Mammals of South America, Volume 2

Mammals of South America, Volume 2

Rodents

Edited by James L. Patton, Ulyses F. J. Pardiñas, and Guillermo D'Elía

The University of Chicago Press

Chicago and London

James L. Patton is professor emeritus of integrative biology and curator of
mammals at the Museum of Vertebrate Zoology, University of California,
Berkeley. Ulyses F. J. Pardiñas is senior scientist at the Centro Nacional
Patagónico, Puerto Madryn, Argentina. Guillermo D'Elía is professor
in the Instituto de Ciencias Ambientales y Evolutivas at the Universidad
Austral de Chile, Valdivia.

The University of Chicago Press, Chicago 60637
The University of Chicago Press, Ltd., London
© 2015 by The University of Chicago
All rights reserved. Published 2015.
Printed in the United States of America

24 23 22 21 20 19 18 17 16 15   1 2 3 4 5

ISBN-13: 978-0-226-16957-6 (cloth)
ISBN-13: 978-0-226-16960-6 (e-book)
DOI: 10.7208/chicago/9780226169606.001.0001

Library of Congress Cataloging-in-Publication Data

Mammals of South America / edited by Alfred L. Gardner.
        p.   cm.
    Includes bibliographical references and index.
    ISBN-13: 978-0-226-28240-4 (cloth : alk. paper)
    ISBN-10: 0-226-28240-6 (cloth : alk. paper)
    1. Mammals—South America. 2. Mammals—South America—
Classification. I. Gardner, Alfred L.
    QL725.AI.M36 2007
    599.098—dc22
            2007017496

♾ This paper meets the requirements of ANSI/NISO Z39.48-1992
(Permanence of Paper).

THIS VOLUME IS DEDICATED TO THOSE SCHOLARS OF MAMMALS IN ACADEMIC institutions, governmental agencies, and nongovernmental organizations in countries of the South American continent. At a pivotal point in the worldwide biodiversity crisis, these individuals have rightly assumed the mantle for the exploration, discovery, description, and conservation of the mammals of their continent through the programs they have developed, the academic institutions and professional organizations they have established and populated, the front rank publication series they have initiated and are supporting, and the students they have trained and who will train those of future generations. There has been remarkable growth in these programs, especially over the past two to three decades, with an energy level and commitment to discovering, archiving, and understanding mammalian diversity in South America that is currently unsurpassed anywhere else in the world.

Contents

Sections and Authors

Introduction

The science of Natural History is of so unbounded an extent, that perhaps I may be allowed, comparatively speaking, to say, that scarcely a day passes without an opportunity being afforded to zoologists of bringing to light unknown instances of its latent treasures.

JOSHUA BROOKES (1829:95), British anatomist and naturalist, in reference to his description of the Plains Viscacha, *Lagostomus*.

IN HIS INTRODUCTION to the first volume in this series, Alfred L. Gardner (2008) detailed the history of the intention that he, Sydney Anderson, and James L. Patton had to produce an updated reference to the mammals of South America, one that would provide a thorough review of knowledge about mammals dating from the mid-1700s through today. The goal was to document the rich nomenclatural history for all taxa (families, genera, and species); describe morphological and other traits that aid in identification, both in the field and museum; delimit species' distributions by listing and mapping marginal records confirmed by voucher specimens and the published literature; and summarize available observations on natural history. Gardner's volume 1, which covered marsupials, xenarthrans, shrews, and bats, set an exemplary standard of scholarship for this effort. This, the second volume in the series, is restricted to the single order Rodentia, the most diverse of all mammalian groups worldwide, accounting for more than 42% of species and 39% of genera (Wilson and Reeder 2005). All Recent and extant taxa are included; those known only from the fossil record or now extinct are not (e.g., *Megaoryzomys* from the Galapagos Islands, Ecuador; *Noronhomys*, from Ilha Fernando de Noronha, Brazil; *Dusimys*, from Curaçao, Caribbean Netherlands), nor are those introduced to the continent (such as beavers, *Castor canadensis*, and muskrats, *Ondatra zibethicus*, from North America, or the Old World rats [*Rattus* spp.] and mice [*Mus* spp.], family Muridae).

The first substantive review of South American rodents was Angel Cabrera's seminal *Catálogo*, the rodent part of which was published in 1961 after his death, thanks to the efforts of Jorge Crespo. In the intervening 50 years, the recognized diversity of South American rodents at both the genus and species levels has increased by approximately 160%. Perhaps not surprisingly, many of the taxa at all categorical levels treated by Cabrera have also been re-ordered in fundamentally different ways. To illustrate, we use taxon diversity trends for the largest group of South American mammals: the rats and mice of the cricetid subfamily Sigmodontinae, which encompass some 62% of the generic diversity and 56% of the species diversity of rodents recognized in the accounts presented in this volume. Only about 34% of the genera and 39% of the South American rodent species we recognize today owe their discovery and description to scholars of the nineteenth century. The first two decades of the twentieth century saw an increase of another 25% to the totals of both genus- and species-group taxa, followed by a 50-year rather quiescent period (from 1921 until 1970) which saw only a modest growth in overall diversity at both taxonomic levels of less than 2% per decade. Tracing these trend lines up to the time when our project was initiated (in the 1970s), one might have surmised that most of the known taxonomic diversity of sigmodontines on the continent had already been discovered and described. Such a conclusion, however, could not have been further from reality. Indeed, in the last 30 years, the generic diversity in South American sigmodontines has increased by 27%, with 24 new genera added, and species diversity by 24%, with 56 new species described. And, in the few months since this book went into production, additional species descriptions have appeared in the literature (e.g., Gonçalves and Oliveira 2014). Fully one quarter of all genera and species of this group of rodents that are recognized today have been described in the terminal three of the 25 decades since Linnaeus (1758) established the earliest currently accepted technical names, and the exploration of the biota of South America began (see the historical reviews of Neotropical mammalogy given by Hershkovitz [1987b] and the Introduction in A. L. Gardner [2008]).

The remarkable rate of recent discovery and description of South American sigmodontines is mirrored by most other rodent groups (as well as for mammals in general worldwide [e.g., Reeder et al. 2007] or those limited to the Neotropical Realm [e.g., B. D. Patterson 2000, 2001]). This increased rate owes its origin to many factors, perhaps the most important of which has been the establishment of exceptionally strong academic programs in systematic and evolutionary sciences, including mammalogy, in most countries on the continent, a development that has accelerated in the past two decades. Naturally coupled to expansion of academic programs has been the development of professional organizations wherein annual or biannual meetings bring together scholars of all academic stages to share knowledge and inspire further research. These two critical components in the development of the professional field of mammalogy have been directly tied to other factors underlying the increase in our knowledge of South America mammals, including but not limited to (1) substantially increased field research, especially into the remote areas of this ecologically diverse continent; (2) an increase in the diversity of survey methods for securing samples (particularly owl pellet analyses in Argentina and pitfall trapping in Brazil); (3) computer-driven analytical methodologies to evaluate characters, compare morphologies in multivariate space, and construct the trees upon which phylogenetic systematics depend; (4) the explosion of molecular-based methodologies, particularly DNA sequencing, which can be applied to historic specimens in museum collections as well as recently collected materials; (5) the requisite shift in systematic philosophy to one based on phylogenetic principles, with taxon definition appropriately based on relationships determined by shared-derived characters rather than overall similarity, and (6) the recognition, through both policy implementation and financial support, by governmental and nongovernmental agencies of the importance of basic systematic research as a key component of biodiversity assessment, which in turn underlies sound environmental policy and establishes conservation priorities. We intend that this volume and the others in our series will identify gaps in our current knowledge and encourage additional field, museum, and laboratory research to fill those gaps.

As noted by A. L. Gardner (2008) in his introduction to volume 1, we selected Cabrera's (1958–1961) catalog as the primary point of departure because of its strong empha-

sis on taxonomy and nomenclature. Our area of coverage includes nuclear South America as well as the continental islands of Trinidad and Tobago and the Caribbean Netherlands off the Venezuelan coast (Map 1). Along with extensive but not exhaustive synonymies, the accounts in this volume include identification keys and descriptions of each order, family, genus, and species. Each species account also includes comments on distribution; lists of selected localities that are plotted on maps; comments on subspecies and, in a few cases only, lists of subspecies and their synonyms and distributions; summaries of natural history information; and discussions of issues related to type localities, taxonomic interpretations, matters of nomenclatural importance, and karyotype, if known. As with volume 1, this one contains its own Gazetteer and Literature Cited sections, with the latter including all citations in the text as well as those cited in the references relating to dates of publication.

Also, as was the case for volume 1, the scope of individual contributions and thus authorship assignments is uneven. The authority and responsibility for a given contribution continue until a new author line is encountered; for example, some authors were responsible for all entries in a single family, while multiple individuals singly or in combination authored the accounts in other families. The Acronyms and Abbreviations, Literature Cited, and Gazetteer sections were compiled from information provided by each account author.

With a few exceptions, the taxonomy of higher categories follows McKenna and Bell (1997), and the sequence of family, genus, and species group taxa is based on the third edition of *Mammal Species of the World*, edited by Wilson and Reeder (2005). Documentation by reference to more recent literature is provided for deviations from either of these two general references, or is noted when based on the personal views of the contributing authors from their own, often unpublished, research. A. L. Gardner (2008:xvi) provided the guidelines followed herein for the synonymies, including citation of the original description and any unique name combination subsequently applied by any author for the first time to that taxon. Interested persons should see his written synopsis for details. We, as did Gardner (see his dedication in volume 1), relied heavily on the "scrapbooks" compiled by Remington Kellogg, the former curator of mammals and director of the U.S. National Museum, in constructing synonymies. Fortunately, much of the early eighteenth and nineteenth century literature is now available in digital form from the Biodiversity Heritage Library (http://www.biodiversitylibrary.org/), a truly remarkable resource. Finally, J. R. Ellerman is credited in this volume with the name combinations cited from his *The Families and Genera of Living Rodents*; however, if a name came from one of the lists and was available before 1937, in many cases either R. W. Hayman or G. W. C. Holt compiled the individual list on which those name combinations are based.

This volume deviates from volume 1 in limited but important ways. First, we asked account authors to provide a description of each species beyond those characters that comprise the keys. Most rodents have not been monographed in recent decades or ever, and many species are known only from their initial description. Moreover, an increasing number of species have been defined in the past few decades by karyotypic data (diploid number and arm number of both autosomes and sex chromosomes) and/or DNA sequences. As important as these characters are, neither the karyotype nor sequence are available for the vast majority of specimens in museum collections nor for live animals captured during field studies. As a consequence, reliance on keys alone for identification in the field or museum can be problematic. We thus believe that descriptions of external and/or craniodental characters of each species is an important addition to this volume, one enhanced by multiauthored accounts for large, geographically widespread, and particularly vexing taxa, such as the sigmodontine genus *Akodon* or the echimyid genus *Proechimys*. Where possible, the authors have provided the standard external measurements of head and body length, tail length, hindfoot length, ear length, and/or mass, taken either from the literature or from labels of museum vouchers they examined as well as characters of the pelage (color, color pattern, details of hair types), appendages, and both soft and craniodental anatomy. However, these descriptions vary greatly in depth of detail owing to both the level of knowledge of individual and geographic variation available and an author's direct familiarity with specimens in museum collections. In some cases, our knowledge of a particular species is so poor that little beyond the original description is available.

Second, we have allowed authors liberty in delineating locality records beyond just those circumscribing the edge of a species' range. Many species of rodents are so poorly known that we believed it valuable to list and map virtually all known records in those cases. However, we caution users of metadata, such as marginal locality maps in global analyses of biodiversity, not to confuse a shaded range bounding locality records with actual occurrence throughout. Many species of rodents are habitat specialists and are certainly not evenly spread across the geographic landscape indicated by that shaded range. Shading surrounding the mapped localities is provided primarily to highlight those locality placements and not to identify actual geographic boundaries, although the maps are drawn with as much precision as possible given their scale. The ranges mapped also include locality records assembled over many decades; as a result, they do not accommodate any range shift or fragmentation that may have resulted from anthropogenic landscape modification or other factors in recent decades. Nor should these range maps be viewed as static into even the near future, as continuing fieldwork in remote areas will invariably expand those ranges now known and global climate change continues to impact both distribution and presence (e.g., Malcolm et al. 2006; Urban et al. 2012). Finally, because the shapes of species' ranges are diverse, often linear or geographically restricted, we have not followed the uniform clockwise listing of localities, starting with N on a compass, employed in volume 1. Rather than explaining deviations from this scheme in so many individual cases, localities are simply listed alphabetically by country, department/province/state, and specific locality in all accounts. Virtually all locality records are based on vouchered specimen records where the identification could be verified, either literature records from revisionary accounts or the actual examination of a museum voucher. We note, however, that the advent of widespread camera traps and archives of verifiable digital photos—which have been used so successfully to document new records for large mammals—are beginning to provide insight even into smaller and often very secretive rodents (e.g., Blake et al. 2010). Although photos are unacceptable replacements for vouchered specimens in systematic research, a few accounts have included localities derived from unambiguous recent photographs of rare taxa with few records. Inset maps depicting the range of the genus are also provided in the maps for each species.

Third, there is also substantial unevenness in coverage of natural history and other information on population biology, ecology, reproduction, and other attributes across accounts. Some of this simply results from the actual quantity of information available, as so many species of rodents have never been studied in their natural habitats, but some also stems from whether existing or relatively recent synopses have already been published (e.g., the *Mammalian Species* accounts produced by the American Society of Mammalogists, http://www.mammalsociety.org/publications/mammalian-species).

Finally, the construction and use of vernacular names is a vexing problem in mammalogy, as there is no standard set of rules that apply to their formation (but see Duckworth and Pine 2003). This issue is exacerbated by the common application of English names (as given, for example, in Wilson and Cole [2000] for mammal species worldwide) rather than use of the larger number of common terms applied to those species by South American biologists, the lay public in each country, and/or indigenous groups. The common names employed herein are those of the account authors and may reflect either or both English and local terms or equivalents. Many of these follow Wilson and Cole (2000) and/or those of account authors in Wilson and Reeder (2005).

Map 1 Political map of the South American continent including the Caribbean Netherlands, Trinidad and Tobago, and the Galapagos Islands.

Acronyms and Abbreviations

The following acronyms and abbreviations are used in the text to identify collectors and museum collections.

AMNH: American Museum of Natural History, New York, New York, United States
AL: Alfredo Langguth field numbers
AN: Andrea Nunes field numbers
APC: Ana Paula Carmignotto field numbers
ARG: Field numbers of uncataloged specimens in the Sam Noble Oklahoma Museum of Natural History, Norman, Oklahoma, United States
ARP: Alexandre R. Percequillo field numbers
BM: The Natural History Museum, London (NHM); formerly the British Museum (Natural History), London, United Kingdom
CAS: California Academy of Sciences, San Francisco, California, United States
CBF: Colección Boliviana de Fauna, Instituto de Ecología y Museo de Historia Natural, La Paz, Bolivia
CEM: Colección Elio Massoia, acquired by the Fundación de Histora Natural Félix de Azara, Buenos Aires, Argentina
CIT: Laboratório de Citogenética de Vertebrados, Universidade de São Paulo, São Paulo, Brazil
CLH: Christine L. Hice field numbers
CM: Carnegie Museum of Natural History, Pittsburgh, Pennsylvania, United States
CML: Colección Mamíferos Lillo, Universidad Nacional de Tucumán, Tucumán, Argentina
CML-PIDBA: Colección Mamíferos Lillo, Programa de Investigaciones de Biodiversidad Argentina, Universidad Nacional de Tucumán, Tucumán, Argentina
CMUFL: Mammal Collection, Universidade Federal de Lavras, Minas Gerais, Brazil
CNP: Colección de Mamíferos, Centro Nacional Patagónico, Puerto Madryn, Chubut, Argentina
CNP-E: Colección de Material de Egagrópilas y Afines "Elio Massoia," Centro Nacional Patagónico, Puerto Madryn, Chubut, Argentina
CONN: University of Connecticut, Storrs, Connecticut, United States
CRB: Cibele R. Bonvicino field numbers
CUVLA: Colección de Vertebrados de la Universidad de Los Andes, Mérida, Venezuela
EBRG: Museo de la Estación Biológica de Rancho Grande, Maracay, Venezuela
EPN: Escuela Politécnica Nacional, Quito, Ecuador
FAG: Valéria Fagundes field numbers
FLMNH: University of Florida Museum of Natural History, Gainesville, Florida, United States
FMNH: Field Museum, Chicago, Illinois, United States
GD: Guillermo D'Elía field numbers
HGB: Helena de Godoy Bergallo field numbers
HTC: Departamento de Microbiología, Universidad del Valle, Cali, Colombia
IADIZA: Colección de Mastozoología del Instituto Argentino de Investigaciones de Zonas Áridas, Mendoza, Argentina
IAvH: Instituto Alexander von Humboldt, Cartegena, Colombia

ICN: Instituto de Ciencias Naturales, Universidad Nacional de Colombia, Bogotá, Colombia

INDERENA: Instituto Nacional de los Recursos Naturales Renovables y del Ambiente, Bogotá, Colombia

INEVH: Instituto Nacional de Enfermedades Virales Humanas, Pergamino, Buenos Aires, Argentina

INPA: Instituto Nacional de Pesquisas da Amazônia, Amazonas, Manaus, Brazil

JAO: João A. de Oliveira field numbers

JPB: Jean-Phillipe Boubli field numbers; voucher specimens in INPA collection.

JPJ: Jorge Pablo Jayat field numbers

JRM: Jay R. Malcolm field numbers

KU: Natural History Museum, University of Kansas, Lawrence, Kansas, United States

LACM: Los Angeles County Museum of Natural History, Los Angeles, California, United States

LAMAQ: Laboratório de Mamíferos Aquáticos, Universidade Federal de Santa Catarina, Florianópolis, Santa Catarina, Brazil

LBCE: Laboratório de Biologia e Controle da Esquistossomose, Fundação Instituto Oswaldo Cruz (FIOCRUZ), Rio de Janeiro, Brazil

LC (and LPC): Leonora Pires Costa field numbers

LCM: Laboratorio de Genómica Evolutiva, Facultad de Medicina, Universidad de Chile, Santiago, Chile

LF: Luiz Flamarion field numbers

LGA: Laboratório de Genética Animal, Departamento de Ciências Biológicas, Universidade Federal do Espírito Santo, Vitória, Espírito Santo, Brazil

LHE: Louise H. Emmons field numbers

LMT: Liliani Marilia Tiepolo field numbers

LPC (and LC): Leonora Pires Costa field numbers

LSUMZ: Louisiana State University Museum of Zoology, Baton Rouge, Louisiana, United States

LTU: Proyecto Localidades Tipo—CONICET, Argentina (vouchers to be deposited in CNP)

LV: Laboratório de Vertebrados, Universidade Federal do Rio de Janeiro, Rio de Janeiro, Brazil

MACN: Colección Nacional de Mastozoología, Museo Argentino de Ciencias Naturales "Bernardino Rivadavia," Buenos Aires, Argentina

MAM: Meika A. Mustrangi field numbers

MARNR: Ministerio del Ambiente y de los Recursos Naturales Renovables, Maracay, Venezuela

MBUCV: Museo de Biología, Universidad Central de Venezuela, Caracas, Venezuela

MCZ: Museum of Comparative Zoology, Harvard University, Cambridge, Massachusetts, United States

MECN: Museo Ecuatoriano de Ciencias Naturales, Quito, Ecuador

MF: Mônica Fonseca field numbers

MFA-ZV-M: Museo Provincial de Ciencias Naturales "Florentino Ameghino," Santa Fe, Argentina

MHNCI: Museu de História Natural Capão de Imbuia, Curitiba, Paraná, Brazil

MHNG: Ville de Genève Muséum d'Histoire Naturelle, Genève, Switzerland

MHNLS: Museo de Historia Natural La Salle, Caracas, Venezuela

MHNSM: Museo de Historia Natural, Universidad de San Marcos (formerly Museo de Historia Natural Javier Prado), Lima, Peru

MHNUC: Museo de Historia Natural, Universidad de Caldas, Caldas, Colombia

MIC: María Inés Carma field numbers

MLP: Museo de La Plata, La Plata, Argentina

MLS: Mel L. Schamberger field numbers

MMD: M. Mónica Díaz field numbers

MNHM: Muséum d'Histoire Naturelle, Neuchâtel, Switzerland

MNHNC: Museo Nacional de Historia Natural de Chile, Santiago, Chile

MNK: Museo de Historia Natural Noel Kempff Mercado, Santa Cruz de la Sierra, Bolivia

MNRJ: Museu Nacional, Rio de Janeiro, Brazil

MPEG: Museu Paraense Emilio Goeldi, Belém, Pará, Brazil

MSB: Museum of Southwestern Biology, University of New Mexico, Albuquerque, New Mexico, United States

MSU: Michigan State University, East Lansing, Michigan, United States

MUSA: Museo de Historia Natural, Universidad Nacional de San Agustín, Arequipa, Peru

MUSM: Museo de Historia Natural, Universidad Mayor de San Marcos, Lima, Peru

MV: Michael Valqui field numbers

MVZ: Museum of Vertebrate Zoology, University of California, Berkeley, California, United States

MW: Marcelo Weksler field numbers

MZUFV: Museu de Zoologia João Moojen, Universidade Federal de Viçosa, Viçosa, Minas Gerais, Brazil

MZUSP: Museu de Zoologia, Universidade de São Paulo, São Paulo, Brazil

NMW: Naturhistorisches Museum Wien, Vienna, Austria

NRM: Naturhistoriska Riksmuseet, Stockholm, Sweden

NUPEM: Núcleo de Pesquisas Ecológicas de Macaé, Universidade Federal do Rio de Janeiro, Brazil

OMNH: Sam Noble Oklahoma Museum of Natural History, University of Oklahoma, Norman, Oklahoma, United States

P: Yves Sbalqueiro laboratory numbers, Universidade Federal do Paraná, Curitiba, Paraná, Brazil

PEO-e: Pablo E. Ortiz field numbers

PMC: field numbers, specimens housed in Museu de História Natural Capão da Imbuia, Curitiba, Paraná, Brazil

PPA: Proyecto Patagonia Agencia—Argentina (voucher specimens to be deposited in CNP)

PRG: Pablo R. Gonçalves field numbers

PVL: Colección de Paleontología de Vertebrados del Instituto Miguel Lillo, Universidad Nacional de Tucumán, Tucumán Province, Argentina

QCAZ: Museo de Zoología, Pontifica Universidad Católica del Ecuador, Quito, Ecuador

RDS: Richard D. Sage field numbers

RMNH: Rijksmuseum van Natuurlijke Historie, Leiden, the Netherlands

RNHMS: Royal Natural History Museum, Stockholm, Sweden

RNP: Natalie Goodall collection, Estancia Harberton, Museo Acatashún, Tierra del Fuego, Argentina

ROM: Royal Ontario Museum, Toronto, Ontario, Canada

SDM: Sterling D. Miller field numbers

SMF: Naturmuseum Senckenburg, Frankfurt, Germany

T: Laboratoire de Paléontologie, Paléobiologie et Phylogénie, Université Montpellier, Montpellier, France

TTU: The Museum, Texas Tech University, Lubbock, Texas, United States

UACH: (formerly IIEUACH), Facultad de Ciencìas, Universidad Austral de Chile, Valdivia, Chile

UFC: Universidade Federal do Ceará, Fortaleza, Ceará, Brazil

UFMG: Universidade Federal de Minas Gerais, Belo Horizonte, Minas Gerais, Brazil

UFPB: Universidade Federal da Paraíba, João Pessoa, Paraíba, Brazil

UFPE: Universidade Federal de Pernambuco, Recife, Pernambuco, Brazil
UFRGS: Universidade Federal do Rio Grande do Sul, Porto Alegre, Rio Grande do Sul, Brazil
UFRJ: Universidade Nacional do Rio de Janeiro, Rio de Janeiro, Brazil
UFROM: Fundação Universidade de Rondônia, Porto Velho, Rondônia, Brazil
UFSC: Universidade Federal de Santa Catarina, Florianópolis, Santa Catarina, Brazil
UMMZ: University of Michigan Museum of Zoology, Ann Arbor, Michigan, United States
UNB: Coleção de Mamíferos, Departamento de Zoologia, Universidade Nacional de Brasília, Brasília, Brazil
UNRC: Universidad Nacional de Río Cuarto, Río Cuarto, Córdoba, Argentina
UP: Ulyses F. J. Pardiñas field numbers
USNM: National Museum of Natural History (formerly the United States National Museum), Washington, D.C., United States
UV: Departamento de Biología, Universidad del Valle, Cali, Colombia
UWBM: University of Washington Burke Museum, Seattle, Washington, United States
VA: Vale das Antas project field numbers
VPF: Valéria Pena-Firme field numbers
VPT: Víctor Pacheco field numbers
YPM: Yale Peabody Museum, Yale University, New Haven, Connecticut, United States
ZINRAS: Zoological Institute, Russian Academy of Sciences, St. Petersburg, Russia
ZMB: Museum für Naturkunde der Humboldt-Universität zu Berlin, Berlin, Germany
ZMFK: Zoologisches Forschungsinstitut und Museum Alexander Koenig, Bonn, Germany
ZMUC (also ZMK): Zoological Museum of the University, Copenhagen, Denmark
ZMUM: Zoological Museum, University of Montana, Missoula, Montana, United States

The following abbreviations are used to identify statistical or other terms.

bp: base pairs (of DNA sequences)
FN: fundamental number
ft: feet
g: grams
ha: hectares
HPD: highest posterior density
IRBP: interphotoreceptor retinoid-binding protein
kg: kilograms
kya: thousands of years ago
m: meters
mm: millimeters
mtDNA: mitochondrial DNA
mya: millions of years ago
nucDNA: nuclear DNA
pg: picograms
s.d.: standard deviation
s.s.: sensu stricto
ybp: years before present

Acknowledgments

No treatise such as this can be completed without the input of many individuals, and not just those who authored the accounts herein. With apologies beforehand for excluding some who deserve mention, we single out several individuals for special recognition for aiding us in so many substantive ways. Specifically, Louise H. Emmons read through most of the completed manuscript in its several iterations, freely offering her vast knowledge, correcting parts, and updating others. Mere mention of her name here does not do justice to her insights and overall contributions. Michael D. Carleton offered expert advice on details of synonymy construction and reviewed large parts of the manuscript with the eye of an editor; his efforts are deeply appreciated. Carola Cañón (Chile), Hugo Mantilla-Meluck and Héctor Ramírez (Colombia), and Julio Torres (Paraguay) reviewed and corrected locality designations in the accounts and gazetteer; we are grateful to each for this very important contribution. Susan Abrams and Christie Henry, both senior editors at the University of Chicago Press over the gestation period of our project, deserve special recognition for their unflagging interest and incredible patience while encouraging us to complete this project, as do Abby Collier, Kyle Adam Wagner, and Amy Krynak, also at the University of Chicago Press, for help with the final stages of manuscript production.

Finally, we are especially grateful to Ronald H. Pine and Robert S. Voss for their careful, critical reviews of the entire manuscript; neither should be held responsible for the final product, but their nomenclatural expertise and especially their personal field and museum experience with many taxa have improved virtually every section.

Jim Patton acknowledges Alfred L. Gardner, whose editorship of volume 1 in this series set a standard of scholarship we can only hope to achieve. Gardner not only initiated Patton's career in mammalogy in the early stages of their respective graduate studies and introduced him to South America and its spectacular mammalian fauna (see Patton 2005c), but he also answered many queries quickly and concisely. Michael Carleton, Judith Chupasco, Robert Fisher, David Flores, Mark Hafner, Yuri Leite, João de Oliveira, Víctor Pacheco, Bruce Patterson, Mario de Vivo, and Robert S. Voss provided Patton with either direct access to the collections under their control, examined critical specimens for him, or sent him crucial specimen records or literature for inclusion in this volume; he thanks each for their professionalism and friendship. James Hanken at the Museum of Comparative Zoology of Harvard University and Mary Sears, librarian of the Mayr Collection of the Museum of Comparative Zoology's library, kindly provided a digital scan of an important historical document.

Ulyses Pardiñas acknowledges the economic support from Agencia Nacional de Promoción Científica y Tecnológica (PICT 2008-547), and Consejo Nacional de Investigaciones Científicas y Técnicas (PIPs 6179 and 2001-164). He is especially grateful to Pablo Teta and Carlos Galliari for many years of collaborative work in the field and museum on sigmodontines, and to Erika Cuéllar, whose love and unconditional support rescued him from unhappiness, and his son Joaquín, a tireless friend in the field, both of whom changed his life.

Guillermo D'Elía is grateful to those colleagues, teachers, and students who shared and discussed their views on the evolutionary history and systematics of South American rodents, and gives special thanks to everyone who made this project possible.

Louise Emmons wishes to thank João Oliveira for providing a photograph of *Oxymycterus talpinus*, essential for her review of *Juscelinomys*.

Claudio Bidau acknowledges Julio R. Contreras, Patricia Mirol, Jeremy B. Searle, Chris G. Faulkes, Enrique Lessa, Andrés Parada, Guillermo D'Elía, Ricardo Ojeda, Thales de Freitas, Yolanda Davies, Alonso Medina, Diego Verzi, Mabel D. Giménez, Claudia Ipucha, Pablo Suárez, Cecilia Lanzone, Joyce Bernardo, Pavel Borodin, Dardo Marti, Juan Luis Santos, and a large number of students during his years at the Universidad Nacional de Misiones, Universidade Estadual de Rio de Janeiro, and Fundação Oswaldo Cruz. He gives his special thanks to Valeria Xinena Rodríguez for her help and endless love.

Scott Steppan and Oswaldo Ramirez thank Jim Patton for substantial contributions to their *Phyllotis* account, and they acknowledge the financial support of National Science Foundation grants DEB-0841447 and DEB-0108422.

Alexandra Bezerra thanks Manoel Santos-Filho and Maria Nazareth F. da Silva for discussions and clarifications regarding the correct identification of a specimen allocated to *Euryzygomatomys spinosus* in previous studies; Ana Paula Carmignotto and Roberta Paresque for sharing their unpublished karyotypic data for *Carterodon sulcidens*; Ulyses F. J. Pardiñas for providing key bibliographic references; and the CNPq (150599/2008-0) for postdoctoral support and Conservation International for grants to visit critical museum collections.

Mario de Vivo wishes to thank Dr. Robert S. Voss and the Mammal Department of the American Museum of Natural History in New York for postdoctoral advisorship and funding, respectively. He and Ana Paula Carmignotto would also like to extend their heartfelt thanks to Jim Patton "for everything, and more."

Taxonomic Accounts

Class Mammalia Linnaeus, 1758
Order Rodentia Bowdich, 1821

NEW WORLD RODENTS

James L. Patton

KEY TO THE SUBORDERS OF RECENT SOUTH
AMERICAN RODENTS:

1. Well-defined zygomatic plate anterior to zygomatic arch present; infraorbital foramen small and either piercing zygomatic plate or positioned on side of rostrum anterior to zygomatic plate; lower jaw sciurognathous, with origin of angular process directly ventral to sheath of lower incisor. 2

1′. No zygomatic plate; infraorbital foramen greatly enlarged; lower jaw hystricognathous, with the root of the angular process deflected lateral to sheath of lower incisor. . . . . Hystricomorpha

2. Zygomatic plate oriented anteriorly; infraorbital foramen small, transmitting only nerves and blood vessels; maxillary teeth number four or five 3

2′. Zygomatic plate oriented more laterally; infraorbital foramen enlarged, narrow, but typically expanded dorsally, transmitting medial masseter muscle in addition to nerves and blood vessels; maxillary teeth number three or fewer . . . . Myomorpha

3. Infraorbital foramen pierces zygomatic plate; cheek teeth number four or five, with molars brachydont, bunodont, with three transversely oriented crests . . . .
 . . . . Sciuromorpha

3′. Infraorbital foramen positioned on side of rostrum anterior to zygomatic plate; cheek teeth number four, either hypsodont, cylindrical, and planar (Geomyidae) or brachydont, bunodont, and cuspidate (Heteromyidae) . . . . Castorimorpha

Suborder Sciuromorpha Brandt, 1855
Infraorder Sciurida Carus, 1868

Family Sciuridae G. Fischer, 1817
Mario de Vivo and Ana Paula Carmignotto

Squirrels are ubiquitous inhabitants of all forest biomes throughout South America except the temperate forests, with all species adapted for life in the canopy, even though some forage on the ground. Specialized adaptations for climbing include elongate bodies, forefeet with four long toes with claws, hindfeet with five long toes with claws, large plantar pads on all feet, broad heads, large eyes, short ears, generally soft fur, long bushy tails, and elongated legs. The ankle joint is flexible, allowing the hindfoot to rotate sufficiently for headfirst descent of vertical tree trunks. Most species have enlarged, long incisors accompanied by correspondingly large jaw muscles, which enable them to gnaw through the hardest nuts. Some species specialize on hard-shelled seeds of tropical trees such as palms, but most are more omnivorous, feeding on a range of food types such as nuts, fruits, insects, fungi, and sometimes leaves, flowers, and bark (Emmons and Feer 1997). All South American squirrels are diurnal; as such, individuals can be easily seen and often readily heard as they move about or gnaw on large nuts. They build round, ball-like nests of leaves and twigs in vine tangles or on tree limbs, or they nest in tree holes. Capuchin monkeys, *Cebus* [now *Sapajus*] *apella* (Galetti 1990), and both medium and large felids (Emmons 1978b) are known predators of South American squirrels.

Body size can be arranged into three broadly separable classes, with one pygmy squirrel (maximum head and body length about 115 mm, tail length 120 mm, and mass 45 g), many dwarf squirrels of intermediate size (head and body length 160–300 mm, tail length 150–280 mm, and mass 130–520 g), and a few large squirrels (head and body length 240–300 mm, tail length 240–340 mm, and mass 570–900 g; Thorington et al. 2012). The skull retains the same generalized shape in all South American species: braincase domed and rounded; interorbital region wide; postorbital processes on both frontals and jugals prominent; rostrum relatively short but broad; zygomatic arches deep and stout; auditory bullae evenly rounded and relatively large but rarely excessively inflated, and divided by several transverse bony septa; incisive foramina short; cheek teeth brachydont and bunodont, with prominent cusps and transverse crests usually on molars; and dental formula I 1/1, C 0/0, PM 2–1/1, and M 3/3 = 20–22, with PM3 reduced or absent. Angular processes of the mandible are slightly inflected. A baculum is present in the phallus of males of all species.

Peterka (1937) and Bryant (1945), among others, have provided detailed descriptions of sciurid morphology. Only four taxa have been karyotyped: *Guerlinguetus brasiliensis brasiliensis* (Lima and Langguth 2002, $2n = 40$, FN = 76), *G. b. ingrami* (Fagundes et al. 2003, $2n = 40$, FN = 74), *Hadrosciurus spadiceus* (Lima and Langguth 2002, $2n = 40$, FN = 76), and *Notosciurus granatensis* (Nadler and Hoffmann 1970, $2n = 42$, FN = 78).

Despite their broad range throughout much of the continent and local ubiquity in all forest types, South American squirrels remain poorly known in nearly all aspects of their life history and, in particular relevance to this volume, their evolutionary relationships and diversification history. There have been no revisionary studies published in the last half-century or longer. Furthermore, and in contrast to virtually all other South American rodent groups, students of molecular phylogenetics have largely bypassed the Sciuridae, except for the inclusion of a few species in the phyletic delineation of the major lineages in this nearly cosmopolitan family (Mercer and Roth 2003; Steppan, Storz, and Hoffmann 2004; Pecnerová and Martínková 2012). These studies have unambiguously identified two lineages among the South American representatives of the family. One of these contains the monotypic *Sciurillus*, which forms a

basal offshoot that diverged very early in the history of the family. Although Moore (1959) did recognize the uniqueness of *Sciurillus* by placing it in its own subtribe Sciurillina in the tribe Sciurini, molecular studies clearly document that *Sciurillus* has no phylogenetic affinity to other tree squirrels. Consequently, this genus is now uniformly placed in its own subfamily, the Sciurillinae (Steppan, Storz, and Hoffmann 2004; Thorington and Hoffmann 2005; Thorington et al. 2012). The basal position of *Sciurillus* in the family, moreover, suggests great age for its lineage, which Roth and Mercer (2008) estimated at about 30 mya, a divergence event only slightly younger than the 36-mya age for the family. In contrast, the entire radiation of all remaining tree squirrels in South America, now placed in the subfamily Sciurinae (Thorington and Hoffmann 2005; Thorington et al. 2012), appears to have descended from a single lineage, one that arrived on the continent subsequent to the closure of the Panamanian land bridge (Roth and Mercer 2008; Pecnerová and Martínková 2012).

The last revision of South American representatives was J. A. Allen's classic monograph *Review of the South American Sciuridae*, published in 1915 and thus now nearly a century old. Allen recognized nine genera, a number reduced to three or four by subsequent reviewers (e.g., Ellerman 1940, three genera [*Sciurillus*, *Microsciurus*, and *Sciurus*, divided into three subgenera]; Moore 1959, four genera [*Guerlinguetus*, *Microsciurus*, *Sciurillus*, and *Syntheosciurus*]; Cabrera 1961, three genera [*Microsciurus*, *Sciurillus*, and *Sciurus*, divided into two subgenera]; R. S. Hoffmann et al. 1993 and Thorington and Hoffmann 2005, three genera [*Microsciurus*, *Sciurillus*, and *Sciurus*, divided into three subgenera]). The number of species recognized in various treatments over the past 50 years has also differed, sometimes substantially. For example, J. A. Allen (1915a) divided South American squirrels into 25 species, Cabrera (1961) into 13, Emmons and Feer (1997) described and mapped 11, and R. S. Hoffmann et al. (1993) and Thorington and Hoffmann (2005; see also Thorington et al. 2012) listed 15. In part, differences in species boundaries recognized by these and other authors attest to the extreme variability often present geographically in color and color pattern. The concordance of both generic membership and categorical level across these treatments is limited, a fact also attesting to both the extensive variation in coloration that characterizes most of these taxa and the lack of critical reviews.

The lack of any comprehensive or even regional analysis of character variation for nearly all species of South American squirrels prompted us to examine most available specimens, including holotypes, to assess both genus- and species-level boundaries. As a result, we recognize seven genera and a total of 19 species. These taxonomic hypotheses differ from previous treatments of South American squirrels, the most recent of which is that of Thorington et al. (2012), and they are ripe for testing by phylogenetic and phylogeographic analyses employing molecular data and methods.

KEY TO THE SUBFAMILIES OF SOUTH AMERICAN SCIURIDAE (FROM MOORE 1959:199–200):

1. Lateral wall of cranium with squamosal extending dorsally to a point less than or equal to halfway from base of zygomatic process of squamosal to base of postorbital process of frontal; typically two transbullar septa; well-developed masseteric tubercle; maxilla contributing much less than half to the lateral side of rostrum; robust zygoma with a well-developed superior process . . . . Sciurinae

1′. Lateral wall of cranium with squamosal extending dorsally notably more than halfway between base of zygomatic process of squamosal and base of postorbital process of frontal; one transbullar septum; no masseteric tubercle; maxilla contributing much more than half of the lateral side of rostrum; slender zygoma lacking superior process. . . . . Sciurillinae

Subfamily Sciurillinae Moore, 1959

The subfamily Sciurillinae contains only the single genus and species *Sciurillus pusillus*. This is the smallest of the New World Sciuridae and one of the two most ancient lineages in the family, the other being the Ratufinae, or giant tree squirrels of southern Asia. In erecting this taxon, Moore (1959:180) delineated 12 diagnostic characters of the skull, including those listed in the key.

Cladistic analyses of both morphological (Roth 1996) and molecular characters (allozymes: M. S. Hafner et al. 1994; mitochondrial and nuclear DNA sequences: Mercer and Roth 2003; Herron et al. 2004; Roth and Mercer 2008) support the basal position of the Sciurillinae in the Sciuridae and refute the hypothesized phyletic link between *Sciurillus* and the Asian (*Nannosciurus*) and African (*Myosciurus*) pygmy squirrels proposed by G. S. Miller and Gidley (1918). Mercer and Roth (2003) estimated the split between the lineage leading to *Sciurillus* and all other sciurids (flying, ground, and tree squirrels) at about 36 mya, in the late Eocene. These authors also argued that the deep phylogenetic separation between *Sciurillus* and other South American sciurids, which formed a monophyletic assemblage, indicated separate invasions of the continent by squirrels. Because *Sciurillus* has no close relatives in the family, its place of origin and timing of entry to South America remain enigmas.

Genus *Sciurillus* Thomas, 1914

The genus *Sciurillus* is a tiny rainforest squirrel with three pairs of mammae, distributed in the Guianas and Amazon Basin. This is the smallest of the New World tree squirrels, with a head and body length averaging 113 mm, tail length 114 mm, hindfoot length 30 mm, and body mass 44 g (N = 5). The ears are small and round, up to 14 mm long. The skull has a strongly convex dorsal profile, with the inflexion point at the middle of the frontals; the rostrum is long, more than one-third the skull length; the frontals are short, as wide or wider than their length; the braincase is relatively large and globose; the temporal part of the orbitotemporal fossa is completely obliterated, with the postorbital process extending posteriorly to almost coincide with a vertical plane at the posteriormost part of the squamosal zygomatic process; the anterior portion of the zygoma is nearly vertical, situated posterior to the first upper molar; a distinct semicircular crest, originating at the infraorbital foramen, runs dorsally along the line of contact between the premaxilla and maxilla on each side; the anterior opening of the infraorbital foramen is positioned midway between the anteriormost part of the zygoma and the incisors, but the infraorbital canal is not conspicuously elongated; the parapterygoid fossa is deep, well delimited, and sometimes perforated by bilateral or unilateral vacuities; the posterior aperture of the alisphenoid canals is small, maximally one-third the width of the foramen ovale; the aperture of the transverse canal is distinctly visible in ventral view, and not covered by the posterior border of the pterygoid encircling the foramen ovale; a sphenopalatine foramen is situated above the third upper molar or, at most, over the region between the second and third molars; and only one bullar septum is present. Very short coronoid processes characterize the mandible. Temporal muscles are correspondingly unusually small for a squirrel, but the anterior deep masseters are enlarged and oriented to assist in retraction of the jaw (Ball and Roth 1995).

The dental formula is I 1/1, C 0/0 PM 2/1, M 3/3 = 22. Upper incisors are proodont. PM3 has one root; all remaining permanent cheek teeth have three roots. In occlusal view, the maxillary toothrows form a more or less rounded to slightly oval arrangement. PM3 is small, with a simple, unicuspidate occlusal surface. PM4 is molariform, only slightly smaller than M1; M2 is the largest cheek tooth. PM4 through M2 lack a paraconule, metaconule, and ectostyle; the trigon of all maxillary teeth is shallow. The lower cheek teeth lack a paraconid and ectostylid, and a trigonid is obsolete or absent.

SYNONYMS:

[*Myoxus*]: Shaw, 1801:171; part (included "*le petit guerlinguet*" of Buffon 1789); not *Myoxus* Zimmerman, 1780.

Sciurus: É. Geoffroy St.-Hilaire, 1803:177; part (inclusion of *pusillus*); not *Sciurus* Linnaeus, 1758.

Sciurus: Desmarest, 1817c:109; part (inclusion of *pusillus*); not *Sciurus* Linnaeus, 1758.

Macroxus: F. Cuvier, 1823b:119; 1823c:161; part (included "*le guerlinguet*" of Buffon, 1789).

[*Funambulus*]: Fitzinger, 1867a:487; part (inclusion of *pusillus*); not *Funambulus* Lesson, 1835.

Sciurus (*Macroxus*): Liais, 1872:503; part (inclusion of *pusillus*); not *Sciurus* Linnaeus, 1758.

Sciurus (*Microsciurus*): E. W. Nelson, 1899b:32; part (inclusion of *pusillus* and *kuhli*); not *Microsciurus* J. A. Allen, 1895.

Microsciurus: E.-L. Trouessart, 1904:329; part (inclusion of *pusillus* and *kuhli*); not *Microsciurus* J. A. Allen, 1895.

Sciurillus Thomas, 1914f:36 [1914f:415]; type species *Sciurus pusillus* Desmarest, 1817 (= *Sciurus pusillus* É. Geoffroy St.-Hilaire, 1803), by original designation.

REMARKS: Buffon (1789:261, plates 65 and 66, respectively) recognized the existence of two "squirrel-like" animals in the Guiana region, "*le grand guerlinguet*" (= *Sciurus aestuans* Linnaeus) and "*le petit guerlinguet*" (= *Sciurillus pusillus* É. Geoffroy St.-Hilaire). Buffon used the vernacular French term "*guerlinguet*" rather than "*écureuil*" because he was not convinced that the animals were "true" squirrels.

When É. Geoffroy St.-Hilaire (1803) named "*le petit guerlinguet*" as *Sciurus pusillus*, he implicitly allied the species with other known tree squirrels. Desmarest (1817c:109) redescribed Geoffroy St.-Hilaire's *Sciurus pusillus* and, because many subsequent authors regarded the taxa proposed by É. Geoffroy St.-Hilaire (1803) as unavailable manuscript names, much of the nineteenth- and twentieth-century literature credited *Sciurus pusillus* to Desmarest (J. A. Allen 1915a; Ellerman 1940; Cabrera 1961; R. S. Hoffman et al. 1993). Desmarest (1822:337) even coined a new common name for the species, calling it "*l'Écureuil nain*" or "dwarf squirrel," probably to emphasize its similarity to the European forms and to abolish the distinctiveness suggested by "*guerlinguet*." However, some authors continued to distinguish between "*guerlinguets*" and squirrels. F. Cuvier, in a series of papers dealing with the dentition of mammals, described and figured the dentition of "*le petit guerlinguet*" (F. Cuvier 1812:277, plate 15, Fig. 1), the distinctive dental characters being the size of the cheek teeth depicted and the presence of a diminutive upper third premolar, a feature present only in *Sciurillus pusillus* and not in *Sciurus aestuans*. Indeed, F. Cuvier did not mention "*le grand guerlinguet*" in his 1812 paper, but in a later publication (F. Cuvier 1819:248), he provided brief descriptions of the "*grand*" and "*petit*" guerlinguets, stating that "leurs

molaires [= cheek teeth] sont tout-à-fait semblables à celles des écureuils," thus making them essentially equal in that respect (although "*le grand guerlinguet*" [= *Sciurus aestuans*] actually has a distinctive dentition that lacks third upper premolars).

Later, when F. Cuvier published a comprehensive account of mammalian dentition (1823c:161–162), he erected the genus *Macroxus*, characterized by (among other features) the presence of two upper premolars, and he included as species "*le guerlinguet*" and "*le toupaye*" (the latter an oriental squirrel). If his diagnosis is interpreted strictly and if reference is made to his 1812 publication, "*le guerlinguet*" would mean only the "*petit guerlinguet*." However, as he clearly considered (incorrectly) that both "*guerlinguets*" had similar dentition (F. Cuvier 1819), it can be safely assumed that by "*guerlinguet*" Cuvier meant both *pusillus* and *aestuans*. Husson (1978:382–383) recognized this possible confusion and specifically designated the specimen described and figured by Buffon (1789:261, plate 66: "*le petit guerlinguet*") as the lectotype of *Sciurus pusillus* É. Geoffroy St.-Hilaire. Earlier, Thomas (1898g:933) had designated *Sciurus aestuans* Linnaeus, 1766 as the type species of *Macroxus* F. Cuvier, 1823, thus making this name unavailable for *Sciurillus*.

The remaining generic names associated with *Sciurillus pusillus* (É. Geoffroy St.-Hilaire) clearly belong either to other rodent groups (e.g., *Myoxus* Zimmermann, Gliridae) or to other groups of sciurids (e.g., *Sciurus* Linnaeus) and need not be discussed further here.

Sciurillus pusillus (É. Geoffroy St.-Hilaire, 1803)
Neotropical Pygmy Squirrel
SYNONYMS:

Sciurus pusillus É. Geoffroy St.-Hilaire, 1803:177; type locality "Cayenne," French Guiana.

Sciurus guianensis Goldfuss, 1809:122; based on Buffon's (1789:261, plate 66) "*le petit guerlinguet*."

Sciurus olivascens Illiger, 1815:69; *nomen nudum*.

Sciurus pusillus Desmarest, 1817c:109; renaming of *Sciurus pusillus* É. Geoffroy St.-Hilaire, 1803; based on Buffon's (1789:261, plate 66) "*le petit guerlinguet*."

Sciurus olivascens Olfers, 1818:208; part; not *olivascens* Illiger, 1815, *fide* Hershkovitz, 1959b:345.

Macroxus pusillus: Lesson, 1842:111; name combination.

[*Funambulus Pucheranii*] *pusillus*: Fitzinger, 1867a:487; part; name combination.

Macroxus kuhlii Gray, 1867:433; type locality "Brazil," restricted by Thomas (1928c) to Pebas, Loreto, Peru.

Sciurus leucotis Gray, 1867:433; unavailable name published as a junior synonym and attributed to the French naturalist François de Caumont LaPorte, Comte de Castelnau (see Hershkovitz 1959b:345).

Sciurus (Macroxus) pusillus: Liais, 1872:503; name combination.

Sciurus aestuans var. *aestuans*: J. A. Allen, 1877:756, part; not *Guerlinguetus aestuans* (Linnaeus, 1766).

Macroxus aestuans: E.-L. Trouessart, 1897:428, part (inclusion of *pusillus* and *kuhlii* as synonyms); not *Guerlinguetus aestuans* (Linnaeus, 1766).

Sciurus (Microsciurus) pusillus: E. W. Nelson, 1899b:32; name combination.

Sciurus (Microsciurus) kuhli: E. W. Nelson, 1899b:32; name combination and incorrect subsequent spelling of *Macroxus kuhlii* Gray.

Microsciurus pusillus: E.-L. Trouessart, 1904:329; name combination.

Microsciurus kuhli: E.-L. Trouessart, 1904:329; name combination and incorrect subsequent spelling of *Macroxus kuhlii* Gray.

Microsciurus kuhlii: J. A. Allen, 1914a:162; name combination.

[*Sciurillus pusillus*] *kuhli* Thomas, 1914b:575; incorrect subsequent spelling or invalid emendation of *Macroxus kuhlii* Gray.

Sciurillus pusillus: Thomas, 1914b:575; first use of current name combination.

Sciurillus pusillus glaucinus Thomas, 1914b:575, type locality "Great Falls of Demerara River, British Guiana," = Ororo Marali, Upper Demerara–Berbice, Guyana.

Sciurillus pusillus pusillus: J. A. Allen, 1915a:197; name combination.

S[*ciurillus*]. *kuhlii*: Thomas, 1928c:291; name combination.

Sciurus pusilus Olalla, 1935:426; incorrect subsequent spelling of *Sciurus pusillus* É. Geoffroy St.-Hilaire.

Sciurillus pusillus kuhlii: Cabrera and Yepes, 1940:192; name combination.

Sciurillus pusillus hoehnei Miranda-Ribeiro, 1941:10; type locality "Rio Teles Pires (São Manoel), northern Mato Grosso," Brazil (page number incorrectly given as 139 by Hershkovitz 1959b:346).

Sciurus aestuans: Rode, 1943:385; part; not *Sciurus aestuans* Linnaeus.

Microsciurus pusillus hoehnei: C. O. da C. Vieira, 1955: 410; name combination.

Sciurillus guajanensis: Hershkovitz, 1959b:345; not *guajanensis* of Gmelin, 1788, nor *guajanensis* of Kerr, 1792.

Sciurillus pusilus Moore, 1959:204; incorrect subsequent spelling of *Sciurus pusillus* É. Geoffroy St.-Hilaire.

DESCRIPTION: As for the genus.

DISTRIBUTION: *Sciurillus pusillus* is known from scattered localities in Guyana, Surinam, French Guiana, Amazonian Brazil in Amapá, Amazonas, Pará, and Mato Grosso states, and northeastern Amazonian Peru. Cuervo Díaz et al. (1986) and Rodríguez-Mahecha et al. (1995) listed this species in their faunal compilations of Colombia but without locality or voucher specimen citations; Alberico et al. (2000) listed *S. pusillus* from the department of Caquetá in southeastern Colombia without providing an exact locality but referring to specimens in the IAvH and ICN in Bogotá. Jessen, Gwinn, and Koprowski (2013) included southeastern Colombia and Venezuela south of the Río Orinoco in the mapped range, without either literature or specimen documentation. The species has not been recorded from the fauna of either Ecuador (Tirira 2007) or Venezuela (O. J. Linares 1998).

SELECTED LOCALITIES (Map 2): BRAZIL: Amapá, Amapá (MPEG 1541), Oiapoque, upper Rio Oiapoque (MZUSP 20353); Amazonas, Andira, near Villa Bella da Imperatriz (AMNH 93154), Estirão do Equador, Rio Javari (MPEG 1557); Mato Grosso, São Manoel, Rio Teles Pires (type locality of *Sciurillus pusillus hoehnei* Miranda-Ribeiro); Pará, Boim, Rio Tapajós (MCZ 30216), Igarapé Amorim (AMNH 95727), Igarapé Brabo (= Bravo), left bank Rio Tapajós (AMNH 94743). FRENCH GUIANA: Cayenne (type locality of *Sciurus pusillus* É. Geoffroy St.-Hilaire), Inini, Arataye River (USNM 548447), Paracou (Voss et al. 2001), Tamanoir, Mana River (FMNH 21788). GUYANA: Upper Demerara–Berbice, Great Falls of Demerara River (type locality of *Sciurillus pusillus glaucinus* Thomas). PERU: Loreto, Estación Biológica Quebrada Blanco (Heymann and Knogge 1997), Pebas (BM 28.7.21.79), Quebrada Orán (M. S. Hafner et al. 1994), Santa Cecília, Río Maniti (FMNH 87182), Sarayacu, Río Ucayali (AMNH 76185). SURINAM: Sipaliwini, Emmaketen [= Emma Keten] (Husson 1978), Frederik Willem IV Falls, Corentyne River (FMNH 48419).

SUBSPECIES: Two (J. A. Allen 1915a) or three (Anthony and Tate 1935; Cabrera 1961; Thorington and Hoffman 2005) subspecies have generally been recognized. These include the nominotypical form (the Guianas and northeastern Brazil), *glaucinus* Thomas (the Amazon Basin of central Brazil and interior Guyana), and *kuhlii* Gray (northeastern Peru and probably western Brazil and southwestern Colombia).

Anthony and Tate (1935), who studied specimens from French Guiana, Peru, and the region of the Rio Tapajós, Brazil, believed that three subspecies could be distinguished on the basis of color, having found no significant differences in cranial measurements, specifically: *S. p. pusillus*—ears black; head decidedly reddish when compared with the grayish dorsum; *S. p. glaucinus*—ears the same color as the head, which was not very distinct in color from the rest of the grayish dorsum; and *S. p. kuhlii*—like *glaucinus* but more "saturated," with "belly even more cinnamon" (p. 8) than in *pusillus* and with a darker back. However, our microscopic examination of the pinnae revealed that the epithelium is pigmented (thus appearing black on dried skins) in specimens from the entire range of the genus. Moreover, the color of the head is variable in the series from the Rio Tapajós, as is the intensity of the orange mixed with gray in the belly. The number of specimens we have examined from Peru, the putative region of occurrence of *kuhlii*, is small, but a good series from Estirão do Equador, Rio Javari, Amazonas, which is geographically close to the Peruvian localities, cannot be distinguished from the larger series from the Rio Tapajós, and which has been ascribed to *glaucinus*. We have not seen the holotype of *Sciurillus pusillus hoehnei* from northern Mato Grosso, but nothing in the description supports it as a plausible valid taxon.

Mercer and Roth (2003) reported substantial sequence divergence in both mitochondrial and nuclear genes between two specimens from opposite ends of the geographic range of this species (USNM 578014 from French Guiana, and LSUMZ 27994 from Loreto department, Peru), positing a late Miocene divergence between these samples. Despite the extensive level of divergence and apparent deep age of the species, we are unable to find significant, geographically partitioned variation among the few samples available from this very large area; we provisionally regard *Sciurillus pusillus* as a monotypic species.

NATURAL HISTORY: Jessen, Gwinn, and Koprowski (2013) synthesized the few data available on ecology, behavior, and life history. Emmons and Feer (1990, 1997), Ball and Roth (1995), Roth (1996), Heymann and Knogge (1997), Voss et al. (2001:76), and Youlatos (2011) detailed observations of solitary individuals in trees in well-drained primary forest, primarily in *Inga* (French Guiana) and *Parkia* (Peru), both in the subfamily Mimosoideae. In both geographic areas, animals were observed gnawing on the bark, probably feeding either on cambium or on exudates (sap or gum), incising pieces of the upper bark layer, and leaving the trunk densely pockmarked. Heymann and Knogge (1997) recorded large amounts of bark chips that accumulated on the ground under one such tree. Emmons and Feer (1990:176) noted that individuals forage on trunks from near the ground to high in the canopy, "hopping and flitting from branch to branch," and traveling from tree to tree high in the canopy.

Their small size makes individuals difficult to spot, but they are not wary of humans and can be easily observed. Youlatos (2011) observed five individuals using wide vertical substrates in the canopy, with locomotion primarily by claw climbing. Velhagen and Roth (1997) suggested that a decrease in bite force with decreasing size might account for a shift in diet of the pygmy squirrels, including *Sciurillus*, from hard fruits and nuts to gleaning bark. Because most sightings were of solitary individuals, Heymann and

Knogge (1997) also suggested that a solitary lifestyle was the dominant mode of sociospatial organization. Peres et al. (2003) reported four sightings, each averaging 1.5 individuals, in the headwaters of the Rio Arapiuns, Pará state, Brazil, and estimated their density at 0.023 individuals/ha. Dalecky et al. (2002) stated that home range size in French Guiana was small, which facilitated individual persistence in habitat islands created by flooding. Olalla (1935), Heymann and Knogge (1997), and Voss et al. (2001) observed this species only in noninundated or *terra firme* forest, and Olalla (1935) reported a litter size of two.

REMARKS: Due to the scarcity of specimens of *S. pusillus* from the time of its discovery in the late eighteenth century until the 1930s, the taxonomic history of this taxon has presented certain persistent problems. Most of the trouble relates to confusion between *Sciurillus pusillus* (or some junior synonym) and several other South American squirrels, usually *Sciurus aestuans* Linnaeus (herein *Guerlinguetus aestuans*). Because both of these species were first known from Guianan specimens, their respective taxonomic histories are entangled. We comment on all names, available or not, that have been associated with *Sciurillus pusillus*; however, new combinations and changes in specific versus subspecific status are not discussed because they can be easily followed in the synonymy provided earlier.

Sciurillus pusillus was the second squirrel to be known from the Guianan region, the first being *Sciurus aestuans*, described by Linnaeus in 1766. Ten years later, Buffon (1776:146–147) published his first report on a Guianan squirrel. The essential parts of Buffon's report are translated as: "M. de la Borde, Physician of the King in Cayenne, *said* that there is *only one species* of squirrel in the Guyanne, that it lives in the forests, that *its fur is reddish*, that *it is not larger than the European rat* . . . I am not quite certain that this animal from the Guyanne, of which M. de la Borde *speaks*, is a true squirrel, because the latter are hardly found in excessively hot climates, such as that of the Guyanne" (emphases ours). This is a very uninformative account, but it is clear that Buffon did not have specimens to describe in 1776, only an oral or written account by de la Borde. It is also clear that Buffon believed that there was a single species of "squirrel-like" mammal in the Guiana region, and that it was not larger than the European *Rattus*. Such reference to size is vague, but it can be interpreted to mean an animal about the size of a Norway or black rat; because *Sciurillus* is so much smaller, we believe this feature would not go unnoticed even by an amateur such as de la Borde. Also, *Sciurillus pusillus* can hardly be described as having "reddish" fur, a character that could roughly be applied to *Sciurus aestuans*. Therefore, it is plausible that the animal reported by de la Borde was *Sciurus aestuans*, by far the most common species found in the Guianas.

Later, Buffon (1789) was finally provided with specimens from French Guiana, which had been collected by M. Sonini de Manoncourt, and he was able to give firsthand information on the entire squirrel fauna of the region. The essential aspects of this later contribution are, first, that Buffon (1789:261–264, plates 45 and 46) believed the animals were not real squirrels (e.g., the European "écureuil" or *Sciurus vulgaris*), confirming his earlier prediction, and he called them "*guerlinguets*"; and, second, that he considered them two species, "*le grand*" and "*le petit guerlinguet*." He then proceeded to reinterpret his own 1776 account of one species to accommodate the two he now had in front of him. Thus, the "*petit guerlinguet*" became the species "not larger than a rat," and the "*grand*" one was compared in size to *Sciurus vulgaris*. Obviously, anything the size of *Sciurillus* is "not larger than a rat," but then neither are its fleas. Buffon's awkward adaptation of his earlier, vague size reference to the actual specimens collected by Sonini de Manoncourt allowed for some misinterpretation, mainly because other naturalists were also busy reporting and cataloging the Guianan sciurid fauna. It is now important to devote attention to a number of publications dealing with Guianan squirrels, which appeared before É. Geoffroy St.-Hilaire (1803) coined the name *Sciurus pusillus* for the "*petit guerlinguet*" of Buffon, because Hershkovitz (1959b) considered some of these names to have priority over *pusillus*.

The first of these is *Sciurus gujanensis*, coined by Gmelin (1788:152) for two squirrels from different reports concerning the Guianan region. One of these reports is that by de la Borde, which we discussed earlier and associated with *Sciurus aestuans* Linnaeus. The other report commented on by Gmelin was Bancroft's (1769:143), taken from his account of the natural history of Guiana, where he states that the animal was close to the common English squirrel in size and morphology, and that it had a pale yellowish brown dorsum, white under parts, small white streaks on the sides of the body, and a tail similar in color to the dorsum but "variegated" with white and black. The presence of lateral white streaks does not fit with the color pattern of any known South American squirrel, and *Sciurus aestuans* does not have white under parts. Whichever squirrel Bancroft had in mind, however, both his description of the color pattern and size completely eliminate the possibility that his animal was *Sciurillus pusillus*.

Hershkovitz (1959b:345) ignored Gmelin's name and its complete lack of correspondence with *Sciurillus pusillus*. He proposed that the valid name for the dwarf squirrel should be *Sciurillus guajanensis* (Kerr) and then organized a synonymy with that in mind. However, Kerr (1792:265)

both acknowledged that his usage derived from Gmelin and decided to keep the name *Sciurus guajanensis* (his different spelling of the name either is an unjustified emendation or a *lapsus calami*) for de la Borde's animal, while creating the new name *Sciurus bancrofti* for the other "species." As documented earlier, these two names have nothing to do with *Sciurillus pusillus* and do not belong in its synonymy. Hershkovitz (1959b:345) also stated that Turton (1802:94) employed the name *Sciurus guajanensis* based on the "*petit guerlinguet*" of Buffon. There is, however, no mention in Turton's work of the French naturalist or of the "*petit guerlinguet*." Indeed, Turton's *General System* is only an updating of Kerr's, which is entirely referable to Gmelin (1788).

Finally, Hershkovitz (1959b:345) cited yet another author (Goldfuss 1809:122) who used the name *Sciurus guianensis* for the "*petit guerlinguet*," a publication that can be viewed in the same perspective as those by Kerr and Turton. However, as neither of us has personally seen the work by Goldfuss, we herein treat his *Sciurus guianensis* as a possible synonym of *Sciurillus pusillus*.

The first scientific name unequivocally associated with "*le petit guerlinguet*" is *Myoxus guerlingus* Var. (?), coined by Shaw (1801:171), which actually encompasses both Guianan squirrels, *Sciurus aestuans* Linnaeus and *Sciurillus pusillus* Thomas. Because we fix the type specimen of *Myoxus guerlingus* as the same as that of *Sciurus aestuans* (*guerlingus* had no type, being a name for both of Buffon's guerlinguets; see the account for *Guerlinguetus aestuans*), Shaw's name cannot be employed as the valid one for "*le petit guerlinguet*" and is not listed in the synonymy of *S. pusillus* as a result.

É. Geoffroy St.-Hilaire (1803:177) created the name *Sciurus pusillus* exclusively for "*le petit guerlinguet*," making it the valid one for the species. He stated, however, that de la Borde had collected the specimen he studied. This made Husson (1978:382) question whether Geoffroy St.-Hilaire's specimen was the same as that seen by Buffon, who obtained his material from Sonini de Manoncourt. The holotype of *Sciurus pusillus* É. Geoffroy St.-Hilaire in the Paris museum (MNHN 312, a mounted skin), however, has an old label bearing de la Borde's name attached to its support. In our opinion, this is probably the same specimen seen by Buffon (1789), and de la Borde's name is on the label because Buffon believed it represented the "*guerlinguet*" he reported. Husson (1978:382) never questioned that the holotype is really what is currently called *Sciurillus pusillus*, in spite of its somewhat deteriorated condition, but he felt the need to select a lectotype, which he indicated (1978:383) to be the same specimen described and figured by Buffon in 1789 (p. 261, plate 46). Husson's act, however, was unnecessary, as the specimen (MNHN 312) is the holotype of *Sciurus pusillus* É. Geoffroy St.-Hilaire, as the name is applied today, and is characterized by the small and rounded pinnae and the small size of the feet. It is not the young of a *Sciurus aestuans* as proposed by Rode (1943:385).

Another early name resurrected by Hershkovitz (1959b:345) was *Sciurus olivascens*, placed in the synonymy of *Sciurillus guajanensis* (Kerr) [= *Sciurillus pusillus*], which he attributed to Olfers, 1818. According to Hershkovitz (1959b:345), Olfers erected the name after a manuscript usage by Illiger, purportedly based on "*le petit Guerlinguet*" of Buffon, but of this Olfers himself was not completely sure because he added a question mark after the French vulgar name. Illiger (1815:69) published the name *Sciurus olivascens* in a list of mammals from South America without diagnosis or description, and he made no subsequent reference to this name in his brief discussion of South American squirrels (pp. 74–75). It thus seems clear that *olivascens* is unavailable from Illiger (1815) in the meaning of the International Code of Zoological Nomenclature (ICZN 1999). The same name was again mentioned in yet another list of South American mammals (Minding 1829:89), also unavailable for the same reasons. Given that Olfers (1818:208) only questionably associated "*Sc. olivascens* Ill." to Buffon's "*le petit guerlinguet*," and that both Illiger's (1815) original usage of this name and the subsequent usage by Minding (1829) are *nomina nuda*, it seems to us that the name continues to be unavailable. But even if Olfers's usage were judged available, *olivascens* would remain a junior synonym of *Sciurus pusillus*, being antedated by E. Geoffroy St.-Hilaire's name by 15 years.

For some decades, no new specimens other than the one(s) in the Paris museum were reported until Gray (1867:433) listed and described *Macroxus pusillus* and *Macroxus kuhlii*, the latter a new species said to have been collected by Castelnau in "Brazil." The name *Sciurus leucotis* was also given by Gray (1867:433) in the synonymy of *kuhlii* and attributed to Castelnau (manuscript); *leucotis*, having never been employed as valid, is not available under the International Code of Zoological Nomenclature (ICZN 1999). Gray's (1867:433) descriptions are not good, but one of us (MdV) has seen the holotype of *Macroxus kuhlii* (BM 51.7.3.7, skin and skull); the skin is in bad shape, with extensive loss of hair in the region immediately posterior to the crown, legs, tail, throat, and chest, and with small losses in other scattered areas. Nevertheless, what is left of the animal is very similar to the specimen from Cayenne (BM 45.5.1.3) that probably served as basis for Gray's account of *Macroxus pusillus*. Thomas (1920f:276) assigned two specimens from Villa Braga on the Rio Tapajós, Pará, Brazil, to *S. pusillus* but noted that these were "quite like" the type of *Macroxus kuhlii* and suggested that the latter likely

came from the same region in Brazil. He later restricted the type locality of *kuhlii* Gray to Pebas, Loreto, Peru (Thomas 1928c:291). Because we consider *Sciurillus pusillus* to be a monotypic species, the matter of the correct type locality of *Macroxus kuhlii*, a junior synonym, is not critical. However, because *kuhlii* has been considered by Anthony and Tate (1935) to be a valid subspecies (and may even be a species separate from *pusillus*, as suggested by the deep molecular divergence of western and eastern Amazonian samples; Mercer and Ross 2003), validation of Thomas's restriction of the type locality to Peru is warranted.

J. A. Allen (1877:756–757), without examining specimens of the dwarf squirrel, placed both *pusillus* and *kuhlii* as junior synonyms of his excessively broad *Sciurus aestuans* var. *aestuans*, believing they could be young specimens of the larger animal. Alston (1878:670) immediately corrected Allen, considering *Sciurus pusillus* a valid species with *Macroxus kuhlii* as a junior synonym. Most subsequent authors were convinced of the validity of *pusillus* but placed the species in the genus or subgenus *Microsciurus* (e.g., E. W. Nelson 1899b; E.-L. Trouessart 1904; J. A. Allen 1914a). Any remaining doubts regarding the distinctiveness of *pusillus* were finally dispelled when Thomas (1914b,f) erected the genus *Sciurillus*, completely divorcing it from the larger squirrels of the genus *Microsciurus*.

Thomas (1914b:575) described *Sciurillus pusillus glaucinus* from Guyana (former British Guiana), thus adding a third species group taxon to the picture. Anthony and Tate's (1935) review of *Sciurillus* took advantage of the first relatively large series of specimens available at any museum; they recognized three subspecies, namely, *Sciurillus p. pusillus*, *S. p. glaucinus*, and *S. p. kuhlii*. They restricted the

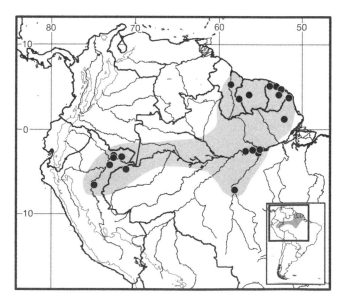

Map 2 Selected localities for *Sciurillus pusillus* (●). Contour line = 2,000 m.

nominotypical subspecies to French Guiana, *glaucinus* to British Guiana (= Guyana) south to the lower Rio Tapajós, Pará, Brazil, and *kuhlii* to northeastern Amazonian Peru. Another subspecies, *Sciurillus pusillus hoehnei*, was posthumously published by Miranda-Ribeiro (1941), from northern Mato Grosso, Brazil. For reasons discussed earlier, we consider *Sciurillus pusillus* to be a monotypic species, with *kuhlii*, *glaucinus*, and *hoehnei* as junior synonyms.

Subfamily Sciurinae G. Fischer, 1817

Species in the subfamily Sciurinae are the "typical" tree squirrels, with members distributed on all continents save Antarctica and Australia. South American representatives include taxa allocated by Moore (1959) to his subtribes Sciurina and Microsciurina. Both groups are now collectively placed in the Sciurinae (Thorington and Hoffmann 2005). Species in some genera (e.g., *Hadrosciurus*, *Microsciurus*, *Notosciurus*) are characterized by an ear patch colored differently than the top of the head, neck, and shoulders; typically, this patch is continuous from the distal part of the medial surface of the pinna to just behind the ear (herein referred to as a postauricular patch). In a few cases, the colored hairs of the patch protrude above the rim of the ear. In species of *Simosciurus*, some hairs may extend from the dorsal edge of the pinnae, but none of these resemble a tuft, such as that characterizing the European *Sciurus vulgaris* (see Thorington et al. 2012).

Phylogenetic relationships among the genera and species we recognize are very incompletely known, as few of these taxa have been included in any of the surprisingly small number of molecular phylogenetic studies to date. The most comprehensive analysis of sciurid phylogeny (Mercer and Roth 2003) included only three South American taxa of "*Sciurus*" and *Microsciurus*. Assuming an entry into South America at or subsequent to the closure of the Panamanian seaway some 3–4 mya, the net rate of diversification of tree squirrels was exceedingly rapid, more than twice that for any other continental assemblage (Roth and Mercer 2008).

KEY TO THE GENERA OF SOUTH AMERICAN SCIURINAE:

1. Mean head and body length of adults <160 mm; orbitotemporal fossa of skull, in dorsal view, unevenly divided by the postorbital process of the frontals, resulting in nearly obliterated temporal fossa; generally with two upper premolars, the anteriormost very small. . . . . *Microsciurus*
1'. Mean head and body length of adults >160 mm; orbitotemporal fossa of skull, in dorsal view, evenly divided by the postorbital process of the frontals, resulting in nearly equal-sized orbital and temporal fossa; always with only one upper premolar 2
2. Mean head and body length of adults between 160 and 180 mm; ears small and rounded . . . . . *Syntheosciurus*
2'. Mean head and body length of adults >160 mm; ears relatively large and pointed. 3
3. Adult females with three pairs of mammae; distinctive postauricular patch of different color than that of top of head usually present . . . . *Notosciurus*
3'. Adult females with four pair of mammae; postauricular patch may or may not be present 4
4. Pelage coarse; distribution west of the Andes in Ecuador and Peru; pinnae hirsute, with somewhat long hairs, but not quite forming a tuft . . . . *Simosciurus*
4'. Pelage normal to soft; distribution east of the Andes; pinnae hirsute, covered with very short hairs 5
5. Size medium, mean head and body length between 175 and 180 mm . . . . *Guerlinguetus*
5'. Size large, mean head and body length >215 mm . . . . *Hadrosciurus*

Genus *Guerlinguetus* Gray, 1821

Guerlinguetus is a tropical tree squirrel genus occurring from eastern Colombia, Venezuela, and the Guianas, south throughout the lowland Amazonian rainforest, then through the Atlantic Forest of eastern and southern Brazil to northeastern Argentina. It is not known to occur in neighboring Paraguay and Bolivia, where the genus is replaced by the morphologically very similar *Notosciurus*. Although widespread in Amazonian Brazil, *Guerlinguetus* is not known from south of the Rio Solimões, between the Rio Javari and Rio Madeira. It is absent from all but the broad gallery forests at the periphery of the Cerrado biome in central Brazil and mesic forest enclaves in the Caatinga of eastern Brazil (Carmignotto et al. 2012). Species of *Guerlinguetus*, as is true of all other South American squirrels, are exclusive forest dwellers. The genus is mostly parapatric with *Notosciurus*, except in the general region of the Río Marañón in Peru.

Guerlinguetus is easily distinguished from other South American squirrels by its size (larger than *Microsciurus*, *Sciurillus*, and *Syntheosciurus*, but smaller than *Hadrosciurus* and *Simosciurus*). In comparison with *Notosciurus*, which has an approximately similar size range, *Guerlinguetus* is unique in the possession of four pairs of mammae instead of three and typically by the absence of a differentially colored postauricular patch. Cranially, the genus can be distinguished by an orbitotemporal fossa that is equally divided by a well-developed postorbital process and an upper second molar with two parallel, buccolingual crests perpendicular to the anteroposterior axis of the skull.

We recognize only two species in *Guerlinguetus*: *G. aestuans*, a Guianan and Amazonian species, and *G. brasiliensis*, from eastern Amazonia to eastern and southern Brazil and northern Argentina. Although there are more publications on aspects of the natural history of these species than for any other South American squirrels, details of ecology, behavior, and other aspects of population biology remain unknown. *Guerlinguetus* appears to rely heavily on small palm fruits, using several palm genera through its distribution. Several regional Brazilian vernacular names are used for these squirrels, without distinction as to species, including *esquilo*, *serelepe*, *caxinguelê*, and *quatipuru*, the last used exclusively in the Brazilian Amazon. In Argentina, where only *G. brasiliensis* occurs, the local name is *ardilla gris* or *ardilla misionera*.

SYNONYMS:

Sciurus: Brisson, 1762:107; part (inclusion of *brasiliensis*); not *Sciurus* Linnaeus, 1758; rejected for nomenclatural purposes (ICZN 1998:Opinion 1894).

Myoxus: Shaw, 1801:171; part (included *"le grand guerlinguet"* of Buffon as species); not *Myoxus* Zimmermann, 1780.

Guerlinguetus Gray, 1821:304; type species *Sciurus guerlinguetus*, by tautonomy (*"le grand guerlinguet"* of Buffon, 1789 = *Myoxus Guerlingus* Shaw, 1801 = *Sciurus guerlinguetus* Gray, 1821 = *Sciurus aestuans* Linnaeus, 1766. Gray [1821] writes "Gerlinguet, Guerlinguetus, sciurus guerlinguetus. Lin.").

Macroxus F. Cuvier, 1823c:161; type species *Sciurus aestuans* Linnaeus designated by Thomas (1898g:933).

[*Funambulus*]: Fitzinger, 1867a:486; part (listing of *aestuans*); not *Funambulus* Lesson, 1835.

[*Sciurus*] (*Guerlinguetus*): Ellerman, 1940:322, 340; as subgenus.

REMARKS: The first available generic name for this group of medium-sized South American squirrels is *Guerlinguetus* Gray, 1821, with type species *Sciurus guerlinguetus*. Gray attributed *S. guerlinguetus* to Linnaeus, a name that is actually a *nomen nudum* because Gray's usage was not followed by any description or plate. However, Gray's generic name *Guerlinguetus* is clearly identifiable to the French vernacular *"guerlinguet,"* which Buffon (1789:261, plates 65, 66) used for the squirrels of the Guianas, for which he recognized two varieties, a larger *"le grand guerlinguet"* [= *Guerlinguetus aestuans* (Linnaeus)], and a smaller *"le petit guerlinguet"* [= *Sciurillus pusillus* (E. Geoffroy St. Hilaire)]. Since Gray (1821), the name *Guerlinguetus* has been either ignored or variously employed for a full genus or subgenus, but at either categorical level it broadly encompassed most medium-sized South American squirrels, some now placed in other genera. J. A. Allen (1915a) was the first to restrict the con-

cept of *Guerlinguetus*, a view that is very close to ours herein.

KEY TO THE SPECIES OF *GUERLINGUETUS*:

1. Ventral coloration orange, or, if mostly orange, only the throat has white or grayish-white hairs; tail color lined with orange . . . . *Guerlinguetus aestuans*
1′. Ventral coloration heavily grizzled with white or grayish-white, either along the entire venter or else only in throat, chest, and inguinal areas; tail color either lined with orange or grayish white . . . . *Guerlinguetus brasiliensis*

Guerlinguetus aestuans (Linnaeus, 1766)

Guianan Squirrel, Quatipuru

SYNONYMS:

[*Sciurus*] *aestuans* Linnaeus, 1766:88; type locality "Surinam."

Sciurus bancrofti Kerr, 1792:265; type locality "Guiana."

Sciurus guajanensis Kerr, 1792:265; type locality "Cayenne."

Myoxus Guerlingus Shaw, 1801:171; type locality "Guiana"; based on Buffon's (1789:261, plate 65) *"le grand guerlinguet."*

Sciurus brasiliensis: Lichtenstein, 1818:216; part; remainder under *Guerlinguetus brasiliensis*.

Sc[*iurus*]. *guianensis* Lichtenstein, 1818:216; type locality not given; name proposed as a synonym of *S. brasiliensis* and based on Bancroft's report of a squirrel from (British) Guiana; name became available when Peters (1864) recognized it as a valid taxon.

Sciurus guerlinguetus Gray, 1821:304; *nomen nudum*; name attributed to Linnaeus.

aesciurus stuans: F. Cuvier, 1823c:255; typographic error.

Macroxus aestuans: Lesson, 1827:238; name combination.

Sc[*iurus*]. *gilvigularis* Wagner, 1843c:43; *nomen nudum*.

Sciurus gilvigularis Wagner, 1845a:148; type locality "nördlichen Brasilien"; name attributed to Johann Natterer (Wagner 1850:283).

S[*ciurus*]. *aestuans*: Cabanis, 1848:778; name combination.

Sciurus aestuans var. *Guianensis* Peters, 1864:655; type locality "Guiana" = Guyana.

[*Funambulus*] *gilvigularis*: Fitzinger, 1867a:486; name combination.

Sciurus gilviventris Fitzinger, 1867a:486; incorrect subsequent spelling of *Sciurus gilvigularis* Wagner; unavailable name cited under [*Funambulus*] *gilvigularis*.

[*Funambulus*] *aestuans*: Fitzinger, 1867a:486; part; name combination.

Macroxus aestuans Var. 1: Gray, 1867:432; name combination.

[*S. aestuans*] var. *aestuans*: J. A. Allen, 1877:672; part; name combination.

Sciurus gilviventris Pelzeln, 1883:59; type locality "Borba," Amazonas, Brazil (see Goeldi 1893:78).

[*Macroxus*] *gilvigularis*: E.-L. Trouessart, 1897:429; name combination.

Sciurus Quelchii Thomas, 1901e:147; type locality "Kanuku Mountains, British Guiana," Upper Takutu–Upper Essequibo, Guyana.

Sciurus Macconnelli Thomas, 1901e:148; type locality "Mt. Roraima, near its base," Cuyuni–Mazaruni, Guyana.

[*Guerlinguetus*] *gilvigularis*: E.-L. Trouessart, 1904:328; name combination.

Sciurus aestuans gilvigularis: J. A. Allen, 1910c:146; name combination.

Sciurus (Guerlinguetus) aestuans gilvigularis: J. A. Allen, 1911:244; name combination.

G[*uerlinguetus*]. *aestuans*: Cabrera, 1912:89; part (also included species of *Notosciurus*); first use of current name combination.

Guerlinguetus aestuans aestuans: J. A. Allen, 1915a:256; name combination.

Guerlinguetus aestuans gilvigularis: J. A. Allen, 1915a:256; name combination.

Guerlinguetus aestuans macconnelli: J. A. Allen, 1915a:259; name combination.

Guerlinguetus aestuans quelchii: J. A. Allen, 1915a:259; name combination.

Guerlinguetus aestuans venustus J. A. Allen, 1915a:260; type locality "Boca Sina (altitude 440 ft.), Rio Cunacunumá (southern base of Mount Duida), [Amazonas,] Venezuela."

Guerlinguetus Aestuans Gilvicularis O. M. O. Pinto, 1931:290; incorrect subsequent spelling of *Sciurus gilvigularis* Wagner.

Sciurus [(*Guerlinguetus*)] *aestuans aestuans*: Ellerman, 1940:342; name combination.

Sciurus [(*Guerlinguetus*)] *aestuans quelchi*: Ellerman, 1940:342; name combination, and incorrect subsequent spelling of *Sciurus quelchii* Thomas.

Sciurus [(*Guerlinguetus*)] *aestuans venustus*: Ellerman, 1940: 342; name combination.

Guerlinguetus gilvigularis gilvigularis: Moojen, 1942:6; name combination.

Guerlinguetus poaiae Moojen, 1942:11; type locality "Mata da Poaia, Tapirapoã, Chapada, Mato-Grosso," Brazil.

G[*uerlinguetus*]. *gilvigularis gilvigularis*: Melo-Leitão, 1943: 356; name combination.

Sciurus (Guerlinguetus) poaiae: W. P. Harris, 1944:11; name combination.

Sciurus gilvigularis gilvigularis: Moojen, 1952b:25; name combination.

Sciurus poaiae: Moojen, 1952b:26; name combination.

Sciurus aestuans quelchii: Moojen, 1952b:27; name combination.

Sciurus (Guerlinguetus) gilvigularis gilvigularis: C. O. da C. Vieira, 1955:407; name combination.

Sciurus [(*Guerlinguetus*)] *aestuans poaiae*: Cabrera, 1961:361; name combination.

Sciurus [(*Guerlinguetus*)] *aestuans quelchii*: Cabrera, 1961:361; name combination.

Sciurus (Guerlinguetus) aestuans georgihernandezi Barriga-Bonilla, 1966:492; type locality "Margen derecha del río Vaupés, cerca de Mitú, Vaupés, Colombia, altura 240 m.s.n.m."

Sciurus (Guerlinguetus) aestuans: Eisenberg, 1989:331; name combination.

Sciurus (Guerlinguetus) gilvigularis: Eisenberg, 1989:331; name combination.

DESCRIPTION: Adult body weight averages 174 g (N=28), head and body length 176 mm (N=235), tail length 177 mm (N=205), hindfoot length 47 mm (N=245), and ear length 22 mm (N=83). Dorsal color uniformly brown to olive brown, heavily streaked with orange throughout, including the upper parts of the legs; ventral color highly variable, with throat orange to grayish white and remaining parts orange, sometimes with dark gray hairs interspersed; inner parts of forelimbs colored similarly to chest and belly; inner parts of hindlimbs darker than belly; area immediately circumscribing mammae not distinct from remainder of venter, and tail color indistinct from dorsum. Postauricular patches absent, except for some specimens from Guianan highlands.

DISTRIBUTION: *Guerlinguetus aestuans* occurs throughout Amazonia, in northeastern Peru, eastern Colombia, Venezuela south of the Río Orinoco, and Brazil north of the Rio Amazonas and west of the Rio Xingu, and throughout the Guianan region.

SELECTED LOCALITIES (Map 3): BRAZIL: Amapá, Cachoeira de Santo Antônio, Rio Jari (MPEG 21804), Mazagão, Rio Vila Nova (MPEG 438), Vila Velha do Cassiporé, Oiapoque (MPEG 6742); Amazonas, Igarapé Auará, Rio Madeira (AMNH 91831), Santo Isidoro, Tefé (AMNH 78949); Mato Grosso, Cachoeira Dardanelos, right bank Rio Aripuanã (MPEG 8786), Estação Ecológica Serra das Araras (MPEG 15272); Pará, Aramapucu, Rio Parú do Oeste (MPEG 10019), Cussary (AMNH 37453), Foz do Rio Curuá, right bank Rio Amazonas (MZUSP 4601), Rio Jamanxim, right bank Rio Tapajós (MPEG 453); Rondônia, Cachoeira Nazaré, left bank Rio Ji-Paraná (MPEG 20825), Ouro Preto d'Oeste (MPEG 20529). COLOMBIA: Vaupés, La Providencia, Río Apaporis (Barriga-Bonilla 1966), right bank Río Vaupés, near Mitú (Barriga-Bonilla 1966). FRENCH GUIANA: River Iracoubo (MNHN 1983/350), River Oyapock (MNHN 1983/346). GUYANA: Essequibo Islands–West Demerara, Supinaam River (BM 10.5.4.12); Upper Takutu–Upper Essequibo, Kanuku Mountains (type locality of *Sciurus Quelchii* Thomas). PERU: Loreto, Orosa, Río Amazonas (AMNH 73917), Sarayacu, Río Ucayali (AMNH 75277). SURINAM: Para, Zanderij (BM 52.1102). VENEZUELA: Amazonas, Monduapo, right bank upper Río Orinoco (BM 99.9.11.29–30); Bolívar, Ciudad Bolívar, Suapure, Río Caura valley (AMNH 16128), El Yagual, lower Río Caura (AMNH 16948).

SUBSPECIES: We recognize two subspecies, namely, *G. a. aestuans* (Linnaeus, 1766) and *G. a. gilvigularis* (Wagner, 1845). Distributions and coat color variation of each are discussed in the Remarks section.

NATURAL HISTORY: There are very few published field observations of the ecology or behavior of this species in the wild. It ranges throughout lowland rainforest and, marginally, into the dry forests of Mato Grosso state, Brazil. In French Guiana, and likely throughout this broad range, seed-caching behavior plays an important role in the demography of understory palms such as *Astrocaryum sciophilum* (see Sist 1989) and other large-seeded plants such as *Carapa procera* (Forget 1996). Forget (1991) reported that this species did not forage at ground level even when seeds were available. In French Guiana, this squirrel may persist on small and medium-sized islands produced by flooding after dam construction, which suggests that their home range size is relatively small (Dalecky et al. 2002). Peres et al. (2003), from a site in the headwaters of the Rio Arapiuns, Pará state, Brazil, estimated density of 0.444 individuals/ha, with sightings of seven solitary individuals. Voss et al. (2001) observed animals perched in trees at heights of 3–30 m above the ground, with most individuals solitary. Individuals these authors captured came primarily from primary forest at both well-drained and swampy sites as well as from somewhat disturbed habitats.

REMARKS: Use of the name *Sciurus aestuans* Linnaeus by almost all authors from the late eighteenth and early nineteenth centuries (e.g., Gmelin 1788:151; Turton 1802:93; Illiger 1815:75; Thunberg 1823:2; Minding 1829:29; Schinz 1840:144; P. Gervais 1841:37, 1854:308; Gray 1843b, 1867; Chenu 1850:34) to the present (e.g., Emmons and Feer 1997; Eisenberg and Redford 1999; Thorington and Hoffmann 2005) includes both *G. aestuans*, as we restrict this species herein, and *G. brasiliensis* (next section) or even species we allocate to *Notosciurus* (e.g., Alston 1878:668).

Geographic variation of coat color across the distribution of *G. aestuans* has led authors to name distinct taxa based on slight variations, all of which names we regard either as applying to subspecies or as synonyms. Here, we summarize the geographic variation of characters to justify our recognition of a single species. There

are no qualitative cranial distinctions along the entire distribution; in fact, the species is cranially indistinct from *G. brasiliensis*.

We recognize two subspecies. The nominotypical form occurs in the Guianas, adjacent Venezuela, Amazonian Brazil, in the states of Roraima, Amazonas (to the north of the Rio Solimões–Amazonas axis), Amapá, Pará (except on the south bank of the Rio Amazonas, to the east of the Rio Xingu, where the other species of *Guerlinguetus* occurs), in northern Peru from scattered localities along the Río Marañón, and in Colombia near its border with Brazil. By contrast, *G. a. gilvigularis* (Wagner, 1845) occurs in Brazil, in the Rio Tapajós and Rio Madeira basins, and from there southward to the states of Rondônia and western Mato Grosso. These two taxa can be distinguished by the color of their ventral surfaces, *aestuans* with a grayish white throat but orange or yellowish chest and belly, and *gilvigularis* with an entirely orange or yellowish venter. Specimens from the Rio Negro region in Amazonas, Brazil, exhibit a ventral color pattern characteristic of both subspecies, suggesting that this area is a zone of intergradation between *aestuans* and *gilvigularis*. Dorsally, all animals are very similarly colored.

One important aspect of morphological variation in the distribution of *G. a. aestuans* is that some animals collected in the highlands of the Guianas and Venezuela possess postauricular patches whereas in the adjacent lowlands such patches are either inconspicuous or absent altogether. The presence of a patch is associated with the names *guianensis* Peters, 1864, and *macconnelli* Thomas, 1901, both of which we regard as synonyms to *Guerlinguetus aestuans aestuans* (Linnaeus, 1766).

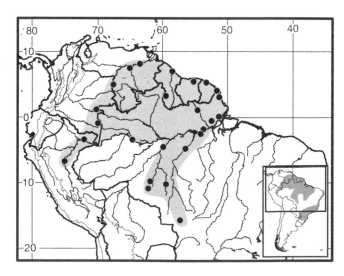

Map 3 Selected localities for *Guerlinguetus aestuans* (●). Contour line = 2,000 m.

Guerlinguetus brasiliensis (Gmelin, 1788)
Brazilian Squirrel, Caxinguelê, Serelepe, Esquilo
SYNONYMS:

[*Sciurus*] *brasiliensis* Gmelin, 1788:151; part; based on "*Sciurus brasiliensis*" of Marcgraf (1648), cited as synonym; type locality "Brasilia et Guiana," but restricted herein to Recife, Pernambuco, Brazil (see Remarks).

Sc[*iurus*]. *guianensis*: Lichtenstein, 1818:216; part; remainder under *G. aestuans*.

Sc[*iurus*]. *brasiliensis* Lesson, 1838:468; type locality "Brésil."

Sciurus Ingrami Thomas, 1901d:368; type locality "Tunnel, Southern Minas Gerais: alt. 1200 m," Brazil.

Sciurus Roberti Thomas, 1903d:463; type locality "S. Lourenço [= São Lourenço], near Pernambuco, Alt. 50 m," Pernumbuco, Brazil.

Sciurus aestuans var. *paraensis* Goeldi and Hagmann, 1904:70; type locality "Mates do Pará," Pará, Brazil.

[*Guerlinguetus*] *ingrami*: E.-L. Trouessart, 1904:328; name combination.

[*Guerlinguetus*] *roberti*: E.-L. Trouessart, 1904:328; name combination.

Sciurus Alphonsei Thomas, 1906c:442, replacement name for *S. Roberti* Thomas, 1903, preoccupied by *Sciurus thaiwanensis roberti* Bonhote, 1901:166.

Guerlinguetus alphonsei alphonsei: J. A. Allen, 1915a:261; name combination.

Guerlinguetus alphonsei paraensis: J. A. Allen, 1915a:261; name combination.

Guerlinguetus ingrami: J. A. Allen, 1915a:262; name combination.

Guerlinguetus Aestuans Paraensis: O. M. O. Pinto, 1931:291; name combination.

Guerlinguetus Aestuans Alphonsei: O. M. O. Pinto, 1931:292; name combination.

Guerlinguetus Aestuans Garbei O. M. O. Pinto, 1931:294; type locality "Villa Collatino, Espírito Santo," Brazil.

Sciurus [(*Guerlinguetus*)] *ingrami*: Ellerman, 1940:343; name combination.

Sciurus [(*Guerlinguetus*)] *aestuans garbei*: Ellerman, 1940:343; name combination.

Sciurus [(*Guerlinguetus*)] *alphonsei alphonsei*: Ellerman, 1940:34; name combination.

Sciurus [(*Guerlinguetus*)] *alphonsei paraensis*: Ellerman, 1940:34; name combination.

Sciurus aestuans henseli Miranda-Ribeiro, 1941:10; type locality "Porto Feliz, Rio Uruguay, Rio Grande do Sul," Brazil.

Guerlinguetus gilvigularis paraensis: Moojen, 1942:9; name combination.

Guerlinguetus ingrami ingrami: Moojen, 1942:14; name combination.

Guerlinguetus ingrami henseli: Moojen, 1942:16; name combination.

G[*uerlinguetus*]. g[*ilvigularis*]. *paraensis*: Melo-Leitão, 1943:356; name combination.

Sciurus ingrami ingrami: Moojen, 1952b:27; name combination.

Sciurus ingrami henseli: Moojen, 1952b:28; name combination.

Sciurus garbei: Moojen, 1952b:28; name combination.

Sciurus (*Guerlinguetus*) *ingrami ingrami*: C. O. da C. Vieira, 1953:131; name combination.

Sciurus (*Guerlinguetus*) *ingrami henseli*: C. O. da C. Vieira, 1955:407; name combination.

Sciurus (*Guerlinguetus*) *garbei*: C. O. da C. Vieira, 1955:407; name combination.

Sciurus (*Guerlinguetus*) *alphonsei*: C. O. da C. Vieira, 1955:407; name combination.

Sciurus (*Guerlinguetus*) *gilvigularis paraensis*: C. O. da C. Vieira, 1955:407; name combination.

Sciurus gilvigularis paraensis: C. T. Carvalho, 1959:461; name combination.

Sciurus [(*Guerlinguetus*)] *aestuans alphonsei*: Cabrera, 1961:359; name combination.

Sciurus [(*Guerlinguetus*)] *aestuans henseli*: Cabrera, 1961:360; name combination.

Sciurus [(*Guerlinguetus*)] *aestuans ingrami*: Cabrera, 1961:360; name combination.

Sciurus [(*Guerlinguetus*)] *aestuans paraensis*: Cabrera, 1961:361; name combination.

Sciurus ingrami sebastiani P. Müller and Vesmanis, 1971:378; type locality "São Sebastião island [= Ilha de São Sebastião], São Paulo, Brazil."

Sciurus aestuans ingrami: Avila-Pires, 1977:23; name combination and typographical error.

Sciurus [(*Guerlinguetus*)] *aestuans*: Honacki, Kinman, and Koeppl, 1982:362; name combination; part.

Sciurus aestuans ingrami: Vaz, 1983:34; name combination.

Guerlinguetus alphonsei: J. A. Oliveira and Bonvicino, 2006:348; name combination.

Guerlinguetus henseli: J. A. Oliveira and Bonvicino, 2006:349; name combination.

Sciurus aestuans sebastiani: Hutterer and Peters, 2010:14; name combination.

DESCRIPTION: Adult mass averages 193 g (N = 33), head and body length 180 mm (N = 175), tail length 183 mm (N = 171), foot length 46 mm (N = 177), and ear length 22 mm (N = 143). Dorsal color uniform brown to olive brown, heavily streaked with orange throughout body, including upper parts of forelegs and hindlegs; ventral color highly variable, with throat always white to grayish white, with white hairs sometimes extending to chest and/or belly and/or inner sides of arms and legs; area immediately around mammae lighter than rest of venter; and dorsal color of tail either indistinct from dorsum or, if distinct, grizzled with whitish bands on hairs. Postauricular patches absent.

DISTRIBUTION: This species is distributed in Amazonian Brazil east of the Iriri and Xingu rivers and south of the lower Amazon in Pará, Maranhão, Tocantins, and Mato Grosso states, and extends throughout northeastern and southwestern Brazil from the states of Ceará to Rio Grande do Sul and from there to Misiones in Argentina. The species appears to occupy two disjunct areas, one corresponding to the Amazonian subspecies *G. b. paraensis* and the other to the two subspecies *G. b. brasiliensis* and *G. b. ingrami*, which occur continuously from northeastern to southern Brazil and Argentina.

SELECTED LOCALITIES (Map 4): ARGENTINA: Misiones, Eldorado (BM 26.2.11.1), Santa Ana (FMNH 49885). BRAZIL: Alagoas, Usina Sinimbú, Mangabeiras (Manimbu) (MZUSP 7532); Bahia, Ilhéus (MZUSP 3501), Macaco Seco, Monte Andaraí, Rio Paraguaçu (FMNH 20424), Senhor do Bonfim [= Villa Nova] (MZUSP 2608); Ceará, Serra do Castelo (BM 20.7.1.7); Espírito Santo, Colatina, right bank Rio Doce (type locality of *Guerlinguetus aestuans garbei* O. M. O. Pinto), Rio Itaúnas (MZUSP 7071), Rive [= Engenheiro Reeve] (BM 3.9.4.44); Maranhão, Aldeia Gurupiúna, Reserva Indígena Alto Turiaçú (MPEG 21981), Estreito, right bank Rio Tocantins (MPEG 2444), São Bento (BM 25.5.21.6); Mato Grosso, Serra do Roncador, 264 km N of Xavantina (BM 79.203–207); Minas Gerais, Rio das Velhas, near Lagoa Santa (FMNH 20733); Pará, Belém (MPEG 710), Gradaús, right bank Rio Fresco (MZUSP 20399), mouth of Rio Bacajá, right bank Rio Xingú (MZUSP 25455), Portel, Rio Procupi (MPEG 606), Rio Riosinho, left bank Rio Fresco (MPEG 1039), Santa Júlia, Rio Iriri (BM 20.7.1.6), Urumajó, Arabóia, east of Bragança, Estrada de Ferro de Bragança (MZUSP 20398); Paraíba, Mamanguape, Camaratuba (MZUSP 8241); Pernambuco, São Lourenço (type locality of *Sciurus Roberti* Thomas); Rio de Janeiro, Sepetiba (NMW B1393), Terezópolis (NMW 1693); Rio Grande do Sul, Colônia do Mundo Novo (Ihering 1892:107); Santa Catarina, Joinville (BM 13.7.8.1); São Paulo, Ilha de São Sebastião (type locality of *Sciurus ingrami sebastiani* P. Müller and Vesmanis), Ilha do Cardoso (MZUSP 3905), Porto Cabral, Rio Paraná (MZUSP 6021), Presidente Epitácio, left bank Rio Paraná (MZUSP 3733), Santos (MZUSP 3699), Serra da Juréia (MZUSP 12835).

SUBSPECIES: *Guerlinguetus brasiliensis* is polytypic: *G. b. brasiliensis* (Gmelin) occurs in northeastern Brazil, in the states of Ceará, Rio Grande do Norte, Paraíba, Per-

nambuco, Alagoas, Sergipe, and Bahia; *G. b. paraensis* (Goeldi and Hagmann) is distributed in eastern Brazilian Amazonia, in the states of Pará, the south of the Rio Amazonas from the Rio Xingu basin eastward to eastern Maranhão, Tocantins, and eastern Mato Grosso states; and *G. b. ingrami* (Thomas), which is found in coastal Brazil from Espírito Santo state south to Rio Grande do Sul, and extending to Misiones in Argentina. See Remarks for comments on the geographic variation of these subspecies.

NATURAL HISTORY: Although the distribution suggests that *G. brasiliensis* inhabits several distinct biomes in eastern South America, from Amazonian rainforest to the central Brazilian savannas (Cerrado), northeastern Brazil dry forests (Caatinga), and Atlantic rainforest, we believe this species is restricted to forest habitats and does not occur in open vegetation communities. In the Amazonian and Atlantic rainforests, this species is common, and specimens are well represented in collections. In contrast, the species appears to be rare in the Cerrado and Caatinga, where it is restricted to marginal gallery forests and isolated pockets of mesic vegetation, respectively. One of us (MdV) has observed solitary individuals or groups of up to four animals in the wild, but there are literature reports of associations of up to eight animals (Solórzano-Filho 2006).

As with other tree squirrels, *G. brasiliensis* is an arboreal specialist, shot, trapped, or observed typically at 5–12 m in the midstratum layer (Grelle 2003) where it builds nests made of moss, twigs, leaves, and plant fibers on the outside and exclusively of fibers internally; nest circumference averages 56 cm, and nests are usually placed at midcanopy (Alvarenga and Talamoni 2005). The diurnal activity pattern was represented by two peaks, one in the morning and other during the afternoon (Bordignon and Monteiro-Filho 2000). Fixed routes between the nests and food resources have been observed, and animals mark territories with urine and head-gland rubbing against the substrate. Foods include lichens, bryophytes, mushrooms, leaves, fruits, insects, and bird eggs (L. F. Ribeiro et al. 2009). At some sites, coconuts may be the most consumed item, comprising up to 70% of the diet; foods also include exotic plant seeds such as *Pinus*, demonstrating this species' ability to occupy altered habitats (Bordignon and Monteiro-Filho 1999; Alvarenga and Talamoni 2006). Several studies have discussed the role of *G. brasiliensis* as either a seed predator and/or disperser, especially of palm species with large seeds and hard endocarps (e.g., Paschoal and Galetti 1995; G. S. Silva and Tabarelli 2001; Pimentel and Tabarelli 2004; Donatti et al. 2009).

Alvarenga and Talamoni (2005) and T. G. Oliveira et al. (2007) presented evidence of reproduction associated with the onset of the rainy season, but Bordignon and Monteiro-Filho (2000) captured pregnant females during two peaks, one in the winter and the other during the summer; the latter investigators also estimated a population density of 0.89 squirrels/ha, with males occupying larger areas during the winter mating period.

REMARKS: The name for the Brazilian Squirrel is *Guerlinguetus brasiliensis*, and its author is Gmelin (1788). The name was published as a synonym of his *Sciurus aestuans* and was attributed to G. Marcgraf (1648). Lesson (1838) recognized the species as valid, stating that F. Cuvier in Buffon's *Supplement*, Tome I, had provided the name; however, in this work, F. Cuvier (1831) does not employ a binomial designation for Marcgraf's squirrel. According to the International Code of Zoological Nomenclature (ICZN 1999:Arts. 11.6.1 and 50.7), the authorship of such a name must be attributed to the first publisher of the name, even if it was given as a junior synonym. Gmelin (1788) gave the type locality as "Brasilia et Guiana," but as other authors have argued for the species first reported by Marcgraf (1648), we restrict it here to Recife, Pernambuco, Brazil.

There is considerable geographic variation in the wide distributional range of this species, and some of this variation has been recognized at the species or subspecies level. The material examined does allow us to recognize three valid subspecies in *G. brasiliensis*. The typical subspecies, *G. b. brasiliensis*, occurs in northeastern Brazil, from the states of Ceará to Bahia, where the color of the venter is generally white on the throat, chest, belly, and inner parts of forelegs and hindlegs, and the tail is grizzled. To the

south of this area, intermediate forms between the nominotypical subspecies and *G. b. ingrami* appear in the state of Espírito Santo (the form to which O. M. O. Pinto [1931] gave the name *garbei*), with specimens having less white and more yellowish under parts but retaining grizzled tails. Farther to the south as well as west of Espírito Santo, animals show white only at the throat and chest, sometimes also at the posterior and inguinal parts of the venter; the tail is never grizzled and is always punctuated with orange or red. The third subspecies, *G. b. paraensis* from eastern Brazilian Amazonia, resembles *ingrami* but has less gray on the belly; individuals also appear to have shorter pelage and are slightly smaller. Current records suggest that *G. b. paraensis* is not contiguous with the other two subspecies because *G. b. brasiliensis* is known only from forest pockets in the otherwise open formation of the Caatinga and *G. b. ingrami* occurs in the Atlantic Forest, which is mostly separated from Amazonia by open savannas of the Cerrado, a biome without any squirrel records (Carmignotto et al. 2012).

Genus *Hadrosciurus* J. A. Allen, 1915

Hadrosciurus contains the largest of the South American squirrels, distributed in Venezuela south of the Orinoco, Colombia, Ecuador, and Peru east of the Andes, western Brazilian Amazonia, and western central Brazil, Bolivia, and northern Paraguay.

This genus can be distinguished from most other South American sciurid genera by their size, except for the similar-sized *Simosciurus*, which differs from *Hadrosciurus* by its distinctly coarser pelage, ears covered by longer hairs, shorter skull with a broader rostrum, and a trans-Andean distribution. Individuals of some species of *Notosciurus* reach sizes that overlap the lower size limits of *Hadrosciurus*, but the two genera can be distinguished because the latter exhibits four pair of mammae instead of three.

Hadrosciurus, like other South American squirrels, exhibits considerable variation in external morphology and especially in color. Mass varies from 450 to 750 g, head and body length from 215 to 320 mm, tail length from 220 to 330 mm, hindfoot length from 50 to 75 mm, and ear length from 20 to 40 mm. A well-developed postorbital process equally divides the orbitotemporal fossa of the skull.

The genus includes three polytypic species, namely, *Hadrosciurus igniventris*, with three subspecies, *Hadrosciurus pyrrhinus*, with two subspecies, and *Hadrosciurus spadiceus*, with three subspecies. These three species present an interesting pattern of sympatry, with their large-scale ranges overlapping along the lower Solimões and lower Purús rivers, the only region where museum specimens document the presence of all three. All remaining

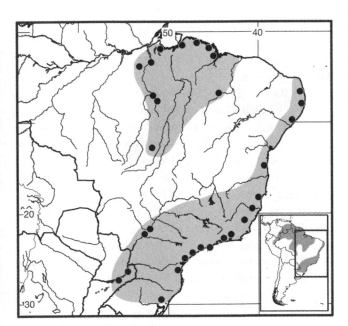

Map 4 Selected localities for *Guerlinguetus brasiliensis* (●). The likely disjunct distributions of Amazonian, Caatinga, and coastal populations are emphasized. Contour line = 2,000 m.

sympatry records are for pairs of species, with *H. igniventris* and *H. spadiceus* broadly sympatric in Ecuador and northern Peru; *H. pyrrhinus* and *H. spadiceus* sympatric along the middle and lower Rio Juruá in western Brazil; and *H. pyrrhinus* and *H. igniventris* sympatric along the lower Rio Negro, also in Brazil. It is somewhat surprising that so few localities have yielded sympatric species of squirrels, if we consider that, at least potentially, the area of sympatry at map scale covers much of central Peru and a vast area of the Brazilian western Amazonia. Patton et al. (1982) and Lemke et al. (1982) documented sympatry for *H. igniventris* and *H. spadiceus* in northern Peru and southern Colombia, respectively.

SYNONYMS:

Sciurus: Olfers, 1818:208; part (listing of *spadiceus*); not *Sciurus* Linnaeus, 1758.

Macroxus: Gray, 1867:428; part (listing of *langsdorffii*); not *Macroxus* F. Cuvier, 1823.

Guerlinguetus: Miranda-Ribeiro, 1914:36; part (listing of *langsdorffii*); not *Guerlinguetus* Gray, 1821.

Mesosciurus (*Histriosciurus*) J. A. Allen, 1915a:213; type species *Sciurus gerrardi* Gray, by original designation; part; all species-level taxa in this genus belong to *Notosciurus* J. A. Allen except for J. A. Allen's *Mesosciurus pyrrhinus*.

Hadrosciurus J. A. Allen, 1915a:265; type species *Sciurus flammifer* Thomas, by original designation.

Urosciurus J. A. Allen, 1915a:267, type species *Sciurus tricolor* Tschudi, by original designation.

Sciurus (*Urosciurus*): Thomas, 1926f:635; name combination.

Sciurus (*Hadrosciurus*): Thomas, 1928b:259; name combination.

[*Sciurus*] (*Guerlinguetus*): Ellerman, 1940:340; as subgenus (among the included species is *Sciurus pyrrhinus* Thomas).

Guerlinguetus (*Hadrosciurus*): Moore, 1959:177; name combination.

REMARKS: Our concept of *Hadrosciurus* mirrors that of Thomas (1928b), who suggested that *Sciurus* (*Hadrosciurus*) encompassed all the largest squirrels east of the Andes, but this concept requires more detailed clarification. As new species of squirrels were added to the roster of taxa from South America, J. A. Allen (1915a), acting as the first true reviewer of the family for the continent, recognized several distinct genera and subgenera, many of which he described as new. For the largest squirrels, Allen described two new genera: *Hadrosciurus*, containing only *H. flammifer*, and *Urosciurus*, with several species, including *U. langsdorffii*, *U. igniventris*, *U. pyrrhonotus*, and others. He also created the genus *Mesosciurus* for many species and subspecies in the *Sciurus granatensis* complex, in which he included *Sciurus*

pyrrhinus Thomas as a species—*Mesosciurus* (*Histriosciurus*) *pyrrhinus*.

However, Thomas (1927a) suggested that Allen's genera were "very intangible" and decided to change the status of many by placing them as subgenera of *Sciurus*. In this same work, Thomas allied *pyrrhinus* with *igniventris*, thus, for all practical purposes, removing *pyrrhinus* from *Mesosciurus*. Shortly thereafter, Thomas (1928b) realized that *Sciurus flammifer*, the type and only species of Allen's *Hadrosciurus*, belonged together with all the other large squirrels in a single taxonomic entity and that, having been described a few pages earlier than *Urosciurus*, the former would be the proper generic name for these squirrels, with *Urosciurus* as a junior synonym. Except for a few specific and subspecific taxa not yet known at the time of Thomas's publication, the content he assigned to *Hadrosciurus* is the same as ours.

Ellerman (1940) did not subscribe to Thomas's views and rearranged the specific contents of Thomas's *Sciurus* (*Hadrosciurus*), placing *pyrrhinus* with most other South American squirrels in his *Sciurus* (*Guerlinguetus*). Moore (1959) arranged all non-dwarf South American squirrels in the genus *Guerlinguetus*, with *Hadrosciurus* as a subgenus. Subsequent authors, for example, Thorington and Hoffmann (2005), believed that Moore (1959) had additionally recognized *Urosciurus* J. A. Allen as a subgenus of his *Guerlinguetus*. However, Moore (1959:177, 198) included *Sciurus igniventris* Wagner, the type species of *Urosciurus* J. A. Allen, as *Guerlinguetus* (*Hadrosciurus*) *igniventris*, thus treating *Urosciurus* as a junior synonym. Because Cabrera's (1961) work was published posthumously, he probably never saw Moore's (1959) arrangement and so employed Thomas's scheme for the genus. The current practice of regarding *Urosciurus* as a valid generic name for the large South American squirrels has resulted from a misreading of Moore's intentions.

Hershkovitz (1959b) expanded the discussion of cranial characters initiated by J. A. Allen (1915a) by recognizing two groups of large South American squirrels, which he treated as distinct species based on a characterization of the skull as either "short and broad" (*Sciurus igniventris*) or "long and narrow" (*Sciurus spadiceus*). Patton (1984) performed a morphometric study of Peruvian samples, agreed with Hershkovitz (1959b) that two distinct skull morphotypes could be recognized in the area, and resurrected *Urosciurus* as a subgenus of *Sciurus*.

At present, we cannot ascertain how the three species and subordinated subspecies that we have recognized in the genus *Hadrosciurus* can be distinguished by the cranial characters described by Hershkovitz (1959b) and Patton (1984), particularly because the presence of these characters has not been tested throughout the entire distribution of the genus. We employ *Hadrosciurus* as a full genus to keep what seems to us a distinct group of species—distinguished mainly by their large size from other South American taxa—and to separate these squirrels from *Sciurus*, a near cosmopolitan taxon. We wish to emphasize that the differences in skull morphology demonstrated by Patton (1984) are quite substantive, contrary to what some previous authors believed (such as Oldfield Thomas). Indeed, we have not found such trenchant character differences in other South American squirrels, and it seems highly unlikely that strongly bimodal distribution is populational or merely geographical. *Hadrosciurus* might, with further study, be justifiably split into distinct taxonomic groups based on their skull morphology alone.

KEY TO THE SPECIES OF *HADROSCIURUS*:

1. Dorsum black or dark brown background streaked with red, orange, or yellow (dark hairs with red, orange, or yellow subterminal bands) 2
1'. Dorsum brownish red to reddish orange with little or no streaking; postauricular patch small, but distinct. . .
. . . . *Hadrosciurus pyrrhinus*
2. Forelimbs uniformly orange or red, differently colored from dorsum; postauricular patch rarely present . . . . .
. . . . *Hadrosciurus igniventris*
2'. Forelimbs similar to that of anterior dorsum; postauricular patch always absent. . . . *Hadrosciurus spadiceus*

Hadrosciurus igniventris (Wagner, 1842)

Northern Amazon Red Squirrel, Quatipuru-açú

SYNONYMS: See under subspecies.

DESCRIPTION: Large, with head and body length 246–300 mm (N=117 adults), tail length 236–310 mm (N=118), hindfoot length 58–75 mm (N=111), ear length 29–40 mm (N=73), and mass 453–700 g (N=12). Dorsum dark, distinctly punctuated by light subapical bands to hairs, with background color from dark brown to black, and light bands from pale yellow to orange and red. Presence of distinct postauricular patches variable geographically; if present, pale orange in color. Forelimbs, and sometimes the hind legs, homogeneously orange to red; ventral color from white to reddish orange or orange yellow; base of tail colored like dorsum, with the terminal portion distinctly orange. Completely or partially melanistic individuals common.

DISTRIBUTION: The species occurs in northwestern Amazonia from the right bank of the Río Orinoco in Venezuela southward to the Rio Negro and Rio Solimões in Brazil and westward to Colombia, Ecuador, and northern Peru east of the Andes. The record of two specimens (MPEG 21807, 21808) from the Brazilian state of Amapá (Cachoeira do Santo Antonio, Rio Jari) is likely in error because there are no other records between this locality and the easternmost known distributional limit of the species

in Roraima, Brazil, about 1,000 km to the west. Cláudia Regina da Silva (pers. comm. to MdV) has conducted extensive mammal inventories in Amapá and has never encountered this species in the wild, nor has she ever heard locals refer to such a squirrel.

SELECTED LOCALITIES (Map 5): BRAZIL: Amazonas, Canabuoca, Paraná do Jacaré (SMF 6918), Igarapé de Alvarães, right bank Rio Solimões (MZUSP 18921), Paraná do Aiapuá, Rio Purus (MZUSP 20357), Paraná do Manhãna, between Rio Japurá and Rio Solimões (MZUSP 6141); Roraima, BR 174, frontier between Brazil and Venezuela, frontier mark 8 (MZUSP 22347), lower Rio Mucajaí, south of Boa Vista (MZUSP 9779); Roraima, Rio Catrimani (MZUSP 13688). COLOMBIA: Boyacá, Río Lengupa (AMNH 38407); Caquetá, Florencia, Río Bodoquera (AMNH 33691); Cundinamarca, Sabana Grande, lowlands near Bogotá (type locality of *Sciurus igniventris taedifer* Thomas); Meta, La Macarena, Río Yerli (FMNH 87954). ECUADOR: Loja, El Porotillo San José (MNHN 1934/1207); Morona–Santiago, Chiguaza (FMNH 54287), Mendéz Sur (MNHN 1960/3790); Orellano, Río Cotapino (FMNH 47600); Pichincha, Quito (MNHN 1937/893). PERU: Cajamarca, Chaupe (AMNH 64047); Loreto, Pebas (AMNH 98426), Santa Elena, Río Samiria (FMNH 87158), Santa Rita, Iquitos (FMNH 87148), Yurimaguas, Río Huallaga (FMNH 19674); San Martín, about 35 mi W of Moyobamba, Río Negro (AMNH 73214). VENEZUELA: Amazonas, Nericagua (Caño Usate), Río Orinoco (BM 99.9.11.26), Raya 32 km SSE of Puerto Ayacucho (USNM 409801); Bolívar, Ciudad Bolívar (AMNH 16127), El Yagual, lower Río Caura (AMNH 16940).

SUBSPECIES: We recognize three subspecies of *H. igniventris*. See Remarks for a description of the geographic variation.

H. i. cocalis (Thomas, 1900)

SYNONYMS:

Sciurus cocalis Thomas, 1900e:138, type locality "Mouth of Coca River, Upper Rio Napo," Orellana, Ecuador.
Urosciurus igniventris cocalis: J. A. Allen, 1915a:273; name combination.
Sciurus (Hadrosciurus) igniventris: Thomas, 1928c:289; part.
Sciurus [(Hadrosciurus)] igniventris cocalis: Ellerman, 1940: 345; name combination.
Sciurus igniventris cocalis: Soukup, 1961:244; name combination.
Sciurus [(Urosciurus)] igniventris [cocalis]: Thorington and Hoffmann, 2005:762; name combination.

This subspecies occurs in Ecuador east of the Andes, and in Peru in the departments of Loreto, Cajamarca, and San Martín.

H. i. igniventris (Wagner, 1842)

SYNONYMS:

Sciurus igniventris Wagner, 1842c:360; type locality "Rio Negro"; restricted by Wagner (1850:277) to "Marabitanos, überhaupt vom Rio negro" [= Marabitanas], right bank of the upper Rio Negro, Amazonas, Brazil.

Sciurus Morio Wagner, 1850:275, type locality "Marabitanos, überhaupt vom Rio negro" [= Marabitanas], right bank of the upper Rio Negro, Amazonas, Brazil.

(?) *Macroxus fumigatus* Gray, 1867:428, type locality "Brazil, Upper Amazons"; identity uncertain as holotype is almost entirely melanistic.

(?) *Macroxus brunneo-niger* Gray, 1867:429, type locality "Brazil"; identity uncertain as holotype is almost entirely melanistic.

Sciurus igniventer Gray, 1867:433; incorrect subsequent spelling of *Sciurus igniventris* Wagner.

Sciurus igniventris niger Fitzinger, 1867a:479; type locality "Am. Brasilien, Rio negro, Marabitanas," Amazonas, Brazil.

Sciurus igniventris taedifer Thomas, 1903b:487, type locality "Sabaña Grande, near Bogota," Cundinamarca, Colombia.

Sciurus tricolor morio: E.-L. Trouessart, 1904:328; name combination.

Sciurus brunneo-niger: E.-L. Trouessart, 1904:328; name combination.

Sciurus duida J. A. Allen, 1914d:594, type locality "Rio Cunucunumá (altitude 700 feet), base of Mount Duida, [Amazonas,] Venezuela."

Sciurus igniventris zamorae J. A. Allen, 1914d:594, type locality "Zamora (altitude 2,000 feet), [Zamora–Chinchipe,] Ecuador."

Urosciurus duida: J. A. Allen, 1915a:270; name combination.

Urosciurus igniventris igniventris: J. A. Allen, 1915a:271; name combination.

Urosciurus igniventris taedifer: J. A. Allen, 1915a:272; name combination.

Urosciurus igniventris zamorae: J. A. Allen, 1915a:274; name combination.

S[ciurus]. (Hadrosciurus) igniventris: Thomas, 1928c:290; name combination.

Hadrosciurus igniventris igniventris: Tate, 1939:176; name combination.

Hadrosciurus igniventris duida: Tate, 1939:176; name combination.

Hadrosciurus igniventris: Cabrera and Yepes, 1940:195; name combination.

Sciurus [(Hadrosciurus)] duida: Ellerman, 1940:344; name combination.

Sciurus [(Hadrosciurus)] igniventris igniventris: Ellerman, 1940:344; name combination.

Sciurus [(Hadrosciurus)] igniventris taedifer: Ellerman, 1940:344; name combination.

Sciurus [(Hadrosciurus)] igniventris zamorae: Ellerman, 1940:345; name combination.

Hadrosciurus igniventris manhanensis Moojen, 1942:24, type locality "Paraná do Nanhana [= Manhana; Paynter and Traylor 1991], entre o Rio Japurá e o Amazonas, Amazonas," Brazil.

Hadrosciurus pyrrhonotus purusianus Moojen, 1942:31, type locality "Lago Aiapuá, Baixo Purús, Estado do Amazonas," Brazil.

Sciurus [(Hadrosciurus)] igniventris manhanensis: W. P. Harris, 1944:12; name combination.

Sciurus [(Hadrosciurus)] igniventris purusianus: W. P. Harris, 1944:12; name combination.

Sciurus igniventris igniventris: Sanborn, 1949b:285; name combination.

Sciurus igniventris manhanensis: Moojen, 1952b:30; name combination.

Sciurus pyrrhonotus purusianus: Moojen, 1952b:31; name combination.

Sciurus (Hadrosciurus) pyrrhonotus purusianus: C. O. da C. Vieira, 1955:408; name combination.

Guerlinguetus (Hadrosciurus) igniventris: Moore, 1959:203; name combination.

Sciurus [(Urosciurus)] igniventris: Patton, 1984:53; name combination.

The nominotypical subspecies occurs in Colombia east of the Andes in Cundinamarca, Meta, and Caquetá departments; along the Río Napo in Loreto, Peru; in Amazonas department of Venezuela; and in Roraima state in Brazil, on the left bank of the Rio Branco, and in Amazonas state, along the Rio Solimões and Rio Negro.

H. i. flammifer (Thomas, 1904)

SYNONYMS:

Sciurus flammifer Thomas, 1904d:33, type locality "La Union, Caura district, Lower Orinoco," Bolívar, Venezuela.

Hadrosciurus flammifer: J. A. Allen, 1915a:266; name combination.

S[ciurus (Hadrosciurus)]. flammifer: Thomas, 1928c:290; name combination.

Guerlinguetus (Hadrosciurus) flammifer: Moore, 1959:203; name combination.

This subspecies is distributed from the Venezuelan department of Bolívar, south of the Orinoco and west of Río Caroní, to the Brazilian border at Serra de Pacaraima.

NATURAL HISTORY: Emmons and Feer (1997) noted that individuals are diurnal, arboreal, territorial, and solitary, although several may feed together in a single fruiting tree. Primary foods include large palm nuts and other tree nuts and fruits. Silvius (2002) reported individuals feeding on beetle larvae as well as the endocarp of the palm *Attalea maripa*. The species is largely limited to closed-canopy rainforest, either mature or disturbed (Emmons and Feer 1997). Youlatos (1999) reported that individuals used all forest heights, preferring tree crowns and lianas, with an ability to walk, leap, and bound along small horizontal supports. Jessen, Palmer, and Koprowski (2013) reported a nest for an unidentifiable melanistic female of either *H. igniventris* or *H. spadiceus* from Loreto, Peru, built at 3.6 m above ground with two offspring.

REMARKS: *Hadrosciurus igniventris cocalis* is a large squirrel with orange forelimbs, which differs from the nominotypical subspecies by a dark to almost black dorsal color streaked with orange or dark red, the presence of conspicuous orange postauricular patches, and a yellowish orange venter.

Hadrosciurus igniventris igniventris is also a large squirrel with orange to red forelimbs, with hindlimbs colored as the forelimbs or dorsum; head blackish above, with an orange muzzle diluted with black and with no conspicuous eye-ring; ears brown to almost black on both sides; postauricular patches, if present, inconspicuous;

dorsum black to dark brown heavily streaked with orange or red; venter red, with same color extending to undersides of legs; and tail dark, mostly black over the proximal third, with some orange distally. J. A. Allen (1915a:267–268) placed *Sciurus igniventris* Wagner in his new genus *Urosciurus*.

Hadrosciurus igniventris flammifer is similar to *H. i. igniventris*, with orange to red limbs and a black dorsum heavily streaked with pale yellow to orange, but the former differs from the latter by having conspicuous pale orange postauricular patches and a white venter. The genus *Hadrosciurus* was created by J. A. Allen (1915a), with *Sciurus flammifer* as its type and only species.

Hadrosciurus pyrrhinus (Thomas, 1898)
Junín Red Squirrel, Quatipuru-açú

SYNONYMS: See under subspecies.

DESCRIPTION: Equivalent in size to *H. igniventris*, with head and body length 220–290 mm (N = 44 adults), tail length 252–305 mm (N = 42), hindfoot length 60–70 mm (N = 40), and ear length 31–35 mm (N = 19). Color of dorsum, forelimbs, and hindlegs uniform dark red to brownish red, without streaking by hairs with light apical or subapical bands. Conspicuous postauricular patches present, generally pale yellow or light orange throughout most of range but inconspicuous in specimens from upper Rio Negro area of Brazil at Colombian border. Venter white to pale red, mixture of both colors in some individuals; when white, transition from venter to torso, limbs, and head demarcated by a red strip.

DISTRIBUTION: This species occurs from the eastern Andean slopes of central Peru to western Brazilian Amazonia, probably limited by the Purus river basin on the east and Negro basin on the north.

SELECTED LOCALITIES (Map 6): BRAZIL: Amazonas, between Cajutuba and Airão, Rio Jaú, Rio Negro (syntype of *Sciurus igniventris* Wagner), Codajás, left bank Rio Solimões (MZUSP 4202), Dejedá, Rio Juruá (type locality of *Urosciurus nigratus* O. M. O. Pinto), Estirão do Equador, right bank Rio Javari (MPEG 1807), Manaus (syntype of *Sciurus igniventris* Wagner), Santo Antônio do Içá, left bank Rio Içá (MZUSP 20384), Tahuapunto (Tauá), Rio Vaupés, left bank (AMNH 78620). PERU: Cusco, San Juan Grande, Quincemil, Quispicanchis (FMNH 75208); Huánuco, Tingo Maria, Río Huallaga (FMNH 24112); Junín, Tarma, 2 mi NW of San Ramón (AMNH 231771); Pasco, Puerto Bermudez (FMNH 24110).

SUBSPECIES: We tentatively recognize two subspecies of *H. pyrrhinus*, which we compare in the Remarks section.

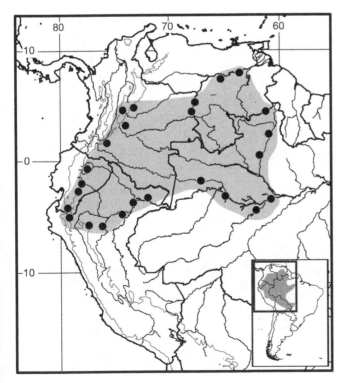

Map 5 Selected localities for *Hadrosciurus igniventris* (●). Contour line = 2,000 m.

H. p. fulminatus (Thomas, 1926)

SYNONYMS:

Sciurus [(*Urosciurus*)] *igniventris fulminatus* Thomas, 1926f:637; type locality "Manacapuru," Rio Solimões, Amazonas, Brazil.

S[ciurus. (Urosciurus)] i[gniventris]. fulminatus: Thomas, 1928c:290; name combination.

Urosciurus nigratus O. M. O. Pinto, 1931:309; type locality "Rio Juruá (Amazonas)," Brazil.

Sciurus [(*Hadrosciurus*)] *nigratus*: Ellerman, 1940:344; name combination.

Sciurus [(*Hadrosciurus*)] *igniventris fulminatus*: Ellerman, 1940:345; name combination.

Hadrosciurus igniventris fulminatus: Moojen, 1942:22; name combination.

Sciurus igniventris fulminatus: Moojen, 1952b:29; name combination.

Sciurus (*Hadrosciurus*) *pyrrhonotus fulminatus*: Avila-Pires, 1964:15; name combination.

This subspecies occurs primarily in Amazonas state, Brazil; it is also known from a single locality in Cusco department, Peru.

H. p. pyrrhinus (Thomas, 1898)

SYNONYMS:

Sciurus pyrrhinus Thomas, 1898f:265; type locality "Garital del Sol, Vitoc, [Junín,] Peru."

Mesosciurus [(*Histriosciurus*)] *pyrrhinus*: J. A. Allen, 1915a:252; name combination.

Sciurus [(*Urosciurus*)] *pyrrhinus*: Thomas, 1927a:369; implicit allocation to subgenus due to accepted resemblance to *S. igniventris*.

Sciurus [(*Guerlinguetus*)] *pyrrhinus*: Ellerman, 1940:342; name combination.

Sciurus (*Hadrosciurus*) *pyrrhinus*: Cabrera, 1961:378; name combination.

The nominotypical subspecies occurs on the eastern slopes of the Andes in the Peruvian departments of Huánuco, Junín, Pasco, and Ucayali.

NATURAL HISTORY: Emmons and Feer (1997) stated that this species inhabits montane forests across an elevational range from 600 to 2,000 m, but it occurs at less than 300 m elevation at localities in the central Amazon.

REMARKS: The taxonomic history of *Hadrosciurus pyrrhinus* (Thomas) is marked by its erroneous allocation to a group of Andean and peri-Andean taxa allied to *Notosciurus granatensis* by J. A. Allen (1915a). Although it is superficially similar to these squirrels, its distribution on the eastern slopes of the central Peruvian Andes only added to the confusion. The alliance of *pyrrhinus* to other large South American squirrels we include in *Hadrosciurus* be-

comes clear when it is considered together with its sister subspecies, as we shall do.

Hadrosciurus pyrrhinus fulminatus is similar to *H. p. pyrrhinus* except that dorsal tones are paler, duller red; pale orange postauricular patches are conspicuous; the venter is uniformly yellow to orange. This taxon occurs sympatrically with two other forms in the genus (namely *H. i. igniventris* and *H. spadiceus pyrrhonotus*) in the region of the lower Rio Solimões and Rio Purus in the Brazilian state of Amazonas. In this area of sympatry, *H. pyrrhinus* can be distinguished from *H. i. igniventris* by its uniform dull dark red dorsal color with very little streaking (which results from a pale band near the tip of the hairs) and the possession of conspicuous postauricular patches. *Hadrosciurus pyrrhinus* is distinguished from *H. spadiceus pyrrhonotus* by presence of distinct postauricular patches, a less vivid red dorsum with less streaking, and yellowish orange as opposed to white venter. Since its description, *fulminatus* has been allied to *Sciurus igniventris*, and since Thomas (1927a) transferred *pyrrhinus* to the *igniventris* complex, both have been usually treated as distinct subspecies under *igniventris*. The similarity in coat color between *pyrrhinus* and *fulminatus* made us consider a probable affinity, here inferred at the subspecific level.

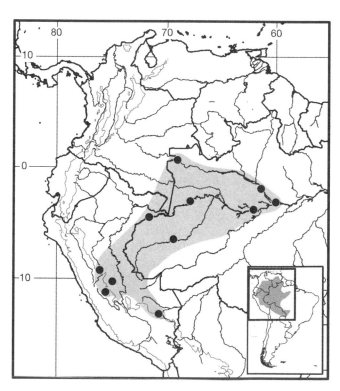

Map 6 Selected localities for *Hadrosciurus pyrrhinus* (●). Contour line = 2,000 m.

Hadrosciurus spadiceus (Olfers, 1818)

Southern Amazon Red Squirrel, Quatipuru-açú

SYNONYMS: See under subspecies.

DESCRIPTION: Slightly larger than other species in the genus, with head and body length 215–316 mm (N = 78 adults), tail length 225–325 mm (N = 78), hindfoot 50–75 mm (N = 76), ear length 20–35 mm (N = 36), and mass 650–750 g (N = 3). Highly variable species, as apparent from several subspecies we recognize (see Remarks for distinctive characters of each subspecies). Dorsum dark, with black background color, strongly streaked with pale yellow, yellow, orange, or red; specimens from Peru have darker mid-dorsal strip. Alternatively, the dorsum may be mostly or entirely red, or divided into relatively distinct anterior and posterior regions, with posterior part uniformly colored orange but anterior portion dark and streaked. Dorsal sides of forelimbs and hindlimbs colored same as adjacent region of upper body; in southeastern portion of distribution, hindlegs divided into paler anterior and darker posterior regions. Differently colored postauricular patches absent. Ventral color variable: white in specimens from Brazilian Amazon and parts of Bolivia, white to yellow in those from Peru, and red in west-central Brazil, central Bolivia, and northern Paraguay.

DISTRIBUTION: The species occurs from southern Colombia, Ecuador, and Peru east of the Andes, eastward into Amazonian Brazil, where it is limited to the region south of the Rio Solimões-Amazonas system, from the Rio Tapajós west to the Peruvian border, and south to west-central Brazil in the states of Mato Grosso and Mato Grosso do Sul as well as across the lowlands of Bolivia to northern Paraguay.

SELECTED LOCALITIES (Map 7): BOLIVIA: La Paz, Chimate (BM 1.2.1.7); Santa Cruz, Santa Cruz de la Sierra (FMNH 21420). BRAZIL: Acre, Oriente (= Seringal Oriente), near Taumaturgo, Rio Juruá (MPEG 725); Amazonas, Eirunepé, left bank Rio Juruá (MZUSP 4634), Lago do Batista, Ilha de Tupinambarama, right bank Rio Amazonas (MZUSP 4567), Lago do Mapixi, east of Rio Purus (FMNH 38881), Manaqueri, Rio Solimões (NMW B 1533), mouth Rio Guariba, left bank Rio Aripuanã (MPEG 8785); Mato Grosso, Mato Grosso (= Vila Bela da Santíssima Trindade) (NMW B 1520), Palmeiras (MZUSP 6338), Villa Maria, NNW of Cáceres (NMW B 1518); Mato Grosso do Sul, Urucum (type locality of *Sciurus langsdorffii urucumus* J. A. Allen); Pará, Flexal, Itaituba–Jacareacanga road, km 212, left bank Rio Tapajós (MPEG 10165), Rio Arapiuns, left bank Rio Tapajós (MPEG 424), Sumaúma, left bank Rio Tapajós (MZUSP 20365); Rondônia, Ouro Preto d'Oeste, Rio Paraíso (MPEG 12411). COLOMBIA: Meta, Cabaña Duda, junction of Río Duda and Río Guayabero (Lemke et al. 1982), Fundo "Guami," Piñalto (Lemke et al. 1982).

ECUADOR: Morona–Santiago, Río San José (AMNH 68264); Napo, mouth of Río Coca, upper Río Napo (BM 0.6.3.2), boca del Río Lagarto Cocha (AMNH 72231); Pastaza, Mera (AMNH 67674). PARAGUAY: Alto Paraguay, Estancia Kambá Aka, Parque Nacional Río Negro (D'Elía, Mora et al. 2008). PERU: Cajamarca, Huarandosa (AMNH 67095); Huánuco, Tingo Maria, Río Huallaga (FMNH 24113); Loreto, boca del Río Curaray (AMNH 72177), Río Apayacu (AMNH 74389), San Fernando, Río Yavari (FMNH 88967); Puno, Pauquiplaya, below San Ignacio, Río Tambopata (FMNH 79903).

SUBSPECIES: We tentatively recognize four subspecies of *H. spadiceus*, as follows.

H. s. castus (Thomas, 1903)

SYNONYMS:

Sciurus castus Thomas, 1903b:488, type locality "Chimate, [La Paz,] Bolivia, . . . on the Upper Rio Beni."

Urosciurus pyrrhonotus castus: J. A. Allen, 1915a:276; name combination.

Hadrosciurus pyrrhonotus castus: Cabrera and Yepes, 1940: 195; name combination.

Sciurus [(*Hadrosciurus*)] *pyrrhonotus castus*: Ellerman, 1940:345; name combination.

Hadrosciurus langsdorffii rondoniae Moojen, 1942:39; type locality "Cabeceiras do Rio Aripuanã, Rondonia, Mato-Grosso," Brazil.

Sciurus [(*Hadrosciurus*)] *langsdorffii rondoniae*: W. P. Harris, 1944:12; name combination.

Sciurus langsdorffii rondoniae: Moojen, 1952b:32; name combination.

Sciurus pyrrhonotus castus: Soukup, 1961:244; name combination.

This subspecies is known from Rondônia state, Brazil, and an isolated locality in La Paz department, Bolivia. It is quite similar morphologically to the nominotypical form, differing primarily by more extensive white on the venter and the lack of a conspicuous eye-ring. The head is black, mixed with orange on the crown; the rostrum also black and orange; the ears brown on both sides; postauricular patches are absent; the dorsum black, finely streaked with vivid yellow; forefeet and limbs the same color as the dorsum, only slightly lighter distally; hindfeet orange streaked with black; bicolored legs—their anterior portion the same color as the sides but darker orange posteriorly; venter entirely white to white with some yellow from throat to chest, white extending to the undersides of fore and hind limbs almost to the feet, with the color near the feet the same as dorsal parts of the legs; and the proximal quarter to third of the tail is dark, and also turning black with an orange tip. *Hadrosciurus langsdorffii rondoniae* Moojen, from west-central Brazil, is identical to *castus* and thus is placed as a synonym.

H. s. pyrrhonotus (Wagner, 1842)

SYNONYMS:

Sciurus pyrrhonotus Wagner, 1842c:360, type locality "Borba," Amazonas, Brazil.

Urosciurus pyrrhonotus pyrrhonotus: J. A. Allen, 1915a: 275; name combination.

Urosciurus pyrrhonotus: O. M. O. Pinto, 1931:303; name combination.

Sciurus [(Hadrosciurus)] pyrrhonotus pyrrhonotus: Ellerman, 1940:345; name combination.

Hadrosciurus pyrrhonotus pyrrhonotus: Moojen, 1942:26; name combination.

Sciurus pyrrhonotus pyrrhonotus: Moojen, 1952b:30; name combination.

Sciurus pirrhonotus Alho, 1982:144; incorrect subsequent spelling of *Sciurus pyrrhonotus* Wagner.

The range of *H. s. pyrrhonotus* is limited to Amazonas state, Brazil, south of the Rio Amazonas–Solimões along the lower and middle courses of the Rio Purus and Rio Madeira. Specimens typically have a black head but orange rostrum, and lack a conspicuous eye-ring. The ears are red on both sides. Postauricular patches are absent. The dorsum is red, slightly darker on the midline. The forelimbs and feet are the same color as the dorsum, only slightly lighter distally. The hindlegs and feet are reddish. The venter is white, with the white extending to the undersides of arms and legs almost to the feet. The color near forefeet and hindfeet is the same as dorsal parts of the legs. The tail is dark over the proximal third of its length, with red predominating in the middle section and black predominating distally. Some specimens of this subspecies have the dorsum partially divided into anterior and posterior sections with slightly distinct hues of red and distinct amounts of banding of the hairs, but the dorsal color is always red, and the venter is white.

H. s. spadiceus (Olfers, 1818)

Sc[iurus]. spadiceus Olfers, 1818:208; type locality restricted by Hershkovitz (1959b:346) to Cuiabá, Mato Grosso, Brasil.

Sciurus langsdorffii Brandt, 1835:89, type locality "Brazil," restricted to Cuiabá, Mato Grosso, Brazil by Wagner (1850:275).

Sciurus Langsdorfii Gray, 1843b:137; incorrect subsequent spelling of *Sciurus langsdorffii* Brandt.

Sciurus langsdorffii Var. *rufa* Wagner, 1844:plate CCXV; preoccupied by *Sciurus rufus* Kerr, 1792.

Sciurus Langsdorffi: E.-L. Trouessart, 1897:429; name combination and incorrect subsequent spelling of *Sciurus langsdorffii* Brandt.

Sciurus Langsdorfi: Thomas, 1900e:137; name combination and incorrect subsequent spelling of *Sciurus langsdorffii* Brandt.

Sciurus langsdorffii urucumus J. A. Allen, 1914d:595; type locality "Urucum, (altitude 400 feet), Rio Paraguay (at mouth of Rio Tacuari), [Mato Grosso do Sul,] Brazil."

Sciurus langsdorffii steinbachi J. A. Allen, 1914d:596; type locality "Santa Cruz de La Sierra, [Santa Cruz,] Bolivia."

Guerlinguetus langsdorffi: Miranda-Ribeiro, 1914:36; name combination and incorrect subsequent spelling of *Sciurus langsdorffii* Brandt.

Urosciurus langsdorffii langsdorffii: J. A. Allen, 1915a:276; name combination.

Urosciurus langsdorffii urucumus: J. A. Allen, 1915a:278; name combination.

Urosciurus langsdorffii steinbachi: J. A. Allen, 1915a:279; name combination.

Urosciurus langsdorffi urucumus: J. A. Allen, 1916d:572; name combination.

Hadrosciurus langsdorffii urucumus: Cabrera and Yepes, 1940:196; name combination.

Hadrosciurus langsdorffii steinbachi: Cabrera and Yepes, 1940:196; name combination.

Sciurus [(Hadrosciurus)] langsdorffi langsdorffi: Ellerman, 1940:345; name combination.

Sciurus [(Hadrosciurus)] langsdorffi urucumus: Ellerman, 1940:345; name combination.

Sciurus [(Hadrosciurus)] langsdorffi steinbachi: Ellerman, 1940:345; name combination.

Hadrosciurus langsdorffii langsdorffii: Moojen, 1942:36; name combination.

Sciurus langsdorffii langsdorffii: Moojen, 1952b:32; name combination.

Guerlinguetus (Hadrosciurus) urucumus: Moore, 1959: 203; name combination.

Guerlinguetus (Hadrosciurus) spadiceus: C. T. Carvalho, 1965:52; name combination.

Sciurus [(Urosciurus)] spadiceus: Patton, 1984:53; name combination.

Sciurus (Hadrosciurus) spadiceus: Eisenberg, 1989:335; name combination.

Sciurus [(Urosciurus)] spadiceus [steinbachi]: Thorington and Hoffmann, 2005:763; name combination.

Urosciurus spadiceus: J. A. Oliveira and Bonvicino, 2006: 350; name combination.

The nominotypical subspecies occurs in the southern part of Mato Grosso and northwestern Mato Grosso do Sul states, Brazil, and extends west to Santa Cruz department, Bolivia. Specimens are characterized by a black head mixed with orange on the crown, a black and orange muzzle, a conspicuous eye-ring, ears brown on both sides; no postauricular patches; dorsum black finely lined with yellow and slightly darker on the midline; forefeet and limbs the same color as dorsum, only slightly lighter distally; black hindfeet streaked with orange; bicolored hindlegs, anteriorly the same pale orange as the sides but posteriorly darker reddish; venter yellowish orange, with some white on the throat and with color extending to the undersides of forelegs and hindlegs, almost to the feet; color near feet same as dorsal parts of legs; and proximal quarter of tail dark, becoming black before terminating in an orange tip.

H. s. tricolor (Tschudi, 1845)

SYNONYMS:

Sc[iurus]. tricolor Tschudi, 1845:156, plate XI; type locality "Peru"; restricted to "Maynas, in the angle between the lower Huallaga and the Marañon [Loreto]" (Thomas 1900e:137).

Urosciurus tricolor: J. A. Allen, 1915a:269; name combination.

Sciurus (Urosciurus) pyrrhonotus taparius Thomas, 1926f: 635; type locality "Uricurituba, Santarem [= Santarém]," Pará, Brazil.

Sciurus [(Urosciurus)] pyrrhonotus juralis Thomas, 1926f:636, type locality "Juruá River," Amazonas, Brazil.

Sciurus (Urosciurus) tricolor: Thomas, 1927f:599; name combination.

Sciurus (Hadrosciurus) tricolor: Thomas, 1927f:599; name combination.

Hadrosciurus tricolor: Cabrera and Yepes, 1940:195; name combination.

Sciurus [(Hadrosciurus)] pyrrhonotus taparius: Ellerman, 1940:345; name combination.

Sciurus [(Hadrosciurus)] pyrrhonotus juralis: Ellerman, 1940:345; name combination.

Hadrosciurus pyrrhonotus taparius: Moojen, 1942:30; name combination.

Hadrosciurus pyrrhonotus juralis: Moojen, 1942:34; name combination.

Sciurus pyrrhonotus taparius: Moojen, 1952b:30; name combination.

Sciurus pyrrhonotus juralis: Moojen, 1952b:31; name combination.

Sciurus [(Urosciurus)] spadiceus [tricolor]: Thorington and Hoffmann, 2005:763; name combination.

This subspecies occurs east of the Andes in Colombia south through Ecuador and Peru to central Bolivia, extending to the east to Acre and Amazonas states in Brazil, and south of the Rio Solimões along the upper course of the Rio Juruá and Rio Purus. There is an apparent gap in the distribution, as the taxon is also known from the Rio Aripuanã,

a tributary on the right bank of the middle Rio Madeira, Amazonas, Brazil, and from there to Pará state west of the Rio Tapajós.

Melanistic individuals of H. s. tricolor are common, with some individuals entirely black, others only partially so. The color pattern for non-melanistic specimens includes a black head with orange rostrum, no conspicuous eye-ring, reddish brown to brownish black ears on both sides; no postauricular patches; dorsum generally black, sometimes darker red posteriorly, but heavily streaked with orange; forefeet and limbs same color as the dorsum, only slightly paler distally; hindfeet and limbs reddish orange, with or without black streaking; venter white, extending to undersides of forelegs and hindlegs almost to the paws, color near hands and feet the same as dorsal parts of the legs; proximal third of the tail dark, with red predominating in the middle section and orange and black predominating distally.

NATURAL HISTORY: Gwinn et al. (2012) compiled information on the natural history of this species. Peres, Barlow, and Haugaasen (2003) reported an estimated density of 0.08 individuals/ha at a site in the headwaters of Rio Arapiuns, Pará state, Brazil; they observed solitary individuals twice. H. Gómez et al. (2003) reported a larger average density for a site in the department of La Paz, Bolivia, with 0.122 individuals/ha. Patton et al. (2000) recorded scrotal males in the headwaters of the Rio Juruá, Amazonas state, Brazil, in September and February. They also collected a lactating female with three placental scars in February. Anderson (1997) reported litter sizes ranging from two to four in August from Bolivia. These very limited data suggest breeding in both wet and dry seasons. According to sources cited in Gwinn et al. (2012), the species is mostly seen at ground level, using the canopy when the forest is flooded. Nuts of four genera of trees (the palms Astrocaryum, Attalea, and Scheelea, and the legume Dipteryx) comprise the majority of its diet. Della-Flora et al. (2013) reported the association of this species with mixed-species bird flocks in western Brazil.

REMARKS: The distributions of three subspecies (H. s. castus, H. s. tricolor, and H. s. spadiceus) approach one another in Bolivia. Here, those specimens we assign to tricolor differ from both spadiceus and castus in dorsal color, with the dorsum black streaked with red or orange, not the pale yellow characteristic of the latter two subspecies. However, Bolivian tricolor also differ from their counterparts east of the Andes in Ecuador and Peru, where specimens typically exhibit a mid-dorsum darker than the sides, a characteristic not present in Bolivian tricolor. Samples from the western Brazilian Amazon, in the states of Acre and Amazonas, also differ from typical tricolor, with the mid-dorsum not markedly darker than the sides and the venter tending to white, with some orange. The dorsum of these Brazilian samples also varies from being uniformly colored from front to back or divided between a more orange posterior section and a black grizzled with orange anterior section; this phenotype has an available name, *Sciurus pyrrhonotus juralis* Thomas, with type locality on the middle Rio Juruá, Amazonas, Brazil.

There are few samples from the east and northeast along the Rio Amazonas, but those from the Rio Aripuanã, in southeastern Amazonas state, as well as those from the left bank of the Rio Tapajós, Pará state, are similar to those from the Juruá region, except that the top of the head is paler. As is usual for most color variants of South American squirrels, there is a name available for these squirrels, namely, *Sciurus (Urosciurus) pyrrhonotus taparius* Thomas. If taken in isolation, specimens from central Peru and from the Rio Tapajós are quite distinct from each other and could be distinguished as valid taxa. We do not do so, however, because we cannot presently establish clear geographical limits to this variation. We believe that the taxon we are recognizing here is a complex and will probably be split into a number of distinct, valid taxa with a deeper study of geographical variation.

Sciurus langsdorffi steinbachi J. A. Allen, described from Santa Cruz de la Sierra, Bolivia, is identical to our concept of *H. s. spadiceus*, and we treat it as a junior synonym. In the southern portion of the geographical distribution of *spadiceus*, specimens from the Serra do Urucum and Cáceres, Mato Grosso do Sul, Brazil, are identical to typical *spadi-*

ceus in color but average distinctly smaller in size. J. A. Allen (1914d) described these specimens as *Sciurus langsdorffii urucumus*; we have placed *urucumus* as a junior synonym of *S. spadiceus* because there is an overlap in size between the largest specimens of *urucumus* and the smallest of *spadiceus*.

Genus *Microsciurus* J. A. Allen, 1895

Microsciurus is a genus of small tree squirrels distributed in Central and South America. Among neotropical squirrels, *Microsciurus* species are intermediate in size between the diminutive *Sciurillus* and the slightly larger *Syntheosciurus*, with mass ranging from 80–120 g and external measurements of head and body length of 112–170 mm, tail length 82–165 mm, hindfoot length 28–45 mm, and ear length 8–19 mm. The ears are small and rounded. Females possess three pairs of evenly spaced mammae. Except for a few taxa, the color pattern is very conservative, usually some shade of brown finely streaked with orange or red on the dorsum, and with a paler ventrum. Several species have colored postauricular patches. High-elevation populations typically have longer and denser fur than lowland ones, and the tail is never very bushy.

The skull has a strongly curved dorsal profile with an inflection point at the middle of the frontals. From the inflection point to the anteriormost portion of skull, the rostrum maintains a straight or even slightly concave profile. The rostrum is short, at most one-third the length of the skull. The frontals are longer than they are wide, and the braincase is globose. The temporal portion of the orbitotemporal fossa is greatly reduced but not obliterated; the postorbital process extends posteriorly almost to a vertical plane passing through the internal (anterior) portion of the squamosal zygomatic process in the fossa. The maxillary process of the zygomatic arch is oblique in lateral view, originating anterior to the cheek teeth and above the infraorbital foramen. The sphenopalatine foramen is situated above the second molar or the contact between the second and third molars. There are two complete bullar septa, and rarely a third incomplete septum. The mandible has a very short coronoid process.

The dental formula is I 1/1, C 0/0, PM 2–1/1, M 3/3 = 20–22; PM3 is absent in some populations, particularly those of *M. flaviventer*. When present, PM3 is simple, small, unicuspidate, and has one root; all remaining permanent cheek teeth possess three roots. PM4 is molariform but varies in size, from one-third smaller to only slightly smaller than M1. M1 and M2 are subequal, but M3 is distinctly smaller in size. Paraconule, metaconule, and ectostyle are absent or vestigial; the trigon is shallow. Only pm4 has a vestigial paraconid, a structure absent in all remaining lower cheek teeth. The trigonid is vestigial.

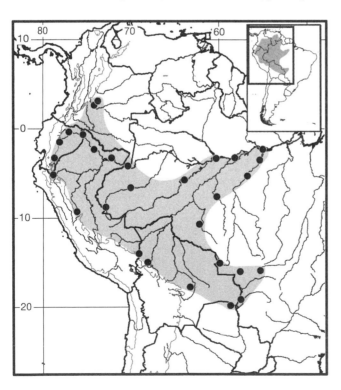

Map 7 Selected localities for *Hadrosciurus spadiceus* (●). Contour line = 2,000 m.

SYNONYMS:

Macroxus: Gray, 1867:432; part (description of *flaviventer*); not *Macroxus* F. Cuvier.

Sciurus: J. A. Allen, 1877:750; part (inclusion of *rufoniger*); not *Sciurus* Linnaeus.

[*Sciurus*] (*Microsciurus*) J. A. Allen, 1895b:332; type species *Sciurus* (*Microsciurus*) *alfari*, by original designation.

Microsciurus: E.-L. Trouessart, 1897:429; elevation to generic status.

Microsciurus (*Microsciurus*): Moore, 1959:179; name combination.

REMARKS: J. A. Allen (1895b) described *Microsciurus* as a subgenus, designating *Sciurus* (*Microsciurus*) *alfari* J. A. Allen as the type species. Being quite easily diagnosable in terms of size and a number of cranial characters from other squirrels, the taxonomic history of *Microsciurus* is relatively unproblematic. The only major taxonomic changes subsequent to Allen's description were E.-L. Trouessart's (1897) elevation to full genus and Thomas's (1914b,f) placement of *Microsciurus pusillus* (*sensu* J. A. Allen 1914a) in its own genus, *Sciurillus*. Moore (1959) used the name in a broader sense, including several genera and subgenera in it, but his subgenus *Microsciurus* is equal to the genus *Microsciurus* of other authors and ourselves.

We recognize 11 species of *Microsciurus*, three in Central America (*M. alfari*, *M. boquetensis*, and *M. venustulus*) and eight in South America (*M. flaviventer*, *M. isthmius*, *M. mimulus*, *M. otinus*, *M. sabanillae*, *M. santanderensis*, *M. similis*, and *M. simonsi*). Only the South American species are detailed herein. Both Hall (1981) and Thorington et al. (2012) extended the range of *boquetensis* into extreme northwestern Colombia (as a subspecies of *M. mimulus*).

KEY TO THE SOUTH AMERICAN SPECIES OF *MICROSCIURUS*:

1. Postauricular patch present 2
1′. Postauricular patch absent 3
2. Dorsum brown, streaked with red . . . . *Microsciurus flaviventer*
2′. Dorsum dark brown, with mid-dorsal dark stripe . . . . *Microsciurus santanderensis*
3. Dorsum dark brown streaked with red to reddish orange tones 4
3′. Dorsum dark brown streaked with yellow to pale yellow tones 5
4. Venter orange, mixed with brown on belly . . . .*Microsciurus isthmius*
4′. Venter washed with red on throat and chest; mixed brown, dark gray, and red on belly . . . .*Microsciurus otinus*
5. Basal third of hairs of venter gray . . . .*Microsciurus mimulus*
5′. Hairs of ventrum without gray basal third 6

6. Venter with predominantly red tones 7
6. Ventral color variable, dark gray, pale gray, or pale yellow on throat, and dark gray to dark brown and yellow on chest and belly . . . . *Microsciurus simonsi*
7. Venter red to mixed red and gray on throat and chest, dark gray to dark brown punctuated with red on belly . . . . *Microsciurus similis*
7′. Venter brown washed with red on throat and chest, dark gray and red on belly . . . .*Microsciurus sabanillae*

Microsciurus flaviventer (Gray, 1867)
Amazon Dwarf Squirrel
SYNONYMS:

Macroxus flaviventer Gray, 1867:432; type locality "Brazil."

Sciurus aestuans: Alston, 1878:668 part (inclusion of *flaviventris* [= *flaviventer*]); not *aestuans* Linnaeus.

Macroxus flaviventris Alston, 1878:668; incorrect subsequent spelling of *Macroxus flaviventer* Gray.

Sciurus chrysurus: Thomas, 1893a:337; part; not *chrysurus* Pucheran.

Sciurus (*Microsciurus*) *peruanus* J. A. Allen, 1897c:115; type locality "Guayabamba, alt. 4000 feet," = Santa Rosa de Huayabamba, Amazonas, Peru.

Sciurus (*Microsciurus*) *peruanus napi* Thomas, 1900f:295; type locality "Mouth of Coca River, Upper Rio Napo," Orellana, Ecuador.

Sciurus peruanus: Ihering, 1904:420; name combination.

Microsciurus peruanus: E.-L. Trouessart, 1904:329; name combination.

Microsciurus peruanus napi: E.-L. Trouessart, 1904:329; name combination.

Microsciurus brevirostris J. A. Allen, 1914a:147; *nomen nudum*; indicated as a new species in a list but with no further description or reference to a description or illustration, likely a *lapsus calami* for *M. rubrirostris*, described in the same paper.

Microsciurus napi: J. A. Allen, 1914a:163; name combination.

Microsciurus rubrirostris J. A. Allen, 1914a:163; type locality "Chanchamayo, [Junín,] central Peru; altitude 2000 m."

Microsciurus florenciae J. A. Allen, 1914a:164; type locality "Florencia (altitude 1000 feet), Caquetá district, Colombia."

Microsciurus avunculus Thomas, 1914b:574; type locality "Gualaquiza; alt. 2500′," Morona–Santiago, Ecuador.

Microsciurus rubricollis Thomas, 1914b:574; incorrect subsequent spelling of *Microsciurus rubrirostris* J. A. Allen.

Microsciurus manarius Thomas, 1920f:275; type locality "Acajutuba, Rio Negro, near its mouth," Amazonas, Brazil.

Microsciurus flaviventer: Thomas, 1920f:275; first use of current name combination.

Microsciurus mannarius: O. M. O. Pinto, 1931:284; inadvertent subsequent spelling of *Microsciurus manarius* Thomas.

Microsciurus [(*Microsciurus*)] *napi*: Moore, 1959:203; name combination.

Microsciurus flaviventer flaviventer: Cabrera, 1961:355; name combination.

Microsciurus flaviventer napi: Cabrera, 1961:356; name combination.

Microsciurus flaviventer peruanus: Cabrera, 1961:357; name combination.

Microsciurus flaviventer rubrirostris: Cabrera, 1961:357; name combination.

DESCRIPTION: Adult mass 80–111 g (N = 16); head and body length 124–170 mm (N = 76), tail length 115–165 mm (N = 79), hindfoot length 30–44 mm (N = 76), and ear length 13–19 mm (N = 65). Dorsal color brown streaked with red; most samples have yellowish to white postauricular patches; venter of mixed colors, with throat and chest yellowish orange and belly slightly darker.

DISTRIBUTION: *Microsciurus flaviventer* occurs widely in western Amazonia, from southern Colombia through eastern Ecuador, Peru, and northeastern Bolivia, then eastward to the Brazilian states of Amazonas, Acre, and Rondônia. It ranges in elevation from 100 to 1,500 m.

SELECTED LOCALITIES (Map 8): BOLIVIA: Pando, left bank Río Madre de Dios near Santa Elena (Salazar-Bravo, Yensen et al. 2002). BRAZIL: Acre, Oriente (= Seringal Oriente), near Taumaturgo, Rio Juruá (MZUSP 20351); Amazonas, Acajutuba, Rio Negro (type locality of *M. manarius* Thomas), Codajás, left bank Rio Solimões (MZUSP 4207), Eirunepé, left bank Rio Juruá (MZUSP 4621), Estirão do Equador, right bank Rio Javari (MZUSP 20350), Lago do Mapixi, east of Rio Purus (FMNH 38880), Manacapuru, Rio Solimões (BM 26.5.5.36), Paraná do Aiapuá, Rio Purus (BM 27.8.11.61); Rondônia, Cachoeira Nazaré, left bank Rio Ji-Paraná (MPEG 20826). COLOMBIA: Caquetá, Florencia (type locality of *M. florenciae* J. A. Allen); Putumayo, La Tagua, Tres Troncos, Río Caquetá (FMNH 71118); Vaupés, Río Vaupés, in front of Tahuapunto (AMNH 78625). ECUADOR: Loja, El Porotillo, San José (MNHN 1934/1210); Morona–Santiago, Mendez Sur, Oriente (MNHN 1987/338); Napo, Archidona (MNHN 1932/2876), San José abajo (AMNH 66792). PERU: Amazonas, Guayabamba (AMNH type locality of *Sciurus* (*Microsciurus*) *peruanus* J. A. Allen), near Huampani, Río Cenepa (MVZ 154929), 12 km E of La Peca Nueva (LSUMZ 21876); Cusco, Quincemil (FMNH 75210); Junin, Chanchamayo (type locality of *Microsciurus rubrirostris* J. A. Allen); Loreto, boca Río Curaray (AMNH

71606), Hacienda Santa Elena, ca. 35 km NE of Tingo Maria (LSUMZ 17720), left bank at mouth of Río Yaquerana, alto Río Yavarí (FMNH 88991), Pebas. Río Amazonas (BM 28.7.21.74–78), Quebrada Pushaga, left bank Río Morona, alto Río Amazonas (FMNH 88984), San Fernando, left bank Río Yavari (FMNH 88990), Santa Rita, Iquitos (FMNH 87178), Sarayacu, Río Ucayali (AMNH 75279); Madre de Dios, Altamira, Manu (FMNH 98067); Ucayali, Lagarto, upper Río Ucayali (AMNH 76538).

SUBSPECIES: We regard *M. flaviventer* as monotypic (but see Remarks). Thorington and Hoffmann (2005) recognized seven subspecies, and Thorington et al. (2012) listed eight, three and four of which, respectively, we regard herein as full species.

NATURAL HISTORY: This species is found only in evergreen rainforest (Emmons and Feer 1997). Surprisingly, few ecological, reproductive, behavioral, or other natural history data are available for this wide-ranging squirrel. Emmons and Feer (1997) stated that individuals search actively over the trunks of large trees, on vines, and in treefalls, apparently searching for insects, but that they also feed on substances scraped from tree bark such as gum (as was suggested by A. C. Smith [1999], who observed two solitary individuals moving through the subcanopy). Youlatos (1999) noted that individuals primarily used the subcanopy or canopy at heights above 5 m. Emmons and Feer (1997) recorded ball nests of leaves lined with fibers in the top of palms. Eisenberg (1989) reported a litter size of two offspring. Buitron-Jurado and Tobar (2007) reported on associations between individuals of this species and bird flocks, with squirrels actively searching for insects at heights of 2–10 m in the company of birds.

REMARKS: *Microsciurus flaviventer* has a wide distribution, from the eastern slopes of the Andes to central Amazonia. In this vast area, authors have distinguished several taxa, all listed in the previous synonymy. Our study of material deposited in many collections revealed that this variation is real, but we have not been able to decide how much of it is of taxonomic value. There are important differences in size (Andean forms such as *napi* are large, for example). And the upper third premolar may be present or absent in different samples. The fur is also longer and denser in Andean samples, and the color, especially that of ventral surfaces, varies regionally. We believe that recognizing a single species for the whole of the samples examined is, in this case, a decision of convenience. It is quite possible that further study of this material will reveal the existence of additional taxa, either to be recognized as valid subspecies or as full species.

Microsciurus flaviventer is one of the two Colombian species of the genus occurring east of the Andes, the other being *M. santanderensis*. Their known distributions are interrupted

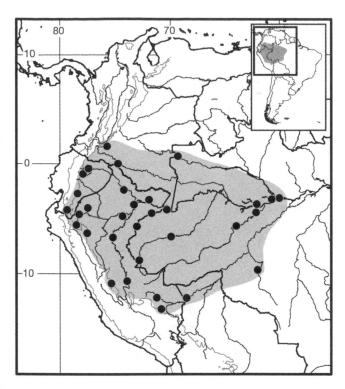

Map 8 Selected localities for *Microsciurus flaviventer* (●). Contour line = 2,000 m.

by a large gap in the eastern Andes, a region well sampled for other squirrels, and this gap likely represents a true disjunction. Salazar-Bravo, Yensen et al. (2002) extended the range of this species into northern Bolivia, reidentifying a specimen that S. Anderson (1997:397) originally assigned to *Sciurus ignitus* (= *Notosciurus pucheranii* herein).

Microsciurus isthmius (E. W. Nelson, 1899)
Isthmian Dwarf Squirrel

SYNONYMS:

(?) *Sciurus aestuans* var. *rufoniger*: J. A. Allen, 1877:757; part.

Sciurus (Microsciurus) isthmius E. W. Nelson, 1899a:77; type locality "Truando River [= Río Truandó], [Chocó,] Colombia."

Microsciurus isthmius: E.-L. Trouessart, 1904:329; name combination.

Sciurus (Microsciurus) palmeri Thomas, 1909:234; type locality "Sipi, Rio Sipi, Rio San Juan, Chocó, western Colombia."

Microsciurus isthmius vivatus Goldman, 1912b:4; type locality "near Cana, (altitude 3,500 feet), in the Pirri range of mountains, [Darien,] eastern Panama."

Microsciurus isthmius isthmius: J. A. Allen, 1914a:57; name combination.

Microsciurus palmeri: J. A. Allen, 1914a:160; name combination.

Microsciurus [(Microsciurus)] isthmius: Moore, 1959:203; name combination.

Microsciurus flaviventer isthmius: Cabrera, 1961:355; part; name combination.

Microsciurus mimulus isthmius: Handley, 1966:778; part; name combination.

DESCRIPTION: Adult mass 120 g (one female); head and body length 120–165 mm (N = 41), tail length 98–150 mm (N = 36), hindfoot length 32–45 mm (N = 39), and ear length 10–19 mm (N = 20). Dorsum dark brown to black, heavily speckled with yellow; postauricular patches absent; venter not sharply distinct from dorsum, generally orange in color but with mixture of orange and brown on belly.

DISTRIBUTION: *Microsciurus isthmius* occurs in rainforests west of the Andes in northwestern Colombia and adjacent Panama; in Colombia, it ranges in elevation from sea level to 800 m.

SELECTED LOCALITIES (Map 9; South America only): COLOMBIA: Antioquia, Alto Bonito, upper Río Sucio (AMNH 37645), Murindó (FMNH 44855); Cauca, Río Saija (FMNH 90210); Chocó, Bagadó (AMNH 34132), Baudó (AMNH 33812), Nóvita, Río Tamaná (BM 9.7.17.26), Río Nuqui, base of Baudó Mts. (USNM 292132), Río Truandó (type locality of *Sciurus [Microsciurus] isthmius* E. W. Nelson), Sipí, Río Sipí, Río San Juan (type locality of *Sciurus [Microsciurus] palmeri* Thomas); Nariño, Buenavista (AMNH 34158), La Guayacana (FMNH 89518); Valle Del Cauca, Mechenquito (ROM 63217), Río Mechengue (FMNH 90201), San José (AMNH 31680).

SUBSPECIES: We treat *M. isthmius* as monotypic.

NATURAL HISTORY: Goldman (1920: 144) reported specimens from about 1,150 m on Cerro Pirre, Darién, Panama (his *M. isthmius vivatus*), collected from lower branches or the trunks of trees in cloud forest.

REMARKS: Variation in the coat color is limited to the amount of black on the crown and rest of dorsum, which is slightly more pronounced in some specimens. J. A. Allen (1877:756–763) recognized five species of squirrels in South America, among which was *Sciurus aestuans* and in which he listed "*aestuans*" and "*rufoniger*" as two "varieties." These together encompassed all of the then-known smaller South American squirrels. This was a very broad concept, and even Allen himself (1914a, 1915) did not succeed in settling all the synonymic implications of his earlier lumping effort. E. W. Nelson (1899a:77–78) subsequently described *Sciurus (Microsciurus) isthmius* based on the material collected at the Río Truandó, Chocó, Colombia.

Cabrera (1961) proposed a new arrangement for the species level taxa in the genus *Microsciurus*, one in which *isthmius* E. W. Nelson and all other South American taxa

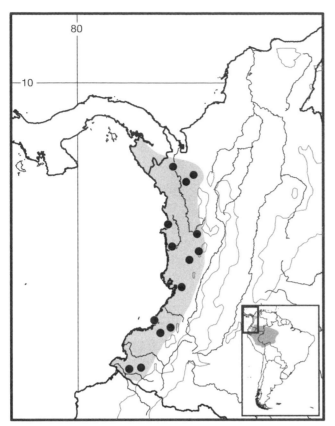

Map 9 Selected localities for *Microsciurus isthmius* (●). Range in Panama from Hall (1981). Contour line = 2,000 m.

were placed as subspecies, or synonyms, of a single species, namely, *M. flaviventer* (Gray). Cabrera's arrangement, however, was not uniformly followed (e.g., see Handley 1966; Hall 1981; and Honacki et al. 1982). Thorington et al. (2012) retained *isthmius* as a subspecies in their concept of *M. mimulus*.

Microsciurus mimulus (Thomas, 1898)

Western Dwarf Squirrel

SYNONYMS:

Sciurus (*Microsciurus*) *mimulus* Thomas, 1898f:266; type locality "Cachavi [= Cachabí], [Esmeraldas,] N. Ecuador, alt. 167 m."

Microsciurus mimulus: E.-L. Trouessart, 1904:329; first use of current name combination.

Microsciurus [(*Microsciurus*)] *mimulus*: Moore, 1959:203; name combination.

Microsciurus flaviventer mimulus: Cabrera, 1961:356; name combination.

DESCRIPTION: Head and body length 114–159 mm (N = 17), tail length 90–125 mm (N = 17), hindfoot length 30–38 mm (N = 16), and ear length 13–16 mm (N = 8). Dorsum dark brown to black, speckled with red or orange, sometimes with mid-dorsal dark stripe; postauricular

patches absent; venter not sharply distinct from dorsum, with color ranging from orange to pale yellow with basal third of hairs gray, with belly orange to pale yellow with some brown.

DISTRIBUTION: *Microsciurus mimulus* occurs in rainforests west of the Andes in northwestern Ecuador and southwestern Colombia, at elevations from sea level to 1,000 m.

SELECTED LOCALITIES (Map 10): COLOMBIA: Nariño, Barbacoas (AMNH 34159), La Guayacana (USNM 309028). ECUADOR: Esmeraldas, Cachabí (type locality of *Sciurus* [*Microsciurus*] *mimulus* Thomas), Carondelet (USNM 113310); Imbabura, Lita (BM 1.1.6.3).

SUBSPECIES: *Microsciurus mimulus* is monotypic.

NATURAL HISTORY: We have found no information about the natural history of this species.

REMARKS: Thomas (1898f:266–267) described *Sciurus* (*Microsciurus*) *mimulus* based on three specimens from northwestern Ecuador. This was the first name applied to dwarf squirrels from South America west of the Andes. Cabrera (1961:356) considered *M. flaviventer* to be the only South American species in the genus, using all other names either for subspecies (including *mimulus* Thomas) or as synonyms. As in many occasions throughout his catalog of South American mammals, this decision was based on the literature and on his own (limited, in this case) experience with the group. Handley (1966:777) considered *M. mimulus* to be a good species, and although he did not provide a reason for this decision, his arrangement

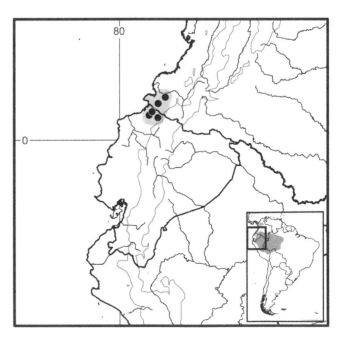

Map 10 Selected localities for *Microsciurus mimulus* (●). Contour line = 2,000 m.

met with wide acceptance and has been adopted by most subsequent authors (e.g., Emmons and Feer 1997; Thorington and Hoffmann 2005). Our concept of *M. mimulus* is more restricted geographically than that espoused by these authors, generally equivalent to that given by Thorington et al (2012:35) for the nominotypical form of their polytypic *M. mimulus*.

Some specimens from Nariño, Colombia, show individual variation in belly color, from almost black punctuated with orange to brown mixed with orange. The type series of *M. mimulus* and other specimens from northwestern Ecuador have a black mid-dorsal stripe continuing to the crown (Thomas 1898f; J. A. Allen 1914a). This character, however, while present in Ecuadorean material, is variable in specimens from farther to the north in Nariño, Colombia, with only one specimen in five presenting it fully.

Microsciurus otinus (Thomas, 1901)
White-eared Dwarf Squirrel
SYNONYMS:

Sciurus (*Microsciurus*) *otinus* Thomas, 1901c:193; type locality "Medellin, [Antioquia,] Colombia."

Microsciurus otinus: E.-L. Trouessart, 1904:329; first use of current name combination.

Microsciurus flaviventer otinus: Cabrera, 1961:357; name combination.

DESCRIPTION: Head and body length 120–141 mm (N = 15), tail length 95–132 mm (N = 12), hindfoot length 35–39 mm (N = 13), and ear length 11–16 mm (N = 6). Dorsum dark brown streaked with pale yellow; postauricular patches absent but ears tipped with white; throat and chest washed with reddish orange, belly mixed brown, dark gray, and red; tail frosted whitish.

DISTRIBUTION: *Microsciurus otinus* occurs west of the Andes, in rainforests in the Colombian departments of Antioquia, Bolívar, and Córdoba, at elevations from 200 to 1,500 m.

SELECTED LOCALITIES (Map 11): COLOMBIA: Antioquia, La Tirana, 25 km S and 22 km W of Zaragoza (USNM 499515), Medellín (type locality of *Sciurus* [*Microsciurus*] *otinus* Thomas), Purí, above Caceres (FMNH 70046), Valdívia, right bank of lower Río Cauca (AMNH 37640); Bolívar, 15 mi W of Simiti, 6 mi above Santa Rosa (USNM 282755); Córdoba, Socorré, upper Río Sinú (FMNH 69034).

SUBSPECIES: *Microsciurus otinus* is monotypic.

NATURAL HISTORY: Rojas-Robles et al. (2008) reported on the importance of a species of *Microsciurus* in the short-distance dispersal (typically <50 m) of the large seeds of *Oenocarpus bataua*, a subcanopy and canopy palm in premontane humid forest. Because their study site

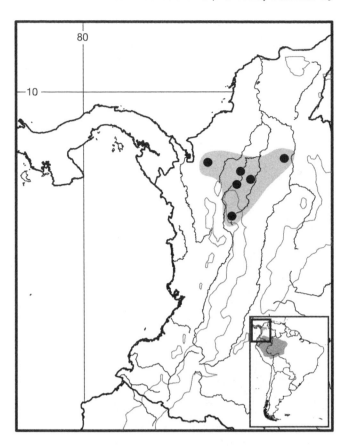

Map 11 Selected localities for *Microsciurus otinus* (●). Contour line = 2,000 m.

was on the Río Porce in the Cordillera Central of Antioquia department, Colombia, the taxon they studied was likely *M. otinus*, not *M. mimulus* as they originally reported. Otherwise, little information is available on this species.

REMARKS: Thomas (1901c) contrasted *M. otinus* with the geographically adjacent *M. isthmius*, noting that the two were similar in all external aspects except that *otinus* had white pinnae. Although we recognize *M. otinus* here as a valid species, this scheme needs to be tested by additional collecting and comparisons of existing specimens. Thorington et al. (2012) included *otinus* as a subspecies of *M. flaviventer*.

Microsciurus sabanillae Anthony, 1922
Sabanilla Dwarf Squirrel
SYNONYMS:

Microsciurus sabanillae Anthony, 1922: 2, type locality "Sabanilla, Prov. de Loja, Ecuador; altitude 5700 feet."

Microsciurus flaviventer sabanillae: Cabrera, 1961:357; name combination.

DESCRIPTION: Head and body length 134–151 mm (N = 3), tail length 140–148 mm (N = 3), and hindfoot length 40–42 mm (N = 3). Dorsum dark brown to black streaked with red; postauricular patches absent; throat and chest washed with red, belly mixed dark gray and red.

DISTRIBUTION: *Microsciurus sabanillae* is known from Ecuador and adjacent northwestern Peru, in midelevation forests on the eastern Andean slope, at an elevational range from 600 to 1,750 m.

SELECTED LOCALITIES (Map 12): ECUADOR: Loja, Sabanilla, near Río Destrozo, a small tributary of Río Zamora (type locality of *Microsciurus sabanillae* Anthony); Napo, Río Napo (BM 54.406). PERU: Amazonas, Santa Rosa (AMNH 69237).

SUBSPECIES: *Microsciurus sabanillae* is monotypic.

NATURAL HISTORY: Nothing has been published on the ecology of this species.

REMARKS: Contra Thorington et al. (2012), we recognize *M. sabanillae* as a species distinct from *M. flaviventer* on the basis of details of ventral coloration, the lack of the postauricular patch, and partial sympatry. However, as in the case of *M. otinus*, this kind of variation has to be considered in the context of the highly variable *M. flaviventer*. The validity of supposed species-level taxa such as *M. sabanillae*, *M. otinus*, and several other forms here treated as synonyms of *M. flaviventer* requires additional study and further collecting efforts.

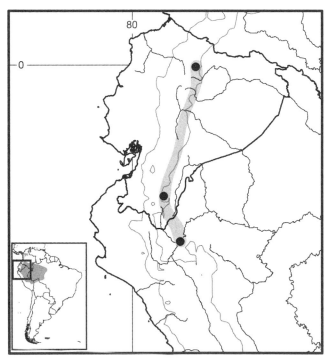

Map 12 Selected localities for *Microsciurus sabanillae* (●). Contour line = 2,000 m.

Microsciurus santanderensis (Hernández-Camacho, 1957)
Santander Dwarf Squirrel
SYNONYMS:

Sciurus pucheranii santanderensis Hernández-Camacho, 1957: 219, type locality "Meseta de los Caballeros, 5 km NW of La Albania, San Vicente de Chururí [= San Vicente de Chucurí], Santander, Colombia."

Microsciurus santanderensis: Hernández-Camacho, 1960: 360; first use of current name combination.

DESCRIPTION: (The following summarized from the original description, which was based on five specimens; no specimens seen by us.) Head and body length 133–160 mm, tail length 136–152 mm, hindfoot length 42–45 mm, and ear length 15–18 mm. Dorsum dark brown, with mid-dorsal black stripe; pale ochraceous postauricular patches present; venter with mixed colors, but predominantly light ochraceous salmon.

DISTRIBUTION: This species occurs in the rainforests on the western slopes of the eastern Cordillera, and from the right bank of the middle Río Magdalena, in the vicinity of Barrancabermeja and San Vicente de Chucurí, Santander, Colombia. Thorington et al. (2012) extended the range west and northwest to include the upper portion of the Cordillera Occidental, but did not cite specific localities.

SELECTED LOCALITIES (Map 13): COLOMBIA: Santander, Barancabermeja (Hernández-Camacho 1960), Meseta de Los Caballeros, 5 km NW of La Albania, San Vicente de Chucurí (type locality of *Sciurus pucheranii santanderensis* Hernández-Camacho).

SUBSPECIES: *Microsciurus santanderensis* is monotypic.

NATURAL HISTORY: Little is known about the ecology of this species, other than its distribution in humid forests in the low to mid elevations in the Cordillera de La Paz, Colombia (Hernández-Camacho 1960). Thorington et al. (2012) included marshlands and montane forests (at an elevational range of 2,700–3,800 m) among habitats this species occupies.

REMARKS: Variation in this species has not been studied beyond what Hernández-Camacho (1957, 1960) presented in his original description and subsequent redescription. It appears to be a valid species because it differs from similarly colored squirrels occurring east of the Andes. For example, *M. santanderensis* differs from *M. flaviventer* by the possession of a mid-dorsal dark stripe and from *M. sabanillae* by its possession of postauricular patches. Although originally described as a subspecies of *Notosciurus pucheranii* (as *Sciurus pucheranii santanderensis*) by Hernández-Camacho, he (1960) subsequently transferred *santanderensis* to *Microsciurus*, as more specimens became available.

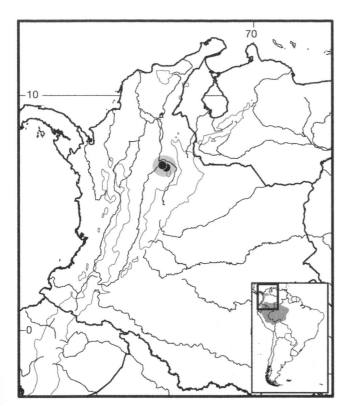

Map 13 Selected localities for *Microsciurus santanderensis* (●). Contour line = 2,000 m.

The remarkable distribution gap between this species and *M. flaviventer*, which happens to be in a region of Colombia well sampled for squirrels in general, and this species' resemblance to *Notosciurus pucheranii*, may indicate that *M. santanderensis* does not belong in the genus *Microsciurus*. Because we have not seen specimens (and, so far as we know, neither has anyone but the author of the name), we keep it among the dwarf squirrels, but with this cautionary note.

Microsciurus similis (E. W. Nelson, 1899)
Chocó Dwarf Squirrel

SYNONYMS:

Sciurus (Microsciurus) similis E. W. Nelson, 1899a:78; type locality "near Cali, Cauca Valley, [Valle de Cauca,] Colombia (alt. 6000 ft.)".

Microsciurus similis: E.-L. Trouessart, 1904:329; first use of current name combination.

Sciurus (Microsciurus) similis fusculus Thomas, 1910d:503; type locality "Juntas, Rio San Juan. Alt. 400′, Chocó, W. Colombia."

Microsciurus similis similis: J. A. Allen, 1914a:153; name combination.

Microsciurus similis fusculus: J. A. Allen, 1914a:154; name combination.

Microsciurus [(*Microsciurus*)] *similis*: Moore, 1959:203; name combination.

Microsciurus flaviventer similis: Cabrera, 1961:357; name combination.

Microsciurus alfari fusculus: Handley, 1966:777; name combination.

DESCRIPTION: Head and body length 112–147 mm (N = 23), tail length 82–133 mm (N = 17), hindfoot length 28–38 mm (N = 21), and ear length 8–16 mm (N = 17). Dorsum dark brown to black punctuated with red; postauricular patches absent; venter not sharply distinct from the dorsum, colored red to mixed red and dark gray on throat and chest but dark gray to dark brown heavily streaked with red on belly.

DISTRIBUTION: *Microsciurus similis* occurs in rainforests west of the Andes in western Colombia, across an elevational range from 130 to 2,200 m.

SELECTED LOCALITIES (Map 14): COLOMBIA: Antioquia, Alto Bonito, upper Río Sucio (AMNH 37643); Cauca, Munchique (FMNH 86788), Rio Munchique (FMNH 90200); Chocó, Baudó (AMNH 33184, 33186), Juntas de Tamaná (type locality of *Sciurus* [*Microsciurus*] *similis fusculus* Thomas), Río Jurubidá, Baudó Mts.

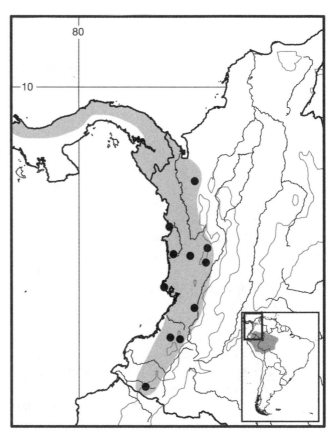

Map 14 Selected localities for *Microsciurus similis* in South America (●). Contour line = 2,000 m.

(USNM 292134), Río San Juan (BM 10.7.16.1); Nariño, Ricaurte (USNM 309027); Risaralda, Siató, Rio Siató, San Juan (FMNH 18877); Valle del Cauca, Albán (= Salencio) (AMNH 32852), Cali (BM 99.9.6.56).

SUBSPECIES: We treat *M. similis* as monotypic, although Thomas (1910d) differentiated specimens from Chocó in western Colombia as *M. s. fusculus.*

NATURAL HISTORY: We found no information about the natural history of this species.

REMARKS: *Microsciurus similis*, on the scale of Maps 9 and 14, appears to broadly overlap the range of *M. isthmius.* The two species, however, occupy separate elevational ranges in the Colombian Andes, with limited overlap. The lowest elevation recorded for *M. similis* is 150 m, and the highest for *M. isthmius* is 800 m. This species is also sympatric with *M. mimulus* in Nariño department in southwestern Colombia.

Microsciurus simonsi (Thomas, 1900)
Simons's Dwarf Squirrel

SYNONYMS:

Sciurus (Microsciurus) simonsi Thomas, 1900f:294; type locality "Porvenir, near Zaparal, Province of Bolivar, Ecuador. Altitude 1500 m."

Microsciurus simonsi: E.-L. Trouessart, 1904:329; first use of current name combination.

Microsciurus similis fusculus: Lönnberg, 1913:26; not *fusculus* Thomas.

Microsciurus flaviventer simonsi: Cabrera, 1961:358; name combination.

DESCRIPTION: Head and body length 126–154 mm (N = 10), tail length 104–124 mm (N = 10), hindfoot length 32–37 mm (N = 11), and ear length 14–16 mm (N = 11). Dorsum dark brown streaked with orange; postauricular patches absent; ventrum mixture of dark gray, pale gray, pale yellow on throat, dark gray to dark brown and yellow on chest and belly.

DISTRIBUTION: *Microsciurus simonsi* occurs west of the Andes in the Ecuadorean provinces of Bolívar and Pichincha, at elevations from 250 to 1,500 m.

SELECTED LOCALITIES (Map 15): ECUADOR: Bolívar, Porvenir (type locality of *Sciurus [Microsciurus] simonsi* Thomas); Pichincha, Ila, 36 mi SW of Santo Domingo (BM 15.1.1.25), Mindo, Río Blanco (BM 13.10.24.36), Santo Domingo (BM 15.1.1.23–24).

SUBSPECIES: *Microsciurus simonsi* is monotypic.

NATURAL HISTORY: Nothing has been published on the ecology or behavior of this species.

REMARKS: *Microsciurus simonsi* is the southernmost species of the genus west of the Andes. Each of the four trans-Andean species can be distinguished mainly by its ventral coloration and the color of the subterminal band of dorsal guard hairs. Thorington et al. (2012) in-

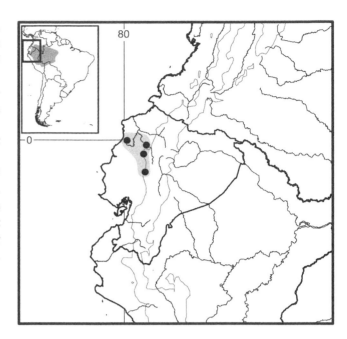

Map 15 Selected localities for *Microsciurus simonsi* (●). Contour line = 2,000 m.

cluded *simonsi* as a subspecies of a highly polytypic *M. flaviventer.*

Genus *Notosciurus* J. A. Allen, 1914

Notosciurus is a genus of medium-sized Central and South American squirrels occurring from eastern Costa Rica through Colombia, Venezuela, Ecuador, Peru, western Amazonian Brazil, and Bolivia, to northernmost Argentina. In South America, its distribution is mostly associated with the Andean Cordillera, but in western Colombia, eastern Peru, and neighboring areas of Brazil, it is also found in lowland evergreen forests. The distribution of the genus is largely parapatric to that of *Guerlinguetus*, but both genera may be found living together at the margins of their ranges, particularly in the Marañón, Solimões, and Madeira river basins of Amazonia and possibly also in northwestern Mato Grosso state in Brazil.

Species of *Notosciurus* can be distinguished from all other South American squirrels except for *Guerlinguetus* by their size (larger than species of *Microsciurus*, *Sciurillus*, and *Syntheosciurus*, but smaller than those of *Hadrosciurus* and *Simosciurus*). Compared with the similarly sized *Guerlinguetus*, *Notosciurus* possess three pairs of mammae instead of four and typically have orange postauricular patches, which are generally lacking in *Guerlinguetus*. *Notosciurus* is highly variable in size and mass, varying from 136–418.5 g (N = 31 adults). The head and body length range from 151–262 mm (N = 469), tail length from 134.5–265 mm (N = 459), hindfoot length from 34–65 mm (N = 455), and ear length from 15–32 mm (N = 359). The skull has the orbitotemporal fossa equally divided by a well-developed postorbital process, and an upper second molar with two parallel, buccolingual crests. These crests are obliquely oriented relative to the anteroposterior axis of the skull.

We recognize two polytypic species: *Notosciurus granatensis* with six subspecies (five in South America), and *N. pucheranii* with four subspecies. Our concept of both species is considerably different from that of previous authors, most recently Thorington et al. (2012). These authorities, for example, recognized 32 subspecies for *granatensis*, 29 in South America. They also treated the trans- and cis-Andean populations of our *N. pucheranii* as separate species.

SYNONYMS:

Sciurus: Pucheran, 1845:336; part (description of *rufoniger* and *chrysurus*); not *Sciurus* Linnaeus, 1758.

Funambulus: Fitzinger, 1867a:487; part (listing of *pucheranii* and *chrysurus*); not *Funambulus* Lesson, 1835.

Macroxus: Gray, 1867:429; part (description of *ignitus*); not *Macroxus* F. Cuvier, 1823.

Sciurus (*Guerlinguetus*): J. A. Allen, 1899b:217; part (description of *quebradensis*); not *Guerlinguetus* Gray, 1821.

Notosciurus J. A. Allen, 1914d:585; type species *Notosciurus rhoadsi* J. A. Allen, 1914, by original designation.

Guerlinguetus: J. A. Allen, 1914d:587; part (description of *pucheranii salentensis*, *hoffmanni quindianus*, *hoffmanni manavi*, and *griseimembra*); not *Guerlinguetus* Gray, 1821.

Leptosciurus J. A. Allen, 1915a:199; type species *Sciurus rufoniger* Pucheran, 1845 (preoccupied by *Sciurus rufonigra* Gray, 1842:263, replaced by *Funambulus pucheranii* Fitzinger, 1867).

Mesosciurus J. A. Allen, 1915a:212; type species *Sciurus aestuans* var. *hoffmanni* Peters, 1864, by original designation.

Histriosciurus J. A. Allen, 1915a:213, 236; as subgenus; type species *Sciurus gerrardi* Gray, 1861, by original designation (p. 236).

Mesosciurus (*Mesosciurus*): J. A. Allen; 1915a:216; name combination.

Guerlinguetus (*Mesosciurus*): J. A. Allen; 1915a:228; name combination, but *lapsus calami*.

Mesosciurus (*Histriosciurus*): J. A. Allen; 1915a:236; name combination.

Sciurus (*Mesosciurus*): Thomas, 1921k:609; name combination.

Sciurus (*Leptosciurus*): Anthony, 1923a:5; name combination.

Sciurus [(*Notosciurus*)]: Ellerman, 1940:344; name combination.

REMARKS: The midsized South American squirrels with three pairs of mammae were organized with all other South American species until J. A. Allen (1915a) created new genera to accommodate the several groups of apparently related species he was able to recognize. Among the new genera were *Leptosciurus* (type species *Sciurus rufoniger* Pucheran, 1845) and *Mesosciurus* (type species *Sciurus aestuan* var. *hoffmanni* Peters, 1864), the latter with two subgenera: the nominotypical one and *Histriosciurus* (type species *Sciurus gerrardi* Gray). Allen did not compare *Leptosciurus* and *Mesosciurus* with each other, except in his key for the genera; in his characterizations, he chose to compare both only to *Guerlinguetus*, and quite superficially. Furthermore, he distinguished between *Leptosciurus* and *Mesosciurus* only on the basis of size, which overlapped slightly in his own key (J. A. Allen 1915a:187).

However, a year earlier, J. A. Allen (1914d) had described a new genus and species based on a single specimen from Ecuador, which he named *Notosciurus rhoadsi*. We have not seen the holotype of this taxon, but Hershkovitz (1947:8) identified it as a young specimen of *Sciurus hoffmanni soederstroemi* Stone. Thorington and Hoffmann (2005:761) also listed Allen's *rhoadsi* as a synonym of *soederstroemi* Stone, which they regarded as a subspecies of a polytypic *Sciurus granatensis* Humboldt. These authors also included *chrysuros* Pucheran and *hoffmanni* Peters as additional subspecies of *S. granatensis*. Because we consider *soederstroemi* Stone to be a synonym of *chrysuros* Pucheran and we treat *granatensis* Humboldt as a species in the genus we discuss here, *Notosciurus* J. A. Allen takes priority over *Leptosciurus* J. A. Allen as the generic name for these squirrels.

We do not recognize subgenera in *Notosciurus*, as all species share the presence of three pairs of mammae; additionally, most of the included species and subspecies possess postauricular patches, both characteristics that distinguish our concept of *Notosciurus* from *Guerlinguetus*. These two genera, as here recognized, are largely parapatric, with marginal sympatry in parts of western Brazil and eastern Peru. See the discussion on the validity of the presently recognized supraspecific groups in the family account.

KEY TO THE SPECIES OF *NOTOSCIURUS*:

1. Size smaller, head and body length approximately 150–210 mm; ventral hairs of various colors but usually with a gray base (except in Bolivia and Argentina) . . . . *Notosciurus pucheranii*

1'. Size larger (but overlapping; in eastern Andean Colombia, where both species occur in sympatry [*N. g. chrysuros* and *N. p. pucheranii*], they diverge greatly in size), head and body length approximately 170–265 mm; ventral hairs of various colors but usually of a single color throughout (species does not occur in Bolivia or Argentina) . . . . *Notosciurus granatensis*

Notosciurus granatensis (Humboldt, 1811)
Red-tailed Squirrel

SYNONYMS: See under subspecies.

DESCRIPTION: Highly variable species, with head and body length 169–262 mm (N = 348 adult specimens), tail length 145–265 mm (N = 342), hindfoot length 40–65 mm (N = 330), ear length 17–32 mm (N = 250), and mass 190–420 g (N = 25). We briefly describe each subspecies following its synonymy.

DISTRIBUTION: This species ranges from Costa Rica to Colombia, Ecuador, and Venezuela. In South America, *N. granatensis* is essentially linked to the Andes, but may also occur in adjacent lowlands and coastal areas at sea level.

SELECTED LOCALITIES (Map 16): COLOMBIA: Antioquia, Puerto Valdivia (type locality of *Mesosciurus* (*Histriosciurus*) *gerrardi valdiviae* J. A. Allen), Sonsón (USNM 239942); Bolívar, Cartagena (type locality of *Sciurus granatensis* Humboldt); Cauca, Cocal (type locality of *Sciurus milleri* J. A. Allen); Cesar, Colonia Agricola de Caracolicito (type locality of *Sciurus granatensis agricolae* Hershkovitz); Chocó, Baudó (type locality of *Mesosciurus* (*Histriosciurus*) *gerrardi baudensis* J. A. Allen), Río Salaqui (type locality of *Sciurus gerrardi salaquensis* J. A. Allen); Córdoba, Catival, upper Río San Jorge (FMNH 69008), Socorré, upper Río Sinú (FMNH 68987); Cundinamarca, Bogotá (type locality of *Sciurus chrysuros* Pucheran and *Sciurus hyporrhodus* Gray); Huila, La Palma (AMNH 33669); La Guajira, Laguna de Junco (USNM 271318), Puerto Estrela (USNM 216696); Magdalena, Don Diego (AMNH 23317); Meta, Villavicencio (USNM 241350); Nariño, Guayacana (USNM 309034); Norte de Santander, El Guayabal, 10 mi from San José de Cucuta (type locality of *Sciurus gerrardi cucutae* J. A. Allen), San Calixto, Río Tarra (type locality of *Sciurus granatensis tarrae* Hershkovitz); Valle del Cauca, Río Frio, Río Cauca (type locality of *Guerlinguetus hoffmanni quindianus* J. A. Allen), Río Raposo, (USNM 334696). ECUADOR: Carchi, Atal, 5 mi SE of San Gabriel (type locality of *Sciurus candelensis carchensis* W. P. Harris and Hershkovitz); Chimborazo, Chunchi (type locality of *Notosciurus rhoadsi* J. A. Allen); Esmeraldas, Panbilar (USNM 113314); Manabí, Manaví, Río de Oro (type locality of *Guerlinguetus hoffmanni manavi* J. A. Allen); Napo, Sumaco abajo (AMNH 68159); Pichincha, Mt. Pichincha (type locality of *Sciurus hoffmanni soederstroemi* Stone). TRINIDAD AND TOBAGO: Tobago, Runnemede (USNM 461886); Trinidad, Carenage (AMNH 14008), Diego Martin (AMNH 188365). VENEZUELA: Anzoátegui, Cantaura (USNM 283879); Apure, Nulita, 3 km N of Nula (USNM 442011), Tama (type locality of *Sciurus griseogena tamae* Osgood); Barinas, Hato Corozal, Caño Tutumito, Arismendi (type locality of *Sciurus* [*Guerlinguetus*] *granatensis llanensis* Mondolfi

and Boher); Carabobo, Urama, E of Yaracuy (AMNH 130289); Guárico, Guárico (AMNH 32877); Mérida, Escorial, Sierra de Mérida (type locality of *Sciurus griseogena meridensis* Thomas); Monagas, near San Agustin, 2 km N and 4 km E of Caripe (USNM 409776); Nueva Esparta, El Valle, Isla de Margarita (type locality of *Sciurus nesaeus* G. M. Allen); Sucre, Cristobal Colón (AMNH 36156), Nevera (AMNH 69911); Vargas, Galiparé, Cerro del Ávila, near Caracas (type locality of *Sciurus griseogena klagesi* Thomas); Zulia, El Panorama, Río Aurare (type locality of *Sciurus granatensis maracaibensis* Hershkovitz), Hacienda Platanal, 18 km N and 49 km W of Maracaibo (USNM 442017), Novita, 3.1 km S and 19 km W of Machiquos (USNM 442013).

SUBSPECIES: We recognize six subspecies of *N. granatensis*, with five in South America. A sixth subspecies, *N. g. hoffmanni*, occurs only in Costa Rica and western Panama. For comparison with South American taxa, specimens of *hoffmanni* possess postauricular patches, a reddish dorsum streaked with orange or yellow but without a dark middorsal area, an orange venter with gray-based hairs, and a homogeneously colored tail fringed and/or streaked with orange.

N. g. chapmani (J. A. Allen, 1899)

SYNONYMS:

Sciurus chapmani J. A. Allen, 1899a:16, type locality "Caparo, Trinidad," Trinidad and Tobago.

Sciurus (*Guerlinguetus*) *aestuans quebradensis* J. A. Allen, 1899b:217, type locality "Quebrada Secca [= Quebrada Seca, = Villarroel], [Sucre,] Venezuela."

Sciurus (*Guerlinguetus*) *chapmanni* E.-L. Trouessart, 1904: 327; name combination, and incorrect subsequent spelling of *Sciurus chapmani* J. A. Allen.

Sciurus tobagensis Osgood, 1910:27, type locality "Tobago Island, Caribbean Sea," Trinidad and Tobago.

Mesosciurus (*Mesosciurus*) *chapmani*: J. A. Allen, 1915a: 230; name combination.

Mesosciurus (*Mesosciurus*) *chapmani tobagensis*: J. A. Allen, 1915a:232; name combination.

Sciurus [(*Guerlinguetus*)] *chapmani chapmani*: Ellerman, 1940:341; name combination.

Sciurus [(*Guerlinguetus*)] *chapmani tobagensis*: Ellerman, 1940:341; name combination.

Sciurus granatensis chapmani: Hershkovitz, 1947:37; name combination.

Sciurus granatensis tobagensis: Hershkovitz, 1947:37; name combination.

Sciurus (*Guerlinguetus*) *granatensis chapmani*: Goodwin, 1953:277; name combination.

Sciurus (*Guerlinguetus*) *granatensis*: Eisenberg, 1989:331; part; name combination.

Notosciurus g. chapmani occurs on the islands of Trinidad and Tobago as well as neighboring coastal Venezuela, including both lowlands and coastal ranges. Specimens on the islands average slightly smaller than those on the mainland but are otherwise indistinguishable. Top of head and dorsum dark brown streaked with orange or pale orange; postauricular patches absent (the only case of this, along with *N. g. morulus*, in the genus); venter differently colored than sides, being washed with orange to reddish orange on the throat, chest, and belly (hairs with a short, gray base and orange on the remainder); and tail colored like the dorsum, but the distal tips of hairs are washed with red.

N. g. chrysuros (Pucheran, 1845)

SYNONYMS:

Sciurus chrysuros Pucheran, 1845:337, type locality "Colombie (Santa-Fé de Bogota)," Cundinamarca, Colombia.

Sciurus chrysurus: Wagner, 1850:272; unnecessary Latinization of *chrysuros* Pucheran.

Sciurus hyporrhodus Gray, 1867:419; type locality "Santa Fé de Bogotá," Cundinamarca, Colombia.

Macroxus griseogena Gray, 1867:429; type locality given as a number of localities from Mexico to Colombia and Venezuela.

[*Funambulus*] *chrysurus*: Fitzinger, 1867a:487; name combination, and incorrect Latinization of *chrysuros* Pucheran.

Sciurus griseogenys Alston, 1878:667; unjustified emendation of *griseogena* Gray.

Sciurus rufo-niger: Alston, 1878:669; part.

Sciurus aestuans: Thomas, 1880:400; part; not *aestuans* Linnaeus, 1766.

Macroxus chrysuros: E.-L. Trouessart, 1897:427; part; name combination.

Sciurus griseogena: W. Robinson and Lyon, 1901:144; name combination.

Sciurus griseogena meridensis Thomas, 1901c:192; type locality "Escorial, Sierra de Merida. Alt. 2500 m," Mérida, Venezuela.

Sciurus griseogena tamae Osgood, 1912:48; type locality "Paramo de Tama, [Norte de Santander,] Colombia. Alt. 6000–7000 ft."

Sciurus hoffmanni: J. A. Allen, 1912:90; part; name combination.

Notosciurus rhoadsi J. A. Allen, 1914d:585; type locality "Pagma Forest, Chunchi (altitude 6300 feet), [Chimborazo,] Ecuador."

Sciurus hoffmanni söderströmi Stone, 1914:14; type locality "Mt. Pichincha," Pichincha, Ecuador.

Sciurus griseogena klagesi Thomas, 1914d:240; type locality "Galifaré, Cerro del Avila [= Pico Ávila], near Caracas, [Vargas,] Venezuela. Alt. 6500′."

Guerlinguetus griseimembra J. A. Allen, 1914d:589; type locality "Buena Vista (altitude 4500 ft.), eastern slope of Eastern Andes, about 50 miles southeast of Bogotá," Meta, Colombia.

Guerlinguetus candelensis J. A. Allen, 1914d:590; type locality "La Candela (altitude 6500 ft.), near San Agustin, Huila, Colombia."

Sciurus gerrardi cucutae J. A. Allen, 1914d:592; type locality "El Guayabal, 10 miles north of San José de Cucuta, [Norte de Santander,] Colombia (near Venezuela boundary)."

Mesosciurus hoffmanni hoffmanni: J. A. Allen, 1915a:216; part; name combination.

Mesosciurus (Mesosciurus) hoffmanni hyporrhodus: J. A. Allen, 1915a:223; name combination.

Mesosciurus (Mesosciurus) griseogena griseogena: J. A. Allen, 1915a:226; name combination.

Guerlinguetus (Mesosciurus) griseogena meridensis: J. A. Allen, 1915a:228; name combination.

Mesosciurus (Mesosciurus) griseimembra: J. A. Allen, 1915a:233; name combination.

Mesosciurus (Mesosciurus) candalensis J. A. Allen, 1915a:235; name combination and incorrect subsequent spelling of *Guerlinguetus candelensis* J. A. Allen.

Mesosciurus (Histriosciurus) gerrardi cucutae: J. A. Allen, 1915a:247; name combination.

Mesosciurus ferminae Cabrera, 1917:49; type locality "Baeza (Ecuador oriental)," Napo, Ecuador.

Mesosciurus candelensis sumaco Cabrera, 1917:51; type locality "San José, al pie del monte Sumaco, [Napo,] Ecuador oriental."

Sciurus gerrardi imbaburae W. P. Harris and Hershkovitz, 1938:1; type locality "Peñaherrera (Intag) western subtropical part of Imbabura Province, Ecuador; altitude, approximately 1500 meters."

Sciurus candelensis carchensis W. P. Harris and Hershkovitz, 1938:3; type locality "Atal, about five miles southeast of San Gabriel, Montúfar, Carchi Province, Ecuador, in the cold temperate rainforests of the western slopes of the eastern cordillera of the Andes. Altitude about 2900 meters."

Mesosciurus griseogena meridensis: Cabrera and Yepes, 1940:194; name combination.

Mesosciurus caudeleusis carchensis: Cabrera and Yepes, 1940:194; name combination and incorrect subsequent spelling of *Guerlinguetus candelensis* J. A. Allen.

Mesosciurus gerrardi imbaburae: Cabrera and Yepes, 1940:195; name combination.

Mesosciurus gerrardi cucutae: Cabrera and Yepes, 1940:195; name combination.

Sciurus [(Guerlinguetus)] hoffmanni hyporrhodus: Ellerman, 1940:340; name combination.

Sciurus [(*Guerlinguetus*)] *hoffmanni soderstromi*: Ellerman, 1940:340; name combination.

Sciurus [(*Guerlinguetus*)] *griseogena griseogena*: Ellerman, 1940:340; name combination.

Sciurus [(*Guerlinguetus*)] *griseogena meridensis*: Ellerman, 1940:340; name combination.

Sciurus [(*Guerlinguetus*)] *griseimembra*: Ellerman, 1940: 341; name combination.

Sciurus [(*Guerlinguetus*)] *candelensis candelensis*: Ellerman, 1940:341; name combination.

Sciurus [(*Guerlinguetus*)] *candelensis sumaco*: Ellerman, 1940:341; name combination.

Sciurus [(*Guerlinguetus*)] *ferminae*: Ellerman, 1940:341; name combination.

Sciurus [(*Guerlinguetus*)] *gerrardi cucutae*: Ellerman, 1940: 342; name combination.

Sciurus [(*Notosciurus*)] *rhoadsi*: Ellerman, 1940:344; name combination.

Sciurus granatensis tarrae Hershkovitz, 1947:26; type locality "Rio Tarra, a small tributary of the upper Catatumbo, San Calixto, Department of Norte de Santander, Colombia; altitude 200 meters."

Sciurus granatensis griseogena: Hershkovitz, 1947:29; name combination.

Sciurus granatensis meridensis: Hershkovitz, 1947:30; name combination.

Sciurus granatensis salaquensis: Hershkovitz, 1947:35; name combination.

Sciurus granatensis chrysurus: Hershkovitz, 1947:36; name combination, and unnecessary Latinization of *chrysuros* Pucheran.

Sciurus granatensis griseimembra: Hershkovitz, 1947:36; name combination.

Sciurus granatensis candalensis: Hershkovitz, 1947:36; name combination.

Sciurus granatensis carchensis: Hershkovitz, 1947:37; name combination.

Sciurus granatensis söderströmi: Hershkovitz, 1947:37; name combination, and incorrect use of diacritical marks (ICZN 1999:Art. 32.5.2.1).

Sciurus granatensis imbaburae: Hershkovitz, 1947:37; name combination.

Sciurus granatensis ferminae: Hershkovitz, 1947:37; name combination.

Sciurus granatensis sumaco: Hershkovitz, 1947:37; name combination.

Sciurus (*Guerlinguetus*) *granatensis griseimembra*: Goodwin, 1953:276; name combination.

Sciurus (*Guerlinguetus*) *granatensis candelensis*: Goodwin, 1953:277; name combination.

Sciurus (*Guerlinguetus*) *granatensis carchensis*: Cabrera, 1961:363; name combination.

Sciurus (*Guerlinguetus*) *granatensis chrysurus*: Cabrera, 1961:363; name combination, and incorrect Latinization of *chrysuros* Pucheran.

Sciurus (*Guerlinguetus*) *granatensis ferminae*: Cabrera, 1961:363; name combination.

Sciurus (*Guerlinguetus*) *granatensis griseogena*: Cabrera, 1961:364; name combination.

Sciurus (*Guerlinguetus*) *granatensis imbaburae*: Cabrera, 1961:365; name combination.

Sciurus (*Guerlinguetus*) *granatensis meridensis*: Cabrera, 1961:366; name combination.

Sciurus (*Guerlinguetus*) *granatensis söderströmi*: Cabrera, 1961:368; name combination.

Sciurus (*Guerlinguetus*) *granatensis sumaco*: Cabrera, 1961: 368; name combination.

Sciurus (*Guerlinguetus*) *granatensis tarrae*: Cabrera, 1961: 369; name combination.

Sciurus (*Guerlinguetus*) *granatensis llanensis* Mondolfi and Boher, 1984:312, type locality "Hato Corozal, Caño Tutumito, Arismendi, Barinas, Venezuela."

Sciurus (*Guerlinguetus*) *granatensis*: Eisenberg, 1989:331, part; name combination.

Sciurus (*Guerlinguetus*) *granatensis soederstroemi*: Thorington and Hoffmann, 2005:761; name combination; required replacement of diacritical marks used in original description (ICZN 1999:Art. 32.5.2.1).

Notosciurus g. chrysuros is the name for the taxon occurring on the eastern slopes of the Andes, from Venezuela to Ecuador. Crown of the head brown to dark brown streaked with pale yellow to dark red; dorsum uniformly colored brown to dark brown streaked with yellow, grayish yellow to red, sometimes with black on midline; pale orange but inconspicuous postauricular patches present; venter reddish orange on the throat, chest, and belly, with white spots on the chest in some specimens. Tail colored like dorsum on its proximal quarter to third, and washed with reddish orange succeeded by a black tip on some specimens; in others, tail may have an intermediate section washed with red and a black distal portion.

N. g. granatensis (Humboldt, 1811)

SYNONYMS:

Sciurus granatensis Humboldt, 1811:8, 13, plate III; type locality "Carthagène [= Cartagena]," Bolívar, Colombia.

Sciurus Grenatensis Illiger, 1815:69; incorrect subsequent spelling of *Sciurus granatensis* Humboldt.

Sciurus variabilis I. Geoffroy St.-Hilaire, 1833:2, plate 4; type locality "Amérique [in reference to both North and South America]." [The volume is dated 1832 but was published in "Mars 1833."]

Sciurus splendidus Gray, 1842:262; no type locality given (see Thomas 1928e:590).

Macroxus variabilis: Lesson, 1842:112; name combination.

Sciurus gerrardi Gray, 1861:92, type locality "New Granada."

Sciurus gerrardii: Gerrard, 1862:214; name combination and incorrect subsequent spelling of *Sciurus gerrardi* Gray.

Macroxus gerrardii: Gray, 1867:430; name combination and incorrect subsequent spelling of *Sciurus gerrardi* Gray.

Sciurus variabilis variabilis: Bangs, 1898b:184; name combination.

Sciurus variabilis saltuensis Bangs, 1898b:185, type locality "Pueblo Viejo [= El Pueblito], [north slope Sierra Nevada de Santa Marta, La Guajira,] Colombia (altitude 8000 ft [2,000 ft; see Paynter 1997:139])."

Sciurus saltuensis bondae J. A. Allen, 1899b:213, type locality "Bonda, Santa Marta District, [Magdalena,] Colombia."

Sciurus saltuensis: J. A. Allen, 1904c:431; name combination.

Sciurus variabilis gerrardi: E.-L. Trouessart, 1904:326; name combination.

Sciurus versicolor zuliae Osgood, 1910:26; type locality "Orope, Zulia, Venezuela."

Guerlinguetus hoffmanni quindianus J. A. Allen, 1914d: 587, type locality "Rio Frio (central Rio Cauca Valley, altitude 3500 feet), western slope of Central (or Quindio) Andes," Magdalena, Colombia.

Sciurus saltuensis magdalenae J. A. Allen, 1914d:593, type locality "Banco (altitude 50–100 feet) Rio Magdalena, a few miles above mouth of Rio Caura," Magdalena, Colombia.

Mesosciurus (Mesosciurus) hoffmanni quindianus: J. A. Allen, 1915a:222; name combination.

Mesosciurus (Histriosciurus) gerrardi gerrardi: J. A. Allen, 1915a:236; name combination.

Mesosciurus (Histriosciurus) gerrardi zuliae: J. A. Allen, 1915a:246; name combination.

Mesosciurus (Histriosciurus) saltuensis saltuensis: J. A. Allen, 1915a:247; name combination.

Mesosciurus (Histriosciurus) saltuensis bondae: J. A. Allen, 1915a:249; name combination.

Mesosciurus (Histriosciurus) saltuensis magdalenae: J. A. Allen, 1915a:251; name combination.

Mesosciurus (Histriosciurus) gerrardi valdiviae J. A. Allen, 1915a:309; type locality "Puerto Valdivia (alt. 360 ft. [600 ft; see Paynter 1997:352]), lower Rio Caura," Antioquia, Colombia.

Mesosciurus hoffmanni quindanus: Cabrera and Yepes, 1940:194; name combination and incorrect subsequent spelling of *Guerlinguetus hoffmanni quindianus* J. A. Allen.

Sciurus [(Guerlinguetus)] hoffmanni quindianus: Ellerman, 1940:340; name combination.

Sciurus [(Guerlinguetus)] gerrardi gerrardi: Ellerman, 1940:341; name combination.

Sciurus [(Guerlinguetus)] gerrardi zuliae: Ellerman, 1940: 342; name combination.

Sciurus [(Guerlinguetus)] gerrardi valdiviae: Ellerman, 1940:342; name combination.

Sciurus [(Guerlinguetus)] splendidus splendidus: Ellerman, 1940:342; name combination.

Sciurus [(Guerlinguetus)] splendidus saltuensis: Ellerman, 1940:342; name combination.

Sciurus [(Guerlinguetus)] splendidus bondae: Ellerman, 1940:342; name combination.

Sciurus granatensis granatensis: Hershkovitz, 1947:12; name combination.

Sciurus granatensis bondae: Hershkovitz, 1947:13; name combination.

Sciurus granatensis saltuensis: Hershkovitz, 1947:15; name combination.

Sciurus granatensis agricolae Hershkovitz, 1947:16; type locality "Colonia Agrícola de Caracolicito, Río Araguaní, on the southern slopes of the Sierra Nevada de Santa Marta, Department of Magdalena, Colombia; altitude 335 meters."

Sciurus granatensis splendidus: Hershkovitz, 1947:17; name combination.

Sciurus granatensis variabilis: Hershkovitz, 1947:19; name combination.

Sciurus granatensis norosiensis Hershkovitz, 1947:21, type locality "Norosí, Mompós, Department of Bolívar, Colombia; altitude 120 meters."

Sciurus granatensis perijae Hershkovitz, 1947:23, type locality "Sierra Negra, above Villanueva, on the western slope of the Sierra de Perijá, Valledupar, Department of Magdalena, Colombia; altitude 1,265 meters."

Sciurus granatensis maracaibensis Hershkovitz, 1947:25, type locality "El Panorama, Río Aurare, a small river emptying into lake Maracaibo, opposite the city of Maracaibo, Zulia, Venezuela; altitude near sea level."

Sciurus granatensis zuliae: Hershkovitz, 1947:28; name combination.

Sciurus granatensis gerrardi: Hershkovitz, 1947:35; name combination.

Sciurus granatensis valdiviae: Hershkovitz, 1947:35; name combination.

Sciurus granatensis quindianus: Hershkovitz, 1947:36; name combination.

Sciurus (Guerlinguetus) granatensis bondae: Goodwin, 1953:275; name combination.

Sciurus (Guerlinguetus) granatensis splendidus: Goodwin, 1953:275; name combination.

Sciurus (Guerlinguetus) granatensis valdiviae: Goodwin, 1953:276; name combination.

Sciurus (Guerlinguetus) granatensis quindianus: Goodwin, 1953:276; name combination.

Sciurus (Guerlinguetus) granatensis agricolae: Cabrera, 1961:362; name combination.

Sciurus (Guerlinguetus) granatensis gerrardi: Cabrera, 1961:363; name combination.

Sciurus (Guerlinguetus) granatensis granatensis: Cabrera, 1961:364; name combination.

Sciurus (Guerlinguetus) granatensis maracaibensis: Cabrera, 1961:365; name combination.

Sciurus (Guerlinguetus) granatensis noroslensis: Cabrera, 1961:367; name combination and incorrect subsequent spelling of *Sciurus granatensis norosiensis* Hershkovitz.

Sciurus (Guerlinguetus) granatensis perijae: Cabrera, 1961:367; name combination.

Sciurus (Guerlinguetus) granatensis saltuensis: Cabrera, 1961:368; name combination.

Seiurus [*sic*] *(Guerlinguetus) granatensis variabilis*: Cabrera, 1961:369; name combination.

Sciurus (Guerlinguetus) granatensis zuliae: Cabrera, 1961: 370; name combination.

Sciurus (Guerlinguetus) granatensis norosiensis: Thorington and Hoffmann, 2005:761; name combination.

Sciurus (Guerlinguetus) granatensis variabilis: Thorington and Hoffmann, 2005:761; name combination.

The nominotypical subspecies is parapatric to *N. g. morulus*, occurring at high elevations in the western Andes of Colombia to the Sierra de Santa Marta, and across the central and eastern cordilleras of Colombia to western Venezuela. This subspecies also extends into the intervening lowlands, including the area around Lake Maracaibo. Average size larger than for all other subspecies; postauricular patches present; dorsal color an "agouti" pattern, mostly brown streaked with orange or red, or uniformly completely orange; posterior dorsum may exhibit dark patterning, ranging from a spot near the base of the tail to larger area; dorsal orange or red tones may appear merely as a scapular patch or extend farther posteriorly. Ventral color ranges from entirely red to orange with white spots (of varying sizes) on chest and belly to entirely white; white ventral spots may also extend as continuous lines across body at midventer. Entirely orange or red individuals with a white venter predominate in the northern parts of the range, on the slopes of the Serra de Santa Marta; the "agouti" dorsal pattern with a mostly or entirely red venter is most common in southern samples.

N. g. morulus (Bangs, 1900)

SYNONYMS:

Sciurus variabilis morulus Bangs, 1900b:43, type locality "Loma del Leon, Panama."

Sciurus versicolor Thomas, 1900g:385: type locality "Cachabi [= Cachabí], Prov. Esmeraldas, N Ecuador, alt. 160m."

Sciurus milleri J. A. Allen, 1912:91; type locality "Cocal (altitude 4000 ft.)," Cauca, Colombia.

Sciurus variabilis choco Goldman, 1913:4, type locality "Cana (altitude 3500 feet). [Darién,] Eastern Panama."

Guerlinguetus hoffmanni manavi J. A. Allen, 1914d:589, type locality "Manavi [= Manabí] (Rio de Oro, near sea-level), Ecuador," [= sea port of Bahía de Caráquez, Manabí province; see Paynter 1993:15].

Sciurus gerrardi salaquensis J. A. Allen, 1914d:592, type locality "Rio Salaqui, [Chocó,] northwestern Colombia."

Mesosciurus (Mesosciurus) hoffmanni manavi: J. A. Allen, 1915a:221; name combination.

Mesosciurus (Histriosciurus) gerrardi milleri: J. A. Allen, 1915a:241; name combination.

Mesosciurus (Histriosciurus) gerrardi versicolor: J. A. Allen, 1915a:242; name combination.

Mesosciurus (Histriosciurus) gerrardi morulus: J. A. Allen, 1915a:243; name combination.

Mesosciurus (Histriosciurus) gerrardi choco: J. A. Allen, 1915a:244; name combination.

Mesosciurus (Histriosciurus) gerrardi salaquensis: J. A. Allen, 1915a:245; name combination.

Mesosciurus (Histriosciurus) gerrardi baudensis J. A. Allen, 1915a:308; type locality "Baudo [= Serranía de Baudó] (alt. 3500 ft.), [Chocó,] coast region of western Colombia."

Sciurus (Guerlinguetus) variabilis choco: Elliot, 1917:19; name combination.

Sciurus (Guerlinguetus) gerrardi choco: Goldman, 1920: 139; name combination.

Sciurus (Guerlinguetus) gerrardi morulus: Goldman, 1920: 140; name combination.

Sciurus gerrardi inconstans Osgood, 1921:40, replacement name for *Sciurus versicolor* Thomas, 1900, said to be preoccupied by *Sciurus versicolor* Zimmermann, 1777; Hershkovitz (1947:36) stated that Zimmermann's name was invalid, but without specifying a reason.

Sciurus (Mesosciurus) gerrardi leonis Lawrence, 1933:369, replacement name for *Sciurus milleri* J. A. Allen, 1912, preoccupied by *Sciurus epomophorus milleri* H. C. Robinson and Wroughton, 1911.

Sciurus [(Guerlinguetus)] hoffmanni manavi: Ellerman, 1940:340; name combination.

Sciurus [(Guerlinguetus)] gerrardi leonis: Ellerman, 1940: 341; name combination.

Sciurus [(Guerlinguetus)] gerrardi inconstans: Ellerman, 1940:341; name combination.

Sciurus [(Guerlinguetus)] gerrardi morulus: Ellerman, 1940: 341; name combination.

Sciurus [(Guerlinguetus)] gerrardi choco: Ellerman, 1940: 341; name combination.

Sciurus [(Guerlinguetus)] gerrardi salaquensis: Ellerman, 1940:342; name combination.

Sciurus [(*Guerlinguetus*)] *gerrardi baudensis*: Ellerman, 1940:342; name combination.

Sciurus granatensis salaquensis: Hershkovitz, 1947:35; name combination.

Sciurus granatensis leonis: Hershkovitz, 1947:36; name combination.

Sciurus granatensis versicolor: Hershkovitz, 1947:36; name combination.

Sciurus granatensis manavi: Hershkovitz, 1947:36; name combination.

Sciurus (*Guerlinguetus*) *granatensis salaquensis*: Goodwin, 1953:275.

Sciurus (*Guerlinguetus*) *granatensis leonis*: Goodwin, 1953:276; name combination.

Sciurus (*Guerlinguetus*) *granatensis manavi*: Goodwin, 1953:276; name combination.

Sciurus (*Guerlinguetus*) *granatensis choco*: G. S. Miller and Kellogg, 1955:257; name combination.

Sciurus (*Guerlinguetus*) *granatensis morulus*: G. S. Miller and Kellogg, 1955:257; name combination.

Sciurus (*Guerlinguetus*) *granatensis versicolor*: Cabrera, 1961:370; name combination.

Notosciurus g. morulus is the oldest available name for the subspecies occurring from the Panama Canal Zone to western Colombia and northwestern Ecuador. Individuals lack postauricular patches over most of this range, except in Ecuador, where traces of a patch are present on specimens. The mid-dorsum slightly to more strongly darker than in other races, a characteristic more pronounced in Colombian specimens than those of central and eastern Panama; venter red, with hairs lacking a grayish base; tail usually divided into three differently colored sections: a dark proximal portion, a large middle red section, and a dark (black) distal terminus.

N. g. nesaeus (G. M. Allen, 1902)

SYNONYMS:

Sciurus nesaeus G. M. Allen, 1902:93; type locality "El Valle, Margarita Island, [Nueva Esparta,] Venezuela."

Mesosciurus (*Mesosciurus*) *nesaeus*: J. A. Allen, 1915a:233; name combination.

Sciurus [(*Guerlinguetus*)] *nesaeus*: Ellerman, 1940:341; name combination.

Sciurus granatensis nesaeus: Hershkovitz, 1947:37; name combination.

Sciurus (*Guerlinguetus*) *granatensis nesaeus*: Cabrera, 1961:367; name combination.

Notosciurus g. nesaeus is an insular subspecies known only from Isla Margarita, Venezuela. It is easily distinguished from mainland samples, but we have decided not to recognize it as a distinct species on the same grounds as Hershkovitz (1947); that is, the color pattern fits a general geograph-

ical trend. Head and back uniformly colored brown, heavily streaked with orange; inconspicuous orange postauricular patches present; ventral coloration sharply set off from the dorsolateral coloration, with uniform reddish orange throat, chest, and belly (hairs are entirely reddish orange). Coloration of tail variable, similar to that of dorsum on its proximal quarter to third but washed with reddish orange distally.

NATURAL HISTORY: Nitikman (1985) summarized the data on form and function, ontogeny and reproduction, ecology, behavior, and genetics of *N. granatensis*, with most studies focused on Panamanian populations (e.g., Heaney 1978; Glanz et al. 1982; Glanz 1984).

REMARKS: Samples referred to *N. granatensis* range from Costa Rica to Panama, and then south to Andean South America, from western Venezuela to northern Ecuador. As the synonymies for each subspecies indicate, dozens of names have been proposed for squirrels from this vast area (Cabrera 1961, for example, listed 28 subspecies in his concept of this polytypic species). One could easily conclude that the diversity of squirrels in this region is extensive simply by reading the original descriptions of these taxa. Indeed, based on our own examination of hundreds of specimens deposited in the major U.S. and European museums as well as most of the type material, both the nongeographic and geographic variation exhibited by these samples is staggering. In fact, many subspecies have been described from localities very close to one another, without even a separation by elevation or river basin. Hershkovitz (1947) personally collected dozens of specimens in Colombia and described several new subspecies, some of them with identical provenance! Other authors have frequently described series in which different

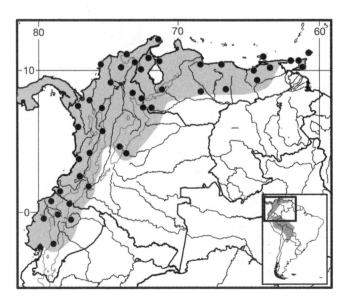

Map 16 Selected localities for *Notosciurus granatensis* (●). Contour line = 2,000 m.

specimens from the same locality could be assigned to separate subspecies.

Here, we offer a brief summary of geographic trends expressed in *N. granatensis*, and we attach names that we believe best represent the major axes of this variation. In so doing, we reduce the number of subspecies to only five for the South American part of the species' distribution. The description we provide above for each one can, with benevolence, be termed "impressionistic." However, we cannot accept a single name for all such squirrels at the risk of obscuring a most baffling but manifest evolutionary phenomenon. In fact, Hershkovitz (pers. comm. to MdV) first devised his "metachromism theory" (Hershkovitz 1968, 1970a) because he wanted to explain the color variation he saw in Colombian squirrels. Keeping at least some of the plethora of names as "valid" will, we hope, encourage researchers to study these squirrels further and discover how and why they vary so much.

Notosciurus pucheranii (Fitzinger, 1867)
Pucheran's Squirrel, Ardija, Quatipuru

SYNONYMS: See under subspecies.

DESCRIPTION: Variable squirrel of medium size, with head and body length 151–212 mm (N = 121 adult specimens), tail length 135–225 mm (N = 117), hindfoot length 34–58 mm (N = 125), ear length 15–29 mm (N = 109), and mass 136–242 g (N = 6). Dorsal color of head and back brown, streaked with orange, sometimes with defined mid-dorsal black stripe, and darker than that of sides; cheeks brownish orange; orange eye-ring present, and orange postauricular patches conspicuous. Ventral surfaces distinct in color from dorsolateral region, with throat white or the same white, yellow, or orange color as chest; belly orange; ventral hairs usually have gray base. Didier (1955) illustrated the baculum for specimens from eastern Peru.

DISTRIBUTION: *Notosciurus pucheranii* is a polytypic species of primarily Andean distribution, ranging from the central and eastern Andean forests of Colombia to the eastern Andes of Peru, and from the Peruvian lowlands to western Amazonian Brazil, Bolivia, and northwestern Argentina. Their ecological range spans upper montane forests, lowland rainforest, and lowland seasonally dry forests, across a broad elevational range from 100 to 3,200 m. A geographic gap in this range is apparent between the Nariño department in southern Colombia and the Loreto department in northern Peru, with no specimens as yet recorded from Ecuador. It is unclear whether this gap is real or the result of inadequate sampling. Most collecting localities are Andean or situated in Andean foothills; there are fewer localities at lower elevations east of the Andean slopes.

SELECTED LOCALITIES (Map 17): ARGENTINA: Jujuy, Higuerilla (type locality of *Sciurus* [*Mesosciurus*] *argentinius* Thomas). BOLIVIA: Beni, Totaisal, 1 km SW of Estación Biologica del Beni (AMNH 261917); Chuquisaca, Sud Cinti, 15.6 km N of El Palmar, above Río Santa Marta (LHE 1310); Cochabamba, Incachaca (FMNH 35224); Río Chapare (FMNH 21547); La Paz, Charuplaya (BM 2.1.1.17), Mapiri (BM 1.1.1.58); Santa Cruz, Buenavista (FMNH 21423), Santiago, Chiquitos (FMNH 105927); Tarija, pié Sierra Santa Rosa, Itau (BM 25.2.1.18–23), Vermejo [= Bermejo] (AMNH 38994). BRAZIL: Acre, Manoel Urbano (MZUSP 11235), Oriente, near Taumaturgo (MPEG 758), Plácido de Castro (type locality of *Sciurus ignitus iquiriensis* Vieira), Rio Branco (AMNH 149145); Amazonas, Igarapé Grande [= Eirunepé], left bank Rio Juruá (MZUSP 4581); Mato Grosso, São João, cabeceiras do Rio Aripuanã (type locality of *Guerlinguetus rufus* Moojen); Rondônia, Ouro Preto d'Oeste (MPEG 20529), Samuel (MPEG, field number 402–8557). COLOMBIA: Antioquia, 4 km NE of Bellavista, above Río Ponce (FMNH 70038), Urrao, Río Aná (FMNH 71106), 15 km E of Río Negrito, Sonsón (FMNH 70043), 10 km S of Valdivia (FMNH 70048); Boyacá, Miraflores (AMNH 32173); Cauca, Salento, west Quindio Andes (AMNH 32850); Cundinamarca, El Roble, above Fusugasugá, east Andes (AMNH 34622), Panamá (AMNH 34680); Huila, Andalucáa, eastern Andes (AMNH 33700), La Candela (AMNH 33699), San Adolfo, Acevedo (FMNH 71110); Nariño, La Laguna (AMNH 42358); Valle del Cauca, vicinity of Palmira, central Andes (AMNH 32175). PERU: Cusco, Cosñipata, Hacienda Cadena (FMNH 65673), Ocabamba, near Cusco (type locality of *Sciurus aestuans cuscinus* Thomas), Río San Miguel (BM 22.1.1.21); Huánuco, Agua Caliente, Río Pachitea (FMNH 55407), Tingo Maria (FMNH 24116); Junín, Perené (AMNH 61814); Loreto, Puerto Punga, Río Tapiche (AMNH 99203), Quebrada Pushaga, Río Morona, upper Río Amazonas (FMNH 88974), San Fernando, Río Yavari (FMNH 88981), Santa Cecilia, Río Maniti (FMNH 87173), Sarayacu, Río Ucayali (AMNH 76442); Madre de Dios, Altamira, Manu (FMNH 98060), Río Inambari (AMNH 16562); Pasco, Rumicruz (AMNH 60599); Puno, Pampa Grande, Sandia (FMNH 79902); Ucayali, Lagarto, upper Río Ucayali (AMNH 78943).

SUBSPECIES: We recognize the four subspecics of *N. pucheranii*.

N. p. argentinius (Thomas, 1921)
SYNONYMS:

Sciurus (*Mesosciurus*) *argentinius* Thomas, 1921k:609, type locality "Higuerilla, 2000 m., in the Department of Valle Grande, about 10 km. east of the Zenta range and 20 km. of the town Tilcara," Jujuy, Argentina.

Sciurus argentinius: Thomas, 1925b:577; name combination.

Leptosciurus argentinius: Cabrera and Yepes, 1940:193; name combination.

Sciarus [sic] [(*Guerlinguetus*)] *ignitus argentinius*: Cabrera, 1961:370; name combination.

This subspecies occurs in extreme southern Bolivia and adjacent northwestern Argentina.

N. p. boliviensis (Osgood, 1921)

SYNONYMS:

Macroxus leucogaster Gray, 1867:430, type locality "Bolivia, Santa Cruz de la Sierra"; preoccupied by *Sciurus leucogaster* F. Cuvier, 1831:300 (= *Sciurus aureogaster* F. Cuvier, 1829c).

Leptosciurus leucogaster: J. A. Allen, 1915a:207; name combination.

Sciurus boliviensis Osgood, 1921:39, replacement name for *Macroxus leucogaster* Gray, 1867, which is preoccupied by *Sciurus leucogaster* F. Cuvier, 1831.

Sciurus [(*Guerlinguetus*)] *boliviensis*: Ellerman, 1940:344; name combination.

Sciurus sanborni Osgood, 1944:191; type locality "La Pampa, between Rio Inambari and Rio Tambopata, about twenty miles north of Santo Domingo, Madre de Dios, Peru. Altitude about 1900 feet."

Sciurus ignitus boliviensis: Sanborn, 1951:18; name combination.

Sciurus [(*Guerlinguetus*)] *ignitus boliviensis*: Cabrera, 1961:371; name combination.

Sciurus [(*Guerlinguetus*)] *sanborni*: Cabrera, 1961:373; name combination.

Notosciurus p. boliviensis is distributed from southern Peru to central Bolivia.

N. p. ignitus (Gray, 1867)

SYNONYMS:

Macroxus ignitus Gray, 1867:429, type locality "Bolivia."

Macroxus irroratus Gray, 1867:431, type locality "Brazil, Upper Ucayali."

Sciurus aestuans cuscinus Thomas, 1899a:40, type locality "Ocabamba, Cuzco," Peru.

S[*ciurus*]. *irroratus*: Thomas, 1899a:40; name combination.

Sciurus cuscinus: Thomas, 1902b:129; name combination.

Sciurus irroratus: Ihering, 1904:420; name combination.

[*Sciurus* (*Guerlinguetus*)] *aestuans cuscinus*: E.-L. Trouessart, 1904:327; name combination.

[*Sciurus* (*Guerlinguetus*)] *irroratus*: E.-L. Trouessart, 1904:328; name combination.

Sciurus cuscinus ochrescens Thomas, 1914a:362, type locality "Bolivia, in upperparts of Beni and Mamoré Rivers. Type from Astillero, [La Paz,] 67°W, 16°S. Alt. 2700 m."

Leptosciurus ignitus ignitus: J. A. Allen, 1915a:204; name combination.

Leptosciurus ignitus irroratus: J. A. Allen, 1915a:206; name combination.

Sciurus irroratus ochrescens: Osgood, 1916:204; name combination.

Sciurus ignitus irroratus: Thomas, 1927f:599; name combination.

Sciurus [(*Guerlinguetus*)] *ignitus ignitus*: Ellerman, 1940:344; name combination.

Sciurus [(*Guerlinguetus*)] *ignitus irroratus*: Ellerman, 1940:344; name combination.

Guerlinguetus rufus Moojen, 1942:14; type locality "S. João [São João] (cabeceiras do Aripuanã) M. [Mato] Grosso," Brazil; preoccupied by *Sciurus rufus* Kerr, 1792.

Sciurus (*Guerlinguetus*) *rufus*: W. P. Harris, 1944:12; name combination.

Sciurus rufus: Moojen, 1952b:28; name combination.

Sciurus ignitus ignitus: Sanborn, 1951:18; name combination.

Sciurus ignitus iquiriensis C. O. da C. Vieira, 1952:28, type locality "Placido de Castro, Rio Abunã, Acre," Brazil.

Sciurius [sic] *ignitus*: Didier, 1955:423; name combination, but typographic error.

Sciurus (*Guerlinguetus*) *ignitus iquiriensis*: C. O. da C. Vieira, 1955:408; name combination.

Sciurus cabrerai Moojen, 1958:50; replacement name for *Guerlinguetus rufus* Moojen, 1942.

Sciurus [(*Guerlinguetus*)] *ignitus cabrerai*: Cabrera, 1961: 372; name combination.

Sciurus [(*Guerlinguetus*)] *ignitus*: R. S. Hoffmann, Anderson, Thorington, and Heaney, 1993:441; name combination.

Guerlinguetus ignitus: J. A. Oliveira and Bonvicino, 2006: 348; name combination.

This subspecies occurs on the eastern Andean slopes and in the Amazon Basin from northern to southern Peru, extending eastward to the states of Acre and Amazonas in Brazil.

N. p. pucheranii (Fitzinger, 1867)

SYNONYMS:

Sciurus rufoniger Pucheran, 1845:336; type locality "Colombie (Santa-Fé de Bogotá)," Cundinamarca, Colombia; preoccupied by *Sciurus rufonigra* Gray, 1842 [see Fitzinger 1867a].

Funambulus Pucheranii Fitzinger, 1867a:487; replacement name for *Sciurus rufoniger* Pucheran [preoccupied by *Sciurus rufonigra* Gray, 1842]; type locality given as "Am. Columbien."

Macroxus griseogena Gray, 1867:429; at least in part; no explicit type locality was given, but "Santa Fé de Bogotá" was listed among the several localities cited.

Macroxus tephrogaster Gray, 1867:431; probably part (?); while no explicit type locality was given, "Bogotá" was listed among the several localities cited. Thorington and Hoffman (2005) treated *tephrogaster* as a synonym of *S. deppei*.

Macroxus medellinensis Gray, 1872:408; type locality "Medellin, Antioquia," Colombia.

S[*ciurus*]. *aestuans* var. *rufoniger*: J. A. Allen, 1877:669; part.

Sciurus tephrogaster: J. A. Allen, 1877:763; name combination.

Sciurus griseogenys: Alston, 1878:667, part; name combination.

Sciurus deppei: Alston, 1878:668, part; name combination.

Sciurus rufo-niger: Alston, 1878:669; name combination.

Sciurus aestuans rufo-niger: True, 1884:595, part; name combination.

S[*ciurus*]. *Pucherani*: Thomas, 1898f:266; name combination and incorrect subsequent spelling of *Funambulus pucheranii* Fitzinger.

S[*ciurus*]. (*Guerlinguetus*) *caucensis* E. W. Nelson, 1899a:79, type locality "Lima River, upper Cauca Valley, [Valle del Cauca?; see Paynter 1997:252,] Colombia (alt. 6000 feet)."

[*Sciurus*] *medellinensis*: E.-L. Trouessart, 1904:327; name combination.

[*Sciurus* (*Guerlinguetus*)] *caucensis*: E.-L. Trouessart, 1904:328; name combination.

Guerlinguetus pucheranii salentensis J. A. Allen, 1914d:587, type locality "near Salento (altitude 9000 ft.), Central Andes, [Cauca,] Colombia."

Leptosciurus pucheranii pucheranii: J. A. Allen, 1915a:200; name combination.

Leptosciurus pucheranii medellinensis: J. A. Allen, 1915a:202; name combination.

Leptosciurus pucheranii caucensis: J. A. Allen, 1915a:203; name combination.

Leptosciurus pucheranii salentensis: J. A. Allen, 1915a:203; name combination.

Sciurus (*Leptosciurus*) *pucheranii medellinensis*: Anthony, 1923a:5; name combination.

Leptosciurus pucherani: Cabrera and Yepes, 1940:192; name combination and incorrect spelling of *pucheranii* Fitzinger.

Sciurus [(*Guerlinguetus*)] *pucherani pucherani*: Ellerman, 1940:343; name combination and incorrect spelling of *pucheranii* Fitzinger.

Sciurus [(*Guerlinguetus*)] *pucherani medellinensis*: Ellerman, 1940:343; name combination and incorrect spelling of *pucheranii* Fitzinger.

Sciurus [(*Guerlinguetus*)] *pucherani salentensis*: Ellerman, 1940:343; name combination and incorrect spelling of *pucheranii* Fitzinger.

Sciurus aestuans rufoniger: Rode, 1943:384; name combination, and required elimination of hyphen used in original description (ICZN 1999:Art. 32.5.2.2).

Sciurus pucherani caucensis: Goodwin, 1953:277; name combination, and incorrect subsequent spelling of *Funambulus pucheranii* Fitzinger.

Sciurus pucheranii salentensis: Goodwin, 1953:277; name combination.

Sciurus pucherani: Didier, 1955:416; name combination and incorrect subsequent spelling of *Funambulus pucheranii* Fitzinger.

Sciurus pucheranii: Didier, 1955:422; name combination.

Sciurus [(*Guerlinguetus*)] *pucheranii santanderensis* Hernández-Camacho, 1957:219, type locality "Meseta de los Caballeros, Santander," Colombia.

Sciurus [(*Guerlinguetus*)] *pucheranii caucensis*: Cabrera, 1961:372; name combination.

Sciurus [(*Guerlinguetus*)] *pucheranii medellinensis*: Cabrera, 1961:372; name combination.

Sciurus [(*Guerlinguetus*)] *pucheranii pucheranii*: Cabrera, 1961:373; name combination.

Sciurus [(*Guerlinguetus*)] *pucheranii santanderensis*: Cabrera, 1961:373; name combination.

Sciurus [(*Guerlinguetus*)] *pucheranii*: Eisenberg, 1989:333; name combination.

The nominotypical subspecies is distributed in the Andean cordilleras of central and southern Colombia. It is geographically separated by a large gap from the cis-Andean distributions of the other three subspecies, as noted previously.

NATURAL HISTORY: Leonard et al. (2009) summarized the natural history, behavior, and other aspects of the biology of this species, restricting their compilation to the trans-Andean Colombian populations. Otherwise, little has been published on this species, surprisingly so given its broad cis-Andean range. Emmons and Feer (1997) provided a few ecological and behavioral details for *argentinius*, *ignitus*, and *sanborni*, which they regarded as separate species.

REMARKS: Both *Funambulus pucheranii* Fitzinger and *Macroxus ignitus* Gray were published in 1867, but Fitzinger's paper appeared in March ("21 März 1867") and Gray's in December. The species name *pucheranii* Fitzinger thus has priority over *ignitus* Gray.

Notosciurus pucheranii is highly variable throughout its range, and many supposed species and subspecies have been described that we treat either as subspecies rather than species or as junior synonyms. All names proposed have been largely based on coat color, which, although variable, presents a reasonably defined geographical trend, as follows. Animals with denser and longer fur are found at higher elevations, while specimens with less dense and shorter pelage come from lower elevations. In northern Colombia

(Antioquia and Quindio departments), most specimens have a well-defined mid-dorsal black stripe (characteristic of *N. p. pucheranii*), but in samples from Cundinamarca, Huila, and Nariño this stripe becomes less pronounced and in some series disappears entirely. In Peru, from Loreto department southward, the *N. p. ignitus* phenotype predominates, with animals having dark brown dorsal parts, without a mid-dorsal stripe, and a grayish white throat and orange chest and belly, with ventral hairs gray at the base. This same phenotype extends to adjacent Brazil, in the states of Amazonas and Acre. Specimens from localities in central and southern Peru have the same general color as typical *N. p. ignitus*, but with less demarcation of the throat from that of the chest and belly, and an overall ventral color ranging from pale yellow to orange.

Southern Peruvian and adjacent Bolivian samples are generally similar, although some Bolivian samples have more white on the venter. By central Bolivia, this general trend of lightening of the venter reaches the extreme, where specimens have white bellies but with hairs still with a gray base; this is the phenotype of *N. p. boliviensis*. The name *Sciurus sanborni* Osgood has been applied to animals with this color pattern; we treat this name as a junior synonym of *boliviensis*. Southern Bolivian and northern Argentinean samples belong to *N. p. argentinius*, with the main distinction between them and *N. p. boliviensis* being in the color of the venter, which is self-colored orange in the former and white with a gray base in the latter. Specimens from Bermejo, Tarija department, Bolivia, have a white venter (characteristic of typical *boliviensis*), but this locality is situated in the middle of the range of *N. p. argentinius*.

The reason that we recognize four subspecies in *N. pucheranii* rather than a single, monotypic species or four separate species is that the characters that would identify distinct taxa do not segregate in a clear-cut manner. Samples frequently contain specimens with mixed characteristics in regions where one subspecies replaces another.

Throughout its distribution, specimens of *N. pucheranii* possess postauricular patches and three pairs of mammae. However, L. H. Emmons (pers. comm. to MdV) collected a female with four pairs of mammae from Sud Cinti, Chuquisaca, Bolivia (LHE 1313). Because all females of *Notosciurus* we have examined have three pairs of mammae, Emmons's specimen may be an anomaly or perhaps has been misidentified.

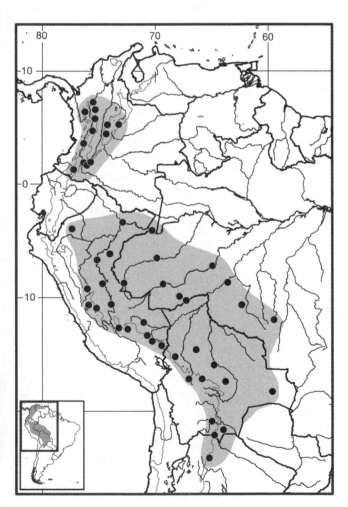

Map 17 Selected localities for *Notosciurus pucheranii* (●); cis- and trans-Andean distributions are mapped as disjunct (see Remarks). Contour line = 2,000 m.

Genus *Simosciurus* J. A. Allen, 1915

Simosciurus is a genus of large, trans-Andean squirrels that occurs in western Ecuador and northwestern Peru. We recognize two species, both of which inhabit dry to moist forests from sea level to about 2,300 m. *Simosciurus* can be distinguished from most South American squirrels by their coarser pelage and larger size (relative to *Sciurillus*, *Microsciurus*, *Guerlinguetus*, *Notosciurus*, and *Syntheosciurus*). Although comparable in size, *Simosciurus* is distinguished from *Hadrosciurus* by its very short and broad rostrum and distinctly coarse fur. The rostral character is difficult to quantify, but it can be visualized most readily in dorsal view comparisons. Not only are the nasals wider but also the premaxillae remain visible distally to the tip of the rostrum. In *Hadrosciurus*, the premaxillae are narrower and disappear from view before the rostral tip. The respective trans- versus cis-Andean distributions of the two genera circumvent the possibility of their misidentification based on geography alone.

J. A. Allen (1915a) erected the genus *Simosciurus* to include *Sciurus stramineus* P. Gervais, with three subspecies (*nebouxii* I. Geoffroy St.-Hilaire; *guayanus* Thomas; and *zarumae* J. A. Allen) in addition to the nominotypical form. Moore (1959) retained *stramineus* in *Simosciurus* (as a subgenus of *Microsciurus*), but Ellerman (1940), Cabrera (1961), and Thorington and Hoffmann (2005) placed the species in *Sciurus*, subgenus *Guerlinguetus*. Allen (1915a:280–281) recognized that two of these taxa had a white nuchal patch (*nebouxii* and *zarumae*) and two did not (*stramineus* and *guayanus*). Emmons and Feer (1997) suggested that animals possessing the nuchal patch were only a highland color variant, a position that supported Allen's initial recognition of a single species. Our study of more than 140 specimens, however, indicates clear geographical separation of specimens with and without a distinct nuchal patch. We interpret this pattern as indicative of species, rather than subspecies, and thus recognize *S. nebouxii* I. Geoffroy St.-Hilaire and *S. stramineus* P. Gervais. No measurements of mass are available in the material we examined, but the two species are of similar size, with an average head and body length of 260 mm, tail of about 294 mm, ear of 35 mm, and hindfoot of 63 mm.

The vernacular name "Guayaquil Squirrel" (or *ardilla de Guayaquil*, Tirira 2004, 2007) has been used in the recent accounts that recognize a single species of these trans-Andean squirrels. However, because the main difference between the two species we recognize lies in the presence (in *S. nebouxii*) or absence (in *S. stramineus*) of a nuchal patch, we refer to the former as the "White-naped Squirrel," the English equivalent of the local Spanish name, *ardilla de nuca blanca* (Tirira 2007).

SYNONYMS:

Macroxus: Lesson, 1842:112, part (listing of *stramineus*); not *Macroxus* F. Cuvier, 1823.

Sciurus: I. Geoffroy St.-Hilaire, 1855:165; part (inclusion of *nebouxii*); not *Sciurus* Linnaeus, 1758.

Echinosciurus E.-L. Trouessart, 1880a:292; part (listing of *stramineus*); proposed as a subgenus of *Sciurus*; type species *Sciurus hypopyrrhus* Wagler, 1831, a synonym of *Sciurus aureogaster* F. Cuvier, 1829.

Simosciurus J. A. Allen, 1915a:280; type species *Sciurus stramineus* P. Gervais, 1841.

[*Sciurus*] (*Guerlinguetus*): Ellerman, 1940:343; name combination.

Sciurus (*Simosciurus*): Goodwin, 1953:278; name combination.

Microsciurus (*Simosciurus*): Moore, 1961a:15, table 2; name combination.

Guerlinguetus (*Simosciurus*): Moore, 1961b:15; name combination.

REMARKS: J. A. Allen (1915a) proposed the genus *Simosciurus* with type species *Sciurus stramineus* P. Gervais, 1841 (a name often attributed to Eydoux and Souleyet, 1841, but see Remarks under the *Simosciurus stramineus*). The name *Echinosciurus* E.-L. Trouessart actually predates *Simosciurus*, but the type species of *Echinosciurus* is *Sciurus hypopyrrhus* Wagler, 1831. E.-L. Trouessart (1880a)

thought that J. A. Allen's (1877) *Sciurus hypopyrrhus* was the same as the Ecuadorean taxon, an error probably resulting from the unusually coarse pelage of *Simosciurus*, which resembles that of some North American species and is completely different from that of any other South American form. Because *Sciurus hypopyrrhus* is firmly in the synonymy of *Sciurus aureogaster* F. Cuvier, 1829 (Thorington and Hoffmann 2005), the name *Echinosciurus* is not available for this genus of squirrels.

KEY TO THE SPECIES OF *SIMOSCIURUS*:

1. Nuchal patch present, colored white to pale gray, slightly tinged with very pale yellow; dorsal color strongly streaked with yellowish gray anteriorly (same color as nuchal patch) and more vivid pale yellow posteriorly; head about the same color as other anterior parts of body; venter mostly pale yellowish gray. . . . . *Simosciurus nebouxii*
1'. No nuchal patch; dorsal color usually brown heavily streaked anteriorly with gray, sometimes yellowish gray posteriorly; head distinctly darker than rest of dorsum, with brownish tones; venter dark brown, sometimes with tinge of gray . . . . *Simosciurus stramineus*

Simosciurus nebouxii (I. Geoffroy St.-Hilaire, 1855)
White-naped Squirrel

SYNONYMS:

Sciurus Nebouxii I. Geoffroy St.-Hilaire, 1855:165, plate 12, a–d; type locality "Pérou, à Payta [= Paita], [Piura]."

Sciurus stramineus: Alston, 1878:664; part; *S. nebouxii* I. Geoffroy St.-Hilaire placed in synonymy.

Macroxus nebouxi: E.-L. Trouessart, 1897:428; name combination, and incorrect subsequent spelling of *Sciurus nebouxii* I. Geoffroy St.-Hilaire.

Sciurus stramineus nebouxii: Thomas, 1900a:151; name combination.

Sciurus stramineus nebouxi: E.-L. Trouessart, 1904:326 ; name combination, and incorrect subsequent spelling of *Sciurus nebouxii* I. Geoffroy St.-Hilaire.

Sciurus nebouxi: Neveu-Lemaire and Grandidier, 1911:8; incorrect subsequent spelling of *Sciurus nebouxii* I. Geoffroy St.-Hilaire.

Sciurus stramineus zarumae J. A. Allen, 1914d: 597; type locality "Zaruma, [El Oro,] southwestern Ecuador; altitude 6000 ft."

Simosciurus stramineus nebouxii: J. A. Allen, 1915a:280; name combination.

Simosciurus stramineus zarumae: J. A. Allen, 1915a:284; name combination.

Sciurus [(*Guerlinguetus*)] *stramineus nebouxi*: Ellerman, 1940:343; name combination, and incorrect subsequent spelling of *nebouxii* I. Geoffroy St.-Hilaire.

Sciurus [(*Guerlinguetus*)] *stramineus zarumae*: Ellerman, 1940:343; name combination.

Simosciurus stramineus zaruma: Cabrera and Yepes, 1940: 196; name combination.

Sciurus (*Simosciurus*) *stramineus zarumae*: Goodwin, 1953: 278; name combination.

Sciurus (*Guerlinguetus*) *stramineus*: Cabrera, 1961:373; part; placed *S. nebouxii* I. Geoffroy St.-Hilaire and *S. s. zarumae* J. A. Allen in synonymy.

Sciurus nebouxiii: Soukup, 1965:360; incorrect subsequent spelling of *Sciurus nebouxii* I. Geoffroy St.-Hilaire; placed in synonymy of *S. stramineus* P. Gervais.

DESCRIPTION: Medium-sized squirrel with head and body length 230–281 mm (N = 52), tail length 264–330 mm (N = 52), hindfoot length 52–70 mm (N = 53), and ear length 26–43 mm (N = 24). Body color light gray, with posterior dorsum and posterior parts of forelegs and hindlegs pale yellow; head, including cheeks, same color as body; ears distinctly darker than head, pinnae rimmed with elongated black hairs; nuchal patch present, white to grayish white; forefeet and distal parts of forelimbs darker than body; hindfeet and distal parts of hindlegs dark brown; venter yellow streaked with gray. Tail black to light gray, except near yellowish base.

DISTRIBUTION: This species ranges from extreme southwestern Ecuador, in El Oro and Loja provinces, to northwestern Peru, in Lambayeque, Tumbes, and Piura departments, from sea level to 2,300 m. Jessen et al. (2010) reported the presence of this species (as *S. stramineus*) in and around Lima, Peru, but considered these animals a probable human introduction.

SELECTED LOCALITIES (Map 18): ECUADOR: El Oro, Arenillas (FMNH 53258), Río Pindo (AMNH 60474), Zaruma (type locality of *Sciurus stramineus zarumae* J. A. Allen); Loja, Guainche [not located; see Paynter 1993] (AMNH 61375), Los Posos (AMNH 67680), Malacatos (FMNH 54285). PERU: Laybayeque, Chongoyape, Cabache (FMNH 80995); Piura, Cerro Prieto, Lancones, Encuentros (FMNH 83437), Hacienda Bigotes, Salitral (FMNH 80908), Le Payta [= Paita] (type locality of *Sciurus nebouxii* I. Geoffroy St.-Hilaire), Monte Grande, 14 km N and 25 km E of Talara (MVZ 135638), near Talara, Pariñas Valley (FMNH 53864), Palambla (AMNH 63701); Tumbes, Huasimo (FMNH 80991).

SUBSPECIES: We treat *Simosciurus nebouxii* as monotypic. J. A. Allen (1915a:280) distinguished *zarumae*, which he named from southwestern Ecuador, from the nominotypical form from adjacent Peru on the basis of dorsal and ventral color differences. Our examination of larger series fails to substantiate this geographic segregation.

NATURAL HISTORY: Little is known about the ecology and population biology of this species. One specimen (FMNH 53864) was found in a grass nest on a canyon wall.

REMARKS: We regard *Simosciurus stramineus* and *S. nebouxii* as distinct species based on their different color patterns and separate geographic ranges. However, since Alston (1878:665) first noted the appearance of "irregular tufts of pure white hairs, . . . and sometimes uniting in larger patches," others have viewed these differences as constituting individual variation rather than species-group characters (e.g., Emmons and Feer 1997). Individuals of both color patterns have been obtained at the same locality in southern Ecuador (Arenillas, El Oro province; specimens in FMNH). Two of the three specimens available possess the nuchal patch characteristic of *S. nebouxii*, but the third has no nuchal patch, as is diagnostic of *S. stramineus*.

Arenillas is the northernmost locality for *S. nebouxii*, and this mixture of characters may be due to unusual polymorphism, may indicate intergradation with the northern *S. stramineus*, or may actually represent an area of distributional overlap between the two species. Nevertheless, a gap of more than 150 km exists in the known ranges of two species between Arenilles in the south (*nebouxii*) and the Guayaquil area in the north (*stramineus*). Collecting efforts directed to this region, including the application of molecular genetic methods, will aid in determining the degree of intergradation, if any, and ultimately the proper number of species to recognize.

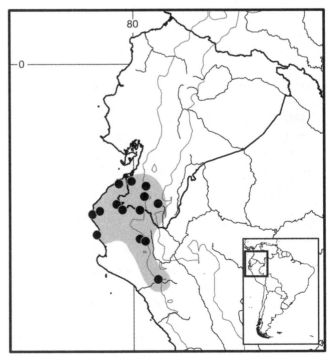

Map 18   Selected localities for *Simosciurus nebouxii* (●). Contour line = 2,000 m.

Simosciurus stramineus (P. Gervais, 1841)

Guayaquil Squirrel

SYNONYMS:

Sciurus stramineus P. Gervais, 1841:37, plate 9; type locality "Omatopé, ou Pérou" [= Amotape, Piura, Peru]; here corrected to Guayaquil, Guayas, Ecuador (see Remarks, below).

Macroxus stramineus: Lesson, 1842:112; name combination.

Macroxus fraseri Gray, 1867:430; type locality given as "Republic of Ecuador."

Sciurus hypopyrrhus: J. A. Allen, 1877:746; part (material from Guayaquil, Ecuador).

Sciurus tephrogaster: J. A. Allen, 1877:763; part (*Macroxus fraseri* Gray treated as a synonym).

Sciurus variabilis: J. A. Allen, 1877:768; part (*Sciurus stramineus* P. Gervais treated as a synonym).

Sciurus (*Echinosciurus*) *stramineus*: E.-L. Trouessart, 1880b:202; name combination.

Sciurus stramineus guayanus Thomas, 1900a:150; type locality "Balzar mountains, upper Río Palenque, [Manabí,] Ecuador."

Sciurus stramineus stramineus: Thomas, 1900a:151; name combination.

Simosciurus stramineus stramineus: J. A. Allen, 1915a:281; name combination.

Simosciurus stramineus guayanus: J. A. Allen, 1915a:283; name combination.

Sciurus [(*Guerlinguetus*)] *stramineus stramineus*: Ellerman, 1940:343; name combination.

Sciurus [(*Guerlinguetus*)] *stramineus guayanus*: Ellerman, 1940:343; name combination.

Simosciurus stramineus: Moore, 1959:204; name combination.

Sciurus (*Guerlinguetus*) *stramineus*: Cabrera, 1961:373; part.

Microsciurus (*Simosciurus*) *stramineus*: Moore, 1961a:15, table 2; name combination.

Guerlinguetus (*Simosciurus*) *stramineus*: Moore, 1961b:15; name combination.

DESCRIPTION: Similar in size to *S. nebouxii*, with head and body length 241–327 mm (N=40), tail length 250–380 mm (N=40), hindfoot length 50–69 mm (N=40), ear length 20–38 mm (N=7). Body color brownish gray, with posterior dorsum and posterior parts of forelegs and hindlegs yellowish; crown, cheeks, and ears distinctly darker than body; no nuchal patch; forefeet and distal parts of forelimbs from darker than body to whitish gray; hindfeet and distal parts of hind legs dark brown; venter brownish gray; tail black and gray, except for the yellowish base.

DISTRIBUTION: Central Ecuador west of the Andes, where it ranges from sea level to about 300 m, in Manabi, Los Ríos, and Guayas provinces.

SELECTED LOCALITIES (Map 19): ECUADOR: Guayas, Chongoncito (AMNH 60548), Daulé (AMNH 34684), Guayaquil (type locality of *Macroxus fraseri* Gray); Los Ríos, Hacienda Pijigal, Vinces (AMNH 62886), Pimocha, Río Babahoyo (MNHN 1932/2893); Manabí, Balzar Mountains, upper Río Palenque (type locality of *Sciurus stramineus guayanus* Thomas); Tama [= Jama] (FMNH 53256); Santa Elena, Cerro Manglar Alto [= Manglaralto, placed at Cerros de Colonche following Painter 1993] (AMNH 66640).

SUBSPECIES: We treat *Simosciurus stramineus* as monotypic. However, J. A. Allen (1915a:281, 284) recognized two subspecies among those geographic samples without a nuchal patch, with *guayanus* Thomas "much lighter colored throughout than true *stramineus*."

NATURAL HISTORY: Tirira (2007) summarized the natural history attributes of this species; however, his map of *S. stramineus* also included the range of *S. nebouxii*, so his account possibly includes both species. *Simosciurus stramineus* inhabits both primary and secondary forests and disturbed areas, including both coffee and citrus plantations where individuals cause damage to crops. The natural habitat is primarily dry forest, but the species can be found in adjoining humid forests as well. This species is diurnal and arboreal, and in contrast with the solitary nature of most other squirrels, this species may be found in groups of three to five individuals. They feed on a variety of seeds taken directly from trees but occasionally eat insects, fungi, green leaves, and tree bark. Their activity is concentrated in the early daylight hours. They take refuge in treetops, in hollow logs, or in foliage. Their nests are made from leaves and twigs.

REMARKS: P. Gervais (1841) described *stramineus* as a member of the genus *Sciurus*. The name, however, has been commonly attributed to the naturalists J. F. T. Eydoux and L. F. A. Souleyet (e.g., J. A. Allen 1915a; Cabrera 1961; Thorington and Hoffmann 2005; Thorington et al. 2012), although these authors explicitly stated that Paul Gervais both identified and described all mammals obtained during the voyage of French corvette *La Bonite* (Eydoux and Souleyet 1841:ii; 3, footnote).

P. Gervais (1841:39) gave the type locality of *Simosciurus stramineus* as "Omatopé, Pérou." There is no locality with this spelling in Peru; the locality referred to is most likely Amotape, Piura, Peru, about 25 km northeast of the port town of Paita (= Payta as given in Eydoux and Souleyet 1841). "Omatopé" (= Amotape) is also the type locality of *Eptesicus innoxius* (P. Gervais 1841), collected on the same expedition (see A. L. Gardner 2008:449). However, Amotape is located in the center of the geographic range of *Simosciurus nebouxii* I. Geoffroy St.-Hilaire, and all specimens that we have examined of this species have the nuchal

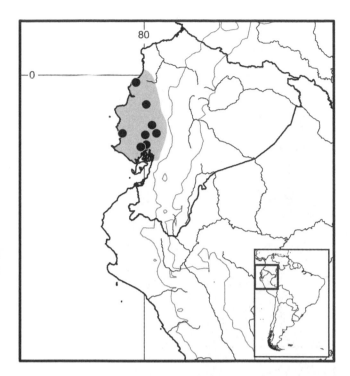

Map 19 Selected localities for *Simosciurus stramineus* (●). Contour line = 2,000 m.

patches. But P. Gervais's (1841:38–39) detailed description of color characteristics of his *Sciurus stramineus* includes no mention of this very distinctive character. As can be seen in Maps 18 and 19, the two species of *Simosciurus* we recognize are allopatric, with their ranges apparently separated by about a 150-km gap. Therefore, we believe that the locality of the type specimen was originally recorded in error or that some other mistake occurred during the voyage and/or after arrival of the specimens in Paris. The ship *Le Bonite* remained anchored at Paita for a few days and then proceeded to Guayaquil, further north, where it stayed for several more days before departing. Because Guayaquil is an area inhabited by *Simosciurus stramineus*, we hereby designate Guayaquil, Provincia de Guayas, Ecuador as the type locality of the species (ICZN 1999:Art. 76A).

P. Gervais (1841:38–39) described the species as if he had an actual specimen in hand, but the holotype could not be located by one of us (MdV) in the Muséum National d'Histoire Naturelle in Paris in 1992, nor has any other author reported on the holotype since the original description. It may have been lost.

Genus *Syntheosciurus* Bangs, 1902

Syntheosciurus is a genus of small squirrels known until now by the single species *S. brochus* from Central America. We report what is likely an undescribed species from San

Martín department, Peru, the first record of the genus for South America. *Syntheosciurus brochus* occurs in evergreen montane forests of the Cordillera Talamanca of Costa Rica and Panama at elevations above 1,700 m. The Peruvian specimen was taken at a single locality in Parque Nacional Río Abiseo, an area covered by montane rainforest at elevations above 1,500 m on the eastern slope of the Andes (S. C. Ramirez and Vela 2003). The large geographic hiatus between the known localities of the Central American *S. brochus* and the Peruvian record appears to be real, because intervening areas in the northern Andes of Ecuador and Colombia with the appropriate habitat and elevations have been extensively collected, with large series of other taxa of squirrels recorded with no *Syntheosciurus* taken.

Syntheosciurus can be distinguished from other South American squirrels by the following combination of characteristics: size intermediate between that of *Microsciurus* and *Guerlinguetus*, with head and body length approximately 180 mm and tail length 150 mm; the single weight available among specimens we examined was that of a adult male, recorded as 244 g. Externally, *Syntheosciurus* is distinguished by its small ears, which are, on average, about 15 mm in length. Females have three pairs of mammae. The temporal portion of the orbitotemporal fossa is similar to that of the larger South American squirrels, with the postorbital process more or less equally dividing it from the orbital portion. This is different from the condition in *Microsciurus*, in which the temporal portion of the orbitotemporal fossa is greatly reduced, and from that in *Sciurillus*, in which it is obliterated. The incisors are clearly proodont, PM3 is present, and the cristae of PM4 to M3 are oriented obliquely. In the original description of the genus, Bangs (1902:25) stated that the incisors have well-developed grooves on the buccal surface; this characteristic is present in the holotype of *S. brochus* but is absent in other specimens examined (Goodwin 1943; Heaney and Hoffmann 1978).

SYNONYMS:

Syntheosciurus Bangs, 1902:25; type species *Syntheosciurus brochus*, by monotypy.

Synthetosciurus Elliot, 1904:91; incorrect subsequent spelling of *Syntheosciurus* Bangs.

Sciurus: Goodwin, 1943:1, part (description of *poasensis*).

Syntheosciurus (*Syntheosciurus*): Moore, 1959:179; name combination.

Syntheosiurus Villalobos and Cervantes-Reza, 2007:31; incorrect subsequent spelling of *Syntheosciurus* Bangs.

REMARKS: Phylogenetic relationships of *Syntheosciurus* to other American squirrels remain unclear, but Mercer and Roth (2003) proposed that *Syntheosciurus* was a sister group to *Microsciurus*, based on DNA sequence analyses. This hypothesis agrees with the earlier, more superficial assessments of Bangs (1902) and Hall (1981). Moore

(1959) included *Syntheosciurus* in his subtribe Microsciurina, together with *Microsciurus* and *Leptosciurus* [= *Notosciurus* herein] (*sensu* J. A. Allen 1915a) and *Simosciurus*. A phylogenetic hypothesis based on 57 morphological characters (Villalobos and Cervantes-Reza 2007) placed *Syntheosciurus* in a group with other Central American taxa, and separate from squirrels with more obvious South American affinities such as *Microsciurus and Notosciurus*.

In keeping with our adoption of most of J. A. Allen's (1915a) arrangement for the genera of South American Sciuridae, we employ *Syntheosciurus* as the generic name for these animals. See Hall (1981) for a list of synonyms and distributional records for the Central American *S. brochus*.

KEY TO THE SPECIES OF *SYNTHEOSCIURUS*:
1. Body color brown punctuated with orange; belly pale orange; eye-ring present and conspicuous; no postauricular patch . . . . *Syntheosciurus brochus*
1'. Body color brown punctuated with ochraceous; belly orange red; no eye-ring; postauricular patch present. . .
 . . . . *Syntheosciurus* sp.

Syntheosciurus sp.
Peruvian Montane Squirrel

SYNONYMS: None.

DESCRIPTION: In comparison to Central American *S. brochus*, the Peruvian specimen has longer palmar pads; belly tinted more reddish than orange; tail hairs fringed with orange; dorsum brown streaked with ochraceous rather than orange; no eye-ring, but postauricular patch present. Skull shorter in length, and bullae have two well-separated rather than closely appressed and parallel septa.

DISTRIBUTION: This species is only known from a single locality in Andean montane forest at 2,800 m in northern Peru.

SELECTED LOCALITIES (Map 20): PERU: San Martín, Vilcabamba del Pajate, 31 km NE of Pataz [La Libertad] (USNM [MML 279]).

SUBSPECIES: None.

NATURAL HISTORY: Nothing is known about the ecology or other aspects of the natural history of this animal. The single known locality is in upper montane forest (S. C. Ramírez and Vela 2003).

REMARKS: This species is here reported thanks to the late Dr. Charles O. Handley Jr., then at the Smithsonian Institution, National Museum of Natural History, Washington D. C., who showed a single specimen to one of us (MdV). Mariella Leo collected this specimen in 1989 during an inventory in the Parque Nacional del Río Abiseo (see S. C. Ramírez and Vela 2003). In that report, made to support a management plan for the park, the specimen was identi-

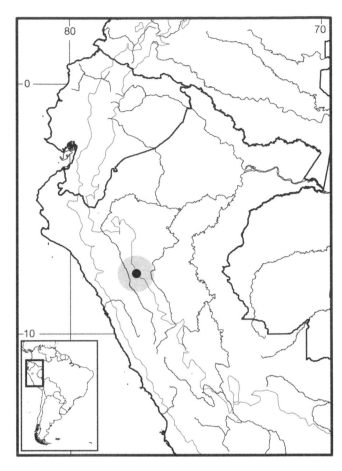

Map 20 Single known locality for *Syntheosciurus* sp. (●). Contour line = 2,000 m.

fied as *Microsciurus* sp., and it was similarly listed in recent compilations of the mammal fauna of Peru (Pacheco et al. 1995, 2009). Several additional specimens may exist in Peruvian collections (V. Pacheco, pers. comm.).

Suborder Castorimorpha Wood, 1955
Infraorder Geomorpha Thaler, 1966

Superfamily Geomyoidea Bonaparte, 1845

James L. Patton

Superficially dissimilar in body form and habits, the pocket gophers (Geomyidae) and kangaroo rats and pocket mice (Heteromyidae) have been united as a recognizable monophyletic assemblage for many decades. All genera share external fur-lined cheek pouches, four cheek teeth (dental formula I 1/1, C 0/0, PM 1/1, M 3/3 = 20) with a bilophodont or more simplified occlusal pattern, and an infraorbital foramen positioned anterior to the zygomatic plate on the side of the rostrum, opening anterolaterally in the Geomyidae, but laterally, completely piercing the rostrum, in the Heteromyidae. The fossil record extends to the Oligocene, with the rapidly expanding diversification of living genera correlated with the general drying trend of the early to mid-Miocene.

McKenna and Bell (1997) grouped geomyids, heteromyids, and the fossil Entotychinae as subfamilies in the Geomyidae, following earlier suggestions of Shotwell (1967) and Lindsay (1972). Although the hierarchical rank of the pocket gophers and kangaroo rats and mice and pocket mice is partly a matter of taxonomic philosophy (Patton 2005a,b), it is also a decision that stems from the phylogenetic placement of the fossil entotychines. The sister relationship between entotychines and living pocket gophers, documented by both Wahlert (1985, 1988) and Korth (1994), supports the traditional separation of the Geomyidae (including the Entotychinae) as a family distinct from the Heteromyidae.

KEY TO THE FAMILIES OF GEOMYOIDEA:

1. Body bauplan adapted for subterranean living, with fusiform body, short powerful limbs, forefeet with elongated and strong claws, small eyes, very short ear pinnae, and short and nearly naked tail; skull blocky with strong zygomatic arches, heavy ridging, broad incisors, and euhypsodont cheek teeth consisting of simple prismatic columns and flattened occlusal surfaces . . . . Geomyidae

1'. Body bauplan adapted for bounding terrestrial gait or bipedal saltatorial locomotion, with elongated hind limbs and feet; forefeet with shortened but strong claws; tail elongated, often with both terminal crest and pencil; eyes large; ear pinna projecting above dorsal surface of skull; hair soft or often with imbedded spines; skull variable, some genera with hugely enlarged auditory bullae expanded onto the dorsal side of the skull; cheek teeth cuspidate, bunodont, and brachydont, or hypsodont and columnar . . . . Heteromyidae

Family Geomyidae Bonaparte, 1845

Mark S. Hafner

The family Geomyidae (pocket gophers) includes six extant genera and approximately 40 extant species of fossorial rodents distributed from southern Canada to northwestern Colombia. Pocket gophers are known from the lower Oligocene of North America (Wahlert 1985); both fossil (Russell 1968; Korth 1994) and molecular (Spradling et al. 2004) evidence suggest that modern genera radiated from a common ancestor approximately 5 mya.

All pocket gophers are characterized by numerous adaptations for subterranean life. Their bodies are fusiform, which facilitates movement in the narrow tunnels they excavate while foraging for roots, tubers, or other plant parts. Their external appendages are short relative to body length, and they use their muscular forelimbs for digging and for pushing excavated soil to the surface. The cranium is robust and features prominent ridges for attachment of powerful temporal muscles. The mouth can be closed behind the large, procumbent incisors to prevent entry of soil when the incisors are used for digging. The incisors and cheek teeth are ever-growing, and the latter have simplified crown patterns consisting of dentine pools surrounded by enamel rings or transverse enamel plates.

Pocket gophers range in size from approximately 100–1,000 g, and they live in a wide variety of soils types. They rarely occupy shallow soils (<0.5 m) or soils prone to flooding. In addition to their principal diet of underground plant foods, many species of pocket gophers also forage opportunistically on surface vegetation near to their burrow entrances. Their preference for certain cultivated crops, including alfalfa, banana roots, and yuca (*Manihot*) make pocket gophers serious agricultural pests over parts of their geographic range. Their subterranean habits make them difficult to control, and they can cause significant damage to the small food plots of subsistence farmers.

Detailed descriptions of geomyid morphology can be found in Merriam (1895a) and Russell (1968), and a review of the taxonomic history of pocket gophers is available in Patton (2005b). Systematic revisions of pocket gopher taxa, appearing since the publication of *Mammal Species of the World* (Wilson and Reeder 2005), include M. S. Hafner et al. (2004, 2005, 2009, 2011), Sudman et al. (2006), D. J. Hafner et al. (2008), M. S. Hafner and Hafner (2009), and Mathis et al. (2013). Only one genus of pocket gopher (*Orthogeomys*) occurs in South America.

Genus *Orthogeomys* Merriam, 1895

Species of the genus *Orthogeomys* are medium to large pocket gophers (200–1,000 g). The skull is heavy and broad, with most of the breadth owing to extreme lateral expansion of robust zygomatic arches. The nasals and rostrum also are relatively broad and heavy, and the angular process of the dentary is short relative to overall length of the mandible. The anterior surface of each upper incisor has a broad longitudinal groove located either medially or near the inner edge of the tooth. The premolars consist of two columns united at their midpoints to yield a figure-8-shaped occlusal surface. The pelage generally is sparse and coarse in species living at low elevations and long and dense in high-elevation forms. Dorsal pelage markings (either a white head spot or a white lumbar belt) are present in all, or nearly all, specimens of some species, while others are characterized only by a low to moderate frequency of individuals with head blazes or lumbar belts, which can occur on either the left, right, or both sides of the body (M. S. Hafner and Hafner 1987). Fossils attributed to the genus *Orthogeomys* (subgenus *Heterogeomys*) are known from Late Pleistocene deposits (San Josecito Cave local fauna) of Nuevo León, Mexico (Russell 1960).

SYNONYMS:

Orthogeomys Merriam, 1895a:172; type species *Geomys scalops* Thomas, 1894, by original designation.

Heterogeomys Merriam, 1895a:179; type species *Geomys hispidus* Le Conte, 1852, by original designation.

Macrogeomys Merriam, 1895a:185; type species *Geomys heterodus* Peters, 1865, by original designation.

REMARKS: Russell (1968) revised the genus and included *Heterogeomys* and *Macrogeomys* as subgenera. The genus includes 10 recognized species, including two in the subgenus *Heterogeomys*, six in *Macrogeomys*, and two in *Orthogeomys*. One species, *O.* (*Macrogeomys*) *dariensis*, occurs in South America.

Orthogeomys dariensis (Goldman, 1912)

Darién Pocket Gopher

SYNONYMS:

Macrogeomys dariensis Goldman, 1912b:8; type from "Cana (altitude 2,000 feet), in the mountains of eastern [Darién] Panama."

Orthogeomys (*Macrogeomys*) *dariensis*: Russell, 1968:532; first use of current name combination.

Orthogeomys thaeleri Alberico, 1990:104; type locality "ca. 7 km S. Bahía Solano, Municipio Bahía Solano, Departamento de Chocó, Colombia, ca. 100 m."

DESCRIPTION: Head and body length 210–278 mm (Goldman 1912b; Alberico 1990). Dorsal color reddish brown to dull chocolate brown, or nearly pure black in fresh pelage in some individuals; venter and inner sides of limbs sparsely haired, with scattered hairs grayish to light brown; vibrissae white to pale brown; upper surfaces of forefeet and hindfeet as well as tail either brownish, dark pinkish, or nearly white. Skull large (greatest length of holotype 71 mm), elongated, and with anteriorly spreading zygomatic arches, narrow braincase, and narrow rostrum (Goldman 1912b; Alberico 1990).

DISTRIBUTION: *Orthogeomys dariensis* occurs in forests and cultivated regions of eastern Panamá and the lowlands and western foothills of the Serranía de Baudó in extreme northwestern Colombia. It also may occur on the eastern slopes of the Serranía de Baudó (Alberico 1990). The species ranges in elevation from 10 to approximately 1,200 m. *Orthogeomys dariensis* is known only from the Recent.

SELECTED LOCALITIES (Map 21): COLOMBIA: Chocó, ca. 7 km S of Bahía Solano (type locality of *Orthogeomys thaeleri* Alberico), Parque Nacional Natural Utría, Ensenada de Utría (LSUMZ 30057), Santa Marta (Alberico 1990).

SUBSPECIES: I treat *Orthogeomys dariensis* as monotypic.

NATURAL HISTORY: *Orthogeomys dariensis* occupies a wide variety of habitat types but is most common in forested areas or in small clearings in forests. Like other subterranean mammals, *O. dariensis* requires well-drained soils that are not prone to flooding and are of sufficient depth to prevent easy access to their burrows by digging predators. Alberico (1990) reported that the tunnels produced by *O. dariensis* in the clayey lateritic soils of the Chocó are shallower (only a few centimeters below the surface) than those typically produced by large pocket gophers. The diet of *O. dariensis* consists primarily of roots and tubers, although surface vegetation in the immediate vicinity of the burrow opening occasionally is consumed.

REMARKS: Intensive search for *O. dariensis* in the foothills of the northern Andes and near the headwaters of the Atrato and San Juan rivers (the most likely path of colonization of the northern Andes by a fossorial mammal) produced no evidence of pocket gophers (Alberico 1990). Thus, it appears that *O. dariensis* has not crossed the Bolivar Trough, which represents the geological (but not political) dividing line between Central and South America.

Alberico (1990:106) recognized specimens of *O. dariensis* from Chocó department, Colombia, as a distinct species, *O. thaeleri*, based on subtle craniodental character differences and because the chewing lice he collected from the type series "do not appear to be specifically related to *Geomydoecus dariensis* reported for *O. dariensis* from Panama" by Price and Emerson 1971. However, Sudman and Hafner (1992) sequenced a portion of two mitochondrial genes (16S rDNA and cytochrome-*b*) in a specimen of *O. thaeleri* from near its type locality in Colombia and compared the sequences with those from a specimen of *O. dariensis* from near its type locality in Panamá. These authors found only three transitional differences between *O. thaeleri* and *O. dariensis* in the 929 base pairs examined (0.3% uncorrected sequence divergence) and concluded that *O. thaeleri* likely represents only a geographic variant of *O. dariensis*.

My comparison of the study skin and skull of a near-topotypic specimen of *O. thaeleri* (LSUMZ 30057) with the skins and skulls of three topotypes (or near topotypes) of *O. dariensis* (LSUMZ 25434–25436) revealed no differences beyond those consistent with normal geographic variation in a species of pocket gopher. Dorsal pelage coloration in the two supposed species was indistinguishable to the naked eye, and two of the three specimens of *O. dariensis* had interorbital constrictions >10 (considered a diagnostic trait of *O. thaeleri* by Alberico 1990). The

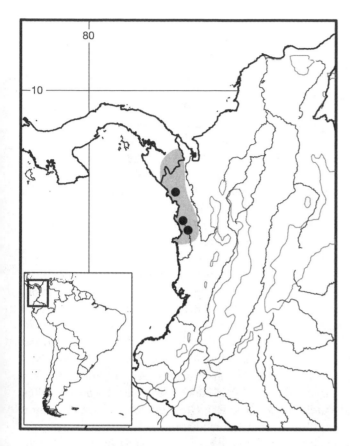

Map 21 Selected localities for *Orthogeomys dariensis* (●). Range in Panama from hall (1981). Contour line = 2,000 m.

height of the external auditory meatus in all three specimens of *O. dariensis* exceeded 2.3 mm, which also was considered diagnostic of *O. thaeleri* by Alberico (1990). The fact that *O. thaeleri* and *O. dariensis* have been collected within 100 km of each other in what appears to be continuous habitat suitable for pocket gophers leads me to conclude that *O. thaeleri* is simply a southward extension of *O. dariensis*.

Family Heteromyidae Gray, 1868
Robert P. Anderson

Three subfamilies comprise the Heteromyidae, but only one of them occurs in South America. Whereas the Dipodomyinae (kangaroo rats and kangaroo mice) and Perognathinae (pocket mice, both silky and coarse-haired) occur exclusively in North America, the Heteromyinae (spiny pocket mice) range from southern Texas, United States, to central Ecuador (Patton 2005a). Heteromyines represent a well-defined monophyletic group distinct from the two other subfamilies (M. S. Hafner 1981; J. C. Hafner and Hafner 1983; Wahlert 1991; J. C. Hafner et al. 2007). Of the two currently recognized genera of

heteromyines, only the genus *Heteromys* inhabits South America.

Modern usage of the genera *Heteromys* and *Liomys* began a century ago. After a rather turbulent taxonomic history of the subfamily during the nineteenth century (see account of the genus *Heteromys*), Merriam (1902) espoused recognition of two genera: *Heteromys* Desmarest, and *Liomys*, which he described. Soon thereafter, Goldman (1911b) accomplished the first comprehensive revision of the subfamily, providing emended diagnoses of the two genera and using these to place previously described species into their appropriate genera. Subsequent twentieth-century authors reviewing the taxonomy of heteromyines (e.g., Genoways 1973; Hall and Kelson 1959; Hall 1981; D. F. Williams et al. 1993; Patton 1993, 2005a) continued to recognize the two genera, which can be identified based on several diagnostic characters and which inhabit markedly different habitats, with species of *Heteromys* generally found in wetter, denser forests (R. P. Anderson et al. 2006).

However, substantial morphological diversity exists in *Liomys* (Genoways 1973), and recent phylogenetic studies have recovered three clades in the subfamily (*Heteromys*; the *adspersus–salvini* clade of *Liomys*; and the *irroratus–pictus–spectabilis* clade of *Liomys*), based on mtDNA sequences (Rogers and Vance 2005; J. C. Hafner et al. 2007) and allozymes and morphology (R. P. Anderson et al. 2006). Although some of these studies indicated the paraphyly of *Liomys*, all documented the monophyly of species previously allocated to *Heteromys*. Whereas some authors have suggested that all heteromyines should be placed in the genus *Heteromys* to avoid recognition of a paraphyletic *Liomys* (J. C. Hafner et al. 2007), I continue to recognize *Liomys* until conclusive studies with multiple independent data sources (e.g., unlinked nuclear markers) are completed, allowing for a comprehensive reevaluation of the generic-level taxonomy of the subfamily (see R. P. Anderson et al. 2006:1227). Notably, available generic-level names exist for each of the three clades recovered in the subfamily, allowing for the possible designation of three genera, which would acknowledge the substantial morphological and ecological distinctiveness of each clade and maximally promote nomenclatural stability in the subfamily (see R. P. Anderson and Gutiérrez [2009] for discussion of *Schaeferia* Lehmann, described in Lehmann and Schaefer [1979]).

Three unique synapomorphies of the genus *Heteromys* are unreversed in the family Heteromyidae: optic foramen small, with posterior border generally formed by a strong bar of bone; anterior margin of posterior loph of permanent upper premolar [PM4] with a long fold; and permanent lower premolar [pm4] with three or more lophids (R. P. Anderson et al. 2006). In addition, the lateral terminations of the lophs of the upper molars and lophids of the lower molars tend to be smooth in species of *Heteromys*, but generally are pointed in all other heteromyids.

Genus *Heteromys* Desmarest, 1817

Species of *Heteromys* are medium-sized heteromyids (approximately 40–130 g in adults) that constitute an easily recognizable element of the mammalian fauna of South America. Like all members of the superfamily Geomyoidea, they possess external fur-lined cheek pouches. Except for pocket gophers of the family Geomyidae (the sister group to the Heteromyidae), species of *Heteromys* represent the only South American mammals with such cheek pouches. In contrast to the subterranean pocket gophers (which evolved short, broad limbs and large claws for digging), *Heteromys* (like other heteromyids) have small claws and long, thin forelimbs; the hindlimbs are characteristically well developed. In addition to their characteristic cheek pouches, most adult *Heteromys* (like other heteromyines) have spiny dorsal pelage, with dark brown or black spines intermixed with soft and generally slightly paler brown or gray hairs, creating a grizzled effect. However, the dorsal pelage of juveniles of all species and even adults of some populations of some species is soft and more uniform in color. As in the rest of the subfamily, the ventral pelage of individuals of all ages lacks spines and is white, sharply demarcated from the dark pelage of the dorsum. A faint, narrow ventral stripe of dark hairs occurs in some individuals of some species in the midsagittal plane of the throat, chest, and/or abdominal regions. The tail is long—as long as or longer than the length of the head and body. As in other heteromyines, reproductively active adult male *Heteromys* possess extremely large testes, with the scrotal sac extending well posterior and lateral to the base of the tail and creating a distinctive outline in dorsal, ventral, and lateral views.

SYNONYMS:

Mus: Thompson, 1815:161; part (description of *anomalus*); not *Mus* Linnaeus, 1758.

Cricetus: Desmarest, 1817e:180; part (listing of *anomalus*); not *Cricetus* Leske, 1779.

Heteromys Desmarest, 1817e:181; type species *Mus anomalus* Thompson, 1815, by monotypy.

Saccomys F. Cuvier, 1823c:186; type species *Saccomys anthophile* F. Cuvier, 1823d:422, by subsequent designation.

Loncheres: Kuhl, 1820:72; part (listing of *anomala*, presumably a mandatory change of spelling of *anomalus* to match gender of genus).

Dasynotus Wagler, 1830:21; type species *Mus anomalus* Thompson, 1815, by monotypy.

Cricetomys: Schinz, 1845:204; part (listing of *anomalus*); not *Cricetomys* Waterhouse, 1840.

Perognathus: Gray, 1868:202; part (description of *bicolor*); not *Perognathus* Wied-Neuwied, 1839.

Xylomys Merriam, 1902:43; type species *Heteromys (Xylomys) nelsoni* Merriam, 1902, by original designation (subgenus).

REMARKS: F. Cuvier (1823d) described *Saccomys anthophile*, with type locality unknown, based on a single fluid-preserved specimen. This specimen, however, is apparently lost, as no holotype for the species appears in Rode (1945). The same specimen may have been the one in the Muséum National d'Histoire Naturelle in Paris studied by Desmarest (1817e) and considered by him to be conspecific with *Mus anomalus* Thompson. Ellerman (1940:472) stated "the type species of *Saccomys* is presumably unidentifiable." The illustrations in F. Cuvier (1823d) show dental and external characters that allow placement of the *Saccomys anthophile* in *Heteromys*, but I concur that neither these illustrations nor the extensive textual description are sufficient to associate the specimen with any particular species of the genus. Given the presently available information, I consider *S. anthophile* to be a *nomen dubium*. Note relevant listings by various subsequent authors: *Saccomys anthophilus* Wagler, 1830:22 (incorrect subsequent spelling of *anthophile* F. Cuvier); *H[eteromys]. anthophilus* Alston, 1880b:119 (name combination and incorrect subsequent spelling of *anthophile* F. Cuvier); and *Saccomys anthopilus* Ellerman, 1940:472 (incorrect subsequent spelling of *anthophile* F. Cuvier).

The species-level taxonomy of *Heteromys* remains highly problematic in Mexico and Central America, but South American species have received recent revisionary study. Just over two decades ago, only six (D. F. Williams et al. 1993) or seven (Patton 1993) species were recognized in the genus. However, karyotypic and allozymic studies indicate that the widespread *Heteromys desmarestianus* (currently considered to range from southern Mexico to northwestern Colombia) is composite (Rogers 1989, 1990; Mascarello and Rogers 1988). In addition, morphological studies have revealed several new species, bringing the number of currently recognized species in the genus to 10 (R. P. Anderson and Jarrín 2002; R. P. Anderson 2003; R. P. Anderson and Timm 2006; R. P. Anderson and Gutiérrez 2009). Whereas the species limits of *Heteromys* in Central America remain unclear (in particular in what currently is considered *H. desmarestianus*), the recent revisionary studies of South American species allow the current accounts, which remain subject to evaluation with molecular data.

Even compared with other genera of rodents, specimens from many populations of *Heteromys* are extremely difficult to identify to species, especially when only one or a few specimens are available from a single locality. Vast age-related cranial variation exists in size and shape, coupled with substantial ontogenetic changes in pelage; hence, age must be taken into account for valid comparisons and reliable identifications (Rogers and Schmidly 1982). Direct comparison with previously identified specimens from large series at single localities (in which age-related variation can be appreciated) is highly desirable and often necessary to ensure correct identification.

KEY TO THE SOUTH AMERICAN SPECIES
OF *HETEROMYS*:

1. Braincase narrow relative to length of skull; interorbital constriction narrow relative to length of skull 2
1'. Braincase wide relative to length of skull; interorbital constriction wide relative to length of skull 3
2. Size diminutive (rostral length <13.4 mm in adults); little or no dark coloration present on dorsal and external surfaces of forelimbs . . . . *Heteromys oasicus*
2'. Size typical for genus (rostral length >13.4 mm in adults); patch of moderately dark coloration present on dorsal and external surfaces of forelimbs . . . .
. . . . *Heteromys anomalus*
3. Rostrum short; dorsal coloration chocolate brown . . . .
. . . . *Heteromys desmarestianus crassirostris*
3'. Rostrum typical in length for genus; dorsal coloration dark slate gray or black 4
4. Braincase distinctly inflated . . . . . . *Heteromys australis*
4'. Braincase not inflated, or only moderately so 5
5. Zygomatic arches wide and bowed . . .*Heteromys teleus*
5'. Zygomatic arches not distinctly bowed, but rather parallel sided, or nearly so . . . . *Heteromys catopterius*

Heteromys anomalus (Thompson, 1815)
Caribbean Spiny Pocket Mouse
SYNONYMS:

Mus anomalus Thompson, 1815:161; type locality "Trinidad . . . [near] St. Anne's barracks," Trinidad and Tobago.
Cricetus anomalus: Desmarest, 1817e:180; name combination.
[*Heteromys*] *anomalus*: Desmarest, 1817e:181; name combination and formation as type species by monotypy; first use of current name combination.
Loncheres anomala: Kuhl, 1820:72; name combination.
Heteromys Thompsonii Lesson, 1827:264; type locality "la Trinité," Trinidad, Trinidad and Tobago.
[*Dasynotus*] *anomalus*: Wagler, 1830:21; name combination.
Cr[icetomys]. anomalus: Schinz, 1845:204; name combination.
Heteromys Thomsonii Schinz, 1845:204; incorrect subsequent spelling of *Heteromys thompsonii* Lesson.
Perognathus bicolor Gray, 1868:202; type locality originally given as "Honduras" but subsequently corrected

to "Venezuela" (Alston 1880b:118–119; see also D. F. Williams et al. 1993:102).
Heteromys thompsoni Gray, 1868:203; incorrect subsequent spelling of *Heteromys thompsonii* Lesson.
Heteromys melanoleucus Gray, 1868:204; type locality originally given as "Honduras" but subsequently corrected to "Venezuela" (Alston 1880b:118–119; see also D. F. Williams et al. 1993:102).
Heteromys bicolor: Alston, 1880b:119; name combination.
Heteromys jesupi J. A. Allen, 1899b:201; type locality "near Minca (at an alt. of 1000 ft. [305 m]), Santa Marta District, [Magdalena,] Colombia."
Heteromys anomalus brachialis Osgood, 1912:54; type locality "El Panorama, Rio Aurare [= Río Anaure], eastern shore of Lake Maracaibo, [Zulia,] Venezuela."
Heteromys anomalus jesupi: Osgood, 1912:54; name combination, listed as subspecies.
Heteromys [(*Heteromys*)] *anomalus anomalus*: Ellerman, 1940:475; name combination, listed as nominotypical subspecies.
[*Heteromys* (*Heteromys*)] *melanoleucas* Ellerman, 1940:475; incorrect subsequent spelling of *Heteromys melanoleucus* Gray.
Heteromys (*Heteromys*) *anomalus hershkovitzi*: Hernández-Camacho, 1956:3; type locality "Volcanes, cerca a la cabecera del corregimiento de Córdoba, Municipio de Caparrapí, Departamento de Cundinamarca; vertiente occidental de la Cordillera Oriental, Colombia. Alt. 250 metros." See also R. P. Anderson (1999:618).

DESCRIPTION: Adults with following combination of characters (R. P. Anderson 2003; R. P. Anderson and Gutiérrez 2009; see also R. P. Anderson 1999): Dorsal pelage typically pale brown, strongly grizzled, with thin ochraceous hairs intermixed with spines, but occasionally almost uniformly dark slate gray, nearly uniform to moderately grizzled (in wet lowlands of Zulia state, Venezuela, in some parts of Cordillera de Mérida, in mesic areas of Monagas and Sucre states, Venezuela, and on islands of Trinidad and Tobago); weak patch of dark coloration present on dorsal and external surfaces of forelimbs; ears distinctively rounded, brown to gray and large relative to body size; orange band on flanks absent; plantar surface of hindfeet naked. Skull variable in size geographically, average to large for genus (occipitonasal length 31.1–39.0 mm in adult specimens of age class 4 of Rogers and Schmidly [1982]); rostrum moderately long and moderately tapered anteriorly, without anterodorsal flare; anterior portion of premaxillary convex (inflated), forming a smooth (not stepped) lateral border of rostrum; interorbital constriction narrow; braincase narrow and not inflated; interparietal variable in size and shape; tubercle or swelling at posteroventral border of infraorbital foramen absent; mesopterygoid fossa formed by long, thin hamular processes of pterygoids; optic foramen especially small, with posterior margin formed by strong bar of bone. PM4 with straight, moderately long fold in anterior margin of posterior loph; pm4 with three lophids.

DISTRIBUTION: This species is found in low elevations along Caribbean coast of Colombia and Venezuela (east of Río Atrato), upper Río Magdalena valley of Colombia, gallery forests of *llanos* (savannas) of Venezuela north of Río Orinoco, and up to middle elevations on adjacent slopes of Sierra Nevada de Santa Marta, Cordillera Central (of Colombia; eastern versant only), Cordillera Oriental (of Colombia; western versant only), Serranía de Perijá, Cordillera de Mérida, Cordillera de la Costa (of Venezuela), and other coastal ranges; it also inhabits the Caribbean islands of Margarita, Trinidad, and Tobago; the known elevational range is from near sea level to 2,430 m, but most occurrences are from sea level to approximately 1,600 m (R. P. Anderson and Soriano 1999; R. P. Anderson 2003; R. P. Anderson and Gutiérrez 2009).

SELECTED LOCALITIES (Map 22): COLOMBIA: Antioquia, Hacienda Barro, 12 km S of Caucasia (R. P. Anderson 2003), Urabá, Río Currulao (R. P. Anderson 2003); Atlántico, Ciénaga de Guájaro, Sabana Larga (R. P. Anderson 2003); Cundinamarca, Volcanes, Municipio de Caparrapí (type locality of *Heteromys* [*Heteromys*] *anomalus hershkovitzi* Hernández-Camacho); Huila, Camp Coscorrón, Hacienda San Diego, 17 km SE of Villavieja (R. P. Anderson 2003); La Guajira, Las Marimondas, Fonseca (R. P. Anderson 2003); Magdalena, Mamatoca (R. P. Anderson 2003); Norte de Santander, Guamalito, near El Carmen (R. P. Anderson 2003), Finca La Palma, Durania (R. P. Anderson 2003). TRINIDAD AND TOBAGO: Tobago, Pigeon Peak, Tobago Forest Reserve, 2.5 km SSW of Charlotteville (R. P. Anderson and Gutiérrez 2009); Trinidad, Botanic Gardens, behind St. Anne's barracks (type locality of *Mus anomalus* Thompson), Mayaro (R. P. Anderson and Gutiérrez 2009). VENEZUELA: Anzoátagui, Paso Los Cocos, Río Caris, S of El Tigre (R. P. Anderson and Gutiérrez 2009), Pekín Abajo, Río Neverí (R. P. Anderson and Gutiérrez 2009); Falcón, Cerro la Danta, Sierra de San Luis, Parque Nacional J. C. Falcón (R. P. Anderson 2003); Guárico, Fundo Pecuario Masaguaral, 45 km (by road) S of Calabozo (R. P. Anderson and Gutiérrez 2009); Miranda, Estación Experimental Río Negro (R. P. Anderson and Gutiérrez 2009); Monagas, Caripito (R. P. Anderson and Gutiérrez 2009); Nueva Esparta, San Juan, Cerro Copey (R. P. Anderson and Gutiérrez 2009); Portuguesa, Cogollal, near Guanarito (R. P. Anderson 2003); Sucre, población de Macuro, Distrito Valdez (R. P. Anderson and Gutiérrez 2009), entre Cariaco y Chacopata (R. P. Anderson and Gutiérrez 2009); Táchira, Urbante,

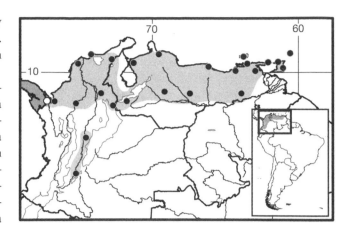

Map 22 Selected localities for *Heteromys anomalus* (●) and *Heteromys desmarestianus crassirostris* (○). Range of *H. desmarestianus* in Panama follows Hall (1981). Contour line = 2,000 m.

Río Potosí (R. P. Anderson 2003); Zulia, El Panorama, Río Anaure (type locality of *Heteromys anomalus brachialis* Osgood).

SUBSPECIES: A thorough review of geographic variation in this relatively widespread species has yet to be undertaken, and I do not recognize subspecies at this time (but see D. F. Williams et al. 1993 and Patton 2005a for recognition of four subspecies).

NATURAL HISTORY: *Heteromys anomalus* inhabits both primary and secondary forests, as well as some low-intensity agricultural settings (R. P. Anderson 1999, 2003; R. P. Anderson and Gutiérrez 2009). Most localities are in evergreen and deciduous forests; the available information for records in landscapes dominated by nonforested vegetation indicates that the species is restricted to areas of denser vegetation (e.g., gallery forests) in such situations (Osgood 1912; August 1984; Soriano and Clulow 1988; R. P. Anderson 2003; R. P. Anderson and Gutiérrez 2009). Mark–recapture studies at Bush Bush Forest on Trinidad documented limited movement for individuals of the species (<300 ft [ca. 90 m]; Worth et al. 1968). Eisenberg (1963) reported some behavioral information for the species.

REMARKS: The karyotype is characterized by $2n = 60$ and $FN = 68$ (Engstrom et al. 1987).

Heteromys australis Thomas, 1901
Southern Spiny Pocket Mouse

SYNONYMS:

Heteromys australis Thomas, 1901c:194; type locality "St. [San] Javier, Lower Cachabi River [Río Cachaví], [Esmeraldas,] N[orthern]. Ecuador. Alt. 20 m."

Heteromys lomitensis J. A. Allen, 1912:77; type locality "Las Lomitas, Cauca [now Valle del Cauca], Colombia . . . Altitude 5000 feet [1,524 m], west slope of Western Andes."

Heteromys australis consicus Goldman 1913:8; type locality "Cana (altitude 2000 feet [610 m]), [Darién,] Eastern Panama."

Heteromys australis australis: Goldman, 1913:8; name combination, listed as nominotypical subspecies; see also Goldman (1913:9).

Heteromys australis lomitensis: Goldman, 1913:8; name combination, listed as subspecies; see also Goldman (1913:9).

Heteromys australis pacificus Pearson 1939:4; type locality "Amagal, [Darién,] eastern Panama, altitude 1000 feet [305 m]".

Heteromys lomitansis Cabrera, 1961:513; incorrect subsequent spelling of *Heteromys lomitensis* J. A. Allen.

DESCRIPTION: Adults characterized by (R. P. Anderson 1999, 2003, R. P. Anderson and Jarrín 2002; see also R. P. Anderson et al. 2006) dorsal coloration of dark slate gray or black, nearly uniform or moderately grizzled with thin ochraceous hairs intermixed with spines; distinctive patch of dark coloration present on dorsal and external surfaces of forelimbs; ears black and small; orange band on flanks absent; plantar surface of hindfeet naked. Skull average to small for genus (occipitonasal length 32.1–36.0 mm in specimens of age class 4 of Rogers and Schmidly [1982] from Ecuador and southwestern Colombia; see also R. P. Anderson [1999]); rostrum unremarkable for genus, but moderately tapered anteriorly, without anterodorsal flare; anterior portion of premaxillary convex (inflated), forming smooth (not stepped) lateral border of rostrum; interorbital constriction wide; braincase wide and distinctly inflated; interparietal wide; tubercle or swelling at posteroventral border of infraorbital foramen absent; mesopterygoid fossa formed by long, thin hamular processes of pterygoids; optic foramen especially small, with posterior margin formed by strong bar of bone. PM4 with straight, moderately long fold in anterior margin of posterior loph; pm4 with three lophids.

DISTRIBUTION: Low elevations along Pacific coast of northwestern South America, from northwestern Ecuador through western Colombia to eastern Panama, and up to middle and high elevations on adjacent slopes of the Cordillera Occidental (of Ecuador), Cordilleras Occidental and Central (of Colombia), Cordillera Oriental (of Colombia; western versant only), and Cordillera de Mérida (Río Uribante drainage only, apparently disjunct) in western Venezuela. The known elevational range is from near sea level to 2,450 m, but most occurrences are found from sea level to approximately 2,000 m (R. P. Anderson 1999; R. P. Anderson and Jarrín 2002).

SELECTED LOCALITIES (Map 23; from R. P. Anderson 1999, except as noted): COLOMBIA: Antioquia, Alto Bonito, Purí, above Cáceres; Chocó, Bagadó,

Unguía, upper Río Ipetí; Córdoba, Socorré, upper Río Sinú; Cundinamarca, Paime; Huila, near San Adolfo, Río Aguas Claras, Río Suaza; Nariño, Barbacoas (R. P. Anderson and Jarrín 2002); Valle del Cauca, Bahía Málaga, Quebrada Valencia, road to Quebrada Alegría, Reserva Forestal Yotoco. ECUADOR: Carchi, El Pailón (Anderson and Jarrín 2002); Esmeraldas, San Javier, Río Cachaví (type locality of *Heteromys australis* Thomas); Pichincha, Cooperativa Salcedo Lindo, on road from Pedro Vicente Maldonado to Encampamento de CODESA (R. P. Anderson and Jarrín 2002). VENEZUELA: Táchira, Presa La Honda, 10 km SSE of Pregonero (R. P. Anderson 2003).

SUBSPECIES: A thorough review of geographic variation in this relatively widespread species has yet to be undertaken, and I do not recognize subspecies (but see D. F. Williams et al. 1993 and Patton 2005a for recognition of three subspecies).

NATURAL HISTORY: *Heteromys australis* inhabits very wet, unseasonal evergreen forests, but very little additional information regarding its natural history is available (R. P. Anderson 1999; R. P. Anderson and Jarrín 2002).

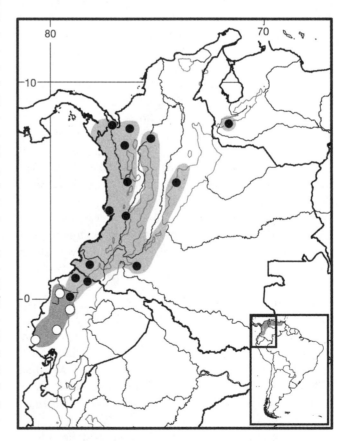

Map 23 Selected localities for *Heteromys australis* (●) and *Heteromys teleus* (○). The disjunct locality of *H. australis* in the Cordillera de Mérida of Venezuela is mapped separately from the remainder of the species' distribution; its range in Panama follows Hall (1981). Contour line = 2,000 m.

REMARKS: The karyotype is unknown. Populations in central Colombia may be specifically distinct from *H. australis* (R. P. Anderson 1999:620).

Heteromys catopterius R. P. Anderson and Gutiérrez, 2009
Overlook Spiny Pocket Mouse

SYNONYMS:

Heteromys anomalus: Tate, 1947:66; part; first name combination applied.

Heteromys catopterius R. P. Anderson and Gutiérrez, 2009:40; type locality "Venezuela: Aragua: Rancho Grande, near Biological Station, 13 km NW Maracay, at 3576 ft [1,090 m]."

DESCRIPTION: Adults with following combination of characters (R. P. Anderson and Gutiérrez 2009): dorsal coloration dark slate gray or black, moderately grizzled with thin ochraceous hairs intermixed with spines; distinctive patch of dark coloration present on dorsal and external surfaces of forelimbs; ears dark gray to black and medium in size; orange band on flanks absent; plantar surface of hindfeet naked. Skull average to large for genus (occipito-nasal length 34.7–38.7 mm in adult specimens of age class 4 of Rogers and Schmidly [1982]); rostrum long and wide, only slightly to moderately tapered anteriorly, without anterodorsal flare; anterior portion of premaxillary convex (inflated), forming a smooth (not stepped) lateral border of rostrum; interorbital constriction wide; braincase wide and moderately inflated; interparietal wide; tubercle or swelling at posteroventral border of infraorbital foramen absent; mesopterygoid fossa formed by long, thin hamular processes of pterygoids; optic foramen especially small, with posterior margin formed by strong bar of bone. PM4 with straight, moderately long fold in anterior margin of posterior loph; pm4 with three lophids.

DISTRIBUTION: Middle-to-high elevations of the Cordillera de la Costa of Venezuela. The known elevational range is from 350 to 2,425 m, but most records are above approximately 700 m (R. P. Anderson and Gutiérrez 2009).

SELECTED LOCALITIES (Map 24; from R. P. Anderson and Gutiérrez 2009). VENEZUELA: Aragua, Campamento Rafael Rangel, Loma de Hierro [= Minas de Niquel], 10 km (by road) WNW [NNE] of Tiara; Carabobo, Campamento La Justa, Río Morón; Miranda, 25 km (by road) N of Altagracia de Orituco, Hacienda Las Planadas; Monagas, 2 km N and 4 km W of Caripe, near San Agustín.

SUBSPECIES: *Heteromys catopterius* is monotypic.

NATURAL HISTORY: *Heteromys catopterius* inhabits wet montane forests, particularly cloud forests with many palms, and has been collected in both primary and second-

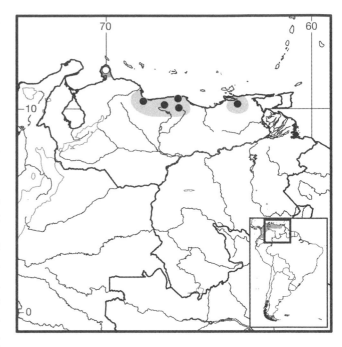

Map 24 Selected localities for *Heteromys catopterius* (●) (note disjunct range, with absence from the lowlands of the Depresión de Unare on the coast) and *Heteromys oasicus* (○). Contour line = 2,000 m.

ary forests (R. P. Anderson and Gutiérrez 2009). Rood (1963) and Rood and Test (1968) provided substantial information regarding the reproduction, development, diet, and behavior for the species.

REMARKS: The karyotype is characterized by $2n = 60$ and FN = 72 (Schmid et al. 1992). However, the FN reported by those authors included the four arms of the female sex chromosomes; thus, the autosomal FN is actually 68 (R. P. Anderson and Gutiérrez 2009). The Y chromosome of this species shows exceptional quinacrine-positive heterochromatin, indicating a preponderance of AT-rich sequences (Schmid et al. 1992).

Heteromys desmarestianus Gray, 1868
Desmarest's Spiny Pocket Mouse

SYNONYMS: [Synonymy for forms from Mexico and elsewhere in Central America not included; see Hall 1981; D. F. Williams et al. 1983.]

Heteromys desmarestianus Gray, 1868:204; type locality "Coban [Cobán]," Departamento de Alta Verapaz, Guatemala.

Heteromys crassirostris Goldman, 1912b:10; type locality "near head of Rio Limon [Río Limón] (altitude 5,000 feet [1,524 m]), Mount Pirri [= Cerro Pirre], [Darién,] eastern Panama."

Heteromys desmarestianus crassirostris: Goldman, 1920:115; name combination, listed as valid subspecies; see also Goldman (1920:117–118).

Heteromys d[esmarestianus]. acutirostris R. P. Anderson, 1999:623; incorrect subsequent spelling of *Heteromys crassirostris* Goldman; editor's error.

DESCRIPTION: (For *H. desmarestianus crassirostris*, see Remarks.) Adults with following combination of characters (R. P. Anderson 1999; see also R. P. Anderson and Timm 2006 and R. P. Anderson et al. 2006): dorsal pelage chocolate brown, only moderately grizzled with thin ochraceous hairs intermixed among spines; distinctive patch of dark coloration present on dorsal and external surfaces of forelimbs; ears small; orange band on flanks absent; plantar surface of hindfeet naked. Skull small for genus (occipitonasal length 32.2–35.1 mm in adult specimens of age classes 4 and 5 of Rogers and Schmidly [1982; published data for age class 4 not available]); rostrum distinctively short, wide, and strongly tapered anteriorly, without anterodorsal flare; anterior portion of premaxillary convex (inflated), forming smooth (not stepped) lateral border of rostrum; interorbital constriction wide; braincase wide and moderately inflated; interparietal moderately wide; tubercle or swelling at posteroventral border of infraorbital foramen absent; mesopterygoid fossa formed by long, thin hamular processes of pterygoids; optic foramen small, with posterior margin formed by strong bar of bone. PM4 with bent, extremely long fold in anterior margin of posterior loph; pm4 with three lophids.

DISTRIBUTION: Definitive distribution of this species of the *H. desmarestianus* species complex is only confirmed for extreme northwestern Colombia and eastern Panama along the Serranía del Darién and associated mountain chains (R. P. Anderson 1999; R. P. Anderson et al. 2006).

SELECTED LOCALITIES (Map 22, in South America): COLOMBIA: Chocó, Alto de Barrigonal, Serranía del Darién (R. P. Anderson 1999).

SUBSPECIES: *Heteromys desmarestianus*, as currently conceived, is polytypic, with 12 subspecies recognized (D. F. Williams et al. 1993; Patton 2005a). As presently understood, only *H. d. crassirostris* Goldman extends into South America (see Remarks).

NATURAL HISTORY: This form is known from cloud forests along the Colombian-Panamanian border, where it is locally common (R. P. Anderson 1999). The contents of cheek pouches have included a variety of seeds, nuts, and other items (R. P. Anderson 1999).

REMARKS: Morphological, karyological, and allozymic data indicate that, as currently conceived, *Heteromys desmarestianus* represents a complex of several species (Mascarello and Rogers 1988; Rogers 1989, 1990; R. P. Anderson and Timm 2006; R. P. Anderson et al. 2006). Detailed specimen-based studies of alpha-level taxonomy are needed to elucidate the species in this complex, and current knowledge does not permit definitive conclusions regarding the valid name for the species present in extreme northwestern Colombia. These populations are provisionally referred to here as *H. d. crassirostris* (following R. P. Anderson 1999; R. P. Anderson et al. 2006). A karyotype for these populations is unknown.

Heteromys oasicus R. P. Anderson, 2003
Paraguaná Spiny Pocket Mouse

SYNONYMS:
Heteromys anomalus: Bisbal-E., 1990:180; first name combination applied.
Heteromys oasicus R. P. Anderson, 2003:9; type locality "Venezuela: Estado Falcón: 49 km N, 32 km W of Coro, Cerro Santa Ana, at 550 m."

DESCRIPTION: Adults have following combination of characters (R. P. Anderson 2003). Dorsal pelage very pale brown, strongly grizzled with thin ochraceous hairs intermixed with spines; little or no dark coloration on dorsal and external surfaces of forelimbs; ears pale brown and large relative to body size; orange band on flanks absent; plantar surface of hindfeet naked. Skull diminutive for genus (occipitonasal length 29.7–31.2 mm in adult specimens of age class 4 of Rogers and Schmidly [1982]); rostrum short and strongly tapered anteriorly, without anterodorsal flare; anterior portion of premaxillary convex (inflated), forming a smooth (not stepped) lateral border of rostrum; interorbital constriction narrow; braincase narrow and not inflated; interparietal moderately wide and rounded laterally; tubercle or swelling at posteroventral border of infraorbital foramen absent; mesopterygoid fossa formed by long, thin hamular processes of pterygoids; optic foramen especially small, with posterior margin formed by strong bar of bone. PM4 with straight, moderately long fold in anterior margin of posterior loph; pm4 with three lophids.

DISTRIBUTION: Low-to-middle elevations of Cerro Santa Ana and Fila de Monte Cano on the Península de Paraguaná in northwestern Venezuela; known elevational range from 90 to 700 m (R. P. Anderson et al. 2012).

SELECTED LOCALITIES (Map 24): VENEZUELA: Falcón, 49 km N and 32 km W of Coro, Cerro Santa Ana (type locality of *Heteromys oasicus* R. P. Anderson).

SUBSPECIES: *Heteromys oasicus* is monotypic.

NATURAL HISTORY: *Heteromys oasicus* inhabits small, isolated areas of forests, especially evergreen vegetation on the upper slopes of Cerro Santa Ana, but it also has been captured in riparian vegetation at lower elevations (i.e., on the lower slopes of Cerro Santa Ana and along the Fila de Monte Cano; R. P. Anderson 2003).

REMARKS: The karyotype is unknown.

Heteromys teleus R. P. Anderson and Jarrín, 2002
Ecuadorean Spiny Pocket Mouse

SYNONYMS:

Heteromys australis: Albuja, 1992:125; first name combination applied.

Heteromys teleus R. P. Anderson and Jarrín, 2002:6; type locality "Ecuador: Provincia Guayas [now Santa Elena]: Cerro Manglar Alto, western slope . . . 2000 ft [610 m]."

DESCRIPTION: Adults with following combination of characters (R. P. Anderson and Jarrín 2002; see also R. P. Anderson et al. 2006): dorsal coloration dark slate gray or black, nearly uniform or moderately grizzled with thin ochraceous hairs intermixed with spines; distinctive patch of dark coloration on dorsal and external surfaces of forelimbs; ears black and small to medium in size; orange band on flanks absent; plantar surface of hindfeet naked. Skull size average to large for genus (occipitonasal length 35.5–38.3 mm in adult specimens of age class 4 of Rogers and Schmidly [1982]); rostrum extremely wide and only slightly to moderately tapered anteriorly, without anterodorsal flare; anterior portion of premaxillary convex (inflated), forming smooth (not stepped) lateral border of rostrum; interorbital constriction moderately wide; braincase moderately wide and not inflated; interparietal narrow and rounded laterally; tubercle or swelling at posteroventral border of infraorbital foramen absent; mesopterygoid fossa formed by long, thin hamular processes of pterygoids; optic foramen especially small, with posterior margin formed by strong bar of bone. PM4 with straight, moderately long fold in anterior margin of posterior loph; pm4 with three lophids.

DISTRIBUTION: Low elevations along the Pacific coast of central-western Ecuador, and up to middle elevations on adjacent slopes of Cordillera Occidental and Cordillera de Chongón-Colonche of Ecuador. The known elevational range is from near sea level to 2,000 m, but most records are from sea level to approximately 600 m (R. P. Anderson and Jarrín 2002).

SELECTED LOCALITIES (Map 23; from R. P. Anderson and Jarrín 2002): ECUADOR: Cotopaxi, San Francisco de las Pampas, Reserva Bosque Nublado Otonga; Esmeraldas, Quinindé; Los Ríos, Estación Biológica Pedro Franco Dávila, Jauneche; Santa Elena, Cerro Manglar Alto, western slope (type locality of *Heteromys teleus* R. P. Anderson and Jarrín).

SUBSPECIES: *Heteromys teleus* is monotypic.

NATURAL HISTORY: *Heteromys teleus* inhabits evergreen but seasonal forests and is tolerant of moderate levels of disturbance (R. P. Anderson and Jarrín 2002). In the region near the type locality, almost all specimens were captured on the banks of small streams (R. P. Anderson and Jarrín 2002). R. P. Anderson and Martínez-Meyer (2004) provided an estimate of the species' potential geographic distribution and a preliminary conservation assessment.

REMARKS: The karyotype is unknown.

Suborder Myomorpha Brants, 1855
James L. Patton

Infraorder Myodonta Schaub, in Grassé and Dekeyser, 1955

Superfamily Muroidea Illiger, 1811

Family Cricetidae G. Fischer, 1817

The muroid rodents of the suborder Myomorpha comprise three subfamilies of the Cricetidae in South America. Two of these, the Neotominae and Tylomyinae, are North or Middle American lineages of which a single genus of the latter and two of the former extend into northwestern South America. The third, the Sigmodontinae, are highly diverse and largely autochthonous to that continent, with a few members extending north into Middle America and reaching a northern limit in the northeastern United States. These three groups, together with the Holarctic Arvicolinae and Old World Cricetinae, are currently included in the Cricetidae (e.g., Musser and Carleton 2005). Initial molecular analyses (e.g., Jansa and Weksler 2004; Steppan, Adkins, and Anderson 2004) failed to recover robust phylogenetic relationships among the three subfamilies that occur in South America. However, three recent taxon and gene-expanded studies (Parada et al. 2013; Schenk et al. 2013; J. F. Vilela et al. 2013) supported a sister relationship between the Tylomyinae and Sigmodontinae; the latter authors posited the split between the two subfamilies at about 18 mya based on a time-calibrated ultrmetric tree analysis.

Musser and Carleton (2005) detailed the long and tortuous history of the concept of the Muroidea, answering the question these two authors had posed in 1984 about the number of families to recognize and the family-group ranks for those lineages recognized in the New World, specifically South America. Although refinements to what is known of phyletic relationships and group membership will continue, particularly with the future expansion of molecular analyses, the three subfamilies of Cricetidae now recognized as part of the native fauna of South America will likely remain stable into the foreseeable future (see Schenk et al. 2013).

The Sigmodontinae comprise the vast majority of cricetid rodents in South America, representing a large radiation (approximately 84 genera and nearly 380 species), with representatives occupying virtually all habitats from sea level to the high Andes, and from evergreen rainforests of the Amazon to some of the world's most xeric communities along the Pacific coast from central Peru to northern Chile. Only single species of *Reithrodontomys* and of *Isthmomys* of the otherwise highly diverse and mostly North or Middle American Neotominae (16 genera, 124 species; Musser and Carleton 2005), and of *Tylomys* of the much less diverse Middle American Tylomyinae (four genera, 10 species; Musser and Carleton 2005), extend into South America. Neotomine and tylomyine taxa are limited to western Colombia and adjacent Ecuador; throughout the remainder of the continent, only members of the Sigmodontinae are found.

Craniodental, skeletal, and soft anatomical traits for representative genera of each of the three subfamilies of Cricetidae that occur in South America are described and illustrated in Carleton (1980), Steppan (1995), Pacheco (2003), and Weksler (2006). Focus should be on those trans-Andean genera in the northwestern part of the continent, the only region where members of the three subfamilies may be found together or in close proximity: *Tylomys*, the sole member of the Tylomyinae in South America, *Isthmomys* and *Reithrodontomys* of the Neotominae, and 25 genera of the Sigmodontinae.

Only a few obvious traits are needed to distinguish *Tylomys* from *Reithrodontomys* and *Isthmomys*, or the latter two from each other. Distinguishing each of these three from the array of trans-Andean sigmodontine genera, however, is more challenging, both because of phyletic diversity (these 25 genera belong to five of the nine currently recognized tribes) and because of disparate morphologies resulting from adaptation to the very wide array of occupied environments. The key emphasizes characters that readily distinguish *Tylomys* from *Reithrodontomys* and *Isthmomys*, and it pairs these comparisons only with the few co-occurring sigmodontine genera that share one or more character sets.

KEY TO THE SOUTH AMERICAN SUBFAMILIES OF CRICETIDAE (LIMITED TO THOSE GENERA IN TRANS-ANDEAN REGION OF ECUADOR AND COLOMBIA):

1. Upper incisors either ungrooved or grooved; lower third molar medium to large (>25% mandibular toothrow length); glans penis lacks dorsal papilla and lateral bacular mounds (cartilaginous baculum with single, medial digit), with crater absent or only incipiently developed; male accessory reproductive glands include no or only single pair of preputials; vesiculars variable but not J-shaped 2
1'. Upper incisors uniformly ungrooved (except for *Sigmodon alstoni*); lower third molar small (<23% mandibular toothrow length); glans penis of males with dorsal papilla, lateral bacular mounds (cartilaginous baculum tridigitate), and deep crater; male accessory reproductive glands include one or two pairs of preputials; vesiculars J-shaped. . . . . Sigmodontinae

2. Body size relatively large (>100 g); tail naked, scales large and conspicuous; anterocone bifurcated by deep anteromedian flexus dividing two equally large cusps; cusps arranged in opposite pattern; upper incisors smooth, ungrooved; supraorbital region with well-developed ledges that extend across the parietals to the lambdoidal crest as pronounced ridges; glans penis short and broad, the bony baculum short and broad, not extending beyond tip of glans, with long cartilaginous tip, crater incipiently developed; male accessory reproductive glands include a single pair of preputials, long and intestiniform vesiculars . . . . Tylomyinae (*Tylomys*)
2'. Body size smaller (<50 g); tail thinly haired, scales not conspicuous; anterocone undivided or bifurcated by weak flexus dividing two small conules; cusps weakly alternate in pattern; upper incisors with or without single groove on anterior face; interorbital region smooth to slightly beaded; temporal ridges absent; glans penis elongated, and baculum long and thin, extending beyond glans tip, without cartilaginous tip, crater absent; male accessory reproductive glands lack preputials but include J-shaped vesiculars 3
3. Body size diminutive (head and body length 70–90 mm); upper incisors with longitudinal groove on anterior face. . . . . Neotominae (*Reithrodontomys*)
3'. Body size larger (head and body length 150–200 mm); upper incisors with smooth anterior face . . . . Neotominae (*Isthmomys*)

REMARKS: *Sigmodon alstoni* is the only sigmodontine in trans-Andean northwestern South America that shares with *Tylomys* the wide supraorbital ledges that extend across the parietals as distinct ridges and end at the lambdoidal crest. These two taxa, however, do not occur in the same area, and they are distinct in habitat (lowland grasslands versus montane forest, respectively). Cranially the two can be easily distinguished by their (respectively) planar versus cuspidate teeth, narrow (< rostral breadth) versus broad (> rostral breadth) interorbital regions, broad zygomatic plates with pronounced spines and deep zygomatic notches (when viewed from above) versus narrow zygomatic plates without spines and thus imperceptible zygomatic notches, among a number of other easily observed craniodental differences.

Externally, *Sigmodon* has a well-furred, dorsoventrally bicolored tail, elongated hindfeet, and coarse, hispid fur. *Tylomys* has a naked, anteroposteriorly bicolored tail, short and broad hindfeet, and soft fur. *Sigmodon alstoni* and *Reithrodontomys* also share grooved upper incisors, but these two are also readily separable externally by size (large versus small) and tail length (shorter than versus longer than head and body), and cranially also by planar versus cuspidate molars, ledged versus smoothly rounded interorbital edges, and zygomatic plates with and without dorsal spines.

Subfamily Neotominae Merriam, 1894

Musser and Carleton (2005:1048) reviewed the concept and taxonomic history of the Neotominae, a group in which Merriam (1894) originally included only the North American woodrats (genus *Neotoma*) and certain South American fossils with high-crowned molars (*Ptyssophorus* and *Tretomys*, both now regarded as synonyms of the extant genus *Reithrodon*; see Pardiñas 2000 and that account here). Today, the subfamily includes all North and Middle American Cricetidae, excluding the Arvicolinae, the four genera of Middle American Tylomyinae, and the few genera of Sigmodontinae that extend north of the Panama-Colombia border. Two genera of neotomines, *Isthmomys* and *Reithrodontomys*, extend into South America.

KEY TO THE SOUTH AMERICAN GENERA
OF NEOTOMINAE:

1. Upper incisors conspicuously sulcate, with a single longitudinal groove on anterior enamel face . . . .
. . . . *Reithrodontomys*
1'. Upper incisors smooth, without conspicuous groove . .
. . . . *Isthmomys*

Genus *Isthmomys* Hooper and Musser, 1964
James L. Patton

This genus includes two species, only one of which extends into extreme northwestern Colombia (*I. pirrensis*). *Isthmomys* was diagnosed originally (Hooper and Musser 1964) to reflect conspicuous penile differences from the remainder of *Peromyscus*. Carleton (1989:125–126) subsequently provided an expanded diagnosis that included the following morphological attributes: Beaded supraorbital edges and weakly developed temporal ridges; subsquamosal fenestra absent or minute, postglenoid foramen small and narrow; sphenopalatine vacuities tiny or absent; M1 strongly ovate, narrow anterocone deeply bifurcate; posteroflexid on m3 isolated as fossetid; entepicondylar foramen absent; stomach with edges of discoglandular zone partially infolded (incipient pouch); phallus long and broad distally, baculum longer then length of glans penis, spines long and heavy, urinary meatus open almost terminally through large mound of nonspinous, soft tissue; preputials and lateral ventral prostates absent, cephalic border of vesiculars with lobulated folds; postaxial pair of mammae absent.

SYNONYMS:
Megadontomys: Bangs, 1902:27; part (description of *flavidus*); not *Megadontomys* Merriam, 1898.
Peromyscus: Osgood, 1909:221 (listing of *flavidus*); part; not *Peromyscus* Gloger, 1841.
Isthmomys Hooper and Musser, 1964:12; type species *Megadontomys flavidus* Bangs, by original designation; as subgenus of *Peromyscus*.

REMARKS: Hooper and Musser (1964) proposed *Isthmomys* as a subgenus of *Peromyscus*. Hooper (1968) maintained this rank, but Carleton (1980, 1989) subsequently provided an emended diagnosis and accorded generic status. Pine et al. (1979:357), citing Article 13:1.1 of the Code, argued that the generic name as proposed by Hooper and Musser (1964) was a *nomen nudum*. Subsequently, Pine (in Honaki, Kinman, and Koeppl 1982:421) wrote that Carleton (1980:124) may have made *Isthmomys* available. Neither Carleton (1989) nor Musser and Carleton (1993, 2005) accepted Pine's position, and continued to list Hooper and Musser (1964) as the author of the name.

Although the placement of *Isthmomys* in the Neotominae is noncontroversial, its phylogenetic relationship to other genera remains ambiguous. Carleton (1980) grouped *Isthmomys* in a clade with *Megadontomys*, based on a number of shared morphological characters. Alternatively, Rogers et al. (2004), using 25 allozyme loci, hypothesized a phyletic relationship between *Isthmomys* and *Reithrodontomys*, a result supported by subsequent analyses of the mtDNA cytochrome-*b* gene (Bradley et al. 2007), although not by other mitochondrial sequences (Engle et al. 1998). Two species (*I. flavidus* and *I. pirrensis*) are currently recognized (Hall 1981; Musser and Carleton 2005). Middleton (2007), in an unpublished M.S. thesis, examined morphometric relationships between them, confirming their phenotypic distinction.

Isthmomys pirrensis (Goldman, 1912)
Mount Pirri Deer Mouse
SYNONYMS:
Peromyscus (*Megadontomys*) *pirrensis* Goldman, 1912b:5; type locality "near head of Rio Limon (altitude 4,500 feet), Mount Pirri, [Darién,] eastern Panama."
[*Peromyscus*] *Isthmomys pirrensis*: Hooper and Musser, 1964:13; name combination.
[*Isthmomys*] *pirrensis*: Carleton, 1980:124; first use of current name combination.

DESCRIPTION: Size large (total length 342–376 mm, tail length 185–204 mm, hindfoot length 36–36.5 mm); upper parts dark brownish cinnamon to cinnamon rufous lined with black, becoming grayish brown on head and more rusty on rump; sides brighter, more rufescent; venter dull buffy white with gray basal color of fur apparent throughout; tail brownish and nearly unicolored.

DISTRIBUTION: Extreme eastern Panama and adjacent Colombia.

SELECTED LOCALITIES (Map 25; South America only): COLOMBIA: Chocó, Alto de Barrigonal, Serranía del Darién (Voss 2014).

SUBSPECIES: *Isthmomys pirrensis* is monotypic.

NATURAL HISTORY: This is a terrestrial species found in evergreen forest (tropical pluvial and tropical rainforest) at an elevational range from 100–500 m (Handley 1966; Alberico et al. 2000; Cuartas-Calle and Muñoz-Arango 2003).

REMARKS: Alberico et al. (2000:63) included *I. pirrensis* in their list of Colombian mammals, citing Cuervo et al. (1986) as their source and identifying the IAvH as the collection presumptively housing one or more vouchers. Neither publication, however, provided a specific locality or catalog number(s) as confirmation. Cuartas-Calle and Muñoz-Arango (2003) included the species, as a probable occurrence, in their list of mammals from Antioquia, Colombia, based on Alberico et al. (2000). Voss (2014) confirmed the presence of this species in Colombia when he examined specimens from the Serranía del Darién in Chocó department, then in the INDERENA collection in Bogotá, during a visit there in 1984.

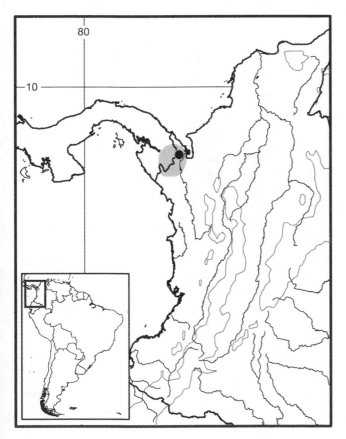

Map 25 Single known South American locality for *Isthmomys pirrensis* (●). Panama distribution from F. A. Reid (2009). Contour line = 2,000 m.

Genus *Reithrodontomys* Giglioli, 1874

Elizabeth Arellano

The genus *Reithrodontomys*, with 20 species (Bradley et al. 2004a; Musser and Carleton 2005), covers an extensive geographic and elevational range from the southwestern portions of Canada to western Colombia and Ecuador (Hall 1981; Eisenberg 1989). Fossils are known from Pleistocene deposits of North America. The elevational range extends from below sea level in Death Valley, California, to over 4,000 m in the mountains of Mexico, Central, and South America (Hooper 1952; Hall 1981). Ecologically, these mice are associated with a variety of habitats and local climates, including grass and shrub lands, both marine and freshwater marshlands, well-drained uplands in temperate climates, arid zones with xerophytic vegetation, tropical forest in humid lowlands, and conifer forests up to the timber line on high mountains (Hooper 1952).

Harvest mice are small rodents with proportionally long tails. They are distinguished from other members of the tribe Reithrodontini (*sensu* Musser and Carleton 2005, including *Habromys, Isthmomys, Megadontomys, Neotomodon, Onychomys, Osgoodomys, Peromyscus,* and *Podomys*) by the possession of grooved upper incisors (Le Conte 1853). This character alone distinguishes these mice from species of similar body size in other genera that co-occur in that part of the range that extends into South America (e.g., *Microryzomys* and *Oligoryzomys*).

The genus is divided into the two subgenera *Reithrodontomys* and *Aporodon,* which differ from each other in characteristics of the molars (A. H. Howell 1914; Hooper 1952) and other features (summarized by Carleton 1980), allozymes (K. Nelson et al. 1984; Arellano et al. 2003), satellite DNA (Hamilton et al. 1990), and DNA sequences (Bell et al. 2001; Bradley et al. 2004b; Arellano et al. 2006).

SYNONYMS:

Reithrodon Le Conte, 1853:413; type species *Mus Le Contii* Audubon and Bachman [= *Mus humulis* Audubon and Bachman]; not *Reithrodon* Waterhouse.

R[eithrodon].: Saussure, 1860:109; part (description of *mexicanus*); not *Reithrodon* Waterhouse.

Reithrodontomys Giglioli, 1874:326; type species *Reithrodon megalotis* Baird, designated by A. H. Howell (1914:13).

Ochetodon Coues, 1874:184; no type species selected among those listed (*humulis* Audubon and Bachman, *longicauda* Baird, *mexicanus* Saussure, *montanus* Baird, and *sumichrasti* Saussure).

Aporodon A. H. Howell, 1914:63; type species *Reithrodontomys tenuirostris* Merriam, by original designation.

Cudahyomys Hibbard, 1944:725; type species *Cudahyomys moorei* Hibbard, by original designation.

REMARKS: See A. H. Howell (1914:13) for why *Mus lecontii* Audubon and Bachman (= *Reithrodontomys humulis*) was incorrectly selected as the type species of the genus *Reithrodontomys* by G. S. Miller and Rehn (1901:95). Merriam (1892:26, footnote) was the first to recognize that *Reithrodontomys* Giglioli had date of publication priority over *Ochetodon* Coues, a position more thoroughly detailed by Hooper (1952:32, footnote).

Only a single species, *R. mexicanus*, is known to occur in South America, but Alberico et al. (2000) listed *R. darienensis* Pearson as a probable resident of Colombia in the region adjacent to Panama (see also Musser and Carleton 2005:1081). Furthermore, analyses of molecular, morphological, and ecological data suggest elevation of at least two of the three subspecies of *mexicanus* in Ecuador to species status (D. Chávez, unpubl. data; see below), and phylogenetically couple these harvest mice to *R. darienensis* rather than to *R. mexicanus*. When published, these analyses will result in substantial changes in the taxonomy and biogeographic history of harvest mice in South America.

Reithrodontomys mexicanus (Saussure, 1860)
Mexican Harvest Mouse

SYNONYMS (only as applied to South American taxa; additional synonyms under subspecies):

Mus tazamaca Gray, 1843a:79; *nomen nudum* (see Alston 1880a:152).

R[*eithrodon*]. *mexicanus*: Saussure, 1860:109; "mts. of Veracruz," restricted to "Mirador, Veracruz, Mexico" (Hooper 1952:140).

Ochetodon mexicanus: Coues, 1874:186; name combination.

Reithrodontomys mexicanus: J. A. Allen, 1894a:319; name combination.

DESCRIPTION: Moderate to large for genus; tail much longer than head and body (about 130–160% of head and body length); body coloration varies geographically from uniform dark brown to blackish, basic color of upperparts tawny or orange-cinnamon, with slight variation in tone; upper parts of intensely pigmented races (such as *mexicanus* and *soederstroemi*) heavily covered with black over tawny ground color; upper parts of paler races nearly pure tawny; venter varies from whitish to pale cinnamon or orange-cinnamon; epidermis of ears fuscous and sparsely covered with black hairs; duskiness of tarsus extends onto foot as a longitudinal stripe of varying width and length, and in northern races covers most of upper surface of hindfoot, leaving only narrow rim of white; in more southern populations, this stripe narrower and wedge shaped, with

apex not extending to base of toes; dusky stripe in some specimens entirely lacking, and upper surface of hindfoot whitish or pale cinnamon; toes in all races whitish (Hooper 1952). Tirira (1999:lamina 21, 7) provided a color photo of a specimen from Ecuador.

DISTRIBUTION: *Reithrodontomys mexicanus* in South America occurs in the western and central Andes of Colombia to as far south as Quito, Ecuador, across an elevational range of 1,200 to nearly 4,000 m (A. H. Howell 1914; Hershkovitz 1941; Hooper 1952; Eisenberg 1989; Voss 2003). Humid tropical broadleaf forests are the typical habitat, but specimens have been collected in dry deciduous forests (Hooper 1952) and at the edge of grassy páramo (Voss 2003). This species exceeds all others in the subgenus *Aporodon* in the extent of geographic (including elevational) range as well as in expressed morphological variability. The South American populations of this species are apparently disjunct from those in Middle America, as a distributional gap in the range extends from northwestern Colombia at least to western Panama (Hall 1981; F. A. Reid 2009).

SELECTED LOCALITIES (Map 26): COLOMBIA: Antioquia, Guapantal (FMNH 71684), 15 km E of Río Negrito (FMNH 70181), Santa Elena (FMNH 69653), Valdivia, La Selva (FMNH 70167); Cauca, Munchique (type locality of *R. m. milleri* J. A. Allen), Río Guachicono (FMNH 90342), Río Mechengue (FMNH 90341), Valle de las Papas (Hooper 1952); Cundinamarca, San Cristóbal (FMNH 71663); Huila, Río Aguas Claras (FMNH 71701); Quindio, El Roble (Hooper 1952); Valle de Cauca, Correa, Farallones de Cali (Universidad de Cali collection), 4 km NW of San Antonio (MVZ 124064). ECUADOR: Imbabura, Intag (Hooper 1952), San Nícolas, near Pimanpiro (type locality of *R. m. eremicus* Hershkovitz); Napo, Río Oyacachi (AMNH 66808); Pichincha, Gualea (AMNH 46800), Mojanda (AMNH 46796), Quito (type locality of *R. m. soederstroemi* Thomas); Tungurahua, San Francisco (AMNH 67551).

SUBSPECIES: Three subspecies of *R. mexicanus* are reported for South America: *R. m. eremicus*, *R. m. milleri*, and *R. m. soederstroemi* (A. H. Howell 1914; Hershkovitz 1941; Hooper 1952). The molecular, morphological, and ecological analyses of D. Chávez (unpubl. data) suggest elevation of at least two of these subspecies to species status.

R. m. eremicus Hershkovitz, 1941
SYNONYM:

Reithrodontomys mexicanus eremicus Hershkovitz, 1941:4; type locality "near Pimanpiro, Imbabura Province, Ecuador; altitude, approximately 6500 feet." Given as "Valle de la Chota, near Pimanpiro, San Nicolas" by Hooper (1952:160).

This race is restricted to light sandy soils supporting arid deciduous forest in Valle del Chota in Ecuador (Hershkovitz 1941:5). Hershkovitz (1941:5) stated that *eremicus* was distinguishable from other subspecies of *R. mexicanus* by its large ears and paler color, which he believed was "obviously correlated with the light sandy soil and open country of its habitat." He also suggested that "although the area occupied by *R. m. eremicus* appears as an island within the range of *söderströmi* [*sic*], it is actually more directly connected with the range of *R. m. milleri* though the Río Chota Valley, which is the interandean trunk of the western Andean Río Mira Valley." Even though Hershkovitz (1941) pointed to ear size as a diagnostic attribute, Hooper (1952) failed to find significant differences in ear size among races of *R. mexicanus* in South America in the samples he examined.

R. m. milleri J. A. Allen, 1912

SYNONYMS:

Reithrodontomys milleri J. A. Allen, 1912:77; type locality "Munchique (alt. 8325 ft), Cauca, Colombia."
Reithrodontomys mexicanus milleri: Hershkovitz, 1941:2; name combination.

This race is the most widely distributed subspecies in South America, inhabiting Andean valleys and slopes at moderate to high elevations (1,800–3,150 m) in Colombia and northwestern Ecuador. It occurs in both the Cordillera Occidental and Cordillera Central but is unknown from the Cordillera Oriental. Hershkovitz (1941) characterized the habitat as subtropical forest, but specimens have been collected in humid, montane evergreen forests.

Hooper (1952:159) regarded this race to be more strongly differentiated from the southern races *eremicus* and *soederstroemi* than it was from those to the north in Central America. He noted a general size increase from north to south, but specimens were too few to know whether the shift was gradual. A. H. Howell (1914) and Hershkovitz (1941) both assigned specimens from Valle de las Papas, 3,050 m, Huila department, Colombia, to *soederstroemi*, but Hooper (1952) considered them most similar to topotypes of *milleri*.

R. m. soederstroemi Thomas, 1898

SYNONYMS:

Reithrodontomys söderströmi Thomas, 1898e:451; type locality "Quito," 9,200 ft, Pichincha, Ecuador (see Hooper 1952:161).
Reithrodontomys mexicanus söderströmi Hershkovitz, 1941:3; name combination.

This race is apparently restricted to the inter-Andean temperate zone of Ecuador, in the vicinity of Quito (Hooper 1952). It is larger than *milleri*, with a broader braincase

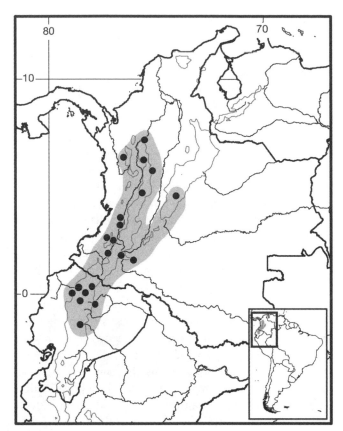

Map 26 Selected localities for *Reithrodontomys mexicanus* in South America (●). Contour line = 2,000 m.

and narrower zygomatic plate, but of similar color. Compared to *eremicus*, *soederstroemi* is similar in size and most cranial details but is distinct in color, with upper parts tan rather than dark reddish brown.

NATURAL HISTORY: Little has been published on the ecology or other aspects of the natural history of this species anywhere in its range in South America. Ludovic Söderström, who collected in Ecuador in the late 1890s, described individuals feeding on flowers and seeds in the gardens of Quito, emerging from climbing plants in the evening (Thomas 1898e).

Subfamily Sigmodontinae Wagner, 1843
Guillermo D'Elía and Ulyses F. J. Pardiñas

Sigmodontines constitute the most diverse South American subfamily of Cricetidae (Reig 1986), with members present in all terrestrial environments, ranging continentwide from the Cape Horn islands in the south to the Caribbean coasts of Colombia and Venezuela, and across an elevational gradient from sea level to nearly 5,000 m. Outside

South America, sigmodontines today range through Central America and southern North America (to about 35°N), and historically occurred on some Caribbean islands. Most of the extant sigmodontine diversity (69 of 86 genera) is endemic to the South American continent (D'Elía 2003b), including neighboring continental islands such as Aruba, Curaçao, Trinidad, and Tobago.

The genus *Nesoryzomys* is endemic to the oceanic Galápagos Archipelago. Two genera (*Oryzomys* and *Sigmodon*) have largely North American distributions and diversities, and they enter South America only in the very northwest of the continent. One genus (*Oligoryzomys*) ranges throughout South America and reaches northward to near the Mexico-United States border. Eleven genera are distributed in Central and South America, with four of these extending barely into Panama (*Ichthyomys*, *Neacomys*, *Oecomys*, and *Rhipidomys*) while the other seven have broader ranges in Central America (*Handleyomys*, as currently delimited to include the "*alfaroi* species group," *Melanomys*, *Nephelomys*, *Sigmodontomys*, *Tanyuromys*, *Transandinomys*, and *Zygodontomys*).

The water mouse *Rheomys* is the only mainland extant genus of sigmodontines not present in South America. Eight genera of sigmodontines became extinct in historic times or during the Pleistocene; four of these inhabited islands that are considered part of South America and thus would have been included in this book had they remained extant: *Megaoryzomys* (Galápagos Archipelago; Steadman and Ray 1982), *Noronhomys* (Fernando de Noronha; Carleton and Olson 1999), *Agathaeromys* (Bonaire; Zijlstra et al. 2010), and *Dushimys* (Curaçao; Zijlstra 2012). Some forms of *Aegialomys* and *Nesoryzomys* have become extinct in the Galápagos Archipelago in recent decades (Patton and Hafner 1983; Hutterer and Oromi 1993; Dowler et al. 2000).

Recent and continuing fieldwork has substantially expanded prior knowledge of the distributional range of many genera and species. For example, L. Luna and Patterson (2003) reported the rare genus *Rhagomys*, known for almost a century only from the Atlantic coastal Brazilian state of Rio de Janeiro, from the eastern slopes of the Peruvian Andes, and more recently Percequillo, Tirelli et al. (2011) documented its presence in central Amazonia.

Nevertheless, distributional patterns of the major sigmodontine groups are reasonably well known. The Oryzomyini, one of three tribes that occur in North, Central, and South America, has by far the largest geographical range, one nearly equivalent to that of the subfamily as a whole (Reig 1986; Weksler 2006). Members of this tribe, however, reach their greatest diversity in the tropical areas of South America. In contrast, the abrotrichine genera are restricted to the Andean region, from the Altiplano in west-central Peru and western mountain ranges in Argentina and Bolivia to southernmost South America (including several islands) and the Patagonian steppes, reaching the Atlantic coast of Argentina (D'Elía et al. 2007).

Sigmodontines inhabit almost all habitat types found in their cumulative distributional range, including deserts, wet tropical and temperate forests, scrublands, wetlands, savannas, steppes, high elevation grasslands, and salt flats (Hershkovitz 1962). Many genera and even some species range across more than one habitat type. For example, *Abrothrix longipilis* can be found in steppe, bunchgrass prairie, Valdivian and subantarctic forests, and sparse shrubland. Many others, however, appear restricted both geographically and to specific habitat types. For instance, *Phyllotis anitae* is known only from the upper belt of the Yungas of northwestern Argentina. Tropical forested areas are roughly characterized by a predominance of oryzomyines and thomasomyines, temperate lowlands by akodontines, Andean piedmont deserts and high Andean areas by phyllotines, and cold arid lands, southern steppes, and temperate and subantarctic forests by abrotrichines (B. D. Patterson 1999).

Sigmodontines are generally small to medium in size, with the smallest species about 12 g in mass and the largest nearly 500 g. The largest living taxon, *Kunsia tomentosus*, does not surpass 300 mm in head and body length. External ears vary from nearly absent or much reduced (e.g., the aquatic rat *Anotomys* or the fossorial mouse *Notiomys*) to large (e.g., the cony rats of the genus *Reithrodon* or the leaf-eared pericotes of the genus *Phyllotis*). Most sigmodontines have a brownish or blackish upper pelage and grayish or whitish under parts. However, *Chinchillula* has a striking contrastingly colored pelage, with buffy or grayish upper parts with black lines, white under parts, and with both hips and rump white banded with black. *Abrothrix jelskii* is another species with a showy, contrastingly patterned pelage, at least for some of its multiple geographic races. Sigmodontine fur may be velvety, soft, woolly, long, thick, harsh, and/or composed of underfur and guard hairs. The genera *Abrawayaomys*, *Neacomys*, and *Scolomys* and the species *Rhagomys longilingua* have some hairs modified into stiffened spines.

In climbing and scansorial sigmodontines, the tail is usually longer than the head and body length; in more generalized terrestrial forms, it is about equal in length to the head and body; and in fossorial species, it is less than half the head and body length. The tail is typically thinly haired, but in some sigmodontines it ends in a penciled tip (e.g., in *Andalgalomys* and in many *Rhipidomys*).

The forefeet are in general relatively small, but those of fossorial forms, such as the giant rat *Kunsia*, are robust and have extremely long claws. The hindfeet of aquatic species (e.g., those in *Holochilus*, *Lundomys*, *Nectomys*, and ichthyomines) are large and webbed, and they may have stiffened natatory hairs along the edges, including the digits (ichthyomyines). Climbers (e.g., *Oecomys* and *Rhipidomys*) typically have shortened but broad hindfeet, elongated toes for grasping, and enlarged plantar pads. The hindfeet of the arboreal mouse *Rhagomys* are unique among sigmodontines in having an elongated digit V and a hallux with a nail instead of a claw.

Sigmodontines possess one pair of upper (rarely grooved) and lower incisors and lack canines, as do all other rodents; as with other cricetids, they lack premolars. Three cheek teeth, or molars, are present on each side, above and below, in all but two sigmodontines that exhibit reduction to two molars (*Neusticomys ferreirai* and *N. oyapocki*). As such, the total number of teeth is typically 16. Supernumerary molars have not been reported, in contrast with their presence in some North American neotomines. The molars exhibit substantial variation in size, morphology, and number of roots (from one to six) and are fairly complex structures. For an idealized sigmodontine molar, Reig (1977) named more than 30 cusps, cuspules, styles, lophs, folds, and islands (see also Hershkovitz 1993 for additional topographical details). Molar crown height is apparently related to diet. Sigmodontines that feed on animals, seeds, fruits, or fungus have low-crowned or brachydont molars. High-crowned or coronal hypsodont molars, which extend well above the gum line, are generally present in sigmodontines that feed on abrasive vegetation (e.g., grass). The hypsodont condition is often associated with simplified occlusal surfaces, with the connection of lophs on each tooth side forming lamina, planation of the crown, and an increase in the number of roots. Two main types of molars are recognized with respect to the number of lophs: tetralophodont and pentalophodont. Hershkovitz (1962; see also Hershkovitz 1993) suggested that the pentalophodont type represents the ancestral state from which a simpler, tetralophodont molar evolved. Remarkably, given the substantial influence that Hershkovitz's hypothesis has had on sigmodontine systematics, the evolution of molar traits has not been studied within a contemporary phylogenetic approach. In light of current phylogenetic hypotheses and the available fossil record, it appears more likely that, among the Sigmodontinae, tetralophodont and pentalophodont molar types represent the plesiomorphic and apomorphic character states, respectively. Nevertheless, the number of character state transformations of the molars that have occurred across the sigmodontine radiation, as well as the timing and the placement of those transformations, remains a mystery.

As is true for nearly all mammals, craniodental morphology has been the basis of sigmodontine taxonomy and systematics, owing to the remarkable variety even among closely related lineages (e.g., genera; Voss and Marcus 1992). Different regions of the skull and mandibles have been minutely scrutinized, with features of the bony palate, supraorbital region, foraminal pattern, and mandible body expression forming the foundation for taxonomic assignment. Several anatomical traits show great variation. For example, the zygomatic plate (despite the myomorph condition of the sigmodontine skull) varies from being a narrow but robust structure in ichthyomyines to broad and high plates in granivorous and grazing members of the tribes Phyllotini and Sigmodontini. The bony palate also displays trenchant differences, including both short versus long and wide versus narrow conditions (Hershkovitz 1962). Another anatomical trait useful in separating major groups is the absence in oryzomyines of a suspensory process of the squamosal bone, which contacts the tegmen tympani (Weksler 2006). The internal carotid artery and its derivatives display three main circulatory patterns (Voss 1988) that have proven valuable in diagnosing taxa and supporting phylogenetic hypotheses. Recent phylogenetic analyses strongly suggest that several morphological state characters have evolved more than once within the sigmodontine radiation, whereas others may be symplesiomorphies.

With a few well-recognized exceptions (i.e., some species of *Abrothrix* and at least one of *Punomys*; Spotorno 1992), sigmodontines have complex penises (Hooper 1959; Hershkovitz 1962). A complex penis has two lateral horns on the cartilaginous distal baculum, making it appear trident-shaped. A simple, saclike stomach (Vorontsov 1979; Carleton 1973) also characterizes all sigmodontines, although the details of morphology vary among genera (Carleton 1973).

Over the last decades of the past century, the contents of the subfamily Sigmodontinae have been the center of much debate (reviewed by Carleton 1980; Reig, 1980; Steppan 1995; D'Elía 2000; Steppan, Adkins, and Anderson 2004; Musser and Carleton 2005). This debate primarily concerns whether to limit the Sigmodontinae, as currently understood, to those predominantly South American cricetids with the tridigitate baculum of the complex penis morphology (e.g., *Sigmodon*, *Oryzomys*, *Akodon*, *Phyllotis*) or to also include the predominantly North American forms with a single-pronged baculum, commonly referred to as the neotomine-peromyscines and currently placed in the cricetid subfamilies Neotominae (e.g., *Neotoma*, *Peromyscus*, *Reithrodontomys*) and Tylomyinae (e.g., *Tylomys*). Early studies based on the muscular system (Rinker 1954), accessory glands of the male reproductive system (Arata 1964; see also Voss and Linzey 1981), glans penis morphology (Hooper 1959; Hooper and Musser 1964), stomach morphology (Carleton 1973), and ectoparasite distributions (Wenzel and Tipton 1966), together with considerations of geographic distribution, led Reig (1980; see also Hershkovitz 1966a, 1972) to argue strongly in favor of the phylogenetic distinction between sigmodontines and neotomine-peromyscines. Under this hypothesis, the Sigmodontinae are restricted to including only the predominantly South American complex penis cricetids. We note, however, that Hershkovitz (1966a) considered some forms (e.g., *Nyctomys*) in his concept of the Sigmodontinae, taxa that Reig (1984) later placed in the Tylomyinae, a new cricetid subfamily (see the emended diagnosis of this subfamily in Musser and Carleton 2005:1186).

The same year that Reig (1980) published his study favoring the restriction of Sigmodontinae to the predominantly South American complex penis cricetids, Carleton (1980) published a landmark and highly influential monograph on muroid morphological variation in which he favored an alternative, broader definition of Sigmodontinae. Even though Carleton's analysis (Fig. 41, p. 117) supported the monophyly of complex penis sigmodontines and their close relationship with Old World murids rather than with neotomine-peromyscines, he stated that "it is premature to taxonomically formalize a division between New World cricetines" (i.e., sigmodontines on the one hand and neotomine-peromyscines on the other).

Despite early molecular work that also supported sigmodontine monophyly (Sarich 1985; Catzeflis et al. 1993), Carleton's suggestion soon became solidified in the most important treatises of muroid rodent taxonomy (Carleton and Musser 1984; Musser and Carleton 1993; McKenna and Bell 1997), in which Sigmodontinae was defined to also include both the neotomine-peromyscines and tylomyines. However, DNA-based phylogenetic analyses published at the end of the 1990s (M. Robinson et al. 1997; Engel et al. 1998; M. F. Smith and Patton 1999; Dubois et al. 1999) strongly supported the monophyly of the Sigmodontinae (*sensu* Reig 1980) to the exclusion of the neotomine-peromyscines and tylomyines. In addition, in some of the trees recovered in these studies, the sigmodontine group was not even recovered as sister to clades of genera now placed in the Neotominae or Tylomyinae (e.g., Musser and Carleton 2005). These studies have reinforced, as presciently recognized by D'Elía (2000), the substantive distinction between sigmodontines and the rest of the New World cricetids. More recent molecular studies (Jansa and Weksler 2004; Steppan, Adkins, and Anderson 2004; Parada et al. 2013; Schenk et al. 2013; J. F. Vilela et al. 2013) have corroborated these earlier phylogenetic results and affirmed the independence of Sigmodontinae relative to both Neotominae and Tylomyinae.

Although the usage of a broadly defined Sigmodontinae has persisted in some recent literature (e.g., Álvarez-Castañeda and Cortés-Calva 2003; Amman and Bradley 2004), Musser and Carleton's (2005) muroid catalog solidified the usage of Sigmodontinae *sensu* Reig (1980), drawing heavily on the increasing depth and breadth of molecular phylogenetic analyses. Thus, as currently defined, the subfamily Sigmodontinae includes 86 living genera and about 400 living species, of which 85 genera and 381 species inhabit South America. These numbers will continue to increase as new genera (see Weksler et al. 2006; Pardiñas, D'Elía, and Teta 2009; Percequillo, Weksler, and Costa 2011) and species (e.g., Jayat et al. 2010; R. G. Rocha, Costa, and Costa 2011; Tavares et al. 2011; C. F. Jiménez et al. 2013; Pardiñas, Teta et al. 2013; Gonçalves and Oliveira 2014) are discovered and described. Also notable, in addition to playing a key role in the discovery of new species, the increasing use of molecular data in taxonomic studies of Sigmodontinae has significantly clarified species boundaries, resulting either in the validation or synonymization of nominal forms (see, for example, D'Elía, Pardiñas, Jayat, and Salazar-Bravo 2008).

Hershkovitz (1966b) showed that the correct name to apply to the clade comprising the predominantly South American complex-penis mice and rats was based on Wagner's (1843a) "Sigmodontes," later Latinized to Sigmodontinae (Thomas 1897a), which has priority over Hesperomyinae Murray, 1866 (and such later variants as Hesperomyes Winge, 1887; Hesperomyidae Ameghino, 1889; or Hesperomyini Simpson, 1945). "Sigmodontes," and thus Sigmodontinae, is based on *Sigmodon* Say and Ord 1825, the oldest sigmodontine generic name. Reig (1980) agreed with Hershkovitz that Sigmodontinae was the correct name to use for this clade because it antedated Hesperomyinae, not because, as Hershkovitz (1966b) argued, the latter was formally unavailable. Hershkovitz (1966b) also ranked the sigmodontine group at the level of tribe (Sigmodontini) because he treated the group of mice and rats with two longitudinal rows of molar cusps as a subfamily, Cricetinae, of a broadly conceived family Muridae.

Historically, genera of Sigmodontinae have been united into various groups, beginning with Oldfield Thomas (1906c, 1916a, i, 1917d; see synopsis in Carleton and Musser 1989:53–55), only some of which have been formalized at the rank of tribe in zoological classifications. The number and content of these groups, as well as of their constituent genera, have varied from authority to authority (e.g., Hershkovitz 1966a), mostly depending on the relative weight given to different trenchant characters. The early use of genital morphology (Hooper and Musser 1964) and karyotypic data (A. L. Gardner and Patton 1976), both to advance hypotheses on the contents of sigmodontine suprageneric groups and delimit relationships, deserves special mention. By the early 1990s, the widespread adoption of phylogenetic methodology applied to morphological character sets (Braun 1993; Steppan 1995; Pacheco 2003), DNA sequence data (M. F. Smith and Patton 1991, 1993, 1999; D'Elía 2003a; Weksler 2003, 2006; D'Elía, Luna et al. 2006; J. J. Martínez et al. 2012; Parada et al. 2013; Salazar-Bravo et al. 2013; Schenk et al. 2013), or a combination of both (e.g., Weksler 2006) has prompted a reconsideration of the distinctiveness of suprageneric groups. The major changes promoted by these studies include the recognition of a previously unnoted group (Abrotrichini), the coalescence of some groups previously recognized (i.e., scapteromyines and oxymycterines into Akodontini), and corroboration of the separation of others (e.g., Reithrodontini, Wiedomyini), as well as changes in the generic composition of most groups. Detailed historical accounts of these changes are given in the tribal accounts that follow herein.

In the most recent classification of sigmodontine rodents, D'Elía et al. (2007) recognized nine tribes plus 11 extant genera (i.e., *Abrawayaomys*, *Andinomys*, *Chinchillula*, *Delomys*, *Euneomys*, *Irenomys*, *Juliomys*, *Neotomys*, *Phaenomys*, *Punomys*, and *Wilfredomys*) they considered individually unique lineages and provisionally treated as Sigmodontinae *incertae sedis*. Since early 2013, the main elements of this classificatory scheme have not changed, and it is the one we follow here. With additional data, however, the most significant departure from this classification may be the establishment of a tribe for a strongly supported clade of mostly Andean genera that would include *Andinomys* and *Punomys* as well as *Euneomys*, *Neotomys*, and *Irenomys* (J. J. Martínez et al. 2012; Salazar-Bravo et al. 2013). Furthermore, the phyletic placement of newly described living genera (e.g., *Neomicroxus* Alvarado-Serrano and D'Elía, 2013) or those already known but yet to be described will need to be ascertained or affirmed.

Perhaps not surprisingly, genus- or species-group taxa are not evenly distributed among tribes. The Akodontini, Oryzomyini, and Phyllotini together account for more than a half of all extant genera and species in the subfamily. The Abrotrichini, Ichthyomyini, Reithrodontini, Sigmodontini, Thomasomyini, and Wiedomyini include one to five extant genera each, and only in a few cases are one or more highly speciose genera included. For example, the Thomasomyini, as currently conceived, encompasses but four living genera, one of which (*Thomasomys*) is the most diverse genus in the subfamily, with 44 species.

As will be clear from the tribal and generic accounts, the alpha taxonomic schemes advanced by Gyldenstolpe (1932), Ellerman (1941), Cabrera (1961), Carleton and Musser (1984), and Musser and Carleton (1993, 2005) have been highly influential and are mandatory sources for those interested in sigmodontine species diversity and generic allocations. Hershkovitz (1962), Reig (1987), and

Voss (1988) each played an equivalent central role for phyllotines, akodontines, and ichthyomyines, respectively. Similarly, the generic nomenclatural historical accounts assembled by Tate (1932a–i) are invaluable sources of information, particularly for tracing the often tortuous use of names and their authors. Finally, the voluminous contributions of Oldfield Thomas in describing new sigmodontine species and genera are unparalleled.

Recent and current sigmodontine taxonomic studies, in general, have been based on multiple sources of evidence. Even though it is not a universal practice (see Tavares et al. 2011), the majority of descriptions of new sigmodontine species now take advantage of the power of the genealogical analysis of DNA sequences to determine species-level lineages. The mtDNA cytochrome-*b* gene has been, by far, the sequence of choice. It is noteworthy, however, that only a handful of studies (e.g., D'Elía and Pardiñas 2004; Jayat, Ortiz et al. 2007; Jayat et al. 2010; D'Elía et al. 2011) have taken full advantage of genealogical analyses of sequence data by providing molecular synapomorphies for species newly defined. Instead, most studies relying on molecular data to delimit species have done so using the principle of monophyly and have ignored molecular characters themselves in the diagnosis of newly circumscribed species.

Although the generic composition of sigmodontine tribes may be relatively stable, phylogenetic relationships among the tribes remain uncertain. At present (mid-2013), only the coupling of the Sigmodontini and Ichthyomyini appears both well corroborated and as a sister pair falling outside of a well-supported clade that includes all other sigmodontines (Weksler 2003; Jansa and Weksler 2004; D'Elía, Luna et al. 2006; Parada et al. 2013; Salazar-Bravo et al. 2013; Schenk et al. 2013). Steppan, Adkins, and Anderson (2004) named the clade comprising all sigmodontines except Sigmodontini and Ichthyomyini the Oryzomyalia. All Sigmodontinae *incertae sedis* genera that have been included in any phylogenetic analysis fall in this oryzomyalid clade. However, within the Oryzomyalia only a few tribal relationships are relatively well corroborated. For example, the Abrotrichini and Wiedomyini are sister to each other, as are Phyllotini and the *incertae sedis* genus *Delomys* (D'Elía, Luna et al. 2006; Salazar-Bravo et al. 2013), although the latter case is based solely on the nucDNA interphotoreceptor retinoid-binding protein (IRBP) locus. As always, data from multiple other mitochondrial but especially nuclear loci will be needed to clarify clade relationships as well as to affirm the membership in each.

Since the 1970s, cytogenetic studies of Sigmodontinae have become widespread, with studies reporting new cytotypes routinely published (e.g., A. L. G. Souza, Corrêa et al. 2011). Hence, chromosomal data are available for a large number of species in the subfamily. Available data on chromosome and autosomal arm numbers ($2n$ and FN, respectively) cover the entire or large portions of the radiation represented by living forms in the tribes Abrotrichini, Akodontini, Oryzomyini, Phyllotini, Reithrodontini, Sigmodontini, and Wiedomyini (see, for example, N. O. Bianchi et al. 1971; A. L. Gardner and Patton 1976; Pearson and Patton 1976; Kasahara and Yonenaga-Yassuda 1984). Data for species of the Thomasomyini are fewer, and those for ichthyomyine species are limited or nonexistent. Banding patterns are known for a much smaller number of sigmodontine species than those for which we have only non-differentially stained karyotypes. Cytogenetic data have been gathered mostly in the course of alpha taxonomic studies and have been used primarily to support recognition of new species (e.g., *Loxodontomys pikumche* Spotorno, Cofre et al. 1998; see comments in the account of *Loxodontomys*). Most taxonomic studies in which cytogenetic information has been used, however, have integrated those data together with morphological and/or molecular character sets (e.g., Patton et al. 2000).

Parallel to taxonomically oriented studies, several reports have focused on sigmodontine chromosomal evolution (e.g., R. J. Baker et al. 1983). In a still influential study, A. L. Gardner and Patton (1976; see also Pearson and Patton, 1976) suggested a general trend of sigmodontine chromosomal evolution toward a decrease in both chromosome and fundamental numbers from an initial complement of 70 to 80 chromosomes. Although no phylogenetically based study has been conducted to test the general applicability of this pioneering scheme of chromosomal evolution, Salazar-Bravo et al. (2001) found support for some parts in the genus *Calomys*. Nevertheless, we suggest that the pattern of sigmodontine chromosomal evolution is much more complex than that suggested by A. L. Gardner and Patton (1976). For example, the oryzomyines are not the stem stock from which all other sigmodontine lineages arose as these authors, and others at that time, envisioned. Furthermore, in a few cases (e.g., *Sigmodon*, where $2n$ ranges from 22 to 82; see Swier et al. 2009), examination of karyotypic diversity in an explicit phylogenetic context supports both an increase and a decrease in chromosome and arm number (see also R. J. Baker et al. 1983). The use of novel molecular cytogenetic techniques (e.g., reciprocal chromosome painting; K. Ventura et al. 2009) is a promising approach to determine putative chromosome homologies as well as the type of chromosome rearrangements involved in chromosome evolution for any given lineage. Such studies, however, require use of an explicit phylogenetic framework to separate homoplasy from homologies due to genealogical descent.

Although extensive, sigmodontine diversity in the fossil record of South America is not as great as at present (for a review, see Pardiñas et al. 2002). The oldest recorded South American sigmodontines come from presumed latest Miocene sediments in La Pampa (Verzi and Montalvo 2008; but see Prevosti and Pardiñas 2009) and Catamarca (Nasif et al. 2010) provinces, Argentina. By the beginning of the Pliocene, about 5 mya, genera representing at least four tribes as currently defined were present in central Buenos Aires province, Argentina. These include the Reithrodontini, Phyllotini, Akodontini, and Abrotrichini. Sigmodontines are rare in Early Pliocene beds (Montehermosan to Sanandresian ages) but become more abundant in Late Pleistocene sediments. Although more than half of the extant South American genera are represented in the fossil record (many from Holocene deposits), seven genera are known only from the fossil record, and an additional four genera have gone extinct in historic times. The assignment of some North American fossils, such as *Abelmoschomys*, *Prosigmodon*, *Bensonomys*, and *Symmetrodontomys* (see Lindsay 2008), is questionable and deserves further scrutiny (Prevosti and Pardiñas, in press). The oldest unequivocal fossil sigmodontines in the North American fossil record are representatives of *Sigmodon* from Early Blancan deposits (Peláez-Campomanes and Martin 2005).

An understanding of sigmodontine evolutionary history has proved to be a complex task and the center of much debate. D'Elía (2000) and Pardiñas et al. (2002) provided the most comprehensive and recent reviews, and we refer the reader to these sources for details and especially for historical aspects of the debate. There is general agreement that even though most sigmodontines are endemic to South America, the immediate ancestor of the group did not originate on the continent. This consensus derives from the lack of a potential ancestor having been discovered in South America, and this implies that, at some point in their history, sigmodontines, or their immediate ancestor, invaded South America, with Central America being the most likely source area. As mentioned previously, the fossil record indicates that this invasion was as late as the latest Miocene (or Early Pliocene).

As of early 2014, two main aspects of sigmodontine historical biogeography remain unresolved: the geographic placement of the basal sigmodontine radiation, including how many sigmodontine lineages invaded South America (see, for example, B. Patterson and Pascual 1968b, 1972; Baskin 1978, 1986; Reig 1984, 1986), and the timing of the sigmodontine invasion of South America (see, for example, Baskin 1978, 1986; Marshall 1979; Simpson 1950; Reig 1984, 1986). Although these two issues have thus far eluded resolution, advances prompted by available phylogenetic analyses offer two hypotheses. These analyses tend to place the Sigmodontinae as sister to the Tylomyinae (Parada et al. 2013; Schenk et al. 2013; J. F. Vilela et al. 2013). Because both groups occur only in the Americas, it is most parsimonious to hypothesize that their common ancestor resided in the New World, thus resolving one of the issues of the sigmodontine historical biogeography debate, namely, a New versus Old World home for the immediate ancestor of the Sigmodontinae (see the different positions taken by B. Patterson and Pascual 1968b, 1972; Hershkovitz 1972; Slaughter and Ubelaker 1984; and L. L. Jacobs and Lindsay 1984).

As to the second issue, the most comprehensive molecular clock-based estimates of the cladogenic events underlying the sigmodontine tribes (Parada et al. 2013; Schenk et al. 2013; J. F. Vilela et al. 2013) have indicated that the split between lineages leading to the Sigmodontini + Ichthyomyini and Oryzomyalia was between 14.70 and 9.28 mya. Furthermore, the base of the Oryzomyalia, and thus the subsequent radiation of the tribal lineages in this larger clade, was between 9.64 to 6.01 mya. Because the Panamanian land bridge was completed about 3 mya (Bartoli et al. 2005), these estimates falsify those historical biogeographic scenarios that posited a sigmodontine radiation in South American only after closure of the Panamanian land bridge (see Simpson 1950).

More importantly, the presence of sigmodontines in southern South America before the origin of the Panamanian Isthmus is now firmly established (e.g., Pardiñas and Tonni 1998), despite the reinsertion of doubts in recent reviews (e.g., S. D. Webb 2006). The number of sigmodontine lines that invaded South America remains to be ascertained (but see the discussion in Steppan, Adkins, and Anderson 2004:548–549; and especially that of Parada et al. 2013), and it also remains unclear whether some North American sigmodontines (i.e., some species of *Sigmodon* and some ichthyomyines) are in situ direct descendants of the first sigmodontines or whether they represent secondary invasions of North America from lineages that initially differentiated in South America. This issue will remain unresolved until a deeper assessment and understanding of presumptive sigmodontine North American fossils are acquired through comparison with extinct and extant South American forms (Prevosti and Pardiñas, in press). Similarly, a comprehensive phylogenetic analysis that incudes all living forms of *Sigmodon* and of the Ichthyomyini is required to infer the geographic distribution of the ancestor of the sigmodontine crown group (D'Elía 2003a). This latter goal seems more feasible, and perhaps also more promising, than a comprehensive review of putative sigmodontine North

American forms, a chaotic and still expanding narrative that includes several unsupported ancestor-descendent hypotheses linking enigmatic fossils to living entities (Lindsay and Czaplewski 2011).

Reig's (1984, 1986) hypothesis regarding sigmodontine historical biogeography deserves special mention. He identified different geographic areas, which he termed "Areas of Original Differentiation, or AOD," and in which sigmodontine tribes would have radiated. He defined "an AOD [as] the geographic space within which a given taxon experienced the main differentiation of its component taxa of subordinate rank" (Reig 1984:411). Thus, for a given tribe, this area may or may not be the same as that where the ancestor of the tribe would have originated. Reig (1984:344–345) identified six tribal AODs, each defined a priori on the basis of highest occurrence of species, genera, and endemic genera. All tribal AODs fell along the Andes. As a corollary, most genus-group lineages originated in the Andes and later colonized the extensive South American lowlands. As will be developed further in the Akodontini account, testing Reig's hypothesis is not straightforward, and no study to date has provided a rigorous test. However, available evidence (i.e., known distributions of extinct and living forms, and current understanding of tribal limits) suggests that the historical biogeography of Sigmodontinae in South America is much more complex than Reig envisioned. Clearly, a significant number of generic lineages originated outside the Andes, with multiple colonizations of both the Andes and the lowlands (see M. F. Smith and Patton 1999; D'Elía, Pardiñas, Jayat, and Salazar-Bravo 2008; Salazar-Bravo et al. 2013).

Despite great progress made to date, there remain important unresolved problems in sigmodontine systematics, issues involving both low and high taxonomic levels. For example, how many sigmodontine species are there? Is the polytomy found at the base of Oryzomyalia hard or soft? Furthermore, these questions relate to both pattern and process: Did the tribal crown groups diversify at about the same time, or did the more diverse ones radiate earlier? Have sigmodontine species and genera accumulated at a constant rate, or have there been diversification pulses? Have chromosomal changes triggered speciation events? Why are akodontines less diverse in Amazonia than in temperate open areas? Some of these questions may be clarified through phylogenetic analyses with dense taxonomic coverage, preferably exhaustive at the species level and based on multiple loci within a molecular clock-based temporal inference (such as Parada et al. 2013; Schenk et al. 2013). The answers to other questions depend on the continuation of collection-based research, with specimens already at hand and/or the capture of additional specimens in nature, including excavation of new and better fossils. Similarly, new areas of study such as breeding experiments coupled with developmental genetics hold great promise for new insights. We are confident that the next few years will bring new understanding of the sigmodontine radiation, one that will substantively increase our comprehension of the biological processes that generated and sustain it.

KEY TO THE TRIBES OF SIGMODONTINAE (NOTE: 12 GENERA DO NOT CLEARLY FIT WITH ANY DEFINED TRIBE AND ARE TREATED SEPARATELY UNDER THE HEADING SIGMODONTINAE *INCERTAE SEDIS*):

1. Pentalophodont molars with persistent mesoloph. . . . 2
1′. Tetralophodont or even simplified molars 4
2. Absence of the posterior suspensory process of squamosal connected to the tegmen tympani . . . . . Oryzomyini
2′. Presence of the posterior suspensory process of squamosal connected to the tegmen tympani 3
3. Bony palate, short and wide . . . . Thomasomyini
3′. Bony palate, long and wide . . . . Wiedomyini
4. Absence of pectoral mammae. 5
4′. Presence of pectoral mammae. 6
5. Stiff mystacial vibrissae ventrally recurved, absence of zygomatic notches, presence of large masseteric tubercles, absence of a pars flaccida, and nuchal ligament attached to the third thoracic vertebra. . . . . Ichthyomyini
5′. Skull with long rostrum and rounded braincase, without frontal or lambdoid crests; interorbital region amphora-shaped; frontal sinuses moderately developed; palate long; anterior border of mesopterygoid fossa square . . . . Abrotrichini
6. m3 strongly sigmoid. 7
6′. m3 not strongly sigmoid. 8
7. Large ears and eyes; double grooved upper incisors; cuniculoid aspect; short tail; posterior suspensory process of squamosal not connected to the tegmen tympani . . . . Reithrodontini
7′. Moderate ears and eyes; upper incisors usually smooth; posterior suspensory process of squamosal connected to tegmen tympani . . . . .Sigmodontini
8. Moderate or large ears; parapterygoid fossa relatively broader than mesopterygoid fossa; very open sphenopalatine vacuities; complete loss of mesoloph; posterior extensions of premaxillaries and nasals subequal; gall bladder present. . . . . Phyllotini
8′. Moderate or small ears; skull with long rostrum and rounded braincase, with frontal sinuses well developed; interorbital region amphora-shaped; persistent mesoloph usually fused to paracone; gall bladder present or absent . . . . Akodontini

Sigmodontinae incertae sedis
Guillermo D'Elía

Attempts to place sigmodontine genera in suprageneric groups have revealed some genera with unclear affinities. In precladistic studies, these genera were mostly those showing otherwise trenchant characters of more than one group. One early example of this conundrum was Thomas's (1906c, 1917d) observation that pentalophodont sigmodontines were not cleanly divided into taxa with short, unpitted palates and with six mammae (e.g., *Thomasomys* and *Rhipidomys*, the basis for the tribe Thomasomyini) versus those with long, pitted palates and eight mammae (e.g., *Oryzomys* and *Oecomys*, tribe Oryzomyini). For example, the southeastern Brazilian *Delomys* and *Phaenomys* have short and unpitted palates but eight mammae, whereas *Rhagomys rufescens* has six mammae but a posterior palate similar to that of *Oryzomys* and *Oecomys*. Musser and Carleton (2005) illustrated this problem with respect to *Abrawayaomys*, another example a taxon displaying key features seen in more than one tribe: "diagnostic traits seem to combine aspects of *Neacomys*, *Oryzomys*, and *Akodon* . . . Certain cranial features of *Abrawayaomys* suggest an archaic 'thomasomyine,' perhaps distantly related to the other endemic genera of SE Brazil."

Analyzing the distribution of character states displayed by pentalophodont sigmodontines, Voss (1993) argued that thomasomyine genera (*sensu* Hershkovitz 1962, 1966a; see also Cerqueira 1982 and Eisenberg and Redford 1999) were united by sharing putative primitive character states and that, therefore, the group was not natural (see the account of the tribe Thomasomyini). Voss (p. 25) added that these genera "could simply be called 'plesiomorphic Neotropical muroids.' " Preferring not to place genera such as *Abrawayaomys*, *Aepeomys*, *Chilomys*, *Delomys*, *Juliomys* (in 1993 *pictipes* was still placed in *Wilfredomys*), *Phaenomys*, *Rhagomys*, *Rhipidomys*, *Thomasomys*, and *Wilfredomys* into monogeneric tribes, Voss (1993) implicitly regarded them as Sigmodontinae *incertae sedis*.

Subsequently, with the adoption of an explicit phylogenetic approach given by the formal construction of phylogenetic trees, several sigmodontines could not be placed into any robust monophyletic group of tribal level. A number of these genera included, as expected, some of those traditionally considered to be of uncertain phylogenetic position (e.g., *Delomys*), whereas others were genera that for most of sigmodontine taxonomic history had been placed in a tribe. This group is well illustrated by genera that, at one time or another, have been placed in the Phyllotini (e.g., *Andinomys*, *Chinchillula*, *Euneomys*, *Irenomys*, *Neotomys*, *Punomys*;

Steppan 1995; M. F. Smith and Patton 1999; D'Elía 2003a; Salazar-Bravo et al. 2013).

Realization that several sigmodontine genera do not belong in any of the groups ranked as tribes invites two classificatory alternatives: either to erect a monogeneric tribe for each, or to treat them as Sigmodontinae *incertae sedis*. In agreement with Voss (1993) and M. F. Smith and Patton (1999), D'Elía (2003a) chose the option to treat these genera as Sigmodontinae *incertae sedis*, because, in this case, the proliferation of taxa (e.g., monogeneric tribes) would not provide any additional information to systematists. Thus, those genera without clear phylogenetic relationships (e.g., *Chinchillula*) or that are sister to well-defined tribes (e.g., *Delomys*, which is sister to Phyllotini) do not have to be forced into tribes. This policy does not exclude the continuing use of existing tribes (i.e., Sigmodontini, Reithrodontini, and Wiedomyini) composed of a single extant genus. Rather, the acceptance of both *incertae sedis* genera and current monogeneric tribes is justified by the availability of existing names. Moreover, these currently recognized tribes are only monogeneric with respect to Recent genera. The Wiedomyini, for example, includes the extinct *Cholomys* in addition to the extant *Wiedomys*; the Sigmodontini also includes the extinct *Prosigmodon* (but see Peláez-Camomanes and Martin 2005); and several extinct genera (i.e., *Ichthyurodon*, *Panchomys*, and *Tafimys*) may be part of the Reithrodontini, which is otherwise composed only of the living *Reithrodon* (but not *Euneomys* or *Neotomys*; D'Elía 2003a; J. J. Martínez et al. 2012).

The list of genera treated as Sigmodontinae *incertae sedis* has varied from author to author. For example, Reig (1986) assigned this status to *Abrawayaomys*, *Punomys*, *Rhagomys*, and *Zygodontomys*, while Musser and Carleton (1993) added a fifth genus, *Pseudoryzomys*, to Reig's list. M. F. Smith and Patton (1999), in their influential molecular phylogenetic study, found that *Delomys*, *Irenomys*, *Juliomys* (as represented by *pictipes*, then placed in *Wilfredomys*), *Reithrodon*, and *Scolomys* lacked clear phylogenetic relationships and referred to them as "additional unique lines" as a way to emphasize their uncertain placement. These authors also listed *Abrawayaomys*, *Microakodontomys*, *Phaenomys*, *Punomys*, and *Rhagomys* as *incertae sedis*, following previous morphologically based assessments, as their molecular analyses did not include them. Those taxa Smith and Patton identified as "additional unique lines" are also best viewed as *incertae sedis* (except *Reithrodon*, for which the tribal name Reithrodontini was already available, and *Scolomys*, for which subsequent broader-based molecular studies have shown a clear linkage in the Oryzomyini; Weksler 2006; D'Elía et al. 2007; Parada et al. 2013). Musser and Carleton (2005) listed nine genera as Sigmodontinae *incertae sedis*: *Abrawayaomys*, *Delomys*,

Irenomys, *Juliomys*, *Phaenomys*, *Punomys*, *Rhagomys*, *Scolomys*, and *Wilfredomys*.

The genera of Sigmodontinae *incertae sedis* listed here comprise 12 Recent taxa following the classification given by D'Elía et al. (2007). Molecular-based phylogenetic analyses (e.g., D'Elía 2003a; J. J. Martínez et al. 2012; Alvarado-Serrano and D'Elía 2013; Parada et al. 2013; Salazar-Bravo et al. 2013; K. Ventura et al. 2013) indicate that *Abrawayaomys*, *Andinomys*, *Delomys*, *Chinchillula*, *Euneomys*, *Irenomys*, *Juliomys*, *Neomicroxus*, *Neotomys*, and *Punomys* fall in the large clade referred to as Oryzomyalia, which groups all Sigmodontinae except for Sigmodontini and Ichthyomyini (Steppan, Adkins, and Anderson 2004). *Abrawayaomys*, in particular, is notable for constituting "the single representative so far known of one of the main lineages of the radiation of Sigmodontinae" (K. Ventura et al. 2013:8). Pacheco (2003), in a morphologically based phylogenetic analysis, recovered *Phaenomys* and *Wilfredomys* as members of a clade otherwise composed of thomasomyine genera (see the account of Thomasomyini); neither of these two genera has been included as yet in a molecular-based phylogenetic analysis.

Andinomys and *Punomys* appear as sisters in analyses encompassing both mtDNA and nucDNA sequences (Parada et al 2013; Salazar-Bravo et al. 2013; Schenk et al. 2013) in a larger clade that somewhat more weakly also includes *Chinchillula* and *Juliomys* with *Irenomys*, *Euneomys*, and *Neotomys*. J. J. Martínez et al. (2012) also recovered *Euneomys*, *Irenomys*, and *Neotomys* as a well-supported clade, which they argued deserved tribal rank. Further studies thus may prompt the proposition of one or more new tribes to encompass some or all of these genera. Alternatively, as emphasized by D'Elía (2000), it is possible that *incertae sedis* genera truly constitute unique members of deep sigmodontine lineages and, as such, that their uncertain position is not due to the amount or quality of the data analyzed to date. If so, sigmodontine phylogenetic diversity is even greater than has been acknowledged previously.

KEY TO THE GENERA OF SIGMODONTINAE INCERTAE SEDIS:

1. Fur soft, without spines; upper incisors opisthodont; cheek teeth pentalophodont or tetralophodont; occlusal surfaces varied 2

1'. Fur with strongly developed, flat spines; upper incisors opisthodont to proodont; cheek teeth pentalophodont; occlusal surfaces crested . . . . *Abrawayaomys*

2. Pelage short, either coarse or soft; interorbital region broad, diverging posteriorly; incisive foramina short, typically not reaching anterior edges of M1s; anteromedian

flexus of M1 prominent; palate wide; sphenopalatine vacuities absent or small, roof of mesopterygoid fossa partially or completely ossified 3

2'. Pelage long, soft; interorbital region narrow, laterally compressed; incisive foramina medium to long, penetrating to or beyond anterior edges of M1s; anteromedian flexus of M1 present or not; palate narrow; sphenopalatine vacuities present, either moderate or widely open, resulting in nearly completely unossified roof of mesopterygoid fossa; cheek teeth pentalophodont or tetralophodont; molar occlusal surface crested, terraced, or planed 6

3. Size small (mass ≤30 g); pelage soft, zygomatic notches shallow; palate short to moderate, wide; rostral tube small or indistinct 4

3'. Size medium (mass ≥45 g); zygomatic notches medium to deep; palate short, wide, or narrow; rostral tube distinct 5

4. Pelage short and dense; tail short, approximately 50% of head and body length; zygomatic plates narrow, sloping backward from base; palate wide, moderate in length, reaching end of molar toothrow; auditory bullae inflated; carotid circulation pattern 1 (*sensu* Voss 1988); capsular projection of mandible inconspicuous. . . . . *Neomicroxus*

4'. Pelage long; tail longer than head and body; zygomatic plates broad; palate wide, short, not reaching end of molar toothrow; auditory bullae small; carotid circulation pattern 2 or 3 (*sensu* Voss 1988); capsular projection of mandible conspicuous. . . . . *Juliomys*

5. Pelage short, coarse; tail length subequal with head and body length, scales conspicuous; nose same color as head and back; zygomatic notches moderately excavated; palate wide, short; rostral tube present; sphenopalatine vacuities absent, roof wholly ossified; auditory bullae small, uninflated; primary cusps alternating. . . . . *Delomys*

5'. Pelage short, soft; tail long, >120% of head and body length, and hairy, but scales visible; nose conspicuously reddish; palate narrow, short; rostral tube absent; sphenopalatine vacuities moderate, elongated; auditory bullae large, inflated; primary cusps opposite . . . . *Wilfredomys*

6. Upper incisors smooth, without groove; carotid circulation pattern 1 (*sensu* Voss 1988); cheek teeth tetralophodont or pentalophodont 7

6'. Upper incisors with longitudinal groove; carotid circulation pattern 1 or 2 (*sensu* Voss 1988); cheek teeth tetralophodont 10

7. Body not uniformly colored, darker on midback, with conspicuous black and white thigh stripes and strongly contrasting whitish venter; body large, very stout, mass >120 g; pelage long, very thick and soft; tail short (<70% head and body length), hairy, with scales not visible; skull with long and narrow palate; sphenopalatine vacuities large, mesopterygoid roof unossified; cheek teeth tetralophodont with terraced molars; anteromedian flexus of M1 absent or very shallow . . . . *Chinchillula*

7'. Body color uniform, without black and white thigh stripes or contrasting ventral fur; body either stout or elongated, medium to large, mass <100 g; pelage long, not thick but soft; tail medium to short, haired, scales either visible or not; palate narrow but either long or short; sphenopalatine vacuities either large or medium, mesopterygoid roof either unossified or with elongated openings. 8

8. Body color brilliant reddish; tail long, >120% head and body length; rostrum broad, tapering anteriorly; zygomatic notch shallow; cheek teeth pentalophodont, crested . . . . *Phaenomys*

8'. Body color yellowish or grayish brown; tail short, <90% head and body length; rostrum broad, parallel sided; zygomatic notch deep, well developed; cheek teeth pentalophodont and crested or tetralophodont and planar . 9

9. Body elongate; dorsum streaked with dark brown; tail about 80% of head and body length, moderately haired, scales visible; zygomatic spine strongly developed; palate short, narrow; cheek teeth tetralophodont; molar cusps alternate; anteromedian flexus of M1 absent or very shallow . . . . *Andinomys*

9'. Body stout; dorsum without dark brown streaking; tail very short, <50% head and body length, well haired, scales not visible; zygomatic spine absent; palate long, narrow; cheek teeth pentalophodont, crested; molar cusps opposite; anteromedian flexus of M1 prominent . . . . *Punomys*

10. Tail long, >120% of head and body length, well haired but scales visible, unicolored; rostrum narrow, tapering anteriorly; zygomatic notch shallow; interorbital region somewhat laterally compressed but divergent posteriorly; upper incisors hyperopisthodont . . . . . . *Irenomys*

10'. Tail short, <70% head and body length, well haired and scales not visible, weakly to strongly bicolored; rostrum broad, parallel sided; interorbital region laterally compressed with raised edges; upper incisors weakly opisthodont 11

11. Pelage long, lax, and very thick; dorsal color gray to dark-gray brown; nose same color as head and body; tail distinctly bicolored; incisive foramina medium, reaching anterior portion of M1; compressed interorbital region very long; palate long, narrow; carotid circulation pattern 1 (*sensu* Voss 1988); occlusal surface of cheek teeth terraced; primary cusps opposite . . . . *Euneomys*

11'. Pelage long, lax, but not as thick; dorsal color dark reddish brown; nose distinctly orange; tail indistinctly bicolored; incisive foramina short, not reaching M1s; compressed interorbital region short; palate short, narrow; carotid circulation pattern 2 (*sensu* Voss 1988); occlusal surface of cheek teeth planar; primary cusps alternate . . . . *Neotomys*

Genus *Abrawayaomys* F. Cunha and Cruz, 1979

Ulyses F. J. Pardiñas, Pablo Teta, and Guillermo D'Elía

Abrawayaomys is a sylvan genus that uniquely combines a spiny pelage with an unusual set of craniodental attributes in the sigmodontine radiation. The genus was regarded as monotypic (e.g., Musser and Carleton 2005) until the recent description of a second species (Pardiñas, Teta, and D'Elía 2009). This genus is one of four sigmodontine taxa with spiny pelage, the others being *Neacomys*, *Scolomys*, and *Rhagomys longilingua*. It is known from the Brazilian coastal forests, from Bahia to Rio de Janeiro and from the Paraná-Paraíba interior forest in northeastern Argentina. No fossils of *Abrawayaomys* are known (G. A. B. Fonseca and Kierulff 1989; Massoia 1993).

Species of this genus are medium-sized rodents (total length 200–290 mm, condylobasal length of the skull 27–29 mm) with a robust head, rounded small ears, and a tail varying from slightly shorter to longer than combined head and body length, with or without a terminal tuft of brown to white hairs. The body pelage is short and close with a moderately spiny texture. The overall appearance is strongly agouti and hispid, with weak distinction between dorsal and ventral colorations. Dorsal and ventral hairs of approximately the same length are either spiny or long and thin. The spiny hairs are flat, have a dorsal longitudinal groove, are rigid, and are broadest at their midpoint, especially on the dorsum. Spines are black-tipped and most abundant on the dorsum and rump, shorter and darker on the head, and slightly shorter, without dark tips, fewer in number, and thinner in structure on the under parts. Both mental and inguinal areas are free of spines. Mystacial and supraorbital vibrissae are short. Rigid tail hairs are arranged in sets composed of five hairs per scale. The pes is long, but the claws are short and covered with dense ungual tufts; digit V is very short; and the plantar surface has six pads, with the hypothenar overlapping the thenar. A mammary count has not been previously reported in the literature, but a specimen from São Paulo state, Brazil, has three pairs, one each in postaxial, abdominal, and inguinal positions.

The skull of *Abrawayaomys* is robust and compact, notable for its extremely short and broad rostrum. The rostrum is broader than the narrowest part of the interorbital region, and has conspicuous nasolacrimal capsules. The

premaxillaries end behind the anterior plane of the incisors. The nasals, with the anterior half considerably expanded, do not extend beyond the anterior ends of the premaxillaries. The interorbital region is smooth, hourglass in shape, weakly beaded, and has its narrowest point positioned anteriad. The coronal suture is U-shaped, and the interparietal is transversally compressed and rhomboid in outline. Anterior borders of the zygomatic plates are straight and lack zygomatic spines; the zygomatic notches are inconspicuous. An alisphenoid strut is usually present. The tegmen tympani contacts the squamosal. The hamular process of the squamosal is long, obliquely oriented, and terminates in a spatula shape; the subsquamosal fenestra and the postglenoid foramen are subequal in size. The carotid circulation is pattern 1 (*sensu* Voss 1988), with a squamosal-alisphenoid groove and sphenofrontal foramen present.

The incisive foramina are short, not reaching the anterior face of M1s. The palate is long, extending slightly beyond the posterior margin of M3s. Posterolateral palatal pits are very small and positioned at the level of the anterior border of the mesopterygoid fossa; the anterior margin of the mesopterygoid fossa is rounded, with its roof typically lacking sphenopalatine vacuities. The middle lacerate foramen is nearly closed. Tympanic bullae are inflated; bony tubes are reduced. Maxillary toothrows are parallel. The mandible of *Abrawayaomys* is short and robust, with an enlarged retromolar fossa; the condyloid, coronoid, and angular processes are subequal in size. The angular notch is shallow, and the capsular projection is very conspicuous. Internally, a well-developed and curved duct for the mandibular nerve is visible through the translucent bone and leads to a large mandibular foramen.

The upper incisors are narrow but noticeably deep, and they vary from slightly opisthodont to proodont. The anterior enamel is orange; the dentine fissure straight and large (*sensu* Steppan 1995:23). The lower incisors are noticeably deep but transversally thin, faced with paler enamel than the uppers. The molars are small and brachydont, slightly crested, and with an "intermediate" main cusp arrangement (*sensu* Steppan 1995:24). M1 is pentalophodont with an anteromedian flexus varyingly visible and a reduced mesoloph fused with the paracone; the procingulum of M1 is anteroposteriorly narrow and labially displaced; enterostyles are present on M1–2. M3 is notably reduced, with an occlusal morphology suggesting persistence of only the proto- and paracone areas, because the posterior portion of this tooth is greatly narrowed. The number of molar roots is M1/m1 = 3/2, M2/m2 = 3/2, and M3/m3 = 1–2/2.

Axial skeletal counts of one Argentinean specimen (*A. chebezi*) totaled 12 ribs and 15 thoracolumbar, three sacral, and 36 caudal vertebrae (Pardiñas, Teta, and D'Elía 2009).

The glans penis is short, stout, and entirely covered with spines; lateral bacular mounds are not visible in external view; and the medial bacular mound is slightly larger than the lateral ones. The proximal baculum is short, stout, and triangular; a laterally flared base has a distinctly deep median notch. The cartilaginous distal baculum is small and tridigitate (Pardiñas, Teta, and D'Elía 2009). The stomach is hemiglandular-unilocular (Finotti et al. 2003).

SYNONYM:

Abrawayaomys F. Cunha and Cruz, 1979:2; type species *Abrawayaomys ruschii* F. Cunha and Cruz, 1979, by original designation.

REMARKS: Two species are currently recognized, one from Brazil (*A. ruschii*) and the other from Argentina (*A. chebezi*), but the few available specimens hamper any assessment of intra- or interpopulation variation (Pardiñas, Teta, and D'Elía 2009). Specimens from Rio de Janeiro may well represent *A. chebezi*. As both samples from Argentina and Rio de Janeiro depart less from the holotype of *A. ruschii* (from Espírito Santo state) than do those from Minas Gerais, the latter may represent a third, as yet undescribed species. To further complicate the existing taxonomy, five skins collected in São Paulo state, Brazil, document important morphological differences between young and adults. For example, young individuals have a proportionally shorter tail (relative to head and body length), a tail tuft is either lacking or very short, the hindfeet are brown above, and there is very limited contrast between the dorsal and ventral colors. Adults, in contrast, have proportionally longer tails that terminate in a long tuft of white hairs (ca. 20 mm), whitish hindfeet, and strikingly demarcated dorsal and ventral colors. These limited comparisons clearly indicate that *Abrawayaomys* is in need of a thorough revision aimed to define both the number of species and their respective geographic distributions.

KEY TO THE SPECIES OF *ABRAWAYAOMYS*:

1. Tail longer than head and body, with a dark brown tuft; interorbital region hourglass in shape; upper toothrow length ≤4 mm; anteromedian flexus on M1 inconspicuous or absent . . . .*Abrawayaomys chebezi*
1'. Tail shorter than head and body and without tuft, or longer and terminating in white tuft; interorbital region with anteriorly convergent sides; upper toothrow length ≥4 mm; anteromedian flexus on M1 conspicuous. . . . .
. . . . *Abrawayaomys ruschii*

Abrawayaomys chebezi Pardiñas, Teta, and D'Elía, 2009
Chebez's Spiny Mouse

SYNONYMS:

Thomasomys (?) *pictipes*: Massoia, 1988c:6; part; not *Thomasomys pictipes* Osgood, 1933.

Abrawayaomys ruschii: Massoia, Chebez, and Heinonen Fortabat, 1991a:39; part; not *Abrawayaomys ruschii* F. Cunha and Cruz, 1979.

Abrawayaomys chebezi Pardiñas, Teta, and D'Elía, 2009; type locality "Argentina: Province of Misiones, Department of Iguazú, conjunction Arroyo Mbocaí and route 12 [25.680115°S, 54.508060°N]."

DESCRIPTION: Smaller than *A. ruschii*, especially in molar measurements (upper toothrow length 3.30–3.96 mm), and characterized by tail longer than head and body, terminating in dark tuft; basally gray ventral hairs with cream to yellowish tips; anterior third of nasals not conspicuously expanded laterally; nasofrontal suture V-shaped, extending posteriorly past level of lacrimals; interorbital region hourglass in shape; alisphenoid strut present; inconspicuous or absent anteromedian flexus on M1, and proodont upper incisors.

DISTRIBUTION: *Abrawayaomys chebezi* is known from the western, forested portions of Misiones province in Argentina (Massoia 1993; Pardiñas, Teta, and D'Elía 2009) that drain towards the Río Paraná.

SELECTED LOCALITIES (Map 27): ARGENTINA: Misiones, Iguazú, junction Arroyo Mbocaí and Rte 12 (type locality of *Abrawayaomys chebezi* Pardiñas, Teta, and D'Elía), Eldorado, Segunda Iglesia Cuadrangular, Barrio Parque Km 11 (Massoia, Chebez, and Heinonen Fortabat 1991a), Montecarlo (Massoia 1993, 1996), Oberá, Escuela Provincial 639 "Rosario Vera Peñaloza," Lote 92, Sección II de Campo Ramón (Massoia 1988c, 1996), Parque Provincial Urugua-í (CNP 3631).

SUBSPECIES: *Abrawayaomys chebezi* is monotypic.

NATURAL HISTORY: Little is known about the natural history of this species. It is certainly preyed upon by the Barn Owl, *Tyto alba*, as three of five records from Misiones are based on remains recovered from the pellets of this owl (Massoia 1993). The holotype (MACN 20253) was captured using a Sherman trap on the southern margin of Arroyo Mbocaí, in secondary forest belonging to the "Rosewood and Assai Palm Forest" unit (*sensu* Martínez-Crovetto 1963), where the predominant tree species are *Matayba elaeagnoides*, *Begonia descoleana*, and *Podostemun comatum*. Some localities are in areas highly disturbed by human activities, where only small patches of the "Laurel and Guatambu Forest" survive, or in ecotonal areas between forest and the open grasslands of the "Campos and Malezales" ecoregion (Pardiñas, Teta, and D'Elía 2009).

C. Galliari collected the second specimen (CNP 3631) in late September 2012 in disturbed secondary forest in Parque Provincial Urugua-í. This animal is an adult male weighing 35 g and had scrotal testes. A large assemblage of ectoparasites collected from this individual included mites, most of the genus *Androlaelaps*, and two fleas, one identi-

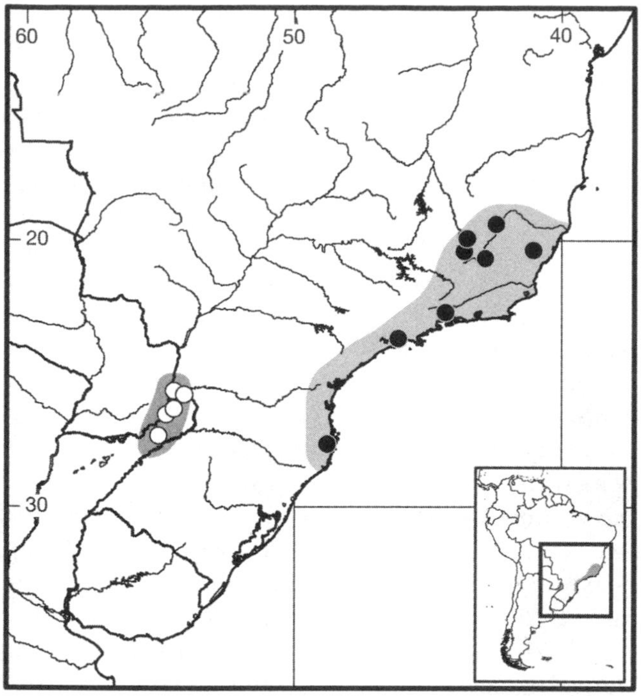

Map 27 Selected localities for *Abrawayaomys chebezi* (○) and *Abrawayaomys ruschii* (●).

fied as *Polygenis* (*Polygenis*) sp. and the other *Adoratopsylla* (*Adoratopsylla*) *antiquorum antiquorum*. The latter may represent an accidental association because fleas of this genus are mainly associated with marsupials (M. Lareschi, pers. comm.).

REMARKS: Massoia (1988c:Figs. 3 and 4) referred to the first specimen (a mandible recovered from owl pellets) of *Abrawayaomys* collected in Argentina as *Thomasomys* (?) *pictipes*. There is an unconfirmed published mention of this species from Puerto Caraguatay, Misiones (Massoia 1993). Specimens from Rio de Janeiro, Brazil, are perhaps referable to this species (Pardiñas, Teta, and D'Elía 2009).

Abrawayaomys ruschii F. Cunha and Cruz, 1979
Ruschi's Spiny Mouse
SYNONYM:

Abrawayaomys ruschii F. Cunha and Cruz, 1979:2; type locality "Forno Grande [= Parque Estadual do Forno Grande (20.5°S, 41.6°W)], município de Castelo, ES [= Espírito Santo], Brasil."

DESCRIPTION: Larger than *A. chebezi*, especially in molar measurements (upper toothrow length of holotype, 5.41 mm); characterized by tail shorter than head and body and without terminal tuft, or longer than head and body with white tuft; yellowish to whitish ventral coloration; rostral third of nasals conspicuously expanded laterally; straight nasofrontal suture at about level of lacrimals; in-

terorbital region with anteriorly convergent sides; alisphenoid strut sometimes present; conspicuous anteromedian flexus on M1; and slightly opisthodont upper incisors.

DISTRIBUTION: *Abrawayaomys ruschii* is known from few localities in the Brazilian states of Espírito Santo, Minas Gerais, Rio de Janeiro, São Paulo, and Santa Catarina.

SELECTED LOCALITIES (Map 27): BRAZIL: Espírito Santo, Castelo, Parque Estadual do Forno Grande (type locality of *Abrawayaomys ruschii* F. Cunhaand Cruz); Minas Gerais, Mata da Prefeitura, 6 km SW of Viçosa (USNM 552416), Ouro Branco (G. Lessa et al. 2009), Parque Estadual do Rio Doce (UFMG 2492), Parque Estadual da Serra do Cipó (Passamani et al. 2011); Rio de Janeiro, Angra dos Reis, Terra Indígena Guarani Sapukai (L. G. Pereiraet al. 2008); São Paulo, Biritiba Mirim (M. J. de J. Silva et al. 2009); Santa Catarina, Caldas da Imperatriz (Cherem et al. 2005).

SUBSPECIES: *Abrawayaomys ruschii* is monotypic.

NATURAL HISTORY: Little is known about the natural history of this species. Observations on diet, derived from a single individual, indicate that *Abrawayaomys ruschii* feeds preferentially on fruits, seeds, and leaves (Finotti et al. 2003). Capture sites of most specimens are from the interior of forests fragments as well as along their margins; most specimens have been caught during the rainy season (G. Lessa et al. 2009). Postcranial anatomy suggests that this species is terrestrial in its mode of locomotion (Coutinho et al. 2013), an observation supported by reported captures in traps set on the ground.

REMARKS: Specimens from Minas Gerais display a long tail with a large white apical tuft, a variably present alisphenoid strut, and an interorbital region with anteriorly convergent borders, character differences from other samples that invite further exploration of their taxonomic status (Pardiñas, Teta, and D'Elía 2009).

The diploid number for specimens from Rio de Janeiro and São Paulo states, Brazil, is 58, with all autosomes predominantly acrocentric; the X is a medium-sized metacentric and the Y a medium-sized submetacentric (L. G. Pereira et al. 2008; M. J. de J. Silva et al. 2009).

Genus *Andinomys* Thomas, 1902
Jorge Salazar-Bravo and Jorge Pablo Jayat

Andinomys occurs between 14°S and 28.5°S, from southern Peru and northernmost Chile through southwestern Bolivia and northwestern Argentina. Individuals have been trapped across an elevational range from 650 to 4,500 m, with most records occurring above 1,500 m. As expected from this broad geographic distribution, the genus occupies multiple habitats, from subtropical mountain forests and semiarid Puna and pre-Puna to high elevation grasslands

(J. P. Jayat, Pacheco, and Ortiz 2009). The following account is abstracted from Osgood (1947), Hershkovitz (1962), Steppan (1995), S. Anderson and Yates (2000), and P. E. Ortiz and Jayat (2007a), which should be consulted for additional details on morphological characters, fossil record, and distribution.

Andinomys is characterized by a large and heavy body (head and body length 134–176 mm for specimens from central Bolivia and northwestern Argentina). The fur is soft and lax; the color to the dorsum and sides is drab, but the under parts are buffy white with hairs basally slate colored. Tail length ranges from about 60% to nearly equal to head and body length; the tail is sharply bicolored, densely furred, but lacks a terminal pencil. Forefeet and hindfeet are well haired on their upper surfaces, covered with white hair above, and with silvery-white ungual tufts that extend well beyond the claws. Digit V of hindfoot (not including claw) reaches at least to the base (often to the middle) of the second phalanx of digit IV; palmar pads large but not fused. The pinnae are comparatively small (24–29 mm long). Eight mammae are present, with two pairs in both pectoral and inguinal positions.

The skull is flattened, with a broad rostrum and nasals that are broader than the interorbital constriction and somewhat expanded at their anterior edges. The premaxillomaxillary suture is oriented dorsoventrally. The supraorbital region is narrow, with squared or slightly raised edges. Infrafrontal fontanelles are often present (S. L. Gardner and Anderson 2001). Zygomatic plates are deep, with their anterior borders slightly concave and with anterodorsal corners projecting as a short spine; zygomatic notches are broad and deeply excavated. The hamular process of the squamosal is long and thin, with a flattened posterior terminus. Incisive foramina are comparatively widely open, with smoothly rounded edges. Posterolateral palatal depressions are pitted and troughlike, continuing as gutters onto the anterolateral portion of the palate. The mesopterygoid fossa is comparatively broad; the parapterygoid fossae are shallow. The carotid circulation is characterized by conspicuous sphenofrontal and stapedial foramina and a squamosal-alisphenoid groove (pattern 1 of Voss 1988). The mandible is robust, with a poorly developed capsular projection, the coronoid process is higher than the articular condyle, and the anterior margin of the masseteric crest extends to the level of the m1 protoconid.

Upper incisors are opisthodont, broad, and asulcate. The upper molar rows diverge posteriorly, and molars above and below are hypsodont, large, prismatic, and flat crowned, with primary cusps having alternating and deeply penetrating flexi and flexids (P. E. Ortiz and Jayat 2007a). A labial root of M1 is present.

Steppan (1995) provided vertebral counts of 13 thoracic, six lumbar, and 28–30 caudal (modal number of 30). The stomach is atypical for most sigmodontine rodents in that it is bilocular, with a deep incisura angularis projecting well past the esophageal opening; the size and position of the glandular epithelium is unknown (Spotorno 1976). The phallus in males is barrel shaped with a slightly ventrally placed crater. The distal process of the baculum is unique among the sigmodontines, with a short and wide central element flanked by longer (at least 60% longer) wing-shaped lateral digits (Spotorno 1986).

SYNONYM:

Andinomys Thomas, 1902g:116; type species *Andinomys edax* Thomas, 1902, by original designation.

REMARKS: The genus and its single living species are known from the Middle-Upper Pleistocene of Argentina and Lower-Middle Pleistocene of Bolivia, to the Recent (Marshall et al. 1984; Marshall and Sempere 1991; Pardiñas et al. 2002; P. E. Ortiz and Jayat 2007a). It was present at seven of nine Late Quaternary micromammal localities in Argentina but was not abundant in any (P. E. Ortiz and Jayat 2007a).

Andinomys edax Thomas, 1902
Andean Rat, Chozchorito

SYNONYMS:

Andinomys edax Thomas, 1902g:116; type locality "El Cabrado, between Potosi and Sucre, [Potosí,] Bolivia. Altitude 3700 metres."

Andinomys edax lineicaudatus Yepes, 1935b:345; type locality "Argentina, Tucumán, Cerro San Javier 2000 m."

Andinomys lineicaudatus: M. M. Díaz and Barquez, 2007: 500; name combination.

DESCRIPTION: As for the genus.

DISTRIBUTION: *Andinomys edax* has been recorded from approximately 60 localities between 500 and 4,500 m in habitats ranging from subtropical mountain forests to the semiarid Puna and pre-Puna of extreme southern Peru, northernmost Chile, Bolivia, and northwestern Argentina.

SELECTED LOCALITIES (Map 28): ARGENTINA: Catamarca, Barranca Larga, Cueva Los Viscos (P. E. Ortiz, Cirignoli et al. 2000), 45 km W of Chumbicha (Thomas 1919f); Jujuy, Cerro Hermoso (MACN 19.554), Rinconada, 6 km N, road to Timón Cruz (M. M. Díaz and Barquez 2007); Salta, La Poma, 3 km E (PEO-e 201), ca. 5 km S of Los Toldos (PEO-e 271); Tucumán, Horco Molle (Capllonch et al. 1997). BOLIVIA: Chuquisaca, 9 km (by road) N of Padilla (S. Anderson 1997); Cochabamba, 16.5 km NW of Colomi (AMNH 268919); Oruro, Finca Santa Helena, approx. 10 km (by road) SW of Pazna (UMMZ 155850); Potosí, 20 mi S of Potosí (S. Anderson 1997); Tarija, Rancho Tambo (AMNH 262771). CHILE: Tarapacá, Arica (FMNH 132651). PERU: Puno, Juli (MCZ 39470); Tacna, 1.5 mi N (10 km by road) of Tarata (MVZ 139480).

SUBSPECIES: Yepes (1935b) recognized two subspecies, *A. edax edax* from southern Bolivia and extreme northwestern Argentina, and *A. edax lineicaudatus* in northwestern Argentina. These two names continue to be used in the literature, both at subspecific and specific levels (see M. M. Díaz and Barques 2007 versus Jayat, Ortiz, and Miotti 2008, and Remarks).

NATURAL HISTORY: Little is known about the natural history of the species. Of 10 specimens collected by one of us (JPJ) in high-elevation grasslands of northwestern Argentina, two females caught in April had signs of reproductive activity (open vaginas). The remainder of the specimens, captured in June, July, and August, showed no signs of reproductive activity. Pearson (1951) reported one (of three collected females) as pregnant with three large embryos (crown-rump length = 34 mm), in mid-December in southern Peru. For Jujuy, M. M. Díaz and Barquez (2007) reported females with closed vaginas in February and March, one lactating female and another with three embryos in February, and females with open vaginas in March (one with two embryos). Young individuals were trapped in February, June, and July. From these observations, it appears that *Andinomys edax* is reproductively active from early December to June.

Molting was found in all individuals collected in the high-elevation grasslands of northwestern Argentina (with the exception of one female taken at the end of August). This observation agrees with M. M. Díaz and Barquez (2007), who observed molting in specimens from Jujuy and trapped in February and March. Two fleas (*Ectinorus uncinatus* and *Neotyphloceras crassispina hemisus*) and a louse (*Hoplopleura zentaensi*) parasitize *Andinomys edax* (Beaucournu and Gallardo 1991; D. Castro and Gonzalez 1997; Autino and Lareschi 1998). No other parasites are known.

Andinomys edax is never an abundant member of the sigmodontine fauna anywhere in its known range, accounting for no more than 2% of the individuals captured at sites in northwestern Argentina. The species is also not abundant in owl pellets, where it occurs in about 2% of specimens examined (JPJ, pers. obs.).

REMARKS: M. M. Díaz (1999) elevated each of the subspecies recognized by Yepes (1935b) to specific status, based on sympatry at localities of the pre-Puna of Jujuy province, Argentina. M. M. Díaz and Barquez (2007) also recognized these two taxa at the species level, stating *A. edax* was restricted to Puna and pre-Puna habitats between 2,000 and 4,800 m, while *A. lineicaudatus* occurred in the Yungas forest below 2,500 m. However,

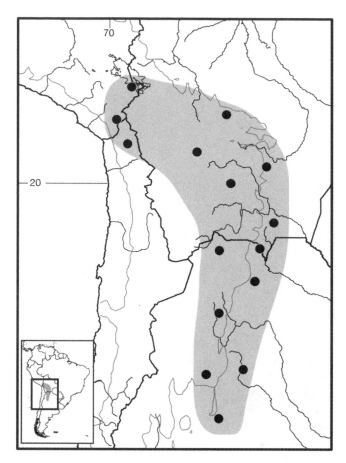

Map 28 Selected localities for *Andinomys edax* (●). Contour line = 2,000 m.

specimens captured by one of us (JPJ) in ecotonal habitats between semiarid highlands and Yungas forests in northwestern Argentina displayed combinations of qualitative (tail coloration) and morphometric (hindfoot length) characters thought to be diagnostic for the two taxa, which suggested the possibility of gene flow between them (Jayat, Ortiz, and Miotti 2008). Given the lack of detailed study of patterns of variation, combined with the lack of differences in karyotypic characters, for the moment at least we maintain the view of Yepes (1935b) and regard *Andinomys* as comprising a single species with two subspecies.

The karyotype is uniform across localities from Argentina (Pearson and Patton 1976), Bolivia (Peurach 1994), and Chile (Spotorno et al. 1994, 2001), with $2n = 56$ and $FN = 56$, an autosomal complement composed of one pair of metacentric and 27 pairs of acrocentric elements and both X and Y chromosomes acrocentric. The Y is about 60% the length of the X. Comparisons of G-banded chromosomes led Spotorno et al. (2001) to suggest homology in 18 chromosomal arms between *A. edax* and *Euneomys chinchilloides*.

Genus *Chinchillula* Thomas, 1898

Jorge Salazar-Bravo

These colorful animals are known from scattered localities between 3,500 and 4,800 m in southeastern Peru, western Bolivia, and northern Chile, where they are often associated with rocky outcrops and stone walls in the highland Puna. Except as noted below, the following account is abstracted from Osgood (1947) and Hershkovitz (1962), who should be consulted for additional details. The genus is not known from the fossil record (McKenna and Bell 1997).

Chinchillula is medium to large in size, with a stout body (total length ranging from 247 to 282 mm; S. Anderson 1997). The tail is short and well-haired (between 50% and 75% of the total length, ranging between 94 and 114 mm; Pearson 1951; S. Anderson 1997), bicolored, with a mid-dorsal brown stripe, and terminating in a thin pencil. The pelage is long, very dense, silky, and distinctly colored buffy to tawny, lined with black over the back. The venter is white, and the hips and rump are white but are crossed by a conspicuous black band. The ears are long, broad, and brown with white preauricular and postauricular tufts. The limbs and feet are white, and ungual tufts almost conceal the claws. The plantar pads are smooth, bare, and have large tubercles. The mammary count is eight, with two pairs in both axillary and inguinal positions.

The skull is flattened in profile. The supraorbital region is narrow and parallel-sided, with edges squared or raised, but with temporal ridges obsolete or absent. The zygomatic arches are evenly bowed, and the zygomatic plates are broad but slant posteriorly from bottom to top, and well-developed zygomatic spines are present. The incisive foramina are long, reaching to the middle of M1s. The bony palate is also long, extending posterior to M3s. Posterolateral palatal depressions are shallow. The mesopterygoid fossa is broad. The width at the anterior base of pterygoid processes is greater than that of the parapterygoid fossa measured on same plane. The parapterygoid fossa is deep in adults. The tympanic bullae are moderately inflated but are often shorter than the length of the molar row (Hershkovitz 1962). Carotid circulation is primitive (pattern 1 of Voss 1988); both stapedial and sphenofrontal foramina and a squamosal-alisphenoid groove are present.

The upper incisors are long, orthodont to slightly opisthodont, with smooth to weakly grooved anterior surfaces. Upper molar rows are parallel or slightly divergent posteriorly. The cheek teeth are highly hypsodont, flat-crowned, and prismatic (Osgood 1947); the cusps are subtriangular or subovate in outline, and the enamel pattern is essentially 8-shaped. The high points of opposing enamel folds never touch, but extend nearly to, or rarely beyond, the midline

of each molar. M1 has a subtriangular procingulum in older specimens; a posteroloph is always absent. Both M2 and M3 have a simplified occlusal pattern, as only the metaflexus and hypoflexus are present (there is no procingulum or posteroloph). The lower first molar has a procingulum with a small anterolingual stylid, and m2–m3 have a simplified occlusal structure without a procingulum or posterolophids.

Didier (1959) described the general morphology of the phallus; Spotorno (1986) extended this description and stressed the unique nature of the structures. Axial skeletal elements consist of 13 thoracic, six lumbar, and 21 or more caudal vertebrate (Steppan 1995). Voss (1991a) noted the presence of a gall bladder, and Dorst (1973b) detailed stomach morphology.

SYNONYM:

Chinchillula Thomas, 1898c:280; type species *Chinchillula sahamae* Thomas, by monotypy.

REMARKS: Thomas (1898c) described *Chinchillula* as a genus distantly allied to *Phyllotis*, a concept followed by most subsequent authors (e.g., Ellerman 1941; Osgood 1947; Olds and Anderson 1990; Braun 1993; Steppan 1993; Steppan 1995). Analyses of mitochondrial and nuclear sequence data have indicated that *Chinchillula* is the sister taxon to a clade formed by *Irenomys*, *Punomys*, and *Andinomys* (Salazar-Bravo et al. 2013) and that these four genera, for the moment, should be considered Sigmodontinae *incertae sedis* as suggested by D'Elía et al. (2007).

Chinchillula sahamae Thomas 1898
Altiplano Chinchilla Mouse, Achallo, Chinchillón, Carpinto

SYNONYM:

Chinchillula sahamae Thomas, 1898c:280; type locality "Esperanza, a 'tambo' in the neighbourhood of Mount Sahama, [La Paz,] Bolivia."

DESCRIPTION: As for the genus.

DISTRIBUTION: *Chinchillula sahamae* is known from fewer than 20 localities in western Bolivia, southern Peru, and northern Chile, all at elevations between 3,500 and 4,800 m.

SELECTED LOCALITIES (Map 29): BOLIVIA: La Paz, Esperanza (type locality of *Chinchillula sahamae* Thomas 1898), 5 km E of Ulla-Ulla (S. Anderson 1997); Oruro, Quebrada Kohuiri (S. Anderson 1997). CHILE: Arica y Parinacota, Parinacota (Mann 1978). PERU: Apurimac, 25 mi SW of Chalhuanca (MVZ 116034); Arequipa, Cabrerías (Zeballos and Carrera 2010), El Rayo (Zeballos and Carrera 2010), 21 mi ENE of Puquio (MVZ 116035); Cusco, Coordillera de Sicuani (Hershkovitz 1962); Puno, 11 km W and 12 km S of Ananea (MVZ 172677), 5 km E of Lago Suche (MVZ 115939), 5 mi SSW of Limbani (MVZ

116183), 36 km SE (by road) of Macusani (MVZ 172675); Tacna, 5 km E of Lago Suche (MVZ 115939), 2.6 mi N of Tarata (MVZ 139484).

SUBSPECIES: *Chinchillula sahamae* is monotypic.

NATURAL HISTORY: *Chinchillula sahamae* is relatively poorly known in most aspects of its ecology, behavior, and reproductive biology. All known localities are in bunchgrass-dominated Altiplano, where it is often associated with rocky outcrops or man-made stone walls (e.g., Pearson 1951). In both Parque Nacional Sajama (Bolivia) and Reserva Nacional de Salinas y Aguada Blanca (Peru), individuals were associated with *Polylepis* patches (Yensen and Tarifa 2002; Zeballos and Carrera 2010). Sympatric species in southern Peru include two species of *Abrothrix*, two species of *Akodon*, *Necromys amoenus*, two species of *Phyllotis*, three of *Auliscomys*, *Neotomys ebriosus*, and a species each of *Galea* and *Lagidium* (Pearson 1951). In west-central Bolivia, *Chinchillula* co-occurs with *Akodon albiventer*, *Abrothrix jelskii*, *Phyllotis chilensis*, and *Lagidium viscaccia* (Yensen and Tarifa 2002).

At one locality in La Paz (Bolivia), the capture rate per 100 traps was 0.9 (Yoneda 1984), but with this rate ranging from 0.3 to 2 in different study years at various sites in Arequipa and Moquegua (Peru; Zeballos and Carrera 2010). The species is herbivorous, feeding on grass or herbs (Pearson 1951; Dorst 1973b; Mann 1978). In southeastern Peru, the diet was composed of 96.57% grass and herbs and 0.01% seeds (H. Zeballos, pers. comm.). Signs of reproductive activity start at the beginning of October (Pearson 1951). Observations of the growth pattern in captive individuals have suggested an exponential growth for the first eight weeks after birth, reaching an asymptote at about week 15 (Aliaga-Rossel et al. 2009).

Ectoparasites recorded from *Chinchillula sahamae* include two species of trombiculid mites of the genus *Crotiscus* (Brennan 1957), one of the genus *Odontacarus* (J. T. Reed and Brennan 1975), one of *Euschoengastia* (misspelled by Hershkovitz 1962 as *Euschonygastia*), five species of fleas (Hershkovitz 1962; Smit 1968), and a beetle (Hershkovitz 1962).

Although thought to be threatened or even at risk in Chile and Peru (e.g., Coffre and Marquet 1999; Zeballos 2001), possibly due to overexploitation for their colorful pelts, no studies have been conducted to assess local population status. The species was deemed "data deficient" by Bolivian authorities (Aguirre et al. 2009).

REMARKS: Pine et al. (1979) and Mann (1978) made a case for the presence of morphological and chromatic differences between Bolivian and Peruvian samples, and one specimen from Chile. The record from "Sumbay, Arequipa" mapped by Hershkovitz (1962) is at odds with current maps and ornithological gazetteers (Stephens 1983),

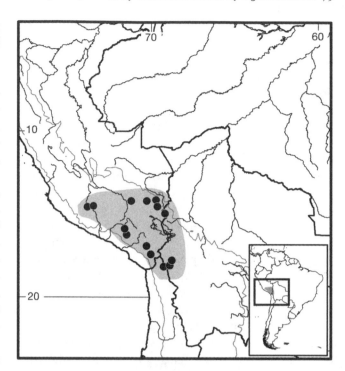

Map 29 Selected localities for *Chinchillula sahamae* (●). Contour line = 2,000 m.

which locate Sumbay to the northeast, not the southwest, of the city of Arequipa. In addition, as mapped, Hershkovitz's Sumbay would be at about 2,000 m, an elevation from which no records of *Chinchillula* currently exist. Eisenberg and Redford (1992) mapped two localities in Chile, but Parinacota is the only one from which a vouchered specimen exists (see Selected Localities).

Pearson and Patton (1976) described a karyotype with $2n = 60$, FN = 64 for Peruvian specimens.

Genus *Delomys* Thomas, 1917
Robert S. Voss

The genus *Delomys* contains two species that inhabit moist forests and perhaps other habitats from sea level to 2,700 m elevation in northeastern Argentina and southeastern Brazil. The only known fossils are from late Pleistocene cave deposits near Lagoa Santa in the Brazilian state of Minas Gerais (Winge 1887; Voss and Myers 1991; Voss 1993). Although Osgood (1933a), Ellerman (1941), and Cabrera (1961) regarded *Delomys* as a junior synonym of *Thomasomys*, subsequent analyses of karyotypic, morphological, and molecular data suggest that these taxa are not closely related (Zanchin, Sbalqueiro et al. 1992; Voss 1993; M. F. Smith and Patton 1999; D'Elía 2003a; Weksler 2003), and the genus remains unclassified at the tribal level (D'Elía et al. 2007). Recent molecular analyses (e.g., Parada et al. 2013; Salazar-Bravo et al. 2013) have recovered *Delomys* as the sister taxon of the tribe Phyllotini, a wholly unexpected and surprising result. Except as noted below, the following account is abstracted from Voss (1993), which should be consulted for additional information about comparative morphology, taxonomy, nomenclature, and natural history.

Species of *Delomys* are small to medium-sized sigmodontines with relatively short, close, grizzled yellowish, brownish, or grayish dorsal pelage. A dark mid-dorsal stripe that extends from the nape to the base of the tail may be present or absent. The ventral pelage is distinctly paler than the dorsal fur, but the ventral pelage is always gray based (never self-colored). Mystacial, superciliary, genal, submental, interramal, and carpal vibrissae are present. The pinnae are large, thin, pliant, and sparsely covered with short hairs. The hindfoot is narrow, with six large plantar pads (thenar, hypothenar, and three interdigitals) and short outer digits; digit V is nonopposable, and its claw does not extend much beyond the first interphalangeal joint of digit IV. Conspicuous ungual tufts of long silvery hairs are present at the bases of the pedal claws. Females have either eight mammae (in inguinal, abdominal, postaxial, and pectoral pairs) or six mammae (in which case the pectoral pair is absent). The naked-appearing tail may be shorter than, equal to, or longer than the combined length of the head and body.

In dorsal view, the skull is characterized by a long tapering rostrum flanked by small but well-defined zygomatic notches; the interorbital region is hourglass in shape, neither conspicuously inflated nor strongly constricted, with smoothly rounded supraorbital margins. The braincase is rounded and smooth, without well-developed ridges or crests. In lateral view, the premaxillae and nasals are produced anteriorly beyond the small incisors to form a short bony rostral tube. Zygomatic plates are moderately broad with approximately vertical anterior margins, and lack anterodorsal spinous processes.

In ventral view, the incisive foramina are long (70–77% of diastemal length, on average), but they do not extend posteriorly between the molar rows. The bony palate is wide and short, without conspicuous ridges, grooves, or complex posterolateral pits. The mesopterygoid fossa is wide, with a square or biconcave anterior margin that extends anteriorly between the last molars; the bony roof of the fossa is complete or perforated by very small sphenopalatine vacuities. Because the alisphenoid strut is absent, the buccinator-masticatory foramen and the foramen ovale are confluent.

The carotid circulation is primitive (pattern 1 of Voss 1988), as evidenced by a large stapedial foramen (for the stapedial artery) and a prominent squamosal-alisphenoid groove and sphenofrontal foramen (for the supraorbital stapedial ramus). The postglenoid foramen is separated from the subsquamosal fenestra by a slender hamular process of the squamosal. An anterior process of the tegmen tympani overlaps a posterior suspensory process of the squamosal above the anterodorsal lip of the bulla. The bullae are small, the tympanic membrane pars flaccida is large, and the orbicular apophysis of the malleus is well developed.

The mandible has a short, falciform coronoid process but lacks a distinct capsular process alveolus. The basihyal lacks an entoglossal process.

The upper incisors are small, narrow, ungrooved, and strongly opisthodont. The molars are brachydont, bunodont, and pentalophodont (with well-developed mesolophs and mesolophids). The principal cusps of the upper teeth are arranged in opposite labial/lingual pairs, but the labial and lingual reentrant folds interpenetrate to create an incipiently lophodont pattern. The anterocone of M1 is divided by a deep anteromedian flexus into distinct anterolabial and anterolingual conules. An ectolophid is always present on m1 and sometimes also on m2. The upper molars usually have three roots each, and the lower molars have two roots each.

The tuberculum of the first rib articulates with the transverse processes of the seventh cervical and first thoracic vertebrae, and the second thoracic vertebra has a greatly elongated neural spine. The entepicondylar foramen of the humerus is absent. Axial skeletal counts include 13 ribs and 19 thoracolumbar, four sacral, and 29 or more caudal vertebrae.

The stomach is unilocular and hemiglandular, without any extension of gastric glandular epithelium into the corpus. The gall bladder is present. Phallic and other male reproductive characters remain undescribed.

SYNONYMS:
Hesperomys: Hensel, 1872b:42; part (description of *dorsalis*); not *Hesperomys* Waterhouse, 1839.
Akodon: E.-L. Trouessart, 1897:537; part (listing of *dorsalis*); not *Akodon* Meyen, 1833.
Oryzomys: Thomas, 1903c:240; part (description of *sublineatus*); not *Oryzomys* Baird, 1857.
Thomasomys: Thomas, 1906c:443; part (listing of *dorsalis* and *sublineatus*); not *Thomasomys* Coues, 1884.
Delomys Thomas, 1917d:196; type species *Hesperomys dorsalis* Hensel, 1872, by original designation.

REMARKS: Species of *Delomys* have at one time or another been referred to *Hesperomys*, *Calomys*, *Akodon*, *Oryzomys*, and *Thomasomys* (none of which are, properly speaking, generic synonyms). This historical confusion reflects both the uncertain morphological diagnoses of neotropical cricetid taxa in general and the genuinely ambiguous relationships of *Delomys* in particular. Although the sister-group relationship between *Delomys* and phyllotines recovered by Parada et al. (2013) seems to be impressively supported by their molecular data, future testing with sequence data from additional loci is much to be desired.

Hershkovitz's (1998:212) comments on morphological variation in *Delomys* imply that the two recognized species (*D. dorsalis* and *D. sublineatus*) cannot be distinguished by the phenotypic characters used in the key that follows here. However, all of the material that I examined (Voss 1993) was unambiguously identifiable by these criteria. Because Hershkovitz did not describe any other diagnostic differences, it is not clear how his species samples were identified if not by the morphological characters employed here.

Voss (1993:32–34) assigned to *Delomys* Winge's (1887) *Calomys plebjus*, described from a large cranial fossil fragment collected by P. W. Lund at Lapa da Serra das Abelhas, Lagoa Santa, Minas Gerais state, Brazil. The connection between *plebjus* to either of the extant species in the genus is problematical, and Voss (1993:34) concluded that "*plebejus* is a *nomen dubium* that should be used only in reference to Winge's hypodigm."

KEY TO THE SPECIES OF *DELOMYS*:

1. Dorsal pelage very soft and dense, usually some shade of dark brown or brownish gray, often with indistinct dark mid-dorsal stripe from nape to base of tail; no band of clear yellow or buff separating dorsal and ventral coloration along flanks; vibrissae long, extending to or beyond tips of pinnae when laid back along head; individual hairs on dorsal surface of hindfoot with indistinct melanic bands (visible only under magnification); claw of digit V extends to or slightly beyond first interphalangeal joint of digit IV; tail subequal to or longer than combined length of head and body; $2n = 82$ chromosomes. . . . .*Delomys dorsalis*
1'. Dorsal pelage coarser, less dense, and paler (usually distinctly grizzled yellowish), the dark mid-dorsal stripe (when present) shorter, usually not extending from nape to tail base; dorsal and ventral pelage almost always separated by lateral band of clear yellow or buff along flanks; mystacial vibrissae short, not extending to tips of pinnae when laid back along head; individual hairs on dorsal surface of hindfoot usually pure white (without any melanic bands); claw of digit V does not extend to first interphalangeal joint of digit IV; tail usually shorter than combined length of head and body; $2n = 72$ chromosomes . . . . *Delomys sublineatus*

Delomys dorsalis (Hensel, 1872)
Striped Delomys
SYNONYMS:

Hesperomys dorsalis Hensel, 1872b:42; type locality "Provinz Rio Grande do Sul," Brazil; restricted to Taquara, Rio Grande do Sul, by Avila-Pires (1994:372).

Hesperomys dorsalis obscura Leche, 1886:696; no type locality given.

Akodon dorsalis: E.-L. Trouessart, 1897:537; name combination.

Akodon dorsalis obscura: E.-L. Trouessart, 1897:537; name combination.

Akodon dorsalis lechei E.-L. Trouessart, 1904:434; replacement name for *obscura* Leche, preoccupied by *Mus obscurus* Waterhouse, 1837.

Thomasomys dorsalis: Thomas, 1906c:443; name combination.

Delomys dorsalis: Thomas, 1917d:196; first use of current name combination.

Delomys dorsalis collinus Thomas, 1917d:197; type locality "Itatiaya, Rio de Janeiro, 4800 ft.," Brazil.

Delomys dorsalis dorsalis: Gyldenstolpe, 1932a:60; name combination.

Delomys dorsalis lechei: Gyldenstolpe, 1932a:61; name combination.

Thomasomys dorsalis dorsalis: Ellerman, 1941:368; name combination.

Thomasomys dorsalis collinus: Ellerman, 1941:369; name combination.

Thomasomys dorsalis lechei: Ellerman, 1941:369; name combination.

Thomasomys lechei: Moojen, 1952b:59; name combination.

Thomasomys collinus: Moojen, 1952b:60; name combination.

Delomys collinus: Avila-Pires, 1960b:32; name combination.

DESCRIPTION: A softer-furred and longer-tailed species than *Delomys sublineatus*, with or without distinct mid-dorsal stripe from nape to base of tail, but consistently lacking lateral line of clear yellow or buff between dorsal and ventral pelage color zones. Mystacial vibrissae long, extending to or beyond tips of adducted pinnae; hindfoot covered dorsally with dark-banded hairs; claw of digit I extending to middle of first phalanx of digit II; claw of digit V extending to or just beyond first interphalangeal joint of digit IV.

DISTRIBUTION: From Rio Grande do Sul (Brazil) and Misiones (Argentina) northward along the humid Atlantic littoral and interior mountains of southeastern Brazil to Minas Gerais. Specimens have been collected near sea level in northern Argentina and southern Brazil (between 25° and 30°S latitude), but most northern collections are from higher elevations (to 2,700 m according to Hershkovitz 1998). *Delomys dorsalis* is known to occur sympatrically with *D. sublineatus* at several localities in the Brazilian states of São Paulo and Rio de Janeiro (Voss 1993).

SELECTED LOCALITIES (Map 30): ARGENTINA: Misiones, Caraguatay, Río Paraná (FMNH 26818). BRAZIL: Minas Gerais, Parque Nacional do Caparaó (Hershkovitz

1998); Paraná, Roça Nova (Voss 1993); Rio de Janeiro, Itatiaya (type locality of *Delomys dorsalis collinus* Thomas), Teresópolis (Voss 1993); Rio Grande do Sul, Taquara (type locality of *Hesperomys dorsalis obscura* Leche); São Paulo, Alto da Serra (Voss 1993), Apiaí (MZUSP 26936), Estação Biológica de Boracéia (MVZ 183055), Casa Grande (Voss 1993), Fazenda Intervales, Base do Carmo (MVZ 183047), Piquete (Voss 1993).

SUBSPECIES: Voss (1993:30–31) reviewed geographic variation in morphology among available samples of *D. dorsalis* and remarked that "the names *collinus* and *dorsalis* could be retained for subspecies distinguished by mammary counts alone, an option that should be evaluated by future researchers with larger samples and material from additional localities in São Paulo state" (see also Remarks).

NATURAL HISTORY: Field observations summarized by Voss (1993) suggest that species of *Delomys* are nocturnal inhabitants of moist-forest habitats. Trapping studies (e.g., Olmos 1991; Cademartori et al. 2008) consistently indicate that these rodents are primarily terrestrial, although there is also some evidence that *D. dorsalis* climbs well and may seasonally occur in subcanopy vegetation (Cademartori et al. 2005). Although montane (or premontane) forests are apparently favored at tropical latitudes, some subtropical collection localities for *D. dorsalis* are closer to sea level. Several reports (e.g., Miranda-Ribeiro 1905; Davis 1945a, 1947) suggest that these rats move among and under the sheltering clutter of logs, organic debris, mossy roots, tangled bamboo, and herbaceous understory vegetation rather than frequenting open areas on the forest floor. In the moist mixed-coniferous forests of Rio Grande do Sul, *D. dorsalis* is more abundant in habitats with denser undergrowth and fewer *Araucaria* pines than in habitats with sparse undergrowth and more abundant *Araucaria* (Cademartori et al. 2004; Dalmagro and Vieira 2005). In the only dietary study published to date, *D. dorsalis* was found to eat a variety of fruits and seeds as well as invertebrates and fungi (E. M. Vieira et al. 2006). Apparently, *D. dorsalis* is reproductively active year round; the gestation period is 21 to 22 days, and the litter size ranges from two to five young (Cademartori et al. 2005).

REMARKS: Zanchin, Sbalqueiro et al. (1992) described the karyotype of *Delomys dorsalis* as $2n = 82$, but fundamental numbers (FN) differ among karyotyped populations of this species.

As recognized herein, *Delomys dorsalis* is geographically variable. Available samples from southern populations (e.g., in Argentina and Rio Grande do Sul) have smaller measurement means and more mammae (eight, in four pairs) than samples from northern populations (e.g., in Rio de Janeiro), which are larger on average and have fewer mammae (six, in three pairs). However, geographically in-

termediate samples (from Paraná and southern São Paulo) are morphometrically intermediate, and adjacent eight- and six-mammate populations (in southern and northern São Paulo, respectively) are not morphometrically divergent. Therefore, although different names are available for the small eight-mammate southern form (*dorsalis*) and the large six-mammate northern form (*collinus*), these appear to represent opposite ends of two independent north-south clines rather than distinct taxa (Voss 1993).

Alternatively, Bonvicino and Geise (1995) argued that *collinus* and *dorsalis* are distinct species based on their analyses of karyotypic data. They found that specimens from the type locality of *collinus* (Itatiaya, in Rio de Janeiro state) and from Caparaó (in Minas Gerais) had a fundamental number (FN) of 86 autosomal arms, whereas the specimens karyotyped by Zanchin, Sbalqueiro et al. (1992) had FN = 80. If this karyotypic difference were correlated with the morphological variation previously described, the case for taxonomic discrimination would be strong. However, the FN = 80 karyotype has been reported from both small-bodied eight-mammate populations (e.g., in Rio Grande do Sul) and large-bodied six-mammate populations (in northern São Paulo), so the taxonomic picture is still unclear. Bonvicino and Geise (1995:125) implied that "morphological analyses" corroborated their recognition of *collinus* and *dorsalis* as distinct species, and they suggested that the absence of heterozygotes (with intermediate fundamental numbers) indicated reproductive isolation. However, no diagnostic morphological criteria were provided in their report, and

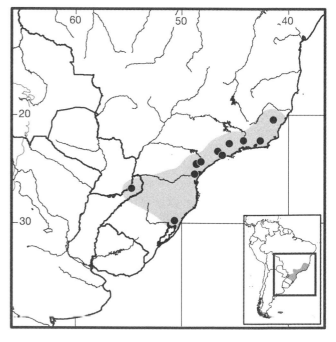

Map 30 Selected localities for *Delomys dorsalis* (●).

geographic variation in fundamental number is an obvious alternative explanation for the scant data at hand. In the absence of other evidence, the hypothesis that *dorsalis* and *collinus* are conspecific (Thomas 1917d; Voss 1993; Hershkovitz 1998) remains to be convincingly falsified.

Delomys sublineatus (Thomas, 1903)
Pallid Delomys

SYNONYMS:

Oryzomys sublineatus Thomas, 1903c:240; type locality "Engenheiro Reeve, Inland of Victoria, Prov. Espirito [= Espírito] Santo, Brazil. Alt. 500 m."

Thomasomys sublineatus: Thomas, 1906c:443; name combination.

Delomys sublineatus: Thomas, 1917d:196; first use of current name combination.

Thomasomys dorsalis sublineatus: Cabrera, 1961:428; name combination.

DESCRIPTION: A coarser-furred, shorter-tailed species than *Delomys dorsalis*, usually lacking a distinct mid-dorsal stripe and almost always with a lateral line of clear yellow or buff separating dorsal and ventral pelage color zones. Mystacial vibrissae short, not extending to tips of pinnae when laid back alongside head; hindfoot covered dorsally with (usually) pure white hairs; digit I short, with claw not extending to middle of first phalanx of digit II; digit V also short, with claw not extending to second interphalangeal joint of digit IV.

DISTRIBUTION: *Delomys sublineatus* occurs from the state of Paraná northward along the Atlantic coast and interior mountains of southeastern Brazil to the state of Espírito Santo. Specimens have been collected from near sea level to about 1,200 m elevation. See the preceding account for records of sympatry with *D. dorsalis*.

SELECTED LOCALITIES (Map 31): BRAZIL: Espírito Santo, Engenheiro Reeve (type locality of *Oryzomys sublineatus* Thomas), Santa Teresa (Voss 1993); Minas Gerais, Conceição do Mato Dentro (Voss 1993); Rio de Janeiro, Teresópolis (Voss 1993); Santa Catarina, Hansa (Voss 1993); São Paulo, Estação Biológica de Boracéia (MVZ 183075), Caucaia do Alto (Püttker et al. 2008a), Iporanga (Voss 1993), Salto Grande (Voss 1993).

SUBSPECIES: *Delomys sublineatus* is monotypic.

NATURAL HISTORY: Unfortunately, the available information is not sufficient to confidently identify natural history differences between this species and *Delomys dorsalis* (see earlier). However, its shorter vibrissae, tail, and fifth pedal digits all suggest that *D. sublineatus* may be less scansorial than *D. dorsalis*, and there are anecdotal indications that the two species may differ in habitat use along moisture and/or successional gradients where they occur sympatrically (Voss 1993).

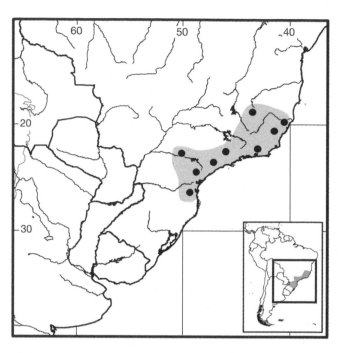

Map 31 Selected localities for *Delomys sublineatus* (●).

Delomys sublineatus is said to be sensitive to forest fragmentation, with local abundance decreasing in smaller or more isolated patches (Pardini et al. 2005; Püttker et al. 2008a; Püttker, Bueno et al. 2013). This species was restricted to native vegetation in a comparison among fragments (Umetsu and Pardini 2007) but seemed to be insensitive to variation in vegetation structure in fragments of native habitat (Püttker et al. 2008b). Püttker, Pardini et al. (2008) examined the relationship between body size, parasite load, and density as a function of vegetation fragment size, and Püttker et al. (2006) examined individual movement patterns based on a capture-recapture study.

REMARKS: No noteworthy geographic variation has been reported for this species. Zanchin, Sbalqueiro et al. (1992) described a karyotype with $2n = 72$, FN $= 90$.

Genus *Euneomys* Coues, 1874
Janet K. Braun and Ulyses F. J. Pardiñas

Members of the genus *Euneomys* range from about 33°S in Argentina and Chile southward to the southernmost part of Cabo de Hornos (Osgood 1943b; Pardiñas, Teta, Chebez et al. 2010; however, see comments for *E. fossor* regarding a single locality north of 35°S). Four species are currently recognized. These occur in dry, open grassland with thickets of *Nothofagus* above the tree line to bare, rocky, windswept scree with limited vegetation (Musser and Carleton 2005). Locality records span elevations from near sea level in the southern part of the distribution to more than 3,300 m in the north. The following generic description is, except as noted, compiled from information presented by Coues (1894), Gyldenstolpe (1932), Ellerman (1941), Mann (1944), Hershkovitz (1962), and Steppan (1995), supplemented by our examination of holotypes and recently collected specimens.

All species of *Euneomys* are externally similar, vole-like animals that are rather large in size and heavy in body. The eyes are large and rimmed in black, a feature reminiscent of *Reithrodon*. The ears are moderate in size (about 20 mm in length) and covered by fine hairs. The fur is thick and soft. The tail is relatively short (<70% of head and body length), bicolored, and lacks a terminal pencil. The hindfeet are broad and unwebbed; the claws are short, and the plantar surface is naked, with six large and well-developed tubercles (thenar, hypothenar, and four interdigitals) present. Eight mammae (in four pairs) are present.

The dorsal contour of the skull is convex in lateral view, the braincase is relatively short, and the rostrum is broad and short. The interorbital region is narrow, parallel-sided, without beading, but is slightly raised dorsally. The zygomata are expanded posteriorly. The anterior border of the zygomatic plates is nearly straight or very slightly concave; their dorsal surfaces rise above the dorsal surface of the rostrum so that the anterior roots of the zygomata insert at the dorsal surface. The premaxillomaxillary suture below the infraorbital foramen is directed forward along a line about parallel to the longitudinal plane of the palate. The incisive foramina are narrow and extend just past the anterior plane of M1. The palate is long, ridged, and has paired and sometimes complicated posterolateral depressions. The parapterygoid fossae are broad and moderately excavated. The mesopterygoid fossa is distinctly narrower than the parapterygoid fossae. A vertical strut of the alisphenoid typically separates the buccinator-masticatory and accessory oval foramina. The hamular process is long. The bullae are not notably inflated, and the lacerate foramina are largely open. The mandible is dorsoventrally deep in proportion to its length, and there is a pronounced capsular process at the base of the coronoid process.

The upper incisors are opisthodont, broad, and possess a single anterior groove. The upper molar rows are slightly to strongly divergent posteriorly. The molars are tetralophodont (mesolophs and mesolophids are absent), hypsodont, and flat-crowned, with their principal cusps and crests forming a sigmoidal shape; folds are narrow. M3 is S-shaped with an open hypoflexus. The procingulum of m1 is typically separated (that is, an anterior murid is absent) from the protoconid and metaconid, and the anterolabial cingulum is absent. Both M1 and M3 are three-rooted.

Axial skeletal elements have counts of 13 thoracic ribs, six lumbar vertebrae, and 22 to 24 caudal vertebrae. The neural spine of the third thoracic vertebra is highly elongated, and the spine of the second cervical vertebra overlaps the third cervical vertebra. The stomach is unilocular and hemiglandular, but the glandular epithelium extends past the esophageal orifice into the corpus (Carleton 1973). A gall bladder is present (Voss 1991).

Information on parasites is available in Fain (1976), Lewis and Spotorno (1984), Smit (1987), Beaucournu and Gallardo (1988), and J. P. Sánchez (2013). Fossil material is known from Holocene deposits in Argentina (Pardiñas et al. 2002; Teta et al. 2005) and Late Pleistocene deposits in Chile (Simonetti and Rau 1989; Latorre 1998).

REMARKS: The genus is in need of systematic of revision; see Yáñez et al. (1987), Reise and Gallardo (1990), Pearson and Christie (1991), and Musser and Carleton (2005) for comments regarding information known to date.

SYNONYMS:

Reithrodon Waterhouse, 1839:68; part (description of *typicus* and *chinchilloides*); not *Reithrodon* Waterhouse, 1837.

Euneomys Coues, 1874:185; as a subgenus of *Reithrodon*; type species *Reithrodon chinchilloides* Waterhouse, 1839, by original designation.

Chelemyscus Thomas, 1925d:585; type species *Reithrodon fossor* Thomas, 1925, by monotypy.

KEY TO THE SPECIES OF *EUNEOMYS*:

1. Known only from the type locality in Salta province, Argentina . . . .*Euneomys fossor*
1'. Not known from Chile and Argentina north of 30°S . 2
2. Area from just dorsal to rhinarium to lower lips covered with short, brown hairs; upper incisors with grooves placed centrally; size larger; dorsally darker and grayer and laterally cinnamon; length of maxillary toothrow generally >6.0 mm; interorbital breadth generally >4.2 mm; posterolateral palatal pits moderately deep. . . . . *Euneomys mordax*
2'. Area from just dorsal to rhinarium to lower lips covered with short, white hairs; upper incisors with grooves placed laterally; size smaller; dorsally browner and laterally buffy; length of maxillary toothrow generally <6.0 mm; interorbital breadth generally <4.2 mm; posterolateral palatal pits deep. .3
3. Known only from Tierra del Fuego and nearby areas . . . .*Euneomys chinchilloides*
3'. Known from areas north of Tierra del Fuego in Chile and Argentina. . . . . *Euneomys petersoni*

Euneomys chinchilloides (Waterhouse, 1839)
Tierra del Fuego Euneomys
SYNONYMS:

Reithrodon chinchilloides Waterhouse, 1839:72; type locality "South shore of the Strait of Magellan, near the

Eastern entrance." The diary of Captain R. FitzRoy (1839:322–323) for the period from February 17–25, 1834, indicated that the HMS *Beagle* was anchored in San Sebastián Bay, Tierra del Fuego, where C. Darwin and others worked the northeastern portion of the island. Therefore, the holotype of *chinchilloides* Waterhouse was probably collected near Bahia San Sebastián on the Argentina side of Tierra del Fuego (see also Osgood 1943b).

Reithrodon (*Euneomys*) *chinchilloides*: Coues, 1874:185; name combination.

Euneomys chinchilloides: Thomas, 1901f:254; name combination.

Euneomys ultimus Thomas, 1916b:185; type locality "St. Martin's Cove, Hermite Island," Hermite Islands, Magallanes y Antártica Chilena, Chile.

Euneomys chinchilloides chinchilloides: Osgood, 1943b: 214; name combination.

Euneomys chinchilloides ultimus: Osgood, 1943b:216; name combination.

Euneomys [(*Euneomys*)] *chinchilloides chinchilloides*: Cabrera, 1961:499; name combination.

Euneomys [(*Euneomys*)] *chinchilloides ultimus*: Cabrera, 1961:499; name combination.

DESCRIPTION: Medium sized, with short tail, dense and soft pelage with rich coloration, and rather small ears for genus; slightly larger that *E. petersoni*. Skull broad and heavy; zygomatic plates with straight anterior border or nearly straight, only slightly inclined backward from bottom to top; posterior palate marked with lateral pits separated from parapterygoid fossae by high ridge; lower half of premaxillomaxillary suture extends forward to point more than halfway between cheek teeth and incisors; front of upper incisors laterally grooved; and cheek teeth hypsodont and with a markedly oblique pattern. Specimens from Tierra del Fuego and from the southern mainland in Chile tend to be darker than those elsewhere (Osgood 1943b; Pearson and Christie 1991).

DISTRIBUTION: *Euneomys chinchilloides* occurs on Isla Grande de Tierra del Fuego, neighboring islands, and southernmost mainland Chile (Magallanes y Antártica Chilena). Fossils of true *chinchilloides* are recorded in the Late Pleistocene-Late Holocene sequence of the archaeological site Tres Arroyos 1, Tierra del Fuego (Latorre 1998).

SELECTED LOCALITIES (Map 32): ARGENTINA: Tierra del Fuego, Bahía Buen Suceso (USNM 482138), Lake Fagnano [= Lago Kami] (Osgood 1943b), southeastern end Lake Fagnano [= Lago Kami] (Osgood 1943b), south shore of the Strait of Magellan, near the eastern entrance (type locality of *Reithrodon chinchilloides* Waterhouse). CHILE: Magallanes y Antártica Chilena, Cerro Cóndor, top, 5 km W of Lapataia (Reise and Venegas 1987), Estancia la Fron-

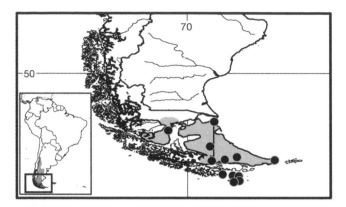

Map 32 Selected localities for *Euneomys chinchilloides* (●). Contour line = 2,000 m.

tera (Reise and Gallardo 1990), Grevy Island (Olrog 1950), Martial Bay, Herschel Island (RNP 1534), vicinity of Punta Arenas (Osgood 1943b), Orange Bay, Hoste Island (Milne-Edwards 1890), St. Martin's Cove, Hermite Island (type locality of *Euneomys ultimus* Thomas), Wollaston Island (Olrog 1950).

SUBSPECIES: Osgood (1943b) treated *ultimus* Thomas as a subspecies of *E. chinchilloides*, based on its larger size and darker coloration.

NATURAL HISTORY: In much of the published literature, data reported for *E. chinchilloides* refers to other species of *Euneomys*, particularly *E. petersoni*. Little information is available for this species as now defined. Reproduction is reported to occur during the summer; a juvenile specimen was captured in April (Pine et al. 1978), and immatures were collected in February and March (Osgood 1943b).

REMARKS: Mitochondrial DNA cytochrome-*b* data suggest that *E. petersoni* should be treated as a synonym of *E. chinchilloides* (see E. P. Lessa et al. 2010). Although the available morphological evidence supports this hypothesis (see Reise and Gallardo 1990; Pearson and Christie 1991), additional research is needed to resolve the systematic status and relationships of the southernmost populations of *Euneomys*. The karyotype with $2n = 36$ is based on one female collected in the vicinity of Ushuaia, Tierra del Fuego, Argentina (Lizarralde et al. 1994).

Euneomys fossor (Thomas, 1899)
Salta Euneomys

SYNONYMS:

Reithrodon fossor Thomas, 1899c:280; type locality "Salta Province, N. Argentina."

E[*uneomys*]. *fossor*: Thomas, 1901f:254; name combination.

[*Chelemyscus*] *fossor*: Thomas, 1925d:585; name combination.

Euneomys [(*Chelemyscus*)] *fossor*: Cabrera, 1961:500; name combination.

DESCRIPTION: Skull larger and heavier than that of *E. chinchilloides*; nasals decidedly expanded anteriorly, just surpassing premaxillae posteriorly; supraorbital edges square, not ridged, their anterior ends marked by slight projections, somewhat like rudimentary postorbital processes; coronal suture evenly curved; interparietal broad; anterior zygomatic root almost as in *E. chinchilloides*, but with slight concavity on front edge rather than deep undercutting in latter species; incisive foramina widely open anteriorly and extend posteriorly to level of M1s; palate ends just behind M3s; posterolateral palatal pits well excavated; and parapterygoid plates shallow (Thomas 1899c:281).

DISTRIBUTION: *Euneomys fossor* is known only from the type locality, the exact position of which is unknown because Salta is a large province in northern Argentina. No additional specimens of this species have been reported, and no other species of *Euneomys* has been recorded from north of about 32°S (Pardiñas, Teta, Chebez et al. 2010).

SELECTED LOCALITIES (Map 33): ARGENTINA: Salta, "Salta Province" (type locality of *Reithrodon fossor* Thomas).

SUBSPECIES: *Euneomys fossor* is monotypic.

NATURAL HISTORY: The natural history of this species is unknown.

REMARKS: The status of this species is uncertain, both because the holotype consists of a mismatched skin and skull, and because the type locality is vague. However, as Pearson and Christie (1991) clearly stated, the name is available for nomenclatorial purposes. The holotype was a fluid-preserved specimen sent to Oldfield Thomas by Francisco P. Moreno, then director of the Museo de La Plata. Although Thomas (1899c:281) wrote that "it [the skull] was extracted [from the fluid] in the Museum [Natural History] on arrival, so that any mistake seems quite impossible," his name was, in fact, based on a *Euneomys* skull and a *Chelemys* skin (see Hershkovitz 1962; Pearson 1984; Pearson and Christie 1991). Thomas must have recognized a problem with the specimen, however, as he also noted (Thomas 1899c:281) that "the skull should be taken as the type if it were hereafter shown not to belong to the skin." Thomas (1925d:585) erected the genus *Chelemyscus* for *fossor* based on the combination of "deeply grooved incisors and large fossorial fore-claws." Galliari et al. (1996) considered *fossor* a *nomen dubium*, an incorrect conclusion because the type of *fossor* presumably contains sufficient character information to resolve its specific status.

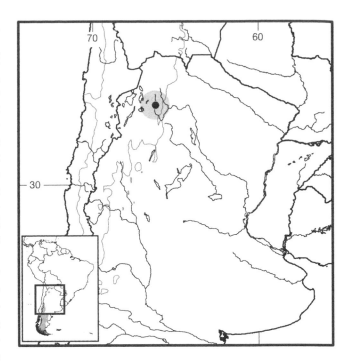

Map 33 Type locality of *Euneomys fossor* (●), mapped without certainty and only to indicate "Salta province." Contour line = 2,000 m.

Euneomys mordax Thomas, 1912
Large Euneomys

SYNONYMS:

Euneomys mordax Thomas, 1912b:410; type locality "Fort San Rafael, Province of Mendoza," Argentina. Fuerte de San Rafael (= Fort San Rafael) is located 24 km WNW San Rafael, Mendoza province, Argentina; however, the geographic provenance of the holotype of *mordax* was questioned by Reise and Gallardo (1990:79), Pearson and Christie (1991:126), and Pearson and Lagiglia (1992:38). According to the last authors "nosotros sugerimos . . . que Bridges colectó los tipos de . . . *Euneomys mordax* . . . durante su cruce de Los Andes [*sic*] cerca del Volcán Peteroa."

Euneomys noei Mann, 1944:95; type locality "Valle de la Junta, Cajón del Río Volcán, a 2.400 m de altura," Metropolitana de Santiago, Chile.

Euneomys [(*Chelemyscus*)] *mordax*: Cabrera, 1961:500; name combination.

Euneomys chinchilloides noei: Mann, 1978:26; name combination.

DESCRIPTION: Larger than that *E. chinchilloides* or *E. petersoni*; fur long, thick, and woolly; hairs on back about 12–13 mm in length; general color dull gray, belly paler, not sharply defined, hairs broadly washed with cream-buff; ears of medium length, well-haired, blackish; manus and pes dull grayish white above; foreclaws not enlarged;

and tail thickly haired, grayish white with indistinct darker line above. Skull stout and heavily built; nasals considerably inflated anteriorly; posterolateral palatal depressions not well developed; supraorbital edges less sharply angular than in *E. fossor*; posterolateral palatal pits comparatively less developed and narrower than in either *E. chinchilloides* or *E. petersoni*; incisors very broad and heavy, with strong groove in more or less central position.

DISTRIBUTION: *Euneomys mordax* is currently known from isolated high Andean (usually above 1,700 m) localities in west central Argentina, Mendoza and Neuquén provinces, and adjacent parts of Chile, including Región Metropolitana in the north and Araucanía in the south (Reise and Gallardo 1990; Pearson and Christie 1991). Pearson and Christie (1991) tentatively associated a large fossil form collected in Latest Pleistocene–Early to Late Holocene strata of Traful I cave, Neuquén, Argentina, with *E. mordax*. Pearson (1995:112) also suggested that *E. mordax* was present in Río Negro province, Argentina: "I have seen what appear to be its runways and fecal pellets in appropriate habitat at high altitude at Laguna Toncek in Nahual Huapi Park."

SELECTED LOCALITIES (Map 34): ARGENTINA: Mendoza, Laguna de la Niña Encantada (Jayat, Ortiz, and Miotti 2006), Valle Hermoso, 28 km E of Volcán Peteroa (A. A. Ojeda et al. 2005); Neuquén, 1.5 km S of Copahue (MVZ 181568). CHILE: Araucanía, 3.5 km W of Paso Pino Hachado (MSU 7460); Biobío, Aguas Calientes, 5 km E of Termas Chillán (Reise and Venegas 1987); Santiago, La Parva (Pine et al. 1979), Valle de la Junta, Cajón del Río Volcán (Mann 1944).

SUBSPECIES: We treat *Euneomys mordax* as monotypic.

NATURAL HISTORY: Little is known of the natural history of this species, except for limited information on habitat preference. In Malleco province, Chile, individuals were captured in ñirre (*Nothofagus*) grassland habitat at 1,740 m (Greer 1965); in Neuquén province, Argentina, Pearson and Christie (1991:123) collected it in "a clearly defined system of runways and tunnels in deep soil in a wet 'alpine' meadow with 100% ground cover, including low bushes." Specimens of *E. petersoni* have been collected at or near most localities where *E. mordax* has been collected (e.g., Pine et al. 1979; Reise and Gallardo 1990; Pearson and Christie 1991; Pearson 1995; A. A. Ojeda et al. 2005). Pearson and Christie (1991) provided information on habitat separation between these two species and suggested that *E. mordax* preferred deeper soils and *E. petersoni* was more often found on bare, windswept rocky scree. Mella (2006) found *E. mordax* to be very uncommon during a population ecology study at El Monumento Natural "El Mo-

rado," Región Metropolitana de Santiago, Chile. Mann (1944, 1978) provided abundant information about the environments occupied by *E. noei* in the canyon of the Río Maipo, as well as adaptations displayed by this rodent in surviving the hostile high Andean conditions.

REMARKS: Resemblances between *E. fossor* and *E. mordax* have not been highlighted since Thomas (1912b); if both nominal forms belong to the same entity, *fossor* antedates *mordax* and should be replace it. For comments on the location of the type locality of *mordax*, see the detailed study by Pearson and Lagiglia (1992). Although A. A. Ojeda et al. (2005) indicated that the type locality was restricted to "cerca del Volcán Peteroa" by Pearson and Lagiglia (1992), it is unclear as to whether this was a formal redesignation of the type locality because it was not followed by Musser and Carleton (2005). Pine et al. (1979), Yáñez et al. (1987), Reise and Gallardo (1990), Pearson and Christie (1991), and A. A. Ojeda et al. (2005) compared morphology and karyology of *E. mordax* and *E. petersoni*.

The status of *noei* Mann remains problematic. Hershkovitz (1962:500) considered this form (not mentioned in Cabrera 1961) as "doubtfully separable from *Euneomys mordax*." Furthermore, Pine et al. (1979) highlighted the composite nature of the type series of *noei* Mann, which includes a small form similar to *chinchilloides* and a large form similar to *mordax*. According to Pearson and Christie (1991:126), "the type specimens of *Euneomys mordax* Thomas and *E. noei* Mann have grooves positioned

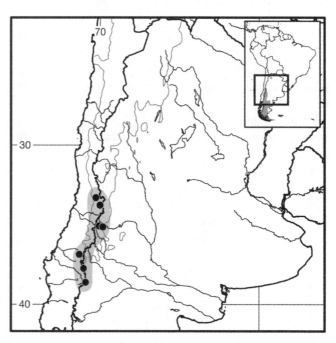

Map 34   Selected localities for *Euneomys mordax* (●). Contour line = 2,000 m.

centrally on the upper incisors and are relatively large . . . There is no evidence that *noei* is specifically distinct from *mordax*."

The karyotype is $2n = 42$, FN $= 66$ (specimens from Paso Pino Hachado, Chile, and Valle Hermoso, Argentina; Reise and Gallardo 1990; A. A. Ojeda et al. 2005).

Euneomys petersoni J. A. Allen, 1903
Peterson's Euneomys
SYNONYMS:

Euneomys petersoni J. A. Allen, 1903b:192; type locality "upper Rio Chico de Santa Cruz, near the Cordilleras, Patagonia," Santa Cruz, Argentina. The holotype was collected by J. B. Hatcher, and our review of Hatcher's (1903) geographical notes, coupled with fieldwork that followed his itinerary, suggests that the holotype was secured near the headwaters of the Río Chico, 4.5 km upstream from Estación El Portezuelo, at about 48.42°S, 71.78°W.

Euneomys dabbenei Thomas, 1919e:127; type locality "Lago Viedma, Santa Cruz, Patagonia," Argentina.

Euneomys dabbbeni Ellerman, 1941:460; incorrect subsequent spelling of *dabbenei* Thomas.

Euneomys [(*Euneomys*)] *petersoni*: Cabrera, 1961:499; name combination.

Euneomys chinchilloides petersoni: Hershkovitz, 1962: 499; name combination.

DESCRIPTION: Smaller than *E. chinchilloides*. Pelage very long and soft, almost woolly, and colored dark gray-brown above variegated with blackish and fulvous streaks; sides much paler and more fulvous, especially along lower border; belly soiled white; ears dark brown on both surfaces and very thinly haired; sides of nose and lower border of cheeks whitish gray, either with faint tinge of yellow or pure white; soles naked except for posterior third; tail one-third or less of total length, well clothed, dusky brown above, with sides and ventral surface white. Skull long and narrow; bullae small and pointed; upper incisors possess lateral grooves.

DISTRIBUTION: *Euneomys petersoni* is widespread in arid lands of southern and central Patagonia, Argentina, from the Atlantic coast to the Andes and from the Straits of Magellan to northwestern Mendoza and southern San Juan provinces, Argentina (Osgood 1943b; Pardiñas, Teta et al. 2003; Pardiñas, Teta, Chebez 2010; Viana et al. 2011). In Chile, recorded localities are restricted to the Andean piedmont from Torres del Paine to Santiago (Osgood 1943b; Pine et al. 1979; Reise and Gallardo 1990). Fossils and subfossils of *E. petersoni* are known from several archaeological sites that range in age from terminal Pleistocene to Recent in southern (e.g., Cueva del Milodón; Simonetti and Rau 1989), western, and central Patagonia (e.g.,

Epullán Grande cave, Traful I cave; Pearson and Pearson 1993; Pardiñas, Teta, D'Elía, and Lessa 2011).

SELECTED LOCALITIES (Map 35): ARGENTINA: Chubut, Cerro El Sombrero (CNP-E 294), Estancia El Guachito (Pardiñas et al. 2000), Estancia Talagapa (CNP-E 204), Pampa de los Guanacos (CNP-E 303), Sierra Apas (Pardiñas, Teta, and Udrizar Sauthier 2008); Mendoza, Laguna de la Niña Encantada (CNP-E 55), RN 40 footbridge over Río Grande (CNP-E 198), Seccional Horcones PP Aconcagua (Pardiñas, Teta, Chebez et al. 2010); Neuquén, Cañadón del Tordillo (Pardiñas, Teta et al. 2003), vicinity of Copahue (Pearson and Christie 1991), 4 km S of Pilolil (Pardiñas, Teta et al. 2003), PN Laguna Blanca, 0.58 km W and 4.20 km N of Co. Mellizo Sud (RDS 18216), Risco Alto, Auca Mahuida (PPA 147); Río Negro, Cerro Corona, Meseta de Somuncura (CNP 2426), 10 km WNW of Comallo (MVZ 165850), Laguna del Molino, Ingeniero Jacobacci (CNP-E 280); San Juan, Pampa larga (Vianna et al. 2011); Santa Cruz, Cerro Fortaleza (CNP-E 500), Cerro Ventana (CNP-E 376), Estancia La Anita (CNP-E 380), Estancia La Dorita, Cañadón Loreley (CNP-E 588), Estancia La María (CNP-E 426), 4 km W of Faro de Cabo Vírgenes (CNP-E 433), Parador Luz Divina, Río La Leona (CNP-E 93), Punta Beagle (CNP-E 434). CHILE: Aysén, Coihaique (Reise and Gallardo 1990), Puerto Ibáñez (Reise and Gallardo 1990); Araucanía, 9.4 km W of Paso Pino Hachado (Greer 1965); Biobío, Chillán (Reise and Gallardo 1990); Magallanes y Antártica Chilena, Cueva del Milodón (Reise and Gallardo 1990), Palli [= Pali] Aike (Yáñez et al. 1987) Sector Lago Toro, Parque Nacional Torres del Paine (Jaksic et al. 1978); O'Higgins, Baños del Flaco (Yáñez et al. 1987); Santiago, El Colorado (LCM 774).

SUBSPECIES: *Euneomys petersoni* is monotypic.

NATURAL HISTORY: *Euneomys petersoni* has been captured on barren, rocky, windswept slopes with little vegetation in Chile (Kelt 1994) and in Neuquén and Río Negro provinces, Argentina (Pearson and Christi 1991). In Malleco province, Chile, *E. petersoni* was captured in ñirre (*Nothofagus*) grassland at 1,555 m (Greer 1965). As noted in the *E. mordax* account, *E. petersoni* has been collected at or near most localities were *E. mordax* has been collected (e.g., Pine et al. 1979; Reise and Gallardo 1990; Pearson and Christie 1991; Pearson 1995; A. A. Ojeda et al. 2005). This species is nocturnal and herbivorous (Kelt 1994; Pearson 1995); Kelt (1994) also reported that the diet includes insects. Individuals collected during the summer (December) in Chile were reproductively active (Kelt 1994); subadults and a female that had been lactating were collected in February in Santa Cruz (J. A. Allen 1903b). Major predators include the Barn Owl (*Tyto alba*) and Burrowing Owl (*Athene cunicularia*) in Patagonia, Argentina (De Santis et al. 1997; Sahores and Trejo 2004), and the Magellanic Horned Owl

(*Bubo magellanicus*) in southern Chile (Reise and Venegas 1974). Blood samples taken from *E. petersoni* in southern Chile contained the protozoan *Trypanosoma cruzi*, which causes Chagas disease in humans (Rozas et al. 2007).

REMARKS: Osgood (1943b) regarded *petersoni* J. A. Allen as a valid species, a view resurrected by Musser and Carleton (2005:1115) based on putative morphometric differences between this taxon and *E. chinchilloides*. Nevertheless, in much of the published literature, this taxon has been included in *E. chinchilloides*. The inclusion of *petersoni*, along with *dabbenei* Thomas, as likely synonyms of *chinchilloides* Waterhouse is supported by the mtDNA cytochrome-*b* analyses of E. P. Lessa et al. (2010), which also delineated an apparently unnamed taxon (if *fossor* Thomas does not apply) of *Euneomys* in northern Patagonia, western Mendoza province, Argentina, and central Chile. The uniqueness of this latter clade is concordant with a different diploid number for animals from Mendoza (see below). Additional research, particularly a morphological restudy of available museum specimens, is needed to resolve the systematic status and relationships of these forms.

A karyotype with $2n = 36$, $FN = 66$ has been reported for specimens from San Carlos de Bariloche, Río Negro, Argentina, and Coyhaique, Chile (Reise and Gallardo 1990). Specimens from Valle Hermoso, Mendoza province, Argentina, had $2n = 34$ (FN not reported; Ojeda et al. 2005; cited as *E. chinchilloides*). Animals from Farellones, Santiago, Chile, had $2n = 36$, $FN = 69$, with an all-biarmed complement, except the smallest autosome pair and the X-chromosome, both of which are acrocentric (Spotorno 1986; reported as *E. noei*, but we refer both specimens, LCM 631 and 774, to *E. petersoni* because they possess characters of this species, most notably a lateral position of incisive grooves).

Genus *Irenomys* Thomas, 1919

Pablo Teta and Ulyses F. J. Pardiñas

The genus *Irenomys* is primarily restricted to cold, forested areas in southern South America (Kelt 1993) and is the only sigmodontine rodent endemic to *Nothofagus* forest (Osgood 1943b). Osgood (1943b), Mann (1978), Kelt (1993, 1994), Steppan (1995), and Pardiñas, Cirignoli et al. (2004) reviewed anatomical characters, distribution, and ecology.

Irenomys is medium-sized, with a tail much longer than the head and body length (about 150%). The pelage is thick and soft. The ears are of medium length and densely haired. The eyes are large. The forefeet and hindfeet are large and broad. The heel is only sparsely furred. The dorsum is grayish cinnamon rufous, streaked with fine dusky lines. The under parts are plumbeous, heavily washed with pinkish cinnamon buff. Ears are brownish black, occasionally with an indistinct subauricular white spot. The forefeet and hindfeet are whitish; toes are white. The tail is unicolored blackish brown, conspicuously clothed with short hairs, and ends in a short (ca. 5 mm) pencil of hairs. The mean and range of external measurements of 37 adults from Chile are: total length, 280 mm (270–326 mm), length of tail, 165 mm (162–196 mm), length of hindfoot, 30 mm (28–32 mm), and length of ear, 22 mm (20–25 mm). The average adult mass of 36 Chilean specimens is 42 g (Kelt 1993, 1994). The adult weight of Argentinean specimens ranged from 30 to 67 g (Pearson 1983). Males and females are of the same weight (Pearson 1983).

The skull is relatively long and narrow. The braincase is long, the interparietal is very large, and the interorbital region is rather parallel-sided but greatly constricted. The coronal suture is V-shaped. The zygomatic plates are high and slightly projected forward. The nasals are rather flat and project beyond the anterior tips of the premaxillae. The incisive foramina are long and wide, extending posteriorly to the protocone of the M1; the posterior end of the palate lies even with the posterior end of the toothrow, defining

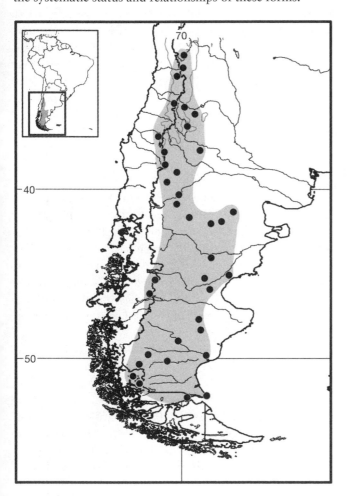

Map 35 Selected localities for *Euneomys petersoni* (●). Contour line = 2,000 m.

a short bony palate. Posterior palatal pits are absent. The parapterygoid fossa is deeply excavated, with its lateral border unusually thickened, flattened, and directly outward. The hamular process is thin along its entire length. The tegmen tympani is present and attached to the squamosal process. A thick alisphenoid strut is present. The circulatory carotid pattern is primitive (*sensu* Voss 1988). The mandible is short and robust with a prominent lower masseteric ridge and an inconspicuous capsular projection. The height of the coronoid process is subequal with the mandibular condyle. The lunar notch is well defined by a long condyloid process (Kelt 1993, 1994; Steppan 1995). Greer (1965) and Kelt (1993, 1994) provided cranial measurements.

The upper incisors are opisthodont, with a sharply defined longitudinal groove on the anterior face (Steppan 1995:Fig. 9D). The molars are hypsodont and laminated, with very deeply dissected surfaces, providing a characteristic prismatic appearance. Hershkovitz (1962:94) described this pattern, unique among sigmodontines, as "partial transverse lamination." Opposite flexi/flexids of all molars meet at midline; there is no posterolophid on m1, nor is there a labial root for either M1/m1; three roots are present on M3, and the molar rows diverge slightly posteriorly (Steppan 1995).

Axial skeletal elements have counts of 13 ribs, 19 thoracolumbar vertebrae, and 34–39 caudal vertebrae. The first rib articulates only with the first thoracic vertebra. The deltoid tuberosity is greater than 59% of the length of the humerus as measured from the condyle (Steppan 1995).

The stomach is unilocular-hemiglandular with the glandular epithelium extending to the corpus (Carleton 1973:13–14). A gall bladder is present (Voss 1991a). The male phallus is complex but without hooks or knobs on the dorsal surface of the lateral mounds (Steppan 1995: Table 4). Preputial glands consist of a single large lateral pair and a medial pair of medium length (Voss and Linzey 1981).

Species of *Oligoryzomys* are of similar color and also have long tails, but they lack grooves on their upper incisors. The eyes of *Irenomys* are larger and are surrounded by dark fur. The tail is thicker and strikingly bicolor. *Phyllotis* have larger ears, proportionally shorter tails, and lack grooves on the upper incisors; their molars are not as highly dissected (Pearson 1995).

SYNONYMS:

Mus: Philippi, 1900:10; part (description of *tarsalis*); not *Mus* Linnaeus, 1758.

Reithrodon: Philippi, 1900:64; part (description of *longicaudatus*), not *Reithrodon* Waterhouse, 1837.

Irenomys Thomas, 1919b:201, type species *Reithrodon longicaudatus* Philippi, 1900, by original designation.

Irenomys tarsalis (Philippi, 1900)

Chilean Tree Rat

SYNONYMS:

Mus tarsalis Philippi, 1900:10, plate 1, Fig. 3; type locality "in provincia Valdivia," stated (p. 11) to have been collected "en el pequeño conservatorio de me jardín en mi fundo San Juan," near La Unión, Los Ríos, Chile.

Reithrodon longicaudatus Philippi, 1900:64, plate XI, Fig. 1; type locality "in litore occidental Patagoniae;" cited as "Melinca, Guaitecas Islands, 44° S, [Los Lagos,] Chile" by Osgood 1943b:219).

Irenomys longicaudatus: Thomas, 1919b:201; generic description and name combination.

Irenomys tarsalis tarsalis: Osgood, 1943b:217; name combination.

Irenomys tarsalis longicaudatus: Osgood, 1943b:219; name combination.

DESCRIPTION: As for the genus.

DISTRIBUTION: *Irenomys tarsalis* is restricted largely to the *Nothofagus* forests of the southern Andes of Argentina and Chile. In Chile, the species is known from forested habitat from Chillán (Ñuble province, VII Region) to at least Puerto Ibáñez (General Carrera province, XI Region) along the Andes and from Nahuelbuta to Isla Grande de Chiloé and the Guaitecas Islands (Osgood 1943b; Kelt 1994) along the coastal range. Although the species is mostly absent from the valleys of south central Chile (Kelt 1993), Saavedra and Simonetti (2000) reported isolated populations in coastal patches of temperate forest and Kelt et al. (2008) captured one individual in a shrubby riparian corridor about 40 km S of and 900 m lower in elevation than the nearest forested region. In Argentina, it is found in *Nothofagus* forests and precordilleran steppes from Neuquén province south to southern Chubut province (Pearson 1983; see Pardiñas, Cirignoli et al. 2004 and G. M. Martin 2010 for records of Argentinean localities). The known fossil record is restricted to latest Pleistocene and Holocene deposits in Río Negro (El Trébol archaeological site; Hajduk et al. 2004) and Neuquén provinces (Cueva Traful I and Estancia Nahuel Huapi; Pearson and Pearson 1993; Rebane 2002), Argentina.

SELECTED LOCALITIES (Map 36): ARGENTINA: Chubut, Lago Fontana (MLP 11.VI.96.10), Lago Futalaufquen (Monjeau et al. 1997), Lago La Plata, west side (Pardiñas, Cirignoli et al. 2004); Neuquén, 3 km NW of Confluencia (MVZ 159421), Pilolil (Pardiñas, Cirignoli et al. 2004); Río Negro, 19 km NNE of El Bolsón (MVZ 152171), Cerro Leones, 15 km ENE of Bariloche (Pearson and Pearson 1982). CHILE: Aysén, 0.25 km W of Puerto Ingeniero Ibáñez (Kelt et al. 2008), Reserva Nacional Coyhaique, 3 km N of Coyhaique (Kelt 1996), Melinka, Islas Guaitecas (Osgood 1943b); Los Lagos, mouth of Río Inio,

Isla Grande de Chiloé (Osgood 1943b), Petrohue, Lago Todos Los Santos (Osgood 1943b); Los Ríos, Fundo San Juan, near La Unión, Valdivia (type locality of *Mus tarsalis* Philippi); Biobío, Aserradero, 7 km below Termas Chillán (Reise and Venegas 1987), Reserva Nacional Los Queules (isolated northern record; Saavedra and Simonetti 2000).

SUBSPECIES: Osgood (1943b) tentatively recognized two subspecies: *tarsalis* Philippi, spread over the mainland range of the species, and *longicaudatus* Philippi, restricted to Isla Melinka, one of the small islands of the Guaiteca group off the southern end of Chiloé Island. However, as noted by Osgood (1943b:200), the series of *longicaudatus* that he examined was composed largely of juvenile specimens, which compromised the taxonomic value of his observations. Thus, the adequacy of these subspecies designations has yet to be tested by either more appropriately aged samples or any modern methodology.

NATURAL HISTORY: *Irenomys tarsalis* is mainly found in dense *Nothofagus* forest with thick stands of *Chusquea* bamboo (Pearson 1983). Population densities in a Coihue (*Nothofagus dombeyi*) forest in southern Argentina varied from 1.4 individuals per hectare in November to 5.1 in May (Pearson and Pearson 1982). The species is nocturnal and scansorial (Pearson 1983; Kelt 1994). Individuals are frequently captured in traps placed along logs, at the base of trees, or in hollows formed by surface roots and boulders covered by lichens and mosses (Kelt 1994). It also has been collected in shrubby habitats in Chile (B. D. Patterson et al. 1990) and in the more arid fringes of cypress trees (*Austrocedrus chilensis*) at the edge of the Argentinean precordillera (Pearson 1983). *Irenomys tarsalis* has been considered a rare or uncommon species (B. D. Patterson et al. 1989, 1990; Saavedra and Simonetti 2000), a view probably resulting from low capture success due to ground placement of traps. It was the second most common mammal observed in the arboreal stratum with camera traps (Amico and Aizen 2000). It climbs well, sometimes jumping a few centimeters from one branch to another, or it may hold on with the hindfeet and reach across a gap with extended front feet (Pearson 1983). This mouse breeds in the spring, although the season may extend to late summer–early autumn (Greer 1965; Pearson 1983). Litter size ranges from three to six (Greer 1965; Pearson 1983; Kelt 1994). The species is primarily granivorous-frugivorous, but also eats green vegetation and fungi (Pearson 1983; Meserve et al. 1988). A captive specimen accepted rolled oats, apple, and tender bamboo shoots but refused carrot, bamboo blossoms, bread, and grubs (Pearson 1983).

Skeletal remains of *Irenomys* have been found in owl pellets of the Magellanic Horned Owl (*Bubo magellanicus*; Massoia, Chebez, and Heinonen Fortabat 1991a), the Rufous-legged Owl (*Strix rufipes*; D. R. Martínez 1993), and the Barn Owl (*Tyto alba*; Trejo and Lambertucci, 2007). These are docile rodents that may be removed from livetraps by hand (Pearson 1983). When released from captivity, they climbed bamboos or trees, scampered away across the forest floor, or disappeared down a burrow (Pearson 1983).

REMARKS: *Mus mochae* Philippi, 1900, included by Gyldenstolpe (1932) in the genus *Irenomys*, is, according to Osgood (1943b), a composite of an *Irenomys tarsalis* skull and an *Abrothrix olivacea* skin. Osgood (1943b:219) noted that the type of Philippi's *tarsalis* has apparently been lost, and that no other specimen of this species has been found subsequently at the type locality or in the region surrounding it. He also stated (1943b:220) that the skull of the type of *longicaudatus* Philippi had been lost, although the skin still remained as a mount.

Irenomys tarsalis has $2n = 64$ chromosomes (FN = 98) and a C-banding pattern characterized by small amounts of centromeric heterochromatin. The autosomal complement consists of five pairs of metacentrics (two large, two medium-sized, and one small), 13 pairs of subtelocentrics (medium- to small-sized), and 13 pairs of medium-sized acrocentrics. The X chromosome is the largest acrocentric

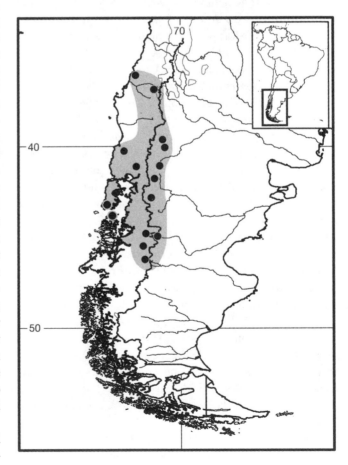

Map 36 Selected localities for *Irenomys tarsalis* (●). Contour line = 2,000 m.

element in the karyotype; the Y chromosome is a medium-sized metacentric (A. A. Ojeda et al. 2004).

Genus *Juliomys* E. M. González, 2000

Enrique M. González, João A. de Oliveira, and Pablo Teta

The genus *Juliomys* includes at least three living and one extinct species of mice endemic to the Atlantic Forest. E. M. González (2000) erected the genus for an odd species of uncertain affinities that was described by Osgood (1933d) as *Thomasomys pictipes*. Prior to the description of *Juliomys*, Musser and Carleton (1993) moved *pictipes* to *Wilfredomys*, mostly based on overall impressions concerning its closeness to *W. oenax* made by Osgood (1933d) and Pine (1980). The other two species were described recently (J. A. Oliveira and Bonvicino 2002; L. P. Costa et al. 2007), but even more recent analyses suggest a greater diversity in this genus than now recognized (Paresque et al. 2009; see also Pavan and Leite 2011). In addition, Cherem et al. (2004) and Cherem (2005) recognized, as *Juliomys* sp., several individuals from coastal localities in the Brazilian state of Santa Catarina, as did D. O. Lima et al. (2010) for ones from the southernmost locality for the genus at the boundary of the Atlantic Forest in Rio Grande do Sul state. It is unclear at present whether these southern Brazilian localities all hold the same entity or different ones (Pavan and Leite 2011). Finally, Pardiñas and Teta (2011a) recorded an apparently extinct species of *Juliomys*, assigning to this genus a Lagoa Santa fossil described by Winge (1887) as *Calomys anoblepas*.

Analyses of the karyotypes of *Juliomys* indicate that centric and tandem fusions plus pericentric inversions have been the most frequent rearrangements involved in the karyotypic differentiation in the genus (Paresque et al. 2009). Phylogenetic analysis using nucDNA IRBP sequences placed *Juliomys* as sister to a clade formed by *Euneomys* and *Irenomys* (D'Elía et al. 2006). Similar analysis based on the mtDNA cytochrome-*b* sequences, or on concatenated nucDNA IRBP and cytochrome-*b* sequences, failed to associate *Juliomys* with any other sigmodontine, with reasonable levels of support (M. F. Smith and Patton 1999; D'Elía 2003a). *Juliomys*, once considered a member of Thomasomyini, is currently regarded as Sigmodontinae *incertae sedis* (D'Elía et al. 2007). Divergence in cytochrome-*b* sequences among extant species is extensive, ranging from 10–14% (L. P. Costa et al. 2007; Pardiñas, Teta et al. 2008; de la Sancha et al. 2009; Pavan and Leite 2011).

Species of *Juliomys* are small sigmodontines, with a tail longer than the head and body. Despite notable differences in the intensity of color, particularly on the muzzle, and in the length of pelage among the three species, all of them have an ochraceous dorsum, orangish top of the head and on the rump, and dusky anterior back and shoulders. The flanks are paler than the dorsum, gradually transitioning toward the venter, which varies between white and pale ochraceous in the different species, contrasting in some regional samples with dark patches resulting from the slaty hair bases. The tail varies from slightly to markedly bicolored in the different species, with the terminal part uniformly dusky. Dorsal surfaces of forefeet and hindfeet are covered with light ochraceous hairs intermixed with a few dark hairs, resulting in a general color varying from pale orange to orange-brown; the toes are covered with whitish hairs, and ungual tufts of white hairs are intermixed with a few dark hairs, giving a general color that varies from self-colored, yellow-white to gray-based. Plantar surfaces of the hindfoot have six plantar pads (thenar, hypothenar, and four interdigital). Digit I of the hindfoot is approximately half the length of digit V. Mammary count varies from eight (one pectoral, one postaxial, and two inguinal pairs), in *J. ossitenuis*, to 10 (one pair each in pectoral, postaxial, thoracic, abdominal, and inguinal positions), in *J. pictipes*. The stomach is unilocular-hemiglandular (L. P. Costa et al. 2007).

The skull is delicate, with a short rostrum (<35% occipitonasal length) and rounded braincase. The nasals become wider anteriorly. The interorbital region is narrow with somewhat squared but not ridged supraorbital edges. The anterior portion of the interfrontal suture is incompletely fused, with a slit-like fontanelle sometimes present. The coronal suture is V-shaped, the interparietal is wide, and the occipital region is reduced. The zygomatic plates are narrow, with their anterior borders nearly vertical; the zygomatic notches are shallow. The palate is short and broad; the mesopterygoid fossa is lyre-shaped and extends anteriorly between the posterior third of the last molars, and is narrower than the parapterygoid fossae. The auditory bullae are small, with the periotic bone externally apparent in basicranial view and forming part of the carotid canal. The mandibles display a curved lateral profile and a well-developed capsular process. The incisors are robust, orthodont to opisthodont. The molars are brachydont with a crested, coronal topography and well-developed flexus/ids and mesolophs/ids; an ectolophid is usually present on m1 and m2.

Species of *Juliomys* are forest dwellers, presumably arboreal mice, although some of the few specimens available were trapped on the ground. Individuals are rather uncommonly collected using standard terrestrial live or kill traps, but may be the most common rodent of the understory or the sole taxon captured in the canopy when pitfall or canopy traps are used (Graipel 2003).

SYNONYMS:

Calomys: Winge, 1887:44; part (description of *anoblepas*); not *Calomys* Waterhouse, 1837.

Oryzomys: E.-L. Trouessart 1897:527; part (listing of *anoblepas*); not *Oryzomys* Baird, 1857.

Thomasomys: Osgood, 1933d:11; part (description of *pictipes*); not *Thomasomys* Coues, 1884.

Oligoryzomys: Massoia, Chebez, and Heinonen-Fortabat, 1991b:17; part (regarded *pictipes* as synonym of *Oligoryzomys flavescens antoniae* Massoia); not *Oligoryzomys* Bangs, 1900.

Wilfredomys: Musser and Carleton, 1993:752; part (listing of *pictipes*); not *Wilfredomys* Avila-Pires.

Juliomys E. M. González, 2000:3; type species *Thomasomys pictipes* Osgood, by monotypy.

KEY TO THE SPECIES OF *JULIOMYS*:

1. Size relatively large; zygomatic notch deep; interorbital region broad, convergent anteriorly; tympanic bullae small; squamosal-alisphenoid groove absent; sphenopalatine vacuities absent or minute; lateral extension of parietals small; posterior projection of condyloid process short; $2n = 36$. . . . .*Juliomys pictipes*

1'. Size relatively small; zygomatic notch shallow; interorbital region narrow, hourglass in shape; tympanic bullae large; squamosal-alisphenoid groove present; sphenopalatine vacuities large to very large; lateral extension of parietals broad; posterior projection of condyloid process long; $2n = 20$. 2

2. Nose brown; dorsal pelage long; ventral pelage pale ochraceous; tail markedly bicolored; interfrontal fontanelle present; posterolateral palatine pits large; ectolophid/ectostylid of m1–m2 small; NF = 34 . . . .*Juliomys rimofrons*

2'. Nose orange; dorsal pelage short; ventral pelage cream-white; tail slightly bicolored; interfrontal fontanelle usually absent; posterolateral palatine pits small; ectolophid/ectostylid of m1–m2 conspicuous; NF = 36 . . . .*Juliomys ossitenuis*

Juliomys ossitenuis L. P. Costa, Pavan, Leite, and Fagundes, 2007

Delicate Red-nosed Tree Mouse

SYNONYM:

Juliomys ossitenuis L. P. Costa, Pavan, Leite, and Fagundes, 2007; type locality "Fazenda Neblina, Parque Estadual da Serra do Brigadeiro, 20 km W Fervedouro, Minas Gerais, Brazil, 20°43′S 42°29′W, elevation 1300 m."

DESCRIPTION: Small, with tail (89–116 mm) longer than head and body (82–97 mm); fur soft and short; dorsal coloration dark orange brown above, venter contrastingly cream white; nose, rump, and inner pinna orange brown; tail faintly bicolored, terminating in short tuft; hands and feet short and broad, their dorsal surfaces covered with pale orange hairs. Skull slender and delicate (occipitonasal length 21.9–26.8 mm); rostrum short and narrow, less than one-third skull length; zygomatic notches very shallow; interorbital region narrow and hourglass in shape, with margins only slightly angled; posterolateral palatine pits small; tympanic bullae large; squamosal-alisphenoid groove present; sphenopalatine vacuities very large; lateral extension of parietals broad; condyloid process of mandible long; m1–m2 with conspicuous ectolophid/ectostylid (J. A. Oliveira and Bonvicino 2002; L. P. Costa et al. 2007).

DISTRIBUTION: *Juliomys ossitenuis* is endemic to the Atlantic rainforest and semideciduous forests of southeastern Brazil in the states of São Paulo, Minas Gerais, and Espírito Santo (L. P. Costa et al. 2007; Pavan and Leite 2011).

SELECTED LOCALITIES (Map 37; from L. P. Costa et al. 2007, except as noted): BRAZIL: Espírito Santo, Casa Queimada, Parque Nacional do Caparaó, Parque Estadual do Forno Grande (Pavan and Leite 2011); Minas Gerais, Fazenda do Itaguaré, 16 km SW of Passa Quatro, Fazenda Neblina, Parque Estadual da Serra do Brigadeiro, 20 km W of Fervedouro (type locality of *Juliomys ossitenuis* L. P. Costa, Pavan, Leite, and Fagundes); São Paulo, Cristo (Pavan and Leite 2011), Estação Ecológica do Bananal, Mulheres, Museros, Quilombo, Reserva Florestal do Morro Grande.

SUBSPECIES: *Juliomys ossitenuis* is monotypic.

NATURAL HISTORY: This species inhabits Atlantic rainforest and semideciduous forests above 800 m, including cloud forests at higher elevations. Apparently, populations from northern latitudes prefer higher elevations (L. P. Costa et al. 2007). Most of the known specimens are from protected areas of mature, continuous forests. The holotype and three paratypes were collected with Sherman traps placed on tree branches or vines, but other specimens were obtained with pitfall traps (L. P. Costa et al. 2007). The only reproductive data available correspond to a lactating female collected in May at the beginning of the dry season (L. P. Costa et al. 2007). This species was found in syntopy with *J. pictipes* at two localities (Mulheres and Museros) in São Paulo state (L. P. Costa et al. 2007).

REMARKS: The karyotype is $2n = 20$, FN = 36 (L. P. Costa et al. 2007). This chromosome complement differs from that of *J. pictipes* by at least eight Robertsonian fusions/fissions plus one pericentric inversion. Although the $2n = 20$ karyotypes of *J. ossitenuis* and *J. rimofrons* appear very similar, several chromosomes are clearly different in size and morphology, which suggests that several complex rearrangements separate their autosomal complements.

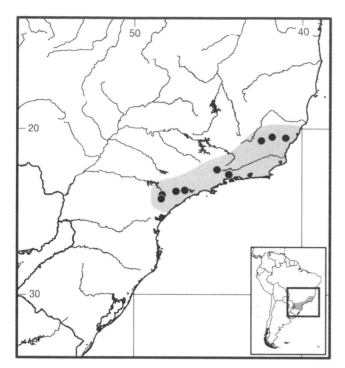

Map 37 Selected localities for *Juliomys ossitenuis* (●).

The Y chromosomes are also very distinctive. Molecular phylogenetic analyses based on the mtDNA cytochrome-*b* gene sequences confirmed that *J. ossitenuis* is a well-supported monophyletic clade, highly divergent from the other two species in the genus (L. P. Costa et al. 2007; Pardiñas, Teta et al. 2008).

Juliomys pictipes (Osgood, 1933)
Small Red-nosed Tree Mouse

SYNONYMS:

Thomasomys pictipes Osgood, 1933d:11; type locality "Caraguatay, Rio Parana, 100 miles south of Rio Iguassu, Misiones, Argentina," restricted by Pardiñas et al. (2007) to Puerto Caraguatay (26.616°S, 54.766°W), Misiones, Argentina, although the holotype was probably obtained at the nearby locality Puerto Benson (Pardiñas, Teta et al. 2008).

Wilfredomys pictipes: Musser and Carleton, 1993:752; name combination.

Juliomys pictipes: E. M. Gonzálcz, 2000:3; first use of current name combination.

DESCRIPTION: Largest species in genus, with tail (82–95 mm) slightly shorter than head-body length (90–110 mm); body color pale orange-brown above and white to cream-white below; tail markedly bicolored (except for dusky terminal half-inch); hindfeet clear ochraceous-tawny above, with whitish toes. Interorbital region of skull broad, with anteriorly convergent borders; lateral

expansion of frontal bones restricted; zygomatic notches deeply to moderately expressed; upper free border of zygomatic plates reduced; incisive foramina short, not reaching anterior plane of M1s; tympanic bullae relatively small; squamosal-alisphenoid grove and sphenofrontal foramen absent; sphenopalatine vacuities absent or minute; condyloid process of mandible short (L. P. Costa et al. 2007; Pardiñas, Teta et al. 2008).

DISTRIBUTION: *Juliomys pictipes* is endemic to the Atlantic rainforest of southeastern Brazil, eastern Paraguay, and northeastern Argentina. Pavan and Leite (2011) suggested that the distribution is disjunct, with those populations in southern Brazil, eastern Argentina, and eastern Paraguay separated from the remainder of the range by a broad expanse of *Araucaria* forest.

SELECTED LOCALITIES (Map 38): ARGENTINA: Misiones, Arroyo de Salamanca, Parque Provincial "Ernesto Che Guevara" (Pardiñas, Teta et al. 2008), Balneario de la Reserva Privada de Usos Múltiples de la Universidad Nacional de La Plata "Valle del Arroyo Cuña Pirú" (Pardiñas, Teta et al. 2008), Parque Nacional Iguazú, Sendero Macuco (Pardiñas, Teta et al. 2008). BRAZIL: Espírito Santo, Alto Alegre, Reserva Biológica Duas Bocas (Pavan and Leite 2011); Minas Gerais, Estação de Pesquisa e Desenvolvimiento Ambiental de Peti (L. P. Costa et al. 2007), Reserva Particular do Patrimônio Natural do Caraça, 25 km SW of Santa Bárbara (Pavan and Leite 2011); Paraná, Fazenda Monte Alegre (Pavan and Leite 2011); Rio de Janeiro, Fazenda Boa Fé (L. P. Costa et al. 2007), Mata do Mamede (L. P. Costa et al. 2007); Rio Grande do Sul, Parque Estadual do Turvo (Pavan and Leite 2011); Santa Catarina, Barragem do Rio São Bento (Graipel 2003), Santo Amaro da Imperatriz (Graipel 2003); São Paulo, Fazenda Intervales (Pavan and Leite 2011), Floresta Nacional de Ipanema, 20 km NW Sorocaba (L. P. Costa et al. 2007), Riacho Grande (L. P. Costa et al. 2007), São Luís do Paraitinga (Pavan and Leite 2011), Mulheres (Pavan and Leite 2011). PARAGUAY: Alto Paraná, Refugio Biológico Limoy, north of Río Limoy (de la Sancha et al. 2009).

SUBSPECIES: *Juliomys pictipes* is monotypic.

NATURAL HISTORY: This is a primarily arboreal rodent that typically inhabits the lower to middle strata in mature and secondary Atlantic rainforest (E. M. Vieira and Monteiro-Filho 2003; Pardini and Umetsu 2006). Three of four Argentinean specimens were captured on the ground near watercourses in areas of dense underbrush of bamboos (Pardiñas, Teta et al. 2008). One individual in Paraguay was trapped with a Sherman live-trap 1.5 m above the ground in primary forest with ferns and low lianas (de la Sancha et al. 2009). Pine (1980) reported young animals caught by hand in nests situated in bamboo and on a bromeliad branch, 5–8 m above the ground. This rodent has been more frequently caught with pitfall traps than with Sherman traps, in forested areas of southeastern Brazil (Pardini and Umetsu 2006; Umetsu et al. 2006). Diets of wild-caught individuals include fruit pulp and small seeds (<10 mm diameter; E. M. Vieira and Monteiro-Filho 2003). One pregnant female was caught in December in Paraguay (de la Sancha et al. 2009) and another in August in Argentina (Pardiñas, Teta et al. 2008); both contained three embryos. Ectoparasites found on specimens from Brazil include the mites *Androlaelaps fahrenholzi*, *Eubrachylaelaps rotundus*, *Gigantolaelaps gilmorei*, *G. oudemansi*, *G. wolffsohni*, *Laelaps castroi*, *L. navasi*, *L. paulistanensis*, *L. thori*, *Mysolaelaps parvispinosus*, the tick *Ixodes loricatus*, and the fleas *Craneopsylla minerva*, *Polygenis (Neopolygenis) atopus*, *P. (N.) pradoi*, and *P. (P.) roberti* (Nieri-Bastos et al. 2004).

REMARKS: Massoia, Chebez, and Heinonen Fortabat (1991b) regarded *"Thomasomys" pictipes* as a synonym of *Oligoryzomys flavescens antoniae* Massoia, which was clearly in error. The observed molecular differentiation among the mtDNA cytochrome-*b* haplotypes of Argentinean, Brazilian, and Paraguayan specimens of *J. pictipes* is moderate, ranging from 0.13% to 2.3%. The pattern of divergence does not conform to an isolation-by-distance pattern, so the variation is not a simple function of geography (Pardiñas, Teta et al. 2008; de la Sancha et al. 2009). The karyotype of *J. pictipes* is 2n = 36, FN = 34, based on specimens from Fazenda Intervales, São Paulo, Brazil (Bonvicino and Otazu 1999).

Juliomys rimofrons J. A. Oliveira and Bonvicino, 2002
Montane Red-rumped Tree Mouse

SYNONYM:

Juliomys rimofrons J. A. Oliveira and Bonvicino, 2002:310; type locality "Brejo da Lapa (22°21′S, 44°44′W, altitude approx. 2000 m), situated to the east of Pedra Furada (2589 m) in the Serra da Mantiqueira, municipality of Itamonte, state of Minas Gerais, Brazil."

DESCRIPTION: Size intermediate among three species, with head and body length 85–93 mm, tail length 99–121 mm; fur long and soft, brownish ochraceous above, dark orangish on rump, pale brown on venter, but with slaty hair base often apparent; dark eye-ring present; hairs on neck and chin whitish with gray bases, and those around mouth totally white; tail bicolored, dark above and whitish below, with distinct longitudinal line of blackish hairs extending midventrally and entire ventral surface becoming darker toward tip, terminating in an inconspicuous pencil of dark and whitish hairs; tail also rather hirsute, covered with dark hairs up to three scales in length. Skull delicate, with a short rostrum (30–34% of occipitonasal length); braincase round and inflated; interorbital region narrow, with gently squared or angled but not ridged supraorbital edges; narrow zygomatic plates, with anterior margins straight; zygomatic notches shallow; incisive foramina long, reaching anterior part of M1s; squamosal-alisphenoid groove present; tympanic bullae larger than in other species, but not obstructing view of periotic in basicranial aspect; single pair of posterolateral palatal pits; sphenopalatine vacuities large to very large; condyloid process of mandible long; ectolophid/ectostylid of m1–m2 reduced. Persistent fontanelle along middle third of frontal suture (basis for specific epithet) present in all known specimens, and in six *J. ossitenuis* from one locality, but absent from *J. pictipes* (J. A. Oliveira and Bonvicino 2002; L. P. Costa et al. 2007).

DISTRIBUTION: This species is known from the type locality in Minas Gerais state and two closely adjacent localities to the immediate south in São Paulo and Rio de Janeiro states, Brazil.

SELECTED LOCALITIES (Map 39): BRAZIL: Minas Gerais, Itamonte, Brejo da Lapa (type locality of *Juliomys rimofrons* J. A. Oliveira and Bonvicino); Rio de Janeiro, Paraty (R. Fonseca et al. 2013); São Paulo, São José do Barreiro (R. Fonseca et al. 2013).

SUBSPECIES: *Juliomys rimofrons* is monotypic.

NATURAL HISTORY: Little is known about the ecology or other aspects of the natural history of this species. All known specimens were collected on the ground, in a trapline extending for nearly 100 m through vegetation varying

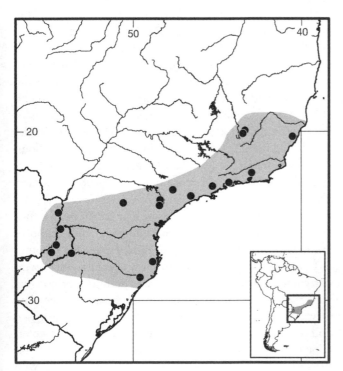

Map 38 Selected localities for *Juliomys pictipes* (●).

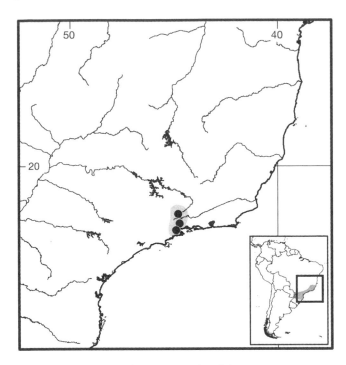

Map 39 Selected localities for *Juliomys rimofrons* (●).

from *Araucaria* forest to patches of grasses and bromeliads (J. A. Oliveira and Bonvicino 2002).

REMARKS: The karyotype of *J. rimofrons* has a reduced diploid number ($2n=20$) relative to specimens of *J. pictipes* ($2n=34$). However, the same number of autosomal arms ($FN=34$) in these two species suggests Robertsonian changes as the mechanism of generating karyotypic divergence (J. A. Oliveira and Bonvicino 2002).

Genus *Neomicroxus* Alvarado-Serrano and D'Elía, 2013

Diego F. Alvarado-Serrano and Guillermo D'Elía

Neomicroxus is a recently established genus for two species, *bogotensis* (Thomas) and *latebricola* (Anthony), both previously placed within *Akodon* although their distinctiveness had been recognized for many years (Reig 1987; J. Ventura et al. 2000; Voss 2003). The genus, which is currently of uncertain position in the sigmodontine radiation, is restricted to the northern Andes, from central Ecuador north to the Cordillera de Mérida of Venezuela.

The taxonomic history of species allocated to *Neomicroxus* has been unstable. Thomas (1895c) described *bogotensis* under *Akodon*, but he later grouped it together with *Akodon mimus* under a "small *Akodon*-like *Oxymycteri* group" (Thomas 1901b:183). Thomas (1909) subsequently formalized this association when he erected *Microxus* (currently a synonym of *Akodon*), placing *bo-*

gotensis in his new genus because it and *mimus* (the type species of *Microxus*) share uniquely narrow and posteriorly slanting zygomatic plates (see also Thomas 1920h). Later, Anthony (1924b) described *latebricola*, also under *Microxus*, based on its strong resemblance to *bogotensis*, although he did not compare it to *mimus*.

The validity of *Microxus* either as a genus or subgenus of *Akodon* has been debated. Gyldenstolpe (1932a) maintained *Microxus* as a distinct genus, placing both *bogotensis* and *latebricola* within it. Tate (1932h) also maintained *Microxus* as a genus but suggested that it was part of an *Akodon* complex and that *bogotensis* was undoubtedly an akodont. Furthermore, Tate (1932h:27) stated that the slope of the zygomatic plates, the principal character used to group species under *Microxus*, was "less important than it appeared to be a first light." Following Tate's (1932h) suggestion, Cabrera (1961), Hooper and Musser (1964), and Hinojosa et al. (1987) placed *Microxus* as a subgenus in *Akodon*; alternatively, Hershkovitz (1966a) placed it in *Abrothrix*. Reig (1987) revalidated *Microxus* as a genus, but his view was restricted to *bogotensis* and not *mimus*, the type species (Patton et al. 1989). Phylogenetic analyses of allozymes (Patton et al. 1989) and both mtDNA (M. F. Smith and Patton 1999) and nucDNA sequences (D'Elía 2003a) have consistently shown that *mimus* falls in *Akodon*, and as a result *Microxus* is currently regarded as a simple subjective synonym of *Akodon* (see that account). Furthermore, the several species that have been historically associated with *Microxus* by various authors are now allocated to other genera, including *Akodon* (*affinis*, *mimus*, and *torques*; D'Elía 2003a; M. F. Smith and Patton 2007), *Brucepattersonius* (*iheringi*; Hershkovitz 1998), and *Abrothrix* (*lanosa*; Feijoo et al. 2010). Until very recently, however, the position of both *bogotensis* and *latebricola* had remained in limbo. Alvarado-Serrano and D'Elía (2013) have now documented the uniqueness of this pair of taxa as well as their distinction from *mimus* and all other *Akodon* species, based on combined morphological, morphometric, and molecular analyses. These authors erected *Neomicroxus* to encompass *bogotensis* and *latebricola*, selecting the later as the type of their new genus.

Both species of *Neomicroxus* are small, delicate mice (total length about 90–100 mm), with a highly similar bauplan and that share a soft dorsal pelage, grizzled to dark chestnut or blackish brown in color, with a slightly paler venter without countershading. The pinnae are dark and project beyond the fur of the head but do not reach the eyes when laid forward. Mystacial vibrissae are proportionally short and do not reach the pinnae when laid back. The overall coloration of the pes, manus, and tail is dark brown. The plantar surfaces of the hindfeet are smooth and have six pads, with the hypothenar separated from the thenar by a distinct gap.

The skull is delicate, with an inflated braincase and a proportionally long, tapering rostrum. Zygomatic notches are indistinct. The interorbital region is hourglass in shape, with smoothly rounded supraorbital edges. The zygomatic arches are parallel and have a distinct jugal that separates the maxillary and squamosal zygomatic processes in lateral view. The zygomatic plates are narrow and slanted posteriorly from their bases, surrounding a wedge-shaped infraorbital foramen; the anterior root of the zygoma forms a very low anterorbital bridge. The anterior margin of the superficial masseter is evident in both species (as a scar in the skull of *latebricola* and as an indistinct tubercle in *bogotensis*). The nasals extend posteriorly past the suture between the frontals, lacrimals, and maxillaries. The incisive foramina are relatively short, extending posteriorly just to the anterior margins of the M1s. The palate is long, with very shallow palatal troughs and reduced or no posterolateral pits. The anterior border of the mesopterygoid fossa is broad, U-shaped, and does not reach the plane formed by the posterior margins of the M3s; the roof of the fossa is perforated by sphenopalatine vacuities. An alisphenoid strut is absent. The carotid circulation pattern (type 1; Voss 1988) consists of a prominent trough for the masticatory-buccinator nerve, a large stapedial foramen, a posterior opening of the alisphenoid canal present but not enlarged, and a prominent anteroposterior squamoso-alisphenoid groove. The subsquamosal fenestra and postglenoid foramen are open, with a significant dorsoventral overlap in *latebricola* but not in *bogotensis*. Auditory bullae are inflated, the stapedial process is conspicuous, and the mastoid fenestra is usually present. The mandible is well developed, with falciform coronoid processes that are subequal in height to the mandibular condyles.

The upper incisors are nearly orthodont, ungrooved on their anterior faces, and with a straight dentine fissure. The molars are hypsodont (more so in *latebricola*); their main cusps are opposite, rounded, and subequal in size. M1s have a well-developed anteromedial flexus that divides the procingulum into anterolabial and anterolingual conules of similar size. An anteroloph (more developed in *bogotensis*) is evident only in barely worn teeth. Both M1 and M2 lack a mesostyle and enterostyle; a mesoloph is visible in *bogotensis* but inconspicuous in *latebricola*; a posteroloph is also present, with a well-developed posteroflexus. M3 is reduced and cylindrical in occlusal, with a rounded interior island of enamel. An anteromedian flexid is present on m1, but only in little worn teeth; no anterolabial cingulum is evident. A mesolophid is visible in m1 and m2 of *bogotensis* but is inconspicuous in *latebricola*; a posterlophid is present in both species. A mesostylid is absent in all lower molars.

Male reproductive accessory glands include a single pair of ventral prostates (for *bogotensis*), which contrasts with the double condition found in many akodontines (Voss and Linzey 1981).

Neomicroxus is endemic to the northern Andes, ranging from central Ecuador through the Colombian Andes to the Cordillera de Mérida in western Venezuela, at elevations above 2,400 m and reaching as high as 3,900 m.

Knowledge of the conservation status of *Neomicroxus* is limited. Currently, *bogotensis* is listed as a species of least concern (Gómez-Laverde and Rivas 2008), and *latebricola* (represented by its holotype until recently [Voss 2003]) is listed as vulnerable (Boada, Gómez-Laverde, and Anderson 2008). However, trapping efforts targeting *latebricola* suggest that the species is more common than previously thought (D. F. Alvarado-Serrano, unpubl. data).

SYNONYMS:

Acodon: Thomas, 1895c:369; part (description of *bogotensis*); not *Akodon* Meyen.

Microxus: Thomas, 1909:237; part (listing of *bogotensis*); but not *Microxus* Thomas.

Neomicroxus Alvarado-Serrano and D'Elía, 2013:1008; type species *Microxus latebricola* Anthony, by original designation.

REMARKS: Although the generic allocation of *bogotensis* and *latebricola* relative to *Akodon*, especially to *A. mimus*, were questioned previously (Patton et al. 1989; Voss 2003), no author had doubted their position in the Akodontini. However, the molecular analyses of Alvarado-Serrano and D'Elía (2013), which included samples of *latebricola*, unambiguously placed this species outside of the Akodontini and underscored *Neomicroxus* as an unexpectedly non-akodontine genus substantially convergent on an akodont morphology, a situation similar to that of *Abrothrix* (tribe Abrotrichini; see that account). Pending future phylogenetic resolution, we place *Neomicroxus* as *incertae sedis* in the Oryzomyalia (*sensu* Steppan, Adkins, and Anderson 2004), as it does not appear closely related to any other lineage.

KEY TO THE SPECIES OF *NEOMICROXUS*:

1. Origin of superficial masseter from indistinct tubercle; visible mesoloph; inflated frontal sinuses . . . .
. . . .*Neomicroxus bogotensis*
1'. Origin of superficial masseter from scar; indistinguishable mesoloph; flatter frontal sinuses . . . .
. . . .*Neomicroxus latebricola*

Neomicroxus bogotensis (Thomas, 1895)
Bogotá Grass Mouse, Bogotá Akodont
SYNONYMS:

Acodon bogotensis Thomas 1895c:369; type locality: "Plains of Bogota," Cundinamarca, Colombia.

[*Akodon* (*Akodon*)] *bogotensis*: E.-L. Trouessart, 1897:537; name combination.

M[*icroxus*]. *bogotensis*: Thomas, 1909:237; name combination.

N[*eomicroxus*]. *bogotensis*: Alvarado-Serrano and D'Elía, 2013:1008; first use of current name combination.

DESCRIPTION: Small, head and body length 78–90 mm, tail length 62–75 mm, ear length 13–20 mm, hindfoot length (without claw) 19–21 mm, greatest skull length 21.1–24.1 mm, and upper molar series length 3.4–3.8 mm. Dorsal fur grizzled to blackish brown, slightly more grayish underneath. Forefeet and hindfeet dark brown above, with short, grayish ungual tufts not surpassing claws. Tail short, about 60–70% of head and body, and indistinctly bicolored (dark above, paler below by presence of brownish and silvery hairs). Mystacial vibrissae short, not extending beyond pinna when laid back. Skull delicate, with inflated braincase and long, tapering rostrum; interorbital region broad (about width of rostrum across nasolacrimal capsules), with narrowest point located midback; supraorbital edges smooth; zygomatic notches very shallow; masseteric tubercles indistinct; incisive foramina short and proportionately wide, extending posteriorly to anterior margins of M1s; mesopterygoid fossa broad, its roof with large sphenopalatine vacuities; auditory bullae inflated with short bony tubes. Details of teeth as for genus.

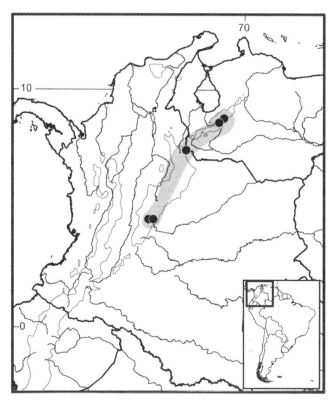

Map 40 Selected localities for *Neomicroxus bogotensis* (●). Contour line = 2,000 m.

DISTRIBUTION: This species is endemic to the northern Andes, in the Cordillera Oriental of Colombia and Cordillera de Mérida and Páramo de Tamá in Venezuela. Soriano et al. (1999) listed this species for Trujillo state, Venezuela, but without documentation. No fossils are known.

SELECTED LOCALITIES (Map 40): COLOMBIA: Cundinamarca, Choachí (AMNH 61567), El Verjón (AMNH 62754). VENEZUELA: Mérida, Mucubají (M. A. Barros and Reig 1979), 8 km SE of Tabay, near La Coromoto (USNM 374614); Táchira, Buena Vista, 41 km SW of San Cristobal, near Páramo de Tamá (USNM 442345).

SUBSPECIES: *Neomicroxus bogotensis* is monotypic.

NATURAL HISTORY: This species is apparently rare, omnivorous, and nocturnal, inhabiting high-elevation evergreen montane forests and shrubby upland meadows. It is most commonly found in *Polylepis* forests or *Espeletia* stands, and is often associated with moss-covered rocks (O. J. Linares 1998; J. Ventura et al. 2000).

REMARKS: M. A. Barros and Reig (1979) described a karyotype from specimens collected at Mucubají, Mérida state, Venezuela, with $2n = 35$–37 and FN = 48.

Neomicroxus latebricola (Anthony, 1924)
Ecuadorean Grass Mouse, Ecuadorean Akodont
SYNONYMS:

Microxus latebricola Anthony 1924b:3; type locality "Hacienda San Francisco, east of Ambato, on Rio Cusutagua, elevation about 8000 feet, Ecuador."

Akodon latebricola: Musser and Carleton, 1993:690; name combination.

N[*eomicroxus*]. *latebricola*: Alvarado-Serrano and D'Elía, 2013:1008; first use of current name combination.

DESCRIPTION: Small, with head and body length 74–103 mm, tail length from 63–98 mm, hindfoot length (without claw) 19–23 mm, ear length 14–17 mm, greatest skull length 22.88–25.63 mm, and upper molar series length 3.32–3.74 mm. Dorsal pelage grizzled dark chestnut brown, somewhat paler underneath, and without visible lateral line separating dorsal and ventral color; dorsal surfaces of both fore- and hindfeet dark brown, covered by gray hairs, and with relatively reduced light gray ungual tufts covering, but not surpassing, claws. Tail about same length as head and body, slightly bicolored, and covered with dark hairs above and silvery hairs below. Vibrissae short, not extending beyond pinna when laid back. Skull delicate with long, thin, tapering rostrum and inflated braincase; interorbital region hourglass in shape, about same width as rostrum at nasolacrimal level, with smooth supraorbital edges; origin of superficial masseter from scar on anterior margin of zygomatic plates; zygomatic notches absent to barely perceptible; alisphenoid struts absent; incisive foramina short and proportionately wide, barely extending posteriorly to anterior

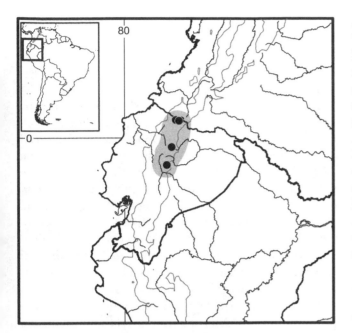

Map 41 Selected localities for *Neomicroxus latebricola* (●). Contour line = 2,000 m.

margins of M1s; auditory bullae inflated; masseteric tubercles of mandible indistinct. See generic description for molar details.

DISTRIBUTION: This species is endemic to high elevations of the eastern Andes in Ecuador, from Tungurahua province in the middle of the country (C. Boada, pers. comm.) to Carchi province near the frontier with Colombia. It is not known from the fossil record.

SELECTED LOCALITIES (Map 41): ECUADOR: Carchi, La Buitrera ravine, Reserva Ecológica el Ángel (EPN 9803), ravine near "Potrerillos," Ipuerán (EPN 9641); Napo, 10.6 km (by road) W of Papallacta (UMMZ 155616); Tungurahua, 2 km SW of Laguna Pisayambo, Pisayambo, Parque Nacional Llanganates (MECN 1739).

SUBSPECIES: *Neomicroxus latebricola* is monotypic.

NATURAL HISTORY: Few data are available on the ecology of this species. It is known only from high Andean environments, including subalpine forests and páramo above 3,000 m. The reproductive biology and other aspects of the behavior or population attributes remain unknown beyond the fact that it is not strictly nocturnal (Voss 2003).

REMARKS: The karyotype of *N. latebricola* is unknown.

Genus *Neotomys* Thomas, 1894

Pablo E. Ortiz and Jorge Pablo Jayat

Neotomys is one of the more distinctive genera of sigmodontine rodents and is restricted to the Central Andes. The single species of *Neotomys* is commonly known as the Andean swamp mouse (M. M. Díaz and Barquez 2007), "ratón de hocico rojo" (D'Elía et al. 2006), or Red-nosed Neotomys (Musser and Carleton 2005), with the latter two names derived from its distinctly rufous muzzle. The genus ranges from about 10°S in central Peru southward throughout most of southern and western Bolivia and northernmost Chile to northwestern Argentina at about 29°S (S. Anderson 1997; Pardiñas and Ortiz 2001; Jayat, Ortiz, and Miotti 2008). Individuals occur in high-elevation grasslands of the Puna environment above 2,700 m elevation as well as in humid grasslands below that elevation and below the limits of the Puna in the southern part of its range. Because *Neotomys* is known from only a few specimens and localities, much remains to be understood about its distribution, natural history, and patterns of geographic variation.

The following account is abstracted mainly from Thomas (1894b), Sanborn (1947a), Pine et al. (1979), Steppan (1995), S. Anderson (1997), and Pardiñas and Ortiz (2001). These sources should be consulted for additional details concerning taxonomy, morphology, distribution, fossil record, and ecology.

Externally, *Neotomys* is characterized by its medium size (total length 176–225 mm), a tail shorter than the body (tail length 72–88 mm), small hindfeet with the plantar surface entirely naked (hindfoot length 22–26 mm), and moderate-sized ears nearly circular in outline (ear length 18–19 mm). The dorsal pelage is dense, soft, long, and colored grizzled grayish brown; long gray hairs with white tips project above the dorsum. A rufous muzzle contrasts with the duller face and head coloration. The venter is grayish white to gray and the tail is bicolored, brownish gray on the dorsum and whitish on the lower surface; an ochraceous patch is present at the tail base. The upper surfaces of both forefeet and hindfeet are white, with a tinge of cinnamon, and the ears are brownish.

The skull is distinctively short and robust, with the nasals greatly expanded anteriorly. The interorbital region is very narrow, narrower than the muzzle, hourglass in shape, and angled in cross-section but not overhanging. The frontals are typically not completely fused along the midline, and a small fontanelle is often present. The zygomatic plates have a pronounced concavity at their anterior margins and prominent spinous anterodorsal processes. Premaxillaries end well anterior of the upper incisive plane. The premaxillomaxillary suture has an acutely angled bend in the middle of the rostrum (Steppan 1995; Pardiñas 1997; Steppan and Pardiñas 1998). The incisive foramina are short and wide, not reaching the anterior plane of M1. The posterior part of the palate has two deep excavations between M2 and M3, and is divided by a posterior palatine ridge. The mesopterygoid fossa reaches the posterior plane of M3 and the parapterygoid fossae are deeply excavated above the level of the bony palate.

The mandible is very deep and robust with an indistinct capsular projection. The coronoid process is small and does not extend to the level of the mandibular condyle. The mandibular ramus has a distinct ventromedial process at the posterior terminus of the symphysis (Steppan 1995).

The upper incisors are extremely broad, and have a smooth anterior surface with a groove on their labial margin. The lower incisors are robust; their anterior surface is broad and flat. The molars are hypsodont and strongly laminated, with numerous roots and planar occlusal crowns. The primary cusps are alternate. M3 is very large, much larger than M2. In young specimens, the procingulum of m1 is slightly separated from the rest of the molar by fusion of the meta- and protoflexids. An anteromedian flexid may also be visible in young individuals. This flexid dissects the procingulum on the lingual side but disappears with very little wear as individuals age (Pardiñas and Ortiz 2001).

SYNONYM:

Neotomys Thomas, 1894b:346; type species *Neotomys ebriosus* Thomas, by monotypy.

REMARKS: The tribal position of *Neotomys* has been strongly debated. Sanborn (1947a) pointed out several similarities between *Neotomys*, *Reithrodon*, and *Euneomys*. Hershkovitz (1955b), on the basis of molar occlusal morphology and palate pattern, placed *Neotomys* and *Reithrodon* together with *Sigmodon* and *Holochilus* in his "Sigmodont" group. However, Reig (1986) considered *Neotomys* to be a transitional form between the Sigmodontini and the Phyllotini. Subsequent research, including phylogenetic analyses of morphological data, placed *Neotomys* in the Phyllotini as part of a restricted generic group (the *Reithrodon* group) together with *Reithrodon*, *Euneomys*, and a few extinct genera (Olds and Anderson 1990; Steppan 1995; Steppan and Pardiñas 1998; P. E. Ortiz, Pardiñas, and Steppan 2000). Currently considered Sigmodontinae *incertae sedis*, *Neotomys* has recently been linked in molecular sequence analyses as sister to *Euneomys*, another *incertae sedis* genus (Parada et al. 2013; Salazar-Bravo et al. 2013; Schenk et al. 2013); these and earlier molecular studies (M. F. Smith and Patton 1999; D'Elía 2003, see also Pardiñas et al. 2002) failed to support any connection between *Neotomys* and phyllotines.

Neotomys ebriosus Thomas, 1894
Andean Swamp Rat

SYNONYMS:

Neotomys ebriosus Thomas, 1894b:346; type locality "Valley of Vitoc, [Junín,] East Central Peru."

Neotomys vulturnus Thomas, 1921k:612; type locality "Sierra de Zenta, 4500 m," Jujuy, Argentina; corrected to Sierra de Tilcara (M. M. Díaz and Barquez 2007).

Neotomys ebriosus ebriosus: Sanborn, 1947a:52; name combination.

Neotomys ebriosus vulturnus: Sanborn, 1947a:54; name combination.

DESCRIPTION: As for the genus.

DISTRIBUTION: *Neotomys ebriosus* is found in the central Andes from central and southern Peru (Ancash, Pasco, Junín, Cusco, and Puno regions) and south through the highlands of Bolivia (La Paz, Cochabamba, Potosí, and Tarija departments) to northernmost Chile (Tarapacá region) and northwestern Argentina (Jujuy, Salta, Catamarca, and San Juan provinces; Sanborn 1947a; Pine et al. 1979; Redford and Eisenberg 1992; Musser and Carleton 1993; S. Anderson 1997; M. M. Díaz et al. 2000; Pardiñas and Ortiz 2001; Jayat, Ortiz, and Miotti 2008). Although *N. ebriosus* has been treated historically as a mammal of the Altiplano, inhabiting elevations above 3,000 m (Pearson 1951), in the southern part of its range it occurs in places outside the limits of Puna and related highland systems down to elevations as low as 2,700 m.

P. E. Ortiz and Pardiñas (2001) and Pardiñas and Ortiz (2001) reported Pleistocene fossils from the province of Tucumán, Argentina, at 1,900 m elevation, where *N. ebriosus* is not known to be part of the Recent fauna. Remains of this species from several Holocene archaeological sites have also been reported in highlands of northwestern Argentina (P. E. Ortiz 2001; Pardiñas and Ortiz 2001; Teta and Ortiz 2002).

SELECTED LOCALITIES (Map 42): ARGENTINA: Catamarca, Alero Los Viscos, Barranca Larga (Pardiñas and Ortiz 2001), Sierra de Aconquija (Thomas 1926h); Jujuy, Sierra de Venta (= Sierra de Tilcara; type locality of *Neotomys vulturnus* Thomas), Yavi, near La Quiaca (Pardiñas and Ortiz 2001); Salta, Valle Encantado (P. E. Ortiz, Pardiñas, and Steppan 2000); San Juan, Vega Agua del Godo (Barquez 1983). BOLIVIA: Cochabamba, Incachaca (S. Anderson 1997), 7.5 mi SE of Rodeo (S. Anderson 1997); La Paz, Lago Viscachani (S. Anderson 1997); Potosí, Lípez (S. Anderson 1997); Tarija, Sama (S. Anderson 1997). CHILE: Arica y Parinacota, 1 km W of Putre (Pine et al. 1979). PERU: Ancash, 12 km W of Huaras (MVZ 150167), Ticapampa (Steppan 1995); Arequipa, 2 km W of Sumbay (MVZ 174043); Cusco, Ccolini, near Marcapata (Pine et al. 1979); Junín, Yauli Valley, Pomacocha (MVZ 138144), Vitoc Valley, Marainiyoc (Sanborn 1947a); Lima, 1 mi E of Casapalca (MVZ 120212); Pasco, Chiquirín (Sanborn 1947a); Puno, Abra Aricoma, 13 mi ENE of Crucero (MVZ 139590), Hacienda Collacachi (Sanborn 1947a), Hacienda Ontave (MVZ 141616).

SUBSPECIES: Both Sanborn (1947a) and Cabrera (1961) recognized two subspecies: *N. e. ebriosus*, occurring from Peru south to central Bolivia, and *N. e. vulturnus*, from southern Bolivia and northern Argentina. A lack of suitable series of specimens from sufficient localities precludes our ability to evaluate this hypothesis.

NATURAL HISTORY: The limited knowledge of the natural history of *N. ebriosus* comes from Thomas (1921k, citing field observations of Emilio Budin), Sanborn (1947a), Pine et al. (1979), and Barquez (1983). Individuals are predominantly herbivorous and have been trapped most commonly in grasslands along streams with dense cover, among reeds on lake margins, and in marshes. The diel activity pattern may be either diurnal or nocturnal, with individuals at higher elevations more active during the warmth of the day and those at lower elevations more nocturnal. Individuals apparently avoid rocky areas, although they may shelter under isolated rocks on level ground. The species is not common at any recorded locality, and local populations are scattered and isolated.

REMARKS: The "Sierra de Zenta" given by Thomas (1921k) as the type locality of his *vulturnus* is actually the Sierra de Tilcara (see M. M. Díaz and Barquez 2007:423–424), some 50km to the south (see also Pardiñas and Ortiz 2001).

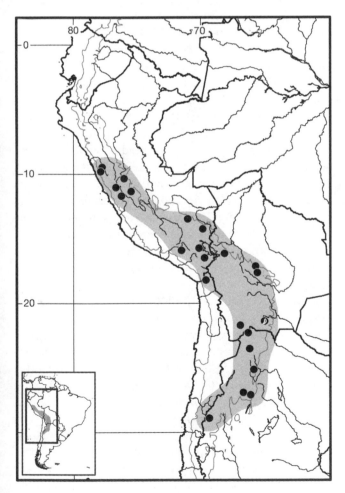

Map 42 Selected localities for *Neotomys ebriosus* (●). Contour line = 2,000 m.

The karyotype is $2n=70$ and $FN=72$, with an autosomal complement of 68 acrocentric pairs ranging in size from large to small and sex chromosomes consisting of a large subtelocentric X and a small metacentric Y (Pearson and Patton 1976).

Genus *Phaenomys* Thomas, 1917
Alexandre R. Percequillo

The monotypic *Phaenomys* is among the rarest of the endemic genera of the Brazilian Atlantic Forest. Currently known from only a few specimens and localities, much remains to be known about its morphology, distribution, ecology, and consequently, phylogenetic relationships. Vaz (2000), Bonvicino, Oliveira et al. (2001), and Passamani et al. (2011) have provided important recent information about some of these matters.

Phaenomys is medium sized (head and body length 143–170 mm), with the tail longer than the head and body (tail length 187–199 mm), short and wide hindfeet (length 31–32 mm), and short and rounded ears (length 17–20 mm). The dorsal pelage is dense, lax, and long (wool hairs 9–10 mm; cover hairs 12–13 mm; guard hairs 14–16 mm). The dorsal color varies from rufous to rust-orange; the ventral color varies from nearly white to pure white, sometimes slightly washed with buff or rufous. The tail is uniformly colored dorsoventrally, gray to dark brown. The hindfeet have a distinctive yellowish to pale rust-colored patch in the medial metatarsal region; the proximal part of the hindfeet is rust, but the distal part and digits are white or dirty white. The ears are brownish, covered externally with short rusty hairs and internally with faded rusty hairs. The mystacial vibrissae are dense, thick, and long, extending backward beyond the ears. Females possess four pairs of mammae, one pair each in the thoracic, pectoral, abdominal, and inguinal positions.

The skull is distinctive, with a long and narrow rostrum and small nasolacrimal capsules flanked by very shallow and narrow zygomatic notches. The interorbital region is narrow, hourglass in shape, and may diverge slightly posteriorly. Its dorsolateral margins have a small but well-defined crest that extends dorsolaterally to the middle part of the braincase; the development of this crest is defined by a shallow depression along the frontal bones. The braincase itself is long and rounded. The interparietal is long and wide. The lambdoidal crest is weakly developed. The zygomatic arches are heavily built and divergent posteriorly, and zygomatic plates are narrow with straight anterior margins. An alisphenoid strut is absent. The incisive foramina are long and wide, occasionally reaching the plane of M1, and they occupy about 79% of the diastema; the paired foramina are widest medially. The palate is short and narrow, with the rounded anterior margin of the mesopterygoid fossa reaching the paracone of M3; the posterolateral palatal pits are small and simple. The mesopterygoid fossa is wide; its roof is not completely ossified, as narrow and long sphenopalatine vacuities are present at the presphenoid-basisphenoid suture. The parapterygoid fossae are wide and flat, with small posterior openings for the alisphenoid channel. The carotid circulatory pattern is pattern 1 (*sensu* Voss 1988). The foramen lacerum medium is very narrow; the bullae are in contact with the squamosal along almost its entire length. The tegmen tympani overlaps the suspensory process of the squamosal, configuring a small postglenoid foramen. The hamular process of the squamosal is long and wide, resulting in a narrow or sometimes absent subsquamosal fenestra. The tympanic bullae are small, flask shaped, and have short bony tubes. The mandible has a strong falciform coronoid process that surpasses the condyloid process in height and defines a deep superior notch; the angular process is weakly projected posteriorly, extends to the level of the condyloid process, and forms a shallow inferior notch. A capsular process is absent.

The upper incisors are asulcate, strongly opisthodont, with a yellow anterior enamel band. The molars are pentalophodont, brachydont, and bunodont, with well-defined cusps arranged in opposite pairs. The main cusps are connected by a flexus(-id) and mure(-id). The molar toothrow is large and wide (M1–M3 length averages 6.36 mm, range: 6.05–6.55 mm; M1 breadth averages 1.85 mm, range: 1.8–1.95 mm). The procingulum of M1 is divided by a deep and well-defined anteromedian flexus; the labial anteroconule is larger and more mesially situated than the lingual one. The anteroloph, paraloph, mesoloph, metaloph, and posteroloph are quite visible in unworn M1 and M2. A distinct enamel island, the mesofossetus, is visible on M1 and M2. The M3 is smaller than M1 and M2, with the posterior cusps, flexi, and lophs coalesced. The anteroconid of m1 is divided by the anteromedian flexid.

SYNOMYMS:

Oryzomys: Thomas, 1894b:352; part (description of *ferrugineus*); not *Oryzomys* Baird, 1857.

Thomasomys: Thomas, 1906c:443; part (listing of *ferrugineus*); not *Thomasomys* Coues, 1884.

Phaenomys Thomas, 1917d:196; type species *Oryzomys ferrugineus* Thomas, by original designation.

Phaenomys ferrugineus (Thomas, 1917)
Rusty Phaenomys
SYNONYMS:

Oryzomys ferrugineus Thomas, 1894b:352; type locality "Rio Janeiro [= Rio de Janeiro]," Brazil.

[*Oryzomys (Oryzomys)*] *ferrugineus*: E.-L. Trouessart, 1904:419; name combination.

[*Thomasomys*] *ferrugineus*: Thomas, 1906c:443; name combination.

Phaenomys ferrugineus: Thomas, 1917d:196; first use of current name combination.

DESCRIPTION: As for the genus.

DISTRIBUTION: *Phaenomys ferrugineus* is endemic to the Atlantic Forest of southeastern Brazil, unambiguously known from a few localities in Minas Gerais, Rio de Janeiro, and São Paulo states, and that range in elevation from about 900 to 1,600 m (see Remarks).

SELECTED LOCALITIES (Map 43): BRAZIL: Minas Gerais, Casa Alpina Hotel (Passamani et al. 2011); Rio de Janeiro, Rio de Janeiro (type locality of *Oryzomys ferrugineus* Thomas), Nova Friburgo, Córrego Grande, 26 km NE of Socavão (Vaz 2000; misspelled as Sacovão by Passamani et al. 2011), Socavão, near Teresópolis (Vaz 2000); São Paulo, Fazenda Califórnia, Serra do Bananal (MZUSP 6004).

SUBSPECIES: *Phaenomys ferrugineus* is monotypic.

NATURAL HISTORY: There is virtually no information on ecology, life history, or behavior for *P. ferrugineus*. Bonvicino, Oliveira et al. (2001) trapped one specimen on the ground in herbaceous vegetation in an agricultural area, part of which was pasture. Humid and dense forests, with abundant ferns and trees festooned with orchids and bromeliads, cover the unaltered areas where this species presumably occurs (Vaz 2000). In Minas Gerais, this species was captured in montane Atlantic Forest, at a site "with abundant leaf litter" and a "continuous and dense canopy formed by young trees" (Passamani et al. 2011:827). Based on the few known localities, *P. ferrugineus* occurs in the higher elevations (between about 900 [Socavão] and 1,550 m [Casa Alpina Hotel, Itamonte]) of the Serra do Mar, Serra do Bocaina, and Serra da Mantiqueira. If the record from the city of Rio de Janeiro were valid (see Remarks), the species would thus range from sea level to 1,550 m.

REMARKS: Thomas (1894b) originally described *P. ferrugineus* as a member of the genus *Oryzomys*, distinguishing his new species from other taxa described earlier, namely, *Mus vulpinus* Brants (= *Holochilus vulpinus*), *Mus leucogaster* Brandt (= *Trinomys* sp.), *Mus cinnamomeus* Pictet (= *Oecomys catherinae*), *Mus physodes* (also = *Euryoryzomys russatus*), and *Mus vulpinus* Lund (= *Cerradomys subflavus*). In 1906, Thomas placed *ferrugineus* in the genus *Thomasomys* based on palatal morphology, especially extension of mesopterygoid fossa between the last molars and the absence of posterior palatal pits. Nevertheless, he noted that *T. ferrugineus* and *T. dorsalis* [= *Delomys dorsalis*] were exceptional among the group of species he placed in *Thomasomys* in possessing "2–2 = 8 mammae" (1906:444). Finally, Thomas (1917d) erected a new genus to accommodate this intriguing species, an action accepted unequivocally by all subsequent authorities (Gyldenstolpe 1932; Ellerman 1941; Cabrera 1961; Musser and Carleton 1993, 2005). However, although the unanimous recognition of *P. ferrugineus* as a monotypic genus has brought nomenclatural stability, this stability has not clarified the phylogenetic position of this apparently unique taxon.

Phaenomys ferrugineus has never been included in any of the morphological phylogenetic analyses of sigmodontine relationships (e.g., Steppan 1995), probably due to the scarcity of specimens in collections. This lack of specimens has also prevented the inclusion of this species in molecular-based phylogenetic studies, which are being widely and wisely employed for phyletic assessment. At present, only 14 specimens are known in museums (CMUFL, MNRJ, MZUSP, BM, ZMB). Only skins were preserved for about half of these (Vaz 2000); when skulls were preserved, most are damaged (all BM specimens, including the holotype). Only recently, in 2001, more than a century after the original description by Thomas (1894b), have new specimens been reported, each providing important data on geographic distribution, morphology, and cytogenetics (Vaz 2000; Bonvicino, Oliveira et al. 2001; Passamani et al 2011). Passamani et al. (2011) provided the first color photograph of a living animal as well as views of a complete skull and mandible.

The specimen from Itamonte, Minas Gerais state, represents an important extension to the distribution of *P. ferrugineus*; it is the first record of the species from the inland Serra da Mantiqueira, a complex of hills and high mountains (some reaching more then 2,000 m) oriented parallel to the Serra do Mar, where most other known localities are found. The reported locality of *P. ferrugineus* in Bahia (see Thomas 1917d) and, especially, the purported type locality in Rio de Janeiro (Thomas 1894b) cannot be confidently regarded as valid records. The specimens bearing these provenances on skin tags (BM 63.11.16.2 and 76.12.8.3) were bought by the BM from dealers such as "Maison Verreaux, Magazin de Zoologie" in Paris. Therefore, the putative collecting localities of both specimens are problematic (Percequillo et al. 2004).

The karyotype consists of $2n = 78$ and $FN = 114$, with an autosomal complement of 19 biarmed and 19 acrocentric pairs that grade in size from large to small and sex chromosomes that include a large submetacentric X and small Y (Bonvicino, Oliveira et al. 2001).

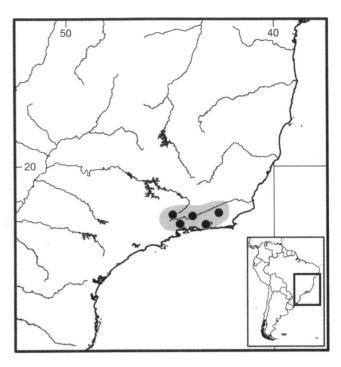

Map 43 Selected localities for *Phaenomys ferrugineus* (●).

Genus *Punomys* Osgood, 1943
James L. Patton

Known only from seven localities in southern Peru (Pacheco and Patton 1995), *Punomys* has the highest elevational range of any species of mammals in the neotropics. This is a medium-sized, stoutly built, vole-like mouse with long and lax fur, a short tail, and relatively small ears. The soles of the hindfeet are hairy for the proximal one quarter of their length, the toes are of medium length, and the claws are uniformly small. Females have eight mammae, paired in pectoral, axillary, abdominal, and inguinal positions. The skull is rather heavy, with a broad and short rostrum, well-developed nasals expanded anteriorly, a squared braincase, a narrow interorbital region with rounded edges, and robust zygomatic arches. The incisive foramina are long and narrow; the palate is complex and dissected by two long and deep sulci; the mesopterygoid fossa extends nearly to the posterior margin of M3; and tympanic bullae are large and inflated. The carotid circulation is pattern 1 (of Voss 1988), characterized by an enlarged carotid canal and stapedial foramen, a squamosal-alisphenoid groove, sphenofrontal foramen, and a groove on the posterolateral margin of the parapterygoid plates. An alisphenoid strut is absent. The tegmen tympani is well developed and overlaps the squamosal, and the mastoid is large with a small to medium-sized fenestra.

The upper incisors are strong and orthodont. The molars combine coronal hypsodonty with high, sharp-angled individual cusps in the unworn condition; molar toothrows diverge posteriorly. The occlusal surface of the molars is complex due to the presence of both labial and lingual styles, with the latter better developed. M1 is strongly curved with the procingulum divergent anteriorly, and with anterolabial and anterolingual conules of subequal size. Four main, distinct cusps are present on M3.

SYNONYM:

Punomys Osgood, 1943a:369; type species *Punomys lemminus* Osgood, by original designation.

REMARKS: *Punomys* is a poorly known and enigmatic taxon of incompletely understood phylogenetic and biogeographic affinities. Vorontsov (1959) placed the genus in the tribe Phyllotini, an action adhered to by Olds and Anderson (1990) and Braun (1993). Reig (1986) suggested that *Punomys* may represent an early descendant of a protophyllotine stock. Alternatively, Hershkovitz (1962) and Reig (1980, 1984) suggested that the genus be maintained as *incertae sedis* in the sigmodontines to emphasize its unique combination of morphological attributes. Steppan (1993, 1995), in a cladistic analysis of craniodental, postcranial skeletal, external, and soft organ characters, supported this latter position and removed *Punomys* from the Phyllotini. His analyses suggested that *Punomys* lies outside the phyllotines, near the base of a phyllotine-akodontine-scapteromyine radiation. Recent mtDNA and nucDNA sequence analyses, however, have supported a sister relationship between *Punomys* and *Andinomys*, in a larger clade that contained *Chinchillula*, *Euneomys*, *Juliomys*, *Irenomys*, and *Neotomys* (Parada et al. 2013; Salazar-Bravo et al. 2013; Schenk et al. 2013).

KEY TO THE SPECIES OF *PUNOMYS*:

1. Pelage of upper parts pale yellow-gray, strongly contrasting with that of under parts; tail <30% of head and body length. . . . .*Punomys lemminus*
1'. Pelage of upper parts overall darker, not contrasting in color with under parts; tail >30% of head and body length . . . . *Punomys kofordi*

Punomys kofordi Pacheco and Patton, 1995

Koford's Puna Mouse

SYNONYM:

Punomys kofordi Pacheco and Patton, 1995:86; type locality "13 mi (20.8 km) ENE Crucero, Lago Aricoma, Department of Puno, Perú, 15,000 ft (4,550 m), approximately 14°17′S, 69°47′W."

DESCRIPTION: Pelage similar to that of *P. lemminus*, but overall darker with under parts and not contrasting

with dorsum; tail short (range: 65–77 mm), but averaging 34% of total length rather than 27%; hindfoot absolutely shorter (mean 27.2 mm versus 28.7 mm). Skull distinguished by parallel as opposed to more anteriorly convergent zygomatic arches; squared anterior root of zygoma; narrower zygomatic plates with straight frontal edges as opposed to broad plates, and more posteriorly an upward slanting edge; spine of zygomatic notches angled laterally rather than directed anteriorly, particularly in ventral view; zygomatic notches broad rather than shallow; nasals not abruptly expanded anteriorly, terminating in deep, posterior V-shaped notch; palatal bridge short (mean 6.6 mm versus 7.4 mm); anterior opening of alisphenoid canal small; and postcingular conules of M1 subequal in size.

DISTRIBUTION: *Punomys kofordi* is known from only three localities in the vicinity of Abra Aricoma and the head of the Limbani Valley, all above 15,000 ft [4,500 m] in the Cordillera Carabaya (of the Cordillera Oriental) in southern Peru, and one locality at 4,770 m in eastern La Paz department, Bolivia.

SELECTED LOCALITIES (Map 44): BOLIVIA: La Paz, cumbre del camino a Yungas (Salazar-Bravo et al. 2011). PERU: Puno, Abra Aricoma, 13 mi ENE of Crucero, Lago Aricoma (type locality of *Punomys kofordi* Pacheco and Patton), 8 mi SSW of Limbani, 15,000 ft (Pacheco and Patton 1995).

SUBSPECIES: *Punomys kofordi* is monotypic.

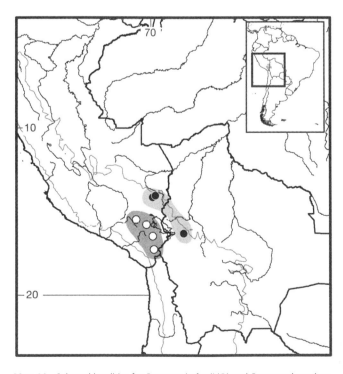

Map 44 Selected localities for *Punomys kofordi* (●) and *Punomys lemminus* (○). Contour line = 2,000 m.

NATURAL HISTORY: This species is known only from the more humid eastern Cordillera bordering the Altiplano. Pregnant or lactating females were taken from June to September (field notes of C. B. Koford, MVZ archives), suggesting that reproduction is timed differently for *P. kofordi* than for *P. lemminus*. Litter size ranged from two to three (summarized by Pacheco and Patton 1995). Specimens from 8 mi SSW Limbani were captured "in barren, broken rock areas and . . . were near fleshy-leaved, pungent *Senecio* plants or piles of *Senecio* cuttings" (Pearson 1957:4). The recent specimen from Bolivia (Salazar-Bravo et al. 2011) was trapped at the entrance of a small, shallow cave fronted by the perennial shrub *Cajophora horrida* (Loasaceae).

REMARKS: Although very similar to and clearly closely related to *P. lemminus*, *P. kofordi* has most likely been geographically separated from its congener for much of the Pleistocene by the expanded glacier of the Quelccaya Ice Cap of the Cordillera Vilcanota, which lies at the northern nexus of the Cordillera Occidental and Cordillera Oriental.

Punomys lemminus Osgood, 1943
Western Puna Mouse

SYNONYM:

Punomys lemminus Osgood, 1943a:369; type locality "San Antonio de Esquilache, Department of Puno, Peru; altitude 15,000 feet."

DESCRIPTION: As for genus. Differences between *P. lemminus* and *P. kofordi* noted under description of latter.

DISTRIBUTION: *Punomys lemminus* is known from only four widely scattered localities, all above 14,600 ft [4,400 m] in the Cordillera Occidental of southern Peru (see map, Pacheco and Patton 1995).

SELECTED LOCALITIES (Map 44): PERU: Arequipa, Huaylarco, 55 mi ENE of Arequipa (Pearson 1957); Puno, Caccachara (Pearson 1951), San Antonio de Esquilache (type locality of *Punomys lemminus* Osgood); Tacna, 12 mi NE of Tarata (Pearson 1957).

SUBSPECIES: *Punomys lemminus* is monotypic.

NATURAL HISTORY: Pearson (1951) noted that *P. lemminus* was neither scarce nor shy at Caccachara, where he observed it during the day feeding or scurrying from the shelter of one rock to another. Here, they fed mostly or entirely on a dwarf, fleshy-leaved tola shrub (*Senecio adenophylloides*) or a low, ground-pine-like herb (*Werneria digitata*) growing in most places. They cut twigs of both plants into lengths as great as 12 inches and stored them under rocks in caches of several dozen twigs. These materials were chewed only slightly when eaten so that stomach contents were a mass of coarsely chopped plant material, and feces were large and irregularly shaped pellets consisting of fragments of undigested

vegetation. The species was always encountered among rocks near yareta (*Azorella yarita*) and the other two food plants mentioned and near surface water. Other species of small mammals in the same local community included *Lagidium peruanum*, *Auliscomys boliviensis*, *Phyllotis xanthopygus*, *Chinchillula sahamae*, and *Abrothrix jelskii*.

Genus *Wilfredomys* Avila-Pires, 1960
Enrique M. González, João A. de Oliveira and Ulyses F. J. Pardiñas

Wilfredomys is a monotypic genus endemic to the southern portion of the Atlantic Forest in Brazil and the gallery and canyon forests of Uruguay (E. M. González and Oliveira 1997). The single species is medium-sized and long-tailed (head and body 110–115 mm; tail 170–220 mm; mass 50–75 g) with striking coloration, including ears covered by very short ochraceous hairs on both sides and a bright ochraceous-rufous area around the nose. The dorsal surface of the head and body is usually pale orange-brown lined with gray due to a mixture of light yellow to orange distal parts of the shorter hairs, and dark brown to black tips of the longer hairs, with gray bases of both types of hair. The rump of some specimens can attain an ochraceous-orange tone due to the prevalence of distally orange hairs. The venter lacks dark hairs and is thus lighter than the dorsum without a clear line of demarcation from it. The distal light-yellow portions of the ventral hairs often obscure their slaty basal halves. Hairs are whitish to their bases only in the throat and inguinal regions. Mixed whitish and ochraceous short hairs cover the upper surfaces of the manus and pes. The tail is very long, brown, slightly paler below, covered by very short ochraceous hairs, and terminates in a short pencil of hairs of 2–3 mm long. The mystacial vibrissae are dark and long, extending to the posterior limit of the pinnae when laid back beside the head. Plantar surfaces of the hindfeet have six tubercles, the thenar, hypothenar, and four interdigitals.

The skull has a relatively short rostrum, wide zygomatic arches, a shallow and narrow zygomatic notch, and a round braincase. The incisive foramina are very long and wide, reaching the anterior face of the protocone of M1, and the interorbital region is narrow, hourglass in shape, with squared (but not ridged) supraorbital margins, and a shallow depression along the frontal suture. Bullae are rather large. The palate is short, as the mesopterygoid fossa reaches to the last upper molar. Large posterolateral palatal pits are located along side the mesopterygoid fossa behind M3. The parapterygoid fossae are wider than the mesopterygoid fossa. An alisphenoid strut is lacking, and the carotid circulation pattern is the generalized pattern 1 (*sensu* Voss 1988). Incisors are orthodont, and molars are moderately lophodont, the main cusps high and well defined. The anterocone bears a deep and obliquely oriented anteromedian flexus defining a lingual anteroconule much smaller than the labial one. An ectolophid is lacking. Mesolophs and mesostyles are well developed. Vaz Ferreira (1960), E. M. González (2000), and Pardiñas and Teta (2011a) have each illustrated several cranial and dental traits.

SYNONYMS:

Thomasomys: Thomas, 1928a:154; part (description of *oenax*); not *Thomasomys* Coues, 1884.

Wilfredomys Avila-Pires, 1960c:4; type species *Thomasomys oenax* Thomas, by original designation.

REMARKS: The name *Wilfredomys* was coined by Avila-Pires (1960c) in the course of clarifying the relationships of several taxa traditionally included in *Thomasomys*. Although ranked as a full genus in most recent treatises (e.g., Musser and Carleton 2005), in prior decades the name *Wilfredomys* was either not mentioned at all or treated as a subgenus (e.g., Pine 1980; Carleton and Musser 1984). The external similarity between *Wilfredomys* and *Wiedomys* often resulted in misidentifications (e.g., Ximénez 1965). These similarities were attributed to convergence by Pine (1980); however, Pacheco's (2003) character analysis added several traits of internal morphology to the known, shared features of the two genera.

Wilfredomys oenax (Thomas, 1928)

Red-nosed Tree Mouse

SYNONYMS:

Hesperomys pyrrhorhinus: Thomas, 1886a:421; part; not *Mus pyrrhorhinos* Wied Neuwied.

Thomasomys oenax Thomas, 1928a:154; type locality "Rio Grande do Sul. Type from San Lorenzo" (Thomas 1928a:155); São Lourenço do Sul, Rio Grande do Sul, Brazil (see Pine 1980:196).

Wilfredomys oenax: Avila-Pires, 1960c:4; first use of current name combination.

Wiedomys pyrrhorhinos: Ximénez, 1965:135; part; not *Mus pyrrhorhinos* Wied Neuwied.

DESCRIPTION: As for the genus.

DISTRIBUTION: *Wilfredomys oenax* is known from a few scattered localities in the Atlantic Forest, in the Brazilian states of São Paulo, Santa Catarina, Paraná, and Rio Grande do Sul and in riparian forest of Uruguay, in the departments of Artigas, Cerro Largo, Durazno, Rivera, Tacuarembó, Treinta y Tres, and Florida (Bonvicino, Oliveira, and D'Andrea 2008; E. M. González and Martínez Lanfranco 2010). Hershkovitz (1998:201, citing K. Blair) reported an unconfirmed record for Caparaó National Park, Minas Gerais, Brasil, from nonvouchered owl pellet remains.

SELECTED LOCALITIES (Map 45): BRAZIL: Paraná: Curitiba (Avila-Pires 1960); Rio Grande do Sul, Itaqui (MZUSP 3181), Uruguaiana (MNRJ 2010); Rio Grande do Sul, São Lourenço do Sul (type locality of *Thomasomys oenax* Thomas); São Paulo, Ubatuba (E. M. González and Oliveira 1997). URUGUAY: Artigas, Rincón de Franquía (E. M. González and Oliveira 1997); Florida, Arteaga, Cerro Colorado (E. M. González and Martínez Lanfranco 2010); Rivera, Minas de Corrales (E. M. González and Martínez Lanfranco 2010); Treinta y Tres: Paso Ancho, Cañada de las Piedras and Rte 8 (E. M. González and Martínez Lanfranco 2010).

SUBSPECIES: We consider *Wilfredomys oenax* to be monotypic. Pine (1980), however, suggested the existence of two subspecies, and Vaz Ferreira (1960) highlighted differences between the type and one specimen from Uruguay. Examination of most of the known material reveals variability in some cranial and pelage features, of which the length of the nasals, the shape of the mesopterygoid fossa, and the extent of an ochraceous-orange tone on the dorsum are the most remarkable. Nevertheless, among the few specimens available in collections, specimens with various degrees of development of the conditions of these characters can be found. In addition, label information (e.g., sex, locality) is often lacking, hampering an evaluation of possible geographic trends as opposed to simple sexual dimorphism.

NATURAL HISTORY: The scarce field data available for this rare arboreal rat, mostly gathered in Uruguay, indicate that it occurs in gallery and montane forests with considerable epiphytic vegetation, particularly bromeliads (*Tillandsia* spp.). A specimen from Imboá, Uruguaiana (Rio Grande do Sul, Brazil), obtained by Emilie Snethlage was taken from a nest in the low forest bordering a river. In Uruguay, *W. oenax* was found nesting in the abandoned globular nests of the furnariid *Anumbius annumbi* on several occasions (R. Vaz Ferreira and F. Achaval, pers. comm.; E. M. González, unpubl. data; G. D'Elía, pers. comm.) and in nests made from *Tillandsia* (Cravino, pers. comm.; Prigioni and Sappa, pers. comm.). It is not clear whether the rats made the second type of nest or they were abandoned bird nests. Four specimens trapped by the American Museum of Natural History expedition to Uruguay in 1962–1963 were obtained in a canyon with mesic subtropical woodland and with a sampling effort of 4,400 trap-nights (Barlow 1969). One specimen (AMNH 206018) was trapped among large rocks in the forest on the north-facing slope of the canyon and was the only rat caught in a line of 50 traps. The remaining three specimens were trapped 20 m or less from the first, in branches from 1.5 to 2 m above the ground. Two of these specimens were males in breeding condition, with testes measuring 10×6 mm and that contained abundant mature sperm in the epididymides and the semi-

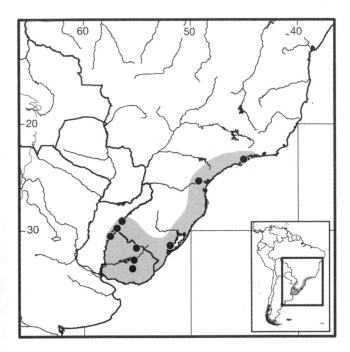

Map 45 Selected localities for *Wilfredomys oenax* (●). Contour line = 2,000 m.

niferous tubules. Stomachs of three of the mice contained about 6 cc of greenish plant material, bait, and purplish berries from the *aruera* (*Lithraea molleoides* [Anacardiaciae]). Vaz Ferreira (1960) captured his specimen during the night while it was walking on a branch. This species has been found in Barn Owl (*Tyto alba*) pellets (Langguth 1965); a semidigested specimen was found in the gut of the Pampas Lancehead (*Bothrops pubescens*; E. M. González and Martínez Lanfranco 2010).

REMARKS: Pine (1980) detailed the history of the joint confusion of synonyms of *Wilfredomys oenax* and *Wiedomys pyrrhorhinos* (see also Hershkovitz 1959a:8).

Tribe Abrotrichini D'Elía, Pardiñas, Teta, and Patton, 2007

Guillermo D'Elía, Ulyses F. J. Pardiñas, and Pablo Teta

One of the most remarkable results of the DNA sequence-based phylogenetic analyses assessing sigmodontine relationships (M. F. Smith and Patton 1999; D'Elía 2003a) was the discovery of a clade formed by five taxa of generic rank (*Abrothrix* [including *Chroeomys*], *Chelemys*, *Geoxus*, *Notiomys*, and *Pearsonomy*) and that was independent of the Akodontini, the assemblage of sigmodontines among which these taxa had been traditionally placed (e.g., Ellerman 1941; Cabrera 1961; Reig 1987; Musser and Carleton 1993). This finding was particularly surprising because some of these taxa (e.g., the genus *Abrothrix* as well as forms currently allocated to it such as *andina*, *olivacea*,

and *xanthorhina*) had been placed in the genus *Akodon* (Musser and Carleton 1993). However, Hershkovitz (1966a) posited that *Abrothrix* (which, at that time was restricted to *A. longipilis* and closely related forms) was part an "oxymycterine" group that he erected, an assemblage of taxa, including the nominotypical genus *Oxymycterus*, now phylogenetically firmly embedded in the Akodontni (see that account). Massoia (1983) followed Hershkovitz's suggestion and gave the Oxymycterini tribal status.

In his classic review of the Akodontini, Reig (1987) provided morphological diagnoses for *Abrothrix* (as a subgenus of *Akodon*), *Chroeomys* (as a valid genus, but now considered a synonym of *Abrothrix*; D'Elía 2003a), *Chelemys*, *Geoxus*, and *Notiomys*, and included a discussion of chromosomal and ecological similarities of these taxa. His treatment of *Chelemys* and *Geoxus* as genera separate from *Notiomys* followed Pearson (1983, 1984).

The first indication that abrotrichines constituted a distinct group came from allozyme studies (Patton et al. 1989; Spotorno et al. 1990; Dickerman 1992; Barrantes et al. 1993) as well as from the initial mtDNA gene sequence analysis (M. F. Smith and Patton 1991). However, the limited taxonomic coverage of these studies precluded a clear picture of the phyletic relationships between abrotrichines and akodontines until the expanded sequence analysis by M. F. Smith and Patton (1999; see also M. F. Smith and Patton 1993). M. F. Smith and Patton (1993, 1999) listed some character states, including anatomical, chromosomal, and allozyme characters, along with distributional and ecological features, that supported the distinctiveness of the abrotrichine group. Furthermore, they recommended (M. F. Smith and Patton 1999:106–107) ranking this group of genera as a tribe, suggesting that an appropriate name for it would be Abrotrichini, although they referred to the group informally as "the Andean Clade" (see also M. F. Smith and Patton 1993:170). References to the group by subsequent authors used this informal name, the incorrect spelling Abrothrichini (e.g., Weksler 2003; Rodríguez-Serrano, Palma, and Hernández 2008), or the informal derivative "abrothrichines" (e.g., D'Elía 2003a). Further phylogenetic studies, which combined mtDNA and nucDNA sequences, corroborated the distinctiveness of the group in the radiation of Sigmodontinae (D'Elía 2003a; D'Elía et al. 2003; D'Elía, Luna et al. 2006; D'Elía, Ojeda et al. 2006). Given the body of molecular evidence, D'Elía et al. (2007) formally described the tribe Abrotrichini, provided a diagnosis based on morphological characters, and established its current contents. The Abrotrichini constitutes the most recently proposed sigmodontine tribe.

The phylogenetic position of Abrotrichini in the sigmodontine radiation remains unresolved. Molecular sequences place the tribe in Oryzomyalia (see Steppan, Adkins, and

Anderson 2004), where the Abrotrichini appear either as sister to Wiedomyini (D'Elía, Luna et al. 2006; D'Elía, Ojeda et al. 2006; Parada et al. 2013; Salazar-Bravo et al. 2013) or in a broader clade containing the Phyllotini, Oryzomyini, and Wiedomyini (D'Elía 2003a). Reig (1986, 1987), who considered abrotrichines to be typical akodonts, also suggested, on morphological grounds, a phylogenetic relationship between the oryzomyines, akodontines, and phyllotines. The analyses of Parada et al. (2013) provided strong support for a phylogenetic linkage between the Abrotrichini and Wiedomyini.

M. F. Smith and Patton (1993, 1999), D'Elía (2003a), D'Elía, Ojeda et al. (2006), Rodríguez-Serrano, Palma, and Hernández (2008), Feijoo et al. (2010), Teta, D'Elía et al. (2011), and Cañón-Valenzuela (2012) have addressed phyletic relationships in the Abrotrichini with DNA sequences. Even without the few species of *Abrothrix* that have yet to be included in any study, results are robust in documenting two main clades in the tribe. One of these includes the terrestrial forms currently referred to *Abrothrix*, while the other encompasses the mainly fossorial, long-clawed genera *Chelemys*, *Geoxus*, *Notiomys*, and *Pearsonomys*. Relationships within both of these clades remain less resolved.

Continued molecular phylogenetic analyses suggest that adjustments of the current genus-group classification of Abrotrichini may be required because neither *Chelemys* nor *Geoxus* appear to be monophyletic (Rodríguez-Serrano, Palma, and Hernández 2008; E. P. Lessa et al. 2010). Alternative classificatory choices might include but are not limited to: erect a new genus for *Chelemys macronyx* and another for southern populations currently referred to as *Geoxus valdivianus*, or return to Osgood's (1925) view of *Notiomys* (*sensu lato*), one that included both *Chelemys* and *Geoxus* (see Pearson 1984 for review). Given current knowledge, which is much more exhaustive than that available to Osgood (see, for example, Pardiñas, Udrizar Sauthier et al. 2008), a scheme that would unite into a single genus such morphologically divergent taxa as *Geoxus valdivianus* and *Notiomys edwardsii* is difficult to support.

Pending publication of a revision of Abrotrichini currently in progress (P. Teta, unpubl. data), we recognize the traditional five abrotrichine genera, with a total of 14 species. This assemblage is distributed from the Altiplano in west-central Peru south through the western mountains of Bolivia to southernmost Argentina and Chile (including Cape Horn and other islands), and the Patagonian steppes to the Atlantic coast of Argentina (Osgood 1943b; Teta, Pardiñas, and D'Elía 2006).

Abrotrichines are small to medium sized (from approximately 20 g in *Notiomys edwardsii* to 75 g in *Chelemys macronyx*) sigmodontine rodents, diagnosed by the following combination of characters (from D'Elía et al. 2007).

Pelage long and soft, generally slate gray or brownish, although some are blackish (*A. sanborni*, *Chelemys*, *Geoxus*) or with bright colors (*A. jelskii*); tail typically short and well haired; feet large and strongly built, with naked plantar surfaces; claws equally robust on forefeet and hindfeet (Gyldenstolpe 1932; Osgood 1943b); skull with long muzzle and rounded braincase, without frontal or lambdoid crests; interorbital region amphora-shaped; nasals and premaxillae slightly projected anterior to the incisors to somewhat trumpet-like; nasals longer than frontals; zygomatic plates typically narrow, with the upper free border reduced or absent; infraorbital foramina wide; frontal sinuses moderately developed; palate long; anterior border of mesopterygoid fossa squared; mandibular ramus generally gracile and elongate, except in *Chelemys* (Gyldenstolpe 1932; Osgood 1943b; Reig 1987).

Upper incisors broad and ungrooved; teeth brachydont to mesodont; upper and lower molars with labial and lingual cusps arranged in opposite or slightly alternate pairs; M1/m1 reduced, fan-shaped, procingulum and shallow or absent anteroflexus(-id); M1 with very shallow to obsolete anteromedian flexus; paraflexus and metaflexus strongly oriented backward to transversely oriented; paracone globose and oriented forward, mesoloph(id) poorly to moderately developed, usually fused to the paracone; M3 reduced and subcylindrical, except in *Chelemys*, with an internal ring-like enamel fossette (Gyldenstolpe 1932; Reig 1987; D'Elía, Ojeda et al. 2006).

The axial skeleton includes 13 ribs, 13 thoracic vertebrae, six lumbar vertebrae, and 18–29 caudal vertebrae (Steppan 1995). Both types of unilocular-hemiglandular stomachs described by Carleton (1973) are present in the Abrotrichini (Carleton 1973; D'Elía, Ojeda et al. 2006; Pardiñas, Udrizar Sauthier et al. 2008). Species of *Abrothrix* have elongated phalli with a much reduced or no distal baculum; the genera *Chelemys*, *Geoxus*, and *Pearsonomys* have a tridigitate distal baculum (the phallus of *Notiomys* is unknown); the phalli of *Pearsonomys* and *Geoxus* have a prominent dorsal hood that extends beyond the terminal crater (Hooper and Musser 1964; Spotorno 1986; D'Elía, Ojeda et al. 2006). *Abrothrix longipilis*, *A. jelskii*, *A. olivacea*, and *Geoxus valdivianus* have two pairs of preputial glands (Voss and Linzey 1981). The number of preputial glands in other genera is unknown.

The karyotype among the Abrotrichini is quite uniform. A diploid complement of 52 chromosomes is present in several species of *Abrothrix*, *Chelemys macronyx*, and *Geoxus valdivianus* (N. O. Bianchi et al. 1971; M. H. Gallardo 1982; Pearson 1984; Liascovich et al.1989; Spotorno et al. 1990), prompting M. F. Smith and Patton (1999) to suggest that 2n = 52 was a synapomorphy for the tribe. Later studies have shown that *Pearsonomys* has 2n = 56 (D'Elía,

Ojeda et al. 2006), that typical *Chelemys macronyx* has $2n=54$ (A. A. Ojeda et al. 2005), and that some populations of *Abrothrix olivacea* from eastern Patagonia have $2n=44$ (V. A. Rodríguez and Theiler 2007). The karyotype of *Notiomys edwardsii* and other nominal forms remains unknown. Further studies are needed to assess the directions and modes of abrotrichine chromosomal evolution, including the ancestral condition for the tribe (i.e., whether the common ancestor had $2n=52$).

The earliest known fossils considered to be abrotrichines are *Abrothrix kermacki* Reig, 1978, and *A. magnus* Reig, 1987, both from the Pliocene of Argentina (Reig 1978, 1987). Pardiñas (1995b) tentatively questioned the generic allocation of *magnus*. Our most recent and unpublished assessment suggests that both taxa are more likely to belong to two different and as yet undescribed Akodontini genera. A third form, referred in the literature as aff. *Abrothrix* and from Pliocene deposits in Jujuy province of northwestern Argentina, represents an extinct, undescribed species of this genus (P. E. Ortiz, García López et al. 2012). No extinct genus has been assigned to the tribe Abrotrichini. The oldest fossils of living abrotrichines are known from the Lujanian and Platan South American Land Mammal Ages (Late Pleistocene–Early Holocene; Pardiñas et al. 2002). Molecular clock dating based on both mtDNA and nucDNA sequences has suggested that the abrotrichine crown group originated approximately 4.9 mya (3.66–6.33), with differentiation of the two main clades beginning about 3.5–4.0 mya (Parada et al. 2013). Rodríguez-Serrano, Palma, and Hernández (2008) found statistical support that the common ancestor of Abrotrichini was fossorial, which would necessitate that fossoriality was lost in the lineage leading to *Abrothrix*. This nonparsimonious directionality of character change presents us with an open question.

KEY TO THE GENERA OF ABROTRICHINI:

1. Tail proportionally long, ≥60% of the combined head and body length; anterior claws <3 mm . . . . *Abrothrix*
1'. Tail proportionally short, ≤60% of the combined head and body length; anterior claws >4 mm 2
2. Pinnae concealed in pelage; dorsal (agouti) and ventral (whitish) coloration sharply demarcated; sides of muzzle brightly colored with orange; rhinarium leathery . . . . *Notiomys*
2'. Pinnae protruding from pelage; dorsal and ventral coloration usually darker, but sometimes slightly differentiated, brown to blackish, paler on the venter; sides of the muzzle without orange color; rhinarium not leathery . . 3
3. Manual claws not flattened at tips; skull robust; toothrow long (>4.5 mm); upper molars with well-developed anteroloph and mesoloph; zygomata heavy and widely flaring . . . . *Chelemys*

3'. Manual claws flattened at tips; skull more delicate; toothrow short (<4.5 mm); upper molars lacking mesolophs; zygomata not widely flaring 4
4. Size small (greatest length of skull usually <29 mm); ears short; pes short (<24 mm); pelage smooth and dense . . . . *Geoxus*
4'. Size large (greatest length of skull usually >30 mm); ears long; pes long (>25 mm); pelage coarse . . . *Pearsonomys*

Genus *Abrothrix* Waterhouse, 1837
Bruce D. Patterson, Margaret F. Smith, and Pablo Teta

The genus *Abrothrix* comprises species of mice that have variously been allocated to *Mus*, *Hesperomys*, *Akodon*, *Oxymycterus*, *Microxus*, *Bolomys*, and *Chroeomys*. Its current contents stem from several DNA sequence analyses, beginning with M. F. Smith and Patton (1991) and followed sequentially by M. F. Smith and Patton (1993, 1999), D'Elía (2003a), Pearson and Smith (1999), M. F. Smith et al. (2001), and Teta, D'Elía, et al. (2011). Those species not yet included in any published molecular analysis are placed in *Abrothrix* by morphological and/or karyotypic characters (see, for example, Spotorno, Zuleta, and Cortes 1990; Spotorno 1992). The genus ranges from the Altiplano of southern Peru through the highlands of Bolivia, northern Chile, and Argentina south through Tierra del Fuego; its range reaches the Pacific at 30°S and the Atlantic at about 40°S, and includes numerous islands where an endemic species occurs.

Abrothrix was named for its long, soft fur, which fairly characterizes all members of the genus. Most species are colored slate gray, although the range includes blackish (*sanborni*) to brightly bicolored (*jelskii*). The tail is short and well haired; Osgood (1943b:195) noted that in at least some *Abrothrix* (i.e., *longipilis* and *sanborni*) the skin of the tail clings tightly to the caudal vertebrae, preventing the preparator from easily removing it without making ventral incisions on the tail. Feet are large and strongly built, with naked palms and soles, and claws equally strong on forefeet and hindfeet.

The skull has a relatively long rostrum and rounded braincase. The nasals and premaxillae usually project anterior to the incisors into a somewhat trumpet-shape, and are longer than the frontals. The zygomatic plates are relatively deep and short, sometimes slanting posteriorly upward; the incisive foramina reach the protocone of M1, and the palate extends well posterior to M3 (except in *A. illutea* and *A. jelskii*). The frontal sinuses are moderately inflated and the bullae are not enlarged (except in *A. jelskii*). The upper incisors are broad and ungrooved, and the molars above and below are broad, rooted, and with hypsodont crowns.

Abrothrix was originally diagnosed by the lack of an antero-median sulcus on M1; although this characterizes the type species *A. longipilis*, other species now allocated to the genus have this trait (especially *olivacea*, *hershkovitzi*, and *andina*; see also Osgood 1943b:196). *Abrothrix* has a shallow anteroflexus, poorly developed mesoloph(-id), strong meso-flexid, and weaker posteroflexid oriented transversely, and entoconids bulging into the buccal cavity; m3 is elongate but smaller than m2 (see Gyldenstolpe 1932; Reig 1987).

A combination of these characters can be used to sep-arate *Abrothrix* from other abrotrichine and akodontine rodents, but members of the genus can be recognized at a glance by their bacula and/or glandes penes. *Abrothrix longipilis* and *A. sanborni* have a simplified baculum that lacks the characteristic (and primitive) tridigitate distal elements of most other muroid lineages, including other sigmodontines (see Musser and Hooper 1964; Spotorno 1992). In contrast, other taxa here allocated to *Abrothrix* have bacula that retain the tridigitate arrangement, albeit greatly reduced so that the apical digits are scarcely visible, little more than vestigial remnants (M. H. Gallardo et al. 1988; Spotorno 1992; Feijoo et al. 2010).

Abrothrix longipilis and *A. jelskii* have two pairs of preputial glands (Voss and Linzey 1981). The ampullary glands of *A. longipilis* are especially large in relation to the male tract as a whole. Dorsal prostates are also unusually large, and are partly covered by extensive lateral prostatic tissue. Vesicular glands are lobed medially and along their greater curvatures, but in some individuals papilla-like pro-cesses rather than rounded lobes are found on the medial surfaces (Voss and Linzey 1981). In *A. jelskii*, the lateral pre-putial glands extend beyond the ventral flexure of the penis, and the vesicular glands are lobed medially and along their curvatures; the subterminal flexures are rough and irregular in appearance. The stomachs of *A. longipilis*, *A. jelskii*, and *A. olivacea* are unilocular-hemiglandular (Carleton 1973).

Axial skeletal elements have counts of 13 ribs, 13 tho-racic vertebrae, six lumbar vertebrae, and 26–29 (*Abrothrix longipilis*), 23–26 (*A. jelskii*), 2–23 (*A. olivacea*, *A. andina*), or 24–27 (*A. sanborni*) caudal vertebrae (Steppan 1995).

Almost all species of *Abrothrix* that have been karyo-typed to date (*andina*, *olivacea* [including *xanthorhina*], *lanosa*, *longipilis*, and *illutea*) share a diploid number of 52, and where details of individual chromosomes are avail-able, an autosomal fundamental number of 56 (Spotorno and Fernandez 1976; M. H. Gallardo 1982; M. Rodríguez et al. 1983; B. D. Patterson et al. 1984; Liascovich et al. 1989; Espinosa et al 1991). Recently, however, V. A. Ro-dríguez and Theiler (2007) described a $2n = 44$ karyotype for 12 *A. olivacea* specimens from Comodoro Rivadavia, in coastal Chubut province, Argentina. The significance of this chromosomal difference is unclear.

The earliest fossils assigned to *Abrothrix* are those of *A. kermacki* from the Chapadmalal Formation, up-per Pliocene–lowermost Pleistocene, near the present city of Mar del Plata, Buenos Aires province, Argentina (Reig 1987). *Abrothrix magna* is known from early Pleistocene (Vorohuean) deposits, also from near Mar del Plata (Reig 1987). These locality records are from almost 1,000 km east of the present distributional limits of the genus (Reig 1987). As noted above, however, Pardiñas (1995b) questioned the generic placement of *magna* and Teta et al. (2012), through inspection of their type series and additional materials, sug-gested that *kermacki* and *magna* probably belong to the Akodontini. P. E. Ortiz, García López et al. (2012) described other fossils, including some Pliocene remains referred as aff. *Abrothrix*, from sites in northwestern Argentina; these perhaps represent the earliest true Abrotrichini.

SYNONYMS:

Mus (*Abrothrix*) Waterhouse, 1837:21; type species *Mus* (*Abrothrix*) *longipilis* Waterhouse, by original designation.

Abrothrix: Gray, 1843b:xxiv, 114; first use as genus.

Habrothrix Wagner, 1843a:519; unjustified emendation of *Abrothrix* Waterhouse.

Acodon Thomas, 1895c:369; part (description of *hirtus*); incorrect subsequent spelling of *Akodon* Meyen.

Akodon: E.-L. Trouessart, 1897:535; part (listing of *lon-gipilis*, *andinus*, *hirtus*, *xanthorhinus*, *brachyotis*, *oliva-ceus*, and *jelskii*); not *Akodon* Meyen, 1833.

Chroeomys Thomas, 1916i:340; type species *Akodon pul-cherrimus* Thomas, by original designation.

Chraeomys Gyldenstolpe, 1932a:122; incorrect subse-quent spelling of *Chroeomys* Thomas.

Microxus: Gyldenstolpe, 1932a:133; part (listing of *pul-cherrimus*, *cayllomae*, *inambarii*, *cruceri*, *inornatus*, *bac-chante*, *sodalis*, *scalops*, *jelskii*, *pyrrhotis*); not *Microxus* Thomas.

Bolomys: Tate, 1932h:23; part (listing of *andinus*, *jucun-dus*, *gossei*); not *Bolomys* Thomas.

REMARKS: We note that *Abrothrix* is based on a femi-nine Greek name (*-trichos*, hair), but *Akodon* is based on a masculine Greek name (*-odon*, tooth). As a consequence of taxonomic history, there has been chaotic variation in the genders of specific epithets as taxa have been transferred between these two genus-groups.

KEY TO THE SPECIES OF *ABROTHRIX*:

1. Coloration generally striking, often with fulvous on nose (sometimes also on feet and tail) and white postau-ricular patches, white (or grayish) venter contrasts with darker dorsum; palate ending at about back edges of last molars; incisors very slender and yellow; bullae en-larged . . . . *Abrothrix jelskii*

1'. Coloration not so striking, some forms with fulvous wash on nose and face; palate ending at about or at some distance behind back edges of last molars; incisors broader and darker; bullae not enlarged 2

2. Length of tail usually >75 mm; condyloincisive length >27 mm. 3

2'. Length of tail usually <75 mm; condyloincisive length <27 mm. 4

3. Dorsal color uniform dark olive gray; ventral surfaceashy gray, with small patches of white under chin and in inguinal area; nasals and premaxillae slightly projected anterior to incisors; nasofrontal sutures not forming acute angle, not reaching line connecting lacrimals; palate ending near back edge of last molars; cartilaginous distal baculum reduced, composed of two apical digits . . . . *Abrothrix illutea*

3'. Dorsal color ranges from gray to brown, sometimes washed with chestnut or reddish brown along back; venter silvery; nasals extend anteriorly into slight trumpet; nasofrontal sutures form acute angle, reaching line connecting lacrimals; palate ending some distance behind back edges of last molars; baculum long and sharply curved near tip, completely lacking apical digits. . . . . *Abrothrix longipilis*

4. Dorsal profile of the skull flat, convex; nasals and premaxillae strongly projected anterior to the incisors into a trumpet shape; baculum long and sharply curved near the tip, without apical digits 5

4'. Dorsal profile of the skull flat to slightly bowed; nasals and premaxillae not strongly projected anterior to the incisors; baculum consists proximally of a straight, blunt rod bearing three apical digits much reduced in size 6

5. Coloration uniformly blackish brown, including hairs of tail and feet, without indication of counter coloration; ears normal length; tail long (70–80% head and body length); nasofrontal sutures form acute angle, surpassing line connecting lacrimals; mandible slender, ramus hardly constricted behind m3; lunar notch shallow . . . . *Abrothrix sanborni*

5'. Dorsal coloration olivaceous, slightly demarcated from venter; dorsal surface of manus and pes covered by short buffy hairs; ears short; tail short (60% to 70% head and body length); nasofrontal suture acute, reaching line connecting lacrimals; mandible slender, with marked constriction of ramus below m3; lunar notch well excavated anteriorly. . . . . *Abrothrix lanosa*

6. Pelage long and dense; color pale, ochraceous; venter gray with strong buffy wash; small white postauricular spots; tail length 47–67 mm; hindfoot length 19–23 mm; ear length 13–16 mm . . . . *Abrothrix andina*

6'. Pelage shorter and sparser; color more saturate, ranging from dark brown or olive brown to ochraceous gray; some populations exhibit a wash of yellow or orange on nose, feet, and tail; length of tail 60–84 mm; length of hindfoot 22–25 mm; length of ear 15–18 mm. 7

7. Size large (adult mass >30 g); tail about two-thirds length of head and body; condylobasal length >23.5 mm; maxillary diastema >6.5 mm. . . . . *Abrothrix hershkovitzi*

7'. Size small (adult mass usually <30 g); tail shorter than 65%; condylobasal length <23.5 mm; maxillary diastema <6.5 mm . . . . *Abrothrix olivacea*

Abrothrix andina (Philippi, 1858)
Andean Soft-haired Mouse
SYNONYMS:

Mus andinus Philippi, 1858:77; type locality "andibus elevatis prov. Santiago," Altos Andes, Metropolitana de Santiago, Chile.

Hesperomys dolichonyx Philippi, 1896:21, plate II, Fig. 1a, c, d, e, f; type locality "vecinidad de Atacama" [= San Pedro de Atacama], Antofagasta, Chile.

Hesperomys dolichonyx cinnamomea Philippi, 1896:22, plate II, Fig. 1b; type locality "la oasis de Leoncito," Antofagasta, Chile.

Akodon andinus: E.-L. Trouessart, 1897:535; name combination.

Mus dolichonyx: Philippi, 1900:58; name combination.

M[us]. dolichonyx cinnamomea: Philippi, 1900:59; name combination.

Akodon (Chelemys) andinus: Wolffsohn, 1910:90; name combination.

Akodon jucundus Thomas, 1913a:140; type locality "Cerro de la Lagunita, E. of Maimara. 4500 m," Jujuy, Argentina.

Akodon gossei Thomas, 1920g:418; type locality "Puente de Inca, Andes of Mendoza. Alt. 10,000'," Mendoza, Argentina.

Chelemys megalonyx: Gyldenstolpe, 1932a; part; not *Hesperomys megalonyx* Waterhouse.

Bolomys andinus: Tate, 1932h:23; name combination.

Bolomys jucundus: Tate, 1932h:23; name combination.

Bolomys gossei: Tate, 1932h:23; name combination.

Akodon [(Akodon)] gossei: Ellerman, 1941:411; name combination.

Akodon [(Akodon)] jucundus: Ellerman, 1941:411; name combination.

Akodon andinus andinus: Osgood, 1943b:177; name combination.

Akodon andinus dolichonyx: Osgood, 1943b:179; name combination.

Akodon andinus polius Osgood, 1944:196; type locality "Salinas, Arequipa, Peru. Altitude 14,000 feet."

Akodon [(*Akodon*)] *andinus*: Honacki, Kinman, and Koeppl, 1982:394; name combination.

"*Akodon*" *andinus*: Patton and Smith, 1992b:93; name combination.

Chroeomys andinus: S. Anderson, 1997:433; name combination.

Abrothrix andinus: Tamayo et al., 1987:5; first use of current name combination but incorrect gender agreement.

Chroeomys andinus dolichonyx: S. Anderson, 1997:434; name combination.

[*Abrothrix andinus*] *cinnamomea*: Musser and Carleton, 2005:1089; name combination and incorrect gender agreement.

[*Abrothrix andinus*] *dolichonyx*: Musser and Carleton, 2005:1089; name combination.

[*Abrothrix andinus*] *cinnamomea*: Musser and Carleton, 2005:1089; name combination and incorrect gender agreement.

[*Abrothrix andinus*] *gossei*: Musser and Carleton, 2005: 1089; name combination.

[*Abrothrix andinus*] *jucundus*: Musser and Carleton, 2005: 1089; name combination and incorrect gender agreement.

[*Abrothrix andinus*] *polius*: Musser and Carleton, 2005: 1089; name combination and incorrect gender agreement.

DESCRIPTION: Smallest species (average head and body length 90 mm; average tail length 65 mm); pelage mainly light buffy and without countershading, with distinct whitish postauricular patches and whitish lips and chin. Skull delicate, with a short and narrow rostrum, rounded braincase, and moderately extended nasals and premaxillae; borders of the mesopterygoid fossa slightly divergent posteriorly; tympanic bullae relatively large and rounded (Osgood 1943b; Mann 1978). Baculum with minute, vestigial apical digits (Spotorno 1992).

DISTRIBUTION: High-elevation grasslands, generally between 2,000 and 3,000 m, from south-central Peru through western Bolivia to northwestern Argentina and northern Chile (to 34°S; Muñoz-Pedreros 2000). In places, *A. andina* ranges to 5,000 m elevation (Mann 1978) and may be found at lower elevations (950 m) in central Chile during extremely cool winters (Iriarte and Simonetti 1986).

SELECTED LOCALITIES (Map 46): ARGENTINA: Catamarca, Minas Capillitas (Mares, Ojeda, Braun et al. 1997); Jujuy, Cerro Lagunita, near Maimara (type locality of *Akodon jucundus* Thomas), Tres Cruces (Osgood 1943b); La Rioja, Pastillos, Reserva Provincial "Laguna Brava" (OMNH 23218); Mendoza, 3 km W of Refugio Militar General Alvarado (OMNH 23623), ca. 3 km SSE of Vallecitos (OMNH 30208); Salta, Cauchari (MSB 75243); San Juan, Tudcum (OMNH 23627); Tucumán, Paso El Infiernillo (MSU 19221). BOLIVIA: Oruro, 1 km SW of Sajama (S. Anderson 1997); Potosí, Lipez (S. Anderson

1997). CHILE: Antofagasta, Ojos San Pedro, 55 km NE of Calama (MVZ 116766), San Pedro de Atacama (type locality of *Hesperomys dolichonyx* Philippi); Arica y Parinacota, Chungará (Rodríguez-Serrano, Hernández, and Palma 2008); Atacama, Leoncitos (Osgood 1943b); Valparaíso, Farellones, 51 km E of Santiago (Spotorno 1992). PERU: Arequipa, Huaylarco, 55 mi ENE of Arequipa (MVZ 116009), Laguna Salinas (Spotorno 1992), 2 km W of Sumbay (M. F. Smith and Patton 1999); Tacna, 2 km NW of Nevado Livine (Spotorno 1992).

SUBSPECIES: Two subspecies have been traditionally recognized in *Abrothrix andina*. The nominotypical race is larger, exceeding 150 mm in total length, is colored reddish brown, and is found in Chile from Santiago to Coquimbo. It is replaced to the north by *A. a. dolichonyx*, which is smaller and paler, and ranges into Peru (Mann 1978) and Bolivia (S. Anderson 1997). However, this species has not been subjected to rangewide analyses of geographic variation, so the validity of these subspecies remains to be substantiated.

NATURAL HISTORY: *Abrothrix andina* is mostly nocturnal, although it is diurnal during autumn to winter (J. R.

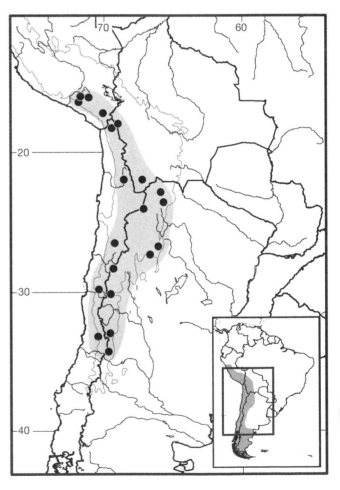

Map 46 Selected localities for *Abrothrix andina* (●). Contour line = 2,000 m.

Contreras and Rosi 1981; Mares, Ojeda, and Kosco 1981). It is a good digger that constructs a system of galleries about 5 cm deep in the soil, winding among the rocks (Mann 1978; J. R. Contreras and Rosi 1981). Its preferred habitat is dense patches of *Lepidophyllum* (Asteraceae), with suitable numbers of refuges. The sparse and low-stature bunchgrass and small shrub community on hard soil pavements contrast with the lusher, denser vegetation occupied by sympatric *Akodon* (Pearson 1951). In Chile, its diet includes arthropods, seeds, and other plant material (Mann 1978; S. I. Silva 2005). In Argentina, males with scrotal testes were trapped in January (Mares, Ojeda, Braun et al. 1997) and lactating females in March (R. A. Ojeda and Mares 1989). *Abrothrix andina* is an important prey item in the diet of the Barn Owl (*Tyto alba*) and foxes (*Lycalopex* spp.) in the Atacama Desert (Jaksíc et al. 1999).

REMARKS: Palma, Marquet, and Boric-Bargetto (2005) examined the phylogeography of pre-Puna and Puna populations from northern Chile. *Abrothrix andina* is known from late Holocene remains at the archaeological site Tebenquiche Chico 1, Catamarca, Argentina (P. E. Ortiz 2001; P. E. Ortiz, Jayat, Nasif et al. 2012).

Abrothrix hershkovitzi (B. D. Patterson, Gallardo, and Freas, 1984)

Hershkovitz's Soft-haired Mouse

SYNONYMS:

Akodon (Akodon) hershkovitzi B. D. Patterson, Gallardo, and Freas 1984:8; type locality "Chile, [Magallanes y Antártica Chilena,] Magallanes Prov., Isla Capitán Aracena, head of Bahía Morris, elev. ca. 60 m; 54°14′S, 71°30′W."

Akodon hershkovitzi: Tamayo et al., 1987:5; name combination.

Akodon (Abrothrix) hershkovitzi: Reig, 1987:366; name combination.

Abrothrix hershkovitzi: Musser and Carleton, 2005; first use of current name combination.

DESCRIPTION: Medium sized, larger than adjacent races of *A. olivacea* (*xanthorhina*, *canescens*), with an absolutely and relatively longer tail (71 mm, 65% versus 55% head and body length) and greater condylobasal length (>23 mm). Pelage pattern similar to other Cape forms, with head, back, and sides agouti, a buffy white chin and venter, rufescent dorsum of feet, tricolored tail, and ochraceous (yellowish-brown) on the sides of nose. Pelage generally coarsely grizzled, forefeet and hindfeet more tan than rufescent, and tail heavily pigmented dorsally, with rufous on lateral margins and whitish below. Rostrum long, with somewhat trumpeted nasals and premaxillae. Baculum with minute, vestigial apical digits (*sensu* Spotorno 1992; authors unpubl. data).

DISTRIBUTION: This species occurs in the Magellanic steppe and coastal forests of outer islands in the Chilean Archipelago (B. D. Patterson et al. 1984). The type locality is on Isla Capitán Aracena, on the Magdalena Channel, south of the Magellanic Straits, and the species also occurs on four islands in the Cabo de Hornos group.

SELECTED LOCALITIES (Map 47; from B. D. Patterson et al. 1984): CHILE: Magallanes y Antártica Chilena, Caleta Lientur, Isla Wollaston, Caleta Toledo, Isla Deceit, Isla Capitán Aracena (type locality of *Akodon (Akodon) hershkovitzi* B. D. Patterson, Gallardo, and Freas), Isla Grevy, Isla Hornos.

SUBSPECIES: *Abrothrix hershkovitzi* is monotypic.

NATURAL HISTORY: This species has not been studied in the field since its initial discovery.

REMARKS: This species was originally described in the subgenus *Akodon* to emphasize its evident affinities with *A. markhami*, *A. olivacea*, and *A. xanthorhina*, which are all now included in *A. olivacea* (B. D. Patterson et al. 1984; Muñoz-Pederos 2000). However, despite phallic differences, this group shows affinity with typical *Abrothrix* in cranial and dental characters, and no consistent or qualitative differences between these two groups have been demonstrated. On the basis of cranial characters, especially of the rostrum, Reig (1987) placed *hershkovitzi* in the subgenus *Abrothrix*, which subsequently was accorded generic rank after its distinctive bacular and other penial characters came to be appreciated (Spotorno et al. 1990). To date, the molecular analyses that confirm the membership of *olivacea* and *xanthorhina* in *Abrothrix* (Pearson and Smith 1999; M. F. Smith and Patton 1999) have not included *A. hershkovitzi*.

Abrothrix hershkovitzi occurs on outer islands in the Chilean coastal archipelago, a broad swath of about 3,000 islands that fringe the Pacific coast of southern Chile, from Chiloé Island at 40°S latitude to Isla Hornos at 56°S. Many of these islands were likely buried under glacial ice as recently as 12,000 ybp (Heusser 2003). Thus, like *A. o. markhami*, *A. hershkovitzi* was probably derived via colonization from adjacent mainland populations now represented by *A. o. xanthorhina*. Because both *xanthorhina* and *markhami* are now regarded as subspecies of *A. olivacea* (Pearson and Smith 1999; Rodríguez-Serrano, Hernández, and Palma 2008), *A. hershkovitzi* cannot logically be considered a distinct species without rendering this enlarged *olivacea* paraphyletic. However, founder effect and genetic drift promote rapid evolutionary divergence, especially in insular forms (Mayr 1963; Wright 1969), and cladistic classifications inevitably break down if biological species concepts are employed. Patton and Smith (1994) presented a particularly well-documented example of this involving the pocket gophers *Thomomys bottae* and *Thomomys townsendii* (Geomyidae). The genetic and morphological differentiation of island forms of *Abrothrix* deserves closer scrutiny.

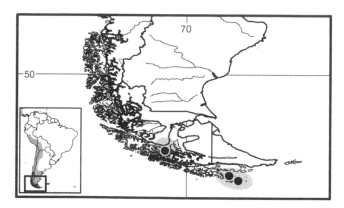

Map 47 Selected localities for *Abrothrix hershkovitzi* (●).

Abrothrix illutea Thomas, 1925
Gray Soft-haired Mouse

SYNONYMS:

Abrothrix illutea Thomas, 1925c:582; type locality originally given as "Concepción, 400 m," Tucumán, Argentina, later modified to "Sierra de Aconquija, 3000–4000 m" in the same province (Thomas 1929:41). Teta, D'Elía et al. (2011), however, reviewed this history and reaffirmed Concepción as the correct type locality.

[*Abrothrix*] *illutea*: Tate, 1932h:24; name combination.

Akodon [(*Abrothrix*)] *illutea*: Ellerman, 1941:416; name combination.

Akodon (*Akodon*) *illutea*: A. L. Gardner and Patton 1976: 30; name combination.

Akodon illuteus: Mares, Ojeda, Braun, and Barquez 1997: 114; name combination.

Abrothrix illuteus: Capllonch, Autino, Díaz, Barquez, and Goytia, 1997:54, 61; incorrect gender agreement.

DESCRIPTION: Dorsal color uniform dark olive-gray, with whitish patches of hairs on chin and hairs white to base in inguinal area (Liascovich et al. 1989); skull robust, with interorbital constriction hourglass in shape and smoothly rounded; muzzle elongated; nasals and premaxillae not strongly flared and project only slightly anterior to incisors (M. M. Díaz 2000); palate ends at about posterior margins of M3s; upper incisors broad and orthodont; molar crowns hypsodont, M1 with well-developed anteroloph and mesoloph. Stomach unilocular-hemiglandular. Terminal crater of phallus ventrally directed; cartilaginous distal baculum reduced, composed of two lateral digits. See Teta, D'Elía, Pardiñas et al. 2011 for emended diagnosis.

DISTRIBUTION: Confirmed records of *A. illutea* are limited to the Catamarca and Tucumán provinces in northwestern Argentina, between 700 and 2,500 m (Teta, D'Elía, Pardiñas et al. 2011). The presence of *A. illutea* at elevations of 3,000 to 4,000 m in the Nevados de Aconquija on the border between Catamarca and Tucumán

(Thomas 1929) is doubtful, and the presence of the species in Jujuy (Olrog 1979) remains to be confirmed (see M. M. Díaz 1999b; Teta, D'Elía, Pardiñas et al. 2011).

SELECTED LOCALITIES (Map 48): ARGENTINA: Catamarca, Andalgalá (MACN 50.434), 5 km S of Las Higuerillas on Hwy 9 (Teta, D'Elía, Pardiñas et al. 2011); Tucumán, Concepción (type locality of *Abrothrix illutea* Thomas), El Naranjal (TTU 32830), 10 km S of Hualinchay (Jayat et al. 2006), Zanjón de Tafí, 2 km SW of Tafí del Valle (Teta, D'Elía, Pardiñas et al. 2011).

SUBSPECIES: *Abrothrix illutea* is monotypic.

NATURAL HISTORY: This species inhabits moist southern pine (*Podocarpus parlatorei* [Podocarpaceae]) and *aliso* (*Alnus acuminata* [Betulaceae]) forests on steep slopes at intermediate elevations, about 1,700 m (Capllonch et al. 1997; Barquez et al. 1991). It is also known from a few records for the "Selva Montana" between 700 and 2.200 m. At its upper elevational limit, it inhabits tall grass and brushy habitats along streams (Fonollat 1984). *Abrothrix illutea* was the second most common item in the diet of Barn Owls (*Tyto alba*) in Tucumán province (P. E. Ortiz and Pardiñas 2001). It is known from late Pleistocene remains from the Tafí Valley, Tucumán, Argentina (P. E. Ortiz and Pardiñas 2001).

REMARKS: Thomas (1929) changed the type locality from Concepción to the Aconquija Mountains, 3,000 to 4,000 m, believing that a temperate-zone taxon like *Abrothrix* could not occur in the subtropical lowlands, where it has been subsequently confirmed to occur (Liascovich et al.1989;

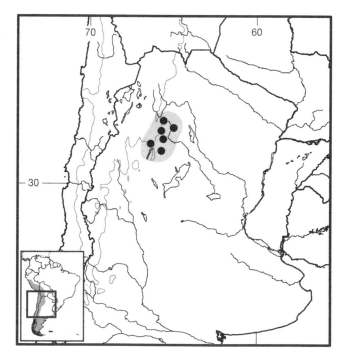

Map 48 Selected localities for *Abrothrix illutea* (●). Contour line = 2,000 m.

Teta, D'Elía, Pardiñas et al. 2011). Osgood (1943b) never examined this species and suspected, from its widely disjunct distribution, that the species was wrongly allocated to *Abrothrix* and might possibly belong in *Hypsimys*. However, both morphology and karyology ally it with *Abrothrix* (Liascovich et al. 1989). Cranially and dentally, it fits revised diagnoses of *Abrothrix*, and its karyotype ($2n = 52$, $FN = 56$) is practically identical to that of *Abrothrix longipilis* and other species in the genus. The report of a $2n = 39$–42 karyotype that N. O. Bianchi et al. (1971) and Dulout et al. (1976) ascribed to *A. illutea* was based on incorrectly identified specimens of *Akodon varius* (see Liascovich et al. 1989). Phylogenetic analyses of nuclear DNA sequences support a clade composed of *A. illutea*, *A. andina*, *A. jelskii*, and *A. olivacea* (Teta, D'Elía, Pardiñas et al. 2011).

Abrothrix jelskii (Thomas, 1894)
Ornate Soft-haired Mouse

SYNONYMS:

Hesperomys (*Habrothrix*) *scalops*: Thomas, 1884; part; not *Oxymicterus scalops* Gay [= *Chelemys megalonyx*].

Acodon Jelskii Thomas, 1894b:360; type locality "Junin, Central Peru."

Acodon Jelskii pyrrhotis Thomas, 1894b:361; type locality "Maraynioc, [Junín,] Central Peru."

Akodon pulcherrimus Thomas, 1897f:549; type locality "Puno, [Puno,] Peru, alt. 4000 metres."

Akodon pulcherrimus cayllomae Thomas, 1901b:185; type locality "Caylloma [= Cailloma]. Alt 4300 m," Arequipa, Peru.

Akodon pulcherrimus inambarii Thomas, 1901b:185; type locality "Limbane [= Limbani], on the Inambari River, Upper Madre de Dios. Alt. 3400 m," Puno, Peru.

Akodon pulcherrimus cruceri Thomas, 1901b:186; type locality "Crucero, on the pass between Puno and the Upper Inambari. Alt. 4550 m," Puno, Peru.

Akodon bacchante Thomas, 1902b:139; type locality "Choro" [= El Choro], upper Río Secure, Cochabamba, Bolivia.

Akodon bacchante sodalis Thomas, 1913a:141; type locality "Cerro de la Lagunita, E. of Maimara," Jujuy, Argentina.

C[*hroeomys*]. *bacchante*: Thomas, 1916i:340; name combination.

Chroeomys inornatus Thomas, 1917g:2; type locality "Ollantaytambo, 13,000 feet," Cusco, Peru.

Chroeomys jelskii: Thomas, 1926b:317; name combination.

Chroeomys jelskii pyrrhotis: Thomas, 1926b:317; name combination.

Chroeomys bacchante sodalis: Thomas, 1926h:195; name combination.

Chraeomys pulcherrimus pulcherrimus: Gyldenstolpe, 1932a:122; name combination.

Chraeomys pulcherrimus cayllomae: Gyldenstolpe, 1932a: 122; name combination.

Chraeomys pulcherrimus inambarii: Gyldenstolpe, 1932a: 122; name combination.

Chraeomys pulcherrimus cruceri: Gyldenstolpe, 1932a:122; name combination.

Chraeomys inornatus: Gyldenstolpe, 1932a:122; name combination.

Chraeomys bacchante bacchante: Gyldenstolpe, 1932a: 122; name combination.

Chraeomys bacchante sodalis: Gyldenstolpe, 1932a:123; name combination.

Chraeomys jelskii jelskii: Gyldenstolpe, 1932a:123; name combination.

Chraeomys jelskii pyrrhotis: Gyldenstolpe, 1932a:123; name combination.

Akodon [(*Chroeomys*)] *bacchante bacchante*: Ellerman, 1941:415; name combination.

Akodon [(*Chroeomys*)] *bacchante sodalis*: Ellerman, 1941: 415; name combination.

Akodon [(*Chroeomys*)] *inornatus*: Ellerman, 1941:415; name combination.

Akodon [(*Chroeomys*)] *jelskii jelskii*: Ellerman, 1941:415; name combination.

Akodon [(*Chroeomys*)] *jelskii pyrrhotis*: Ellerman, 1941: 415; name combination.

Akodon [(*Chroeomys*)] *pulcherrimus pulcherrimus*: Ellerman, 1941:415; name combination.

Akodon [(*Chroeomys*)] *pulcherrimus cayllomae*: Ellerman, 1941:416; name combination.

Akodon [(*Chroeomys*)] *pulcherrimus cruceri*: Ellerman, 1941:416; name combination.

Akodon [(*Chroeomys*)] *pulcherrimus inambarii*: Ellerman, 1941:416; name combination.

Akodon (*Chroeomys*) *jelskii inornatus*: Sanborn, 1947b: 108; name combination.

Akodon (*Chroeomys*) *jelskii ochrotis*: Sanborn, 1947b:133; type locality "Huacullani, Department of Puno, Peru."

Akodon (*Chroeomys*) *jelskii bacchante*: Sanborn, 1947b: 137; name combination.

Akodon (*Chroeomys*) *jelskii inambarii*: Sanborn, 1947b: 139; name combination.

Akodon (*Chroeomys*) *jelskii pulcherrimus*: Sanborn, 1947b: 139; name combination.

Akodon (*Chroeomys*) *jelskii cruceri*: Sanborn, 1947b:140; name combination.

Akodon (*Chroeomys*) *jelskii sodalis*: Sanborn, 1947b:142; name combination.

Akodon jelskii bacchante: Cabrera, 1961:459; name combination.

Akodon jelskii cruceri: Cabrera, 1961:459; name combination.

Akodon jelskii inambarii: Cabrera, 1961: 459; name combination.

Akodon jelskii inornatus: Cabrera, 1961:460; name combination.

Akodon jelskii jelskii: Cabrera, 1961:460; name combination.

Akodon jelskii ochrotis: Cabrera, 1961: 460; name combination.

Akodon jelskii pulcherrimus: Cabrera, 1961:460; name combination.

Akodon jelskii pyrrhotis: Cabrera, 1961:461; name combination.

Akodon jelskii sodalis: Cabrera, 1961:461; name combination.

Abrothrix jelskii: Spotorno, 1992:509; first use of current name combination.

[*Abrothrix jelskii*] *bacchante*: Musser and Carleton, 2005:1089; name combination.

[*Abrothrix jelskii*] *cayllomae*: Musser and Carleton, 2005:1089; name combination.

[*Abrothrix jelskii*] *cruceri*: Musser and Carleton, 2005: 1089; name combination.

[*Abrothrix jelskii*] *inambarii*: Musser and Carleton, 2005: 1089; name combination.

[*Abrothrix jelskii*] *inornatus*: Musser and Carleton, 2005: 1089; name combination and incorrect gender agreement.

[*Abrothrix jelskii*] *ochrotis*: Musser and Carleton, 2005: 1089; name combination.

[*Abrothrix jelskii*] *pulcherrimus*: Musser and Carleton, 2005:1089; name combination and incorrect gender agreement.

[*Abrothrix jelskii*] *pyrrhotis*: Musser and Carleton, 2005: 1089; name combination.

[*Abrothrix jelskii*] *scalops*: Musser and Carleton, 2005: 1089; name combination.

[*Abrothrix jelskii*] *sodalis*: Musser and Carleton, 2005:1089; name combination.

DESCRIPTION: Color pattern striking and variable over range, generally strongly countershaded, with venter whitish or grayish; contrasting patches of white or buff hair behind ears and fulvous or ochraceous hairs on muzzle, feet, tail dorsum, flanks, or back. Skull with enlarged tympanic bullae, short rostrum, and nasals and premaxillae that do not form trumpet-like tube; incisors unusually slender, with pale yellow anterior surfaces. Baculum minute, with vestigial apical digits (Spotorno 1992).

DISTRIBUTION: High-elevation grasslands above about 3,400 m, from Ancash and Pasco regions in central Peru (Arana-Cardó and Ascorra 1994) south to western Bolivia and northwestern Argentina.

SELECTED LOCALITIES (Map 49): ARGENTINA: Jujuy, Cerro de la Lagunita, east of Maimará (type locality of *Akodon bacchante sodalis* Thomas); Salta, Santa Victoria range, 13 km NW of Lizoite (Jayat et al. 2013). BOLIVIA: Cochabamba, Choro (type locality of *Akodon bacchante* Thomas); La Paz, Laguna Viscachani (S. Anderson 1997), Pelechuco (S. Anderson 1997); Oruro, Quebrada Kohuiri (S. Anderson 1997); Potosí, Kirkari Mountains (S. Anderson 1997), Lipez (S. Anderson 1997). PERU: Ancash, 3 km S and 12 km W of Huaras (MVZ 135680); Arequipa, 5 km W of Cailloma (MVZ 174286), Salinas, 22 mi E of Arequipa (Spotorno 1992); Ayacucho, 4 km W (by road) of Pampamarca (MVZ 174290), 15 km WNW of Puquio (MVZ 137986), 6 mi. NNE of Tambo (AMNH 208088); Cusco, Ollantaytambo (type locality of *Chroeomys inornatus* Thomas); Huancavelica, Huancavelica (Sanborn 1947b); Junín, 7 km E of Concepción (LSUMZ 14405); Lima, 1.5 km W of Casapalca (MVZ 119941); Pasco, 10 mi NE of Cerro de Pasco (MVZ 119942); Puno, Limbani (type locality of *Akodon pulcherrimus inambarii* Thomas), 6.5 km SW of Ollachea (MVZ 173080); Tacna, 1 mi NE of Challapalca (MVZ 141357).

SUBSPECIES: *Abrothrix jelskii* varies remarkably over its range. Subspecific variation was reviewed by Sanborn (1947b) and has not been analyzed since; gene flow among

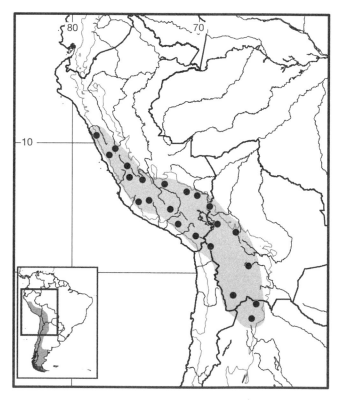

Map 49 Selected localities for *Abrothrix jelskii* (●). Contour line = 2,000 m.

the geographic races has not been studied or demonstrated in this species. Patton and Smith (1992b) noted that the sharp chromatic differences between northern and southern populations might indicate that they belong to different species. Pearson (1982) noted sharp discontinuities in morphology that coincided with the Río Acarí (southeast of Nazca, Peru, a western drainage) and the Río Tambo (formed by the confluence of the Ene and Perené, draining to the Urubamba in the east).

NATURAL HISTORY: Pearson (1982) considered *Abrothrix jelskii* to be a species of broad distribution found throughout the entirety of the Altiplano. It occupies a variety of habitats "such as grassy places, rocks, vacant huts, and even occupied houses" (Pearson 1951:140), and therefore is possibly sympatric with every other mammal of the Altiplano. It may be active day and night (Dorst 1971). In Peru, stomachs were found to contain forbs, insects, grass, and seeds (Pizzimenti and DeSalle 1980). Fleas, lice (*Hoplopleura affinis*), and mites (*Atricholaelaps glasgowi*) were collected from *Abrothrix jelskii* at Caccachara (Pearson 1951).

Abrothrix lanosa (Thomas, 1897)
Woolly Soft-haired Mouse
SYNONYMS:

Oxymycterus lanosus Thomas, 1897d:218; type locality "Monteith Bay, Straits of Magellan," equated to Seno Monteith, in the Madre de Dios Archipelago, Magallanes y Antártica Chilena, Chile by Teta, Pardiñas, and D'Elía (2006).

Microxus lanosus: Thomas, 1909:237; name combination.

Akodon (Abrothrix) lanosus: Osgood, 1943b:197; name combination.

Akodon longipilis lanosus: Mann, 1978:157; name combination.

Akodon lanosus: Tamayo, Núñez, and Yáñez, 987:5; name combination.

Abrothrix lanosus: Lozada, Monjeau, Heinemann, Guthmann, and Birney, 1996:1; first use of name combination but incorrect gender agreement.

DESCRIPTION: One of smaller members of genus; reduced eyes and ears barely visible through dense, woolly pelage; reduced countershading of venter (especially as compared to sympatric *A. olivacea xanthorhina*); and tail length about 65% head and body length (Feijoo et al. 2010). Dorsal pelage brown to cinnamon brown; under parts usually heavily washed with fulvous; feet white, and tail bicolored (Osgood 1943b). Skull delicate, with enlarged and pointed rostrum, gracile zygomatic arches, and rounded braincase; nasals project beyond anterior plane of incisors and together with premaxillae form well-defined trumpet-like tube; nasofrontal suture acuminate (reaching level of small lac-

rimal bones); masseteric tubercle present; palate long, with anterior border of mesopterygoid fossa posterior to plane of M3s (Feijoo et al. 2010). Baculum "simple" (*sensu* Spotorno 1992); glans penis described and figured by Feijoo et al. (2010).

DISTRIBUTION: *Abrothrix lanosa* is known along a narrow strip extending from 48° to 55°S in northwestern Santa Cruz province of Argentina, south through Última Esperanza and Magallanes in southern Chile to Tierra del Fuego, Argentina (see Galliari and Pardiñas 1999; Feijoo et al. 2010). Osgood (1943b) suspected that either *A. sanborni* or *A. lanosa* would be found living in the coastal archipelago between Chiloé and Tierra del Fuego, where *A. hershkovitzi* and *A. olivacea* (*markhami*) were later discovered. Late Pleistocene to Late Holocene fossils have been recorded from three archaeological deposits, two within the current range of *A. lanosa* on Isla Grande de Tierra del Fuego about 150 km north of the nearest known extant population (summarized by Feijoo et al. 2010). The species is known from Late Pleistocene and Late Holocene remains collected at Cueva del Milodón, Magallanes y Antártica Chilena, Chile (Simonetti and Rau 1989); Rockshelter 1, Tres Arroyos, Tierra del Fuego, Chile; and Cerro Casa de Piedra 5, Santa Cruz, Argentina (Feijoo et al. 2010).

SELECTED LOCALITIES (Map 50): ARGENTINA: Santa Cruz, Cerro Casa de Piedra (Galliari and Pardiñas 1999), Valle del Río Tucu-Tuco (J. A. Allen 1905); Tierra del Fuego, Bahía Buen Suceso (Pine et al. 1978), Lapataia (Reis and Venegas 1987), Lago Fagnano (Osgood 1943b). CHILE: Magallanes y Antártica Chilena, Bahía Parry (Feijoo et al. 2010), Parque Nacional Torres del Paine (Jaksic et al. 1978), vicinity of Punta Arenas (Osgood 1943b), Seno Monteith, Isla Madre de Dios (type locality of *Oxymycterus lanosus* Thomas).

SUBSPECIES: *Abrothrix lanosa* is monotypic.

NATURAL HISTORY: Osgood (1943b:197) encountered *A. lanosa* "in or near deep forest on Tierra del Fuego and in the vicinity of Punta Arenas where it shows preference for cool, damp habitat." Pine et al. (1978) recorded this species from dense, humid forest dominated by *Nothofagus betuloides* [Nothofagaceae] and *Drimys winteri* [Winteraceae] (see Massoia and Chebez 1993). Feijoo et al. (2010) found this mouse in coastal localities with well-developed soils and dense vegetation cover of *matanegra* (*Junielia tridens* [Verbenaceae]), although very close to the edge of nearby Fuegian forest. *Abrothrix lanosa* has been found in low numbers in Magellanic Horned Owl (*Bubo magellanicus*) pellets from Magallanes y Antártica Chilena, Chile (Jaksic et al. 1978). Captive individuals are docile and easy to manage with bare hands. Times of capture during the summer reveal that the species is basically nocturnal but also exhibits some diurnal activity (Feijoo et al. 2010).

REMARKS: The affinities of *A. lanosa* with *Abrothrix* are now clear. Although initially allied with *Oxymycterus* and *Microxus* (Gyldenstolpe 1932), this species was later linked to the subgenus *Abrothrix* of *Akodon* by Osgood (1943b; see also Reig 1987). Mann (1978) considered *lanosa* a subspecies of *A. longipilis*, but Osgood (1943b) and Yáñez et al. (1978) affirmed its specific rank. The lack of recently collected material for comparative studies of non-traditional characters have shielded this species from the most recent cytogenetic and molecular analyses of group relationships. However, Feijoo et al. (2010) reported new records as well as phylogenetic analyses of mtDNA cytochrome-*b* sequences that indicate relationships with *A. jelskii*, and of nucDNA IRBP sequences showing its affinities with *A. longipilis*, the latter seeming more reasonable on geographic grounds. Future analyses with expanded data sets can be expected to clarify the phylogenetic position of *A. lanosa*.

Musser and Carleton (1993) erroneously gave the type locality as "Argentina, Tierra del Fuego Prov., Bahía Monteith, Straits of Magellan." Osgood (1943b) was apparently unable to locate this locality. However, Teta, Pardiñas, and D'Elía (2006), based on historical research, equated the type locality to Seno Monteith, in the Madre de Dios Archipelago, Magallanes y Antártica Chilena region, Chile (see also Feijoo et al. 2010).

Abrothrix longipilis (Waterhouse, 1837)
Long-haired Soft-haired Mouse
SYNONYMS:

Mus longipilis Waterhouse, 1837:16; type locality "Coquimbo," Coquimbo, Chile.

Hesperomys longipilis: Wagner, 1843a:466; name combination.

Mus porcinus Philippi, 1858:78; type locality "in planitie prov. Santiago prope locum Angostura," Metropolitana de Santiago, Chile.

Hesperomys (Habrothrix) longipilis: Thomas, in Milne-Edwards, 1890:28; name combination.

Acodon hirtus Thomas, 1895c:370; type locality "Fort San Rafael, Mendoza," Argentina; but see Pearson and Lagiglia (1992) for doubts about this locality.

Mus brachytarsus Philippi, 1900:37, plate XV, Fig. 2; type locality "Prope Santiago loco dicto Quinta Normal," Metropoolitana de Santiago, Chile.

Mus fusco-ater Philippi, 1900:45, plate XIX, Fig. 1; type locality not specified.

Mus melampus Philippi, 1900:49, plate XX, Fig. 4; type locality "Ad vicum Cartajena ad austrum urbis Valparaiso," vicinity of Cartagena, Valparaiso, Chile.

Akodon suffusus Thomas, 1903c:241; type locality "Valle del Lago Blanco, Southern Chubut (Cordillera Region)," Argentina.

Akodon francei Thomas, 1908:496; type locality "Santa Maria [= Estancia Santa María], Tierra del Fuego," Magallanes y Antártica Chilena, Chile.

Abrothrix francei: Thomas, 1916i:340; name combination.

Abrothrix hirtus: Thomas, 1916i:340; name combination but incorrect gender agreement.

Abrothrix longipilis: Thomas, 1916i:340; first use of current name combination.

Abrothrix suffusus modestior Thomas, 1919b:202; type locality "Maiten [= El Maitén]," Chubut, Argentina.

Abrothrix suffusus moerens Thomas, 1919b:203; type locality "Beatriz, [Península Quetrihué,] Nahuel Huapi. 800 m," Río Negro, Argentina.

Chelemys angustus Thomas, 1927d:654; type locality "Bariloche, E. of Lake Nahuel Huapi. Alt. 800 m," Río Negro, Argentina (see Pearson 1984).

Abrothrix hirta nubila Thomas, 1929:40; type locality "Estancia, Alta Vista, Lago Argentino, 600m," Santa Cruz, Argentina.

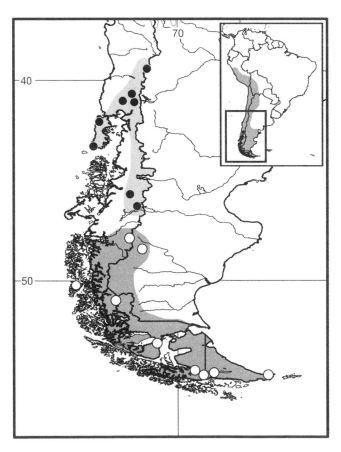

Map 50 Selected localities for *Abrothrix lanosa* (○) and *Abrothrix sanborni* (●). Contour line = 2,000 m.

Abrothrix hirta hirta: Thomas, 1929:40; name combination.

Abrothrix hirta moerens: Thomas, 1929:40; name combination.

Abrothrix hirta suffusa: Thomas, 1929:40; name combination.

Akodon [(*Abrothrix*)] *francei*: Ellerman, 1941:416; name combination.

Akodon [(*Abrothrix*)] *hirta hirta*: Ellerman, 1941:416; name combination.

Akodon [(*Abrothrix*)] *hirta moerens*: Ellerman, 1941:416; name combination.

Akodon [(*Abrothrix*)] *hirta nubila*: Ellerman, 1941:417; name combination.

Akodon [(*Abrothrix*)] *hirta suffusa*: Ellerman, 1941:416; name combination.

Akodon (*Abrothrix*) *longipilis apta* Osgood, 1943b:188; type locality "Piedra de Aguilas, Sierra Nahuelbuta, Malleco, [Araucanía,] Chile. Altitude about 4,000 feet."

Akodon (*Abrothrix*) *longipilis castaneus* Osgood, 1943b:189; type locality "Mocha Island, coast of southern Chile, Province of Arauco."

Akodon (*Abrothrix*) *longipilis francei*: Osgood, 1943b:193; name combination.

Akodon (*Abrothrix*) *longipilis hirta*: Osgood, 1943b:191; name combination.

Akodon (*Abrothrix*) *longipilis longipilis*: Osgood, 1943b: 184; name combination.

Akodon (*Abrothrix*) *longipilis moerens*: Osgood, 1943b: 190; name combination.

Akodon (*Abrothrix*) *longipilis suffusa*: Osgood, 1943b: 192; name combination.

Akodon (*Abrothrix*) *longipilis nubila*: Osgood, 1943b:193; name combination.

Akodon longipilis aptus: Cabrera, 1961:455; name combination.

Akodon longipilis castaneus: Cabrera, 1961:455; name combination.

Akodon longipilis francei: Cabrera, 1961:455; name combination.

Akodon longipilis hirtus: Cabrera, 1961:456; name combination.

Akodon longipilis longipilis: Cabrera, 1961:456; name combination.

Akodon longipilis moerens: Cabrera 1961:456; name combination.

Akodon longipilis nubilus: Cabrera, 1961:457; name combination.

Akodon longipilis suffusus: Cabrera, 1961:457; name combination.

Akodon (*Abrothrix*) *longipilis casteneus* Yáñez et al., 1978:188, 189, 196; incorrect subsequent spelling of *Akodon* (*Abrothrix*) *longipilis castaneus* Osgood.

Abrothrix longipilis hirtus: Pearson and Lagiglia, 1992:37; name combination and incorrect gender agreement.

[*Abrothrix longipilis*] *angustus*: Musser and Carleton, 2005:1090; name combination and incorrect gender agreement.

[*Abrothrix longipilis*] *apta*: Musser and Carleton, 2005: 1090; name combination.

[*Abrothrix longipilis*] *brachytarsus*: Musser and Carleton, 2005:1090; name combination and incorrect gender agreement.

[*Abrothrix longipilis*] *castaneus*: Musser and Carleton, 2005:1090; name combination and incorrect gender agreement.

[*Abrothrix longipilis*] *francei*: Musser and Carleton, 2005: 1090; name combination.

[*Abrothrix longipilis*] *fusco-ater*: Musser and Carleton, 2005: 1090; name combination.

[*Abrothrix longipilis*] *melampus*: Musser and Carleton, 2005: 1090; name combination.

[*Abrothrix longipilis*] *modestior*: Musser and Carleton, 2005:1090; name combination.

[*Abrothrix longipilis*] *moerens*: Musser and Carleton, 2005:1090; name combination.

[*Abrothrix longipilis*] *nubila*: Musser and Carleton, 2005: 1090; name combination.

[*Abrothrix longipilis*] *porcinus*: Musser and Carleton, 2005: 1090; name combination and incorrect gender agreement.

[*Abrothrix longipilis*] *suffusus*: Musser and Carleton, 2005:1090; name combination and incorrect gender agreement.

DESCRIPTION: Large and robust (head and body length up to 130 mm); tail bicolored; pelage with long guard hairs. Color varies greatly geographically but most races retain reddish-brown markings on back bordered by more uniformly grayish flanks and venter. Skull robust, with sharply trumpeted nasals; nasofrontal suture markedly acuminate (extending posteriad beyond line drawn between lacrimal bones); masseteric tubercle absent; palate long, extending past posterior borders of M3s. Baculum "simple" (*sensu* Spotorno 1992); strongly curved at the tip, lacking discrete apical elements.

DISTRIBUTION: *Abrothrix longipilis* ranges from Coquimbo, Chile, and Mendoza, Argentina, in the north to Tierra del Fuego in the south (Osgood 1943b; Mann 1978; Pardiñas, Teta et al. 2003). The species is known from Late Pleistocene and Holocene deposits in Chile (Simonetti and Rau 1989; Simonetti and Saavedra 1998; Saavedra et al. 2003) and from several Holocene archaeological sites in extra-Andean Patagonia of Argentina (Pearson 1987; Pearson and Pearson 1993; Pardiñas 1998, 1999b; Teta et al. 2005).

SELECTED LOCALITIES (Map 51): ARGENTINA: Chubut, Estancia San Pedro (Pardiñas, Teta et al. 2003),

Paso del Sapo (Pardiñas, Teta et al. 2003), Río Arrayanes, Parque Nacional Los Alerces (MVZ 188127); Mendoza, Fort San Rafael (type locality of *Acodon hirtus* Thomas), Laguna de la Niña Encantada, Los Molles (Massoia et al. 1994b); Neuquén, Cañadón del Tordillo (Pardiñas, Teta et al. 2003), Parque Nacional Laguna Blanca (Pardiñas, Teta et al. 2003); Río Negro, Cerro Somuncurá Chico (Pardiñas, Teta et al. 2003), Estancia Maquinchao (Pardiñas, Teta et al. 2003); Santa Cruz, Lago Cardiel (Osgood 1943b), Seccional Glacier Moreno (MVZ 160117), mouth of Río Coig (Osgood 1943b); Tierra del Fuego, Bahía San Sebastián (Marconi 1988). CHILE: Aysén, Río Coihuique [*sic*] Station (Osgood 1943b); Araucanía, Parque Nacional Nahuelbuta (MVZ 140000); Coquimbo, Coquimbo (type locality of *Mus longipilis* Waterhouse), Parque Nacional Fray Jorge (FMNH 130902); Los Lagos, Maicolpue, 2 km S of Bahía Mansa (FMNH 130584), Petrohué, Lago Todos Santos (Osgood 1943b); Los Ríos, 6 km S of Puerto Fuy (MVZ 173700); Magallanes y Antártica Chilena, El Torcido (TTU 30136), Lago Pehoe, Torres del Paine (MVZ 175129), Santa Maria, near Porvenir (type locality of *Akodon francei* Thomas); Santiago, Angostura (type locality of *Mus porcinus* Philippi); Valparaíso, Los Molles, 30 km N of La Ligua (MVZ 150038).

SUBSPECIES: Osgood (1943b), Cabrera (1961), and Mann (1978) regarded *A. longipilis* as polytypic with eight subspecies, four limited to Chile (from north to south *longipilis*, *apta*, *castanea*, and *francei*) and four either restricted to or largely restricted to Argentina (*hirta*, *moerens*, *suffusa*, and *nubila*). Palma et al. (2010) analyzed variation in the mtDNA cytochrome-*b* and nucDNA beta-fibrinogen sequences across the species' full latitudinal range in Chile. They found reasonable concordance between clades and subspecific taxonomy, and they recovered the northernmost subspecies (*longipilis*) as the basal split in *A. longipilis*. In a broader study, including 41 Argentinean and Chilean populations, Sierra-Cisternas (2010) tested the agreement between classical taxonomy and the recovered molecular phylogeographic pattern. This author found eight major clades strongly supported, structured geographically, and with significant genetic differentiation. In contrast with Palma et al. (2010), Sierra-Cisternas (2010) found little correspondence between clades and traditional subspecific classification. Both morphological (M. H. Gallardo et al. 1988) and molecular (Palma et al. 2010; Sierra-Cisternas 2010) data strongly suggest that, in its current meaning, *A. longipilis* includes two or more species. This complex is ripe for a comprehensive analysis of the geographic variation in both morphological and molecular characters, both to validate the existing infraspecific taxonomy

and to delineate the processes underlying the impressive diversification observed.

NATURAL HISTORY: Pearson (1983) and Kelt (1994) have provided notes on the natural history of *Abrothrix longipilis* in Argentina and southern Chile, respectively. This species is abundant in dense forests of *Nothofagus* (Meserve, Lang et al. 1991), but also occurs in most other habitats, such as marshes, shrubby steppes, tussock grass, or rocky areas (Kelt 1994; Pearson 1995). Although these animals burrow well, they can also climb trees (Pearson 1983). They may be active in the morning, afternoon, and at night (Pearson 1983). In Argentina, the diet included berries, seeds, fern spores, insects, fungi, worms, and slugs (Pearson 1983). In Chile, the species has been alternately characterized as either herbivorous-insectivorous or highly insectivorous, and fungivorous-insectivorous or strongly fungivorous (Meserve et al. 1988; S. I. Silva 2005).

Abrothrix longipilis breeds in spring and summer months in Río Negro province, Argentina, and in southern Chile (Pearson 1983; L. A. González and Murúa 1985; Kelt 1994; Guthmann et al. 1997). Pearson (1992) provided details of the reproductive cycle of both sexes in Río Negro and Neuquén provinces, Argentina. He found that males were equal in length to but 8% heavier than females and that sexually active males were longer and heavier than inexperienced males. Few wild mice survived two winters, and most females underwent several infertile ovulation cycles before becoming pregnant, shedding 4.66 ova and giving birth to litters averaging 3.85 young. According to Pearson (1983) and Kelt (1994), by late October most overwintering males had entered breeding condition, and by November and December essentially all adult males were breeding. Testes and vesicular glands of breeding individuals each exceeded 9 mm in length (Pearson 1983). Pregnant and lactating females were caught in late October, November, December, and February (Pearson 1983; Kelt 1994), and reproduction ceased before April (Pearson 1983). In northern Chile, reproductive activity peaked much earlier than in Argentina (Meserve and Glanz 1978). Embryo counts averaged 3.78 (range: 2–5) in Río Negro (Pearson 1983), and 3.7 (range: 2 to 5) in Malleco, Chile (Greer 1965). In the northern Chilean scrub, females had embryo counts of four and six (Meserve and Le Boulengé 1987).

The age structure in Río Negro and southern Chile remains relatively constant over the year, although populations include more juveniles in the fall (Pearson 1983; Kelt 1994). High numbers of old individuals, especially in spring samples, suggested that populations contain significant numbers of individuals that have survived two winters (Pearson 1983; Guthmann et al. 1997). Longitudinal studies suggest that *A. longipilis* populations are less eruptive than sympatric *A. olivacea* or *Oligoryzomys longicaudatus* (Meserve et al. 1999). Population peaks of *A. longipilis* occurred at the end of summer and beginning of autumn (Guthmann et al. 1997). Home-range size in forest agroecosystems of central Chile varied between 1,636 m² in winter to 2,758 m² in spring (Muñoz-Pedreros et al. 1990). Recaptures of this species at Malleco were in areas no greater than 40 to 241 m² (Greer 1965). Densities in the *Nothofagus* forest of Río Negro, Argentina, ranged from 0.4 to 4.8 individuals per hectare in the spring to 2.8 to 10.8 in the autumn (Pearson and Pearson 1982). In southern Chile, *Abrothrix longipilis* was found to be one of the most abundant rodents in pristine temperate rainforests, and occurred across all sampled elevations (B. D. Patterson et al. 1989, 1990). Pearson (2002) and Sage et al. (2007) described a population outbreak of this species in northwestern Argentinean Patagonia.

Abrothrix longipilis is a main prey item of both Barn (*Tyto alba*) and Magellanic Horned Owls (*Bubo magellanicus*) in Argentina and Chile (Jaksic et al. 1981; Trejo and Grigera 1998; Pillado and Trejo 2000; Teta et al. 2001; Trejo and Ojeda 2004; Nabte et al. 2006). Other important predators are the Rufous-legged Owl (*Strix rufipes*) in forests of Chile and Argentina (I. Díaz 1999; D. R. Martínez and Jaksic 1997; Udrizar Sauthier et al. 2005), the White-tailed Kite (*Elanus leucurus*) in central and southern Chile (Jaksic et al. 1981; González-Acuna et al. 2009), and the South American Gray Fox (*Lycalopex griseus*) in the forests of Valdivia, Chile (Rau et al. 1995). In the Mediterranean region of Chile, *A. longipilis* is an important reservoir of Andes Virus, which is the predominant etiological agent of hantavirus cardiopulmonary syndrome, in which it is second only to *Oligoryzomys longicaudatus* (R. A. Medina et al. 2009).

REMARKS: This species was long thought to be closely related to and perhaps conspecific with *A. sanborni* (see Osgood 1943b; Pine et al. 1979) and *A. lanosus* (Mann 1978). Based on both mtDNA and nucDNA sequence analyses, Palma et al. (2010) suggested a sister relationship between *A. longipilis* and *A. jelskii* rather than with *A. sanborni*, whereas Teta, D'Elía, Pardiñas et al. 2011) recovered *A. longipilis* as sister to *A. lanosa*. Pearson (1984) assigned *Chelemys angustus* to the synonymy of *A. longipilis*; other synonyms follow Osgood (1943b) and Reig (1987). Osgood (1943b) and Muñoz-Pedreros (2000) delineated and mapped subspecies boundaries in Chile. As yet, the patterns of character variation that might support subspecies recognition across the broad species range to the east in Río Negro and Chubut provinces, Argentina (Teta et al. 2002), have yet to be evaluated.

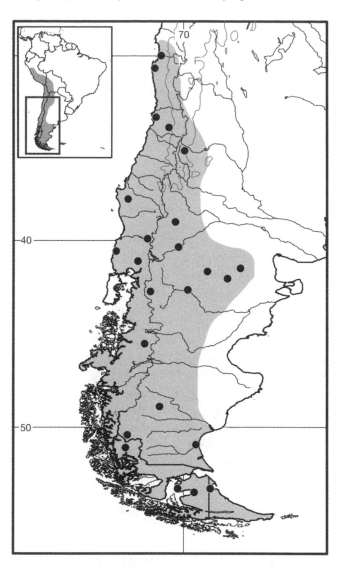

Map 51 Selected localities for *Abrothrix longipilis* (●). Contour line = 2,000 m.

Abrothrix olivacea (Waterhouse, 1837)
Olive Soft-haired Mouse
SYNONYMS:

Mus olivaceus Waterhouse, 1837:16; type locality "Valparaiso," Valparaíso, Chile.

Mus brachiotis Waterhouse, 1837:17; type locality "in insula parvulâ apud Midship Bay, Chonos Archipelago" subsequently fixed to "Islet off E. coast of [Isla] Chiloé," Los Lagos, Chile, by Thomas (1927b:551).

Mus xanthorhinus Waterhouse, 1837:17; type locality "Santa Cruz," Santa Cruz, Argentina, subsequently fixed to "Hardy Peninsula, Tierra del Fuego [= Isla Hoste, Magallanes, Chile]," by Thomas (1927b:551).

Mus canescens Waterhouse, 1837:17; type locality "Port Desire" [= Puerto Deseado], Santa Cruz, Argentina.

Mus Renggeri Waterhouse, 1839:51; type locality "Valparaiso and Coquimbo," subsequently fixed to "Valparaíso," Valparaíso, Chile, by Thomas (1927b:550).

Mus brevicaudatus Philippi, 1872:446; type locality "Puerto Montt," Los Lagos, Chile.

Hesperomys [*Calomys*] *canescens*: Burmeister, 1879:227; name combination.

Hesperomys (*Habrothrix*) *xanthorhinus*: Thomas, 1881b: 5; name combination.

Hesperomys (*Habrothrix*) *olivaceus*: Thomas, 1882:110; name combination.

A[*codon*]. *olivaceus*: Thomas, 1894b:364; name combination.

Akodon olivaceus: J. A. Allen, 1897c:117; name combination.

Hesperomys (*Acodon*) *xanthorhinus*: Matschie, 1898:7; name combination.

Mus brachyotis Philippi, 1900:12; incorrect subsequent spelling of *Mus brachiotis* Waterhouse.

Mus lepturus Philippi, 1900:17, plate IV, Fig. 2; type locality "provincia Santiago ad Peine," Antofagasta, Chile.

Mus psilurus Philippi, 1900:17, plate IV, Fig. 3; type locality "provincia Colchagua," O'Higgins, Chile.

Mus trichotis Philippi, 1900:18, plate V, Fig. 1; type locality "Andibus provinciae Santiago," Andes, Metropolitana de Santiago, Chile.

Mus Foncki Philippi, 1900:20, plate V, Fig. 4; type locality "prope Puerto Montt in provincia Llanquihue," Los Lagos, Chile.

Mus chonoticus Philippi, 1900:24, plate VII, Fig. 2; type locality "in insulis 'Chonos'," Chonos Archipelago, Los Lagos, Chile.

Mus vinealis Philippi, 1900:24, plate VII, Fig. 3; type locality "provincia Santiago in vineis sarmenta adscendens, uvas consumens," Metropolitana de Santiago, Chile.

Mus (*Oxymycterus*) *Landbecki* Philippi, 1900:26, plate VIII, Fig. 2; type locality "prope Illapel et Choapa," Coquimbo, Chile.

Mus (*Oxymycterus*) *senilis* Philippi, 1900:27, plate VIII, Fig. 3; type locality "Andibus provinciae Santiago loco dicto 'Valle de Yeso'," Metropolitana de Santiago, Chile.

Mus Germaini Philippi, 1900:32, plate XII, Fig. 2; type locality "provincia Santiago," Metropolitana de Santiago, Chile.

Mus nasica Philippi, 1900:38, plate XV, Fig. 3; type locality not specified but presumed to be Chile.

Mus ruficaudus Philippi, 1900:40, plate XVII, Fig. 1; type locality "Chile."

Mus macronychos Philippi, 1900:40, plate XVII, Fig. 2; type locality "ex provinciis centralibus" [= from central provinces], Chile.

Mus infans Philippi, 1900:41, plate XVII, Fig. 3; type locality "provinciis centralibus?" [= central provinces?], Chile.

Mus xanthopus Philippi, 1900:41, plate XVII, Fig. 4; type locality "prope Osorno," Los Lagos, Chile.

Mus Mochae Philippi, 1900:42, plate XVII, Fig. 5; type locality "isla Mocha," Arauco, Biobío, Chile.

Mus pencanus Philippi, 1900:46, plate XIX, Fig. 2; type locality "prope Concepción," Concepción, Biobío, Chile.

Mus nemoralis Philippi, 1900:49, plate XX, Fig. 3; type locality "prope oppidum Valdivia," Valdivia, Los Ríos, Chile.

Mus longibarbus Philippi, 1900:51, plate XXI, Fig. 4; type locality "provincia Valdivia," Valdivia, Los Ríos, Chile.

Mus atratus Philippi, 1900:57, plate XXV, Fig. 3; type locality "provincia Maule," Maule, Chile.

Akodon beatus Thomas, 1919b:204; type locality "Beatriz, [Península Quetrihué,] Nahuel Huapi. 800 m," Río Negro, Argentina.

Abrothrix brachiotis: Thomas 1927b:551; name combination.

Akodon arenicola beatus: Gyldenstolpe, 1932a:103; name combination.

Akodon [(*Akodon*)] *olivaceus*: Ellerman, 1941:412; name combination.

Akodon [(*Akodon*)] *xanthorhinus*: Ellerman, 1941:432; name combination.

Akodon [(*Akodon*)] *brachiotis*: Ellerman, 1941:416; name combination.

Akodon olivaceus beatus: Osgood, 1943b:176; name combination.

Akodon olivaceus brachiotis: Cabrera, 1961:444; name combination.

Akodon olivaceus mochae: Cabrera, 1961:445; name combination.

Akodon olivaceus olivaceus: Cabrera, 1961:445; name combination.

Akodon olivaceus pencanus: Cabrera, 1961:446; name combination.

Akodon xanthorhinus canescens: Cabrera, 1961:450; name combination.

Akodon xanthorhinus xanthorhinus: Cabrera, 1961:450; name combination.

Akodon (*Akodon*) *markhami* Pine, 1973:423; type locality "1.2 km NWW [or 1.2 km NNW] Puerto Edén, Isla Wellington, ca. 600 km NNW, aprox. de Punta Arenas, Magallanes, [Magallanes y Antártica Chilena,] Chile."

Akodon (*Akodon*) *llanoi* Pine, 1976:63; type locality "head of Bahía Capitán Canepa (54°51′S, 64°27′W), Isla de los Estados (Staten Island), Territorio Nacional de la Tierra del Fuego [–Terra del Fuego, Antártida e Islas del Atlántico Sur], Argentina."

Akodon (*Abrothrix*) *mansoensis* De Santis and Justo, 1980:121; type locality "Río Manso superior, Estación Aforo, Bariloche," Río Negro, Argentina.

Akodon (*Akodon*) *brachiotis*: Reig, 1987:366; name combination.

Akodon (*Akodon*) *olivaceus*: Reig, 1987:366; name combination.

Akodon (*Abrothrix*) *xanthorhinus*: Reig, 1987:367; name combination.

Abrothrix olivaceus: Tamayo, Núñez, and Yáñez, 1987:5; first use of current name combination but incorrect gender agreement.

Abrothrix xanthorhinus: Tamayo Núñez, and Yáñez, 1987:5; name combination and incorrect gender agreement.

Abrothrix xanthorhinus xanthorhinus: Massoia and Chebez, 1993:98; name combination and incorrect gender agreement.

Abrothrix xanthorhinus llanoi: Massoia and Chebez, 1993:102; name combination.

[*Abrothrix olivaceus*] *atratus*: Musser and Carleton, 2005:1000; name combination and incorrect gender agreement.

[*Abrothrix olivaceus*] *beatus*: Musser and Carleton, 2005: 1000; name combination and incorrect gender agreement.

[*Abrothrix olivaceus*] *brachiotis*: Musser and Carleton, 2005:1000; name combination.

[*Abrothrix olivaceus*] *brevicaudatus*: Musser and Carleton, 2005:1000; name combination and incorrect gender agreement.

[*Abrothrix olivaceus*] *canescens*: Musser and Carleton, 2005:1000; name combination.

[*Abrothrix olivaceus*] *chonoticus*: Musser and Carleton, 2005:1000; name combination and incorrect gender agreement.

[*Abrothrix olivaceus*] *foncki*: Musser and Carleton, 2005: 1000; name combination.

[*Abrothrix olivaceus*] *germaini*: Musser and Carleton, 2005: 1000; name combination.

[*Abrothrix olivaceus*] *infans*: Musser and Carleton, 2005: 1000; name combination.

[*Abrothrix olivaceus*] *landbecki*: Musser and Carleton, 2005:1000; name combination.

[*Abrothrix olivaceus*] *lepturus*: Musser and Carleton, 2005:1000; name combination and incorrect gender agreement.

[*Abrothrix olivaceus*] *longibarbus*: Musser and Carleton, 2005:1000; name combination and incorrect gender agreement.

[*Abrothrix olivaceus*] *macronychos*: Musser and Carleton, 2005:1000; name combination and incorrect gender agreement.

[*Abrothrix olivaceus*] *mansoensis*: Musser and Carleton, 2005:1000; name combination.

[*Abrothrix olivaceus*] *mochae*: Musser and Carleton, 2005: 1000; name combination.

[*Abrothrix olivaceus*] *nasica*: Musser and Carleton, 2005: 1000; name combination.

[*Abrothrix olivaceus*] *nemoralis*: Musser and Carleton, 2005:1000; name combination.

[*Abrothrix olivaceus*] *pencanus*: Musser and Carleton, 2005:1000; name combination and incorrect gender agreement.

[*Abrothrix olivaceus*] *psilurus*: Musser and Carleton, 2005: 1000; name combination and incorrect gender agreement.

[*Abrothrix olivaceus*] *renggeri*: Musser and Carleton, 2005:1000; name combination.

[*Abrothrix olivaceus*] *ruficaudus*: Musser and Carleton, 2005:1000; name combination and incorrect gender agreement.

[*Abrothrix olivaceus*] *senilis*: Musser and Carleton, 2005: 1000; name combination.

[*Abrothrix olivaceus*] *trichotis*: Musser and Carleton, 2005: 1000; name combination.

[*Abrothrix olivaceus*] *vinealis*: Musser and Carleton, 2005: 1000; name combination.

[*Abrothrix olivaceus*] *xanthopus*: Musser and Carleton, 2005:1000; name combination and incorrect gender agreement.

Abrothrix olivaceus tarapacensis Rodríguez-Serrano, Cancino, and Palma, 2006:977; type locality "Quebrada de Tarapacá, región I Tarapacá, Chile."

DESCRIPTION: Small; pelage color varies greatly geographically, ranging from dark brown or olive brown to gray or pale ochraceous; some forms with yellow or orange patches on nose, feet, and tail. Skull small (condyloincisive length <27mm) and delicate, with short and narrow rostrum, bowed dorsal profile, and moderately extended nasals and premaxillae; palate long, extending past posterior borders of M3s. Distal baculum minute, with vestigial apical digits (Spotorno 1992).

DISTRIBUTION: *Abrothrix olivacea* ranges from northernmost Chile through southernmost Chile and Argentina. A disjunct low-elevation population lives in lomas of the Atacama Desert, and the species ranges exclusively west of the Andes until the latitude of Concepción, where it extends onto the eastern slopes of the Andes, reaching the Atlantic near Península Valdés at about 43°S and spreading across much of Patagonia. Its northern range in Argentina corresponds to that of the Patagonian Botanical Province (Osgood 1943b; Lozada et al. 1996; Pardiñas, Teta et al. 2003). Isolated historical records, perhaps reflecting a past wider distribution, are known from Monte Desert localities in northern Patagonia (i.e., Choele Choei [MACN 28.111], Laguna del Barro [Thomas 1929], Puerto Madryn [Mahnert 1982], and Rawson [Thomas 1929]). *Abrothrix olivacea* also occurs on many of the larger islands in the coastal archipelago, including Isla

Grande de Tierra del Fuego and Isla Wellington (represented by the taxon *markhami*), but appears to be contiguously allopatric with the insular endemic *Abrothrix hershkovitzi* on Isla Capitán Aracena and the Cape Horn island group.

SELECTED LOCALITIES (Map 52): ARGENTINA: Chubut, 4 km N of El Maitén (MVZ 151006), 30 km W of José de San Martín (MVZ 206338), Rawson (Pardiñas, Teta et al. 2003); Mendoza, Laguna de la Niña Encantada, Los Molles (Massoia et al. 1994b); Neuquén, Parque Nacional Laguna Blanca (Pardiñas, Teta et al. 2003); Río Negro, Cerro Somuncurá Chico (Pardiñas, Teta et al. 2003); Santa Cruz, Cape Fairweather (Osgood 1943b), 52 km WSW of El Calafate (MVZ 151011), 4.4 km E of Los Antiguos, Estancia La Aurora (MVZ 206341), Puerto Deseado (type locality of *Mus canescens* Waterhouse); Tierra del Fuego, Estancia Cullen (Osgood 1943b), Estancia Via Monte (Osgood 1943b), head of Bahía Capitán Canepa, Isla de los Estados (type locality of *Akodon (Akodon) llanoi* Pine), Lake Yerwin (Osgood 1943b). CHILE: Aysén, Archipiélago de los Chonos (type locality of *Mus chonoticus* Philippi), Puerto Aysén (Osgood 1943b); Antofagasta, Paposo (MVZ 119157); Araucanía, Parque Nacional Nahuelbuta (MVZ 137886); Arica y Parinacota, Quebrada de Camarones (Rodríguez-Serrano, Hernández, and Palma 2008); Atacama, Caldera (Osgood 1943b); Biobío, Isla Mocha (type locality of *Mus Mochae* Philippi); Coquimbo, Romero (Osgood 1943b); Los Lagos, Puerto Montt (type locality of *Mus Foncki* Philippi), mouth of Río Inio, Isla Grande de Chiloé (Osgood 1943b); Magallanes y Antártica Chilena, Hardy Peninsula, Isla Hoste (type locality of *Mus xanthorhinus* Waterhouse), Isla Wellington, Puerto Eden (type locality of *Akodon (Akodon) markhami* Pine), Lake Sarmiento (Osgood 1943b), Puerto Natales, Última Esperanza (Osgood 1943), Punta Arenas (Osgood 1943b), east end of Riesco Island (Osgood 1943b); Tarapacá, mouth of Río Loa (Rodríguez-Serrano, Hernández, and Palma 2008); Valparaíso, Valparaíso (type locality *Mus olivaceus* Waterhouse), San Carlos de Apoquindo (Osgood 1943b).

SUBSPECIES: Multiple subspecies had been recognized for both *olivacea* and *xanthorhina* before their amalgamation into a single species (see Remarks). For example, Cabrera (1961) listed five geographic races for *olivacea*, four in Chile (from north to south along the coast, the nominotypical form, *pencana*, and *brachiotis*, with *mochae* from Isla Mocha) and *beata* in Argentina (see also Osgood 1943b; Mann 1978). Cabrera (1961) included two subspecies in his concept of *xanthorhina*, the nominotypical form from forests of extreme southern Chile and *canescens* from the steppes of Patagonian Argentina. Since then, two insular taxa described as species

have been synonymized with *A. olivacea*, namely, *llanoi* and *markhami* (B. D. Patterson et al. 1984; Rodríguez-Serrano, Hernández, and Palma 2008). Rodríguez-Serrano, Hernández, and Palma (2008) described *tarapacensis* as a subspecies from northern Chile. A thorough understanding of patterns of morphological variation as well as comprehensive phylogeographic analyses are needed to gauge the validity of each of these geographic forms and their relationships to the insular endemic *A. hershkovitzi*.

NATURAL HISTORY: In central Chile, *A. olivacea* lives in semiarid scrub habitat; in Patagonia, it occupies semiarid bushy steppes and bunchgrass habitats (Kelt 1994; Pearson 1995); and in the Andes, it is found in moister habitats such as meadows, thick grasses, and dense *Nothofagus* forests from western Neuquén, western Río Negro, and northwestern Chubut, Argentina, to southern Chile and Tierra del Fuego (B. D. Patterson et al. 1984; Pearson 1995; Lozada et al. 1996). In the Chilean archipelago, the species occurs in dense coastal forest of *Nothofagus betuloides* [Nothofagaceae], *Drimys winteri* [Verbenaceae], and *Lomatia ferruginea* [Proteaceae] and in grassy meadows with sparse shrub cover (the type series of *markhami* Pine).

Pearson (1983), Kelt (1994), and Lozada et al. (1996) summarized the natural history of *Abrothrix olivacea*. It is partly diurnal and climbs well (Pearson 1983; Tapia 1995). It builds simple nests of grass, underground or sheltered in roots or rocks (Mann 1978), and, when living in dense grass, creates runways (Pearson 1983). It can also dig, and in Chile has been found living in burrows of other rodents such as *Spalacopus*. The diet includes berries, seeds, arthropods, green vegetation, fungi (Pearson 1983; Lozada et al. 1996), and invertebrates (see S. I. Silva 2005). These mice reproduce from early spring through summer (Meserve and Glanz 1978; Pearson 1983; Murúa et al. 1987; Kelt 1994; Guthmann et al. 1997), but the length of the reproductive season may vary (Kelt 1994). Pregnant and lactating females have been caught in late October, November, December, February, March, and April (Pearson 1983; Kelt 1994). The average litter size ranged from 5.0 (in steppe habitat) to 5.1 (in forest habitat) in Río Negro (Pearson 1983; Lozada et al. 1996), and 5.5 in Malleco, Chile (Greer 1965), and Tierra del Fuego (Marconi 1988). Females produced two or three litters per season (Lozada et al. 1996; Muñoz-Pedreros 2000). Older females tended to have large litters (Pearson 1983). Reproductively active males had testes 10–13 mm long and vesicular glands 10 mm in length (Pearson 1983).

The population structure clearly reflects summer recruitment to the autumn population (Pearson 1983; Kelt 1994). Adult males weighed a significant 9.2% more than adult females (Pearson 1983). Murúa and González (1985) documented seasonal fluctuations in numbers and body mass, and described multiannual cycles with temporally different cohorts characterized by different age structures and survival rates. These multiannual fluctuations are variable both in space and time, and local populations may remain stable for long periods (Meserve, Kelt, and Martínez 1991; Meserve et al. 1999). El Niño events can result in increased densities by an order of magnitude or more (Meserve et al. 1995).

Factors involved in regulating numbers may include age composition of the breeding stock, differential survival of cohorts, and presaturation dispersal (Murúa and González 1985). Maximum density can vary both seasonally and geographically. In Chilean shrubland-grassland, maximum density was attained in the fall, with 37.7 individuals per hectare; in temperate Chilean woodlands, density peaked in winter at 37.9 individuals per hectare (L. C. González et al.1982). At a semiarid site in northern Chile, Fulk (1975) estimated density at 30.3 individuals per hectare in August and 97.0 individuals per hectare in November, with an average home range diameter of 54.0 m. Densities in *Nothofagus* forest of Río Negro province, Argentina, ranged from 0 to 1.9 individuals per hectare in spring and increased to 4.2 to 17.9 in autumn (Pearson and Pearson 1982). Lozada et al. (1996) reported a maximum of 40 individuals per hectare in steppe habitat in Río Negro province, and Marconi (1988) recorded an even higher density (63 individuals per hectare) in southern Tierra del Fuego province, Argentina. Over four consecutive years, densities in steppe habitat in Río Negro, Argentina, peaked consistently in March, with lows in September (Lozada et al. 1996). Jaksic and Lima (2003) reviewed this history of population outbreaks.

In forest agroecosystems of central Chile, the home range varied between 1,779 m² in winter to 2,776 m² in spring (Muñoz-Pedreros et al. 1990). In Chilean temperate rainforest and shrubland-grassland, males increased their home range at the onset of the reproductive period while females maintained the same home range size throughout the year (L. C. González et al. 1982). The average home range in steppe areas of Río Negro varied from 548 m² for females and 598 m² for males during the nonreproductive season, and increased to 610 m² and 1,307 m² in reproductively active females and males, respectively (K. M. Heinemann et al. 1995). Home range was reduced substantially during outbreaks in southern Chile (L. A. González et al. 2000). B. D. Patterson et al. (1989, 1990) examined the distribution, abundance, and habitat association along an elevational transect in a temperate rainforest in Chile; *A. olivacea* was the most frequently captured of the 10 small mammals sampled and occurred at all elevations as well as exhibited modal habitat preferences.

Barn Owls (*Tyto alba*) and Magellanic Horned Owls (*Bubo magellanicus*) prey upon *A. olivacea* throughout its range (Jaksic et al. 1981; Torres-Mura and Contreras 1989; Trejo and Grigera 1998; Pillado and Trejo 2000; Teta et al. 2001; Trejo and Ojeda 2004; Nabte et al. 2006). Additional predators include by the Burrowing Owl (*Athene cunicularia*) in the Mediterranean ecosystems of Chile (Fulk 1977; Meserve et al. 1987) and in northern Patagonia, Argentina (A. Andrade, Udrizar Sauthier et al. 2004), the Short-eared Owl (*Asio flammeus*) in Chile's Lake District (Figeroa-R. et al. 2009) and in the agricultural landscapes of southern Chile (D. R. Martínez et al. 1998), the Rufous-legged Owl (*Strix rufipes*) in forests of Chile and Argentina (I. Díaz 1999; Udrizar Sauthier, Andrade et al. 2005), the White-tailed Kite (*Elanus leucurus*) in central and southern Chile (Jaksic et al. 1981; González-Acuna et al. 2009), and the South American Gray Fox (*Lycalopex griseus*) in the Valdivian forests of Chile (Rau et al. 1995).

This species is known from Holocene remains in Chile (Simonetti and Saavedra 1998; Saavedra, Quiroz, and Iriarte 2003) and from several Holocene archaeological sites in the extra-Andean Argentinean Patagonia (Pearson 1987; Pearson and Pearson 1993; Pardiñas 1998, 1999b; Teta et al. 2005).

REMARKS: For most of the twentieth century, *olivacea* Waterhouse and *xanthorhina* Waterhouse were regarded as separate species, and locally they can appear and behave quite differently (see Osgood 1943b; B. D. Patterson et al. 1984; Pearson 1995). Yet the two merge clinally in cranial and body measurements and in molar characters, which caused Yáñez et al. (1978) to consider them conspecific. More recently, genetic studies of contact zones between the two indicate that they are reproductively confluent. Pearson and Smith (1999) studied what were then regarded as *A. olivacea brachiotis* and *A. xanthorhina canescens* where their ranges are contiguous in northern Patagonia. Morphological analyses demonstrated that intermediates predominated across a 30-km zone near Bariloche, and DNA analyses confirmed their genetic similarity. Similar results were obtained in studies of contact between these same nominal taxa east of Aísen, 500 km south of Bariloche (M. F. Smith et al. 2001). Haplotypes of "*canescens*" had not yet achieved reciprocal monophyly relative to those of "*brachiotis*," thus supporting *olivacea* and *xanthorhina* as conspecific, in keeping with morphometric appraisals (Yáñez et al. 1979). Studies of additional contact zones are needed to resolve the specific or subspecific rank of other *Abrothrix* taxa, especially *canescens-xanthorhina* and *xanthorhina-hershkovitzi* (see also Musser and Carleton 2005).

Both Rodríguez-Serrano, Hernández, and Palma (2008) and Abud (2011) used comparable molecular analyses to show that the insular endemics *markhami* Pine and

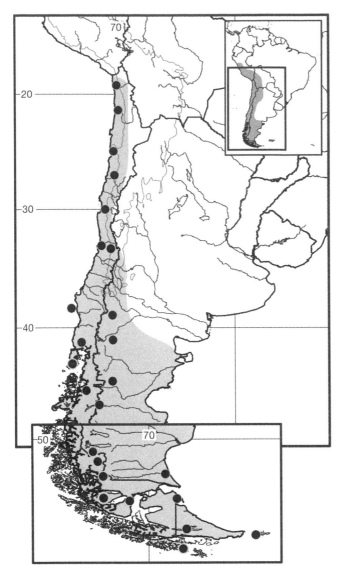

Map 52 Selected localities for *Abrothrix olivacea* (●). Contour line = 2,000 m.

hershkovitzi B. D. Patterson, Gallardo, and Freas, respectively, were independent and recent derivatives of *A. olivacea*. The southern Andes and outlying islands were extensively glaciated until the end of the Pleistocene (Heusser 2003). Thus, territory currently occupied by *A. olivacea* and these two insular derivatives would only have been available for colonization after glacial retreat, either by expanding low-latitude populations or by high-latitude relicts that managed to persist in refugia (Rodríguez-Serrano, Hernández, and Palma 2008; E. P. Lessa et al. 2010). The apparent paraphyly of *A. olivacea* with respect to both *markhami* and *hershkovitzi* is to be expected, given their recent, post-Pleistocene origin. Although we regard *hershkovitzi* as a species, this phylogenetic placement could support its potential synonymy with *A. olivacea*, as we treat *markhami* Pine.

Abrothrix sanborni Osgood, 1943
Sanborn's Soft-haired Mouse

SYNONYMS:

Akodon (*Abrothrix*) *sanborni* Osgood, 1943b:194; type locality "mouth of Rio Inio, south end of Chiloe Island [= Isla Grande de Chiloé]," Los Lagos, Chile.

Akodon sanborni: Cabrera, 1961:457; name combination.

Akodon longipilis sanborni: Mann, 1978:157; name combination.

Abrothrix sanborni: Tamayo, Núñez, and Yáñez, 1987:5; first use of current name combination.

DESCRIPTION: Small; coloration uniformly blackish-brown, including tail and feet, without indication of countershading; small, scantily haired ears; head and body length averaging 105 mm, tail about 76 mm (Osgood 1943b). Skull delicate with narrow and elongate rostrum, trumpeted nasals, acuminate nasofrontal suture extending posteriad beyond line drawn between lacrimals, gracile zygomatic arches, and rounded braincase; infraorbital plates narrow and sloping, with conspicuous masseteric tubercles at base. Baculum "simple" (*sensu* Spotorno 1992) and strongly curved at tip, lacking discrete apical elements.

DISTRIBUTION: *Abrothrix sanborni* inhabits the coastal rainforests of southern Chile, including Isla Grande de Chiloé but not, insofar as known, the Guaitecas Archipelago (Osgood 1943b; B. D. Patterson, unpubl. data). Pearson (1995) recorded this species at Lago Quillén, Neuquén, Argentina. At La Picada, Chile, it was significantly more common at low elevations, and was not recorded at elevations above 820 m (B. D. Patterson et al. 1989).

SELECTED LOCALITIES (Map 49): ARGENTINA: Neuquén, NW shore of Lago Quillén (Pearson 1995). CHILE: Aysén, Lago Atravesado (Figueroa et al. 2001), 4 km NW of Puerto Ibáñez (FMNH 130197); Los Lagos, Anticura, Parque Nacional Puyehue (MVZ 173501), 15 km E of Chepu, Isla Grande de Chiloé (Reis and Venegas 1987), Peulla, Parque Nacional Vicente Perez Rosales (FMNH 124050), mouth of Río Inio, Isla Grande de Chiloé (Osgood 1943b), Valle de la Picada (B. D. Patterson 1989).

SUBSPECIES: *Abrothrix sanborni* is monotypic.

NATURAL HISTORY: *Abrothrix sanborni* is active both day and night (Meserve et al. 1982). Its diet includes high proportions of fungi, moderate amounts of larval and adult arthropods, plant foliage, seeds, and fruits (Meserve et al. 1988). This species breeds in spring and summer (Figueroa et al. 2001). B. D. Patterson (1989, 1990) studied distribution, abundance, and habitat associations along an elevational transect in a temperate rainforest of Chile. *Abrothrix sanborni* was caught at sites with dense canopy cover, tall trees, sparse shrub cover, deep litter, ground cover of bryophytes and bamboo, and gentle slopes (B. D. Patterson et al. 1990).

REMARKS: The species is dark, almost blackish in color and has smaller, more sparsely haired ears than *Abrothrix longipilis*. However, there are persistent reports of putative hybrids between them (see Pine et al. 1979; M. H. Gallardo et al. 1988; Pearson 1995). Morphometric analyses by Yáñez et al. (1978) showed greater quantitative differences between *sanborni* and *longipilis* than among geographic samples of the latter. M. H. Gallardo et al. (1988), citing gross similarities in phenetic and immunodiffusion characters, suggested that *sanborni* was merely another geographic race of *Abrothrix longipilis*. However, Palma et al. (2010) documented that *A. sanborni* and *A. longipilis* were nonsister species, based on phylogenetic analyses of mtDNA and nucDNA sequences. More recently, G. D'Elía, P. Teta, N. S. Upham, and B. D. Patterson (pers. obs.) have each uncovered morphometric and mtDNA sequence evidence for two distinct groupings in *A. sanborni* as treated in this account.

Genus *Chelemys* Thomas, 1903

Pablo Teta, Ulyses F. J. Pardiñas, and Guillermo D'Elía

The genus *Chelemys* includes two species of medium-sized sigmodontine rodents of fossorial habits. It ranges from central Chile and west-central Argentina through the Andes south to the vicinity of the Straits of Magellan. Originally described as a subgenus of *Akodon*, *Chelemys* was later considered to be a distinct genus (J. A. Allen 1905; Thomas 1919b, 1927d; Gyldenstolpe 1932) or was subsumed under *Notiomys* (Osgood 1925, 1943b; Ellerman 1941; Cabrera 1961). Pearson (1984) and Reig (1972, 1987) established the basis for the current classification, treating *Chelemys* as a distinct genus while also according generic rank to *Geoxus* and *Notiomys*. B. D. Patterson (1992a) made additional morphological and morphometric comparisons among *Chelemys* and the other long-clawed Abrotrichini.

Chelemys is a medium-sized, stout, short-tailed mouse. The fur is short, soft, and dense. Ears are small and project above the fur only slightly, and the tail is short and very thickly clothed with hairs that completely cover the scales. The pollex possesses a distinct claw; forefeet are strong and digits II through V possess long and sharply pointed claws. Hindfeet are short and broad, also furnished with strong claws and possess a fringe of hairs along their lateral margins, although these are less developed than in *Notiomys*. The palms and soles are naked; the plantar skin of the hindfeet is squamate, and six large, rounded, and prominent pads are present; pedal digit V has a well-developed claw that reaches the second phalanx of dIV (Gyldenstolpe 1932).

The skull is robust, with a relatively short and broad rostrum, nearly quadrangular braincase, noninflated frontals, an arcuately biconcave interorbital region with non-squared supraorbital edges, a comparatively well-developed although narrow interparietal, broad and nearly vertical to slightly inclined forward zygomatic plates, a palate that extends to (*C. macronyx*) or slightly posteriorly past the molars (*C. megalonyx*), small or indistinct posterolateral palatal pits, a deep and broad mesopterygoid fossa with a rounded to square anterior border, and comparatively deep parapterygoid fossae (Gyldenstolpe 1932; Reig 1987). The carotid circulation pattern is type 1 (*sensu* Voss 1988), characterized by an enlarged carotid canal and stapedial foramen and the supraorbital branch of the stapedial artery visible as a groove along the inferior surface of the squamosal and alisphenoid bones and exiting through a sphenofrontal foramen or anterior alar canal. An alisphenoid strut is absent. The mandible is strongly built, with the ramus much deeper than in *Geoxus*, *Notiomys*, or *Pearsonomys*. The coronoid process is long and strong, and the capsular projection is of medium size (Reig 1987).

The upper incisors are fairly thick and robust, ungrooved, and proodont (*C. megalonyx*) or orthodont (*C. macronyx*). Molars are moderately high, with a greater degree of hypsodonty than in other members of the tribe, rather large and broad in *C. macronyx* and smaller in *C. megalonyx*. M1 has a moderately reduced procingulum without an anteromedian flexus; a parastyle is well developed. Upper molars have the para- and metaflexus strongly oriented backward; there is no mesoflexus on any teeth, as the para- and mesolophs are completely fused; a mesostyle is normally absent, and an enterostyle is totally missing. M3 is relatively well developed, about two-thirds the length of M2. The lower first molar (m1) has an anterior-posteriorly compressed procingulum without an anteromedian flexid in adult individuals and a well-developed prostylid. All lower molars have oblique ento- and posteroflexids, a more transverse metaflexid, coalesced mesoloph and paraloph, and lack both a mesostylid and ectostylid. The third lower molar is sigmoid-shaped, not reduced, more than three-quarters the length of m2 (Reig 1987).

Vertebral counts in nine individuals were 13 thoracic, six lumbar, and 20–22 caudal; the rib count was 13 (Steppan 1995).

Unlike *Notiomys* and *Geoxus*, *Chelemys* has a very large cecum (Pearson 1984). The stomach is unilocular-hemiglandular (Pardiñas, Udrizar Sauthier et al. 2008). *Chelemys macronyx* has a gall bladder (Voss 1991a); its presence is unknown for *C. megalonyx*.

SYNONYMS:

Hesperomys: Waterhouse, 1845:154; part (description of *megalonyx*).

Oxymicterus Gay, 1847:108; part (description of *scalops*); incorrect subsequent spelling of *Oxymycterus* Waterhouse.

Oxymycterus: Philippi, 1872:445; part (inclusion of *megalonyx* and *scalops*).

Acodon Thomas, 1894b:362; part (description of *macronyx*).

Akodon: E.-L. Trouessart, 1897:538; part (listing of *megalonyx* and *macronyx*).

Mus: Philippi, 1900:57; part (description of *microtis*).

Akodon (*Chelemys*) Thomas, 1903c:242; type species *Hesperomys megalonyx* Waterhouse, as subgenus.

Chroeomys: Thomas, 1916i:340; part (described as subgenus, included *scalops* Gay).

Chelemys: Thomas, 1919b:207; first use intended to represent a valid genus.

Notiomys: Osgood, 1925:121; part (listed *megalonyx*, *macronyx*, and *vestitus* as species; described *connectens* as species and *alleni* as subspecies of *vestitus*).

REMARKS: *Oxymycterus delfini* Cabrera 1905 (type locality: "Punta Arenas") is known only by the type specimen, which is lost (Osgood 1943b:166; Cabrera 1961:471). This form was treated as allied to *Notiomys vestitus alleni* (= *Chelemys macronyx*) (Osgood 1925), *C. megalonyx* (Mann 1978; Tamayo and Frassinetti 1980), or as a distinct species under the genera *Microxus* (Tate 1932h) or *Chelemys* (W. E. Johnson et al. 1990; Musser and Carleton 2005). Our evaluation of the original description casts doubt on the generic allocation of *delfini*. First, even though the external measurements of *delfini* are broadly consistent with those reported for large series of *Chelemys*, their combination seems unlikely in the same individual, as first noted by Osgood (1943b:166). Second, the greatest length of the skull and upper molar length are much less than those of typical *C. macronyx* (e.g., upper toothrow length = 4.5 mm in *delfini* vs. 5.4–5.9 in *C. macronyx* [B. D. Patterson 1992a]). According to the original description, *delfini* has a nearly uniform coloration, and its tail annulations are clearly visible (Cabrera 1905:15). By contrast, both known *Chelemys* species generally show countershading and have very thickly haired tails in which the hairs hide the scale annulations (B. D. Patterson 1992a). Finally, as Cabrera (1905:16) stated, "El tercer molar inferior es casi como una tercera parte del que le precede" [the third lower molar is almost a third of the preceding one (translation ours)]. This dental feature is especially different from the condition in *Chelemys*, where m3 is almost equal in size to m2. Thus, the inclusion of *delfini* Cabrera in *Chelemys* lacks convincing support.

Our comparisons with other southern abrotrichines, including *Abrothrix lanosa*, *A. olivacea*, and *Geoxus valdivianus*, suggest to us a connection between *delfini* and *Geoxus*, especially southern populations usually included in the subspecies *G. v. michaelseni*. Both *delfini* and *michaelseni* have a uniformly dark color and, most importantly, small third lower molars. We thus consider *Oxymycterus delfini* Ca-

brera a junior synonym of *G. v. michaelseni*, a decision that has two major taxonomic consequences. First, *Chelemys* has only two living species, not three as has been listed in recent compendia (e.g., Musser and Carleton 2005). Second, the southernmost limits of *Chelemys* are at about 50°30'S, not at a point farther to the south.

As currently understood, *Chelemys* is not monophyletic, based on phylogenetic analysis of both mitochondrial and nuclear DNA sequences, with *C. megalonyx* the sister to *Notiomys edwardsi* rather than to its present congener *C. macronyx* (Rodríguez-Serrano, Palma, and Hernández 2008; Feijoo et al. 2010; Teta, D'Elía, Pardiñas et al. 2011). If the current abrotrichine generic classification is to be maintained (i.e., *Chelemys* is to be considered a genus separate from *Notiomys*), a new generic name to contain *C. macronyx* is needed.

KEY TO THE SPECIES OF *CHELEMYS*:

1. Skull robust; incisors orthodont; anterior borders of zygomatic plates nearly vertical; anterior border of mesopterygoid fossa even with posterior face of M3; bullae small; width of manus about 5 mm . . . . *Chelemys macronyx*

1'. Skull more delicate; incisors proodont; anterior borders of zygomatic plates moderately slanted posteriorly upward; anterior border of mesopterygoid fossa posterior to the M3; bullae larger; width of manus about 7 mm . . . . *Chelemys megalonyx*

Chelemys macronyx (Thomas, 1894)
Andean Long-clawed Mouse

SYNONYMS:

Acodon macronyx Thomas, 1894b:362; type locality "East side of the Andes, near Fort San Rafael, Province of Mendoza," Argentina. Pearson and Lagiglia (1992; see also Reise and Gallardo 1990:79; Pearson and Christie 1991:126) questioned the designated type locality, arguing that the actual provenance was probably near Volcán Peteroa, on the Argentinean-Chilean border.

Akodon (*Chelemys*) *vestitus* Thomas, 1903c:242; type locality "Valle del Lago Blanco, Cordillera region of Southern Chubut Territory, Patagonia," Argentina; further restricted to "Estancia Valle Huemules (45.94°S, 71.91°W, 593 m, Río Senguerr, Chubut)" by Pardiñas et al. (2007:406).

Chelemys vestitus: Thomas, 1919b:207; name combination.

Notiomys connectens Osgood, 1925:120; type locality "Villa Portales [38.45°S, 71.37°W], Province of Cautin [Araucanía], Chile. Altitude 3,300 ft."

Notiomys macronyx: Osgood, 1925:122; name combination.

Notiomys vestitus: Osgood, 1925:123; name combination.

Notiomys vestitus alleni Osgood, 1925:124; type locality "upper Río Chico, Santa Cruz, southern Argentina." Our reconstruction of the itinerary followed by J. B. Hatcher (1903), the collector of the type specimen, indicates that this locality is Estancia Tucu Tucu (ca. 48.42°S, 71.78°W), Río Chico, Santa Cruz, Argentina.

Chelemys vestitus fumosus Thomas, 1927d:654; type locality "Sierra de Pilpil, 2000m," about 15 km S San Martín de Los Andes (40.20°S, 71.25°W), Lácar, Neuquén, Argentina.

Chelemys macronyx: Gyldenstolpe, 1932a:126; first use of current name combination.

Chelemys vestitus vestitus: Gyldenstolpe, 1932a:126; name combination.

Chelemys connectens: Gyldenstolpe, 1932a:127; name combination.

Chelemys vestitus alleni: Gyldenstolpe, 1932a:127; name combination.

Notiomys macronyx macronyx: Osgood, 1943b:159; name combination.

Notiomys macronyx vestitus: Osgood, 1943b:162; name combination.

Notiomys macronyx alleni: Osgood, 1943b:165; name combination.

Notiomys vestitus fumosus: Ellerman, 1941:425; name combination.

DESCRIPTION: Large and robust, with relatively short ears and tail (total length 182–209 mm; tail length 52–63mm); claws on forefeet long (ca. 6–7mm); fur thick and short, dark olive brown to dark coffee dorsally and buffy brown to grayish or whitish on venter. Ventral color often extends onto sides, sharply demarcated from dorsum in most individuals. Average mass 66.7g (range 43–86 g; Pearson 1983). Skull characterized overall by robustness (greatest length 30.4–32.2 mm), squared outline, and broad rostrum; zygomatic plates high, with anterior borders nearly vertical and without upper free borders; incisive foramina long, reaching protocones of M1s; anterior border of mesopterygoid fossa even with posterior faces of M3s; bullae small; upper incisors orthodont, faced with pale orange enamel; molars large and heavy (upper toothrow 5.4–5.9mm) with developed mesolophs usually fused to paracones. Both B. D. Patterson (1992a) and Kelt (1994) provided external and cranial measurements.

DISTRIBUTION: *Chelemys macronyx* occurs in the southern Andean portions of Argentina and Chile, from about 33.5°S south to 52°S. Most of its range is in Argentina, where the species is mainly found in a narrow band of high-elevation open areas and forested environments, from southwestern Mendoza to southwestern Santa Cruz provinces (Osgood 1943b; Pardiñas, Teta et al. 2003; Pardiñas,

Udrizar Sauthier et al. 2008; V. L. Roldán 2010). Scattered, isolated populations occur through central Patagonia in suitable environments (Teta et al. 2002; Pardiñas, Teta et al. 2003). In Chile, *C. macronyx* is known from mainly along the Andes, but in the southernmost portion of its distribution it reaches lowlands close to the Pacific Ocean. The northernmost record for Chile is based on a specimen (LCM 1761) collected with the type series of *Loxodontomys pikumche* (Spotorno, Cofre et al. 1998; Teta, Pardiñas, and D'Elía 2011). Remarkably, *C. macronyx* is absent from several well-studied communities in the Valdivian forest, such as those at La Picada, Osorno (B. D. Patterson et al. 1989), and Fundo San Martín (Murúa et al. 1987, in which article the name "*Chelemys* spp." [pg. 730] refers to *Pearsonomys*).

The known fossil record of *C. macronyx* is restricted to the Holocene of Argentinean Patagonia (Pearson and Pearson 1993, Pardiñas 1998, 1999a; Teta, Andrade, and Pardiñas 2005; Udrizar Sauthier 2009). This rodent has been extralimitally recorded in Late Holocene assemblages both in northwestern Patagonia (Pilcaniyeu; Teta, Andrade, and Pardiñas 2005) and the southernmost mainland (Pali Aike; Pardiñas, Teta, Formoso, and Barberena 2011). However, the species is absent in the Late Pleistocene-Holocene sequence of Cueva del Milodón (Simonetti and Rau 1989; U. F. J. Pardiñas, unpubl. data), an area where it is present today. Both situations suggest that this species experienced one or more expansion-retraction events, involving hundreds of kilometers (Teta, Andrade, and Pardiñas 2005; Pardiñas, Teta, Formoso, and Barberena 2011). This dynamic is supported by phylogeographic analyses of mtDNA cytochrome-*b* gene sequences (Alarcón et al. 2011).

SELECTED LOCALITIES (Map 53): ARGENTINA: Chubut, 41km W of Alto Río Senguerr (CNP-E 473), Cabaña Arroyo Pescado (CNP-E 327), Cholila (CNP-E 135), Establecimiento La Ollada (CNP-E 291), Estancia San Pedro (Teta et al. 2002), Estancia Santa María, Puesto El Chango (CNP-E 328), Paso del Sapo (Pardiñas, Teta et al. 2003), Sierra de Tepuel, Cañadón de la Madera (CNP 2374), 4km S of Tres Banderas (CNP-E 75); Mendoza, Cerro Colorado (Alarcón et al. 2011), Laguna del Diamante (OMNH 23520), Laguna de la Niña Encantada (Massoia et al. 1994a; CNP-E 55); Neuquén, Arroyo Covunco and RNN°40 (CNP-E 87), Cueva Traful I (CNP-E 595), 12km NE of Refugio Parque Provincial Tromen (CNP-E 529); Río Negro, Cañadón Quetrequile (CNP-E 284), Puesto de Hornos, Estancia Maquinchao (Teta et al. 2002); Santa Cruz, Alero Destacamento Guardaparque (CNP-E 103), 8km N of El Chaltén (CNP-E 418), Cerro Comisión (CNP-E 413), Cueva de las Manos (CNP-E 47). CHILE: Aysén, 3km N of Coyhaique, Reserva Nacional

Coyhaique, Laguna Venus (Kelt 1996), Osorno, Paso Puyehue (Reise and Venegas 1987), Sector el Manzano (Alarcón et al. 2011); Araucanía, Paso de Las Raíces, 58 km W of Curacautín (Reise and Venegas 1987); Biobío, 7 km below Termas Chillán (Reise and Venegas 1987); Magallanes y Antártica Chilena, Cueva del Milodón (U. F. J. Pardiñas, unpubl. data), Laguna Lazo, Lake Sarmiento, Última Esperanza (Osgood 1943b); Maule, Baño San Pedro, Romeral (MNHNC 298 JCT); Santiago, Las Melosas (Pardiñas, Teta, Formoso, and Barberena 2011).

SUBSPECIES: Although four trinomials have been used for *C. macronyx*, the reality of these supposed taxa has never been evaluated by a rigorous assessment of morphological variation. A phylogeographic analysis based on mtDNA sequences retrieved a shallow genealogy for *C. macronyx* that is geographically structured into two well-supported and allopatric clades. One of these is distributed at high-Andean localities in the Argentinean provinces of Mendoza and northern Neuquén, and the other clade occupies most of the distributional range of the species at medium- to low-elevation localities in northwestern Neuquén, Chubut, and Santa Cruz provinces in Argentina as well as Aysén and Magallanes y Antártica Chilena regions in Chile. The northern clade appears to have been demographically stable, while the southern clade shows a signal of recent demographic expansion. Divergence between the two main clades found is limited (i.e., 2.3% at the mtDNA cytochrome-*b* gene), supporting a relatively recent origin, with expansion from late Pleistocene refugia. A study in progress, based on cranial anatomy and metrics, indicates two distinct morphotypes in *C. macronyx* that are congruent with the phylogroups uncovered by Alarcón et al. (2011). Northern samples are characterized by a mesopterygoid fossa with a rounded anterior border and both large and nearly circular foramina ovale; central and southern samples have a squared anterior border to the mesopterygoid fossa and ovate to piriform foramina ovale. One of us (PT) believes that two subspecies are recognizable, the nominotypical *C. m. macronyx* for northern samples and *C. m. vestitus* for central and southern ones. Under this treatment, *alleni* Osgood, for which a haplotype from a topotypic specimen was included in the Alarcón et al. (2011) study, is a junior synonym of *vestitus* Thomas. The same would likely apply to *fumosus* Thomas, for which haplotypes recovered from specimens collected near its type locality belong to the central-southern clade. Further analyses incorporating larger geographic coverage, especially in Chile, combined with variation at nuclear loci are needed to test this hypothesis.

NATURAL HISTORY: *Chelemys macronyx* lives primarily in *Nothofagus pumilio* forest, from timberline to the edge of the precordilleran steppe (Pearson 1983). It also occurs in moist habitats or under shrubs in semiarid steppes or in areas of lush grasses and loose soils with scattered canopy of *N. pumilio* or *N. antarctica* (Pearson 1983; Kelt 1994). In Malleco, Chile, it prefers deep soils in forested to shrubby areas (Greer 1966). This mouse is primarily subterranean, and can be both diurnal and nocturnal. During winter, *Chelemys* constructs networks under the snow, especially in areas with a dense cover of amancay lilies (*Alstroemeria aurantiaca*) in *Nothofagus* forests. Such networks may cover almost one-third of the surface over an area of 10–30 m² (Pearson 1984). Stomach contents have included arthropods, green vegetable matter, and fungi (Pearson 1983).

Population densities in mixed *Nothofagus betuloides–N. pumilio* forest and in pure stands of *N. antarctica* in Río Negro province, Argentina, varied between 1.1 (November) and 14.7 (May) and 0 (November) and 1.3 (May) individuals per hectare, respectively (Pearson and Pearson 1982). In a high elevation shrubby area (>2,000 m) in central Chile, Mella (2006) found densities that varied between 1.6 (spring) and 17.2 (summer). Testes in reproductively active males ranged from 11 to 14 mm long, and vesicular glands were 10 mm or longer (Pearson 1983). Both Pearson (1983) and Kelt (1994) described the age structure and reproductive cycles. The breeding season seems to begin abruptly when the ground is still covered by snow and continues for a considerable period during spring and summer. Individuals of both sexes may begin breeding before reaching full adult size (Pearson 1983). One pregnant female carried four embryos, another five (Pearson 1983). The autumn population included two age groups, one composed of young of the year and the other of adults >1 year in age; the spring population was composed almost entirely of adults born the preceding spring or summer (Pearson 1983). Individuals born at the beginning of the reproductive season do not venture from their burrows until late spring (Kelt 1994).

REMARKS: Both Mann (1978) and Tamayo Frassinetti (1980) considered *Chelemys macronyx* to be a subspecies of *C. megalonyx*. According to Osgood (1943b), *Notiomys connectens* Osgood (1923) is a composite of a *Chelemys* skin and an *Abrothrix* skull; to fix the name, Osgood (1943b) selected the skin as the holotype of *connectens* and placed the name in his synonymy of *Notiomys macronyx vestitus*. *Chelemys angustus* Thomas, 1927, was based on a specimen of *Abrothrix longipilis* (see Pearson 1984).

Specimens from southwestern Mendoza and considered to be topotypes have $2n = 54$, $FN = 62$, and four microchromosomes (A. A. Ojeda et al. 2005). Specimens from Bariloche (Argentina), Paso de Pino Hachado, and Coyhaique

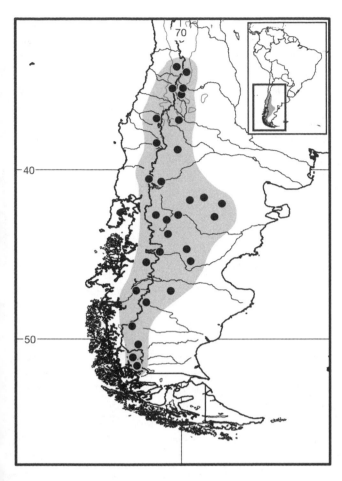

Map 53 Selected localities for *Chelemys macronyx* (●). Contour line = 2,000 m.

(Chile) show the same autosomal complement but differ in the number of microchromosomes (unpubl. data cited by A. A. Ojeda et al. 2005). It is noteworthy that Pearson (1984) reported a $2n = 52$ complement for individuals from Bariloche and Castaño Overo, northwestern Patagonia. The biological and taxonomic significance of these differences remain unclear.

Chelemys megalonyx (Waterhouse, 1845)
Chilean Long-clawed Mouse

SYNONYMS:

Hesperomys megalonyx Waterhouse, 1845:154–155; type locality "Chile . . . Lake of Quintero," Valparaíso, Chile.

Oxymicterus scalops Gay, 1847:108; type locality "campos de las provincias centrales," Chile.

A[codon] megalonyx: Thomas, 1894b:362; name combination.

Akodon megalonyx: E.-L. Trouessart, 1897:538; name combination.

Mus microtis Philippi, 1900:57; type locality "E provincia Maule," Chile.

Akodon (Chelemys) megalonyx: Thomas, 1903c:242; name combination.

Notiomys megalonyx: Osgood, 1925:12; name combination.

Chraeomys scalops: Gyldenstolpe, 1932a:123; name combination.

Chelemys megalonyx: Gyldenstolpe 1932a:126; first use of current name combination.

Notiomys megalonyx megalonyx: Osgood, 1943b:157; name combination.

Notiomys megalonyx microtis: Osgood, 1943b:158; name combination.

DESCRIPTION: Medium-sized, mole-like mouse, with elongated front claws (<6 mm), thick pelage, and short tail (total length <180 mm; tail <60 mm); fur long, soft, nearly uniform grayish brown above and grayish white below; chest with brown mark; tail uniformly brown. Skull more elongated and gracile than that of *C. macronyx*; frontal sinuses well inflated; nasals narrow with parallel borders; nasofrontal suture U-shaped and extends posteriad to level of lacrimals; braincase globose; zygomatic plates without upper free border, with anterior margins convex and inclined backward with noticeable masseteric-like tubercle; incisive foramina do not reach beyond anterior faces of M1s; palate long and flat; mesopterygoid roof fully ossified or with vacuities reduced to narrow slits; bullae well inflated; incisors proodont, with pale orange to whitish enamel; upper molar series slightly divergent anteriorly; molars comparatively small and narrow (upper toothrow <5 mm); M3 more reduced than in *C. macronyx*.

DISTRIBUTION: This species is known from only a few scattered localities in an apparently disjunct distribution in coastal central Chile (Osgood 1943b; Mann, 1978). In the north, it is limited to the Coquimbo to Santiago regions and in the south to Araucanía, and presumably Maule regions, with no records from intervening regions.

SELECTED LOCALITIES (Map 54): CHILE: Araucanía, Comuna de Victoria, cuenca del Río Quino, near Curacautín (V. Quintana 2009), Temuco (Osgood, 1943b); Coquimbo, Parque Nacional Fray Jorge (UACH 1462, 2732); Santiago, Rinconada de Maipú (MNHN 1545, 1546; J. Yáñez, pers. comm.); Valparaíso, La Ligua (USNM 391799), Lake of Quintero (Waterhouse 1845), Olmué (Osgood 1925), Valparaíso (Osgood 1943b).

SUBSPECIES: Osgood (1943b) recognized two subspecies, the nominotypical *C. m. megalonyx* (Waterhouse), found around Valparaíso in the north, and *C. m. microtis* (Philippi), with type locality unspecified but somewhere in Maule province, in the southern portion of the distribution. As noted by Osgood (1943b), the inclusion of *microtis*

Philippi in *C. megalonyx* is questionable, as some measurements of the holotype (tail length, hindfoot length) are much smaller than typical of *megalonyx*. Apparently, the only known specimen collected at a precisely known locality, and referable without question to Philippi's *microtis*, is BM 8.3.1.15 from Temuco (identified by Osgood 1943b:159). We provisionally assign specimens reported by V. Quintana (2009) from a forest fragment near Curacautín to Philippi's *microtis*, although no voucher was preserved and both an available photograph and measurements taken in the field are equivocal. It is possible that this record is of *C. macronyx*, not *C. megalonyx*. The presence of *C. megalonyx* in the lowlands of south-central Chile requires adequate documentation. No recent study has assessed the distinction between the two subspecies recognized by Osgood (1943b).

NATURAL HISTORY: Almost nothing is known about the ecology of this rodent. It is semifossorial and digs tunnels, preferably in humid soils, with burrow openings usually protected by rocks, shrubs, or fallen logs (Mann 1978). Koford (1955) mentioned the capture of one individual between shrub and cactus at the foot of a long steep slope. Mann (1978) reported this rodent in dense mesophitic forest and shrubby coastal areas. *Chelemys megalonyx* appears to be rare, or at least very difficult to trap. Recently collected specimens come from Parque Nacional Fray Jorge, a protected

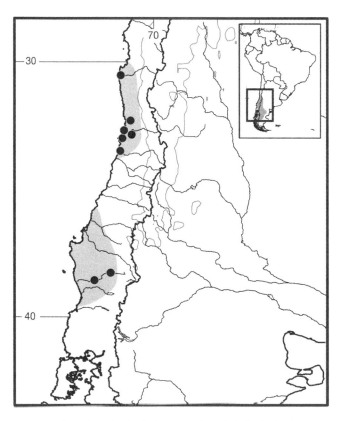

Map 54 Selected localities for *Chelemys megalonyx* (●); distribution is mapped as disjunct (see Subspecies). Contour line = 2,000 m.

area of arid Mediterranean thorn-scrub with remnants of fog forests at higher elevation near the city of Coquimbo, and from disturbed forest along the Río Quino in the municipality of Victoria, Malleco province (V. Quintana 2009).

REMARKS: Mann (1978) and Tamayo and Frassinetti (1980) listed *delfini* Cabrera (see Remarks under genus), along with *macronyx* Thomas, as subspecies of *C. megalonyx*. ?*Oxymycterus niger* Philippi, 1872:445 has been considered a synonym of this species by some authors (e.g., Gyldenstolpe 1932:126), presumably following the tentative remarks made by Osgood (1925:122). However, the holotype of *niger* is lost, and, as recognized by Osgood (1925, 1943b), the allocation of *niger* to *megalonyx* is uncertain.

Genus *Geoxus* Thomas, 1919

Pablo Teta, Ulyses F. J. Pardiñas, and Guillermo D'Elía

The monotypic genus *Geoxus* is mostly restricted to humid habitats, such as forest, shrublands, and wetlands; in Chile from about 36°S south to Fuerte Bulnes (53°40'S) and in Argentina from 39°S to Santa Cruz province (47°S; Osgood 1943b; Pearson 1995). Osgood (1925, 1943b) reviewed the genus as part of his broad concept of *Notiomys*. Thomas (1927d) was the first to defend generic status, a position later promoted by Pearson (1984; see Reig 1987). Maximum parsimony and likelihood analyses of mtDNA cytochrome-*b* and nucDNA sequences suggested that *Geoxus* is closely related to *Pearsonomys* (M. F. Smith and Patton 1999; D'Elía, Ojeda et al. 2006). However, as currently conceived, the genus is non-monophyletic because northern populations form a clade sister to *Pearsonomys* with southern populations sister to that clade (E. P. Lessa et al. 2010; see further discussion in the tribal account). Except as noted, the following account is based on Osgood (1943b) and Pearson (1983, 1984), each of whom should be consulted for further details (see also B. D. Patterson 1992a). Our descriptions of the skull and molars are taken from Reig (1987).

The body shape is highly modified for fossorial life. The size is small (head and body length about 96 mm, tail length 45 mm, and mass 25 g). The fur is velvety and mole-like; the upper parts vary from blackish brown to olive brown, and the under parts are only slightly paler than the upper parts, slaty gray to brown. The ears are small but pinnae are visible externally; the tail is shorter than half the length of head and body, may be unicolored or indistinctly bicolored, and is very thickly clothed with hairs that conceal the scales. The pollex and other digits of the forefeet have long and strong claws; those of the hindfeet are much shorter, without a fringe of hairs along their lateral margins. Both palms and soles are naked, with pads smooth and rounded (Gyldenstolpe 1932; Reig 1987).

The skull is slender and rather delicately built, with a long and narrow rostrum. The premaxillary bones extend to or end slightly anterior to the nasals; the frontal region is inflated; the interorbital region has smooth and rounded sides so that supraorbital edges are not square shaped; and the interparietal is narrow and short. In lateral view, the zygomatic plates are reduced, narrow, and slanted posteriorly from bottom to top; the alisphenoid is anteriorly expressed to hide the (sometimes incomplete) sphenofrontal foramen and the anterior lacerate foramen. The incisive foramina are long, extending to or between the anterior borders of M1; the palate is long, extending slightly behind the level of the third molars, and lacks posterior palatal pits. The pterygoid region is long; the mesopterygoid fossa has nearly parallel sides and a squared anterior border. The stapedial process of the bulla is fused to the parapterygoid. The mandible is particularly slender, with a smooth diastema, short coronoid process, and small capsular projection; the condyloid process not inclined inward (Reig 1987).

The incisors are rather slender, and the upper ones orthodont. The molars are small, narrow, simple, and brachydont. The upper molars have transversely oriented para- and metaflexi, a mesoflexus is absent, and the para- and mesoloph are coalesced; both mesostyle and anterostyle are missing. M1 has a reduced procingulum, shallow anteroflexus, and no anteromedian flexus. |M3 is greatly reduced, less than two-thirds the length of M2, without a metacone and round in occlusal view. The lower molars have transverse lophids and flexids with little infolding. The procingulum of m1 is narrow and lacks an anteromedian flexid, and a protostylid is present; m3 is also reduced, less than three-fourths the length of M2, and with a simplified enamel pattern (Reig 1987).

The axial skeletal elements in *Geoxus* have modal counts of 13 ribs and 13 thoracic, six lumbar, and 20–23 caudal vertebrae (Steppan 1995).

The stomach is unilocular hemiglandular. The antrum, which is covered by glandular epithelium, is larger than the corpus, which in turn is mostly covered by cornified epithelium; glandular epithelium from the antrum extends to the corpus and surpasses to the left of the esophageal orifice (Carleton 1973). A caecum is absent (Reig 1987), but a gall bladder is present (Voss 1991a).

The penis of *Geoxus* has a prominent dorsal hood extending beyond the terminal crater; there are large urethral flaps, with each bearing at its distal end a medially directed spine, and three cartilaginous digits, with the central one larger than both laterals and with its tip protruding from the crater. The baculum is ventrally keeled and has a rounded and prominent head. Its base resembles an equilateral triangle with a notch on its proximal face (Hooper and Musser 1964). *Geoxus* has two pairs of preputial glands, a large pair lying lateral to the glans penis and a much smaller pair near the midventral margin of the prepuce; these glands do not exceed the prepuce in length. The bulbourethral, dorsal and anterior prostates, and ampullary glands are unusually large and well developed (Voss and Linzey 1981).

SYNONYMS:

Oxymycterus: Philippi, 1858:303; part (description of *valdivianus*).

[*Acodon*]: Thomas, 1894b:362; part (description of *macronyx*).

Hesperomys (*Acodon*): Matschie, 1898:5; part (description of *michaelseni*).

Mus (*Oxymycterus*): Philippi, 1900:21; part (listing of *valdivianus*).

Acodon (*Chelemys*): J. A. Allen, 1905:80; part (listing of *michaelseni*).

Geoxus Thomas, 1919b:208; type species *Geoxus fossor* Thomas, by original designation.

Notoxus Thomas, 1919b:209; *lapsus calami* for *Geoxus* Thomas.

Notiomys: Osgood, 1925:115; part (listing of *valdivianus*, *fossor*, and *michaelseni*; description of *auraucanus* and *chiloensis*).

Geoxus valdivianus (Philippi, 1858)

Valdivian Long-clawed Mouse

SYNONYMS:

Oxymycterus valdivianus Philippi, 1858:303; type locality "Valdivia," Los Ríos, Chile.

[*Acodon*] *valdivianus*: Thomas, 1894b:363; name combination.

Hesperomys (*Acodon*) *michaelseni* Matschie, 1898:5; type locality "Süd-Patagonien, Punta Arenas," Magallanes y Antártica Chilena, Chile.

Mus (*Oxymycterus*) *valdivianus*: Philippi, 1900:21; name combination.

Oxymycterus microtis J. A. Allen, 1903b:189; type locality "Pacific slope of the Cordilleras, near the head of the Rio Chico de Santa Cruz," Argentina. Our reconstruction of the itinerary followed by J. B. Hatcher and O. A. Peterson, the collectors of the type specimen, indicates that this locality corresponds to a point in front of Estancia Tucu Tucu (ca. 48.46°S, 71.98°W), Río Chico, Santa Cruz, Argentina.

Akodon (*Chelemys*) *michaelseni*: J. A. Allen, 1905:80; name combination.

Notiomys michaelseni: E.-L. Trouessart, 1904:436; name combination.

Notiomys microtis: E.-L. Trouessart, 1904:436; name combination.

Geoxus fossor Thomas, 1919b:208; type locality "Maiten [= El Maitén, Cushamen], W. Chubut. 700 m," Argentina.

Notoxus fossor: Thomas, 1919b:209; *lapsus calami.*

N[*otoxus?*]. *microtis*: Thomas, 1919b:209; name combination.

[*Geoxus*] *michaelseni*: Thomas, 1919b:209; name combination.

N[*otoxus?*]. *michaelseni*: Thomas, 1919b:209; name combination.

[*Geoxus*] *valdivianus*: Thomas, 1919b:209; first use of current name combination.

Notiomys valdivianus: Osgood, 1925:115; name combination.

Notiomys valdivianus araucanus Osgood, 1925: 117; type locality "Tolhuaca, Province of Malleco, [Araucanía,] Chile."

Notiomys valdivianus chiloensis Osgood, 1925:117; type locality "Quellon [= Quellón], Chiloe Island [Isla Grande de Chiloé, Los Lagos,] Chile."

Notiomys fossor: Osgood, 1925:118; name combination.

Geoxus valdivianus valdivianus: Gyldenstolpe, 1932a:124; name combination.

Geoxus valdivianus araucanus: Gyldenstolpe, 1932a:125; name combination.

Geoxus valdivianus chiloensis: Gyldenstolpe, 1932a:125; name combination.

Notiomys valdivianus bullocki Osgood, 1943b:154; type locality "Mocha Island, coast of southern Chile, Province of Arauco [Biobío]."

Notiomys valdivianus valdivianus: Osgood, 1943b:151; name combination.

Notiomys valdivianus bicolor Osgood, 1943b:155; type locality "Casa Richards, Rio Nirehuao [= Río Ñirehuao], [Aysén,] Chile. Lat. 45°3′S [45.260°S, 71.705°W]."

Notiomys valdivianus michaelseni: Osgood, 1943b:156; name combination.

Notiomys valdivianus fossor: Cabrera, 1961:473; name combination.

G[*eoxus*]. v[*aldivianus*]. *fossor*: Pearson, 1984:232; name combination.

DESCRIPTION: As for the genus.

DISTRIBUTION: Chile from about 36°S, including Mocha and Chiloé islands, to the Straits of Magellan; and southwestern Argentina, in *Nothofagus* forests and precordilleran steppes, from central Neuquén province south to Santa Cruz province (Osgood 1943b; Pearson 1983, Pardiñas, Teta et al. 2003). Saavedra and Simonetti (2001) reported isolated populations in coastal patches of temperate forest in central Chile. The fossil record is restricted to Holocene deposits in the Argentinean Patagonia (Pearson and Pearson 1993; Pardiñas 1998, 1999a).

SELECTED LOCALITIES (Map 55): ARGENTINA: Chubut, El Maitén (type locality of *Geoxus fossor* Thomas), 6.5 km W of bridge over Río Tecka, on RPN°17

(CNP-E 323), Capilla El Triana (CNP-E 449); Neuquén, Cueva Traful I (Pearson and Pearson 1993), south margin of Lago Caviahue (CNP 812), bridge on RPN°13 over Río Carreri (Pardiñas, Teta et al. 2003); Río Negro, Cerro Leones (Pearson and Pearson 1993), Cabaña Cacique Foyel (Udrizar Sauthier, Andrade, and Pardiñas 2005); Santa Cruz, near Estancia Tucu Tucu (type locality of *Oxymycterus microtis* J. A. Allen). CHILE: Aysén, ca. 15 km E of Puerto Chacabuco, 15 km S of Aysén, NW end of Lago Riesco (Kelt 1996), El Manzano (E. P. Lessa et al. 2010); Araucanía, Reserva Nacional Malalcahuello (E. P. Lessa et al. 2010), Villarica (UACH 2450); Biobío, Isla Mocha (Osgood 1943b), vicinity refugio Club Andino Alemán on W side of Chillán Volcano (Reise and Venegas 1987); Los Lagos, Isla Grande de Chiloé, Río Inio (Osgood 1943b), La Picada (UACH 3488), Puyehue (UACH 5428); Los Ríos, Costa Río Caunahue (E. P. Lessa et al. 2010), Fundo San Martín, San José, Valdivia (UACH 700); Magallanes y Antártica Chilena, Cueva del Milodón (U. F. J. Pardiñas, unpubl. data), Fuerte Bulnes (http://www.flickr.com/photos/martinezpina/5331904118/in/photostream/), Punta Arenas (type locality of *Hesperomys* (*Acodon*) *michaelseni* Matschie); Maule, Reserva Nacional Los Queules (Saavedra and Simonetti 2001).

SUBSPECIES: Osgood (1943b), Cabrera (1961), and Mann (1978) each recognized six subspecies in *G. valdivianus* (*G. v. bicolor, G. v. bullocki, G. v. chiloensis, G. v. fossor, G. v. michaelseni,* and *G. v. valdivianus*). This hypothesis has yet to be tested using modern morphological and/or molecular methodologies (also see comment under Remarks).

NATURAL HISTORY: *Geoxus valdivianus* is a forest dweller that occurs in pure and mixed stands of *Nothofagus*, although it may be caught occasionally in lush meadows or in marshes bordering forests, in areas with heavy undergrowth of cane and *Berberis* shrubs, in small clumps of *Nothofagus pumilio* surrounded by pasture, or at timberline in tussock grass with patches of *Araucaria araucana* or low bushes of *Nothofagus* (Greer 1965; Reise and Venegas 1974; Pine et al. 1979; Pearson 1983, 1984). More recently, K. García et al. (2011) caught one individual in an exotic plantation of *Pinus contorta*. Population densities in *Nothofagus* forests of Río Negro, Argentina, varied between 0.4 and 0.5 animals per hectare (Pearson and Pearson 1982). This species is active both during the day and night, and eats small invertebrates (e.g., adult and larval arthropods and annelids), vegetation, and fungi (Pearson 1983; Meserve et al. 1988). It is partially subterranean but often is found in dense litter or herbaceous vegetation in runaways alongside logs (Pearson 1983, 1984; Kelt 1994). It may use tunnels made by other rodents (e.g., *Aconaemys*), or it may make its own burrows in soft soil rich in humus (Pearson 1983, 1984; see also Greer 1965).

Males with descended testes were caught in October, November, and December in northwestern Patagonia, Argentina (Pearson 1983); and in March and November in the Aysén region, Chile. Females in breeding condition were caught in October, November, and December in northwestern Patagonia (Pearson 1983), with an average of 3.5 embryos. Pearson (1983) documented reproduction, age structure, sex ratio, and body size in Patagonian Argentina. B. D. Patterson et al. (1989, 1990) examined the distribution, abundance, and habitat selection along an elevational gradient in a temperate rainforest of southern Chile.

REMARKS: Pearson (1984) noted the uncertainty of the generic association and specific status of *Hesperomys* (*Acodon*) *michaelseni*, and considered it perhaps allied to *Notiomys* or *Geoxus*. Pardiñas, Udrizar Sauthier et al. (2008), however, rejected the link of Matschie's *michaelseni* with *Notiomys*. Reig (1987) considered *michaelseni* as a probably valid species of *Geoxus*. Although the genus is currently viewed as monotypic, the available, although limited, molecular data indicate that samples of *G. valdivianus* are paraphyletic with respect to *Pearsonomys* (E. P.

Lessa et al. 2010), suggesting that at least two species are contained in our current concept. If so, names are available from the current list of subspecies epithets.

The diploid number of a specimen of *Geoxus* from Río Negro, Argentina, was $2n = 52$, with one pair of large subtelocentric autosomes, two pairs of very small metacentrics, remaining autosomes acrocentric, a large submetacentric X chromosome, and a very small metacentric Y chromosome (Pearson 1984). Chilean specimens apparently have a $2n = 56$ karyotype (M. H. Gallardo, unpubl. data, cited in D'Elía, Ojeda et al. 2006).

Genus *Notiomys* Thomas, 1890

Pablo Teta and Ulyses F. J. Pardiñas

Notiomys is the only endemic sigmodontine genus of the Patagonian steppes and tablelands of southern Argentina. Currently considered to be monotypic, containing the single species *N. edwardsii*, several other taxa were placed in *Notiomys* during the last century (see Osgood 1925, 1943b; Cabrera 1961). Pearson's (1984) revision established the current contents of the genus, clarifying the status of *angustus* Thomas (1927d; a synonym of *Abrothrix longipilis*) while placing *macronyx* Thomas (1894b; including *delfini* Cabrera [but see Remarks for *Chelemys*]), and *megalonyx* Waterhouse (1845) in *Chelemys*, and *valdivianus* Philippi (1858) in *Geoxus*. Keystone contributions that document the taxonomic history and changing contents of *Notiomys* are those of Osgood (1925), Pearson (1984), Reig (1987), and B. D. Patterson (1992a). Phylogenetic analyses based on nucDNA sequences have placed *Notiomys* as sister to *Geoxus* + *Pearsonomys* (D'Elía, Ojeda et al. 2006). However, Rodríguez-Serrano, Hernández, and Palma (2008) recovered, with low support, a sister relationship between *Notiomys* and *Chelemys macronyx*. This account is based mostly on Pardiñas, Udrizar Sauthier et al. (2008), who included detailed anatomical descriptions as well as data on distribution, natural history, and both genetic and morphological variation.

Notiomys is a small mouse with a short tail (head and both length about 86 mm, tail length about 40 mm, mass about 20 g). Its fur is not mole-like, as has sometimes been reported, but has bright colors. The dorsal pelage is dense, soft, but not long (mid-dorsal hairs about 7–8 mm in length). The dorsal head and body coloration shows an agouti effect, produced by almost entirely gray hairs with ochraceous tips; the ventral color is gray cream to white, with individual hairs basally gray and distally white. A line of hairs with their distal half orange separates the dorsal and ventral colors. The muzzle has bright orange spots laterally; the rhinarium is tipped with a pinkish, leathery button. The eyes are reduced in size (about 4 mm in diameter).

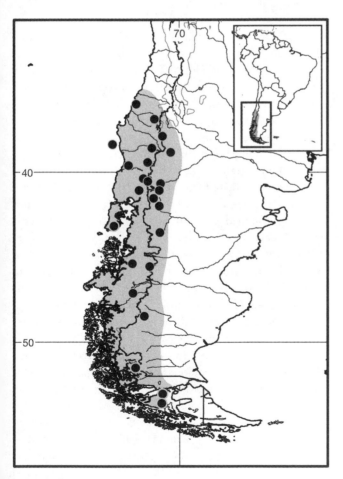

Map 55 Selected localities for *Geoxus valdivianus* (●). Contour line = 2,000 m.

The longest mystacial vibrissae (about 30 mm) are white in color and surpass the pinnae margin when appressed against the side of the head. Pinnae are small and extremely thin, edged with long and silky white hairs that make it difficult to detect their margins. The tail is whitish, slightly brownish above, and well haired. The hindfeet are short and broad, with a lateral fringe of stiff hairs extending well beyond the plantar surfaces. The fore claws are long (about 4 mm) but more weakly developed than those of *Geoxus*; the claws of the hindfeet are shorter (about 3 mm) than in *Geoxus*. Palms and soles are naked, except for the heel.

The skull is heavily built (greatest length 23.5–24.1 mm; zygomatic breadth 11.8–13.5 mm). The rostrum is short, wide, and conical. The interorbital region is especially broad (4.7–5.1 mm), amphoral or hourglass in shape, with inflated frontal sinuses and sharply squared supraorbital ridges. The interparietal is small and narrow. The braincase is short and round. The nasals and premaxillae extend anteriorly beyond the incisors, giving the rostrum a trumpet-like appearance. The zygomatic plates are tall and narrow, moderately slanting posteriorly from bottom to top to almost vertical and almost without upper free borders. The infraorbital foramina are strikingly large and flared. Incisive foramina are short, typically not reaching the level of M1; the palate is broad and long, extends well past the last molars, and has minute to large posterior palatal pits. The pterygoid region is short, with the anterior border of the mesopterygoid fossa rounded. An alisphenoid strut is absent, and the foramen ovale is confluent with the alisphenoid canal. A tunnel for a secondary arterial connection between the internal carotid and orbital-maxillary circulation is present in some individuals (e.g., MVZ 163067). A trough for the masticatory-buccinator nerve is broad but very shallow, and the squamosal-alisphenoid groove is clearly visible with a medial perforation, leading anteriorly to a medium-sized sphenofrontal foramen.

The mandible is gracile, although not as slender as in *Geoxus*, with a large and lateralized mental foramen, long and robust coronoid process, small capsular projection, and short angular process (Pearson 1984; Reig 1987). The masseteric crest is moderately developed; its inferior ridge meets the upper ridge, forming a unique crest at the level of the mental foramen.

The upper incisors are robust and orthodont, and the molars are brachydont, with some tendency to mesodonty, and small, fairly narrow, and elongated (toothrow 3.15–3.40 mm). Coronal topography is crested (in juvenile specimens) to planate (in adults); the main cusps are arranged in an opposite pattern. M1 is trilophodont; the procingulum is fan-shaped and slightly compressed transversely with a shallow but consistently present anteromedian flexus and a mesoflexus. M2 is bilophodont with a short anteroloph directed forward, with the paraflexus, hypoflexus, and mesoflexus each almost obsolete; a short posteroloph is directed posteriorly. M3 is simple and greatly reduced, round in outline, and less than one-half the length of M2. On the lower molars, the meso- and posteroflexids are not deeply infolded, and the lophids are oriented rather transversely; the procingulum of m1 is narrow, with a shallow anteromedian flexid, and mesolophid remnants are visible on m1; m3 is greatly reduced, rounded, and simple, less than one-half the length of M2, with a shallow hypoflexid. Both mesostyle/mesostylid and enterostyle/ectostylid are missing (Reig 1987). M1 has two large and circular roots (one anterior, one posterior); a small accessory labial root may be present under the paracone. M2 has three roots, one lingual and elliptical in outline, and two labials that are subequal and circular. M3 has a single circular root. The m1 has two large roots located along the midline of the tooth, and one reduced root on the labial side; m2 has two roots; m3 has a single circular root.

Vertebral counts of two specimens (CNP 1, MVZ 163065) are 13 thoracic, six lumbar, and 18 caudal; there are 13 ribs (three free) (Steppan 1995; Pardiñas, Udrizar Sauthier et al. 2008).

The stomach is unilocular hemiglandular (Carleton 1973). The antrum, which is covered by glandular epithelium, is larger than the corpus, which in turn is mostly covered by cornified epithelium. The glandular epithelium of the antrum extends to the corpus, surpassing the esophageal orifice (Pardiñas, Udrizar Sauthier et al. 2008). A gall bladder is present. Spotorno (1986) described the male phallus as short and barrel shaped, with tips of the distal baculum clearly protruding from the crater. The distal processes of the baculum are small with three very similar digits, tips somewhat hooked, and laterals with faint hooks. The proximal baculum has a short tip with a very well-developed base. The rounded urethral processes are proportionally large for a penis of this size.

Chelemys and *Geoxus* are larger than *Notiomys* and have mole-like fur. The claws of *Geoxus* are stouter than those of *Notiomys* and much longer that those on the hindfeet. Both *Notiomys* and *Geoxus* have tiny, almost cuspless teeth, but the skull of *Notiomys* is much shorter and wider than that of *Geoxus*. The skull of *Chelemys* is much larger than that of *Notiomys*, and its teeth are much larger and heavier (Pearson 1995). *Notiomys* also lacks a well-developed cecum like that of *Chelemys* (Pearson 1984).

SYNONYMS:

Hesperomys (*Notiomys*) Thomas, 1890.24; as a subgenus of *Hesperomys* Waterhouse; type species *Hesperomys* (*Notiomys*) *edwardsii*, by original designation.

Notiomys: Thomas, 1897a:1020; elevation to genus.

Notiomys edwardsii (Thomas, 1890)
Edwards's Long-clawed Mouse

SYNONYM:

Hesperomys (*Notiomys*) *edwardsii* Thomas, 1890:24; type locality "la Patagonia, au sud de Santa Cruz, vers 50° degré de latitude Sud" (Thomas in Milne-Edwards 1890:A.25); restricted to Puerto Santa Cruz (province of Santa Cruz, Argentina) by Pardiñas, Udrizar Sauthier et al. (2008).

DESCRIPTION: As for the genus.

DISTRIBUTION: This species occurs in shrub and herbaceous steppes and basaltic plateaus in the Argentinean provinces of Santa Cruz, Chubut, and Río Negro. The fossil record is restricted to four Holocene deposits in northern Patagonia (Pardiñas 1999a; Pardiñas, Udrizar Sauthier et al. 2008). Two of these records fall in areas where this species is not currently found, suggesting at least a wider distribution during part of the last 10 kya (Pardiñas, Udrizar Sauthier et al. 2008).

SELECTED LOCALITIES (Map 56): ARGENTINA: Chubut, Campo de Pichiñán (Pardiñas, Udrizar Sauthier et al. 2008), Colonia Nahuel Pan (G. M. Martin and Archangelsky 2004), Establecimiento La Ollada (Pardiñas, Udrizar Sauthier et al. 2008), Estancia Talagapa (Pardiñas, Udrizar Sauthier et al. 2008), Estancia Valle Huemules (Thomas 1919b), 30 km W of José de San Martín (Pardiñas and Galliari 1998a), Paso del Sapo (Pardiñas and Galliari 1998a), near Salina Grande (Pardiñas, Udrizar Sauthier et al. 2008), Subida del Naciente (CNP-E 27), 4 km S of Tres Banderas on RP 11 (Pardiñas, Udrizar Sauthier et al. 2008); Río Negro, Arroyo Pinturas (A. Andrade 2008), Campo Anexo Pilcaniyeu (Pearson 1984), Cerro Puntudo (A. Andrade 2008), Estancia Calcatreo (A. Andrade et al. 2002); Santa Cruz, Cueva de la Manos (Pardiñas, Udrizar Sauthier et al. 2008), Estancia La Ascención (G. R. Cueto et al. 2008), Estancia Laguna Manantiales (Pardiñas, Udrizar Sauthier et al. 2008), Estancia La María (CNP-E 442), Meseta del Lago Buenos Aires, nacimiento Río Ecker (CNP, not cataloged), N end of Lago Cardiel and RN 40 (Pardiñas, Udrizar Sauthier et al. 2008), Laguna del Diez (Jayat et al. 2006), Piedra Clavada (Thomas 1929), Puerto Santa Cruz (type locality of *Hesperomys* (*Notiomys*) *edwardsii* Thomas), Puesto El Cuero (CNP-E 426).

SUBSPECIES: As currently understood, *N. edwardsii* is monotypic, but see Remarks for comments on geographic variation.

NATURAL HISTORY: Little is known about this fossorial Patagonian rodent. The low trapping success in areas where owls are apparently successful at catching them suggests trap avoidance may be related to the use of traditional baits and traps. Pearson (1984) stated that *Notiomys* is found on arid steppes with sandy soil and a mixture of low shrubs and bunchgrasses. He reported a single individual obtained in a *Ctenomys haigi* burrow (Pearson 1984). It is insectivorous, but also may eat some seeds (Pearson 1984; G. M. Martin and Archangelsky 2004). In least at some localities on the Somuncurá Plateau, it is a moderately abundant species in open grassy steppes of *Poa ligularis* and *Festuca pallescens* above 1,000 m. Two young females (CNP collection) with imperforate vaginas were recorded during the summer (February and March).

REMARKS: The publication year of Thomas's description of this genus and species is sometimes cited as 1891 (e.g., Tate 1932h) or 1892 (Thomas 1929), but see Pardiñas, Udrizar Sauthier et al. (2008) who gave 1890 as the correct year of publication.

Some geographic variation in both molecular sequences and morphometric characters is known. For example, partial mtDNA cytochrome-*b* gene sequences are available for three specimens of *N. edwardsii* captured at three different localities, each with a different haplotype. Haplotypes from Campo Anexo Pilcaniyeu (41°S) and Estancia Laguna Manantiales (47.5°S), some 760 km apart, are 0.64% divergent.

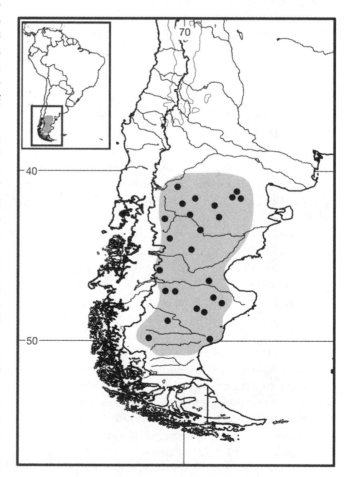

Map 56 Selected localities for *Notiomys edwardsii* (●). Contour line = 2,000 m.

A third haplotype, found at Laguna Blanca (41.4°S) diverges 0.75% from that of Pilcaniyeu (300 km) and 0.74% from the specimen from Manantiales (700 km) (Pardiñas, Udrizar Sauthier et al. 2008). Specimens from south of 46°S have large to very large posterolateral palatal pits, in some cases with internal subdivisions, whereas in samples from north of this latitude those pits are small to absent. There are also significant differences in some cranial measurements (zygomatic plate length, and both maxillary and mandibular molar alveolar lengths) between northern (Somuncurá) and southern (Santa Cruz) samples. Externally, specimens from southern populations are more distinctly bicolored, with more pure white on the belly and dorsal surface of the manus and pes, than those from northern populations, which generally have grayish cream venters. The systematic implications of the genetic and morphological variation, if any, remain to be explored in detail with larger and more densely distributed samples.

Until the end of the 1990s, this rodent was known from only eight localities widely dispersed across Patagonia (Pardiñas and Galliari 1998a). Today, as a result of intensified mammal surveys in the arid lands of southern Argentina, more than 50 localities are now known, based on trapped specimens or on remains recovered from owl pellets (Pardiñas and Galliari 1998a; Teta et al. 2002; Pardiñas, Teta et al. 2003; Pardiñas, Udrizar Sauthier et al. 2008).

Genus *Pearsonomys* B. D. Patterson, 1992
Guillermo D'Elía

Pearsonomys is the most recent abrotrichine genus described, but it remains poorly known. Only two reports (D'Elía, Ojeda et al. 2006; Figeroa et al. 2012) have been published since the genus was described two decades ago.

Pearsonomys comprises the single living species *P. annectens*, which was described on the basis of one specimen from the vicinity of Mehuín in the southern Chilean province of Valdivia. *Pearsonomys* has the smallest distributional range of any abrotrichine genus, being known from only five localities in a small portion of the Cordillera de la Costa, primarily in southern Chilean Valdivian temperate forest. This forest type is characterized by evergreen trees mixed with a few deciduous species that grow under conditions of high annual rainfall (1,500–3,000 mm), wet and frost-free winters, and short dry summers (see details in Murúa et al. 1987).

The description of external features that follows is mostly taken from B. D. Patterson (1992a). *Pearsonomys annectens* is of moderate size, with measurements of the holotype including total length 201 mm; tail length

77 mm; hindfoot length 27 mm; ear length 19 mm; and mass 58.5 g. D'Elía, Ojeda et al. (2006) provided external dimensions and weights of additional specimens. Males are larger than females (Figueroa et al. 2012). The tail is densely furred, unicolored, and about 60% of the head and body length. The pinnae are much larger than those of all other fossorial abrotrichines. The fur is short. Specimens housed at the Mammal Collection of the Universidad Austral de Chile display substantial variation of pelage color. Some of these, as noted by B. D. Patterson (1992a), are blackish on the dorsum and buffy on the venter; others are uniformly blackish or brown.

B. D. Patterson (1992a) described the skull of *Pearsonomys* as delicate and elongate, with nasals and premaxillae forming a trumpet. The frontal sinuses are inflated; the interorbital constriction is smoothly concave; the zygomatic plates are narrow and strongly inclined backward; the zygomatic arches are weakly developed, with the breadth across them scarcely greater than that of the braincase; and the hamular processes are well developed and separate large subsquamosal and postglenoid fenestrae. The incisive foramina are long, reaching the level of the M1s; the palate is long, extending past the posterior border of the M3s; and the bullae are longer than wide. The mandible is elongate, with the coronoid process parallel to the condyloid process and the angular process directed posteriorly.

The original description (B. D. Patterson 1992a) included few details of dental morphology, because the holotype is an old adult male with advanced tooth wear. D'Elía, Ojeda et al. (2006) subsequently described dental characteristics based on multiple specimens, noting that the teeth of *Pearsonomys* resembled those of *Geoxus*. The incisors are delicate; the upper ones orthodont, pale orange, and without grooves on the anterior surface; the lower incisors are much paler. The molar toothrows are long, narrow, and with simple occlusal surfaces that become even more simplified with wear. The upper molars are crested, with labial cusps slightly taller than lingual cusps. The principal cusps are arranged opposite one another. Flexi from opposite sides do not reach each other and are perpendicular to the main molar axis. M1 is particularly large, with a distinct and large procingulum attached to the primary cusps by a wide connecting mure, but lacking an anteroflexus; it has a shallow anteromedian flexus that disappears as tooth wear advances. The anteromedian flexus divides the procingulum into two conules; the anterolabial conule is slightly smaller than the anterolingual one. The protocone and paracone are equal in size, as the hypocone and the metacone are to each other as well. In young individuals, a small mesoloph fused to the paracone can be distinguished. Both mesostyle and enterostyle are missing. The posteroloph is short and fused with the metacone; there is no postero-flexus. M2 has two pairs of main cusps, with the proto-cone and paracone slightly larger than the hypoconc and metacone. M3 is reduced (less than two-thirds the length of M2), simple, and round in outline; the mesoflexus persists as a small fossette. The lower molars are crested and transversally compressed, with flexids oriented slightly posteriad. A shallow anteromedian flexid is present on m1; the anterolabial and anterolingual conulids are subequal and define a well-developed and quadrate procingulum; an anterolabial cingulum is present; the metaconid is anterior to the protoconid; the hypoflexid is wide; the entoflexid is large; a mesolophid is absent; the entoconid is anterior to the hypoconid; and the posterolophid is well developed and slightly transverse. In m2, the protoconid is clearly larger than the hypoconid; the posterolophid is well developed; and there is no mesolophid. The m3 is shorter than m2, with a large protoconid and wide hypoflexid (D'Elía, Ojeda et al. 2006).

D'Elía, Ojeda et al. (2006) described the penile morphology. A prominent dorsal hood extends beyond the terminal crater, and the urethral flaps are relatively large, each bearing at their distal end a medially directed spine. The baculum has three distal, cartilaginous digits; the central one is larger than both laterals, and its tip protrudes from the crater. The bony baculum is ventrally keeled and has a rounded and prominent head, the base of which resembles an equilateral triangle with a notch on its proximal face. Overall, the penis of *Pearsonomys* resembles that of *Geoxus* (reported on as *Notiomys valdivianus* by Hooper and Musser 1964). The presence of a dorsal hood is especially noteworthy; this structure was previously known only from *Geoxus* (but see Spotorno 1986:116), and as such Hooper and Musser (1964:27) considered it to be a diagnostic feature the later species. Spotorno (1986) studied the glans penis of several abrotrichine taxa and, in regard to *Notiomys*, made no mention of the presence of a dorsal hood. Therefore, the presence of a dorsal hood may be considered a synapomorphy of the *Pearsonomys-Geoxus* clade (D'Elía, Ojeda et al. 2006). Among abrotrichines for which male phalli have been described, that of *Pearsonomys* is tridigitate (or of the complex type described by Hooper and Musser 1964), as are those of *Chelemys*, *Geoxus*, and *Notiomys* (Spotorno 1986).

D'Elía, Ojeda et al. (2006) also described the stomach of *Pearsonomys*. It is of the unilocular-hemiglandular type (*sensu* Carleton 1973), with a single chamber and the incisura angularis barely extending beyond the opening of the esophagus. The corpus is slightly larger than the antrum. Glandular and cornified epithelia coincide almost in distribution with the antrum and the corpus, respectively. However, the glandular epithelium does cover a small area of the corpus, because the bordering fold recurves sharply and passes to the left of the esophageal orifice. *Pearsonomys* shares the same type of unilocular-hemiglandular stomach with *Geoxus* and *Notiomys*.

SYNONYM:

Pearsonomys B. D. Patterson (1992a:132); type species *Pearsonomys annectens*, by original designation.

REMARKS: *Pearsonomys* and *Geoxus* share many morphological attributes, including a dorsal hood in the glans penis as well as details of the skull, molars, and stomach, and $2n = 56$ chromosome complement (B. D. Patterson 1992a; D'Elía, Ojeda et al. 2006; M. H. Gallardo, unpubl. data, cited in D'Elía, Ojeda et al. 2006; but see also Pearson 1984). They are also sister genera in phylogenetic analyses based on molecular sequences (M. F. Smith and Patton 1999; D'Elía, Ojeda et al. 2006). D'Elía, Ojeda et al. (2006) thus questioned the placing of *annectens* in its own genus, suggesting that *annectens* may merely represent a large-sized species of *Geoxus*. Lessa et al. (2010) supported this hypothesis in a study with a larger geographic sampling, as they found *Geoxus* paraphyletic with respect to *Pearsonomys*. However, none of these authors made a formal taxonomic change, pending revision of the tribe Abrotrichini.

Pearsonomys annectens B. D. Patterson, 1992
Pearson's Long-clawed Mouse

SYNONYM:

Pearsonomys annectens B. D. Patterson, 1992a:137; type locality "Near Mehuín, 42 km N and slightly E of Valdivia, in the Provincia de Valdivia, Región de Los Lagos [now Región de Los Ríos], Chile (39°26'S, 73°10'W; Paynter, 1988), at 100 m elevation."

DESCRIPTION: As for the genus.

DISTRIBUTION: As for the genus, known from only five localities in a small fraction of the Coastal Cordillera of southern Chile. For a decade and a half following its description, *P. annectens* was known only from its type locality. Subsequently, D'Elía, Ojeda et al. (2006) reported four additional localities. No fossils are known.

SELECTED LOCALITIES (Map 57; from D'Elía, Ojeda et al. 2006, except as noted): CHILE: Araucanía, Comuy; Los Lagos, Bahía San Pedro; Los Ríos, Fundo San Martín, Fundo Santa Rosa, near Mehuín (type locality of *Pearsonomys annectens* B. D. Patterson).

SUBSPECIES: *Pearsonomys annectens* is monotypic.

NATURAL HISTORY: All known specimens were trapped in Valdivian temperate forest or in forest plantations with native vegetation under the canopy and deep soils (Figueroa et al. 2012). The species has been recorded in syntopy with the sigmodontines *Abrothrix longipilis*, *A. olivacea*, *Geoxus valdivianus*, and *Oligoryzomys longicaudatus*, and the microbiotheriid *Dromiciops gliroides*. A male collected in April 2011 at Fundo San Martín had scrotal testes, but another secured at the same place in June 2011 had undescended testes. *Pearsonomys annectens* is apparently scarce, as Murúa et al. (2005) reported that they collected only seven specimens over more than 20 years and nearly 213,000 trap-nights of effort at Fundo San Martín, where most specimens have been taken.

REMARKS: No etymology for *annectens* is given in the original description of the species. However, Bruce Patterson (pers. comm.) explained that *annectens* (meaning joined to, connected to, annectent) refers to the intermediate morphology of this species, which combines the longer tail and large pinnae of scansorial abrotrichines of the genus *Abrothrix* with the large claws and other synapomorphies of the semifossorial *Geoxus*, *Chelemys*, and *Notiomys*.

As a result of geopolitical redistricting of the higher administrative units of Chile in 2007, the type locality is now in the Región de Los Ríos, which was split from the Región de Los Lagos.

The karyotype of *P. annectens* is $2n = 56$, FN = 62, with an autosomal complement composed of one pair of large subtelocentrics, 23 pairs of medium to small-sized acrocentrics, and three pairs of small metacentrics; the X is submetacentric, and the largest element of the complement; the Y is a small subtelocentric (D'Elía, Ojeda et al. 2006).

Tribe Akodontini Vorontsov 1959
Guillermo D'Elía and Ulyses F. J. Pardiñas

The Akodontini ranks behind the Oryzomyini as the second most diverse sigmodontine tribe in numbers of both genera and species. Members of the Akodontini range over most of South America, being absent only from southern Patagonia and most of Chile and with low diversity in Amazonia. Akodontines also constitute a diverse ecomorphological group, encompassing typical vole-like cursorial forms such as those of the genera *Akodon* and *Necromys*, the shrew-like fossorial *Blarinomys*, large semiaquatic rats of the genus *Scapteromys*, and the largest known living sigmodontine, the fossorial genus *Kunsia*. Remarkably, no akodontines have adapted to the scansorial or arboreal niches.

The concept of an akodontine group can be traced to Thomas (1916i, 1918b) and Tate (1932h), both of whom recognized the morphologic resemblances between *Akodon* and some other taxa ranked as genera (e.g., *Abrothrix*, *Chroeomys*, *Deltamys*, *Hypsimys*, *Necromys* [= *Bolomys*], *Thalpomys*, *Thaptomys*). Four decades later, Vorontsov (1959) coined the term Akodontini for this assemblage. In turn, Reig (1972, 1978, 1984, 1987) made some of the most important contributions toward delimiting the group and clarifying its evolutionary history. During this period, much of the debate on akodontine systematics centered on issues at low taxonomic levels, including species boundaries in almost all genera, the generic allocation of some species, and the distinction of several genus-level taxa, mostly those related to *Akodon*. At the peak of typological taxonomy, several taxa were named to encompass species that were morphologically similar to *Akodon* but distinguishable from it only by one or two trenchant characters. For example, Thomas (1918b) erected *Hypsimys* to contain a species characterized by its high-crowned molars, separated *Microxus* from *Akodon* based on its narrow and posteriorly sloped (from base to top) zygomatic plates (Thomas 1909), and separated *Thalpomys* from *Akodon* by its ridged interorbital region (Thomas 1916i). These taxa, together with *Deltamys*, *Necromys*, and *Thaptomys* have been considered to be either synonyms of *Akodon*, subgenera of *Akodon*, or genera depending on the weight given to these and other characters (see Gyldenstolpe 1932; Tate 1932h; Ellerman 1941; Cabrera 1961; Reig 1987; Hershkovitz 1990a, 1998). *Podoxymys* has been aligned morphologically with either or both *Oxymycterus* and *Microxus* (Anthony 1929; Ellerman 1941; Reig 1987; Pérez-

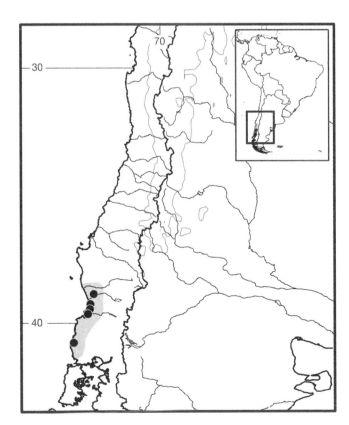

Map 57 Selected localities for *Pearsonomys annectens* (●). Contour line = 2,000 m.

Zapata et al. 1992) or has been placed in an "oxymycterine" assemblage of genera (Hershkovitz 1966a:86–87, footnote). The position of *Podoxymys* in the Akodontini, however, has not been challenged.

Aside from these examples, contents of the Akodontini remained more or less stable until molecular-based phylogenetic studies prompted substantive changes in generic membership. The most radical of these was the allocation of *Abrothrix*, *Chelemys*, *Geoxus*, *Notiomys*, and *Pearsonomys* to the newly diagnosed tribe Abrotrichini (D'Elía et al. 2007; see that account earlier). Molecular data also challenged the grouping of the "oxymycterine" genera delineated by Hooper and Musser (1964) and Hershkovitz (1966a) into the Oxymycterini (Massoia 1981a, 1983), genera that Reig (1980, 1987) placed in the Akodontini. Massoia (1979b, 1980c), Reig (1980), and Pardiñas (1996) also placed *Bibimys*, *Kunsia*, and *Scapteromys* in a separate tribe, the Scapteromyini, based on craniodental traits and ecological attributes. Molecular sequences (both mtDNA [M. F. Smith and Patton 1999; D'Elía et al. 2005] and nucDNA [D'Elía 2003a]), however, while confirming a sister relationship between *Kunsia* and *Scapteromys*, aligned these two genera in a subclade of the Akodontini that otherwise included *Blarinomys*, *Lenoxus*, and *Brucepattersonius*. Furthermore, D'Elía (2003a) and D'Elía et al. (2005) found that *Bibimys* was also an akodont, although its relationship to either *Scapteromys* or *Kunsia* was unclear (our unpublished studies do support the phyletic connection between *Bibimys* and/or *Scapteromys* and *Kunsia*). None of these analyses supported the monophyly of the Scapteromyini with respect to other akodonts, as originally envisioned.

The last important formal change at the genus level within the Akodontini was the erection of two new genera for species previously allocated to known groups. *Brucepattersonius* was proposed by Hershkovitz (1998) to include *Oxymycterus iheringi* (together with several new species), and Pardiñas, D'Elía, and Teta (2009) coined *Gyldenstolpia* for *Kunsia fronto*. Furthermore, D'Elía (2003a) and D'Elía et al. (2003) have shown that *"Akodon" serrensis* traditionally placed in the genus *Akodon*, does not belong to that radiation but is sister to *Thaptomys*. An ongoing morphological assessment of this taxon supports the placement of *serrensis* in its own genus. As no name is available, a new one should be erected. Similarly, Alvarado-Serrano and D'Elía (2013) have shown recently that a pair of northern Andean species, *bogotensis* and *latebricola*, that were also previously considered part of *Akodon* fall outside the akodontine radiation. These authors erected a new genus, *Neomicroxus*, to encompass these species, and excluded their new genus-group taxon from the Akodontini, placing it in the *incertae sedis* group of sigmodontines.

We treat the Akodontini as delineated by M. F. Smith and Patton (1999). The tribe is composed of 15 living genera: *Akodon*, *Bibimys*, *Blarinomys*, *Brucepattersonius*, *Deltamys*, *Gyldenstolpia*, *Juscelinomys*, *Kunsia*, *Lenoxus*, *Necromys*, *Oxymycterus*, *Podoxymys*, *Scapteromys*, *Thalpomys*, and *Thaptomys*. As noted previously, a new genus for *"Akodon" serrensis* should be added. The fossil genus *Dankomys* Reig has also been considered a member of the tribe (Reig 1978; but see discussion). Until now, no published phylogenetic analysis has included all akodontine genera, as all such studies have lacked representatives of *Gyldenstolpia* and *Podoxymys*. Of these two, the phylogenetic position of *Podoxymys* appears more uncertain, as *Gyldenstolpia* is closely related to *Kunsia* and *Scapteromys* (Pardiñas, D'Elía, and Teta 2009). What follows is a discussion of akodontine phylogenetic relationships at the generic level, based mostly on D'Elía (2003a). M. C. Barros et al. (2009) posited some different relationships, but limited taxon sampling precludes confidence in their results.

Akodontine genera group into five strongly supported major clades, referred to as divisions by D'Elía (2003a). Relationships among these, however, are poorly resolved (i.e., it is not possible to confidently delimit larger clades than these divisions in the Akodontini). The five divisions are an *Akodon* Division (including *Akodon*, *Deltamys*, *Necromys*, *Thalpomys*, *Thaptomys*, and *"A."* serrensis), which is equivalent to the akodontine tribe *sensu stricto*; the *Blarinomys* Division (*Blarinomys*, *Brucepattersonius*, and *Lenoxus*); the *Oxymycterus* Division (*Oxymycterus* and *Juscelinomys*); the *Scapteromys* Division (*Scapteromys* and *Kunsia*, and probably *Gyldenstolpia* on morphological grounds); and a monotypic *Bibimys* Division (*Bibimys*). Conclusions as to the number and contents of each division seem robust but may need adjustment once multiloci sequence analyses become available that include all genera. In particular, the distinction of the *Bibimys* Division from the *Scapteromys* Division, mostly found in analyses of mtDNA cytochrome-*b* sequences, requires further testing.

Reig (1972), in his doctoral thesis and published later in separate papers (Reig 1978, 1987, 1994), covered most of the fossil record of the tribe. These fossils were recovered from strata exposed on the coast of southern Buenos Aires province, Argentina, between Mar del Plata and Bahía Blanca, but their assignment to the Akodontini is questionable. The oldest known akodontine is argued to be *Necromys bonapartei*, based on a single mandible from the classic fossiliferous locality of Farola de Monte Hermoso (Reig 1978). The exact stratigraphic provenance of this fossil has been debated (Tonni et al. 1992; Marshall and Sempere 1993). Pardiñas and Tonni (1998) stated that the holotype of *N. bonapartei* was found in rocks of the upper part of the Monte Hermoso formation, lower Chapadmalalan in age (about 4–3.5 mya), a fact regrettably overlooked in recent accounts (e.g., S. D. Webb 2006). However, studies in progress regarding the generic affiliation of *bonapartei* suggest that it is neither a *Necromys* nor even an akodontine (see also Pardiñas 2010). *Dankomys* is an extinct genus described by Reig (1978) and based on a combined fragment of the skull and mandible exhumed from the Atlantic cliffs between Mar del Plata and Miramar, southeastern Buenos Aires province. Reig (1994), in a posthumous publication, described a second species of *Dankomys* from younger beds in the same area. *Dankomys* has not been adequately revised; Reig (1978) stated it was close to *Necromys*. Morphological differences between the two species of *Dankomys* (the type species, *D. simpsoni*, and *D. vorohuensis*) are so marked that it is hard to allocate both to the same genus. For example, the former has the anterior border of the zygomatic plate rounded, but the latter shows a conspicuous spine projected forward and a sharply bent premaxillomaxillary suture, both conditions that strongly resemble those in phyllotines and reithrodontines rather than those in akodontines.

We believe that the oldest fossil unambiguously assignable to the Akodontini is *Akodon lorenzinii* Reig, found in exposures of the Vorohué and San Andrés formations, Late Pliocene in age. Reig (19878) believed this species was related to the living *Akodon iniscatus*. In this paper Reig also described *Akodon johannis*, from Ensenadan beds (Late Pliocene–Middle Pleistocene), and considered it close to *Thaptomys*. The oldest scapteromyine is *Scapteromys hershkovitzi* Reig from the Vorohué Formation (Late Pliocene). However, as with some of the fossil taxa already mentioned, the generic affiliation of this species needs to be reevaluated (see Pardiñas, D'Elía, and Teta 2009). The type series of Reig's (1994) *S. hershkovitzi*, in addition to the holotype, includes an additional specimen that belongs to *Abrothrix magnus*. In fact, *hershkovitzi* seems to belong to an undescribed akodontine fossil genus more closely related to *Gyldenstolpia* and *Kunsia*. The remaining genera of the tribe have their fossil records restricted to strata no older than the Ensenadan or Lujanian. Fossil *Bibimys*, *Oxymycterus*, and *Scapteromys* belong to species that are the same, or at least ones morphologically very close to living ones (Pardiñas et al. 2002). Finally, several genera, such as *Deltamys*, *Lenoxus*, and *Podoxymys*, have no known fossil record. This may be because of the particular environments inhabited by these taxa not favoring fossilization or the limited amount of paleontological prospecting in areas that are currently forested.

Reig (1984, 1986), in his general scenario of sigmodontine historical biogeography, suggested that most akodontine genera originated in the central Andes in the Late Miocene or Early Pliocene, an area that includes present-day southwestern Bolivia, northern Chile, and northwestern Argentina. Subsequent to the origin of the basal akodontine, new genera then spread into the eastern lowlands as well as into both the northern and southern Andes to reach the current tribal range. As noted above, Reig's concept of the Akodontini differs from the one employed here; in fact, Reig (1986) envisioned an akodontine secondary *Area of Original Diversification* in the southern Andes to accommodate those taxa now placed in the Abrotrichini. As such, testing Reig's hypothesis is not straightforward, and no study has been designed to do so yet. For instance, the geographic distribution of the common ancestor of the akodontine crown group has not been inferred in any study. However, available evidence (i.e., known distributions of extinct and living forms and results of phylogenetic analyses) has indicated that the historical biogeography of Akodontini is more complex than envisioned by Reig. First, as previously reviewed, the oldest known akodontine comes from the Atlantic coast of central Argentina, a fact consistently ignored in discussions of historical biogeography because it is difficult to reconcile with the generally accepted scenario—that sigmodontines entered South America from the north and, as such, that the oldest members would have been Andean—while appealing to the incompleteness of the fossil record. Second, most genera of Akodontini are non-Andean (M. F. Smith and Patton 1999), with only one, *Lenoxus*, exclusively limited to the Andes. It is also noteworthy that *Lenoxus* is closely related to *Blarinomys* and *Brucepattersonius*, two genera distributed in eastern South America, rather than to genera with Andean species (i.e., *Akodon*, *Necromys*, and *Oxymycterus*). This suggests that akodontines reached and/or dispersed from the Andes multiple times. Relationships among species of *Akodon*, *Necromys*, and *Oxymycterus* suggest similar scenarios. In summary, we are still far from understanding akodontine biogeographic history, and no study has as yet dated most important cladogenic events, but an emerging complex picture points to the South American lowlands as an important, if not the most important, area for akodontine diversification (M. F. Smith and Patton 1999; see also Parada et al. 2013).

The extreme diversity of its members makes diagnosing the Akodontini a substantial challenge, one that, until now, has not been undertaken. Overall, the akodontine bauplan seems to be simply that of a generalized sigmodontine. Akodontines are small to medium sized, averaging 40 g in weight, although some forms are much larger (e.g., *Kunsia* reaches 500 g). The typical condition is an unspecialized, fusiform body with small, rounded ears; short legs; dark coloration with noncontrasting under parts; and a tail shorter than the combined head and body lengths (with the exception of *Lenoxus*). Eyes tend to be small or of medium size (*Blarinomys*, *Podoxymys*, *Deltamys*), and claws are moderately to well developed (*Gyldenstolpia*, *Kunsia*). The digestive system is usually without specializations for herbivory; rather, most species are omnivorous to insectivorous and have a unilocular-hemiglandular stomach. *Blarinomys*, *Brucepattersonius*, *Kunsia*, *Lenoxus*, *Oxymycterus*, and *Scapteromys* have the unilocular-discoglandular condition (the same may be also the case in *Juscelinomys* and *Gyldenstolpia*, but stomach morphology for these genera has not been described). Large intestines are short, usually less than 15% the length of small intestines; the caecum is small or absent.

Upper incisors are never grooved. Akodontine molars are mesodont to subhypsodont; the coronal surfaces are crested or terraced with main cusps in an alternate pattern, or exceptionally planate (*Gyldenstolpia*, *Kunsia*). Mesolophs and mesolophids are moderate sized (*Bibimys*, *Scapteromys*, *Gyldenstolpia*) to reduced or, more typically, are vestigial (e.g., *Akodon*, *Necromys*, *Thaptomys*), often fully or partially coalesced with paralophs or entolophids and present only as terminal remnants; posterolophs are coalesced with metalophs and a posteroflexus is usually obsolete; the upper third molar is reduced and circular in outline. Interorbital regions are typically hourglass in shape, exceptionally with sharply divergent borders (*Thalpomys*, some species of *Necromys*); frontal sinuses are inflated; zygomatic plates are little to moderately developed, rarely high or projecting before the antorbital bridge (e.g., *Scapteromys*, *Necromys*), and never with a developed spinous process. Incisive foramina usually reach posteriorly beyond the anterior edges of the first upper molars; the palate is broad and flat. Preputial glands are usually single; a medial ventral prostate usually is reduced or absent. Diploid complements vary widely, from 10 and 70 chromosomes (after Reig 1987, with modifications).

REMARKS: Vorontsov (1959) diagnosed the Akodontini in his review of cricetid rodents, along with most other tribes currently recognized in the subfamily Sigmodontinae. Cockerell et al. (1914:359), however, were apparently the first to coin the term Akodontini, as well as Ichthyomyini and Sigmodontini, as tribes of the Sigmodontinae. Although these authors neither defined nor diagnosed these family-groups, their use of the names probably emerged from a more substantive work then in progress but apparently never published, as suggested when they wrote "in the course of some work on the classification and relationships of rodents" (Cockerell et al. 1914:347).

As mentioned previously, the current concept of the tribe Akodontini encompasses the type genus and other genera earlier placed in the tribe Oxymycterini as well as all members of the tribe Scapteromyini. As such, both family-group names, Oxymycterini and Scapteromyini, are synomyns of Akodontini.

The taxonomic history of the oxymycterine group is more complex than that of the scapteromyines, given that more genera were included in it and that some of these genera are relatively diverse. Hershkovitz (1966a), in a footnote, recognized the group; listed as members the genera *Oxymycterus*, *Podoxymys*, *Lenoxus*, and *Abrothrix* (in which he subsumed *Microxus*, now included in the synonymy of *Akodon*; M. F. Smith and Patton 1993); and provided a set of cranial and dental character states that characterize it. Two decades later, Reig (1987) reinforced the hypothesis that the oxymycterines belong to the akodontine radiation and explicitly questioned the unity of Hershkovitz's grouping, in part because Reig considered *Abrothrix* a subgenus of *Akodon*. The same year, Hinojosa et al. (1987) added to the group the genera *Blarinomys* and *Geoxus* and reevaluated the supposedly diagnostic character states provided by Hershkovitz, demonstrating that several, if not most, of these were also present in other akodontines or not present in all oxymycterines. More significantly, given these results and a preliminary assessment of electrophoretic data, Hinojosa et al. (1987:9) cast doubts on the monophyly of the oxymycterine group. Hershkovitz (1994:14–15) accepted their conclusions and, in addition to arguing against the oxymycterine affinities of *Blarinomys* and *Geoxus*, agreed that what once seemed to be a natural group actually comprised a set of unrelated akodontine lineages united by morphological convergences. As noted previously, phylogenetic analyses based on DNA sequences strongly corroborated the presumption of Hinojosa et al. (1987) as to the polyphyly of the oxymycterine group. Core "oxymycterine" genera fall in two divisions of Akodontini (the *Oxymycterus* and *Blarinomys* divisions, *sensu* D'Elía 2003a) as well as in the tribe Abrotrichini (M. F. Smith and Patton 1999; D'Elía 2003a).

The taxonomic history of the scapteromyine group is relatively simple. A scapteromyine concept began with Peters (1861), who placed *Mus tomentosus* Lichtenstein together with *Hesperomys tumidus* Waterhouse under *Scapteromys* as a subgenus of *Hesperomys*. Later, Fitzinger (1867b) elevated *Scapteromys* to the genus level, although the usage of *Scapteromys* as a subgenus of *Hesperomys* continued (e.g., Thomas 1884). In the following years, new fossil and living species were added to *Scapteromys* (Gyldenstolpe 1932; Miranda-Ribeiro 1914; Winge 1887). In a key contribution, Hershkovitz (1966a) removed several taxa previously assigned to *Scapteromys* (e.g., *S. tomentosus*) and placed them in the newly created genus *Kunsia*. Hershkovitz also suggested that both genera be combined into an informal group he referred to as the scapteromyine group, which he diagnosed by a set of external, cranial, dental, and phallic characters. Hershkovitz also doubted that some other taxa belonged to his scapteromyine group, including *Scapteromys labiosus* Winge, which up to that time had been placed in *Scapteromys*. Finally, he suggested that the scapteromyines were part of the akodont radiation. Avila-Pires (1972) expanded the contents of the group by describing a new subspecies of *Kunsia fronto* (see the account of *Gyldenstolpia*). The same year Reig (1972) used the name Scapteromyini and fully diagnosed the tribe. Massoia (1979b) was the first author to make the name Scapteromyini available by actual publication. He also expanded the group with the description of a new genus and species, *Bibimys torresi*. The following year, Massoia (1980c) allocated both *Akodon chacoensis* Shamel and *Scapteromys labiosus* Winge to *Bibimys*, species that Hershkovitz (1966a) had explicitly excluded from the group. As stated earlier, phylogenetic analyses of molecular sequences (M. F. Smith and Patton 1999; D'Elía 2003a; D'Elía et al. (2005) firmly place all "scapteromyine" genera in the Akodontini and fail even to support them as a monophyletic unit in the tribe.

KEY TO THE GENERA OF AKODONTINI:

1. Tail length equal to or greater than head and body length and with a white tip that strongly contrasts in color with the basal portion . . . . *Lenoxus*
1′. Tail length equal to or shorter than head and body length and without distal white tip 2
2. Size very large, reaching 500 g in adults; pes bicolored . . . .*Kunsia*
2′. Size small to large, not exceeding 250 g in adults; pes unicolored 3
3. Size small, reduced eyes, dark colored above and below 4
3′. Size small to large, eyes moderately reduced, underparts contrastingly colored 6
4. Pelage smooth, dense, and blackish; very short tail; ears totally obscured by fur . . . . *Blarinomys*
4′. Pelage velvety and dark; tail moderately long; ears quite visible 5
5. Snout elongated . . . . *Brucepattersonius*
5′. Snout not elongated . . . .*Deltamys*
6. Snout elongated 7
6′. Snout not elongated 8
7. Tail hairless and short . . . . *Oxymycterus*
7′. Tail sparsely haired and long . . . .*Podoxymys*
7″. Tail well-haired and conical . . . .*Juscelinomys*
8. Conspicuously pink and enlarged upper lip . . . .*Bibimys*
8′. Upper lips normal in size, not pink 9
9. Large size >100 g, powerful claws on forefeet 10
9′. Small to medium sized, <100 g 11
10. Tail length >80% of head and body length . . . . *Scapteromys*

10′. Tail length <80% of head and body length . . . . *Gyldenstolpia*
11. Ears rounded and well haired, color dark 12
11′. Ears moderately large, color not dark . . . *Thalpomys*
12. Tail length >50% of head and body length 13
12′. Tail length >50% of head and body length . . . . *Thaptomys*
13. Length of fifth digit not extending beyond base of fourth digit . . . . *Necromys*
13′. Length of fifth digit extending beyond base of fourth digit . . . .*Akodon*

Genus *Akodon* Meyen, 1833
Ulyses F. J. Pardiñas, Pablo Teta, Diego Alvarado-Serrano, Lena Geise, Jorge Pablo Jayat, Pablo E. Ortiz, Pablo R. Gonçalves, and Guillermo D'Elía

The contents of *Akodon* have fluctuated greatly since Meyen first designated the genus in 1833. J. A. Allen (1905:70–71) was the first to highlight the necessity of dividing the genus, by that time composed of more than 100 species, although he pointed out that "in cranial and dental characters there is great uniformity of structure, and no very evident lines of division." Thomas (1916i) attempted that task by recognizing seven genera: *Zygodontomys*, *Akodon* (with two subgenera, *Akodon* and *Chalcomys*), *Thalpomys*, *Thaptomys*, *Bolomys*, *Chroeomys*, and *Abrothrix*. It is interesting to note the importance of molar morphology in Thomas's arrangement; for example, when erecting *Zygodontomys*, he remarked, "The recognition of *Zygodontomys* as a genus distinct from *Akodon* is somewhat provisional, and can only be finally settled when specimens with unworn molars of many more species are available for examination . . . species as *Akodon lenguarum* undoubtedly tend to connect the two groups" (1916i:338).

Reworking Thomas's diagnosis of *Akodon*, Gyldenstolpe (1932) was the first to view the genus in a way similar to its current concept. He included 36 species divided into two subgenera and employed 23 trinomials to cover *Akodon* diversity. Gyldenstolpe's (1932) understanding of *Akodon*, however, included both many species now placed in this genus as well as taxa now in *Abrothrix* (*beatus*, *olivaceus*, and *gossei*), in *Bibimys* (*chacoensis*), or even in the phyllotine genus *Calomys* (*pusillus*, a putative although improbable placement).

Ellerman's (1941) comprehensive study of living rodents represented for sigmodontines the beginning of the systematics school that favored a polytypic species concept and taxonomic organization into species-groups and subgenera instead of genera. Ellerman (1941:410ff.) thus subsumed several previously proposed genera as subgenera under

Akodon, and recognized about 85 species divided among eight subgenera (*Abrothrix*, *Akodon*, *Bolomys*, *Chroeomys*, *Deltamys*, *Hypsimys*, *Thalpomys*, and *Thaptomys*). Within the subgenus *Akodon* he created the *boliviensis* and *urichi* species groups that were basically equivalent to the contents of subgenera *Akodon* and *Chalcomys*, respectively, as proposed by Gyldenstolpe (1932). This "lumping" trend reached a peak with Cabrera (1961:437), who explained his rationale "sigo la opinión de Ellerman al incluir en este género [*Akodon*], como subgéneros, los diversos géneros en que fue disgregado por Thomas, así como *Abrothrix*; pero no veo motivo para conservar separado a *Microxus*" (freely translated: "I follow the opinion of Ellerman to include in this genus [*Akodon*], as subgenera, the several genera as was partitioned by Thomas, and also *Abrothrix*; but I see no reason to keep separate *Microxus*"). Cabrera (1961) partitioned *Akodon* into nine subgenera, with the nominotypical subgenus the most diverse with 21 species, many of which are now allocated to three other genera (*Abrothrix*, *Bibimys*, and *Necromys*). Cabrera also used trinomials extensively in several complex taxa, such as *A. boliviensis*, *A. mollis*, *A. orophilus*, *A. urichi*, and *A. varius*.

As he did for *Necromys* (see that account), Osvaldo Reig (1972) established the roughly current concept of *Akodon*. This is not surprising because the dissociation of *Necromys* from *Akodon* was the first fundamentally historical point in establishing the modern concept of *Akodon* (see also Reig 1978, 1987). Reig's (1987:358) concept of *Akodon* included "a central genus *Akodon* subdivided into five subgenera, namely *Abrothrix*, *Akodon* s.s., *Chroeomys*, *Deltamys*, and *Hypsimys*." However, the second fundamental departure in the taxonomic history of *Akodon* was the segregation of *Abrothrix* as not only a valid genus but one unrelated to *Akodon* s.s., a phyletic understanding only possible with the application of molecular data (e.g., Patton et al. 1989; Barrantes et al. 1993; M. F. Smith and Patton 1991, 1993, 1999; see the Abrotrichini account). The confused placement of *Akodon* and *Abrothrix* is perhaps unique among sigmodontines because it involved two genera from separate tribes that are so similar in external and craniodental morphology that misidentification of both specimens and species was commonplace. Although some *Abrothrix* such as *A. longipilis* were early recognized as distinct from *Akodon* s.s. (*longipilis* is the type species of *Abrothrix*), others such as *A. andinus* and *A. olivaceus* were regarded as typical *Akodon* until molecular phylogenetic studies firmly established their correct allocation to *Abrothrix* (see the historical summary by Musser and Carleton 2005).

We include 38 species in *Akodon*, plus one still undescribed ("*Akodon* sp. $2n = 10$" of several authors), totaling 39 extant species. Some authors have claimed an additional undescribed species, based on a karyotype of $2n = 34$,

from grassland-Araucaria forest ecotones in southern Brazil (e.g., Pedó, Freitas, and Hartz 2010). Our concept of *Akodon* largely agrees with Musser and Carleton's (2005) treatment of the genus but incorporates recent taxonomic hypotheses as well as newly described species (*philipmyersi* and *polopi*; Pardiñas, D'Elía et al. 2005; Jayat et al. 2010; C. F. Jiménez et al. 2013), synonyms (e.g., *oenos* as junior synonym of *A. spegazzinii*, *viridescens* as junior synonym of *A. polopi*, and *aliquantulus* as a junior synonym of *A. caenosus*; Jayat et al. 2010; D'Elía et al. 2011; Pardiñas, Teta, D'Elía, and Díaz 2011), and resurrections (*caenosus* as a species different from *A. lutescens*; Jayat et al. 2010). There are also three described extinct species of *Akodon*, *A. clivigenis* (Winge, 1887), *A. johannis* Reig, 1987, and *A. lorenzinii* Reig, 1987 (see below).

The most recent change in *Akodon* alpha diversity involved the species *bogotensis* Thomas and *latebricola* Anthony. The non-*Akodon* condition of both was briefly highlighted by Voss (2003:21–22) on morphological grounds. Alvarado-Serrano and D'Elía (2013) have now coined *Neomicroxus* as a new genus to allocate these forms and demonstrated by phylogenetic analyses of DNA sequences that their lineage is not even part of Akodontini (see that account under Sigmodontinae *incertae sedis*). We are also confident that *serrensis* Thomas does not belong to *Akodon*, as D'Elía (2003a; see also M. F. Smith and Patton 2007; Jayat et al. 2010; Coyner et al. 2013) recovered this species outside of *Akodon* based on molecular markers. This taxon represents an undescribed genus, one in the Akodontini. The remaining 38 extant species of *Akodon* we treat herein position the genus as one the few muroids with an alpha diversity greater than 30 recognized species (Musser and Carleton 2005), and second in diversity only to the 44 species of *Thomasomys* within the Sigmodontinae.

Several of the recognized species of *Akodon* are very poorly known, and their status remains uncertain. For example, *A. pervalens*, described from the Yungas of Tarija, Bolivia, has never been reviewed although its validity has been questioned (Myers 1990; Jayat, Ortiz et al. 2007). A similar lack of data characterizes *A. affinis*, *A. dayi*, *A. sanctipaulensis*, and *A. surdus*, so the status of these as valid species should be considered provisional. In contrast, *Akodon* also includes many forms that have been thoroughly studied by multiple character sets; some of these are among the most reviewed of sigmodontine taxa. For example, detailed field and laboratory analyses of *A. azarae* or *A. cursor* have detailed diet, ecophysiology, behavior, reproduction, sex determination, habitat selection, parasites, and predation, among other attributes (e.g., Pearson 1967; Dalby 1975; Zuleta and Bilenca 1992; Gentile et al. 1997; Kittlein 1997; O. Suárez and Kravetz 1998a,b; Bergallo and Magnusson 1999; Macdonald 2001; O. Suárez and Bonaventura 2001; C. Busch and Del Valle 2003; Cerqueira et al. 2003; Moraes et al. 2003; Olifiers et al. 2004; De Conto 2007, and the references therein). Furthermore, general knowledge of the genus has increased significantly, especially within the past two to three decades, thanks to the revision of several species or group of species. Starting with Myers' (1990) seminal study of the species related to *Akodon varius*), additional contributions focused on Andean species distributed in Peru, Bolivia, and Argentina, including the description of new entities (*siberiae*, Myers and Patton 1989a; *kofordi*, Myers and Patton 1989b; *juninensis* and *arequipae*, Myers et al. 1990) and a general revision of the *boliviensis* group (Myers et al. 1990). These contributions promoted similar efforts in non-Andean regions, and many species widespread in the Atlantic Forest and neighboring grasslands were studied, including a definition of the *Akodon cursor* species group (Rieger et al. 1995), phylogenetic hypotheses based on molecular and chromosomal data (e.g., Geise et al. 1998, 2001; Pardiñas, D'Elía et al. 2005; P. R. Gonçalves et al. 2007; Hass et al. 2008), descriptions of additional new species (*lindberghi* and *sanctipaulensis*, Hershkovitz 1990b; *mystax*, Hershkovitz 1998; *reigi*, L. A. González et al. 2000; *paranaensis*, Christoff et al. 2000; *philipmyersi*, Pardiñas, D'Elía et al. 2005), and karyological and morphological studies (e.g., Christoff 1997; Fagundes et al. 1998; Pardiñas, D'Elía et al. 2003; Geise et al. 2005; P. R. Gonçalves et al. 2007). More recently, and mostly associated with an important effort to collect fresh materials, revisionary studies reached west central and northwestern Argentina where several new nominal forms have been described (*aliquantulus*, M. M. Díaz et al. 1999; *oenos*, Braun et al. 2000; *polopi*, Jayat et al. 2010; *viridescens*, Braun et al. 2010; see also D'Elía et al. 2011) or restudied (*sylvanus*, Jayat, Ortiz et al. 2007; *oenos*, Pardiñas, Teta, D'Elía 2011), and the *boliviensis* group fully addressed (Jayat et al. 2010). A few additional contributions focused on species that inhabit geographic regions previously poorly known (e.g., Ecuador, Alvarado-Serrano 2005; Patagonia, Pardiñas 2009).

Reig (1978, 1980, 1981, 1984, 1986, 1987), based on both the analysis of geographic occurrences of the species of *Akodon* as well as fossil and karyological evidence, favored an "original differentiation" of the genus in the "area of the present altiplano" (Reig 1987:393). According to Reig, "The frequency of localities from which species of *Akodon* are reported shows that 82% of the occurrences belong to Andean environment . . . The ancestral akodontine may have been a generalized *Akodon*-like form . . . This ancestral form may have encountered adequate conditions in the southern proto puna, and from here local differentiation may have developed" (1987:393). Excluding "*A.*" *serrensis*, 21 of the remaining 37 living species are

Andean in distribution, eight occur in middle elevation and central semiarid lowlands, and the remaining eight characterize the forested or grassy eastern, mainly lowland, portions of the subcontinent. Of the Andean taxa, over 70% inhabit the eastern versant of the Andes and contiguous arid lands, with a latitudinal concentration between 8°S to 25°S. Only one species, *A. affinis*, has its entire distribution north of the Equator in the Cordillera Oriental. As Voss and Emmons (1996:12) highlighted, *Akodon* is definitively a non-Amazonian genus. The few records of *Akodon* sp. $2n = 10$ in the Brazilian states of Mato Grosso (M. J. de J. Silva and Yonenaga-Yassuda 1998) and Pará (T. D. Lambert et al. 2005) are restricted to enclaves of heterogeneous or nonforested habitats. Members of the genus are absent from the Guiana shield (e.g., Tate 1939; Voss 1991) and Patagonia, with the exception of *A. iniscatus*, the southernmost representative of the genus, which reaches about 47.5°S (Pardiñas 2009). Neither *Akodon* nor any akodontine occupies the western southern slope of the Andes to the Pacific coast in Chile; rather, this region is the domain of *Abrothrix* and other abrotrichines. The roughly "dumbbell" distributional pattern of species alpha diversity concentrated in the central Andes and Atlantic Forest is shared by *Oxymycterus*, another diverse akodontine genus (e.g., Hershkovitz 1994).

The oldest *Akodon* is *A. lorenzinii* Reig, 1987, whose remains were found in the strata of Vorohué and San Andrés formations (Upper Pliocene) in the coastal cliffs exposed south of Mar del Plata, Buenos Aires province, Argentina. This small *Akodon* was stated as similar to *A. iniscatus* and *A. caenosus* (the later referred to as *puer*; Reig 1987). Another extinct species from Buenos Aires province is *A. johannis*, described by Reig (1987) from several cranial fragments of a single individual excavated from the outcrops of Miramar Formation (Middle Pleistocene). According to Reig, *A. johannis* "seems to be more closely related to *A.* [= *Thaptomys*] *nigrita* . . . Doubtless it should be allocated in the subgenus *Akodon s.s.*" (1987:380). Reig (1987) also mentioned two other living species, *Akodon* cf. *A. iniscatus* from the Vorohué Formation and *A.* cf. *A. montensis* (cited as *cursor*) from the Miramar Formation. The fragmentary nature of the available material of the former, however, does not allow a confident specific assignation (Pardiñas 1995b). The later probably represents the oldest record of *A. montensis* or a morphologically related species in the *cursor* group. These were also recovered in association with a complex molar sigmodontine that Reig (1972, 1987) referred to *Nectomys*, which suggests a very different sigmodontine assemblage by that time than the one currently observed in the Pampean region of Argentina. Remains referred to *A. azarae* have been described from Middle Pleistocene (about 0.7 mya) deposits in northeastern and southeastern Buenos Aires province, Argentina (Pardiñas 1993; Voglino and Pardiñas 2005). A Middle Pleistocene owl pellet fossil assemblage from north of Mar del Plata contained an indeterminate *Akodon*, one similar in size to *A. iniscatus* or *A. azarae* (Pardiñas 2004). During the Late Pleistocene, an assemblage also produced by owls in the Andean ranges of Catamarca province, Argentina, shows a dominance of *A. spegazzinii* and a single record of *A. simulator*; in contrast, *A. caenosus*, abundant in this region today, was absent (P. E. Ortiz et al. 2011a). *Akodon spegazzinii* has a rich fossil record in northwestern Argentina (P. E. Ortiz et al. 2011b). Late Quaternary Brazilian deposits at Lagoa Santa, Minas Gerais (Winge 1887), are also rich in *Akodon* representatives, including *A. cursor*, *A. angustidens* (= *A. serrensis*; see that account), and the extinct *A. clivigenis*. The taxonomic status of the later has not been reviewed, however, and plausible synonymies with extant Brazilian forms described during recent decades are likely. *Akodon* is also registered as fossil in the Late Pleistocene of Uruguay (Ubilla 1996) and Ecuador (Fejfar et al. 1993).

Species of *Akodon* live in wide range of habitats, including the Puna, Páramo, montane tropical and subtropical forests, grassy pampas, dry montane Andean valleys, semidesert Patagonian tablelands, and Cerrado shrubby areas. However, all species are characterized by a single external morphological bauplan, small to medium in size with short legs, claws, and ears, nonreduced eyes, inconspicuous vibrissae, brownish to grayish coloration, poorly contrasting back and ventral colors, and tail moderate in length but always shorter than the combined head-body length. Not surprisingly, there are exceptions to this general pattern, but these are few. For example, *A. albiventer* has a snowy white venter that, combined with bullar enlargement and other cranial features, supported its earlier placement in *Bolomys* (= *Necromys*), and *A. montensis* has moderately large foreclaws.

Craniodental morphology in *Akodon* follows the same pattern of interspecific similarity (Reig 1987). The skull is normally built and usually somewhat elongated behind. The nasals are either longer or slightly shorter than the frontals. Anterior borders of the zygomatic plates are usually vertical and are never projected to form dorsal processes or points. The braincase is moderately long, usually not broadened, and either as wide as it is long or slightly shorter than half of the condylobasal length. The interorbital region is typically fairly narrow, without supraorbital ridges. Anterodorsal frontal sinuses are evident although not specially inflated. Bullae are usually not enlarged. Incisive foramina are long, typically reaching the protocone of M1. The anterior border of the mesopterygoid fossa is usually even with or slightly behind the posterior face of M3.

Zygomatic arches are delicate. The squamosal-alisphenoid region is characterized by a foramen ovale divided by a consistently present alisphenoid strut and a reduced anterior opening of the alisphenoid canal. A well-developed trough for the masticatory-buccinator nerve is present, partly configuring the primitive, pattern 1 carotid circulatory system described by Voss (1988), the widespread condition in the genus (Myers et al. 1990). The mandible is relatively low and stout, more slender in the smaller species, with the masseteric crest normally developed and reaching the middle of m1. The capsular projection is usually not defined as a distinct tubercle. Deviations from this standard in cranial morphology are infrequent. Thomas (1920h) remarked that the narrowed and slanted zygomatic plate of *A. mimus* (and to a lesser extent, *A. torques*), which he termed "microxine," was more similar to that of *Oxymycterus* than to typical *Akodon*. The very shallow zygomatic notches, partially related to the zygomatic plate expression, also distinguished the skulls of *A. mimus* from those of most *Akodon* species, all of which have typically deeper zygomatic notches.

Myers et al. (1990:22) vividly wrote that "*Akodon* teeth are frustrating to study because they wear rapidly, and as they wear they lose much of their occlusal topology." The upper incisors are frequently opisthodont, more rarely orthodont, always smooth and well pigmented. The molars are neither markedly elongated nor narrowed, and have a moderately developed tubercular hypsodonty; as an exception, crowns are noticeably higher in *A. budini* and *A. siberiae*. The molar crowns are usually bilevel or terraced, and the main cusps alternate in position. M1 always has an anteromedian flexus (except in "*A.*" *serrensis*), but in several members this structure disappears with wear in adult individuals (e.g., the *varius* group, *sensu* Myers 1990). Remnants of the mesoloph are usually united to the mesostyle, typically on both M1 and M2. M3 tends to be a simple cylinder (Reig 1987).

Studies of soft anatomy show that all species of *Akodon* studied to date share the widespread unilocular-hemiglandular condition, with the bordering fold bisecting the stomach on a line from the incisura angularis to a point opposite it on the greater curvature (Carleton 1973). The presence of a gall bladder is another common trait in the genus (Voss 1991; Geise, Weksler, and Bonvicino 2004); however, at least one species, *A. montensis*, lacks this structure. A single pair of large preputial glands is present that typically extends well beyond the ventral flexure of the penis (Voss and Linzey 1981).

Montes (2007) summarized the major trends across the extensive karyological literature of *Akodon*. Karyotypes have been commonly used to distinguish many species of *Akodon*, especially those that inhabit eastern and central

South America (e.g., *cursor* group taxa). Even more, several changes of taxonomic status of different nominal forms have been made on the basis of karyotypic information, either to establish them at the species level (e.g., *montensis*) or regard them as synonyms (e.g., *arviculoides*). N. O. Bianchi and Contreras (1967), in describing polymorphism of sex chromosomes in *A. azarae*, published the first cytogenetic work on *Akodon*. Later, N. O. Bianchi et al. (1971) recognized at least three karyological assemblages in *Akodon*, one of which included the species currently referred to *Abrothrix*. G and C-banding techniques were also used in early papers, starting with the analysis of N. O. Bianchi et al. (1973, 1976) and followed by many others (e.g., Yonenaga-Yasuda 1979; M. A. Barros et al. 1990; Fagundes and Yonenaga-Yasuda 1998; Geise et al. 1998). Fagundes et al. (1997) published the initial work on fluorescent hybridization in *Akodon* species. During the period from 1970 to 1990, most of the currently recognized species in the genus were karyotyped; these included individual studies that often combined cytogenetic and morphological data (see review by Montes 2007) that supported species-groups in *Akodon* (e.g., the *varius* group, *sensu* Myers 1990; *boliviensis* group of Myers et al. 1990; *fumeus* group of Myers and Patton 1989b; and *cursor* group of Rieger et al. 1995).

Outstanding contributions on karyological patterns in *Akodon*, and by extension of the Akodontini as a whole, were made by N. O. Bianchi, O. A. Reig, and their collaborators, in a series of 10 papers published between 1971 and 1984, all under the general title of "Cytogenetics of South American Akodont Rodents (Cricetidae)," as well as by Y. Yonenaga-Yasuda and her collaborators (e.g., Yonenaga 1972b; Yonenaga-Yasuda et al. 1976; Yonenaga-Yasuda 1979). In both cases, these authors dealt almost exclusively with species living in open lowland to forested areas of South America, primarily in Argentina and Brazil. Some of these works included laboratory cross-breeding experiments of putative species, such as between *A. dolores* and *A. molinae* (e.g., Merani et al. 1978). Andean forms were also the subject of several studies, most of them focused on Argentinean, Bolivian, and Peruvian forms (Barquez et al. 1980; Myers and Patton 1989a,b; Myers et al. 1990; Patton et al. 1990; Patton and Smith 1992b; Blaunstein et al. 1992). With respect to the most recent phylogenetic analyses (see below), *Akodon* species groups can be karyologically characterized as follows (Montes 2007): the *aerosus* group (*sensu* M. F. Smith and Patton 2007), plus *A. albiventer* either have $2n = 22–26$ (*A. mollis*, *A. torques*, *A. orophilus*, some *A. aerosus*) or $2n = 38–40$ (other *A. aerosus*, *A. albiventer*, *A. budini*, *A. mimus*, *A. siberiae*), although are all characterized by FN = 40. Members of the

varius and *dolores* groups (*sensu* Jayat et al. 2010) have $2n = 33–44$, sharing a high FN (42–44). Those of the *boliviensis* group (*A. boliviensis*, *A. spegazzinii*, *A. lutescens*, *A. subfuscus*, *A. juninensis*, *A. kofordi*, *A. fumeus*, *A. polopi*) all have the same $2n = 40$, FN = 40 karyotype. Finally, a fourth group of species that occupies open to forested areas in lowland terrains from east-central Argentina to central and eastern Brazil, some of uncertain affinities (e.g., *A. azarae*) but others closely related (e.g., *A. montensis*, *A. reigi*, *A. paranaensis*; see M. F. Smith and Patton 2007), are characterized by substantial variation, with $2n$ ranging from 14–16 (*A. cursor*), 24–26 (*A. montensis*), 36 (*A. philipmyersi*), 37–38 (*A. azarae*), 42 (*A. lindberghi*), to 44 (*A. paranaensis*, *A. reigi*).

Remarkably, besides normal XX females, in several species of *Akodon* females possess karyotypes indistinguishable from those of males, with XY females known for eight different species with at least six independent origins (Hoekstra and Edwards 2000). The XY females, known since the 1960s (N. O. Bianchi and Contreras 1967), are fertile. Cytogenetic and molecular studies have shown that these females are not Xx, with one X resulting from an arm deletion (i.e., an extreme case of dosage compensation), but XY*, where the *Sry* gene differentially expressed (see review in N. O. Bianchi 2002). Recently, A. Sánchez et al. (2010) showed that, at least for *A. azarae* and *A. boliviensis*, *Sry* mutations are not the basis of sex reversal XY* given that the *Sry* gene sequences of XY* do not significantly differ from those of males. Thus, the basis of the inactivation of the *Sry* gene in XY* remains unclear.

Beginning in the early 1990s with M. F. Smith and Patton's (1991, 1993) pioneering studies, DNA sequence data began to be extensively used in phylogenetic studies of *Akodon*. Most of these studies are based on mtDNA cytochrome-*b* gene sequences, with sequential papers offering an increasingly broader taxonomic coverage. Jayat et al. (2010) and Coyner et al. (2013) are the most recent approaches with broad taxonomic coverage of *Akodon*.

Phylogenetic studies of *Akodon* have had their major impact in two critical systematic areas: membership in the genus and delineation of species group relationships. By assessing the phylogenetic position of the type species of genus-group taxa related to *Akodon*, these molecular studies (M. F. Smith and Patton 1999; D'Elía 2003a; D'Elía et al. 2003) have both set the limits of *Akodon* and demonstrated that taxa such as *Deltamys*, *Thalpomys*, and *Thaptomys* deserved generic status, given that their respective type species, *kempi*, *lasiotis*, and *nigrita*, fell outside the crown-group clade of *Akodon* s.s. (i.e., the least inclusive clade containing "non-problematic" *Akodon* taxa and

A. boliviensis, which is the type species of *Akodon*). It is relevant to note here, however, that most of the topologies supporting the distinction of *Deltamys* are either weakly supported or, in several topologies gathered by M. F. Smith and Patton (2007) and Jayat et al. (2010), *Akodon s.s.* is paraphyletic with respect to *Deltamys*. Clearly, the generic distinction of *Deltamys* needs further scrutiny. In addition, these phylogenetic analyses have shown that *serrensis* Thomas, whose placement in *Akodon* had never been questioned, falls outside the *Akodon s.s.* clade and that, as result, this species should be given generic status. On the other hand, these phylogenetic studies have also shown that *mimus* and *budini*, type species of *Microxus* and *Hypsimys*, respectively, fall in the clade of *Akodon s.s.*, rendering these two genus-group taxa synonyms of *Akodon*. As is the case of relationships among *Akodon* species groups (see below), the phylogenetic position of some of these taxa (notably *mimus* and *kempi*) varies with the taxonomic coverage, gene(s) sequenced, and analytical method employed (M. F. Smith and Patton 2007; Jayat et al. 2010; Coyner et al. 2013). Therefore, to obtain a stable delimitation of *Akodon*, analysis of a taxonomically dense matrix using multiple mtDNA and nucDNA sequences of different loci remains a necessity.

Traditionally, species of *Akodon* have been divided into a number of species groups. Some of these groupings (e.g., the *boliviensis* versus *mollis* groups of Hershkovitz 1990b) simply define size-based associations. Others, however, were assumed to represent taxa that were part of a common evolutionary lineage "explicitly based on the overall similarity of the taxa and their apparently contiguous geographic distributions" (Myers 1990:35, in his definition of the *varius* group). Myers (1990) defined the *varius* group on morphological grounds alone; Myers et al. (1990) based their *boliviensis* group both on morphology and allozymes; and Rieger et al. (1995) delineated the *cursor* group mostly on karyotypic data. More recent studies, built primarily around a phylogenetic rather than a phenetic framework, have reviewed the contents of these earlier groups while defining new ones. In particular, M. F. Smith and Patton (2007) and Jayat et al. (2010) deserve special mention (see also Pardiñas, D'Elía et al. 2005; Jayat, Ortiz et al. 2007; Montes 2007; Coyner et al. 2013), given their dense taxonomic sampling and the overall relevance of their findings. What follows in this account is mostly based on M. F. Smith and Patton (2007) and Jayat et al. (2010); we also make some comments about the results recently published by Coyner et al. (2013), based on Coyner (2010), although these two studies have more limited taxonomic coverage and do not include several major changes proposed in recent years. However, and importantly, Coyner et al. (2013) presented a phylogenetic hypothesis derived from concatenated mtDNA and nucDNA sequences, expanding the available molecular data for *Akodon*.

Currently, five species groups are recognized in *Akodon*. (1) The *Akodon aerosus* group is minimally composed of *A. aerosus*, *A. albiventer*, *A. affinis*, *A. mollis*, *A. orophilus*, *A. surdus* (varyingly regarded as a synonym of *aerosus* [Patton and Smith 1992a,b] or as a species, such as herein [see also Coyner et al. 2013]), and *A. torques*. M. F. Smith and Patton (2007) recovered *A. budini* and *A. siberiae* as part of their *A. aerosus* species group, but Jayat et al. (2010), in a more taxon dense study, found that both species fell outside of a core *aerosus* clade. (2) The *Akodon boliviensis* group contains *A. boliviensis*, type species of the genus, plus *A. caenosus*, *A. fumeus*, *A. juninensis*, *A. kofordi*, *A. lutescens*, *A. polopi*, *A. spegazzinii*, *A. subfuscus*, and *A. sylvanus*. (3) The *Akodon cursor* species group includes *A. cursor*, *A. montensis*, *A. reigi*, and *A. paranaensis* as well as the forms referred in the literature as *A.* aff. *cursor* and *Akodon* sp. $2n = 10$ (see M. J. de J. Silva et al. 2006). M. F. Smith and Patton (2007) posited that *A. sanctipaulensis* was a member of this group, although to date it has not been included in any molecular analysis. Preliminary assessments also place *A. mystax* and *A. lindberghi* together as a small body-size subgroup in the *cursor* group, one from eastern Brazil that shares a $2n = 42$ karyotype (P. R. Gonçalves et al. 2007; Jayat et al. 2010; Coyner et al. 2013). (4) The recently delimited *Akodon dolores* group (Jayat et al. 2010; see also Coyner et al. 2013), which includes *A. dayi*, *A. dolores* (including *molinae*), *A. iniscatus*, and *A. toba*. (5) An *Akodon varius* group that is now restricted to *A. simulator* and *A. varius* (see Jayat et al. 2010). Coyner et al. (2013) proposed to transfer these species to the *aerosus* group, given they appear as sister to that clade. We do not favor this proposition, given the large morphological gap between members of the two groups. *Akodon pervalens*, for which no DNA sequence data are yet available, likely belongs to the *varius* species group. Finally, there is a set of *Akodon* species whose phylogenetic position is currently unresolved, including *A. azarae*, *A. budini*, *A. mimus*, *A. philipmyersi*, and *A. siberiae*. Relationships among species groups as well as sister-pair relationships within the groups also vary greatly among analyses. More robust analyses using nuclear in addition to mitochondrial gene sequences are required to clarify further the contents of each species group as well as the relationships both among and within them.

Phylogenetic analyses of *Akodon* species have been also used in recent new species descriptions with the aim of correctly placing the new forms in the *Akodon* radiation and then to guide species comparisons (e.g., *philipmyersi*, Pardiñas, D'Elía et al. 2005; *polopi*, Jayat et al. 2010). Similarly, phylogenetic analyses have been used in taxonomic studies designed to assess the distinction of different nominal forms as, for example, cases where either recognition at the species level was supported (e.g., *sylvanus*, Jayat, Ortiz et al. 2007) or was falsified, resulting in names becoming junior synonyms (e.g., *oenos*, Pardiñas, Teta, D'Elía, and Diaz 2011; *viridescens*, D'Elía et al. 2011, see also Coyner et al. 2013:9). Finally, phylogenetic analyses have recently begun to tackle aspects of *Akodon* evolutionary history in addition to traditional questions of systematics and taxonomy, notably the independent origin of XY* females (Hoekstra and Edwards 2000), the direction of chromosomal evolution (M. C. Barros et al. 2009), or more general historical biogeography (M. F. Smith and Patton 2007; M. C. Barros et al. 2009).

SYNONYMS:

Mus: J. B. Fischer, 1829:313; part (description of *azarae*); not *Mus* Linnaeus.

Akodon Meyen, 1833:599; type species *Akodon boliviense* Meyen, by original designation.

Hesperomys: Wagner, 1843a:519; part (listing of *arenicola*); not *Hesperomys* Waterhouse.

Acodon Tschudi, 1844:177; incorrect subsequent spelling of *Akodon* Meyen.

Axodon Giebel, 1855:48; incorrect subsequent spelling of *Akodon* Meyen.

Habrothrix: Winge, 1887:27; part (description of *cursor*); not *Habrothrix* Wagner.

Hesperomys (*Habrothrix*): J. A. Allen, 1891b:210; name combination.

[*Akodon* (*Akodon*)]: E.-L. Trouessart, 1897:535; name combination.

Microxus Thomas, 1909:237; type species *Oxymycterus mimus* Thomas, by original designation.

Akodon (*Chalcomys*) Thomas, 1916i:338; type species *Akodon ærosus* Thomas, by original designation.

Hypsimys Thomas, 1918b:190; type species *Hypsimys budini* Thomas, by original designation.

Bolomys: Thomas, 1919f:131; part (listing of *albiventer*); not *Bolomys* Thomas.

[*Akodon*] (*Akodon*): Ellerman, 1941:410; name combination.

[*Akodon*] (*Hypsimys*): Ellerman, 1941:414; name combination.

[*Akodon*] (*Bolomys*): Ellerman, 1941:415; part (listing of *albiventer*, *berlepschii*, and *leucolimnaeus*); not *Bolomys* Thomas.

Akodon (*Bolomys*): Sanborn, 1950:13; name combination.

Plectomys Borchert and Hansen, 1983:237; *nomen nudum* (see Hershkovitz 1990b:21–23).

KEY TO THE SPECIES OF *AKODON*: [NOTE THAT MANY SPECIES ARE SO SIMILAR MORPHOLOGICALLY, AND SPECIES BOUNDARIES ARE SO POORLY UNDERSTOOD, THAT IT IS DIFFICULT TO PRODUCE A DEFINITIVE KEY. THE ONE PROVIDED HERE IS UNCONVENTIONAL IN THAT IT IS NOT STRICTLY DICHOTOMOUS. SUBSTANTIAL WORK IS REQUIRED TO IMPROVE ON THIS ATTEMPT.]

1. Anteromedian flexus absent . . . . *Akodon serrensis*
1'. Anteromedian flexus present 2
2. Found in Andes, Bolivian lowlands, dry Chaco, Monte, and Patagonia. 3
2'. Found in Atlantic forest, Cerrado, humid Chaco, and temperate eastern grasslands. 15
3. Molars strongly hypsodont; braincase rounded; rostrum narrow and pointed 4
3'. Molars slightly to moderately hypsodont; braincase squared; rostrum broad and moderately short 5
4. Found in eastern Andes of Cochabamba department, Bolivia; zygomatic plate narrow; tail length >45% of total length . . . . *Akodon siberiae*
4'. Found in eastern Andes of northwestern Argentina and southernmost Bolivia; zygomatic plate moderately broad; tail length <45% of total length. . . . .
. . . . *Akodon budini*
5. Size large (maxillary tooth-row length >4.5 mm); supraorbital region squared or beaded, often with edges divergent posteriorly; rostrum short and broad; interorbital region broad; $2n = 34$–42, $FN = 40$–44 6
5'. Size small to medium (maxillary toothrow length <4.5 mm); supraorbital region smooth, without edges divergent posteriorly; rostrum moderately long and narrow; interorbital region narrow 7
6. Found in northern and central Bolivia; tail unicolored dark; foot dark . . . . *Akodon dayi*
6'. Found in central and southern Bolivia; zygomatic plate narrower. . . . . *Akodon varius*
6''. Found in Yungas of southern Bolivia; rostrum especially broad. . . . . *Akodon pervalens*
6'''. Found in Yungas of southern Bolivia and northwestern Argentina; dorsally gray or brown; conspicuous buffy eye rings . . . . *Akodon simulator*
6''''. Found in dry Chaco; skull proportionately broader; shorter nasals . . . . *Akodon toba*
6'''''. Found in montane Chaco and Monte; colour uniform pale brown; skull strongly bowed. . . . *Akodon dolores*
7. Found in Patagonia; $2n = 33$–34 . . . . *Akodon iniscatus*
7'. Found in Andes and Andean ranges, $2n = 40$, $FN = 40$.
. . . . 8
8. Zygomatic plate of "microxine" type, inclined backward and narrow; zygomatic notches very shallow. . . .
. . . . *Akodon mimus*

8'. Zygomatic plate of "akodont" type, straight and moderately broad; zygomatic notches distinct 9

9. Body pelage pattern distinctly countershaded; bullae especially enlarged; upper incisors slightly proodont . . . .
. . . . *Akodon albiventer*

9'. Body pelage pattern indistinctly or not countershaded; bullae normal sized; upper incisors opisthodont to slightly orthodont 10

10. Size small to medium; interorbit narrow; 2n = 34 or 40, FN = 40 11

10'. Size medium; interorbit broad; 2n variable in number
. . . . 14

11. Size small; tail length <70 mm 12

11'. Size medium; tail length >70 mm 15

12. Zygomatic notches and plates broad; braincase moderately broad; strongly bilophodont third molars . . . . . .
. . . . *Akodon boliviensis*

12'. Zygomatic notches shallow; zygomatic plates narrow; 2n = 34 . . . . *Akodon caenosus*

12''. Zygomatic notches narrow; teeth relatively large; M3 usually oval . . . .*Akodon subfuscus*

12'''. . . . Zygomatic notches small and shallow; zygomatic plates narrow; M3 with a distinct metaflexus . . . . . .
. . . . *Akodon juninensis*

12''''. Hindfoot <21 mm; length of maxillary toothrow <3.8 mm . . . . *Akodon lutescens*

12'''''. Found on eastern Andean slopes of northwestern and central Argentina; coloration variable; rostrum well developed . . . .*Akodon spegazzinii*

13. Found isolated in central Argentina in elevations >1,500 m . . . . *Akodon polopi*

13'. Found isolated in Sierra de Santa Bárbara, Jujuy province, Argentina . . . .*Akodon sylvanus*

13''. Tail bicolored; mesopterygoid fossa broad . . . .
. . . .*Akodon fumeus*

13'''. Tail unicolored; mesopterygoid fossa lyre shaped . . .
. . . . *Akodon kofordi*

14. Found in the Cordillera Occidental, western Colombia
. . . . *Akodon affinis*

14'. Found in upper eastern Andean slopes, central Peru; 2n = 26, FN = 40 . . . .*Akodon orophilus*

14''. Found from northwerstern Peru to northern Ecuador; 2n = 22–23, FN = 40 . . . . *Akodon mollis*

14'''. Found in upper montane forests of eastern Andean slopes, 1,000–2,000 m, in Ecuador, Peru, and Bolivia; 2n = 22, 38, or 40, FN = 40 . . . . *Akodon aerosus*

14''''. Found in eastern Andean forests, 2,900–3,800 m, of Huánuco department, central Peru; 2n = 22, FN = 40 . .
. . . . *Akodon josemariarguedasi*

14'''''. Found on eastern Andean slopes of Ayacucho and Cusco departments, Peru; tail nearly naked and very long; zygomatic plate resembling "microxine"

type . . . .
. . . . *Akodon torques*

14''''''. Found in cloud forests, eastern flank of southeastern Andes in Peru; fur short, crisp, and dark; cranial profile relatively flat . . . . *Akodon surdus*

15. Size large; tail length >70 mm; maxillary toothrow length >4.5 mm 16

15'. Size small to medium; tail length <70 mm; maxillary toothrow length <4.3 mm 17

16. 2n = 10 . . . . *Akodon* sp. 2n = 10

16'. 2n = 14–16; gall bladder present . . . . . *Akodon cursor*

16''. 2n = 44, FN = 42 . . . .*Akodon reigi*

16'''. 2n = 44, FN = 44 . . . . *Akodon paranaensis*

16''''. 2n = 24–25; gall bladder absent . . *Akodon montensis*

17. Zygomatic plates narrow 18

17'. Zygomatic plates broad; found in nonforested habitats of southern Brazil, eastern Paraguay, south to Uruguay and central Argentina . . . . *Akodon azarae*

18. Isolated distribution in southern Misiones province, Argentina; 2n = 36 . . . .*Akodon philipmyersi*

18'. Isolated on Mt. Caparaó, Brazil; 2n = 42–44; masseteric scar distinct . . . .*Akodon mystax*

18''. Found in Cerrado and Atlantic forest of Brazil; 2n = 42; masseteric scar indistinct . . *Akodon lindberghi*

18'''. Isolated in São Paulo state, Brazil; interorbit broad
. . . . *Akodon sanctipaulensis*

Akodon aerosus Thomas, 1913
Yungas Grass Mouse, Yungas Akodon
SYNONYMS:

Hesperomys (Habrothrix) caliginosus: J. A. Allen, 1891b: 210; part; not *Hesperomys caliginosus* Tomes.

[*Acodon*] *caliginosus*: Thomas, 1894b:356; part; not *Hesperomys caliginosus* Tomes.

Akodon aerosus Thomas, 1913b:406; type locality "Mirador, Baños, Ecuador. Alt. 1500 m.," Baños (ca. 2.924°S, 79.066°W), Azuay, Ecuador.

Akodon aerosus baliolus Osgood, 1915:192: type locality "Inca Mines, Inambari River, Peru," Santo Domingo (ca. 13.85°S, 69.683°W), Puno, 1,689 m, on the Río Inambari, general area of gold mines near town called Inca Mine belonging to the Inca Mining Co. (see Stephens and Traylor 1983).

Akodon (Chalcomys) aerosus: Thomas, 1916i:338; name combination.

Akodon (Chalcomys) aerosus aerosus: Gyldenstolpe, 1932a: 110; name combination.

Akodon (Chalcomys) aerosus baliolus: Gyldenstolpe, 1932a: 110; name combination.

Akodon urichi aerosus: Cabrera, 1961:448; name combination.

Oryzomys (*Melanomys*) *caliginosus caliginosus*: S. Anderson, 1985:12; part; not *Hesperomys caliginosus* Tomes.

DESCRIPTION: Described originally as large dark species almost indistinguishable externally from *Melanomys caliginosus*. Pelage dark brown above and slightly paler below, with ears, forefeet, hindfeet, and tail blackish brown. Patton and Smith (1992b) characterized species by its large body size (total length ca. 190 mm), dense but short fur, moderately long and nearly naked tail, and nearly uniform coloration (deep blackish or reddish brown dorsally and paler on the venter). Skull strongly built (condyloincisive length ca. 25.6 mm), with broad interorbital region; rostrum short and broad; smooth and nonbeaded supraorbital ridges; evenly convex dorsal profile; anterior edges of zygomatic plates straight to slightly concave; incisive foramina long, widely open, and with smoothly rounded edges; broad mesopterygoid fossa; and long maxillary toothrow (ca. 4.6 mm). S. Anderson (1997) provided external and cranial measurements for Bolivian specimens.

DISTRIBUTION: *Akodon aerosus* occurs in lower montane forests along the eastern Andean slopes, between 1,200–2,400 m, from southern Ecuador, through Peru, to central Bolivia (Patton and Smith 1992a,b; S. Anderson 1997). In Peru (Patton and Smith 1992a,b), this species is found in a series of disjunct populations in Amazonas, San Martín, Huánuco, Junín, Ayacucho, Cusco, and Puno departments. It is apparently absent from the Urubamba drainage in Cusco department, Peru, where it is replaced by *A. surdus* (see Remarks for that species). In Bolivia, it is found along a narrow band of Yungas forest between Cochabamba and La Paz departments (S. Anderson 1997). No fossils are known for *A. aerosus*.

SELECTED LOCALITIES (Map 58): BOLIVIA: Cochabamba, Alto Palmar, Chapare (S. Anderson 1997); La Paz, 17 km N of Apolo (Anderson 1997), base of Mt. Sorata (S. Anderson 1997). ECUADOR: Azuay, Baños (type locality of *Akodon aerosus* Thomas). PERU: Amazonas, Cordillera del Cóndor, Valle Río Comaina, Puesta Vigilancia 3, Alfonso Ugarte (USNM 581938); Ayacucho, Hacienda Luisiana [= Luisiana], Río Apurimac (Patton and Smith 1992a,b); Cusco, 72 km NE of Paucartambo (Patton and Smith 1992a), Hacienda Cadena [= Cadena], Marcapata (Patton and Smith 1992a,b), 2 km SW of Tangoshiari (USNM 588097); Huánuco, 35 km NE of Tingo Maria on Carretera Central (LSUMZ 12600); Junín, 10 km WSW (by road) of San Ramón (MVZ 172873); Puno, 14 km W of Yanahuaya (Patton and Smith 1992a,b), Inca Mines, Río Inambari (= Santo Domingo) (type locality of *Akodon aerosus baliolus* Osgood), Ocancque, 10 mi N of Limbani (MVZ 116110), 11 km NNE of Ollachea (Patton and Smith 1992a,b).

SUBSPECIES: S. Anderson (1997) recognized *baliolus* as a subspecies for Bolivian populations, and Pacheco et al. (2012) treated *baliolus* as a species. As yet comparisons of available samples have not substantiated either hypothesis of this latitudinally widely distributed and karyotypically variable species.

NATURAL HISTORY: There is no published information of the natural history of this species. Known records are restricted to forested environments of the Yungas along the eastern versant of the Andes, between 1,200–2,400 m. Pacheco et al. (2011; referred to as *A. baliolus*) found high frequencies of *A. aerosus* in premontane forests between 1,200 and 1,985 m in the Río Tambopata drainage of Puno department, Peru. Limited reproductive data from labels of specimens in the MVZ document pregnant and lactating females in the months of July and August in southern Peru, with litter size (based on embryo counts or placental scars) ranging from two to seven.

REMARKS: A member of the *Akodon aerosus* group, as defined by M. F. Smith and Patton (2007), which includes the nominal taxa *aerosus*, *affinis*, *baliolus*, *budini*, *mollis*, *orophilus*, *siberiae*, *torques*, and putatively *albiventer*. Thomas (1916i), mostly based on the dark-colored appearance and velvety fur of this rodent, established the subgenus *Akodon* (*Chalcomys*), in which he also included *urichi*, *venezuelensis*, and *meridensis* (all three names now allocated to *Necromys*; see that account). Cabrera (1961) considered *aerosus* a synonym of *urichi*, but A. L. Gardner and Patton (1976) listed the two as separate species based on a pronounced difference in diploid number. Phylogenetic analysis of mtDNA sequences failed to recover a monophyletic *aerosus*, suggesting that more than one species is encompassed within the current concept of this taxon (see Patton et al. 1990; Patton and Smith 1992a,b; M. F. Smith and Patton 2007). Patton and Smith (1992b) included *surdus* under *aerosus*, based on their examination of mtDNA sequences of a single specimen from Machu Picchu, which is within the range of *surdus* as defined by Thomas (1917g, 1920h). Pacheco et al. (2012) treated *baliolus* as a species, based on karyotype differences noted by Patton and Smith (1992a,b) and shorter nasals of Puno samples relative to those from central Peru north to Ecuador. A thorough revision of *A. aerosus*, as delineated herein, is badly needed, both to define the number of taxa included and delimit their respective geographic ranges. Myers (1990) documented the morphological distinctiveness of *A. aerosus* as compared with *A. varius* and related forms.

Patton et al. (1990) and Patton and Smith (1992a,b) reported considerable variation in karyotype among populations in southern Peru that shared identical morphology, with 2*n* = 40 for specimens from Ayacucho department,

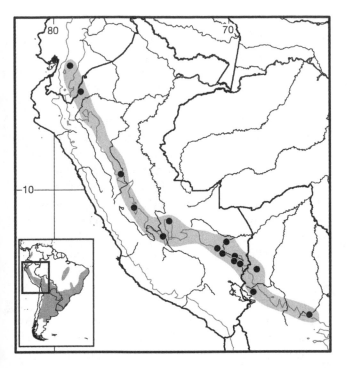

Map 58 Selected localities for *Akodon aerosus* (●). Contour line = 2,000 m.

$2n = 38$ for specimens from Puno department, and $2n = 22$ for those from Cusco department.

Akodon affinis (J. A. Allen, 1912)
Colombian Grass Mouse, Colombian Akodon
SYNONYMS:

Microxus affinis J. A. Allen, 1912:89; type locality "San Antonio (near Cali, alt. 8000 ft.), Cauca, Colombia"; San Antonio (3.2993°N, 76.5192°W), Santiago de Cali, Colombia.

Akodon tolimae J. A. Allen, 1913a:480; type locality "Rio Toché, Quindio Andes, Tolima, Colombia; altitude, 7000 feet."

Microxus (?) *affinis*: Gyldenstolpe, 1932a:134; name combination.

Akodon affinis: Goodwin, 1953:320; first use of current name combination.

DESCRIPTION: Medium sized (total length = 160 mm; tail = 70 mm; greatest length of skull = 25.8 mm; upper toothrow = 5 mm [measurements of holotype, from J. A. Allen 1912]), characterized by uniform dusky brown upper parts, with tips of hairs yellowish, giving general olivaceous effect; venter dark grayish brown, with tips of the hairs olivaceous, giving the general effect of dark grayish brown with faint olive tinge; ears dull brown, nearly naked, rather short and broad; feet grayish brown, hairs dark with lighter tips; tail brown, not appreciably bicolored, nearly naked, shorter than head and body.

DISTRIBUTION: *Akodon affinis* is known from the Cordillera Occidental of western Colombia (Musser and Carleton 2005). No fossils are known.

SELECTED LOCALITIES (Map 59): COLOMBIA: Antioquia, Vereda Puente Peláez, Finca Cañaveral (Voss et al. 2002); Nariño, Galera (J. A. Allen 1913a), La Florida (J. A. Allen 1913a); Risaralda, La Pastora, Reserva Ucumarí (Quiceno 1993), Vereda Siató (Voss et al. 2002); Santander, 10 km W of Santander (MVZ 124065); Valle del Cauca, Bosque San Antonio, Cordillera Occidental (Quiceno 1993), Corregimiento Baragán, Hacienda La Esperanza (Quiceno 1993), Hacienda La Sirena, carretera Palmira-La Nevera (Quiceno 1993), Hacienda Los Alpes (Quiceno 1993), Laguna Sonso (Quiceno 1993), Municipio de Roldanillo (Quiceno 1993), 4 km NW of San Antonio (MVZ 124066).

NATURAL HISTORY: Few ecological data are available for this species. It is apparently common, especially in secondary and altered habitats (R. P. Anderson and M. Gómez-Laverde, pers. comm.). Voss et al. (2002) captured 10 specimens in the valley of the Río Siato, Risaralda department, at 1,520–1,620 m in 405 trap-nights. It has been found in both primary cloud forests and adjacent anthropogenically modified vegetation in relatively cool and very humid environments (Voss et al. 2002). This mouse is terrestrial, mostly diurnal and crepuscular, and probably feeds on insects, seeds, and vegetation (Quiceno 1993).

REMARKS: J. A. Allen originally allocated this species to *Microxus*, possibly influenced by its external resemblance to *A. bogotensis*. Anthony (1924:4), in his description of *A. latebricola*, suggested that *affinis* was unlike other *Microxus* and perhaps was more appropriately placed in *Akodon*. Goodwin (1953:320) later formalized Anthony's opinion, a treatment followed by Cabrera (1961:440). One individual sequenced by M. F. Smith and Patton (2007), putatively referred to this species and collected about 180 km NNE of the type locality, was found to be part of the *Akodon aerosus* group (see Remarks under *A. aerosus*). *Akodon tolimae* was treated as a full species by several authors (e.g., Reig 1987) but later was subsumed under *affinis* (Musser and Carleton 1993:688; see also Reig, Kiblisky, and Linares 1971:162). *Akodon tolimae* was briefly diagnosed as a taxon similar to *aerosus* but much smaller and darker. Although the conspecifity of *affinis* and *tolimae* is plausible, given their geographic proximity and similar morphology, their actual relationship needs clarification. Quiceno (1993) found two different cytotypes, $2n = 24$ and $2n = 26$, both with FN = 40 in a survey of *Akodon* populations in Valle del Cauca and Risaralda departments, Colombia. He associated the former karyotype with the name *affinis* and the latter with *tolimae*, but regardless of these assignments, the chromosomal data suggest systematic complexity among *Akodon* populations in the northern Andes of Colombia.

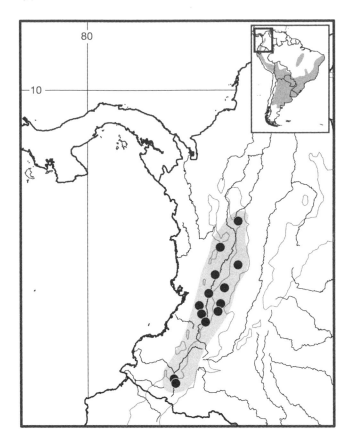

Map 59 Selected localities for *Akodon affinis* (●). Contour line = 2,000 m.

Akodon albiventer Thomas, 1897
White-bellied Grass Mouse, White-bellied Akodon
SYNONYMS:

Akodon albiventer Thomas, 1897d:217; type locality "Lower Cachi," Salta, Argentina; it is likely that Thomas referred to the lower course of the Río Cachi, which passes through the town of Cachi (25°07′11.93″S, 66°09′47.00″W, 2,341 m, Salta province), close to the junction with the Río Calchaquí (see Jayat et al. 2010).

Akodon Berlepschii Thomas, 1898c:281; type locality "Esperanza, a 'tambo' in the neighbourhood of Mount Sahama, [La Paz,] Bolivia."

B[*olomys*]. *albiventer*: Thomas, 1916i:339; name combination.

B[*olomys*]. *berlepschi*: Thomas, 1916i:340; name combination and incorrect subsequent spelling of *Akodon berlepschii* Thomas.

Akodon [(*Bolomys*)] *albiventer*: Ellerman, 1941:415; name combination.

Akodon [(*Bolomys*)] *berlepschii*: Ellerman, 1941:415; name combination.

Akodon albiventer berlepschii: Pine et al., 1979:347; name combination.

DESCRIPTION: Small to moderate sized with total body length from 162–179 mm, tail length 69–77 mm, ear length 13–13.5 mm, and hindfoot length (with claw) 20–22 mm; fur short and coarse, slightly woolly in texture; general color pale grizzled grayish, tips of darker hairs brown and of paler ones whitish buff, all slate gray at their base; venter snowy white, contrasting strongly with the dorsum; eyes with whitish eye ring; ears short, with a few whitish hairs that form an indistinct postauricular spot; dorsal surfaces of both forefeet and hindfeet pure white; tail short, well haired, brown above and white on sides and below. Skull with short rostrum, shallow and narrow zygomatic notches, zygomatic arches not flared laterally, and broad interorbital region with rounded edges; zygomatic plates broad, with anterior margins straight; incisive foramina extend to hypoflexus of M1; mesopterygoid fossa broad, with rounded anterior margin and parallel sides; auditory bullae very large for the genus, with short and broad bony tubes. Mandibular ramus low, with short symphysis. Upper incisors orthodont. M1 with well-developed anteromedian flexus and conspicuous anteroloph; major cups nearly parallel; mesoloph present but reduced. M2 similar to M1 but with reduced procingulum. M3 even more reduced and oval in occlusal view. Lower m1 with well-developed anteromedian flexid and anterolabial cingulum; both ectostylid and ectolophid present but reduced; m2 with very shallow protoflexid visible only in young individuals; m3 bilophodont and sigmoid in shape. Condyloincisive length averages 23.7 mm and upper molar series 4.1 mm.

DISTRIBUTION: *Akodon albiventer* is widely distributed in the Andean Altiplano and other highlands, from southern Peru in Arequipa and Puno departments, through southwestern Bolivia and northern Chile, to northwestern Argentina in Jujuy and Salta provinces. Most localities are between 3,000 to 4,500 m, but a few are from as low as 2,350 m (Thomas 1897d; Pearson 1951; Pine et al. 1979; S. Anderson 1997; M. M. Díaz and Barquez 2007). The species is known from a late Holocene archaeological site at Inca Cueva, Jujuy, Argentina (Teta and Ortiz 2002; P. E. Ortiz et al. 2011b).

SELECTED LOCALITIES (Map 60): ARGENTINA: Jujuy, El Toro, 55 km W of Susques (IADIZA 3336), 0.5 mi E of Tilcara (MVZ 119934); Salta, Cauchari (MSB 75252), Lower Cachi (type locality of *Akodon albiventer* Thomas). BOLIVIA: La Paz, Salla (S. Anderson 1997); Oruro, 7 km S and 4 km E of Cruce Ventilla (S. Anderson 1997), Luca (S. Anderson 1997), 40 km S of Oruro (MVZ 119935); Potosí, Quetena Chica on Río Quetena (S. Anderson 1997), 4 mi E of Uyuni (MVZ 119936), 5 mi [= 8 km] E of Uyuni (S. Anderson 1997), 5 mi N of Villazon

(MVZ 119938); Tarija, 4.5 km E of Iscayachi (S. Anderson 1997). CHILE: Arica y Parinacota, Caritaya, 75 mi SE of Arica (MVZ 116770). PERU: Arequipa, 15 km S of Cailloma (Pearson 1951); Moquegua, 5 mi NW of Toquepala (MVZ 145543); Puno, Hacienda Calacala, 7 mi SW of Putina (MVZ 116650); Puno, 1 mi SW of Ancomarca (MVZ 141330), Hacienda Calacala, 7 mi SW of Putina (MVZ 116650), Hacienda Pairumani, 24 mi S of Ilave (MVZ 114651), 5 mi N of Mazocruz (MVZ 139552); Tacna, 0.5 mi W of Challapalca (MVZ 141335), Río Tarata (MVZ 115700).

SUBSPECIES: Thomas (1902c) noted that *albiventer* and *berlepschii* might be conspecific and that color differences between the two probably intergraded. Pine et al. (1979) unequivocally stated that the two were conspecific but that fresh pelage of individuals from within the range of *albiventer* may turn out to be more buffy-brown and less gray than in individuals from within the range of what has been called *berlepschii*. Considering the geographical distance between the two forms as currently known and keeping in mind that supposed pelage differences could have some validity, these authors provisionally recognized the two as subspecies. However, this hypothesis has not as yet been tested through geographic analysis of any character set.

NATURAL HISTORY: *Akodon albiventer* primarily occupies semiarid Puna and high Andean environments, but at their southern distributional limit the species also extends into the Monte desert. Although it has been trapped in open grassy places, *A. albiventer* seems to prefer more sheltered habitats such as dense tola shrubland and man-made rock walls that provide protection (Pearson 1951). In both Bolivia and Peru, *A. albiventer* lives together with *A. boliviensis* (Pearson 1951; S. Anderson 1997). In northern Chile, this species inhabits meadows with dense grass and rock walls, especially near watercourses (Pine et al. 1979). Here it was found in association with *Abrothrix andina*. In Argentina, *A. albiventer* occurs sympatrically with *A. spegazzinii* (M. M. Díaz and Barquez 2007). This species has been reported as mostly or entirely diurnal by several authors (Pearson 1951; Mann 1978, Pine et al. 1979). M. M. Díaz and Barquez (2007) commented on an individual feeding on grass in Jujuy province, Argentina.

There are few published observations on reproduction and molting for this species. S. Anderson (1997) reported two males with enlarged testes (8 and 9 mm long) caught in July, but 10 other large specimens caught at the same time were still sexually undeveloped. A female caught in July had an open vagina, but no females contained embryos, so the breeding season must not begin until after July. M. M. Díaz and Barquez (2007) observed young individuals in Jujuy province between January and June. Males with nonscrotal testes were captured from February to June and in November, and males with scrotal testes in February, March, and December. These authors also mentioned females with closed vaginas in February, March, and June, and females with open vaginas in February, March, and December. Lactating and pregnant females were trapped in February, March, and December. Molting animals were collected in February, March, June, and December. Beaucournu et al. (2011) described a new species of flea in the genus *Agastyopsylla* from specimens collected from in Tarapacá, Chile.

REMARKS: *Akodon albiventer* was once considered a member of the genus *Bolomys* [= *Necromys*] (Thomas 1916i; Gyldenstolpe 1932; N. O. Bianchi et al. 1971; A. L. Gardner and Patton 1976; E. Massoia, in Massoia and Pardiñas 1993). However, the inclusion of *albiventer* in *Akodon* is supported on molecular and, to a more limited extent, morphological grounds (Pine et al. 1979; Reig 1987; D'Elía et al. 2003; M. F. Smith and Patton 2007). Phylogenetic analyses based on mtDNA cytochrome-*b* gene sequences supported a sister relationship of *A. albiventer* and several Andean species of *Akodon*, including *A. aerosus*, *A. affinis*, *A. orophilus*, and *A. torques* (Jayat et al. 2010).

N. O. Bianchi et al. (1971) reported a $2n = 40$ chromosomal complement, represented by 38 acrocentric and two small metacentric elements.

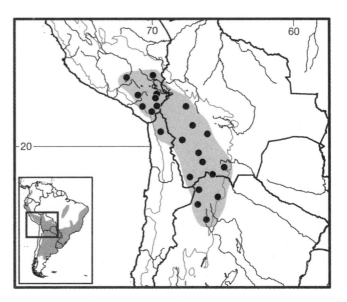

Map 60 Selected localities for *Akodon albiventer* (●). Contour line = 2,000 m.

Akodon azarae (J. B. Fischer, 1829)
Azara's Grass Mouse, Azara's Akodon

SYNONYMS:

Ratton agreste Brants, 1827:184: based on Azara's (1802:94) "*el agreste*"; ruled by the International Commission (ICZN 1982:Opinion 1232) to be a vernacular name and unavailable for use in zoological nomenclature.

M[*us*]? *Azarae* J. B. Fischer, 1829:325; based on Azara's (1802:94) "*el agreste*" with type locality given as "In Paraguaya" although Azara wrote (1802:94) "en los 30 ½ grados de latitude," erroneously interpreted as "30½° (= latitude of Entre Ríos) Argentina" by Tate (1932:26; see also Cabrera 1961:440); Pardiñas et al. (2007:401) reviewed the available evidence and corrected the type locality to "São Gabriel (30°19′S, 54°19′W, 118 m) . . . in the Brazilian State of Rio Grande do Sul."

Mus arenicola Waterhouse, 1837:18; type locality "Maldonado," Maldonado, Uruguay.

Hesperomys arenicola: Wagner, 1843a:466; name combination.

[*Akodon (Akodon)*] *arenicola*: E.-L. Trouessart, 1897:536; name combination.

Akodon arenicola hunteri Thomas, 1917c:97; type locality "Isla Ella, in the delta of the Rio Parana, at the top of the La Plata Estuary"; restricted by Pardiñas et al. (2007:402) to a "small island (ca. 1 km²) surrounded by the streams Espera and Esperita (34°22′S, 58°38′W, Primera Sección Delta del Paraná, Tigre, Buenos Aires . . .)."

[*Akodon arenicola*] *arenicola*: Thomas, 1917c:97; name combination.

Akodon azarae: Tate, 1932:5; first use of current name combination.

Akodon azarae azarae: Cabrera, 1961:440; name combination.

Akodon azarae hunteri: Cabrera, 1961:440; name combination.

Akodon azarae bibianae Massoia, 1971b:131; type locality "Estación Experimental El Colorado (INTA), Departamento de Pirané, Provincia de Formosa, República Argentina."

DESCRIPTION: Medium-sized akodont, with head and body length 90–114 mm and tail length 69–90 mm; fur moderately soft and long (9–13 mm); dorsal color olive brown to deep brown, usually tinged buffy on sides and cheeks; ventral hairs gray based and buffy tipped, and thus moderately demarcated from dorsum; ears clothed in short brown hairs; hindfeet tan above; bicolored tail shorter than head and body. Skull delicate; rostrum narrow and pointed; interorbital region hourglass shaped with rounded edged; temporal and mastoid ridges weak; interparietal small; zygomatic arches weak, not flaring laterally; zygo-matic plates narrow; palate short; incisors delicate (Myers 1990). Massoia and Fornes (1964a) and Hershkovitz (1990b) provided external and cranial measurements. Sierra de Soriano (1969) detailed the external morphology of Uruguayan specimens.

DISTRIBUTION: *Akodon azarae* ranges from Rio Grande do Sul in southern Brazil, and eastern Paraguay south through Uruguay to central Argentina (Reig 1964; Barlow 1969; Massoia 1971b; Dalby 1975; Myers and Wetzel 1979; Gamarra de Fox and Martin 1996). Known populations in Argentina are mostly restricted to the eastern part of the country (Buenos Aires, Chaco, Corrientes, Entre Ríos, Formosa, and Santa Fe provinces), extending to the southern tip of Buenos Aires province and westward to La Pampa and Córdoba provinces. There is an unconfirmed record for the Parque Nacional Teniente Enciso, in the western Chaco of Paraguay (Gamarra de Fox and Martin 1996:537). Several fossils collected from Pleistocene deposits of Buenos Aires province, Argentina, have been referred to *A. azarae* (Ameghino 1889; Pardiñas 1999a). The oldest records for the species are Middle Pleistocene (Ensenadan), from Ramallo, Mar del Plata, and the vicinity of Bahía Blanca (see Pardiñas 1993, Pardiñas and Deschamps 1996, Voglino and Pardiñas 2005).

SELECTED LOCALITIES (Map 61): ARGENTINA: Buenos Aires, Bahía Blanca (Pardiñas, Abba, and Merino 2004), Balneario San Antonio (Pardiñas, Abba, and Merino 2004), Bañado de Flores, Buenos Aires (Massoia and Fornes 1967b), Cabaña San José (Pardiñas, Abba, and Merino 2004), Estancia El Abra (Pardiñas, Abba, and Merino 2004), Monte Hermoso (Pardiñas, Abba, and Merino 2004), Santa Clara del Mar (Reig 1964); Chaco, Capitán Solari (Massoia et al. 1995b); Córdoba, La Nacional (CNP-E 653-1), Marull (CNP-E 59-2), Río Cuarto (E. Castillo et al. 2003); Corrientes, Santo Tomé (CNP-E 650-1); Formosa, Cooperativa de Villa Dos Trece [= Kilómetro 213] (Pardiñas and Teta 2005); La Pampa, Estancia Los Guadales (Siegenthaler et al. 1993), Laguna El Chañar, Puesto El Chato (Siegenthaler et al. 1993); Santa Fe, Estación Santa Margarita (Massoia et al. 1995a), Jacinto L. Arauz (Teta and Pardiñas 2010). BRAZIL: Rio Grande do Sul, Alegrete (G. Gonçalves, pers. comm.), Esmeralda (Sbalqueiro, Mattevi, Oliveira, and Freitas 1982), Guaíba (G. Gonçalves, pers. comm.), Reserva Ecológica do Taim (Sbalqueiro, Mattevi, Oliveira, and Freitas 1982). PARAGUAY: Canindeyú, Reserva Natural Privada Morombí (J. Torres, pers. comm.); Presidente Hayes, 15.5 km by road NNW of Chaco-i (Myers and Wetzel 1979), 83 km NW of Puerto Falcón (Gamarra de Fox and Martin 1996). URUGUAY: Maldonado, Maldonado (type locality of *Mus arenicola* Waterhouse); San José, Kiyú (GD 327).

SUBSPECIES: Molecular and morphological data support three subspecies in our current concept of *A. azarae*:

A. a. azarae, *A. a. bibianae*, and *A. a. hunter*. The nominate race is restricted to southern Brazil and Uruguay; the other two are in central-eastern Argentina and northeastern Argentina and eastern Paraguay, respectively.

NATURAL HISTORY: *Akodon azarae* inhabits open grasslands, wet grasslands along rivers and marshes, open thorn woodlands, and borders of cultivated fields (Crespo 1966; Pearson 1967; Dalby 1975; Teta, González-Fischer et al. 2010). It utilizes shallow holes and will occasionally burrow; Massoia and Fornes (1965b) and K. Hodara et al. (1997) described the structure of the burrow systems. This animal may be either nocturnal or diurnal (J. R. Contreras 1979b).

Akodon azarae has a polygynous mating system (Bilenca and Kravetz 1998; O. Suárez and Kravetz 1998a; Zuleta and Bilenca 1992); the breading season lasts about eight months, a period that roughly coincides with spring through autumn (Crespo 1966; Pearson 1967; Dalby 1975; Teta, González-Fischer et al. 2010). Although delayed implantation may occur, gestation in the laboratory usually lasts 22.7 to 24.5 days; the average litter size varies between 4.7 and 5.7 (range: 3–7 in Uruguay; 3–10 in Argentina); and neonates are altricial with an average birth weight of 2.2 g (De Villafañe 1981; Mills, Ellis, McKee et al. 1992; C. Busch and Del Valle 2003). Mothers can control the sex ratio of their litters (Zuleta and Bilenca 1992). O. Suárez and Kravetz (1998a,b) described copulatory behavior, mating system, and transmission of food selectivity from mothers to offspring. De Villafañe (1981) examined development under laboratory conditions. Pearson (1967) and Zuleta et al. (1988) documented age structure and population dynamics of Argentinean populations. Life span varied between 7 and 8 months in individuals born in the spring, and from 10 and 12 months for those born in the autumn (K. Hodara et al. 2000).

Diet varied geographically and seasonally, but includes variable proportions of invertebrates and plant material, such as green vegetation, leaves, fruits, and seeds (Barlow 1969; Bilenca and Kravetz 1998; C. Busch and Del Valle 2003; Macdonald 2001; O. Suárez and Bonaventura 2001; O. Suárez and Kravetz 1998b). Dalby (1975), Zuleta and Bilenca (1992), and Priotto and Steinmann (1999) reported on home range use and size. Owls are common predators of this mouse (e.g., Kittlein 1997; Pardiñas 1999a; González-Fischer 2011). Navone et al. (2009) reported on parasites. This species is known to carry the Maciel and Pergamino hantavirus strains (O Suárez et al. 2003).

REMARKS: J. B. Fischer (1829) based his *M? Azarae* on "*el agreste*" of Azara (1802:94), but his name was generally ignored in favor of *A. arenicola* Waterhouse for much of the early nomenclatural history of this species. Tate

(1932c) and Cabrera (1961) established the current concept of *A. azarae*, including *arenicola* as a subjective junior synonym, and restricted the type locality to northern Entre Ríos province, Argentina. However, as detailed by Pardiñas et al. (2007) in their study of Felix de Azara's voyages, the Spanish naturalist apparently captured this mouse in Brazil, in the vicinity of the city of São Gabriel, Rio Grande do Sul.

Ratton agreste Brants is a senior synonym of *Mus azarae* J. B. Fisher, but had not been used in the literature since its publication, including major taxonomic synopses (e.g., E.-L. Trouessart 1897, 1904; Tate 1932c; Cabrera 1961; Musser and Carleton 2005), except for a single mention by Hershkovitz (1966a:106; see also Hershkovitz 1990b:5). To stabilize the use of *azarae* J. B. Fisher for this species, Langguth (1966a, 1978) twice petitioned the International Commission on Zoological Nomenclature that Brants's name be rejected. In 1982 (Opinion 1232), the Commission ruled that *Ratton* Brants was suppressed for the purpose of the Principle of Priority (but not for Homonymy) and placed it on the Official Index of Rejected and Invalid Generic Names in Zoology with the Name Number 2131. The Commission further ruled that *agrestes* Brants, as published in combination with *Ratton*, was a vernacular name and unavailable for use in zoolological nomenclature (see ICZN 1987).

Karyotypes with $2n = 37$–38, FN = 40–44, are known for specimens from central Argentina (*hunteri* from Buenos Aires and La Pampa provinces, *bibianae* from Chaco province), Uruguay, and Brazil (*arenicola* from Maldonado and *azarae* from Rio Grande do Sul; e.g., N. O. Bianchi and Contreras 1967; Vitullo et al. 1984; Kasahara and

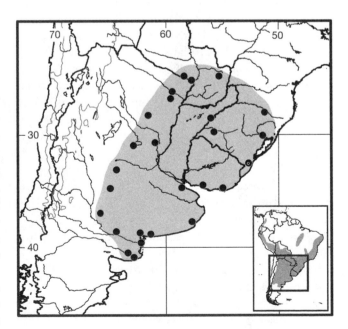

Map 61   Selected localities for *Akodon azarae* (●). Contour line = 2,000 m.

Yonenaga-Yassuda 1984; Tiranti 1999a). Two types of Y chromosomes have been reported from Brazilian specimens: acrocentric in populations in the northern part of Rio Grande do Sul, and submetacentric in the southern part of that state (Mattevi et al. 1981; Sbalqueiro, Mattevi, Oliveira, and Freitas 1982).

Akodon boliviensis Meyen, 1833
Bolivian Grass Mouse, Bolivian Akodon

SYNONYMS:

Akodon boliviense Meyen, 1833:600; type locality "Auf der Hochebene von Hochperu, in dem Indianer-dorfe Pichu-pichun, auf einer Höhe von 14000 Fuss gefangen"; Pichu-Pichún, 14,000 ft., Chucuito Province, Puno Department, Peru (as clarified by Myers et al. 1990:49–50).

Acodon boliviense: Tschudi, 1844:177; name combination.

[*Akodon* (*Akodon*)] *boliviensis*: E.-L. Trouessart, 1897: 535; name combination.

Akodon pacificus Thomas, 1902b:135; type locality "La Paz, [La Paz,] 4000 m," Bolivia.

DESCRIPTION: Small, with total body length averaging 155.7 mm, tail length 64.6 mm, ear length 13.7 mm, and hindfoot length 21.1 mm; dorsal fur pale, brown-tinged olivaceous, uniformly colored across back and crown, lightly streaked with black; flanks similar to dorsum in color and pattern; narrow but conspicuous yellowish eye-ring present; venter paler, with hairs buffy tipped; inguinal area especially dark, almost rufous, in some individuals; small patch of white fur often occurs at apex of chin; tail sharply bicolored and heavily clothed with hairs. Skull with comparative broad zygomatic notches, with zygomatic arches tending to flare laterally more than other members of group; interorbital region narrow, with posterior margins tending to be squared; temporal and mastoid ridges unusually strongly developed; zygomatic plates relatively broad, with anterior faces vertically oriented; posterior ascending process of alisphenoid projects at least to squamosoalisphenoid groove; tympanic hook relatively stout; mesopterygoid fossa narrow, with anterior margin usually squared and often with a small median spine; M1 with well-formed parastyle, mesostyle present but tiny, and metalophule occasionally present; M2 with anterior cingulum and mesostyle always present; M3 relatively large and complex for *boliviensis* group species, averaging about half length of M2; metaflexus and hypoflexus of M3 partly divide tooth, giving occlusal surface a distinctive figure-8 shape. Lower m1 with distinguishable, but sometimes small, protostylid; both m1 and m2 usually with ectostylid and mesostylid present but always tiny; m3 long and conspicuously S-shaped. Condyloincisive length averages 22.4 mm, upper molar series length 4.0 mm.

DISTRIBUTION: *Akodon boliviensis* is widely distributed in highland grasslands of the Puna and western Andean slopes from southern Peru through central Bolivia to northwestern Argentina in Salta province. Most specimen records are from elevations above 3,500 m, but it has been found as low as 2,400 m (Myers and Patton 1989a; Myers et al. 1990; S. Anderson 1997; Salazar-Bravo, Yensen et al. 2002; Tarifa et al. 2007; Jayat, Ortriz et al. 2008, 2010). No fossils are known.

SELECTED LOCALITIES (Map 62): ARGENTINA: Salta, ca. 3 km SE of Abra de Ciénaga Negra (MACN 23499), Azul Cuesta, ca. 9 km S of Nazareno (MACN 23503). BOLIVIA: Chuquisaca, Río Limón (S. Anderson 1997); Cochabamba, 4.4 km (by road) N of Tablas Monte (S. Anderson 1997); La Paz, Pelechuco (Tarifa et al. 2007), Río Aceromarca (S. Anderson 1997); Oruro, 10 km by road SW of Pazña, Finca Santa Helena (Myers and Patton 1989a), 1.5 km SW of Sajama (S. Anderson 1997); Potosí, 20 km S of Potosí (Myers et al. 1990), Serranía Siberia, 11 km (by road) NW of Torrecillas (S. Anderson 1997); Tarija, Sama (S. Anderson 1997). PERU: Moquegua, 19 km NE of Torata (Myers et al. 1990); Puno, 11 km W and 12 km S of Ananea (MVZ 173033), 13 mi W and 2 mi N of Crucero (Myers et al. 1990), Hacienda Calacala, 7 mi SW of Putina (MVZ 116653), 4 km E Juli (MVZ 115670), Mazocruz (MVZ 115671), 5 km W of Puno (MVZ 115673), Río Huanque (MVZ 136259), 12 km S of Santa Rosa [de Ayaviri] (Myers et al. 1990), Tincopalca, 50 mi W of Puno (MVZ 114649); Tacna, Tarata (MVZ 141313).

SUBSPECIES: Cabrera (1961) considered *A. spegazzinii* and *A. tucumanensis* as subspecies of *A. boliviensis*, and *A. alterus* as a synonym of *A. boliviensis tucumanensis*. Myers et al. (1990), in their review of geographic variation in several craniodental, soft anatomical, and other traits, concluded that *spegazzinii*, *tucumanensis*, and *alterus* were best excluded from *A. boliviensis*. These authors, however, noted differences between specimens of *A. boliviensis* from the extreme northwestern (Moquegua department, Peru) and southeastern (Tarija, Bolivia) segments of the species' range, relative to a rather homogeneous set of geographically intermediate samples, but did not recommend any formal recognition of these end points.

NATURAL HISTORY: *Akodon boliviensis* is found in the drier environments of the Puna in southern Peru but extends into humid grasslands adjacent to the eastern Andes in central Bolivia. It is common in dense bunch grasses dominated by *Stipa* and *Festuca*, and in *Polylepis* woodlands fragments, and is readily trapped where rocks or shrubs are interspersed with grasses, and along rock walls of corrals or houses where there are clumps of grasses (Myers et al. 1990; S. Anderson 1997; Salazar-Bravo, Yensen et al. 2002; Tarifa et al. 2007). In Argentina, most records come from relatively humid areas of the eastern Andean slopes, in the transition between uppermost Yungas and high Andean grasslands (Jayat et al. 2010). Along this broad distributional range, *A. boliviensis* maybe syntopic or contiguously allopatric with several other congeneric species, including *A. aerosus, A. albiventer, A. budini, A. lutescens, A. fumeus, A. caenosus, A. mimus, A. pervalens, A. siberiae, A. simulator, A. subfuscus, A. toba,* and *A. varius* (Myers et al. 1990; S. Anderson 1997; Salazar-Bravo, Yensen et al. 2002; Tarifa et al. 2007; Jayat, Ortiz, and Miotti 2008; Jayat et al. 2010). S. Anderson (1997) reported pregnant females between the months of January and September, with a mean of 4.5 embryos. In northwestern Argentina, Jayat et al. (2010) reported molting individuals in winter (July and August), but only a single individual was reproductively active; some specimens caught in spring (November) were both molting and reproductively active.

REMARKS: In the first revision of the small *Akodon* from the central Andes, Myers et al. (1990) defined a *boliviensis* group, characterized the morphological and distributional limits of *A. boliviensis,* and considered *A. spegazzinii* a valid species. Jayat et al. (2010) reidentified specimens previously referred to *A. boliviensis* from Salta province, Argentina, as *A. spegazzinii,* and noted plausible occurrence of *A. boliviensis* in Jujuy province, based on animals cited as *A. alterus* by M. M. Díaz and Barquez (2007).

Karyotyped individuals of *A. boliviensis* have a 2n=40 karyotype, with slight geographic variation in autosomal arm number and/or sex chromosome morphology. In Puno

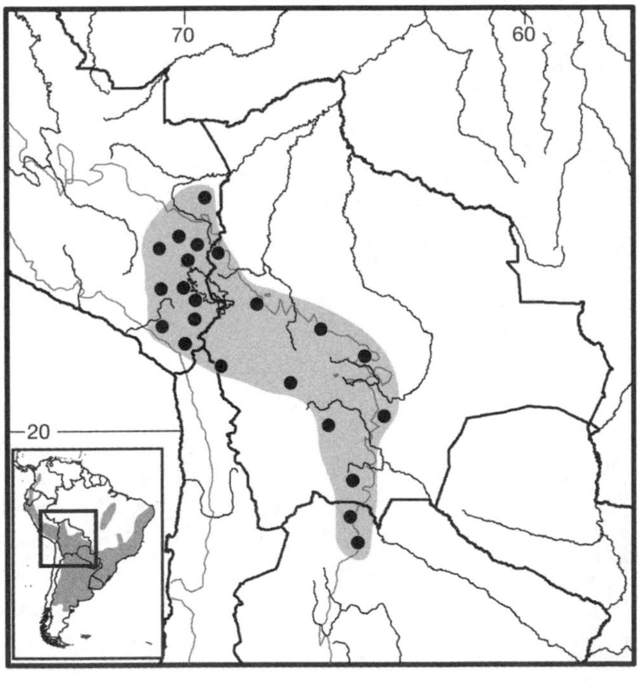

department, Peru, individuals are FN=40, with 18 pairs of uniarmed autosomes (one distinctly larger than the remainder, which grade in size evenly from medium to small) and one very small pair of biarmed elements. The X chromosome is a medium-sized subtelocentric, and the Y is a small submetacentric, but larger than the smallest uniarmed autosome (Myers et al. 1990). Specimens from Tacna department, Peru, have a FN=42, with 17 pairs of uniarmed autosomes, one pair of large subtelocentrics, and one very small pair of biarmed elements. The X chromosome in this karyotype is a large subtelocentric, and the Y is a small submetacentric (A. L. Gardner and Patton 1976).

Akodon budini (Thomas, 1918)
Budin's Grass Mouse, Budin's Akodon
SYNONYMS:

Hypsimys budini Thomas, 1918b:190; type locality "Leon [= León], Jujuy, 1500 m"; Argentina.

Hypsimys deceptor Thomas, 1921k:613; type locality "Higuerilla, 2000 m"; Higuerilla= Pampichuela (23°32′S, 65°02′W, 1,735 m, Valle Grande, Jujuy, Argentina; see Pardiñas et al. 2007).

Akodon [(*Hypsimys*)] *budini*: Ellerman, 1941:414; name combination.

Akodon [(*Hypsimys*)] *deceptor*: Ellerman, 1941:414; name combination.

Akodon budini: Musser and Carleton, 1993:689; name combination.

DESCRIPTION: Large, with total body length 197–221 mm, tail length 52–99 mm, ear length 12–22 mm, and hindfoot length (with claw) 22–29 mm; general coloration olivaceous brown, finely sprinkled with black hairs; dorsal color uniform from head to rump, and onto flanks; ventral color paler, but contrasts only slightly with dorsum; conspicuous patch of white hairs covers chin and, in some specimens, throat; tail bicolored, dorsally brown or blackish brown and ventrally whitish or buffy, but not densely furred. Skull large; rostrum elongated; zygomatic notches narrow and deep; and frontal sinuses slightly swollen. Interorbital region hourglass in shape, with rounded margins and without overhanging borders; braincase broad and inflated, with poorly developed mastoid and temporal ridges; zygomatic plates relatively narrow and sharply slanted backward from bottom to top; incisive foramina long, generally extending to protocone of M1 or slightly beyond; mesopterygoid fossa broad, straight or lyre shaped, and anterior margin with median spine. Mandibular ramus relatively deep; anterior end of masseteric crest situated well behind level of anterior border of m1, generally reaching protoconid; capsular projection weakly developed and placed posterior to coronoid process. Upper incisors orthodont

Map 62 Selected localities for *Akodon boliviensis* (●). Contour line=2,000 m.

or slightly proodont; molars strongly hypsodont for an *Akodon*; M1 with well developed procingulum but with only inconspicuous anteromedian flexus; both anteroloph and mesoloph present; M2 with remnant anteroloph and mesoloph; M3 large and complex, with well-developed paraflexus and metaflexus in most specimens and with anteroloph and mesoloph small but clearly visible. Lower m1 with well-developed anteromedian flexid and antero-labial cingulum, tiny ectostylid, and vestigial mesolophid; m2 also with ectostylid and mesolophid, but less developed than in m1; m3 very large, about 50% size of m2. Greatest skull length 26.8–29.8 mm; upper molar series length 4.7–5.4 mm.

DISTRIBUTION: All the records of *A. budini* are along eastern Andean slopes from south-central Bolivia in Chuquisaca department to northernmost Argentina, in Jujuy and Salta provinces, with most at elevations between 1,000 and 3,000 m (Emmons and Feer 1997; Pardiñas, D'Elía et al. 2006; M. M. Díaz and Barquez 2007; Jayat, Ortiz, and Miotti 2008; Jayat et al. 2010). Because the distribution of *A. budini* in northwestern Argentina is limited to the eastern slopes of mountain ranges (the Santa Victoria, Zenta, Calilegua, and Chañi ranges), uniformly above 1,200 m, records from Jujuy well below this elevation (see M. M. Díaz and Barquez 2007) are questionable and need to be corroborated. In this regard, restudy of a specimen from Caimancito (MACN 20282), elevation 600 m, confirms it is an *Akodon simulator*, not *A. budini*. M. M. Díaz and Barquez (2007) listed *A. budini* as occurring at several localities around the Santa Bárbara range, eastern Jujuy province, but their specimens may actually be *A. sylvanus*, a species with similar external traits. No fossils are known.

SELECTED LOCALITIES (Map 63): ARGENTINA: Jujuy, Cerro Morado, sobre Río Morado, 11 km NW of San Antonio (CML 4604), Mesada de las Colmenas (Heinonen and Bosso 1994); Salta, Pampa Verde, ca. 8 km WSW of Los Toldos and S Cerro Bravo (MACN 23505), Parque Nacional Baritú, Finca Jakulica, Los Helechos (MACN 20666). BOLIVIA: Chuquisaca, Cerro Bufete (Emmons 1997b; this record needs confirmation), Horcas, 80 km SE of Sucal [= Sucre?] (MVZ 134648).

SUBSPECIES: We treat *Akodon budini* as monotypic.

NATURAL HISTORY: *Akodon budini* is mainly an inhabitant of the upper elevational belts of the Yungas forest. There are few published observations on reproduction and molting. In Jujuy province, Argentina, M. M. Díaz and Barquez (2007) reported young individuals between February and July, subadults in June, males with nonscrotal testes between June and September, males with scrotal testes in February and July, and females with closed and open vaginas in June and August. Individuals in active molt were found in June. No specimen caught in northernmost

Salta province in July ($N = 3$) or May ($N = 43$) was reproductively active (P. Jayat and P. E. Ortiz, unpubl. data). At least four other *Akodon* species may be found in sympatry, including *A. boliviensis*, *A. caenosus*, *A. fumeus*, and *A. simulator* (Barquez, Díaz, and Ojeda 2006; M. M. Díaz and Barquez 2007; Jayat, Ortiz, and Miotti 2008; Jayat et al. 2010).

REMARKS: *Akodon budini* is the type species of the genus *Hypsimys*, which was diagnosed from *Akodon* by the very hypsodont molars, as emphasized by Thomas: "molars [are] quite unique in this group, highly hypsodont, almost as much so as in *Chinchillula*, though of so different a type to that as to make comparison difficult. They are just what *Akodon* teeth might be expected to become if made very hypsodont, high, narrow, with the vertical grooves extending far down towards the roots, of simple sectional pattern" (1918b:190). Three years later, Thomas (1921k) described *Hypsimys deceptor*, based on its larger size in comparison with *H. budini*. Dismissing the metric differences reported by Thomas, Cabrera (1961:451) placed *deceptor* Thomas in the synonymy of *budini* Thomas, an action followed by subsequent students of sigmodontines (e.g., Reig 1987; M. M. Díaz and Barquez 2007). Cabrera (1961), Reig (1987), and M. F. Smith and Patton (1999) recognized *Hypsimys* as a subgenus of *Akodon*, but Galliari et al. (1996) considered it a valid genus. The allocation of *budini* to *Akodon* has been reinforced by the similarity of its karyotype to those of other species of *Akodon* (Apfelbaum et al. 1993; Vitullo et al. 1986), its close re-

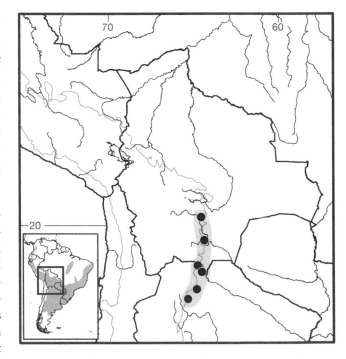

Map 63 Selected localities for *Akodon budini* (●). Contour line = 2,000 m.

semblance in morphology and molecular characters to *A. siberiae* (Myers and Patton 1989a), and its phylogenetic position nested in the *Akodon* radiation (D'Elía 2003a; D'Elía et al. 2003; M. F. Smith and Patton 2007).

The karyotype of *A. budini* is $2n = 38$, FN = 40 (Vitullo et al. 1986; Myers and Patton 1989a). The autosomal complement consists of 18 pairs, including a single pair of large submetacentrics and two pairs of medium-sized metacentrics. The remaining autosomes are acrocentrics that decrease evenly in size from medium to small. The X chromosome is telocentric and intermediate in size between those of autosome pairs 4 and 5; the Y is telocentric and the smallest chromosome of the complement (Vitullo et al. 1986).

Akodon caenosus Thomas, 1918
Unicolored Grass Mouse, Unicolored Akodon

SYNONYMS:

Akodon puer caenosus Thomas, 1918b:189; type locality "León, Jujuy, 1500 m," Argentina.

Akodon caenosus: Thomas, 1920a:192; name combination.

Akodon puer: Vitullo et al., 1986:70; part; not *Akodon puer* Thomas.

Akodon (*Akodon*) *puer*: Apfelbaum and Reig, 1989:260; part; not *Akodon puer* Thomas.

Akodon diminutus Barquez, Díaz, and Goytia, 1994; *nomen nudum*.

Akodon lutescens caenosus: S. Anderson, 1997:421; name combination.

Akodon aliquantulus M. M. Díaz, Barquez, Braun, and Mares, 1999:788; type locality "Argentina: Tucumán Province: Departamento Tafí Viejo: Las Aguitas, Cumbres del Taficillo, 1,700 m, 26°42′S, 65°22′W."

Akodon lutescens: Pardiñas, D'Elía, Teta, Ortiz, Jayat, and Cirignoli, 2006:151; part; not *Akodon lutescens* J. A. Allen.

Akodon lutescens puer: M. M. Díaz and Barquez, 2007:488 part; neither *Akodon lutescens* J. A. Allen nor *Akodon puer* Thomas.

DESCRIPTION: Small, with total body length 124–169 mm, tail length 46–75 mm, ear length 12–15 mm, and hindfoot length (with claw) 20–6 mm; dorsal color uniform along body but may be highly variable within and among populations, typically ochraceous brown but with yellowish, rufous, or olivaceous casts, and almost always dark; venter whitish gray, buffy gray, yellowish or even ruddy, clearly contrasting with dorsum; few white hairs may be present on chin but do not form conspicuous patch; tail strongly bicolored, dorsally blackish-brown and ventrally whitish or buffy. Skull with short rostrum, very narrow and shallow zygomatic notches, and weakly developed frontal sinuses; interorbital region comparatively broad, hourglass in shape, and with rounded or slightly squared margins;

zygomatic arches not flared; braincase small and inflated, with undeveloped temporal and lambdoid crests; zygomatic plates narrow, with anterior border straight or slightly concave and generally sloping gently backward from bottom to top; incisive foramina typically extend to protocone of M1; mesopterygoid fossa very narrow, its anterior margin rounded or slightly squared and lateral margins straight to slightly divergent posteriorly; parapterygoid fossae generally broader than mesopterygoid, with convex border and slight backward divergence. Mandibular ramus very delicate; masseteric crest extends slightly behind anterior margins of m1; and angular process terminates anterior to condyle. Upper incisors orthodont to slightly opisthodont. M1 with well-developed anteromedian flexus, both anteroloph and mesoloph always present; M2 with conspicuous mesoloph; M3 variable, from completely oval without distinct features or with well-developed anteroflexus, metaflexus, and hypoflexus. Lower m1 with well-developed anteromedian flexid and both protostylid and ectostylid; m2 usually retains only labial stylid although some specimens have very small ectostylid; m3 with deep labial and lingual flexids. Greatest skull length 21.3–24.2 mm; upper molar series length 3.5–3.8 mm.

DISTRIBUTION: *Akodon caenosus* has a wide distribution, with records from Chuquisaca and Tarija departments in southern Bolivia south to Catamarca province in northwestern Argentina. Elevational range is from 400 to 3,100 m, but most records are from below 2,500 m (Myers et al. 1990; S. Anderson 1997; Jayat, Ortiz, and Miotti 2008; Jayat et al. 2010). No fossils are known.

SELECTED LOCALITIES (Map 64): ARGENTINA: Catamarca: ca. 2 km SE of Huaico Hondo, on highway 42, E of Portezuelo (Jayat et al. 2010), Las Chacritas, ca. 28 km NNW of Singuil (Jayat et al. 2010); Jujuy, El Duraznillo, Cerro Calilegua (CML 1734), La Herradura, 12 km SW of El Fuerte on Rte 6 (Jayat et al. 2010), Termas de Reyes, Mirador, on Rte 4 (Jayat et al. 2010); Salta, ca. 5 km NW of Campo Quijano, Km 30 on Rte 51 [Quebrada del Toro] (Jayat et al. 2010), Cañadón Ojo de Agua, 10 km S of Rosario de la Frontera (Jayat et al. 2010), Pampa Verde, ca. 8 km WSW of Los Toldos and S of Cerro Bravom (MACN 23506), 1 km ENE of Rodeo Pampa, km 59 on Rte 7 (MACN 23508); Tucumán, Hualinchay, on road to Cafayate (MACN 23451). BOLIVIA: Chuquisaca, 2 km E of Chuhuayaco (S. Anderson 1997), 11 km N and 16 km W of Padilla (S. Anderson 1997); Tarija, 1 km E of Iscayachi, Río Tomayapo (S. Anderson 1997), Tapecua (S. Anderson 1997).

SUBSPECIES: We treat *Akodon caenosus* as monotypic (but see Remarks, below).

NATURAL HISTORY: In Argentina, most of the records for this species come from Yungas environments, from the lowest elevational belt to high-elevation grasslands (Mares, Ojeda, Braun et al. 1997; Jayat, Ortiz, and Miotti 2008; Jayat et al. 2010). Jayat et al. (2010) extended the species' range into Chacoan environments near the ecotone with Yungas forests, and into the lower elevational limits of the High Andean grasslands. Although individuals in reproductive condition have been recorded throughout the year in northwestern Argentina, most were reproductively active between November and January (M. M. Díaz and Barquez 2007; Jayat et al. 2010). The highest proportion of molting animals in this area was found in the fall and winter (Mares, Ojeda, Braun et al. 1997; M. M. Díaz and Barquez 2007; Jayat et al. 2010). Across its range, *A. caenosus* may coexist with several other species of *Akodon*, depending on latitude and elevation. In the ecotone between highland grasslands and upper belts of Yungas forest in northernmost Argentina, at 1,400 m, *A. caenosus* was captured alongside *A. budini* and *A. sylvanus*. At the same latitude in grasslands above 2,000 m, *A. caenosus* is sympatric with *A. boliviensis*. In both Yungas areas and Chaco-Yungas ecotone of Tucumán province, the species was caught together with *A. spegazzinii*. It may be sympatric with *A. simulator*, another broadly distributed species in northwestern Argentina, over most of the latter's distribution range (Capllonch et al. 1997; Mares, Ojeda, Braun et al. 1997; M. M. Díaz et al. 2000; M. M. Díaz and Barquez 2007; Jayat, Ortiz, Miotti 2008; Jayat et al. 2010). In Bolivia, *A. caenosus* may be found at localities also inhabited by one or more of five *Akodon* species, including *A. albiventer*, *A. boliviensis*, *A. fumeus*, *A. simulator*, and *A. varius* (S. Anderson 1997).

REMARKS: Although Thomas (1918b) originally described *caenosus* as a subspecies of *Akodon puer* (= *A. lutescens* J. A. Allen), two years later (Thomas 1920a), he elevated *caenosus* to species status, a decision followed by Cabrera (1961), Barquez et al. (1980), and Mares, Ojeda, and Kosco (1981). Vitullo et al. (1986) and Apfelbaum and Reig (1989) considered specimens from León, Jujuy province (the type locality of *caenosus*) as *A. puer*, and Myers et al. (1990) accepted this conclusion (but see Hershkovitz 1990b) and retained *caenosus* as a subspecies of *A. puer* inhabiting northwestern Argentina (see also M. A. Barros et al. 1990). S. Anderson (1997) also recognized three subspecies for *A. puer* but noted the priority of the name *lutescens* over *puer*. Since then, the nomenclature used for this species has been constantly changing (see Capllonch et al. 1997; Mares, Ojeda, Braun et al. 1997; M. M. Díaz 1999; M. M. Díaz et al. 1999, 2000; Musser and Carleton 2005; Pardiñas, D'Elía et al. 2006; M. M. Díaz and Barquez 2007). M. M. Díaz et al. (1999) described *A. aliquantulus*, based on a few specimens from the ecotone between upper Yungas forest and highland grassland in Tucumán province, and assigned it

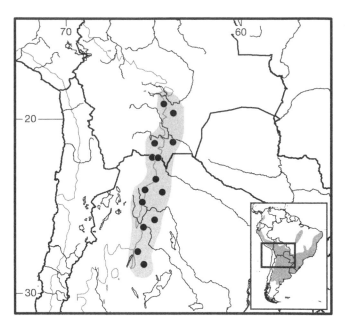

Map 64   Selected localities for *Akodon caenosus* (●). Contour line = 2,000 m.

to the *boliviensis* group, based on its morphology. Jayat et al. (2010) subsequently placed the name *aliquantulus* in synonymy with *A. caenosus*.

The karyotype of this species consists of $2n = 34$, FN = 40, based on specimens from León, Jujuy province, and El Cadillal, Tucumán province, in Argentina (Barquez et al. 1980; Vitullo et al. 1986, as *A. puer*).

Akodon cursor (Winge, 1887)

Cursorial Grass Mouse, Cursorial Akodon

SYNONYMS:

Habrothrix cursor Winge, 1887:25; type locality "Lagoa Santa," Minas Gerais, Brazil.

[*Akodon* (*Akodon*)] *cursor*: E.-L. Trouessart, 1897:536; name combination.

[*Akodon*] *arviculoides cursor*: Thomas, 1913b:406; name combination.

A[*kodon*]. c[*ursor*]. *cursor*: Reig, 1987:382; name combination.

DESCRIPTION: Medium sized and sexually dimorphic, with males averaging larger than females: mean head and body length 120.40 ± 9.30 mm s.d. (N = 212) for males and 115.15 ± 10.02 mm (N = 198) for females, tail length 93.16 ± 11.11 mm (N = 206) versus 89.7 ± 9.21 mm (N = 196), and mass 54.20 ± 11.70 g (N = 201) versus 43.23 ± 10.57 g (N = 180). Tail always shorter than head and body, averaging about 77% of that length in both sexes; ears sparsely covered by short agouti fur; dorsal pelage homogeneous dark to golden brown, sides slightly paler, and venter grayish or yellowish with unicolored hairs; white patch present in gular region in some specimens.

Skull supraorbital region with rounded margins, without crests; zygomatic plates with straight anterior borders, not reaching maxillary; jugals small; incisive foramina with rounded lateral margins, relatively short, only occasionally reaching anterior margin of M1; distinct depressions present on posterior palate; palatal bridge short and narrow; mesopterygoid fossa narrow, beginning at posterior border of M3, anterior border smoothly rounded, occasionally with small median projection; foramen ovale larger than lacerate foramen; stapedial and sphenofrontal foramina and squamosal groove present; and upper incisors opisthodont. Condyloincisive length averages 28.10±s.d. 1.27 mm (N=254) in males and 27.14±1.19 mm (N=223) in females; upper molar series length 4.49±0.17 mm (N=266) in males and 4.46±0.18 mm (N=228) in females. Data for four syntypes measured by L. Geise (unpubl. data), condyloincisive length 29.64±0.40 mm, upper molar series 4.51±0.07 mm, and least interorbital breadth 5.03±0.11 mm. Gall bladder present, distinguishing *A. cursor* from similar and often sympatric *A. montensis* (Geise, Weksler, and Bonvicino 2004).

DISTRIBUTION: *Akodon cursor* occurs in the Atlantic Forest in the Brazilian states of Paraíba, Pernambuco, Bahia, Espírito Santo, Minas Gerais, Rio de Janeiro, São Paulo, and Paraná. A record of the species from Misiones province, Argentina is unconfirmed (Pardiñas, D'Elía, and Cirignoli 2003). Fossils of *A. cursor* are known from Quaternary deposits of the Lagoa Santa caves, Minas Gerais (Winge 1887; Voss and Myers 1991).

SELECTED LOCALITIES (Map 65; localities of specimens for which molecular and/or chromosomal data are available; see Remarks, below): BRAZIL: Bahia, Chapada Diamantina, Morro da Torre da TeleBahia, Lençóis (L. G. Pereira and Geise 2009), Ilhéus (Geise and Pereira 2008), Nova Viçosa (MNRJ 47922), Una (FAG 002); Espírito Santo, Cariacica (LGA 325); Minas Gerais, Estação Ecológica de Acauã, 17 km N of Turmalina (LC 81), Rio das Velhas, Lagoa Santa (type locality of *Habrothrix cursor* Winge); Paraíba: Fazenda Alagamar, 9 km S and 6 km E of Mamanguape (UFPB 4366), João Pessoa (UFPB 112); Paraná, Ilha Rasa, Guaraqueçaba (MHNCI [P 682]); Pernambuco, Bom Conselho (PMN 165; see Furtado 1981), Mata do Macuco, Sítio Bituri, Brejo da Madre de Deus (UFPE 1096), São Lourenço da Mata (UFPB 3523); Rio de Janeiro, Casimiro de Abreu (Geise et al. 2005); São Paulo, Fazenda Intervales, Capão Bonito (MVZ 182072), Ubatuba (MNRJ 55728).

SUBSPECIES: *Akodon cursor* is monotypic.

NATURAL HISTORY: Geise (2012) summarized the literature on ecology, behavior, and other aspects of the natural history of this species. *Akodon cursor* is a member of the forest and open area (grassland) fauna in the Atlantic Forest of eastern Brazil; it also occurs in contact areas between the eastern *campo cerrado* and inland forest isolates, such as in the Chapada Diamantina in Bahia state (L. G. Pereira and Geise 2007) and in Caatinga (F. F. Oliveira and Langguth 2004). Males typically have larger home ranges than females. In one long-term study at Restinga de Barra de Maricá (Maricá, Rio de Janeiro state), Gentile et al. (1997) documented home ranges from 0.12 to 0.68 ha (mean 0.28 ha, s.d. 0.14, N=24). At the same site, food preferences varied according to environmental conditions (Cerqueira et al. 2003). Sexually active females on Ilha do Cardoso (São Paulo state) were present throughout the year, although there was a decrease during the coldest months (June through August), with rainfall affecting pregnancy rates (Bergallo and Magnusson 1999). The gestation period in laboratory colonies ranged from 21 to 24 days (mode, 23 days; Mello and Mathias 1987), and a postpartum estrus was observed in 53% of females in a sample of 389 litters, occurring between the first and fifth days following birth. Litter size ranged from 1 to 10 (mean 4.64) with no skew in sex ratio. Birth weight averaged 3.95 g for males and 3.88 g for females (De Conto 2007).

This species inhabits both primary forest and degraded forest fragments (Olifiers et al. 2004). At localities from Rio de Janeiro state, *A. cursor* has been found from sea level to elevations above 1,000 m (Geise et al. 2005). G. A. B. Fonseca and Kierulff (1989) and Stallings (1989) regarded *A. cursor* as an insectivore-omnivore. Ectoparasites from specimens taken at Itatiaia include a rich assemblage of fleas, including *Adorapsylla antiquorum*, *Craneopsylla minerva minerva*, *Polygenis atopus*, *P. dentei*, *P. pradoi*, and *P. rimatus* (Moraes et al. 2003).

REMARKS: Earlier synoptic treatments of *Akodon* regarded *cursor* as a subspecies of *A. arviculoides* (Wagner), along with *montensis* Thomas (Gyldenstolpe 1932; Ellerman 1941) and *tapirapoanus* J. A. Allen (Cabrera 1961). The names *arviculoides* and *tapirapoanus* are now placed in synonymy with *lasiurus* and *lenguarum*, respectively, in *Necromys* (see *Necromys* account, and Reig 1978, 1987), and *montensis* has been elevated to species status in *Akodon* (see that account).

The widespread distribution of *A. cursor* was originally recognized by Thomas (1902a:60) based on several Brazilian specimens, but this species is not easily distinguishable from the largely sympatric *A. montensis* (see below), based on external morphological features. This great similarity has led to substantial confusion concerning these taxa and, therefore, the boundaries of their geographic ranges. Karyotypic differences became one of the most commonly used ways to distinguish the two, once it was evident that two nonoverlapping chromosomal groups comprised this complex in southern Brazil. We now regard those animals with a $2n = 14$, 15, or 16 as *A. cursor*, and those with $2n = 24$ or 25 as *A. montensis*. Cerqueira et al. (1994) were the first to apply the name *A. cursor* to specimens with 14, 15, and 16 chromosomes, an action supported by Rieger et al. (1995) who documented a specimen from near the type locality of *A. cursor* as having the $2n = 14$ karyotype. Subsequent authors uniformly began to associate this karyotype with *A. cursor* (e.g., Sbalqueiro and Nascimento 1996; Geise et al. 1998; Geise et al. 2001; Fagundes and Nogueira 2007). Within its range of diploid numbers, however, is a remarkably variable karyotype, with 28 cytotypes now described (see Fagundes et al. 1998).

Following the realization that *cursor*-like animals of the Atlantic Forest were divisible into two karyotypic groups, additional character complexes were found to distinguish between *A. cursor* and *A. montensis*. These include cranial features, the most useful being upper molar series length and least interorbital breadth (Geise et al. 2005; P. R. Gonçalves et al. 2007), and also mtDNA cytochrome-*b* sequences (Geise et al. 2001). More recently, Pereira (2006) and Geise et al. (2007) documented two phylogeographic groups of cytochrome-*b* sequences through broad sampling across the species' range, with the division between them close to the Rio Jequitinhonha in southern Bahia state. These two groups differed by an average molecular distance (Kimura 2-parameter) of 3.6%,

further demonstrating that *A. cursor* exhibits a substantial degree of both cytological and molecular polymorphism. Population analysis of nuclear microsatellites (A. Cunha 2005) also indicated strong differentiation among populations, without indication of isolation-by-distance among sampled areas.

Akodon dayi Osgood, 1916
Day's Grass Mouse; Dusky Akodon

SYNONYMS:

Akodon dayi Osgood, 1916:208; type locality "Todos Santos, Chaparé River, Bolivia" Todos Santos (16.804°S, 65.143°W), Cochabamba, Bolivia.

[*Akodon* (*Chalcomys*)] *dayi*: Tate, 1939:187; name combination.

Akodon tapirapoanus dayi: Cabrera, 1961:447; name combination but not *Zygodontomys tapirapoanus* J. A. Allen [= *Necromys lenguarum* (Thomas)].

Akodon urichi dayi: S. Anderson, 1985:12; name combination.

DESCRIPTION: Large, with moderately short tail; total length averages about 200 mm, tail length 80 mm, greatest length of skull 29.5 mm, and upper toothrow 4.8 mm. Dorsal coloration rich blackish brown, slightly paler on sides and rufous ventrally; ears blackish; fore- and hindfeet diagnostically brownish black above; moderately short tail uniformly blackish; no eye-ring present. Skull large and robust, with interorbital region hourglass in shape, with squared but not beaded edges; temporal ridges well defined; and zygomatic notches moderately broad. Upper incisors opisthodont. Procingulum of M1 with well-developed anteromedian flexus. Gall bladder present. Myers (1990:29) described type series in detail, and S. Anderson and Olds (1989) illustrated molar and other anatomical traits.

DISTRIBUTION: Lowland areas and up into the Yungas to at least 2,450 m elevation, from Pando to west-central Santa Cruz in Bolivia, including the departments of Cochabamba, Santa Cruz, and La Paz (Anderson 1997). No fossils are known for *A. dayi*.

SELECTED LOCALITIES (Map 66): BOLIVIA: Beni, El Consuelo (B. D. Patterson 1992b); Cochabamba, 25 km W of Comarapa (S. Anderson 1997), Todos Santos (Osgood 1916); La Paz, Chulumani (B. D. Patterson 1992b), Ixiamas (S. Anderson 1997); Pando, Bella Vista (S. Anderson 1997), Remanso (S. Anderson 1997); Santa Cruz, 6 km W of Ascención [= Ascención de Guarayos] (S. Anderson 1997), Ayacucho (S. Anderson 1997), 2.5 km NE of El Refugio, Parque Nacional Noel Kempff Mercado (USNM 584503; Emmons et al. 2006), Las Lomitas, 15 km S of Santa Cruz (S. Anderson 1997).

SUBSPECIES: *Akodon dayi* is monotypic.

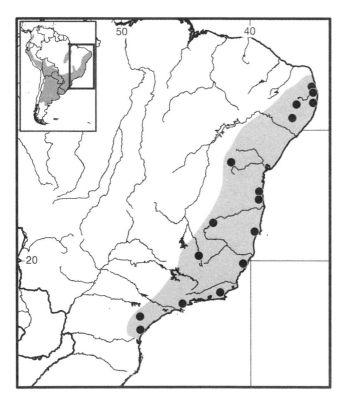

Map 65 Selected localities for *Akodon cursor* (●). Contour line = 2,000 m.

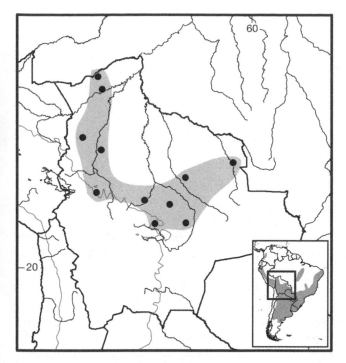

Map 66 Selected localities for *Akodon dayi* (●). Contour line = 2,000 m.

NATURAL HISTORY: Little is known about the population ecology or behavior of this species. *Akodon dayi* occupies mesic areas between 250 and 2,450 m, including savannas, lowland dry forests, and moist forests (Myers 1990; S. Anderson 1997). Embryo counts range from 2 to 5 (mean: 3) for five pregnant females, with pregnancies recorded in March, June, and July. Nonpregnant females were recorded in June, July, and August, and two juveniles were collected in March.

REMARKS: S. Anderson and Olds (1989) and S. Anderson (1997) presented external and cranial measurements. Braun et al. (2008) reported high divergence values of molecular variation between two individuals referred to this species, suggesting the possibility that the current concept of *A. dayi* is composite. These authors found a sister relationships between *A. dayi* and *A. toba* and included both in a lowland clade of the *varius* group (see also Jayat et al. 2010). Cabrera (1961:447) regarded *dayi* as a subspecies of *Akodon tapirapoanus* (J. A. Allen), but S. Anderson (1997:418) correctly noted that *tapirapoanus* should be placed in *Bolomys* (= *Necromys*), not *Akodon*.

Akodon dolores Thomas, 1916
Dolores Grass Mouse, Córdoba Akodon
SYNONYMS:

Akodon dolores Thomas, 1916h:334; type locality "Yacanto, near Villa Dolores, south-western slopes of the Sierra de Cordova. Alt. 900 m," Yacanto (32.063°S, 65.024°W), San Javier, Córdoba, Argentina.

Akodon neocenus Thomas, 1919c:213; type locality "Neuquen, Rio Limay, Upper Rio Negro, Patagonia," clarified as Neuquén (38°57′S, 68°04′W, 259 m, Confluencia, Neuquén; Argentina by Pardiñas et al. (2007).

Akodon varius neocenus: Gyldenstolpe, 1932a:103; name combination.

Akodon molinae J. R. Contreras, 1968:10; type locality "Vivero del Ministerio de Asuntos Agrarios, Laguna Chasicó, Partido de Villario, Provincia de Buenos Aires," = Ex vivero Humboldt, junction of Arroyo Chasicó and Laguna Chasicó (ca. 38.645°S, 63.015°W), Villarino, Buenos Aires, Argentina.

Akodon varius molinae: Massoia, Morici, and Lartigau, 1993:34; name combination.

Akodon varius dolores: Massoia, Morici, and Lartigau, 1993:34; name combination.

DESCRIPTION: One of larger-bodied species, with head and body length >110 mm, tail >80 mm, greatest length of the skull >30 mm, and upper toothrow >4.5 mm. Dorsal color pale olive brown to olive gray, some individuals with reddish tint; venter pale, drab white to gray; distinction between dorsal and ventral color moderately abrupt. Pale eye-ring present. Fur long and dense; ears large, up to 18 mm long; tail usually short, bicolored, and densely furred. Skull robust and bowed, with strongly flared zygomatic arches, especially broad zygomatic plates with round and long upper free borders, short nasals, wide interorbital constriction with squared supraorbital region, well-developed temporal and mastoid ridges, and posterolateral palatal pits at level of anterior border of mesopterygoid fossa. Capsular projection of mandible pointed but lies below coronoid process. Upper incisors opisthodont and robust; procingulum of M1 labially compressed; anteromedian flexi/ids usually absent in adult individuals; and anterior portion of M2 more transversally expanded than posterior part.

DISTRIBUTION: As is currently understood, the range of *A. dolores* is broadly coincident with the lowland semiarid areas of the Monte Desert and Espinal in central Argentina. Recorded localities are distributed from Tucumán and Santiago del Estero provinces south to southwestern Buenos Aires, reaching the Ventania hill system, and to northeastern Chubut province in Patagonia. The western boundary of this distribution roughly coincides with the Monte-Huayquerías limit in Mendoza province (J. R. Contreras and Rosi 1981; Myers 1990; Braun et al. 2008; Pardiñas 2009; Nabte et al. 2009; Jayat, Ortiz, Pacheco, and González 2011). The known fossil record is restricted to the Holocene of southeastern Buenos Aires province and northern Patagonia (Pardiñas 1999a; Udrizar Sauthier 2009).

SELECTED LOCALITIES (Map 67): ARGENTINA: Buenos Aires, Arroyo de los Loros (Pardiñas, Abba, and Merino 2004), Bahia Blanca (Pardiñas, Abba, and Merino 2004), Cabaña San José (Pardiñas, Abba, and Merino. 2004), Puesto El Chara (Pardiñas, Abba, and Merino 2004); Catamarca, Trampasacha (Jayat, Ortiz, Pacheco, and González 2011); Chubut, Estancia La Irma (Nabte et al. 2009), Estancia Los Nogales (CNP-E 402), Laguna La Blanca (De Santis and Pagnoni 1989); Córdoba, Chucul (Wittouck et al. 1995), Villa de María del Río Seco (Wittouck et al. 1995); La Pampa, 15 km SW of Chamaicó, Loma Loncovaca (Tiranti 1998), Colonia Gobernador Ayala (Sieghentaler et al. 1993), Estancia El Puma, Chicalcó (Sieghentaler et al. 1990), 10 km SW of Santa Rosa, Chacra la Lomita (Tiranti 1998); La Rioja, Ambil (Massoia et al. 1999); Mendoza, La Pega (J. R. Contreras and Rosi 1980b), Rincón del Atuel (J. R. Contreras and Rosi 1980b); Neuquén, Barda Negra, Parque Nacional Laguna Blanca (Massoia and Pastore 1997), 3–5 km upstream Chos Malal, along Río Neuquén (Tiranti 1996b); Río Negro, Cerro Castillo [= Cerro Guacho] (Pardiñas and Massoia 1989), El Rincón [= Establecimiento El Rincón] (Pardiñas, Teta et al. 2003), Estancia Liempi (Nabte et al. 2009), San Carlos (Pardiñas 2009); Santiago del Estero, INTA "La María" Research Station, 2.9 km W of station entrance (Jayat, Ortiz, Pacheco, and González 2011), ca. 30 km N of Pozo Hondo, along Rte 34 (Jayat, Ortiz, Pacheco, and González 2011); Tucumán, ca. 4 km NW of Las Cajas (Jayat, Ortiz, Pacheco, and González 2011).

SUBSPECIES: We treat *Akodon dolores* as monotypic (but see Remarks).

NATURAL HISTORY: *Akodon dolores* is nocturnal (J. R. Contreras 1979b). This species has a broad habitat range, having been collected in Espinal forests of *Prosopis caldenia* (Tiranti 1998a), in halophytic shrub communities (Tiranti 1998a), mixed shrublands (Teta, Pereira, Fracassi et al. 2009), borders of grasslands and cultivated fields (Polop et al. 1985; Tiranti 1998a), clumps of grass near watercourses (J. R. Contreras 1966), Chaco Serrano transition with palms (Tiranti 1998a), rocky hilly environments (Pardiñas, Abba, and Merino 2004; Teta et al. 2005), and deforested areas where there is adequate shrub or litter cover (Tiranti 1998a; Corbalán 2005). Some populations also occur in urbanized areas (E. Castillo et al. 2003). Polop et al. (1985), Polop and Sabattini (1993), Bonaventura et al. (1998), Corbalán and Ojeda (2004), Corbalán (2005), Corbalán et al. (2006), Tabeni et al. (2006), and Teta et al. (2005) addressed habitat use, microhabitat selection, and/or abundance by habitat. R. A. Ojeda (1989) studied response to natural fires.

Corbalán and Ojeda (2005) and Teta et al. (2005) determined mean home range size from 785–924 m^2 in central and west-central Argentina, respectively, with those of males larger than females. Highest population densities occurred during the autumn (J. R. Contreras 1979b; Polop et al. 1993), following a prolonged reproductive season from August to March or April (M. C. Navarro 1991; Teta, Pereira, Fracassi et al. 2009). Piantanida (1981) examined fertility, postnatal development, and growth in a laboratory population, and determined a mean embryo count of 7.3 but a reduced mean litter size of 4.6. Apfelbaum and Blanco (1985) examined levels of allozyme variation in central Argentina populations.

This species is omnivorous in its diet, with local and seasonal differences in the degree of insectivory or herbivory (J. R. Contreras 1979b; R. L. Martínez et al. 1990; C. M. Campos, Ojeda et al. 2001; Castellarini et al. 2003; Gianonni et al. 2005). Ectoparasites from animals captured in northwestern Patagonia (from specimens referred to as *A. neocenus*) include the fleas *Barreropsylla excelsa*, *Ectinorus onychiusonychius*, *Plocopsylla wolffsohni*, *Sphinctopsylla ares*, *Tetrapsyllus* (*Tetrapsyllus*) *rhombus*, and *T.* (*T.*) *tantillus* (Autino and Lareschi 1998; Lareschi and Mauri 1998) and the sucking lice *Hoplopleura aitkeni* and *H. varia* (D. Castro and Cicchino 1998). The ectoparasite assemblage recorded from populations in Mendoza province was also diverse, including mites (*Ornithonyssus bacoti*), fleas (*Ectinorus* (*Ectinorus*) *barrerai*, *Craneopsylla minerva wolffhuegeli*, *Polygenis* (*Polygenis*) *bohlsi bohlsi*, *P.* (*P.*) *platensis*, *P.* (*P.*) *rimatus*, *P.* (*Neopolygenis*) *puelche*), and lice (*Hoplopleura aitkeni*) (Autino and Lareschi 1998; Lareschi and Mauri 1998; Lareschi and Linardi 2009).

REMARKS: Our concept of *A. dolores* includes *molinae* J. R. Contreras and *neocenus* Thomas as synonyms; these names are available for trinomial usage as might prove to be appropriate in the future. The close relationship between *A. dolores* and *A. molinae* was initially suggested by karyotypic and cross-breeding studies (Myers 1990, and the references therein), with conspecificity proposed by Hershkovitz (1990b), and later documented by phylogenetic analysis of mtDNA sequences (Braun et al. 2008). Some authors maintain the validity of both *dolores* and *molinae* by attributing to the former a diploid complement of 34–40 (e.g., Tiranti 1998a; Musser and Carleton 2005) instead of the 2n = 42–44 of typical *molinae* (N. O. Bianchi et al. 1971). However, topotypes of *A. dolores* have a 2n = 42–44 karyotype (Wittouck et al. 1995). This supports the proposition that *dolores* Thomas is a senior synonym of *molinae* J. R. Contreras. If further studies demonstrate that the 2n = 34–40 cytotype is specifically distinct, a new name may have to be coined for that taxon. Populations of *Akodon dolores* with $2n = 34–40$ are mainly distributed along the eastern flank of the Sierra de Córdoba (N. O. Bianchi et al. 1971; N. O. Bianchi and Merani 1984; Wittouck et al. 1995), extending to at least southern Catamarca province (N. O. Bianchi et al. 1979). As has been discussed by several authors, those with this distributional range are apparently allopatric relative to those with $2n = 42–44$ (see Tiranti 1998a). Although the two cytotypes do not overlap, or even contact one another, they are completely interfertile in experimental crossbreeding studies (Wittouck et al. 1995). Furthermore, Wittouck et al. (1995) found no differences in cranial measurements between populations of the two cytotypes from opposite sides of Sierra de Córdoba.

As for supporting the inclusion of *neocenus* in the synonymy of *A. dolores*, we note that Thomas (1916h), in his original concept of *neocenus*, included specimens from populations that J. R. Contreras (1968) later used in his description of *molinae*. Myers (1990) supported the validity of *neocenus*, but he examined only a small sample of specimens from central Argentina—one of these he referred to *A. dolores*, a second to *A. molinae*, and the remaining 11 to *A. neocenus*. Myers (1990) also reported sympatry between *neocenus* and *molinae* in southwestern Buenos Aires province. The conspecificity of *neocenus* and *molinae*, however, is supported by karyotypic, molecular, and morphological data. First, one near topotype of *neocenus*, caught at Cipoletti, Río Negro, had a $2n = 42$ (N. O. Bianchi et al. 1971). Second, a study in progress that includes topotypes of *dolores*, *molinae*, and *neocenus* has indicated very low levels of molecular divergence among these specimens and a lack of reciprocal monophyletic lineages that would support species status (G. D'Elía and U. F. J. Pardiñas, unpubl. data). Finally, both external as well as craniodental morphology, despite some variation in overall color, is very similar among populations referred to the different supposed species. The hypothesis of only one large species of *Akodon* living in central Argentina does make biogeographic sense as well because the distribution of the entire complex is broadly coincident with the xerophytic shrublands and woodlands of the Monte Desert and Espinal.

The separation of *A. dolores* from to *A. toba* requires further study because the latter has a similar $2n = 42–44$ karyotype (Myers 1990) and also occupies similar environments in the Chacoan region (see that account). Phylogenetic analyses of DNA sequences indicate that *A. dolores* and *A. toba* represent reciprocally monophyletic lineages, but available specimens within the range of *toba* are relatively few, and levels of genetic divergence among lineages are low (<3%; Braun et al. 2008).

Akodon fumeus Thomas, 1902
Smoky Grass Mouse, Smoky Akodon

SYNONYMS:

Akodon fumeus Thomas, 1902b:137; type locality "Choro, 3500 m," Cochabamba, Bolivia.

Akodon mollis fumeus: Gyldenstolpe, 1932a:106; name combination.

DESCRIPTION: Medium sized, with average total length 171.2 mm, tail length 76.2 mm, ear length 15.9 mm, and hindfoot length 21.4 mm. Dorsal pelage dark brown with rufous overtones, not sharply differentiated from venter; pale ring present around each eye. Tail sparsely furred, not sharply bicolored. Skull with relatively flat dorsal profile in lateral view, narrow and pointed rostrum, distinctively swollen frontal sinuses, moderately developed zygomatic arches, relatively broad zygomatic plates, unusually shallow zygomatic notches, and smoothly rounded to squared but never beaded interorbital region. Squamosoalisphenoid groove conspicuous on lateral wall of braincase; mesopterygoid fossa broad and usually with rounded anterior margin that lacks median spine; parapterygoid fossae deeply excavated with lateral margins slightly or moderately convex. Upper incisors orthodont. M1 with pronounced anteromedian flexus and prominent mesoloph and mesostyle; M2 with well-developed anteroloph and parastyle that projects to edge of tooth; M3 relatively small, with outline that varies from oval to figure-8 shaped. Lower m1 also with well-

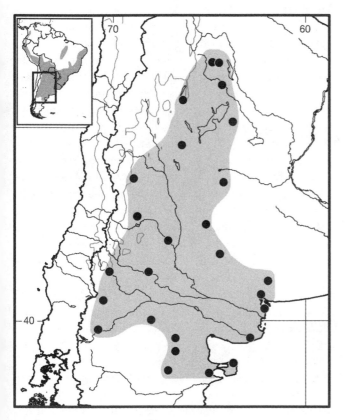

Map 67   Selected localities for *Akodon dolores* (●). Contour line = 2,000 m.

developed anteromedian flexid and conspicuous antero-labial cingulum; m2 with conspicuous anterolabial lophid and well-developed posterolophid; and m3 relatively simple, about two-thirds length of m2. Mean condyloincisive length 23.48 mm, mean upper molar series length 4.07 mm.

DISTRIBUTION: *Akodon fumeus* occurs on eastern Andean slopes from Puno department in southeastern Peru through Bolivia to northwestern Argentina, in Salta and Jujuy provinces (Myers and Patton 1989b; S. Anderson 1997; M. F. Smith and Patton 2007; Tarifa et al. 2007; Jayat et al. 2010). No fossils are known.

SELECTED LOCALITIES (Map 68): ARGENTINA: Jujuy, Río Las Capillas, 15 km N of Las Capillas on Rte 20 (Jayat et al. 2010); Salta, Finca Alto Verde (Jayat et al. 2010), Parque Nacional Baritú, near Río Baritú (M. M. Díaz et al. 2000). BOLIVIA: Cochabamba, Palmar, 101 km by road SW of Epizana, Siberia cloud forest, Cordillera Oriental (Myers and Patton 1989b), Yungas de Cocha-bamba (Myers and Patton 1989b); La Paz, Bellavista (S. Anderson 1997), 35 km by road N of Caranavi, Serranía Bellavista (S. Anderson 1997), Huaraco (S. Anderson 1997), Khallutaka [near Laja] (S. Anderson 1997), 10 km by road N of Sorata, Moyabaya, Río Challapampa (S. Anderson 1997); Santa Cruz, Guadalupe, 10 km S of Vallegrande (S. Anderson 1997); Tarija, Caiza (S. Anderson 1997), Erquis (S. Anderson 1997). PERU: Puno, Sandia (Myers and Patton 1989b).

SUBSPECIES: *Akodon fumeus* is monotypic.

NATURAL HISTORY: *Akodon fumeus* lives mostly in forested regions of the eastern Andean slopes, from 700 m to at least 3,800 m at the ecotone with Puna-like grass-lands. Across this extensive elevational range, *A. fumeus* occupies a diversity of forest types, from high-elevation *Polylepis* woodlands, cloud forest, and *Podocarpus* and *Alnus* forests (Myers and Patton 1989b; S. Anderson 1997; M. F. Smith and Patton 2007; Tarifa et al. 2007; Jayat et al. 2010). S. Anderson (1997) reported Bolivian fe-males with embryos in October and November, and non-pregnant females in June, July, and September. None of the 75 specimens captured in May 2006 in northernmost Salta province, Argentina (J. P. Jayat et al., unpubl. data), were reproductively active. In August 2006, three females still had a closed vagina, but five males had semiscrotal testes. *Akodon fumeus* may be sympatric or contiguously allopat-ric with several other *Akodon* species, including *A. aerosus*, *A. albiventer*, *A. boliviensis*, *A. caenosus*, *A. dayi*, *A. mimus*, *A. lutescens*, *A. kofordi*, *A. pervalens*, *A. siberiae*, *A. simula-tor*, *A. subfuscus*, and *A. varius* (Myers and Patton 1989b; Patton and Smith 1992b; S. Anderson 1997; S. Anderson and Yates 2000; Salazar-Bravo, Yensen et al. 2002; Salazar-Bravo and Yates 2007; Tarifa et al. 2007). In northernmost Argentina, *A. fumeus* was captured at the same localities

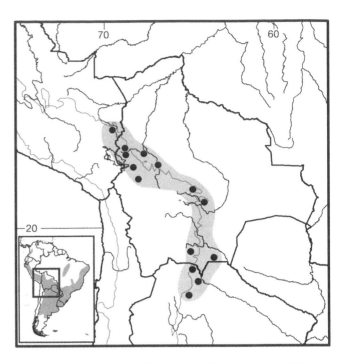

Map 68 Selected localities for *Akodon fumeus* (●). Contour line = 2,000 m.

and in the same habitats as *A. budini*, *A. caenosus*, and *A. simulator* (Jayat and Ortiz 2010; Jayat et al. 2010).

REMARKS: Gyldenstolpe (1932), Cabrera (1961), and Hershkovitz (1990b) regarded *fumeus* as a subspecies of *A. mollis*, but Myers and Patton (1989b) established its specific status and allied it to *A. kofordi*, grouping the two together in an informal *fumeus* group. Recent molecular studies confirmed the close relationship between *A. fumeus* and *A. kofordi* but placed both species in the *boliviensis* group (M. F. Smith and Patton 2007). These two species have been confused in the recent literature, as S. Anderson (1997) assigned Bolivian specimens to *A. fumeus* but that actually represent *A. kofordi* (Salazar-Bravo, Yensen et al. 2002). M. M. Díaz and Barquez (1999) added this species to the fauna of Argentina on the basis of one record for Ju-juy province. M. M. Díaz and Barquez (2007) reported ad-ditional Argentinean localities, but most of the specimens assigned by these authors to *A. fumeus* were reidentified by Jayat et al. (2010) as *A. sylvanus* or *A. budini*. How-ever, all the specimens from Salta province, as well as a few from Jujuy province, assigned to *A. sylvanus* by Jayat, Ortiz et al. (2007) also were misidentified, and represent *A. fumeus* (see Jayat et al. 2010).

Akodon iniscatus Thomas, 1919

Intelligent Grass Mouse; Patagonian Akodon

SYNONYMS:

Akodon iniscatus Thomas, 1919b:205; type locality "Valle de Lago Blanco, Koslowsky region, Patagonia,

46°S"; restricted to Estancia Valle Huemules (45°57'S, 71°31'W, 593 m, Río Senguerr, Chubut, Argentina (Pardiñas et al. 2007:406).

Akodon iniscatus collinus Thomas, 1919b:206; type locality "Maiten, W. Chubut. 700 m." El Maitén (42.050°S, 71.160°W), Cushamen, Chubut, Argentina.

Akodon nucus Thomas and St. Leger, 1926b:636; type locality "Chos Malal, 800 m." Chos Malal (37.376°S, 70.270°W), Chos Malal, Neuquén, Argentina.

Akodon iniscatus iniscatus: Cabrera, 1961:442; name combination.

Akodon iniscatus nucus: Cabrera, 1961:443; name combination.

DESCRIPTION: Small, with head and body length 80–110 mm, tail length 50–80 mm, greatest length of skull <27 mm, and upper toothrow <4.5 mm. Dorsal color darkly grizzled olivaceous brown to uniformly brown; venter gray and moderately demarcated from dorsum; white spot variably present on chin. Tail short, well haired, and bicolored. Skull delicate and bowed. Samples from northwestern part of species' range larger than those to south, with broader rostrum, wide zygomatic plates with long upper free borders, subparallel zygomatic arches, narrower mesopterygoid fossa, larger pterygoids, and moderately opisthodont upper incisors. M1 and m1 in all samples with persistent anteromedian flexus/flexid and greatly reduced M3 and m3.

DISTRIBUTION: *Akodon iniscatus* inhabits northern Argentinean Patagonia, ranging from southwestern Buenos Aires, La Pampa, and probably southern Mendoza provinces south through Río Negro, Neuquén, and Chubut provinces to at least the northeastern portion of Santa Cruz province (Roig 1965; Pardiñas and Galliari 1999; Pardiñas, Teta et al. 2003; Pardiñas 2009), and extends, barely, into Aysén region, Chile. The occurrence of *A. iniscatus* in southern Mendoza province, first mentioned by Cabrera (1961; see also Massoia et al. 1994a, who recorded this species from Laguna de la Niña Encantada [about 35°09'S]), needs confirmation; these records may be of misidentified *A. spegazzinii* (Pardiñas, Teta, Formoso, and Barberena 2011). The northeastern portion of the historic range has apparently retracted contracted considerably because of anthropogenic agroecosystem expansion, as recent surveys in this region have failed to encounter *A. iniscatus* (see Teta, Pereira, Fracassi et al. 2009). Thus, the historic record of this species from Estación Perú is indicated as disjunct from the main portion of the species range (Map 66). Fossils attributed to *A. iniscatus* have been described from the Middle Pleistocene of Mar del Plata (Reig 1987) and Bahía Blanca (Pardiñas and Deschamps 1996), Buenos Aires province. A rich Holocene record for this rodent is also available from several caves in Chubut and Neuquén

provinces, Patagonia (Pardiñas 1999a; Udrizar-Sauthier 2009; F. J. Fernández et al. 2012).

SELECTED LOCALITIES (Map 69): ARGENTINA: Buenos Aires, Bahía San Blas (Pardiñas and Galliari 1999), Carmen de Patagones (Thomas 1919b), Estación Perú (Thomas 1919b); Chubut, Astra (Nabte et al. 2006), Buenos Aires Chico (CNP-E 49), Estancia Valle Huemules (Thomas 1919b), Isla Escondida (UP 283), 20 km S of Leleque (M. A. Barros et al. 1990), Los Altares (Pardiñas, Teta et al. 2003), 36 km W of Los Altares (CNP-E 51), Paso del Sapo (Pardiñas, Teta et al. 2003), Puerto Lobos (CNP-E 190), Punta Delgada (Pardiñas, Teta et al. 2003); La Pampa, La Humada (Tiranti 1989); Mendoza, General Alvear (Roig 1965); Neuquén, Barrancas (F. J. Fernández et al. 2012), Estancia Corcel Negro (CNP-E 88), Paraje La Querencia (Pardiñas, Teta et al. 2003), Riscos Bayos (Pardiñas, Teta et al. 2003); Río Negro, Barda Esteban, Pilcaniyeu (CNP-E 417), Choele Choel (MACN 28.110); Santa Cruz, Cañadón Minerales (CNP-E 365), Puerto Deseado (Thomas 1919b). CHILE: Aysén, 1 km E of Coyhaique Alto (M. F. Smith and Patton 1999; FMNH 129843).

SUBSPECIES: Currently, *A. iniscatus* is regarded as monotypic. However, the taxon *nucus* Thomas and Saint Leger, sometimes listed as a full species (Galliari et al. 1996; Hershkovitz 1990b; Reig 1987), was retained as a subspecies of *A. iniscatus*, pending critical revision, by M. A. Barros et al. (1990) and Pearson (1995). M. F. Smith and Patton (2007) provided mtDNA cytochrome-*b* sequence from a specimen of *nucus* to support the conspecificity of this taxon with typical *A. iniscatus*. However, typical *nucus* populations have several diagnostic morphological and morphometric features, as discussed by M. A. Barros et al. (1990) and Pardiñas (2009). Whether these character differences justify subspecies or species status remains to be determined.

NATURAL HISTORY: This is the most common but also one of the most poorly known *Akodon* species in Patagonia (Pardiñas 2009). It has been regularly confused with *Abrothrix olivacea* (see discussion in Thomas 1919b; B. D. Patterson et al. 1984). Many ecological data concerning this species come from coastal areas in northeastern Chubut and have been presented in meeting abstracts but not formally published. *Akodon iniscatus* primarily inhabits the Monte Desert biome and sub-Andean steppe grasslands, generally from sea level to 900–1,000 m. This rodent is particularly abundant in the eastern lowlands of the Monte Desert where it is apparently replaced by *A. olivacea* toward the central basaltic plateau (Pardiñas 2009). It is a common prey species for several species of owls (see Pardiñas 1999a; Nabte et al. 2008, Trejo and Lambertucci 2008). Antinuchi and Busch (1999) examined renal morphology and studied the physiology of a population near Puerto Madryn, Chubut. Historical data and current collections suggest that its range, especially at the northern and eastern boundary, has recently retracted, in part due to the expansion of agroecosystems such as those in Buenos Aires province (Pardiñas, Abba, and Merino 2004). Ectoparasites reported include the fleas *Agastopsylla boxi boxi*, *Craneopsylla minerva wolffhuegeli*, *Ectinorus ixanus*, *E. levipes*, *E. onychius onychius*, *Eritranis andricus*, *Hectopsylla (Hectopsylla) gracilis*, *Listronius fortis*, *Neotyphloceras crassispina*, *Polygenis (Polygenis) platensis*, *P. (P.) rimatus*), *Tetrapsyllus rhombus*, and *T. tantillus*, and the ticks *Amblyomma tigrinum* and *Ixodes* sp. (Smit 1987; Autino and Lareschi 1998; Lareschi and Mauri 1998).

REMARKS: M. F. Smith and Patton (2007) placed *A. iniscatus* in their *varius* group, based on mtDNA cytochrome-*b* sequences. With an expanded mtDNA analysis, Jayat et al. (2010) included *iniscatus* in a subclade (termed the *dolores* group) of Myers's (1990) *varius* group, composed of *A. dolores*, *A. molinae*, *A. toba*, and *A. dayi*. B. D. Patterson et al. (1984:14) argued that *Mus infans* Philippi might be a synonym of *nucus* (and thus of *A. iniscatus*) and not of *Abrothrix olivacea* (cited as *xanthorhina*), as Osgood (1943b) had proposed. This hypothesis has not as yet been examined adequately (see the *Abrothrix olivacea* account herein).

M. A. Barros et al. (1990) reported a $2n = 33$–34 karyotype from specimens collected in Neuquén (*nucus*) and Chubut (*iniscatus*) provinces. Tiranti (1999) reported the same karyotype for animals obtained at Parque Nacional Laguna Blanca, Neuquén province.

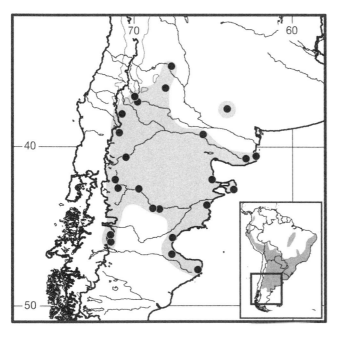

Map 69 Selected localities for *Akodon iniscatus* (●). The apparently isolated population on the northeastern edge of the range is shown as disjunct from the main distribution (see Distribution). Contour line = 2,000 m.

Akodon josemariarguedasi Jiménez, Pacheco, and Vivas, 2013

Arguedas's Grass Mouse, Arguedas's Akodont

SYNONYMS:

Akodon orophilus: Pacheco, Córdova, and Velásquez, 2012:107; part; not *Akodon orophilus* Osgood.

Akodon josemariarguedasi Jiménez, Pacheco, and Vivas, 2013:232; type locality "Galloganán, Ambo, Huánuco, Peru, 3420 m. 10° 09′S, 76° 08′W."

DESCRIPTION: Large *Akodon*, with mean total body length 192.3 mm, tail length 89 mm, ear length 15.6 mm, and hindfoot length 22.8 mm; greatest skull length ranges from 26.1–27.5 mm and upper molar series length 3.9–4.4 mm. Distinguished from congeners by the following combination of characters: dorsal pelage coloration uniformly brownish olive, strongly contrasting with smoky gray to whitish venter; fur hairs with paler gray bases; tail dark and moderately countershaded; nasals extend posterior to premaxillofrontal suture, posterior end blunt with serrated contour; lacrimals large; zygomatic plates with anterior margin less slanting, with either convex or straight anterior edges; antorbital bridge moderately wide; zygomatic notches somewhat deep; palate narrow; maxillary toothrows somewhat short; mandible delicate with thin and large condylar process, and deeper lunar notch; internal fossettus between paracone and protocone present on M1.

DISTRIBUTION: *Akodon josemariarguedasi* is only known from the montane cloud forest between 2,900 to 3,800 m in Huánuco department, south of the Río Huallaga, Peru (C. F. Jiménez et al. 2013). We extend the species range to Junín department, based on the suggestion by C. F. Jiménez et al. (2013) and examination of specimens in the MVZ (J. L. Patton, pers. comm.). It is not known from the fossil record.

SELECTED LOCALITIES (Map 70): PERU: Huánuco, Galloganán (type locality of *Akodon josemariarguedasi* Jiménez, Pacheco, and Vivas), Hacienda Exito (FMNH 24495), Hatuncucho (C. F. Jiménez et al. 2013), Ichocán (C. F. Jiménez et al. 2013), mountains, 15 mi NE Huánuco (FMNH 23657), Palmapampa (C. F. Jiménez et al. 2013), Unchog [= Bosque Unchog], pass between Churrubamba and Had. Paty, NNW Acomayo (LSUMZ 27952); Junín, 16 km NNE (by road) Palca (MVZ 173057).

SUBSPECIES: *Akodon josemariarguedasi* is monotypic.

NATURAL HISTORY: *Akodon josemariarguedasi* is a montane forest inhabitant, co-occurring with the sigmodontines *Thomasomys kalinowskii*, *T. incanus*, *Microryzomys minutus*, *M. altissimus*, and *Oligoryzomys* sp. (C. F. Jiménez et al. 2013). This species is largely insectivorous, consuming primarily adult beetles (as *A. orophilus*, see Noblecilla and Pacheco 2012).

REMARKS: C. F. Jiménez et al. (2013) separated *A. josemariarguedasi* from *A. orophilus*, supporting the hypothesis that the latter was a complex of species (see that account). Both morphological and karyological data suggest that *josemariarguedasi* is a member of the *aerosus* group. *Akodon josemariarguedasi* has a $2n=22$ and $FN=40$ karyotype (originally referred to *A. orophilus* by Pacheco, Córdova, and Velásquez 2012:107); all autosomes are biarmed (8 metacentric and 2 submetacentric pairs), but both sexchromosomes are uniarmed, with the Y chromosome about a third of the length of the X chromosome.

C. F. Jiménez et al. (2013:234) highlighted the potential occurrence of *Akodon josemariarguedasi* in Junín department based on data in Patton and Smith (1992b) for specimens collected east of Palca. Both M. F. Smith and Patton (1993) and Coyner et al. (2013) had referred these specimens to *A. orophilus* in their respective molecular phylogenetic analyses. However, J. L. Patton examined the series of specimens in the MVZ, using the character descriptions and illustrations in C. F. Jiménez et al. (2013), and determined that they do represent *A. josemariarguedasi*, as suggested by these authors. The holotype of *josemariarguedasi* is an adult individual (MUSM 22754; C. F. Jiménez et al. 2013:Fig. 5), regrettably too old to exhibit details of occlusal morphology.

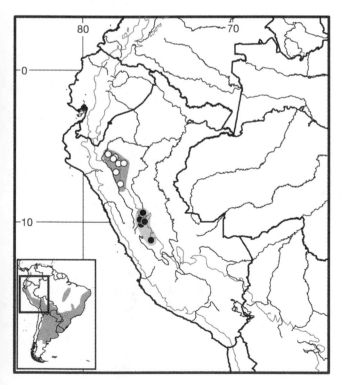

Map 70 Selected localities for *Akodon josemariarguedasi* (●) and *Akodon orophilus* (○). Contour line = 2,000 m.

Akodon juninensis Myers, Patton, and Smith, 1990
Junín Grass Mouse, Junín Akodon

SYNONYM:

Akodon juninensis Myers, Patton, and Smith, 1990:41; type locality "22 km N La Oroya (by road), Depto. Junín, Peru, 4040 m (at the junction of roads to Junín [Hwy 31] and Tarma [Hwy 201]."

DESCRIPTION: Medium-sized *Akodon* with body length averaging 149.4 mm, tail length 61.8 mm, ear length 12.2 mm; and hindfoot length 19.5 mm. Dorsal coloration brownish to pale olivaceous, heavily lined with black; hairs of ventral region with whitish or buffy tips; hairs on chin and sometimes upper throat monocolored and white; yellowish eye-ring present but inconspicuous compared with other members of *boliviensis* group; tail densely furred and sharply bicolored. Skull relatively narrow, with especially narrowed zygomatic notches; interorbital region hourglass in shape, with sides rounded anteriorly to squared or nearly squared on posterior part of orbit; both postglenoid and subsquamosal foramina smaller than in other members of *boliviensis* group; tympanic hook slender to medium sized; mesopterygoid fossa broad with anterior border usually gently rounded or square and lacking median spine; foramen ovale very small and sometimes completely enclosed by bone. Most molar features as described for other species of the *boliviensis* group, but some differences include M1 with anteromedian flexus unusually shallow and always lacking metalophule, and small M3 with shallow metaflexus. Mean condyloincisive length 20.06 mm, upper molar series length 3.74 mm.

DISTRIBUTION: *Akodon juninensis* is known only from the central Peruvian departments of Ancash, Junín, and Lima, on both eastern and western slopes of the Andes, and on the Pacific slope in Huancavelica and Ayacucho departments, at elevations above 2,700 m (Myers et al. 1990; Pacheco 2002). This species is not known from the fossil record.

SELECTED LOCALITIES (Map 71): PERU: Ancash, 25 mi S of Huaras (MVZ 137958); Ayacucho, 18 km E (by road) of Puquio (MVZ 174261); Huancavelica, Hacienda Piso, Locroja (FMNH 75561); Junín, 22 km N of La Oroya (type locality of *Akodon juninensis* Myers, Patton, and Smith), Pomacocha, Yauli Valley (MVZ 137967), near San Blas (UMMZ 120285); Lima, Zarate, 6 mi E of Pueblo San Bartolomé (MVZ 119931); Pasco, 10 mi NE of Cerro de Pasco (MVZ 119933).

SUBSPECIES: We treat *Akodon juninensis* as monotypic, although Myers et al. (1990) documented slight geographic variation in craniodental morphometric traits.

NATURAL HISTORY: *Akodon juninensis* primarily occupies dense bunchgrass in well-drained and fine-grained soils, and sparse perennial forests at elevations between 2,700 m and 3,800 m. Other sigmodontines coexisting with *juninensis* are *Abrothrix jelskii*, *Auliscomys pictus*, *Phyllotis andium*, *P. xanthopygus*, *Calomys sorellus*, *Neotomys ebriosus*, and an undescribed species of *Oligoryzomys* (see Carleton and Musser 1989).

REMARKS: In their original description, Myers et al. (1990) assigned *A. juninensis* to the *boliviensis* group. Their analyses of 26 allozyme loci suggested a close relationship with *A. boliviensis*, *A. lutescens*, and *A. subfuscus*, three other members of the *boliviensis* group in southern Peru. Both M. F. Smith and Patton (2007) and Jayat et al. (2010) recovered *A. juninensis* as a member of a clade containing *A. fumeus* and *A. kofordi* at the base of the remainder of *boliviensis* group taxa, based on mtDNA cytochrome-*b* sequences. Because of this relationship, the similarity of the karyotypes, and the small genetic distances among these species, M. F. Smith and Patton (2007) expanded the *boliviensis* group as advocated by Myers et al. (1990) to include both *A. kofordi* and *A. fumeus*.

The karyotype of *A. juninensis* consists of a 2*n*=40, FN=40. The autosomal complement consists of 18 pairs of acrocentrics, one of which is significantly larger than the others, which grade evenly in size from medium to small, and one pair of small metacentrics. The X chromosome is an acrocentric, and the morphology of the Y chromosome is unknown (Myers et al. 1990).

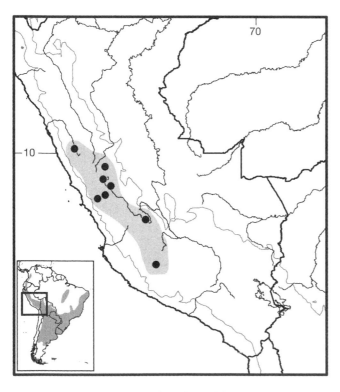

Map 71 Selected localities for *Akodon juninensis* (●). Contour line = 2,000 m.

Akodon kofordi Myers and Patton, 1989
Koford's Grass Mouse, Koford's Akodon

SYNONYM:
Akodon kofordi Myers and Patton, 1989b:14; type locality "Agualani, 9 km N by road Limbani, Depto. Puno, Peru, elevation 2840 m."

DESCRIPTION: Medium sized, with average total length 173.9 mm, tail length 77.6 mm, hindfoot length 22.5 mm, and ear length 15.3 mm. Overall color dark olivaceous brown without patches of contrasting colored fur; no trace of eye-ring; tail well haired to unaided eye, unicolored or faintly bicolored; hindfeet clothed with banded hairs. Skull with narrow and pointed rostrum, distinctively swollen frontal sinuses, and moderately developed and laterally flared zygomatic arches; zygomatic notches relatively broad; interorbital area neither remarkably constricted nor broad, with rounded sides that become increasingly squared with age; zygomatic plates relatively broad, with ventral roots directly below dorsal roots; relatively inconspicuous squamosoalisphenoid groove present; incisive foramina broad but of average length for *Akodon*, reaching anteroloph of M1; mesopterygoid fossa distinctly broad, usually lyre-shaped and with well-developed median spine present on anterior border. Dentally, similar to *A. fumeus* but paraflexus of M1 remarkably deep, extending almost to posterior limit of paracone; small section of paraflexus becomes isolated with wear, forming distinctive island. Condyloincisive length of type series averages 23.15 mm, upper molar series length 3.95 mm.

DISTRIBUTION: *Akodon kofordi* is known from the Marcapata and Limbani drainages on the eastern Andean slope in Cusco and Puno departments of southern Peru, south to Cochabamba department in central Bolivia. Recorded localities range in elevation between 1,800 and 3,700 m (Myers and Patton 1989b; Salazar-Bravo, Yensen et al. 2002; Tarifa et al. 2007). No fossils are known.

SELECTED LOCALITIES (Map 72): BOLIVIA: Cochabamba, Cocapata (Tarifa et al. 2007), 4.4 km (by road) N of Tablas Monte (Salazar-Bravo, Yensen et al. 2002). PERU: Cusco, Amacho (FMNH 75483); Puno, Agualani, 9 km N by road of Limbani (type locality of *Akodon kofordi* Myers and Patton).

SUBSPECIES: *Akodon kofordi* is monotypic.

NATURAL HISTORY: This species occupies moist bunch grass, disturbed shrubby areas, and forest (including *Alnus* cloud forest and *Polylepis* woodlands) at the contact between high-elevation elfin forest and Puna (Myers and Patton 1989b; Salazar-Bravo, Yensen et al. 2002; Tarifa et al. 2007). At the type locality in southeastern Peru, *A. kofordi* was common along rock walls of old terraces and buildings and in thick clumps of grass and shrubs in the relatively more open and disturbed areas (Myers and Patton

1989b). In Bolivia, this species was captured in open *Polylepis* woodland (with trees 2–3 m tall) and at the base of steep shale slopes, in secondary cloud forest dominated by *Alnus* (40% cover), *Cecropia* shrubs (40–90% cover), with 10% to 80% forb cover in tree-fall clearings, and with moss covering the rocks, vines, and tree trunks (Salazar-Bravo, Yensen et al. 2002). In both areas, up to 70% of specimens were captured adjacent to streams. Other *Akodon* found in sympatry, or near sympatry, with *A. kofordi* included *A. aerosus*, *A. fumeus*, *A. lutescens*, *A. mimus*, and *A. subfuscus* (Myers and Patton 1989b; Patton and Smith 1992b; Salazar-Bravo, Yensen et al. 2002; Tarifa et al. 2007).

REMARKS: Myers and Patton (1989b) allied *A. kofordi* with *A. fumeus* because of similar morphological attributes, grouping both into a *fumeus* group. MtDNA sequence analyses (M. F. Smith and Patton 2007) confirmed this sister relationship but showed that both taxa were part of their broader *boliviensis* group.

The karyotype of *A. kofordi*, as described by Myers and Patton (1989b) for specimens from Peru, is composed of a $2n=40$, $FN=40$. The autosomal complement includes 18 pairs of uniarmed elements, one of which is notably larger than the others, which grade evenly in size from medium to small, and a single pair of very small metacentric elements. The X chromosome is a medium-sized subtelocentric with very short second arms, and the Y is a very small acrocentric equal in size to one of the small biarmed autosomes. This karyotype is identical, or nearly so, to those of other members of the *boliviensis* complex, particularly *A. boliviensis*, *A. lutescens*, and *A. subfuscus* (Myers and Patton 1989b; Myers et al. 1990).

Akodon lindberghi Hershkovitz, 1990
Lindbergh's Grass Mouse, Lindbergh's Akodon

SYNONYMS:

Plectomys paludicola Borchert and Hansen, 1983:237; *nomen nudum* (see Hershkovitz 1990b:21–23).

Akodon lindberghi Hershkovitz 1990b; type locality "Matosa, a former *fazenda* now part of the Parque Nacional de Brasília, about 20 km NW of Brasília, Distrito Federal, Brazil; altitude about 1,100 m."

DESCRIPTION: Small, with head and body length 67–93 mm, tail length 49–74 mm, hindfoot length (without claw) 17–20 mm, and ear length 11–16 mm. Dorsal pelage thick, long, and lax, uniformly dark olivaceous agouti in color; sides of body lighter, but without distinct lateral line; underparts from chin to tail base not sharply defined from the sides, hairs buffy to ochraceous terminally and slaty basally. Vibrissae short, not extending beyond pinna when laid back; tail dark brown above, slightly paler beneath, thinly haired, with scales visible; hindfeet thinly covered above by grayish or buffy hairs, but brown scaly skin exposed; claws comparatively thick, moderately curved, and only limitedly covered by short ungual hairs. Eight mammae distributed in pectoral, postaxial, abdominal, and inguinal pairs. Skull delicate, with relatively short rostrum and globular braincase; interorbital region symmetrically biconcave with squared supraorbital margins, and relatively broad in relation to zygomatic arches (least interorbital breadth >38% of zygomatic breadth); nasals broad and short, not extending posteriorly beyond lacrimals; frontoparietal suture U-shaped and colinear with frontosquamosal suture; frontal never contacts squamosal dorsally; zygomatic plates very narrow, with posterior margin barely reaching level of M1 protocone; zygomatic notches shallowly incised so that anterodorsal edge of plates weakly projected in relation to dorsal zygomatic root; incisive foramina wide and long, extending posteriorly beyond protocone of M1 and angled outward at their posterior ends; auditory bullae comparatively small; mesopterygoid fossa comparatively wide, roof with wide sphenopalatine vacuities; parapterygoid fossae wide proximally, becoming wider distally. Greatest skull length 21.7–24.7 mm; upper molar series length 3.6–4.1 mm.

DISTRIBUTION: The scattered records for this species suggest a wide distribution from the central portion of the Brazilian Cerrado in Goiás state and the Distrito Federal to the Atlantic Forest in the states of Minas Gerais and Rio de Janeiro. No fossils are known.

SELECTED LOCALITIES (Map 73): BRAZIL: Distrito Federal, Matosa, Parque Nacional de Brasília (type locality

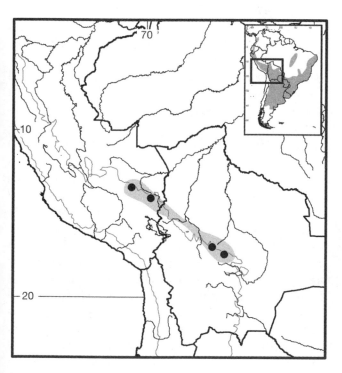

Map 72 Selected localities for *Akodon kofordi* (●). Contour line = 2,000 m.

of *Akodon lindberghi* Hershkovitz); Minas Gerais, Juiz de Fora (Queirolo and Granzinolli 2009), Parque Nacional da Serra da Canastra (Queirolo and Granzinolli 2009), Sítio Maglândia, Simão Pereira (MNRJ 48026).

SUBSPECIES: *Akodon lindberghi* is monotypic.

NATURAL HISTORY: *Akodon lindberghi* has a marked preference for grassland-dominated habitats in the Cerrado and Atlantic Forest, being captured more frequently in open areas, especially in wet grasslands or near swamps, with few found in forested patches (Borchert and Hansen 1983; Hershkovitz 1990b; Geise et al. 1996; Queirolo and Granzinolli 2009). Records from the Atlantic Forest, therefore, include localities where the forest cover has been significantly altered by human activities, suggesting that this species might be extending its original range, owing to the fragmentation of semideciduous and evergreen forests (Queirolo and Granzinolli 2009). In the Cerrado of central Brazil, local densities appear to be low in relation to those of other sympatric sigmodontines; however, at the few localities where this species has been recorded in the Atlantic Forest, it is as common as the other local rodents. Breeding activity is markedly higher during the rainy season, and the highest densities are observed during the dry season, the period of greatest production of grass seeds (Queirolo and Granzinolli 2009). De Conto and Cerqueira (2007) studied reproduction and life history in captives. A litter size varied from 1 to 4, with a modal number of 3 and mean of 2.72, and was negatively correlated with birth weight. The gestation period was 23

days, with a postpartum estrus within a short time following parturition. The sex ratio at birth was not significantly different from 1:1, nor was there significant dimorphism in neonatal birth weight. Males and females attain sexual maturity in 26 to 61 days and 30 to 56 days, respectively, with females mating successfully at 47 to 54 days of age. The mortality of neonates raised by a female in the male's absence was lower, suggesting that only the female is with the young under natural conditions (De Conto and Cerqueira 2007).

REMARKS: Borchert and Hansen (1983) referred to mice collected from what is now the type locality of *A. lindberghi* under the name *Plectomys paludicola*. Despite the lack of any voucher specimens or formal description of *P. paludicola*, Hershkovitz (1990b:21–23) linked this name, which he treated as a *nomen nudum*, to the specimens he subsequently collected and described as *A. lindberghi*.

The karyotype of *A. lindberghi* is 2n = 42, FN = 42 chromosomes (Svartman and Almeida 1994; Geise et al. 1996).

Akodon lutescens J. A. Allen, 1901
Altiplano Grass Mouse, Altiplano Akodon

SYNONYMS:

Hesperomys (*Habrothrix*) *xanthorhinus*: Thomas, 1884: 450; part; not *Abrothrix xanthorhina* Waterhouse.

Akodon lutesens J. A. Allen, 1901b:46; type locality "Tirapata, [Puno,] Peru, (about 15,000 ft)"; spelling emended to *lutescens* by J. A. Allen (1901b:46).

Akodon puer Thomas, 1902b:136; type locality "Choquecamate, 4000 m," Cochabamba, Bolivia.

Akodon andinus lutescens: Sanborn, 1949c:315; name combination.

Akodon puer lutescens: Myers, Patton, and Smith, 1990: 70; name combination.

Akodon puer puer: Myers, Patton, and Smith, 1990: 67; name combination.

Akodon lutescens lutescens: S. Anderson, 1997:422; name combination.

Akodon lutescens puer: S. Anderson, 1997:422; name combination.

DESCRIPTION: Small species with total body length averaging 139.2 mm, tail length 59.0 mm, hindfoot length 18.5 mm, and ear length 12.4 mm. Dorsal coloration pale grayish or olive brown, moderately to heavily lined with black; ventral pelage grayer than dorsum, with few white hairs on chin; moderately conspicuous eye-ring present; tail sharply bicolored, with narrow dark dorsal stripe. Skull with shallow and narrow zygomatic notches, weakly flared zygomatic arches, relatively broad interorbital region, and small braincase; zygomatic plates slender, with anterior borders that slope gently backward from bottom to top; tympanic hook particularly narrow and delicate (illustrated in Myers et al. 1990:Fig. 16); posterior ascending process of alisphe-

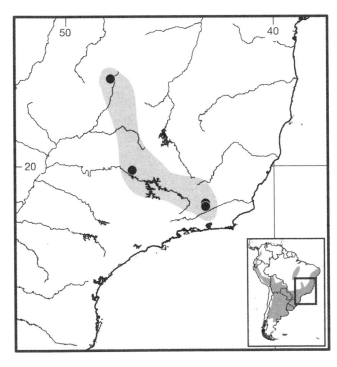

Map 73 Selected localities for *Akodon lindberghi* (●).

noid not reaching level of squamosoalisphenoid groove; incisive foramina short in comparison with most other members of *boliviensis* group, but still extend to anterior part of M1; mesopterygoid fossa very narrow, with strongly squared anterior margin, usually with median spine; mandible very slender. Cheek teeth resemble those of *A. boliviensis*, but molar series shorter and M3 generally oval in outline, never figure-8 shaped, lacking deep hypoflexus and deep metaflexus. Condyloincisive length averages 20.7 mm, upper molar series length 3.6 mm.

DISTRIBUTION: *Akodon lutescens* inhabits the Altiplano and eastern Andean slopes from Puno department, Peru to Chuquisaca department in south-central Bolivia, at elevations between 1,130 m to 4,560 m (Myers and Patton 1989a; Myers et al. 1990; Patton and Smith 1992b; S. Anderson 1997; Salazar-Bravo, Yensen et al. 2002; Tarifa et al. 2007). Argentinean records of this species (see Pardiñas, D'Elía et al. 2006; Jayat, Ortiz, and Miotti 2008; P. E. Ortiz et al. 2011a) actually apply to *A. caenosus*. M. M. Díaz and Barquez (2007) recorded both *lutescens* (as *Akodon lutescens puer*) and *Akodon caenosus* from Yungas environments in Jujuy province, but the broadly comparative study by Jayat et al. (2010) failed to confirm any records of *A. lutescens* from Argentina. No fossils are known.

SELECTED LOCALITIES (Map 74): BOLIVIA: Cochabamba, Ayopaya (Myers et al. 1990), 15 mi ENE of Tiraque (MVZ 119923); La Paz, 12 km by road SW of Jesús de Machaca, Río Desaguadero (S. Anderson 1997), Pelechuco (S. Anderson 1997), Río Aceromarca (S. Anderson 1997); Oruro, 1 km W of Huancaroma (S. Anderson 1997); Santa Cruz, above Cocapata on old Cochabamba to Santa Cruz road (S. Anderson 1997), 1 km N and 8 km W of Comarapa (Myers et al. 1990). PERU: Cusco, 20 km N (by road) of Paucartambo, km 100 (MVZ 173788); Puno, 11 km W and 12 km S of Ananea (MVZ 173073), 36 km SE by road of Macusani (Myers et al. 1990), 6 km S (by road) of Pucara (MVZ 173398), 5 km W of Puno (Myers et al. 1990), 12 km S of Santa Rosa [de Ayaviri] (Myers et al. 1990).

SUBSPECIES: In addition to *caenosus* Thomas, now regarded as a valid species, both Myers et al. (1990) and S. Anderson (1997) recognized the subspecies *lutescens* J. A. Allen from southern Peru and extreme northern Bolivia and *puer* from west-central Bolivia.

NATURAL HISTORY: In Peru, *Akodon lutescens* has been found exclusively in high Puna grasslands, almost always associated with the coarse and spiny *Festuca* bunchgrass that dominates well-drained and hard soils (Myers et al. 1990). In Bolivia, records for this species also come primarily from high-elevation Puna grasslands and eastern Andean slopes, but the species is also known from Yungas

cloud forest (S. Anderson 1997; Myers et al. 1990). A single female collected in March had four embryos; all others obtained between March and September were not pregnant (S. Anderson 1997). *Akodon lutescens* may be sympatric or contiguously allopatric with several other *Akodon* species, including *Akodon aerosus*, *A. albiventer*, *A. boliviensis*, *A. fumeus*, *A. kofordi*, *A. mimus*, *A. siberiae*, *A. subfuscus*, and *A. varius* (Myers et al. 1990; Patton and Smith 1992b; S. Anderson 1997; Salazar-Bravo, Yensen et al. 2002; Tarifa et al. 2007).

REMARKS: J. A. Allen (1901b:46) initially spelled the species epithet *lutesens*, but amended the name to *lutescens* later on the same page, a spelling that has been in subsequent use in virtually all publications, from Trouessart (1904:435) to Musser and Carleton (2005:1096). Myers et al. (1990:71–73) encapsulated the confusing history of this species relative to that of other named *Akodon* from the Altiplano of Peru and Bolivia, although they erred in ignoring the priority of J. A. Allen's (1901b) *A. lutescens* over Thomas's (1902b) *A. puer* (see S. Anderson 1997:421–422). Osgood (1944), in his description of *Akodon boliviensis subfuscus*, noted that *lutescens* might be a synonym of *boliviensis*. In the same paper, Osgood described *Akodon andinus polius*, and later Sanborn (1949c) stated that Osgood believed *polius* might be a synonym of *lutescens*, thus associating Allen's *lutescens* with *Akodon* (now *Abrothrix*) *andinus*, a decision followed by Cabrera (1961). Both Myers et al. (1990) and Hershkovitz (1990b) recognized

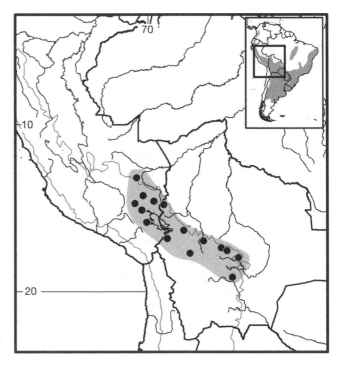

Map 74 Selected localities for *Akodon lutescens* (●). Contour line = 2,000 m.

that the true affinities of Allen's *lutescens* were not with *andinus* but with the *Akodon boliviensis* group of mice. Furthermore, Myers et al. (1990), using allozyme as well as morphological comparisons, documented broad sympatry without interbreeding between *lutescens* and both *boliviensis* and *subfuscus*, thus falsifying Osgood's (1944) hypothesis and validating the specific status of *A. lutescens*.

The karyotype of *A. lutescens* is $2n = 40$, $FN = 40$ (Myers et al. 1990). As with other species of the *boliviensis* group, the autosomal complement consists of 18 pairs of uniarmed elements, with one pair notably larger and the others grading evenly from medium sized to small, and with one pair of small metacentrics about one-half the size of the smallest acrocentric autosome. The X chromosome is a medium-sized subtelocentric and the Y is a small acrocentric, considerably smaller than the smallest uniarmed autosome.

Akodon mimus (Thomas, 1901)

Thespian Grass Mouse, Hocicudo-like Akodon

SYNONYMS:

Oxymycterus mimus Thomas, 1901b:183; type locality "Limbane [= Limbani], Dept. Puno. Alt. 2600 m," Peru.

Microxus mimus: Thomas, 1909:237; name combination, type species of *Microxus* Thomas.

Akodon (Abrothrix) mimus: Sanborn, 1950:15; name combination.

Akodon (Microxus) mimus: Cabrera, 1961:458; name combination.

[*Abrothrix*] *mimus*: Hershkovitz, 1966:86; name combination.

Akodon mimus: Patton and Smith, 1992b:91; name combination.

DESCRIPTION: Moderately large with average total length 190.9 mm, tail length 94.1 mm, hindfoot length (with claw) 24 mm, and ear length 18.3 mm. Overall color uniformly gray-brown, with dorsum and venter only weakly contrasting; no eye-ring; whitish spot on chin present in many specimens. Tail proportionally long, equal to head and body length, and unicolored dark. Skull notable for elongated and narrow rostrum, very shallow and narrow zygomatic notches, broad interorbital region, nonflared zygomatic arches, swollen braincase, and, especially, very narrow zygomatic plates that slant posteriorly from bottom to top. Mesopterygoid fossa conspicuously broadened, with rounded anterior margin. Condyloincisive length averages 24.51 mm, the upper molar series 4.50 mm.

DISTRIBUTION: *Akodon mimus* inhabits upper elfin forests on the eastern Andean slopes from Puno department in southern Peru to Santa Cruz department in Bolivia, at elevations between 2,000 and 3,700 m (Patton and Smith 1992b; S. Anderson 1997). No fossils are known.

SELECTED LOCALITIES (Map 75): BOLIVIA: Cochabamba, 25 km W of Comarapa, Siberia (UMMZ 155962), Corani (S. Anderson 1997), 17 km E of Totora, Tinkusiri (S. Anderson 1997), Yungas del Palmar (S. Anderson 1997); La Paz, Cocapunco (S. Anderson 1997), Saynani (S. Anderson 1997), 10 km by road N of Sorata, Moyabaya (S. Anderson 1997), 30 km by road N of Zongo (S. Anderson 1997). PERU: Puno, 4 mi N of Limbani (MVZ 116109), 14 km W of Yanahuaya (MVZ 171752).

SUBSPECIES: *Akodon mimus* is monotypic.

NATURAL HISTORY: *Akodon mimus* is a dweller of the more humid parts of the Yungas, from cloud forests to the transition between these and the high-elevation grasslands. Specimens from near Yanahuaya in southern Peru were obtained in forest dominated by epiphyte-laden trees, cecropias, bamboos, and tree ferns. In the Siberia Cloud Forest spanning the boundary of Cochabamba and Santa Cruz departments, Bolivia, *A. mimus* was trapped on the mossy, fern-laden, and humid forest floor (Hinojosa et al. 1987). *Akodon mimus* was the most common species of rodent in the Yungas forests of La Paz department, Bolivia (J. Vargas et al. 2007). There, it was trapped in humid primary and secondary forest. A single individual was obtained in the moist, high-elevation grassland bordering forest communities, near a stream. Across its latitudinal and elevational range, *A. mimus* may co-occur with several other *Akodon* species, including *A. aerosus*, *A. boliviensis*, *A. fumeus*, *A. kofordi*, *A. lutescens*, *A. siberiae*, and *A. subfuscus*. S. Anderson (1997) recorded 38 nonpregnant females through the autumn to the end of winter (May, June, July, and September), with pregnant females obtained only in October (with a mean number of embryos = 2.0) and one lactating female in June.

REMARKS: *Akodon mimus* is the type species of *Microxus*, a nominal taxon that has, at times, been treated either as a subgenus of *Akodon* (Cabrera 1961) or as a valid akodontine genus (Gyldenstolpe 1932; Ellerman 1941; Reig 1987). Allozymes as well as both nuclear and mtDNA sequences support the inclusion of *mimus* in the genus *Akodon* (Patton et al. 1989; M. F. Smith and Patton 1991, 1993, 2007; D'Elía 2003a). However, the skull and dental morphology of *mimus* are very distinctive compared with that of other *Akodon*, and especially its narrowed zygomatic plates that slant posteriorly from bottom to top. As Thomas (1920h:241) noted, "the long head, especially the long muzzle, and the small eyes, give the *Microxus* quite a different aspect to that of the *Akodon* with its blunt snout and normal eyes, and I now feel no hesitation in considering them as belonging to different genera." This differentiation is partially supported by several phylogenetic reconstructions based on DNA sequences, which place *mimus* as sister to *Deltamys* (M. F. Smith and Patton 2007).

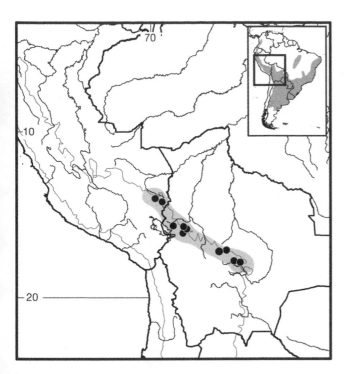

Map 75   Selected localities for *Akodon mimus* (●). Contour line = 2,000 m.

Akodon mollis Thomas, 1894
Soft-furred Grass Mouse, Soft-furred Akodont

SYNONYMS:

Hesperomys olivaceus: Thomas, 1882:110; part; not *Mus olivaceus* Waterhouse.

Acodon mollis Thomas, 1894b:363; type locality "Tumbez, [Tumbes,] N Peru."

[*Akodon (Akodon)*] *mollis*: E.-L. Trouessart, 1897:538; name combination.

Akodon mollis altorum Thomas, 1913b:404; type locality "Cañar. Alt. 2600 m," Cañar, Ecuador.

Akodon mollis mollis: Gyldenstolpe, 1932a:106; name combination.

Akodon mollis fulvescens Hershkovitz, 1940b:2; type locality "The Río Mira Valley, parish of La Carolina, Ibarra, Imbabura Province, Ecuador; subtropical forest of the western slope of the western Andes; altitude about 900 meters."

DESCRIPTION: Medium-sized with head and body length 81–113 mm, tail length 55–92 mm, hindfoot length (without claw) 18–26 mm, ear length 8–18 mm, greatest skull length 23.71–29.37 mm, and upper molar series length 3.5–4.7 mm. Dorsal pelage soft and variable in color, from dark olive to yellowish brown, contrasting with paler, usually grayish or dull buff under parts; lateral line, if present, narrow and weakly developed; dorsal surfaces of both forefeet and hindfeet pale brown to gray, covered with basally dark banded hairs; ungual tufts grayish or light brown,

with distal margin of most hairs ending at or just surpassing claws. Tail short, moderately to well furred, distinctly to indistinctly bicolored, with darker hairs above. Vibrissae short, not extending beyond pinna when laid back. Eight mammae in pectoral, postaxial, abdominal, and inguinal pairs. Skull compact with fairly robust rostrum; interorbital region broader than muzzle, with narrowest point centrally located and with smooth supraorbital edges; nasals narrow and extend posteriorly beyond suture between maxillary, frontal, and lacrimal; parietal and occipital bones in contact, isolating narrow interparietal from squamosal; alisphenoid strut present; zygomatic plates broad with slightly, posteriorly inclined anterior margin and posterior border reaching protocones of M1s; zygomatic notches moderately to well developed; maxillary and mandibular diastemata long, as are incisive foramina, which extend posteriorly past anterior margin of M1s, reaching to level protocones in some specimens; and carotid circulation pattern 1 (*sensu* Voss 1988).

DISTRIBUTION: *Akodon mollis* is distributed from northwestern Peru to northern Ecuador (and likely occurs in southern Colombia) and across an extensive elevation range from near sea level to above 4,900 m. An isolated population occurs on the coast of southern Peru in Tacna department. In central and northern Peru, this species occurs along the Pacific coast and adjacent western slopes of the Andes, west of the Río Marañón (see C. F. Jiménez et al. 2013), but its distribution farther north becomes restricted to the higher elevations on both the western and eastern slopes of the Andes and throughout the inter-Andean valleys to near the snow line. Fossil *Akodon* from the Late Pleistocene and Holocene of Ecuador (Fejfar et al. 1993) probably belong to this species.

SELECTED LOCALITIES (Map 76): ECUADOR: Azuay, Molleturo, Cuenca (AMNH 61990); Carchi, Páramo del Artesón, Comuna La Esperanza (EPN 9641); Cotopaxi, Hacienda Sr. Cepeda, 10 km NW of Chugchilán (QCAZ 11281); Imbabura, sector "Los Cedros," Manduroyacu, Reserva Ecológica Cotacachi-Cayapas (EPN 7583); Loja, Laguna Negra, Bosque Protector Colomobo-Yacurí (QCAZ 11281), Puyango, Alamor (AMNH 47358); Morona-Santiago, Cerro Bosco, forest surrounding the Pacifictel Transmitting/Receiving Antenna (QCAZ 4812); Napo, Oyacachi, El Chaco County (AMNH 63842). PERU: Ancash, Huari, Yanacocha (MUSM 24072), 1 km N and 12 km E of Pariacoto (MVZ 135676), 4 km S and 8 km E of Recuay (MVZ 137975); Cajamarca, Contumaza, Bosque Cachil, entre Cascas y Contumaza (VPT 1673), Cutervo, San Andrés de Cutervo (UMMZ 176689), Hacienda Taulis (MVZ 121157); Huánuco, Chinchuragra, Punchao (Jiménez et al. 2013); La Libertad, 10 mi WNW of Santiago de Chuco (137980), 5 km NE of Pacasmayo (MVZ 137679), 5 mi SW of Otuzco (MVZ 137961); Piura, Cerro

Chinguela, Huancabamba (Jiménez et al. 2013), Cerro Chinguela, ca. 5 km NE of Zapalache (LSUMZ 27149), Morropón, Chalaco (MUSM 21691), 2 km W of Porculla Pass (MVZ 137969); Tacna, Morro de Sama, 65 km W of Tacna (MVZ 143697).

SUBSPECIES: Hershkovitz (1940b) reviewed geographic variation in morphology and delineated three subspecies based primarily on differences in dorsal color and minor craniodental traits: *A. m. mollis* from northwestern Peru, *A. m. altorum* from the intermontane valleys in Ecuador, and *A. m. fulvescens* from the eastern Andean slopes in Ecuador.

NATURAL HISTORY: *Akodon mollis* occupies a wide variety of habitats across its large latitudinal and elevational range, including dry coastal areas on the Pacific coast, cloud forest and evergreen to semideciduous forests on upper slopes on both sides of the Andes, and grasslands at high elevations (D. F. Alvarado-Serrano 2005, pers. comm.). It is, however, most commonly found in grassland and shrubby habitats at elevations above 2,000 m (Alvarado-Serrano 2005; Voss 2003). As noted by Barnett (1999), *A. mollis* may be the most common mouse in high-elevation grasslands of Ecuador, often comprising >60% of all captures; it is one of the few sigmodontine species in the northern Andes that can be regularly found in highly disturbed environments such as introduced *Pinus* or *Eucalyptus* plantations. Little is known about the ecology, but based on an analysis of specimens from Parque Nacional El Cajas, *A. mollis* is a predominantly nocturnal generalist, with at least two reproductive cycles per year and two to four embryos per pregnancy (Barnett 1999).

REMARKS: Originally considered to be a northern representative of *Hesperomys* (now *Abrothrix*) *olivaceus* (Thomas 1882:110, 1884:456), this species is currently grouped with four other *Akodon* species (*A. aerosus*, *A. torques*, *A. orophilus*, and *A. affinis*) in a weakly supported clade of Andean-slope *Akodon* based on mtDNA sequences (M. F. Smith and Patton 1993, 1999, 2007). With the exception of *A. mollis*, which was described from the northwestern coast of Peru and occurs at low to high elevations, all other species in this clade are restricted to montane habitats in the central and northern Andes. To date, however, all samples of *A. mollis* included in the published molecular analyses are from high-elevation populations and are in a complex set of paraphyletic relationships with samples of other species in the clade (M. F. Smith and Patton 1993; Alvarado-Serrano 2005, unpubl. data). Therefore, until samples from coastal areas at or near the type locality of *mollis* and additional data that can resolve phyletic relationships become available, the proposition that *A. mollis*, as construed herein and by earlier authors, is a species complex with currently undefined boundaries and an unknown number of included taxa remains a hypothesis

to be tested. Certainly, the conspecific status of each described subspecies needs adequate assessment (Voss 2003).

Myers and Patton (1989b) suggested that *A. mollis* was allied to *A. fumeus*, *A. kofordi*, also Andean species of *Akodon* based on commonality in some morphological traits. However, the phyletic propinquity of both *A. fumeus* and *A. kofordi* with species in the broader *boliviensis* group (M. F. Smith and Patton 2007) falsifies Myers and Patton's hypothesis. Alvarado-Serrano et al. (2013) examined the extent of phenotypic differentiation across the species' range in Peru using geometric morphometric and ecomorphological analyses of skull shape. This ecologically widespread species exhibited little phenotypic specialization across their range, with the exception of samples from the high-elevation puna. From molecular genetic analyses, geographic isolation alone did not explain this localized phenotype.

Karyotypes with $2n = 22$ have been recorded from Ancash department, Peru (Patton and Smith 1992b), and central Ecuador (Lobato et al. 1982), and $2n = 36-38$ for specimens from Piura department, Peru (Patton and Smith 1992b). The Ecuadorean specimens had a fundamental number of 43–44 (Lobato et al. 1982). This geographic diversity of chromosomal complements underscores the earlier comment that *A. mollis* likely represents a complex of separate species.

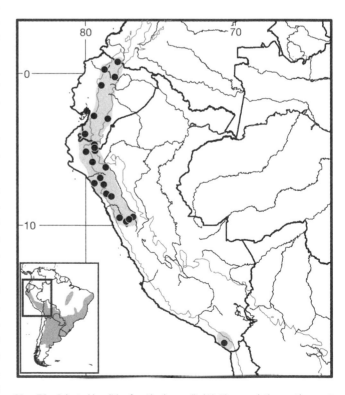

Map 76 Selected localities for *Akodon mollis* (●). The population on the coast of SW Peru is presumed to be disjunct from the main distribution. Contour line = 2,000 m.

Akodon montensis Thomas, 1913

Montane Grass Mouse, Montane Akodont

SYNONYMS:

Akodon arviculoides montensis Thomas, 1913b:405; type locality "Paraguay. Type from Sapucay [= Sapucaí], [Paraguarí]."

Akodon cursor montensis: Massoia, 1980b:22; name combination.

Akodon montensis: Geise, Canavez, and Seuánez, 1998: 158; first use of current name combination.

DESCRIPTION: Medium sized similar to but somewhat smaller than *A. cursor*. Males usually larger than females, with head and body length averaging 115.21 ± 8.13 mm (N = 46) versus 108.83 ± 11.10 mm (N = 38), tail length 93.01 ± 10.13 mm (N = 48) versus 88.41 ± 8.44 mm (N = 35), and mass 43.31 ± 11.14 mm (N = 48) versus 36.43 ± 12.83 mm (N = 35; Geise et al. 2005). Tail in both sexes always shorter than head and body length, approximately 81%. Greatest skull length averages 27.15 ± 1.06 mm (N = 52) mm in males and 25.71 ± 1.21 mm (N = 39) in females; the upper molar series length averages 4.23 ± 0.18 mm (N = 91) for both sexes combined. Very similar to *A. cursor* in external morphology, but two species differ in craniodental morphometry (Christoff 1997; Geise et al. 2005), in distribution of allozyme variants (Rieger et al.1995), are reciprocally monophyletic in mtDNA sequences (Geise et al. 2001; D'Elía et al. 2003), and have distinct karyotypes (Fagundes and Yonenaga-Yassuda 1998; Geise et al. 1998). *Akodon montensis* also lacks a gall bladder, which is present in *A. cursor* (Geise, Weksler, and Bonvicino 2004).

DISTRIBUTION: *Akodon montensis* is widespread from eastern Paraguay (Myers 1982; Gamarra de Fox and Martin 1996), northeastern Argentinain gallery forests along rivers or wetlands in the provinces of Formosa, Chaco, Corrientes, and Misiones (Massoia and Fornes 1962; Pardiñas, D'Elía, and Cirignoli 2003), and along the southern coast of Brazil from Rio Grande do Sul state in the south to Rio de Janeiro and eastern of Minas Gerais states in the north (Geise et al. 2001, 2005). Reig (1987) reported *A. montensis* from the Middle Pleistocene (Ensenadan) of southeastern Buenos Aires province, which is noteworthy because this locality is extralimital to the present range of the species (Pardiñas, D'Elía, and Cirignoli 2003).

SELECTED LOCALITIES (Map 77): ARGENTINA: Chaco, Estación Experimental Colonia Benítez, INTA (Massoia and Fornes 1962); Corrientes, El Sombrero (J. R. Contreras et al. 2003), Santa Tecla, Rte 12, km 1287 (Pardiñas, D'Elía, and Cirignoli 2003); Formosa, Estación Experimental El Colorado, INTA (Massoia 1971b), Estancia Guaycolec (Massoia, Heinonen Fortabat, and Diéguez

1997); Misiones, Arroyo Itaembé Miní, bridge over Rte 12 (Massoia 1983), Concepción de la Sierra (Pardiñas, D'Elía, and Cirignoli 2003), Reserva Privada UNLP "Valle del Arroyo Cuña Pirú" (Pardiñas, D'Elía, and Cirignoli 2003). BRAZIL: Mato Grosso do Sul, Fazenda Maringá, 54 km W of Dourados (MVZ 197459), Fazenda Princesinha (Cáceres et al. 2007); Minas Gerais, Estação Biológica de Caratinga, Caratinga (MNRJ 31450), Jambreiro, Nova Lima (MNRJ [LG 207]); Paraná, Mananciais da Serra, Piraquara (MNRJ [JAO 965]); Rio Grande do Sul, Rota do Sol, Tainhas–Terra de Areia (Valdez and D'Elía 2013); Rio de Janeiro, Sítio Xitaca, Debossan, Nova Friburgo (MNRJ 35925); São Paulo, Estação Ecológica do Bananal, Bananal (MZUSP [EEB 563]), Fazenda Passes, Santo Antônio do Aracanguá (MNRJ [EDH 61]). PARAGUAY: Amambay, Parque Nacional Cerro Corá (Gamarra de Fox and Martin 1996); Paraguarí, Sapucaí (type locality of *Akodon arviculoides montensis* Thomas); San Pedro, Ganadera Jejui (UMMZ 174897).

SUBSPECIES: *Akodon montensis* is monotypic.

NATURAL HISTORY: Less is known about the natural history of *A. montensis* in comparison to its close relative *A. cursor*. The species apparently prefers closed and dense forest understory and more pristine habitats at elevations above 800 m in Rio de Janeiro state (Geise et al. 2001, 2005; Geise, Pereira et al. 2004). Where *A. montensis* and *A. cursor* are sympatric, the two species are elevationally segregated, with *A. montensis* at higher elevations (Geise et al. 2001). In Santa Catarina state, where *A. montensis* was also among the most frequently trapped species (Cherem and Perez 1996), it is found at sea level in the mangrove–coastal forest interface (as on Ilha Santa Catarina; Graipel et al. 2001). In Paraguay, individuals exhibited a preference for either forest with little understory or dense vegetative cover on or near the ground (Goodin et al. 2009). And in Misiones, Argentina, this species was the dominant member of the small mammal assemblage in both secondary forest and other anthropogenic disturbed habitats (Cirignoli et al. 2011). At most trapping sites in Brazil, *A. montensis* was also among the most frequently trapped species, typically comprising more than 70% of all captures (Davis 1945b; Cherem and Perez 1996). Based on long-term studies at Teresopolis and other sites in Rio de Janeiro state, home ranges were large, with individual movements between trapping sessions between from 100 and 370 m (Davis 1945b). Püttker, Bueno et al. (2013) described survival rates, population change rates, and capture probability for populations in different forest cover landscapes. Fontes et al. (2007) reported large and sexually dimorphic home ranges, with the space used by males larger ($1,460 \pm 934$ m^2) than that of females ($1,092 \pm 1007$ m^2). Reproduction is seasonal, with a peak during the rainy season (Couto and Talamoni 2005), with litter sizes ranging between 3 and 5 (Graipel et al. 2001). Graipel et al. (2003) found these animals to be most active at the beginning and end of each night, although they recorded 23% of captures during daylight hours. Horn (2005) provided evidence that *A. montensis* functions as a seed disperser for at least two important native plants, *Ficus organensis* and *Piper* cf. *solmisianum*. *Akodon montensis* harbors a hantavirus (Goodin et al. 2009).

REMARKS: The validity of specific status for *A. montensis* and its separation from *A. cursor* has been the subject of interest for many years. Cytogenetic studies showed two distinct sets of diploid complements, 24–25 and 14–15, making clear the existence of two synmorphic species (see Fagundes and Yonenaga-Yassuda 1998; Geise et al. 1998). Because the $2n=14–15$ karyotype was found at Lagoa Santa, type locality of *A. cursor* Winge, these diploid complements were assigned to that species; the $2n=24–25$ was then assigned to *A. montensis*. DNA phylogenetic analyses clearly show that both species are valid because both sets of karyomorphs form reciprocally monophyletic groups (Geise et al. 2001). In a more extensive DNA phylogenetic analysis, G. D'Elía (unpubl. data) found one topotype of *montensis* part of the $2n=24–25$ clade identified by Geise et al. (2001), supporting the association of *montensis* Thomas with these karyotypes (Pardiñas, D'Elía, and Cirignoli 2003). Cestari and Imada (1968) incorrectly used the name *A. arviculoides cursor* for specimens with $2n=24$ and collected in São Paulo state, Brazil. Phylogenetic reconstruction based on mtDNA cytochrome-*b* sequences

documented that *A. montensis* is part of a well-supported clade also containing *A. cursor*, *A. mystax*, *A. paranaensis*, *A. reigi*, and an undescribed species sometimes referred to as "*A.* aff. *cursor*" (M. F. Smith and Patton 2007). Valdez and D'Elía (2013) detailed phylogeographic structure in relation to Pleistocene refugia hypotheses for the Atlantic Forest.

Akodon mystax Hershkovitz, 1998
Caparaó Grass Mouse, Caparaó Akodont
SYNONYM:

Akodon mystax Hershkovitz 1998; type locality "Arrozal, Pico da Bandeira, western slope Mt. Caparaó, Minas Gerais, Brazil, elevation 2300 m."

DESCRIPTION: Small, very similar externally to *A. lindberghi*, with head and body length 66–101 mm, tail length 48–76 mm, hindfoot length (without claw) 17–20 mm, and ear length 11–18 mm. Dorsal fur soft and long, and of overall grayish-brown tone, slightly grayer than *A. lindberghi*; sides somewhat paler than dorsum with phaeomelanic bands generally very pale yellow color; ventral pelage weakly delimited from dorsum, but also grayish in tone due to hair with dark gray basal bands and terminal whitish bands; small cluster of short melanic bands above rhinarium visible in some specimens, providing what looks like dark rostral patch. Tail well furred and bicolored, with blackish hairs dorsally and whitish hairs ventrally; short hairs with dark basal and silvery apical bands, resulting in overall light gray tone, cover dorsal surfaces of both forefeet and hindfeet; well-developed ungual tufts present, with most hairs extending past tips of claws. Vibrissae short, not extending beyond pinna when laid back. Eight mammae distributed in pectoral, postaxial, abdominal, and inguinal pairs. Skull appears delicate with relatively short rostrum and globular braincase; interorbital region symmetrically biconcave with rounded supraorbital edges and relatively broad in relation to zygomatic arch breadth (least interorbital breadth >38% of zygomatic breadth); nasal bones broad and short, not extending posteriorly beyond lacrimals; incisive foramina wide and long, as in *A. lindberghi*, extending posteriorly past M1 protocone and expanded laterally at posterior end; frontoparietal suture wide, U-shape and not colinear with frontosquamosal suture, producing narrow area of dorsal contact between squamosal and frontal; zygomatic plates narrow with posterior margin barely reaching level of M1 protocone, and anterodorsal edge weakly projected in relation to dorsal zygomatic root; zygomatic notches shallowly incised; masseteric crest evident at base of each zygomatic plate. Greatest skull length 22.62–25.58 mm, upper molar series length 3.63–4.1 mm.

DISTRIBUTION: *Akodon mystax* is known only from the highest elevation zone (1,800–2,900 m) in the vicinity

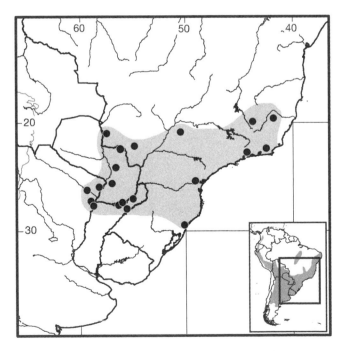

Map 77 Selected localities for *Akodon montensis* (●). Contour line = 2,000 m.

of Pico da Bandeira, on the border between Minas Gerais and Espírito Santo states in southern Brazil (Hershkovitz 1998). This species is not known in the fossil record.

SELECTED LOCALITIES (Map 78): BRAZIL: Minas Gerais, Pico da Bandeira, Parque Nacional do Caparaó (type locality of *Akodon mystax* Hershkovitz).

SUBSPECIES: *Akodon mystax* is monotypic.

NATURAL HISTORY: This species occurs in the cool-humid, high-elevation grasslands called *campos de altitude* found above 1,800 m on Mt. Caparaó (Bonvicino et al. 1997; P. R. Gonçalves et al. 2007). The highest densities have been found in bamboo patches (*Chusquea* spp.) and bunchgrasses (*Cortadeira*) that grow on shallow hygrophilic soils and among rocky outcrops. No *A. mystax* were captured in the gallery montane forest at 2,500 m. One female contained four embryos. Hershkovitz (1998) stated that the blackish rostral patch was more frequent in males than in females.

REMARKS: P. R. Gonçalves et al. (2007) clarified the identification of a specimen treated as *A. mystax* by Geise et al. (2001) and by Pardiñas, D'Elía, and Cirignoli (2003), both of whom had suggested a possible sister relationship between *A. mystax* and *A. lindberghi*. M. F. Smith and Patton (2007) placed *A. mystax* in their *cursor* group, as the well-supported sister of *A. paranaensis*. Two karyotypes, one with 2n=42, FN=42 (P. R. Gonçalves et al. 2007), and a second with 2n=44 (Bonvicino et al. 1997), have been reported for topotypes.

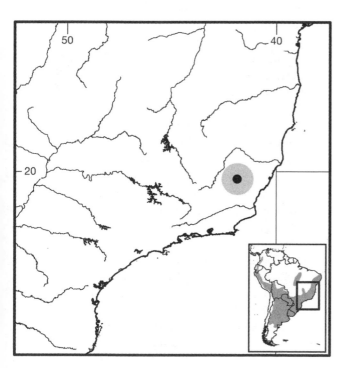

Map 78 Single known locality for *Akodon mystax* (●).

Akodon orophilus Osgood, 1913
El Dorado Grass Mouse, Utcubamba Akodont

SYNONYMS:

Akodon mollis orophilus Osgood, 1913:98; type locality "six miles west of Leimabamba [= Leymebamba], [Amazonas,] Peru (in mountains near headwaters of Utcubamba River)."

Akodon mollis orientalis Osgood, 1913:99; type locality "Poco Tambo, between Chachapoyas and Rioja, [San Martín,] Peru. Altitude about 6000 ft."

Microxus orophilus: Thomas, 1926e:615; name combination.

Microxus orophilus orientalis: Thomas, 1926e:616; name combination.

[*Akodon*] *orophilus*: Thomas, 1927a:370; name combination.

[*Akodon orophilus*] *orientalis*: Osgood, 1943b:197; name combination.

Akodon orophilus orophilus: Cabrera, 1961:446; name combination.

DESCRIPTION: Medium-sized species (measurements of holotype: total length 192 mm, head and body length 107 mm, tail length 85 mm, hindfoot length 23 mm, greatest length of skull 26.8 mm, basal length 24.5 mm, zygomatic breadth 13.3 mm, interorbital breadth 4.9 mm, nasals 10.8×2.8 mm, incisive foramina 5.8 mm, diastema 6.8 mm, and upper molar series 4.4 mm). Similar to *A. mollis altorum* from Ecuador, but averages slightly larger and more fulvous in color; two species, however, exhibit marked cranial differences. Skull in *A. orophilus* with laterally compressed and attenuate rostrum; nasals slender and elongate; zygomatic plates narrow, with anterior edge convex, slanting posteriorly from base; braincase broader and more smoothly rounded than that of *A. mollis* or *A. m. altorum*, without temporal ridges.

Akodon o. orientalis slightly larger, with longer tail and much darker overall coloration than *A. o. orophilus*. External measurements of holotype include total length 201 mm, head and body length 116 mm, tail length 85 mm, and hindfoot length 24 mm. Craniodental measurements include zygomatic breadth 13.8 mm, breadth of braincase 13 mm, interorbital breadth 5.5 mm, nasals 10.6×3.4 mm, palatine foramina 5.7 mm, diastema 7 mm, and upper molar 4.6 mm. Upper parts deep mummy brown in general appearance; hairs annulated, with dark umber and tipped blackish; under parts heavily washed with tawny russet; tail and feet entirely blackish. Skull of *orientalis* with same general form of slender rostrum and narrow, receding zygomatic plates as in *orophilus* but braincase larger (Osgood 1913).

Akodon orophilus distinguishable from other members of *aerosus* group by uniform olive green general coloration; dorsal and ventral pelage not strongly contrasting; contours of posterior edge of nasals smooth, tapering, and

V-shaped; lacrimals small; zygomatic plates slanted, with anterior edge sigmoid or slightly convex; antorbital bridge moderately wide; zygomatic notches wide and intermediate in depth; incisive foramina penetrate between first upper molars, but comparatively short; diastema intermediate in length; and jaw moderately robust, with delicate capsular projection (C. F. Jiménez et al. 2013).

DISTRIBUTION: This species occurs in wet elfin forests on eastern Andean slopes east of the Río Marañón in Amazonas and northern San Martín departments of northern Peru, between 1,900 to 2,860 m (C. F. Jiménez et al. 2013). Specimens that Patton and Smith (1992b) assigned to this species from Cajamarca department are *Akodon mollis*; those originally referred to *orophilus* from Huánuco and Junín departments (see M. F. Smith and Patton 1993, 1999, 2007) are *Akodon josemariarguedasi* (see that account and C. F. Jiménez et al. 2013). Osgood (1943b:197) remarked that in northern Peru, *A. mollis altorum* and *A. orophilus* were separated only by the canyon of the Río Marañón (see also C. F. Jiménez et al. 2013:238), and, contrary to Thomas's supposition (1920h), these two taxa are not "respectively lowland and highland" forms, but forms of the western, central, and eastern Cordilleras. We treat the reported occurrences of *orophilus* in Ecuador (Barnett 1999; P. Moreno and Albuja 2005) as belonging to *Akodon mollis*, following C. F. Jiménez et al. (2013:240).

SELECTED LOCALITIES (Map 70): PERU: Amazonas, Cordillera Colán, E of La Peca (LSUMZ 21859), Leimabamba, ca. 20 km W by road (FMNH 129234), 6 road km SW of Lake [= Laguna] Pomacochas (LSUMZ 19236), 5 km N and 5 km E of Pomacocha [= Florida] (MVZ 135672), Tambo Ventija [= Ventilla], near Molinopampa (Osgood 1914b); San Martín, Poco Tambo [= Puca Tambo], 50 mi E of Chachapoyas (type locality of *Akodon mollis orientalis* Osgood), Puerto del Monte, ca. 30 km NE of Los Alisos (Patton and Smith 1992b).

SUBSPECIES: Two subspecies are currently recognized, *A. o. orophilus* and *A. o. orientalis*, both originally described as subspecies of the widespread *A. mollis*. According to Osgood (1914b:163), "as a species, *Akodon mollis* has continuous distribution from the arid west coast up to an altitude of at least 12,000 feet and thence across the mountains and down into the upper part of the eastern forest or montagna . . . Four well-marked subspecies are recognizable. Their more important characters may be summarized as follows: Rostrum shorter and broader, zygomatic plate upright in front. Pelage short; color paler: *A. mollis*; Pelage long; color darker: *A. m. altorum*. Rostrum longer and slenderer, zygomatic plate with receding front edge. Feet grayish; tail bicolor: *A. m. orophilus*. Feet and tail wholly blackish: *A. m. orientalis*." Osgood (1913) reported intergradation between *orophilus* and *orientalis* in the region of

Puca Tambo and related color differences to environmental conditions.

NATURAL HISTORY: Collections at Utcubamba were made in a swampy mountain glade near the source of a western branch of the Río Utcubamba, about six miles west of Leimebamba, elevation 8,000 ft (Osgood 1914b). Specimens identified as *orophilus* were obtained in the "heavily wooded canyons, which increased in number after we crossed the Marañon, associated with *Thomasomys*, *Rhipidomys*, and *Oryzomys albigularis* [= *Nephelomys albigularus*] and it generally out-numbered any of these. It was found also in open swamps at high altitudes living in long grass or rushes quite after the manner of northern voles of the genus *Microtus*. In certain of these places, the labyrinthine runways, open burrows, and fresh grass cuttings so familiar to the northern collector were found in great numbers. In other places, as for example heavy woods or rocky stream beds, the Akodons seemed to lead wandering lives and have as retreats only natural openings in or near the ground" (Osgood 1914b:164). Osgood (1914b) stated that *orientalis* was found in "the dense, humid, but relatively cool forest which forms the practically unbroken cover of the lower slopes of the eastern Andes. In this region it is possible that its range may overlap that of *Akodon aerosus* which is found slightly lower down." *Akodon orophilus* has been trapped with *Hylaeamys yunganus*, *Microryzomys minutus*, *Nelphelomys albigularis*, *Thomasomys notatus*, and *T. ischyrus* in montane forests of Amazonas department. The species may be very common, especially in corn and potato agricultural plots, comprising 46% of all animals captured in a 12-night survey during July and August 2007 (C. F. Jiménez et al. 2013).

REMARKS: Thomas (1920h, 1926e) discussed the generic allocation of *orophilus* and *orientalis* when he described *surdus* and *torques*, which were based on the collections obtained by Edmund Heller from Machu Picchu and its surroundings. Thomas (1920h:237) wrote "[*Akodon surdus*] differs from its nearest ally, *A. mollis*, by its stouter build and smokier color, without tinge of buffy. The animal is of interest in connection with its strong external resemblance to the species here called *Microxus torques*. The two are found, respectively, at low (5,000–6,000 feet) and high (8,000–14,000 feet) altitudes and apparently represent the lowland and highland forms of northern Peru, distinguished by Osgood merely as subspecies of *A. mollis* [i.e., *orophilus* and *orientalis*]." Thomas (1920h:240) further added "[*torques*] is distinguishable from *M. mimus* by its broader and less characteristically Microxine zygomatic plate, but is otherwise very similar to that animal. Osgood's *Akodon mollis orophilus* is said to be 'more fulvous than *A. m. altorum*,' which is anything but the case with the present animal; and his *A. m. orien-

talis is a low-land form and much darker in color . . . But, on the other hand, I quite admit that while the zygomatic plate of the earlier described species [*mimus*] was strongly and characteristically different from that of *Akodon*, that of *M. torques* (and I presume of *Akodon mollis orophilus* and *orientalis*) is more or less intermediate between the two. We have therefore to decide whether *Microxus* shall be amalgamated with *Akodon*, ignoring its peculiar zygomatic plate, or whether we shall recognize *Microxus* and put *torques*, *orophilus*, and *orientalis* into it, where they would form a group of species annectent with *Akodon*." Osgood (1943b:197) later treated *orophilus* as a distinct species and also discussed its generic position, saying "a re-examination of my original collections, together with series of *torques* and much additional material, leads to the conclusion that *orophilus* is specifically but not generically distinct from *mollis*, at least not in northern Peru." Cabrera (1961) placed all these forms in *Akodon* and treated *torques* as a subspecies of *orophilus*.

C. F. Jiménez et al. (2013) described a karyotype for specimens from Amazonas department with $2n=26$, $FN=40$, with 8 pairs of biarmed and 3 pairs of uniarmed autosomes, a large biarmed X, and a small uniarmed Y. Specimens with other karyotypes originally assigned to *A. orophilus* ($2n=26$, from Ayacucho department [Hsu and Benirschke 1973], and $2n=22$, from Huánuco department [Pacheco, Córdova, and Velásquez 2012]) are referable instead to *A. torques* (Patton and Smith 1992a,b) and *A. josemariarguedasi* (C. F. Jiménez et al. 2013), respectively.

Akodon paranaensis Christoff, Fagundes, Sbalqueiro, Mattevi, and Yonenaga-Yassuda, 2000
Paraná Grass Mouse, Paraná Akodont

SYNONYM:

Akodon paranaensis Christoff, Fagundes, Sbalqueiro, Mattevi, and Yonenaga-Yassuda, 2000:844; type locality "Piraquara (Estação Ecológica Canguiri), Paraná, Brazil."

DESCRIPTION: Similar externally and cranially to *A. cursor* and *A. montensis* but distinguishable by karyotype and cranial dimensions (particularly least interorbital breadth and molar series length). The head and body length 88–125 mm, tail length 60–92 mm, ear length 15–19 mm, and hindfoot length (without claw) 22–24 mm. Dorsal pelage soft and long, with overall gray olivaceous tone and without clear distinction from either lateral or ventral pelage; venter grayish or creamy; forefeet and hindfeet grayish brown above, covered by banded hairs with dark base and white tip, and with long ungual tufts that extend beyond tip of each claw. Vibrissae short, not extending past pinnae when laid back. Tail bicolored, especially near base, with scales exposed and clearly visible to eye among short

bristles. Eight mammae distributed in pectoral, postaxial, abdominal, and inguinal pairs. Skull elongated with pronounced rostrum and globular braincase and symmetrically biconcave and relatively narrow (least interorbital breadth <5.1 mm or 34% of zygomatic breadth) interorbital region, especially in comparison to *A. cursor*; nasals pointed and long, but do not extend posterior to lacrimals in most specimens; frontoparietal suture U-shaped and collinear with frontosquamosal suture so that frontal never contacts squamosal dorsally; zygomatic plates wide with straight anterior borders in most specimens; anterodorsal edges of plates square and projected anteriorly in relation to dorsal zygomatic root, forming deeply incised zygomatic notches; incisive foramina long, usually reaching level of the protocones of M1s. Greatest skull length 26–31 mm, upper molar series length 4.4–4.9 mm.

DISTRIBUTION: *Akodon paranaensis* is found in southern and southeastern Brazil, extending westward to northeastern Argentina (Misiones province) and eastern Paraguay. Northeastern populations occur on isolated mountaintops in the Serra da Mantiqueira (states of Minas Gerais and Rio de Janeiro), those in the south are at lower elevations (P. R. Gonçalves et al. 2007). No fossils are known, but *Habrothrix clivigenis* Winge, a fossil *Akodon* from Capão Secco Cave in Minas Gerais, Brazil, is very similar to *A. paranaensis*. These two taxa require careful comparison, not just to extend a temporal range but also in regard to name priority.

SELECTED LOCALITIES (Map 79): ARGENTINA: Misiones, Parque Provincial Islas Malvinas (Liascovich and Reig 1989). BRAZIL: Minas Gerais, Serra do Papagaio, Itamonte (Gouveia 2009); Paraná, Estação Ecológica do Canguiri (Christoff et al. 2000), Mangueirinha (Hass 2001), Salto Caxias Dam, lower Rio Iguaçu (Casella and Cáceres 2006), Tijucas do Sul (Hass 2001); Rio de Janeiro, Campos do Itatiaia, Parque Nacional do Itatiaia (P. R. Gonçalves et al. 2007); Rio Grande do Sul, Cambará do Sul, Parque Nacional da Serra Geral (MNRJ [LMT 294]), Itapeva, Torres (Tiepolo 2007), Pelotas (Tiepolo 2007), Santa Maria (T. G. Santos et al. 2008), Três Barras, margins of Rio Uruguai, Aratiba (Christoff et al. 2000), Venancio Aires (Schleiber and Christoff 2007); Santa Catarina, Urubici, Morro da Igreja, Parque Nacional de São Joaquim (MNRJ [LMT 304]). PARAGUAY: Alto Paraná, Reserva Biológica Limoy (de la Sancha 2010); Canindeyú, Reserva Natural del Bosque de Mbaracayú (de la Sancha 2010), Reserva Natural Privada Morombí (de la Sancha 2010); Itapúa, Estancia Parabel, 0.3 km E of houses (D'Elía, Mora et al. 2008), Reserva de Recursos Manejados San Rafael (de la Sancha 2010).

SUBSPECIES: *Akodon paranaensis* is monotypic.

NATURAL HISTORY: This species inhabits primary and secondary tropical forests as well as grasslands and shrubby transitional habitats (Christoff et al. 2000; Schleiber and Christoff 2007). In southern Brazil, the species is broadly distributed in humid montane forests dominated by *Araucaria angustifolia*, a tree that occurs in temperate regions above 900 m (Christoff et al. 2000). In southeastern Brazil, it appears to be limited to isolated high-elevation habitats such as the *campos de altitude* and montane forests, which occur above 1,800 m in the Serra da Mantiqueira and Serra do Mar ranges (P. R. Gonçalves et al. 2007). In Paraguay, this is primarily an insectivorous-omnivorous species, with stomach contents containing both coleopterans and a substantial proportion of fruit (de la Sancha 2010). Casella and Cárceres (2006) suggested, based on the high proportion of intact seeds found in the stomach contents, that *A. paranarensis* might play an important role as a seed disperser. R. C. Oliveira et al. (2012) isolated the Jabora strain of hantavirus (JABV) from specimens of *A. paranaensis* collected in Santa Catarina state, Brazil.

REMARKS: The karyotype of *A. paranaensis* is now understood to be 2*n* = 44, FN = 44 (Christoff et al. 2000; Geise et al. 2001). Specimens with this karyotype, however, were previously identified as *A. serrensis* (Liascovich and Reig 1989; Sbalqueiro 1989) until Christoff et al. (2000) corrected this assignment. Previous authors had also identified *Akodon* populations with 2*n* = 44 from Mt. Itatiaia, in southeastern Brazil, and from Misiones, Argentina, as allied to *A. mystax* (Geise et al. 2001; Geise, Pereira et al. 2004; Pardiñas, D'Elía, and Cirignoli 2003—specimens referred to as *Akodon* sp. 2), and Geise, Pereira et al. (2004)

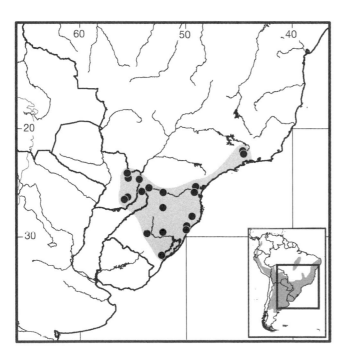

Map 79 Selected localities for *Akodon paranaensis* (●). Contour line = 2,000 m.

incorrectly referred to specimens of this species as *A. reigi*. Morphological and molecular comparisons, including the type series, showed that these samples actually belong to *A. paranaensis* (P. R. Gonçalves et al. 2007; D'Elía, Mora et al. 2008). Nevertheless, *A. paranaensis* is a member of the *cursor* species group identified by M. F. Smith and Patton (2007), and is sister to *A. mystax* in this group.

Akodon pervalens Thomas, 1925
Tarija Grass Mouse, Tarija Akodont

SYNONYMS:

Akodon sylvanus pervalens Thomas, 1925b:579; type locality "Carapari [= Caraparí], 1000 m.," Tarija, Bolivia.

?*Akodon tartareus*: Thomas, 1926c:322; part; not *Akodon tartareus* Thomas.

Akodon varius pervalens: Cabrera, 1961:449; name combination.

Akodon pervalens: Myers, 1990:19; first use of current name combination.

DESCRIPTION: Large with total body length averaging 209 mm, tail length 93 mm, ear length 19.4 mm, and hindfoot length 24.6 mm. Dorsal coloration dark grayish or brownish olive; ventrally fur with very dark bases and buffy tips; eye-ring lacking; no white spot on either chin or throat; tail unicolored dark above and below. Skull broad and heavy, with broad but smoothly rounded interorbital region with no traces of supraorbital ridges; zygomatic notches deep and broad; frontal sinuses inflated; zygomatic arches slightly expanded laterally; braincase not inflated; zygomatic plates low and broad, with straight anterior borders; incisive foramina extend posteriorly to protocone of M1; mesopterygoid fossa relatively broad, with anterior border rounded, without medial process, and straight and parallel lateral margins; parapterygoid fossae slightly narrower than mesopterygoid fossa, with convex lateral borders that diverge slightly posteriorly; auditory bullae large and globular. Upper incisors orthodont but with slight tendency toward proodonty. M1 with well-developed anteromedian flexus and mesoloph that extends nearly to lateral tooth margin; M2 also has conspicuous mesoloph and penetrating paraflexus; M3 with poorly developed paraflexus and weakly developed hypoflexus and metaflexus. The m1 with well-developed anteromedian flexid, anterolabial cingulid, and mesolophid; m2 lacks ectolophid and mesolophid; m3 S-shaped, with well-developed hypoflexid and mesoflexid. Mean condyloincisive length 28 mm, upper molar series 4.70 mm.

DISTRIBUTION: *Akodon pervalens* is known from only six localities in Chuquisaca and Tarija departments in south-central Bolivia, between 900 and 2,100 m (S. Anderson 1997). However, S. Anderson (1997) himself indicated the dubious taxonomic status of all the specimens he referred to this species, thus the need for a comprehensive restudy. Yepes (1933) assigned specimens from northernmost Argentina to this species, but Pardiñas, D'Elía et al. (2006) concluded, based on a reexamination of voucher specimens, these did not represent *A. pervalens*. No fossils are known.

SELECTED LOCALITIES (Map 80): BOLIVIA: Chuquisaca, Tihumayu (S. Anderson 1997), Tola Orko, 40 km from Padilla (S. Anderson 1997); Tarija, Caraparí (type locality of *Akodon sylvanus pervalens* Thomas), 8 km by road N of Cuyambuyo (S. Anderson 1997), Pino [= Pinos] (S. Anderson 1997).

SUBSPECIES: *Akodon pervalens* is monotypic.

NATURAL HISTORY: Virtually nothing is known about the natural history of this species. It has been captured in humid Yungas forest, Chaco woodland, and semiarid inter-Andean valleys. It may co-occur with *A. boliviensis*, *A. fumeus*, *A. simulator*, *A. toba*, and *A. varius* (S. Anderson 1997).

REMARKS: *Akodon pervalens* was originally described as a subspecies of *A. sylvanus*, as Thomas (1925b:579) referred to his new taxon as "like *A. sylvanus* of Jujuy in all essential respects, but larger, with longer tail, and the colour inconspicuously less obscure. Skull larger, heavier throughout, with broader interorbital region, longer braincase, and slightly larger bullae." Cabrera (1961) was the first to separate *pervalens* from *A. sylvanus*, but treated *pervalens* as a subspecies of *A. varius*. Myers (1990), how-

ever, tentatively treated *pervalens*, *sylvanus*, and *varius* as three separate species. The status of *A. pervalens* as a full species, supported by some authors (Myers 1990; S. Anderson 1997; Musser and Carleton 2005; M. M. Díaz and Barquez 2007), and the proper delineation of its distribution require further studies focused on topotypical material (see Jayat, Ortiz et al. 2007) and the reexamination of the specimens assigned to this species by S. Anderson (1997). Both Thomas (1925b) and Myers (1990) mentioned a superficial resemblance between *A. pervalens* and *A. montensis*.

Akodon philipmyersi Pardiñas, D'Elía, Cirignoli, and Suárez, 2005
Philip Myers's Grass Mouse, Jesuita Akodont

SYNONYM:

Akodon philipmyersi Pardiñas, D'Elía, Cirignoli, and Suárez, 2005:465; type locality "ARGENTINA: Province of Misiones, Department of Posadas, Estancia Santa Inés, Ruta No. 105 km 10 (27°31′32″S, 55°52′19″W, 95 m)."

DESCRIPTION: Small species distinguished by condyloincisive length <24 mm, zygomatic breadth <1 3 mm, maxillary toothrow length <4.10 mm), tail short (40% of head and body length), short rostrum (mean incisive foramen length 5.6 mm), short and wide nasals almost covering rostrum in dorsal view, zygomatic plates very narrow almost without free upper border, mesopterygoid fossa wide with median palatine process on anterior border, interorbital constriction wide (4.3–4.6 mm), and auditory bullae medium in size. Gall bladder present.

DISTRIBUTION: This species is known from only two localities in southern Misiones province, Argentina. It is not known from the fossil record.

SELECTED LOCALITIES (Map 81): ARGENTINA: Misiones, Estancia Santa Inés (type locality of *Akodon philipmyersi* Pardiñas, D'Elía, Cirignoli, and Suárez), Parada Leis (Pardiñas, D'Elía et al. 2005).

SUBSPECIES: *Akodon philipmyersi* is monotypic.

NATURAL HISTORY: *Akodon philipmyersi* is found in natural grasslands of southern Misiones Province, some of which have been converted to yerba mate (*Ilex paraguayensis*) fields. Specimens were obtained at night in a portion of a field covered by tall (2 m) grasses. Of 12 captured in March and for which reproductive data are available, four females were pregnant (with three or more embryos) and six males had scrotal testes (≥ 5 mm in length). The topotypical series of 13 specimens, obtained in May 2009, included nine males (five with scrotal testes) and four females (three pregnant). These data suggest that this species is reproductively active at least until the middle of the autumn. *Akodon philipmyersi* is the type host of the recently

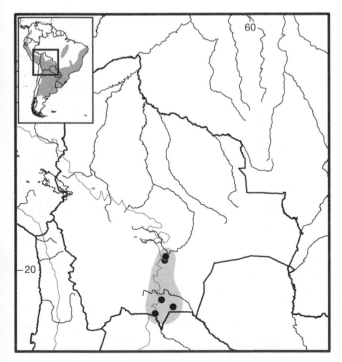

Map 80 Selected localities for *Akodon pervalens* (●). Contour line = 2,000 m.

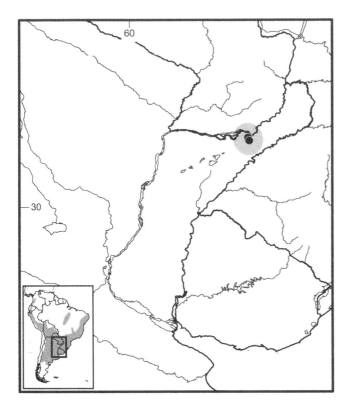

Map 81 Selected localities for *Akodon philipmyersi* (●).

described laelapid mite *Androlaelaps ulysespardinasi* (Lareschi 2011).

REMARKS: Pardiñas, D'Elía, and Cirignoli (2003) originally discovered this species in an owl pellet sample collected at the type locality, and referred to it as *Akodon* sp. 1. Specimens trapped subsequently allowed Pardiñas, D'Elía et al. (2005) to confirm that the owl pellet materials represented a previously undescribed species. Molecular and morphological analyses suggested that *A. philipmyersi* may be the sister taxon to the *A. mystax*–*A. lindberghi* complex (Pardiñas, D'Elía et al. 2005; P. R. Gonçalves et al. 2007). The karyotype has a $2n = 36$, $FN = 42$ (Pardiñas, D'Elía et al. 2005).

Akodon polopi Jayat, Ortiz, Salazar-Bravo, Pardiñas, and D'Elía, 2010
Polop's Grass Mouse, Polop's Akodont
SYNONYMS:

Akodon polopi Jayat, Ortiz, Salazar-Bravo, Pardiñas, and D'Elía, 2010:34; type locality "Pampa de Achala, 6 km E (by highway 34) from antena repetidora La Posta, 2200 m (31°36′44.5″S, 64°48′48.7″W), San Alberto Department, Córdoba Province, Argentina."

Akodon viridescens Braun, Mares, Coyner, and Van Den Bussche, 2010:393; type locality "Argentina: San Luis: Coronel Pringles, 9 km N Paso del Rey, elevation 4,800 feet (1,455 m)."

DESCRIPTION: Medium-sized with total body length 150–188 mm, tail length 60–76 mm, ear length 14–16 mm, and hindfoot length (with claw) 21–25 mm. Fur dense and soft; dorsal coloration uniform buffy brown lightly spattered with black hairs; venter contrasts with dorsum with buffy or tawny tinged hairs separating them; chin with small but clearly evident white patch; tail bicolored, blackish brown above and whitish below. Claws on forefeet and hindfeet somewhat longer than in other species of *boliviensis* group. Skull heavily constructed relative to other members of group; rostrum relatively short and broad; frontal sinuses clearly inflated; zygomatic notches broad and deep; interorbital region hourglass in shape, with rounded to slightly squared borders and with posterior margins with greater tendency to be sharply squared than typical for other species; both temporal and lambdoid ridges well developed; zygomatic plates relatively broad, with anterior margin straight and vertical in most individuals; hamular process relatively robust but variable in extent; incisive foramina long, reaching posteriorly to hypoflexus of M1; mesopterygoid fossa narrow, a rounded anterior margin and slightly divergent lateral margins; parapterygoid fossae only weakly excavated, broader than mesopterygoid fossa, with generally straight lateral margins that diverge posteriorly. Mandible robust for a *boliviensis* group member, with deeper horizontal ramus and broader coronoid process; angular process ends just anterior to condyloid process; masseteric crest reaches level of anterior margin of m1 or slightly beyond. Upper incisors essentially orthodont, but somewhat proodont in many individuals. M1 with well developed procingulum and anteromedian flexus, and with both small anteroloph and mesoloph present, and, on lingual side, visible tiny enteroloph in some specimens; M2 with reduced procingulum, vestigial mesolophs, and weakly developed posteroflexus; M3 with paraflexus and metaflexus always present in young specimens, but hypoflexus present only in a few individuals. The m1 with deep anteromedian flexid and clearly defined anterolabial cingulum; young individuals bear well-developed ectostylid, and, on lingual side, young individuals show well-developed metastylid and relatively robust mesolophid; m2 with anterolabial cingulum and small ectostylid, but reduced mesolophid; m3 large and S-shaped. Axial skeleton with 13–14 thoracic ribs and with 13–14 thoracic, 8 lumbar, and 26–27 caudal vertebrae (N = 6). Greatest skull length 24.0–28.0 mm, upper molar series 4.14–4.64 mm.

DISTRIBUTION: *Akodon polopi* is known from fewer than 10 localities in Córdoba and San Luis provinces in central Argentina. All localities are in the Central Pampean Ranges, between elevations of 1,300 to 2,250 m (Polop 1989, 1991; Priotto et al. 1996; Kufner et al. 2004; Braun et al. 2010; Jayat et al. 2010). No fossils are known for this species.

SELECTED LOCALITIES (Map 82): ARGENTINA: Córdoba, Cerro de Oro (Priotto et al. 1996), Pampa de Achala (type locality of *Akodon polopi* Jayat, Ortiz, Salazar-Bravo, Pardiñas, and D'Elía); San Luis, 9 km N of Paso del Rey (type locality of *Akodon viridescens* Braun, Mares, Coyner, and Van Den Bussche).

SUBSPECIES: *Akodon polopi* is monotypic.

NATURAL HISTORY: This species is found in highland grasslands dominated by *Festuca* and *Stipa*, intermingled with scattered patches of woodland and rocky outcrops (Polop 1989, 1991; Braun et al. 2010; Jayat et al. 2010). None of the specimens captured by Jayat et al. (2010) at the end of the winter in 2008 (August) nor most of those collected by Braun et al. (2010) in April and May showed signs of reproductive activity. This suggests that the species is reproductively active in late spring and summer and agrees with previous studies in which the greatest number of pregnancies occurred in November and December, and in which the average embryo counts was 4.7 (range: 3 to 7; Polop 1989). Few specimens captured in fall and winter were molting (Braun et al. 2010; Jayat et al. 2010). Other sigmodontines captured with *A. polopi* and in the same habitat were *Calomys musculinus, Oxymycterus rufus* (given as *O. paramensis* in Polop [1989]), *Oligoryzomys flavescens, Phyllotis xanthopygus,* and *Reithrodon auritus* (Braun et al. 2010; Jayat et al. 2010). *Akodon polopi* was the dominant cricetid at the sites where it was collected, typically making up more than 70% of all animals captured (Kufner et al. 2004; Jayat et al. 2010), although Polop (1989, 1991) recorded a smaller relative abundance (34–53%).

REMARKS: Polop (1989) originally thought that Akodon *polopi* was an undescribed species, but before it was formally described this taxon was subsequently treated in the literature as *A. boliviensis* (Polop 1991; Morando and Polop 1997), *A. alterus* (Priotto et al. 1996), or *A. spegazzinii* (D'Elía 2003a; D'Elía et al. 2003; Kufner et al. 2004; Pardiñas, D'Elía et al. 2005; P. R. Gonçalves et al. 2007; M. F. Smith and Patton 2007). Based on morphological similarities, possession of the same karyotype, its paraphyletic position relative to *A. polopi,* and the allocation of a specimen (UP AC008 from Pampa de Achala, Córdoba, Argentina) to both taxa in their original descriptions, D'Elía et al. (2011) demonstrated that *A. viridescens* (Braun et al. 2010) is a junior synonym of *A. polopi.*

The karyotype of *A. polopi* is $2n = 40$, with 18 acrocentric and one small metacentric autosomal pair, a chromosomal complement similar to many other members of the *boliviensis* group. The X chromosome is a large subterminal, and the Y is a small metacentric (Polop 1989; Pinna-Senn et al. 1992; Braun et al. 2010).

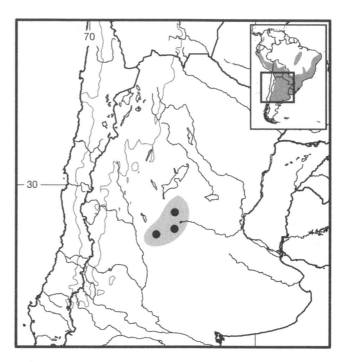

Map 82 Selected localities for *Akodon polopi* (●). Contour line = 2,000 m.

Akodon reigi E. M. González, Langguth, and Oliveira, 1998
Reig's Grass Mouse, Reig's Akodont

SYNONYM:

Akodon reigi E. M. González, Langguth, and Oliveira, 1998:2; type locality "Paso Averías, Departamento de Lavalleja, Uruguay (Lat. 33°60′S; Long. 54°40′W)."

DESCRIPTION: Medium to large species, morphologically very similar to *A. paranaensis,* with commonalities between them including traits useful in distinguishing both from *A. cursor* and *A. montensis.* Externally, *A. reigi* shares with *A. paranaensis* the same overall gray olivaceous tone of dorsal pelage, bicolored tail, and grayish hairs covering dorsal surfaces of both forefeet and hindfeet. Shared cranial characters between these species include relatively narrow interorbital region (least interorbital breadth <5.1 mm), long upper molar series (>4.5 mm), and wide zygomatic plates with straight anterior border. Condyloincisive length 25–27 mm; upper molar series length 4.5–4.8 mm. Differs from *A. montensis* in karyotype, with $2n = 44$ rather than $2n = 24$–25.

DISTRIBUTION: *Akodon reigi* occurs in central and eastern Uruguay, including the departments of Tacuarembó, Durazno, Cerro Largo, Maldonado, and Lavalleja, and in the southernmost portion of Rio Grande do Sul state, Brazil (E. M. González et al. 1998; E. M. González and Martínez Lanfranco 2010). It is not known as a fossil.

SELECTED LOCALITIES (Map 83): BRAZIL: Rio Grande do Sul, Reserva Ecológica do Taim (E. M. González et al. 1998). URUGUAY: Durazno, Arroyo Cordobés y Río Negro (E. M. González and Martínez Lanfranco 2010); Lavalleja, Paso Averías (E. M. González et al. 1998); Maldonado, Hostería La Laguna, N de Maldonado (E. M. González and Martínez Lanfranco 2010); Río Negro, Paso de las Piedras (E. M. González et al. 1998); Tacuarembó, Tacuarembó (E. M. González and Martínez Lanfranco 2010).

SUBSPECIES: *Akodon reigi* is monotypic.

NATURAL HISTORY: In Uruguay and southern Brazil this species has been trapped in subtropical gallery forests at sites with dense litter or at the interface of forested habitats and grasslands. It is especially common in swampy areas and avoids open grassland habitats. Substantial variation in year-to-year capture rates suggests irregular fluctuations in population density (E. M. González and Martínez Lanfranco 2010). Breeding probably occurs from the spring to autumn, with autumn being when the number of juvenile dispersers greatly increases (E. M. González et al. 1998).

REMARKS: Morphological and cytogenetic evidence suggested that the populations of *Akodon* from southern Brazil, northeastern Argentina, Paraguay, and Uruguay with $2n=44$ comprised a single morphologically diagnosable unit. Some of these samples had been previously allocated to *A. cursor montensis* (e.g., Ximenez and Langguth 1970). Phylogenetic analyses of mtDNA cytochrome-*b* sequences, however, indicated moderate

levels of divergence between Brazilian and Uruguayan populations (Pardiñas, D'Elía, and Cirignoli 2003; P. R. Gonçalves et al. 2007; D'Elía, Mora et al. 2008). D'Elía, Mora et al. (2008) reported that haplotypes recovered from populations assigned to *A. paranaensis* (i.e., those from Brazil, northern Argentina, and Paraguay) and to *A. reigi* (those from Uruguay) form two reciprocally monophyletic groups, diverging on average by 5.9% genetic distance. This level of divergence indicates that these samples comprised two genetically divergent lineages of *Akodon*, despite the similarities in morphology and chromosome number. Therefore, we provisionally maintain *A. paranaensis* and *A. reigi* as distinct species, although we contend that any morphological differences between these species are unclear and should be tested by analysis of geographically intermediate populations, especially those from the southernmost part of Rio Grande do Sul state where the ranges of both species interdigitate (P. R. Gonçalves et al. 2007).

The karyotype of *A. reigi* has a $2n=44$ and FN = 44 (E. M. González et al. 1998).

Akodon sanctipaulensis Hershkovitz, 1990
São Paulo Grass Mouse, São Paulo Akodont
SYNONYM:

Akodon sanctipaulensis Hershkovitz, 1990b:23; type locality "Primeiro Morro, São Paulo, Brazil," about 24°22′S and 47°49′W (Hershkovitz 1990b:3).

DESCRIPTION: Medium sized with markedly darker pelage than potentially co-occurring *A. cursor*, *A. montensis*, and *A reigi*. Head and body length 73–100 mm, tail length 58–75 mm, ear length 12–16 mm, and hindfoot length (without claw) 23–24 mm. Both dorsal and lateral pelage very dark in tone owing to reduction in ochraceous bands on guard hairs and expanded melanic bands; dorsal, flank, and ventral pelage color not distinct; tail weakly bicolored and covered with short hairs that leave scales exposed; hairs with dark basal bands and white tips cover both forefeet and hindfeet, and ungual tufts long, extending beyond distal limits of claws; vibrissae short, not extending beyond pinna when laid back. Skull slender, with distinctively narrow, long, and pointed rostrum and globular braincase (greatest skull length 23.24–24.61 mm, the upper molar series length from 4.0–4.5 mm); interorbital region symmetrically biconcave and relatively broad (least interorbital breadth >40% of zygomatic breadth); nasals long, terminating posteriorly at level of lacrimals in most specimens; frontoparietal suture U-shaped and colinear with frontosquamosal suture so that frontal never contacts squamosal dorsally; zygomatic platesmoderately wide with straight or slightly reclined anterior borders; anterodorsal edge of plate projected in relation to dorsal zygomatic root,

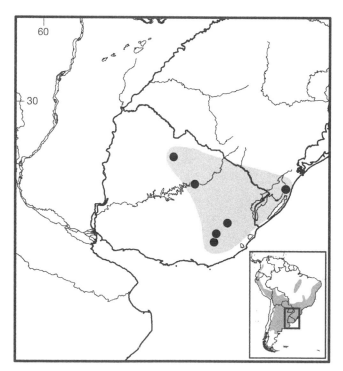

Map 83 Selected localities for *Akodon reigi* (●). Contour line = 2,000 m.

forming a moderately incised zygomatic notch; incisive foramina long, usually reaching level of protocones of M1s.

DISTRIBUTION: *Akodon sanctipaulensis* is known from only four localities on the north bank the Rio Juquiá and Rio Etá on the Atlantic coastal plain in São Paulo state, Brazil (Hershkovitz 1990b). This species has no fossil record.

SELECTED LOCALITIES (Map 84): BRAZIL: São Paulo, Iporanga (Hershkovitz 1990b), Morretinho (Hershkovitz 1990b), Primeiro Morro (type locality of *Akodon sanctipaulensis* Hershkovitz), Quadro Penteado (Hershkovitz 1990b).

SUBSPECIES: *Akodon sanctipaulensis* is monotypic.

NATURAL HISTORY: Little is known of the ecology of this species. The few localities are in the Parque Estadual Turístico do Alto Ribeira (PETAR), one of the largest remaining remnants of montane and submontane dense evergreen forest in the Atlantic Forest domain of coastal Brazil.

REMARKS: This is the least known species of *Akodon* in southeastern Brazil. Data on its karyotype are lacking, and no additional specimens have been collected at or near the type locality since A. M. Olalla obtained the original series in 1961. The series examined by Hershkovitz (1990b) is composed of young individuals and a few subadults, and most with damaged skulls, limiting any comparisons with adults of other *Akodon*. One of the specimens included in the original description (FMNH 94513) lacks an anteromedian flexus on the procingulum of M1, a condition also

found in *Akodon serrensis*. The similarity of this specimen to *A. serrensis* suggests that the original series of *A. sanctipaulensis* may actually be composite. Nevertheless, the skull of the holotype (FMNH 94516), a subadult, is different from those of *A. serrensis* and adults of *A. cursor* and *A. montensis*, especially in the rostral and interorbital regions (Christoff et al. 2000; P. R. Gonçalves et al. 2007). Because these portions of the skull are considerably influenced by age, further analysis of young and subadults of other co-occurring *Akodon* species from southeastern Brazil is warranted to reassess the status of *A. sanctipaulensis*.

Akodon serrensis Thomas, 1902
Serra do Mar Grass Mouse, Serra do Mar Akodont

SYNONYMS:

Akodon serrensis Thomas, 1902a:61; type locality "Roça Nova, situated at an altitude of about 1000 metres . . . in the Serro do Mar of the Province of Paraná, and on the railway between Paranagua and Curitiba" (Thomas 1902a:59). Roça Nova, Serra do Mar, Paraná, Brazil, 25°28'S 49°01'W (see Emmons, Leite et al. 2002:24).

A[kodon]. *serrensis* var. *leucogula* Miranda-Ribeiro, 1905: 188; type locality "Retiro de Ramos," Serra de Itatiaia, Rio de Janeiro, 2,200 m, Brazil.

Akodon serrensis serrensis: Gyldenstolpe, 1932a:104; name combination.

Akodon serrensis leucogula: Gyldenstolpe, 1932a:104; name combination.

DESCRIPTION: Moderately large, with total length ca. 170 mm, tail length ca. 80 mm; similar to *A. cursor* and other related *Akodon* species from southeastern Brazil (i.e., *A. montensis*, *A. paranaensis*) in size, general bauplan, and color pattern. Fur thick and woolly (Thomas 1902a:61) with hairs of back 8 to 9 mm; general color above finely grizzled olivaceous by fine mixture of blackish and yellowish hairs; under surface ochraceous and line of demarcation on flanks not sharply marked; anal area and cheeks more rich ochraceous. Ears of moderate size, dark brown; manus and pes dull brown above. Tail long, finely ringed, almost naked, covered by minute hairs brown above, dull white below. Skull strongly built, greatest length ca. 27 mm, upper toothrow ca. 5 mm, with pointed rostrum, broad interorbital region usually without sharp borders, flat profile, and unridged braincase. Incisive foramina especially broad, not extending backwards to M1 protocone and with very well-developed palatal process of premaxillary; mesopterygoid fossa also broad and mesopterygoid roof perforated by paired sphenopalatine vacuities. Molars unique among *Akodon*, with occlusal features wearing rapidly with wear, coronal surface crested, main cusps almost in opposite pattern, M1 procingulum without anteromedian flexus, M3 large, m1 procingulum narrow, and m3 subequal in length to m2.

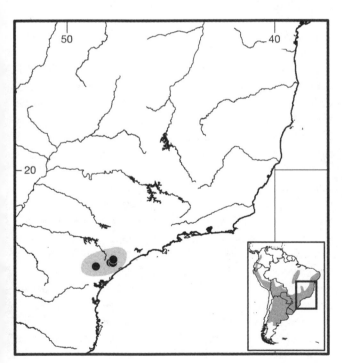

Map 84 Selected localities of *Akodon sanctipaulensis* (●). Contour line = 2,000 m.

DISTRIBUTION: Originally thought to be restricted to the Serra do Mar of coastal Brazil, *A. serrensis* is now known across the Atlantic Forest, from Espírito Santo state in the north (ca. 20°30′S) to Rio Grande do Sul state in the south (29°45′S) (P. P. D. Oliveira and Geise 2006; Gislene Gonçalves, pers. comm. to Lena Geise), and extending into Misiones province, Argentina (J. Pereira et al. 2005). In the north, *A. serrensis* appears restricted to elevations above 1,200 m, reaching 2,700 m on Pico da Neblina, Caparaó. In the south, this species is found at elevations as low as 400 m (Mocaná area, Misiones, Argentina; U. F. J. Pardiñas, unpubl. data). Justo and De Santis (1977:47) were the first to report this species in Argentina, based on a specimen from "río San Antonio, 32 km al norte de Bernardo de Irigoyen." Although dismissed by subsequent authors (Galliari et al. 1996; Pardiñas, D'Elía, and Cirignoli 2003; J. Pereira et al. 2005), this record likely represents an actual occurrence of this species. J. Pereira et al. (2005) documented the first corroborated record from Misiones, based on a specimen from Reserva Privada de Vida Silvestre Uruguá-í, and Testoni et al. (2012) extended the species's range into Santa Catarina state, Brazil. It is likely that Quaternary fossils from Lagoa Santa, described as *Habrothrix angustidens* by Winge (1887), are of this species (see Remarks).

SELECTED LOCALITIES (Map 85): ARGENTINA: Misiones, Reserva Privada de Vida Silvestre Uruguá-í (J. Pereira et al. 2005), RP 2, 6 km NE of Arroyo Paraíso (CNP-E 449). BRAZIL: Espírito Santo, 4 km N of Castelinho (MNRJ 32631), Parque Nacional do Caparaó (L. Geise, pers. comm.); Minas Gerais, Fazenda da Onça (UFMG 1857), Mata do Banco (MZUFV 1900), Mata do Sossego (MNRJ 35910), 13 km SE of Posses (UFMG 1855), Serra das Cabeças, Parque Estadual da Serra do Brigadeiro (MZUFV 1267); Paraná, Fazenda Panagro (P 566), Represa de Guaricana (MZUSP 29128), Roça Nova (P 888); Rio de Janeiro, Fazenda Marimbondo (HGB 457), Parque Nacional da Serra dos Órgãos, base da Pedra do Sino, Campo das Antas (MNRJ 69806), Sítio Xitaca, Debossan (MNRJ 35912); Rio Grande do Sul, São Francisco de Paula (G. Gonçalves, pers. comm. to L. Geise); Santa Catarina, Alta da Boa Vista (Testoni et al. 2012), Fazenda Gateados (Testoni et al. 2012), Reserva Biológica Estadual do Sassafrás (Testoni et al. 2012); São Paulo, Boracéia (MZUSP 9469), Fazenda Intervales (MZUSP 27226), Parque Nacional da Bocaina (HGB [DB 15]), Serra da Fartura (L. Geise, pers. comm.).

SUBSPECIES: Miranda-Ribeiro (1905) described *A. serrensis* var. *leucogula* based on subtle color differences between a specimen obtained at Retiro de Ramos, Rio de Janeiro, and Thomas's (1902a) description of *serrensis*. Although Christoff et al. (2000:849) found no morphologi-cal reason to treat *leucogula* as valid, there is a moderate degree of divergence in mtDNA cytochrome-*b* sequences (G. D'Elía, unpubl. data) that suggests more than one evolutionary unit in the species. It remains for future analyses to determine whether these lineages deserve recognition as subspecies, and, if so, whether *leucogula* Miranda-Ribeiro would apply to one of them.

NATURAL HISTORY: *Akodon serrensis* replaces *A. cursor* at elevations above 1,600 m in the Serra do Brigadeiro, Minas Gerais state, although both species were the most abundant rodents along the gradient between 1,200 and 1,800 m that spans forest and upper elevation grasslands (J. C. Moreira et al. 2009). Here, adults of both sexes were statistically monomorphic, although males averaged slightly larger than females in several cranial measures (Rodarte and Lessa 2010). Musser and Carleton (1993) stated that the wide variation in morphometry, described by Kosloski (1997) from specimens identified as *A. serrensis* from Paraná state, may have included other *Akodon* species (*A. paranaensis* or *A. reigi*) in the sample. Barros-Battesti et al. (1998) listed a wide range of ectoparasites for specimens collected at Tijucas do Sul, Paraná state, Brazil, including the mites *Androlaelaps fahrenholzi* and *A. rotundus*, the sucking louse *Hoplopleura imparata*, the staphylinid beetle *Amblyopinus* sp., and the fleas *Craneopsylla minerva*, *Polygenis pradoi*, *P. pygaerus*, *P. rimatus*, and *P. tripus*. *Akodon serrensis* shares most of its ectoparasite fauna with *A. montensis*. Specimens collected in the Serra da Bocaina (São Paulo state) and the Serra de Itataia (Rio de Janeiro) were infested with the fleas *C. minerva*, *Polygenis atopus*, *P. dentei*, *P. frustratus*, *P. pradoi*, and *P. rimatus* (Moraes et al. 2003).

REMARKS: On morphometric grounds, Christoff et al. (2000) identified karyotyped specimens with 2*n* = 46 as *A. s. leucogula*. These authors also clarified the karyological diversity among species of *Akodon* in southeastern Brazil and attributed the 2*n* = 44 diploid complement to their newly described *A. paranaensis* rather than *A. serrensis*, as previous authors had done (Liascovich and Reig 1989; Sbalqueiro 1989; see the account for *A. paranaensis*).

Robert S. Voss (pers. comm.) has concluded that the name *Habrothrix angustidens*, as applied by the fossil species from Lagoa Santa described by Winge (1887), is a senior synonym of *A. serrensis*. Therefore, after justification for this conclusion is published, *serrensis* Thomas will become a subjective junior synonym of *angustidens* Winge, and *angustidens* will become the proper name for this species.

The allocation of *A. serrensis* to the genus *Akodon* has been questioned as the result of several molecular genetic analyses, the first by D'Elía (2003a) who obtained a clade composed of *Thaptomys* + *serrensis* and well apart from *Akodon* (see also Geise et al. 2001). Additional stud-

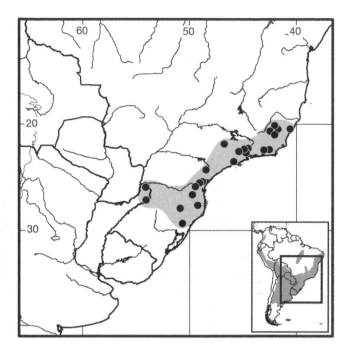

Map 85 Selected localities for *Akodon serrensis* (●). Contour line = 2,000 m.

ies using other molecular markers and somewhat different taxonomic coverage have supported this result (e.g., D'Elía et al. 2003; M. F. Smith and Patton 2007; P. R. Gonçalves 2006; P. R. Gonçalves et al. 2007). D'Elía's (2003a) phylogenetic hypothesis supported early inferences based on chromosomal data as to the distant position of *serrensis* vis-à-vis other *Akodon* species of southeastern Brazil (e.g., Geise et al. 1998; Christoff et al. 2000). More recent studies on chromosomal evolution in the Akodontini (e.g., Hass 2006; Hass et al. 2008), as well as a reevaluation of morphology (U. F. J. Pardiñas et al., unpubl. data) strongly suggest that *serrensis* Thomas represents an undescribed and monotypic akodontine genus.

Akodon siberiae Myers and Patton, 1989
Cochabamba Grass Mouse, Cochabamba Akodont

SYNONYM:

Akodon siberiae Myers and Patton, 1989a:4; type locality "28 km by road W of Comarapa, Cochabamba Dept., Bolivia, elevation 2800 m, 17°51′S, 64°40′W."

DESCRIPTION: One of largest species, with average body length 191.1 mm, tail length 89.1 mm, ear length 18.5 mm, and hindfoot length 24.1 mm. Overall color very dark, with dorsum medium to dark brown heavily lined with black; sides slightly paler but with same black streaking; venter paler, clothed with hairs basally slate gray but pale buff or tan tips; small and rather inconspicuous area of contrasting white hairs on the chin; ears clothed on anterior surfaces with agouti hairs near pinna base but mostly

black or dark brown hairs near edges; forefeet and hindfeet sparsely furred with mixture of whitish and agouti hairs. Claws rather long, slightly curved, and strongly built. Tail unicolor or slightly bicolor, appearing naked to unaided eye. Skull with long and narrow rostrum; narrow and shallow zygomatic notches; zygomatic arches flared posteriorly and midsections converging anteriorly; broad interorbital region with sides rounded anteriorly but squared, never beaded, posteriorly; conspicuously swollen frontal sinuses; and exceptionally broad and inflated braincase. Zygomatic plates narrow, with dorsal root slightly posterior to ventral root; incisive foramina long, ending at level of paracones of M1s; mesopterygoid fossa broad, with rounded anterior margin that sometimes with small median spine; parapterygoid fossae relatively broad, with slightly convex lateral margins. Upper incisors relatively delicate and nearly orthodont. Molars moderately hypsodont although not especially high crowned; M3 unusually elongate, with distinctive figure-8 shape due to penetrating mesoflexi and hypoflexi. Average condyle-incisive length 24.88 mm, upper toothrow length 4.81 mm.

DISTRIBUTION: *Akodon siberiae* is known only from the vicinity of the type locality in the Siberia Cloud Forest of Cochabamba and western Santa Cruz departments, central Bolivia, between 1,800 and 3,100 m (Myers and Patton 1989a; S. Anderson 1997). No fossils are known.

SELECTED LOCALITIES (Map 86): BOLIVIA: Cochabamba, 4.4 km N of Tablas Monte (S. Anderson 1997), Colomi (S. Anderson 1997); Santa Cruz, 28 km by road W of Comarapa (type locality of *Akodon siberiae* Myers and Patton), 1 km N and 8 km W of Comarapa (MSB 210746).

SUBSPECIES: *Akodon siberiae* is monotypic.

NATURAL HISTORY: All records of *Akodon siberiae* are from cloud forest on the eastern Andean slopes of the Bolivian Andes (Myers and Patton 1989a; S. Anderson 1997). S. Anderson (1997) reported four females with embryos in September. This species has been recorded in sympatry with *A. boliviensis*, *A. fumeus*, *A. mimus*, *A. lutescens*, and *A. varius* (S. Anderson 1997).

REMARKS: Myers and Patton (1989a) believed that *Akodon siberiae* was related to *Akodon budini*, the type species of *Hypsimys*, because of similarities in morphology and karyotype. M. F. Smith and Patton (2007) and Jayat et al. (2010) found this relationship well supported by mtDNA cytochrome-*b* sequences, with the two species being related to other members of their *aerosus* group of *Akodon*, which also included *A. aerosus*, *A. affinis*, *A. albiventer*, *A. mollis*, *A. orophilus*, and *A. torques*. The karyotype of *A. siberiae*, $2n = 38$, $FN = 40$, is virtually identical to that of *A. budini* (Myers and Patton 1989a). The overall chromosome complement of *A. siberiae* consists of 16 pairs of evenly graded uniarmed elements, one pair of very large

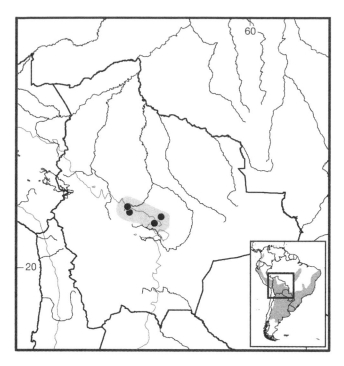

Map 86 Selected localities for *Akodon siberiae* (●). Contour line = 2,000 m.

subtelocentrics, one pair of medium-large submetacentrics, and a single pair of very small metacentrics. The morphology of the sex chromosomes is unknown because only a female has been karyotyped, but comparisons to other *Akodon* suggest that the X chromosomes of this individual are likely the medium-sized submetacentric pair (Myers and Patton 1989a).

Akodon simulator Thomas, 1916

Gray-bellied Grass Mouse, White-throated Akodont

SYNONYMS:

Akodon simulator Thomas, 1916h:335; type locality "Villa Nouges, San Pablo. Alt. 1200 m," Tucumán, Argentina.

Akodon glaucinus Thomas, 1919a:116; type locality "Chumbicha . . . about 60 kilometres due south-west of the town of Catamarca . . . 600 m," Catamarca, Argentina (Thomas 1919a:115).

Akodon tartareus Thomas, 1919g:155; type locality "Tartagal, 600 m.," Salta, Argentina (Thomas 1919g:154).

Akodon varius simulator: Thomas, 1926d:604; name combination.

Akodon varius glaucinus: Gyldenstolpe, 1932a:102; name combination.

Akodon simulator simulator: Myers, 1990:21; name combination.

Akodon simulator glaucinus: Myers, 1990:23; name combination.

Akodon simulator tartareus: Myers, 1990:25; name combination.

DESCRIPTION: Large, with average total length 192.7 mm, tail length 81.6 mm, ear length 18.8 mm, and hindfoot length (with claw) of 24.1 mm. General coloration brownish gray, suffused on back with yellowish or reddish highlights, and darker or paler depending on individuals and localities; yellowish or reddish highlights on back become more intense posteriorly so that rump more ochraceous tawny; flanks and venter grayish white or washed buffy, with change from dorsal to ventral pattern abrupt; diffuse white patch on chin, and prominent eye-ring present; tail bicolored, dorsally brown or blackish brown and ventrally whitish or buffy; dorsal surface of hindfeet covered with whitish hairs. Skull large and heavily built (greatest skull length ranges from 27.5–29.0 mm, upper molar series length from 4.7–4.8 mm); rostrum relatively short and comparatively narrow; zygomatic notches shallow; interorbital region hourglass in shape, with squared supraorbital edges along posterior half of orbit; zygomatic arches not strongly flared; zygomatic plates broad, vertically oriented, and with straight anterior margin; incisive foramina generally extend posteriorly to protocones of M1s; mesopterygoid fossa relatively broad, with anterior margin either slightly rounded or squared. Upper incisors with tendency to proodonty and molars exhibit slight hypsodonty. M1 with weakly developed anteromedian flexus and anteroloph, with small mesoloph present and partially coalesced with paracone, and poorly developed posteroflexus; M2 with remnant of anteroloph, weak mesoloph, and relatively shallow protoflexus and hypoflexus; M3 greatly reduced, with paraflexus isolated as deep pit. Anteromedian flexid and anterolabial cingulum well developed on m1, which possesses well-defined mesolophid and mesostylid, and prominent posterolophid; m2 with large hypoconid, shallow protoflexid, and tiny ectolophid; m3 with well-defined mesoflexid and hypoflexid, but protoflexid very small.

DISTRIBUTION: *Akodon simulator* is distributed along the lower Andean slopes of northwestern Argentina, in La Rioja, Catamarca, Tucumán, Salta, and Jujuy provinces, and in southern Bolivia, in Tarija department, over an elevational range of 400 to 3,000 m (Myers 1990; S. Anderson 1997; Capllonch et al. 1997; Mares, Ojeda, Braun et al. 1997; M. M. Díaz et al. 2000, Pardiñas, D'Elía et al. 2006; M. M. Díaz and Barquez 2007; Jayat, Ortiz, and Miotti 2008). This species is known from the Late Pleistocene of Catamarca province and the Late Holocene of Tucumán province (P. E. Ortiz et al. 2011a). Several specimens morphologically very similar to *A. simulator* have been collected in recent field trips to La Rioja and San Juan provinces, west-central Argentina, in unusual environments for the species. If these are determined to represent *A. simulator*, its distributional range and ecological parameters will be significantly extended.

SELECTED LOCALITIES (Map 87): ARGENTINA: Catamarca, access to the town of Catamarca by Rte 38 (PEO-e 193), El Espinillo, Campo del Pucará, Las Estancias (PEO-e 101); Jujuy, El Simbolar, 25 km SW of Palma Sola (CML 1992), 1.5 km E of Tiraxi on Rte 29 Catamarca (CML 4898); La Rioja, Cuesta La Cébila, 1 km from Chumbicha by Rte 60 (CML 3751); Salta, Aguaray (MACN 3223), 27 km W of Aguas Blancas (MACN 17517), Arroyo Salado, 7 km E de Rosario de la Frontera, al lado del ACA (PEO-e 33), Cafayate (CML 852), El Corralito, ca. 23 km SW of Campo Quijano by Rte 51 (Jayat, Ortiz, and Miotti 2008), Embarcación (MACN 16439); Tucumán, 2 km S of Gobernador Garmendia by Rte 34 (PEO-e 72). BOLIVIA: Tarija, 1 km S of Camatindi (S. Anderson 1997), Tablada (S. Anderson 1997).

SUBSPECIES: Once treated as a subspecies of *A. varius* (Thomas 1926d; Cabrera 1961), Myers (1990) raised *A. simulator* to full species as a member of his *varius* group. He also considered *glaucinus* Thomas and *tartareus* Thomas to be subspecies of *A. simulator*; these latter two are regarded herein as species.

NATURAL HISTORY: *Akodon simulator* occupies the ecotone of Chaco and Yungas forests and all elevational belts of the Yungas, including highland grasslands (Pardiñas, D'Elía et al. 2006; Jayat, Ortiz, and Miotti 2008). In these areas, it has been recorded in forests, secondary growth, grasslands, cutover areas, and along rivers and streams. Few data are available on its reproductive biology or other population attributes. M. M. Díaz and Barquez (2007) reported females in Jujuy province with closed vaginas in January, May, June, and July, and females with open vaginas and/or lactating in May. They recorded males with scrotal testes in January and May, and with abdominal testes in May, June, and July. In Catamarca province, Mares, Ojeda, Braun et al. (1997) noted reproductively inactive males and females in July, and males with large scrotal testes and females with open vaginas in February and December. Embryo counts ranged from five to eight. These authors also recorded molting individuals during July and December. S. Anderson (1997) reported two females with no embryos in August. *Akodon simulator* is nocturnal and common throughout its range (Jayat, Ortiz, and Miotti 2009; Jayat and Ortiz 2010). Many sigmodontine species, such as *Akodon caenosus*, *A. spegazzinii*, *Oligoryzomys* cf. *O. flavescens*, *Calomys musculinus*, and *Andinomys edax*, share much of its range (S. Anderson 1997; M. M. Díaz et al. 2000; Barquez, Díaz, and Ojeda 2006; Jayat, Ortiz, and Miotti 2008, 2009; Jayat et al. 2010; Jayat and Ortiz 2010). In highland grasslands, this species has been trapped along with *A. boliviensis*, *Grao-*

mys edithae, *Necromys lactens*, *N. lasiurus*, *Phyllotis osilae*, *P. xanthopygus*, and *Reithrodon auritus* (Dalby and Mares 1974; Mares, Ojeda, Braun et al. 1997; M. M. Díaz et al. 2000; P. E. Ortiz, Cirignoli et al. 2000; Jayat, Ortiz, and Miotti 2008). It has been reported as a reservoir of hantavirus in Jujuy and Salta provinces (Pini et al. 2003; Puerta et al. 2006).

REMARKS: Braun et al. (2008) recovered, based on mtDNA sequences, a monophyletic *Akodon varius* group, placed *A. simulator* (plus *A. varius*) in a Yungas forest clade, and recommended recognizing *glaucinus*, *simulator*, and *tartareus* as full species. However, in more recent and broader taxonomic studies, the *A. varius* species group (*sensu* Myers 1990) was not recovered as a monophyletic unit in any analysis (Jayat et al. 2010). The primarily Yungas species, *A. simulator* and *A. varius*, formed a clade only distantly related to the one containing the mostly lowland species *A. dayi*, *A. dolores*, *A. iniscatus*, *A. molinae*, and *A. toba*. In this analysis, the Yungas clade formed part of a larger central Andean clade also composed of the *A. aerosus* species group plus *A. albiventer*.

Barquez et al. (1980) reported the karyotype of *A. simulator* (as *A. varius*), based on specimens from Catamarca and Tucumán provinces. This karyotype has $2n = 42$–41 and $FN = 42$, with polymorphism in the largest autosomes and X chromosome; the Y chromosome is a small submetacentric or subtelocentric. Liascovich et al. (1990) recorded eight different karyomorphs, with diploid numbers 38, 39, 40, 41, and 42, but a constant $FN = 42$ for specimens from Tucumán province.

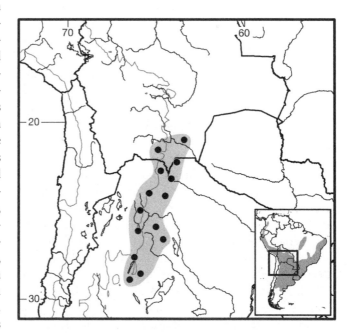

Map 87 Selected localities for *Akodon simulator* (●). Contour line = 2,000 m.

Akodon spegazzinii Thomas, 1897
Spegazzini's Grass Mouse, Spegazzini's Akodont
SYNONYMS:

Akodon spegazzinii Thomas, 1897d:216; type locality "Lower Cachi," Salta, Argentina; it is likely that
Thomas referred to the lower course of the Río Cachi,
which passes through the town of Cachi (25°07'11.93"S,
66°09'47.00"W, 2,341 m, Salta province), close to the
junction with the Río Calchaquí (see Jayat et al. 2010).

Akodon tucumanensis J. A. Allen, 1901c:410; type locality "Tucuman, Argentina, alt. 450 m.," San Miguel de
Tucumán, Tucumán, Argentina; the type was collected by
L. Dinelli (misspelled as Dionelli in J. A. Allen, 1901c), a
famous naturalist who worked in northwestern Argentina.

Akodon alterus Thomas, 1919d:496; type locality "Otro
Cerro," Catamarca, Argentina (= Otro Cerro, 28°45'S,
66°17'W, 2,023 m, Capayán, Catamarca, about 4 km
SSE of Cerro Catalán; Pardiñas et al. 2007).

Akodon leucolimnaeus Cabrera, 1926:320; type locality "Laguna Blanca, Catamarca (3,100 m de altura),"
Argentina.

Bolomys (?) *leucolimnaeus*: Gyldenstolpe, 1932a:119;
name combination.

Akodon [(*Bolomys*)] *leucolimnaeus*: Ellerman, 1941:415;
name combination.

Akodon boliviensis spegazzinii: Cabrera, 1961:441; name
combination.

Akodon boliviensis tucumanensis: Cabrera, 1961:441;
name combination.

Akodon lactens leucolimnaeus: Cabrera, 1961:454; name
combination.

Akodon minoprioi J. R. Contreras and Rosi, 1981; *nomen
nudum*.

[*Akodon*] *spegazzinii spegazzinii*: Myers, Patton, and
Smith, 1990:62; name combination.

[*Akodon*] *spegazzinii tucumanensis*: Myers, Patton, and
Smith, 1990:62; name combination.

Akodon oenos Braun, Mares, and Ojeda, 2000:218; type
locality "Argentina: Mendoza Province: Departamento
La Valle, La Pega (32°48'S, 68°40'W)," an old winery
NE Ciudad de Mendoza.

DESCRIPTION: Medium sized with total length 93–
196 mm, tail length 46–83 mm, ear length 12–21 mm, and
hindfoot length (with claw) 18–25 mm. Overall coloration
markedly variable among both individuals and populations, ranging from ochraceous brown to ruddy brown,
fulvous brown, or buffy brown, all in darker or paler tones;
dorsal coloration uniform from head to rump with base
color interrupted by scattered black or dark brown hairs;
yellow eye-ring always present but may be variable in expression; ventral color ranges from buffy to ruddy gray or
ochraceous gray, contrasting only slightly with dorsum;

few isolated white hairs rather than conspicuous patch of
white present on chin; tail conspicuously bicolored, dorsally brown or blackish brown and ventrally whitish or
buffy. Skulls with stout rostrum; zygomatic notches relatively narrow; frontal sinuses lightly swollen; interorbital
region hourglass in shape, with rounded or slightly squared
margins but without overhanging borders; breadth of zygomatic plates highly variable, with anterior margin straight
to slightly concave; incisive foramina generally extend to
anterior border of hypocones of M1s; mesopterygoid fossa
intermediate in breadth relative to other members of *boliviensis* group, with anterior margin slightly rounded or
squared and with straight to slightly divergent lateral borders; parapterygoid fossae similar in breadth or slightly
broader than mesopterygoid fossa, relatively shallow, and
with straight or slightly convex lateral margins diverging
posteriorly. Mandibular ramus delicate; anterior end of
masseteric crest situated just behind level of anterior border of m1; delicate coronoid process delicate extends just
above the condyle; condyle extends behind posterior margin of angular process. Upper incisors orthodont. M1 with
well-developed anteromedian flexus, conspicuous anteroloph, short mesoloph, small enteroloph, and weakly developed posteroflexus; M2 with remnant of anteroloph, relatively well-developed paraflexus, weak mesoloph, and very
shallow protoflexus and posteroflexus; M3 with paraflexus
and metaflexus clearly visible in most specimens, although
tooth lacks figure-8 shape because hypoflexus is vestigial;
m1 with well-developed anteromedian flexid and anterolabial cingulum, tiny ectostylid, and vestigial mesolophid;
m2 with very shallow protoflexid and tiny ectolophid; m3
retains remnant of protoflexid, mesoflexid, and transverse
and conspicuous hypoflexid, giving this tooth an S-shape.
Vertebral column with 13–14 thoracic ribs, 7–8 lumbar,
and 23–26 caudal vertebrae (N = 19). Greatest skull length
22.24–27.00 mm, upper molar series 3.80–4.60 mm.

DISTRIBUTION: *Akodon spegazzinii* is distributed
along the eastern slopes of the Andes in northwestern and
central Argentina, with reliable records from the provinces
of Salta, Tucumán, Catamarca, and Mendoza, and at elevations from 400 to 3,500 m (Jayat et al. 2010, 2011; Pardiñas, Teta, D'Elía, and Diaz 2011). Records from La Rioja
province (Thomas 1920g; cited as *A. alterus*) need confirmation; we have not seen these specimens. The fossil record for
A. spegazzinii is from two sites in the Tafí valley, Tucumán
province, of both Middle-Upper Pleistocene (P. E. Ortiz
and Pardiñas 2001) and Lower Holocene ages (P. E. Ortiz
and Jayat 2007a), and from additional Upper Pleistocene and
Upper Holocene localities in Catamarca and Tucumán provinces (P. E. Ortiz 2001; P. E. Ortiz et al. 2011a,b).

SELECTED LOCALITIES (Map 88): ARGENTINA:
Catamarca, El Bolsón (JPJ 2058), ca. 2 km SE of Huaico

Hondo, on Rte 42, E of Portezuelo (MACN 23430), Laguna Blanca (type locality of *Akodon leucolimnaeus* Cabrera), Loma Atravesada, ca. 3 km NW of Leandro Vega ranch, NW of Chumbicha (JPJ 1158); La Rioja, Cueva de Díaz, Sierra de Famatina (MACN 24040); Mendoza, Laguna Llancanelo (Pardiñas, Teta, D'Elía, and Diaz 2011), Villavicencio, 2 km S of Rte 32 (Braun et al. 2000); Salta, Arroyo Salado, 7 km E of Rosario de la Frontera, on the side of ACA (PEO-e 34), ca. 2 km NNE of Cachi Adentro, on the road to Las Pailas (MACN 23495), Campo Quijano, ca. 5 km NW, Km 30 on RN 51, Quebrada del Toro (JPJ 139), Santa Rosa de Tastil (MACN 15643); Tucumán, La Tranquera, northern border Los Chorrillos farm, on highway 205 (JPJ 1773), San Miguel de Tucumán (type locality of *Akodon tucumanensis* J. A. Allen).

SUBSPECIES: Cabrera (1961) treated *A. spegazzinii* and *A. tucumanensis* as subspecies of *A. boliviensis* and *A. alterus* as a synonym of *A. boliviensis tucumanensis*. Myers et al. (1990) excluded *A. spegazzinii* and *A. tucumanensis* from the synonymy of *A. boliviensis*. These authors viewed *spegazzinii* as a valid species, listed *tucumanensis* as a subspecies of *spegazzinii*, and regarded *alterus* as "properly allied" to *spegazzinii*. Blaustein et al. (1992) studied *alterus* and *tucumanensis* and found minor qualitative and morphometric differences and identical cytogenetic characteristics. The sample of *A. spegazzinii* from southern Mendoza (representing *oenos* Braun, Mares, and Ojeda) was both the most geographically and genetically distant of those studied by Jayat et al. (2010). Whether this represents a simple isolation-by-distance pattern or reflects a step cline that might support the trinomial status of *oenos* needs further testing.

NATURAL HISTORY: *Akodon spegazzinii* occupies varied habitats, from the ecotone between Chacoan and Yungas forests to semiarid Monte, pre-Puna, Puna, and high Andean environments. In more xeric areas, this species occurs only in grassy microhabitats usually along watercourses (Jayat, Ortiz, and Miotti 2008; Jayat et al. 2010). There are few published observations on reproduction and molting for this species, but Mares et al. (1997) recorded pregnant and lactating females in December and February in the Monte Desert of central Catamarca province. The number of embryos ranged from five to eight. All males caught in these months had large scrotal testes, and two were molting. Jayat et al. (2010) stated that reproduction appears to occur throughout the year, although with a clear peak between November and April in Yungas environments. In this environment, most of these mice were molting in fall and winter (April to August). In northern Mendoza province, *A. spegazzinii* occupies low-elevation halophytic desert scrub at the base of the Andean and pre-Andean ranges. Males with abdominal or subscrotal testes

and females with closed vaginas were caught in August and September: males with scrotal testes in February, October, and December, and pregnant females (embryo range: 3–5) in October and December. Population density at La Pega (the type locality of *oenos* Braun, Mares, and Ojeda) was estimated as 21 animals/ha, with a biomass of 461 g/ha, although only eight individuals were captured (J. R. Contreras and Rosi 1981). The home range of a single individual was calculated to be less than 300 m². *Akodon spegazzinii* is a common species throughout its distribution. It is chiefly nocturnal although it can be caught during the day (Mares, Ojeda, Braun et al. 1997). At localities above 3,000 m in Puna and high Andean habitats, *A. spegazzinii* is sympatric with *Calomys lepidus*, *Eligmodontia* sp., *Graomys edithae*, *Neotomys ebriosus*, *Phyllotis xanthopygus*, and *Reithrodon auritus* (Dalby and Mares 1974; P. E. Ortiz, Cirignoli et al. 2000; Jayat, Ortiz, and Miotti 2008; Jayat et al. 2010). In grasslands, it was caught with *Necromys lactens*, *N. lasiurus*, and *Phyllotis osilae*. In forested areas of the Yungas, *A. spegazzinii* coexists with *Abrothrix illutea*, *Oligoryzomys* sp., *Oxymycterus paramensis*, *O. wayku*, and *Phyllotis anitae* (M. M. Díaz et al. 2000; Jayat, Ortiz, and Miotti 2008; Jayat et al. 2010). In the ecotone between Yungas and Chaco, *A. spegazzinii* was caught with *Calomys fecundus*, *Calomys musculinus*, *Graomys chacoensis*, *Holochilus chacarius*, and *Necromys* sp. Other species broadly distributed in the region, such as *Akodon caenosus*, *A. simulator*, *Oligoryzomys* cf. *flavescens*, *C. musculinus*, and *Andinomys edax*, have also been found in sympatry with *A. spegazzinii*. In addition, in Monte Desert environments *Akodon albiventer*, *A. dolores*, *Andalgalomys olrogi*, *Eligmodontia bolsonensis*, *E. moreni*, *Graomys griseoflavus*, and *Phyllotis xanthopygus* all co-occur with *A. spegazzinii* (Mares, Ojeda, Braun et al. 1997; M. M. Díaz et al. 2000; P. E. Ortiz, Cirignoli et al. 2000; Pardiñas, Teta, D'Elía, and Diaz 2011).

REMARKS: Until recently, treatments of the nominal forms *spegazzinii* Thomas, *tucumanensis* J. A. Allen, *alterus* Thomas, and *leucolimnaeus* Cabrera has been variable, with each hypothesis lacking a defensible basis (e.g., M. M. Díaz et al. 2000; Musser and Carleton 2005; Pardiñas, D'Elía et al. 2006; M. M. Díaz and Barquez 2007). *Akodon leucolimnaeus*, considered a synonym of *Necromys lactens* for many decades (e.g., Cabrera 1961; Reig 1987; Musser and Carleton 1993), was recently placed in the *boliviensis* group of *Akodon* (Galliari et al. 1996) and treated as full species (Musser and Carleton 2005; Pardiñas, D'Elía et al. 2006) in *Akodon*. Following a comprehensive study of thousands of specimens of *Akodon* trapped from Jujuy to Catamarca provinces, Jayat (2009) produced the first detailed analysis of the *boliviensis* group of *Akodon* in Argentina. He included in *A. spegazzinii* other putative species whose taxonomic status had been debated, including *alterus, leucolimnaeus*, and *tucumanensis* (see Myers et al. 1990:60–62). Jayat (2009; see also Blaustein et al. 1992) documented that *A. spegazzinii* is a medium-sized species that exhibits substantial phenotypic variation, especially in pelage color, which ranges from darker in lowland populations (previously referred to *tucumanensis*) to brownish and reddish in medium- to high-elevation populations (e.g., *alterus* and *leucolimnaeus*). Subsequently, Jayat et al. (2010) corroborated this morphologically based hypothesis with an expanded morphometric and molecular sequence analyses based on extensive samples, which included both topotypes and samples representing of all nominal forms. They concluded that there were two pairs of nominal taxa, *A. spegazzinii* + *A. alterus* and *A. tucumanensis* + *A. leucolimnaeus*, with each pair consisting of taxa that were indistinguishable between from each other. However, there were significant differences in several measurements between pairs. These authors also demonstrated that specimens from type localities (Cachi, Laguna Blanca, vicinity of Otro Cerro, and Yungas forest in Tucumán) exhibited no clear or consistent morphological differences from each other. Furthermore, topotypes of the nominal taxon (*spegazzinii, tucumanensis, alterus*, and *leucolimnaeus*) all formed part of a single monophyletic assemblage of haplotypes. Pardiñas, Teta, D'Elía, and Diaz (2011) extended these analyses to include specimens representing *oenos* Braun, Mares and Ojeda, documenting that this taxon was part of the same morphological and molecular assemblage. This combination of a broad array of studied morphological characters and molecular analyses lends strong support to the hypothesis that *tucumanensis* J. A. Allen, *alterus* Thomas, *leucolimnaeus* Cabrera, and *oenos* Braun, Mares and Ojeda are best considered junior synonyms of *A. spegazzinii* Thomas.

The taxonomic status of *A. spegazzinii* with respect to *A. boliviensis*, however, remains a debatable issue. Based on a small series of specimens, Myers et al. (1990) proposed that *spegazzinii* was a valid species, and much of the data available to date support this hypothesis. For example, these two allopatric taxa are reciprocally monophyletic with respect to mtDNA sequences, and each is clearly identifiable in morphometric multivariate space and by qualitative morphological traits (Jayat et al. 2010). The two taxa also have largely different habitat preferences and distinctly different karyotypes, including heteromorphism in the largest autosomal pair and the X chromosome in *A. spegazzinii* (A. L. Gardner and Patton 1976; Barquez et al. 1980; Myers et al. 1990).

Barquez et al. (1980; as *A. boliviensis*) described the karyotype of *A. spegazzinii* based on 10 specimens from Catamarca, Salta, and Tucumán provinces, with $2n = 40$, FN = 40–41, and an autosomal complement consisting of

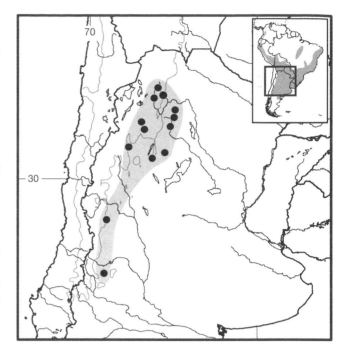

Map 88   Selected localities for *Akodon spegazzinii* (●). Contour line = 2,000 m.

one pair of large and heteromorphic chromosomes (acrocentric or subtelocentric), 17 pairs of acrocentrics that grade in size from medium to small, and a single pair of very small metacentrics. The X chromosome is large and may be subtelocentric or acrocentric; the Y is a small submetacentric. Karyotypes of samples allocated to *A. varius neocenus* (N. O. Bianchi et al. 1971) or *A. varius* (N. O. Bianchi and Merani 1984) are of *A. spegazzinii*, as were some of those grouped as *A. neocenus* in the analysis of allozyme variation (Apfelbaum and Reig 1989).

Akodon subfuscus Osgood, 1944
Puno Grass Mouse, Puno Akodont

SYNONYMS:

Akodon boliviensis subfuscus Osgood, 1944:195; type locality "Limbani, Puno, Inambari drainage, Peru. Altitude about 9,000 feet."

A[kodon]. subfuscus: Myers and Patton, 1989a:28; first use of current name combination.

Akodon subfuscus subfuscus: Myers, Patton, and Smith, 1990:80; name combination.

Akodon subfuscus arequipae Myers, Patton, and Smith, 1990:86; type locality "15 km S Callalli, Depto. Arequipa, Peru, elevation 4150 m."

DESCRIPTION: Small, with mean total length 159.6 mm, tail length 68.2 mm, hindfoot length 20.6 mm, and ear length 13.8 mm. Dorsal coloration uniform from head to rump, olivaceous to grayish brown and heavily lined with black; faint buffy or yellowish eye-ring present; venter clearer than dorsum, with tip of hairs dark buff to whitish; most specimens with small patch of all-white hairs on chin; tail sharply bicolored and moderately furry, with middorsal stripe. Skull with weakly flared zygomatic arches; narrow, shallow zygomatic notches; interorbital region with sides usually rounded or squared posteriorly; weakly developed temporal and mastoid ridges; narrow zygomatic plates; slender tympanic hook; and narrow to moderately open mesopterygoid fossa with clearly squared anterior margin typically bearing median spine. M1 with well-developed parastyle and mesostyle, and metalophule usually present but tiny; M2 with anterior cingulum and conspicuous mesostyle; M3 small and usually oval in outline. Both m1 and m2 with ectostylid usually present, and anterior cingulid generally present on m3.

DISTRIBUTION: *Akodon subfuscus* is widely distributed in the high-elevation Puna grasslands of southern Peru, from the western Andean slopes in Ayacucho and Arequipa departments, through the highlands of Apurimac and Cusco departments and the eastern Andean slopes in Cusco and Puno departments, and then south to at least Cochabamba department in Bolivia. Most records for this species are from elevations above 2,500 m, but the species may occur at elevations as low as 1,900 m (Myers and Patton 1989a; Myers et al. 1990; S. Anderson 1997; Salazar-Bravo, Yensen et al. 2002; R. Vargas 2005; Solari 2007; Tarifa et al. 2007). Fossils of this species are unknown.

SELECTED LOCALITIES (Map 89): BOLIVIA: Cochabamba, Cocapata (Tarifa et al. 2007), Cuesta Cucho (S. Anderson 1997); La Paz, Alaska Mine (Myers et al. 1990), Ayane [= Yani] (S. Anderson 1997), Pelechuco (S. Anderson 1997). PERU: Apurimac, 36 km S (by road) of Chalhuanca (Myers et al. 1990); Arequipa, 28 km NE (by road) of Arequipia (MVZ 136253), 15 km S of Callalli (type locality of *Akodon subfuscus arequipae* Myers, Patton, and Smith), Salinas, 22 mi E of Arequipa (MVZ 116074), 2 km W of Sumbay (MVZ 174078); Ayacucho, 2 mi SE of Huanta (MVZ 141321), 32 km NE (by road) of Pampamarca (MVZ 174248), 10 mi WNW of Puquio (Myers et al. 1990), San Miguel Tambo (Myers et al. 1990); Cusco, 55.4 km N (by road) of Calca (MVZ 166720), 5 km N of Huancarani (MVZ 172369), 32 km NE (by road) of Paucartambo, km 112 (MVZ 151572), 90 km SE (by road) of Quillabamba (MVZ 166721), 16 km SW of Yauri (MVZ 174222); Puno, Limbani (type locality of *Akodon boliviensis subfuscus* Osgood), 6.5 km SW of Ollachea (MVZ 172984).

SUBSPECIES: In their review of the *boliviensis* group, Myers et al. (1990) recognized two subspecies, one of which they described as new: *A. s. subfuscus*, characterized by larger size, darker pelage, a broader mesopterygoid fossa, and a distribution largely across the northern Altiplano and eastern Andean slopes; and *A. s. arequipae*, a smaller and paler taxon predominantly limited to the western side of the Altiplano and upper slopes of the western Andes. The difference in size is most noticeable in adult animals, for which several craniodental measurements are 5–10% larger in *subfuscus* than in *arequipae*. There are several dental differences, including an absent or tiny mesostylid on m1 in *arequipae* versus an unusually large mesostylid in *subfuscus*. Data from 26 allozyme loci, however, failed to separate populations assigned to the two subspecies by the morphological criteria of their respective diagnoses.

NATURAL HISTORY: In Peru, *A. subfuscus* inhabits bunchgrass at high Andean elevations, where it is easily trapped in grass clumps mixed with shrubs and in areas of grass and large rocks. It is also common along the base of stone walls bordering agricultural plots (Myers et al. 1990; Solari 2007). In Bolivia, this species occurs in fragmented *Polylepis* woodlands between 3,200 and 3,800 m elevation in Puna and Yungas environments, and treeless grassland Puna created by repeated human-caused fires (Salazar-Bravo, Yensen et al. 2002; R. Vargas 2005; Tarifa et al. 2007). Solari (2007) recorded high proportions of arthropods (adults and larvae) in stomach contents (with a mean of 59.3%), and which showed little variation seasonally or between the sexes in southern Peru. Plant material was also important in the diet of this species, comprising 26% of the total volume analyzed. Solari (2007) stated that *A. subfuscus* exhibited only a slight specialization for insectivory, with a niche breadth (following the Levins metric) of 4.53, which overlapped extensively (0.98) with that of the sympatric but typically habitat-segregated *A. torques*. *Akodon subfuscus* occurs, or is contiguously allopatric, with several other *Akodon* in high areas of Peru and Bolivia, including *A. aerosus*, *A. boliviensis*, *A. fumeus*, *A. juninensis*, *A. kofordi*, *A. lutescens*, *A. torques*, and *A. varius* (Myers et al. 1990; Patton and Smith 1992b; S. Anderson 1997; Salazar-Bravo, Yensen et al. 2002; R. Vargas 2005; Solari 2007; Tarifa et al. 2007).

REMARKS: Although it was originally described as a subspecies of *A. boliviensis*, Myers et al. (1990) raised *A. subfuscus* to specific level based on substantive craniometric, soft anatomical (including male phallus and palatal rugae), and allozyme differences. They included *A. subfuscus* in their *boliviensis* group, a hypothesis since corroborated by phylogenetic analyses of mtDNA cytochrome-*b* sequences (M. F. Smith and Patton 1993, 2007; Jayat et al. 2010).

The karyotype has $2n = 40$, $FN = 40$. The autosomal complement is identical to that described for *A. boliviensis* and *A. lutescens*. The X chromosome is a medium-sized subtelocentric, as in *A. lutescens*, but the Y is a very small acrocentric (Myers et al. 1990).

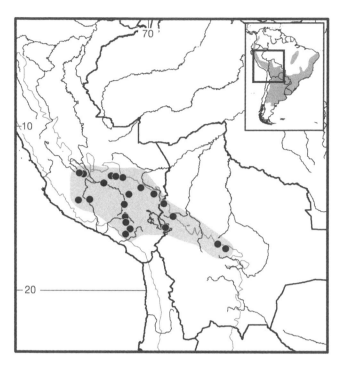

Map 89 Selected localities for *Akodon subfuscus* (●). Contour line = 2,000 m.

Akodon surdus Thomas, 1917
Silent Grass Mouse, Slate-bellied Akodont

SYNONYM:

Akodon surdus Thomas, 1917g:2; type locality "Huadquiña, 5,000 feet"; Hacienda Huadquiña is on the southern margin of the Río Urubamba, Santa Teresa, La Convención, Cusco, Peru (ca. 13.13°S, 72.59°N; see route map of the Peruvian expedition in Bingham 1916).

DESCRIPTION: A dark-colored species resembling *A. mollis*; comparatively large, with head and body length 110 mm, tail length 80 mm; stout and heavily built; fur rather coarse; dorsum dark olivaceous gray dorsal and venter soiled grayish, but only vaguely separable; medium length ears; forefeet and hindfeet grayish brown above but blackish on soles; tail fairly long, finely haired, and essentially unicolored but only slightly paler below. Skull robust (greatest length 28 mm, upper molar series 4.9 mm), with domed profile; interorbital region broad with squared edges; zygomatic plates broad, project forward from bottom to top, with straight anterior border; incisive foramina widely open; auditorybullae moderately sized. Myers (1990) distinguished *surdus* by its short, crisp, dark fur, relatively flat cranial profile, generally rounded interorbital region, small interparietal, and very shallow but moderately broad zygomatic notches.

DISTRIBUTION: This species is known only from the vicinity of Machu Picchu in the mid-elevation montane forest of the Urubamba Valley in Cusco department, Peru,

at elevations between 600 and 2,000 m (Thomas 1917g, 1920h). No fossils are known for *A. surdus*.

SELECTED LOCALITIES (Map 90): PERU: Cusco, Huadquiña (type locality of *Akodon surdus* Thomas), Idma (Thomas 1920h), Machu Picchu (ZMUM 16986; Patton and Smith 1992a), Paltaybamba [= Paltaibamba] (Thomas 1920h), Santa Ana (Thomas 1920h).

SUBSPECIES: *Akodon surdus* is monotypic.

NATURAL HISTORY: Nothing is known about the ecology of this species. The few specimens available were trapped in very wet mid-montane forest on the steep slopes of the Urubamba Valley.

REMARKS: *Akodon surdus* has been treated as a species by most authors since its description (Reig 1987; Myers 1990; Hershkovitz 1990b; Musser and Carleton 2005), but Patton and Smith (1992b) suggested it *surdus* was best regarded as a local variant of *A. aerosus* confined to the Urubamba Valley. Specimens from localities to the west and south of the Urubamba Valley along the eastern flank of the southern Peruvian Andes, and in the same moist forest habitat, are uniformly treated as *A. aerosus* (see that account), and mtDNA sequences obtained from skin samples of museum specimens of *A. surdus* from Machu Picchu were placed in the same monophyletic clade as all geographic samples of *A. aerosus* from southern Peru (Patton and Smith 1992a). Samples of both *A. surdus* and *A. aerosus* in southern Peru do share the same dark color tones

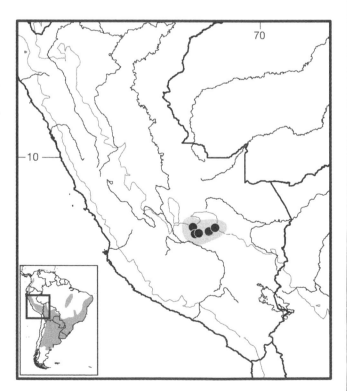

Map 90 Selected localities for *Akodon surdus* (●). Contour line = 2,000 m.

and general cranial attributes, but careful and direct comparisons among them have yet to be made. Therefore, M. F. Smith and Patton (2007:846), in establishing the *aerosus* group of *Akodon* based on mtDNA sequences, concluded "while data are as yet unavailable, *A. surdus* . . . is likely a member of this group, either as a distinct species or perhaps conspecific with *aerosus*."

Akodon sylvanus Thomas, 1921
Forest Grass Mouse, Woodland Akodont

SYNONYMS:

Akodon sylvanus Thomas, 1921d:184; type locality "Sierra de Santa Barbara, S.E. Jujuy. Type from Sunchal, 1200 m"; clarified by Jayat, Ortiz et al. (2007) as Sunchal (24°15′S, 64°26′W, 1,454 m), Santa Bárbara department, Jujuy province, Argentina.

Akodon sylvanus sylvanus: Gyldenstolpe, 1932a:104; name combination.

Akodon azarae sylvanus: Cabrera, 1961:441; name combination.

DESCRIPTION: Medium-sized species with body length 151–203 mm, tail length from 68–89 mm, ear length from 14–21 mm, and hindfoot length (with claw) from 22–27 mm. Dorsal color uniform along body, olivaceous brown finely streaked with black hairs, generally not contrasting with ventral color; ears same color as dorsum; eye-ring either absent or weakly developed; white spotting on chin barely developed; tail only slightly bicolored, with somewhat paler venter. Skull relatively narrow and elongated, with prominent rostrum, average zygomatic notches in breadth and depth, slightly expanded zygomatic arches, and uninflated braincase; interorbital region with rounded margins; mesopterygoid fossa relatively narrow, with rounded or squared anterior border without medial process. Upper incisors orthodont. Molars more hypsodont than other species of *boliviensis* group. M1 with well-developed anteroloph-parastyle, mesoloph may extend to buccal edge of tooth; mesoloph and paraflexus of M2 less developed than those structures on M1; M3 with poorly developed paraflexus but conspicuous hypoflexus. Protostylid of m1 well developed but metastylid, mesolophid, and mesostylid tiny; m2 without evidence of either ectolophid (ectostylid) or mesolophid; m3 with well-developed hypoflexid and mesoflexid. Greatest skull length 25.30–28.30 mm; upper molar series length 4.04–4.70 mm. Axial skeleton with 7 cervical, 13 thoracic, 7 lumbar, and 27–29 caudal vertebrae, and 13 ribs (N = 10).

DISTRIBUTION: *Akodon sylvanus* is known from only a few localities in the Sierra de Santa Bárbara and the surrounding area, all in southeastern Jujuy province, Argentina, and between 900 and 2,400 m (M. M. Díaz and Barquez 2007; Jayat, Ortiz et al. 2007, 2008; Jayat et al. 2010). No fossils are known.

SELECTED LOCALITIES (Map 91): ARGENTINA: Jujuy, El Simbolar, 25 km SW of Palma Sola (CML 2029), Finca El Piquete, margin of Río Volcán, ca. 5 km from junction of Río Tamango and logging road (MACN 23490), La Antena, Sierra del Centinela, S of El Fuerte (MACN 23434), La Herradura, 12 km SW of El Fuerte, on Rte 6 (CNP-E 1485).

SUBSPECIES: *Akodon sylvanus* is monotypic.

NATURAL HISTORY: *Akodon sylvanus* occupies forest and high-elevation grasslands of the Yungas ecoregion as well as grasslands in the Chacoan ecoregion (M. M. Díaz and Barquez 2007; Jayat, Ortiz et al. 2007; Jayat, Ortiz, and Miotti 2008; Jayat et al. 2010). Reproductive activity begins in early spring (late August), with all individuals caught during winter (June) reproductively inactive. Almost all individuals trapped in grasslands between June and October were molting (Jayat, Ortiz et al. 2007; Jayat et al. 2010). *Akodon sylvanus* has been caught sympatrically with *A. caenosus*, *A. budini*, and *A. simulator*. It is the most common species in its restricted range, comprising 40–60% of all sigmodontines caught in the La Herradura (N = 50) and Sierra del Centinela (N = 46) areas (Jayat, Ortiz et al. 2007; Jayat et al. 2010).

REMARKS: Thomas (1921d) described *A. sylvanus*, based on specimens from southeastern Jujuy province, Argentina. Four years later (Thomas 1925b), he erected *A. sylvanus pervalens*, from Tarija department, Bolivia, a taxon now recognized as unrelated to *A. sylvanus* and probably allied to the *varius* group (see that account). Thomas

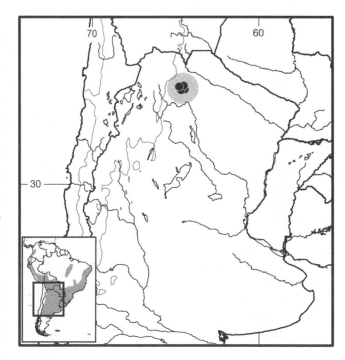

Map 91 Selected localities for *Akodon sylvanus* (●). Contour line = 2,000 m.

(1921d:184–185) characterized *sylvanus* as "A large species of the *arenicola* [i.e., *azarae*] group . . . this Santa Barbara *Akodon* has a curious resemblance to the *A. arenicola hunteri* of the Parana Delta, with which it agrees closely in size and colour." Cabrera (1961) treated *sylvanus* as a valid subspecies of *A. azarae*. Myers (1990) regarded *azarae*, *sylvanus*, and *pervalens* as different species, as did Musser and Carleton (2005) and Pardiñas, D'Elía et al. (2006). Jayat, Ortiz et al. (2007) and Jayat et al. (2010), based on both molecular and morphological character analyses, validated *A. sylvanus* as a species and placed it in the *boliviensis* group. Jayat et al. (2010) reidentified, as *Akodon fumeus*, specimens wrongly referred to *A. sylvanus* by Jayat, Ortiz et al. (2007), from the Baritú area (northern Salta province) and Las Capillas (Jujuy province), and thereby redefined the distribution and morphologic characteristics of *A. sylvanus*.

Akodon toba Thomas, 1921

Toba Grass Mouse, Toba Akodont

SYNONYMS:

Akodon toba Thomas, 1921b:178; type locality "Jesematathla, Northern Chaco. Alt. 100 m."; according to Pine and Wetzel (1976:651), "This place is apparently the same as 'Estancia Yesamathasla' . . . 76–80 k E Juan de Zalazar," at roughly 23.07°S and 58.48°W, Presidente Hayes, Paraguay.

Akodon varius toba: Cabrera, 1961:450; name combination.

DESCRIPTION: Large with total body length 183–203.7 mm, tail length 87–85.1 mm, average hindfoot length 25.5 mm, and ear length 19–20 mm. Dorsal pelage strongly olivaceous, finely ticked with gray; eye-ring present but not clearly defined; venter frosted grayish, white, or buffy, color that contrasts strongly with dorsum; some specimens with white spot on chin less conspicuous than in most members of *varius* group; tail sharply bicolored. Skull strongly bowed in lateral view and with notably well-developed and widely expanded zygomatic arches; zygomatic notches unusually broad and deep; supraorbital and temporal ridges well developed; interorbital region with squared edges that become beaded in older individuals; zygomatic plates notably broad with slightly concave anterior margin; incisive foramina reach posteriorly to level of M1 protoflexus; both mesopterygoid and parapterygoid fossae proportionally narrow. Mandible strongly built for genus, with coronoid process clearly surpassing condyle height, which extends slightly posterior to vertical plane of angular process. Upper incisors generally opisthodont. Upper molars not particularly hypsodont but posses marked anteromedian flexus on M1, tiny mesoloph and poorly developed posteroflexus poorly developed on both M1 and M2, remnant anteroloph on M2, and well-developed paraflexus and metaflexus on

M3. Of the lower molars, m1 with well-developed anteromedian flexid and anterolabial cingulum; m2 with shallow protoflexid; and m3 proportionally large, with well-developed mesoflexid and transverse hypoflexid. Greatest skull length 28.5–29.5 mm; upper molar series length from 4.7–5.2 mm.

DISTRIBUTION: *Akodon toba* occurs primarily in Chaco environments of southeastern Bolivia, western Paraguay, southern Brazil, and northern Argentina, at elevations below 1,000 m (Massoia 1971c; Myers 1990; S. Anderson 1997; Jayat et al. 2006; Bonvicino, Oliveira, and D'Andrea 2008; Braun et al. 2008; Cáceres et al. 2008). A few records extend into the forested Yungas habitats, reaching elevations of 1,500 m (see Myers 1990; M. M. Díaz et al. 2000; M. M. Díaz and Barquez 2007). Specimens referred to *A. toba* from Yungas environments in Jujuy and Salta provinces, especially those located some distance from Chaco environments and at elevations above 700 m (such as those mentioned by Myers 1990; M. M. Díaz et al. 2000; and M. M. Díaz and Barquez 2007) should be reexamined. Jayat, Ortiz, and Miotti (2009) suggested that several of these specimens are, in fact, *A. simulator*. No fossils are known for *A. toba*.

SELECTED LOCALITIES (Map 92): ARGENTINA: Formosa, Parque Nacional Río Pilcomayo (Massoia 1971c); Jujuy, Villa Carolina, Río Lavallén (M. M. Díaz and Barquez 2007); Santiago del Estero, Bandera (Jayat et al. 2006). BOLIVIA: Chuquisaca, 2 km S and 10 km E of Tiquipa, Laguna Palmar (S. Anderson 1997); Santa Cruz, 4 km N and 1 km W of Santiago de Chiquitos (S. Anderson 1997), Tita (S. Anderson 1997); Tarija, 3 km WNW of Caraparí (S. Anderson 1997). BRAZIL: Mato Grosso, Corumbá, 7 km WSW of Urucum, São Marcus road (Myers 1990). PARAGUAY: Alto Paraguay, 50 km WNW of Fortín Madrejón, Cerro León (Myers 1990); Boquerón, Estancia Iparoma, 19 km N of Filadelfia (Myers 1990); Presidente Hayes, 8 km NE of Juan de Salazar (Braun et al. 2008), 69 km by road NW of Villa Hayes (Myers 1990).

SUBSPECIES: *Akodon toba* is monotypic.

NATURAL HISTORY: *Akodon toba* occupies several habitats in the Chacoan ecoregion, including patches of forest intermingled with palm savanna, marshland, thorn forest, semiarid scrublands, and gallery forest (Massoia 1971c; Myers 1990; Pardiñas and Teta 2005). In Paraguay, Yahnke et al. (2001) recorded a strong preference for grassland habitats, although *A. toba* was also captured in other habitats, such as croplands, forest, and thorn scrub. S. Anderson (1997) reported females with no embryos for August and October. One female with a closed vagina and one male with semiscrotal testes were caught in November in eastern Jujuy province (P. Jayat, unpubl. data). One of these specimens was molting.

Akodon toba is sympatric with many of the sigmodontines documented for dry Chaco environments (Massoia 1971c; Myers 1990; Gamarra de Fox and Martin 1996; S. Anderson 1997; Yahnke et al. 2001). In eastern Argentina, this species was found with the widely distributed *A. azarae*, and there are isolated populations in humid Chaco environments, where it coexists with species typically associated with the floodplain and gallery forests of the Río Paraguay (Pardiñas and Teta 2005).

Known ectoparasites reported for *A. toba* are the mite *Androlaelaps rotundus* and the flea *Craneopsylla minerva minerva* (Gettinger and Owen 2000; Lareschi et al. 2003). Whitaker and Abrell (1978) reported on ectoparasites for the Paraguayan specimens they referred to *A. varius* but which are most likely *A. toba*.

REMARKS: *Akodon toba* was treated as a subspecies of *A. varius* by Cabrera (1961), likely following Thomas (1921b:178–9), who noted that "[*A. toba*] skull is, on the whole, most like that of *A. simulator* [regarded by Cabrera as a subspecies of *varius*] and its allies of North Argentina west of the Chaco." Myers (1990) treated *A. toba* as a full species, a decision supported by subsequent authors (Musser and Carleton 2005; Pardiñas, D'Elía et al. 2006). Recent mtDNA molecular-based analyses placed *A. toba* close to *A. dolores* and in a clade that also included other lowland species such as *A. dayi* and *A. iniscatus* (including *nucus*) rather than to the Yungas forest clade of which *A. varius* is a member (Braun et al. 2008; Jayat et al. 2010). Jayat et al. (2010) referred to the lowland assemblage of species as the *dolores* group. Although more limited in taxon coverage, this group

was also recovered by D'Elía's (2003a) combined mtDNA and nucDNA analyses, although he incorrectly identified a specimen of *A. toba* as *A. varius*.

A. L. Gardner and Patton (1976) reported a karyotype of *A. toba* from Paraguayan specimens as $2n=40$ and FN$=40$, but Myers (1990) reported $2n=42$ or 43 and FN$=44$ for specimens from the central and northern Chaco of Paraguay. In the $2n=42$ complement, the first autosomal pair is heteromorphic, composed of two large metacentrics or one metacentric and one subtelocentric. The autosomes form a size-graded series of uniarmed elements, with the smallest pair being very small metacentrics. The sex chromosomes are acrocentric (Myers 1990), but A. L. Gardner and Patton (1976) reported a subtelocentric X and a submetacentric Y.

Akodon torques (Thomas, 1917)

Cloud Forest Grass Mouse, Cloud Forest Akodont

SYNONYMS:

Microxus torques Thomas, 1917g:3; type locality "Matchu [*sic*] Picchu, 10,000 feet," Machu Picchu (13.163°S, 72.544°W), Cusco, Peru.

[*Akodon*] *torques*: Thomas, 1927a:370; first use of current name combination.

Microxus (?) *torques*: Gyldenstolpe, 1932a:134; name combination.

Akodon orophilus torques: Cabrera, 1961:446; name combination.

DESCRIPTION: Somewhat larger than *A. mimus* (total length about 195 mm), with close, soft, and woolly fur, smoky to olivaceous gray dorsally and slightly paler ventrally. Ears short and colored like dorsum; forefeet and hindfeet grayish brown above, with soles flesh colored, not blackish; tail proportionally long (averaging about 94 mm), finely haired, and brown above. Skull slightly bowed in lateral profile; condyloincisive length averages about 24.4 mm, upper toothrow about 4.3 mm; rostrum somewhat elongated, but not to degree found in *A. mimus*; interorbital region broad, with smooth or only slightly squared edges; zygomatic plates intermediate in width, not as narrow as in *A. mimus* nor broad as in *A. aerosus*; anterior border slants posteriorly from base to top, but not to degree as *A. mimus*; incisive foramina quite open and elongated, extending to anterior one-third of M1s.

DISTRIBUTION: *Akodon torques* occurs in the upper elfin forests on the eastern slope of the Andes, from Junín, Ayacucho, and Cusco departments, Perú, at elevations between 2,000 and 3,500 m (Patton and Smith 1992b:92). This species is replaced in similar habitats by *A. orophilus* to the north and by *A. mimus* to the south. In lower elevation montane forest, it is replaced by *A. aerosus* and in the bunch grass habitats above tree line by *A. subfuscus*. No fossils are known.

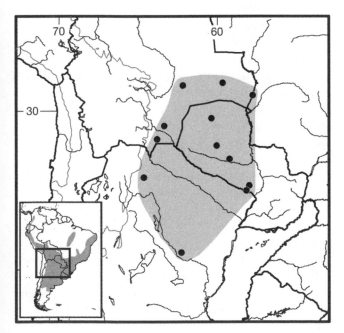

Map 92 Selected localities for *Akodon toba* (●). Contour line = 2,000 m.

SELECTED LOCALITIES (Map 93): PERU: Ayacucho, Puncu, ca. 30 km NE of Tambo (Patton and Smith 1992a); Cusco, Amacho (Patton and Smith 1992a), La Convención (USNM 582129), Lucma, Cosireni Pass, Quirapata (USNM 194679), Machu Picchu (type locality of *Microxus torques* Thomas), 1 km below of Marcapata (LSUMZ 19337), 32 km NE (by road) of Paucartambo, km 112 (MVZ 166758), 54 km NE (by road) of Paucartambo, km 134 (MVZ 171736), 90 km SE (by road) of Quillabamba (MVZ 166723), Tres Cruces, 18 km N of Paucartambo (MVZ 115749); Junín, Satipo, Cordillera de Vilcabamba (USNM 582178).

SUBSPECIES: *Akodon torques* is monotypic.

NATURAL HISTORY: *Akodon torques* is syntopic with, but segregated by habitat from, *A. subfuscus*, being found commonly in the dense moss mats under canopies formed by trees or large shrubs. In the highland grasslands ("pajonal") of Parque Nacional Manu, east of Paucartambo in the humid Puna of the Cordillera Oriental, Cusco department, Peru (3,300 to 3,800 m), *A. subfuscus* and *A. torques* represent 35% of total captures. In this same area, *A. torques* is primarily insectivorous, consuming mostly arthropod larvae (Solari 2007). In large series collected in May and July in Cusco department, most females were nulliparous, none were pregnant, but parous adults had embryo scar counts ranging from two to four. Old males collected at the same time were scrotal with enlarged accessory glands, although most males were not in reproductive condition (data from specimen labels in the MVZ collection).

REMARKS: Thomas (1917g, 1920h) placed *torques* with *Microxus* instead of *Akodon* after comparing it with the type species of his genus *M. mimus*, although he recognized than the former has "[a] less characteristically Microxine zygomatic plate" (Thomas 1920h:240). Ellerman (1941) maintained the validity of *Microxus* although he placed *torques*, without explanation but probably following Thomas (1927a), in *Akodon*. Later, Osgood (1943b:197) wrote that *torques* was probably a subspecies of *A. orophilus*, stating "it is difficult to find any external distinction between *torques* and *orientalis* [which, in 1913, he had described as a subspecies of *orophilus*] and the only cranial character appears to be the somewhat wider braincase of *torques*." Cabrera (1961) followed Osgood and treated *torques* as a subspecies of *A. orophilus*. Subsequently, both allozyme (Patton et al. 1989) and mtDNA cytochrome-*b* sequences (Patton and Smith 1992a,b; M. F. Smith and Patton 1991, 1993, 2007) provided support for species status of *torques*. The most recent phylogenetic reconstructions based on cytochrome-*b* favored a close relationship between *A. torques* and both *A. orophilus* and *A. mollis* (M. F. Smith and Patton 2007). Patton et al. (1990)

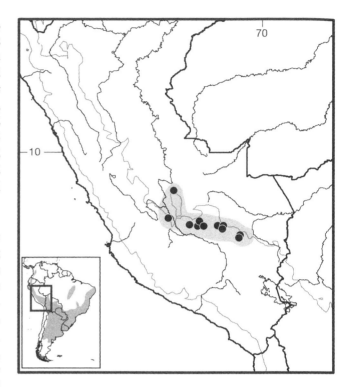

Map 93 Selected localities for *Akodon torques* (●). Contour line = 2,000 m.

illustrated striking differences in rostral depth and length as well as features of the zygomatic plates in comparisons among *A. torques*, *A. mimus*, and *A. aerosus*. Although *A. torques* shares more narrowed and posteriorly upward slanting zygomatic plates with *A. mimus*, but not to the degree found in the latter, it is more similar to *A. aerosus* in its somewhat shorter and deeper rostrum.

Diploid numbers range from 26 in Ayacucho department (specimens called *Akodon orophilus* by Hsu and Benirschke 1973) to 22 in the Urubamba drainage or 24 in the Cosñipata drainage, both in Cusco department, Peru (Patton et al. 1990; Patton and Smith 1992b).

Akodon varius Thomas, 1902
Variable Grass Mouse, Variable Akodont
SYNONYMS:
Akodon varius Thomas, 1902b:134; type locality "Cochabamba, [Cochabamba,] 2400 m.," Bolivia.
Akodon varius varius: Gyldenstolpe, 1932a:102; name combination.

DESCRIPTION: One of larger species, with average total length ca. 193.5 mm. Dorsal pelage medium brown; individual hairs with yellow phaeomelanin band 1–3 mm long and dark tips 0.1–0.5 mm long; saturate eumelanic guard hairs abundant; combination of dark-tipped underfur and black guard hairs results in heavily streaked dorsal appearance; top of head, shoulders, middorsal region, and rump same hue, tending gradually to paler coloration on sides; distinct pale eye-ring present; chin white or gray; throat and ventral hairs gray basally with white to buffy tips; dorsal surfaces of forefeet and hindfeet whitish. Individual rump hairs 11–13 mm long. Tail relatively long (44% of the total length), bicolored, and densely furred. Skull bowed in lateral view; interorbital region sharply squared, not beaded, weakly to moderately divergent to hourglass in shape; zygomatic arches moderately flared; nasals long; zygomatic plates proportionally narrow; incisors slightly opisthodont, less proodont than other related species (Myers 1990).

DISTRIBUTION: *Akodon varius* occurs along the eastern Andean slopes in Bolivia over a broad elevational range from 365 m (SE of Santa Cruz) to at least 3,200 m (Serranía Sama; S. Anderson 1997). No fossils are known for *A. varius*.

SELECTED LOCALITIES (Map 94): BOLIVIA: Cochabamba, Chaparé (S. Anderson 1997), 15 mi E of Tapacari (S. Anderson 1997); Potosí, Río Cachimayo (S. Anderson 1997); Santa Cruz, Cerro Colorado (S. Anderson 1997), Palmarito, Río San Julián (S. Anderson 1997); Tarija, 3 km WNW of Caraparí (S. Anderson 1997), 11.5 km N and 5.5 km E of Padcaya (S. Anderson 1997).

SUBSPECIES: *Akodon varius* is monotypic.

NATURAL HISTORY: S. Anderson (1997) provided data on 21 females, none with embryos, taken in May (1), July (14), August (4), and September (2). Ectoparasites from Paraguayan specimens identified as *A. varius* (but that are probably *A. toba*) are the mites *Androlaelaps rotundus, A. fahrenholzi), Andalgalomacarus paraguayensis, Paratrombicula enciscoensis,* and *Paraguacarus callosus,* fleas of the genus *Polygenis* sp., and sucking lice of the genus *Hoplopleura* sp. (Whitaker and Abrell 1987).

REMARKS: Cabrera (1961) treated *neocenus, pervalens, simulator,* and *toba* as subspecies of *A. varius,* but Myers (1990) ranked each of these as full species and placed all in his *varius* group, which he based on morphological similarity, not on determined phyletic propinquity. Jayat et al. (2010) failed to recover as monophyletic the assemblage of species Myers (1990) placed in his *varius* group; rather, the predominantly Yungas species *A. simulator* and *A. varius* formed a clade that was not sister to one that included the lowland species *A. dayi, A. dolores, A. iniscatus* (including *nucus*), and *A. toba.* Jayat et al. (2010) termed this latter clade the *dolores* group. The Yungas clade (i.e., *A. varius* group *sensu stricto*) is also part of a larger, central Andean clade connected to both the *A. aerosus* species group and to *A. albiventer.* Braun et al. (2008), in a study focused on the systematics of the *A. varius* group, also provided a molecular-based phylogeny in which two clades (i.e., *A. varius* group *sensu stricto* and *A. dolores*

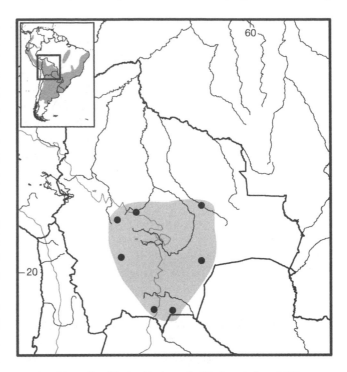

Map 94 Selected localities for *Akodon varius* (●). Contour line = 2,000 m.

group) were recovered, but the likely polyphyly of an *A. varius* species group *sensu lato* was not testable because of the design of their study.

Akodon sp. 2n = 10
Mato Grosso Grass Mouse, Mato Grosso Akodont
SYNONYMS:

Akodon sp.: M. J. de J. Silva and Yonenaga-Yassuda, 1998:47.

Akodon sp. nov.: T. D. Lambert, Malcolm, and Zimmerman, 2005:984.

Akodon sp. n.: M. J. de J. Silva, Patton, and Yonenaga-Yassuda, 2006:469.

DESCRIPTION: Large species morphologically similar to *A. cursor* and *A. montensis*: external measurements of two adult males include total length ?/221 mm, tail length ?/90 mm, hindfoot length, with and without claw, 23/25 and 26/27 mm, respectively; and mass 52/51 g. Pelage slightly crisp and markedly agouti, brownish above and somewhat paler gray below; forefeet and hindfeet covered by yellowish hairs and claws by dense, ungual tufts. Skull strongly built (condyloincisive length 27.6/28.7 mm), with relatively broad interorbital region (5.0/5.2 mm) and short rostrum; anterior edges of zygomatic plates straight; incisive foramina long and widely open; mesopterygoid fossa U-shaped; maxillary toothrow long (4.57/4.72 mm); alisphenoid strut variably present.

DISTRIBUTION: This species is restricted to central Brasil, with known localities in Mato Grosso (M. J. de J.

Silva and Yonenaga-Yassuda 1998) and Pará states (T. D. Lambert et al. 2005). Specimens reported by Pinc et al. (1970) as *Akodon* sp. from Serra do Roncador may belong to this form. No fossils are known for this species.

SELECTED LOCALITIES (Map 95): BRAZIL: Mato Grosso, Vila Rica (M. J. de J. Silva and Yonenaga-Yassuda 1998), Gaúcha do Norte (M. J. de J. Silva and Yonenaga-Yassuda 1998); Pará, Programa de Assentamento Benfica I, Itupiranga (MNRJ 69785), Pinkaití Research Station, Kayapó Indigenous Area (T. D. Lambert et al. 2005).

SUBSPECIES: Species presumably monotypic.

NATURAL HISTORY: This species apparently occurs in a transitional and highly seasonal area between the Amazon rain forest and the Cerrado dry forest (M. J. de J. Silva and Yonenaga-Yassuda 1998). T. D. Lambert et al. (2005) noted the abundance of this *Akodon* in the Kayapó Indigenous Area, where the forest structure, despite the lack of human disturbance, is very heterogeneous, with numerous open areas and dense understory. They trapped *Akodon* sp. exclusively on the ground; it was the third most common of 12 sigmodontines, after *Euryoryzomys emmonsae* and *Hylaeamys megacephalus*.

REMARKS: First detected by a unique karyotype with the lowest diploid number known to date for a sigmodontine rodent (M. J. de J. Silva and Yonenaga-Yassuda 1998), this species remains undescribed. The diploid number is 10

in most individuals, although one female was found with $2n=9$ and a presumptive XO sex chromosome system. In samples from Mato Grosso, individuals may be homozygous for a biarmed or uniarmed autosomal pair, with heterozygotes identified (M. J. de J. Silva 1999). GTG-banded metaphase comparisons between *Akodon* sp. ($2n=10$) and *A. cursor* ($2n=16$) indicated that the karyotypic differentiation between these species involved several complex chromosomal rearrangements, including tandem fusions, Robertsonian rearrangement, and pericentric inversions (M. J. de J. Silva 1999). Phylogenetic analyses of the mtDNA cytochrome-*b* gene (M. J. de J. Silva et al. 2006) also supported species status for this $2n=10$ *Akodon*. Animals from Gaúcha do Norte and Vila Rica shared almost identical haplotypes and were grouped into a strongly supported monophyletic group recovered as the sister to *A. cursor* cytomorphs with $2n=14$, 15, and 16. Fagundes and Nogueira (2007) confirmed the absence of shared species-specific haplotypes among *A.* sp. $2n=10$, *A. cursor*, and *A. montensis* using polymerase chain reaction restriction fragment length polymorphism analysis of the mtDNA cytochrome-*b* gene.

Genus *Bibimys* Massoia, 1979
Ulyses F. J. Pardiñas, Guillermo D'Elía, and Pablo Teta

Dramatic pink and enlarged upper lips make representatives of this genus among the most easily distinguishable small sigmodontine rodents in the field; unfortunately, this trait is poorly preserved in museum specimens. *Bibimys* is known from isolated localities in tropical and subtropical eastern South American lowlands, where it inhabits grasslands close to forested areas in Argentina, Brazil, and Paraguay. Winge (1887), Massoia (1979b), Pardiñas (1996), P. R. Gonçalves, Oliveira et al. (2005), and D'Elía et al. (2005) have been the major contributors to our knowledge of this genus. The information provided here is largely based on the conclusions reached by D'Elía et al. (2005), supplemented by our unpublished data obtained from recently trapped specimens.

Bibimys is a small sigmodontine (head and body length <130 mm; tail <80 mm) with soft and dense pelage. The anterior part of the snout, from just below the nostrils to the opening of the mouth, is strikingly swollen, making it bulbous. The entire area from just dorsal to the rhinarium to the lower lips is covered with remarkably short, bristle-like, all-white hairs that extend stiffly and straight outward from the lips, the skin underlying these hairs is distinctively reddish in living/fresh specimens, although this color becomes lost in both skins and fluid-preserved specimens. Mystacial vibrissae are 2.0 to 2.5 cm long, dark

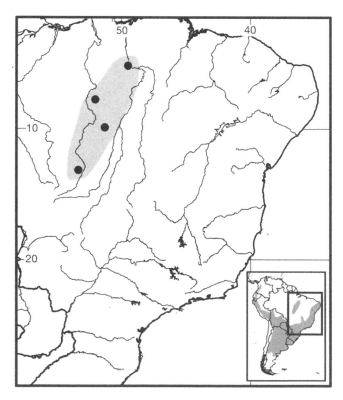

Map 95 Selected localities for *Akodon* sp. $2n=10$ (●).

brown proximally, and whitish distally. Hairs in the submental region are longer than those on the swollen muzzle, less dense, and entirely white, as opposed to the remaining ventral hairs, which are a dark slate at their base. The ears are small and rounded and finely haired, and the tail is bicolored. The forefeet are dorsally covered with hair, either proximally gray and distally whitish or entirely whitish. There is a greatly reduced claw-bearing pollex, and there are five palmar pads, including three reduced interdigital and two larger metacarpal pads. There are no carpal vibrissae. The hindfeet are skin colored, with ungual tufts mainly whitish but with some dark brown hairs with white tips, the longest extending beyond the tips of the claws. The hindfoot claws are long; those on digits 2–4 are longer than one-third of the length of the digits, and are paler in color than the skin of the pes. The hallux has a claw similar in size to that of digit V and about half the length of the other three claws. The plantar surface has two metatarsal pads, including a reduced hypothenar and thenar, and four interdigital pads (P. R. Gonçalves, Oliveira et al. 2005; D'Elía et al. 2005).

The skull is robust and domed in lateral profile. The premaxilla has a well-developed gnathic process forming a projecting plate anterior to the incisors. The nasals project anteriorly beyond the premaxillae, and they are distally partly inflected dorsally. The interorbital region is squared, not beaded or ledged, and is hourglass in shape. Zygomatic notches are moderately deep and broad; the anterior margins of the zygomatic plates are nearly vertical. The anterior half of the zygomatic arch is flattened and expanded dorsoventrally. The braincase is inflated, and its width is subequal to the zygomatic breadth.

The incisive foramina are large and reach the level of the protocones of the M1s. The palate is long, extending well past the posterior faces of the M3s. The mesopterygoid fossa is narrower than either parapterygoid fossa; its anterior margin is squared and lacks a median spine. The alisphenoid strut is present, and a crest connects the squamosal root of the zygomatic arch to a slender hamular process. The subsquamosal foramen is small; the tegmen tympani is enlarged and reaches the squamosal. The stapedial foramen and posterior opening of the alisphenoid canal are large; both squamosal-alisphenoid groove and sphenofrontal foramen are absent. The mandible is short, with an inconspicuous capsular projection, a distally slender and high coronoid process, a dorsoventrally expanded condyloid process (internally excavated), and a short angular process. The inferior ridge of the masseteric crest is sharp and well marked.

Upper incisors are strongly opisthodont. The molars are terraced and moderately hypsodont. The anteromedian flexus/-id is present in M1–2/m1–2; mesolophs/-ids in

M1–2/m1–2 are fused with paralophule/entolophulid. The M3 is rounded in occlusal outline and has a persistent hypofossette and posterofossette, but there is no evidence of a hypoflexus. The m3 is rectangular with a large, oblique hypoflexid; its length is slightly less than that of m2. A full description of molar occlusal and root patterns is available in Pardiñas (1996).

External and cranial measurements of *Bibimys* species are available in Massoia (1979b), P. R. Gonçalves, Oliveira et al. (2005), and D'Elía et al. (2005). Pardiñas (1996) provided dental measurements.

Skeletal element counts from four *B. chacoensis* specimens are 13 ribs, 13 thoracic vertebrae, six lumbar vertebrae, three sacral vertebrae, and 26–29 caudal vertebrae. A gall bladder is present. The stomach is unilocular and hemiglandular. Geise et al. (2006) noted a pair of paravaginal mammae in female specimens, a condition unique for Sigmodontinae. A specimen (CNP 1891) from Las Palmas, a topotype of *B. chacoensis*, shows the same condition. Massoia (1979b) and P. R. Gonçalves, Oliveira et al. (2005) described glans and bacular morphology. According to P. R. Gonçalves, Oliveira et al. (2005), the glans penis is an elongated, laterally compressed, rod-like structure uniformly covered by small spines except on the crater rim. A major median groove extends along the length of its ventral surface. The bacular mound is an expanded structure, covered by spinous epidermis with three distally directed lobules on its extremity but without traces of lateral digits. The proximal end of the bacular base is paddle-like and without a conspicuous notch. The dorsal papilla is a finger-like structure, flattened dorsoventrally or slightly divided into three lobes. The urethral process is a bifurcated projection that extends beyond the ventral opening.

Dyzenchauz and Massarini (1999) reported the karyotype of *B. torresi*, and P. R. Gonçalves, Oliveira et al. (2005) reported the same for *B. labiosus*. Both species have $2n = 70$, but vary in FN: 76 in *B. torresi* and 72 in *B. labiosus*. In addition, *B. torresi* shows C-bands on the X chromosome and in pair 3; *B. labiosus* has C-bands only on the X chromosome. A new karyomorph, reported for a specimen referred to *B. labiosus* collected in Minas Gerais state, Brazil, had a $2n = 68$, $FN = 80$ (Geise et al. 2006).

Massoia (1979b) put *Bibimys* in the Scapteromyini based on molar morphology, a hypothesis generally accepted by subsequent authors (Reig 1980; Musser and Carleton 1993). However, when envisioning and delineating the scapteromyine group, Hershkovitz (1966a) explicitly excluded *B. labiosus* (*Scapteromys labiosus* at that time), stating that *labiosus* morphologically resembled *Akodon azarae*. His conclusions were based mainly on the cranial and molar morphology of *B. labiosus* as portrayed by Winge (1887). Recent phylogenies based on mtDNA cytochrome-*b* and nucDNA

IRBP sequences (D'Elía 2003a; D'Elía et al. 2005) support Hershkovitz's (1966a) conclusion. *Bibimys* is not closely related to either *Scapteromys* or *Kunsia* but rather appears as the sole member of one of the five main clades in the Akodontini. However, in topologies obtained by analysis of IRBP sequences alone tend to place *Bibimys* closer to *Scapteromys* (D'Elía, Luna et al. 2006) or *Kunsia* (G. D'Elía, unpubl. data). Thus, for the present, the phylogenetic position of *Bibimys* in the Akodontini remains unclear.

Currently, *Bibimys* contains three living species, *B. chacoensis*, *B. labiosus*, and *B. torresi*, each characterized by complex nomenclatorial histories and a scarcity of available specimens for study. In addition, these taxa were erected over the course of more than a century and without cross-comparisons. As a result, the taxonomic status of all nominal forms remains uncertain, a problem exacerbated by the general lack of specimens and little knowledge of their natural history. The level of variation of the mtDNA cytochrome-*b* gene among the species of *Bibimys* is extremely low (D'Elía et al. 2005), surprisingly so considering the geographic distances among studied populations and their high degree of isolation. Observed variation is much lower than that reported among species in other akodontine genera and even lower than interpopulation variation seen in some akodontine species (summarized in M. F. Smith and Patton 1991, 1993). Pardiñas (1996) and D'Elía et al. (2005) addressed the degree of morphological variation. These authors, working with small samples, failed to find definitive anatomical traits of or morphometric ranges that reliably delimit the named species of *Bibimys*. They thus claimed that recognition of a single species composed of several subspecies better represented the known diversity in the genus. Newly available specimens of *Bibimys* may allow a much deeper understanding of variation within the genus. For example, at the time of the redescription made by P. R. Gonçalves, Oliveira et al. (2005), *B. labiosus* was known only from six specimens, one of them the lectotype. Subsequently, however, Geise et al. (2006) reported the capture of 17 individuals of this taxon from a single locality, Poços de Caldas, Minas Gerais state, Brazil.

Bibimys has been recorded from several paleontological and archaeological sites in the Pampean region of Argentina. The oldest record comes from the late Pleistocene of Constitución, southeastern Buenos Aires province (Pardiñas, Cione et al. 2004). Several late Holocene occurrences from across the eastern Pampean region indicate a wider distributional range for this rodent until at least the eighteenth century (Pardiñas 1996; M. Silveira et al. 2010; Teta et al. 2013). In Brazil, fossil specimens have come from the Late Pleistocene-Holocene beds of Lapa da Escrivânia Nr. 5, Minas Gerais (Winge 1887).

REMARKS: *Bibimys* is hard to trap using Sherman or kill traps and traditional baits. Most of the data regarding distribution in Argentina have come from the analysis of owl pellets (Massoia 1983, 1988b; Massoia, Chebez, and Heinonen Fortabat 1989a, 1989b, 1989c, 1989d; Massoia, Tiranti, and Torres 1989). Its frequency in owl pellets, although moderate, is great enough to indicate that its apparent rarity is perhaps a sampling artifact. The use of pitfall traps in Brazil has resulted in several new localities in the last decade (e.g., Casado et al. 2006).

SYNONYMS:

Scapteromys: Winge, 1887:39; part (description of *labiosus*); not *Scapteromys* Waterhouse.

Akodon: Shamel, 1931:427; part (description of *chacoensis*); not *Akodon* Meyen.

Bibimys Massoia, 1978:56; *nomen nudum*.

Bibimys Massoia, 1979b:2; type species *Bibimys torresi* Massoia, by original designation.

KEY TO THE SPECIES OF *BIBIMYS*

1. Length of M1–3 usually >4 mm; range restricted to east-central Argentina . . . . *Bibimys torresi*
1'. Length of M1–3 usually <4 mm; range outside of east-central Argentina 2
2. Dorsal coloration strongly olivaceous; condylobasal length usually ≤ 23.8 mm; range northeastern Argentina and eastern Paraguay . . . . *Bibimys chacoensis*
2'. Dorsal coloration less olivaceous, almost brownish; condylobasal length usually ≥ 23.8 mm; southeastern Brazil . . . . *Bibimys labiosus*

Bibimys chacoensis (Shamel, 1931)
Pink-lipped Mouse, Swollen-nosed Mouse, Chacoan Akodont
SYNONYMS:

Akodon chacoensis Shamel, 1931:427; type locality "Las Palmas, Chaco, Argentina." According to the itinerary of Alexander Wetmore (1926:3), the collector of the type, he arrived at Puerto Las Palmas on July 12, 1920, and immediately traveled to "the little village of Las Palmas, headquarters for a large estancia, that covered 60 leagues of land." There, Wetmore worked until August 2. Thus, the type specimen of *chacoensis* was secured somewhere around the current town of Las Palmas (ca. 27.0485°S, 58.6889°W), Bermejo, Chaco province, Argentina.

Bibimys chacoensis: Massoia, 1980c:285; first use of current name combination.

DESCRIPTION: General characters as in generic description. Overall coloration olivaceous with some buff around eyes, on sides of head, and flanks; darker middorsal line present, almost black from shoulders to the tail; underparts whitish, with buffy tinge; both dorsal and ventral hairs with slate graybases (guard hairs ca. 11 mm; cover hairs ca. 8 mm); feet appear dark brown (foreclaws ca. 2.6 mm), with some hairs whitish at their tips; toes white. Compared with *B. torresi*, *B. chacoensis* has a more whitish venter, head and flanks are paler, and tail only slightly haired.

DISTRIBUTION: This species occurs in a narrow band from eastern Chaco province (Argentina), through eastern Paraguay, to southern Misiones and northwestern Corrientes provinces (Argentina). For many years after its original description, *B. chacoensis* was known only from the type locality and vicinity in Chaco province (Shamel 1931; J. R. Contreras 1984c). D'Elía et al. (2005), on morphological and geographical grounds, referred the Argentinean records from Misiones province, formerly regarded as of *B. labiosus* (see Massoia 1983, 1993), to *B. chacoensis*. Then several new localities were discovered in eastern Paraguay, filling the gap between records from Chaco and Misiones provinces (D'Elía, Mora et al. 2008). Even more recently, this species was recorded in northwestern Corrientes (C. Galliari and U. F. J. Pardiñas, unpubl. data). The geographic range displayed by *B. chacoensis* resembles that of *Chacodelphys formosa*, a small opossum (Teta and Pardiñas 2007). Most of the Argentinean records for *B. chacoensis* come from owl pellets. No fossils are known.

SELECTED LOCALITIES (Map 96): ARGENTINA: Chaco, Las Palmas (type locality of *Akodon chacoensis* Shamel); Corrientes, Santo Tomé (CNP-E uncataloged); Misiones, El Dorado (D'Elía et al. 2005), Estación Experimental INTA Cuartel Río Victoria (Massoia 1980b), Los Helechos (Massoia, Chebez, and Heinonen Fortabat 1989b), Santa Inés (D'Elía et al. 2005). PARAGUAY: Itapuá, Estancia Parabel, 0.3 km E of house (D'Elía, Mora et al. 2008), Parque Nacional San Rafael (D'Elía, Mora et al. 2008).

SUBSPECIES: *Bibimys chacoensis* is monotypic.

NATURAL HISTORY: Trapping data suggest that *B. chacoensis* inhabits perisylvan grasslands, including severely disturbed habitats, from humid Chaco to northern Campos ecoregions. The type specimen was secured in a small patch of marsh grasses in savanna. Field notes available for some Paraguayan specimens indicate that they were trapped in small patches of humid grassland near or within remnants of evergreen forest. Four males collected from July to September had scrotal testes. A female caught in January had a closed vagina, while another, taken in November, was lactating and had the vagina open (D'Elía, Mora et al. 2008).

REMARKS: Shamel (1931:427) stated that A. Wetmore acquired the holotype of *B. chacoensis* on June 20, 1920. However, this must be an error because Wetmore did not arrive in Argentina until June 21, 1920, and he worked the Las Palmas area between July 12 and August 2 (Wetmore 1926:2–3); the correct date of collection is probably July 20, 1920.

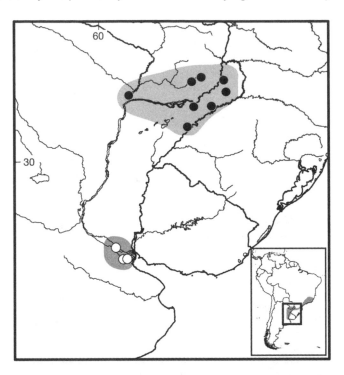

Map 96 Selected localities for *Bibimys chacoensis* (●) and *Bibimys torresi* (○).

Bibimys labiosus (Winge, 1887)
Lagoa Santa Pink-lipped Mouse, Lagoa Santa Swollen-nosed Mouse, Lagoa Santa Akodont

SYNONYMS:

Scapteromys labiosus Winge, 1887:39; type locality "Lagoa Santa [19°39′S, 43°54′W]," Minas Gerais, Brazil.

Bibimys labiosus: Massoia, 1980c:285; first use of current name combination.

DESCRIPTION: General characters as in the generic description. Pelage soft, long, and dense; color of dorsum homogeneous brownish, becoming somewhat paler on sides; some guard hairs with yellow-orange subapical band, producing black-lined pattern in combination with adjacent dark guard hairs; underparts whitish and sharply demarcated in color from dorsum. Area around mouth, from nostrils down (including lower lip) covered with densely distributed, short, white hairs that form round, white, velvety area on of muzzle in preserved skins. Mystacial vibrissae 2 to 2.5 cm long, dark brown proximally and whitish distally. Tail densely haired, with individual hairs extending over about two rows of scales, entirely dark brown on the upper surface of tail but with proximal dark brown and distal whitish halves on ventral surface (see P. R. Gonçalves, Oliveira et al. 2005).

DISTRIBUTION: This species was long known exclusively from animals, both extant and fossil, secured by Peter Lund in 1834 at the type locality and that were the basis of Winge's description. Recent trapping efforts in Brazil, however, have recorded the species at several additional localities that collectively expand the known range from northern Rio

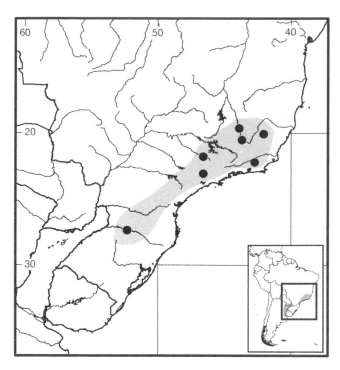

Map 97 Selected localities for *Bibimys labiosus* (●).

Grande do Sul northeast to Minas Gerais and Rio de Janeiro states in Brazil (Paglia et al. 1995; P. R. Gonçalves, Oliveira et al. 2005; Bonvicino, Oliveira, and D'Andrea 2008). Regrettably, most of these locality details remain unpublished (see Casado et al. 2006; Geise et al. 2006).

SELECTED LOCALITIES (Map 97): BRAZIL: Minas Gerais, Lagoa Santa (type locality of *Scapteromys labiosus* Winge), Viçosa, Mata do Paraíso (Paglia et al. 1995), Ouro Branco (G. Lessa et al. 2009), Poços de Caldas (Geise et al. 2006); Rio de Janeiro, Teresópolis, Vieira (Bonvicino, Oliveira, and D'Andrea 2008); São Paulo, Atibaia (Geise et al. 2006); Rio Grande do Sul, Aratiba (Casado et al. 2006).

SUBSPECIES: *Bibimys labiosus* is monotypic.

NATURAL HISTORY: Specimens from Viçosa were trapped in pitfall buckets placed in a disturbed patch of Atlantic Forest (Paglia et al. 1995). No other information on the ecology, behavior, or life history of this species is known.

Bibimys torresi Massoia, 1979
Torres's Pink-lipped Mouse, Torres's Swollen-nosed Mouse, Torres's Akodont

SYNONYMS:

Bibimys torresi Massoia, 1978:56; *nomen nudum*.

Bibimys torresi Massoia, 1979b:3; type locality "borde de camino de tierra y a orillas del canal de endicado de la Estación Experimental Agropecuaria Delta del Paraná del INTA [34.175061°S, 58.868071°W]," Campana, Buenos Aires, Argentina.

DESCRIPTION: General characters as in generic description. Large, with mass up to 40+ g (comparative measurements in D'Elía et al. 2005). Pelage long (hairs ca. 9 mm) and silky; individual hairs gray-based with ochraceous tips, giving general agouti appearance, especially yellowish on flanks and cheeks; hairs around snout, axillae, and genitalia with some orange tinge; ears small and rounded, externally and internally covered by short, delicate, cream-colored hairs; underparts cream, with the fur gray at base; dorsal and ventral colorations contrast sharply; tail bicolored, brown above and orange below, covered by short hairs; scale annulations quite visible; mystacial vibrissae short, not reaching pinna when laid backward; ungual tufts shorter than claws.

DISTRIBUTION: *Bibimys torresi* is restricted to the islands and margins of Delta del Paraná in northern Buenos Aires and southernmost Entre Ríos provinces of Argentina. It is known from only six localities, half of which have produced only materials recovered from owl pellets. The prehistoric record of the Late Holocene, however, indicates a larger range, one reaching to 38°S and including coastal and hilly wet environments (Pardiñas 1995b, 1999b; Teta et al. 2004; Teta et al. 2013).

SELECTED LOCALITIES (Map 96): ARGENTINA: Entre Ríos, Isla Ibicuy (Massoia 1983); Buenos Aires, Estación Experimental INTA Delta del Paraná (Massoia 1979b), San Fernando, Establecimiento Carabelas (E. M. González 1997).

SUBSPECIES: *Bibimys torresi* is monotypic.

NATURAL HISTORY: The few specimens obtained by trapping (<10) came from long-term (>1 year) efforts. These data suggest a nocturnal activity pattern and presence in habitats near water bodies or flooded areas covered by dense tall grass and hydrophilic vegetation. *Bibimys torresi* is a secondary prey item of the Barn Owl (*Tyto alba*) in the Delta del Paraná, Argentina (Massoia 1983). Pardiñas (1996) hypothesized fossorial habits. *Bibimys torresi* is the type-host of *Hoplopleura massoiai* (see D. Castro and González 2003), a louse also shared by *B. chacoensis* (see Gugliemone and Nava 2011).

REMARKS: The original version of the type locality as given by Massoia (1979b) is somewhat ambiguous, a fact that may explain Musser and Carleton's (2005:1104) mixture of two separate localities ("confluence of Arroyo Las Piedras with Arroyo Cucarachas" and "Estación Experimental del INTA, Canal 6," which are separated by ca. 6 km).

Genus *Blarinomys* Thomas, 1896
Pablo Teta and Ulyses F. J. Pardiñas

The monotypic *Blarinomys*, with its adaptations for fossorial living, is one of the more distinctive genera of sigmodontine rodents (Hershkovitz 1966a:95). The single species, *Blarinomys breviceps*, is known from a limited number of localities in the Atlantic Forest of southeastern Brazil and extreme northeastern Argentina. Matson and Abravaya (1977) reviewed the morphology, fossil record, and ecology of the genus; Geise et al. (2008) added karyological information, new morphological data, and provided a gazetteer of known localities. K. Ventura et al. (2012) provided the first phylogeographic assessment of *Blarinomys* and reported eight distinct karyomorphs, indicating impressive karyotypic diversity in this fossorial rodent. The present account is largely based on these three contributions.

In general appearance, this small sigmodontine resembles the North American short-tailed shrews of the genus *Blarina*, hence the name *Blarinomys* (Thomas 1896). The body form is highly modified for fossorial life, with a short and conical head, extremely reduced eyes hidden in the fur, very small and well haired pinnae, and a short, unicolored, and sparsely haired tail (usually <50% of head and body length). The manus has long, pointed claws and a strongly reduced fifth digit. The pes has short and curved claws. The palms and soles of the feet are naked, except for the heel. The pelage is short, crisp, and velvety, and more or less uniform dark slate gray throughout, with reddish or brownish tips to the hairs. The dorsum is slightly iridescent with a ruby tinge. Under parts are gray-brown, scarcely differentiated from the dorsum. The tip of the muzzle and chin are whitish. The hands and feet are brown above (Gyldenstolpe 1932; Matson and Abravaya 1977). The head and body length ranges from 91–122 mm, tail length 30–59 mm, hindfoot length 16– 21 mm, and ear length 8–10 mm. Both Abravaya and Matson (1975) and Geise et al. (2008) documented reverse sexual dimorphism, with females larger than males.

The skull is very distinctive among sigmodontines, short but conical in the rostral region and broadened posteriorly. The braincase is short, laterally rounded, and almost flat dorsally. The interorbital region is broad, tubular throughout, and smooth, with an almost undistinguishable constriction. Frontal sinuses are well inflated. Nasals are subequal in length to the frontals and extend beyond the anterior end of the premaxillae. Zygomatic plates are narrow and reduced, their anterior bordersposteriorly slanting from bottom to top. Masseteric tuberclesare well developed. The interparietal is extremely reduced or absent. Lambdoidal crests are well-developed, even in juveniles. The infraorbital foramina are large. Incisive foramina are expanded in their posterior half, and extend posteriorly to the level of the procingula of M1s. The palate is wide and short. The mesopterygoid fossa is remarkably broad and parallel sided, and extends to the posterior roots of M3s. Each parapterygoid fossa is narrower than the mesoptery-

goid fossa, and well excavated. The hamular process of the squamosal is thin, defining subequal postglenoid foramen and subsquamosal fenestra. An alisphenoid strut is present. The tegmen tympani is well developed and contacts the squamosal. The mandibular ramus is slender; the capsular projection absent; the coronoid process well developed, with a broad base and slightly backward oriented; and the angular process delicate.

The upper incisors are orthodont with a straight dentine fissure. The molar rows are slightly convergent posteriorly. The molars are hypsodont, with high, crowns; main cusps are opposite to each other; and enamel folds wear quickly with age to leave simple indented surfaces in subadults and adults. M1 has a reduced procingulum with a shallow median anteroflexus. M3 is small but shows a central hypofossetus. Both m1 and m2 have reduced mesolophids.

Geise et al. (2008:7) described and figured a unilocular-hemiglandular stomach. However, these authors failed to note that the glandular epithelium is contained in a pouch-like diverticulum that apparently opens into the main chamber of the stomach by a minute aperture. In addition, Geise et al. (2008) stated that they observed a bordering fold, but, from the inspection of their figure, there is no clear indication of this structure. The stomach of *Blarinomys* seems to be very similar to those of *Brucepattersonius* (Hershkovitz 1998:229–230) and *Oxymycterus* (Vorontsov 1960:365; Carleton 1973:15; Hershkovitz 1994:7 and 9).

SYNONYMS:

Oxymycterus: Winge, 1887:34; part (description of *breviceps*); not *Oxymycterus* Waterhouse.

Blarinomys Thomas, 1896:310; type species *Oxymycterus breviceps* Winge, by original designation.

Blarinomys breviceps (Winge, 1887)

Atlantic Forest Burrowing Mouse, Blarinine Akodont

SYNONYMS:

Oxymycterus breviceps Winge, 1887:34; type locality "Lapa do Capão Secco [= Lapa do Capão Seco]," Lagoa Santa, Minas Gerais, Brazil (name appears first in list on p. 4).

Blarinomys breviceps: Thomas, 1896:310; generic description and first use of current name combination.

DESCRIPTION: As for the genus.

DISTRIBUTION: *Blarinomys breviceps* is found mostly in forested areas from sea level to about 1,600 meters in the Brazilian states of Bahia, Espírito Santo, Minas Gerais, Rio de Janeiro, and São Paulo, and the Argentinean province of Misiones (Massoia 1993; C. R. Silva et al. 2003; Massoia et al. 2006). Despite a few localities near sea level (S. F. Reis et al. 1996; see also Moojen 1952b and Avila-Pires 1960), most collections from the northern portion of the range are from middle to high elevations (750 to 1,500 m;

Geise et al. 2008). Matson and Abravaya (1977) regarded
the species as chiefly montane. This species is primarily an
Atlantic Forest dweller, but a few records are from localities transitional between it and the Cerrado (e.g., Serra do
Ouro Branco; V. B. Pereira et al. 2010). The only known
fossils are from the Quaternary cave deposits at Lagoa
Santa, Minas Gerais, the type locality (Winge 1887; Voss
and Myers 1991). The species is apparently not found in
that area today (C. R. Silva et al. 2003). The nearest known
modern locality to the type locality is Estação de Proteção
e Desenvolvimento Ambiental de Peti, about 60 km ESE of
Lagoa Santa (Paglia et al. 2005).

SELECTED LOCALITIES (Map 98): ARGENTINA:
Misiones, Puerto Esperanza (Massoia 1993), R. N. E. San
Antonio (Massoia et al. 2006), San Martín (Massoia et al.
2006). BRAZIL: Bahia, Fazenda Imbaçuaba, 30 km N of
Prado (Geise et al. 2008), Ilhéus (Moojen 1952b), RPPN da
Serra do Teimoso (Geise et al. 2008); Espírito Santo, Castel-
ingho (Abravaya and Matson 1975), Reserva Florestal
Nova Lombardia (Abravaya and Matson 1975); Minas
Gerais, Boca da Mata (Avila-Pires 1960b), Jacutinga (Oehl-
meyer et al. 2010), Juiz de Fora (Granzinolli and Motta-
Junior 2006), Lapa do Capão Seco (type locality of Oxy-
mycterus breviceps Winge), Mata do Paraíso (Geise et al.
2008), Parque Estadual da Serro do Ouro Branco (V. B.
Pereira et al. 2010); Rio de Janeiro, Fazenda Marimbondo
(Geise et al. 2008); São Paulo, Fazenda João XXIII (C. R.
Silva et al. 2003), Parque Estadual da Cantareira (Nieri-
Bastos et al. 2004), Parque Estadual Turístico do Alto Ri-
beira (C. R. Silva et al. 2003), Reserva Florestal do Morro
Grande (Pardini and Umetsu 2006).

SUBSPECIES: *Blarinomys breviceps* is monotypic (but
see Remarks).

NATURAL HISTORY: Little is known about this rare
sigmodontine rodent. It has been caught in both primary
and secondary growth forests (C. R. Silva et al. 2003; Paglia et al. 2005), and in moderately disturbed habitats such
as forest fragments in agricultural areas (Oehlmeyer et al.
2010). Abravaya and Matson (1975) reported catching five
individuals with snap traps baited with corn and placed in
leaf litter, but most Brazilian specimens have been obtained
in pitfall buckets (Pardini and Umetsu 2006; Geise et al.
2008; V. B. Pereira et al. 2010). *Blarinomys* uses galleries
under the leaf litter at depths up to 25 cm (V. B. Pereira et al.
2010).

Males with scrotal testes were collected in January and
February; pregnant females were trapped in September, January, and February. Litter size varied from 1 to 2 (Abravaya
and Matson 1975). The diet consists primarily of invertebrates, with stomachs containing arthropods of six different
orders, mostly insects (90%), but also arachnids (S. F. Reis
et al. 1996). Davis (1944) and Abravaya and Matson (1975)

reported that captive animals were docile and accepted a
wide variety of insects, mainly orthopterans, roaches, and
lepidopteran pupae and larvae, but refused fruit and seeds
except for those of oranges (Geise et al. 2008).

REMARKS: Tate (1932d, 1932h) believed that *Mus
fossorius* and *Mus talpinus*, two extinct species named
by Lund (1840b, 1841c), might be species of *Blarinomys*.
Paula Couto (1950) included *M. fossorius*, a taxon described on the basis of an isolated humerus, in the genus
Nectomys, while Winge (1887) and Hershkovitz (1966a)
believed that its affinities were indeterminable. Paula
Couto (1950) also referred *M. talpinus* to *Blarinomys*, although Winge (1887) included this species in *Oxymycterus*. Moojen (1965), on the other hand, considered *talpinus* a species of *Juscelinomys*. João A. Oliveira (in Pardiñas
et al. 2002:222) believed that *Mus talpinus* was referable
to *Brucepattersonius* (see that account and the one for *Juscelinomys*), and, after the examination of the holotype,
Pardiñas and Teta (2013a) formally placed *talpinus* Lund
in the synonymy of *Brucepattersonius*.

Recently described geographic variation in karyotypic,
morphological, and molecular characters supports the possibility that more than one species is contained in our present concept of *B. breviceps*. For example, Fagundes and
Costa (2008) described a $2n=45$, $FN=51$ karyotype from
a specimen caught at Reserva Biológica de Duas Bocas,
Cariacica, Espírito Santo, which is substantially different
from the $2n=28$, $FN=48$ known for specimens from Rio
de Janeiro (Geise et al. 2008). Gudinho and Ximenez (2010)
reported variation in several morphological traits (e.g., foramen magnum morphology, anterior expansion of the nasals)
between northern and southern populations in Brazil. Fi-

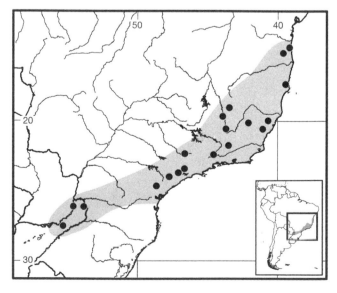

Map 98 Selected localities for *Blarinomys breviceps* (●).

nally, mtDNA cytochrome-*b* sequence analyses (K. Ventura et al.2012; also K. Ventura 2009) have identified two evolutionary lineages, one found north and one south of the Rio Doce fluvial system in eastern Brazil. Pairwise divergences between members of these clades ranged from 4.9–8.4%. These authors also identified eight distinct karyomorphs, including $2n = 52$ (50A, XX), $2n = 52$ (48A, XY+2Bs), $2n = 45$ (42A, XY+1B), $2n = 43$ (37A, XX+4Bs), $2n = 37$(34A, XY+1B), $2n = 34$ (32A, XX), $2n = 31$ (27A, XX+2Bs), and $2n = 28$ (26A, XY), all with the same number of autosomal arms (FN = 50).

Genus *Brucepattersonius* Hershkovitz, 1998

*Júlio Fernando Vilela, Pablo R. Gonçalves,
and João A. de Oliveira*

The genus *Brucepattersonius* was recently proposed for a clade of shrew-like akodonts from southeastern-southern Brazil and northeastern Argentina. Museum records have been obtained from both coastal and interior portions of the Atlantic Forest domain, from the Brazilian states of Minas Gerais and Espírito Santo to Rio Grande do Sul, and westward to the Argentinean province of Misiones. In addition to *B. iheringi* (Thomas 1896), which was originally described as an *Oxymycterus*, seven other recently described forms (Hershkovitz 1998; Mares and Braun 2000a) have been assigned to the genus, namely, *B. soricinus* (type species of *Brucepattersonius*), *B. griserufescens*, *B. igniventris*, *B. albinasus*, *B. paradisus*, *B. misionensis*, and *B. guarani*, the last four based on single specimens, and the last three from closely adjacent localities in the department of Guaraní, Misiones province, Argentina. The type specimen of *B. albinasus* is indistinguishable, both in mtDNA cytochrome-*b* sequence and cranial morphology, from *B. griserufescens*, which was described from a nearby locality (J. F. Vilela et al. 2006). The apparently extinct *Oxymycterus talpinus* Winge, 1887, described from the Pleistocene-Holocene cave faunal assemblage of Lapa da Serra das Abelhas, Lagoa Santa, Minas Gerais, is here assigned to the genus *Brucepattersonius* based on comparisons with *Oxymycterus iheringi* (= *B. iheringi*) by Thomas (1896), J. A. Oliveira (1998), and Pardiñas and Teta (2013a).

The species of *Brucepattersonius* are small akodontine rodents (total length 194–221 mm), with small eyes, relatively large ears (16–19 mm), long and tapered rostrum, moderately developed forefoot and hindfoot claws with whitish ungual tufts, and tail varying from shorter to a fourth longer (82–112 mm) than the combined head and body length (93–128 mm). The hindfoot, with claw, averages 25 mm in adults of all species, and usually between 24 and 26 mm. The dorsal fur is long and soft, owing to guard

hairs that are proximally thin and distally thickened. Dorsal coloration is generally dark brown to grayish brown, but some forms have a reddish-brown dorsum (e.g., *B. igniventris*). The sides vary from slightly paler to almost as dark as the dorsum, and the venter can be weakly to strongly set off from the lateral coloration. In the latter case, the ventral pelage is usually paler and grayish, sometimes with a touch of ochraceous. Despite the supposed taxonomic significance of chromatic variation in the genus (see Hershkovitz 1998; Mares and Braun 2000a), current museum holdings seem to show that there are high levels of intraspecific variation in pelage color, preventing a clear-cut distinction among nominal forms based on pelage color alone (J. F. Vilela et al. 2006).

The skull is delicately built, with a narrow and elongated rostrum and smooth, subglobular braincase. The nasals have rounded or bluntly pointed tips that extend anteriorly beyond the incisors but do not form a distinct tube as in *Oxymycterus*. The lateral premaxillary walls become free from their contact with the nasals at or slightly posterior to the plane of incisors. The interorbital region is very broad and wider than the rostrum and is slightly biconcave or amphoral in shape, without crests or ridges. The supraorbital edges of frontals are rounded, and their antorbital portions are considerably inflated due to expanded frontal sinuses. The interparietal is very reduced and quite narrowed anteroposteriorly. Incisive foramina can be long, extending posteriorly to the level of the paraflexi of M1s, or short, not reaching the level of the anterior edges of M1s. The bony palate is short and wide, and the mesopterygoid fossa extends anteriorly to the posterior limits of the third molars. A large stapedial foramen is present, coalesced with a deep petrotympanic fissure. A sphenofrontal foramen and an anterior opening in the alar fissure are also present, providing evidence of superior and inferior branches of the stapedial artery, conforming to the basic or primitive pattern of carotid circulation (Bugge 1970; Carleton 1980; Voss 1988). There is no bony alisphenoid strut, and the buccinator-masticatory and accessory oval foramina are coalesced to form a unique foramen. Auditory bullae barely touch the squamosal anteriorly through the tegmen tympani, which overlaps with a reduced posterior suspensory process. A second point of attachment of the auditory complex to the squamosal is provided by a slender hamular process, which shows a median ridge that extends from the posterior tip of the hamular process to coalesce with the squamosal process of the zygomatic arch. The mandible is slender and smooth, and without a defined incisor root capsule.

Molar rows are very nearly parallel sided, aside from greater width of first molars as compared with third molars; opposing cusps are oblique. Molars are tetralophodont, quadritubercular, and hypsodont; cusps of unworn teeth are narrow and pointed; the occlusal surface is crested; upper and lower first molars have anteromedian flexi and flexids; anteroloph present; mesolophs (and -ids) are absent; upper and lower third molars are rounded and simplified with paracone and protocone as the sole pronounced cusps; and upper incisors are short, narrow, and usually orthodont.

The stomach is bilocular-discoglandular, and the glandular epithelium is contained in a pouch-like diverticulum that opens into the main chamber of the stomach by a minute orifice (Hershkovitz 1998:Fig. 20), a condition found in insectivorous rodents (Carleton 1973). Although Hershkovitz (1998) suggested that the gall bladder is present in *Brucepattersonius*, Geise, Weksler, and Bonvicino (2004) did not find it in specimens from Itatiaia, Rio de Janeiro state, Brazil. Females have six mammae, one thoracic and two inguinal pairs (Thomas 1896).

The baculum is long, a distally thin shaft that becomes gradually wider to an enlarged cuneiform base. Two lateral indentations are present, one near the distal end and another near the base; in lateral view, the baculum is markedly curved, and the lateral indentations do not form a conspicuous lateral enlargement of the base. Lateral condyles are located almost in the proximal extremity on each side of the base and are separated by a shallow median depression. The distal end is rounded, but the tip more squared. Both lateral and medial cartilaginous digits are lacking (J. A. Oliveira 1998).

Part of the information that follows is based on our examination of newly available specimens that have not been previously published and on comparisons with type specimens, particularly of Brazilian taxa. Unlike many other sigmodontine genera in which the number of species currently recognized underestimates true diversity, the number of those recognized for *Brucepattersonius* likely overrepresents that diversity. For example, the three supposed species of *Brucepattersonius* from Misiones province in Argentina (Mares and Braun 2000a) were described from single individuals captured at localities separated by only a few kilometers. Our inability to examine the type specimens has prevented us from making unequivocal decisions regarding their taxonomic status. We thus follow herein the taxonomic structure, diagnostic characters, and key for species originally provided by Mares and Braun (2000a). Morphological and molecular comparisons of larger samples from Misiones province, comparisons that include the holotypes of *B. paradisus*, *B. misionensis*, and *B. guarani*, are critically required to evaluate the true species diversity in the Misiones region as well as the relationships of these forms to previously described species from Brazil (see J. Pereira et al. 2005; Cirignoli et al. 2011).

SYNONYMS:

Hesperomys: Hensel, 1872b:39; part (listing of *nasutus*); not *Hesperomys* Waterhouse.

Hesperomys (*Oxymycterus*): Ihering, 1893:109; part (listing of *nasutus*); not *Oxymycterus* Waterhouse.

Oxymycterus: Winge, 1887:36; part (description of *talpinus*); not *Oxymycterus* Waterhouse.

Microxus: Thomas, 1909:237; part (listing of *iheringi*); not *Microxus* Thomas.

?*Microxus*: Gyldenstolpe:134; part (listing of *iheringi*); not *Microxus* Thomas.

Akodon (*Microxus*): Cabrera, 1961:458; part (listing of *iheringi*); neither *Akodon* Meyen nor *Microxus* Thomas.

Juscelinomys: Moojen, 1965:284; part (listing of *talpinus*); not *Juscelinomys* Moojen.

Brucepattersonius Hershkovitz, 1998:227; type species *Brucepattersonius soricinus* Hershkovitz, by original designation.

KEY TO THE SPECIES OF *BRUCEPATTERSONIUS* (ADAPTED FROM HERSHKOVITZ 1998 AND MARES AND BRAUN 2000A):

1. Length of incisive foramina 5.8 mm or more 2
1'. Length of incisive foramina less than 5.8 mm 4
2. Tail length < 90% of head and body length; under parts grayish with ochraceous wash . . . .
. . . .*Brucepattersonius soricinus*
2'. Tail length ≥90% to head and body length. 3
3. Under parts primarily grayish . . . .
. . . .*Brucepattersonius griserufescens*
3'. Under parts primarily reddish or orange . . . .
. . . . *Brucepattersonius igniventris*
4. Dorsum uniform gray, venter scarcely paler . . . .
. . . .*Brucepattersonius iheringi*
4'. Dorsum reddish-brown or grayish brown 5
5. Dorsal and ventral colorations not sharply demarcated laterally; dorsum reddish-brown; chin ochraceous; pterygoid wings robust and with distinctive knobs at tips; zygomatic arches thick; rostrum wider (>5 mm), with well-developed nasolacrimal capsules . . . .
. . . . *Brucepattersonius paradisus*
5'. Dorsal and ventral colorations sharply demarcated laterally; dorsum dark-brown to grayish brown; chin whitish; pterygoid wings delicate and lacking knobs at tips; zygomatic arches thin; rostrum narrower (<5 mm), with poorly developed nasolacrimal capsules 6
6. M3 one-half the length of M2; tail bicolored . . . .
. . . .*Brucepattersonius misionensis*
6'. M3 < one-half the length of M2; tail unicolored. . . . . .
. . . .*Brucepattersonius guarani*

Brucepattersonius griserufescens Hershkovitz, 1998
Gray-bellied Brucie

SYNONYMS:

Brucepattersonius griserufescens Hershkovitz, 1998:233; type locality "Terreirão, Parque Nacional do Caparaó, Minas Gerais, Brazil, elevation 2400 meters."

Brucepattersonius albinasus Hershkovitz, 1998:235; type locality "Pico da Bandeira, Parque Nacional de Caparaó, Minas Gerais, Brazil, elevation, 2700 meters."

DESCRIPTION: Stated to be largest species in genus by Hershkovitz (1998), but mean head and body length 103 mm (range: 93–109 mm) similar or even smaller than other species described by him. Tail length for 10 specimens from Caparaó, including holotype, average 103 mm (range 97–112 mm). Dorsum uniformly brownish or dark brown from rostrum to tail base; pelage silky and fluffy, with hairs about 10 mm long that possess narrow subterminal ochraceous orange band and gray hair base entirely concealed beneath long lax fur; sides of trunk and limbs like back; underparts predominantly to entirely grayish, variably washed pale ochraceous and more or less defined from sides; tail uniformly brown, dorsal hairs about 1 scale long, and ventral hairs, plus tip and terminal pencil, whitish and about three scales long; scales visible to eye; ears brown, partly hidden in fur; digital and facial vibrissae whitish, rostral vibrissae when laid back reach ear bases; hands and feet pale brown to grayish above, palms pigmented or unpigmented, soles brown; manual claws 2 mm long; pedal claws 4 mm long.

Skull smooth, without crests or ridges; rostrum long, slender, and tapered, with nasal tips rounded and extended slightly beyond incisors, but without expansion or formation of trumpet; zygomatic plates little exposed when viewed from above; interorbital region wide, with rounded edges; incisive foramina short, with posterior edges that do not reach level of M1 protoflexi, with posterior portions narrower than anterior and median portions; sphenopalatine vacuities laterally expanded and oblong in shape, mastoid perforated by conspicuous foramen or fenestra; palatal bridge extends to posterior plane of M3s; capsular process only weakly pronounced.

DISTRIBUTION: Known from the higher elevations of the Caparaó, Mantiqueira, and Serra dos Órgãos massifs, in the Atlantic Forest of southeastern Brazil, and at elevations from 1,800 to 2,400 m.

SELECTED LOCALITIES (Map 99): BRAZIL: Espírito Santo, Santa Clara do Caparaó, Parna Caparaó, vertente nordeste (MNRJ 69581); Minas Gerais, Alto Caparaó, Parna Caparaó, vertente sudoeste (Hershkovitz 1998), Itamonte, Brejo da Lapa (MNRJ 60602), Fazenda do Itaguaré, Passa Quatro (UFMG 1891); Rio de Janeiro, Macaé de Cima, Pirineus (MNRJ 69762), Teresópolis, Vale das Antas (MNJR 77118).

SUBSPECIES: We treat *Brucepattersonius griserufescens* as monotypic.

NATURAL HISTORY: On the Caparaó Massif in southeastern Brazil, Bonvicino et al. (1997) obtained individuals in submontane secondary forest, montane scrub, but primarily in undisturbed humid montane forest. Other details of population of natural history are lacking.

REMARKS: J. F. Vilela et al. (2006) placed *albinasus* Hershkovitz, the "white-nosed Brucie," for which there is only the holotype (MNRJ 32017), in synonymy with *B. griserufescens* based on molecular and morphological characters (J. F. Vilela et al. 2006). The supposedly diagnostic tuft of hair on the tip of the tail was found to be indistinguishable from the condition found in some individuals in a larger series assignable to *B. griserufescens*, and the white "nose," actually the region around the rhinarium, was interpreted as an additional case of incomplete penetrance of genotypes and intraspecific variation *sensu* Axenovich et al. (2004). Specimens from Itamonte in the Serra da Mantiqueira, assigned to *B. griserufescens* by Bonvicino et al. (1998), as well as those from Serra dos Órgãos may constitute a separate taxonomic unit (J. F. Vilela 2005).

The karyotype consists of $2n = 52$, $FN = 52$ (Svartman and Cardoso de Almeida 1993b), with 24 pairs of acrocentric chromosomes that vary in size from large to small, plus a small pair of metacentric chromosomes; the X and Y chromosomes are medium and small acrocentrics (J. F.

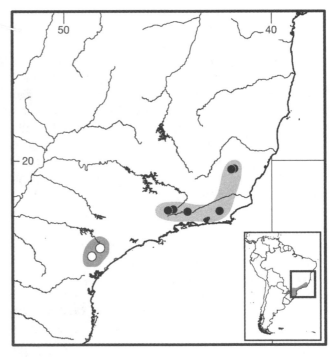

Map 99 Selected localities for *Brucepattersonius griserufescens* (●) and *Brucepattersonius igniventris* (○).

Vilela 2005). Bonvicino et al. (1998) recorded an inversion polymorphism resulting in an FN of 53 in specimens from Itamonte, Minas Gerais state.

Brucepattersonius guarani Mares and Braun, 2000
Guaraní Brucie

SYNONYM:

Brucepattersonius guarani Mares and Braun, 2000a:9; type locality "Argentina: Misiones Province: Departamento Guaraní, 6 km NE by Highway 2 of Jct. Highway 2 and Arroyo Paraíso. The elevation at the type locality is 360 m (1,180 ft.)"; ca. 2 km W Parque Provincial Moconá.

DESCRIPTION: Medium sized, with head and body length 101 mm, and tail 91 mm; ears originally described as small and rounded, but reported length (18 mm) slightly longer than the mean of all Brazilian species. Dorsum dark brown, with numerous and long guard hairs dark at their tips and becoming gradually paler basally; banded hairs with dark gray base along most of length, narrow pale-colored middle band (about 0.5 mm in length), and dark tip (about 0.5 mm); sides similar to dorsum in color, gradually lightening toward venter, which is washed with ochraceous, but hairs gray basally and with cinnamon-buff tips; general coloration of throat smoky gray; chin within distinct patch of white hairs; tip of muzzle with very small patch of white hairs; ears dark, outer surfaces gray; hindfeet brown, covered with hairs whitish basally, light brownish in middle, and whitish at tip; tail described as unicolored, but hairs brownish black above and bicolored below (brownish basally with whitish tips), extending over two to three scale rows; about 19 scales per cm at base of tail.

Skull slender, with delicate rostrum, and blunt posterior ends of nasals that extend beyond premaxillofrontal suture; nasolacrimal capsules not well developed; zygomatic plates markedly inclined posteriorly from bottom to top; zygomatic arches delicate, slightly expanded laterally; incisive foramina long, extending posteriorly to antero-lingual conules of M1s, and with prevomerine process of premaxilla at least three-fourths the length of the opening; pterygoid wings delicate, tips without distinct knob. Mandible slender, with posterior extension of condyloid process about equal to that of angular process, which itself is not especially delicate and lacks a distinct lateral hook. M3 less than or equal to one-half length of M2.

DISTRIBUTION: *Brucepattersonius guarani* is known only from its type locality.

SELECTED LOCALITIES (Map 100): ARGENTINA: Misiones, 6 km NE by Hwy 2 of junction Hwy 2 and Arroyo Paraíso (type locality of *Brucepattersonius guarani* Mares and Braun).

SUBSPECIES: *Brucepattersonius guarani* is monotypic.

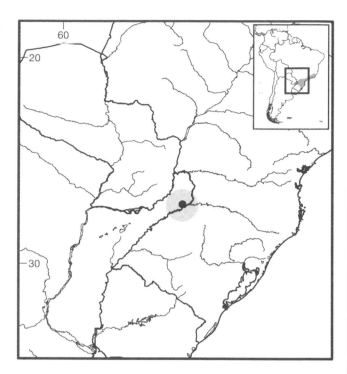

Map 100 Single known locality for *Brucepattersonius guarani* (●).

NATURAL HISTORY: The holotype, taken in November 1990, had a closed vagina and was molting. Unidentified roundworms were found in the intestines (Mares and Braun 2000a). Nothing else is known about the natural history of this species.

REMARKS: The holotype appears to be a subadult, based on the relatively unworn molars, its nonreproductive status, and molt condition.

Brucepattersonius igniventris Hershkovitz, 1998
Red-bellied Brucie

SYNONYM:

Brucepattersonius igniventris Hershkovitz, 1998:232; type locality: "Iporanga (Petar) State Park [= Parque Estadual Turístico do Alto Ribeira], southwestern São Paulo, Brazil."

DESCRIPTION: Mean and range values for four specimens include head and body length 116 mm (114–118 mm), tail length 93 mm (86–99 mm), hindfoot with claw 24 mm (22–25 mm), and ear length 17 mm (16–18 mm). Dorsal pelage fine, soft, appressed, 7–8 mm long; coloration from snout to rump reddish brown, sides slightly more reddish and merging into reddish orange of cheeks, chin, throat, arms, and belly; broad orange lateral line separates reddish brown dorsum from reddish venter. Tail shorter than head plus body, uniformly brownish in color, thinly clothed with scales clearly visible, and terminating in short, thin pencil of whitish to brownish hairs; digital tufts of three middle toes of forefeet and hindfeet whitish, those on the outer toes sparse or absent; manual claws small and thin; vibrissae fine, longest barcly reaching ear base when laid back. Skull slender; rostrum attenuated; nasal tips project well beyond incisors; palate long, extending to posterior plane of M3s; incisive foramina long, with posterior limits near the level of protoflexi of M1s, or, if not reaching this plane, posterior portion of incisive foramina wider than median portion.

DISTRIBUTION: Known only from two localities in the Atlantic Forest of southeastern São Paulo state, Brazil.

SELECTED LOCALITIES (Map 99): BRAZIL: São Paulo, Capão Bonito (MZUSP [AB107]), Iporanga (Petar) State Park (type locality of *Brucepattersonius igniventris* Hershkovitz).

SUBSPECIES: *Brucepattersonius igniventris* is monotypic.

NATURAL HISTORY: The few known specimens have been collected in secondary forest. Little else is known about the natural history of this species.

REMARKS: Despite being easily distinguished by pelage color from other Brazilian species, preliminary molecular analyses (J. F. Vilela et al., unpubl. data) which include the specimen from Capão Bonito suggest that this form is not distinct from *B. soricinus*. See also M. F. Smith and Patton (1999) and Remarks for *B. soricinus*.

Brucepattersonius iheringi (Thomas, 1896)
Ihering's Brucie

SYNONYMS:

Hesperomys nasutus: Hensel, 1872b:43; part; not *Mus nasutus* Waterhouse.

Oxymycterus talpinus Winge, 1887:36; type locality "Lapa da Escrivania [= Lapa da Escrivânia] Nr. 5," Lagoa Santa, Minas Gerais, Brazil.

Hesperomys (Oxymycterus) nasutus: Ihering, 1893:109; part; not *Mus nasutus* Waterhouse.

Oxymycterus Iheringi Thomas 1896:308; type locality "Taquara, Rio Grande do Sul," Brazil.

M[*icroxus*]. *iheringi*: Thomas 1909:237; name combination.

Microxus (?) *iheringi*: Gyldenstolpe, 1932a:134; name combination.

Microxus iheringii C. O. da C. Vieira, 1953:145; incorrect subsequent spelling of *Oxymycterus Iheringi* Thomas.

Akodon (Microxus) iheringi: Cabrera, 1961:45; name combination.

[*Oxymycterus rutilans*] *iheringi*: Hershkovitz, 1966a:86; name combination.

Brucepattersonius iheringi: Hershkovitz, 1998:240; first use of current name combination.

DESCRIPTION: Medium sized, with mean and ranges for four specimens, including holotype, of head and body length 100 mm (89–113 mm), tail length 100 mm (94–107 mm), hindfoot length 23 mm (22–26 mm), and ear length 17 mm (15–18 mm); fur soft and thick; upper parts and sides overall grayish, under parts uniform grizzled brown, scarcely paler than sides. Skull slender; incisive foramina short, with posterior limits not reaching level of M1 protoflexus, with posterior portions narrower than anterior and median portions; sphenopalatine vacuities absent or reduced to longitudinal slits; mastoid completely ossified without distinct foramina.

DISTRIBUTION: The core distribution of this species consists of the *Araucaria* forest on the Brazilian Meridional Plateau and the coastal forests of southern Brazil. Its elevation range is from near sea level to 1,000 m in the mountainous regions of the Aparados da Serra (Rio Grande do Sul state) and Piraquara (Paraná state). C. O. da C. Vieira (1955) suggested that this species occurs in Uruguay, a possibility that has not been corroborated, including by voucher specimens in collections (Massoia 1963b).

SELECTED LOCALITIES (Map 101): ARGENTINA: Misiones, Aristóbulo del Valle, Cuña Piru (MLP 16.VII.02.09), Dos de Mayo (Massoia and Fornes 1969), Puerto Gisela (Massoia 1963b), Sierra de La Victoria (MACN-BR18950), Tobunas, Rte 14, Km 352 (Massoia 1963b). BRAZIL: Rio Grande do Sul, Aratiba (MNRJ [CRB 1944]), Cambará do Sul, Parque Nacional da Serra Geral (MNRJ 78511), Osório, Morro Osório (MNRJ 49798), Sapiranga, Alto Ferrabraz, Morro Ferrabraz (MNRJ 49800), Taquara (type locality of *Oxymycterus iheringi* Thomas), Tôrres, Faxinal Norte Lagoa Itapeva (MNRJ 49802); Santa Catarina, Caldas da Imperatriz, Parque Estadual Serra do Tabuleiro (UFSC [ZOOL] 735), Jaborá (MNRJ 69785).

SUBSPECIES: *Brucepattersonius iheringi* is monotypic.

NATURAL HISTORY: Field notes indicate that this mouse has been most frequently captured in forests with dense litter and herbaceous cover; captive animals have tried to burrow under the litter placed in cages. Specimens recorded by Massoia (1963b) were trapped at night in wooded areas, one on the border of a rocky stream, the other in a low brushy area on the edge of a path.

REMARKS: Note that the fossil species *talpinus* Winge is herein tentatively placed in the synonymy of *iheringi* Thomas. Should careful analyses of available specimen confirm this hypothesis, Winge's name would have priority over that of Thomas.

Hershkovitz (1998) reported on the mean and extremes of measurements from at least 11 specimens from Misiones province, Argentina, including ones from Tobuna and Puerto Gisela, referred by Massoia (1963b) and Massoia and Fornes (1969) to *O. iheringi*, but which Hershkovitz (1998) simply called *Brucepattersonius* sp. Mares and Braun (2000a) did not assign any specimens collected

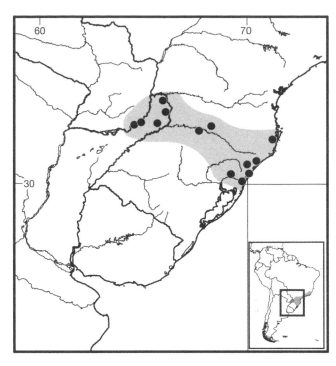

Map 101 Selected localities for *Brucepattersonius iheringi* (●).

previously in Misiones to the species they described. J. F. Vilela (2005) examined 16 specimens from Misiones, except holotypes. He identified a single evolutionary lineage in that province, one that extends into adjacent Rio Grande do Sul, Brazil, based on external morphology and craniodental morphometric analysis.

Brucepattersonius iheringi shares the same chromosomal complement, including morphologies of autosomes and sex chromosomes, as *B. griserufescens*, with $2n = 52$, FN = 52 (J. F. Vilela 2005).

Brucepattersonius misionensis Mares and Braun, 2000
Misiones Brucie

SYNONYM:

Brucepattersonius misionensis Mares and Braun, 2000a:7; type locality "Argentina: Misiones Province: Departamento Guaraní, jct. Hwy 21 and Arroyo Oveja Negra, approx. 2 km W Parque Provincial Moconá."

DESCRIPTION: Dorsal pelage brownish; dark guard hairs numerous and long, banded with dark gray base (about 8 mm), yellowish middle (about 1 mm in length), and dark tip (<1 mm); sides and flanks similar in color to dorsum, although slightly paler, gradually merging into color of venter; ventral hairs whitish basally (about 2 mm), banded neutral gray in middle (about 6 mm in length), and tipped ivory yellow; ears rounded, their outer surfaces edged with hairs colored similarly to those of dorsum; tail bicolored, olive brown above and lighter

below, clothed in short hairs that have brownish bases and whitish tips; tail with 16 scales per cm, measured at base; forefeet whitish and covered with hairs entirely whitish or that have light brownish bases and whitish tips; ungual tufts surround nails of forefeet, but do not extend beyond tips. Skull slender; rostrum delicate; tips of nasals bluntly pointed and deflected slightly downward; posterior ends of nasals blunt and extend beyond premaxillofrontal suture; nasolacrimal capsules not well developed; zygomatic arches delicate; incisive foramina long, with prevomerine process of premaxilla extending less than three-fourths their length; pterygoid wings delicate, with tips lacking distinct knob. Mandible slender; condylar process extends beyond angular process; and angular process has delicate hook. M3 longer than one-half length of M2.

DISTRIBUTION: This species is known only from its type locality in Misiones province, Argentina.

SELECTED LOCALITIES (Map 102): ARGENTINA: Misiones, junction Hwy 21 and Arroyo Oveja Negra, approximately 2 km W of Parque Provincial Moconá (type locality of *Brucepattersonius misionensis* Mares and Braun).

SUBSPECIES: *Brucepattersonius misionensis* is monotypic.

NATURAL HISTORY: The holotype, caught under a log uphill from the nearby river in December 1990, was molting and had testes that measured 13×6 mm.

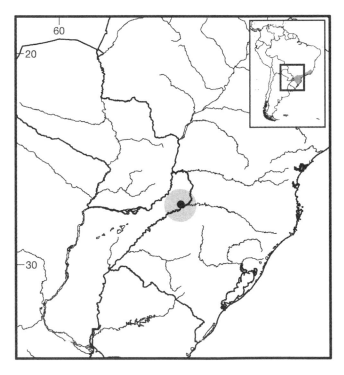

Map 102 Single known locality for *Brucepattersonius misionensis* (●).

REMARKS: Together with *B. guarani*, the holotype of which is an old adult and molting specimen, *B. misionensis* was described primarily on the basis of pelage characters. These two taxa may prove to be synonyms of *B. paradisus*, the first of the three species described by Mares and Braun (2000a) from closely spaced localities, or even of *B. iheringi* (see earlier and J. F. Vilela 2005).

Mares and Braun (2000a) described a $2n = 52$, $FN = 52$ karyotype with large to small uniarmed to subtelocentric autosomes and X and Y chromosomes indistinguishable from the autosomes.

Brucepattersonius paradisus Mares and Braun, 2000
Paradise Brucie

SYNONYM:

Brucepattersonius paradisus Mares and Braun, 2000a:3; type locality "Argentina: Misiones Province: Departamento Guaraní, jct. Hwy 2 and Arroyo Paraíso. The elevation at the type locality is 197 m."

REMARKS: Size medium, with head and body length 108 mm and tail 90 mm. Ears <19 mm in length, and rounded; hindfeet long, narrow, base of middle digits slightly webbed; manual digits 3 mm in length, pedal digits 5 mm. Overall dorsal coloration reddish brown; dorsal fur long and rather lax, about 10 mm in length on midrump, with dark gray basal band (about 8 mm), followed by yellowish-buffy band (about 1 mm) and dark tip (<1 mm); sides and flanks similar in color to dorsum, although slightly paler, with gradual transition from dorsum to venter, which is strongly washed with bright ochraceous; ventral hairs grayish basally, for about 50% of length; hairs of check, throat, and chin pinkish cinnamon; tip of muzzle has small patch of white hairs; outer surfaces of ears furred with hairs like those of dorsum; hindfeet dark gray and scantily clothed with hairs with brownish bases and whitish tips, with skin visible; ungual tufts extend beyond nails of hindfeet; forefeet covered with reddish brown hairs, ungual tufts surround nails, but generally do not extend beyond tips; tail essentially unicolored, overall dark gray in color, but caudal hairs unicolored dorsally and with dark bases and whitish tips ventrally; caudal hairs sparsely distributed, extending over fewer than two scale rows, underlying epidermal scales generally visible by eye, about 17 scales per cm at base of tail; tip of tail ends in short tuft of hairs 2 mm long. Skull with robust rostrum that gradually diverges posteriorly due to well-developed nasolacrimal capsules; gnathic process well developed but does not extend beyond nasals; nasal tips blunt when viewed in lateral profile, upturned, posterior ends blunted and not extending past premaxillofrontal suture; interorbital region moderately broad and hourglass in shape, with no ridges or ledges; zygomatic arches thick, slightly convergent anteriorly and slightly expanded laterally; zy-gomatic plates narrow and inclined posteriorly from bottom to top, anterior borders convex; zygomatic notches shallow; incisive foramina long, prevomerine processes of premaxilla extend at least three-fourths length of opening and wider than one-half of greatest width; superior masseteric ridges present; pterygoid wings robust, tips with distinct knob; palate with very shallow grooves in which oblong posterior palatine foramina are located; small posterolateral palatal pits present; mesopterygoid fossa "heart-shaped," extending anterior to posterior root of M3. Mandible slender, with coronoid process well developed and condyloid process extending just slightly beyond delicate angular process with distinct lateral hook. Molar rows parallel sided or slightly convergent posteriorly; molars tetralophodont, quadritubercular, and hypsodont; M3 greater than one-half length of M2. Upper incisors orthodont.

DISTRIBUTION: *Brucepattersonius paradisus* is known only from its type locality.

SELECTED LOCALITIES (Map 103): ARGENTINA: Misiones, junction Hwy 2 and Arroyo Paraíso (type locality of *Brucepattersonius paradisus* Mares and Braun).

SUBSPECIES: *Brucepattersonius paradisus* is monotypic.

NATURAL HISTORY: The single known specimen was collected in rainforest, among rocks, 6 m uphill from the nearby river. It was molting and pregnant with two embryos,

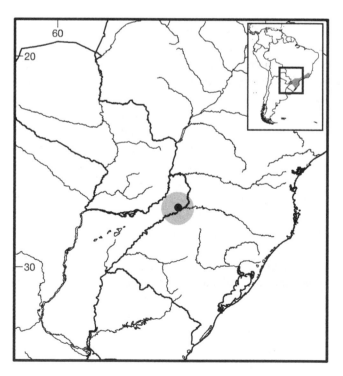

Map 103 Single known locality for *Brucepattersonius paradisus* (●).

one in each uterine horn. The stomach was full of hair that was the same color as the animal.

REMARKS: The possibility that *B. paradisus* is conspecific with *B. iheringi* requires appropriate comparison of the holotype with series of the latter species (see J. F. Vilela 2005).

Brucepattersonius soricinus Hershkovitz, 1998
Short-tailed Brucie, Soricine Brucie

SYNONYM:

Brucepattersonius soricinus Hershkovitz, 1998:232; type locality "Ribeirão Fundo, São Paulo, Brazil."

DESCRIPTION: Described as one of smallest species, *B. soricinus* has relatively shortest tail (Hershkovitz 1998). Mean values for 26 specimens include head and body length 99 mm, tail length 88 mm, hindfoot length (with claw) 25 mm, and ear length 16 mm. Pelage moderately appressed, fur about 8 mm long, insufficient to conceal pinnae; dorsum brownish to dark brownish, crown to rump brown, hairs with tips minutely tipped blackish, followed by ochraceous orange band and dark gray base; guard hairs entirely blackish; sides paler than back, with wider ochraceous bands, basal portions gray; side color merges into grayish chest and belly that may be washed reddish, with hairs broadly banded pale ochraceous with bases dark gray; throat and chin dominantly gray, hairs directed forward; coloration of forelimbs and hindlimbs like that of trunk; digital vibrissae sparse, facial vibrissae short, hardly reaching ear base when laid back. Skull broad; nasal bones long, tapered, extending anterior of incisors by 1–2 mm, tips either rounded or blunt, not squared, and trumpet shaped; premaxillary bones do not reach nasal tips; zygomatic arches weak, barely spreading beyond greatest width of braincase; zygomatic plates narrow, markedly reclined posteriorly from bottom to top, hardly visible when viewed from above; interorbital region smooth, wide; braincase smooth and subglobular in shape; interparietal only 1 or 2 mm in depth; incisive foramina long and narrow, terminating slightly behind anterior plane of M1s; palate extends to or slightly behind posterior plane of M3; sphenopalatine vacuities absent. Molars tetralophodont and hypsodont; molar rows parallel sided or slightly convergent posteriorly; both M1 and m1 with median fold; enamel of anterior margin of m1 more or less crenulated, mesoloph(id)s may be present or absent; cusps ovate to subtriangular, cuspids subprismatic, opposing pairs in echelon. M3 about two-thirds size of m2.

DISTRIBUTION: This species occurs in southeastern and southern Brazil, in the states of São Paulo and Paraná, reaching the Mantiqueira massif at its northern limits.

SELECTED LOCALITIES (Map 104): BRAZIL: Minas Gerais, Itamonte, Brejo da Lapa (MNRJ 48018); Paraná,

Ortigueira (MZUSP 31623), Piraquara, Mananciais da Serra (MNRJ 78421), Telêmaco Borba (MNRJ 68342); São Paulo, Bananal, Reserva Ecológica do Bananal (MZUSP 33713), Cotia, Reserva Morro Grande (MNRJ 78678), Estação Biológica de Boracéia (MVZ 183036), Tapiraí (MZUSP [AB 201]).

SUBSPECIES: *Brucepattersonius soricinus* is monotypic.

NATURAL HISTORY: This species has been collected in the interior of well-preserved relicts of moist broadleaf forest (*floresta ombrófila densa*), under ground litter (Serra de Paranapiacaba; Umestu and Pardini 2007), in native vegetation in early stages of regeneration, in mosaics of forest remnants, and in anthropogenic habitats (Caucaia do Alto; Umestu and Pardini 2007). At Reserva Florestal de Morro Grande, this species was caught with a relative frequency, among 22 other species of small mammals (592 captured in all) of (0.9%) in secondary forest and 1.2% in mature, primary forest, with an overall frequency 1% (Pardini and Umetsu 2006). A specimen (MNRJ 48018) from Itamonte, Minas Gerais state, was captured in *floresta ombrófila densa*. Known localities range in elevation from sea level to more than 2,000 m.

REMARKS: We suggested earlier that *B. igniventris* might not be specifically distinct from *B. soricinus*, based on the high similarity in their mtDNA cytochrome-*b* sequences. M. F. Smith and Patton (1999), in fact, in their analysis of generic relations among sigmodontine rodents,

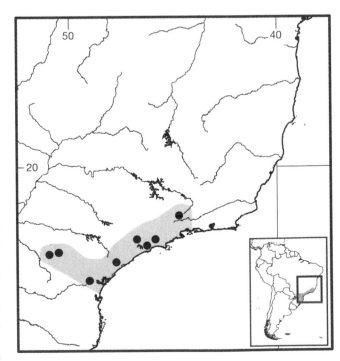

Map 104 Selected localities for *Brucepattersonius soricinus* (●).

allocated specimens from Base do Carmo, Fazenda Inter-vales, Capão Bonito (*igniventris*), and Estação Biológica de Boracéia (*soricinus*) to *Brucepattersonius* "sp. 1" be-cause they formed part of a single clade with very simi-lar cytochrome-*b* haplotypes. J. F. Vilela et al. (2006) also noted that the haplotypes from these two localities formed a strongly supported clade relative to *B. griserufescens* (in-cluding *albinasus*).

Genus *Deltamys* Thomas, 1917
Ulyses F. J. Pardiñas and Pablo Teta

The monotypic genus *Deltamys* lives in humid grasslands, gallery forest, and marshes of central-eastern Argentina, Uruguay, and southern Brazil. Since the original description, this taxon has been considered to be a full genus (Gylden-stolpe 1932; Massoia 1980a) or a subgenus of *Akodon* (Ellerman 1941; Cabrera 1961; Massoia 1964; Reig 1987; Musser and Carleton 1993). The morphological analyses of Bianchini and Delupi (1994) and E. M. González and Massoia (1995), supplemented by chromosomal data (Gen-tile de Fronza et al. 1979, 1981; E. C. Castro et al. 1991), supported the generic status of *Deltamys*. Massoia (1964), Bianchini and Delupi (1994), E. M. González and Massoia (1995), E. M. González and Pardiñas (2002), and Teta, Cueto, and Suárez (2007) provided summaries of taxonomy, morphology, and/or natural history of *Deltamys*. However, we stress that the genus is still poorly known, and the extent and pattern of geographic variation deserve further explora-tion (see Montes et al. 2008). As noted in the introduction to the genus *Akodon*, the phyletic relationship of *Deltamys* vis-à-vis *Akodon* remains unresolved.

Deltamys kempi is a small mouse (head and body length <100 mm), with conspicuously small eyes and dark-colored and velvety pelage. The dorsum is shiny blackish brown, and the venter is dull brownish gray. The guard hairs have dark tips and pale bases, and the under fur is gray-ish (Massoia 1964). A white patch on the chin is present in some populations (Massoia 1964; Bianchini and Delupi 1994). The tail, only faintly bicolored, is slightly shorter than the head plus body and is finely haired throughout. Mean Based upon 39 males and 29 females from the Ciudad Autónoma de Buenos Aires, Argentina, the ratio of tail length to head and body length was 0.8, and the ear length did not exceed 12 mm (Teta, Cueto, and Suárez 2007:Table 1). Males were significantly larger and heavier than females (Teta, Cueto, and Suárez 2007). The hindfeet are narrow and have six plantar pads and moderately long claws (ca. 2.5 mm in adults); ungual tufts are whitish or brown and are typically shorter than claws. At least in the Argentinean literature, inability to distinguish between *D. kempi* and

Akodon azarae has often clouded the utility of anatomical (Bianchini and Delupi 1994) and parasitological (Lares-chi 2000) data. *Deltamys* differs from *A. azarae* in having darker coloration, more conspicuous tail scales, and a nar-rower and more gracile skull, especially in regard to the rostrum, anterior portions of the frontal bones, and the zygomatic arches.

The skull is relatively long and narrow, and with a char-acteristic tubular-shaped interorbital region. Nasals and premaxillae project slightly beyond the incisors, and the premaxillae have moderately developed gnathic processes. The braincase is slender and inflated, while the interpari-etal is exceedingly slender. Zygomatic plates are narrow, with straight anterior borders and greatly reduced upper free borders. In the squamosal region, the very weak and slender descending hamular process divides the subequal subsquamosal foramen and postglenoid fenestra. A stout alisphenoid strut is present; the carotid circulatory pat-tern is type 1 (*sensu* Voss 1988). The tegmen tympani and the suspensory process of the squamosal are both well de-veloped. Incisive foramina are relatively narrow, sharply pointed posteriorly, and end at the level of the M1 proto-cones. The postdiastemal portion of the palate is long and wide. The anterior border of the mesopterygoid fossa is wide, not reaching past the posterior edges of the M3s. The parapterygoid plates are flat and narrow. Auditory bullae are flattened, and their ventral borders do not extend be-low a plane defined by the molar surfaces. Periotic capsules are small, and paroccipital processes are little developed.

The mandible is elongated, slender, and low, with the mas-seteric crest scarcely developed. The coronoid process is rela-tively long, projected backward, and the condyloid process is elongated. The capsular projection, which lies at the level of the middle part of the sigmoid notch, is weakly developed.

The upper incisors are orthodont, ungrooved, and fronted by orange enamel. The molars have a moderately developed tubercular hypsodonty with usually crested crowns. As in many akodontines, in *D. kempi* the teeth wear rapidly, los-ing their occlusal pattern. The M1 usually exhibits a mod-erately developed anteromedian flexus, but one shorter than in *Akodonazarae*. As in M2, the anteroloph and mesoloph are absent. M3 is subcylindrical. The lower molars are en-larged and simplified. The m1 is slightly larger than m2 in young individuals, or subequal to shorter in adults and senile animals; its procingulum is wide but anteroposteriorly com-pressed with a shallow, but distinguishable, anteromedian flexid in juveniles. The m1 appears to be divided into two subequal lobules by the opposition of the hypoflexid and the mesoflexid; m2 shows a weakly developed protoflexid; and m3 is transversely compressed, with a short hypoflexid. E. M. González and Pardiñas (2002) provided mean cranial, mandibular, and dental measurements.

Steppan (1995:Table 5) reported, based on two individuals from Uruguay, that *D. kempi* has 13 thoracic, 6 lumbar, and 29–30 caudal vertebrae. Teta, Cueto, and Suárez (2007) provided information, based on individuals from northeastern Buenos Aires province, Argentina, on previously unknown details of the soft anatomy. There are two complete diastemal and four incomplete interdental palatal rugae, resembling the pattern described for the *boliviensis* group of *Akodon* (Myers et al. 1990). Dissection of two individuals revealed the existence of a gall bladder, the typical condition in the tribe. The stomach is unilocular-hemiglandular, with a nearly semicircular bordering fold and shallow incisura angularis. The glans penis is short and stocky, and covered by small epidermal spines imbedded in individual pits. The cartilaginous distal baculum is well developed, with the lateral digits about three-fourths the length of the medial digit. Only the medial bacular mound extrudes distally from the crater rim and is thus visible externally. Urethral flaps are long, taper distally, with well-separated tips, and extend beyond the crater rim so as to be visible externally. The osseous baculum is short and stout with a laterally flared base. In general, male genital morphology of *Deltamys* is similar to that of *Akodon* spp.

SYNONYMS:

Deltamys Thomas, 1917c:98; type species *Deltamys kempi* Thomas, by original designation.

[*Akodon*] (*Deltamys*): Ellerman, 1941:414; as subgenus, with *kempi* Thomas listed as sole species.

REMARKS: Although *Deltamys* has been uniformly regarded as monotypic, containing the single species *D. kempi*, both chromosomal and molecular sequence diversity suggest at least two, possibly more species than currently recognized (e.g., K. Ventura et al. 2011; see below).

Deltamys kempi Thomas, 1917

Delta Mouse or Kemp's Mouse (Ratón Aterciopelado)

SYNONYMS:

Deltamys kempi Thomas, 1917c:98; type locality: "Isla Ella, in the delta of the Rio Parana, at the top of the La Plata Estuary" (Thomas 1917:95). The exact location of this island had been uncertain until, after a detailed survey of historical and geographic sources, Pardiñas et al. (2007) tentatively located Isla Ella at 34°22′S, 58°38′W, Delta del Paraná, Buenos Aires, Argentina.

Akodon ([*Deltamys*]) *kempi*: Ellerman, 1941:414; name combination.

Akodon (*Deltamys*) *kempi*: Cabrera, 1961:451; name combination.

Akodon obscurus obscurus: Barlow, 1963: 22; name combination but not *Akodon* [= *Necromys*] *osbcurus* Waterhouse.

Deltamys kempi langguthi Massoia, 1980a:179; *nomen nudum*.

Deltamys kempi kempi: E. M. González and Massoia, 1995:3; name combination.

Deltamys kempi langguthi E. M. González and Massoia, 1995:3; type locality: "Uruguay, Montevideo, Parque Lecocq" (ca. 34°47′S, 56°22′W, 20 m), Montevideo, Uruguay.

DESCRIPTION: As for the genus.

DISTRIBUTION: *Deltamys kempi* occurs in a small portion of southeastern South America, from northeastern Buenos Aires province (ca. 36°30′S) and southern Entre Ríos province, through Uruguay to the Brazilian state of Rio Grande do Sul (ca. 29°). In Argentina, it is known from Delta del Paraná and from several localities bordering the Río de La Plata estuary, to at least the southernmost tip of Samborombón Bay (Udrizar Sauthier, Abba et al. 2005; Teta, Cueto, and Suárez 2007). In Uruguay, it is known from the departments of Colonia, San José, Montevideo, Canelones, Maldonado, Rocha, Lavalleja, Treinta y Tres, Cerro Largo, Durazno, Tacuarembó, and Rivera (E. M. González 2001; E. M. González and Pardiñas 2002; E. M. González, unpubl. data). Brazilian records are from several coastal localities in Rio Grande do Sul state (Montes et al. 2008). A fossil record for *D. kempi* is unknown.

SELECTED LOCALITIES (Map 105): ARGENTINA: Entre Ríos, Arroyo Brazo Largo and Arroyo Brazo Chico (Massoia 1983); Buenos Aires, Canal 6 and Paraná de las Palmas (Massoia and Fornes 1964a), Reserva Ecológica Costanera Sur (Teta, Cueto, and Suárez 2007), Reserva El Destino (Udrizar Sauthier, Abba et al. 2005), Arroyo Las Tijeras (CNP 2377). BRAZIL: Rio Grande do Sul, Tôrres (E. C. Castro et al. 1991), Osório (Montes et al. 2008), Charqueadas (Montes et al. 2008), Tapes (E. C. Castro et al. 1991), Taim (Sbalqueiro et al. 1984). URUGUAY: Rivera, Minas de Corrales (E. M. González and Massoia 1995); Tacuarembó, Paso Baltasar, Arroyo Tres Cruces (E. M. González and Massoia 1995), Treinta y Tres, Arrozal 33 (E. M. González and Massoia 1995), Lavalleja, Paso Averías, Río Cebollatí (E. M. González and Massoia 1995); Rocha, Parque Nacional Santa Teresa (E. M. González and Massoia 1995), Laguna de Castillos (Gambarotta et al. 1999); Colonia, Playa Ferrando, 2 km E Cnia. Sacramento (E. M. González and Massoia 1995), Puente del arroyo Pereira sobre RN 1 (E. M. González and Massoia 1995); Canelones, Laguna del Cisne (E. M. González and Massoia 1995); Montevideo, Parque Lecoq (E. M. González and Massoia 1995).

SUBSPECIES: E. M. González and Massoia (1995) segregated populations from Uruguay and Brazil, as *D. k. langguthi*, from the nominotypical race represented by Argentinean samples. However, neither D'Elía et al. (2003)

nor Montes et al. (2008) found support for the recognition of subspecies. We currently regard *D. kempi* as monotypic.

NATURAL HISTORY: *Deltamys kempi* inhabits marshy environments, particularly borders of wetlands, flooded grassland, and places with reeds and straws usually without trees (E. M. González and Pardiñas 2002). Less commonly, it is also found in woodland with a high diversity of trees and shrubs (E. M. González 1996). Most of the documented captures in Argentina were made in swamps dominated by the sedges *Scirpus giganteus* and *Schoenoplectus californicus*, relatively pure stands of the Pampa grass *Cortaderia selloana*, or in areas of dense grassy cover in humid, wooded environments (Massoia 1964; Bianchini and Delupi 1994; Teta, Cueto, and Suárez 2007). In Uruguay, E. M. González and Pardiñas (2002) collected *D. kempi* in traps placed on floating vegetation. Despite a predilection for wet environments, this species exhibits no special external adaptations for aquatic life and has no greater swimming ability than other mice of similar size and morphology.

Bianchini and Delupi (1994) considered that the long rostrum, morphology of both manus and pes, and claw length were indicative of digging habits and an animalivorous diet. L. M. Miller and Anderson (1977) found that body proportions of *D. kempi* correspond to those of fossorial species. Animals kept in captivity sometimes dug but usually foraged on the ground and looked for resting sites on the surface (Massoia 1964). The nest of *D. kempi*, made with plant fibers, is spherical and sometimes placed in hollow logs (Massoia 1964). In Uruguay, one nest with three young was found in a fallen palm (*Syagrus capitata*), with the cavity entrance 30 cm above ground level (E. M. González and Pardiñas 2002). Trapping data indicate that *D. kempi* is nocturnal. Little is known of the diet; the few stomachs examined contained insects, a few pieces of seeds, and limited remains of green plants. Barn (*Tyto alba*), Burrowing (*Athene cunicularis*), and Great Horned Owls (*Bubo virginianus*) are known predators (Massoia 1983; E. M. González and Saralegui 1996; Teta, Malzof et al. 2006; see E. M. González and Pardiñas 2002 for review).

In northeastern Buenos Aires province, males with scrotal testes and females with open vaginas were caught mostly in late winter, spring, and summer; pregnant and lactating females were recorded in spring and summer (Teta, Cueto, and Suárez 2007). In Uruguay, Gonzalez (2001) found lactating females in November (late spring) and with embryos in December and February (summer). Thus, the reproductive season seems to start at the late winter/early spring and to last until the end of summer/early autumn. Weights of reproductive individuals were 19–29 g in males and 17–28 g in females (Teta, Cueto, and Suárez 2007). Population structure in northeastern Buenos Aires province suggested

significant early spring recruitment to the late spring and summer population. Animals of intermediate age constituted the majority of individuals in autumn and winter and to a lesser extent in spring and summer. Very old animals were found mostly in spring and summer. The life span was at least 18 months in one individual, and 9 and 15 months in two others (Teta, Cueto, and Suárez 2007). Overall sex ratio, in this same study area, was 55.1% males and 44.9% females.

Autino and Lareschi (1998) and Lareschi and Mauri (1998) studied ectoparasites in Argentinean populations, including dermanyssoid mites and fleas. Individuals from coastal localities on the Río de La Plata in Argentina (Navone et al, 2009) were parasitized by the fleas *Polygenis* (*Neopolygenis*) *atopus* and *P.* (*P.*) *bohlsi bohlsi*, the mite *Androlaelaps mauri*, the flukes *Levinseniella* (*Monarrhenos*) *cruzi*, the tapeworm *Rodentolepis* sp., and the nematode *Stilestrongylus* sp. Other parasites found in *D. kempi* from Buenos Aires province, Argentina, included the mites *Eulaelaps stabularis* and *Laelaps paulistanensis*, and the fleas *P.* (*Polygenis*) *platensis* and *P.* (*Polygenis*) *rimatus* (Autino and Lareschi 1998; Lareschi and Mauri 1998). *Deltamys kempi* is the type host of *Androlaelaps mauri*; this mite appears to be host specific and is ubiquitous on this rodent throughout its distribution in Argentina and Uruguay (Lareschi and Gettinger 2009). Reca et al. (1996) regarded this species as rare in Argentina, but it is likely that the scarcity of records for this country is an artifact produced by misidentification of *Deltamys* specimens as *Akodon azarae*.

REMARKS: The recent phylogeographic analysis by Montes et al. (2008), based on mitochondrial and nuclear gene sequences, obtained a topology composed of two reciprocally monophyletic clades roughly north and south of Lagoa dos Patos (Brazil). Thus, a strict application of a phylogenetic species concept would support the hypothesis that each clade constitutes to a separate species. Montes et al. (2008) also claimed that there are differences in skull proportions between these two groups, but the small size of the samples analyzed means that these results should be taken with caution. Clearly, the potential existence of more than one species in what is now understood as *D. kempi* requires further exploration.

Barlow (1969:23), following personal communication with S. Anderson and his own observations, considered this species conspecific with *Necromys obscurus* (given as *Akodon obscurus obscurus*). Because of this, and taking into account the locations of the collecting localities, his observations concerning the natural history of "*A. o. obscurus*" in Uruguay may have been based on a mixture of data from these two different species. Ximénez et al. (1972) partly clarified this situation, stating that at least the specimens housed at the AMNH were in fact referable to *Deltamys*. A critical reexamination of the entire series referred by Barlow (1963) to *Akodon obscurus obscurus* is needed to resolve the uncertainty (see E. M. González and Massoia 1995).

The karyotype of *Deltamys kempi* is highly variable, with $2n = 35$, 36, 37, and 38, but $FN = 38$ (E. C. Castro et al. 1991). Gentile de Fronza et al. (1981) recorded a somatic complement of 36 acrocentric chromosomes and a single submetacentric chromosome. Chromosome studies on 28 specimens collected from two Brazilian populations revealed seven different karyotypes, due to two autosomal centric fusions (chromosomes 2 with 3 and 9 with 15) in homozygous and heterozygous states, and a Y-autosome translocation present in all males (E. C. Castro et al. 1991). Four autosomal centric fusions were found in 44% of specimens, with each rearrangement restricted to a different locality (E. C. Castro et al. 1991). In parts of its range, the species exhibits an uncommon multiple sex-chromosome determining mechanism of the type $X_1X_1X_2X_2/X_1X_2Y$ (Sbalqueiro et al. 1984), unique among sigmodontines. However, specimens from Esmeralda, Rio Grande do Sul, southern Brazil, had a $2n = 40$, $FN = 40$ karyotype, with a normal XX/XY sex chromosome system. These are allopatric to the main distribution of *D. kempi*. Phylogenetic analyses based on mtDNA grouped these specimens as sister to other *D. kempi* samples, and at a divergence up to 12% (K. Ventura et al. 2011). Both chromosomal and molecular differences prob-

ably signify separate species status and highlight the need for a revision of the genus.

Genus *Gyldenstolpia* Pardiñas, D'Elía, and Teta, 2009

Ulyses F. J. Pardiñas and Alexandra M. R. Bezerra

This genus was recently erected for species that had previously been placed under the name *Kunsia fronto* and its relatives, a group of sigmodontines poorly known from a few, scattered fossil and Recent specimens from the Chaco and Cerrado biomes of central South America (Hershkovitz 1966a; Pardiñas, D'Elía, and Teta 2009). *Gyldenstolpia* includes two species, *G. fronto* (Winge) and *G. planaltensis* (Avila-Pires), known collectively from five localities (Bezerra 2011). There are indications that these sigmodontines were widespread in the past (Pardiñas, D'Elía, and Teta 2009).

The history of *Gyldenstolpia* started with Winge (1887) naming *Scapteromys fronto*, based on a fossil consisting of a fragment of the anterior part of a skull found in Lapa da Escrivânia Nr. 5, one of the richest fossil sites excavated by Peter Lund in the vicinity of Lagoa Santa, Minas Gerais state, Brazil. Gyldenstolpe (1932b) added a second piece to the puzzle when he described *Scapteromys chacoensis* on the basis of a single specimen, a broken skull and an imperfect skin, from the Río de Oro in the Argentinean Chaco. Gyldenstolpe (1932b:2) discussed the resemblance between the two supposedly separate species, stating that "the present species [*S. chacöensis*] seems . . . most closely allied to the subfossil form named *Scapteromys fronto* by Winge. This species is, however, only known from some rather incomplete skull fragments found in caves, and external comparison between the two forms is therefore impossible. It is rather unlikely, that they are identical as their respective type-localities are situated far from each other." Hershkovitz (1966a) erected *Kunsia* to contain *chacoensis* and *fronto* as well as the giant forms *tomentosus* Litchtenstein, 1830, *principalis* Lund, 1840, and *gnambiquarae* Miranda-Ribeiro, 1914. Moreover, Hershkovitz (1966a) subsumed *chacoensis* Gyldenstolpe into *fronto* Winge while retaining it as a valid subspecies. Hershkovitz, though recognizing that both taxa were known from only a single specimen from their respective type localities, hypothesized larger distributional ranges for both. Based on a larger series of specimens secured during the construction of the city of Brasília, Distrito Federal, Brazil, Avila-Pires (1972) proposed *planaltensis* as a new subspecies for *K. fronto*. That same year Reig (1972) referred to *K. fronto* a fossil mandibular fragment from the Quaternary deposits in Tarija department, Bolivia, thus extending the prehistoric distribution of the genus to that country. Finally, Pardiñas,

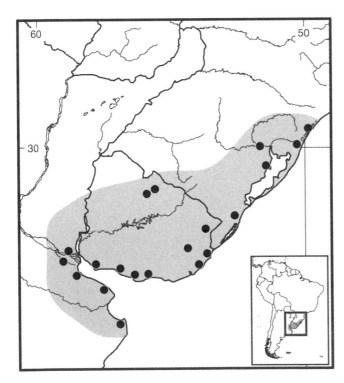

Map 105 Selected localities for *Deltamys kempi* (●).

D'Elía, and Teta (2009) revised the "scapteromyine" genera (see the introduction to the Akodontini), based on morphological characters, separating the large and smaller forms of *Kunsia* and placing the latter in their new genus *Gyldenstolpia*. These authors also preliminarily discussed part of the fossil record associated with this group, a matter that requires more thorough investigation.

Morphologically, *Gyldenstolpia* appears to be close to *Scapteromys* and *Kunsia* and probably can be placed in the *Scapteromys* division in the Akodontini (*sensu* D'Elía 2003a), pending investigation beyond reported similarities among the included genera (Pardiñas, D'Elía, and Teta 2009). Karyological and molecular data are totally lacking for this group of taxa.

Gyldenstolpia is characterized by a unique combination of traits, including large size, short tail (ca. 55% of combined head and body length), hispid dorsal hair, and small tail scales, each covered by three hairs; small and rounded ears; unicolored manus and pes; large nasals with acute posterior ends reaching the level of the lacrimals; distinctive enlargement of the zygomatic arches; large incisive foramina reaching the level of the protocones of M1s; a denticulate and open U-shaped coronal suture; a carotid circulatory pattern with an enlarged stapedial foramen and carotid canal, with evidence of the supraorbital branch of the stapedial artery on the inferior surface of the squamosal and alisphenoid bones (pattern type 1, *sensu* Voss 1988); conspicuous coronal hypsodonty; well-developed mesolophid, protolophid, and protoconulid on both first and second lower molars; and first lower molar with three roots.

Species of *Gyldenstolpia* are large sigmodontines (total length about 300 mm, weight about 150 g) with a semi-fossorial bauplan exemplified by moderately short (ca. 10% of head and body length), rounded, and semihidden ears and short and somewhat hispid dorsal pelage; small eyes, thick and short tail (tail about 110 mm), and well-developed and powerful claws on both forefeet and hindfeet (about 4.5 and 3.7 mm in length, respectively). Coloration of the back, pes, manus, and tail is dark wood brown to grayish-brown; upper parts of the body with a glistening greenish sheen visible under certain lighting conditions; the ventral surface is a not sharply contrasting grayish-white. The hindfoot has a naked plantar surface and six pads: an elongated and distally rounded thenar, a very reduced hypothenar, and four interdigital pads subequal in size. The unicolored tail has rigid hairs and subrhomboidal scales. As currently understood, there are four pairs of mammae.

The skull is solidly built with a somewhat triangular outline when viewed dorsally, especially in *G. fronto*; the skull of *G. planaltensis* is more elongated. The rostrum is moderately long, and the zygomatic notches are both well developed and concealed by large upper free borders of the zygomatic plates. The nasofrontal suture is U- or V-shaped and extends posteriorly to the point of contact of the premaxilla with the frontal. The interorbital region is hourglass in shape, but not especially constricted, with the lateral frontal borders varying from smoothly to more sharply divergent backward. The frontal sinuses are well inflated. The zygomatic plates are broad and medium in height; the zygomatic arches are robust, flare outward, and are dorsoventrally enlarged along their anterior one third. Postorbital processes are present and conspicuous. The incisive foramina are narrow but long, reaching posteriorly to the level of the M1 protocones. The palate is short and narrow; the anterior border of the mesopterygoid fossa is at the level of the posterior edges of the M3s. Sphenopalatine vacuities are reduced to lateral slits. The auditory capsules are medium sized and flat. A well-developed alisphenoid strut divides the foramen ovale. The subsquamosal fenestra is typically closed in adults by an enlarged hamular process, and an expanded tegmen tympani partially closes the postglenoid foramen.

The upper incisors are orthodont to slightly opisthodont; the enamel band is orange, and the dentine fissure is straight. The molars are quite hypsodont with alternating main cusps and usually planate occlusal surfaces. The procingulum of M1 lacks an anteromedian flexus; the mesolophs are greatly reduced to absent in M1–2, and M3 is noticeably cylindrical in shape. Features that characterize the lower molars include a reduced, subcircular procingulum on m1 without an anteromedian flexid, well-developed mesolophids on m1–3, and noticeable protoconulids associated with protoconids on m1. The mandible is robust, short, and moderately high, with a well-developed capsular projection.

The highly fragmented distribution of *Gyldenstolpia* is also shared by other sigmodontines in Chaco and Cerrado biomes. For example, *Juscelinomys*, a Cerrado specialist, was originally described from Brasília, Distrito Federal (Moojen 1965), and was more recently recorded from northeastern Bolivia, about 1,500 km to the west (Emmons 1999b). The known range of *Kunsia* consists of a number of isolated localities distributed across 3,000 km from the Pampas del Heath in northern Bolivia to Lagoa Santa in east-central Brazil (Bezerra, Carmignotto et al. 2007; Terán et al. 2008; Pardiñas, D'Elía, and Teta 2009). Other sigmodontines with larger and more continuous distributions also show disjunct areas of occupation, separated by large gaps (e.g., *Cerradomys*; Percequillo et al. 2008). It is unclear at present if this fragmented distributional pattern is an artifact of limited knowledge or reflects a disjunction of suitable habitats across the arid diagonal that includes the Chaco, Cerrado, and Caatinga biomes of central South America.

All nominal taxa of *Gyldenstolpia* are classified as Endangered, both in the 2011 IUCN Red List (Pardiñas, D'Elía et al. 2008) and the official list of Brazilian species at risk of extinction (A. B. M. Machado et al. 2008). The few and old records of *G. planaltensis*, in addition to its fragmented distribution, suggest that this species is already extinct or headed for extinction due to the loss of its specialized habitat (Bezerra 2011). The conservation status of *G. fronto* is presumably worse because this species is known only from the type series.

SYNONYMS:

Scapteromys: Winge, 1887:44; part (description of *fronto*); not *Scapteromys* Waterhouse.

Kunsia: Hershkovitz, 1966a:112; part (listing of *fronto*); not *Kunsia* Hershkovitz.

Gyldenstolpia Pardiñas, D'Elía, and Teta, 2009:552; type species *Scapteromys fronto* Winge, by original designation.

REMARKS: Although the date printed on the journal issue in which *Gyldenstolpia* was published reads "jul./dez.2008" (i.e., July/December 2008), this issue of the *Arquivos do Museu Nacional* was not released until 2009, and thus that year must be taken as the year of publication for nomenclatural purposes. Recognized species and subspecies in *Gyldenstolpia* are based on few and fragmentary specimens of different ages and geographic provenances. Morphological features that putatively distinguish these entities are subtle and perhaps merely reflect poorly assessed intrapopulational variation. Soft anatomy is almost unknown for any representative of *Gyldenstolpia*, despite the capture of several specimens in early 1990; no specimen as yet has been preserved in fluid. The classification used here is that of Pardiñas, D'Elía, and Teta (2009).

KEY TO THE SPECIES OF *GYLDENSTOLPIA*:

1. Skull robust; anterior border of the zygomatic plates moderately projected forward; upper toothrow >7.5 mm
. . . .*Gyldenstolpia fronto*
1'. Skull more delicate; anterior border of the zygomatic plates rounded; upper toothrow <7.5 mm. . . . .
. . . . *Gyldenstolpia planaltensis*

Gyldenstolpia fronto (Winge, 1887)
Fossorial Giant Rat

SYNONYMS:

Scapteromys fronto Winge, 1887:44; type locality "Lapa da Escrivania [= Lapa da Escrivânia] Nr. 5," a cave-chamber excavated by Peter W. Lund near Lagoa Santa, Minas Gerais, Brazil. The exact location of this cave is unknown, but was probably not far from the city of Lagoa Santa, 20 km N Belo Horizonte, Minas Gerais.

Scapteromys chacoensis Gyldenstolpe, 1932b:1; type locality "Argentine [*sic*], Rio de Oro, Chaco Austral" (caught in 'Lagunas de aqua [*sic*] dulce' "; Gyldenstolpe, 1932b:2). According to Pardiñas et al. (2007), the holotype of *chacoensis* was taken somewhere along the river where it crosses the northeastern part of Chaco province, Argentina.

Kunsia fronto fronto: Hershkovitz, 1966a:116; name combination.

Scapteromys fronto chacoensis: Hershkovitz, 1966a:116; *lapsus calami*.

G[yldenstolpia]. fronto fronto: Pardiñas, D'Elía, and Teta, 2009:554; name combination.

G[yldenstolpia]. fronto chacoensis: Pardiñas, D'Elía, and Teta, 2009:554; name combination.

DESCRIPTION: General characters as for the genus. Largest of two species, with total length ca. 330 mm, length of upper toothrow ca. 8.2 mm. Skull robust with more constricted interorbital region, with lateral frontal borders moderately sharp and divergent posteriorly, and projected anterior upper border of zygomatic plates.

DISTRIBUTION: The known present-day range of *G. fronto* is limited to a single locality in the northeastern corner of Chaco province, Argentina. Winge (1887) recorded subfossils from cave deposits near Lagoa Santa, Minas Gerais, Brazil.

SELECTED LOCALITIES (Map 106): ARGENTINA: Chaco, Río de Oro (type locality of *Scapteromys chacoensis* Gyldenstolpe).

SUBSPECIES: Pardiñas, D'Elía, and Teta (2009) recognized two subspecies, *G. fronto chacoensis* and †*G. fronto fronto*, both restricted to their respective type localities. *Gyldenstolpia f. chacoensis* is slightly larger than *G. f. fronto* and characterized by a U-shaped nasofrontal suture (V-shaped in *fronto*), a slightly acute anteriormost point of the nasals (condition unknown in *fronto*), the anterior border of zygomatic plates with distinct dorsal spines (less conspicuous in *fronto*), sharp lateral edges of the frontals (smooth in *fronto*), and a straight jugal-maxillary suture (condition unknown in *fronto*).

NATURAL HISTORY: No details of ecology have been reported.

REMARKS: *Gyldenstolpia f. fronto* is known from only four fossils of Quaternary age recovered in two caves from the Lagoa Santa area, Lapa da Escrivânia Nr. 5, the type locality, and Lapa da Serra das Abelhas (Pardiñas, D'Elía, and Teta 2009). According to Winge (1887:4), *fronto* was not represented in the recent fauna sampled by Lund from Minas Gerais state, Brazil. The holotype and unique specimen of *G. f. chacoensis* was collected in 1896, and there have been no new records since then. It is highly likely that *G. fronto* is now extinct.

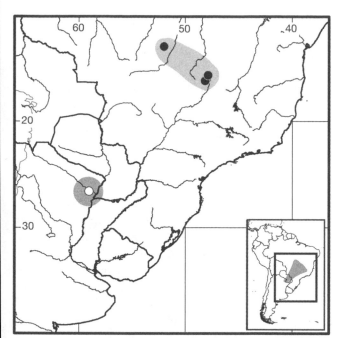

Map 106   Selected localities for *Gyldenstolpia fronto* (○) and *Gyldenstolpia planaltensis* (●). Contour line = 2,000 m.

Gyldenstolpia planaltensis (Avila-Pires, 1972)
Cerrado Giant Rat

SYNONYMS:

Kunsia fronto planaltensis Avila-Pires, 1972:421; type locality "Parque Zoobotânico de Brasília, Distrito Federal," Jardim Zoológico de Brasília Sgt. Sílvio Delmar Hollembach [15°51'S, 47°56'W], Brasília, Distrito Federal, Brazil (see Pardiñas, D'Elía, and Teta 2009:566).

Gyldenstolpia planaltensis Pardiñas, D'Elía, and Teta, 2009:564; first use of current name combination.

DESCRIPTION: General characters as for the genus; smaller than *G. fronto*, with total length ca. 275 mm, length of upper toothrow ca. 7.3 mm. Skull more delicate, with more elongated interorbital region, frontals with smooth lateral edges, and anterior edges of zygomatic plates not specially projected as spines.

DISTRIBUTION: Bezerra (2011) reviewed known localities; *G. planaltensis* is endemic to the central plateau of west-central Brazil, with two occurrences near Brasília, Distrito Federal, and one in the Serra do Roncador, Mato Grosso state.

SELECTED LOCALITIES (Map 106): BRAZIL: Distrito Federal, Planaltina, Águas Emendadas Ecological Station (Marinho-Filho et al. 1998, cited as *Kunsia tomentosus*, see Bezerra, Carmignotto et al. 2007; Pardiñas, D'Elía, and Teta 2009), Sgt. Silvio Delmar Hollembach Brasília Zoological Garden (type locality of *Kunsia fronto planaltensis* Avila-Pires); Mato Grosso, Ribeirão Cascalheira, R3 Serra do Roncador (Bezerra 2011).

SUBSPECIES: *Gyldenstolpia planaltensis* is monotypic.

NATURAL HISTORY: The semifossorial bauplan of *G. planaltensis* suggests specialization for life in palustrine habitats and a digging search strategy for food (see Pardiñas, D'Elía, and Teta 2009). There is no indication of burrow use or construction by this species. Available data suggest that this rodent lives in grasslands that are probably at least seasonally flooded, and is associated with marshes and/or other fresh water bodies (Avila-Pires 1972; Langguth, Limeira, and Franco 1997; Marinho-Filho et al. 1998). The specimen from Serra do Roncador, Mato Grosso state, was caught in a transect of 7 km × 20 m, through a vegetation type described as "cerrado *sensu stricto*" (Bezerra 2011), a plant association of the Cerrado biome that is dominated by dense savanna in which trees and shrubs, ca. 3–8 m in height, comprise up to 40% of the arboreal cover. Data concerning feeding habits, reproduction, parasitology, predators, or other aspects of the natural history are lacking.

REMARKS: The specimen from Serra do Roncador departs slightly in cranial morphology from the series of *planaltensis* collected in Brasília, Distrito Federal, particularly in shape of the zygomatic plates. Until samples of some size are available from multiple localities, the significance of this apparent difference will remain obscure.

Genus *Juscelinomys* Moojen, 1965
Louise H. Emmons

Species of this genus are medium-sized, soft-furred rats with an akodontine body shape very like that of *Oxymycterus* spp. The pelage is reddish or olivaceous, and the short tail is 60–70% of the head and body length, thick at the base, tapering to the tip, and completely clothed above with stiff blackish or reddish hairs. The ears are short and well clothed with hair on both sides. The toes of the forefoot are short, with greatly elongated claws on the middle three digits. The genus can be distinguished from all other Akodontini by the combination of broad and unusually long mesopterygoid fossa and narrow and extremely long parapterygoid plates with smoothly convex posterior margins. It can be distinguished from *Oxymycterus* by the absence of the trumpet-shaped forward projection of the nasals and premaxillaries beyond the incisors that is typical of that genus (description largely from Moojen 1965; and Emmons 1999b). The lower first molar has an anteromedian flexus. In *Juscelinomys* spp., the mandible is slender and dorsoventrally narrow; the sigmoid notch between the coronoid and condyloid processes is long, shallow, and nearly horizontal, and the coronoid process is a short, slender, posteriorly pointing hook, extending about a third of the way across the notch (measured to the edge of the articular surface of the condyloid process). The angular process is slender below a deep notch and slender coronoid process (illustrated in Emmons 1999b). In *Oxymycterus* spp., the mandible is deep, with the sigmoid notch short, and the coronoid process generally strongly hooked so as to reach higher than the robust condyloid process. The angular process is short and blunt, below a shallow notch (illustrated in Hershkovitz 1994), but not always (Jayat, D'Elía et al. 2008). The fully haired, thick tail is distinctive. *Kunsia* has such a tail, but does not otherwise resemble *Juscelinomys*.

Jayat, D'Elía et al. (2008:35) supported a close relationship between *Juscelinomys* and *Oxymycterus* but showed that *Juscelinomys huanchacae* may fall in the clade of *Oxymycterus* spp.: "in the ML [maximum likelihood] analysis *Oxymycterus* appears paraphyletic with respect to *Juscelinomys* or, depending on the placement of the root of the tree, to a clade formed by the remaining outgroups. As this study was not designed to test akodontine generic limits nor relationships among genera, we do not further discuss this issue." Pending a more thorough analysis, Moojen's genus is retained for these species. The two known species are from the Cerrado biome in grass-dominated habitats.

This genus was erected in 1965 to include one living species from central Brazil and provisional inclusion of Winge's (1887:36) *Oxymycterus talpinus*, a Lund fossil from Lagoa Santa. Moojen (1965) admitted that Winge's figures of *Oxymycterus talpinus* were not detailed enough for certain assignment. I exclude it from the genus because the cheek teeth of the specimen from which Winge's figure of the holotype was drawn are not the correct shape for the genus (Winge 1887:plate II, Fig. 13; L. H. Emmons, pers. obs. of a recent photo; see also *Brucepattersonius* account, which includes *Oxymycterus talpinus* Winge in the synonymy of *B. iheringi*). "*Juscelinomys vulpinus* (Winge, 1887)," listed by G. A. B. Fonseca et al. (1996:28) among living species of Brazil, is an apparent confusion with *talpinus* Winge. Lund's *Mus vulpinus* is considered a synonym of *Cerradomys subflavus* (Musser and Carleton 2005; see that account) and is not near *Juscelinomys*. Emmons (1999b) described two additional species of *Juscelinomys* from Bolivia, each from a single specimen. Subsequently, Emmons and Patton (2012), based on a larger series of specimens, and including molecular comparisons, reduced *J. guaporensis* to a subjective synonym of *J. huanchacae*. Because Moojen (1965) did not explicitly designate a type for the genus, in which he included two species, I here select *J. candango* Moojen as the type species of *Juscelinomys*.

SYNONYM:

Juscelinomys Moojen, 1965:281; type species *Juscelinomys candango* Moojen, by subsequent designation, herein.

KEY TO THE SPECIES OF *JUSCELINOMYS*:

1. Upper parts ochraceous orange lined with black; ear length <15 mm; M1 with anteromedian flexus . . . .
. . . . *Juscelinomys candango*
1'. Upper parts olivaceous (hairs agouti); ear length ≥16 mm; M1 lacking anteromedian flexus . . . .
. . . . *Juscelinomys huanchacae*

Juscelinomys candango Moojen, 1965
Candango Akodont

SYNONYM:

Juscelinomys candango Moojen, 1965:281; type locality "terrenos da Fundação Zoobotánica de Brasília (alt. 1,030 m), Distrito Federal," Brazil.

DESCRIPTION (from Moojen 1965 and Emmons 1999b): Medium sized, with head and body length 128–155 mm, with short tail about 70% of head and body length. Upper parts strongly tinged rusty orange, with individual hairs gray-based and with either orange or black tips, so overall impression of pelage is reddish, streaked with black; mystacial area, cheeks, and eye-rings pure orange; nose completely hairy except between nostrils; under parts orange-buff, with sharp transition from dorsal color on sides; ventral hairs pale based, but with some individuals having sooty patches on under parts, which may be due to soiling. Ears small (12 to 15 mm) and fully haired on back and on internal, lateral surface; tail short and broad at the base, fully clothed with hairs, black basally with orange tips; generally blackish above and more orange below. Feet dusky washed with ochraceous; ungual tufts sparse, on hind toes few hairs do not reach tips of claws. Vibrissae fine and inconspicuous; mystacials reach to just behind eye when pressed back; no genals evident; two inconspicuous superciliary vibrissae.

Skull slightly convex dorsally in lateral view, with frontal bones adjacent to nasals (frontal sinus) slightly swollen; each zygomatic arch in lateral view forms smooth, downward curve to meet squamosal; incisive foramina reach to centers of M1s, are broad and parallel sided posteriorly, but narrow with slight step for anterior one-quarter of their length; parapterygoid fossa broad anteriorly with lateral edge forming nearly straight line from behind toothrow; supraorbital edges smooth and without beading, and viewed dorsally, narrowest part of interorbital constriction rounded and does not form sharp angle. Both upper and lower first molars have anteromedian flexi; anterolophs/ids and mesolophs/ids prominent in M1/m1 and M2/m2.

DISTRIBUTION: *Juscelinomys candango* is known only from its type locality in the Distrito Federal, Brazil.

SELECTED LOCALITIES (Map 107): BRAZIL: Distrito Federal, Fundação Zoobotánica de Brasília (type locality of *Juscelinomys candango*).

SUBSPECIES: *Juscelinomys candango* is monotypic.

NATURAL HISTORY: This species is known from *campo cerrado* habitat in the Cerrado biome of central Brazil, an area of grasslands with sparse trees (see photos in Hershkovitz 1998). It uses burrows, to which lead bare access paths packed with dirt excavated from these burrows. Two tunnels, oval in cross-section and 9 cm at the widest diameter, leave the surface and meet at a nest chamber about 80 cm belowground. The nests are poorly furnished with pieces of grass and fine roots. Moojen (1965) included photos of living animals and of a burrow. Stomachs contained unidentified fibrous plant material and 10 to 20 large ants. Moojen (1965) noted that the tail is fragile and that several specimens have short or missing tails.

REMARKS: In the original description, Moojen (1965) designated the holotype as female MNRJ 22.807. The specimen has been renumbered, and it is now catalogued as MNRJ 23870. I am confident that the short ear-length given in Moojen's (1965) original table (there are no measurements on the skin labels) was measured from the notch because he states of the holotype "orelha interna, 15 mm." Although he was not the collector and may not have made the measurements himself, his manual for preparation of specimens shows the ear as measured from the internal notch (Moojen 1943b), which I assume was then the standard in his research group. This species is known from

only nine specimens collected in 1960 by three collectors or projects (Pedro Brito, Ary, and O. P. Belem). Despite intensive efforts to capture additional individuals (e.g., Hershkovitz 1998), no others have ever been found. I have captured *J. huanchacae* quite easily with standard bait and large Sherman or Tomahawk traps, so the rarity of *J. candango* is likely to be genuine and not just a result of low trapability. Because it may be critically endangered or extinct, efforts should be made to document its presence and location.

Juscelinomys huanchacae Emmons, 1999
Huanchaca Akodont

SYNONYMS:

Juscelinomys huanchacae Emmons, 1999b:3; type locality "Bolivia: Santa Cruz; Provincia Velasco, Serranía de Huanchaca, Parque Nacional Noel Kempff Mercado, Campamento "Huanchaca II." [S] 14°31′25″; W 60°44′22″, elev. 700 m."

Juscelinomys guaporensis Emmons, 1999b:4; type locality "Bolivia: Santa Cruz; Prov. Velasco, Parque Nacional Noel Kempff Mercado, Flor de Oro; 13°33.25′S; 61°00.51′W, elev. 210 m."

DESCRIPTION: General color uniform olivaceous, with color derived from mixture of yellow- and black-tipped hair; pelage 10–12 mm long at midback, with pale gray base, darkening to blackish at midlength, either tipped black or broadly ochraceous yellow with or without a narrow terminal black tip; color of sides similar to that of dorsum. Ventral pelage uniformly gray-based, tinged with buff or pale gray; sides of muzzle paler, washed with ochraceous yellow; chin and circumoral region whitish or yellowish; ears well-clothed inside and out with short, pale-tipped hairs, rim with slight, pale fringe; hair of forefeet and hindfeet with dusky bases and white tips; ungual tufts silvery. Claw of central digit of manus 5–6 mm in length. Tail completely covered with stiff hairs, predominately black, contrasting with body color above; paler, washed with pale buff below, palest near base, but not conspicuously bicolored. Four pairs of mammae, one pectoral, one axillary, one lower abdominal, and one inguinal. Vibrissae short, fine, and inconspicuous; mystacial vibrissae reach to behind eye; no genal vibrissae evident; and two superciliary vibrissae difficult to distinguish from pelage hairs; no vibrissae evident on forelegs.

Skull narrow, and rostrum broad; posterior part of supraorbital ridge sharp and not beaded. When viewed laterally, forward projection of frontal bones beside nasals (frontal sinus) slightly to markedly inflated dorsally, most prominently in younger individuals, producing marked convexity of skull. Anterior premaxillary bones and nasals not projected forward and upward into trumpet shape beyond incisors. Incisive foramina long and wide;

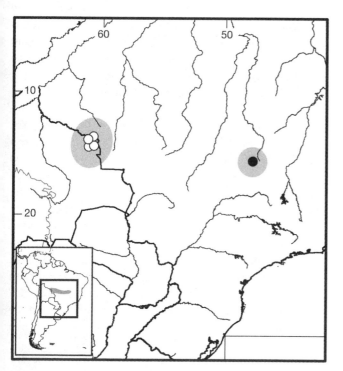

Map 107 Selected localities for *Juscelinomys candango* (●) and *Juscelinomys huanchacae* (○). Contour line = 2,000 m.

bony palate shorter than toothrows, with anterior margin of mesopterygoid fossa reaching forward of posterior edges of M3s; sphenopalatine vacuities present; carotid arterial circulation characterized by prominent squamosalalisphenoid groove and sphenofrontal foramen; foramen ovale opens lateral to parapterygoid processes, nearly in vertical plane, and covered with broad alisphenoid strut; trough for maxillary-buccinator nerve above foramen ovale broad and prominent auditory bullae small, appressed to basioccipital bone; stapedial foramen and carotid canal large and prominent. Posterior tip of jugal-squamosal suture on internal face of zygomatic arch does not extend onto bend on squamosal portion. Mandible slender; small, sharply hooked coronoid process at same height as condyloid process, with wide, shallow notch between them; angular process of mandible slender. An anteromedian flexus present on m1, but absent on M1. Toothwear rapidly eliminates most occlusal detail. Incisors faced with orange enamel.

DISTRIBUTION: *Juscelinomys huanchacae* is known from eastern Bolivia, only from the type locality and three other localities in Parque Nacional Noel Kempff Mercado. The maximum distance between known localities is 117 km.

SELECTED LOCALITIES (Map 107): BOLIVIA: Santa Cruz, Campamento Huanchaca II, Parque Nacional Noel Kempff Mercado (type locality of *Juscelinomys huanchacae* Emmons), Campamento Los Fierros, Parque Nacional Noel Kempff Mercado (USNM 584512), Campamento Mangabalito, Parque Nacional Noel Kempff Mercado (USNM 584508), Puesto Flor de Oro, Parque Nacional Noel Kempff Mercado (type locality of *Juscelinomys guaporensis* Emmons).

SUBSPECIES: I treat *Juscelinomys huanchacae* as monotypic.

NATURAL HISTORY: The type locality (Huanchaca II) is on the Huanchaca or Caparuch tableland massif (meseta) at 700 m in a patch of wet open savanna with scattered short trees. Other specimens were captured in well-drained grasslands interspersed with woody shrubs and trees in the lowlands at the foot of the tableland (200 m elevation). This species was readily captured and recaptured in large Sherman traps with oatmeal-based baits. A male and female were often successively captured in a given trap. Stomach contents included adult ants and insect larvae as well as small amounts of plant tissue. Many females were pregnant or lactating in early October (late dry–early rainy season), but few other months were sampled. Embryo numbers were 1 to 3 (N = 3). Burrows were not identified.

The tail-skin breaks easily and slips off at any point along its length, and many individuals have short or entirely missing tails. Study skins are extremely greasy, and exude oil. Over 12 years of trapping a standard plot,

Emmons (2009, and unpubl. data) observed the population of this species go to apparent extinction (0 individuals captured) for three successive years, followed by partial recovery, two more years of absence, then presence again in 2012. This suggests that this species may be susceptible to population swings that lead to local extinction of fragmented demes, a possible scenario also in the case of *J. candango*. Agroindustry and cattle raising on grasslands planted with exotic species have already destroyed much of the natural savannas of the Cerrado, particularly in the region between the localities of the two *Juscelinomys* species. Biofuel production threatens nearly all of the rest. *Juscelinomys candango* is not currently known to occur at any locality, and *J. huanchacae* is known only from sporadic presence in savanna fragments. Targeted conservation action may be needed to ensure a future for this genus.

Genus *Kunsia* Hershkovitz, 1966

Alexandra M. R. Bezerra

The genus *Kunsia* contains a single species, *K. tomentosus* (Litchtenstein, 1830), distributed in tropical open grasslands and savannas at elevations from 60 to 750 m in central Brazil and northern Bolivia (Musser and Carleton 2005; Bezerra, Carmignotto et al. 2007; Terán, Ayala, and Hurtado 2008). Fossil specimens of *Kunsia* are known from Brazilian cave deposits (Winge 1887) at Lagoa Santa in Minas Gerais state, some of which may date from the Late Pleistocene. Except as noted otherwise, the following account is based on Hershkovitz (1966a), Bezerra, Carmignotto et al. (2007), and Pardiñas, D'Elía, and Teta (2009), which should be consulted for additional details concerning taxonomy, morphology, and geographic distribution.

Kunsia tomentosus is a rare, large, semifossorial rodent, the largest of the extant Sigmodontinae (head and body length 185–287 mm and mass 241–654 g). It has harsh and dense fur; proportionally short limbs with large and powerful feet and claws; small and round ears; and a unicolored, short tail (less than half of the head plus body length) sparsely covered with short, unbanded hair and large subrectangular scales, with five to seven hairs per scale. The dorsum is dark gray, and the venter is grayish, without a sharp demarcation laterally between dorsal and ventral colors. The pinnae are covered with short hairs colored like the dorsal pelage. There are four pairs of mammae.

There are five digits on the manus; digits II to V have long claws (almost equal in length to the digit), and the pollex is reduced with a pointed nail. The plantar surface of the manus has five pads: three interdigital pads (II, III,

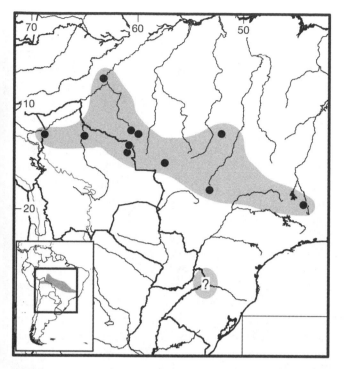

Map 108 Selected localities for *Kunsia tomentosus* (●); ? = approximate position of type locality, as restricted by Hershkovitz (1966a). Contour line = 2,000 m.

Genus *Lenoxus* Thomas, 1909

James L. Patton

This is a medium-sized rat, generally dull blackish washed with gray above, with paler grayish sides, and gray-brown venter tinged slightly with buff. The ears are small but exposed, and thinly haired. The tail is moderately long (equal to or greater than head and body length), uniformly gray-brown above and below, and with a white tip that may extend as much as a quarter of its length and that strongly contrasts with the basal portion. The tail is thinly haired with the scale annuli quite visible. The plantar surfaces of the hindfeet are naked, except for that of the heel; the dorsal surface is gray contrasting with white toes. The claws of both manus and pes are short, not stoutly developed as in *Oxymycterus*. The skull is long and narrow with a basally broad but elongated snout, long and tapering nasals, a broad interorbital region with rounded margins, obsolete parietal ridges, weak and posteriorly diverging zygomatic arches, narrow and posteriorly upwardly slanting zygomatic plates, shallow zygomatic notches, long but posteriorly broad incisive foramina, broad mesopterygoid fossa penetrating anteriorly to the third molars, no posterior-lateral palatal pits, and small tympanic bullae. Reig (1987) regarded the genus as closely related to *Oxymycterus* on general morphological grounds. However, as Thomas (1909:237) noted, "this striking species has none

of the peculiar appearance characteristic of *Oxymycterus*, and there is no doubt it should form a distinct genus."

I have found that, based on specimens in the Museum of Vertebrate Zoology, *Lenoxus* possesses a unilocular-discoglandular stomach, with the glandular epithelium confined to a diverticulum located on the greater curvature, a morphology more similar to that of *Oxymycterus* than to *Scapteromys* (as described by Carleton 1973).

Lenoxus contains the single species, *L. apicalis* J. A. Allen.

SYNONYM:

Lenoxus Thomas, 1909:236; type species *Oxymycterus apicalis* J. A. Allen, 1900, by original designation.

Lenoxus apicalis (J. A. Allen 1900)
White-tailed Akodont

SYNONYMS:

Oxymycterus apicalis J. A. Allen 1900a:224; type locality "Juliaca, Peru, altitude 6000 feet," corrected to "Inca Mines [= Santo Domingo Mine], about 200 miles northeast of Juliaca, on the east side of the Andes, on the Inambary River [Río Inambarí]," Puno, Peru, by J. A. Allen (1901b:41).

Lenoxus apicalis: Thomas, 1909:236; first use of current name combination.

Lenoxus apicalis apicalis: Sanborn 1950:15; name combination.

Lenoxus apicalis boliviae Sanborn 1950:16; type locality "Nequejahuira, Unduavi River. Altitude, 8000 feet," La Paz, Bolivia.

DESCRIPTION: As for the genus.

DISTRIBUTION: Known from cloud forest and upper montane forest on the eastern Andean slopes of southern Peru and northern Bolivia, at elevations between 1,500 and 2,500 m.

SELECTED LOCALITIES (Map 109): BOLIVIA: La Paz, Ñequejahuira, Río Unduavi (type locality of *Lenoxus apicalis boliviae* Sanborn), Okara (S. Anderson 1997), Serranía Bella Vista (S. Anderson 1997), Ticunhuaya (S. Anderson 1997). PERU: Puno, Inca Mines, Río Inambari (type locality of *Oxymycterus apicalis* J. A. Allen), 6 mi N of Limbani (MVZ 116666), Ocaneque, 10 mi N of Limbani (MVZ 116112), 14 km W of Yanahuaya (MVZ 171512).

SUBSPECIES: Sanborn (1950) regarded Bolivian samples as constituting a subspecies (*boliviae* Sanborn) distinct from the nominotypical form from southern Peru, and S. Anderson (1997) used this name for all samples from Bolivia. My own limited comparison of samples from the Limbani Valley, which contains the type locality of *apicalis*, with those from the San Juan Valley near the Bolivian border and those from Bolivia, however, suggests that the species is monotypic.

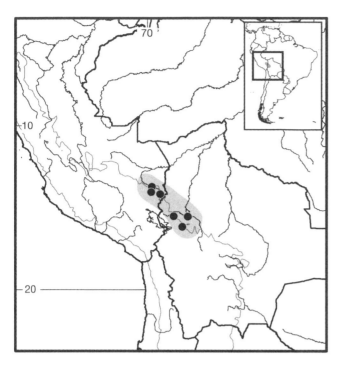

Map 109 Selected localities for *Lenoxus apicalis* (●). Contour line = 2,000 m.

NATURAL HISTORY: Peruvian specimens were trapped in runways along moss-covered tree roots or among dense ferns on the floor of elfin forest in southern Peru (C. B. Koford and J. L. Patton field notes, MVZ archives). Sanborn (1950) noted that the Bolivian localities of Nequehajira and Okara are drier and less heavily forested than the area around the type locality at Inca (= Santo Domingo) Mine in southern Peru. No other published data on natural history are available.

Genus *Necromys* Ameghino, 1889

Ulyses F. J. Pardiñas, Pablo Teta, Pablo E. Ortiz, Jorge Pablo Jayat, and Jorge Salazar-Bravo

Necromys is one of the few highly polytypic sigmodontine genera that remain unrevised. With a midcontinental distribution, encompassing more than 5,400 km, from Trinidad and Tobago to central Argentina, and up to 30 included nominal forms, these mice have presented a continuous challenge for students of rodent systematics (Musser and Carleton 2005). Most contributions on *Necromys* taxonomy are based on a single kind of evidence and focused on a small geographic area or have limited species coverage. Some exemplars deserving special mention are studies based on dimensions and morphology of Bolivian populations of *N. amoenus*, *N. lactens*, and *N. lenguarum* (S. Anderson and Olds 1990), the morphometric analyses of geographic variation in *N. lasiurus* (Macêdo and Mares 1987)

and *N. urichi* (J. Ventura et al. 2000), and the clarification of species boundaries between *N. benefactus* and *N. obscurus* from central Argentina and Uruguay (Galliari and Pardiñas 2000). Two substantive studies established the baseline for our current understanding of the genus and this account of *Necromys* systematics. The first, which remains largely unpublished, was Reig's (1972) morphological analyses, with a shortened version of his conclusions appearing in Reig (1987) and partial results published posthumously in Reig (1994). The second is the molecular phylogenetic analyses of D'Elía, Pardiñas, Jayat, and Salazar-Bravo (2008) based on mtDNA cytochrome-*b* sequences with broad taxonomic and geographic coverage (see below).

The complex taxonomic and nomenclatural history of *Necromys* is explained in part by the large geographic distribution of the genus but mostly by the morphological similarity of species in the genus to several species of *Akodon* (Akodontini) and *Zygodontomys* (Oryzomyini). In fact, the confusion among these three genera has persisted despite the fact that rodent systematists have reached important taxonomic conclusions, some of which have been partially overlooked. For example, Tate (1939:188) was clear in removing *lasiurus* from *Zygodontomys* when he wrote, "the forms from south of the Amazon typified by *tapirapoanus* are not *Zygodontomys* but *Akodon*. They can be distinguished easily by their blackish instead of whitish feet and by the extension of the palatal foramina backward as narrowed, pointed slits" (see also Voss 1991a:24, footnote). However, Hershkovitz (1962:205) reinserted the notion of a "Southern Group" of *Zygodontomys* to include *lasiurus* and several allied nominal forms. Thinking along the same lines as Tate, Massoia and Fornes (1967a) correctly removed these southern representatives from *Zygodontomys* and erected *Cabreramys* to contain *benefactus*, *obscurus*, and *lenguarum*.

Most of the nominal forms of *Necromys* were named during two time periods: the second half of the nineteenth century for those of Brazil, and the first decades of the twentieth century for those from the Andes. The generic allocation of the entities described during these times was initially to a growing yet vague concept of *Akodon*, particularly with *A. varius* and putatively related species. Gradually, however, Thomas (1902b, 1910c) and Gyldenstolpe (1932) moved several species of *Necromys* to *Zygodontomys*, a genus widespread in northern South America and southern Central America. This is not surprising because *Zygodontomys* is an unusual oryzomyine that has many features of the akodontine bauplan. In fact, the first solid clues for differentiating *Akodon* from *Zygodontomys* arose from chromosome and genital data, not from skull and molar morphology. By the time that Tate (1932h) assembled his review of "Akodont" rodents, species of *Necromys* were intermingled with those of *Akodon* and *Zygodontomys*.

Meanwhile, a poorly known Andean form (*amoenus*) remained in the nominal genus *Bolomys*, erected by Thomas (1916i) to contain forms with contrasting dorsal and ventral coloration and large inflated bullae. Three decades after Tate's treatment, Hershkovitz (1962) made an attempt to order this complex of taxa. He subsumed many Brazilian names under *lasiurus* and allocated this species to his "Southern Group"of the genus *Zygodontomys*.

Many genus-group taxa have a major systematic architect; for *Necromys*, this was Osvaldo Reig. Reig tested, by direct comparison with type material, the numerous taxonomic insights scattered in several papers of Oldfield Thomas. Also, for the southern nominal forms (*benefactus* and *obscurus*), he built on the taxonomic contributions made by Massoia and Fornes (1967a). Reig (1972:119–120) arrived at five fundamental conclusions with regard to the context and content of *Necromys*: (1) *Bolomys*, as redefined on the basis of its type species *Akodon amoenus*, is a valid and distinctive genus; (2) *A. amoenus* and *A. obscurus*, genotype of *Cabreramys*, are inseparable at the generic level, thus *Cabreramys* is a junior synonym of *Bolomys*; (3) *Akodon albiventer* does not belong to *Bolomys*; (4) *A. lactens* is a valid species of *Bolomys*; and (5) Hershkovitz's "Southern Group" of *Zygodontomys* and *Akodon arviculoides* are also to be included in *Bolomys*. In 1987, Reig published his main findings concerning this new consolidated and clarified *Bolomys*. Several years before, however, Massoia (1985b) had found a critical piece of this complex puzzle. In revising materials in the collections made by the famous paleontologist Florentino Ameghino and housed at the Museo Argentino de Ciencias Naturales "Bernardo Rivadavia" in Buenos Aires, Massoia realized that the specimens on which Ameghino had based the name *Necromys conifer* were morphologically indistinguishable from *Akodon benefactus*, a realization that made *Necromys* Ameghino, 1889 a senior synonym of *Bolomys* Thomas. Although Reig (1989) quickly accepted this nomenclatural conclusion, other contemporaries (Voss 1991a,b, 1992; S. Anderson 1997; M. M. Díaz 1999) were reluctant to use *Necromys*. As a result, both *Necromys* and *Bolomys* were used interchangeably for many years. Even after Massoia and Pardiñas (1993) fully documented the correspondence between the two generic names, the priority of *Necromys* was not firmly established until the treatise of Musser and Carleton (2005). In the intervening period, the *urichi* group from northern South America, which had been traditionally placed in the subgenus *Chalcomys* of *Akodon*, was found to belong with *Necromys*, as based on molecular phylogenetic analyses (M. F. Smith and Patton 1991, 1993; D'Elía 2003a). None of the authors, from Thomas (1916i) to Reig, Kiblisky, and Linares (1971), who had viewed *urichi* from a morphological viewpoint, had recognized the similarities between it and species of *Necromys*.

Seven living and one fossil species are currently recognized in *Necromys* (Musser and Carleton 2005; D'Elía, Pardiñas, Jayat, and Salazar-Bravo 2008). Isolated in northern South America (i.e., Venezuela, Colombia, and Trinidad and Tobago), the *N. urichi* complex inhabits forests, grasslands, shrublands, and montane areas. Along eastern Andean slopes and up to Puna habitats are three species that replace one another from north to south: *N. punctulatus*, *N. amoenus*, and *N. lactens*. Only a few Ecuadorean and Colombian specimens, all with ambiguous geographic provenance, are known for *N. punctulatus* (Voss 1991b). *Necromys amoenus* and *N. lactens* have widespread and parapatric Andean ranges, collectively extending from 12°S in southern Peru to 29°S in Argentina. A third group of species extends throughout the eastern lowlands south of the Rio Amazonas. This group comprises an endemic Chacoan representative, *N. lenguarum*, and the very similar and more widespread *N. lasiurus* that is distributed primarily in Cerrado, Caatinga, and other lowlands in tropical, subtropical, and temperate areas of the east-central midcontinent but also extends into the Chaco. Finally, a coastal grassland-shrubland specialist, *N. obscurus*, exhibits a disjunct distribution in southern Uruguay and eastern Buenos Aires province, Argentina. Most of the species of *Necromys* have small (e.g., *N. obscurus*, ca. 15,000 km²) to medium-sized (e.g., *N. lactens*, ca. 100,000 km²) ranges, with that of *N. lasiurus* contrasting strongly, as it covers more than 4 million km², from fragmented grasslands in the Amazon Basin to the northern boundary of Patagonia.

A comprehensive morphology-based assessment of *Necromys* diversity is still needed. Reig's (1972) attempt to make one was restricted to molar comparisons, and S. Anderson and Olds (1989) focused on only a few species. The extensive phenotypic variation seen in some species, especially in color but also in size, has contributed to the confusion regarding species boundaries and thus geographic ranges (Macêdo and Mares 1987; J. Ventura et al. 2000).

Members of *Necromys* are generalized akodontines in external morphology, with heavy bodies, medium-sized eyes and ears, short limbs and tails, and unspecialized forefeet and hindfeet that cause them to be easily confused with several species of *Akodon* (especially those of the *A. varius* species group in the case of the southern *Necromys* and with *saturatus* of the northern *N. urichi* complex). Size in *Necromys* varies from small in Andean forms, with combined head and body length of about 100 mm, to large in lowland forms, in which it reaches 130 mm (the holotype of *N. tapirapoanus*) or 133 mm (syntype of *N. obscurus*). *Necromys obscurus* has been stated to be the largest species (Massoia and Fornes 1967a; Galliari and Pardiñas 2000), but some Amazonian and Chacoan species are of equal or larger size (Macêdo and Mares 1987; S. Anderson and Olds 1990). In addition, there is significant variation in size within species; for example, in *N. lasiurus*, populations assigned to the nominal form *temchuki* are markedly larger than the southernmost populations of the nominal form *benefactus* (Massoia and Fornes 1967a; Massoia 1980c, 1982; J. R. Contreras 1982a). The length of the tail is always shorter than that of the head and body; *Necromys* is considered to be a short-tailed taxon, especially in comparison with *Akodon* (S. Anderson and Olds 1990). Although this is true for the many forms of *Necromys* that have tails that range in length between 50–65% of the head and body length (e.g., *benefactus*, *lasiurus*, *punctulatus*, *negrito*, *orbus*, *obscurus*, *tapirapoanus*), Andean forms such as *N. amoenus* have longer tails (about 75% of the head and body length), and the tail is even longer in the *urichi* complex (about 80% in *meridensis*). Typically, in *Necromys* the hindfoot is relatively narrower than in any species of *Akodon*, and digit V is much shorter in relation to IV. In *Akodon*, the claw of digit V extends to the most proximal interphalangeal joint of IV, but in *Necromys* the claw of V extends either only about one-half or two-thirds the length of the proximal phalanx of IV or does not extend past the anteriormost interdigital pad (Massoia and Fornes 1967a; Voss 1991a). Claws of both forefoot and hindfoot are large (except in *N. urichi*; Reig, Kiblisky, and Linares 1971) and curved; in the hindfoot, the claw of digit III is greater than 50% of the length of this digit itself (Voss 1991a).

Necromys has a short rostrum, deep and often wide zygomatic notches, a posteriorly divergent interorbital region with ridged edges, relatively narrow incisive foramina, narrow parapterygoid fossae, reduced or absent pterygoid bridges, and anterior face of upper incisors that extends anterior to the premaxillae, a reduced anteromedian flexus on M1, and lacks an anteromedian flexid on m1 (S. Anderson and Olds 1990:15). The Andean forms *N. amoenus* and *N. lactens* are characterized by paler, proodont to orthodont upper incisors, broad interorbits, and short free upper edges of the zygomatic plates. In contrast, lowland *Necromys* populations now referred to as *N. lasiurus* and *N. lenguarum* have orthodont, saturated orange upper incisors; larger nasals; interorbits with posteriorly divergent edges; and zygomatic plates with well-developed upper free borders. *Necromys obscurus* differs somewhat from other species owing to its inflated frontal sinuses, smooth edges of the frontal bones, U-shaped frontoparietal suture, and a well-developed capsular process that is centrally located between the coronoid and condyloid processes. Apparently, all *Necromys* species share a well-developed alisphenoid strut partially covering the anterior opening of the alisphenoid canal, although this strut is weaker in *N. amoenus* and *N. lactens*. Many species of *Necromys* lack the pterygoid bridge, a delicate bony bar

that crosses the posterior opening of the alisphenoid canal (S. Anderson and Olds 1990:Fig. 6).

Reig (1978) described the extinct species *Necromys bonapartei*, based on a fragmentary mandible with two molars and found by José Bonaparte in Monte Hermoso, southwestern Buenos Aires province, Argentina. The available information attached to this fossil indicates that it came from the upper part of the Lower Pleistocene Monte Hermoso Formation (Pardiñas and Tonni 1998). However, a study in progress of the holotype and only known specimen of *N. bonapartei* casts doubt on its generic allocation. Several traits shown by this specimen are more similar to those of phyllotines, especially the absence of mesolophids, the robustness of the masseteric crest, and the proportions of M2 as compared to M1. Thus, if *bonapartei* were excluded from the genus, the oldest remains attributed to *Necromys* would be those reported from Pliocene deposits exposed near Miramar, Buenos Aires province (Reig 1972, 1994, cited as "*Bolomys* sp. A"). The referred material, a fragmentary left dentary, makes it difficult to differentiate this taxon from generalized akodontines such as *Akodon*. Voglino and Pardiñas (2005) recorded *Necromys* cf. *N. lasiurus* (cited as *benefactus*) in a micromammal assemblage from northeastern Buenos Aires province with an age close to the Brunhes-Matuyama paleomagnetic boundary (ca. 0.78 mya). The genus is apparently also present in the Middle-Late Pleistocene of the Tarija Basin, Bolivia (Steppan 1996; referred to *Bolomys*). Late Pleistocene and Holocene records of *Necromys* are widespread, documenting its presence in central (Pardiñas, Teta, and D'Elía 2010) and northwestern Argentina (P. E. Ortiz et al. 2011b), southern Bolivia (Hoffstetter 1968; Pardiñas and Galliari 1998c), and southeastern Brazil (Winge 1887; Salles et al. 2006). The Late Pleistocene type species of the genus, *N. conifer*, is from northeastern Buenos Aires province (Ameghino 1889). *Necromys* spp. have undergone significant changes in abundance and distribution during the last millennia or recent centuries according to fossil evidence from the Pampean region of Argentina, which conclusively documents regional extinctions and population fragmentation (Pardiñas 1999b; Galliari and Pardiñas 2000).

Most species of *Necromys* possess a karyotype with 34 chromosomes (Reig 1987; Tiranti 1996a). Vitullo et al. (1986) described the diploid autosomal complement as consisting of acrocentrics except for a single metacentric pair, the smallest in the series. A $2n = 34$ (FN 34) has been reported for the species and nominal forms *N. amoenus*, *N. obscurus*, *N. benefactus*, *N. lasiurus*, and *N. temchuki* (N. O. Bianchi et al. 1971; A. L. Gardner and Patton 1976; Vitullo et al. 1986; Tiranti 1996a; Fagundes and Yonenaga-Yassuda 1998); a minor variant, $2n = 33$, was found in animals referred to *N. lasiurus* and from Serra dos Cavalos, Pernambuco, Brazil (Maia and Langguth 1981; Svartman and Cardoso de Almeida 1993c). In contrast, animals published under the names *N. urichi meridensis* and *N. urichi venezuelensis* have a diploid number of 18, with the autosomal complement consisting of entirely metacentric chromosomes (N. O. Bianchi et al. 1971; Reig, Kiblisky, and Linares 1971; Reig, Olivo, and Kiblisky 1971).

Akodon and *Necromys* have been recognized as closely related genera, according to morphological, electrophoretic, and mtDNA sequence data (Reig 1987; M. F. Smith and Patton 1999). Phylogenetic reconstructions place *Necromys* sister to the clade *Akodon-Thaptomys* (M. F. Smith and Patton 1993, 1999), to *Thaptomys* (D'Elía et al. 2003), or to *Thalpomys* (D'Elía 2003a), depending upon the gene sequence studied. D'Elía, Pardiñas, Jayat, and Salazar-Bravo (2008) conducted the most recent phylogenetic analysis of *Necromys* phylogeny. These authors found a three-lineage polytomy at the base of *Necromys*: one clade comprised *N. urichi* and *N. amoenus*, a second was formed by only *N. lactens*, and a third included all lowland species occurring south of Amazonia. Given this polytomy, an alternative classificatory option would be to recognize three genera: *Necromys* for the morphologically homogeneous lowland forms; *Bolomys* for *amoenus* and *urichi* (and probably *punctulatus*); and a third unnamed genus for *lactens*.

SYNONYMS:

Ratton Brants, 1827:187; part (inclusion of *Ratton colibreve*, which was based on Azara's [1802:86] "*el colibreve*"; suppressed for the purpose of Principle of Priority [but not Homonomy] by the International Commission [ICZN 1982:Opinion 1232]).

Mus: Waterhouse, 1837:16; part (description of *obscurus*); not *Mus* Linnaeus.

Habrothrix: Winge, 1887:31; part (listing of *lasiurus*); not *Habrothrix* Wagner.

Necromys Ameghino, 1889:120; type species *Necromys conifer* Ameghino, by original designation.

Abrothrix: J. A. Allen and Chapman, 1893:217; part (series used in subsequent description of *urichi* J. A. Allen and Chapman referred to as *Abrothrix caliginosus* [now = *Melanomys caliginosus*]); not *Abrothrix* Waterhouse.

Acodon: Thomas, 1894b:361; part (description of *punctulatus*); not *Acodon* Agassiz nor *Akodon* Meyen.

Akodon: J. A. Allen and Chapman, 1897a:19; part (description of *urichi*); not *Akodon* Meyen.

Zygodontomys: Thomas, 1902a:61; part (listing of *brachyurus*); not *Zygodontomys* J. A. Allen.

Akodon (Chalcomys): Thomas, 1916i:338; part (listing of *urichi*, *venezuelensis*, and *meridensis*); not *Akodon* Meyen nor *Chalcomys* Thomas.

Bolomys Thomas, 1916i:337; type species *Akodon amoenus* Thomas, by original designation.

Akodon [(*Akodon*)]: Ellerman, 1941:410; part (listing of *arviculoides*, *lenguarum*, *obscurus*, *chapmani*, *meridensis*, *urichi*, and *venezuelensis*); not *Akodon* Meyen.

Akodon [(*Bolomys*)]: Ellerman, 1941:415; part (listing of *amoenus*, *lactens*, *negrito*, and *orbus*); name combination.

Cabreramys Massoia and Fornes, 1967a:418; type species *Mus obscurus* Waterhouse, by original indication.

KEY TO THE SPECIES OF *NECROMYS*:

1. $2n = 18$; length of tail about 80% of head and body length; range in Venezuela, Colombia, and Trinidad and Tobago . . . . *Necromys urichi*
1'. $2n = 34$; length of tail <80% of head and body length 2
2. Isolated populations in Colombia and Ecuador. . . . . *Necromys punctulatus*
2'. Andean ranges of Peru, Bolivia, and Argentina; and lowlands of Bolivia, Brazil, Paraguay, Argentina, and Uruguay 3
3. Upper incisors pale, distinctly proodont; range in Andes of Peru, Bolivia, and Argentina. 4
3'. Upper incisors shiny orange, orthodont; range in lowlands of Bolivia, Brazil, Paraguay, Argentina, and Uruguay 5
4. Head and body length ca. 100 mm; auditory bullae enlarged, rounded; sphenopalatine vacuities present . . . . *Necromys amoenus*
4'. Head and body length ca. 120 mm; auditory bullae small, flattened; sphenopalatine vacuities absent. . . . . *Necromys lactens*
5. Head and body length ca. 110 mm; foreclaws as large as corresponding digits; interorbital region without marked edges; frontal sinuses well inflated . . . . *Necromys obscurus*
5'. Head and body length ca. 115–130 mm; foreclaws shorter than corresponding digits; interorbital region with marked constriction point and posteriorly divergent frontal edges; frontal sinuses not well inflated. . . 6
6. Populations mostly restricted to Boreal Chaco; nasals long, obscuring incisors when skull viewed from above . . . . *Necromys lenguarum*
6'. Populations mostly restricted to arid to mesic lowlands of Paraguay, Brazil, and Argentina; nasals shorter . . . . *Necromys lasiurus*

Necromys amoenus (Thomas 1900)
Pleasant Akodont
SYNONYMS:

Akodon amoenus Thomas, 1900h:468; type locality "Calalla, rio Colca, near Sumbay, [Arequipa,] Peru, elevation 3500 m."

Bolomys amoenus: Thomas, 1916i:340; name combination.

Akodon [(*Bolomys*)] *amoenus*: Ellerman, 1941:415; name combination.

Necromys amoenus: Massoia, 1985b:4; first use of current name combination.

DESCRIPTION: Small, with head and body length 95–103 mm (N = 11); upper parts yellowish brown, finely streaked with black hairs; sides more clear and richly colored, yellowish buff on muzzle, side of head, and side of rest of the body, forming well-defined lateral line in some specimens; under parts whitish, sharply contrasting with rest of body; hairs on chin and, in some cases, throat completely white. Vibrissae relatively short, and ears densely furred with hairs of same general hue as muzzle. Most specimens show well-developed eye-ring. Forefeet and hindfeet with ochraceous dorsal hue; claws covered with dense tuft of hairs. Tail short (63–79 mm; N = 11), bicolored, dorsally ochraceous brown or blackish brown, and ventrally white or ochraceous. Skull with short rostrum (even shorter than *N. lactens*), zygomatic notches proportionally broad and deeply excavated, zygomatic arches bowed outward, and interorbital region with squared and posteriorly divergent dorsolateral margins; nasals short, not surpassing anterior plane of upper incisors; zygomatic plates relatively broad, with straight or slightly concave anterior margins that slant slightly backward from bottom to top; temporal crests relatively weak, but lambdoidal crests well developed; mesopterygoid fossa relatively broad and parapterygoid fossae narrow; bullae large for genus. Mandible delicate but body relatively deep; masseteric ridges extend to anterior face of m1 and end at level of mental foramen; capsular projection conspicuous, located under sigmoid notch and slanted backward. Upper incisors notably proodont, faced with whitish or yellowish enamel. Molars robust, broad, and with almost completely transversely oriented principal cusps; M1 with relatively small procingulum, shallow anteromedian flexus in young specimens, tiny anteroloph, and well-developed mesoloph; M2 with trace of mesoloph; M3 with hypoflexus and mesoflexus, the latter tending to form island. Lower m1 with deep anteromedian flexid, anterolabial cingulum, ectolophid, and tiny mesolophid; m2 conserves labial structures but is reduced; m3 S-shaped.

Axial skeleton with 13 thoracic, 6 lumbar, and 23–26 caudal vertebrate (Steppan 1995); phallus and baculum described by Spotorno (1986), stomach by Dorst (1973b); estimates of heart size in relation to body size reported by Dorst (1973a).

DISTRIBUTION: *Necromys amoenus* is known from relatively few localities in Altiplano grasslands above 3,200 m in southern Peru and western Bolivia and from high grasslands of the eastern Andean slopes in Salta province, Argentina, at 3,100 m (S. Anderson and Olds 1990; S. Anderson 1997; Salazar-Bravo, Yensen et al. 2002; Musser and Carleton 2005; Jayat et al. 2006; Jayat, Ortiz, and Miotti 2008).

SELECTED LOCALITIES (Map 110): ARGENTINA: Salta, 1 km ENE Rodeo Pampa, km 59 on Rte 7 (MACN 24849). BOLIVIA: Chuquisaca, Potolo (= Estancia Pupayoj), 22 km S and 13 km E of Icla (Salazar-Bravo, Yensen et al. 2002); Cochabamba, Colomi (BM 266), 9.5 km by road SE of Rodeo, then 2.5 km on road to ENTEL antenna (AMNH 260890); La Paz, 2.5 km S of Tacacoma (Salazar-Bravo, Yensen et al. 2002); Tarija, Serranía del Sama (MSB 67194). PERU: Arequipa, 15 km S of Callalli (MVZ 174058); Cusco, 20 km N (by road) of Paucartambo, km 100 (MVZ 171563); Puno, 11 km W and 12 km S of Ananea (MVZ 172894), Hacienda Ontave (MVZ 141329), 36 km SE (by road) of Macusani (MVZ 173202), Puno (AMNH 213552), 12 km S of Santa Rosa (MVZ 172878).

SUBSPECIES: We treat *Necromys amoenus* as monotypic (but see Remarks).

NATURAL HISTORY: Of three specimens collected in August in high-elevation grasslands of northwestern Argentina, one male had semiscrotal testes, and all were molting. These specimens were caught in rocky places intermingled with sparse grasses. Salazar-Bravo, Yensen et al. (2002) reported two specimens caught at the base of rock outcrops, on rocky soils among ferns and mosses in the *Bosque húmedo-montano subtropical* life zone. An additional specimen mentioned by Salazar-Bravo, Yensen et al. (2002) came from a barley field surrounded by *Polylepis* woodlands in the *Bosque húmedo-montano templado*. This species is apparently mostly insectivorous but with some variation between localities (Pizzimenti and DeSalle 1980; Dorst 1973b). Individuals are diurnal and often associated with open grassy places, dominated by *Stipa ichu* or Tola

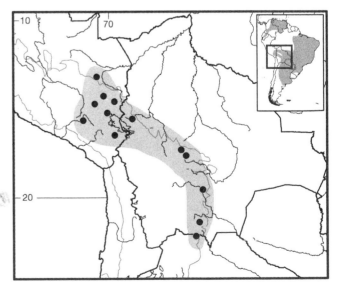

Map 110 Selected localities for *Necromys amoenus* (●). Contour line = 2,000 m.

(*Parastrephia lepidophylla*) (Pearson 1951). Reynafarja and Morrison (1962) and Morrison et al. (1963) examined hematocrit values and myoglobin levels with respect to high-elevation adaptation.

REMARKS: *Necromys amoenus* is the type species of *Bolomys*, which was later considered to be a junior synonym of *Necromys* (Massoia 1985b; Reig 1989; Massoia and Pardiñas 1993).

D'Elía, Pardiñas, Jayat, and Salazar-Bravo (2008) suggested that *N. amoenus* might be a composite of more than one species. They found a deep phylogeographic break between haplotypes in Cusco and Puno departments, southern Peru, and ones from Tarija, in southern Bolivia, and from Salta, in northwestern Argentina, with large genetic divergence between clades.

Necromys lactens (Thomas 1918)
White-chinned Akodont

SYNONYMS:

Akodon lactens Thomas, 1918b:188; type locality "Leon [= León], Jujuy, Argentina, elevation 1500 m."

Akodon orbus Thomas, 1919d:497; type locality "Otro Cerro" Catamarca, Argentina (= Otro Cerro, 28°45′S, 66°17′W, 2,023 m, Capayán, Catamarca, about 4 km SSE of Cerro Catalán; Pardiñas et al. 2007).

Bolomys negrito Thomas, 1926a:312; type locality "Las Paras [= Las Pavas], about 4000 m" Tucumán, Argentina.

Bolomys lactens: Thomas, 1926d:605; name combination.

Bolomys orbus: Gyldenstolpe, 1932a:119; name combination.

Akodon [(Bolomys)] lactens: Ellerman, 1941:415; name combination.

Akodon [(Bolomys)] negrito: Ellerman, 1941:415; name combination.

Akodon [(Bolomys)] orbus: Ellerman, 1941:415; name combination.

Necromys lactens: Massoia, 1985b:4; first use of current name combination.

DESCRIPTION: Medium size, with head and body length 98–124 mm (N = 29, age class 3 from northwestern Argentina); upper parts buffy brown, finely lined with black; sides paler and richly colored, fulvous or cinnamon in some specimens; under parts often cinnamon, but with varying intensity. All examined individuals from high-elevation grasslands of northwestern Argentina with contrasting white spot on chin and/or throat. Vibrissae relatively short, and ears densely furred with hairs of same color as rest of dorsum. Some specimens with well-developed eye-ring. Claws relatively long and covered with tuft of white hairs. Tail short (57–77 mm, N = 30 specimens of age class 3 from northwestern Argentina), bicolored, dorsally blackish brown and whitish or buffy in ventral view. Skull with short rostrum, zygomatic notches relatively narrow and shallow, and interorbital region ridged with posteriorly divergent dorsolateral margins; nasals short, not surpassing anterior plane of upper incisors; zygomatic plates relatively broad, with slightly concave anterior margins and slanted slightly backward, from bottom to top; incisive foramina end posteriorly at about level of protoflexus or anterior face of protocone of M1; mesopterygoid fossa relatively broad, with anterior margin rounded or slightly squared, with or without median spine, and lateral sides diverge slightly backward; parapterygoid fossae narrow, with straight lateral edges that diverge slightly posteriorly. Mandible delicate but with relatively deep body; masseteric ridge extends to anterior face of m1 and ends at same level as mental foramen; capsular projection conspicuously expressed as backwardly slanting knob located under sigmoid notch. Upper incisors proodont, with anterior face whitish or yellowish. Molars robust, broad, and with principal cusps almost completely transverse; M1 with well-developed, fan-shaped procingulum with shallow anteromedian flexus in young specimens; some individuals with a small mesostyle; posteroflexus tending to form enamel island in young specimens; M2 occasionally with trace of mesoloph and posteroflexus tending to form island; M3 large, with deeply penetrating mesoflexus and hypoflexus. Lower m1 with well-developed procingulum with deep anteromedian flexid (forming island in some young specimens), anterolabial cingulum, and trace of ectolophid and mesolophid in some individuals; m2 with anterolabial cingulum; m3 large and S-shaped.

DISTRIBUTION: *Necromys lactens* occurs between 1,400 and 4,000 m in the eastern Andean highlands of south-central Bolivia and northwestern Argentina (S. Anderson and Olds 1990; S. Anderson 1997; P. E. Ortiz, Cirignoli et al. 2000; Musser and Carleton 2005; Jayat and Pacheco 2006; Jayat et al. 2006; Jayat, Ortiz, and Miotti 2008; M. M. Díaz and Barquez 2007). The fossil record for *N. lactens* is restricted to two localities in the Tafí valley, Tucumán, Argentina, representing the Middle-Upper Pleistocene and the Pleistocene-Holocene boundary (P. E. Ortiz and Pardiñas 2001; P. E. Ortiz and Jayat 2007b; P. E. Ortiz et al. 2011b).

SELECTED LOCALITIES (Map 111): ARGENTINA: Catamarca, ca. 2 km SE of Huaico Hondo, on Rte 42, E of Portezuelo (MACN 23484), Otro Cerro (type locality of *Akodon orbus* Thomas); Jujuy, La Antena, Sierra del Centinela, S of Fuerte (MACN 23492), León (type locality of *Akodon lactens* Thomas); Salta, ca. 15 km W of Escoipe, on Rte 33 (Jayat, Ortiz, and Miotti 2008); Tucumán, Las Paras (type locality of *Bolomys negrito* Thomas). BOLIVIA: Chuquisaca, 14 km N of Monteagudo, Cañon de Herida, Río Bañado (S. Anderson 1997); Cochabamba,

14 km SE of Rodeo (S. Anderson and Olds 1993); Tarija, Rancho Tambo, 61 km by road E of Tarija (S. Anderson 1997), Serranía del Sama (S. Anderson 1997).

SUBSPECIES: We treat *Necromys lactens* as monotypic.

NATURAL HISTORY: In northwestern Argentina, *N. lactens* principally inhabits high-elevation grasslands between 1,500 and 3,100 m, in the highest vegetation belts of the *Yungas* ecoregion (M. M. Díaz et al. 2000; Jayat and Pacheco 2006; M. M. Díaz and Barquez 2007; Jayat, Ortiz, and Miotti 2008). M. M. Díaz and Barquez (2007) reported specimens caught near water at Cerro Hermoso and Chilcayoc, Jujuy province, Argentina. These authors registered females with closed vaginas and males with abdominal testes in July and September, and a female with developed mammae in July. Additional unpublished data suggest that *N. lactens* reproduces mainly between November and April and molts primarily during fall and winter (May to July). In this region, the species is mainly diurnal.

REMARKS: Capllonch et al. (1997) resurrected *orbus* (as *Bolomys orbus*), based on a single specimen from Tucumán province (northwestern Argentina), but Pardiñas and Galliari (1998b) criticized this decision. Phylogenetic studies conducted by D'Elía, Pardiñas, Jayat, and Salazar-Bravo (2008) recovered *N. lactens* as one of the three main lineages in the genus. The geographic structure of the mtDNA variation these authors observed was consistent with a history molded by the high-altitude grasslands of the central Andes. Their study included haplotypes obtained from near the type localities of *lactens* Thomas, *negrito* Thomas, and *orbus* Thomas, all found to be closely related with molecular divergence about 2%.

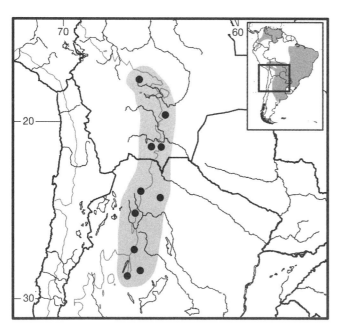

Map 111 Selected localities for *Necromys lactens* (●). Contour line = 2,000 m.

Necromys lasiurus (Lund, 1840)

Hairy-tailed Akodont

SYNONYMS:

Mus lasiurus Lund, 1838:50; *nomen nudum*.

Mus lasiotis Lund, 1838:50; *nomen nudum*.

[*Mus*] *lasiolis* Lund, 1839a:233; *nomen nudum*; incorrect subsequent spelling of *Mus lasiotis* Lund.

Mus lasiuro Lund, 1839b:208; *nomen nudum*.

Mus lasiurus Lund, 1840b:50 [1841c:280]; type locality "Rio das Velhas's Floddal," Lagoa Santa, Minas Gerais, Brazil.

Mus lasiotis Lund, 1840b:50 [1841c:280]; type locality "Rio das Velhas's Floddal"; refined to "Curvelo, una localidad ubicada al WNW de Lagoa Santa en el valle del Rio das Velhas a los 18°45'S, 44°25'W, Brasil" by Langguth (1975:45).

Hesperomys arviculoides Wagner, 1842a:361; type locality "Brasilia" = the country of Brazil, not the capital city of Brasília (see comments in Ximénez and Langguth 1970:3).

Hesperomys orobinus Wagner, 1842a:361; type locality "Brasilia" = Brazil (see comments in Ximénez and Langguth 1970:3).

Mus renggeri Pictet and Pictet, 1844:76, plates 21, 22, and 23, Fig. 7; type locality "Bahia," Brazil; not *renggeri* Waterhouse (= *Abrothrix olivacea*).

Hesperomys brachyurus Wagner, 1845:147; type locality "Itararé," São Paulo, Brazil.

Akodon fuscinus Thomas, 1897c:496; type locality "Source, Marajó Island"; interpreted as a misspelling of Soure (0.721°S, 48.504°W), Ilha Marajó, Pará, Brazil by Hershkovitz (1962:205).

Necromys conifer Ameghino, 1899:120; type locality "Mercedes, Olivera y Luján, en la provincia de Buenos Aires," Argentina.

Akodon benefactus Thomas, 1919c:214; type locality "Bonifacio, South-west Buenos Ayres, Province; alt. 50"; Laguna Alsina (= Estación Bonifacio; 36.809°S, 62.245°W), Argentina.

Zygodontomys arviculoides: Tate, 1932h:17; name combination.

Zygodontomys brachyurus: Tate, 1932h:17; name combination.

Zygodontomys fuscinus: Tate, 1932h:17; name combination.

Zygodontomys lasiurus: Tate, 1932h:17; name combination.

Zygodontomys orobinus: Tate, 1932h:17; name combination.

Zygodontomys pixuna Moojen, 1943a:8; type locality "Crato, Ceará," Brazil.

Zygodontomys lasiurus brachyurus: Hershkovitz, 1962:206; name combination.

Zygodontomys lasiurus fuscinus: Hershkovitz, 1962:205; name combination.

Zygodontomys lasiurus lasiurus: Hershkovitz, 1962:206; name combination.

Zygodontomys lasiurus pixuna: Hershkovitz, 1962:206; name combination.

Cabreramys benefactus: Massoia and Fornes, 1967a:421; name combination.

Cabreramys temchucki Massoia, 1980a:179; *nomen nudum*.

Cabreramys temchuki Massoia, 1982:91; type locality "Costa del Arroyo Zaimán, Villa Miguel Lanús, Departamento Capital, Provincia de Misiones, República Argentina"; Arroyo Zaimán (ca. 27.418°S, 55.891°W).

Bolomys temchuki temchuki: J. R. Contreras, 1982a:174; name combination.

Bolomys temchuki elioi J. R. Contreras, 1982a:174; type locality "Mantilla Cué, frente a Laguna Paiva, Las Lomas, Departamento Capital, Provincia de Corrientes"; Laguna Paiva (ca. 27.480S° 58.740W°), Argentina.

Bolomys temchuki liciae J. R. Contreras, 1982a:175; type locality "Colonia Río Tragadero, 2 km al norte de Resistencia, Departamento San Fernando, Provincia del Chaco"; Río Tragadero (ca. 27.390°S, 58.939°W), Argentina.

Necromys lasiurus: Massoia, 1985b:4; first use of current name combination.

Necromys pixuna: Massoia, 1985b:4; name combination.

Necromys temchuki: Massoia, 1985b:4; name combination.

[*Necromys*] *lasiotis*: Musser and Carleton, 2005:1130; incorrect subsequent spelling of *Mus lasiotis* Lund.

DESCRIPTION: Medium sized, with head and body length 118–138 mm, tail length 66–96 mm. Considerable variation in pelage color throughout distributional range; dorsal color olivaceous gray to dark brown, usually with well-defined agouti pattern; rump and sides generally paler and more ochraceous tawny; venter whitish to gray, sometimes washed with buff, with individual hairs dark gray at their bases; limit between dorsal and ventral color poorly defined; eye-ring either absent or ill-defined by buffy or cinnamon hairs. Ears small, rounded, and sparsely haired; tail bicolored and covered by moderately dense, short hairs. Manus and pes broad, dorsally brownish, with well-developed nails; ungual tufts whitish. Skull strongly built; rostrum short; zygomatic notches deep; interorbital region converges anteriorly and with weakly beaded supraorbital margins; incisive foramina long, reaching protocone of M1, more or less parallel sided but with anterior part wider than posterior; palate short, without conspicuous posterolateral pits. Upper incisors orthodont with orange enamel. Main cusps of molars slightly alternate anteroposteriorly. Massoia and Fornes (1967a), Crespo (1969), and Macêdo and Mares (1987) provided external and cranial measurements.

DISTRIBUTION: As here understood (i.e., to include the nominal taxa *benefactus* Thomas, *conifer* Ameghino, and *temchuki* Massoia), *N. lasiurus* has one of the largest distributions of any sigmodontine. Geographic limits are not totally understood, but specimens assigned to this species are known from the Ilha do Arapiranga in the Brazilian state of Pará (approximately 1°20′S), from southern Buenos Aires province (38°40′S), from close to the Brazilian Atlantic coast, and from the Sierra de Ambato (66°W) in the Argentinean Andes. The species occurs widely in eastern and central Brazil, from the Atlantic coast of the state of Pará west to Rondônia state. In Paraguay, it is primarily restricted to the eastern portion of the country, but some populations also are found west of the Río Paraguay where overlaps occur with the range of *N. lenguarum*. In Argentina, *N. lasiurus* is nearly continuously distributed throughout the northeastern portion of the country although it is absent east of the Río Paraná in Entre Ríos province and the southern portion of Corrientes province. West of the Río Paraná, *N. lasiurus* has a continuous distribution to at least southern Santa Fe province (J. R. Contreras 1982a). Isolated populations occur in east-central Argentina, in Córdoba, La Pampa, and Buenos Aires provinces (Galliari and Pardiñas 2000; Teta, González-Fischer et al. 2010).

SELECTED LOCALITIES (Map 112): ARGENTINA: Buenos Aires, Bonifacio (type locality of *Akodon benefactus* Thomas), Fulton (Galliari and Pardiñas 2000), Laguna Chasicó (LTU 85), Monte Hermoso (Galliari and Pardiñas 2000), Olavarría (CNP 2333), Ramallo (Galliari and Pardiñas 2000); Catamarca, Las Chacritas, ca. 28 km NNW of Singuil (D'Elía, Pardiñas, Jayat, and Salazar-Bravo 2008); Chaco, Colonia Río Tragadero, 2 km N of Resistencia (type locality of *Bolomys temchuki liciae* Contreras); Córdoba, Río Ceballos (Galliari and Pardiñas 2000); Corrientes, Santo Tomé (CNP-E 650–2); La Pampa, 20 km SE de Luán Toro, Estancia la Florida (J. R. Contreras and Justo 1974), Quehué, Estancia Los Molinos (Tiranti 1996), Laguna Guatraché (Galliari and Pardiñas 2000); Santa Fe, Berna (CNP 531). BRAZIL: Alagoas, Anadia (Libardi 2013), Capela (Libardi 2013); Bahia, Feira de Santana (Libardi 2013), Mucugê, Parque Nacional da Chapada Diamantina, Rio Cumbuca (D'Elía, Pardiñas, Jayat, and Salazar-Bravo 2008); Ceará, Crato (type locality of *Zygodontomys pixuna* Moojen), Serra Baturité, Pacoti (Libardi 2013); Maranhão, Ribeirãozinho, Imperatriz (Libardi 2013), UTE Ponta Madeira, São Luis (Libardi 2013); Mato Grosso, Casa de Pedra, Chapada dos Guimarães (Libardi 2013), Posto Jacaré (Libardi 2013), Rodovia 7 Placas, Cabeceira, Rio Cuiabá (Libardi 2013); Mato Grosso do Sul, Corumbá (Godoi et al. 2010); Minas Gerais, Caratinga, Fazenda Montes Claros (D'Elía, Pardiñas, Jayat, and Salazar-Bravo 2008),

Lagoa Santa (type locality of *Mus lasiurus* Lund); Pará, Alter do Chão (Lima Francisco et al. 1995), Ilha de Arapiranga (D'Elía, Pardiñas, Jayat, and Salazar-Bravo 2008), Souré (type locality of *Akodon fuscinus* Thomas); Pernambuco, Caruaru (Maia and Langguth 1981), Parque Nacional do Catimbau (Geise et al. 2010); Rio Grande do Sul, Rondinha, Parque Estadual Florestal de Rondinha (D'Elía, Pardiñas, Jayat, and Salazar-Bravo 2008); São Paulo, Itararé (type locality of *Hesperomys brachyurus* Wagner). PARAGUAY: Alto Paraguay, Palmar de Las Islas (D'Elía, Pardiñas, Jayat, and Salazar-Bravo 2008); Presidente Hayes, 8 km NE of Juan de Zalazar (D'Elía, Pardiñas, Jayat, and Salazar-Bravo 2008).

SUBSPECIES: Macêdo and Mares (1987) assessed geographic variation in *N. lasiurus*, retaining *fuscinus* Thomas and *lasiurus* Lund as subspecies. J. R. Contreras (1982a) described *elioi* and *liciae* as subspecies of *Bolomys temchuki* Massoia. Divergence in the mtDNA cytochrome-*b* gene for specimens studied by D'Elía, Pardiñas, Jayat, and Salazar-Bravo (2008) ranged from 0–3.9% and from 1.3–6.3% at the intrapopulation and interpopulation levels, respectively. These authors noted two clades in *N. lasiurus*, one from localities as far from each other as the Ilha do Arapiranga (in Pará, Brazil) and Paraguay, and the other occupying all remaining localities, including specimens from near the type locality at Lagoa Santa, Minas Gerais, Brazil. Despite its large geographic distribution and great range of morphological variation, *N. lasiurus* exhibits little phylogeographic structure. When all haplotypes were analyzed together, tests of neutrality supported a recent increase in population size, likely leading to range expansion. In combination with the lack of phylogeographic structure, these results are consistent with the hypothesis that *N. lasiurus* has recently invaded several regions where it is currently found (D'Elía, Pardiñas, Jayat, and Salazar-Bravo 2008).

NATURAL HISTORY: *Necromys lasiurus* is found in grasslands, along the borders of cultivate fields, abandoned croplands, or even in secondary or gallery forests (Crespo 1969; Dietz 1983; Ellis et al. 1997; Geise, Paresque et al. 2010). It is basically a diurnal rodent with more pronounced crepuscular and nocturnal activity in the dry season than in the wet season (E. M. Vieira et al. 2010). In grassland areas, it is a moderately abundant species, reaching densities up to 19.02 individuals/ha in central Brazil (Becker et al. 2007) and 33.8 individuals/ha in east-central Argentina (De Villafañe et al. 1973). Diet varies both geographically and seasonally to include variable quantities of seeds, green vegetation, and arthropods (Ellis et al.1998; Talamoni et al. 2008).

The reproductive season in east-central Argentina extends through the spring and summer, from October to February (Mills, Ellis, McKee et al. 1992). Parreira and Cardoso (1993) determined seasonal variation in spermatogenic activity in southeastern Brazil. The average number of embryos varies from 4.2 (central Brazil; Dietz 1983) to 6.2 (east-central Argentina; Mills, Ellis, McKee et al. 1992). C. R. de Almeida et al. (1982) and Piantanida and Nani (1993) studied its reproduction and growth under laboratory conditions. Home range size varied both seasonally and by sex from 0.02 to 0.52 ha (Santos Pires et al. 2008). J. A. Oliveira et al. (1998) examined relative age and age structure in natural populations in northeastern Brazil, and Cangussu et al. (2002) studied sexual dimorphism of submandibular glands.

Necromys lasiurus is a reservoir for at least two hantavirus strains: Macielin, from the Pampas region of Argentina, which has not yet been associated with human disease (Enría and Levis 2004), and Araraquara, from the central plateau and southeastern portion of Brazil, which appears to be the most virulent strain infecting humans in South America (L. T. M. Figueiredo et al. 2009; G. G. Figueiredo et al. 2010). A long list of ectoparasites has been associated with this species, including mites (*Androlaelaps fahrenholzi*, *A. rotundus*, and *Eulaelaps stabularis*), fleas (*Polygenis adelus*, *P. axius*, *P. bohlsi*, *P. pradoi*, *P. pygaerus*, *P. frustratus*, *P. massoiai*, *P. platensis*, *P. rimatus*, *P. tripus*, *P. occidentalis*, *P. tripopsis*, *Adoratopsylla antiquorum*, *A. intermedia*, *Craneopsylla minerva minerva*, *Ctenocephalides felis felis*, *Xenopsylla cheopis*, and *X. brasiliensis*), and lice (*Hoplopleura imparata* and *H. misionalis*) (Autino and Lareschi 1998; D. Castro and Cicchino 1998; Lareschi and Mauri 1998; Linardi and Guimarães 2000). Endoparasitic nematodes are abundant, and include *Syphacia alata*, *Protospirura numidica criceticola*, *Pterigodermatites* (*Paucipectines*) *zygodontomys*, *Litomosoides carinii*, *L. silvai*, *Stilestrongylus freitasi*, *Trichuris laevitestis*, and *Strongyloides venezuelensis* (Quentin, 1967, 1968; Quentin et al. 1968; M. Robles 2008, 2010; and Suriano and Navone 1994).

REMARKS: Langguth (1975) and Hershkovitz (1990a) discussed the convoluted history of the use of Lund's *lasiurus* and *lasiotis*, with the latter often causing confusion with *Thalpomys lasiotis* Thomas (see that account). Paula Couto (1950: footnote 66) regarded *Mus lasiotis* Lund as a synonym of *Hesperomys* (= *Calomys*) *expulsus* Lund.

The current concept of *N. lasiurus* (see D'Elía, Pardiñas, Jayat, and Salazar-Bravo 2008) includes several nominal forms with complex taxonomic histories, such as *Hesperomys arviculoides*, *Akodon benefactus*, and *Cabreramys temchuki*. Cabrera (1961) and Reig (1978) placed Pampean populations of *N. lasiurus*, previously treated as *N. benefactus*, in part or totally in *N. obscurus*. Alternatively, Massoia and Fornes (1967a) and Galliari and Pardiñas (2000) regarded both *benefactus* and *obscurus* as valid species, although their comparisons were generally limited to

Argentinean samples. Even so, Galliari and Pardiñas (2000) recognized the similarities between *benefactus* and *lasiurus*, placing both in the same morphological group of species. According to Massoia and Pardiñas (1993) and Galliari and Pardiñas (2000), the fossil species *N. conifer* Ameghino is a senior synonym of *benefactus* Thomas, although the latter authors preferred to retain *benefactus* as the name for this species for reasons of familiarity (but see Musser and Carleton 2005). Hershkovitz (1962) revised Brazilian populations of *N. lasiurus* as part of his "Southern Group" of the genus *Zygodontomys*. However, A. L. Gardner and Patton (1976), Maia and Langguth (1981), and Voss and Linzey (1981) supported the inclusion of these populations in either *Akodon* or *Bolomys* (= *Necromys*) on karyological and morphological grounds. Macêdo and Mares (1987) regarded *N. lenguarum*, currently recognized as a valid species, as a synonym of *N. lasiurus*. Indeed, the morphological discrimination and distributional limits of these two species require refinement (see below). *Hesperomys arviculoides* Wagner, sometimes listed as a valid species (Gyldenstolpe 1932; Cabrera 1961; Ximénez and Langguth 1970), was considered related to *N. obscurus* (Galliari and Pardiñas 2000) or as a synonym of *N. lasiurus* (Macêdo and Mares 1987; Reig 1978, 1987). Musser and Carleton (2005) demonstrated that *Cabreramys temchucki* Massoia, 1980, is a *nomen nudum*; the availability of *temchuki* actually dates from Massoia's (1982) explicit diagnosis of the species.

Libardi (2013) addressed nongeographic variation in Brazilian populations based on morphometrics of 1,572 specimens; this author also constructed detailed maps of the distribution of *N. lasiurus* in Brazil. Reported specimens

of *N. lasiurus* from the Pampas del Heath in southeastern Perú (Pacheco et al. 1995; Emmons, Romo et al. 2002) have been tentatively reassigned to *N. lenguarum* (Pacheco et al. 2009).

Necromys lenguarum (Thomas, 1898)
Paraguayan Akodont

SYNONYMS:

Akodon lenguarum Thomas, 1898f:271; type locality "Waikthlatingwaialwa, Northern Chaco of Paraguay," Presidente Hayes, Paraguay (an old Chacoan mission located at about 23°25′S and 58°10′W [see Paynter 1989, under the spelling Waikthlatingwayalwa]).

Zygodontomys tapirapoanus J. A. Allen, 1916c:528; type locality "Tapirapoan [= Tapirapuã], Rio Sepotuba, Mato Grosso, Brazil."

[*Akodon*] *tapirapoanus*: Tate, 1939:188; name combination.

Akodon (*Akodon*) *tapirapoanus tapirapoanus*: Cabrera, 1961:447; name combination.

Akodon (*Akodon*) *obscurus lenguarum*: Cabrera, 1961:444; name combination.

"*Zygodontomys* [? *lasiurus*] *lenguarum*": Hershkovitz:1962: 206; name combination.

"*Zygodontomys* [? *lasiurus*] *tapirapoanus*": Hershkovitz: 1962:207; name combination.

Cabreramys lenguarum: Massoia and Fornes, 1967a:422; name combination.

Bolomys lenguarum: Reig, 1978:167; name combination.

Bolomys lenguarum tapirapoanus: S. Anderson and Olds, 1989:19; name combination.

Necromys lenguarum: Massoia, 1985b:4; first use of current name combination.

DESCRIPTION: Medium sized, with total body length of holotype 191 mm, tail 76 mm, ear length 14 mm, and hindfoot length 21 mm. Dorsal pelage dark gray, with more yellowish on rump; sides paler; venter whitish, with tips of hairs white or with slight yellowish tinge; forefeet and hindfeet uniformly pale gray above. Tail bicolored, blackish above and whiter below. Skull heavily built; nasals short, supraorbital region broad and beaded, and zygomatic plates low and broad, with straight anterior margin; incisive foramina wide and short, not constricted posteriorly. Upper incisors slightly proodont. Greatest skull length 28.5 mm, upper molar series length 4.7 mm (dimensions of holotype).

DISTRIBUTION: *Necromys lenguarum* occurs in isolated grassland patches in southeastern Peru (Pacheco et al. 2009), eastern Bolivia, western Paraguay, southwestern Brazil, and possibly northern Argentina (Galliari et al. 1996), inhabiting Chacoan lowlands below 1,300 m elevation. Galliari et al. (1996) cited the first possible specimens for Argentina, from Pampa Bolsa, Chaco province

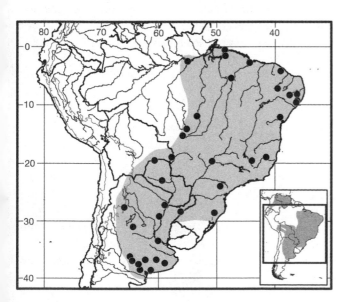

Map 112 Selected localities for *Necromys lasiurus* (●). Contour line = 2,000 m.

(PROBBAS, numbers 800 to 832). M. M. Díaz and Barquez (2007) mentioned a possible record from Jujuy province, Argentina (Ingenio La Esperanza) in the collection of Elio Massoia. These Argentinian specimens should be reexamined in the course of a comprehensive taxonomic review of the genus. Similarly, many of the localities in Bolivia, given by S. Anderson (1997), should be treated with caution because they could be *N. lasiurus*, a species with a much wider distribution and ecological amplitude (D'Elía, Pardiñas, Jayat, and Salazar-Bravo 2008).

SELECTED LOCALITIES (Map 113): ARGENTINA: Chaco, Pampa Bolsa (Galliari et al. 1996); Jujuy, Ingenio La Esperanza (M. M. Díaz and Barquez 2007). BOLIVIA: Chuquisaca, Río Limón (S. Anderson 1997; D'Elía, Pardiñas, Jayat, and Salazar-Bravo 2008); Santa Cruz, 4.5 km N and 1.5 km E of Cerro Amboró, Rio Pitisama (S. Anderson 1997), Estancia San Marcos, 6 km W of Ascención (D'Elía, Pardiñas, Jayat, and Salazar-Bravo 2008), 4 km N and 1 km W of Santiago de Chiquitos (S. Anderson 1997), Velasco, Parque Nacional Noel Kempff Mercado, Campamento Los Fierros (D'Elía, Pardiñas, Jayat, and Salazar-Bravo 2008). BRAZIL: Mato Grosso, Tapirapoã, Rio Sepotuba (type locality of *Zygodontomys tapirapoanus* J. A. Allen). PARAGUAY: Alto Paraguay, Laguna Placenta (D'Elía, Pardiñas, Jayat, and Salazar-Bravo 2008); Presidente Hayes, Waikthlatingwaialwa (type locality of *Akodon lenguarum* Thomas). PERU: Madre de Dios, Pampas del Heath (Pacheco et al. 2009).

SUBSPECIES: S. Anderson (1997) maintained *tapirapoanus* J. A. Allen, with type locality in Mato Grosso, western Brazil, as a subspecies, but the validity of this nominal form at either specific or subspecific levels requires further evaluation.

NATURAL HISTORY: All the localities reported by D'Elía, Pardiñas, Jayat, and Salazar-Bravo (2008) for *N. lenguarum* are situated in seasonally dry environments of the *Chaco Seco* ecoregion. Tapirapoã, the type locality of *tapirapoanus* J. A. Allen, is in the Cerrado of west-central Brazil. The record in southern Peru (Pacheco et al. 2009) is from the Pampas del Heath, which spans the border with Bolivia, a moist grassland isolated in the lowland rainforest of the Río Madre de Dios Basin. This species has not been studied in regard to other aspects of its ecology or behavior.

REMARKS: The taxonomic status of *N. lenguarum* is unresolved. It has been considered as either conspecific with *N. lasiurus* (Macêdo and Mares 1987; Pardiñas, D'Elía et al. 2006; Cáceres et al. 2008) or a separate species (Reig 1987; S. Anderson and Olds 1990; S. Anderson 1997; Galliari and Pardiñas 2000; D'Elía, Pardiñas, Jayat, and Salazar-Bravo 2008). The reciprocally monophyletic relationship of this taxon as compared with a broad array of geographic samples of *N. lasiurus*, derived from mtDNA analyses, offers

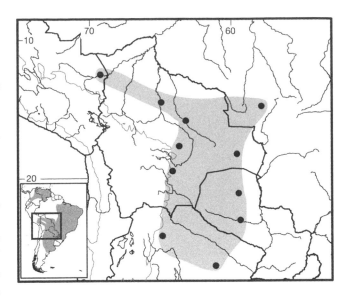

Map 113 Selected localities for *Necromys lenguarum* (●). Contour line = 2,000 m.

strong support for the specific separation of these two taxa (D'Elía, Pardiñas, Jayat, and Salazar-Bravo 2008). However, these authors remarked that the valid application of the name *lenguarum* to the sister clade to *N. lasiurus* was not properly tested because no topotypical specimen was analyzed. Rather, specimens of *Necromys* collected in the area of the type locality of *N. lenguarum* were all referable to *N. lasiurus*. Thus, D'Elía, Pardiñas, Jayat, and Salazar-Bravo (2008) argued that several taxonomic issues must be resolved before the taxonomic status of *N. lenguarum* is firmly established. Therefore, we restrict the distributional range of *N. lenguarum* to those localities listed by D'Elía, Pardiñas, Jayat, and Salazar-Bravo (2008), the type localities of *lenguarum* Thomas and of its junior synonym, *tapirapoanus* J. A. Allen, and that from southern Peru, which is most likely of this species (see Pacheco et al. 2009).

Necromys obscurus (Waterhouse, 1837)
Dark-furred Akodont

SYNONYMS:

Ratton colibreve Brants, 1827:186; based on Azara's (1802:86) "*el colibreve*"; ruled by the International Commission (ICZN 1982:Opinion 1232) to be a vernacular name and unavailable for use in zoological nomenclature.

Mus obscurus Waterhouse, 1837:16; type locality "Maldonado," Maldonado, Uruguay.

Hesperomys obscurus: Wagner, 1843a:516; name combination.

Akodon obscurus: E.-L. Trouessart, 1897:536; name combination.

Zygodontomys obscurus: Reig, 1964:211; name combination.

Cabreramys obscurus: Massoia and Fornes, 1967a:419; name combination, as the type species of *Cabreramys* Massoia and Fornes.

Bolomys obscurus: Reig, 1978:167; name combination.

Necromys obscurus: Massoia, 1985b:4; first use of current name combination.

Necromys obscurus bonapartei: Massoia, 1985b:4; name combination.

"*Bolomys* sp. C": Reig, 1994:104; name combination.

Necromys obscurus scagliarum Galliari and Pardiñas, 2000:219; type locality "mouth of Arroyo Corrientes, Mar del Plata (38°01'S, 57°31'30"W, General Pueyrredón county, Buenos Aires Province, Argentina."

N[ecromys]. obscurus obscurus: Galliari and Pardiñas, 2000:214; name combination.

DESCRIPTION: Largest species in genus, with head and body length averaging ca. 110mm and tail ca. 75mm. Fur rather long and somewhat glossy; dorsum varies from dark chestnut brown to blackish brown; individual hairs blackish at base and tip and yellowish in middle, giving general agouti appearance; guards hairs entirely blackish and long (ca. 13mm); flanks and cheeks tinged with orange or buffy; ventral hairs gray based with yellowish to orange tips; in some individuals, a white spot on the chin. Ears small and rounded, dark brown skin covered by short buffy hairs; tail bicolored, dark brown above and grayish below, with scales evident to naked eye. Manus and pes both dark, but pes covered dorsally by short buffy hairs; second interdigital pad usually larger than the first and third; claws on digits I, III, and IV longer than the corresponding digit. Skull with short nasals, U-shaped frontoparietal suture, slightly marked supraorbital borders, and U-shaped mesopterygoid fossa. Mandible with short, robust, and slightly backwardly curved coronoid process; well-developed capsular projection covers base of sigmoid notch. Procingulum of M1 fan shaped. Massoia and Fornes (1967a) and Galliari and Pardiñas (2000) provided external and cranial measurements. Carleton (1973), Voss and Linzey (1981), and Hooper and Musser (1964) described stomach morphology, male accessory glands, and male genitalia, respectively.

DISTRIBUTION: This species has a disjunct distribution with two main geographic centers, one in southern Uruguay and the other in east-central Argentina (Galliari and Pardiñas 2000). Uruguayan populations are mostly restricted to the coast in San José, Maldonado, Montevideo, and Canelones departments (E. M. González and Martínez Lanfranco 2010). There is an unconfirmed report for Rivera department (Mones et al. 2003), which would provide an extension to the north of about 350km. Argentinean occurrences are from the southeastern corner of Buenos Aires province, including coastal environments as well as grasslands and hilly areas (Reig 1964; Massoia and Fornes

1967a; Galliari and Pardiñas 2000; Teta, González-Fischer et al. 2010). The fossil record of *N. obscurus* is restricted to the Late Holocene in Buenos Aires province (Pardiñas 1999b). The reference by Teta et al. (2004) to *Necromys* sp. from two archaeological sites in northeastern Buenos Aires is to this species. These past occurrences for *N. obscurus* indicate a wider distribution than at present and that partially connected Argentinean and Uruguayan populations. Sympatric occurrence of *N. lasiurus* and *N. obscurus*, not presently known, has been recorded in Late Holocene coastal assemblages (Pardiñas and Tonni 2000; Pardiñas 1999b).

SELECTED LOCALITIES (Map 114): ARGENTINA: Buenos Aires, Arroyo de las Brusquitas (Fornes and Massoia 1965), Estación Deferrari (Teta, González-Fischer et al. 2010), Estación San José (Teta, González-Fischer et al. 2010), INTA Balcarce (Galliari and Pardiñas 2000), Laguna Mar Chiquita (Bó et al. 2002), Punta Negra, Necochea (Galliari and Pardiñas 2000). URUGUAY: Canelones, Arroyo Tropa Vieja (MACN 17442), Laguna del Cisne (E. M. González and Fregueiro 1999); Maldonado, Maldonado (type locality of *Mus obscurus* Waterhouse); Montevideo, Bañados de Carrasco (Galliari and Pardiñas 2000), Parque Lecocq (E. M. González 1996), Villa Colón (Massoia and Fornes 1967a); San José, Bañados del Arazatí (E. M. González and Martínez Lanfranco 2010), Barra del Río Santa Lucía (Galliari and Pardiñas 2000).

SUBSPECIES: Two geographically disjunct subspecies are currently recognized, based on morphology. The nominotypical form ranges along a narrow band bordering the La Plata River and the Atlantic Coast of Uruguay, whereas *N. o. scagliarum* is known from a few localities in southeastern Buenos Aires province (Galliari and Pardiñas 2000). These morphologically defined subspecies are not molecularly divergent (D'Elía, Pardiñas, Jayat, and Salazar-Bravo 2008). Massoia (1985b) treated the fossil form *bonapartei* Reig as a subspecies of *N. obscurus*.

NATURAL HISTORY: *Necromys obscurus* is mostly diurnal and cursorial, although it exhibits some tendency toward fossoriality (Fornes and Massoia 1965). It has been captured in low to moderate numbers in grasslands, the borders of cultivated fields, moist areas near lagoons and streams along the coast, and among rocks in hilly environments (Reig 1964; E. M. González and Martínez Lanfranco 2010). It is omnivorous, including such arthropods as beetles and crickets and plants in its diet (E. M. González and Martínez Lanfranco 2010). This mouse is a minor component of the diet of the Barn Owl (*Tyto alba*) in southeastern Buenos Aires province (Leveau et al. 2006; Teta, González-Fischer et al. 2010). E. M. González and Martínez Lanfranco (2010) recorded reproductive activity in spring and winter months, and Reig (1965) noted

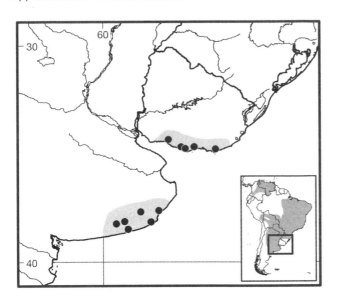

Map 114 Selected localities for *Necromys obscurus* (●); those in Argentina and Uruguay are mapped as disjunct (see Distribution). Contour line = 2,000 m.

all dorsal color black finely lined with gray and yellow that turns cheeks, sides of neck, and flanks to olive; under parts dirty yellowish white. Ears small, well haired, and brown; tail short and bicolored. Both manus and pes brownish white above; digit V of pes short, barely reaching the base of digit IV. Skull dorsally convex in lateral profile with short muzzle and long, narrow braincase; strongly beaded to slightly overhanging supraorbital edges; and long palatal foramina nearly twice length of molar series (Thomas 1894b).

DISTRIBUTION: Thomas (1894b:362) gave the holotype as probably from Pallatanga, an Ecuadorean town on the Pacific slope of the western Andean cordillera. Voss (1991b:6–7) discussed in detail the geographic provenance of the few specimens of *N. punctulatus* then available, summarizing "at least two independent records from Ecuador and one rather more dubious record from eastern Andean Colombia."

SELECTED LOCALITIES (Map 115): ECUADOR: Chimborazo, Pallatanga (type locality of *Acodon punctulatus* Thomas). COLOMBIA: Cundinamarca, Bogotá (Voss 1991b).

SUBSPECIES: *Necromys punctulatus* is monotypic.

NATURAL HISTORY: Nothing is known about the ecology or behavior of this species.

REMARKS: After a review of all available specimens of this species, Voss (1991b) removed *punctulatus* from *Zygodontomys* and noted its resemblance to, if not conspecifity with, *N. lasiurus*. Thomas (1894b:362) also recognized a possible connection to *N. lasiurus*, although he later (Thomas 1900h:469) allied *punctulatus* with *N. amoenus*, writing in reference to the latter "an isolated species, apparently only allied to *A. punctulatus*, from which it is distinguishable by its very different colour." This potential relationship, overlooked by Voss (1991b) and incorrectly referenced by Musser and Carleton (2005:1130), deserves further exploration. The geographic provenances of the known specimens of *N. punctulatus* make its being a northern Andean expression of *N. amoenus* biogeographically defensible. If it were such, then this could explain the mensural differences between the holotype of *punctulatus* (perhaps partially a result of age) as compared with *N. lasiurus*, specifically its smaller size (the length of the upper molar series of *punctulatus* is 4.2 mm [Thomas 1894b:362] versus >4.5 mm for samples of *lasiurus* [Macêdo and Mares 1987]). However, one specimen figured by Voss (1991b:Fig. 1) and identified as *N. punctulatus* (AMNH 36312) appears more like *N. lasiurus*, especially with respect to the orientation of the upper incisors and the nature of the interorbital constriction. Clearly, additional specimens of this rare mouse are needed to assess both its systematic status and relationships to other members of the genus.

that sperm production was most prevalent in November and January. Navone et al. (2010) provided a complete list of known parasites. D'Elía, Pardiñas, Jayat et al. (2008) placed *N. obscurus* in the IUCN Near Threatened category.

REMARKS: Brants (1827) based his *Ratton colibreve*, a probable synonym of *N. obscurus* (Langguth 1966a), on the "*colibreve*" in Azara's Spanish edition (1802), which was described from an individual acquired in Montevideo, Uruguay. Brants's name, however, was rejected by the International Commission (ICZN 1982:Opinion 1232).

The diploid number is $2n = 34$, based on specimens from southeastern Buenos Aires (N. O. Bianchi et al. 1971).

Necromys punctulatus (Thomas, 1894)
Ecuadorean Akodont

SYNONYMS:

Acodon punctulatus Thomas, 1894b:361; type locality "Ecuador (probably Pallatanga)."

Akodon punctulatus: E.-L. Trouessart, 1897:536; name combination.

Zygodontomys punctulatus: Gyldenstolpe, 1932a:114; name combination.

Zygodontomys punctulatus punctulatus: Cabrera, 1961:464; name combination.

Zygodontomys brevicauda punctulatus: Hershkovitz, 1962:204; name combination.

Bolomys punctulatus: Voss, 1991b:4; name combination.

N[ecromys]. punctulatus: Galliari and Pardiñas, 2000:226; first use of current name combination.

DESCRIPTION: Medium sized, with total length ca. 200 mm; tail ca. 71 mm. Dorsal pelage short and crisp; over-

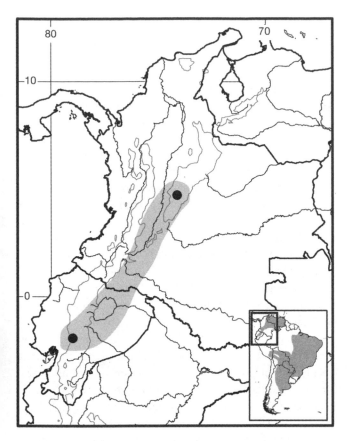

Map 115 Selected localities for *Necromys punctulatus* (●). Contour line = 2,000 m.

Necromys urichi (J. A. Allen and Chapman, 1897)
Northern Grass Mouse, Ratón campestre, Ratón mochilero

SYNONYMS:

Akodon urichi J. A. Allen and Chapman, 1897a:19; type locality "Caparo, Trinidad," Trinidad and Tobago.

Akodon venezuelensis J. A. Allen, 1899b:203; type locality "Quebrada Seca, near Cumana, [Sucre,] Venezuela."

Akodon meridensis J. A. Allen, 1904b:329; type locality "Merida (alt. 1630 m.), [Mérida,] Venezuela."

Akodon chapmani J. A. Allen, 1913c:600; type locality "Chipaque (altitude 8500 feet), Eastern Andes, [Cundinamarca,] Colombia."

Chalcomys aerosus saturatus Tate, 1939:187; type locality "Mt. Auyan-tepui plateau, middle R[ío]. Caroni, [Bolívar,] Venezuela, 6000 ft."

Akodon urichi tobagensis Goodwin, 1962:6; type locality "Speyside, St. John Parish, eastern Tobago, the West Indies, at an altitude of 300 feet," Trinidad and Tobago.

Necromys urichi: Musser and Carleton, 2005:1131; first use of current name combination.

DESCRIPTION: Medium sized (smaller than *N. lasiurus* and *N. lenguarum*, but slightly larger than *N. amoenus*).

Pelage grizzled and short, with hairs about 3–4 mm long over middle rump; dorsal color dark, rusty chestnut finely intermixed with black; venter washed with yellowish, or pale to medium gray tips, but always with dark-based hairs; hair tips of chin, throat, and around inguinal area gray in most specimens; dorsal-ventral pelage transition undefined; eye-ring present. Ears dark brown, thinly covered with short, dark rusty brown hairs; tail slightly more than one-third of total length, uniformly dark, sparsely haired, and without tuft. Manus and pes covered dorsally with short, dark ochraceous hairs; ungual tufts present on both front and hindfeet, but more developed in latter; hindfoot about 20% of head and body length, narrow, with digits I and V much shorter than digits II to IV; claw of digit V about two-thirds length of first phalanx of digit IV in holotype (AMNH 7725/6110). Skull with short rostrum and relatively broad interorbital region; zygomatic notches comparatively (to other species in genus) narrow and shallow; supraorbital shelves present, converging toward middle of orbits; ledge over orbit heavy in some specimens, approaching incipient bead; zygomatic arches widest across squamosal roots and therefore convergent anteriorly; zygomatic plates relatively broad; rostrum conical in lateral view, tapering forward from zygomatic root; nasal bones long (ca. 40% of the condylobasilar length), projecting anteriorly such that distal portion of premaxillary bones and incisors hidden beneath nasals and cannot be seen from above; incisive foramina long and broad, posteriorly reaching protocone of M1, widest at, or near to, premaxillomaxillary suture and tapering slightly posteriorly; hard palate basically flat, with barely conspicuous palatal gutters; palate of intermediate length, with mesopterygoid fossa broad, wide, approximately parallel, and with U-shaped anterior margin, extending between maxillae and reaching level of posterior border of M3s; roof of mesopterygoid fossa without sphenopalatine vacuities (holotype and various individuals from northeastern Venezuela with tiny perforations but never large vacuities, but some individuals from southern populations with large vacuities); distinct posterior palatal pits absent; parapterygoid fossae slightly concave dorsally (not flat, as in other species), with straight lateral margins, posteriorly divergent, narrower than mesopterygoid fossae; foramen ovale about size of M3 and often partially hidden by pterygoid bridge of variable width. Upper incisors orthodont, enamel colored dull yellow-orange to saturated orange (in holotype). Molars moderately short-crowned, robust, and with main cusps alternating along longitudinal plane. Procingulum broad and conspicuously divided into anterolingual and anterolabial conules by deep anteromedian flexus (lost in very old individuals). Both M1–2 with paralophules and well-developed anterolophs and mesolophs; M3 about half size of M2, with well-developed hypoflexus and metaflexus (clearly visible in young individuals, but not holotype). The m1 lacks anteromedian flexid but does possess anterolabial cingulum, ectolophid, and mesolophid; m2 with same structures as m1 but about two-thirds its size; m3 about two-thirds size of m2, elongated and figure-8 shaped. Mandible with mental foramen opening dorsally at diastema; capsular process distinct and slightly slanted backward; superior and inferior masseteric ridges meet below m1; coronoid process slightly higher than condyloidprocess, which extends dorsally slightly beyond angular process. Vertebral column with 13 thoracic, six lumbar, and 24–27 caudal elements (Steppan 1995). Structure of phallus and baculum similar to those of *Akodon surdus* and *Akodon boliviensis* (Hooper and Musser 1964).

DISTRIBUTION: *Necromys urichi* is known from over 70 localities in northern South America, including confirmed reports from Colombia, Venezuela, and Trinidad and Tobago; reports of the species from northern Brazil are based on the several Venezuelan localities near the Brazilian-Venezuelan border. As currently understood, *N. urichi* occurs in moist habitats from near sea level to over 3,600 m in elevation and appears to be able to tolerate a range of climatic and ecological zones. The range of the species appears to be somewhat disjunct with a large number of localities forming a NE-SW arc from Trinidad and Tobago, through the lowlands of northern Venezuela, adjacent Colombia, and across the Cordillera de Mérida Andes into southwestern Colombia. It appears again in southern and western Venezuela and perhaps northernmost Brazil, in areas associated with or, close to, the foothills of the tepuis, where it is also found at several elevations (J. A. Allen and Chapman 1897a; Handley 1976; O. J. Linares 1998; M. F. Smith and Patton 1999; Musser and Carleton 2005; D'Elía et al. 2008).

SELECTED LOCALITIES (Map 116): COLOMBIA: Cauca, La Gallera (AMNH 32467); Cundinamarca, Chipaque (type locality of *Akodon chapmani* J. A. Allen); La Guajira, Las Marimondas, Fonseca (USNM 280805); Meta, Finca El Buque, Villavicencio (NMNH 507258). TRINIDAD AND TOBAGO: Tobago, Speyside (AMNH 184789); Trinidad, Caparo (AMNH 6110), Princestown (J. A. Allen and Chapman 1897a). VENEZUELA: Amazonas, Cerro Duida, Cabecera Del Cano Culebra, 40 km NNW of Esmeralda (USNM 406120), Cerro Neblina, Camp XI (USNM 560866); Apure, Nulita, 29 km SSW of Santo Domingo, Selvas De San Camilo (USNM 418446); Bolívar, Arabopo, Mt. Roraima (AMNH 75640); Carabobo, La Copa, 4 km NW of Montalban (USNM 442343), Auyan-tepui (type locality of *Chalcomys aerosus saturatus* Tate), 85 km SSE of El Dorado, km 125 (USNM 387978), 45 km NE of Icabarú, Santa Lucia de Surukun (USNM 495663); Falcón, Cerro Socopo, 84 km NW of Carora

(USNM 442354); Guárico, Parque Nacional Guatopo, 15 km NW of Altagracia (USNM 387980); Mérida, Mérida (type locality of *Akodon meridensis* J. A. Allen), Mucubaji, 3.25 km ESE of Apartaderos (USNM 579557); Monagas, San Agustín, 5 km NW of Caripe (USNM 409941); Sucre, Manacal, 26 km ESE of Carupano (USNM 416718), Quebrada Seca (AMNH 14743); Vargas, Pico Ávila, 5 km NNE of Caracas, near Hotel Humboldt (USNM 371213).

SUBSPECIES: As currently understood, *Necromys urichi* may be a composite of several taxa, involving at least the three nominal ones (*saturatus*, *meridensis*, and *venezuelensis*), in addition to the nominate form. Morphometric analyses of Venezuelan populations and the analyses of mtDNA cytochrome-*b* gene sequences indicated that these taxa deserve equal taxonomic rank (M. F. Smith and Patton 1999; J. Ventura et al. 2000; D'Elía, Pardiñas, Jayat, and Salazar-Bravo 2008). Clearly, more comprehensive studies are needed to build consensus as to hierarchical rank (species or subspecies) appropriate for these taxa.

NATURAL HISTORY: Specimens in Venezuela have been trapped in savannas as well as evergreen, cloud, and deciduous forests, but almost always on the ground, near streams, and in rocky, grassy areas with thick herbaceous vegetation (Handley 1976; Aagaard 1982; O. J. Linares 1998; Voss 1992). A study of over nine months in the páramos of the Corodillera de Mérida showed that animals trapped in the wet season were larger, on average, than those trapped in the dry season. Also, males were larger and heavier than females (Aagaard 1982). In the premontane humid forest of the Coastal Range in Venezuela, a study over 22 months found a mean population density of 2.9 animals per hectare (range: 0.6–5.9) and a mean residence time on the study area of just over three months for males and four months for females. This population was slightly female biased (1:1.22), and the estimated juvenile survivorship was 55%. Animals were active day and night. Average litter size was 5, and females averaged 3.6 litters per year. Sexual maturity was reached at 2.7 months; reproduction occurred throughout the year but with a peak from May and June, at the onset of the rainy season (O'Connell 1989). Goff and Brennan (1978), Fain and Lukoschus (1982), and P. T. Johnson (1972) detailed arthropod ectoparasites, Arcay (1982) protozoan parasites, and Vogelsang and Espin (1949) nematodes.

REMARKS: J. A. Allen and Chapman (1893:217) provisionally assigned specimens from Trinidad to *Abrothrix caliginosus* (now = *Melanomys caliginosus*), but later used the same series in their description of *Akodon urichi* (J. A. Allen and Chapman 1897a:19). Tatc (1939:187), in his description of *Chalcomys aerosus saturatus* from Mt. Auyan-tepui, also reported on a series of specimens he called "*Chalcomys aerosus* near *chapmani*" from the lower slopes of the same tepui. These are paler than *saturatus* and similar in coloration to J. A. Allen's *chapmani* from Colombia.

The phylogenetic position of *N. urichi* in the genus requires thorough reevaluation, a process that may refine the content and context of *Necromys*. Phylogenetic studies based on mitochondrial and nuclear gene data support the hypothesis that *N. urichi* is closely related to other species in *Necromys* (M. F. Smith and Patton 1999; D'Elía 2003a; D'Elía, Pardiñas, Jayat, and Salazar-Bravo 2008) and that this relationship is best characterized by a three-group basal polytomy (see above). As currently understood on morphological grounds, *N. urichi* is the most divergent *Necromys*. In fact, combinations of characters currently presumed to be diagnostic for *Necromys* (e.g., size and shape of zygomatic notches, length of rostrum, narrow parapterygoid and mesopterygoid fossae, size and shape of the foramen ovale, lack of an anteromedian flexus in M1, size and shape of the hypo-, meso-, and metaflexi) are not present in *N. urichi*. Only a single character diagnostic of *Necromys* is present in the holotype of *N. urichi* (AMNH 7725); that is that digit V, including claw, does not extend beyond half of the proximal phalanx of digit IV. In combination, these are trenchant character differences that need to be evaluated in a phylogenetic context when more nuclear sequence data are available for species in the genus.

Reig and Kiblisky (1968b) and Reig, Olivo, and Kiblisky (1971) described a karyotype with $2n = 18$, composed of eight pairs of metacentric autosomes of decreasing size and heteromorphic sex chromosomes. N. O. Bianchi et al. (1971) reported a lack of X0 females in this species, an observation supported by Hoekstra and Edwards (2000).

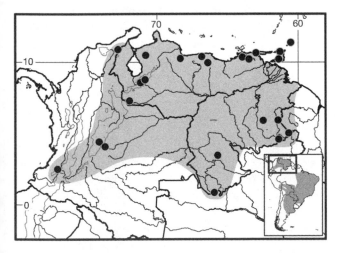

Map 116 Selected localities for *Necromys urichi* (●). Contour line = 2,000 m.

Genus *Oxymycterus* Waterhouse, 1837

João A. de Oliveira and Pablo R. Gonçalves

The genus *Oxymycterus*, as treated here, includes 15 species: *O. amazonicus*, *O. caparaoe*, *O. dasytrichus*, *O. delator*, *O. hiska*, *O. hucucha*, *O. inca*, *O. josei*, *O. juliacae*, *O. nasutus*, *O. nigrifrons*, *O. paramensis*, *O. quaestor*, *O. rufus*, and *O. wayku*. Records are from localities ranging from the southern bank of the Rio Amazon in eastern Brazil and western Brazilian and southern Peruvian Amazonia to southern Buenos Aires province in Argentina. Within this range, records are more densely concentrated in eastern Brazil and the Bolivian-Peruvian Andeans and eastern slopes, but also throughout Paraguay and central Brazil. Pleistocene and Recent material has been reported from Lagoa Santa, southeastern Brazil, but of the fossil species described from there and referred to *Oxymycterus* by Winge (1887), only *Oxymycterus cosmodus*—based on a fragment of maxilla with the first upper molar and recovered from Lapa da Serra das Abelhas (Hansen 2012:67)—is currently included in the genus (Voss and Myers 1991). As for the fossil nominal forms proposed by Ameghino (1889) in *Oxymycterus*, none are currently assigned to the genus (Massoia and Pardiñas 1993; Pardiñas 2000), despite fossils referable to extant species from Argentinean sites (P. E. Ortiz and Pardiñas 2001; Pardiñas et al. 2002).

The smallest species of *Oxymycterus* is *O. hucucha*, whereas the largest specimens recorded have been assigned to *O. quaestor*. Adult head and body lengths in the genus range from 99–197 mm, and tail length varies from 70–156 mm. Among the different species, tail length varies from shorter than (about 60%) to approximately equal to head and body length. Hindfoot length (with claw) varies from 21–42 mm, and ear length (from notch) ranges from 14–27 mm.

Species of this genus are quite variable in size and pelage length and color, but the genus can be easily recognized by the long muzzle, small eyes, and by the long and strong claws both in forefeet and hindfeet. The tail is shorter than the head and body; the ears usually protrude above the fur of the head but vary in length among species, so the combination of ear and hindfoot lengths is useful in distinguishing among sympatric forms of similar body size.

The dorsal pelage of *Oxymycterus* is usually dense, and sometimes hispid and shiny. As in most sigmodontines, dorsal guard hairs have an expanded stiffer distal portion and thinner and more flexible base. The boundary between these two sections, located, in all species of the genus, near the midlength of the hair, is usually a constricted region that causes the expanded distal portion of the hair to bend backward. Guard hairs of the back and sides are usually of two kinds, both cylindrical and thin at their bases and with a grayish proximal portion: a longer type, 10–18 mm in length among species, with a thickened, dark brown to black distal segment, and a shorter, thinner type 7.5–15 mm in length, tipped with a yellow-orange to reddish-brown 0.3–4.5 mm subapical band. Both kinds of guard hairs are also present laterally, but the first type gradually becomes less common progressing ventrad, and the subapical band in the second type gets longer toward the venter. Ventral hairs have grayish bases becoming yellowish distally. Very often the hair bases of the chin, and rarely those of the chest, are pure white.

Skulls of *Oxymycterus* are relatively long and narrow, mainly owing to the nasal tube that projects beyond the upper incisors, the anterior extremity of which is contributed equally to by nasals and premaxillaries. Prenasal ossification (the *os rostri*) is present in all species, although usually being lost during preparation, providing a more pointed appearance to an otherwise squared anterior end of the nasals. The nasal tube may be trumpet-shaped with a high anterior nasal aperture, owing to an upward flaring of the nasals, or the aperture may below and inconspicuous. Zygomatic plates are relatively narrow, not forming obvious zygomatic notches, and have markedly curved anterior margins, a morphology approached only by *Brucepattersonius*, *Blarinomys*, *Juscelinomys*, and *Lenoxus*.

Considerable variation is found among species in the width of the inferior zygomatic root and in the curvature formed by the anterior margin of the zygomatic plates, which may be smoothly to strongly curved anteroposteriorly, without any remarkable anterodorsal knob just anterior to their upper zygomatic roots. The incisive foramina are very long, usually reaching the anterior third or half of the first molars. The anterior end of the mesopterygoid fossa is often at the level of the third molars, usually with their posterior edges, but consistent deviation from this is found among certain populations or species. The interparietal is reduced in comparison with that of other sigmodontines, but also shows considerable variation in shape and size in some species. All species possess the primitive muroid pattern of carotid arterial circulation, including large stapedial and sphenofrontal foramina and a squamosal-alisphenoid internal groove (pattern 1 of Voss 1988).

Compared with other akodont skulls, the rostrum is rather elongated in *Oxymycterus*, notwithstanding that in some species (*O. rufus*, *O. amazonicus*, and *O. delator*) this character is less conspicuous. In cranial morphology, the genera *Juscelinomys*, *Lenoxus*, and *Brucepattersonius* are the most similar to *Oxymycterus*. However, *Oxymycterus* is unique among this group in having prenasal ossifications (*os rostri*) and in the structure of the nasal tube, which is formed by equal contributions of the nasal and premaxillary bones.

Upper incisors are opisthodont, and molars are narrow, moderately hypsodont, and cusps are highly crested. M1 is

marked by a well-incised anteromedian flexus and is twice the length of M2, which is twice as long as M3. The major cusps and internal dentine structures conform to the tetralophodont pattern of akodontines, with an anteroloph(id) and a reduced median loph(id) instead of a well-developed mesoloph. These structures become rather worn in most subadult and adult individuals, so the teeth commonly lose much of the enamel folding complexity of the occlusal surfaces. Although dental morphology has been regarded as conservative in the genus (Hershkovitz 1994), F. G. Hoffmann et al. (2002) reported that young individuals of *O. josei* differ from *O. nasutus* in having a protostyle and a well-marked enterostyle on the M1 and by analogous differences in the m1.

Vertebral counts are 7 cervical, 13 thoracic, 6 lumbar, and 26–33 caudal elements in complete tails. The tail scales are annularly arranged, 13–16 rings per centimeter, and usually have three bristles about 2–3 mm long extending from beneath each scale. The tail ranges from moderately to fairly hirsute, the scales unconcealed to almost concealed in some species. The tail can be unicolored to moderately bicolored, darker above and lighter below.

Forefeet and hindfeet are relatively long and narrow (hindfoot length with claw varies from 16–32% of the head and body length in adults among different species), and covered dorsally with short hairs to the distal phalanges. The palmar surfaces are naked, with two conspicuous metacarpal pads and three smaller interdigital pads. The plantar surfaces bear four interdigital pads, with the two middle ones larger. The hypothenar pad is slightly smaller than the fourth interdigital pad, and is situated halfway between it and the thenar pad, which is as large as the two middle interdigital pads. Ungual tufts are usually short, but can be as long as the claws in the hindfeet. Digits II through IV bear long, strong, and keeled claws, in both the manus and pes; those of the manus are usually slightly larger. The claw on digit III is the longest, ranging from 2.5–4.5 mm in the different species, as measured in a straight line from the tip of the phalanx in ventral view. Claws of digits II and IV are usually subequal in size, and both are slightly shorter than the third digit's claw, hardly surpassing 4 mm in the largest animals. As in other akodontine rodents, the pollex is atrophied but bears a small claw, 1–2 mm long, that hardly reaches beyond the distal ends of the metacarpals. The fifth manual digit is reduced, with a small to medium-sized claw (1–3 mm long), often reaching the level of the distal phalanx of the fourth digit. The hallux and the fifth pedal digit are of subequal lengths, reaching the level of the distal edges of the interdigital pads, the former bears a larger claw (1.5–2.5 mm long versus 1–1.5 mm for digit V).

Four pairs of mammae, one each in pectoral, thoracic, abdominal, and inguinal positions, are present in all species of the genus. Hershkovitz (1994) reported only six pairs of nipples for *O. amazonicus*, but close inspection of specimens housed in the Museu Nacional (MNRJ 32896) and Museu Goeldi (MPEG 8994, 8995) indicates that the mammae count for this species is the same as in others.

Hooper and Musser (1964) described the glans penis of *Oxymycterus* as having a crenate crater rim with essentially no bordering band of spineless tissue, dissimilar bacular mounds, and a robust spinous urethral flap. The baculum is long with a straight shaft and a broadened base. Despite considerable age-related variation, the baculum varies among species in *Oxymycterus* in the shape of the proximal end, developmental extent of lateral condyles, and shape of the distal end (J. A. Oliveira 1998). Soft parts of the glans also vary among species and populations, including size of the dorsal papilla and the spinous ornamentation of the urethral flap (P. R. Gonçalves 2006).

The stomach is bilocular-discoglandular with glandular epithelium contained in a pouch-like diverticulum, characteristics found in insectivorous rodents (Hershkovitz 1994). The gall bladder is usually present, but Geise, Weksler, and Bonvicino (2004) detected intrapopulation variation with respect to this character in *O. dasytrichus*. Specimens referred to *O. dasytrichus* from Ilha Grande, off the southwestern coast of Rio de Janeiro state, lack a gall bladder (Geise, Weksler, and Bonvicino 2004).

SYNONYMS:

M[us].: G. Fischer, 1814:71; part (description of *rufus*, based on "*rat cinquième ou rat roux*" of Azara); not *Mus* Linnaeus.

M[us].: Olfers, 1818:209; part (description of *rutilans*, based on "*rat cinquième ou rat roux*" of Azara); not *Mus* Linnaeus.

Mus: Desmarest, 1819c:62; part (description of *rufus*, based on "*rat cinquième ou rat roux*" of Azara); not *Mus* Linnaeus.

Mus: Schinz, 1821:288; part (description of *dosytrichos*); not *Mus* Linnaeus.

H[ypudaeus].: Wied-Neuwied, 1826: 425; part (listing of *dasytrichos*); not *Hypudaeus* Illiger.

L[emmus].: J. B. Fischer, 1829:293; part (listing of *dasytrichos*); not *Lemmus* Link.

Oxymycterus Waterhouse, 1837:21; proposed as a subgenus; type species *Mus nasutus* Waterhouse, by original designation.

Arvicola: Lesson, 1842:147; part (listing of *dasytrichos*); not *Arvicola* Lacépède.

Hesperomys: Wagner, 1842a:361; part (description of *rostellatus*); not *Hesperomys* Waterhouse.

Oxymycterus: Pictet, 1843a:211; first use as full genus.

Holochilus: Gray, 1843b:114; part (listing of *nasutus*); not *Holochilus* Brandt.

Oxymycteris Schinz, 1845:179; incorrect subsequent spelling of *Oxymycterus* Waterhouse.

Oxymyctorus Tomes, 1860a:221; incorrect subsequent spelling of *Oxymycterus* Waterhouse.

Oxymicterus Tomes, 1861 [1862]:285; incorrect subsequent spelling of *Oxymycterus* Waterhouse.

Oxymcterus Vorontsov, 1979:17; incorrect subsequent spelling of *Oxymycterus* Waterhouse.

Oxymeterus Vorontsov, 1979:27; incorrect subsequent spelling of *Oxymycterus* Waterhouse.

Oxymycetrus Vorontsov, 1979:198; incorrect subsequent spelling of *Oxymycterus* Waterhouse.

Oxymyceterus Vorontsov, 1979:269; incorrect subsequent spelling of *Oxymycterus* Waterhouse.

Oximycterus Marinho-Filho, Reis, Oliveira, Vieira, and Paes, 1994:151; incorrect subsequent spelling of *Oxymycterus* Waterhouse.

REMARKS: J. A. Oliveira (1998) provided detailed histories and descriptions, along with photographs of the skulls, of the type specimens of most nominal taxa of *Oxymycterus*.

KEY TO THE SPECIES OF *OXYMYCTERUS*:

1. Hindfoot (with claw) about 30 mm long or less, not exceeding 32 mm; ear length (from notch) <20 mm in length; upper molar series usually <5.3 mm, not exceeding 6 mm. 2

1'. Hindfoot (with claw) usually >30 mm; ear length (from notch) >20 mm; upper molar series usually >5.3 mm, and not shorter than 5 mm 12

2. Ratio between length of nasal tube (measured from anterior border of incisive alveolus to anteriormost tip of premaxillary) to that of molar series <0.5 3

2'. Ratio between length of nasal tube (measured from anterior border of incisive alveolus to anteriormost tip of premaxillary) to that of molar series >0.5 5

3. Nasals short, not reaching level of upper zygomatic root, interparietal reduced (ratio of interparietal breadth/braincase breadth usually <0.5); dorsal pelage bright rufous. . . . . *Oxymycterus rufus*

3'. Nasals long, reaching level of upper zygomatic root; interparietal relatively larger (ratio of interparietal breadth/braincase breadth usually >0.5); dorsal pelage blackish or dark brown 4

4. Palate shorter, not extending posteriorly beyond M3, leaving anterior end of presphenoid visible in ventral view; frontal sinuses less inflated; pelage dark brown . . . . *Oxymycterus amazonicus*

4'. Palate longer, extending posteriorly beyond M3 and covering anterior end of presphenoid so that it is not visible in ventral view; frontal sinuses more inflated; pelage blackish . . . . *Oxymycterus delator*

5. Length of hindfoot with claw <or about 23 mm . . . . . .
. . . . *Oxymycterus hucucha*
5′. Length of hindfoot with claw >23 mm 6
6. Found on Atlantic coast of Brazil or Uruguay 7
6′. Foundin eastern Andes of Peru, Bolivia, or Argentina . 9
7. Zygomatic plate wide, greatest width equal to or >2.2 mm . . . . *Oxymycterus josei*
7′. Zygomatic plate narrow, width generally <2 mm, rarely exceeding 2.2 mm 8
8. Incisive foramina long, extending well beyond posterior limit of inferior root of zygomatic plate; ectotympanic portion of bulla less expanded anterolaterally, resulting in smaller and conical bullae . . . .
. . . . *Oxymycterus nasutus*
8′. Incisive foramina short, not extending beyond posterior limit of inferior root of zygomatic plate; ectotympanic portion of bulla more expanded anterolaterally, resulting in more quadrangular bullae . . . *Oxymycterus caparaoe*
9. Pelage showing contrast in color between anterior and posterior parts of dorsum or homogeneously dark or blackish brown; interorbital region and nasal tube relatively narrower; zygomatic plate relatively larger (ratio between the interorbital constriction and zygomatic breadth <0.42; ratio between width of zygomatic plate and width of nasal tube >0.49) 10
9′. Pelage color paler and homogeneously brown throughout dorsum; interorbital region and nasal tube relatively larger; zygomatic plate relatively narrower (ratio between interorbital constriction and zygomatic breadth >0.42; ratio between width of zygomatic plate and width of nasal tube usually <0.49) 11
10. Head and anterior part of dorsum of general olive-grayish color, strongly lined with black, becoming more brownish or ochraceous brown toward rump; narrower braincase and zygomatic notches; claws on forefeet and hindfeet moderately developed; less expanded zygomatic arches; incisive foramina longer and with straight edges; lunar notch of mandible symmetrically excavated . . . . *Oxymycterus paramensis*
10′. Head and dorsal pelage blackish brown with slight ochre or reddish hue; claws on forefeet and hindfeet strong and long, those of digit III reaching more than 6 mm on forefoot and 5 mm on hindfoot; well-expanded zygomatic arches; incisive foramina shorter; asymmetrical due to more excavated ventral portion . . . .
. . . . *Oxymycterus wayku*
11. Size averages larger (hindfoot with claw >25 mm); skull longer, with occipitonasal length usually >31 mm; if smaller, ratio between zygomatic breadth and occipitonasal length usually >0.46 . . . *Oxymycterus nigrifrons*
11′. Size averages smaller (hindfoot with claw ≤25 mm); skull shorter, with occipitonasal length usually <31 mm;

ratio between zygomatic breadth and occipitonasal length ≥0.46 . . . . *Oxymycterus hiska*
12. Andean-Amazonian distribution 13
12′. Atlantic distribution (west to Río Paraguay) 14
13. Ratio between width of zygomatic plate and width of nasal tube >0.53 . . . . *Oxymycterus inca*
13′. Ratio between width of zygomatic plate and width of nasal tube usually ≤0.53 . . . . *Oxymycterus juliacae*
14. Interorbital region relatively broad and nasal tube less elongated, ratio between interorbital width and occipitonasal length generally >0.17; if lower, ratio between length of nasal tube (measured distally from anterior margin of upper incisors) and length of upper molar row <0.6; dorsal pelage varying from olivaceus to dark brown, not markedly lined with black . . . .
. . . . *Oxymycterus dasytrichus*
14′. Interorbital region relatively narrow and nasal tube more elongated, ratio between interorbital width and occipitonasal length generally <0.17; if higher, ratio between length of nasal tube and length of upper molar row >0.6; dorsal pelage varying from dark yellow gray to dark orange, markedly lined with black, in some specimens providing a metallic iridescent sheen . . . . . .
. . . . *Oxymycterus quaestor*

Oxymycterus amazonicus Hershkovitz, 1994
Amazonian Hocicudo

SYNONYM:

Oxymycterus amazonicus Hershkovitz, 1994:23; type locality "Fordlândia, right bank, lower Rio Tapajóz [*sic*], Pará, Brazil, 3°40′S, 55°30′W."

DESCRIPTION: Dorsal and lateral coloration of generally dark-brown from head to base of tail, with some specimens paler and more reddish-brown; ventral color varies from orange to reddish orange, with hairs of throat generally paler or even whitish; hindfeet covered dorsally by dark-brown hairs. Head and body length of adults 140–148 mm, tail length 86–95 mm, hindfoot length (with claw) 28–29 mm, and ear length 16–20 mm. Skull with comparatively short and wide rostrum and robust braincase, similar in structure and size to *O. delator* but with frontal sinuses less inflated; nasals longer than in *O. delator* but do not project anteriorly beyond perpendicular plane of anterior incisor border, hardly forming the typical trumpet-shaped nasal tube of genus; zygomatic plates quite variable in width, but their curvature less pronounced than in most species; zygomatic notches relatively broad, correlated with wide infraorbital foramina; incisive foramina relatively wide, specially along maxillary region, and robust palatine septum present; bony palate shorter than *O. delator* not extending posteriorly beyond level of M3, leaving anterior end of presphenoid visible in ventral view.

DISTRIBUTION: Lower Amazon, including regions along the lower parts of Tocantins, Xingú, and Tapajós rivers, southwestward along middle and upper tributaries of Aripuanã and Tapajós rivers to northwestern Mato Grosso state, at least to the southern edge of Serra do Norte (= Serra dos Parecis); to the west, *Oxymycterus amazonicus* has been recorded on the Rio Jamari, an upper affluent of Rio Madeira Basin, in the state of Rondônia.

SELECTED LOCALITIES (Map 117): BRAZIL: Mato Grosso, Aripuanã, Alto do Rio Madeira (MNRJ 46705), Apiacás (MZUSP 35159), Campos Novos (AMNH 37101), Juruena (MZUSP 35162); Pará, Agrovila União, 18 km S and 19 km W of Altamira, Rio Xingu (USNM 521485), BR 165, Santarém to Cuiabá, Santarém (USNM 544637), Fazenda São Raimundo, km 42 (estreito), São João do Araguaia, Marabá (MPEG 10140), Fordlândia, Rio Tapajós (type locality of *Oxymycterus amazonicus* Hershkovitz), [near] Jatobal, 73 km N and 45 km W of Marabá (USNM 519787), Piquiatuba, Rio Tapajós (AMNH 94809); Rondônia, Alto Paraíso (MNRJ 79724).

SUBSPECIES: *Oxymycterus amazonicus* is monotypic.

NATURAL HISTORY: Specimens of *O. amazonicus* from near Altamira (Pará) have been obtained in secondary forest and in *capoeiras* (areas of secondary growth brush) intermixed with rice or manioc plantations, as documented on specimen labels. A specimen from Apiacás, Mato Grosso state, was found from the stomach of a *Bothrops*.

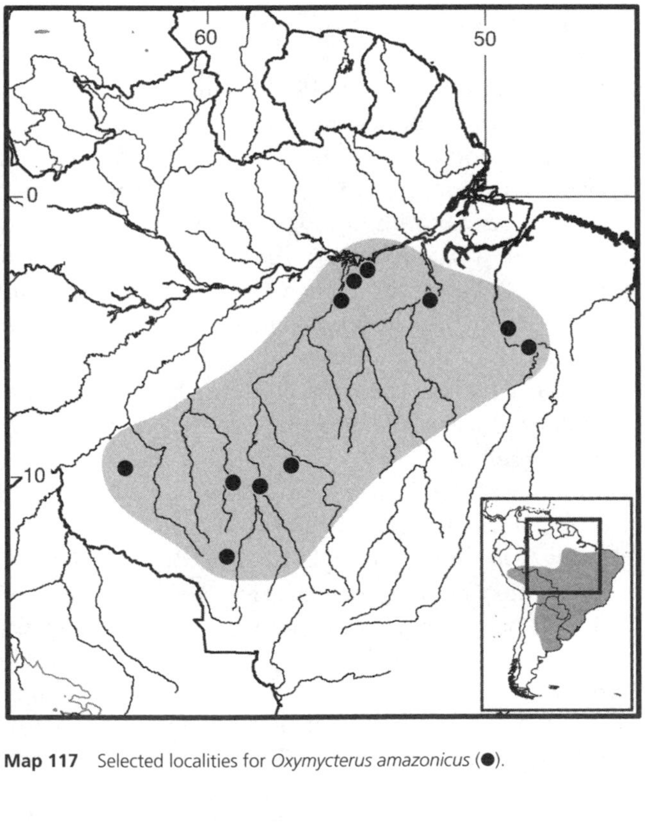

REMARKS: *Oxymycterus amazonicus* is closely related to *O. delator*, based on phylogenetic analysis of mtDNA cytochrome-*b* sequences (F. G. Hoffmann et al. 2002), differing by only 2.9% p-distance. See also remarks under *O. delator*.

Oxymycterus caparaoe Hershkovitz, 1998
Mt. Caparaó Hocicudo

SYNONYMS:

Oxymycterus caparaoe Hershkovitz, 1998:244; type locality "Arrozal, Parque Nacional do Caparaó, Minas Gerais, Brazil, elevation 2400 m."

O[*xymycterus*]. *caparaonense* F. G. Hoffmann, Lessa, and Smith, 2002:415; incorrect subsequent spelling of *caparaoe* Hershkovitz.

Oxymycterus caparaoe Musser and Carleton, 2005:1157; incorrect subsequent spelling of *caparaoe* Hershkovitz.

DESCRIPTION: Overall dorsal coloration predominantly dark brown (individual hairs with blackish tips, narrow orange subterminal band, and slate gray base); sides of body slightly more orange due to wider phaeomelanin band; venter orange but with slaty base showing through; ears dark brown and hindfeet covered dorsally by grayish brown hairs (gray basally and white apically). Pelage long and lax. Head and body length of adults 110–139 mm, tail length 80–104 mm, hindfoot length (with claw) 27–31 mm, and ear length 18–20 mm. Rostrum markedly elongated and narrow, with nasals and premaxillary bones parallel sided and combined to form well projected trumpet-shaped nasal tube; nasals with caudal limits anterior to line linking upper zygomatic roots; zygomatic plates slender, notably curved anteroposteriorly and not considerably arched laterally, delimiting narrow infraorbital foramina and reduced zygomatic notches; interorbital region broad due to laterally inflated supraorbital margins and frontal sinuses, which remain visible in ventral view up to level of M2; incisive foramina relatively short, not extending beyond posterior margin of inferior root of zygomatic plates; bony palate relatively short, not extending posteriorly beyond M3; bullae large and quadrangular, with ectotympanic anteriorly expanded.

DISTRIBUTION: This species is known only from the western slope and altiplano of the Pico da Bandeira, Parque Nacional do Caparaó, along the boundary between the states of Espírito Santo and Minas Gerais, at elevations from 2,100 to 2,400 m.

SELECTED LOCALITIES (Map 118): BRAZIL: Espírito Santo and Minas Gerais: Arrozal, Parque Nacional do Caparaó, 3.0 km N and 0.1 km W of Pico da Bandeira (type locality of *Oxymycterus caparaoe* Hershkovitz).

SUBSPECIES: *Oxymycterus caparaoe* is monotypic.

Map 117 Selected localities for *Oxymycterus amazonicus* (●).

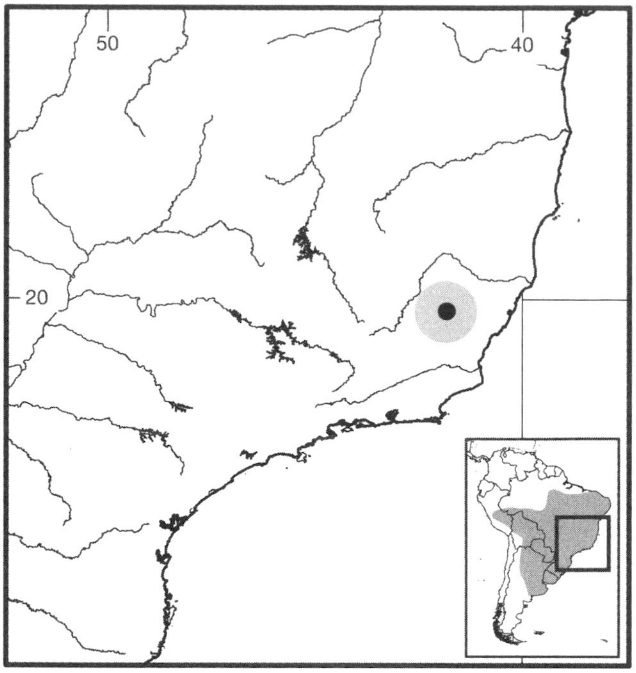

Map 118 Single known locality for *Oxymycterus caparaoe* (●).

NATURAL HISTORY: This species has been encountered only at high elevations (>2,100 m) on Mt. Caparaó, in close association with the *campos de altitude*, a series of cool-humid grasslands that occur atop the highest summits and altiplano of southeastern Brazil (Bonvicino et al. 1997; Safford 1999). Trapping records of topotypical specimens deposited in MNRJ indicate higher abundance in areas with hygrophilic soils dominated by bunchgrasses (*Cortadeira* spp.) and Andean bamboo (*Chusquea* spp.), which form a cluttered and dense herbaceous cover that is probably used for nesting places.

REMARKS: Hershkovitz (1998) considered *O. caparaoe* closely related to *O. nasutus*, as both two species share an extremely long and narrow rostrum, relatively narrow and curved zygomatic plates, and relatively small hindfeet. Despite these morphological similarities, however, *O. caparaoe* can be distinguished from *O. nasutus* by its overall darker dorsal color, relatively larger size, shorter incisive foramina, broader interorbit, and anteroposteriorly inflated and quadrangular bullae.

Oxymycterus dasytrichus (Schinz, 1821)

Atlantic Forest Hocicudo, or "rato-bubo" (Bahia), "rato-porco" (Pernambuco), and "trioréu" (São Paulo)

SYNONYMS:

Mus dosytrichos Schinz, 1821:288; type locality "Camamú unweit Bahia," Brazil (as given by Wied-Neuwied [1826] for the "Rato Bubo" of Schinz [1821]); modified

by Cabrera (1961) to "Rio Mucuri, Bahia, Brazil," and further to "lower Rio Mucuri" by Avila-Pires (1965), the locality of provenance of the second specimen described by Wied-Neuwied (1826).

H[ypudaeus]. dasytrichos Wied-Neuwied, 1826:425; unjustified emendation of *dosytrichos* Schinz, name combination, and redescription of *Mus dosytrichos* Schinz; Avila-Pires (1965) designated a lectotype.

L[emmus]. dasytrichos: J. B. Fischer, 1829:293; name combination.

Mus dasytrichos: J. B. Fischer, 1829:293; name combination.

Arvicola dasytrichos: Lesson, 1842:147; name combination.

Hesperomys [Oxymycterus] rostellatus Wagner, 1842a: 361; type locality "Brasilia"; restricted by Gyldenstolpe (1932) to "Eastern Brazil, Bahia" and further restricted by Hershkovitz (1994) to "São Salvador, or Salvador, Bahia, Brazil, 12°59′S, 38°31′W, sea level."

H[esperomys]. rostellatus: Wagner, 1843a:514; name combination.

Oxymycterus hispidus Pictet, 1843a:211, plates IV and V; no type locality given, but listed by Baud (1977) as "Iles Eos" [= Ilhéus?], Bahia, Brazil (see J. A. Oliveira 1998 for further details concerning the type locality).

O[xymycterus]. rostellatus: Pictet, 1843a:213; name combination.

M[us]. hispidus: Schinz, 1845:179; name combination; not *Mus hispidus* Lichtenstein.

M[us]. hispidulus Schinz, 1845:179; unjustified emendation of *hispidus* Pictet.

Oxymycteris hispidus: Schinz, 1845:180; name combination.

M[us]. rostellatus: Schinz, 1845:189; name combination.

Hesperomys rufus: Burmeister, 1854:183; name combination; part; not *Mus rufus* G. Fischer.

O[xymycterus]. rufus: Burmeister, 1855a:10; name combination: part; not *Mus rufus* G. Fischer.

Hesperomys (Oxymycterus) rufus: E.-L. Trouessart, 1880b:142; name emendation (see Remarks) and name combination; listed as synonym of but not *rufus* G. Fischer.

Hesperomys (Oxymycterus) dasytrichos: E.-L. Trouessart, 1880b:142; name combination, listed as synonym of *O. rufus nasutus*.

Oxymycterus rufus: Winge, 1887:36, plate 1, Figs. 10–11; plate 2, Fig. 14; part; not *rufus* G. Fischer.

[Oxymycterus] dasytrichus: E.-L. Trouessart, 1897:539; emendation of *dosytrichos* Schinz and first use of current name combination; listed as a synonym of *O. rufus*.

Oxymycterus Roberti Thomas, 1901g:530; type locality "Rio Jordão, in the district of Araguary [= Araguari], S.W. Minas Geraes [= Minas Gerais]," Brazil.

Oxymycterus angularis Thomas, 1909:237; type locality "São Lourenço [= São Lourenço da Mata], near Pernambuco. Alt. 30m," Pernambuco, Brazil.

Hypudaeus dasytrichus: Tate, 1932h:17; name combination.

Oxymycterus rufus dasytrichus: Cabrera, 1961:468; name combination.

Oxymycterus hispidus hispidus: Cabrera, 1961:466; name combination.

Oxymycterus quaestor: Avila-Pires and Gouveia, 1977:25–26; part.

Oxymycterus rosettellatus Vorontsov, 1979:95, Fig. 47; incorrect subsequent spelling of *rostellatus* Wagner.

Oxymycteris angularis: Mares, Willig, Streilen, and Lacher, 1981:4, 117; name combination.

Oxymycterus rufus dasytrichos: Hershkovitz, 1987:37; name combination.

DESCRIPTION: Dorsal color dark-brown to paler brownish red, without strong lining of black; ventral color dark gray to paler, cinnamon buff or pinkish buff. Head and body length of adults 125–197mm, tail length 90–156mm, hindfoot length 30–42mm, and ear length 18–26mm. Skull robust, proportionally larger in width dimensions of braincase, interorbital region, and zygomatic plates, with relatively longer molar tooth rows (mean 5.6mm) than most other species of Atlantic coast; nasals long and wide, slightly flared upward at distal ends, with posterior limits extending to line connecting upper zygomatic roots; zygomatic plates wide, with slightly curved anterior margins delimiting broad zygomatic notches and wide infraorbital foramina; premaxillary part of septum of incisive foramina relatively large, approximately two-thirds of septal length, and maxillary portion often thick and sometimes expanded; interparietal characteristically very large.

DISTRIBUTION: *Oxymycterus dasytrichus* occurs throughout eastern Brazil in the states of Pernambuco, Alagoas, Sergipe, Bahia, Minas Gerais, Espírito Santo, Rio de Janeiro, São Paulo, and Paraná, in both coastal forests and isolated interior forest remnants.

SELECTED LOCALITIES (Map 119): BRAZIL: Alagoas, Matriz de Camaragibe, Fazenda Santa Justina, 6km SSW of Matriz de Camaragibe (UFPB 6960), São Miguel dos Campos, Mangabeiras (MZUSP 7551); Bahia, Palmeiras, Chapada Diamantina, Gerais da Cachoeira da Fumaça (MNRJ 67798), Reserva Biológica Pau Brasil, 15km NW of Porto Seguro (UFPB 576), Rio Mucuri (AMNH 559), Três Braços, 37km N and 34km E of Jequié (USNM 545060); Espírito Santo, Cachoeiro de Itapemirim, 4mi N of Castelinho (MNRJ 32880), Reserva Florestal Nova Lombardia (MNRJ 32874); Goiás, Anápolis (MNRJ 3357[skull]/AMNH 134881[skin]); Minas Gerais, Além Paraíba (MNRJ 5375), Araguari, Rio Jordão (BM 1.11.3.51), Boca da Mata, km 104–105 on the road from Lagoa Santa to Conceição do

Mato Dentro (MNRJ 13406), Fazenda Esperança, Serro (USNM 282591), Lagoa Santa (ZFMK 397), Passos (MNRJ 12853); Pernambuco, Garanhuns (MNRJ 27218), Macaparana: Fazenda Água Fria (UFPB 6961), São Lourenço da Mata (BM 3.10.1.56); Rio de Janeiro, Angra dos Reis (MNRJ 2566), Fazenda Velha, Tijuca (MNRJ 32857); São Paulo, Barra do Guaraú, Cananéia (MZUSP 24887), Fazenda Intervales (Base do Carmo), Capão Bonito (MVZ 183125), Salto de Pirapora (MZUSP 22693), Varjão, Bertioga (USNM 484400); Sergipe, Fazenda Cruzeiro, 13km SSE of Cristinápolis (MNRJ 66570).

SUBSPECIES: We treat *Oxymycterus dasytrichus* as monotypic.

NATURAL HISTORY: This species occurs both in lowland and mountainous regions up to elevations of 1,800m, in seasonally flooded *várzea* forest and humid open areas near or in forests; in agricultural crop fields, such as sugar cane, coffee, corn, and bananas; and in the tall, pioneer grasses characteristic of secondary formations in the Atlantic Forest of eastern Brazil. M. V. Vieira et al. (2009) documented higher abundances in forest fragments surrounded by rural properties than in those near urban areas in Rio de Janeiro state. *Oxymycterus dasytrichus* has been found in sympatry with other species of the genus in a number of localities in eastern Brazil, throughout the states of Bahia (*O. delator*), Minas Gerais (*O. rufus, O. delator*) and São Paulo (*O. judex*) (J. A. Oliveira 1998; P. R. Gonçalves and Oliveira 2004; Geise, Pereira et al. 2004). On the eastern slopes of the Serra deItatiaia (on the border of Rio de Janeiro and Minas Gerais states), *O. dasytrichus*is more abundant in submontane forests, becoming increasingly rare at higher elevations (>2,000m) where *O. delator* becomes more frequent (Geise, Pereira et al. 2004). Habitat information by collector A. Robert on the label of the type specimen of *O. roberti* reads "Foret/Endroit humide." Also based on label comments, at Anápolis in Goiás state, C. Lako and R. M. Gilmore in 1937 collected individuals amid coffee, corn, and sugar cane fields apparently adjacent to a forest or at the margins of creeks. Pregnant females with two to four embryos have been obtained in August and September at Caruaru, Pernambuco, and from April to November at Viçosa, Alagoas. Both laelapine mites (*Androlaelaps*) and staphylinid beetles (*Amblyopinodes*) parasitize animals on Ilha Grande, Rio de Janeiro state (Martins-Hatano et al. 2002; Bittencourt and Rocha 2002).

REMARKS: Schinz (1821) spelled the specific epithet of his new species *dosytrichos*, and Wied-Neuwied (1826) subsequently gave the variant spelling *dasytrichos*. E.-L. Trouessart (1897) later Latinized Wied-Neuwied's name to *dasytrichus*, the spelling that has been in prevailing use since then. We thus regard Trouessart's spelling *dasy-*

trichus as a justified emendation of both *dosytrichos* Schinz and *dasytrichos* Wied-Neuwied (ICZN 1999:Art. 33.2.3.1). Avila-Pires (1965) designated a lectotype for *H[ypudaeus]. dasytrichos* Wied-Neuwied, 1862, a subjective junior synonym of Schinz's (1821) *Mus dosytrichos*, without commenting on the applicability of his action to Schinz's name.

Molecular analyses (F. G. Hoffmann et al. 2002; P. R. Gonçalves and Oliveira 2004) have confirmed the separation of this species from other Atlantic forms, such as *O. quaestor* (including *O. judex*) and *O. rufus*, as had been previously suggested on the basis of morphological comparisons (J. A. Oliveira 1998). The concept of *O. dasytrichus* adopted here includes samples from the northern limits of the range (in Alagoas and Pernambuco states) that have been traditionally referred to as *O. angularis* Thomas, as well as samples from the western (inland) localities such as the "forests bordering the upper Rio Paranaiba," which formed the basis for Thomas' (1901g) description of *O. roberti*, and those from Parnaguá, Piauí, tentatively regarded as representatives of *O. hispidus* by J. A. Oliveira (1998). This concept is based on cranial morphometric patterns, which indicate that these samples represent extremes of a continuum that encompasses the entire range of the "*dasytrichus* and *angularis* groups" (*sensu* J. A. Oliveira 1998) rather than separate and isolated units. Moderate levels of sequence divergence have been found between samples from Bahia state and those from southeastern Brazil, suggesting well-defined geographic structure uncorrelated with the smooth morphological cline revealed so far (P. R. Gonçalves and Oliveira 2004; P. R. Gonçalves 2006). More comprehensive molecular and morphological sampling of the entire geographic range of *O. dasytrichus* is needed to test the taxonomic scheme presented here.

Oxymycterus delator Thomas, 1903
Spy Hocicudo, Rato-da-Vereda, Hocicudo-Purpúreo de Paraguay

SYNONYMS:

Oxymycterus delator Thomas, 1903b:489; type locality: "Sapucay [= Sapucaí], [Paraguarí,] Paraguay."

Oxymycterus roberti: Myers, 1982:85; part; not *Roberti* Thomas, 1903b.

Oximycterus roberti: Marinho-Filho, Reis, Oliveira, Vieira, and Paes, 1994:151; part; not *Roberti* Thomas, 1903b.

DESCRIPTION: Dorsal color of dark tones, with black predominating because of quite short subapical phaeomelanin bands in banded hairs; samples from Paraguay and central Brazil darkest, those from northeastern and southeastern Brazil less marked blackish; marked patch of paler hairs present above cheeks in most specimens. Ventral pelage distinct from sides and dorsum, generally buffy but darkened in some regions due to visibility of slate-colored hair bases. Tail with relatively long hairs, spanning 3–4 scales, with scales visible to eye in most specimens; tail unicolored and blackish, similar in color to hindfeet and ears. Head and body length of adult specimens 115–174 mm, tail length 70–120 mm, hindfoot length 23–31 mm, and ear length 16–20 mm. Skull with proportionally larger zygomatic plates and basicranial length than other species; most distinctive features inflated anterior part of frontal bones, rounded and heightened braincase, and very angular supraorbital edges. Rostrum reduced and markedly reclined downward, with nasals less projected anteriorly and less flared, not forming typical trumpet-shaped nasal tube, a condition approached only by *O. amazonicus* and *O. rufus*. Incisive foramina wide with septum usually elevated and with thick maxillary part, sometimes expanded at contact with premaxillary part; palate long, extending posteriorly beyond M3 and covering anterior end of presphenoid.

DISTRIBUTION: *Oxymycterus delator* ranges from eastern Paraguay to northeastern Brazil, along a belt of open formations that extends across the states of Mato Grosso, Mato Grosso do Sul, Paraná, São Paulo, Minas Gerais, Bahia, Goiás, Tocantins, Piauí, the western part of Ceará, and the Distrito Federal.

SELECTED LOCALITIES (Map 120): BRAZIL: Bahia, Abaíra, Mata do Tijuquinho, Chapada Diamantina

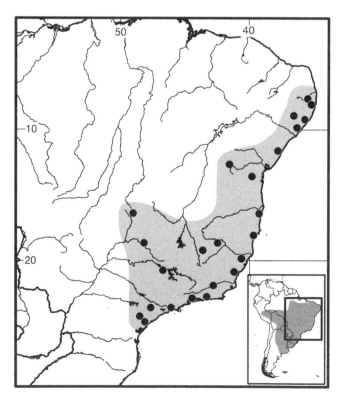

Map 119 Selected localities for *Oxymycterus dasytrichus* (●).

(MNRJ 67832), Cocos, Fazenda Sertão Formoso (MNRJ 66037), Lençóis, Chapada Diamantina (MNRJ 67558); Ceará, Guaraciaba do Norte (MNRJ 33162); Goiás, Rio das Almas, Jaraguá (MZUSP 3913); Mato Grosso, UHE Manso (UNB 777), 264 km N of Xavantina, Serra do Roncador (BM 80.475); Minas Gerais, Itamonte (MNRJ 46651), Parque Nacional Grande Sertão Veredas (MZUSP 35167); Paraná, Jaguariaiva (MHNCI [PMC 1705]); Piauí, Estação Ecológica de Uruçuí-Una (MZUSP 30349); São Paulo, Itapetininga (USNM 460550), Paranapiacaba (MZUSP 35157); Tocantins, Peixe (MZUSP 35158), Ponte Alta do Tocantins, Estação Ecológica Serra Geral do Tocantins (MZUSP 35163). PARAGUAY: La Cordillera, 1.6 km by road S of Tobati (UMMZ 125953); Misiones, Refugio Faunistico "Atingy" (TK 60856, to be deposited in MNHNP; R. D. Owen, pers. comm.); Paraguarí, Sapucay (BM 3.4.7.18).

SUBSPECIES: We treat *Oxymycterus delator* as monotypic.

NATURAL HISTORY: *Oxymycterus delator* is a common member of the small mammal community of the Brazilian Cerrado, being frequently trapped in ecological studies and inventories in central Brazil, most of which have treated this species as *O. roberti* (Alho 1982; Ernest and Mares 1986; Mares et al. 1986; Nitikman and Mares 1987; Gettinger and Ernest 1995; Marinho-Filho et al. 1998; Lacher and Alho 2001). Among the Cerrado physiognomies, the wet grasslands known locally as *veredas* and *campos* are the preferred habitat of *O. delator*. The easternmost populations of the species are associated with montane grasslands (*campos de altitude*) and Cerrado enclaves in Minas Gerais, São Paulo, and Paraná states. The species is apparently sensitive to impacts on herbaceous vegetation cover, as local abundances seem to be greatly reduced after fires. Borchert and Hansen (1983) note the prevalence of earthworms (74%) and insects, especially termites, in stomachs. Predation experiments by Redford (1984) revealed that *O. delator* discriminates between types of termites, preferring species with mechanically based defenses to those with chemically based defenses. These rats generally find termite nests by gleaning the ground surface and uncovering hidden trails. Few reproductive data are available. Males are apparently more territorial than females, and mark their territory with urine and feces (Alho 1982). Pregnant females with up to four embryos were collected in November in Parque Nacional da Chapada dos Veadeiros, Goiás state, and in February and July in Serra do Roncador, Mato Grosso state, Brazil (Bonvicino, Lindbergh, and Maroja 2002; Bonvicino, Lemos, and Weksler 2005).

REMARKS: This species was previously considered to be restricted to Paraguay and possibly adjacent Brazil (Moojen 1952b), but J. A. Oliveira (1998) documented an expansive distribution across the wide belt of open forma-

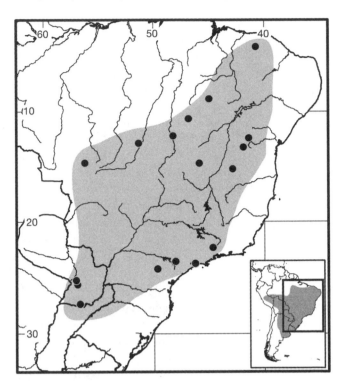

Map 120 Selected localities for *Oxymycterus delator* (●).

tions from Paraguay to northeastern Brazil. F. G. Hoffmann et al. (2002) detected low sequence (cytochrome-*b*) divergence between *O. delator* and *O. amazonicus*. P. R. Gonçalves (2006), in a more comprehensive phylogeographic analysis based on both mitochondrial and nuclear genes, revealed that samples of *O. amazonicus* are nested within an expanded *O. delator* clade, resulting in a paraphyletic *O. delator*. Nevertheless, samples representing both *O. amazonicus* and *O. delator* are diagnosable for a number of morphological attributes. Consequently, we continue to recognize the two as separate species. The karyotype is $2n = 54$, FN = 62 (Svartman and Cardoso de Almeida 1993b; Bonvicino, Lemos, and Weksler 2005).

Oxymycterus hiska Hinojosa, Anderson, and Patton, 1987
Small Yungas Hocicudo

SYNONYMS:

Oxymycterus hiska Hinojosa, Anderson, and Patton, 1987: 14; type locality "14 km W of Yanahuaya, department of Puno, Peru, at 2210 m elev.; lat. 14°19′S and long. 69°21′W," near Abra Marracunca, on the ridge separating the drainages of the Río Huari Huari and Río Tambopata.

Oxymycterus paramensis nigrifrons: S. Anderson, 1997: 439; part; not *Oxymycterus paramensis nigrifrons* Osgood.

DESCRIPTION: Small, with head and body length 97–126 mm, tail length 77–91 mm, hindfoot length (with claw) 23–25 mm, ear length 13–16 mm, and mass of adult specimens 30–44 g. Dorsal pelage short and fine, glossy in appearance, and colored blackish gray with hairs on both dorsal and ventral surfaces tipped ochraceous, especially in older individuals, so that younger animals appear darker, almost black. Tail uniformly blackish brown from base to tip above and below; caudal scales readily visible to eye. Skin of dorsal and palmar surfaces of pes dark, with sparsely distributed hairs on dorsum blackish; ungual hairs of hindfeet do not reach claw tips. Skull also small, with condyloincisive length 23.5–25.9 mm; narrow (1.46–1.76 mm) zygomatic plates slant distinctly backward from bottom to top; relatively long and parallel-sided incisive foramina reach protocone of M1s; relatively long nasal tube (ratio of tube length to maxillary toothrow length, 0.57) flared distally to form slight to well-developed trumpet; short tooth rows (4.79–5.01 mm); and anteroloph of M1 divided by shallow anteromedian flexus into equal-sized conules.

DISTRIBUTION: This species is known from only a few specimens obtained at elevations from 600 to about 3,500 m, in the Yungas of the southern part of Cordillera de Carabaya (Puno, Peru), the Cordillera Real (La Paz, Bolivia), and the Siberia Cloud Forest of the Cordillera Oriental (Cochabamba department, near the border with Santa Cruz department, Bolivia).

SELECTED LOCALITIES (Map 121): BOLIVIA: Cochabamba, Yungas near Pojo, Carrasco (ZFMK 92403); La Paz, Mapiri, Larecaja (AMNH 72889), Okara (AMNH 72750), Tacacoma (AMNH 91601). PERU: Puno, 14 km W of Yanahuaya (type locality of *Oxymycterus hiska* Hinojosa, Anderson, and Patton).

SUBSPECIES: *Oxymycterus hiska* is monotypic.

NATURAL HISTORY: The holotype and other specimens obtained at the type locality were collected in mossy runways, under logs, or around tree roots in low montane humid forest dominated by epiphyte-laden trees, cecropias, bamboos, and tree ferns. The specimen from Yungas near Pojo (Cochabamba), obtained by J. Niethammer in August 1951, has the information "in urwald" (= in primary forest, or jungle) written on the label. A female was taken at 2.5 km S Tacacoma, La Paz department, Bolivia, ca. 3,500 m, in humid subtropical montane forest and in mixed *Polylepis* woodland, composed of dense, epiphyte-covered trees of short stature (5–6 m) on a 30–40° slope (Salazar-Bravo, Yensen et al. 2002). The holotype and other females obtained at the type locality showed signs of recent reproductive activity in late July and early August, suggesting a breeding season at the end of the rainy season (March or April) (Hinojosa et al. 1987).

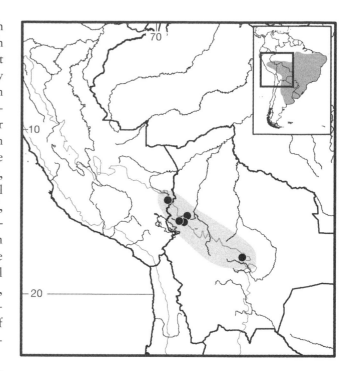

Map 121 Selected localities for *Oxymycterus hiska* (●). Contour line = 2,000 m.

REMARKS: J. A. Oliveira (1998) assigned Bolivian specimens from La Paz department, Bolivia, previously referred to *Oxymycterus paramensis nigrifrons* by S. Anderson (1997), to *O. hiska*. Salazar-Bravo, Yensen et al. (2002) corroborated the presence of *O. hiska* in Bolivia. The record from Mapiri is based on an incomplete specimen, but if confirmed would suggest possible sympatry with *O. inca*. *Oxymycterus hiska* is morphometrically similar to *O. nigrifrons*, with which it is possibly sympatric in the upper Río Inambari drainage (Puno department, Peru).

Oxymycterus hucucha Hinojosa, Anderson, and Patton, 1987

Quechuan Hocicudo, Hucucha

SYNONYMS:

Oxymycterus hucucha Hinojosa, Anderson, and Patton, 1987:15; type locality "28 km by road W of Comarapa (Santa Cruz) but in the department of Cochabamba, Bolivia, at 2800 m elev.; lat. 17°51′S and long. 64°40′W."

Oxymycterus hucuca: Salazar-Bravo and Emmons, 2003:147; incorrect subsequent spelling of *hucucha* Hinojosa, Anderson, and Patton.

DESCRIPTION: Smallest species in genus (only slightly smaller than *O. hiska*), with head and body length 99–109 mm, tail length 71–75 mm, hindfoot length 21 mm, ear length 14–15 mm, and mass 26–36 g. Coloration similar to that of *O. hiska*, but tips of dorsal hairs conspicuously paler and more reddish. Ungual tufts also longer, some

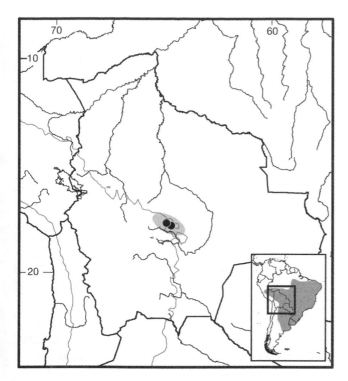

Map 122 Selected localities for *Oxymycterus hucucha* (●). Contour line = 2,000 m.

reaching claw tips on pes. Skull also similar to that of *O. hiska*, but narrower across rostrum, interorbital region, and braincase, and smaller in most other dimensions (see Hinojosa et al. 1987).

DISTRIBUTION: *Oxymycterus hucucha* is known from only a few localities in the vicinity of the Siberia Cloud Forest, Cochabamba and Santa Cruz departments, Bolivia.

SELECTED LOCALITIES (Map 122): BOLIVIA: Cochabamba, 28 km W of Comarapa (type locality of *Oxymycterus hucucha* Hinojosa, Anderson, and Patton), 20 mi E of Totora (MVZ 119948).

SUBSPECIES: *Oxymycterus hucucha* is monotypic.

NATURAL HISTORY: The holotype and two other specimens were collected in the Siberia Cloud Forest, a humid forest with trees 6 m in height, trunks covered with lichens, and dense ground cover comprising lichens, ferns, and mushrooms in central Bolivia (Hinojosa et al. 1987). There is no other information on this species.

REMARKS: This very distinctive small species is craniometrically closest to *O. paramensis* (J. A. Oliveira 1998).

Oxymycterus inca Thomas, 1900
Inca Hocicudo, Hocicudo Castaño del Perú
SYNONYMS:

Oxymycterus inca Thomas, 1900:298; type locality "Perené, Ucayali watershed, Department of Junin, E. Peru. Altitude 800 m."

Oxymycterus iris Thomas, 1901a:183; type locality "San Ernesto, near Mapiri, Upper Beni River, [La Paz,] Bolivia. Alt. 1000 m."

Oxymycterus incae Reig, 1987:361; incorrect subsequent spelling of *inca* Thomas.

Oxymycterus inca inca: Cabrera, 1961:467; name combination.

Oxymycterus inca iris: Cabrera, 1961:467; name combination.

DESCRIPTION: Moderate sized, with adult head and body length 135–184 mm, tail length 85–122 mm, hindfoot length 29–35 mm, ear length 19–23 mm. Overall dorsal color tones brown, notably grizzled with black on crown and along midline, becoming paler on sides. Ventral pelage predominantly dark orange, slate-colored base of hairs completely covered by orangish tips, with characteristically pale, banded hairs extending along undersurface of forelimbs, with clear strip of yellowish orange hairs extending to the wrists. Forefeet, hindfeet, and tail dark brown; tail unicolored and moderately hairy, with hairs spanning four tail scales in length near base and becoming longer toward tip. Skull with proportionally long upper toothrows (5.2–5.9 mm), wide rostrum (4.32–5.25 mm), and relatively short orofacial dimensions. Cranial morphology resembles that of Atlantic coast *O. dasytrichus*, from which it differs by the relatively narrower bullae (5.2–5.8 mm), braincase (13.3–14.9 mm), and interorbital constriction (5.9–6.9 mm).

DISTRIBUTION: This species occurs along affluents of the upper Río Ucayali in Junín department, Peru, and in the lowlands in Santa Cruz department, Bolivia, extending eastward into the western Amazon of Acre state, Brazil. Records from La Paz, Bolivia, are from lower montane affluents of the Río Beni system; the sole record from western Brazil is from lowland tropical rainforest. All records are from localities at 1,000 m or below.

SELECTED LOCALITIES (Map 123): BOLIVIA: La Paz, Guanay (AMNH 72747), San Ernesto, Mapiri (type locality of *Oxymycterus iris* Thomas); Santa Cruz, 6 km by road W of Ascención (AMNH 262258), Ayacucho (AMNH 263344), Buenavista, Ichilo (BM 26.12.4.62), Estancia Cachuela Esperanza (AMNH 260603), Santa Cruz de la Sierra, Andrés Ibáñez (MNRJ 22787). BRAZIL: Acre, Sena Madureira (USNM 546292). PERU: Junín, Perené (type locality of *Oxymycterus inca* Thomas).

SUBSPECIES: We recognize two subspecies, *O. i. inca* from Peru and Santa Cruz department, Bolivia, and *O. i. iris* from La Paz department Bolivia.

NATURAL HISTORY: Few data are available on habitat, behavior, food, or reproduction (S. Anderson 1997)—available information is restricted to collectors' notes recorded on specimen labels. Specimens from San

Ernesto (Mapiri, La Paz department, Bolivia) were captured in grasses during the daytime (P. O. Simmons, on specimen labels). Females captured in June 1964, near La Merced, Junín, Peru, were either lactating or had placental scars from four to six embryos; males collected in the same period had scrotal testes (label notes by T. B. Seifert, N. E. Coon, F. J. Myers, D. R. Seidel, and D. Castanon). This species was found only at sites with the highest diversity of rodents and marsupials, which was probably related to the presence of bamboo in forest habitats of the upper Río Urubamba valley in southeastern Peru (Solari et al. 2001). Jiménez-Ruiz and Gardner (2003) reported parasitic nematodes in Bolivian specimens.

REMARKS: Cabrera (1961) treated the nominal forms *inca*, *iris*, *juliacae*, and *doris* as subspecies of *O. inca*, an arrangement followed in most subsequent taxonomic lists. Morphometric analyses by J. A. Oliveira (1998) revealed that samples from Junín, Peru (including the holotype of *O. inca*), La Paz, Bolivia (including the type series of *O. iris*), and Santa Cruz, Bolivia, are different from other samples from Peru and Bolivia and that included the type specimens of *O. juliacae* and *O. doris*. Therefore, we treat *juliacae* and *doris* as belonging to a separate species, for which the name *O. juliacae* is the older (see *O. juliacae* account). J. A. Oliveira (1998) interpreted the morphometric variation among the samples from Junín (Peru), La Paz (Bolivia), and Santa Cruz (Bolivia) as suggestive of subspecific differentiation. Therefore, the samples from Junín and Santa Cruz are assigned to

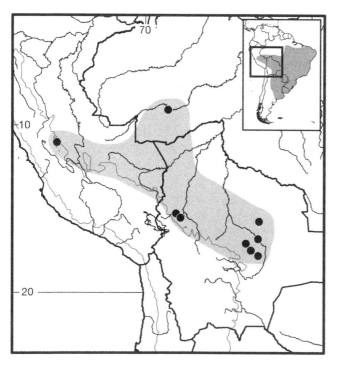

Map 123 Selected localities for *Oxymycterus inca* (●). Contour line = 2,000 m.

O. inca inca, and the sample from La Paz is recognized as *O. inca iris*. Morphological and genetic studies of additional samples of *O. inca* are pending in order to properly evaluate the taxonomic status of the subspecies we recognized.

Oxymycterus josei F. G. Hoffmann, Lessa, and Smith, 2002
Cook's Hocicudo

SYNONYM:

Oxymycterus josei F. G. Hoffmann, Lessa, and Smith, 2002:411; type locality "West margin of Arroyo Tarariras, Balneario Las Flores, Departamento de Maldonado, Uruguay."

DESCRIPTION: Medium sized with reddish to dark-brown pelage and occasional white spots on venter; dorsal pelage of most specimens long and resembles that of *O. rufus* from Córdoba province, Argentina, with light ochraceous buffy color strongly lined with black. Adult head and body length 122–172 mm, tail length 75–104 mm, hindfoot length 24–29 mm, ear length 15–18 mm, and mass 36–125 g. Skull with relatively short rostrum, inflated frontal sinuses, and downward dorsal profile without distal flaring of nasal tube, conditions also seen in *O. rufus* and *O. delator*. Zygomatic plates relatively wide, approaching conditions also in *O. rufus* and *O. delator* when compared with sympatric *O. nasutus* (García-Olaso 2008), but narrower than those of other large-sized species such as *O. dasytrichus*. Interorbital region with rounded margins, braincase walls parallel and not rounded, both supraoccipital and lambdoidal ridges well developed, and interparietal reduced. Ventrally, incisive foramina elongated and ovate, extending beyond anterior conule of M1s. Baculum resembles that of *O. rufus*, with longitudinally enlarged proximal end and reduced lateral condyles with small lateral spikes, characters not present in sympatric *O. nasutus* (J. A. Oliveira 1998).

DISTRIBUTION: This species is known only from southwestern Uruguay, south of the Río Negro, in the departments of Soriano, Colonia, San José, Canelones, and Maldonado (F. G. Hoffmann et al. 2002).

SELECTED LOCALITIES (Map 124): URUGUAY: Maldonado, Las Flores (MVZ 183266); Soriano, 3 km E of Cardona (AMNH 206205).

SUBSPECIES: *Oxymycterus josei* is monotypic.

NATURAL HISTORY: F. G. Hoffmann et al. (2002) reported sympatry between *O. josei* and *O. nasutus* at two localities in Maldonado, Uruguay. They stated that Barlow's (1969) data on the natural history of *O. nasutus* needed reevaluation because *O. josei* was not recognized at the time of his studies.

REMARKS: *Oxymycterus josei* was described as very similar in morphology to *O. nasutus*, differing by larger size, broader zygomatic plates, and a few dental traits

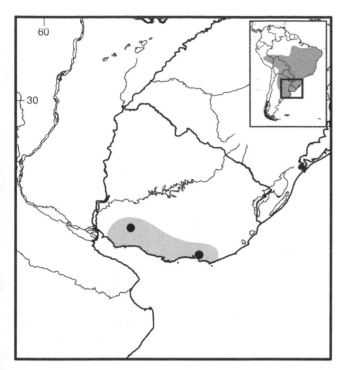

Map 124 Selected localities for *Oxymycterus josei* (●).

(F. G. Hoffmann et al. 2002). The discriminating value of the dental characters was later questioned by García-Olaso (2008) based on comparisons with sympatric *O. nasutus*. J. A. Oliveira (1998) remarked on the similarities of dorsal pelage color pattern and baculum between *O. josei* and *O. rufus*. Phylogenetic analyses of mtDNA cytochrome-*b* sequences placed *O. josei* as sister to *O. rufus*, but differing only by 1.7–1.9% corrected pairwise sequence divergence (F. G. Hoffmann et al. 2002; P. R. Gonçalves and Oliveira 2004; Jayat, D'Elía et al. 2008).

Oxymycterus juliacae J. A. Allen, 1900
Upper Yungas Inca Hocicudo

SYNONYMS:

Oxymycterus juliacae J. A. Allen, 1900a:223; type locality "Juliaca, [Puno,] Peru"; corrected to "Inca Mines [= Santo Domingo Mine], about 200 miles northeast of Juliaca, on the east side of the Andes, on the Inambary River [= Río Inambarí]," Puno, Peru, by J. A. Allen (1901b:41).

Oxymycterus doris Thomas, 1916j:478; type locality "Charuplaya, Upper Mamoré, 65°5′W., 16°S′, [Beni,] Bolivia, Alt. 1350m."

Oxymycterus inca juliacae: Cabrera, 1961:467; name combination.

Oxymycterus inca doris: Cabrera, 1961:467; name combination.

Oxymycterus juliacae doris: J. A. Oliveira, 1998:284; name combination.

DESCRIPTION: Moderate sized, head and body length 125–155 mm, tail length 95–140 mm, hindfoot length 27–36 mm, and ear length 19–24 mm. Dorsal color overall dark brown, slightly darker on posterior part of back and rump; venter drab brown. Dark hairs cover dorsal surfaces of forefeet and hindfeet. Short hairs, extending over 2–3 scales in length, uniformly cover the tail. Skull with proportionately long toothrows (5.0–5.8 mm), wide zygomatic breadth (15.1–17.8 mm), and wide interorbital constriction (6.3–6.9 mm), contrasting with all species except *O. dasytrichus*.

DISTRIBUTION: *Oxymycterus juliacae* occurs in upper montane and cloud forests in tributaries of the Río Inambari (Cusco and Puno departments, Peru) and Río Chaparé (Cochabamba department, Bolivia), at elevations between 1,000 and 2,700 m.

SELECTED LOCALITIES (Map 125): BOLIVIA: Cochabamba, Charuplaya, Río Securé (type locality of *Oxymycterus doris* Thomas), Incachaca (AMNH 38641), Yungas, N of Locotal (AMNH 38646). PERU: Cusco, Hacienda Cadena, Marcapata FMNH 65706); Puno, Inca Mines (type locality of *Oxymycterus juliacae* J. A. Allen).

SUBSPECIES: We treat *Oxymycterus juliacae* as monotypic.

NATURAL HISTORY: Based on data from specimen labels, this species is a terrestrial inhabitant of the upper montane and cloud forests.

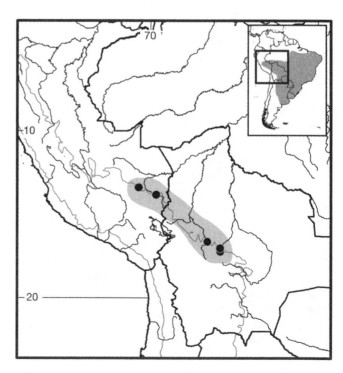

Map 125 Selected localities for *Oxymycterus juliacae* (●). Contour line = 2,000 m.

REMARKS: The type locality, originally given as Juliaca (15°30′S, 70°00′W), situated west of Lake Titicaca on the Andean Altiplano, was corrected to Inca Mines [= Santo Domingo Mine] on the eastern Andean slope (J. A. Allen 1901b). Following Cabrera (1961), most authors have listed *Oxymycterus juliacae* as either a valid subspecies or a synonym of *O. inca* Thomas. However, J. A. Oliveira (1998) documented the morphometric distinctness of *O. juliacae* relative to *O. inca* and the conspecificity of *doris* Thomas with the former.

Oxymycterus nasutus (Waterhouse, 1837)
Darwin's Hocicudo, Long-nosed Hocicudo

SYNONYMS:

Mus nasutus Waterhouse, 1837:16; type locality "Maldonado," Maldonado, Uruguay.

Mus (Oxymycterus) nasutus: Waterhouse, 1837:21; name combination.

O[xymycterus]. nasutus: Pictet, 1843a:211; first use of current name combination.

H[esperomys]. nasutus: Wagner, 1843a: 514; name combination.

Holochilus nasutus: Gray, 1843b:114; name combination.

M[us]. nasutus: Schinz, 1845:179; name combination.

Hesperomys (Oxymycterus) nasutus: Thomas, 1884:450; name combination.

Oxymycterus rufus nasutus: Cabrera, 1961:468; name combination.

DESCRIPTION: Adult specimens with head and body length 123–131 mm, tail length 87–97 mm, hindfoot length 27–28 mm, and ear length 17–18 mm. Dorsal pelage color quite variable; specimens from Uruguay vivid dark orange, heavily lined with black (particularly on head and middorsum), but paler on sides; those from Brazil much darker and duller, without black lined characteristic, or intermediate between two chromatic extremes. Venter in both extremes paler than dorsum, but of same overall hue. Forefeet and hindfeet well-haired, light to more strongly brown above. Tail slightly bicolored to unicolored; hairs about 2–3 scales in length. Skull with protruded, narrow, and trumpet-shaped rostrum; braincase elongated. Overall appearance very similar to that of *O. caparaoe* but with relatively longer incisive foramina, narrower interorbital region, and smaller bullae.

DISTRIBUTION: *Oxymycterus nasutus* ranges through Uruguay and the southern Brazilian states of Rio Grande do Sul, Santa Catarina, and Paraná. It occurs at sea level in the southernmost localities of Uruguay and Rio Grande do Sul, but occupies progressively higher elevations (up to 1,700 m) at more northern localities.

SELECTED LOCALITIES (Map 126): BRAZIL: Paraná, Castro (MZUSP 2497), Estação Ecológica Canguiri,

Piraquara (MNRJ 78463); Rio Grande do Sul, Cambará do Sul (MNRJ 46696), Eldorado do Sul, Charqueadas (MNRJ 46689), Estação Ecológica do Taim (MNRJ 46664), São Lourenço do Sul (BM 85.6.26.19). URUGUAY: Canelones, Arroyo Tropa Vieja (MNRJ 42435); Maldonado, 15 km N of San Carlos (FMNH 29249); Rocha, Cerro Largo, 6 km SE of Melo (AMNH 206172), 22 km SE of Lascano (AMNH 206189).

SUBSPECIES: *Oxymycterus nasutus* is monotypic.

NATURAL HISTORY: This species inhabits wet grasslands and steppes (*campos sulinos*) that dominate much of the plains of Uruguay and southern Brazil, and that are scattered throughout the southern Brazilian plateau. *Oxymycterus nasutus* can be the most abundant small mammal species in these habitats in the high-elevation grasslands in the states of Rio Grande do Sul and Santa Catarina. It occurs in sympatry with the larger *O. judex judex* in Brazil, with *O. josei* in Uruguay, and probably with *O. quaestor* in Paraná state in its northernmost distributional limits. Some ecological observations made by Barlow (1969) in his study of Uruguayan rodent populations are worth mentioning, although they should be interpreted with caution because he probably included *O. josei* in the animals he studied (F. G. Hoffmann et al. 2002). Barlow reported that *O. nasutus* was most frequently trapped in wet meadows with stands of bunch grass and in tall grass near streams and rivers, but never occurred in inundated marshes where more hydrophilic cricetids (e.g., *Holochilus* and *Scapteromys*) were present. These rats commonly used trails of *Cavia* and *Hydrochoerus* as pathways through the tall grass. Stomach contents revealed a predominantly arthropod-based diet, but also included small quantities of earthworms (Oligochaeta), slugs (Gastropoda), and plant material. Barlow's detailed description of the habitats and habits of Uruguayan rodents is one of the first to note that specimens of *Oxymycterus* have a distinct and strong scent when caught, similar to acrylic aldehyde, to which he attributed a putative warning function. The species exhibited diurnal activity pattern, with peaks during dusk and dawn hours, in southern Brazil (Paise and Vieira 2006). Barlow (1969) reported litter sizes ranging from one to six embryos.

REMARKS: The type specimen was acquired by Charles Darwin during his stay in Maldonado, Uruguay, and is the type species of the genus. Cabrera (1961) questioned the species status of *O. nasutus*, regarding it as a subspecies of *O. rufus*. Since then, J. A. Oliveira (1998), F. G. Hoffmann et al. (2002), and P. R. Gonçalves and Oliveira (2004) have presented morphological and/or molecular evidence for its status as a species. Congeneric species sympatric with *O. nasutus* differ consistently from it in size, external morphology, and craniodental traits (F. G. Hoffmann et al. 2002).

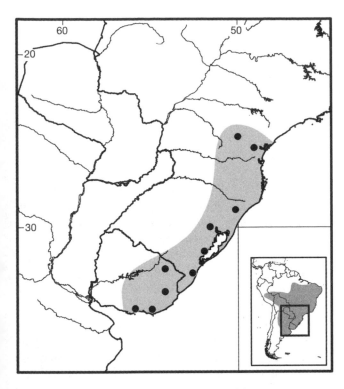

Map 126 Selected localities for *Oxymycterus nasutus* (●).

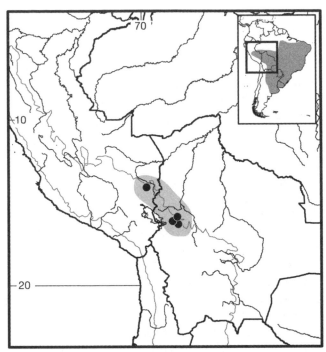

Map 127 Selected localities for *Oxymycterus nigrifrons* (●). Contour line = 2,000 m.

Oxymycterus nigrifrons Osgood, 1944
Elfin Forest Hocicudo

SYNONYMS:

Oxymycterus paramensis nigrifrons Osgood, 1944:197; type locality: "Limbani, Puno, Inambari drainage, Peru. Altitude about 9,000 feet."

Oxymycterus nigrifrons: J. A. Oliveira, 1998:254; first use of current name combination.

DESCRIPTION: Adults with head and body length 117–133 mm, tail length 79–96 mm, and hindfoot length 23–27 mm. Overall dorsal color light brownish, neither particularly darker in any region nor lined with black; ventral color paler yellowish wash, but with dark gray bases of hairs apparent. Hairs of forefeet and hindfeet almost totally nonpigmented, and those of tail short (2–3 scales in length), dark above and nonpigmented below, but still providing overall unicolored aspect and notably light brownish tone. Skull relatively long (occipitonasal length >31 mm), narrow, and with sharper angles; toothrow length (4.3–5.2 mm) comparatively short.

DISTRIBUTION: *Oxymycterus nigrifrons* is known from the type locality, in the elfin forests of the southern Cordillera Carabaya (Puno department, Peru) at 3,350 m, and from the Cordillera Real in La Paz department, Bolivia, at elevations from 2,000 to 2,700 m.

SELECTED LOCALITIES (Map 127): BOLIVIA: La Paz, Ñequejahuira (AMNH 72751), 18 km by road N of Zongo, Cuticucho (UMMZ 156084), 30 km by road N of Zongo, Cement Mine (UMMZ 155943). PERU: Puno, Limbani (type locality of *Oxymycterus paramensis nigrifrons* Osgood).

SUBSPECIES: *Oxymycterus nigrifrons* is monotypic.

NATURAL HISTORY: Habitat descriptions (field notes of G. G. H. Tate, cited in Boom 1981) of collecting localities along the Río Aceromarca, Bolivia, indicate that the species inhabits cold temperate forests, second-growth brush, and fields in valleys edged by abrupt cliffs with Puna vegetation on the tops.

REMARKS: *Oxymycterus nigrifrons* was described as a subspecies of *O. paramensis* on the basis of pelage characters, and was named after a supranarial blackish patch regarded as distinctive. J. A. Oliveira (1998), however, documented that the type series from Limbani (Puno department, Peru) was craniometrically different from *O. paramensis* from the Altiplano and was related to samples from the elfin forests of the upper Yungas of La Paz, Bolivia, here regarded as conspecific, as well as different from *O. hiska*, a smaller form from similar elevation and habitat.

Oxymycterus paramensis Thomas, 1902
Páramo Hocicudo, Huacucho

SYNONYMS:

Oxymycterus paramensis Thomas, 1902b:139; type locality "Choquecamate [= Choquecamata], [Cochabamba,] 4000 m," Bolivia.

Oxymycterus akodontius Thomas, 1921k:615; type locality "Higuerilla.—2000 m., in the Department of Valle Grande, about 10 km. east of the Zenta range and 20 km. of the town of Tilcara," Jujuy, Argentina. Pardiñas et al. (2007) equated Higuerilla (an abandoned ranch) to the modern village of Pampichuela (23°32′S, 65°02′W, 1,735 m).

Oxymycterus paramensis jacentior Thomas, 1925b:580; type locality "Carapari [= Caraparí], 1000 m., about 35 kilometres north of Yacuiba, on the way toward the town of Tarija, [Cochabamba,] Bolivia."

Oxymycterus paramensis paramensis: Gyldenstolpe, 1932a: 130; name combination.

Oxymycterus rutilans paramensis: Hershkovitz, 1966a: 127; name combination.

Oxymycterus inca doris: S. Anderson, 1997:438; name combination; part, not *doris* Thomas.

Oxymycterus akodontinus Mares, Ojeda, Braun, and Barquez 1997:124; incorrect subsequent spelling of *akodontius* Thomas (see Remarks).

DESCRIPTION: Adult individuals with head and body length 105–160 mm, tail length 75–124 mm, hindfoot length 25–32 mm, and ear length 15–23 mm. Dorsal color predominantly deep olive buff, strongly lined with black on anterior half of body, and brownish with less contrasting black lines over rump; ventral color pale buff to yellowish gray, with gray base of individual hairs visible; tail bicolored along length and well haired, with individual hairs extending over 4–5 scales; scales not visible. Dorsal surfaces of both hands and feet covered by ochraceous or buff colored hairs. Young specimens generally darker, with noticeably darker feet and tail, and relatively more limited contribution of yellowish distal parts of hairs to general ventral color. Skull slender, with proportionally large values for basisphenoid and premaxilla lengths than in other species; toothrow length longer and rostral tube wider, but both incisive foramina and palatal length shorter than in sympatric *O. nasutus*.

DISTRIBUTION: *Oxymycterus paramensis* occurs throughout the Andean Altiplano from Cusco department (Peru), south through La Paz, Cochabamba, and Tarija departments, Bolivia, and on the eastern Andean slopes in Chuquisaca and Tarija departments, Bolivia, south to Jujuy and Salta provinces, Argentina.

SELECTED LOCALITIES (Map 128): ARGENTINA: Jujuy, Caimancito (FMNH 41280), León (BM 18.1.1.33); Salta, Sierra de Aguaray, Tartagal (FMNH 35251). BOLIVIA: Chuquisaca, 2 km E of Chuhuayacu (AMNH 263889), Tola Orco, 40 km from Padilla, Tomina (USNM 276603); Cochabamba, Arani (FMNH 46151), Chapare (BM 34.9.2.166), Choquecamata (type locality of *Oxymycterus paramensis* Thomas), Pocona (BM 34.9.2.167),

17 km E of Totora, Tinkusiri (AMNH 264209), Vinto (AMNH 38625); La Paz, Caracato (AMNH 248999); Tarija, Caraparí (type locality of *Oxymycterus paramensis jacentior* Thomas), 3 km SE of Cuyambuyo (AMNH 264206), 4.5 km E of Iscayachi (AMNH 244203), Tapecua (AMNH 264201). PERU: Cuzco, Chospyoc, Urubamba (USNM 194699), Paso Ocobamba (BM 22.1.1.990), W of Pilcopata, Consuelo (FMNH 123960).

SUBSPECIES: We regard *Oxymycterus paramensis* as monotypic (but see Remarks).

NATURAL HISTORY: Brief microhabitat descriptions from localities distributed throughout the range are written on specimen labels, such as "rounded hills covered with grass" (specimens from Ocobamba Paso, Ollantaytambo, Peru); "in stone pile," "in grassy place," "in grass hill side," "in rock wall," or "rocky hillside at base of *Puya raimondii*" (specimens from Cochabamba, Bolivia, labels by P. O. Simmons and D. C. Schmitt); "tiene sus cuevas en tierra humeda, como *Ctenomys*, raro" (type series of *O. akodontius*, labels by E. Budin); or "cazado en un cerco de ramas," "habita las barrancas humadas entre los pinos," and "habita entre ramas de pinos" (holotype of *jacentior*, label also by E. Budin). Malygin and Rosmiaret (2005) included *O. paramensis* in a group of species inhabiting transitional biotopes, boundary areas between forest and open spaces, where individuals refuged in burrows, consumed seeds and grasses, and maintained low densities despite an average of five embryos per litter.

A single specimen obtained in Tucumán province, Argentina, came from the most humid part of a transition forest (Barquez et al. 1991) related to the Yungas, from the basal stratum to the highland pastures (P. E. Ortiz and Pardiñas 2001). R. A. Ojeda and Mares (1989) also collected specimens in lower montane and transitional humid forested habitats and on stream banks in Salta province, Argentina, where the species seems to be restricted to humid forests and secondary vegetation (Mares, Ojeda, and Kosco 1981; Mares, Ojeda, and Barquez 1989). Jiménez-Ruiz and Gardner (2003) reported parasitic nematodes in Bolivian specimens, and Pia et al. (2003) noted predation by the Culpeo (*Lycalopex culpaeus*).

REMARKS: Although considering *O. paramensis* to be distributed along a wide latitudinal range in the Andean Altiplano, Thomas (1925b) recognized *O. p. jacentior* as a subspecies of lower elevations in northern Argentina. Osgood (1944) described *O. paramensis nigrifrons* from Limbani, Puno department, Peru, on the basis of pelage characters. Cabrera (1961) also regarded *nigrifrons* as a subspecies of *O. paramensis* and extended the range of *O. p. jacentior* to extreme southern Bolivia and to the provinces of Salta and Jujuy in Argentina, an arrangement that was followed by S. Anderson (1997) for Bolivian samples. However, J. A. Oliveira (1998), using principal component and discriminant analyses of craniometric characters, showed that samples referred to O. p. paramensis and O. p. jacentior by Anderson (1997), as well as the type specimen of O. akodontius, were indistinguishable but were divergent from the type series and other samples identified as O. nigrifrons, which is here recognized as a distinct species. J. A. Oliveira (1998) also documented that young and subadult individuals of O. paramensis have a darker pelage, similar to the type series of O. akodontius. Furthermore, Jayat, D'Elía et al. (2008) found that mtDNA cytochrome-b haplotypes referred to O. paramensis paramensis and O. paramensis jacentior are not reciprocally monophyletic. Thus, O. paramensis is here regarded as monotypic, pending further analyses including topotypical specimens, a survey of diagnostic characters, and a more precise definition of geographic limits among putative species or subspecies in this group.

The supposed record of O. "akodontinus" from Andalgalá in Catamarca province, Argentina, cited by Mares, Ojeda, Braun et al. (1997), is based on a specimen (MACN 50.434) acquired by J. A. Crespo in 1950. Pardiñas, D'Elía et al. (2006) reidentified this specimen as Abrothrix illutea.

Kajon et al. (1984) reported a karyotype with $2n = 54$, FN = 64 for specimens they identified as O. akodontius and from San León de Jujuy in northwestern Argentina.

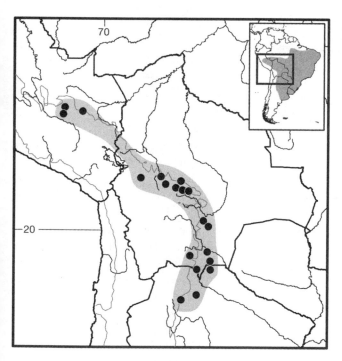

Map 128 Selected localities for Oxymycterus paramensis (●). Contour line = 2,000 m.

Oxymycterus quaestor Thomas, 1903
Quaestor Hocicudo, Rato-Quati, Rato-Mineiro
SYNONYMS:

Hesperomys rufus: Burmeister, 1854:183; part; not Mus rufus G. Fischer.

Oxymycterus rufus: Pelzeln, 1883:75; part; not Mus rufus G. Fischer, 1814.

Oxymycterus quaestor Thomas, 1903a:226; type locality "Roça Nova, Paraná, altitude 1000m," Brazil.

Oxymycterus judex Thomas, 1909:238; type locality "Joinville, Santa Catarina," Brazil.

Oxymycterus misionalis Sanborn, 1931b:1; type locality "Rio Paranay [= Río Paranay Guazú], an affluent of Rio Parana, near Caraguatay, about 100 miles south of Rio Iguassu, Misiones Territory, Argentina." Pardiñas et al. (2007:397) restricted the type locality to the "confluence of the río Paranay-Guazú with the río Paraná (26°41′S, 54°49′W, 122m), Libertador General San Martín, Misiones."

Oxymycterus questor Gyldenstolpe, 1932a:131; incorrect subsequent spelling of quaestor Thomas.

Oxymycterus hispidus judex: Cabrera, 1961:466; name combination.

Oxymycterus hispidus quaestor: Cabrera, 1961 467; name combination.

Oxymycterus hispidus misionalis: Cabrera, 1961:467; new name combination.

Oxymycterus rutilans: Roguin, 1986:1018; not Mus rutilans Olfers [= Oxymycterus rufus].

DESCRIPTION: Large, with adult head and body length 135–180mm, tail length 97–143mm, hindfoot length 34–40mm, and ear length 21–26mm. Overall dorsal coloration dark orange brown strongly lined with black, sometimes more intensely black on crown and middorsum, to reddish brown without black. Ventral pelage dark yellowish gray to dark orange. Skull elongated, with largest dimensions of occipitonasal, mandibular, incisive foramina, and nasal lengths in genus. Zygomatic plates, palate, and rostrum proportionally narrow; maxillary in area of toothrow uninflated. Proportionally smaller interparietal and narrower interorbital region distinguish this species from O. dasytrichus, with which it may overlap broadly in distribution.

DISTRIBUTION: Oxymycterus quaestor is known from eastern Paraguay and northeastern Argentina east through northern Rio Grande do Sul to southern São Paulo states, Brazil; an isolated population is present in the Serra dos Órgãos in Rio de Janeiro state.

SELECTED LOCALITIES (Map 129): ARGENTINA: Misiones, Dos de Mayo, Guarany (FMNH 122696), Puerto Gisela, San Ignacio (BM 24.6.6.50). BRAZIL: Paraná, Roça Nova (type locality of Oxymycterus quaestor Thomas), Rio Paranapanema, Fazenda Caioá, Cambará

(MZUSP 1275), Usina Hidroeléctrica de Guaricana, Morretes (MNRJ 79073); Rio Grande do Sul, Usina Hidroeléctrica de Itá, right margin Rio Uruguay (MNRJ 62143), Victor Graeff (MNRJ 79072); Rio de Janeiro, Teresópolis (FMNH 26585), Serra de Macaé (MZUSP 2770); Santa Catarina, Florianópolis (BM 14.1.26.33), Joinville (type locality of *Oxymycterus judex* Thomas), Pinheiros, Anitápolis (UFSC 473); São Paulo, Fazenda Intervales (Base do Carmo), Capão Bonito (MZUSP 29011), Posto Indígena Icatu, Braúna (MZUSP 3740), Teodoro Sampaio (USNM 309164). PARAGUAY: Alto Paraná, Puerto Bertoni (BM 21.4.21.4).

SUBSPECIES: We treat *Oxymycterus quaestor* as monotypic.

NATURAL HISTORY: Graipel et al. (2006) recorded individuals on the ground in humid and disturbed vegetations, and at low densities relative to other species of rodents and marsupials on Ilha Santa Catarina, Brazil. At this locality, reproduction was apparently seasonally bimodal, with young specimens found in both late spring-summer (November and February) and winter months (April to July). Reproductive females (as judged by perforated vaginae) were captured in July and January, and lactating females in October. All adult males exhibited scrotal testes. Individuals are apparently short lived, as the average residence period was about 76 days; a higher percentage of females was considered resident, with one female exhibiting the maximum persistence in the area of 180 days.

REMARKS: Thomas (1909) described *O. judex* to distinguish larger specimens with a more robust braincase from *O. quaestor*, which he had described six years earlier (1903). Sanborn (1931) described *O. misionalis*, distinguishing it from *O. judex* by its larger size. Subsequently, Cabrera (1961) treated *quaestor* Thomas as a subspecies of *O. hispidus*, together with *judex* Thomas and *misionalis* Sanborn, in an arrangement that has been followed by most subsequent authors. As understood here, following J. A. Oliveira (1998) and P. R. Gonçalves and Oliveira (2004), *O. hispidus* is a junior synonym of *O. dasytrichus*, a widely distributed species from eastern Brazil (see that account) and distinct from other large forms from southern Brazil and the Misiones region of Argentina. J. A. Oliveira's (1998) morphometric analyses showed that the holotype, paratype, and one topotype of *O. quaestor* were included in a morphological continuum together with the types and samples assigned to *O. judex* and *O. misionalis*. Musser and Carleton (2005) treated *O. judex* and *O. misionalis* as junior synonyms of *O. quaestor*, based on the molecular results of F. G. Hoffmann et al. (2002).

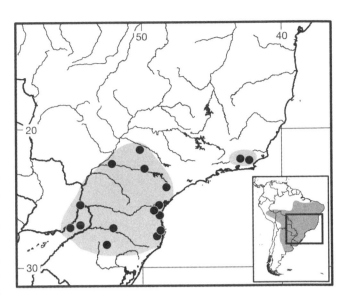

Map 129   Selected localities for *Oxymycterus quaestor* (●). The disjunct populations in the Serra dos Órgãos, Rio de Janeiro state, Brazil, are mapped separately.

Oxymycterus rufus (G. Fischer, 1814)

Red Hocicudo, Rat Roux, Ratón Hocicudo de Azara, Hocicudo Común, Hocicudo Rojuzochico, Yaguarasapá, Anguya Pihtâ

SYNONYMS:

Mus rufus G. Fischer, 1814:71; based on Azara's (1801:94) "*rat cinquième ou rat roux*"; type locality "Paraguay"; restricted by Rengger (1830:231) to "von Asuncion." Hershkovitz (1994:36) wrote that the type locality "was said to have been found at 32°30′S in the Río Paraná drainage system; this restricts the type locality to northern Entre Rios, Argentina." Galliari et al. (1996) criticized Hershkovitz's (1994) restriction of the type locality and argued that the type locality of Azara's animal should be restricted to San Ignacio Guazú, Paraguay, as this was the place of residence of the priest Noséda, who obtained the specimen Azara (1801) described. J. R. Contreras and Teta (2003) noted, however, that Noséda had actually resided at three, closely adjacent places in the Paraguayan department of Misiones (Santa María de Fe, Santiago, and San Ignacio Guazú), all approximately 250 km SE of Asunción; Azara's "*rat roux*" could thus have come from any of these three localities. Here, we select a neotype for Fischer's *Mus rufus* and, by that action, restrict the type locality of *O. rufus* (G. Fischer, 1814) to "Estancia San Juan Poriahú, Depto. San Miguel, Provincia Corrientes, Argentina (27.71667°S, 57.19389°W" (see Remarks).

Mus ?rutilans Illiger, 1815:108; *nomen nudum.*

M[us]. rutilans Olfers, 1818:209; based on Azara's (1801) "*rat cinquième ou rat roux.*"

H[esperomys]. rufus: Wagner, 1843a:595; name combination.

O[xymycterus]. rufus: Burmeister, 1855a:10; part; first use of current name combination.

[Hesperomys (Oxymycterus)] rufus: E.-L. Trouessart, 1880b:142; name combination.

Oxymycterus platensis Thomas, 1914d:244; type locality "Ensenada, La Plata. Sea Level," Buenos Aires, Argentina.

Oxymycterus nasutus: Bertoni, 1914b:10; part; not *Mus nasutus* Waterhouse.

Oxymycteris rutilans: Hershkovitz, 1959b:339; name combination.

Mus nasutus: Hershkovitz, 1959b:339; part; not *Mus nasutus* Waterhouse.

Oxymycterus rufus platensis: Cabrera, 1961:469; name combination.

Oxymycterus rufus rufus: Cabrera, 1961:469; name combination.

Oxymycterus rutilans: Echave Llanos and Vilchez, 1964:187; name combination.

Oxymycterus rutilans rutilans: Hershkovitz, 1966a:127; name combination.

Oxymycterus rutilans platensis: Massoia and Fornes, 1969:316; name combination.

Oxymycteris [sic] *rutilans*: A. L. Gardner and Patton 1976: 28, Table 2; name combination.

DESCRIPTION: Moderate sized, with adult head and body length 129–161 mm, tail length 69–118 mm, hindfoot length 27–32 mm, ear length 14–20 mm, and mass 61–130 g. Dorsal coloration varies as function of prevalence of one of two types of guard hairs (banded and unbanded), but also varying intensity and length of ochraceous subapical band of banded hair. General dorsal color thus cinnamon brown, strongly lined with black over head and middorsum and ochraceous tawny on rump and sides, to paler dorsum in which black component is less conspicuous; some individuals strongly reddish, notably those from Buenos Aires coast and region of La Plata Delta, but also populations from Minas Gerais state, Brazil, where entire pelage, especially on ventral surface, ochraceous orange; specimens from Córdoba province, Argentina, among palest, with general color of entire dorsum pale ochraceous buff, but strongly lined with black, resembling some specimens of *O. josei*. Skull medium-sized, with relatively wide braincase and moderately protruded rostrum; nasal tube not markedly flared, with weak dorsal inflexion when viewed laterally, and dorsal outline of nasals not forming parallel plane in relation to diastemal region. Palatine septum delicate, with maxillary portion particularly thin, providing keeled appearance in ventral view. Interparietal triangular and greatly reduced in relation to other species. Lacrimal laterally reduced and rectangular, generally conforming to antorbital wall. Posterior extension of bony palate variable, short (reaching posterior border of M3s) in Minas Gerais specimens or long (extending beyond M3s) in most Argentinean specimens.

DISTRIBUTION: *Oxymycterus rufus* occurs in Argentina, from southern Buenos Aires province, at least from Bahía Blanca and the Atlantic coast to Ensenada, La Plata, Buenos Aires province, and to the lower Río Paraná, including the Delta region and the provinces of Corrientes (Goya) and Entre Rios (Gualegauychu). It has been recorded from as far west as Villa Dolores, Córdoba province. P. R. Gonçalves and Oliveira (2004) recorded isolated populations in Minas Gerais state, Brazil.

SELECTED LOCALITIES (Map 130): ARGENTINA: Buenos Aires, Abra del Hinojo, Sierras de Curamalal (TTU 64489), Arroyo Brusquitas, Miramar (UMMZ 115501), Bahía Blanca (MHNG 611.52), Ensenada, La Plata (type locality of *Oxymycterus platensis* Thomas), Mar del Tuyú, General Lavalle (FMNH 122697), Tandil (TTU 33153), Valeria del Mar (MVZ 134244); Córdoba, Candonga (ZSM 1955/236), Espinillo, Río Cuarto (TTU 66588), Noetinger (BM 17.1.25.36), Villa Dolores (BM 16.1.6.32); Corrientes, Goya (BM 98.12.3.23), San Miguel, San Juan Poriahú (MLP 26.XII.01.5); Entre Rios, Gualeguaychu (TTU 33155), Yuqueri (MZUSP 573). BRAZIL: Minas Gerais, Ouro Preto (MNRJ 14568), Viçosa (MNRJ 65522).

SUBSPECIES: We treat *Oxymycterus rufus* as monotypic.

NATURAL HISTORY: In Argentina, *O. rufus* avoids woodlands, preferentially inhabiting grasslands and steppes with dense herbaceous cover, such as the *pastizales altos de los médanos* (tall dune grasses), along margins of small creeks and lagoons, and the rocky outcrops in the sierras (Bonaventura et al. 1991). The same habitat preference seems to hold for Brazilian representatives of the species, which have also been frequently trapped in exotic and native grasslands near water bodies (Paglia et al. 1995; G. Lessa et al. 1999). Temporal activity patterns, however, seem to be more variable, with both diurnal and nocturnal activity reported for Argentinean populations (Massoia 1961; Kravetz 1972). As are other species of the genus, *O. rufus* feeds primarily on small soil invertebrates, with 70–85% of its diet composed of arthropods, annelids, and nematodes, with plant mass accounting for only 15–20% (Kravetz 1972; O. V. Suárez 1994). Reproductive intensity is higher during the spring and summer months (November to March), with a mean litter size of four embryos (range: 2–5) (Reig 1965; V. R. Cueto et al. 1995c). Mark-recapture data suggest a polygynous mating system because males have larger home ranges than females and stronger territoriality during the mating season (V. R. Cueto et al. 1995b). Both longevity and local residence of individuals are relatively long, probably spanning two breeding seasons, as suggested by the persistence of marked individuals for more than 24 months (V. R. Cueto et al. 1995b). Individuals exhibit considerable jumping capability, readily escaping from 40 liter buckets used in pitfall traps (P. R. Gonçalves, pers. obs.)

REMARKS: Musser et al. (1998) discussed the validity of the binomials proposed by Gotthelf Fischer (1814) for the sigmodont rodents described by Azara (1801). Their conclusions with respect to the availability of most of these names, which support Langguth's (1966b) original petition to the International Commission, are followed here with respect to the priority of *Mus rufus* (G. Fischer) over the names provided by Illiger (1815), Olfers (1818), and Desmarest (1819c) for Azara's (1801) "*rat cinquième ou rat roux.*" One of the two primary arguments for disregarding Fischer's names, the inconsistent usage of binomial nomenclature (Sabrosky 1967), does not apply to the combination *Mus rufus* as used both in the index and in the main text of Fischer's work.

Brants (1827), Rengger (1830), Burmeister (1854), and Winge (1887) each concluded that specimens from eastern Brazil represented Azara's "*rat cinquième ou rat roux.*" Importantly, Burmeister (1854) explicitly regarded *Hypudaeus dasytrichos* Wied-Neuwied and *Hesperomys* (*Oxymycterus*) *rostellatus* Wagner as synonyms of *O. rufus*. Cabrera (1961) adhered to this view in his treatment of subspecies and most subsequent authors followed his position. However, recent morphological and molecular analyses (P. R. Gonçalves and Oliveira 2004) have supported the validity of *O. rufus* and *O. dasytrichus* as distinct species, as they are treated herein. Furthermore, these analyses did not support recognition of subspecies of *O. rufus*. Hence, as did Gonçalves and Oliveira, we use the name *O. rufus* to refer to the species that is widespread in Buenos Aires, Córdoba, Corrientes, and Entre Rios provinces in Argentina and known from a few localities in Minas Gerais state, Brazil.

As yet, no museum specimens are known from Paraguay, the supposed origin of the specimen Azara (1801) referred to as the "*rat cinquième ou rat roux.*" A series of *O. rufus* from Estancia San Juan Poriahú, Corrientes province, Argentina, housed in the Museo de La Plata (MLP) is the geographically nearest sample to the region of probable provenance of Azara's animal. This Argentinean locality is located some 75 km SW, 92 km SW, and 105 km SW from Santiago, San Ignacio Guazú, and Santa María de Fe, respectively, the three localities in Misiones department, Paraguay, suggested by J. R. Contreras and Teta (2003) as putative potential provenances for Azara's specimen. Therefore, we select as a neotype for *O. rufus* (G. Fischer 1814) the specimen MLP 26.XII.01.05 (original number 191), an adult male collected by Mariano L. Merino and Augustín

Abba on September 23, 1999. This specimen is represented by a skull, mandible, postcranial skeleton, and stuffed skin missing the left forefeet and hindfeet, which were cleaned and preserved as part of the postcranial skeleton. The skull is complete and still preserves the prenasal ossification. External measurements are total length = 252 mm, tail length = 105 mm, hindfoot length (with claw) = 30.5 mm (27 mm without claw), ear length = 17 mm, and mass = 110 g; cranial measurements (as defined in J. A. Oliveira 1998) include occipitonasal length = 38.5 mm, incisive foramina length = 8.4 mm, upper toothrow length = 5.01 mm, zygomatic plate width = 2.7 mm, zygomatic breadth = 16.4 mm, nasal length = 15.1 mm, least interorbital breadth = 6.0 mm, braincase breadth = 14.3 mm, mandible length = 19.6 mm. The skin of the neotype has a dark cinnamon brown dorsal color, strongly lined with black toward the midline and less in the laterals, without strong reddish tones. The ventral pelage is generally ochraceous with the gray bases of hairs showing through, except on the chin, where a small patch of white hairs is present. The tail is moderately haired and mostly dark brown, except on its ventral surface around the tail base, where the hairs are less pigmented. Claws are well developed and keeled, measuring 5.1 mm in length on the longest digit of the manus and 5.8 mm on the longest digit of the pes. The skull has the typical morphology of *O. rufus*, with the nasal tube moderately protruded but

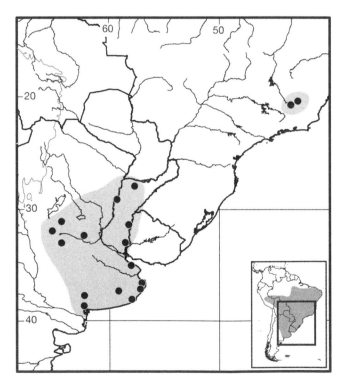

Map 130 Selected localities for *Oxymycterus rufus* (●). The disjunct populations in Minas Gerais state, Brazil, are mapped separately. Contour line = 2,000 m.

not markedly flared in lateral view, wide zygomatic plates, palatine septum of incisive foramina with a keeled maxillary part, long bony palate, reduced interparietal, and alisphenoid strut absent on both sides of cranium. Molars are worn, with enamel folds and islets obliterated, as typical of adult individuals of *Oxymycterus* spp.

Five additional specimens collected at the same locality and on the same date are preserved (MLP 26.XII.01.6, 26.XII.01.7, 26.XII.01.8, 26.XII.01.9, 26.XII.01.10). P. R. Gonçalves and Oliveira (2004) included this series and the neotype in a craniometric analysis of *O. rufus* and showed that the samples from Argentina form a morphological cline spanning samples from Corrientes province in the northwest to Buenos Aires in the southeast.

Oxymycterus wayku Jayat, D'Elía, Pardiñas, Miotti, and Ortiz, 2008
Ravine Hocicudo

SYNONYM:

Oxymycterus wayku Jayat, D'Elía, Pardiñas, Miotti, and Ortiz, 2008:40; type locality "Argentina, Province of Tucumán, Department of Trancas, 10 km by road south of Hualinchay on the trail to Lara (26 19'20.2"S, 65 36'45.5"W, 2316 m)."

DESCRIPTION: Intermediate in size relative to other hocicudos, with adult head and body length 124–144 mm, tail length 77–97 mm, hindfoot length 28–32 mm (with claw), ear length 17–21 mm, and mass 44.5–76.5 g. Pelage blackish brown with slight ochre or reddish hue (juvenile specimens darker, almost black), with dorsal hairs dark gray for most of length but tips ochraceous; guard hairs completely black and extend well beyond underfur, up to 3.5 mm on rump. Coloration of flanks similar to that of dorsum. Venter paler, with ochre color intermingled with ash-gray hair bases; chin with conspicuous white spot that, in some specimens, extends onto throat. Tuft of black hairs with ochre tips immediately anterior to ear extends posteriorly to the middle of the pinnae. Forefeet and hindfeet dark gray, almost black; claws long and robust, with forefoot and hindfoot digit III 6.4 and 5.1 mm long, respectively, covered by white ungual tufts that contrast sharply with darkness of the body, thighs, and feet. Tail almost completely black, only slightly paler ventrally, and densely covered by hairs. Skull long and robust (mean condyloincisive length of adults 30.37 mm), with rounded braincase and expanded zygomatic arches. Rostrum wide (mean rostral width 5.93 mm) and robust, with frontal sinuses inflated; zygomatic plates narrow and delicate; incisive foramina markedly short and oval in shape, their posterior margins not reaching level of the protoflexus of M1. Mandible delicate, low, and elongated, with lunar notch well developed but asymmetrically excavated.

DISTRIBUTION: *Oxymycterus wayku* is distributed along the eastern humid slopes of the Sierra del Aconquija and Cumbres Calchaquíes ranges, between 800 and 2,400 m elevation, in Tucumán province, Argentina. The suitable habitat occurs along a narrow strip about 200 km long, from the northeastern end of the Cumbres Calchaquíes in southern Salta province to the Sierra de Narváez on the boundary between Tucumán and Catamarca provinces.

SELECTED LOCALITIES (Map 131; from Jayat, D'Elía et al. 2008, except as noted): ARGENTINA: Tucumán, 10 km by road S of Hualinchay (type locality of *Oxymycterus wayku* Jayat, D'Elía, Pardiñas, Miotti, and Ortiz), La Angostura, Los Sosa, Reserva Provincial La Florida.

SUBSPECIES: *Oxymycterus wayku* is monotypic.

NATURAL HISTORY: *Oxymycterus wayku* appears to be rare across its known, limited range. Only five of the more than 400 total animals collected in the vicinity of the type locality belong to this species, which is otherwise known from only two additional complete specimens and the cranial remains of a few individuals recovered from Barn Owl (*Tyto alba*) pellets. Most specimens were trapped in the ecotone between montane forest and highland grasslands, at the upper elevational limit of Yungas forests (*Bosque Montano, sensu* A. D. Brown et al. 2001). This vegetation belt develops on wet slopes between 1,500 and 3,000 m and represents the landscape with maximum heterogeneity in the Yungas. The holotype was caught in alder forest (*Alnus acuminata* [Betulaceae]) with an

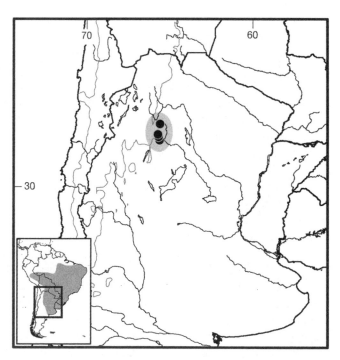

Map 131 Selected localities for *Oxymycterus wayku* (●). Contour line = 2,000 m.

understory of grasses dominated by *Festuca, Deyeuxia,* and *Stipa,* with deep soils, a well-developed organic horizon, abundant leaf detritus and fallen log cover, and on steep slopes interrupted by rocky outcroppings. Two additional specimens were collected at a lower elevation in montane rainforest (*Selva Montaña*), a vegetation belt between 700 and 1,500 m and dominated by evergreen trees. One female collected in the fall (late April to early May) exhibited signs of reproductive activity; males collected in July or early spring were reproductively inactive.

REMARKS: Molecular phylogenetic analysis did not recover a consistent sister relationship between *O. wayku* and other species, but placed it in a large clade that included *O. nasutus, O. paramensis, O. hiska, O. josei, O. judex, O. misionalis, O quaestor,* and *O. rufus* (see Jayat, D'Elía et al. 2008).

Genus *Podoxymys* Anthony, 1929
Ulyses F. J. Pardiñas and Pablo Teta

This genus is perhaps the least known among living akodontines. Reported specimens total six, five from the original series used by H. E. Anthony (1929) in his description of *Podoxymys roraimae,* and an additional young female obtained by an expedition organized by the La Salle National History Museum of Caracas, Venezuela, in 1989. All specimens came from the summit of the Roraima Tepui, the place that inspired Sir Arthur Conan Doyle's novel *The Lost World,* on the borders between Guyana, Venezuela, and Brazil.

Podoxymys was described as "a rodent which appears somewhat intermediate in character between *Akodon* and *Oxymycterus*" and with the "external appearance that of a dark-colored, long-tailed *Akodon,* with long slender claws (which suggest the generic name); skull with long, slender rostrum, narrow zygomatic plate and general appearance of *Oxymycterus*" (Anthony 1929:4). *Podoxymys* is small (total length = 196 mm; tail length = 95 mm; measurements from the holotype). Its pelage is long and lax (hairs on the back 10–11 mm in length), dorsally blackish slate at base and for most of the length of the hair, with only the tip being colored; sides of head and under parts slightly lighter in tone than back. Dorsal surfaces of the manus and pes are clove-brown. The tail is about half of the total length, very sparsely haired, and unicolored brown. Ears are partially hidden in the long pelage. Eyes are rather small. Claws of the forefeet are long (that on digit III extending 3 mm beyond the fleshy tip of the digit), slender, strongly compressed laterally, and moderately curved; those of hindfeet are slightly shorter. Palmar and plantar morphology have not been described, but based on the drawings in Hinojosa et al. (1987:Fig. 6B), the pollex has a developed claw; dig-

its I and V of the hindfoot are reduced, the former reaching the second phalanx of digit II in a pattern roughly like that of *Akodon mimus.*

The skull is long and slender (greatest length of skull = 27.5 mm), with the rostrum elongated and the zygomatic notches inconspicuous. The nasals are long and project forward, totally covering the incisors in dorsal view and combine with the premaxillaries to for a moderately developed trumpet. A noticeable gnathic process is present. The interorbital region is broad, greater than the rostral width, with rounded edges but without a definite constriction point. Zygomatic arches scarcely flare laterad beyond the level reached by the braincase, and are weak and thread-like at midpoint. The braincase is moderately inflated and evenly rounded, with a U-shaped coronal suture and a reduced interparietal. The zygomatic plates are narrow with upward and backwardly sloping anterior edges. The incisive foramina are long, widely open posteriorly and parallel sided, and reach posteriorly to the protocones of M1s. The palate is short, terminating in a broad mesopterygoid fossa. Auditory bullae are moderate sized and only slightly inflated. The basicranial region is proportionally long. Anthony (1929) did not describe the dentition of *Podoxymys,* recoding only the length of the maxillary toothrow (4.4 mm for the holotype) and providing a general indication that its molars are somewhat weaker than in *Akodon aerosus.*

Pérez-Zapata et al. (1992), on the basis of a newly collected juvenile specimen, reported additional morphological traits and the karyotype. They stated that *Podoxymys* has a tail as long as or only slightly longer than the head and body, claws of hindfeet as long as in *Akodon,* and ears covered by dense pelage. Cranially, the nasals reach posteriorly well behind the maxillofrontal suture (see also Hinojosa et al. 1987:Fig. 7B). The mandible is slender and elongated, with a short coronoid process and a small capsular projection. The upper incisors are opisthodont and the upper molars brachydont and bunodont, and with the main cusps in opposite pairs. The procingulum of M1 is slightly reduced and has a well-developed anteromedian flexus; the mesoloph is well developed in M1 and M2, reaching the labial margin in both. In the lower dentition, there is a reduced procingulum on m1, with a distinct anteromedian flexid (overlooked by Pérez-Zapata et al. 1992:218, but clearly visible in their figure); mesolophids are present on both m1 and m2; and posterolophids are well developed and directed backward.

The stomach of *Podoxymys* is unilocular-hemiglandular, with a reduced area of glandular epithelium and with some cornified epithelium occupying part of the antrum to the right of the esophageal opening (Carleton 1973:Fig. 5A). The caecum is well developed but very short, only 4.4% of the total intestine length; the large intestine is also short

(Pérez-Zapata et al. 1992). A gallbladder is present (Voss 1991a).

SYNONYMS:

Podoxymys Anthony, 1929:4; type species *Podoxymys roraimae* Anthony, by original designation.

Podoxomys Carleton, 1973:15; incorrect subsequent spelling of *Podoxymys* Anthony.

Podoximys O. J. Linares, 1998:273, incorrect subsequent spelling of *Podoxymys* Anthony.

Microxus: O. J. Linares, 1998:274, part (listing of *roraimae*); not *Microxus* Thomas.

Podoxymys roraimae Anthony, 1929
Roraima Akodont

SYNONYMS:

Podoxymys roraimae Anthony, 1929:4; type locality "Mt. Roraima, British Guiana; altitude 8600 feet." Interpreted as "Guyana, Mazaruni-Potaro Dist., summit of Mt Roraima, 8600 ft (2621 m)" by Musser and Carleton (2005:1164), but restricted by O. J. Linares (1998:274) to the southwestern corner of Tepui Roraima (ca. 5.144°N,60.762°S), between 2,580 and 2,700 m, Bolívar, Venezuela.

Podoxomys roraimae: Carleton, 1973:15; name combination.

Podoxymys roraimi Hinojosa, Anderson, and Patton, 1987:12; incorrect subsequent spelling of *roraimae* Anthony.

Podoximys roraimae: O. J. Linares, 1998:273; name combination.

Microxus roraimae: O. J. Linares, 1998:274; name combination.

DESCRIPTION: As for the genus.

DISTRIBUTION: *Podoxymys roraimae* is known only from the vicinity of its type locality.

SELECTED LOCALITIES (Map 132): VENEZUELA: Bolívar, summit of Roraima Tepuí (type locality of *Podoxymys roraimae* Anthony).

NATURAL HISTORY: All known specimens come from an area dominated by small trees of the genus *Bonnetia* [Clusiaceae], now highly altered due to wood extraction and tourism (O. J. Linares 1998). Pérez-Zapata et al. (1992) reported the capture of one individual, together with four specimens of *Rhipidomys macconnelli*, in a carpet of *Sphagum* moss in which there were various small cavities.

REMARKS: The putative presence of *Podoxymys roraimae* in Brazil (see Bonvicino, Oliveira, and D'Andrea 2008) or in adjoining tepuis of Venezuela and Guiana (see O. J. Linares 1998) needs confirmation by vouchered specimens.

The karyotype consists of a $2n = 16$, $FN = 26$, with five pairs of large biarmed and three pairs of telocentric chromosomes; each chromosome pair is very distinctive in size and morphology (Pérez-Zapata et al. 1992).

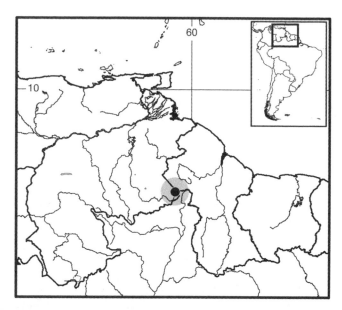

Map 132 Single known locality for *Podoxymys roraimae* (●). Contour line = 2,000 m.

Genus *Scapteromys* Waterhouse, 1837
Guillermo D'Elía and Ulyses F. J. Pardiñas

Swamp rats of the genus *Scapteromys* are among the largest members of the tribe Akodontini, and are easily distinguished in the field. They occur through east-central and northern Argentina, the southern coast of Brazil, southern Paraguay, and Uruguay, living near watercourses, including large rivers, small creeks, ponds, and swamps (Hershkovitz 1966a; D'Elía and Pardiñas 2004). Charles Darwin, the collector of the holotype of *Scapteromys tumidus*, wrote, "This rat was caught in so wet a place amongst the flags bordering a lake, that it must certainly be partly aquatic in its habits" (Waterhouse 1839:58). *Scapteromys* is an excellent swimmer, propelling itself by horizontal undulations of the tail and by means of all four limbs (Massoia and Fornes 1964c). *Scapteromys* is also able to climb trees, a behavior considered an adaptation for living in seasonally inundated areas (Barlow 1969; Sierra de Soriano 1969). *Scapteromys* is mainly nocturnal and feeds primarily on insects and oligochaetes, but also on leeches and vegetation (Barlow 1969; Massoia 1961).

Massoia and Fornes (1964c), Hershkovitz (1966a), Sierra de Soriano (1969), D'Elía and Pardiñas (2004), and Pardiñas, D'Elía, and Teta (2009) have described external and craniodental morphology. These are large rats (head and body length about 180 mm, tail length about 85 mm; mean values of adult males) with their external form adapted for palustrine and subaquatic life. The pelage is thick, soft, and glossy; the upper parts of head and body are somewhat blackish; and the underside of the head, chest, and belly are more or less gray washed with buffy to ochraceous. The tail is well developed but not thick, covered by stiff but not hispid dark brown pelage; each tail scale has three hairs, with ventral keel hairs usually long; a terminal pencil is present. The ears are small, rounded, and covered by hairs on both surfaces. The forefeet are well developed but not markedly enlarged, digits II–V are strong, and claws are slender and of equal length. Hindfeet are strong and elongate, their length (including claws) approximately 24% of the head and body length; the claws are slender. There are narrow bands of webbing between the first phalanges of the three middle digits and along the inner side of each outer digit.

The skull of *Scapteromys* is mostly without crests or ridges. The rostrum is slender and the nasals are rounded distally, taper proximally, and extend well behind the maxillofrontal sutures. The zygomatic plates are strong, with their posterior edges reaching to the level of the alveoli and strongly inclined laterally. Lacrimal bones are conspicuous. The zygomatic arches are not enlarged at their midpoints. The braincase is globular with a well-developed coronal suture. The hamular process is slender, bisecting the subsquamosal and postglenoid openings so as to make them subequal in size. The alisphenoid strut is typically absent. The carotid circulatory pattern conforms to pattern 1 of Voss (1988), with a large stapedial foramen and both squamosalalisphenoid groove and sphenofrontal foramen present. The palate is flat, short, and narrow (*sensu* Hershkovitz 1962), typically extending behind the posterior faces of M3s. The mandibles are rather weak and elongated with comparatively low horizontal rami and ascending processes; the capsular projection of the incisor does not form a tubercular process (Reig 1972). The incisors in *Scapteromys* are relatively weak and ungrooved. Molars are hypsodont, terraced in juveniles and young adults but planate in full adults, with the main cusps alternating. *Scapteromys* has a molar pattern similar to those of *Kunsia*, *Gyldenstolpia*, and *Bibimys*, characterized by a general akodontine bauplan slightly more complex than that observed in *Akodon* and related genera (e.g., *Necromys*, *Thaptomys*). Mesolophs and mesolophids are typically well developed and fused with mesostyles/stylids. The occlusal surface of each M3 is rounded in outline but retains a central enamel ring like that of abrotrichines. The first lower molar has a moderately developed anteromedian flexid and a labiolophulid. Vertebral counts for *Scapteromys* typically include 13 thoracic, 6 lumbar, 3 sacral, and 29 caudal elements (Steppan 1995).

The stomach, described for specimens of *S. tumidus*, is unilocular and discoglandular; the wall of the antrum is more thickly muscular than that of sigmodontines, having a hemiglandular stomach (Carleton 1973). A gall bladder is present (Voss 1991). The male phallus was described by Hooper and Musser (1964) and further detailed by Hershkovitz (1966a). The glans is nearly twice as long as wide and laterally compressed, almost entirely covered with spines; the tapered distal portion is furrowed longitudinally. Hershkovitz (1966a) reported a limited degree of variability, and Hooper and Musser (1964) detailed differences between the male phallus of *S. aquaticus* and *S. tumidus*. Hooper and Musser (1964:29) stated that the phallus of *Scapteromys* was "set well apart . . . from all other South American examples at our disposal" in morphological details, a view strongly criticized by Hershkovitz (1966a:131) who considered the characters as typically sigmodontine. Males possess bulbourethral, ampullary, vesicular, and prostate accessory glands; paired preputial glands are large (Hershkovitz 1966a; Voss and Linzey 1981).

Hershkovitz (1966a) and Massoia (1980c) established the basis of the current contents of *Scapteromys* by removing some species previously allocated to the genus and placing them into the new genera *Kunsia* and *Bibimys*. One species Hershkovitz assigned to *Kunsia*, *fronto* Winge, is now the type species of its own genus (see the account for *Gyldenstolpia*). Therefore, as currently understood, *Scapteromys* includes two living species (but, see Remarks under both species), whose taxonomic history, together with that of the genus, we summarize next.

Waterhouse (1837:15) described *Mus tumidus* from Maldonado, Uruguay, and five pages later (i.e., p. 20) erected *Scapteromys* as a new subgenus of *Mus* with *tumidus* designated as its type species. Therefore, though when *tumidus* and *Scapteromys* were described in the same work and *tumidus* as the then sole species of *Scapteromys*, *tumidus* was originally placed in *Mus* and, as such, as indicated by the use of parentheses to contain the authority and year of description, it is considered to have been only secondarily treated as a *Scapteromys*. Subsequently, Waterhouse (1839:57) again described *tumidus* but included it (on p. 77) in his new comprehensive genus *Hesperomys* while ignoring his earlier *Scapteromys*. Fitzinger (1867b) was the first to use *Scapteromys* at the genus level, although some authors continued to consider *Scapteromys* as a subgenus of *Hesperomys* (e.g., Thomas 1884). Thomas (1917c) referred a series of *Scapteromys* collected at Isla Ella, Argentina, to the species *S. tomentosus* (a species currently allocated to the genus *Kunsia*), stating that those specimens differed from *S. tumidus* only in color. Three years later, Thomas (1920b) noticed that his previous assignment of the specimens from Isla Ella to *tomentosus* was wrong and described a new species, *S. aquaticus*, based on these specimens. Remarkably, Thomas made no comment regarding his previous statement about the similarity of his new form *aquaticus* to *tumidus*. During the following four decades, taxonomic references to *Scapteromys sensu stricto* were limited to a handful of citations in general treatises on mammal taxonomy (e.g., Gyldenstolpe 1932; Ellerman 1941; Cabrera 1961), mammal lists (e.g., Devicenzi 1935), or accounts of nomenclatural history (e.g., Tate 1932d).

Massoia and Fornes (1964c) analyzed variation in skull, teeth, and external morphology within and between Argentinean and Uruguayan populations of *Scapteromys*, stating that morphological differences between *aquaticus* and *tumidus* were minor and limited to the shape of the frontoparietal suture and the width of the mesopterygoid fossa. These authors also mentioned that individuals of *S. tumidus* are brownish and seemed slightly larger than the blackish *S. aquaticus*. Massoia and Fornes (1964c) considered the two forms as only subspecifically distinct on the basis of individuals of intermediate phenotypes. However, Massoia and Fornes indicated that the correspondence between the trenchant characters they used to define and diagnose the two supposed subspecies and their geographic distributions was not absolute. Subsequently, Hershkovitz (1966a) argued that the characters enumerated by Massoia and Fornes (1964c) were highly variable, dismissed recognition of two different forms, and stated that *Scapteromys* consisted of a single species, *S. tumidus*, without subspecies.

Scapteromys exhibits considerable karyotypic diversity that appears to be geographically structured. Samples from Argentina (Fronza et al. 1976; Brum-Zorrilla et al. 1986), Paraguay (see D'Elía and Pardiñas 2004), and one Brazilian population (Bonvicino et al. 2013) have a $2n = 32$ complement, while those in Uruguay are $2n = 24$ (Brum-Zorrilla 1965; Brum-Zorrilla et al. 1972, 1986). Diploid numbers of 24, 34, and 36 have also been reported from Brazilian populations (T. R. O. Freitas et al. 1984). Brum-Zorrilla et al. (1972, 1986) proposed that individuals with $2n = 32$ correspond to *S. aquaticus* (distributed in Argentina) and those with $2n = 24$ to *S. tumidus* (distributed in Uruguay). T. R. O. Freitas et al. (1984) also suggested the existence of an undescribed species in Brazil. In spite of all this, Hershkovitz's (1966a) view of a single species in the genus continued to prevail among several taxonomists and ecologists (e.g., Musser and Carleton 1993; V. R. Cueto et al. 1995a; Eisenberg and Redford 1999). Finally, D'Elía and Pardiñas (2004) analyzed populations of *Scapteromys* from Argentina, Paraguay, and Uruguay, and showed that DNA haplotypes fell into two well differentiated and strongly supported reciprocally monophyletic clades that were geographically segregated and mostly congru-

ent with patterns of morphological and chromosomal variation. In light of these results, D'Elía and Pardiñas (2004) suggested that *S. aquaticus* be considered a species distinct from *S. tumidus*. We follow this scheme here, but see Remarks under both species regarding the status of Brazilian populations. A third species, *Scapteromys meridionalis*, was described by Quintela et al. (2014) after this volume went into production; it extends the geographic and habitat range of the genus to the *Araucaria angustifolia* biogeographic province in the Brazilian states of Rio Grande do Sul, Santa Catarina, and Paraná.

REMARKS: Reig (1994) described *S. hershkovitzi* from the Upper Pliocene (Vorohué Formation) of southeastern Buenos Aires province, Argentina, which constitutes the only known fossil species of the genus. However, restudy of this form suggests that it belongs to an undescribed extinct genus more closely related to *Kunsia* and *Gyldenstolpia* (Pardiñas, D'Elía, and Teta 2009).

SYNONYMS:

Mus (Scapteromys) Waterhouse, 1837:20; type species *Mus (Scapteromys) tumidus* Waterhouse, by original designation.

Hesperomys Waterhouse, 1839:75, 77; part (inclusion of *Mus tumidus*).

Hesperomys: Wagner, 1843a:515; part (listing of *tumidus*); not *Hesperomys* Waterhouse.

KEY TO THE SPECIES OF *SCAPTEROMYS*:

1. Frontoparietal suture more or less resembling an open U or V, anterior border of mesopterygoid fossa square with a bluntly pointed median palatine process; $2n = 32$ *Scapteromys aquaticus*
1'. Frontoparietal suture resembling a W, anterior border of mesopterygoid fossa rounded, without a median palatine process; $2n$ not 32 . . . . *Scapteromys tumidus*

Scapteromys aquaticus Thomas, 1920
Argentinean Swamp Rat
SYNONYMS:

Scapteromys tomentosus: Thomas, 1917c:96; initial reference to specimens from Isla Ella, Buenos Aires, Argentina; not *Scapteromys tomentosus* E.-L. Trouessart.

Scapteromys aquaticus Thomas, 1920b:477; type locality "Isla Ella" (see Thomas 1917c:95, who gave locality as "Isla Ella, in the delta of the Rio Parana, at the top of the La Plata Estuary"); according to Pardiñas et al. (2007:402) "Isla Ella is a small island (ca. 1 km²) surrounded by the streams Espera and Esperita (34°22′S, 58°38′W, Primera Sección Delta del Paraná, Tigre, Buenos Aires, Argentina)."

DESCRIPTION: Slightly smaller than *S. tumidus*. Dorsal coloration gray or dark brown and belly grayish white.

Cranially, frontoparietal suture typically U- or V-shaped, and mesopterygoid fossa narrower than in *S. tumidus* and with pointed median palatine process.

DISTRIBUTION: Wetlands in east-central and northern Argentina in Buenos Aires, Corrientes, Chaco, Entre Ríos, Formosa, and Santa Fe provinces, in southern Paraguay (D'Elía and Pardiñas 2004; Pardiñas, D'Elía, and Teta 2009), and in the Brazilian state of Rio Grande do Sul close to the border with Argentina (Bonvicino et al. 2013). Reig (1994:113) mentioned a locality "in the vicinity of Santa Clara del Mar," citing a personal communication with J. R. Contreras, in southeastern Buenos Aires province, Argentina. This locality is about 170 km south of the southernmost record reported here; we do not include it as we have been unable to verify its validity.

SELECTED LOCALITIES (Map 133): ARGENTINA: Buenos Aires, General Lavalle (D'Elía and Pardiñas 2004), Isla Ella, Delta del Paraná (type locality of *Scapteromys aquaticus* Thomas), Isla Martín García (CNP-E 652–2), Punta del Indio (D'Elía and Pardiñas 2004), Ramallo (D'Elía and Pardiñas 2004); Chaco, Selvas del Río de Oro (D'Elía and Pardiñas 2004); Corrientes, Ahoma Sur, Empedrado (D'Elía and Pardiñas 2004), Paraje Uguay, Esteros del Iberá (CNP 2521), Santo Tomé (CNP-E 650–4); Formosa, 17 km W of Colonia Villafañe (D'Elía and Pardiñas 2004), Laguna Naick Neck, INTA IPAF-NEA (LTU 582), Misión Franciscana Tacaaglé (CNP-E, not cataloged); Santa Fe, Estero La Zulema, 4 km NE of Estación Guaycurú (J. R. Contreras 1966), La Matilde, 16 km N of Alejandra (J. R. Contreras 1966), Puerto Ocampo (D'Elía and Pardiñas 2004). BRAZIL: Rio Grande do Sul, São Borja (Bonvicino et al. 2013). PARAGUAY: Caaguazú, Estancia San Ignacio, 24 km NNW of Carayao (D'Elía and Pardiñas 2004); Misiones, Isla Yaciretá (D'Elía and Pardiñas 2004); Paraguarí, Costa del Río Tebicuary (D'Elía and Pardiñas 2004); Presidente Hayes, 24 km W of Villa Hayes (D'Elía and Pardiñas 2004).

SUBSPECIES: *Scapteromys aquaticus* is monotypic.

NATURAL HISTORY: V. R. Cueto et al. (1995a) studied demographic structure of a population in the Delta del Paraná, Argentina, by capture-mark-recapture methods over a two-year period. Density variation was limited despite seasonal reproduction, with major intensity during the spring and summer. Individuals persisted in the sampling area for longer than 20 months, which supports a minimal longevity of two years. Density, residence, and reproduction were higher in microhabitats with a more extensive vegetative cover with a height of at least 0.50 m. Massoia and Fornes (1964c; see also Massoia 1961) also studied the natural history of *S. aquaticus* in the Delta del Paraná and coastal wetlands of Buenos Aires province, Argentina. The species was frequently encountered in flooded places dominated by stands of *Cortaderia* sp. and *Scirpus giganteus*.

A variety of vocalizations were recorded, including a sharp cry by nestling young and a sonorous adult call. In contrast with *S. tumidus* (see below), *S. aquaticus* apparently feeds preferentially on oligochaetes rather than beetles. The parasites of *S. aquaticus* have been well surveyed based on animals collected in the Delta del Paraná and several Argentinean coastal localities (see Navone et al. 2009). Lareschi (2006) found no host sexual dimorphism in ectoparasite abundances, except for the mite *Laelaps manguinhosi* being significantly more common on male rats. The staphylinid beetle *Amblyopinodes gahani gahani* was also found on *S. aquaticus*. Known internal parasites include helligmondllid nematodes of the genus *Malvinema* (Digiani et al. 2003) and the digenean nematodes *Conspicuum minor* (Dicrocoeliidae), *Echinoparyphium scapteromae* (Echinostomatidae), *Echinostoma platensis* (Echinostomatidae), and *Levinseniella* (*Monarrhenos*) *cruzi* (Microphallidae) (Lunaschi and Drago 2007 and references therein). M. Robles, Bain, and Navone (2012) recently described a new capillariine nematode, *Capillaria alainchabaudi*, with *S. aquaticus* as the type host.

REMARKS: On the basis of mtDNA variation, D'Elía and Pardiñas (2004) assigned specimens collected at Las Cañas, Río Negro department, Uruguay, to this species despite a *tumidus*-like karyotype that had been reported for the same locality (Brum-Zorrilla et al. 1972). E. M. González and Martínez Lanfranco (2010) dismissed this assignment, however, and placed all Uruguayan popu-

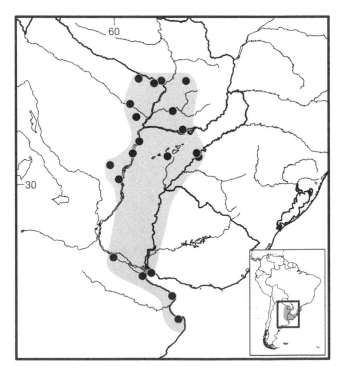

Map 133 Selected localities for *Scapteromys aquaticus* (●).

lations of *Scapteromys* in *S. tumidus*. More recent work (G. D'Elía, unpubl. data) substantiates the presence of specimens with *aquaticus*-like mtDNA but *tumidus*-like karyotype at Las Cañas as well as at an additional nearby locality. Further studies are needed to clarify the basis for discrepancy between the mitochondrial gene tree and the diploid complement, especially the possible role of reticulation through hybridization. Until this issue is adequately resolved, we take a conservative posture and assign all Uruguayan populations of *Scapteromys* to *S. tumidus*. The karyotype has $2n=32$, $FN=40$; autosomes include three large pairs of submetacentrics or subtelocentrics, two small pairs of submetacentrics, and 10 pairs of acrocentrics grading in size from medium to small; the X chromosome is a medium acrocentric, the Y a small subtelocentric (Bonvicino et al. 2013).

Scapteromys tumidus (Waterhouse, 1837)
Uruguay Swamp Rat

SYNONYMS:

Mus tumidus Waterhouse, 1837:15; type locality "Maldonado," Maldonado, Uruguay.

Mus (*Scapteromys*) *tumidus*: Waterhouse, 1837:21; name combination.

[*Hesperomys*] *tumidus*: Waterhouse, 1839:75, 77; name combination (assignment of *Mus tumidus* Waterhouse to *Hesperomys* Waterhouse).

Hesperomys tumidus: Wagner, 1843:515; name combination.

Scapteromys tumidus: Fitzinger, 1867b:80; first use of current name combination.

DESCRIPTION: Slightly larger and heavier than *S. aquaticus* (especially populations in Uruguay and southernmost Brazil). Dorsal coloration brownish to ochraceous brown and venter ivory white. Skull with frontoparietal suture typically W-shaped, mesopterygoid fossa broad and without median palatine process on anterior border.

DISTRIBUTION: This species ranges across most of Uruguay and along a narrow coastal strip in the Brazilian states of Rio Grande do Sul, Santa Catarina, and Paraná (T. R. O. Freitas et al. 1984; J. González 1994; D'Elía and Pardiñas, 2004).

SELECTED LOCALITIES (Map 134): BRAZIL: Paraná, Nova Tirol (T. R. O. Freitas et al. 1984); Rio Grande do Sul, Bagé (T. R. O. Freitas et al. 1984), Cambará do Sul (T. R. O. Freitas et al. 1984), Esmeralda (T. R. O. Freitas et al. 1984), Porto Alegre (Hensel 1872b), Taim (T. R. O. Freitas et al. 1984). URUGUAY: Artigas, La Isleta, Colonia Artigas (Massoia and Fornes 1964c); Colonia, Arroyo Artilleros, Santa Ana (D'Elía and Pardiñas 2004); Maldonado, Barra del Arroyo Maldonado (type locality of *Mus tumidus* Waterhouse); Rivera, Estancia La Quemada (D'Elía and

Pardiñas 2004); Rocha, Arroyo La Palma, Ruta 15 km 10, La Paloma (D'Elía and Pardiñas 2004).

SUBSPECIES: *Scapteromys tumidus* is monotypic.

NATURAL HISTORY: Little is known about the ecology of this species. Field observations in Uruguay (Barlow 1969:27) indicate that this rat is usually trapped in marshy or low places where *Eryngium* sp. and *Cortaderia selloana* are the dominant plants. Swamp rats seem not to make runways of their own, but frequently use runways constructed by *Cavia aperea*; they also dig to obtain food, but there is no evidence that they excavate burrows. Vocalizations are often heard when the animals are in distress. Most have been trapped at dusk or during the night; diurnal activity is limited to juveniles and subadults and may be correlated with high population levels. *Scapteromys tumidus* is largely insectivorous; stomach contents of 11 individuals trapped in April and May were 85% invertebrate, mostly Coleoptera and particularly scarabaeid beetles (Barlow 1969:28). In Uruguayan populations, males were reproductively active every month of the year except January, June, and July, with a peak of testis size in April and May; pregnant females were taken in the spring and summer (Barlow 1969). These rats are good swimmers despite limited interdigital webbing (Sierra de Soriano 1969:485). The tick *Ixodes longiscutatum* (see Venzal et al. 2001) and the nematode *Ansiruptodera scapteromi* (Ganzorig et al. 1999) parasitize this species.

REMARKS: Uruguayan and lowland Brazilian populations have a $2n=24$, $FN=40$ karyotype; highland Brazilian

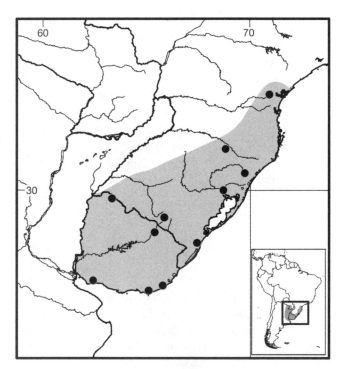

Map 134 Selected localities for *Scapteromys tumidus* (●).

populations comprise two karyomorphs, $2n=34$ (northern Rio Grande do Sul state) and $2n=36$ (Paraná state). The latter karyotypic forms likely belong to at least one undescribed species (T. R. O. Freitas et al. 1984; see also Tiepolo 2007, Pedó et al. 2010). These individuals exhibit striking morphological and molecular differences from both *S. aquaticus* and *S. tumidus* (Tiepolo 2007; G. D'Elía and U. F. J. Pardiñas, unpubl. data) as well as extensive karyotypic diversity. The treatment of all Brazilian populations of *Scapteromys* as *S. tumidus*, with the exception of the single record assigned to *S. aquaticus* (see above), is a conservative position followed here and requires a comprehensive study of Brazilian specimens and the naming of new taxa. According to the few published data, animals with $2n=34$ and 36 are restricted to localities above 800 m, in the coastal Serra do Mar.

Genus *Thalpomys* Thomas, 1916

Ulyses F. J. Pardiñas and Pablo Teta

The genus *Thalpomys* is endemic to the Brazilian Cerrado and includes two living species, *T. cerradensis* Hershkovitz and *T. lasiotis* Thomas. For many years, *Thalpomys* was both very poorly known and constituted a highly confusing generic concept. Its complex taxonomic and nomenclatural history, paradoxically tortuous considering the limited information available for the genus, has been addressed by several authors (e. g., Langguth 1970; Reig 1987) but not fully resolved until Hershkovitz's (1990a) treatment. The present account is largely based on the results of the last and with recent additions provided by A. F. B. Andrade, Bonvicino et al. (2004).

Both species of *Thalpomys* are small (head and body length 80–101 mm; mass <30 g) with the general bauplan of *Akodon*. The overall pale or bright phaeomelaninic color contrasts sharply with the dusky olivaceous or blackish coat of other akodontines. The tail is of medium length (<65% of head and body length) and well haired, which obscures the scales. The pinnae are of medium length, rounded, and covered by short hairs both surfaces. The pes and manus have white hairs on the upper surface and the ungual tufts are shorter than the claws (Bonvicino, Oliveira, and D'Andrea 2008). Hershkovitz (1990a) noted that the coloration of *Thalpomys* is unique among akodontines, resembling that of some oryzomyines such as *Oligoryzomys*.

The skull of *Thalpomys* is characterized by strongly developed supraorbital ledges that extend posteriorly as crests or beading along the length of parietals (thus resembling the condition in *Graomys* [Phyllotini]). This condition is unusual in the Akodontini, although some species of *Necromys* have a roughly similar configuration (a re-

semblance noted by Winge [1887] when he described *T. lasiotis*). The nasals are tapered distally and with rounded tips, proximally extending beyond the premaxillary sutures. The rostral width is subequal to the least interorbital width. The frontoparietal suture is U- to V-shaped. Zygomatic arches are more or less bowed, and converge anteriorly. The braincase is moderately inflated, with its dorsal contour rounded. Zygomatic plates are medium to high, broad anteroposteriorly, similar to those in many *Akodon* species, and have straight anterior borders. Incisive foramina are well developed and reach posteriorly to the protocones of the M1s. The palate is wide, flat, and extends to slightly behind the posterior faces of the M3s; posterolateral palatal pits range in size from small to large. The mesopterygoid fossa (subequal in width to each parapterygoid fossa) is parallel sided and with its anterior border square; sphenopalatine vacuities are conspicuous. Auditory bullae are well inflated, and the distance between them at the basisphenoid-basioccipital suture is about the same as the greatest width of the mesopterygoid fossa.

Upper incisors are orthodont to slightly proodont, and do not extend beyond the nasal tips when viewed from above; their anterior faces are covered with pale orange enamel. Molars are parallel sided, moderately hypsodont, and possess terraced crowns and a simplified occlusal topology. The anteromedian flexus is obvious in unworn M1s but rapidly decays with wear. Upper molars have their major cusps subtriangular in outline and possess a moderate-sized hypoflexus and a wide transverse mesoflexus, both reaching the molar midline. Cusps of opposite sides are subequal in size; flexi are nearly transverse, especially on M1 and M2. The occlusal view of M3 is nearly oval. Lower molars display a characteristic occlusal pattern with a reduced and compressed procingulum on m1, lack an anteromedian flexid, and have crown flexids that nearly touch. The anterolabial conulid of m1 is reduced in size, subequal to the entoconid in size, and narrowly oval to subtriangular in shape. Minute mesolophids are present on both m1 and m2.

Thalpomys has three pairs of mammae, one each in pectoral, abdominal, and inguinal positions. Bacular and penis morphology are similar to other akodontines, based a single individual of *T. cerradensis*. Geise, Weksler, and Bonvicino (2004) reported the presence of the gall bladder in one specimen of *T. lasiotis*. No other data about soft anatomy are available.

The two species currently recognized overlap in distribution, with sympatry or near-sympatry at one locality in Bahia state (Fazenda Sertão do Formoso, Jaborandi [A. F. B. Andrade, Bonvicino et al. 2004]) and two in the Distrito Federal (Parque Nacional de Brasília [Hershkovitz 1990a] and Reserva Ecológica do IBGE [A. F. B. Andrade, Bonvicino et al. 2004]). A. F. B. Andrade, Bonvicino et al. (2004) documented unique karyotypes and reciprocally monophyletic mtDNA lineages that are consistent with the two named taxa.

SYNONYMS:

[?] *Hesperomys*: Burmeister, 1854:177; part (included *lasiotis* Lund [= *Mus lasiurus* Lund,= *Necromys lasiurus*] as species); not *Hesperomys* Waterhouse.

Habrothrix: Winge, 1887:29; part; (description of *lasiotis*); not *Habrothrix* Wagner.

Thalpomys Thomas, 1916i:339; type species *Habrothrix lasiotis* Winge, by original designation.

[*Akodon*] (*Thalpomys*): Ellerman, 1941:409; name combination.

Akodon: Langguth, 1975:47; part (*reinhardti* as unnecessary replacement name for *lasiotis* Winge); not *Akodon* Meyen.

KEY TO THE SPECIES OF *THALPOMYS*:

1. Size larger, mean head and body length >90 mm, greatest adult skull length >25 mm; postpalatal pits large; sphenoidal sinus large; diploid number 36 . . . . *Thalpomys cerradensis*
1'. Size smaller, mean head and body length <90 mm, greatest adult skull length <25 mm; postpalatal pits small; sphenoidal sinus small; diploid number 38 . . . . *Thalpomys lasiotis*

Thalpomys cerradensis Hershkovitz, 1990
Cerrado Akodont

SYNONYM:

Thalpomys cerradensis Hershkovitz, 1990a:777; type locality "Parque Nacional de Brasília, Distrito Federal, Brazil" (elevation ca. 1,100 m).

DESCRIPTION: Larger than *T. lasiotis* with only marginal overlap in external and some cranial measures. Pelage dense and stiff; general coloration of dorsal surface of head and body tawny; individual hairs tricolored with tips black, middle orange-buffy, and base blackish; guard hairs blackish; sides of body paler than back; chin pale buff; check and eye-ring ochraceous orange; vibrissae short, extending to base of ears; tail brownish above and pale buff beneath; hindfoot small, pale buff above, with outer toes short; claws (1 mm) barely extend to base of adjacent toes.

DISTRIBUTION: Cerrado biome in central Brazil, in the states of Bahia, Goiás, and Mato Grosso, and in Distrito Federal.

SELECTED LOCALITIES (Map 135; from A. F. B. Andrade, Bonvicino et al. 2004, except as noted): BRAZIL: Bahia, Cocos, Sertão do Formoso, Jaborandi; Distrito Federal, Parque Nacional de Brasília (type locality of *Thalpomys cerradensis* Hershkovitz); Goiás, Baliza; Mato Grosso, Cuiabá, Xavantina, 260 km N of Serra do Roncador (Hershkovitz 1990a).

SUBSPECIES: *Thalpomys cerradensis* is monotypic.

NATURAL HISTORY: Specimens of *T. cerradensis* have been caught in open grasslands (*campo limpio*), wet prairies with palms (*veredas*), and sparsely wooded savannas in the Brazilian Cerrado (Hershkovitz 1990a; Bonvicino, Cerqueira, and Soares 1996). According to E. M. Vieira and Baumgarten (2009), *T. cerradensis* is more active immediately after sunset and in the last three hours before sunrise. This species is one of the first to recolonize burned grasslands in the Cerrado, reaching population peaks <2 years after fires (Henriques et al. 2006). This rodent represents 10.7% of the diet of the Barn Owl (*Tyto alba*), based on pellet analyses from the state of Bahia; however, no specimens were caught with traps in this same area (Bonvicino and Bezerra 2003). *Thalpomys cerradensis* is parasitized by the botfly *Megacuterebra apicalis* (E. M. Vieira 1993).

REMARKS: Specimens from Serra do Roncador are paler throughout than ones from elsewhere, especially the underparts (Hershkovitz, 1990a). Specimens recorded as *Akodon* sp. nov.? by Pine et al. (1970:669) and as *Akodon* sp. 1 by Mares, Braun, and Gettinger (1989:13) represent *T. cerradensis*.

The karyotype consists of $2n = 36$, FN = 34 (A. F. B. Andrade, Bonvicino et al. 2004).

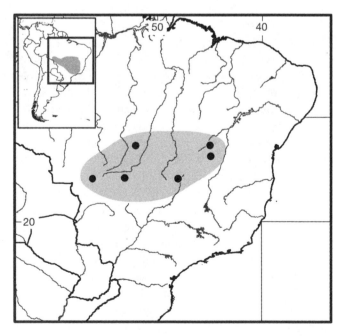

Map 135 Selected localities for *Thalpomys cerradensis* (●).

Thalpomys lasiotis Thomas, 1916

Hairy-eared Akodont

SYNONYMS:

[?] *Hesperomys lasiotis*: Burmeister, 1854:177; name combination.

Habrothrix lasiotis Winge, 1887:29; type locality "Lagoa Santa," Minas Gerais, Brazil.

T[*halpomys*]. *lasiotis* Thomas, 1916i:339; redescription of *Habrothrix lasiotis* Winge, with the same type locality.

Akodon [(*Thalpomys*)] *lasiotis*: Ellerman, 1941:414; name combination.

Akodon reinhardti Langguth, 1975:47; unnecessary replacement name for *lasiotis* Thomas.

DESCRIPTION: A smaller version of *T. cerradensis* but with longer and more lax fur, more saturate color throughout, and slightly less complex molar occlusal pattern. Hershkovitz (1990a) provided measurements of craniodental variables.

DISTRIBUTION: This species is restricted to Cerrado formations in central Brazil, in the states of Bahia, Minas Gerais, Rondônia, and São Paulo, and the Distrito Federal. Hershkovitz (1990a) considered the skin-only specimen recorded by C. O. da C. Vieira (1955) from São Paulo state as of dubious provenance.

SELECTED LOCALITIES (Map 136; from A. F. B. Andrade, Bonvicino et al. 2004, except as noted): BRAZIL: Rondônia, Vilhena); Bahia, Sertão do Formoso, Jaborandi; Distrito Federal, Parque Nacional de Brasília (Hershkovitz 1990a); Minas Gerais, Lagoa Santa (type locality of *Thalpomys lasiotis* Thomas); São Paulo, Estação de Ferro Noroeste, Braúna (C. O. da C. Vieira 1955); Minas Gerais, Parque Nacional Serra da Canastra, São Roque de Minas, Serra da Canastra, Poços de Caldas.

SUBSPECIES: *Thalpomys lasiotis* is monotypic.

NATURAL HISTORY: Capture sites of *T. lasiotis* in the Cerrado have mostly been in open grasslands (*campo limpio*), open grasslands mixed with low shrubs (*campo cerrado*), open grasslands with scattered shrubs and trees (*campo sujo*), wet grasslands with water-saturated soil (*campo umido*), and wild grasslands with small natural earthen mounds without trees or underbrush (*campo de murumdus*) (Dietz 1983; Mares, Braun, and Gettinger 1989; Hershkovitz 1990a; Bonvicino and Bezerra 2003; R. Ribeiro et al. 2011; C. R. Rocha et al. 2011). Deitz (1985) documented a unimodal reproductive period during the wet season at Serra da Canastra, Minas Gerais, but R. Ribeiro et al. (2011) found breeding during the dry season at a locality in the northeastern part of the Distrito Federal. Near Brasília, Alho and Pereira (1985) recorded males with scrotal testes from August to October and pregnant females from September to May. Finally, at a site in northeastern Distrito Federal, reproductive activity began at the end of the dry season (August) and lasted through the end of the rainy season (March), with pregnant females recorded from November through March (R. Ribeiro et al. 2011). Pregnant females carried two to three embryos (reviewed by Hershkovitz 1990a). Mean biomass varied between 1.2 and 19.6 g/ha and mean home range size for 26 individuals was 2,278 m² (Alho and Pereira 1985). Densities per hectare expressed as the minimum number of individuals known alive for the months of July and August (winter) at two sites in the vicinity of Brasília were 0.5 and 0.6 (Lacher et al. 1990) and 0.5 to 1.6 (R. Ribeiro et al. 2011). R. Ribeiro et al. (2011) provided data on home range size and dispersal distances. L. A. Pereira (1982, fide Armada et al. 1984) found *T. lasiotis* to be the second most abundant rodent, after *Necromys lasiurus*, in the Cerrado. At least three species of laelapid mites (*Androlaelaps foxi*, *A. pachyptilae*, and *A. fahrenholzi*) and some undetermined Acariformes are external parasites of this species (Mares, Braun, and Gettinger 1989).

REMARKS: Both Langguth (1975) and Hershkovitz (1990a) documented that *Thalpomys lasiotis*, the type species of *Thalpomys* Thomas, is not Lund's *Mus lasiotis* (1840b:50, 1841c:280), as has often been assumed (e.g., Cabrera 1961). Rather, it is the species Winge (1887:29) described and figured as *Hesperomys lasiotis*, and which Winge mistakenly attributed to Lund. Lund's *lasiotis* is a junior synonym of *Mus lasiurus* Lund, which is now regarded as a valid species of *Necromys* Ameghino (see that account). Langguth (1975:47) unnecessarily renamed Winge's *Hesperomys lasiotis* as *Akodon reinhardti*, placing it in a sub-

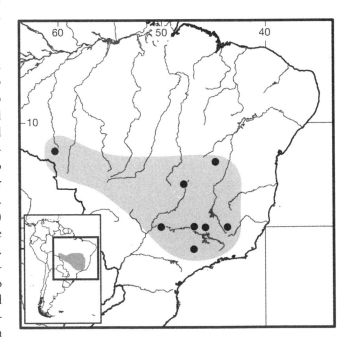

Map 136 Selected localities for *Thalpomys lasiotis* (●).

genus, *Thalpomys*. Thomas (1916i:339) becomes the author of *Thalpomys lasiotis*, even though he gave the type species as the misidentified "*T. lasiotis* (*Mus lasiotis*, Lund; *Habrothrix lasiotis*, Winge)" (ICZN 1999:Art. 11.10).

Armada et al. (1984) and A. F. B. Andrade, Bonvicino, et al. (2004) recorded a karyotype with $2n = 38$, FN = 38, with apparently geographically fixed morphological differences in the X chromosome.

Genus *Thaptomys* Thomas, 1916

Pablo Teta, Ulyses F. J. Pardiñas, and Guillermo D'Elía

In current usage, the genus *Thaptomys* includes only the single species *Thaptomys nigrita*, which is restricted to moist tropical forest and second-growth forest in southeastern Brazil, eastern Paraguay, and northeastern Argentina (Massoia 1963a; Hershkovitz 1998). Ellerman (1941), Cabrera (1961), and Musser and Carleton (1993) all listed *Thaptomys* as a subgenus of *Akodon*. Reig (1987) included *nigrita* in the genus Akodon, subgenus *Akodon* (*Akodon*). Hershkovitz (1990b; see also Hershkovitz 1998) reintroduced the concept of *Thaptomys* as a valid genus based on morphological and karyotypic data. Subsequent molecular-based phylogenetic analyses recovered *nigrita* outside of *Akodon* (*sensu stricto*). These analyses recovered, with weak support, *Thaptomys* as sister either to *Akodon* s.s. (M. F. Smith and Patton 1999), to "*Akodon*" *serrensis* (D'Elía 2003; D'Elía et al. 2003), or to an *Akodon-Deltamys-"A." serrensis* clade (M. F. Smith and Patton 2007).

Thaptomys is modified for a semifossorial life. The fur is velvety and short; eyes are very small; ears are short; and the tail is short, usually equal to or less than 50% of the head and body length, and very thinly haired (the short stiff hairs brown above, whitish below, not concealing scales). Manual claws are long and fine; pedal claws are shorter. Palms and soles are naked, with prominent pads. The dorsum is dark brown or glistening olive brown to reddish brown, finely grizzled with ochraceous brown and intermixed with some longer blackish brown hairs. The under parts are ochraceous to dull brownish gray washed with buff, the hairs dark slaty at their bases. The mean and ranges of external measurements of 12–17 adults from Brazil are: total length, 96 mm (82–116 mm), length of tail, 45 mm (30–50 mm), length of hindfoot, 19 mm (16–20 mm), and length of ear, 11 mm (10–12 mm). Females have three to four pairs of mammae (Gyldenstolpe 1932; Hershkovitz 1998).

The skull is rather strongly built, with a short and thick rostrum and short nasals that do not extend to the level of the incisors. The interorbital region is wide, with a moderate constriction and marked supraorbital edges. The braincase is smooth and square; the supraoccipital bone is inclined forward; the interparietal is small; and the parietals are moderately ridged. Zygomatic arches are delicate. Zygomatic plates are broad, with almost vertical anterior borders, very short upper free borders, and an excrescence at the base, at the point of origin of the superficial masseter. Incisive foramina are longer than the molar rows, parallel sided, narrow, and extend to the level of the protocones of M1s. The palate is short, flat, and wide. The mesopterygoid fossa is broad; sphenopalatine vacuities are absent (Gyldenstolpe 1932; Hershkovitz 1998). The carotid circulation pattern is pattern 1 (*sensu* Voss 1988); a thick alisphenoid strut is present. Massoia (1963a) and Hershkovitz (1990b, 1998) provided cranial measurements. The mandible is robust, with a prominent capsular projection directed toward the condyle and with a robust coronoid process; the mental foramen is large and clearly visible in lateral view.

Upper incisors are long, opisthodont to slightly proodont. Molars are hypsodont, moderately high crowned, and small; the unworn crowns are terraced. An anteromedian flexus of M1 is well defined; the paralophule in both M1 and M2 is fused to the mesoloph, and both are coalescent with the paracone; M3 is reduced, but with an enamel ring usually present. The first lower molar has a conspicuous anterolabial cingulum with a well-developed anteromedian flexid, oblique main cusps, and a short posterolophid. The mesolophid in both m1 and m2 is reduced, but both teeth have a well-developed ectolophid. The third lower molar is bilophodont, lacking the posteroflexid (details from Hershkovitz 1998).

The vertebral count is 7 cervical, 19 thoracolumbar, 2 sacral, and 23–24 caudal elements (Steppan 1995; Hershkovitz 1998).

Thaptomys possess a single pair of prostates (Voss and Linzey 1981). A gall bladder is absent, both in Brazilian (Geise, Weksler, and Bonvicino 2004) and Argentinean specimens (LTU 846, 858, 877, 879). The stomach is hemiglandular-unilocular, with a reduced glandular area (our dissection of two specimens, LTU 877, 879 from Misiones, Argentina).

SYNONYMS:

Mus: Lichtenstein, 1829:7; part (description of *nigrita*); not *Mus* Linnaeus.

H[*esperomys*].: Wagner, 1843a:523; part (listing of *nigrita*); not *Hesperomys* Waterhouse.

Habrothrix: Fitzinger, 1867b:81; part (listing of *nigrita*); not *Habrothrix* Wagner.

Hesperomys (*Habrothrix*): Ihering, 1894:20; part (listing of *fuliginosus*); not *Hesperomys* Wagner nor *Habrothrix* Wagner.

Akodon: E.-L. Trouessart, 1897:536; part (listing of *nigrita*); not *Akodon* Meyen.

Thaptomys Thomas, 1916i:339; type species *Hesperomys subterraneus* Hensel, by original designation.

[*Akodon*] (*Thaptomys*): Ellerman, 1941:409, 414; name combination as subgenus.

Akodon (*Abrothrix*): Paula Couto, 1950:272; part (listing of *nigrita*); not *Abrothrix* Waterhouse.

Thaptomys nigrita (Lichtenstein, 1829)
Ebony Akodont
SYNONYMS:

Mus nigrita Lichtenstein, 1829:7; type locality "in der Gegend von Rio de Janeiro gefunden," region of Rio de Janeiro, Brazil (Cabrera 1961).

[*Mus*] *orycter* Lund, 1839b:208; *nomen nudum*.

M[*us*]. *orycter* Lund, 1840b [1841c:281]; type locality "Rio das Velhas's Floddal" (Lund 1841c:292, 294), restricted herein to "Lapa da Serra das Abelhas," Minas Gerais, Brazil (see Remarks).

H[*esperomys*]. *nigrita*: Wagner, 1843a:523; name combination.

Hesperomys fuliginosus Wagner, 1845:148; type locality "Ypanema" (= Floresta Nacional de Ipanema, 20 km NW Sorocaba, São Paulo, Brazil, 23°26′7″S 47°37′41″W, 701 m; L. P. Costa et al. 2003).

Habrothrix nigrita: Fitzinger, 1867b:82; name combination.

Hesperomys subterraneus Hensel, 1872b:44; type locality "Provinz Rio Grande do Sul," Brazil.

Hesperomys subterraneus var. *henseli* Leche, 1886:697; type locality "Taquara do Mundo Novo," Taquara, Rio Grande do Sul, Brazil (see discussion by Musser et al. 1998:184).

Habrothrix orycter: Winge, 1887:27; name combination.

Hesperomys (*Habrothrix*) *fuliginosus*: Ihering, 1894:20; name combination.

Akodon nigrita: E.-L. Trouessart, 1897:537; name combination.

Akodon subterraneus: E.-L. Trouessart, 1897:537; name combination.

Akodon fuliginosus: E.-L. Trouessart, 1897:537; name combination.

T[*haptomys*]. *subterraneus*: Thomas, 1916i:339; name combination.

Thaptomys nigrita: Gyldenstolpe, 1932a:117; first use of current name combination.

Akodon [(*Thaptomys*)] *nigrita*: Ellerman, 1941:414; name combination.

Akodon [(*Thaptomys*)] *subterraneus*: Ellerman, 1941:414; name combination.

Akodon (*Abrothrix*) *nigrita*: Paula Couto, 1950:272; name combination.

Akodon (*Thaptomys*) *nigrita nigrita*: Cabrera, 1961:452; name combination.

Akodon (*Thaptomys*) *nigrita subterraneus*: Cabrera, 1961:453; name combination.

A[*kodon*]. (*Ak*[*odon*].) *nigrita*: Reig, 1987:351; name combination.

DESCRIPTION: As for the genus.

DISTRIBUTION: *Thaptomys nigrita* is found in southeastern Brazil, from Bahia south through Minas Gerais, São Paulo, Paraná, Santa Catarina, and Rio Grande do Sul states into southeastern Paraguay and the Argentinean province of Misiones (Massoia 1963a; Myers and Wetzel 1979; Gamarra de Fox and Martin 1996; Hershkovitz 1998; J. C. Moreira and Oliveira 2011).

SELECTED LOCALITIES (Map 137; from J. C. Moreira and Oliveira 2011, except as noted): ARGENTINA: Misiones, San Ignacio, San Ignacio (Massoia 1993), Candelaria, Arroyo Yabebyri (Massoia 1993), Oberá, Campo Ramón (Massoia 1993). BRAZIL: Bahia, Fazenda Almada, Fazenda Ribeirão da Fortuna, Vila Brasil; Espírito Santo, Santa Teresa; Minas Gerais, Boca da Mata, Conceição do Mato Dentro (Avila-Pires 1960b); Rio Grande do Sul, Taquara (type locality of *Hesperomys subterraneus* var. *henseli* Leche); São Paulo, São João da Boa Vista, Fazenda Intervales, Capão Bonito; Paraná, Fênix; Santa Catarina, Corupá. PARAGUAY: Itapúa, Ape Aime (TK 66003), 8 km N of San Rafael (Myers and Wetzel 1979).

SUBSPECIES: Two subspecies, *T. n. nigrita* and *T. n. subterraneus*, have been generally recognized since Cabrera (1961), a view supported by Massoia's (1963a) morphological assessment (but see Remarks).

NATURAL HISTORY: Little is known about this semifossorial rodent. Specimens have been collected in both primary and second-growth forest and in grasslands near forest (Davis 1947; Myers and Wetzel 1979). In Paraguay, they have been trapped in wet tropical forest along runways through grass, in woodpiles, and next to fallen logs (Myers and Wetzel 1979). In Brazil, they were found under logs and tree roots (Davis 1947). *Thaptomys nigrita* is strongly terrestrial and mostly diurnal (Davis 1947). This species makes tunnels in the leaf litter or in the soft earth (Davis 1947; Hershkovitz 1998). In São Paulo state, Brazil, females with perforated vaginas were trapped in April-May and September-October, while males with descended testes were found in May and October. Breeding between April and May is consistent with the trapping of juveniles between May and August (Olmos 1992). The observed number of embryos in four females from Brazil ranged from three to five (Davis 1947). Several specimens were collected with a single trap in a small area on consecutive days, suggesting some degree of sociality or at least that a single burrow system is used by more than one individual (U. F. J. Pardiñas, unpubl. data on captures made in Misiones, Argentina).

REMARKS: Thomas (1902a:62; see also Tate 1932h:2), when describing specimens from São Paulo, referred to at that time as *Akodon subterraneus*, was the first to associate *orycter* Lund with that form and with *nigrita*. Paula Couto (1950) directly equated *Mus orycter* with *Akodon* (*Abrothrix*) *nigrita*. Later, Avila-Pires (1960b:39) listed *orycter* in the synonymy of *Thaptomys nigrita*. However, none of these authors indicated explicitly that they had examined the original material referred by Lund to *orycter*. The holotype (fixed by monotypy; ICZN 1999:Art. 73.1.2) of *Mus orycter* is a large skull fragment housed in the Natural History Museum of Denmark (ZMK 1/1845:13239, figures at http://www.zmuc.dk/VerWeb/lund/lund_mammals .html) and illustrated by Winge (1887) in plate II, Fig. 8 and by Hansen (2012:56–57). Accordingly, the type locality of *orycter* Lund is fixed at "Lapa da Serra das Abelhas," where this fossil originated (Winge 1887:27; Hansen 2012:56). From inspection of the holotype, the morphology (including nasal length, zygomatic plates with short upper free borders, wide interorbital region with well-inflated frontal sinuses, presence of a visible excrescence at the point of origin for the superficial masseter, and large incisive foramina that are parallel sided and narrow) is in agreement with that of *Thaptomys*, which validates Thomas's (1902a) early impression. However, further studies are needed to determine whether *orycter* Lund represents a distinct (perhaps extinct?) species of *Thaptomys* or whether it is a synonym of *T. nigrita* as we treat it here.

Karyological, molecular, and morphometric data combine to suggest greater taxonomic diversity in the currently recognized single species. The diploid complement of *T. nigrita* across most of its range is $2n = 52$, $FN = 52$ (Yonenaga-Yassuda 1975; K. Ventura et al. 2010). This karyomorph is composed of 24 pairs of acrocentric autosomes that gradually decrease in size, a minute metacentric pair, a medium-sized metacentric X chromosome, and a subtelocentric medium-sized Y (Paresque et al. 2004). Ventura et al. (2004) found a second, $2n = 50$, $FN = 48$, karyomorph in animals from Una, Bahia state, and suggested that this population represents an undescribed species. Phylogeographic analyses (K. Ventura et al. 2010) based on mtDNA cytochrome-*b* sequences recovered two main clades whose distribution broadly corresponded with those of the two karyomorphs, one restricted to a small area in central-eastern Brazil ($2n = 50$), and another widely distributed in southeastern Brazil and perhaps Argentina and Paraguay ($2n = 52$). The latter clade is composed of three geographically structured subclades. Silva Gómez (2008) obtained similar results with broader geographic sampling that included Argentinean populations and referred Bahia as well as São Paulo populations to separate undescribed species, while placing samples from Rio de Janeiro, Es-

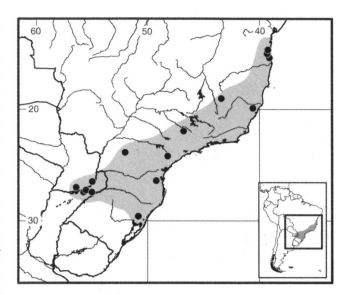

Map 137   Selected localities for *Thaptomys nigrita* (●).

pírito Santo, and Minas Gerais in *T. nigrita* and those from Paraná, Santa Catarina, Rio Grande do Sul, and Misiones to *T. subterraneus*. J. C. Moreira and Oliveira (2011), using morphometrics and qualitative morphological character analyses, failed to find differences between northeastern and southeastern populations, reporting only subtle metric differences in cranial descriptors. Thus, these authors were unable to reject the hypothesis that the differences found between northern and southern populations of *Thaptomys* simply constitute those of a widely distributed polytypic species. The possibility that *T. nigrita* might be composite, comprising two or more species, thus remains an open question.

Tribe Ichthyomyini Vorontsov, 1959
Robert S. Voss

The tribe Ichthyomyini includes five genera and 17 species, of which four genera and 13 species are known to occur in South America. In Central America, the tribe ranges from southern Mexico through Guatemala, El Salvador, and Costa Rica, to eastern Panama. In South America, ichthyomyines occur from Colombia southward to Bolivia and eastward to the Guianas and Brazil. Although ichthyomyines share numerous derived morphological traits that are not observed in other New World cricetids (Voss 1988), currently available molecular results do not support ichthyomyine monophyly (J. J. Martínez et al. 2012; Parada et al. 2013), a noteworthy anomaly. Ichthyomyines are unknown from any island fauna, and there is no fossil record. Except as noted otherwise, this account is abstracted from Voss (1988), which should be consulted for additional details concerning taxonomy, comparative morphology, geographic distributions, and natural history.

Ichthyomyines are small to medium-sized (35–155 g) semiaquatic carnivorous rodents that are morphologically distinctive in many external, skeletal, and visceral characters. The eyes and pinnae are small (*Anotomys* lacks visible external ears altogether), and the blunt rostrum is provided with numerous ventrally recurved mystacial vibrissae; genal vibrissae (present in most other Sigmodontinae) are absent. There are six mammae (in postaxial, abdominal, and inguinal pairs), the hindfeet are provided with a more or less distinct fringe of stiff hairs along the plantar margins, and the tail is well haired (lacking visible epidermal scales). The cranium is distinctive in dorsal view, lacking any trace of zygomatic notches; the supraorbital margins are smoothly rounded, without beads or sharp edges; and the braincase is smooth in lateral aspect, without any temporal ridges. In lateral view, a well-developed gnathic process of the premaxillae (projecting between the upper incisors) is invariably present, and a distinct bony tubercle for the origin of the superficial masseter is usually present at the base of the inferior zygomatic root. The infraorbital foramen is large and oval in cross-section (not narrowed ventrally as in most other muroids). The bony palate is smooth and produced posteriorly behind the toothrows; the posterolateral palatal pits are small, simple, and inconspicuous. The parapterygoid fossae are narrow and not deeply excavated, and the bony roof of the mesopterygoid fossa is not perforated by large sphenopalatine vacuities. The buccinator-masticatory and accessory oval foramina are widely separated by a stout vertical strut of the alisphenoid. The small auditory bullae are firmly anchored to the skull by broad overlap between the tegmen tympani and a posterior suspensory process of the squamosal. The tympanic membrane pars flaccida (present in all other Sigmodontinae) is absent, and there is no subsquamosal fenestra.

Ichthyomyine molars are tetralophodont (lacking well-developed mesolophs and mesolophids) with the principal cusps arranged in opposite labial-lingual pairs. The unworn anterocone of M1 has two functional cusps, but the anteromedian flexus is shallow, indistinct, or absent. The upper and lower third molars (M3/m3) are small or absent. The first and second upper molars have three roots each, whereas the lower first and second molars each have two roots; third molars (M3/m3) have variable root morphologies.

Counts of axial skeletal elements include 13–14 ribs, 19–20 thoracolumbar vertebrae, 4–5 sacral vertebrae, and 30–33 caudal vertebrae. The tuberculum of the first rib articulates with the transverse processes of the seventh cervical and first thoracic vertebrae, and the third (not the second) thoracic vertebra has a greatly elongated neural spine. The entepicondylar foramen of the humerus is absent.

The glandular epithelium of the stomach is variously reduced from the hemiglandular condition (Carleton 1973), the caecum is simple and very small, and a gall bladder is either present or absent. Male accessory secretory organs conform to the widespread condition among other Sigmodontinae (Voss and Linzey 1981). The glans penis is complex but taxonomically variable, with a terminal crater containing one to three bacular mounds, one or two dorsal papillae, two lateral papillae, and a bifurcate urethral process.

KEY TO THE GENERA OF ICHTHYOMYINI:

1. Carotid circulation primitive (pattern 1 of Voss 1988); five plantar pads on manus (hypothenar and third interdigital pads separate); stomach with glandular epithelium between esophagus and pyloric sphincter 2
1'. Carotid circulation derived (patterns 2 or 3 of Voss 1988); five or fewer plantar pads on manus; stomach with glandular epithelium restricted to oval patch on greater curvature 3
2. Hindfoot narrow, with weakly developed fringe of stiff hairs; claw of pedal digit V extends to but not beyond first interphalangeal joint of dIV; ears clearly visible above fur of head; philtrum present; tail < head and body length. . . . . *Neusticomys*
2'. Hindfoot broad, with well-developed fringe of stiff hairs; claw of pedal digit V extends well beyond first interphalangeal joint of dIV; ears buried in fur of head; philtrum present or absent; tail > head and body length . . . . *Chibchanomys*
3. Adult dorsal pelage dull gray-black when fresh (fading to dull rust color in old museum skins); pinnae visibly absent; tufts of pure white fur over external auditory canals; philtrum absent; superciliary vibrissae present; carotid circulation pattern 2 . . . . *Anotomys*
3'. Adult dorsal pelage glossy grizzled-brownish; pinnae visibly present; ear region without conspicuous tuft of white fur; philtrum present or absent; superciliary vibrissae absent; carotid circulation pattern 3 4
4. Manus with five plantar pads (hypothenar and third interdigital pads separate); nasal bones short, revealing nasal orifice in dorsal view; supraorbital foramina on dorsal surface of frontals in large adults; occipital condyles produced posteriorly well behind the rest of the occiput; gall bladder absent . . . . *Ichthyomys*
4'. Manus with four or fewer plantar pads (hypothenar and third interdigital pads always fused); nasal bones long, concealing nasal orifice in dorsal view; supraorbital foramina opening laterally within the orbits; occipital condyles not produced posteriorly behind the rest of the occiput; gall bladder present . . . . *Rheomys*

REMARKS: The key above includes *Rheomys*, a genus currently known only from Central America, but which is almost certainly present on the Colombian side of the mountainous divide (Serranía de Pirre, Serranía del Darién) that forms the political boundary between that country and Panama.

Genus *Anotomys* Thomas, 1906

The genus *Anotomys* contains a single species, *A. leander* Thomas, 1906, which is known to occur only in northern Ecuador between 2,800 and 4,000 m.

Anotomys leander is a medium-sized ichthyomyine (recorded adult weights range from 40 to 60 g) with body pelage that is dull gray-black dorsally and abruptly silvery white ventrally; there are tufts of pure white fur over the ear region (where no pinnae are visible), and the tail is unicolored-brownish. The philtrum (a naked median groove that extends from the rhinarium to the upper lip in most Sigmodontinae) is absent, supraorbital vibrissae (absent in other ichthyomyines) are present, and the external ears are reduced to low ridges of skin that are buried beneath the fur of the head. The manus has only four plantar pads because the hypothenar and third interdigital pads are fused. The large hindfeet (averaging about one-third of head and body length) are conspicuously webbed and provided with a dense fringe of silvery hairs, the plantar surface is heavily pigmented (blackish), the hypothenar pad is absent, and the claw of digit V extends to the base of the claw on digit IV. The tail is longer than the combined length of head and body.

The nasal bones are long, concealing the nasal orifice from dorsal view. The rostrum is narrow (with dorsally exposed nasolacrimal capsules), and the interorbital region is also narrow in proportion to the inflated braincase. The supraorbital foramina open laterally (within the orbits), not onto the dorsal surface of the frontals. The condyles are not produced posteriorly behind the rest of the occiput and are not dorsally visible. In lateral view, the dorsal profile of the skull is distinctively flat over the rostrum and orbits, but abruptly convex over the braincase. The posterior edge of the inferior zygomatic root of each zygomatic plate is located above (not anterior to) the anterocone of M1, and the tip of the masseteric process never projects ventrally below the plane of the molar alveoli. The carotid arterial supply conforms to pattern 2 (of Voss 1988), with a large stapedial foramen but without a squamosal-alisphenoid groove for the supraorbital stapedial ramus. The lower third molar (m3) is small, without a large posterior lobe corresponding to the entoconid-hypoconid cusp pair.

There are 14 ribs, 20 thoracolumbar vertebrae, and 5–6 sacral vertebrae. Gastric glandular epithelium is restricted to a broad band that crosses the greater curvature on the right-hand side of the stomach, and a gall bladder is present. The penis has a single bacular mound bearing two large dorsolateral spines and is supported by a single cartilaginous digit; two dorsal crater papillae are present. Other skeletal, visceral, and reproductive characters conform to those of the tribe.

SYNONYMS:

Anotomys Thomas, 1906a:86; type species *Anotomys leander* Thomas, 1906, by original designation.

Anotomys leander Thomas, 1906
Earless Water Mouse
SYNONYM:

Anotomys leander Thomas, 1906a:87; type locality "Mount Pichincha," Pichincha, Ecuador.

DESCRIPTION: As for the genus.

DISTRIBUTION: *Anotomys leander* is known only from three identifiable localities (two of which are so close together as to be indistinguishable on a large-scale map) at recorded elevations of 2,800 and 4,000 m in northern Ecuador (see Remarks).

SELECTED LOCALITIES (Map 138): ECUADOR: Napo, 6.9 km by road W of Papallacta (AMNH 244605); Pichincha, Mount Pichincha (type locality of *Anotomys leander* Thomas).

SUBSPECIES: *Anotomys leander* is monotypic.

NATURAL HISTORY: The only available natural history information about *Anotomys leander* comes from 12 specimens that were trapped along the banks of a torrential

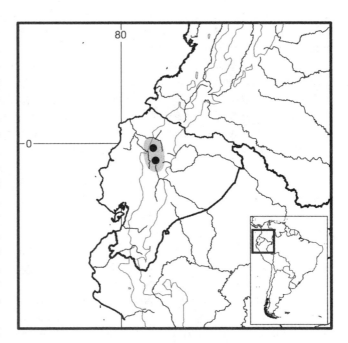

Map 138 Selected localities for *Anotomys leander* (●). Contour line = 2,000 m.

mountain stream between 3,600 and 3,800 m near Papallacta, Provincia de Napo, Ecuador. The bordering vegetation at this locality (where another ichthyomyine, *Neusticomys monticolus*, was also trapped) included grassy páramo, mossy elfin forest, and a shrubby ecotone at the forest edge. Dissected stomachs from these specimens contained the remains of numerous small aquatic arthropods, principally the immature stages (larvae or nymphs) of caddisflies (Trichoptera: Hydropsychidae, Limnephilidae, Rhyacophilidae), mayflies (Ephemeroptera: Baetidae), stoneflies (Plecoptera: Gripopterygidae), net-winged midges (Diptera: Blephariceridae) and other benthic insects. An adult female caught in the late dry season (March) had a single large embryo in utero.

REMARKS: A published record of *A. leander* from the western Andes of Colombia (Cuervo Díaz et al. 1986) appears to be unvouchered and cannot be confirmed (M. Gómez-Laverde, pers. comm.).

Genus *Chibchanomys* Voss, 1988

The genus *Chibchanomys* contains two species that inhabit high elevations in the Andes of Colombia, Ecuador, and Peru. A juvenile specimen from Bolivia (unassignable to species) has also been referred to this genus (S. Anderson 1997).

Species of *Chibchanomys* are small ichthyomyines (recorded adult weights range from 35 to 50 g) with dull gray-black (not glossy brownish) dorsal body pelage and silvery or pale-gray venters. Supraorbital vibrissae (present in *Anotomys*) are absent, and the pinnae (visibly absent in *Anotomys*) are small and concealed in the fur of the head. The manus has five separate plantar pads (thenar, hypothenar, and three interdigitals). The pes is provided with a well-developed fringe of stiff, silvery hairs, and the claw of pedal digit V extends beyond the first interphalangeal joint of digit IV. The tail is longer than the combined length of the head and body.

The nasal bones are long, concealing the nasal orifice from dorsal view, and the rostrum is narrow with dorsally exposed nasolacrimal capsules. The interorbital region is narrow in proportion to the inflated braincase. The supraorbital foramina open laterally (within the orbits), not onto the dorsal surface of the frontals. The occipital condyles are not produced posteriorly behind the rest of the occiput, so they are not dorsally visible. In lateral view, the dorsal cranial profile is more or less flat over the rostrum and interorbit but convex over the braincase. The posterior edge of the inferior zygomatic root of each zygomatic plate is located above (not anterior to) M1, and the tip of the masseteric tubercle does not project below the plane of the molar alveoli. The carotid arterial morphology conforms to pattern 1 (of Voss 1988), with a large stapedial foramen and a prominent squamosal-alisphenoid groove for the supraorbital stapedial

ramus. The lower third molar (m3) is small, without a large posterior lobe corresponding to the entoconid-hypoconid cusp pair.

There are 13 to 14 ribs, 19 thoracolumbar vertebrae, and 4 sacral vertebrae. The stomach is nearly hemiglandular, with gastric glandular epithelium present between the esophagus and the pyloric sphincter; and a gall bladder is present. The penis has three bacular mounds supported by a tridigitate bacular cartilage, and a single dorsal crater papilla. (Postcranial, visceral, and genital characters are based on *Chibchanomys trichotis* because relevant anatomical preparations are not available from *C. orcesi*.)

SYNONYMS:

Ichthyomys: Thomas, 1897e:220; part (description of *trichotis*); not *Ichthyomys* Thomas.

Rheomys: Thomas, 1906b:421; part (listing of *trichotis*); not *Rheomys* Thomas.

Anotomys: Handley, 1976:53; part (listing of *trichotis*); not *Anotomys* Thomas.

Chibchanomys Voss, 1988:321; type species *Ichthyomys trichotis* Thomas, by original designation.

KEY TO THE SPECIES OF *CHIBCHANOMYS*:
1. Philtrum (naked median groove from rhinarium to upper lip) present; rhinarium unpigmented (brownish on dry skins); metatarsal formula (relative lengths of bones) III > IV > II >> V > I; orbicular apophysis of malleus present
. . . . *Chibchanomys orcesi*
1′. Philtrum absent; rhinarium pigmented (blackish on dry skins); metatarsal formula IV > III > II = V > I; orbicular apophysis of malleus absent . . . .*Chibchanomys trichotis*
REMARKS: A single immature Bolivian specimen referred to *Chibchanomys* by S. Anderson (1997) is not referred to either of the species recognized here.

Chibchanomys orcesi Jenkins and Barnett, 1997
Orces's Andean Water Mouse
SYNONYM:

Chibchanomys orcesi Jenkins and Barnett, 1997:124; type locality "Lake Luspa, Las Cajas, Provincia Azuay, Ecuador, 02°50′S 79°30′W, altitude 3700 m."

DESCRIPTION: See keys (above) and Jenkins and Barnett (1997).

DISTRIBUTION: *Chibchanomys orcesi* is known from 2,400 to 3,700 m in southern Ecuador and central Peru.

SELECTED LOCALITIES (Map 139): ECUADOR: Azuay, Las Cajas, Lake Luspa (type locality of *Chibchanomys orcesi* Jenkins and Barnett). PERU: Huánuco, Cordillera Carpish (LSUMZ 14406).

SUBSPECIES: *Chibchanomys orcesi* is monotypic.

NATURAL HISTORY: Everything that is known about the ecology and behavior of *C. orcesi* was summarized by Barnett (1999), whose observations were based on six specimens trapped along small mountain streams bordered by grassy páramo vegetation or pastures near Las Cajas, Provincia Azuay, Ecuador. Dissected stomachs from two kill-trapped specimens contained the remains of aquatic insects (mayfly naiads and caddisfly larvae) and unidentified fish. A single live-trapped animal from the same locality and maintained in captivity for four months was almost exclusively nocturnal, emerging from its burrow at night to capture small fish in an artificial stream.

REMARKS: The Peruvian specimen (LSUMZ 14406) here referred to *Chibchanomys orcesi* was originally reported as *Anotomys leander* by A. L. Gardner (1971) and subsequently identified as *C. trichotis* by Voss (1988). However, the qualitative morphological characters of LSUMZ 14406 more closely resemble those of *C. orcesi* than of *C. trichotis*. External measurements reported by Jenkins and Barnett (1997:Table 1) suggest that *C. orcesi* has much smaller hindfeet (19 to 24 mm) than do *C. trichotis* and the Peruvian specimen (30 to 33 mm), but some of this disparity may be an artifact of different measurement protocols; measured by the American method (heel to tip of longest claw), the dry hindfoot of BM 82.816 (the holotype of *orcesi*) is 28 mm.

A. L. Gardner (1971) karyotyped LSUMZ 14406 and reported a diploid number ($2n$) of 92 chromosomes and a fundamental number (FN) of 98 or 100 autosomal arms; the X chromosomes were thought to be the largest biarmed pair.

Chibchanomys trichotis (Thomas, 1897)
Northern Andean Water Mouse

SYNONYMS:

Ichthyomys trichotis Thomas, 1897e:220; type locality allegedly "W. Cundinamarca, in low country near to Magdalena R[iver]" in Colombia, but the holotype was probably collected in the Cordillera Oriental near Bogotá (Voss 1988).

Rheomys trichotis: Thomas, 1906b:421; name combination.

Anotomys trichotis: Handley, 1976:53; name combination.

Chibchanomys trichotis: Voss, 1988:321; first use of current name combination.

DESCRIPTION: See keys (above), Thomas (1897e), and Voss (1988: 322–324).

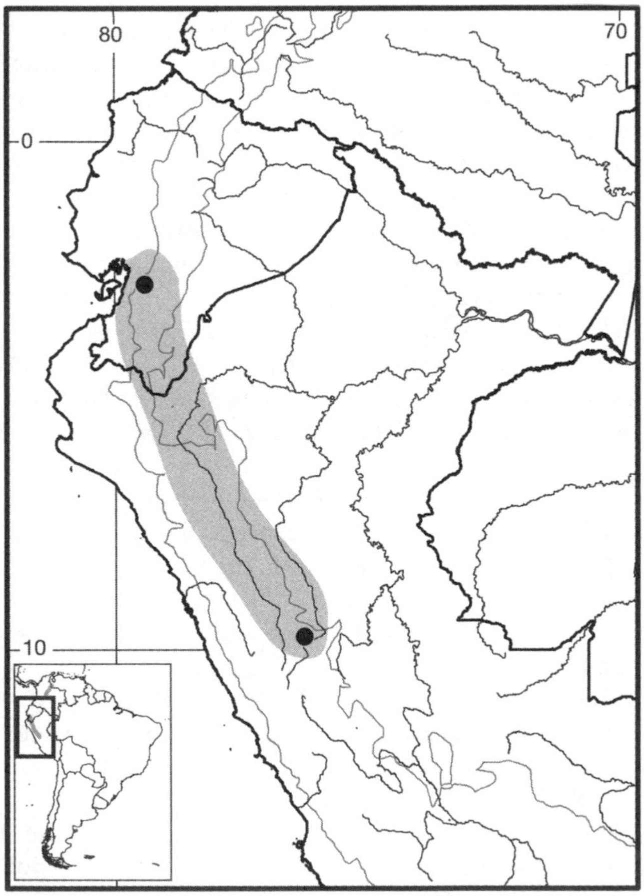

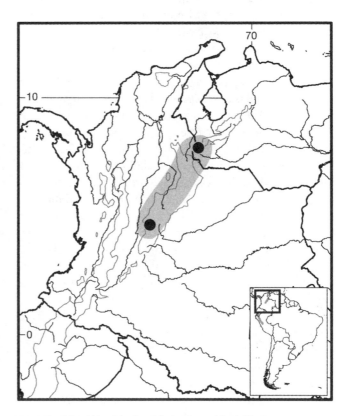

Map 139 Selected localities for *Chibchanomys orcesi* (●). Contour line = 2,000 m.

Map 140 Selected localities for *Chibchanomys trichotis* (●). Contour line = 2,000 m.

DISTRIBUTION: *Chibchanomys trichotis* is known from elevations between 2,400 and 2,900 m in the Táchira Andes of western Venezuela and in the Cordillera Oriental near Bogotá, Colombia.

SELECTED LOCALITIES (Map 140): COLOMBIA: Cundinamarca, San Cristobal (FMNH 71226). VENEZUELA: Táchira, Buena Vista (USNM 442606).

SUBSPECIES: *Chibchanomys trichotis* is monotypic.

NATURAL HISTORY: Five specimens of *Chibchanomys trichotis* were trapped along the banks of a small mountain stream bordered by cloud forest near Buena Vista, Estado Táchira, Venezuela. Dissected stomachs from four of these specimens contained only aquatic insects, including larval beetles (Helodidae), larval crane flies (Tipulidae), larval caddisflies (Leptoceridae), and nymphs of mayflies (Leptophlebiidae, Trichorythidae) and stoneflies (Perlidae). An adult female *C. trichotis* trapped in the middle of the dry season (January) at the same locality contained a single near-term embryo. Two specimens of *C. trichotis* collected at other localities were trapped along streams, but no additional ecological information was recorded about them.

Genus *Ichthyomys* Thomas, 1893

The genus *Ichthyomys* contains at least four valid species that occur in lowland and montane habitats from central Panama southward to central Peru and eastward to the coastal cordilleras of northern Venezuela.

Species of *Ichthyomys* are large ichthyomyines (recorded adult weights range from 85 to 155 g) with glossy brownish dorsal fur and countershaded (whitish or silvery) under parts. The philtrum is always present and broad, supraorbital vibrissae are absent, and the small external ears are always visible above the fur of the head. The manus has five separate plantar pads, including the thenar, hypothenar, and three separate interdigital pads. The hindfoot varies in size (its average length ranging from about one-fifth to almost one-third of head and body length in different species) but is always broad, with pigmented plantar epithelium (blackish on dried skins), and a well-developed natatory fringe of stiffened (usually silvery) hairs; the claw of digit V always extends beyond the first interphalangeal joint of digit IV. The tail may be shorter than, equal to, or longer than the combined length of the head and body; it is unicolored (all dark) in some species and bicolored (dark above, white below) in others.

The skull is highly distinctive in several respects. The nasal bones are short and truncated anteriorly so as to expose the nasal orifice in dorsal view, and the rostrum is broad with expanded lateral margins that conceal the nasolacrimal capsules from above. The supraorbital foramina open laterally (near the orbital margins) in young animals, but migrate dorsally as the interorbital region constricts in most adult specimens. The zygomatic processes of the maxilla and squamosal are strongly developed in old adults, flaring widely from the rostrum and braincase, respectively. The lambdoidal crests are likewise well developed in large adults, and the occipital condyles are produced posteriorly behind the rest of the occiput. In lateral view, the skull is diagnostically flattened from the tips of the nasals to the occiput, without any conspicuous convexity over the braincase. The posterior edge of the inferior zygomatic root of each zygomatic plate is always located anterior to M1, and the tip of the masseteric tubercle sometimes projects ventrally to or below the plane of the molar alveoli. The incisive foramina have straight, subparallel lateral margins. The carotid arterial circulation conforms to pattern 3 (of Voss 1988), with or without a tiny stapedial foramen, and lacking any grooves or foramina for the stapedial rami; instead, the ophthalmic and internal maxillary circulations are provided by an endocranial anastomosis of the internal carotid. In most species (*Ichthyomys pittieri* is the only exception), m3 is a distinctly bilobed tooth with subequal protoconid/metaconid and entoconid/hypoconid cusp pairs.

There are 13 ribs, 19 thoracolumbar vertebrae, and 4 sacral vertebrae in all examined skeletons. The gastric glandular epithelium is restricted to a broad band that crosses the greater curvature on the right-hand side of the stomach, and the gall bladder is absent. The penis has three bacular mounds (supported by a tridigitate bacular cartilage) and a single dorsal crater papilla. Other skeletal, visceral, and reproductive characters conform to those of the tribe.

SYNONYMS:

Habrothrix: Winge, 1891:20; part (description of *hydrobates*); not *Habrothrix* Wagner.

Ichthyomys Thomas, 1893a:337; type species *Ichthyomys stolzmanni* Thomas, by original designation.

KEY TO THE SPECIES OF *ICHTHYOMYS*:

1. Tail sharply bicolored (dark above, white below); hindfoot very large (>39 mm); in Andean foothills of eastern Ecuador and Peru . . . . *Ichthyomys stolzmanni*
1′. Tail unicolored (dark above and below); hindfoot smaller (<38 mm) 2
2. Tail (measured from dorsal flexure to fleshy tip) distinctly shorter than head and body length (obtained by subtracting length of tail from total length); upper molars very small (crown length M1–M3 <3.7 mm); lower third molar (m3) not bilobed (without a distinct entoconid-hypoconid cusp pair); in Serranía de Aroa and coastal cordilleras of northern Venezuela . . . . *Ichthyomys pittieri*

2′. Tail about as long as combined length of head and body; upper molars larger (crown length >4.0 mm); m3 bilobed 3
3. Incisors, rostrum, zygomatic plates, zygomatic arches, and lambdoidal crests massively developed in older adults; in Pacific lowlands and foothills of western Ecuador . . . . *Ichthyomys tweedii*
3′. Incisors, rostrum, zygomatic plates, and zygomatic arches much narrower, even in old adults; lambdoidal crests usually inconspicuous; in high Andes and adjacent foothills of Ecuador, Colombia, and western Venezuela . . . . *Ichthyomys hydrobates*

REMARKS: In the absence of many qualitative character differences, species of *Ichthyomys* are less easily keyed out than species of other ichthyomyine genera. Indeed, *I. tweedii* and *I. hydrobates* are principally distinguished by ontogenetically variable morphometric traits such as size and correlated allometric proportions that are hard to reduce to simple ratio comparisons, hence the ambiguity of the last key couplet. However, where the ranges of these two species are juxtaposed on the Andean slopes of western Ecuador, the local form of *I. hydrobates* can be identified by the absence of distinct lateral parapterygoid margins (Voss 1988:Fig. 37B), which are always well developed in *I. tweedii*.

Ichthyomys hydrobates (Winge, 1891)
Silver-bellied Ichthyomys

SYNONYMS:

Habrothrix hydrobates Winge, 1891:20; type locality "Sierra de Mérida," Mérida, Venezuela.

Ichthyomys hydrobates: Thomas, 1893a:338; first use of current name combination.

Ichthyomys söderströmi de Winton, 1896:512; type locality "Rio Machangara," near Quito, Pichincha, Ecuador.

Ichthyomys nicefori Thomas, 1924a:165; type locality "Neighborhood of Bogotá," Cundinamarca, Colombia.

Ichthyomys hydrobates hydrobates: Cabrera, 1961:509; name combination.

Ichthyomys hydrobates nicefori: Cabrera, 1961:510; name combination.

Ichthyomys hydrobates soderstromi: Cabrera, 1961:510; name combination.

DESCRIPTION: See keys (above) and Voss (1988:334).

DISTRIBUTION: *Ichthyomys hydrobates* is known to occur in the Andes of western Venezuela, in the eastern and western Andes of Colombia, and in the western Andes of northern Ecuador; recorded elevations range from about 1,000 to 2,800 m.

SELECTED LOCALITIES (Map 141): COLOMBIA: Cauca, Cerro Munchiquito (USNM 294985); Cundina-marca, Paime (BM 23.11.13.9). ECUADOR: Pichincha, Guápulo (AMNH 39593). VENEZUELA: Mérida, La Mucuy (UMMZ 156375); Táchira, Paso Hondo, Río Potosí (Ochoa and Soriano 1991).

SUBSPECIES: Cabrera (1961) recognized three subspecies as did Voss (1988:334), who commented "each name is the descriptor of a single local population" present in Venezuela (nominotypical form), Colombia (*nicefori*), and Ecuador (*soderstromi*).

NATURAL HISTORY: All specimens of *Ichthyomys hydrobates* for which the ecological circumstances of capture have been recorded were trapped or caught by hand along the banks of streams or irrigation canals; although most were taken in agricultural landscapes, the original vegetation throughout the range of this species was probably cloud forest. *Ichthyomys hydrobates* has been collected sympatrically with other ichthyomyines, including *Neusticomys monticolus* (at Guarumal, Provincia Pichincha, Ecuador) and *N. mussoi* (at Paso Hondo, Estado Táchira, Venezuela). Unfortunately, no dietary data are available from this species.

REMARKS: As currently recognized, *Ichthyomys hydrobates* is very widely distributed across dissected mountainous terrain and includes diagnosably different local populations. Whether the three nominal taxa currently synonymized under this binomen are really conspecific remains to be convincingly demonstrated.

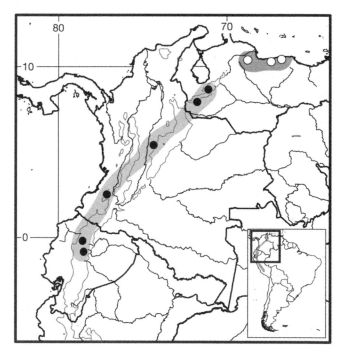

Map 141 Selected localities for *Ichthyomys hydrobates* (●) and *Ichthyomys pittieri* (○). Contour line = 2,000 m.

Ichthyomys pittieri Handley and Mondolfi, 1963
Pittier's Ichthyomys

SYNONYM:

Ichthyomys pittieri Handley and Mondolfi, 1963:417; type locality "near the head of the Río Limón, Parque Nacional de Rancho Grande, Estado Aragua, Venezuela."

DESCRIPTION: See keys (above), Mondolfi and Handley (1963), Voss (1988:334–336), and F. J. García, Machado et al. (2012).

DISTRIBUTION: *Ichthyomys pittieri* is known from a few scattered localities in the coastal cordilleras and adjacent hills of northern Venezuela, at elevations ranging from 700 to 1,750 m (F. J. García, Machado et al. 2012).

SELECTED LOCALITIES (Map 141): VENEZUELA: Aragua, Rancho Grande (type locality of *Ichthyomys pittieri* Handley and Mondolfi); Vargas, Naiguatá (MBUCV I-2803). Yaracuy, Finca El Jaguar, 21 km NW of Aroa (MHNLS 8114).

SUBSPECIES: *Ichthyomys pittieri* is monotypic.

NATURAL HISTORY: All known specimens of *Ichthyomys pittieri* have been trapped or captured by hand in small streams or drainage canals bordered by lower montane forest. Captive animals consumed a wide range of small aquatic animals but evinced a marked preference for pseudothelphusid crabs, which were killed and eaten in a stereotypical manner (Voss et al. 1982). The stomach contents of one specimen contained aquatic insects (Trichoptera and Ephemeroptera; F. J. García, Machado et al. 2012). An old adult female trapped at Finca El Jaguar, Estado Yaracuy, Venezuela, contained four embryos (field notes of H. G. Castellanos, 26 July 1987).

REMARKS: Schmid et al. (1988) karyotyped a female specimen of *Ichthyomys pittieri* and reported a diploid number (2*n*) of 92 chromosomes and a fundamental number (FN) of 98 autosomal arms, apparently identical to the karyotype that A. L. Gardner (1971) reported for *Chibchanomys orcesi*.

Ichthyomys stolzmanni Thomas, 1893
Stolzmann's Ichthyomys

SYNONYMS:

Ichthyomys stolzmanni Thomas, 1893a:339; type locality "Chanchamayo," Junín, Peru.

Ichthyomys orientalis Anthony, 1923b:7; type locality "near Rio Napo," Napo, Ecuador.

Ichthyomys stolzmanni orientalis: Cabrera, 1961:510; name combination.

Ichthyomys stolzmanni stolzmanni: Cabrera, 1961:510; name combination.

DESCRIPTION: See keys (above), Thomas (1893a), Anthony (1923b), Voss (1988:336–337), and Pacheco and Ugarte-Núñez (2011).

DISTRIBUTION: *Ichthyomys stolzmanni* is known only from one locality in the eastern Andean piedmont of Ecuador and from four localities in Peru, ranging in elevation from 2,800 to 3,400 m.

SELECTED LOCALITIES (Map 142): ECUADOR: Napo, Río Napo [probably near Tena] (AMNH 62382). PERU: Ayacucho, Arizona (Pacheco and Ugarte-Núñez 2011), El Bagrecito (Pacheco and Ugarte-Núñez 2011), Vischongo, Río Ponacochas (Pacheco and Ugarte-Núñez 2011); Junín, Chanchamayo (type locality of *Ichthyomys stolzmanni* Thomas).

SUBSPECIES: Although Voss (1988) recognized *I. s. orientalis* (in Ecuador) and *I. s. stolzmanni* (in Peru) as valid subspecies, Pacheco and Ugarte-Núñez (2011) found that these nominal taxa could not be distinguished by any phenotypic criterion and concluded that *I. stolzmanni* is monotypic.

NATURAL HISTORY: According to Pacheco and Ugarte-Núñez (2011), *Ichthyomys stolzmanni* is common along Andean streams in the Puna biome of south-central Peru, where this species is locally known as *mayo sonso*, a derogatory term that applies to a riverine animal that causes damage (*I. stolzmanni* is apparently a major predator of non-native trout in local fish farms). Standard trapping methods are ineffective for this species, most individuals having been captured by hand or in fishing nets.

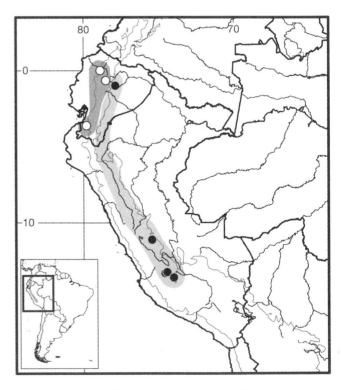

Map 142 Selected localities for *Ichthyomys stolzmanni* (●) and *Ichthyomys tweedii* (○). Contour line = 2,000 m.

Ichthyomys tweedii Anthony, 1921

Tweedy's Ichthyomys

SYNONYMS:

Ichthyomys tweedii Anthony, 1921b:1; type locality "Portovelo, Prov. del Oro, Ecuador; altitude, 2000 ft."

Ichthyomys caurinus Thomas, 1924d:541; type locality "below Gualea," Pichincha, Ecuador.

Ichthyomys hydrobates caurinus: Cabrera, 1961:509; name combination.

Ichthyomys hydrobates tweedii: Cabrera, 1961:510; name combination.

DESCRIPTION: See keys (above), Anthony (1921b), Thomas (1924d), and Voss (1988:337–338).

DISTRIBUTION: *Ichthyomys tweedii* is known from the lowlands and Andean foothills (near sea level to about 1,350 m) of western Ecuador and central Panama. It presumably also occurs in geographically intermediate regions, such as the Pacific lowlands of western Colombia.

SELECTED LOCALITIES (Map 142): ECUADOR: El Oro, Portovelo (type locality of *Ichthyomys tweedii* Anthony); Imbabura, Paramba (BM 99.11.3.4); Pichincha, Mindo (UMMZ 126300).

SUBSPECIES: I treat *Ichthyomys tweedii* as monotypic.

NATURAL HISTORY: The only natural history information about South American populations of *Ichthyomys tweedii* comes from eight specimens that were trapped between 1,290 and 1,330 m near Mindo, Provincia Pichincha, Ecuador. All were taken along small streams bordered by dense secondary vegetation. Stomachs dissected from these specimens contained the remains of fish (Characidae), crabs (Pseudothelphusidae), roaches (Blattidae), dobsonfly larvae (Corydalidae), and other aquatic insects. A female specimen collected in the late wet season (May) at this locality contained two newly implanted embryos in utero.

Genus *Neusticomys* Anthony, 1921

As currently constituted (by Voss 1988), *Neusticomys* includes *Daptomys* Anthony, 1929, as a synonym and contains the most plesiomorphic (least aquatically adapted) ichthyomyine species. Based on published records, the genus is known to occur in montane habitats in Colombia and Ecuador, and from lower montane and lowland habitats in Venezuela, Brazil, the Guianas, and Peru.

Species of *Neusticomys* are small to medium-sized (30–75 g) ichthyomyines with dull gray-black or glossy-brownish dorsal body pelage; the ventral pelage is colored more or less like the dorsum, slightly paler in some specimens but never abruptly white or silvery. Supraorbital vibrissae are absent, but a philtrum is always pres-

ent, and the small pinnae are distinctly visible above the fur. The manus has five separate plantar pads including the thenar, hypothenar, and three interdigitals. The hindfoot is always small (its length consistently less than one-fourth of the head and body length) and narrow, with inconspicuous interdigital webs and weakly developed fringing hairs; the plantar surface is unpigmented or weakly pigmented (pink in fresh specimens, brownish on dried skins), and the claw of digit V extends just to or slightly beyond the first interphalangeal joint of digit IV. The tail is less than or equal to the combined length of the head and body, and is invariably unicolored (dark above and below).

The nasals are long, concealing the nasal orifice from dorsal view. The rostrum is moderately to very broad, with nasolacrimal capsules that are partially or completely concealed from dorsal view. The interorbital region is broad in proportion to the uninflated braincase, and the supraorbital foramina open laterally (within the orbits), not onto the dorsal surface of the frontals. The condyles are concealed from dorsal view (in *N. monticolus*) or are produced slightly behind the rest of the occiput (in other congeners). The dorsal cranial profile is almost flat, without a conspicuous bulge over the braincase. The posterior edge of the inferior zygomatic root is anterior to M1 in some species and above the anterocone of that tooth in others; the tip of the masseteric tubercle is above or below the plane of the molar alveoli. The incisive foramina are subparallel or have weakly convex lateral margins. The carotid arterial supply conforms to pattern 1 (of Voss 1988), the osteological traces of which include a large stapedial foramen (for the stapedial artery) and a prominent squamosal-alisphenoid groove and sphenofrontal foramen (which transmit the supraorbital stapedial ramus). The third mandibular molar (m3), when present, lacks a distinct posterior lobe corresponding to the entoconid-hypoconid cusp pair.

There are 13 to 14 ribs, 19 thoracolumbar vertebrae, and 4 sacral vertebrae. The stomach is nearly hemiglandular, with gastric glandular epithelium between the esophagus and the pyloric sphincter, and the gall bladder is present. The penis has three bacular mounds (supported by a tridigitate bacular cartilage) and a single dorsal papilla. Other external, skeletal, and visceral characters conform to those of the tribe.

SYNONYMS:

Neusticomys Anthony, 1921b:2; type species *Neusticomys monticolus* Anthony, by original designation.

Daptomys Anthony, 1929:1; type species *Daptomys venezuelae* Anthony, by original designation.

REMARKS: Voss (1988:339) synonymized *Daptomys* with *Neusticomys* for reasons that now seem less compelling.

Although all the species here included in *Neusticomys* can be diagnosed from other ichthyomyines by a number of shared primitive attributes, there appears to be no evidence that they form a monophyletic group. Instead, *N. monticolus* shares at least one conspicuous apomorphy (dull gray-black adult pelage) with *Anotomys* and *Chibchanomys*, whereas all of the lowland forms (*ferreirai, mussoi, oyapocki, peruviensis,* and *venezuelae*) share other derived traits inter se (notably the dorsal exposure of the occipital condyles; Ochoa and Soriano 1991). Future analyses of ichthyomyine relationships as based on molecular sequence data and other characters not analyzed by Voss (1988) may justify the recognition of *Daptomys* as a valid taxon.

KEY TO THE SPECIES OF *NEUSTICOMYS*:

1. Adult dorsal fur dull gray-black (when fresh, foxing with age to brownish-gray in old museum specimens); rostrum narrow, with dorsally exposed nasolacrimal capsules; occipital condyles not projecting posteriorly beyond rest of occiput, not exposed dorsally . . . . *Neusticomys monticolus*
1'. Adult dorsal fur glossy brownish; rostrum broad, nasolacrimal capsules not dorsally exposed; occipital condyles projecting posteriorly behind rest of occiput, exposed dorsally. 2
2. Molars 2/2 or 3/2 (M3 may be absent, m3 always absent) 5
2'. Molars 3/3 (M3/m3 present) 3
3. Ears and feet covered with dark hairs; posterior edge of inferior zygomatic root of zygomatic plates above M1 anterocone . . . . *Neusticomys venezuelae*
3'. Inside of ears and dorsal surface of feet covered with pale (cream-colored) hairs; posterior edge of inferior zygomatic root anterior to M1. 4
4. Hindfoot about 30 mm, crown length of M1–M3 about 3.8 mm . . . . *Neusticomys peruviensis*
4'. Hindfoot about 21 mm, crown length of M1–M3 range from 3.3–3.4 mm . . . . *Neusticomys mussoi*
5. Upper third molar absent (molars 2/2) . . . . *Neusticomys oyapocki*
5'. Upper third molar present (molars 3/2) . . . . *Neusticomys ferreirai*

Neusticomys ferreirai Percequillo, Carmignotto, and de Silva, 2005
Ferreira's Ichthyomyine

SYNONYM:

Neusticomys ferreirai Percequillo, Carmignotto, and de Silva, 2005:874; type locality "near Juruena, 20 km W of the west bank of the Rio Juruena, Matto Grosso, Brazil."

DESCRIPTION: See keys (above) and Percequillo et al. (2005).

DISTRIBUTION: *Neusticomys ferreirai* is an Amazonian lowland species known from two localities between the Rio Madeira and the Rio Tapajós and another between the Rio Xingu and the Rio Tocantins (C. L. Miranda et al. 2012).

SELECTED LOCALITIES (Map 143): BRAZIL: Mato Grosso, Aripuanã (C. L. Miranda et al. 2012), Juruena (type locality of *Neusticomys ferreirai* Percequillo, Carmignotto, and de Silva); Pará, Floresta Nacional Tapirapé-Aquiri, Marabá (C. L. Miranda et al. 2012).

SUBSPECIES: *Neusticomys ferreirai* is monotypic.

NATURAL HISTORY: All of the five known specimens of *Neusticomys ferreirai* were taken in pitfall traps in primary lowland forest; none were taken in streams (Percequillo et al. 2005; C. L. Miranda et al. 2012). The female holotype was pregnant with five large embryos, the most yet reported from any ichthyomyine. The stomach of the paratype contained the remains of unidentified crustaceans.

REMARKS: *Neusticomys ferreirai* is phenotypically similar to *Neusticomys oyapocki* (see Percequillo et al. 2005:Table 2), notably in lacking m3, and it seems plausible that these species are allopatric sister taxa separated by the lower Amazon. The karyotype of *N. ferreirai* (2*n* = 92, FN = 98; Percequillo et al. 2005) closely resembles those of other ichthyomyines, as previously described in the literature (e.g., *Chibchanomys orcesi, Ichthyomys pittieri*).

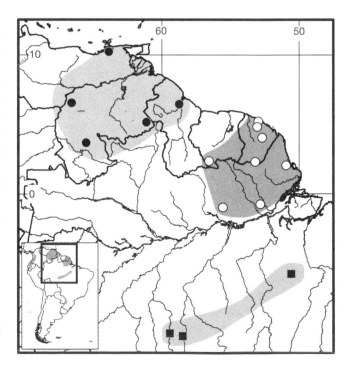

Map 143 Selected localities for *Neusticomys ferreirai* (■), *Neusticomys oyapocki* (○), and *Neusticomys venezuelae* (●).

Neusticomys monticolus Anthony, 1921
Montane Ichthyomyine

SYNONYM:

Neusticomys monticolus Anthony, 1921b:2; type locality "Nono Farm, 'San Francisco' near Quito," Pichincha, Ecuador.

DESCRIPTION: See keys (above), Anthony (1921b), and Voss (1988:343–344).

DISTRIBUTION: *Neusticomys monticolus* is known from elevations between about 1,800 and 3,800 m in Andean Ecuador and Colombia. In Ecuador, *N. monticolus* has been collected in both the eastern and western cordilleras from about 2°S northward to the equator. In Colombia, *N. monticolus* has been collected in the western Andes (Cordillera Occidental) to about 6°N, and in the Central Andes (Cordillera Central) to about 5°N (Gómez-Laverde 1994; Velandia-Perilla and Saavedra-Rodríguez 2013).

SELECTED LOCALITIES (Map 144): COLOMBIA: Antioquia, Santa Barbara (FMNH 71221); Huila, San Antonio (FMNH 71224); Risaralda, La Pastora (ICN 12118); Valle del Cauca, Pichindé (HTC 3365). ECUADOR: Bolívar, Sinche (AMNH 66848); Chimborazo, Pauchi (AMNH 62920); Napo, 8.2 km W of Papallacta (UMMZ 155605); Pichincha, Guarumal (UMMZ 126298); Tunguragua, San Antonio (AMNH 67548).

SUBSPECIES: *Neusticomys monticolus* is monotypic.

NATURAL HISTORY: Most specimens of *Neusticomys monticolus* have been trapped in small streams bordered

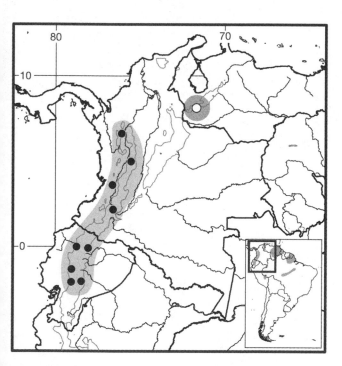

Map 144 Selected localities for *Neusticomys monticolus* (●) and *Neusticomys mussoi* (○). Contour line = 2,000 m.

by páramo vegetation or cloud forest, but a few have been taken at moist sites that were not adjacent to running water. This species occurs sympatrically with other ichthyomyines, as documented in the preceding accounts for *Anotomys leander* and *Ichthyomys hydrobates*. Dissected stomachs from eight specimens trapped at Guarumal, Pichincha province, Ecuador, contained only aquatic insects, including larval water beetles (Helodidae), larval crane flies (Tipulidae), mayfly nymphs (Baetidae, Leptophlebiidae), stonefly nymphs (Perlidae), and larval caddisflies (Calamoceratidae, Hydropsychidae). A single female specimen collected in the early wet season (mid-May) near Papallacta, Provincia Napo, Ecuador, contained two large embryos.

REMARKS: The few available specimens from the western Andes (Cordillera Occidental) of Colombia exhibit morphological differences from Ecuadorean material and may represent a distinct species.

Neusticomys mussoi Ochoa and Soriano, 1991
Musso's Ichthyomyine

SYNONYM:

Neusticomys mussoi Ochoa and Soriano, 1991:97; type locality "Paso Hondo, Rio Potosí, 14 km SE Pregonero, Táchira, Venezuela (7°57′N, 71°39′W), 1,050 m elevation."

DESCRIPTION: See keys (above) and Ochoa and Soriano (1991). Percequillo et al. (2005) tabulated qualitative character comparisons between *N. mussoi* and other lowland congeners.

DISTRIBUTION: *Neusticomys mussoi* is known only from the type locality.

SELECTED LOCALITIES (Map 144): VENEZUELA: Táchira, Paso Hondo, Río Potosí (Ochoa and Soriano).

SUBSPECIES: *Neusticomys oyapocki* is monotypic.

NATURAL HISTORY: The two known specimens of *Neusticomys mussoi* were trapped in a stream bordered by semideciduous forest, where another ichthyomyine species, *Ichthyomys hydrobates*, was also collected.

REMARKS: Although Ochoa and Soriano (1991) focused their taxonomic comparisons on the problem of distinguishing *Neusticomys mussoi* from another Venezuelan congener, *N. venezuelae*, *N. mussoi* more closely resembles a third species, *N. peruviensis*, with which it appears to differ principally in size. Noteworthy similarities between *mussoi* and *peruviensis* include their pale (cream-colored) ears and feet, together with the anterior position of the inferior zygomatic root (see key). Future phylogenetic research should test the hypothesis that *mussoi* and *peruviensis* are sister taxa, sharing a common ancestor more recently than either does with other lowland congeners (*N. ferreirai*, *N. oyapocki*, and *N. venezuelae*).

Neusticomys oyapocki (Dubost and Petter, 1979)
Guianan Ichthyomyine

SYNONYMS:

Daptomys oyapocki Dubost and Petter, 1979:436; type locality "Trois Sauts, French Guyana."

Neusticomys oyapocki: Voss, 1988: 344; first use of current name combination.

DESCRIPTION: See keys (above), Dubost and Petter (1979), Voss et al. (2001:99–102), and R. N. Leite et al. (2007:Table 1). Qualitative character comparisons of *N. oyapocki* with other lowland congeners were tabulated by Percequillo et al. (2005).

DISTRIBUTION: *Neusticomys oyapocki* is currently known from just eight lowland localities (<500 m elevation) in Surinam, French Guiana and the neighboring Brazilian states of Amapá and Pará (R. N. Leite et al. 2007; Lim and Joemratie 2011; C. L. Miranda et al. 2012).

SELECTED LOCALITIES (Map 143): BRAZIL: Amapá, Fazenda Itapoã (MPEG 24252); Pará, Castanhal, Monte Dourado, Rio Jari (R. N. Leite et al. 2007), Floresta Estadual de Trombetas (C. L. Miranda et al. 2012). FRENCH GUIANA: Les Nouragues (C. L. Miranda et al. 2012), Paracou (AMNH 267597), St.-Eugène (MNHN 1995–3234), Trois Sauts (type locality of *Daptomys oyapocki* Dubost and Petter). SURINAM: Nickerie, Kutari River (Lim and Joemratie 2011).

SUBSPECIES: *Neusticomys oyapocki* is monotypic.

NATURAL HISTORY: Two French Guianan specimens of *Neusticomys oyapocki* were captured in pitfall traps that were about 5 m from small streams in primary rainforest (Voss et al. 2001), and another was captured by hand under a native house (Catzeflis, 2012). Pitfall traps were also used to collect Brazilian specimens in a managed forestry landscape (R. N. Leite et al. 2007), in rainforest (C. L. Miranda et al. 2012), and in *campo* (savanna) vegetation located at least 1 km from any forested watercourse (Nunes 2002). The single known Surinam specimen was live-trapped at a site with both upland forest and swamps (Lim and Joemratie 2011). The ecological circumstances of other collected specimens are not recorded in the literature, but the type was said to have been found "non loin des rives de l'Oyapock" (Dubost and Petter 1979:436), implying that it was not actually taken in a river or stream. Therefore, there is abundant evidence that this ichthyomyine species is not restricted to aquatic habitats.

Neusticomys peruviensis (Musser and Gardner, 1974)
Peruvian Ichthyomyine

SYNONYMS:

Daptomys peruviensis Musser and Gardner, 1974: 7; type locality "Balta (10°08′S, 17 [*sic*, 71]°13′W), at the point where the streams known to the local Cashinahua Indi-

ans as the Inuya and the Xumuya enter the Río Curanja, elevation 300 meters, Departamento Loreto (now Región de Ucayali), Perú."

Neusticomys peruviensis: Voss, 1988:344; first use of current name combination.

DESCRIPTION: See keys (above) and Musser and Gardner (1974). Percequillo et al. (2005) tabulated qualitative character comparisons of *N. peruviensis* with other lowland congeners.

DISTRIBUTION: *Neusticomys peruviensis* is currently known from just two localities in the Amazonian lowlands of southeastern Peru.

SELECTED LOCALITIES (Map 145): PERU: Madre de Dios, Pakitza (Pacheco and Vivar 1996); Ucayali, Balta (type locality of *Daptomys peruviensis* Musser and Gardner).

SUBSPECIES: *Neusticomys peruviensis* is monotypic.

NATURAL HISTORY: The specimen from Balta was trapped in a small stream (Musser and Gardner 1974), but the specimen from Pakitza was trapped at least 200 m from water (Pacheco and Vivar 1996). The surrounding habitat of both captures was primary evergreen lowland forest.

REMARKS: Pacheco and Vivar (1996) drew attention to several character differences between their young adult specimen and the fully adult holotype that could reflect either ontogenetic or taxonomic variation. Additional specimens are needed to evaluate the possibility that more than one species of *Neusticomys* occurs in eastern Peru.

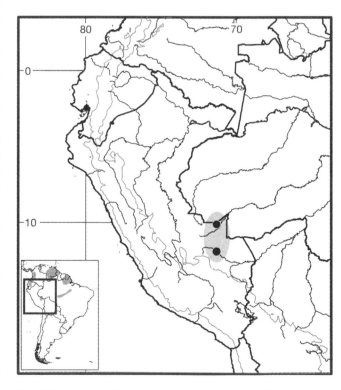

Map 145 Selected localities for *Neusticomys peruviensis* (●). Contour line = 2,000 m.

Neusticomys venezuelae (Anthony, 1929)
Venezuelan Ichthyomyine

SYNONYMS:

Daptomys venezuelae Anthony, 1929:1; type locality "Neverí [= Neverí], about 15 miles west of Cumanacoa," Sucre, Venezuela.

Neusticomys venezuelae: Voss, 1988:344; first use of current name combination.

DESCRIPTION: See keys (above), Anthony (1929), Musser and Gardner (1974), and Voss et al. (2001:Table 25). Qualitative character comparisons of *N. venezuelae* with other lowland congeners tabulated by Percequillo et al. (2005).

DISTRIBUTION: *Neusticomys venezuelae* is currently known from only five localities, ranging in elevation from near sea level to 1,400 m, in Venezuela and Guyana.

SELECTED LOCALITIES (Map 143): GUYANA: Cuyuni-Mazaruni, Kartabo (AMNH 142818). VENEZUELA: Amazonas, Cerro Duida (USNM 406123); Bolívar, Serranía de los Pijiguaos (EBRG 11939), San Ignacio Yuruaní (AMNH 257344); Sucre, Neverí (type locality of *Daptomys venezuelae* Anthony).

SUBSPECIES: *Neusticomys venezuelae* is monotypic.

NATURAL HISTORY: All of the specimens for which habitat information has been published were trapped along small streams bordered by evergreen lowland rainforest, semideciduous forest, or montane vegetation (Anthony 1929; Musser and Gardner 1974; Handley 1976; Ochoa and Soriano 1991). Unpublished field notes (by R. S. Voss, archived in the AMNH Department of Mammalogy) additionally record that three specimens collected in 1987 near San Ignacio Yuruaní, Bolívar state, Venezuela, were trapped among wet rocks behind a waterfall; the surrounding vegetation at this locality was primary lowland rainforest. Two stomachs dissected from specimens trapped at Neverí, Sucre state, Venezuela, contained crustacean gill lamellae and unidentified insect fragments (R. S. Voss, unpublished notes). Two additional stomachs from specimens trapped at Serranía de los Pijiguaos, Bolívar state, Venezuela, contained prey fragments that Ochoa and Soriano (1991) identified as insects (Coleoptera, Orthoptera, Dictyoptera), pseudothelphusid crabs, and leptodactylid frogs.

Tribe Oryzomyini Vorontsov, 1959
Marcelo Weksler

The oryzomyines form the most taxonomically diverse tribe of sigmodontine rodents, with 130 recognized extant species in 34 genera. Members of the tribe are distributed in the Neotropical and Nearctic (southeastern section) regions from Tierra del Fuego to the southern and eastern United States, in the Galapagos Archipelago, and on Trinidad and Tobago. Oryzomyines are found in almost all major biomes in South America, including forests, savannas, swamps, scrublands, and semiarid environments, and at elevations ranging from sea level to above 4,000 m in the high Andes above treeline. In many of these habitats, they are among the most speciose and abundant small mammals (Voss and Emmons 1996; Eisenberg 1999). J. R. Prado and Percequillo (2013) have provided a gazetteer of localities for each genus of oryzomyine rodents in South America, based on museum records, analyzed both the geographic and habitat distribution for clades of genera identified by Weksler (2006) and Percequillo, Weksler, and Costa (2011), and presented a hypothesis for areas of original differentiation of each genus. Two years earlier, Valencia-Pacheco et al. (2011) had examined latitudinal diversification patterns in the tribe, developed a source-sink ecological hypothesis for that pattern, and argued that the Amazon Basin was the center of tribal diversity. L. F. Machado et al. (2013) have examined the phylogeny and biogeography of tetralophodont members of the tribe.

Most oryzomyines are predominantly cursorial, but some species display marked arboreal (e.g., *Oecomys*) or semiaquatic (e.g., *Holochilus*, *Lundomys*, and *Nectomys*) specializations. Oryzomyines serve as primary or secondary host and reservoirs to a wide range of disease-carrying organisms, including hantaviruses, arenaviruses, digeneans (*Schistosoma*), and trypanosomes (Fulhorst et al. 1997; Bharadwaj et al. 1997; Powers et al. 1999; Calderon et al. 1999; V. L. Rodrigues and Ferraz Filho 1984; Mello 1979; Picot 1992; A. C. Ribeiro et al. 1998; D'Andrea et al. 2000).

Oryzomyines are known in the fossil record of South America from Early to Middle Pleistocene deposits in Bolivia and Argentina (Steppan 1996; Pardiñas et al. 2002; Pardiñas 2008). In North America, oryzomyines are found in Middle to Late Pleistocene deposits of Florida, Georgia, Kansas, and Texas (reviewed by Weksler 2006). Oryzomyines are also found in Pleistocene and Recent deposits on all major Caribbean islands, in the Caribbean Netherlands, and Fernando de Noronha Island (Brazil). Each of these insular taxa are now extinct, some becoming so in historic times (G. M. Allen 1942; McFarlane and Debrot 2001; Carleton and Olson 1999; Turvey et al. 2010, 2012). Eight extinct genera are currently recognized: *Agathaeromys* Zijlstra, Madern, and van den Hoek Ostende; *Carletonomys* Pardiñas; *Dushimys* Zijlstra; *Megalomys* E.-L. Trouessart; *Noronhomys* Carleton and Olson; *Pardinamys* P. E. Ortiz, Jayat, and Steppan; *Pennatomys* Turevey, Weksler, Morris, and Nokkert; and *Reigomys* Machado, Leite, Christoff, and Giugliano.

The tribe can be diagnosed by seven putative synapomorphies (Voss and Carleton 1993; Steppan 1995; Weksler

2006; Weksler and Percequillo 2011): presence of a long palate with prominent posterolateral pits; absence of an alisphenoid strut; absence of a posterior suspensory process of the squamosal attached to tegmen tympani; absence of a gall bladder; 12 thoracic vertebrae; absence of a hemal arch on the first caudal vertebra; and fewer than 36 caudal vertebrae. The last two synapomorphies are reversed in several oryzomyine taxa. Members of the tribe can also be recognized by the soft fur (except in *Neacomys* and *Scolomys*, which have spiny fur); small, unkeeled manual claws (except in *Lundomys*, with long, ventrally keeled manual claws); a mammary complement of eight teats in inguinal, abdominal, postaxial, and pectoral pairs (except in *Handleyomys*, *Scolomys*, and *Oligoryzomys griseolus*, which have six mammae, lacking the pectoral pair); a sparsely haired tail covered with more or less conspicuous epidermal scales and lacking a terminal tuft of long hairs (the well-haired tail of *Nesoryzomys* does not appear scaly, and species of *Oecomys* and *Tanyuromys* may have weakly to prominently tufted tails); zygomatic plates without an anterodorsal spinous process (except in *Pseudoryzomys*, *Lundomys*, and *Holochilus*, which have a spinous process); nasal bones with rounded or squared posterior margins (except in *Nectomys*, *Scolomys*, and *Sigmodontomys*, which have acutely angled posterior nasal margins); a smooth posterior wall of the orbit (except in *Holochilus*, which has a well-developed postorbital ridge); a bony palate between the molar rows smooth or weakly sculpted (except in *Holochilus* and *Lundomys*, which have a well-developed median keel flanked by deep lateral gutters); an alisphenoid canal with a large anterior opening (the anterior opening of the alisphenoid canal is absent or very small in *Scolomys*); upper incisors with smoothly rounded enamel bands (the upper incisor enamel is distinctly faceted in *Holochilus*); low-crowned and bunodont or terraced molars (except in *Holochilus*, which has high-crowned, planar molars); labial flexi enclosed by a cingulum (the labial flexi are not enclosed in *Holochilus* and *Lundomys*); parallel maxillary toothrows (*Holochilus* and *Lundomys* have anteriorly convergent toothrows); a median mure connected to the protocone on M1 (except in *Holochilus*, which has the median mure connected to the paracone); a unilocular-hemiglandular stomach; and a male accessory reproductive gland complement that includes one pair each of bulbourethral, dorsal prostate, anterior prostate, vesicular, and ampullary glands, and two pairs of ventral prostate glands (except *Nesoryzomys*).

KEY TO THE GENERA OF ORYZOMYINI (MODIFIED FROM WEKSLER AND PERCEQUILLO 2011):

1. Dorsal and ventral fur with grooved spines 2
1′. Dorsal and ventral fur without grooved spines 3
2. Six mammae in inguinal, abdominal, and postaxial pairs . . . .*Scolomys*
2′. Eight mammae in inguinal, abdominal, postaxial, and pectoral pairs . . . . *Neacomys*
3. Hypothenar pad absent or vestigial 4
3′. Hindfeet with developed hypothenar pad 11
4. Hindfeet without natatory fringes (including continuous combs of stiff hairs along the plantar margins and/or webbing sometimes between the digits) 5
4′. Hindfeet with natatory fringes 8
5. Dorsal surface of hindfeet covered with dark hairs, feet appear brown 6
5′. Dorsal surface of hindfeet sparsely covered with short silvery hairs, feet appear grayish white or pale tan . . . 7
6. Hindfeet with interdigital webs . . . . . . *Sigmodontomys*
6′. Hindfeet without interdigital webs . . . . . . *Tanyuromys*
7. Interorbital region weakly beaded even in adults, mesoloph small on M1, M2, absent in M3 . . *Pseudoryzomys*
7′. Interorbital region with strongly developed supraorbital crest in adults, mesoloph present and well developed on M1–M3 . . . . *Oryzomys*
8. Nasals with acutely pointed posterior terminus; molars brachydont, cingula closing labial folds 9
8′. Nasals with blunt posterior terminus; molars hypsodont or planar, labial folds open 10
9. Body pelage with strong countershading; upper incisors orthodont . . . . *Amphinectomys*
9′. Body pelage with weak countershading; upper incisors opisthodont . . . . *Nectomys*
10. Alisphenoid strut present; postorbital ridge present; accessory labial root on M1 present . . . . *Holochilus*
10′. Alisphenoid strut absent; postorbital ridge absent; accessory labial root on M1 absent . . . . *Lundomys*
11. Plantar pads on hindfeet highly developed, large and fleshy, interdigitals 1–4 set close together, often in contact; mystacial vibrissae long and abundant 12
11′. Pads smaller, interdigitals 1–4 displaced proximally relative to 2–3; mystacial vibrissae shorter and sparser 13
12. Venter without gular patch of self-colored hairs; palate long; supraorbital margins squared, strongly beaded or with distinct crests . . . .*Oecomys*
12′. Venter with gular patch of self-colored hairs; dorsal surface of hindfeet with distinct dark patch; palate short; supraorbital margins lightly beaded . . . .*Drymoreomys*
13. Tail much shorter than head and body 14
13′. Tail subequal to or longer than head and body 15
14. Dorsal pelage grizzled or light brown . . *Zygodontomys*
14′. Dorsal pelage dark brown . . . . *Melanomys*
15. Sphenofrontal foramen absent 16
15′. Sphenofrontal foramen present 24

16. Small mice (head and body length of adults rarely >110 mm) 17

16'. Medium or large rats (head and body length of young rarely <110 mm) 19

17. Anteromedian flexus present 18

17'. Anteromedian flexus absent . . . .
. . . . *chapmani* group [of "*Handleyomys*"]

18. Jugal absent or vestigial; alisphenoid strut absent; stapedial foramen present (pattern 2; Voss, 1988; except in *O. rupestris*); mesoloph and mesolophid usually present (absent in *O. rupestris* and *O. fornesi*) . . *Oligoryzomys*

18'. Jugal present; alisphenoid strut always present; stapedial foramen absent (pattern 3; Voss, 1988); mesoloph and mesolophid always absent . . . . *Microakodontomys*

19. Stapedial foramen present . . . . *Hylaeamys*

19'. Stapedial foramen absent or vestigial 20

20. Tail densely furred, scales not visible. . . *Nesoryzomys*

20'. Tail sparsely furred, appearing impression of nakedness, scales visible 21

21. Mystacial vibrissae very long; interorbital region hourglass in shape with squared margins . . . . . . *Sooretamys*

21'. Mystacial vibrissae short; interorbital region anteriorly convergent with beaded margins 22

22. First upper and lower molars without accessory roots; capsular process absent or indistinct. . . *Eremoryzomys*

22'. First molars with accessory roots; capsular process well developed 23

23. Anterocone of M1 undivided (anteromedian flexus absent); baculum bifid . . . . *Cerradomys*

23'. Anterocone of M1 divided by anteromedian flexus; baculum trifid . . . . *Aegialomys*

24. Very small mice (head and body length < 100) with tail much longer than head and body length 25

24'. Medium-sized and large rats (head and body length > 100) with tail as long as or longer than head and body length 26

25. Pelage distinctly countershaded; foramen magnum oriented caudally; anteroconid of m1 undivided (anteromedian flexid absent) . . . . *Oreoryzomys*

25'. Pelage not countershaded; foramen magnum oriented posteroventrally; anteroconid of m1 divided by anteromedian flexid . . . . *Microryzomys*

26. Superciliary vibrissae extending posteriorly beyond pinnae. 27

26'. Superciliary vibrissae not extending posteriorly beyond pinnae 28

27. Zygomatic notch indistinct; zygomatic plates small. . .
. . . . *Mindomys*

27'. Zygomatic notch deep; zygomatic plates broad . . . . .
. . . . *Transandinomys*

28. Anterocone of M1 divided by anteromedian flexus . . .
. . . . *Nephelomys*

28'. Anterocone of M1 undivided (anteromedian flexus absent) 29

29. M1 without labial accessory root; m2 with two roots
. . . . *Euryoryzomys*

29'. M1 with labial accessory root; m2 with three roots . 30

30. Six mammae in inguinal, abdominal, and postaxial pairs . . . . *Handleyomys*

30'. Eight mammae in inguinal, abdominal, postaxial, and pectoral pairs . . . . . *alfaroi* group [of "*Handleyomys*"]

REMARKS: The generic composition of the tribe, corroborated by phylogenetic analyses of morphological and molecular data (Voss and Carleton 1993; Steppan 1995; Weksler 2003; 2006; Turvey et al. 2010; Percequillo, Weksler, and Costa 2011; Pine et al. 2012), includes pentalophodont and tetralophodont sigmodontines, as first advanced by Voss (1991) and Voss and Carleton (1993). For historical shifts in the content of Oryzomyini, see Hershkovitz (1944, 1955b, 1962), A. L. Gardner and Patton (1976), Reig (1984, 1986), Carleton and Musser (1989), Voss (1991a), Voss and Carleton (1993), and Weksler (2006).

Systematic work on the tribe, at all levels, has been very active in recent years, and work in progress indicates that the current taxonomy is not yet stable. The generic organization of the accounts follows Weksler et al. (2006), with the addition of taxa described thereafter (Percequillo, Weksler, and Costa 2011; Pine et al. 2012). The position of the following taxa is provisional: *Microakodontomys* is considered here as a valid genus based on its distinctive morphology (Weksler and Percequillo 2011) and preliminary molecular data (R. Paresque, pers. comm.; J. D. Hanson, pers. comm.), but validation awaits a published analysis. The ex-"*Oryzomys alfaroi*" group was previously included in *Handleyomys* (Weksler et al. 2006), but at least one new genus should be erected to contain the *alfaroi-chapmani-melanotis* clade (Weksler 2006; Weksler et al. 2006; Weksler and Percequillo 2011). Members of this group that reside in South America are treated herein as "*Handleyomys*."

Genus *Aegialomys* Weksler, Percequillo, and Voss, 2006
Alexandre R. Percequillo

The genus *Aegialomys* is a trans-Andean endemic lineage of Oryzomyini, occurring from northern Peru to northern Ecuador as well as two islands in the Galapagos Archipelago. Two species are recognized herein, but five nominal taxa have been associated with *Aegialomys*: *bauri* J. A. Allen and *galapagoensis* Waterhouse for the insular populations, with the latter name currently assigned to all

Galapagos specimens; and *baroni* J. A. Allen, *ica* Osgood, and *xanthaeolus* Thomas, with the last name provisionally applied to all mainland populations of the genus. Patton and Hafner (1983) suggested a close phylogenetic linkage between *A. galapagoensis* and *A. xanthaeolus*, based on both morphological and molecular (protein gel electrophoresis) comparisons, a hypothesis that was supported by shared karyotypes (A. L. Gardner and Patton 1976).

The genus has never been revised and is possibly more diverse than currently understood (see Musser and Carleton 2005; Weksler et al. 2006). The most important studies dealing with this genus, aside from the original descriptions, are those of Patton and Hafner (1983) and Weksler et al. (2006). The majority of this account is based on those works, supplemented by my personal examination of voucher specimens in several North American museums. The recent distributional and ecological studies of Dowler and colleagues (Dowler and Carroll 1996; Dowler et al. 2000) have added substantially to our knowledge of the Galapagos taxa.

The genus is quite distinct from other oryzomyine genera, as apparent from the following description (largely based on Weksler et al. 2006). The dorsal pelage is very long, lax, and dense, coarsely grizzled yellowish or grayish brown; the ventral pelage is abruptly paler (superficially whitish or pale yellow), but the ventral hairs are always gray-based. Pinnae are small, do not reach the eye when laid forward, and are densely covered by short hairs. Mystacial and superciliary vibrissae do not extend posteriorly beyond the pinnae when laid back. The hindfeet are short and wide, with conspicuous tufts of ungual hairs at the base of claws of all digits; the plantar surfaces are densely covered with distinct squamae distal to the thenar pad; the hypothenar pad is large; the claw of digit I extends beyond the middle of phalanx 1 (almost to the first interphalangeal joint) of digit II, and the claw of digit V extends just beyond the first interphalangeal joint of digit IV. The tail is shorter than or about as long as the head and body in *A. galapagoensis* (head and body length 130–170 mm [mean 149.0 mm], tail length 131–165 mm [mean 149.1 mm]), and manifestly longer than the head and body in *A. xanthaeolus* (head and body length 107–183 mm [mean 142.4 mm], tail length 107–200 mm [mean 154.6 mm]). The tail is weakly to distinctly bicolored.

The skull is medium sized (greatest skull length range for the genus: 29.2–38.9 mm [mean 33.9 mm]); the rostrum is stout, and flanked by deep zygomatic notches. The interorbital region converges anteriorly, with strongly beaded supraorbital margins or developed supraorbital crests. The braincase is oblong, usually with well-developed temporal crests continuous with supraorbital crests; the lambdoidal and nuchal crests are well developed in older adults.

The posterior margin of the zygomatic plates is dorsal to the alveoli of M1s in some specimens, anterior to the M1 alveoli in others, regardless of the species. Jugals are small, and thus the maxillary and squamosal zygomatic processes broadly overlap (in lateral view) but do not contact each other. The nasals are long and tapered, extending posteriorly behind the lacrimals in *A. galapagoensis*, while extending to but usually not behind the lacrimals in *A. xanthaeolus* (shorter nasals). The lacrimals usually have longer maxillary sutures than frontal sutures. The frontosquamosal suture is usually colinear with the frontoparietal suture. The parietals have broad lateral expansions. In ventral view, the skull also shows distinctive features, with long incisive foramina typically extending posteriorly to or between the M1 alveoli, almost parallel sided (in *A. galapagoensis*) or widest at midlength and tapering symmetrically anteriorly and posteriorly (in *A. xanthaeolus*). The palate is wide and is short to long in length: the mesopterygoid fossa penetrates anteriorly between the maxillae in *A. galapagoensis* but only occasionally in *A. xanthaeolus*; the bony roof of the mesopterygoid fossa is perforated by very large sphenopalatine vacuities; the posterolateral palatal pits are large, complex, and recessed in deep fossae. The alisphenoid strut is absent, thus the buccinator-masticatory foramen and accessory foramen ovale are confluent. The stapedial foramen and posterior opening of the alisphenoid canal are small, the squamosal-alisphenoid groove and sphenofrontal foramen are absent, and the secondary anastomosis of the internal carotid crosses the dorsal surface of the pterygoid plate, configuring the carotid circulatory pattern 3 of Voss (1988). The postglenoid foramen is large and rounded, and the subsquamosal fenestra is small but distinct in most forms (vestigial or absent in an unnamed species from coastal Ecuador; Weksler et al. 2006). The periotic is exposed posteromedially between the ectotympanic and basioccipital, but usually does not extend anteriorly to the carotid canal. The mastoid is unfenestrated or has a small but distinct posterodorsal fenestra (in specimens from coastal Ecuador). The mandible is not strongly built and has shallow superior and inferior notches; the coronoid process is falciform to weakly falciform, and is lower than the condyloid process; the capsular process is well developed in most fully adult specimens; the superior and inferior masseteric ridges are conjoined anteriorly as a single crest below m1.

Upper incisors are opisthodont, fronted with a distinctly orange enamel band. The molar series are parallel; labial and lingual cones are arranged in opposite pairs; and labial and lingual flexi of M1 and M2 do not interpenetrate along the molar midline. The anterocone of M1 is divided into anterolabial and anterolingual conules by a distinct anteromedian flexus in all examined specimens of *A. gala-*

pagoensis and in some young specimens of *A. xanthaeolus* (most adults of this species have a small internal fossette that seems to represent a vestigial anteromedian flexus; samples from coastal Ecuador also have an anteromedian flexus); the anteroloph is well developed and fused with the anterostyle on the labial cingulum and is separated from the anterocone by a persistent anteroflexus in *A. xanthaeolus* or fused to the anterocone (anteroflexus reduced or absent) in *A. galapagoensis*; the protostyle is absent; and the paracone is usually connected by an enamel ridge to the posterior moiety of the protocone. M2 has a distinct protoflexus; the mesoloph is present in most individuals of *A. xanthaeolus*, but is more frequently reduced in *A. galapagoensis*, being narrow and short; the mesostyle is reduced or absent in some specimens; the protocone is connected only to the paracone (not to a median mure), producing a long and single mesoflexus (with wear the mesoflexus is present as a single internal fossette). M3 has a posteroloph and diminutive hypoflexus (the latter tending to disappear with moderate to heavy wear). The accessory labial root of M1 is often present.

The anteroconid of m1 is usually without an anteromedian flexid; the anterolabial cingulum is present on all lower molars; the anterolophid is present on m1 but is absent on m2 and m3; the ectolophid is absent on m1 and m2 but is not observable on m3 due to the diminutive size of this tooth. On m2, the mesolophid shows great individual and taxonomic variation: in *A. galapagoensis*, it is more frequently reduced and sometimes partially fused to the entoconid; in *A. xanthaeolus*, it is usually distinct on an unworn m1 but more frequently is reduced on m2; m2 has a small hypoflexid. The m3 is reduced and morphologically similar to m2. Both accessory lingual and labial roots of m1 are present; m2 and m3 each have two small anterior roots and one large posterior root.

The fifth lumbar (17th thoracolumbar) vertebra has a well-developed anapophysis. The hemal arches of the second and third caudal vertebrae have a posterior spinous process. Patton and Hafner (1983) detailed structures of the stomach, glans penis and baculum, and male reproductive organs. The stomach is hemiglandular and unilocular, without an extension of glandular epithelium into the corpus. The glans penis is of the complex type (*sensu* Hooper and Musser 1964); nonspinous tissue on the crater rim does not conceal bacular mounds, the dorsal papilla is spineless, and the urethral processes is without subapical lobules. The baculum is a short and slender central bony digit capped by a small, trifid cartilaginous tip. Only one pair of preputial glands is present.

SYNONYMS:

Mus: Waterhouse, 1839:66; part (description of *galapagoensis*); not *Mus* Linnaeus.

Hesperomys: Wagner, 1843a:517; part (listing of *galapagoensis*); not *Hesperomys* Waterhouse.

Oryzomys: J. A. Allen, 1892:48; part (description of *bauri*); not *Oryzomys* Baird.

Aegialomys Weksler, Percequillo, and Voss, 2006:5; type species *Oryzomys xanthaeolus* Thomas, by original designation.

KEY TO THE SPECIES OF *AEGIALOMYS*:

1. Tail equal to or shorter than head and body length; nasals extending posteriorly behind lacrimals; lateral margins of incisive foramina almost parallel sided; mesopterygoid fossa penetrates anteriorly between maxillae . . . . *Aegialomys galapagoensis*

1'. Tail longer than head and body length; nasals not extending behind lacrimals; lateral margins of incisive foramina widest at midlength and tapering symmetrically anteriorly and posteriorly; mesopterygoid fossa rarely penetrates anteriorly between maxillae . . . .
. . . . *Aegialomys xanthaeolus*

Aegialomys galapagoensis (Waterhouse, 1839)
Galapagos Aegialomys

SYNONYMS:

Mus galapagoensis Waterhouse, 1839:66; type locality: "Chatham Island [Isla San Cristóbal], Galapagos Archipelago, Pacific Ocean," Ecuador.

Hesperomys galapagoensis: Wagner, 1843a:517; name combination.

Oryzomys bauri J. A. Allen, 1892:48; type locality: "Barrington Island [Isla Santa Fe]," Galapagos Islands, Ecuador.

O[ryzomys]. galapagoensis: Thomas, 1894b:354; name combination.

[*Oryzomys*] *Bauri*: E.-L. Trouessart, 1897:527; name combination.

[*Oryzomys (Oryzomys)*] *galapagoensis*: E.-L. Trouessart, 1904:419; name combination.

[*Oryzomys (Oryzomys)*] *bauri*: E.-L. Trouessart, 1904:419; name combination.

Oryzomys galapagoensis Gyldenstolpe, 1932a:23; incorrect subsequent spelling of *Mus galapagoensis* Waterhouse.

[*Aegialomys*] *galapagoensis*: Weksler, Percequillo, and Voss, 2006:5; first use of current name combination.

DESCRIPTION: Dorsal pelage very long, dense, and lax, light yellowish to buffy grizzled with brown; ventral pelage grayish. Tail equal in length or shorter than head and body; unicolored (in the holotype from Isla San Cristóbal) or bicolored (specimens, including holotype of *bauri* J. A. Allen, from Isla Santa Fe) and densely covered with short hairs. Hindfoot short, densely covered by short white hairs, with thick and very long ungual tufts. Pinnae small and rounded densely covered by short hairs. Mystacial vibrissae dense and thick but very short, not reaching pinnae when laid back. Skull medium sized (skull length 32.4–35.9 mm); rostrum long and narrow; zygomatic notch narrow and moderately deep; nasals generally with distinct narrow anterior projection and extending posteriorly beyond lacrimals; incisive foramina long (length range 6.2–7.7 mm) with nearly parallel-sided lateral margins (slightly wider medially; width range 2.2–2.9 mm); posterior margins of foramina reaching level of anterocones of M1s.

Phallus elongate, with length/diameter ratio of 1.8. Male accessory reproductive glands consist of one pair of preputial, bulbourethral, ampullary, and vesicular glands and four pairs of prostates, similar to the situation in other oryzomyines (Patton and Hafner 1983).

DISTRIBUTION: This species is restricted to the Galapagos Archipelago of Ecuador, where it is known from only two islands (San Cristóbal and Santa Fe).

KNOWN LOCALITIES (Map 146): ECUADOR: Galapagos, Cerro Brujo, Bahia d' Esteban, Isla San Cristóbal (= Chatham Island; type locality of *Mus galapagoensis* Waterhouse), Isla Santa Fe (= Barrington Island), Barrington Cove (MVZ 145372).

SUBSPECIES: I treat *Aegialomys galapagoensis* as monotypic.

NATURAL HISTORY: There are few data available on the natural history of this species. The population from Isla San Cristóbal is probably now extinct (Osgood 1929; Heller 1904; Patton and Hafner 1983). As a consequence, except for the limited information about the specimen from that island that Charles Darwin secured in 1835, all available data are from the population on Isla Santa Fe. Darwin (in Waterhouse 1839:66) commented that the species was "abundant," and exclusive to Chatham Island (= Isla San Cristóbal). He also stated that *A. galapagoensis* "frequents the bushes, which sparingly cover the rugged streams of basaltic lava, near the coast, where there is no fresh water, and where land is extremely sterile." Darwin did not visit Isla Santa Fe during the visit of the HMS *Beagle* to the Galapagos. J. A. Allen (1892) mentioned that the species was common on Isla Santa Fe, where it was found "between the bushes near the shore, and also high up between grass and the lava rocks." Based on data of other researchers, Patton and Hafner (1983) reported that total population size ranged from 1,000 to 2,000 individuals and that the species "breeds during the rainy season, from January through April." Clark (1980), who studied the ecology of this species, positively correlated abundance with vegetation density, noting that food was a limiting resource. Clark (1980) also showed that males have larger home ranges than females and stated that reproduction takes place in March and May, with embryo pro-

duction limited by the quantity of rainfall. This species also shows considerable longevity, with maximum life spans of nearly two years as compared with continental Oryzomyini, which barely "survive for more than a year" (D. B. Harris and MacDonald 2007). Clifford and Hoogstraal (1980) reported on tick (*Ixodes*) parasitism.

REMARKS: It is unclear where on Chatham Island [= Isla San Cristóbal] C. Darwin collected the holotype of *galapagoensis*. Heller (1904) suggested that the specimen came from the eastern end of the island. However, itineraries of the HMS *Beagle* (FitzRoy 1839; Grant and Estes 2009) indicate that, although the island was circumnavigated from September 16–22, 1835, the only night Darwin spent on the island (and when he most likely would have obtained this animal) was that of September 21–22, near Cerro Brujo at the eastern end of Stephens Bay on the north-central coast, where the *Beagle* was anchored.

Conspecificity of the two putative taxa, *galapagoensis* Waterhouse (from Isla San Cristóbal) and *bauri* J. A. Allen (from Isla Santa Fe) has been suggested at least since Osgood (1929; see also Gyldenstolpe 1932; Cabrera 1961). Patton and Hafner (1983) provided convincing support for this hypothesis, based on qualitative and quantitative analysis of the skull and soft anatomy. So far, the only observable difference between the specimen from Isla San Cristóbal and those from Isla Santa Fe involves countershading of the tail (unicolored in the former and bicolored on the latter); this trait shows considerable individual variation in several species of sigmodontines and this may also be the case in this species.

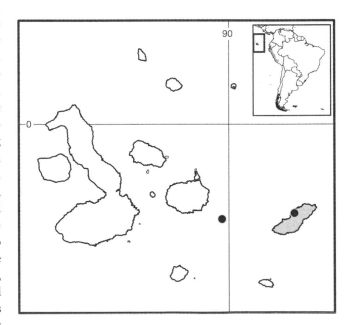

Map 146 Selected localities for *Aegialomys galapagoensis* (●). The species is believed extinct from the eastern-most island (Isla San Cristóbal [= Chatham Island]).

The karyotype of specimens from Isla Santa Fe is $2n = 56$ and $FN = 58$ (A. L. Gardner and Patton 1976); that for animals from Isla San Cristóbal is unknown.

Aegialomys xanthaeolus (Thomas, 1894)
Yellowish Aegialomys

SYNONYMS:

Oryzomys xanthaeolus Thomas, 1894b:354; type locality: "Tumbez [= Tumbes, Depto. Tumbes], N. Peru."

[*Oryzomys*] *xantholaeus* E.-L. Trouessart, 1897:527; incorrect subsequent spelling of *Oryzomys xanthaeolus* Thomas.

Oryzomys baroni J. A. Allen, 1897c:117; type locality: "Malca, Cajabamba, alt. 8000 feet," [on road from Cajabamba to San Marcos; Stephens and Traylor 1983] Cajamarca, Peru.

[*Oryzomys (Oryzomys)*] *xantholaeus* E.-L. Trouessart, 1904:419; name combination and incorrect subsequent spelling of *xanthaeolus* Thomas.

[*Oryzomys (Oryzomys)*] *baroni*: E.-L. Trouessart, 1904: 419; name combination

Oryzomys xanthaeolus baroni: Osgood, 1914b:157; name combination.

Oryzomys xanthaeolus xanthaeolus: Gyldenstolpe, 1932a: 23; name combination.

Oryzomys xanthaeolus ica Osgood, 1944:192; type locality "Hacienda San Jacinto, near Ica, Province of Ica, southwestern Peru."

Oryzomys xantheolus Cabrera, 1961:396; incorrect subsequent spelling of *xanthaeolus* Thomas.

Oryzomys xantheolus ica: Cabrera, 1961:396; name combination and incorrect subsequent spelling of *xanthaeolus* Thomas.

Oryzomys xantheolus xantheolus: Cabrera, 1961:397; name combination and incorrect subsequent spelling of *xanthaeolus* Thomas.

[*Aegialomys*] *xanthaeolus*: Weksler, Percequillo, and Voss, 2006:5; first use of current name combination.

DESCRIPTION: Dorsal fur long, dense, and lax, but comparatively shorter than in *A. galapagoensis*. In some Ecuadorean samples, ventral pelage with gular and/or pectoral patch of self-colored white or cream hairs. Tail varies from bicolored (most Peruvian samples) to unicolored or weakly bicolored (several Ecuadorean samples), and longer than head and body length. Hindfoot medium in length (range: 27–38 mm), with sparse to dense (in some Peruvian specimens) ungual tufts, and short to equal to claw length (in few specimens tufts exceed claws). Skull highly variable in size (skull length 29.2–38.9 mm), but shape conserved across geographic samples. Anterior margins of nasals rounded (without distinct process observed in *A. galapagoensis*), with posterior margins not extending behind lacrimals; incisive foramina long (length range: 5.3–8.6 mm), posterior margins usually reaching anterocones of M1s, anteroposterior margins rounded, and lateral margins markedly widest at midlength (width range: 1.7–3.3 mm), tapering symmetrically both anteriorly and posteriorly; palate devoid of bony excrescences; mesopterygoid fossa rarely penetrating anteriorly between maxillae; auditory bullae very inflated, and external auditory meatus wide; orbicular apophysis large and rounded.

DISTRIBUTION: *Aegialomys xanthaeolus* is restricted to the arid coastal and montane areas from southwestern Ecuador to southern Peru, at elevations from sea level to above 2,500 m (in the upper Río Marañón valley of northern Peru). Most of this range is covered by xerophytic vegetation, similar to other savanna-like landscapes in South America. On the coastal region of Peru and Ecuador, *A. xanthaeolus* occurs in the fog-shrouded "lomas," where vegetation is supported by mist from the Pacific Ocean in otherwise barren coastal desert (Hueck 1972; Ellenberg 1959).

SELECTED LOCALITIES (Map 147): ECUADOR: Guayas, Huerta Negra, 20 km ESE of Baláo, east of Tenguel (USNM 534361), Río Chongón, 1.5 km SE of Chongón (USNM 513544); Loja, Loja (USNM 461652); Los Ríos, Hacienda Santa Teresita (Abras de Mantequilla), ca. 12 km NE of Vinces (USNM 534364). PERU: Amazonas, 8 km WSW of Bagua (MVZ 135667); Ancash, 4 km by road NE of Chasquitambo, km 51 (UMMZ 155915); Arequipa, Chavina, on coast near Acari, Lomos River (USNM 277572); Cajamarca, Cajabamba, Malca (type locality of *Oryzomys baroni* J. A. Allen); Huánuco, Chinchao, Hacienda Buena Vista (USNM 304533); Ica, Hacienda San Jacinto (type locality of *Oryzomys xanthaeolus ica* Osgood); La Libertad, Pacasmayo (FMNH 19461), Trujillo (FMNH 19459); Lambayeque, 8 km S of Morrope (MVZ 135670); Lima, 10 km ENE of Pucusana (MVZ 137594), Cerro Azul, Río Cañete Valley, 100 m from ocean (UMMZ 161219), Lomas de Lachay, 22 km N and 11 km W of Cancay (MVZ 135665); Piura, Catacaos (USNM 304523); Tumbes, Tumbes (type locality of *Oryzomys xanthaeolus* Thomas).

SUBSPECIES: Cabrera (1961) listed *ica* Osgood as a valid subspecies, limiting its range to southwestern Peru, with the nominotypical form distributed from the central Peruvian coast from Lima department northward. However, pending a review of the actual patterns of geographic variation, I regard *A. xanthaeolus* as monotypic.

NATURAL HISTORY: Few data are available on the natural history of *A. xanthaeolus*. The species is subject to producing *ratadas*, or local population explosions, at several different localities in Peru, where population density may range above 250 individuals per hectare. These numbers are among the largest observed for several species of sigmodontine rodents as well as marsupials in South America (Jaksic and Lima 2003). *Aegialomys xanthaeolus* is host to the flagellate protozoan *Trypanosoma rangeli* (C. M. Pinto et al. 2006).

REMARKS: This species exhibits notable variation in size throughout its range, variation that stimulated the recognition of supposed new species (*O. baroni* J. A. Allen) or subspecies (*xanthaeolus ica* Osgood). Several authors (e.g., Osgood 1914b; Cabrera 1961; Musser and Carleton 1993, 2005) have questioned the validity of these taxa at any level, specific or subspecific, although Weksler et al. (2006) suggested that *A. xanthaeolus* may be a composite. In particular, these latter authors noted that animals from the Ecuadorean coast are smaller in size and possess several distinctive features, such as the absence of a subsquamosal fenestra, a mastoid with a posterodorsal fenestra, and the presence of an anteromedian flexus on M1. Although these features appear diagnostic, further comparisons with larger series are needed to ascertain whether these Ecuadorean animals represent an undescribed lineage of *Aegialomys*. Thorough analyses can also determine whether other biological entities are hidden in what is now called *A. xantheolus*, and to which the specific epithets *baroni* and *ica* could be applied. Preferably, future research will include both karyotypic and molecular analyses in addition to morphological comparisons.

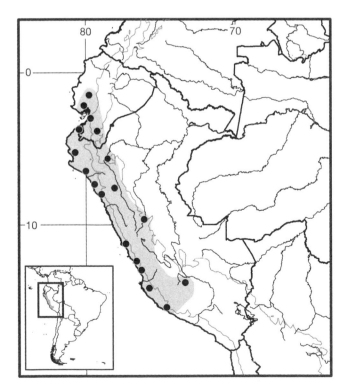

Map 147 Selected localities for *Aegialomys xantheolus* (●). Contour line = 2000 m.

Genus *Amphinectomys* Malygin, 1994

Marcelo Weksler and Michael Valqui

Amphinectomys is a monotypic genus found thus far in only two localities in the lowland Amazon forests of the Ucayali-Amazon-Yavarí interfluvial region of eastern Peru. Very little is known about its ecology, systematics, or conservation status, and more collecting effort targeted at this taxon is needed. The following description is based on a translation of Malygin et al. (1994) and the accompanying illustrations.

These semiaquatic rats are large (head and body length 187–190 mm) and heavy (mass 214–225 g), with the tail length similar to the combined head and body length (tail length 173–206 mm), long and robust hindfeet (hindfeet 53–54 mm), and short ears (ear length 24 mm). The body pelage is very soft, shiny, and glossy, with long dorsal hair (length 22 mm) and dense underfur. The dorsal color is dark brown to fulvous brown, sides are more ochraceous, and the venter is white-grayish, with hairs gray at their bases (except the chin and throat, where hairs are completely white). The limits of the lateral and ventral colors are sharply delineated, creating a conspicuous countershading effect. The mystacial vibrissae are dense and very long, extending beyond the pinnae when laid back. The pinnae are small, only slightly hair, appear superficially naked, and are dark brown. The manus and pes are covered dorsally with short, pale hairs. Digits II–IV of the hindfeet are much longer than I and V. Large webs of skin are present on the manus and pes between digits II, III, and IV, extending to the base of the claws. Natatory fringes are present along medial plantar margins of the metatarsus and digits I and II, and along lateral plantar margins of digits IV and V. The tail is dark, unicolored, and sparsely haired, with dark brown hairs forming a small brush, but not a well-developed pencil, at the tip. There are eight pairs of mammae, one each in inguinal, abdominal, postaxial, and pectoral positions.

The skull is robust, and deep zygomatic notches flank the rostrum. The nasals have squared or gently rounded posterior margins, extending posteriorly to approximately the same level as, or slightly beyond, the lacrimals, which usually have much longer maxillary than frontal sutures. The interorbital margins converge anteriorly and possess weakly developed, overhanging supraorbital crests. The frontosquamosal suture is colinear with the frontoparietal suture (the dorsal facet of the frontal is not in contact with the squamosal). The parietal has a broad lateral expansion, a large portion dipping below the temporal ridge. The interparietal is narrow and wedge shaped. The zygomatic arches are slightly convergent anteriorly and widest posteriorly; the jugal is robust and not constricted at its middle, so the maxillary and squamosal processes of the zygoma are not in contact. The zygomatic plates lack an anterodorsal spinous process, and their posterior margins are level with the alveoli of the M1s. Incisive foramina are broad, short, and teardrop shaped, terminating well anterior to the alveolar line of the M1s. The palatal bridge is long, extending well behind the third molars and possessing conspicuous posterolateral palatal pits. The bony roof of the mesopterygoid fossa is perforated by reduced and narrow sphenopalatine vacuities. The postglenoid foramen is large and rounded, but the subsquamosal fenestra is absent or vestigial. The mandible is short and deep, with a large falciform coronoid process slightly higher than the condyloid process. The capsular process is not pronounced, and the anterior masseteric ridge reaches the level of the procingulum of m1s. Upper incisors are orthodont.

SYNONYM:

Amphinectomys Malygin, 1994:198; type species *Amphinectomys savamis* Malygin, 1994, by original designation.

REMARKS: Based both on nucDNA IRBP sequences and discrete morphological characters, Weksler (2006) found *Amphinectomys* as sister to *Nectomys squamipes* and *Nectomys apicalis*, but analyses that include a comprehensive sampling of *Nectomys* are necessary for phylogenetic corroboration of *Amphinectomys* as a genus distinct from *Nectomys*.

Amphinectomys savamis Malygin, 1994
Ucayali Water Rat

SYNONYM:

Amphinectomys savamis Malygin, 1994:198; type locality "7 km east Jenaro Herrera, right bank of Río Ucayali (04°55′S, 73°45′W) in the province Requena, Department of Loreto, Peru" [translation from the original in Russian].

DESCRIPTION: As for the genus.

DISTRIBUTION: *Amphinectomys savamis* is known from only two localities in the lowland Amazon rainforest of the Ucayali-Amazon-Yavarí interfluvial region of eastern Peru.

SELECTED LOCALITIES (Map 148): PERU: Loreto, 7 km E of Jenaro Herrera, right bank Río Ucayali (type locality of *Amphinectomys savamis* Malygin); San Pedro, 80 km NE of Jenaro Herrera (MV 970045).

SUBSPECIES: We treat *Amphinectomys savamis* as monotypic.

NATURAL HISTORY: The two collecting sites of *Amphinectomys* are part of the uninterrupted lowland tropical forest of the Ucayali-Amazon-Yavarí interfluvial region. The holotype was collected in 1991 (Malygin et al. 1994). The second known specimen was trapped in June 1997 in the Tahuayo valley, a right tributary of the Río Amazonas northeast of the Jenaro Herrera biological station (Valqui 2001).

The climate of the region is typical of rainforest close to the equator, with a mean temperature around 26.5°C, a mean rainfall for San Pedro around 2,500 mm, and somewhat drier climate at Jenaro Herrera. Locals recognize a short "dry season," in which the monthly rainfall is sometimes below 100 mm; however, rain falls throughout the year. Both places are close to true *várzea* forests, although they are beyond the direct influence of seasonal flooding that, depending the relative water height, begins in November–December and ends in May–June. Deforestation is very localized and scattered, and the area still retains a very diverse mammal community (Valqui 2001). At both localities, *A. savamis* was trapped close to a small creek in primary forest. The San Pedro specimen was caught in the streambed of a small permanent creek (water 0.5–1 m across, 0.5 m deep), connected by a gentle slope to a 3–4 m higher lying terrain. Although streams inside the forest are often covered by a dense canopy of evergreen trees, as in this case, poorly drained areas are dominated by palms (*Scheelea* sp., *Jessenia* sp.), and the canopy is not as dense. The morphological adaptations shown by *Amphinectomys* (extensive interdigital webbing, long dense hair) suggest a high specialization for amphibious life.

REMARKS: Malygin et al. (1994) posited that the closest relative of *Amphinectomys savamis* were the water rats of the genus *Nectomys*, which are similar in size and adaptations. Compared to *Nectomys*, the pelage is longer, darker,

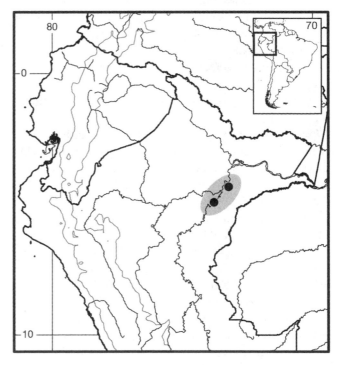

Map 148 Selected localities for *Amphinectomys savamis* (●). Contour line = 2,000 m.

shinier, denser, and more colorful; the feet are whiter, and the webbing of the hindfeet extends farther up the digits; and the ear is longer. See, however, the earlier remarks regarding the need for additional molecular information to establish the phylogenetic identity of *Amphinectomys*. The coordinates given in Malygin's description also place the type locality about 8 km west of Jenaro Herrera, not east of the station. Malygin et al. (1994) reported the karyotype as $2n = 50$, FN = 56.

Genus *Cerradomys* Weksler, Percequillo, and Voss, 2006

Alexandre R. Percequillo

Species of *Cerradomys* occur in the open vegetation belt that crosses South America from northeastern Brazil to southeastern Bolivia and northwestern Paraguay. The Caatinga, Cerrado, and Chaco biomes form this area, also known as the "dry diagonal" (see L. P. Costa 2003). This is the one of the few genera of Oryzomyini, along with *Aegialomys*, *Holochilus*, *Pseudoryzomys*, and a few others, that occupies these open and drier habitats. Nevertheless, the species of *Cerradomys* are predominantly associated with forested habitats that occur throughout this region, including gallery forests, moist semideciduous forest patches (*brejos*), and woodland savannas (*cerradão*). There are two exceptions to this pattern: in northeastern and south-central Brazil, three species penetrate the coastal Atlantic Forest, including the distinctive *restinga* vegetation on sandy coastal soils, with the extension of two of these possibly due to anthropogenic habitat alteration. In the central Brazilian Cerrado, one species inhabits the *cerrado/campo cerrado* ecotone, a more open habitat.

The present account is based primarily on Percequillo et al. (2008), supplemented by the studies of Bonvicino et al. (1999), Bonvicino and Moreira (2001), Langguth and Bonvicino (2002), Bonvicino (2003), and Tavares et al. (2011). Additional details on morphological variation, taxonomy, and molecular-based phylogenetic relationships can be found in these publications.

Species of *Cerradomys* are large (head and body length 119–185 mm) with the tail much longer than the head and body (tail length 129–210 mm). The hindfeet are robust but vary considerably in length (20–43 mm); ungual tufts are short and dense but do not conceal the claws. The dorsal fur is long and dense, ranging in color from buffy-yellow grizzled with dark brown to buffy-orange grizzled with black, and contrasting greatly with the paler ventral pelage. In some species, the coloration of the head contrasts with that of the body, being grayish to grizzled reddish versus orangish to brown. The ventral coloration

grades from whitish or weakly buffy grizzled to an almost pure white, buffy, or buffy-yellow. The length of hair on the tail varies from short to long among species; in some species, the tail is unicolored, but *C. langguthi* has a distinctly bicolored tail. Dense, brown hairs with buffy or orange tips cover both surfaces of the pinnae. The mystacial vibrissae are very dense and long, extending past the ears when laid back; the most dorsal vibrissae are dark brown with golden tips, and the most ventral are entirely white.

The skull of all species is robust, with the average total length from 29.4–40.4 mm. The rostrum is short and narrow, and flanked by deep and wide zygomatic notches; rostral fossae (the dorsal margins of the nasolacrimal foramina that are confluent with the zygomatic notches; see Percequillo et al. 2008) are moderately to deeply excavated. The zygomatic plates range from narrow (2.8 mm) to wide (4.4 mm), with their anterior margins straight or slightly concave; zygomatic spines are present in some species but are predominantly rounded. The interorbital region diverges strongly posteriorly; the supraorbital margins form a distinctly projecting shelf. The braincase is long, with a well-developed temporal crest. Incisive foramina are very long (6.1–8.3 mm), their lateral margins broadest at midlength, and both ends acutely pointed . The palate is long and wide, typically with multiple posterolateral palatal pits located in very deep and wide depressions. The roof of the mesopterygoid fossa is perforated by small to very large sphenopalatine vacuities, except in *C. marinhus* and *C. maracajuensis*. The anterior border of the mesopterygoid fossa reaches the level of the alveoli of the M3s in younger individuals, a condition that persists in some adults. The stapedial foramen is minute or absent, and both the squamosoalisphenoid groove and sphenofrontal foramen are absent, configuring carotid circulatory pattern 3 of Voss (1988). The buccinator-masticatory and accessory oval foramina are confluent, thus the alisphenoid strut is absent, in all but one species (*C. scotti*). The postglenoid foramen is narrow and the subsquamosal fenestra a small aperture in most species, being obliterated in *C. scotti*. The suspensory process of the squamosal is absent, and the tegmen tympani is short and may or may not overlap the squamosal. Auditory bullae are small and globose, with a distinct stapedial process that may or may not overlap the squamosal.

The mandible has a low coronoid process that is nearly triangular and equal in height to the condyloid process. The angular process does not reach past the condyloid process posteriorly. Both superior and inferior notches are shallow, and a capsular projection of the lower incisor is present in adult specimens.

The upper incisors are opisthodont. The molars are pentalophodont or tetralophodont, and low crowned; the main cusps are arranged in opposite pairs, and the labial and lingual flexi interpenetrate only slightly at the median molar plane. The paracone is connected medially to the protocone on both M1 and M2, not to the median mure, configuring a long and obliquely oriented parafossette. The posteroloph is long, fusing to the metacone only after extensive wear. One species (*C. scotti*) exhibits various degrees of reduction of both mesolophs and mesolophids, these structures being absent in some individuals. The M1 has four roots, one anterior, one posterior, and two accessory rootlets, and M2 and M3 each have two roots.

The phallus is very elongate. The distal cartilaginous baculum is extremely reduced and bidigitate due to the absence of a central digit; the cartilaginous baculum extends distal to the body of the glans penis. The bony baculum is extremely elongated, with its cartilaginous portion about one-eighth the length of its osseous portion. Urethral ventral flaps and dorsal papilla are absent. Small spines densely ornament the phallus's epidermis. The stomach is unilocular and hemiglandular.

The genus *Cerradomys* currently includes seven species: *C. goytaca*, *C. langguthi*, *C. maracajuensis*, *C. marinhus*, *C. scotti*, *C. subflavus*, and *C. vivoi*. However, based on recent morphological, karyological, and molecular evidence (Percequillo 1998, 2003; Bonvicino et al. 1999; Bonvicino and Moreira 2001), the genus is clearly more diverse, with several additional forms requiring formal description. Tavares et al. (2011) recognized two informal groups of these species based on shared external and cranial characters. One group, comprising *C. maracahuensis*, *C. marinhus*, and *C. scotti*, is characterized by a homogeneously olivaceous dorsal pelage that does not contrast sharply with the yellowish buff ventral color, a completely ossified bony roof of the mesopterygoid fossa, and a proportionally larger opening to the alisphenoid canal (larger than the outline of M3). The second group, including *C. goytaca*, *C. langguthi*, *C. subflavus*, and *C. vivoi*, has shorter dorsal and ventral fur, a sharper contrast between the dorsal and ventral colors, medium-sized to large sphenopalatine vacuities, and a smaller opening of the alisphenoid canal (equal to or less than the outline of M3). The limited molecular sequence data available (Percequillo et al. 2008), while not including all species, are at least consistent with these morphological groupings.

SYNONYMS:

Mus: Lund, 1840:279; part (description of *vulpinus*); not *Mus* Linnaeus.

Hesperomys: Wagner, 1842c:362; part (description of *subflavus*); not *Hesperomys* Waterhouse.

Calomys: Fitzinger, 1867b:86; part (listing of *vulpinus*); not *Calomys* Waterhouse.

Oryzomys: E.-L. Trouessart, 1897:526; part (listing of *vulpinus* as synonym of *vulpinoides* Schinz); not *Oryzomys* Baird.

Holochilus: C. O. da C. Vieira, 1953:134; part; not *Holochilus* Brandt.

Cerradomys Weksler, Percequillo, and Voss, 2006:8; type species *Hesperomys subflavus* Wagner, by original designation.

KEY TO THE SPECIES OF *CERRADOMYS*:

1. Roof of mesopterygoid fossa perforated by medium-sized to large sphenopalatine vacuities 2
1′. Roof of mesopterygoid fossa completely ossified, unperforated by vacuities or perforated by very small vacuities. 6
2. Head grayish and back orangish to reddish; tail moderately covered with hairs, uniformly colored or bicolored; hindfeet large and covered by sparse brown and brown/white hairs; alisphenoid strut absent; subsquamosal fenestra present; mesoloph and mesolophid well developed 3
2′. Head and back grayish; tail well-furred and sharply bicolored; hindfeet small, entirely and densely covered by short white hairs; alisphenoid strut present; subsquamosal fenestra absent in adults; mesoloph reduced; mesolophid reduced or absent . . . . *Cerradomys scotti*
3. Sphenopalatine vacuities long and wide, extending through pre- and basisphenoid, exposing the orbitosphenoid; 2n not 50, or if so then FN 60 or greater 4
3′. Sphenopalatine vacuities short and narrow, restricted to the presphenoid, barely exposing the orbitosphenoid; bony tube with distinct medial projection; very deep posterolateral palatal pits, recessed in deep palatal fossa; tail bicolored; 2n = 50, FN = 56 . . . . *Cerradomys langguthi*
4. Sphenopalatine vacuities long but narrow; ratio of diastemal length to condylobasilar length <28.9%; M1 broad, width >1.5 mm; 2n = 54, FN = 62 . . . . *Cerradomys subflavus*
4′. Sphenopalatine vacuities long but wide; ratio of diastemal length to condylobasilar length >30%; M1 narrow, width >1.45 mm; 2n = 50 or 54, FN = 62–63 or 66 . . . 5
5. Tail with sparse hairs but large scales (14 per cm at midtail); lacrimals broad and posterolaterally expanded; supraorbital and lambdoidal crests pronounced; 2n = 54, FN = 66 . . . . *Cerradomys goytaca*
5′. Tail more hirsute and with small scales (16–18 per cm at midtail); lacrimals diminutive and sometimes absent; supraorbital and lambdoidal crests slender and small; 2n = 50, FN = 62–63 . . . . *Cerradomys vivoi*
6. Mesolophid reduced or absent and posterolophid absent; 2n = 58, FN = 54 . . . . *Cerradomys marinhus*
6′. Mesolophid and posterolophid well developed; 2n = 58, FN = 58 . . . . *Cerradomys maracajuensis*

Cerradomys goytaca Tavares, Pessôa,
and Gonçalves, 2011
Goytacá Cerradomys

SYNONYM:

Cerradomys goytaca: Tavares, Pessôa, and Gonçalves,
2011:647: type locality "Parque Nacional Restinga de
Jurubatiba, municipality of Carapebus, Rio de Janeiro,
Brazil, 22°15′28″S, 41°39′49″W, 1 m altitude."

DESCRIPTION: Small, with head and body length
116–166 mm, tail length 130–181 mm. Dorsal pelage
short, overall color orange grizzled with brown; head
color grayish, especially around eyes and on cheeks; venter whitish and covered by short and sparse hairs with
pale gray to brownish gray bases. Tail markedly bicolored, with completely unpigmented hairs and scales on
ventral surface along proximal portion, and uniformly
dark above; scales large, with modal number 14 per cm
near base. Ungual hair tufts over claws rarely exceed claw
length. Skull large and robust; rostrum broad, lacrimals
broad; zygomatic arches wide; supraorbital ridges that
extend almost to occipital crest along temporal region
prominent; palatal fossae deep with complex and broad
posteropalatal pits; mesopterygoid roof with wide and
large sphenopalatine vacuities extending broadly along
presphenoid and alisphenoid (exposing the orbitosphenoid); no alisphenoid strut; small posterior opening of
alisphenoid canal; and relatively narrow upper molars.
Cerradomys goytaca larger in most craniodental measurements than *C. langguthi* and *C. vivoi*, and significantly
larger than samples of *C. subflavus* in length of diastema
and breadth of rostrum.

DISTRIBUTION: *Cerradomys goytaca* is known from
only five localities in coastal Espírito Santo and Rio de Janeiro states, Brazil.

SELECTED LOCALITIES (Map 149): BRAZIL: Espírito Santo, Praia das Neves, Presidente Kennedy (MNRJ
67543); Rio de Janeiro, Parque Nacional da Restinga
de Jurubatiba, Carapebus (type locality of *Cerradomys
goytaca* Tavares, Pessôa, and Gonçalves, 2011), Restinga
de Iquipari-Grussai, Grussai, São João da Barra (MNRJ
73261), Restinga do Faroizinho, Farol de São Tomé, Campos dos Goytacazes (MNRJ 73276), Sítio Santana, Beira
de Lagoa, Quissamã (MNRJ 73198).

SUBSPECIES: *Cerradomys goytaca* is monotypic.

NATURAL HISTORY: All known collecting localities are in the narrow sandy plains covered by distinctive
shrubby and open vegetation, locally known as *restingas*, along the Atlantic coast of south-central Brazil (see
D. Araújo 1992; Oliveira-Filho and Fontes 2000). *Cerradomys goytaca* is one of the most abundant small mammals in the Parque Nacional da Restinga de Jurubatiba,
occurring in most plant formations but at higher densities

in the open shrublands that are distributed as vegetation
islands in sandy plains. It is most common in patches of
Clusia hilariana (Clusiaceae), the largest tree in these shrubland patches. Specimens have also been taken in forests
that develop in more humid terrain and near lagoons. This
species plays an important role as a seed predator of the
palm *Allagoptera arenaria* (Grenha et al. 2010). Specimens
were captured in traps placed on limbs up to 2 m above the
ground, suggesting some degree of arboreal activity. Nests
were found on tree branches and tangled bromeliads (Bergallo et al. 2005). Dense litter is also likely used as shelter.
Pregnant females were captured in April and September,
each with four embryos. In the laboratory, litters consisted
of four to five pups; eyes and external ears opened four to
nine days following birth, with different growth rates in
the two sexes. Both macronyssid (*Ornithonyssus* sp.) and
laelapid mites (*Laelaps manguinhosi*, *Gigantolaelaps goyanensis*, and *G. vitzthumi*) parasitize this species (Bergallo
et al. 2004). One individual was found with a botfly larva
(Diptera, Cuterebridae).

REMARKS: Among the diagnostic features given by
Tavares et al. (2011) in their description of *C. goytaca*
is a unique karyotype with $2n = 54$ and $FN = 66$, including a single large acrocentric pair and four small biarmed
pairs in the autosomal complement, a large subtelocentric X chromosome, and a medium-sized acrocentric Y
chromosome.

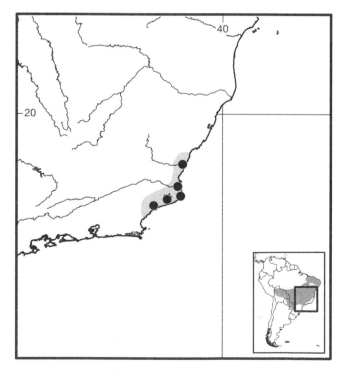

Map 149 Selected localities for *Cerradomys goytaca* (●).

Cerradomys langguthi Percequillo, Hingst-Zaher, and Bonvicino, 2008

Langguth's Cerradomys

SYNONYM:

Cerradomys langguthi: Percequillo, Hingst-Zaher, and Bonvicino, 2008:13; type locality "Corredor São João-Fazenda Pacatuba, Sapé, State of Paraíba, Brazil, at ca. 07°02'S, 35°09'W."

DESCRIPTION: Small, with head and body length 119–153 mm, tail length 144–195 mm. Dorsal pelage short and dense; body color orange grizzled with brown; head color contrastingly grayish; ventral body color grayish or slightly yellowish. Skull with short and narrow sphenopalatine vacuities, restricted to presphenoid and thus partially exposing orbitosphenoid; no alisphenoid strut; palatal fossae deep (complex posterolateral palatal pits).

DISTRIBUTION: *Cerradomys langguthi* occurs from the left bank of the Rio São Francisco through the Brazilian states of Pernambuco, Paraíba, Ceará, and Maranhão. In Pernambuco, Paraíba, and Ceará, records extend from the coastal lowlands to the interior highlands and mountain ranges; in Maranhão, the species is associated with lowlands of the central portion of the state.

SELECTED LOCALITIES (Map 150): BRAZIL: Ceará, Sítio Friburgo, Serra de Baturité, Pacoti (MZUSP [ARP 11]); Maranhão, Fazenda Lagoa Nova (MPEG 23482); Paraíba, João Pessoa (UFPB 31), Fazenda Pacatuba, Sapé, Corredor São João (type locality of *Cerradomys langguthi* Percequillo, Hingst-Zaher, and Bonvicino); Pernambuco, Buíque, Sítio Mata Verde (MZUSP 20604), Exu (MZUSP 18900), Fazenda Saco IBA, 6.6 km NNE of Serra Talhada (MZUSP 18905), São Vicente Ferrer (UFPB 19).

SUBSPECIES: *Cerradomys langguthi* is monotypic.

NATURAL HISTORY: *Cerradomys langguthi* inhabits several vegetation types, including the coastal lowland humid Atlantic Forest (locally called *zona da mata*), the open and relatively dry forests (locally called *agreste*) in the zone between the more humid and dense coastal forest and the more open and drier Caatinga, the Caatinga, and the forests restricted to humid slopes of mountain ranges in areas of Caatinga (*brejos*).

This species is associated with sugarcane plantations, near humid and mesic areas in Ceará and Pernambuco states (Paiva 1973; Mares, Willig et al. 1981); with natural and cultivated fields, where the rats construct nests in the wetter grass patches (Karimi et al. 1976); and with secondary forests and shade-grown coffee plantations on the slopes of Serra de Baturité (a typical *brejo*) in Ceará state (Percequillo et al. 2008).

REMARKS: In the genus, *C. langguthi* is nested in a monophyletic group along with *C. vivoi* and *C. subflavus*

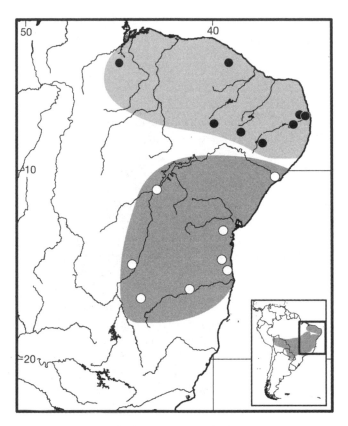

Map 150 Selected localities for *Cerradomys langguthi* (●) and *Cerradomys vivoi* (○).

(Percequillo et al. 2008), as based on mtDNA cytochrome-*b* sequences. Cranially, *C. langguthi* exhibits short and narrow sphenopalatine vacuities that barely expose the orbitosphenoids; *C. subflavus* and *C. vivoi*, in contrast, have long and narrow or long and wide vacuities, respectively. There are also noticeable quantitative differences among these three species, with *C. subflavus* being significantly larger and more robust, *C. vivoi* of intermediate size, and *C. langguthi* the smallest.

The karyotype of *C. langguthi* is variable in diploid number ($2n=46$ or 48–50) but constant in fundamental number (FN=56), with the variation in diploid number due to multiple centric fusions that are polymorphic in some populations (Maia and Hulak 1981; Bonvicino et al. 1999; Percequillo et al. 2008).

Cerradomys maracajuensis (Langguth and Bonvicino, 2002)

Maracaju Cerradomys

SYNONYMS:

Oryzomys maracajuensis Langguth and Bonvicino, 2002: 292; type locality "Brazil, Mato Grosso do Sul, Municipality of Maracaju (approx. 21°38'S, 55°09'W) Fazenda da Mata."

[*Cerradomys*] *maracajuensis*: Weksler, Percequillo, and Voss, 2006:8; first use of current name combination.

DESCRIPTION: Large, head and body length 140–185 mm, and tail length 171–227 mm; hindfeet long and robust, 34–43 mm; dorsal body color coarsely grizzled, buffy brown to orange brown, and ventral body color grayish to buffy to yellow gray. Skull with shallow rostral fossae; roof of mesopterygoid fossa with small and narrow sphenopalatine vacuities or fully ossified; palatal fossae shallow, with large posterolateral palatal pits; palatal excrescences developed; all lower molars with well-developed mesolophids (but narrow and reduced in some individuals from Bolivia). Phallus lacks central cartilaginous digit of distal baculum.

DISTRIBUTION: *Cerradomys maracajuensis* occurs in open vegetation habitats of central South America, including the Cerrado in Brazil, the eastern Chaco and grasslands of Paraguay (see D'Elía, Pardiñas, Jayat, and Salazar-Bravo 2008), the Cerrado and Chaco Boreal of Bolivia, and the islands of open, Cerrado-like vegetation in southern Peru. Elevational range is from 100 m (Tacuatí, Paraguay) to 1,750 m (Pitiguaya, Bolivia), with most records from the lowland.

SELECTED LOCALITIES (Map 151): BOLIVIA: Beni, Boca del Río Baures (AMNH 210227), Pampa de Meio, Rio Itenéz (AMNH 210024); La Paz, Pitiguaya, Río Unduavi (AMNH 72641). BRAZIL: Mato Grosso, 264 km N of Xavantina, Serra do Roncador (BM 81.450); Mato Grosso do Sul, Fazenda Primavera, Bataiporã (MZUSP 28766); Minas Gerais, Reserva do Jacob, Nova Ponte (UFMG 1970). PARAGUAY: Paraguarí, Sapucay (BM 3.2.3.9). PERU: Puno, Aguas Claras camp, Río Heath (USNM 579688).

SUBSPECIES: *Cerradomys maracajuensis* is monotypic.

NATURAL HISTORY: *Cerradomys maracajuensis* has been collected in "woods," "bush and grass," "brush pile in grass at edge of rice field," "brush-swampy," "forest," and "brush and grass" (R. M. Gilmore, from AMNH and MNRJ museum labels on specimens collected near the type locality), dense shrubby and grassy fields (cleared forest) and adjacent dense forest (G. Schmitt, AMNH labels), gallery forest (Langguth and Bonvicino 2002), and seasonally flooded pampas (Emmons et al. 2006). Pregnant females were caught in Bolivia during June and September, with two and four embryos (S. Anderson 1997).

REMARKS: A recently described species, *C. maracajuensis* is readily distinguishable from other members of the genus (with the exception of *C. marinhus*) by its larger size and long tail, longer and more robust hindfeet, shallower rostral fossae, narrower zygomatic notches, developed palatal excrescences, and fully ossified roof of mesopterygoid fossa (small sphenopalatine vacuities are present in some Bolivian and Paraguayan specimens). Another diagnostic feature of *C. maracajuensis* is the $2n = 56$ and $FN = 54$ karyotype (Bonvicino et al. 1999). Bonvicino

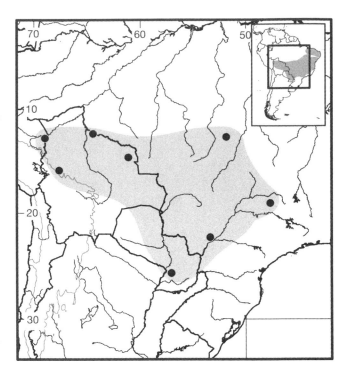

Map 151 Selected localities for *Cerradomys maracajuensis* (●). Contour line = 2,000 m.

(2003) hypothesized that *C. marinhus* and *C. maracajuensis* form the most basal clade of the genus.

Cerradomys marinhus (Bonvicino, 2003)
Marinho's Cerradomys

SYNONYMS:

Oryzomys marinhus Bonvicino, 2003:84; type locality "Fazenda Sertão do Formoso (known formerly as Fazenda Jucurutu, 14°40′20″S 45°49′71″ [*sic*] W, altitude around 775 m), Jaborandi municipality, Goiás, Brazil" (corrected to Bahia state, at coordinates 14°48′S, 45°57′W; see comment under Remarks).

[*Cerradomys*] *marinhus*: Weksler, Percequillo, and Voss, 2006:8; first use of current name combination.

DESCRIPTION: Large, with head and body length 153–179 mm; tail longer than body (length 198–212 mm); hindfeet long and robust (length 38–43 mm). Dorsal body color coarsely grizzled, buffy brown to orange brown; ventral body color grayish to buffy to yellowish gray. Skull with shallow rostral fossae; roof of mesopterygoid fossa either fully ossified or with small and narrow sphenopalatine vacuities; palate with shallow fossae and large posterolateral palatal pits; mesolophid of m3 reduced or absent.

DISTRIBUTION: *Cerradomys marinhus* is known only from two localities, one each in Bahia and Minas Gerais states, Brazil.

SELECTED LOCALITIES (Map 152): BRAZIL: Bahia, Fazenda Sertão do Formoso (type locality of *Oryzo-*

mys marinhus Bonvicino); Minas Gerais, Parque Nacional Grande Sertão Veredas (MZUSP [APC 764]).

SUBSPECIES: *Cerradomys marinhus* is monotypic.

NATURAL HISTORY: Specimens of *C. marinhus* have been captured in seasonally flooded semideciduous forests (Percequillo et al. 2008) and in *vereda*, the periodically flooded grasslands with scattered palm trees of the genera *Mauritia* and *Mauritiella* and that usually occur along headwater streams in the Cerrado (Bonvicino 2003). Pregnant females were captured in both dry and rainy seasons, suggesting that breeding takes place throughout the year; embryo counts range from two to four, with mode of four (Bonvicino 2003:87). Bonvicino also stated that individuals of *C. marinhus* were infested with mesostigmatid mites and ticks, *Polygenis tripus* fleas, and flies of the family Hippoboscidae.

REMARKS: The type locality specified in the original description is incorrect. As Bonvicino (2003:79) has noted, Fazenda Sertão do Formoso is "located in Jaborandi and Cocos municipalities, Bahia state" and not in Goiás state. Moreover, all specimens were captured in one particular habitat type, the geographical coordinates of which were specified by Bonvicino (2003:79). Thus, the correct type locality for this species is Fazenda Sertão do Formoso (formerly known as Fazenda Jucurutu), 14°48′S, 45°57′W, elevation about 775 m, municipality of Jaborandi, Bahia, Brazil.

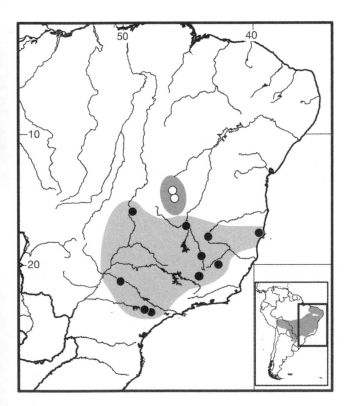

Map 152 Selected localities for *Cerradomys marinhus* (○) and *Cerradomys subflavus* (●).

Cerradomys marinhus and *C. maracajuensis* were recovered as sister taxa in a molecular phylogenetic analysis with high bootstrap support (Bonvicino and Moreira 2001; Bonvicino 2003), and perhaps not surprisingly because the two species are remarkably similar in body and skull size, overall pelage color, and qualitative skull characters (e.g., shallow rostral fossae, ossified mesopterygoid fossa roof, and presence of palatal excrescences; see Percequillo et al. 2008). The karyotype of *C. marinhus* consists of $2n=56$ and $FN=54$, with an entirely uniarmed autosomal set whereas that of *C. maracajuensis* is $2n=56$ and $FN=58$, with two metacentric pairs (Bonvicino 2003). The absence of a mesolophid on m1 is the only diagnostic feature separating *C. marinhus* from *C. maracajuensis*.

Cerradomys scotti (Langguth and Bonvicino, 2002)
Lindbergh's Cerradomys

SYNONYMS:

Oryzomys scotti Langguth and Bonvicino, 2002:290; type locality "Brazil, Goiás: municipality of Corumbá de Goiás (approx. 15°54′S, 48°48′W), Morro dos Cabeludos [= Morro de Cabeludo]."

Oryzomys andersoni Brooks, Baker, Vargas, Tarifa, Aranibar, and Rojas, 2004:3; type locality "Pozo Mario, Estancia Las Conchas, Santa Cruz, Bolivia; 17°35′46.9″S; 59°30′20.5″W."

[*Cerradomys*] *scotti*: Weksler, Percequillo, and Voss, 2006:8; first use of current name combination.

DESCRIPTION: Medium sized, with head and body length 125–181 mm and tail length 141–198 mm; hindfeet small (length 24–27 mm). Dorsal body color coarsely grizzled buffy brown, ventral color grayish. Tail bicolored and noticeably hirsute. Skull with deep rostral fossae and mesopterygoid fossa roof with large and wide sphenopalatine vacuities that expose the orbitosphenoid; alisphenoid strut present; palatal fossae deep (complex posterolateral palatal pits). M2 with reduced mesoloph, and mesolophid absent or reduced on m1 and m2. Central cartilaginous digit of distal baculum absent.

DISTRIBUTION: *Cerradomys scotti* occurs in the highlands of central South America, with records from the Cerrado of Brazil and eastern Bolivia and the Chaco Oriental of Paraguay. The species ranges in elevation from 250 m to about 1,180 m.

SELECTED LOCALITIES (Map 153): BOLIVIA: Santa Cruz, El Refugio Pampa, 3 km NE from camp (USNM 584583), Santa Rosa de la Roca (AMNH 263872). BRAZIL: Bahia, Fazenda Sertão do Formoso (MNRJ 61667); Maranhão, Estiva (MPEG 22670); Mato Grosso, 264 km N of Xavantina, Serra do Roncador (BM 81.439); Minas Gerais, Conquista, Usina Hidroelétrica de Igarapava (UFMG 29), Lagoa Santa (BM 88.1.9.5), Serra Azul

(COPASA, Mateur Leme) (UFMG 21); Piauí, Estação
Ecológica de Uruçuí-Una (MZUSP 137); São Paulo, Estação Ecológica de Santa Bárbara (MZUSP [APC 1157]).
PARAGUAY: Paraguarí, Sapucaí (BM 3.4.7.13).

SUBSPECIES: I treat *Cerradomys scotti* as monotypic.

NATURAL HISTORY: *Cerradomys scotti* is most frequently captured in open habitats of the Cerrado and through
the ecotone between forested and open areas (voucher data,
journal notes, and collector information), in habitats locally
referred to as *cerrado* and *pampa arbolada* (Emmons et al.
2006), *cerrado sensu stricto* (grassland densely covered
with bushes and small trees), *cerrado rupestre* (similar to
the previous, but with rocky outcrops), *campo úmido* (seasonally flooded grassland), and *campo cerrado* (grassland
sparsely covered with bushes; Bonvicino, Lemos, and Weksler 2005). This is in contrast to other members of the genus
that are associated more with forested habitats, such as gallery forest and woodland savanna, or *cerradão* (Percequillo
et al. 2008). Langguth and Bonvicino (2002) stated that
this species is also found in *veredas* and gallery forests but
less frequently than in more open habitats of the Cerrado.
Cerradomys scotti is scansorial (Alho and Villela 1984).
This species may be sympatric with three other species:
C. maracajuensis in the Brazilian states of Minas Gerais,
Mato Grosso, and Mato Grosso do Sul, the Paraguayan
state of Paraguarí, and in Santa Cruz department, Bolivia;
C. marinhus in Bahia state, Brazil; and *C. subflavus* in
Minas Gerais and São Paulo states, Brazil.

REMARKS: *Cerradomys scotti* can be easily distinguished from its congeners by the presence of an alisphenoid
strut and most notably by reduction to loss of the mesolophid in almost all individuals and of the mesoloph in many
(Percequillo et al. 2008). *Cerradomys scotti*, along with
Microakodontomys transitorius, *Akodon lindberghi*, and
Thalpomys cerradensis, belongs to a fringe zone fauna that
collectively inhabits transitional communities between open
and forested habitats. Each of these species is characterized
by reduction or loss of a mesoloph(id) (Hershkovitz 1993).

Based on molecular evidence, Emmons et al. (2006)
suggested that *Cerradomys andersoni* and *C. scotti* are
conspecific. Brooks et al. (2004) also highlighted the close
relationship between the haplotype of the holotype of *C.
andersoni* and specimens of *C. scotti*. The only specimen
ever assigned to *C. andersoni* shares some similarities with
specimens of *C. scotti* such as a bicolored tail, presence of
the alisphenoid strut, and reduction of the mesoloph on
M2. Additionally, this specimen is nearly identical to the
holotype of *C. scotti* as well to some other Bolivian specimens of this latter species from Santa Cruz department
and to some Paraguayan specimens from Sapucay, Tobati,
Altos, and Concepción departments (Percequillo et al.
2008). Thus, I feel confident in regarding *andersoni* as a

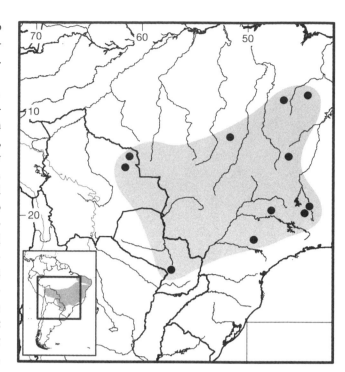

Map 153 Selected localities for *Cerradomys scotti* (●). Contour line = 2,000 m.

synonym of *C. scotti*. The karyotype of *C. scotti* is $2n = 58$
and FN = 70 to 72 (Bonvicino et al. 1999).

Cerradomys subflavus (Wagner, 1842)
Flavescent Cerradomys

SYNONYMS:

Mus vulpinus Lund, 1839a:233; *nomen nudum.*

Mus vulpinus Lund, 1840b [1841c:279]; type locality "Rio
das Velhas's Floddal" (Lund 1841c:292), Lagoa Santa,
Minas Gerais, Brazil; part; preoccupied by *Mus vulpinus* Brants.

Hesperomys subflavus Wagner, 1842c:362; type locality
"Brasilia"; restricted to "Lagoa Santa, estado de Minas
Gerais," Brazil by Cabrera (1961:396).

Mus subflavus: Schinz, 1845:190; name combination.

Mus vulpinoides Schinz, 1845:193; renaming of *Mus vulpinus* Lund.

Hesperomys laticeps: Burmeister, 1854:171; part; not *Mus
laticeps* Lund.

Calomys vulpinus: Fitzinger, 1867b:86; part; name combination.

Calomys laticeps: Winge, 1887:51; part; not *Mus laticeps*
Lund (see Percequillo 1998 and Musser et al. 1998:263,
298, concerning this usage).

Oryzomys subflavus: Thomas, 1901g:528; name combination.

[Oryzomys (Oryzomys)] subflavus: E.-L. Trouessart, 1904:
421; name combination.

[*Oryzomys*] *vulpinoides*: Tate, 1932e:18; name combination.

Holochilus physodes physodes: C. O. da C. Vieira, 1953: 134; part.

Oryzomys laticeps intermedius: C. O. da C. Vieira, 1953: 137–138; part.

Oryzomys subflavus subflavus: C. O. da C. Vieira, 1955: 411; part; name combination.

[*Cerradomys*] *subflavus*: Weksler, Percequillo, and Voss, 2006:8; first use of current name combination.

DESCRIPTION: Medium sized to large, with head and body length 120–179 mm and tail length 150–210 mm. Hindfeet small (length 31–34 mm). Dorsal body color coarsely grizzled buffy to orange brown, head contrastingly grayish, and ventral color also grayish. Skull with deep rostral fossae; roof of mesopterygoid fossa with large and wide sphenopalatine vacuities exposing orbitosphenoid; palatal fossae long, deep, and narrow, with multiple posterolateral palatal pits; bony tubes of bullae short, with distinct medial lamina dorsal to carotid canal. Central cartilaginous digit of distal baculum present but reduced in size.

DISTRIBUTION: *Cerradomys subflavus* is endemic to central Brazil, in the states of São Paulo, Minas Gerais, Bahia, and Goiás.

SELECTED LOCALITIES (Map 152): BRAZIL: Bahia, Nova Viçosa (LV [MW 13]); Goiás, Anápolis (MNRJ 4345); Minas Gerais, Fazenda Triangulo Formoso, Buritizeiros (UFMG 3), Fazenda Esmeralda, 30 km E and 4 km N (by road) from Rio Casca (UFMG 1200), Lagoa Santa (type locality of *Hesperomys subflavus* Wagner), Parque Estadual do Rio Preto, 15 km S of São Gonçalo do Rio Preto (UFMG 2852), Prados (UFMG 448); São Paulo, Avanhandava, Rio Tieté (MZUSP 2872), Itapetininga (UFMG [MAM 182]), Salto de Pirapora, Bairro da Ilha (MZUSP 24599).

SUBSPECIES: I treat *Cerradomys subflavus* as monotypic.

NATURAL HISTORY: *Cerradomys subflavus* occurs in mesic habitats in the Cerrado, including gallery forest and woodland savanna (*cerradão*), and in patches of semideciduous forest of the interior Brazilian highlands. It is primarily terrestrial (Stallings 1989), although individuals have been caught in trees (G. A. B. Fonseca and Kierulff 1989).

REMARKS: The karyotype consists of $2n = 54$ and $FN = 62$ or 64 (E. J. C. Almeida and Yonenaga-Yassuda 1974; Bonvicino and Moreira 2001; J. C. Moreira et al. 2009).

Cerradomys vivoi Percequillo, Hingst-Zaher, and Bonvicino, 2008
Vivo's Cerradomys

SYNONYM:

Cerradomys vivoi Percequillo, Hingst-Zaher, and Bonvicino, 2008:24; type locality "Parque Zoobotânico da Comissão Executiva do Plano da Lavoura Cacaueira (CEPLAC), situated 6 km E of Itabuna by road, state of Bahia, Brazil, at 14°48′S, 39°16′W, on sea level."

DESCRIPTION: Size intermediate for genus, with head and body length 127–160 mm and tail length 153–200 mm; hindfeet rather small (length 30–36 mm). Dorsal pelage short and dense, colored orange grizzled with brown, head contrastingly grayish, and venter grayish or slightly yellowish. Skull with long and wide sphenopalatine vacuities; alisphenoid strut absent; palatal fossae deep, with multiple posterolateral palatal pits.

DISTRIBUTION: *Cerradomys vivoi* occurs through the Brazilian states of Minas Gerais, Bahia, and Sergipe. In Minas Gerais, most records are restricted to northern basins of the Rio Jequitinhonha and Rio São Francisco, including its right bank tributaries. In Bahia, records of *C. vivoi* are concentrated in the coastal region on the right (eastern) bank of Rio São Francisco, while in Sergipe, the only record is from the coast, near the mouth of the Rio São Francisco.

SELECTED LOCALITIES (Map 150): BRAZIL: Bahia, Parque Zoobotânico CEPLAC, 6 km E of Itabuna (type locality of *Cerradomys vivoi* Percequillo, Hingst-Zaher, and Bonvicino), Lagoa de Itaparica (MZUSP 28889), Fazenda Bolandeira, 10 km S of Una (UFMG, not cataloged), Valença, Rio Unamirim (UFPB 540); Minas Gerais, Riacho Mocambinho, Jaíba (MNRJ 34433), Fazenda Canoas, 36 km NE and 12 km W of Montes Claros (MNRJ 61662), Jequitinhonha (UFMG 1458); Sergipe, Fazenda Capivara, 7 km SE of Brejo Grande (MNRJ 30587).

SUBSPECIES: *Cerradomys vivoi* is monotypic.

NATURAL HISTORY: *Cerradomys vivoi* is associated with the drier areas of the Cerrado and Caatinga as well as the transition between these biomes and the wetter Atlantic Forest (Percequillo et al. 2008). In the drier biomes, this species inhabits secondary semideciduous and gallery forests as well as arboreal Caatinga. In the Atlantic Forest of southern Bahia (Pardini 2004), *C. vivoi* was considered absent or very rare in mature forests (including their edges) but occurred in disturbed forests, particularly those in initial stages of regeneration or more commonly in open areas (Percequillo et al. 2008). Diet consists mainly of plant material and arthropods. The only ectoparasite thus far found on *C. vivoi* is the mite *Gigantolaelaps vitzthum*: (Hingst et al. 1997).

REMARKS: See comparisons in the discussion of *C. langguthi*. There are some karyotypic similarities between *C. vivoi* ($2n = 50$, $FN = 64$; Bonvicino et al. 1999, as "*O. subflavus* variant 3") and *C. langguthi*, which is characterized by population polymorphism of $2n = 48$ to 50 and $FN = 56$ (Percequillo et al. 2008), but the overlap is restricted to the diploid number, not to the fundamental number. Morphological differences in the autosomal complements of these two species suggest that hybrids, if possible, would suffer meiotic incompatibilities, thus supporting recognition of both forms as separate species.

Genus *Drymoreomys* Percequillo, Weksler, and Costa, 2011

Alexandre R. Percequillo and Marcelo Weksler

Drymoreomys is a monotypic genus found in the Atlantic Forest biome of eastern Brazil, known from fewer than 20 specimens from four localities ranging in elevation from 650 m to 1,200 m. Its sole species, *D. albimaculatus* Percequillo, Weksler, and Costa (2011), first collected in 1992, is the sister group to *Eremoryzomys polius*, a taxon endemic to the dry forests of the Río Marañón basin in the northern Andes of Peru, according to phylogenetic analyses of molecular and morphological data (Percequillo, Weksler, and Costa 2011; Suárez-Villota et al. 2013).

Drymoreomys are medium-sized rodents, with moderate head and body size (length 109–149 mm) and mass (44–102 g). The tail is longer than combined head and body length (tail length 150–176 mm), the pinnae are small, and the hindfeet are wide and short (hindfoot length 26–30 mm). The dorsal pelage consists of long and dense underfur and longer and lax overfur (length of aristiforms 14–17 mm). Dorsal color ranges from dull orange to fulvous buffy weakly grizzled with reddish-brown; the ventral pelage is grayish, with distinctive white patches present in gular, thoracic, and inguinal regions (the inguinal patch is sometimes absent), and distinctively paler than the dorsal pelage. The ears are rounded and small, covered internally with short golden hairs and externally densely covered with reddish-brown hairs. The mystacial vibrissae usually extend posteriorly for a few millimeters beyond the caudal margins of the pinnae when laid back. The tail is unicolored, densely covered with short and stiff brown hairs on dorsal and ventral surface and with very reduced scales. Forefeet are small and wide, covered by brown hairs proximally with the digits covered with white/silvery hairs. Hindfeet are short and moderately wide, densely covered above with a patch of brown hairs over the metatarsus surrounded by white/silvery hairs; conspicuous ungual tufts of long white to silvery hairs are present at the base of the claws on digits II to V; the plantar pads are very large and fleshy, and the interdigital pads are set close together.

The skull is medium sized and moderately robust, its condyloincisive length ranging from 27.1–32.4 mm. The rostrum is long (nasal length 11.25–13.51 mm) and wide (breadth of rostrum range from 4.82–5.98 mm), tapering distally; nasals and premaxillae are slightly projected anteriorly, forming a short bony rostral tube. Shallow zygomatic notches flank the rostrum; the zygomatic plates are not projected forward, their anterior margins are straight or slightly concave, without dorsal free margins and zygomatic spines. The interorbital region is long and narrow (least interorbital breadth ranges from 5.01–5.68 mm), anteriorly convergent, with squared to lightly beaded supraorbital margins. The braincase is oblong, with or without weakly developed temporal crests. The incisive foramina are long (averaging about 71% of the diastema length) and widest posteriorly, sometimes extending to the level of the alveoli or anterocones of M1s; lateral margins are abruptly indented anteriorly near the premaxillomaxillary suture or sharply constricted posteriorly, near the posterior border. The palate is short and wide, without palatal excrescences, and with small to moderately large posterolateral palatal pits, single to multiple and recessed in shallow fossae. The mesopterygoid fossa reaches the level of the posterior borders of M3s (penetrating to the level of the hypocones of M3s in some specimens); the bony roof of the mesopterygoid fossa is completely ossified, or perforated only by small sphenopalatine vacuities restricted to the presphenoid (only marginally in the basisphenoid). The stapedial foramen, squamosoalisphenoid groove, and sphenofrontal foramen are absent; the posterior opening of alisphenoid canal is large and conspicuous but lacks a posterior groove or depression; and the secondary anastomosis of internal carotid artery crosses the dorsal surface of parapterygoid plate, all traits that define carotid circulatory pattern 3 (*sensu* Voss 1988). An alisphenoid strut is present as a robust bony bar separating the buccinator-masticatory and accessory oval foramina. Auditory bullae are globose, with short bony tubes; the stapedial process is long and narrow, overlapping the lateral margin of the squamosal. The posterior suspensory process of the squamosal is absent; a bony projection dorsolateral to the stapedial process is present, overlapping the squamosal; and the periotic is exposed posteromedially between the ectotympanic and the basioccipital, but does not extend anteriorly to the carotid canal in most specimens. Each mastoid is perforated by a conspicuous posterodorsal fenestra; the postglenoid foramina are large and rounded; and the subsquamosal fenestrae are large and patent.

The mandible is long and shallow; the coronoid process is large, falciform or triangular, nearly equal in height to the condyloid process; the angular process is short, not extending past the condyloid process posteriorly. Both superior and inferior notches are shallow. The capsular process is not noticeable.

Upper incisors are opisthodont, and molars are pentalophodont and high-crowned. The maxillary toothrows are subparallel or weakly convergent anteriorly; the molar series is robust and long, 4.93–5.43 mm. Labial and lingual flexi of upper molars shallowly penetrate at the molar midline; both are arranged in opposite pairs. The labial and lingual cusps are high (molars nearly hypsodont) and compressed anteroposteriorly; labial cusps are quite distinct and project laterally. The occlusal surfaces of the labial cusps of the upper molars are oriented posteriorly; the occlusal surfaces of the lingual cusps are labially oriented. The anterocone of M1 is divided into labial and lingual conules (anteromedian flexus present); the anteroloph is well developed and fused with an anterostyle on the labial cingulum, and separated from the anterocone by a persistent anteroflexus. A protostyle is present (in most specimens) as an isolated accessory cuspule or fused to an anterior mure. An enamel bridge connects the paracone to the middle or the anterior portion of the protocone. The protoflexus of M2 is variably present; the mesoflexus is long and obliquely or transversely oriented, divided into two (labial and lingual) or three (labial, medial, and lingual) fossettes. M3 has a mesoloph, posteroloph, and persistent hypoflexus; the paracone and protocone are distinguishable on the occlusal surface; the hypocone is reduced, and the metacone is greatly reduced and completely fused with the posteroloph. The labial accessory root of M1 is possibly absent (not visible in lateral view).

Labial cusps of the lower molars are slightly anterior to the lingual cusps; the occlusal surfaces of the lingual cusps are oriented anteriorly, and those of the labial cusps are lingually oriented. The anteroconid of m1 is divided by an anteromedian flexid; the anterolabial cingulum is developed in m1 and m2, and variably present in m3; the anterolophid is present in all lower molars; the ectolophid is variably present on m1 only; and the posteroflexid on m3 is large and persistent. All lower molars have two roots, one each in anterior and posterior positions.

The axial skeleton includes 12 ribs and 19 thoracolumbar, 4 sacral, and 36–38 caudal vertebrae. The hemal arch at the boundary of the second and third caudal vertebrae hasa posterior spinous process. The trochlear process of the calcaneus is long and shelf-like, positioned immediately posterior to the articular facet. The entepicondylar foramen is absent. The supratrochlear is absent in adults but is observable in young individuals.

The stomach is unilocular and hemiglandular, with the gastric glandular epithelium restricted to the antrum. The glans penis does not have lateral bacular mounds; the distal cartilaginous apparatus is tridigitate, with the central digit more robust than the lateral ones; the dorsal papilla has one apical spine; and the urethral process has a large lateral lobule. The preputial glands are very large.

SYNONYM:

Drymoreomys Percequillo, Weksler, and Costa, 2011:360; type species *Drymoreomys albimaculatus* by original designation and monotypy.

Drymoreomys albimaculatus Percequillo, Weksler, and Costa, 2011
White-throated Montane Forest Rat

SYNONYM:

Drymoreomys albimaculatus Percequillo, Weksler, and Costa, 2011:368; type locality "Brazil, Estado de São Paulo, Município de Ribeirão Grande, Parque Estadual Intervales, Base do Carmo, 700 m; 24°20'S, 48°25'W."

DESCRIPTION: As for the genus.

DISTRIBUTION: All known specimens referred to *Drymoreomys* are from the eastern slopes of the Serra do Mar, in the coastal Brazilian Atlantic Forest, from São Paulo to Santa Catarina states, at elevations ranging from 650 m to 1,200 m.

SELECTED LOCALITIES (Map 154): BRAZIL: São Paulo, Estação Ecológica do Bananal (MZUSP 33782), Estação Biológica de Boracéia (MZUSP [BO 24]), Parque Estadual Intervales, Base do Carmo (type locality of *Drymoreomys albimaculatus*); Santa Catarina, Parque Estadual da Serra do Tabuleiro (UFSC 860).

SUBSPECIES: *Drymoreomys albimaculatus* is monotypic.

NATURAL HISTORY: Available evidence (see Percequillo, Weksler, and Costa 2011) suggests that *D. albimaculatus* is an Atlantic Forest endemic, inhabiting dense and humid montane and premontane forests. Apparently, this species does not exhibit preferences for pristine forests, as it has also found in secondary and disturbed forests, and even in areas that were completely deforested a few decades ago. Additional evidence (Pardini and Umetsu 2006; Bueno 2008; R. Pardini, pers. comm.), however, suggests that *D. albimaculatus* inhabits areas of both pristine and disturbed forest. Females were reproductively active in June and from November to December, and males possessed scrotal testes in December, suggesting that reproduction may take place throughout the year (Percequillo, Weksler, and Costa 2011).

REMARKS: This recently discovered rat comes from one of the most densely populated areas in South America. Furthermore, the unexpected close phylogenetic relationship between *Drymoreomys* and *Eremoryzomys*, as recovered by Percequillo, Weksler, and Costa (2011) and Suárez-Villota et al. (2013), provides one of the most noteworthy sister taxon duos in South America biogeography. *Drymoreomys* is endemic to lower montane areas in the Atlantic Forest of eastern Brazil, whereas *Eremoryzomys* is an Andean endemic (760 to 2,100 m) from the valley of the lower Río Marañón in northern Peru. Each taxon is restricted to well-known Neotropical centers of endemism (P. Müller 1973; Cracraft 1985), with *D. albimaculatus* part of the Serra do Mar center and *E. polius* from the Marañón center. A closely parallel case illustrating a biogeographic connection

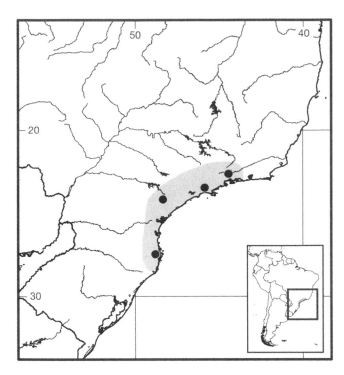

Map 154   Selected localities for *Drymoreomys albimaculatus* (●).

between the Brazilian Atlantic Forest and central Andes is that of *Rhagomys*. It is likely that additional discoveries of rodents or many other organismal groups will fill the large geographic gap between these two distributional areas.

Suárez-Villota et al. (2013) described a karyotype with $2n = 62$, $FN = 62$; the autosomal complement consists of 29 pairs of acrocentrics grading evenly from large to small plus a single pair of small metacentrics; the X chromosome is a large subtelocentric, the Y a slightly small submetacentric.

Genus *Eremoryzomys* Weksler, Percequillo, and Voss, 2006

Alexandre R. Percequillo

The genus *Eremoryzomys* is monotypic, with the single species, *E. polius*, limited to the upper Río Marañón valley in northern Peru. Since it was described, the validity of this species has never been questioned (see Cabrera 1961; Musser and Carleton 1993, 2005), although its phylogenetic relationships to other oryzomyines have remained unclear. Osgood (1913) emphasized the distinctness of *E. polius* in comparison with *Aegialomys xanthaeolus*, and Musser and Carleton (2005) declared its affinities to be obscure.

The morphology of *Eremoryzomys* is quite distinctive both in external and cranial traits. This rat is medium sized, with the head and body length 125–161 mm. The

tail is longer than the head and body (135–188 mm). The dorsal pelage is dense and harsh, coarsely grizzled grayish to brownish or yellowish gray; the ventral pelage is noticeable paler, with gray-based hairs tipped very light buff. The tail is distinctly bicolored. Ears are small and rounded, and do not reach the eye when laid forward. The hindfoot has conspicuous and dense tufts of long, white ungual hairs at the base of the claws; the hypothenar pad is large and distinct, as are other plantar pads.

The skull is medium-sized, with its condyloincisive length 29.9–36.8 mm. The rostrum is long and stout, flanked by moderately deep zygomatic notches. The interorbital region is divergent posteriorly, with beaded supraorbital margins. The braincase is rounded, with distinct temporal crests. Posterior margins of the zygomatic plates are located above the alveoli of the M1s. The zygomatic arches are strong and flared, widest at the squamosal root; jugals are large (the maxillary and squamosal zygomatic processes are widely separated, not overlapping, in lateral view). The nasals are short and do not extend posterior to the lacrimals, which are equally sutured to the maxillary and frontal bones. The frontosquamosal suture usually is collinear with the frontoparietal suture. Incisive foramina are very long, usually extending posteriorly between the anterocones or even protocones of the M1s; the lateral margins are subparallel and widest medially. The palate is short and wide, with the anterior margin of the mesopterygoid fossa penetrating anteriorly to, or slightly between, the molar rows; the bony roof of mesopterygoid fossa is perforated by large sphenopalatine vacuities. The palate has large and complex posterolateral palatal pits recessed in deep fossae. The alisphenoid strut is usually present so that the buccinator-masticatory foramen and accessory foramen ovale are separate, but the strut is unilaterally absent on some skulls. The stapedial foramen and the posterior opening of alisphenoid canal are small, the squamosal-alisphenoid groove and the sphenofrontal foramen are absent, and the secondary anastomosis of the internal carotid crosses dorsal surface of the pterygoid plate, all features that definecarotid circulatory pattern 3 (sensu Voss 1988). The postglenoid foramen is large and rounded, and the subsquamosal fenestra is large and conspicuous. The periotic bone is exposed posteromedially between the ectotympanic and the basioccipital but does not extend anteriorly to the carotid canal; each mastoid is perforated by a small or large posterodorsal fenestra. The capsular process is indistinct or absent, and both superior and inferior masseteric ridges are usually conjoined anteriorly below m1 as a single crest.

Labial and lingual flexi of M1 and M2 do not interpenetrate. The anterocone of M1 is not divided into labial and lingual conules (but a small internal fossette obviously derived from the anteromedian flexus is present); the anteroloph usually is well developed, fused with an anterostyle on the labial cingulum, and separated from the anterocone by a persistent anteroflexus; a protostyle is absent; the paracone is connected by an enamel bridge to the middle or posterior part of the protocone. The protoflexus of M2 is present but shallow; the mesoflexus is present as one or more internal fossettes (both conditions occurring on opposite sides of some specimens: e.g., AMNH 64054). Both a posteroloph and diminutive hypoflexus are present on M3, the latter tending to disappear with moderate to heavy wear. A labial accessory root of M1 is absent. The anteroconid of m1 lacks an anteromedian flexid; the anterolabial cingulum and anterolophid are present on all lower molars; an ectolophid is absent on m1 and m2; a mesolophid is variably developed on m1 and m2, large and distinct in some specimens but much reduced or absent in others; and the hypoflexid of m2 is short. Accessory roots are absent on m1; m2 and m3 each has one large anterior and posterior root.

SYNONYMS:

Oryzomys: Osgood, 1913:97; part (description of *polius*); not *Oryzomys* Baird.

Eremoryzomys Weksler, Percequillo, and Voss, 2006:10; type species *Oryzomys polius* Osgood, by original designation.

REMARKS: According to Weksler (2003, 2006), *Eremoryzomys* is the most basal lineage of a clade (Clade D) containing *Oryzomys*, *Pseudoryzomys*, *Lundomys*, *Holochilus*, *Sooretamys*, *Cerradomys*, *Nesoryzomys*, *Aegialomys*, *Amphinectomys*, *Nectomys*, *Sigmodontomys*, and *Melanomys*.

Eremoryzomys polius (Osgood, 1913)

Gray Eremoryzomys

SYNONYMS:

Oryzomys polius Osgood, 1913:97; type locality "Tambo Carrizal, mountains east of Balsas, [Amazonas], Peru. Altitude about 5000 ft."

Oryzomys [(*Oryzomys*)] *polius*: Cabrera, 1961:394; name combination.

[*Eremoryzomys*] *polius*: Weksler, Percequillo, and Voss, 2006:10; first use of current name combination.

DESCRIPTION: As for the genus.

DISTRIBUTION: Known only from seven localities in the upper Río Marañón valley of northern Peru, in Amazonas and Cajamarca departments.

SELECTED LOCALITIES (Map 155): PERU: Amazonas, Condechaca, Río Utcubamba (BM 26.5.3.16), Tambo Carrizal, mountains E of Balsas (type locality of *Oryzomys polius* Osgood), Tingo, 30 km S and 41 km E of Bagua, Río Utcubamba (MVZ 135658); Cajamarca, Chaupe (AMNH 64054), San Ignacio (AMNH 64055).

SUBSPECIES: *Eremoryzomys polius* is monotypic.

NATURAL HISTORY: *Eremoryzomys polius* inhabits the upper Marañón Valley westward to the Cordillera Central of northern Peru, in a rain-shadow area characterized by hot and arid climate (Cracraft 1985; Osgood 1914b:145). Nothing is known about the behavior or population biology of this species. What little is known of the habitat was summarized by Osgood (1914b:145), who stated that the type series was obtained at "a dilapidated tambo on the side of the mountains directly east of and overlooking Balsas and the [Río] Marañón. A tiny spring here is surrounded by a clump of trees . . . and the steep slope below has some moderately extensive thickets of low bushes which moisture from the spring permits to flourish. Elsewhere conditions are arid with occasional bunches of grass, cactuses, or small bushes." Thomas (1926g:157) stated that Condechaca, situated a few kilometers south of Chachapoyas but north of Tambo Carrizal in the Marañón valley, "is surrounded by relatively arid foothills."

REMARKS: As with species of *Aegialomys*, *Eremoryzomys polius* appears to be an open habitat specialist. As such, members of these two genera share some resemblances

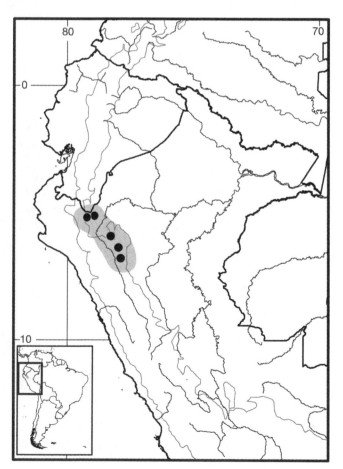

Map 155 Selected localities for *Eremoryzomys polius* (●). Contour line = 2,000 m.

(grayish dorsal coloration, long incisive foramina, mesoloph/id reduction) that are apparently convergences owing to adaptations to open, dry areas (Weksler 2003, 2006). Weksler et al. (2006:11) noted that "*Eremoryzomys* differs from *Aegialomys* by its much grayer dorsal pelage (the dorsal fur is distinctly yellowish or brownish in *Aegialomys*); larger jugal (the jugal of *Aegialomys* is much smaller); longer incisive foramina (these openings never extend posteriorly between the M1 protocones in *Aegialomys*); shorter palate (the mesopterygoid fossa never extends anteriorly to the molar rows in *Aegialomys*); presence of an alisphenoid strut separating the buccinator-masticatory and accessory oval foramina (the strut is invariably absent and the foramina are confluent in *Aegialomys*); absence of a distinct capsular process (although well developed in *Aegialomys*); absence of accessory roots on M1/m1 (accessory roots are normally present on these teeth in *Aegialomys*); and presence of the anterolophid on m2 and m3 (the anterolophid is absent on these teeth in *Aegialomys*)."

Genus *Euryoryzomys* Weksler, Percequillo, and Voss, 2006

Alexandre R. Percequillo

According to recent revisionary work (Weksler 1996, 2006; Musser et al. 1998; Percequillo 1998, 2003), the genus *Euryoryzomys* includes six valid species: *E. emmonsae*, *E. lamia*, *E. legatus*, *E. macconnelli*, *E. nitidus*, and *E. russatus*. These studies also emphasize that some members of the genus exhibit variation in morphology, karyology, and/or molecular phylogeography that may merit taxonomic recognition. This is especially so for *E. russatus* from northeastern Brazil (Weksler 1996; Percequillo 1998; M. J. de J. Silva et al. 2000b), *E. macconnelli* from western Amazonia (Musser et al. 1998; see also Patton et al. 2000; Voss et al. 2001), and *E. emmonsae* from south central Amazonia (L. P. Costa 2003).

The species of *Euryoryzomys* are widely distributed throughout South America (hence the name, from the Greek *eurus*, meaning "far reaching or far-spread"), occurring in lowland and lower montane tropical rainforest of the Amazon Basin; Brazilian and Argentinean Atlantic Forest; Bolivian and Argentinean yungas; the semideciduous forests of Paraguay, Brazil, and Bolivia; isolated patches of rainforest (*brejos*) in the northeastern Brazilian Caatinga; and gallery forests in the Brazilian Cerrado and Paraguayan Chaco.

Species of this genus exhibit a considerable range in size (head and body length 90–96 mm; tail length 100–185 mm). The tail is longer than the head and body, except in *E. nitidus* and *E. lamia*, which have a tail as long as head and body (head and body length 113–160 mm versus tail length 108–163 mm, and head and body length 135–165 mm versus tail length 135–165 mm, respectively). The dorsal pelage is soft and dense, varying from long in most species (fur length ranges from 12–16 mm) to very long in *E. macconnelli* (fur length from 17–20 mm). The dorsal and ventral pelage coloration is variable among species, but always shows strong countershading; the dorsal color ranges from buffy-brown to reddish cinnamon brown, and the ventral color varies from grayish-white to almost pure white. The tail is always bicolored. The hindfoot is long and narrow (length 25–44 mm), and has very dense and long ungual tufts that conceal the claws. The ungual tufts are usually white.

The skull ranges from medium sized to large (mean greatest skull length varies among species from 28.8–39.2 mm), and is of lightly to moderately heavily build. The rostrum is long and narrow, flanked by narrow and shallow to deep (in *E. lamia* and *E. legatus*) zygomatic notches. Incisive foramina range from short to long (4.0–7.0 mm) and are variable in shape, with their lateral margins either nearly parallel, convex, or divergent posteriorly (width 1.9–3.2 mm); their posterior margins never reach the level of the alveoli of M1s. The interorbital region is narrow (4.4–6.5 mm in least breadth) with lateral margins divergent posteriorly; the supraorbital margins are weakly to strongly beaded, and form discrete crests that extend posteriorly onto the dorsolateral margins of braincase, which is long and narrow. The zygomatic arches project laterally and are widest at their squamosal roots; jugals are absent in almost all specimens of *E. lamia*. The zygomatic plates span a considerable range in width, from 2.6–5.2 mm; the free dorsal margins project anteriorly, with their anterior margins sloping backward from top to bottom in *E. lamia* and *E. legatus* and either straight to slightly concave. A stapedial foramen, squamosal-alisphenoid groove, and sphenofrontal foramen are present, defined as carotid circulatory pattern 1 (*sensu* Voss 1988). An alisphenoid strut, present in all specimens of *E. legatus*, occurs at low to moderate frequencies in *E. russatus*, *E. lamia*, and *E. nitidus*, but is absent in individuals of *E. macconnelli* and *E. emmonsae*. The palate is long and narrow, with puntiform to small posterolateral palatal pits placed either at palatal level or in a shallow palatal depression; palatal excrescences are present in most species. The mesopterygoid fossa is narrow with its anterior margin rounded; the roof of the fossa is completely ossified or perforated only by small sphenopalatine vacuities. Parapterygoid fossae are flat or only moderately concave. Auditory bullae are small and rounded, with both the bony tubes and stapedial processes long and narrow. The tegmen tympani is small, rounded, squared, or triangular, and may or may not overlap the squamosal; the suspensory process of the squamosal is absent.

The mandible is deep and robust. The coronoid and condyloid processes are equal in height, and separated by a shallow superior notch. The angular process does not extend posteriorly past the condyloid process. The capsular process is a distinct projection in most species but is absent in *E. macconnelli* and *E. emmonsae*.

Upper incisors are opisthodont, fronted by distinct orange enamel bands. Toothrows are parallel and of medium length (length 4.3–5.6 mm). Main cusps are arranged in opposite pairs and are low to high crowned. The anterocone of M1 is undivided; M1 and M2 have two distinct enamel islands between the paracone and metacone; and M3 possesses a distinct anterolabial cingulum in some species such as *E. lamia*. The first lower molar is not divided by an anteromedian flexid; m2 has a distinct ectostylid or even a narrow and delicate ectolophid. M1 has three main roots, anterior and posterior labial and lingual, as well as an accessory rootlet between the two labials and beneath the paracone; M2 and M3 each has three main roots. All lower molars have two roots, one anterior and one posterior.

The axial skeleton is composed of 7 cervical, 12–13 thoracic, 6–7 sacral, and 25–33 caudal vertebrae; thoracic vertebrae bear 12–13 ribs, or rarely 11 ribs plus a vestigial one. Hemal arches on caudal elements at the junction of the first and second and/or second and third caudal vertebrae have a distinct medial posterior process.

The stomach is unilocular and hemiglandular, with a shallow angular notch. The glans penis is of the complex type (Hooper and Musser 1964), with a well-developed, tridigitate cartilaginous baculum. Both a single dorsal papilla and urethral flap are present, and three bacular mounds are visible at the terminal crater. The basal portion of the bony baculum is wide and spatulate, barely concave.

SYNONYMS:

Mus: Brants, 1827:139; part (description of *physodes*); not *Mus* Linnaeus.

Hesperomys: Wagner, 1848:312; part (description of *russatus*); not *Hesperomys* Waterhouse.

Calomys: Winge, 1887:51; part (description of *coronatus*); not *Calomys* Waterhouse.

Oryzomys: E.-L. Trouessart, 1897:525; part (inclusion of *intermedius* and *nitidus* as subspecies of *laticeps*); not *Oryzomys* Baird.

Oryzamys: Avila-Pires, 1968:162; *lapsus calami* (description of *kelloggi* and *ratticeps moojeni*).

Euryoryzomys Weksler, Percequillo, and Voss, 2006:8; type species *Oryzomys macconnelli* Thomas, by original designation.

REMARKS: In 1960 (and subsequently in 1966a), Hershkovitz committed one of the most controversial and influential nomenclatural acts in the convoluted history of oryzomyine species-group taxa (see Pine 1973; A. L. Gardner and Patton 1976; Musser et al. 1998; Weksler 2006). In a footnote comment, apparently not based on the examination of voucher specimens (especially type material) and clearly without any extensive discussion of morphological variation, Hershkovitz (1960:544) lumped under *Oryzomys laticeps/capito* (see Musser et al. 1998:10) the species of the genus *Euryoryzomys* (*boliviae*, *legatus*, and *macconnelli*) as well as those now belonging to other oryzomyine genera such as *Hylaeamys* (*goeldii*, *modestus*, *oniscus*, *perenensis*, *saltator*, and *velutinus*), *Nephelomys* (*caracolus*), and *Transandinomys* (*casteneus* [*sic*; Hershkovitz attributed the name authorship of *castaneus* to Oldfield Thomas, but this author never coined either *castaneus* or *casteneus* for any sigmodontine rodent genus according to J. E[dwards]. Hill 1990; the name *castaneus* was proposed by J. A. Allen in 1901; see Cabrera 1961], *magdalenae*, *medius*, *mollipilosus*, *rivularis*, *sylvaticus*, and *talamancae*). Moreover, Hershkovitz also mentioned that "a few other" species were also synonyms of *Oryzomys laticeps/capito*, but he never clarified the composition of this group in subsequent papers (e.g., Hershkovitz 1966a).

Through the following decades, Hershkovitz's footnote taxonomy was followed, partially or entirely, by other authorities (most notably Cabrera [1961]), until the publication of two important but temporally widely separated contributions. A. L. Gardner and Patton (1976), in their pioneering and influential paper on the cytogenetics of Neotropical Oryzomyini, paved the way for subsequent authors by showing that, under the name *capito*, Hershkovitz (1960, 1966a) had hidden extensive morphologic, karyologic, and taxonomic diversity. And, 20 years later in their massive revision, Musser et al. (1998) defined the limits of several species historically assigned to *Oryzomys* "*capito*" (*sensu* Hershkovitz 1960), recognizing 11 species grouped into six species groups based on extensive analysis of geographic variation and, critically, on their examination of almost all available type material. Combined, these two efforts plus several subsequent contributions have been responsible for the substantial advances in the classification and nomenclature of the Oryzomyini, as now presented throughout this volume.

KEY TO THE SPECIES OF SOUTH AMERICAN *EURYORYZOMYS*:

1. Lateral face of mandibular ramus without trace of capsular process 2
1′. Lateral face of mandibular ramus with prominent capsular process 3
2. Dorsal pelage very dense, lax, and long (aristiform length 17–20 mm); dorsal pelage cinnamon brown or orange brown grizzled with dark brown or black; inci-

sive foramina short (length 4.1–6.2 mm, mean 5.4 mm) but widest posteriorly (teardrop in shape; see Musser et al. 1998:196, Fig. 82; palatal excrescences frequently present; upper molars with high cusps; enamel islands (fossettes) on M1 and M2, specially the labial fossette, quickly obliterated by wear; m1 with ectolophid frequently present; diploid number of western Brazilian Amazon and Peruvian populations 2n = 64 and fundamental number FN = 70 and FN = 64, respectively; of Venezuelan populations 2n = 76 and FN = 85, and of Brazilian central Amazon 2n = 58 and NF = 90 . . . .
. . . . *Euryoryzomys macconnelli*
2′. Dorsal pelage soft and short (aristiform length 12–15 mm); dorsal pelage ochraceous brown grizzled with dark brown; incisive foramina short (length 4.0–6.0 mm, mean 5.2 mm) but widest medially; palatal excrescences absent; upper molars with lower cusps, more flat-crowned; enamel islands more persistent, well resistant to wear; m1 without noticeable ectolophid; 2n = 80 and FN = 86 . . . . *Euryoryzomys emmonsae*
3. Zygomatic plates very wide (3.7–5.2 mm), defining extremely deep zygomatic notches 4
3′. Zygomatic plates narrower (3.2–4.6 mm), defining moderately deep zygomatic notches 5
4. Dorsal fur long (aristiform length 15–16 mm) and dense; head and body length (135–165 mm, mean 149.4 mm) equal to tail length (135–165 mm, mean 149.2 mm) in most specimens (few specimens have tail longer or shorter than head and body length); hindfoot more robust and longer (32–37 mm, mean 34 mm); zygomatic plates very wide (4.0–5.2 mm); zygomatic notches narrow and very deep; jugal absent; alisphenoid struts present in few specimens; palate without palatal excrescences . . . . *Euryoryzomys lamia*
4′. Dorsal fur very long (aristiform length 17–19 mm) and dense; tail length (130–180 mm, mean 146.5 mm) exceeds head and body length (110–159 mm, mean 136.3 mm) in all specimens; hindfoot shorter (30–36 mm, mean 32.9 mm); zygomatic plates wide (3.7–4.6 mm); zygomatic notches narrow and deep; jugal present; alisphenoid struts present; palate ornate with small palatal excrescences . . . . . . *Euryoryzomys legatus*
5. Tail (112–185 mm, mean 151.7 mm) longer than head and body (105–196 mm, mean 143.2 mm); restricted to eastern South America, in Atlantic Forest from Bahia to Rio Grande do Sul states, Brazil, west to Missiones province, Argentina and eastern Paraguay . . . .
. . . . *Euryoryzomys russatus*
5′. Tail (108–163 mm, mean 136.1 mm) equals head and body length (113–160 mm, mean 136.2 mm); restricted to western Amazon Basin south to western Mato Grosso state, Brazil. . . . . *Euryoryzomys nitidus*

Euryoryzomys emmonsae (Musser, Carleton, Brothers, and Gardner, 1998)

Emmons's *Euryoryzomys*

SYNONYMS:

Oryzomys emmonsae Musser, Carleton, Brothers, and Gardner, 1998:233; type locality "east [= right] bank of Rio Xingu, 52 km SSW Altamira (03°39′S/52°22′W), below 100 m in Estado de Pará, Brazil."

[*Euryoryzomys*] *emmonsae*: Weksler, Percequillo, and Voss, 2006:11; first use of current name combination.

DESCRIPTION: Characterized by long tail relative to head and body length (average about 114%); small and delicate cranium, with long but slender rostrum; narrow zygomatic plates; narrow interorbital region with smooth dorsolateral margins; and narrow occiput. Incisive foramina short and wide; bony palate correspondingly long.

DISTRIBUTION: *Euryoryzomys emmonsae* is currently known from the south bank of the Rio Amazonas, throughout the Amazonian interfluvial forest between the left bank of the Rio Tocantins to the right bank of the Rio Xingu, in Pará state, Brazil. Two records from the Rio Teles Pires (a headwater tributary of the Rio Tapajós) in Mato Grosso state are included here, although these might represent an undescribed species (see below and L. P. Costa 2003). All localities range in elevation from 80 and 200 m.

SELECTED LOCALITIES (Map 156): BRAZIL: Mato Grosso, Cláudia (L. P. Costa 2003), Reserve Ecológica Cristalino, 40 km N of Alta Floresta (MVZ 197523), Vila Rica (L. P. Costa 2003), 264 km N of Xavantina, Serra do Roncador (BM 81.436); Pará, 52 km SSW of Altamira, right bank Rio Xingu (type locality of *Oryzomys emmonsae* Musser, Carleton, Brothers, and Gardner), Floresta Nacional Tapirapé-Aquiri (Patton et al. 2000), Serra Norte (MPEG 8188).

SUBSPECIES: *Euryoryzomys emmonsae* is monotypic.

NATURAL HISTORY: Data summarized by Musser et al. (1998:236) suggests that this species is typically captured on the ground and under logs in viney forests and bamboo thickets. Most collecting localities are from nonflooded (*terra firme*) forest, although some are on riverbanks where the habitat may be influenced by seasonal flooding.

REMARKS: *Euryoryzomys emmonsae* is very similar to both *E. nitidus* and *E. russatus* in morphological and karyological traits, although the size and proportions of both body and cranial parts distinguish these three species (see Musser et al. 1998). The most important diagnostic feature of this species is the absence of a capsular process on the mandibular ramus. The karyotype of *E. emmonsae* is very similar to those described for *E. nitidus* (A. L. Gardner and Patton 1976) and for *E. russatus* (M. J. de J. Silva et al. 2000b), with 2n = 80 and FN = 86. Specimens from two localities in the drainage of the Rio Teles Pires

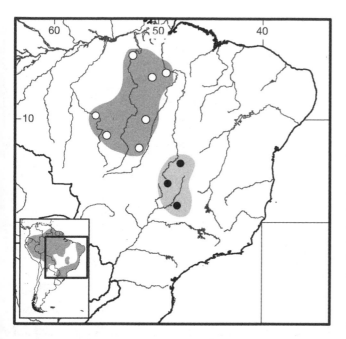

Map 156 Selected localities for *Euryoryzomys emmonsae* (○) and *Euryoryzomys lamia* (●).

in Mato Grosso state share similar morphological traits with *E. emmonsae* but are divergent from it in mtDNA cytochrome-*b* sequence (L. P. Costa 2003: localities 20 and 21). These likely represent an undescribed species but are included here under *E. emmonsae* because they are placed with this species in phylogenetic analyses of sequence data.

Euryoryzomys lamia (Thomas, 1901)
Buffy-sided Euryoryzomys

SYNONYMS:

Oryzomys lamia Thomas, 1901g:528; type locality "Rio Jordão near Araguary," Minas Gerais, Brazil.

Oryzomys (Oryzomys) lamia: Tate, 1932e:18; name combination.

[*Euryoryzomys*] *lamia*: Weksler, Percequillo, and Voss, 2006:11; first use of current name combination.

DESCRIPTION: Dorsal pelage long and dense, with aristiforms 15–16 mm long and bicolored tail equal to or shorter than head and body length (average about 92%). Skull long and robust; rostrum wide; zygomatic notches deeply excavated; zygomatic plates very wide with straight anterior free margin; incisive foramina medium in length (mean 6.2 mm; range: 5.5–7.0 mm) with parallel lateral margins; palatal excrescences absent; roof of mesopterygoid fossa completely ossified; and M2 with well-developed anterolabial cingulum.

DISTRIBUTION: *Euryoryzomys lamia* is known from three localities in the semideciduous forest of the Brazilian central plateau in western Minas Gerais and southern Goiás states (Percequillo 1998) and in Cerrado habitats of northern Goiás state (Bonvicino et al. 1998).

SELECTED LOCALITIES (Map 156): BRAZIL: Goiás, Anápolis (MNRJ 4352), Fazenda Fiandeira, Parque Nacional da Chapada dos Veadeiros, 65 km SSW of Cavalcante (MNRJ 46826); Minas Gerais, Araguari, Rio Jordão (type locality of *Oryzomys lamia* Thomas).

SUBSPECIES: *Euryoryzomys lamia* is monotypic.

NATURAL HISTORY: Available evidence (Weksler 1996; Percequillo 1998) suggests that *E. lamia* is a forest-dweller, inhabiting gallery forests (along the Rio Jordão) and a luxuriant semideciduous forest locally known as *Mato Grosso de Goiás* (at Anápolis; Faissol 1952). More recent published data suggest that this species also occurs in Cerrado habitats, such as *cerradão* (a dry woodland savanna), *cerrado sensu stricto*, and *campo úmido* (both open habitats), although, in the last, two specimens were captured in the ecotone with gallery forest (Bonvicino et al. 1998).

REMARKS: Although regarded as a synonym of *Euryoryzomys russatus* by Musser et al. (1998), *E. lamia* is now recognized as a separate species (e.g., Musser and Carleton 2005; Weksler et al. 2006). These two species differ in several morphological attributes (Weksler 1996; Percequillo 1998), notably, the tail length of *E. lamia* is equal to or shorter than the head and body length while that of *E. russatus* exceeds the head and body length; *E. lamia* has much deeper zygomatic notches; and *E. lamia* has a longer palate (palatal bridge length 7.2–8.2 mm) rarely ornamented by palatal excrescences whereas *E. russatus* possesses a shorter palate (palatal bridge length 5.6–7.6 mm). Univariate and multivariate analyses also differentiate the two taxa (Weksler 1996; Percequillo 1998; Musser et al. 1998). The two species also have different karyotypes ($2n = 58$, FN = 82 for *E. lamia* [Bonvicino et al. 1998] and $2n = 80$, FN = 86 for *E. russatus* [Musser et al. 1998]) and display divergent mitochondrial and nuclear sequences (Bonvicino and Moreira 2001; Weksler 2003).

The assignment of specimens from Cavalcante, Goiás state (reported by Bonvicino et al. 1998), to *E. lamia* deserves further investigation because there are noticeable morphometric differences between these specimens and series from Anápolis and Rio Jordão (Percequillo 1998).

Euryoryzomys legatus (Thomas, 1925)
Tarija Euryoryzomys

SYNONYMS:

Oryzomys legatus Thomas, 1925b:577; type locality "Carapari [= Caraparí], 1000 m., about 35 kilometres north of Yacuiba, on the way towards Tarija," Tarija, Bolivia.

Oryzomys (Oryzomys) legatus: Tate, 1932e:18; name combination.

Oryzomys laticeps: Hershkovitz, 1960:544, footnote; part; not *Mus laticeps* Lund (= *Hylaeamys laticeps* [Lund]).

Oryzomys capito legatus: Cabrera, 1961:386; name combination.

[*Euryoryzomys*] *legatus*: Weksler, Percequillo, and Voss, 2006:11; first use of current name combination.

DESCRIPTION: Most informative traits include very dense and long dorsal pelage where aristiform hairs reach 17–19 mm in length; distinctly bicolored tail longer than combined head and body length (average about 108%); small skull (greatest length 32.9–36.8 mm), more delicate than *E. lamia* (length 34.2–38.1 mm) but more robust than other species; zygomatic notches narrow and deep, similar to *E. lamia* but less excavated; zygomatic plates wide; incisive foramina short (length 4.9–6.6 mm, mean 5.5 mm) and narrow, with parallel lateral margins; bony palate with small palatine excrescences; roof of mesopterygoid fossa completely ossified; and alisphenoid strut universally present.

DISTRIBUTION: The collecting localities of *E. legatus* are in premontane and montane forests along the eastern Andean slope in extreme southern Bolivia and northernmost Argentina, at elevations from 500 to 2,100 m; there are no records of this species in the Bolivian or Argentinean lowlands.

SELECTED LOCALITIES (Map 157): ARGENTINA: Jujuy, 8.4 km E of El Palmar, Sierra Santa Bárbara (Musser et al. 1998); Salta, Santa Victoria, Parque Nacional Baritú, Arroyo Santa Rosa (MACN 20708), Piquirenda (Musser et al. 1998). BOLIVIA: Chuquisaca, 2 km E of Chuhuayaco (AMNH 263886), Tola Orco, 40 km from Padilla (USNM 271584); Santa Cruz, 1 km NE of Estancia Las Cuevas (AMNH 264182), Mataracú (FMNH 129267); Tarija, 1 km S of Camatindi (AMNH 264186), 3 km SE of Cuyambuyo (AMNH 264280).

SUBSPECIES: *Euryoryzomys legatus* is monotypic.

NATURAL HISTORY: This species is found in mesic and transitional forests of Salta province, Argentina (Mares, Ojeda, and Kosco 1981:174), where specimens were commonly acquired in "littered forested areas supporting little undergrowth," but also caught in "dense second growth vegetation along streams and roads, and along the rocky banks of water courses." Mares, Ojeda, and Kosco (1981) also stated that this species inhabited burrows in the forest floor, but the notes on specimen tags (MACNBR) also indicate that it frequents the canopy (*habita la selva sobre los arboles*). *Euryoryzomys legatus* feeds on vegetation and insects. S. Anderson (1997:401) summarized available information on the ectoparasites found on Bolivian specimens, including the ticks *Eulaelaps halleri* and *Schistolaelaps mazzai* as well as the flea *Polygenis* (*Polygenis*) *typus*.

REMARKS: Musser et al. (1998) treated *E. legatus* as a synonym of *E. russatus*, but it is currently recognized as a

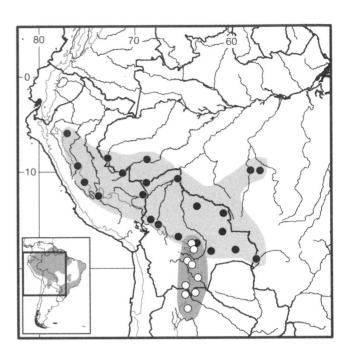

Map 157 Selected localities for *Euryoryzomys legatus* (○) and *Euryoryzomys nitidus* (●). Contour line = 2,000 m.

separate species (Musser and Carleton 2005; Weksler et al. 2006). The two are demonstrably different by morphologic and morphometric comparisons (Weksler 1996; Percequillo 1998), with the most distinctive morphological traits of *E. legatus* being the presence of a deep and narrow zygomatic notch and an alisphenoid strut present in all specimens. Limited mtDNA sequence analyses, however, suggest that *E. legatus* is nested within samples of *E. nitidus* (Patton et al. 2000). While not refuting the species distinctness of *E. legatus*, these data do suggest that the origin of this species has been relatively recent, possibly as a peripheral isolate of a more widespread *E. nitidus*. These data also strongly support phylogenetic separation from *E. russatus*.

Euryoryzomys macconnelli (Thomas, 1910)
McConnell's Euryoryzomys

SYNONYMS:

Oryzomys macconnelli Thomas, 1910b:186; type locality "River Supinaam [= Supenaam], a tributary of the lower Essequibo, Demerara, British Guiana," Pomeroon-Supenaam, Guyana (see Musser et al. 1998:278).

Oryzomys incertus J. A. Allen, 1913c:598; type locality "La Murelia (altitude 600 feet), Rio Bodoquera, Caquetá, Colombia."

Oryzomys mureliae J. A. Allen, 1915b:630; replacement name for *Oryzomys incertus* J. A. Allen, preoccupied by *Oryzomys alfaroi incertus* J. A. Allen.

Oryzomys (*Oryzomys*) *macconnelli*: Tate, 1932e:17; name combination.

Oryzomys (*Oryzomys*) *mureliae*: Tate, 1932e:17; name combination.

Oryzomys macconnelli macconnelli: Cabrera, 1961:392; name combination.

Oryzomys macconnelli mureliae: Cabrera, 1961:392; name combination.

[*Euryoryzomys*] *macconnelli*: Weksler, Percequillo, and Voss, 2006:11; first use of current name combination.

DESCRIPTION: Medium sized, with average head and body length 143 mm; tail equal to or longer than head and body (average 104%); closely resembles *E. nitidus* but differs in very dense, lax, long (aristiforms 17–20 mm), and orange to reddish brown fur in most samples. Skull with narrower zygomatic plates with correspondingly shallower zygomatic notches; wider interorbital region; shorter and wider incisive foramina (teardrop in shape); longer bony palate; alisphenoid struts absent in majority of specimens; m1 with ectolophid frequently present; capsular process not evident.

DISTRIBUTION: *Euryoryzomys macconnelli* is distributed throughout the Amazon Basin, from eastern Ecuador to eastern Pará state, Brazil, and from northern Guyana to southern Peru (Musser et al. 1998:176, Fig. 78). Over this extensive area, elevational records range from near sea level in the eastern Amazonian lowlands and Guianan region to about 1,500 m in the tepuis of Venezuela and Andean foothills of Peru.

SELECTED LOCALITIES (Map 158): BRAZIL: Amapá, Serra do Navio (MZUSP 20521); Amazonas, Barro Vermelho, left bank Rio Juruá (Patton et al. 2000), Macaco, left bank Rio Jaú (Patton et al. 2000), PD-BFF, 82 km N of Manaus (Patton et al. 2000), alto Rio Urucu (Patton et al. 2000); Mato Grosso, Aripuanã (L. P. Costa 2003), Juruena (L. P. Costa 2003); Pará, 54 km S and 150 km W of Altamira (Musser et al. 1998), Flexal, Itaituba-Jacareacanga, km 212 (USNM 545293), Floresta Nacional Tapirapé-Aquiri (INPA 2796), Igarapé-Assu (BM 4.7.4.65), 19 km S of Itaituba (Musser et al. 1998), Serra Norte (MPEG 15101). COLOMBIA: Caquetá, La Morelia [= La Murelia], Río Bodoquera (AMNH 33756). ECUADOR: Orellana, San José Abajo (Musser et al. 1998). FRENCH GUIANA: Cayenne, Arataye (USNM 548449). GUYANA: Pomeroon-Supenaam, Supenaam River (type locality of *Oryzomys macconnelli* Thomas). PERU: Amazonas, headwaters Río Kagka (MVZ 154972); Cusco, Río San Miguel (USNM 194564); Huánuco, Hacienda Éxito, Río Cayumba (Musser et al. 1998); Junín, Perené (BM 0.7.7.31); Puno, Pampa Grande, below San Ignacio on the Río Tambopata (FMNH 79900). SURINAM: Para, Finisanti, Saramacca River (FMNH 95595). VENEZUELA: Amazonas, Cerro Duida, Cabecera del Caño Culebra, 40 km NW of Esmeralda (USNM 406044); Bolívar, Auyán-tepuí (AMNH 131119).

SUBSPECIES: As currently understood, *E. macconnelli* is monotypic, but the considerable geographic variation in karyotype and mtDNA diversity (as described here) could signal greater specific and/or subspecific diversity than currently recognized.

NATURAL HISTORY: This uncommon terrestrial species is found exclusively in primary tropical evergreen rainforest (Musser et al. 1998) and is considered a good indicator of pristine forests (Patton et al. 2000), being negatively affected by forest fragmentation (Malcolm 1991). Malcolm (1990) reported trapping density of 0.387 individuals/ha and nocturnal census density of 0.095 individuals/ha in the central Brazilian Amazon, near Manaus; these values are smaller than those reported for other sympatric oryzomyines, especially *Hylaeamys megacephalus*. Individuals of both sexes with limited to moderate wear on the upper molars were sexually mature, with pregnant or lactating females found in both the rainy and dry seasons. Embryo counts ranged from two to four (modal number three) in the western Brazilian Amazon (Patton et al. 2000).

REMARKS: *Euryoryzomys macconnelli* has been consistently considered a valid species by most authors (Gyldenstolpe 1932; Tate 1932e; Ellerman 1941; Cabrera 1961; Pine 1973; A. L. Gardner and Patton 1976; Husson 1978; Musser and Carleton 1993; Musser et al. 1998), except for Hershkovitz (1960), who incorrectly included this species in the synonymy of *Hylaeamys megacephalus* (then *Oryzomys capito*; see Musser et al. 1998).

Analyses of geographic variation in external and cranial traits exhibited no significant differences among samples (Musser et al, 1998), but karyological data (summarized by Musser et al. 1998 and Patton et al. 2000) and molecular evidence (Patton et al. 2000) suggest that *E. macconnelli* is

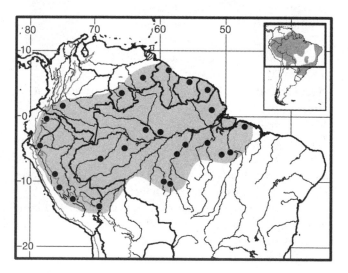

Map 158 Selected localities for *Euryoryzomys macconnelli* (●). Contour line = 2,000 m.

composite. Populations north of the Amazon River are quite distinct, with the Rio Negro isolating two groups, one on the eastern and northeastern (left) bank with $2n = 76$, FN = 85 and another on the west (right) bank with $2n = 58$, FN = 90. Samples from Para state, Brazil, south of the Amazon River, and those from the western reaches of Amazonia in Ecuador, Peru, and western Brazil have a similar karyotype with $2n = 64$, FN = 70. The very limited mitochondrial sequence data suggest that each karyotype, and thus geographic region, represents reciprocally monophyletic lineages.

The status of *Oryzomys mureliae* J. A. Allen as a synonym of *E. macconnelli* was pointed out early by Tate (1932e) and confirmed by Musser et al. (1998). If, however, the geographic variants in this species merit taxonomic recognition, this is an available name (Voss et al. 2001).

Euryoryzomys nitidus (Thomas, 1884)
Elegant Euryoryzomys

SYNONYMS:

Hesperomys (Calomys) laticeps: Thomas, 1882:102; part (assignment of specimens from Peru); not *Mus laticeps* Lund.

Hesperomys (Oryzomys) laticeps: Thomas, 1884:452; part (assignment of specimens from Peru); not *Mus laticeps* Lund.

Hesperomys laticeps, var. *nitidus* Thomas, 1884:452; type locality "situated between the streams of Chanchamayo and Anamayo, at a little distance from the river Tutamayo [*sic*]" (Thomas 1884:447), and designated as "Amable Maria" based on the selection of a lectotype (Thomas 1927b:549); type locality further refined to "Amable María, valley of Río Tulumayo, 10 km south of San Ramón, Departamento de Junín, Peru, 2000 ft" (A. L. Gardner and Patton 1976:42).

[*Oryzomys*] *laticeps nitidus*: E.-L. Trouessart, 1897:525; name combination.

Oryzomys boliviae Thomas, 1901h:536; type locality: "Mapiri, Upper Beni River, (about 68°W. and 15°S.). Alt. 800 m," La Paz, Bolivia.

Oryzomys nitidus: Thomas, 1920h:228; name combination.

Oryzomys (Oryzomys) nitidus nitidus: Tate, 1932e:17; name combination.

Oryzomys capito nitidus: Cabrera, 1961:386; name combination.

[*Euryoryzomys*] *nitidus*: Weksler, Percequillo, and Voss, 2006:11; first use of current name combination.

DESCRIPTION: Medium sized, with mean head and body length 136.1 mm (range: 108–163 mm) and tail nearly as long (mean length 136.2 mm; range: 113–160); short and close dorsal pelage; upper parts bright tawny or tawny brown with ochraceous highlights along sides of head and body; venter whitish gray; unicolored tail,

mottled on ventral surface near tip; dorsal surfaces of forefeet and hindfeet covered with dense white hairs, claws nearly hidden by long and dense ungual tufts; all plantar pads present; superciliary, genal, and mystacial vibrissae moderately long, not reaching past ear. Skull elongate (mean skull greatest length 33.3 mm; range: 25.5–36.3 mm) with narrow interorbital region (compared with *E. macconnelli*) bounded by dorsolateral ridges; alisphenoid struts present in about half of specimens examined; incisive foramina long (mean 5.7 mm; range: 4.6–6.8 mm) and narrow; bony palate short; zygomatic plates wide, zygomatic notches correspondingly moderately deep. M2 with short paraflexus and labial as well as medial fossettes; m2 with moderately deep entoflexid and short hypoflexid extending only halfway across tooth, with fossetid present.

DISTRIBUTION: *Euryoryzomys nitidus* is restricted to the western Amazon Basin, east of Andes in Peru and Bolivia and extending into adjacent western Brazil, as far east as the left bank of the Rio Teles Pires. It ranges in elevation from about 200 m in Brazil to 1,500 m on the eastern Andean slopes of Peru.

SELECTED LOCALITIES (Map 157): BOLIVIA: Beni, Guayaramerín, Rio Mamoré (S. Anderson 1997), Lago Victoria (S. Anderson 1997); Cochabamba, 2 km E of Villa Tunari (AMNH 247779); La Paz, La Reserva (AMNH 264722), Mapirí (S. Anderson 1997); Pando, Río Nareuda (S. Anderson 1997); Santa Cruz, Parque Noel Kempff Mercado, 17 km S of Los Fierros (S. Anderson 1997), San Ignacio de Velasco (Musser et al. 1998), San Rafael de Amboró (Musser et al. 1998), 4 km N and 1 km W of Santiago de Chiquitos (S. Anderson 1997), 8 km SE of Tita (AMNH 260377). BRAZIL: Acre, Igarapé Porongaba, right bank Rio Juruá (Patton et al. 2000), Manuel Urbano, Sena Madureira, BR 364 km 8 (MPEG 10609); Mato Grosso, Apiacás (L. P. Costa 2003), Juruena (L. P. Costa 2003); Mato Grosso do Sul, Urucum, 22 km S of Corumbá (USNM 390109). PERU: Ayacucho, San José, Río Santa Rosa above Hacienda Luisiana (LSUMZ 16693); Huánuco, Chinchavito, at the mouth of the Río Chiraco (Musser et al. 1998); Junín, Amable Maria (type locality of *Hesperomys laticeps* var. *nitidus* Thomas); Madre de Dios, Reserva Cusco Amazónico (Woodman et al. 1991); San Martín, Moyobamba (USNM 259606); Ucayali, Balta, Río Curanja (A. L. Gardner and Patton 1976).

SUBSPECIES: I treat *Euryoryzomys nitidus* as monotypic.

NATURAL HISTORY: Based on information compiled by Musser et al. (1998:188–189; see also Patton et al. 2000), *E. nitidus* is a terrestrial inhabitant of humid tropical evergreen forests, largely restricted to Amazonian uplands on the western and southwestern border of the Amazon Basin.

The species also occurs in mature and secondary deciduous forests and gallery forests through palm savannas and Cerrado, floodplain forest (including that adjacent to seasonally inundated swamp forest), and near anthropogenic areas (such as soccer fields, old vegetable gardens, and pastures). Emmons and Patton (2005) stated that *E. nitidus* was also captured in dry forests and grassland habitats in the Parque Nacional Noel Kempff Mercado of eastern Bolivia.

The outlier locality of Urucum in Mato Grosso do Sul (also called Serra do Urucum) is an isolated massif on the western border of the Brazilian Pantanal covered with deciduous forest. This area is characterized by the presence of various Amazonian mammalian taxa, such as the marmoset *Callithrix melanura* (Vivo 1991), the titi monkey *Callicebus donacophilus*, and the tree squirrel *Hadrosciurus spadiceus* (see that account).

REMARKS: My concept of *E. nitidus* is more restricted than that of Musser et al. (1998), because I believe that samples west of the Rio Teles Pires to the west bank of Rio Xingu belong to an undescribed species (see L. P. Costa 2003), but specimens from localities on the east bank of the Rio Xingu are *E. emmonsae* (see above). Therefore, it is likely that samples from the Serra do Roncador, assigned by Musser et al. (1998) to *E. nitidus*, belong to this undescribed form. Furthermore, specimens from Paraguayan and Argentinean localities, as reported by Musser et al. (1998), are not *E. nitidus* but rather *E. russatus*. A. L. Gardner and Patton (1976) suggested that *Oryzomys boliviae* was a synonym of *E. nitidus*, based on their examination of the holotype, a hypothesis subsequently confirmed by the morphological and morphometric analyses of Musser et al. (1998). The karyotype of *E. nitidus* consists of a $2n = 80$ and FN = 86 (A. L. Gardner and Patton 1976; Patton et al. 2000).

Euryoryzomys russatus (Wagner, 1848)
Russet Euryoryzomys

SYNONYMS:

Mus physodes Brants, 1827:139; type locality "Provinz San Paulo," São Paulo, Brazil; preoccupied by *Mus physodes* Olfers, a species of *Reithrodon* (see Musser et al. 1998:284).

Hesperomys russatus Wagner, 1848:312; no type locality given in original description, but the selection by Musser et al. (1998) of a lectotype for *E. russatus*, from among specimens collected by J. Natterer, unequivocally established the type locality as Ipanema, São Paulo, Brazil (= Floresta Nacional de Ipanema, 20 km NW Sorocaba, São Paulo, Brazil, 23°26'7"S, 47°37'41"W, 701 m; L. P. Costa et al. 2003).

Hesperomys physodes: Burmeister, 1854:167; name combination.

Holochilus physodes: Fitzinger, 1867b:90; name combination.

Hesperomys Darwinii?: Hensel, 1872b:48; part; not *Mus darwini* Waterhouse.

Hesperomys laticeps var. *intermedia* Leche, 1886:693; type locality "Rio dos Linos [= Rio dos Sinos], Taquara do Mundo Novo [= Taquara]," Rio Grande do Sul, Brazil.

Calomys coronatus Winge, 1887:51; type locality "Lapa da Serra das Abelhas," Lagoa Santa, Minas Gerais, Brazil.

Hesperomys (Oryzomys) laticeps var. *intermedia*: Ihering, 1893:15; name combination.

Hesperomys leucogaster: Ihering, 1897:150; not *Hesperomys leucogaster* Wagner [= *Sooretamys angouya*].

[Oryzomys] laticeps intermedia: E.-L. Trouessart, 1897:525; name combination.

Oryzomys intermedius: Thomas, 1901g:528; name combination.

Oryzomys physodes: Thomas, 1904b:142; name combination.

Oryzomys kelloggi Avila-Pires, 1959a:2; type locality "Brasil, Minas Gerais, Além Paraíba, Fazenda São Geraldo."

Oryzomys ratticeps moojeni Avila-Pires, 1959b:3; type locality "Brasil, São Paulo, Cananéia, Morro de São João."

Oryzomys capito intermedius: Cabrera, 1961:385; name combination.

[Euryoryzomys] russatus: Weksler, Percequillo, and Voss, 2006:11; first use of current name combination.

DESCRIPTION: Closely similar to *E. nitidus* but with differences that allow specific recognition based on greater average external, cranial, and dental dimensions. Large (mean head and body length 151.7 mm; range: 112–185 mm) with tail equal to or longer (mean length 143.2 mm; range: 105–196). Skull robust and large (mean greatest length 34.6 mm; range: 28.3–39.2 mm); molar toothrow long (mean length 5.1 mm; range: 4.5–5.5 mm; in *E. nitidus*, average length 4.8; range: 3.9–5.6 mm). Other differences between two species include moderately long and generally brighter dorsal pelage in *E. russatus*; usually shorter incisive foramina relative to occipitonasal length in most specimens (Musser et al. 1998); presence of alisphenoid strut in less than half of examined specimens; and presence of small sphenopalatine vacuities in several specimens.

DISTRIBUTION: *Euryoryzomys russatus* is distributed along the eastern Brazilian coastal region from Bahia (lowland) to Rio Grande do Sul (montane areas), including coastal islands in São Paulo and Rio de Janeiro states, and from there to the western highlands of São Paulo state, reaching Paraguay (Caaguazú) and Argentina (Misiones), and limited to the west by the Rio Paraná in São Paulo and by the Río Paraguay in Paraguay. *Euryoryzomys russatus* is also a part of the fossil fauna of Lagoa Santa (see Remarks). Specimens from Ceará and Paraíba states are provisionally assigned to this species (see Remarks).

SELECTED LOCALITIES (Map 159): ARGENTINA: Misiones, Puerto Gisela, San Ignacio (Musser et al. 1998). BRAZIL: Bahia, Ilhéus, Almada, Rio do Braço (MNRJ 9032); Ceará, São Benedito (MNRJ 37049), Sitio Friburgo, Serra de Batuité, Pacoti (MZUSP [ARP 15]); Espírito Santo, Castelo, 3 km NE Forno Grande (MNRJ 32813), Fazenda Santa Terezinha, 33 km NE of Linhares (UFMG [LPC 27]); Minas Gerais, Lagoa Santa (ZMUC uncatalogued specimen); Paraíba, Mata do Pau Ferro, 6 km from Areia (UFPB 2046); Rio de Janeiro, Centro de Primatologia, Magé (L. P. Costa 2003); Rio Grande do Sul, Fazenda Aldo Pinto, São Nicolau (MPEG 22249), Morro do Osório (UFRGS 1080); Santa Catarina, Florianópolis, Rio Tavares, CASAN, Rodovia Estadual SC 405 (LAMAQ 168); São Paulo, Ariri (MZUSP [ARP 5]), Estação Biológica de Boracéia (MVZ 183114), Lins, Campestre (MZUSP 6160), Teodoro Sampaio (MZUSP 8859). PARAGUAY: Caaguazú, Sommerfeld Colony #11 (MNRJ 32764).

SUBSPECIES: I treat *Euryoryzomys russatus* as monotypic.

NATURAL HISTORY: *Euryoryzomys russatus* is predominantly associated with forested coastal plains and with the evergreen and semideciduous forests of the coastal highlands. This species apparently avoids open vegetation types, such as the Cerrado and Chaco, preferring primary forests (Bonvicino et al. 1997; R. Pardini, pers. comm.). In northeastern Brazil, this species is found in montane enclaves covered by moist semideciduous forest in the Caatinga (locally called *brejos*).

The abundance of this species in the Brazilian Atlantic Forest varies at both large and small geographical scales. In some areas of southeastern Brazil, *E. russatus* is the most common terrestrial species (Bergallo 1994) while in others it is less common (Olmos 1992). In northeastern Brazil where the small mammal community is dominated by *Hylaeamys laticeps*, this species is uncommon (Laemmert et al. 1946; Pardini 2004).

Bergallo (1994) reported population and reproductive data for the species in southern Brazil. The home range varied from 0.16 to 1.12 ha, with a mean size of 0.46 ± 0.31 ha; population density, averaging 5.26 ± 1.02 individuals/ha, was stable throughout the year, with no correlation between rainfall and birth frequencies. Apparently this species breeds year-round, because juveniles were captured in every month except July, November, and January, and pregnant females were recorded from March to July, August, and November. Females attained reproductive maturity by day 115 following birth and could produce more than six litters in a single year (Bergallo 1995). Puttker, Bueno et al. (2013) described survival rates, population change rates, and capture probability for populations in different forest cover landscapes.

Euryoryzomys russatus builds egg-shaped nests (15 cm long and 10 cm wide), with dry leaves and grass, under or inside fallen logs, and among the aerial roots of various species of palms (Briani et al. 2001).

REMARKS: The record of *E. russatus* from Lagoa Santa (as *Calomys coronatus* Winge; see synonymy above) is based on specimens obtained from cave deposits. Several other species described from the cave material of Lagoa Santa (e.g., *Protopithecus brasiliensis*, *Caipora bambuirorum*, *Blarinomys breviceps*, *Delomys* sp., *Sooretamys angouya*; see Harttwig and Cartelle 1996; Voss and Myers 1991; C. R. Silva et al. 2003) are members of a wet forest community that is no longer found in the region. Although the vegetation of the Lagoa Santa region today is a mosaic of Cerrado and semideciduous forest habitats, no recent specimens of any of these species have been found at or in the vicinity of Lagoa Santa. Therefore, it is reasonable to assume that *E. russatus* is now presently extinct in this part of interior Brazil.

In the early nineteenth to middle twentieth centuries, several specific names were coined for this species. Each of these is available, but *russatus* Wagner has priority (see Musser et al. 1998). Throughout history, this species was considered to be a subspecies of *Hylaeamys laticeps* (Leche 1886; E.-L. Trouessart 1897), *Hylaeamys megacephalus* (Cabrera 1961), or even a synonym of *E. nitidus* (A. L. Gardner and Patton 1976). However, the validity of *Euryoryzomys russatus* as a full species is supported by morphological (Weksler

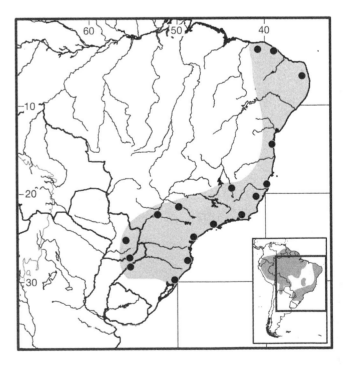

Map 159 Selected localities for *Euryoryzomys russatus* (●). Contour line = 2,000 m.

1996; Musser et al. 1998; Percequillo 1998) and molecular (Patton et al. 2000; L. P. Costa 2003) analyses.

Euryoryzomys russatus is the only species in its distributional range to have an orange to orange-rufous dorsal pelage sharply contrasting with a grayish-white to almost pure white ventral coloration; a white foot with very long and dense ungual tufts; an M2 with two enamel islands; and a distinct capsular process on the mandibular ramus.

While I assign samples of genus *Euryoryzomys* from *brejos* in northeastern Brazil (Paraíba and Ceará states; see Weksler 1996; Percequillo 1998; M. J. de J. Silva et al. 2000b) to *E. russatus*, these are likely to represent an undescribed species, based on morphological and karyological data. These samples have smaller and more delicate skulls, with shorter and narrower incisive foramina, an anterolingual cingulum on M2, and a karyotype with $2n = 76$ rather than the $2n = 80$, $FN = 86$ that characterizes *E. russatus* throughout the majority of its range (E. J. C. Almeida and Yonenaga-Yassuda 1974, as "*Oryzomys capito*"; Musser et al. 1988).

Genus *Handleyomys* Voss, Gómez-Laverde, and Pacheco, 2002

Marcela Gómez-Laverde, Robert S. Voss, and Víctor Pacheco

The genus *Handleyomys* includes two species: *H. fuscatus*, endemic to the western Andes of Colombia, and *H. intectus*, endemic to the central Andes of Colombia; the elevational range of the genus is from about 1,700 to 2,800 m. No fossil material is known. The following account is abstracted from Voss et al. (2002), with additional information based on newly examined specimens.

Handleyomys fuscatus and *H. intectus* are both small (17–34 g) mice with tails that are about as long as the combined length of the head and body (length of head and body 80–130 mm; length of tail 75–110 mm). The adult body pelage is fine and soft, uniformly dull brownish gray dorsally, but usually darker (sometimes almost blackish) middorsally than on the flanks; the ventral pelage is dark gray frosted with paler gray or buff and is not sharply countershaded. The tail appears naked to the eye (a sparse caudal pelage is visible under magnification) and unicolored (dark above and below). Mystacial, superciliary, genal, submental, interramal, and carpal vibrissae are present; mystacial vibrissae are neither very short nor very long, extending posteriorly to (but not beyond) the caudal margins of the pinnae when laid back against cheeks. The pinnae are not large, but they are clearly visible above the fur of the head and are sparsely covered with short dark hairs. The manus is sparsely covered dorsally with short

pale hairs; the plantar surface is unpigmented, with two carpal and three interdigital pads; and the claws are short, neither conspicuously elongated nor unusually curved. The pes is long (21–26 mm) and narrow, with the outer digits much shorter than the middle three (the claw of digit I extends to the middle of the first phalanx of digit II, and the claw of digit V extends just beyond the first interphalangeal joint of digit IV); conspicuous ungual tufts of long silvery hairs are rooted at the bases of the claws on digits II to V, but the pedal dorsum otherwise is only sparsely covered with short pale (whitish or silvery) or dark-banded hairs; the plantar surface (including the heel) is naked, weakly pigmented (grayish in life), with two metatarsal and four interdigital pads; indistinct squamae (scale-like tubercles) are sparsely distributed along the outer distal plantar surfaces, but not in the center of the sole. Unusually for an oryzomyine, there are only six mammae (in inguinal, abdominal, and postaxial pairs).

The skull has a long, tapering rostrum flanked by shallow but distinct zygomatic notches; the interorbital region is hourglass shaped, neither greatly inflated nor unusually constricted, with rounded supraorbital margins; the braincase is moderately inflated and rounded, without prominent temporal crests, ridges, or beads. The zygomatic plates are moderately broad, with their anterior edged vertical or nearly so, with a rounded (never angular or spinous) anterodorsal contour. Premaxillae are short (not produced anteriorly beyond the incisors to form a rostral tube with the nasals). Incisive foramina are neither very short nor greatly elongated, averaging about 60% of the diastemal length (not extending posteriorly between the molar rows) and are widest near the premaxillary-maxillary suture. The palatal bridge is long and wide, without a median ridge or deep lateral gutters; the posterolateral pits are usually large and often complex. The mesopterygoid fossa does not penetrate anteriorly between the molar rows, and the bony roof of the fossa is complete or perforated only by narrow slits (it is never conspicuously fenestrated). The alisphenoid strut is absent, and the buccinator-masticatory foramen and the foramen ovale are therefore confluent. The carotid arterial morphology is primitive, with the orbitofacial circulation supplied by separate supraorbital and infraorbital branches of a large stapedial artery (= pattern 1 of Voss, 1988); the course of the supraorbital stapedial ramus is marked by a prominent squamosalisphenoid groove and a sphenofrontal foramen. The postglenoid foramen is separated from the large subsquamosal fenestra by a slender hamular process of the squamosal. The tegmen tympani either does not overlap the squamosal at all, or the tegmen tympani-squamosal overlap is shallow (not involving a distinct posterior suspensory process of the latter bone). The bullae are small, the pars flaccida of the tympanic membrane is present and large, and the orbicular apophysis of the malleus is well developed.

The mandible has a well-developed, falciform coronoid process, and a distinct capsular process on the lateral mandibular surface is absent. The basihyal bone (known only in *H. fuscatus*) is not strongly arched and lacks an entoglossal process; the thyrohyals are shorter than the basihyal.

The incisors are ungrooved and opisthodont. The maxillary molar rows are parallel. The principal molar cusps, arranged in opposite labial/lingual pairs, are bunodont when unworn but are quickly eroded with age to the same level as other enamelled occlusal structures (the occlusal surfaces of most adult teeth are more or less planar); the labial and lingual re-entrant folds are interpenetrating (incipiently lophodont *sensu* Voss 1993); the anterocone of M1 is not divided into labial and lingual conules (an anteromedian flexus is absent); the anteroflexus is very deep, extending lingually beyond the dental midline on M1 and M2; anterolophs and mesolophs are large, fused with corresponding (antero- and meso-) styles on the labial margins of M1 and M2; posterolophs are distinct on M1 and M2, persisting with moderate to heavy wear; M3 is subtriangular in occlusal view, smaller than more anterior teeth, with most of the same occlusal elements but usually without a distinct hypocone or posteroloph. M1 has one accessory labial root (four roots total), but M2 and M3 are each three-rooted.

The anteroconid of m1 is undivided by a median flexid and is fused with the protolophid and/or the anterolophid to enclose a persistent internal fold of uncertain homology (anteroflexid and/or protoflexid); the anterolophid is absent on m2 and m3; the anterolabial cingulum is absent or indistinct on m2, and consistently absent on m3; the mesolophids and posterolophids are large and well developed on all mandibular teeth; the ectolophids are consistently absent. The m1 has an accessory labial root and occasionally also an accessory lingual root (three to four roots total); m2 and m3 each have two small anterior roots and one large posterior root (three roots total).

Axial skeletal elements include 12 ribs, 19 thoracolumbar vertebrae, 4 sacral vertebrae, and 28–30 caudal vertebrae; hemal arches appear to be present on the caudal vertebrae of *H. fuscatus*, but not on those of *H. intectus*. The tuberculum of the first rib articulates with the transverse processes of the seventh cervical and first thoracic vertebrae, and the second thoracic vertebra has a greatly elongated neural spine. The entepicondylar foramen of the humerus is absent.

The stomach in both species is unilocular and hemiglandular, without any extension of glandular epithelium into the corpus; the bordering fold crosses the lesser curvature slightly to the right of the incisura angularis (between that flexure and the pylorus), and it crosses the greater curvature opposite the incisura angularis. The gall bladder is absent. The morphology of the glans penis and other male reproductive structures is unknown.

SYNONYMS:

Aepeomys: J. A. Allen, 1912:89; part (description of *fuscatus*); not *Aeepomys* Thomas.

Oryzomys: Thomas, 1921j:356 (description of *intectus*); part; not *Oryzomys* Baird.

Thomasomys: Ellerman, 1941:369; part (listing of *fuscatus*); not *Thomasomys* Coues.

?*Nectomys*: Ellerman, 1941:351 (provisional listing of *intectus*); part; not *Nectomys* Peters.

Handleyomys Voss, Gómez-Laverde, and Pacheco, 2002:5; type species *Aepeomys fuscatus* J. A. Allen, by original designation.

KEY TO THE SOUTH AMERICAN SPECIES OF *HANDLEYOMYS*:

1. Hindfeet covered dorsally with dark-banded hairs (the melanic bands distinctly visible under low magnification); nasals short, truncated posteriorly at or near the premaxillomaxillary suture; incisive foramina anteriorly constricted, with lateral margins abruptly narrowed at or near the premaxillomaxillary suture; interparietal broad (transverse dimension) relative to its depth (anteroposterior dimension)...... *Handleyomys fuscatus*

1'. Hindfeet covered dorsally with pure white or indistinctly pigmented hairs; nasals long, extending posteriorly well beyond the premaxillomaxillary suture; incisive foramina smoothly tapering, with evenly rounded lateral margins; interparietal narrow (transverse dimension) relative to its depth (anteroposterior dimension) ....*Handleyomys intectus*

Handleyomys fuscatus (J. A. Allen, 1912)

Colombian Western Andes Cloud Forest Mouse

SYNONYMS:

Aepeomys fuscatus J. A. Allen, 1912:89; type locality "San Antonio (near Cali, alt. 7000 ft), Cauca [= Valle del Cauca], Colombia."

Thomasomys fuscatus: Ellerman, 1941:369; name combination.

Thomasomys lugens fuscatus: Cabrera, 1961:431; name combination.

Handleyomys fuscatus: Voss, Gómez-Laverde, and Pacheco, 2002:23; first use of current name combination.

DESCRIPTION: See the Key.

DISTRIBUTION: *Handleyomys fuscatus* is known only from the western Andes (Cordillera Occidental) of Colombia between 1,800 and 2,530 m in the departments of Antioquia and Valle del Cauca.

SELECTED LOCALITIES (Map 160): COLOMBIA: Antioquia, Vereda La Soledad, Finca La Reina (ICN 16510–16523); Valle del Cauca, Finca La Playa (Voss et al. 2002), San Antonio (type locality of *Aepeomys fuscatus* J. A. Allen).

SUBSPECIES: *Handleyomys fuscatus* is monotypic.

NATURAL HISTORY: This apparently terrestrial and nocturnal species inhabits both primary cloud forest and adjacent secondary growth, usually in cool and very humid environments (12–18°C mean ambient temperature; >2,000 mm annual precipitation) corresponding to the vegetation zone that Espinal and Montenegro (1963) called *Bosque muy húmedo montano bajo*. Four of six known pregnant females (each with two to three embryos) were collected during the rainy season (April or June), but another was taken during the late dry/early rainy season (September), and the fourth in the dry season (February); one pregnant female was also lactating in June. The only additional recorded lactating female was captured during the rainy season (September). Juveniles have been caught during the rainy season and the late dry/early rainy seasons.

REMARKS: The karyotype is $2n=54$, FN=62; the X and Y chromosomes are both submetacentrics (A. L. Gardner and Patton 1976).

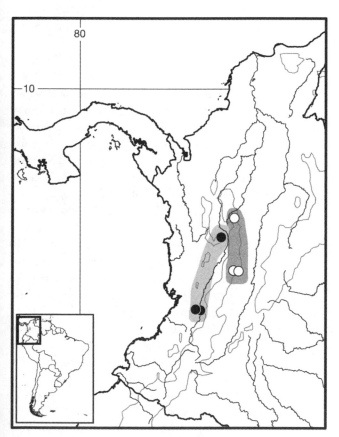

Map 160 Selected localities for *Handleyomys fuscatus* (●) and *Handleyomys intectus* (○). Contour line = 2,000 m.

Handleyomys intectus (Thomas, 1921)
Colombian Central Andes Cloud Forest Mouse

SYNONYMS:

Oryzomys intectus Thomas, 1921j:356; type locality "Medellin. Santa Elena," Antioquia, Colombia.

?*Nectomys intectus*: Ellerman, 1941:351; suggested name combination.

Oryzomys (Oryzomys) intectus: Tate, 1932e:16; name combination.

Handleyomys intectus Voss, Gómez-Laverde, and Pacheco, 2002:24; first use of current name combination.

DESCRIPTION: See the Key.

DISTRIBUTION: *Handleyomys intectus* is known only from the central Andes (Cordillera Central) of Colombia between 1,700 and 2,800 m, in the departments of Antioquia, Quindío, and Risaralda.

SELECTED LOCALITIES (Map 159): COLOMBIA: Antioquia, Santa Elena (type locality of *Oryzomys intectus* Thomas), Ventanas (Voss et al. 2002); Quindío, Salento (Voss et al. 2002).

SUBSPECIES: *Handleyomys intectus* is monotypic.

NATURAL HISTORY: This apparently terrestrial and nocturnal species occurs in the same types of cloud forest and secondary habitats as those inhabited by *H. fuscatus*. The only recorded pregnant female was caught during the rainy season (April) and gave birth to two young in the trap; one lactating female was captured during the dry season (January). Juveniles have been caught during both rainy and dry seasons.

REMARKS: The karyotype is $2n=54$, FN=68–70; the morphology of the X and Y chromosomes is undetermined (M. L. Bueno, M. Gómez-Laverde and C. A. Delgado-V., unpubl. data).

Genus *"Handleyomys"*

"Handleyomys" alfaroi (J. A. Allen, 1891)
Alfaro's Rice Rat
Marcelo Weksler

This species is treated here independently because of its provisional position as a member of *Handleyomys*. The ex-"*Oryzomys alfaroi*" group was included in *Handleyomys* by Weksler et al. (2006); as they explained (Weksler et al. 2006:2), however, "[pending] the description of other new genera (by M. D. Carleton and G. G. Musser, pers. com.), we provisionally transfer members of the '*alfaroi* group' (herein understood to include *alfaroi, chapmani, melanotis, rhabdops, rostratus*, and *saturatior*) to *Handleyomys*, a suboptimal but phylogenetically defensible nomenclatural option previously discussed by Weksler (2006)." Unfortunately, no new genus to encompass these taxa has yet been described. The taxon is included in the Key to genera of the Oryzomyini, but not in the Key to species of *Handleyomys*.

SYNONYMS:

Hesperomys (Oryzomys) alfaroi J. A. Allen, 1891b:214; type locality "San Carlos," Alajuela Province, Costa Rica.

Oryzomys alfaroi: J. A. Allen, 1894b:36; name combination.

Oryzomys gracilis Thomas, 1894b:358; type locality "Concordia, Medellin, Colombia."

Oryzomys palatinus Merriam, 1901:290; type locality "Teapa, Tabasco, Mexico."

Oryzomys alfaroi incertus J. A. Allen, 1908:655; type locality "Rio Grande," Río Grande de Matagalpa, south of Tuma, Nicaragua.

Oryzomys palmirae J. A. Allen, 1912:83; type locality "a few miles east of Palmira, eastern slope of Central Andes," Valle del Cauca, Colombia.

Oryzomys alfaroi dariensis Goldman, 1915:128; type locality "Cana, eastern Panama (altitude 2,000 feet)."

Oryzomys alfaroi alfaroi: Goldman, 1918:59; name combination.

Oryzomys alfaroi palatinus: Goldman, 1918:65; name combination.

Oryzomys (Oryzomys) alfaroi intagensis Hershkovitz, 1940a:78; type locality "Hacienda Chinipamba near Peñaherrera, Intag, in the subtropical forest of the western slope of the western cordillera of the Andes, Imbabura Province, northwestern Ecuador. Altitude, about 1500 meters."

Oryzomys alfaroi gloriaensis Goodwin, 1956:8; type locality "La Gloria, rain forest, about 2500 feet elevation, 10 kilometers south of Santa Maria Chimalapa, District of Juchitan, Oaxaca, Mexico."

Oryzomys alfaroi agrestis Goodwin, 1959:7; type locality "12 miles southeast of Tapachula, Chiapas, Mexico; altitude 1000 feet."

Oryzomys alfaroi gracilis: Cabrera, 1961:383; name combination.

Oryzomys alfaroi palmirae: Cabrera, 1961:383; name combination.

[*Handleyomys*] *alfaroi*: Weksler, Percequillo, and Voss, 2006:2; name combination.

DESCRIPTION: Small sized; adult head and body length 90–106 mm, tail length 89–101 mm, hindfoot length 25–28 mm, ear length 14–17, mass 20–32 g; dorsal pelage color dull yellow-brown to dark reddish-brown; sides ochraceous to orangish; venter grayish white. Fur short (<5 mm); ears usually thinly lined with blackish or orangish hairs, sometimes appearing almost hairless. Tail either uniformly dark or weakly bicolored over proximal half, then dark to tip. Skull small with short rostrum; relatively broad and rather low braincase; slender zygomata; temporal ridges slightly developed; interorbital region broad, lateral margins slightly elevated; incisive foramina short and wide and not reaching plane of M1s; molariform teeth small.

DISTRIBUTION: This species ranges predominantly in evergreen forests of tropical Middle America, from southern Mexico through Panama, extending through low and middle elevation forests of western Colombia and northwestern Ecuador in South America. Elevational range is from near sea level to about 2,500 m.

SELECTED LOCALITIES (in South America; Map 161): COLOMBIA: Antioquia, Concordia (type locality of *gracilis* Thomas); Boyacá, Muzo (AMNH 62784); Cauca, Sabanetas (AMNH 181451); Cesar, Villanueva, Sierra Negra (USNM 280562); Cundinamarca, Volcanes, Municipio de Caparrapí (USNM 282116); Huila, La Candela (AMNH 33867); Nariño, Barbacoas (AMNH 34219); Valle de Cauca, Palmira (type locality of *palmirae* J. A. Allen), Río Frío (AMNH 32858). ECUADOR: Chimborazo, Puente de Chimbo (AMNH 62305); El Oro, 1 km SW of Puente de Moromoro (USNM 513561); Esmeraldas, Esmeraldas (AMNH 33205); Guayas, Bucay (AMNH 61343); Imbabura, Hacienda Chinipamba, Intag (type locality of *intagensis* Hershkovitz); Los Rios, Limón (AMNH 66948); Manabí, Cerro of Pata de Pajaro (AMNH 64771); Santo Domingo de los Tsáchilas, Santo Domingo de los Colorados (AMNH 213506).

SUBSPECIES: Musser and Carleton 2005:1145) listed eight nominal taxa in their concept of "*Oryzomys*" *alfaroi*. Five of these are Middle American forms, four of which were regarded as subspecies by Hall (1981). I treat South American representatives of this species as monotypic. Pending a thorough analysis of character trends, it is unclear if the South American populations would warrant subspecific designation separate from the eastern Panamanian "*H.*" *alfaroi dariensis* (see Hall 1981).

NATURAL HISTORY: In Costa Rica, Goodwin (1946) reported that "*H.*" *alfaroi* occurred on heavily forested mountain slopes in the humid lower tropical zone. In Nicaragua, individuals were captured in a variety of habitats, including grasslands, woodlands, and along rivers, but appeared to be uncommon (J. K. Jones and Engstrom 1986). These same authors reported a pregnant female in March with three embryos. In Belize, specimens were captured in both pitfalls and traps on the forest floor, usually in damp sites with intact overstory; one pregnant female had five embryos. "*Handleyomys*" *alfaroi* was reported as prey for the Blue-crowned Motmot (*Momotus momota*) in Costa Rica (J. L. Reid and Sanchez-Gutierrez 2010). One of two individuals collected in southern Mexico tested positive for Venezuelan equine encephalitis virus antibodies (Estrada-Franco et al. 2004). Both fleas (*Jellisonia amadoi*; Ponce-Ulloa 1989)

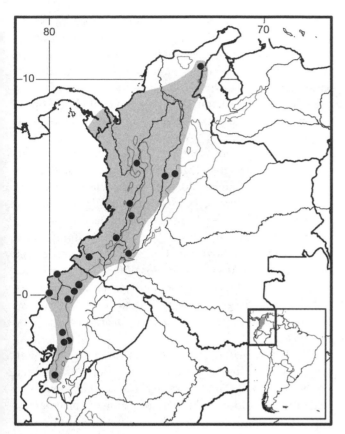

Map 161 Selected localities for *"Handleyomys" alfaroi* (●) in South America; distribution in eastern Panama from Hall (1981) and F. A. Reid (2009). Contour line = 2,000 m.

and mites (*Gigantolaelaps trapidoi*; D. Lee and Furman 1970) have been recovered in the fur of some individuals.

REMARKS: Haiduk et al. (1979) reported a karyotype of $2n = 60$, FN = 104 for specimens collected in Veracruz, México. Almendra et al. (2014) recently documented the phylogenetic relationship of specimens of *alfaroi* from Ecuador relative to Middle American taxa based on mtDNA and nucDNA sequences.

Genus *Holochilus* Brandt, 1835

Pablo R. Gonçalves, Pablo Teta, and Cibele R. Bonvicino

Marsh rats of the genus *Holochilus* are large rodents with numerous morphological specializations for an herbivorous diet and semiaquatic mode of life (Hershkovitz 1955b). The mass of adult individuals is 160–455 g, head and body length from 130–220 mm, tail length 115–183 mm, and ear length 14–22 mm (Massoia 1976b; Barreto and Garcia-Rangel 2005). The genus is broadly distributed throughout riparian or marshy habitats in tropical lowlands (Hershkovitz 1955b). It ranges from northern Venezuela and Guyana south to the temperate grasslands

and steppes of Uruguay and east-central Argentina, including through the open habitats in the Cerrado and Chaco ecoregions of central Brazil, Bolivia, and Paraguay.

The fur is short, dense, and soft; most species are tawny to brownish dorsally, grizzled by darker guard hairs; the ventral pelage varies from ochraceous to white, with the basal hair bands always grayish; ears are relatively small, rounded, and well furred; the muzzle is short; vibrissae are short and fine, none usually reaching to the ear tip; the tail is unicolored, sparsely haired so that scales are visible, and varies in length from shorter than to as long as the head and body length; hindfeet are long and have webbing between the digits, most pronounced between dII and dIII, and dIII and dIV, where the membrane extends to the distal ends of the first phalanges; natatory fringes are also well developed along the plantar margins; ungual tufts are absent or very short and do not conceal the claws on either forefeet or hindfeet; the plantar surface of the hindfoot has four reduced interdigital pads, one thenar pad, and a hypothenar that is variably present across populations and species of the genus; mammae number varies from 8 (pectoral, postaxial, abdominal, and inguinal pairs) to 10 (with the addition of a thoracic pair) among populations (Gyldenstolpe 1932; Hershkovitz 1955b; Massoia 1976b; Barreto and Garcia-Rangel 2005).

The skulls of *Holochilus* are heavily built, with a short and broad rostrum and squared braincase; the interorbital region is narrower than the rostrum and has a typical amphoral morphology (hourglass shaped) with symmetrically biconcave lateral margins; supraorbital margins have developed supraorbital beads; the zygomatic plates are broad with typically concave anterior profiles, sometimes marked by anterodorsal spines in adult specimens, and with their anterior margins reaching the nasolacrimal capsules; zygomatic notches are, consequently, well incised; the palate is moderately long, with the mesopterygoid fossa rarely extending anteriorly to the level of the third molars; the palatal bridge has a complex morphology, with a median ridge and two deep lateral sulci that coalesce caudally in a well-excavated pair of posterolateral palatal pits; auditory bullae are globular and predominantly composed of the ectotympanic, which almost completely obscures the periotic in ventral view; the stapedial foramen is absent, and the correlated absence of the sphenofrontal foramen and alisphenoid vascular groove identifies the derived carotid circulation pattern 3 (*sensu* Voss 1988); and a bony alisphenoid strut separates the buccinator-masticatory foramen from the anterior opening of the foramen ovale. The mandible is robust and deep, with the coronoid process nearly vertical and extending well above the condyloid process; the sigmoid notch is broad and oval (Gyldenstolpe 1932; Massoia 1976b; Voss and Carleton 1993; Carleton and Olson 1999; Voglino et al. 2004).

Upper incisors are opisthodont and are faced with angular enamel bands that form a distinct lateral bevel. Molars are flat crowned, moderately hypsodont, with cusps and connecting lophs in a single occlusal plane. Major cusps are ovate or subtriangular in occlusal outline, nearly opposite (*brasiliensis* group) or alternating (*sciureus* group). An anteroloph is absent and a reduced mesoloph(id)-like structure may be present in some species, partially fused to a paralophule (Pardiñas 2008); M1 has four roots; m1 has two large roots (anterior and posterior) and two accessory rootlets (lingual and labial); m2 and m3 are both two-rooted. Hershkovitz (1955b), Massoia (1971a), Voss and Carleton (1993) and Pardiñas and Galliari (1998c) described tooth morphology in great detail.

The stomach is unilocular-hemiglandular, with glandular epithelium covering an extensive area of the corpus because the bordering fold recurves sharply and extends beyond the esophageal orifice (Carleton 1973). The caecum is large (68 mm in a Uruguayan specimen of *H. brasiliensis*). A gall bladder is absent (Voss 1991a).

Hooper and Musser (1964) described the anatomy of the glans penis, and both Massoia (1971a) and Díaz de Pascual and Péfaur (1982) the baculum. *Holochilus* has a single pair of preputial glands that reach anteriorly slightly beyond the ventral flexure of the penis; the medial ventral prostates are longer than the lateral ventral prostates; vesiculars glands are lobed medially and along their greater curvatures, and the subterminal flexures are smooth (Voss and Linzey 1981).

In the first taxonomic revision and hypothesis of intergeneric relationships of *Holochilus*, Hershkovitz (1955b) included the genus among a group of dentally high-crowned sigmodontines with sigmoidal molars lacking well-developed mesolophs (e.g., with *Sigmodon*, *Neotomys*, and *Reithrodon*). Shortly thereafter, however, Hooper and Musser (1964) disaggregated these taxa, based on phallic morphology. More recent analyses of morphological and molecular characters strongly support *Holochilus* as a member of the tribe Oryzomyini (Voss and Carleton 1993; Steppan 1995; M. F. Smith and Patton 1999; Weksler 2006).

Hershkovitz (1955b) allocated 13 nominal forms to the genus, but recognized only two species, *H. brasiliensis* and *H. magnus*. The first species included 10 nominal forms as subspecies (*amazonicus*, *balnearum*, *berbicensis*, *brasiliensis*, *guianae*, *incarum*, *leucogaster*, *vulpinus*, *nanus*, and *venezuelae*) with *sciureus* and *chacarius* as junior synonyms of *H. brasiliensis brasiliensis*. This arrangement was sustained more on the basis of convenience than as a result of geographic variation (Hershkovitz 1955b:661). Massoia (1971, 1976, 1981b) was the first to validate *chacarius* and *sciureus* as full species, and Voss and Carleton

(1993) coined *Lundomys* to contain *H. magnus*, the sister-genus of *Holochilus* in some phylogenetic hypotheses (e.g., Weksler 2006). Subsequent studies have suggested a larger number of species, revealing either morphological differentiation among synonymized nominal forms (e.g., Marques 1988; Voss and Carleton 1993; Voglino et al. 2004) or extensive cytogenetic variation (e.g., T. R. O. Freitas, Mattevi, Oliveira et al. 1983; Aguilera and Pérez-Zapata 1989; Nachman and Myers 1989; Nachman 1992a, b). The latest taxonomic accounts (Musser and Carleton, 2005; Barreto and García-Rangel 2005) recognize at least three valid species (*H. brasiliensis*, *H. chacarius*, and *H. sciureus*) in the genus, which is clearly an underestimate.

Based on external, cranial, and dental features, Massoia (1981b) and Voss and Carleton (1993) recognized two morphological groups in *Holochilus*: the *brasiliensis* group, which included *darwini*, *leucogaster*, and *vulpinus*; and the *sciureus* group, containing *amazonicus*, *balnearum*, *berbicensis*, *chacarius*, *guianae*, *incarum*, *nanus*, and *venezuelae*.

The *brasiliensis* group comprises large animals with dense and luxurious fur, a tail about as long as the head and body or longer, no hypothenar pad on the plantar surface of the hindfoot, eight pairs of mammae, pronounced postorbital ridges, a long palate, distinct and subsquamosal fenestrae, masseteric crests that become closely approximated anteriorly to form a single crest that extends beyond the anterior root of m1 and approaches the mental foramen, somewhat medially flattened upper incisors with an indistinct labial bevel, and opposite or slightly alternating main molar cusps with the posterior faces of protoconid and hypoconid oriented 45° inward and typically persistence of mesoloph(id)-like structures. Axial skeletal elements in specimens referred to in the literature as *H. b. brasiliensis* and *H. b. vulpinus* have modal counts of 12 ribs, 19 thoracolumbar vertebrae, and 33–35 caudal vertebrae.

The *sciureus* group includes medium-sized animals with short and close fur, a tail shorter than the head and body, a very small hypothenar pad occasionally present, ether 8 or 10 pairs of mammae, weak to pronounced postorbital ridges, a mesopterygoid fossa extending to or slightly between the third molars, subsquamosal fenestrae that are often obscured by an expanded hamular process or by an internal crest or septum of the periotic, upper, and lower masseteric crests closely approximated anteriorly to form a single crest in some populations or coalescing at the level of the anterior root of m1 in others, laterally flattened upper incisors with a distinct labial bevel, lower incisors flattened laterally with a lingual bevel, and alternating main molar cusps with the posterior faces of both protoconid and hypoconid transversally oriented and mesoloph(id)-

like structures typically absent. Axial skeletal elements have modal counts of 12 ribs, 19 thoracolumbar vertebrae, and 29–32 caudal vertebrae (in *H. chacarius*).

Holochilus has a rich fossil record, especially in middle latitudes of southern South America, ranging from Middle Pleistocene to Holocene (Ameghino 1889; Voss and Carleton 1993; Pardiñas 1999a, 2004, 2008; P. E. Ortiz 2001; Teta et al. 2004; Pardiñas and Teta 2011b). Steppan (1996) described *Holochilus primigenus*, an extinct species known from Tarija, Bolivia. Although Steppan placed *primigenus* in *Holochilus*, he considered it transitional between *Lundomys* and *Holochilus*. Carleton and Olson (1999), Pardiñas (2008), and Pardiñas and Teta (2011b) questioned the generic assignment, and L. F. Machado et al. (2013) removed *primigenus* to their newly described *Reigomys*, as a member of a clade that included *Pseudoryzomys* and *Holochilus* along with the extinct *Noronhomys* and *Carletonomys*.

In this account, we recognize *H. brasiliensis*, *H. chacarius*, *H. sciureus*, *H. venezuelae*, *H. vulpinus*, and the recently described *H. lagigliai* as valid species. Despite extensive intrapopulation cytogenetic variability (Nachman 1992a), known karyotypes can be arranged into four groups with discrete ranges of autosomal fundamental and diploid numbers (FN/2*n* ratio). Species distinctions in the *brasiliensis* and *sciureus* groups are thus largely based on cytogenetic variation. Unnamed forms believed to constitute cytogenetically different species, such as the one with $2n=50$, $FN=58$ karyotype and from Colombia (A. L. Gardner and Patton 1976), are not addressed in this account.

SYNONYMS:

Mus: Desmarest, 1819a:62; part (description of *brasiliensis*); not *Mus* Linnaeus.

[*Mus*] *Holochilus* Brandt, 1835:428; as subgenus; type species *Holochilus sciureus* Wagner, designated under the plenary power of the International Commission (ICZN 2001:Opinion 1984).

Holochyse Lesson, 1842:137; incorrect subsequent spelling of *Holochilus* Brandt.

Hesperomys Wagner, 1843a:535; part; not *Hesperomys* Waterhouse.

Halochilus Schinz, 1845:192; incorrect subsequent spelling of *Holochilus* Brandt.

Mus (*Holochilus*): P. Gervais, 1854:411; as subgenus; *brasiliensis* included as species.

Holochilomys Brandt, 1855:304; proposed as a replacement name for *Holochilus* Brandt (see Remarks).

Holocheilus Coues, 1874:177; incorrect subsequent spelling of *Holochilus* Brandt.

[*Hesperomys*] *Holochilus*: Burmeister, 1879:210; as subgenus; included *vulpinus* as species and *brasiliensis* as synonym.

Sigmodon: Winge, 1887:12; part; not *Sigmodon* Say and Ord.

Hesperomys (*Holochilus*): Ihering, 1894:19; part (listing of *brasiliensis*); name combination.

Hesperomys (*Nectomys*): Ihering, 1894:19; incorrect assignment of *sciureus* Wagner to *Nectomys* Peters.

REMARKS: Brandt (1835:428) erected *Holochilus* as a subgenus of *Mus*, and named two species for it to contain: *Mus* (*Holochilus*) *leucogaster* (on p. 428) and *Mus* (*Holochilus*) *anguya* (on p. 430), the latter a misspelling of *Mus angouya* Desmarest (a homonym of *Mus angouya* G. Fischer, which today equals *Sooretamys angouya*; see that account). The type species of Brandt's *Holochilus* was subsequently designated by Miller and Rehn (1902:89) as *Mus* (*Holochilus*) *leucogaster*. Wagner (1842c) described *Holochilus sciureus* and provided an expanded diagnosis of Brandt's genus while noting that the teeth of both *leucogaster* and *anguya* were unknown but were likely the same as in his *sciureus*.

As detailed by Voss and Abramson (1999:256–257), however, the holotype of Brandt's *Mus leucogaster* is an hystricomorphous rodent, not a myomorphous rat to which the name *Holochilus* has been connected since 1835 (e.g., Palmer 1904; Gyldenstolpe 1932; Hershkovitz 1955b; Cabrera 1961; Musser and Carleton 1993). Brandt (1855:304, 315), in his monographic description of the major variants of rodent jaw anatomy, recognized the crucial distinction between these jaw morphologies and realized that his *Mus leucogaster* had the hystricomorph condition. Further, in admitting his mistake concerning the identity of Desmarest's *Mus angouya* (a myomorph), Brandt (1855:304) proposed the name *H. langsdorffii* for the taxon that he had previously called *H. anguya*, and classified *Holochilus* in the subfamily Echinomyina of his suborder Hystricomorphi, along with *Loncheres*, *Echinomys*, *Nelomys*, and *Mesomys*, all now in the Echimyidae. And he proposed *Holochilomys*, which he placed in the family Myoides of his suborder Myomorphi, to contain the myomorphous species referred by others to his *Holochilus* (e.g., Wagner 1842a, b, c; 1843a; Burmeister 1854). Among subsequent authors, only Peters (1861) used the name *Holochilomys* as Brandt had intended (for a myomorphous rodent), but without bibliographic reference; consequently, Cabrera (1961:503), for example, gave Peters credit for the name and listed *Holochilomys* Peters, 1861 as a junior synonym of *Holochilus* Brandt, without comment.

Voss and Abramson (1999) examined the holotypes of Brandt's *leucogaster* and *langsdorffii*, noting that the former was a specimen of *Trinomys*, the genus of spiny rats occurring in the Atlantic Forest of coastal Brazil, while the latter was referable to a *Proechimys*, the spiny rats of Amazonia and Middle America. Voss and Abramson (1999) thus petitioned the International Commission to conserve *Holochilus* Brandt for the myomorphous marsh rats and both *Proechimys* J. A. Allen, 1899 and *Trinomys* Thomas, 1921 for the hystricomorphous spiny rats, name associations that had stood in the literature for nearly a century or longer. In Opinion 1984, the International Commission (ICZN 2001) used its plenary power to designate *Holochilus sciureus* Wagner, 1842 as the type species of *Holochilus*, an action that maintained *Holochilus* as a name for the group of myomorphous rodents for which it had long been used. The name associations that follow from this action, beyond the fixation of Brandt's *Holochilus* to a sigmodontine rodent include: (1) Brandt's *anguya*, which was a misspelling of *angouya* Desmarest, is appropriately associated with the sigmodontine rodent now called *Sooretamys angouya* (see that account), and not with a species of *Holochilus* (contra Musser and Carleton 2005:1119, who inadvertently misspelled the name *anguyu*). (2) Brandt's *leucogaster* should be associated with the echimyid *Trinomys*, and not with the sigmodontine *Holochilus* where it is often listed in synonymies (e.g., Musser and Carleton 2005:1119), although *leucogaster* cannot be equated with any of the currently recognized species (see the *Trinomys* account). (3) Brandt's *langsdorffii* is correctly associated with another echimyid, the genus *Proechimys*, but also cannot be equated to an extant species (see the *Proechimys* account).

KEY TO THE SPECIES OF *HOLOCHILUS*:
1. Shorter tail (<150 mm, <23 caudal vertebrae); M3 hypoflexus and hypocone absent ...... *Holochilus lagigliai*
1'. Longer tail (>150 mm, >25 caudal vertebrae); M3 hypoflexus and hypocone always present 2
2. Hindfoot shorter (<46 mm); hypothenar pad absent; tail shorter than head and body; upper and lower masseteric ridges usually coalesce at the level of the posterior root of m1; M1 with expanded paracones; mesoloph(id)-like structure absent on M1–M2 and m1–m2 3
2'. Hindfoot larger (>47 mm); hypothenar pad present; tail equal in length, or slightly longer, than head and body; upper and lower masseteric ridges usually coalesce at the level of the anterior root of m1; M1 with unexpanded paracones; mesoloph(id)-like structure present on M1–M2 and m1–m2 5
3. Lophs(ids) compressed with acute outer margins; distributed throughout the lowlands of lower Paraguay Basin in Brazil (Mato Grosso do Sul state), Argentina, and Paraguay; characterized by 2n = 48–56, FN = 57–63 ... . *Holochilus chacarius*
3'. Lophs(ids) compressed with strongly acute and more prismatic outer margins; range elsewhere than Paraguay Basin of Argentina, Brazil, and Paraguay; 2n 44–46 or 55–56, FN = 56 4

4. Restricted to Venezuela; 2n = 44–46 and FN = 56 .... *Holochilus venezuelae*
4'. Distributed throughout the Amazon Basin and northeastern Brazil; 2n = 55–56, FN = 56. .... *Holochilus sciureus*
5. Size smaller (condyloincisive length 34.6–36.9 mm, length of upper molar rows 6.3–7.5 mm); distributed throughout the Atlantic forest and adjoining areas of southeastern and southern Brazil; 2n = 40, FN = 56. .... *Holochilus brasiliensis*
5'. Size larger (condyloincisive length 36.3–45.9 mm, length of upper molar rows 7.3–8.1 mm); distributed in Uruguay and northeastern Argentina, 2n = 35–39, FN = 57–61. .... *Holochilus vulpinus*

Holochilus brasiliensis (Desmarest, 1819)
Brazilian Marsh rat
SYNONYMS:
Mus brasiliensis Desmarest, 1819a:62; type locality "Brésil"; restricted to Lagoa Santa, Minas Gerais, by Hershkovitz (1955b:662), based on the itinerary of August St.-Hilaire.
Holochilus brasiliensis: Gray, 1843b:114; part (listing of *Mus brasiliensis* Desmarest); first use of current name combination.
Holochilomys brasiliensis: Hensel, 1872b:32; name combination.
Hesperomys [*Holochilus*] *brasiliensis*: Ihering, 1894:19; name combination.
Holochilus wagneri Moojen, 1952:65; type locality "Matas de Ipanema, São Paulo," Brazil.
Holochilus brasiliensis brasiliensis: Hershkovitz, 1955b:662; name combination.
Holochilus brasiliensis nanus: Hershkovitz, 1955b:666; part, not *Holochilus nanus* Thomas.

DESCRIPTION: Intermediate in size relative to other species recognized herein, with head and body length 167–211 mm, tail length 183–214 mm, and hindfoot length between 51–56 mm (Bonvicino, Oliveira, and D'Andrea 2008). General coloration cinnamon above, becoming brightly orange on sides and pale orange on venter, with throat, chest, and inguinal region entirely white.

DISTRIBUTION: *Holochilus brasiliensis* occurs in southeastern and southern Brazil, from Espírito Santo state to coastal Rio Grande do Sul state, including the eastern border of Minas Gerais, and west into the Humid Chaco of Paraguay. Most sites are located in the Atlantic Forest ecoregion, including both coastal and interior localities. Distributional limits in relation to those of *H. vulpinus* are unclear in southern Brazil, with the southernmost locality vouchered by a karyotyped specimen caught in the swampy grasslands of Pelotas, in Rio Grande do Sul.

SELECTED LOCALITIES (Map 162): BRAZIL: Minas Gerais, Lagoa Santa (type locality of *Mus brasiliensis* Desmarest), Pirapora (MNRJ 4271), Viçosa (ROM 77592); Paraná, Castro (MHNCI 552); Rio Grande do Sul, Pelotas (T. R. O. Freitas, Mattevi, Oliveira et al. 1983); Rio de Janeiro, Carapebus (NUPEM 237), Maricá (MNRJ 28809), Silva Jardim (MNRJ 61802); São Paulo, Floresta Nacional Ipanema ("Ipanema," type locality of *H. wagneri*), UHE de Ourinhos, Rio Ribeira (Tiepolo 2007).

SUBSPECIES: We treat *Holochilus brasiliensis* as monotypic.

NATURAL HISTORY: This species is nocturnal, amphibious, and herbivorous, with capture sites associated with riparian and swampy habitats in the Brazilian Atlantic Forest. Despite its infrequent capture in live-traps, it is a common prey item of the Barn Owl (*Tyto alba*) and small carnivores. Rocha-Mendes et al. (2010) recorded it in the diet of three felid species in Paraná (*Leopardus pardalis*, *L. tigrinus*, and *Puma yagouaroundi*). *Holochilus brasiliensis* has been documented in *ratadas* (rapid and extreme local population increases), along with other sigmodontines, in Paraná state, Brazil (Hershkovitz 1955b). It is a reservoir of *Schistosoma mansoni* in São Paulo (Dias et al. 1978).

REMARKS: Desmarest's *Mus brasiliensis* was based on a specimen received by the Muséum National d'Histoire Naturelle in Paris from Aguste St.-Hilaire, and referred to as "*le rat du Brézil*." Some nominal forms treated under *H. brasiliensis* (e.g., Musser and Carleton 2005:1119) have proven to belong to other rodent genera. For example, *Mus leucogaster* Brandt, 1835:428, once considered to be a junior synonym of *H. brasiliensis*, and once the type species of *Holochilus* (Miller and Rehn 1902), is actually a *Trinomys* (Voss and Abramson 1999). *Holochilus canellinus* Wagner and *Hesperomys leucogaster* Wagner are synonyms of *Sooretamys angouya*, and *Mus physodes* Brants is a synonym of *Euryoryzomys russatus* (see Musser et al. 1998 and the accounts for *Sooretamys* and *Euryoryzomys*). Our delimitation of *H. brasiliensis* is also considerably more restricted than those of recent taxonomic treatments of large *Holochilus* with mesoloph(id)-like structures on the molars (e.g., Musser and Carleton 2005). Several populations previously identified as *H. brasiliensis*, mainly in Argentina, Paraguay, and Uruguay, are herein considered to belong to *H. vulpinus*.

The karyotype is $2n=40$ and $FN=56$, with eight large to medium-sized and one small biarmed autosomes, and 10 small uniarmed autosomes. The X chromosome is a medium-sized acrocentric; the Y is a small biarmed chromosome (T. R. O. Freitas, Mattevi, Oliveira et al. 1983). This karyotype has been recorded in *H. brasiliensis* populations from southern (T. R. O. Freitas, Mattevi, Oliveira

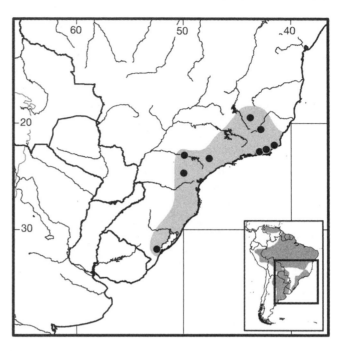

Map 162 Selected localities for *Holochilus brasiliensis* (●). Contour line = 2,000 m.

et al. 1983) and southeastern Brazil (MNRJ 61802; P. R. Gonçalves et al., unpubl. data).

Holochilus chacarius Thomas, 1906
Chacoan Marsh Rat
SYNONYMS:

Holochilus chacarius Thomas, 1906c:447; type locality "Chaco 1 league N.W. of Concepcion, [Concepción,] Paraguay."

Holochilus balnearum Thomas, 1906c:447; type locality "Bañado de S. Felipe, Tucuman [= Tucumán]. Alt. 435 m," Argentina.

Holochilus brasiliensis vulpinus: Hershkovitz, 1955b:662; name combination; part, not *Mus vulpinus* Brants.

Holochilus brasiliensis balnearum: Hershkovitz, 1955b: 665; name combination.

Holochilus brasiliensis chacarius: Massoia, 1971a:13; name combination.

Holochilus chacarius chacarius: Massoia, 1976b:12; name combination.

Holochilus chacarius balnearum: Massoia, 1976b:56; name combination.

DESCRIPTION: Small for genus, with head and body length 140–201 mm, tail length 148–183 mm, and hindfoot length 37–46 mm (Massoia 1976b; Bonvicino, Oliveira, and D'Andrea 2008). General coloration above reddish to cinnamon, intermixed with long black hairs; sides of body buffy, brightening to ochraceous buff along lower edge;

throat, chest, and inguinal regions entirely white but remainder of ventral surface buff or grayish, with hairs white basally; some individuals with venter entirely white.

DISTRIBUTION: *Holochilus chacarius* occupies lowlands of the Chaco ecoregion from western Brazil and Paraguay south into Argentina. The range in Argentina includes a narrow band in the provinces of Jujuy, Salta, Santiago del Estero, and Tucumán, and a second, wider region in the Humid Chaco from Formosa province south to northeastern Buenos Aires. Although S. Anderson (1997) failed to record this species from Bolivia, Pardiñas and Teta (2011b) recently referred specimens from Santa Cruz department to *H. chacarius*. Distributional relationships between *H. chacarius* and *H. sciureus* need clarification, especially in Bolivia. Pardiñas and Teta (2011b) summarized the fossil record.

SELECTED LOCALITIES (Map 163): ARGENTINA: Buenos Aires, San Pedro (Voglino et al. 2004); Chaco, Roque Saenz Peña (Massoia 1976b); Corrientes, Manantiales (Massoia 1976b); Entre Ríos, Gualeguay (Voglino et al. 2004), Strobel (P. Teta, unpubl. data); Formosa, Bañado La Estrella (P. Teta, unpubl. data); Jujuy, Ledesma (Massoia 1976b); Salta, El Talar (Massoia 1976b); Santa Fe, Jacinto L. Aráuz (Teta and Pardiñas 2010), Puerto Gaboto (Voglino et al. 2004); Tucumán, Concepción (Massoia 1976b). BOLIVIA: Santa Cruz, Bañados del Izozog (MNK 2038). BRAZIL: Mato Grosso do Sul, Corumbá (MNRJ [LBCE 5409]), Fazenda Rio Negro, Aquidauna (MNRJ [LBCE 4401]). PARAGUAY: Cordillera, 12 km N (by road) of Tobatí (A. L. Gardner and Patton 1976).

SUBSPECIES: Massoia (1976b) viewed *H. chacarius* as polytypic, with two subspecies: *H. c. balnearum* in southern Bolivia and northwestern Argentina in the lower Yungas, and *H. c. chacarius* in the Paraguayan Chaco and northeastern Argentina south to Buenos Aires province. These taxa can be differentiated by color and measurements as well as by several molecular markers (J. D. Hanson, unpubl. data).

NATURAL HISTORY: *Holochilus chacarius* inhabits swamps, flooded grasslands, and cultivated fields in open, mostly nonforested habitats; it swims, dives, and climbs well (Massoia 1976b). In the Paraguayan Chaco, *H. chacarius* was strongly associated with pasturelands, in areas with large amounts of litter and deep herbaceous ground cover (Yahnke 2006). Gestations under laboratory conditions last 27.8 ± 1 days, the mean number of embryos was 3.3 ± 1.2, and sexual maturity was reached at the age of 60 days in both sexes (Piantanida 1993); the average number of embryos in western Paraguay was seven (range: 4–9). This species makes grass nests measuring 20–30 cm in diameter (weight = 25–95g) in sugarcane and banana plantations, rice fields, or flooded grasslands, up to 1.4 m above the ground (Massoia 1976b). Nests found in north-

western Argentina were built on the ground in sugarcane plantations, with rounded living quarters constructed of interlaced strips and scraps of cane (Llanos 1944). This species also uses subterranean galleries, with one to three openings (Massoia 1971a). *Holochilus chacarius* is strictly herbivorous and can cause extensive damage to rice, banana, sugarcane, and other crops (Massoia 1974, 1976b). This species is a frequent prey item of the Barn Owl (*Tyto alba*) in northeastern Argentina (Voglino et al. 2004; Pardiñas, Teta, and Heinonen Fortabat 2005), and probably throughout its range.

REMARKS: Massoia (1971a, 1976b) and Pardiñas and Galliari (1998c) described external, cranial, and dental morphology. *Holochilus balnearum* Thomas has been considered a valid species (Reig 1986), a subspecies, or a junior synonym (at the subspecies level) of *H. chacarius* (Massoia 1976b; Musser and Carleton 2005).

The karyotype of *H. chacarius* is highly variable, with 2n ranging from 48 and 56 although the FN is more stable, varying between 57 and 60 (Vidal-Rioja et al. 1976; Nachman and Myers 1989; Nachman 1992a b). The karyotype described by Vidal et al. (1976), which was based on specimens from Salta province, is nearly identical to that of typical *H. sciureus*; however, specimens from the same general area, collected in Jujuy province, were recovered as sister to other *H. chacarius* as based on several gene sequences (J. D. Hanson et al., unpubl. data). Sequences

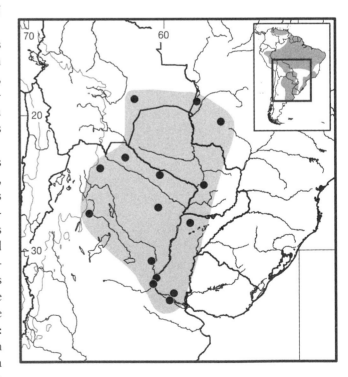

Map 163 Selected localities for *Holochilus chacarius* (●). Contour line = 2,000 m.

of the mtDNA ND3 gene of this species and of *H. vulpinus* differ by 18.9% (Kennedy and Nachman 1998).

Holochilus lagigliai Pardiñas, Teta, Voglino, and Fernández, 2013
Lagiglia's Marsh Rat

SYNONYM:

Holochilus lagigliai Pardiñas, Teta, Voglino, and Fernández, 2013; type locality not precisely known; the label of the holotype states only "El Nihuil–San Rafael–Mendoza–Argentina." El Nihuil is a dam constructed in 1947 on the middle of the Río Atuel, at 35°04'S, 68°43'W, 1,300 m.

DESCRIPTION: Large, characterized by strikingly short tail, especially broad braincase, wide and anteriorly rounded zygomatic plates, bony palate without excrescences, M1 with subelliptical procingulum and persistent parastyle and protostyle, well-developed mesolophostyles on M1; M3 shorter than M2, and, uniquely for genus, with hypoflexus and hypocone absent.

DISTRIBUTION: *Holochilus lagigliai* is known only from the type locality and from Holocene fragments obtained at Gruta del Indio, both localities in Mendoza province, west-central Argentina.

SELECTED LOCALITIES (Map 164): ARGENTINA: Mendoza, EL Nihuil (type locality of *Holochilus lagigliai* Pardiñas, Teta, Voglino, and Fernández).

SUBSPECIES: *Holochilus lagigliai* is monotypic.

NATURAL HISTORY: No aspects of the natural history of *H. lagigliai* are known. The holotype was collected half a century ago nearby a dam (El Nihuil) along the Río Atuel at 1,300 m, the highest elevation recorded for the genus. The other three specimens referred to this species correspond to Holocene-aged mandibular remains from the Gruta del Indio archaeological site. *Holochilus lagigliai* is biogeographically isolated in the northwestern corner of Patagonia, with the nearest populations of living *Holochilus* about 500 km to the east (*H. brasiliensis* in Buenos Aires Province) and 850 km to the north (*H. chacarius* in Tucumán province; Massoia 1976). Along with other amphibious mammals, such as *Lutreolina crassicaudata*, *Hydrochoerus hydrochaeris*, and *Myocastor coypus*, *H. lagigliai* forms part of a group of subtropical mammals that have patchy records in Mendoza province. The conservation status of the species is unknown, as the southern corner of Mendoza province has been poorly prospected for mammals (Pardiñas et al. 2013).

REMARKS: The tail of *H. lagigliai* is strikingly short compared with other similarly sized marsh rats and may influence its swimming capacity. Amphibious rats use the tail to change direction during swimming and as a point of rest against the bottom surface (Sierra de Soriano 1969). Mor-

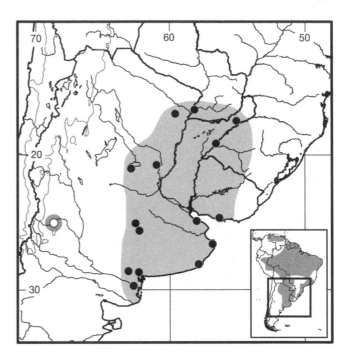

Map 164 Selected localities for *Holochilus lagigliai* (○) and *Holochilus vulpinus* (●). Contour line = 2,000 m.

phologically, *H. lagigliai* possesses a mosaic of traits present in species of the *sciureus* and *brasiliensis* groups, including a large cranium and molars, presence of a mesolophostyle, and small hindfeet and short tail. It does, however, share the unique M3 morphology with other members in the genus, with the hypoflexus and hypocone absent (Pardiñas et al. 2013).

Holochilus sciureus Wagner, 1842
Amazonian Marsh Rat

SYNONYMS:

Holochilus sciureus Wagner, 1842c:17; type locality "Rio Francisco [= Rio São Francisco] in Brasilien," regarded to be in Minas Gerais, Brazil (Hershkovitz 1955b).

Hesperomys squamipes: Burmeister, 1854:165; part (listing of *sciureus*); not *Mus squamipes* Brants.

Mus (Holochilus) brasiliensis: P. Gervais, 1854:411; part (reference to "*le rat du Brésil*"); not *Mus brasiliensis* Desmarest.

Hesperomys (Nectomys) sciureus: Ihering, 1894:19; name combination but incorrect assignment of Wagner's *sciureus* to the genus *Nectomys*.

Holochilus nanus Thomas, 1897c:495; type locality "Source [= Soure], Island of Marajó [= Ilha de Marajó], Mouth of the Amazon [= Rio Amazonas]," Pará, Brazil.

Holochilus guianae Thomas, 1901e:149; type locality "Kanuku Mountains, about 59° W. and 3° N., and on the savannahs near their base, at altitudes varying from 240 to 2000 feet," Upper Takutu–Upper Essequibo, Guyana.

Holochilus amazonicus Osgood, 1915:188; type locality "Itacoatiara, Amazon River, [Amazonas,] Brazil."

Holochilus incarum Thomas, 1920h:227; type locality "Santa Ana," Cusco, Peru.

Holochilus sciureus berbicensis Morrison-Scott, 1937:20; type locality "Blairmont Plantation, Berbice, coastal British Guiana," Upper Demerara-Berbice, Guyana.

Holochilus brasiliensis nanus: Hershkovitz, 1955b:666; name combination.

Holochilus brasiliensis amazonicus: Hershkovitz, 1955b: 667; name combination.

Holochilus brasiliensis incarum: Hershkovitz, 1955b: 668; name combination.

Holochilus brasiliensis guianae: Hershkovitz, 1955b:668; name combination.

Holochilus brasiliensis berbicensis: Hershkovitz, 1955b: 669; name combination.

Holochilus sciureus berbicensis: Barreto and García-Rangel, 2005:2; name combination.

Holochilus sciureus sciureus: Barreto and García-Rangel, 2005:2; name combination.

DESCRIPTION: Smallest species in genus, head and body length 123–193 mm, tail length 115–178 mm, and hindfoot 35–46 mm. General coloration above coarsely mixed brownish and rufous, darkest at midline and lighter and more ochraceous on sides; ventral surface whitish, washed with buff, hairs with narrow slate gray bases.

DISTRIBUTION: *Holochilus sciureus* occurs in the lowlands of the Guyanas, Bolivia, Peru, and Brazil. It may extend to northwestern Argentina (Vidal-Rioja et al. 1976; but see Remarks for *H. chacarius*).

SELECTED LOCALITIES (Map 165): BOLIVIA: Beni, Totaisal, 1 km SW Estación Biológica del Beni (S. Anderson 1997), Río Tijhamuchi (S. Anderson 1997); Santa Cruz, 3.5 km W of Estación El Pailón (S. Anderson 1997), San Rafael de Amboró (S. Anderson 1997), 9 km SE of Tita (MSB 55299). BRAZIL: Alagoas, Penedo (J. A. Oliveira et al. 2003); Amazonas, Altamira, right bank Rio Juruá (Patton et al. 2000), Itacoatiara (type locality of *Holochilus amazonicus* Osgood); Bahia, Bom Jesus da Lapa (MNRJ 4145); Ceará, Fortaleza (J. A. Oliveira et al. 2003), Ipu (J. A. Oliveira et al. 2003); Mato Grosso, Taiamã Island, Cáceres (T. R. O. Freitas, Mattevi, Oliveira et al. 1983); Minas Gerais, Jaíba (MNRJ 43825), Pirapora (MNRJ 4208); Pará, Soure, Ilha de Marajó (type locality of *Holochilus nanus* Thomas), Utinga (MNRJ 20928); Pernambuco, Bodocó (J. A. Oliveira et al. 2003), Pesqueira (J. A. Oliveira et al. 2003); Roraima, Caracaraí, Parque Nacional do Viruá (MNRJ [JAO 2059]). GUYANA: Mahaica-Berbice, Blairmont Plantation (type locality of *Holochilus sciureus berbicensis* Morrison-Scott); Potaro-Siparuni, Paramakatoi (USNM 565655); Upper Takutu–Upper Esse-

quibo, Kanuku Mountains (type locality of *Holochilus guianae* Thomas). PERU: Ayacucho, Hacienda Luisiana (A. L. Gardner and Patton 1976); Cusco, Santa Ana, upper Río Urubamba Valley (Thomas 1920h); Loreto, Genaro Herrera (Aniskin 1994); San Martín, Rioja (FMNH 19853); Ucayali, Yarinacocha (LSUMZ 16709). SURINAM: Sipaliwini, Sipaliwini airstrip (R. J. Baker et al. 1983).

SUBSPECIES: We regard *Holochilus sciureus* as being monotypic, although Barreto and García-Rangel (2005) recognized two subspecies (*H. s. sciureus* from the Orinoco Basin [including *venezuelae* J. A. Allen, herein regarded as a distinct species] and *H. s. berbicensis* from the Amazon Basin). This hypothesis needs to be evaluated by an adequate review of museum specimens.

NATURAL HISTORY: Barreto and García-Rangel (2005) summarized general characters, distribution, form and function, ontogeny and reproduction, ecology, behavior, and genetics in a treatment that also included *H. venezuelae*. *Holochilus sciureus* is found primarily in the Amazon drainage and in the savannas of Guyana (Twigg 1962; Barreto and García-Rangel 2005). It occurs in open areas, inundated grass patches along river banks, or in small agricultural plots close to river banks in lowland rainforest (Emmons and Feer 1997; Patton et al. 2000). It feeds mainly on leaves and seeds of savanna grasses but is also a major crop pest in sugarcane fields. In Guyana, it constructs spherical, dual-chambered nests, usually placed 0.5–1.5 m above the ground in sugarcane and rice fields. Sometimes, nests consist only of a mass of fine leaves placed in cracks in the soil, under dense grass or tussocks, or on the soil surface beneath dense litter (Twigg 1962, 1965).

Embryo counts for pregnant females ranged from one to eight in the coastal region of Guyana (Twigg 1965), and four to six in northeastern Brazil (Moojen 1952). In the Amazonian forests of western Brazil, Patton et al. (2000) caught males with scrotal testes and pregnant or lactating females in the months of August through November. In Bolivia, a lactating female was taken in March; nonpregnant females were recorded in May, September, and October (S. Anderson 1997). In Guyana, males with scrotal testes made up 70–98% of the population, and >27% of females were reproductively active (Twigg 1962). Litter size was affected by environmental variables and increased with the size and age of the females (Twigg 1965). In captivity, gestation lasted about 28–30 days, and the mean number of embryos was 3.1 (range: 1–6); males reached maturity at the age of 2–3 months and females within 2–5 months (Mello 1986). This species is a reservoir of *Leishmania* (*Viannia*) *brasiliensis* at Amajari, Pernambuco state, Brazil (Brandão-Filho et al. 2003).

REMARKS: Massoia (1981b) raised *sciureus* Wagner to species rank, distinguishing between this taxon and members of the *brasiliensis* group of species. Our deductions about the concept of this species stem from the analysis of samples from northeastern Brazil (Bahia, Alagoas, Pernambuco, and Ceará states) and Amazonia, which appear to form a morphologically cohesive entity lacking mesoloph-like structures on M1–M2 and with upper and lower mandibular masseteric ridges fused at the level of the anterior portion of m1 in most specimens. In addition, Pardiñas and Galliari (1998c) and Pardiñas and Teta (2011b) found morphological differences between *H. sciureus* and *H. chacarius*, particularly in their respective lower dentitions, including the form of the lophids (compressed with acute outer margins in *H. chacarius* versus compressed with strongly acute outer margins and more prismatic in *H. sciureus*); form and size of the anteromedian fossetid (subcircular and large in *H. chacarius* versus transversally elongated and small in *H. sciureus*); and occlusal development of the metaflexid (scarcely developed, not reaching the midline of the tooth in *H. chacarius* versus well developed, freely connected with the protoflexid in subadults of *H. sciureus*; see also Percequillo 2006). Petter and Tostain (1981) described morphological variation in M3 for French Guiana specimens.

This species is characterized by $2n = 55$–56 and FN = 56, with a marked predominance of acrocentrics in the autosomal complement, as based on specimens from Argentina (Vidal-Rioja et al. 1976), Brazil (T. R. O. Freitas, Mattevi, Oliveira et al. 1983), Peru (A. L. Gardner and Patton 1976; Aniskin 1994), and Surinam (R. J. Baker et al. 1983). Karyotypes from specimens identified as *H. balnearum*

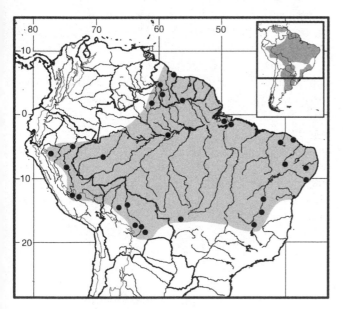

Map 165 Selected localities for *Holochilus sciureus* (●). Contour line = 2,000 m.

(= *H. chacarius* herein) from northwestern Argentina are nearly identical to those of *H. sciureus* (Vidal-Rioja et al. 1976; but see Remarks under *H. chacarius*), leaving some question as to whether *H. sciureus* may extend south of 20°S latitude.

Holochilus venezuelae J. A. Allen, 1904
Venezuelan Marsh Rat
SYNONYMS:

Holochilus venezuelae: J. A. Allen, 1904b:330; type locality "El Llagual [= El Yaguel], [Bolívar,] Venezuela."

Holochilus brasiliensis venezuelae: Hershkovitz, 1955b: 670; name combination.

Holochilus sciureus: Voss and Carleton, 1993:12; part (*venezuelae* listed as synonym); not *Holochilus sciureus* Wagner.

DESCRIPTION: Second largest species, with head and body length 130–226 mm, tail length 124–194 mm, and hindfoot 36–48 mm (O. J. Linares 1998). General color reddish brown above, strongly streaked with black middorsally, lighter and more fulvous on sides; and reddish fulvous on lower back and rump; venter buffy gray, hairs gray basally with yellowish white to deep buff tips; feet thinly haired, grayish flesh-colored; claws whitish with subapical dusky ring, and fringed with whitish hairs at base; tail dark brown, not appreciably bicolored, with short black bristles increasing in length and abundance apically; terminal fifth of tail well clothed with blackish bristly hairs that conceal underlying scales (J. A. Allen 1904b).

DISTRIBUTION: *Holochilus venezuelae* occurs in Venezuela, from the lower Orinoco Basin to Lago Maracaibo (Fuentes and Poleo 2005). The distributional limits of *H. venezuelae* in comparison with those of *H. sciureus* need clarification.

SELECTED LOCALITIES (Map 166): VENEZUELA: Bolívar, El Yagual (type locality of *Holochilus venezuelae* J. A. Allen); Guárico, Calabozo (Eiris and Barreto 2009); Portuguesa, Piritu (FLMNH 13323); Zulia, Misión Tokuko (USNM 448570).

SUBSPECIES: *Holochilus venezuelae* is monotypic.

NATURAL HISTORY: *Holochilus venezuelae* is considered an important agricultural pest in Venezuelan rice plantations, with densities measured between 5 and 713 rats/ha (Cartaya and Aguilera 1985). Its diet is mostly grasses, especially *Leptochloa scabra* and *Oryza sativa*, with small amounts of Cyperaceae, dicots, and invertebrates (A. M. G. Martino and Aguilera 1993). The mean home range size for 11 individuals inhabiting one rice field was 0.278 ha (Cartaya and Aguilera 1984). Eiris and Barreto (2009) examined home range, territoriality, individual movement, and dispersal. Gestation lasts about 29 days, and the mean number of embryos ranged between 5.8 (Venezuelan Llanos) and 7.4 (Acarigua, Venezuela) (Agüero 1978; Cartaya 1983; Poleo and Mendoza 2001). The mass of pregnant females increased about 48% (Aguilera 1987). In Venezuelan savannas dominated by *Paspalum fasciculatum*, marsh rats account for 0.2–1.3% of all rodents captured. Ectoparasites found on *H. venezuelae* include mites and ticks (*Amblyomma ovale*, *Androlaelaps fahrenholzi*, *Eutrombicula alfreddugesi*, *E. batatas*, *Gigantolaelaps canestrinii*, *G. mattogrossensis*, and *Ornithonyssus bacoti*), lice (*Hoplopleura contigua* and *H. quadridentata*), and a flea (*Polygenis dunni*). Known endoparasites include the nematodes *Physaloptera* sp., *Stilestrongylus* sp., and *Strongyloides* sp. (Guerrero 1985).

REMARKS: *Holochilus venezuelae* has been considered a full species (Aguilera et al. 1993), a subspecies of *H. sciureus* (O. J. Linares 1998), or a junior synonym of *H. s. sciureus* (Barreto and García-Rangel 2005). This taxon, however, is characterized by a karyotype with $2n=44$, $FN=56$ (Aguilera and Pérez-Zapata 1989), and thus with a relatively higher $FN/2n$ ratio than *H. sciureus*. A. L. Gardner and Patton (1976) described a karyotype from Villavicencio, Colombia, with $2n=50$, $FN=58$, one outside the known karyotypic variation for *H. venezuelae* and *H. sciureus* and which may represent an undescribed form. Further morphological and molecular comparisons are needed to test the affinities between the Colombian form and *H. venezuelae*. If these two forms prove to be conspecific, the range of *H. venezuelae* will extend westward into the headwaters of the Meta-Orinoco fluvial system in Colombia.

Holochilus vulpinus (Brants, 1827)
Crafty Marsh Rat
SYNOMYNS:

Mus vulpinus Brants, 1827:137; type locality "Brasil"; regarded as Rio Uruguay, southeastern Brazil, by Lichtenstein (1830; see Hershkovitz 1955b:660); restricted to Maldonado, Uruguay, by Cabrera (1961), but Cerqueira (1975) convincingly placed the type locality in Brazil, on the banks of Rio Uruguay, between Itaqui and Uruguay, Rio Grande do Sul.

Mus Braziliensis Waterhouse, 1839:58; type locality "Bahia Blanca," Buenos Aires, Argentina; not *Mus brasiliensis* Desmarest (see Remarks).

Holochilus brasiliensis: Gray, 1843b:114; part (listing of *brasiliensis* Waterhouse); not *Mus brasiliensis* Desmarest.

H[olochilus]. vulpinus: Wagner, 1843a:554; first use of current name combination.

Hesperomys vulpinus: Burmeister, 1854:163; name combination.

Hesperomys [(Holochilus)] vulpinus: Burmeister, 1879:210; name combination.

Holochilus multannus Ameghino, 1889:117; type locality "río Luján cerca de la estación Olivera en la provincia de Buenos Aires," Argentina.

Holochilus Darwini Thomas, 1897c:496; replacement name for *Mus braziliensis* Waterhouse, which is preoccupied by *Mus brasiliensis* Desmarest.

Holochilus chrysogaster Cabrera, 1912:102; *nomen nudum*, erroneously attributed to Waterhouse.

Holochilus brasiliensis vulpinus: Hershkovitz, 1955b:662; name combination.

DESCRIPTION: Largest species in genus, with range in head and body length 141–238 mm, tail length 160–238 mm, and hindfoot 41–55 mm (Massoia 1976b; Bonvicino, Oliveira, and D'Andrea 2008). General coloration above deep orangish to reddish brown strongly intermixed with long, glistening blackish brown hairs; sides bright ochraceous buff; ventral surface entirely white (northern populations), with narrow band of grayish to orangish hairs, or golden yellowish with throat and chest covered by patches of entirely white hairs (Massoia 1976b).

DISTRIBUTION: *Holochilus vulpinus* ranges from Paraguay and east-central Argentina throughout Uruguay to Rio Grande do Sul state in Brazil (Massoia 1976b, 1981b; Marques 1988). Extralimital fossils, referred to *H. brasiliensis* by Pardiñas and Teta (2011b), suggest that this species had a wider distribution during the Late Holocene,

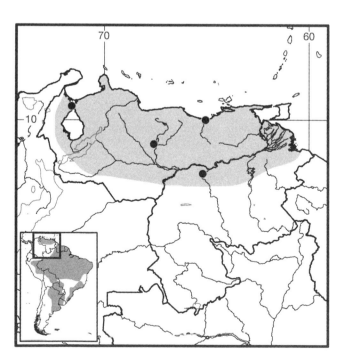

Map 166 Selected localities for *Holochilus venezuelae* (●). Contour line = 2,000 m.

reaching the northern border of Patagonia and the hill systems of central Argentina.

SELECTED LOCALITIES (Map 164): ARGENTINA: Buenos Aires, Arroyo Brusquitas (Massoia 1976b), Arroyo Chasicó (Massoia 1976), Bahía Blanca (type locality of *Holochilus brasiliensis* Waterhouse), Estancia La Providencia (Teta, González-Fischer et al. 2010), Mar del Tuyú (Massoia 1976b), Marucha (Teta, González-Fischer et al. 2010), 20 km S of Pedro Luro on Hwy 3 (Formoso et al. 2010), Punta Lara (Massoia 1976); Chaco, Parque Nacional Chaco (Teta, González-Fischer et al. 2010); Córdoba, Marull (Jayat et al. 2006 [erroneously cited as for *H. chacarius*]); Misiones, Oberá, Escuela Provincial 639 "Rosario Vera Peñaloza," Lote 92, Sección II de Campo Ramón (Massoia 1993); Santa Fe, Jacinto L. Aráuz (Teta and Pardiñas 2010). BRAZIL: Rio Grande do Sul, Itaqui (Tiepolo 2007). PARAGUAY: Estancia Yacaré (GD 071). URUGUAY: Montevideo, Parque Lecocq (E. M. González 1996).

SUBSPECIES: We regard *Holochilus vulpinus* as monotypic, but Massoia (1976b), based on morphometrics, treated *darwini* Thomas as a subspecies separate from a more northern nominal form.

NATURAL HISTORY: *Holochilus vulpinus* constructs spherical and unlined nests of leaves, sticks, and grasses that average 22–27 mm in diameter or excavates galleries at the water level. Nests are often in tangles of vines and terminal forks of tree branches, up to 2–3 m above the ground (Barlow 1969) or in hygrophilous vegetation such as the bulrushes *Schoenoplectus californicus* or *Scirpus giganteus* (Cyperaceae) as high as 40–50 cm above the water level (Massoia 1976b). In Buenos Aires province, Argentina, up to seven young individuals were found in each nest (Massoia 1976b). In Uruguay, males and females found in reproduction condition were caught in May and April; pregnant females had two to four embryos (Barlow 1969). Population densities in southeastern Buenos Aires reached 4.3 individuals/ha (range 0–12.5; Dalby 1975). In Uruguay, Barlow (1969) found 45 nests in an area of 50 ha. Plants, including aquatic, riparian, and cultivated species, comprise the diet. Barlow (1969) described the molt cycle in Uruguayan specimens. This species is infested by the mites *Laelops manguinhosi*, *Androlaelaps fahrenholzi*, and *Gigantolaelaps wolffsohni*, and the fleas *Craneopsylla minerva*, *Polygenis* (N.) *atopus*, *P.* (N.) *frustratus*, *P.* (P.) *acodontis*, *P.* (P.) *platensis* (Nava et al. 2003; Navone et al. 2010), the mite *Notoedres muris* (Klompen and Nachman 1990), the digenean *Urotrema scabridum*, *Cladorchis pyriformis*, and the nematode *Tapironema coronatum* (Durette-Desset et al. 1997).

REMARKS: Waterhouse (1839) described *Mus brasiliensis* based on a specimen collected by Charles Darwin at Bahía Blanca, on the coast of Buenos Aires province, Argentina. Although Waterhouse equated his animal to A. Saint-Hilaire's "*le rat du Brézil*," which formed the basis for Desmarest's earlier description of *Mus brasiliensis*, Waterhouse's taxon is not the same species although his *Mus brasiliensis* is pre-occupied by Desmarest's name. Thomas (1897c) recognized this problem and proposed *Holochilus darwini* as a replacement for Waterhouse's *brasiliensis*.

Massoia (1971a, 1976b) and Pardiñas and Galliari (1998c) described external, cranial, and dental morphology of *H. vulpinus* but under the name of *H. brasiliensis*. Specimens from owl pellets in central and eastern Argentina and previously identified as *H. brasiliensis* (e.g., Formoso et al. 2010) probably represent *H. vulpinus*. Specimens of *H. vulpinus* exhibit larger dimensions than *H. brasiliensis*, especially in its cranial measurements, such as the condyloincisive length (36.3–45.9 mm in *H. vulpinus* and 34.6–36.9 mm in *H. brasiliensis*) and length of the upper molar series (7.3–8.1 mm in *H. vulpinus* and 6.3–7.5 mm in *H. brasiliensis*).

This species is characterized by $2n = 35$–39 and $FN = 57$–61 in Argentina and Paraguay (Nachman 1992a, b), with a higher proportion of biarmed autosomes than in *H. brasiliensis* from coastal Rio Grande do Sul (Riva et al. 1977; T. R. O. Freitas, Mattevi, Oliveira et al. 1983; Nachman 1992a,b). Better documentation of the morphological distinction between *H. brasiliensis* and *H. vulpinus* is needed to clarify the geographic distribution of both taxa in Uruguay, Paraguay, and Argentina. *Holochilus vulpinus* differs from *H. chacarius* by a genetic distance of 18.9% in sequence comparisons of the mtDNA ND3 gene (Kennedy and Nachman 1998).

Genus *Hylaeamys* Weksler, Percequillo, and Voss, 2006

Alexandre R. Percequillo

Species of *Hylaeamys*, previously known as the *capito* complex or *megacephalus* group of the genus *Oryzomys* (*sensu lato*), are widely distributed throughout cis-Andean South America, from Colombia south to Paraguay and east to coastal Brazil. In this large area, they inhabit lower slopes of the Andes, extending to an elevation of 1,000 m, Amazonian lowlands and floodplains, lowlands and highlands of the Atlantic Forest, and forested habitats in Cerrado and Chaco, among the drier biomes of the continent. There are no known records from the Caatinga or the central portion of the Brazilian Pantanal.

I recognize seven species of *Hylaeamys*: *H. acritus* Emmons and Patton, *H. laticeps* (Lund), *H. megacephalus*

(G. Fischer), *H. oniscus* (Thomas), *H. perenensis* (J. A. Allen), *H. tatei* (Musser, Carleton, Brothers, and Gardner), and *H. yunganus* (Thomas). I include *seuanezi* Weksler, Geise, and Cerqueira as a synonym of *H. laticeps*, but see Brennard et al. (2013) who characterized this taxon morphologically and chromosomally, and mapped its range, as a species distinct from other coastal Brazilian *Hylaeamys* (*laticeps*, *megacephalus*, and *oniscus*).

Morphological characters that diagnose the genus include a tail length equal or subequal to head and body length; short, teardrop-shaped incisive foramina; and carotid circulatory pattern 2 (*sensu* Voss 1988), a trait that is shared only with the genus *Oligoryzomys* among members of the Oryzomyini. The pelage is usually short and harsh, with viliform, setiform, and aristiform hairs homogeneous in texture and similar in length (aristiforms 9–14 mm). The dorsal color ranges from a dark grayish to a rich buffy brown; ventral color is uniformly gray. The tail is shorter than the head and body; its surface is usually darker than the ventral one, but the tail is not conspicuously bicolored. Hindfeet are long and narrow, with ungual tufts just reaching the tips of the claws; and with four interdigital pads and two plantar pads, the thenar and hypothenar (with polymorphism in the presence of the hypothenar in some samples of *H. yunganus*; see Musser et al. 1998:59, Table 7).

The skull is moderately to heavy built, 27.5–37.9 mm in greatest length. The rostrum has weakly projected nasolacrimal capsules; zygomatic notches are narrow and either shallow to moderately deep; the free dorsal margins of the zygomatic plates are reduced, with their anterior margins either straight or weakly concave. The zygomatic arches are slender and nearly parallel; the jugal is small. The interorbital region is slightly divergent posteriorly; supraorbital margins are squared or possess small dorsolateral projections, rarely strongly beaded. The braincase is elongate with a discrete temporal crest. Incisive foramina are short (3.5–5.8 mm long), and their lateral margins diverge posteriorly, making them teardrop in shape. The palate is long and wide, with small to large posterolateral palatal pits in shallow to deep palatine depressions; palatal excrescences are present in most specimens (Musser et al. 1998:74, Table 12). The anterior margin of the mesopterygoid fossa does not reach the level of the alveoli of the M3s; the roof of the fossa is either completely ossified or perforated only by small and narrow sphenopalatine vacuities on the presphenoid or at the presphenoid-basisphenoid suture. A modified cephalic arterial circulation is present (pattern 2 of Voss 1988), as indicated by absence of sphenofrontal foramina and squamosoalisphenoid grooves and presence of large stapedial foramina and a deep furrow in the posterolateral portion on the ventral surface of each

pterygoid plate. The buccinator-masticatory and accessory oval foramina are confluent, so an alisphenoid strut is absent in all specimens. Auditory bullae are small, stapedial processes are long and narrow, and bony tubes are short. The suspensory process of the squamosal is absent, and the tegmen tympani varies from small and triangular to large and squared, and does not extend anteriorly to overlap the squamosal; the postglenoid foramen is long and wide, and the subsquamosal fenestra is absent. The mandible is shallow; the posterior margin of the angular process is at the same vertical plane as that of the condyloid process; the coronoid process reaches the same height as the condyle; inferior and superior notches are shallow; and a capsular projection is absent.

Upper incisors are opisthodont. Molars are pentalophodont and moderately high-crowned; labial and lingual cusps are arranged in opposite pairs. Both labial and lingual flexi are moderately deep and narrow, overlapping on each molar at midplane. Musser et al. (1998) described and figured occlusal surfaces of *H. laticeps*, *H. megacephalus*, *H. oniscus*, *H. perenensis*, *H. tatei*, and *H. yunganus*. Young and a few adult specimens of *H. megacephalus* have m2s bisected by a long hypoflexid (Musser et al. 1998:79, Figs. 31 and 32). M1 has three main roots, anterior and posterior labial and lingual; M2 and M3 also have three main roots. Each lower molar has two roots, one each in anterior and posterior positions.

The axial skeleton is composed of 7 cervical, 12–13 thoracic, 6–7 lumbar, 4 sacral, and 27–30 caudal vertebrae; there is some variation in the number of ribs, but the modal count is 12. Hemal processes and hemal arches (rarely observed in *H. megacephalus*) are present at the junctions of the first and second and/or second and third caudal vertebrae.

The stomach is unilocular and hemiglandular, with a shallow angular notch. The glans penis is of the complex type (Hooper and Musser 1964), with a tridigitate and well-developed cartilaginous baculum. The three bacular mounds are visible in the terminal crater, and the basal portion of the bony baculum is wide and spatulate, barely concave. A single dorsal papilla and one urethral flap are present.

All species are very similar in external and cranial morphology, more easily differentiated by morphometric and karyological criteria and geographic ranges (see Percequillo 1998; Musser et al. 1998; Emmons and Patton 2005). Consequently, my Key employs all useful characters in the recognition of these species. Brennand et al. (2013) provided details of locality distribution and morphological comparisons among four species (*H. laticeps*, *H. megacepahlus*, and *H. oniscus*, including the recognition of *H. seuanezi*) that occur in coastal Brazil.

SYNONYMS:

Mus: G. Fischer, 1814:71; part (description of *megacephalus*); not *Mus* Linnaeus.

H[*esperomys*].: Wagner, 1843a:542; part (listing of *cephalotes*); not *Hesperomys* Waterhouse.

Calomys: Fitzinger, 1867b:87; part (listing of *cephalotes*); not *Calomys* Waterhouse.

Oryzomys: Thomas, 1894b:354; part (listing of *laticeps*); not *Oryzomys* Baird.

Hylaeamys Weksler, Percequillo, and Voss, 2006:8; type species *Mus megacephalus* G. Fischer, by original designation.

REMARKS: The species-group taxa assigned to the genus *Hylaeamys* were profoundly affected by Hershkovitz's (1960, 1966a) revisionary acts in footnotes. I discussed the consequences of these acts in the Remarks section for the genus *Euryoryzomys*.

KEY TO THE SOUTH AMERICAN SPECIES OF *HYLAEAMYS*:

1. Second upper molar with distinct medial mesofossette and short paraflexus 2
1'. Second upper molar without mesofossette and with long paraflexus 3
2. Pelage long (fur length 9–12 mm; Musser et al. 1998:101) and dense, uniformly dark brown; upper molar toothrow long (5.5–5.7 mm) and wide (1.7–1.9 mm); incisive foramina short relative to greatest skull length (4.2–4.9 mm and 30.9–33.9 mm, respectively) . . . .
. . . . *Hylaeamys tatei*
2'. Pelage short (fur length 6–10 mm) and less dense, from grayish brown to buffy brown; upper molar toothrow short (3.9–5.5 mm); incisive foramina long relative to greatest skull length (3.9–5.8 mm and 27.5–35.3 mm, respectively); 2*n*=58–60 and FN=62–66 in western populations (Brazil and Peru), and 2*n*=52–59 and FN=66–67 in eastern samples (Surinam; see Musser et al. 1998:80–81, Table 13 for further details) . . . .
. . . . *Hylaeamys yunganus*
3. Head and body (80–158 mm) and tail (90–138 mm) long; hypothenar pad large or very small; skull smaller and delicate (greatest length 27.2–34.9 mm); molar toothrow shorter (3.7–5.2 mm) and molars narrower; sphenopalatine vacuities variably present; lower second molar without distinct entoflexid or entofossetid and with long hypoflexid nearly bisecting the molar in most specimens; 2*n*=54 4
3'. Head and body (81–174 mm) and tail (94–162 mm) long; hypothenar pad medium sized; skull large and robust (greatest length 29.5–37.9 mm); molar toothrow longer (4.5–5.6 mm) and molars robust; roof of mesopterygoid fossa completely ossified in most specimens (>90% of

specimens from Rio Juruá, Amazonas and Acre states, and eastern Atlantic Forest, Brazil); lower second molar with distinct entoflexid or entofossetid and short hypoflexid in most specimens; 2*n*=48 and 52 5
4. Head and body (80–158 mm) and tail (90–138 mm) shorter; hypothenar pads small; skull smaller (greatest length 27.2–34.9 mm) and delicate; molar toothrow shorter (3.7–5.2 mm) and molars delicate; sphenopalatine vacuities small and narrow (vacuities present in 50–59% of Paraguayan specimens); multiple posterolateral palatal pits set in shallow to deep palatal notches; palatal excrescences present, from small projections to large bone struts fused to maxilla; 2*n*=54 and FN=62 . . . . .
. . . . *Hylaeamys megacephalus*
4'. Head and body (131–152 mm) and tail shorter (110–130 mm); hypothenar pads large; skull smaller (greatest length 33.9–34.6 mm) and delicate; molar toothrow shorter (4.7–5.2 mm) and molars delicate; roof of mesopterygoid fossa infrequently perforated by sphenopalatine vacuities (when present, very small and narrow); multiple posterolateral palatal pits set in shallow to deep palatal notches; palatal excrescences present as large bone struts fused to maxilla . . . *Hylaeamys acritus*
5. Head and body 81–169 mm and tail 94–162 mm; 2*n*=52, FN=62 6
5'. Head and body 122–174 mm and tail 100–157 mm; 2*n*=48, FN=60; pelage sparser and coarser; dorsal color dark, grayish or blackish brown; skull larger (greatest length 31.8–37.6 mm); molars more robust (toothrow length 4.6–5.5 mm) . . . . *Hylaeamys laticeps*
6. Head and body (131–162 mm), tail (135–162 mm), and skull (33.7–37.6 mm slightly larger; occurs in northeastern Brazil, north of Rio São Francisco . . . .
. . . . *Hylaeamys oniscus*
6'. Head and body (81–169 mm), tail (94–154 mm), and skull (29.5–37.9 mm smaller; occurs in Amazon Basin of Colombia, Ecuador, Peru, Bolivia, and western Brazil . . . . *Hylaeamys perenensis*

Hylaeamys acritus (Emmons and Patton, 2005)
Bolivian Hylaeamys
SYNONYMS:

Oryzomys capito: S. Anderson, 1997:399; part (listing of specimens from MNK); not *Mus capito* Olfers (= *Hylaeamys megacephalus* [G. Fischer]).

Oryzomys megacephalus: Musser, Carleton, Brothers, and Gardner, 1998:32; part; not *Mus capito* Olfers (= *Hylaeamys megacephalus* [G. Fischer]).

Oryzomys acritus Emmons and Patton, 2005:14; type locality "Bolivia: Departamento de Santa Cruz, Provincia Velasco, Parque Nacional Noel Kempff Mercado, El Refugio Huanchaca, an outpost with a few buildings and an airstrip on private property, but within the park (14°42.553'S, 061°2.034'W [WGS 84]; elev. 170 m)."

[*Hylaeamys*] *acritus*: Weksler, Percequillo, and Voss, 2006: 14; first use of current name combination.

DESCRIPTION: Characterized externally by tail much shorter than head and body length (average about 81%); long dorsal (9–10 mm) and ventral hairs (3–4 mm); whitish venter; large and fleshy hypothenar pad; one ring of squamae on digit I of pes. Skull robust and with strongly developed palatal excrescences; very large foramen ovale; M2 with short paraflexus and single labial fossette (mesoflexus); m2 with short hypoflexid.

DISTRIBUTION: *Hylaeamys acritus* is an endemic species of northeastern Bolivia, limited to the west bank tributaries of the Río Iténez (= Rio Guaporé in Brazil), in the departments of Santa Cruz and Beni.

SELECTED LOCALITIES (Map 167): BOLIVIA: Beni, Curicha (USNM 551653), San Joaquín, Monto Río Machupo (Emmons and Patton 2005); Santa Cruz, El Refugio Huanchaca, Parque Nacional Noel Kempff Mercado (type locality of *Oryzomys acritus* Emmons and Patton), Lago Caiman, Parque Nacional Noel Kempff Mercado (MNK 1949), Urubicha, Río Negro (MNK [NR 10]).

SUBSPECIES: *Hylaeamys acritus* is monotypic.

NATURAL HISTORY: All natural history data are from populations in Parque Nacional Noel Kempff Mer-

cado (Emmons and Patton 2005). There, individuals were captured on the ground in evergreen riverine (gallery) forests, deciduous and semideciduous *terra firme* forests, and seasonally flooded forests on the edge of seasonally flooded pampa. These forests are characterized by two physiognomies: a "seasonally deciduous emergent canopy, sparse understory and midstory varying in deciduousness depending both on the location and on the dryness of the particular year," or liana forests, exhibiting "low canopies and dense lianas at all levels" (Emmons and Patton 2005:15). Elevational range is 170–700 m.

REMARKS: This recently recognized species is thoroughly diagnosed and described by Emmons and Patton 2005), who provided molecular and both metric and qualitative morphological characters in support of species status. Phenotypically, the most similar species is *H. megacephalus*, with which *H. acritus* shares similarities in body color, body and skull size, and molar topography. Despite these common features, *H. acritus* is the sister of *H. perenensis*, a western Amazonian *Hylaeamys* that is more closely related to the Atlantic Forest *H. laticeps* than to the eastern Amazonian-Cerrado *H. megacephalus*. *Hylaeamys acritus* is sympatric with *H. yunganus* (Emmons and Patton 2005).

Hylaeamys laticeps (Lund, 1840)
Atlantic Forest Hylaeamys

SYNONYMS:

Mus laticeps Lund, 1839a:233; *nomen nudum*.

Mus laticeps Lund, 1840b [1841c:279]; type locality "Rio das Velhas's Floddal" (Lund 1841c:292, 294), Lagoa Santa, Minas Gerais (see also Musser et al. 1998).

Hesperomys laticeps: Burmeister, 1854:171; part (excluding *Hesperomys subflavus* Wagner [= *Cerradomys subflavus*]); name combination.

Calomys saltator Winge, 1887:48, plate I, Figs. 16 and 17; plate III, Fig. 7; renaming of *Mus laticeps* Lund.

Oryzomys laticeps: Thomas, 1894b:354; name combination.

[*Oryzomys* (*Oryzomys*)] *laticeps laticeps*: Tate, 1932e:18; part; name combination.

[*Oryzomys* (*Oryzomys*)] *saltator*: Tate, 1932e:18; name combination.

Oryzomys laticeps laticeps: Gyldenstolpe, 1932a:17; name combination.

Oryzomys oniscus: Moojen, 1952b:47; part; not *Oryzomys oniscus* Thomas (= *Hylaeamys oniscus*).

Oryzomys [(*Oryzomys*)] *capito laticeps*: Cabrera, 1961: 386; name combination.

Oryzomys capito: Hershkovitz, 1966a:137, footnote; part; not *Mus capito* Olfers (= *Hylaeamys megacephalus* [Fischer]).

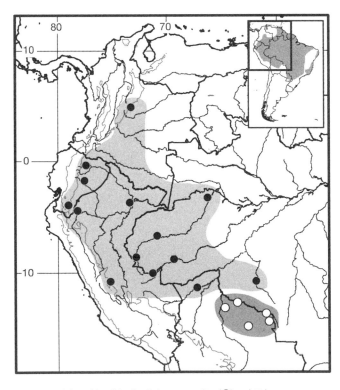

Map 167 Selected localities for *Hylaeamys acritus* (○) and *Hylaeamys perenensis* (●). Contour line = 2,000 m.

Oryzomys seuanezi Weksler, Geise, and Cerqueira, 1999: 454; type locality "Fazenda União, Casimiro de Abreu municipality, Rio de Janeiro state (22°25′S, 42°02′W; 50 m."

[*Hylaeamys*] *laticeps*: Weksler, Percequillo, and Voss, 2006: 14; first use of current name combination.

DESCRIPTION: Large, with head and body length 122–174 mm; tail short (100–157 mm); hindfeet robust, with fleshy hypothenar pads and sparse ungual tufts shorter than claws. Dorsal fur very short and harsh, colored buffy, yellowish, or grayish brown densely grizzled with dark brown; ventral fur shorter and predominantly grayish in color; tail bicolored, basally and uniformly colored distally; ears sparsely covered with reddish-brown hairs. Skull robust and large (greatest skull length 31.8–37.6 mm), with short, narrow, and teardrop-shaped incisive foramina (length 4.3–5.9 mm; width 2.1–2.9 mm); interorbital region diverges slightly posteriorly, with squared or weakly to moderately beaded supraorbital margins; palate with complex, multiple posterolateral palatal pits recessed in shallow palatine depressions and developed palatal excrescences. Toothrows robust (length 4.6–5.5 mm; M1 width 1.4–1.8 mm).

DISTRIBUTION: *Hylaeamys laticeps* is endemic to the eastern Brazilian Atlantic Forest. Known collecting localities are distributed along coastal lowland forests and eastern forested foothills of the Serra do Mar, from southern Bahia to northern Rio de Janeiro states. Consequently, the elevational range of this species is limited, extending from near sea level (Casimiro de Abreu, Rio de Janeiro, at 17 m) to approximately 100 m at Buerarema, Bahia, and about 300 m in Parque Estadual do Rio Doce, Minas Gerais.

SELECTED LOCALITIES (Map 168): BRAZIL: Bahia, Almada, Rio do Braço (MNRJ 9314); Espírito Santo, Linhares, Fazenda Santa Terezinha, 33 km NE of Linhares (UFMG MAS 4); Minas Gerais, Lagoa Santa (type locality of *Mus laticeps* Lund), Parque Estadual do Rio Doce (UFMG JRS 382); Rio de Janeiro, Casimiro de Abreu, Reserva Biológica de Poço das Antas (MNRJ 29406).

SUBSPECIES: I treat *Hylaeamys laticeps* as monotypic.

NATURAL HISTORY: This terrestrial species inhabits lowland evergreen forest and subtropical deciduous forest along the Atlantic coast, where it may be found in swampy, disturbed, and primary forests (G. A. B. Fonseca and Kierulff 1989; Stallings 1989), but most frequently in pristine habitats (Laemmert et al. 1946). In one southern Bahia community, *H. laticeps* was the most common small mammal, where it was encountered in forest fragments of all sizes as well as in the surrounding matrix, although it was more abundant in mature forests (Pardini 2004).

REMARKS: The southeastern Atlantic Forest populations of genus *Hylaeamys*, called *H. laticeps* by Musser et al. (1998), were only recently recognized as specifically distinct from other populations of this group, particularly those from the Cerrado (allocated to *H. megacephalus*) and northeastern Atlantic Forest in Brazil (*H. oniscus*; see Percequillo 1998; Musser et al. 1998; Weksler et al. 1999; Andrades-Miranda, Zanchin et al. 2002).

Throughout much of its recent history, the specific epithet *laticeps* Lund was not applied to any species, regarded only as a synonym of *H. megacephalus* (Musser and Carleton 1993; but see Hershkovitz 1960:544, footnote, who recognized *laticeps* but included in his concept a large number of names now allocated to species in five different genera). Musser et al. (1998), however, stated that *laticeps* Lund was the valid and available name that should be applied to the large Atlantic Forest *Hylaeamys* and regarded *oniscus* Thomas to be a junior synonym. Almost simultaneously, Weksler et al. (1999) proposed *Oryzomys seuanezi* for the morphologically and karyologically divergent populations from the southeastern Atlantic Forest (Weksler et al. 1999), regarding *Oryzomys oniscus* Thomas as a species restricted to northeastern Atlantic Forest. Here, I treat *seuanezi* Weksler, Geise, and Cerqueira as a junior synonym of *H. laticeps*, but I also regard *H. oniscus* as a valid species (see that account). Brennard et al. (2013)

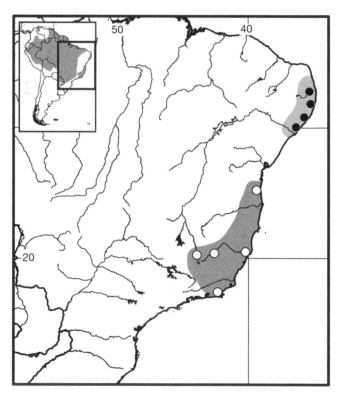

Map 168 Selected localities for *Hylaeamys laticeps* (○) and *Hylaeamys oniscus* (●).

regarded *seuanezi* as a species distinct from other Atlantic Forest *Hylaeamys*, including *H. laticeps*, *H. megacephalus*, and *H. oniscus*.

Although most recent records would suggest that *H. laticeps* is a coastal Atlantic Forest dweller, the type locality at Lagoa Santa, Minas Gerais (where this species occurs in sympatry with *H. megacephalus*) is situated on the western slope of Serra do Espinhaço in a region mostly dominated by open vegetation of the Cerrado, although in a mosaic with patches of semideciduous forest. Lagoa Santa contrasts elevationally with most locality records for this species, because it is about 760 m above sea level.

Hylaeamys laticeps can be differentiated from *H. megacephalus* by its sparser and coarser pelage; dark, grayish or blackish brown dorsal body coloration; larger skull; more robust molars; larger and wider palatal excrescences; diploid number $2n = 48$ and $FN = 60$ (in contrast to $2n = 54$ and $FN = 62$ in *H. megacephalus*), and geographical distribution (while the former is restricted to eastern Brazilian Atlantic Forest, the latter has a more widespread range that extends from the Cerrado throughout eastern Amazonia).

Hylaeamys megacephalus (G. Fischer, 1814)
Azara's Hylaeamys

SYNONYMS:

Mus megacephalus G. Fischer, 1814:71; based on Azara's (1801, 1802) "*rat second ou rat a grosse tête*" and "*del cola igual al cuerpo*"; type locality "Paraguay east of the Río Paraguay, Departamento de Canendiyu [= Canindeyú], 13.3 km (by road) N Curuguaty (24°31'S/55°42'W, 255 m," based on neotype designation (Musser et al. 1998:251–2).

Mus ? Capito Illiger, 1815:70; *nomen nudum*.

? *Mus capito*: Olfers, 1818:209; redescription of *Mus Capito* Illiger; based on Azara's (1801, 1802) "*rat second ou rat a grosse tête*" and "*del cola igual al cuerpo*."

Mus cephalotes Desmarest, 1819d:63; based on Azara's (1801, 1802) "*rat second ou rat a grosse tête*" and "*del cola igual al cuerpo*."

H[esperomys]. cephalotes: Wagner, 1843a:542; name combination.

Calomys cephalotes: Fitzinger, 1867b:87; part; name combination.

Oryzomys velutinus J. A. Allen and Chapman, 1893:214; type locality "Princestown [= Princes Town], Trinidad," Trinidad and Tobago.

Oryzomys Goeldi Thomas, 1897c:494; type locality "Itaituba, Rio Tapajós," Pará, Brazil.

[*Oryzomys*] *cephalotes*: E.-L. Trouessart, 1897:526; name combination.

Oryzomys modestus J. A. Allen, 1899b:212; type locality "Campo Alegre, Sucre, Venezuela."

[*Oryzomys (Oryzomys)*] *goeldi*: Tate, 1932e: 7; part; name combination.

[*Oryzomys (Oryzomys)*] *cephalotes*: Tate, 1932e:18; part; name combination.

[*Oryzomys (Oryzomys)*] *saltator*: Tate, 1932e:18; part; name combination.

Oryzomys angouya: C. O. da C. Vieira, 1953:137; part; not *Mus angouya* G. Fischer (= *Sooretamys angouya* [G. Fischer]).

Oryzomys capito: Hershkovitz, 1959b:339; part; name combination.

Oryzomys [(*Oryzomys*)] *capito capito*: Cabrera, 1961:385; name combination.

Oryzomys [(*Oryzomys*)] *capito goeldii*: Cabrera, 1961: 385; name combination and incorrect subsequent spelling of *Oryzomys goeldi* Thomas.

Oryzomys (Oryzomys) laticeps cf. *modestus*: C. T. Carvalho, 1962:290; part; name combination.

Oryzomys capito goeldi: C. T. Carvalho and Toccheton, 1969:290; name combination.

[*Oryzomys*] *megacephalus*: Langguth, 1966b:287; name combination.

Oryzomys oniscus: Alho, 1981:225; part; not *Oryzomys oniscus* Thomas (= *Hylaeamys oniscus* [Thomas]).

[*Hylaeamys*] *megacephalus*: Weksler, Percequillo, and Voss, 2006:14: first use of current name combination.

DESCRIPTION: Small to medium sized (head and body length 80–158 mm; mean 121.9 mm, N = 151); tail shorter than head and body (range: 90–138 mm; mean 115.4 mm, N = 145); hindfeet short with small plantar pads (especially hypothenar), and sparse ungual tufts shorter than claws. Dorsal fur short, dense, and slightly harsh, colored overall with ochraceous, yellowish, or orangish weakly to moderately ticked with dark brown; ventral fur shorter and predominantly gray, often with small white gular and inguinal patches; tail uniform in color or weakly to completely bicolored; ears sparsely covered with either entirely brown or banded hairs, brown basally but white or golden distally. Skull small and delicate (greatest length 27.2–34.8 mm); incisive foramina very short and broad, teardrop in shape (length 3.5–5.4 mm; width 1.9–3.1 mm); interorbital region diverges slightly posteriorly, with supraorbital margins rounded to weakly beaded; palate with complex and multiple posterolateral palatal pits recessed in shallow palatine depressions and with palatal excrescences varying from small projections to large struts fused to maxilla; short and narrow toothrows (length 3.7–5.2 mm; M1 width 1.1–1.6 mm).

DISTRIBUTION: *Hylaeamys megacephalus* has the largest known distribution of all congeners, occurring throughout forested areas of the Cerrado and Chaco in central Paraguay, Cerrado and semideciduous forests in southeastern and central Brazil, and Amazonian forests of eastern Brazil, the Guianan region, and southern Venezuela, as well as forested areas in the Venezuelan Llanos; and on the coastal island of Trinidad. *Hylaeamys megacephalus* ranges in elevation from the Amazonian lowlands, such as Itaituba, Pará state, at 15 m, to the central Brazilian highlands, such as at Brasília, Distrito Federal, at 1,100 m.

SELECTED LOCALITIES (Map 169): BRAZIL: Amapá, Terezinha, Rio Amapari, Serra do Navio (MPEG 15024); Amazonas, Tambor, left bank Rio Jaú (INPA 2014); Ceará, Serra de Maranguape (UFC 74); Maranhão, Alto Parnahyba [= Alto Parnaíba] (FMNH 26447); Mato Grosso, Cidade Laboratório de Humboldt, Aripuanã (MPEG 12647), Juruena (MZUSP M 97.179); Mato Grosso do Sul, Fazenda Califórnia, Morraria do Sul, Bodoquena (MZUSP [APC 731]), Porto Faya (MZUSP 1700); Minas Gerais, Formoso, Parque Nacional Grande Sertão Veredas (MZUSP M 97.140); Pará, Bragança, Santa Maria, Tracuateua [= Tracuatena] (USNM 394180), Itaituba, Rio Tapajós (type locality of *Oryzomys Goeldi* Thomas); São Paulo, Fazenda Sete Lagoas, Mogi Guaçu (MZUSP 25759). FRENCH GUIANA: Cayenne, Paracou (Voss et al. 2001). GUYANA: Barima-Waini, Santa Cruz (ROM 98827). PARAGUAY: Caaguazú, 22.5 km N of Coronel Oviedo by road (UMMZ 124499); Canindeyú, 13.3 km by road N of Curuguaty (type locality of *Mus megacephalus* G. Fischer); San Pedro, Tacuati, Aca-Poi (USNM 293148). SURINAM: Saramacca, La Poule (FMNH 95623). TRINIDAD AND TOBAGO: Trinidad, Princestown (type locality of *Oryzomys velutinus* J. A. Allen and Chapman). VENEZUELA: Amazonas, Cerro Neblina Base Camp, left (west) bank Río Baria [= Río Mawarinuma] (USNM 559193), Tamatama, Río Orinoco (USNM 409873); Sucre, Campo Alegre (type locality of *Oryzomys modestus* J. A. Allen).

SUBSPECIES: As currently understood, *Hylaeamys megacephalus* is monotypic, although molecularly divergent samples from south of the Amazon River (from eastern Pará state, Brazil, to Paraguay), and those from north of the Amazon in central Brazil and the Guianan region (Patton et al. 2000; L. P. Costa 2003) may represent either valid infraspecific groups or require specific recognition.

NATURAL HISTORY: This species inhabits a range of biomes across its distribution but is always restricted to forested habitats (Mares et al. 1986; Ochoa et al. 1993). Specimens have been obtained in well-drained, swampy, or creekside primary forest as well as secondary forest. This species can also be common near human dwellings and rock outcrops (Handley 1976) and in overgrown orchards (Barnett and Cunha 1994). Sites of trap captures include under fallen logs, under roots and buttresses of fallen trees, in tangled branches of fallen trees, at the base of buttressed trees, and in hollow logs (Voss et al. 2001). *Hylaeamys megacephalus* is a predominantly terrestrial species (Nitikman and Mares 1987; Voss et al. 2001) but can be scansorial (Marinho-Filho et al. 1998; Voss et al. 2001), with arboreal captures always in the lower forest strata (<2.0 m; Malcolm 1988, 1991).

In the gallery forests of the Cerrado, Mares and Ernest (1995) reported an increase in population density during the dry season, when densities reached 7.5 individuals/ha, whereas in central Brazilian Amazonia, Malcolm (1990) estimated densities at 0.404 individuals/ha. Pregnant females and reproductively active males have been observed in both dry (May and June) and wet (from September to November) seasons in Cerrado vegetation (Mares and Ernest 1995). In French Guiana, Henry (1994) recorded births and pregnant females year-round, with peaks coincident with the rainy season (April–May). Here, captive animals exhibited a gestation period of 26 days and litter sizes of one to six. Barnett and Cunha (1994) trapped pregnant females in December (dry season) in the transition between the Llanos and Amazon forest in Brazil.

REMARKS: Most of the nomenclatural issues involving this species have been clarified by Musser et al. (1998) and Percequillo (1998), especially the allocation of problematic names such as *Mus capito* Olfers, *Mus cephalotes* Desmarest, *Oryzomys velutinus* J. A. Allen and Chapman, *Oryzomys goeldi* Thomas, and *Oryzomys modestus* J. A. Allen, as well as the priority of *megacephalus* Fischer over that of *capito* Olfers. The names *megacephalus* (Fischer), *capito* (Olfers) and *cephalotes* (Desmarest; see also Desmarest 1804:24) were coined for the species Azara (1801, 1802) referred to as the "*rat second ou rat a grosse téte*" and "*del cola igual al cuerpo*" (respectively).

The concept of *H. megacephalus* as a widely distributed species throughout Amazonia as presented by Musser et al. (1998) in their revision of the complex has been modified by more recent analyses. Molecular sequence studies, for example, documented reciprocal monophyly for samples from eastern and western Amazonia, and to which the names *H. megacephalus* and *H. perenensis* are now applied, respectively (Patton et al. 2000; Voss et al. 2001; Bonvicino and Moreira 2001; L. P. Costa 2003; Musser and Carleton 2005). Morphometric, qualitative morphological, and karyological data also support this hypothesis. *Hylaeamys megacephalus* is smaller in size (head and body length 80–158 mm); has a smaller skull (greatest skull length 27.2–34.9 mm); has short, teardrop-shaped incisive foramina (length 3.4–5.4 mm); has a shorter maxillary toothrow (length 3.7–5.2 mm), and a karyotype with $2n = 54$ and $FN = 62$ (see account of *H. perenensis* for comparisons). However, even within this re-

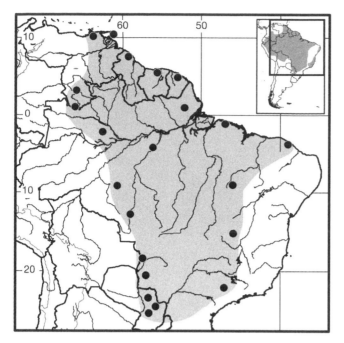

Map 169 Selected localities for *Hylaeamys megacephalus* (●). Contour line = 2,000 m.

stricted concept of *H. megacephalus*, molecular clades separated by the Amazon River are reciprocally monophyletic, suggesting the possibility that further taxonomic divisions of this species may be warranted.

There are few records of this species from eastern coastal Brazil, except for those from Amapá, Maranhäo, and Ceará states. The eastern edge of the range coincides with the limits of the Cerrado biome. There are no definitive records of this species in western Amazonia (see also Patton et al. 2000), and the western distributional limits of this common species remain unclear as well. Musser et al. (1998) suspected that contact between *H. perenensis* and *H. megacephalus* might be found along the lower Rio Tapajós, and also stated that more samples from this region should be examined to clearly evaluate the possibility that this river may represent the boundary between these forms. Given the degree of overall morphological similarity between *H. megacephalus* and *H. perenensis*, additional collections should provide for both karyotypic and molecular data. The karyotype of *H. megacephalus* consists of 2*n* = 54, FN = 62 (Musser et al. 1998:Fig. 11); that of *H. perenensis* is 2*n* = 52, Fn = 62 (see that account).

Hylaeamys oniscus (Thomas, 1904)
Northern Atlantic Forest Hylaeamys

SYNONYMS:

Oryzomys oniscus Thomas, 1904b:142; type locality "São Lourenço [= São Lourenço da Mata]," Pernambuco, Brazil.

[*Oryzomys (Oryzomys)*] *oniscus*: Tate, 1932e:18; name combination.

Oryzomys laticeps: Hershkovitz, 1960:544, footnote; part (*oniscus* Thomas listed as synonym); not *Mus laticeps* Lund (= *Hylaeamys laticeps*).

Oryzomys [(*Oryzomys*)] *capito oniscus*: Cabrera, 1961: 387; name combination.

Oryzomys capito: Hershkovitz, 1966a:137, footnote; part; not *Mus capito* Olfers (= *Hylaeamys megacephalus* [G. Fischer]).

[*Hylaeamys*] *oniscus*: Weksler, Percequillo, and Voss, 2006: 14; first use of current name combination.

DESCRIPTION: Large, head and body 140–162 mm (mean 147.2, N = 8); tail equal to or shorter than head and body (range: 135–162 mm; mean 146.8, N = 8); hindfeet robust, with fleshy and large plantar pads and sparse ungual tufts shorter than claws. Dorsal fur short and harsh, color overall yellowish to buffy heavily ticked with dark brown; ventral fur short and predominantly grayish; tail bicolored basally and uniformly colored distally; ears sparsely covered with reddish-brown hairs (more so dorsally) or yellowish (more so ventrally). Skull large and robust (skull length 33.7–37.6 mm); teardrop-shaped incisive foramina long and wide (length 4.9–5.8 mm; width 2.3–3.4 mm); slightly and posteriorly divergent interorbital region with squared to beaded supraorbital margins; palate with complex and multiple posterolateral palatal pits recessed in shallow to moderately deep palatine depressions, and palatal excrescences varying from small projections to large and robust bone struts fused to maxillae; and toothrows long and wide (length 4.9–5.3 mm; M1 width 1.4–1.6 mm).

DISTRIBUTION: *Hylaeamys oniscus* is endemic to the northeastern Atlantic Forest of Brazil, with all known localities north of the Rio São Francisco in the states of Alagoas, Pernambuco, and Paraíba (see Brennand et al. 2013).

SELECTED LOCALITIES (Map 168): BRAZIL: Alagoas, Fazenda do Prata, 13 km SSW of São Miguel dos Campos (MNRJ 30595), Fazenda Santa Justina, 6 km SSE of Matriz de Camaragibe (UFPB 978); Paraíba, Fazenda Pacatuba, 10 km NE of Sapé (UFPB 1931); Pernambuco, São Lourenço da Mata (type locality of *Oryzomys oniscus* Thomas).

SUBSPECIES: *Hylaeamys oniscus* is monotypic.

NATURAL HISTORY: No data are available on the natural history of *H. oniscus*, although all known localities are in typical coastal Atlantic rainforest.

REMARKS: *Hylaeamys oniscus* has been considered as either a valid taxon, as a species or subspecies, or as a synonym of *H. megacephalus*. More recently, this species was considered to be morphologically, morphometrically (Musser at al. 1998), and karyologically (Andrades-Miranda, Zanchin et al. 2002) indistinguishable from the eastern populations of *H. laticeps* (as defined previously), and therefore as a junior synonym because *laticeps* Lund, 1840, has priority over *oniscus* Thomas, 1904. However, I regard *Hylaeamys oniscus* as a valid species, distinguishable from *H. laticeps* by its larger size (mean head and body length 147 mm versus 136 mm, respectively; mean greatest skull length 35.8 mm versus 34.3 mm; mean diastema length 9.8 mm versus 8.8 mm), by shorter toothrows and narrower molars (mean maxillary toothrow length 5.08 mm versus 5.14 mm; mean M1 width 1.45 mm versus 1.53 mm), and by a $2n = 52$, $FN = 62$ karyotype (specimens from the type locality; Maia 1990). Specimens with a karyotype of $2n = 48$ and $FN = 60$–64, and identified by Andrades-Miranda, Zanchin et al. (2002) as "*Oryzomys oniscus*," I assign to *Hylaeamys laticeps* (see account earlier).

Hylaeamys perenensis (J. A. Allen, 1901)
Western Amazonian Hylaeamys

SYNONYMS:

Oryzomys perenensis J. A. Allen, 1901c:406; type locality "Perené, Department of Junin, Peru; altitude 800 m," defined as "Valle Perené, Colonia del Perené; a coffee plantation at junction of ríos Paucartambo and Chanchamayo, 1000 m" (Stephens and Traylor 1983:161).

[*Oryzomys* (*Oryzomys*)] *perenensis*: E.-L. Trouessart, 1904: 419; name combination.

Oryzomys laticeps perenensis: Gyldenstolpe, 1932a:18; part; name combination.

Oryzomys [(*Oryzomys*)] *capito nitidus*: Cabrera, 1961:386; part; but neither *Mus capito* Olfers (= *Hylaeamys megacephalus* [G. Fischer]) nor *Hesperomys laticeps*, var. *nitidus* Thomas (= *Euryoryzomys nitidus*).

Oryzomys megacephalus: Musser, Carleton, Brothers, and Gardner, 1998:251; part (inclusion of samples from western Amazonia).

[*Hylaeamys*] *perenensis*: Weksler, Percequillo, and Voss, 2006:14; first use of current name combination.

DESCRIPTION: Medium sized (head and body length 81–169 mm; mean 135.7, N = 337); tail shorter than head and body length (tail 94–154 mm; mean 122.9, N = 339); hindfeet moderately robust, with medium-sized hypothenar pads. Coloration of dorsal pelage ochraceous buffy weakly to intensely ticked with dark brown or reddish brown; ventral pelage predominantly gray; dorsal surfaces of hindfeet grayish to whitish; ungual tufts short, whitish in color, but sparse; tail unicolored or only weakly bicolored. Skull large and robust (greatest length 29.5–37.9 mm); incisive foramina long, narrow, teardrop-shaped (length 3.6–6.0 mm;

width 1.5–2.9 mm); molars long and robust (maxillary toothrow length 4.5–5.6 mm); roof of mesopterygoid fossa completely ossified (in more than 90% of specimens from Rio Juruá in Amazonas and Acre states, Brazil; Patton et al. 2000). Lower second molars with distinct entoflexid, entofossetid, and short hypoflexid in most specimens.

DISTRIBUTION: *Hylaeamys perenensis* ranges through lowland and lower montane rainforests of the western Amazon Basin, from eastern Colombia south through Ecuador and Peru to northeastern Bolivia, and extending into western Brazil. Extent of the range to the east is uncertain, but it is possibly bounded by the Río Putumayo in the north and the Rio Madeira–Rio Guaporé in the south. The Andean Cordillera forms the western limits of the range for *H. perenensis*; there are no known trans-Andean records. The species ranges in elevation from about 65 m (Lago Vai Quem Quer, Amazonas, Brazil) to 1,000 m (Mámbita, Cundinamarca, Colombia).

SELECTED LOCALITIES (Map 167): BOLIVIA: Pando, La Cruz (AMNH 262937). BRAZIL: Acre, Sena Madureira, km 8 on route BR 364 between Sena Madureira and Manuel Urbano (USNM 545295), Sobral, left bank Rio Juruá (MPEG 28312); Amazonas, Colocação Vira-Volta, left bank Rio Juruá (MVZ 190530), Seringal Condor, left bank Rio Juruá (MVZ 190485); Rondônia, Cachoeira Nazaré (MPEG 20811). COLOMBIA: Cundinamarca, Mámbita (AMNH 70526). ECUADOR: Orellana, San José Abajo (AMNH 68050); Pastaza, Sarayacu, Río Bobonaza (AMNH 67352); Zamora-Chinchipe, Zamora (AMNH 36577). PERU: Amazonas, Huampami, Río Cenepa (Patton et al. 2000); Junín, Perené (type locality of *Oryzomys perenensis* J. A. Allen); Loreto, Reserva Nacional Allpahuayo-Mishana (Hice and Velazco 2012); Ucayali, Balta, Río Curanja (A. L. Gardner and Patton 1976, as *Oryzomys capito*).

SUBSPECIES: *Hylaeamys perenensis* is monotypic.

NATURAL HISTORY: *Hylaeamys perenensis* is found in seasonally inundated floodplain forests (*várzea*), nonflooded upland forests (*terra firme*), river beach grass, *Cecropia* stands, flooded and nonflooded grasslands, *Heliconia* thickets, active and abandoned garden plots, and second-growth forests (Woodman et al. 1995; Patton et al. 2000). Exclusively terrestrial, this species, where it occurs, may be the most abundant rodent both in *terra firme* and *várzea* forests, but has been trapped more commonly in *várzea* habitat (Patton et al. 2000). On the Rio Juruá in western Brazil, *H. perenensis* was reproductively active throughout the year "spanning both dry and rainy seasons" (Patton et al. 2000:153), where modal litter size was four, with embryo counts two to five. In Bolivia, the modal number was three embryos, also with a range of two to five (S. Anderson 1997).

REMARKS: *Hylaeamys perenensis* has been treated as a subspecies of what is now called *Hylaeamys laticeps* (Gyldenstolpe 1932; Ellerman 1941), as a synonym of *Euryoryzomys nitidus* (Cabrera 1961), or as part of *Hylaeamys megacephalus* (Musser and Carleton 1993). Musser et al. (1998) did demonstrate differences between *H. perenensis* and *H. megacephalus*, but adopted a conservative taxonomic posture and maintained both western (*H. perenensis*) and eastern Amazonian samples under the same species *H. megacephalus*. Their decision was based on the absence of evidence of a geographical barrier precluding gene flow between populations (Musser et al. 1998:41–42). However, Patton et al. (2000; see also L. P. Costa 2003) presented molecular evidence to sustain the morphological and karyological differentiation recognized by Musser et al. (1998), and I treat *H. perenensis* as a valid species based on these morphometric and molecular analyses.

Specimens of *H. perenensis* can be distinguished from those of *H. megacephalus* by their larger size, larger skull, longer teardrop-shaped incisive foramina, more robust molars, and a $2n=52$, FN$=62$ karyotype (A. L. Gardner and Patton 1976, as *Oryzomys capito*; Patton et al. 2000). Moreover, DNA sequences (Patton et al. 2000; L. P. Costa 2003; Emmons and Patton 2005) support a sister relationship between *H. perenensis* and *H. acritus*, which in turn are coupled to *H. laticeps* and then, finally, to *H. megacephalus*.

Hylaeamys tatei (Musser, Carleton, Brothers, and Gardner, 1998)

Tate's Hylaeamys

SYNONYMS:

Oryzomys tatei Musser, Carleton, Brothers, and Gardner, 1998:100; type locality "Palmera (01°25′S/78°12′W; . . .), at 4000 ft (1220 m), Provincia del Tungurahua, Ecuador."

[*Hylaeamys*] *tatei*: Weksler, Percequillo, and Voss, 2006:14; first use of current name combination.

DESCRIPTION: Large (mean head and body length 130.2 mm, range: 118–137 mm); pelage dark; incisive foramina short and narrow relative to size of cranium; toothrows long and wide; molars robust; coronoid process blunt and triangular in shape. Distinguishable from congeners by having long and deep mesoflexus on M1 and M2 (moderate wear forms either long medial mesofossette or small labial and long medial mesofossettes), and robust ectolophid on m1 and m2.

DISTRIBUTION: *Hylaeamys tatei* is known from only three, closely adjacent localities along the upper Río Pastaza on the eastern slope of the Ecuadorean Andes (Musser et al. 1998:106, Fig. 47).

SELECTED LOCALITIES (Map 170): ECUADOR: Tunguruhua, Palmera (type locality of *Oryzomys tatei* Musser, Carleton, Brothers, and Gardner).

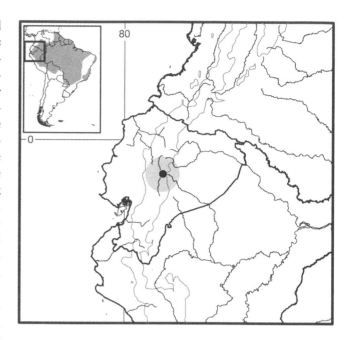

Map 170　Selected locality for *Hylaeamys tatei* (●). Contour line = 2,000 m.

SUBSPECIES: *Hylaeamys tatei* is monotypic.

NATURAL HISTORY: The region where all specimens of *H. tatei* were captured is characterized by a mosaic of "paramo or temperate plants" and "subtropical vegetation" (G. H. H. Tate notes cited in Musser et al. 1998:103), cultivated lands, second-growth forest, and primary forest (Rageot and Albuja 1994) between 1,159 and 1,524 m. Musser et al. (1998) also suggested that *H. tatei* could be a wet and cool premontane evergreen forest inhabitant of the eastern slopes of Ecuador and Peru, because four other broadly distributed species of this general habitat (*Nephelomys auriventer*, *Oreoryzomys balneator*, *Chilomys instans*, and *Akodon aerosus*) were collected with *H. tatei*.

REMARKS: *Hylaeamys tatei* is easily distinguished from congeners, but especially from *H. yunganus*, by its large body size and cranial dimensions, shorter incisive foramina (length 4.2–4.9 mm in *H. tatei*; length 3.9–5.8 mm in *H. yunganus*), and longer and wider maxillary toothrow (length 5.5–5.7 mm and breadth 1.7–1.9 mm in *H. tatei*; length 3.9–5.5 mm and breadth 1.3–1.7 mm in *H. yunganus*).

Hylaeamys yunganus (Thomas, 1902)

Amazonian Hylaeamys

SYNONYMS:

Oryzomys yunganus Thomas, 1902b:130; type locality "Charuplaya, 1350 m," stated by Thomas (1902b:126) as being "on the [Río] Securé, just north of 16 S.," Cochabamba, Bolivia.

[*Oryzomys* (*Oryzomys*)] *yunganus*: E.-L. Trouessart, 1904: 419; name combination.

Oryzomys [(*Oryzomys*)] *capito yunganus*: Cabrera, 1961: 387; name combination.

[*Hylaeamys*] *yunganus*: Weksler, Percequillo, and Voss, 2006:14; first use of current name combination.

DESCRIPTION: Moderate sized with tail equal to or shorter than head and body length (head and body length 115–149 mm; tail length 107–128 mm). Dorsal fur dense and velvety; hairs longer than sympatric *H. megacephalus* (mean sample length 7.1–8.4 mm versus 5.4–6.6 mm); upper parts dark brownish tawny to brownish black; under parts dark grayish white; tail monocolored or with ventral basal half either unpigmented or mottled; dorsal surfaces of front and hindfeet whitish; claws sparsely covered by ungual tufts; plantar surfaces of hindfeet lack hypothenar pads in some specimens; superciliary, genal, and mystacial vibrissae not exceptionally long, extending only to ears. Cranial size and shape similar to that of *H. megacephalus* and *H. perenensis*, but zygomatic plates much wider and lengths of bony palate, incisive foramina (4.3–5.6 mm), and molar rows (4.8–5.5 mm) each longer relative to occipitonasal length. Posterior margins of dentary shallowly concave between condyloid and angular processes (compared with more deeply concave margins in *H. megacephalus* and *H. perenensis*). Molars chunky and wide (M1 width 1.5–1.6 mm); M2 with short paraflexus, labial and medial fossettes present; m2 with conspicuous entoflexid, short hypoflexid that extends only to medial plane of tooth, and single fossetid.

DISTRIBUTION: *Hylaeamys yunganus* is widely distributed, with a range nearly concordant with the lowland Amazonian rainforest in Brazil, Bolivia, Peru, Ecuador, Colombia, Venezuela, Guyana, Surinam, and French Guiana (Musser et al. 1998:47, Fig. 15). Across this broad region, *H. yunganus* also occupies a wide elevational range, from sea level in French Guiana (Île de Cayenne) to nearly 2,000 m on the eastern Andean slopes of Peru (Yambrasbamba).

SELECTED LOCALITIES (Map 171): BOLIVIA: Cochabamba, Charuplaya (BM 2.1.1.39); Pando, Bella Vista, 10 km SSW of Mapiri (BM 1.1.1.65); Santa Cruz, 4.5 km N and 1.5 km E of Cerro Amboró, Río Pitasama (AMNH 262079). BRAZIL: Amapá, Serra do Navio, Macapá (USNM 392065); Mato Grosso, Rio Saueniná (AMNH 37112), 264 km N of Xavantina, Serra do Roncador (BM 81.470); Pará, Agrovila da União, 18 km S and 19 km W of Altamira (USNM 521519); Rondônia, Rio Roosevelt, left bank upper Rio Aripuanã (AMNH 37113). COLOMBIA: Cundinamarca, Guaicáramo (AMNH 71328); Putumayo, Río Macaya, at Río Caquetá (FMNH 72067). ECUADOR: Napo, Volcán Sumaco (AMNH 68118); Zamora-Chinchipe, Zamora (AMNH 47830). French Guiana: Île de Cayenne (MNHN 1986-322). Guyana: Barima-Waini,

Santa Cruz (ROM 98719). PERU: Amazonas, Yambrasbamba (BM 26.8.6.14); Huánuco, Chinchao (FMNH 23721); Puno, Río Inambari, probably near Oroya (BM 1.1.1.26); Ucayali, San José, on Río Santa Rosa (LSUMZ 16685). SURINAM: Marowijne, Powakka, ca. 50 km S of Paramaribo (CM 54048). VENEZUELA: Amazonas, Cerro Duida, Agüita Camp (AMNH 77314); Bolívar, Auyántepuí (AMNH 130906).

SUBSPECIES: Despite a wide geographic range and demonstrable variation (see Remarks), *Hylaeamys yunganusis* currently regarded as monotypic.

NATURAL HISTORY: *Hylaeamys yunganus* inhabits flooded (*várzea*) and nonflooded (*terra firme*) forests (Musser at al. 1998; Patton et al. 2000; Voss et al. 2001). There is some apparent geographical variation in habitat preferences. In the western Amazon of Brazil, Patton et al. (2000) trapped specimens of *H. yunganus* in primary and secondary flooded and nonflooded forests as well as in abandoned and active garden plots and *Heliconia* thickets. In French Guiana, Voss et al. (2001) reported that the species was limited to primary forest, most frequently in moister sites such as swampy and creekside forest formations where specimens were captured under or beside logs, around the bases of trees, in dense undergrowth unsheltered by woody objects, in tangled dead branches, among fallen palm fronds, and among the stilt roots of standing trees. This difference in habitat utilization may involve competitive interactions with congeners, for in western Brazil, Patton et al. (2000) found *H. yunganus* more frequently associated with *terra firme* forests than was sympatric *H. perenensis* (see above). *Hylaeamys yunganus* has been captured only in traps set at ground level and never in lianas above the ground as has *H. megacephalus*. Capture data also suggest that this species is less abundant than sympatric *H. perenensis* or *H. megacephalus* in the western and eastern Amazon Basin, respectively (Patton et al. 2000; Voss et al. 2001).

Reproductively active males (those with scrotal testes) and females (pregnant, lactating, or postlactating) have been recorded throughout the year in the western Amazon; the modal litter size was two embryos (range: 1–4; Patton et al. 2000). When compared with the sympatric and syntopic *H. perenensis*, *H. yunganus* had smaller modal litter sizes, lower pregnancy rates, and apparently delayed parturition, on average. These features may be related to differences in abundance observed in the field among sympatric species of *Hylaeamys*.

REMARKS: *Hylaeamys yunganus* has been recognized as a valid form since its description (Thomas 1902b), being only provisionally treated as a subspecies of *Hylaeamys megacephalus* by Cabrera (1961). In external morphology, *H. yunganus* is very similar to sympatric congeners, namely, *H. megacephalus* in eastern Amazonia and *H. perenensis* in western Amazonia. The most conspicuous external character distinguishing *H. yunganus* is the reduction in size or absence of the hypothenar pad on the hindfoot. However, given substantial individual variation of size and presence, this feature lacks utility as an absolutely diagnostic trait (Musser et al. 1998:63, Table 7). In cranial morphology, *H. yunganus* can be told readily from both *H. megacephalus* and *H. perenensis* by the presence of two enamel islands and a short paraflexus on M2, the most reliable character state that distinguishes these species.

Musser et al. (1998:269–271) discussed problems associated with the placement of the type locality. In Brazil, distributional records are widely scattered in the lowland forests of the Amazon Basin. In Bolivia, Peru, and Ecuador the species ranges from lowland rainforest to montane forests in the foothills of the Andes. In Colombia, all records are associated with the narrow area of lowland forest along the eastern base of the Andes, in the headwaters of left bank tributaries of the Río Orinoco. There are no known records of *H. yunganus* in eastern Colombia along the middle and lower reaches of Orinocan tributaries. All collecting sites in Venezuela are from the table mountains (*tepuis*) in the southern and eastern parts of the country.

Hylaeamys yunganus exhibits marked geographical variation in cranial dimensions, with samples from western Amazonia and from the Venezuelan *tepuis* having larger

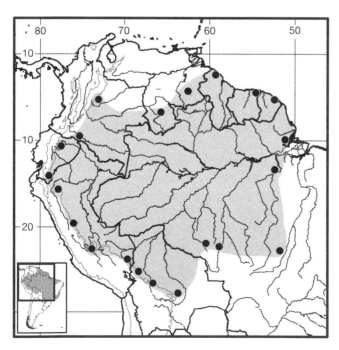

Map 171 Selected localities for *Hylaeamys yunganus* (●). Contour line = 2,000 m.

skulls than those from eastern Amazonia and elsewhere in the Guianan region (Musser et al. 1998). This species is also characterized by chromosomal polymorphism both within and between western ($2n = 58$–60; FN = 62–66; A. L. Gardner and Patton 1976; Patton et al. 2000) and eastern populations ($2n = 52$–59; FN = 64–67; Musser et al. 1998). Whether these differences warrant taxonomic recognition must remain for future analyses.

Genus *Lundomys* Voss and Carleton, 1993
Robert S. Voss

The genus *Lundomys* contains a single species, *L. molitor*, which now occurs only in southern Brazil (Rio Grande do Sul) and Uruguay. Previously known as *Holochilus magnus* and thought to be a "sigmodont" (closely related to *Sigmodon*; Hershkovitz 1955b), this taxon was referred to the tribe Oryzomyini by Voss and Carleton (1993) on the basis of derived cranial and visceral characters. Recent phylogenetic analyses (e.g., Weksler 2003, 2006; Parada et al. 2013) unequivocally support the hypothesis that *Lundomys* and *Holochilus* are oryzomyines (not closely related to *Sigmodon*), although they are not consistently recovered as sister taxa (contra Pardiñas and Teta 2011b). Pleistocene and/or early Holocene fossils have been reported from central Brazil (Minas Gerais; Voss and Carleton 1993), Uruguay (Ubilla et al. 2004), and northern Argentina (Lezcano et al. 1992; Pardiñas and Lezcano 1995; Pardiñas and Deschamps 1996; Teta and Pardiñas 2006; Pardiñas and Teta 2011b). Except as noted otherwise here, this account is based on Voss and Carleton (1993), which should be consulted for additional details concerning morphological characters, taxonomy, distribution, and natural history.

Lundomys molitor is a large rat (mature adults probably weigh >300g) with long, dense, soft, yellowish-brown dorsal pelage and buffy under parts. The ears are small and well provided with short brownish or yellowish fur (not contrasting in color with the fur of the head). Mystacial, superciliary, genal, submental, interramal, and carpal vibrissae are present; the mystacial hairs are short and do not extend posteriorly behind the pinnae when laid back against the head. The manus is unremarkable, but the huge (58–68 mm) hindfoot has naked claws (not concealed by long tufts of ungual hairs), conspicuous interdigital webbing, and dense fringes of long silvery hairs along the plantar margins; the plantar surface is hairless and densely squamate, with one metatarsal and four very small interdigital pads (the hypothenar pad is absent); the claw of pedal digit V extends to the end of phalanx 1 of digit IV. There are eight mammae in inguinal, abdominal, postaxial, and pectoral pairs. The sparsely haired tail (with clearly visible epidermal scales) is unicolored and much longer than the combined length of head and body.

The skull in dorsal view has a blunt, massive rostrum flanked by deep zygomatic notches; the interorbital region is narrow with sharp (but not beaded) supraorbital margins; and the braincase lacks well-developed temporal crests. The zygomatic plates are very broad, with concave anterior margins and spinous anterodorsal processes; the jugals are present and consistently separate the maxillary and squamosal zygomatic processes; and there is a well-developed vertical ridge along the frontosquamosal suture in the back of the orbit. Incisive foramina are long, extending posteriorly to or between the first molar alveoli; the palatal bridge is long and narrow, with prominent posterolateral pits; the parapterygoid fossae are about as wide as the mesopterygoid fossa; and the roof of the mesopterygoid fossa is completely ossified (or perforated by very small sphenopalatine vacuities). Diagnostic features of the basicranium and ear region include the absence of an alisphenoid strut (the buccinator-masticatory foramen and the foramen ovale are confluent); a derived carotid arterial circulation (pattern 3 of Voss 1988); and a tegmen tympani that is not connected to a posterior suspensory process of the squamosal. The auditory bullae are globular, with short and narrow bony tubes, and although the petrosal is broadly exposed posteromedially between the ectotympanic and the basioccipital, it does not extend anteriorly to the carotid canal; above each auditory bulla, the lateral surface of the braincase is perforated by a large postglenoid foramen and a subequally sized subsquamosal fenestra. The coronoid process of the mandible is about level with the condyle; the tip of the angular process does not extend posteriorly much behind the condylar process; the superior and inferior masseteric crests converge anteriorly as an open chevron (not tightly compressed to form a single anterior ridge); and the capsular process is indistinct.

The upper incisors are opisthodont, with smoothly rounded (not faceted) yellow-orange enamel bands. The upper molars are bunodont (not planar), with rounded (rather than angular) labial and lingual cusp margins. The anterocone of M1 is weakly divided (when unworn) into subequal labial and lingual lobes by a shallow anteromedian flexus; an anteroloph is present but small on unworn M1 and well developed on M2 and M3; a shallow protoflexus is present on unworn M2. A small mesoloph is always present on M1 and M2, but it is never fused with a mesostyle on the labial cingulum; no mesoloph is present on M3. All upper molars have three roots. The anteroconid of m1 is entire (not divided by an anteromedian flexid), but it encloses a large enamel pit; minute anterolophids are present on all unworn lower molars; small mesolophs are present on m1 and m2

but not on m3; and discrete posterolophids are present and persistent on m1 and m2. The lower first molar has three or four roots, but m2 and m3 have two roots each.

The axial skeleton includes 12 ribs, 19 thoracolumbar vertebrae, 4 sacral vertebrae, and 35–36 caudal vertebrae. The tuberculum of the first rib articulates with the transverse processes of the seventh cervical and first thoracic vertebrae, and the second thoracic vertebra has a greatly elongated neural spine. The entepicondylar foramen of the humerus is absent.

The stomach is unilocular and hemiglandular, and the gall bladder is absent. Other visceral and genital characters are unknown.

SYNONYMS:

Hesperomys: Winge, 1887:14; part (description of *molitor*); not *Hesperomys* Waterhouse.

Holochilus: Hershkovitz, 1955b:657; part (description of *magnus*); not *Holochilus* Brandt.

Lundomys Voss and Carleton, 1993:5; type species *Hesperomys molitor* Winge, by original designation.

Lundomys molitor (Winge, 1887)
Lund's Water Rat

SYNONYMS:

Hesperomys molitor Winge, 1887:14; type locality given originally as "Lapa do Capão Secco [= Lapa do Capão Seco], da Escrivania [= Escrivânia] Nr. 5 og da Serra das Abelhas"; fixed by lectotype designation (Voss and Myers 1993) to "Lapa da Escrivania [= Escrivânia] Nr. 5, "a cave excavated by P. W. Lund near Lagoa Santa, Minas Gerais, Brazil.

Oryzomys molitor: E.-L. Trouessart, 1897:528; name combination.

Holochilus magnus Hershkovitz, 1955b:657; type locality "Paso de Averías, Río Cebollati, about 40 km south of Trienta y Tres [Lavalleja], eastern Uruguay."

[*Calomys*] *molitor*: Hershkovitz, 1962:123; name combination.

Holochilus molitor: Massoia, 1980c:282; name combination.

Lundomys molitor: Voss and Carleton, 1993:5; first use of current name combination.

DESCRIPTION: As for the genus.

DISTRIBUTION: All known Recent specimens of *Lundomys molitor* have been collected in Uruguay and in the adjacent Brazilian state of Rio Grande do Sul.

SELECTED LOCALITIES (Map 172): BRAZIL: Rio Grande do Sul, Tupanciretã (T. R. O. Freitas, Mattevi, Oliveira et al. 1983). URUGUAY: Canelones, Bañado de Tropa Vieja (AMNH 206363); Lavalleja, Paso Averías, Río Cebollatí (type locality of *Holochilus magnus* Hershkovitz); Soriano, 3 km E of Cardona (AMNH 206368); Treinta y Tres, 8 km E of Treinta y Tres (FMNH 29260).

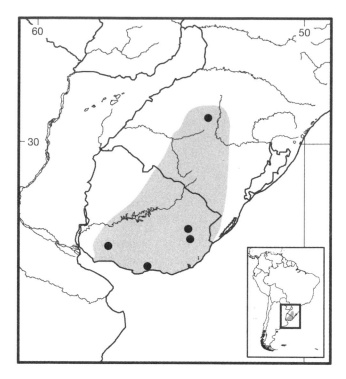

Map 172 Selected localities for *Lundomys molitor* (●).

SUBSPECIES: I treat *Lundomys molitor* as monotypic.

NATURAL HISTORY: *Lundomys molitor* has been collected in freshwater marshes and along streams bordered by grass and brush. Throughout its range, *L. molitor* is sympatric (and perhaps syntopic) with *Holochilus brasiliensis*, an externally similar semiaquatic species with which it was formerly confused (Hershkovitz 1955b). Scant information from several sources suggests that *L. molitor* is nocturnal, herbivorous, and builds spherical elevated nests supported by and constructed from reeds (Barlow 1969; Sierra de Soriano 1969; Voss and Carleton 1993).

REMARKS: Extralimital fossil remains of *Lundomys molitor* (from Argentina and central Brazil) have been interpreted as evidence for geographic range shifts correlated with Quaternary climatic fluctuations—northward into the tropics during cool episodes (Voss and Carleton 1993) and southward into temperate regions during warmer periods (Teta and Pardiñas 2006; Pardiñas and Teta 2011b).

Genus *Melanomys* Thomas, 1902

Marelo Weksler and Simone Lóss

Members of the genus *Melanomys*, generally known as dusky rice rats, are found in lowland or lower montane forested habitats in Central America from Honduras to Panama (Hall 1981; F. A. Reid 1997), and south to north-

western South America, in Colombia, Ecuador, Peru, and westernmost Venezuela. We recognize six species in the genus: *M. caliginosus*, *M. columbianus*, *M. robustulus*, *M. zunigae*, *M. chrysomelas*, and *M. idoneus*. The last two species are currently found only in Central America and not included in this account. The genus is poorly known, as the last review was that of J. A. Allen (1913b). The following morphological description is based on the analyses of specimens deposited in the AMNH and USNM collections, including most holotypes, but also includes information from J. A. Allen (1913b), Goldman (1918), Hooper and Musser (1964; glans penis), Carleton (1973; stomach), Voss and Linzey (1981; male accessory glands), Voss (1991a; gall bladder), Weksler (2006), and Lóss and Weksler (unpubl. data).

Melanomys are medium-sized (head and body length, 100–150 mm), vole-like rodents with dorsal pelage dark brown, and ventral pelage grayish or yellowish brown; the limits of dorsal and ventral colors are subtle or indistinct. The body pelage is short and close in coastal (lowland) populations, or long, dense, and soft in interior (mountain) forms. Pinnae are small and slightly hairy, not contrasting in color with the dorsal fur. The mystacial vibrissae are short and sparse, extending only to base of pinnae when laid back against cheek. The manus and pes are covered dorsally with short and dark hairs, and the manual claws are long and keeled. The hindfeet do not have natatory fringes or interdigital webs, and have short and sparse tufts of ungual hairs at bases of claws on digit II through digit V; the plantar surface is densely covered with distinct squamae distal to the thenar pad; the hypothenar pad is absent or vestigial, and the interdigital pads are small and fleshy, with pads 1 and 4 displaced proximally relative to 2 and 3. The claw of digit I extends to the middle of phalanx 1 of digit II, and claw of digit V extends to the middle of phalanx 1 of digit IV. The tail is unicolored and dark, sparsely haired, and covered with more or less conspicuous epidermal scales; it is much shorter than the combined length of head and body and does not have a tuft of terminal long hairs. The mammary complement consists of eight teats in inguinal, abdominal, postaxial, and pectoral pairs.

The skull has a robust rostrum flanked by moderate to deep zygomatic notches. The interorbital region is convergent anteriorly, with weakly beaded supraorbital margins. The braincase is rounded, without well-developed temporal crests; the lambdoidal and nuchal crests are weakly developed in older adults. The posterior margins of the zygomatic plates are situated anterior to the M1 alveoli in most specimens, but approximately even with M1 in others. The zygomatic plates lack anterodorsal spinous processes. Jugals are present, but they are very small (maxillary and squamosal zygomatic processes often overlap in lateral view but rarely are in contact). Lacrimals have a longer maxillary than frontal suture. The frontosquamosal suture is placed either anterior to or continuous with the frontoparietal suture. Parietals have broad lateral expansions. The incisive foramina are short, not extending beyond the M1 alveoli, either with slightly posteriorly divergent or parallel margins. The bony palate between the molar rows is smooth and long; the mesopterygoid fossa penetrates anteriorly between maxillae but not between molar rows. Posterolateral palatal pits are variable in size, number, and morphology but are usually contained within deeply recessed fossae. The bony roof of mesopterygoid fossa is completely ossified or perforated by extremely reduced (slit-like) sphenopalatine vacuities. An alisphenoid strut is absent (buccinator-masticatory foramen and accessory foramen ovale are confluent). The alisphenoid canal has a large anterior opening. The stapedial foramen and posterior opening of the alisphenoid canal are small; a squamosoalisphenoid groove and sphenofrontal foramen are absent; and a secondary anastomosis crosses the dorsal surface of pterygoid plate (characters of carotid circulatory pattern 3 of Voss 1988). A posterior suspensory process of the squamosal is absent. The postglenoid foramen is large and rounded, and the subsquamosal fenestra is absent or vestigial. The periotic is exposed posteromedially between the ectotympanic and basioccipital, usually not extending anteriorly to the carotid canal; the mastoid usually is completely ossified, not fenestrated. A capsular process is present but not well developed in most fully adult specimens. The basihyal is without entoglossal process.

Upper incisors have smoothly rounded enamel bands. The maxillary toothrows are parallel; the molars are bunolophodont and with labial flexi enclosed by a cingulum. Labial and lingual flexi of M1 and M2 meet at the midline, with overlapping enamel. The anterocone of M1 is not divided into labial and lingual conules (anteromedian flexus absent), the anteroloph is fused with the anterocone labially, and an anteroflexus is present as a small fossette. M1 lacks a protostyle. Mesolophs are present on all upper molars, the paracone is connected by an enamel bridge to an anterior moiety of the protocone, and the median mure is not connected to the protocone. The protoflexus is absent from M2, but a mesoflexus is present, either as a single internal fossette or divided into labial and medial fossettes. The paracone of M2 lacks an accessory loph. M3 also lacks a posteroloph, but it possesses a diminutive hypoflexus (which tends to disappear with moderate to heavy wear). An accessory labial root of M1 is present. The anteroconid of m1 is usually without an anteromedian flexid. An anterolabial cingulum is present on all lower molars.

An anterolophid is present on m1 but absent on m2 and m3, an ectolophid is absent on m1 and m2, the mesolophid is distinct on unworn m1 and m2, and a posteroflexid is present on m3. Accessory lingual and labial roots are present on m1; m2 and m3 each have one large anterior root and one large posterior root.

The first rib has a double articulation. The humerus lacks an entepicondylar foramen, but has a supratrochlear foramen. There are 12 ribs. The fifth lumbar (17th thoracolumbar) vertebra has a vestigial anapophysis. A posterior spinous process is present on the hemal arch at the junction of the second and third caudal vertebrae.

The stomach is unilocular-hemiglandular, without extension of glandular epithelium into the corpus. The gall bladder is absent. One pair of preputial glands is present. The distal bacular cartilage of the glans penis is small and trifid (with a short and slender central digit); the nonspinous tissue on the rim of the terminal crater does not conceal the bacular mounds; the dorsal papillae are spineless; and the urethral processes are without subapical lobules.

SYNONYMS:

Hesperomys: Tomes, 1860b:263; part (description of *caliginosus*); not *Hesperomys* Waterhouse.

Hesperomys (*Habrothrix*): J. A. Allen, 1891b:210; part (listing of *caliginosus*); not *Hesperomys* Waterhouse nor *Habrothrix* Waterhouse.

Oryzomys: Thomas, 1894b:355; part (description of *phaeopus*); not *Oryzomys* Baird.

Akodon: E.-L. Trouessart, 1897:537; part (listing of *caliginosus*); not *Akodon* Meyen.

Oryzomys (*Zygodontomys*): Bangs, 1900a:95; part (listing of *obscurior* as subspecies of *phaeopus*); not *Oryzomys* Baird nor *Zygodontomys* J. A. Allen.

Oryzomys (*Melanomys*) Thomas, 1902f:248; type species of new subgenus *Oryzomys phaeopus* Thomas (= *Hesperomys caliginosus* Tomes).

Melanomys: J. A. Allen, 1913b:537; first use as generic designation.

REMARKS: The genus has no fossil record. *Melanomys* can be distinguished from other oryzomyine genera by the following combination of characters: dark dorsal coloration with minimal ventral countershading, tail shorter than head and body, reduction of ungual tufts, robust rostrum, lacrimals with longer maxillary than frontal sutures, long palate, carotid circulatory pattern 3 (Voss, 1988), and bunolophodont molars.

Monophyly of *Melanomys* is corroborated by cladistic analysis of morphological characters (Pine et al. 2012), an unsurprising result given the morphological distinctiveness of *Melanomys* within the Oryzomyini, but recent molecular phylogenetic analyses (Hanson and

Bradley 2008; Pine et al. 2012) recovered *Sigmodontomys alfari* within *Melanomys*, rendering the genus polyphyletic. Pine et al. (2012) proposed that incomplete lineage sorting of sequence haplotypes could be causing the non-monophyly of *Melanomys*. Herein, both *Melanomys* and *Sigmodontomys* are retained as separate genera, but additional analyses are required to confirm or reject this arrangement.

KEY TO THE SOUTH AMERICAN SPECIES
OF *MELANOMYS*:

1. Dorsal pelage dark brown, with tips slightly reddish or dark brown; nasals long (mean 12 mm) 2
1'. Dorsal pelage streaked with ochraceous and black, paler than in other species; nasals short (mean 11 mm); known only from the region of Sierra Nevada de Santa Marta in Colombia and westernmost Venezuela. . . . . . .
. . . .*Melanomys columbianus*
2. Incisive foramina short (3.7–5.6 mm) 3
2'. Incisive foramina long (5.6–6.6 mm); known only from the Lomas de Atocongo, Peru. . . . .*Melanomys zunigae*
3. Dorsal color coppery brown; interorbital region and rostrum wide; size large . . . . . . *Melanomys robustulus*
3'. Dorsal color olivaceous; interorbital region and rostrum narrow; size smaller . . . . . *Melanomys caliginosus*

Melanomys caliginosus (Tomes, 1860)

Dusky Melanomys

SYNONYMS:

Hesperomys caliginosus Tomes, 1860b: 263; type locality "Ecuador" (stated by J. A. Allen [1913b:537] to be "doubtless Esmeraldas [Esmeraldas] (near sea level), Ecuador").

Hesperomys (Habrothrix) caliginosus: J. A. Allen, 1891b: 210; name combination.

Oryzomys phaeopus Thomas, 1894b:355; type locality "Pallatanga [Chimborazo], Ecuador."

Oryzomys phaeopus obscurior Thomas, 1894b:356; type locality "Concordia, Medellin, [Antioquia,] Colombia."

Akodon caliginosus: E.-L. Trouessart, 1897:537; name combination.

Oryzomys (Zygodontomys) phaeopus obscurior: Bangs, 1900a:95; part; name combination.

Oryzomys phaeopus olivinus Thomas, 1902f:247; type locality "Zaruma, Southern [El Oro] Ecuador."

Oryzomys (Melanomys) phaeopus: J. A. Allen, 1912:87; name combination.

Oryzomys (Melanomys) obscurior affinis J. A. Allen, 1912:88; type locality "San José, (altitude 200 ft), Cauca [= Valle del Cauca], Colombia."

Melanomys caliginosus caliginosus: J. A. Allen, 1913b:537; name combination.

Melanomys caliginosus oroensis J. A. Allen, 1913b:538; type locality "Rio de Oro (altitude about 1500 feet), Manavi [Manabí] Province, Ecuador."

Melanomys affinis affinis: J. A. Allen, 1913b:539; name combination.

Melanomys affinis monticola J. A. Allen, 1913b:540; type locality "Gallera [= La Gallera] (altitude 5700 ft), west slope of Western Andes," Cauca, Colombia.

Melanomys phaeopus phaeopus: J. A. Allen, 1913b:541; name combination.

Melanomys phaeopus olivinus: J. A. Allen, 1913b:543; name combination.

Melanomys phaeopus vallicola J. A. Allen, 1913b:544; type locality "Rio Frío (altitude 3500 feet), Cauca Valley, [Valle del Cauca,] Colombia."

Melanomys phaeopus tolimensis J. A. Allen, 1913b:545; type locality "Rio Toche (altitude 6800 feet), Tolima Province, Colombia."

Melanomys lomitensis J. A. Allen, 1913b:545; type locality "Las Lomitas [= Lomitas] (altitude 5000 feet), Western Andes, [Valle del Cauca,] Colombia."

Melanomys obscurior: J. A. Allen, 1913b:546; name combination.

Melanomys buenavistae J. A. Allen, 1913b:547; type locality "Buenavista (altitude 4500 feet), Eastern Andes, about 50 miles, in a straight line, southeast of Bogota, [Meta,] Colombia."

Melanomys affinis monticolor J. A. Allen, 1916b:214; incorrect subsequent spelling of *Melanomys affinis monticola* J. A. Allen.

Oryzomys (Melanomys) caliginosus caliginosus: Tate, 1932f: 11; name combination.

Oryzomys (Melanomys) phaeopus phaeopus: Tate, 1932f: 11; name combination.

Oryzomys (Melanomys) phaeopus obscurior: Tate, 1932f: 11; name combination.

Oryzomys (Melanomys) affinis affinis; Tate, 1932f:11; name combination.

Oryzomys (Melanomys) caliginosus oroensis: Tate, 1932f: 11; name combination.

Oryzomys (Melanomys) affinis monticola: Tate, 1932f:11; name combination.

Oryzomys (Melanomys) phaeopus vallicola: Tate, 1932f: 11; name combination.

Oryzomys (Melanomys) phaeopus tolimensis: Tate, 1932f: 11; name combination.

Oryzomys (Melanomys) lomitensis: Tate, 1932f:11; name combination.

Oryzomys (Melanomys) buenavistae: Tate, 1932f:11; name combination.

Oryzomys (Melanomys) caliginosus phaeopus: Cabrera, 1961:401; name combination.

Oryzomys (Melanomys) caliginosus obscurior: Cabrera, 1961:401; name combination.

Oryzomys (Melanomys) caliginosus olivinus: Cabrera, 1961:401; name combination.

Oryzomys (Melanomys) caliginosus monticola: Cabrera, 1961:401; name combination.

Oryzomys (Melanomys) caliginosus buenavistae: Cabrera, 1961:401; name combination.

DESCRIPTION: Geographically variable species, but overall very dark dorsally, almost black flecked with ochraceous or reddish, with dark reddish-brown sides, and orange-brown under parts. Nominotypical form more brownish washed with pale yellow and finely lined with black, especially along midback; other geographic populations range from much darker with hairs tipped ochraceous brown to heavily washed with chestnut red. Most similar to *M. columbianus*, distinguished from it primarily by darker overall color and longer nasals.

DISTRIBUTION: *Melanomys caliginosus* ranges from west-central Colombia to southwestern Ecuador, from sea level to 2,300 m. It is distributed mostly along the Andean Cordillera, but also in the lowland rainforest of the Pacific versant, including the Chocó (see Cadena et al. 1998), and in the Cauca Valley. Hanson and Bradley (2008) assigned Central American records ascribed to this species by previous workers (e.g., Hall 1981; Musser and Carleton 2005) to either *M. chrysomelas* or *M. idoneus*.

SELECTED LOCALITIES (Map 173): COLOMBIA: Antioquia, Bellavista (FMNH 70356), Concordia (type locality of *Oryzomys phaeopus obscurior* Thomas), Santa Bárbara, Río Urrao (FMNH 71836); Boyacá, Muzo (FMNH 71838); Caldas, Samana, Río Hondo (FMNH 71816); Cauca, Munchique (AMNH 32406); Huila, Andalucía (AMNH 33783), San Adolfo (FMNH 71823); Meta, Buenavista (type locality of *Melanomys buenavistae* J. A. Allen); Nariño, Candelilla (FMNH 89550); Tolima, Río Toche (type locality of *tolimensis*; AMNH 32976); Valle del Cauca, Río Frio (type locality of *Melanomys phaeopus vallicola* J. A. Allen), Zabaletas (USNM 507245). ECUADOR: Chimborazo, Pallatanga (type locality of *Oryzomys phaeopus* Thomas); El Oro, Salvias (AMNH 47766), Santa Rosa (AMNH 61296); Esmeraldas, Esmeraldas (type locality of *Hesperomys caliginosus* Tomes); Loja, Celica (USNM 461644), Guayas, Los Pozos (AMNH 67470); Manabí, Cuaque (AMNH 66338); Pichincha, Las Máquinas (AMNH 66326), San Tadeo (FMNH 53249); Santa Elena, Cerro Manglar Alto (AMNH 66352).

SUBSPECIES: *Melanomys caliginosus* is clearly polytypic, displaying extensive variation over its known range. The 10 names we currently treat as synonyms of *M. caliginosus* may represent either valid subspecies or even species,

but to date we have been unable to detect external or cranial features that consistently separate any into diagnosable units (Weksler and Lóss, unpubl. data). Thus, for the moment we see no value in trying to delineate infraspecific units.

NATURAL HISTORY: Cadena et al. (1998) trapped this species in virtually intact forest with many palms and epiphytes. They found a male with scrotal testes and a lactating and pregnant female during March. The pregnant female gave birth to two young a few hours after capture and proceeded to eat them over the following days. Previous studies describing the natural history of *M. caliginosus* actually refer to populations from Costa Rica and Panama that we recognize as *M. chrysomelas* and *M. idoneus*, respectively. We include this information here because the genus seems to be ecologically conservative. T. H. Fleming (1970) reported pregnant or lactating females in May, July, August, and November in Panama (= *M. idoneus*). A. L. Gardner (1983b) found pregnant or lactating females in February, March, and August in Costa Rica (= *M. chrysomelas*); 11 pregnant females from Costa Rica averaged 3.5 embryos (range: 1–6); one female had postpartum estrus. Laboratory specimens were eager to eat insects, and some specimens were captured

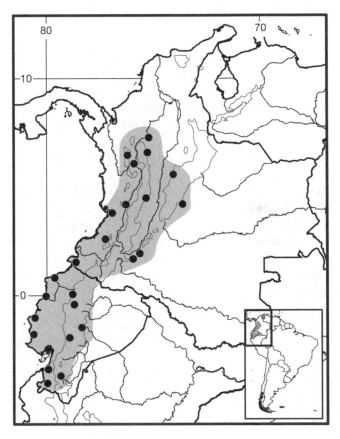

Map 173 Selected localities for *Melanomys caliginosus* (●); range in Panama from Hall (1981). Contour line = 2,000 m.

during the day in Costa Rica, around leaves and other debris associated with logs, buttress of large trees, and bases of banana foliage (A. L. Gardner 1983b).

REMARKS: Hanson and Bradley (2008) described extensive mitochondrial variation in what had been treated as *M. caliginosus*, which led them to recognize *M. chrysomelas*, *M. idoneus*, and *M. columbianus* as distinct species. We follow their arrangement based on our analyses of morphological and morphometric characters of specimens housed in the AMNH and USNM (Weksler and Lóss, unpubl. data). This complex of species is in need of revision.

Specimens from eastern Colombia (Norte de Santander and Arauca) in the Field Museum that are listed as *Melanomys caliginosus* in the FMNH database (http:// http:// emuweb.fieldmuseum.org/mammals/Query.php) are not included in our analysis. These localities are outliers to the defined range of the species, and we have not been able to examine them to verify this identification.

A. L. Gardner and Patton (1976) described the chromosomal complement if *M. caliginosus* (now *M. chrysomelas*) from Costa Rica as $2n = 56$ and $FN = 58$. The karyotype of *M. caliginosus*, as we restrict this species herein, is unknown.

Melanomys columbianus (J. A. Allen, 1899)
Colombian Melanomys

SYNONYMS:

Akodon columbianus J. A. Allen, 1899b:203; type locality "Manzanares (alt. 3000 ft), Santa Marta District, [Magdalena,] Colombia."

Oryzomys (Zygodontomys) phaeopus obscurior: Bangs, 1900a:95; part (listing of specimens from Chirua, La Guajira); not *phaeopus obscurior* Thomas.

Melanomys columbianus: J. A. Allen, 1913b:550; first use of current name combination.

Oryzomys (Melanomys) caliginosus columbianus: Cabrera, 1961:400; name combination.

DESCRIPTION: Distinguished of *M. caliginosus* and *M. robustulus* by paler dorsal pelage, dark brown grizzled with strong yellowish brown varied with black, with tendency for black dorsal stripe. Venter uniform rusty brown, with rusty tips concealing grayish undertones; tail dusky, unicolored, nearly naked, with scanty short hairs not concealing scale annuli. Rostrum broader than in *M. caliginosus*; otherwise, cranial measurements similar for two species.

DISTRIBUTION: Known only from the region of Sierra Nevada de Santa Marta, northern Colombia, and adjacent northwestern Venezuela, at elevations between 1,000 m and 2,500 m. Extensive collecting below 1,000 m in the Sierra Nevada de Santa Marta yielded no specimens (J. A. Allen 1913b).

SELECTED LOCALITIES (Map 174): COLOMBIA: La Guajira, Chirua (AMNH 36825); Magdalena, Don Diego (AMNH 23291), Manzanares (AMNH 15336). VENEZUELA: Mérida, La Azulita (FMNH 22159); Zulia, Misión Tukuko (USNM 448569).

SUBSPECIES: *Melanomys columbianus* is monotypic.

NATURAL HISTORY: There is no published information on any aspect of the ecology or behavior of this species.

REMARKS: In his original description, J. A. Allen (1912) regarded *M. columbianus* as a member of the genus *Akodon*, comparing it only to *Akodon venezuelensis* (now *Necromys venezuelensis*; see Musser and Carleton 2005) and *Akodon bogotensis*. A year later, in his revision of *Melanomys*, J. A. Allen (1913b) treated *columbianus* as a species of *Melanomys*. Cabrera (1961), however, placed it as a subspecies of *M. caliginosus* without comment. Hanson and Bradley (2008), based on sequences from the mtDNA cytochrome-*b* gene, recommended that *M. columbianus* be recognized as a distinct species, pending further morphological review. We regard this species as distinct from *M. caliginosus* based on as yet unpublished morphometric analyses (Weksler and Lóss, unpubl. data), geographic isolation from the other species, and a few qualitative characters (see above). The karyotype is unknown.

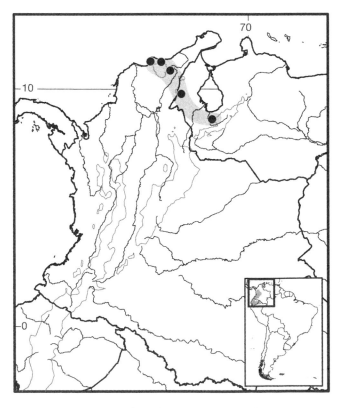

Map 174 Selected localities for *Melanomys columbianus* (●). Contour line = 2,000 m.

Melanomys robustulus Thomas, 1914
Robust Dark Rice Rat

SYNONYMS:

Melanomys robustulus Thomas, 1914d:243; type locality "Gualaquiza, Oriente of Ecuador [Morona-Santiago]. Alt. 250′."

Oryzomys (Melanomys) robustulus: Tate, 1932f:12; name combination.

DESCRIPTION: Largest species in genus, with head and body length of holotype = 122 mm. Skull large, heavily built, with large and convex braincase, broad interorbital region (6.8 mm for holotype), and overhanging supraorbital ledges. Distinguished from both *M. caliginosus* and *M. columbianus* by coppery brown rather than olivaceous dorsal color. Forefeet and hindfeet and tail uniformly black.

DISTRIBUTION: Southeastern Ecuador and northwestern Peru, in tropical lowland rainforest between elevations of 400 to 1,200 m.

SELECTED LOCALITIES (Map 175): ECUADOR: Morona-Santiago, Gualaquiza (type locality of *Melanomys robustulus* Thomas); Napo, San Javier (FMNH 29454); Pastaza, Montalvo (FMNH 41475), Río Pindo Yacu (FMNH 43228); PERU: Loreto, Boca del Río Curaray (AMNH 71555; see LeCroy and Sloss 2000:43 for position of this locality).

SUBSPECIES: *Melanomys robustulus* is monotypic.

NATURAL HISTORY: Little is known of this species. It has been reported as nocturnal and terrestrial (Tirira and Boada 2008).

REMARKS: Thomas (1914d:243) regarded this species as more similar to *M. chrysomelas* from Central America than to the neighboring *M. caliginosus*. The karyotype is unknown.

Melanomys zunigae Sanborn, 1949
Zuniga's Dark Rice Rat

SYNONYMS:

Oryzomys (Melanomys) zunigae Sanborn, 1949d:2; type locality "Lomas de Atocongo, Department of Lima, Perú."

Melanomys zunigae Musser and Carleton, 1993:708; first use of current name combination.

DESCRIPTION: Palest species in genus, with general dorsal color dark brown becoming paler on the sides; under parts paler brown mixed with grayish black but with pale tips; both forefeet and hindfeet brown rather than black; tail blackish brown above and slightly paler below. Skull distinguished from congeneric species by long incisive foramina that extend posteriorly to between first molars.

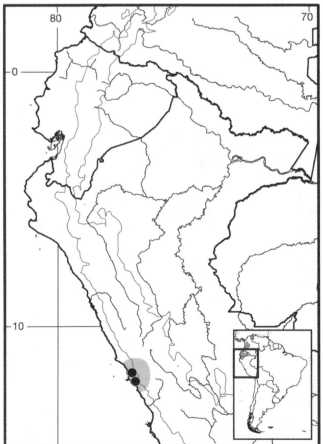

Map 176 Selected localities for *Melanomys zunigae* (●). Contour line = 2,000 m.

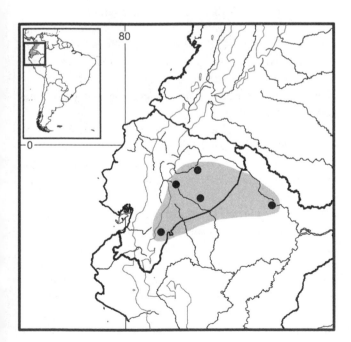

Map 175 Selected localities for *Melanomys robustulus* (●). Contour line = 2,000 m.

DISTRIBUTION: This species is known only from the western coast of central Peru, between 600 and 800 m.

SELECTED LOCALITIES (Map 176): PERU: Lima, Cerro San Jerónimo (FMNH 65578), Lomas de Atocongo (type locality of *Oryzomys* [*Melanomys*] *zunigae* Sanborn).

SUBSPECIES: *Melanomys zunigae* is monotypic.

NATURAL HISTORY: This species is known only from the Lomas de Atocongo, a group of low lying hills close to Lima in west central Peru (Musser and Carleton 2005). This area is characterized by desert or dryland habitat, with mesic vegetation along streams and on hilltops as a product of sporadic precipitation from fog during the winter season (Mena et al. 2007). The species is terrestrial and probably nocturnal (Zeballos and Vivar 2008).

Melanomys zunigae is listed as Critically Endangered (Possibly Extinct) by the IUCN (2011) because it has not been seen despite extensive searching over the past 50 years, and because there is a continuing, severe decline in the extent and quality of any remaining habitat (Zeballos and Vivar 2008).

REMARKS: Sanborn (1949d) regarded *M. zunigae* as most closely related to populations of *M. caliginosus phaeopus* from western Ecuador. No phylogenetic analyses of any character set, however, are available comparing this species to any other in the genus. The karyotype is unknown.

Genus *Microakodontomys* Hershkovitz, 1993

Roberta Paresque and J. Delton Hanson

Microakodontomys is a monotypic genus (Hershkovitz 1993) with a restricted distribution in the Brazilian Cerrado biome. The single species is one of the smallest sigmodontines (mean head and body length = 35.3 ± 7.7 mm), with a long, conspicuously scaled tail that is much longer than head and body (mean 93 ± 6.1 mm). It has a relatively long and narrow hindfoot (mean 24 ± 0.8 mm) with small interdigital and hypothenar plantar tubercles and an elongate thenar tubercle. Digit I is much shorter than digits II–IV, which are about the same length, and digit V extends just to the base of the third phalanx of digit IV. Claws are thin, and moderately long with sparse pale ungual hairs.

Dorsally *Microakodontomys* is buffy-ochraceous tinged blackish in color; the venter is creamy. Tricolored hairs, dark brown at the base and black at the tip, with an intermediate and broad buffy band, define the dorsal coloration. Guard hairs on the dorsum are black, forming a dark dorsal strip compared with the fewer, shorter guard hairs on the flanks, which appear more ochraceous. The snout has a black band reaching from the tip to the corner of the eye where it becomes a black eye-ring. The mystacial vibrissae are relatively short, with the longest only extending to the ears. The proximal half of the tail is bicolored, and on the bicolored half of the tail the hair on the dorsal surface is black, whereas on the ventral surface the hair grades from completely ochraceous (proximal portion), to basally black with ochraceous tips (distal portion). The tip of the tail has a slight tuft.

The skull is small (greatest length 19.3–22.8 mm) with a moderately long, slender rostrum (the nasals are about 35% of the skull length). The lacrimals are slightly inflated, leading into a narrow interorbital region with a slight medial constriction and comparatively little posterior divergence. The zygomatic plates project slightly and partially overlap the capsule of the incisor root. The braincase is flat above and moderately curved behind, with the inflated portion of the supraoccipital bone dorsally visible and the interparietal long and wide. The incisive foramina are narrow and extend to the anterior margins of the M1s, whereas the large posterolateral pits border the palatal bridge, which extends beyond the posterior plane of the M3s. *Microakodontomys* lacks stapedial or sphenofrontal foramina and a squamosal-alisphenoid groove (i.e., it conforms to carotid circulatory pattern 3 of Voss 1988).

Microakodontomys is tetralophodont, with mesolophs and mesolophids absent in all molars. The brachydont upper molar row is short (3.1–3.4 mm) and tuberculate. The paraflexus in all upper molars is isolated from the margin, as is the metaflexus in the first and third upper molars. In M1, the anterocone has well-defined anterolabial and anterolingual conules; an anterior fossette is absent, as are the posterostyle and posteroloph. M3 is less than half the size of M2, but has an enlarged median fossette and is subtriangular in occlusal outline. The lower molars are subhypsodont; m1 and m2 has isolated mesoflexids and postereoflexids and lacks ectolophids; m1 is also missing an anterolophid and stylid; and m3 is subtrapezoidal in occlusal outline and has isolated lingual flexids. Upper incisors are opisthodont.

Nearly all cranial and dental information for this species was gleaned from Hershkovitz's (1993) description of the type specimen. Recent material has been collected from the Cerrado of Brazil and is discussed here.

SYNONYM:

Microakodontomys Hershkovitz, 1993:2; type species *Microakodontomys transitorius*, by original designation.

REMARKS: In his description, Hershkovitz (1993) discussed affinities of the genus based on size as well as on cranial measurements. He viewed this taxon as a derived member of the Sigmodontinae, suggesting it was intermediate between akodontines and oryzomyines: resembling *Akodon* in its molar pattern and certain cranial characteristics and *Oligoryzomys* or *Microryzomys* in overall cranial similarity. Although Hershkovitz (1993:4) acknowledged the possibility that his new taxon was only a "well-differentiated species of *Oligoryzomys*," he believed that based on dental characteristics, facial coloration, and tail length *Microakodontomys* was easily distinguishable from any other sigmodontine rodent. However, the lack of mesolophs and mesolophids in some specimens of *Oligoryzomys rupestris* as well as in *O. fornesi* and the general characteristics of small body, long tail, and long hindfeet have called into question the status of the *Microakodontomys*. Musser and Carleton (2005) recognized *Microakodontomys* but recommended additional studies to determine its relationships with *Oligoryzomys*. In contrast, Weksler et al. (2006) considered *M. transitorius* an aberrant form of *Oligoryzomys* and subsumed *Microakodontomys* within that genus.

During a recent collecting effort at the type locality of *M. transitorius*, R. Paresque and colleagues (unpubl. data) trapped four additional specimens and compared them with sympatric individuals of two *Oligoryzomys* species (*O. nigripes* and *O. fornesi*) to test whether *Microakodontomys* was a valid genus. Their parsimony analyses of DNA sequence data from *Microakodontomys* (including from the holotype) and other sigmodontine genera supported *Microakodontomys* as a valid genus within the Oryzomyini, and one with a closer relationship to *Sigmodontomys alfari*, *Melanomys caliginosus*, and *Oryzomys couesi* than to *Oligoryzomys*.

Microakodontomys transitorius Hershkovitz, 1993
Transitional Colilargo

SYNONYMS:

Microakodontomys transitorius Hershkovitz, 1993:2; type locality "Parque Nacional de Brasília [Distrito Federal], about 20 km NW of Brasília, D. F., [Brazil] about 1100 m."

Oligoryzomys transitorius: Weksler, Percequillo, and Voss, 2006:4; name combination.

DESCRIPTION: As for the genus.

DISTRIBUTION: This species has an extremely restricted distribution, known only from two sites in the Parque Nacional de Brasília in central Brazil.

SELECTED LOCALITIES (Map 177): BRAZIL: Distrito Federal, Parque Nacional de Brasília (type locality of *Microakodontomys transitorius* Hershkovitz).

SUBSPECIES: *Microakodontomys transitorius* is monotypic.

NATURAL HISTORY: *Microakodontomys transitorius* is found in edge habitat in the Cerrado biome of Brazil that is transitional between *campo limpo* (wet grassland) and *campo sujo* (grass and shrub assemblage). With its long tail and moderately long feet, this mouse is presumably adapted to both sylvan environs as well as more pastoral habitats

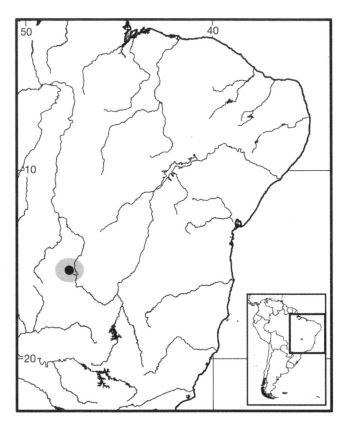

Map 177 Selected locality for *Microakodontomys transitorius* (●).

(Hershkovitz 1993). Additionally, its external morphology is similar to that of *Oligoryzomys fornesi* and *O. nigripes*, both of which are also found syntopically in edge habitats. The restricted distribution, combined with the low incidence of capture despite frequent collection efforts over the years, suggests that *M. transitorius* is a rare species.

REMARKS: The karyotype has $2n = 38$, $FN = 46$ (R. Paresque et al., unpubl. data). The autosomal complement includes 13 acrocentric pairs, gradually decreasing in size, and five pairs of small, biarmed chromosomes. The X chromosome is large submetacentric, and the Y chromosome is a small submetacentric.

Genus *Microryzomys* Thomas, 1917
Michael D. Carleton

Species of *Microryzomys* are small oryzomyines (10–15 g) that are externally recognizable by their soft, luxuriant fur and a tail notably longer than head and body. The hindfeet are short and slender, the plantar surface bears six fleshy pads (thenar and hypothenar pads are large), and the fifth digit (V) is nearly as long as the central three (II–IV). Females possess eight mammae, arranged in four pairs and anatomically positioned as in other Oryzomyini (Carleton and Musser 1989; Voss and Carleton 1993).

The skull is delicately constructed, characterized by a short, attenuate rostrum, hourglass-shaped interorbit, and smooth, rounded braincase devoid of ridges. The zygomatic plate is conspicuously narrow and the dorsal notch rudimentary. The cranial wall above the otic capsule is perforated by a spacious subsquamosal fenestra and postglenoid foramen, which together delineate a narrow hamular process. For an oryzomyine, the bony palate is short, extending only slightly beyond the end of the third molars, and the posterolateral palatal pits are correspondingly small and simple; parapterygoid fossae are shallow and narrow. An alisphenoid strut is uniformly absent, the masticatory-buccinator foramen and foramen ovale accessorius being broadly confluent. The co-occurrence of certain cranial foramina and osseus grooves indicates a primitive cephalic arterial supply (Carleton and Musser 1989; pattern 1 of Voss 1988): large stapedial foramen, large posterior opening of the alisphenoid canal, and presence of an squamosoalisphenoid groove leading to the sphenofrontal foramen. The ectotympanic bulla is small, exposing a broad wedge of the periotic in ventral view, and the tegmen tympani does not overlap the squamosal (Voss 1993).

Species of *Microryzomys* possess diminutive, cuspidate molars with brachydont crowns and a pentalophodont occlusal pattern. The upper and lower third molars are noticeably smaller than the second molars. The anterocone(-id) is deeply bifurcated and wide, imparting a rectangular shape to the first molars; the well-developed mesoloph(id) reaches the buccal margin of the tooth; and an anteroloph(id) is usually present and distinct. Three roots anchor each upper molar; each lower molar has two. The upper incisors are narrow, asulcate, and typically orthodont. See Carleton and Musser (1989) and accounts herein for additional external, cranial, and dental descriptions and comparisons with *Oligoryzomys* and *Oryzomys sensu stricto*; see Musser and Carleton (2005) and Weksler et al. (2006) for contrasts with the externally similar Andean species *Oryzomys balneator* Thomas (1900), now placed in the newly described genus *Oreoryzomys* Weksler, Percequillo, and Voss (2006).

Stomach morphology conforms to the unilocular-hemiglandular plan, but glandular epithelium is more extensive over the lesser curvature than in *Oryzomys sensu stricto* (Carleton and Musser 1989). No gall bladder is present (Voss 1991a). A full complement of male accessory reproductive glands, including preputials, has been reported based on single examples of each species (Voss and Linzey 1981; Carleton and Musser 1989).

SYNONYMS:

Hesperomys: Tomes, 1860a:215; part (description of *minutus*); not *Hesperomys* Waterhouse.

Oryzomys: E.-L. Trouessart, 1897:527; part (listing of *minutus*); not *Oryzomys* Baird.

Oryzomys (*Microryzomys*) Thomas, 1917g:1; type species *Oryzomys minutus* (Tomes), by original designation.

Microryzomys: Thomas, 1926b:314; first usage as genus.

Oryzomys (*Oligoryzomys*): Thomas, 1926e:613: part (allocation of *Microryzomys* as full synonym, invalid as subgenus); not *Oligoryzomys* Bangs.

Oligoryzomys: Gyldenstolpe, 1932a:26; part (listing of *Microryzomys* as full synonym); not *Oligoryzomys* Bangs.

Thallomyscus Thomas, 1926e:612; type species *Oryzomys dryas* Thomas, by original designation.

Oryzomys (*Thallomyscus*): Tate, 1932f:5; name combination.

Oryzomys (*Microryzomys*): Osgood, 1933c:1; name combination (*Thallomyscus* allocated as full synonym, invalid as subgenus).

REMARKS: *Microryzomys* was originally designated as a subgenus of *Oryzomys* and subsequently retained as such (Osgood 1933c; Ellerman 1941; Cabrera 1961) or placed in synonymy with *Oligoryzomys* (Thomas 1926e; Gyldenstolpe 1932; Tate 1932f). The junior synonym *Thallomyscus* has been formally retained as a subgenus (Tate 1932f) or allocated as a full synonym (Osgood 1933c; Carleton and Musser 1989). *Microryzomys*, along with *Oligoryzomys*, was once considered to be a taxonomic grade that artificially united merely small-bodied species of *Oryzomys* (Goldman 1918; Tate 1932f; Hershkovitz 1944); however, Carleton and Musser (1989) consolidated discrete morphological traits that differentiate species of *Microryzomys* from both *Oligoryzomys* and typical *Oryzomys* and recast its diagnosis as genus. *Microryzomys* is a member of Oryzomyini (*sensu* Voss and Carleton 1993; Musser and Carleton 2005; Weksler 2003, 2006) and shares certain morphological traits with *Neacomys*, *Oligoryzomys*, and *Oreoryzomys* (Carleton and Musser 1989; Weksler 2006; Weksler et al. 2006). In phylogenetic interpretations using allozymic or mitochondrial sequence data, *Microryzomys* variously falls in a clade that includes *Neacomys* and *Oligoryzomys* (Dickerman and Yates 1995; Myers et al. 1995; Patton and da Silva 1995; M. F. Smith and Patton 1999). In taxonomically broad studies of oryzomyines (Weksler 2003, 2006; Weksler et al. 2006; Percequillo, Costa, and Weksler 2011; Pine et al. 2012), using morphology in combination with nuclear and/or mitochondrial genes, the genus emerges as sister to *Oreoryzomys*, usually in the cladistic hierarchy (*Neacomys* (*Oligoryzomys* (*Microryzomys* + *Oreoryzomys*))).

Both forms of *Microryzomys* may prove to be species composites. Carleton and Musser (1989) drew attention to the strong morphometric differentiation of their isolated samples of *M. altissimus* in central Colombia and questioned the apparent morphological homogeneity of both *M. altissimus* and *M. minutus* across the Huancabamba Depression in northwestern Peru. The high-elevation occurrence of the genus in different Andean cordilleras elegantly lends itself to a phylogeographic study using molecular data.

KEY TO THE SPECIES OF *MICRORYZOMYS*:

1. Pelage principally ochraceous-tawny; dorsal and ventral body fur faintly contrasted; tail basically unicolored and longer (usually >110 mm); dusky markings present on tops of forefeet and hindfeet; skull with shorter, ovate incisive foramina (usually terminating short of or even with the level of the M1s) and shorter toothrow (usually <3.0 mm); dentary with distinct incisor tubercle . . . . . .
. . . . *Microryzomys minutus*

1'. Dorsal pelage more buffy, and under parts grayish buff; dorsum and venter clearly demarcated; tail bicolored and shorter (usually <110 mm); tops of forefeet and hindfeet whitish; skull with longer, narrower incisive foramina (usually penetrating between the M1s) and longer toothrows (usually >3.0 mm); dentary with inconspicuous incisor tubercle . . . . *Microryzomys altissimus*

Microryzomys altissimus (Osgood, 1933)
Páramo Colilargo

SYNONYMS:

Oryzomys minutus: Osgood, 1914b:158; part.

Oryzomys (Microryzomys) minutus altissimus Osgood, 1933c:5; type locality "La Quinua, mountains north of Cerro de Pasco, [Pasco,] Peru. Alt. 11,600 ft."

M[*icroryzomys*]. a[*ltissimus*]. *altissimus*: Hershkovitz, 1940a:81; name combination.

Microryzomys altissimus hylaeus Hershkovitz, 1940a:81; type locality "Atal, within the confines of the Hacienda Indújel del Vincula, near San Gabriél, Montúfar, Carchi Province, Ecuador. Altitude, about 2900 meters."

Microryzomys altissimus chotanus Hershkovitz, 1940a:82; type locality "southern slopes of the arid Chota Valley, parish of Pimanpiro, Imbabura Province, Ecuador. Altitude, about 2000 meters."

Oryzomys altissimus altissimus: Cabrera, 1961:397; name combination.

Oryzomys altissimus hylaeus: Cabrera, 1961:398; name combination.

Oryzomys altissimus chotanus: Cabrera, 1961:398; name combination.

DESCRIPTION: *Microryzomys altissimus* is characterized by more buff in the upper body fur and grayish buff in the under parts such that the dorsum and venter are clearly demarcated; grayish head, tail bicolor and usually shorter than 110 mm, and tops of forepaws and hindpaws whitish; hindfeet and metatarsal pads narrower; skull slightly

heavier with broader braincase; longer incisive foramina, and longer (usually greater than 3.0 mm), more robust toothrows; and incisor tubercle of dentary small (see Carleton and Musser 1989:68).

DISTRIBUTION: *Microryzomys altissimus* occurs in the high elevations of the central Andes, including the Cordillera Central of central Colombia; Cordilleras Occidental and Oriental of Ecuador; Cordilleras Occidental and Oriental of northern Peru, and southward to the vicinity of Junín in central Peru. Its known elevational range is from 2,000 to 4,300 m, but most occurrences are from 2,500 to 4,000 m.

SELECTED LOCALITIES (Map 178; from Carleton and Musser 1989, except as noted): COLOMBIA: Quindio, Finca La Cubierta, 6 km N of Salento-Cocora road; Tolima, Nevado de Ruiz. ECUADOR: Azuay, Bestión; Bolívar, Sinche; Cañar, Chical, Naupan Mountains; Carchi, Atal, near San Gabriél, Montúfar (type locality of *Microryzomys altissimus hylaeus* Hershkovitz); Napo, Antisana; Pichincha, Pichincha; Tungurahua, San Francisco. PERU: Amazonas, mountains E of Balsas; Ancash, Huaras, Quilcahuanca; Huánuco, Huánuco; Junín, Maraynioc, 45 km NE of Tarma; Libertad, mountains NE of Otuzco; Pasco, La Quinua, mountains N of Cerro de Pasco (type locality of *Oryzomys (Microryzomys) minutus altissimus* Osgood); Piura, Huancambamba.

SUBSPECIES: Carleton and Musser (1989) provisionally regarded *Microryzomys altissimus* as monotypic.

NATURAL HISTORY: *Microryzomys altissimus* inhabits grassy páramo and assorted subalpine wet forest ecotones, where it has been obtained in terrestrial settings (Voss 1978–1980 [in Carleton and Musser 1989]; Barnett 1999; Voss 2003). Most instances of sympatry with *M. minutus* have been documented where upper montane rainforest and páramo intermix, within an elevational belt of 2,500 to 3,500 m (Osgood 1933c; Carleton and Musser 1989; Barnett 1999). Its diet in Peruvian montane forest was characterized as omnivorous, including grass stems and seeds, insect larvae and adults, and fruits (Noblecilla and Pacheco 2012). Basic aspects of behavior, ecology, and reproduction require much illumination.

REMARKS: Osgood (1933c) originally diagnosed *altissimus* as a subspecies of *Oryzomys minutus*, but Hershkovitz (1940a) elevated it to species and created two additional subspecies. Cabrera (1961) recognized the two subspecies named by Hershkovitz (1940a) from Ecuador, but Carleton and Musser (1989) considered neither valid. Carleton and Musser (1989:21), however, noted the cranial morphometric distinctiveness of samples from Colombia, a separation that is not reflected in the distribution of the named subspecies. The karyotype is characterized as $2n = 57$ and $FN = 58$, based on a single female judged to be "a heterozygote for a Robertsonian translocation" (A. L. Gardner and Patton 1976).

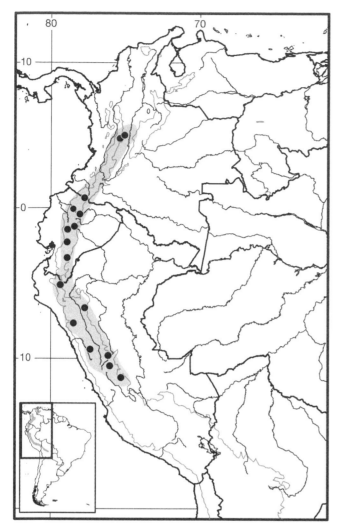

Map 178 Selected localities for *Microryzomys altissimus* (●). Contour line = 2,000 m.

Microryzomys minutus (Tomes, 1860)
Montane Colilargo

SYNONYMS:

H[*esperomys*]. *minutus* Tomes, 1860a:215; type locality unknown, probably near Pallatanga, on the western slope of the Cordillera, Chimborazo, Ecuador (Carleton and Musser 1989:65).

Oryzomys minutus: E.-L. Trouessart, 1897:527; name combination.

Oryzomys dryas Thomas, 1898f:267; type locality "Pallatanga, [Chimborazo,] Ecuador."

Oryzomys dryas humilior Thomas, 1898f:268; type locality "Plains of Bogotá," Cundinamarca, Colombia.

Oryzomys (Oligoryzomys) dryas humilior: Bangs, 1900a:95; name combination.

Oryzomys (Oligoryzomys) fulvirostris J. A. Allen, 1912:86; type locality "Munchique (alt. 8325 ft)," Cauca, Colombia.

O[*ryzomys (Oligoryzomys)*]. *humilior*: J. A. Allen, 1916b:214; name combination.

Oryzomys (Microryzomys) minutus: Thomas, 1917g:1; name combination.

Oryzomys (Microryzomys) aurillus Thomas, 1917g:1; type locality "Torontoy," Cusco, Peru.

Microryzomys aurillus: Thomas, 1926b:314; name combination.

Thallomyscus aurillus: Thomas, 1926e:612; name combination.

Thallomyscus dryas: Thomas, 1926e:613; name combination.

Oligoryzomys minutus minutus: Gyldenstolpe, 1932a:29; name combination.

Thallomyscus dryas dryas: Gyldenstolpe, 1932a:30; name combination.

Thallomyscus dryas humilior: Gyldenstolpe, 1932a:31; name combination.

Oligoryzomys fulvirostris: Gyldenstolpe, 1932a:30; name combination.

Oryzomys (Microryzomys) [minutus] minutus: Osgood, 1933c:3; name combination.

Oryzomys (Microryzomys) [minutus] minutus: Osgood, 1933c:4; name combination.

Oryzomys (Microryzomys) minutus humilior: Osgood, 1933c:5; name combination.

Oryzomys (Microryzomys) minutus fulvirostris: Osgood, 1933c:5; name combination.

Oryzomys (Microryzomys) minutus aurillus: Osgood, 1933c:4; name combination.

Microryzomys minutus: Hershkovitz, 1940a:83; name combination.

Microryzomys m[inutus]. humilior: Hershkovitz, 1940a:83; name combination.

Microryzomys m[inutus]. fulvirostris: Hershkovitz, 1940a:83; name combination.

Microryzomys m[inutus]. aurillus: Hershkovitz, 1940a:83; name combination.

DESCRIPTION: This species of *Microryzomys* is characterized by the prevalence of ochraceous-tawny pelage tones, faint or no contrast from the ventral body fur, tail basically unicolored and usually longer than 110 mm, and dusky markings on tops of the forefeet and hindfeet; hindfeet slightly wider, and plantar pads more robust; skull more delicate with narrower braincase, shorter and ovate incisive foramina, and shorter toothrow (usually less than 3.0 mm); dentary with distinct tubercle (see Carleton and Musser 1989:65–66).

DISTRIBUTION: This species is known from middle to high elevations in the northern and central Andes, including the Caribbean Coastal Ranges and Mérida Andes of Venezuela; Sierra Nevada de Santa Marta, Cordilleras

Oriental, Central, and Occidental of Colombia; Cordilleras Oriental and Occidental of Ecuador; Cordillera Occidental of northwestern Peru and Cordillera Oriental and its eastern flanks in central Peru; into west-central Bolivia as far as westernmost Santa Cruz department. This range excludes the Altiplano region and coastal ranges of southwestern Peru. Known elevational occurrence is from 800 to 4,265 m, with over 80% of collecting localities between 2,000 and 3,500 m.

SELECTED LOCALITIES (Map 179; from Carleton and Musser 1989, except as noted): BOLIVIA: La Paz, 30 km N of Zongo; Santa Cruz, Siberia, 25 and 30 km W of Comarapa. COLOMBIA: Antioquia, Santa Bárbara, Río Urrao, Ventanas, Valdivia; Cauca, Cerro Munchique (type locality of *Oryzomys (Oligoryzomys) fulvirostris* J. A. Allen), Valle de la Papas; Cundinamarca, Guasca, Río Balcones; Magdalena, Macotama; Quindío, Salento. ECUADOR: Azuay, Molleturo, Bestión; Carchi, Montúfar, near San Gabriél; Chimboraza, Pallatanga (type locality of *Oryzomys dryas* Thomas); El Oro, Taraguacocha, Cordiller de Chilla; Napo, Baeza; Pichincha, Pacto. PERU: Amazonas, Cordillera Colán, NE of La Peca, Uchco, Tambo Almirante; Ayacucho, Puncu, 30 km NE of Tambo; Cajamarca, Taulís; Cusco, Marcapata, Torontoy (type locality of *Oryzomys (Microryzomys) aurillus* Thomas); Huánuco, below Carpish Pass, trail to Hacienda Paty; Junín, Acobamba, 45 km NE of Cerro de Pasco (incorrectly given as Pasco department), Yano Mayo, Río Tarma; Piura, Cerro Chinguela, 5 km NE of Zapalache; Puno, Agualani, 9 km N of Limbani (MVZ 171476). VENEZUELA: Mérida, Páramo Tambor; Miranda, Alto de Nuevo León, 33 km WSW of Caracas; Sucre, Cerro Negro, 10 km NW of Caripe; Tachira, Buena Vista, 41 km SW of San Cristóbal; Trujillo, Hacienda Misisí, 15 km E of Trujillo.

SUBSPECIES: Carleton and Musser (1989) treated *Microryzomys minutus* as monotypic.

NATURAL HISTORY: The species is found in various Andean moist forest environments, from lower montane and upper montane rainforest, through various subalpine forest and shrubby habitats, into the lower fringes of páramo (Osgood 1933c; Carleton and Musser 1989; Barnett 1999; Voss 2003). In Venezuela, *M. minutus* was most frequently encountered in upper montane rainforest (cloud forest as per Handley 1976), in both disturbed and pristine habitats (Cabello et al. 2006), and most collecting localities with known elevation fall within this vegetational zone (Carleton and Musser 1989). Cabello et al. (2006) considered the species to be the most abundant rodent in Venezuelan cloud forests. Diet in the Peruvian montane forest was characterized as omnivorous, including grass stems and seeds, insect larvae and adults, and fruits (No-

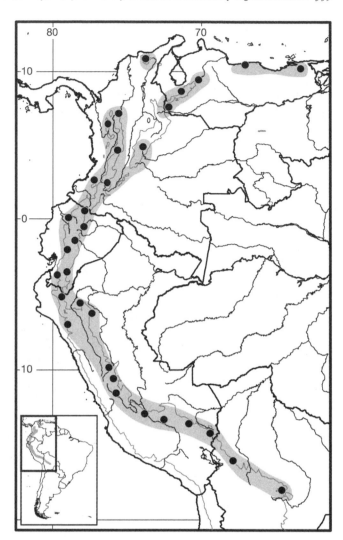

Map 179 Selected localities for *Microryzomys minutus* (●). Contour line = 2,000 m.

blecilla and Pacheco 2012). *Microryzomys minutus* has been collected usually on the ground (Handley 1976; Voss 1978–1980 [in Carleton and Musser 1989]; Cabello et al. 2006) like its congener *M. altissimus*, but Carleton and Musser (1989) suggested that the longer tail and larger plantar pads of *M. minutus* indicate greater scansorial or arboreal proclivities. Possible differences in climbing behavior, along with other natural history information, need detailed autecological study where the two species occur in syntopy.

REMARKS: *Microryzomys minutus* formerly included *altissimus* as a subspecies as described by Osgood (1933c), who also treated *aurillus*, *humilior*, and *fulvirostris* as valid subspecies. Carleton and Musser (1989) found no geographic units that they deemed diagnosable as subspecies. Diploid chromosomal count reported as $2n = 58$ (Kiblisky 1969).

Genus *Mindomys* Weksler, Percequillo, and Voss, 2006

Alexandre R. Percequillo

Known only from a handful of specimens from the Ecuadorean Andes, *Mindomys* is a monotypic genus with its single species, *M. hammondi*, characterized by a distinctive morphology but uncertain phylogenetic relationships (Weksler 2006). The body size of all specimens is large (head and body length 173–293 mm), the tail is much longer than head and body (tail length 222–251 mm), and the hindfeet are robust (length 38–42 mm). The dorsal body pelage is long (aristiform length 14–16 mm), compact, dense, and harsh; aristiforms are extremely abundant. The dorsal body coloration varies from grayish to buffy yellow grizzled with brown; the ventral coloration is light buff densely grizzled with grayish, markedly gray grizzled with white in some specimens. Long hairs that partially conceal large scales cover the tail surface; these caudal hairs increase in length from the base to the tip of the tail. The tail scales and hairs are blackish or dark brown, and there is no dorsoventral countershading. Both external and internal surfaces of the ears are well furred with short, entirely brown hairs. The forefeet are robust and completely covered above by short brown hairs. The hindfeet are densely covered by short hairs above that are overall whitish or grayish, but brown basally and white apically. Ungual tufts are short and sparse but extend to the tips of the claws. The mystacial vibrissae are thick, very dense, long, and reach considerably past the ears when laid back; supraorbital vibrissae are very long, largely surpassing the tips of the pinnae.

The skull is very large and robust, ranging from 37.6–44.3 mm in length. The rostrum is long (nasal length 13.4–16.6 mm) and wide (rostral breadth 6.4–7.2 mm), with weakly projected nasolacrimal capsules. The zygomatic notches are narrow and very shallow; the free dorsal margins of the zygomatic plates are very reduced, with their anterior margins exhibiting straight profiles. The zygomatic arches are robust and convergent anteriorly; jugals are small. The interorbital region is wide (interorbital constriction 6.1–8.0 mm) and slightly divergent posteriorly; supraorbital margins are squared or have small dorsolateral projections. The braincase is elongate and has distinct temporal crests. Incisive foramina are medium to long (length 5.7–6.7 mm) and laterally convex, forming an elongated oval. The palate is long and wide, with small posterolateral palatal pits set in shallow palatine depressions; no palatal excrescences are present. The anterior margin of the mesopterygoid fossa may reach the alveoli of the M3s; its roof is completely ossified or has only small and narrow sphenopalatine vacuities in the presphenoid or at the presphenoid-basisphenoid suture. The stapedial foramen is large, and the squamosoalisphenoid groove and sphenofrontal foramen are present, features that define carotid circulatory pattern 1 (*sensu* Voss 1988). The buccinator-masticatory and accessory oval foramina are confluent, thus an alisphenoid strut is absent in all specimens. Auditory bullae are small, with a robust, long, and flattened stapedial process; the bony tubes are short and flat. The bullae are in close contact with the squamosal, resulting in a small and narrow postglenoid foramen; the subsquamosal fenestra is absent. The suspensory process of the squamosal is absent, and the tegmen tympani is short, overlapping the squamosal.

The mandible is deep and robust; the angular process surpasses the condyloid process posteriorly, and the coronoid process is shorter than the condyloid. Inferior and superior notches are shallow. A small but discrete capsular process is present in older specimens.

Upper incisors are opisthodont. The molars are pentalophodont and moderately high-crowned; cusps are compressed anteroposteriorly, high in profile, and with labial and lingual cusps arranged in opposite pairs. Both labial and lingual flexi are deep and narrow, overlapping at the molar midplane. Molars are extremely robust, with the upper series 6.1–6.8 mm in length. The paracone of M1 is connected medially to the protocone and is also linked to the mesoloph by an oblique paralophule, which defines a narrow and oblique medial fossette. M3 is long, with all main elements discernible on its occlusal surface, including the metacone and posteroloph.

SYNONYMS:

Nectomys: Thomas, 1913c:570; part (description of *hammondi*); not *Nectomys* Peters.

Oryzomys (*Macruroryzomys*) Hershkovitz, 1948b:56; type species *Nectomys hammondi* Thomas; *nomen nudum*.

[*Oryzomys*] (*Macruroryzomys*): Cabrera, 1961:410; name combination.

Oryzomys: Musser and Carleton, 1993:722 (listing of *hammondi*); part; not *Oryzomys* Baird.

Mindomys Weksler, Percequillo, and Voss, 2006: 16; type species *Nectomys hammondi* Thomas, by original designation.

REMARKS: The uniqueness of this species with respect to other oryzomyine rodents prompted Hershkovitz (1948b) to erect the genus-group name *Macruroryzomys* for *N. hammondi*. However, as Hershkovitz (1970b) later admitted and as reinforced by Pine and Wetzel (1976), *Macruroryzomys* is a *nomen nudum* since no characters were given for it. Hence, this distinctive taxon lacked a valid generic name until the descriptions of Weksler et al. (2006).

Mindomys hammondi (Thomas, 1913)
Hammond's Mindomys

SYNONYMS:

Nectomys hammondi Thomas, 1913c:570; type locality "Mindo, N.W. of Quito. Alt. 4213 ft.," Pichincha, Ecuador.

[*Oryzomys*] (*Macruroryzomys*) [*hammondi*]: Hershkovitz, 1948b:56; name combination.

Oryzomys hammondi: Musser and Carleton, 1993:722; name combination.

[*Mindomys*] *hammondi*: Weksler, Percequillo, and Voss, 2006:16; first use of current name combination.

DESCRIPTION: As for the genus.

DISTRIBUTION: *Mindomys hammondi* is known from only three localities, all in the Ecuadorean Andes. Two of these are in the vicinity of Mindo on the western slope (elevation 1,260 to 1,330 m) and other in the Amazonian lowlands (300–500 m). The latter record, however, is problematic because it poses an unusual distributional pattern among Andean small mammals: no species that is distributed below 1,500 m in the Ecuadorean Andes, as is *M. hammondi*, is known to occur on both the east and west slopes. Rather, species below this elevation occur either on one or the other side of the Andes (Weksler et al. 2006:16, footnote 5; see McCain et al. 2007:135 for further discussion).

SELECTED LOCALITIES (Map 180): ECUADOR: Pichincha, Mindo (type locality of *Nectomys hammondi*), Concepción (MCZ 52543).

SUBSPECIES: *Mindomys hammondi* is monotypic.

NATURAL HISTORY: No data other than those dealing with general habitat are available for this species. Voss (1988) provided good habitat descriptions for the Mindo region (where the most recent specimen of *M. hammondi* was captured), which is drained by the Mindo, Canchupi, and Saguambi rivers. This general area is covered by montane rainforest (cloud forest), except in river valleys in which the forest is similar to late successional lowland rainforest because these valleys were deforested some decades ago. The canopy is of moderate height (15 to 20 m), and the trees are covered by abundant epiphytes (orchids, bromeliads, Aracea, and lianas) as well as by a thin moss layer. The most common emergents are palms and giant bamboo. The understory is dense and formed by woody dicots (mainly Piperaceae and Melastomataceae), Musaceae, Zingiberaceae, Cuclanthaceae, small Arecaceae and a giant *Equisetum*, this last commonly along riverbeds. The litter layer is formed by logs, branches, leaves, flowers, and fruits, and covers a thin humus layer. The small rivers in the Mindo region are shallow and sluggish, with muddy or sandy bottoms, and the larger rivers are deeper and faster, with sandy, gravel, or cobble-bottomed. During July, midwinter, the

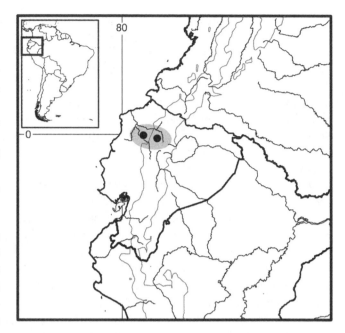

Map 180 Selected localities for *Mindomys hammondi* (●). Contour line = 2,000 m.

minimum and maximum daily air temperatures average 14 and 22° C, respectively.

REMARKS: *Mindomys hammondi* has always been viewed as a valid species, regardless of the genus or subgenus to which it has been allocated (*Nectomys* [Thomas 1913], *Macruroryzomys* [Hershkovitz 1948b], or *Oryzomys sensu lato* [Musser and Carleton 1993]). In fact, the combination of external and cranial traits exhibited by this species clearly distinguishes it from all oryzomyine genera and species, notably its large size, short and harsh dark grayish pelage, long unicolored tail, large and robust skull and molars, narrow and slightly diverging but beaded interorbital region, and carotid circulatory pattern 1, among other features. Species of the genus *Nectomys* superficially resemble *M. hammondi* but are separable by cranial (carotid circulatory pattern; interorbital region) and dental traits (general molar occlusal pattern). While the phylogenetic position of *M. hammondi* within the Oryzomyini remains problematic (Weksler 2006), its status as a unique lineage led Weksler et al. (2006) to erect the genus *Mindomys*.

Genus *Neacomys* Thomas, 1900
Marcelo Weksler and Cibele R. Bonvicino

Members of the genus *Neacomys*, called bristly or spiny mice (e.g., Musser and Carleton 2005), occur throughout the Amazon Basin in Brazil, Colombia, Peru, Ecuador,

Bolivia, Venezuela, Guyana, Surinam, and French Guiana, and west of the Andes from easternmost Panama (Darién province) to Ecuador and Venezuela, where they are present in lowland, premontane, and lower montane rainforest below about 1,500 m elevation. Seven species are currently recognized from South America: *Neacomys dubosti*, *N. guianae*, *N. minutus*, *N. musseri*, *N. paracou*, *N. spinosus*, and *N. tenuipes*. An eighth species, *N. pictus*, is presently known only from easternmost Panama (see Hall 1981).

Species of *Neacomys* are small oryzomyines (head and body length 64–100 mm) with short and grooved spines in the coarsely grizzled yellowish-, reddish-, or grayish-brown dorsal fur and on the contrastingly pale or white ventral fur, with small ears, moderately long vibrissae, and sparsely haired tails as long as or slightly longer than the combined head and body length. The hindfeet are narrow and without interdigital webs; the three middle digits (II, III, and IV) are much longer than the outer two (I and V); the plantar surface is entirely naked (from heel to toes) and scaly, with six fleshy tubercles (thenar, hypothenar, and four interdigitals). Mammae number eight, positioned in inguinal, abdominal, postaxial, and pectoral pairs (not six as reported by Gyldenstolpe 1932 and Steppan 1995; see Weksler 2006).

The skull is delicate, with nasal bones blunt their posterior margins and extending posteriorly behind the lacrimals; lacrimals have subequal maxillary and frontal suture borders. The posterior wall of the orbit is smooth. The frontosquamosal suture is colinear with the frontoparietal suture. The parietals are mostly restricted to the dorsal surface of the skull. The zygomatic plates lack anterodorsal spinous processes, and their posterior borders lie anterior to the M1s. Jugals are present but small (the maxillary and squamosal zygomatic processes broadly overlap but are not in direct contact). Incisive foramina are short, not extending posteriorly to or between the M1 alveoli. The bony palate is smooth or weakly sculptured, and long (the mesopterygoid fossa does not penetrate anteriorly between maxillary bones). Posterolateral palatal pits are present but are not recessed in deep fossae. The bony roof of the mesopterygoid fossa is perforated by small sphenopalatine vacuities. An alisphenoid strut is almost always absent (buccinator-masticatory foramen and accessory foramen ovale are confluent); the alisphenoid canal has a large anterior opening; and the stapedial foramen is large (= carotid circulatory patterns 1 or 2 of Voss 1988). A posterior suspensory process of the squamosal is absent. The postglenoid foramen is large and rounded, and the subsquamosal fenestra is present. The periotic is exposed posteromedially between the ectotympanic and basioccipital, extending anteriorly to the carotid canal. The capsular process is present in most fully adult specimens; superior and inferior masseteric ridges join anteriorly at a point so as to form a chevron.

The upper incisors are opisthodont, with smoothly rounded enamel bands. The maxillary toothrows are parallel. Molars are bunodont and with labial flexi enclosed by a cingulum. The anterocone of M1 is divided into anterolabial and anterolingual conules by a distinct anteromedian flexus in some forms but undivided in others. The anteroloph is well developed and can be joined or not with the anterocone by a labial cingulum; a protostyle is absent; mesolophs are well developed on M1 and M2. An enamel bridge connects the paracone to the anterior moiety of the protocone. The protoflexus of M2 is present but poorly developed; the mesoflexus is present as a single internal fossette; the paracone lacks an accessory loph. M3 lacks a posteroloph and usually also a hypoflexus (the latter, if present, disappears with moderate to heavy wear). An accessory labial root of M1 may be present or absent. The anteroconid of m1 lacks an anteromedian flexid; an anterolabial cingulum is usually present on all lower molars (absent in m3 of some species); an anterolophid is present on m1 but absent (or weakly expressed) on m2 and m3; an ectolophid is absent on m1 and m2; a mesolophid is distinct on unworn m1 and m2; and a posteroflexid is well developed on m3. Lower molars each have two roots.

Axial skeletal elements have modal counts of 12 ribs, 19 thoracolumbar vertebrae, 4 sacral vertebrae, and 29–35 caudal vertebrae (Steppan 1995). The tuberculum of the first rib articulates with the transverse processes of the seventh cervical and the first thoracic vertebrae, and the second thoracic vertebra has a greatly elongated neural spine. The humerus lacks an entepicondylar foramen.

The stomach is unilocular and hemiglandular (Carleton 1973; Weksler 2006), and the gall bladder is absent (Voss 1991a; Weksler 2006). Male accessory secretory organs consist of one pair each of preputial, bulbourethral, anterior and dorsal prostate, ampullary, and vesicular glands, and two pairs of dorsal prostate glands (Voss and Linzey 1981). The glans penis is complex, with a tridigitate bacular cartilage and a deep terminal crater containing three bacular mounds, one dorsal papilla, and a deeply bifurcated urethral flap (Hooper and Musser 1964; Weksler 2006).

SYNONYMS:

Hesperomys (*Calomys*): Thomas, 1882:105; part (description of *spinosus*); not *Hesperomys* Waterhouse or *Calomys* Waterhouse.

Hesperomys (*Oryzomys*): Thomas, 1884:448; part (listing of *spinosus*); not *Hesperomys* Waterhouse or *Oryzomys* Baird.

Oryzomys: E.-L. Trouessart, 1897:528; part (listing of *spinosus*); not *Oryzomys* Baird.

Neacomys Thomas, 1900a:153; type species *Hesperomys* (*Calomys*) *spinosus* Thomas, by original designation.

REMARKS: Despite the recent description of several new species (Patton et al. 2000; Voss et al. 2001), considerable revisionary work is still needed for the genus. Patton et al. (2000), for example, detected three clades of highly divergent mtDNA haplotypes that probably correspond to yet unnamed small species, one from northern Peru and eastern Ecuador (referred to as *N.* cf. *minutus* by Tirira 2007; see also Hice and Velazco 2012), a second from the Imerí region between the Rio Solimões and Rio Negro in central Brazil, and a third on the lower Rio Xingu in eastern Brazil. *Neacomys spinosus* also may comprise more than one species across its wide range (Aniskin 1994; Patton et al. 2000). Most of the following account is based on Patton et al. (2000) and Voss et al. (2001), and should be viewed as provisional.

KEY TO THE SOUTH AMERICAN SPECIES
OF *NEACOMYS*:

1. Stapedial foramen present, squamosoalisphenoid groove and sphenofrontal foramen absent (carotid circulatory pattern 2; Voss 1988) . . . . *Neacomys musseri*
1'. Stapedial foramen, squamosoalisphenoid groove, and sphenofrontal foramen present (carotid circulatory pattern 1; Voss 1988). 2
2. Anterocone divided by anteromedian flexus 3
2'. Anterocone undivided 4
3. Interorbital region hourglass in shape; supraorbital beads absent or weakly pronounced . . . . *Neacomys tenuipes*
3'. Interorbital region strongly convergent; supraorbital beads developed . . . . *Neacomys minutus*
4. Size large, total length of adults more than 165 mm . . . . *Neacomys spinosus*
4'. Size smaller, total length of adults less than 165 mm. . 5
5. Tail distinctly bicolored . . . . *Neacomys guianae*
5'. Tail usually unicolored (rarely indistinctly bicolored at base). 6
6. Long outer digits of hindfeet, claw of fifth digit (dV) extends almost to end of first phalanx of dIV, claw of dI extends to middle of first phalanx of dII; capsular part of auditory bullae narrow gradually to merge with bony tubes, bullae are more or less flask shaped . . . .*Neacomys dubosti*
6'. Very short outer digits of hindfeet, claw of dV extends ca. two-thirds length of first phalanx of dIV, claw of dI barely extends beyond base of first phalanx of dII; bullae abruptly constricted anteromedially with sharper transition between a more globular tympanic capsule and narrower bony tubes . . . . *Neacomys paracou*

Neacomys dubosti Voss, Lunde, and Simmons, 2001
Dubost's Spiny Mouse
SYNONYM:
Neacomys dubosti Voss, Lunde, and Simmons, 2001:78; type locality "Paracou [= Domaine Experimental Paracou]," French Guiana.

DESCRIPTION: Small (head and body length from 64–81 mm), distinguished from other diminutive congeners by short (range: 70–85 mm) and usually unicolored tail; moderately short rostrum flanked by relatively shallow zygomatic notches; broad and strongly convergent interorbital region with highly developed; shelf-like supraorbital beads; broad and distinctly inflated braincase; short and convex-sided incisive foramina; carotid circulation pattern 1 (*sensu* Voss 1988); flask-shaped auditory bullae; M2 with undivided anterocone; mesoloph of M1 with more or less symmetrical connections to protocone and hypocone; and persistently tubercular molar cusps.

DISTRIBUTION: *Neacomys dubosti* occurs in French Guiana, northeastern Brazil, and southeastern Surinam. All localities are situated in lowland and premontane moist broadleaf forest in the Guianan subregion of Amazonia.

SELECTED LOCALITIES (Map 181; from Voss et al. 2001, except as noted): BRAZIL: Amapá, Serra do Navio; Pará, Monte Dourado (Cardoso 2011). FRENCH GUIANA: Cacao, Camopi, Iracoubo, Paracou (type locality of *Neacomys dubosti* Voss, Lunde, and Simmons), Piste de Saint Élie, Saül Trois Sauts. SURINAM: Sipaliwini, Oelemarie, Sipaliwini airstrip.

Map 181 Selected localities for *Neacomys dubosti* (●). Contour line = 2000 m.

SUBSPECIES: *Neacomys dubosti* is monotypic.

NATURAL HISTORY: Few data are available beyond the fact that the holotype of *N. dubosti* was taken in a pitfall trap in creekside primary forest (Voss et al. 2001). This species occurs in sympatry with *N. paracou* in the center of French Guiana (Säul; Mauffrey et al. 2007), and in several other localities in French Guiana, Surinam, and Brazil (Catzeflis and Tilak 2009; Cardoso 2011). It also occurs in sympatry with *N. guianae* in the Nickerie district of Surinam (Voss et al. 2001).

REMARKS: The karyotype of Amapá specimens is $2n = 62$ (Voss et al. 2001). Phylogeographic studies showed that *N. dubosti* has very little genetic variation, containing only half of the nucleotide diversity found in the co-distributed *N. paracou* (Catzeflis and Tilak 2009). The record from Pará state, Brazil was originally reported to be *N. guianae* (R. N. Leite 2006; see Cardoso 2011).

Neacomys guianae Thomas, 1905
Guianan Spiny Mouse

SYNONYM:

Neacomys guianae Thomas, 1905b:310; type locality "Demerara River, British Guyana. Alt. 120 feet," probably Demerara-Mahaica, Guyana.

DESCRIPTION: Small, with head and body length 64–85 mm; tail about as long as head and body (range: 67–81 mm), distinctly bicolored in some specimens, and with small caudal scales; hindfoot with short outer digits (claw of digit V extends only about three-fourths length of first phalanx of digit IV, claw of dI extends to less than one-half length of first phalanx of digit II). Cranially, rostrum moderately long and flanked by shallow zygomatic notches; interorbital region narrow, hourglass in shape or weakly convergent anteriorly; zygomatic plates narrow; incisive foramina short; and bullae flask shaped.

DISTRIBUTION: *Neacomys guianae* occurs in eastern Venezuela, Guyana, Surinam, and probably adjacent parts of northeastern Brazil, although confirmed specimens from this country are lacking. The record of this species from Monte Dourado, Pará state (R. N. Leite 2006) is a misidentified specimen of *N. dubosti* (Cardoso 2011). All localities are situated in lowland and premontane moist broadleaf forest in the Guianan subregion of Amazonia.

SELECTED LOCALITIES (Map 182; from Voss et al. 2001, except as noted): GUYANA: Cuyuni-Mazaruni, Kartabo; Demerara-Mahaica, Demerara River (type locality of *Neacomys guianae* Thomas). SURINAM: Sipaliwini, Sipaliwini airstrip, Tafelberg. VENEZUELA: Bolívar, 45 km NE of Icabarú.

SUBSPECIES: *Neacomys guianae* is monotypic.

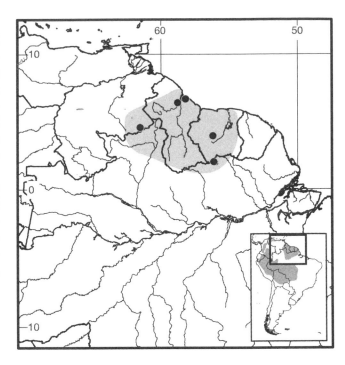

Map 182 Selected localities for *Neacomys guianae* (●). Contour line = 2000 m.

NATURAL HISTORY: *Neacomys guianae* occurs in sympatry with *N. dubosti* and *N. paracou* over much of the Guianan subregion (Voss et al. 2001).

REMARKS: Arenavirus Amaparí was isolated from *Neacomys* specimens collected in Serra do Navio and identified as *N. guianae* (F. P. Pinheiro et al. 1966; Cajimat et al. 2007; Charrel and de Lamballerie 2010); however only *N. paracou* and *N. dubosti* occur in this region (Voss et al. 2001). Specimens from Baramita (Barima-Waini region, Guyana) identified as *N. guianae* (Patton et al. 2000:Table 22) are actually *N. paracou* (Voss et al. 2001). Several other specimens identified as *N. guianae* in earlier publications (e.g., Carleton 1973; C. T. Carvalho 1962; Genoways et al. 1981; Hooper and Musser 1964; Husson 1978; Malcolm 1990; Voss and Emmons 1996; Patton et al. 2000; R. N. Leite 2006) are referable to *N. dubosti*, *N. paracou*, or *N. tenuipes* (see Voss et al. 2001; Cardoso 2011).

Neacomys minutus Patton, da Silva, and Malcolm, 2000
Minute Spiny Mouse

SYNONYM:

Neacomys minutus Patton, da Silva, and Malcolm, 2000: 105; type locality "Altamira, left bank Rio Juruá, Amazonas, Brazil, 06°35′S, 68°54′W."

DESCRIPTION: Another diminutive species, with head and body length 65–76 mm; relatively long tail (tail length 70–84 mm); dark orange dorsal coloration strongly but finely streaked with black; short ears (13 mm or less);

small and delicate skull with short maxillary toothrow (<2.75 mm); primitive carotid arterial system (pattern 1 of Voss 1988), with squamosoalisphenoid groove and sphenofrontal foramen; teardrop-shaped incisive foramina; and weakly developed anteromedian flexus(id) on both upper and molar first molars.

DISTRIBUTION: *Neacomys minutus* is known from the lowland, moist broadleaf rainforest of the western Amazon Basin, from the central and lower sections of the Rio Juruá, Amazonas state, Brazil, and from the Río Gálvez, Loreto department, in northeastern Peru.

SELECTED LOCALITIES (Map 183; from Patton et al. 2000, except as noted): BRAZIL: Amazonas, Altamira, right bank Rio Juruá (type locality of *Neacomys minutus* Patton, da Silva, and Malcolm), Barro Vermelho, left bank Rio Juruá, Colocação Vira-Volta, left bank of Rio Juruá, Igarapé Nova Empresa, left bank Rio Juruá, Penedo, right bank Rio Juruá, Sacado, right bank Rio Juruá. PERU: Loreto, Nuevo San Juan, Río Gálvez (AMNH 272867).

SUBSPECIES: *Neacomys minutus* is monotypic.

NATURAL HISTORY: *Neacomys minutus* was caught both in upland (*terra firme*) and seasonally flooded (*várzea*) forest along the Rio Juruá (Patton et al. 2000). Pregnant females collected during dry and wet season, and litter size of three young. One lactating female was also pregnant, suggesting a postpartum estrus in this species. Reproductively active individuals of both sexes still in partial juvenile pelage but with completely erupted but unworn teeth suggest that breeding commences at an early age. This species is sympatric with *N. spinosus* and is apparently replaced by the very similar *N. musseri* in the headwaters of the Rio Juruá (Patton et al. 2000), but the two species are sympatric on the Río Gálvez in northeastern Peru (Catzeflis and Tilak 2009).

REMARKS: Substantial molecular divergence (8% in mtDNA cytochrome-*b* sequences) divide this species into upriver versus downriver clades along the Rio Juruá in western Brazil (Patton et al. 2000). Concordant nonoverlap in discriminate morphometric scores across the same clades and geography may signal species-level divergence in this taxon. If so, and as the holotype of *N. minutus* is a member of the upriver clade, populations from the lower Rio Juruá would lack an available name.

The karyotype of *N. minutus* is characterized by $2n = 35$–36, FN $= 40$, the uneven diploid number due to a Robertsonian fusion in the autosomal complement (Patton et al. 2000). Specimens of small *Neacomys* collected elsewhere in Peru, especially north of the Río Marañón, and in Ecuador are morphologically similar to *N. minutus* but are extremely divergent in mtDNA cytochrome-*b* sequences (Patton et al. 2000). These are best considered a separate and as yet undescribed species. Aniskin (1994) reported a karyotyped *Neacomys* sp. from Genaro [= Jenaro] Her-

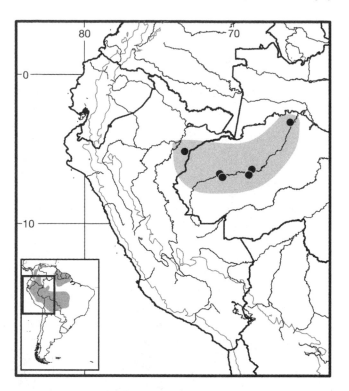

Map 183 Selected localities for *Neacomys minutus* (●). Contour line = 2000 m.

rera, Peru, with $2n = 30$–32 and FN $= 38$, plus from 1 to 6 B chromosomes. This karyotype, although clearly different from those reported by Patton et al. (2000) for *N. minutus*, shows some resemblance, especially the morphology of the three largest acrocentric pairs. In addition, this *Neacomys* sp. has a large submetacentric that might be homologous to the heterozygous pair due to a centric fusion observed in some specimens of *N. minutus* (Patton et al. 2000).

Neacomys musseri Patton, da Silva, and Malcolm, 2000
Musser's Spiny Mouse

SYNONYM:

Neacomys musseri Patton, da Silva, and Malcolm, 2000:98; type locality "72 km NE Paucartambo (by road), at km 152, Manu Biosphere Reserve, Departamento de Cusco, Peru, 1460 m."

DESCRIPTION: Another small species, with head and body length 65–79 mm; tail only slightly longer (68–90 mm), slightly bicolored dark brown above, paler below, with scales visible but clothed in short hairs. Cranially similar to other small forms, but unique in having derived carotid pattern 2 (of Voss 1988), without squamosoalisphenoid groove and sphenofrontal foramen but with persistence of large stapedial foramen. M1 with deep anteromedian flexus that divides procingulum into two subequal conules, and incisive foramina evenly rounded rather than teardrop in shape.

DISTRIBUTION: *Neacomys musseri* is known from lowland and lower montane moist broadleaf forest in western Amazonia Brazil and eastern Peru. The recent record of *N. musseri* from San Fermín on the Río Tambopata in extreme eastern Puno department (Pacheco et al. 2012), Peru, suggests that the species might range into adjacent Bolivia.

SELECTED LOCALITIES (Map 184): BRAZIL: Acre, left bank Rio Juruá opposite Igarapé Porongaba (Patton et al. 2000). PERU: Cusco, 72 km NE of Paucartambo (type locality of *Neacomys musseri* Patton, da Silva, and Malcolm); Loreto, Nuevo San Juan, Río Gálvez (AMNH 272676); Madre de Dios, Quebrada Aguas Calientes, left bank Río Alto Madre de Dios, 2.75 km E of Shintuya (Solari et al. 2006); Puno, San Fermín (Pacheco et al. 2012).

SUBSPECIES: *Neacomys musseri* is monotypic.

NATURAL HISTORY: Patton et al. (2000) reported a pregnant subadult female with two embryos, collected on the ground in *várzea* habitat during the rainy season (February) in Brazil. Conversely, none of the specimens collected in Peru during the dry season (July) were in reproductive condition (Patton et al. 2000). *Neacomys musseri* is found in sympatry with *N. spinosus* (Patton et al. 2000; Solari et al. 2006) and with *N. minutus* in northeastern Peru (Catzeflis and Tilak 2009).

REMARKS: The karyotype of *N. musseri* is characterized by $2n = 34$, $FN = 64$ or 68 (Patton et al. 2000). This complement is similar in diploid number to that reported

by Aniskin (1994; see account for *N. minutus*), but differs markedly in the number of chromosome arms. Specimens that Woodman et al. (1991) referred to *N. tenuipes* from the Río Madre de Dios in southeastern Peru are likely this species. Because the range of *N. musseri* extends almost to the Bolivian border, the species may occur in that country.

Neacomys paracou Voss, Lunde, and Simmons, 2001
Paracou Spiny Mouse

SYNONYM:

Neacomys paracou Voss, Lunde, and Simmons, 2001:81; type locality type locality "Paracou [= Domaine Experimental Paracou]," French Guiana.

DESCRIPTION: Distinguished from other small-sized congeners by very short outer pedal digits (claw of dV extends only two-thirds length of phalanx 1 of dIV, claw of dI barely extends beyond base of phalanx 1 of dII); short and usually unicolored tail (sometimes weakly bicolored basally) with large scales; short rostrum flanked by relatively deep zygomatic notches; broad and usually strongly convergent interorbital region with well-developed and often shelf-like supraorbital beads; narrow and uninflated braincase; long and parallel-sided incisive foramina; carotid circulatory pattern 1 (of Voss 1988); globular auditory bullae; M1 with narrow and undivided anterocone, stout mesoloph often curving from and disproportionately connected to the hypocone; and principal molar cusps that wear quickly to enamel loops, not persistently tubercular.

DISTRIBUTION: *Neacomys paracou* occurs in French Guiana, Surinam, Guyana, eastern Venezuela (Bolívar state), and northern Brazil (Amapá, Amazonas, and Pará states). All localities are situated in lowland and premontane moist broadleaf forest in the Guianan subregion of Amazonia (Voss et al. 2001).

SELECTED LOCALITIES (Map 185; from Voss et al. 2001, except as noted): BRAZIL: Amapá, Serra do Navio; Amazonas, 80 km N of Manaus; Pará, Cachoeira Porteira. FRENCH GUIANA: Arataye, Cayenne, Mont St. Michel (Catzeflis and Tilak 2009), Paracou (type locality of *Neacomys paracou* Voss, Lunde, and Simmons). GUYANA: Barima-Waini, Baramita; Cuyuni-Mazaruni, Kartabo; Potaro-Siparuni, Kurupukari; Upper Takutu–Upper Essequibo, Nappi Creek. SURINAM: Marowijne, Perica; Para, Locksie Hattie; Sipaliwini, Oelemarie. VENEZUELA: Bolívar, San Ignacio de Yuruani.

SUBSPECIES: *Neacomys paracou* is monotypic.

NATURAL HISTORY: Voss et al. (2001) collected *N. paracou* in primary forest, either well drained or swampy, and in secondary vegetation, always at or near ground level. In Brazil, this is a rare species occurring mainly in primary forest (R. N. Leite 2006). *Neacomys paracou* occurs in sym-

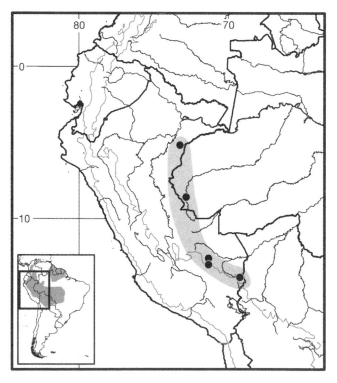

Map 184 Selected localities for *Neacomys musseri* (●). Contour line = 2000 m.

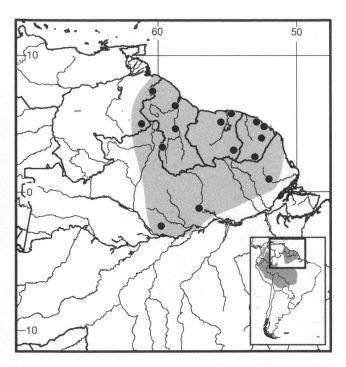

Map 185 Selected localities for *Neacomys paracou* (●). Contour line = 2000 m.

patry with *N. dubosti* and *N. guianae* (Voss et al. 2001; R. N. Leite 2006).

REMARKS: Voss et al. (2001) recorded a karyotype with $2n = 56$ for specimens from French Guiana and eastern Venezuela.

Neacomys spinosus (Thomas, 1882)
Large Spiny Mouse
SYNONYMS:

Hesperomys (*Calomys*) *spinosus* Thomas, 1882:105; type locality "Huambo—a plantation in the forest of the same name, to the east of Chachapoyas and Chirinoto, 3700 feet in altitude, on the banks of the river Huambo, a tributary of the Huallaga," Amazonas, Peru.

Hesperomys (*Oryzomys*) *spinosus*: Thomas, 1884:448; name combination.

Oryzomys spinosus: E.-L. Trouessart, 1897:528; name combination.

N[*eacomys*]. *s*[*pinosus*]. *typicus* Thomas, 1900a:153; *nomen nudum*, in reference to the type species of *Neacomys*, *Oryzomys spinosus* Thomas.

Neacomys spinosus amoenus Thomas, 1904a:239; type locality "Santa Ana da Chapada, a village situated at an altitude of about 800 m., on the Serra do Chapada [= Chapada dos Guimarães], some thirty miles N.E. of Cuyabá," Mato Grosso, Brazil.

Neacomys spinosus spinosus: Gyldenstolpe, 1932a:34; name combination.

Neacomys spinosus carcelini Hershkovitz, 1940c:1; type locality "Llunchi, an island on the northern side of the Río Napo, west of the mouth of the Río Jivino, latitude and longitude approximately 0°37'S., 76°46'W.; parish of La Coca, Napo-Pastaza, Ecuador; altitude about 250 meters."

DESCRIPTION: Largest species in genus, with head and body length 75–105 mm, almost nonoverlapping other species, and large skull, nearly 10% larger than that of other species. Specimens from eastern Ecuador to southern Peru and western Brazil rather uniform in body dimensions, but exhibit variation in darkness of dorsal pelage, from paler yellow-reddish brown mixed with black to darker reddish-brown. Thomas (1904a), in his description of *amoenus* from Mato Grosso state, Brazil, noted brighter ochraceous coloration in comparison with western Amazonian samples. Skull long with relatively narrow braincase, differing only in size but not proportions from other species compared by Patton et al. (2000). Carotid circulation pattern of primitive type, retaining well-developed squamosoalisphenoid groove and sphenofrontal foramen indicative of presence of supraorbital branch of stapedial artery (pattern 1 of Voss 1988). Maxillary toothrow long, averaging >3 mm; molar occlusal morphology similar to that of other species except that procingulum of both M1 and m1 either entire, or only weakly divided into anterolabial and anterolingual conules by anteromedian flexus.

DISTRIBUTION: *Neacomys spinosus* is the most widely ranging species, occurring throughout the Amazon Basin from west-central Brazil to the Andean foothills and lowlands of southeastern Colombia, eastern Ecuador, eastern Peru, and northern and central Bolivia. It can be found commonly in both lowland evergreen rainforest and in midmontane forest on the eastern Andean slopes, to elevations of nearly 2,000 m.

SELECTED LOCALITIES (Map 186): BOLIVIA: Chuquisaca, 2 km E of Chuhuayaco (S. Anderson 1997); Cochabamba, El Palmar, Río Cochi Mayu (S. Anderson 1997); La Paz, 8 km from mouth of Río Madidi (S. Anderson 1997), Río Zongo (S. Anderson 1997); Pando, La Cruz (S. Anderson 1997); Santa Cruz, Estancia Cachuela Esperanza (S. Anderson 1997). BRAZIL: Acre, Sobral, left bank Rio Juruá (Patton et al. 2000); Amazonas, Penedo, right bank Rio Juruá (Patton et al. 2000), Seringal Condor, left bank Rio Juruá (Patton et al. 2000); Goiás, Baliza (UFPB [CRB 70]); Mato Grosso, Ponte Branca (UNB 1193), São José do Xingu (MNRJ [CRB 2794]); Pará, Castelo dos Sonhos (MNRJ [SVS/PA 282]). COLOMBIA: Meta, Finca El Buque, Villavicencio (A. L. Gardner and Patton 1976); Putumayo, Río Mecaya (FMNH 71792). ECUADOR: Orellana, Parque Nacional Yasuní (Patton et al. 2000);

Pastaza, Tinguino, ca. 130 km S of Coca (USNM 574567); Zamora-Chinchipe, Zamora (AMNH 47795). PERU: Amazonas, Huambo (type locality of *Hesperomys* (*Calomys*) *spinosus* Thomas), Huampani, Río Cenepa (Patton et al. 2000); Ayacucho, Río Santa Rosa (A. L. Gardner and Patton 1976); Loreto, Reserva Nacional Allpahuayo-Mishana (Hice and Velazco 2012); Madre de Dios, Albergue Cusco Amazonica, Río Madre de Dios, ca. 12 km E of Puerto Maldonado (MVZ 157811), Cocha Cashu Biological Station (Solari et al. 2006); Puno, 11 km NNE of Ollachea (MVZ 172650); Ucayali, Balta, Río Curanja (A. L. Gardner and Patton 1976), Pucallpa (Aniskin 1994).

SUBSPECIES: Cabrera (1960) recognized three subspecies: the nominotypical form from the Andean slopes and Amazon Basin of Peru, *N. s. amoenus* from southwestern Brazil, and *N. s. carceloni* from the Andean foothills and Amazon Basin of Colombia and Ecuador. The validity of these taxa, however, must await an analysis of character variation over the range of the species, something that has yet to be accomplished.

NATURAL HISTORY: In Peru, Hice and Velazco (2012) recorded pregnant females in June, November, and December, with an average litter size of 3.1 (N = 10, range 2–4), suggesting year-round breeding, but perhaps with a peak at the beginning of the rainy season in December. On the Rio Juruá in Brazil, pregnant females were captured both in the wet (February and March) and dry (August and September) seasons, also suggesting year-round breeding (Patton et al. 2000). Here, embryo counts ranged from two to four. In Peru, individuals were frequently captured in secondary growth and cultivated land, only rarely in forested habitats (Hice and Velazco 2012). Along the Rio Juruá, specimens were collected in primary rainforest and secondary growth in both *terra firme* and *várzea* habitats (Patton et al. 2000). Bonvicino, Cerqueira, and Soares (1996) also encountered this species in conserved forested formations elsewhere in Brazil. *Neacomys spinosus* occurs in sympatry with one or more small species over much of its range in western Amazonia, specifically with both *N. minutus* and *N. musseri* in western Brazil and Peru, and with other as yet unnamed species in northern Peru and eastern Ecuador (Patton et al. 2000). Some recent publications still refer to *Neacomys spinosus* as the reservoir of the Rio Mamoré hantavirus (e.g., Bi et al. 2008), but the role of this species as a reservoir is questionable (Young et al. 1998). The only verified reservoir of this hantavirus is *Oligoryzomys microtis* (Richter et al. 2010). Gettinger and Gardner (2005) described the mite *Laelaps neacomydis* from the pelage of *Neacomys spinosus* collected in Bolivia.

REMARKS: A uniform karyotype with $2n = 64$, FN = 68 has been recorded for samples from eastern Colombia, throughout eastern Peru, and from western Brazil (A. L.

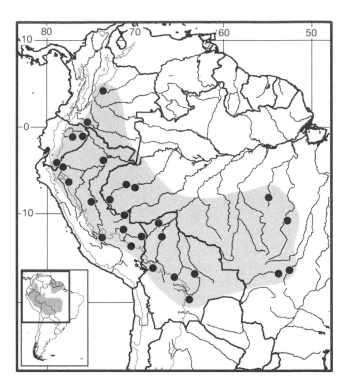

Map 186 Selected localities for *Neacomys spinosus* (●). Contour line = 2000 m.

Gardner and Patton 1976; Aniskin 1994; Patton et al. 2000).

The mtDNA cytochrome-*b* sequence data are geographically limited for such a widely distributed species, with about 2% divergence between samples from eastern Ecuador versus those from western Brazil (Patton et al. 2000). However, unpublished data (J. L. Patton) document substantial mtDNA cytochrome-*b* divergence (up to 8.5%) between a sample from the middle Andean slopes in southern Peru and specimens from eastern Ecuador, northern Peru, and western Brazil, perhaps signaling species-level divergence.

Neacomys tenuipes Thomas, 1900
Slender Foot Spiny Mouse

Neacomys spinosus tenuipes Thomas, 1900a:153; type locality "Guaquimay, near Bogotá," Cundinamarca, Colombia.

Neacomys pusillus J. A. Allen, 1912:81; type locality "San José [near Barbacoas], Cauca, Colombia."

Neacomys tenuipes: Lawrence, 1941b:425; first use of current name combination.

DESCRIPTION: Another small species, with head and body length 72–97 mm. Tail longer than head and body (about 115%, length 74–108 mm), contrasting with subequal tails of other small species; distinctly bicolored, at least near base, with small caudal scales. Hindfoot with long outer digits (claw of dV extending to end of first phalanx of dIV, claw of dI to at least middle of first phalanx of dII). Skull with long rostrum flanked by shallow zygomatic notches; interorbital region narrow and either hourglass in shape or weakly convergent anteriorly; zygomatic plates narrow; incisive foramina short; and auditory bullae flask shaped. M1 approximately rectangular in occlusal view, with broad, often divided anterocone; labial and lingual cusps subequal and persistently tubercular; and slender mesoloph with symmetrical attachments to protocone and hypocone.

DISTRIBUTION: *Neacomys tenuipes* is distributed in western and north-central Colombia and in northern Venezuela. Tirira (2007) extended the distribution into northwestern Ecuador under the species name *N. pictus* but provided no specific localities. All known localities are situated in premontane or lower montane forests, between 400 and 1,750 m (Voss et al. 2001; Musser and Carleton 2005; Tirira 2007).

SELECTED LOCALITIES (Map 187; from Voss et al. 2001): COLOMBIA: Antioquia, 11 km S and 30 km E of Cisneros, Quebrada del Oro, Sonsón, 25 km S and 22 km W of Zaragoza; Boyacá, Muzo; Caldas, Samana; Cundinamarca, Paime; Huila, Acevedo. VENEZUELA: Aragua, Rancho Grande; Falcón, Cerro Socopo; Vargas, Los Venados; Yaracuy, Minas de Aroa.

SUBSPECIES: We treat *Neacomys tenuipes* as monotypic.

NATURAL HISTORY: Virtually nothing is known about the ecology or behavior of *Neacomys tenuipes*, al-

though, like other species in the genus, it is assumed to be nocturnal, solitary, and terrestrial. This species is considered vulnerable in Ecuador (Tirira 2001). Laelapine mites are associated with *N. tenuipes* in Venezuela (Furman 1972a).

REMARKS: Specimens from eastern Ecuador allocated to this species by Lawrence (1941b) are not *N. tenuipes* but likely represent the unnamed small species east of the northern Andes that Patton et al. (2000) identified as *Neacomys* sp. Clade 3 and Tirira (2007) referred to as *N.* cf. *minutus*. Neither the karyotype nor any analysis of geographic variation, in morphology or molecular sequences, has been recorded for *N. tenuipes*.

Genus *Nectomys* Peters, 1861

Cibele R. Bonvicino and Marcelo Weksler

Commonly called water rats, members of the genus *Nectomys* are found in most lowland rainforest biomes in South America, and in the riverine (gallery) forests of open habitats, such as the Cerrado, Llanos, and other grasslands. We recognize five species: *N. apicalis*, *N. grandis*, *N. palmipes*, *N. rattus*, and *N. squamipes*, but this number will likely change in the future due to continuing taxonomic research on the genus (which needs comprehensive revision). These rats are endemic to South America, found in all countries (except Chile) and on the continental shelf islands of Trinidad and Margarita, from sea level to 2,200 m (Hershkovitz 1944). Pardiñas et al. (2002) recorded fossil *Nectomys* from Middle to Late Pleistocene deposits in Argentina and Recent deposits in both Argentina and Brazil.

All species of the genus *Nectomys* share similar integumental, cranial, and dental characters. Water rats are of large size (head and body length 150–254 mm) with partially webbed hindfeet and long tails (tail length 151–255 mm). The pelage is thick and long, with well-developed underfur and long guard hairs. The dorsal pelage is dark, varying from brownish gray to a paler gray; the lateral pelage is paler than the back, and the venter is whitish or ochraceous with a gray base. The hindfeet are large, long, and have interdigital webbing and natatory fringes along the medial margins of digits I and II and lateral margins of digits IV and V. The plantar surface is naked and densely set with epidermal squamae (the heel is smooth); the hypothenar pad may be absent or reduced, and the four interdigital pads are very small. The tail is longer than head and body length, naked in appearance (small triads of hair are visible under magnification), and unicolored. There are eight pairs of mammae, one each in inguinal, abdominal, postaxial, and pectoral positions.

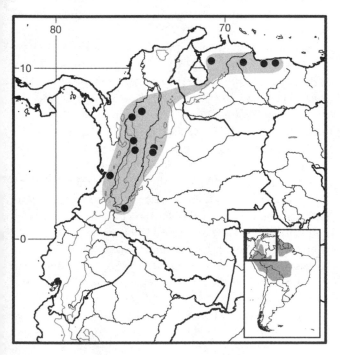

Map 187 Selected localities for *Neacomys tenuipes* (●). Contour line = 2000 m.

The skull is robust, with a broad and short rostrum flanked by deep zygomatic notches; the interorbital margins converge anteriorly and possess strongly beaded supraorbital edges; the braincase is squared and relatively narrow, with well-developed temporal, lambdoidal, and nuchal crests. The zygomatic plates lack anterodorsal spinous processes, and their posterior margins lie level to the alveoli of the M1s. The nasal bones have sharply convergent posterior margins. The posterior wall of the orbit is smooth. The frontosquamosal suture is colinear with the frontoparietal suture, and the parietals have large lateral expansions. The bony palate between the molar rows is smooth and long, and the mesopterygoid fossa extends anteriorly between the maxillae but does not extend to the level of the third molars. Posterolateral palatal pits are recessed in deep fossae, and the bony roof of mesopterygoid fossa is completely ossified. An alisphenoid strut is absent (buccinator-masticatory foramen and accessory foramen ovale are confluent), and the alisphenoid canal has a large anterior opening. The stapedial foramen and the posterior opening of the alisphenoid canal are small, and the squamosoalisphenoid groove and sphenofrontal foramen are absent; a secondary anastomosis of the internal carotid crosses the dorsal surface of the pterygoid plate (= carotid circulatory pattern 3 of Voss 1988). The postglenoid foramen is large and rounded, but the subsquamosal fenestra is absent or vestigial. The mastoid is completely ossified or has only a small perforation. In the mandible, the capsular process is absent or poorly developed, and the superior and inferior masseteric ridges are conjoined anteriorly as a single crest below m1. The basihyal lacks the entoglossal process.

Hershkovitz (1944:19) characterized the molars as large, pentalophodont, and bunodont in occlusal configuration but with with high crowns, low crests, their dentine exposed at an early stage of wear, and fully-rooted. The labial and lingual flexi of M1 and M2 approach the midline of the tooth but do not interpenetrate. The anterocone of M1 is not divided, and the anteroloph is well developed and connected to the anterocone by the labial cingulum. The protostyle is absent, and the paracone is usually connected to the posterior moiety of the protocone. There is no protoflexus on M2, but the mesoflexus is present as single internal fossette, as the mesoloph is connected to the paracone by a labial cingulum. M3 has a posteroloph. The m1 anteroconid lacks an anteromedian flexid, an anterolophid is present on m1 but absent from m2 and m3, and ectolophids are absent from m1 and m2. Both M1 and m1 have accessory roots, and m2 and m3 each has two roots.

The stomach is unilocular-hemiglandular, but the glandular epithelium is extensive and reaches the corpus (Hersh-

kovitz 1962; Carleton 1973). The distal bacular cartilage of the glans penis is large and trifid; the central digit is robust.

SYNONYMS:

Mus: Brants, 1827:138: part (description of *squamipes*); not *Mus* Linnaeus.

Nectomys Peters, 1861:151; type species *Mus squamipes* Brants, by subsequent designation (G. S. Miller 1912: 180).

Holochilus: Fitzinger, 1867b:90; part (listing of *squamipes*); not *Holochilus* Brandt.

Potamys Liais, 1872:505; type species *Potamys brasiliensis*, by original designation (= *N. squamipes*); a homonym of *Potamys* Larranhaga, 1823:83, a junior synonym of *Myocastor* (see Hershkovitz 1944:26; McKenna and Bell 1997:207).

Hesperomys: Hensel, 1872b:28; part (description of *rattus*); not *Hesperomys* Waterhouse.

Sigmodontomys: Bonvicino, 1999:254; not *Sigmodontomys* J. A. Allen.

REMARKS: Hershkovitz (1944, 1948b) divided *Nectomys* into two subgenera, *Nectomys* and *Sigmodontomys* J. A. Allen, and considered *Nectomys squamipes* (Brants) the single species of the subgenus *Nectomys*, and within which he recognized 17 subspecies. Cabrera (1961) proposed a similar taxonomic arrangement, but A. L. Gardner and Patton (1976), on the basis of karyology and bacular morphology, considered *Sigmodontomys* to be a subgenus of *Oryzomys* Baird, 1857, and postulated that *Nectomys* was polytypic. Musser and Carleton (1993) regarded *Sigmodontomys* as a full genus and included three species in *Nectomys* (*N. squamipes*, *N. palmipes*, and *N. parvipes*).

Bonvicino (1994) argued that chromosomal diversification played an important role in speciation of water rats; eight basic diploid numbers have been described for *Nectomys*, including $2n = 16$, $2n = 17$, $2n = 34$, $2n = 38$, $2n = 40$, $2n = 42$, $2n = 52$, and $2n = 56$ (Yonenaga 1972a; Yonenaga-Yassuda et al. 1976; A. L. Gardner and Patton 1976; T. R. O. Freitas 1980; Furtado 1981; Sbalqueiro, Mattevi, Freitas, and Oliveira 1982; Sbalqueiro et al. 1986, 1987; Bossle et al. 1988; Maia et al. 1984; Zanchin 1988; M. A. Barros et al. 1992; Bonvicino, D'Andrea et al. 1996; Goméz-Laverde et al. 1999; Patton et al. 2000; Bonvicino and Gardner 2001). Bonvicino (1994) proposed that *Nectomys* comprised 11 species, and placed *N. parvipes* as a species in the genus *Sigmodontomys* (see also Bonvicino 1999); we include *N. parvipes* as a synonym of *N. rattus* (see below).

Most recently, Musser and Carleton (2005) retained *Sigmodontomys* as a valid genus and recognized five species of *Nectomys* (*N. apicalis*, *N. magdalenae* [= *N. grandis*], *N. palmipes*, *N. rattus*, and *N. squamipes*). Weksler's

(2006) phylogenetic analysis corroborated the recognition of *Nectomys* and *Sigmodontomys* as different genera. Our taxonomic arrangement here follows that of Musser and Carleton (2005) but should be viewed as provisional. Notably, we consider the different karyotypes as diagnostic traits of subspecies in some cases, such as *N. grandis grandis* and *N. palmipes tatei*, differentiating them from *N. grandis magdalenae* and *N. palmipes palmipes*, respectively, to reinforce the need for further studies on these taxonomic entities. Furthermore, despite chromosome variation found across the geographic range of *N. apicalis*, we retained all karyomorphs in a single species due to ongoing phylogeographic analyses of mtDNA cytochrome-*b* sequences that document genetic cohesion throughout this distribution for the karyomorphs with $2n = 38$, $2n = 40$, and $2n = 42$. Finally, the identity of the $2n = 56$ cytotype from Surinam (R. J. Baker et al. 1983) also needs further study, as it likely belongs to an undescribed species.

KEY TO THE SPECIES OF *NECTOMYS*:

1. Interparietal deep (anteroposterior dimension) relative to its width (transverse dimension); exoccipital occupying most of the dorsolateral surface of occiput 2
1′. Interparietal much wider than deep; dorsolateral exposition of the exoccipital reduced 4
2. Dorsal color light brown; distributed in Trinidad and northeastern Venezuela; diploid number <20 . . . . *Nectomys palmipes*
2′. Dorsal color brown; distributed in Colombia, Peru, Ecuador, Brazil, and probably Bolivia; diploid number >30 3
3. Dorsal color overall grayish-ochraceous without a dark middorsal band; distributed in northern Andean valleys of Colombia, in Cauca and Magdalena rivers basins; $2n = 32$ and 34 . . . . *Nectomys grandis*
3′. Distinct dark middorsal band covering less than one third of dorsal breadth; distributed in western Amazonia (Ecuador, Peru, Brazil, and probably Bolivia); $2n = 38–42$ . . . . *Nectomys apicalis*
4. Distributed in the basins of the Río Uruguay, Río Paraná, Rio São Francisco, and in independent river basins south of São Lourenço da Mata, Pernambuco, in the Atlantic Forest of Brazil, Uruguay, and Argentina; $2n = 56$ . . . .*Nectomys squamipes*
4′. Distributed in the Amazonian region of Colombia and Venezuela (upper and middle Río Orinoco Basin), eastern Ecuador, Peru (Ucayali Basin), and Brazil (parts of Amazon Basin, Río Paraguay Basin, and independent river basins of the Atlantic Forest north of São Lourenço da Mata, Pernambuco; $2n = 52–54$ . . . . *Nectomys rattus*

Nectomys apicalis Peters, 1861
Western Amazonian Water Rat
SYNONYMS:

Nectomys apicalis Peters, 1861:152; type locality "Guayaquil"; subsequently restricted to Tena, Napo-Pastaza Province, Ecuador (Hershkovitz 1944:26).

Nectomys fulvinus Thomas, 1897c:499; type locality "Believed to be Quito," Ecuador. Hershkovitz (1944:26–27), however, convincingly argued that "it is certain that [the type of] *fulvinus* was collected by Jameson somewhere east of the Andes during his excursion from Quito to the Río Napo."

Nectomys saturatus Thomas, 1897f:546; type locality "Ibarra, [Imbaburra], N. Ecuador, alt. 2225 metres."

Nectomys garleppii Thomas, 1899a:41; type locality "Occobamba, Cuzco," Peru; subsequently restricted to "a station somewhere in the tropical or subtropical zone of the Río Ocobamba Valley, department of Cuzco, Perú" by Hershkovitz (1944:59).

Nectomys garleppi Thomas, 1902b:129; incorrect subsequent spelling of *Nectomys garleppii* Thomas.

Nectomys squamipes napensis Hershkovitz, 1944:56–57; type locality "site 'San Francisco', left bank of Río Napo, above the mouth of the Río Challuacocha; latitude and longitude approximately 0°47′S, 76°25″W, Parish of La Coca, Napo-Pastaza Province, Ecuador; altitude, about 200 meters."

Nectomys squamipes garleppii: Hershkovitz, 1944:59; name combination.

Nectomys squamipes vallensis Hershkovitz, 1944:61; type locality "Santa Ana, a semiarid, tropical pocket in the Río Urubamba Valley, department of Cuzco, Peru, altitude 3480 feet."

Nectomys squamipes saturatus: Hershkovitz, 1948b:51; name combination.

DESCRIPTION: Dorsum ochraceous buff with mixture, rarely with heavy overlay, of cinnamon brown or dark brown; dark middorsal band narrowly defined (ocassionally absent); coloration from nose to crown on head not markedly different from that of middorsal region of back; under parts pallid neutral gray moderately washed with ochraceous; inner surface of forelimbs from wrist to elbow usually brown with mixture of ochraceous; tail brown, keel entirely brown or mixed with gray to entirely gray; hindfoot usually with five plantar tubercles.

DISTRIBUTION: *Nectomys apicalis* occurs in eastern Ecuador, eastern Peru, and northwestern Brazil (southwestern Amazonas state), and is probably also found in western Bolivia and Acre state in Brazil.

SELECTED LOCALITIES (Map 188): BRAZIL: Amazonas, Barro Vermelho (MVZ 190374), Seringal Condor (INPA 3070/MVZ 193781). ECUADOR: Pastaza, Sarayacu

(AMNH 67327); Pichincha, Nanegal (AMNH 47009); Sucumbíos, Limoncocha (USNM 513585); Zamora-Chinchipe, 4 km ENE of Los Encuentros (USNM 513584). PERU: Amazonas, vicinity of Huampami, Río Cenepa (MVZ 153533); Ayacucho, Hacienda Luisiania, left bank of Río Apurimac (LSUMZ 16712); Cusco, Kiteni, Río Urubamba (MVZ 166700); Junín, 3 mi SW of San Ramón (AMNH 231190); Loreto, Boca del Río Curaray (AMNH 71594); Ucayali, Balta, Río Curanja (MVZ 136641), Lagarto, upper Río Ucayali (AMNH 76512).

SUBSPECIES: We regard *Nectomys apicalis* as monotypic. However, the degree to which the karyological variation found in this species ($2n = 38$, 40, or 42) might map onto patterns of regional morphological discontinuity needs to be investigated.

NATURAL HISTORY: This species has not been studied in the field, so data on ecology or behavior are scarce. Patton et al. (2000) noted that individuals with $2n = 42$ collected along the Rio Juruá were trapped mainly along streams in undisturbed or second-growth forests, or in garden plots adjacent to such forest. Juveniles were collected between August and November, and adult females taken over the same months were either pregnant (three embryos; N = 1) or lactating (three placental scars; N = 1).

REMARKS: Diploid and autosomal fundamental numbers in this species are variable. Populations from Ecuador have $2n = 40$, FN = 40 (Bonvicino and Gardner 2001);

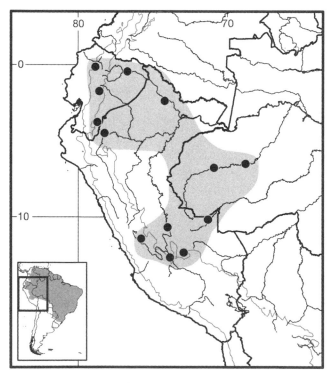

Map 188 Selected localities for *Nectomys apicalis* (●). Contour line = 2,000 m.

those from the Alto Marañón drainage in northern Peru (Patton et al. 2000), the headwaters of the Río Purus in Peru (A. L. Gardner and Patton 1976, as *N. squamipes* variant 2), and the Rio Juruá in western Brazil (Patton et al. 2000) were $2n = 42$, FN = 40–41; those from southeastern Peru were $2n = 38$, FN = 40 (A. L. Gardner and Patton 1976, as *N. squamipes* variant 3). These karyotypes are easily derived from one another by centric fusion and pericentric inversions. Note that populations of *Nectomys rattus* ($2n = 52$) are found along the central Río Ucayali, between those of *Nectomys apicalis* population with $2n = 40$ and those with $2n = 38$ (A. L. Gardner and Patton 1976, as *N. squamipes* variant 1). Unpublished analysis of molecular variation in these forms reveals a genetic cohesion across the range of the species, but the significance of the karyotypic variation in *N. apicalis* requires further study.

Nectomys grandis Thomas, 1897
Magdalena-Cauca Water Rat
SYNONYMS:

Nectomys grandis Thomas, 1897c:498; type locality "Concordia, Medellin, Colombia"; emended to "Concordia, western slope of the Río Cauca Valley, Medellín, Antioquía, Colombia; altitude 1790 meters" (Hershkovitz 1944:62).

Nectomys magdalenae Thomas, 1897c:499; type locality "W. Cundinamarca, in lowlands near Magdalena R.," Colombia.

Nectomys squamipes grandis: Hershkovitz, 1944:62; name combination.

Nectomys squamipes magdalenae: Hershkovitz, 1944:64; name combination.

DESCRIPTION: Dorsal pelage color overall coarsely mixed grayish ochraceous, with ochraceous tones predominant; dark middorsal stripe normally absent. Skull heavy with upper profile evenly convex throughout, and frontal and parietal regions markedly swollen; uniform convexity markedly different from frontally flattened skulls of other species. Nasals fairly broad, slightly compressed about their middle, pointed behind, but not markedly extended posteriorly; supraorbital ridges curve slightly as they diverge, continuous with temporal ridges; frontoparietal suture almost transverse; interparietal longer anteroposteriorly and narrower transversely than in other species; incisive foramina widely open, parallel sided, and evenly rounded at both ends; posterior palate with unusually large and deep lateral pits; palatal margin of mesopterygoid fossa squared.

DISTRIBUTION: *Nectomys grandis* is found throughout the Río Magdalena and the Río Cauca basins in north-central Colombia.

SELECTED LOCALITIES (Map 189): COLOMBIA: Antioquía, Bellavista (FMNH 70110), Concordia (type lo-

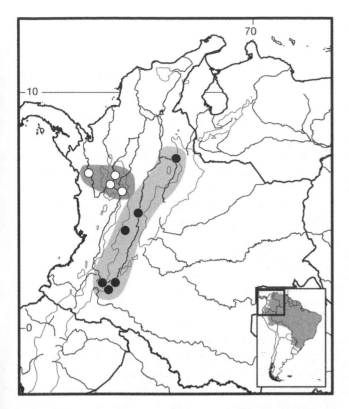

Map 189 Selected localities for *Nectomys g. grandis* (○) and *Nectomys g. magdalenae* (●). Contour line = 2,000 m.

cality of *Nectomys grandis* Thomas), La Pintada, S of Medellín (USNM 544410), San Jerónimo, about 35 km NW of Medellín (FMNH 70104); Cundinamarca, lowlands near Río Magdalena (type locality of *Nectomys magdalenae* Thomas); Huila, Andalucía (AMNH 33729), San Adolfo, Acevedo (FMNH 71640), San Agustín (AMNH 33736); Norte de Santander, Río de Oro, 91 km N of Tibú (AMNH 255816); Tolima, Río Chilí (AMNH 69174).

SUBSPECIES: *Nectomys grandis* includes two subspecies: *N. g. magdalenae*, characterized by $2n = 34$ and autosomal FN = 40 (Goméz-Laverde et al. 1999) and restricted to the Río Magdalena basin; and *N. g. grandis*, with $2n = 32$ (M. L. Bueno, pers. comm.) and restricted to the Río Cauca Basin.

NATURAL HISTORY: No information is available on the ecology or behavior of this species.

REMARKS: We tentatively recognize *grandis* and *magdalenae* as subspecies due to their allopatric distributions and different karyotypes. Further studies with a broader sample are necessary to clarify this taxonomic arrangement.

Nectomys palmipes J. A. Allen and Chapman, 1893
Trinidad Water Rat
SYNONYMS:

Nectomys palmipes J. A. Allen and Chapman, 1893:209; type locality "Princestown [= Princes Town], Trinidad."

Nectomys squamipes tatei Hershkovitz, 1948b:52; type locality "San Antonio, about 15 km. east of Mount Turumiquire, Sucre [actually Monagas], northern Venezuela; altitude about 1,800 feet."

DESCRIPTION: Pelage soft, dense, and glossy; overall dorsal color pale yellowish brown but middorsally strongly blackish from from nose to tail; sides grayish buffy brown, sparingly varied with black-tipped hairs; top of head, from muzzle to behind eyes, blackish varied with gray; under parts whitish with wash of pale buff and base grayish; line of demarcation between coloration of dorsal and ventral surfaces very indistinct; ears oval in shape, evenly rounded above, flesh colored at their base, dusky apically, and sparsely haired; lateral surfaces of limbs grayish brown; feet scaly, so thinly haired as to be nearly naked; palms and soles scaly; plantar surface of pes with five tubercles; tail slightly shorter than head and body, blackish, nearly unicolored, basal 10 mm or more heavily furred, distal part scantily clothed with short bristly hairs, which form very slight pencil at tip.

DISTRIBUTION: *Nectomys palmipes* occurs on the island of Trinidad, and in the Río Orinoco delta and other independent basins of northern Venezuela.

SELECTED LOCALITIES (Map 190): TRINIDAD AND TOBAGO: Trinidad, Bush Bush Forest (AMNH 185474), Las Cuevas, Maraba Bay (M. A. Barros et al. 1992), Princestown (type locality of *Nectomys palmipes* J. A. Allen and Chapman), Savanna Grande (FMNH 8772). VENEZUELA: Anzoátegui, Cueva del Agua (M. A. Barros et al. 1992); Delta Amacuro, La Horqueta, Tucupita (M. A. Barros et al. 1992); Monagas, Cachipo (M. A. Barros et al. 1992), Caripito (AMNH 142608), San Antonio, 15 km E of Mount Turumiquere (type locality of *Nectomys squamipes tatei* Hershkovitz); Sucre, Dos Ríos (M. A. Barros et al. 1992), Latal (AMNH 69900).

SUBSPECIES: *Nectomys palmipes* includes two subspecies: *N. p. palmipes* with $2n = 17$ from Trinidad, and *N. p. tatei* with $2n = 16$ from Venezuela.

NATURAL HISTORY: Like all other species of this genus, *N. palmipes* is adapted for semiaquatic existence. Jonkers et al. (1968) reported a Caraparu-like virus, which causes Venezuelan equine encephalitis, in *N. p. palmipes* from Trinidad. Known ectoparasites of *N. p. tatei* include the flea *Polygenis punni* (Tipton and Machado-Allison 1972). Endoparasites of *N. p. palmipes* include the nematode *Litomosoides taylori* (Guerrero et al. 2011).

REMARKS: The name *palmipes* was first used by J. A. Allen and Chapman (1893) to describe a rat from Trinidad, later reported in Sucre state, northern Venezuela (Pittier and Tate 1932), and with a distribution extending to other northern Venezuelan states. Hershkovitz (1948b), based on morphologic and biogeographic data, designated the Venezuelan populations as *tatei* and those from Trinidad as *palmipes*. The allopatric distribution of the two forms, combined with morphometric and karyological analyses, indicate that they constitute two different subspecies.

Voss et al. (2001) compared *N. palmipes* with *N. squamipes* and *N. melanius* (treated as a synonym of *N. rattus* herein). Besides differences in karyotype, *N. palmipes* has nasal margins that abruptly taper behind the premaxillae, nasolacrimal capsules that are partially concealed in lateral view, a tegmen tympani that always has a large anterior process, and a deeper (relative to width) interparietal.

The diploid and autosomal fundamental numbers of *palmipes* are $2n = 17$ and FN = 29 (M. A. Barros et al. 1992); those for *tatei* are $2n = 16$ and FN = 25–28, with variation in fundamental number due to a pericentric inversion polymorphism in pairs 3 and 4 (M. A. Barros et al. 1992). These two karyotypes differ primarily in the morphology of chromosome pair 1. Both taxa share a pale dorsal and ventral coloration, one trait in which they differ from all other species of the genus. Hershkovitz (1944:66–67) referred to *N. p. tatei* as *N. squamipes* subspecies II.

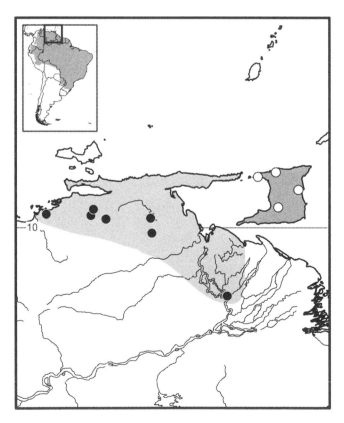

Map 190 Selected localities for *Nectomys p. palmipes* (○) and *Nectomys p. tatei* (●).

Nectomys rattus Pelzeln, 1883
Common Water Rat
SYNONYMS:

Hesperomys rattus Pelzeln, 1883:73; type locality "Marabitanas [= Marabitanos]," upper Rio Negro, Amazonas, Brazil.

Nectomys squamipes mattensis Thomas, 1904a:238; type locality "Santa Ana da Chapada, a village situated at an altitude of about 800 m., on the Serra do Chapada, some thirty miles N.E. of Cuyabá," Mato Grosso, Brazil.

Nectomys squamipes melanius Thomas, 1910b:185; "Lower Essequibo River, 12 miles from mouth. Alt. 40 feet," probably Demerara-Mahaica, Guyana.

Nectomys squamipes pollens Hollister, 1914b:104; type locality "Sapucay [= Sapucaí], Paraguay"; Hershkovitz (1944:42) states that Sapucaí is "a station on the railroad line about halfway between Villa Rica and Asunción, in the region drained by the Tevicuarhy, a tributary of the lower Paraguay"; see Remarks.

Nectomys squamipes amazonicus Hershkovitz, 1944:47; type locality "Tauary, on the east bank of the lower Rio Tapajóz [= Tapajós] about fifteen kilometers below Aveiros, state of Pará, Brazil."

Nectomys squamipes montanus Hershkovitz, 1944:57; type locality "Hacienda Exito, on the Río Cayumbá, a small stream which enters the Huallaga next below the Río Chinchao, department of Huanuco, Peru; altitude, 3000 feet."

Nectomys squamipes tarrensis Hershkovitz, 1948b:51; type locality "Río Tarra, upper Río Catatumbo, department of Norte de Santander, Colombia; altitude, 250 meters."

Nectomys melanius: Voss, Lunde, and Simmons, 2001:93; name combination.

Nectomys parvipes Petter, 1979:507; type locality "Cacao au bord de la rivière Comté (4°35′N, 52°28′W)," French Guiana.

Sigmodontomys parvipes Bonvicino, 1999:254; name combination.

DESCRIPTION: Color similar to *Rattus rattus*, but sides of body, chest, and upper half of abdomen pale with whitish tips; iris very dark brown; eye-ring blackish; nose purple-gray; incisors yellow-ochre; ear very rounded, blackish, with grayish skin color at root; top of tarsus grayish, with very few short hairs and thus appears almost naked; claws tan with whitish tips; forefeet with distinct rudiment of thumb with nail; and tail scaly but moderately haired (approximate translation of Pelzeln 1883:73).

DISTRIBUTION: *Nectomys rattus* occurs throughout riverine forests of eastern Colombia, Venezuela (Orinoco Basin), Peru (Ucayali Basin), Guyana, French Guiana, Paraguay, and Brazil (parts of the Amazon and Paraguay basins, and in several independent rivers basins of the northernmost Atlantic Forest).

SELECTED LOCALITIES (Map 191): BRAZIL: Amazonas, Estirão do Equador (MPEG 1685), Marabitanas (type locality of *Hesperomys rattus* Pelzeln); Goiás, Anápolis (MNRJ 1505), Posse (MNRJ 2150); Maranhão, Bacabal (MPEG [CZ 1346]); Mato Grosso, Tapirapoan (MNRJ 2079); Mato Grosso do Sul, Salobra (MZUSP 2010); Pará, Belém (MPEG 622), Cachoeira da Porteira, Rio Trombetas (MPEG 10084); Pernambuco, Exu (PMN 404; see Furtado 1981), São Lourenço da Mata (PMN 252; see Furtado 1981); Piauí, Lagoa Alegre (MPEG [CZ 1380]). COLOMBIA: Meta, Villavicencio (AMNH 136324); Norte de Santander, Río Tarra (USNM 279730). French Guiana: Awala-Yalimapo (T-4420), Kaw (T-4578). GUYANA: Essequibo Islands-West Demerara, lower Essequibo River, 12 mi above mouth (type locality of *Nectomys squamipes melanius* Thomas). PARAGUAY: Paraguarí, Sapucaí (FMNH 18162). PERU: Loreto, Lagunas (FMNH 19648); Ucayali, Yarinacocha (LSUMZ 14371). VENEZUELA: Bolívar, 46 km S and 7 km E of Ciudad Bolívar (USNMN 406048), El Lagual [= El Yagual] (AMNH 16964), Auyan-Tepuí (AMNH 130731); Mérida, Fundo Vista Alegre, San Isidro de Bejuquero (M. A. Barros et al. 1992); Zulia, Misión Tokuko (USNM 448572).

SUBSPECIES: Despite the number of names applied to different geographic segments of a very large range (e.g., Hershkovitz 1944), we regard *Nectomys rattus* as monotypic. However, the taxonomic status of populations from Surinam (Brokopondo, Rudi Kappelvligveld and Marowijme, Olea Marie) with a $2n=56$ karyotype need to be revaluated.

NATURAL HISTORY: In the Cerrado region of Brazil, *Nectomys rattus* is found primarily along streams in gallery forest (Paula 1983; Mares and Ernest 1995; Hannibal and Cáceres 2010; Talamoni and Dias 1999). Paula (1983) recorded a home range size of 2,200 m², and Mares and Ernest (1995) estimated population densities of two to six individuals per hectare. In Venezuela, this species was trapped in dry forest, moist forest, and wet forest, always on the ground and near watercourses (Handley 1976). Ectoparasites of this species include the flea *Polygenis tripus* (L. R. Guimarães 1972) and the tick *Amblyomma* sp. (E. K. Jones et al. 1972). *Nectomys rattus* is also infected by the trematode *Schistosoma mansoni*, with an ability to eliminate viable eggs to complete the transmission cycle of the parasite (A. C. Ribeiro et al. 1998), and can be a reservoir of the protozoan genus *Leishmania* (Lainson and Shaw 1969).

REMARKS: Thomas (1904a) included *Hesperomys rattus* Pelzeln (1883) in *Nectomys*. Subsequently, Hershkovitz (1944) excluded *Hesperomys rattus* (Pelzeln) from the genus *Nectomys*, although he did not examine the holotype

but rather based his decision only on Pelzeln's original description, which did not mention interdigital membranes. However, interdigital membranes are not readily apparent in young specimens of *Nectomys*, including the skin of the holotype of *N. rattus* (NMW 471 housed in the Naturhistorisches Museum Wien in Vienna). Examination of the holotype (Alfredo Langguth, pers. comm.) and of photographs confirmed that Pelzeln's type is a *Nectomys*.

Nectomys rattus includes forms previously described as *N. squamipes mattensis* Thomas (1904a), *N. s. amazonicus* Hershkovitz (1944), *N. s. montanus* Hershkovitz (1944), *N. s. tarrensis* Hershkovitz (1948b), *N. s. pollens* Hollister (1914b), and *N. s. melanius* Thomas (1910b). Voss et al. (2001) used the name *N. melanius* for the species of water rats in the Guianan region but included in their concept karyotypes ranging from $2n=52$ to 56. They also suggested that *rattus* Pelzeln could be a senior synonym of *melanius* Thomas pending an examination of the holotype. However, as construed herein, *N. rattus* has a $2n=52$, with this basic diploid number supplemented in some populations by the presence of up to two supernumerary chromosomes, giving a $2n=54$ (Bonvicino, D'Andrea et al. 1996). For the moment, the systematic significance of the numerical variability in diploid number reported by Voss et al. (2001) for Guianan specimens is unclear. Nevertheless, we regard *melanius* Thomas as a junior synonym of *rattus* Pelzeln based on: (1) our unpublished molecular analysis that grouped specimens from Guyana and French Guiana with those of *N. rattus* with $2n=52$; (2) the morphological assessment of Voss et al. (2001) that regarded water rats from the Guianas and Amazon Basin to be the same; and (3) the inclusion of a $2n=52$ karyotype among the French Guianan specimens reported by Voss et al. (2001).

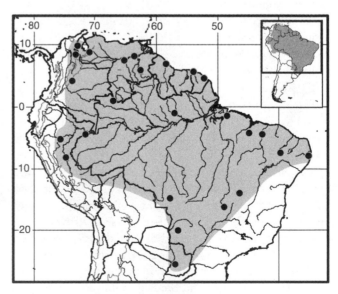

Map 191 Selected localities for *Nectomys rattus* (●). Contour line = 2,000 m.

Bonvicino (1999) treated *parvipes* Petter as a species of *Sigmodontomys*, but Voss et al. (2001:94–95), after examination of the holotype of *parvipes*, stated that "none of its qualitative characters diverge from the range of variation exhibited by other specimens of French Guiana *Nectomys* with substantially larger measurements." We concur, based on our examination of photographs of Petter's holotype, and include *parvipes* Petter as junior synonym of *N. rattus*.

Nectomys squamipes Brants, 1827
Atlantic Water Rat

SYNONYMS:

M[*us*]. *squamipes* Brants, 1827:138; type locality "Brasilien"; subsequently restricted to São Sebastião, São Paulo State, Brazil (Hershkovitz 1944).

Mus aquaticus Lund, 1839a:233; *nomen nudum*.

M[*us*]. *aquaticus* Lund, 1840a [1841b:240]; type locality "Rio das Velhas's Floddal" (Lund 1841b:264, 266), Lagoa Santa, Minas Gerais, Brazil.

H[*esperomys*]. *Squamipes*: Wagner, 1843a:540; name combination.

Hesperomys (*Holochilus*) *robustus* Burmeister, 1854:164; type locality "Brazil"; see Hershkovitz (1944) for remarks about type locality.

Hesperomys (*Holochilus*) *aquaticus*: Burmeister, 1855a:6; name combination.

Nectomys squamipes: Peters, 1861:152; first use of current name combination.

Potamys brasiliensis Liais, 1872:507; renaming of *Mus aquaticus* Lund.

Hesperomys (*Nectomys*) *squamipes*: Ihering, 1893:14; name combination.

Nectomys squamipes squamipes: Gyldenstolpe, 1932a:66; name combination.

Nectomys squamipes aquaticus: Hershkovitz, 1944:40; name combination.

Nectomys squamipes olivaceus Hershkovitz, 1944:41; type locality "five miles north of Therezopolis [= Teresópolis], Rio de Janeiro, Brazil."

DESCRIPTION: Color of back ochraceous orange mixed with brown; underfur, basal parts of cover hairs, and guard hairs neutral gray; rump colored like back but with tips of guard hairs frequently gray rather than ochraceous; upper parts of sides like back but with less dark brown, lower halves nearly uniformly ochraceous; hairs of chest and belly neutral gray basally, paling subterminally to gray, tips ochraceous orange; head from nose to crown cinnamon brown or dark brown, lightly mixed with ochraceous; narrow orbital ring present but poorly defined; cheeks colored like sides; area around mouth, chin, throat, and ventral surface of forelegs and hindlegs pallid neutral gray to nearly white with faint to medium-heavy ochraceous wash;

ears brown, thinly haired externally, and nearly naked internally; vibrissae black at their base, terminally black, brown, or gray; forefeet and hindfeet covered thinly above with white to brown hairs; brown plantar surfaces naked; hindfoot fringed with white; tail uniformly brown except for scattering of gray keel hairs, especially over terminal one-fourth, in some specimens.

DISTRIBUTION: *Nectomys squamipes* occurs in the São Francisco, Paraíba do Sul, and Paraná river basins of eastern Brazil (from Pernambuco state in the north to Rio Grande do Sul in the south), and in the Cerrado gallery forests bordering the Atlantic Forest in the Brazilian states of Minas Gerais, São Paulo, and Mato Grosso do Sul, Misiones in Argentina, and probably in Uruguay, although no specimens are known from that country.

SELECTED LOCALITIES (Map 192): ARGENTINA: Misiones, Caraguatay (FMNH 26734). BRAZIL: Alagoas, Quebrângulo (MNRJ 12408); Bahia, Prado (UFPB [AL 1319]), Valença (UFPB [AL 3467]); Mato Grosso do Sul, Maracaju (MNRJ 3827); Minas Gerais, Araguari (MNRJ 7520), Barra do Paraopeba, Rio Extrema (MNRJ 2058), Conceição do Mato Dentro (MNRJ 13349); Pernambuco, São Lourenço da Mata (PMN 299, see Furtado 1981); Rio Grande do Sul, Sapiranga (UFRGS [LF 80]); Santa Catarina, Florianópolis (UFRGS [LF 2003]); São Paulo, Ilha de São Sebastião (type locality of *Mus squamipes* Brants), Itapura (MZUSP 10184); Sergipe, Cristinápolis (MNRJ 30574).

SUBSPECIES: We regard *N. squamipes* as monotypic.

NATURAL HISTORY: *Nectomys squamipes* inhabits primary forest vegetation near streams (Briani et al. 2001; Prevedello et al. 2010). Atlantic water rats are nocturnal and omnivorous, feeding on fungi, invertebrates, leaves, fruit, and small vertebrates (Crespo 1982b). D'Andrea et al. (1996) recorded population densities between 1.2 and 3.4 individuals per hectare with home range sizes between 2.2 to 12.0 per hectare. Litter sizes ranged from one to seven (mode = 4 and 5), the gestation period was 30 days, and the minimum age to weaning was between 20 and 25 days. On Ilha do Cardoso (São Paulo state), *N. squamipes* reproduced seasonally and where survival rate was tied to fruit availability (Bergallo and Magnusson 1999), but at Sumidouro (Rio de Janeiro state), *N. squamipes* reproduced throughout the year but mostly during rainy periods, and survivorship increased with the effects of the dry periods (Bonecker et al. 2009). Briani et al. (2001) studied *N. squamipes* nests and found them to be egg shaped, about 15 cm long and 10 cm wide, without any apparent entrance, and composed of dry leaves and grass.

Endoparasites of *N. squamipes* include the trematodes *Schistosoma mansoni* (see A. V. Martins et al. 1955) and *Canaania obese* (see Maldonado et al. 2010) and the nematodes *Aspidorera raillicti*, *Physaloptera getula* (see Vicente et al. 1982), and *Syphacia venteli* (see M. Robles and Navone 2010). The species is highly susceptible to infection by *Schistosoma mansoni* and is able to maintain the transmission cycle of this parasite (Maldonato Jr. et al. 1994; Gentile et al. 2006). Ectoparasites include the flea *Polygenis pradoi* (Botelho and Linardi 1980), and the mites *Eubrachylaelaps rotundus*, *Gigantolaelaps goyanensis*, *Laelaps manguinhosi*, and *Laelaps mazzai* (Botelho et al. 1981). *Nectomys squamipes* is preyed upon by the Barn Owl (*Tyto alba*; R. P. Martins et al. 1980).

REMARKS: Hershkovitz (1944:65–66) included all names applicable to *Nectomys* in his highly polytypic *N. squamipes*, but we restrict this name only to those populations throughout the Atlantic Forest of coastal Brazil. Bonvicino, D'Andrea et al. (1996) substantiated the species-level status of the Amazonian *N. rattus* and *N. squamipes* through experimental hybrid crosses that produced sterile males with abnormal spermatogenesis, thus indicating that the two represent independent genetic lineages. Moreover, karyological data confirmed that both species occur in sympatry in Pernambuco state where no natural hybrids have been found through karyotyping (Furtado 1981).

The diploid and autosomal fundamental numbers ($2n=56$, $FN=56$) in this species are variable due to the presence of up to three supernumerary chromosomes (Bonvicino 1994). The upper toothrow of *N. squamipes* is significantly shorter than that of *N. rattus*, *N. apicalis*, *N. grandis*, or *N. palmipes* (Bonvicino 1994).

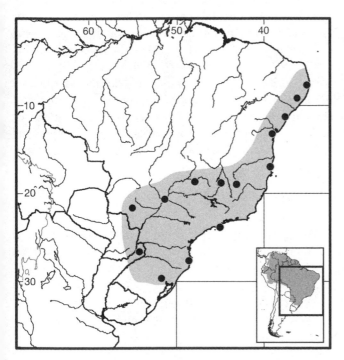

Map 192 Selected localities for *Nectomys squamipes* (●). Contour line = 2,000 m.

Genus *Nephelomys* Weksler, Percequillo, and Voss, 2006

Alexandre R. Percequillo

Species of the genus *Nephelomys* in South American occurd along the Andes from Bolivia to Colombia and in the coastal ranges of Venezuela, extending northward into the cordilleras of Panama and Costa Rica in Central America. These rats are found in montane cloud and elfin forests, and wooded *páramo*, across an elevational range from 900–3,500 m. Although there are no records of *Nephelomys* on the western dry western slopes of the Andes in Peru or southwestern Ecuador, more northern species may be present on both versants.

The genus has not been revised, and many nominal taxa are known primarily from their original descriptions. I have, however, examined most material in major museum collections and herein provisionally recognize 13 species: *N. albigularis*, *N. auriventer*, *N. caracolus*, *N. childi*, *N. devius*, *N. keaysi*, *N. levipes*, *N. maculiventer*, *N. meridensis*, *N. moerex*, *N. nimbosus*, *N. pectoralis*, and *N. pirrensis*. Two species are restricted to Central America (*N. devius* and *N. pirrensis*), although Voss (2014) provided a record from the Serranía de Darién on the Colombian side of the international border with Panama that most likely represents *N. pirrensis*. For details of these two species, see also Goldman (1913, 1918, 1920), Hall (1981), and F. A. Reid (2009).

Externally, species of this genus are defined by a tail longer than head and body length; long and dense mystacial vibrissae, surpassing the ears when laid backward; and long and narrow hindfeet, with the dorsal surfaces weakly furred and the ventral surfaces naked and with well-developed interdigital and plantar (thenar and hypothenar) pads. The pelage is soft and dense, varying from long to very long; the ventral pelage may have patches of self-colored hairs in the gular, thoracic, abdominal, and inguinal regions that vary in size and position depending upon the species; the ventral pelage coloration varies from grizzled white to rich buffy ochraceous tones.

The skull is very distinctive in comparison to those of other oryzomyine genera. In dorsal view, the rostrum is long and robust; the nasolacrimal capsules project weakly; the zygomatic notches are usually shallow and wide; and the interorbital region is hourglass in shape, with dorsolateral margins diverging or converging posteriorly and with the supraorbital border varying from rounded, squared, beaded, or strongly beaded. Incisive foramina may be short, medium, or long, with their lateral margins widest posteriorly (short to long teardrop shape) or widest medially (egg shaped). The palate is typically long and wide (*sensu* Hershkovitz 1962); in some species, the anterior margin of the mesopterygoid fossa reaches the alveoli or the hypoflexus of the M3s, resulting in a shorter palate. Posterolateral palatal pits are located in shallow to very deep palatal depressios and range from simple at the palatal level to complex. An enlarged stapedial foramen, squamosoalisphenoid groove, and sphenofrontal foramen are present, configuring pattern 1 of the carotid circulatory system (*sensu* Voss 1988). An alisphenoid strut, separating the buccinator-masticator and accessory oval foramina, is present in some species. Auditory bullae are large and triangular in shape. The tegmen tympani is weakly developed, only rarely in contact with the squamosal in individuals of some species.

The mandible has a small and weakly falciform coronoid process, and the angular process extends posteriorly past the condyloid process. The capsular process is inconspicuous in most species and is frequently absent.

The molars are pentalophodont and hypsodont, with the main cusps arranged in opposite pairs. These cusps are connected transversely by lophs(ids) and anteroposteriorly by mures(ids). M1 and m1 both have a deep anteromedian flexus and flexid, respectively.

The glans penis is complex, with a tridigitate and well-developed cartilaginous baculum; there is a single dorsal papilla, one short and bifurcate urethral flap, and three bacular mounds visible in the terminal crater (Hooper and Musser 1964). Male accessory glands consist of one pair each of ventral prostate, ampullary, vesicular, and bulbo-urethral glands, and one pair each of medial and lateral prostate glands; preputial glands are absent (Voss and Linzey 1981). The stomach is unilocular-hemiglandular (Carleton 1973).

SYNONYMS:

Hesperomys: Tomes, 1860b:264; part (description of *albigularis*); not *Hesperomys* Waterhouse.

Hesperomys (*Calomys*): Thomas, 1882:103; part (listing of *albigularis*); not *Calomys* Waterhouse.

Hesperomys [*Oryzomys*]: Thomas, 1884:448; part (listing of *albigularis*); not *Oryzomys* Baird.

Oryzomys: E.-L. Trouessart, 1897:524; part (listing of *albigularis* and *meridensis*); not *Oryzomys* Baird.

Zygodontomys: Gyldenstolpe, 1932a:114; part (listing of *obtusirostris*); not *Zygodontomys* J. A. Allen.

Nephelomys Weksler, Percequillo, and Voss, 2006:18; type species *Hesperomys albigularis* Tomes, by original designation.

REMARKS: Several species-group nominal taxa historically and currently included in the genus *Nephelomys* (namely, *devius* Bangs, *meridensis* Thomas, *pirrensis* Goldman, *childi* Thomas, *o'connelli* [*sic*] J. A. Allen, *pectoralis* J. A. Allen, *maculiventer* J. A. Allen, *auriventer* Thomas, *keaysi* J. A. Allen, *obtusirostris* J. A. Allen, and *levipes* Thomas) were listed as synonyms of *Oryzomys albigularis* (= *N. albigularis*) in Hershkovitz's (1944:72) unfortunate footnote revision. Later, in another footnote, Hershkovitz

(1960:544) inadvertently listed *caracolus* Thomas, another name historically associated with *Oryzomys albigularis*, among synonyms of "*Oryzomys laticeps*" (now in *Hylaeamys*; see also that account) but subsequently corrected his error (Hershkovitz 1966a:137, footnote 11) by treating *caracolus* along with *devius* and *villosus* J. A. Allen (see Musser and Williams 1985) as "additional representatives" of *O. albigularis*. In my opinion, Hershkovitz's (1966a) concept of *Oryzomys* (=*Nephelomys*) *albigularis* is the same he developed previously (Hershkovitz 1944), with the only changes being the addition of other species-group names. These "additional representatives" were listed as junior synonyms without being given a taxonomic categorical rank. However, Hershkovitz (1966a) did recognize two taxa as distinct subspecies of *O. albigularis* (*meridensis* Thomas and *caracolus* Thomas).

From the 1940s through the 1960s, Hershkovitz treated several nominal taxa (now in the genera *Euryoryzomys*, *Hylaeamys*, *Nephelomys*, and *Transandinomys*) in synonymy with a small number of species-group names without the benefit of morphological diagnoses or apparently the examination of name-bearing types or other specimens. Unfortunately, Cabrera (1961) and others followed these nomenclatural acts, actions that have obscured the current application of many species-group names.

Species limits in the genus are poorly understood, and, as noted earlier, many named forms have been regarded as either synonyms or valid subspecies of a widespread *Nephelomys albigularis*. However, I have found both the frequency and extent of the white ventral patches to be valuable diagnostic features for species recognition in *Nephelomys*, contrary to past opinion (J. A. Allen 1904c; Osgood 1912, 1913, 1914b; Eisenberg 1989). Once segregated on this basis and by the limited karyotypic data, the 11 species I recognize herein are diagnosable by additional external and craniodental characters. My current hypothesis of species limits must await more thorough geographic sampling as well as more extensive morphological, karyotypic, and molecular analyses.

KEY TO THE SOUTH AMERICAN SPECIES OF *NEPHELOMYS*:

1. Ventral patches of self-colored white hairs absent 2
1′. Ventral patches of self-colored white hairs present on variable frequencies 5
2. Alisphenoid strut present in >75% of specimens; interorbital region hourglass in shape, with rounded or weakly beaded supraorbital margins; ventral pelage grayish to grayish yellow. 3
2′. Alisphenoid strut present in <25% of specimens; skull robust and long (mean length 37.9 mm; range: 36.2–39.8 mm); incisive foramina very long (mean length

6.26 mm; range: 5.6–6.7 mm), teardrop in shape; interorbital region with strongly beaded supraorbital margins, diverging posteriorly; coloration of ventral pelage strongly ochraceous yellow; $2n = 70$, FN = 84 . . . . *Nephelomys auriventer*

3. Alisphenoid strut small and delicate, a dorsoventrally oriented bony bar, defining the oval accessory and the buccinator-masticatory foramina 4

3′. Alisphenoid strut very robust, varying from a wide bar oriented anteroposteriorly (defining the oval accessory and the buccinator-masticatory foramina) to an almost complete alisphenoid bony wall, with small individual foramina for the nerves and vessels; incisive foramina very short (mean length 4.56 mm; range: 4.1–5.1 mm), teardrop in shape . . . . *Nephelomys moerex*

4. Incisive foramina long (mean length 6.0 mm; range: 5.2–7.2 mm) with lateral margins widest medially; posterior margins of incisive foramina close to alveoli of M1s; palate short (mean length 6.6 mm; range: 5.9–8.1 mm), with anterior margin of mesopterygoid fossa frequently slightly penetrating between toothrows; posterolateral palatal pits elongated and situated in oblique, narrow and deep palatal fossae; $2n = 76$ and FN = 88 . . . . *Nephelomys levipes*

4′. Incisive foramina short (mean length 5.4 mm; range: 4.1–6.5 mm) with lateral margins widest posteriorly; posterior margins of incisive foramina more distant from alveoli of M1s; palate long (mean length 7.5 mm; range: 6.7–8.7 mm), with anterior margin of mesopterygoid fossa rarely penetrating between toothrows; posterolateral palatal pits situated in more rounded and deep palatal fossae; $2n = 80$ and FN = 92 to 94 . . . . *Nephelomys keaysi*

5. Ventral pelage grayish or whitish washed with gray; white patches in gular, pectoral, and/or inguinal regions 6

5′. Ventral pelage ochraceous yellow with a golden tone, with very small white gular patches; incisive foramina short (mean length 5.0 mm; range 4.6–5.1 mm), with lateral margins slightly rounded in outline; alisphenoid strut present in about 30% of specimens; posterolateral palatal pits small to large, in shallow palatal fossae; interorbital region hourglass in shape, with supraorbital margins rounded or squared . . . *Nephelomys nimbosus*

6. Ventral white patches present in moderate to high frequencies (>50%) 7

6′. Ventral white patches present in low frequencies (ca. 12%); posterolateral palatal pits small and numerous, situated in shallow or moderately deep palatal fossae. . . . . *Nephelomys childi*

7. Ventral white patches large and wide, occupying a large area on gular and pectoral regions 8

7′. Ventral white patches small and narrow, restricted to the gular and pectoral regions in about 60% of specimens; incisive foramina long (mean length 6.1 mm; range: 5.3–6.9 mm) and narrow, with nearly parallel lateral margins; interorbital region divergent posteriorly, with supraorbital margins beaded to strongly beaded; posterolateral palatal pits small, varying in depth from palatal level or in shallow palatal fossae; $2n = 66$, FN = 90 . . . . *Nephelomys caracolus*

8. Ventral white patches large, present in more than 90% of specimens; alisphenoid strut absent 9

8′. Ventral white patches present in about 50% of specimens; delicate alisphenoid strut present in >60% of specimens; interorbital region convergent posteriorly, with rounded supraorbital margins in its anterior portion and squared margins in posterior portion . . . . *Nephelomys albigularis*

9. Posterolateral palatal pits vary from few to many, deep, and situated in shallow to moderately deep palatal fossae 10

9′. Posterolateral palatal pits numerous and deep, situated in sharply delimitated, very deep and complex palatal fossae; white ventral patches very large, occupying gular and pectoral regions, rarely inguinal region; $2n = 66$, FN = 94. . . . . *Nephelomys pectoralis*

10. White ventral patches extremely long and wide, occupying gular and pectoral regions, reaching middle of abdomen or, more frequently, inguinal region (very few specimens have venter entirely white); few posterolateral pits, placed at palatal level or in shallow palatal fossae; shorter rostrum (mean nasal length 12.1 mm; range 10.85–13.65 mm); $2n = 66$, FN = 112 . . . . *Nephelomys maculiventer*

10′. White ventral patches very large, occupying gular and pectoral regions; numerous posterolateral palatal pits, situated in shallow to deep palatal fossae; longer rostrum (mean nasal length 13.01 mm; range: 11.78–14.3 mm); $2n = 66$, FN = 104 . . . . *Nephelomys meridensis*

Nephelomys albigularis (Tomes, 1860)
White-throated Nephelomys

SYNONYMS:

Hesperomys albigularis Tomes, 1860b:264; type locality: "taken en camino [*sic*] on my return from Pallatanga"; restricted to "Pallatanga, [Chimborazo,] Ecuador" (Tate 1939:189).

Hesperomys (*Calomys*) *albigularis*: Thomas, 1882:103; name combination.

[*Hesperomys* (*Oryzomys*)] *albigularis*: Thomas, 1884:448; name combination.

[*Oryzomys*] *albigularis*: E.-L. Trouessart, 1897:524; name combination.

[*Oryzomys (Oryzomys)*] *albigularis*: E.-L. Trouessart, 1904:
417; name combination.

[*Oryzomys (Oryzomys)*] *albigularis albigularis*: Tate, 1932e:
16; name combination.

Oryzomys albigularis albigularis: Gyldenstolpe, 1932a:14;
name combination.

[*Nephelomys*] *albigularis*: Weksler, Percequillo, and Voss,
2006:18; first use of current name combination.

DESCRIPTION: Recognized by combination of following traits: medium body size (head and body length
102–162 mm) with tail longer than head and body (138–
184 mm); tail bicolored or weakly bicolored (few specimens with unicolored tails); long, dense and soft dorsal
fur, ranging from buffy orange grizzled with dark brown
to ochraceous intensely grizzled with dark brown; whitish
or yellowish ventral pelage, with or without small patches
of self-colored hairs in gular region (rarely in pectoral region); skull with long incisive foramina, nearly parallel
lateral margins widening posteriorly; posterolateral palatal
pits small and numerous, on level of palate or situated in
moderately deep palatal fossae; alisphenoid strut variably
present; interorbital region hourglass in shape or convergent posteriorly, with supraorbital margins rounded or
slightly squared; sphenopalatine vacuities small and variably present.

DISTRIBUTION: *Nephelomys albigularis* occurs on
the eastern and western slopes of the Andean Cordillera
Oriental, from northwestern Peru to central Ecuador. The
species is present on both sides of the arid Huancabamba
Depression, and in isolated forested patches in this area.
The elevational range is from 900 m (Hacienda Éxito,
Huánaco, Peru) to 3,100 m (Taraguacocha, Cordillera de
Chilla, El Oro, Ecuador).

The southernmost record of this species is Hacienda
Éxito, in Huánuco department, Peru, at about 900 m. This
site is located at the left margin of Río Cayumba, a left margin tributary of the Río Huallaga, which drains the upper
elevations of the Cordillera Oriental. Farther to the north
in Peru, most samples are distributed along the eastern
slope of the Cordillera Oriental (in the upper Río Huallaga
drainage) or on both banks of rivers (e.g., Río Cutervo,
Río Utcubamba, and Río Huancabamba, all tributaries of
the Río Marañón) between the central and eastern Andean
cordilleras. Only specimens from Taulis (Cajamarca) and
Seques (Lambayeque) represent unequivocal records of this
species on the western slope in northern Peru. All remaining localities in the Cordillera Occidental, such as Cutervo
(on the Río Chatano) and Tambo (on the Río Huancabamba), are in drainages of the Amazon Basin.

North of the Huancabamba Depression, in Ecuador,
records of *N. albigularis* are predominantly from the western slopes of Cordillera Occidental. The species is known
from only two localities on the eastern slopes that drain
to the Amazon Basin, near Loja in the headwaters of the
Río Zamora (a Río Marañón tributary) and Mazán (in the
inter-Andean Cuenca basin). The northernmost record of
this species is Carmen, in Bolívar province, central Ecuador.

SELECTED LOCALITIES (Map 193): ECUADOR:
Azuay, Mazán (BM 84.306), Molleturo (AMNH 61898);
Bolívar, Carmen, near Sinchic (AMNH 66944); Cañar,
Chical (AMNH 63333); Chimborazo, Pallatanga (type lo-
cality of *Hesperomys albigularis* Tomes); El Oro, El Chiral
(AMNH 46481); Loja, Celica (AMNH 47813), 8 km W of
Loja (UMMZ 164889). PERU: Amazonas, Uscho, about
50 km E Chachapoyas (FMNH 19698); Cajamarca, Taulis
(AMNH 73135); Huánuco, Hacienda Éxito, Río Cayumba
(FMNH 24771); La Libertad, Utcubamba, on trail to
Ongón (LSUMZ 24790); Piura, Cerro Chinguela, ca. 5 km
NE of Zapalache (LSUMZ 26992).

SUBSPECIES: *Nephelomys albigularis* is monotypic.

NATURAL HISTORY: Throughout its latitudinal and
elevational range, *N. albigularis* inhabits the humid slopes
of the Peruvian and Ecuadorean Andes (Chapman 1926),
generally in montane and cloud/elfin forests. Barnett
(1999) trapped *N. albigularis* in both primary and secondary cloud forests between 2,400 m and 2,700 m in Ecuador.
Over a five-year period from June to October, he observed
no pregnant females and captured one lactating female; of
16 males trapped, only four were in breeding condition
(Barnett 1999). In northern Peru, *N. albigularis* inhabits
very humid forests from 1,525 to 2,440 m, dwelling under
logs, roots, and debris. Females captured in Amazonas,
Peru, during August showed no evidence of embryos, although one lactating female was collected (data from museum tags).

REMARKS: *Nephelomys albigularis* was previously
considered to be widely distributed, occurring from northwestern Peru to northern Colombia, northwestern Venezuela, and eastern Panama (Eisenberg 1989; Corbet and Hill
1991; Musser and Carleton 1993; Eisenberg and Redford
1999). This concept included populations that exhibit
substantial karyological variation and morphological differences in external features such as the presence and extension of ventral white patches and also in several skull
characters. Herein, I recognize these morphologically and
chromosomally distinct geographic units as separate species (see following accounts).

My more restricted concept of *N. albigularis* includes
samples that share the following morphological traits:
medium frequency (50%) of gular and/or pectoral white
patches; bicolored tail present in >70% of specimens; interorbital region hourglass in shape, with dorsolateral
margins converging posteriorly; supraorbital margins
rounded or only slightly beaded; incisive foramina long,

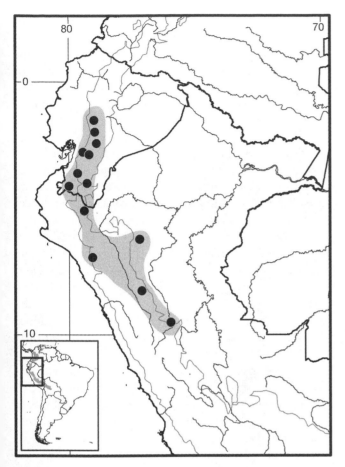

Map 193 Selected localities for *Nephelomys albigularis* (●). Contour line = 2,000 m.

wide, and teardrop in shape; palate long and ornate, with shallow to deep posterolateral palatal pits at the palatal level or in deep palatal notches; and alisphenoid strut present in >60% of examined specimens.

Nephelomys auriventer (Thomas, 1899)
Greater Golden-bellied Nephelomys

SYNONYMS:

Oryzomys auriventer Thomas, 1899d:379; type locality "Mirador, below Baños, on the Upper River Pastasa [= Río Pastaza], E Equador [= Tunguruhua, Ecuador]. Altitude 1500 m."

Oryzomys (*Oryzomys*) *auriventer auriventer*: Tate, 1932e: 16; name combination.

[*Oryzomys*] *albigularis*: Hershkovitz, 1944:72, footnote; name combination; part; not *Hesperomys albigularis* Tomes, 1860 (= *Nephelomys albigularis* [Tomes]).

Oryzomys albigularis auriventer: Cabrera, 1961:381; name combination.

[*Nephelomys*] *auriventer*: Weksler, Percequillo, and Voss, 2006:18; first use of current name combination.

DESCRIPTION: Body long and robust (head and body length 140–173 mm), tail very long (range: 160–191 mm). Dorsal pelage short and harsh, colored orange yellow to orange golden densely grizzled with brown; ventral pelage light yellowish to intensely golden; white gular patches absent (except for one specimen from Peru with a small patch); tail uniformly colored or weakly bicolored. Skull long and robust (greatest length 36.2–39.9 mm); incisive foramina long and wide posteriorly (teardrop in shape); interorbital region slightly to strongly convergent anteriorly, with sharply angled supraorbital margins; palate long, with large and complex posterolateral palatal pits recessed in shallow fossae; alisphenoid strut rarely present; medial lacerate foramen narrow, and postglenoid foramen small and narrow.

DISTRIBUTION: *Nephelomys auriventer* is known from three closely adjacent localities in the upper Río Pastaza drainage of the Cordillera Oriental of Ecuador and from one locality in the Apurímac drainage in the Cordillera Oriental of Peru, at elevations between 1,140 m (Mera) to 1,660 m (Huanhuachayo).

SELECTED LOCALITIES (Map 194): ECUADOR: Pastaza, Mera (AMNH 67351), Tunguruhua: Mirador (type locality of *Oryzomys auriventer* Thomas). PERU: Ayacucho, Huanhuachayo (LSUMZ 16667).

SUBSPECIES: *Nephelomys auriventer* is monotypic.

NATURAL HISTORY: The few collecting localities for *N. auriventer* are characterized by "paramo or temperate plants" and "subtropical vegetation" (G. H. H. Tate notes cited in Musser et al. 1998:103), cultivated lands, second-growth forest, and primary forest (Rageot and Albuja 1994).

REMARKS: Known from fewer than 20 specimens, *N. auriventer* is readily distinguished from congeners by the following combination of characters: ochraceous yellow to nearly golden ventral pelage color; large size (mean skull length 37.9 mm; range: 36.2–39.8 mm); robust molar series (mean length 6.22 mm; range: 6.03–6.5 mm); strongly beaded supraorbital ridges; and extremely narrow lacerate foramen and medium foramen postglenoid. A. L. Gardner and Patton (1976) reported karyotypic data from a Peruvian specimen with $2n = 70$ and FN = 84; although there is no karyological information available for Ecuadorean specimens, these share the same unique combination of morphological traits as the Peruvian specimen.

The considerable distributional hiatus (about 1,300 km) between Ecuadorean and Peruvian localities may be due to a truly disjunct range or to a sampling artifact. Few recent collecting efforts have been made across the wide region between these distributional end points since the Godman-Thomas expeditions of 1926 and 1927 (see, for example, Thomas 1926b,e,g, 1927a,f; Thomas and Leger 1926a).

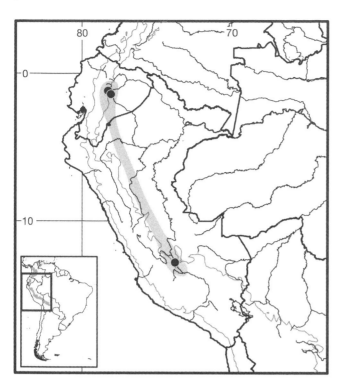

Map 194 Selected localities for *Nephelomys auriventer* (●). Contour line = 2,000 m.

Nephelomys caracolus (Thomas, 1914)

Coastal Cordilleran Nephelomys

SYNONYMS:

Oryzomys caracolus Thomas, 1914d:242; type locality "Galiparé [= Galipán or Pichacho de Galipán], Cerro del Avila, near Caracas. Alt. 6000'," Vargas, Venezuela.

Oryzomys (Oecomys) caracolus: Tate, 1939:190; name combination.

Oryzomys laticeps: Hershkovitz, 1960:544, footnote; part; not *Mus laticeps* Lund (= *Hylaeamys laticeps* [Lund]).

Oryzomys capito velutinus: Cabrera, 1961: 387; part; not *Oryzomys velutinus* J. A. Allen and Chapman (= *Hylaeamys megacephalus* [G. Fischer])

Oryzomys albigularis caracolus: Hershkovitz, 1966a:137, footnote; name combination.

[*Nephelomys*] *caracolus*: Weksler, Percequillo, and Voss, 2006:8; first use of current name combination.

DESCRIPTION: Medium body size (head and body length 120–165 mm) and tail longer than head and body combined (tail length 144–176 mm). Dorsal pelage long and dense, yellowish intensely grizzled with brown in color; ventral pelage grayish to whitish, with narrow white pectoral patches variably present (in some specimens patches especially long, extending to gular region). Skull of medium size (greatest length 31.1–38.4 mm); incisive foramina long and

narrow, with acutely pointed anterior and posterior margins; palate with few, small posterolateral palatal pits at palate level or recessed in very shallow fossae, lacking palatine crests or excrescences; interorbital region divergent posteriorly, with supraorbital margins strongly angled and beaded.

DISTRIBUTION: *Nephelomys caracolus* is distributed throughout the Venezuelan Cordillera de la Costa, in Aragua, Carabobo, and Miranda states as well as in the Distrito Federal. Most records are associated with the Caribbean (or northern) slope; only a few samples are from the southern slope. Elevation ranges from 1,050 to 2,300 m. See R. P. Anderson and Raza (2010), who modeled the climatic niche and potential distribution of the species.

SELECTED LOCALITIES (Map 195): VENEZUELA: Aragua, Rancho Grande, Estación Biologica de Rancho Grande, 13 km NW of Maracay (USNM 405988); Carabobo, La Cumbre de Valencia (AMNH 31541); Miranda, Curupao, 19 km E of Caracas (AMNH 135170), Hacienda Las Planadas (R. P. Anderson and Raza 2010); Vargas, Galiparé [= Galipán], Cerro de Ávila (type locality of *Oryzomys caracolus* Thomas).

SUBSPECIES: *Nephelomys caracolus* is monotypic.

NATURAL HISTORY: This species has been captured in mature and disturbed humid evergreen and cloud forests with a rich subcanopy of ferns and vines and covered with abundant moss and epiphytes (Handley 1976). Specimens have also been taken in drier evergreen and cloud forests, with limited epiphytic cover, on the inland slopes of the Cordillera de la Costa (Handley 1976). O. J. Linares (1998) summarized the natural history of "*Oryzomys albigularis*" in Venezuela, but his concept included specimens and localities now assigned to both *N. caracolus* and *N. meridensis*.

REMARKS: *Nephelomys caracolus* has been listed as a species of the genus *Oecomys* (Tate 1939), as a synonym of *Hylaeamys megacephalus* (= *Oryzomys capito, sensu* Hershkovitz 1960; or *Oryzomys capito velutinus*, see Cabrera 1961), and as a synonym or subspecies of *Oryzomys* (= *Nephelomys*) *albigularis* (Hooper and Musser 1964; Musser and Carleton 1993; Handley 1976; Hershkovitz 1966a). Tate (1947:66) included some specimens of this species in his concept of *Oryzomys* [= *Nephelomys*] *meridensis*. More recently, Aguilera, Pérez-Zapata, and Martino (1995) and Márquez et al. (2000) established the species validity of *N. caracolus* based on karyological (2n = 66, FN = 90) and morphometric data, respectively. This species is morphologically distinct from other Venezuelan populations, treated here as *N. meridensis*, by the following combination of characters: lower frequency of ventral patches of self-colored white hairs, which are smaller and restricted to the gular region; interorbital region divergent posteriorly, with supraorbital margins beaded; incisive foramina longer with lateral margins wider medially (convex); and postero-

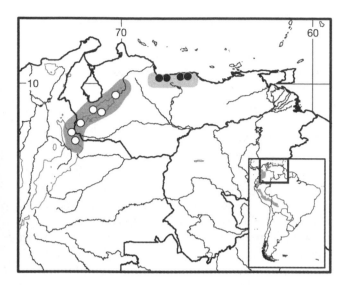

Map 195 Selected localities for *Nephelomys caracolus* (●) and *Nephelomys meridensis* (○). Contour line = 2,000 m.

lateral palatal pits smaller and placed at the palatal level or in shallow palatal notches.

Nephelomys childi (Thomas, 1895)
Child's Nephelomys

SYNONYMS:

Oryzomys childi Thomas, 1895b:59; type locality "Bogota," Cundinamarca, Colombia."

Oryzomys meridensis: Bangs, 1900a:93; part; *childi* Thomas given as a synonym of *Oryzomys meridensis* Thomas.

O[*ryzomys*]. *Childi*: Thomas, 1901b:188; name combination.

[*Oryzomys* (*Oryzomys*)] *childi*: E.-L. Trouessart, 1904:419; name combination.

Oryzomys o'connelli J. A. Allen, 1913c:597; type locality "Buenavista (altitude 4500 feet), about 50 miles southeast of Bogota," Meta, Colombia.

[*Oryzomys* (*Oryzomys*)] *o'connelli*: Tate, 1932e:17; name combination.

[*Oryzomys*] *albigularis*: Hershkovitz, 1944:72 footnote; part; not *Hesperomys albigularis* Tomes, 1860 (= *Nephelomys albigularis* [Tomes]).

Oryzomys albigularis childi: Cabrera, 1961:381; name combination.

Oryzomys albigularis o'connelli: Cabrera, 1961:82; name combination.

[*Oryzomys*] *oconnelli*: Márquez, Aguilera, and Corti, 2000:85; corrected spelling of species epithet (ICZN 1999:Art.32.5.2.3).

[*Nephelomys*] *childi*: Weksler, Percequillo, and Voss, 2006:18; first use of current name combination.

DESCRIPTION: Distinguished from congeners by following traits: medium body size (head and body length 120–184 mm); tail longer than head and body (tail length 110–194 mm); dorsal pelage varying in length and texture geographically, with some samples presenting dense and shorter fur and other exhibiting very dense and longer hairs; ventral pelage of few individuals with small white patches in gular region. Skull with interorbital region hourglass in shape or convergent posteriorly, with supraorbital margins rounded; rounded braincase, with discrete temporal border; long incisive foramina, with lateral margins wider posteriorly (teardrop in shape); palate with numerous posterolateral palatal pits recessed in shallow to moderately deep fossae; and developed palatal excrescences situated on palatine level or on palatine crests.

DISTRIBUTION: *Nephelomys childi* is endemic to Colombia, occurring in all three cordilleras of the northern Andes. In the Cordillera Oriental, this species is distributed on both the eastern and western forested slopes from the headwaters of Río Magdalena (Huila department), in the Sierra de Macarena (Meta department), and in the plains of Bogotá (Cundinamarca department). In the Cordillera Central, isolated populations occur in upper Cauca and Magdalena drainages (Huila and Cauca departments, respectively) and on the eastern slope in the departments of Antioquia, Caldas, Quíndio, and Tolima. The only known populations of *N. childi* in the Cordillera Occidental are isolated on the extreme northwestern slope at Santa Barbara and Guapantal (Antioquia department). Elevational range is 1,140 to 3,400 m.

SELECTED LOCALITIES (Map 196): COLOMBIA: Antioquia, San Pedro (FMNH 72101), Santa Barbara, RíoUrrao (FMNH 71882); Cauca, Almaguer (BM 13.5.27.42); Cundinamarca, Anolaima (BM 99.10.3.23), Guasca, Río Balcones (FMNH 71896), Paime (AMNH 71340); Huila, Andalucía, eastern Andes (AMNH 33755); Meta, Buenavista (type locality of *Oryzomys oconnelli* J. A. Allen), Macarena Mt. (AMNH 142160); Tolima, El Éden, E Quindio Andes (AMNH 32983).

SUBSPECIES: *Nephelomys childi* is monotypic.

NATURAL HISTORY: Data on ecology or other aspects of the natural history of *N. childi* are lacking.

REMARKS: The holotypes and topotypes of *N. childi* and *N. oconnelli* are extremely similar in external and cranial traits, and, as a result, I regard these taxa as conspecific. Previous authors have considered *N. childi* a synonym of *Nephelomys meridensis* based on the presence and size of ventral white patches (Bangs 1900a; Thomas 1901b; Osgood 1914b) or as a synonym (Hershkovitz 1944) or subspecies (Cabrera 1961) of *N. albigularis*. Furthermore, many specimens referrable to *N. childi* have been erroneously identified as *N. auriventer* (J. A. Allen 1916b; Cuervo Díaz et al. 1986) or *N. pectoralis* (J. A. Allen 1916b; Anthony 1923a).

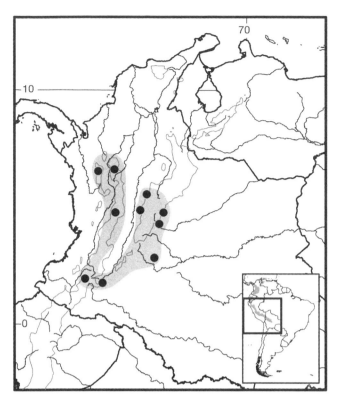

Map 196 Selected localities for *Nephelomys childi* (●). Contour line = 2,000 m.

This species can be readily recognized from its Colombian congeners *N. pectoralis* and *N. maculiventer* by the combination of a low frequency of small ventral white patches (about 12%) and shallow and small posterolateral palatal pits placed in shallow to moderately deep palatal fossae.

Nephelomys keaysi (J. A. Allen, 1900)

Keays's Nephelomys

SYNONYMS:

Oryzomys keaysi J. A. Allen, 1900a:225; type locality "Juliaca, Peru, altitude 6000 ft," corrected to "Inca Mines [= Santo Domingo Mine], about 200 miles northeast of Juliaca, on the east side of the Andes, on the Inambary River [= Río Inambarí]," Puno, Peru, by J. A. Allen (1901b:41).

Oryzomys obtusirostris J. A. Allen, 1900a:225; type locality "Juliaca, Peru, altitude 6000 ft"; corrected to "Inca Mines [= Santo Domingo Mine], about 200 miles northeast of Juliaca, on the east side of the Andes, on the Inambary River [= Río Inambarí]," Puno, Peru, by J. A. Allen (1901b:41).

Oryzomys (Oligoryzomys) obtusirostris: Thomas, 1901b: 188; name combination.

[*Oryzomys (Oryzomys)*] *keaysi*: E.-L. Trouessart, 1904:419; name combination.

[*Oryzomys (Oryzomys)*] *obtusirostris*: E.-L. Trouessart, 1904: 419; name combination.

Zygodontomys obtusirostris: Gyldenstolpe, 1932a:114; name combination.

[*Oryzomys*] *albigularis*: Hershkovitz, 1944:72, footnote; part; not *Hesperomys albigularis* Tomes (= *Nephelomys albigularis* [Tomes]).

Oryzomys albigularis keaysi: Cabrera, 1961:381; name combination.

[*Nephelomys*] *keaysi*: Weksler, Percequillo, and Voss, 2006: 18; first use of current name combination.

DESCRIPTION: Medium sized (head and body length 125–165 mm); tail much longer than head and body (length 146–207), not bicolored in most specimens; dorsal pelage dense, long, and lax, colored ochraceous to buffy ochraceous intensely grizzled with brown; ventral pelage grayish, whitish, or rarely yellowish, without white pectoral or gular patches; interorbital region of skull slightly to strongly divergent posteriorly, with supraorbital margins squared to strongly beaded; incisive foramina long, with lateral margins wider posteriorly and posterior margins not reaching M1s; palate long, with large posterolateral pits at palate level or recessed in shallow and narrow fossae; palatal excrescences small and delicate; alisphenoid strut present in most specimens.

DISTRIBUTION: *Nephelomys keaysi* occurs on the forested eastern slopes of the Andes in central Peru (Santa Cruz, Pasco) south to central Bolivia (Chaparé, Cochabamba), at elevations between 1,000–2,600 m; most records are from 1,200–1,500 m).

SELECTED LOCALITIES (Map 197): BOLIVIA: Cochabamba, Chaparé (BM 34.9.2.128); Yungas (AMNH 38557); La Paz, Pitiguaya, Río Unduavi (AMNH 72647). PERU: Ayacucho, Yuraccyacu (LSUMZ 15693); Cusco, Machu Picchu, San Miguel Bridge (BM 22.1.1.28), Marcapata, Camante (FMNH 68628), 72 km NE of Paucartambo (MVZ 166682); Junín, Acobamba, 45 mi NE of Cerro de Pasco (BM 27.11.1.91), Satipo, Cordillera de Vilcabamba (USNM 582120); Pasco, Santa Cruz, ca. 9 km SSE of Oxapampa (LSUMZ 25894); Puno, Sagrario (FMNH 52705), Inca Mines (type locality of *Oryzomys keaysi* J. A. Allen and *Oryzomys obtusirostris* J. A. Allen).

SUBSPECIES: *Nephelomys keaysi* is monotypic.

NATURAL HISTORY: *Nephelomys keaysi* inhabits montane and cloud forests (Patton et al. 1990). Pregnant females were captured in November, with three embryos, at Santa Cruz, Peru, and females with uterine scars (four to five) were observed in July, east of Paucartambo, Peru. Stomachs of specimens from Bosque Aputinye, Cusco, Peru contained arthropods and vegetable matter (diet and reproductive data from museum tags and field notes).

REMARKS: *Nephelomys keaysi* was resurrected from the synonymy of *N. albigularis* on the basis of karyological (A. L. Gardner and Patton 1976) and molecular studies (Patton et al. 1990; see also Musser and Carleton 1993). The karyotype of *N. keaysi* is 2*n* = 80, FN = 92–94, based on specimens from localities in Junín, Ayacucho, and Cusco departments, Peru (A. L. Gardner and Patton 1976, "*Oryzomys albigularis* variant 3"; J. L. Patton, pers. comm.).

The separate specific status of *N. keaysi* and *N. levipes*, which replace one another on the elevational gradient in southeastern Peru, was demonstrated by karyological (J. L. Patton, pers. comm.) and allozyme analyses (Patton 1986; Patton et al. 1990). Morphologically, *N. keaysi* differs from *N. albigularis* by absence of self-colored ventral patches, divergent interorbital margins with sharp or beaded supraorbital margins, and higher frequency of an alisphenoid strut. For comparisons with *N. levipes*, see Remarks under that species.

Nephelomys obtusirostris, here recognized as a synonym of *N. keaysi*, has been previously assigned to the *longicaudatus-stolzmanni* group of *Oryzomys* (= genus *Oligoryzomys*; Thomas 1901b) or to *Zygodontomys* (Gyldenstolpe 1932; Ellerman 1941). However, Tate (1939) correctly noted that the holotype of *Oryzomys obtusirostris* J. A. Allen was a young specimen of *Nephelomys keaysi*, with immature pelage and unerupted third upper molars.

Nephelomys keaysi is parapatric to *N. levipes* throughout much of its distribution (see account) and with *N. auriventer* at Huanhuachayo, Ayacucho, Peru (see account).

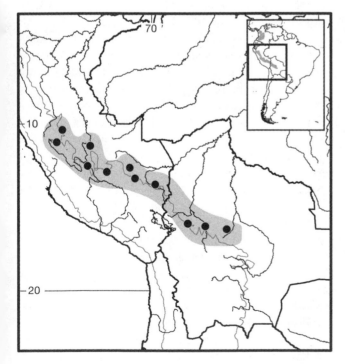

Map 197 Selected localities for *Nephelomys keaysi* (●). Contour line = 2,000 m.

Nephelomys levipes (Thomas, 1902)
Nimble-footed Nephelomys

SYNONYMS:

Oryzomys levipes Thomas, 1902b:129; type locality "Limbane [= Limbani], 2,000 m," Puno, Peru.

[*Oryzomys* (*Oryzomys*)] *levipes*: E.-L. Trouessart, 1904:419; name combination.

[*Oryzomys*] *albigularis*: Hershkovitz, 1944:72 footnote; part; not *Hesperomys albigularis* Tomes (= *Nephelomys albigularis* [Tomes]).

Oryzomys [(*Oryzomys*)] *albigularis keaysi*: Cabrera, 1961:381; part; not *Oryzomys keaysi* J. A. Allen (= *Nephelomys keaysi* [J. A. Allen]); name combination.

[*Nephelomys*] *levipes*: Weksler, Percequillo, and Voss, 2006:18; first use of current name combination.

DESCRIPTION: Medium sized (head and body length 113–161 mm); tail longer than head and body (length 143–185 mm). Dorsal pelage very dense, very long, and lax, colored from yellowish to ochraceous grizzled with brown; ventral pelage shorter and predominantly yellowish, without patches of self-colored hairs; tail frequently weakly to strongly bicolored, rarely unicolored. Skull length 33.5–39.8 mm; interorbital region slightly to strongly divergent posteriorly, with squared to beaded supraorbital margins; incisive foramina long, wider in middle, with posterior margins almost reaching M1 alveoli; palate short, with anterior border of mesopterygoid fossa reaching alveoli or even M3 hypocones; posterolateral palatal pits deep, recessed at moderately deep to deep oblique and lateral fossa positioned at maxillopalatine suture; palate with small excrescences; sphenopalatine vacuities long and wide; alisphenoid strut present in all examined specimens.

DISTRIBUTION: *Nephelomys levipes* inhabits cloud and elfin forests on the eastern slopes of the Andes in northwestern Bolivia and southwestern Peru, at elevations from 1,800 to 3,200 m. This species replaces *N. keaysi* at higher elevations on the eastern Andean slope.

SELECTED LOCALITIES (Map 198): BOLIVIA: La Paz, Cocopunco (AMNH 72118); 15 km (by road) NE of Unduavi (UMMZ 156180); Santa Cruz, Serranía Siberia, 11 km NW by road of Torrecillas (AMNH 264192). PERU: Cusco, 32 km NE of Paucartambo (MVZ 166679); Puno, Limbani (type locality of *Oryzomys levipes* Thomas), 11 km NNE of Ollachea (MVZ 172637), 14 km W of Yanahuaya (MVZ 172639).

SUBSPECIES: *Nephelomys levipes* is monotypic.

NATURAL HISTORY: *Nephelomys levipes* is predominantly terrestrial but has also been captured in trees. Notes on specimen tags state that the species dwells in "humid, dark, mossy forest," "high forest," on "steep slopes," and in "gorges near streams." Pregnant females with one or three embryos were collected in July and with two, four, or

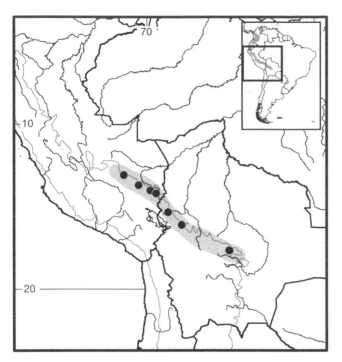

Map 198 Selected localities for *Nephelomys levipes* (●). Contour line = 2,000 m.

five embryos during June–August at localities in Puno and Cusco departments, Peru, respectively (specimen labels and field notes).

REMARKS: *Nephelomys levipes* has been considered a junior synonym of either *N. keaysi* (Gyldenstolpe 1932; Cabrera 1961) or *N. albigularis* (Hershkovitz 1944, 1966a; Corbet and Hill 1991). The molecular data of Patton et al. (1990), however, support *N. levipes* as specifically distinct from both *N. keaysi* and *N. albigularis*, an arrangement followed by subsequent authors (Musser and Carleton 1993, 2005; Eisenberg and Redford 1999). *Nephelomys levipes* can be differentiated from *N. keaysi* by a denser and longer dorsal pelage; longer incisive foramen, with lateral margins wider medially; posterolateral palatal pits situated in deeper and narrower palatal slits; and a higher frequency in the presence of an ectolophid on m1. The two species also differ in karyotype, with a 2*n* = 76 and FN = 88 for *N. levipes* and 2*n* = 80, FN = 92 to 94 in *N. keaysi* ("*Oryzomys albigularis* variant 3" of A. L. Gardner and Patton 1976; J. L. Patton, pers. comm.).

Nephelomys maculiventer (J. A. Allen, 1899)
Santa Marta Nephelomys
  SYNONYMS:
Oryzomys maculiventer J. A. Allen, 1899b:204; type locality: "Sierra El Libano (alt. 6000 ft.), Santa Marta District," Magdalena, Colombia.
Oryzomys meridensis: Bangs, 1900a:92; part; *maculiventer* listed as a synonym of *Oryzomys meridensis* Thomas.

[*Oryzomys* (*Oryzomys*)] *maculiventer*: E.-L. Trouessart, 1904:418; name combination.
Oryzomys meridensis maculiventer: Osgood, 1912:49; name combination.
Oryzomys albigularis maculiventer: Osgood, 1914b:159; name combination.
[*Oryzomys*] *albigularis*: Hershkovitz, 1944:72 footnote; part; not *Hesperomys albigularis* Tomes (= *Nephelomys albigularis* [Tomes]).
[*Nephelomys*] *maculiventer*: Weksler, Percequillo, and Voss, 2006:18; first use of current name combination.

DESCRIPTION: Medium to large body size (head and body length 127–180 mm); tail longer than head and body (length 148–194); long, dense, and lax dorsal pelage; ventral pelage with patches of self-colored hairs present in almost all examined specimens, with large patches in gular, pectoral (including ventral surface of forelimbs), abdominal, and inguinal areas; tail usually unicolored, rarely weakly bicolored; small and delicate skull (greatest skull length 32.7–35.9 mm); interorbital region hourglass in shape or slightly convergent posteriorly, with rounded supraorbital margins; short incisive foramina, wider posteriorly (teardrop in shape); long palate, with few but deep posterolateral pits either recessed in shallow fossae or at palate level.

DISTRIBUTION: All known localities of *N. maculiventer* are from the slopes of Sierra Nevada de Santa Marta and Serranía de Perijá, in northern Colombia, where this species ranges from 900–2,600 m in montane and cloud forests.

SELECTED LOCALITIES (Map 199): COLUMBIA: Cesar, Villenueva, Sierra Negra (USNM 280641); La Guajira, San Miguel (AMNH 3894); Magdalena, Sierra El Libano (type locality of *Oryzomys maculiventer* J. A. Allen).

SUBSPECIES: *Nephelomys maculiventer* is monotypic.

NATURAL HISTORY: Nothing is known about the ecology, population biology, or other aspects of the natural history of this species.

REMARKS: A. L. Gardner and Patton (1976) regarded the population of *Nephelomys* from the Sierra Nevada de Santa Marta as a subspecies of *N. albigularis* based on karyological and morphological evidence. Subsequent authors (Musser and Williams 1985; Musser and Carleton 1993, 2005) treated *maculiventer* simply as a synonym of *N. albigularis*. The specimens from Santa Marta, however, are sharply differentiated morphologically from other species in the genus and have a different karyotype (2*n* = 66, FN = 112; "*Oryzomys albigularis* variant 1" of A. L. Gardner and Patton 1976). As a result, I recognize *N. maculiventer* as a valid species. Among the morphological features that diagnose this species are large ventral patches of self-colored hairs in most specimens and a skull with simple and few posterolateral palatal pits situated at the palatal level or in shallow palatal fossae.

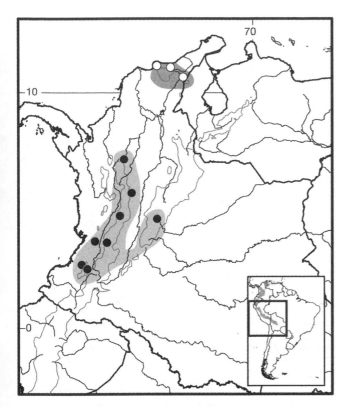

Map 199 Selected localities for *Nephelomys maculiventer* (○) and *Nephelomys pectoralis* (●). Contour line = 2,000 m.

Nephelomys meridensis (Thomas, 1894)
Mérida Nephelomys

SYNONYMS:

Oryzomys meridensis Thomas, 1894b:351; type locality "Merida, Venezuela," Mérida, Venezuela.

[*Oryzomys (Oryzomys)*] *meridensis*: E.-L. Trouessart, 1904: 418; name combination.

[*Oryzomys (albigularis)*] *meridensis*: Osgood, 1914b:159; name combination.

[*Oryzomys*] *albigularis*: Hershkovitz, 1944:72 footnote; part; not *Hesperomys albigularis* Tomes (= *Nephelomys albigularis* [Tomes]).

O[*ryzomys*]. *albigularis meridensis*: Hershkovitz, 1966a:137, footnote; part; not *Hesperomys albigularis* Tomes (= *Nephelomys albigularis* [Tomes]); name combination.

Oryzomys [(*Oryzomys*)] *albigularis*: Eisenberg, 1989:351; part.

[*Nephelomys*] *meridensis*: Weksler, Percequillo, and Voss, 2006:18; first use of current name combination.

DESCRIPTION: Medium body size (head and body length 135–160 mm); long tail that exceeds head and body (length 170–172 mm), typically unicolored but rarely weakly bicolored; dorsal pelage dense, lax, and very long; ventral pelage with large pectoral patches of self-colored hairs (gular, axillary, and inguinal patches rarely pres-

ent). Skull of medium size (greatest length 32.4–37 mm); interorbital region slightly convergent posteriorly or symmetrically constricted (hourglass in shape), with rounded or slightly beaded supraorbital margins posteriorly; palate with multiple and deep posterolateral palatal pits recessed in shallow depressions; palatine crest present, with small to very large palatal excrescences (if large, may be fused to maxilla).

DISTRIBUTION: This species occurs on forested slopes from the northeastern Cordillera Oriental of Colombia in Boyacá department, through the Páramo de Tamá on the Colombian-Venezuelan border, to the northern portion of the Sierra de Mérida, in Trujillo state, Venezuela, at elevations from 1,100 to 4,000 m. See R. P. Anderson and Raza (2010), who modeled the climatic niche and potential distribution of the species.

SELECTED LOCALITIES (Map 195): COLUMBIA: Boyacá, La Primavera (FMNH 92408); Norte de Santander, Páramo de Tamá (FMNH 18648). VENEZUELA: Mérida, Montes de Los Nevados (AMNH 33163), Páramo Tambor (BM 29.3.17.55); Táchira, Táchira (FMNH 7044); Trujillo, Hacienda Misisí, 14 km E of Trujillo (USNM 374310).

SUBSPECIES: *Nephelomys meridensis* is monotypic.

NATURAL HISTORY: Osgood (1912) secured specimens in ground litter and under logs and roots in the dense, cold, forested areas of the Páramo de Tamá, on the Venezuelan/Colombian border.

REMARKS: *Nephelomys meridensis* was recognized as a valid taxon throughout much of its history, either as a species (E.-L. Trouessart 1904; Goldman 1918; Gyldenstolpe 1932; Tate 1939; Ellerman 1941; Aguilera, Pérez-Zapata, and Martino 1995; Soriano and Ochoa 1997; Márquez et al. 2000; Musser and Carleton 2005) or subspecies (Osgood 1914b; Cabrera 1961; Hershkovitz 1966a; O. J. Linares 1998; Rivas and Péfaur 1999a). Other authorities simply included this name in the synonymy of *N. albigularis* (Hershkovitz 1944; Handley 1976; Péfaur and Díaz de Pascual 1985; Eisenberg 1989; Corbet and Hill 1991; Musser and Carleton 1993; Rivas and Péfaur 1999b).

Aguilera, Pérez-Zapata, and Martino (1995) and Márquez et al. (2000) separated populations from the Mérida Andes (= *N. meridensis*; $2n = 66$, FN = 104) from those from the Cordillera de la Costa (= *N. caracolus*; $2n = 66$, FN = 90) of Venezuela on karyological and morphometric grounds, respectively. These authors, as well as Percequillo (2003), also distinguished samples from the Páramo de Tamá on the Colombian/Venezuelan border ($2n = 66$, FN = 92) from those of the Mérida Andes. Morphologically, specimens from Mérida and the Páramo de Tamá are quite similar in the extent of their ventral white patches and in cranial traits, differing only slightly in the morphology of posterolateral palatal pits, which are deeper on *N. meridensis*. The specimens from the Páramo de Tamá are included in my concept of *N. meridensis* until the validity of this decision is rejected by subsequent analyses. If differences between specimens from the Páramo de Tamá and the Mérida Andes prove to warrant taxonomic recognition, a new name will have to be erected for the former because none is currently available.

Nephelomys moerex (Thomas, 1914)
Gray-bellied Nephelomys

SYNONYMS:

Oryzomys albigularis moerex Thomas, 1914d:241; type locality "Mindo, N. W. of Quito. Alt. 4200′," Pichincha, Ecuador.

[*Oryzomys*] *albigularis maerex* Anthony, 1926a:4; incorrect subsequent spelling of *moerex* Thomas.

[*Oryzomys*] *moerex*: Tate, 1939:189; name combination.

Oryzomys albigularis: Boyce et al., 1982:435; part; not *Hesperomys albigularis* Tomes (= *Nephelomys albigularis* [Tomes]).

[*Nephelomys*] *moerex*: Weksler, Percequillo, and Voss, 2006: 18; first use of current name combination.

DESCRIPTION: Small to medium body (head and body length 114–160); tail longer than head and body (length 139–169 mm) and usually bicolored; dorsal pelage dense, lax, and short, overall colored ochraceous brown finely grizzled with dark brown; ventral pelage grayish, without patches of self-colored hairs. Skull with slightly convergent posteriorly or symmetrically constricted (hourglass in shape) interorbital region, with rounded or squared supraorbital margins; incisive foramina very short (length 4.1–5.1 mm; mean 4.6 mm), with sides widest medially; alisphenoid strut present in high frequency but ranges from narrow strut oriented parallel to lateral border of parapterygoid plate to wide bone covering almost completely ovale accessorius and buccinator-masticatory foramina.

DISTRIBUTION: *Nephelomys moerex* is endemic to Ecuador, with all known localities from a small area on the northwestern and southwestern faces of Volcán Pichincha and the Mindo river drainage, at elevations from 1,200 to 2,800 m.

SELECTED LOCALITIES (Map 200): ECUADOR: Pichincha, Las Máquinas, San [= Santo] Domingo trail (AMNH 64702), Río Tulipe, Gualea (FMNH 94979).

SUBSPECIES: *Nephelomys moerex* is monotypic.

NATURAL HISTORY: No data on natural history are available.

REMARKS: Thomas (1914d) originally described *moerex* as a subspecies of *N. albigularis*, a treatment followed by many subsequent authors (Anthony 1926a; Gyldenstolpe 1932; Ellerman 1941; Cabrera 1961). Others, however, regarded *moerex* as a full species (e.g., Tate 1939). Nevertheless, the name *moerex* was, until recently (Percequillo

Map 200 Selected localities for *Nephelomys moerex* (●) and *Nephelomys nimbosus* (○). Contour line = 2,000 m.

2003), lost in the synonymy of *N. albigularis* (Hershkovitz 1966a; Boyce et al. 1982; Corbett and Hill 1991; Musser and Carleton 1993, 2005).

Known from only a few specimens (about three dozen deposited in the BM, AMNH, MCZ, and UMMZ collections), *N. moerex* is considerably different from typical *N. albigularis* or any other nominal form of the genus, and thus I recognize it here as a valid species (see also Percequillo 2003). This species is diagnosed by the combination of absence of white ventral patches (in all but one specimen), very short and oval incisive foramina, and, most conspicuously, presence of a robust and obliquely oriented alisphenoid strut in all specimens (in some specimens the strut almost obliterates the alisphenoid wall, leaving only small foramina for the passage of nerves and vessels).

Nephelomys nimbosus (Anthony, 1926)
Lesser Golden-bellied Nephelomys

SYNONYMS:

Oryzomys auriventer nimbosus Anthony, 1926a:4; type locality "San Antonio, on Rio Ulva, northeastern slope of Tunguragua, [Tunguruhua,] Ecuador, altitude 6700 feet."

[*Oryzomys (Oryzomys)*] *auriventer nimbosus*: Tate, 1932e: 16; name combination.

[*Oryzomys*] *nimbosus*: Tate, 1939:189; name combination.

Oryzomys [(*Oryzomys*)] *albigularis nimbosus*: Cabrera, 1961:382; part; name combination.

O[*ryzomys*]. *albigularis*: Corbet and Hill, 1991:151; part; not *Hesperomys albigularis* Tomes (= *Nephelomys albigularis* [Tomes]); name combination.

Oryzomys auriventer: Musser and Carleton, 1993:720; part; not *Oryzomys auriventer* Thomas (= *Nephelomys auriventer* [Thomas]); name combination.

[*Nephelomys*] *nimbosus*: Weksler, Percequillo, and Voss, 2006:18; first use of current name combination.

DESCRIPTION: Small species, with head and body length 122–144 mm, and tail length 147–170 mm. Dorsal pelage very dense and very long, colored yellow orange to golden orange densely grizzled with brown; ventral pelage pale ochraceous yellow to intensely golden, with small white gular patches in most specimens. Skull small and delicate, length 31.5–35.4 mm; interorbital region narrow and either convergent posteriorly or amphoral in shape, with rounded or slightly squared supraorbital margins; incisive foramina short and widest posteriorly (teardrop in shape; mean length 4.98 mm; range: 4.6–5.1 mm); palate with small and simple posterolateral palatal pits located at palatal level or recessed in shallow depressions; alisphenoid strut found in about 30% of specimens.

DISTRIBUTION: *Nephelomys nimbosus* is endemic to Ecuador, with the four known collecting localities on the eastern Andean slope in Napo and Tunguruhua provinces, at elevations from 2,000–2,440 m. Along the Río Pastaza drainage, this species is parapatric to *N. auriventer*, replacing it at higher elevations.

SELECTED LOCALITIES (Map 200): ECUADOR: Napo, Baeza (AMNH 63840), Chaco, Rio Oyacachi (AMNH 66807); Tunguruhua, San Antonio (type locality of *Oryzomys auriventer nimbosus* Anthony), San Francisco, E of Ambato (AMNH 67354).

SUBSPECIES: *Nephelomys nimbosus* is monotypic.

NATURAL HISTORY: There is no published information on the natural history of *N. nimbosus*.

REMARKS: Anthony's *nimbosus* was considered either a subspecies of *N. auriventer* or of *N. albigularis*, or a synonym of the latter, since its description until Weksler et al. (2006) treated it as a species. Specimens of *N. nimbosus* are rare in museum collections, with fewer than 15 known, and few specialists have examined them. Based on my examination of the available material, I recognize *N. nimbosus* as a valid species because it can be distinguished from *N. auriventer* and other species of the genus by the combination of the characters mentioned above.

Nephelomys pectoralis (J. A. Allen, 1912)
Western Colombian Nephelomys
SYNONYMS:
Oryzomys pectoralis J. A. Allen, 1912:83; type locality "crest of Western Andes (alt. 10, 340 ft.), 40 miles west of Popayán, Cauca, Colombia."

[*Oryzomys albigularis*] *meridensis*: Osgood, 1914b:159; part; not *meridensis* Thomas.

[*Oryzomys* (*Oryzomys*)] *pectoralis*: Tate, 1932e:17; name combination.

[*Oryzomys*] *albigularis*: Hershkovitz, 1944:72 footnote; part; not *albigularis* Tomes.

Oryzomys [(*Oryzomys*)] *albigularis pectoralis*: Cabrera, 1961:382; part; name combination.

[*Nephelomys*] *pectoralis*: Weksler, Percequillo, and Voss, 2006:18; first use of current name combination.

DESCRIPTION: Differentiated from congeners from Colombia or generally northern Andes by medium head and body (length 124–180 mm); tail longer than head and body (length 110–196); long, dense, and lax dorsal pelage, colored yellowish to ochraceous grizzled with brown; ventral pelage grayish, with large and conspicuous white gular and pectoral patches that rarely reach abdominal region; and usually unicolored tail. Distinguishing cranial characteristics include medium-sized skull (greatest length 32.0–37.5 mm); robust and long rostrum; interorbital region hourglass in shape or posteriorly convergent, with slightly squared posteriorly supraorbital margins; incisive foramina short, with lateral margins widest posteriorly (teardrop in shape); numerous and complex posterolateral palatal pits recessed in very deep and well-defined depressions; roof of mesopterygoid fossa frequently ossified so that sphenopalatine vacuities, when present, small and narrow.

DISTRIBUTION: *Nephelomys pectoralis* is endemic to Colombia, occurring in all three Andean cordilleras. It is restricted to the western slope of the Cordillera Oriental, occurs along the length of the Cordillera Central, and is present on the forested western slope and upper regions of the eastern slope of the Cordillera Occidental.

SELECTED LOCALITIES (Map 199): COLUMBIA: Antioquia, La Bodega, N side Río Negrito (USNM 293778), Ventanas, Valdívia (FMNH 70417); Cauca, Cocal (AMNH 32539), Río Mechengue (FMNH 90217); Cundinamarca, Cuchillas del Carnicero, near Bogotá (BM 99.10.3.16); Quindío, El Roble (AMNH 32985); Valle del Cauca, Las Lomitas (AMNH 32200), Miraflores, near Palmira (AMNH 32185).

SUBSPECIES: *Nephelomys pectoralis* is monotypic.

NATURAL HISTORY: No details have been published.

REMARKS: The validity of the name *pectoralis* has been questioned throughout its history (Osgood 1914b; Hershkovitz 1944; Musser and Carleton 1993), although Cabrera (1961) listed it as a subspecies of *N. albigularis*. Specimens from Valle del Cauca (Peñas Blancas, Río Pichindé) have a $2n=66$, FN=94 karyotype (A. L. Gardner and Patton 1976, as "*Oryzomys albigularis* variant 2"), which differs from those of all other forms of *Nephelomys*. The limited geographic reprsentation by specimens and lack of karyotypic data from topotypes, however, precluded these authors from associating this karyotype with the name *pectoralis*.

Nevertheless, my examination of these specimens and others from the Andes of Colombia supports the hypothesis that *N. pectoralis* should be distinguished from other members of the genus, and I recognize it as a valid species. Among its distinctive features are large and wide white ventral patches, numerous and deep posterolateral pits placed in deep palatal fossae, the low frequency of small sphenopalatine vacuities, and a karyotype of $2n = 66$ and $FN = 94$ (A. L. Gardner and Patton 1976).

Genus *Nesoryzomys* Heller, 1904

Robert C. Dowler

Members of the genus *Nesoryzomys* are restricted to the Galápagos Islands, Ecuador, with Recent species known from only four of the major islands in the archipelago: Isla Baltra, Isla Fernandina, Isla Santa Cruz, and Isla Santiago. Undescribed species, known from fragmentary subfossil remains, have been reported from Isla Isabela and Isla Rábida (Steadman and Zousmer 1988). Of the five described species, two are now considered extinct. Much of the following account is based on Weksler's (2006) morphological analysis.

The five species of *Nesoryzomys* fall in two size classes. The smaller species (*N. darwini* and *N. fernandinae*) have total lengths of about 200 mm and masses about 32 g, whereas the larger (*N. indefessus*, *N. narboroughi*, and *N. swarthi*) are all greater than 250 mm in total length and have masses around 100 g. The ears are medium-sized, well haired proximally and less so distally. Mystacial vibrissae are long, with the longest projecting posterior to the ears. The tail is shorter than the head and body and well haired, with scales rarely visible. Fur is medium-length and color varies among species, usually medium brown or dark brown approaching black. The ventral pelage is paler than that of the dorsum. A white subauricular patch is usually present. Both manus and pes have ungual tufts. The plantar surfaces of the hindfeet are naked with thenar, hypothenar, and four interdigital pads present. There are eight mammae.

The skull is elongate with a long rostrum and a very narrow, symmetrical interorbital region that has little, if any, supraorbital bead. The nasals are short, terminating bluntly even with the maxillary-frontal-lacrimal junction. The zygomatic plates are broad with distinct zygomatic notches. The zygomatic arches converge slightly anteriorly with the widest portion at the squamosal root. The lacrimal is primarily in contact with the maxillary and has relatively little contact with the frontal. The incisive foramina are long, with the posterior margins at same level as or posterior to the anterior margins of the M1s. The palate is flat with posterolateral palatal pits. There are large sphenopalatine vacuities that reach the basisphenoid. The anterior alisphenoid canal is large, and an alisphenoid strut is absent. The carotid circulation is the derived pattern 3 (*sensu* Voss 1988), with a small stapedial foramen and both squamosoalisphenoid groove and sphenofrontal foramen absent. The ectotympanic bullae are large, covering most of the periotic bone viewed ventrally. The mandible has a sharply curved coronoid process equal in height to the condyle. The posterior border of the angular process lies directly below the tip of the condyle. The mandible has an obvious capsular process.

The molars are brachydont and bunodont, with lateral elements higher than central ones. The anterocone of M1 is divided into labial and lingual conules. All upper molars have mesolophs, and a mesoloph is present on m1 and m2. The M2 lacks a protoflexus. A posteroloph is present on M3. Both M1 and m1 have four roots.

The axial skeleton has 12 ribs. The first rib has a dual articulation with the seventh cervical and first thoracic vertebrae. A supratrochlear foramen is present in the humerus. The stomach is unilocular and hemiglandular (Patton and Hafner 1983), and a gall bladder is absent. Male accessory reproductive glands include a single pair of preputial glands, a pair of bulbourethral glands, and one pair of ampullary and vesicular glands. There is a single pair of dorsal prostate glands (Patton and Hafner 1983). The glans penis is elongate with a very shallow terminal crater. The osseous baculum is long and slender with a very small tridigitate cartilaginous apparatus (Patton and Hafner 1983).

SYNONYMS:

Oryzomys: Thomas, 1899c:280; part (description of *indefessus*); not *Oryzomys* Baird.

Nesoryzomys Heller, 1904:241; type species *Oryzomys indefessus* Thomas, by original designation.

[*Oryzomys*] (*Nesoryzomys*): Ellerman, 1941:343; usage as subgenus.

REMARKS: Patton and Hafner (1983) reviewed soft and craniodental characters in *Nesoryzomys* relative to those of other subgenera (now genera) of "*Oryzomys*" (*sensu lato*) and supported the generic status originally proposed by Heller. This opinion has been followed in recent reviews (e.g., Musser and Carleton 1993, 2005). The genus *Aegialomys* is considered sister to *Nesoryzomys* based on both molecular and morphological phylogenetic analyses (Patton and Hafner 1983; Weksler 2003, 2006; Pine et al. 2012). Patton and Hafner (1983) also suggested that all large forms of *Nesoryzomys* (*indefessus*, *narboroughi*, and *swarthi*) could be considered subspecies of one species, with *N. indefessus* having priority. Musser and Carleton (2005) concurred for *N. narboroughi* and *N. indefessus*, but retained *N. swarthi* as a distinct species based on craniodental features presented in the original description of *N. swarthi* (Orr 1938). Recent molecular studies (Weksler

2003) verified the divergence of *N. swarthi* and *N. narboroughi*. Additional molecular and chromosomal studies under way by the author further support the recognition of different species on separate islands and argue for a conservative retention of the original names for all *Nesoryzomys*.

KEY TO THE SPECIES OF *NESORYZOMYS*:

1. Hindfeet dark; adult skull length usually less than 36 mm; premaxillae short, terminating anterior to posterior margins of nasals 2
1′. Hindfeet white, adult skull length usually greater than 36 mm; premaxillae longer, terminating at the level of posterior margins of nasals. 3
2. Pelage dark brown with few yellow-tipped hairs; dorsal hairs of hindfoot brown; interparietal about half as wide as posterior margin of squamosal. . . . . *Nesoryzomys darwini*
2′. Pelage brown suffused with yellow-tipped hairs; dorsal hairs of hindfoot silver; interparietal wider than half the width of the posterior margin of squamosal . . . . *Nesoryzomys fernandinae*
3. Pelage dark brown to black; subsquamosal fenestra absent. . . . . *Nesoryzomys narboroughi*
3′. Pelage cinnamon brown; subsquamosal fenestra present 4
4. Posterior margins of the incisive foramina in adults projecting well between M1s; mastoid usually completely ossified; palate long with mesopterygoid fossa extending to posterior margin of maxillae; length of tail <122 mm . . . . *Nesoryzomys indefessus*
4′. Posterior margin of incisive foramina in adults at anterior margin of first molars; mastoid usually not fully ossified but having distinct fenestra; palate shorter with mesopterygoid fossa extending between maxillae; length of tail >122 mm . . . . *Nesoryzomys swarthi*

Nesoryzomys darwini Osgood, 1929
Darwin's Nesoryzomys

SYNONYMS:

Nesoryzomys darwini Osgood, 1929: 23; type locality "Academy Bay, Indefatigable Island [= Isla Santa Cruz], Galapagos Islands," Ecuador.

Oryzomys [(*Nesoryzomys*)] *darwini*: Ellerman, 1941:359; name combination.

DESCRIPTION: One of two small species in genus, total length 204–222 mm; tail shorter than head and body (70–80%); dorsal color reddish brown streaked with black but providing overall bright fulvous tone; sides slightly paler than back; and under parts rufous with gray-based hairs. Skull slender and without sharp ridges or angles. Distinguished from sympatric *N. indefessus* by much smaller size and cinnamon-brown overall color.

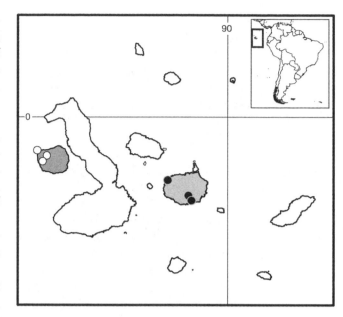

Map 201 Selected localities for *Nesoryzomys darwini* (●) and *Nesoryzomys fernandinae* (○).

DISTRIBUTION: *Nesoryzomys darwini* was known only from the island of Santa Cruz, Galápagos Islands, Ecuador; it is now considered extinct (Clark 1984; Dexter et al. 2004).

SELECTED LOCALITIES (Map 201): ECUADOR: Galápagos, Isla Santa Cruz (= Indefatigable Island), Academy Bay (type locality of *Nesoryzomys darwini* Osgood), Conway Bay (Osgood 1929), Fortuna, near Bellavista (AMNH 99922).

SUBSPECIES: *Nesoryzomys darwini* is monotypic.

NATURAL HISTORY: This species is presumed extinct, with the last specimens known collected in April 1930 (Patton and Hafner 1983). The species likely occurred across most of the island of Santa Cruz as the few specimens are from two coastal sites and two upland sites at 800 and 1,200 feet elevation, respectively. Nothing has been reported on the natural history of this species, although Osgood (1929) commented that it was much less numerous than sympatric *N. indefessus* based on trapping records of the specimens in the type series.

Nesoryzomys fernandinae Hutterer and Hirsch, 1979
Small Fernandina Nesoryzomys

SYNONYM:

Nesoryzomys fernandinae Hutterer and Hirsch, 1979:278; type locality "Insel [= Isla] Fernandina, 300 m, Galápagos, Ecuador."

DESCRIPTION: Second small species, with alveolar length in type series 4.9–5.6 mm, overlapping only minimally with that of *N. darwini* at lower end of distribution and *N. narboroughi* at the upper end (see Hutterer and Hirsch 1979). Distinguished from its sympatric congener *N. narboroughi* by smaller size (mass averaging 32.3 g versus 77.5 g, respectively) and darker brown dorsal pelage with dark feet, rather than paler grayish black and white feet (Dowler et al. 2000).

DISTRIBUTION: *Nesoryzomys fernandinae* is known only from the island of Fernandina, Galápagos Islands, Ecuador. All specimen records are from the western half of the island.

SELECTED LOCALITIES (Map 201): ECUADOR: Galápagos, Isla Fernandina (type locality of *Nesoryzomys fernandinae* Hutterer and Hirsch), Cabo Douglas (Dowler et al. 2000), caldera rim (Dowler et al. 2000).

SUBSPECIES: *Nesoryzomys fernandinae* is monotypic.

NATURAL HISTORY: *Nesoryzomys fernandinae* has been collected at coastal sites with saltbush (*Cryptocarpus pyriformis* [Nyctaquinaceae] up to the highest elevation on Fernandina at approximately 1,360 m. At the higher areas, habitat is dominated by large *Scalesia microcephala* (Asteraceae), *Darwiniothamnus tenuifolius* (Asteraceae), and mixed grasses. This species occurs sympatrically with *N. narboroughi*, but it is rare at coastal sites and much more common at high elevations where vegetation is dense (Dowler and Carroll 1996). This species has not been collected in areas dominated by mangroves, where *N. narboroughi* may be abundant. Specimens taken in July and August were not in reproductive condition.

Nesoryzomys indefessus (Thomas, 1899)

Santa Cruz Nesoryzomys

SYNONYMS:

Oryzomys indefessus Thomas, 1899c:280; type locality "Indefatigable Island [= Isla Santa Cruz]," Galápagos Islands, Ecuador.

Nesoryzomys indefessus: Heller, 1904:241; name combination.

Oryzomys [(*Nesoryzomys*)] *indefessus*: Ellerman, 1941:359; name combination.

DESCRIPTION: Large bodied, with total length 260–297 mm; tail length 108–117 mm; Orr 1938); palest species in genus in dorsal color tones (Patton and Hafner 1983). Orr (1938) noted that the skull was virtually indistinguishable from that of *N. narboroughi*, although the two are slightly separable in multivariate discriminate space based on craniodental measurements (Patton and Hafner 1983).

DISTRIBUTION: *Nesoryzomys indefessus* was known only from the islands of Santa Cruz and Baltra, Galápagos Islands, Ecuador, but is now considered extinct (Clark 1984; Dexter et al. 2004).

SELECTED LOCALITIES (Map 202): ECUADOR: Galápagos, Isla Santa Cruz (= Indefatigable Island), Academy Bay (Patton and Hafner 1983), Conway Bay (Patton and Hafner 1983), Isla Baltra (= South Seymour Island; Patton and Hafner 1983).

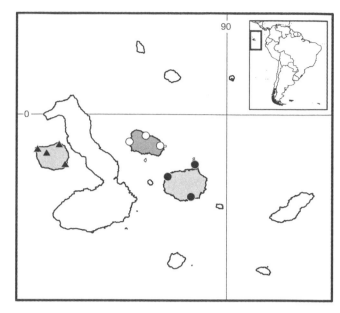

Map 202 Selected localities for *Nesoryzomys indefessus* (●), *Nesoryzomys narboroughi* (▲), and *Nesoryzomys swarthi* (○).

SUBSPECIES: *Nesoryzomys indefessus* is monotypic.

NATURAL HISTORY: This species is presumed extinct, with the last specimens collected in 1941 on the island of Baltra. On Santa Cruz, A. Rambech, an early settler, reported native rats to be abundant from the coast to the wet summit in the late 1920s before the introduction of *Rattus rattus*. Collections by J. P. Chapin in March of 1935 at Conway Bay revealed only *Rattus rattus* (Clark 1984). Feral cats also may have played a role in the extinction of this species (Dexter et al. 2004*)*. The species likely was common on both Santa Cruz and Baltra based on the number of specimens in collections. Recent surveys on both islands have failed to document an extant population (Patton and Hafner 1983; Dowler et al. 2000; Dexter et al. 2004). Heller (1904) reported that individuals inhabited burrows or rock crevices beneath bushes and was nocturnal.

Nesoryzomys narboroughi Heller, 1904

Large Fernandina Nesoryzomys

SYNONYMS:

Nesoryzomys narboroughi Heller, 1904:242; type locality "Mangrove Point, Narborough Island [= Isla Fernandina]," Galápagos Islands, Ecuador.

Oryzomys [(*Nesoryzomys*)] *narboroughi*: Ellerman, 1941:359; name combination.

Nesoryzomys indefessus: Patton and Hafner, 1983:562; part (*narboroughi* assigned as synonym); name combination.

DESCRIPTION: Large, with total length 264–300 mm; tail length 127–140 mm; Orr 1938); darkest color tones of any species in genus (Patton and Hafner 1983). Readily separable from sympatric *N. fernandinae* by both size and overall color, as noted in that account.

DISTRIBUTION: *Nesoryzomys narboroughi* is known only from the island of Fernandina, Galápagos Islands, Ecuador. It occurs at every locality that has been sampled on the island, from the coast to the summit of the caldera.

SELECTED LOCALITIES (Map 202): ECUADOR: Galápagos, Isla Fernandina (= Narborough Island), Cabo Douglas (Dowler et al. 2000), caldera rim (Dowler et al. 2000), Mangrove Point (type locality of *Nesoryzomys narboroughi* Heller), Punta Espinosa (Patton and Hafner 1983).

SUBSPECIES: *Nesoryzomys narboroughi* is monotypic.

NATURAL HISTORY: *Nesoryzomys narboroughi* is common in all vegetated habitats on the island of Fernandina, including coastal sites with saltbush (*Cryptocarpus pyriformis* [Nyctaginaceae]) and mangrove up to high elevations where *Scalesia microcephala* (Asteraceae), *Darwiniothamnus tenuifolius* (Asteraceae), and mixed grasses dominate. *Nesoryzomys narboroughi* occurs sympatrically with the smaller species *N. fernandinae*, but it is far more common at coastal sites and less common at high elevations. Population density in a saltbush community at Cabo Douglas was estimated at 100 per hectare, though the actual population at this site was restricted to relatively small habitat patches (Dowler and Carroll 1996). This species is nocturnal, though occasionally individuals may be active during daylight hours. These rodents show little fear of humans and may damage tents of researchers by gnawing. They probably have a catholic diet composed of both vegetation and invertebrates. Heller (1904) reported red material in the stomach that likely represented remains of crabs (*Grapsus grapsus*). Both the Galápagos Hawk (*Buteo galapagoensis*) and the Barn Owl (*Tyto alba*) prey on *N. narboroughi*.

REMARKS: I consider *Nesoryzomys narboroughi* to be a species separate from *N. indefessus*, contra Patton and Hafner (1983). The karyotype is $2n=32$, $FN=50$, with acrocentric X and Y chromosomes (A. L. Gardner and Patton 1976).

Nesoryzomys swarthi Orr, 1938
Santiago Nesoryzomys, Swarth's Nesoryzomys
SYNONYMS:

Nesoryzomys swarthi Orr, 1938:304; type locality "Sulivan [= Sullivan] Bay, James Island [= Isla Santiago]", Galápagos Islands, Ecuador.

O[ryzomys].(*Nesoryzomys*) *swarthi*: Corbet and Hill, 1980: 144; name combination.

Nesoryzomys indefessus: Patton and Hafner, 1983:562; part (*swarthi* assigned as subspecies); name combination.

DESCRIPTION: Largest species (total length 310–317 mm; tail length 124–140 mm; Orr 1938), but only slightly more so than *N. narboroughi*. Most similar in body size, length of tail, and hindfoot to *N. narboroughi* but indistinguishable from *N. indefessus* in color tones. Patton and Hafner (1983), using colorimetry, verified the similarity in color tones between *N. swarthi* and *N. indefessus* and also demonstrated complete cranial separation between it and other large species.

DISTRIBUTION: *Nesoryzomys swarthi* is known only from the island of Santiago, Galápagos Islands, Ecuador. Specimens exist from only James Bay, Sullivan Bay, and La Bomba on the north coast (Dowler et al. 2000). Recent surveys revealed a population restricted to a 14-km stretch of north-central coastline centered on La Bomba (D. B. Harris et al. 2006); these surveys have failed to record the species at the other two historical localities.

SELECTED LOCALITIES (Map 202): ECUADOR: Galápagos, Isla Santiago (= James Island), La Bomba (Dowler et al. 2000; D. B. Harris et al. 2006), Sullivan Bay (type locality of *Nesoryzomys swarthi* Orr), James Bay (Peterson 1966).

SUBSPECIES: *Nesoryzomys swarthi* is monotypic.

NATURAL HISTORY: *Nesoryzomys swarthi* is currently restricted to mature cactus thorn-scrub dominated by the cactus *Opuntia galapageia* and tree species *Bursera graveolens* (Burseraceae) and *Croton scouleri* (Euphorbiaceae). The abundance of this species at La Bomba was positively correlated with cactus density, the proportion of mature cacti, and the presence of the shrub species *Lantana peduncularis* (Verbenaceae) (D. B. Harris et al. 2006). D. B. Harris and Macdonald (2007) reported median home ranges (minimum convex polygon) to be 1.76 ha for males and 0.68 for females. This *Nesoryzomys* is the only one to be found sympatrically with the invasive rodent species *Rattus rattus* and *Mus musculus*. *Nesoryzomys swarthi* is usually nocturnal with peak activities between 20:00 and 03:00 hours; however, in areas where high populations of *R. rattus* occur, activity may extend to crepuscular periods (D. B. Harris et al. 2006). Reproduction coincides with the rainy season from December through May and two litters of two to four young may be produced annually. Annual survival rates for *N. swarthi* are high (23.2%) relative to most continental oryzomyine species (D. B. Harris and Macdonald 2007).

REMARKS: As noted previously, this species was regarded as an insular race of *Nesoryzomys narboroughi* by Patton and Hafner (1983) but as a valid species by Musser and Carleton (1993, 2005), a decision with which I concur.

Genus *Oecomys* Thomas, 1906
Michael D. Carleton and Guy G. Musser

Oecomys is a moderately speciose genus of Sigmodontinae, containing at least 16 valid species according to current understanding (Musser and Carleton 2005; Carleton et al.

2009). The generic distribution generally adheres to tropical lowland rainforest, where the greatest species diversity is concentrated in the broad reaches of the Amazon Basin, including lower Atlantic-facing slopes of the Andean cordilleras. Marginal occurrences are found to the north of Amazonia, in coastal forests of northern Venezuela and trans-Andean forest of southernmost Central America, and to the south, in the Atlantic Forest of southeastern Brazil. The genus has been recorded from the Chaco in northern Argentina, but the species allocation of these records remains uncertain (see account for *O. mamorae*).

Species of *Oecomys* are small to medium-sized oryzomyines (average weights from 15–20 g in *O. rutilus* to 80–90 g in *O. superans*), externally characterized by soft fur, short and broad hindfeet with relatively long toes, and a tail moderately longer than the head and body (tail length about 110–120% of head and body length). The short, wide hindfoot bears thick, bright-white ungual tufts on digits II–V; pedal digit V is nearly as long as digits II–IV; and the plantar surface is smooth with six large, closely positioned pads, including a well-defined hypothenar pad. The tail in most species is basically unicolored, medium to dark brown above and below, or slightly paler underneath in some but never distinctly bicolored; caudal hairs are short, scarcely visible to the unaided eye and not obscuring the fine scale pattern; a rudimentary pencil or tuft of hairs is expressed at the tail tip in several species. The fur is soft, never spinose, and short to moderately long in length. Dorsal coloration varies from grayish brown to tawny to reddish brown, dull to bright in tone; ventral pelage color ranges from pure white to dark gray. Females possess eight mammae, including the pectoral pair, anatomically positioned as in most oryzomyines (Voss and Carleton 1993; Weksler 2006).

The skull is moderately robust and features a short and broad rostrum, a broad interorbital region, and slender zygomatic plate with a shallow or indistinct dorsal notch. The supraorbital edges are ridged or beaded, convergent anteriorly and notably divergent posteriorly (interorbital constriction cuneate), and extend across the parietals as temporal ridges in older individuals. A posteroventral flange of the parietal extends onto the lateral wall of the braincase; the interparietal is large, nearly as wide as the posterior border of the frontals. The incisive foramina are moderately long and generally wide, and in most species posteriorly terminate in front of the first molars or even with them. The bony palate is flat, long, and wide in conformation (*sensu* Hershkovitz 1962), and projects slightly or distinctly beyond the end of the third molars; posterolateral palatal pits are uniformly present but vary in development; parapterygoid fossae are moderately excavated above the plane of the bony palate; the roof of the mesopterygoid fossa is solid (sphenopalatine vacuities absent) or weakly perforated (va-

cuities present as narrow slits). The ectotympanic bullae are small and expose a broad wedge of the periotic in all species except *O. mamorae*, in which the broader ectotympanic capsule conceals more of the periotic; the tegmen tympani is reduced (see Voss 1993:18), occasionally touching but not overlapping the squamosal in all species except *O. rex*. Members of *Oecomys* possess different carotid circulatory patterns: complete in most species (stapedial foramen present, and large, posterior opening of alisphenoid canal large, and squamosal-alisphenoid groove leading to sphenofrontal foramen present; character state 0 of Carleton [1980] or pattern 1 per Voss [1988]); or derived in a few (stapedial foramen absent, or minute, squamosal-alisphenoid groove and sphenofrontal foramen absent; character state 3 of Carleton [1980] or pattern 3 per Voss [1988]; also see those different patterns as illustrated for oryzomyines by Carleton and Musser 1989). The occurrence of alisphenoid struts is variable in the genus, but they tend to be consistently present or absent (buccinator-masticatory and accessory oval foramina confluent) in a particular species. The hamular process of the squamosal is generally short and wide, the subsquamosal fenestra small or absent; a postglenoid foramen is uniformly present and moderately developed in most species. No significant sexual dimorphism is evident in craniodental measurements of those species that have been analyzed (*O. bicolor*, *O. catherinae*, *O. roberti*, *O. sydandersoni*—Patton et al. 2000; Carleton et al. 2009; Asfora et al. 2011). Explanation and/or illustration of the above characters and additional morphological description of *Oecomys* are provided by Hershkovitz (1960), Carleton and Musser (1989), Patton et al. (2000), Voss et al. (2001), Weksler (2006), and Carleton et al. (2009).

Most members of *Oecomys* possess opisthodont upper incisors, nearly orthodont in some species, and their maxillary toothrows are parallel. Molars are brachydont and cuspidate, the major lingual and labial cusps positioned oppositely; three roots anchor the upper molars and two roots the lowers. Labial and lingual flexi are broadly open and meet approximately at the tooth's midline, not interpenetrating. The M1 bears a broad anterocone with anterolingual and anterolabial conules joined medially, not cleft by an anteromedian fold; the conules are distinct in juveniles and young adults but lose definition with wear. A discrete anteroloph is present on M1 and a well-formed mesoloph on M1–2; the M3 retains an irregularly formed mesoloph and weak posteroloph. A mesolophid is consistently present on m1–2, but an ectolophid is variably developed, absent in most species. Hershkovitz (1960) and Weksler (2006) provide additional dental descriptions and character state definitions of *Oecomys*.

Few species of *Oecomys* have been surveyed for soft anatomical traits of the alimentary canal and reproductive tract. Those that have been studied exhibit a complex glans penis, with a large medial bacular mound and smaller lateral bacular mounds; have a complete assortment of male accessory reproductive glands, including one pair of preputial glands; possess a unilocular-hemiglandular gastric morphology; and lack the gall bladder (Hooper and Musser 1964; Carleton 1973; Voss and Linzey 1981; Voss and Carleton 1993; Weksler 2006).

No long-term, autecological study is available for any species of *Oecomys*. Furthermore, correspondence of past natural history and ecological reports on *O. "bicolor"* or *O. "concolor," sensu* Hershkovitz (1960), to a currently recognized species must be cautiously interpreted. In general, species are nocturnal, and individuals are typically associated with tropical lowland rainforest and numerous microhabitats in that broad classification, whether primary or secondary. Hershkovitz (1960) viewed *Oecomys* as the oryzomyine most specialized for arboreal life, a characterization sustained by collecting surveys, canopy-trapping, and nocturnal head-lamp observations, all of which attest to the superior scansorial abilities of its members (Handley 1976; Malcolm 1991; Patton et al. 2000; Voss et al. 2001). Linares (1998) characterized all species of Venezuelan *Oecomys* as frugivorous and granivorous; Voss et al. (2001) provisionally assigned *Oecomys* to the arboreal granivore/frugivore guild, but admitted that no detailed dietary investigations have been published to support their conjecture.

Species of *Oecomys* and the thomasomyine *Rhipidomys* superficially resemble one another and may be difficult to distinguish in the field where similar-sized members of the two genera co-occur. Several external differences serve to separate adults of the two genera in most instances, notably coloration of the dorsal surface of the hindfeet (darker over the metatarsum in *Rhipidomys*), length of the tail pencil (caudal pilosity and terminal pencil more strongly expressed in *Rhipidomys*), and number of mammae (three pairs in *Rhipidomys*, which lacks the pectoral pair found in *Oecomys*) (Thomas 1917d; Voss 1993; Patton et al. 2000; Voss et al. 2001).

SYNONYMS:

Mus: Pictet and Pictet, 1844:64; part (description of *cinnamomeus*); not *Mus* Linnaeus.

Hesperomys: Wagner 1845a:147; part (description of *concolor*); not *Hesperomys* Waterhouse.

Oryzomys: J. A. Allen and Chapman, 1893:212; part (descriptions of *speciosus* and *trinitatis*); not *Oryzomys* Baird.

Rhipidomys: E.-L. Trouessart, 1897:520; part (listing of *bicolor*); not *Rhipidomys* Tschudi.

Oryzomys (*Oecomys*) Thomas, 1906c:444; type species *Rhipidomys benevolens* Thomas, 1901 [= *Hesperomys bicolor* Tomes, 1860], by original designation.

Oecomys: Thomas, 1909:379; first usage as genus.

REMARKS: Thomas (1906c) erected *Oecomys*, as a subgenus, to segregate arboreal, pencil-tailed oryzomyines with a long palate from species of *Rhipidomys*, the externally similar genus under which many *Oecomys* taxa were first described. *Oecomys* was afterward treated as a valid genus by some systematists (Thomas 1909, 1917d; Gyldenstolpe 1932), but most regarded it as a subgenus of *Oryzomys* (Goldman 1918; Ellerman 1941; Hershkovitz 1960; Cabrera 1961; Hall 1981; Honacki et al. 1982; Corbet and Hill 1986). The karyotypic study of A. L. Gardner and Patton (1976) was pivotal in establishing the modern usage of *Oecomys* as a monophyletic genus (Carleton and Musser 1984; Reig 1984; Voss and Carleton 1993; Musser and Carleton 1993, 2005), a rank vindicated by broad phylogenetic evaluations of Sigmodontinae or Oryzomyini (M. F. Smith and Patton 1999; A. F. B. Andrade and Bonvicino 2003; D'Elía 2003a; Weksler 2003, 2006; Percequillo, Weksler, and Costa 2011; Pine et al. 2012). Within Oryzomyini, *Oecomys* belongs to a large clade that includes other genera that co-occur with it in lowland rainforest (*Euryoryzomys, Hylaeamys, Transandinomys*) and some that inhabit Andean montane forest (*Handleyomys, Nephelomys*).

Hershkovitz's (1960) revision consolidated the taxonomic contents of *Oecomys* but swept some 25 species and immense morphological variety into just two polytypic species, *O. bicolor* and *O. concolor*. His gross underestimation of specific diversity has been exposed by regional field studies in which three to five species of *Oecomys* have been documented in sympatry (Carleton et al. 1986, 2009; Woodman et al. 1991; Patton et al. 2000; Voss et al. 2001). Morphological definitions and distributional limits of the species recognized herein conform to the taxonomy presented in Musser and Carleton (1993, 2005), as based on the authors' personal examination of some 1,600 museum specimens and of all type specimens except that of *O. cleberi* Locks (1981). Identification of valid species and refinement of their distributions still require much systematic research and basic biological inventory. The treatment of all species as monotypic is a pro forma concession to this imperfect understanding of *Oecomys* systematics.

Species of *Oecomys* may be distinguished based on unique combinations of morphological traits, the most important of which include size; pelage color, especially distinctiveness of the dorsal-ventral contrast; fur texture and length; relative length of tail and development of a modest caudal pencil or tuft; pronouncement of supraorbital and temporal ridging; relative width of the zygomatic plate and expression of the dorsal notch; length and shape of the incisive foramina; presence and size of the subsquamosal and postglenoid foramina; condition of the tegmen tympani; presence or absence of alisphenoid struts; and the carotid arterial pattern. Chromosomal differences promise to be a useful tool for discriminating species, but many more forms need to be surveyed (A. L. Gardner and Patton 1976; Patton et al. 2000; Andrades-Miranda, Oliveira, Zanchin et al. 2001; A. F. B. Andrade and Bonvicino 2003; Langguth et al. 2005; C. C. Rosa et al. 2012). Descriptions in the species accounts do not repeat traits that are generally characteristic of the genus, as reviewed earlier. "Smaller than" and "larger than" comparative statements in the following descriptions reflect the ordering of select external and cranial dimensions (Table 1).

KEY TO THE SPECIES OF *OECOMYS* (VARIABLE ABBREVIATIONS: CLM, CORONAL LENGTH OF MAXILLARY TOOTHROW, M1–M3; HBL, LENGTH OF HEAD AND BODY; HFL, LENGTH OF HIND FOOT INCLUDING CLAW; IOB, LEAST INTERORBITAL BREADTH; ONL, OCCIPITONASAL LENGTH; TL, LENGTH OF TAIL; ALSO SEE TABLE 1):

1. Carotid circulatory pattern derived (sphenofrontal foramen and squamosoalisphenoid groove absent, stapedial foramen minute or absent) 2
1′. Carotid circulatory pattern complete (sphenofrontal foramen and squamosoalisphenoid groove present, stapedial foramen small but present) 4
2. Dorsal pelage predominantly gray to grayish buff, with a buffy to ochraceous lateral stripe; pelage moderately long for the genus (≈7–9 mm over middle rump); alisphenoid struts typically absent; interorbit narrower (IOB ≤5.1 mm), supraorbital shelves weakly defined for the genus; palatal bridge short, terminating about the level of the posterior M3s, posterolateral palatal pits simple . . . . *Oecomys mamorae*
2′. Dorsal pelage ochraceous brown to rufous brown, without distinct lateral stripe; pelage shorter (≤5–7 mm over middle rump); alisphenoid struts typically present; interorbit broader (IOB ≥5.1 mm), supraorbital shelves moderately developed as per genus; palatal bridge long, projecting notably beyond M3s, posterolateral palatal pits pronounced 3
3. Size larger (HFL ≈ 26–30 mm, ONL ≈ 31–34 mm, CLM ≈ 4.8–5.2 mm); tail notably longer than head and body; dorsal pelage more uniformly rufous brown and darker in tone, shorter in length (4–5 mm over middle rump); ventral pelage usually dull white, dorsal-ventral contrast more apparent; incisive foramina evenly curving, widest near their middle with the anterior and posterior ends acutely pointed . . . . *Oecomys concolor*
3′. Size smaller (HFL ≈ 22–26 mm, ONL ≈ 28–31 mm, CLM ≈ 4.4–4.8 mm); tail slightly longer than head and

body; dorsal pelage ochraceous brown to pale tawny generally bright in tone and showing more grayish over forequarters, slightly longer (5–7 mm over middle rump); ventral pelage grayish; incisive foramina noticeably widest toward the rear, the posterior ends blunt and anterior ends sharp. . . . . . *Oecomys sydandersoni*
4. Size large (HFL ≈ 27–29 mm, ONL ≈ 33–35 mm, CLM ≈ 5.1–5.5 mm); pelage thick, long, and richly luxuriant; interorbital region wide (IOB ≥ 6 mm), with conspicuous supraorbital shelves and distinct postorbital processes; tegmen tympani well developed, securely overlapping squamosal . . . .*Oecomys rex*
4′. Size small to large; pelage short to long, coarse to soft in texture; interorbital region narrower (IOB usually ≤ 6 mm), supraorbital shelves moderate to strongly developed, without postorbital processes; tegmen tympani reduced, sometimes contacting but not overlapping squamosal, postglenoid foramen medium to large 5
5. Size small (HFL ≈ 18–23 mm, ONL ≈ 23–28 mm , CLM ≈ 3.2–4.0); underparts typically all white, bright in tone, dorsal-ventral pelage contrast well defined; tail relatively short (TL ≈ 105–110% of HBL), caudal hairs long, obscuring distal scale rows and forming a penicillate or tufted tip 6
5′. Size medium to large (HFL usually ≥ 23 mm, ONL usually ≥ 27 mm, CLM ≥ 4.0); under parts variable, dull white to some shade of grayish white, dorsal-ventral contrast variably delineated 7
6. Cranium and toothrow diminutive (ONL ≤ 25.0 mm, CLM ≈ 3.2–3.4 mm); dorsal pelage moderately long for body size (6–8 mm over the middle rump); terminal tail tuft longer, better developed; alisphenoid struts usually present . . . . *Oecomys rutilus*
6′. Cranium and toothrow more robust (ONL ≥ 25.0 mm, CLM ≈ 3.6–4.0 mm); dorsal pelage short for body size (3–6 mm over the middle rump); terminal tail tuft shorter, less conspicuous; alisphenoid struts usually absent. . . . . *Oecomys bicolor/Oecomys cleberi*
7. Size medium (HFL ≈ 22–26 mm, ONL ≈ 27–31 mm, CLM ≈ 4.0–4.6) 8
7′. Size medium large to large (HFL ≈ 26–31 mm, ONL ≈ 30–37 mm, CLM ≈ 4.6–5.6) 11
8. Dorsal pelage soft, dense, and long (9–12 mm over middle rump), caudal hairs long, obscuring scale pattern and forming a conspicuous terminal tuft 9
8′. Dorsal pelage shorter (5–8 mm over the middle rump), texture coarser; under parts typically pure white, sharply contrasting with bright orange-brown dorsum and ochraceous flanks; tail absolutely and relatively long (TL ≈ 118% of HBL), with short caudal hairs and indistinct terminal pencil . . . . *Oecomys speciosus*

9. Tail relatively short (TL ≈ 107% of HBL); alisphenoid struts typically present; subsquamosal fenestra occluded. . . . .*Oecomys auyantepui*
9'. Alisphenoid struts typically absent; subsquamosal fenestra open. 10
10. Under parts typically grayish white; tail absolutely and relatively long (TL ≈ 117% of HBL). . . . .
. . . . *Oecomys paricola*
10'. Under parts typically dull white; tail relatively short (TL ≈ 105% of HBL) . . . .*Oecomys phaeotis*
11. Dorsal pelage shorter (5–10 mm over middle rump) and moderately soft 12
11'. Dorsal pelage longer (10–14 mm over middle rump), very dense and palpably soft to touch. . . . .
. . . . *Oecomys trinitatis* complex
12. Size large (HFL ≥ 28 mm, ONL ≥ 34 mm, CLM ≥ 5.2 mm), dorsal pelage medium in length (8–10 mm over middle rump), pelage somber in tone with dark grayish venter, tail moderately longer than head and body (TL ≈ 118% of HBL). . . . . . . *Oecomys superans*
12' Size medium large (HFL ≤ 28 mm, ONL ≤ 34 mm, CLM ≤ 5.1 mm), dorsal pelage short (5–7 mm over middle rump), pelage brighter in tone with white to grayish white venter, tail notably longer than head and body (TL ≈ 123% of HBL) . . . .
. . . .*Oecomys roberti*
13. Size medium large (HFL ≈ 25–27 mm, ONL ≈ 30–32 mm, CLM ≈ 4.6–4.8 mm), confined to lower elevations in mountains of northeastern Colombia and northern Venezuela . . . . *Oecomys flavicans*
13'. Size medium large (HFL ≈ 26–29 mm, ONL ≈ 31–33 mm, CLM ≈ 4.8–5.2 mm), distribution in greater Amazonia and trans-Andean forests of lower Central America and northwestern South America. . . . .
. . . .*Oecomys trinitatis*
13''. Size large (HFL ≈ 28–30 mm, ONL ≈ 32–34 mm, CLM ≈ 5.0–5.2 mm), confined to Atlantic Forest region of southeastern Brazil. . . . . *Oecomys catherinae*

Oecomys auyantepui Tate, 1939
Guianan Oecomys

SYNONYMS:

Oecomys auyantepui Tate, 1939:193; type locality "south slopes of Mt. Auyan-tepui, Caroni River, [Bolívar,] Venezuela, 3500 feet."
Oryzomys auyantepui: Goodwin, 1953:302; name combination.
Oryzomys concolor speciosus: Hershkovitz, 1960:553; part (*auyantepui* allocated as full synonym, invalid as subspecies).
Oryzomys trinitatis auyantepui: Cabrera, 1961:407; name combination.

Oecomys paricola: Musser and Carleton, 1993:716; part (*auyantepui* allocated as synonym without indication of rank).

DESCRIPTION: Similar in body size to *O. paricola* and *O. phaeotis* (Table 1), larger than *O. rutilus* and *O. bicolor*, the two smallest-bodied species in genus, and appreciably smaller than all other congeners. Luxuriant pelage soft, dense, and long (10–12 mm over middle rump), with demarcation between dorsal and ventral pelage sharp. Dorsal fur rich ochraceous tawny (nearly orange brown) in most specimens, but darker—brownish tawny—in others; tops of hind feet dark brown to tip of digits. Ventral pelage typically grayish white (hairs with gray bases and white tips), or uncommonly with rich ochraceous overwash of gray bases; in some specimens, chin and throat pure white, in others white strip extends along midline from throat to groin. Tail only slightly longer than combined head and body (TL ≈ 106% of HBL) and dark brown on all surfaces; caudal hairs long, typically obscuring distal scale rows and forming conspicuous tuft at tip (6–10 mm long per Voss et al. 2001).

Skull small and compact, with relatively short rostrum and narrow interorbit. Conspicuous supraorbital ridges outline interorbital and postorbital regions and sweep caudad to join distinct temporal beading. Zygomatic arches narrower rostrally and wider toward braincase; zygomatic plates relatively narrow and dorsal notches shallow. Incisive foramina moderately long and broad, their posterior margins situated about even with anterior rim of first molars. Pair of prominent posterolateral palatal pits and smaller foramina adorn bony palate behind molars. Mesopterygoid fossa wide and U-shaped at anterior margin, its roof entirely osseous. Strong alisphenoid struts present (Voss et al. 2001), and carotid circulatory plan primitive (pattern 1 of Voss 1988). Posterolateral wall of braincase bears postglenoid foramen but no subsquamosal fenestra. Ectotympanic bullae small, in ventral view exposing much of medial periotic.

DISTRIBUTION: Southeastern Venezuela (Bolívar), the Guianas, and north-central Brazil (Amazonas, Amapá)—the Guiana subregion of Amazonia; known elevational range from sea level to about 1,100 m.

SELECTED LOCALITIES (Map 203; from Voss et al. 2001, except as noted): BRAZIL: Amapá, Serra do Navio; Amazonas, 80 km N of Manaus. FRENCH GUIANA: Paracou; Trois Sauts. GUYANA: Cuyuni-Mazaruni, Kartabo, Cuyuni River; Upper Takutu–Upper Essequibo, 5 km SE of Surama. SURINAM: Sipaliwini, Avanavero Falls, Kabalebo River (RMNH 21733); Brokopondo, Brownsberg Nature Park, 7 km S and 18.5 km W of Afobakka (CM 54055). VENEZUELA: Bolívar, Auyan-tepuí, south slope, Río Caroni (type locality of *Oecomys auyantepui* Tate).

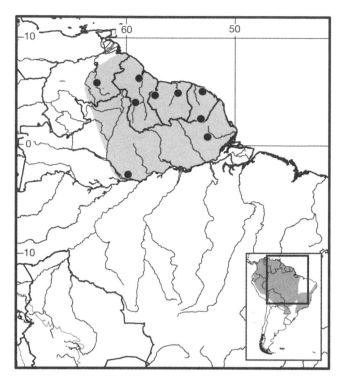

Map 203 Selected localities for *Oecomys auyantepui* (●). Contour line = 2,000 m.

SUBSPECIES: In resurrecting *O. auyantepui* to species, Voss et al. (2001) provided no evidence for definable subspecies; we provisionally regard the species as monotypic.

NATURAL HISTORY: At the type locality in Venezuela (Auyantepui, 3500 ft), Tate (1939) merely noted the habitat as humid forest. In French Guiana, specimens were obtained in primary lowland forest, both in well-drained sites and in creekside or swampy settings, mostly in arboreal trap-sets (lianas, canopy platforms) but rarely on the ground (Voss et al. 2001).

REMARKS: Populations of *auyantepui* are clearly distinct from those of *O. concolor* or *O. trinitatis*, the species under which it was earlier synonymized (Cabrera 1961; Hershkovitz 1960), and the taxon was returned to specific status by Voss et al. (2001). However, its level of relationship and nomenclatural status with regard to *O. paricola* (a synonym per Musser and Carleton 1993) and to *O. phaeotis* require further investigation. The geographic range of *O. auyantepui*, as so far known, is limited to the Guiana subregion of Amazonia, its distribution congruent with the ranges of *O. rex* and *O. rutilus*.

Oecomys bicolor (Tomes, 1860)
White-bellied Oecomys
SYNONYMS:

Hesperomys bicolor Tomes, 1860a:217; type locality "Gualaquiza," Rio Gualaquiza, 885 m, Morona-Santiago, Ecuador.

Hesperomys (Rhipidomys) bicolor: Thomas, 1884:448; name combination (tentative subgeneric referral of species).

Rhipidomys bicolor: E.-L. Trouessart, 1897:520; name combination.

Rhipidomys dryas Thomas, 1900c:271; type locality "Paramba, [Imbabura,] N. Ecuador. Alt. 1100 m."

Rhipidomys benevolens Thomas, 1901d:369; type locality "Chimate, Upper Beni River, [La Paz,] Bolivia, 68° W., 15° S.; alt. 700 m."

Rhipidomys rosilla Thomas, 1904d:35; type locality "La Union, Lower Orinoco," Bolívar, Venezuela.

[*Oryzomys (Oecomys)*] *bicolor*: Thomas, 1906c:445; name combination.

Oryzomys (Oecomys) benevolens: Thomas, 1906c:445; name combination.

Oryzomys (Oecomys) dryas: Thomas, 1906c:445; name combination.

[*Oryzomys (Oecomys)*] *rosilla*: Thomas, 1906c:445; name combination.

Oecomys nitedulus Thomas, 1910d:505; type locality "Lower Essequibo, 13 miles from mouth," Essequibo Islands-West Demerara, Guyana.

Oecomys bicolor: Osgood, 1914b:161; first use of current name combination.

Oecomys milleri J. A. Allen, 1916c:523; type locality "Barão Malgaço [Barão de Malgaço], Matto Grosso [now Rondônia], Brazil."

Oecomys florenciae J. A. Allen, 1916c:524; type locality "Florencia (altitude 675 feet), Rio Caquetá drainage, [Caquetá,] southwestern Colombia."

Oecomys trabeatus G. M. Allen and Barbour, 1923:262; type locality "Rio Jesusito, [Darién,] eastern Panama."

Oecomys bicolor: Gyldenstolpe, 1932a:39; name combination (*dryas* allocated as full synonym without indication of rank).

Oecomys rosilla: Gyldenstolpe, 1932a:39; name combination.

Oecomys benevolens: Gyldenstolpe, 1932a:40; name combination.

Oecomys endersi Goldman, 1933:525; type locality "Barro Colorado Island, Canal Zone [now Panamá province], Panama."

Oecomys phelpsi Tate, 1939:194; type locality "south slope Mt. Auyan-tepui, 3500 feet," Bolívar, Venezuela.

Oryzomys [(*Oecomys*)] *benevolens*: Ellerman, 1941:357; name combination.

Oryzomys [(*Oecomys*)] *florenciae*: Ellerman, 1941:358; name combination.

Oryzomys [(*Oecomys*)] *milleri*: Ellerman, 1941:358; name combination.

Oryzomys [(*Oecomys*)] *nitedulus*: Ellerman, 1941:358; name combination.

Oryzomys [(*Oecomys*)] *rosilla*: Ellerman, 1941:359; name combination.

Akodon (*Chalcomys*) *aerosus*: Tate, 1945:316; part (restriction of type specimen, *phelpsi* allocated as synonym).

Oryzomys (*Oecomys*) *bicolor trabeatus*: Hershkovitz, 1960: 533; name combination (*endersi* allocated as full synonym, invalid as subspecies).

Oryzomys (*Oecomys*) *bicolor occidentalis* Hershkovitz, 1960: 533; replacement name for *Rhipidomys dryas* Thomas, 1900, preoccupied by *Oryzomys dryas* Thomas (now = *Microryzomys minutus*).

Oryzomys bicolor bicolor: Hershkovitz, 1960:535 name combination (*florenciae, milleri, nitedulus,* and *rosilla* allocated as full synonyms, invalid as subspecies).

Oryzomys bicolor phaeotis: Hershkovitz, 1960:540; part (*benevolens* allocated as full synonym, invalid as subspecies).

Oryzomys bicolor bicolor: Cabrera, 1961:403; name combination (*dryas* allocated as full synonym, invalid as subspecies).

Oryzomys bicolor benevolens: Cabrera, 1961:403; name combination.

Oryzomys bicolor florenciae: Cabrera, 1961:403; name combination.

Oryzomys bicolor milleri: Cabrera, 1961:403; name combination.

Oryzomys bicolor nitedulus: Cabrera, 1961:404; name combination.

Oryzomys bicolor rosilla: Cabrera, 1961:404; name combination.

Oryzomys marmosurus guianae: Cabrera, 1961:405; part (*phelpsi* questionably listed as synonym).

Oecomys bicolor: Musser and Patton, 1989:6; name combination (restriction of type specimen, provisional allocation of *phelpsi* as synonym without indication of rank).

Oecomys bicolor: Musser and Carleton, 1993:716; name combination (*benevolens, dryas* [of Thomas, 1900], *endersi, florenciae, milleri, nitedulus, occidentalis, rosilla,* and *trabeatus* listed as synonyms without indication of rank).

DESCRIPTION: Recognizable chiefly by small size (averaging only larger than *O. rutilus*), combined with white under parts and relatively short tail (TL ≈ 105–110% of HBL) that bears moderate terminal tuft. Pelage dense and felt-like, somewhat crisp to touch, and short compared with body size (3–6 mm over middle rump); demarcation between dorsal and ventral pelage sharp. Dorsal fur ranges from rich ochraceous tawny (nearly orange brown) to rufous brown, finely mixed with dark brown over middle dorsum and becoming slightly paler along lower sides of head and body, a bright buffy gray. Darker dorsum sharply set off from bright ventral pelage, which is uniformly white (hairs self-colored) in the majority of specimens examined; variations include ventral hairs with white bases and buff to ochraceous tips, imparting pale wash to white ground color, or white over belly and chest with varying suffusion of gray-based hairs toward sides. Short tail all brown, and caudal hairs notably long for small size of species, typically obscuring scale rows and forming short but discrete tuft at tip (hairs extend 5–7 mm beyond tip).

Skull stoutly built for its compact size, with short rostrum, narrow interorbit with finely beaded supraorbital ledges, and relatively inflated braincase. Temporal ridges weakly defined and become obsolete over posterior braincase. The zygomatic arches slightly narrower rostrally and slightly wider toward braincase; zygomatic plates narrow and dorsal notches shallow. Incisive foramina relatively long and narrow, their posterior margins situated about even with anterior margin of first molars. Pair of prominent posterolateral palatal pits adorns bony palate; mesopterygoid fossa wide and U-shaped at anterior margin; roof entirely osseous. Postglenoid foramen and subsquamosal fenestra present and spacious. Alisphenoid struts commonly absent, but occurrence highly variable within species as currently defined; carotid circulatory plan complete (pattern 1 of Voss 1988). Ectotympanic bullae small, exposing much of medial periotic.

DISTRIBUTION: Lowlands of trans-Andean rainforests (eastern Panama to western Ecuador) and greater Amazonia (eastern Colombia and central Venezuela, through the Guianas, eastern Peru and Ecuador, to eastern Bolivia and south-central Brazil); known elevational range from sea level to 1,537 m, most locality records falling between sea level and 800 m.

SELECTED LOCALITIES (Map 204): BOLIVIA: Beni, Las Peñas (Anderson 1997); La Paz, Chimate (type locality of *Rhipidomys benevolens* Thomas); Santa Cruz, Ayacucho (Anderson 1997), 2.5 km NE of El Refugio, Parque Nacional Noel Kempff Mercado (Carleton et al. 2009). BRAZIL: Acre, Nova Vida, right bank Rio Juruá (Patton et al. 2000); Amapá, Serra do Navio (Voss et al. 2001); Amazonas, Colocação Vira-Volta, left bank Rio Juruá (Patton et al. 2000); Distrito Federal, 20 km S of Brasília (OMNH 17482, 17483); Mato Grosso, 264 km N of Xavantina, Serra do Roncador Base Camp (BM 81.385–81.388); Mato Grosso do Sul, Urucum de Corumbá (FMNH 26806); Pará, Capim (AMNH 188963), Gonotire, Rio Fresco at confluence with Rio Xingu (MZUSP 1298, 1301); Rondônia, Barão de Malgaço (type locality of *Oecomys milleri* J. A. Allen). COLOMBIA: Amazonas, Río Apaporis, Ina Gaje, Río Pacoa (FMNH 57237, 57239); Boyacá, Arauca, Río Covaria (FMNH 92524); Caquetá, Florencia, Río Bodoquera (type locality of *Oecomys florenciae* J. A. Allen). ECUADOR: Imbabura, Paramba, Río Mira (type locality of *Oryzomys* (*Oecomys*) *bicolor occidentalis* Hershkovitz, replacement name for *Rhipidomys dryas* Thomas); Morona-Santiago, Gualaquiza (type locality of *Hesperomys bicolor* Tomes); Sucumbios, Santa Cecilia (KU 135116, 135153). FRENCH GUIANA: Nouragues (Voss et al. 2001). GUYANA: Essequibo Islands-West Demerara, 13 mi from mouth of Lower Essequibo (type locality of *Oecomys nitedulus* Thomas); Upper Takutu–Upper Essequibo, 5 km SE of Surama (Voss et al. 2001). PERU: Junín, Chanchamayo (BM 5.11.2.16, FMNH 29453); Loreto, San Jacinto (KU 158191), Boca Río Peruate, Río Amazonas (FMNH 88940–88943); Madre de Dios, Reserva Cuzco Amazonico, 14 km E de Puerto Maldonado (KU 144302–144304); San Martín, Yurac Yacu (BM 27.1.1.122–27.1.1.124). SURINAM: Sipaliwini, Oelemarie (CM 76862–76866); Paramaribo, Paramaribo (USNM 319963–319967). VENEZUELA: Amazonas, Tamatama, Río Orinoco (Handley 1976); Aragua, Estación Biológica Rancho Grande (AMNH 144851, KU 135127); Bolívar, La Unión (type locality of *Rhipidomys rosilla* Thomas), south slope Auyán-tepuí (type locality of *Oecomys phelpsi* Tate).

SUBSPECIES: As here defined, *O. bicolor* encompasses many junior synonyms and appreciable size and color variation. Whether such variation represents definable geographic races or the presence of additional valid species requires further investigation. In lieu of such studies, we provisionally treat *O. bicolor* as monotypic. Linares (1998) retained four subspecies for Venezuelan populations (*O. b. bicolor*, *O. b. nitedulus*, *O. b. phelpsi*, *O. b. rosilla*).

NATURAL HISTORY: *Oecomys bicolor* inhabits tropical evergreen forest, both primary and secondary and in seasonally inundated vegetation (várzea) and unflooded, upland forest (terra firme) (Handley 1966; Carleton et al. 1986; Patton et al. 2000). Along the Rio Juruá, western Brazil, Patton et al. (2000) collected most specimens of *O. bicolor* in undisturbed várzea forest but some in terra firme; a majority were captured in canopy traps (74%), some in traps placed in viny tangles about 2 m above ground and very few in ground sets. Pregnant females (litter size mode = 2, range: 1–4) have been recorded throughout the dry season and beginning of the rainy season (August–February; Patton et al. 2000). Perhaps more than any other *Oecomys*, individuals of *O. bicolor* have been documented from commensal settings, such as orchards, overgrown fields, and rustic dwellings (Hershkovitz 1960; Hutterer et al. 1995; Emmons et al. 2006).

REMARKS: Of the species-group taxa that Hershkovitz (1960) associated under his polytypic view of *O. bicolor*, three have been resurrected to species (*O. paricola*, *O. phaeotis*, *O. rutilus*—see Musser and Carleton 1993, 2005; Voss et al. 2001). The contents presented here also may be a complex of species. The status of taxa described from the Trans-Andean region (*trabeatus*, *endersi*, and

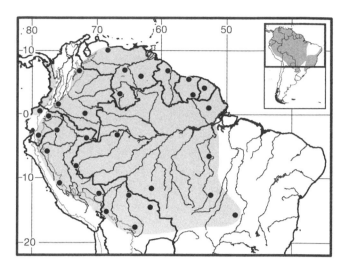

Map 204 Selected localities for *Oecomys bicolor* (●). Contour line = 2,000 m.

occidentalis) in particular warrant attention. As recorded by Hershkovitz (1960) and affirmed by our own museum research, a wide geographic hiatus separates these trans-Andean forms, described from Panama and northwestern Ecuador, from the core distribution of *O. bicolor* in Amazonia (Map 204), in spite of ample inventory in northern Colombia and northwestern Venezuela where other *Oecomys* have been recovered (e.g., *O. flavicans*, *O. speciosus*). In addition, Patton et al. (2000) identified a small, white-bellied form in sympatry with *O. bicolor* in western Brazil; the status of this indeterminate form invites resolution.

The karyotype has a diploid number of 80, with a high proportion of biarmed chromosomes (FN = 140; A. L. Gardner and Patton 1976; Patton et al. 2000).

Oecomys catherinae Thomas, 1909
Atlantic Forest Oecomys
SYNONYMS:

Mus cinnamomeus Pictet and Pictet, 1844:64; type locality, Brazil, Bahia (name preoccupied by *Mus cinnamomeus* Lichtenstein, 1830, a *Trinomys* [Echimyidae]).

Oecomys catherinae Thomas, 1909:234; type locality "Joinville, Santa Catherina [Santa Catarina], S. Brazil."

Oryzomys [(*Oecomys*)] *catherinae*: Ellerman, 1941:357; name combination.

Oecomys cinnamomeus: Moojen, 1952b:53; name combination.

Oryzomys subflavus: Hershkovitz, 1960:543; part (*catherinae* allocated as full synonym); not *subflavus* Wagner.

Oryzomys concolor bahiensis Hershkovitz, 1960:561; type locality "Fazenda Almada, Ilhéus, Bahia, Brazil."

Oryzomys catherinae catherinae: Cabrera, 1961:404; name combination.

Oecomys trinitatis: Musser and Carleton, 1993:717; part (*bahiensis*, *catherinae*, and *cinnamomeus* listed as synonyms without indication of rank); not *trinitatis* J. A. Allen and Chapman.

Oecomys catherinae: Musser and Carleton, 2005:1137 (*bahiensis* and *cinnamomeus* listed as synonyms without indication of rank).

DESCRIPTION: Specimens suggest slightly larger version of *O. trinitatis*. Dorsal pelage lush and thick (12–14 mm), sleek in texture, brownish-orange to tawny chestnut in color, and bright in tone. General impression of under parts dark grayish-white, ventral hairs with dark to slate gray bases, usually tipped with white; variations include streaks or patches of buffy-tipped hairs on chest and/or abdomen and creamy hairs confined to throat. Demarcation between dorsum and ventrum weakly defined. Tail dusky brown all around for most of length, paler beneath toward base; caudal hairs short, revealing the fine scale rows and not forming discrete tuft.

Robust skull features relatively long and narrow rostrum, pronounced supraorbital shelves with free edge reflected dorsally, and conspicuous temporal ridges that extend across braincase to occiput. Postglenoid foramen small, and subsquamosal fenestra tiny or absent. Incisive foramina relatively narrow and long, reaching level of anterior roots of upper first molars. Roof of mesopterygoid fossa usually fully ossified, sphenopalatine vacuities absent or minute. Alisphenoid struts commonly absent; carotid circulatory plan complete (pattern 1 of Voss 1988). Ectotympanic bullae small, exposing much of medial periotic.

DISTRIBUTION: Atlantic coast forests in southeastern Brazil (Paraíba southward to northern Santa Catarina) and along riverine forest into south-central Brazil (Goiás, Federal District, Mato Grosso do Sul) (see Fig. 3 in Asfora et al. 2011); known elevational range sea level to 700 m.

SELECTED LOCALITIES (Map 205; from Asfora et al. 2011, except as noted): BRAZIL: Paraíba, Mamanguape; Alagoas, Ibateguara; Bahía, Fazenda Almada, Ilhéus (type locality of *Oryzomys concolor bahiensis* Hershkovitz); Goiás, Barra do Rio São Domingos (MZUSP 4012, 4020), Minaçu; Distríto Federal, 20 km S of Brasília (OMNH 17487, 17488); Goiás, Caldas Nova; Espírito Santo, Águia Branca; Minas Gerais, Pirapitanga; Espírito Santo, Viana; Mato Grosso do Sul, Brasilândia; Rio de Janeiro, Cambuci, Mangaratiba; São Paulo, Parque Estadual Intervales; Santa Catarina, Joinville (type locality of *Oecomys catherinae* Thomas).

SUBSPECIES: Although *bahiensis* Hershkovitz has been recognized as a subspecies, knowledge of intraspecific variation is too poor to defend usage of subspecies at this stage of understanding; *O. catherinae* is provisionally regarded as monotypic (also see Asfora et al. 2011).

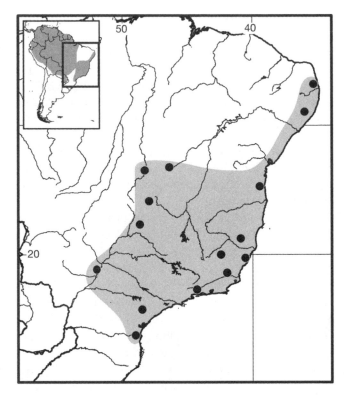

Map 205 Selected localities for *Oecomys catherinae* (●).

NATURAL HISTORY: Asfora et al. (2011) recorded the species as occurring in dense rainforest, open rainforest, and semideciduous forest in a savanna-forest mosaic (i.e., gallery forest settings in the Cerrado biome).

REMARKS: Hershkovitz (1960) erected his new subspecies *Oryzomys concolor bahiensis* because he viewed *Oecomys catherinae* Thomas (1909) as well as *Oecomys rex* Thomas (1910) as a synonym of *Oryzomys subflavus* (now *Cerradomys subflavus sensu* Weksler et al. 2006). With correction of this misidentification (Musser and Carleton 2005) and because *Mus cinnamomeus* Pictet and Pictet is a junior homonym, *O. catherinae* assumes priority for Atlantic Forest populations.

Samples of *O. catherinae* from the Atlantic Forest and Cerrado riparian formations are closely related to, but genetically well differentiated from, samples of *O. trinitatis* in the Amazon Basin (L. P. Costa 2003). The karyotype uniformly consists of $2n=60$ and displays minor variation in the number of autosomal arms (FN$=62/64$) as based on broad geographic survey of chromosomal numbers (Asfora et al. 2011).

Oecomys cleberi Locks, 1981
Cleber's Oecomys

SYNONYM:

Oecomys cleberi Locks, 1981:1; type locality "Fazenda Agua Limpa, da Universidade de Brasília, Distrito Federal, Brasil. Longitude: 45°54′W. Latitude: 15°57′S."

DESCRIPTION: Small species, slightly larger than examples of *O. bicolor* and slightly smaller than those of *O. paricola* (Table 1). In general appearance, it recalls *O. bicolor*, having pure white to pale cream under parts sharply contrasted to orange-brown upper parts and relatively short tail with modest terminal tuft. Cranial architecture also resembles *O. bicolor*, including possession of primitive carotid circulatory plan (pattern 1 of Voss 1988); alisphenoid struts usually absent. R. G. Rocha et al. (2012) provide additional description and comparisons with *O. paricola* and *O. roberti*, other species that cohabit Cerrado of south-central Brazil.

DISTRIBUTION: Known from the southern Cerrado biome of Brazil based on four localities documented to date.

SELECTED LOCALITIES (Map 206; from R. G. Rocha et al. 2012, except as noted): BRAZIL: Mato Grosso, Fazenda São Luís, 30 km N of Barra do Garças; Distrito Federal, Fazenda Água Limpa, Universidade Federal de Brasília (type locality of *Oecomys cleberi* Locks); Minas Gerais, Fazenda Cafundó, 13 km NE of Nova Ponte; Mato Grosso do Sul, Fazenda Maringá, 54 km of Dourados.

SUBSPECIES: Based on the few specimens (15) and few localities (4) so far known, *O. cleberi* is regarded as monotypic (R. G. Rocha et al. 2012).

NATURAL HISTORY: Locks (1981) reported the habitat as riparian forest.

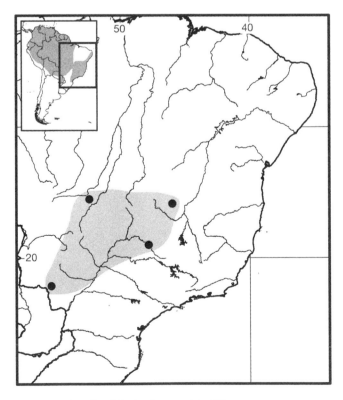

Map 206 Selected localities for *Oecomys cleberi* (●).

REMARKS: Locks (1981) critically contrasted her new form to *O. bicolor* and *O. milleri* (here considered another synonym of *O. bicolor*). Morphological evidence and analysis of mtDNA cytochrome-*b* sequences affiliate *O. cleberi* and *O. bicolor* as cognate species, the former apparently endemic to the Cerrado and a vicariant relative to the latter distributed across greater Amazonia (R. G. Rocha et al. 2012). R. G. Rocha et al. (2012) cautioned that past reports of *O. bicolor* from Cerrado localities might instead represent *O. cleberi*. Morphological discrimination of the two species and the distributional extent of *O. cleberi* require further study.

Oecomys concolor (Wagner, 1845)
Natterer's Oecomys

SYNONYMS:

Hesperomys concolor Wagner, 1845a:147; type locality "Flusse *Curicuriari* im nordwestlichen Brasilien" [= Brazil, Amazonas, Rio Curicuriari, a tributary of the upper Rio Negro, below São Gabriel (da Cachoeira), fide Hershkovitz 1960; also see Carleton et al. 2009:24].

Rhipidomys marmosurus Thomas, 1899d:378; type locality "Maipures, Upper Orinoco," Vichada, Colombia.

[*Oryzomys (Oecomys)*] *marmosurus*: Thomas, 1906c:445; name combination.

O[*ecomys*]. *marmosurus*: Thomas, 1910b:187; name combination.

[*Oryzomys (Orzyomys)*] *concolor*: Tate, 1932e:3; name combination, taxonomic history.

Oryzomys [*(Oecomys)*] *marmosurus*: Ellerman, 1941:358; name combination.

Oryzomys (Oecomys) concolor concolor: Hershkovitz, 1960: 545–546; name combination (*marmosurus* allocated as full synonym, invalid as subspecies).

Oecomys concolor: A. L. Gardner and Patton, 1976:13; first use of current name combination.

Oecomys concolor: Musser and Carleton, 1993:716; name combination (*marmosurus* listed as synonym without indication of rank).

DESCRIPTION: Characterized by combination of medium-large size, relatively long tail (TL ≈ 118% of HBL), and short pelage (5–7 mm over middle rump). Dorsal pelage varies from dull fulvous brown to brighter ochraceous-tawny, especially over shoulders and cheeks. Venter also varies, wholly dull white in most specimens, or some with strong overwash of buff-tipped to ochraceous-tipped hairs, or some with small to broad expanses of gray over middle abdomen; dorsal-ventral pelage contrast weakly defined in the last two conditions. Pinnae dark brown, usually contrasting with adjacent upper parts. Tail brown to dark brown all around for most of length, mottled beneath near base; short caudal hairs expose scale rows for most of tail length, without formation of terminal pencil.

Skull with moderately prominent supraorbital ridges, with distinct interorbital constriction and posteriorly diverging supraorbital shelves (cuneate shape). Incisive foramina relatively narrow, their anterior and posterior ends acute, gently curving along lateral edges and widest near middle. Bony palate noticeably projecting beyond posterior margins of M3s, as in most oryzomyines; posterolateral palatal pits well developed, typically consisting of one large pit with interior perforations plus one or two supernumerary foramina. Alisphenoid struts typically present (Carleton et al. 2009), and carotid circulatory plan derived (pattern 3 of Voss 1988).

DISTRIBUTION: Lowland rainforest to the north of the Rios Amazonas-Solimões in northwestern Brazil and to the south of the Río Orinoco in eastern Colombia and southern Venezuela; known elevational range sea level to 400 m.

SELECTED LOCALITIES (Map 207; from Carleton et al. 2009, except as noted): BRAZIL: Amazonas, Ilha das Onças, left bank Rio Negro, Macaco, left bank Rio Jaú, Rio Curicuriari below São Gabriel da Cachoeira (type locality of *Hesperomys concolor* Wagner), Rio Uaupés, opposite Tahuapunta, Yavanari, right bank Rio Negro; Roraima, Rio Uraricoera. COLOMBIA: Meta, Los Micos, 18 km SW of San Juan de Arama; Vichada, Maipures, middle Río Orinoco (type locality of *Rhipidomys marmosurus* Thomas). VENEZUELA: Amazonas, Belén, Río Cunucunuma, Boca Mavaca, 68 km SE of Esmeralda, 12 mi W of Río Jawasu, left bank Río Casiquiare, San Juan, Río Manapiare, Tamatama, Río Orinoco; Apure, Hato Caribén, 60 km NE of Puerto Páez, Río Cinaruco.

SUBSPECIES: Carleton et al. (2009) provided no evidence for subspecific divisions within *O. concolor*.

REMARKS: The epithet *concolor* Wagner (1845a) was a forgotten name in oryzomyine taxonomy until Tate (1932e) tentatively identified it as a species of *Oryzomys* and Hershkovitz (1960) recognized it as a member of *Oecomys*. The highly variable morphology and pantropical range of *O. concolor* as defined by Hershkovitz reflected its composite nature, which embraced entirely or partly at least nine valid species among his synonyms (*O. auyantepui*, *O. catherinae*, *O. concolor*, *O. flavicans*, *O. mamorae*, *O. roberti*, *O. speciosus*, *O. superans*, *O. trinitatis*, as per Musser and Carleton, 2005). Among species of *Oecomys*, the derived circulatory plan of *O. concolor* is shared only with *O. mamorae* and *O. sydandersoni* and offers a morphological basis to hypothesize their closer relationship (Weksler 2006; Carleton et al. 2009). The geographic occurrence of *O. concolor* proper is confined to the Rio Negro West subregion of northern Amazonia, an area of endemism delineated from phylogeographic and distributional studies of neotropical birds (Cracraft and Prum 1988; Stotz et al. 1996).

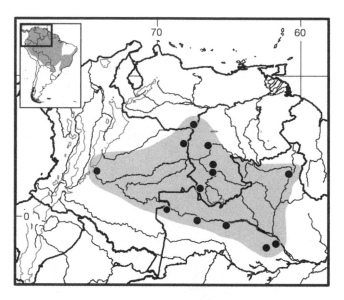

Map 207   Selected localities for *Oecomys concolor* (●). Contour line = 2,000 m.

Oecomys flavicans (Thomas, 1894)
Tawny Oecomys
SYNONYMS:

Oryzomys flavicans Thomas, 1894b:351; type locality "Merida, [Mérida,] Venezuela."

Oryzomys flavicans illectus Bangs, 1898:164; type locality "Pueblo Viejo, [Sierra Nevada de Santa Marta, Magdalena,] Colombia, altitude 8000 ft."

Oecomys mincae J. A. Allen, 1913c:603; type locality "Minca (altitude 2000 feet), Santa Marta region, [Magdalena,] Colombia."

Oryzomys flavicans flavicans: Gyldenstolpe, 1932a:22; name combination.

Oecomys illectus: Gyldenstolpe, 1932a:42; name combination, elevation to species.

Oryzomys [(Oecomys)] mincae: Ellerman, 1941:358; name combination.

O[ryzomys] flavicens: Goodwin, 1953:310; spelling lapsus for *flavicans*.

Oryzomys (Oecomys) concolor concolor: Hershkovitz, 1960:545–547; name combination (*flavicans*, *illectus*, and *mincae* allocated as full synonyms, invalid as subspecies).

Oryzomys trinitatis flavicans: Cabrera, 1961:407; name combination.

Oryzomys trinitatis illectus: Cabrera, 1961:408; name combination.

Oecomys flavicans: Musser and Carleton, 1993:716; first use of current name combination (*illectus* and *mincae* allocated as synonyms without indication of rank).

DESCRIPTION: Medium large in body size (e.g., larger than *O. paricola* and *O. specious* but smaller than *O. trinitatis* and *O. superans*). Dorsal pelage soft, dense, and moderately long (8–10 mm over middle rump); head and back bright ochraceous tawny that transforms to rich ochraceous lateral line along sides of body. Ventral pelage typically dull white spattered with buff in some individuals, with encroachment of basally gray hairs over middle abdomen; demarcation between bright lateral line and whitish venter sharp, with ochraceous tones wash along borders of ventral fur in some specimens. Tail longer than combined head and body (tail length ≈ 115% of head and body length), brown on top, sides and all around distal one-fourth but pale brown below; caudal hairs short, revealing scale rows, without formation of tuft at tip.

Skull chunky, with relatively short rostrum and narrow interorbit. Distinct supraorbital ridges outline interorbital and postorbital margins and sweep caudad to join prominent temporal ridges. Zygomatic arches converge little toward rostrum; zygomatic plates broad and dorsal notches moderately incised. Incisive foramina moderately long and impressively broad, their posterior margins situated anterior to molar rows. A pair of large posterolateral palatal pits marks each side of bony palate behind toothrows; mesopterygoid fossa broad, U-shaped at anterior margin, with roof entirely osseous. Postglenoid foramen moderately large but subsquamosal fenestra typically closed (hamular process fused with squamosal). Alisphenoid struts typically present (ca. 70% of sample), and carotid circulatory plan primitive (pattern 1 of Voss 1988). Ectotympanic bullae small, revealing much of medial periotic.

DISTRIBUTION: *Oecomys flavicans* occurs across northeastern Colombia and northern Venezuela; known elevation range from sea level to 2,000 m.

SELECTED LOCALITIES (Map 208): COLOMBIA: Cesar, Colonia Agrícola de Caracolicito (USNM 280595), Pueblo Viejo, Sierra Nevada de Santa Marta (type locality of *Oryzomys flavicans illectus* Bangs); Huila, 5 km N of Villavieja (MVZ 113378); Magdalena, Minca, Santa Marta region (type locality of *Oecomys mincae* J. A. Allen); Norte de Santander, San Calixto, Río Tarrá (USNM 280593, 280594). VENEZUELA: Aragua, Camp Rafael Rangel (USNM 317713–317715); Carabobo, El Trompillo (BM 14.9.1.43–14.9.1.45), San Esteban, near Venezuela Hills (BM 11.5.25.145); Mérida, Mérida (type locality of *Oryzomys flavicans* Thomas); Miranda, Hacienda La Guapa, 6 km S of Río Chico (USNM 387867); Portuguesa, La Hoyada, near Guanarito (AMNH 266915, 266916); Trujillo, Isnotú, 10 km WNW of Valera (USNM 371165); Zulia, Misión Tukuko (USNM 448580–448582).

SUBSPECIES: *Oecomys flavicans* is provisionally regarded as monotypic.

NATURAL HISTORY: Linares (1998) recorded *O. flavicans* as an inhabitant of dense, multistrata forest with abundant bromeliads, lianas, and epiphytes, in both sub-

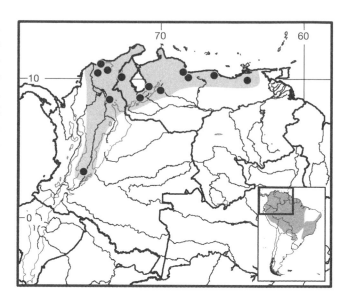

Map 208 Selected localities for *Oecomys flavicans* (●). Contour line = 2,000 m.

montane and montane zones; it is occasionally numerous in agricultural settings such as cafetales.

REMARKS: Cabrera (1961) arranged *flavicans* as a subspecies of *trinitatis*, but the two forms occur in sympatry or altitudinal parapatry in the coastal ranges of northern Venezuela. Morphological differences between the two are slight: specimens of *O. flavicans* typically have a brighter tawny dorsum and flanks (usually darker brown in *O. trinitatis*), with a whitish venter (usually some shade of gray in *O. trinitatis*), and average smaller in external and cranial dimensions (Table 1), in particular length of the molar row (CLM usually < 5 mm in *O. flavicans*; CLM usually ≥ 5 mm in *O. trinitatis*).

Oecomys mamorae (Thomas, 1906)

Mamore Oecomys

SYNONYMS:

Oryzomys (Oecomys) mamorae Thomas, 1906c: 445; type locality "Mosetenes, Upper Mamoré, Yungas, [Cochabamba,] Bolivia"; but see Remarks.

Oecomys mamorae: Osgood, 1916b:206; first use of current name combination.

Oryzomys [(Oecomys)] mamorae: Ellerman, 1941:358; name combination.

Oryzomys mamorae mamorae: Cabrera, 1961:405; name combination.

Oryzomys concolor roberti: Hershkovitz, 1960:559; part (*mamorae* allocated as full synonym, invalid as subspecies).

Oecomys concolor roberti: S. Anderson, 1985:12; part (*mamorae* listed as full synonym, invalid as subspecies).

DESCRIPTION: Combination of medium-large size, absolutely and relatively long tail (Table 1), and moderately long pelage (7–9 mm) characterizes examples of *O. mamorae*. Dorsal pelage generally medium buffy brown, dominated by gray hues in some series that range from gray to grayish buff; ventrum appears dull white from chin to inguen in most specimens, although some individuals exhibit encroachment of gray-based hairs over middle abdomen. Dorsal-ventral pelage contrast tends to be sharply marked, in most specimens accentuated by buff to bright ochraceous lateral strip that demarcates upper and under parts. Long tail (≈ 115–118% of head and body length) brown to dusky brown for most of its length, proximal section slightly paler beneath; short caudal hairs weakly obscure scale rows, without expression of terminal pencil.

Within the genus, cranium exhibits weakest expression of supraorbital ridging, which is confined more to rear of orbit and weakly diverges posteriorly. Interorbital shape thus appears more amphoral and its span narrow compared with constriction observed in other large-bodied species. Zygomatic plates broad for genus, with dorsal notches clearly incised. Incisive foramina relatively long, more nearly parallel sided, and relatively narrow without pronounced outward bowing. Rear termination of hard palate more or less even with caudal margins of M3s; perhaps in correlation with shorter palate, pit construction simpler, usually consisting of single opening. Alisphenoid struts typically absent and carotid circulatory plan derived (pattern 3 of Voss 1988; Carleton et al. 2009). Development of auditory bullae is exceptional for the genus: ectotympanic shell slightly more inflated, and less of posteromedial periotic exposed.

DISTRIBUTION: Subhumid and gallery forests in savanna and Chaco zones of central and eastern Bolivia, contiguous west-central Brazil, and northern and eastern Paraguay. The known elevational range is from sea level to 2,100 m, with most localities between 200–500 m.

SELECTED LOCALITIES (Map 209; from Carleton et al. 2009, except as noted): BOLIVIA: Beni, Baures, Puerto Salinas, Río Beni, Río Mamoré, Río Tijamuchi; Chuquisaca, Ticucha, Río Capirenda, Tola Orko, 40 km from Padilla; Cochabamba, Mosetenes, upper Río Mamoré (type locality of Oryzomys (Oecomys) mamorae Thomas); La Paz, 1 mi W of Puerto Linares; Santa Cruz, Camiri, 5 km S of Choreti, Punta Rieles, Santa Ana. BRAZIL: Mato Grosso, Caiçara; Mato Grosso do Sul, Urucum de Corumbá. PARAGUAY: Amambay, Colonia Sargento Duré, 3 km E of Río Apa; Chaco, Cerro León, 50 km WNW of Fortín Madrejón; Misiones, 40 km S of San Ignacio.

NATURAL HISTORY: A specimen from the Chaco of Paraguay was captured on the ground in thorn forest (Myers and Wetzel 1979, reported as Oryzomys concolor roberti).

SUBSPECIES: Carleton et al. (2009) offered no data that indicate geographic variation, but the possibility of recognizable subspecies should be revisited with the fresh redefinition of O. mamorae; we provisionally regard it as monotypic.

REMARKS: Systematists recognized this distinctive species as valid until Hershkovitz (1960) confused matters by treating it as a full synonym of Oryzomys concolor roberti. The ranges of O. mamorae and O. roberti approach one another in eastern Bolivia, where they are strikingly different in external and cranial morphology, but sympatry has yet to be documented (Carleton et al. 2009). Specimens of Oecomys have been recovered from owl pellets in northeastern Argentina, from the provinces of Chaco (Massoia and Fornes 1965b, as O. concolor) and Formosa (Pardiñas and Ramírez-Llorens 2005, as O. sp.). By geographic proximity, these samples are plausibly referable to O. mamorae, but critical review of variation within nominal O. mamorae and comparisons with the Argentinean populations are required to verify their specific assignment.

Ulyses Pardiñas (pers. comm.) kindly informed me that the type locality "Mosetenes" actually refers to an indigenous people distributed principally within the Yungas of central Bolivia, not a specific placename. The origin of the type specimen from somewhere within Cochabamba department, specifically the Cordillera de Mosetenes (Anderson 1997; Carleton et al. 2009), has been presumed since Hershkovitz's (1960) interpretation of the locality, which regional placement plausibly followed Thomas's (1906c) approximation of Mosetenes as the "Upper Mamoré." However, the collector, Italian naturalist Luigi Balzan, also descended the Rio Beni, La Paz department, and had visited Mosetenes settlements along that river (Balzan 1893,

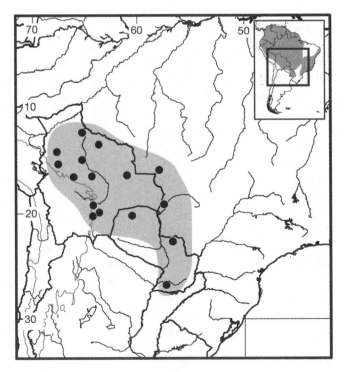

Map 209 Selected localities for *Oecomys mamorae* (●). Contour line = 2,000 m.

2008), an area where the type may conceivably have been collected. Additional archival research is required to justify formal amendment of the type locality.

Oecomys paricola (Thomas, 1904)
Brazilian Oecomys

SYNONYMS:

Rhipidomys paricola Thomas, 1904e:194; type locality "Igarapé-Assu, near Pará, [Para, Brazil]. Alt. 50 m."

[*Oryzomys (Oecomys)*] *paricola*: Thomas, 1906c:445; name combination.

Oecomys paricola: Gyldenstolpe, 1932a:39; first use of current name combination.

Oryzomys [(*Oecomys*)] *paricola*: Ellerman, 1941:358; name combination.

Oryzomys bicolor bicolor: Hershkovitz, 1960:535; part (*paricola* allocated as full synonym).

Oryzomys bicolor paricola: Cabrera, 1961:404; name combination.

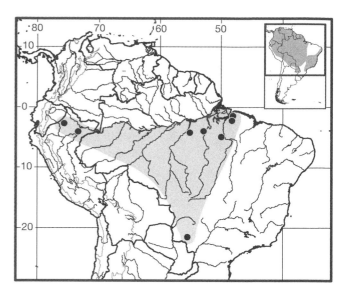

Map 210   Selected localities for *Oecomys paricola* (●). Contour line = 2,000 m.

DESCRIPTION: Like *O. auyantepui* and *O. phaeotis*, dorsal coat of *O. paricola* thick, silky in texture, and long (8–10 mm) for size of body. Dorsal color ranges from ochraceous brown to rich dark brownish red; flanks tend to be paler and faint ochraceous line may occur between upper and under parts. Venter generally dull grayish white, showing more white on throat, neck, and chest in some specimens, or having strong buffy wash across belly and chest in others. Tail dark brown over its circumference; caudal hairs lengthen over distal half of tail, obscuring scale rows and forming modest terminal pencil.

Skull resembles that of *O. auyantepui*—stockily built with blunt rostrum, well-defined supraorbital ridges, and typical cuneate interorbital constriction. Zygomatic arches converge anteriorly from their widest expanse toward braincase, and zygomatic plates relatively narrow with shallow dorsal notches. Moderately long and wide incisive foramina terminate anterior to front rims of M1s in most specimens, or extend even with anterior borders in some. Mesopterygoid fossa short and wide, with U-shaped anterior margin; roof either entirely osseous or perforated by short, narrow sphenopalatine vacuities. In contrast to *O. auyantepui*, cranium of *O. paricola* typically lacks alisphenoid struts and retains short, narrow subsquamosal fenestrae. Carotid circulatory pattern primitive (pattern 1 of Voss 1988), and ectotympanic bullae small, as typical of genus.

DISTRIBUTION: Central Brazil, south of the Amazon River, and northeastern Peru; range limits uncertain.

SELECTED LOCALITIES (Map 210): BRAZIL: Pará, 52 km SSW of Altamira, right bank Rio Xingu (USNM 549531, 549532), Capim (AMNH 203391, 203392, 203394), Igarapé-Assu, near Pará (type locality of *Rhipidomys paricola* Thomas), Km 217, BR 165, Santarem to

Cuiabá (USNM 544601), Porto Jatobal, 73 km N and 45 km W of Marabá (USNM 521453, 521528); Mato Grosso, 264 km N of Xavantina, Serra do Roncador Base Camp (BM 81.380, 81.595, 86.1127); Mato Grosso do Sul, Maracaju (AMNH 134510, 134511). PERU: Loreto, Estación Biológica Allpahuayo, 25 km SW of Iquitos, on Iquitos-Nauta Highway (TTU [CLH 2348, 2846]), San Jacinto (KU 158190).

SUBSPECIES: *Oecomys paricola* is provisionally regarded as monotypic.

NATURAL HISTORY: Carleton et al. (1986) recorded *O. paricola* from primary lowland forest along the Rio Xingu, central Brazil.

REMARKS: The geographic range of *O. paricola* (south of the Amazon River) is allopatric to that of *O. auyantepui* (north of the Amazon River) and to that of *O. phaeotis* (western Amazonia); the status of the three as valid species needs verification.

Chromosomal variation ($2n = 68, 70$; FN = 72, 76) has been reported for geographically close populations of *O. paricola* located at the mouth of the Amazon River (C. C. Rosa et al. 2012); the authors suspected the existence of cryptic species, although accompanying morphological or molecular differentiation was not demonstrable.

Oecomys phaeotis (Thomas, 1901)
Dusky Oecomys

SYNONYMS:

Rhipidomys phaeotis Thomas, 1901b:181; type locality "Segrario, 13°5′S., 70°5′W., Upper Inambari, [Puno,] S.E. Peru. Alt. 1000 m."

[*Oryzomys (Oecomys)*] *phaeotis*: Thomas, 1906c:445; name combination.

Oecomys phaeotis: Gyldenstolpe, 1932a:40; first use of current name combination.

Oryzomys [(Oecomys)] phaeotis: Ellerman, 1941:358; name combination.

Oryzomys bicolor phaeotis: Hershkovitz, 1960:540; name combination.

DESCRIPTION: Medium-sized species—larger than *O. bicolor*, about same as *O. auyantepui* and *O. paricola*, and smaller than all other species (Table 1). Pelage luxuriant: soft, dense, and long (10–12 mm over middle rump). Dorsal fur rich tawny brown in most specimens, darker along top of head and back than on sides. Ventral pelage typically dull white (hairs white from tips to roots); demarcation between dorsal and ventral pelage coloration sharp and usually accentuated by bright ochraceous strip. Tail longer than combined head and body (≈110% of HBL) and dark brown on all surfaces; caudal hairs relatively long, typically obscuring scale rows and forming noticeable tuft at tip (6–8 mm).

Skull compactly built, with relatively short rostrum and narrow interorbit. Conspicuous supraorbital ridges outline interorbital and postorbital regions and sweep caudad to join distinct temporal beading. Zygomatic arches narrower rostrally and wider toward braincase; zygomatic plates relatively narrow and dorsal notches shallow. Incisive foramina moderately long and broad, with posterior margins situated about even with anterior faces of M1s or extend just beyond them. A pair of prominent posterolateral palatal pits and smaller vascular perforations adorn bony palate behind molars; mesopterygoid fossa wide and U-shaped at anterior margin, its roof entirely osseous. Postglenoid foramen and subsquamosal fenestra large. Alisphenoid struts typically absent (present in 1 of 7 specimens), and carotid circulatory plan primitive (pattern 1 of Voss 1988). Ectotympanic bullae small, exposing much of medial periotic.

DISTRIBUTION: Eastern Andean slopes from Peru to northern Bolivia; limits unresolved.

SELECTED LOCALITIES (Map 211): BOLIVIA: Cochabamba, Alto Palmar, Chaparé (AMNH 119406). PERU: Cusco, Hacienda Villa Carmen, Cosñipata (FMNH 84314); Junín, 10 km WSW of San Ramón (MVZ 172649); Loreto, Quebrada Oran, 5 km N of Río Amazonas, 85 km NE of Iquitos (LSUMZ 27975); Puno, Segrario, upper Río Inambari (type locality of *Rhipidomys phaeotis* Thomas).

SUBSPECIES: Too few specimens of *O. phaeotis* exist to allow rigorous analysis of intraspecific variation and delineation of subspecies.

NATURAL HISTORY: There is no published information on the ecology or behavior of this species. Most collecting localities appear to be in lower montane forest.

REMARKS: See comments in accounts of *O. auyantepui* and *O. paricola*.

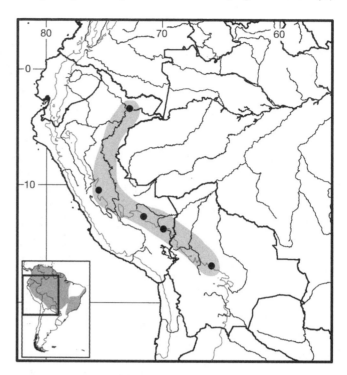

Map 211 Selected localities for *Oecomys phaeotis* (●). Contour line = 2,000 m.

Oecomys rex Thomas, 1910
Regal Oecomys

SYNONYMS:

Oecomys rex Thomas, 1910d:504; type locality "R. Supinaam [= Supenaam], Lower Essequibo, Demerara [= Georgetown]," Demerara-Mahaica, Guyana.

Oryzomys [(Oecomys)] rex: Ellerman, 1941:359; name combination.

Oryzomys subflavus regalis Hershkovitz, 1960:543; unnecessary replacement name for *Oecomys rex* Thomas, 1910, mistakenly viewed as a junior synonym of *Oryzomys subflavus* [= *Cerradomys subflavus* (Wagner)].

Oryzomys catherinae rex: Cabrera, 1961:405; name combination.

DESCRIPTION: Large, morphologically distinctive species that exhibits several features whose expression is unique within genus. In size, *O. rex* is surpassed only by individuals of *O. superans* and approximates those of *O. catherinae* (Table 1). Dorsal pelage thick and long (10–12 mm), its texture very fine and more luxurious than that characteristic of long-furred *O. trinitatis*. Upper parts generally dark in tone, ranging from dull ochraceous brown to deep chestnut, finely intermixed with black-tipped hairs over middle dorsum and grading to brighter on flanks. Ventral pelage uniformly dark grayish white (hairs with gray bases and white tips) in most specimens, although some display overwash of buff to ochraceous on belly and dingy white on gular and inguinal regions. Tail brown all around, caudal hairs short and inconspicuous without formation of terminal pencil or tuft. Tops of forefeet and hindfeet fuscous brown, becoming dingy white over phalanges.

Robust skull bears exaggerated supraorbital shelves and strong temporal ridges that extend across braincase and reach supraoccipital. An angular postorbital process at lateral union of frontal and parietal accentuates orbital-temporal ridging, and flaring supraorbital ridges account for extreme width of interorbit (Table 1). Nasolacrimal opening slit-like, capsules more nearly flat than bulging; rostrum consequently appears narrow relative to cranial size. Unlike other species of *Oecomys*, tegmen tympani well developed, adnate to squamosal and essentially occluding postglenoid foramen; subsquamosal fenestra absent or persists as short, narrow slit. Incisive foramina terminate noticeably anterior to M1s, except subequal with the anterior roots in old individuals. Carotid circulatory plan complete (pattern 1 of Voss 1988).

DISTRIBUTION: Southeastern Venezuela (Bolívar), the Guianas, and northeastern Brazil (Amapá, Amazonas); no elevational records.

SELECTED LOCALITIES (Map 212): BRAZIL: Amapá, Rio Amapari de Macapá (MPEG 6704, 6725), Serra do Navio, Rio Amapari (USNM 394254–394256); Amazonas, Fazenda Dimona, 80 km N of Manaus (INPA [JRM 2107]). FRENCH GUIANA: Riviere Approuague (MNHN 1983.389, 1983.390), Trois Sauts (MNHN 1982.621, 1983.391). GUYANA: Demerara-Mahaica,

Demerara, Supinaam River (type locality of *Oecomys rex* Thomas); Potaro-Siparuni, 40 km NE of Surama (ROM 98033). SURINAM: Marowijne, 3 km SW of Albina (CM 76867); Sipaliwini, Avanavero (CM 68459). VENEZUELA: Bolívar, Serranía Imataca (MHNLS [JO 562]).

SUBSPECIES: *Oecomys rex* is provisionally regarded as monotypic.

NATURAL HISTORY: In southeastern Venezuela, Linares (1998) associated *O. rex* with humid, evergreen forest that harbors a lush undergrowth of bromeliads, epiphytes, and lianas.

REMARKS: Like *O. auyantepui* and *O. rutilus*, the geographic range of *O. rex* adheres to the Guiana subregion of Amazonia.

Oecomys roberti (Thomas, 1904)
Robert's Oecomys
SYNONYMS:

Rhipidomys roberti Thomas, 1904a:237; type locality "Santa Ana de Chapada, a village situated at an altitude of about 800 m, on the Serra do Chapada, some thirty miles N.E. of Cuyabá [Cuiabá]," Mato Grosso, Brazil.

[*Oryzomys (Oecomys)*] *roberti*: Thomas, 1906c:445; name combination.

Oecomys tapajinus Thomas, 1909:378; type locality "Santa Rosa, R. Tamauchim [= Rio Jamanxim], right bank of the Upper Tapajoz [= Tapajós]," Pará, Brazil.

Oecomys guianae Thomas, 1910b:187; type locality "River Supinaam, Demerara," Demerara–Mahaica, Guyana.

Oecomys roberti: Gyldenstolpe, 1932a:40; first use of current name combination.

Oecomys guianae guianae: Tate, 1939:193; name combination.

Oryzomys [(*Oecomys*)] *roberti*: Ellerman, 1941:358; name combination.

Oryzomys [(*Oecomys*)] *guianae*: Ellerman, 1941:358; name combination.

Oryzomys [(*Oecomys*)] *tapajinus*: Ellerman, 1941:359; name combination.

Oryzomys (Oecomys) concolor concolor: Hershkovitz, 1960:546; part (*tapajinus* allocated as full synonym); not *concolor* Wagner.

Oryzomys concolor speciosus: Hershkovitz, 1960:553; part (*guianae* allocated as full synonym); not *speciosus* J. A. Allen and Chapman.

Oryzomys concolor roberti: Hershkovitz, 1960:559; name combination.

Oryzomys marmosurus guianae: Cabrera, 1961:405; name combination.

Oryzomys marmosurus tapajinus: Cabrera, 1961:406; name combination.

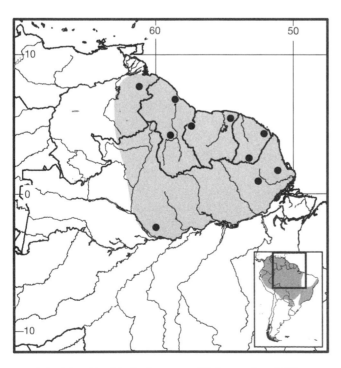

Map 212 Selected localities for *Oecomys rex* (●). Contour line = 2,000 m.

Oecomys roberti: Musser and Carleton, 1993:716; name combination (*guianae* and *tapajinus* allocated as synonyms without indication of rank).

DESCRIPTION: Moderately large species that possesses absolutely and relatively long tail ($\approx$ 123% of HBL) and relatively short fur for its body size (5–7 mm over middle rump). Dorsal pelage texture soft, thought slightly harsh to touch. Upper parts bright in tone, tawny to fulvous brown, grading to ochraceous on flanks, although without obvious lateral line. Appearance of venter ranges from all white (hairs self-colored) to grayish white (hairs with gray bases and white tips), with varying degrees of intermediacy depending upon extent of gray-based hairs, or sometimes tinted with buffy or ochraceous overwash on chest and belly; dorsal-ventral pelage contrast sharp in individuals with mostly white under parts but only moderately defined in examples with grayish white or grayish buff venters. Tail wholly dark brown; caudal hairs so short that tail appears naked macroscopically, revealing fine scale rows; no tuft formed at tip. Pinnae relatively short and brown; tops of hindfeet white or tinged with ochraceous.

Chunky skull characterized by relatively short rostrum and wide interorbit. Supraorbital shelves distinctive but moderately developed compared with *O. trinitatis* and have fairly straight edges that do not diverge as strongly posteriorly; temporal ridges of low relief present, becoming poorly defined over parietals. Nasolacrimal capsules conspicuously bulge laterally and contribute to short-broad appearance of rostrum. Zygomatic arches converge little from braincase toward rostrum; zygomatic plates relatively narrow, and dorsal notches shallow. Incisive foramina relatively short, terminating well anterior to M1 roots and expanding widely over their middle section. Posterolateral palatal pits well expressed; mesopterygoid fossa broad, V- or U-shaped at anterior margin, and roof entirely osseous in some specimens, broken by slits in others. Tegmen tympani may touch rear edge of squamosal, but postglenoid foramen remains open, larger than tiny or absent subsquamosal fenestra (hamular process fused with the squamosal). Occurrence of alisphenoid struts variable, but typically absent; carotid circulatory plan complete (pattern 1 of Voss 1988). Ectotympanic bullae small, revealing much of medial periotic.

DISTRIBUTION: Southern Venezuela (Amazonas), Guyana, Surinam, northern and west-central Brazil, eastern Colombia, eastern Peru, and extreme eastern Bolivia; known elevational range sea level–800 m.

SELECTED LOCALITIES (Map 213): BOLIVIA: Pando, Remanso (Anderson 1997); Santa Cruz, 2.5 km NE of El Refugio, Parque Nacional Noel Kempff Mercado (Carleton et al. 2009). BRAZIL: Amazonas, Colocação Vira-Volta, left bank Rio Juruá (Patton et al. 2000),

Manacapuru, Rio Solimões (BM 1920.7.1.15), Sacado, right bank Rio Juruá (Patton et al. 2000); Pará, Campos do Ariramba, Igarapé Jaramacaru (MZUSP 69145), Capim (MZUSP 9896, 9897), Santa Rosa, Rio Jamanxim, upper Rio Tapajós (type locality of *Oecomys tapajinus* Thomas); Mato Grosso, Santa Ana do Chapada, Serra do Chapada, 30 mi NE of Cuiabá (type locality of *Rhipidomys roberti* Thomas), 264 km N of Xavantina, Serra do Roncador Base Camp (BM 81.412–81.417); Rondônia, Barão de Melgaço, Rio Commemoração, upper Rio Ji-Paraná (AMNH 37115, 37116, 37118). COLOMBIA: Caquetá, Tres Troncas, Río Caquetá (FMNH 72024). GUYANA: Demerara-Mahaica, Demerara, Supinaam River (type locality of *Oecomys guianae* Thomas); Upper Takutu–Upper Essequibo, Quarter Mile Landing, 5 km S of Annai (ROM 98125). PERU: Amazonas, Río Cenepa, vicinity of Huampami (MVZ 153528, 155005); Loreto, Boca Río Peruate (FMNH 88937); Madre de Dios, Itahuania, right bank Río Madre de Dios (FMNH 84303); Pasco, San Pablo (AMNH 231133); Puno, La Pampa, Limbani-Asterillos road (MCZ 39499). SURINAM: Paramaribo, Paramaribo (USNM 319968, 319969); Sipaliwini, Sipaliwini airstrip (CM 64562). VENEZUELA: Amazonas, Cerro Neblina, Base Camp (USNM 560822); Bolívar, Río Yuruani, 12 mi from mouth (AMNH 30726).

SUBSPECIES: Although broad in distribution, no evaluation of variation has been conducted to rigorously delimit infraspecific units; we provisionally treat *O. roberti* as monotypic.

NATURAL HISTORY: Patton et al. (2000) obtained samples of *O. roberti* primarily in várzea forest (88%) along the Rio Juruá, western Brazil, and rarely at the edges of second-growth forest; individuals were captured in equal numbers on the ground and in the canopy. Along the Rio

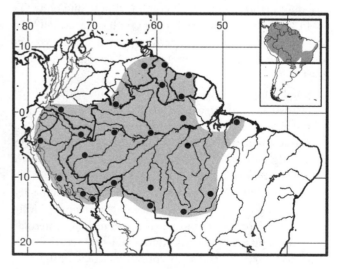

Map 213 Selected localities for *Oecomys roberti* (●). Contour line = 2,000 m.

Xingu, central Brazil, trap success was notable in ecotones of dense growth that proliferated around natural tree-falls within primary forest and in viney forest with low canopy (Carleton et al. 1986). Within the Cerrado biome of eastern Bolivia, examples of *O. roberti* were captured in semideciduous forest (Emmons et al. 2006). Litter size, as based on fetal counts and embryo scars, is only two or three (Patton et al. 2000).

REMARKS: Hershkovitz (1960) retained *roberti* as a subspecies of *O. concolor*, but submerged the otherwise dissimilar *mamorae* within his construct of *O. concolor roberti*.

Patton et al. (2000) reported the karyotype to consist of a high diploid count ($2n = 80$) and intermediate number of biarmed chromosomes (FN = 114).

Oecomys rutilus Anthony, 1921
Reddish Oecomys

SYNONYMS:

Oecomys rutilus Anthony, 1921a:4; type locality "Kartabo, [Cuyuni-Mazaruni,] British Guiana" [Guyana].

Oryzomys [(*Oecomys*)] *rutilus*: Ellerman, 1941:359; name combination.

Oryzomys bicolor bicolor: Hershkovitz, 1960:535; part (*rutilus* allocated as full synonym); not *bicolor* Tomes.

Oryzomys bicolor nitedulus: Cabrera, 1961:404; part (*rutilus* allocated as full synonym); not *nitedulus* Thomas.

DESCRIPTION: Smallest species of *Oecomys* so far recognized, suggesting dainty facsimile of *O. bicolor*. Pelage soft and dense, lax to touch, and moderately long compared with body size (6–8 mm over middle rump). Dorsal coat rich ochraceous tawny (nearly orange brown), and slightly paler along lower sides of head and body, bright buffy gray. Ventral pelage pure white, hairs white from root to tip, and sharply demarcated from dorsum. Tail only slightly longer than combined head and body (TL ≈ 111% of HBL) and dark brown on all surfaces; tail hairs relatively long, typically concealing distal scale rows and forming conspicuous terminal tuft (5–11 mm per Voss et al. 2001).

Small skull delicately but compactly built with relatively short rostrum and narrow interorbit. Conspicuous supraorbital ridges (narrowly ledge-like) outline interorbital and postorbital regions and sweep caudad to join distinct temporal ridges. Zygomatic arches converge slightly toward rostrum; zygomatic plates relatively narrow and dorsal notches shallow. Short and narrow incisive foramina terminate posteriorly about even with anterior faces of M1s. Rear of bony palate contains pair of prominent posterolateral palatal pits; mesopterygoid fossa narrow and U-shaped at anterior margin, and roof entirely osseous. Postglenoid foramen and subsquamosal fenestra spacious. Alisphenoid

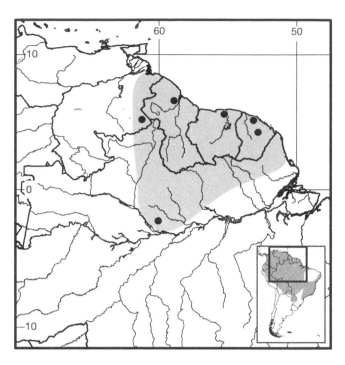

Map 214 Selected localities for *Oecomys rutilus* (●). Contour line = 2,000 m.

struts typically present, and carotid circulatory plan primitive (pattern 1 of Voss 1988). Ectotympanic bullae small, exposing much of medial periotic.

DISTRIBUTION: Extreme eastern Venezuela, Guyana, Surinam, and French Guiana, to Amazonas, Brazil; single elevational record at 60 m.

SELECTED LOCALITIES (Map 214; from Voss et al. 2001, except as noted): BRAZIL: Amazonas, 80 km N of Manaus. FRENCH GUIANA: Nouragues, Paracou. GUYANA: Cuyuni-Mazaruni, Kartabo, lower Essequibo (type locality of *Oecomys rutilus* Anthony). SURINAM: Paramaribo, Carolina Kreek. VENEZUELA: Bolívar, San Ignacio Yuruaní, 5.2 km NE via La Toma de Agua.

SUBSPECIES: *Oecomys rutilus* is provisionally regarded as monotypic.

NATURAL HISTORY: In French Guiana (Voss et al. 2001), nearly all examples of *O. rutilus* were captured in primary forest, one in roadside secondary growth; settings in primary vegetation included well-drained sites as well as swampy forest and along creeks; specimens issued about equally from terrestrial (pitfall lines) and arboreal (lianas, canopy platform) trapping stations. Adler et al. (2012) also captured this species most commonly on lianas, especially at lower heights in the forest at Paracou, French Guiana.

REMARKS: As so far documented, the geographic range of *O. rutilus* adheres to the Guiana subregion of Amazonia, along with those of *O. auyantepui* and *O. rex*.

Oecomys speciosus (J. A. Allen and Chapman, 1893)
Savanna Oecomys

SYNONYMS:

Oryzomys speciosus J. A. Allen and Chapman, 1893:212; type locality "Princestown, Trinidad," Trinidad and Tobago.

Oryzomys trichurus J. A. Allen, 1899b:206; type locality "El Libano plantation, near Bonda (alt. 500 ft.), Santa Marta District," Magdalena, Colombia.

Oecomys caicarae J. A. Allen, 1913c:603; type locality "Caicara, Rio Orinoco, [Bolívar,] Venezuela."

Oecomys guianae caicarae: Tate, 1939:193; name combination.

Oryzomys [(Oecomys)] caicarae: Ellerman, 1941:357; name combination.

Oryzomys (Oecomys) concolor concolor: Hershkovitz, 1960:546–547; part (*caicarae* and *trichurus* allocated as full synonyms); not *concolor* Wagner.

Oryzomys concolor speciosus: Hershkovitz, 1960:553; name combination.

Oryzomys marmosurus marmosurus: Cabrera, 1961:406; part (*caicarae* allocated as full synonym); not *marmosurus* Thomas.

Oryzomys trinitatis illectus: Cabrera, 1961:408; part (*trichurus* allocated as full synonym), not *illectus* Bangs.

Oecomys speciosus: Musser and Carleton, 1993:716; first use of current name combination (*caicarae* and *trichurus* allocated as synonyms without indication of rank).

DESCRIPTION: Medium-sized, orange-brown mouse with bright ochraceous sides and pure white venter. Larger than examples of *O. bicolor* or *O. auyantepui* but appreciably less robust than those of *O. concolor* or *O. trinitatis* (Table 1). Dorsal fur soft, dense, and medium in length (5–8 mm over middle rump). Dorsal coloration usually rich ochraceous tawny, grading to bright ochraceous orange along sides of body from nose to rump. Individual variation occurs, some specimens showing redder dorsal coat and others darker upper parts, approaching brownish tawny; in many examples, bright ochraceous hue confined to cheeks and attenuates as distinctive lateral line bordering ventral pelage. Venter typically pure white and contrasts sharply with dorsum. Feet buffy, ears dark brown. Tail relatively long (≈ 118% of HBL) and brown throughout; caudal hairs short over most of tail, lengthen toward tip to end in barely perceptible tuft.

Although medium in size, skull presents sturdy conformation and moderately long rostrum. Low supraorbital shelves extend posteriorly to merge with temporal ridges. Long and narrow incisive foramina extend between M1s, and bony palate projects beyond third molars to form moderately wide shelf. Large postglenoid fissure and small but obvious subsquamosal foramen penetrate lateral wall of braincase. Alisphenoid struts typically absent; carotid circulatory plan primitive (pattern 1 of Voss 1988).

DISTRIBUTION: Northeastern Colombia, northern Venezuela, and Trinidad; known elevational range from sea level to 1,250 m; most site records below 400 m.

SELECTED LOCALITIES (Map 215): COLOMBIA: Cesar, Valledupar, Río Guaimaral (USNM 280521); Magdalena, El Libano plantation, near Bonda (type locality of *Oryzomys trichurus* J. A. Allen). TRINIDAD AND TOBAGO: Trinidad, Princestown (type locality of *Oryzomys speciosus* J. A. Allen and Chapman). VENEZUELA: Apure, Hato Cariben, 60 km NE of Puerto Paez, Río Cinaruco (USNM 374314–374316), Hato El Frio, 30 km W of El Samán (MHNLS 7923; USNM 448583); Bolívar, Caicara, Río Orinoco (type locality of *Oecomys caicarae* J. A. Allen), Hato San José, 20 km W La Paragua (USNM 406009); Falcón, Capatárida (USNM 442297–442300); Guárico, San José de Tiznados, 52 km NNW of Calabozo (USNM 545116–545119); Monagas, Hato Mato de Bejuco, 54 km SE of Maturín (USNM 442141–4421450); Sucre, near Hacienda Tunantal, 21 km E of Cumaná (USNM 405999), Ensenada Cauranta, 7 km N and 5 km E of Guira (USNM 409856–409860); Trujillo, Agua Santa, 23 km NW of Valera (USNM 371152, 371153); Yaracuy, 19 km NW of Urama, Km 40 (USNM 372514, 372515).

SUBSPECIES: Substantial variation in pelage color is apparent among the samples referred to *O. speciosus*, but whether this variation reflects broad geographic patterns that concord with any of the available junior synonyms as subspecies invites further study. Among Venezuelan populations, Linares (1998) recognized two subspecies, *O. s. speciosus* and *O. s. caicarae*, but we provisionally view *O. speciosus* as monotypic in lieu of studies across the entire distribution.

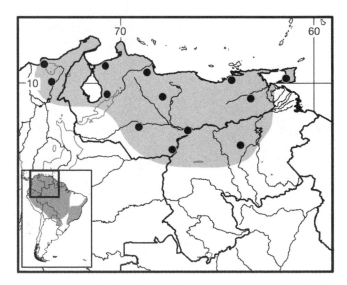

Map 215 Selected localities for *Oecomys speciosus* (●). Contour line = 2,000 m.

NATURAL HISTORY: Linares (1998) associated *O. speciosus* with palm and woodland savannas, semideciduous forest, and gallery forest.

REMARKS: Hershkovitz's (1960) concept of the subspecies *O. concolor speciosus* was an amalgam of four species (*O. speciosus* proper plus *O. auyantepui*, *O. roberti*, and *O. trinitatis*).

Oecomys superans Thomas, 1911

Large Oecomys

SYNONYMS:

Oecomys superans Thomas, 1911c:250; type locality "Canelos, Rio Bobonaza, [Pastaza,] Oriente of Ecuador. Alt. 2100′."

Oecomys palmeri Thomas, 1911c:251; type locality "Canelos, [Pastaza,] Oriente of Ecuador. Alt. 2100′."

Oecomys melleus Anthony, 1924b:4; type locality "Zamora, [Zamora-Chinchipe,] eastern Ecuador, altitude about 3250 feet."

Oecomys superans: Gyldenstolpe, 1932a:42; name combination (*palmeri* allocated as synonym without indication of rank).

Oryzomys [(*Oecomys*)] *melleus*: Ellerman, 1941:358; name combination.

Oryzomys [(*Oecomys*)] *superans*: Ellerman, 1941:359; name combination (*palmeri* allocated as synonym without indication of rank).

Oryzomys (*Oecomys*) *superans*: Sanborn, 1951:21; name combination.

Oryzomys concolor superans: Hershkovitz, 1960:556; name combination (*melleus* and *palmeri* allocated as full synonyms, invalid as subspecies).

Oryzomys mamorae superans: Cabrera, 1961:405; part (*palmeri* allocated as full synonym).

Oecomys superans: Musser and Carleton, 1993:716; name combination (*melleus* and *palmeri* allocated as synonyms without indication of rank).

DESCRIPTION: As attested by the majority of dimensions measured, largest species of *Oecomys* (Table 1). Fur soft and fine in texture and moderately long (8–10 mm over middle rump); dorsal-ventral pelage transition moderately defined, usually without ochraceous lateral line. Overall, upper parts somber in tone, dark ochraceous brown, generally stronger concentration of blackish-brown over middle dorsum and paler toward sides. Under parts dull grayish white to dirty gray, tinged or washed with buff in some specimens whose venters appear dark grayish buff; individual hairs gray basally and tipped with buff. Tail moderately longer than combined head and body (TL ≈ 111% of HBL) and uniformly dark brown above and below; caudal hairs short and expose fine scale rows, without formation of terminal tuft or pencil. Pinnae appear small and closely rounded for size of species.

Skull large and stoutly built, with moderately long and broad rostrum and relatively narrow interorbit. Supraorbital ledges well developed, distinctly diverging to sweep caudad and join prominent temporal ridges. Zygomatic arches noticeably expand toward rear and strongly taper rostrally; zygomatic plates relatively broad and dorsal notches well defined. Incisive foramina relatively short and narrow, slightly oval in shape, their posterior margins situated well anterior to molar rows. Posterolateral-palatal pits are well expressed; the mesopterygoid fossa broad, V- or U-shaped at anterior border, roof entirely osseous in some specimens or broken by narrow slits in others. Postglenoid foramen small; subsquamosal fenestra usually tiny or absent (hamular process fused with squamosal). Alisphenoid struts typically absent, and carotid circulatory plan primitive (pattern 1 of Voss 1988). Tegmen tympani contacts squamosal but does not overlap its posterior margin. Ectotympanic bullae small, revealing much of medial periotic.

DISTRIBUTION: Lower Andean slopes and foothills in south-central Colombia, eastern Ecuador and Peru, and westernmost Brazil (Amazonas); known elevational range from sea level to 1,800 m; most localities below 400 m.

SELECTED LOCALITIES (Map 216): BRAZIL: Amazonas, Altamira, right bank Rio Juruá (Patton et al. 2000), Penedo, right bank Rio Juruá (Patton et al. 2000). COLOMBIA: Caquetá, Florencia, Río Bodoquera (FMNH 72013), Tres Troncas, La Tiqua, Río Caquetá (FMNH 72015, 72025). ECUADOR: Pastaza, Canelos, Río Bobonaza (type locality of *Oecomys superans* Thomas and *Oecomys palmeri* Thomas), Río Tigre (BM 54.473); Sucumbios, Santa Cecilia (KU 135117, 135150), Lagarto Yaco (MCZ 52609); Zamora-Chinchipe, Zamora (type locality of *Oecomys melleus* Anthony). PERU: Amazonas, vicinity of Huampami, Río Cenepa (MVZ 155010–155013); Cusco, Quincemil, Río Marcapata (FMNH 68636; LSUMZ 19276); Huánuco, Tingo Maria (BM 27.11.1.131–27.11.1.137); Junín, Satipo (MCZ 39676); Loreto, Boca del Río Curaray (AMNH 71589–71591), Santa Elena, Río Samiria (FMNH 87195, 87196); Madre de Dios, Reserva Cuzco Amazonico, 14 km NE of Puerto Maldonado (Woodman et al. 1991); San Martín, Yurac Yacu (BM 27.1.1.117–27.1.1.121); Ucayali, Balta, Río Curanja (LSUMZ 12330–12332), Sarayacu, Río Ucayali (AMNH 76175, 76298, 76456).

SUBSPECIES: *Oecomys superans* is provisionally regarded as monotypic.

NATURAL HISTORY: Although generally associated with lowland tropical forest, little detailed habitat information is available for *O. superans*. Along the Rio Juruá, western Brazil, the few specimens were collected on the ground in highly disturbed secondary growth (Patton et al. 2000).

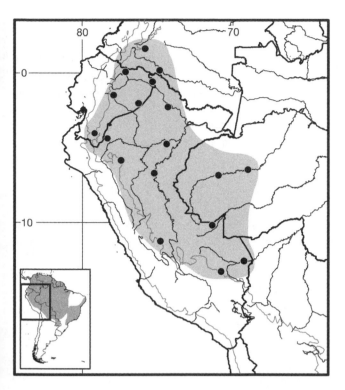

Map 216 Selected localities for *Oecomys superans* (●). Contour line = 2,000 m.

REMARKS: Hershkovitz (1960) viewed *superans* as the largest and darkest of the five subspecies he identified within *O. concolor*. His morphological definition and distribution of the subspecies largely conform to our understanding of the species *O. superans*, including the recent documentation of the species in westernmost Brazil (Patton et al. 2000). The karyotype has been uniformly recorded as $2n = 80$, FN = 108 (A. L. Gardner and Patton 1976; Patton et al. 2000; A. F. B. Andrade and Bonvicino 2003).

The geographic ranges of *O. superans* and *O. roberti* overlap along the Andean foothills, where both have been trapped at the same localities, and the two are superficially similar and often misidentified in collections. *Oecomys roberti* is smaller in external and craniodental dimensions (Table 1), its ventral pelage is grayish white without any buffy overlay (although a few specimens from elsewhere in the range of *O. roberti* show a buffy tinge), and the incisive foramina are relatively shorter and broader, more tear-dropped shaped.

Oecomys sydandersoni Carleton, Emmons, and Musser, 2009
Anderson's Oecomys
SYNONYMS:

Oecomys concolor: Musser and Carleton, 1993:716; part (provisional identification and range extension).

Oecomys sydandersoni Carleton, Emmons, and Musser, 2009:8; type locality "Bolivia, Departamento de Santa

Cruz, Provincia Velasco, El Refugio Huanchaca, 210 m; 14°46′01″S/61°02′02″W."

DESCRIPTION: Medium-size for genus (e.g., larger than *O. bicolor* or *O. paricola*, smaller than *O. roberti* or *O. mamorae*). Pelage soft and fine in texture and moderately short (about 5–7 mm over middle rump). Dorsal color ochraceous brown to pale tawny, generally bright in tone with more grayish showing on head and flanks; ventral pelage pale to medium gray, with hairs of chin, throat, and inguinum entirely white to base in most specimens; dorsal-ventral pelage transition moderately defined, without an ochraceous lateral line. Tail only slightly longer than head and body (TL ≈ 106% of HBL), brown to dark brown all around for most of length; short caudal hairs scarcely visible to unaided eye, revealing fine scale pattern, and rudimentary pencil expressed at tip.

Skull ruggedly built for its size, with short rostrum and relatively broad interorbit. Supraorbital shelves strongly convergent anteriad, with free ledge approaching incipient bead. Zygomatic arches noticeably expand toward rear and strongly taper rostrally; zygomatic plates relatively broad and dorsal notches well defined. Medium-long incisive foramina very broad toward rear, posterior ends obtuse (blunt) and anterior ends acute. Posterolateral-palatal pits well expressed; mesopterygoid fossa broad, bluntly U-shaped at anterior margin, and roof entirely osseous. Alisphenoid struts typically present, and carotid circulatory plan derived (pattern 3 of Voss 1988). Ectotympanic bullae small, revealing much of medial periotic.

DISTRIBUTION: Extreme eastern Bolivia, departments of Beni and Santa Cruz; single elevational record at 210 m.

SELECTED LOCALITIES (Map 217): BOLIVIA: Beni, 4 km above Costa Marques, Río Iténez (Carleton et al. 2009), Bahía de los Casara, 20 km W of Larangiera, Río Iténez (Carleton et al. 2009); Santa Cruz, El Refugio Huanchaca (type locality of *Oecomys sydandersoni* Carleton, Emmons, and Musser).

SUBSPECIES: *Oecomys sydandersoni* is monotypic.

NATURAL HISTORY: Most specimens have been captured in small clumps of woody vegetation surrounded by seasonally flooded grasslands, typically in traps placed above ground on vines, trunks, or branches within arm's reach (Carleton et al. 2009). One individual was captured in closed riverine forest, taken on a vine snaking through a shrub along the river's edge.

REMARKS: Specimens of *O. sydandersoni* were tentatively reported as *O. concolor* based on their superficial similarity (Musser and Carleton 1993). As in *O. concolor* and *O. mamorae*, *O. sydandersoni* possesses a derived carotid circulatory pattern (Carleton et al. 2009), a condition that may indicate closer relationship among the three relative to other species of *Oecomys*. The species has been collected

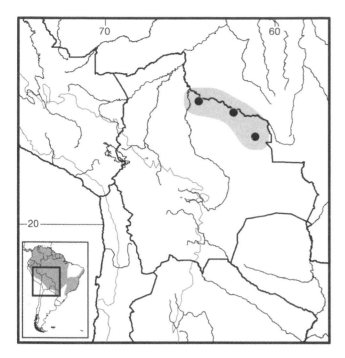

Map 217 Selected localities for *Oecomys sydandersoni* (●). Contour line = 2,000 m.

near to but not in syntopy with *O. bicolor*, *O. roberti*, and *O. trinitatis* in the Parque Nacional Noel Kempff Mercado, eastern Bolivia (Emmons et al. 2006; Carleton et al. 2009).

Oecomys trinitatis (J. A. Allen and Chapman, 1893)
Long-furred Oecomys

SYNONYMS:

Oryzomys trinitatis J. A. Allen and Chapman, 1893:213; type locality "Princestown, Trinidad," Trinidad and Tobago.

Oryzomys flavicans subluteus Thomas, 1898f:268; type locality "W. Cundinamarca," western slope of Cordillera Oriental, Cundinamarca, Colombia.

Oryzomys palmarius J. A. Allen, 1899b:210; type locality "Quebrada Secca, [Sucre,] Venezuela."

Oryzomys fulviventer J. A. Allen, 1899b:212; type locality "Quebrada Secca, [Sucre,] Venezuela (altitude about 3000 feet)."

Oryzomys tectus Thomas, 1901f:251; type locality "Bogava [Bugaba], Chiriqui, Panama. Altitude 250 m."

Oryzomys klagesi J. A. Allen, 1904b:327; type locality "El Llagual, [Bolívar,] Venezuela."

Oryzomys frontalis Goldman, 1912a:6; type locality "Corozal, Canal Zone, Panama."

Oryzomys helvolus J. A. Allen, 1913c:597; type locality "Villa Vicencio [Villavicencio] (altitude 1600 feet), about 50 miles southeast of Bogota, [Meta,] Colombia."

Oryzomys vicencianus J. A. Allen, 1913c:598; type locality "Villa Vicencio [Villavicencio] (alt. 1600 ft, about 50 miles southeast of Bogota), at base of eastern Andes, [Meta,] Colombia."

Oryzomys tectus tectus: Goldman, 1918:84; name combination.

Oryzomys tectus frontalis: Goldman, 1918:85; name combination.

Oecomys osgoodi Thomas, 1924f:287; type locality "Moyobamba, [Amazonas,] N. Peru. Alt. 2700'."

Rhipidomys(?) *klagesi*: Gyldenstolpe, 1932a:50; name combination (provisional generic allocation).

Oecomys splendens Hayman, 1938:381; type locality "Mayaro, S.E. Trinidad," Trinidad and Tobago.

Oryzomys trinitatis: Tate, 1939:190; name combination (*palmarius* and *fulviventer* listed as synonyms without indication of rank).

Oryzomys [(*Oecomys*)] *osgoodi*: Ellerman, 1941:358; name combination.

Oryzomys (*Oecomys*) *concolor concolor*: Hershkovitz, 1960: 546; part (*frontalis*, *helvolus*, *klagesi*, *subluteus*, *tectus*, and *vicencianus* allocated as full synonyms, invalid as subspecies); not *concolor* Wagner.

Oryzomys concolor speciosus: Hershkovitz, 1960:553; part (*fulviventer*, *palmarius*, *splendens*, and *trinitatis* allocated as full synonyms, invalid as subspecies); not *speciosus* J. A. Allen and Chapman.

Oryzomys concolor superans: Hershkovitz, 1961:556; part (*osgoodi* allocated as full synonym, invalid as subspecies); not *superans* Thomas.

Oryzomys trinitatis frontalis: Cabrera, 1961:408; name combination.

Oryzomys trinitatis helvolus: Cabrera, 1961:408; name combination (*vicencianus* allocated as full synonym, invalid as subspecies).

Oryzomys trinitatis klagesi: Cabrera, 1961:408; name combination.

Oryzomys trinitatis trinitatis: Cabrera, 1961:409; name combination (*fulviventer* and *palmarius* allocated as full synonyms, invalid as subspecies).

O[*ryzomys*]. *c*[*oncolor*]. *tectus*: Handley, 1966:781; name combination (*frontalis* indicated as full synonym, invalid as subspecies).

Oecomys trinitatis: Musser and Carleton, 1993:716–717; first use of current name combination (*frontalis*, *fulviventer*, *helvolus*, *klagesi*, *osgoodi*, *palmarius*, *splendens*, *subluteus*, *tectus*, and *vicencianus* allocated as synonyms without indication of rank).

DESCRIPTION: Medium-large species, similar in size to *O. roberti* and *O. concolor*. Dorsal fur soft and dense, palpably luxuriant and lustrous, and deep (10–13 mm over middle rump). Head and back dark tawny brown, commonly grading to brighter ochraceous-brown over cheeks and shoulders and along flanks; some individuals exhibit

Table 1. Size variation among *Oecomys* species, as indexed by select external and craniodental variables and roughly graded from smaller to larger.

Species	HBL	TL	HFL	MASS	ONL	ZB	IOB	CLM
O. rutilus	84	93	20	17	24.2	12.8	4.4	3.3
	82–86	88–98	18–21	16, 18	23.4–25.0	12.2–13.4	4.2–4.6	3.2–3.4
O. bicolor	100	109	22	31	27.0	14.3	4.7	3.9
	94–107	99–119	21–23	25–37	25.9–28.1	13.6–15.0	4.5–4.7	3.8–4.0
O. cleberi	105	111	23	—	28.2	14.6	4.9	4.1
	92–117	99–123	21–29	—	26.7–30.1	14.3–16.0	4.6–5.3	3.9–4.3
O. auyantepui	111	118	24	39	27.9	14.9	5.3	4.1
	101–121	108–128	23–25	30–48	26.8–29.0	14.2–15.6	5.1–5.5	4.0–4.2
O. paricola	109	128	24	41	28.5	15.0	5.2	4.2
	101–117	117–139	22–25	32–50	27.2–29.7	14.3–15.7	4.9–5.5	4.1–4.3
O. phaeotis	110	115	23	48	29.2	15.5	4.9	4.3
	102–118	107–123	22–24	43, 54	28.3–30.1	15.0–16.0	4.8–5.1	4.2–4.4
O. speciosus	117	138	25	40	29.3	15.5	5.0	4.4
	107–127	126–150	24–27	30–50	27.6–30.8	14.5–15.5	4.7–5.3	4.2–4.6
O. sydandersoni	125	132	24	44	29.8	16.5	5.3	4.6
	113–137	124–140	23–26	37–52	28.6–31.0	15.9–17.1	5.1–5.5	4.4–4.7
O. flavicans	123	141	26	60	31.1	16.1	5.3	4.7
	106–140	132–150	25–27	50–70	29.8–32.4	15.6–16.8	5.0–5.6	4.6–4.8
O. roberti	123	151	27	56	32.2	17.1	5.6	4.9
	111–135	138–164	26–28	46–66	30.8–33.6	16.2–18.0	5.3–5.9	4.7–5.1
O. trinitatis	127	150	27	61	32.3	16.6	5.3	5.0
	117–137	135–165	25–29	48–74	30.9–33.7	15.7–17.5	5.1–5.5	4.8–5.2
O. concolor	125	148	27	58	32.2	17.7	5.5	4.9
	119–131	141–155	26–28	49–67	31.3–33.1	17.1–18.3	5.3–5.7	4.8–5.0
O. mamorae	136	159	27	61	32.4	17.0	4.9	5.0
	127–145	146–172	26–28	49–73	31.0–33.8	16.2–17.8	4.7–5.1	4.8–5.2
O. catherinae	130	154	29	59	33.5	17.4	5.8	5.1
	121–139	143–165	28–30	47–71	32.5–34.5	16.8–18.0	5.6–6.0	5.0–5.2
O. rex	128	155	28	59	34.0	17.6	6.5	5.3
	121–135	142–168	27–29	47–71	32.8–35.2	16.8–18.4	6.0–7.0	5.1–5.5
O. superans	150	167	30	85	35.8	18.3	5.9	5.4
	135–165	150–184	28–31	66–104	34.3–37.2	17.4–19.2	5.5–6.3	5.2–5.6

Abbreviations: CLM, coronal length of maxillary toothrow (M1–M3); HBL, length of head and body; HFL, length of hindfoot (including claw); IOB, least interorbital breadth; ONL, occipitonasal length; TL, length of tail; WT, weight; ZB, zygomatic breadth.
Notes: Sample statistics include the mean and range based on ±1 standard deviation, which encompasses about 70% of observed cases; units of external and craniodental variables are in millimeters and mass in grams. Species samples were coarsely defined to convey broad limits of measurement variation as follows: *O. auyantepui*—Amapá, Brazil, French Guiana, Guyana, and Surinam (N=26–48); *O. bicolor*—Amazonas and Madre de Dios, Peru (N=36–46) ; *O. cleberi*—south-central Brazil (N=14; R. G. Rocha et al. 2012); *O. catherinae*—Federal District and Minas Gerais, Brazil (N=23–27); *O. concolor*—Amazonas, Brazil (N=23–24); *O. flavicans*—Venezuela (N=7–55); *O. mamorae*—Beni and Santa Cruz, Bolivia (N=13–49); *O. paricola*—Para, Brazil (N=36–61); *O. phaeotis*—Peru (N=2–12); *O. rex*—Amapá, Brazil, French Guiana, Guyana, and Surinam (N=8–21); *O. roberti*—Amazonas, Mata Grosso, and Para, Brazil (N=87–104); *O. rutilus*—French Guiana and Guyana (N=2–16); *O. speciosus*—Venezuela (N=26–51); *O. superans*—Colombia, Ecuador, and Peru (N=20–86); *O. sydandersoni*—Beni and Santa Cruz, Bolivia (N=21–24); *O. trinitatis*—Venezuela (N=33–76). In general, sample sizes are larger for the craniodental variables, smaller for external measurements, and smallest for weight.

bright buffy orange dorsum; dorsal surfaces of forefeet and hindfeet dark brownish. Ventral pelage chromatically ranges from dark grayish white, through grayish white tinged with buff, to rich ochraceous gray. Demarcation between dorsal and ventral fur not well defined, occasionally marked by ochraceous lateral line in some specimens, whereas in others ochraceous tones wash along borders with grayish ventral fur. Tail longer than combined head and body (TL ≈ 118% of HBL), dark brown and unicolor or slightly paler below; caudal hairs short, scalation thus visible to eye, and form slight pencil at tip.

Robust skull with relatively long rostrum, narrow interorbit, and strongly opisthodont upper incisors. Distinct supraorbital ridges or shelves delineate interorbital edges and strongly diverge caudad to join prominent temporal ridges. Zygomatic arches narrow toward rostrum but widen toward braincase; zygomatic plates broad and dorsal notches relatively deep. Incisive foramina moderately long and narrow to moderately broad, posterior margins situated anterior to molar rows. Large posterolateral palatal pits mark each side of bony palate beyond toothrows; mesopterygoid fossa broad, U-shaped at anterior margin, and its roof entirely osseous. Postglenoid foramen medium sized, but subsquamosal fenestra either closed (hamular process fused with squamosal) or very small in most specimens. Alisphenoid struts present in about 40% of sample, and carotid circulatory plan primitive (pattern 1 of Voss 1988). Tegmen tympani usually contacts rear edge of squamosal. Ectotympanic bullae small, revealing much of medial periotic.

DISTRIBUTION: Trinidad and Tobago, eastern, central and southern Venezuela, Guyana and Surinam, eastern Ecuador and Peru, northern and central Brazil; known elevational range from sea level to 2,150 m.

SELECTED LOCALITIES (Map 218): BOLIVIA: Santa Cruz, El Refugio, Parque Nacional Noel Kempff Mercado (Carleton et al. 2009). BRAZIL: Acre, opposite Igarapé Porongaba, left bank Rio Juruá (Patton et al. 2000); Amazonas, Villa Bella da Imperatriz, Boca Rio Andirá (AMNH 93532, 93533), Colocação Vira Volta, left bank Rio Juruá (Patton et al. 2000); Goiás, Annapolis (AMNH 134571, 134659, 134676); Mato Grosso, Humboldt Laboratory, Rio Roosevelt, Aripuanã (USNM 545275–545277), 264 km N of Xavantina, Serra do Roncador Base Camp (BM 81.418–81.420); Mato Grosso do Sul, Maracaju (AMNH 134904); Pará, Porto Jatobal, 73 km N and 45 km W of Marabá (MPEG 15360, 15361; USNM 521532). COLOMBIA: Cauca, Sabanetas (FMNH 90281); Córdoba, Socorre, upper Río Sinu (FMNH 69197–69199); Cundinamarca, western slope of Cordillera Oriental (type locality of *Oryzomys flavicans subluteus* Thomas); Meta, Villa Vicencio, about 50 mi SE of Bogotá (type locality of *Oryzomys helvolus* J. A. Allen and *Oryzomys vicencianus*

J. A. Allen). GUYANA: Upper Demerara-Berbice, Comackpea, Rio Cunerara (BM 6.1.1.14). PERU: Amazonas, Moyobamba (type locality of *Oecomys osgoodi* Thomas); Huánuco, Río Cayamba, Hacienda Exito (FMNH 24574, 24576); Loreto, San Jacinto (KU 158192). SURINAM: Para, Carolina Kreek (FMNH 95577–95579). TRINIDAD AND TOBAGO: Tobago, Pigeon Peak, Tobago Forest Preserve (USNM 538101); Trinidad, Mayaro (type locality of *Oecomys splendens* Hayman), Princestown (type locality of *Oryzomys trinitatis* J. A. Allen and Chapman). VENEZUELA: Amazonas, Cerro Neblina, Camp V (USNM 560648–560650); Aragua, Rancho Grande Biological Station, 13 km NW of Maracay (USNM 517572–517576); Bolívar, El Llagual, near Maripa, lower Río Caura (type locality of *Oryzomys klagesi* J. A. Allen), San Ignacio de Yuruaní (USNM 448577–448579); Sucre, Quebrada Secca (type locality of *Oryzomys palmarius* J. A. Allen and *Oryzomys fulviventer* J. A. Allen).

SUBSPECIES: *Oecomys trinitatis* is provisionally regarded as monotypic. In Venezuela, however, Linares (1998) retained *O. t. klagesi* along with *O. t. trinitatis* as recognizable subspecies.

NATURAL HISTORY: In Panama, Handley (1966) considered the species (reported as *Oryzomys concolor*) as semiarboreal and associated with tall grass and herbaceous growth in savannas and forest openings; another Panamanian example was obtained on a log in lowland forest characterized as Dry Tropical Forest (Fleming 1970). In Venezuela, Linares (1998) considered *O. trinitatis* an inhabitant of dense, evergreen forests with lush undergrowth. In western Brazil, specimens were taken in *terre firme* and at the edge of flooded forest (Patton et al. 2000).

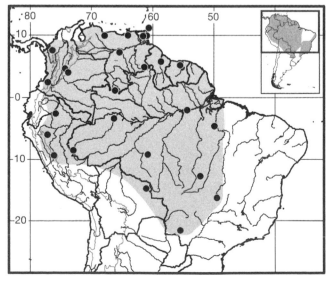

Map 218 Selected localities for *Oecomys trinitatis* (●). Contour line = 2,000 m.

REMARKS: This nominal species forms the core of Hershkovitz's (1960) *O. concolor* in terms of distributional extent and the large number of synonyms included. Even after peeling away clearly diagnosable species (see accounts of *O. auyantepui*, *O. catherinae*, *O. concolor*, *O. flavicans*, *O. mamorae*, *O. roberti*, *O. speciosus*, and *O. superans*), based on morphological differences and/or evidence of sympatry, the residuum now comprising *O. trinitatis* remains a composite. For example, the relationship of the senior taxon *trinitatis*, described from Trinidad, to mainland forms, the status of trans-Andean populations (*tectus* and *frontalis*), and level of differentiation of populations to the north and south of the Amazon River will all require attention in unraveling the heterogeneity evident within *O. "trinitatis."*

Patton et al. (2000) reported a karyotype of $2n = 58$, $FN = 96$ for populations of *O. trinitatis* in western Amazonia.

Genus *Oligoryzomys* Bangs, 1900

Marcelo Weksler and Cibele R. Bonvicino

Members of the genus *Oligoryzomys*, commonly known as Colilargos or Pygmy Rice Rats, are found from northeastern Mexico to extreme southern Chile and Argentina, and across an elevation from sea level to around 3,500 m. Species of the genus are present in lowland and montane forests (humid and dry), such as the Atlantic Forest, lowland Amazonian rainforest, and Andean cloud forest, open-vegetation habitats (grasslands and wetlands) such as the Cerrados, Pampas, Llanos, Pantanal, Chaco, and in drier environments such as Patagonia, Caatinga, and the Peruvian Pacific coast. We recognize 19 species of Colilargos for South America: *O. andinus*, *O. arenalis*, *O. brendae*, *O. chacoensis*, *O. delicatus*, *O. destructor*, *O. flavescens*, *O. fornesi*, *O. griseolus*, *O. longicaudatus*, *O. magellanicus*, *O. mattogrossae*, *O. messorius*, *O. microtis*, *O. moojeni*, *O. nigripes*, *O. rupestris*, *O. stramineus*, and *O. utiaritensis*. The genus contains four additional species extralimital to the continent, *O. costaricensis*, *O. vegetus*, and *O. fulvescens* in Central America, and *O. victus* from Saint Vincent Island and Martinique in the Lesser Antilles, now probably extinct (Musser and Carleton 2005; Hanson et al. 2011). An undescribed species, also extinct, is reported from Guadeloupe (Lorvelec et al. 2001). The earliest known fossils are from the Ensenadense (middle Pleistocene) in Argentina (Pardiñas et al. 2002).

Species of *Oligoryzomys* are very small (head and body ranges between 70 and 110 mm) cursorial mice. The dorsal pelage is coarsely or homogeneous yellowish to orangish-brown; the ventral pelage is paler (superficially whitish or pale yellow). The fur is soft, without subauricular patches (as in *Microryzomys*, with which *Oligoryzomys* species have been historically confused; see Carleton and Musser 1989).

The hindfeet lack well-developed natatory fringes and interdigital webs, and have short tufts of ungual hairs at bases of claws on digits II through V. The plantar surface is densely covered with distinct squamae distal to thenar pad (the heel is smooth); the hypothenar pad is large, and interdigital pads are small and fleshy (pads 1 and 4 displaced proximally relative to 2 and 3). The claw of digit I extends to the middle of the first phalanx of digit II, and the claw of digit V extends just beyond first interphalangeal joint of digit IV. The tail is usually longer than the combined head and body length, sparsely haired, and covered with more or less conspicuous epidermal scales; it lacks a long tuft of terminal hairs, and is weakly to distinctly bicolored (dark above, pale below). Four pairs of mammae are present in most species, one each in inguinal, abdominal, postaxial, and pectoral positions, except in *O. griseolus* with three pairs.

The skull has a delicate rostrum flanked by moderate zygomatic notches; the interorbital region is symmetrically constricted (hourglass or amphoral in shape), with rounded supraorbital margins; the braincase is squared or rounded, without well-developed temporal, lambdoidal, or nuchal crests. The zygomatic plates lack an anterodorsal spinous processes, and their posterior margins lie anterior to the M1 alveolus. Jugals are absent or reduced to slivers of bone (maxillary and squamosal processes are in contact). The nasal bones have blunt posterior margins and do not extend posteriorly beyond the lacrimals, which are equally sutured to the maxillary and frontal bones. The posterior wall of the orbit is smooth. The frontosquamosal suture is colinear with the frontoparietal suture, and the parietals lack or have small lateral expansions. The bony palate between the molar rows is smooth and long, and the mesopterygoid fossa does not extend anteriorly between the third molars. The posterolateral palatal pits are variable in size, number, and morphology but are usually not recessed in fossae. The bony roof of the mesopterygoid fossa is perforated by very large sphenopalatine vacuities. The alisphenoid strut is absent (buccinator-masticatory foramen and accessory foramen ovale are confluent), and the alisphenoid canal has a large anterior opening. The stapedial foramen and the posterior opening of the alisphenoid canal are large, but the squamosoalisphenoid groove and the sphenofrontal foramen are absent (carotid circulatory pattern 2 of Voss 1988) in all species except *O. rupestris*, which either lack or have only a vestigial stapedial foramen (= pattern 3). The posterior suspensory process of the squamosal is absent. The postglenoid foramen is large and rounded, and the subsquamosal fenestra is always patent. The periotic is exposed posteromedially between the ectotympanic and the basioccipital, and usually extends anteriorly to the carotid canal. The mastoid is perforated by conspicuous posterodorsal fenestra. In the mandible, the capsular process is well developed in most adults, and the superior and inferior masseteric ridges converge anteriorly as an open chevron below m1. The basihyal lacks the entoglossal process.

The upper incisors have smoothly rounded enamel bands, and the maxillary toothrows are parallel. The molars are bunodont, with labial flexi enclosed by a cingulum. The labial and lingual flexi of M1 and M2 do not (or shallowly) interpenetrate. The anterocone of M1 is divided into anterolabial and anterolingual conules by a distinct anteromedian flexus; the anteroloph is well developed and separated from the anterocone by a persistent anteroflexus; a protostyle is absent; mesolophs are present on all upper molars in most species, but absent in some specimens of *O. rupestris*; the paracone is connected by an enamel bridge to the anterior moiety of protocone; a median mure is connected to the protocone. The protoflexus of M2 is present in most species; a mesoflexus is present as a single internal fossette; the paracone is without an accessory loph. M3 has a posteroloph; hypoflexus is diminutive (tending to disappear with moderate to heavy wear) in some species, but persistent in others. An accessory labial root of M1 is absent. The anteroconid of m1 is usually without an anteromedian flexid; an anterolabial cingulum is present on all lower molars; the anterolophid is present on m1 but absent on m2 and m3; an ectolophid is absent on m1 and m2 in most species; the mesolophid is distinct on unworn m1 and m2 in most species, but reduced or absent in specimens of *O. mattogrossae*; the posteroflexid is present on m3. Accessory roots are absent for m1, but both m2 and m3 have one large anterior root and one large posterior root.

The stomach is unilocular-hemiglandular (Carleton 1973), and the glandular epithelium does not reach the corpus; the gall bladder is absent. The distal bacular cartilage of the glans penis is large and trifid, with a robust central digit; the nonspinous tissue on rim of the terminal crater does not conceal the bacular mounds.

SYNONYMS:

Mus: Olfers, 1818: 209; part (description of *nigripes*); not *Mus* Linnaeus.

Hesperomys: Waterhouse, 1839:74; part (description of *flavescens*).

Hesperomys (*Calomys*): Wagner, 1843a:528; part (listing of *flavescens* and *longicaudatus*); not *Hesperomys* Waterhouse.

Hesperomys (*Oryzomys*): Milne-Edwards, 1890:27; part (listing of *longicaudatus*); not *Hesperomys* Waterhouse.

Oryzomys: Thomas: 1894b:357; part (description of *stolzmanni*); not *Oryzomys* Baird.

Oligoryzomys Bangs, 1900a:94; type species *Oryzomys navus* Bangs (= *Oligoryzomys delicatus*), by original designation.

Oryzomys (*Oligoryzomys*): J. A. Allen, 1910b:100; name combination.

REMARKS: *Oligoryzomys* was variously considered as a full genus, subgenus, or synonym of *Oryzomys*, and its status as full genus was established only recently (Carleton and Musser 1989). Monophyly of *Oligoryzomys* is supported by phylogenetic analyses of allozymes (Dickerman and Yates 1995), mitochondrial and nuclear gene sequences (Myers et al. 1995; Weksler 2003; Agrellos et al. 2012), and morphological characters (Weksler 2006). The taxonomy of the genus, however, is in much disarray, and comprehensive studies on this widely distributed and specious genus are lacking. The taxonomic arrangement employed here is based on Carleton and Musser (1989), Musser and Carleton (1993, 2005), Bonvicino and Weksler (1998), Weksler and Bonvicino (2005), and Agrellos et al. (2012); it should be viewed as provisional. *Microakodontomys* was included as a junior synonym of *Oligoryzomys* by Weksler et al. (2006), but molecular and genetic data have indicated that this taxon might merit generic recognition (see account of *Microakodontomys transitorius* for additional information).

Carleton and Musser (1989) identified three undescribed species ("sp. A" from Caparaó, "sp. B" from Bolivian and Peruvian Andes, and "sp. C" from the Altiplano of southern Peru). The "sp. A" from Caparaó is identified here as *O. flavescens*, the other two cannot be allocated to any species, despite the karyological data from "sp. B" ($2n = 68$, $FN = 74$, A. L. Gardner and Patton 1976). Another undescribed species reported by Agrellos et al. (2012) from the Brazilian state of Amapá and French Guiana, is here considered as *O. messorius* (see that account).

KEY TO THE SOUTH AMERICAN SPECIES OF *OLIGORYZOMYS*:

1. Large opening of stapedial foramen (carotid circulation pattern 2) 2
1'. Small or vestigial opening of stapedial foramen (carotid circulation pattern 3); undivided anterocone; endemic to high elevations in central Brazil; $2n = 44–46$, $FN = 52–53$ . . . . *Oligoryzomys rupestris*
2. With eight mammae 3
2'. With six mammae; fur thick and long (length almost 1 cm); found only in the Táchira Andes of Venezuela, and Cordillera Oriental of easternmost Colombia . . . . .
. . . . *Oligoryzomys griseolus*
3. Tail shorter, equal or slightly longer than head and body length 4
3'. Tail considerably longer than head and body length. . 5
4. Ventral color whitish; tail length approximately equal to head and body length; found only south of 45°S in Tierra del Fuego; $2n = 54$, $FN = 56$ . . . .
. . . . *Oligoryzomys magellanicus*

4'. Ventral color whitish or buffy with perigenital and gular patches of pure white hairs; found in forested habitats in Amazonian basin; $2n = 64$, $FN = 66$ . . . .
. . . . *Oligoryzomys microtis*

5. Limit between ventral and dorsal coloration well defined 6

5'. Limit between ventral and dorsal coloration scarcely defined or absent 13

6. Ventral coloration creamy or yellowish white 7

6'. Ventral coloration whitish 9

7. Tail clearly bicolored on proximal half 8

7'. Tail unicolored, slightly bicolored near the base; dorsal color reddish brown; found in the Andes at elevations between 600–3,350 m; $2n = 60$, $FN = 76$ . . . .
. . . .*Oligoryzomys destructor*

8. Dorsal color pale ochraceous darkened by numerous fine dusky lines, head color grayish; found in Andes; $2n = 60$, $FN = 70$ . . . . *Oligoryzomys andinus*

8'. Dorsal color fawn-colored or pale rufous, head color similar to body color; distributed along eastern and western flanks of Andes in Chile and Argentina; $2n = 56$, $FN = 66$ . . . .*Oligoryzomys longicaudatus*

9. Orange or yellowish pectoral band present 10

9'. Pectoral band absent 11

10. Smaller zygomatic plate, darker overall dorsal coloration, head color similar to body color; found in eastern Brazil, Paraguay, Uruguay, and Argentina; $2n = 62$, $FN = 82$ . . . . *Oligoryzomys nigripes*

10'. Larger zygomatic plate, paler overall dorsal coloration, head color gray, especially in sides; endemic to central and northeastern Brazil; $2n = 52$, $FN = 68$ . . . . .
. . . . *Oligoryzomys stramineus*

11. Hair on skin and throat regions with gray base; distinctive tufts of orangish hairs anterior to the ears absent. .
. . . . 12

11'. Hair on skin and throat regions all white, including on the base; distinctive tufts of orangish hairs anterior to the ears present; occurs in northern Argentina, eastern Bolivia, southwestern Brazil, and Paraguay; $2n = 58$, $FN = 74$. . . . . *Oligoryzomys chacoensis*

12. Tail unicolored or slightly bicolored; $2n = 72$, $FN = 76$
. . . . *Oligoryzomys utiaritensis*

12'. Tail sharply bicolored; $2n = 58$, $FN = 74$ . . . .
. . . . *Oligoryzomys brendae*

13. Dorsal coloration darker or orangish brown 14

13'. Dorsal coloration clay-colored; found only in hydromorphic vegetation patches in arid and semiarid coastal plains of Peru . . . . *Oligoryzomys arenalis*

14. Dorsal coloration grayish, ventral coloration whitish with large gray basal band; found in montane regions of northernmost Brazil, southern Venezuela, and Guyana; $2n = 66$, $FN = 74$ . . . . *Oligoryzomys messorius*

14'. Dorsal coloration orangish brown or dark brown, ventral coloration buffy or cream 15

15. Smaller incisive foramen, never extending posteriorly beyond alveolar line of M1 16

15'. Longer incisive foramen, extending posteriorly beyond alveolar line of M1; $2n = 64–66$, $FN = 66–68$ . . . .
. . . . *Oligoryzomys flavescens*

16. Very small size, head and body length <85 mm. . . . 17

16'. Small size, head and body >84 mm; $2n = 70$, $FN = 74$.
. . . . *Oligoryzomys moojeni*

17. Dorsal pelage strongly lined with dark hairs, general heterogeneous coloration, head color gray 18

17'. Dorsal pelage coloration homogeneous, head color similar as the body color, orange hair around the nose; sides lack guard hairs; distributed in northern South America, $2n = 60$, $FN = 74$. . . . .*Oligoryzomys delicatus*

18. Distributed in Brazil and Paraguay (and possibly Bolivia); $2n = 62$, $FN = 64$ . . . *Oligoryzomys mattogrossae*

18'. Restricted to Formosa province, Argentina . . . .
. . . . *Oligoryzomys fornesi*

Oligoryzomys andinus (Osgood, 1914)

Andean Colilargo

SYNONYMS:

Oryzomys andinus Osgood, 1914b:156; type locality "Hacienda Llagueda [= Llaugueda, Stephens and Traylor 1983], upper Rio Chicama, [Libertad,] Peru. Altitude 6,000 ft."

Oligoryzomys andinus: Gyldenstolpe, 1932a:29; first use of current name combination.

DESCRIPTION: Medium-sized mouse allied to *O. destructor* but with longer hindfoot and very long tail. Coloration of upper parts pale ochraceous darkened by numerous fine dusky lines; slight ochraceous lateral line present; face and head somewhat grayish; and under parts creamy white, sharply differentiated, hairs with slaty bases. Ears thinly haired with ochraceous basally and internally, brownish distally and externally. Dorsal surfaces of both forefeet and hindfeet white. Tail thinly clothed with very short hairs, brownish above, whitish below for proximal half, gradually becoming dusky toward tip. Skull similar in general form to that of *O. destructor* but relatively narrower with smaller teeth and bullae; incisive foramina compressed; braincase angled and truncate rather than rounded behind; and interorbital region at base of nasals slightly concave between sharply angled edges.

DISTRIBUTION: *Oligoryzomys andinus* occurs on the dry Pacific slopes of the Peruvian Andes and the western margins of the Altiplano of Peru and Bolivia, between 1,700 and 4,000 m. An isolated population occurs in the eastern Peruvian Andes in Cusco department (Carleton and Musser 1989). The precise distributional limits of this species are unknown (Musser and Carleton 2005).

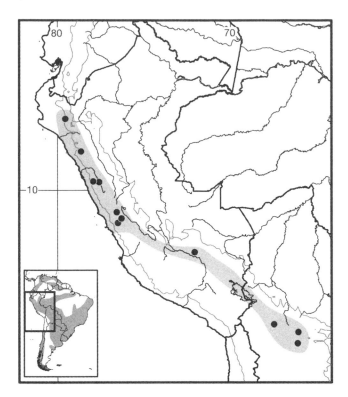

Map 219 Selected localities for *Oligoryzomys andinus* (●). Contour line = 2,000 m.

SELECTED LOCALITIES (Map 219; from Carleton and Musser 1989, except as noted): BOLIVIA: Oruro, 1 km W of Huancaroma; Potosí, 31 km SW Acacio on road to Uncia (S. Anderson 1997), 3 km SE of Pocoata. PERU: Ancash, I km N and 12 km E of Pariacoto; Callon, 2 km S and 11 km W of Huarás; Cusco, 94 km W of Cuzco on road to Abancay; La Liberdad, Hacienda Llaugueda (type locality of *Oryzomys andinus* Osgood); Lima, below Huaros, 6.3 mi W of Casapalca, Zarate, 6 km E of Pueblo San Bartolomé; Piura, Porculla Pass.

SUBSPECIES: *Oligoryzomys andinus* is monotypic.

NATURAL HISTORY: *Oligoryzomys andinus* is used as prey by the Barn Owl (*Tyto alba*) in Bolivia (S. Anderson 1997). No other aspect of ecology, behavior, reproductive biology, or other life history attributes is known.

REMARKS: Samples of this species used in morphometric (Olds and Anderson 1987:269) and allozyme analyses (Dickerman and Yates 1995:188) were misidentified as *Oryzomys longicaudatus* and *Oligoryzomys destructor*, respectively. The karyotype of *Oligoryzomys andinus* is characterized by 2*n* = 60 and FN = 70 (A. L. Gardner and Patton 1976). The species was recorded sympatrically with *O. arenalis* in Lima department, Peru (Arana-Cardó and Ascorra 1994, in Musser and Carleton 2005).

Oligoryzomys arenalis Thomas, 1913
Sandy Colilargo
SYNONYMS:
Oryzomys arenalis Thomas, 1913c:571; type locality "Eten [= Etén]; sea-level," Lambayeque, Peru.
Oligoryzomys arenalis: Gyldenstolpe, 1932a:27; first use of current name combination.

DESCRIPTION: Smaller than any allied species. General dorsal color clay over underlying buffy, lined with dark brown; head slightly grayer and less buffy; sides buffy, with somewhat buffy line separating dorsal from ventral pelage; under surfaces creamy white, with slight buffy tint and with slaty bases to hairs; ear only slightly darker than dorsum; chin white; dorsal surfaces of forefeet and hindfeet white; tail long and bicolored, brown above and creamy white below. Skull similar to but smaller than that of *O. destructor*; zygomatic plates well developed and projecting forward; braincase not especially enlarged; interorbital space narrow, narrow part long and evenly curved, not especially divergent posteriorly, and supraorbital edges sharply square.

DISTRIBUTION: *Oligoryzomys arenalis* is found in the arid and semiarid coastal plains of Peru (Musser and Carleton 2005).

SELECTED LOCALITIES (Map 220): PERU: Arequipa, Atiquipa (Zeballos et al. 2000), Lomas, 7 km E of Matarani (MVZ 150165); Cajamarca, 35 mi WNW of

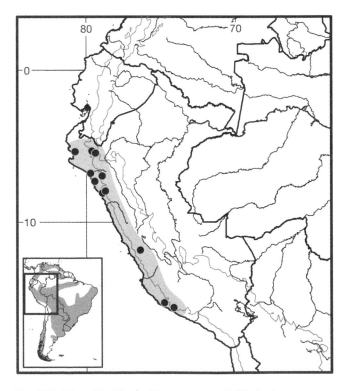

Map 220 Selected localities for *Oligoryzomys arenalis* (●). Contour line = 2,000 m.

Cajamarca (MVZ 137918); La Libertad, Menocucho (FMNH 19422), Pacasmayo (Osgood 1914b), Trujillo (Osgood 1914b); Lambayeque, Etén (type locality of *Oryzomys arenalis* Thomas); Lima, Matucana (FMNH 53058); Piura, 15 km by road E of Canchaque (LSUMZ 19261), Faique, Los Potreros (FMNH 84442), Palambla (AMNH 63712), 33 road km SW of Huancabamba, W slope (LSUMZ 19256).

SUBSPECIES: *Oligoryzomys arenalis* is monotypic.

NATURAL HISTORY: *Oligoryzomys arenalis* is a rare species with an insectivorous-granivorous diet (Zeballos et al. 2000). This species is found in the dry coast of Peru in hydromorphic vegetation patches close to rivers and lagoons, at the edges of small swamps or irrigating ditches (Osgood 1914b), but it is not found in gallery forest (H. Zeballos, pers. comm.).

REMARKS: *Oligoryzomys arenalis* was recorded sympatrically with *O. andinus* in Lima department (Arana-Cardó and Ascorra 1994, in Musser and Carleton 2005).

Oligoryzomys brendae Massoia, 1998
San Javier's Colilargo

SYNONYM:

Oligoryzomys brendae Massoia, 1998:243; type locality "Cerro San Javier, Dpto. Tafí Viejo (aprox. 1000m. de altura)," Tucumán, Argentina.

DESCRIPTION: Size large for genus (mean head and body length 89.7 mm; tail length 126.0 mm; hindfoot length [with claw] 26.5 mm; condyloincisive length of skull 22.9 mm; maxillary toothrow length 3.9 mm); dorsal color orangish brown, somewhat grizzled in appearance; ventral hairs with gray bases and white to ochraceous tips; ears short (mean 18.2 mm in length), rounded, and dark brown; and tail longer than head and body (140% of length), sharply bicolored. Skull relatively robust with short, broad rostrum; well-expanded zygomatic arches; inflated and broad braincase; narrow and hourglass-shaped interorbital region with slightly defined supraorbital ridges; and relatively long incisive foramina extending anterior margins of M1s (emended diagnosis from Teta, Jayat et al. 2013).

DISTRIBUTION: *Oligoryzomys brendae* is found mainly in forested and highland grassland environments of the Yungas from northernmost Salta province southward to Catamarca province, Argentina, between 700 and 2,900 m. It extends farther to the south in La Rioja province where it is restricted to isolated humid ravines at the ecotone between the xeric Chaco Seco and Monte desert formations (Teta, Jayat et al. 2013).

SELECTED LOCALITIES (Map 221; from Teta, Jayat et al. 2013, except as noted): ARGENTINA: Catamarca, Las Chacritas, ca. 28 km NNW of Singuil, Las Juntas (MIC 203), Mogote Las Trampas, ca. 15 km NW of Chumbicha;

Jujuy, ca. 3 km S Bárcena, La Herradura, 12 km SW of El Fuerte, Quyegrada Alumbriojo, ca. 8 km NE of Santa Ana, Parque Nacional Calilegua; La Rioja, Pampa de la Viuda, km 19 Rte 73; Salta, El Corralito, ca. 25 km SW of Campo Quijano, 1.6 km W of Los Toldos, Metán (CML-PIDBA 986); Tucumán, Alberdi, ca. 10 km S of Hualinchay, Cerro San Javier (type locality of *O. brendae*).

SUBSPECIES: *Oligoryzomys brendae* is monotypic.

NATURAL HISTORY: Published information on the natural history of this species is lacking.

REMARKS: Recent phylogenetic analyses of mtDNA sequences (Gonzalez-Ittig et al. 2010; Palma et al. 2010; Teta, Jayat et al. 2013) suggested that *O. brendae* is a member of a *destructor* species group, and we recognize it as a valid species based on these data. Nevertheless, both the taxonomic status and geographic distribution of *brendae* needs to be re-evaluated in a comprehensive revision of *destructor* species group, particularly with respect to those southern Bolivian specimens from Potosi and Tarija departments assigned herein to *O. destructor* (see also Teta, Jayat et al. 2013). Palma et al. (2010) considered *brendae* as a *nomen nudum* because Massoia's work was published in a meeting presentation. However, we agree with Musser and Carleton (2005) that *brendae* is a valid name because the International Code of Zoological Nomenclature (ICZN 1985, 1999) does not exclude this type of publication as valid. Furthermore, although the description is very short, it contains all formal requirements, such as holotype and

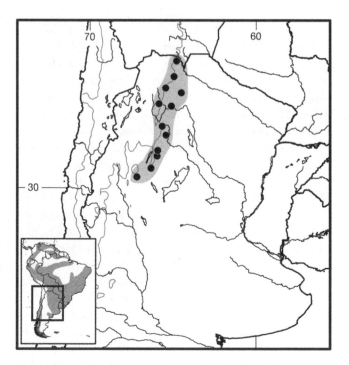

Map 221 Selected localities for *Oligoryzomys brendae* (●). Contour line = 2,000 m.

type locality designations, and comparisons to other species. Selected localities are from Gonzalez-Ittig et al. (2010; specimens identified as *O. destructor*); Palma et al. (2010; specimens identified as *Oligoryzomys* sp.1), and Teta, Jayat et al. (2013), who recently reviewed this species. Specimens with a $2n=58$, FN$=74$ karyotype described by Espinosa and Reig (1991) as *Oligoryzomys* cf. *longicaudatus* belong to *O. brendae* (see Teta, Jayat et al. 2013).

Oligoryzomys chacoensis (Myers and Carleton, 1981)
Chacoan Colilargo

SYNONYMS:

Oryzomys (*Oligoryzomys*) *chacoensis* Myers and Carleton, 1981:19; type locality "419 km by road NW Villa Hayes (alongside the Trans Chaco Highway), Dept. Boquerón, Paraguay."

Oligoryzomys chacoensis: Carleton and Musser, 1989: 72; first use of current name combination.

DESCRIPTION: Medium sized, distinguished from congeners by combination of whitish under parts, with hairs white to base on chin and throat, relatively long ears with hairs on inner surface with unusually short or absent dark basal bands, small but distinctive tufts of orange hairs anterior to ears, and karyotype with $2n=58$, FN$=74$.

DISTRIBUTION: *Oligoryzomys chacoensis* occurs in the Cerrado, Chaco, and semidry forests of northern Argentina, eastern Bolivia, southwestern Brazil, and Paraguay (Weksler and Bonvicino 2005).

SELECTED LOCALITIES (Map 222): ARGENTINA: Formosa, Riacho Pilagá (USNM 236243); Jujuy, San Salvador (INEVH [JY 1332]); Salta, Orán (INEVH [OR 22498]). BOLIVIA: Chuquisaca, Porvenir (AMNH 262120); Santa Cruz, 7 km E and 3 km N of Ingeniero Mora (AMNH 247769), San Ignacio de Velasco (USNM 391524); Tarija, Entre Rios (USNM 271412), Taringuiti (Weksler and Bonvicino 2005). BRAZIL: Mato Grosso, Cáceres (USNM 390126); Mato Grosso do Sul, Corumbá, 10 km NE Urucum (USNM 390124). PARAGUAY: Alto Paraguay, Agua Dulce (Weksler and Bonvicino 2005); Boquerón, Fortín Guachalla (Weksler and Bonvicino 2005), Fortín Teniente Pratts Gil (Weksler and Bonvicino 2005), 419 km by road NW of Villa Hayes (type locality of *Oryzomys* [*Oligoryzomys*] *chacoensis* Myers and Carleton); Presidente Hayes, Estancia Laguna Porá (Weksler and Bonvicino 2005), 8 km NE of Juan de Zalazar (UMMZ 134358), Puerto Pinasco (USNM 236293).

SUBSPECIES: *Oligoryzomys chacoensis* is monotypic.

NATURAL HISTORY: Specimens of *O. chacoensis* have been caught in bushes and low trees in the Paraguayan Chaco (Myers and Carleton 1981); the same authors reported an average of 4.6 embryos (range: 2–5) in 10 pregnant females from Paraguay, and suggested that reproduction occurred in January, February, July, with few births

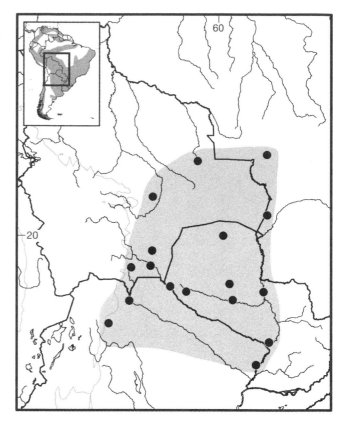

Map 222 Selected localities for *Oligoryzomys chacoensis* (●). Contour line = 2,000 m.

in June (winter). S. Anderson (1997) reported two females with three embryos each, and one with five embryos, from Bolivia. Andean and Bermejo hantaviruses are associated with *O. chacoensis* (González Della Valle et al. 2002; Fulhorst et al. 2004).

REMARKS: Thomas (1925b:578) referred to specimens of this species as *Oryzomys* sp., *flavescens* group. The karyotype of *O. chacoensis* is characterized by $2n=58$, FN$=74$ (Myers and Carleton 1981). Two specimens from northeastern Brazil (USNM 528416, 304583) identified by Carleton and Musser (1989) as *O. chacoensis* are actually *O. stramineus* (see Bonvicino and Weksler 1998). The distribution of *O. chacoensis* given by Weksler and Bonvicino (2005) mistakenly included specimens from Tucumán and Jujuy provinces, Argentina, as *O. chacoensis* based on a wrong interpretation of karyotype provided by Espinosa and Reig (1991); these specimens are herein considered as *O. brendae* (see that account).

Oligoryzomys delicatus (J. A. Allen and Chapman, 1897)
Delicate Colilargo

SYNONYMS:

Oryzomys delicatus J. A. Allen and Chapman, 1897a:19; type locality "Caparo, Trinidad," Trinidad and Tobago.

Oryzomys navus Bangs, 1899:9; type locality "Pueblo Viejo [= El Pueblito; Paynter 1997], Sierra Nevada de Santa Marta, [La Guajira,] Colombia. Altitude, 8000 feet."

Oryzomys tenuipes J. A. Allen, 1904b:328; type locality "Merida (alt. 1630 m.), [Mérida,] Venezuela."

Oryzomys (Oligoryzomys) munchiquensis J. A. Allen, 1912:85; type locality "La Florida (alt. 7700 ft.)," Cauca, Colombia.

Oligoryzomys navus navus: Gyldenstolpe, 1932a:27; name combination.

Oligoryzomys delicatus: Gyldenstolpe, 1932a:28; first use of current name combination.

Oligoryzomys tenuipes: Gyldenstolpe, 1932a:28; name combination.

Oligoryzomys munchiquensis: Gyldenstolpe, 1932a:28; name combination.

Oligoryzomys delicatulus Tate, 1939:191; incorrect subsequent spelling of *Oryzomys delicatus* J. A. Allen and Chapman.

Oryzomys delicatus navus: Cabrera, 1961:388; name combination.

Oryzomys delicatus tenuipes: Cabrera, 1961:389; name combination.

Oligoryzomys fulvescens: Carleton and Musser, 1989:70; part, specimens from northern South America.

DESCRIPTION: Small with yellowish brown fur above, darker and more rufus brown medially, mixed sparingly with blackish-tipped hairs; rump clear yellowish rufus; sides paler, yellowish buff; under parts clear buff; legs like adjoining parts of body; and dorsal surfaces of both forefeet and hindfeet yellowish white.

DISTRIBUTION: *Oligoryzomys delicatus* is distributed in South America throughout Colombia, northernmost Ecuador, northern Venezuela, and Trinidad and Tobago, and the coastal region of the Guianas. In Venezuela and the Guianas, the species is limited to coastal lowlands and llanos.

SELECTED LOCALITIES (Map 223): COLOMBIA: Cauca, Almaguer (AMNH 33860), Cocal (AMNH 32600); Cundinamarca, Paime (AMNH 71367); Huila, La Palma (AMNH 33858), near Villavieja (MVZ 113373); La Guajira, San Miguel (AMNH 38920); Magdalena, Palomino (AMNH 38917); Meta, 7 km NE of Villavicencio (AMNH 207936), Puerto López (ROM 90530); Quindio, Salento (AMNH 32925); Valle del Cauca, Pichindé (USNM 483980). ECUADOR: Napo, Santa Cecilia (KU 112879). SURINAM: Paramaribo, Paramaribo (USNM 319970). TRINIDAD AND TOBAGO: Trinidad, Caparo (type locality of *Oryzomys delicatus* J. A. Allen and Chapman), Nariva, Nariva Swamp (AMNH 189372). VENEZUELA: Apure, Hato El Frio (AMNH 257241); Aragua, Rancho Grande Biological Station (USNM 517581); Falcón, Río Socopo, 80 km NW Carora (USNM 442175); Guárico,

Hato Las Palmitas (USNM 442154); Lara, near El Tocuyo (USNM 495473); Mérida, Mérida (type locality of *Oryzomys tenuipes* J. A. Allen), Timotes (USNM 270814); Miranda, Río Chico (USNM 387881); Monagas, Hato Mata de Bejuco (USNM 442192), San Agustín (USNM 416707); Portuguesa, La Arenosa (AMNH 269143); Sucre, Cumaná, 2 km E (USNM 406024), Manacal (USNM 409911); Yaracuy, near Urama (USNM 374693); Vargas, Los Venados (USNM 371167); Zulia, near Cerro Azul (USNM 442176).

SUBSPECIES: Cabrera (1961) regarded three of the names we list above in the synonymy of *O. delicatus* as valid subspecies (*navus* Bangs and *tenuipes* J. A. Allen, in addition to the nominotypical form), with another considered a valid species (*munchiquensis* J. A. Allen). However, because none of these taxa have been evaluated, we believe it is premature to comment on the validity of any names as recognizable geographic races.

NATURAL HISTORY: *Oligoryzomys delicatus* was one of the most frequently captured rodents of the Llanos of Venezuela, being trapped in *Paspalum* savanna and lowland savanna (Utrera et al. 2000, as *O. fulvescens*). This species is a natural host of Maporal hantavirus in Venezuela (Bayard et al. 2004; Fulhorst et al. 2004; Hanson et al. 2011).

REMARKS: We recognize *O. delicatus* as a valid species distinct from *O. fulvescens* based on phylogenetic assessments of Rogers et al. (2009) and Hanson et al. (2011), which show that *Oligoryzomys* specimens from northern South America (i.e., *O. delicatus*) form a clade separate from *O. fulvescens* from Mexico and Honduras. Therefore, we considered most taxa from South America previously treated as junior synonyms of *O. fulvescens* (e.g., Musser and Carleton, 2005) as belonging to *O. delicatus*, with *O. fulvescens* restricted to Mexico and Central America. We note, however, that neither Hanson et al. (2011) nor we have examined relevant type material to support of

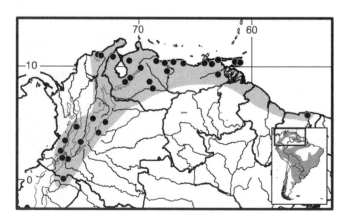

Map 223 Selected localities for *Oligoryzomys delicatus* (●). Contour line = 2,000 m.

this taxonomic arrangement. Karyotypes of *O. delicatus* are reported as $2n = 62$, FN = 74, 76 (Venezuela, Bolívar, as *Oryzomys* [*Oligoryzomys*] *longicaudatus*, variant 3, A. L. Gardner and Patton 1976) and $2n = 60$, FN = 72 (as *O. fulvescens* from Miranda, Venezuela [Kiblisky 1969] and Surinam [R. J. Baker et al. 1983]). We also recognize *O. messorius*, listed as a subspecies of *O. delicatus* by Cabrera (1961), as a valid species based on the distinct karyotype and morphology of specimens from Mt. Roraima and elsewhere in southern Guyana (see that account), and phylogenetic results (G. B. Miranda et al. 2009; Agrellos et al. 2012).

Oligoryzomys destructor (Tschudi, 1844)
Tschudi's Colilargo

SYNONYMS:

Hesperomys destructor Tschudi, 1844:182; type locality "im Oststriche" (= east region), Peru; restricted by Cabrera, 1961:390 to "las haciendas junto al río Chinchao, departamento de Huánuco, entre los 900 y los 1000 m de altitude" based on Hershkovitz (1940a:81).

Hesperomys melanostoma Tschudi, 1844:182; type locality "im Oststriche" (= east region), Peru; restricted by Cabrera, 1961:393 to "Departamento de Huánuco" based on Hershkovitz (1940a:81).

Oryzomys stolzmanni Thomas, 1894b:357; type locality "Huambo, [Amazonas,] N. Peru, 3700 feet."

Oryzomys destructor: E.-L. Trouessart, 1897:528; name combination.

Oryzomys melanostoma: E.-L. Trouessart, 1897:528; name combination.

Oryzomys stolzmanni maranonicus Osgood, 1914b:155; type locality "Hacienda Limon [= Hacienda Limón], near Balsas, Marañon River," upper Río Marañón, Cajamarca, Peru.

Oryzomys longicaudatus destructor: Thomas, 1928b:261; name combination.

Oryzomys (*Oligoryzomys*) *destructor*: Tate, 1932f:9; name combination.

Oryzomys (*Oligoryzomys*) *stolzmanni stolzmanni*: Tate 1932f:10; name combination.

Oligoryzomys stolzmanni: Gyldenstolpe, 1932a:28; name combination.

Oligoryzomys minutus maranonicus: Gyldenstolpe, 1932a:29; name combination.

Oryzomys (*Oligoryzomys*) *spodiurus* Hershkovitz, 1940a:79; type locality "Hacienda Chinipamba near Peñaherrera, Intag, in the subtropical forest of the western slope of the western cordillera of the Andes, Imbabura Province, Ecuador. Altitude, about 1500 meters."

Oryzomys (*O*[*ligoryzomys*].) *longicaudatus destructor*: Hershkovitz, 1940a:81; name combination.

Oryzomys (*O*[*ligoryzomys*].) *longicaudatus stolzmanni*: Hershkovitz, 1940a:81; name combination.

Oryzomys stolzmanni stolzmanni: Sanborn, 1950:2; name combination.

Oryzomys longicaudatus maranonicus: Cabrera, 1961:391; name combination.

Oryzomys longicaudatus stolzmanni: Cabrera, 1961:392; name combination.

Oryzomys spodiurus: Cabrera, 1961:395; name combination.

Oligoryzomys destructor: Carleton and Musser, 1989:73; first use of current name combination.

DESCRIPTION: Upper body reddish brown, interspersed with numerous black hairs and some gray with light tips; sides of body and outer part of hindlimbs reddish yellow; hair color behind ears lighter, while on head black hair more frequent; throat and chest whitish gray, while belly yellowish white; dorsal surfaces of feet covered by short silver-gray hair, with hairs of nail base very long; soles light brown, and nails are yellowish brown; tail brown and slightly shorter than the head and body length; internal ear surfaces with short, light brown hair, external surface covered with longer, blackish hairs; vibrissae black on basal-half and white toward tip; nose hair silver white.

DISTRIBUTION: *Oligoryzomys destructor* occurs in the Andes from southernmost Colombia to Bolivia, through Ecuador (on both eastern and western slopes) and Peru (eastern slopes and Amazonian drainage), over an elevational range from 600 to 3,350 m (Carleton and Musser 1989).

SELECTED LOCALITIES (Map 224): ARGENTINA: Jujuy, San Salvador (Rivera et al. 2007); Salta, Metán (Rivera et al. 2007); Tucumán, Monteros (OMNH 18656). BOLIVIA: Chuquisaca, Tola Orko (S. Anderson 1997); Cochabamba, Incachaca (S. Anderson 1997); La Paz, La Reserva (S. Anderson 1997), Ñequejahuira (S. Anderson 1997), Sorata (S. Anderson 1997), Ticunhuaya (S. Anderson 1997); Oruro, Poopo (S. Anderson 1997); Potosí, Tupiza (S. Anderson 1997); Santa Cruz, 1 km N and 8 km W of Comarapa (S. Anderson 1997), Vallegrande (S. Anderson 1997); Tarija, Tarija, 3 km SE of Cuyambuyo (S. Anderson 1997). COLOMBIA: Nariño, Ricaurte (AMNH 34233). ECUADOR: Chimborazo, Pallatanga (Carleton and Musser 1989); El Oro, Pinas (AMNH 61356); Loja, Puyango, Alamor (AMNH 61358); Pichincha, Gualea (AMNH 36297), Mojanda region (Carleton and Musser 1989), near Mount Illiniza (Carleton and Musser 1989); Zamora-Chinchipe, Zamora (AMNH 60546). PERU: Amazonas, 6 road km SW of Lake Pomacochas (LSUMZ 19265), Uscho (AMNH 73197); Apurimac, 10 km SSE of Abancay (MVZ 115650); Ayacucho, Huanhuachayo (A. L. Gardner and Patton 1976); Cajamarca, Hacienda

Limón (type locality of *Oryzomys stolzmanni maranonicus* Osgood); Cusco, directly below Marcapata (LSUMZ 19252), Bosque Aputinye above Huyro (LSUMZ 19285); Huánuco, E slope Cordillera Carpish, Carretera Central (LSUMZ 14361), Río Chinchao (type locality of *Hesperomys destructor* Tschudi); Junín, near San Ramón, 22 mi E of Tarma (AMNH 231676), Perené (AMNH 61817); Puno, 3 mi N of Limbani (MVZ 116065), 11 km NNE of Ollachea (MVZ 172610), 14 km W of Yanahuaya (MVZ 173423).

SUBSPECIES: Several subspecies have been recognized in the literature (Hershkovitz 1940a; Cabrera 1961), but the validity of these must wait an appropriate evaluation of geographic variation based on adequate samples of museum specimens.

NATURAL HISTORY: Most Bolivian records of *O. destructor* are from the Yungas forests (S. Anderson 1997:394). Three pregnant females were recorded from Bolivia in May and April with three, four, and five embryos each (S. Anderson 1997). Recorded ectoparasites include the mites *Gigantolaelaps wolffsohni*, *Laelaps paulistanensis*, and *Mysolaelaps microspinosus*, and the flea *Craneopsylla minerva* (Lareschi et al. 2003).

REMARKS: The karyotype of *O. destructor* is characterized by $2n = 60$ and $FN = 76$ (A. L. Gardner and Patton 1976, as *Oryzomys* (*Oligoryzomys*) *longicaudatus*, variant 4). Olds and Anderson (1987), in their review of *Oligoryzomys*, included specimens of this species under *O. longicaudatus*. Both the taxonomy and distributional limits are poorly understood (Musser and Carleton 2005:1140).

Oligoryzomys flavescens (Waterhouse, 1837)
Flavescent Colilargo

SYNONYMS:

Mus flavescens Waterhouse, 1837:19; type locality "Maldonado," Maldonado, Uruguay.

Hesperomys flavescens: Waterhouse, 1839:74; name combination.

H[esperomys]. (*Calomys*) *flavescens*: Wagner, 1843a:530; name combination.

Oryzomys longicaudatus flavescens: E.-L. Trouessart, 1897: 527; name combination.

Oryzomys flavescens: Thomas, 1926d:603; name combination.

Oryzomys flavescens: J. R. Contreras and Rosi, 1980a:157; first use of current name combination.

Oligoryzomys flavescens occidentalis J. R. Contreras and Rosi, 1980a: 158; type locality "Colonia Alto del Algarrobal, Departamento San Rafael, Provincia de Mendoza," Argentina.

Oligoryzomys flavescens antoniae Massoia, 1983; type locality "República Argentina, provincia de Misiones, departamento de Capital, arroyo Itaembé Mini, debajo del puente sobre la Ruta Nacional N° 12."

Oligoryzomys microtis: Carleton and Musser, 1989:71; part; not *Oryzomys* (*Oligoryzomys*) *microtis* J. A. Allen.

Oligoryzomys sp. A: Carleton and Musser, 1989:72; name combination.

DESCRIPTION: Small species characterized by dorsal color bright brownish-orange finely intermixed with dark hairs and lateral color of brighter orange. Skull with long incisive foramina that usually reach M1s, and short mesopterygoid fossa that does not reach M3s.

DISTRIBUTION: *Oligoryzomys flavescens* occurs in northern Argentina, eastern Paraguay, throughout Uruguay, and eastern Brazil, from sea level to about 1,800 m (Bonvicino et al. 1997).

SELECTED LOCALITIES (Map 225; from Weksler and Bonvicino 2005, except as noted): ARGENTINA: Buenos Aires, Balcarce, Berazategui, Diego Gaynor, Ezeiza, Punta Lara, General Lavalle, Monte Hermoso; Córdoba, Río Cuarto;

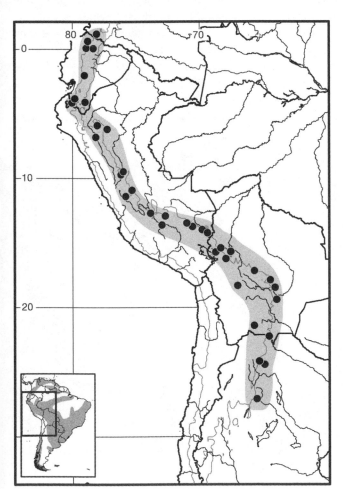

Map 224 Selected localities for *Oligoryzomys destructor* (●). Contour line = 2,000 m.

Jujuy, Maimará; Misiones, Arroyo Itaembé Mini (type locality of *Oligoryzomys flavescens antonia* Massoia); Tucumán, Concepción. BOLIVIA: Tarija, Tarija. BRAZIL: Bahia, Rio Una, 10 km ESE of São José); Minas Gerais, Itamonte, Viçosa; Paraná, Piraquara, Ponta Grossa; Rio Grande do Sul, Esmeralda, Sapiranga, Taím, Tôrres; Santa Catarina, Itá (MNRJ [CRB 1932]); São Paulo, Americana, Caçapava, Casa Grande. PARAGUAY: Canendiyú, 6.3 km by road N of Curuguaty (Myers and Carleton 1981); Misiones, San Pablo; Presidente Hayes, La Golondrina. URUGUAY: Artigas, Artigas; Colonia, Colonia del Sacramento; Maldonado, Maldonado (type locality of *Mus flavescens* Waterhouse); Montevideo, Montevideo; Río Negro, Fray Bentos.

SUBSPECIES: Musser and Carleton (2005) regarded *Oligoryzomys flavescens* to be monotypic.

NATURAL HISTORY: *Oligoryzomys flavescens* is found in the Pampas, Chaco, Atlantic Forest (pristine and second-growth communities), and gallery forests in the Cerrado near the limits with Atlantic Forest. This species is also found along cropland borders in Argentina (M. Busch and Kravetz 1992). Population levels of *O. flavescens* dramatically increased during the flowering of *taquara-lixa* bamboo (*Merostachys skvortzovii*) in a possible *ratada* event (Galiano et al. 2007). Ectoparasites of *O. flavescens* include the mites *Gigantolaelaps wolffsohni*, *Laelaps paulistanensis*, and *Mysolaelaps microspinosus*, and the sucking louse *Hoplopleura travassosi* (Lareschi et al. 2003; Lareschi, 2010). This species is also the principal carrier of the Andean hantavirus in northern Argentina (Gonzalez Della Valle et al. 2002), and is a natural host of the Lechiguanas hantavirus in both Argentina and Uruguay (Fulhorst et al. 2004).

REMARKS: The basic chromosomal complement of *Oligoryzomys flavescens* is $2n = 64$, $FN = 66$, but its diploid and fundamental numbers are variable ($2n = 64$–67, $FN = 66$–70) due to the presence of acrocentric or metacentric supernumerary chromosomes (Espinosa and Reig 1991; Sbalqueiro et al. 1991). Myers and Carleton (1981) erroneously regarded their specimens with this karyotype to be *Oligoryzomys fornesi* (here recognized as *O. mattogrossae*; see that account).

Oligoryzomys fornesi (Massoia, 1973)
Fornes's Colilargo

SYNONYM:

Oryzomys fornesi Massoia, 1973a:22; type locality "Ceibo 13, Naineck, Departamento de Río Pilcomayo, provincia de Formosa, República Argentina."

DESCRIPTION: Another small species with small and pale ears covered on outer surface with ochre hairs, orangish brown or grayish brown dorsal pelage, with paler venter varying from pale ochre to yellowish white.

DISTRIBUTION: By current understanding, *O. fornesi* is limited to the vicinity of the type locality in Argentina.

SELECTED LOCALITIES (Map 226): ARGENTINA: Formosa, near Laguna Blanca (CEM 3436), Naineck (type locality of *Oryzomys fornesi* Massoia).

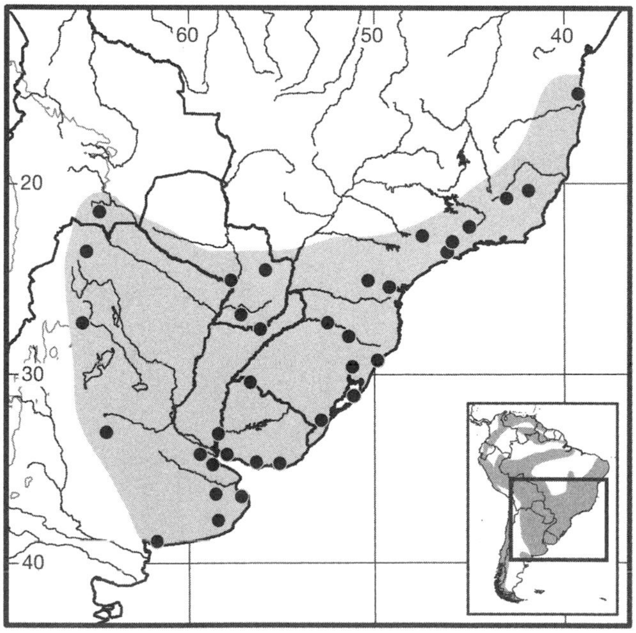

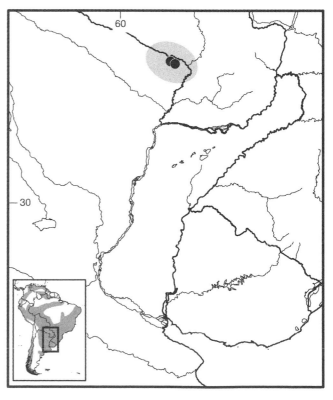

Map 225 Selected localities for *Oligoryzomys flavescens* (●). Contour line = 2,000 m

Map 226 Selected localities for *Oligoryzomys fornesi* (●). Contour line = 2,000 m.

SUBSPECIES: *Oligoryzomys fornesi* is monotypic.

NATURAL HISTORY: *Oligoryzomys fornesi* has been collected in open vegetation areas used as pasture for domestic livestock; numerous nests were found in banana trees (Massoia 1973a).

REMARKS: The taxonomic status of this species needs to be reassessed in relation to *O. mattogrossae*, with particular attention paid to study of the type specimen. See Remarks for *O. mattogrossae*.

Oligoryzomys griseolus (Osgood, 1912)
Grizzled Colilargo

SYNONYMS:

Oryzomys griseolus Osgood, 1912:49; type locality "Paramo de Tama, head of Tachira River, Venezuela. Alt. 6,000–7,000 ft."

Oligoryzomys griseolus: Gyldenstolpe, 1932a:27; first use of current name combination.

Oryzomys delicatus griseolus: Cabrera, 1961:388; name combination.

DESCRIPTION: Overall ground color of upper parts pale clay, much duller than ochraceous or ochraceous buff of related forms, with abundant mixture of black-tipped hairs producing somewhat grizzled effect which is most pronounced on forehead and sides of face with fulvous minimized and gray predominates; small preauricular tuft of ochraceous-tipped hairs usually present; under parts mostly between clay color and ochraceous buff, almost or quite concealing slaty bases of hairs; middle of chin and throat white or whitish to bases of hairs; forefeet and hindfeet white above, with outer sides of tarsal joints dusky; tail bicolored, dusky above and dull whitish below. Only three pairs of mammae present. Skull with slender rostrum, well-developed zygomatic plates project forward farther than in related forms, and elongated incisive foramina that extend posteriorly beyond plane of M1s.

DISTRIBUTION: *Oligoryzomys griseolus* occurs in the Táchira Andes of westernmost Venezuela and in the Cordillera Oriental of easternmost Colombia (Musser and Carleton 2005).

SELECTED LOCALITIES (Map 227): VENEZUELA: Táchira, Buena Vista, 41 km SW of San Cristóbal (USNM 442155), Páramo de Tamá (type locality of *Oryzomys griseolus* Osgood), Buena Vista (USNM 442155).

SUBSPECIES: *Oligoryzomys griseolus* is monotypic.

NATURAL HISTORY: Specimens comprising the type series were collected in a small grassy swamp (Osgood 1912); others were found in lower montane very humid forest (Handley 1976).

REMARKS: No karyotypic data are available. Osgood (1912) and Musser and Carleton (2005) stated that *O. griseolus* sharply contrasts with *O. fulvescens* and resembles

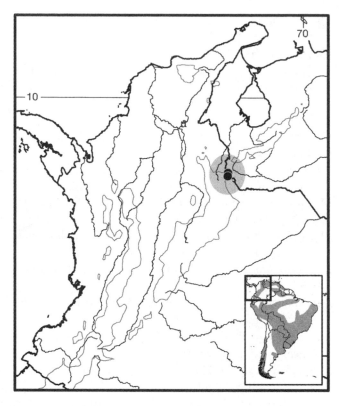

Map 227 Selected localities for *Oligoryzomys griseolus* (●). Contour line = 2,000 m.

Central American *O. vegetus*, especially "in the extent of white on the throat and in the forwardly projecting zygomatic plate" (Osgood 1912:50). In the description of the species, Osgood (1912) stated that *O. griseolus* had six pairs of mammae, but all *Oligoryzomys* species examined to date have eight pairs (Steppan 1995; Weksler 2006); the only Oryzomyini taxa with six mammae are *Handleyomys* and *Scolomys* (Patton and da Silva 1995; Voss et al. 2002; Weksler 2006). A recent examination of the type series (B. D. Patterson, pers. comm.) confirms Musser and Carleton's (2005) assignment of *griseolus* to the genus *Oligoryzomys*. Further examination is necessary to confirm Osgood's mammae count. Although Musser and Carleton (2005) included the Cordillera Oriental of easternmost Colombia within the range of the species, we are unable to confirm the allocation of specimens from this region from our personal examination of specimens. Hence, possible records from Colombia are neither listed nor mapped as selected localities here.

Oligoryzomys longicaudatus (Bennett, 1832)
Long-tailed Colilargo

SYNONYMS:

Mus longicaudatus Bennett, 1832a:2; type locality: "Chiliæ"; restricted to Valparaíso, Chile by Osgood (1943b:143).

H[esperomys]. (Calomys) longicaudatus: Wagner, 1843a:529; name combination.

Mus philippii Landbeck, in Philippi, 1857:361; type locality "Valdivia," Chile.

Hesperomys (Calomys) coppingeri Thomas, 1881b:4; type locality "Cockle Cove, Trinidad Channel, Madre de Dios Island, W. Patagonia," Magallanes y Antártica Chilena; fixed by lectotype selection (Thomas 1927b: 549).

Oryzomys longicaudatus: E.-L. Trouessart, 1897:527; name combination.

Oryzomys coppingeri: E.-L. Trouessart, 1897:529; name combination.

Mus dumetorum Philippi, 1900:14; type locality "provincia Valdivia," Chile.

Mus exiguus Philippi, 1900:19; type locality "Andibus provinciae Santiago," [= Andes of Santiago,] Chile."

Mus commutatus Philippi, 1900:25; type locality "Valdivia," Chile.

Mus macrocercus Philippi, 1900:30; type locality "provincia Colchagua," Chile.

Mus nigribarnis Philippi, 1900:31; type locality " Andibus de Talcaregüe," [= Talcaregue Andes,] Chile.

Mus amblyrrhynchus Philippi, 1900:36; type locality "provincia Valdivia," Chile.

Mus melanizon Philippi, 1900:39; type locality "Chile."

Mus diminutivus Philippi, 1900:43; type locality "Illapel et provincia O'Higgins," Chile.

Mus agilis Philippi, 1900:44; type locality "Illapel," Chile.

Mus (Rhipidomys) araucanus Philippi, 1900:46; type locality "Conceptione," [= Concepción,] Chile.

Mus pernix Philippi, 1900:48; type locality "La Ligua," Chile.

Mus glaphyrus Philippi, 1900:51; type locality "provincia Maule," Chile.

Mus peteroanus Philippi, 1900:56, type locality "Andibus de Peteroa," Peteroa, Curicá Province, Chile.

Mus malaenus Philippi, 1900:62; type locality "provincia Maule," Chile.

Oryzomys magellanicus mizurus Thomas, 1916b:186; type locality "Koslowsky Valley, 46°S., 71°W., Central Patagonia," Chubut, Argentina.

Oryzomys longicaudatus pampanus Massoia, 1973a:34; type locality "Estación Experimental Hilario Ascasubi (INTA), partido de Villarino, provincial de Buenos Aires, República Argentina."

O[ryzomys]. l[ongicaudatus]. *philippii*: M. H. Gallardo and Patterson, 1985:49; name combination.

Oligoryzomys longicaudatus: Carleton and Musser, 1989: 74; first use of current name combination.

DESCRIPTION: Extremely long tail, nearly double length of head and body (head and body length of holotype given as 3 inches [76.2 mm], tail length 5½ inches [139.7 mm]). Fur soft, smooth, and luxuriant; hairs deep ashy gray at base; those of the upper surface fawn-colored or pale rufous toward their points, tip frequently black; those of under surface tipped with white slightly tinged with fawn; face covered with short hairs of mingled fawn and black; lips nearly white; vibrissae extremely long, black at their base and silvery at tip; ears rounded and of moderate size, covered on inside with short hairs same color as those of face, and on outside with very short whitish hairs scarcely discernible on blackish skin; color of back mixed fawn and black, black disappearing on almost purely fawn-colored sides; front of forelegs and the outside of hind legs fawn-colored; tail scaly, furnished with numerous very short bristly hairs, brownish above and nearly white beneath; hairs of upper surface of tarsi short, very pale fawn approaching white; those of toes still more white; and lengthened bristles covering claws almost silvery.

DISTRIBUTION: *Oligoryzomys longicaudatus* occurs along the eastern and western flanks of the Andes in Chile and Argentina, from Tierra del Fuego to Atacama (Chile) and Tucumán (Argentina) in the north; isolated populations are reported from the lowland areas in the provinces of Buenos Aires, La Pampa, and Rio Negro.

SELECTED LOCALITIES (Map 228): ARGENTINA: Buenos Aires, Bahía San Blas (Palma, Rivera-Milla et al. 2005), Villarino, Estación Experimental INTA Hilario Ascasubi (type locality of *Oryzomys longicaudatus pampanus* Massoia); Chubut, Cushamen (LSUMZ 16882), Parque Nacional Los Alerces (MVZ 188421); La Pampa, Parque Nacional Lihue Calel (MVZ 182026); Neuquén, 2 km SE of La Rinconada (MVZ 163775), 5 km N of Las Coloradas (MVZ 159332); Río Negro, Chimpay (FMNH 50909), Choele Choel (FMNH 129276), Pichi Leufú (Rivera et al. 2007); Santa Cruz, 52 km WSW of El Calafate (MVZ 150983); Tierra del Fuego, Bahía Lapataia (Espinosa and Reig 1991). CHILE: Aysén, 20 km S of Chile Chico (FMNH 133384), Coyhaique (TTU 30185), Isla Gunther (FMNH 127745), Parque Nacional Queulat (FMNH 133358), Río Aysén (FMNH 22561); Araucanía, 7 km NNW of Los Álamos (MSU 6403), 1.7 km W of Paso Pino Hachado (MSU 6737), 11 km SE of Temuco (MSU 6739), Termas de San Luis (UMMZ 155906), 2 km NE of Tirúa (MSU 6708), 11 km N of Villa Ranquil (MSU 6402); Atacama, Ramadilla (FMNH 22660), 40 km E of Vallenar (TTU 30170); Biobío, Concepción (FMNH 22624), 13 km E of Recinto (FMNH 119489); Coquimbo, Conchalí (TTU 30180), Parque Nacional Fray Jorge (FMNH 133210), La Serena (FMNH 133211), Paihuano (FMNH 22644); Los Lagos, 2 km S of Bahía Mansa (FMNH 133313), Cuesta El Moraga (FMNH 133808), Isla Grande de Chiloé, Quellón (FMNH 22574), Parque Nacional Puyehue (FMNH 133264); Los Ríos, Valdivia (type locality of *Mus philippii* Landbeck); Magallanes y Antártica Chilena, Cockle Cove [= Caleta Cockle], Isla Madre de Dios (type locality of *Hesperomys* [*Calomys*] *coppingeri* Thomas), Puerto del Hambre (Belmar-Lucero et al. 2009), Reserva Nacional Magallanes (Belmar-Lucero et al. 2009), San Martín (M. H. Gallardo and Palma 1990), Torres del Paine (Palma, Rivera-Milla et al. 2005); Maule, Pilén Alto (FMNH 22615); O'Higgins, Baños de Cauquenes (FMNH 22631); Santiago, La Parva (TTU 30183); Valparaíso, Valparaíso (type locality of *Mus longicaudatus* Bennett).

SUBSPECIES: We regard *Oligoryzomys longicaudatus* to be monotypic, although an adequate assessment of geographic trends for any character set that might indicate valid races has not been achieved to date.

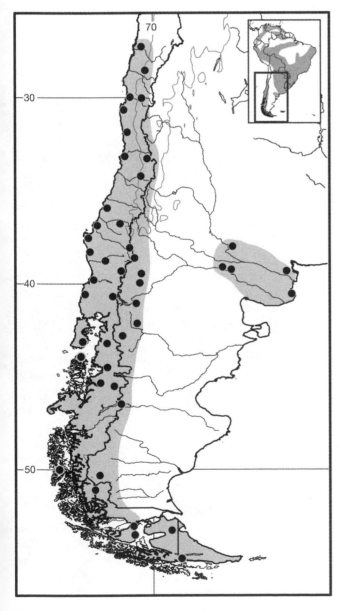

Map 228 Selected localities for *Oligoryzomys longicaudatus* (●); the disjunct populations in eastern Argentina are mapped separately from the main distribution. Contour line = 2,000 m.

NATURAL HISTORY: Large-scale outbreaks (*ratadas*) of *O. longicaudatus* have been observed in both Chile (Gonzalez et al. 2000; Murúa et al. 1986) and Argentina (Sage et al. 2007), both associated with cyclical masting of the arborescent bamboo *Chusquea quil* and *C. culeou*, respectively. This species was found to select microsites with high foliage density and thick understory, which offer protection from horizontal view of predators (Gonzalez et al. 2000). *Oligoryzomys longicaudatus* have a single, limited reproduction period each year in Chile (Murúa et al. 1996). Ectoparasites of *O. longicaudatus* include the mites *Gigantolaelaps wolffsohni* and *Mysolaelaps microspinosus* (Lareschi et al. 2003). This species is also host to the nematode *Litomosoides pardinasi* (Notarnicola and Navone 2011) and to Andes hantavirus in the Chilean Andes and Oran hantavirus in Argentina (Belmar-Lucero et al. 2009; Fulhorst et al. 2004).

REMARKS: The karyotype of *O. longicaudatus* is $2n = 56$, $FN = 66$ (Espinosa and Reig 1991). M. H. Gallardo and Palma (1990) provided the morphological basis for the current classification of colilargos from Chile, including the lack of differentiation of previously identified *philippii* populations (e.g., Osgood 1943b). *Mus philippii* is usually assigned to Landbeck and Philippi (1858), but this species was actually described by Luis Landbeck within the article by Philippi (1857). Palma, Rivera-Milla et al. (2005) presented a phylogeographic study using mtDNA sequences of the species and discussed its relationship with *O. magellanicus*. Thomas (1916b:186) noted that his original designation of the type locality of *Hesperomys* (*Calomys*) *coppingeri* to the Straits of Magellan was incorrect, and that Cockle Cove (= Caleta Cockle) is "in the Trinidad Channel, at the north end of Madre de Dios Island, West Patagonia, in 50° S. lat." Belmar-Lucero et al. (2009) showed that the distribution of *O. longicaudatus* extends into Tierra del Fuego, where it is sympatric with *O. magellanicus* (see also the account of this species).

Oligoryzomys magellanicus (Bennett, 1836)
Patagonian Colilargo
SYNONYMS:

Mus magellanicus Bennett, 1836:191; type locality "apud Portum Famine dictum, in fretu Magellanicus," Port Famine [= Puerto del Hambre], Magallanes y Antártica Chilena, Chile.

Hesperomys (*Oryzomys*) *longicaudatus*: Milne-Edwards, 1890:27; part; not *Mus longicaudatus* Bennett.

Oryzomys magellanicus: J. A. Allen, 1905:47; name combination.

Oryzomys longicaudatus magellanicus: Gyldenstolpe, 1932a: 11; name combination.

Oryzomys (*Oligoryzomys*) *magellanicus magellanicus*: Tate, 1932f:10; name combination.

DESCRIPTION: Head and body length similar to tail length; dorsal color saturated with grayish buff; ventral color whitish; and dorsal surfaces of forefeet and hindfeet white, ears of moderate size, rounded, and hairy.

DISTRIBUTION: *Oligoryzomys magellanicus* is limited to the Patagonian region of Chile and Argentina, including Tierra del Fuego.

SELECTED LOCALITIES (Map 229): ARGENTINA: Terra del Fuego, Estancia Via Monte (Osgood 1932), Lago Fagnano (= Lago Cami; Osgood 1932), Lago Yehuin (= Lago Yerwin; Osgood 1932). CHILE: Magallanes y Antártica Chilena, Harrison Island (M. H. Gallardo and Patterson 1985), Puerto del Hambre (type locality of *Mus magellanicus* Bennett), Punta Arenas (M. H. Gallardo and Patterson 1985), Lago Sarmiento, Parque Nacional Torres del Paine (M. H. Gallardo and Palma 1990).

SUBSPECIES: *Oligoryzomys magellanicus* is monotypic.

NATURAL HISTORY: Osgood (1932) found this species to be fairly common at some localities on Tierra del Fuego, but in the vicinity of Punta Arenas and northward in western Patagonia it was found with difficulty and only in small numbers. It co-occurs with *Abrothrix olivacea* in forest communities but not with this species in the brush or open pampa.

REMARKS: *Oligoryzomys magellanicus* is characterized by 2n = 54, FN = 66 (M. H. Gallardo and Patterson 1985). M. H. Gallardo and Palma (1990) described external, cranial, bacular, chromosomal, and allozyme characters that differentiate *O. magellanicus* from *O. longicaudatus*. Palma, Rivera-Milla et al. (2005) and Belmar-Lucero et al. (2009), however, showed, based on molecular and karyological characters, that specimens from southern Chile (including from Puerto Hambre, type locality of *O. magellanicus*) and Tierra del Fuego, in the range of *O. magellanicus* as proposed by M. H. Gallardo and Palma (1990), actually belong to *O. longicaudatus*. Palma, Rivera-Milla et al. (2005) also showed an outlier haplotype from Rio Penitente, which they regarded to be *O. magellanicus*. We recognize *O. magellanicus* as a valid species based on its distinctive short tail (same length as head and body, compared to the much longer tail of *O. longicaudatus*), its different karyotype, and the molecular results of Palma, Rivera-Milla et al. (2005). Our assessment is that *O. magellanicus* and *O. longicaudatus* are sympatric in southernmost Chile and Tierra del Fuego (based on molecular and karyological results), the first being rare (and thus not collected by Belmar-Lucero et al. 2009). Nevertheless, Palma et al. (2012) argued, based on their molecular data, that *O. magellanicus* is a subspecies of *O. longicaudatus*; the taxonomic and distributional limits of these species need re-evaluation, especially in a more integrative approach that combines molecular, karyotypic, and morphological data.

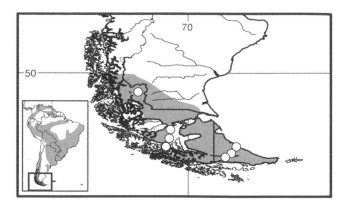

Map 229 Selected localities for *Oligoryzomys magellanicus* (O). Contour line = 2,000 m.

Oligoryzomys mattogrossae (J. A. Allen, 1916)
Mato Grosso Colilargo

SYNONYMS:

Oryzomys (*Oligoryzomys*) *mattogrossae* J. A. Allen, 1916c: 528; type locality "Utiarity, [Rio Papagaio, 2,500 ft,] Matto Grosso [= Mato Grosso], Brazil."

Oligoryzomys mattogrossae: Gyldenstolpe, 1932a: 29; first use of current name combination.

Oryzomys [(*Oligoryzomys*)] *mattogrossae*: Ellerman, 1941: 355; name combination.

Oryzomys utiaritensis: Cabrera 1961:396; part; not *Oligoryzomys utiaritensis* J. A. Allen.

Oligoryzomys microtis: Carleton and Musser, 1989:71; part; not *Oryzomys* (*Oligoryzomys*) *microtis* J. A. Allen.

Oryzomys (*Oligoryzomys*) *fornesi*: Myers and Carleton, 1981:25; not *Oryzomys fornesi* Massoia.

Oligoryzomys fornesi: Bonvicino and Weksler, 1998:90; not *Oryzomys fornesi* Massoia.

DESCRIPTION: Characterized by rufous tone, especially on rump; under parts pale ochraceous buff instead of grayish white; tail relatively long.

DISTRIBUTION: *Oligoryzomys mattogrossae* occurs in the open-vegetation belt throughout Paraguay, and central and northeastern Brazil; this species also possibly occur in Bolivia (e.g., Santa Cruz specimens listed as *O. microtis* by Olds and Anderson [1987] and S. Anderson [1997]) and Argentina, but no voucher material was analyzed to confirm identification.

SELECTED LOCALITIES (Map 230): BRAZIL: Alagoas, Matriz de Camaragibe (UFPB 977); Bahia, Fazenda Sertão do Formoso, Jaborandi (MNRJ 62637); Distrito Federal, Brasília (UNB 279); Goiás, Aporé (LBCE 8579), Corumbá de Goiás (MNRJ 34440), Fazenda Vão dos Bois, Teresina de Goiás (MNRJ [CRB 674]), Mambaí (LBCE 10851), Serranópolis (LBCE 8509); Mato Grosso, Utiariri (type locality of *Oryzomys* [*Oligoryzomys*] *mattogrossae*

J. A. Allen); Mato Grosso do Sul, Cassilândia (LBCE 12111), Corumbá (LBCE 5718); Paraíba, Mamanguape (UFPB [MPS 72]); Pernambuco, Bom Conselho (UFPB-PMN 60), Buíque (UFPB 1893), Macaparana (UFPB [MPS 34]). PARAGUAY: Caaguazú, 24 km NNW of Carayaó, Estancia San Ignacio (UMMZ 133818), Canindeyú, 6.3 km NE by road of Curuguaty (UMMZ 124218).

SUBSPECIES: *Oligoryzomys mattogrossae* is monotypic.

NATURAL HISTORY: *Oligoryzomys mattogrossae* is an inhabitant of open-vegetation biomes such as the Cerrado, Caatinga, and Chaco, but can also be found in forest formations in the transition with Amazonian forest. It can occur in sympatry with *O. nigripes*, *O. stramineus* (see Weksler and Bonvicino 2005), *O. utiaritensis* (see J. A. Allen 1916c), and maybe with *O. flavescens* (see Myers and Carleton, 1981, and below). *Oligoryzomys mattogrossae* (specimens identified as *O. fornesi*) was the reservoir of Anajatuba hantavirus in the Brazilian state of Maranhão (E. S. T. Rosa et al. 2005).

REMARKS: *Oligoryzomys mattogrossae* was described by J. A. Allen (1916c) based on two specimens, the holotype from Utiariti and one paratype from "Guatsué" (not located by Painter and Traylor 1991, but presumably on middle Rio Papagaio), both in Mato Grosso state. Later, Cabrera (1961:396) considered this taxon a junior synonym of *O. utiaritensis*. However, morphologic and phylogenetic analyses (C. R. Bonvicino and M. Weksler, in prep.) showed that *O. mattogrossae* is an independent lineage; in particular, *O. mattogrossae* has a yellow-ochraceous ventral coloration that contrasts with the whitish venter of *O. utiaritensis* (as previously noted by Myers and Carleton 1981). Some authors (e.g., Carleton and Musser 1989) also considered *O. mattogrossae* a junior synonymous of *O. microtis*, but these two species differ in a number of attributes, including distinct karyotypes ($2n=64$, FN$=66$ for *O. microtis*; A. L. Gardner and Patton 1976 [as *Oryzomys* (*Oligoryzomys*) *longicaudatus*, variant 2]; Aniskin 1994; Patton et al. 2000), nonsister status in molecular phylogenetic analyses (C. R. Bonvicino and M. Weksler, in prep.), and distinct habitats (*O. mattogrossae* is found mostly in open-vegetation biomes, whereas *O. microtis* is found only in the forested environments).

We recognize *Oligoryzomys mattogrosae* as a valid species based on the examination of the holotype and the discovery of a specimen from Sapezal (same municipality as Utiariti, type locality of *O. mattogrossae*) that shares the same traits of the holotype and that is a member of a well-supported clade within *Oligoryzomys*; specimens of this clade were previously referred to as *O. fornesi* (e.g., Weksler and Bonvicino 2005), a name that we restrict here to the type locality and vicinity (see that account). Therefore, the small body size and yellow belly of specimens, with $2n=62$, FN$=64$, found in the Cerrado and Caatinga domains, pre-

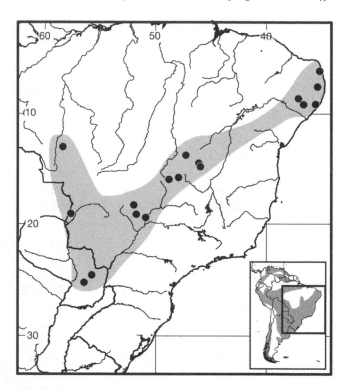

Map 230 Selected localities for *Oligoryzomys mattogrossae* (●). Contour line = 2,000 m.

viously identified by us (e.g., Bonvicino and Weksler 1998; Weksler and Bonvicino 2005) as *O. fornesi*, are here considered as *O. mattogrossae*. The karyotype of this species is frequently mistaken as a variant of the karyotype of *O. flavescens* ($2n=64$–66, FN$=66$–68; e.g., Myers and Carleton 1981; Sbalqueiro et al. 1991). However, *O. mattogrossae* ($2n=62$, FN$=64$) and *O. flavescens* ($2n=64$–66, FN$=66$–68) karyotypes are discontinuous because a $2n=63$ karyotype has not been reported. Furthermore, these two taxa appear to differ in morphological traits, such as the incisive foramen length (longer in *O. flavescens*). The $2n=62$, FN$=64$ karyotype has also been attributed to *O. eliurus* (Svartman 1989; Andrades-Miranda, Oliveira, Lima-Rosa et al. 2001), but examination of the karyotyped specimens listed by these authors confirmed them as *O. mattogrossae*.

Oligoryzomys messorius (Thomas, 1901)
Hairy Colilargo
SYNONYMS:

Oryzomys navus messorius Thomas, 1901e:151; type locality "Kanuku Mountains, about 59° W. and 3° N., and on the savannas near their base, at altitudes varying from 240 to 2000 feet," Upper Takutu–Upper Essequibo, Guyana.

Oligoryzomys navus messorius: Gyldenstolpe, 1932a:27; name combination.

Oryzomys delicatus messorius: Cabrera, 1961:388; name combination.

Oligoryzomys fulvescens: Carleton and Musser, 1989:70; part, specimens from northern South America.

DESCRIPTION: Color above grizzled grayish fawn, varying towards rufous; rump more rufous than back; face grayer; ears comparatively short, but little darker than general color of head, very different therefore to blackish ears of *O. fulvescens*; under surface dull whitish, with scarcely trace of buff, not sharply defined laterally, hairs all grayish basally; dorsal surfaces of hands and feet white. Tail short for group, thinly haired, brown above, lighter beneath; not sharply contrasted as in *O. fulvescens*.

DISTRIBUTION: *Oligoryzomys messorius* is distributed primarily in higher elevations of southern Venezuela, Guyana, Surinam, and northeastern Brazil.

SELECTED LOCALITIES (Map 231): BRAZIL: Amapá, Tartarugalzinho (MNRJ 35967); Roraima, Parque Nacional do Viruá (MNRJ 70475), Rio Cotinga, Limão (AMNH 75411). GUYANA: Upper Takutu–Upper Essequibo, near Awarawaunowa (ROM 38522), near Dadanawa (ROM 36021), Kanuku Mountains (type locality of *Oryzomys navus messorius* Thomas), Nappi Creek (ROM 31529), 5 km SE of Surama (ROM 103348). SURINAM: Sipaliwini, Sipaliwini airstrip (CMNH 76892). VENEZUELA: Amazonas, Pozón, 50 km NE of Puerto Ayacucho (ROM 107871); Bolivar, Auyán-Tepuí (AMNH 130883), Mount Roraima, Arabopo (AMNH 75662).

SUBSPECIES: *Oligoryzomys messorius* is monotypic.

NATURAL HISTORY: Nothing is known about the ecology, population biology, or behavior of this species.

REMARKS: We recognize *O. messorius* as a valid species distinct from *O. delicatus* based on morphological differences of specimens of both taxa in the AMNH and

MNRJ collections (M. Weksler and colleagues, unpubl. data), karyotypic data ($2n=66$, FN = 74) of specimens identified as *Oligoryzomys* sp. by Andrades-Miranda, Oliveira, Lima-Rosa et al. (2001; see also R. J. Baker et al. 1983 for specimens from Surinam with $2n=66$), and the phylogenetic position relative to other species in the genus (G. B. Miranda et al. 2009; Agrellos et al. 2012). Based on our ongoing morphological, karyotypic, and molecular analyses, the specimen reported as *Oligoryzomys* cf. *messorius* by Andrades-Miranda, Oliveira, Lima-Rosa et al. (2001), with karyotype of $2n=56$, FN = 58, is distinct from the concept of *O. messorius* we espouse herein.

Oligoryzomys microtis (J. A. Allen, 1916)
Small-eared Colilargo

SYNONYMS:

Oryzomys (Oligoryzomys) microtis J. A. Allen, 1916c:525; type locality "Lower Rio Solimoens [= Solimões] (fifty miles above mouth)," Amazonas, Brazil; restricted to Manacapuru, north bank of lower Rio Solimões, 70 km WSW Manaus (Voss et al. 2001:118–119, footnote).

Oryzomys chaparensis Osgood, 1916:205; type locality "Todos Santos, Chaparé River, [Cochabamba,] Bolivia. Altitude about 1200 feet."

Oligoryzomys microtus Gyldenstolpe, 1932a:27; incorrect subsequent spelling of *Oryzomys (Oligoryzomys) microtis* J. A. Allen.

Oryzomys [(Oligoryzomys)] microtis: Ellerman, 1941:355; name combination.

Oligoryzomys microtis: Carleton and Musser, 1989:71; first use of current name combination.

DESCRIPTION: Similar in size and general coloration to *O. delicatus* but very much less rufous above and clearer white below, and with very much smaller ears. Upper parts dull yellowish brown, finely lined with black-tipped hairs; under parts white (in holotype, but faintly tinged with buff in some specimens); upper surfaces of feet thinly clothed with light buffy brown hairs; soles dark brown; tail shorter than head and body, brown, nearly naked, finely annulated, with tip not distinctly tufted; ears small, brownish, and nearly naked. Patton et al. (2000:Table 33) provided mean, standard error, and range for external and cranial measurements for large series of adults from Rio Juruá in western Amazonian Brazil; total length 168–233 mm, tail length 80–125 mm, hindfoot length 21–26, and ear length 11–15 mm; series as small-bodied, long-tailed, and short-haired mouse, yellowish brown above and grayish white below; skull with elongated braincase, broad but long rostrum, and hourglass-shaped interorbital region with squared edges.

DISTRIBUTION: *Oligoryzomys microtis* occurs broadly in the Amazon Basin of eastern Peru, northeastern

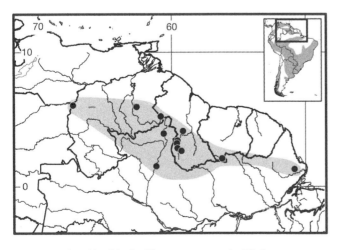

Map 231 Selected localities for *Oligoryzomys messorius* (●). Contour line = 2,000 m.

Bolivia, and western Brazil, extending eastward along the floodplain of the Rio Amazonas in Pará state.

SELECTED LOCALITIES (Map 232): BOLIVIA: Beni, El Triunfo (USNM 391298), La Penas (USNM 460741); Cochabamba, Todos Santos (type locality of *Oryzomys chaparensis* Osgood); La Paz, Mapiri (AMNH 72697); Pando, La Cruz (AMNH 262901), Pemanso (AMNH 262912), Río Nareuda (AMNH 248982); Santa Cruz, El Refugio (AMNH 268971). BRAZIL: Acre, Igarapé Porongaba, right bank Rio Juruá (Patton et al. 2000); Amazonas, Jainú, right bank Rio Juruá (Patton et al. 2000), Manacaparu, north bank Rio Solimões, 70 km WSW of Manaus (type locality of *Oryzomys* [*Oligoryzomys*] *microtis* J. A. Allen), Parintins (AMNH 92899), Rio Madeira, Rosarinho, Lago Miguel (AMNHN 92705), Seringal Condor, left bank Rio Juruá (Patton et al. 2000); Pará, Belém (USNM 461069), Capim, 150 mi SE of Belém (AMNH 188964), Marabá, Serra do Norte (USNM 543345), Villarinho do Monte, Rio Xingu (AMNH 95984). PERU: Loreto, Jenaro Herrera (Aniskin 1994), Mishana Allpahuayo (Aniskin 1994); Madre de Dios, Río Manu, 57 km above mouth (USNM 559399), Río Tambopata, 30 km above mouth (USNM 530925); Pasco, ca. 10 km N of Puerto Bermudez (AMNH 245551); Ucayali, Balta, Río Curanja (LSUMZ 14360).

SUBSPECIES: We treat *Oligoryzomys microtis* as monotypic.

NATURAL HISTORY: *Oligoryzomys microtis* is a terrestrial inhabitant of edge habitats within the lowland rainforest of the Amazon Basin. Population genetic analyses of *O. microtis* from the Brazilian Amazon (Patton et al. 1996, 2000) showed weak isolation by distance occurring only over large geographical distances. Migration rates (*Nm*) were high, and most (about 80%) of haplotype variation was observed within local populations. Specimens of *O. microtis* captured in Bolivia and Peru were positive for the Río Mamoré hantavirus (Carroll et al. 2005; Powers et al. 1999; Richter et al. 2010).

Pregnant females were recorded in March, May, August, and September in Bolivia; the number of embryos vary from three to five (S. Anderson 1977). Along the Rio Juruá in the western Amazon of Brazil, individuals reached reproductive maturity quickly, some while still in juvenile pelage, pregnancy rates were high (with litter size, based on embryo counts, ranging from two to eight [modal number = 4]), and reproduction was recorded throughout the dry season. They inhabit mainly seasonally flooded habitats along river margins (*várzea*) exposed during the dry season, but also occur in *terra firme* forest (Patton et al. 2000).

REMARKS: A. L. Gardner and Patton (1976) reported a karyotype of $2n = 64$, $FN = 66$ for specimens from eastern Peru (as *Oryzomys* (*Oligoryzomys*) *longicaudatus*, variant

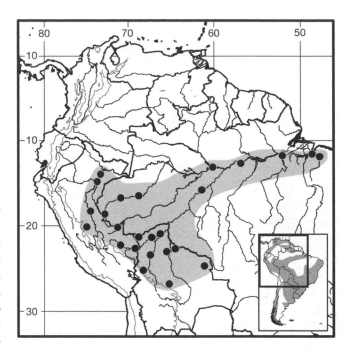

Map 232 Selected localities for *Oligoryzomys microtis* (●). Contour line = 2,000 m.

2). Aniskin and Volobouev (1999) and Patton et al. (2000) both reported the same karyotype from additional localities in the Peruvian and the western Brazilian Amazon, respectively. Other karyotypes are found in colilargos from the Brazilian Amazon within the geographic range of *O. microtis* (C. R. Bonvicino, unpubl. data), and it is likely that this is a karyotypically polytypic species.

Voss et al. (2001:118–119, footnote) discussed the location of the type locality, north of the Rio Solimões, and the lack of morphological differentiation between the type series and topotypes of *O. fulvescens*. *Oligoryzomys* species, however, are very similar to one another, and we cannot discard that such is the case for *O. microtis* and *O. fulvescens*. Additional morphological comparisons are necessary to resolve this problem, and we keep the name *microtis* in this account. This species needs an in-depth taxonomic assessment.

Oligoryzomys moojeni Weksler and Bonvicino, 2005
Moojen's Colilargo

SYNONYM:

Oligoryzomys moojeni Weksler and Bonvicino, 2005:116; type locality "Fazenda Fiandeira in 'Morro do Chapéu' region, in the lowest part of the Chapada dos Veadeiros National Park, 65 km SSW Cavalcante (14°04'S 47°45'W, altitude ranging from 550 to 740m), Goiás State, Brazil."

DESCRIPTION: Medium sized, characterized by brown-orange dorsal pelage that does not contrast sharply with creamy ventral pelage, notably short incisive foramina, and second highest diploid number ($2n = 70$) for genus.

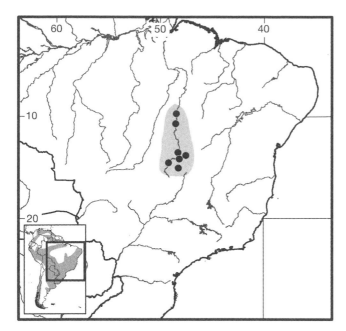

Map 233 Selected localities for *Oligoryzomys moojeni* (●).

DISTRIBUTION: Central Brazil, in the Cerrado of Goiás and Tocantins states, at elevations below 800 m.

SELECTED LOCALITIES (Map 233; from Weksler and Bonvicino 2005, except as noted): BRAZIL: Goiás, Cavalcante, Colinas do Sul, Mimoso de Goiás, Minaçu, Uruaçu; Tocantins, Lajeado (J. F. S. Lima et al. 2003), Porto Nacional (J. F. S. Lima et al. 2003).

SUBSPECIES: *Oligoryzomys moojeni* is monotypic.

NATURAL HISTORY: *Oligoryzomys moojeni* occurs at elevations between 500 and 800 m in the Brazilian Cerrado, primarily in open vegetation formations such as *cerrado sensu stricto* and *campo úmido*, but also in open gallery forest (Weksler and Bonvicino 2005). Two pregnant females were captured at the type locality in August (dry season), each with three embryos.

REMARKS: The karyotype of *Oligoryzomys moojeni* is characterized by $2n=70$ and FN$=74$–76; variation in the autosomal fundamental number is due to a pericentric inversion affecting a medium chromosome pair (J. F. S. Lima et al. 2003). In their respective karyotypic analyses, Andrades-Miranda, Oliveira, Lima-Rosa et al. (2001), and F. J. S. Lima et al. (2003) referred to this species as *Oligoryzomys* sp.

Oligoryzomys nigripes (Olfers, 1818)
Black-footed Colilargo
SYNONYMS:

Mus nigripes Olfers, 1818:209; type locality fixed by neotype designation to "Ybycuí National Park, Department Paraguarí, approximately 85 km SSE Atyrá, Paraguay" (Myers and Carleton 1981:14).

Mus longitarsus Rengger, 1830:232; type locality "Ufer des Paraguaystromes, nördlich von Villa-Real."

Hesperomys eliurus Wagner, 1845a:147; type locality "Ytarare [= Itararé]," São Paulo, Brazil.

Hesperomys pygmaeus Wagner, 1845a:147; type locality "Ypanema" (= Floresta Nacional de Ipanema, 20 km NW Sorocaba, São Paulo, Brazil, 23°26′7″S 47°37′41″W, 701 m; L. P. Costa et al. 2003).

Hesperomys (Calomys) pygmaeus: Burmeister, 1854:173; name combination.

Oryzomys pygmaeus: E.-L. Trouessart, 1897:527; name combination.

Oryzomys nigripes: E.-L. Trouessart, 1897:528; name combination.

Oryzomys eliurus: E.-L. Trouessart, 1897:528; name combination.

Oryzomys longitarsus: E.-L. Trouessart, 1897:528; name combination.

Oryzomys delticola Thomas, 1917c:96; type locality "Isla Ella, in the delta of the Rio Parana, at the top of the La Plata Estuary," Buenos Aires, Argentina.

Oryzomys (Oligoryzomys) delticola: Tate, 1932f:10; name combination.

Oryzomys (Oligoryzomys) longitarsus: Tate, 1932f:10; name combination.

Oryzomys (Oligoryzomys) nigripes: Tate, 1932f:10; name combination.

Oryzomys (Oligoryzomys) eliurus: Tate, 1932f:10; name combination.

Oryzomys (Oligoryzomys) pygmaeus: Tate, 1932f:10; name combination.

Oligoryzomys eliurus: Carleton and Musser, 1989:73; name combination.

Oligoryzomys nigripes: Carleton and Musser, 1989:73; first use of current name combination.

Oligoryzomys delticola: Carleton and Musser, 1989:75; name combination.

DESCRIPTION: Large, characterized by dark-brown to dark-yellowish dorsal pelage color, with sharply defined transition to whitish ventral coloration, and often with orange pectoral band; long ears; $2n=62$, FN$=78$–82 karyotype.

DISTRIBUTION: *Oligoryzomys nigripes* occurs in eastern Brazil (from Pernambuco in the north to Rio Grande do Sul in the south along the coast, and in the southern portion of the Cerrado in Minas Gerais and São Paulo states and the Distrito Federal), Paraguay (east of the Río Paraguay), Uruguay (Salto, Durazno, Colonia, and Maldonado departments), and Argentina (Buenos Aires, Misiones, and Chaco provinces).

SELECTED LOCALITIES (Map 234; from Weksler and Bonvicino 2005, except as noted): ARGENTINA: Buenos

Aires, Isla Ella, delta of Río Paraná (type locality of *Oryzomys delticola* Thomas); Chaco, Las Palmas; Misiones, Caraguatay. BRAZIL: Bahia, Rio Una, 10 km ESE of São José; Distrito Federal, Brasília; Espírito Santo, Venda Nova; Goiás, Flores de Goiás; Minas Gerais, Fazenda Canoas, Juramento, Passos, Peirópolis; Paraíba, Pirauá; Pernambuco, Bom Conselho, Buíque; Rio de Janeiro, Itaguaí, Nova Friburgo; Rio Grande do Sul, Faxinal, Pontal do Morro Alto; Santa Catarina, Florianópolis, Itá; São Paulo, Araraquara, Guaratuba, Iguapé, Itapetininga, Pedreira. PARAGUAY: Amambay, 4 km (by road) SW of Cerro Corá (Myers and Carleton 1981), 28 km SW of Pedro Juan Caballero (Myers and Carleton 1981); Caaguazú, Sommerfield Colony; Canindeyú, 6.3 km by road N of Curuguaty; Central, Asunción; Itapúa, Encarnación; Misiones, San Francisco. URUGUAY: Durazno, Durazno; Maldonado, Punta del Este; Salto, Salto.

SUBSPECIES: *Oligoryzomys nigripes* is monotypic.

NATURAL HISTORY: *Oligoryzomys nigripes* had an average of 4.7 embryos (range: 4–6) in 21 pregnant females from the Brazilian states of Rio de Janeiro and São Paulo collected in September and November (C. R. Bonvicino, unpubl. data). Myers and Carleton (1981) reported an average of 3.57 embryos (range 2 to 5) in 32 pregnant females from Paraguay collected around June and August (with a hiatus in July). *Oligoryzomys nigripes* is the most habitat-generalist of all Brazilian *Oligoryzomys* species, occurring in primary and secondary vegetation in Atlantic Forest and Cerrado. It can be sympatric with *O. stramineus* (see Bonvicino and Weksler 1998), though the two species have never been caught in the same trapline. It can be syntopic with *O. flavescens* and *O. mattogrossae*. *Oligoryzomys nigripes* is more commonly trapped on the ground, but Myers and Carleton (1981) mentioned specimens collected in trees and suggested some arboreal activity. Püttker, Pardini et al. (2008) found the species at sites with a low canopy and a dense understory. In southern Brazil, population levels peaked in the cold and dry months and reached null values during the warmest months, with the population increase connected to greater availability of food resources, especially of grass seeds (Antunes et al. 2009). This colilargo can reach a density of 33 animals per hectare during population explosions (study population in Buenos Aires province, Argentina; Jaksic and Lima 2003). In the Rio de la Plata delta (and probably elsewhere), *O. nigripes* is prey of the Barn Owl (*Tyto alba*) and host to the ectoparasitic mites *Haemolaelaps glasgowi*, *Gigantolaelaps mattogrossensis*, *Laelaps paulistanensis*, and *Mysolaelaps microspinosus* (see Massoia and Fornes 1964b). The species is also host to the nematodes *Stilestrongylus lanfrediae* in Rio de Janeiro, Brazil (J. G. R. Souza et al. 2009) and *Syphacia kinsellai* in Misiones, Argentina (M. Robles and Navone 2007), and individuals are a reservoir for the Araucaria hantavirus (R. C. Oliveira et al. 2011).

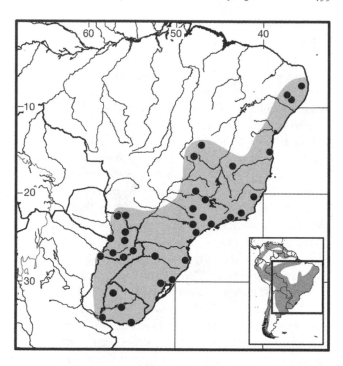

Map 234 Selected localities for *Oligoryzomys nigripes* (●). Contour line = 2,000 m.

REMARKS: The name *nigripes* was first given to Azara's (1801:98) *le rat à tarse noir* by Olfers (1818:209; see Desmarest 1819f; Hershkovitz 1959b; Myers and Carleton 1981). As no type specimen was selected in the original or in subsequent publications, Myers and Carleton (1981:14) designated a neotype from Ybycuí National Park, 85 km SSE Atyrá (which is approximately where Azara obtained his specimens), in the department of Paraguaí, Paraguay.

Oligoryzomys nigripes has a diploid number of 62 chromosomes but a variable autosome fundamental number ranging from 79 to 82 due to pericentric inversions (Bonvicino, D'Andrea, and Borodin 2001). Because Bonvicino and Weksler (1998) and Weksler and Bonvicino (2005) were unable to separate samples of *O. nigripes* from those of *delticola* Thomas or *eliurus* Wagner on morphological and karyological grounds, we consider both *delticola* and *eliurus* as equivalent to *O. nigripes* herein. Recent analyses of mtDNA sequence data (D-loop and cytochrome-*b*) corroborate that *delticola* is best considered a synonym of *O. nigripes* (Francés and D'Elía 2006; Rivera et al. 2007). L. F. Machado et al. (2011) presented further characters that differentiate sympatric *O. nigripes* from *O. flavescens*.

Oligoryzomys rupestris Weksler and Bonvicino, 2005
Highlands Colilargo

SYNONYM:

Oligoryzomys rupestris Weksler and Bonvicino, 2005:119; type locality "Pouso Alto (14°01′S 47°31′W), in the highest part of the Chapada dos Veadeiros National Park, at 1,500m altitude, 14km NNW of Alto Paraíso, Goiás State, Brazil."

DESCRIPTION: Small sized, characterized by gray head that contrasts with lighter yellow-brownish dorsal body coloration; small tufts of whitish hairs anterior to base of pinna; reduced or absent stapedial foramen, squamosal-alisphenoid groove, and sphenofrontal foramen that characterizes carotid circulatory pattern 3 (*sensu* Voss 1988); and lowest known diploid number ($2n=44$–46) among species in the genus.

DISTRIBUTION: *Oligoryzomys rupestris* occurs in central Brazil, where it is known from only three localities, one each in Goiás, Minas Gerais, and Bahia states, at elevations above 1,000m.

SELECTED LOCALITIES (Map 235): BRAZIL: Bahia, Pico das Almas (MZUSP 29015); Goiás, Pouso Alto (type locality of *Oligoryzomys rupestris* Weksler and Bonvicino); Minas Gerais, Serra do Cipó (MZUSP 27423).

SUBSPECIES: *Oligoryzomys rupestris* is monotypic.

NATURAL HISTORY: This species occurs only in high-elevation areas characterized by *campos rupestre* vegetation in the Cerrado Biome (Weksler and Bonvicino 2005). No female was found pregnant in November (end of dry season) at the type locality (Bonvicino, Lemos, and Weksler 2005). No other natural history information is available.

REMARKS: *Oligoryzomys rupestris* has the lowest diploid number of all species in the genus, with $2n=44$ to 46 and an autosomal fundamental number varying from 52 to 53 (M. J. de J. Silva and Yonenaga-Yassuda 1997; as *Oligoryzomys* sp. 1 and sp. 2). The derived carotid circulation

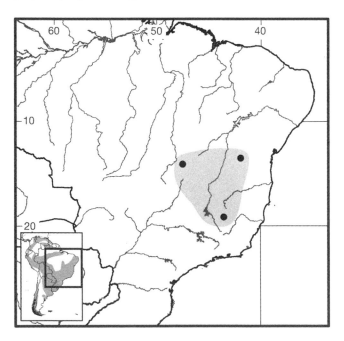

Map 235 Selected localities for *Oligoryzomys rupestris* (●). Contour line = 2,000 m.

pattern (pattern 3 of Voss 1988) of *O. rupestris* is unique among *Oligoryzomys* species.

Oligoryzomys stramineus Bonvicino and Weksler, 1998
Straw-colored Colilargo

SYNONYM:

Oligoryzomys stramineus Bonvicino and Weksler, 1998: 98; type locality "Fazenda Vão dos Bois (13°34′29″S 47°10′57″W, 424 m), Terezina de Goiás, Goiás State, Brazil, 24 km N of Terezina de Goiás, 15 km SW of Rio Paranã, a tributary of the upper Rio Tocantins, road GO-118, km 275."

DESCRIPTION: Large species characterized by pale dorsal color, with defined transition between lateral and whitish ventral pelage, long incisive foramina, broad zygomatic plates, and $2n=52$, FN = 68–70 karyotype.

DISTRIBUTION: *Oligoryzomys stramineus* occurs in central and northeastern Brazil, from Goiás and Minas Gerais states in the south to Ceará, Piauí, Paraíba, and Pernambuco states in the north, in both Cerrado and Caatinga vegetational domains.

SELECTED LOCALITIES (Map 236): BRAZIL: Bahia, Caetité (MNRJ 63416); Ceará, Russas (Fernandes et al. 2012), Santanopole (USNM 304583); Goiás, Fazenda Vão dos Bois, Teresina de Goiás (type locality of *Oligoryzomys stramineus* Bonvicino and Weksler), Mambaí (Weksler and Bonvicino 2005); Minas Gerais, Fazenda Canoas, Juramento (Weksler and Bonvicino 2005); Paraíba, Pirauá (UFPB [LFS 49]); Pernambuco, Agrestina (UFPE [PMN 62]), Buíque (UNPE [PMN 147]), Correntes (Weksler and Bonvicino 2005), Exu (Weksler and Bonvicino 2005).

SUBSPECIES: *Oligoryzomys stramineus* is monotypic.

NATURAL HISTORY: *Oligoryzomys stramineus* is primarily found in gallery forests within the Cerrado. Young juveniles were collected in August, suggesting that reproduction occurred around June–July; one pregnant female was captured in September with four embryos (Weksler and Bonvicino 2005). This species is sympatric with *O. fornesi* (with both obtained in the same trapline) and *O. nigripes* (but never in the same trapline) at some localities in central Brazil (Weksler and Bonvicino 2005).

REMARKS: The karyotype of *Oligoryzomys stramineus* is characterized by $2n=52$, FN = 68 to 69 (Furtado 1981; Maia et al. 1983; both as *Oligoryzomys* aff. *eliurus*). Differences in autosomal fundamental number are due to an inversion in a small acrocentric pair (Bonvicino and Weksler 1998). See remarks in the account for *Oligoryzomys chacoensis* for the confusion over the distributional range of that species due to misidentification with specimens of *O. stramineus* (e.g., Carleton and Musser 1989:72).

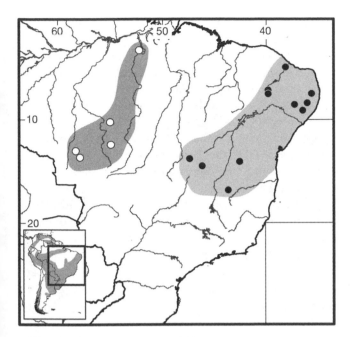

Map 236 Selected localities for *Oligoryzomys stramineus* (●) and *Oligoryzomys utiaritensis* (○). Contour line = 2,000 m.

Oligoryzomys utiaritensis J. A. Allen, 1916
Utiariti Colilargo

SYNONYMS:

Oryzomys (*Oligoryzomys*) *utiaritensis* J. A. Allen, 1916c: 527; type locality "Utiarity [= Utiariti], Rio Papagaio, Matto Grosso, Brazil."

Oligoryzomys utiaritensis: Gyldenstolpe, 1932a:29; first use of current name combination.

Oryzomys [(*Oligoryzomys*)] *utiaritensis*: Ellerman, 1941: 356; name combination.

Oligoryzomys eliurus: Carleton and Musser, 1989:73; part; not *Hesperomys eliurus* Wagner.

DESCRIPTION: Upper parts clay-colored, finely lined with black-tipped hairs over back, nearly pure clay-color on sides; front of head and nose darker and slightly grayish; under parts grayish white, hairs darker at base; upper surfaces of forefeet and hindfeet paler, about flesh color, soles dark brown; ears large, light brown, nearly naked; and tail longer than head and body, finely annulated, light brown, and nearly naked, apical portion with short, bristly grayish hairs. Agrellos et al. (2012) diagnosed *O. utiaritensis* by the following characters: grizzled yellowish-brown dorsal pelage contrasting with whitish ventral pelage and tail weakly bicolored; long incisive foramina with posterior borders reaching or almost reaching alveoli of first upper molars, but never extending posteriorly; highest diploid number for genus ($2n = 72$); and three putative synapomorphies in mtDNA cytochrome-*b* and one in nuclear intron 7 of beta-fibrinogen gene sequences.

DISTRIBUTION: *Oligoryzomys utiaritensis* is distributed across the northwest of Mato Grosso and southwest of Pará states, an area that includes the Chapada dos Parecis, a massive plateau in northwestern Mato Grosso that marks the transition between the Cerrado and Amazonian domains in central Brazil.

SELECTED LOCALITITES (Map 236): BRAZIL: Pará, Castelo dos Sonhos, Altamira (MNRJ 75609); Mato Grosso, Campo Novo do Parecis (MNRJ 75616), Feliz Natal (MNRJ [SVS 819]), Peixoto de Azevedo (MNRJ 75600), Utiariti, Rio Papagaio (type locality of *Oryzomys* [*Oligoryzomys*] *utiaritensis* J. A. Allen).

SUBSPECIES: *Oligoryzomys utiaritensis* is monotypic.

NATURAL HISTORY: Specimens of *Oligoryzomys utiaritensis* were captured between 110 and 570 m elevation, in secondary semideciduous forest at the limits of corn and soy plantations, in eucalypt plantations with herbaceous vegetation, and in very altered vegetation near plantations (Agrellos et al. 2012). This species is sympatric with *O. mattogrossae* at the type locality (J. A. Allen 1916c). Individuals serve as a reservoir of the Castelo dos Sonhos hantavirus (E. S. T. Rosa et al. 2011).

REMARKS: Recent karyotypic, morphologic, and molecular analyses (Agrellos et al. 2012) demonstrated that *O. utiaritensis* and *O. nigripes* (= *O. eliurus*; see the species account) were different species. *Oligoryzomys utiaritensis* had a karyotype with $2n = 72$ and FN = 76, unlike *O. nigripes* with $2n = 62$, FN = 78–82. Analyses of mtDNA and nucDNAmarkers indicated that *O. nigripes* and *O. utiaritensis* were not sister taxa.

Genus *Oreoryzomys* Weksler, Percequillo, and Voss, 2006
Alexandre R. Percequillo

Oreoryzomys balneator, the sole species in the genus, is endemic to high-elevation montane forests in Ecuador and northern Peru. It is currently known from very few specimens and localities.

Externally, *O. balneator* has a small body size, with head and body length ranging from 75–94 mm; the tail is longer than head and body (tail length 95–120 mm); hindfeet are narrow and delicate (length 23.5–27 mm); and ears are short and rounded. The dorsal pelage is dense and compact, with short hairs (aristiforms reaching 12 mm in length), and darkly colored, ochraceous densely grizzled with dark brown. The ventral pelage is shorter, sharply distinct from the dorsum and sides of the body in color, grayish to buffy gray. The tail is whitish along its basal half, with the distal half uniformly dark brown; the tail tip may be white in some specimens. Hindfeet are covered dorsally with short, bicolored (brown proximally and white distally), and self-colored hairs (entire white); ungual tufts are short, and do not conceal the claws. Mystacial vibrissae are dense, thick, and long, surpassing the tip of ears when laid back.

The skull of *O. balneator* is very small and delicate (greatest length range: 22.2–25.7 mm). The rostrum is long and narrow, flanked by small and discrete nasolacrimal foramina, and has very shallow and narrow zygomatic notches. The zygomatic plates are narrow, with their anterior margins straight; zygomatic arches are slightly divergent posteriorly, and jugals are absent. The interorbital region is anteroposteriorly short and wide, with an hourglass shape; the anterior half has rounded supraorbital margins, and the posterior half shows squared margins. The braincase is rounded and globose, without temporal crests; in occipital view, the braincase is well rounded, and both the foramen magnum and occipital condyles are oriented more ventrally. Incisive foramina are short (length 2.8–3.8 mm), with convex lateral margins and rounded anterior and posterior margins. The palate is long and wide, with or without small and inconspicuous palatal excrescences; the posterolateral palatal pits are large and deep. The anterior border of mesopterygoid fossa is rounded; its roof is perforated by long and narrow sphenopalatine vacuities present at the presphenoid-basisphenoid suture. An alisphenoid strut is absent. The stapedial foramen, squamosoalisphenoid groove, and sphenofrontal foramen are present, configuring carotid circulatory pattern 1 (*sensu* Voss 1988). Both the subsquamosal fenestra and postglenoid foramen are large. Auditory bullae are small but globose, with long and flat bony tubes; the stapedial process is small; the tegmen tympani varies in size, and may or may not marginally overlap the squamosal.

The mandible is shallow; the coronoid process is very short and triangular, and positioned lower than the condyloid process, thus defining an extremely shallow superior notch. The angular process is shorter than the condyloid process, so the inferior notch is also shallow. A capsular process is well developed.

Upper incisors are orthodont. Molars are pentalophodont, although some specimens may have tetralophodont lower molars, with a reduced to absent mesolophid. Molars have low crowns with the main cusps arranged in opposite pairs. Labial and lingual flexi interpenetrate moderately at the median molar plane; the labial flexus is deep and wide, while the lingual flexus is relatively shorter. The anterocone of M1 is divided by a deep anteromedian flexus; the paracone is connected to the protocone by its medial portion, configuring a very long mesoflexus and thus a long and narrow mesoloph (the anteroloph is also very long); with wear, it forms a long mesofossette. M2 is similar in shape and occlusal details, except for the absence of anterocone, but the anteroloph is well developed. M3 is

smaller, but similar to the other teeth. A deep anteroflexid is present on m1 as is a long and narrow mesolophid (although this latter structure is absent in one specimen). The second lower molar exhibits individual variation in the presence of a mesolophid, conspicuous in most specimens, but others lack a noticeable trace, even in unworn molars.

SYNONYMS:

Oryzomys: Thomas, 1900c:273; part; not *Oryzomys* Baird.
Oreoryzomys Weksler, Percequillo, and Voss, 2006:21; type species *Oryzomys balneator* Thomas, by original designation.

REMARKS: *Oreoryzomys* is recovered as the sister taxon to *Microryzomys* in phylogenetic analyses of oryzomyine rodents based on molecular and morphological data (Weksler 2003, 2006). In fact, the two genera are morphologically similar (as first observed by M. D. Carleton, pers. comm.); both share a deep anteromedian flexus on M1, a short, convexly shaped incisive foramina, type 1 carotid circulatory pattern, a rounded interorbital region hourglass in shape, a rounded brain case, and a posteroventral oriented foramen magnum.

Oreoryzomys balneator (Thomas, 1900)
Ecuadorean Oreoryzomys

SYNONYMS:

Oryzomys balneator Thomas, 1900c:273; type locality "Mirador," Baños, Tungurahua, Ecuador.
Oryzomys ([*Oryzomys*]) *balneator*: E.-L. Trouessart, 1904: 418; name combination.
Oryzomys balneator hesperus Anthony, 1924b:7; type locality "El Chiral, Western Andes, Provincia del Oro, Ecuador, elevation 5350 ft."
Oryzomys balneator balneator: Gyldenstolpe, 1932a:13; name combination.
[*Oreoryzomys*] *balneator*: Weksler, Percequillo, and Voss, 2006:21; first use of current name combination.

DESCRIPTION: As for the genus.

DISTRIBUTION: This species is known from a few localities on both western and eastern slopes of the Andes in Ecuador and northern Peru, at intermediate elevations (from 1,500 to 1,586 m).

SELECTED LOCALITIES (Map 237): ECUADOR: El Oro, El Chiral (*Oryzomys balneator hesperus* Anthony); Napo, Baeza (BM 15.7.12.9); Tunguruhua, Mirador (type locality of *Oryzomys balneator*); Zamora-Chinchipe, 4 km E of Sabanilla (USNM 513570). PERU: Cajamarca, 4 km W of Chaupe (AMNH 268144).

SUBSPECIES: I treat *Oreoryzomys balneator* as monotypic.

NATURAL HISTORY: This species occurs in Andean montane and cloud forest. According to Musser et al. (1998), Mirador, the type locality of *O. balneator*, has a

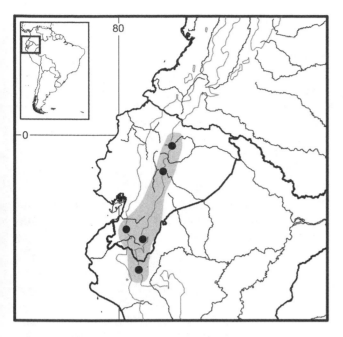

Map 237 Selected localities for *Oreoryzomys balneator* (●). Contour line = 2,000 m.

matrix of vegetation with sharp contrasts, including *páramo* mixed with temperate plants in the valleys interdigitating with extensive areas of subtropical forest. The climate is rainy, cold, and windy. There are no published details on the habitat where specimens were caught, but museum skin labels (BM) indicate that individuals were "trapped amidst the roots on a tree," found "at old stump," "among dead logs," and "in woods," suggesting that they are terrestrial. No other information on natural history is available.

REMARKS: The validity of *Oryzomys balneator hesperus* Anthony remains to be determined, pending examination of the type material. Until that definition has been achieved, I provisionally recognize Anthony's name as a synonym of *Oreoryzomys balneator*. It is noteworthy that the type and only known locality of *hesperus* represents the single western slope site for *O. balneator*, which is otherwise distributed along the eastern Andean gradient.

Genus *Oryzomys* Baird, 1857

Alexandre R. Percequillo

Until very recently, the genus *Oryzomys* was one of the most taxonomically diverse of the many genera within the subfamily Sigmodontinae, comprising some 43 species distributed from the southern United States in North American to northern Argentina in South America (Musser and Carleton 2005). However, the extensive taxonomic rearrangement proposed by Weksler et al. (2006), based on morphological (Percequillo 1998, 2003; Weksler 2006) and

molecular analyses (Weksler 2003, 2006), restricted the genus *Oryzomys* in its taxonomic content and thus substantially narrowed its geographic range, especially in South America. Currently five species are attributed to this genus (Musser and Carleton 2005; Voss and Weksler 2009). Three of these (*O. antillarum*, *O. dimidiatus*, and *O. palustris*) are restricted to North and Central America (see Hall 1981; Weksler et al. 2006) and are not considered further here. Two species either extend into northwestern South America (*O. couesi*) or are limited to this continent (*O. gorgasi*).

The South American species of genus *Oryzomys* (based on Hall 1981; H. Sánchez et al. 2001) are medium-sized oryzomyine rats, with the tail ranging from shorter to longer than head and body length, with conspicuous epidermal scales, and with ventral caudal hairs longer than dorsal ones and a small pencil of terminal hairs. Hindfeet are long and narrow, with webs of skin connecting digits II, III, and IV, and a fringe of stiff and short hairs along the metatarsal margins; ungual tufts are sparse and short, and do not conceal the claws; plantar surfaces are covered with squamae, and the hypothenar pads are small and inconspicuous. Ears are small and rounded. The pelage is harsh and dense, usually short; dorsal pelage is grizzled grayish brown to ochraceous tawny, mixed with black; color of the sides is paler and with less black; ventral pelage is white to ochraceous buff. The tail is distinctly bicolored at its base or throughout most of its length.

The skull is strongly built although not very long. The rostrum is short and strong, with deep and wide zygomatic notches and well-projected and inflated nasolacrimal capsules. The interorbital region is broad and strongly divergent posteriorly with heavily beaded supraorbital margins or with well-developed supraorbital crests; temporal crests are well developed and collinear with the supraorbital margins. Incisive foramina are very long and narrow, with their posterior margins penetrating between the first molars; lateral margins are widest medially. The palate is long and wide (*sensu* Hershkovitz 1962); posterolateral palatal pits are complex, numerous, and recessed in deep and rounded depressions. The anterior border of the mesopterygoid fossa is rounded, and its roof is completely ossified without a trace of sphenopalatine vacuities (except in *O. palustris*, which has very large and wide vacuities). The stapedial foramen is minute or absent, and both the squamosoalisphenoid groove and sphenofrontal foramen are absent, configuring carotid circulatory pattern 3 (*sensu* Voss 1988). The buccinator-masticatory and accessory oval foramina are confluent, thus an alisphenoid strut is absent. The postglenoid foramen is narrow, and the subsquamosal fenestra is a small aperture in most species, due to a broad hamular process of the squamosal, obliterated in both South American species. The suspensory process of the squamosal is absent and the tegmen tympani is short, and may or may not overlap the squamosal. Auditory bullae are small but globose, with a distinct dorsostapedial process that may or may not overlap the squamosal; bony tubes are abruptly constricted, and ectotympanic capsules are globose.

The mandible is strong, with or without the capsular projection of lower incisors; superior and inferior notches are shallow. Upper incisors are strongly opisthodont. Molars are pentalophodont, and brachydont; toothrows are parallel and appear smaller in relation to the remaining parts of the skull; main cusps are arranged in opposite pairs; and labial and lingual flexi interpenetrate only slightly at the median molar plane. Carleton and Musser (1989) described occlusal topology in detail. The first upper and lower molars have accessory roots.

SYNONYMS:

Mus: Harlan, 1837:385; part (description of *palustris*); not *Mus* Linnaeus.

[*Hesperomys*] *Oryzomys* Baird, 1857:458; as subgenus; type species *Mus palustris* Harlan, by monotypy.

Hesperomys: Alston, 1877:756; part (description of *couesi*); not *Hesperomys* Waterhouse.

Oryuzomys: Coues, 1890:4165; elevation to generic rank.

Nectomys: Thomas, 1905a:586; part (description of *dimidiatus*); not *Nectomys* Peters.

Micronectomys: Hershkovitz, 1948:55; *nomen nudum*.

REMARKS: The number of species within the genus *Oryzomys* is in flux. Weksler et al. (2006:4,Table 1) recognized five (*O. antillarum* Thomas, *O. couesi* Alston, *O. dimidiatus* Thomas, *O. gorgasi* Hershkovitz, and *O. palustris* Harlan). Carleton and Arroyo-Cabrales (2009) reviewed the *couesi* complex from western Mexico, and recognized three other species in addition to *O. couesi* (*O. albiventer* Merriam, *O. nelsoni* Merriam, and *O. peninsulae* Thomas). Most recently, Hanson et al. (2010), based on molecular phylogenetic analysis of samples from throughout the North and Middle American range, increased this number by four. These authors split *O. texensis* J. A. Allen from *O. palustris* and *O. mexicanus* J. A. Allen from *O. couesi* while identifying two undescribed species from Costa Rica and Panama. To date, no samples from South America have been included in any molecular phylogenetic study.

KEY TO THE SOUTH AMERICAN SPECIES OF *ORYZOMYS*:

1. Rostrum long and slender; incisive foramina wider posteriorly; hamular process of squamosal slender, configuring a large subsquamosal fenestra . . . .*Oryzomys couesi*
1′. Rostrum stout and blunt; incisive foramina wider medially, tapering posteriorly; hamular process of squamosal broad, configuring a small subsquamosal fenestra . . . .*Oryzomys gorgasi*

Oryzomys couesi (Alston, 1877)
Coues's Marsh Rice Rat
SYNONYMS:

Hesperomys couesi Alston, 1877:756; type locality "Cobán, Guatemala."

Oryzomys aquaticus J. A. Allen, 1891c:289; type locality "Brownsville, Cameron Co., Texas."

H[*esperomys*]. (*O*[*ryzomys*].) *couesi*: J. A. Allen, 1891c:290; name combination.

[*Oryzomys*] *Couesi*: Thomas, 1893b:403; first use of current name combination.

Oryzomys couesi: J. A. Allen, 1893:240; name combination.

Oryzomys fulgens Thomas, 1893b:403; type locality "Mexico."

Oryzomys mexicanus J. A. Allen, 1897b:52; type locality "Had. San Marcos, 3550 ft., Tonila, Jalisco," Mexico.

Oryzomys bulleri J. A. Allen, 1897b:53; type locality "Valle de Banderas, Terro Tepic, Jalisco, Mexico."

Oryzomys peninsulae Thomas, 1897:548; type locality "Santa Anita, Baja California," Mexico.

Oryzomys jalapae J. A. Allen and Chapman, 1897b:206; type locality "Japala, 4400 ft., Veracruz," Mexico.

Oryzomys cozumelae Merriam, 1901:103; type locality "Cozumel Island, Quintana Roo," Mexico.

Oryzomys albiventer Merriam, 1901:279; type locality "Ameca, 4000 ft., Jalisco," Mexico.

Oryzomys crinitus Merriam, 1901:281; type locality "Tlalpan, Distrito Federal, Mexico."

Oryzomys crinitus aztecus Merriam, 1901:282; type locality "Yautepec, Morelos," Mexico.

Oryzomys peragrus Merriam, 1901:283; type locality "Río Verde, San Luis Potosi," Mexico.

Oryzomys richmondi Merriam, 1901:284; type locality "Escondido River, 50 mi, above Bluefields, Nicaragua."

Oryzomys jalapae rufinus Merriam, 1901:285; type locality "Catemaco, 1000 ft., Veracruz," Mexico.

Oryzomys goldmani Merriam, 1901:288; type locality "Coatzacoalcos, Veracruz," Mexico.

Oryzomys rufus Merriam, 1901:287; type locality "Santiago, 200 ft., Nayarit," Mexico.

Oryzomys teapensis Merriam, 1901:286; type locality "Teapa, Tabasco," Mexico.

Oryzomys zygomaticus Merriam, 1901:285; type locality "Nenton, Guatemala."

Oryzomys molestus Elliot, 1903:145; type locality "Ocotlan, State of Jalisco, Mexico."

Oryzomys japalae apatelius Elliot, 1904a:266; type locality "San Carlos, State of Vera Cruz, Mexico."

[*Oryzomys* (*Oryzomys*)] *couesi*: E.-L. Trouessart, 1904:416; name combination.

Oryzomys richardsoni J. A. Allen, 1910b:99; type locality "Peña Blanca, Nicaragua."

Oryzomys gatunensis Goldman, 1912a:7; type locality "Gatún, Canal Zone, Panama."

Oryzomys couesi regillus Goldman, 1915:129; type locality "Los Reyes, Michoacán," Mexico.

Oryzomys couesi pinicola A. Murie, 1932:1; type locality "pine ridge, 12 mi. S El Cayo, Belize."

Oryzomys couesi lambi Burt, 1934:107; type locality "San José de Guaymas, Sonora," Mexico.

Oryzomys azuerensis Bole, 1937:165; type locality "Paracoté, 1½ mi. S mouth Río Angulo, Mariato-Suay lands, Veraguas, Panama."

Oryzomys palustris couesi: Hall, 1960:173; name combination.

Oryzomys [(Oryzomys)] palustris couesi: Hall, 1981:609; name combination.

DESCRIPTION: Large and robust skull, with short and wide rostrum; incisive foramina penetrate between procingula of first upper molars; posteriorly strongly divergent interorbital region with strongly beaded to shelved supraorbital margins; prominent posterolateral palatal pits recessed in medium to deep palatal depressions; and completely ossified roof of mesopterygoid fossa (see Weksler 2006). Hershkovitz (1987a) noted typical ochraceous orange color of body (dorsal, lateral, and ventral) throughout species' range and entirely whitish mystacial facial region.

DISTRIBUTION: This species is widespread throughout North and Central America (Hall 1981; Musser and Carleton 2005), with only a single record from South America, in northwestern Colombia (Hershkovitz 1987a).

SELECTED LOCALITIES (Map 238): COLOMBIA: Córdoba, Montería (Hershkovitz 1987a).

SUBSPECIES: Substantial subspecies diversity is currently recognized within the North and Central American range of *O. couesi* (Hall 1981, as *Oryzomys palustris*).

NATURAL HISTORY: Nothing is known about the natural history of this species in South America, aside from the comment made by Hershkovitz (1987a) that the single specimen he obtained was collected in a rice field.

REMARKS: Long considered a subspecies of a very wide-ranging *Oryzomys palustris* (e.g., Hall 1960, 1981), *O. couesi* is now widely recognized as a valid species based on a number of morphological, karyotypic, and molecular geographic analyses (see summaries in Musser and Carleton 2005; Hanson et al. 2010). Hershkovitz (1971) regarded *O. couesi* as distinct from *O. gorgasi*, based on comparisons he made in the original description of the latter. H. Sánchez et al. (2001) also compared *O. couesi* to *O. gorgasi*, pointing out their differences. However, none of theses authors directly compared the voucher of *O. couesi* (FMNH 127250) from Montería, Colombia, to the holotype or the Venezuelan series of *O. gorgasi*. Thus, there remains a remote possibility that only one species of *Ory-*

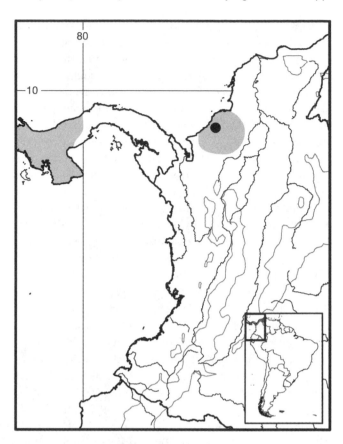

Map 238 Single known South American locality for *Oryzomys couesi* (●); range in Panama from Hall (1981). Contour line = 2,000 m.

zomys is present at South America. If this is correct, the appropriate name applied to the South American species would be *O. gorgasi* (see H. Sánchez et al. 2001).

The diploid number of the Colombian specimen is $2n = 56$ (Hershkovitz 1987a), the same as that recorded for specimens from Middle American localities (Haiduk et al. 1979; R. J. Baker et al. 1983; Burton et al. 1987).

Oryzomys gorgasi Hershkovitz, 1971
Gorgas's Marsh Rice Rat
SYNONYMS:

Oryzomys gorgasi Hershkovitz, 1971:700; type locality "Loma Teguerre (7°54′N, 77°W), Departamento Antioquia, northwestern Colombia, between Río Atrato and mouth of a *caño* (channel) of the east bank *ciénaga* just below and opposite Sautatá (Chocó), Interoceanic Canal Survey Route 25 (Golfo de Urabá [*sic*] to Bahía Humboldt, via Ríos Atrato-Truandó); altitude, 1 meter above sea level."

Oryzomys curasoae McFarlane and Debrot, 2001:182; type locality "Fissure, 30 m below edge of north face of Tafelberg Santa Barbara, (UTM coordinates: 19–518570, 13–34570)," Curaçao, Caribbean Netherlands.

DESCRIPTION: Same external and craniodental traits typical of genus *Oryzomys*, as presented earlier. H. Sánchez et al. (2001:210) stated that *O. gorgasi* differed from other species (specifically *O. couesi*, *O. dimidiatus*, and *O. palustris*) by presence of stout and blunt rostrum, posteriorly tapering incisive foramina, absence of sphenopalatine vacuities, and broad hamular process of squamosal, which reduces subsquamosal fenestra to small opening. However, Weksler (2006) scored the roof of mesopterygoid fossa of *O. couesi* as completely ossified, the same condition as *O. gorgasi*, thus suggesting that this trait may be polymorphic in samples of *O. couesi*. Consistency in the lack of sphenopalatine vacuities and thus the utility of an unossified mesopterygoid fossa roof as diagnostic of *O. gorgasi* warrant further evaluation by examination of more extensive series of specimens of all species in the genus.

DISTRIBUTION: *Oryzomys gorgasi* is known from two localities in the trans-Andean portion of northern South America, one each in Colombia and Venezuela. This species is not known to extend into Panama, despite the close proximity of the single Colombian locality to that country. It is apparently extinct from Curaçao, off the northwestern Venezuelan coast.

SELECTED LOCALITIES (Map 239): COLOMBIA: Antioquia, Loma Teguerre (Hershkovitz 1971). VENEZUELA: Zulia, El Caimito, Refugio de Fauna Silvestre y Reserva de Pesca Ciénaga de Los Olivitos, 40 km NE of Maracaibo (H. Sánchez et al. 2001).

NATURAL HISTORY: In Colombia, the only known specimen was collected in the "swamp, or ciénaga, forests in the Río Atrato basin" (Hershkovitz 1971:701) "at least 60 km upriver from the gulf of Urubá in an area of high annual rainfall, . . . a freshwater swamp" (H. Sánchez et al. 2001:210). The series from Venezuela were obtained at El Caimito, a small coastal island covered by xerophytic vegetation and by mangroves along swampy lagoon margins (H. Sánchez et al. 2001). Here, the species was omnivorous, feeding on crustaceans, insects, seeds, and other plant tissues. Endoparasites observed included the onchocercid filariod *Litomosoides sygmodontis* and the recitulariid genus *Pterygodermatites* (subgenus *Paucipectines*). Despite being common in Neotropical cricetids, neither oxyurid nor trychostrongylid nematode endoparasites were found in examined stomachs.

Oryzomys gorgasi is a species apparently uncommon in Venezuela, as considerable trapping effort at three localities near El Caimito has produced no other specimens (H. Sánchez et al. 2001). Instead, the black rat (*Rattus rattus*) was commonly captured, suggesting that this more aggressive exotic species may be competing with and displacing *O. gorgasi*. Similarly, Voss and Weksler (2009), based on paleontological (McFarlane and Debrot 2001) and ecological data (H. Sánchez et al. 2001) for *O. gorgasi*, suggested that black rats were possibly responsible for the extinction of this species on the island of Curaçao.

REMARKS: Since its description by Hershkovitz (1971) and, at the time, only known from the holotype, *O. gorgasi* has been considered a valid species, closely related to *O. palustris* and *O. couesi* (Honacki et al. 1982; Musser and Carleton 1993). In 2001 H. Sánchez et al. reported on a series of specimens from the coastal mangroves of northwestern Venezuela, supplemented Hershkovitz's description of *O. gorgasi*, and solidified its specific status relative to other species of *Oryzomys*. They noted, however, that the markedly wide rostrum of the holotype might be due to allometric growth of the facial region during captivity.

More recently, *Oryzomys curasoae*, an extinct species based on remains obtained from limestone fissures, was described from the continental shelf island of Curaçao off the northern Venezuelan coast (McFarlane and Debrot 2001). In the original description, these authors suggested that their species was close to *Oryzomys*, subgenus *Oecomys*, but refrained from addressing phylogenetic relationships. Subsequently, Voss and Weksler (2009), based on an extensive comparative analysis of several members of tribe Oryzomyini, constructed a phylogenetic hypothesis in which *O. curasoae* was nested with other species of *Oryzomys* (*sensu* Weksler et al. 2006), specifically as sister to *O. gorgasi*.

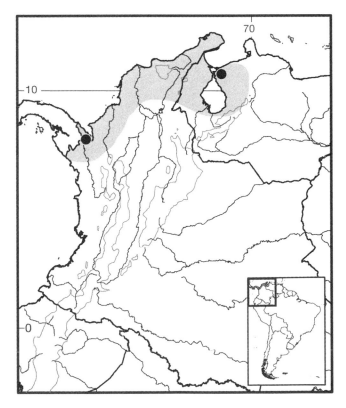

Map 239 Selected localities for *Oryzomys gorgasi* (●). Contour line = 2,000 m.

Because quantitative traits also highlight the similarity between these two forms, these authors considered both taxa as "strict synonyms," with priority for the name *O. gorgasi*. They limited the name *curasoae* "solely to refer to the subfossil material from Curaçao" (Voss and Weksler 2009:5).

Musser and Carleton (2005) incorrectly gave the type locality of *O. curasoae* as "Venezuela, Curaçao Isl." This island is, however, part of the Caribbean Netherlands, although on the continental shelf just north of the Venezuelan mainland (map, Fig. 1).

Genus *Pseudoryzomys* Hershkovitz, 1962
Robert S. Voss

The genus *Pseudoryzomys* contains a single species, *P. simplex*, which occurs in tropical and subtropical grassland habitats below 1,000 m elevation in Brazil, northern Argentina, western Paraguay, eastern Bolivia, and extreme southeastern Peru. Although Hershkovitz (1962), Reig (1984, 1986), and Braun (1993) classified *Pseudoryzomys* as a phyllotine, other morphological researchers have suggested that the genus is more closely related to oryzomyines (Olds and Anderson 1990; Voss and Myers 1991; Langguth and Neto 1993; Voss and Carleton 1993; Steppan 1995). Recent phylogenetic analyses (e.g., Weksler 2003, 2006; Percequillo, Weksler, and Costa 2011; Parada et al. 2013) unequivocally support the hypothesis that *Pseudoryzomys* is an oryzomyine closely related to *Holochilus*. Subfossil material of *Pseudoryzomys* is known from Brazilian cave deposits (Winge 1887)—some of which may date from the late Pleistocene—and from an archaeological site in Buenos Aires province, Argentina (Pardiñas 1995a). As far as known, the genus is not a member of any island fauna. Except as noted otherwise below, the following account is abstracted from Voss and Myers (1991), which should be consulted for additional details concerning taxonomy, morphology, and geographic distribution.

Pseudoryzomys simplex is a medium-sized (30–60 g; Pine and Wetzel 1976; J. R. Contreras and Berry 1982) sigmodontine with coarsely grizzled-brownish dorsal fur and gray-based yellowish or buffy ventral fur. The pinnae are small, covered with short hairs that are colored like the fur of the head, and do not appear naked. The hindfeet are long and narrow with the middle three digits (II–IV) much longer than the outer two (I and V); the heel is smooth and hairless, but the rest of the naked plantar surface is densely covered with small tubercles, among which one or two metatarsal pads (the hypothenar may be present or absent) and four very small interdigital pads can be distinguished. The tail, distinctly bicolored (dark above and pale below) is about as long as the combined length of the head and body; it is covered with short hairs, but the underlying

epidermal scales are clearly visible, and there is no terminal tuft of distinctly longer hairs. There are eight mammae in inguinal, abdominal, postaxial, and pectoral pairs.

The skull is distinctively proportioned in dorsal view, with a short rostrum, deep zygomatic notches, and a narrow interorbital region. The posterior supraorbital margins are anteriorly convergent and either sharp-edged or beaded; the zygomatic arches are widest across their squamosal roots and converge anteriorly. In lateral view, the zygomatic plates are very broad, with straight or concave anterior margins, and sometimes bearing blunt anterodorsal processes. The incisive foramina are narrow, parallel sided, and long, usually extending posteriorly to or between the anterocones of the first molars. The palate is smooth, without deep lateral gutters or median keels, and long (extending posteriorly well behind the third molars); prominent posterolateral palatal pits are present. The bony roof of the mesopterygoid fossa is perforated by large sphenopalatine vacuities; the parapterygoid fossae are shallow (not deeply excavated) and narrow (without expanded lateral margins). An alisphenoid strut separating the buccinator-masticatory foramen from the foramen ovale accessorius may be either present or absent. The carotid arterial pattern is derived (pattern 3 of Voss 1988). The auditory bullae are small, and the tegmen tympani does not overlap the squamosal. A very large postglenoid foramen and a smaller subsquamosal fenestra perforate the lateral surface of the braincase above the bulla on either side of the skull. The mandible is short and deep, with a large falciform coronoid process; the tip of the angular process is below or slightly behind the condyle. The lower incisor alveolus terminates in a well-developed capsular process below or just behind the base of the coronoid process.

The upper incisors are narrow, strongly opisthodont, and smooth (ungrooved), with yellow-orange enamel bands. The upper molar rows are parallel, and the teeth themselves are low-crowned and bunodont (even when unworn). The principal labial and lingual cusps of the upper molars are arranged in opposite pairs connected by transverse lophs. The anterocone of M1 is undivided, as is the anteroconid of m1. Small mesolophs are present on M1 and M2 but do not connect with mesostyles on the labial cingulum; mesolophids are absent on all of the lower molars. M1 and m1 each have four roots, but all the remaining molars have only three roots each.

Axial skeletal elements include 12 ribs, 19–20 thoracolumbar vertebrae, 3–4 sacral vertebrae, and 29 caudal vertebrae. The tuberculum of the first rib articulates with the transverse processes of the seventh cervical and first thoracic vertebrae, and the second thoracic vertebra has a greatly elongated neural spine. Each ring-shaped hemal arch on the undersides of the caudal vertebrae has a distinct spinous process on its posterior margin (Steppan 1995). The entepicondylar foramen of the humerus is absent.

The stomach is unilocular and hemiglandular, and the gall bladder is absent. One pair of preputial glands is present, but other male accessory secretory organs remain undescribed. The glans penis (illustrated by Langguth and Neto 1993) conforms to the complex bauplan, with a spinous exterior and a deep terminal crater containing three bacular mounds (supported by a tridigitate bacular cartilage), a dorsal papilla, and a bifurcate urethral flap.

The karyotype consists of 56 diploid chromosomes, all of which appear to be acrocentric; the fundamental number (FN) is 54.

SYNONYMS:

Hesperomys: Winge, 1887:11; part (description of *simplex*); not *Hesperomys* Waterhouse.

Oryzomys: E.-L. Trouessart, 1897:528; part (listing of *simplex*); not *Oryzomys* Baird.

Oecomys: Moojen, 1952b:55; part (listing of *simplex*); not *Oecomys* Thomas.

Pseudoryzomys Hershkovitz, 1959a:8; *nomen nudum* (see Pine and Wetzel 1976).

Calomys: Hershkovitz, 1962:123; part; (listing of *simplex*); not *Calomys* Waterhouse.

Pseudoryzomys Hershkovitz, 1962: 208; type species *Oryzomys wavrini* Thomas, by original designation.

REMARKS: Hershkovitz (1959a) originally proposed the name *Pseudoryzomys*, but the name was not made formally available until three years later, when Hershkovitz (1962) provided a generic diagnosis (Pine and Wetzel 1976).

Pseudoryzomys simplex (Winge, 1887)

False Oryzomys

SYNONYMS:

Hesperomys simplex Winge, 1887:11; type locality (fixed by lectotype selection; Voss and Myers, 1991) "Lapa da Escrivania [= Escrivânia] Nr. 5," a cave excavated by P. W. Lund near Lagoa Santa, Minas Gerais, Brazil.

Oryzomys simplex: E.-L. Trouessart, 1897:528; name combination.

Oryzomys wavrini Thomas, 1921b:177; type locality "Jesematathla [= Jesamatathla], west of Concepción," Presidente Hayes, Paraguay.

Oecomys simplex: Moojen, 1952b:55; name combination.

Pseudoryzomys wavrini: Hershkovitz, 1959a:8; name combination, but *Pseudoryzomys* a *nomen nudum* with this date (see Remarks, above).

Calomys simplex: Hershkovitz, 1962:123; name combination.

Pseudoryzomys wavrini: Hershkovitz, 1962:208; first valid use of this name combination.

Pseudoryzomys wavrini reigi Pine and Wetzel, 1976:649; type locality "Pampa de Meio, Río Itenez, Depto. Beni, Bolivia."

Pseudoryzomys simplex: Massoia, 1980c:282; first use of current name combination.

DESCRIPTION: As for the genus.

DISTRIBUTION: *Pseudoryzomys simplex* is known from low-lying areas below 900 m from northeastern Argentina, western Paraguay to eastern Bolivia and the Pampas del Heath in extreme southeastern Peru, and from there eastward through Brazil to the eastern margin of the Amazon Basin and the northeastern Atlantic coast. The species is largely confined to lowland areas with strongly seasonal rainfall, such as the grasslands and wetlands (*campo úmido*) found throughout the Chaco, Cerrado, and Caatinga domains (Voss and Myers 1991; L. G. Lessa and Talamoni 2000; F. F. Oliveira and Langguth 2004; Pardiñas, Cirignoli, and Galliari 2004; Bezerra et al. 2009; Prada and Percequillo 2013). Subfossils document a Holocene distribution that once extended southward to the Argentinean province of Buenos Aires (Pardiñas 1995a).

SELECTED LOCALITIES (Map 240): ARGENTINA: Chaco, Predisencia Roque Sáenz Peña (Prado and Percequillo 2013); Formosa, Estancia Guayacolec (Massoia, Heinonen Fortabat, and Diéguez 1997); Santa Fe, Pedro Gómez Cello (Pardiñas, Cirignoli, and Galliari 2004), Santa Margarita (Massoia et al. 1995b). BOLIVIA: Beni, Estación Biológica del Beni (AMNH 262048); Santa Cruz, Pampa de Meio, Río Itenéz (Prado and Percequillo 2013), Parque Nacional Noel Kempff Mercado (Prado and Percequillo 2013), Santa Ana (Prado and Percequillo 2013). BRAZIL: Amapá, Fazenda Asa Branca, Tartarugalzinho (Prado and Percequillo 2013); Amazonas, Humaitá, Escola Agrotécnica (Prado and Percequillo 2013); Bahia, Mata do Zé Leandro (L. G. Pereira and Geise 2009); Goiás, Moro da Baleia, Alto Paraíso, Parque Nacional da Chapada dos Veadeiros (Prado and Percequillo 2013); Mato Grosso, Posta Leonardo (Prado and Percequillo 2013); Mato Grosso do Sul, Gruta São Miguel (Prado and Percequillo 2013), Rio Sucuriú (Prado and Percequillo 2013); Minas Gerais, Lagoa Santa (Voss and Myers 1991); Pernambuco, 40 km W of Recife (UMMZ 164995), Vila Feira Nova (Prado and Percequillo 2013); São Paulo, Estação Ecológico Jataí (L. G. Lessa and Talamoni 2000); Tocantins, Parque Nacional do Araguaia (Bezerra et al. 2009), Peixe (Prado and Percequillo 2013). PARAGUAY: Presidente Hayes, Fortín Juan de Zalazar (CONN 16487). PERU: Madre de Dios, Pampas del Heath (MUSM 11732).

SUBSPECIES: I treat *Pseudoryzomys simplex* as monotypic.

NATURAL HISTORY: All known collection localities for *Pseudoryzomys simplex* are from tropical and subtropical landscapes characterized by open vegetation (e.g., in the Cerrado, Chaco, and Pantanal biomes). The few specimens

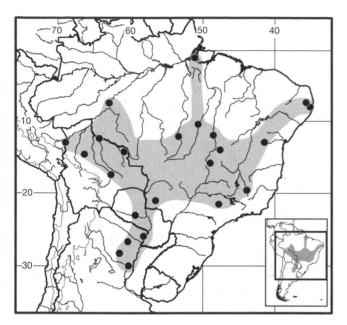

Map 240 Selected localities for *Pseudoryzomys simplex* (●). Contour line = 2,000 m.

accompanied by explicit habitat information were captured on the ground in grassy habitats, usually seasonally flooded savannas, marshes, and river floodplains. For some reason, *P. simplex* is hard to trap, and most known specimens have been recovered from Barn Owl (*Tyto alba*) pellets (e.g., by Pardiñas, Cirignoli, and Galliari 2004; Teta, Pereira, Muschetto, and Fracassi 2009). Nothing has been recorded in the literature concerning diet, reproduction, or other natural history topics.

Genus *Scolomys* Anthony, 1924

James L. Patton

Described in 1924 by H. E. Anthony from the eastern slope of the Ecuadorean Andes, the genus *Scolomys* remained an enigmatic taxon for the next 60 years, known in the published literature only from the original series of six specimens collected by G. H. H. Tate that comprised the single, and during that time, only known species, *S. melanops* Anthony. Following in rapid succession, and based on surveys at several sites in the western Amazon Basin, Pacheco (1991) described a second species (*S. ucayalensis*) from northern Peru, Patton and da Silva (1994) described a third species (*S. juruaense*) from western Brazil, and Gómez-Laverde et al. (2004) recorded specimens from southern Colombia and reviewed the systematics of the genus. These latter authors extended the range of *S. melanops* from the upper Río Pastaza and the Río Napo in Ecuador southeast to include three localities in northern Peru to the left (= north) bank of

the Río Amazonas near Iquitos. They assigned newly collected specimens from the Río Caquetá and Río Rumiyaco to the Peruvian taxon *S. ucayalensis* Pacheco, and extended the range of this species to a second locality in northeastern Peru. Finally, they argued that *S. juruaense* Patton and da Silva was not specifically distinct from *S. ucayalensis*.

Species of *Scolomys* are small (head and body length range: 80–90 mm), short-tailed (tail length 55–77 mm), and short-eared (pinna height 15–17 mm) mice with a short and broad head and distinctly spiny fur both above and below (Patton and da Silva 1995; Gómez-Laverde et al. 2004). The body pelage is short and close, colored grizzled pale reddish-black to nearly totally black dorsally and gray ventrally. Dorsal hairs are of two types: long (averaging 12 mm), stout, flat, and broad (averaging 0.6 mm) spines with a medial trough on both surfaces, with the terminal one-third to one-quarter increasingly dark to the tip and proximal portion clear; and long, thin hairs equal in length to the spines and with tips reddish or blackish. Ventral hairs of both types are uniformly gray from base to tip. Mystacial, superciliary, genal, submental, interramal, and carpal vibrissae are present. Pinnae are small, appearing somewhat thickened and thus stiff, and, while appearing naked from a distance, are clothed externally and internally with short reddish-brown hairs. The manus has five large, fleshy plantar pads (two carpal and three interdigital), with toes pale in color. Digit I is reduced but has a small nail; digits II through V are long and well developed with short, stout, and curved claws. Hindfeet are rather short and broad, although the metatarsus is nearly twice as long as digit III; the heel is haired, and the naked sole begins at about one-quarter the length of the plantar surface (not including the digits). The outer digits are shorter than the middle three (with the claw of digit I extending to or just past the base of digit II and that of digit V extends to the proximal phalanx of digit IV). Conspicuous tufts of long, silvery hairs are present at dorsal base of each claw, extending past their tips, but the claw is visible from above. The claws are short, stout (about twice as long as deep) and strongly curved along their dorsal surface. There are five to six plantar pads, with the thenar and four interdigital pads large, fleshy, always present; the hypothenar pad is either absent or only weakly developed. The tail is shorter than head and body length, is sparsely haired, and lacks any terminal tuft or pencil or long hairs. There are 15–18 scale annuli per centimeter at midlength of the tail, with annual hairs broad, blackish, and 2.5–4 scale rows in length, but sparsely distributed so that the tail scales are conspicuous. Females possess only three pairs of mammae (one thoracic, one abdominal, and one inguinal), in contrast the four pairs typical of other genera of the tribe Oryzomyini.

The skull, in dorsal view, has a short and broad or tapering rostrum flanked by shallow, barely perceptible zygomatic notches, laterally expanded nasolacrimal capsules (especially so in *S. melanops*), a broad interorbital region hourglass in shape, and with well-developed beaded ledges overhanging the margins from the middle of the frontals and continuing along the posterior margins of the orbit onto the braincase just above the squamosoparietal suture as moderately developed temporal ridges. The braincase is distinctly rounded and globular, dominating the dorsal aspect of the skull (length of braincase is about one-half the length of the skull). Nasals are somewhat expanded and taper posteriorly to a median point that terminates well behind the premaxillofrontal sutures. The interparietal is large, one-half to one-third as deep as wide.

In lateral view, the nasals extend onto or just beyond the anterior curvature of the incisors. The zygomatic plates are narrow, vertical to slightly angled posteriorly from its base, and without a distinct, free dorsal edge (thus, the zygomatic notches are markedly shallow when viewed from above). The zygomatic arch is thin, with a reduced jugal. The postglenoid foramen is moderate to small, hamular processes of the squamosal are stout, the subsquamosal foramen is reduced to totally occluded, and the mastoid fenestra is very small to nonexistent. The tegmen tympani of the periotic either does not contact or abuts but does not overlap the squamosal. Tympanic bullae are small and inflated ventrally only to the level of the molar series.

In ventral view, the incisive foramina are moderate in size (occupying about 60% of the diastemal distance), distinctly teardrop in shape, and pointed anteriorly with diverging sides and expanded, rounded posterior margins. The premaxillary-vomerine septum is greatly swollen, nearly filling the entire cavity of the incisive foramina. The bony palate is long (*sensu* Hershkovitz 1962), without a median ridge or palatal excrescences, with only weakly evident lateral folds, but with large and complex posterolateral pits. The mesopterygoid fossa is wide with parallel sides and a rounded or squared anterior margin ending well behind the third upper molars. The bony roof of the fossa is complete or perforated only by barely perceptible sphenopalatine vacuities along the presphenoid. Parapterygoid fossae are well developed, with their lateral margins straight to slightly convex and strongly divergent toward the bullae, devoid of vacuities except for a small foramen ovale, and moderately excavated, not flat in appearance. An alisphenoid strut is absent, but only the foramen ovale is present laterally, without an anterior opening of the alisphenoid canal. A shallow trough indicating the passage of the masticatory-buccinator branch of the maxillary nerve is visible; it emanates from the anterior margin of the foramen ovale and obliquely crosses the alisphenoid onto the squamosal. Facial circulation is apparently derived from the internal carotid artery (pattern 3, *sensu* Voss 1988), as indicated by a greatly reduced to absent stapedial foramen, no squamosoalisphenoid groove along the internolateral wall of the braincase, and no sphenofrontal foramen.

The mandible is short and stout, with a short coronoid process with a weakly to moderately curved posterior projection. A capsular process is weakly developed. Lower incisors are thin, elongate, and with enamel essentially devoid of pigment. Upper incisors are asulcate and faced with yellow to pale yellow enamel. These are small, deeper than wide, and proodont (in *S. melanops*) to orthodont (in *S. ucayalensis*).

Maxillary toothrows converge slightly posteriorly and are angled obliquely downward and outward at about 40°. Upper and lower molars are small and form a graded series with the third molar greatly simplified. Upper teeth are pentalophodont with their principal cusps arranged transversely and slightly obliquely. Labial and lingual reentrant folds do not interdigitate or contact, with the major labial folds of M1 and M2 (paraflexus and metaflexus) deep, extending at least two-thirds across the tooth, lingual folds reduced, with protoflexus only evident as a shallow lateral indentation in M1 and not visible at all in M2. The procingulum of M1 and m1 is well developed but not divided into separate anterolabial and anterolingual conules (no anteriormedial flexus[id] present). An anteroflexus on M1 is absent, so the anteroloph is not separated from the labial anteroconule. The anteroloph of M2 is well developed. Distinct mesolophs are present and extend to the labial margin of all three molars. The posteroloph is well developed on M1 but barely perceptible on M2 and absent on M3. The paracone and metacone of M1 and M2 is tall and well developed, with the protocone and hypocone proportionately reduced in size and much lower in topography. Only the paracone and weakly developed protocone are present on M3.

The stomach is unilocular and hemiglandular. The male phallus (of *S. ucayalensis*, this structure has not been described for *S. melanops*) is cylindrical with an incomplete crater rim, terminally exposed urethral flaps, lateral mounds of distal baculum hidden by tissue of the crater rim, and an epidermis with widely spaced small spines.

SYNONYM:

Scolomys Anthony, 1924b:1; type species *Scolomys melanops*, by original designation.

REMARKS: *Scolomys* possesses each of the seven putative synapomorphies that diagnose the tribe Oryzomyini (Weksler 2006; see above), but of the remaining characters shared by most members of the tribe, *Scolomys* has only six rather than eight mammae in females (Patton and da Silva 1995) and acutely angled posterior nasal margins rather than rounded or squared ones (Weksler 2006).

Initial molecular sequence analyses of both mitochondrial and nuclear genes failed to support the placement of *Scolomys* as a member of the Oryzomyini (M. F. Smith and Patton 1999; D'Elía 2003). However, more taxon dense molecular analyses (L. F. García 1999; Weksler 2003), especially when combined with a cladistic analysis of morphological characters (Weksler 2006), firmly placed *Scolomys* within the Oryzomyini as the sister to *Zygodontomys* but near the base of the radiation. The apparent sister relationship between these two genera was unexpected, as noted by Weksler (2006:75): "*Scolomys* and *Zygodontomys* are two of the most distinctive clades of oryzomyines, and they are ecologically and morphologically dissimilar from one another. Whereas species of *Scolomys* are rainforest specialists (Patton and da Silva 1995; Gómez-Laverde et al. 2004), species of *Zygodontomys* are highly specialized for savannas and other open vegetation formations (Voss 1991a). The IRBP sequence divergence of both lineages was also high, suggesting an early split within oryzomyine evolution. Although the clade lacks IRBP synapomorphies, analyses using faster-evolving mitochondrial genes have also recovered the same grouping (L. F. García 1999). This suggests an early cladogenetic event between the *Scolomys* and *Zygodontomys* lineages after the appearance of the *Scolomys-Zygodontomys* ancestor."

I follow Gómez-Laverde et al. (2004) and recognize two species in the genus.

KEY TO THE SPECIES OF *SCOLOMYS*:
1. Incisors proodont; rostrum short and blunt; nasolacrimal capsules expanded when viewed from above; zygomatic arches rounded; subsquamosal fenestra open; mandible short and deep; coronoid process short; hypothenar pad usually present; $2n=60$ . . . .*Scolomys melanops*
1'. Incisors orthodont; rostrum longer and more tapering; nasolacrimal capsules not expanded; zygomatic arches straight, converging anteriorly; subsquamosal fenestra occluded; mandible gracile, with elongated coronoid process; hypothenar pad reduced to absent; $2n=50$ . . .
. . . . *Scolomys ucayalensis*

Scolomys melanops Anthony, 1924
Gray Spiny Mouse
SYNONYM:
Scolomys melanops Anthony, 1924b:2; type locality "Mera, about 3800 feet elevation, eastern [Pastaza] Ecuador."
DESCRIPTION: Karyologically and molecularly well differentiated from *S. ucayalensis*, although externally both species share same small body, short and nearly naked tail, distinctly spiny fur, and grizzled pale reddish-brown to dark reddish-black finely streaked with black color above and below (Anthony 1924; Patton and da Silva 1994; Gómez-

Laverde et al. 2004). Cranially, however, two species readily distinguished. *Scolomys melanops* with short, wide, and blunt rostrum; wide and rounded zygomatic arches circumscribing wide and short orbital openings; laterally expanded nasolacrimal capsules; an open subsquamosal fenestra; proodont upper incisors; and short and deep mandible with short coronoid process. Hypothenar pad on plantar surface of hindfoot also well developed.

DISTRIBUTION: *Scolomys melanops* is known from three localities in eastern Ecuador and four in northeastern Peru, all between the Río Pastaza and Río Napo and north of the Río Amazonas. Elevation ranges from about 150 m to 1200 m.

SELECTED LOCALITIES (Map 241): ECUADOR: Orellana, Parque Nacional Yasuní, 38 km S of Pompeya Sur (ROM 104537); Pastaza, Mera (type locality of *Scolomys melanops* Anthony), Puyo (YPM 823); Sucumbios, Limoncocha (USNM 513581). PERU: Loreto, Estación Biológica Allpahuayo (Hice 2001), Quebrada Orán, ca. 5 km N of Río Amazonas, 85 km NE of Iquitos (LSUMZ 27927), San Jacinto (KU 158212), 1.5 km N of Teniente Lopez (KU 158214).

SUBSPECIES: *Scolomys melanops* is monotypic.

NATURAL HISTORY: This species has been trapped in cultivated fields and at the edge of mature forest (Rageot and Albuja 1994) and in patches of remnant forest (Gómez-Laverde et al. 2004) in eastern Ecuador, and in

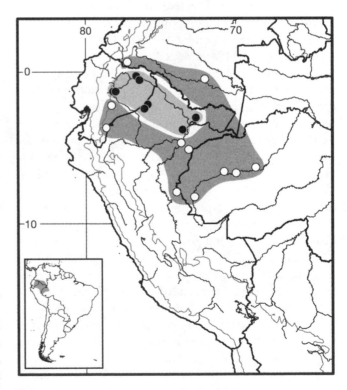

Map 241 Selected localities for *Scolomys melanops* (●) and *Scolomys ucayalensis* (○). Contour line = 2,000 m.

undisturbed forest as well as in selectively logged forest types at localities in northeastern Peru (Hice 2001; Hice and Velazco 2012). Except for two individuals caught on fallen logs, all others were on the ground, where the species was obtained on pitfall lines as commonly as in traps (Hice and Schmidly 2002). Hice (2001) recorded this species in all months of the wet season but never captured individuals in the four drier months of the year. She noted that reproductively active males were captured in March, October, and November, and pregnant females in March and April. The average litter size was 2.5 embryos. Anthony (1924b) reported pregnant females in March, with a litter size of three. Rengifo and Aquino (2012) described the nest, noting in particular the use of fiber taken from the palm *Lepidocaryum tenue*. Little else is known of the ecology or behavior of *S. melanops*.

REMARKS: Patton and da Silva (1995) described the karyotype of specimens from the Río Napo, Ecuador, with $2n = 60$, $FN = 78$, a large submetacentric X chromosome and large subtelocentric Y chromosome.

Scolomys ucayalensis Pacheco, 1991
Ucayali Spiny Mouse

SYNONYMS:

Scolomys ucayalensis Pacheco, 1991:1; type locality "Centro de Invesgicationes 'Jenaro Herrera', 2.8 km E of Jenaro Herrera, department of Loreto, right bank of the Ucayali river; at 135 m elev.; 73°39′W, 04°52′S."

Scolomys juruaense Patton and da Silva, 1994:324; type locality "Seringal Condor, left bank Rio Juruá, Amazonas, Brazil 70°51′W, 6°45′S."

DESCRIPTION: Small mouse with short, nearly naked tail (83% of head and body length); short and broad head with distinctly pointed snout; short and relatively broad hindfeet; greatly reduced to absent hypothenar pad but well-developed thenar and interdigital pads; small and rounded ears; and dorsal color varying from grizzled pale reddish-brown to dark reddish-black finely streaked with black. Skull with rounded and inflated braincase; short and basally broad rostrum that tapers distally; narrowed and straight zygomatic arches; narrow but long orbital openings; subsquamosal fenestra totally occluded by stout hamular process of squamosal; short and distally broad incisive foramina; and wide mesopterygoid fossa with parallel sides and squared, as opposed to rounded, anterior margin.

DISTRIBUTION: *Scolomys ucayalensis* occurs largely in lowland rainforest from extreme southern Colombia to northeastern Peru, western Brazil, and southern Ecuador, across an elevational range from 200 to 1,400 m. The distribution circumscribes that of its congener *S. melanops*, and the two are likely to be found in sympatry with additional collecting north of the Río Marañón-Río Amazonas axis.

SELECTED LOCALITIES (Map 241): BRAZIL: Acre, Sobral, left bank Rio Juruá (INPA 2485); Amazonas, Seringal Condor, left bank Rio Juruá (type locality of *Scolomys juruaense* Patton and da Silva), Penedo, right bank Rio Juruá (MVZ 183165), Barro Vermelho, left bank Rio Juruá (MVZ 183170). COLOMBIA: Amazonas, Corregimiento Puerto Santander, cerca de la Quebrada Bocaduché, margen sur Río Caquetá (ICN 16233); Nariño, cuenca alta del Río Rumiyaco (IAvH 6203). ECUADOR: Morona-Santiago, Domono (M. Brito and Arguero 2012); Zamora-Chinchipe, Alto Machinaza (M. Brito and Arguero 2012). PERU: Loreto, 2.8 km E of Jenaro Herrera (type locality of *Scolomys ucayalensis* Pacheco), Nuevo San Juan, Río Gálvez (AMNH 272668); Ucayali, Reserva Nacional Sierra del Divisor (H. Quintana et al. 2009).

SUBSPECIES: *Scolomys ucayalensis* is monotypic, with little expressed variation in mensural characters among the known samples (Gómez-Laverde et al. 2004).

NATURAL HISTORY: Pacheco (1991) obtained his two specimens at the type locality in dense undergrowth at the edge of primary forest artificially cut for regeneration studies, suggesting that this species naturally occurs in disturbed or secondary forest. Patton et al. (2000), however, collected their specimens along the Rio Juruá in western Brazil in undisturbed *terra firme* forest, though where there were often areas of local, natural disturbances (such as tree falls) and thus with a dense understory. There, pregnant females were captured from August through March, which spans both the wet and dry seasons. Embryo counts ranged from one to three. In contrast, the two Ecuadorean localities are both on the eastern Andean slope, one in mature subtropical rainforest dominated by the *cauchillo* tree (*Ficus* [Moraceae]) and the second in cloud forest with trees that reach 20 m in height and covered by bryophytes and bromeliads.

REMARKS: Gómez-Laverde et al. (2004), with the availability of much larger and geographically more expansive samples, showed that the presumptive distinctive characters differentiating *S. juruaense* to *S. ucayalensis* were primarily age related. Combined with very limited mitochondrial sequence divergence (about 1%), they argued that the two taxa were conspecific, a decision followed here. M. Brito and Arguero (2012) provided cranial measurements for the four known Ecuadoran specimens of *S. ucayalensi* and contrasted the morphological characteristics of this species with those of *M. melanops*.

Patton and da Silva (1995) described a karyotype of western Brazilian specimens as $2n = 50$, $FN = 68$, with a large acrocentric X chromosome and small acrocentric Y chromosome. These authors also figured and described the male phallus.

Genus *Sigmodontomys* J. A. Allen, 1897
Marcelo Weksler

Sigmodontomys is a monotypic genus found in lowland and lower montane trans-Andean forests in northwestern South America (Colombia, Venezuela, and Ecuador) and from Honduras to Panama in Central America. Until recently (e.g., Musser and Carleton 2005), *Sigmodontomys* included the two species *S. alfari* J. A. Allen, 1897 and *S. aphrastus* (W. P. Harris 1932), but phylogenetic analyses did not recover the two taxa as sister species (Weksler 2006; Pine et al. 2012), and *aphrastus* is now placed in its own genus: *Tanyuromys* Pine, Timm, and Weksler, 2012 (see *Tanyuromys* account). The following morphological description is based on AMNH and USMN specimens supplemented by information extracted from Hershkovitz (1944), Arata (1964; male accessory reproductive glands), Hooper and Musser (1964, glans penis), and Voss (1991a; gall bladder).

Sigmodontomys are medium-sized rodents, with moderate head and body size (range 120–152 mm). The tail is longer than the combined head and body length (length 149–190 mm), and the hindfeet are long and robust (length 34–37 mm). The body pelage is long, dense, and glossy; coloration of the upper parts ranges from dark ochraceous-orange to grizzled-yellowish brown, and the under parts from gray, white, or ochraceous, but hairs always with a dark gray base. The pinnae are small, slightly hairy, and dark, contrasting with paler dorsal coloration. The manus and pes are covered dorsally with short, dark hairs. Digits II–IV of the hindfeet are much longer than dI and dV (claw of dI reaches the medial portion of first phalanx of dII; claw of dV extends to interphalangeal joint of dIV). The plantar surface of the hindfoot is naked, densely covered with small epidermal tubercles distally, and has one or two small metatarsal pads (the thenar pad is always present, but the hypothenar pad is either absent or extremely reduced) and four very small interdigital pads. Small webs of skin are present between the digits II, III, and IV, but seldom extending more than about half the length of the first phalanx. Both forefeet and hindfeet have sparse ungual tufts covering but seldom extending beyond the tips of the claws (the tufts are absent on the hallux); the hindfeet lack natatory fringes. The tail is unicolored and sparsely haired, without a terminal pencil or tuft of distinctly longer hairs. Females have four pairs of mammae, one each in inguinal, abdominal, postaxial, and pectoral positions.

The skull is medium sized and robust. The rostrum is short and wide, and flanked by deep zygomatic notches. The nasals have squared or gently rounded posterior margins that extend posteriorly approximately to the same level as or slightly beyond the lacrimals; lacrimals usually have a longer maxillary than frontal suture. The interorbital margins converge anteriorly and possess strongly beaded supraorbital edges; the braincase is squared and relatively narrow, with well-developed temporal, lambdoidal, and nuchal crests. The posterior wall of the orbit is smooth. The frontosquamosal suture is anterior to the frontoparietal suture (the dorsal facet of the frontal is in broad contact with the squamosal). The parietal has a broad lateral expansion, a large portion of which dips below the temporal ridge posteriorly. The zygomatic arches are rounded or slightly convergent anteriorly, widest across middle portion of the arch or more posteriorly. The zygomatic plates lack anterodorsal spinous processes, and their posterior margins lie level to the alveolus of M1. Incisive foramina are broad and short, terminating well anterior to the alveoli of the M1s. The palatal bridge lacks deep furrows or median ridges, and extends slightly behind the third molars; conspicuous posterolateral palatal pits are present. The bony roof of the mesopterygoid fossa is perforated by reduced and narrow sphenopalatine vacuities. The postglenoid foramen is large and rounded, but the subsquamosal fenestra is absent or vestigial. The mastoid is completely ossified, or has only a small perforation (mastoid fenestra absent). The mandible is robust, with a large falciform coronoid process slightly higher than the condyloid process, and a rounded angular process. The capsular process is developed, situated below or just posterior to the base of the coronoid process.

The upper incisors are ungrooved and opisthodont. The molars are pentalophodont and bunodont. Maxillary toothrows are parallel; the molar series is robust and long, ranging from 5.3–6 mm in length; the labial and lingual flexi of upper molars shallowly penetrate at molar midline; the labial and lingual flexi are arranged in opposite pairs. Labial cingula are present, obliterating the lateral view of labial folds. The anterocone of M1 is undivided (anteromedian flexus absent); the anteroloph is well developed and fused with the anterocone by a labial cingulum; a protostyle is absent; an enamel bridge connects the paracone to the posterior portion of the protocone. M2 usually lacks a protoflexus and the mesoflexus is present as single fossette. M3 has a mesoloph, posteroloph, and shallow hypoflexus (that disappears with wear); the paracone and protocone are distinguishable on the molar surface, the hypocone is reduced, and the metacone is greatly reduced and completely fused with the posteroloph. The labial accessory root of M1 is present (four roots total).

Lower molars have the labial and lingual cusps in an opposite orientation. The anteroconid of m1 is undivided but contains a large internal fossetid; an anterolabial cingulum is developed in m1 but is absent on m2 and m3; an anterolophid is absent on m2, and the mesolophid is present in all lower molars; the ectolophid is absent, but an ectostylid is observed in the m1 of some specimens; the posteroflexid on m3 is large and persistent. There are four roots on m1, one anterior, one posterior, one lingual (below the metaconid), and one labial (below the metaconid); both m2 and m3 possess three roots each (two anterior and one posterior).

The axial skeleton consists of 12 ribs and 7 cervical, 19 thoracolumbar, 4 sacral, and 33–36 caudalvertebrae. The hemal arch between the second and third caudal vertebrae rarely forms a posterior spinous process. The entepicondylar foramen is absent, and the supratrochlear is present. The trochlear process of the calcaneus is noticeably posterior to the articular facet.

The gall bladder is absent. The glans penis has a deep terminal crater with three bacular mounds containing a tridigitate distal cartilaginous apparatus, with the central digit more slender than lateral ones; the dorsal papilla exhibits one apical spine, and a spineless urethral process without lateral lobule. The preputial glands are large.

SYNONYMS:

Sigmodontomys J. A. Allen, 1897a:38; type species *Sigmodontomys alfari* J. A. Allen, by original designation.

Nectomys: Thomas, 1897f:547; part (description of *russulus*); not *Nectomys* Peters.

Oryzomys: J. A. Allen, 1908:635; part (description of *ochraceous*); not *Oryzomys* Baird.

Nectomys (Sigmodontomys): Hershkovitz, 1944:21; part (listing of *alfari*, *efficax*, *russulus*, and *emeraldarum*); subgeneric designation.

Oryzomys (Sigmodontomys): A. L. Gardner and Patton, 1976:11; subgeneric designation.

REMARKS: Hershkovitz (1944, 1948b) included *Sigmodontomys* as a subgenus of *Nectomys*, and considered *Sigmodontomys alfari* as the single species, with four subspecies. Cabrera (1961) proposed a similar taxonomic arrangement, but A. L. Gardner and Patton (1976), on the basis of karyology and bacular morphology, considered *Sigmodontomys* to be a subgenus of *Oryzomys* (Baird 1857). Musser and Carleton (1993, 2005) regarded *Sigmodontomys* as a full genus, an arrangement corroborated by Weksler's (2003, 2006) phylogenetic analyses.

Cadena et al. (1998) reported a distinct species from Nariño in the Colombian Chocó as "*Sigmodontomys* sp.," but ongoing research (M. Pinto and M. Weksler, unpubl. data) suggests that this specimen belongs to a new taxon phylogenetically close to *Mindomys hammondi* (see Pine et al. 2012).

Bonvicino (1999) placed *Nectomys parvipes* (Petter 1979) in the genus *Sigmodontomys*, but following Voss et al. (2001: 94–95), I treat this taxon as a synonym of *Nectomys rattus* (see *Nectomys* account).

Recent molecular phylogenetic analyses (Hanson and Bradley 2008; Pine et al. 2012) recovered *Sigmodontomys alfari* as nested within *Melanomys*, rendering the latter taxon polyphyletic. Pine et al. (2012) suggested that the

non-monophyly of *Melanomys* was due to incomplete lineage sorting as their cladistic analysis of morphological characters corroborated the monophyly of this genus relative to other Oryzomyini. Herein, I retain both *Melanomys* and *Sigmodontomys* as separate genera, but additional analyses are required to confirm this arrangement.

Sigmodontomys alfari J. A. Allen, 1897
Alfaro's Water Rat

SYNONYMS:

Sigmodontomys alfari J. A. Allen, 1897a:39; type locality "Jimenez (altitude, 700 feet), [Limón,] Costa Rica."

Nectomys russulus Thomas, 1897f:547; type locality "Valdivia, [Antioquia,] Colombia, alt. 1200 metres."

Nectomys esmeraldarum Thomas, 1901f:250; type locality "St. Javier, Esmeraldas Prov., N.W. Ecuador. Altitude 20 m."

Sigmodontomys alfaroi Thomas, 1901f:251; incorrect subsequent spelling of *Sigmodontomys alfari* J. A. Allen.

Oryzomys ochraceus J. A. Allen, 1908:655; type locality "Rio Grande, [León,] Nicaragua."

Nectomys alfari alfari: Goldman, 1913:7; name combination.

Nectomys alfari efficax: Goldman, 1913:7; type locality "Cana (Altitude 1,800 feet), [Panamá,] Eastern Panama."

Oryzomys barbacoas J. A. Allen, 1916a:85; type locality "Barbacoas, [Nariño,] Altitude 75 feet, Southwestern Colombia."

Oryzomys barbacoas ochrinus Thomas, 1921f:449; type locality "Ecuador, [Esmeraldas,] west of Quito (Hershkovitz [1948b:54] wrote: "The original specimen of *ochrinus* 'collected' by Söderström from 'west of Quito' could very well have originated in the Río Mira region on the western slope of the Cordillera Occidental."

Nectomys esmeraldorum Gyldenstolpe, 1932a:68; incorrect subsequent spelling of *Nectomys esmeraldarum* Thomas.

Nectomys [(Sigmodontomys)] alfari alfari: Hershkovitz, 1944:74; name combination.

Nectomys [(Sigmodontomys)] alfari efficax: Hershkovitz, 1944:75; name combination.

Nectomys [(Sigmodontomys)] alfari russulus: Hershkovitz, 1944:77; name combination.

Nectomys [(Sigmodontomys)] alfari esmeraldarum: Hershkovitz, 1944:78; name combination.

Oryzomys (Sigmodontomys) alfari: A. L. Gardner and Patton, 1976:11; name combination.

DESCRIPTION: As for the genus.

DISTRIBUTION: *Sigmodontomys alfari* is found throughout trans-Andean lowland and lower montane forests in Colombia, Venezuela, and Ecuador; most collecting localities range from sea level to 1,300 m; this species also occurs in Central American lowland forests from Panama to Honduras.

SELECTED LOCALITIES (Map 242): COLOMBIA: Antioquia, La Tirana (USNM 499611), Valdivia (type locality of *Nectomys russulus* Thomas); Caldas, Samaná, Río Hondo (FMNH 71639); Cauca, Güengüé (AMNH 32183), Sabanetas (FMNH 90282); Chocó, Teresita, Río Truando (FMNH [B-185]); Córdoba; Socorré, upper Río Sinú (FMNH 69192); Nariño, Barbacoas (type locality of *Oryzomys barbacoas* J. A. Allen), Guayacana (FMNH 89563); Norte de Santander, San Calixto, Río Tarrá (USNM 279742); Valle del Cauca, Buenaventura, Palmares del Pacifico (USNM 483981), Virology Field Station, Río Raposo (USNM 334703). ECUADOR: Esmeraldas, Río Blanco, near Mindo (BM 34.9.10.214), San Javier (type locality of *Nectomys esmeraldorum* Gyldenstolpe); Guayas, Bucay (AMNH 61355). VENEZUELA: Zulia, Novito, 19 km WSW of Machiques (USNM 442251).

SUBSPECIES: I treat *Sigmodontomys alfari* as monotypic.

NATURAL HISTORY: In Panama, Handley (1966) recorded *Sigmodontomys alfari* as a common species, occurring in marshes, abandoned cane fields, and other forest openings. In Venezuela, Handley (1976) secured specimens on the ground under low bushes in a banana patch, both near streams and at dry sites. The species is said to be semiaquatic (F. A. Reid 1997; Lord 1999), but no factual report on its swimming ability was found; analysis of morphological characters (e.g., interdigital webbing reduced, natatory fringes absent), however, indicate that it has fewer semiaquatic adaptations than *Nectomys* (Thomas 1897f; Weksler 2006). Handley (1966) reported S. *alfari* at lower elevations in Panama, but Goodwin (1946: 402) stated that in Costa Rica, "[S.] *alfari* occupies higher regions"; most South American records of S. *alfari* are from localities between 200 and 1,300 m (Musser et al. 1998; McCain et al. 2007), but one locality (Sabanetas) is situated at 1,900 m (Musser et al. 1998). Recorded parasites include the flea *Polygenis roberti*, the chiggers *Crotiscus desdentatus*, *Eutrombicula goeldii*, and *Pseudoschoengastia bulbifera*, the laelapids mites *Echinolaelaps lowei* and *Gigantolaelaps* spp., the tick *Teratothrix higae*, and the nematodes *Hypocristata anguillula*, *Stilestrongylus barusi*, and *Syphacia alata* (Quentin 1969; Durette-Desset and Guerrero 2006; Digiani and Durette-Desset 2007; Goff 1981; Wenzel and Tipton 1966).

REMARKS: A. L. Gardner and Patton (1976) reported a $2n = 56$ FN = 54 karyotype for Costa Rican specimens.

Genus *Sooretamys* Weksler, Percequillo, and Voss, 2006
Alexandre R. Percequillo

The only known species of this genus, *Sooretamys angouya*, is widely distributed in eastern South America, from the Brazilian Atlantic rainforest to the humid forests of eastern Argentina and Paraguay.

Sooretamys angouya is easily distinguished from other sympatric oryzomyines (such as *Euryoryzomys russatus*, *Hylaeamys laticeps*, *Hylaeamys megacephalus*, and *Nectomys squamipes*) by its larger size (head and body length 125–215 mm), a tail much longer than head and body (range: 160–240 mm), and nonwebbed, robust, and wide hindfeet (length 25–42 mm). The body pelage is soft, long, and very dense; wool hairs are thin, long, and wavy; cover hairs are longer and thicker on their distal half; and guard hairs are much longer, stiffer, and thicker over their distal third. The dorsal body color varies from buffy yellow grizzled with dark brown to buffy orange grizzled with black. Ventral color grades from a whitish or buff weakly grizzled with gray to an almost pure white, buff, or buffy yellow. The tail is covered with short hairs, which do not conceal large scales; tail scales and hairs are blackish or dark brown, and the tail is unicolored. Internal and external surfaces of ears are well

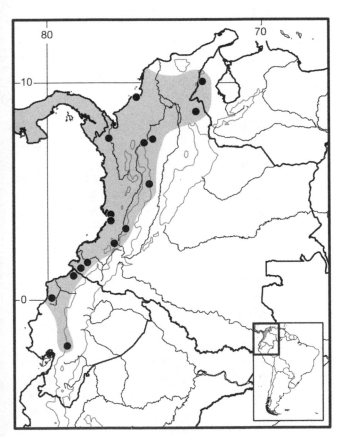

Map 242 Selected localities for *Sigmodontomys alfari* (●); range in Panama from F. A. Reid (2009). Contour line = 2,000 m.

furred, with hairs entirely brown. Forefeet and hindfeet are covered above by short hairs, and the ungual tufts of the toes are dense and long, almost concealing the claws. Brown hairs typically cover the dorsal surface of the hindfeet although some specimens have only a dark brown line or diffuse patch along the length of the foot. Ungual tuft hairs are proximally brown but white at their tips. Mystacial vibrissae are very dense and long, surpassing the ears when laid back; the most dorsal vibrissae are dark brown with golden tips, and the most ventral are entirely white.

The skull of *Sooretamys* is unique among genera of the Oryzomyini, easily recognized by its larger size (skull length 35.6–44.3 mm); longer incisive foramina (length 7.3–9.5 mm) that reach the alveolus of M1 in many specimens; long and narrow interorbital region (breadth 4.9–6.0 mm), hourglass in shape, with squared supraorbital margins; and very robust molars (toothrow length 5.6–6.5 mm; M1 breadth 1.5–1.8 mm). The rostrum is long and wide, flanked by well-inflated capsular projections of the nasolacrimal foramen. The zygomatic plates are wide, with straight or concave free anterior margins well projected anteriorly. The zygomatic arches are expanded laterally and wider near their squamosal roots. The braincase has a rounded profile, with the temporal border squared but without crests or beads. The palate is long with deep and complex posterolateral pits; palatal excrescences are absent. The roof of the mesopterygoid fossa is perforated by long and wide sphenopalatine vacuities that expose the orbitosphenoid bone. The stapedial foramen is minute or absent and, with the lack of both a squamosoalisphenoid groove and sphenofrontal foramen, configures the derived carotid circulatory pattern 3 (*sensu* Voss 1988). The buccinator-masticatory and accessory oval foramina are confluent, so all specimens lack an alisphenoid strut. Both the postglenoid foramen and subsquamosal fenestra are large apertures. Auditory bullae are large (breadth of one bulla range: 4.9–5.9 mm), with short stapedial processes and well-developed dorsostapedial processes overlapping the squamosal. The suspensory process of the squamosal is absent, and the tegmen tympani is short and may or may not overlap the squamosal.

The mandible has a low coronoid process, nearly triangular in shape and equal in height to the condyloid process; the angular process does not exceed the condyloid process. Both superior and inferior notches are shallow.

Upper incisors are opisthodont. Molars are pentalophodont and low-crowned with the main cusps arranged in opposite pairs, and with labial and lingual flexi interpenetrating only slightly at the median molar plane. On M1 and M2, the paracone is connected medially to the protocone, not to the median mure, defining a long and obliquely oriented parafossette that fuses to the metacone only with heavy wear. M1 has four roots, one anterior, one posterior, and two accessory rootlets; M2 and M3 each has two roots.

The axial skeleton comprises 7 cervical, 12–13 thoracic, 6–7 lumbar, 4 sacral, and 31–36 caudal vertebrae. Hemal arches have a distinct median posterior process.

The stomach is unilocular and hemiglandular; the bordering fold between gastric and glandular epithelium is well developed and oriented at 45° to the right of stomach. The gall bladder is absent. The glans penis is complex (terminology of Hooper and Musser 1964). The proximal bony baculum has a wide and flattened basal portion and long and rounded apical portion. The cartilaginous terminal baculum is tridigitate and well developed, about half the length of the bony baculum; the central cartilaginous digit is longer than the laterals, with its distal apex extending past the glans body.

SYNONYMS:

Mus: G. Fischer, 1814:71; part (description of *angouya*; not *Mus* Linnaeus.

Holochilus: Brandt, 1835:96; part (listing of *anguya* [=*angouya*]); not *Holochilus* Brandt.

Hesperomys: Wagner, 1843a:534; part (listing of *anguya* [=*angouya*]); not *Hesperomys* Waterhouse.

Hesperomys (*Oryzomys*): Thomas, 1884:448; part (listing of *angouya*); not *Oryzomys* Baird.

Calomys: Winge, 1887:50; part (description of *rex*); not *Calomys* Waterhouse.

Sooretamys Weksler, Percequillo, and Voss, 2006:23; type species *Mus angouya* G. Fischer, by original designation.

Sooretamys angouya (G. Fischer, 1814)

Angouya Sooretamys

SYNONYMS:

Mus angouya G. Fischer, 1814:71; no type locality given; based on Azara's (1801:86) "*rat troisième ou rat angouya*"; type locality "Paraguay east of the Río Paraguay, Departamento de Misiones, 2.7 km (by road) N of San Antonio," based on neotype designation (Musser et al. 1998:300).

Mus ? buccinatus Illiger, 1815:70; *nomen nudum.*

M[*us*]. *buccinatus* Olfers, 1818: 209; validation of *Mus buccinatus* Illiger 1815; based on Azara's (1801, 1802) "*rat troisième ou rat angouya*" and "*del anguyá*," respectively. .

Mus angouya Desmarest, 1819b:62; based on Azara's (1801, 1802) "*rat troisième ou rat angouya*" and "*del anguyá*," respectively; a homonym of *Mus angouya* G. Fischer.

mus angouya: Desmarest, 1820:305; name combination.

Mus Anguya Rengger, 1830:229; unjustified emendation of *Mus angouya* G. Fischer.

Mus (*Holochilus*) *Anguya* Brandt, 1835:428; name combination and unjustified emendation of *Mus angouya* Desmarest (= *Mus angouya* G. Fischer).

Hesperomys Anguya: Wagner, 1843a:534; name combination and unjustified emendation of *Mus angouya* G. Fischer.

H[*olochilus*]. *canellinus* Wagner, 1843a:552; substitute name for *Mus anguya* Brandt.

Mus canellinus: Schinz, 1845:192; name combination.

Hesperomys leucogaster Wagner, 1845a:147; type locality "Ypanema" (= Floresta Nacional de Ipanema, 20 km NW Sorocaba, São Paulo, Brazil, 23°26'7"S 47°37'41"W, 701 m; L. P. Costa et al. 2003).

Hesperomys ratticeps Hensel, 1872b:36, plate 1, Figs. 25a and 25b; plate 2, Figs. 15a and 15b; type locality "Rio Grande do Sul, Brasil."

H[*esperomys*. (*Oryzomys*)] *angouya*: Thomas, 1884:448; name combination.

Calomys rex Winge, 1887:50, plate 3, Fig. 8; type locality "Rio das Velhas, Lagoa Santa, Minas Gerais, Brasil."

Hesperomys (*Calomys*) *ratticeps*: Ihering, 1893:15 (108); name combination.

[*Oryzomys*] *anguya*: E.-L. Trouessart, 1897:525; name combination and unjustified emendation of *Mus angouya* G. Fischer.

[*Oryzomys*] *ratticeps*: E.-L. Trouessart, 1897:525; name combination.

[*Oryzomys* (*Oryzomys*)] *anguya*: E.-L. Trouessart: 1904: 420; name combination and unjustified emendation of *Mus angouya* G. Fischer.

[*Oryzomys* (*Oryzomys*)] *ratticeps*: E.-L. Trouessart, 1904: 420; name combination.

Oryzomys angouya: J. A. Allen, 1916d:570; name combination, reference to *Mus angouya* Desmarest, 1919; not *Mus angouya* G. Fischer.

O[*ryzomys*]. *angouya*: Thomas, 1921b:177; name combination.

Oryzomys ratticeps ratticeps: Thomas, 1924e:143; name combination.

Oryzomys ratticeps tropicius Thomas, 1924e:143; type locality "Piquete, São Paulo, Brasil."

Oryzomys ratticeps paraganus Thomas, 1924e:144; type locality "Sapucay, Paraguari, Paraguai," Paraguay.

Oryzomys [(*Oryzomys*)] *angouya*: Tate, 1932e:18; name combination.

Oryzomys [(*Oryzomys*)] *ratticeps ratticeps*: Tate, 1932e: 18; name combination.

Oryzomys [(*Oryzomys*)] *ratticeps tropicius*: Tate, 1932e: 18; name combination.

Oryzomys [(*Oryzomys*)] *ratticeps paraganus*: Tate, 1932e: 18; name combination.

Oryzomys buccinatus: Hershkovitz, 1955b:660; name combination.

[*Holochilus*] *anguyu*: Musser and Carleton, 2005:1119; *lapsus calami.*

[*Sooretamys*] *angouya*: Weksler, Percequillo, and Voss, 2006: 23; first use of current name combination.

DESCRIPTION: As for the genus.

DISTRIBUTION: *Sooretamys angouya* is distributed along the forested foothills and slopes of the Serra do Mar, Serra da Mantiqueira, and eastern hillsides of the Serra do Espinhaço (all in the Atlantic Forest), and coastal lowland forests (*restinga*) from Espírito Santo to Rio Grande do Sul states, Brazil, including some of the coastal islands in Santa Catarina and São Paulo states. To the west, this species reaches the Argentinean provinces of Misiones, Corrientes, and Entre Ríos, on the right (= east) bank of the Río Paraguay and the Argentinean provinces of Formosa and Chaco and the Paraguayan state of Presidente Hayes, on the left (= west) bank of the Río Paraguay.

SELECTED LOCALITIES (Map 243): ARGENTINA: Corrientes, Santa Tecla, Rte 12, km 1287 (MLP 1.X.94.5); Entre Ríos, Isla Chapetón, Río Paraná (Teta, Pardiñas et al. 2007); Formosa, Estancia Guaycolec (Teta, Pardiñas et al. 2007). BRAZIL: Espírito Santo, Hotel Faz Monte Verde, 24 km SE of Venda Nova do Imigrante (UFPB 334); Rio São José (MZUSP 6210); Minas Gerais, Lagoa Santa (type locality of *Calomys rex* Winge), Passos (MNRJ 32745); Rio de Janeiro, Fazenda do Tenente, São João de Marcos (MNRJ 5768); Rio Grande do Sul, Capão do Leão, Mostardas (UFRGS 2114), Fazenda Aldo Pinto, São Nicolau (MPEG 23275), Tôrres (UFRGS 8); Santa Catarina, Florianópolis (UFRGS 2042); São Paulo, Iguapé (MZUSP 26799), Ilha do Mar Virado (MZUSP AUC 165), Rio Feio (MZUSP 1910), Teodoro Sampaio (MZUSP 25324). PARAGUAY: Amambay, 4 km by road SW of Cerro Corá (UMMZ 125457); Paraguarí, Sapucay (type locality of *Oryzomys ratticeps paraganus* Thomas); Presidente Hayes, 24 km NW of Villa Hayes (UMMZ 133795).

SUBSPECIES: I treat *Sooretamys angouya* as monotypic.

NATURAL HISTORY: *Sooretamys angouya* is a scansorial species that can be captured both on the ground, on vines, and in trees (data from skin tags and collector field journals). Olmos (1992) reported that *S. angouya* is an aggressive species that, when caught in the same trap with another species, always killed and ate the individual, and that fights were commonplace when conspecifics were together. Olmos (1992:561) also reported that this species moved on the ground by meter-long leaps, and that individuals were good climbers of both trees and rock walls, as well good diggers, producing "extensive tunnels, even in hard, rocky ground."

In the eastern part of its range, *S. angouya* inhabits mature and undisturbed portions of the Atlantic Forest, but is more commonly observed in secondary forests (Pardini and Umetsu 2006). In westernmost regions (Formosa, Argentina, and Presidente Hayes, Paraguay), the species occurs in patches and remnants of semideciduous and gallery forests amid Chaco vegetation (G. D'Elía, pers. comm.).

Females were reproductively active during October and November (Olmos 1992), with juveniles being trapped in June (dry season).

REMARKS: The taxonomic history of *S. angouya* is long and contorted, confounded especially by the several names proposed for the rat described by D. Felix de Azara (1801, 1802) as the "*rat troisième ou rat angouya*" and "*del anguyá*," the specimens of which have been lost since the nineteenth century (see also Desmarest 1804, who referred to Azara's "*rat angouya*"). Some authors have suggested that the epithet "*angouya*" should refer to a Paraguayan species of the genus *Cerradomys* (specifically *Cerradomys maracajuensis*; e.g., Avila-Pires 1960a; Myers et al. 1995; Percequillo 1998) or was best considered a *nomen dubium* (Weksler 1996). Musser et al. (1998), on the other hand, resolved the conundrum by defining a neotype for *Mus angouya* G. Fischer and associating it to the species then formerly known as *Oryzomys ratticeps* Hensel. This nomenclatural act brought stability to the taxon, although the official availability of G. Fischer's names has not as yet been accepted by explicit action of the International Commission on Zoological Nomenclature (see Langguth 1966b; Sabrosky 1967; Musser et al. 1998).

Winge (1887) named as *Calomys rex* those specimens of *S. angouya* obtained by Peter W. Lund, preserved as fossil or subfossil material from the caves of Lagoa Santa, Minas Gerais. There are, however, no recent records of *S. angouya* from this region. The specimen Avila-Pires (1960b) recorded from the "Lagoa Santa region" actually came from Conceição do Mato Dentro, which is located nearly 110 km northeast of Lagoa Santa on the eastern slope of Serra do Espinhaço, a more humid area predominantly covered by forests. Today, the region of Lagoa Santa is covered mostly by Cerrado vegetation or is transitional between semideciduous forest and Cerrado. The same distributional pattern of typical forest-dwelling species registered in the caves of Lagoa Santa as fossils and subfossils is also observed for many other sigmodontine and echimyid genera, such as *Blarinomys breviceps*, *Brucepattersonius*, *Delomys* sp., *Euryoryzomys russatus*, and *Callistomys pictus* (Voss 1993; Emmons and Vucetich 1998; Voss and Carleton 1993; Musser et al. 1998; C. R. Silva et al. 2003; Pardiñas and Teta 2013a; see also Voss and Myers 1991 for a summary of the Lagoa Santa cave fauna).

C. O. da C. Vieira (1953:134) identified specimens of this species from São Paulo state, Brazil, as *Holochilus physodes physodes*, a name combination confusingly applied histori-

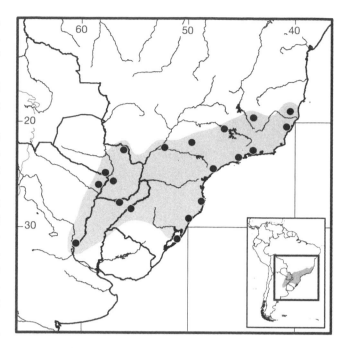

Map 243 Selected localities for *Sooretamys angouya* (●). Contour line = 2,000 m.

cally to both *Holochilus* and *Reithrodon* (see Musser et al. 1998:280–284, and accounts of those genera).

The karyotype of *S. angouya* is polymorphic, with diploid numbers ranging from 57–60 and fundamental numbers from 60–64 arms (E. J. C. Almeida 1980; Zanchin 1988; M. J. de J. Silva 1994; Geise 1995).

Genus *Tanyuromys* Pine, Timm, and Weksler, 2012
Marcelo Weksler

Tanyuromys is a monotypic genus found at a few midelevation forest sites in Costa Rica, Panama, and Ecuador. To date, only eight specimens are known for the genus, and thus our distributional and ecological knowledge of this taxon is quite sparse (e.g., we do not know whether the lack of collecting sites in Colombia is a sampling artifact or real). Until recently (Musser and Carleton 2005), *aphrastus* W. P. Harris was included in *Sigmodontomys*. Phylogenetic analyses (Weksler 2006; Pine et al. 2012), however, demonstrated that *aphrastus* was not closely related to *S. alfari* J. A. Allen, and Pine et al. (2012) thus erected *Tanyuromys* to contain *aphrastus*. The following morphological description is a synthesis of the recent work on this taxon by Weksler (2006), McCain et al. (2007), and Pine et al. (2012).

Tanyuromys are medium-sized rodents (adult head and body length range: 116–152 mm). The tail is much longer than the combined head and body length (length

176–235 mm), and the hindfeet are long and robust (length 35–40 mm). The dorsal pelage is long, thick, and soft. Coloration of the upper parts is grizzled dark brown with orange highlights, and that of the under parts is paler gray with pronounced ochraceous highlights but always with a dark gray base. The dorsoventral countershading is subtle, not as strong as in many other oryzomyines. The pinnae are small and sparsely to moderately covered with blackish, dark brown, or reddish-brown hairs both internally and externally. The mystacial and superciliary vibrissae are very long (extending posteriorly beyond caudal margins of the pinnae). The manus and pes have a sparse to moderate coverage of pale brown to dark brown hairs dorsally. Digits II–IV of the hindfeet are longer than dI and dV (claw of dI reaches the proximal portion of first phalanx of dII; claw of dV extends to the distal portion of second phalanx of dIV). The plantar surface of the hindfoot is naked, densely covered with small epidermal tubercles distally, and has one or two small metatarsal pads (the thenar pad is always present, but the hypothenar pad may be absent or extremely reduced) and four large and fleshy interdigital pads. The hindfeet lack well-developed natatory fringes and interdigital webs, and have sparse ungual tufts covering, but seldom extending beyond, the tips of claws (the tufts are absent on the hallux). The tail is unicolored brown and sparsely haired, terminating in a short tuft of hairs but without a distinctly longer terminal pencil. Four pairs of mammae are present, one each in inguinal, abdominal, postaxial, and pectoral positions.

The skull is medium sized and robust. The rostrum is short and stout, and flanked by very shallow zygomatic notches. The nasals have acutely angled posterior margins extending posteriorly beyond the premaxillae and behind lacrimals, nearly reaching the interorbital constriction level; the lacrimals have longer maxillary than frontal sutures. The interorbital margins converge anteriorly and possess overhanging supraorbital edges; the braincase is rounded and broad, with well-developed temporal, lambdoidal, and nuchal crests. The frontosquamosal suture is anterior to the frontoparietal suture (the dorsal facet of the frontal is in broad contact with the squamosal). The parietal has a broad lateral expansion, a large portion dipping below the temporal ridge posteriorly. The zygomatic arches are convergent anteriorly, relatively unbowed, and widest at the squamosal root; the jugal is large, and the maxillary and squamosal zygomatic processes are widely separated, not overlapping in lateral view. The zygomatic plates lack anterodorsal spinous processes, and their posterior margins lie anterior to the alveoli of the M1s. The incisive foramina are short and terminate far anterior of the alveoli of the M1s. The palatal bridge lacks deep furrows or median ridges, extends only slightly behind the M3s, and is perforated by small and simple posterolateral palatal pits. The bony roof of the mesopterygoid fossa is usually perforated by narrow sphenopalatine vacuities. The stapedial foramen and posterior opening of alisphenoid canal are very small, and the squamosoalisphenoid groove and sphenofrontal foramen are absent, with secondary anastomosis of the internal carotid across the dorsal surface of the pterygoid plate (= carotid circulatory pattern 3 of Voss 1988). The postglenoid foramen is large and rounded, but the subsquamosal fenestra is absent or vestigial. The mastoid is completely ossified, or has only a small perforation (mastoid fenestra absent). The mandible is robust, with a large falciform coronoid process slightly higher than the condyloid process, and with a rounded angular process. The capsular process is usually absent, or present as a slight rounded elevation not protruding above the level of the coronoid-condylar notch. Superior and inferior masseteric ridges join anteriorly as a single crest below m1.

The upper incisors are ungrooved and opisthodont. The molars are pentalophodont and bunodont, with an apomorphic and complex occlusal pattern having extensive, deep, steep-sided flexi and fossettes (enamel islands) with irregular and jagged borders. The maxillary toothrows are mostly parallel, with a slight divergence posteriorly. The molar series is robust and long, ranging from 5.6–6.2 mm in length. Labial and lingual flexi of the upper molars are deeply interpenetrating; labial flexi are convoluted and enclosed by a cingulum. The anterocone of M1 is undivided, the anteroloph is well developed and fused with the anterocone by a labial cingulum, and the protostyle absent; an enamel bridge connects the paracone to the posterior portion of the protocone. The M2 protoflexus is usually absent, the mesoflexus is usually present as a single fossette. M3 has a mesoloph, posteroloph, and shallow hypoflexus (which disappears with wear); the paracone and protocone are distinguishable on the molar surface; and the hypocone is reduced, and the metacone is greatly reduced and completely fused with the posteroloph. The labial accessory root of M1 is present (four roots total).

On the lower molars, the labial and lingual cusps are opposite. The anteroconid of m1 is undivided but contains a large internal fossetid; the anterolabial cingulum and a small anterolophid are present on all lower molars, the latter disappearing with moderate to heavy wear. The mesolophid is present in all lower molars, and the ectolophid is absent. The posteroflexid on m3 is large and persistent. The first lower molar has four roots, one anterior, one posterior, one lingual (below the metaconid), and one labial (also below the metaconid).

The gall bladder is absent. The stomach is unilocular-hemiglandular with glandular epithelium extending into corpus.

SYNONYMS:

Oryzomys: W. P. Harris, 1932:5; part (description of *aphrastus*); not *Oryzomys* Baird.

Sigmodontomys: Musser and Carleton 1993:748; part (listing of *aphrastus*); not *Sigmodontomys* J. A. Allen.

Tanyuromys Pine, Timm, and Weksler, 2012:858; type species *Oryzomys aphrastus* W. P. Harris, by original designation.

REMARKS: Phylogenetic analyses (Weksler 2006; Pine et al. 2012) indicate that *Tanyuromys* is a member of an oryzomyine clade that also includes *Aegialomys*, *Nesoryzomys*, *Melanomys*, and *Sigmodontomys*, all trans-Andean taxa distributed in southern Central America and northern South America (especially west of the Andes), and the Galapagos Islands.

Tanyuromys aphrastus (W. P. Harris, 1932)
Long-tailed Montane Rat

SYNONYMS:

Oryzomys aphrastus W. P. Harris, 1932:5; type locality "Joaquin de Dota, Costa Rica. This locality is southeast of Santa Maria de Dota in the Pacific rain forest at an altitude of about 4,000 feet"; [= San Joaquín de Dota, Cordillera de Talamanca, San José, Costa Rica, 1,220 m; 09°35′N, 83°59′W].

Sigmodontomys aphrastus: Musser and Carleton 1993:748; name combination.

Tanyuromys aphrastus: Pine, Timm, and Weksler, 2012: 859; first use of current name combination.

DESCRIPTION: As for the genus.

DISTRIBUTION: *Tanyuromys aphrastus* is known only from six middle-elevation (700–2,000 m) localities in north-central Costa Rica, western Panama, and northwestern Ecuador.

SELECTED LOCALITIES (Map 244; South American only): ECUADOR: Imbabura, 10 km E of Santa Rosa (QCAZ 10427); Pichincha; Guarumos (MCZ 50396), Mindo (UMMZ 155808).

SUBSPECIES: *Tanyuromys aphrastus* is monotypic.

NATURAL HISTORY: McCain et al. (2007) provided almost all known information about the natural history of this species, a synthesis of which is presented here. All eight known specimens were collected in traps set on the ground. In Costa Rica and Pichincha (Ecuador), specimens were taken in intact, mature montane forest; in Imbabura (Ecuador), one specimen "was collected in a mixed forest and tall-grass area within 10 m of a stream" (T. E. Lee et al. 2010:10); and in Panama, two specimens were captured "in disturbed, fairly open, relatively dry habitat, which included mixed grass, weeds, brush, oaks, and other trees" (R. H. Pine, pers. comm., cited in McCain et al. 2007:131). It is unclear from the few specimens available despite substantial trapping effort whether *T. aphrastus* occurs in very

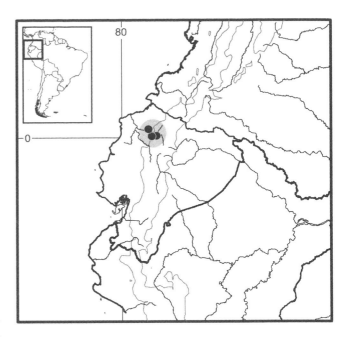

Map 244 Selected localities for *Tanyuromys aphrastus* in South America (●). Contour line = 2,000 m.

low densities or is simply difficult to trap using standard methods.

Of the eight available specimens, three are females, and five are males. Six specimens are adults; one adult male caught at Monteverde, Costa Rica, had descended testes measuring 7 × 5 mm and another from Chiriquí, Panama, had testes measuring 5 mm in length. The remaining two specimens include a juvenile female caught in Monteverde Cloud Forest Reserve and a young adult male from Chiriquí. One species of tick (*Ixodes sinaloa*) was collected from an adult at Monteverde.

REMARKS: The karyotype of *Tanyuromys aphrastus* is unknown. *Tanyuromys* differs from all other extant Oryzomyini (*sensu* Weksler 2003; Weksler et al. 2006; Weksler and Percequillo 2011) in the degree of lophodonty and the complicated enamel folding pattern of flexi and flexids of the molars (Pine et al. 2012).

Genus *Transandinomys* Weksler, Percequillo, and Voss 2006
Michael D. Carleton

Transandinomys contains two species (*bolivaris* J. A. Allen, 1901; *talamancae* J. A. Allen, 1891) that inhabit lowland to premontane wet forests in Central America and northwestern South America. They are medium sized (head and body length ranges from 100 to 140 mm) and possess a nearly naked, finely scaled tail that is as long as to slightly longer than the head and body. Conformation of the hindfoot indicates scampering, terrestrial habits: relatively long (about 26 to 32 mm) and narrow; digits I and V short relative to II–IV; ungual tufts ample; plantar pads six in number, 1 and 4 displaced proximally from 2 and 3, hypothenar small but discrete. The long arrays of cephalic vibrissae constitute the most definitive external trait, notably the sweeping superciliary vibrissae that extend posteriorly well behind the pinnae. Eight teats are present.

The cranium is moderate in build and size (occipitonasal length about 28 to 32 mm) and elongate in conformation, displaying a long, attenuate rostrum, nearly parallel-sided zygomatic arches, moderately incised zygomatic notch, and a rectangular, boxy braincase. Distinct supraorbital shelves with beaded edges mark the frontal borders and impart a cuneate shape to the interorbital constriction. The subsquamosal fenestra penetrates the posterior wall of the braincase as a narrow cleft, notably smaller than the companion postglenoid foramen. The incisive foramina are intermediate in length (i.e., not as long as in *Oryzomys* proper but longer than *Hylaeamys*), but do not extend between the alveoli of the M1s. The bony palate is narrow and long, as characteristic of most oryzomyines, and posterolateral palatal pits are well defined. The anterior border of the mesopterygoid fossa is evenly rounded, and sphenopalatine vacuities are consistently patent as short, narrow slits. The ectotympanic bulla is small, revealing a narrow posteromedial wedge of the periotic. The carotid arterial pattern of both species is complete (pattern 1 of Voss 1988: stapedial foramen, squamosal-alisphenoid groove, and sphenofrontal foramen present. Alisphenoid struts are absent).

The molar rows in both species of *Transandinomys* (coronal length of M1–3 range: 4.3–4.7 mm) are brachydont and cuspidate, the primary cusps positioned as opposite pairs, and the occlusal configuration is fundamentally pentalophodont (anterolophs, mesolophs, mesolophids present). The anterocone of the M1 is broad and undivided (anteromedian flexus absent). M2 has a long paraflexus and single median fossette, and the m2 has a long hypoflexid that crosses the midline of the tooth. The upper molars possess three roots (labial accessory root absent), and the lowers have two. The upper incisors are moderately robust, asulcate, and opisthodont.

Most taxonomic, nomenclatural, morphological, and distributional information on the component species is contained in the studies of Goldman (1918), Pine (1971), Musser and Williams (1985), and Musser et al. (1998), all as reported under *Oryzomys* in its formerly all-inclusive sense.

SYNONYMS:

Oryzomys: J. A. Allen, 1891a:193; part (description of *talamancae*); not *Oryzomys* Baird.

Transandinomys Weksler, Percequillo, and Voss 2006:25; type species *Oryzomys talamancae* J. A. Allen, by original designation.

REMARKS: No special relationship between *bolivaris* and *talamancae* was perceived in the early taxonomic literature on oryzomyine rodents. Goldman (1918) segregated the two into monotypic species groups, *talamancae* and *bombycinus* (= *bolivaris*), within the genus *Oryzomys sensu lato*. The identity of each was lost following Hershkovitz's (1960) precipitous allocation of forms believed to be synonymous with *Oryzomys capito*, a footnoted opinion formalized by Cabrera (1961) in his influential classification of South American mammals. Musser et al. (1998) differentiated *bolivaris* and *talamancae* in detail because of their close morphological resemblance and overlapping trans-Andean distributions, and compared the two to members of the *megacephalus* and *nitidus* complexes. In separate and combined analyses of morphological and molecular data, *Transandinomys talamancae* regularly emerged in a large oryzomyine clade that also included species of *Euryoryzomys*, *Handleyomys*, *Hylaeamys*, *Nephalemys*, and *Oecomys* (Weksler 2003, 2006; Percequillo, Weksler, and Costa 2011; Pine et al. 2012). See Weksler et al. (2006, especially Table 3) for morphological traits that discriminate these groups. As noted by those authors, the exceptionally long superciliary vibrissae offer the most persuasive synapomorphy for associating *bolivaris* with *talamancae* in a monophyletic genus.

KEY TO THE SPECIES OF *TRANSANDINOMYS*:

1. Dorsal pelage longer, dark brown and somber, under parts dark gray; mystacial and superciliary vibrissae extremely long (>45 mm); tail as long as head and body; supraorbital ridges well defined; parietal rarely contributes to the lateral braincase wall below the temporal ridge; incisive foramina long relative to bony palate, nearly reaching level of M1s . . . . *Transandinomys bolivaris*
1'. Dorsal pelage shorter, tawny and brightly colored, under parts bright whitish-gray; mystacial and superciliary vibrissae long (<40 mm); tail slightly longer than head and body; supraorbital ridges weak; parietal contributes substantially to the lateral braincase wall below the temporal ridge; incisive foramina short relative to bony palate, terminating conspicuously anterior to level of M1s . . . . *Transandinomys talamancae*

Transandinomys bolivaris (J. A. Allen, 1901)

Long-whiskered Oryzomys

SYNONYMS:

Oryzomys bolivaris J. A. Allen, 1901c:405; type locality "Porvenir, Bolivar, Ecuador, altitude 1800."

Oryzomys castaneus J. A. Allen, 1901c:406; type locality "San Javier, [Esmeraldas,] northern Ecuador, altitude 60 ft."

Oryzomys rivularis J. A. Allen, 1901c:407; type locality "Rio Verde, [Pichincha,] northern Ecuador, altitude 3200 ft."

Oryzomys bombycinus Goldman, 1912a:6; type locality "Cerro Azul (altitude 2500 feet), near the headwaters of the Chagres River, [Panamá,] Panama."

Oryzomys nitidus alleni Goldman, 1915:128; type locality "Tuis (about 35 miles east of Cartago), [Cartago,] Costa Rica."

Oryzomys bombycinus bombycinus: Goldman, 1918:77; name combination.

Oryzomys bombycinus alleni: Goldman, 1918:78; name combination.

[*Oryzomys (Oryzomys)*] *bolivaris*: Tate, 1932e:16; name combination.

[*Oryzomys (Oryzomys)*] *castaneus*: Tate, 1932e:16; name combination.

[*Oryzomys (Oryzomys)*] *rivularis*: Tate, 1932e:16; name combination.

Oryzomys bombycinus orinus Pearson, 1939:2; type locality "Mount Pirre, [Darién,] eastern Panama, at an altitude of 4700 feet along the Rio Limón."

Oryzomys capito bolivaris: Cabrera, 1961:385; name combination.

Oryzomys capito castaneus: Cabrera, 1961:385; name combination (*rivularis* allocated as full synonym, invalid as subspecies).

Oryzomys talamancae: Musser and Williams, 1985:14; part (*castaneus* included as synonym).

Oryzomys bolivaris: Musser and Carleton (1993:720); name combination (*alleni*, *bombycinus*, *castaneus*, *orinus*, and *rivularis* allocated as synonyms without indication of rank).

[*Transandinomys*] *bolivaris*: Weksler, Percequillo, and Voss, 2006:25; name combination (*alleni*, *bombycinus*, *castaneus*, *orinus*, and *rivularis* listed as synonyms without indication of rank).

DESCRIPTION: Individuals remarkable for extraordinary length of superciliary and genal vibrissae (see the Key) and for luxuriant dark-brown fur, very soft in texture and long (about 9–11 mm over middle dorsum); ventral pelage dark gray but clearly demarcated from upper parts. Tail appears nearly naked and approximates head and body in length. Moderately elongate skull characterized by wider interorbit and more prominent supraorbital shelves with beaded edges; minimal or no projection of parictal onto lateral braincase wall; and moderately long incisive foramina that end just anterior to molar rows. Most easily confused with specimens of its congener *T. talamancae* or with those of "*Handleyomys*" *alfaroi*, particularly where they may occur in sympatry (see the Key; Musser et al. [1998:125–143] supplied detailed comparisons with both taxa).

DISTRIBUTION: *Transandinomys bolivaris* occurs at low to middle elevations from southeastern Honduras, through eastern Nicaragua, Costa Rica, and Panama, to western Colombia and west-central Ecuador (as far as Manabí province), west of the Andes; elevational range in South America, about sea level to 1,800 m.

SELECTED LOCALITIES (Map 245; from Musser et al. 1998, except as noted): COLOMBIA: Cauca, Río Mechenque; Chocó, Condoto, mouth Río Condoto; Nariño, Barbacoas, on Río Telembí; Valle del Cauca, 6 km N of Buenaventura (USNM 554232). ECUADOR: Bolívar, Hacienda Porvenir, west of Hacienda Talahua (type locality of *Oryzomys bolivaris* J. A. Allen); Esmeraldas, San Javier (type locality of *Oryzomys castaneus* J. A. Allen); Manabí, Cerro de Pata de Pájaro; Pichincha, Gualea, Río Tulipe.

SUBSPECIES: I treat *Transandinomys bolivaris* as monotypic (see Musser et al. 1998:143–149).

NATURAL HISTORY: *Transandinomys bolivaris* is a denizen of lowland evergreen rainforest and lower montane rainforest (cloud forest), typically in regions of exceptional annual rainfall and rich floristic diversity (summary in Musser et al. 1998). The species is terrestrial, typically collected in very moist microenvironments within the forest (e.g., under logs, around the base of large trees, among mossy boulders along small forest streams, and in low terrain subject to flooding [Handley 1966; Pine 1971; Timm et al. 1989; González-M. and Alberico 1993]).

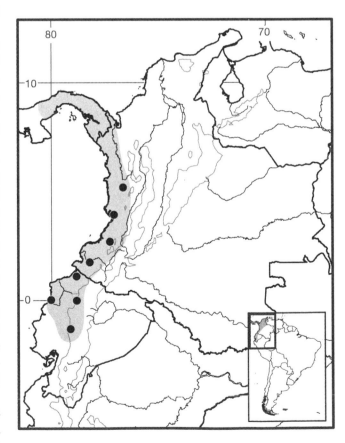

Map 245 Selected localities for *Transandinomys bolivaris* (●); range in Panama follows Musser et al. (1998). Contour line = 2,000 m.

REMARKS: Other names historically applied to this species include *bombycinus*, as reviewed by Pine (1971), and *rivularis*, as discussed by A. L. Gardner and Patton (1976). Musser and Carleton (1993) established the priority of *bolivaris*, and Musser et al. (1998) addressed the morphological definition, variation, geographic distribution, and synonymy of the species. Pine (1971) retained subspecies (under the name *bombycinus*), but Musser et al. (1998) judged intraspecific variation too insubstantial and population samples too small to critically delimit geographic races. The chromosome complement consists of $2n = 58$, $FN = 80$ (A. L. Gardner and Patton 1976; based on Costa Rican individuals reported as *bombycinus*).

Transandinomys talamancae (J. A. Allen, 1891)
Transandean Oryzomys
SYNONYMS:

Oryzomys talamancae J. A. Allen, 1891a:193; type locality "Talamanca," Limón, Costa Rica.

Oryzomys mollipilosus J. A. Allen, 1899b:208; type locality "Valparaiso [= Cincinati] (alt. 4500 ft.), Santa Marta District," Magdalena, Colombia.

Oryzomys magdalenae J. A. Allen, 1899b:209; type locality "Minca (alt. 2000 ft.), Santa Marta District of Magdalena Province, Colombia."

Oryzomys villosus J. A. Allen, 1899b:210; type locality "Valparaiso [= Cincinati] (alt. 4500 ft.), Santa Marta District, [Magdalena,] Colombia."

Oryzomys sylvaticus Thomas, 1900c:272; type locality "Santa Rosa, [El Oro,] Southern Ecuador. Altitude 10 m."

Oryzomys panamensis Thomas, 1901f:252; type locality "Panama (City of)," Panama, Panama.

Oryzomys medius W. Robinson and Lyon, 1901:142; type locality "San Julián, 8 miles east of La Guaira, [Vargas,] Venezuela."

Oryzomys carrikeri J. A. Allen, 1908:656; type locality "Rio Sicsola, Talamanca, [Limón,] Costa Rica."

Oryzomys talamancae: Goldman, 1918:73; *carrikeri* and *panamensis* allocated as full synonyms, invalid as subspecies.

[*Oryzomys (Oryzomys)*] *mollipilosus*: Tate, 1932e:16; name combination.

[*Oryzomys (Oryzomys)*] *magdalenae*: Tate, 1932e:16; name combination.

[*Oryzomys (Oryzomys)*] *villosus*: Tate, 1932e:16; name combination.

[*Oryzomys (Oryzomys)*] *sylvaticus*: Tate, 1932e:16; name combination.

[*Oryzomys (Oryzomys)*] *medius*: Tate, 1932e:16; name combination.

Oryzomys laticeps magdalenae: Gyldenstolpe, 1932a:18; name combination.

Oryzomys talamancae talamancae: Goodwin, 1946:392; name combination.

Oryzomys talamancae carrikeri: Goodwin, 1946:392; name combination.

Oryzomys talamancae talamancae: Hall and Kelson, 1959:564; name combination (*panamensis* included as full synonym, invalid as subspecies).

Oryzomys capito mollipilosus: Cabrera, 1961:386; name combination.

Oryzomys capito magdalenae: Cabrera, 1961:386; name combination

Oryzomys capito sylvaticus: Cabrera, 1961:387; name combination.

Oryzomys capito velutinus: Cabrera, 1961:387; part (*medius* listed as full synonym, invalid as subspecies).

Oryzomys capito talamancae: Handley, 1966:780; name combination (*panamensis* included as full synonym, invalid as subspecies).

O[*ryzomys*]. *c*[*apito*]. *carrikeri*: Handley, 1966:780; name combination.

Oryzomys talamancae: Musser and Williams, 1985:13; specific revision (*carrikeri*, *magdalenae*, *medius*, *mollipilosus*, *sylvaticus*, *panamensis*, and *villosus* allocated as synonyms without indication of rank).

[*Transandinomys*] *talamancae*: Weksler, Percequillo, and Voss, 2006:25; name combination (*carrikeri*, *magdalenae*, *medius*, *mollipilosus*, *sylvaticus*, *panamensis*, and *villosus* listed as synonyms without indication of rank).

DESCRIPTION: Resembles *T. bolivaris* in general size but with longer tail, distinctly longer than head and body, and smaller hindfoot. Fur more brightly colored (tawny over dorsum, tawny-buff on sides, and grayish-white on ventrum), shorter (7–9 mm over middle dorsum), and palpably coarser in texture. Superciliary and genal vibrissae long compared with most oryzomyines, but significantly shorter than those of *T. bolivaris* (see the Key). Skull with narrower interorbit and weaker supraorbital ridging; shorter incisive foramina that terminate appreciably in front of molar rows and correspondingly longer bony palate; and substantial contribution of parietal to lateral braincase wall. See Musser et al. (1998) for additional comparisons as well as diagnostic contrasts with larger *Hylaeamys megacephalus* and smaller "*Handleyomys*" *alfaroi*.

DISTRIBUTION: *Transandinomys talamancae* occupies low to middle elevations of northwestern Costa Rica, through Panama, to western and northern Colombia, northwestern Venezuela, western Ecuador, and extreme northwestern Peru; known elevational range in South America is from sea level to 1,385 m.

SELECTED LOCALITIES (Map 246; from Musser et al. 1998, except as noted): COLOMBIA: Antioquia,

24–26 km S and 21–22 km W of Zaragoza; Bolívar, San Juan Nepomuceno; Boyacá, Muzo, Río Cobaria Fatima; Caldas, Samaná; Cauca, Río Mencheque; Cesar, San Alberto; Chocó, Gorgas Memorial Laboratory, Teresita, Serranía de Baudó; Córdoba, Socorré; La Guajira, Las Marimondas; Magdalena, Cincinati (type locality of *Oryzomys mollipilosus* J. A. Allen and *Oryzomys villosus* J. A. Allen); Norte de Santander, Río Tarrá. ECUADOR: Chimborazo, Ríos Chimbo-Coco; El Oro, Salvias; Esmeraldas, 3 km W Majua; Santa Elena, Cerro de Manglaralto. PERU: Tumbes, Quebrada Los Naranjos, Zarumilla (Pacheco et al. 2009). VENEZUELA: Aragua, Rancho Grande; Miranda, Parque Nacional Guatopo; Yaracuy, Finca El Jaguar, 12 km NW of Aroa, Palmichal, 23 km N of Bejuma; Zulia, Hacienda El Tigre, 17 km N and 55 km W of Maracai; Vargas, San Julián, 8 mi E of La Guaira (type locality of *Oryzomys medius* W. Robinson and Lyon); Zulia, Misión Tukuko.

SUBSPECIES: I treat *Transandinomys talamancae* as monotypic.

NATURAL HISTORY: *Transandinomys talamancae* is a terrestrial oryzomyine that inhabits evergreen, semideciduous, and deciduous tropical forests, including both moist and dry associations and within primary or secondary growth (Handley 1966, 1976; T. H. Fleming 1971; O. J. Linares 1998; Musser et al. 1998; F. J. García, Delgardo-Jaramillo et al. 2012). Knowledge of ecology, reproduction, population dynamics, and life history of the species is mostly derived from T. H. Fleming's (1971) long-term field study

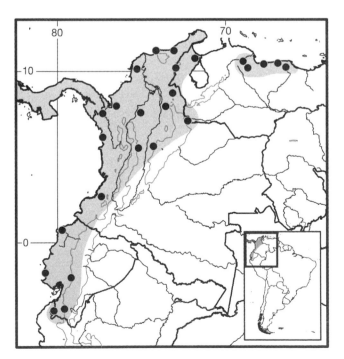

Map 246 Selected localities for *Transandinomys talamancae* (●). For Central American records, see Musser et al. (1998). Contour line = 2,000 m.

of Panamanian populations (reported as *Oryzomys capito*). O. J. Linares (1998) reported its diet as granivorous and omnivorous.

REMARKS: Recognition of this distinctive species was long hampered by its association as a form of *Oryzomys* "*capito*" (Hershkovitz 1960; Cabrera 1961; Handley 1966, 1976). Musser and Williams (1985) distilled the taxonomy, morphological identity, and distribution of the species; Musser et al. (1998) amplified those topics and refined the specific contents. See Musser et al. (1998:275) for commentary on the ambiguous connotation of the type locality "Talamanca," whether a specific village, the eastern lowlands adjacent to the Cordillera Talamanca, or the Cordillera Talamanca itself. Substantial karyotypic variation in diploid (2*n* = 34, 36, 40, 54) and fundamental numbers (FN = 64, 66, 67) has been encountered in samples from Venezuela and Ecuador (Pérez-Zapata et al. 1986; Musser et al. 1998), unaccompanied by appreciable morphological differences. The possible taxonomic significance of this variation requires study.

Genus *Zygodontomys* J. A. Allen, 1897
Robert S. Voss

Zygodontomys is a genus of two species that occur in unforested landscapes below about 1,300 m elevation from southeastern Costa Rica throughout Panama to Colombia, Venezuela, the Guianas, and northern Brazil. Populations also occur on adjacent continental-shelf islands, including Isla Margarita, Trinidad, and Tobago, where they may occupy forest as well as nonforest habitats. *Zygodontomys* is known from skeletal material found in caves on the island of Aruba (Hooijer 1967), but it is not a member of the extant fauna of that island. Hershkovitz (1962) thought that *Zygodontomys* was a phyllotine, but morphological evidence that *Zygodontomys* belongs to the tribe Oryzomyini was discussed inter alia by Voss (1991a), Voss and Carleton (1993), and Steppan (1995). Subsequent phylogenetic analyses of sequence data from the nuclear IRBP gene strongly support the hypothesis that *Zygodontomys* is an oryzomyine (Weksler 2003, 2006). Except as noted otherwise here, the following account is based on Voss (1991a), which should be consulted for additional details about taxonomy, morphology, distribution, and natural history.

Species of *Zygodontomys* have grizzled-brownish dorsal pelage, short vibrissae (not extending behind the pinnae when laid back alongside the head), and relatively small ears and eyes. The hindfeet are short and narrow, with middle digits (II, III, and IV) that are much longer than the outer digits (I and V); distinctively long tufts of ungual hairs cover the claws dorsally, and the plantar surface is provided with two metatarsal and four interdigital pads.

The tail is short, about three-quarters of the combined length of head and body, sparsely haired, and bicolored (dark above, pale below). Females have eight mammae in inguinal, abdominal, postaxial, and pectoral pairs.

The skull is unremarkable in dorsal view, with a moderately long rostrum flanked by well-developed zygomatic notches; the interorbital region is convergent and distinctly beaded, and the supraorbital bead on each side is continuous with a well-developed temporal crest on the dorsolateral surface of the braincase; the stout zygomatic arches are convergent (widest across the squamosal roots). In lateral view, the zygomatic plates are broad, with straight or convex anterior margins, but without anterodorsal spinous processes. The incisive foramina (in ventral view) are long and usually extend posteriorly to or between the first molar alveoli. The palatal bridge is broad and smooth, without conspicuous ridges, deep furrows, or a spine-like median posterior process; one or more prominent posterolateral palatal pits are present on each side behind M3. The mesopterygoid fossa does not extend anteriorly between the toothrows, and its bony roof is either complete or perforated by sphenopalatine vacuities; the flanking mesopterygoid fossae are narrow and not deeply excavated. An alisphenoid strut is absent (the buccinator-masticatory foramen and the foramen ovale are confluent). A large stapedial foramen is almost always present, but other osteological traces of the stapedial circulation are highly variable (most specimens exhibit either pattern 1 or pattern 2 of Voss 1988). The tegmen tympani does not overlap the posterior margin of the squamosal above the auditory bulla. The subsquamosal fenestra is usually present but often small; it is absent from occasional specimens in most large samples. The lower incisor root is contained in a small but usually distinct capsular process below the base of the coronoid process on the lateral surface of each mandible.

The upper incisors are smooth (never grooved), narrow, and opisthodont, with yellow-orange enamel bands. The upper molar rows are parallel, and the teeth themselves are bunodont with labial and lingual cusps arranged in opposite pairs; the anterocone of M1 is undivided, and mesolophs are absent from all of the upper teeth, each of which has only three roots. The anteroconid of m1 is likewise undivided, and mesolophids are correspondingly absent from all of the lower molars; m1 has two roots, m2 has three roots, and m3 has either two or three roots.

Modal counts axial skeletal elements include 12 ribs and 19 thoracolumbar, 4 sacral, and 24 to 26 caudal vertebrae. The tuberculum of the first rib articulates with the transverse processes of the seventh cervical and first thoracic vertebrae, and the second thoracic vertebra has a hypertrophied neural spine. The entepicondylar process of the humerus is absent.

The stomach is unilocular and hemiglandular (Carleton 1973), and the gall bladder is absent (see Remarks). Male accessory secretory organs consist of one pair each of preputial, bulbourethral, anterior and dorsal prostate, ampullary, and vesicular glands, and two pairs of lateral prostate glands (Voss and Linzey 1981). The glans penis is complex, with a tridigitate bacular cartilage and a deep terminal crater containing three bacular mounds, one dorsal papilla, and a bifurcate urethral flap (Hooper and Musser 1964).

REMARKS: Absence of a gall bladder in *Zygodontomys* was first reported by Voss (1991a), who dissected 22 specimens of *Z. brevicauda* from several localities scattered throughout the range of that species (no fluid-preserved specimens of *Z. brunneus* were available for dissection). A gall bladder is likewise absent in other oryzomyines, and this state optimizes as a tribal synapomorphy on phylogenetic trees (Steppan 1995; Weksler 2006). In a paper on the digestive adaptations of Venezuelan rodents, however, Domínguez-Bello and Robinson (1991) reported that a gall bladder was observed in two specimens identified by them as *Z. microtinus* (= *Z. brevicauda*, see below). Unfortunately, no voucher material was preserved in that study (M. G. Domínguez-Bello, in litt.), so the observations in question cannot be checked. (Fluid-preserved specimens dissected by Voss [1991a] are listed by museum catalog number in appendix 2 of that report.)

SYNONYMS:

Oryzomys: J. A. Allen and Chapman, 1893:215; part (description of *brevicauda*); not *Oryzomys* Baird.

Akodon: J. A. Allen and Chapman, 1897a:20; part (description of *frustrator*); not *Akodon* Meyen.

Zygodontomys J. A. Allen, 1897a:38; type species *Oryzomys cherriei* J. A. Allen, by original designation.

Oryzomys (*Micronectomys*): Hernández-Camacho, 1957: 223; part (description of *borreroi*); not *Oryzomys* Baird; *Micronectomys* Hershkovitz is a *nomen nudum* (Musser and Carleton 2005:1144).

KEY TO THE SPECIES OF *ZYGODONTOMYS*:

1. Length of hindfoot >27 mm, upper molar toothrow >4.6 mm, breadth of braincase (measured just above squamosal zygomatic roots) >12.2 mm; in side-by-side comparisons with sympatric *Z. brevicauda*, with slightly more hypsodont molars, more oblique mures and murids, less opisthodont upper incisors, shallower zygomatic notches, and less inflated nasolacrimal capsules . . . . *Zygodontomys brunneus*

1'. Measurements and qualitative characters overlapping those of *Z. brunneus* except in the upper Magdalena valley, where length of hindfoot <28 mm, upper molars <4.5 mm, and breadth of braincase <12.1 mm; with slightly less hypsodont molars, less oblique mures and murids, more opisthodont upper incisors, deeper zygomatic notches, and more inflated nasolacrimal capsules . . . . Zygodontomys brevicauda

Zygodontomys brevicauda (J. A. Allen and Chapman, 1893)

Cane Mouse

SYNONYMS:

Oryzomys brevicauda J. A, Allen and Chapman, 1893:215; type locality "Princestown [= Princes Town], Trinidad," Trinidad and Tobago.

Oryzomys microtinus Thomas, 1894b:358; type locality "Surinam."

Oryzomys cherriei J. A. Allen, 1895b:329; type locality "Boruca, Costa Rica."

Zygodontomys brevicauda: J. A. Allen, 1897a:38; first use of current name combination.

Akodon frustrator J. A. Allen and Chapman, 1897a:20; type locality "Caura, Trinidad," Trinidad and Tobago.

Zygodontomys cherriei: J. A. Allen, 1897a:38; name combination.

Zygodontomys microtinus: Thomas, 1898f:270; name combination.

Oryzomys sanctaemartae J. A. Allen, 1899b:207; type locality "Bonda, Santa Marta District, [Magdalena,] Colombia."

Zygodontomys stellae Thomas, 1899d:380; type locality "Maipures, Upper Orinoco" [Amazonas, Venezuela].

Zygodontomys brevicauda tobagi Thomas, 1900c:274; type locality Richmond, "Island of Tobago, W. Indies," Trinidad and Tobago.

Zygodontomys thomasi J. A. Allen, 1901a:39; type locality "Campo Alegre, [Sucre,] Venezuela, 90 miles south of Cumana."

Zygodontomys seorsus Bangs, 1901:642; type locality "San Miguel Island," Panama.

Zygodontomys cherriei ventriosus Goldman, 1912a:8; type locality "Tabernilla, Canal Zone, Panama."

Zygodontomys thomasi sanctaemartae: Osgood, 1912:52; name combination.

Zygodontomys griseus J. A. Allen, 1913c:599; type locality "El Triumfo [*sic*,= El Triunfo] (altitude 600 feet), Magdalena Valley, [Tolima,] Colombia."

Zygodontomys fraterculus J. A. Allen, 1913c:599; type locality "Chicoral (altitude 1800 feet), Coello River, Tolima, Colombia."

Zygodontomys brevicauda brevicauda: Gyldenstolpe, 1932a:112; name combination.

Zygodontomys microtinus stellae: Tate, 1939:188; name combination.

Zygodontomys microtinus fraterculus: Cabrera, 1961:464; name combination.

Zygodontomys punctulatus griseus: Cabrera, 1961:464; name combination.

Zygodontomys punctulatus sanctaemartae: Cabrera, 1961:464; name combination.

Zygodontomys punctulatus thomasi: Cabrera, 1961:465; name combination.

Zygodontomys brevicauda cherriei: Hershkovitz, 1962:203; name combination.

Zygodontomys brevicauda ventriosus: Hershkovitz, 1962:204; name combination.

Zygodontomys brevicauda seorsus: Hershkovitz, 1962:204; name combination.

Zygodontomys brevicauda sanctaemartae: Hershkovitz, 1962:204; name combination.

Zygodontomys brevicauda microtinus: Hershkovitz, 1962:205; name combination.

Zygodontomys brevicauda stellae: Hershkovitz, 1962:205; name combination.

Zygodontomys brevicauda thomasi: Hershkovitz, 1962:205; name combination.

Zygodontomys brevicauda soldadoensis Goodwin, 1965:2; type locality "Soldado Rock, latitude 10°4′25″ N., longitude 62°0′56″ W., an island 6 miles west of the southwestern tip of Trinidad, the West Indies, and 7 miles from the nearest point on the Venezuelan coast."

Zygodontomys microtinus thomasi: Reig, 1986:407; name combination.

Zygodontomys reigi Tranier, 1976:1202; type locality "aux environs de Cayenne," French Guiana.

DESCRIPTION: See the Key, and Voss (1991a) for thorough descriptions and illustrated characters.

DISTRIBUTION: As recognized herein, *Zygodontomys brevicauda* occurs in unforested habitats below about 1,300 m from southeastern Costa Rica throughout Panama to Colombia, Venezuela, the Guianas, and northern Brazil. This species is sympatric with *Z. brunneus* in the upper Magdalena valley of Colombia.

SELECTED LOCALITIES (Map 247): BRAZIL: Amapá, Tartarugalzinho (Mattevi et al. 2002); Amazonas, Castanhal (Bonvicino, Gonçalves et al. 2008), Igarapé Tucunaré (Bonvincino, Maroja et al. 2003), Serra do Aracá (Bonvicino, Gonçalves et al. 2008), Tabocal (AMNH 79401); Pará, Serra do Tumucumaque (USNM 392078); Roraima, Serra da Lua (FMNH 20051), Sunumú (Bonvicino, Gonçalves et al. 2008). COLOMBIA: Antioquia, Cisneros (Voss 1991a); Atlántico, Ciénaga de Guájaro (USNM 280383); Casanare, Finca Balmoral (UVM 3820); Chocó, Unguía (FMNH 70219); Córdoba, Montería (UVM 3452), Socorré (FMNH 69130); Huila, Valle de Suaza (USNM 541900), Villavieja (MVZ 113379); Magdalena, Bonda (type locality of *Oryzomys sanctaemartae* J. A. Allen); Meta, Puerto Lleras (Voss 1991a); Tolima, Chicoral (type locality of *Zygodontomys fraterculus*

J. A. Allen), El Triunfo (type locality of *Zygodontomys griseus* J. A. Allen); Vichada, Maipures (BM 99.9.11.39). FRENCH GUIANA: Cacao (MNHN 1980–244), Kourou (MNHN 1986–904). GUYANA: Demerara-Mahaica, Hyde Park (USNM 172939); Mahica-Berbice, Tauraculi (ROM 31476); Rupununi, Dadanawa (ROM 36096). SURINAM: Marowijne, Wiawia Bank (RMNH 21612); Paramaribo, Paramaribo (RMNH 21588). TRINIDAD AND TOBAGO: Tobago, Richmond (type locality of *Oryzomys brevicauda* J. A. Allen and Chapman); Trinidad, Princestown (type locality of *Zygodontomys brevicauda tobagi* Thomas). VENEZUELA: Amazonas, Esmeralda (AMNH 77316); Barinas, Santa Bárbara (Voss 1991a); Bolívar, Boca de Parguaza (Voss 1991a), Hato La Florida, 44 km ESE Caicara (USNM 406147), 55 km NE Icabarú (USNM 442423); Carabobo, Patanemo (MARNR 8191); Falcón, Mirimire (USNM 496830); Miranda, Río Chico (USNM 388026); Monagas, Isla Guara (Voss 1991a); Sucre, Cristobal Colón (AMNH 36188), Cumaná (USNM 406135); Táchira, Estación Experimental (Voss 1991a); Zulia, El Panorama, Río Aurare (Voss 1991a).

SUBSPECIES: See Remarks.

NATURAL HISTORY: On the Central and South American mainland, *Zygodontomys brevicauda* is invariably found in unforested lowland habitats such as savannas, shrublands, pastures, orchards, and grassy roadsides. Favored microhabitats usually include an herbaceous groundcover interspersed with woody plants growing in small thickets, usually on moist soils. It is among the commonest nonvolant mammals taken in anthropogenic landscapes (where it is sometimes an agricultural pest; A. M. G. Martino and Aguilera 1993), and its geographic range has doubtless increased with the widespread deforestation of northern South America. Other sigmodontines that are often collected syntopically with mainland populations of *Z. brevicauda* include *Sigmodon alstoni*, *S. hirsutus*, and one or more species of *Oligoryzomys*. Island populations of *Zygodontomys brevicauda* are ecologically distinctive because they inhabit closed-canopy rainforest, a habitat from which *Zygodontomys* is consistently absent on the Central and South American mainland.

Information from many sources (e.g., Enders 1935; Aldrich and Bole 1937; Ibáñez 1980; August 1981, 1984; O'Connell 1981, 1989; Vivas 1984; Vivas et al. 1986) indicates that *Zygodontomys brevicauda* is nocturnal, terrestrial, and omnivorous, differing in one or more of these traits from other commonly syntopic taxa such as *Sigmodon* (which is diurnal and herbivorous) and *Oligoryzomys* (scansorial). Reproductive studies (e.g., Aguilera 1985; Voss et al. 1992) have shown that *Z. brevicauda* is a spontaneous ovulator that gives birth to litters of 1 to 11 offspring (4 and 5 are modal litter sizes in most studies; Voss 1991a) after a gestation of about 25 days; there is a postpartum estrus, and implantation is not delayed by lactation. The neonates, which typically weigh 3–4 g each, are sparsely haired and appear naked; among other postnatal developmental events, the eyes open on day 6, the molars begin erupting on day 8, and the young are weaned by day 16. Females reach sexual maturity at about three to four weeks of age, males at six to eight weeks. Apparently, reproduction is continuous in wild populations, even those in conspicuously seasonal habitats (subject to seasonal cycles of flooding and draught; August 1981, 1984; O'Connell 1981, 1989; Vivas 1984). Due to its numerical abundance, *Z. brevicauda* is probably an important item in the diet of many sympatric predators; to date it has been recorded from vomitus, scats, or stomach contents of the Barn Owl (*Tyto alba*), Striped Owl (*Asio clamator*), White-tailed Kite (*Elanus leucurus*), Crab-eating Fox (*Cerdocyon thous*), and Jaguarundi (*Puma yagouaroundi*). Fifty-five species of arthropod ectoparasites have been collected from *Z. brevicauda* in Panama and Venezuela (Voss 1991a:Table 29).

Zygodontomys brevicauda is the primary reservoir of Guanarito virus, the etiologic agent of Venezuelan hemorrhagic fever (Fulhorst, Bowen et al. 1999; Fulhorst, Ksiazek et al. 1999; contra Tesh et al. 1993). Given the widespread geographic distribution of *Z. brevicauda*, it seems likely that this dangerous disease could break out sporadically throughout eastern Central America and northern South America, despite the fact that human cases have only been reported from Venezuela (Manzione et al. 1998).

REMARKS: As suggested by the many names and name combinations listed as synonyms above, this species exhibits geographic variation in size, coat color, and other morphological traits. Several geographic patterns of morphological variation are noteworthy: (1) trans-Andean populations (in Central America, Colombia, and northwestern Venezuela) differ from cis-Andean populations (in central, southern, and eastern Venezuela, and the Guianas) in frequencies of alternative carotid circulatory morphologies; (2) distinct sphenopalatine vacuities are usually present in specimens from Central America, Colombia, and Venezuela, but they are usually absent in specimens from coastal Guyana, Surinam, and French Guiana; and (3) insular populations (e.g., on Isla Cebaco, Trinidad, and Tobago) are consistently larger-bodied than adjacent mainland populations.

Voss (1991a) recognized three subspecies of *Zygodontomys brevicauda* based on the first and second patterns described here. These were diagnosed as follows: *Z. b. brevicauda* (including *frustrator*, *soldadoensis*, *stellae*, *thomasi*, and *tobagi*) usually has large sphenopalatine vacuities and a derived pattern of carotid circulation; *Z.*

b. cherriei (including *fraterculus, griseus, sanctaemartae, seorsus,* and *ventriosus*) usually has large sphenopalatine vacuities and the primitive pattern of carotid circulation; and *Z. b. microtinus* (including *reigi*) usually lacks distinct sphenopalatine vacuities and has a derived pattern of carotid circulation. Island gigantism was hypothesized to have evolved independently within each subspecies, and the notion that all large-bodied forms of *Zygodontomys* are closely related to one another (Tate, 1939) was rejected.

However, neither morphometric analyses (discussed by Voss 1991a) nor karyotypic data (Mattevi et al. 2002; Bonvicino, Maroja et al. 2003; Bonvicino, Gonçalves et al. 2008) corroborate this subspecies classification, whose heuristic value now seems questionable. Chromosomal complements with diploid numbers ranging from 84 to 88 and fundamental numbers varying from 116 to 118 have been reported for this species (A. L. Gardner and Patton 1976; Pérez-Zapata et al. 1984; Reig et al. 1990; Mattevi et al. 2002; Bonvicino et al. 2003, 2008). Three of these karyomorphs ($2n=86$, FN=96–98; $2n=84$, FN=96–98; and $2n=82$, FN=94), each from savanna systems along the southern margins of the species range in northern Brazil and French Guiana, were reported to form reciprocally monophyletic cytochrome-*b* haplotype groups that Bonvicino, Gonçalves et al. (2008) interpreted as distinct species. However, a subsequent analysis of cytochrome-*b* sequence data that included Bonvicino's (2008) sequences together with others from Central American and Venezuelan specimens (P. González et al. 2010) did not recover these karyomorphs as reciprocally monophyletic clades. The latter analysis additionally revealed surprisingly high levels of sequence divergence among some Central American populations and, overall, a lack of clear phylogeographic structure (P. González et al. 2010).

Clearly, a comprehensive assessment of geographic variation in morphology, karyotypes, and DNA sequences (ideally including one or more nuclear genes in addition to the single mitochondrial locus hitherto analyzed) is needed to make sense of these discrepant results. The effort involved in such an enterprise would be biogeographically rewarding because mainland populations of *Zygodontomys* are strongly associated with savannas and other nonforest lowland vegetation, the late Cenozoic history of which might be reconstructed from evolutionary analyses of the vertebrate fauna that inhabits it (Voss 1991a). Pending such an integrative analysis of morphological, karyotypic, and molecular variation, a conservative nomenclatural option is to recognize a monotypic *Z. brevicauda,* with the caveat that this might represent a complex of closely related but evolutionarily independent taxa.

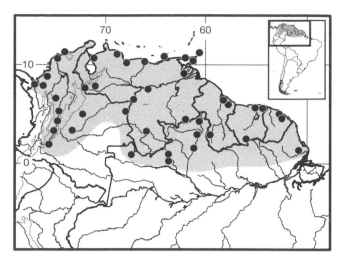

Map 247 Selected localities for *Zygodontomys brevicauda* (●). For Central American records, see Voss (1991a). Contour line=2,000 m.

Zygodontomys brunneus Thomas, 1898
Colombian Cane Mouse

SYNONYMS:

Zygodontomys brunneus Thomas, 1898f:269; type locality "El Saibal, W. Cundinamarca," Colombia.

Zygodontomys brevicauda brunneus: Gyldenstolpe, 1932a: 112; name combination.

Oryzomys (*Micronectomys*) *borreroi* Hernández-Camacho, 1957:223; type locality "Ladera norte del Cerro de San Pablo, hacienda Montebello, Municipio de Betulia, Valle del Río Chucurí, Departamento de Santander, Colombia."

DESCRIPTION: For details, see Thomas (1898f), Voss (1991a), and the Key.

DISTRIBUTION: *Zygodontomys brunneus* inhabits the intermontane valleys of Colombia, including those of the upper Río Magdalena, the upper Río Cauca, the upper Río Patía, and the upper Río Dagua.

SELECTED LOCALITIES (Map 248): COLOMBIA: Antioquia, San Jerónimo, 35 km NE of Medellín (FMNH 70205); Boyacá, Muzo (FMNH 71258); Cundinamarca, Volcanes, Municipio de Caparrapí (USNM 282114); Huila, Andalucía (AMNH 33773); Nariño, Finca Arizona, 4.5 km S of Remolino (UV 3099); Santander, Hacienda Montebello, near Cerro San Pablo (type locality of *Oryzomys* [*Micronectomys*] *borreroi* Hernández-Camacho); Valle de Cauca, Atuncela (UV [Voss 1991a]), Hacienda Formosa (UV 3579).

NATURAL HISTORY: Nothing has been published about the natural history of *Zygodontomys brunneus,* which, however, probably resembles *Z. brevicauda* in most ecological and behavioral traits. Interspecific habitat divergence, if any, might be expected where these taxa are sympatric in the upper Magdalena valley.

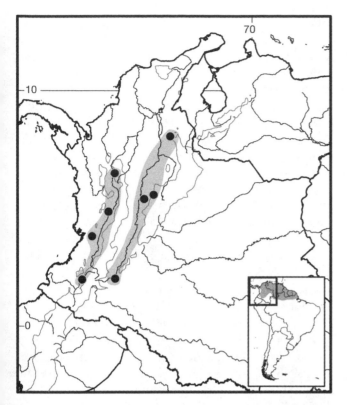

Map 248 Selected localities for *Zygodontomys brunneus* (●); those in the Magdalena and Cauca valleys are mapped as disjunct. Contour line = 2,000 m

REMARKS: The taxonomic status of populations in western Colombia (in the Cauca, Dagua, and Patía valleys) is uncertain and merits future research.

Tribe Phyllotini Vorontsov, 1959
Jorge Salazar-Bravo

The tribe Phyllotini includes 11 extant genera of small to medium-sized rodents widely distributed in south-central and southern South America, often associated with Andean, Patagonian, and Chacoan bioregions. The maximum species diversity for the tribe occurs in the Puna habitats of Bolivia, Peru, northern Chile, and northern Argentina, an area suggested to be the region of original differentiation (Reig 1986). Species boundaries, relationships among species, and the content of the tribe have been studied by Hershkovitz (1962), Pearson (1958), Pearson and Patton (1976), Olds and Anderson (1990), Braun (1993), Steppan (1993), Steppan (1995), M. F. Smith and Patton (1999), and Pardiñas et al. (2014); these authors should be consulted for further reference. What follows is a summary of the more recent taxonomic history of the phyllotines. Additional details can be found in Olds and Anderson (1990), Braun and Mares (1995), Steppan (1995), Pardiñas (1997), Steppan and Pardiñas (1998), M. F. Smith and

Patton (1999), S. Anderson and Yates (2000), Mares and Braun (2000b), P. E. Ortiz, Pardiñas, and Steppan (2000), Spotorno et al. (2001), Steppan et al. (2007), Steppan and Sullivan (2000), P. E. Ortiz, Jayat, and Steppan (2012), and Salazar-Bravo et al. (2013), who collectively documented the sometimes acrimonious debates regarding the taxonomy of the tribe.

Olds and Anderson (1990) were the first to use morphological characters to formally diagnose and summarize the contents of the tribe Phyllotini. They included 14 genera: *Andalgalomys*, *Andinomys*, *Auliscomys*, *Calomys*, *Chinchillula*, *Eligmodontia*, *Euneomys*, *Galenomys*, *Graomys*, *Irenomys*, *Neotomys*, *Phyllotis*, *Punomys*, and *Reithrodon*. In addition, they recognized and diagnosed a "*Reithrodon* group" that included *Euneomys* and *Neotomys* along with *Reithrodon*. Braun's (1993) concept of the Phyllotini was rather similar to that of Olds and Anderson (1990), except that she also included *Pseudoryzomys*. Important taxonomic conclusions from her analyses included the description of a new genus (*Maresomys*) for *Auliscomys boliviensis*, and the resurrection of *Loxodontomys* for *Auliscomys micropus* and *Paralomys* for *Phyllotis gerbillus* and *P. amicus*. Almost simultaneously, Steppan (1993) presented a cladistic analysis of mostly discrete morphological characters for the suite of taxa within Olds and Anderson's (1990) concept of the Phyllotini plus several outgroups. He concluded that *Punomys* should be removed from the tribe. In addition, his analyses did not support the distinction of *Andalgalomys* with respect to *Graomys*, suggested the paraphyly of both *Phyllotis* and *Calomys*, and also found strong support for a *Reithrodon* group composed of *Euneomys*, *Neotomys*, and *Reithrodon*. Pardiñas (1997), Steppan and Pardiñas (1998), and P. E. Ortiz, Pardiñas, and Steppan (2000) provided further support for an expanded *Reithrodon* group. In a subsequent analysis, Steppan (1995) confirmed the majority of his earlier conclusions with regard to the content of the Phyllotini. However, he removed *Pseudoryzomys* from the tribe and found no support for a monotypic *Maresomys*. He also defined three subtribes, the Phyllotina (*Phyllotis* s. s. + *Graomys* [including *Andalgalomys*]), the Reithrodonina (*Auliscomys* [including *Maresomys*], *Andinomys*, and *Reithrodon* groups), and the Calomyina including only *Calomys*. His phylogenetic analysis supported a nested set of subtribe relationships of (Phyllotina, Reithrodonina (Calomyina)). Both Mares and Braun (2000b) and Steppan and Sullivan (2000) agreed with retaining *Andalgalomys* at the generic level, although the latter authors made it clear that their decision was provisional pending a full resolution of the phylogeny of *Graomys*. This same period saw the description of two phyllotine genera, *Salinomys* (Braun and Mares 1995) and *Tapecomys* (S. Anderson and Yates 2000) as well as the expansion of molecular studies to study the composition of the Phyllotini and that of other suprageneric groups within the Sigmodontinae. Most recently, *Calassomys* has been added as another new member of the tribe, a taxon that appears as the basal member in DNA sequence results (Pardiñas et al. 2014).

Two pioneering studies shed light on the composition of the Phyllotini. Engel et al. (1998) supported earlier claims based on DNA-DNA hybridization studies (Catzeflis et al. 1993) that *Sigmodon* was the sister group to the rest of the Sigmodontinae. These authors also posited that *Reithrodon* did not belong within the Phyllotini. Using a larger sampling of sigmodontine species and mtDNA cytochrome-*b* gene sequences, M. F. Smith and Patton (1999) supported the hypothesis that *Reithrodon* as well as *Irenomys* were best treated as Sigmodontinae *incertae sedis* as they did not cluster within the Phyllotini. Steppan, Adkins, and Anderson (2004) also supported the exclusion of *Reithrodon* from the Phyllotini based on both nuclear and mtDNA analyses. Their analysis consistently placed *Reithrodon* as sister to the remaining Oryzomalia, rather that within the Phyllotini. Steppan, Adkins, and Anderson (2004) also showed that *Andinomys* and *Irenomys* were sister taxa to a clade formed by *Calomys*, *Auliscomys*, and *Phyllotis*, although the sampling was not designed to test the composition of the Phyllotini. Additional molecular studies further showed that *Euneomys* often formed a sister-taxon relationship with *Irenomys* and that both were outside of the Phyllotini (D'Elía 2003; D'Elía et al. 2006a; D'Elía et al. 2006b). In a later analysis explicitly directed to the genus *Phyllotis* and its close relatives, Steppan et al. (2007) found no support for *Paralomys* and moved *Phyllotis wolffsohni* to *Tapecomys*. None of these molecular phylogenetic studies, however, were sufficiently taxon dense to test adequately either the generic composition of the tribe or the limits of genera that had been questioned in prior studies.

Salazar-Bravo et al. (2013), in an expanded taxon and gene sequence study, revised the contents of the Phyllotini using both mtDNA and nucDNA gene sequences. Their analyses supported the generic separation of *Andalgalomys* from *Graomys* and the exclusion of *Andinomys*, *Chinchillula*, *Neotomys*, and *Punomys* from the tribe. They found no support for a *Reithirodon* group as proposed by Olds and Anderson (1990). These results strongly suggested the occurrence of multiple instances of putative morphological convergence among distinct sigmodontine lineages— for example, between *Euneomys* and *Neotomys* relative to *Reithrodon* and between the Phyllotini (as they restricted the tribe) and other Andean taxa. Salazar-Bravo et al. (2013) also concluded that fossil forms such as *Panchomys*, *Tafimys*, and *Olimipicomys*, once associated with the *Reithrodon* group

sensu Olds and Anderson (1990), and by implication included within the tribe Phyllotini, required additional study to determine whether their affinities lie with *Reithrodon*, with *Euneomys–Neotomys*, or with another group, including a restricted Phyllotini.

Here, I follow Salazar-Bravo et al. (2013) and Pardiñas et al. (2014) and recognize 11 extant genera in the tribe Phyllotini: *Andalgalomys*, *Auliscomys*, *Calassomys*, *Calomys*, *Eligmodontia*, *Galenomys*, *Graomys*, *Loxodontomys*, *Phyllotis*, *Salinomys*, and *Tapecomys*. The oldest known sigmodontine, and phyllotine, is *Auliscomys formosus* from the Montehermosense of Argentina, which dates to between 5 and 4 mya (Pardiñas et al. 2002; Pardinas and Tonni 1998). Whether or not fossil taxa (e.g., *Bensonomys*) from the late Miocene to early Pliocene of North America (Baskin 1978; Lindsay and Cazplewski 2011) are also referable to the Phyllotini is both controversial and a question beyond the scope of this volume. The description of *Pardinamys humahuaquensis* from late Pliocene deposits (between 3 and 2.5 mya) from the Uquia formation in central Jujuy province, Argentina (P. E. Ortiz, Jayat, and Steppan 2012), suggests an important radiation before the establishment of modern communities in the central Andean region. Another new fossil genus is currently being described (U. F. J. Pardiñas et al., in litt.).

Members of the tribe are ubiquitous in pastoral habitats and forest fringes throughout the southern half of South America, from the highlands of central Ecuador to the Straits of Magellan along the main axis of the Andes, and from the Pacific coast of Peru and Chile east through Patagonia to southeastern Brazil. *Calomys hummelincki*, the sole representative of the tribe north of the Amazon Basin, occurs in the savannas of northern Venezuela and northeastern Colombia and some of the continental islands off the Venezuelan coast.

In general, members of the Phyllotini have a body and limbs adapted for terrestrial life with normal ambulatory, slightly saltatorial, or scansorial activity. Notable specializations for burrowing, climbing, or swimming are absent. Body size varies from very small (mass of 12 g in adult *Calomys laucha*) to the approximate average for sigmodontine rodents (about 100 g in *Graomys*, some populations of *Phyllotis*, and *Tapecomys* [Pearson 1951; Hershkovitz 1962; S. Anderson 1997]). The pelage is normally thick, fine, and soft to moderately coarse, but never hispid or spiny. Dorsal coloration contrasts sharply to moderately with the ventral color, with or without a delimiting lateral line. Tail length is variable, ranging from less than one-half of the combined head and body to about the same as or slightly longer than the head and body (generally no more than 1.4 times, except in *Salinomys*, which can reach 1.6 times that of the head and body [Braun and Mares

1995]). The tail is moderately to well haired along the entire length; it may be either bicolored or unicolored; and a terminal pencil is present in some taxa. The eyes are normal in size and may or may not be circumscribed by a pale eye-ring. Preauricular and/or postauricular pale patches are usually present; the pinnae are moderate to large in size, and always more than 60% of the hindfoot length. The hindfoot is not markedly broadened or elongated; it is well haired above, the heel beneath is hairy, and the plantar surface may be either bare or hairy. The pollex posses a nail, all other digits have unspecialized claws (Pearson 1958; Hershkovitz 1962; Olds and Anderson 1990).

The skull is variable in both shape and form, and may appear either delicate or robust. The distal edge of the nasals varies from roughly square, rounded, or obtusely triangular, terminating from slightly in front to slightly behind the frontomaxillary suture, but never with a long pointed process extending deeply between frontals. Supraorbital edges are square, ridged, beaded, or produced as ledges, never evenly rounded. The interparietal is well developed, at least transversely, and has a distinct suture separating each of its borders from at least two-thirds of the width of parietal and occipital, respectively. The zygomatic arches are complete. The zygomatic notches are deep and broad; the zygomatic plates are broad, often wider than one-half of the least interorbital breadth, and their anterior borders are straight or slightly convex, but not deeply excised. The palate extends behind the posterior plane of last molars; its posterior border may be rounded or square, with or without a short median spine. Posterolateral palatine pits are often marked by distinct perforations. The mesopterygoid fossa is more or less U-shaped, and slightly divergent posteriorly. The parapterygoid fossa is shallow and often ovate in outline; its lateral wall is not parallel to the inner wall but gradually merges into dorsal surface and anterior wall. Incisive foramina are long, often narrow, pointed behind, and, with few individual exceptions, extend posteriorly to or behind the anterior plane of the M1s. The rostrum is not markedly tapered dorsoventrally from its base to tip. The anterior process of the premaxillary is projected slightly beyond the vertical plane of the incisors, but even if united with the anterolateral border of the nasals it never forms a well-developed tubular projection or trumpet. The premaxillomaxillary suture on the side of the rostrum is oriented essentially dorsoventrally, forming an angle of 90–135° (but never <90°) with the ventral edge of the rostrum as it passes around to the ventral side (Hershkovitz 1962; Steppan 1995). The inferred carotid circulatory pattern is primitive (*sensu* Voss 1988), with a functional stapedial foramen, sphenofrontal foramen, and squamosal groove all present (Steppan 1995).

The incisors are strong but never extremely broad, and orthodont to opisthodont. Their frontal surfaces are pigmented yellowish to dark orange and are usually asulcate, except in *Auliscomys* where fine striae (requiring magnification) may be present. Molars are well developed; the procingulum of M1 may or may not be divided; a mesoloph(id), enteroloph, and posteroflexus are absent; the mesostyle of M1 is often absent, except in *Calomys*; the parastyle of M1 and anteroflexus are present only in some species of *Calomys* and in *Loxodontomys*. The first upper molar has four roots, three along the main axis of the tooth and a large root laterally placed (except in *Auliscomys* where the fourth root is small, and *Loxodontomys* where there are two lateral roots); lower m1 has two medial and one lateral roots. The occlusal surface of M2 is square or broadly rectangular in outline, its width of the posterior half is approximately two-thirds or more than the greatest length of the tooth; it has two or three roots. M3 is one-half to three-fourths the length of M2, with two or three roots. The procingulum of m1 is not isolated from the cusps due to the confluence protoflexids and metaflexids; the hypoflexid, mesoflexid, and posteroflexid are well developed; and the mesoflexid of m3 is reduced or absent.

The axial skeleton is composed of 12 to 14 thoracic, 6 to 7 lumbar, and 17 to 36 caudal vertebrate (Steppan 1995). A gall bladder is present in all genera examined (Voss 1991a). Mammae range from 8 to 14 (Hershkovitz 1962; Olds 1988). Diploid numbers range from $2n=20$ in *Auliscomys boliviensis* to $2n=78$ in *Andalgalomys pearsoni* (Pearson and Patton 1976; Olds et al. 1987; see *Auliscomys* account).

KEY TO THE GENERA OF PHYLLOTINI (FROM PEARSON 1951; HERSHKOVITZ 1962):

1. Upper first and second molars lack mesolophs and mesostyles 2
1'. Upper first and second molars with persistent but vestigial mesolophs and mesostyles . . . . *Calassomys*
2. Size relatively small (toothrow length <4.3 mm); pelage countershaded, preauricular and postauricular patches present; ear, measured from notch, never longer than hindfoot; toes of hindfoot variable in length; upper incisors orthodont or opisthodont, their anterior surface never marked by vertical grooves or striae; molars with relatively low crowns; procingulum of M1 divided by anteromedian flexus; small mesostyle present or absent, when present usually fused with paracone; mammae 6 to 14. 3
2'. Relative body size medium to large (often toothrow length >4.3 mm); pelage countershade weak or strong, with or without preauricular and/or postauricular patches; ears shorter or longer than hindfoot; upper

incisors orthodont or opisthodont, with anterior surface smooth or striated; upper molars high or low crowned; procingulum of M1 divided or not, its occlusal surface flat in outline; mesostyle absent; mammae never more than eight 5

3. Soles with three middle postdigital tubercles more or less united, forming a thickly haired cushion; first postdigital tubercle small, naked; fifth postdigital tubercle reduced or absent; tarsal tubercle small, naked; fifth hind toe long, its tip, less claw, reaching distal end of first phalanx of fourth toe or beyond. . . . . . . *Eligmodontia*

3′. Soles with six well-defined, unfused, and naked plantar pads; fifth hind toe short, not including the claw—this toe barely reaches the first phalanx of fourth toe 4

4. Dorsal color grayish with venter fully white; tail >56% of body length; tail with tufted (or pencilled) tip; white or somewhat discolored area above eyes; skull with supraorbital region posteriorly divergent but without sharp ridges; skull when seen on lateral view rather bowed and with the braincase inflated; karyotype $2n = 18$ in females or $2n = 19$ in males; currently known only from desert-salt marsh environment in the temperate Monte Desert and Monte-Chaco ecotonal areas of central west Argentina . . . . *Salinomys*

4′. Dorsal color often brownish or buffy, with distinct countershade and often an ochraceous delimiting line in between; ventral coloration either pale with white-tipped hairs or fully white, but tail less than 56% of total body length and no longer than 80 mm; skull often flat when seen on dorsal profile, braincase narrow, not inflated; bullae not particularly inflated, its anterior-posterior length (not including tube) less than the length of molar row; known chromosomal complements ranges between $2n = 36$ to 66; widespread . . . . *Calomys*

5. Dorsal color brownish or grayish; tail long (>90 mm), coequal with head and body length, conspicuously bicolored; supraorbital region comparatively broad, sides posteriorly diverging from slightly behind angle of maxillofrontal suture; interorbital breadth at midfrontal plane greater than greatest width of rostrum; supraorbital edges square or beaded, with or without projecting ledges 6

5′. Dorsal color olive to buffy, hair long or short; tail as long as or shorter than head and body length; supraorbital region narrow, the sides either parallel or convergent from slightly behind angle of maxillofrontal suture to form constriction at approximately midfrontal plane; interorbital breadth at midfrontal plane equal to or, usually, less than greatest width of rostrum; supraorbital edges square, pinched, or, sometimes, weakly beaded 8

6. Dorsal color agouti brownish, venter gray to whitish with individual hairs gray-based with white tips; hindfeet brownish with dark dorsal hairs near the base of the toes; interorbital region with sharp edges, but not overhanging ledges; nasals long, extending caudad behind premaxillae; zygomatic plate relatively broad and without a zygomatic spine . . . . *Tapecomys*

6′. Dorsal color grayish to brown, venter white with hairs on chin and throat fully white to base; hindfeet whitish, no patch of dark hairs at base of toes; interorbital region markedly divergent posteriorly, with sharp overhanging edges; nasals short, very rarely extending caudad beyond premaxillae; zygomatic plate relatively narrow, with concave anterior border and well-developed zygomatic spine 7

7. Smaller (average head and body length <102 mm; range: 75–119 mm), with shorter hindfeet (length <25 mm); ventral pelage white, sharply demarcated from dorsal color; ear with subauricular white spot; teeth low-crowned, tending to form lophs; anterocone may be divided; alisphenoid strut absent . . . . *Andalgalomys*

7′. Larger (average head and body length >131 mm; range: 111–159 mm), with longer hindfeet (length >25 mm); ventral pelage gray; no conspicuous subauricular white spot; teeth high-crowned, tending not to form lophs; anterocone entire; alisphenoid strut present . . . . . *Graomys*

8. Size medium (average head and body length 106 mm), body stout, tail short (range: 30–44 mm); dorsal color buffy; venter with throats and chests fully white, elsewhere gray-based; incisors slender; molars brachydont; zygomatic plate nearly as wide as the least interorbital breadth, with the anterior border rounded, slanted backward; forefeet and hindfeet white above; plantar surface of hindfoot well haired and hypothenar pad absent; skull markedly vaulted middorsally . . . . *Galenomys*

8′. Body size variable, with tails measuring >44 mm and >25% of the total length; incisors broad; molars hypsodont or at least subhypsodont; plantar surface of hindfoot naked; six unfused plantar pads; skulls not markedly vaulted 9

9. Tail relatively short (36–46% of total length), body stocky built and size small to medium (head and body length <150 mm); both buffy or white preauricular and postauricular patches present; upper incisors with grooves near center of anterior face (requires magnification); interorbital region narrow (often narrower than the width of nasals); anteromedian flexus of M1 absent . . . . *Auliscomys*

9′. Tail length variable, body size variable; without white postauricular and/or preauricular patches; no grooves on the anterior surface of upper incisors (even under magnification); interorbital region narrow but often not less than width of nasals; anteromedian flexus of M1 present but only as a shallow groove. 10

10. Body robust with thick, lax, and lusterless pelage; dorsal color varies from gray to brownish; venter lighter, often washed with white or yellow; ears small, length from notch 20–50% smaller than hindfoot; fifth hind toe long, reaching to the end of the first phalanx of the fourth toe; tail short in relation to total length; molar toothrow posteriorly divergent; tympanic bullae relatively small (anteroposterior length, exclusive of tubes, is greater than or equal to alveolar length of molar row); upper incisors broad; M2 and M3 about same size, M3 in young adults with a complex pattern; lives in forests, meadows, and mesic brushy habitats in the Andean foothills of southern Chile and southwestern Argentina, from sea level to 3,000 m . . . .
. . . . .Loxodontomys

10′. Body form variable, often robust with long and fine pelage; dorsal color variable, many species with much lighter venter; ears medium to large (many species with ears about as long as hindfoot); fifth hind toe short (tip, less claw, not reaching distal end of first phalanx of fourth toe); tail subequal in length to combined head and body; maxillary toothrows parallel or posteriorly convergent; tympanic bullae relatively large; M2 distinctly larger than M3; widespread in rocky and brushy habitats from sea level to nearly 5,000 m, from northern Ecuador to southern Patagonia. . . . . Phyllotis

REMARKS: Vorontsov (1959) erred in his original construction of the tribal name Phyllotiini because a family-group name is to be based on the stem of the type genus, which in the case of Latin or Greek roots is the genitive singular to which the appropriate suffix is to be added (ICZN 1999:Arts. 29 and 29.3.1). In this case, the type genus is Phyllotis, a compound Greek work derived from phyll- (Greek phyllon, meaning leaf) and -ot (Greek ous, genitive otos, the ear). The stem of the family-group name is thus Phyllot-, to which the suffix -ini, when added, would give Phyllotini as the correct spelling. Reig (1980) was apparently the first to use Phyllotini.

Genus *Andalgalomys* D. F. Williams and Mares, 1978

Janet K. Braun

The genus *Andalgalomys*, commonly known as Chaco mice, ranges from southeastern Bolivia to western Paraguay, and in a narrow band throughout four provinces in west-central Argentina. Three species currently are recognized, where they occur in arid and semiarid habitats below about 1,000 m elevation. The following generic description is, except as noted, compiled from information presented by D. F. Williams and Mares (1978), Olds et al. (1987), Steppan (1995), and Mares and Braun (1996); data are not available for all characters for all species.

All three species of *Andalgalomys* are externally similar, with large ears, and whitish subauricular spots and postauricular patches. The dorsum is yellowish brown to brownish olive, and the venter is immaculate white. The tail is sparsely to moderately haired, bicolored, penicillate, and exceeds the combined length of the head and body. The hindfeet are unwebbed; the three middle digits (II, III, and IV) are longer than the outer two (I and V); the plantar surface is entirely naked (from heel to toes) with six tubercles (thenar, hypothenar, and four interdigitals).

The skull has a long, slender rostrum; relatively deep, open zygomatic notches (about equal to width of zygomatic plate); and posteriorly divergent, overhanging supraorbital margins. The zygomatic arches are only slightly convergent. Zygomatic plates have slightly concave anterior margins and moderately developed anterodorsal spinous processes. The incisive foramina are relatively long, extending to about the level of the anterior conules of M1. The palatines have large, slit-like foramina. Auditory bullae are inflated, and the bullar tubes are short and wide. Parapterygoid fossae are moderately broad and shallowly to moderately excavated. A vertical strut of the alisphenoid generally is absent. The mandible has an angular notch that does not extend anterior to the posterior edge of the capsular process. The coronoid process is small and delicate, and the angular process terminates anterior to the condyle. The lower incisor root is contained in a prominent capsular process posteroventral to the base of the coronoid process.

The molar rows are more or less parallel. The incisors are opisthodont and ungrooved. The molars are tetralophodont (lacking any trace of a mesoloph or mesolophid), brachydont, and the crowns tuberculate and slighted crested. Cusps are about opposite in arrangement. The anterocone of M1 generally has an anterior median fold and an anteromedian style; the postcingulum has a slight notch or fold. The anteroloph of M2 is small and directed nearly anteriorly. The major fold of m3 is deep, separating the metaconid-hypoconid from the protoconid-paraconid by a double enamel wall. M3, m2, and m3 have three roots.

There are 12 ribs plus 18–19 thoracolumbar and 31–33 caudal vertebrae. The second thoracic vertebra has a greatly elongated neural spine. The epicondylar foramen of the humerus is absent.

The gall bladder is present. Male accessory secretory organs consist of two pairs of preputial glands. The glans penis is complex, with a tridigitate bacular cartilage; bacular mounds are simple and unmodified, with the lateral mounds longer than the medial mound; the urethral flap is bifurcate. G. B. Diaz and Ojeda (1999) reported indices of renal relative medullary area, ratio of medulla to cortex, and relative medullary thickness.

Fossil material of *Andalgalomys* has been reported from southern Mendoza province, Argentina from about 8,900 ybp (Neme et al. 2002).

SYNONYMS:

Graomys: Myers, 1977:1; part (description of *pearsoni*); not *Graomys* Thomas.

Andalgalomys D. F. Williams and Mares, 1978:197; type species *Andalgalomys olrogi* Williams and Mares, by original designation.

KEY TO THE SPECIES OF *ANDALGALOMYS*:

1. Dorsum yellowish brown; length of tail ≥124 mm; bullae greatly inflated, >6.4 mm in length, and >4.8 mm in depth, width of bullae >5.8 mm; m3 length >0.80 mm; length of interparietal usually <2.5 mm; known only from the isolated Bolsón de Pipanaco near Andalgalá, Catamarca Province, Argentina . . . .
. . . . *Andalgalomys olrogi*
1'. Dorsum not yellowish brown; length of tail usually <125 mm; bullae only moderately inflated, measurements less than indicated above; m3 length <0.80 mm; length of interparietal usually >2.5 mm 2
2. Dorsum reddish brown; length of skull generally >29.0 mm; breadth of mesopterygoid fossae from 1.0–1.4 mm; rostrum generally longer and broader; palatilar length generally >13.0 mm; width of zygomatic plate >2.7 mm; length of mandible generally >15.0 mm; known from the Chaco of Bolivia and Paraguay . . . . . .
. . . .*Andalgalomys pearsoni*
2'. Dorsum grayish brown; length of skull generally <29.0 mm; breadth of mesopterygoid fossae <1.0 mm; rostrum shorter and more slender; palatilar length generally <13.0 mm; width of zygomatic plate <2.9 mm; length of mandible generally <15.0 mm; known from the western Chaco and the Chaco-Monte Desert ecotone in central Argentina . . . .*Andalgalomys roigi*

Andalgalomys olrogi D. F. Williams and Mares, 1978
Olrog's Chaco Mouse

SYNONYM:

Andalgalomys olrogi D. F. Williams and Mares, 1978:203; type locality "West Bank Rio Amanao, about 15 km W (by road) Andalgalá, Catamarca Province, Argentina."

Graomys olrogi: Steppan, 1995:79; name combination.

G[*raomys*]. (*Andalgalomys*) *olrogi*: Steppan and Sullivan, 2000:267; name combination.

DESCRIPTION: Largest species in genus, with mean head and body length 96.3 mm (range: 86–113 mm) and hindfoot length 24.2 mm (range: 23–25 mm). Tail longest proportionally (130% of head and body length) and absolutely (mean 125 mm; range: 124–126 mm) among the three species. Overall dorsal color yellowish-brown, with small whitish subauricular and postauricular patches present; ventral surface and upper surfaces of both forefeet and hindfeet white. Tail terminates in moderate pencil. Body of glans penis shorter and proportionally broader and baculum much shorter than both structures of *A. pearsoni* (D. F. Williams and Mares 1978).

DISTRIBUTION: *Andalgalomys olrogi* is known only from a few adjacent localities in the Bolsón de Pipanaco, an isolated valley in the Monte Desert of northwestern Argentina at an elevation of about 1,000 m.

SELECTED LOCALITIES (Map 249): ARGENTINA: Catamarca, west bank of Río Amanao, about 15 km W (on Rte 62) of Andalgalá (type locality of *Andalgalomys olrogi* Williams and Mares).

SUBSPECIES: *Andalgalomys olrogi* is monotypic.

NATURAL HISTORY: *Andalgalomys olrogi* occurs in or near creosote brush (*Larrea*, Zygophyllaceae) flats found at the northern end of the Bolsón de Pipanaco (D. F. Williams and Mares 1978). Unpublished field notes indicate nocturnality and little to no reproductive activity during the winter and some activity during the summer.

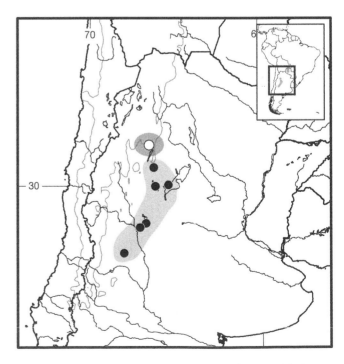

Map 249 Selected localities for *Andalgalomys olrogi* (○) and *Andalgalomys roigi* (●). Contour line = 2,000 m.

REMARKS: The chromosome complement consists of a $2n=60$, $FN=116$, with all autosomes biarmed, a moderately large submetacentric X and small metacentric Y chromosome (D. F. Williams and Mares 1978).

Andalgalomys pearsoni (Myers, 1977)
Pearson's Chaco Mouse
SYNONYMS:

Graomys pearsoni Myers, 1977:1; type locality "410 km NW Villa Hayes by road, Departamento Boquerón, Paraguay."

Andalgalomys pearsoni: D. F. Williams and Mares, 1978: 210; first use of current name combination.

Andalgalomys pearsoni dorbignyi Olds, Anderson, and Yates, 1987:9; type locality "Bolivia: Department of Sant Cruz; 29.5 km W of Roboré, 475 m (18°19′S, 60°02′W)."

Graomys pearsoni: Steppan, 1995:79; name combination.

Graomys pearsoni dorbignyi: S. Anderson, 1997:461; name combination.

G[raomys]. (Andalgalomys) pearsoni: Steppan and Sullivan, 2000:267; name combination.

DESCRIPTION: Medium sized (head and body length 86–113 mm) with elongated tail (110% head and body; range: 97–124 mm) and moderate hindfeet (22–25 mm). Dorsal coloration brown streaked with gray; fur on rump about 10 mm long and slate gray at base; hairs on venter white to their base; small postauricular buffy patch present; both forefeet and hindfeet white on top; tail bicolored, dorsal third brown, ventral two-thirds buffy, and sparsely covered with hair. Tail nonpenicillate, with terminal hairs extending <3 mm beyond tip. Skull with same general confirmation as all species in genus, but averages larger in most length (e.g., condylobasal, rostral, diastema, palatal, incisive foramina) and breadth (e.g., braincase, interorbital, rostral, nasal) dimensions than other species.

DISTRIBUTION: *Andalgalomys pearsoni* occurs in the Chaco of western Paraguay and southeastern Bolivia below 500 m.

SELECTED LOCALITIES (Map 250): BOLIVIA: Santa Cruz, 29.5 km W of Roboré (Olds et al. 1987). PARAGUAY: Boquerón, Loma Plata (MSB 80512), 410 km NW of Villa Hayes by road (type locality of *Graomys pearsoni* Myers), Teniente Enciso (UCONN 17562).

SUBSPECIES: Olds et al. (1987) described Bolivian samples as a subspecies (*A. p. dorbignyi*) separate from the nominotypical form from northern Paraguay, based primarily on larger size, more pronounced anterior zygomatic spines, broader mesopterygoid fossa, and different diploid number.

NATURAL HISTORY: *Andalgalomys pearsoni* is found in dry grasslands that occur as isolated microhabitats in the Chaco of western Paraguay (Myers 1977). In south-

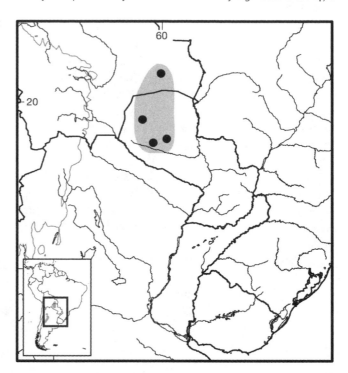

Map 250 Selected localities for *Andalgalomys pearsoni* (●). Contour line = 2,000 m.

eastern Bolivia, it has been collected in "arid, scrubby, low, partly open vegetation" (Olds et al. 1987:9) at the edge of the Chaco. The diet may include some insects (Myers 1977). Limited data suggest that some reproductive activity occurs during the winter (Myers 1977). Goff and Whitaker (1984) summarized information on arthropod ectoparasites.

REMARKS: Myers (1977) compared *A. pearsoni* only with those species of *Graomys* Thomas. Mares and Braun (1996) provided more explicit comparisons among the three species of *Andalgalomys* currently recognized. The karyotype varies from a $2n=76$ in Bolivian populations (*A. p. dorbignyi*) to $2n=78$ in those from Paraguay (*A. p. pearsoni*), with the two cytotypes varying in autosomal count from 10 and 12 pairs of uniarmed and 28 or 27 pairs of biarmed chromosomes. Both X and Y chromosomes are acrocentric (Olds et al. 1987).

Andalgalomys roigi Mares and Braun, 1996
Roig's Chaco Mouse
SYNONYMS:

Andalgalomys roigi Mares and Braun, 1996:929; type locality "Argentina: San Luis Province: 8 km W La Botija, Pampa de Las Salinas, 36°12′27″S, 66°39′35″W. 510 m elevation."

G[raomys]. (Andalgalomys) roigi: Steppan and Sullivan, 2000:267; name combination.

DESCRIPTION: Similar in size to *A. pearsoni*, with head and body length 75–113 mm and hindfeet 21–25 mm, but tail proportionally longer (119% of head and body; range: 85–135 mm). Overall dorsal coloration light brownish olive, with banded hairs neutral gray at their base and cream buff tips; guard hairs dark gray basally with black tips; sides paler than dorsum, with lateral hairs lacking black tips; venter white to base and sharply demarcated from sides and back. Tail bicolored, brownish above and whitish below, and moderately penciled, with hairs extending about 5 mm beyond tip; basal two-thirds moderately haired, although scales remain visible to eye; distal one-third well haired, and scales not visible; tail hairs increase in length distally, giving darker appearance to tip than base. Both subauricular and postauricular patches as well as tops of forefeet and hindfeet white. Skull similar in shape to other two species but smallest in length and breadth dimensions.

DISTRIBUTION: *Andalgalomys roigi* is known from the Chaco and Chaco-Monte ecotone of west-central Argentina at elevations between about 300–1,000 m.

SELECTED LOCALITIES (Map 249): ARGENTINA: Catamarca, Chumbicha, 0.5 km E of Rte 38 along Rte 60 (CML 3693); La Rioja, Salar La Antigua, 45 km NE of Chamical (R. A. Ojeda et al. 2001), 26 km SW of Quimilo (OMNH 30016); Mendoza, Reserva Ñacuñán (Borghi and Eguaras 2000); San Luis, Pampa de las Salinas, 1.3 km N and 7.4 km W of La Botija (OMNH 23794), Parque Provincial Sierra de las Quijadas, 6 km W of Hualtaran (OMNH 23795), Lomas Blancas (D. Castro, Gonzales, and Vidal 1998).

SUBSPECIES: *Andalgalomys roigi* is monotypic.

NATURAL HISTORY: *Andalgalomys roigi* has been collected in arid and semiarid habitats in the narrow ecotone between the Monte Desert and Chaco that extends from Catamarca to Mendoza province. Specimens have been trapped most commonly in low thorn scrub vegetation bordering salt flats (Mares and Braun 1996; R. A. Ojeda et al. 2001). Molting has been reported to occur in the months of May and July (Mares and Braun 1996). Reproductive activity (including young/juvenile individuals) occurs in March, May, and November (Mares and Braun 1996; unpubl. data). Unpublished field notes suggest that this species is nocturnal. D. Castro, Gonzales, and Vidal (1998) and Borghi and Eguaras (2000) summarized information on arthropod ectoparasites. Mares and Braun (1998) documented mammalian associates at three localities where this species was caught.

REMARKS: The chromosome complement consists of a $2n = 60$, $FN = 116$, with both all autosomes and both X and Y chromosomes biarmed. This karyotype is similar to that of *A. olrogi* but differs from that of *A. pearsoni* in both diploid and fundamental numbers.

Genus *Auliscomys* Osgood, 1915

Jorge Salazar-Bravo

Members of the genus *Auliscomys* range from just north of 8°S (La Libertad in central Peru) southward through the Altiplano of Peru and Bolivia to northern Argentina and Chile, occurring in various highland habitats at localities ranging between 3,500 and above 5,000 m, where they occupy rock slides, *Polylepis* woodlands, and both bunchgrass and shrub dominated communities. This account is based on information extracted primarily from Gyldenstolpe (1932), Osgood (1947), Pearson (1951, 1958), Hershkovitz (1962), and Braun (1993), who should be consulted for further reference.

Species of the genus *Auliscomys* are small to medium-sized rats (head and body size <150 mm [Pearson 1958]) with long, lax fur, stout bodies, and tails either as long as or shorter than head and body length. The tail is well haired, bicolored, and lacks a tuft or pencil of hairs at the tip. Dorsal coloration varies among and within species, from uniformly grayish in *A. boliviensis* to rich brown in *A. pictus* to buffy yellow mixed with black in *A. sublimis*. Fur on the underparts is overall gray in tone with hair tips gray, dirty white, or pure white. Ears vary from small to large, between 21 and 23 mm in *A. pictus* and *A. sublimis* to 30 mm in *A. boliviensis*, often with buffy or white postauricular patches. Feet are stout, generally heavily fringed; digit V of the pes is long, reaching to the end of the first phalanx of digit IV (Gyldenstolpe 1932). Palms and soles are naked with six large, low pads. There are four pairs of mammae.

A narrow interorbital region (often narrower than the width of their nasals), with smooth and rounded supraorbital margins that lack ridges or beads, and widely expanded zygomatic arches, characterize the skull of *Auliscomys*. The margins of the supraorbital region diverge anteriorly; the nasals are short, barely extending beyond the maxillofrontal suture at its contact with the lacrimal bone (Steppan 1995); lacrimals are well developed; zygomatic notches are deep and with the anterodorsal margins of the zygomatic plates smoothly rounded (no zygomatic spines); the anterior margin of the orbit is positioned approximately even with the alveolus of M1. Incisive foramina are long, projecting distally between the M1s; the mesopterygoid fossa is narrow (much narrower than the parapterygoid fossae) and does not penetrate anteriorly between the M3s; parapterygoid fossae are broad and posteriorly divergent; the foramen ovale is small (except in some populations of *A. sublimis*); sphenopalatine vacuities are well developed; the carotid circulation is type 1 of Voss (1988), where the stapedial foramen and the posterior opening of the alisphenoid canal are large and the squamosal-alisphenoid groove and the sphenofrontal foramen are present. The internal carotid canal is bounded by both the auditory bullae and the occipital bone. A capsular process is noticeably indistinct.

Upper incisors are orthodont with fine striae or shallow grooves, and with enamel colored whitish yellow to deep orange (see below). Molar rows are parallel. The molars are moderately low-crowned, robust, with main cusps alternating along the longitudinal plane but seldom reaching the midline. The anterocone of M1 is narrow and, except in *A. pictus*, undivided. A small labial root of M1 is set medially. Both M1 and M2 lack a mesostyle, and all three upper teeth lack parastyles. An anteromedian flexid in m1–3 is weakly developed and lost with wear, a posterolophid of m1 is distinct at all wear classes, the procingulum of M2 is moderately to well developed, and m3 has a highly reduced mesoflexid.

SYNONYMS:

Hesperomys: Waterhouse, 1846:9; part (description of *boliviensis*); not *Hesperomys* Waterhouse.

Rheithrodon Thomas, 1884:457; part (description of *pictus*); not *Reithrodon* Waterhouse, and incorrect subsequent spelling of *Reithrodon* Waterhouse.

Rhithrodon Thomas, 1893a:337; part (listing of *pictus*); not *Reithrodon* Waterhouse, and incorrect subsequent spelling of *Reithrodon* Waterhouse.

Phyllotis: E.-L. Trouessart, 1897:534; part (listing of *boliviensis*); not *Phyllotis* Waterhouse.

Euneomys: Thomas, 1901f:254; part (listing of *pictus*); not *Euneomys* Coues.

Auliscomys Osgood, 1915:190; as subgenus of *Phyllotis*, type species *Reithrodon pictus* Thomas, by original designation.

Euneomys (Auliscomys): Thomas, 1916a:141; name combination as a valid subgenus.

Maresomys Braun, 1993:40; type species *Hesperomys boliviensis* Waterhouse, by original designation.

Auliscomys (Maresomys): S. Anderson, 1997:443; name combination.

REMARKS: Originally diagnosed as a subgenus of *Phyllotis* (Osgood 1915), members of *Auliscomys* were subsequently recognized either at that rank (Osgood 1947; Pearson 1958) or as a full genus (Thomas 1926b; Gyldenstolpe 1932; Musser and Carleton 2005). The monophyly of *Auliscomys* is supported by karyotypic (Couve Montane 1975; Pearson and Patton 1976), morphological (Braun 1993; Steppan 1993, 1995), and molecular analyses (Salazar-Bravo et al. 2013). Some molecular data indicate that *Auliscomys* and *Galenomys* share a most recent common ancestor (e.g., Steppan et al. 2007; Salazar-Bravo et al. 2013). *Maresomys* was described by Braun (1993) to include *A. boliviensis* based a presumptive sister-taxon relationship between *boliviensis* and *Galenomys garleppi*. However, this hypothesis did not receive support from other morphological (Steppan 1995; Steppan and Sullivan 2000) or molecular analyses (Salazar-Bravo et al. 2013). Therefore, I follow Steppan and colleagues and treat *Maresomys* as a synonym of *Auliscomys*. Anderson (1997) included *Maresomys* as a subgenus of *Auliscomys*.

Three recent species are currently recognized: *Auliscomys pictus*, *A. sublimis*, and *A. boliviensis*. *Auliscomys formosus*, a fossil form from the Montehermosan age deposits (ca. 5–4 mya) of Buenos Aires province, Argentina, is currently the oldest known sigmodontine fossil in South America (Pardiñas and Tonni 1998). Two other fossil forms associated with the genus (*Auliscomys fuscus* and *Auliscomys osvaldoreigi*) are apparently better arranged under the extinct genera *Olimpicomys* and *Panchomys*, respectively (Teta, Pardiñas et al. 2009). Some authors (e.g., Simonetti and Spotorno 1980; L. I. Walker and Spotorno 1992) assigned *Loxodontomys micropus* to *Auliscomys*, a hypothesis rejected by cladistic analyses of morphology (Braun 1993; Steppan 1993 1995; Spotorno, Cofre et al. 1998; Steppan and Sullivan 2000). Furthermore, analyses of combined mtDNA and nucDNA sequences have failed to support a close relationship between *Auliscomys* and *Loxodontomys*, regardless of the method of analysis (Salazar-Bravo et al. 2013).

KEY TO THE SPECIES OF *AULISCOMYS*:

1. Tail shorter than 70 mm; upper incisors sometimes faintly grooved, somewhat proodont, and white to pale yellow . . . . *Auliscomys sublimis*
1'. Tail longer than 70 mm; upper incisors slightly yellow 2
2. Upper incisors with longitudinal grooves, yellow or orange, not proodont; head plus body shorter than 125 mm; ear length from notch shorter than 25 mm; soles of feet pale . . . . *Auliscomys pictus*
2'. Upper incisors without longitudinal grooves, pale yellow, distinctly proodont; head plus body of adults longer than 125; ear length from notch longer than 25 mm, well clothed with hair and with tuft of long hair at base; soles of hindfeet blackish . . . . . *Auliscomys boliviensis*

Auliscomys boliviensis (Waterhouse, 1846)

Bolivian Pericote, Achulla, Achohalla, Pericote Orejón Boliviano

SYNONYMS:

Hesperomys Boliviensis Waterhouse, 1846:9; type locality "Bolivia, near Potosi," "a few leagues south of Potosi, at an elevation of 12,000 feet" (see Pearson 1958:452).

[*Hesperomys (Hesperomys)*] *Waterhousii* E.-L. Trouessart, 1880b:138; replacement name for *boliviensis* Waterhouse under the assumption that it was a secondary homonym of *Akodon boliviensis* Meyen (see Hershkovitz 1962:417).

H[esperomys]. (Phyllotis) boliviensis: Thomas, 1884:449; name combination.

[Phyllotis] boliviensis: E.-L. Trouessart, 1897:534; name combination.

Phyllotis boliviensis flavidior Thomas, 1902f:248; type locality "Bateas, Cayulloma [= Cailloma], [Arequipa,] Peru. Altitude 4500 m."

P[hyllotis]. (Auliscomys) boliviensis: Osgood, 1915:191; name combination.

P[hyllotis]. (Auliscomys) b[oliviensis]. flavidior: Osgood, 1915:191; name combination.

Euneomys (Auliscomys) boliviensis: Thomas, 1916a:143; name combination.

Euneomys (Auliscomys) boliviensis flavidior: Thomas, 1916a: 143; name combination.

Auliscomys boliviensis boliviensis: Gyldenstolpe, 1932a:94; first use of current name combination as species.

Auliscomys boliviensis flavidior: Gyldenstolpe, 1932a:95; name combination.

Phyllotis (Auliscomys) boliviensis flavidior: Sanborn, 1950: 9; name combination.

Phyllotis boliviensis boliviensis: Pearson, 1958:452; name combination.

Auliscomys (Maresomys) boliviensis: Braun, 1993:40; name combination.

DESCRIPTION: Pale, long-eared species (ear length often >25 mm), with tail shorter than head and body (57 to 95%) and hindfeet with blackish soles (Pearson 1958). Considerable variation in head and body length, with some populations in southern Bolivia averaging 125 mm and others 112 mm. In general, stout-bodied with long and soft fur; upper surface of head and body with coarse mixture of gray, brown, and buff or ochraceous, sides paler, and muzzle whitish; ochraceous lateral line not always well developed; under parts whitish or pale gray, more or less washed with buff, and with ochraceous pectoral streak sometimes present (Hershkovitz 1962). Dorsal surface of tail often ochraceous with terminal portion darker, underside whitish becoming ochraceous terminally; tail not tufted at tip. Hindfeet comparatively larger than in congeners, with silvery-white, gray, or buff dorsal surfaces, sometimes with ochraceous metatarsal patch; soles coarsely scutellated, naked, and blackish. Ears long (>25 mm), covered internally with short yellow hairs and externally with much longer, rusty yellow hairs. Prominent ochraceous and white preauricular and buffy cream to whitish postauricular ear patches present.

Skull with strongly bowed zygomatic arches, very narrow interorbital region, and deep and broad zygomatic notches. Skull with interorbital region anteriorly divergent, narrowest point located behind orbit; rostrum wider than interorbital region (Steppan 1995); braincase not globular but squared in profile; tympanic bullae large and thin walled. Hershkovitz (1962) indicated that anteroposterior length of tympanic bullae minus their tubes is subequal to alveolar length of molar row. Incisive foramina long and broad, posteriorly reaching almost to protocones of M1s; foramina widest at, or near to, premaxillomaxillary suture and taper slightly posteriorly. Dorsal edge of mesopterygoid fossae squared (in most individuals) and does not reach M3 alveolus. Parapterygoid fossae diverge dorsally, much broader than mesopterygoid fossa. Sphenopalatine vacuities well developed. Most adult specimens exhibit well-developed lacrimals.

Upper incisors distinctly proodont, ungrooved (or if present, sulci only faint, shallow striae usually not visible to unaided eye), and with pale yellow enamel. Molar rows parallel sided or slightly bowed. Molars details similar to those of *A. pictus*, but cusps not flexed forward, protoflexus of M2 more reduced, anterior cingulum of M3 and vestige of mesoloph larger, and protoflexid of m1 obsolete (Hershkovitz 1962). A persistent parastyle present on M2 (Mann 1978).

Vertebral column with 13 thoracic, 6 lumbar, and between 25 and 27 caudal elements (Steppan 1995). Spotorno (1987) provided anatomical details of male phallus and baculum. In comparison with both *A. pictus* and *A. sublimis*, glans of *A. boliviensis* with short rather than long urethral process, and baculum without basal median notch. Caecum coiled and long, followed by complex colon (Mann 1978).

DISTRIBUTION: *Auliscomys boliviensis* occurs across the Altiplano of southern Peru, northern Chile, and west-central Bolivia, between 3,500 and 5,000 m.

SELECTED LOCALITIES (Map 251): BOLIVIA: Chuquisaca, Sucre (Hershkovitz 1962); Oruro, 7 km S and 4 km E of Cruce Ventilla (MSB 57101), Estancia Agua Rica, 22 km S and 40 km SE of Sajama (MSB 57102), Livichuco (BM 1902.2.2.16); Potosí, Potosí (Hershkovitz 1962). CHILE: Arica y Parinacota, Guallatiri (Pine et al. 1979), 6 mi E of Putre (USNM 541735). PERU: Arequipa, Huancarama, Orcopampa (MUSA 512), Huaylarco, 54 mi ENE of Arequipa (MVZ 115892), San Ignacio (MCZ 39494), San Juan de Tarucani (MUSA 1403); Moquegua, Caccachara (MCZ 42922); Puno, Pampa de Ancomarca, 123 km S of Ilave (MVZ 115896); Tacna, Lago Suche (MVZ 115893), 13 km NE of Tarata (MVZ 141600).

SUBSPECIES: Both Pearson (1958) and Hershkovitz (1962) recognized two subspecies, *Auliscomys boliviensis boliviensis* in central Bolivia and northern Chile and *Auliscomys boliviensis flavidior* in southwestern Peru. However,

Pearson (1958:452) noted that "the boundary between the two subspecies could be drawn almost anywhere" because the differences between the two are so slight and abundant intermediates exist.

NATURAL HISTORY: The species occurs in the Altiplano, usually in open country with sparse vegetation; it has been observed living among boulder-strewn slopes among yareta (*Azorella compacta* [Apiaceae]), stone walls, and even within abandoned tuco-tucos (*Ctenomys*) burrows (Pearson 1951). In Bolivia, it was found occupying active burrows of *Ctenomys opimus* and sharing burrows with *Galea musteloides*. In northern Chile, it was trapped in tola (*Parastrephia lepidophylla* [Asteraceae]) dominated flatlands (Mann 1978). Individuals are diurnal (Pearson 1951; Mann 1978); although they are not reported to be gregarious, at one locality in Oruro, Bolivia (7 km S and 4 km E Cruce Ventilla), 7 and 10 individuals occupied a small (150 m²) area within active tuco-tuco burrows; these rodents would stick their heads up from inside the burrows as people approached (J. Salazar-Bravo, pers. comm.). They are reported to be fairly tame and can be handled and fed readily (Mann 1945); they interact with vizcachas, a species of much larger size, and cue in to its alarm calls (Pearson 1951). They are reported to feed on a variety of foods, including lichens, forbs, and other vegetation (S. I. Silva 2005). Bozinovic and Rosenmann (1988a) reported basal metabolic rate and thermal conductance values.

Males of different populations apparently enter reproductive activity at different times of the summer, some in September and others in early December (Pearson 1951). Pregnant females with three to four embryos were recorded in October and December in southern Peru. Reported ectoparasites and endoparasites for the species include a flea of the genus *Cleopsylla* as well as lice (*Hoplopleura affinis*), mites (*Atricholaelaps glasgowi*), two species of chiggers in the genus *Trombicula*, and nematodes of the genus *Stilestrongylus* (Colombetti et al. 2008; Diaw 1976; Pearson 1951). The tapeworm *Taenia talicei* was found in two of eight Bolivian animals examined (Scott L. Gardner, pers. comm.).

REMARKS: The record for this species from La Cumbre (La Paz, Bolivia) by Mercado and Miralles (1991) was based on a misidentified specimen of *Punomys* (Salazar-Bravo et al. 2011).

Auliscomys boliviensis has a diploid number of $2n=22$ and $FN=30$ in Peru and northern Chile (Couve Montane 1975; Pearson and Patton 1976; L. I. Walker and Spotorno 1992). In contrast, two males and a female from Oruro, Bolivia, had a diploid number of $2n=20$ (Peurach 1994). The question of whether these regional karyotypic differences might support the current subspecies allocations or even reflect species-level boundaries underscores the need for further taxonomic work.

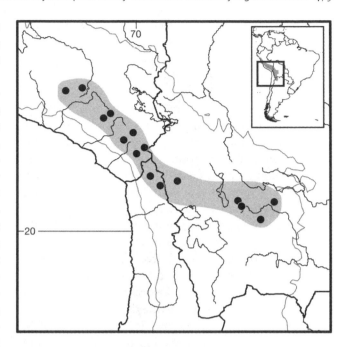

Map 251 Selected localities for *Auliscomys boliviensis* (●). Contour line = 2,000 m.

Auliscomys pictus (Thomas, 1884)
Colorful Pericote, Hucucha Guanaco, Ratón Orejón Pintado

SYNONYMS:

Rheithrodon [*sic*] *pictus* Thomas, 1884:457, plates 19–21; type locality "Junin," given as "Junín, 13,700 feet, Department of Junín, Peru" by Pearson (1958).

Rhithrodon [*sic*] *pictus*: Thomas, 1893a:337.

E[*uneomys*]. *pictus*: Thomas, 1901f:254; name combination.

P[*hyllotis*]. (*Auliscomys*) *pictus*: Osgood, 1915:191; name combination.

Phyllotis (*Auliscomys*) *decoloratus* Osgood, 1915:191; type locality "Tirapata, Dept. Puno, Peru."

Auliscomys pictus: Thomas, 1926b:316; first use of current name combination.

Phyllotis (*Auliscomys*) *pictus*: Pearson, 1958:448; name combination.

Phyllotis pictus: Hershkovitz, 1962:404; name combination.

DESCRIPTION: Medium sized with head and body length 100–133 mm; back and sides of trunk brownish; rump, outer sides of thigh, and base of tail tawny to ochraceous. Head and shoulders often grizzled, and lateral line, when present, ochraceous. Under parts gray to dirty white with dark basal portion of hairs showing through; small ochraceous pectoral streak present in some populations (Hershkovitz 1962). Tail bicolored, typically more than 68 mm long but <91% of head and body length; never conspicuously tufted. Ears <27 mm, with inner posterior surface ochraceous and pale postauricular tufts usually present. Hindfeet short (average 25.8 mm near Lake Titicaca in south to 27 mm in Junín department in north [Pearson 1958]), with upper surfaces gray, lightly washed with ochraceous; soles finely scutellated.

Skull characterized by very narrow interorbital region, with sides either concave or parallel sided, and with squared but never ridged or beaded edges; rostrum short and wide; zygomatic arches moderately expanded; interparietal well developed; posterolateral palatal pits present; and bullae rather noninflated with anteroposterior length, not including bony tubes, shorter than alveolar length of molar row (Hershkovitz 1962).

Upper incisors grooved, moderately heavy, and orthodont. Molar rows parallel sided or slightly divergent posteriorly. Both upper and lower first molars with four roots; all cheek teeth subhypsodont with cusps more or less ovate in outline (Hershkovitz 1962). Procingulum of M1 divided in young individuals; old individuals or heavily worn M1s ovate in outline. Procingulum of M2 well developed, with well-developed paraflexus and obsolete protoflexus. Vestigial mesoloph often present in M1–2. Posteroloph of m2 relatively well developed but weakly developed in m3; m2 with persistent protoflexus even in well-worn tooth; occlusal pattern of m3 sigmoid in shape.

Axial skeleton with mode of 13 thoracic and six lumbar vertebrae and between 22 and 26 caudal vertebrae (Steppan 1995). Gall bladder present (Voss 1991a); Dorst (1973b) described stomach structure; and Hooper and Musser (1964) detailed anatomy phallus and baculum.

Color and size noticeably variable throughout range. Larger individuals (ca. 10% longer in head and body and 4% longer in greatest length of the skull) occur in northern portion of range near type locality relative to specimens from Bolivian-Peruvian border. In general, animals from more humid drainage systems (e.g., populations on eastern escarpment of Altiplano from above Limbani, Puno, Peru) with longer-tailed and darker color tones than those from drier regions of Titicaca drainage (Pearson 1958).

DISTRIBUTION: *Auliscomys pictus* occurs throughout the highlands of Bolivia and Peru, very likely extending into extreme northern Chile, although there are no confirmed records from that country. Known localities range from La Libertad department in central Peru to La Paz department in Bolivia, at elevations from 2,590 to 4,820 m (Pearson 1958; Hershkovitz 1962; S. Anderson 1997).

SELECTED LOCALITIES (Map 252): BOLIVIA: La Paz, Limani (CBF 851), Pongo (AMNH 72727), Quiswarani (CBF 6431). PERU: Ancash, 1 km N and 12 km E of Pariacoto (MVZ 138126); Arequipa, Toccra Pampa, Caylloma (MUSA 881); Ayacucho, Tucumachay Cave (UMMZ

122625), 15 mi WNW of Puquio (MVZ 138133); Cusco, 55.5 km by road N of Calca (UMMZ 160541); La Libertad, 10 mi WNW of Santiago de Chuco (MVZ 138142); Lima, Pacomanta, km 120 on Lima to Huarochirí highway (KU 143594); Pasco: Chipa (AMNH 60580); Puno, 1 mi S of Limbani (MVZ 116150), Santa Rosa (MCZ 42806).

SUBSPECIES: *Auliscomys pictus* is monotypic. Neither Pearson (1958) nor Hershkovitz (1962), in their respective reviews of this species, recognized formal geographic units despite recording variation in size and color across the species's range.

NATURAL HISTORY: *Auliscomys pictus* has been trapped in relatively mesic habitats in the Altiplano containing an abundance of grasses (mostly *Stipa ichu*) and other herbaceous vegetation at high elevations (Pizzimenti and DeSalle 1980) and also in stone walls and places far from water (Pearson 1951). Animals were active day and night (Pearson 1951). In southern Peru, the species used a wide range of food items (Pizzimenti and DeSalle 1980), including both plant material from forbs (93% of diet) and insects (7%). A report from Chile stating that this species feeds on mushrooms, seeds, and fruits is likely incorrect and probably refers to *Loxodontomys micropus* (S. I. Silva 2005).

Pearson (1951) reported that breeding commenced between September and December in southern Peru. Hershkovitz (1962) noted a pregnant female with five embryos from Cusco department taken in April; females, some with up to five embryos and six uterine scars were trapped in January in Arequipa (H. Zeballos, pers. comm.). S. Anderson (1997) reported a female with three embryos in January from La Paz, Bolivia.

Together with *Auliscomys sublimis*, *Auliscomys pictus* composed more than 30% of the diet of owls at Pairumani, Peru (Pearson 1951), and remains of this species were reported in the scats of Culpeo (*Lycalopex culpaeus*) near La Paz, Bolivia (Mercado and Miralles 1991), and the highlands of Abiseo National Park in Peru (Romo 1995). Pearson (1951) mentioned that the ears of *Auliscomys pictus* are often "nibbled" away as a result of infestation by orange mites, and Sarmiento et al. (1999) recorded two genera of nematodes (*Parastrongylus* and *Trichuris*) for the species. Red blood cell volumes and hemoglobin concentrations were within the range of variation of other rodents at both high and low elevations (Morrison et al. 1963).

REMARKS: *Auliscomys decoloratus* was described as a distinct species (Osgood 1915) based on the small size and pale coloration of an individual from Tirapata in southern Peru. Subsequent reviewers (Sanborn 1950; Pearson 1958; Hershkovitz 1962) have consistently regarded *decoloratus* as a synonym of *A. pictus*, and unworthy of even subspecific recognition.

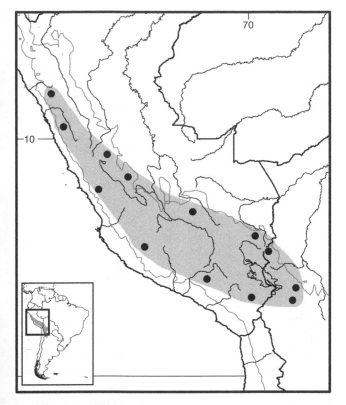

Map 252 Selected localities for *Auliscomys pictus* (●). Contour line = 2,000 m.

Pearson (1972) and Peurach (1994) described a karyotype with $2n = 28$, $FN = 30$ from Peruvian and Bolivian localities, respectively.

Auliscomys sublimis (Thomas, 1900)
Lofty Pericote, Pericote de la Puna, Pericote Andino
SYNONYMS:

Phyllotis sublimis Thomas, 1900h:467; type locality "Rinconado Malo pass, above Caylloma [= Cailloma], on the Sumbay road, Peru. Altitude 5500 metres."

E[uneomys]. sublimis: Thomas, 1901f:254; name combination.

Euneomys sublimis: Thomas, 1902b:134; name combination.

Phyllotis (Auliscomys) sublimis: Osgood, 1915:190–191; name combination.

[Euneomys (Auliscomys)] sublimis: Thomas, 1916a:143; name combination.

Euneomys (Auliscomys) leucurus Thomas, 1919f:129; type locality "La Lagunita, Maimara, Jujuy," Argentina.

Auliscomys sublimis: Gyldenstolpe, 1932a:95; first use of current name combination.

Auliscomys leucurus: Gyldenstolpe, 1932a:95; name combination.

Phyllotis (Auliscomys) sublimis leucurus: Sanborn, 1950:8; name combination.

Phyllotis (Auliscomys) sublimis: Pearson, 1958:446; name combination.

Phyllotis sublimis sublimis: Pearson, 1958:447; name combination.

Phyllotis sublimis leucurus: Pearson, 1958:448; name combination.

DESCRIPTION: Fluffy-furred, short-tailed species with widely bowed zygomatic arches, pinched interorbital region, and small bullae typical of genus (Pearson 1958). Adult body size (head and body length) averages 110 mm (Pearson 1951). Dorsal pelage long, extremely soft, and fine. Overall dorsal color buffy yellow mixed with some black hairs; head pelage somewhat more grizzled than rest of body; no eye-ring present; sides of body paler turning to white on cheeks; although lateral line usually indistinct, under parts sharply defined whitish to pale gray with base of fur often plumbeous; upper surfaces of forefeet and hindfeet silvery white; hindfeet naked below except on heel. Tail length always <68 mm; tail itself covered with fine white hair, often distinctly bicolored, but lacking tuft at tip. Ears are short (average 22.2 mm) and covered with fine yellowish hair inside; distinct postauricular patches often present.

Skull quite similar to that of *A. pictus*. Interorbital region narrow, parallel sided in middle, divergent anteriorly, and with rounded margins; zygomatic arches evenly bowed, which in combination with globular braincase give skull more "rounded" profile than other species of highland phyllotines; rostrum narrower than in *A. pictus*; interparietal small; bullae only slightly smaller than those of *A. pictus* and pear rather than flask shaped; foramen ovale about size of M3 and in many specimens completely unobstructed.

Upper incisors orthodont with enamel colored pale yellow to almost white and with anterior faces marked by faint, shallow grooves (sometimes hard to see by unaided eye). Molar rows parallel. Molars hypsodont, cusps subovate in outline, with flexi from opposite sides not reaching midline (Hershkovitz 1962). Procingulum of M1 broad but undivided; metaloph and paraloph well developed; base of mesoflexus and hypoflexus not alternating, which defines an almost straight median mure. M1 with 4 roots. M2 with small procingulum and weakly developed paraflexus and protoflexus, both becoming lost with moderately wear. Vestige of mesoloph sometimes present on M1–2. Overall, outline of M2 sigmoid when worn. Procingulum of m1 weakly divided by anteromedian flexid, which becomes lost in worn teeth; m1 with 3 roots; and base of the protoflexid opposes that of metaflexid, forming a short anterior murid in medial position (Hershkovitz 1962).

Steppan (1995) reported 13 thoracic, 6 lumbar, and between 24 and 25 caudal vertebrate. Phallus and baculum similar to *A. pictus* (Spotorno 1987).

Both Pearson (1958) and Hershkovitz (1962) called attention to large variability within and among samples, especially in skull size, head and body length, and dorsal and ventral coloration. Specimens from the more humid eastern Andean slope are darker than those from comparatively dry Altiplano of southern Peru (Pearson 1962).

DISTRIBUTION: *Auliscomys sublimis* occurs in the Altiplano highlands from southern Peru through west-central Bolivia and northern Chile to northern Argentina, between 3,200 to above 5,000 m.

SELECTED LOCALITIES (Map 253): ARGENTINA: Jujuy, ca. 3 km S of Guairazul (PEO-e 301); Salta, 3 km E of La Poma (PEO-e 234). BOLIVIA: Cochabamba, 9.5 km by road SE of Rodeo, then 2.5 km on road to EN-TEL antenna (AMNH 260915); La Paz, 20 mi S of La Paz (MVZ 120168), Ulla-Ulla (MSB 75283); Potosi, Lipez (BM 26.6.12). CHILE: Antofagasta, Toconce, 60 mi ENE of Calama (MVZ 116793); Arica y Parinacota, Choquelimpie, 114 km NE of Arica (LCM 1194). PERU: Arequipa, 5 km W of Cailloma (MVZ 174013), Salinas, 4,316 m (MUSA 1141), Toccra Pampa (MUSA 1140); Puno, 8 mi SSW of Limbani (MVZ 114735); Ayacucho, 15 mi WNW of Puquio (MVZ 138143); Tacna, Pampa de Titire, 29 km NE of Tarata (MVZ 115914).

SUBSPECIES: Pearson (1958) recognized two subspecies, the nominotypical form from southern Peru south through northern Chile and north-central Bolivia, and *A. s. leucurus* from extreme southern Bolivia, adjacent Antofagasta, Chile, and northernmost Argentina. The southern race differs primarily in absolutely and proportionally longer toothrows and grayer venters. Pearson (1951) also noted that specimens from the more humid eastern Andean slopes in southern Peru were noticeably darker than those from the drier Altiplano, but did not recommend a formal distinction from typical *A. s. sublimis*. Hershkovitz (1962:430), while also delineating two subspecies, argued that "differences in molar row length and color of underparts between northern and southern populations of the species are not significantly greater than differences between local populations in Arequipa Department, Peru." S. Anderson (1997) retained Pearson's subspecific division for Bolivian populations.

NATURAL HISTORY: *Auliscomys sublimis* reaches record elevations of up to 5,500 m (Thomas 1900h; Pearson 1958). Reported as "one of the commonest mice of the Altiplano of southern Peru" (Pearson 1951), this species is often present in areas with abundant vegetative cover, in pastures, rocky outcrops, and *Stipa*-dominated scrublands, or among boulders and yareta (*Azorella compacta*

[Apiaceae]). It may also be found in high-elevation meadows (called *bofedales*), where it excavates its own burrows or occupies old tuco-tuco or guinea pig burrows (Mercado and Miralles 1991). Pearson (1951) and Hershkovitz (1962) suggested that the species might aestivate, but no direct evidence has been reported. Pizzimenti and DeSalle (1980) argued that *A. sublimis* was omnivorous, as the diet included 0–79% insects, but Mercado and Miralles (1991) found the species to be a specialized herbivore. Pearson (1951) described the habits as gregarious and nocturnal. Some females were sexually precocious (Pearson 1958), and females with up to five embryos were found in Arequipa department in Peru (H. Zeballos, pers. comm.). Pearson (1951) recorded scrotal males at the end of July in southern Peru (Pearson 1951), and a young individual was trapped in May at Sierra de Tilcara, Argentina (M. M. Díaz and Barquez 2007).

Together with *Auliscomys pictus*, *Auliscomys sublimis* comprised more than 30% (by volume) of the prey of owls in southern Peru (Pearson 1951), and it was reported to be the dominant species from owl pellet samples in Jujuy, northern Argentina (Jayat, Ortiz, Pacheco, and González 2011). P. E. Ortiz, Cirignoli et al. (2000) recovered the remains of at least 10 individuals in pellets of the Short-eared Owl (*Asio flammeus*). The species was present in the diet

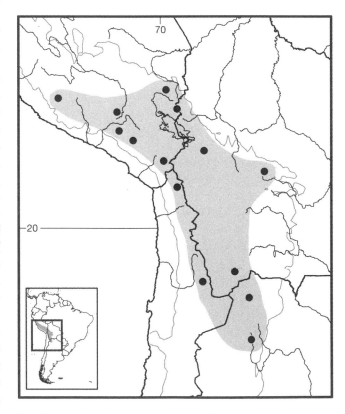

Map 253 Selected localities for *Auliscomys sublimis* (●). Contour line = 2,000 m.

of both the Andean cat (*Leopardus jacobitus*) and Pampas cat (*L. colocolo*) in Bolivia (L. Villalba, pers. comm.).

REMARKS: Teta and Ortiz (2002) reported Holocene records from archaeological sites in Jujuy, Argentina. Mitochondrial and nuclear sequence analyses support a sister relationship between *A. sublimis* and *A. pictus* (Steppan et al. 2007; Salazar-Bravo et al. 2013). Pearson (1972) described a karyotype with $2n = 28$, FN = 30, identical to that of *A. pictus*.

Genus *Calassomys* Pardiñas, Lessa, Salazar-Bravo, and Câmara, 2014

Ulyses F. J. Pardiñas and James L. Patton

This recently described genus has the smallest geographic range of any within the tribe Phyllotini, as the single species (*C. apicalis*) is known from only three, closely adjacent localities in the Cerrado ecoregion of Minas Gerais state, Brazil. Although the general external morphology, especially the abundant and long mystacial vibrissae combined with its long tail, resembles some medium-sized oryzomyine and thomasomyine genera, details of the skull and molars clearly link *Calassomys* with other phyllotines. Morphological attributes suggest an affinity with *Calomys*, but molecular sequence data place *Calassomys* basal to all other members of the tribe. Pardiñas et al. (2014) provide detailed comparisons to *Calomys* and several nonphyllotine genera (*Delomys*, *Wiedomys*, *Zygodontomys*, and *Necromys*) with which *Calassomys* shares some features.

The single species is medium sized (head and body length ranges from 103–125 mm) with a long (tail length 130–158 mm; >125% of head and body length) and bicolored tail, brown above and white below but with the terminal 25–50 mm completely white and ending in a sparse apical tuft 5–6 mm in length. The tail is covered by sets of three delicate hairs per scale, each dorsal hair covering two scale rows. Ears are moderately large, rounded, and partially naked, covered on both surfaces by short and delicate brown hairs but with concha and tragus strikingly yellow. Ear length (from notch) averages >18.2 mm. Dorsal fur is characterized by dark gray base and brown-ochraceous overtones; guard hairs are long (ca. 15 mm) and soft. Ventral hairs have gray bases and white tips; the muzzle and cheeks are whitish gray. Mystacial vibrissae are long, surpassing the ears when laid back alongside the head. A conspicuous black ring surrounds eyes. Forefeet are covered dorsally by short, whitish hairs; digits II and dIV are subequal in size, and dII and dV are shorter, with dII clearly longer than dV; dI is vestigial but has a small nail; dII–V possess inconspicuous claws (about 1.8 mm long), dorsally covered by whitish ungual hairs. The ventral surface of the forefeet is naked and unpigmented, with two carpal and three interdigital large and fleshy pads. The hindfeet are covered dorsally by whitish hairs; the plantar surface is naked and unpigmented, slightly squamated, and has four large, nearly ovate interdigital pads plus a hypothenar pad subequal in size to the first interdigital and located intermediate between the first interdigital and thenar pad; the thenar pad is enlarged and narrow. Long hairs on the sides of the heel partially cover the calcaneal area.

The skull is domed in lateral profile with the braincase generally dominating the dorsal aspect. The rostrum is long and moderately broad, its sides tapering gradually but with the premaxillary bones visible for almost their entire length along the nasal margins. The nasals are parallel sided and broad, projecting anteriorly to form an incipient rostral tube or "trumpet," completely obscuring the upper incisors in dorsal view. Lacrimals are large. The interorbital region is wide and dorsally flat, with the constriction placed immediately posterior to inconspicuous frontal sinuses; margins of the frontal bones are sharp and posteriorly divergent but do not form supraorbital shelves. The coronal suture is V-shaped. The braincase is large and round, and lacks both temporal and lambdoidal ridges; the suture between the parietals and squamosals is concealed by a slightly developed bony ridge in adult specimens. The zygomatic plates are narrow and high, with a moderate forward extension forming a free anterodorsal margin and defining shallow zygomatic notches. The zygomatic arches are delicate; their midpoints dip ventrally to remain visible above the level of the orbital floor. A short jugal distinctly separates maxillary and squamosal portions of the arch. Parietals lack extensions in lateral view. The hamular process of the squamosal is long, thick, and has an unexpanded posterior end; the subsquamosal fenestra is reduced with respect to the postglenoid foramen. Incisive foramina are wide and long, compressed anteriorly, with the posterior ends rounded and extending slightly past the anterior roots of M1s; the premaxillary process occupies about three-fourths the length of the foramina. The palate is wide and short, with diminutive posterolateral palatal pits placed on the sides of the mesopterygoid fossa. That fossa is especially wide, with its anterior border reaching the posterior face of M3s; the anterior portion is slightly lyre-shaped due to the presence of a median palatine process. Sphenopalatine vacuities on the roof of the mesopterygoid fossa are typically present as narrow slits. Pterygoids are short, small, and parallel; the parapterygoid plates are flat and reduced in size with respect to the mesopterygoid fossa. Auditory bullae are medium in size, flat in profile, and subtriangular in outline, with a short and wide bony tube. The carotid circulation is primitive, or pattern 1 (*sensu* Voss

1988), characterized by a large stapedial foramen and posterior opening of alisphenoid canal, squamosal-alisphenoid groove, and sphenofrontal foramen.

The mandible is enlarged with the ascendant portion subtriangular in outline. The anterior point of the diastema is located slightly below the alveolar plane; the posterior face of the diastema is smooth; and the mental foramen is moderately visible in labial view. The masseteric crest is evident, with its upper and lower ridges meeting to produce a thick edge running from the plane defined by the posterior face of m1 to slightly above the level of the mental foramen. The capsular projection is conspicuous, lying near the base of the coronoid process, which is reduced and slightly inflected backward. The condyloid process is wide and short; the angular process is short. The lunar notch is scarcely developed. The mandibular foramen is compressed and located above the mylohyoid line; the retromolar fossa is moderately enlarged.

Upper incisors are faced with bright orange enamel; they are opisthodont, smooth, and narrow, with juxtaposed tips showing a straight cutting edge; the dentine fissure is a long, straight slit (see Steppan 1995). Upper molars are crested, with their labial cusps higher than lingual ones, markedly opposite in pattern, brachydont, and small with respect to the skull. M1 is trilophodont, subrectangular in outline, with a well-developed procingulum usually without an anteromedian flexus; mures are longitudinally oriented; the hypoflexus and protoflexus are transverse; protocones and hypocones are subequal in size; minute mesolophs are consistently present; and both parastyle and mesostyles are usually present as thickenings of the labial cingula. The anteroloph and mesoloph exhibit a tendency to coalesce with the paracone and metacone, respectively. The hypoflexus is wide and posteriorly directed. M2 is square in outline and bilophodont; the anteroloph is reduced or absent; a mesoloph is typically present; and a wide mesoflexus is opposite the hypoflexus. M3 is cylindroform and moderately reduced in size with respect to M2; the hypoflexus is vestigial, but the mesoflexus persists as a small fossette. M1 and M2 appear to have three roots, and M3 one or two. Lower molars are also crested, transversally compressed, with the main cusps moderately alternating and both mesolophids and mesostyles absent. The trilophodont m1 has a well-developed and fan-shaped procingulum, also usually without a anteromedian flexid; the mesoflexid is especially wide; protoconid and hypoconid areas are subtriangular in outline; the entoconid is enlarged and bulbous; and the posterolophid is reduced. The m2 is bilophodont, with a reduced posterolophid. And m3 is subtriangular in outline, sigmoid, and has a reduced mesoflexid. All lower molars lack mesolophids.

The axial skeleton consists of 12 thoracic ribs and 12 thoracic, 7 lumbar, and 30 caudal vertebrae. The neural spine of the second cervical vertebra is enlarged, has a distinctive knob, and does not overlap with the third cervical vertebra. The neural spine of the second thoracic vertebra is conspicuously enlarged, twice or more so than the adjacent spines. Hemal arches are absent from the caudal vertebral, but hemal processes are present, starting between the second and third caudal vertebrae, becoming more pronounced between the fourth and fifth, then decreasing in size until they disappear. The humerus lacks the entepicondylar foramen, and the supratrochlear fossa is perforated.

Little information is available on soft anatomy. There are two diastemal and five interdental palatal rugae. A gall bladder is present, and the stomach is unilocular-hemiglandular (see Carleton 1973), with the antrum slightly smaller than the corpus. The bordering fold crosses the lesser curvature at the apex of the incisura angularis and the greater curvature at a locus opposite the incisura angularis; the glandular epithelium extends past the esophageal orifice and protrudes as a wide bulge into the corpus.

SYNONYM:

Calassomys Pardiñas, Lessa, Salazar-Bravo, and Câmara, 2014:202; type species *Calassomys apicalis*, by original designation.

REMARKS: *Calassomys* appears as basal to all other extant genera of phyllotines in phylogenetic analyses of mtDNA and nucDNA sequences (Pardiñas et al. 2014), a position consistent with the persistence of vestigial mesolophs and mesostyles on both M1 and M2, unique features among the Phyllotini. This phylogenetic hypothesis suggests that the ancestor of the phyllotine crown group ranged outside of the Andes, the area usually credited as the ancestral home of the tribe (Reig 1986).

Calassomys apicalis Pardiñas, Lessa, Salazar-Bravo, and Câmara, 2014
Calaça's White-tailed Mouse
SYNONYM:

Calassomys apicalis Pardiñas, Lessa, Salazar-Bravo, and Câmara, 2014:203; type locality "Brazil, Minas Gerais, Sempre Vivas National Park, 3.25 km by road NW Macacos, Pedreira do Gaio (19°57′50″S, 43°47′18″W, 1251 m asl)."

DESCRIPTION: As for the genus.

DISTRIBUTION: *Calassomys apicalis* is known only from Cerrado habitats within Parque Nacional Sempre Vivas in Minas Gerais state, Brazil, at elevations between 1,100 and 1,450 m.

SELECTED LOCALITIES (Map 254): BRAZIL Minas Gerais, Arrenegado, about 46 km N of Macacos (Pardiñas et al. 2014), 3.25 km NW by road of Macacos (type locality

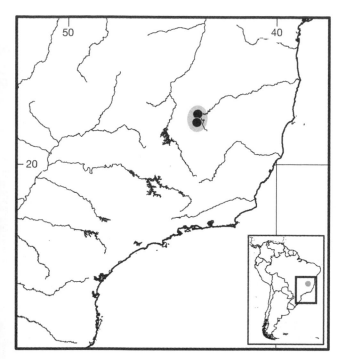

Map 254 Selected localities for *Calassomys apicalis* (●).

of *Calassomys apicalis* Pardiñas, Lessa, Salazar-Bravo, and Câmara).

SUBSPECIES: *Calassomys apicalis* is monotypic.

NATURAL HISTORY: *Calassomys apicalis* inhabits rocky outcrops within *campos rupestres*, an open physiognomy that characterizes plateaus and mountaintops. Here, vegetation displays characteristic marked xerophytic adaptations and occurs on acidic or lithosolic, poorly developed soils, with exposed rock often covered by lichens and rupestrial plants. This species co-occurs with the sigmodontines *Calomys* sp. and *Cerradomys subflavus*, the echimyids *Thrichomys apereoides* and *Trinomys albispinus*, and the marsupial *Monodelphis domestica*.

REMARKS: The reported karyotype is $2n=62$, FN$=116$ and consists of 28 pairs of biarmed and two pairs of uniarmed autosomes, a medium uniarmed X chromosome, and a uniarmed Y chromosome.

Genus *Calomys* Waterhouse, 1837

Jorge Salazar-Bravo

The genus *Calomys* is one of the most widespread genera of the Neotropical rodent fauna. Members of this genus—colloquially known as vesper mice, lauchas, or *rato bolinhas*—are small to very small, mostly granivorous rodents, abundant to common in most of the habitats where they occur.

The genus has a disjunct geographic distribution that is unique among sigmodontine rodents, with the exception of

Necromys. One species occurs in northern South America, in the llanos of Colombia and Venezuela and islands off the Venezuelan coast (Aruba, Curaçao, and Trinidad); all other species are associated with grasslands, savannas, and forest fringes in parts of Brazil, Bolivia, Peru, Argentina, Paraguay, Uruguay, and northern Chile. Although *Calomys* was the subject of a major revision in the early 1960s (Hershkovitz 1962) and an unpublished revision in the late 1980s (Olds 1988), the systematic status and both geographic and taxonomic limits of several species remain uncertain. Fortunately, the past decade has seen a renovated interest on the taxonomy of the genus and consequently the building of consensus on the identity and number of species that should be included in the genus, as well as a more comprehensive attempt (in techniques used and specimens analyzed) to understand the diversity of the genus. Nonetheless, much work in the field and museum remains to be done.

Until the early 2000s, much of what was known or inferred about species boundaries and relationships within the genus was based on morphological, chromosomal, and biochemical (protein electrophoresis) data. Most of these studies recognized species groups based on body size, tail length, skull morphology, or chromosomal counts of diploid numbers (e.g., Olds 1988). This approach was successful only at a very coarse level; at the heart of the problem is that many species in *Calomys* share a uniform morphology (in coloration, size, body proportions, etc.) yet occupy many ecological habitats and show high (in some cases, extreme) levels of chromosomal diversity. Interestingly, cladistic analysis of molecular sequence data only weakly supported some of the groups or species proposed earlier and indicated that, like in other muroid rodents, the burst of diversification in *Calomys* was accompanied by ecomorphological or chromosomal divergence only at a very coarse level (Salazar-Bravo et al. 2001; F. C. Almeida et al. 2007; Haag et al. 2007).

In the accounts that follow, I have been guided by criteria postulated by Musser and Carleton (2005), with more recent additions to the diversity of the genus (Bonvicino et al. 2010). It is very likely that this arrangement will change in the near future; nonetheless, these authors provide a solid base from which to start.

All species of *Calomys* are small to medium-sized mice well deserving of their Latinized Greek name (καλός = beautiful, μυς = mouse). Head and body length ranges from 70 to 150 mm (Hershkovitz 1962), usually with small to medium-sized and well-haired ears (between 13 and 19 mm [Olds 1988]). Most species have white or light buffy postauricular patches with some also presenting preauricular patches of the same color. Dorsal and ventral colors are often well differentiated with the venter paler. Pelage coloration, length, and appearance vary greatly among species and sometimes among populations of the same species; often, high-elevation species or populations of widespread ones show particularly long, thick, and smooth pelages. Dorsal color ranges from bright buffy or pale yellowish-brown to dark brown or gray. Ventral color ranges from pure white to white-tipped gray hairs to medium gray, but it is never the same color as the dorsum. Hindfeet are small (always <23 mm [Olds 1988; Bonvicino et al. 2010]) and covered with white hairs above; soles are naked, except for the heel, and possess six small plantar pads. The tail is less than or equal to the head and body length, covered with short fine hairs, and lacks a tuft at the tip; it may be either weakly to strongly bicolored, darker above. Mammae usually number 8 to 10 (4 to 5 pairs), but can range from 6 to 14 (Hershkovitz 1962; Olds 1988).

The skull is delicate to robust with parallel-sided to posteriorly divergent interorbital regions that may have either smooth edges or sharply beaded ridges. Zygomatic arches are expanded in adults, usually convergent anteriorly, rarely parallel sided. The anterior borders of the zygomatic plates are slightly concave, but never to the point of defining zygomatic spines; zygomatic notches are deep and broad. Incisive foramina are long, extending to the middle of the M1s; the palate is also long, extending beyond the M3s, and may or may not present posterolateral palatal depressions but often possess posterolateral palatal pits. The parapterygoid fossae are shallow and always much broader than the mesopterygoid fossa. Sphenopalatine vacuities are large. The carotid circulation is primitive or pattern 1 (*sensu* Voss 1988), with a large stapedial foramen and posterior opening of alisphenoid canal, squamosal-alisphenoid groove, and sphenofrontal foramen present (Olds 1988; Steppan 1995). An alisphenoid strut may or may not be present.

Upper incisors are opisthodont to nearly proodont, ungrooved, and faced with smooth enamel colored yellow to deep orange. Molar rows are generally parallel sided; individual cheek teeth are brachydont, tetralophodont, and crested; labial and lingual cusps alternate along the anteroposterior axis of the toothrow. The anterocone of M1 is always divided by an anteromedian flexus (except in very old individuals), and the anterolabial conule is often larger than the anterolingual one. M2 is square shaped and slightly smaller than M1; M3 is about 50% the size of M2 and more roundly shaped (Olds 1988). There is no reduction in the mesoflexus of M3 relative to that of M2 (Steppan 1995). The mandible is delicate but relatively deep; the coronoid process is slightly higher than the condylar process; a capsular process is not well developed and is located below the anterior portion of the sigmoid notch and under the coronoid. The masseteric crest is variable in both presence and development, from a barely visible chevron to a distinct knob.

Hooper and Musser (1964) described the male reproductive structures of two species of *Calomys* (*C. callosus* and *C. laucha*); in general, these were similar to each other and equally similar to those of *Eligmodontia* and *Graomys* (Hooper and Musser 1964:48). Voss and Linzey (1981) detailed the male reproductive tracts of the same species and concluded that each represented the plesiomorphic character state of glandular polarity for the Sigmodontinae. The stomach is unilocular and hemiglandular (Carleton 1973), and a gall bladder is present (Voss 1991a). Current karyotypic data on *Calomys* document remarkable diversity, with the number of known karyotypes higher than the number of recognized species and ranging from $2n = 36$ to $2n = 66$ and FN = 48 to FN = 72 (see Bonvicino et al. 2010 for a recent compilation of diploid and fundamental numbers for recognized species of *Calomys*).

Calomys fossils are known from as early as the lower Pleistocene of Bolivia and Argentina (Pardiñas et al. 2002) and as late as the Pleistocene-Holocene boundary of Patagonia and northern Argentina (P. E. Ortiz and Jayat 2007b; Pardiñas, Teta, D'Elía, and Lessa 2011).

SYNONYMS:

Mus: Olfers, 1818:209; part (description of *laucha*); not *Mus* Linnaeus.

Calomys Waterhouse, 1837:21; as subgenus; type species *Mus bimaculatus* Waterhouse (= *Mus laucha* G. Fischer), by original designation.

Hesperomys Waterhouse, 1839:75; type species *Mus bimaculatus* Waterhouse, by subsequent designation (Coues 1874:177); objective junior synonym of *Mus* (*Calomys*) Waterhouse, 1837.

Mus (*Hesperomys*): Wagner, 1843a:545; part (listing of *expulsus*); neither *Mus* Linnaeus nor *Hesperomys* Waterhouse.

Callomys Lesson, 1842:133; incorrect subsequent spelling of *Calomys* Waterhouse.

Eligmodontia: Lesson, 1842:133; part (listing of *bimaculatus* and *gracilipes*); not *Eligmodintia* F. Cuvier.

Hesperomys (*Calomys*): Burmeister, 1879:225; part (listing of *bimaculatus*); not *Hesperomys* Waterhouse.

Hesperomys (*Akodon*): E.-L. Trouessart, 1800:142; part (listing of *pusillus*); neither *Hesperomys* Waterhouse nor *Akodon* Meyen.

Oryzomys: Thomas, 1894a:359; part (description of *venustus*); not *Oryzomys* Baird.

Zygodontomys: E.-L. Trouessart, 1904:423; part (listing of *expulsus*); not *Zygodontomys* J. A. Allen.

Baiomys: Husson, 1960:34; part (description of *hummelincki*); not *Baiomys* True.

REMARKS: Waterhouse (1837) described *Calomys* as a subgenus of *Mus*, designating *bimaculatus* Waterhouse as the type species. Two years later, he (Waterhouse 1839) defined *Hesperomys*, without specifying a type species, to contain New World muroid rodents with bunodont and brachydont teeth. Coues (1874:177) designated *Mus (Calomys) bimaculatus* Waterhouse as the type species of *Hespermys* and, on the previous page, equated *Calomys* with *Eligmodontia* F. Cuvier. In his 1877 monograph on rodents, Coues (1877:43–44) reiterated his selection of *bimaculatus* as the type of *Hesperomys*, again equated *Calomys* with *Eligmodontia*, and inexplicably asserted that Waterhouse, in 1837, had "established the subgenus *Calomys* upon *C. elegans*" [= *Eligmodontia elegans*, a junior synonym of *E. typus* F. Cuvier]. Notwithstanding this incorrect statement, *Calomys* Waterhouse and *Hesperomys* Waterhouse are based on the same type species (*Mus bimaculatus*), the former by Waterhouses's unequivocal original designation ("Subgenus 4. Calomys . . . Type *Mus (Calomys) bimaculatus*;" Waterhouse 1837:21) and the latter by Coues's equally uniquivocal subsequent designation ("we may properly therefore elect the latter [*M. bimaculatus*] as technically the type;" Coues 1874:177). The two names are, thus, objective genus-group synonyms and the older name *Calomys* Waterhouse, 1837, assumes priority over *Hesperomys* Waterhouse, 1839.

Two different studies have questioned the monophyly of *Calomys*, one using mtDNA sequences (Engel et al. 1998) and the other morphology (Steppan 1995). Various lines of evidence suggest that the claim by Engel et al. (1998) was based either on a misidentified specimen or on a misread cryovial label (Mares and Braun 2000; Steppan and Sullivan 2000); their hypothesis thus requires no additional comment. On the other hand, most characters suggested by Steppan (1995) to support a closer relationship between *C. sorellus* and other phyllotines instead of a monophyletic *Calomys* can arguably be considered polymorphic and/or symplesiomorphic and thus do not support his hypothesis for the paraphyly of *Calomys*. Cladistic analyses of both mtDNA cytochrome-*b* (Salazar-Bravo et al. 2001; F. C. Almeida et al. 2007) and nucDNA IRBP sequences (Salazar-Bravo et al. 2013) have supported a monophyletic *Calomys* that includes *Calomys sorellus*.

KEY TO THE SPECIES OF *CALOMYS*:

1. Tail length always less than head and body length, tail less than 45% of total length; edges of supraorbital region divergent posteriorly and with distinct ledges in adults 2
1'. Tail length variable, greater or less than 45% of total length; edges of interorbital region parallel or divergent posteriorly, occasionally with slight beading, but never heavy ridges 9
2. Greatest length of skull of adults generally less than 24 mm; total length generally less than 170 mm, hindfoot shorter than 19 mm, maxillary toothrow between 3.1 and 3.9 mm; venter often dark, and dorsum frequently with a reddish or chestnut color; alisphenoid strut usually present; occurring in Cerrado and altered Atlantic Forest tracts from central and eastern Brazil south to northeastern Argentina (Misiones province), eastern Paraguay, and northeastern Santa Cruz department, Bolivia; $2n = 66$, FN = 66 . . . . *Calomys tener*
2'. Overall larger than above; venter gray to whitish, dorsal coloration brown, olive-brown or grayish (no reddish or chestnut); alisphenoid strut absent; widespread and occupying several types of habitats 3
3. Larger overall, maxillary toothrow averaging above 4.45 mm (range: 4.2–4.9 mm), average breath of braincase above 11.5 mm (11.3–12.5 mm) 4
3'. Overall, size smaller than above; maxillary toothrow averaging below 4.3 mm, braincase breath below 11.5 mm 5
4. Dorsal coloration brownish gray with some black and reddish hairs mixed in. Total length (head and body plus tail) of adults generally greater than 210 mm; greatest length of skull above 29 mm; maxillary toothrow generally greater than 4.30 mm (range: 4.37–4.59 mm); occurring in the Espinal and southern Humid Chaco in central Argentina, in the provinces of Cordoba, San Luis, and Santiago del Estero; $2n = 56$, FN = 66 . . . . *Calomys venustus*
4'. Dorsal coloration grayish somewhat washed with dull yellow, especially toward the rump; slightly smaller than above; occurring in the Yungas of central and southern Bolivia to northern Argentina (and associated habitats, above and below the Yungas vegetation belt), between 600 and 2,700 m; $2n = 54$, FN = 66 . . . . *Calomys boliviae*
5. Diploid chromosomal count of $2n = 50$ or above 6
5'. Diploid chromosomal count below $2n = 50$ 7
6. Breath of palatal bridge below 4.32 mm; length of the orbital fossa above 9.2 mm and length of the molar row above 4.1 mm; tail only slightly bicolor; occurring in central and northeastern Brazil, in habitats associated with Cerrado and Caatinga vegetation (and related ecotones); $2n = 66$, FN = 68 . . . . *Calomys expulses*
6'. Breadth of palatal bridge above 4.26 mm; length of the orbital fossa below 9.6 and length of the molar row below 4.1 mm; tail sharply bicolored; occurring in southwestern Brazil, eastern and northeastern Bolivia, western Paraguay, and northeastern Argentina,

often associated with Chacoan vegetation (and human altered habitats); $2n = 50$; FN $= 66$ . . . .
. . . . *Calomys callosus*

7. Dorsal coloration cinnamon-brown; head paler than dorsum; $2n = 36$ or 38, FN $= 66$ . . . . . . *Calomys cerqueirai*

7'. Dorsal coloration drab olive; head of the same color as the dorsum; $2n$ greater than 36 or 38 8

8. Ratio of length of molar row to greatest length of skull 0.15 or below; occurs in nonflooded grasslands of the Argentinean Mesopotamian and enclaves of grassy vegetation and old fields of western Brazil, from Rondônia in the north to Mato Grosso do Sul in the south; $2n = 48$; FN $= 66$ . . . .*Calomys callidus*

8'. Ratio of length of molar row to greatest length of skull 0.16 or above; occurs in central Brazil, in the upper basin of the Rio Araguaia; $2n = 46$; FN $= 66$ . . . .
. . . . *Calomys tocantinsi*

9. Tail shorter than 60 mm 10

9'. Tail longer than 60 mm, may be equal to head and body length 12

10. Dorsum bright buffy ochraceous; occurs in grasslands and other nonforested habitats, below 700 m, often on sandy soils in northern, mainland Venezuela and Colombia, and on the islands of Curaçao and Aruba . . . .
. . . .*Calomys hummelincki*

10'. Dorsum yellow-brown, gray or buffy; occurs in the grasslands and other nonforested habitats (0 to 4,600 m) south of the Equator. 11

11. Dorsal pelage marbled in appearance, fluffy; hairs of venter gray based; tail whitish or pale gray (not distinctly bicolor), very short, less than 55% of head plus body length; rostrum narrow; maxillary toothrow longer than 3.2 mm; occurring at elevations above 2,500 m in the Andes of northern Argentina, extreme northeast Chile, the central plateau of Bolivia and the southern half of the Altiplano of Peru . . . . . .
. . . . *Calomys lepidus*

11'. Dorsal pelage not marbled, short hairs; venter white; tail bicolored, more than one-half of head plus body length; rostrum not narrowed; maxillary toothrow shorter than 3.3 mm; occurring in grasslands and forest fringes up to 950 m in Argentina, Paraguay, southeast Bolivia, Uruguay, and southern Brazil. . . . .
. . . . *Calomys laucha*

12. Tail approximately equal to head plus body length; dorsum golden brown lined with black hairs; sides of interorbital region divergent posteriorly; occurring at moderate to high elevations (2,000 to 3,500 m) in the eastern slopes of the southern Bolivian and northern Argentinean Andes as well as lowlands below 1,000 m in the dry Chaco of Paraguay and the Pampas and agricultural landscapes of central Argentina, south to at least

the Patagonian province of Santa Cruz; $2n = 38$ . . . . . .
. . . . *Calomys musculinus*

12'. Tail shorter than head plus body; dorsum brown; sides of interorbital region parallel; occurring at high elevations in the Andes of Peru and likely Bolivia; $2n = 62$ or 64 . . . .*Calomys sorellus*

Calomys boliviae (Thomas, 1901)

Bolivian Laucha

SYNONYMS:

Eligmodontia callosa boliviae Thomas, 1901f:253; type locality "Rio Solocame, 67°W., 16°S. [La Paz,] Bolivia. Altitude 1200 m."

[*Hesperomys callosus*] *boliviae*: Thomas, 1916a:141, name combination

Hesperomys fecundus Thomas, 1926c:321; type locality "Tablada, 2000 m," Tarija, Bolivia.

Calomys callosus boliviae: Cabrera, 1961:477; name combination.

Calomys boliviae: Musser and Carleton, 1993:696; first use of current name combination.

DESCRIPTION: Medium sized (measurements for series of specimens from near the type locality in B. D. Patterson [1992b]). General color above grayish, more or less washed with dull yellowish turning richer toward rump; sides buffy and indistinctly separated from under surface by buffy line; under surface uniformly pale buff, with base of hairs slate colored; color of face like back but cheeks paler, similar to sides. Ears of medium size (18–19 mm), thinly haired, covered internally by grayish hairs and externally by olive brown hairs, overall same color as head; inconspicuous whitish postauricular patches noticeable, but no preauricular ear patch present. Arms and legs grayish white; hands and feet clear buffy white. Tail very finely scaled, finely haired, and bicolored, brown above and whiter below. Hindfeet short and narrow; plantar surface naked, except for heel; interdigital pads small, hypothenar smaller than interdigital pads, thenar pad enlarged and narrow. Adult females from type locality in Yungas of La Paz department, Bolivia, with five pairs of mammae; those from southern Bolivia (Tarija department) and Argentina with seven pairs, the increase perhaps related to larger body size of latter populations (Olds 1988). Gall bladder present (J. Salazar-Bravo, pers. obs.), and four individuals examined by Steppan (1995) had 12 thoracic, 7 lumbar, and about 25 caudal vertebrae.

Skull with straight lateral profile; rostrum heavy and short, sides tapering gradually; nasals broad and slightly turned down at tip, overhanging and thus obscuring the incisors when viewed from above. Dorsal margins of nasals not extending beyond suture junction of maxillary, frontal, and lacrimal bones; lacrimals present but not especially enlarged. Interorbital region flat, with sharp and posteriorly divergent frontal margins that produce supraorbital shelves in most adults. Coronal suture broadly U-shaped. Braincase relatively narrow, with temporal and lambdoidal ridges present in very old individuals. Zygomatic plates broad, without zygomatic spines; zygomatic notches shallow. Zygomatic arches project forward and not very broad. Parietals with only small extensions onto squamosal; postglenoid foramen reduced with respect to subsquamosal fenestra. Incisive foramina long, extending past anterior roots of M1s; premaxillary process occupies three-fourths of foraminal opening. Palate long and wide (*sensu* Hershkovitz 1962); mesopterygoid fossa slightly wider than that in other species of large *Calomys*, with anterior portion slightly broader than remainder of fossa; median palatine process lacking. Sphenopalatine vacuities present and well developed. Carotid circulation primitive (pattern 1 of Voss, 1988). Alisphenoid strut absent.

Mandible with ascending portion of coronoid initiated below gum line of teeth; upper ridge of masseteric crest weakly developed, lower ridges well developed; capsular projection present but not prominent, located between base of coronoid process and sigmoid notch; coronoid process falciform, smaller than condyloid process, unremarkable except in younger individuals where it forms less acute angle with main axis of mandible than in older individuals.

Upper incisors opisthodont with bright orange enamel; upper molars crested (*sensu* Hershkovitz 1962), with labial and lingual cusps of about same height, brachydont, and with cusps opposite in pattern. M1 subrectangular in outline with well-developed procingulum and anteromedian flexus fused at base of cusp; hypoflexus and protoflexus broad and parallel, especially in older individuals; in younger individuals, protoflexus somewhat comma shaped; anterolabial style usually present as thickening of labial cingula but without trace of anteroflexus. M2 square in outline; protoflexus present even old individuals; mesoflexus alternate to hypoflexus. M3 triangular and moderately reduced with respect to M2; hypoflexus evident; mesoflexus persists; protoflexus small but present. Lower molars, like uppers, crested. Lower m1 with well-developed procingulum with anteromedian flexid in some individuals (e.g., holotype), but in others this structure not easily seen; mesoflexid broad and deep, almost isolating the metaconid; protoconid and hypoconid subtriangular in outline; entoconid bulbous; and posterolophid reduced. Lower m2 squared in outline, with well-developed flexids, including protoflexid; m3 Z-shaped in outline, with mesoflexid and hypoflexid almost touching at midline.

DISTRIBUTION: *Calomys boliviae* is known from the eastern slopes of the Andes, between ca. 600 and 2,700 m, from west-central Bolivia to northwestern Argentina.

SELECTED LOCALITIES (Map 255): ARGENTINA: Jujuy, Santa Bárbara (M. M. Díaz and Barquez 2007), on Hwy 9 at border with Salta, at campground on the way to El Carmen (M. M. Díaz and Barquez 2007); Salta, 48.8 km NW of junction of Rtes 50 and 18, on road to Isla de Cañas, sobre Rte 18 (OMNH 35262), 5 km WSW of Pulares (OMNH 35260); Santiago del Estero, Virgen del Valle picnic area on Hwy 64 (Dragoo et al. 2003); Tucumán, Faimalla (De Catalfo and Wainberg 1974). BOLIVIA: 2 km E of Chuhuayacu (Salazar-Bravo, Dragoo et al. 2002), Horcas, 80 km SE of Sucre (S. Anderson 1997), 2 km SE of Monteagudo (Salazar-Bravo et al. 2001); Cochabamba, ca. 2 km NW of Peña Blanca (S. Anderson 1997); La Paz, 1 mi W of Puerto Linares (S. Anderson 1997), Río Solocame (type locality of *Eligmodontia callosa boliviae* Thomas); Tarija, 3 km WNW of Caraparí (Salazar-Bravo, Dragoo et al. 2002), 4 km by road N of Cuyambuyo, Fábrica de Papel (UMMZ 156322), Porvenir (Salazar-Bravo, Dragoo et al. 2002), 1 km E of Tucumilla (Salazar-Bravo, Dragoo et al. 2002).

SUBSPECIES: *Calomys boliviae* is monotypic, although southern populations (to which the name *fecundus* Thomas would apply) are somewhat larger in body size than northern samples that represent the nominotypical form.

NATURAL HISTORY: This species has been trapped in diverse habitats, ranging from the Yungas and Chaco (M. M. Díaz and Barquez 2007) to humid Puna (Jayat, Ortiz, and Miotti 2008), to old fields and cultivated areas (J. Salazar-Bravo, pers. obs.) in central Bolivia and northern Argentina. Little is known about its ecology in Bolivia (S. Anderson 1997), but Barquez et al. (1991), M. M. Díaz et al. (2000), and M. M. Díaz and Barquez (2007) provided data on Argentina populations. Pregnant females (one with 10 embryos) or enlarged nipples were captured toward the end of the summer (February) in Jujuy, with scrotal males caught in February and May and young individuals trapped from April to August (M. M. Díaz and Barquez 2007). In the Yungas of La Paz, Bolivia, individuals serve as a reservoir for *Leishmania* (J. Vargas Matos, pers. comm.). Archeological samples probably assignable to this species were described from Ruinas Jesuíticas de San José de Lules in Tucumán province, Argentina, and dated to about 350 to 150 ybp (P. E. Ortiz et al. 2011b).

REMARKS: Both the geographic and taxonomic limits of this species require further work. Originally described as a species of *Eligmodontia*, *boliviae* was reassigned to *Hesperomys* (viewed was equivalent to *Calomys*) by Thomas (1916a). Since then, the status of *C. boliviae* has fluctuated repeatedly from a subspecies of *Calomys callosus* (Hershkovitz 1962) or *Calomys venustus* (Olds 1988; S. Anderson 1997) to a valid species (Musser and Carleton 1993). Salazar-Bravo, Dragoo et al. (2002) and Dragoo et al.

(2003), based on molecular characters (analyses of mtDNA D-loop and cytochrome-*b* sequences and AFLP fragments) and chromosomal data, recognized *Calomys fecundus* as a valid species but were unable to comment of the status of *Calomys boliviae* because of the lack of available material. Subsequently, Salazar-Bravo et al. (2003) suggested that *boliviae* Thomas might be a senior synonym of *fecundus* Thomas, a position earlier advocated by Musser and Carleton (1993) and subsequently supported by M. M. Díaz et al. (2006) and M. M. Díaz and Barquez (2007). Although I follow this hypothesis herein, I stress that the studies advocated by both Salazar-Bravo et al. (2003) and Musser and Carleton (2005) have yet to be conducted, so the taxonomic composition of *C. boliviae* should be viewed with caution.

B. D. Patterson (1992b) reported measurements of 23 specimens of *C. boliviae* from the Bolivian department of La Paz under the name of *C. venustus*. His series of specimens were part of a lot collected by Alfonso Olalla and sold to the Royal Natural History Museum in Stockholm in the late 1930s or early 1940s. Although all the measurements are those of adult *Calomys boliviae*, the figured skull said to be a representative of the species (his Fig. 11) is not;

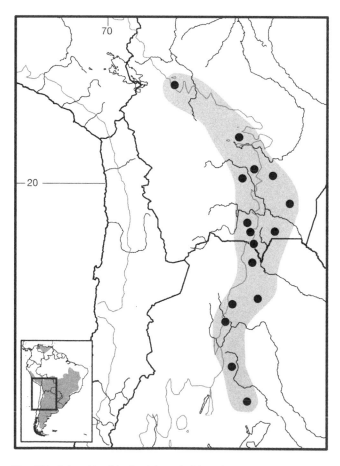

Map 255 Selected localities for *Calomys boliviae* (●). Contour line = 2,000 m.

two of Olalla's specimens had the field identification number AMO 334, one corresponding to *Calomys boliviae*, one to a specimen of *Hylaeamys megacephalus*; during the last steps of the editorial process, the *Hylaeamys* skull was photographed by mistake (B. D. Patterson, pers. comm.).

De Catalfo and Wainberg (1974) and Salazar-Bravo, Dragoo et al. (2002) reported a karyotype with $2n = 54$ and $FN = 66$ for specimens from northern Argentina and southern Bolivia.

Calomys callidus (Thomas, 1916)

Crafty Vesper Mouse, Crafty Laucha, Reclusive Laucha

SYNONYMS

Hesperomys venustus callidus Thomas, 1916b:182; type locality "Goya, Corrientes. 600'," Argentina.

Calomys venustus callidus: Cabrera, 1961:481; name combination.

Calomys callosus callosus: Hershkovitz, 1962:172; part; *callidus* treated as synonym.

Calomys callidus: Vitullo, Kajon, Percich, Zuleta, Merani, and Kravetz, 1984:486; first use of current name combination.

DESCRIPTION: One of large species; pelage short and appressed, dorsal coloration grayish, buffy wash on flanks, and venter pale gray (hairs slaty basally, white or grayish terminally). Ears large, with brown inner and black outer surfaces; inconspicuous postauricular patch present. Tail shorter than head and body, thinly haired, bicolored (pale brownish above and grayish white on sides and below), and without tuft of hair at tip. Mammae usually number 10 (Thomas 1916b).

Skull robust, slightly bowed in lateral profile, and with its greatest length averaging 25 mm in females and 26.7 mm in males (Bonvicino et al. 2010). Interorbital region flat, posteriorly divergent, and with sharp supraorbital edges that continue as visible temporal crests. Nasals with raggedy posterior margins; do not extend dorsally beyond triple-point suture between maxillary, frontal, and lacrimal. Zygomatic plates with straight dorsal edges, without zygomatic spines, posterior margins anterior to alveoli of M1s; zygomatic notches moderately narrow but deep. Incisive foramina long, extending between protocones of both M1s. Mesopterygoid fossa with squared anterior border that does not reach alveoli of M3s; sphenopalatine vacuities present. Alisphenoid strut absent. Upper incisors orthodont, with bright orange enamel band. Upper molar series parallel, with teeth terraced. M1 divided by shallow groove, M2 squared, and M3 small and rounded. Mandible strong; capsular projection straight and located ventral to coronoid process.

No information available on axial elements of skeleton or soft anatomical structures.

DISTRIBUTION: *Calomys callidus* occurs in grasslands and grassland/monte ecotone of the Mesopotamian alluvial system of Argentina as well as in enclaves of grassy vegetation and old fields of western Brazil, from Rondônia in the north to Mato Grosso do Sul in the south.

SELECTED LOCALITIES (Map 256): ARGENTINA: Chaco, Parque Nacional "Chaco" (Vitullo et al. 1984); Corrientes, Estancia San Juan Poriahú (Lareschi et al. 2005), Goya (type locality of *Hesperomys venustus callidus* Thomas); Entre Rios, Parque Nacional "El Palmar" (Vitullo et al. 1984), Diamante (Massa 2010), Puerto Ruiz (Massa 2010). BRAZIL: Matto Grosso, Barão de Melgaço (Bonvicino et al. 2010), Campo Novo do Parecis (Bonvicino et al. 2010); Matto Grosso do Sul, Municipalities of Dois Irmãos de Buriti and Terenos (Caceres et al. 2011); Rondônia, Pimienta Bueno (Mattevi et al. 2005).

SUBSPECIES: *Calomys callidus* is monotypic.

NATURAL HISTORY: Little is known about the ecology of this species; apparently, it prefers microhabitats with nonflooded, tall grass in areas of the Argentinean Mesopotamia. Smaller-sized individuals are preferentially taken in pitfall traps in Brazil (Caceres et al. 2011). Massa (2010) found signs of Barn Owl (*Tyto alba*) predation over *C. callidus* in almost 60% of the localities she studied in southern Entre Rios province (Argentina). V. L. Hodara et al. (1994) reported on blood, serum, and urine chemistry while Percich and Hodara (1990) examined growth and development.

REMARKS: Both taxonomic and geographic limits of *C. callidus*, especially with respect to its relationship to *C. callosus*, require attention. J. R. Contreras et al. (2003) suggested that *callidus* Thomas may be a synonym of *C. callosus* (Rengger), since specimens from 60 km S Pilar (near the type locality of *Calomys callosus*, as amended by J. R. Contreras 1992) had a chromosomal complement of $2n = 48$, $FN = 70$, similar to the $2n = 48$, $FN = 68$ for specimens examined from Entre Ríos and assigned to *C. callidus* (Vitullo et al. 1984). Support for this argument included the apparent restriction of the $2n = 48$ cytotype of *Calomys* to Argentinean Mesopotamia (e.g., Entre Ríos province). However, recent studies have reported populations of large *Calomys* with a $2n = 48$, but with an $FN = 66$ from western Brazil (Mattevi et al. 2005; Bonvicino et al. 2010), suggesting that *C. callidus* is present throughout eastern Paraguay and into western Brazil. One implication of the linkage of *callidus* to *callosus* would be that the widely distributed, large, and lowland mice with $2n = 50$, for which the name *C. callosus* is now used (e.g., Bonvicino et al. 2010; see that account) would require a new name combination, as this taxon represents an independent monophyletic clade (Salazar-Bravo, Dragoo et al. 2002). This taxonomic hypothesis requires work beyond the scope of this volume.

Scapteromys modestus (Miranda-Ribeiro 1914) was described from Caceres in the Brazilian state of Mato Grosso based on two young adult specimens. In the original description, Miranda-Ribeiro compared his two samples to *Scapteromys labiosus* [= *Bibimys labiosus*], *Oryzomys* [= *Oligoryzomys*] *longicaudatus*, and *Mus musculus*, but concluded that they deserved specific status. Since the original description, the identity of these animals has only been cursorily examined: Winge (1941:145) commented on the quality of the "indistinct photographs of the skull" yet did not hesitate to suggest that this form may be a synonym of "*Hesperomys expulsus*" [= *Calomys expulsus*]. Hershkovitz (1966:97) likewise noted that "judg[ing] from the published figure of its skull '. . . . *S. modestus*' corresponds to *Calomys callosus*." Neither Winge nor Hershkovitz, however, examined any of Miranda-Ribeiro's specimens before assigning them to *Calomys*. To compound the problem, Avila-Pires (1968) in effect synonymized *Scapteromys modestus* with [*Oligoryzomys*] *utiaritensis*, a species recently shown to be a valid form of *Oligoryzomys* (Agrellos et al. 2012). The two specimens of *S. modestus* are not members of *Scapteromys*, *Bibimys*, *Mus*, or *Oligoryzomys* and indeed belong to the genus *Calomys* (U. F. J. Pardiñas, pers. comm.; M. Weksler, pers. comm.). Furthermore, morphological characters (e.g., the lack of alisphenoid struts) and morphometric data (J. A. Oliveira, pers. comm.) indicate that these specimens belong to either *Calomys callosus* or *Calomys callidus*. Regrettably, nowadays the only useful marker to separate these two species

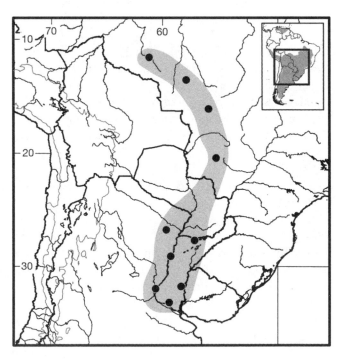

Map 256 Selected localities for *Calomys callidus* (●). Contour line = 2,000 m.

is the diploid chromosome count (see the Key), and obviously this information is not available for the specimens that Miranda-Ribeiro had on hand while describing *S. modestus*. On biogeographic and distribution grounds, it is plausible —or even likely—that *modestus* might be a senior synonym of *callidus*, but data at hand are too limited to support this hypothesis strongly, especially as there is still much doubt regarding the position of *C. callidus* with respect to other species of lowland *Calomys*. For the present, therefore, I treat *modestus* Miranda-Ribeiro as a *nomen dubium*, and list it neither in the synonymies of either *C. callidus* or *C. callosus*.

While the position I advocate here with respect to *C. callidus* and its relationships with other species of the genus is unsatisfactory, it is the most defensible hypothesis at present. What is required to clarify this systematic and taxonomic tangle is a multiloci, inclusive systematic work that includes topotypic material for *callidus*, *boliviae*, and close relatives of the other Brazilian, eastern Bolivian, and northern Argentinean forms of the genus.

Calomys callosus (Rengger, 1830)

Large Vesper Mouse, Big Laucha

SYNONYMS:

Mus Callosus Rengger, 1830:231; type locality "am Ufer des Paraguaystromes, ungefähr unter dem sieben und zwanzigsten Breitengrade" [= on the banks of the Paraguay river, below roughly 27°]; interpreted as "opposite mouth of Río Bermejo, Department of Villa del Pilar" by Hershkovitz (1962:172); amended by J. R. Contreras (1992:2) to vicinity of Ciudad Pilar, Corrientes, Argentina.

Oryzomys callosus: Thomas, 1898i:3; name combination.

Hesperomys callosus: Thomas, 1916a:141; name combination.

Hesperomys muriculus Thomas, 1921l:623; type locality "San Antonio, Parapiti, on 20°S, about 250 km. south of Santa Cruz de la Sierra. Alt. 600 m," Santa Cruz, Bolivia.

Calomys callosus: Cabrera, 1961:477; first use of current name combination.

Calomys hildebrandii Enría, Mills, Bausch, Shieh, and Peters, 2011:453; *nomen nudum*.

DESCRIPTION: Medium sized (head and body length 99–122 mm; tail length 78–110 mm; S. Anderson 1997) with long and moderately lax pelage (or short and appressed in wetter habitats). Upper parts of body ochraceous to tawny and often with fine mixture of black or dark brown; rump and sides generally paler; weakly defined ochraceous lateral band sometimes present; and venter whitish to gray, sometimes washed with buff, individual hairs dark gray at their bases. Eye-ring defined by buffy hairs sometimes evident. Tail bicolored, grayish brown above and whitish below. Ears small (range: 16 to 19 mm), rounded and lightly haired; pale-colored postauricular patches present. Both forefeet and hindfeet white above, with whitish ungual tufts; hindfeet narrow. Mammae count 10 to 14.

Skull strongly built, dorsal aspect flat in lateral profile, with short rostrum and deep zygomatic notches. Anterior edges of zygomatic plates slightly concave, delimiting small zygomatic spines; posterior borders terminate anterior to frontal plane of M1s. Borders of supraorbital region of skull divergent posteriorly, strongly beaded, and form broad ledges in fully mature individuals; midfrontal width always greater than maximum width of rostrum. Interparietal well developed. Zygomatic arches converge anteriorly. Alisphenoid strut absent. Incisive foramina long, reaching protocones of M1s, but with anterior part narrower than posterior. Palate short, somewhat narrow, and rugose, with conspicuous palatine foramina and deep posterolateral pits. Bullae triangular in shape, their bony tubes long and narrow. Middle lacerate foramen large, and foramen ovale small but distinct. Mesopterygoid fossa parallel sided, with anterior border not reaching line connecting posterior edges of M3s; sphenopalatine vacuities present. Upper incisors orthodont, each fronted by orange enamel. Main cusps of molars alternate in anteroposterior position, with molars bileveled. M1 with deeply divided anterocone, and no sign of anterolabial style or anteroflexus. Bonvicino et al. (2010) provided external and cranial measurements.

Modal counts of thoracic, lumbar, and caudal vertebrae 12, 7, and 22–25, respectively (Steppan 1995). Gall bladder present (Voss 1991a); Hooper and Musser (1964) described male phallus of specimen from Matto Grosso; and Voss and Linzey (1981) detailed morphology of male accessory glands from Paraguayan specimens.

DISTRIBUTION: As herein understood *C. callosus* occurs from northern Argentina, throughout Paraguay, eastern and northeastern Bolivia, and western Brazil.

SELECTED LOCALITIES (Map 257): ARGENTINA: Corrientes, Pilar (J. R. Contreras et al. 2003); Formosa, 13 km S of Clorinda (Pardiñas and Teta 2005), Reserva El Bagual (Pardiñas and Teta 2005); Salta, Aguaray (M. M. Díaz et al. 2000), 31 km SSW of Dragones, along Río Bermejo (M. M. Díaz et al. 2000); Santiago del Estero, San Antonio (MACN 17612); Tucumán, Concepción (MACN 997–34). BOLIVIA: El Beni, Nueva Calama (S. Anderson 1997), Mangal, 25 km NW of Exaltación (S. Anderson 1997), El Consuelo (B. D. Patterson 1992b), San Ignacio (S. Anderson 1997); Santa Cruz, El Refugio, Parque Nacional Noel Kempff Mercado (Emmons et al. 2006), La Hoyada, 30 km S of Valle Grande (Bonvicino and Almeida 2000), Las Cruces (Salazar-Bravo, Dragoo et al. 2002), San Ramón (Salazar-Bravo, Dragoo et al. 2002). BRAZIL:

Mato Grosso, Aquidauana (Bonvicino et al. 2010), Si-brolândia (Bonvicino et al. 2010). PARAGUAY: Amambay, Parque Nacional Cerro Corá (Salazar-Bravo et al. 2001).

SUBSPECIES: *Calomys callosus* is monotypic. The polytypic *C. callosus* envisioned by both Cabrera (1961) and Hershkovitz (1962) included taxa regarded as distinct species herein (e.g., *boliviae* Thomas and *expulsus* Lund). Nevertheless, specimens from more humid localities (eastern Santa Cruz department, Bolivia and enclaves of Cerrado and Pantanal vegetation in Paraguay) are both smaller in size and darker in color, both dorsally and ventrally, than those from more xeric geographic areas.

NATURAL HISTORY: The natural history of the species is entrenched in the morass that currently encompasses the taxonomy of the large body species of lowland *Calomys*. Because this species shares many morphological characters with other sympatric and quasi-sympatric species, it is not always possible to ascertain in the published literature which species was actually studied in the field. Based on the reported localities recognized here, the species occupies a range of habitats, from dry Chaco to grassy enclaves, secondary growth, and both old fields and agriculture-dominated landscapes (Myers 1982). For example, Yahnke (2006) found the species almost exclusively in agricultural habitats where pregnant females had an average of 6.25 embryos (N=4, range: 4–11) and were captured in the months of May, August, and December. Rengger (1830:232) noted that individuals were found in pairs within burrows about a foot long and close to water. He saw them active during the day feeding on seeds and roots.

Calomys callosus was identified as the reservoir for Latino virus (reviewed in Salazar-Bravo, Ruedas, and Yates 2002) as well as the reservoir for Laguna Negra hantavirus (Levis et al. 2004).

REMARKS: See Remarks under *C. callidus* for the possible relationship between these two species and possible assignment of *modestus* Miranda-Ribeiro. As currently understood, the karyotype of *C. callosus* is $2n=50$, FN=66, and composed of nine pairs of metacentric, with one pair much smaller than the others, and 15 pairs of acrocentric autosomes, a large and metacentric X chromosome, and a small acrocentric Y chromosome. This karyotype is uniform in specimens from Bolivia (Salazar-Bravo, Dragoo et al. 2002) and Brazil (Bonvicino et al. 2010) and is identical to that reported for a specimen identified as *C. fecundus* from Beni, Bolivia (Pearson and Patton 1976).

Molecular data (Salazar-Bravo, Dragoo et al. 2002) suggest the presence of an undescribed taxon of *Calomys* in the Bolivian department of Beni; individuals of this form serve as reservoirs to Machupo arenavirus, the etiological agent of Bolivian hemorrhagic fever. Because individuals in these populations are morphologically, morphometrically,

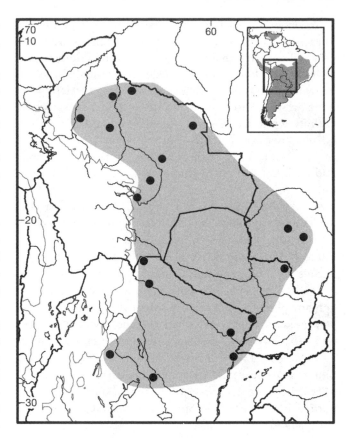

Map 257 Selected localities for *Calomys callosus* (●). Contour line = 2,000 m.

and chromosomally ($2n=50$, FN=66) indistinguishable from *C. callosus*, I treat them here as members of the same species. Enría et al. (2011) referred to the Beni samples as *Calomys hildebrandii*, a *nomen nudum* because this taxon was not formally described in that paper, or elsewhere.

Calomys cerqueirai Bonvicino, Oliveira, and Gentile, 2010

Cerqueira's Vesper Mouse, Rato-bolinha de Cerqueira

SYNONYM:

Calomys cerqueirai Bonvicino, Oliveira, and Gentile, 2010: 27; type locality "Margins of 'Buraco do Cachorro' stream, situated near the head of 'Café' stream, close to Capitaõ Andrade city center (19°04′35″S, 41°52′30″W), Capitaõ Andrade Municipality, state of Minas Gerais, Southeastern Brazil."

DESCRIPTION: Medium sized (mean head and body length 99 mm, greatest length of skull 26 mm) with tails shorter than head and body (75–87%), moderately long and narrow hindfeet, and small, rounded pinnae (16–18% of head and body length). Dorsal pelage soft and dense, colored cinnamon-brown overall; head paler than dorsum; the rostrum covered with yellow to ochraceous-tawny hairs; sides of body slightly paler than dorsum due to reduction in black guard hairs; ventral pelage well delineated from sides, covered with whitish hairs with gray bases. Pinnae internally covered by entirely orange hairs or hairs with long orange bands, and externally by sparse brownish hairs; postauricular patch of completely white hairs present. Forefeet and hindfeet covered dorsally with entirely white hairs; silvery ungual tufts present on distal phalanxes of digits II–V of the manus and digits I–V of pes. Tail sharply bicolored, with white hairs ventrally and dark hairs dorsally. Mammae count in adult females unknown.

Skull robust, with rostrum about 36% of greatest skull length and somewhat bowed lateral profile. Interorbital region divergent posteriorly, with sharp supraorbital edges; interorbital breadth wider than length of upper molar row. Nasals with blunt posterior margins that do not extend beyond suture between maxillary, frontal, and lacrimal. Zygomatic plates project slightly anteriorly, lacking any trace of zygomatic spines; zygomatic notches moderately deep and broad. Carotid circulation pattern primitive (pattern 1, *sensu* Voss 1988). Incisive foramina longer than upper molar row and reach imaginary line across protoflexi of M1s. Mesopterygoid fossa with squared dorsal edge; does not reach alveoli of M3s. Sphenopalatine vacuities present. Alisphenoid strut absent.

Upper incisors opisthodont, and upper series (M1–M3) parallel. Shallow anteromedian flexus in M1 divides anterocone into subequal anterolingual and anterolabial conules; M2 squared and M3 small and rounded. Mandible strongly built; capsular projection present as subtle elevation and located ventral to coronoid process. No data available on axial elements of skeleton.

DISTRIBUTION: *Calomys cerqueirai* is known only from riparian grasslands in southeastern Minas Gerais, Brazil.

SELECTED LOCALITIES (Map 258): BRAZIL: Minas Gerais, Capitão Andrade (type locality of *Calomys cerqueirai* Bonvicino, Oliveira, and Gentile), Lagoa Santa (F. C. Almeida et al. 2007).

SUBSPECIES: *Calomys cerqueirai* is monotypic.

NATURAL HISTORY: Bonvicino et al. (2010) captured specimens of *Calomys cerqueirai* in rural areas along streams, in grassy vegetation, less than 2 m from stream margins. These authors recorded both tapeworms (*Taenia* sp.) and nematodes (probably *Hassalstrongylus epsilon*) in individuals they trapped.

REMARKS: F. C. Almeida et al. (2007) referred to this species as *Calomys sp. nov.*

Chromosomes reported as $2n = 38$ and FN = 66, with 15 pairs of biarmed and three pairs of uniarmed autosomes for specimens of the type locality, and $2n = 36$, FN = 66 for specimens from Lagoa Santa (Bonvicino et al. 2010).

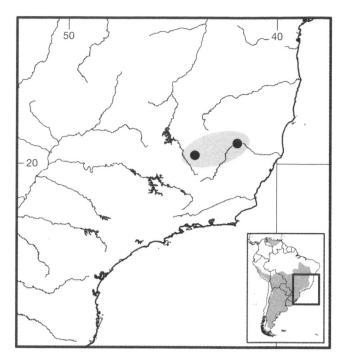

Map 258 Selected localities for *Calomys cerqueirai* (●).

Colombi et al. (2011) reported a $2n = 36$, FN = 66 animal from Nova Venécia, Espírito Santo, Brazil, that is probably best referred to *C. cerqueirai*, based both on karyotype and similarity in mtDNA cytochrome-*b* sequences. I refrain from accepting this assignment until a proper comparison of the voucher specimens provides confirmation.

Calomys expulsus (Lund, 1840)
Caatinga Laucha, Rato-bolinha da Caatinga
SYNONYMS:
Mus expulsus Lund, 1839a:233; *nomen nudum.*
Mus expulsus Lund, 1840b:50 [1841c:280]; type locality "Rio das Velhas's Floddal," Lagoa Santa, Minas Gerais, Brazil.
Mus [*Hesperomys*] *expulsus*: Wagner, 1843a:545; name combination.
Hesperomys expulsus: Burmeister, 1854:175; name combination.
[*Oryzomys*] *expulsus*: E.-L. Trouessart, 1897:528; name combination.
[*Zygodontomys*] *expulsus*: E.-L. Trouessart, 1904:423; name combination.
Calomys expulsus: Cabrera, 1961:478; first use of current name combination.
Calomys callosus expulsus: Hershkovitz, 1962:174; name combination.
DESCRIPTION: Size average for a large, lowland species of genus (mean head and body length 99.7 ±14 mm in

21 specimens from central Brazil; Bonvicino and Almeida 2000); tail short, averaging about 74% of head and body length; upper parts of body yellowish to olive brown, finely streaked with black hairs; sides of body tawny, not as dark as dorsum, sharply countershaded relative to venter; no lateral line separating side and ventral coloration; head same color as dorsum; venter covered with white-tipped but gray-based hairs throughout but completely white around chin. Moderate to well-developed eye-ring present. Forefeet and hindfeet covered dorsally with short, white hairs; claws tufted. Tail only weakly bicolored (Bonvicino et al. 2010), slightly darker above; ears average length for genus (ca. 17 mm) and covered by darker hairs.

Skull with comparatively (but not statistically) shorter rostrum than other species of large bodied, lowland *Calomys*, but with slightly narrower interorbital region (Bonvicino et al. 2010) and more heavily bowed in lateral view. Nasals long, projecting beyond anterior plane of upper incisors. Zygomatic plates broad, with slightly to markedly concave anterior margins; zygomatic spines comparatively more developed than in other species of lowland, large *Calomys*, even in young specimens; zygomatic notches broad and deeply excavated. Zygomatic arches convergent anteriorly. Interorbital region with sharply squared and posteriorly divergent dorsolateral margins. Temporal crests relatively weak, but lambdoidal crest rather strongly developed, especially in older individuals. Incisive foramina long, extending between M1s, and slender, with widest point at about premaxillomaxillary suture. Palate long, without any special marks except for pair of palatine foramina and pair of deep posterolateral palatal pits. Mesopterygoid fossa slightly narrower at midpoint; anterior edge rounded or, in many cases, bullet shaped, and does not reach end of M3s; sphenopalatine vacuities well developed. Bullae of average size for genus. Alisphenoid strut absent. Mandible strongly built with relatively deep body; capsular projection present, slanted backward and located below sigmoid notch.

Upper incisors opisthodont; enamel bright orange. Molars robust, broad, and with principal cusps alternating. Maxillary toothrow averages about 4.06 mm (Bonvicino et al. 2010). Procingulum of M1 deeply divided by anteromedian flexus; both paraflexus and metaflexus deep, bending backward; anterolabial style (*sensu* Hershkovitz 1962) present but anteroflexus is not; protoflexus and hypoflexus laterally oriented and straight; M2 about 70% size of M1, squared, and with well-developed paraflexus, metaflexus, and hypoflexus; M3 squared in profile, with well-developed flexi; many individuals have enamel island in middle of M3. Anteromedian flexid of m1 undivided; cusps arranged in alternating pattern; entoflexid much smaller than either mesoflexid or hypoflexid; all labial struc-tures conserved in m2, and m3 S-shaped and about 80% on size of m2.

DISTRIBUTION: *Calomys expulsus* occurs throughout the Caatinga and Cerrado biomes, and ecotones between them, of the central "Planalto" region of Brazil, from the states of Tocantins and Piauí in the north and northeast to Bahia and Goiás in the south (Bonvicino, Oliveira, and D'Andrea 2008; Nascimento et al. 2011).

SELECTED LOCALITIES (Map 259): BRAZIL: Bahia, Caetité (F. C. Almeida et al. 2007), Corentyne (F. C. Almeida et al. 2007), Fazenda Jucurutu, Jaborandi (Bonvicino, Lima, and Almeida 2003), Ideate (Nascimento et al. 2011); Goiás, Campo Alegre de Goiás (Nascimento et al. 2011), Fazenda Flandeiras, 65 km SSW of Cavalcante (Bonvicino, Lima, and Almeida 2003), Fazenda Vão dos Bois (Bonvicino, Lima, and Almeida 2003), Ipameri (Haag et al. 2007), Mambaí (Haag et al. 2007), Mimoso de Goiás (F. C. Almeida et al. 2007), Mineiros, Emas National Park (F. C. Almeida et al. 2007), Morro dos Cabeludos, Corumbá de Goiás (F. C. Almeida et al. 2007); Minas Gerais: Juramento (Nascimento et al. 2011); Piauí, Coronel José Dias (Nascimento et al. 2011), São João do Piauí (Bonvicino, Lima, and Almeida 2003); Tocantins, Buriti (Nascimento et al. 2011).

SUBSPECIES: *Calomys expulsus* is monotypic. However, Nascimento et al. (2011) found patterns and levels of mtDNA sequence divergence supporting the Rio São Francisco as a biogeographic barrier to gene flow, thus marking the boundary between subpopulations that may warrant formal recognition.

NATURAL HISTORY: This species is practically indistinguishable morphologically from three (perhaps four) congeners of large, lowland *Calomys* with which it may overlap substantially in central Brazil (but see Cordeiro-Estrela et al. 2006, who identified size-free shape differences between this species and *C. tocantinsi*). Therefore, much of the older (e.g., Mello and Teixeira 1977) and some of the newer (e.g., R. A. L. Santos and Henriques 2010) published reports on the ecology, population biology, and/or epidemiology of the species are impossible to assign to *C. expulsus* or any other species with certainty. Nonetheless, in the last several years, various aspects of the biology of this species have been documented. For example, Araripe et al. (2006) studied and compared life-history characteristics (fertility, growth, and rate of postnatal development) in laboratory-raised populations. Hingst-Zaher et al. (2000) detailed changes in patterns of cranial shape through ontogeny. Predators include the Barn Owl (*Tyto alba*; Magrini and Facure 2008).

REMARKS: The karyotype of *C. expulsus* is 2n=66, FN=68, with two pairs of biarmed and 30 uniarmed pairs (Bonvicino and Almeida 2000).

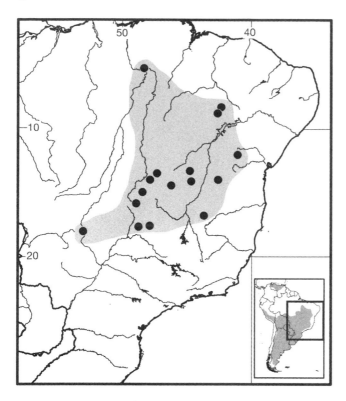

Map 259 Selected localities for *Calomys expulsus* (●).

Calomys hummmelincki (Husson, 1960)
Hummelinck's Vesper Mouse, Hummelinck's Laucha, Dwergwitvoetmuis

SYNONYMS:

Baiomys hummelincki Husson, 1960:34; type locality "Klein Santa Martha, N.W. Curaçao, Netherlands Antilles [now = Caribbean Netherlands]."

Calomys hummelincki: Handley, 1976:53; first use of current name combination.

DESCRIPTION: Small (head and body length 55 mm or less) with soft and silky pelage often brightly colored ochraceous-buff, and traversed mid-dorsally by band of hairs varying from buffy brown to dark brown. Sides colored sandy orange and clearly marked from venter by ochraceous line; venter white, but individual hairs (except those from throat and chin) gray based with white tips. Mystacial vibrissae short. Buffy preauricular and pale white postauricular patches present. Forefeet and hindfeet white above; soles of hindfeet naked, with six small plantar pads. Tail weakly bicolored, averaging <35% of total length (Olds 1988), pale above, lighter below, and terminates in short pencil (Husson 1960). Four pairs of mammae.

Skull small but robust (mean greatest length 20.87 ± 0.54 mm s.d.; Olds 1988), and flat in lateral profile; rostrum relatively short but nasals still obscure incisors when viewed from above; supraorbital region divergent posteri-

orly but lacks ridges; interorbital breadth wider than length of upper molar row. Anterior border of zygomatic plates straight; zygomatic spines lacking; zygomatic notches broad but not deep. Incisive foramina longer than molar series and extend posteriorly between M1 alveoli. Anterior border of mesopterygoid fossa squared and does not reach alveoli of M3s; sphenopalatine vacuities present; and parapterygoid fossae flat and often lightly perforated. Alisphenoid strut absent (Olds 1988). Tympanic bullae small.

Mandible delicate, with body relatively shallow; capsular projection conspicuous but not well developed, located at dorsal base of the sigmoid notch. Upper incisors orthodont with enamel deep orange in color. Molar rows parallel and teeth delicate, narrow, and with principal cusps alternating along anteroposterior axis of toothrow, which is <3.3 mm in length (Olds 1988). M1 with anterocone deeply divided by anteromedian flexus, defining anterolingual and anterolabial conules of unequal size, the latter larger; anteromedian style rarely present (<10%; Olds 1988); M2 about 60% size of M1, with well-developed anterocone; M3 small (ca. 70% of M2), circular, and with paraflexus and mesoflexus evident in young individuals, but no hypoflexus; in older individuals, all flexi of M3 tend to be lost. First lower molar with deep anteromedian flexid; m2 about 75% of m1; and m3 with Z-shaped occlusal surface.

Axial skeletal elements vary in number. Steppan (1995) recorded five individuals with 12 thoracic, 7 lumbar, and 21 to 23 caudal vertebrates, two individuals with 12 thoracic, 6 lumbar and 21–22 caudal, and one individual with 13 thoracic and 6 lumbar but without information on the number of caudal elements. Pérez-Zapata et al. (1987) described sperm morphology and compared it to that of *Calomys laucha*.

DISTRIBUTION: *Calomys hummelincki* occurs in grasslands and other nonforested habitats (below 700 m) on sandy soils in northern Venezuela and Colombia, and on the islands of Curaçao and Aruba. Eshelman and Morgan (1985) reported this species in Pleistocene deposits from Tobago, where it is unknown in the Recent fauna.

SELECTED LOCALITIES (Map 260): CARIBBEAN NETHERLANDS: Aruba, Sero Blanco (Bergmans 2011); Curaçao, Klein Santa Marta (type locality of *Baiomys hummelincki* Husson). COLOMBIA: La Guajira, Cabo de la Vela, Arroyo Cerrejón, Cerro Pilón de Azucar (ICN 19177). VENEZUELA: Anzoátegui, El Merey (A. M. G. Martino et al. 2002); Apure, Puerto Paez (A. M. G. Martino et al. 2002), Río Cinaruco, 38 km NNW of Puerto Paez (Handley 1976); Bolívar, Sipao (A. M. G. Martino et al. 2002); Falcón, Isiro (Salazar-Bravo et al. 2001); Lara, Curarigua (A. M. G. Martino et al. 2002); Monagas, 55 km SSE of Hato Mata de Bejuco (Handley 1976), 8 km NNW of Los Barrancos (KU 135126); Zulia, near Cojoro,

34 km NNE of Paraguaipoa (Handley 1976), confluence of Río Lajas with El Palmar (EBRG 22817).

SUBSPECIES: *Calomys hummelincki* is monotypic. However, the ventral pelage of some individuals from Aruba is completely white, lacking the gray-based hairs that are present in the holotype and other individuals from Curaçao and mainland populations (Husson 1960). Moreover, animals from Zulia and Guárico states on mainland Venezuela are darker and grayer on the dorsum, contrasting with the yellowish or buffy coloration of populations from insular or more northern mainland, dryer habitats in Colombia and Venezuela. Finally, insular samples average slightly smaller but have broader interorbital regions than their mainland conspecifics (comparison of measurements in Husson 1960 and Olds 1988).

NATURAL HISTORY: *Calomys hummelincki* was always trapped on the ground often associated with dry and/ or thorny forests and in localities below 700 m in elevation (Handley 1976). It is present but apparently rare in the diet of the Barn Owl (*Tyto alba*) in Curaçao (Debrot et al. 2001); in Aruba, *C. hummelincki* is a preyed upon by rattlesnakes (Goode et al. 1990), the Giant centipede (*Scolopendra gigantea*), and the American Kestrel (*Falco sparverius*) (Bekker 1999). Reproduction information is limited: Olds (1988) reported six pregnant females with two to five embryos each (no collection date was provided) among specimens from Guárico state, Venezuela. One female with four embryos was captured in early August in Monagas state (EBRG 14946). Bekker (1999) studied Aruba populations and presented the most complete ecological study to date. Estimates of abundance, averaged from 0.4 animals per 100 trap-nights, were that the home

range was 4,100 m² for males and 713 m² females, and a bimodal polyestrous cycle with youngsters was present in early January and late August, indicating births at the end December and middle of August.

REMARKS: Pérez-Zapata et al. (1987) reported a karyotype with $2n = 60$ and $FN = 64$. A. M. G. Martino and Capanna (2002) provided data on differential staining and chromosomal banding patterns, demonstrating a close homology of chromosomal arms between this species and those of *Calomys laucha* as well as some marked differences. A. M. G. Martino et al. (2000) inferred relatively low levels of gene flow among populations using protein electromorphic characters.

Calomys laucha (G. Fisher, 1814)
Small Vesper Mouse, Laucha Chica, Lauchita
SYNONYMS:

Mus Laucha et *Lauchita* G. Fischer, 1814:71; the local common name cited by F. d'Azara (1801:102) for his "*rat septieme ou rat laucha.*"

[*Mus*] ? *Laucha* Illiger, 1815:108; *nomen nudum*.

M[*us*]. *laucha* Olfers, 1818:209; based on Azara's (1801: 102) "*rat laucha*"; type locality "Paraguay," restricted to the "neighborhood of Asunción, Paraguay" (Hershkovitz 1962:153).

Mus (?) *dubius* J. B. Fischer, 1829:326; type locality "Paraguaya," amended by Pardiñas et al. (2007) to "Saõ Gabriel (30°19′S, 54°19′W, 118 m, Rio Grande do Sul, Brazil); see Tate (1932c) for discussion of the applicability of this name to the synonymy of *C. laucha*.

Mus bimaculatus Waterhouse, 1837:18; type locality "Maldonado," Maldonado, Uruguay.

Mus gracilipes Waterhouse, 1837:19; type locality "Bahia Blanca," Buenos Aires, Argentina.

Mus (*Calomys*) *bimaculatus*: Waterhouse, 1837:21; name combination.

Mus (*Calomys*) *gracilipes*: Waterhouse, 1837:21; name combination.

Eligmodontia bimaculatus: Lesson, 1842:133; name combination.

Eligmodontia gracilipes: Lesson, 1842:133; name combination.

H[*esperomys*]. *laucha*: Wagner, 1843a:543; name combination.

Mus pusillus Philippi and Landbeck, 1858:79; type locality "Valparaiso," Valparaíso, Chile (see Osgood 1943b, for discussion of the applicability of this name to the synonymy of *C. laucha*).

Hesperomys bimaculatus: Burmeister, 1861:415; name combination.

Hesperomys [*Calomys*] *bimaculatus*: Burmeister, 1879:225; name combination.

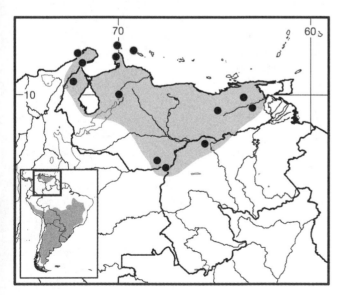

Map 260 Selected localities for *Calomys hummelincki* (●). Contour line = 2,000 m.

[*Hesperomys* (*Calomys*) *laucha*] *gracilipes*: E.-L. Trouessart, 1880b:138; name combination.

[*Hesperomys* (*Akodon*)] *pusillus*: E.-L. Trouessart, 1880b: 142; name combination.

H[*esperomys*]. (*Calomys*)] *gracilipes*: Thomas, 1884:449; name combination.

Oryzomys laucha: Thomas, 1897d:215; name combination.

[*Akodon*] *pusillus*: E.-L. Trouessart, 1897:538; name combination.

[*Eligmodontia*] *laucha*: E.-L. Trouessart, 1904:428; name combination.

Eligmodontia laucha: Thomas, 1910a:242; name combination.

[*Hesperomys*] *laucha*: Thomas, 1916a:141; name combination.

[*Hesperomys*] *gracilipes*: Thomas, 1916a:141; name combination.

Hesperomys laucha: Thomas, 1916b:182; name combination.

Akodon pusillus: Gyldenstolpe, 1932a:101; name combination.

Hesperomys bimaculatus bonariensis Osgood, 1933d:14; type locality "Torrecita, province of Buenos Aires, Argentina."

Calomys laucha laucha: Hershkovitz, 1959b:339; first use of current name combination.

Calomys dubius bonariensis: Cabrera, 1961:477; name combination.

Calomys dubius dubius: Cabrera, 1961:477; name combination.

DESCRIPTION: Smallest species in genus (mean total length 125 mm, greatest skull length 20.46 mm; Olds 1988). Dorsal pelage gray mixed with brown to golden brown; in populations with grayer coloration, upper parts with marked concentration of blackish-tipped hairs forming broad, dark dorsal band from forehead to rump; sides always paler but often with some mix of gray and brown hairs; buffy line often separates dorsal from ventral coloration; venter always washed with white; in some populations, chin and throat have fully white hairs. Buffy hair tufts occur anterior to ears in populations from drier habitats, but all specimens throughout range have conspicuous white postauricular patches. Tail bicolored, brown above, white below; short, no more than 43% of total body length. Mammae count in adult females 8–10 (Olds 1988).

Skull with straight lateral profile; rostrum short and moderately narrow, with sides tapering gradually; nasals narrow and projecting forward above incisors; lacrimals of moderate size; interorbital region wide and posteriorly divergent but lacking supraorbital shelves or ridges; braincase large and rounded; temporal and lambdoidal ridges absent; parietals with small expansion onto lateral surface on squamosal, forming shallow deflection at level of squamosal zygomatic root in mature specimens; interparietal large for size of skull. Zygomatic plates broad, with upper free borders enclosing poorly developed zygomatic notches; no sign of zygomatic spines. Zygomatic arches delicate, not expanded and mostly parallel. Alisphenoid strut variably present. Hamular process robust, dividing large subsquamosal fenestra and relatively smaller postglenoid foramen. Incisive foramina broad and long, slightly surpassing anterior roots of M1s; premaxillary part of septum occupies about three-fourths length of foramina. Palate wide and long, rather flat, with small posterolateral palatal pits on sides of mesopterygoid fossa. Mesopterygoid fossa wide for genus but never wider than parapterygoid fossae measured at same level, and with round anterior border; sphenopalatine vacuities small but present. Parapterygoid plates broad and excavated. Auditory bullae globular of medium size and with short and wide bony tubes.

Mandible delicate; masseteric crest forms well-developed knob, more evident in older individuals; capsular projection inconspicuous; coronoid process long, narrow, strongly curved backward, and ends at plane higher than short and wide condyloid process; angular process short; sigmoid notch deep and well developed.

Upper incisors hyperopisthodont and faced with light orange enamel (Steppan 1995). Upper molars brachydont and small with respect to skull. M1 subrectangular in outline; procingulum well developed with anterocone deeply divided by anteromedian flexus (in 77% of specimens; Olds 1988); anteromedian style present (Olds 1988); mures longitudinally oriented; hypoflexus and protoflexus parallel; protocone and hypocone about same size; paracone and metacone subequal in size and both about 60% smaller than protocone and hypocone; small parastyle (and accompanying anteroflexus) present (Steppan 1995); and hypoflexus narrow and laterally directed. M2 square in outline, trilophodont, and about 60% smaller than M1; both anteroloph and mesoloph absent; mesoflexus opposite to hypoflexus present but very narrow; and posteroflexus small but present (Olds 1988). M3 cylindroform, moderately reduced with respect to M2 (from 64–67%; Olds 1988); hypoflexus reduced; and mesoflexus persists as small fossette. Lower molars have their main cusps moderately alternating. Lower first molar tetralophodont, with procingulum divided by anteromedian flexid, small but present anterolabial cingulum, transverse and well-developed protoflexids and hypoflexids opposite metaconid and entoconid, respectively, and with conspicuous posterolophid. Second lower molar trilophodont, with large proto-, hypo-, meso-, and posteroflexids and well-developed posterolophid. Third lower molar subtriangular in outline, with complex dental pattern (almost right-facing "w"), well-developed hypoflexid opposite small entoconid, and also with well-developed posterolophid.

Axial skeleton composed of 13 thoracic, 6 lumbar, and 21–23 caudal vertebrae (Steppan 1995). Stomach unilocular-hemiglandular (Carleton 1973). Only one pair of preputial glands present among male accessory reproductive glands (Voss and Linzey 1981). E. R. S. Roldán et al. (1985) described sperm structure.

DISTRIBUTION: *Calomys laucha* inhabits grasslands and forest fringes (up to an elevation of 950 m) in Argentina, Paraguay, southeastern Bolivia, northeastern Argentina, Uruguay, and east into Rio Grande do Sul state, Brazil. Unverified records may extend the range to Paraná state, Brazil (L. T. M. D. Souza et al. 2002). Several authors have suggested that *C. laucha* (like *Calomys musculinus*) may have increased in abundance and expanded its range to otherwise forest wetlands and agricultural fields and adjacent habitats (e.g., in Misiones province, Argentina) only in the last few centuries, likely as a direct result of anthropogenic influences (Pardiñas et al. 2000; Pardiñas and Teta 2005; P. E. Ortiz et al. 2011b). Fossils assigned to *Calomys* cf. *laucha* were reported from the late Pliocene–early Pleistocene and the Holocene of southern Bolivia (Hoffstetter 1986; Pardiñas and Galliari 1998c) and from the same time period in northern Argentina (Pardiñas et al. 2002).

SELECTED LOCALITIES (Map 261): ARGENTINA: Buenos Aires, Bahia Blanca (Waterhouse 1837), José C. Paz (Massoia et al.1968), near Otamendi (MSU 19000), 30 road km SW of La Plata, Punta Blanca (UMBW 72276); Córdoba, Colonia Tirolesa (TTU 32992); Jujuy, Yuto (AMNH 167856); Misiones, Candelaria, Río Yabebyrí (CFA 9551), Oberá, Campo Viera (CFA 10052); Río Negro, Chimpay (Hershkovitz 1962), Villa Regina (MACN 14584); Salta, Corralito (TTU 32996); San Luis, Naschel (TTU 32997); Santiago del Estero, Lavelle (TTU 32998). BOLIVIA: Santa Cruz, Boyuibe, near Camiri (S. Anderson 1997), Cerro Colorado (S. Anderson 1997); Tarija, 8 km S and 10 km E of Villa Montes (S. Anderson 1997). BRAZIL: Rio Grande do Sul, Faxinalzinho (Badzinski et al. 2012), Fazenda Marcelina (A. O. Rosa and Vieira 2010), Praia do Cassino (C. R. Camargo et al. 2006). PARAGUAY: Boquerón, Parque Nacional Teniente Enciso (TK 67304); Canendiyú, Estancia Salazar, Río Verde (TK 66816); Ñeembucú, Estancia San Felipe (TK 61623). URUGUAY: Maldonado, Maldonado (type locality of *Mus bimaculatus* Waterhouse).

SUBSPECIES: Geographical variation is extensive in pelage coloration (Waterhouse 1837; Osgood 1933d), diploid chromosomal counts (Brum Zorrilla et al. 1990), the proportion of individuals with an alisphenoid struts in samples (Olds 1988), and the number of mammae (Olds

1988). Osgood (1933d) described *Hesperomys bimaculatus bonariensis* from samples of *C. laucha* from around Buenos Aires and based on the smaller size and "somewhat darker-colored" phenotypes; he also commented on the variability observed in the amount of white hairs on ventral surfaces of this and other species, suggesting that this character was variable and likely of little taxonomic value.

Based on the patterns of pelage coloration, the ratio of tail to total length, the proportion of white hairs in the venter, and the presence/absence of alisphenoid strut in individuals per population, Olds (1988) recognized two species of *Calomys* within our current understanding of *Calomys laucha*. She used the name *C. bimaculatus* for populations of small bodied *Calomys* from Uruguay, southern Brazil, and various localities in Buenos Aires province (e.g., Torrecitas) based on her observation that this form tended to have a dorsal pelage coloration that was overall grayer, often with a darker band of black hairs from the muzzle to the base of the tail, and white hairs on the venter restricted to the chin and throat; one of the 12 specimens she examined of this form had an alisphenoid strut. Olds (1988) restricted the name *C. laucha* for populations from dry biotopes from south central Bolivia, north and northeastern Argentina, and western Paraguay. These populations are characterized by having golden-brown dorsal and fully white ventral colorations (including chins and throats); in addition about 91% of the 35 animals she examined had alisphenoid struts. However, craniodental dimensions of the two taxa Old's recognized overlapped extensively in multivariate space.

Recent work on captive populations of *C. laucha* have elegantly demonstrated that pelage color change in this species is controlled by an endogenous circannual rhythm wherein individuals molt from an orange-colored pelage during summer to a dark gray color during winter (C. R. Camargo et al. 2006). The hypothesis, therefore, is that populations living in warmer, dryer biotopes of Argentina, Bolivia, Paraguay, and parts of Brazil have golden-brown pelage similar to that of the summer pelage of populations living in more seasonal environments on Uruguay, Argentina, and southernmost Brazil.

Despite apparent marked geographic variation in some external phenotypic characters, divergence in mtDNA cytochrome-*b* gene sequences from Bolivian, Paraguayan, Argentinean, and Brazilian populations indicate a pattern of genetic structure coincident with isolation by distance (R. González-Ittig, pers. comm.).

NATURAL HISTORY: The natural history of this species is relatively well known (e.g., Kravetz, Busch et al. 1981; Kravetz, Manjon et al. 1981; Mills, Ellis, McKee, Maiztegui, and Childs 1991; K. Hodara et al. 1997; Bilenca and Kravetz 1999; Yahnke 2006; Courtalon and Busch 2010). In the grassland of Buenos Aires province, Mills, Ellis, McKee, Maiztegui, and Childs (1991) reported a seasonal pattern of population fluctuation with high densities in the summer and autumn, declining to almost nil during winter months and expanding again in the spring. Individuals were more abundant in the center rather than along the edges of field, possibly due to competition with *Akodon azarae* (Courtalon and Busch 2010). In the Argentinean Pampas, *C. laucha* preferentially occupied cornfields in current agricultural production, in contrast with sympatric *C. musculinus*, which was present only in adjacent old fields (Villafañe et al. 1992). In the Chaco region of central Paraguay, *Calomys laucha* was trapped in pastures, cropland, thorn scrub, and thorn forest, being most abundant in the first three types of habitats; it was more highly associated with less herbaceous ground cover than other small mammal species (Yahnke 2006). Reproduction is seasonal, occurring from September to May in east-central Argentina (Mills, Ellis, McKee et al.1992) and from April to August in the central Chaco (Yahnke 2006). The mean number of embryos ranged from 5.15 (in the Chaco) to 6.2 in the pampas of Argentina (Mills, Ellis, Childs et al. 1992). This may be a monogamous species (Laconi and Castro-Vazquez 1999). The diet is reported as omnivorous (Castellarini et al. 2003). In Bolivia and Paraguay, *C. laucha* is the natural reservoir of Laguna Negra hantavirus (A. M. Johnson et al. 1997; Yahnke et al. 2001), of a *Syphacia* nematode in Buenos Aires province, Argentina (E. J. R. Herrera et al. 2011), a *Physaloptera* nematode in Uruguay (Sutton 1989b), and may be infected with *Trypanosoma cruzi* in Córdoba province, Argentina (Moretti et al. 1980). *Calomys laucha* is preyed upon by at least two species of owls (Bellocq and Kravetz 1994).

REMARKS: Musser and Carleton (1993:698; 2005:1107) incorrectly cited "[G.] Fischer, 1814. *In* Eschwege, J. Brasilien, Neue Bibliothek. Reisenb., 15(2):209" as the author and source of *Calomys laucha*. W. L. Eschwege's *Journal von Brasilien, oder vermischte Nachrichten aus Brasilien, auf wissenschaftlichen Reisen gesammelt* was published in 1818, and I. von Olfers was the author of chapter 10, the description and list of mammals contained within, with Olfer's citation of *M. laucha* given on p. 209. Fischer had earlier published *Mus laucha* on p. 71 in his 1814 treatise *Zoognosia tabulis synopticis illustrata*. Both Fischer's and Olfer's *laucha* were based directly on F. d'Azara's (1801) "*rat septieme ou rat laucha*." Hershkovitz (1959b:339, 1962:150) identified Olfers (1818) as the author of *laucha* but did not cite Fischer (1814). Further discussion on the acceptability of Fischer's names can be found in Musser et al. (1998: 28ff.).

Olds (1988) suggested the presence of *Calomys laucha* in the Brazilian state of Mato Grosso, based on a specimen collected by Leo Miller in 1914 at Utiariy on the Rio Papagaio during the Roosevelt Brazilian Expedition. The specimen in question (AMNH 37543) was originally identified as *Oryzomys* (*Oligoryzomys*) *microtis* and at some later unknown date reidentified as *Calomys laucha* (as can be seen on the museum catalogue entry for the specimen). Currently, there are no records of this species in the states of Rondônia, Mato Grosso, or Mato Grosso do Sul in Brazil, even at localities where longitudinal trapping programs (with both live and pitfall traps) have been conducted (e.g., Cáceres et al. 2010; Cáceres et al. 2011; Mallmann et al. 2011). Furthermore, the species has not been recorded at Bolivian localities near the Brazilian border (e.g., Emmons et al. 2006), where a similar composition of small mammal fauna to that of Utiariy is currently found, and where equally long-term studies have taken place. A parsimonious explanation is that AMNH 37543 represents another species of *Calomys*, most likely *Calomys tener*, a species found in other inventories in Mato Grosso and Matto Grosso do Sul (Cáceres et al. 2008) that shares a handful of morphological and morphometric characteristics with *C. laucha*.

Thomas (1898h) reported *Eligmodontia gracilipes* (a junior synonym of *Calomys laucha*) from near Rawson in the Patagonian province of Chubut; this report was based on two specimens collected in 1878 by Henry Durnford, who explored and collected along the basin of the Chubut River between 1876 and 1878 (Durnford 1878). Discussing the identity of the two specimens, Thomas clearly assigned them to the same species represented by "Azara's Laucha" (*Calomys laucha*). Based on the fact that there are no confirmed records of *Calomys laucha* south of the Río Colorado in Patagonia—but those of *C. musculinus* are abundant (Pardiñas, Teta, D'Elía, and Lessa 2011)—and that Thomas did not recognize (and would eventually describe) *musculinus* as a biological entity distinct from *laucha* until 13 years after his 1898 publication, it is safe to assume that the specimens trapped by Durnford in March 1878 are early records of *C. musculinus* in Patagonia. Steppan (1995: 104) reidentified as *C. laucha* two specimens from Chimay (Rio Negro province, south of the Colorado River) originally assigned to *C. murillus* (= *C. musculinus*) by Olds (1988). I disagree with Steppan's reidentification; in fact, extensive and intensive studies by U. F. J. Pardiñas and collaborators over the last decade indicate that currently *Calomys laucha* does not range south of the Río Colorado. The only species of *Calomys* found in northern Patagonia, at least since the late Holocene, is *Calomys musculinus* (Udrizar Sauthier 2009).

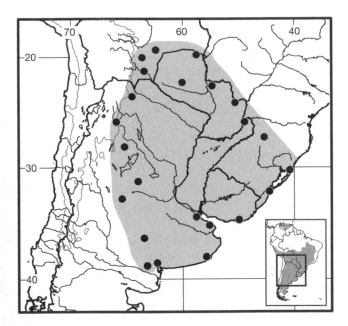

Map 261 Selected localities for *Calomys laucha* (●). Contour line = 2,000 m.

Some karyotypic variability is known for specimens assigned to this species from Argentina and Uruguay, with the most common karyotype being $2n=64$, $FN=68$ (Brum Zorrilla et al. 1990). This karyotype is the only one found in Brazilian samples (Mattevi et al. 2005). The single report of animals from Artigas (Uruguay) with $2n=56$, $FN=59$ (Brum 1962), has never been independently confirmed. M. L. Ortiz et al. (2007) reported the distribution of telomeric signals in various internal positions of the karyotype, suggesting multiple chromosomal rearrangements in the evolutionary history of the species. Ciccioli and Poggio (1993) reported on genome size.

Calomys lepidus (Thomas, 1884)

Andean Vesper Mouse, Graceful Laucha, Hiska Achaco

SYNONYMS:

Hesperomys (Calomys) bimaculatus var. *lepidus* Thomas, 1884:454; type locality "Junin," Peru.

E[ligmodontia]. lepida: Thomas, 1900f:298; name combination.

Eligmodontia ducilla Thomas, 1901b:182; type locality "San Anton [= San Antón], Lake Titicaca, [Puno,] S.E. Peru. Alt. 3800 m."

Eligmodontia carilla Thomas, 1902b:133; type locality "Choro, 3500 m," Cochabamba, Bolivia.

[Hesperomys] carilla: Thomas, 1916a:141; name combination and incorrect gender concordance.

[Hesperomys] ducilla: Thomas, 1916a:141; name combination and incorrect gender concordance.

[Hesperomys] lepidus: Thomas, 1916a:141; name combination.

Hesperomys carillus marcarum Thomas, 1917g:1; type locality "Lauramarca [= Labramarca]," Cusco, Peru.

Hesperomys carillus argurus Thomas, 1919f:130; type locality "Abrapampa, 3500 m," Jujuy, Argentina.

Hesperomys lepidus lepidus: Sanborn, 1950:3; name combination.

Hesperomys lepidus ducillus: Sanborn, 1950:4; name combination.

Hesperomys lepidus montanus Sanborn, 1950: 4: type locality "El [La] Cumbre, [La Paz,] Bolivia. Altitude 15,200 feet."

Hesperomys ducillus: Pearson, 1951:141; name combination.

Calomys lepidus: Cabrera, 1961:479; first use of current name combination.

Calomys lepidus marcarum: Cabrera, 1961:480; name combination.

Calomys lepidus ducillus: Cabrera, 1961:480; name combination.

DESCRIPTION: Small (total length <140 mm) with very short tail (<50 mm), averaging less than <40% of total length (Olds 1988). Fur long, soft, and silky; dorsal coloration dull fawn or grayish fawn, somewhat darker on center of back because of intermixture of numerous longer black hairs; dorsum with marbled appearance, nearly uniform in tone from head to rump (Olds 1988); ventral coloration white, but hairs with gray bases. Ears large, pale brown, each with conspicuous white postauricular spot. Tail uniformly hairy, weakly bicolored, sandy white above, clear white below. Forefeet and hindfeet light above; hindfeet with six small plantar pads, heel and proximal portion of sole hairy. Mammae number 8 to 10 (Hershkovitz 1962).

Frontal outline of skull markedly convex; interorbital region parallel sided and supraorbital edges squared but not ridged (Gyldenstolpe 1932); rostrum short and moderately broad; nasals short but relatively broad, especially first third, covering incisors when skull seen from above and terminating at site of triple juncture of maxillary, frontal, and lacrimal; coronal suture shaped as open U; zygomatic plates about 7% of condyloincisive length, with slightly concave anterior borders and very shallow zygomatic notches; zygomatic arches delicate and parallel at midpart, broader near parietal root; midpoint dips ventrally but remains discernibly above level of orbital floor; jugal narrow but distinctly separates maxillary and squamosal portions of arch; braincase small and somewhat globular; parietals with small extensions in lateral view; interparietal wide across braincase but frontodorsally depressed; postglenoid foramen vestigial, obscured internally by lateral border of tegmen tympani. Incisive foramina long, borders parallel, and surpass only slightly anterior roots of M1s; premaxillary part of the septum occupies about half of opening length. Palate long and wide (*sensu* Hershkovitz 1962), with small posterolateral palatal pits. Mesopterygoid fossa short, narrow, and divergent posteriorly (strongly V-shaped; Steppan 1995), its anterior border behind posterior face of M3; roof of fossa perforated by well-developed sphenopalatine vacuities. Auditory bullae somewhat inflated, pear-shaped in outline.

Upper incisors orthodont and faced with bright orange enamel. Upper molars brachydont, with rows posteriorly divergent (Braun 1993). M1 subrectangular in outline with well-developed procingulum and anteromedian flexus dividing anterocone into anterolingual and anterolabial conules in 95% of specimens (Olds 1988); anteromedian style present in about 33% of specimens examined (Olds 1988); mures longitudinally oriented; both hypoflexus and protoflexus transverse; protocone and hypocone are about same size; and hypoflexus wide and posteriorly directed. M2 square in outline, about 60% smaller than M1; anteroloph persistent, a mesoloph is absent; mesoflexus and hypoflexus well developed. M3 squared, about 67% of the transverse length of M2 (Olds 1988); both hypoflexus and mesoflexus well developed; and anteroloph present. Lower molars crested, with main cusps clearly alternating; both mesolophids and ectolophids absent, but mesostylids and ectostylids present. Procingulum of m1 well developed and divided by anteromedian flexid; mesoflexid narrow and directed backward; posterolophid well developed; m2 with intercalating cusps, well-developed mesoflexids and hypoflexids, and well-developed posterolophid; m3 complex, Z-shaped, with small but well-defined protoflexid; posteroflexid lost with age but otherwise present. Mandible with capsular projection of ramus obsolete or absent (Hershkovitz 1962; Braun 1993).

Vertebral count of 13 thoracic, 6 lumbar, and 20–22 caudal elements (Steppan 1995); the same author also reported a single large lateral pair of preputial glands. Sperm morphology identical to that of *C. musculinus* (Espinosa et al. 1997).

DISTRIBUTION: *Calomys lepidus* occurs in high-elevation, rocky grasslands above 2,900 m from central Peru south to northern Chile, Bolivia, and into northwestern Argentina.

SELECTED LOCALITIES: (Map 262): ARGENTINA: Catamarca, Belén (Jayat, Ortiz, Pacheco, and González 2011); Jujuy, Abrapampa (type locality of *Hesperomys carillus argurus* Thomas), El Toro, 50 km W of Susques (MSB 75238); Tucumán, Huaca Huasi (Ferro and Barquez 2008), La Junta, naciente del Río Pavas (Ferro and Barquez 2008). BOLIVIA: Chuquisaca, 2 km N of Tarabuco (CBF 1090); Cochabamba, Ayopaya (S. Anderson 1997), Choro (type locality of *Eligmodontia carilla* Thomas), ca. 13 km N of Co-

lomi (S. Anderson 1997), 101 km by road SE of Espinaza, Siberia Cloud Forest (AMNH 246798); La Paz, La Cumbre (type locality of *Hesperomys lepidus montanus* Sanborn), Reserva de Fauna Ulla-Ulla (Salazar-Bravo et al. 2001). CHILE: Antofagasta, Ojos de San Pedro (Hershkovitz 1962); Tarapacá, Parinacota (MSB 210378). PERU: Ancash, Rachacoco, 45 km S of Huaraz (MVZ 135689); Arequipa, ca. 48 km by road E of Arequipa (LSUMZ 27888); Ayacucho, 23 km W (by road) of Pampamarca (MVZ 174337), Tucumachay Cave (UMMZ 122623); Cusco, Lauramarca [= Labramarca] (type locality of *Hesperomys carillus marcarum* Thomas); Junín, Carhuamayo (Hershkovitz 1962); Puno, 4 km E of Juli (Olds 1988), San Antón (type locality of *Eligmodontia ducilla* Thomas).

SUBSPECIES: *Calomys lepidus* is polytypic, with extensive morphological and morphometric variation evident from the number of formal taxa recognized across its range, either as separate species or subspecies. Hershkovitz (1962) reviewed patterns of diversification, and delineated four subspecies: the nominotypical form in the Andes of central Peru south to Cusco department; *C. l. ducillus* Thomas from the Altiplano of southern Peru, northeastern Chile, and northern Bolivia; *C. l. carillus*, from La Paz and Cochabamba departments, Bolivia; and *C. l. argurus*, restricted to its type locality in Jujuy province, Argentina.

NATURAL HISTORY: The species is widespread in Puna grassland and ecotonal habitats, at elevations between 2,950 m to about 4,850 m, and from central Peru to Catamarca in Argentina (Hershkovitz 1962; S. Anderson 1997; Yensen and Tarifa 2002; Zeballos and Carrera 2010; Jayat, Ortiz, Pacheco, and González 2011). Pregnant females have been trapped throughout the year, with a mode of four embryos (range: 1 to 6) from Lima department, Peru, to Tarija department, Bolivia (Olds 1988; S. Anderson 1997). In northern Argentina, M. M. Díaz and Barques (2007) recorded scrotal males in March and young individuals in February and December. Pearson and Ralph (1978) reported a density of 0.81 individuals per hectare in bunch grass habitat at 3,900 m in Puno (Peru), and Zeballos and Carrera (2010) recorded abundances of about 0.3 animals/100 trapnights. The Culpeo (*Lycalopex culpaeus*; Mercado and Miralles 1991) and owls (Pearson 1951; P. E. Ortiz, Cirignoli et al. 2000) are known predators.

REMARKS: It is highly likely that at least two species are masquerading within the concept of *C. lepidus* presented here. Patterns of morphometric and morphological diversity found along the almost 1,500 km of the distribution added to the difference in chromosome numbers in populations at the extremes of the distribution, and genetic distances among haplotypes of the mtDNA cytochrome-*b* for individuals from Puno (Peru) and Tarija (Bolivia),

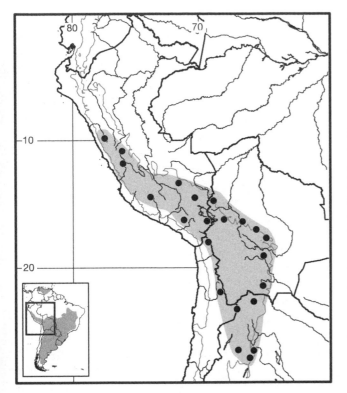

Map 262 Selected localities for *Calomys lepidus* (●). Contour line = 2,000 m.

which average 3.35%, support this hypothesis (J. Salazar-Bravo, unpubl. data).

The karyotype is geographically variable, with $2n = 36$, $FN = 68$, reported for Peruvian samples (Pearson and Patton 1976), but $2n = 44$, $FN = 68$, in specimens from northwestern Argentina (Espinosa et al. 1997).

Calomys musculinus (Thomas, 1913)

Drylands Vesper Mouse, Drylands Laucha, Corn Mouse

SYNONYMS:

Eligmodontia laucha musculina Thomas, 1913a:138; type locality "Maimara, [Jujuy, Argentina]. 2230 m."

[*Hesperomys laucha*] *musculinus*: Thomas, 1916a:141; name combination.

Hesperomys murillus Thomas, 1916b:183; type locality "La Plata City," Buenos Aires, Argentina.

Hesperomys murillus cordovensis Thomas, 1916b:184; type locality "Yacanto, [Córdoba, Argentina,] 900 m.

Hesperomys musculinus: Thomas, 1920d:116; name combination.

Hesperomys musculinus cortensis Thomas, 1920a:190; type locality "Jujuy, [Jujuy, Argentina,] 1,258 m.

Hesperomys murillus murillus: Yepes, 1935a:227; name combination.

H[*esperomys*]. *murillus cordobensis* Yepes, 1935a:227; name combination and incorrect subsequent spelling of *cordovensis* Thomas.

Calomys laucha cordovensis: Cabrera, 1961:478; name combination.

Calomys laucha musculinus: Cabrera, 1961:479; name combination.

Calomys musculinus: Massoia and Fornes, 1965b:6; first use of the current name combination.

DESCRIPTION: Medium sized (total length 140–200 mm) with tails as long as head and body (about 50% of total length) and clear countershading of dorsal and ventral coloration (in most populations). Dorsal color sandy brown lined with black hairs to gray all over with usually darker rumps; head about same color as dorsum; sides of body often bit lighter and contrast sharply with white appearance of ventral region; often hairs on ventral area white only at tips (gray at base), but some populations with fully white hairs on throat. Populations from Paraguay, southern Bolivia, and northern Argentina often with buffy line demarcating sides from venter. Ears about average for small forms of genus (ca. 14 mm long), rounded, covered internally by short, delicate brownish yellowish hairs, and contrasted from head color with concha and tragus white. Hindfeet long and narrow (ca. 18 mm), covered dorsally by small white hairs. Tail often bicolored, dark brown to black above and clear to creamy-white below. Plantar surfaces of hindfeet largely naked but with some hairs covering calcaneus and extending to thenar pad; six plantar pads present. Mammae number 10 to 12 (Olds 1988).

Skull with straight profile when viewed laterally; rostrum moderately short and narrow, with nasals projecting forward so that incisors covered when skull viewed from top. Interorbit about 30% of zygomatic breath; supraorbital region posteriorly divergent but lacks marked ridges or ledges (Olds 1988). Coronal suture broadly open U-shape. Braincase somewhat elongated, without temporal and lambdoidal ridges; parietals with very poorly developed lateral extensions onto squamosal; interparietal expanded across braincase. Zygomatic plates broad and high, without zygomatic spines, and defining poorly developed zygomatic notches; frontal edge of zygomatic plates variable, either somewhat rounded or straight. Hamular process narrow, separating reduced postglenoid foramen with respect to subsquamosal fenestra. Incisive foramina long and narrow, reaching between anterior roots of M1s; premaxillary part of septum occupies about 50% of its length. Palate broad and long (*sensu* Hershkovitz 1962), with very small posterolateral palatal pits. Mesopterygoid fossa narrow, with squared anterior edge and without trace of median palatine process; sphenopalatine vacuities present. Auditory bullae slightly inflated (Olds 1988), and carotid circulatory pattern primitive (pattern 1 of Voss 1988).

Mandible small but robust; mental foramen not visible from labial view; masseteric crest evident, its upper and lower ridges producing thickened chevron running from point of contact (below about second lateral root of m1) to slightly above level of mental foramen; capsular projection very conspicuous, lying at about middle point of sigmoid notch. Coronoid process strong, falciform (with blunt edge in some adults), and inflected backward; condyloid process long and narrow; sigmoid notch well defined.

Upper incisors opisthodont and faced with pale yellow (old adults) to orange enamel; dentine fissure long straight slit (*sensu* Steppan 1995). Upper molars crested and brachydont, markedly opposite in pattern, and small with respect to skull. M1 tetralophodont; an anteromedian flexus divides relatively narrow anterocone; hypoflexus and protoflexus straight and transversal; protocone and hypocone about equal in size; both parastyle and mesostyle present as thickening of labial cingulum; and no trace of a shallow anteroflexus (Steppan 1995). M2 square in outline, bilophodont, and about 60% the length of M1 (Olds 1988); wide mesoflexus opposite to hypoflexus. M3 trilophodont, moderately reduced with respect to M2 (between 72% and 74%); hypoflexus, mesoflexus, and paraflexus all persist, even in old individuals, as small fossettes. Lower molars crested with alternating main cusps. Lower first molar with anteromedian flexid fused at base of cusp; the hypoconid is larger and broader than the protoconid; the entoconid is bulbous, well-developed posterolophid, and posterostylid subequal in size to hypoconid; m2 trilophodont, with posterolophid and metaflexids, mesoflexids, and hypoflexids well developed; m3 sigmoid in outline, with hypoflexid deep and curved upward, but mesoflexid persistent as fossette.

Vertebral elements highly variable in number, with thoracic 12–14, lumbar 5–7, and caudal 24–29 (Steppan 1995). E. R. S. Roldán et al. (1985) described sperm morphology; spermatozoa from mature males with hooked heads with roughly polygonal nucleus, similar in morphology to those present in males of *C. lepidus* and *C. hummelincki* (Espinosa et al. 1997). R. A. Cutrera et al. (1992) described female genital morphology, and Buzzio and Castro-Vazquez (2002) detailed gross morphology of male accessory glands. Stomach unilocular-hemiglandular (Rouaux et al. 2003), with digestive tract structures described and compared with those of other Pampean species of rodents by Ellis et al. (1994). Bee de Speroni and Peregrini de Gstaldo (1988) detailed allometry of brain growth, and Cavia et al. (2008) examined hair microstructure.

DISTRIBUTION: *Calomys musculinus* is known from sea level to 3,300 m in habitats ranging from the Pampas, Espinal, and Monte of central Argentina, to the Yungas and Puna of northern Argentina and south and central Bolivia, and dry Chaco of Paraguay.

SELECTED LOCALITIES (Map 263): ARGENTINA: Buenos Aires, La Plata (type locality of *Hesperomys murillus* Thomas), Mar del Plata (Salazar-Bravo et al. 2001); Catamarca, 13 km NNW of Andalgalá (Salazar-Bravo et al. 2001); Chaco, General Pinedo (Olds 1988); Chubut, Esquel (CFA 02166), 8 km ESE (by road) of Puerto Madryn (Salazar-Bravo et al. 2001), Valle Hermoso, Comodoro Rivadavia (CFA 09070); Córdoba, Yacanto (type locality of *Hesperomys murillus cordovensis* Thomas); Entre Ríos, Gualeguay (MACN-MA 21236); Jujuy, Maimara (type locality of *Eligmodontia laucha musculina* Thomas), Santa Catalina (Olds 1988); La Pampa, 40 km N of Anzoátegui (TTU 64423); La Rioja, El Pesebre (Jayat, Ortiz, González et al. 2011); Mendoza, 10.5 km W of Old Rte. 40, along road to Lago Diamante (OMNH 15360); Neuquén, 5 km N of Las Coloradas (MVZ 159401); Salta, 3 km E of La Poma (Jayat, Ortiz, Pacheco, and González 2011); San Juan, 4 km W of Complejo Astroneomico "El Leondito" (OMNH 20023), Tudcum, Nacedero (OMNH 30022); Santa Cruz, Laguna del Diez, 40 km SW of Monumento Natural Bosques Petrificados (Jayat et al. 2006), Estancia La Julia (CNP-E 489), Estancia Cerro Ventana (PNG 830); Santa Fe, San Javier (MACN-Ma 15734); Santiago del Estero, 1 km S and 2 km E of Pampa de los Guanacos (OMNH 35124); Tucumán, Paso El Infiernillo (MSU 19219). BOLIVIA: Chuquisaca, 4 km N of Tarabuco (S. Anderson 1997); Cochabamba, Cochabamba (S. Anderson 1997); La Paz, Mecapaca (S. Anderson 1997); Potosí, 10 km SSE of Otavi (FMNH 162892); Santa Cruz, 1 km N and 8 km W of Comarapa (S. Anderson 1997), Tarija, Cariazo (S. Anderson 1997). PARAGUAY: Boquerón, Fortín Toledo (MSB 80553), Loma Plata (MSB 80556).

SUBSPECIES: Although several taxa originally described as species or subspecies are included within the synonymy of *C. musculinus*, for the present this species is regarded as monotypic. Cabrera (1961) included *musculinus* Thomas and *cordovensis* Thomas as a valid subspecies of his polytypic *C. laucha*, but Hershkovitz (1962) included all names herein associated with *C. musculinus* as simple subjective junior synonyms of his *C. l. laucha*. Geographic variation in size and color is, however, present, prompting Olds (1988) to divide geographic segments of *C. musculinus* into separate species (see Remarks).

NATURAL HISTORY: Studies on the ecology and natural history of this species have been conducted primarily in three foci of its distribution: central Argentina (e.g., Mills, Ellis, McKee, Maiztequi, and Childs 1991; Mills, Ellis, Childs et al.1992; Villafañe et al. 1992), in the Monte Desert of Mendoza province (recent review by R. A. Ojeda et al. 2011), and the central region of the Paraguayan Chaco (Yahnke 2006).

In central Argentina, *C. musculinus* is abundant in agricultural areas, especially corn and wheat fields, and along linear habitats (fencerows, roadsides, and railroad beds). Population densities were described as relatively high in spring and remaining so through summer and early autumn from November to April (Mills, Ellis, McKee, Ksiazek et al. 1991). Ellis et al. (1997) found that *C. musculinus* was statistically associated with linear habitats (roadsides and fence lines), which contrasted with the results by M. Busch et al. (1997), who found no habitat preference in any season, overlapping with both *C. laucha* and *Akodon azarae* in fields and linear habitats. In the Argentinean Pampas, Villafañe et al. (1992) found *C. musculinus* to occupy preferentially old fields that had been out of agricultural production for several decades. Gonnet and Ojeda (1998) found this species to be one of the two most dominant in the Andean foothills of the Monte Desert of Argentina, where it preferred grassy microhabitats and where population densities were rather constant throughout the year (R. A. Ojeda et al. 2011). In contrast, Yahnke (2006) found *C. musculinus* to occupy several habitats types in the Paraguayan Chaco but with a preference for microhabitats with low herbs, grassy vegetation, and cropland. The species is apparently highly adapted to xeric environments as it appears independent of free water (Mares 1977a) and shows daily torpor (Bozinovic and Rosenmann 1988b).

Calomys musculinus is omnivorous in agroecosystems in Argentina, with the relative proportions of seeds, insects, and vegetation consumed varying seasonally (Ellis et al. 1998). In contrast, animals in the Monte Desert are granivorous (Giannoni et al. 2005). Buzzio and Castro-Vasquez (2002) summarized known reproduction biology, with high seasonality but a long duration from September to June in Argentinean agroecosystems (Mills, Ellis, McKee et al. 1992). The number of embryos ranged from average of 6.07 in the Chaco (Yahnke 2006) to 7.5 in central Argentina (Mills, Ellis, McKee et al. 1992). Both Laconi and Castro-Vazquez (1999) and Yunes et al. (1991) examined nesting and parental behavior. Villafañe (1981) examined reproduction and grown in a laboratory colony, where females reached sexual maturity at age 72.5 days, the gestation period was estimated at 24.5 days and estrous cycle at 5.7 days, average litter size was 5.4 young, a postpartum estrous occurred in 50–64% of females, and males reached sexual maturity at 82 days.

Known predators include several species of owls (García Esponda et al. 1998; P. E. Ortiz, Cirignoli et al. 2000; Pardiñas, Teta et al. 2003) and the Geoffroy's cat (*Leopardus geoffroyi*; Bisceglia et al. 2011). Ticks (Nava et al. 2006; Sanchez et al. 2010; Guglielmone and Nava 2011), lice (Abrahamovich et al. 2006), fleas (Lareschi et al. 2004), and trypanosomes (Basso et al. 1977) are known parasites.

Calomys musculinus also serves as a reservoir for the Junín arenavirus, the etiological agent of the Argentine hemorrhagic fever (Massoia and Fornes 1965b; Salazar-Bravo, Ruedas, and Yates 2002).

REMARKS: In her unpublished doctoral thesis, Olds (1988) suggested that our present understanding of *C. musculinus* includes two species. She used the name *C. musculinus* for populations ranging from 1,000 to 3,000 m on the eastern slopes of the Andes in Bolivia and northern Argentina and *C. murillus* for those populations below 1,200 m elsewhere in Argentina. She differentiated between these two on the basis of size and pelage coloration, although she acknowledged that these differences were "difficult to distinguish." Morphometric analyses conducted by Provensal and Polop (1993) suggested no geographic variation among populations from central Argentina, but did

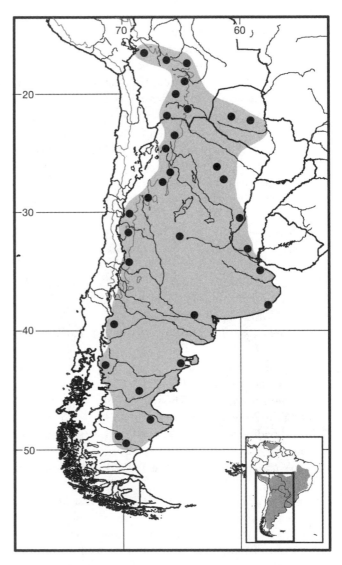

Map 263 Selected localities for *Calomys musculinus* (●). Contour line = 2,000 m.

document significant differences between individuals collected in different years. Since these two reports, evidence has accumulated supporting a recent population expansion of populations in central Argentina (González-Ittig et al. 2007), perhaps as a result of either or both anthropogenic activities (Pardiñas et al. 2000) and climate change (P. E. Ortiz, Madozzo Jaen, and Jayat 2012). In addition, molecular evidence has supported that individuals from populations assigned to either binomial appear to belong to a single species, for which the name *C. musculinus* has priority (Salazar-Bravo et al. 2001).

Myers (1982) mentioned the presence of *Calomys musculinus* in eastern Paraguay, but extensive collections in various localities east of the Paraguay river have recorded both *Calomys tener* and *C. callosus* but not *C. musculinus*. Examination of the two specimens at the Smithsonian currently identified as *C. musculinus* (USNM 390144 and USNM 531236) indicate that these specimens may represent *Calomys tener* (S. Peurach, pers. comm.).

A karyotype of $2n = 38$ and $FN = 48$ (Forcone et al. 1981; Ciccioli 1991) has been described for central Argentinean populations. The autosomal complement is composed of one large metacentric pair, seven pairs of medium-sized metacentric and submetacentric chromosomes, five pairs of medium subtelocentrics, three pairs of small subtelocentrics, and two pairs of small metacentric chromosomes. The X chromosome is a large submetacentric, and the Y an extremely small subtelocentric. No karyotype is known for populations in Bolivia or Paraguay.

Calomys sorellus (Thomas, 1900)

Peruvian Vesper Mouse, Peruvian Laucha

SYNONYMS:

Eligmodontia sorella Thomas, 1900f:297; type locality "eight miles south of Huamachuca, [La Libertad,] N.W. Peru. Altitude 3500 m."

[*Hesperomys*] *sorella*: Thomas, 1916a:141, name combination.

Hesperomys frida Thomas, 1917g:1, type locality "Chospyoc, [Cusco, Peru] 10,000 feet."

Hesperomys frida miurus Thomas, 1926b:314; type locality "Yano Mayo," Río Tarma, Junín, Peru, 8,500 ft.

Calomys frida frida: Cabrera, 1961:478; name combination.

Calomys lepidus sorellus: Cabrera, 1961:481; name combination.

Calomys sorellus: Hershkovitz, 1962:137; first use of current name combination.

DESCRIPTION: Medium sized (mean total length 148.05 ± 13.8 mm, maxillary toothrow 3.61 ± 0.13 mm; Olds 1988) with short tails (<45% of total length) and long, smooth, and fine fur. General dorsal color grayish

brown, sometimes dark brown, finely lined with black; tip of muzzle usually ochraceous (Hershkovitz 1962); body sides paler than dorsum, separated from venter by well-defined buffy line; ventral color whitish, with individual hairs slate-gray basally. Ears moderately long (mean 17.0 ± 1.8 mm), brownish when folded, colored like head, and with white postauricular patch. Upper surface of hands and feet pale to white; proximal third of soles thinly haired. Tail sharply bicolored, brownish above, darkening toward tip, white below and on the sides. Adult females with four pairs of mammae.

Skull moderate to large for genus (mean greatest length 23.4 ± 1.29 mm; Olds 1988), with straight profile in lateral view. Rostrum moderately short and deep throughout, although sides taper gradually toward front. Nasals broad, almost covering breadth of rostrum throughout; when seen from above nasals cover incisors. Lacrimals minute. Interorbital region parallel sided or slightly divergent posteriorly, with squared but not beaded edges; coronal suture U-shaped; braincase rounded and lacks well-defined lambdoidal ridges; parietals only slightly expanded over lateral surface; interparietal depressed dorsoventrally but covers entire back of the braincase. Zygomatic arches delicate, almost straight or convergent toward front, not bowed; zygomatic plates broad, about 10% of skull length, with straight anterior borders; zygomatic spines absent; zygomatic notches visible but neither deep nor broad. Alisphenoid strut absent (Olds 1988). Hamular process long and narrow, subdividing reduced postglenoid foramen and larger subsquamosal fenestra. Incisive foramina long and narrow, compressed anteriorly, and extending posteriorly between procingula of M1s; premaxillary part of septum occupying about half of opening. Palate broad and long (*sensu* Hershkovitz 1962), with small but clearly present palatal posterolateral pits positioned just dorsal to mesopterygoid fossa. Mesopterygoid fossa very narrow; sphenopalatine vacuities present; parapterygoid plates excavated, lyre shaped, and very broad. Auditory bullae small (Olds 1988). Carotid circulatory pattern primitive, pattern 1 (*sensu* Voss 1988; see Steppan 1995).

Mandible small but robust, but lacks well-developed masseteric crest; capsular projection conspicuous and located near base of coronoid process, which is short and inflected backward; condyloid process long and narrow; angular process short and almost as wide as condyloid process.

Upper incisors opisthodont (Steppan 1995), with bright orange enamel and smooth surfaces. Upper molars crested, with markedly alternating labial and lingual cusps along major axis of toothrow; molars brachydont, and moderate in size with respect to skull. M1 subrectangular in outline; anterocone usually with well-developed anteromedian flexus; anteromedian style present in about 44% of specimens (Olds 1988); hypoflexus and protoflexus transverse; protocone slightly larger than hypocone; no trace of parastyle/anteroflexus (Steppan 1995); and hypoflexus posteriorly directed. M2 square in outline, trilophodont, and about 62% length of M1; anteroloph well developed; mesoflexus wide and posteriorly directed, alternating with hypoflexus. M3 moderately bilophodont, rounded in shape, and about 70% of M2 in length (Olds 1988); hypoflexus and mesoflexus vestigial and opposite each other. Lower molars crested, with main cusps alternating; mesolophids and mesostylids absent; m1 trilophodont with well-developed, fan-shaped procingulum with deep anteromedian flexid, well-developed anterolabial cingulum, both ectostylid and mesostylid present, a mesoflexus narrow and posteriorly directed, both entoconid and metaconid subequal in size, and a reduced posterolophid; m2 trilophodont, with well-developed posterolophid; m3 subrectangular in outline, sigmoid, with reduced mesoflexid and deep hypoflexid.

Number of vertebral elements variable, with 13–14 thoracic and 27–31caudal vertebrae, but six lumbar elements consistently present (Steppan 1995). Dorst (1973b) described stomach morphology, and heart weight relative to mass at about 0.85% (Dorst 1973a). Steppan (1995) reported two pairs (one very small, less than 1 mm in length) of preputial glands.

DISTRIBUTION: *Calomys sorellus* is largely confined to Puna vegetation at high elevation (typically above 2,500 m) in Peru, although it likely extends into Bolivia in the vicinity of Lake Titicaca. A few specimens from the *loma* vegetation of coastal Arequipa department, Peru (specimens in the MVZ and MUSA) deserve closer scrutiny.

SELECTED LOCALITIES (Map 264): PERU: Ancash, Callon, 2 km S and 11 km W of Huaras (MVZ 137998), 25 km S of Huaras (MVZ 137999), Rachacoco, 45 km S of Huaras (MVZ 135684); Apurímac, 10 km N of Abancay (MVZ 119957), 36 km S (by road) of Chalhuanca (MVZ 174022); Arequipa, ca. 33 road km E of Arequipa (LSUMZ 27863), 2 mi E of Atiquipa (MVZ 145615), 1 km N of Chivay (MVZ 145616); Ayacucho, 2 mi SE of Huanta (MVZ 141362), 11 mi NW of Puquio (MVZ 116030); Cusco, Chirapata, Cossireni Pass (Hershkovitz 1962), Chospyoc (type locality of *Hesperomys frida* Thomas), 5 km N of Huancarani (MVZ 171549), 20 km N (by road) of Paucartambo, km 100 (MVZ 171544); Huancavelica, Huancavelica (Hershkovitz 1962), Lircay (Hershkovitz 1962), Río Mantaro (MVZ 119959); Huánuco, Cullcui (Hershkovitz 1962), Panao mountains (Hershkovitz 1962); Junín, 17 mi WNW of Huancayo (MVZ 141361), 16 km NNE (by road) of Palca (MVZ 172670), Pomacocha, Yauli Valley (MVZ 138001); La Libertad, 8 mi S of Huamachuca (type

locality of *Eligmodontia sorella* Thomas), mountains NE of Otuzco (Hershkovitz 1962), 10 mi WNW of Santiago de Chuco (MVZ 138022); Lima, 1 mi W of Canta (MVZ 119960), 1 mi E of Casapalca (MVZ 119961); Pasco, Huariaca (Hershkovitz 1962); Puno, 3 mi NE of Arapa (MVZ 116669), 6 km N of Putina (MVZ 171559), 4.5 km NE (by road) of San Antón (MVZ 173429), 12 km S of Santa Rosa [de Ayaviri] (MVZ 171555).

SUBSPECIES: I regard *Calomys sorellus* as monotypic.

NATURAL HISTORY: Dorst (1971,1973b), Pearson and Ralph (1978), Pizzimenti and DeSalle (1980), and Pacheco et al. (2007) have reported on various aspects of the natural history. This species may account for about 4–41% of the biomass at trapsites (Dorst 1971), with population densities ranging from eight to nine animals per hectare (Dorst 1973b) and abundances between 0.24 and 0.36 per 100 trap-nights in the basin of the Río Apurimac (Pacheco et al. 2007). The species is primarily nocturnal, with an activity peak between 7:00 PM and 10:00 PM (Dorst 1971). Individuals may be mainly insectivorous (Pizzimenti and DeSalle 1980) or granivorous (Dorst 1971). A tick (*Ixodes andinus*) is the only published parasite of the species (Hershkovitz 1962). The limited reproductive data indicate breeding active from July to September (Dorst 1971; Olds 1988). Morrison et al. (1963) and Reynafarje and Morrison (1962) examined hematocrit values and myoglobin levels with respect to high-altitude adaptations.

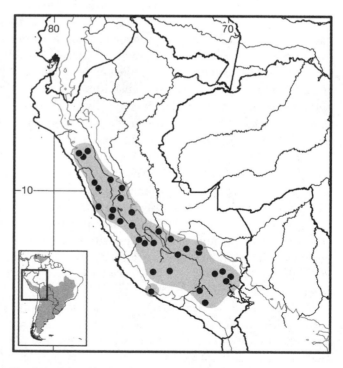

Map 264 Selected localities for *Calomys sorellus* (●). Contour line = 2,000 m.

REMARKS: Ample morphological and chromatic variation has been reported for *Calomys sorellus* across its range, with specimens from northern Peru paler, more grizzled, and smaller throughout, while those from further south are darker, browner, and slightly larger (Olds 1988). Slight chromosomal polytypy is also known (Pearson and Patton 1976), with $2n = 64$, $FN = 68$, in populations ranging from Ancash department in the north to Puno in the south, with $2n = 62$ for a specimen from Apurimac. Lacking detailed analyses of phenotypic variation and/or molecular phylogeographic structure, it is premature to determine whether infraspecific units warrant formal recognition.

Calomys tener (Winge, 1887)

Delicate Vesper Mouse, Delicate Laucha

SYNONYMS:

Hesperomys tener Winge, 1887:15; type locality "Lagoa Santa," Minas Gerais, Brazil; given as "Rio das Velhas, Lagoa Santa" by Gyldenstolpe (1932:75) and as "Lapa da Escrivania [= Escrivânia] Nr. 5" by Hansen (2012:48).

Eligmodontia tener: Shufeldt, 1926:503; name combination.

Calomys tener: Cabrera, 1961:481; first use of current name combination.

Calomys laucha tener: Hershkovitz, 1962:157; name combination.

DESCRIPTION: Size average for genus, with head and body length 74–88 mm (N = 10, from central and eastern Brazil; Bonvicino and Almeida 2000). Both external and cranial measurements smaller when compared with sympatric congeners. Upper parts of body are yellowish to dark brown, with reddish hue in some specimens, but hairs gray at base; venter pale to dark gray, finely streaked with black hairs; sides softer color but easily differentiable from venter; ventral region grayish to whitish with base of hairs gray; no lateral line separating dorsal and ventral colors; some specimens with fully white hairs on chin. Most specimens with eye-ring. Ears average 14.2 mm (Bonvicino and Almeida 2000), covered with short hairs about same color as rest of head. Tail proportionally short, about 45% of total length, mean length 60.6 mm (range: 38 to 77 mm; Bonvicino and Almeida 2000), and slightly bicolored, dorsally with grayish brown, pale below. Pale postauricular patches and ochraceous preauricular spots present. Five pairs of mammae usually present, three pectoral and two inguinal, but Olds (1988) reported one female with only three pairs, two pectoral and one inguinal.

Skull domed in lateral profile, higher above tympanic bullae, with moderately broad and long rostrum (averaging 34% of greatest length; Olds 1988, Bonvicino and Almeida 2000). Nasals parallel sided, protrude beyond plane of incisors such that they cannot be seen from above, and

dorsally terminate bluntly at same level as premaxillaries. Lacrimals minute. Interorbital region wide and flat, with supraorbital margins distinctly ledged and divergent posteriorly. Zygomatic plates broad, with dorsal edges mostly straight and upper free borders enclosing poorly developed zygomatic notches, posterior margin anterior to alveolus of M1. Zygomatic arches delicate, not expanded, and widest at squamosal root of zygomatic arch. Incisive foramina long and narrow, with parallel borders, posterior ends rounded and reaching at least to level of M1 protocones; premaxillary part of septum occupying about 60% of length of opening. Anterior border of mesopterygoid fossa rounded, without median palatine process, and not reaching posterior face of M3s; roof of fossa with large sphenopalatine vacuities. Alisphenoid strut typically present in most specimens (Olds 1988). Auditory bullae of medium size, with short but wide bony tubes; carotid circulation primitive (pattern 1 of Voss 1988).

Mandible delicate, with conspicuous capsular projection lying at base of coronoid process; coronoid process small and inflected backward; condyloid process broad and short; angular process short; sigmoid notch scarcely developed. Upper incisors orthodont and faced with bright orange enamel band. Lingual and labial molar cusps alternate on both upper and lower teeth. Upper molar rows parallel, maxillary toothrow length 3.2–3.6 mm. Anterocone of MI divided by anteromedian flexus, and with anteromedian style (*sensu* Hershkovitz 1962) complete most specimens; weakly developed anterolabial style and anteroflexus also potentially present; metaflexus and mesoflexus deep and directed backward; hypoflexus deep and opposite to paracone. M2 squared in shape, about 67% of M1 length; anteroloph reduced. M3 with well-defined hypoflexus, mesoflexus, and paraflexus, especially in young individuals. Anterocone of m1 sometimes divided by shallow groove but not by deep anteromedian flexid; m2 conserves all structures but in reduced state; m3 S-shaped. One specimen had 12 thoracic, 7 lumbar, and 24 caudal vertebrate (Steppan 1995).

DISTRIBUTION: *Calomys tener* occurs mainly in open vegetative formations in the Cerrado of central Brazil (e.g., grasslands, gallery forests, woodland savanna, or *cerradão*) and secondary vegetation of both the Amazon and the Atlantic Forest adjacent to the Cerrado, from central and eastern Brazil to Misiones province in eastern Paraguay and Santa Cruz department in southeastern Bolivia.

SELECTED LOCALITIES (Map 265): BOLIVIA: Santa Cruz, Santa Rosa de la Roca (Salazar-Bravo et al. 2001). BRAZIL: Bahia, Jaborandi (F. C. Almeida et al. 2007); Goiás, Serra de Mesa (Haag et al. 2007); Mato Grosso do Sul, Nova Alvorada do Sul (Cáceres et al. 2008); Minas Gerais, Viçosa (USNM 541497); Rio Grande do Sul,

Quintão (Haag et al. 2007); São Paulo, Campinas (F. C. Almeida et al. 2007); Tocantins, Aliança do Tocantins (Haag et al. 2007). PARAGUAY: Misiones, San Francisco, 36 km NE of San Ignacio (USNM 390144).

SUBSPECIES: *Calomys tener* is monotypic.

NATURAL HISTORY: The ecology of central Brazilian populations is relatively well known, but nothing is known about the biology of those from elsewhere in the species' range. In the Cerrado of the Distrito Federal, several short-term studies (about 1 year in length) revealed little to no seasonal population fluctuation but high levels of habitat specificity; abundance estimates were about five times higher in areas with a more developed secondary structure and higher proportion of woody plants than in areas fully dominated by grasses and herbs (R. Ribeiro and Marinho-Filho 2005; C. R. Rocha et al. 2011). In addition, the species is described as often caught around agricultural areas or in areas around buildings (Umetsu and Pardini 2007). Patterns of daily activity were reported by E. M. Vieira and Baumgarten (1995). Several aspects of the life history of the species were reported by Araripe et al. (2007) based on captive colonies. Litter size averaged 3.6 embryos, and the interval between reproductive bouts was about 22 days. Lactating females were field-trapped in November, and pregnant females (with four embryos each) were recorded in November and February (Dietz 1983). The diet is reported as that of an herbivore-omnivore, rich in plant material and fruits (V. N. Ramos and Facure 2009). In São Paulo state, *C. tener* may be infected with Araquara hantavirus (G. G. Figueiredo et al. 2010). *Calomys tener* is preyed upon by at least two species of owls (A. Bueno and Motta-Junior 2008; Magrini and Facure 2008) and Maned wolves (Dietz 1983). Mites parasitize this species in Brazil (Gettinger 1992a).

REMARKS: *Calomys tener* is currently viewed as monotypic, but limited karyotypic variation has been recorded. Bonvicino and Almeida (2000) identified a karyotype of $2n = 66$ and $FN = 66$ in five specimens from São Paulo state, Brazil, with an autosomal complement that included a medium-sized pair of metacentrics and 31 acrocentric pairs varying gradually in size. The X chromosome was a large subtelocentric and the Y chromosome a small acrocentric. Mattevi et al. (2005) reported a chromosomal complement for specimens from Rondônia state, Brazil, with $2n = 64$ and $FN = 64$. Furthermore, F. C. Almeida et al. (2007) identified reciprocal monophyletic mtDNA clades grouping specimens from Brazil versus those from Bolivia, one that may also be associated with different karyomorphs. Further work is needed to determine both the geographic concordance of the cytotypes and molecular clades as well as whether formal recognition of either might be warranted.

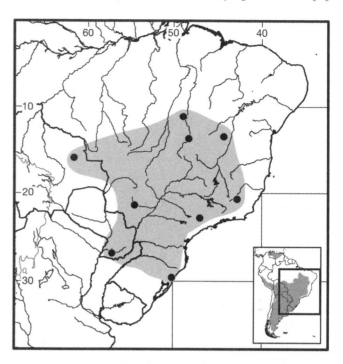

Map 265 Selected localities for *Calomys tener* (●). Contour line = 2,000 m.

Several reports indicate the presence of *Calomys tener* in Pernambuco state in northeastern Brazil (e.g., J. A. Oliveira et al. 2003). An examination of skull morphology of samples of *C. tener* and *C. expulsus*, however, suggested that specimens from Pernambuco are best assigned to *C. expulsus* (Cordeiro-Estrela et al. 2006). Records of the species from Tarija department in southern Bolivia (S. Anderson 1997) are instead *Calomys musculinus*.

Calomys tocantinsi Bonvicino, Lima, and Almeida, 2003
Tocantins Vesper Mouse, Rato-bolinha do Tocantins

SYNONYM:

Calomys tocantinsi Bonvicino, Lima, and Almeida, 2003:301; type locality "Brazil, Tocantins: Formoso do Araguaia (Rancho Beira Rio, 33 km SW, 11°47'S and 49°45'W)."

DESCRIPTION: Medium size (mean head and body 90 mm, mean greatest skull length 25 mm) with tails shorter than head and body. Dorsal pelage soft, brownish-gray, and darker middorsally from muzzle to base of tail. Sides of body slightly paler that back; ventral coloration clearly demarcated from sides and washed with white, individual hairs white-tipped, but gray at base. Tail bicolored, dark above, white below; hands and hindfeet covered with small white hairs; ungual tufts present. Ears small (ca. 17 mm in length), covered with small dark hairs; preauricular and postauricular patches present. Number of mammae unknown.

Skull robust; dorsum bowed in lateral profile; rostrum comparatively short although nasals cover incisors when viewed from above; molar row long; palatal bridge broad (Bonvicino, Lima, and Almeida 2003). Interorbital region diverges posteriorly with sharp, somewhat beaded, edges. Nasals terminate anterior to premaxillofrontal suture. Zygomatic arches not extremely bowed, broader at their root and thus somewhat convergent anteriorly, and jugal small but robust. Zygomatic plates with concave anterior edges forming relatively well developed zygomatic spines. Incisive foramina long (ca. 80% of diastema length), extending posteriorly between procingula of both M1s; opening widest just behind premaxillomaxillary suture; premaxillary part of septum occupies about three-fourths of opening. Braincase round and somewhat globular, with weak lambdoid and temporal ridges. Upper incisors strongly opisthodont. Upper molars crested, with labial cusps in markedly opposite pattern, brachydont, and rather large with respect to skull. M1 tetralophodont, rectangular in outline; procingulum well developed with deep anteromedian flexus defining anterolingual and anterolabial conules; no trace of antero or protoloph; paraflexus and hypoflexus almost connect opposite each other at central mure; metaflexus deep and laterally oriented; posteroloph not quite evident. M2 and M3 both with well developed hypo- and metaflexus that connect opposite to each other at median mure, so both molars appear almost divided into two parts. M2 about 70% of M1, M3 about 65% of M2. Mandible deep and robust; capsular projection not very well defined, located ventral to coronoid process; coronoid process short, only slightly higher than condyloid process. No data available on axial skeleton or characters of soft anatomy.

DISTRIBUTION: *Calomys tocantinsi* occurs in central Brazil, in the upper basin of the Rio Araguaia, a tributary of the Rio Tocantins.

SELECTED LOCALITIES (Map 266): BRAZIL: Mato Grosso, Cocalinho (Fagundes et al. 2000), São José do Xingu (Bonvicino et al. 2010), Vila Rica (Fagundes et al. 2000); Tocantins, Formoso do Araguaia (type locality of *Calomys tocantinsi* Bonvicino, Lima, and Almeida), Lagoa da Confusão (Bonvicino et al. 2010).

SUBSPECIES: *Calomys tocantinsi* is monotypic.

NATURAL HISTORY: This species has been trapped in gallery forest at the ecotone between Amazonian rainforest and Cerrado vegetation, a well as in various formations of the Cerrado proper (Bonvicino, Lima, and Almeida 2003; Bezerra et al. 2009). It was reported to be an important prey item for the Barn Owl (*Tyto alba*; R. G. Rocha, Ferreira, Leite et al. 2011). Nine of 12 females captured in October and November in 2008 were reproductively active: three were pregnant, with five, seven, and 10 embryos each; five had open vaginas, and one was lactating (R. G. Rocha, Ferreira, Costa et al. 2011). No other data on ecology or behavior are available.

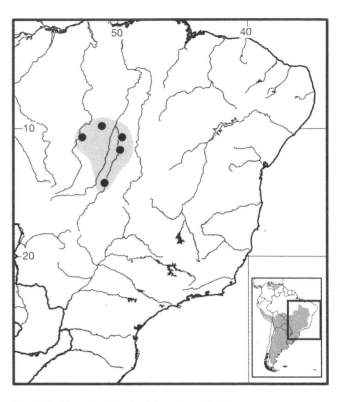

Map 266   Selected localities for *Calomys tocantinsi* (●).

REMARKS: The karyotype consists of a $2n=46$, $FN=66$ (J. F. S. Lima and Kasahara 2001; Bonvicino, Lima, and Almeida 2003), with an equal number of biarmed and uniarmed pairs of autosomes, a medium metacentric X, and a small acrocentric Y chromosome.

Calomys venustus (Thomas, 1894)

Córdoba Laucha

SYNONYMS:

Oryzomys (?) *venustus* Thomas, 1894b:359; type locality "Cosquin [= Cosquín], Cordova [= Córdoba], Argentina."

[*Hesperomys*] *venustus*: Thomas, 1916a:141; name combination.

Calomys venustus: Cabrera, 1961:481; first use of current name combination.

Calomys venustus venustus: Cabrera, 1961:481; name combination.

Calomys callosus: Hershkovitz, 1962:172; part; not *callosus* Rengger.

DESCRIPTION: Largest species in genus (greatest length of skull ca. 28.7 mm; Polop and Provensal 2000). Dorsal pelage soft and long, with dull brownish or yellowish gray colored hairs with black-lined tips; sides sandy buff (Thomas 1894b). Indistinct buffy line separates sides and venter; underparts pale gray with individual hairs gray at base and white at tips. In many specimens, hairs of chin are white to base. Ears medium to large and almost always with buffy tuft at base and buffy postauricular spot, although it may not be as obvious as in other species. Hands and feet well haired and white. Tail shorter than head and body length, well haired, and distinctly bicolored (brown above and white on sides and below). Number of mammae varies from five to seven pairs.

Skull large; rostrum broad, especially in older individuals; interorbital region with strongly ridged edges that diverge posteriorly. Details of dentition are similar to other large lowland species; an anteromedian style frequently present and anteromedian flexus less well developed. No anteroflexus or anterolabial style visible. Elements of axial skeleton somewhat variable, with 12 or 13 thoracic, 7 lumbar, and 22 to 25 caudal vertebrae (Steppan 1995).

DISTRIBUTION: *Calomys venustus* occurs in central Argentina, in the provinces of Córdoba, San Luis, and Santiago del Estero. Because morphological characters in this and other species of large *Calomys* are often insufficient (or at least unsatisfactory) to differentiate among them, in the list of selected localities below I have only included individuals identified as *C. venustus* based on chromosomal and molecular and/or other biochemical data. It is possible that the species may be found in Buenos Aires province (e.g., F. Fernández et al. 2012).

SELECTED LOCALITIES (Map 267): ARGENTINA: Córdoba, Cosquín (type locality of *Oryzomys venustus* Thomas), Espinillo (Tiranti 1996a); San Luis, 3 km W of Hualtaran, Parque Provincial Sierra de las Quijadas (OMNH 35371), 1.3 km N and 5.8 km W of La Toija, Pampas de las Salinas (OMNH 35370), San Francisco del Monte de Oro (OMNH 35369), Papagayos, camp on left bank Río Papagayos (PY 13, specimen to be deposited at CNP); Santiago del Estero, Buena Vista, 15 km NE of Villa Ojo de Agua on Hwy 13 (Dragoo et al. 2003).

SUBSPECIES: *Calomys venustus* is monotypic.

NATURAL HISTORY: Several aspects of the ecology and natural history of the species in the agroecosystems of the pampas of central Argentina are relatively well known. *C. venustus* is abundant in areas with relatively high vegetation cover including crop field edges, roadsides, railway banks, and remnant areas of native vegetation (Priotto and Polop 1997; Polop and Sabbatini 1993). Population densities fluctuate seasonally, with a peak in May-June (Provensal and Polop 2008), and multiannually in response to the previous year's spring and summer rainfall and spring mean temperature (Andreo et al. 2009). Individuals are apparently capable of facultative torpor (Caviedes-Vidal et al. 1990). *Calomys venustus* is reported to be an omnivore with a tendency to folivory in spring and autumn and to gra-

nivory in summer (Castellarini et al. 1998). Reproduction follows a cyclical seasonal pattern characterized by a period of repose and a variable length of the period of sexual activity between August to September and May to June, where three pregnancy peaks were observed (spring, summer, and late summer), each with a different average litter size (Polop et al. 2005). The species is sexually dimorphic (Polop and Provensal 2000); a promiscuous-polygynous mating system was suggested for the species (Priotto et al. 2002). Yunes et al. (1991) examined nesting and digging behavior in the species. There are no known predators reported in the literature and only one tick (*Amblyomma tigrinum*) is known to parasitize individuals (Nava et al. 2006; Guglielmone and Nava 2011). Additional aspects of the biology of the species are treated in Polop and Busch (2010).

REMARKS: The geographic and species limits of *C. venustus* as well as its relationship to other large-body *Calomys* from southern Bolivia and north-central Argentina require detailed study (see remarks under *Calomys boliviae*).

The karyotype of this species has a modal $2n = 56$, $FN = 66$, composed of six pairs of metacentric or submetacentric autosomes with the remainder acrocentric, a large submetacentric X, and a small subtelocentric Y (Tiranti 1996a; Lisanti et al. 1976; M. I. Ortiz et al. 2007). Lisanti et al. (1976) reported the presence of XO females.

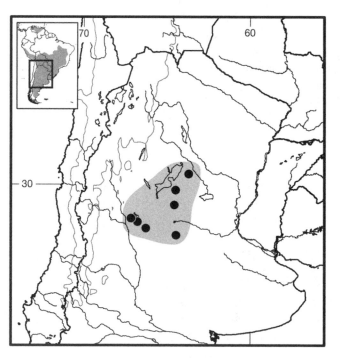

Map 267 Selected localities for *Calomys venustus* (●). Contour line = 2,000 m.

Genus *Eligmodontia* F. Cuvier, 1837

*Cecilia Lanzone, Janet K. Braun, James L. Patton,
and Ulyses F. J. Pardiñas*

Members of the genus *Eligmodontia*, commonly known as gerbil mice, range from southern Peru southward to southernmost Argentina and mainland Chile. This area embraces arid and semi-arid habitats linked to the Andean landscape, from sea level in the southern part of the distribution to above 4,000 m in the north. These rodents are highly desert adapted, and exhibit anatomical, physiological, and behavior specializations that enable their xeric existence (Mares 1977b; G. B. Diaz and Ojeda 1999; Giannoni et al. 2001).

The genus is polytypic, and a decades-long confused taxonomy is now becoming stabilized, primarily through the application of karyotypic and molecular sequence analyses. Nevertheless, several systematic issues remain unclear and others controversial. At the intra-generic level, species, or major groups of species, are associated only with weak statistical support in most molecular analyses, and form a polytomy at the base of the *Eligmodontia* tree (Mares et al. 2008; Lanzone, Ojeda et al. 2011; Spotorno et al. 2013). At the intra-tribal level, the position of *Eligmodontia* is also uncertain, appearing in some analyses near the base of phyllotine radiation but with very low statistical support (Steppan et al. 2007). Species of *Eligmodontia* also may be highly variable geographically in morphological, molecular, and others characters, and poorly delineated in morphology, even in multivariate morphometric space. Linear craniodental or even external variables of one species typically overlap in their extremes with those of others, and qualitative characters are also highly variable, especially with respect to specimen age. In this account, we seven species, but on-going studies are likely to expand this number as taxonomic boundaries become better delineated. Zambelli et al. (1992), Hillyard et al. (1997), Sikes et al. (1997), Lanzone, Ojeda, and Gallardo (2007) and Mares et al. (2008) documented sympatry without hybridization between *E. moreni* and *E. typus*, between *E. moreni* and *E. bolsonensis*, and between *E. typus* and *E. morgani*, supporting their respective status as species. The following generic description is, except as noted, largely compiled from information presented by Hershkovitz (1962), Steppan (1995), Lanzone, Ojeda, and Gallardo (2007), Mares et al. (2008) and Lanzone (2009). Mean external measurements provided for each species are compiled from measured samples reported by Sikes et al. (1997), Lanzone, Ojeda, and Gallardo (2007), Mares et al. (2008), Lanzone, Ojeda et al. (2011), and Spotorno et al. (2013), supplemented by nearly 270 specimens in the MVZ collections.

All species of *Eligmodontia* are externally similar small rodents without marked size differences among them. Ears are large, and the dorsal fur color ranges from yellowish to yellowish brown, grayish brown, or dark brown. Ventral hairs vary at their base from completely white to gray to black, but are always white at their tips. A more or less evident yellow to ochraceous lateral line demarcates the transition from dorsum to venter. Pelage around the mouth and behind the eyes is whitish. The tail is moderately haired, bicolored to unicolored (wholly pale), with or without a pencil, and with a length slightly less than or greater than that of the head and body. The hindfeet are elongated and digits I and V are subequal in length. Interdigital pads 2 to 4 are fused to form a hairy cushion, an external autapomorphic trait that easily distinguishes *Eligmodontia* from other similar-sized sigmodontine genera. Eight mammae (in four pairs) have been reported.

The skull is delicate, with a slender rostrum, and posteriorly divergent supraorbital margins that are neither beaded nor raised. Zygomatic arches are generally parallel-sided and little expanded. Zygomatic plates are straight or with slightly concave anterior margins. The bony palate extends beyond the posterior plane of the M3s. The incisive foramina are relatively long and extend at least to the anterior margins of M1s. The parapterygoid fossae are 1.5 to 2.5 times the mesopterygoid fossae width. An alisphenoid strut is present. The bony tubes of the bullae do not reach the posterior edge of the pterygoid processes. Both the stapedial foramen and posterior opening of the alisphenoid canal are large, and the squamosal-alisphenoid groove and sphenofrontal foramen are present, all characteristics of carotid circulation pattern 1 (of Voss 1988). On the mandible, the knob of the anterior masseteric ridge exceeds the dorsal edge of the diastema. The lower incisor root is contained in a capsular process posteroventral to the base of the sigmoid notch, and the coronoid process is short. Molar rows are parallel to slightly convergent posteriorly. The upper incisors are opisthodont and ungrooved. The molars are tetralophodont (lacking any trace of a mesoloph or mesolophid), brachydont, and the crowns tuberculate and slightly crested. Cusps are approximately alternate in arrangement and mures are obliquely positioned. The anterocone of M1 has an anterior median fold that becomes weak or obsolete when worn. M1 has four roots; m3 is usually two-rooted (Pearson 1995).

Axial skeletal elements have counts of 13 ribs plus 19 thoracolumbar and 25–26 caudal vertebrae. The neural spine of the second cervical vertebra is enlarged into a distinct knob that does not overlap the third cervical vertebra. The second thoracic vertebra has a greatly elongated neural spine.

The stomach is unilocular and hemiglandular (Carleton 1973) and a gall bladder is present (Voss 1991). Male accessory secretory organs consist of two pairs of preputial glands, a single large lateral pair and a medium length lateral pair (Voss and Linzey 1981). The glans penis is complex, with a tridigitate bacular cartilage; the glans is short, without parotoid lobes, and is covered externally by large epidermal spines; the medial mound is longer than lateral mounds; the urethral flap is deeply bifurcate; and the bony baculum is long (equal to glans in length; Hooper and Musser 1964). Spotorno, Zuleta et al. (1998), G. B. Diaz and Ojeda (1999), and Al-kahtani et al. (2004) provided data on renal structure and function, which varies among species. The three species for which diet was studied (*E. typus*, *E. hirtipes*, and *E. moreni*) are considered omnivorous or herbivorous, with leaves the main food item consumed in all three (C. M. Campos, Ojeda et al. 2001; S. I. Silva 2005; Giannoni et al. 2005; Lanzone et al. 2012).

Fossils of *Eligmodontia* are uncommon but have been reported from Chubut, Argentina, from about 1,830 ybp (Pardiñas et al. 2000), the Holocene of Buenos Aires province, Argentina (Pardiñas et al. 2000; Pardiñas 2001), and southern Mendoza province, Argentina, from about 8,900 ybp (Neme et al. 2002). The earliest known material dates from the Middle Pleistocene (Ensenadense, about 1.8 mya) of Argentina (Pardiñas et al. 2002). Latest Pleistocene specimens from the Chilean side of Tierra del Fuego, recovered from the lower levels of the archaeological site Tres Arroyos 1, support the presence of this genus on that island (U. F. J. Pardiñas, unpubl. data), a noteworthy record since the genus today reaches its southern limit at the Straits of Magellan. Uquian deposits in Jujuy province, Argentina, dated at about 2.5 mya lack *Eligmodontia* but contain a morphologically close fossil genus (P. E. Ortiz, García López et al. 2012).

SYNONYMS:

Eligmodontia F. Cuvier, 1837:168; type species *Eligmodontia typus* F. Cuvier, by monotypy (page number incorrectly cited as 169 by Cabrera 1961:482).

Mus: Waterhouse, 1837:19; part (description of *elegans*); not *Mus* Linnaeus.

Mus (*Cal[omys].*): Waterhouse, 1837:21; part (listing of *elegans*); neither *Mus* Linnaeus, 1758, nor *Calomys* Waterhouse.

Elygmodovtia Wiegmann, 1838:388; incorrect subsequent spelling of *Eligmodontia* F. Cuvier.

Elimodon Wagner, 1841:125; incorrect subsequent spelling of *Eligmodontia* F. Cuvier (incorrectly written as *Eligmodon* by Cabrera 1961:482).

Heligmodontia Agassiz, 1846:5; 136, 175; unjustified emendation of *Eligmodontia* F. Cuvier.

Hesperomys [(*Calomys*)]: Burmeister, 1879:220; part (listing of *elegans*); neither *Hesperomys* Waterhouse nor *Calomys* Waterhouse.

Eligmodon Thomas, 1896:307; part (see Remarks).

Phyllotis: Thomas, 1902c:225; part (description of *hirtipes*); not *Phyllotis* Waterhouse.

REMARKS: Cabrera (1961:482) listed Wagner (1841) as the author of *Eligmodon* in his generic synonymy of *Eligmodontia*, but Wagner (1841:125) spelled his name *Elimodon*, and in a simple list without further explanation. Rather, the spelling *Eligmodon* stems from Thomas (1896:307), who both coined the name for the species he therein described, *Eligmodon moreni*, and made no reference to Wagner's similar name. Moreover, in a footnote, Thomas (1896:307) explicitly equated his *Eligmodon* to the combined "*Calomys*, Waterh. (1837) nec Is. Geoff. (1830). *Eligmodontia*, F. Cuvier, 1837. *Hesperomys* Waterh., 1839," while stating in the text (p. 308) that "*Eligmodon Moreni* is most nearly allied to *E. elegans* Waterh. (*E. typus*, F. Cuv. [= the type species of *Eligmodontia*])." Thomas's *Eligmodon* is thus not exactly equivalent to Cuvier's *Eligmodontia*. In his review of rodent genera the following year, Thomas (1897a:1020) listed *Eligmodontia*, F. Cuv., with "*Calomys*, Waterh., nec Geoff. *Hesperomys* (s. s.), Waterh." as synonyms but did not list *Eligmodon*.

Cabrera (1961) recognized two species, *E. typus* (including *elegans* Waterhouse, *morgani* J. A. Allen, and *pamparum* Thomas as either valid subspecies or synonyms) and *E. puerulus* Philippi (with *moreni* Thomas, *hirtipes* Thomas, *marica* Thomas, *jacunda* Thomas, and *tarapacensis* Mann). Hershkovitz (1962) regarded each of these 10 nominal taxa then allocated to *Eligmodontia* as part of a single species, *E. typus* F. Cuvier, dividing that broadly distributed taxon into two subspecies: *E. typus typus* (with a southern range and including *elegans* Waterhouse, *morgani* Allen, and *pamparum* Thomas) and *E. typus puerulus* (with a northern range and including *puerulus* Philippi, *moreni* Thomas, *hirtipes* Thomas, *marica* Thomas, *jacunda* Thomas, and *tarapacensis* Mann). Braun (1993) listed six species (*hirtipes* Thomas, *marica* Thomas, *moreni* Thomas, *morgani* J. A. Allen, *puerulus* Philippi, and *typus* F. Cuvier), although her treatise was not directed to documenting species-level taxonomy. Recent revisionary research has overwhelming demonstrated greater species diversity than recognized by either Cabrera (1961) or Hershkovitz (1962), with karyotypic, mtDNA sequences, and morphological comparisons delineating the seven species we recognize herein (Ortells et al. 1989; Kelt et al. 1991; Zambelli et al. 1992; Spotorno et al. 1994, 2001; Hillyard et al. 1997; Sikes et al. 1997; Tiranti 1997; Spotorno, Zuleta et al. 1998; Lanzone and Ojeda 2005; Lanzone, Ojeda, and Gallardo 2007; Mares et al. 2008; Lanzone, Ojeda et al.

2011; M. C. Da Silva 2011; Spotorno et al. 2013). Tiranti (1997) offered a Plio-Pleistocene biogeographic interpretation to explain phyletic diversification within the genus. Spotorno et al. (1994, 2001) suggested that both genetic and geographical factors have played a complementary role in the process of speciation and divergence in *Eligmodontia*.

We have excluded some of the natural history literature from the accounts because we are unable to verify the species identity given. And, we preferentially cite records of specimens in the list of selected localities for each species for which either, or both, karyotype and mtDNA sequences have been determined.

KEY TO THE SPECIES OF *ELIGMODONTIA*:

1. Tail proportionally short (<100% head and body length) 2

1'. Tail proportionally long (>110% head and body length) 5

2. Dorsal color overall yellowish brown; venter immaculate white; $2n = 50$, FN = 48 3

2'. Dorsal color overall grayish brown; venter with variable color at the based (white, gray, or black); $2n = 31$–37, FN = 32 or 48 4

3. Body size large (mean head and body length >90 mm); sharply distinct fulvous lateral line; short tail tuft (2–4 mm); zygomatic arches slightly divergent posteriorly; northern Puna . . . . *Eligmodontia hirtipes*

3'. Body size very small (mean head and body length ~70 mm); indistinct fulvous lateral line; no tail tuft; zygomatic arches parallel; southwestern of the Atacama Desert, northern Chile . . . . *Eligmodontia dunaris*

4. Body size medium (mean head and body length >85 mm); venter with gray or black hairs at the base; tail strongly bicolored; tail tuft short; ear length short (mean length ~16 mm); anterior border of mesopterygoid fossa squared, often with median spine; $2n = 32$–34, FN = 32; Patagonian steppe . . . . *Eligmodontia morgani*

4'. Body size small (mean head and body length <85 mm); venter with white, gray or black hairs at the base; tail generally weakly bicolored; tail tuft short; ear length medium (mean length ~18 mm); anterior border of mesopterygoid fossa pointed; $2n = 31$–37, FN = 48; southern Puna . . . . *Eligmodontia puerulus*

5. Tail tuft lacking, or very short; tail unicolored or weakly bicolored; ear short (length <18 mm); bullae uninflated, interbullar distance wide, petrotympanic fossae widely open, bullar tube elongated, stapedial spine protuberant, $2n = 44$, FN = 44 6

5'. Tail tuft long (>5 mm); tail bicolored; ear long (generally length >19 mm); bullae inflated, interbullar distance narrow, petrotympanic fossae nearly closed, bullar tube

short, stapedial spine very short; $2n = 52$, FN = 50 . . . . *Eligmodontia moreni*

6. Fur long, lax; venter immaculate white; X chromosome uniarmed; anterior border of mesopterygoid fossa rounded; anterior border of maxilla with developed tubercles; northern Monte Desert . . . . *Eligmodontia bolsonensis*

6'. Fur short; venter white, with gray or white-based hairs; X chromosome biarmed; anterior border of mesopterygoid fossa squared, sometimes with median spine; anterior border of maxilla with tubercles slightly developed or absent; southern and central Monte Desert and Patagonian steppe . . . . *Eligmodontia typus*

Eligmodontia bolsonensis Mares, Braun, Coyner, and van den Bussche, 2008

Bolsón Gerbil Mouse, Bolsón Laucha

SYNONYM:

Eligmodontia bolsonensis Mares, Braun, Coyner, and van den Bussche, 2008:15; type locality "ARGENTINA: Catamarca Province: Pomán: Establecimiento Río Blanco, 28 km S, 13.3 m W Andalgalá 27°51′01″S, 66°18′17″W."

DESCRIPTION: Small, with mean head and body length 82.1 mm (range 73–94 mm), tail length 92.3 mm (74–105 mm), hindfoot length 23.5 mm (21–26 mm), and ear length 17.1 mm (15–19 mm). Tail somewhat longer than head and body, averaging about 113%. Pelage long and lax, very pale yellowish brown above with light ochraceous buff hairs distally but grayish proximally, and dark guard hairs providing slightly darker overall appearance; under parts immaculate, with hairs white to base; transition from dorsum to venter well defined, delineated by slightly brighter ochraceous buff lateral line; nose and mouth areas surrounded by white. Whitish patches behind eyes, and ears with moderately developed pale postauricular patches. Both forefeet and hindfeet covered with whitish hairs, soles only slightly covered with hairs. Tail unicolored, pale white above and below, and without terminal pencil of hairs. Skull delicate; nasals do not extend beyond premaxillofrontal suture; and zygomatic arches delicate, parallel, and little expanded; anterior border of zygomatic plates slightly concave; auditory bullae moderately developed, wider than long (averaging 4.89 mm in width, 4.16 in length) and with long bullar tubes, protuberant stapedial spines, and wide petrotympanic fissures; maxillary toothrow length relatively short for genus, averaging 3.88 mm. Anterior border of maxilla with developed and rounded tubercles. M1 with anteromedian flexus slightly marked or absent; anterolingual style absent or slightly developed, and mesostyle absent; no contact between labial (hypoflexus) and secondary lingual fold (metaflexus); M3 with anteromedian flexid absent; posterolophid present. In most dimensions, *E. bolsonensis* quite similar to means and ranges of *E. typus* but smaller than those of *E. moreni*. Lanzone, Ojeda, and Gallardo (2007) and Mares et al. (2008) provided mensural comparisons between *E. bolsonensis* (karyotype 1 sample in Lanzone, Ojeda, and Gallardo [2007]), *E. moreni* (karyotype 2 sample in the same paper), and *E. typus*.

DISTRIBUTION: *Eligmodontia bolsonensis* is known from Catamarca and Salta provinces, Argentina, in the area north and west of the Sierra de Ambato and Sierra de Manchao, at elevations largely above 1,800 m, with one locality as low as 700 m. It is sympatric with *E. moreni* at localities in Catamarca province, and with *E. puerulus* also at a locality in Catamarca province (Lanzone, Ojeda, and Gallardo 2007; Mares et al. 2008). However, *E. puerulus* and *E. moreni* at some localities of Catamarca are indistinguishable in cytochrome-*b* mtDNA (Lanzone, Ojeda et al. 2011); as this was the character used by Mares et al. (2008) to differentiate the two taxa, some of their data need further confirmation.

SELECTED LOCALITIES (Map 268): ARGENTINA: Catamarca, Agua de Dionisio (Mares et al. 2008), Campo Arenal, Los Nacimientos (Lanzone, Ojeda, and Gallardo 2007); 5.2 km S of El Bolsón (Mares et al. 2008), 21 km SW of El Desmonte (Mares et al. 2008), 1.5 km S of El Peñón (Mares et al. 2008), Establecimiendo Río Blanco (type locality of *Eligmodontia bolsonensis* Mares, Braun, Coyner, and van den Bussche); Salta, Cafayate (Lanzone, Ojeda, and Gallardo 2007).

SUBSPECIES: *Eligmodontia bolsonensis* is monotypic.

NATURAL HISTORY: Little is known about the natural history. Mares et al. (2008) reported that reproductively active individuals were captured in the spring months of October and November with inactive individuals sampled in March and September. Pregnant females contained embryo counts varying from four to seven. General observations indicated that areas where *E. bolsonensis* and *E. moreni* were sympatric, the former preferred soft, sandy soils with the latter more common on dense clay and gravelly soils; in the absence of *E. bolsonensis*, *E. moreni* occupied sandy substrates (Mares et al. 2008). The universality of these observations requires more explicit study of habitat segregation or preferences.

REMARKS: This species is weakly separable morphologically from *E. typus*, which generally occurs at lower elevations and has a more southern distribution. Apportionment of mtDNA cytochrome-*b* sequences of the two species is equivocal, as these sort into reciprocally monophyletic units but with limited support and low molecular distances (Lanzone, Ojeda and Gallardo, 2007; Mares et al. 2008) or those of *E. bolsonensis* fall within the variation presented by a broader sampling of *E. typus* (E. P. Lessa et al. 2010). Lanzone, Ojeda, and Gallardo (2007)

reported a 2*n* = 44, FN = 44, karyotype with an acrocentric X chromosome for specimens from the Salar de Pipanaco, Campo Arenal (both in Catamarca province), and Cafayate (Salta province), which they referred to as "*Eligmodontia* 1." Mares et al. (2008) regarded these specimens to represent *E. bolsonensis*. Lanzone, Ojeda, and Gallardo (2007) regarded the acrocentric X chromosome to be the ancestral condition for the genus because by mtDNA cytochrome-*b* sequence analysis this karyotype was sister to the 2*n* = 44, FN = 44, biarmed X-chromosome karyotype characteristic of *E. typus*. These authors suggested that *marica* Thomas was the appropriate name to apply to this taxon because the type locality of that taxon (Chumbicha, Catamarca province) was only 80 km from the Salar de Pipanaco. DNA sequences, however, place specimens from Chumbicha within the *E. typus* clade and not with *E. bolsonensis* (Mares et al. 2008; E. P. Lessa et al. 2010). Recently, M. C. Da Silva (2011) reported cytochrome-*b* haplotypes she referred to *E. bolsonensis* from two localities in Chubut province, sympatric with *E. typus*. Hence, either *E. bolsonensis* has a larger distribution than previously suspected or cytochrome-*b* haplotypes are not always optimal markers to distinguish the two species. Because the specimens of *E. bolsonensis* from Chubut has not been examined karyotypically or morphologically as yet, neither locality is included in the list of selected localities for this species.

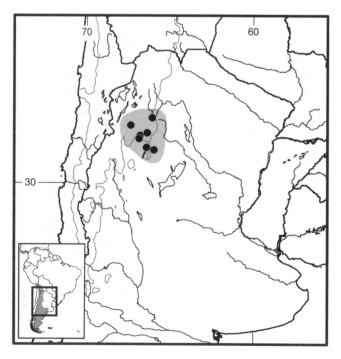

Map 268 Selected localities for *Eligmodontia bolsonensis* (●). Contour line = 2,000 m.

Teta, D'Elía, and Pardiñas (2010) indicated that *hypogaeus* Cabrera was a junior synonym of *E. moreni*, and noted that Mares et al. (2008) assigned two specimens to *E. bolsonensis* that Massoia (1976a) had previously stated were indistinguishable from Cabrera's holotype. Teta, D'Elía, and Pardiñas (2010) thus emphasized the need to consider all available names when making taxonomic and nomenclatural decisions, including the description of presumptively new species. Although *E. bolsonensis* shares with *E. moreni* an immaculate white venter with hairs white to their base, the two differ in most external and craniodental dimensions, have distinctly different karyotypes, and belong to markedly divergent mtDNA clades (Mares et al. 2008; Lanzone, Ojeda et al. 2011).

Eligmodontia dunaris Spotorno, Zuleta, Walker, Manriquez, Valladares, and Marín, 2013
Dune Gerbil Mouse, Dune Laucha
SYNONYM:

Eligmodontia dunaris Spotorno, Zuleta, Walker, Manriquez, Valladares, and Marín, 2013:385; type locality "CHILE: Coquinbo Region: comuna La Higuera, Sector Playa Los Choros (29°14′S; 71°18′W, 15 masl), 55 km N of La Serena city."

DESCRIPTION: Smallest species in genus, mean head and body length 70.6 mm (range: 64–77 mm), tail length 67.1 mm (57–76 mm), hindfoot length 18.9 mm (17–20 mm), and ear length 13.5 mm (10–18 mm). Tail shorter than head and body length (about 95%); both ears and hindfeet absolutely shortest relative to other species. Dorsal color very pale yellowish brownish gray; head with slightly darkish band extending from nose to between ears; ears with whitish preauricular tufts reaching proximal part of pinna; venter immaculately white, sharply demarcated from dorsum with imperceptible fulvous lateral line; tail whitish above and below with slight differentiation; tail lacks a terminal tuft; hindfeet white above; hypothenar pad absent. Skull delicate; nasals not extending beyond premaxillary-frontal suture; zygomatic arches parallel; interorbital region with slightly divergent edges; anterior border of zygomatic plates slightly concave; auditory bullae relatively uninflated with wide interbullar distance, elongated bony tubes (stated to be short in the original description, but photographs of holotype reveal tubes equivalent in length to those of *E. bolsonensis* [Lanzone, Ojeda, and Gallardo 2007; Mares et al. 2008], *E. typus* [Lanzone, Ojeda, and Gallardo 2007], and *E. morgani* [Lanzone 2009]), widely open petrotympanic fissures, and thinly protuberant stapedial spines; M1 lacking anteromedian flexus.

DISTRIBUTION: This species is restricted to semidesert habitats south of the Atacama Desert in northern Chile,

occurring in coastal dunes in Coquimbo Region north to the central plains of Atacama Region.

SELECTED LOCALITIES (Map 269): CHILE: Atacama, Bahia Salado, 55 km S of Caldera (Spotorno et al. 2013), 45 km S of Copiapó (Spotorno et al. 2013), Playa Rodillo, 1.5 km N of Caldera (Spotorno et al. 2013); Coquimbo, comuna La Higuera, Sector Playa Los Choros (type locality of *Eligmodontia dunaris* Spotorno et al. 2013).

SUBSPECIES: *Eligmodontia dunaris* is monotypic.

NATURAL HISTORY: The known habitat is the sand dune belt and central plains along the arid coast of northern Chile. Plant cover is xerophytic low matorral, with sparsely distributed small shrubs and trees, semidry most of the year but with erratic rains resulting from El Niño–Southern Oscillation (ENSO) events. Reproductively active individuals were caught in June, September, and February to March; litter size varied between two and five. Stomach contents of one individual included plant parts, insects, and seeds, but a more representative sample of the species and broader spatial and temporal studies are needed to document a general dietary strategy. The small body size was hypothesized to be an adaptation to poor-quality habitat, in which the construction of sandy galleries and possible torpor were hypothesized to be energy saving strategies. Population density was low but considered opportunistic in response to unpredictable rains resulting from local El Niño–Southern Oscillation (ENSO) episodes (data from Spotorno et al. 2013).

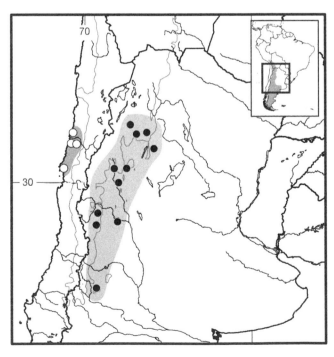

Map 269 Selected localities for *Eligmodontia dunaris* (○) and *Eligmodontia moreni* (●). Contour line = 2,000 m.

REMARKS: The karyotype is similar to that of *E. hirtipes*, with $2n = 50$, $FN = 48$, and both X and Y chromosomes uniarmed. Relationships are uncertain, as *E. dunaris* forms a weakly supported association with *E. puerulus* and *E. moreni*, not with the karyotypically similar *E. hirtipes*, in mtDNA cytochrome-*b* sequence analysis. This species can be distinguished from geographically adjacent *E. hirtipes* and *E. puerulus* by its much smaller size, paler dorsal coloration, shorter hindfoot, and much smaller bullae (Spotorno et al. 2013). However, as is true for all species in the genus, the largest specimens *E. dunaris* overlap in external and cranial dimensions with the smallest samples of *E. puerulus* and *E. hirtipes* (Lanzone, Ojeda et al. 2011; Spotorno et al. 2013).

Eligmodontia hirtipes (Thomas, 1902)
Hairy-footed Gerbil Mouse, Hairy-footed Laucha
SYNONYMS:

Phyllotis hirtipes Thomas, 1902c:225; type locality "Challapata, 3750m," Oruro, Bolivia.

Eligmodontia hirtipes: Thomas, 1916a:141; first use of current name combination.

Eligmodontia hirtipes hirtipes: Gyldenstlope, 1932:71; name combination.

Eligmodontia puerulus hirtipes: Mann, 1945:78; name combination.

Eligmodontia puerulus tarapacensis Mann, 1945:75; type locality "Churiguaya," Arica y Parinacota, Chile.

DESCRIPTION: Characterized by Thomas (1902c) as small, but larger than other species in genus, with mean head and body length 94.7mm (range: 87–103mm), tail length 89.4mm (82–100mm), hindfoot length 25.7mm (24–27mm), and ear length 18.7mm (18–20mm; N=25, specimens from Peru in MVZ collection). Tail always shorter than head and body length (average 94.6%). Fur long, soft, and lax. Dorsal color yellowish brown to sandy buff with gray-based hairs providing darker streaking on back; very distinct buffy lateral line present in adults but less well-developed in younger individuals; undersurface generally immaculate white with hairs white to base; some individuals have very narrow gray-based fur, which typically does not show through from below. Tail weakly bicolored, slightly darker above and paler below, with some individuals having rather dark dorsal stripe and others nearly unicolored; short pencil of 2–4mm long hairs terminates tail in most, but not all, specimens. Forefeet and hindfeet dirty white above and well clothed with stiff whitish hairs on plantar surfaces. Ears dark brown on external surface; inner surfaces clothed with paler, somewhat buffy hairs; small, pale subauricular patches present on most specimens. Skull unremarkable but slightly larger than most species (mean greatest skull length 25.7mm, range: 24.3–26.5mm); auditory bullae notably small (mean length 18.8% that of skull), interbullar distance thus wide, bullar tubes elongated, petrotympanic fissures wide, and stapedial spines protuberant; anteromedian flexus absent on M1 of adults, weakly visible in very young individuals.

DISTRIBUTION: Geographic limits of *E. hirtipes* are unclear, especially with respect to *E. puerulus*. As currently understood, this species occupies dry Puna habitats of the Altiplano of northern Chile, western and southwestern Bolivia, and southern Peru generally at elevations above 3,000m. Apparently, the range of *E. hirtipes* does not overlap with those of any other species as currently understood, but more samples are needed in the possible contact zones among *E. hirtipes*, *E. puerulus*, and *E. dunaris* to delimit these geographic ranges.

SELECTED LOCALITIES (Map 270): BOLIVIA: La Paz, 8.5km W of San Andrés de Machaca (S. Anderson 1997); Oruro, 37km SW of Camino de Oruro, 3.5km NE of Toledo (S. Anderson 1997), Challapata (type locality of *Phyllotis hirtipes* Thomas), Estancia Agua Rica, 22km S and 40km E of Sajama (S. Anderson 1997); Potosí, 2km E of ENDE camp, Laguna Colorada (Kelt et al. 1991), 1mi E of Uyni (MVZ 119973), 5mi N of Villazón (MVZ 119974). CHILE: Arica y Parinacota (Choquelimpie, 114km NE of Arica (Spotorno et al. 1994); Tarapacá, Suricayo, Colchane (Mares et al. 2008). PERU: Puno, Mazocruz (MVZ 115772), Pampa de Ancomarca, 123km W of Ilave (MV 115774); Tacna, 5km E of Lago Suche (MVZ 115771), 2km N of Nevado Livine (MVZ 115776).

SUBSPECIES: We treat *Eligmodontia hirtipes* as monotypic.

NATURAL HISTORY: This species is nocturnal (Pearson 1951) and terrestrial. In Peru, *E. hirtipes* occurs in sandy areas where tola (*Lepidophyllum* and *Baccharis* [Asteraceae]) is present with scattered clumps of bunch grass (Pearson 1951). Pearson and Ralph (1978) found it the most abundant small mammal in the tola habitat where individuals reached densities exceeding 1/ha and had an average home range of 92m. S. Anderson (1997) captured a pregnant female with four embryos in Bolivia in May; little to no reproductive activity was found in July and August in Peru (Pearson 1951). S. I. Silva (2005), in a review of trophic position of small mammals in Chile, described *E. hirtipes* (under the name *E. typus*) as herbivorous.

REMARKS: Previously considered a synonym of *Eligmodontia typus puerulus* (Hershkovitz 1962) or *E. puerulus* (Musser and Carleton 2005), we recognize *E. hirtipes* as a valid species based on its karyotype and well-supported monophyly of mtDNA cytochrome-*b* sequences (Lanzone, Ojeda, and Gallardo 2007; Lanzone, Ojeda et al. 2011; Mares et al. 2008; Spotorno et al. 2013). This species has a $2n=50$, FN = 48, karyotype with an acrocentric X chromosome and small biarmed Y (Pearson and Patton 1976 [reported as *E. typus*]; Ortells et al. 1989; Kelt et al. 1991; and Spotorno et al. 1994 [each reported as *E. puerulus*]), similar in all characteristics to that of *E. dunaris* (Spotorno et al. 2013).

Musser and Carleton (2005) listed *tarapacensis* Mann as a synonym of *E. puerulus*, but we align this name with *E. hirtipes*, in part because the type locality is on the border with Bolivia and near that with Peru. Mann's (1945) description of *tarapacensis* lacks sufficient detail, but he did note that the tail of his new form was only weakly bicolored, a characteristic of *E. hirtipes*, while contrasting it with the dark dorsal stripe of most *E. puerulus* specimens. Nevertheless, we caution that examination of the holotype and/or determination of the karyotype of topotypes is necessary to verify our hypothesis.

Lanzone and Ojeda (2005) mapped the purported range of this species but clearly indicated its questionable limits in Bolivia and Chile. S. Anderson (1997) included all Bolivian specimens in his concept of *E. puerulus*, but it is unclear whether these records are of *E. hirtipes* or whether both species occur in that country. Three specimens in the MVZ collection from Potosí department (MVZ 199771, 199772, 199774) are smaller in body size that Peruvian *E. hirtipes* but have the immaculate white venter characteristic of this species.

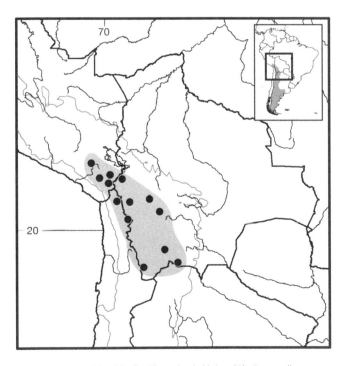

Map 270 Selected localities for *Eligmodontia hirtipes* (●). Contour line = 2,000 m.

Eligmodontia moreni (Thomas, 1896)
Monte Gerbil Mouse, Monte Laucha
SYNONYMS:

Eligmodon Moreni Thomas, 1896:307; type locality "Chilecito, Prov. Rioja [= La Rioja], Argentina, alt. 1200 metres."

Eligmodontia moreni: Thomas, 1916a:141; first use of current name combination.

Graomys hypogaeus Cabrera, 1934:124; type locality "Corral Quemado, Catamarca (2250 m. de alt.)," Argentina.

Eligmodontia morenoi Yepes, 1935a:226; incorrect subsequent spelling of *Eligmodon moreni* Thomas.

Eligmodontia typus moreni: Cabrera, 1961:483; name combination.

DESCRIPTION: Body small but tail absolutely and proportionately longest in genus; mean head and body length 79.4 mm (range: 69–89 mm), mean tail length 106.2 mm (range: 92–125 mm), mean hindfoot length 24.4 mm (range: 22.1–27 mm), and mean ear length 19.2 mm (range: 16–21.5 mm). Tail much longer than head and body length (about 134% greater). Fur long and lax. Dorsal color either brownish (Lanzone, Ojeda, and Gallardo 2007) or coarsely mixed fawn and brown (Thomas 1896); under parts white, with hairs white to base, although Thomas (1896) noted immature specimens with gray-based hairs; junction of dorsal and ventral colors marked by yellowish lateral line. Tail markedly bicolored and moderately haired, dark above and white below and on sides; terminal pencil of hairs extends 5 mm or longer. Area around mouth white; white postoccular patch present. Hindfeet notably large; upper surfaces of forefeet and hindfeet clothed in silvery white hairs, and soles likewise covered by short hairs. Skull with interorbital region slightly divergent posteriorly and with edges angled for approximately half of length and then expanded to their anterior margins; zygomatic arches parallel or slightly divergent posteriorly, narrow, and slender; deep zygomatic notches with oval internal borders; anterior border of maxillae with developed and rounded tubercles; anterior margin of mesopterygoid fossa rounded; small maxillary foramina located at anterior half of M1; teardrop-shaped foramen ovale; pterygoid plates expanded; auditory bullae inflated, with short to very short bullar tubes, distinctly narrowed petrotympanic fissures, and very short but expanded at the base stapedial spines; m1 with anteromedian flexus marked to slightly marked; anterolingual and mesostyle present; m2 with labial fold (hypoflexus) in contact with secondary lingual fold (metaflexus); m3 with two conspicuous folds; M1 with weakly developed or absent anteromedian flexus; M2 with first primary fold well developed; M3 with lingual fold present; condyloid process narrower than angular process.

This species is readily characterized by its long bicolored tail and distinctive structure of the auditory bullae, in which bullar inflation occludes the petrotympanic fissures, the interbullar distance is greatly reduced, and the stapedial spine is narrowed but with a robust base. Lanzone, Ojeda, and Gallardo (2007) compared craniodental measurements of *E. moreni*, *E. bolsonensis* (as *E*. sp. 1), and *E. typus*, as did Mares et al. (2008); Lanzone, Ojeda et al. (2011) compared *E. moreni* with *E. puerulus*. The latter authors also studied the morphometry of two samples of *E. moreni* from Catamarca and Mendoza provinces, recording significant differences in most external and cranial measures between them.

DISTRIBUTION: *Eligmodontia moreni* occurs in Argentina from Salta province south to Mendoza province along the eastern slopes of the Andes in Monte Desert and pre-Puna habitats at elevations from about 500 to 3,900 m. However, as mentioned below, specimens assigned by morphology to both *E. moreni* and *E. puerulus* from one high-elevation locality in Catamarca province share the same mtDNA cytochrome-*b* sequences (Lanzone, Ojeda et al. 2011). Hence, the actual presence of *E. moreni* at high elevation in Salta needs further confirmation. Lanzone, Ojeda, and Gallardo (2007) and Mares et al. (2008) recorded sympatry between *E. moreni* and *E. bolsonensis* in Catamarca province and between *E. moreni* and *E. typus* in both Mendoza and San Juan provinces.

SELECTED LOCALITIES (Map 269): ARGENTINA: Catamarca, Corral Quemado (type locality of *Graomys hypogaeus* Cabrera), 5.2 km S of El Bolsón (Mares et al. 2008), 21 km SW of El Desmonte (Mares et al. 2008), 1.5 km S of El Peñón (Mares et al. 2008); La Rioja, Chilecito (type locality of *Eligmodon Moreni* Thomas), 14.5 km N of Villa Unión (Mares et al. 2008); Mendoza, Malargüe, 70 km NE of Barrancas (Mares et al. 2008), Reserva Telteca (Lanzone, Ojeda et al. 2011), ca. 7 km S of Uspallata (Mares et al. 2008); San Juan, Estancia Leoncito, 1 km W of Observatorio Astronómico (Mares et al. 2008), Ischigualasto, Valle Fértil (Lanzone, Ojeda, and Gallardo 2007).

SUBSPECIES: *Eligmodontia moreni* is monotypic.

NATURAL HISTORY: *Eligmodontia moreni* has been collected in sandy areas with scattered, sparse vegetation in the Monte Desert and pre-Puna of Argentina. Reproduction occurs in the fall (March, April, and May) and spring (October and November). Unpublished observations indicate that this species is nocturnal and terrestrial. Gonnet and Ojeda (1998) examined habitat use of a small mammal community at a midelevation locality in northwestern Mendoza province, Argentina; although identified as *E. typus*, it is probable that the species the studied was *E. moreni*. Traba et al. (2010) described macrohabitat and microhabitat selection by *E. moreni* during both wet and dry seasons, regarding the species as a "fine-grained generalist" because of a broad ecological amplitude that spanned creosote bush scrub, mesquite woodland, columnar cactus slopes, barrens, chical, and saltbush habitats. The diet of *E. moreni* during a dry season in a hyperarid region of the Monte desert was omnivorous with a tendency to herbivory. Most food items were plant leaves, followed by arthropods, with seeds the least consumed. However, there was high interindividual variability in the proportion of these categories, which suggests high flexibility in the dietary strategy of this species (Lanzone et al. 2012).

REMARKS: Thomas (1896) regarded *E. moreni* to be most closely allied to *E. elegans* Waterhouse (= *E. typus* F. Cuvier). Lanzone, Ojeda, and Gallardo (2007) and Mares et al. (2008) provided limited mtDNA cytochrome-*b* sequence support for *E. moreni* and *E. puerulus* as sister species. In a more geographically expansive analysis, however, Lanzone, Ojeda et al. (2011) failed to recover reciprocal monophyly in cytochrome-*b* haplotypes from specimens assigned to both species by karyotype and morphology. Hershkovitz (1962) recognized the link between *moreni* and *puerulus*, listing the latter as a subspecies of his *E. typus* with *moreni* Thomas as a synonym. Cabrera (1961), on the other hand, regarded *moreni* as a valid subspecies of *E. typus*. The species status of *E. moreni* relative to *E. puerulus* is strongly supported by the former's unique 2n = 52, FN = 50, karyotype (karyotype 2 in Lanzone, Ojeda, and Gallardo 2007) and by multivariate morphometric analyses (Lanzone, Ojeda, and Gallardo 2007; Lanzone, Ojeda et al. 2011). The lack of reciprocal monophyly in cytochrome-*b* sequences is thus presumably due to recent splitting of both *moreni* and *puerulus* from a common ancestor, but this hypothesis requires further study. Spotorno et al. (1994) initially assigned the 2n = 34, NF = 48, karyotype to *E. moreni*, but later Spotorno and colleagues examined specimens from the type locality of *E. puerulus* and reassigned this cytotype to that species (Spotorno, Zuleta et al. 1998; Spotorno et al. 2001; Lanzone and Ojeda 2005; Lanzone, Ojeda et al. 2011). Lanzone, Ojeda, and Gallardo (2007) documented qualitative and quantitative cranial character differences between *E. moreni* (their karyotype 2 sample) and both *E. bolsonensis* (their karyotype 1 sample) and *E. typus*.

Cabrera (1961) referred *Graomys hypogaeus* Cabrera (1934) as a full synonym of *Phyllotis griseoflavus medius*, Massoia (1976a) placed it in synonymy under *E. typus sensu lato*, and D. F. Williams and Mares (1978) questioned whether the type was a composite specimen. Teta, D'Elía, and Pardiñas (2010), however, examined the holotype and concluded that *hypogaeus* Cabrera was a junior synonym of the currently restricted concept of *E. moreni*.

Eligmodontia morgani J. A. Allen, 1901

Morgan's Gerbil Mouse, Western Patagonian Laucha

SYNONYMS:

Eligmodontia morgani J. A. Allen, 1901c:409; type locality "Arroya [*sic*] Else, Patagonia"; J. A. Allen (1905:53)
referred to this vague reference as "Balsaltic Cañons, 50
miles southeast of Lake Buenos Aires, Patagonia ('Arroyo
Else' on the labels and in the original description)." There
is no "Arroyo Else" in Patagonia; according to the notes
and itinerary followed by J. B. Hatcher and A. E. Colburn (see Hatcher 1903); the type specimen was collected
near the confluence of the Río Ecker and Río Pinturas,
about 47°07′S and 70°51′W, Santa Cruz, Argentina.

Eligmodontia elegans morgani: Osgood, 1943b:199; name
combination.

Eligmodontia typus morgani: Cabrera, 1961:484; name combination.

DESCRIPTION: One of largest species of genus with regard to length of head and body, with mean head and body
length 86.4 mm (range: 75–90 mm), tail length 78.9 mm
(range: 63–99 mm), hindfoot length 22.5 mm (range: 21–
25 mm), and ear length 16.2 mm (range: 14–19 mm; data
from Sikes et al. 1997 and MVZ collection). Tail usually
shorter than head and body length (about 92%); bicolored
with narrow dark dorsal stripe; terminating in short pencil
about 2 mm long. Dorsal fur dark to grayish brown with
yellowish mottled; intensity of yellow increases laterally
and caudally, but most specimens darker relative to those
of other species. Dorsum separated from whitish venter by
narrow ochraceous lateral line; venter with gray to black
bases of hairs visible, tips white. Ears dark brown and
noticeably short (about <20% of head and body length),
without evidence of preauricular or postauricular patches
of paler color. J. A. Allen (1901c:409) remarked that pelage was "very full, long, and soft," but specimens in the
MVZ collections from Río Negro province have noticeably
shorter dorsal fur than other species. Dorsal surfaces of
forefeet and hindfeet clothed in short, white hairs; plantar
surfaces sparsely covered by short hairs. Cranially similar to *E. typus*, but with longer incisive foramina (mean
5.25 mm, range: 4.7–6.0 mm). The nasals of *E. morgani*
longest among all species in genus (mean 10.2 mm); zygomatic arches rounded when viewed from above, but delicate, thinner in middle and posterior parts relative to other
species; anterolateral border of zygomatic plates slightly
concave; braincase small; tubercles on anterior border of
maxillae poor developed or absent; parapterygoid plates
long, wide, and more elevated in lateral portion than in
others species; mesopterygoid fossa relatively long and narrow, with median spine on anterior margin; auditory bullae uninflated (mean length × width: 3.61 × 3.75 mm), with
elongated bony tubes terminating in serrations of which

one projection longer than others, very thin protuberant
stapedial spines without conspicuous base, and wide petrotympanic fissures; maxillary toothrow length notably
short (mean 3.60, range: 3.3–3.9; Sikes et al. 1997; Lanzone 2009); M1 with anteromedian flexus variable present; mandible with wide angular process with an internal
concavity, and condyloid process narrow but longer than
in other species.

DISTRIBUTION: *Eligmodontia morgani* occurs in the
Patagonian steppe of southern Chile and southern Argentina in Mendoza, Neuquén, Río Negro, Chubut, and Santa
Cruz provinces, at moderate to high elevations along the
slopes and eastern base of the Andes. This species is sympatric with *E. typus* at multiple localities from Mendoza
south to Santa Cruz province where the Patagonian steppe
interfaces with Monte or coastal desert habitats (Zambelli
et al. 1992; Hillyard et al. 1997; Sikes et al. 1997; Mares
et al. 2008; M. C. Da Silva 2011).

SELECTED LOCALITIES (Map 271): ARGENTINA:
Chubut, Estancia La Escondida (Hillyard et al. 1997), Estancia Los Manantiales (E. P. Lessa et al. 2010), Estancia Talagapa (E. P. Lessa et al. 2010), 27 km NW of Pampa de Agnia (Sikes et al. 1997), 3 km N of Tecka along Hwy 4 (Mares
et al. 2008); Mendoza, 2 mi E of Gendarmería Cruz de Piedra (Mares et al. 2008), 2 km N of junction Hwy 40 and El
Manzano Road along Hwy 40 (Mares et al. 2008), 3 km W
of Refugio Militar General Alvarado (Mares et al. 2008);
Neuquén, 8.8 km S of Lonco Luan (Mares et al. 2008),
Parque Nacional Laguna Blanca (Tiranti 1997), 21 km NW
and 1 km N of junction Hwys 60 and 23 (Mares et al. 2008);
Río Negro, Cerro Corona (Lessa et al. 2010), Las Victo-
rias, 4.2 km E of Bariloche (E. P. Lessa et al. 2010), 15 km
NE of Mencué (Sikes et al. 1997), Meseta de Somuncurá
(Sikes et al. 1997), Tembrao (Hillyard et al. 1997); Santa
Cruz, Cabo Vírgenes (UP 1201), Estancia Cerro Ventana (E.
P. Lessa et al. 2010), Estancia La Vizcaina (Hillyard et al.
1997), Río del Deseado, Estancia Cerro del Paso (M. C. Da
Silva 2011), Río Santa Cruz, 4 km W of Punta Quilla s/RP
288 (M. C. Da Silva 2011), RPG Laguna Azul (CNP-E 425).
CHILE: Aysén, Chile Chico, 2 km S and 1 km W of Aeró-
dromo (Kelt et al. 1991); Magallanes y Antártica Chilena,
Lake Sarmiento, Ultima Esperanza (Osgood 1943b).

SUBSPECIES: *Eligmodontia morgani* is monotypic.
Mares et al. (2008) detected a shallow gene-genealogy with
a phylogeographic break between southern and northern
populations, but this geographic division has not been examined adequately as yet.

NATURAL HISTORY: *Eligmodontia morgani* is terrestrial and nocturnal. It is primarily restricted to the semiarid
steppes in western Patagonia, where it may be the most
abundant species (Monteverde et al. 2011). Population
ecology, microhabitat selection, space use, population dy-

namics, reproduction, and longevity have been studied in northwestern Patagonia (Pearson et al. 1987; Guthmann et al. 1997; Lozada and Guthmann 1998; Lozada et al. 2001). Pearson et al. (1987) characterized the reproductive season as October through April, with an average litter size of 5.9; individuals reached sexual maturity at 1.5 months, and the life span was about nine months. Abundance was highest in the fall (3.5 individuals/ha) and lowest in the spring (0.4/ha). Monteverde et al. (2011) recorded a breeding season from August to March, larger home range sizes for females (656.72 ± 90.55 m²) than for males (439.58 ± 114.52 m²), and abundances ranging from 12 to 71 individuals per hectare. Individuals of this species are prey to both the Magellanic Horned Owl (*Bubo magellanicus*; Trejo et al. 2005; Trejo and Guthmann 2003) and the Great Horned Owl (*Bubo virginianus*; Trejo and Grigera 1998). J. P. Sánchez et al. (2009), J. P. Sánchez et al. (2010), and J. P. Sánchez (2013) described tick and flea ectoparasites. Udrizar Sauthier (2009) enlarged the knowledge of the past distribution and abundance of this species through the Holocene of central Patagonia.

REMARKS: J. A. Allen (1901c) described *morgani* as a species, but this name, as with so many others in the genus, became subsumed as a synonym of *E. typus typus* in Herskowtiz's (1962) revision or as a synonym of *E. typus elegans* in Cabrera's (1961) compilation. The recognition of *E. morgani* as a species distinct from *E. typus* was first supported by its unique $2n = 32–34$, FN $= 32$, karyotype (Ortells et al. 1989; Kelt et al. 1991; Zambelli et al. 1992; Tiranti 1997) and has been more recently confirmed by reciprocal monophyly of mtDNA cytochrome-*b* sequences (Hillyard et al. 1997; Mares et al. 2008; Lanzone, Ojeda et al. 2011). At the chromosome level, *E. morgani* possesses both polymorphic and polytypic Robertsonian (Rb) populations. In some localities, the $2n = 32$ and $2n = 34$ cytotypes are fixed, while in others these diploid numbers are part of a polymorphic Rb system where the $2n = 33$ heterozygote is also present. Mares et al. (2008) described two mtDNA cytochrome-*b* clades and suggested that these corresponded to northern and southern divisions concordant with the chromosome differentiation. However, the known karyotype diversity pattern does not exhibit a simple dichotomous distribution, as populations fixed for the $2n = 32$ karyotype are found both in the north and south of the distribution range of the species, with both polymorphic and $2n = 34$ populations in the center (Ortells et al. 1989; Kelt et al. 1991; Zambelli et al. 1992; Tiranti 1997; Lanzone 2009). Furthermore, Lessa et al. (2010) failed to recover any phylogeographic break across their broad sampling throughout Patagonia in their expanded analysis of cytochrome-*b* sequences. Additional studies are needed to understand distributional relationship between cytotypes and mtDNA se-

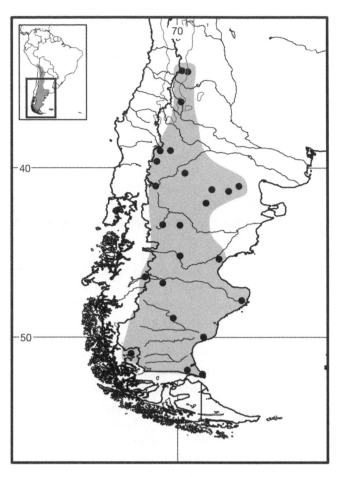

Map 271 Selected localities for *Eligmodontia morgani* (●). Contour line = 2,000 m.

quences. Both Hillyard et al. (1997) and Mares et al. (2008) discussed historical biogeography with regard to *E. typus*, elaborating upon the complementarity of their present-day distributions to biotic provinces within Patagonia.

Eligmodontia puerulus (Philippi, 1896)
Andean Gerbil Mouse, Altiplano Laucha

SYNONYMS:

Hesperomys puerulus Philippi, 1896:20; type locality "cerca de Atacama," [= San Pedro de Atacama,] Antofagasta, Chile (see Osgood 1943b:198; Hershkovitz 1962:187).

Mus puerulus: Philippi, 1900:79; name combination.

Eligmodontia hirtipes jacunda Thomas, 1919f:131; type locality "Abrapampa [= Abra Pampa], 3500 m," Jujuy, Argentina.

Eligmodontia puerulus: Osgood, 1943b:198; first use of current name combination.

Eligmodontia puerulus puerulus: Mann, 1945:78; name combination.

DESCRIPTION: Size moderate for genus, with mean head and body 82.6 mm (range: 67–95 mm); mean tail length 73.8 mm (range: 59–92 mm), mean hindfoot length 23.6 mm (range: 21–26 mm), and mean ear length 17.9 mm (range: 15–21 mm). Fur long and lax; dorsally colored grayish brown to light brown with yellow; distinctly wide buffy lateral line divides dorsal from ventral fur. Ventral hairs white to gray to black at their base but always white at tip; frequency of these ventral colorations varies among populations. Eyes outlined by darker line if variable length that begins on posterior margin, less marked or absent in specimens from some populations. Tail absolutely and proportionally shortest among species in genus, averaging 89.4% of head and body length; coloration varies from markedly bicolor to almost completely white, although slightly bicolored in most specimens, more pronounced at tip, pale brown above and white below, terminating in short pencil of hairs. Both forefeet and hindfeet covered dorsally with short, dirty white hairs; soles clothed in dense, long white hairs. Skull with nasals relatively large, but shorter than those of *E. morgani*; posterior part of interorbital region relatively angular; zygomatic arches relatively thick and rounded, most robust and least delicate among species; well-developed rounded or ovoid tubercles on anterior border of maxilla, but generally smaller than those of *E. moreni*; short incisive foramina (shortest among species of genus), reaching the anterior border of m1; mesopterygoid fossa relatively long, broad with rounded anterior margin; parapterygoid plate relatively narrow and little elevated in lateral portion; oval foramen large and rounded; petrotympanic fissure narrow (narrower than in *E. typus* but wider than in *E. moreni*), although rather variable within and among populations; bullar tubes short and flat, somewhat irregular anteriorly; stapedial spines relatively short and triangular, wide at insertion point on bullae and thin and detached at anterior tip; M1 with weakly developed anteromedian flexus in young individuals but without flexus in adults; M2 with first primary fold absent or poorly developed; M3 with only second primary fold well developed (labial) and deep, more so than in other species, lingual fold, in general, absent or underdeveloped; m1 with marked anteromedian flexus; no contact between labial fold (hypoflexus) and second major lingual fold in m2; and m3 with two conspicuous folds. Skull differs from geographically adjacent (to southeast) *E. moreni* by much smaller bullae (average length and width, 4.67 × 5.07 mm), presence of very open petrotympanic fissures, and distinct stapedial spine (Lanzone 2009; Lanzone, Ojeda et al. 2011); craniodental dimensions completely nonoverlapping in multivariate principal component space (Lanzone, Ojeda et al. 2011).

DISTRIBUTION: *Eligmodontia puerulus* occurs in the Puna of the Altiplano of northeastern Chile and northwestern Argentina in the provinces of Jujuy, Salta, and Catama-

rca, at elevations generally above 3,500 m. The possibility that this species is present in southwestern Bolivia requires careful consideration. This species had been recorded in sympatry only with *E. bolsonensis* in Catamarca province (Mares et al. 2008), but the high genetic similarity between *E. puerulus* and *E. moreni* in that region requires a detailed study of multiple taxonomic characters to identify properly species in this province.

SELECTED LOCALITIES (Map 272): ARGENTINA: Catamarca, Antofagasta de la Sierra, Paycuqui (CML [Arg 5296]), Pastos Largos (Lanzone, Ojeda et al. 2011); Jujuy, Abra Pampa (type locality of *Eligmodontia hirtipes jacunda* Thomas), 8 km E of Saladillo (Lanzone, Ojeda et al. 2011), 18 km E of Salinas Grandes (Lanzone, Ojeda et al. 2011), 26 km W of Susques (Lanzone, Ojeda et al. 2011); Salta, Cauchari (Spotorno et al. 1994), Vega Cortadera (OMNH 34756). CHILE: Antofagasta, Ojos de San Pedro (MVZ 116774), San Pedro de Atacama (type locality of *Hesperomys puerulus* Philippi), 40 km SE of Tonocao (Spotorno et al. 2001).

SUBSPECIES: We treat *Eligmodontia puerulus* as monotypic; Lanzone, Ojeda et al. (2011) detected only very slight differences among three sampled populations in five of 29 mensural variables. However, there is high molecular diversity (more than 5% in mtDNA cytochrome-*b* sequences) among Argentinean Puna samples. Additionally, both Spotorno et al. (1994; see also Spotorno, Zuleta et al. 1998) and Lanzone, Ojeda et al. (2011) have described a complex Robertsonian chromosome system in samples from the Puna of Argentina and Chile. The relationship between molecular and chromosomal diversity as well as the taxonomic implications of both requires further study.

NATURAL HISTORY: Like other species of *Eligmodontia*, *E. puerulus* is nocturnal and terrestrial. It is restricted to the high elevations of the Altiplano. In northern Chile, this species is found in "*tolares arenosos*" (Spotorno, Zuleta et al. 1998); near the type locality, it was captured in association with the saltbush *Atriplex etumatacamensii* (Amaranthaceae). Unpublished data indicate reproductive activity occurs in the fall (March).

REMARKS: The karyotype of *E. puerulus* is characterized by a variable $2n$ ranging from 31 to 37, but with a stable FN of 48 signaling Robertsonian polymorphism (Spotorno et al. 2001; Lanzone, Ojeda et al. 2011). Specimens from the type locality of *E. puerulus* (San Pedro de Atacama, Antofagasta, Chile) have $2n = 34$ and FN = 48 (Spotorno, Zuleta et al. 1998). There are different populations with fixed 32 and 34 chromosomes, and polymorphic ones with $2n = 31, 32, 33$ and $2n = 32, 34, 35, 36, 37$. Meiosis of a few samples of heterozygotes exhibited normal segregation, with no imbalances detected (Lanzone, Ojeda et al. 2011). Thomas (1919f) described *jacunda* as a subspecies of *E. hirtipes*, but specimens from the type locality (Abra Pampa, Jujuy, Argentina) have a $2n = 31$, 32, and 33, at the lower end of the range characteristic of *E. puerulus*, not the $2n = 52$ of *E. hirtipes* (Lanzone, Ojeda et al. 2011). Karyotypic data also support the species status of *E. puerulus* with respect to the geographically adjacent *E. hirtipes* (to the north), *E. dunaris* (to the southwest), and *E. moreni* (to the southeast). Three or all four species form a major clade in some phylogenetic analyses (Lanzone, Ojeda, and Gallardo 2007; Mares et al. 2008; Lanzone, Ojeda et al. 2011; Spotorno et al. 2013). *E. puerulus* is characterized by deep molecular variability and divergences with a complex geographic distribution (Lanzone, Ojeda et al. 2011). Specimens allocated to *E. puerulus* and *E. moreni* are presumptively recently derived species, as their respective mtDNA cytochrome-*b* haplotypes have yet to reach reciprocal monophyly, presenting a polyphyletic pattern. Nevertheless, the two species are markedly distinct in karyotype and differ significantly in 21 of the 29 morphometric variables compared as well as in qualitative traits, such as differences in bullar size and openness of the petrotympanic fossae (Lanzone, Ojeda et al. 2011). All four species have allopatric ranges, but it is possible that ranges overlap between some of them in the Altiplano of northern Chile, western Bolivia, and northwestern Argentina. Additional collections accompanied by karyotypes and molecular sequences are needed to refine the range limits of all four species.

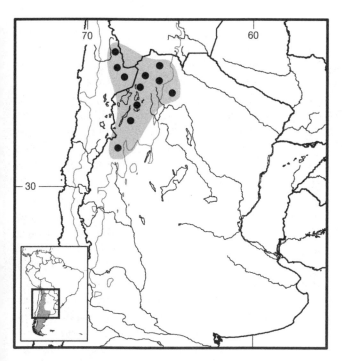

Map 272 Selected localities for *Eligmodontia puerulus* (●). Contour line = 2,000 m.

Eligmodontia typus F. Cuvier, 1837
Highland Gerbil Mouse, Eastern Patagonian Laucha

SYNONYMS:

Eligmodontia typus F. Cuvier, 1837:168; type locality "environs de Buenos-Ayres," Buenos Aires, Argentina (see Remarks for discussion of the actual type locality).

Mus elegans Waterhouse, 1837:19; type locality "Bahia Blanca [= Bahía Blanca]," Buenos Aires, Argentina.

Eligmodontia morgani pamparum Thomas, 1913c:572; type locality "Peru Station, F.C.P. [= Perú]; about 200 kilometres N.W. of Bahia Blanca," La Pampa, Argentina.

Eligmodontia marica Thomas, 1918c:483; type locality "Chumbicha, Catamarca. Alt. 600 m," Argentina.

Eligmodontia typus elegans: Cabrera, 1961:483; name combination.

Eligmodontia typus marica: Cabrera, 1961:483; name combination.

Eligmodontia typus: Cabrera, 1961:483; first use of current name combination.

Eligmodontia typus typus: Cabrera, 1961:484; name combination.

DESCRIPTION: Size similar to *E. bolsonensis*, *E. moreni*, and *E. puerulus*, slightly smaller on average than *E. morgani*, much smaller than *E. hirtipes*, but not as small as *E. dunaris*. Mean head and body length 76.21 (Lanzone, Ojeda, and Gallardo 2007) to 84.03 mm (Mares et al. 2008), with range across both studies 61–88 mm; averages from same two studies for tail length 92.1 mm and 94.67 mm, respectively (combined range: 72–110 mm), hindfoot length 21.26 and 22.01 mm (range: 16–25 mm), and ear length 17.21 and 17.83 mm (range: 11–23 mm). Tail proportionally long, 110–121% of head and body length; slightly bicolored with dark and especially wide dorsal stripe; usually either lacking penciled tip or, if present, one less than 5 mm. Fur short, as in *E. morgani*, with overall brownish dorsal tones; upper parts separated from venter by buffy lateral line of varying distinctness; white under parts with hairs gray or white at their base throughout or confined mainly to margins. Forefeet and hindfeet white above; soles appear naked or nearly so, clothed in sparse, short white hairs. Ears moderately long, 20–23% of head and body length. Skull unremarkable except for overall small size, with mean greatest cranial length varying from 22.75 mm (Mares et al. 2008) to 23.85 mm (Lanzone, Ojeda, and Gallardo 2007), combined range from both studies 20.9–25.7 mm; tubercles on anterior border of maxilla absent or slightly developed and rounded; auditory bullae uninflated, with mean length by width 4.16 mm × 4.31 mm, somewhat larger than parapatric *E. morgani* (Sikes et al. 1997) and distinctly smaller than *E. moreni* (Lanzone, Ojeda, and Gallardo 2007); stapedial spine development and enlarged petrotympanic fissures similar to *E. bolsonensis*, distinct from conditions in *E. moreni* (Lanzone, Ojeda, and Gallardo 2007); anteromedian flexus of M1 weakly visible in very young individuals but not in adults.

DISTRIBUTION: *Eligmodontia typus* occurs east of the Andes from Catamarca south to Santa Cruz province, and east to Buenos province, Argentina, primarily at elevations below 1,000 m. It has the broadest range across the ecoregions of Argentina, ranging from Espinal and Pampa to Monte Desert and Patagonian steppe. *Eligmodontia typus* is sympatric with *E. moreni* at localities in Mendoza and San Juan provinces (Lanzone, Ojeda, and Gallardo 2007; Mares et al. 2008) and with *E. morgani* at several sites in Chubut, Mendoza, Río Negro, and Santa Cruz provinces (Zambelli et al. 1992; Hillyard et al. 1997; Sikes et al. 1997; Mares et al. 2008).

SELECTED LOCALITIES (Map 273): ARGENTINA: Buenos Aires, Bahía Blanca (type locality of *Mus elegans* Waterhouse), Pechuén-có (Massoia and Fornes 1964d), Pirovano (J. R. Contreras 1972); Catamarca, Chumbicha (type locality of *Eligmodontia marica* Thomas), Quimilo (Mares et al. 2008); Chubut, Estancia Los Manantiales (E. P. Lessa et al. 2010), Istmo Ameghino (Sikes et al. 1997), 30 km NW of Pampa de Agnia (Sikes et al. 1997), Paraje Fofocahuel, Campo Netchovit (E. P. Lessa et al. 2010), Paso de Indios, 18 km W of junction Hwys 29 and 27 along Hwy 27 (E. P. Lessa et al. 2010), Rawson (Teta, D'Elía, and Pardiñas 2010); La Pampa, Lihué Calel (Teta, D'Elía, and Pardiñas 2010), Loventué (Teta, D'Elía, and Pardiñas 2010), Peru Station (type locality of *Eligmodontia morgani pamparum* Thomas); Mendoza, Costa de Araujo (Lanzone, Ojeda, and Gallardo 2007), 35 km S of Pareditas by Hwy 40 and 3 km E (Mares et al. 2008), Reserva Telteca (Mares et al. 2008), Salinas del Diamante (Mares et al. 2008); Neuquén, 20 km E of Zapala (Tiranti 1997); Río Negro, Aguada Cecilio (Hillyard et al. 1997), 10 km S of Comallo (E. P. Lessa et al. 2010), Meseta de Somuncurá (Sikes et al. 1997), 18 km SW of Viedma (Hillyard et al. 1997); San Luis, Pampa de las Salinas (Lanzone, Ojeda, and Gallardo 2007), 12 km N of Varela (by road) (E. P. Lessa et al. 2010); Santa Cruz, Estancia Cerro del Paso (E. P. Lessa et al. 2010), Meseta El Pedrero (Sikes et al. 1997).

SUBSPECIES: We treat *Eligmodontia typus* as monotypic. Although this species occupies a broader ecological range than any other in the genus, no study has as yet determined the patterns of morphological variation that might substantiate geographic races.

There are high levels of protein heterozygosity, polymorphism, and genetic identity within and among populations of Patagonia (G. B. Sousa et al. 1996). Limited cladal structure in mtDNA cytochrome-*b* sequences is present, but this structure has not as yet been placed within a geo-graphic context (Hillyard et al. 1997; Mares et al. 2008; E. P. Lessa et al. 2010), nor has it been related to morphological patterns of variation.

NATURAL HISTORY: Most information on the natural history of this species is based on studies of populations in the Monte Desert of Mendoza province, Argentina. Here, and presumably elsewhere, *E. typus* is nocturnal and terrestrial. The diet is omnivorous, with leaves and seeds predominating (Giannoni et al. 2005), and seed hoarding has been reported (Giannoni et al. 2001; C. M. Campos et al. 2007; Tababorelli et al. 2009). The species is resistant to water deprivation (Bozinovic et al. 2007), and is highly efficient at concentrating urine as a means to minimize water loss (G. B. Diaz and Ojeda, 1999). The species is a quadrupedal saltator, but locomotion varies according to the openness of the habitat (Taraborelli et al. 2003). In the Monte Desert, this species prefers areas of low cover in sandy areas, but higher cover where creosote bush (*Larrea* [Zygophyllaceae]) or mesquite (*Prosopis* [Fabaceae]) is present (Corbalán and Ojeda 2004; Corbalán 2006; Corbalán and Debandi 2006; Corbalán et al. 2006). This species does well in areas disturbed by cattle grazing (Tabeni and Ojeda 2005). Known predators include Geoffroy's Cat (*Leopardus geoffroyi*; Bisceglia et al. 2011), the Culpeo (*Lycalopex culpaeus*; Corley et al. 1995), and the Barn Owl (*Tyto alba*; De Santis et al. 1994; García Esponda et al. 1998). Despite possible bipedal locomotion and an erratic escape behavior, individuals were consumed in a greater proportion than their availability, likely due to the open habitat the species utilizes (Tabani et al. 2012).

REMARKS: Although *E. typus* has been recognized as a species by virtually all authorities over the past century and a half, the current and considerably more geographically constrained concept of this species did not coalesce until the application of karyotypic and molecular sequence data within the past two decades. Ortells et al. (1989), Kelt et al. (1991), Zambelli et al. (1992), and Tiranti (1997) each documented the unique karyotype of $2n = 44$, FN = 44, for specimens of this species. More recently, mtDNA cytochrome-*b* sequences have substantiated specimens with this chromosome complement as a monophyletic lineage relative to other species in the genus (Hillyard et al. 1997; Lanzone, Ojeda, and Gallardo 2007; Lanzone, Ojeda et al. 2011; Mares et al. 2008; E. P. Lessa et al. 2010; M. C. Da Silva 2011), except possibly *E. bolsonensis* (E. P. Lessa et al. 2010).

The application of Cuvier's name *typus*, however, to the $2n = 44$ karyotypic form is not straightforward because the type locality of Cuvier's mouse ("environs de Buenos-Ayres") is both imprecise and in doubt. Hershkovitz (1962:185), Massoia and Fornes (1964d), and Ortells et al. (1989) have each discussed the uncertainty of the type locality, noting that, at the time of Cuvier's work, the "environs de Buenos-Ayres" included a large area that cannot be equated to the surroundings of the city of Buenos Aires today. Thomas (1929:39), reflecting on the problem, stated "no doubt Cuvier's giving of Buenos Ayres as the original locality was merely due to this latter term being then used in a general significance—just as if we said Argentina now." He and more recent authors (Cabrera 1961:484; Honacki et al. 1982:410) concluded that the type locality should be "Corrientes," citing d'Orbigny and Gervais (1847). But Alcicde d'Orbigny, who apparently collected the specimen upon which Cuvier's figure of *E. typus* was drawn, simply stated (in d'Orbigny and Gervais 1847:24) "nous avons rencontré cette espèce dans la province de Corrientes, où elle est peu commune" (roughly translated, "we encountered this species in the province of Corrientes, where it is uncommon"); he made no specific reference to where the specimen he provided to Cuvier had originated. To our knowledge, there is no confirmed record of any *Eligmodontia* from Corrientes province, the Mesopotamian region of Argentina; the nearest known locality for the genus is about 500 km distant. If the type of *E. typus* was, in fact, collected by Alcides d'Orbigny, it was most likely obtained in southernmost Buenos Aires province, as he spent several months traveling around Carmen de Patagones and San Blas. The issue of the type locality of Cuvier's *typus*, and thus the correct application of this name to the $2n=44$ karyomorph, remains unresolved, but needs to be "formally restricted in order to settle the morphological identification of *E. typus* and a critical definition of the species within the genus" (Musser and Carleton 2005:1114).

Waterhouse's *elegans*, published in the same year as Cuvier's *typus*, has been uniformly regarded a synonym of *typus* Cuvier (Waterhouse 1839; Lesson 1842; J. A. Allen 1905). Furthermore, all specimens of gerbil mice from the broad vicinity of Bahía Blanca, Buenos Aires province, the type locality of *elegans* Waterhouse, share the $2n=44$ karyotype allocated to *E. typus*, and specimens from nearby Villarino are part of the *E. typus* mtDNA clade (E. P. Lessa et al. 2010). Thus, unless future research is able to contradict the long-accepted association of *typus* Cuvier with *elegans* Waterhouse, we follow those before us and continue to use Cuvier's name for this species.

Thomas (1913c) originally described *pamparum* as a subspecies of *E. morgani*. However, as noted by Mares et al. (2008), the type locality of Peru Station, La Pampa province, is well removed from the currently known range of *E. morgani* and is within the Monte Desert ecoregion that characterizes *E. typus* but not *E. morgani*, and other specimens from this province possess the $2n=44$ karyotype of *E. typus* (e.g., Tiranti 1997). We thus agree with

Mares et al. (2008), who followed both Thomas (1929) and Musser and Carleton (2005), and assign *pamparum* Thomas to *E. typus*.

M. M. Díaz et al. (2006) regarded *marica* Thomas as a valid species, but this taxon is allocated as a synonym of *E. typus* herein based on mtDNA cytochrome-*b* sequences of specimens from near the type locality (Chumbicha, Catamarca province; Mares et al. 2008; E. P. Lessa et al. 2010). However, the low mtDNA cytochrome-*b* sequence divergence between *E. bolsonensis* and *E. typus* (less than 3%) and thus a potential polyphyletic relationship between both taxa cannot be ruled out at present. It is thus possible that Thomas's *marica* is actually a senior synonym of *bolsonensis*. This hypothesis needs to be examined in detail, especially because Thomas (1918c:483) explicitly noted that the type series of *marica* had the "whole of the under surface pure sharply defined white, all the hairs, even laterally, white to their bases." Emilio Budin, who collected the type series of *marica*, noted that the species was "evidently very rare" (Thomas 1918c:484). Given the extensive anthropogenic modification of the

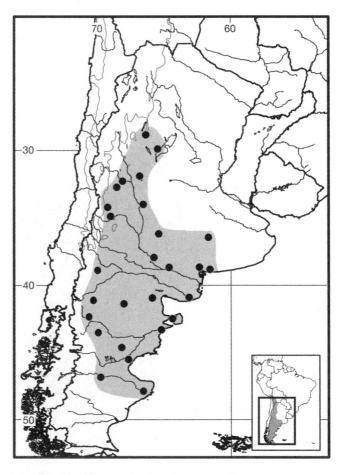

Map 273 Selected localities for *Eligmodontia typus* (●). Contour line = 2,000 m.

habitat around Chumbicha, the potential replacement of *marica* by *typus* also remains a valid, if difficult to falsify, hypothesis.

Genus *Galenomys* Thomas, 1916
Jorge Salazar-Bravo

This monotypic genus ranges between about 16°S and 18°S latitude in southern South America. The single species occurs in arid and semiarid Puna, at elevations between 3,720 and 4,675 m in the departments of La Paz and Oruro in Bolivia and Puno in Peru. Except as noted, the following account is abstracted from Steppan (1995) and Hershkovitz (1962), which should be consulted for additional details on morphological characters. The genus is not known from the fossil record. Phylogenetic analysis based on morphological characters placed *Galenomys* as the sister to *Auliscomys* (Steppan 1995). This topology was generally supported by Braun (1993), but she erected *Maresomys* for *Auliscomys boliviensis* based on its sister-group relationship with *Galenomys*, thus creating an otherwise paraphyletic *Auliscomys* (see Steppan 1995, and account for *Auliscomys*).

Galenomys is characterized by the following combination of characters: medium size (head and body length ranging from 105 to 132 mm) with a stout body, short tail (30 to 44 mm [Pearson 1957; S. Anderson 1997]), and pale color. The upper parts are buffy, thinly lined with brown; the sides are clearer; and the under parts and legs are sharply defined white, the individual hairs wholly white on legs, feet, and hands, and from the chin to the anterior part of chest, but basally gray elsewhere. The ears are large, well haired on their outer sides. The forefeet and hindfeet are white above. The plantar surface of the hindfeet is thinly covered with long white hairs except on the tubercles and terminal phalanges; the thenar pad is absent, and the remaining plantar pads are unfused. The tail is relatively short, only one-third to one-half of the combined head and body length. It is thickly haired white above and below, the scales are not visible, and a terminal tuft is not conspicuous.

The dorsal shape of the skull in lateral profile is strongly arched in the supraorbital region with the rostrum inflected downward; the interorbital region is narrow with squared but not beaded or ridged edges; zygomatic arches are delicate with a pronounced downward inflection; the zygomatic plates are nearly as wide as is the least interorbital breadth, with their anterior borders straight and slanted backward; the palate is arched; the incisive foramina are long, reaching to the middle of the M1s; the mesopterygoid fossa is narrow; and the bullae are moderately in-

flated, with its length approximating the alveolar length of the maxillary toothrow. A large stapedial foramen is present, an internal groove is present on the alisphenoid and squamosal bones, and a sphenofrontal foramen is present (all indicators of pattern 1 [of Voss 1988] to the carotid circulation), and an alisphenoid strut is present. The upper incisors are delicate, asulcate, and orthodont; the lower incisors are highly procumbent. M1 and m1 both have four roots, m3 has three. The crowns of each tooth are low and slightly crested, with ovate cusps. M2 and M3 are square in shape (as long as wide). A distinct anteromedian flexus is present on M1, a reduced hypoflexus on M3, and a highly reduced mesoflexid on m3. The axial skeletal elements consist of 13 ribs, 19 thoracolumbar vertebrae, and 17–19 caudal vertebrae.

SYNONYMS:

Galenomys Thomas, 1916b:143; type species *Euneomys* (*Galenomys*) *garleppii* Thomas, by original designation, as subgenus.

[*Galenomys*]: Thomas, 1926b:317; elevation to genus.

REMARKS: Thomas (1916a) originally regarded *Galenomys* as a subgenus of *Euneomys* but later (Thomas 1926b) raised the taxon to generic status. Ellerman (1941), Osgood (1947), and Corbet and Hill (1991) treated it as a subgenus of *Phyllotis*. Hershkovitz (1962) and Steppan (1995) revised *Galenomys*, considering it to be a distinct genus. Pearson (1958) likewise regarded *Galenomys* separate from *Phyllotis* in his revision of the latter.

Galenomys garleppii (Thomas, 1898)
Garlepp's Pericote

SYNONYMS:

Phyllotis (?) *Garleppii* Thomas, 1898c:279; type locality "Esperanza, a 'tambo' in the neighbourhood of Mount Sahama," La Paz, Bolivia.

Euneomys (*Galenomys*) *garleppi*: Thomas, 1916a:143; name combination.

Galenomys garleppii: Gyldenstolpe, 1932a:96; first use of current name combination.

Phyllotis [(*Galenomys*)] *garleppii*: Ellerman, 1941:455; name combination.

DESCRIPTION: As for the genus.

DISTRIBUTION: *Galenomys garleppii* is known from seven localities in the Altiplano of northern Bolivia and southern Peru, all between 3,300 and 4,650 m.

SELECTED LOCALITIES (Map 274): BOLIVIA: La Paz, Esperanza (type locality of *Phyllotis Garleppii* Thomas); Oruro, Huancaroma (S. Anderson 1997). PERU: Puno, Hacienda Pichupichuni, 6 km NW of Huacullani (Pearson 1957), Pampa de Ancomarca, 123 km W of Ilave (Pearson 1957).

SUBSPECIES: *Galenomys garleppii* is monotypic.

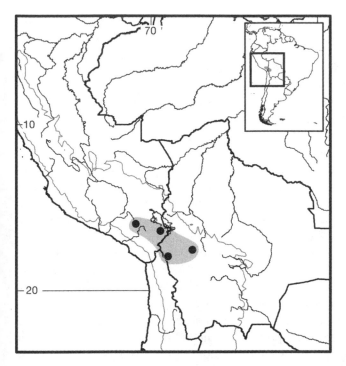

Map 274 Selected localities for *Galenomys garleppii* (●). Contour line = 2,000 m.

NATURAL HISTORY: *Galenomys garleppii* is very poorly known in all aspects of its ecology and behavior. All known localities are in bunch grass dominated Altiplano. Pearson (1957) obtained the two Peruvian specimens in heavily grazed pampa dominated by dwarf grass and prostrate forbs with scattered thorn bushes about 1 foot tall. His unpublished field notes (O. P. Pearson, MVZ archives) suggest that this species might be diurnal. No information is available on food habits. In Bolivia, one female had four embryos in December, and four young were taken in December (S. Anderson 1997).

REMARKS: Both Hershkovitz (1962) and Musser and Carleton (2005) included northern Chile in the range of this species, but no localities in this country from which voucher species exist have been reported. However, given the proximity of both Bolivian and Peruvian localities to the Altiplano of northeastern Chile, it is quite possible that *Galenomys garleppii* will be found there.

Genus *Graomys* Thomas, 1916

Janet K. Braun and James L. Patton

Members of the genus *Graomys* range from about 17°S (south central Bolivia) southward through western Paraguay and adjacent Brazil to about 49°S in southern Argentina (Hershkovitz 1962; Udrizar Sauthier, Formoso et al. 2011); the genus is not known to occur west of the Andes in Chile. Species in the genus generally occupy arid and semi-arid habitats, and occasionally transitional forest, from sea level to about 3,100 m. The following generic description is, except as noted, compiled from information presented by Gyldenstolpe (1932), Ellerman (1941), Hershkovitz (1962), and Steppan (1995).

All species of *Graomys* are externally similar, with large ears, long, lax fur, and well-haired, slightly penicillate, bi-colored tails that are usually longer than head and body. The venter is white with hairs white terminally, and white or gray basally. The hindfeet are unwebbed, with digit I distinctly shorter than digit V; the plantar surface is naked (from heel to toes) with six small, but well-defined tubercles (thenar, hypothenar, and four interdigitals). There are eight mammae (in four pairs).

The skull, in dorsal view, has a relatively broad and characteristic interorbital region that is evenly convergent V-shaped; supraorbital margins are generally beaded and project as ledges. Lacrimals are conspicuous. Zygomatic arches are usually parallel sided or slightly convergent. Zygomatic plates have concave anterior margins, resembling those of *Reithrodon*, and zygomatic notches are deep and well developed. Frontals are relatively broad and the interparietal is well developed. The orbital wings of the presphenoid are posterior to the maximum constriction. The roof of the mesopterygoid fossa has large sphenopalatine vacuities. The parapterygoid fossae are flat and about even with the bony palate. A vertical stout strut of the alisphenoid is present. Bullae are moderately to well developed; the stapedial spine is laterally appressed to the bullae. The tegmen tympani contacts a posterior suspensory process of the squamosal; the hamular process of the squamosal is well developed, long, and distally spatulate; postglenoid foramen is at least twice as large as the subsquamosal fenestra. The coronoid process of the mandible is not well developed. The lower incisor root is contained in a moderately developed capsular process. The condyle is well projected and the angular notch is characteristically excised forward.

Incisors are ungrooved and opisthodont. The molars are tetralophodont (lacking any trace of a mesoloph or mesolophid), moderately hypsodont, and flat-crowned to slightly raised. Molar rows are parallel sided. Cusps are opposite in position. The procingulum of M1 lacks an anteromedian flexus. The anterolabial cingulum of m1 is distinct and large, and the mesoflexid of m3 is not shifted anteriorly relative to m2. M1 and m1 have four roots.

Axial skeletal elements have counts of 12–13 ribs, 19 thoracolumbar vertebrae, and 31–35 caudal vertebrae. The tuberculum of the first rib articulates with the transverse processes of the seventh cervical and the first thoracic vertebrae, and the second thoracic vertebra has a greatly elongated neural spine. The entepicondylar foramen of the humerus is present.

The stomach is unilocular and hemiglandular (Carleton 1973), and a gall bladder is present, the widespread condition for the tribe. Male accessory secretory organs consist of one pair each of bulbourethral, anterior, and dorsal prostate, ampullary, and vesicular glands, and two pairs each of preputial and dorsal prostate glands (Voss and Linzey 1981). The glans penis is complex; three bacular mounds are present, the medial one truncate and the lateral pair slightly double-peaked; hooks are absent. Two pairs of preputial glands are present, a single large lateral pair and a medium length medial pair.

The genus *Graomys* is known from the late Pliocene of Argentina and early Holocene of Bolivia. Reig (1978) described *G. dorae*, a fossil species he considered similar to *G. domorum*, from the Chapadmalal Formation (Upper Pliocene) of southeastern Buenos Aires province. Fossils referred to the living species *G. griseoflavus* have been identified from Middle to Late Pleistocene (Ensenadan, Bonarian and Lujanian) deposits in Buenos Aires province (Burmeister 1879, as *Hesperomys bravardi*; see Hershkovitz 1962; Reig 1978; Pardiñas 1995c; Pardiñas et al. 1996; and Voglino and Pardiñas 2005).

SYNONYMS:

Mus (*Phyllotis*): Waterhouse, 1837; part (description of *griseoflavus*); neither *Mus* Linnaeus nor *Phyllotis* Waterhouse.

Hesperomys: Waterhouse, 1839:76; part (listing of *griseoflavus*); not *Hesperomys* Waterhouse.

Hesperomys (*Phyllotis*): Wagner, 1843a:510; part (listing of *griseoflavus*); neither *Hesperomys* Waterhouse nor *Phyllotis* Waterhouse.

Hesperomys [(*Calomys*)]: Burmeister, 1879:219; part (listing of *griseoflavus*); neither *Hesperomys* Waterhouse nor *Calomys* Waterhouse.

Bothriomys Ameghino, 1889:118; type species *Bothriomys catenatus*, by original designation; *nomen oblitum* (see Remarks).

Eligmodontia: Thomas, 1898h:210; part (listing of *griseoflava*); not *Eligmodontia* F. Cuvier.

Graomys Thomas, 1916a:141; type species *Mus* (*Phyllotis*) *griseoflavus* Waterhouse, by original designation; *nomen protectum* (see Remarks).

REMARKS: Hershkovitz (1962) revised the genus *Graomys*, assigning nine nominal taxa to two species, the monotypic and questionably valid *G. edithae* Thomas and the highly polytypic and widespread *G. griseoflavus*, with two subspecies, the nominotypical form plus *domorum* Thomas. Cabrera (1961) also recognized two species, *G. domorum* (with three subspecies, including *lockwoodi* Thomas and *taterona* Thomas in addition to the nominotypical form) and *G. griseoflavus* (with five subspecies [*cachinus* J. A. Allen, *centralis* Thomas, *chacoensis* J. A. Allen, *griseoflavus* Waterhouse, and *medius* Thomas], with *edithae* Thomas listed as a synonym of the latter).

The composite nature of Hershkovitz's concept of *G. griseoflavus*, however, has been demonstrated by a series of studies documenting karyotypic polytypy that, together with cross-breeding experiments, allozymes, mtDNA sequences, and morphometric analyses diagnose at least three geographically segregated species: *G. chacoensis*, *G. domorum*, and *G. griseoflavus*. We provisionally include *G. edithae* Thomas as a fourth species, pending an adequate review of the holotype and other specimens from the vicinity of the type locality. We emphasize that, although the geographic boundaries of these taxa have become reasonably well understood in Argentina (e.g., J. J. Martínez, Krapovickas, and Theiler 2010; J. J. Martínez and Di Cola 2011; Udrizar Sauthier, Formoso et al. 2011), few of the recent studies substantiating species boundaries in the genus have treated samples from Bolivia or Paraguay. Most localities provided in the following accounts are based on specimens for which karyotype and/or DNA sequence data are available.

Pardiñas (1995c) reviewed the available fossil material of *Bothriomys catenatus* Ameghino, from Pleistocene deposits in Córdoba province, Argentina and identified its synonymy with *Graomys* Thomas. Hershkovitz (1962) had earlier listed *Bothriomys* as a junior synonym of *Euneomys* Coues without having seen the referred specimens (pg. 496, 498). Citing Article 23(b) of the International Code of Zoological Nomenclature (ICZN 1985), Pardiñas (1995c:177) argued that *Bothriomys* be suppressed in favor of *Graomys* because (1) *Bothriomys* was based on fragmentary material (a partial left mandible with incisor and molars in place), with limited useful characters, and (2) *Bothriomys* had not been used as a valid name in the prior 50 years whereas *Graomys* had been cited by a minimum of five authors in 10 publications during the same period. Pardiñas thus concluded that replacement of *Graomys* by Ameghino's earlier but unknown *Bothriomys* would result in nomenclatural instability. The rationale provided by Pardiñas (1995c:177) satisfies both Articles 23.9.1 and 23.9.2 of the current Code (ICZN 1999) in his choice of the younger name *Graomys* as valid relative to *Bothriomys*. The latter would thus become a *nomen oblitum*, the former a *nomen protectum* (see also Musser and Carleton 2005:1116–1117), although a formal declaration from the Commission is still required.

KEY TO THE SPECIES OF *GRAOMYS*:

1. Tail without a well-defined tuft of hairs; dorsal coloration olivaceous to rufous; hairs of venter dark basally; bullae smaller, length <6 mm . . . .*Graomys domorum*

1'. Tail with a well-defined tuft of hairs; dorsal coloration buffy to tawny; hairs of venter white to whitish basally; bullae larger, length >6 mm 2

2. Size small, averaging about 230 mm and always less than 265 mm; supraorbital edges without beading . . . .
. . . . .Graomys edithae

2'. Size larger, averaging about 260 mm and often more than 300 mm; supraorbital edges with beading. 3

3. Found in Monte desert and Patagonian steppe habitats from northwestern to southern Argentina; bullae larger; length of maxillary toothrow longer; width of M1 wider; 2n = 34–38 . . . . Graomys griseoflavus

3'. Found in Chaco and associated habitats from southern Bolivia (and adjacent Brazil), Paraguay, central Argentina; bullae smaller; length of maxillary toothrow shorter; width of M1 narrower; 2n = 42 . . . .
. . . . Graomys chacoensis

Graomys chacoensis (J. A. Allen, 1901)
Chaco Pericote

SYNONYMS:

Phyllotis chacoënsis J. A. Allen, 1901c:408; type locality "Waikthlatingwayalwa, Chaco boreal, [Presidente Hayes,] Paraguay." According to Thomas (1898f:272), "Waikthlatingwayalwa" is an old Chacoan mission, located at about 23°25'S and 58°10'W (Paynter 1989, as Waikthlatingwayalwa).

E[ligmodontia]. chacoensis: Thomas, 1902b:132; name combination.

Eligmodontia griseoflavus centralis Thomas, 1902d:240; type locality "Cruz del Eje," Córdoba Province, Argentina.

Graomys chacoensis: Thomas, 1916a:142; first use of current name combination.

Graomys griseoflavus centralis: Thomas, 1916a:142; name combination.

Graomys medius Thomas, 1919d:494; type locality "Chumbicha, Catamarca," Argentina.

Graomys griseoflavus chacoënsis: Gyldenstolpe, 1932a:91; name combination.

Phyllotis griseoflavus centralis: Cabrera, 1961:495; name combination.

Phyllotis griseoflavus chacoensis: Cabrera, 1961:495; name combination.

Phyllotis griseoflavus medius: Cabrera, 1961:495; name combination.

[Graomys] centralis: Tiranti, 1998:35; name combination.

Graomys centralis: Theiler, Gardenal, and Blanco, 1999:971; name combination.

DESCRIPTION: Similar in size to G. domorum, with dimensions of holotype including head and body length 142 mm, tail length 185 mm, hindfoot (with claw) 33 mm, and ear length 24 mm. Pelage thick and soft, dorsal color overall varying from yellowish brown to grayish drab, varied with black-tipped hairs; nose and top of head more grayish; and ventral color pure white to base of hairs. Ears uniformly brown, thinly hairs on both surfaces. Tail much longer than head and body, well haired, bicolored brown above and white below, and terminates in well-developed pencil or tuft of elongated hairs. Dorsal surfaces of forefeet and hindfeet white. Skull with bullae only slightly larger than those of G. domorum but conspicuously smaller than G. griseoflavus (mean length by width measurements 6.35 × 4.48 mm; J. J. Martínez, Krapovickas, and Theiler 2010). Cheek teeth notably small, and especially narrower than those of G. domorum (M1 breadth in holotype, for example, 1.5 mm compared with 1.8 mm in G. domorum); mean length of maxillary toothrow 5.35 mm (J. J. Martínez, Krapovickas, and Theiler 2010). Supraorbital ridges well marked, with rudimentary postorbital projections, and overhanging points of zygomatic plates well defined. Thomas (1902d) noted that his centralis was readily distinguished from G. griseoflavus by its small bullae and from G. domorum by its small molars.

DISTRIBUTION: Graomys chacoensis occurs through the Dry and Humid Chaco ecoregions, secondarily within the Espinal, of northern Argentina and Paraguay (see J. J. Martínez and Di Cola 2011). Hershkovitz (1962) assigned the single record of Graomys from Brazil (Mato Grosso do Sul state; Kühlhorn 1954) to his G. griseoflavus, but this specimen is likely G. chacoensis on geographic grounds and is so mapped here. We also include those specimens from the Chaco of southeastern Bolivia that S. Anderson (1997) allocated to G. g. griseoflavus under the assumption that only a single species of Graomys occured within the Chaco ecoregion (see Lanzone, Novillo et al. 2007; J. J. Martínez and Di Cola 2011).

SELECTED LOCALITIES (Map 275): ARGENTINA: Catamarca, Chumbicha, 0.5 km E of Rte 38 over Rte 60 (J. J. Martínez and Di Cola 2011), La Carrera (Zambelli et al. 1994); Chaco, 1 km E of Puente General Belgrano, Río Paraná (J. R. Contreras 1982b); Córdoba, Laguna Larga (Zambelli and Vidal-Rioja 1995), 10 km N of Santiago Temple (J. J. Martínez and Di Cola 2011), Yacanto (J. J. Martínez and Di Cola 2011); Formosa, 7 km N of junction of Rte 81 and 95 (J. J. Martínez and Di Cola 2011), 35 km S and 5 km E of Ingeniero Guillermo Juárez, Puesto Divisadero (J. J. Martínez and Di Cola 2011), Parque Nacional Río Pilcomayo (Heinonen Fortabat 2001), Reserva El Bagual (Pardiñas and Teta 2005); La Rioja, Guayapa, Patquía (J. J. Martínez and Di Cola 2011); San Luis, 23 km N of Rte 20, Pampa de las Salinas, near La Botija (J. J. Martínez and Di Cola 2011); Salta, Cabeza de Buey, Campo La Peña (J. J. Martínez and Di Cola 2011); Santa Fe, Berna (J. J. Martínez,

González-Ittig et al. 2010), Jacinto L. Arauz (Teta and Pardiñas 2010), Recreo (Pautasso 2008); Santiago del Estero, Buena Vista, 15 km NE of Villa Ojo de Agua on Hwy 13 (J. J. Martínez and Di Cola 2011), 6 km S and 2 km E of Pampa de los Guanacos (J. J. Martínez and Di Cola 2011); Tucumán, San Pedro de Colalao (J. J. Martínez and Di Cola 2011). BOLIVIA: Chuquisaca, Tomina (S. Anderson 1997); Santa Cruz, 7 km E and 3 km N of Ingeniero Mora (S. Anderson 1997); Tarija, 35 km by road SE of Villa Montes, Taringuiti (S. Anderson 1997). BRAZIL: Mato Grosso do Sul, Ivinheima (Hershkovitz 1962). PARAGUAY: Alto Paraguay, Puerto Casado [= Puerto La Victoria] (Hershkovitz 1962); Boquerón, 420 km NW (by road) of Villa Hayes (MVZ 145247); Presidente Hayes, Waikthlatingwaialwa (type locality of *Phyllotis chacoensis* J. A. Allen).

SUBSPECIES: We treat *Graomys chacoensis* as monotypic.

NATURAL HISTORY: *Graomys chacoensis* occupies the Chacoan and Espinal ecoregions (Myers 1982; Tiranti 1998b; J. J. Martínez and Di Cola 2011) where it may be trapped in a variety of habitats such as woodland, grassland, monte, and marginal areas. It is generally uncommon everywhere (Polop et al. 1985; Polop 1991; Polop and Sabattini 1993). Like most species of *Graomys*, *G. chacoensis* is scansorial, nocturnal, and predominately herbivorous, feeding on grasses and seeds (Hershkovitz 1962). Theiler, Ponce et al. (1999) summarized information on reproductive barriers between *G. centralis* and *G. griseoflavus*. In La Rioja province, pregnant females and reproductively active males were collected in the fall; males and females were not reproductively active in the winter (J. K. Braun, unpubl. data). Because the distributional limits for this species are unclear, it is not possible to determine whether some published research on ecology and natural history pertains to this species or to *Graomys griseoflavus* (e.g., Polop et al. 1982). However, a large study on development and reproduction based on a lab colony surely applies to this species, as it is now understood (Puig and Nani 1981; cited as *G. griseoflavus*). Venzal et al. (2012) recorded the tick *Ornithodoros quilinensis* from specimens identified as *G. centralis* from the Argentine Chaco.

REMARKS: Recognition of this taxon as a species distinct from *G. griseoflavus* is substantiated by allozymes, karyotypes, mtDNA sequences, multivariate morphometrics, and cross-breeding experiments (Wainberg and Fronza 1974; Theiler and Gardenal 1994; Zambelli et al. 1994; Zambelli and Vidal-Rioja 1995; Theiler and Blanco 1996a,b; Theiler, Gardenal, and Blanco 1999; Theiler, Ponce et al. 1999; Catanesi et al. 2002, 2006; Lanzone, Novillo et al. 2007; Ferro and Martínez 2009; J. J. Martínez, González-Ittig et al. 2010; J. J. Martínez, Krapovickas, and Theiler 2010;

J. J. Martínez and Di Cola 2011). Most of these studies employed *centralis* Thomas as the name for this taxon (see also M. M. Díaz et al. 2006) because topotypes from its type locality (Cruz del Eje, Córdoba province) had the 2n = 42 karyotype (e.g., Lanzone, Novillo et al. 2007; J. J. Martínez, Krapovickas, and Theiler 2010); Tiranti (1998:35) was the first to connect the 2n = 42 karyomorph with the name *centralis*. However, Lanzone, Novillo et al. (2007) suggested that *chacoensis* J. A. Allen was the oldest name applicable because specimens from near its type locality in Paraguay had the same 2n = 42 karyotype. Ferro and Martínez (2009) and J. J. Martínez and Di Cola (2011) showed that *centralis* Thomas was best considered a synonym of *chacoensis* J. A. Allen, based both on molecular and morphometric analyses.

The assignment of *medius* Thomas as a synonym of *G. chacoensis* is also based on both karyotypic and morphometric grounds (Lanzone, Novillo et al. 2007), a decision supported by the descriptive details of color and smaller cranial features provided by Thomas (1919d; compare with data in J. J. Martínez, Krapovickas, and Theiler 2010). However, in southern La Rioja province, where the type locality (Chumbicha) of *medius* is placed, both 2n = 42 and 2n 36–38 diploid complements have been recorded and overlap

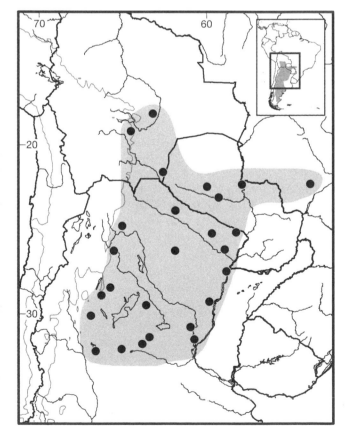

Map 275 Selected localities for *Graomys chacoensis* (●). Contour line = 2,000 m.

in transitional Chaco-Monte areas (e.g., Theiler and Blanco 1996a,b; Tiranti 1998). Moreover, the Chumbicha local environment sampled by E. Budin (the collector of the type series of both *edithae* and *medius*) has changed sharply in the last century due to overgrazing and wood extraction. Consequently, and especially considering the substantial variability among the series of *Graomys* from this locality recognized by Thomas (1919d:494–495), the hypothesis that *medius* Thomas is junior synonym of *chacoensis* J. A. Allen must be confirmed by direct study of the holotypes. The reasonable assumption that only a single species of *Graomys* occupies the Chacoan biome, and thus our assignments of both Bolivian and Brazilian localities to *G. chacoensis*, also requires verification.

The karyotype is $2n = 41-42$, $FN = 44-46$, with 17–18 pairs of uniarmed and two to three pairs of biarmed autosomes, a large metacentric X and small acrocentric Y chromosome. *Graomys chacoensis* has a significantly larger genome size than the $2n = 34-38$ karyotypes of *G. griseoflavus* (P. B. Ramirez et al. 2001).

Graomys domorum (Thomas, 1902)

Pale Pericote

SYNONYMS:

Eligmodontia domorum Thomas, 1902b:132; type locality "Tapacari. Altitude 3000 m," Cochabamba, Bolivia.

E[ligmodontia]. g[riseoflavus]. domorum: Thomas, 1902d:240; name combination.

Graomys domorum: Thomas, 1916a:142; first use of current name combination.

Phyllotis (Graomys) domorum: Osgood, 1916:207; name combination.

Graomys lockwoodi Thomas, 1918b:187; type locality "Manuel Elordi, Vermejo, Salta, alt. 500 m," Argentina.

Graomys taterona Thomas, 1926c:320; type locality "Tablada, 2000 m," Tarija, Bolivia.

Graomys griseoflavus domorum: Gyldenstolpe, 1932a:91; name combination.

Phyllotis domorum domorum: Cabrera, 1961:494; name combination.

Phyllotis domorum lockwoodi: Cabrera, 1961:494; name combination.

Phyllotis domorum taterona: Cabrera, 1961:494; name combination.

Phyllotis griseoflavus domorum: Hershkovitz, 1962:458; name combination.

Phyllotis griseoflavus griseoflavus: Hershkovitz, 1962:452; part; inclusion of *taterona* Thomas as synonym.

Graomys domorum taterona: Reig, 1978:180; name combination.

Graomys domorum domorum: Reig, 1978:182; name combination.

Graomys domorum lockwoodi: Reig, 1978:182; name combination.

DESCRIPTION: Similar in size to *G. chacoensis*, with head and body length 102–150 mm, tail length 140–170 mm, hindfoot length 26–32 mm, and ear length 20–27 mm (Hershkovitz 1962:Table 62). Overall dorsal color olivaceous (as in juveniles of *G. griseoflavus*) to rufous; sides clear buffy; hairs of ventral surface dark basally, but hairs of throat white to their base. Ears finely haired, grayish brown. Upper surfaces of forefeet and hindfeet white as in other species of genus. Tail long and prominently bicolor, dark brown above and white beneath; tail tip differs from other species by being much less tufted terminally, with pencil rarely longer than 4 mm. Skull similar to congeners except that bullae conspicuously small, with bullar length (excluding bony tubes) 6.1 mm or less, approximately equal to or less than length of upper molar toothrow.

DISTRIBUTION: *Graomys domorum* occurs in Yungas and transitional forests at elevations from 600 to 2,700 m east of the Andes in south-central Bolivia and northwestern Argentina.

SELECTED LOCALITIES (Map 276): ARGENTINA: Jujuy, Laguna La Brea (M. M. Díaz and Barquez 2007), Maimará (M. M. Díaz and Barquez 2007); Salta, 20 km W of General Ballivián, on Puerto Baulés road (Mares, Ojeda, and Kosco 1981), 25 km SE of La Viña (OMNH 33334), Manuel Elordi, Vermejo (type locality of *Graomys lockwoodi* Thomas); Tucumán, El Cadillal, Estación de Piscicultura (Barquez et al. 1991), Las Tipas, Parque Biológico Sierra de San Javier (Capllonch et al. 1997). BOLIVIA: Cochabamba, Tapacari (S. Anderson 1997); Potosí, 30 mi WNW of Cotagaita (MVZ 119979), Yuruma (MVZ 119981); Santa Cruz, Comarapa (S. Anderson 1997); Tarija, 2 km S and 5 km E of Palos Blancos (S. Anderson 1997), Tablada (type locality of *Graomys taterona* Thomas).

SUBSPECIES: S. Anderson (1997) distinguished two subspecies in Bolivia, the nominotypical form from Cochabamba, Chuquisaca, northern Potosí, and Santa Cruz departments, and *G. d. taterona* from southern Potosí and Tarija, and presumably extending south into northern Argentina. Capllonch et al. (1997) regarded populations from northern Argentina, especially those from Tucumán and Salta province, to represent *G. d. lockwoodi*. To date, however, there has been no evaluation of character variation of morphological or other traits to establish the validity of any of these races.

NATURAL HISTORY: Little natural history information is available for this species. In Bolivia, it occurs in transitional forest at low to middle elevations. Reproductive activity apparently spans the summer (S. Anderson 1997). In Argentina, animals were captured in Yungas and transitional forest and were both reproductively active and inactive during the spring (Mares, Ojeda, and Kosco 1981). S. Anderson (1997) summarized data on ectoparasites from Bolivian specimens.

REMARKS: Hershkovitz (1962) regarded *domorum* Thomas as a valid subspecies of *G. griseoflavus* (Waterhouse), allocating both *lockwoodi* Thomas and *taterona* Thomas to the nominotypical race based on the intermediate size of the bullae and geographic range. S. Anderson (1997), following Cabrera (1961), placed *taterona* with *G. domorum* and recognized two subspecies, but commented that additional research was needed to assess their systematic relationship. Cabrera (1961) and S. Anderson (1997) also allocated *lockwoodi* Thomas as a synonym of *G. domorum taterona*. Capllonch et al. (1997), however, regarded *lockwoodi* Thomas as a valid subspecies of *G. domorum*. Pending an adequate revision of this species, one that should include examination of holotypes and topotypic series, both *lockwoodi* Thomas and *taterona* Thomas are included herein within the synonymy of *G. domorum*, following Capllonch et al. (1997) and M. M. Díaz et al. (2006). It is interesting to note that Thomas (1902b:133) remarked on the "extraordinary resemblance" between *G. domorum* and *Tapecomys wolffsohni* in external morphology.

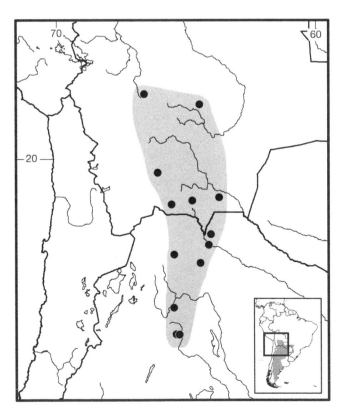

Map 276 Selected localities for *Graomys domorum* (●). Contour line = 2,000 m.

Pearson and Patton (1976) described a karyotype with $2n=28$, FN=46, consisting of 10 pairs of biarmed and three of uniarmed autosomes, and both uniarmed X and Y chromosomes for specimens from Cochabamba, Bolivia. They concluded (p. 341) that this karyotype was so different from the $2n=36$–38 karyotype of *G. griseoflavus* they described that "there is little doubt that Thomas's (1902b) separation of *domorum* as a separate species on the basis of less hairy tail, less inflated bullae, and underparts with gray-based hairs was warranted, and lumping by Hershkovitz (1962:448) unjustified." J. J. Martínez and Di Cola (2011) included a small series (three specimens) of *G. domorum* in their analysis of geometric morphometric differentiation of *Graomys* samples, documenting that *domorum* was indeed phenetically distinct from both *G. chacoensis* and *G. griseoflavus*. Although the species status of *G. domorum* relative to the more broadly distributed *G. griseoflavus* seems incontrovertible, given their morphological and karyological differences, delimitation of the ranges of each in southern Bolivia and northern Argentina remains to be firmly understood.

Graomys edithae Thomas, 1919
Otro Cerro Pericote

SYNONYMS:

Graomys edithae Thomas, 1919d:495; type locality "Otro Cerro," Catamarca Province, Argentina (= Otro Cerro, 28°45′S, 66°17′W, 2,023 m, Capayán, Catamarca, about 4 km SSE of Cerro Catalán; Pardiñas et al. 2007).

Graomys griseoflavus medius: Cabrera, 1961:495; *edithae* Thomas listed as synonym.

DESCRIPTION: Supposedly smallest species in genus; measurements of holotype head and body length 108 mm, tail length 127 mm, hindfoot 25 mm, ear length 20 mm, greatest skull length 28.5 mm, condyloincisive length 26.5 mm, zygomatic breadth 15.0 mm, nasal length 10.5 mm, interorbital breadth 4.5 mm, braincase breadth 13.5 mm, palatal length 12.8 mm, incisive foramina length 6.7 mm, bullar length 6.0 mm, and upper molar series 4.7 mm (Thomas 1919d). Dorsal color similar to that of *G. griseoflavus*, but without buffy wash on sides of latter. Under parts white, as in *G. centralis*, lacking slaty base characteristic of *G. domorum* and some populations of *G. griseoflavus*. Tail absolutely and proportionately shorter than *G. griseoflavus*, less heavily haired terminally but bicolored, brown above and white below. Skull a miniature of those of other species but notable for absence of beading on supraorbital edges.

DISTRIBUTION: *Graomys edithae* is currently known from two nearby localities in Catamarca province, Argentina, at elevations of 400 and 2,000 m.

SELECTED LOCALITIES (Map 277): ARGENTINA: Catamarca, Chumbicha (Thomas 1919d), Otro Cerro (type locality of *Graomys edithae* Thomas).

SUBSPECIES: *Graomys edithae* is monotypic.

NATURAL HISTORY: Nothing is known of the natural history of this species. However, the habitat around Otro Cerro, the type locality, is grassland at the top of the Cerro de Ambato.

REMARKS: Although *G. edithae* is currently recognized as distinct species, its taxonomic status should be considered uncertain pending systematic revision, including the review of the type series and the collection of additional specimens from the type locality and vicinity (Musser and Carleton 2005; Pardiñas et al. 2007). *Graomys edithae* has, judging from the holotype, an almost perfect coincidence in cranial traits with the other species in the genus, as Thomas (1919d) initially remarked. In contrast, in external morphology, the holotype of *G. edithae* embraces several traits that depart from congeners, including a very short ear and very long vibrissae. Additional diagnostic traits may include more opisthodont upper incisors, a wide U-shaped anterior border of the mesopterygoid fossa that lacks a median process, and a unique morphology of the presphenoid.

Thomas (1919d:495) wrote "this interesting little *Graomys* agrees with the larger species in all the essential characters of the group." The type is an old male with worn teeth. In that paper, Thomas recognized three species of *Graomys* from Catamarca province, one he referred to *G. cachinus* J. A. Allen (junior synonym of *G. griseoflavus*), the largest of the three with a skull length between 33.5 and 35.0 mm; the second he named *G. medius* (also now junior synonym of *G. chacoensis*; see Remarks in that account), similar to *G. cachinus* but slightly smaller in size (skull length of holotype, an "adult male" 31.2 mm), and differing from it by less development of buffy band on sides, tail shorter and less tufted, smaller skull with shorter nasals, sharply angular supraorbital edges but with less distinct beads, and smaller bullae; and the third *G. edithae* based on a single old adult male with skull length 28.5 mm.

Cabrera (1961:495) placed *edithae* Thomas as a junior synonym of *Graomys griseoflavus medius* Thomas. Hershkovitz (1962:455, 461) retained *edithae* as a valid species, being impressed by both the small size and lack of supraorbital beading of the holotype. D. F. Williams and Mares (1989:216) regarded the holotype as notable for "the narrow procingulum of M1 . . . which is deeply divided from the protocone-metacone by deep anterior median folds, and the deep major fold on the M3" but noted that "the paratypes of *G. edithae* (from near Chumbicha, Catamarca, at an elevation nearly 2,000 m lower than the type locality), represent, in our opinion, *G. griseoflavus medius*" (= *G. chacoensis* herein). Pardiñas et al. (2007) clarified the location of the type locality as Otro Cerro, 28°45'S,

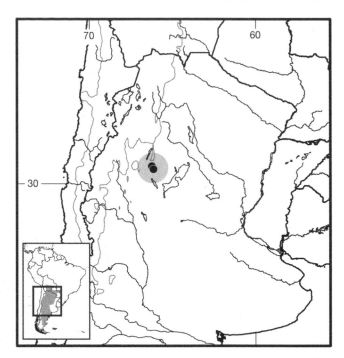

Map 277 Selected localities for *Graomys edithae* (●). Contour line = 2,000 m.

66°17'W, 2,023 m, about 4 km SSE of Cerro Catalán, Capayán, Catamarca province, Argentina.

Graomys griseoflavus (Waterhouse, 1837)
Common Pericote
SYNONYMS:

Mus (*Phyllotis*) *griseo-flavus* Waterhouse, 1837:28; type locality "Rio Negro," Argentina; restricted to mouth of the Río Negro, Río Negro Province, Argentina (see Hershkovitz 1962:453).

Mus griseo-flavus: Waterhouse, 1839:62; name combination.

Phyllotis griseoflavus: Gray, 1843b:112; name combination.

Eligmodontia griseoflava: Thomas, 1898h:210; name combination.

Phyllotis cachinus J. A. Allen, 1901c:409; type locality "Upper Cachi River, Salta Province, Argentina."

Graomys griseoflavus: Thomas, 1916a:142; first use of current name combination.

Graomys cachinus: Thomas, 1916a:142; name combination.

Graomys griseoflavus cachinus: Gyldenstolpe, 1932a:91; name combination.

Graomys griseoflavus griseoflavus: Gyldenstolpe, 1932a:91; name combination.

Phyllotis griseoflavus cachinus: Cabrera, 1961:494; name combination.

Phyllotis griseoflavus griseoflavus: Cabrera, 1961:495; name combination.

DESCRIPTION: Waterhouse (1837) provided brief review of relevant morphological details in original description, and wrote a more expansive description two years later (Waterhouse 1839). Although Waterhouse noted (1839:62) "tail rather shorter than the head and body taken together," Hershkovitz (1962:453) argued that the "original measurements for combined head and body length (6" 8'" = 169 mm) and for tail length (5" 6'" = 140 mm)" of the holotype were incorrect, likely transposed, and in specimens he examined the "tail is always longer than combined head and body length." Of samples for which Hershkovitz (1962:Table 61) provided external measurements, those clearly applicable to *G. griseoflavus* as understood herein range in head and body length 118–169 mm, tail length 134–171 mm, hindfoot length 27–31 mm, and ear length 23–25 mm. General dorsal color grayish yellow, less bright and buffy than *G. chacoensis*; ventral color white, but belly hairs typically gray at base rather than white throughout. Tail, like that of *G. chacoensis*, longer than head and body, bicolor, dark above and pale below, and well haired, with hairs increasing in length toward end, forming slight to bushy pencil at tip. Most craniodental dimensions average larger than their counterparts in *G. chacoensis* (J. J. Martínez, Krapovickas, and Theiler 2010), most notably in basilar length, incisive foramina length, post-palatal length, and bullar dimensions. Maxillary toothrow ranges 5.0–6.1 mm (Hershkovitz 1962:Table 61). Rossi (1983) reported morphometric data for confirmed *G. griseoflavus* populations from Mendoza province, and Carrizo and Díaz (2011) described postcranial skeletal anatomy.

DISTRIBUTION: *Graomys griseoflavus* occurs east of the Andes primarily in Monte, Espinal, and Patagonian steppe ecoregions of Argentina. J. J. Martínez and Di Cola (2011) mapped point localities for northwestern Argentina (Catamarca, Mendoza, Salta, San Juan, San Luis, and Tucumán provinces) and Udrizar Sauthier, Formoso et al. (2011) detailed more than 150 locality records in Patagonia (southern Buenos Aires, Chubut, Neuquén, Río Negro, and Santa Cruz provinces), where *G. griseoflavus* deeply penetrates the Patagonian Province along the major watercourses that dissect the region, but is absent from highland areas.

SELECTED LOCALITIES (Map 278; from Udrizar Sauthier, Formoso et al. 2011, except as noted): ARGENTINA: Buenos Aires, Bahía San Blas, Médanos (Theiler and Blanco 1996a); Catamarca, 4.6 km SW, 0.7 km N of Belén (OMNH 34926); Chubut, Comodoro Rivadavia (V. A. Rodríguez 2005), Escuela N 859, Fofo Cahuel, Península Valdez (J. J. Martínez and Di Cola 2011), Puerto Piojo, Rawson (Hershkovitz 1962), 17 km W of Sarmiento; Jujuy, La Quiaca (MACN 36.802); La Pampa, 12 km NNE of Naicó, Estancia Los Toros (Tiranti 1998b); Mendoza, near Divisadero Largo (J. J. Martínez and Di Cola 2011); Neuquén, 1 km SE of bridge RNN 840 on Río Neuquén, Sierra del Portezuelo; Río Negro, El Espigón, 29 km S of El Cóndor on RPN 1, Las Grutas; Salta, ca. 3 km N of Cachi Adentro (OMNH 33335); San Juan, Castano Nuevo, 9 km NW of Villa Nueva (J. J. Martínez and Di Cola 2011), Tudcum, Nacedero (J. J. Martínez and Di Cola 2011); San Luis, 11 km W of Buena Esperanza (IADIZA 5439), Papagayos (J. J. Martínez and Di Cola 2011); Santa Cruz, Estancia La María, Piedra Clavada Sur, Port Desire (Hershkovitz 1962); Tucumán, Hualinchay (J. J. Martínez and Di Cola 2011).

SUBSPECIES: We regard *Graomys griseoflavus* as monotypic, despite a large geographic range that encompasses several different ecoregions within Argentina. Of the five subspecies recognized by Cabrera (1961), three are names now regarded as synonyms of *G. chacoensis* (*centralis* Thomas, *chacoensis* J. A. Allen, and *medius* Thomas). If a thorough review of geographic trends in morphology eventually supports the recognition of races, *cachinus* Thomas would be an available name for northern populations with the nominotypical form applicable to southern ones.

NATURAL HISTORY: *Graomys griseoflavus* has been collected in diverse habitats, including Espinal, Monte Desert, Patagonian steppe, and areas of cultivation and secondary vegetation (e.g., Mares, Ojeda, and Kosco 1981; Rosi 1983; Mares, Ojeda, Braun et al. 1997; Monjeau et al. 1997; Yahnke 2006). This species is nocturnal, with individuals using both horizontal terrestrial and vertical arboreal dimensions (Albanese et al. 2011). In the Monte Desert, this species differentially selected microhabitats with high cover of litter, subshrubs, or shrubs and trees in mesquite forest, and those with high cover in creosote bush communities (Corbalán 2006). Reproductive activity generally occurs during the spring, summer, and fall (Mares, Ojeda, and Kosco 1981; Mares, Ojeda, Braun et al. 1997; Yahnke 2006). In the Monte Desert, this species has been categorized as an omnivore with a strong tendency toward folivory (C. M. Campos, Ojeda et al. 2001) and a herbivore (Giannoni et al. 2005). Giannoni et al. (2011) reported food caching behavior. *Graomys griseoflavus* is scansorial and exhibits quadrupedal saltation as an escape mode (Taraboreli et al. 2003). Individuals are prey of Geoffroy's Cat (*Leopardus geoffroyi*; Bisceglia et al. 2011), the Burrowing Owl (*Athene cunicularia*; Nabte et al. 2008), and the Barn Owl (*Tyto alba*; Pardiñas, Teta et al. 2003). S. Anderson (1997) summarized information about ectoparasites for this species in Bolivia; Rojas and Digiani (2003) and Digiani and Durette-Desset (2003) described nematodes; and Schramm and Lewis (1988) and Lareschi et al. (2003, 2004) reported on fleas and mites for this species in Argentina.

REMARKS: As noted by J. A. Allen (1901c:409), "doubtless *griseoflavus*, *chacoensis*, and *cachinus* will prove to be closely related forms of the same species, but they show features of difference worthy of note, and, moreover, inhabit quite different faunal districts. The first two are chaco forms, widely separated geographically, while the latter is from the mountain district northwest of Salta."

Cabrera (1961:495) included *Graomys hypogaeus* Cabrera, 1934, in the synonymy of *Phyllotis griseoflavus medius* (Thomas), but Teta, D'Elía, and Pardiñas (2010) convincingly demonstrated that Cabrera's *hypogaeus* is a junior synonym of *Eligmodontia moreni* (see that account).

Pardiñas's (1995c) reallocation of the late Pleistocene taxa *Bothriomys catenatus* and *Oxymicterus* [sic] *impexus*, described by Ameghino (1889), as full synonyms of *G. griseoflavus* broadens the historic range of this species further to northeastern Buenos Aires province relative to its current distribution. However, the reference of these fossils to *G. griseoflavus* was made prior to the recognition of *G. chacoensis* as a valid species (see that account).

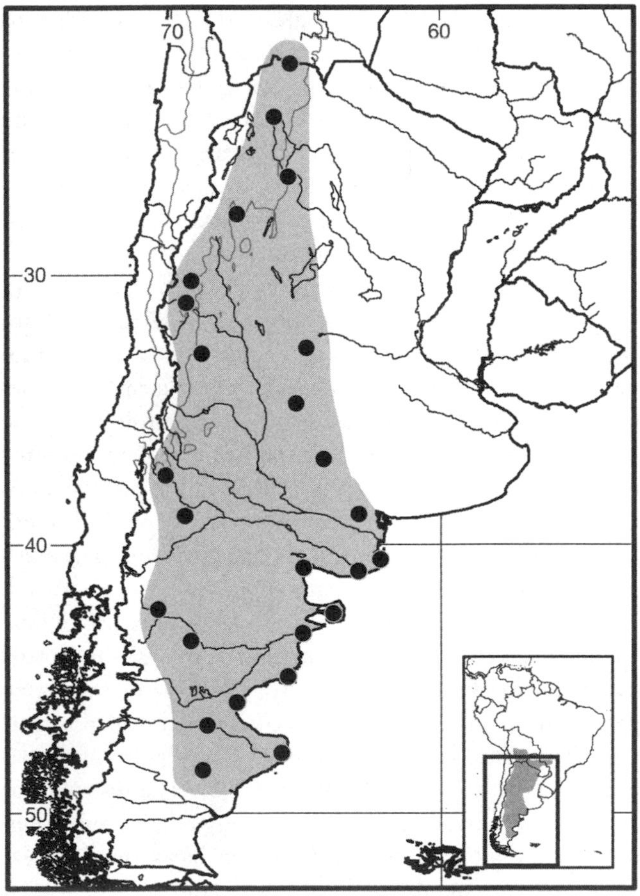

Wainberg and Fronza (1974) and Pearson and Patton (1976) described a karyotype with a variable diploid number of $2n=34$ to 38 but constant FN=44, analyses that have been extended geographically in multiple additional publications (Zambelli et al. 1994, 2003; Catanesi et al. 2002; V. A. Rodríguez 2005). These cytotypes differed by two centric fusions. Theiler and Blanco (1996a) and Thaeiler et al. (1999b) documented reproductive isolation between these karyomorphs and a $2n=42$ cytotype originally allocated to *G. griseoflavus* (e.g., Zambelli et al. 1994) and now regarded as the separate species *G. chacoensis*.

Genus *Loxodontomys* Osgood, 1947

Pablo Teta, Ulyses F. J. Pardiñas, Jorge Salazar-Bravo, and Guillermo D'Elía

Members of the genus *Loxodontomys*, commonly known as pericotes, range from the Central Cordillera of Santiago, Chile (about 33°30′S) south to Punta Arenas in southern Chile (latitude 53°30′S) and to Lago Argentino in southwestern Argentina (latitude 50°S). These mice occur at localities between sea level and 2,500 m and in a variety of habitats, from Patagonian steppe to forests, meadows, and scrub-dominated valleys in the Andes of central Chile (Osgood 1943b; Hershkovitz 1962; Spotorno, Cofre et al. 1998). Except as noted, the following account is based on Teta, Pardiñas et al. (2009), which should be consulted for further details.

Hershkovitz (1962:391–392) summarized the long and meandering taxonomic and nomenclatorial history of *Loxodontomys*, which changed from placement within *Phyllotis sensu stricto* to *Auliscomys* and *Euneomys*. Briefly, Osgood (1947) erected *Loxodontomys* as a subgenus of *Phyllotis* to allocate *micropus* Waterhouse. Later, based on phenetic analyses of karyotypic and morphological data, Simonetti and Spotorno (1980) moved *micropus* from *Phyllotis* to *Auliscomys*. The generic status of *Loxodontomys* was subsequently supported by the morphology based phylogenetic analyses of Braun (1993) and Steppan (1993, 1995). Phylogenetic hypotheses constructed using molecular data suggested a close association with *Auliscomys* (L. I. Walker and Spotorno 1992; M. F. Smith and Patton 1999; Spotorno et al. 2001; D'Elía 2003a). Steppan et al. (2007) using both mtDNA and nucDNA gene sequences recovered a *Phyllotis* group composed of *Auliscomys*, *Loxodontomys*, *Phyllotis*, *Tapecomys*, and possibly the *Andalgalomys/*

Map 278 Selected localities for *Graomys griseoflavus* (●). Contour line = 2,000 m.

Salinomys clade. A more recent and more taxon dense analysis, also based on mtDNA and nucDNA markers, retrieved ambiguous relationships among these same genera (Salazar-Bravo et al. 2013).

Recent taxonomic compilations (e.g., Musser and Carleton 2005) recognize two valid species of *Loxodontomys*, *L. micropus* (Waterhouse) and *L. pikumche* Spotorno, Cofre, Manriquez, Vilina, Marquet, and Walker. Our inspection of the skull, mandibles, and skin of the holotype of *L. pikumche* (LCM 1759, Spotorno, Cofre et al. 1998:Fig. 5) indicated their composite nature, with important taxonomic implications (Teta, Pardiñas, and D'Elía 2011). Although the skin belongs to a specimen of *Loxodontomys*, the skull and mandibles belong to a subadult individual of the genus *Phyllotis*. The main anatomical traits that supported this conclusion included the degree of hypsodonty, size of M3, molar toothrow orientation, bullar morphology, structure of the nasal bones, and the orientation of the mandibular condyle. Given these facts, Teta, Pardiñas, and D'Elía (2011) restricted the name *pikumche* to the skin, an act that maintains the linkage of the name to *Loxodontomys*. Nonetheless, the status of *pikumche*, whether a valid species or a subjective synonym, cannot be addressed until the genus is properly revised. Because analysis of the mtDNA cytochrome-*b* sequences failed to recover *pikumche* as a reciprocally monophyletic lineage (Cañón et al. 2010), for the present we treat *Loxodontomys* as a monotypic genus.

Loxodontomys is a medium-sized, robust sigmodontine rodent (head and body length 83–157 mm) with thick, lax, and lusterless pelage. General color dorsally varies from gray to brownish, grizzled with brownish yellow and intermixed with dusky black dorsally; the ventral surface is paler, often washed with white or yellow. Ears are small, with their length from notch about one-fifth to one-half less than the length of the hindfoot, which is also short. The tail is about as long as the combined head and body length. Postauricular patches are reduced or absent. Soles of the hindfeet are naked and lightly scutellated, with six small tubercles. The scale pattern of dorsal guard hairs is lanceolate (Chehébar and Martín 1989). Specimens from northern populations have a body pelage distinctly countershaded and tails more distinctly bicolored (Pine et al. 1979; Spotorno, Cofre et al. 1998).

The skull in *Loxodontomys* is moderately elongated, with a narrow interorbital region with either concave or parallel sides, sometimes raised, but never ridged nor beaded and lacking supraorbital knobs. The shape of the frontoparietal suture is rounded, and the edge of the frontals is convex. The interparietal is well developed with meandering sutures all around. The zygomatic arches converge anteriorly, forming a triangular outline when seen from above (especially obvious in adults), without temporal ridges. The nasals are long, with their distal ends slightly to moderately expanded. The anterior borders of the zygomatic plates are either vertical or sometimes slightly concave, with the upper zygomatic roots high on the rostrum; zygomatic spines and the posterior margin of the plates are located at, or anterior to, the alveolus of M1. Incisive foramina are comparatively wide, with their borders almost parallel and terminating at the level of protocone of M1. The mesopterygoid fossa has slightly divergent borders and is distinctly narrower than the adjacent, deeply excavated, parapterygoid fossae. The palate is long, posterior palatal pits are minute, and palatal vacuities are large. Tympanic bullae are moderately inflated. The carotid arterial supply is the primitive pattern 1 (of Voss 1988), with a squamosal alisphenoid groove and sphenofrontal foramen. The alisphenoid strut is absent. The mandible is short and high, with a falciform coronoid process; this process ends at a plane slightly higher than the mandibular condyle; the capsular projection is well defined and lies below the coronoid; the angular process is short and blunt. Both Pearson (1958, 1983) and Hershkovitz (1962) provided external and cranial measurements.

Upper incisors are broad, ungrooved, and opisthodont to hyperopisthodont. Molars are tetralophodont, planar, and hypsodont, with principal cusps ovate, and with crests exhibiting a tendency to laminate obliquely at about 45° along main axis of teeth. Molar tooth rows are slightly divergent posteriorly. Paracone and metacone of unworn and moderately worn M1–M2 are inflected forward; M1 is elongated with subeliptical procingulum; anteromedian flexus of M1 is absent in adult individuals; parastyle/anteroflexus of M1 is present but indistinct; anterolophs of M2 and M3 are well-developed; and M3 has an enamel island. Lower molars have alternating primary cusps; m1 lacks an anteromedian flexid and has a rounded procingulum projected to the protoconid by well-defined anterolabial cingulum; the protoflexid of m2 is present as a groove; and the mesoflexid of m3 is reduced and shifted anteriorly relative to m2. The m3 has an obvious S-shape, subequal in length to m2 (Hershkovitz 1962; Braun 1993; Steppan 1995).

Axial skeletal elements have modal counts of 13 ribs, 13 thoracic vertebrae, 6 or 7 lumbar vertebrae, and 26 to 29 caudal vertebrae. Male accessory glands consist of one pair each of bulbourethral, anterior, and dorsal prostate, ampullary, and vesicular glands and two pairs of dorsal prostate glands (Voss and Linzey 1981). The distal bacular cartilage is tridigitate, with the medial digit longer than the lateral digits, and each lateral digit has a small medial hook. The proximal baculum is long and robust, with a central notch on the base. The urethral process of the phallus is long and unilobed (Braun 1993).

SYNONYMS:

Mus: Waterhouse, 1837:17; part (description of *micropus*); not *Mus* Linnaeus, 1758.

Mus (*Ab*[*rothrix*].): Waterhouse, 1837:21; part (listing of *micropus*); neither *Mus* Linnaeus nor *Abrothrix* Waterhouse.

Hesperomys (*Habrothrix*): Burmeister, 1879:217; part; neither *Hesperomys* Waterhouse nor *Habrothrix* Wagner.

Habrothrix: Fitzinger, 1867b:81; part (listing of *micropus*); not *Habrothrix* Wagner.

Akodon (*Akodon*): E.-L. Trouessart, 1897:536; part (listing of *micropus*); not *Akodon* Meyen.

Phyllotis: J. A. Allen, 1905:60; part (listing of *micropus*); not *Phyllotis* Waterhouse.

Euneomys (*Auliscomys*): Thomas, 1916a:143; part (listing of *micropus*); neither *Euneomys* Coues nor *Auliscomys* Osgood.

Euneomys: Thomas, 1919b:201; part (description of *alsus*); not *Euneomys* Coues.

Auliscomys: Gyldenstolpe, 1932a:95; part (listing of *micropus* and *alsus*); not *Auliscomys* Osgood.

[*Phyllotis*] (*Auliscomys*): Ellerman, 1941:452, 455; part (listing of *micropus* and *alsus*); neither *Phyllotis* Waterhouse nor *Auliscomys* Osgood.

Phyllotis (*Loxodontomys*) Osgood, 1947:172; as subgenus; type species *Mus micropus*, Waterhouse.

Loxodontomys: Massoia, 1982:45; first use as genus.

REMARKS: Pine et al. (1979) proposed that the name *Loxodontomys* Osgood was a *nomen nudum*, an opinion discussed and dismissed by Steppan (1995).

Loxodontomys micropus Waterhouse, 1837

Southern Pericote

SYNONYMS:

Mus micropus Waterhouse, 1837:17; type locality: "Santa Cruz." A note written by Charles Darwin and transcribed in Waterhouse (1839:62) added that the holotype was "caught in the interior plains of Patagonia in lat. 50°, near banks of the Santa Cruz."

Mus (*Ab*[*rothrix*].) *micropus*: Waterhouse, 1837:21; name combination.

Habrothrix micropus: Fitzinger, 1867b:81; name combination.

Hesperomys (*Habrothrix*) *micropus*: Burmeister, 1879:217; name combination.

[*Akodon* (*Akodon*)] *micropus*: E.-L. Trouessart, 1897:536; name combination.

Phyllotis micropus: J. A. Allen, 1905:60; name combination.

Euneomys (*Auliscomys*) *micropus*: Thomas, 1916a:143; name combination.

Euneomys micropus alsus Thomas, 1919b:202; type locality "Maiten [= El Maitén (42.05°S, 71.16°W)], W. Chubut. 700 m," Cushamen, Chubut, Argentina.

Auliscomys micropus micropus: Gyldenstolpe, 1932a:95; name combination.

Auliscomys micropus alsus: Gyldenstolpe, 1932a:95; name combination.

Phyllotis (*Auliscomys*) *micropus micropus*: Ellerman, 1941: 455; name combination.

Phyllotis (*Auliscomys*) *micropus alsus*: Ellerman, 1941: 455; name combination.

Phyllotis (*Auliscomys*) *micropus fumipes* Osgood, 1943b: 214; type locality "Quellon [= Quellón (43.136°S, 73.663°W)], Chiloe Island [= Isla Grade de Chiloé], [Los Lagos,] Chile."

[*Phyllotis* (*Loxodontomys*)] *micropus*: Osgood, 1947:172; name combination.

Loxodontomys micropus: Massoia, 1982:45; first use of current name combination.

Loxodontomys pikumche Spotorno, Cofre, Manríquez, Vilina, Marquet, and Walker, 1998:362; type locality "Cajón del Río Maipo, sector Cruz de Piedra (34°10'S 69°58'W, 2.450 msnm), a 55 km S de la Central Hidroeléctrica de Las Melosas . . . en la Cordillera de Región Metropolitana," Chile.

DESCRIPTION: As for the genus.

DISTRIBUTION: Southern Chile and southwestern Argentina from the Andean area near Santiago in central Chile and western Mendoza province, Argentina, south to islands bordering the Straits of Magellan (at least one record from Isla Riesco; see Markham 1970; but absent from Tierra del Fuego), at elevations from sea level to 2,500 m (Osgood 1943b; Hershkovitz 1962; Spotorno, Cofre et al. 1998; Novillo et al. 2009). The known ecological distribution of *L. micropus*, traditionally considered a denizen of Andean forests (Pearson 1983), has been extended significantly as isolated populations in the basaltic tablelands of Río Negro (e.g., Maquinchao), Chubut (Sierra de Chacays), and Santa Cruz (Bosques Petrificados, Cerro Fortaleza) provinces (Heinonen Fortabat and Haene 1994; Teta et al. 2002; Pardiñas, Teta et al. 2003; Udrizar Sauthier, Teta et al. 2008; V. L. Roldán 2010). This species has been found in several archeological and paleontological deposits of the latest Pleistocene-Holocene in Argentinean Patagonia (e.g., Cueva Traful I, Cueva Epullán Grande, Alero Santo Rosario; Pearson 1987; Pearson and Pearson 1993; Pardiñas 1998, 1999a; Rebane 2002; A. Andrade and Teta 2003; Teta et al. 2005; Pardiñas and Teta 2013b) and Chile (Cueva del Milodón, La Batea 1; Simonetti and Rau 1989; Simonetti and Saavedra 1994). Some of these records, especially those spanning the last 3,000 years, are extralimital to the current range of the species, suggesting that *L. micropus* had a more extended distribution in steppe areas during the cold-humid phases of the Late Holocene (Pardiñas 1999a; Teta et al. 2005; Udrizar Sauthier 2009).

SELECTED LOCALITIES (Map 279): ARGENTINA: Chubut, El Maitén (type locality of *Euneomys micropus alsus* Thomas), 8.4 km NW of Estancia El Valley, Valle de Lanzaniyeu (CNP-E 521), Estancia Santa María, Puesto El Chango (CNP-E 290), Paso del Sapo (Pardiñas, Teta et al. 2003), 4 km S of Tres Banderas (Udrizar Sauthier, Teta et al. 2008); Mendoza, Laguna del Diamante (A. Novillo, pers. comm.), 6 km NW (by road) of Las Leñas (Jayat et al. 2006); Neuquén, Cajón Grande, A° Curi Leuvú (CNP-E 632), SW end of Lago Hui Hui, 8 km W and 2 km S of Cerro Quillén (MVZ 163400), Paso Puyehue, 32 km WNW of Villa La Angostura (MVZ 181566), 2.6 km NW of Puente Carreri (CNP-E 186); Río Negro, entrance to Estación Calcatreo (CNP-E 287), Estancia Maquinchao, Puesto de Hornos (Teta et al. 2002), Estancia Pilcañeu (Teta et al. 2001); Santa Cruz, Cerro Comisión (CNP-E 413), Cerro Fortaleza (CNP-E 420), Lago Cardiel, La Península, Estación Las Tunas (Massoia et al. 1994b), laguna de los Cisnes (J. A. Allen 1905), Monumento Natural Bosques Petrificados (Heinonen, Fortabat, and Haene 1994), Río Ecker, 500 m aguas abajo Estación Casa de Piedra, margen norte (CNP-E 377). CHILE: Aysén, Fin del Fiordo Ofhidro (J. Guzmán, pers. com.); Biobío, Laguna del Laja, Río Petronquines (Spotorno, Cofre et al. 1998), Sierra de Nahuelbuta (Hershkovitz 1962); Los Lagos, Quellón, Isla Grande de Chiloé (type locality of *Phyllotis* (*Auliscomys*) *micropus fumipes* Osgood); Los Ríos, Fundo San Martín, San José, Valdivia (UACH 3115); Magallanes y Antártica Chilena, Lago Pehoe, Torres del Paine (MVZ 175150), Puerto Natales (Osgood 1943b), Punta Arenas (Osgood 1943b), vicinity of Puerto Henry, Isla Riesco (Markham 1970); Santiago, El Colorado, Farellones (Spotorno, Cofre et al. 1998).

SUBSPECIES: Three subspecies have been described, mostly based on subtle differences of metrics and external coloration (Thomas 1919b; Osgood 1943b). Gyldenstolpe (1932) and Ellerman (1941) recognized *micropus* and *alsus* as different; later, Osgood (1943b) included *alsus* under *micropus* and at the same time coined the new subspecies *fumipes* based on immature specimens from Chiloé Island. Both Pearson (1958) and Hershkovitz (1962) dismissed the differences between these nominal forms and considered *micropus* as monotypic. Cañón et al. (2010; see also E. P. Lessa et al. 2010) addressed the phylogeography of *Loxodontomys* with mtDNA cytochrome-*b* gene sequences from 24 Argentinean and Chilean localities, and documented a shallow but geographically structured genealogy. Two main evolutionary lineages, replacing one another geographically at about 44°S, gave support to the recognition of subspecies, the nominotypical *L. m. micropus* (from southern populations) and *L. m. alsus* (from northern populations). No specimen referable to *fumipes* Osgood was analyzed. The northern clade included haplotypes recovered from specimens with $2n = 32$ referred to the nominal form *pikumche* by Novillo et al. (2009) and from several Chilean localities ascribed by Spotorno, Cofre et al. (1998) to the range of this putative species as well as from Argentinean and Chilean populations within the range of *L. micropus* (*sensu* Spotorno, Cofre et al. 1998). In light of these results, Cañón et al. (2010) proposed that *pikumche* could be a junior synonym of *L. micropus*. As stated above, if the northern clade deserves taxonomic distinction, *alsus* would have priority given that the haplotype recovered from a topotype falls within the northern clade. Finally, Cañón et al. (2010) suggested a historical scenario whereby ancestral populations of *Loxodontomys* contracted their ranges into a minimum of two glacial refugia (although their location was not stated) to account for the origin of the geographic pattern of current genetic diversity they documented. Most of the Argentinean and Chilean populations lack karyological data, and thus the geographic structure of the chromosome variation is unknown.

NATURAL HISTORY: *Loxodontomys micropus* has been caught in *Nothofagus* forests, precordilleran shrubby steppes, humid dense grasslands (Mann 1978; Pearson 1983; Kelt 1994), and open habitats of spiny shrubs and dry grasslands on hard soil pavements (Spotorno, Cofre et al. 1998). It is nocturnal and both burrows and climbs skillfully (Pearson 1983). This rodent is primarily herbivorous, with stomach contents including green leaves, fungi, grass seeds, and fruits (Pearson 1983; Meserve et al. 1988; Monjeau 1989). In humid dense grasslands of northwestern Patagonia, *L. micropus* excavates tunnel systems, with numerous entrance openings and food chambers (Monjeau 1989). The reproductive season starts at the beginning of the spring and lasts through the end of the summer (Pearson 1983) or early Fall (Kelt 1994). The average number of embryos ranges between 4.1 in Argentina (Pearson 1983) to 4.8 in Chile (Greer 1965; Pine et al. 1979). Both sexes are reproductively active after reaching 48 g of mass (Pearson 1983). Teta, Pardiñas et al. (2009) summarized available information about ontogeny and reproduction, population characteristics, space use, diet, diseases and parasites, and interspecific interactions, mostly based on the original investigations of Pearson and Pearson (1982), Pearson (1983) and Kelt (1994). Specimens collected at different localities from Argentina and Chile were seropositive for hantavirus (Spotorno et al. 2000; Cantoni et al. 2001; Cerda and Sandoval 2004; J. C. Ortiz et al. 2004; Padula et al. 2004). Goff and Gettinger (1995) and D. Castro et al. (2001) described new species of chiggers and lice, respectively, from specimens identified as *L. micropus*.

REMARKS: Charles Darwin collected the type of *Mus micropus* Waterhouse at some point between the mouth of the Río Santa Cruz (= "Santa Cruz," as stated by Waterhouse 1837) and the Andean Cordillera (see Waterhouse 1839). According to Hershkovitz (1962:392) this specimen was collected "perhaps in the vicinity of the present locality of La Argentina, province of Santa Cruz, southern Argentina." However, Darwin did not reach the longitude of La Argentina, as judged by Hershkovitz's (1962:Fig. 89, p. 393) placement of this locality on the south shore of Lago Argentino. The westernmost point reached by Darwin was about 240 km west of the mouth of the Río Santa Cruz, about 30 km from the eastern side of Lago Argentino (G. R. Cueto et al. 2008; Teta, Pardiñas et al. 2009).

Spotorno, Cofre et al. (1998) registered two diploid numbers for *L. micropus*. Chilean specimens south of 37°30'S have $2n = 34$, $FN = 36$, with most chromosomes telocentric. They found the second diploid complement ($2n = 32$, $FN = 34$) north of this latitude, formally describing these specimens under the name *L. pikumche*. In this karyotype, all elements were telocentric except a single small metacentric pair (Spotorno, Cofre et al. 1998). Novillo et al. (2009) reported the same karyotype for four

specimens collected in Mendoza province, Argentina, close to the border with Chile. Cañon et al. (2010) misunderstood the data provided by Spotorno, Cofre et al. (1998) and Novillo et al. (2009) about the fundamental numbers of Chilean and Argentinean populations assigned by them to *L. pikumche* and stated that the Argentinean population differed from those of northern Chile and thus represented a third karyomorph for the genus. Spotorno and Walker (1979) provided a banded karyotype from Chilean representatives of *L. micropus*.

Genus *Phyllotis* Waterhouse, 1837
Scott J. Steppan and Oswaldo Ramirez

The leaf-eared mice of the genus *Phyllotis* include 15 named extant species and one undescribed species that live from the highlands (páramo) of Ecuador throughout the Andes and adjacent xeric habitats to the southern tip of continental South America. Although there has been some fluctuation in generic status and placement for several species, the genus as recognized here includes 15 speecies: *P. alisosiensis, P. amicus, P. andium, P. anitae, P. bonariensis, P. caprinus, P. darwini, P. definitus, P. gerbillus, P. haggardi, P. limatus, P. magister, P. osgoodi, P. osilae,* and *P. xanthopygus*. There have been two complete revisions of the genus (Pearson 1958; Hershkovitz 1962) and a review as part of a revision of the tribe Phyllotini (Steppan 1995). Some phylogenetic studies have suggested that the genus is paraphyletic, but these have not provided a robust framework for a revision of the taxonomy. Braun (1993) recommended removing *P. gerbillus* and *P. amicus* to the genus *Paralomys* Thomas, 1926, a recommendation followed in part by Musser and Carleton (2005), but that change was both weakly supported by morphological data (Steppan 1995) and weakly contraindicated by mtDNA cytochrome-*b* sequence data (Steppan 1998). The evidence for exclusion of *P. wolffsohni* came from both morphology and molecules (Steppan 1995; Steppan 1998), but neither study provided a robust alternative hypothesis. More detailed taxon sampling with cytochrome-*b* sequences, however, supported the removal of *wolffsohni* to *Tapecomys* and, along with nuclear RAG1 sequences, the inclusion of *P. gerbillus* and *P. amicus* within *Phyllotis* (Steppan et al. 2007).

Many taxa lumped into a polytypic *P. darwini* (particularly by Hershkovitz 1962, but also by Pearson 1958) have been removed more recently to separate species: *P. caprinus, P. definitus, P. limatus, P. magister, P. xanthopygus,* and (*Tapecomys*) *wolffsohni*. L. I. Walker et al. (1984) demonstrated reproductive isolation between *P. darwini* and the subspecies *vaccarum*, and associated *vaccarum* with the elevated *P. xanthopygus*. Many studies still

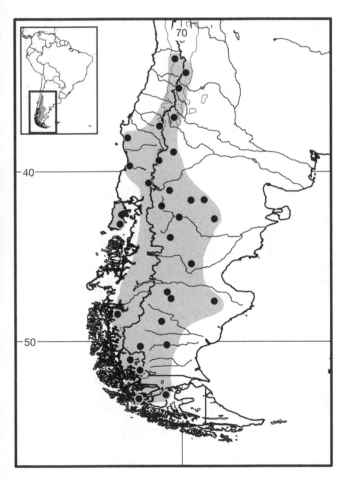

Map 279 Selected localities for *Loxodontomys micropus* (●). Contour line = 2,000 m.

incorrectly refer individuals of *P. xanthopygus* to *P. darwini*, the latter being restricted to central Chile. Musser and Carleton (1993) misinterpreted L. I. Walker et al. (1984) by including northern populations of *P. xanthopygus* in *P. darwini*, although these authors later corrected this assignment (Musser and Carleton 2005). Steppan (1995) elevated *P. xanthopygus chilensis* to species status based on the absence of intermediate phenotypes along elevational transects with *P. xanthopygus rupestris* and due to a putative apomorphy. Subsequent study indicated that it was the Pacific slope populations of *P. x. rupestris* rather than *P. x. chilensis* that possessed a clear apomorphic feature and that could be distinguished from nearby *P. xanthopygus*, leading Steppan (1998) to place the Pacific slope populations of *P. x. rupestris* within *P. limatus* while returning, at least temporarily, *chilensis* to *P. xanthopygus*.

Species of *Phyllotis* live in open, mostly rocky and brushy habitats from sea level to nearly 5,000 m. Fossils attributable to *P. andium* and *P. xanthopygus* have been found from the Early Pleistocene to the Recent (Pardiñas and Dechamps 1996; P. E. Ortiz, Cirignoli et al. 2000) of Ecuador, Peru, Bolivia, Chile, and Argentina, but specific determination is difficult given the relatively undifferentiated molar dentition among the species and the nearly complete absence of cranial characters (U. F. J. Pardiñas and S. J. Steppan, unpubl. data). Ancient DNA assignable to *P. limatus* has been identified from midden remains in the Atacama desert of Chile, dated to near the Pleistocene/Holocene boundary (Kuch et al. 2002).

In addition to the species described here, the genus includes one undescribed species from Peru (V. Pacheco, pers. comm.) that is genetically distinct from other named species, as well as one population currently assigned to *P. amicus* and possibly one to three more that are currently attributed to *P. xanthopygus* (Steppan et al. 2007; S. J. Steppan, unpubl. data).

Phyllotis species are small to medium-sized phyllotine mice with long, fine, and variably colored pelage, the ventrum usually much lighter than the dorsum. The tail is subequal in length to the combined head and body; ears are medium to large, and typically lack postauricular patches. The hindfeet are moderate in length, with a reduced fifth hind toe; the soles are naked with six plantar pads. Females have eight mammae, in four pairs.

The molars are tetralophodont with no trace of a mesoloph or mesolophid; incisors are without grooves, opisthodont to nearly orthodont, and without tripartite dentine fissure. M1 has labial roots; there are two or three roots on M3 and two roots on m3. The anteromedian flexus of M1 is either absent or limited to a shallow groove; mesostyle and parastyle/anteroflexus of M1 are absent; the anterolabial cingulum m1 is long and distinct, fusing with protoconid and leaving the protoflexid as a lake (with wear); m1 possesses a posterolophid/stylid, at least in juveniles; the hypoflexid of M3 is pinched to form a lake; the mesoflexid m3 is shifted anteriorly relative to m2; the hypoflexid of m3 shows no posterior shift; and the opposing flexi in M3 do not meet.

The dorsal surface of the skull is slightly rounded; the supraorbital region is narrow with convex sides. The zygomatic notches are moderate to deep; a zygomatic spine is absent (anterior border is either flat or vertical) or weakly developed. The antorbital bridge of the zygomata is positioned slightly below the rostrum. The narrowest portion of the interorbital ridge is centrally or anteriorly located; the frontoparietal suture forms an obtuse angle; the interparietal is well developed, >0.45% the length of the parietal measured along the median. The incisive foramina are long, reaching the anterolabial and anterolingual conules of M1. The palate is long and broad, with no medial process on its posterior border. Maxillary toothrows are parallel or posteriorly convergent. The mesopterygoid fossa is distinctly narrower than adjacent parapterygoid fossae, with the orbital wings of the presphenoid anterior to maximum constriction; a sphenopalatine foramen is present, small to moderate in size. The tegmen tympani contacts a suspensory process of the squamosal. The stapedial spine of each auditory bulla is circular to ovoid in cross-section, and not laterally appressed. Both the basioccipital and the ectotympanic portion of auditory bullae bound the internal carotid canal. An alisphenoid strut is absent or filamentous. The medioventral process of the mandibular ramus is absent or weakly developed.

The vertebral column has a modal count of 13 thoracic, six lumbar, four sacral, and 28–36 caudal vertebrae. The tuberculum of the first rib articulates with the transverse process of the seventh cervical and first thoracic vertebrae; the second thoracic vertebra has an elongated neural spine; hemal arches are absent. The entepicondylar foramen of the humerus is absent, but the supertrochlear foramen is present.

The gall bladder is present; the stomach is hemiglandular. Two distinct preputial glands are present, with a third, smaller gland found in some populations. The baculum is complex with short to long distal bacular elements. Hershkovitz (1962) detailed bacular variation within and among taxa.

SYNONYMS:

Mus (*Phyllotis*) Waterhouse, 1837:28; part; as subgenus; type species *Mus* (*Phyllotis*) *darwini* Waterhouse, by subsequent designation (Thomas 1884:449).

Hesperomys: Waterhouse, 1839:76; part (listing of *xanthopygus* and *darwini*).

Phyllotis: Gray, 1843b:xxiii; listed as genus equivalent to *Mus* Linnaeus and *Hesperomys* Waterhouse, the two genera under which most sigmodontine names were at that time allocated.

Hesperomys (*Phyllotis*): Thomas, 1884:449; part (listing of *darwini*), as subgenus; not *Hesperomys* Waterhouse.

Phyllotis: Thomas, 1897a:1020; elevated to genus.

Paralomys Thomas, 1926b:315; type species *Paralomys gerbillus* Thomas, by original designation.

KEY TO SPECIES OF *PHYLLOTIS*:

1. Small bodied, maxillary molar row <4.7 mm. 2
1'. Medium to large bodied, maxillary molar row >4.7 mm 3
2. Tail length <95 mm 4
2'. Tail length >95 mm. 5
3. Maxillary molar row between 4.7 and 5.7 mm 6
3'. Maxillary molar row >5.7 mm 11
4. Tail shorter than head and body; upper toothrows posteriorly convergent . . . . *Phyllotis haggardi*
4'. Tail subequal to or longer than head and body; upper toothrows parallel sided, ventrum entirely white, dorsum and sides light sandy color. . . . . *Phyllotis gerbillus*
5. Upper incisors opisthodont, ears as long or longer than hindfoot . . . . *Phyllotis amicus*
5'. Upper incisors orthodont, ears shorter than hindfoot. . . . . *Phyllotis andium*
6. Mesoflexus of M3 highly reduced relative to M2, or absent. 7
6'. Mesoflexus of M3 reduced relative to M2 10
7. Ratio of incisor depth to width across both teeth >0.9 . . . . *Phyllotis limatus*
7'. Ratio of incisor depth to width across both teeth <0.9 8
8. Ratio of M3 length to alveolar length of molar toothrow<0.205 14
8'. Ratio of M3 length to alveolar length of molar toothrow>0.205 9
9. Small swellings or knobs on anterior supraorbital region posterior to lacrimal . . . . *Phyllotis darwini*
9'. Supraorbital swelling or knobs absent . . . . *Phyllotis caprinus*
10. Anteroflexus of M2 distinct; if protoflexus present, then procingulum poorly developed as broad, shallow projection with concave anterior edge; if proflexus not present, then distinct anteroflexus or paraflexus present; mesoflexus of M3 reduced relative to M2. . . . . *Phyllotis andium*
10'. Procingulum of M2 completely absent; posteropalatal pits small, lateral to mesopterygoid fossa, usually even with or behind posterior margin of palate (occasionally slightly anterior); mesopterygoid fossa broad,

with sides parallel or diverging anteriorly, the anterior border broad and frequently interrupted by a median posterior projection of the palate. . . . . *Phyllotis osilae*
11. Nasals broad, clearly greater than minimum interorbital distance 12
11'. Nasals narrow, subequal to or less than minimum interorbital distance. 14
12. Interorbital region very constricted, <3.7 mm, uniformly cinnamon-colored ventrum . . . . *Phyllotis alisosiensis*
12'. Interorbital breadth >3.8 mm, ventrum with grayregion 13
13. Pectoral spot or streak present and well defined; distributed west of the Andean cordillera 14
13'. Pectoral spot or streak absent or weakly defined; confined to Sierra de la Ventana, Argentina . . . . *Phyllotis bonariensis*
14. Mesopterygoid fossa with anterior border square or slightly rounded . . . . *Phyllotis magister*
14'. Mesopterygoid fossa with anterior border pointed. . . . . *Phyllotis definitus*
15. Incisor enamel white to yellowish-white; dorsal pelage dark gray; tail slightly bicolored . . . . . . *Phyllotis anitae*
15'. Incisor enamel yellow to orange; dorsal pelage light to medium ochraceous or brown; tail bicolored 16
16. Karyotype $2n = 38$ . . . . *Phyllotis xanthopygus*
16'. Karyotype $2n = 40$ . . . . *Phyllotis osgoodi*

Phyllotis alisosiensis Ferro, Martínez, and Barquez, 2010
Los Alisos Leaf-eared Mouse

SYNONYM:

Phyllotis alisosiensis, Ferro, Martínez, and Barquez, 2010: 526; type locality "El Papal, 2175 m (27°11′S 65°57′W), Parque Nacional Campo de Los Alisos, Departamento Chicligasta, Tucumán, Argentina."

DESCRIPTION: Large, total length 243–342 mm (mean 271 mm), tail length 121–139 mm (mean 131 mm), hindfoot length 29–32 mm, ear length 20–24 mm, mass 35–70 g. Characterized by long and fluffy fur, cinnamon-brown dorsal color flecked with black hairs, and ventral color strongly cinnamon. Mean hair length 16–17 mm on rump; individual hairs tricolored with short basal white band followed by broad plumbeous band and brownish to cinnamon tip; guard hairs entirely black and project 4–5 mm beyond the level of the underfur. Transition between dorsal and ventral colors gradual, not sharply contrasting. Ears dark, scarcely haired except for conspicuous patch on anterior edge; inner surface covered with dark brown hairs. Inconspicuous grayish postauricular patch present. Tail strongly bicolored, dark gray above and whitish below, about same length as head and body. Both manus and pes covered dorsally with short but dense white hairs; ungual tuft covers each claw. Ventral mystacial vibrissae short and white; dorsal ones longer, black basally and white distally.

Skull robust and wide, with enlarged rostrum. Condyloincisive length 27.1–32.5 mm; zygomatic breadth 14.8–18.8 mm; rostral length 12.1–13.4 mm. Interorbital constriction notably narrow, 3.6 mm or less, especially in comparison to other, codistributed species (*P. anitae, P. osilae,* and *P. xanthopygus*). Anterior end of zygomatic arch high and, from lateral view, antorbital bridge flat and barely visible. In fully developed adults, width of zygomatic plates equal to or greater than width of interorbital constriction. Molars above and below hypsodont, wide, with simplified occlusal pattern. M1 trilophodont, with major cusps alternating; labial flexus oblique and lingual flexus transverse to medial plane; procingulum well developed, wide, and without anteromedian flexus. M2 S- or Z-shaped, with hypoflexus anteriorly oriented and metaflexus transverse. M3 with deep hypoflexus surpassing medial plane of tooth. Lower m1 trilophodont, m2 bilophodont, and m3 also bilophodont with hypoflexid more developed than metaflexid.

DISTRIBUTION: *Phyllotis alisosiensis* is known from only two localities in Tucumán province, Argentina, at the southernmost end of the Yungas ecoregion across an elevational range from 1,234 to 2,175 m.

SELECTED LOCALITIES (Map 280): ARGENTINA: Tucumán, El Papal (type locality of *Phyllotis alisosiensis*, Ferro, Martínez, and Barquez), Reserva Provincial Los Sosa, Rte 307, km 35, campamento de Vialidad Provincial (CML 4106).

SUBSPECIES: *Phyllotis alisosiensis* is monotypic.

NATURAL HISTORY: Little is known about the natural history, reproductive biology, or other aspects of the population ecology of this species. Specimens from the type locality were obtained in upper montane forest (*Bosques Montanos*) where the landscape is dominated by deciduous forest of Andean alder (*Alnus acuminata* [Betulaceae]), intermixed with bamboo (*Chusquea* sp.), ferns (*Blechnum occidentalis* [Blechnaceae]), scattered thickets of queñoa (*Polylepis australis* [Rosaceae]), and native grasses (*Cortadeira* sp.) on steep, rocky slopes. The only other known locality is within cloud forest at the upper part of the Lower Montane Forest (Hueck 1978), habitats dominated by species of *Myrcianthes* (Myrtaceae) with epiphytes, lianas, and grasses.

REMARKS: Ferro et al. (2010) provided detailed morphological and multivariate morphometric comparisons among *P. alisosiensis* and the three other species of *Phyllotis* that occur in Tucumán province, *P. anitae, P. osilae,* and *P. xanthopygus. Phyllotis alisoiensis* can be easily distinguished from all other species in the genus by its strong cinnamon color on the ventrum and strongly constricted interorbital

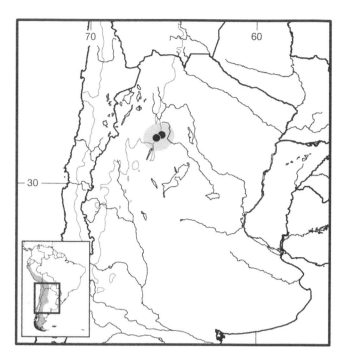

Map 280 Selected localities for *Phyllotis alisosiensis* (●). Contour line = 2,000 m.

region. Based on comparative mtDNA cytochrome-*b* sequences, *P. alisosiensis* is closely related to *P. anitae*, with divergence only 1.5% divergent (a measure exceeded by many conspecific populations; Steppan et al. 2007), with these two species forming a sister group with *P. osilae*.

Phyllotis amicus Thomas, 1900
Friendly Leaf-eared Mouse

SYNONYMS:

Phyllotis amicus Thomas, 1900d:355; type locality "Tolon [= Tolón], coast district, Province Cajamarca, N.W. Peru. Altitude 100 m."

Phyllotis amicus maritimus Thomas, 1900f:296; type locality "Eten, [Lambayeque,] coast of N.W. Peru. Altitude 20 m."

Phyllotis amicus montanus Thomas, 1900f:297: type locality "Uramarca [= probably Yuramarca], [Ancash,] near Pallasca, N.W. Peru. Altitude 1200 m."

DESCRIPTION: Characterized by small size, short hindfeet, especially short molar toothrows (3.5–4.2 mm), and relatively large ears, usually as long or longer than hindfeet. Interorbital region broad and flat, frequently with square or sharp edges; upper incisors opisthodont; auditory bullae large and globular; posteropalatal pits located anterior to mesopterygoid fossa; M2 with two outer entrant angles in unworn teeth. Northernmost coastal specimens small and brightly colored; those inland and from south, even along coast, larger and duller.

Phyllotis amicus meets both *P. limatus* and *P. andium*, and may possibly contact *P. definitus* and *P. gerbillus*. Much smaller in all dimensions relative to *P. definitus* and *P. limatus*, especially smaller teeth, opisthodont incisors, more conspicuous development of two outer entrant angles of M2, shorter fur, and more naked tail. *Phyllotis amicus* distinguished from *P. andium* by smaller dimensions, slightly paler color, more conspicuous development of second out entrant angle on M2, and proportionally larger ears. *Phyllotis gerbillus* with pure white venter, white tail white above and below, and smaller ears.

DISTRIBUTION: *Phyllotis amicus* is limited to western Peru, from Piura south into Arequipa department. In the north it occurs on both slopes of the Cordillera Occidental, but is restricted to the western slope of the Andes further south (Pearson 1972:Fig 2), over an elevational range from sea level to about 2,800 m.

SELECTED LOCALITIES (Map 281): PERU: Ancash, 29 km S (by road) of Casma (MVZ 139263), Uramarca [= Yuramarca], near Pallasca (type locality of *Phyllotis amicus montanus* Thomas); Arequipa, Atiquipa (FMNH 107392); Ayacucho, 35 mi ENE of Nazca (Pearson 1972); Huancavelica, 2 km E of Ticrapo, Pisco Valley (Pearson 1972); Lambayeque, 12 km S and 8 km E of Bayovar (MVZ 135370), Etén (type locality of *Phyllotis amicus maritimus* Thomas); Lima, 19 mi N of Cañete (MVZ 119990), 20 km N and 6 km W of Chancay (MVZ 135723), 1 mi W of Surco (MVZ 119996); Piura, 6.5 mi E (by road) of Canchaque (MVZ 141837), Cerro Amotape, 10 km N and 40 km W of Sullana (MVZ 135736), 45 km W and 15 km N of Chulucanas (MVZ 135731).

SUBSPECIES: Pearson (1958) recognized three subspecies: the nominotypical form ranging from the foothills of Cajamarca south to southern Lima department, at elevations from sea level to 1,800 m; *maritimus* Thomas, known only from the vicinity of the type locality; and *montanus* Thomas, occupying the western slopes of the mountains in La Libertad and Ancash, Peru, between 1,050 and 2,750 m. Hershkovitz (1962) argued that retention of formal subspecies was unwarranted and regarded *P. amicus* as monotypic.

NATURAL HISTORY: *Phyllotis amicus* inhabits sand and rock formations with sparse cover on the coast and Pacific slope of northern and central Peru. It is found in *Tillandsia* (Bromeliaceae) desert, rocky places with sparse cactus vegetation, and dry washes (Pearson 1972), where it may be with *P. limatus* and *P. andium*. Its feeding habits appear broadly omnivorous. Leafy vegetation constituted about half of its diet, which is typical of the genus, but it consumed more insect and less seed than most other species of *Phyllotis* (Pizzimenti and DeSalle 1980). This species maintains low density, highly dispersed populations in a harsh environment, and can reproduce at any time of the year in central Peru

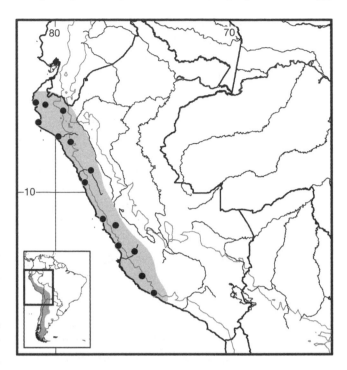

Map 281 Selected localities for *Phyllotis amicus* (●). Contour line = 2,000 m.

(Arana et al. 2002). Movement distances calculated from six individuals in Lachay Lomas was 44.5 m; litter size varied between one and three pups (Arana et al. 2002).

REMARKS: *Phyllotis amicus* appears most closely related to *P. gerbillus* in mtDNA cytochrome-*b* sequences (Steppan et al. 2007), with this pair of species sister to *P. andium* and an undescribed taxon. A sample from Puno department, Peru, far south of the previously known range and attributed to *P. amicus* on morphological grounds may represent a distinct species (Steppan et al. 2007).

The $2n = 38$ karyotype consists of 18 pairs of metacentric and submetacentric autosomes grading from large to small, a large acrocentric or submetacentric X and a small metacentric Y chromosome (Pearson 1972). The morphology of the X chromosome is apparently geographically structured, metacentric in all individuals and localities sampled in the Región de Lima but acrocentric in those further to the south in Huancavelica. The complement with the metacentric X chromosome is shared with *P. caprinus*, *P. darwini*, *P. gerbillus*, *P. magister* (except for the morphology of the Y chromosome), *P. haggardi*, and *P. xanthopygus* (Pearson 1972; Pearson and Patton 1976).

Phyllotis andium Thomas, 1912
Andean Leaf-eared Mouse
SYNONYMS:
Phyllotis andium Thomas, 1912b:409; type locality "Cañar, [Cañar,] Ecuador. Alt. 2600 m."

Phyllotis melanius Thomas, 1913b:407; type locality "Porvenir, Bolivar, Ecuador. Alt. 1800 m"; corrected to Cañar, Cañar, Ecuador by Pearson (1958:440).

Phyllotis andium stenops Osgood, 1914b:165; type locality "Rio Utcubamba, 15 miles above Chachapoyas, [Amazonas,] Peru."

Phyllotis tamborum Osgood, 1914b:165; type locality "Tambo Carrizal, mountains east of Balsas, [Amazonas,] Peru. Altitude 5,000 ft."

Phyllotis fruticicolus Anthony, 1922:1; type locality "Guachanamá, [Loja,] southern Ecuador; altitude, 9050 ft."

Phyllotis andium tamborum: Thomas, 1926e:614; name combination.

Phyllotis lutescens andium: Osgood, 1944:194; name combination, as subspecies of what is now *Phyllotis osilae* J. A. Allen.

DESCRIPTION: Medium-sized, rich-brown species, with delicate vibrissae, lightly haired tail, relatively small ears, pale gray venter with distinct buffy pectoral streak in some specimens, and occasionally entire belly tinged with buff. Comparing *P. andium* to four other species within range, *P. amicus* has a smaller body size with proportionately larger ears (usually as long as hindfeet), small molars, and opisthodont upper incisors; *P. definitus* is much larger in almost all dimensions, especially toothrow; *P. haggardi* is a shorter-tailed species (tail <95 mm and shorter than head and body length) with longer palate, posteropalatal pits usually sunk in depressions, toothrows tending to converge posteriorly, zygomatic arches more widely flaring, and interorbital edges rather sharp; and *P. xanthopygus* has fluffier fur, hairier tail, heavier vibrissae, and larger ears. In Lima department, Peru, where *P. andium* contacts *P. xanthopygus*, there is almost no overlap between them in ear length (19–25 mm versus 24–29 mm, respectively.

DISTRIBUTION: *Phyllotis andium* occurs on both slopes of the Andes, although mostly on the western slope, from Lima department, Peru, north to Tungurahua province in Ecuador, mostly at elevations between 1,500 and 4,000 m but extending down the west slope in northern Peru (Lambayeque department) to about 300 m (Pearson 1958).

SELECTED LOCALITIES (Map 282): ECUADOR: Azuay, Yunguilla Valley (FMNH 53240); Cañar, Cañas (type locality of *Phyllotis andium* Thomas); Loja, Guachanamá (type locality of *Phyllotis fruticicolus* Anthony); Tungurahua, Baños (Pearson 1958). PERU: Amazonas, Río Utcubamba, 15 mi above Chachapoyas (type locality of *Phyllotis andium stenops* Osgood), Tambo Carrizal, mountains east of Balsas (type locality of *Phyllotis tamborum* Osgood); Ancash, Huaraz (FMNH 81202), Macate (FMNH 20935); Cajamarca, Cajamarca (FMNH 19465), El Arenal, 1 km S and 6 km W of Pomahuaca (MVZ 135774); Huánuco, Ambo (FMNH 24724); Lam-

bayeque, Chongoyape (Hershkovitz 1962); La Libertad, 5 mi SW of Otuzco (MVZ 138045); Lima, Huaros (MVZ 120015), Lomas de Lachay, 22 km N and 11 km W of Chancay (MVZ 139269), 5 mi E of Yauyos (MVZ 137638); Piura, Aisayacu (FMNH 84206), Porculla Pass (MVZ 138078).

SUBSPECIES: Despite the association as synonyms of four named taxa, both Pearson (1958) and Hershkovitz (1962) failed to document geographic trends worthy of recognition and thus regarded *P. andium* as monotypic.

NATURAL HISTORY: This is a species of brushy habitats on both sides of the Andes (Pearson 1972), where it replaces *P. xanthopygus* in central Peru. Specimens from the Rímac Valley in Lima department illustrate the contact between the two species. The brushy zone and the small areas of forest between these elevations are occupied by *P. andium*. On the south side at 4,000 m near Casapalca, traps among rock and grass with many woody lupine bushes (the upper limit of this lupine) caught about equal numbers of *P. xanthopygus* and *P. andium*. Among lupine bushes at 3,700 m, only *P. andium* was taken, and only *P. andium* among brush between 2,000 and 2,900 m. *Phyllotis andium* is sympatric with *P. haggardi* in southern Ecuador, with *P. limatus* and *P. x. posticalis* in central Peru, with *P. definitus* on the Pacific slope of northern Peru, and with *P. magister* and *P. amicus* in central and southern Peru (Pearson 1958). *Phyllotis andium* utilizes more leafy vegetation and less seed than most other species in the genus (43% grass, 31% forbs, 4% seeds, and 22% insects; Pizzimenti and DeSalle 1980). Along the Peruvian coast near Lima, *P. andium* show clear seasonal reproductive activity between July and October (Arana et al. 2002). Based on data from three individuals, the movement distance between captures varied between 40.8 and 68.8 m; the gestation period was 24 days, and the litter size varied between one and four young (Arana et al. 2002). Although *P. andium* is sympatric with several other congeneric species, the regions of overlap are always restricted to small contact areas, suggesting local spatial segregation.

REMARKS: Pearson (1958:440) concluded that the holotype of *melanius* Thomas was composite, consisting of a skin of *Akodon aerosus* (the dark color of which was, unfortunately, the basis for the name *melanius*) and a skull of a *Phyllotis*. Because Thomas (1913b) attached the name *melanius* to a *Phyllotis* and not *Akodon*, Pearson selected the skull as the type while noting that a skin-only specimen of a *Phyllotis andium* (also collected by P. O. Simons and bearing the same field number) should be associated with the type skull. This *Phyllotis* skin bears the locality designation of Cañar, Ecuador, which Pearson (1958) argued should be regarded as the correct type locality of *melanius* Thomas.

The karyotype of *P. andium* consists of a $2n = 64$, $FN = 72$, with five pairs of medium and small metacentrics and sub-

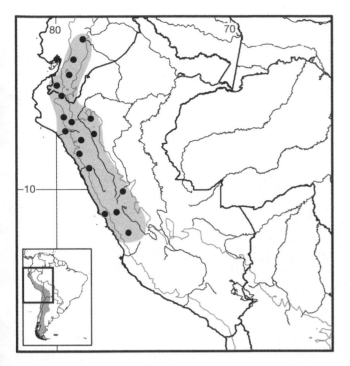

Map 282   Selected localities for *Phyllotis andium* (●). Contour line = 2,000 m.

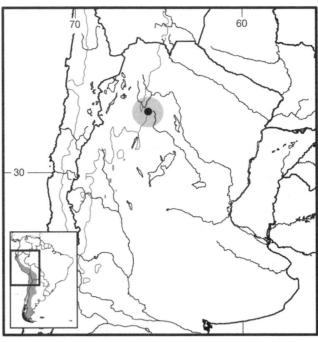

Map 283   Single known locality for *Phyllotis anitae* (●). Contour line = 2,000 m.

metacentrics, a graduated series of 26 pairs of acrocentrics, a large submetacentric X chromosome, and a medium-sized submetacentric Y chromosome (Pearson 1972).

Phyllotis anitae Jayat, D'Elía, Pardiñas, and Namen, 2007
Anita's Leaf-eared Mouse

SYNONYM:

Phyllotis anitae Jayat, D'Elía, Pardiñas, and Namen, 2007: 778; type locality "Argentina: Province of Tucumán, Department of Trancas, 10 km by road south of Hualinchay on the trail to Lara (26°19′20.2″S, 65°36′45.5″W, 2316 m."

DESCRIPTION: Distinguishable from congeners externally by large size (head and body length 89–123 mm, tail length 94–112 mm); dark gray dorsal coloration; light ochraceous venter; ears covered internally and externally by dark brown hairs; manus with digits and distal dorsum covered with white hair, proximal dorsum of both manus and pes dark; tail slightly bicolored and haired but not penicillate; markedly hypsodont molars; and narrow interorbital region hourglass in shape.

DISTRIBUTION: *Phyllotis anitae* is known only from the type locality in the Yungas forest in northwestern Argentina.

SELECTED LOCALITIES (Map 283): ARGENTINA: Tucumán, 10 km by road S of Hualinchay (type locality of *Phyllotis anitae* Jayat, D'Elía, Pardiñas, and Namen).

SUBSPECIES: *Phyllotis anitae* is monotypic.

NATURAL HISTORY: Little is known about this species. The type was trapped in the ecotone between montane forest and high-altitude grasslands, a sparse cloud forest habitat (Jayat, D'Elía et al. 2007), near isolated and large rocks. This species is believed to be a rare specialist of the high-elevation *Alnus* (alder, Betulaceae) forest.

REMARKS: *Phyllotis anitae* is sympatric with *P. osilae* and occurs near to some *P. xanthopygus* populations. It can be distinguished by its darker coloration, especially dorsally. The karyotype is unknown.

Phyllotis bonariensis Crespo, 1964
Buenos Aires Leaf-eared Mouse

SYNONYMS:

Phyllotis darwini bonariensis Crespo, 1964:99; type locality "Abra de la Ventana, part. de Tornquist, prov. de Buenos Aires, a 500 m s.m.," Parque Provincial Ernesto Tornquist, Buenos Aires, Argentina.

Phyllotis bonariensis: Reig, 1978:180: first use of current name combination.

DESCRIPTION: One of larger species (head and body length ranges from 127 to 151 mm), similar in size to *P. magister*, but with ears and hind legs relatively small, vibrissae well developed, dorsal coat light in color (similar to *P. xanthopygus rupestris*), and ventral pelage gray white (similar to *P. xanthopygus xanthopygus*), flanks and belly very well defined, and tail bicolored with scarce hairs and paler than the body. Pectoral buffy band very slightly marked.

Ear length (23–25 mm) shorter than hindfoot length (25–28 mm). Skull large (greatest length 31.4–35.1 mm); minimum interorbital width 4.1–4.5 mm, with sides diverging backward slightly; large posterolateral palatal pits located anterior to mesopterygoid fossa; distal portion of nasals markedly wide, 30–35% of length of nasal suture line; auditory bullar length (4.8–5.3 mm) shorter than upper alveolar molar length (5.4–6.1 mm). Upper incisors orthodont and anteriorly pigmented; upper molar rows parallel; M2 lacks procingulum.

DISTRIBUTION: *Phyllotis bonariensis* is restricted to the Sierra de la Ventana in Buenos Aires province, eastern Argentina.

SELECTED LOCALITIES (Map 284): ARGENTINA: Buenos Aires, Abra de la Ventana (type locality of *Phyllotis darwini bonariensis* Crespo).

SUBSPECIES: *Phyllotis bonariensis* is monotypic.

NATURAL HISTORY: *Phyllotis bonariensis* lives in rocky places at about 400 m elevation in a range of hills north of Bahía Blanca, in Buenos Aires department. There are no published data on ecology or behavior of this species.

REMARKS: Originally described as a subspecies of *P. darwini* (= *P. xanthopygus* herein), Reig (1978) and Galliari et al. (1996) elevated *P. bonariensis* to species status, a decision followed by some authors (e.g., Musser and Carleton 1993, 2005) but not by others (e.g., M. M. Díaz et al. 2006). This species is known only from the eight specimens used in the original description and a few collected subsequently. As noted by Pearson (1972), this taxon is geographically isolated from other species of *Phyllotis* by broad expanses of pampa. Molecularly, *P. bonariensis* is included within a widespread clade otherwise composed of

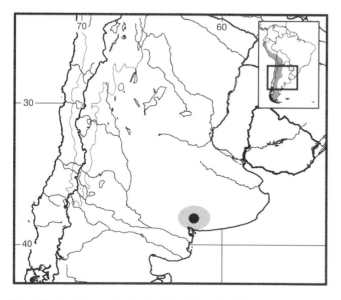

Map 284 Single known locality for *Phyllotis bonariensis* (●). Contour line = 2,000 m.

P. limatus, *P. x. xanthopygus*, *P. x. vaccarum*, and some *P. x. rupestris* (Steppan et al. 2007). A more thorough geographic sampling of these taxa is required to resolve species-level boundaries within this clade. The karyotype is unknown.

Phyllotis caprinus Pearson, 1958
Capricorn Leaf-eared Mouse
SYNONYMS:

Phyllotis caprinus Pearson, 1958:435; type locality "Tilcara, 8,000 feet, Jujuy, Argentina."
Phyllotis darwini caprinus: Hershkovitz, 1962: name combination.

DESCRIPTION: Large, relatively short-furred, with flat, sharp-edged, long-waisted interorbital region, long frontal bones, V-shaped frontoparietal suture, and heavy rostrum. Zygomatic plates sharply vertical or project forward slightly above, their anterior margins straight or concave; posteropalatal pits anterior or lateral to mesopterygoid fossa; nasals even with or project posteriorly behind the premaxillae. M3 small with or without an island; M2 with one outer entrant angle and usually second smaller notch.

In color, *P. caprinus* similar to sympatric or nearly sympatric *P. xanthopygus* but larger in body size (head and body length 102–140 mm, mean 118 mm, as opposed to 87–130 mm, mean of 103 mm), with thicker tail, and typically with buffy pectoral streak that is absent in latter.

DISTRIBUTION: *Phyllotis caprinus* occurs on the eastern Andean slopes in northern Argentina and southern Bolivia, between 2,100 and 4,500 m.

SELECTED LOCALITIES (Map 285): ARGENTINA: Jujuy, Humahuaca (Pearson 1958), Sierra de Zenta (Pearson 1958), Tilcara (type locality of *Phyllotis caprinus* Pearson). BOLIVIA: Chuquisaca, Camargo (MVZ 120121); Potosí, Yuruma (MVZ 120211); Tarija, Rancho Tambo, 61 km by road E of Tarija (AMNH 262987).

SUBSPECIES: *Phyllotis caprinus* is monotypic.

NATURAL HISTORY: *Phyllotis caprinus* occurs in brushy habitats in southern Bolivia and northern Argentina. At the type locality in northern Argentina, *P. caprinus* lives in brushy hedgerows and stonewalls, but just east of Tilcara, at Alfarcito where there is a mixture of brushy and open habitats, *P. caprinus* is sympatric with *P. xanthopygus*. Pearson (1958) collected large numbers of individuals among scattered bushes and *Opuntia* at 2,600 m in southern Bolivia.

REMARKS: Hershkovitz (1962:335–339) disputed Pearson's (1958) conclusion that *P. caprinus* was a species distinct from others in the genus and regarded it merely as a geographic race of *P. darwini* (= *P. xanthopygus* herein). Most other authors, however, have accepted the morphological and distributional data provided by Pearson as substan-

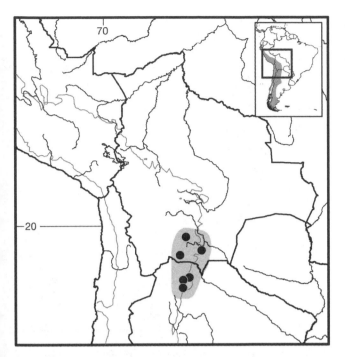

Map 285 Selected localities for *Phyllotis caprinus* (●). Contour line = 2,000 m.

tiating his hypothesis of species-level status (e.g., Cabrera 1961; Musser and Carleton 1993, 2005; Steppan 1998). Pearson (1958) suspected that *P. caprinus* was most closely related to "*P.*" *wolffsohni* (now *Tapecomys wolffsohni*), but Steppan's (1993) phylogenetic analysis of morphological traits placed the species as the sister to *P. darwini* (*sensu stricto*). Molecular phylogenetic analyses, to date, have not included *P. caprinus* (e.g., Steppan et al. 2007). The karyotype is $2n=38$, FN$=72$, with 36 size-graded biarmed autosomes, a large submetacentric X, and a small metacentric Y chromosome (Pearson and Patton 1976). It appears identical to that found in most populations of *P. xanthopygus*.

Phyllotis darwini (Waterhouse, 1837)
Darwin's Leaf-eared Mouse
SYNONYMS:

Mus (Phyllotis) darwinii Waterhouse, 1837:28; type locality "Coquimbo," Coquimbo, Chile.

Hesperomys darwini: Thomas, 1884:449; name combination.

Mus melanonotus Philippi and Landbeck, 1858:78; type locality "prov. Santiago prope locum [= near the place of] Angostura," Chile.

Mus dichrous Philippi, 1900:14, plate III, Fig. 2; type locality "provincia Santiago, prope [= near] Peine," Chile.

Mus megalotis Philippi, 1900:15, plate III, Fig. 3; type locality "provincia Santiago," Chile.

Mus mollis Philippi, 1900:23, plate VII, Fig. 1; type locality "provincia Santiago," Chile.

Mus illapelinus Philippi, 1900:28, plate IX, Fig. 1; type locality "prope [= near] Illapel," Coquimbo, Chile.

Mus Segethi Philippi, 1900:30, plate XI, Fig. 2; type locality "provinciae Santiago loco dicto Peine," Chile.

Mus campestris Philippi, 1900:38, plate XVI, Fig. 1; type locality "Choapa," Coquimbo, Chile.

Mus melanotis Philippi, 1900:39, plate XVI, Fig. 3; no type locality given, but certainly in Chile (Hershkovitz 1962:321).

Mus platytarsus Philippi, 1900:47, plate XIX, Fig. 4; type locality "La Ligua," Aconcagua, Chile.

Mus Boedeckeri Philippi, 1900:53, plate XXII, Fig. 2; type locality "provincia Maule; in praedio Coroney [= Coronel]," near Quirihue, Maule, Chile (see Osgood 1943b:202).

Mus griseoflavus Philippi, 1900:55, plate XXIV, Fig. 1; preoccupied by *Mus (Phyllotis) griseoflavus* Waterhouse; type locality "La Serena in provincia Coquimbo," Chile.

Phyllotis darwini: Thomas, 1902b:131; first use of current name combination.

Phyllotis darwini boedeckeri: Osgood, 1943b:202; name combination.

Phyllotis darwini fulvescens Osgood, 1943b:204; type locality "Sierra Nahuelbuta, west of Angol, Malleco, Chile. Altitude about 4,000 feet."

DESCRIPTION: Relatively short-furred, dark-colored coastal mouse with very large ears and occasional buffy pectoral streak. In comparison to *P. x. vaccarum*, which occurs at higher elevations in Chilean Andes, *P. darwini* has smaller body (mean head and body length, 119.5 versus 124.2 mm), tail (mean length, 119.2 versus 129.6 mm), skull (mean greatest length, 29.63 mm versus 31.90 mm), and especially molars (mean toothrow length, 5.26 versus 5.74 mm). Darker and browner in dorsal color, with paler lateral line; fur shorter; and ears more naked. In comparison to *P. x. rupestris*, which replaces *P. darwini* to north in coastal Chile, *P. darwini* somewhat smaller (mean head and body length, 105.0 versus 112.3 mm), colored darker brown, and with slightly larger (mean height, 26 versus 25 mm) and also more naked ears. Under parts may be nearly white but often creamy or even pale buffy rather than white (Osgood 1943b).

DISTRIBUTION: This species is limited to the coastal region and western slopes of the Andes in central and northern Chile, from Antofagasta south to Cautín regions, at an elevation range from sea level to 2,200 m (Osgood 1943b; Pearson 1958; Spotorno, Zuleta et al. 1998).

SELECTED LOCALITIES (Map 286): CHILE: Antofagasta, Paposo (MVZ 119184); Atacama and Parinacota, 10 mi N of Caldera (MVZ 150056), Vallenar (MVZ 118659); Araucanía, Galvarino (USNM 93254); Biobió, Sierra de Nahuelbuta (type locality of *Phyllotis darwini fulvescens* Osgood); Coquimbo, Cerro Potrerillos, 4 km E of

Guanaqueros (MVZ 150059), Las Palmas, 95 km N of Los Vilos (MVZ 150060), 10 km N of Pte. Los Molles (FMNH 119508); Santiago, Farellones (LCM 746); Valparaíso, Valparaíso (Pearson 1958).

SUBSPECIES: Osgood (1943b) listed three subspecies for *Phyllotis darwini* (*sensu stricto* as currently understood, not the broader view espoused by Osgood [1943b], Pearson [1958], and Hershkovitz [1962]) along the central coast of Chile: the broadly distributed nominotypical form was well as *P. d. boedeckeri* and *P. d. fulvescens*, both known only from their respective type localities near the coast. Pearson (1958) retained both *boedeckeri* Philippi and *fulvescens* Osgood as valid, the former as a short-tailed coastal form and the latter as a darkly colored inhabitant of *Araucaria* forests. Hershkovitz (1962) regarded *boedeckeri* Philippi as a synonym of *P. d. darwini* but retained *fulvescens* Osgood as a valid subspecies. As yet, lacking appropriate samples and analyses, no hypothesis regarding infraspecific groups within *P. darwini* has been tested adequately.

NATURAL HISTORY: *Phyllotis darwini* occurs in semiarid thorn scrub and sparse desert habitats. It is a granivorous-folivorous rodent with an average adult weight of about 50 g (Meserve and Le Boulengé 1987). Large numerical fluctuations and temporal variation in demographic variables has been found (M. Lima and Jaksic 1999; M. Lima et al. 2001), with population size varying between less than 10 (Meserve et al. 1995) to 230 individuals ha–[1] during El Niño events (M. Lima and Jaksic 1999; M. Lima et al. 1999). The mean home range length varied from 36.3 m (± 2.42 m s.d.) to 76. 9 m (± 9.00 m s.d.) (Fulk 1975). Breeding is markedly seasonal, starting in September and lasting until January or February (Fulk 1975; Meserve and Le Boulengé 1987). Females produce two or three litters per reproductive season (Fulk 1975: Meserve and Le Boulengé 1987). The mean litter size was calculated as 4.21 (L. I. Walker et al. 1984) and 5.2 (Meserve and Le Boulengé 1987).

REMARKS: As noted above, *P. darwini* has been traditionally viewed as a highly polytypic species of immense range (from central Peru to southern Argentina; Osgood 1943b; Pearson 1958; Hershkovitz 1962). Both Osgood (1943b) and Pearson (1958) recorded what they believed to be intergrades between *P. d. darwini* and *P. d. rupestris* (now *P. x. rupestris*) in northern Chile and between *P. d. darwini* and *P. d. vaccarum* (now *P. x. vaccarum*) where they replace one another elevationally in central Chile. However, L. I. Walker et al. (1984; see also L. I. Walker et al. 1999) demonstrated very limited hybrid production in laboratory crossings, with male hybrids exhibiting testicular sterility and female hybrids with greatly reduced fecundity, supporting reproductive isolation in the wild. Nevertheless, the failure of some subsequent authors to recognize the taxonomic restriction of *P. darwini* to central Chilean populations (the

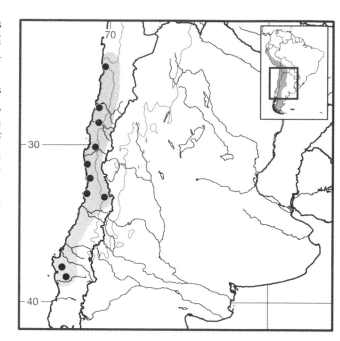

Map 286 Selected localities for *Phyllotis darwini* (●). Contour line = 2,000 m.

recommendation of L. I. Walker et al. 1984) has resulted in many studies and some GenBank sequences incorrectly attributing observations to *P. darwini* rather than correctly those of *P. xanthopygus*. *Phyllotis darwini* does not occur in Peru, Bolivia, Argentina, or high elevations in Chile (>2,200 m), and all studies from these areas are more correctly referred to *P. xanthopygus* or other species.

The karyotype consists of a $2n = 38$, FN = 72, with 18 pairs of metacentric and submetacentric autosomes plus a large, slightly submetacentric X chromosome and a small metacentric Y chromosome (Pearson 1972). This complement is identical to that of *P. xanthopygus* but with much larger blocks of centromeric heterochromatin on all chromosomes (L. I. Walker et al. 1984). It is also identical to *P. magister* except for the latter's subtelocentric Y chromosome (L. I. Walker et al. 1999). Hybrids between *P. darwini* and *P. magister* conform to Haldane's Rule, with fertile females but sterile males L. I. Walker et al. 1999).

Phyllotis darwini has been the focus of numerous studies on physiology and metabolism (e.g., Bozinovic and Nespolo 1997; Bacigalupe et al. 2012) and the quantitative genetic characterization of these traits.

Phyllotis definitus Osgood, 1915
Ancash Leaf-eared Mouse

SYNONYMS:

Phyllotis definitus Osgood, 1915:189; type locality "Macate, 50 miles northeast of Chimbote, [Ancash,] Peru. Altitude 9,000 ft."

Phyllotis magister definitus: Pearson, 1958:431; name combination.

Phyllotis darwini definitus: Hershkovitz, 1962:296; name combination.

DESCRIPTION: Large bodied (mean head and body length 126.3 mm, range: 111–146 mm) with tail about equal in length (mean 120.6 mm, range: 109–136 mm). Overall dorsal color tones deep and rich, cinnamon brown mixed liberally with dusky, but with rump slightly paler than back and shoulders and nape inclining to grayish. Nose, forehead, and sides of face decidedly grayish, in definite contrast to body. Venter buffy ochraceous, paler and more whitish on throat and inguinal region, with definite buffy band across pectoral region; all hairs slate gray at base. Forefeet and hindfeet white, with proximal part of pes more or less ochraceous. Inner surface of ears rufous. Tail bicolored, dusky above and buffy white below.

Skull large (mean length, 31.6 mm, range: 29.4 to 35.0 mm) and stoutly built, with broad nasals, heavy zygomata, and broad teeth as part of long toothrow (>5.8 mm). As noted by Pearson (1958, 1972), cranially *P. definitus* quite similar to *P. magister*.

DISTRIBUTION: *Phyllotis definitus* is confined to the upper slopes of the western Andes in Ancash department, Peru, with the four known localities ranging in elevation from 2,590 to 3,800 m (Pearson 1972).

SELECTED LOCALITIES (Map 287): PERU: Ancash, Macate (type locality of *Phyllotis definitus* Osgood), 1 km N and 12 km E of Pariacoto (MVZ 135794).

SUBSPECIES: *Phyllotis definitus* is monotypic.

NATURAL HISTORY: This species inhabits the brushy zone on the western Andean slopes where habitat loss may have reduced population sizes. It is sympatric with *P. andium*. No other data on reproduction, behavior, or ecology has been published.

REMARKS: *Phyllotis definitus* was described originally as a species (Osgood 1915), but was relegated by both Pearson (1958) and Hershkovitz (1962) to geographic races of *P. magister* or *P. darwini* (*sensu lato*), respectively. Pearson (1958:431) pointed out that *magister* and *definitus* were similar in most characters except color and occupied the same elevational zones on the western Andean slope. He noted, however, the lack of specimens from intervening localities between their respective ranges. In a subsequent study based on more specimens, Pearson (1972) reevaluated his 1958 opinion and agreed with Osgood's (1915) original hypothesis that *P. definitus* was a valid species, distinguishable from *P. magister* by both a richer coloration and karyotype. Steppan's (1993; see also Steppan and Sullivan 2000) cladistic analysis of morphological traits placed *P. definitus* either as the sister to *P. magister*, to the

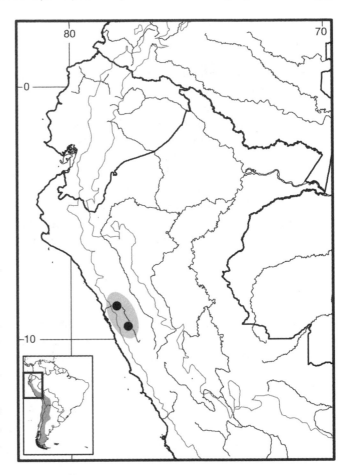

Map 287 Selected localities for *Phyllotis definitus* (●). Contour line = 2,000 m.

exclusion of *P. darwini* + *P. xanthopygus* although with low support, or as part of an unresolved polytomy that formed the base of *Phyllotis* (Steppan and Sullivan 2000). More thorough geographic sampling and molecular data are critically needed.

The karyotype is $2n = 54$, FN = 72, with 20 biarmed and 32 acrocentric autosomes, a large submetacentric X chromosome, and a smaller acrocentric that is probably the Y chromosome (Pearson 1972).

Phyllotis gerbillus Thomas, 1900
Peruvian Leaf-eared Mouse, Gerbil Leaf-eared Mouse
SYNONYMS:

Phyllotis gerbillus Thomas, 1900a:151; type locality "Piura, [Piura,] N.W. Peru, alt. 50 m."

Paralomys gerbillus: Thomas, 1926b:315; name combination, *Phyllotis gerbillus* Thomas, 1900 given as genotype of *Paralomys* Thomas.

Hesperomys [(*Paralomys*)] *gerbillus*: Ellerman, 1941:448; name combination.

P[hyllotis]. (*Paralomys*) *gerbillus*: Osgood, 1947:167; name combination.

DESCRIPTION: Smallest member of genus (mean head and body length 83.2 mm, range: 77–96 mm) with tail subequal in length relative to body (mean length 78.1 mm, range: 62–90 mm). Overall, upper parts of head and body ochraceous or salmon-colored, finely peppered with dark brown; upper half of sides of body like back but hairs without dark tips; lower half wholly white like under parts. Hairs of limbs, cheeks, and sides of rostrum white. Tail thinly haired throughout, overall white in color, but sometimes with occasional dark hair or patch of hairs, with integument of terminal one-fourth to one-third of tail dark brown, contrasting sharply with whitish basal portion; pencil rudimentary. Hindfoot broad and short (19 to 21 mm), sole sparsely set with whitish bristles except on six well-developed plantar pads. Ear comparatively small (17 to 18 mm); white preauricular spot usually present.

DISTRIBUTION: This species is restricted to the Sechura desert on the coast of northwestern Peru.

SELECTED LOCALITIES (Map 288): PERU: Lambayeque, 2 mi SE of Morrope (MVZ 138023); Piura, 10 km E of Piura (MVZ 135706), Reventazon, 64 km S and 19 km W of Sechura (MVZ 135710), 30 km SSE of Sechura (MVZ 135709), 10 km ENE of Talara (Pearson 1972).

SUBSPECIES: *Phyllotis gerbillus* is monotypic.

NATURAL HISTORY: *Phyllotis gerbillus* is nocturnal, spending the day underground, and does not need to drink free water (Koford 1968). The habitat is mounded with loose sand, from drifts around dead bushes to dunes several meters high and half covered with low green shrubs, principally thick-leaved evergreens, *Capparis scabrida*, *C. avicennifolia* (Capparaceae), and the deciduous legume *Prosopis juliflora* (Fabaceae); under many bushes, the ground was littered with dry fruits and seeds (Koford 1968). Habitat loss in the Sechura desert may threaten the remaining populations of this most morphologically distinctive *Phyllotis*. It may be caught together with the introduced *Mus musculus* but does not contact the ranges of any other member of the genus. The number of embryos in eight pregnant females ranged from one to four (average three; Pearson 1972).

REMARKS: The name refers to the gerbil-like and contrasting pale sandy dorsal versus white ventral coloration, a combination unique among species of *Phyllotis*. Pearson (1958) regarded *Paralomys* a valid genus closely related to *Phyllotis* but later changed his mind (Pearson 1972). Hershkovitz (1962) placed *gerbillus* within *Phyllotis*. Cladistic analysis of morphological characters led Braun (1993) to remove *gerbillus* from *Phyllotis* and resurrect *Paralomys*, including *amicus* within that genus. Steppan (1993; 1995) found support for removing *gerbillus* but felt the removal was premature given the weak signal in the data. More recent molecular data indicated that *P. gerbillus* is most closely related to *P. amicus* (Steppan et al. 2007),

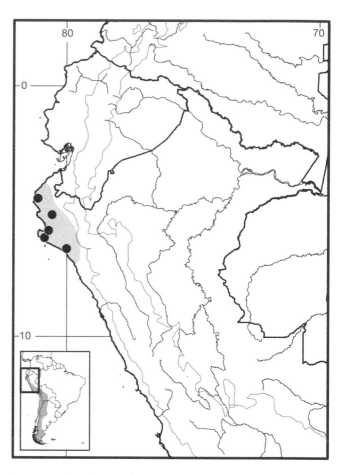

Map 288 Selected localities for *Phyllotis gerbillus* (●). Contour line = 2,000 m.

with these two species together with *P. andium* forming a well-supported clade within *Phyllotis*.

The karyotype is $2n = 38$, $FN = 72$, and consists of 18 pairs of metacentric and submetacentric autosomes, a large submetacentric X chromosome, and a small metacentric Y chromosome (Pearson 1972).

Phyllotis haggardi Thomas, 1898
Haggard's Leaf-eared Mouse

SYNONYMS:

Phyllotis Haggardi Thomas, 1898f:270; type locality "Mount Pichincha, [Pichincha,] Ecuador, altitude 3400–4000 metres."

H[esperomys]. elegans: Tomes, 1860a:213: part; not *Mus elegans* Waterhouse [= *Eligmodontia typus* F. Cuvier].

Phyllotis elegantulus Thomas, 1913a:139; type locality "Pallatanga, [Chimborazo,] Ecuador."

Phyllotis fuscus Anthony, 1924a:1; type locality "Contrayerbas, Provincia del Azuay, 11,000 feet, Western Andes," Ecuador.

Phyllotis haggardi haggardi: Pearson, 1958:442; name combination.

Phyllotis haggardi fuscus: Pearson, 1958:442; name combination.

Phyllotis haggardi elegantulus: Pearson, 1958:443; name combination.

DESCRIPTION: Small species (head and body length 99–115 mm), with proportionally short tail (less than head and body length, 72–89 mm), short ears, colored buffy to dark brown above with head and rump not markedly different from back; venter gray, well defined from sides, and usually lacking even hint of pectoral streak. Tail bicolored, clothed with short hairs and with virtually no terminal pencil. Palate relatively long for a *Phyllotis*, large posteropalatal pits tending to be sunk in depressions, upper toothrows tending to converge posteriorly, and nasals pointed at posterior ends and extending behind premaxillae. In comparison with *P. andium*, with which it overlaps in range, *P. haggardi* with more abruptly flared zygomatic arches, shorter maxillary toothrow (range: 4.0–4.7 mm versus 4.2–5.0 mm for Ecuadorean samples of *P. andium*), and shorter tails (length <95 mm versus >100 mm).

DISTRIBUTION: *Phyllotis haggardi* occurs on the upper slopes on both sides of the Andes through Ecuador, typically between 2,500 and 4,500 m, but may extend to as low as 2,000 m in Pichincha province.

SELECTED LOCALITIES (Map 289): ECUADOR: Chimborazo, Mt. Chimborazo (FMNH 53306), Urbina (Pearson 1958); Napo, Papallacta (AMNH 46824); Pichincha, Pichincha (type locality of *Phyllotis haggardi* Thomas), Solaya (FMNH 53309), Tanda (Pearson 1958).

SUBSPECIES: Pearson (1958) recorded substantial geographic variation within *P. haggardi*, with increasing duskiness in coloration from north to south, together with increased length of the skull and a narrowing and lengthening of the nasals. He recognized this trend by distinguishing a broadly ranging *P. h. fuscus* from the nominotypical form, which he restricted to the northern part of the range of the species in the vicinity of Pichincha and Quito to Mt. Mojanda. Pearson also recognized *P. h. elegantulus* Thomas, restricted to its type locality and based on only two specimens, although he did so with reluctance: "I have retained *elegantulus* as a subspecies primarily to avoid having to place the clearly defined *P. h. fucsus* into synonymy under the almost unknown *elegantulus*" (p. 443). Hershkovitz (1962) regarded *P. haggardi* as monotypic.

NATURAL HISTORY: This species usually lives at higher elevations than *P. andium* at the same latitude, although *P. haggardi* itself descends to relatively low elevation at the northern end of its range. Almost nothing is known about of its ecology and reproductive traits.

REMARKS: Pearson (1958:443) gave the type locality of *elegantulus* Thomas as "Nadonbi, Pallatanga, about 6,5000

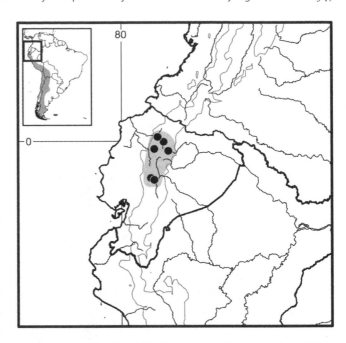

Map 289 Selected localities for *Phyllotis haggardi* (●). Contour line = 2,000 m.

feet" but suggested that this locality was likely incorrect. As noted by Tomes (1860a:211), the collection of which the holotype was a part was "believed to have been collected at Pallatanga, on the western slope of the Cordillera; but the exact locality is not certain, from the specimens having been unfortunately mixed together." Pearson suggested that the two known specimens might have actually come from near Quito or from the vicinity of Riobamba, where the collector L. Frazer had also worked.

The karyotype is $2n = 38$, FN = 72, with both autosomal and sex chromosome complements similar to those of *P. amicus*, *P. caprinus*, *P. darwini*, *P. gerbillus*, *P. magister*, and *P. xanthopygus* (Pearson 1972; Pearson and Patton 1976).

Phyllotis limatus Thomas, 1912
Lima Leaf-eared Mouse

SYNONYMS:

Phyllotis darwini limatus Thomas, 1912b:407; type locality "Chosica, near Lima, [Lima,] Peru. Alt. 850 m."

Phyllotis limatus: Steppan, 1998:582; first use of current name combination.

DESCRIPTION: A member of *darwini* species group, distinguishable from others by deep and narrow upper incisors (depth at midarc >90% of width across both incisors at cutting edge), short to moderate maxillary toothrow (4.2–5.8 mm), light coloration, venter frequently white, and two pairs of preputial glands. Steppan (1998) provided comparative measurements for *P. limatus*, *P. xanthopygus* (subspecies *chilensis*, *posticalis*, and *rupestris*), *P. darwini*, and *P. magister*.

DISTRIBUTION: This species is restricted to the arid coast and Pacific slopes of the Andes from Lima in central Peru south to Tarapacá in northern Chile, at an elevational range from sea level to 2,500 m in the north and sea level to 4,000 m in the south.

SELECTED LOCALITIES (Map 290): CHILE: Antofagasta, San Pedro de Atacama (MVZ 150069); Arica y Parinacota, Putre (FMNH 133844); Tarapacá, mouth of Río Loa (LCM 511). PERU: Arequipa, 43 km E of Arequipa (LSUMZ 27858), 5 mi ENE of Camaná (MVZ 143701), Chala (MVZ 139311); Ayacucho, 35 mi ENE of Nazca (MVZ 138092); Ica, 10 km SSE of Pisco (MVZ 136328); Lima, 4 km ENE of Pucusana (MVZ 137658), Rímac Valley (MVZ 120067), 1 mi W of Surco (MVZ 120023), Yangas (MVZ 136332); Tacna, Morro Sama, 65 km W of Tacna (MVZ 143722).

SUBSPECIES: Steppan (1998) documented a sharp step in the clinal length of the auditory bullae, with the break between samples in northern and central coastal Arequipa, Peru. He thus informally designated a "northern *limatus*" (from Lima south along the coast to extreme northwestern Arequipa) and a "southern *limatus*" (from central Arequipa south to Antofagasta, Chile). Steppan (1998) went on to note that, were one to recognize these geographic units formally, the southern group currently lacked a name.

NATURAL HISTORY: *Phyllotis limatus* is an inhabitant of arid foothills, rocky outcrops, foggy lomas among a pure stands of *Tillandsia* as well as in more varied habitat supporting cacti and scattered weeds (Pearson 1958) on the west slope of the Andes. Diet is primarily insects and forbs, consuming more of them than nearby *P. x. chilensis* of southern Peru (54% versus 35%) and a lower proportion of seed and grass (6% versus 28%; Pizzimenti and DeSalle 1980). The breeding season seems to coincide largely with the wet season (November to April), but some females were pregnant in every month. Litter size is modest, embryo counts in 29 females from the mountains of southern Peru averaged 3.69, but at Morro Sama on the Tacna coast to southern Peru, litters were larger than in any of the other populations during an El Niño event (embryo counts of 6, 6, 7, 2; Pearson 1976).

REMARKS: Pearson (1958) and Hershkovitz (1962) both regarded *limatus* to be a subspecies of *P. darwini* (*sensu lato*) and mapped its range to include Lima, Ica, and Arequipa regions in Peru. These authors regarded populations from southern Peru (southern Arequipa, Moquegua, and Tacna) as well as those in northern provinces of Antofagasta and Tarapacá in Chile to represent *P. darwini rupestris*. Steppan (1998), however, documented both the commonality of morphology of these more southern populations to *limatus* (especially in the width and depth of the upper incisors) and the composite nature of the samples

from this area that Osgood (1943b), Pearson (1958), and Hershkovitz (1962) had referred to *P. d. rupestris*. As a consequence, Steppan extended the range of *P. limatus* to Tarapacá and Antofagasta, Chile, while solidifying its status as a species separate from *P. xanthopygus* (*P. darwini* of these earlier authors). As noted in the account for *P. xanthopygus*, *Mus rupestris* P. Gervais might more correctly apply to *P. limatus* than to that species (see Steppan 1998). If so, it would be the senior name for what we, and earlier authors, herein call *P. limatus*.

Cytochrome-*b* sequences for the few specimens of *P. limatus* examined supported its monophyly, but these are nested within those of *P. xanthopygus*, including representatives of *P. x. chilensis*, *P. x. posticalis*, *P. x. rupestris*, and *P. x. vaccarum* as well as *P. bonariensis* (Steppan 1998; Steppan et al. 2007; Riverón 2011). Incomplete lineage sorting of nuclear RAG1 sequences (Steppan et al. 2007) due to recency of ancestry has likely resulted in the lack of complete reciprocal monophyly of these taxa. Mitochondrial affinity to nearby *P. x. rupestris* might be due either to mitochondrial introgression or very recent morphological divergence and speciation. Both Kuch et al. (2002) and Steppan et al. (2007) have argued that *P. limatus* was derived recently, perhaps during the most recent ice age.

Pearson (1972) recorded a karyotype for specimens from Lima department, Peru with $2n = 38$, $FN = 72$ and the same autosomal and sex chromosome morphology as representatives of three subspecies of *P. xanthopygus* (*chilensis*, *posticalis*, and *rupestris*).

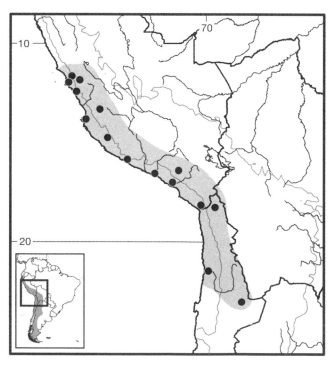

Map 290 Selected localities for *Phyllotis limatus* (●). Contour line = 2,000 m.

Phyllotis magister Thomas, 1912
Master Leaf-eared Mouse

SYNONYMS:

Phyllotis magister Thomas, 1912b:406; type locality "Arequipa, [Arequipa,] Peru. Alt. 2300 m."

Phyllotis magister magister: Pearson, 1958:430; name combination.

Phyllotis darwini magister: Hershkovitz, 1962:288; name combination.

DESCRIPTION: Large (head and body length 108–145 mm), coarse-haired species with buffy pectoral streak, long tail (averaging longer than head and body length, range: 117–158 mm), unusually large molars (toothrow length 5.5–6.5 mm), broad rostrum, sharp interorbital edges, nasals that seldom reach posterior to premaxillae, large posteropalatal pits anterior to mesopterygoid fossa, and bullae that usually taper evenly toward bullar tubes. Dorsal color buffy or ochraceous finely lined with black; head and shoulders sometimes contrastingly paler than back; under parts grayish, with gray basal color of hairs visible. Tail bicolored, thinly to moderately haired.

DISTRIBUTION: *Phyllotis magister* is distributed from west-central Peru to northern Chile on the western slope of the Andes. The northern limit appears to be Pacomanta, near Huarochiri, in Lima department (Arana-Cardó and Ascorra 1994), while the southern limit is the basin of the Río Loa near Calama, Antofagasta, Chile (Spotorno, Zuleta et al. 1998). It has been collected from sea level near the mouths of rivers draining the high elevations, but the normal habitat is between 2,300 and 4,000 m.

SELECTED LOCALITIES (Map 291): CHILE: Antofagasta, Calama (LCM 1817); Arica y Parinacota, Putre (USNM 541790). PERU: Arequipa, Arequipa (type locality of *Phyllotis magister* Thomas), 1 km N of Chivay (MVZ 174035); Huancavelica, 2 km E of Ticrapo, Pisco Valley (MVZ 136335); Lima, Pacomanta (Arana-Cardó 1994), 8 mi NE of Yauyos (MVZ 137663); Moquegua, 5 mi NE of Toquepala (MVZ 145610); Tacna, Morro Sama, 65 km W of Tacna (MVZ 143708), 6 km NE of Tarata (MVZ 141556).

SUBSPECIES: *Phyllotis magister* is monotypic.

NATURAL HISTORY: *Phyllotis magister* inhabits brushy zones on the Pacific slope of the Andes. It is omnivorous, but about 60% of its diet was based on five species of locally abundant forbs in southern Peru (Pizzimenti and DeSalle 1980). This species may be sympatric with *P. andium*, *P. limatus*, *P. x. posticalis*, *P. x. chilensis*, and *P. amicus*. Pearson (1976) recorded an average litter size of 2.3, based on embryo counts; L. I. Walker et al. (1999) reported two litters with a mean neonate count of 3.5.

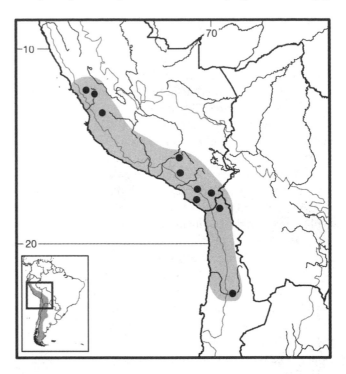

Map 291 Selected localities for *Phyllotis magister* (●). Contour line = 2,000 m.

REMARKS: *Phyllotis magister* has been almost uniformly treated as a valid species since Thomas's (1912b) original description (e.g., Ellerman 1941; Pearson 1958), with Pearson demonstrating morphological separation from sympatric or near-sympatric *P. darwini rupestris* in southern Peru (now *P. limatus*; see that account) and with *P. andium* in central Peru. Hershkovitz (1962) interpreted Pearson's data differently and regarded *magister* as a subspecies of *P. darwini* (*sensu lato*). Pearson and Ralph (1978) documented the ecological separation between these two species when in sympatry at several localities on their elevational transect in Tacna department, Peru. *Phyllotis magister* forms a well-supported clade in mtDNA cytochrome-*b* sequences completely separate from (although sister to) one that contains the various subspecies of *P. xanthopygus* as well as *P. limatus* and *P. bonariensis*, thus confirming its species status (Steppan et al. 2007).

The karyotype is $2n = 38$, $FN = 72$, with the same evenly graded series of metacentric and submetacentric autosomes as other members of the *darwini* complex. However, it appears to have a uniquely metacentric X chromosome and small acrocentric Y (Pearson 1972).

Phyllotis osgoodi Mann, 1945
Osgood's Leaf-eared Mouse

SYNONYMS:

Phyllotis osgoodi Mann, 1945:81; type locality "en los alrededores de Parinacota," Arica y Parinacota, Chile."

Phyllotis darwini rupestris: Hershkovitz, 1962:304; part; neither *darwini* Waterhouse nor *rupestris* Gervais.

DESCRIPTION: As indicated in the Key to species and noted in the Remarks here, with present information this species can be diagnosed definitively only by its karyotype. Mann (1945) described *P. osgoodi* from a single individual (a young male) captured among rocks near Parinacota; specimen described as a mouse with dorsal buffy-brown coloration and pale neutral-gray belly, long tail (53% of total length), auditory bullae of regular size, and interorbital region posteriorly extended. However, Spotorno (1976) published a redescription of *P. osgoodi* based on evaluation of five specimens that emphasized the following features: medium sized, cryptic and sympatric with *P. xanthopygus chilensis*, with variable coat color (predominantly gray on brown). Skull might be distinguishable from *P. xanthopygus chilensis* by molar cusps and folds, more symmetrical and rounded in adults; both hypoflexus and metaflexus of M2 tend to be symmetrical. In addition to central constriction, stomach with prepyloric constriction visible both external and internally. Base of baculum with clear notch in dorsal-ventral view. Karyotype with 40 chromosomes (NF = 72); X chromosome submetacentric, Y metacentric.

DISTRIBUTION: *Phyllotis osgoodi* is known only from its type locality in the Puna of northern Chile.

SELECTED LOCALITIES (Map 292): CHILE: Arica y Parinacota, Parinacota (type locality of *Phyllotis osgoodi* Mann).

SUBSPECIES: *Phyllotis osgoodi* is monotypic.

NATURAL HISTORY: This species has not been studied in the field, so no data on ecology or behavior are available. Little is known about this species because of the difficulty in identifying it.

REMARKS: The status of *P. osgoodi* as a species, or even as a valid subspecies of another *Phyllotis*, is in doubt, largely owing to the fact that no morphological traits have been identified to distinguish it from sympatric *P. xanthopygus*. Indeed, Pearson (1958:415) listed *osgoodi* Mann as a synonym of *chilensis* Mann, reasoning that a "single immature specimen of *osgoodi* and the few specimens of *chilensis* available from the type locality could hardly have been adequate to define two similar species from the same area" (p. 416). He went on to note that, while *osgoodi* had page priority over *chilensis*, he believed "it is wiser to base this form on an old specimen at hand [= the type of *chilensis*] than on an immature specimen [= type of *osgoodi*]." Hershkovitz (1962) placed *osgoodi* Mann as a synonym of *Phyllotis darwini* (= *xanthopygus*) *rupestris*, agreeing with Pearson that *osgoodi* did not warrant recognition while disagreeing with him as to the status of *chilensis* vis-à-vis *rupestris*. Nev-

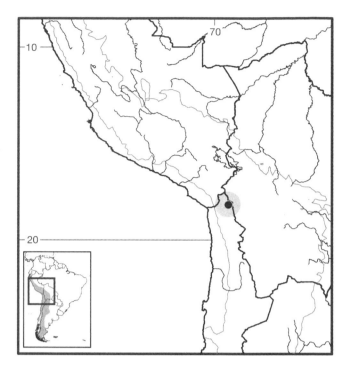

Map 292 Single known locality for *Phyllotis osgoodi* (●). Contour line = 2,000 m.

ertheless, Musser and Carleton (1993, 2005) elevated *osgoodi* Mann to species status based on Spotorno (1976) and Spotorno and Walker (1979, 1983). Samples these authors identified as *P. osgoodi* had a 2n = 40 karyotype, as opposed to 2n = 38, and were basal to *P. darwini* (*sensu stricto*) and *P. xanthopygus* in electrophoretic phenograms. Unfortunately, the morphometric analyses that Spotorno and Walker (1983) presented to distinguish the latter two species did not include their specimens of *P. osgoodi*, nor did they apparently compare either to Mann's type. Consequently, it is not possible as yet to diagnose *P. osgoodi* with reliable morphological characters or even to determine whether the specimens examined by Spotorno and colleagues are correctly assigned. Steppan's (1998) initial mtDNA cytochrome-*b* sequence analysis, which included a specimen (LCM 638) from the type locality at Parinacota, seemed to support the hypothesis of Spotorno and colleagues, although this sequence was subsequently found to be composite (Steppan et al. 2007). Resequencing of a small fragment of that specimen, however, suggested it was a basal member of the *darwini* species group, the sister species to either the *P. xanthopygus* complex or to *P. magister* (S. J. Steppan, unpubl. data). Clearly, additional series from the vicinity of the type locality, with all specimens karyotyped and sequenced and properly compared morphologically to other taxa from the general area, will be necessary before the status of *osgoodi* Mann can be firmly established.

Phyllotis osilae J. A. Allen, 1901
Bunch Grass Leaf-eared Mouse

SYNONYMS:

Phyllotis osilae J. A. Allen, 1901b:44; type locality "Osila" [= Asillo, Puno], Peru (alt. about 12,000 feet)."

Phyllotis lutescens Thomas, 1902b:131; type locality "Choro [= El Choro,] 3500 m," Cochabamba, Bolivia.

Phyllotis darwini tucumanus Thomas, 1912b:408; type locality "Cunbre [= Cumbre] de Mala-mala, Sierra de Tucuman. Alt. 3300 m," Tucumán, Argentina.

Phyllotis tucumanus: Thomas, 1919d:492; name combination.

Phyllotis nogalaris Thomas, 1921k:611; type locality "Higuerilla, 2000 m., in the Department of Valle Grande, about 10 km. east of the Zenta range and 20 km. of the town of Tilcara," Jujuy, Argentina (Thomas 1921k:609). Pardiñas et al. (2007) equated Higuerilla (an abandoned ranch) to the modern village of Pampichuela (23°32'S, 65°02'W, 1,735 m).

Phyllotis phaeus Osgood, 1944:193; type locality "Limbani, Puno, Inambari drainage, Peru. Altitude about 9,000 feet."

Phyllotis osilae osilae: Pearson, 1958:426; name combination.

Phyllotis osilae tucumanensis: Pearson, 1958:427; name combination.

Phyllotis osilae nogalaris: Pearson, 1958:427; name combination.

Phyllotis osilae phaeus: Pearson, 1958:427; name combination.

DESCRIPTION: Both Pearson (1958) and Hershkovitz (1962) characterized *P. osilae* in detail and noted numerous external and cranial characters that separate it from sympatric *P. xanthopygus* (*P. darwini* to them). Head and body length slightly longer in *P. osilae* than *P. xanthopygus* in southern Peru (*P. o. osilae* versus *P. x. chillensis*) but reverse in Catamarca, Argentina (*P. o. tucumanensis* versus *P. x. vaccarum*). Tail of *P. osilae* varies from about 80% to 130% of body length and ears comparatively small, rarely exceeding 25 mm and usually less than 23% of combined head and body length. Upper parts of body buffy or ochraceous to dark brown, more or less mixed with blackish; animals from more humid eastern Andean slopes in southern Peru and northern Bolivia (*P. o. phaeus*) noticeably darker and cocoa colored than those from adjacent Altiplano (*P. o. osilae*). Under parts rarely sharply defined, usually dark gray with plumbeous basal portions of hairs visible. Pectoral streak, patch, or midventral line typically present. Inner and outer sides of ears thinly haired but with dark-brown edging. Tail clothed with short hairs and terminates in weakly developed pencil; sharply bicolored or with terminal one-half to two-thirds or more uniformly brown.

Hershkovitz (1962:Fig. 86) provided idealized drawings of ear, ventral body, ventral cranium, and baculum illustrating differences between *P. osilae* and *P. darwini* (*sensu lato*).

DISTRIBUTION: *Phyllotis osilae* occurs from southern Peru to northwestern Argentina, along the eastern Altiplano and Andean slopes, from 2,700 to 4,000 m in the north and down to 500 m in the south.

SELECTED LOCALITIES (Map 293): ARGENTINA: Catamarca, Chumbicha (Pearson 1958); Jujuy, Higuerilla (type locality of *Phyllotis nogalaris* Thomas), 1 mi W of León (MVZ 120164); Tucumán, Cunbre [= Cumbre] de Mala-mala, Sierra de Tucumán (type locality of *Phyllotis darwini tucumanus* Thomas). BOLIVIA: Cochabamba, El Choro (type locality of *Phyllotis lutescens* Thomas), 6 mi W of Parotani (MVZ 141579), 10 mi NE of Punata (MVZ 120175); La Paz, Sorata (AMNH 91504); Potosí, Lagunillas (Hershkovitz 1962); Tarija, 10 mi NE of Tarija (MVZ 120165). PERU: Cusco, Machu Picchu (USNM 194576), 90 km SE (by road) of Quillabamba (MVZ 166793); Puno, Asillo (type locality of *Phyllotis osilae* J. A. Allen), Hacienda Ontave (MVZ 141570), Huacullani (FMNH 52596), Limbani (type locality of *Phyllotis phaeus* Osgood), 6 km N of Putina (MVZ 171540), Santa Rosa [de Ayaviri] (MVZ 120159).

SUBSPECIES: Both Pearson (1958) and Hershkovitz (1962) recognized four subspecies of *P. osilae*: *P. o. osilae* from Cusco department in southern Peru south across Bolivia to northern Argentina, in bunch grass habitats between 2,750 and 4,200 m; *P. o. tucumanensis*, on the eastern slope of the Andes in Tucumán and Catamarca provinces, Argentina, across a broad elevation amplitude from 500 to 5,000 m; *P. o. nogalaris*, known only from its type locality in Jujuy province, Argentina; and *P. o. phaeus*, from the eastern slope of the Andes in Puno department, Peru south to La Paz department, Bolivia between 3,400 and 4,300 m.

NATURAL HISTORY: *Phyllotis osilae* is largely restricted to *Stipa*-dominated bunch grass habitats on the Altiplano, notably in thick stands of *Stipa ichu* (Pearson 1958). It was caught on grassy hillsides and on the grassy banks of gullies at Pairumani, Peru. It differs, therefore, from the rock-inhabiting *P. xanthopygus*. No pregnant females or breeding males were caught at Pairumani in July. The diet of *P. osilae* resembles both *P. xanthopygus* and *P. magister* in being omnivorous and including a wide variety of items. It differs from both in that it relies most heavily on grass and grass seeds (Pizzimenti and DeSalle 1980). Dorst (1971, 1973b) reported *P. osilae* ate mostly leafy vegetation plus some insects and seeds.

REMARKS: *Phyllotis osilae* is another species very similar to *P. xanthopygus* in many respects, although not closely related phylogenetically (Steppan 1998; Steppan et al. 2007). West of Lake Titicaca, convergence between

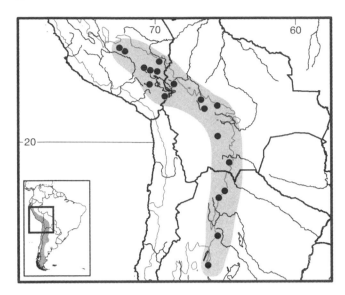

Map 293 Selected localities for *Phyllotis osilae* (●). Contour line = 2,000 m.

these two is most dramatic, and no single morphological character is diagnostic. Hershkovitz (1962) provided detailed summaries of several transects, including the means of distinguishing *P. xanthopygus* (= *P. darwini* to him) from *P. osilae*. Some of Hershkovitz's difficulties in identification probably stemmed from the composite nature of his *P. darwini*, but individuals can be difficult to assign (S. J. Steppan, unpubl. data). There is no genetic evidence for hybridization (Steppan et al. 2007; S. J. Steppan, unpubl. data).

The karyotype is $2n = 70$, FN = 68, with 39 pairs of acrocentric autosomes grades in size from large to small and acrocentric X and Y chromosomes (Pearson 1972). It is markedly distinct from the 38-chromosome complement that characterizes *P. xanthopygus* and other members of the *darwini* complex of species.

Phyllotis xanthopygus (Waterhouse, 1837)
Yellow-rumped Leaf-eared Mouse, Patagonian Leaf-eared Mouse

SYNONYMS:

Mus (*Phyllotis*) *xanthopygus* Waterhouse, 1837:28; type locality "Santa Cruz," Santa Cruz, Argentina.

[*Hesperomys*] *zanthopygus* Waterhouse, 1839:76; incorrect subsequent spelling of *xanthopygus* Waterhouse.

Mus rupestris P. Gervais, 1841:51; type locality "un trou de rocher des hautes montagnes de Cobija," Antofagasta, Chile (see Pearson 1958).

Mus capito Philippi, 1860:159; type locality "Hueso parado," [= Hueso Parado, near Taltal,] Antofagasta, Chile (Osgood:1943b:205); preoccupied by *Mus capito* Olfers [= *Hylaeamys megacephalus* (G. Fischer)].

Hesperomys Xanthopygus: Burmeister, 1879:225; name combination.

Hesperomys glirinus Philippi, 1896:19, plate VII, Fig. 3; type locality "Atacama," [= San Pedro de Atacama,] Antofagasta, Chile.

Hesperomys lanatus Philippi, 1896:19, plate VII, Fig. 2; type locality "Atacama," [= San Pedro de Atacama,] Antofagasta, Chile.

Phyllotis xanthopygus: E.-L. Trouessart, 1897:534; first use of current name combination.

[*Akodon* (*Akodon*)] *rupestris*: E.-L. Trouessart, 1987:535; name combination.

Phyllotis arenarius Thomas, 1902c:224; type locality "Uyuni. Alt. 3670 m," Potosí, Bolivia.

Mus glirinus: Philippi, 1900:59; name combination.

Mus lanatus: Philippi, 1900:59; name combination.

Phyllotis darwini posticalis Thomas, 1912b:406; type locality "Galéra, W. of Oroya, Department of Junin, Peru. Alt. 4800 m."

Phyllotis darwini vaccarum Thomas, 1912b:408; type locality "Las Vacas [= Punta de Vacas; see Paynter 1995], Argentine slope of Cordillera opposite Mendoza. Alt. 2500 m," Mendoza, Argentina.

Euneomys (*Auliscomys*) *xanthopygus*: Thomas, 1916a:143; name combination.

Phyllotis ricardulus Thomas, 1919d:493; ; type locality "'Otro Cerro,' North-eastern Rioja," [actually Catamarca,] Argentina (see Pardiñas et al. 2007 for determination of this locality).

Phyllotis oreigenus Cabrera, 1926:319; type locality "Laguna Blanca, Catamarca (3,100 m. de altura)," Argentina.

Phyllotis abrocodon Thomas, 1926b:316; type locality "Oroya [= La Oroya]. 12,000'," Junín, Peru.

Phyllotis darwini rupestris: Osgood, 1943b:205; name combination.

Phyllotis darwini xanthopygus: Osgood, 1943b:208; name combination.

Phyllotis wolffhuegeli Mann, 1944:102, 108; type locality "Bocatoma, Lo Valdés, Cajón del Río Volcán, a 1.800 m. de altura," Santiago, Chile.

Phyllotis arenarius chilensis Mann, 1945:84; type locality "los alrededores de Parinacota," Tarapacá, Chile.

Phyllotis darwini chilensis: Pearson, 1958:415; name combination.

Phyllotis darwini ricardulus: Pearson, 1958:418; name combination.

Phyllotis chilensis: Steppan, 1995:79; name combination.

DESCRIPTION: Highly variable geographically, as noted below under Subspecies. As a consequence, difficult to encapsulate morphology of *P. xanthopygus* in short and simplistic manner, a task made more difficult since revisions of Pearson (1958) and Hershkovitz (1962), both of which contain detailed descriptions across geography, include two nominal taxa (*P. darwini* and *P. limatus*) that have since been elevated to species status.

Body size average for genus, with head and body length almost always longer than 90 mm (range: 87–148 mm). Tail relatively long, almost always longer than 90 mm and usually longer than 90% of head and body length (range: 89–165 mm). Ears large, sometimes as long as hindfoot (ear length 22–32 mm; hindfoot length 19–31 mm). Dorsal fur denser at higher elevations, less fluffy at lower ones, a typical characteristic for nearly any species distributed across wide elevational band that encompasses warm to cold climates. Vibrissae elongated and luxuriant. Dorsal color varies greatly geographically pale buff through gray and darker tones, depending upon elevation and latitude. Ventral color whitish, grayish, to buffy; pectoral streak variably present both within and among populations. Skull with long and slender rostrum, even in old specimens; palate extends beyond plane of last molars; posteropalatal pits anterior to anterior border of mesopterygoid fossa; bullae inflated, with tubes short and usually emerging abruptly rather than tapering evenly away from each bulla; interorbital edges square or slightly rounded, only occasionally sharp; and frontoparietal suture usually broadly U-shaped.

Most easily confused with *P. andium* near northern edge of range, and with *P. magister*, *P caprinus*, *P. osilae*, and *Tapecomys wolffsohni* near center of range. One should consult Pearson (1958) in particular but also Hershkovitz (1962) for details of distinguishing any single pair of these species in areas of overlap or near contact.

DISTRIBUTION: *Phyllotis xanthopygus* has the largest range of any member of the genus, with a distribution in the high Andes, and adjacent brush habitats to the east, from central Peru (Junín department) to the southern tip of continental South America (the provinces of Magellanes in Chile and Santa Cruz in Argentina), over an elevational range from 1,900 to 5,030 m in the north (and, like *P. limatus* and *P. magister*, isolated sea level populations at river mouths) and from sea level to 2,000 m in the south. This 5,000 m elevational spread is one of the greatest among mammals. East of the Andes in Argentina, populations extend to lower elevations than to the west in Chile and Peru, where instead *P. xanthopygus* is replaced by *P. darwini* or *P. limatus* below about 2,000 m.

SELECTED LOCALITIES (Map 294): ARGENTINA: Catamarca, Otro Cerro (type locality of *Phyllotis ricardulus* Thomas), 22 km NE of Tinogasta (MVZ 134257); Chubut, 4 km W of Lago Blanco (MVZ 151026), 5 km W of Leleque (MVZ 152166), Pico Salamanca, Comodoro Rivadavia (Pearson 1958); Córdoba, Pampa de Achala, 14 km E of Cura Brochero (MVZ 165848); Jujuy, Tilcara (MVZ 141417); Mendoza, 35 km WNW of 25 de Mayo,

Cerro Medio (MVZ 181569); Neuquén, 3 km W of Rahue (MVZ 173703), 7 km NE of Zapala (MVZ 134261); Río Negro, Cerro Leones, 15 km ENE of Bariloche (MVZ 171145), Huanuluan (MCZ 19029); San Juan, Reserve Ischigualasto (MNC 1107); Santa Cruz, 4.4 km E of Los Antiguos, Estancia la Aurora (MVZ 206375), Monumento Nacional Bosque Petrificado (MVZ 160119), Santa Cruz (type locality of *Mus* (*Phyllotis*) *xanthopygus* Waterhouse). BOLIVIA: Chuquisaca, Tarabuco (AMNH 263899); La Paz, 20 mi S of La Paz (MVZ 120054); Oruro, 40 mi S of Oruro (MVZ 120100); Potosí, 20 mi S of Potosí (MVZ 120106), 5 mi N of Villazón (MVZ 120108), Uyuni (type locality of *Phyllotis arenarius* Thomas); Tarija, 12 mi NW of Iscayachi (MVZ 141583). CHILE: Antofagasta, Ojos San Pedro, 55 mi NE of Calama (MVZ 118660), 5 mi S of San Pedro de Atacama (MVZ 150069); Arica y Parinacota, Timar (USNM 391808); Atacama, Salar de Pedernales (LCM 405); Magallanes y Antártica Chilena, Laguna Lazo, Última Esperanza (FMNH 50539); Santiago, Farellones (LCM 730). PERU: Apurimac, 36 km S (by road) of Chalhuanca (MVZ 174041); Arequipa, 5 km W of Cailloma (MVZ 174039); Ayacucho, 2 mi SE of Huanta (MVZ 141587), 15 mi WNW of Puquio (MVZ 138113); Cusco, Marcapata (FMNH 75464), Occobamba Pass (USNM 194577); Huancavelica, 3 mi SE of Izcuchaca (MVZ 141593), Río Mantaro (MVZ 120006); Junín, 1 mi E of Casapalca (MVZ 120081), 10 km W of Jauja (MVZ 139310); Pasco, Huariaca (Hershkovitz 1962); Puno, Abra Aricoma, 13 mi ENE of Crucero (MVZ 139561), 11 km W and 12 km S of Ananea (MVZ 172774), 4 km NE of Pomata (MVZ 115873); Tacna, Morro Sama, 65 km W of Tacna (MVZ 143748), 5 km SE of Tarata (FMNH 107577).

SUBSPECIES: Excluding those nominal taxa regarded herein as valid species (*P. darwini*, *P. limatus*), and their synonyms, Pearson (1958) recognized, diagnosed, and mapped the ranges of six subspecies, including: (1) *P. x. chilensis*, a dark gray, short-tailed taxon intermediate in size and color between *P. x. posticalis* and *P. x. rupestris*, distributed on the Altiplano west and southwest of Lake Titicaca in Peru and Bolivia, and the high Andes between Tarapacá, Chile, and Bolivia above 3,300 m; (2) *P. x. posticalis*, a large and long-tailed subspecies with gray back and sides, larger than *P. x. chilensis*, which it replaces in the mountains of central and southern Peru, on both slopes of the Andes to elevations as low as 1,800 m; (3) *P. x. ricardulus*, a richly colored animal with a bright buffy lateral line, buffy under parts, large ears, and relatively hairy tail, that is distributed along the central and eastern Andes in northwestern Argentina at elevations from 500 to 4,800 m; (4) *P. x. rupestris*, a small, short-tailed, and pale race that ranges across the Bolivian plateau, extending down the eastern slope into Jujuy, Argentina, and at intermediate elevations on the west slope in northern Chile and southern Peru; (5) *P. x. vaccarum*, a large-bodied, long-tailed, pale with more gray on the head and neck form that occurs on both sides of the Andes from Atacama (Chile) and Catamarca (Argentina) in the north to Talca (Chile) and Neuquen (Argentina) in the south, between 800 and 4,000 m; and *P. x. xanthopygus*, a large animal with a proportionately short tail, relatively dark coloration with a distinctly buffy venter that ranges from Neuquen province, Argentina, south almost to the Straits of Magellen and reaching the coast in Santa Cruz, Argentina, with one locality in Chile at the southern end of its range.

NATURAL HISTORY: This is a mouse of rocky microhabitats on the arid Andean slopes where boulder fields, rock slides, cliffs, small shale outcroppings, stone walls, and stone huts are all satisfactory habitations. The species is also found in rocky outcrops in the monte and Patagonian steppes of Argentina and at higher elevations in the Sierras de San Luis and Córdoba, Argentina. *Phyllotis xanthopygus* is sympatric with (and at times difficult to distinguish from) *P. osilae* (see that account), and may be sympatric in places with *P. magister*, *P. andium*, or *P. limatus* on the western Andean slopes. Like *P. magister* and *P. limatus*, populations of *P. xanthopygus chilensis* exist near sea level along the mouths of rivers, giving the species an elevational range of nearly 5,000 m, one of the largest for any mammal. As a consequence, the species has been the subject of physiological studies regarding adaptation and acclimatization to high altitude (e.g., Bozinovic and Marquet 1991).

Phyllotis xanthopygus is strictly nocturnal. Several wild-caught individuals had green vegetable pulp in their stomachs, but a caged specimen ate assorted grains and ignored the tola bush (*Senecio adenophylloides*; Pearson 1951). In the southern Peruvian Altiplano, pregnant females were caught at Santa Rosa on July 28 and August 1, but at Caccachara none were caught until October 5. Pregnant females and breeding males were caught frequently between this date and December 19, and immature specimens were caught on December 9. The average number of embryos was four. In all six cases where the necessary data were recorded, there were more embryos in the right horn than in the left. The ratio of sexes was 64 males to 44 females (Pearson 1951). Kramer et al. (1999) summarized additional details, although some studies referenced were actually conducted on *P. darwini*, as currently understood.

REMARKS: *Phyllotis xanthopygus* represents the residuum from earlier concepts of *P. darwini*, particularly that of Pearson (1958), after removal of *P. bonariensis* (Musser and Carleton 1993; Steppan 1995), *P. limatus* (Steppan 1998), and true *P. darwini* (L. I. Walker et al. 1984). This

remains by far the most widespread species in the genus, and it was suggested to be the parental species for others through a process of peripheral isolate formation and successive range expansions across the Andes onto the Pacific slopes (Steppan 1998). It is paraphyletic with respect to *P. limatus* (Kuch et al. 2002; Riverón 2011). Additional DNA sequence data suggest that this is a composite taxon, containing at least one and possibly three additional species (Steppan et al. 2007; S. J. Steppan et al., unpubl. data), and the taxonomy will likely change pending ongoing revisionary studies. Riverón (2011) detected multiple deep divergences and strong geographic structuring among mitochondrial lineages of southern populations. It seems likely that northern, Altiplano populations currently allocated to the subspecies *posticalis*, *chilensis*, and parts of *rupestris* will be removed to one or more additional species.

The correct attribution of Philippi's names *Mus capito*, *Hesperomys glirinus*, and *Hesperomys lanatus* to this species, which Osgood (1943), Pearson (1958), and Hersh-

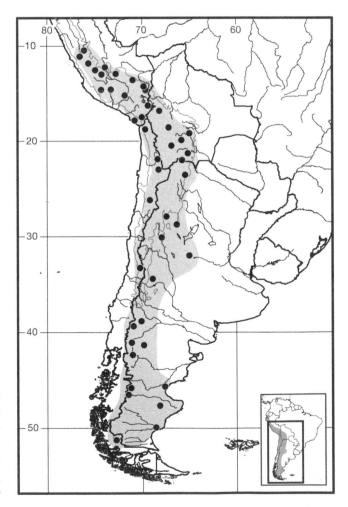

Map 294 Selected localities for *Phyllotis xanthopygus* (●). Contour line = 2,000 m.

kovitz (1962) all listed as synonyms of *P. darwini rupestris* (= *P. xanthopygus rupestris*, herein) is provisional. As noted by Steppan (1998), the types have been lost, so it not possible to determine whether these names are most appropriately applied here or more correctly to *P. limatus* instead.

Genus *Salinomys* Braun and Mares, 1995
Cecilia Lanzone and Janet K. Braun

This monotypic genus ranges between about 30°S and 34°S in southern South America, with the single species *Salinomys delicatus* having both a restricted and patchy distribution, with local populations of small size. Only a few studies of the ecology of the species have been published since its description (Braun and Mares 1995). The species occurs in the arid and semiarid landscape of west-central Argentina, especially in halophytic habitats, with individuals exhibiting morphological and physiological adaptations to these extreme environments (G. B. Diaz and Ojeda 1999). The few studies available indicate special ecological, physiological, and genetic characteristics that warrant further investigation. The following generic description is abstracted from Braun and Mares (1995). External measurements are from D. Rodriguez et al. (2012a), who examined the largest available sample and included data on the holotype and other specimens used in the original description.

Salinomys has large ears, whitish postauricular and supraorbital patches, soft grayish olive fur, a white venter, and a tufted bicolored tail that exceeds the combined length of the head and body. The hindfeet are long, with unwebbed digits; the plantar surface is naked but has six small tubercles (thenar, hypothenar, and four interdigitals). These rodents are small, with a mean head and body length of 71.2 mm (range: 50.0–84.0 mm), tail length 102.1 mm (range: 85.0–123.0 mm), hindfoot length 21.3 mm (range: 18.0–23.0 mm), and ear length 17.0 mm (range: 15.0–21.0 mm).

The skull is delicate, with a long and slender rostrum, slightly convergent supraorbital margins, and an inflated braincase. The foramen magnum is dorsoventrally flattened and oriented slightly posteroventrally. The zygomatic arches are parallel to slightly convergent anteriorly (widest across the squamosal roots). The anterior margins of the zygomatic plates are more or less straight; zygomatic notches are deep (>50% of zygomatic plate width). Incisive foramina are long (>90% of the diastemal length); palatal foramina are often long slits. The stapedial process of each bulla extends nearly to or is in contact with the pterygoid plate. The bullae are not inflated, and the ectotympanic does not completely cover the periotic bone. A vertical

strut of the alisphenoid is present. The coronoid process of the mandible is small. A capsular process is present. The angular notch does not extend anteriorly beyond the capsular process.

The incisors are orthodont; the molars are tetralophodont, brachydont, and slightly crested. The anterocone of M1 is divided; an anteromedian style or shelf is present. The posterior border of M1 has a small notch.

Axial skeletal elements have modal counts of 12 ribs, 19 thoracolumbar vertebrae, 3–4 sacral vertebrae, and 31–33 caudal vertebrae. The entoglossal process of the hyoid is absent. The entepicondylar foramen of the humerus is absent.

Kidneys have a single, very elongate renal papilla, the most extreme condition among all murid rodents inhabiting the Monte Desert. Renal indices (relative medullary thickness, ratio of inner medulla to cortex and relative medullary area) are also high, as is the ability to concentrate urine to minimize water loss (G. B. Diaz and Ojeda 1999).

SYNONYM:

Salinomys Braun and Mares, 1995:505; type species *Salinomys delicatus* Braun and Mares, 1995, by original designation.

Salinomys delicatus Braun and Mares, 1995
Delicate Salt Flat Mouse

SYNONYM:

Salinomys delicatus Braun and Mares, 1995:514; type locality "Province of San Luis, Departamento de Ayacucho, ca. 23 km N Route 20, Pampa de Las Salinas, cerca La Botija, at 1,300 feet elevation," Argentina.

DESCRIPTION: As for the genus.

DISTRIBUTION: *Salinomys delicatus* occurs between 400 and 600 m elevation in several isolated populations at the edge of salt flats (or *salares*) in west-central Argentina, in the provinces of Catamarca, La Rioja, San Juan, Mendoza, and San Luis (Braun and Mares 1995; R. A. Ojeda et al. 2001; Lanzone et al. 2005; D. Rodríguez, Lanzone, Chillo et al. 2012). No fossils have been reported.

SELECTED LOCALITIES (Map 295): ARGENTINA: Catamarca, Bolsón de Pipanaco (Lanzone et al. 2005), Quimilo (ARG 5147); La Rioja, 26 km SW of Quimilo (OMNH 30224), Salar La Antigua, 45 km NE of Chamical (CML 3795; R. A. Ojeda et al. 2001); Mendoza, Estancia El Tapón (D. Rodríguez, Lanzone, Chillo et al. 2012), Laguna del Rosario (D. Rodríguez, Lanzone, Chillo et al. 2012); San Juan, 15 km ESE of José de Martí on road to Chañar Seco (OMNH 23603); San Luis, Pampa de Las Salinas, ca. 23 km N of Rte 20, near La Botija (type locality of *Salinomys delicatus* Braun and Mares), 15 km SSE of Salinas de Bebedero (OMNH 23602).

SUBSPECIES: *Salinomys delicatus* is monotypic.

NATURAL HISTORY: *Salinomys delicatus* has been collected in xeric and semiarid habitats on the edge of salt flats, and in small patches of mesquite trees (*Prosopis* [Fabaceae]) with a hardpan substrate (Braun and Mares 1995; R. A. Ojeda et al. 2001; D. Rodríguez, Lanzone, Chillo et al. 2012). Unpublished field notes indicate that this terrestrial species is scansorial and nocturnal (R. A. Ojeda and Tabeni 2009). Reproductive activity apparently is highest in late summer and early fall (Braun and Mares 1995), although a pregnant female was recorded in spring (D. Rodriguez et al. 2012a). Data from births in captivity and from field-caught pregnant females indicate litter sizes ranging from two to six. The diet of *S. delicatus* is omnivorous, consisting in similar proportions of seeds, arthropods, and halophytic plants (D. Rodriguez et al. 2012a).

REMARKS: The diploid number is the lowest among phyllotine species, but the complement is sexually dimorphic in *S. delicatus*, $2n = 18$ for females and $2n = 19$ for males. The number of autosomal arms (FN) is 32, and all autosomes biarmed. The sex chromosome system is complex due to an X-autosomal translocation, with XY_1Y_2 sex chromosomes in males. Multiple sex chromosomes are uncommon in mammals and *S. delicatus* seems to have undergone a major chromosome restructuring during its karyotypic evolution (Lanzone et al. 2005; Lanzone, Rodríguez et al. 2011).

Salinomys delicatus is also unusual among rodents (and mammals more generally) morphologically, as females are larger than males in head and body length and other ex-

ternal and cranial measurements (D. Rodríguez, Lanzone, Chillo et al. 2012). Field data for this species are very limited, but in most sampling efforts females were more abundant than males, suggesting that either the primary sex ratio or survivorship is biased in favor of females (D. Rodríguez, Lanzone, Chillo et al. 2012). The species is considered rare and vulnerable to extinction, based on its restricted and patchy distribution, few known occurrences, low local abundances, and specialization to halophytic habitats (Lanzone et al. 2005; D. Rodríguez, Lanzone, Teta et al. 2012).

Genus *Tapecomys* S. Anderson and Yates, 2000
Jorge Salazar-Bravo

Mice of the genus *Tapecomys* include only two species living, both from Andean localities in central Bolivia and northern Argentina. Until recently, the genus *Tapecomys* was considered monotypic (S. Anderson and Yates 2000; Musser and Carleton 2005), but morphological (Steppan 1993) and molecular data (Steppan et al. 2007) strongly supported the inclusion of *Phyllotis wolffsohni* Thomas within *Tapecomys*. I thus recognize two species: *T. primus* and *T. wolffsohni*. These mice live at moderate to high elevations (1,200 to 3,900 m) on the eastern versant of the Andes, from Cochabamba (Bolivia) to Salta (Argentina) where they occupy Yungas forests (*T. primus*) or brush and scrub in ecotonal localities above Yungas forests (*T. wolffsohni*). Except as noted below, the following account is abstracted from Pearson (1958), Hershkovitz (1962), Steppan (1993, 1995), S. Anderson (1997), and Barquez et al. (2006), which should be consulted for additional details.

Tapecomys is characterized by the following combination of characters: medium to large with dorsal pelage agouti brownish, bright ochraceous lateral line, and ventral pelage whitish gray (but individual hairs always gray based); a well-defined pectoral streak may or may not be present; tail distinctly bicolored, not penicillate, moderately furred, slightly longer than head and body (6–15% longer); manus white; relatively large pes, covered with white hairs, a darker spot on top near base of each toe may or may not be present; genal and superciliary vibrissae present, extending to the edge of the pinnae when laid back; hindfoot with conspicuous tufts of ungual hairs at the base of the claws on digits II–V, digit I with ungual tufts present but much less conspicuous; hypothenar pad present and large.

The skull has a robust rostrum flanked by deep zygomatic notches. The interorbital region is posteriorly diver-

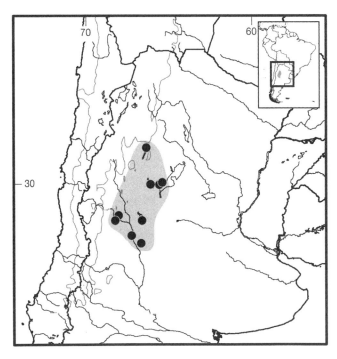

Map 295 Selected localities for *Salinomys delicatus* (●). Contour line = 2000 m.

gent in *T. primus* but mostly parallel in *T. wolffsohni*, but always with sharp edges. The braincase is oblong or globular, usually with clear temporal ridges; lambdoid ridges are not present (if so, poorly developed). The zygomatic plates are broad (*sensu* Voss 2003), their posterior margins anterior to each M1. The nasals extend posteriorly behind the lacrimals in *T. wolffsohni* but are shorter (extending to but usually not behind the lacrimals) in *T. primus*. The frontoparietal suture is U-shaped; the incisive foramina are very long, extending posteriorly between molar alveoli, widest at their midlength; and the bony palate extends slightly behind the M3s. The mesopterygoid fossa is parallel sided, with its bony roof pierced by sphenopalatine vacuities. An alisphenoid strut is absent (buccinator-masticatory foramen and accessory foramen ovale are confluent) and the carotid circulatory pattern is primitive (pattern 1 of Voss 1988).

The upper incisors are opisthodont and fronted by orange enamel. The molars are hypsodont, appear somewhat prismatic, and have planed occlusal surfaces; upper and lower second molars possess a deep hypoflexus(id) and protoflexus(id); and M3s have a hypoflexus(id) that alternates with metaflexus(id) and posteroflexus(id) so that when worn this molar acquires a distinct S- or Z-shape.

SYNONYMS:

Phyllotis: Thomas, 1902b:131; part (description of *wolffsohni*); not *Phyllotis* Waterhouse.

Tapecomys S. Anderson and Yates, 2000:21; type species *Tapecomys primus* S. Anderson and Yates, 2000, by monotypy and original designation.

REMARKS: Initial mitochondrial sequence analyses (S. Anderson and Yates 2000) suggested a phyletic relationship between *Tapecomys* and both *Salinomys* and *Andalgalomys* among the limited set of phyllotine taxa these authors examined. Steppan et al. (2007), however, supported a sister relationship between *Tapecomys* as part of a triad of genera also including *Auliscomys* and *Loxodontomys* that are sister to *Phyllotis* rather than to *Salinomys* and *Andalgalomys*. Of the two more expansive and recent molecular studies, Salazar-Bravo et al. (2013) failed to resolve relationships among the non-*Calomys* phyllotine genera, but Schenk et al. (2013) recovered a grouping of *Tapecomys* with, successively, *Auliscomys*, *Loxodontomys*, and *Phyllotis*, and within a clade that otherwise contained *Andalgalomys* and *Graomys*.

KEY TO THE SPECIES OF *TAPECOMYS*:

1. Skull with sharp, posteriorly divergent interorbital region; nasals short, extending to lacrimals, and posteriorly not pointed; hindfoot larger (range: 29–34 mm; mean 32 mm); molar toothrow longer (range: 6.4–6.7 mm; mean 6.6 mm); 2*n* = 56; found between 1,200 and 1,500 m in Yungas forests of southern Bolivia and northern Argentina . . . . *Tapecomys primus*

1′. Skull with sharp, evenly constricted interorbital region; nasals long, extending posterior to lacrimals, and posteriorly pointed; hindfoot smaller (range: 26–30 mm; mean 27.5 mm); molar toothrow shorter (range: 5.1–6.5 mm; mean 5.7 mm); 2*n* = 54; usually caught in scrub and brush on eastern slope of the Andes from central Bolivia to northern Argentina . . . . *Tapecomys wolffsohni*

Tapecomys primus S. Anderson and Yates, 2000
Tapecua Leaf-eared Mouse

SYNONYM:

Tapecomys primus S. Anderson and Yates, 2000:21; type locality "Tapecua, 1,500 m, 21°26′S, 63°55′W, Department of Tarija, Bolivia."

DESCRIPTION: Medium sized with head and body length 124–139 mm; tail longer than head and body (length 143 mm in single specimen with complete tail), and ears long (33–34 mm). Pelage brownish agouti overall, only slightly paler and with ochraceous hue on venter. In contrast, tail strongly bicolored. Manus white on dorsal surface; pes with darker hairs at base of each toe but paler more proximally. Skull notable for posteriorly divergent supraorbital edges, sharply angled, slightly beaded, and continuing across parietal bone as distinct temporal ridges.

DISTRIBUTION: *Tapecomys primus* is known only from three localities, one in southern Bolivia and two in northern Argentina.

SELECTED LOCALITIES (Map 296): BOLIVIA: Tarija, Tapecua (type locality of *Tapecomys primus* S. Anderson and Yates). ARGENTINA: Jujuy Province, Finca Las Capillas (Barquez et al. 2006), Rte 83, on road to Valle Grande, 9 km N of San Francisco (M. M. Díaz et al. 2009).

SUBSPECIES: *Tapecomys primus* is monotypic.

NATURAL HISTORY: The four known specimens were obtained in lower Yungas Forest vegetation, at elevations ranging from 1,200 to 1,500 m. In southern Bolivia and northern Argentina, this type of forest is often referred to as *Bosque Boliviano-Tucumano* (G. Navarro and Maldonado 2002) or *selva montana* (Barquez et al. 2006), and is characterized by the juxtaposition of montane elements with an abundance and dominance of Myrtaceae intermixed with xeric adapted elements more characteristic of the Chaco at lower elevations to the east (G. Navarro and Maldonado 2002). At the type locality in southern Bolivia, this combination creates a mosaic of patches of columnar cacti and leguminous trees in open areas and broadleaf canopy trees and thick brush on hillsides and in ravines (Anderson and Yates 2000). Barquez, Ferro, and Sanchez (2006) noted large, moss-covered rocks and dense litter at one Argentinean site,

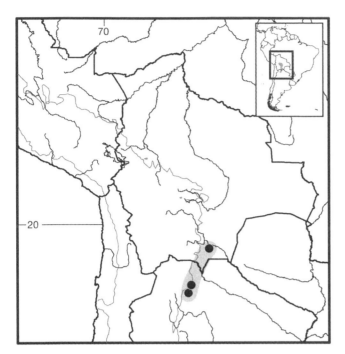

Map 296 Selected localities for *Tapecomys primus* (●). Contour line = 2000 m.

and M. M. Díaz et al. (2009) reported myrtaceous trees (including myrtle, horco molle, and others) at a second. All known specimens are males: one adult collected 14 July 1991 in Bolivia had small, undescended testis measuring 9 × 5 mm; no reproductive data are known for the other three individuals. Lareschi et al. (2010) reported the flea *Craneopsylla minerva minerva* from one Argentinean specimen.

REMARKS: The karyotype of *T. primus* consists of $2n = 56$, $FN = 76$, with 11 pairs of metacentric-submetacentric and 16 pairs of acrocentric autosomes; the X is a large submetacentric, and Y is a small submetacentric (S. Anderson and Yates 2000).

Tapecomys wolffsohni (Thomas, 1902)
Wolffsohn's Leaf-eared Mouse

SYNONYMS:

Phyllotis wolffsohni Thomas, 1902b:131; type locality "Tapacari, 3000 m," west of Cochabamba, Cochabamba, Bolivia.

Phyllotis darwini wolffsohni: Hershkovitz, 1962:339; name combination.

Tapecomys wolffsohni: Steppan, Ramirez, Banbury, Huchon, Pacheco, Walker, and Spotorno, 2007:799; first use of current name combination.

DESCRIPTION: Somewhat larger (head and body length 110–150 mm, mean 132 mm) than *T. primus*, with slightly shorter tail (length 118–160 mm, mean 138 mm) and much smaller ear (height 23–29 mm, mean 25.8 mm).

Overall dorsal color yellowish brown streaked with black; hairs bicolored, with elongated basal gray band and narrow phaeomelan in tip. Venter white but gray base of hairs show through; pectoral streak may or may not be present. Tail moderately bicolored, dark yellowish brown above and paler yellow to white below. Dorsal surfaces of both forefeet and hindfeet white. Ears darker than body. Skull possesses similar sharply edged and divergent supraorbital ledges, but these do not extend across parietals as notable ridges, unlike in *T. primus*.

DISTRIBUTION: *Tapecomys wolffsohni* ranges from central Bolivia to northern Argentina along the eastern slopes of the Andes between 1,300 and 3,875 m.

SELECTED LOCALITIES (Map 297): ARGENTINA: Jujuy, on road between San Francisco and Pampichuela (Jayat et al. 2006). BOLIVIA: Cochabamba, Cuchicancha (Hershkovitz 1962), 15 mi E of Tapacari (Hershkovitz 1962), Ucho Ucho (B. D. Patterson 1992b); Chuquisaca, 9 km by road N of Padilla (S. Anderson 1997), 1.3 km W of Jamachuma (S. Anderson 1997); Santa Cruz, Comarapa (S. Anderson 1997), Guadalupe (Hershkovitz 1962).

SUBSPECIES: *Tapecomys wolffsohni* is monotypic, although Pearson (1958) noted that specimens from more humid habitats near Comarapa, on the border of Cochabamba and Santa Cruz departments in Bolivia, are darker and smaller that those from the type locality.

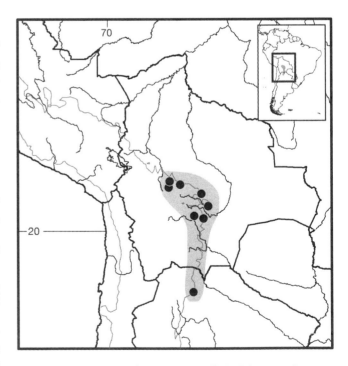

Map 297 Selected localities for *Tapecomys wolffsohni* (●). Contour line = 2000 m.

NATURAL HISTORY: This species inhabits brush and scrub habitat at moderate to high elevations (2,200 to 3,500 m; Pearson 1958). Little is known about the natural history of this species. S. Anderson (1997) recorded a single pregnant female (with two embryos) in March, a lactating female in July, and nonpregnant females in March, July, and September.

REMARKS: *Tapecomys wolffsohni* was originally described as a species of *Phyllotis*, and affirmed so by Pearson (1958), but was treated as a subspecies of *P. darwini* by Hershkovitz (1962). Steppan (1993) and Steppan et al. (2007) allocated *wolffsohni* Thomas to *Tapecomys* based on morphological and molecular phylogenetic analyses, respectively. The identity of some Argentinean specimens assigned to this species is controversial (Jayat et al. 2006).

The karyotype consist of a $2n=54$, $FN=72$, with 16 pairs of acrocentric and 10 pairs of metacentrics or submetacentrics autosomes, a large metacentric X chromosome, and a small metacentric Y chromosome (Pearson and Patton 1976).

Tribe Reithrodontini Vorontsov, 1959
Ulyses F. J. Pardiñas, Carlos A. Galliari, and Pablo Teta

The history of, and justification for the tribe Reithrodontini is given in the general account of the Sigmodontinae, and the characteristics of the sole member of the tribe, the genus *Reithrodon*, are in the following account.

Genus *Reithrodon* Waterhouse, 1837

Representatives of *Reithrodon* are among the most beautiful and widespread sigmodontine rodents of the open lands in the South American Southern Cone. With two recognized living species (*R. auritus* and *R. typicus*) but at least 10 nominal forms in current synonymy, *Reithrodon* occurs in crop fields, grasslands, steppes, and semideserts of Argentina, Chile, Uruguay, and adjoining southernmost Brazil (Musser and Carleton 2005). It is the most frequently encountered genus in the sigmodontine fossil record of Argentina, and several names have been used to describe Plio-Pleistocene remains (Reig 1978; Pardiñas 1995b).

Reithrodon has been placed as a member of the Sigmodontini (see Hershkovitz 1955b) or Phyllotini (see Olds and Anderson 1990; Braun 1993; Steppan 1993, 1995), closely related to other extant and fossil genera that comprised a "*Reithrodon* group" (Steppan 1995; Pardiñas 1997; Steppan and Pardiñas 1998; P. E. Ortiz, Cirignoli et al. 2000). However, phylogenetic analyses of molecular sequences have indicated the distinctiveness of the genus within the sigmodontine radiation (Engel et al. 1998; M. F. Smith and Patton 1999; D'Elía 2003a) and have thus supported Vorontsov's (1959) monotypic tribe Reithrodontini. Several contributions have documented the status, diversity, and morphological traits of *Reithrodon* species (Osgood 1943b; Hershkovitz 1955b; Reig 1978; Ortells et al. 1988; Pearson 1988, 1995; Steppan 1995). The present account is largely based on an extensive summary of the available information compiled for *R. auritus* (Pardiñas and Galliari 2001) and for *R. typicus* (Barlow 1969).

The two species are among the largest sigmodontines in present-day communities in Patagonia and Pampas biomes (head-body length >130 mm; mass about 90 g). Herbivorous in diet and nocturnal in habits, these animals reach high densities under suitable environmental conditions where they are heavily preyed upon by raptorial birds. They live in moderately complex burrow systems and construct runways in short grassland vegetation; their presence is easily recognized by accumulations of cylindrical feces with one edge rounded and the other pointed (Pearson 1988). They share none of the traditional morphological features typical of semifossorial rodents; their common name "conyrat" belies a convergence with the leporid genus *Ochotona* (Lagomorpha; Ortells et al. 1988).

Reithrodon has one of the longest temporal ranges within the sigmodontine radiation (Reig 1972). The oldest known remains are from Monte Hermoso deposits (in southwestern Buenos Aires province, Argentina) with an approximate age of 4–5 mya (Pardiñas 2000). During the Chapadmalalan (Early to Middle Pliocene, 3–4 mya) the genus was well represented in fossil assemblages from central Argentinean latitudes, being one of the most common sigmodontines present. In fact, Florentino Ameghino coined the genus *Proreithrodon*, with two species, to accommodate these fossils; both are recognized today as full synonyms of *R. auritus*. However, an important temporal gap is present in the fossil record, one corresponding to the Sanandresian (Late Pliocene, 2.5 mya). Based on the caviomorph fossil record, Verzi and Quintana (2005) suggested that hyperarid conditions, presumably coupled with strong glaciation, might have depressed the abundance of *Reithrodon*. Subsequently, the genus recovered its earlier dominance to characterize Middle Pleistocene micromammal assemblages, especially those around 1 mya (Ensenadan) or between 0.3–0.15 mya ago (Lujanian). From Lujanian (Late Pleistocene) deposits another two supposedly extinct genera, *Tretomys* and *Ptyssophorus*, were erected to allocate *Reithrodon* fossil remains. P. E. Ortiz et al. (2011b) recorded the genus from Late Pleistocene deposits at several places outside the current geographic range (e.g., around Córdoba city in central Argentina and in the lowlands of northwestern Argentina; see also P. E. Ortiz and Jayat 2012b), and Ubilla et al. (2004), Hadler and Ferigolo (2004), and R. P. Lopes and Buchmann (2011) documented late Pleistocene fossils from Uruguay and Brazil. Over the last 10,000 years, the history of *Reithrodon* has been one of gradual decline in the Pampean region, with a precipitous drop during the Late Holocene. As seems to be the case with several other mammal species, *Reithrodon* populations were probably deeply affected by human-induced landscape changes over the last five centuries, most importantly by massive introduction of cattle and agricultural practices, both of which have affected fire-regime and associated vegetation-structure changes (Pardiñas, Teta, and D'Elía 2010).

Reithrodon species are easily distinguishable from other sigmodontine rodents by their large heads, ears, and eyes, massive bodies, moderately short and haired tails (<105 mm), and short and blunt claws of both manus and pes. Conspicuous bands of lighter pelage form a mask around the eyes and ears that usually decays in stored museum skins. The general color varies from dark brown to pale buffy gray dorsally, whereas the ventral surface is white to brownish ochraceous; flanks and belly have a variable amount of yellow and white (Gyldenstolpe 1932). The fur is soft, luxurious, and dense; regular and rhomboidal cuticular scales of the dorsal guard hairs clearly distinguish *Reithrodon* from other sympatric sigmodontines (M. Busch 1986). The tail is bicolored and densely haired but without an apical tuft. Long whitish hairs cover the dorsal surfaces of both manus and pes and ungual tufts surpass the tip of each claw. Plantar pads on the hindfeet are reduced to four in number, with two large interdigital pads for both digit II and IV embedded in a strongly squamate surface. *Reithrodon* has four pairs of mammary glands, one each in pectoral, postaxial, abdominal, and inguinal positions (Gyldenstolpe 1932; Steppan 1995).

Hooper (1962) described the glans penis and baculum of a specimen from Tierra del Fuego, Argentina, as short and blunt, covered externally by conical tubercles, and divided externally into four indistinct lobes; the three bacular mounds plus the baculum resembled a fleur-de-lis in ventral view. The baculum is shorter than the glans, with a wide bony base divided by a deep medial notch and a thin flat tip; three distal segments are cartilaginous with a more elongated middle segment, ending in a point. The stomach (based on a specimen from Buenos Aires province, Argentina) is unilocular-hemiglandular with the glandular epithelium reaching the corpus (Carleton 1973). Barlow (1969:37) noted, in reference to *R. typicus*, that the single chambered stomach is somewhat sacculated in the fundic region and an enlarged caecum is present at the junction of the large and small intestines. A gall bladder is present (Voss 1991a;

based on a specimen of *R. typicus* from Uruguay, and the same condition was detected in several *R. auritus* from Argentina [authors, unpubl. data]).

Cranially, *Reithrodon* is characterized by a unique combination of traits including a very domed skull in lateral profile, a narrow and parallel-sided interorbital region, a denticulate V-shaped coronal suture, a very large interparietal bone, deep and narrow zygomatic plates with upper well-projected spines, an arched premaxillomaxillary suture, molar toothrows that converge anteriorly, an excavated, narrow, and long palate, small auditory bullae, very large parapterygoid plates, and the absence of the tegmen tympani. The mandible is deep and robust and has a long condyloid process internally excavated by a spoon-like fossa (Pearson 1995; Steppan 1995). Carotid circulatory pattern is the derived pattern 3 (*sensu* Voss 1988).

Upper incisors are opisthodont, faced with orange enamel and two frontal grooves (Steppan 1995:Fig. 9e); molars are noticeably hypsodont, with alternating main cusps and flat crowns (*sensu* Hershkovitz 1962); mesolophs/lophids (including mesostyle/stylids) and enterostyle/stylids are absent. The M1 is trilophodont with the procingulum compressed anteroposteriorly, and there is no evidence of an anteromedian flexus; protoconid and hypoconid areas are subequal. M2 and M3 are similar in occlusal design, the latter is slightly smaller, with the metaflexus more penetrating and directed posteriorly; m1 is elongated and tetralophodont, with a subtriangular procingulum coalescent with anterolabial cingulum at an early stage of wear; m2 has a reduced protoflexid, and lingual lophids are not equal in extent; m3 is simplified relative to m2, with a clearly sigmoidal pattern. Number of roots per molar is M1, 5–4; M2, 3; M3, 4–3; m1, 4; m2, 3; and m3, 3–2 (Pearson 1995).

Number of vertebrae varies in *R. auritus*: each of 10 specimens had 12 thoracic, 7 lumbar, and 24 caudal vertebrae whereas one specimen had 13, 6, and 25–26 (Steppan, 1995). All specimens had 12 ribs.

The alpha taxonomy of *Reithrodon* is still unclear, and revision of the genus is clearly needed. Unfortunately, the few series of specimens preserved in museum collections make this desirable goal presently difficult. The first significant approach to delineating *Reithrodon* diversity was sketched by Thomas (1920b), who recognized three informal groups based on "colour and by the degree of hairiness of the soles." According to Thomas, southern nominal forms in Argentina and Chile (*cuniculoides*, *hatcheri*, *flammarum*) have the plantar hairs extending forward to the anterior pair of sole pads, whereas central Argentinean forms (*auritus*, *pampanus*, *marinus*) show the main mass of the hairs ceasing at the posterior pair of pads. In contrast, specimens from Corrientes and Uruguay (*currentium*, *typicus*) but also those from northern Argentina (*caurinus*) are characterized by a naked pad region. Osgood (1943b) subsumed all available names under *auritus* and treated *Reithrodon* as monotypic although he recognized the potential existence of several subspecies: *cuniculoides* and *pachycephalus* for southern populations, and *auritus*, *typicus*, and *caurinus* for central and northern ones. Ortells et al. (1988) resurrected *R. typicus* as valid species based primarily on chromosomal data.

A preliminary exploration of mtDNA cytochrome-*b* sequences (E. P. Lessa et al. 2010; E. P. Lessa et al., unpubl. data) covering almost the known *Reithrodon* geographic distribution supported the recognition of *auritus* and *typicus* but also indicated high genetic distances between isolated northern Argentina *auritus* populations and those in Patagonia.

SYNONYMS:

Mus: G. Fischer, 1814:71; part (description of *auritus*); not *Mus* Linnaeus.

M[us].: Illiger, 1815:108; part (listing of *physodes*); not *Mus* Linnaeus.

M[us].: Olfers, 1818:209; part (description of *physodes*); not *Mus* Linnaeus.

Mus: Desmarest, 1819e:64; part (description of *auritus*); not *Mus* Linnaeus.

M[us].: Brants, 1827:145; part (listing of *pyrrhogaster*); not *Mus* Linnaeus.

Reithrodon Waterhouse, 1837:29; type species *Reithrodon typicus* Waterhouse, by original designation (ICZN 1999:Art. 68.2.2).

Rithrodon Agassiz, 1846:327; incorrect subsequent spelling of *Reithrodon* Waterhouse.

Rheitrodon Roger, 1887:102; incorrect subsequent spelling of *Reithrodon* Waterhouse.

Rheithrodon Thomas, 1884:457; incorrect subsequent spelling of *Reithrodon* Waterhouse.

Ptyssophorus Ameghino, 1889:111; type species *Ptyssophorus elegans* Ameghino, by monotypy and original designation.

Tretomys Ameghino, 1889:119; type species *Tretomys atavus* Ameghino, by monotypy and original designation.

Rhithrodon Flower and Lydekker, 1891:464; incorrect subsequent spelling of *Reithrodon* Waterhouse.

Proreithrodon Ameghino, 1908:424; type species *Proreithrodon chapalmalense* Ameghino, by original designation.

KEY TO THE SPECIES OF *REITHRODON*:

1. Hindfeet with airy sole pads; diploid complement $2n = 34$ with all acrocentric autosomes . . . .
. . . . *Reithrodon auritus*
1'. Hindfeet with naked sole pads; diploid complement $2n = 28$ with four pairs of metacentric autosomes . . . . .
. . . . *Reithrodon typicus*

Reithrodon auritus (G. Fischer, 1814)
Hairy-soled Conyrat

SYNONYMS:

Mus auritus G. Fischer, 1814:71; type locality "les Pampas, au Sud de Buenos-Ayres" (Azara 1801:91), equated to "south bank of the Río de la Plata," Buenos Aires Province, Argentina, by Hershkovitz (1959b:349). After a study of Azara's itinerary, particularly the monograph by Mones and Klappenbach (1997), we conclude, however, that Azara most likely made his observations on this rodent in the current Pila county, on the south side of the Río Salado, approximately 150 km S Buenos Aires.

M[us]. *physodes* Illiger, 1815:108; *nomen nudum*.

M[us]. *physodes* Olfers, 1818:209; name based on bibliographic reference to Azara's (1801) "*rat quatrième* or *rat oreillard*."

Mus auritus Desmarest, 1819e:64; name based on bibliographic reference to Azara's (1801) "*rat quatrième* or *rat oreillard*."

M[us]. *pyrrhogaster* Brants, 1827:145; in synonymy of *Mus auritus* (fide Hershkovitz, 1959b:348).

Reithrodon cuniculoides Waterhouse, 1837:30; type locality "Santa Cruz." Santa Cruz, in the meaning of Charles Darwin, the collector, must have been the mouth of the Santa Cruz river where the HMS *Beagle* was anchored on the southern bank, near the present locality of Puerto Santa Cruz (ca. 50.117°S, 68.415°W), Santa Cruz, Argentina.

Mus pachycephalus Philippi, 1900:42; type locality "Freto Magellanico," restricted, although with weak support, to the vicinity of Punta Arenas by Osgood (1943b:221–222).

Reithrodon cuniculoides obscurus J. A. Allen, 1903b:190; type locality "Punta Arenas, Patagonia," Magallanes y Antártica Chilena, Chile.

Reithrodon hatcheri J. A. Allen, 1903b:191, type locality "Pacific slope of the Cordilleras, head of the Rio Chico de Santa Cruz," restricted to Perito Moreno National Park by Galliari and Pardiñas (1999:119). However, this restriction is incorrect. Our review of the geographical notes made by J. B. Hatcher (1903), coupled with fieldwork following his itinerary in this portion of the Patagonian territory, suggests that O. A. Peterson and Hatcher collected the type of *hatcheri* close to the course of the Río Tucu Tucu (ca. 48.47°S, 71.87°W, Río Chico county, Santa Cruz province, Argentina), about 8 km downstream from its headwaters.

Reithrodon cuniculoides flammarum Thomas, 1912b:411; type locality "Tierra del Fuego . . . from Spring-hill [= Springhill], in the north of the island," Tierra del Fuego, Chile.

Reithrodon cuniculoides pampanus Thomas, 1916g:305; type locality "Southern pampas of Buenos Ayres Province . . . Peru [= Estación Perú], F.C.P., about 200 kilometres N.W.

of Bahia Blanca [= Bahía Blanca]," La Pampa Province, Argentina.

Reithrodon caurinus Thomas, 1920b:473; type locality "Otro Cerro, Catamarca. Alt. 3000 m." Pardiñas et al. (2007) recently delineated the history of this frequently mentioned Argentinean type locality, and located Otro Cerro at about 28°45′S, 66°17′W in the top of Sierra de Ambato, Catamarca Province (see also Cabrera 1961:501).

[*Reithrodon*] *auritus*: Thomas, 1920b:474; first use of current name combination.

Reithrodon auritus marinus Thomas, 1920b:474; type locality "Mar del Plata, on the south-eastern sea-coast of Buenos Ayres [= Buenos Aires] Province," Argentina.

Reithrodon cuniculoides evae Thomas, 1927d:652; type locality "Zapala," Neuquén Province, Argentina.

Reithrodon physodes caurinus: Cabrera, 1961:501; name combination.

Reithrodon physodes cuniculoides: Cabrera, 1961:501; name combination.

Reithrodon physodes evae: Cabrera, 1961:501; name combination.

Reithrodon physodes pachycephalus: Cabrera, 1961:501; name combination.

Reithrodon physodes physodes: Cabrera, 1961:502; name combination.

DESCRIPTION: General characters as in generic description. Soles more densely haired than in *R. typicus*; tail bicolored; diploid complement = 34.

DISTRIBUTION: *Reithrodon auritus* occurs from sea level to ca. 3,000 m (Pardiñas and Galliari 2001), with a relatively uniform range across the open areas of Patagonian region of Argentina and Chile, from Tierra del Fuego island to about 36°S. North of 36°S, this species is found more patchily or restricted to a few high-elevation (>2,000 m) localities in the Sierras Pampeanas in central and northern Argentina (P. E. Ortiz and Jayat 2012b). Records obtained from Tierra del Fuego are restricted to the northern steppe portion of the island, north of Lago Fagnano (J. A. Allen 1905; Osgood 1943b). In continental Patagonia, *R. auritus* is widespread, but a detailed survey of its distribution indicates several gaps; for example, it is very scarce or virtually absent in the high portions of the basaltic plateaus that characterize the central region (e.g., Somuncura). In central latitudes of Argentina, *R. auritus* is restricted to the Pampean region and the eastern portion of the Monte desert but completely absent from the arid diagonal. A few records are known from piedmont and Andean ranges in Mendoza province, most of them associated with *Payunia* environments (Pardiñas, Teta, and Udrizar Sauthier 2008). Contrary to the prediction of Dalby and Mares (1974), surveys conducted in northwestern Argentina over the last 20 years (Mares, Ojeda, and Barquez 1989; Barquez et al. 1991; Mares, Ojeda, Braun et al. 1997; Jayat

2009; P. E. Ortiz and Jayat 2012b) have reported few new localities for this rodent. In fact, *R. auritus* is known from only five isolated high-elevation, grassland localities in the Argentinean provinces of Jujuy, Tucumán, Catamarca, and Córdoba. Two specimens (MACN 33.246, from Salí, Tucumán, and MACN 31.37, from Maimará, Jujuy) trapped by Emilio Budin and listed by Dalby and Mares (1974:205) as vouchers of the occurrence of *R. auritus* in northwestern Argentina are misidentified (MACN 33.246 is a *Loxodontomys micropus*, and MACN 31.37 is a *Phyllotis caprinus*, not a *Graomys domorum* as indicated by M. M. Díaz and Barquez (1999:331). Ihering (1927) provided an unsubstantiated report of the genus from the Malvinas (Falkland) Islands.

SELECTED LOCALITIES (Map 298): ARGENTINA: Buenos Aires, Granja 17 de Abril (Massoia 1988d), Laguna Alsina (J. R. Contreras 1972); Puesto El Plátano (Leveau et al. 2006), Saladillo (Massoia 1988b); Catamarca, Barranca Larga (P. E. Ortiz, Cirignoli et al. 2000), Otro Cerro (type locality of *Reithrodon caurinus* Thomas); Córdoba, Pampa de Achala (Ortells et al. 1988), Washington (UNRC 8273); Jujuy, Cochinoca (Pardiñas and Galliari 2001); La Pampa, General Pico (MACN 37.100), Laguna Guatraché (Massoia, Tiranti, and Diéguez 1997); Mendoza, Caverna de las Brujas (Gasco et al. 2006), Volcán Carapacho (incorrectly given as "Cerro Morocho" by Jayat et al. 2006 and Pardiñas, Teta, and Udrizar Sauthier 2008); Neuquén, 3–5 km upstream of Chos Malal (Tiranti 1996); Río Negro, Cerro Leones (Pearson 1987); Santa Cruz, Alero Destacamento Guardaparque (Pardiñas 1998), Río Tucu Tucu, about 8 km downstream from headwaters (type locality of *Reithrodon hatcheri* J. A. Allen); Tierra del Fuego, north end of Lake Fagnano (Osgood 1943b), Estancia Via Monte (Osgood 1943b); Tucumán, Paso El Infiernillo (Dalby and Mares 1974). CHILE: Aysén, 3 km N of Coyhaique, Reserva Nacional Coyhaique, Laguna Verde (Kelt 1996); Magallanes y Antártica Chilena, Cueva del Milodón (U. F. J. Pardiñas, unpubl. data), Punta Arenas (type locality of *Reithrodon cuniculoides obscurus* J. A. Allen), Río Verde, east end of Skyring Water (Osgood 1943b).

SUBSPECIES: Although many taxa have been proposed as subspecies, recent evaluations using modern data or methods to document even general geographic trends are lacking. In his revision, Osgood (1943b) retained *cuniculoides* and *pachycephalus* as valid geographic units within the continuously distributed southern populations, with *R. a. cuniculoides* distributed along steppes, prairies, and sparse shrublands (Massoia and Chebez 1993) of the Magellan Straits and Tierra del Fuego, and *R. a. pachycephalus* restricted to the southernmost tip of South America (Osgood 1943b; Reise and Venegas 1987; Tamayo and Frassinetti 1980), in Aysén (XI Region) as well as Última Esperanza, Magallanes y Antártica Chilena, and Tierra del Fuego (XII Region; Mann

1978) of Chile. Cabrera (1961) recognized five subspecies (under the combination of *R. physodes*): *R. a. caurinus* in northwestern Argentina; *R. a. auritus*, in the Pampean region of Buenos Aires; *R. a. evae* for southern Mendoza and Neuquén; *R. a. cuniculoides*, from Chubut and Santa Cruz to the Magellan Straits; and *R. a. pachycephalus* in the western Andean ranges from Neuquén to Tierra del Fuego and adjoining parts of Chile. Molecular sequences (E. P. Lessa et al. 2010; E. P. Lessa et al., unpubl. data) failed to recover structure among austral populations that might support subspecies recognized by Osgood (1943b).

NATURAL HISTORY: In Patagonia, reproductively active males and pregnant females are found mostly in the spring (Kelt 1994; Pearson 1988), but the reproductive season can extend to the end of summer (Guthmann et al. 1997). Litter size varied from 1–8, with a mean of 4.53 in utero fetuses (Pearson 1988; Pine et al. 1979). Juveniles (<50 g) were recorded in spring and in higher numbers and proportions during the summer and autumn. Spring populations are composed of a few old overwintering individuals, some middle-aged overwintering individuals, and numerous young individuals, which mature in summer. In autumn, the oldest individuals disappear and are replaced by middle-aged individuals (Kelt 1994; Pearson 1988). *Reithrodon auritus* reaches its population maxima at the end of austral autumn (Guthmann et al. 1997). From very low minimum abundances, *R. auritus* can reach peaks of ca. 10–15 individuals (minimum known alive) per hectare. Longest recapture times were eight months, but the study of cohorts suggests maximum longevity of 15 months. Nests are made of plant materials, or, in Tierra del Fuego, sheep's wool. *Reithrodon* is most common in open habitats, such as steppes and prairies, usually near green grass (Pearson 1988). However, individuals may inhabit a wide range of environments, from high-elevation bunchgrass prairie in northern Argentina (Dalby and Mares 1974) to southern beech forest (*Nothofagus pumilio*) or shrublands in Tierra del Fuego (Pine et al. 1979). At Pampa de Achala (Córdoba province, Argentina), *R. auritus* was founded on the banks of streams, prairies, and bluffs from 1,850 to 2,170 m (Polop 1991). In Patagonia, Pearson (1988) found animals only in soils with minimum degree of humidity, but overly wet conditions cause individuals to abandon tunnels. In southeastern Buenos Aires province, stomach contents of *R. auritus* consisted only of grasses, especially *Poa* spp. and *Lolium multiflorum*. In captivity, they ate an amount of fresh green vegetation equal to their mass nightly (Pearson 1988). In Chilean winters, individuals remain active in burrows under the snow. Activity began in the evening and lasted until the early hours of the morning (Pearson 1988). Feeding occupies most of the active period, with short foraging periods interspersed. Individuals are not strongly at-

tracted to common baits and traps (Reig 1964), one reason for the paucity of specimens in museum collections. Animals excavate tunnel systems 4–7 cm in diameter, usually vertically through the turf, with grouped entrance openings, but without evidence of earth plugs (Pearson 1988). Sometimes *R. auritus* uses tunnels dug by *Ctenomys* (Dalby and Mares 1974; Pine et al. 1979) and even coinhabits the tunnels with these and *Abrothrix longipilis*, *Loxodontomys micropus*, and *Oligoryzomys longicaudatus* (Pearson 1988). Feces are common near tunnel openings. Feces are fibrous, thick (Pearson 1988), short, and greenish, similar to those of *Euneomys* (Mann 1978) but pointed at one end (Pearson 1988).

A number of ectoparasites can be found on individual *R. auritus* (Autino and Lareschi 1998; Lareschi and Mauri 1998), including the fleas *Tetrapsyllus bleptus*, *Craneopsylla minerva wolffhuegeli*, *and Neotyphloceras crassispina*; and the mites *Androlaelaps fahrenholzi*, *Eulaelaps stabularis*, *Haemolaelaps reithrodontis*, *Laelaps mazzai*, and *L. paulistanensis*. J. P. Sánchez (2013) recovered fleas from specimens collected in Chubut province: *Agastopsylla boxi boxi*, *Craneopsylla minerva wolffhuegeli*, *Ectinorus (Ectinorus) hapalus*, *Ectinorus (Ectinorus) ixanus*, *Ectinorus (Ectinorus) levipes*, *Ectinorus (Ichyonius) onychius*, *Hectopsylla (Hectopsylla) gracilis*, two undescribed species of *Neotyphloceras*, *Polygenis (Polygenis) platensis*, *Tetrapsyllus (Tetrapsyllus) tantillus*, and *Tiarapsylla argentina*. Lice of the genus *Hoplopleura* have been recovered from Patagonian specimens (species *H. argentina*; D. Castro and Cicchino 1987; Durden and Musser 1994) and those from Catamarca province (*H. serrulata*; D. Castro 1997). Specimens from Buenos Aires province (Balcarce) contained the nematode endoparasite *Stilestrongylus aureus* (Durette-Desset and Sutton 1985; Sutton 1989a). This species is a major prey item for several birds of prey and mammalian carnivores (Jaksic et al. 1978; De Santis et al. 1983, 1994, 1996; García Esponda et al. 1998; Massoia 1983, 1985a, 1988b, 1988d, 1988e; Massoia and Pardiñas 1988a, 1988b; Pearson 1987; Tiranti 1992, 1996; Travaini et al. 1997; Trejo and Grijera 1998; M. M. Martínez et al. 1996).

REMARKS: As was noted previously, the taxonomy of *R. auritus* as well as the entire genus needs a fresh review. It is likely that northwestern Argentinean populations deserve full specific rank; if so, *caurinus* Thomas would be an available name. Recent compilations of localities (e.g., Hercolini 2007; Teta, González-Fischer et al. 2010) indicate the existence of a major discontinuity in distribution in northeastern Buenos Aires province where this species occurs patchily (e.g., Massoia 1988d). This fact indicates a potential gap of about 200 km long between the ranges of *R. auritus* (south of the Río Salado) and *R. typicus* (north of the Río Paraná). However, Late Pleistocene and Holocene fossil samples show that this gap is perhaps very recent, probably due to land-cover changes

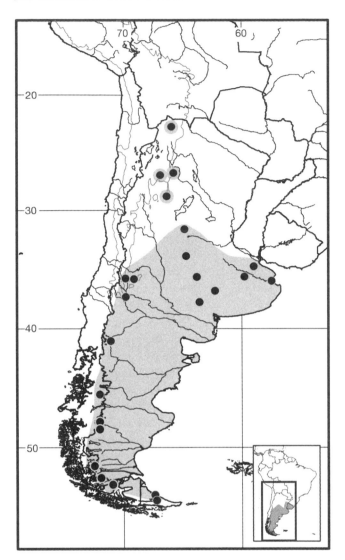

Map 298 Selected localities for *Reithrodon auritus* (●); northwestern localities are shown as disjunct from main distribution (see Distribution). Contour line = 2,000 m.

resulting from the spread of agriculture and urbanization that have occurred in this densely populated area.

Ortells et al. (1988) described a karyotype with $2n = 34$, FN = 32, for specimens from Buenos Aires and Córdoba provinces, Argentina. Using comparative G-banding, these authors demonstrated that this karyotype differs from that of *R. typicus* by at least four centric fusions, including one each involving both X and Y chromosomes, one paracentric inversion in a large autosome, and either a gain or loss of chromatin in another autosomal pair.

Reithrodon typicus Waterhouse, 1837
Naked-soled Conyrat

SYNONYMS:

Reithrodon typicus Waterhouse, 1837:30; type locality "Maldonado," Maldonado, Uruguay.

Reithrodon typicus currentium Thomas, 1920b:475; type locality "Corrientes. Type from Goya," Argentina.

Reithrodon typicus typicus: Gyldenstolpe, 1932a:77; name combination.

Reithrodon auritus typicus: Osgood, 1943b:223; name combination.

Reithrodon physodes typicus: Cabrera, 1961:502; name combination.

DESCRIPTION: General characters as in generic description. Soles of forefeet and hindfeet naked or tending to be less haired than those of *R. auritus*; tail unicolored; and diploid complement = 28.

DISTRIBUTION: *Reithrodon typicus* occurs in natural or modified grasslands in the Argentinean provinces of Entre Ríos and Corrientes, throughout Uruguay (Barlow 1969; Mones et al. 2003; E. M. González and Martínez Lanfranco 2010), and in a small portion of southeastern Brazil close to the Uruguayan border (T. R. O. Freitas, Mattevi, and Oliveira 1983; Bonvicino, Oliveira, and D'Andrea 2008).

SELECTED LOCALITIES (Map 299): ARGENTINA: Corrientes, Goya (type locality of *Reithrodon typicus currentium* Thomas). Estancia Rincón de Animas, 20 km NE of Sauce (MFA-ZV-M 1176); Entre Ríos, Estación Paranacito (J. R. Contreras 1972), Villa Ramírez (Massa 2010). BRAZIL: Rio Grande do Sul, 8 km E of Aceguá (T. R. O. Freitas, Mattevi, and Oliveira 1983). URUGUAY: Cerro Largo, 10 km NW of Paso del Dragón (= Plácido Rosas; Barlow 1969); Rocha, 22 km SE of Lascano (Barlow 1969); Maldonado (type locality of *Reithrodon typicus* Waterhouse).

SUBSPECIES: Thomas (1920b:475) distinguished within *R. typicus* a "dull sea-coast form," that is, the nominotypical *R. t. typicus* originally collected at Maldonado, Uruguay, and an "inland form . . . [with] well-contrasted markings," which he named as *R. t. currentium* based on a single specimen collected by R. Perrens at Goya, Corrientes, Argentina. Osgood (1943b) dismissed the distinction between these putative geographical forms based on pelage color. Ortells et al. (1988; see also T. R. O. Freitas, Mattevi, and Oliveira 1983) indicated chromosomal differences between Brazilian and Uruguayan populations, although both shared the same diploid number, which may warrant subspecific recognition.

NATURAL HISTORY: Very little is known about this species. Most information stems from Barlow (1969) and is based on 16 trapped specimens and observations made throughout Uruguay. Barlow (1969:35–36) stated that *R. typicus* was "usually found in overgrazed pasture, among rocky outcrops or on well-drained slopes with scanty vegetation . . . Coney rats dig burrows in soil varying from sandy to hard-baked clay, or utilize with or without modification abandoned burrows . . . or natural crevices and holes among rocks . . . The nearness of burrows to each other suggests that this cony rat may be variously gregarious or solitary . . . The burrow entrances averaged 5 cm in diameter. From each a passageway descended vertically an average of 25 cm, at which depth the tunnels became level and followed a course parallel to the surface. Active systems were marked by fresh cuttings and droppings near the entrances." Barlow (1969:36) also excavated an extensive burrow system near the Río Olimar Chico and noted "all burrows were less than 2 m in length and had two entrances. They were occasionally branched and tortuous . . . The burrow entrances were from 3 to 5 cm in diameter. The tunnels were 5 to 7 cm in diameter and were 10 to 25 cm beneath the surface. Some burrows contained central, oval chambers up to 30 cm in diameter. In one such chamber we found a nest platform composed of a fine dry grass . . . Defecation sites were just outside of the entrances of the burrows or within a radius of 9 m from them. No distinct runways led to any of these burrows." Most individuals secured by Barlow (1969) were obtained by shooting; E. M. González (2001) noted the same, reporting capture of individuals directly by hand when tracing burrows or locating nests. T. R. O. Freitas, Mattevi, and Oliveira (1983) reported captures using nets. Pregnant females were taken in January (one with four embryos) and May (three); males with scrotal testes activity were recorded in January, March, April, and May (Barlow 1969). Like *R. auritus*, *R. typicus* appears to be strictly nocturnal. Preferred plants consumed by this rodent in Uruguay include corms of *Oxalis* sp. (Oxalidaceae) and rhizomes and roots of the grass *Digitaria* sp.; Barlow (1969:37) also noted that they do not store food within their burrows.

REMARKS: The geographic boundary between *R. auritus* and *R. typicus* is not clear (Pardiñas and Galliari 2001). In Argentina, specimens from Goya, Corrientes province, were referred to *R. typicus* (see Thomas 1920b), whereas specimens from Villa Paranacito and Concordia, both in Entre Ríos province and north of the Río Paraná, were identified as *R. auritus* (see J. R. Contreras 1972; Crespo 1982a). According to unpublished mtDNA cytochrome-*b* sequence data, one specimen from Entre Ríos (MLP 26.VIII.01.17) was sister to Uruguayan representatives of *R. typicus* although diverging at 3%. This finding supports the reference of Argentinean Corrientes and Entre Ríos populations of conyrats to *R. typicus* and also the potential recognition of *currentium* Thomas as a valid subspecies. The "*orejón*," described by Azara (1802:82), was tentatively included in the list of synonyms of *R. auritus* by Pardiñas and Galliari (2001), in part because the type locality of Azara's taxon was

considered to be in Entre Ríos province, Argentina (Tate 1932c; Hershkovitz 1959b). We are now confident that Azara made his observations on the "*orejón*" as well as other rodents (e.g., *Akodon azarae*) around San Gabriel de Batovi (= São Gabriel), Río Grande do Sul, Brazil (see Pardiñas et al. 2007 for discussion). Consequently, the geographical context Azara's "*orejón*" should be considered a "synonym" of *R. typicus*. However, in the Spanish version of his work, Azara (1802) also included observations made at the southern margin of the Río La Plata, which suggests that his "*orejón*" may also be based on both *R. auritus* and *R. typicus*. A similar situation arises around his "*hocicudo*," a supposed synonym of *Oxymycterus rufus*, based on populations today referred to *O. rufus* and *O. nasutus* (J. R. Contreras and Teta 2003; see those accounts).

Brum-Zorrilla (1965) recorded a karyotype with $2n = 28$, $FN = 26$, for specimens from Uruguay but both T. R. O. Freitas, Mattevi, and Oliveira (1983) and Ortells et al. (1988) described a karyotype with $2n = 28$ and $FN = 40$ from specimens collected both in Uruguay and Rio Grande do Sul, Brazil. The apparent arm number differences between these two karyotypes were believed to be technical artifacts (see Ortells et al. 1988) rather than real geographic variation.

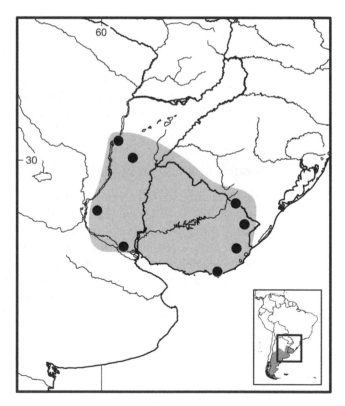

Map 299 Selected localities for *Reithrodon typicus* (●).

Tribe Sigmodontini Wagner, 1843
Robert S. Voss

This tribe includes only a single Recent genus, *Sigmodon*. Although other Recent genera were once considered to be closely related to *Sigmodon*, all are now referred to other tribes or unnamed clades. Hershkovitz (1955), for example, placed *Holochilus*, *Neotomys*, and *Reithrodon* together with *Sigmodon* in his "sigmodont group," primarily on the basis of molar morphology and characters of the facial skeleton associated with the masticatory musculature. Subsequent research based on different anatomical characters and/or molecular sequence data, however, have provided convincing evidence that *Holochilus* is an oryzomyine, and that *Neotomys* and *Reithrodon* are not closely related either to *Sigmodon* or to *Holochilus* (Hooper and Musser 1964; Voss and Carleton 1993; Steppan 1995; Weksler 2003, 2006; J. J. Martínez et al. 2012; Parada et al. 2013; Schenk et al. 2013). Although earlier molecular phylogenetic results suggested that *Sigmodon* was an isolated basal lineage within the subfamily Sigmodontinae (Engel et al. 1998; Patton and Smith 1999; D'Elía 2003a), more taxon-dense analyses supported an ancient sister-group relationship with ichthyomyines, with the two tribes together forming the basal clade to all other Sigmodontinae (Weksler 2003; Parada et al. 2013; Schenk et al. 2013). Whether or not any fossil taxa (e.g., *Prosigmodon*, from the late Tertiary of North America; McKenna and Bell 1997) properly belong to the tribe Sigmodontini is a question beyond the scope of this volume.

Because the tribe Sigmodontini is essentially monotypic, its morphological characteristics and geographic distribution are the same as those described below for the genus *Sigmodon*.

Genus *Sigmodon* Say and Ord, 1825

Species of *Sigmodon*, commonly known as cotton rats, range from about 41°N (in the United States) southward throughout most of Mexico and Central America to northern South America. Four species are currently recognized from South America, where they occur in natural grasslands and deforested landscapes below about 3,800 m elevation from Colombia southward to coastal Peru and eastward to the Guianas and northeastern Brazil. Except as noted below, the following account is abstracted from Voss (1992), which should be consulted for additional details about taxonomy, morphology, distribution, and natural history.

All species of *Sigmodon* are externally similar, with small ears, short vibrissae, coarsely grizzled brownish fur, and a sparsely haired tail that never exceeds the combined length of the head and body. The hindfeet are narrow and unwebbed, with the three middle digits (II, III, and IV) much longer than the outer two (I and V); the plantar surface is entirely naked (from heel to toes) with six small tubercles (thenar, hypothenar, and four interdigitals). There are usually either eight mammae (in four pairs) or 10 (in five pairs).

The skull is distinctive in dorsal view, with a short, blunt rostrum; deep zygomatic notches; and convergent, beaded supraorbital margins. The supraorbital beads are continuous with prominent temporal crests, and the zygomatic arches are usually convergent anteriorly (widest across the squamosal roots). The zygomatic plates are very broad, with concave anterior margins and prominent spinous anterodorsal processes. The incisive foramina are relatively long (exceeding 70% of diastemal length), and the mesopterygoid roof is always fenestrated, with large sphenopalatine vacuities. The parapterygoid fossae are broad and deeply excavated. A vertical strut of the alisphenoid separates the buccinator-masticatory and accessory oval foramina. The tegmen tympani overlaps a posterior suspensory process of the squamosal, and a slender hamular process of the squamosal separates the postglenoid foramen from the subsquamosal fenestra.

The mandible is dorsoventrally deep in proportion to its length, with a strong, falciform coronoid process and an angular process that extends posteriorly well behind the condyle. The lower incisor root is contained in a prominent capsular process posteroventral to the base of the coronoid process.

The molars are tetralophodont (lacking any trace of a mesoloph or mesolophid), hypsodont, and flat crowned, with the principal cusps and crests connected in a more or less continuous sigmoidal pattern (hence the generic name). The anterocone of M1 is undivided, as is the anteroconid of m1. M1 always has four roots, and m1 has either three or four roots; all remaining molars are three rooted.

Axial skeleton elements have modal counts of 12 ribs, 19 thoracolumbar vertebrae, 4 sacral vertebrae, and 23–25 caudal vertebrae. The tuberculum of the first rib articulates with the transverse processes of the seventh cervical and the first thoracic vertebrae, and the second thoracic vertebra has a greatly elongated neural spine. The entepicondylar foramen of the humerus is absent.

The stomach is unilocular and hemiglandular (Carleton 1973), and a gall bladder is present. Male accessory secretory organs consist of one pair each of bulbourethral, anterior prostate, dorsal prostate, ampullary, and vesicular glands, and two pairs each of preputial and dorsal prostate glands (Voss and Linzey 1981). The glans penis is complex (*sensu* Hooper and Musser 1964), with a tridigitate bacular cartilage and a deep terminal crater containing three bacular mounds, one dorsal papilla, two lateral papillae, and a bifurcate urethral flap (Hooper 1962).

SYNONYMS:

Sigmodon Say and Ord, 1825:354; type species *Sigmodon hispidus*, by monotypy.

Lasiomys Burmeister, 1855b:16; type species *Lasiomys hirsutus* Burmeister, by monotypy.

Lasiuromys: Giebel, 1857:caption to plate 5, Fig. 1; inadvertent misuse of *Lasiuromys* Deville for *Lasiomys* Burmeister (Winge 1941:153).

Deilemys Saussure, 1860:98; type species *Hesperomys toltecus* Saussure, by monotypy.

Isothrix: E.-L. Trouessart, 1880b:178; part (listing of *hirsutus*); not *Isothrix* Wagner.

Reithrodon: Thomas, 1881a:691; part (description of *alstoni*); not *Reithrodon* Waterhouse.

Sigmomys Thomas, 1901e:150; type species *Reithrodon alstoni* Thomas, by original designation.

REMARKS: Thomas (1901e:150) erected *Sigmomys* in recognition of the many morphological characters that distinguish *S. alstoni* from *Reithrodon* (the genus to which he originally referred this species) noting, however, that "it seems probable that this form is rather a groove-toothed *Sigmodon* than any close relation to *Reithrodon*." Hershkovitz (1955, 1966a), Cabrera (1961), and Voss (1992) concluded that *alstoni* was only what Thomas suggested—a species of *Sigmodon* with grooved upper incisors—and treated *Sigmomys* as a subjective junior synonym. Other authors have disagreed, continuing to recognize *Sigmomys* as a valid genus (e.g., Husson 1978; S. L. Williams et al. 1983) or subgenus (Musser and Carleton, 2005). Continued recognition of *Sigmomys* as a valid taxon at any rank, however, is unjustified. The notion of uniquely "important" morphological characters—such as the incisor grooving emphasized by Husson (1978) and S. L. Williams et al. (1983)—is obsolete, and current phylogenetic results (Henson and Bradley 2009) do not support the hypothesis that *Sigmodon alstoni* is the sister taxon to other members of the genus (contra the previous results of Peppers et al. 2002).

Because neither the South American species of *Sigmodon* nor the North American species appear to form monophyletic groups (Henson and Bradley 2009), the biogeographic history of the genus (discussed, inter alia, by Reig [1986] and Voss [1992]) must have included multiple intercontinental movements. Future phylogenetic research on this genus should make a particular effort to obtain molecular sequence data from *S. inopinatus*, the only living species that remains unrepresented in recent analyses. Additional nuclear gene sequences would also be helpful for resolving currently problematic interspecific relationships.

KEY TO THE SOUTH AMERICAN SPECIES OF *SIGMODON*:

1. Upper incisors deeply grooved and strongly opisthodont; m1 always with three well-developed roots . . . . *Sigmodon alstoni*

1'. Upper incisors ungrooved, strongly opisthodont or procumbent; m1 with three or four roots. 2

2. Stapedial foramen, squamosal-alisphenoid groove, and sphenofrontal foramen present (carotid arterial pattern 1; Voss 1988); palatal bridge short, narrow, and prominently grooved, with deep posterolateral sulci; auditory bullae large; molars with alternating, acutely angled cusps; m1 usually with three well-developed roots . . . . *Sigmodon peruanus*

2'. Stapedial foramen, squamosal-alisphenoid groove, and sphenofrontal foramen indistinct or absent (carotid arterial pattern 3; Voss 1988); palatal bridge long, broad, and smooth, without prominent grooves or deep posterolateral sulci; auditory bullae small; molars with opposite, rounded cusps; m1 usually with four well-developed roots 3

3. Nasal bones short (exposing incisors to dorsal view; interorbital region constricted (least interorbital breadth <23% of zygomatic breadth); posterior palatal foramina usually bordered by maxillary and palatine bones; upper incisors procumbent (but still weakly opisthodont). . . . . *Sigmodon inopinatus*

3'. Nasal bones long (concealing incisors from dorsal view); interorbital region broad (least interorbital breath >26% of zygomatic breadth); posterior palatal foramina usually enclosed by palatines; upper incisors not procumbent (strongly opisthodont) . . . . *Sigmodon hirsutus*

Sigmodon alstoni (Thomas, 1881)
Groove-toothed Cotton Rat
SYNONYMS:

Reithrodon alstoni Thomas, 1881a:691; type locality "Venezuela," subsequently restricted to Cumaná, Estado Sucre, Venezuela by Thomas (1901e:150).

Sigmomys alstoni: Thomas, 1901e:150; name combination.

Sigmomys savannarum Thomas, 1901e:150; type locality "on the savannas near their base," Kanuku Mountains, Upper Takutu–Upper Essequibo District, Guyana.

Sigmomys venester Thomas, 1914e:412; type locality "El Trompillo, near Lake Valencia, [Carabobo,] N. Venezuela. Alt. 1300'."

Sigmodon alstoni: Cabrera, 1961:508; first use of current name combination.

Sigmodon alstoni alstoni: Cabrera, 1961:508; name combination, as subspecies.

Sigmodon alstoni savannarum: Cabrera, 1961:508; name combination, as subspecies.

Sigmodon alstoni venester: Cabrera, 1961:508; name combination, as subspecies.

DESCRIPTION: Small (see Voss [1992] for measurement data), distinguishable from other congeners by broad, strongly opisthodont, and deeply grooved upper incisors. Additionally, *S. alstoni* differs from other South American species by its long nasal bones; unconstricted interorbital region; incisive foramina that usually extend to or between molar alveoli; unconstricted palatal bridge; posterior palatal foramina always bordered by both maxillary and palatine bones; derived stapedial circulation (pattern 3 of Voss 1988); small auditory bullae; upper molars with opposite, obtusely rounded cusps; and first mandibular molars with only three well-developed roots.

DISTRIBUTION: *Sigmodon alstoni* occurs below about 1,300 m elevation from northeastern Colombia eastward across the Maracaibo Basin, through the llanos, and into the deforested foothills of the Venezuelan coastal cordilleras. Another large and possibly isolated series of populations (surrounded by forests) inhabits the Gran Sabana of southeastern Venezuela, the contiguous Rupununi savannas of Guyana, and the adjacent Rio Branco savannas of Brazil. A few scattered collections document the presence of this species in the coastal and interior savannas of Surinam and Amapá state, Brazil (Husson 1978; S. L. Williams et al. 1983; C. T. Carvalho 1962). Pleistocene fossils have been reported from Tobago (Eshelman and Morgan 1985), but *S. alstoni* is not known to be part of any Recent insular fauna.

SELECTED LOCALITIES (Map 300): BRAZIL: Amapá, Macapa (C. T. Carvalho 1962); Roraima, Serra da Luna (AMNH 27039). COLOMBIA: La Guajira, Villanueva (USNM 280382); Meta, Los Micos (FMNH 88015). GUYANA: Upper Takutu–Upper Essequibo, Kanuku Mountains (type locality of *Sigmomys savannarum* Thomas). SURINAM: Para, Zanderij (RMNH 19620); Sipaliwini, Sipaliwini-savanne-vliegeveld (RMNH 17217). VENEUELA: Amazonas, Paria Grande (USNM 415034), San Juan Manapiare (USNM 415027); Anzoátegui, Río Oritupano (MHNLS 3823); Bolívar, Arabupu (AMNH 75653), La Llagual (AMNH 16971), Vetania (USNM 442581); Sucre, Manacal (USNM 415020); Zulia, Empalado Savannas (FMNH 18685), Río Cogollo (FMNH 27038).

SUBSPECIES: I treat *Sigmodon alstoni* as monotypic.

NATURAL HISTORY: *Sigmodon alstoni* is usually trapped in open, grassy habitats such as savannas, airfields, pastures, overgrown gardens, and weedy roadsides (Hand-ley 1976; Husson 1978; O'Connell 1981, 1982; Ibáñez and Moreno 1982; S. L. Williams et al. 1983; Vivas 1986; Soriano and Clulow 1988). Strictly terrestrial, *S. alstoni* constructs runways through low vegetation and is active both in the daytime and at night (Osgood 1912; Tate 1939; O'Connell 1981, 1982; Vivas et al. 1986). Like other cotton rats, *S. alstoni* is predominantly herbivorous and exhibits a strong preference for grass in feeding experiments (O'Connell 1981), but populations in agricultural fields eat large quantities of grain (A. M. G. Martino and Aguilera 1993). Reproductive autopsies suggest that this species breeds throughout the year, even in habitats with seasonal cycles of flooding and drought (Ibáñez and Moreno 1982). Voss (1992) summarized information about predators and arthropod ectoparasites. Originally suspected to be the principal rodent reservoir of Guaranito virus, the etiologic agent of Venezuelan hemorrhagic fever (Tesh et al. 1993), *S. alstoni* was subsequently shown to carry the related Piri-tal virus that is not known to cause any human disease (Fulhorst, Bowen et al. 1999; Fulhorst et al. 2008).

REMARKS: As noted here, *Sigmodon alstoni* was previously referred to the genus or subgenus *Sigmomys* (e.g., by Thomas 1901e; Husson 1978; S. L. Williams et al. 1983), a biologically arbitrary distinction based on the supposed importance of incisor grooving for rodent classification. Recent analyses of chromosomal and DNA sequence data (Peppers et al. 2002; Carroll and Bradley 2005; Bradley et al. 2008; Henson and Bradley 2009) have failed to convincingly resolve the phylogenetic position of *S. alstoni* with respect to congeners.

A historical record of *Sigmodon alstoni* from French Guiana (Menegaux 1902; discussed by Voss et al. 2001) was based on an old specimen in the National Histoire Naturelle, Paris (1902–44), only the skull of which can now

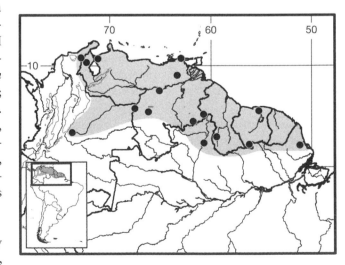

Map 300 Selected localities for *Sigmodon alstoni* (●). Contour line = 2,000 m.

be found (F. Catzeflis, pers. comm.). The specimen in question, a young adult example of *Hylaeamys megacephalus*, was obviously misidentified by Menegaux. Although the presence of *S. alstoni* in Surinam and Amapá (Brazil) suggests that the species could occur in French Guiana, it remains formally undocumented from that country.

Reig et al. (1990) stated that Reig (1987) reported karyotypes with diploid counts of 80 and 82 chromosomes. Reig (1987), however, made no reference to *Sigmodon* karyotypes, although Reig (1986) reported $2n = 82$ for *S. alstoni* without noting the geographic origin of his specimen(s). Voss (1992) reported modal diploid counts of 78 to 82 chromosomes from Venezuelan material.

Sigmodon hirsutus (Burmeister, 1854)
Burmeister's Cotton Rat

SYNONYMS:

Lasiomys hirsutus Burmeister, 1855b:16; type locality "von Maracaibo," Zulia, Venezuela.

Lasiuromys hirsutus: Giebel, 1857:caption to plate 5, Fig. 1; name combination.

Isothrix villosus hirsutus: E.-L. Trouessart, 1880b:178; name combination.

Sigmodon bogotensis J. A. Allen, 1897d:121; type locality "Plains of Bogota, [Cundinamarca] Colombia."

Sigmodon sanctaemartae Bangs, 1898c:189; type locality "Pueblo Viejo, [Cesar,] Colombia," "in the center of the Sierra Nevada [de Santa Marta], not far from the source of Rio Ancho" (p. 189, footnote).

Sigmodon hirsutus: Thomas, 1914e:413; first use of current name combination.

Sigmodon hispidus bogotensis: Cabrera, 1961:508; name combination.

Sigmodon hispidus hirsutus: Cabrera, 1961:508; name combination.

DESCRIPTION: Large (see Voss [1992] for measurement data), distinguished from South American congeners by long nasal bones; broad interorbital region; long incisive foramina; broad and long palatal bridge without conspicuous longitudinal grooves; posterior palatal foramina usually enclosed by palatine bones; derived stapedial circulation (pattern 3 of Voss 1988); small auditory bullae; narrow, strongly opisthodont, ungrooved upper incisors; upper molars with opposite, obtusely rounded cusps; and first mandibular molars usually with four well-developed roots.

DISTRIBUTION: In South America, *Sigmodon hirsutus* occurs below about 2,600 m from the Caribbean coast of Colombia southward throughout the valley of the Río Magdalena and eastward to northern Venezuela. The range of *S. hirsutus* overlaps that of *S. alstoni* in northwestern Venezuela, where both species have been collected sympatrically at several localities. Pleistocene fossils from Aruba,

Netherland Antilles (Hooijer 1967), are possibly referable to this species, but no Recent insular populations of *S. hirsutus* are known. According to Bradley et al. (2008), the North American distribution of *S. hirsutus* extends from Panama northward to southwestern Mexico, excluding Belize, eastern Guatemala, and northern Honduras.

SELECTED LOCALITIES (Map 301): COLOMBIA: Cundinamarca, near Bogotá (type locality of *Sigmodon bogotensis* J. A. Allen); Huila, Valle de Suaza (USNM 542102); Magdalena, Bonda (FMNH 15281); Norte de Santander, Guamalito (USNM 280877); Tolima, Mariquita (BM 12.4.2.6). VENEZUELA: Barinas, near Altamira (USNM 416758); Cojedes, Hato Itabana (MHNLS 3851); Guárico, San Juan de los Morros (USNM 314995); Mérida, near Mesa Bolívar (USNM 388120); Miranda, Curupao (USNM 388131).

SUBSPECIES: As currently recognized (e.g., by Peppers and Bradley 2000; Musser and Carleton 2005), *Sigmodon hirsutus* is monotypic.

NATURAL HISTORY: Like other cotton rats, *Sigmodon hirsutus* is always trapped on the ground, usually in grassy or weedy habitats such as savannas, pastures, and agricultural fields (Handley 1976). In the Venezuelan llanos, the diet of this species consisted principally of the green (vegetative) tissues of grasses and herbaceous dicots (Vivas and Calero 1988). Llanoan populations breed throughout the annual cycle of flooding and drought in this seasonal landscape, although reproductive activity is perhaps most intense in the rainy season when population densities are highest (Vivas and Calero 1985, 1988). Voss (1992) summarized information about predation and arthropod ectoparasites.

REMARKS: *Sigmodon hirsutus* was long considered to be a subspecies or synonym of *S. hispidus* (e.g., by Cabrera 1961; Honaki et al. 1982; Voss 1992; Musser and Carleton 1993), but phylogenetic analyses of DNA sequence data have consistently failed to recover *hirsutus* and *hispidus* as sister taxa (Peppers and Bradley 2000; Peppers et al. 2002; Carroll and Bradley 2005; Bradley et al. 2008). Although these species are phenotypically similar in most respects, the morphology of the posterior palate provides a potentially useful criterion for sorting museum series unaccompanied by molecular data. In typical examples of *S. hispidus* (from the southeastern United States), the palate is marked by deep posterolateral sulci flanking a well-defined keel that is produced posteriorly as a strong median palatal spine. By contrast, South American specimens of *S. hirsutus* have a smooth palate that lacks posterolateral sulci and never develops a distinct median keel or spine.

Venezuelan specimens of *Sigmodon hirsutus* appear to have the same diploid number ($2n = 52$) and autosomal chromosome morphology as specimens of *S. hispidus* from North and Central America (Voss 1992).

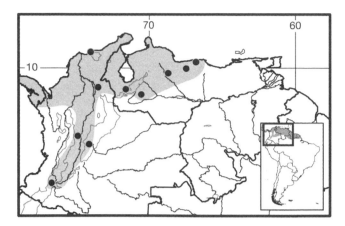

Map 301 Selected localities for *Sigmodon hirsutus* (●). Contour line = 2,000 m.

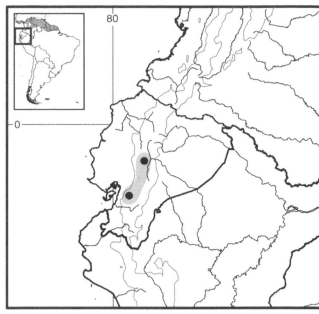

Map 302 Selected localities for *Sigmodon inopinatus* (●). Contour line = 2,000 m.

Sigmodon inopinatus Anthony, 1924
Ecuadorean Cotton Rat

SYNONYMS:

Sigmodon inopinatus Anthony, 1924a:3; type locality "Urbina, slopes of Chimborazo, [Chimborazo,] Ecuador; altitude 11,400 feet."

Sigmodon hispidus inopinatus: Cabrera, 1961:509; name combination.

DESCRIPTION: Large (see Voss [1992] for measurement data), distinguished from other congeners by short nasal bones (exposing incisors to dorsal view); very narrow interorbital region; long incisive foramina (usually, but not always extending between molar alveoli); unconstricted palatal bridge; posterior palatal foramina usually bordered both by maxillary and palatine bones; derived stapedial circulation (pattern 3 of Voss 1988); small auditory bullae; procumbent (but still weakly opisthodont), ungrooved upper incisors; upper molars with opposite, obtusely rounded cusps; and first mandibular molars usually with four well-developed roots.

DISTRIBUTION: *Sigmodon inopinatus* is currently known from just two localities in the high Andes of Ecuador.

SELECTED LOCALITIES (Map 302): ECUADOR: Azuay, Las Cajas National Park (Voss 1992); Chimborazo, Urbina (type locality of *Sigmodon inopinatus* Anthony).

SUBSPECIES: *Sigmodon inopinatus* is monotypic.

NATURAL HISTORY: All known specimens of this species were trapped in treeless vegetation between 3,500 and 3,800 m. At the type locality, Anthony (1924a) reported 12 captures among tufts of coarse grass, presumably on well-drained slopes, but five specimens from Las Cajas were all trapped near water (riverbanks, marshes, and bogs; Barnett 1999). No other information is available on the ecology, natural history, or behavior of this poorly known species.

REMARKS: The karyotype of *Sigmodon inopinatus* is unknown.

Sigmodon peruanus J. A. Allen, 1897
Peruvian Cotton Rat

SYNONYMS:

Sigmodon peruanus J. A. Allen, 1897c:118; type locality "Trujillo, [Libertad,] Peru."

Sigmodon simonsi J. A. Allen, 1901a:40; type locality "Eten, coast region of northwestern [Lambayeque,] Peru."

Sigmodon puna J. A. Allen, 1903a:99; type locality "Puná, Puná Island, [Guayas,] Ecuador, . . . altitude 10 m."

Sigmodon chonensis J. A. Allen, 1913a:479; type locality "Chone, Manavi [= Manabí] Province, Ecuador (altitude less than 100 feet)."

Sigmodon lönnbergi Thomas, 1921f:448; type locality "Quevado, lowlands of Western Ecuador, due north of Guayaquil," Los Ríos province.

Sigmodon hispidus chonensis: Cabrera, 1961:508; name combination.

Sigmodon hispidus peruanus: Cabrera, 1961:509; name combination.

Sigmodon hispidus puna: Cabrera, 1961:509; name combination.

Sigmodon hispidus simonsi: Cabrera, 1961:509; name combination.

Sigmodon peruanus peruanus: Reig, 1986:407; name combination.

Sigmodon peruanus simonsi: Reig, 1986:407; name combination.

Sigmodon peruanus chonensis: Reig, 1986:407; name combination.

Sigmodon peruanus puna: Reig, 1986:407; name combination.

DESCRIPTION: Large (see Voss [1992] for measurement data), distinguished from all congeners by complete, primitive stapedial circulation (pattern 1 of Voss 1988). From other South American species, *S. peruanus* distinguished by long nasal bones; unconstricted interorbital region; short incisive foramina (seldom extending to or between molar alveoli); very narrow, short, and grooved palatal bridge with deep posterolateral sulci; posterior palatal foramina always bordered by both maxillary and palatine bones; large auditory bullae; broad, strongly opisthodont, ungrooved upper incisors; upper molars with alternate, acutely angled cusps; and first mandibular molars usually with only three well-developed roots.

DISTRIBUTION: *Sigmodon peruanus* occurs below about 1,600 m elevation in the Pacific littoral and adjacent Andean foothills of western Ecuador and northwestern Peru.

SELECTED LOCALITIES (Map 303): ECUADOR: Azuay, Tunguilla Valley (FMNH 53375); Chimborazo, Puente de Chimbo (AMNH 63027); Loja, Malacatos (FMNH 53381); Los Ríos, near San Carlos (MSU 32638); Manabí, Chone (type locality of *Sigmodon chonensis* J. A. Allen), Río de Oro (AMNH 34295). PERU: Lambayeque, Trujillo (type locality of *Sigmodon peruanus* J. A. Allen); Piura, Jilili (MCZ 41136).

SUBSPECIES: I treat *Sigmodon peruanus* as monotypic.

NATURAL HISTORY: Specimens of *Sigmodon peruanus* have been collected in xeric and semiarid open habitats, as well as in agricultural fields and secondary vegetation, generally in regions of western Ecuador and northwestern Peru with a long and severe annual dry season. Unpublished field notes indicate that this species, like other cotton rats, is both diurnal and nocturnal, and constructs runways through grass or other low herbaceous vegetation. No other information of ecology, natural history, or behavior of this species is known.

REMARKS: Phylogenetic analyses based on cytochrome-*b* sequence data that suggest that *Sigmodon peruanus* is more closely related to North American cotton rats of the *S. fulviventer* group (Peppers et al. 2002) than to South American or Central American congeners are contradicted by recent molecular results that do not convincingly resolve the relationships of this species (Henson and Bradley 2009). The karyotype of *S. peruanus* is unknown.

Tribe Thomasomyini Steadman and Ray, 1982
Víctor Pacheco, James L. Patton, and Guillermo D'Elía

The concept of a thomasomyine group of genera among the Sigmodontinae has been enigmatic since Thomas (1906c) originally recognized that taxa with pentalophodont molars could be divided into those with short palates without posterolateral pits (e.g., *Thomasomys* and *Rhipidomys*) and those with long palates and well-developed pits (e.g., *Oryzomys*). A separate group related to *Thomasomys*, including *Aepeomys*, *Chilomys*, *Delomys*, *Phaenomys*, *Rhagomys*, and *Rhipidomys*, was varyingly placed into an all-inclusive oryzomyine complex (Tate 1932g), formalized with all pentalophodont taxa into the Oryzomyini (Vorontsov 1959; see also Reig 1986), or informally recognized as "thomasomyines" (Hershkovitz 1962, 1966a). Hershkovitz envisioned thomasomyines as a sylvan pentalophodont grade at the base of the sigmodontine radiation from which one lineage gave rise to the oryzomyines (*sensu stricto*) and another to what he termed the pastoral tetralophodont representatives, such as the Akodontini and Phyllotini. He conceived (Hershkovitz 1966a:125) a

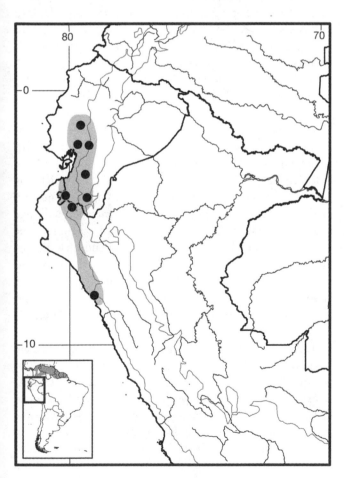

Map 303 Selected localities for *Sigmodon peruanus* (●). Contour line = 2,000 m.

Pacheco's Thomasomyini, notably *Delomys*, *Juliomys*, and *Wiedomys* (M. F. Smith and Patton 1999; Weksler 2003; D'Elía et al. 2005, 2006; Salazar-Bravo et al. 2013; Ventura et al. 2013). In these analyses, *Delomys* is placed as sister to Phyllotini, and *Juliomys* is inconsistently placed within the Sigmodontinae, although in general it is associated with Andean taxa previously considered to be phyllotines.

Unfortunately, sequence data are unavailable to date for *Phaenomys* and *Wilfredomys*, but mitochondrial and nuclear gene sequences (Ventura et al. 2013) for *Abrawayaomys* suggest it is more closely related to the Akodontini rather than to the Thomasomyini, where it was placed in Pacheco's morphological analysis. All molecular studies support *Wiedomys* as a lineage sister to Abrotrichini, one to be retained in the small tribe Wiedomyini originally diagnosed by Reig (1980). Excluding *Wiedomys*, all of these genera are regarded herein as Sigmodontinae *incertae sedis* (see the *incertae sedis* account). We emphasize, however, that our view of the Thomasomyini is provisional and acknowledge that no morphological synapomorphy is currently known to support the group as we use it herein. The composition of the Thomasomyini, therefore, is highly likely to change as future molecular studies expand both taxon representation and number of genes examined.

Our restricted set of five thomasomyine genera includes three with limited species diversity and narrow ranges (*Aepeomys*, *Chilomys*, and *Rhagomys*, with two species each) and two highly speciose genera with cumulatively very broad ranges (*Rhipidomys*, currently 23 species; *Thomasomys*, currently 44 species). Pacheco (2003) questioned the generic status of *Aepeomys*, although it is included herein as valid, and he noted that *Thomasomys*, as currently understood, was polyphyletic while *Rhipidomys* might be paraphyletic. In his revision of *Rhipidomys*, Tribe (1996) tentatively recognized its monophyly but did not support this opinion with a phylogenetic analysis; however, Pacheco (2003:278, 378), with a denser taxon sampling, recovered a monophyletic *Rhipidomys* and provided an emended diagnosis. Substantial systematic work, at all levels, will be necessary to resolve these vexing problems, and it is likely that future work will divide both of these genera to maintain monophyletic taxa, much the way Weksler et al. (2006) did recently for *Oryzomys* (*sensu lato*). These unresolved issues make the construction of a suitable key for identifying the five genera currently recognized difficult, at best.

KEY TO THE GENERA OF THE THOMASOMYINI:
1. Palate long, with mesopterygoid fossa not extending anteriorly to level of M3s..........................2
1'. Palate short, with mesopterygoid fossa extending anterior to or between M3s.......................3

2. Dorsal aperture of ectotympanic ring closed, extending posteriorly to contact or closely approach the petrosal; anterodorsal margin of zygomatic plate deep (depth greater than half width of notch); simple posterolateral palatal pits placed distinctly anterior to mesopterygoid fossa............................*Rhagomys*
2'. Dorsal aperture of ectotympanic ring open, not extending posteriorly to contact petrosal but bordered by a well-exposed tegmen tympani; anterodorsal margin of zygomatic plate shallow (depth less than half width of notch); simple posterolateral palatal pits placed at level with or posterior to anterior margin of mesopterygoid fossa.............................*Chilomys*
3. Gall bladder absent; dark metatarsal patches always present; no gap present between hypothenar and thenar pads of hindfeet; hypothenar pad level or overlapping with interdigital pad 4; claw of hallux long, extending to first interphalangeal joint of dII or beyond; mystacial vibrissae very long, extending well past posterior margin of pinna; body pelage moderately to distinctly countershaded; tail penicillate, typically with long tuft of hairs extending posteriorly from tip; anterior margin of zygomatic plate usually vertical; squamosal roots of zygomatic arch to not expand laterally, joining braincase at oblique angle; zygomatic arches typically parallel; interorbital region convergent and usually distinctly beaded when viewed in cross section; subsquamosal fenestra placed in low position, distinctly below the squamosal root of the zygomatic arch; mesopterygoid fossa convergent anteriorly ....*Rhipidomys*
3'. Gall bladder present; dark metatarsal patches variably present, or absent; gap present between hypothenar and thenar pads usually present; distinct gap present between hypothenar and interdigital pad 4; claw of hallux variable in length; mystacial vibrissae short to variable in length; body pelage usually indistinctly or not countershaded; tail typically nonpenicillate, or with very short extension of hair tuft posterior to tip; anterior margin of zygomatic plate typically slopes backward; squamosal roots of zygomatic arch laterally expanded, joining the braincase more perpendicularly; zygomatic arches typically converge anteriorly; interorbital region hourglass in shape, with supraorbital region smoothly rounded when viewed in cross section or weakly squared or edged; subsquamosal fenestra in high position, level with the squamosal root of the zygomatic arch; mesopterygoid fossa typically parallel sided.......................4
4. Dark metatarsal patches absent; gap between hypothenar and thenar pads always present; claw on hallux short, not extending beyond metatarsal of dII; genal I vibrissae absent; mystacial vibrissae short, not extending to the posterior margin of the pinna; masseteric tubercle

prominent; anteroflexus of M1 small to indistinct . . . .
. . . . *Aepeomys*
4′. Dark metatarsal patches usually present; gap between hypothenar and thenar pads usually present; claw on hallux variable, short to long; genal I vibrissae variably present; mystacial vibrissae variable, from short to very long; masseteric tubercle typically absent, or reduced in size; anteroflexus of M1 typically distinct . . . .
. . . . *Thomasomys*

Genus *Aepeomys* Thomas, 1898

Víctor Pacheco

Aepeomys comprises two species, *A. lugens* and *A. reigi*. This South American endemic occurs in premontane and montane tropical forests along the Venezuelan Andes, including the Sierra de Mérida, at elevations above 1,000 m in western Venezuela (O. J. Linares 1998; Soriano et al. 1999; Ochoa et al. 2001). The following description is based mainly on Thomas (1896), Ochoa et al. (2001), Voss (1991), Voss et al. (2002), and Pacheco (2003).

Species of *Aepeomys* are medium-sized sigmodontines distinguished by a slender tail slightly longer than length of head and body (head and body 100–125 mm, tail 114–142 mm), narrow and slender hindfeet (25–30 mm), and moderate mass (32.0–44.5 g). The tail is thinly haired, uniformly colored above and below, and lacks a white tip. The ears are large (18–21 mm), brown, and furred inside and out. The body pelage is dense, soft, and velvety. Dorsal coloration varies from dark gray-brown to reddish gray–brown in *A. reigi* or olive brown in *A. lugens*, and the ventral pelage is slightly paler than dorsum. Ungual tufts on manus are absent. Digit I of the pes (hallux) is short, its first phalanx (p1) not extending beyond the metatarsal of dII, and the claw does not extend more than half the length of p1 of dII. Claws on manus are medially keeled. Legs, ankles, and dorsal surface of the pes are sparsely covered by brown hair. Dark metatarsal patches are absent. A gap between the hypothenar and thenar pads is present. The mammary complement consists of three pairs, one each in inguinal, abdominal, and thoracic positions (terminology of Pacheco 2003). Genal vibrissae are lacking. Mystacial vibrissae are short, not extending to the posterior margin of the pinna. A prominent protuberant anus is present.

Cranially, the rostrum is long, tapering, and the nasals and premaxillae form a prominent bony tube that projects well beyond incisors. Posterodorsal extensions of the premaxillae are very short, terminating anterior to the nasals and the zygomatic notch. The lumen of the infraorbital foramen is moderately open. The zygomatic notches are shallow, zygomatic plates are narrow and slope posterodorsally,

and the masseteric tubercle is comparatively prominent. The incisive foramina are large but do not reach the level of the M1s. The interorbital region is broad and hourglass in shape, with rounded supraorbital margins. The ethmoid foramen is placed dorsal or posterodorsal to the M2/M3 contact. An alisphenoid strut is absent. The braincase is rounded and inflated; the palate is short to relatively long but does not extend posteriorly much behind the last molar; posterolateral palatal pits are poorly developed; and the mesopterygoid fossa is wide and flanked by narrow, shallow, and unperforated parapterygoid fossae. Carotid circulation is pattern 1 of Voss (1988) with a large stapedial foramen, squamosal-alisphenoid groove, and sphenofrontal foramen present. The tegmen tympani overlaps a posterior suspensory process of the squamosal. The anterodorsal edge of ectotympanic is short and does not extend posteriorly to contact the petrosal. The postglenoid foramen is very small and subequal in size to the tegmen tympani. The parietosupraoccipital suture is intermediate, about one-third the posterior margin of the parietal. Bullae are noninflated. The malleus lamina is square shaped; and the thoracic basket has 13 pairs of ribs (Steppan 1995). The mandible is very slender, and the capsular processes are absent. Upper incisors are narrow, orthodont, and moderately developed; upper and lower molars have comparatively high crowns with well-developed mesolophs and mesolophids; internal segments of paraflexus and metaflexus are oriented anteroposteriorly along the midline of M1–M2; protoflexus and hypoflexus are both narrow and shallow; M1 anterocone exhibits a small or indistinct anteromedian flexus, but anteroconid of m1 is undivided.

Steppan (1995) counted 13 ribs and 13 thoracic, 6 lumbar, and about 38 caudal vertebrae in the postcranial skeleton of a single specimen of *A. lugens*. The stomach is unilocular (single-chambered) and discoglandular (gastric glandular epithelium restricted to a small pouch-like structure on the greater curvature); a gall bladder is present. The glans penis body, lateral mounds, urethral flaps, and dorsal papilla lack spines. The genus is unknown from the fossil record.

SYNONYMS:

Oryzomys: Thomas, 1896:306; part (description of *lugens* Thomas); not *Oryzomys* Baird.

Aepeomys Thomas, 1898e:452; type species *Oryzomys lugens* Thomas, by original designation.

Thomasomys: Anthony, 1932:1; part (description of *ottleyi* Anthony); not *Thomasomys* Coues.

REMARKS: Osgood (1933a) considered *Aepeomys* to be a synonym of *Thomasomys*, an arrangement followed by Ellerman (1941), Cabrera (1961), and Handley (1976). Alternatively, Gyldenstolpe (1932), A. L. Gardner and Patton (1976), and more recent authors (e.g., Aguilera et al. 1994, 2000; Musser and Carleton 1993, 2005) argued for generic status.

Neither of these options, however, is sustained by a phylogenetic hypothesis or a thorough comparison between species of both genera. Much of the support for recognizing *Aepeomys* as a distinct genus came from A. L. Gardner and Patton (1976) who reported a karyotype of "*Aepeomys fuscatus* J. A. Allen" that differed appreciably from those of species in the thomasomyine genera *Thomasomys* and *Rhipidomys* and was more similar to that of oryzomyine rodents. However, after Voss et al. (2002) convincingly removed *fuscatus* to *Handleyomys* Voss, Gómez-Laverde, and Pacheco, *Aepeomys* became restricted to *A. reigi* and *A. lugens*. Although *Aepeomys* needs to be further contrasted with *Thomasomys*, it is tentatively retained here as a separate genus based on cranial characteristics that distinguish it from *Thomasomys sensu stricto* (Ochoa et al. 2001; Pacheco 2003), and by the karyological differences reported by Aguilera et al. (2000). Pacheco's (2003) phylogenetic assessment based on morphological characters suggested that *Aepeomys* is nested within a clade of *Thomasomys* spp., but the support for this clade was weak.

Thomas (1898e) described *Aepeomys vulcani* based on a specimen with a partially crushed skull from Pichincha, Ecuador. Philip Hershkovitz, Charles Handley, and Víctor Pacheco independently found that this specimen shared more characteristics with *Thomasomys sensu stricto* than with *Aepeomys* (as determined by Voss et al. 2002, and herein). *Aepeomys vulcani* Thomas (1898e) was then referred to *Thomasomys* (Voss et al. 2002; see also the account for the genus *Thomasomys*).

Aepeomys lugens was not available to M. F. Smith and Patton (1999) for their phylogenetic assessment based on mtDNA cytochrome-*b* sequence data, but they suggested that *Aepeomys* might belong to a clade composed of *Thomasomys*, *Chilomys*, and *Rhipidomys*. *Aepeomys lugens* was included for the first time in a phylogenetic analysis based on nucDNA IRBP sequences (D'Elía, Luna et al. 2006) where *Aepeomys* appeared as the sister to a *Rhagomys-Thomasomys* clade.

KEY TO THE SPECIES OF *AEPEOMYS*:

1. Head and body length (average 110.1 mm, range: 100–119 mm); fur on head and body shorter and rougher; legs, ankles, and dorsal surface of pes densely haired; digit V of pes moderately long, extending beyond the interphalangeal joint of digit IV; posterior margin of zygomatic ramus of the maxilla lacking a distinct notch; incisive foramina broad and oval; interparietal of moderate size, with anteroposterior midline between 30–40% of parietal length; palate extending to the posterior border of M3 or barely beyond; posterior margin of palate with small medial process; maxillary toothrow (average 4.3 mm, range: 4.0–4.4 mm); M3 small and round in occlusal view; paraflexus of M1 and M2 un-divided; coronoid and condylar processes not notably broad nor large, sigmoid and angular notches not markedly deep; protolophid of m1 reduced or absent; karyotype $2n = 28$, $FN = 48$ . . . .*Aepeomys lugens*

1′. Head and body length longer (average 113.6 mm, range: 104–125 mm); fur on head and body shorter and rougher; legs, ankles, and dorsal surface of pes sparsely haired; digit V of pes shorter, not extending beyond the interphalangeal joint of digit IV; posterior margin of zygomatic ramus of the maxilla with a distinct notch; incisive foramina broader with more convergent anterior margins; interparietal longer in most specimens with anteroposterior midline near half of the parietal length; palate longer, extending behind M3; posterior margin of palate shallow and lacking a medial process in most specimens; maxillary toothrow longer (average 4.5 mm, range: 4.3–4.8 mm); M3 larger and triangular in occlusal view; M1 and M2 with paraflexus divided by an enamel bridge; coronoid and condylar processes broader and larger, producing deeper sigmoid and angular notches; protolophid in m1 usually present; karyotype $2n = 44$, $FN = 46$ . . . . *Aepeomys reigi*

Aepeomys lugens (Thomas, 1896)
Mérida Aepeomys

SYNONYMS:

Oryzomys (?) *lugens* Thomas, 1896:306; type locality "La Loma del Morro, near Merida, Venezuela, alt. 3000 metres."

[*Aepeomys*] *lugens*: Thomas, 1898e:452; first use of current name combination.

Thomasomys ottleyi Anthony, 1932:1; type locality "Paramo de los Conejos, about fifteen miles north of Merida, Venezuela; altitude 9600 feet."

[*Thomasomys*] *lugens*: Osgood, 1933a:161; name combination.

DESCRIPTION: Similar to *A. reigi*, differing in slightly smaller size (mean head and body length 110.1 mm, tail length 121.7 mm, hindfoot length 27.0 mm, and condyloincisive length of skull 26.6 mm), proportionately longer digits I and V of hindfoot, broad and oval incisive foramina, moderate-sized interparietal with its midline 30–40% of parietal length, no distinct notch on posterior margin of zygomatic ramus of maxilla, palate extending to posterior border of M3 or slightly beyond, and other details of cranium. Karyotype distinctly different, $2n = 28$, $FN = 48$, versus $2n = 44$, $FN = 46$; see Ochoa et al. (2001) and the *A. reigi* account.

DISTRIBUTION: Mérida Andes of Venezuela (states of Táchira and Mérida), between 1,990 and 3,500 m.

SELECTED LOCALITIES (Map 304): VENEZUELA: Mérida, La Montana, 4.1 km SE of Mérida (USNM

579550), Paramito, 3 km W of Timotes (Handley 1976), Páramo de los Conejos, 15 mi N of Mérida (Anthony 1932), Santa Rosa, 2 km N of Mérida (Handley 1976); Táchira, Páramo el Zumbador, 8 km SSW El of Cobre (Ochoa et al. 2001).

SUBSPECIES: I treat *Aepeomys lugens* as monotypic.

NATURAL HISTORY: *Aepeomys lugens* is nocturnal, may be semiarboreal (O. J. Linares 1998) or strictly terrestrial (Aagaard 1982). These are solitary and largely insectivorous mice, eating primarily arthropods, larvae, and worms (Díaz de Pascual 1993). Individuals can breed year-round but with a preference on the rainy season, with pregnant females recorded in September and lactating females between May and November. Litter size was usually two, but seven young have been found. *Aepeomys lugens* is found in primary cloud forest and rarely in secondary habitats (Díaz de Pascual 1993), and also in seasonal forests and páramos (Soriano et al. 1999). Microhabitat preferences are apparently similar to those of *A. reigi*. Handley (1976) recorded captures both species of *Aepeomys* beside logs and at the base of trees and tree ferns, among boulders covered with moss and lichen, in thick cover of herbs and ferns on the ground, and on logs; usually in damp habitats and rarely in dry places, all within Lower Montane very humid forest (bmh-MB), Montane rain forest (bp-M), Lower Montane humid forest (bh-MB), and Montane very humid forest (bmh-M1), in the life zone categories of Ewel and Madriz (1968). Aagaard (1982) found this species most abundant in the lower montane wet forest and montane wet forest in the Holdridge system. Males averaged 3.6% heavier than females (Aagaard 1982). O. J. Linares (1998) probably combined data for both *A. lugens* and *A. reigi*, as the latter species was not described until after Linares's book was published.

Aepeomys lugens is the host for several genera of mites, including the acarid genus *Eutropicola* (Brennan and Reed 1974a) and laelapid genus *Hymenolaelaps* (Furman 1972a). Brennan and Reed (1975) documented additional chiggers, including species in the genera *Crotiscus*, *Hoffmannina*, and *Adontacarus*. P. T. Johnson (1972) described anopluran lice in the genus *Hoplopleura*, E. K. Jones et al. (1972) reported ticks of the genus *Ixodes*, and Tipton and Machado-Allison (1972) described fleas in the genera *Ctenidiosomus*, *Neotyphloceras*, *Plocopsylla*, and *Cleopsylla*. All records are from Mérida state, with the host originally identified as *Thomasomys lugens*.

REMARKS: Osgood (1933a) synonymized *Thomasomys ottleyi* Anthony with *A. lugens*, a conclusion supported by Voss et al. (2002). Cuervo Díaz et al. (1986) and Alberico et al. (2000) reported *A. lugens* for Colombia, but these records appear to have been based on misidentified material (Voss et al. 2002). The reported distribution in

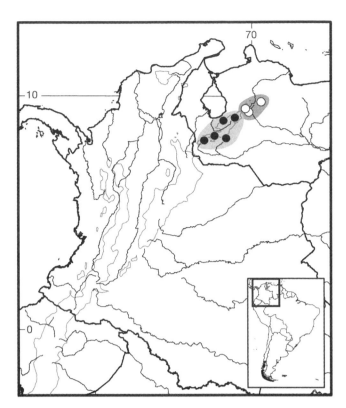

Map 304 Selected localities for *Aepeomys lugens* (●) and *Aepeomys reigi* (○). Contour line = 2,000 m.

Andean Ecuador (Eisenberg 1989; Musser and Carleton 1993; Eisenberg and Redford 1999) was apparently based on the erroneous inclusion of *Aepeomys vulcani* Thomas, 1898 (= *Thomasomys vulcani*) as a synonym of *A. lugens* by Cabrera (1961).

Aguilera et al. (2000) reported the karyotype (2n = 28, FN = 48) and noted its strong divergence from the karyotype (2n = 44, FN = 46) of an unnamed population subsequently described as *Aepeomys reigi* (Ochoa et al. 2001).

Aepeomys reigi Ochoa, Aguilera, Pacheco, and Soriano, 2001

Reig's Aepeomys

SYNONYMS:

Thomasomys lugens: Handley, 1976:51; part; not *Oryzomys* (?) *lugens* Thomas (= *Aepeomys lugens* [Thomas]).

Aepeomys lugens: O. J. Linares, 1998:309; part; not *Oryzomys* (?) *lugens* Thomas (= *Aepeomys lugens* [Thomas]).

Aepeomys reigi Ochoa, Aguilera, Pacheco, and Soriano, 2001:230; type locality "El Blanquito, Parque Nacional Yacambú, 17 km SE Sanare, Lara State, Venezuela, 1600 m (approx. 9°40′N; 69°37′W)."

DESCRIPTION: Larger of two known species in genus, but only slightly so (mean head and body length 113.6 mm, tail length 127.1 mm, hindfoot length 27.9 mm, condyloincisive length of skull 27.8 mm). Differs from *A. lugens* in that digits I and V of hindfoot do not extend beyond commissure of digits II and III and first interphalangeal of digit IV; posterior margin of zygomatic ramus of maxilla with distinct notch; palate extends to posterior border of M2 or even behind it; interparietal length nearly half parietal length; and enamel bridge, crossing from paracone to base of anteroloph, divides paraflexus of M1 and M2.

DISTRIBUTION: *Aepeomys reigi* occurs in the extreme northeastern part of the Venezuelan Andes (states of Lara and Trujillo), between 1,600 and 3,230 m.

SELECTED LOCALITIES (Map 304): VENEZUELA: Lara, El Blanquito, Parque Nacional Yacambu, 17 km SE of Sanare (type locality of *Aepeomys reigi* Ochoa, Aguilera, Pacheco, and Soriano); Trujillo, Hacienda Misisí, 14 km E of Trujillo (Ochoa et al. 2001), Macizo de Guaramacal, 9 km ESE of Boconó (Ochoa et al. 2001).

SUBSPECIES: *Aepeomys reigi* is monotypic.

NATURAL HISTORY: Handley (1976) and Ochoa et al. (2001) wrote that *A. reigi* inhabited primary cloud forest and small patches of páramo, and that it was not rare in cloud forest habitats. Specimens have been trapped on the ground in densely forested areas (beside logs, at the base of trees and tree ferns, in rocky places, along trails, or near small streams) or in open areas (close to the ecotone between páramo and forest) covered by shrubs and herbaceous vegetation (Ochoa et al. 2001). Brennan and Reed (1974a) described three new species of acarid mites in the genus *Eutrombicula* from specimens of this species, J. T. Reed and Brennan (1975) described three new species of chiggers in the genus *Odontacarus*, and Tipton and Machado-Allison (1972) reported a flea in the genus *Cleopsylla* from *A. reigi* (all reported as *Thomasomys lugens*).

REMARKS: Specimens from Trujillo were formerly recorded as *Thomasomys lugens* (Handley 1976) or as *Aepeomys* sp. (Aguilera et al. 1994, 2000; Soriano et al. 1990). Ochoa et al. (2001) differentiated *A. reigi* from *A. lugens* on the basis of larger size, shorter fur, more sparsely haired legs, ankles, and hindfeet, deeper zygomatic notches, and other cranial and dental features in addition to the distinctive karyotype (2*n*=44, FN=46), which differs substantially from that of *A. lugens* (2*n*=28, FN=48).

Genus *Chilomys* Thomas, 1897
Víctor Pacheco

Chilomys comprises two species, *C. fumeus* and *C. instans*. This South American endemic occurs in premontane and montane tropical forests along the Andes from the Cordillera of Mérida in western Venezuela (O. J. Linares 1998;

Soriano et al. 1999) south through the three main cordilleras in Colombia to central Ecuador (Musser et al. 1998). The following description is based primarily on Thomas (1895c), Osgood (1912), and Pacheco (2003).

Species of *Chilomys* are small-sized sigmodontines distinguished by a long and slender tail that exceeds the length of head and body (head and body 72–102 mm, tail 102–137 mm) and narrow and relatively long hindfeet (hindfoot length 22–26 mm). The dorsum varies from dark gray to gray-brown with the venter not countershaded. The pelage hair is short and woolly. The tail is covered with very short hairs, uniform in color, and often exhibits a white tip. The ears are medium sized and also covered with short hairs (ear length 14–18 mm). The soles of the hindfeet lack imbrications, but the dorsal surface is unusually scaly. The mammary complement consists of three pairs, one each in inguinal, abdominal, and thoracic positions (terminology of Pacheco 2003). Genal vibrissae are typically present. Mystacial vibrissae are of moderate length, extending to the posterior margin of the pinna, but not far beyond. A protuberant anus is not prominent.

The skull has a short and delicate rostrum that contrasts with a large, rounded, and deep braincase; the nasals are shorter than the premaxillae and exhibit a distinct notch on their anterior margin; the interorbital region is broad; the anterodorsal margin of the zygomatic arch is shallow (depth less than half width of notch); the ventral margin of the zygomatic arch is placed distinctly above the orbital floor; the jugal bone is long; the optic foramen is small and placed posterior to M3; the parietosupraoccipital suture is distinctly broad, with the suture at least half the posterior margin of the parietal breadth; carotid circulation is pattern 3 (*sensu* Voss 1988), with a small stapedial foramen and no squamosoalisphenoid groove or sphenofrontal foramen; the incisive foramina are short and do not extend to the protocone of M1; the diastema exhibits a pair of swollen ridges lying on either side of the posterior half of the incisive foramina; the masseteric tubercle is distinctly placed anterior to the root of the zygomatic plate; the palate is long, extending posteriorly beyond M3; sphenopalatine vacuities are absent or expressed only as narrow slits; the condition of the alisphenoid strut is variable; parapterygoid fossae exhibit a sharp triangular outline (terminology of Pacheco 2003); the parapterygoid vacuity is variable; the mesopterygoid fossa has parallel borders and is subequal to or broader than the adjacent parapterygoid fossae; the dorsal aperture of the ectotympanic ring is open and does not extend posteriorly to contact the petrosal; the tegmen tympani overlaps the posterior process of the squamosal; and the malleus lamina is rectangular in shape.

The upper incisors are long, slender, and proodont (*sensu* Hershkovitz 1962:103); molars are small, bunodont, brachydont, and lack a distinct labial cingulum; the antero-median flexus on M1 is indistinct, and the anteromedian flexid on m1 is absent. The mandibular diastema is flat and almost horizontal; a distinct capsular process is present. The condylar process is well developed, contrasting with a short angular process, producing a shallow angular notch.

Steppan (1995) recorded 12–13 thoracic, 6–7 lumbar, and more than 33 caudal vertebrae for two specimens of *C. instans*. The gall bladder is present (Voss 1991), and the stomach conforms to the unilocular-hemiglandular type (Carleton 1973). Voss and Linzey (1981) reported one pair of preputials and larger medial and ventral prostates than lateral ones. The genus is not known from the fossil record.

SYNONYMS:

Oryzomys: Thomas, 1895c:368; part (description of *instans*); not *Oryzomys* Baird.

Chilomys Thomas, 1897c:500; type species *Oryzomys instans* Thomas, by original designation.

REMARKS: *Chilomys* has been viewed as an enigmatic genus of peculiar morphology with uncertain affinities. The genus was previously considered an oryzomyine (Tate 1932g; Cabrera 1961; Voss and Linzey 1981; Reig 1984, 1986), but Musser and Carleton (1993) suggested it might be closely related to thomasomyines. Voss and Carleton (1993) subsequently defined and delimited the contents of the tribe Oryzomyini but omitted *Chilomys*. M. F. Smith and Patton (1999) based on mtDNA cytochrome-*b* sequences confirmed the thomasomyine affinities of *Chilomys*, suggesting that it might be sister to *Thomasomys*. Pacheco (2003), based on a broad array of morphological characters and a large representation of thomasomyine rodents, also found *Chilomys* a member of the thomasomyine *sensu stricto* clade, one that contained members of *Thomasomys*, *Aepeomys*, and *Rhipidomys*. More recently, D'Elía, Luna et al. (2006) presented a phylogenetic analysis of nucDNA IRBP sequences that supported the inclusion of *Aepeomys* as well as *Rhagomys* as thomasomyines; unfortunately, *Chilomys* was not included in their analysis.

KEY TO THE SPECIES OF *CHILOMYS*:

1. Nasals short (average 6.96, range: 6.41–7.39), diastema short (average 6.66 mm, range: 6.14–6.96 mm), breadth of zygomatic plate narrow (average 1.75 mm, range: 1.34–1.96 mm), maxillary toothrow short (<3.15 mm) . . . . *Chilomys fumeus*
1'. Nasals long (average 8.05, range: 7.4–8.76), diastema long (average 7.51 mm, range: 7.0–7.91 mm), breadth of zygomatic plate broad (average 2.21 mm, range: 2.05–2.41 mm), maxillary toothrow large (>3.18 mm) . . . . *Chilomys instans*

Chilomys fumeus Osgood, 1912
Smoky Chilomys

SYNONYMS:

Chilomys fumeus Osgood, 1912:53; type locality "Paramo de Tama, head of Tachira River, Santander [= Norte de Santander], Colombia. Alt. 6,000–7,000 ft."

Chilomys instans fumeus: Cabrera, 1961:435; name combination.

DESCRIPTION: *Chilomys fumeus* and *C. instans* very similar in general appearance and size. As detailed by Pacheco (2003), however, skull of *C. fumeus* with especially narrowed zygomatic plates (less than length of M1) and narrowed hypoflexus of M2 (distinctly narrower than mesoflexus). In postcranial skeleton, hemal arches of anterior caudal vertebrae of *C. fumeus* present, with left and right elements fused, and with spinous process on border; distinct gap between trochlear process of calcaneus and posterior articular facet also present.

DISTRIBUTION: *Chilomys fumeus* is known only from the Cordillera de Mérida of Venezuela in the states of Mérida and Táchira, and the Cordillera Oriental of Colombia, in the department of Norte de Santander, at elevations between 1,830 and 2,700 m.

SELECTED LOCALITIES (Map 305): COLOMBIA: Norte de Santander, Páramo de Tamá (type locality of *Chilomys fumeus* Osgood). VENEZUELA: Mérida, Páramo Tambor (FMNH 22172), 6 km ESE of Tabay, Middle Refugio (USNM 387964); Táchira, Páramo El Zumbador, 8 km SSW of El Cobre (CUVLA 5743).

SUBSPECIES: *Chilomys fumeus* is monotypic.

NATURAL HISTORY: This apparently nocturnal, terrestrial, and insectivorous species primarily inhabits cloud forest, rarely extending into secondary growth vegetation, where it feeds on plant material, insects, and worms (O. J. Linares 1998). *Chilomys fumeus* was captured in cloud forests of Venezuela at the base of rotting moss-covered trees, under moss-covered logs and fallen limbs, or under lichen- and moss-covered tree roots; in openings and in dense tangles of vines and bamboo in moist fern, moss, and lichen-laden cloud forests; in the lower montane wet forest and montane rainforest in the Holdridge system (Handley 1976). A female (USNM 579558) with well-developed teats and likely lactating was captured in July, during the local wet season. No other details of reproductive biology are known. O. J. Linares (1998) considered this taxon as rare, in agreement with Osgood (1912) who remarked to have collected only two specimens during some 20 days of trapping. This species hosts chiggers in the genera *Androlaelaps* (Furman 1972a), *Eutrombicula* (Brennan and Reed 1974a), and *Odontacarus*, *Peltoculus* (Brennan and Reed 1975). Tipton and Machado-Allison (1972) reported the flea *Cleopsylla*. *Chilomys instans* was

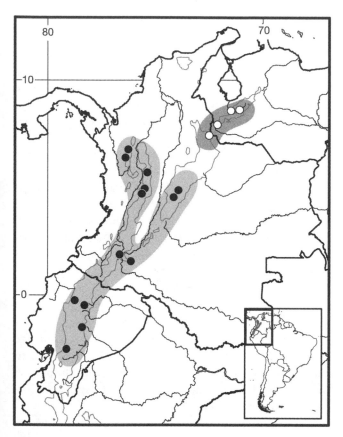

Map 305 Selected localities for *Chilomys fumeus* (○) and *Chilomys instans* (●). Contour line = 2,000 m.

originally indicated as the host species for each of these ectoparasitic records, but all are from *C. fumeus* as currently understood.

REMARKS: Osgood (1912:53) described *Chilomys fumeus* as a species distinct from *C. instans*. Cabrera (1961), without justification, relegated *fumeus* to subspecific status, and the genus has thereafter been considered monotypic (O. J. Linares 1998; Musser and Carleton 1993, 2005). However, based on my examination of the type specimens of *fumeus* and *instans*, and specimens from other available localities (Pacheco 2003), I concur with Osgood (1912) and recognize *fumeus* as a valid species.

Osgood (1912) reported the species at "Paramo de Tama," but according to his field notes (B. D. Patterson, pers. comm.) the specimens FMNH 18690 and 18689 were collected at the "Hd. of Tachira River." Paynter (1982) gave an elevation of 3,329 m for Páramo de Tamá, but he explicitly stated that Osgood's camp between 10 February and 6 March 1910 was in temperate forest (not páramo) at 1,825 to 2,450 m. Therefore, *C. fumeus* is most likely present in the middle montane humid forests and not in páramo habitats. The karyotype is unknown.

Chilomys instans (Thomas, 1895)
Andean Chilomys, Colombian Forest Mouse

SYNONYMS:

Oryzomys instans Thomas, 1895c:368; type locality "Bogota," Cundinamarca, Colombia.

[*Chilomys*] *instans*: Thomas, 1897:501; first use of current name combination.

DESCRIPTION: As noted in description of *C. fumeus*, *C. instans* similar in general external and cranial appearance, but differs from its congener in broader zygomatic plates (subequal in size to length of M1), broader hypoflexus of M2 (similar in width to mesoflexus), no hemal arch on anterior caudal vertebrae, and trochlear process of calcaneus level to or overlapping with posterior articular facet.

DISTRIBUTION: *Chilomys instans* is distributed from the central Andes of Ecuador through the three Andean cordilleras of Colombia between 1,100 and 4,000 m. This species exhibits an allopatric distribution with *Chilomys fumeus*.

SELECTED LOCALITIES (Map 305): COLOMBIA: Antioquia, 7 km E of Sonsón (FMNH 70343), Páramo Frontino (FMNH 71610), Santa Barbara (FMNH 71614); Caldas, Manizales, Río Termales (Castaño et al. 2003); Cundinamarca, San Cristóbal (FMNH 71632), Guasca, Río Balcones (FMNH 71638); Huila, San Adolfo (FMNH 71621), San Antonio (FMNH 71495); Risaralda, La Pastora, Reserva Ucumarí (ICN 16556). ECUADOR: Cañar, Chical (AMNH 62922); Napo, Papallacta (Voss 2003); Pichincha, Volcán Pichincha (Musser et al. 1998); Tungurahua, San Francisco, E of Ambato (AMNH 67542).

SUBSPECIES: *Chilomys instans* is monotypic.

NATURAL HISTORY: O. J. Linares (1998) considered *C. instans* (including *C. fumeus*) to be a nocturnal, terrestrial, and insectivorous species that fed on plant material, insects, and worms. It is not clear if his comment applied to both species or just to *C. fumeus*. Voss (2003) reported a specimen of *C. instans* from Papallacta, Ecuador, trapped on a mossy log over a stream in subalpine rainforest, another on the ground in a narrow runway beneath mossy shrubs in the same habitat, and a third in a mossy tunnel beneath bunch grass in the páramo/forest ecotone. Albuja (1991) listed the species for Ecuador as occurring in the temperate region almost countrywide. Alberico et al. (2000) reported additional specimens of *C. instans* from Valle del Cauca and Quindío, Colombia, but without reference to voucher specimens. Trapido and Sanmartín (1971) processed one specimen of *C. instans* from Munchique for Pichindé virus, but the virus was not detected. Seevers (1955) described three new species of staphylinid beetles in the genus *Amblyopinus* from *C. instans* (reported as *Chilomys* sp.), all from localities in Cundinamarca and Huila,

Colombia. Tipton and Machado-Allison (1972) reported fleas of the genus *Sphinctopsilla*, also from Colombian specimens.

REMARKS: The type locality was given by Thomas (1895c) as simply "Bogota," but in the label on the skin of the holotype reads as "La Selva Estate, plains of Bogota." Voss (2003) reported that the Papallacta specimens closely resembled the holotype of *Chilomys instans* in qualitative characters, except that they have slightly broader interorbits and shorter molar rows, indicating intraspecific variation and the necessity of an updated revision of the genus. Pacheco (2003) also described polymorphism in qualitative characters for this species. Colombian populations from Urrao, Antioquia, and Munchique, Cauca, might represent an undescribed species (V. Pacheco, unpubl. data). The species occurs from lower montane forests to páramo/forest ecotone, a wider elevational and ecological range than that of *C. fumeus*. The karyotype is unknown.

Genus *Rhagomys* Thomas, 1886

Lucía Luna

The genus *Rhagomys* is poorly known and, having no clear systematic affinities, has been regarded as *incertae sedis* in recent accounts of the Sigmodontinae (Reig 1986; M. F. Smith and Patton 1999; McKenna and Bell 1997; Percequillo et al. 2004; Musser and Carleton 2005) or as part of an assemblage that Voss (1993) termed "plesiomorphic Neotropical murids." Relying solely on cranial morphology, Thomas (1917d) associated the single, then-recognized species *R. rufescens* with both the "*Oryzomys-Oecomys* series" (Oryzomyini) and "*Thomasomys-Rhipidomys* series" (Thomasomyini). Initial molecular sequence analyses (Percequillo et al. 2004) failed to provide an unequivocal position for *Rhagomys* within the sigmodontine radiation. These authors (p. 254) thus concluded that *Rhagomys* "would constitute an additional divergent lineage interposed among more supported suprageneric groups" of the Sigmodontinae. However, more recent molecular analyses using both mitochondrial and nuclear gene sequenced placed *Rhagomys* within the Thomasomyini, together with *Thomasomys*, *Aepeomys*, and *Rhipidomys* (D'Elía, Luna et al. 2006). While this hypothesis is based on relatively low support, it remains the best currently available.

The genus has only two species, each known from only a few localities limited to relatively small geographic areas. Their distributions are disjunct and link two regions known for the remarkable degree of endemicity of their respective faunas: *Rhagomys rufescens* (Thomas 1886a), known from a handful of specimens collected in the Atlantic Forest of coastal Brazil (Thomas 1886a, 1917d; Percequillo et al.

2004; P. S. Pinheiro et al. 2004), and *R. longilingua* (L. Luna and Patterson 2003), known from only three specimens collected on the eastern slope of the Andes in Peru (L. Luna and Patterson 2003) and Bolivia (Villalpando et al. 2006). Both Luna Wong (2002) and Pacheco (2003) corroborated the phylogenetic placement of the recently described *R. longilingua* within *Rhagomys* by a cladistic analysis of morphological characters.

Species of *Rhagomys* are small mice with small ears and long mystacial, supraorbital, and genal vibrissae that reach the pinnae. The hindfeet are broad with unusually fleshy plantar pads (Ellerman 1941; Moojen 1952b) and naked soles (Moojen 1952b). Among the Sigmodontinae, the genus is uniquely diagnosed by the presence of a nail (as opposed to a claw) on the hallux and an elongated digit V, which reaches the middle of the second phalanx of digit IV. The tail is slightly shorter than the head and body together and terminates in a small tuft of hairs (Moojen 1952b). *Rhagomys* has six mammae arranged in three pairs, one postaxial, one abdominal, and one inguinal (L. Luna and Patterson 2003; see also Thomas 1917d). Stomach morphology is unilocular and hemiglandular, with the antrum and corpus subequal in area (L. Luna and Patterson 2003; Percequillo et al. 2004).

The ventral and dorsal pelages are strongly countershaded, vividly reddish above in *R. rufescens* and olive brown in *R. longilingua*. Notably, *R. rufescens* has soft hair, whereas the hair of *R. longilingua* is spiny, making *Rhagomys* one of four spiny genera of sigmodontines, together with *Neacomys*, *Scolomys*, and *Abrawayaomys*.

L. Luna and Patterson (2003), Percequillo et al. (2004), and P. S. Pinheiro et al. (2004) each provide expanded morphological descriptions of the genus based on their respective recently collected specimens. Diagnostic cranial characters include the combination of opisthodont upper incisors, prominent incisor capsules, and molars with very conspicuous cusps and thin ridges connecting the conules with the median mure. The skull is delicate and characterized by a short but stout rostrum; nasals strongly tapering posteriorly but extending past the anterior line of the premaxillary bones; a rounded braincase; posteriorly diverging supraorbital ledges; relatively short and narrow incisive foramina; a shallow and narrow zygomatic notch; long and wide palate, with puntiform posterolateral palatal pits; a narrow mesopterygoid fossa and narrow parapterygoid fossae perforated by long and wide fenestrae; a derived carotid circulation (pattern 3 of Voss 1988) that includes a small stapedial foramen, no groove on the internal surfaces of the squamosal and alisphenoid, and no sphenofrontal foramen; a very wide and robust alisphenoid strut; small auditory bullae; an absence of the suspensory process of the squamosal; and a completely planed ventral surface to the dentary. The genus is easily identified by the presence of a nail instead of a claw on digit V of the pes, a character also present in some Old World Muridae (*Dendromus* and *Chiropodomys*) but unique among the Sigmodontinae.

SYNONYM:

Hesperomys: Thomas, 1886a:250; part (description of *rufescens*); not *Hesperomys* Waterhouse.

Rhagomys Thomas, 1917d:192; type species *Hesperomys rufescens* Thomas, by original designation.

REMARKS: Thomas (1886a:250) originally placed his new species within the then very large and inclusive genus *Hesperomys*, but remarked that *H. rufescens* resembled superficially the smaller species of the "*Oryzomys* section" of *Hesperomys*, with *H. bicolor* (= *Oecomys bicolor*) its only close ally. By 1906, however, Thomas (p. 442) expressed doubt regarding his earlier association of *rufescens* with *Oryzomys* and formalized this doubt in his 1917 erection of *Rhagomys* to contain this enigmatic species.

Percequillo, Tirelli et al. (2011) reported on four molars taken from the scat of an ocelot (*Leopardus pardalis*) at Alta Floresta in northern Mato Grosso, Brazil, that were unambiguously identified as *Rhagomys*. Given the lack of diagnostic occlusal characters distinguishing the two known species, however, these authors were unable to assign their sample to either, or even possibly to a third, undescribed form. Nevertheless, this locality is intermediate geographically between that of *R. longilingua* from the eastern Andean slope in southern Peru and northern Bolivia and *R. rufescens* from the Atlantic Forest of coastal Brazil. As such, this record supports the hypothesis of L. Luna and Patterson (2003) that the apparent large distributional hiatus in the ranges of the two known species was possibly a sampling artifact and provides evidence for a biogeographic connection between the western Amazon–eastern Andean slope and the Atlantic Forest.

KEY TO THE SPECIES OF *RHAGOMYS*:

1. Pelage spiny; distribution in eastern Andes . . . .
. . . .*Rhagomys longilingua*
1'. Pelage soft; distribution in coastal Brazil . . . .
. . . . *Rhagomys rufescens*

Rhagomys longilingua L. Luna and Patterson, 2003
Long-tongued Rhagomys

SYNONYM:

Rhagomys longilingua L. Luna and Patterson 2003:3; type locality "Peru: Departamento de Cuzco, Provincia de Paucartambo, below 'Suecia,' a roadside settlement in Manu Biosphere Reserve along the Río Cosñipata; coordinates 13°06.032'S, 71°34.125'W; elevation approximately 1900 m."

DESCRIPTION: Small to medium sized (mass 25–35 g) spiny mouse with small ears and tail subequal in length to head and body. Pelage short and close with markedly spiny texture, dorsal fur more spiny than ventral fur. Color above olive brown but with strongly hispid pattern resulting from both spines and long, thin hairs. Ventral pelage from chin to interior sides of forefeet and hindfeet ochraceous-buff, contrasting sharply with dorsal coloration. Tail weakly bicolored, only slightly paler below, and terminated with small tuft of hairs; scales arranged in annular pattern, with hairs grouped in triplets of same size, shorter at tail base and becoming longer toward tip. Five manual digits unique in being long with broad, squared, narrow, and short claws that minimally project beyond conspicuously blunt tips. Pes with five digits, each with blunt tip bearing calloused surface; hallux noticeably shortened and bearing nail rather than claw. Hindfoot short (length 17 to 20 mm) and broad with well-developed plantar pads. Skull strongly built with short, blunt, and narrow rostrum; short and strongly diverging interorbital region; and large braincase, all conditions that also generally characterize *R. rufescens*. Nevertheless, the two species differ only in subtle ways. For example, *R. longilingua* has strongly beaded supraorbital edges, narrowest anteriorly; somewhat more shallow zygomatic notches; a more conspicuous postorbital process; a smaller posterior opening of alisphenoid canal; and a larger undifferentiated heel on M3 that gives this tooth an inverted triangle shape.

DISTRIBUTION: *Rhagomys longilingua* is known from only four specimens collected on the eastern slope of the Andes of southeastern Peru and northern Bolivia. The maximum distance between collection sites is approximately 520 km (Villalpando et al. 2006). Specimens of *R. longilingua* have been collected over a wide elevational range, from 450 m (in Peru) to 1,900 m (in Peru and Bolivia).

SELECTED LOCALITIES (Map 306): BOLIVIA: La Paz, Parque Nacional y Area Natural de Manejo Integrado Cotapata (Villalpando et al. 2006). PERU: Cusco, Paucartambo, below Suecia, Manu Biosphere Reserve (type locality of *Rhagomys longilingua* L. Luna and Patterson); Madre de Dios, Maskoitania, 13.4 km NNW of Atalaya, left bank Río Alto Madre de Dios (FMNH 175218).

SUBSPECIES: *Rhagomys longilingua* is monotypic.

NATURAL HISTORY: The stomach contents of the holotype of *R. longilingua* contained Diptera, Formicidae, and Lepidoptera, suggesting an insectivorous diet. Insectivory was also indicated by the presence of a relatively short intestine and caecum as well as crested teeth and mandibular characteristics suggesting strong mastication muscles to pierce chitin. A number of characteristics indicate that this species is arboreal, including pedal morphology

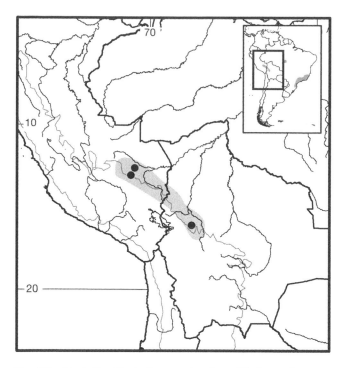

Map 306 Selected localities for *Rhagomys longilingua* (●). Contour line = 2,000 m.

(conspicuously fleshy plantar pads, broad feet, long digit V, grooved plantar surface, and minuscule claws) as well as the forward position of the eyes and a short muzzle (L. Luna and Patterson 2003). Three of the four specimens were collected in pitfall traps, and one was collected at 1 m above ground in a bamboo thicket. These findings suggest that *R. longilingua* can easily exploit both ground and above ground environments (L. Luna and Patterson 2003). The long tongue, from which its specific epithet is derived (*longus* = long, *lingua* = tongue), may serve as a probing device to collect insects from crevices (L. Luna and Patterson 2003). No other data are known of the natural history of this species, including behavior, reproductive biology, or population attributes.

Rhagomys rufescens (Thomas, 1886)
Rufescent Rhagomys

SYNONYMS:

Hesperomys rufescens Thomas, 1886a:250; type locality "Rio Janeiro [*sic*]," Rio de Janeiro, Brazil.

Rhagomys rufescens: Thomas, 1917d:193; first use of current name combination.

DESCRIPTION: Also small to medium in size, similar to *R. longilingua* but differing from that species in having soft and lax rather than spiny fur, reddish dorsal and reddish-gray ventral coloration. Tail unicolored brownish, with small scales, covered with blackish hairs becoming progressively longer toward tip. Hindfeet short and broad,

about 24% of head and body length, with short hallux reaching interdigital pad at base of digit II. Plantar pads large and well developed, as in *R. longilingua*. Skull quite similar to that of *R. longilingua*; characters that distinguish the two species, based on few available specimens of each, detailed in the account of *R. longilingua*.

DISTRIBUTION: *Rhagomys rufescens* is known only from the Atlantic Forest of Minas Gerais, São Paulo, and Rio de Janeiro states in southeastern Brazil, at elevations ranging from sea level to 650 m.

SELECTED LOCALITIES (Map 307): BRAZIL: Espírito Santo, Santa Teresa (Percequillo, Tirelli et al. 2011); Minas Gerais, Parque Estadual da Serra do Papagaio (Passamani et al. 2011), Poços de Caldas (Percequillo, Tirelli et al. 2011), Viçosa, Mata do Paraíso (MNRJ 66056); Rio de Janeiro, Rio de Janeiro (type locality of *Hesperomys rufescens* Thomas); Santa Catarina, Mono, Parque Natural Municipal Nascentes do Garcia (Percequillo, Tirelli et al. 2011); São Paulo, Reserva Floresta Morro Grande, Caucaia do Alto (Percequillo, Tirelli et al. 2011), Ribeirão Grande (Percequillo, Tirelli et al. 2011), Ubatuba, Núcleo Picinguaba, Parque Estadual da Serra do Mar (MNRJ 65545).

SUBSPECIES: *Rhagomys rufescens* is monotypic.

NATURAL HISTORY: Little is known about the ecology, reproductive biology, behavior, or other life history aspects of this species. Preliminary analysis of the stomach contents of one specimen revealed three unidentified species of ants (P. S. Pinheiro et al. 2004). As in the case of *R. longilingua*, limb and foot morphologies suggest arboreal and scansorial habits (Thomas 1917d; P. S. Pinheiro et al. 2004). Also like *R. longilingua*, this species has recently been collected in pitfall traps (Percequillo et al. 2004; P. S. Pinheiro et al. 2004). The specimen from Viçosa, Minas Gerais, was obtained in second growth semideciduous tropical forest (*floresta estacional semidecidual*; Percequillo et al. 2004; P. S. Pinheiro al. 2004). The recent specimen from the Parque Estadual da Serra do Mar was taken at a site characterized by transition between coastal vegetation on Quaternary sandy plains of marine origin (*restinga*) and tropical forest (*floresta ombrófilo densa*; P. S. Pinheiro et al. 2004).

REMARKS: Some published records of this poorly known species are in error, either due to questionable locality provenance or incorrect identifications. Records of *R. rufescens* from Além Paraíba and Passos in Minas Gerais state mentioned by Eisenberg and Redford (1999) are misidentified specimens of *Hylaeamys*. The locality of "Reserve Estadual Itapetinga, Atibaia, São Paulo" given to a specimen in the collection at MZUSP is probably in error and "no precise geographical information can be assigned to it" (Percequillo et al. 2004:241).

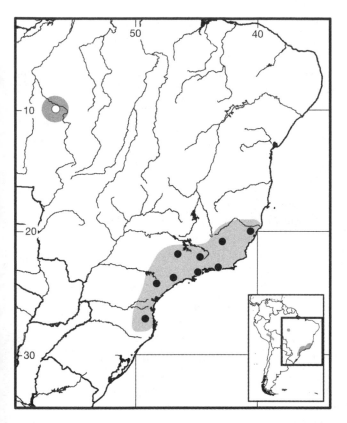

Map 307 Selected localities for *Rhagomys rufescens* (●) and *Rhagomys* sp. (○; see Percequillo, Tirelli et al. 2011).

Testoni et al. (2010) described the karyotype of *R. rufescens* from Santa Catarina state with a $2n = 36$, $FN = 50$, including GTG, CBG, Ag-NOR, and CMA(3)/DAPI banding profiles. Both X and Y chromosomes are acrocentric.

Genus *Rhipidomys* Tschudi, 1845

Christopher J. Tribe

Known in English as South American climbing mice and rats, and generally as *ratos arborícolas* ("tree rats") in Portuguese and *ratas arborícolas* in Spanish, species of *Rhipidomys* are found in forested habitats throughout most of tropical South America. The range of the genus extends from nearly 12°N on the Paraguaná peninsula in northern Venezuela, to 25°S in Salta province in northwestern Argentina and 24°S in Canindeyú department in eastern Paraguay and São Paulo state in southeastern Brazil. It also occurs in eastern Panamá and on certain offshore islands, including Trinidad. Within these latitudes it occupies habitats ranging from lowland rainforest to forest-páramo ecotones at over 3,000 m; no representatives have been found in the arid coastal regions of Peru and Chile, or in the cold, treeless zones of the high Andes. Individuals of several species have been caught in the roofs of buildings. Much of

the following account is drawn from the author's revision of the genus (Tribe 1996), which deals with morphology and distribution in greater detail.

Adult *Rhipidomys* vary in size from roughly that of a house mouse (*Mus musculus*) to that of a black rat (*Rattus rattus*), with known adult head and body lengths ranging from less than 90 mm to over 210 mm, and mass from under 30 g to nearly 200 g. Several aspects of their morphology are associated with their arboreal habits (Rivas-Rodríguez et al. 2010). These animals have large, protruding eyes and long, stiff mystacial vibrissae, extending well beyond the pinnae when pressed back against the head. The long tail (observed range: 93–180% of head and body length) is covered with short, stiff hairs and tipped with a distinct, often long, pencil. The broad hindfeet have relatively short metatarsals and long digits; digits I and V are longer than those of most other sigmodontine rodents. Digit V is partly opposable. There are six large plantar pads on each hindfoot, the hypothenar lying very close to the 4th interdigital. The dorsal surface of the metatarsus is covered along its midline by a narrow or broad dark patch, which may be distinctly or diffusely outlined. Female *Rhipidomys* have six mammae, in inguinal, abdominal, and thoracic positions (terminology according to Pacheco 2003). In lowland and lower montane forms, the adult body pelage is short and moderately fine to medium-coarse in texture; usually sleek, but short, woolly, juvenile-like fur is retained ventrally in many adult *Rhipidomys leucodactylus* and some other species. At higher elevations (above 1,500 to 2,000 m), the pelage tends to be finer and longer, but not to the extremes found in certain *Thomasomys* species or in phyllotines, and juveniles at those elevations tend to resemble adults. Dorsal color ranges from a fairly bright reddish-orange to a duller grayish-brown agouti; ventral color is white to pale yellow, sometimes suffused with orange. *Rhipidomys macconnelli* is exceptional in its generally chocolate-brown dorsum and grayish venter.

Sexual dimorphism in skull dimensions is negligible in most *Rhipidomys* but can be apparent in the larger species, in which males may be 2–3% larger than females on average. Skulls range from about 26–46 mm in occipitonasal length. In dorsal view, skulls have a short, blunt rostrum and shallow zygomatic notches. The zygomatic arches are not greatly bowed and are broadest across their squamosal roots (sometimes across the jugals in older specimens). The squared or beaded supraorbital margins, which may or may not form the broadest part of the interorbital region, diverge posterolaterally and continue as parietal ridges across the large, oval braincase. The braincase profile is domed in smaller species, becoming flatter in adult specimens of larger species. The interparietal bone is long (in the sagittal plane), usually exceeding 40% of its width. A robust hamular process of the squamosal separates the short postglenoid foramen from the subsquamosal fenestra, the dorsal margin of which is lower than the squamosal root of the zygomatic arch. Ventrally, the skull has relatively long incisive foramina, which may just reach the level of the M1s. The short palatal bridge does not extend beyond the M3s; its posterior margin is never concavely rounded but is roughly squared or biconcave with a small, caudally projecting median process. Small posterolateral palatal pits flank the anterior part of the mesopterygoid fossa. The parapterygoid fossae are triangular and generally shallow. Dorsal to them, an alisphenoid strut separates the buccinator-masticatory foramen from the foramen ovale accessorium. The bony roof of the mesopterygoid fossa is usually complete, but in some specimens it may be pierced by short, narrow sphenopalatine vacuities. The relatively small auditory bullae are suspended anteriorly by contact between the posterior suspensory process of the squamosal and the tegmen tympani, which also partly occludes the postglenoid foramen and sometimes even the subsquamosal fenestra. The mandible is neither very slender nor very robust. Relative to the molar occlusal plane, the condyloid process rises somewhat higher than the small, slightly falciform coronoid process, and extends posteriorly about as much as the angular process, or sometimes a little more.

The upper incisors are slightly opisthodont. The molars are pentalophodont, with distinct mesoloph(id)s on M1/m1 and M2/m2, and brachydont, with clearly separated main cusps. The upper molars each have three roots, the lowers two, although M1/m1 sometimes exhibit a small accessory root. The anterocone of M1 consists of two subequal conules only slightly smaller than the main cusps. An anteromedian style may be present. The principal cusps of M1 and M2 are arranged in opposite, rather than alternate, pairs. The lingual cusps are flat even when unworn, their occlusal surfaces lying at a shallow angle to the occlusal plane, and are triangular in occlusal view; they are separated by broad hypoflexi and interdental notches. The labial cusps are sugarloaf-shaped when unworn, with rounded cross-sections in the occlusal plane. The paraflexus and metaflexus each describe a quarter-circle around their respective cusps, and are blocked at their outer (labial) ends by a cingular ridge that connects the labial cusps and styles. The lingual ends of the lower molar flexids are similarly obstructed. A distinct anterolabial cingulid is present on m2 and m3.

Axial skeletal counts include 7 cervical, 19 thoracolumbar, 4 sacral and 35–44 caudal vertebrae, and 12 or 13 pairs of ribs. The first ribs articulate with the transverse processes of the 7th cervical and 1st thoracic vertebrae, while the transverse processes of the 6th and sometimes the 5th cervical vertebrae buttress those of the 7th. The neural spine

of the 2nd thoracic vertebra is only slightly elongated. The entepicondylar and supratrochlear foramina of the humerus are absent (Pacheco 2003).

Rhipidomys species lack a gall bladder (Voss 1991a) and have a unilocular-hemiglandular stomach (Carleton 1973; Pacheco 2003). The glans penis has long, deep grooves on the dorsal and ventral surfaces; the lateral bacular mounds are triple-curved (except in *R. macconnelli*) and extend beyond the crater rim (Hooper and Musser 1964; Pacheco 2003).

The known *Rhipidomys* karyotypes (all from the "leucodactylus" and "macconnelli" sections, see below) have a diploid number of $2n = 44$, with the exception of those allocated herein to *R. nitela* ($2n = 48$ or 50). The number of autosomal arms for specimens with $2n = 44$ displays a roughly bimodal distribution, some taxa clustering at $FN = 46$–52 and others at $FN = 70$–80, although a few intermediate numbers have been found: $FN = 56$ (A. L. G. Souza et al. 2007b), $FN = 61$ (M. J. de J. Silva and Yonenaga-Yassuda 1999), and $FN = 64$ (C. R. Bonvicino, pers. comm.). *Rhipidomys nitela* specimens have $FN = 67$–72. Thomazini (2009) listed the published karyotypes for the genus and added new data, and A. H. Carvalho et al. (2012) provided further discussion on comparative karyology in the genus.

Morphologically and ecologically, species of *Rhipidomys* fall into three sections, within which there are few discrete morphological characters by which the constituent species may be distinguished from each other (Tribe 1996; Pacheco 2003). (1) The "fulviventer" section consists of montane forms from the cordilleras of Venezuela, Colombia, and southeastern Peru as well as the Guiana Shield. These have longer, softer, less agouti pelage extending over the ankles and wrists, and the juvenile pelage resembles that of the adults. Their skulls are broader and more rounded, with less pronounced supraorbital-parietal ridges. They lack a protoloph and protostyle in M1, and m1 has a transverse lophid in the procingulid, anterior to the anterolophid. All member species except *R. caucensis* display the primitive carotid circulation pattern, with a large stapedial foramen and an internal groove across the squamosal and alisphenoid, leading to the "sphenofrontal" foramen in the alisphenoid-orbitosphenoid suture (the frontal is rarely involved) (Bugge 1970; pattern 1 of Voss 1988). (2) The "leucodactylus" section, the most widespread of the three, consists of lowland and lower montane species with short agouti adult pelage covering the ankles and wrists; the juvenile pelage is grayer and somewhat woolly. Their skulls are generally less rounded than those in the "fulviventer" section, and their carotid circulation pattern is derived, with a small stapedial foramen, no internal groove across the squamosal and alisphenoid, and no sphenofrontal foramen (pattern 3 of Voss 1988). A protoloph and protostyle are generally present in M1, as is a transverse lophid in the procingulid of m1. (3) The third section comprises only *Rhipidomys macconnelli*, from the sandstone inselbergs (*tepuis*) of the Guiana shield. This species is less specialized for arboreal life, as reflected in its longer and narrower metatarsi, its slightly smaller and more separate plantar pads, and its lack of body fur on ankles and wrists. Its long, soft dorsal pelage is dark chocolate-brown, while ventrally the cream or buffy tips of the hairs do not disguise their dark gray bases. *Rhipidomys macconnelli* resembles members of the "leucodactylus" section in its skull morphology and its derived carotid circulation pattern (Bugge 1970; pattern 3 of Voss 1988 [not pattern 1 as misstated by Voss et al. 2001:125–126]). A protoloph and protostyle are present in M1, and, uniquely, m1 lacks a transverse lophid in the procingulid, anterior to the anterolophid.

A cladistic analysis of morphological characters in thomasomyine rodents, including seven species of *Rhipidomys* representing all three sections of the genus, found a monophyletic *Rhipidomys* but provided no intrageneric resolution, apart from a close relationship between *R. fulviventer* and *R. caucensis* (Pacheco 2003). Another analysis using only a limited subset of the same character data yielded a highly paraphyletic basal arrangement of these *Rhipidomys* species with respect to *Thomasomys* and other genera, but this phylogeny had poor jackknife and bootstrap support values (Salazar-Bravo and Yates 2007). In a molecular analysis of sigmodontine rodents based on the mtDNA cytochrome-*b* gene, six species drawn from all three sections of *Rhipidomys* formed a single clade, within which *R. wetzeli* (of the "fulviventer" section) was the most basal branch, followed by *R. macconnelli* as the sister taxon to a subclade comprising four members of the "leucodactylus" section (M. F. Smith and Patton 1999). A constrained molecular clock based on cytochrome-*b* sequences suggested that the three sections split apart over 6 million years ago (L. P. Costa 2003). Parasitological evidence also suggests some unity of the genus: the anopluran louse *Hoplopleura angulata* Ferris, 1921, described from *Rhipidomys venezuelae*, has also been reported on seven other *Rhipidomys* species from all three sections of the genus, but otherwise only on *Thomasomys cinereus* (see Durden and Musser 1994). Should future studies demonstrate the monophyly of each of these three sections diagnosed, it would be appropriate to grant them subgeneric status within a monophyletic *Rhipidomys*. No genus-group names are currently available for either the "fulviventer" or the "macconnelli" sections.

The morphological variation exhibited by *Rhipidomys* specimens is not easily arranged into discrete species-level units, and in some cases there is apparently greater individual and geographic variation within species than phylogenetically informative variation among them (Tribe 1996). The uncertainty of specimen allocation based solely on morphological evidence is reflected in the key to species given below. The limited but rapidly increasing amount of karyotype and molecular data, so far available mainly for Brazilian forms, has proven invaluable for distinguishing certain taxa and allying others, often in unexpected ways (e.g., Zanchin, Langguth, and Mattevi 1992; Andrades-Miranda, Oliveira et al. 2002; L. P. Costa 2003; B. M. de A. Costa 2007; Thomazini 2009; A. H. Carvalho et al. 2012). Indeed, three new species were described in 2011 alone based on the combination of renewed fieldwork coupled with molecular and karyotypic data (R. G. Rocha, Costa, and Costa 2011; B. M. de A. Costa et al. 2011). Both of these works also identified an apparently undescribed species from central Brazil allied to *R. cariri* on the basis of molecular data (which they respectively referred to as "*R.* sp. 1" and "*Rhipidomys* sp.," both citing UFMG 2934 as a member). Clearly, much more work in these fields, particularly analyses that indisputably link molecular evidence to both karyotypes and morphology, is needed to test the hypotheses of species distinction and specimen allocation presented here.

SYNONYMS:

Mus: Lund, 1840a:24 [1841b:240]; part (description of *mastacalis*); not *Mus* Linnaeus.

Rhipidomys Tschudi, 1843:252; *nomen nudum*.

Rhipidomys Tschudi, 1844:16; *nomen nudum*.

Rhipidomys Tschudi, 1845:183; as subgenus of *Hesperomys*; type species *Hesperomys leucodactylus* Tschudi, by monotypy.

Hesperomys: Wagner, 1845:147; part (description of *leucodactylus* [= *Rhipidomys micrurus* P. Gervais]); not *Hesperomys* Waterhouse.

Calomys: Burmeister, 1855a:7; as subgenus of *Hesperomys*; part (listing of *mystacalis* [= *Rhipidomys mastacalis* Lund]); not *Calomys* Waterhouse.

Holochilus: Giebel, 1855:542; as subgenus of *Hesperomys*; part (listing of *mystacalis* [= *Rhipidomys mastacalis* Lund]); not *Holochilus* Brandt.

Myoxomys Tomes, 1861 [1862]:284; as subgenus of *Hesperomys*; type species [*Hesperomys* ([*Myoxomys*])] *salvini* Tomes, a junior synonym of *Nyctomys sumichrasti* (de Saussure, 1860), designated by Hunt, Morris, and Best 2004:1; included species: *salvini*, *latimanus* (a *Rhipidomys*), and *bicolor* (an *Oecomys*).

Nyctomys: Fitzinger, 1867b:88; part (listing of *mastacalis* and *leucodactylus*); not *Nyctomys* de Saussure.

Cricetus (*Rhipidomys*): Thomas, 1888b:133; name combination.

Rhipidomys: Winge, 1887 [1888]:54; raised to rank of genus.

Tylomys: J. A. Allen and Chapman, 1893:211; part (description of *couesi*); not *Tylomys* Peters.

Rhipodomys J. A. Allen, 1899b:206; incorrect subsequent spelling of *Rhipidomys* Tschudi.

Oryzomys: J. A. Allen, 1899b:211; part (description of *tenuicauda*); not *Oryzomys* Baird.

Oecomys: J. A. Allen, 1916c:525; part (description of *emiliae*); not *Oecomys* Thomas.

Thomasomys: Thomas, 1917d:196; part (listing of *macconnelli*); not *Thomasomys* Coues.

Rhipidomus Hopkins, 1949:470; incorrect subsequent spelling of *Rhipidomys* Tschudi.

Rhidomys Vorontsov, 1959:135; incorrect subsequent spelling of *Rhipidomys* Tschudi.

Phipidomys Cabrera, 1961:423; incorrect subsequent spelling of *Rhipidomys* Tschudi.

Rhipdomys Saunders, 1975:85; incorrect subsequent spelling of *Rhipidomys* Tschudi.

Rhipodmys Saunders, 1975:85; incorrect subsequent spelling of *Rhipidomys* Tschudi.

Rhyppidomis Guitton, Araújo Filho, and Sherlock, 1986:233; incorrect subsequent spelling of *Rhipidomys* Tschudi.

Rhidpidomys Stallings, 1989:187; incorrect subsequent spelling of *Rhipidomys* Tschudi.

Ripidomys Stallings, 1989:190; incorrect subsequent spelling of *Rhipidomys* Tschudi.

REMARKS: There has been some confusion regarding the year in which the name *Rhipidomys* Tschudi became available. Tschudi first published it in December 1843 (Sherborn 1922:cxxiv) in his article "Mammalium conspectus" (p. 252), as a subgenus of *Hesperomys* Waterhouse: "2. Subg. Rhipidomys. Wagn. in litt. 1844. Tsch.," below which he listed the only included species: "87. 1. *H. leucodactylus*. Tsch. Fauna peruan. cum fig." The reference was to his *Untersuchungen über die Fauna peruana*, published in parts between 1844 and 1846. However, the part of the latter work in which he described *Hesperomys leucodactylus* appeared only in 1845 (Sherborn 1922). Because the "Mammalium conspectus" contained neither description nor definition of the species, nor a reference to a previously published description or definition, the species name *leucodactylus* as used therein is not available (ICZN 1999:Art. 12). Consequently, because the subgeneric name *Rhipidomys* as used in Tschudi's "Mammalium conspectus" was not accompanied by a definition, a description, or an available species-group name, it too is not available from that work under the terms of Article 12. On page 16 of the first part of the *Fauna peruana*, published in 1844, the names *Rhipidomys* and *H. leucodactylus* are again listed without definition, and are therefore also not available according to Article 10.1.1 of the Code, on interrupted publication. These names become available only from page 183 of the *Fauna peruana*, published in 1845. Here Tschudi first briefly defines the subgenus—"Subgenus RHIPIDOMYS. Wagn. Cauda corpore longior apice penicillata" (tail longer than body, with tufted tip)—and then defines and describes the species *H. leucodactylus* at length. The animal is also figured in an engraving (plate XIII, Fig. 2; i.e., the left-hand rodent illustrated).

KEY TO THE SPECIES OF *RHIPIDOMYS*:

1. Adult dorsal pelage dark chocolate brown, never yellowish agouti; ventral pelage with cream or buffy tips not disguising long, dark gray hair bases; tail bicolored, with the ventral surface paler than the dorsal throughout most of its length; heel and wrist bare, not fur covered; hind foot relatively narrow, with long metatarsus; lateral extension of parietal bone (lateral to parietal ridge) variable in shape but usually much narrower posteriorly; m1 procingulid lacking transverse lophid anterior to the anterolophid; lateral bacular mounds of phallus simple, not complexly curved (Pacheco 2003) . . . .*Rhipidomys macconnelli*

1'. Adult dorsal pelage from reddish orange to yellowish gray agouti, not chocolate brown; ventral pelage appearing white, cream, or pale orange when unruffled, with or without gray hair bases; tail unicolored, or if paler ventrally then only in proximal one-third; heel and wrist well furred; hindfoot broad with short metatarsus; lateral extension of parietal bone (lateral to parietal ridge) approximately rectangular or only slightly narrower posteriorly; m1 procingulid displaying one or two transverse lophids anterior to the anterolophid; lateral bacular mounds of phallus "triple curved" (Pacheco 2003) 2

2. Pelage soft, long (body hairs 8–14 mm on mid-dorsum, 6–9 mm midventer, guard hairs 2–5 mm longer); dorsally bright reddish-orange; ventrally gray based, even on throat; juvenile pelage similar to adult in texture and color, though less bright; ears darker than dorsal pelage; skull with broad braincase; interorbital region rounded, narrowest in the middle, with squared or beaded supraorbital margins diverging rearward; these continue as weak parietal ridges curving around the braincase; carotid circulation pattern primitive or derived; protostyle and protoloph absent on M1; protoflexus absent on M2 . . . . ("fulviventer" section) 3

2'. Pelage coarser, short (body hairs typically 5–8 mm on mid-dorsum, 3–5 mm midventer, guard hairs 2–5 mm longer), though rather longer in highland specimens; dorsal pelage brownish-gray to yellowish-gray agouti, occasionally redder; throat and (usually) medial abdomen covered with white or cream hairs without gray bases; juvenile pelage gray and woolly in texture; ear color paler than, similar to, or occasionally darker than dorsal pelage; interorbital region of skull waisted anteriorly, with beaded or slightly ledged supraorbital margins diverging rearward; these continue as evident parietal ridges crossing the sides of the braincase in a straighter line; carotid circulation pattern derived (small stapedial foramen and no internal squamosal-alisphenoid groove or sphenofrontal foramen: carotid arterial pattern 3; Voss 1988); protostyle and (usually) protoloph present on M1; protoflexus usually present on M2. . . . . ("leucodactylus" section) 9

3. Derived carotid circulation pattern, with small stapedial foramen and no internal squamosal-alisphenoid groove or sphenofrontal foramen (carotid arterial pattern 3; Voss 1988); body size very small (adult head and body length usually <110 mm); upper molar row <4.5 mm . . . . *Rhipidomys caucensis*

3'. Primitive carotid circulation pattern, with large stapedial foramen and evident internal squamosal-alisphenoid groove and sphenofrontal foramen (carotid arterial pattern 1; Voss 1988); body size from very small through very large; upper molar row 3.5–6.5 mm. . . . 4

4. Body size very small (adult head and body length usually <105 mm); upper molar row <4.0 mm; tip of paroccipital process bifurcated . . . . *Rhipidomys wetzeli*

4'. Body size larger (adult head and body length usually >105 mm); upper molar row >4.1 mm; tip of paroccipital process not distinctly bifurcated 5

5. Body size large (adult head and body length >150 mm); tail very long (>190 mm); upper molar row >6 mm. . . . . *Rhipidomys ochrogaster*

5'. Body size smaller (adult head and body length usually <150 mm); tail length < 200 mm; upper molar row <6 mm. 6

6. Body size small to medium (adult head and body length <130 mm); upper molar row <4.7 mm in Venezuela or <5.0 mm in Colombia. 7

6'. Body size medium to moderately large (adult head and body length usually >125 mm); upper molar row >4.7 mm in Venezuela or >5.0 mm in Colombia 8

7. Dorsal pelage bright chestnut or orange-brown; ventral pelage with cream hair tips; 1–2 pairs of palatal foramina; found in NE Venezuela (Serranía de Turimiquire) . . . .*Rhipidomys tenuicauda*

7'. Dorsal pelage duller chestnut brown; ventral pelage with orange suffusion; two or more pairs of palatal foramina; found in Eastern Cordillera of Andes. . . . .*Rhipidomys fulviventer*

8. Dorsal pelage reddish to orangey brown; ventral pelage with cream or orange tips; tail much longer than head and body; relatively broad skull; supraorbital ridges only slightly developed; upper molar row averaging 5.25 mm (range: 5.0–5.4 mm), rather large in comparison with skull length; found in SW Colombia (Western and Central Cordilleras). . . . . *Rhipidomys similis*

8'. Dorsal pelage bright chestnut brown; ventral pelage with white or cream tips; tail slightly longer than head and body; skull of average breadth; supraorbital ridges a little more developed than in (8); upper molar row averaging 4.95 mm (range: 4.8–5.2 mm), average in comparison with skull length; found in the mountains of northwestern and northern Venezuela . . . . *Rhipidomys venustus*

9. Size large; upper molar row >5.4 mm (>5.0 mm in Peru; >5.6 mm in Venezuela; >5.9 mm in eastern and southern Brazil and Paraguay); adult occipitonasal length >35 mm (>33 mm in Argentina and Bolivia; >36 mm in northern Colombia and Venezuela; >39 mm in Brazil); adult hindfoot length (with claw) ≥28 mm. 10

9'. Size generally smaller; upper molar row and occipitonasal lengths smaller than in couplet 9; adult hindfoot length (with claw) ≤30 mm (≤32 in the southern Atlantic Forest) 14

10. Tail abundantly furred distally, with long, bushy pencil (extending >15 mm beyond tail tip); adult body pelage sleek or somewhat woolly, especially ventrally; size large to very large; upper molar row 6.2–7.2 mm (>5.8 mm in the Brazilian and Ecuadorian Amazon); skull large, robust, with well-developed supraorbital ridges. . . . .*Rhipidomys leucodactylus*

10'. Tail less well furred, with shorter, sparser pencil (<15 mm beyond tail tip); adult body pelage sleek or somewhat woolly; size moderate to very large; upper molar row 5.0–6.6 mm; skull moderate to large; supraorbital ridges well developed or not 11

11. Adult body pelage somewhat woolly; moderate size; adult hindfoot length (with claw) 28–30 mm; skull with broad interorbit; upper molar row 5.0–5.8 mm; found in eastern Andean valleys of Peru. . . . .*Rhipidomys modicus*

11'. Adult body pelage sleek; moderate to very large size; skull with medium to narrow interorbit; adult hindfoot length (with claw) usually ≥30 mm. 12

12. Skull long and rather narrow, with shallow, flattened braincase and well-developed supraorbital ridges. . . . . *Rhipidomys couesi*

12'. Skull with slightly broader and more inflated braincase; supraorbital ridges well developed or not. 13

13. Adult hindfoot length usually ≤33 mm; skull with narrow interorbit ranging from hourglass in shape with very slight supraorbital ridges in northwestern Bolivia

to more angular with moderate supraorbital ridges in southern Bolivia and Argentina; upper molar row usually ≤6.0 mm. . . . . *Rhipidomys austrinus*

13′. Adult hindfoot length usually ≥33 mm, sometimes shorter especially in northwestern Bolivia; skull with somewhat broader interorbit and moderate supraorbital ridges, more developed in northwestern Bolivia; upper molar row usually ≥5.9 mm . . . .*Rhipidomys gardneri*

14. Combination of small size and small molars; adult hind foot length (with claw) ≤28 mm; adult occipitonasal length ≤32 mm (occasionally up to 34 mm in Estado Bolívar, Venezuela); upper molar row ≤4.7 mm (≤4.9 mm in northern Estado Bolívar); tail with distinct pencil; karyotype 2*n* = 48 or 50; found north of the Amazon and east of the Orinoco and Negro or in Trinidad and Tobago. . . . . *Rhipidomys nitela*

14′. Size small to large; adult hindfoot length (with claw) usually ≥25 mm; karyotype 2*n* = 44; found outside of the region described in (14) 15

15. Dorsal pelage usually bright orange-brown (duller, more olive-brown in Huila, Colombia); ears small; hindfoot usually rather slender (broader in northern Peru and eastern Panama), with narrow, dark metatarsal patch; dark tail with pencil of medium length (usually ca. 10 mm, shorter in Huila); skull moderate to slender with straight supraorbital ridges; found in northern Andean foothills and valleys and in extreme eastern Panama. . . . . *Rhipidomys latimanus*

15′. Dorsal pelage redder or grayer; hindfoot usually broader, with variable metatarsal patch; skull usually broader; tail pencil short to moderate. 16

16. Dorsal pelage varying from rich reddish brown (in mesic areas) to duller yellowish brown (in more arid areas); ears large; hindfoot with proportionately short digits and faint, indistinctly outlined metatarsal patch; tail usually medium brown with short pencil (usually ≤6 mm); skull elongated and becoming flattened with age; braincase not inflated; found in northern Colombia and northern Venezuela . . . . . . *Rhipidomys venezuelae*

16′. Combination of characters not as in (16); found south of the Amazon in Brazil and eastern Paraguay. 17

17. Body size small (mean head and body length ≤120 mm, maximum <140 mm); dorsal pelage yellowish to orange brown or gray brown; interorbital region constricted anteriorly, with or without supraorbital beading; molars small to medium (mean upper molar row 4.7–4.8 mm, range: 4.5–5.0 mm). 18

17′. Body size larger (mean head and body length >130 mm, maximum >150 mm); dorsal pelage color ranges from gray-brown to yellowish or reddish brown; interorbital region from hourglass in shape with rounded sides (no beading) to constricted anteriorly with beaded edges; molars variable, small to moderately large 19

18. Tail pencil short, <8 mm; skull small, mean occipitonasal length <32 mm; supraorbital region rounded, without beading. . . . . *Rhipidomys tribei*

18′. Tail pencil medium to long, >10 mm; skull moderate, mean occipitonasal length >33 mm; supraorbital region with moderately developed beading . . . . *Rhipidomys ipukensis*

19. Interorbital region hourglass in shape (constricted centrally), with slight or no supraorbital beading (rather more prominent in Mato Grosso do Sul and Paraguay); braincase rounded and slightly inflated; incisive foramina bullet shaped. 20

19′. Interorbital region constricted anteriorly, with moderate to prominent supraorbital beading; braincase narrower and flatter, especially in older and larger specimens; incisive foramina elliptical to diamond shaped 21

20. Tail proportionally medium to long, mean 124% of head and body length (range: 107–148%); tail pencil short to moderate, ≤10 mm in length; dorsal pelage yellowish brown to dark reddish brown; braincase and interorbit broad; supraorbital edges with slight beading; molars medium size, mean toothrow length 5.2 mm (range: 4.9–5.4 mm) . . . . *Rhipidomys itoan*

20′. Tail proportionally short, mean 113% of head and body length (range: 96–133%, longer in some eastern populations); tail pencil short to moderate 5–15 mm (sometimes longer in southwestern population, i.e., Mato Grosso do Sul and Paraguay); dorsal pelage dull, reddish gray-brown; braincase average breadth; interorbit narrow; supraorbital region rounded, without beading, or beaded in southwestern population; molars medium to large, mean toothrow length 5.1 mm (range: 4.8–5.5 mm) in central and eastern Brazil, and 5.6 mm (range: 5.3–5.9 mm) in southwestern population . . . . *Rhipidomys macrurus*

21. Body size moderate to large (adult head and body length range: 120–190 mm); dorsal color yellowish-gray brown with conspicuous agouti effect; skull larger and more robust than in other eastern Brazilian species (mean occipitonasal length 35.5 mm, range: 33.6–36.7 mm); interorbit rather narrow, occiput broad; palatal bridge shorter than upper toothrow; bullae large for the genus; molars moderate in size (mean upper toothrow length 5.3 mm, range: 5.0–5.7 mm). . . . .*Rhipidomys cariri*

21′. Body size moderate (adult head and body length 105–160 mm); dorsal color gray-brown or more intense reddish brown; interorbit broad; palatal bridge often longer than upper toothrow; bullae moderate in size; average toothrow length smaller than in (21) 22

22. Dorsal color dull grayish brown; skull slightly smaller on average, mean occipitonasal length 33.3 mm (observed range: 28.2–36.9 mm); molars rather small, upper toothrow rarely exceeding 5 mm (mean 4.8 mm, range: 4.5–5.1 mm); hypocone of upper M3 usually greatly reduced; karyotype with low FN = 46–52 . . . . .
. . . .*Rhipidomys emiliae*

22'. Dorsal color grayish brown to reddish brown; skull slightly larger on average, mean occipitonasal length 34.2 mm (observed range: 29.8–38.5 mm); molars a little larger than (22) (mean upper toothrow length 5.0 mm, range: 4.6–5.5 mm); hypocone of upper M3 slightly reduced; karyotype with high FN = 70–80 . . . .
. . . . *Rhipidomys mastacalis*

Although the species of *Rhipidomys* are treated alphabetically in the accounts that follow, their membership in the three sections recognized herein is as follows: (1) "fulviventer" section, including *R. caucensis*, *R. fulviventer*, *R. ochrogaster*, *R. similis*, *R. tenuicauda*, *R. venustus*, and *R. wetzeli*; (2) "leucodactylus" section, including *R. austrinus*, *R. cariri*, *R. couesi*, *R. emiliae*, *R. gardneri*, *R. itoan*, *R. ipukensis*, *R. latimanus*, *R. leucodactylus*, *R. macrurus*, *R. mastacalis*, *R. modicus*, *R. nitela*, *R. tribei*, and *R. venezuelae*; and (3) "macconnelli" section, including only *R. macconnelli*. The majority of these assignments follow Tribe (1996); of the five species described since then, only *R. ipukensis* was explicitly assigned by its authors (R. G. Rocha, Costa, and Costa 2011) to the "leucodactylus" section. Although Patton et al. (2000), Tribe (2005), and B. M. de A. Costa et al. (2011) made no such assignments for *R. gardneri*, *R. cariri*, *R. itoan*, or *R. tribei*, each of these species are best considered part of the same section based both on the diagnostic characters mentioned herein and on their phylogenetic linkage to other species in that complex. It remains for future analyses of morphological or molecular characters, or both, to determine whether these sections reflect true phylogenetic relationships.

Rhipidomys austrinus Thomas, 1921
Southern Andean Rhipidomys

SYNONYMS:

Rhipidomys austrinus Thomas, 1921d:183; type locality "Sierra de Santa Barbara, S.E. Jujuy [. . .] Sunchal, 1200 m," Argentina.

Rhipidomys collinus Thomas, 1925b:578; type locality "Sierra Santa Rosa, 1000 m," near Itaú, Tarija, Bolivia.

Rhipidomys leucodactylus austrinus: Cabrera, 1961:420; name combination.

Rhipidomys leucodactylus collinus: Cabrera, 1961:421; name combination.

Rhipidomys auritus M. M. Díaz, Braun, Mares, and Barquez, 1997:8; incorrect subsequent spelling of *Rhipidomys austrinus* Thomas.

DESCRIPTION: Size moderate to large (head and body length up to 170 mm) with soft and dense grayish-brown to bright orange-brown agouti dorsal pelage and sharp lateral transition line; ventral pelage also soft with creamy white tips and medium to dark gray hair bases. Hindfeet medium to large (28–33 mm in length), broad, with distinctly outlined dark patch extending to bases of digits II to V (less distinct and more extensive in northwestern samples). Tail long (110–135% of head and body length), medium to dark brown, with short hairs on shaft and short to medium terminal pencil. Ears large and oval. Skull moderate in size and robustness, with short, deep rostrum that does not taper greatly; nasals narrow sharply behind tips, then taper gently to their posterior ends; interorbital region narrow, either hourglass in shape with squared edges or waisted anteriorly with slightly raised supraorbital ridges; braincase rather flattened or only very slightly inflated; auditory bullae prominent; carotid circulation pattern derived (small stapedial foramen, no squamosal-alisphenoid groove, no sphenofrontal foramen); molars generally large compared with skull length, with maxillary toothrow length ranging from 5.3–6.3 mm. See Massoia (1989) for further details of skull, and Carrizo and Díaz (2011) for comparison of postcranial skeleton with that of *Graomys griseoflavus*.

DISTRIBUTION: *Rhipidomys austrinus* occurs in the eastern Andean foothills and valleys and adjacent forested plains at elevations from 360 to 2,000 m, from southern Salta province, Argentina, northward to Santa Cruz department, Bolivia; farther northwest in La Paz department it has been found in the same valley system as *R. gardneri*, but at higher elevations (1,500 to 1,770 m).

SELECTED LOCALITIES (Map 308): ARGENTINA: Jujuy, Cerro Calilegua (Olrog 1979); Salta, Aguaray (Ojeda and Mares 1989), Anta, Parque Nacional El Rey (Jayat, Ortiz, Teta et al. 2006), Río de las Conchas, 5.7 km W of Metán (Carrizo and Díaz 2011). BOLIVIA: Chuquisaca, Cerro Bufete (Emmons 1997b), Porvenir (AMNH 262259); La Paz, Pitiguaya (AMNH 72632); Santa Cruz, Ingenio Mora, 7 km E and 3 km N (AMNH 247782), Río Ariruma, 7 km by road SE of Ariruma (AMNH 264199); Tarija, Cuyambuyo, 4 km N by road, Fábrica de Papel, Río Sidras (UMMZ 155864).

SUBSPECIES: *Rhipidomys austrinus* is monotypic.

NATURAL HISTORY: *Rhipidomys austrinus* occurs in the densely forested valleys of the Yungas on the eastern Andean slopes. For example, the collecting locality at Villa Montes (8 km S, 10 km E), Tarija department, Bolivia, 467 m, was described on the specimen tag as "E. bank Río Pilcomayo. Habitat: dense forest 15 m height, dense understory of 6 m shrubs, 86% ground cover of forbs, grasses and bromeliads." At the southern limit of its range

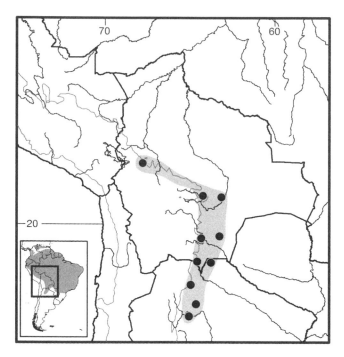

Map 308 Selected localities for *Rhipidomys austrinus* (●). Contour line = 2,000 m.

in Argentina, *R. austrinus* inhabits lowland, montane, and transitional forests; Ojeda and Mares (1989) provide general habitat information for that area. The two specimens from the Sierra Santa Bárbara, Sunchal, Jujuy, Argentina, were collected in tall nut trees (*nogales*), and the two from the Chimasi Yungas, La Paz, Bolivia, were found "in corn sack" (specimen tags).

REMARKS: This member of the "leucodactylus" section of *Rhipidomys* is the most southerly of the species that replace each other (with a certain amount of overlap) along the eastern Andean piedmont (from north to south: *R. couesi*, *R. leucodactylus*, *R. modicus*, *R. gardneri*, and *R. austrinus*). The name *R. couesi* has sometimes been misapplied to this species (e.g., Emmons and Feer 1990:184). Specimens from Pitiguaya are distinguished by their rather more slender, less robust skulls from those from Chijchipani, some 800 to 900 m lower in the same valley, here referred to *R. gardneri*. This distinction requires confirmation from molecular and karyotype evidence.

Rhipidomys cariri Tribe, 2005
Cairi Rhipidomys

SYNONYMS:

Rhipidomys cariri Tribe, 2005:137; type locality "BRAZIL–CEARÁ: Crato (07°14′S 39°23′W), Sítio Caiana."

Rhipidomys cariri baturiteensis Tribe, 2005:141; type locality "BRAZIL–CEARÁ: Pacoti (04°13′S 38°56′W), Sítio Cebola."

DESCRIPTION: Moderate to large bodied, with adult head and body length usually 130–160 mm (observed range: 120–190 mm). Tail length equals 110–140% of head and body length. Dorsal pelage yellowish gray brown, often somewhat grayer, sometimes redder, but always with conspicuous flecking from dark guard hairs and dark tips to body hairs; dorsal body hairs with mid-gray bases (about 62% of length), orange subterminal bands (about 29%), and dark tip (8–9%). Ventral pelage white or pale cream; some specimens with pale to mid-gray hair bases especially toward sides and occasionally across abdomen, as well as midline pectoral spot. Pelage texture slightly coarse and woolly, especially on ventral surface. Mystacial vibrissae noticeably long and coarse. Tail uniformly colored, varying from pale to rather dark; caudal hairs are dark and short (1–2 mm), especially along proximal half of tail, but may be longer at tip forming terminal tuft up to 15 mm in length. Hindfeet broad and relatively long compared with other species of similar size, averaging 27–30 mm (with claw). Dark patch on dorsal surface of metatarsals varies from narrow to broad, often poorly defined; ungual tufts usually about same length as claws. Ears seem larger than in other similar-sized Brazilian species, but average only a millimeter longer. Mystacial vibrissae long (>50 mm) and coarse; genal and supraorbital vibrissae sparse.

Skull large (occipitonasal length averages 35.5 mm) and robust with well-developed, straight supraorbital ridges diverging from front of interorbital region; rostrum moderately long and zygomatic plates broad, generally with shallow zygomatic notches visible in dorsal view; braincase large and angular rather than rounded, with broad occiput relative to greatest breadth across squamosals; lateral process of parietal rectangular; hamular process of squamosal long and almost horizontal; roughly elliptical incisive foramina terminate between anterior roots of M1s, while mesopterygoid fossa penetrates between posterior roots of M3s, making palatal bridge shorter than molar toothrow; sphenopalatine vacuities absent; auditory bullae relatively large for genus and slightly inflated; carotid circulation pattern derived (pattern 3 of Voss 1988), with small stapedial foramen, no translucent internal groove across the squamosal and alisphenoid, and no sphenofrontal foramen. Upper incisors robust and more opisthodont than in most other *Rhipidomys*; molars moderate in size but generally larger than in other eastern Brazilian species, with maxillary toothrow averaging 5.1–5.5 mm in length; M1 and m1 both have well-defined anteromedian flexus/flexid; oblique paralophules present in M1–M3; and M3 not greatly reduced in size or occlusal structure.

DISTRIBUTION: *Rhipidomys cariri* occurs in mesic enclaves surrounded by semiarid caatinga thorn scrub in northeastern Brazil, including the Serra de Baturité and Chapada do Araripe in Ceará state, the Parque Nacional do Catimbau in Pernambuco state, and the Morro do Chapéu in central Bahia state.

SELECTED LOCALITIES (Map 309): BRAZIL: Bahia, Morro do Chapéu (MNRJ 67779); Ceará, Sítio Caiana, Crato (type locality of *Rhipidomys cariri* Tribe), Pacoti, Serra de Baturité (type locality of *Rhipidomys cariri baturiteensis* Tribe); Pernambuco, Buíque, Parque Nacional do Catimbau (Geise, Paresque et al. 2010).

SUBSPECIES: See comment under Remarks.

NATURAL HISTORY: *Rhipidomys cariri* occurs within the semiarid Caatinga in isolated mesic forests (*brejos*), which are maintained by orographic rainfall (Andrade-Lima 1982). Individuals have been caught in forests, palms, and crops, particularly sugar cane and coffee. Two females from Pacoti each had four embryos in late July; six out of seven males captured between mid-August and early September had enlarged testes (Tribe 2005). The specimen from Buíque was captured in wet forest along a dry seasonal riverbed within the Caatinga.

REMARKS: Specimens of this recently described species were included in prior studies under a variety of different name combinations, such as *Rhipidomys cearanus* (Moojen 1943a:10), *Holochilus sciureus* (C. A. de Freitas 1957:131), and *Rhipidomys mastacalis* (Mares, Willig et al. 1981:84).

Rhipidomys cariri is a member of the "leucodactylus" section of the genus. Mitochondrial DNA studies using cytochrome-*b* (B. M. de A. Costa 2007; B. M. de A. Costa

et al. 2011) and cytochrome oxidase I (Thomazini 2009) sequences place specimens of *R. cariri* from Crato (southern Ceará) and Buíque (central Pernambuco) as the sister group to a clade containing certain specimens currently included in *R. macrurus* from central Bahia, eastern Tocantins, northern Goiás, and northern and eastern Minas Gerais states. In view of this and the expansion of *R. cariri* to include specimens from Pernambuco and Bahia, the subdivision of the species into the nominotypical subspecies and *R. cariri baturiteensis*, based on distinctions occurring solely within the state of Ceará (Tribe 2005), requires re-evaluation. Together, the above-mentioned clade and *R. cariri* form the sister clade to *R. macrurus* (*sensu stricto*) (B. M. de A. Costa 2007; Thomazini 2009; B. M. de A. Costa et al. 2011). The karyotype of *R. cariri* from Crato (type locality) and Buíque is $2n=44$, $FN=50$ (Thomazini 2009; Geise, Paresque et al. 2010), whereas that of a specimen from Pacoti (type locality of *R. cariri baturiteensis*) is $2n=44$, $FN=48$ (M. J. de J. Silva et al. 2000a).

Rhipidomys caucensis J. A. Allen, 1913
Lesser Colombian Rhipidomys

SYNONYM:

Rhipidomys caucensis J. A. Allen, 1913c:601; type locality "[Cerro] Munchique (altitude 8225 feet), Western Andes, Cauca, Colombia."

DESCRIPTION: Very small (about 100 mm in head and body length) with bright orange-brown dorsal pelage and ventral pelage with cream or pale orange hair tips over dark-gray bases. Tail longer than head and body (130 mm), medium to dark brown in color, with hairs lengthening over the distal 25% and terminating in moderately long pencil (10–15 mm). Ears very dark. Hindfeet short and proportionately narrower than in most other *Rhipidomys* species, with poorly defined dorsal dark patch that, nevertheless, contrasts sharply with pale digits. Skull with hourglass-shaped interorbit, very slight supraorbital ridges, globose braincase, and derived carotid circulation pattern (small stapedial foramen, no groove on the inner surface of the squamosal and alisphenoid, and no sphenofrontal foramen). Mandible with pronounced incisor root capsule. Molars small (maxillary toothrow length 3.8–4.3 mm), even in comparison with skull length; first two upper molars with strong oblique paralophules paralleling mesoloph; part of third upper molar posterior to hypoflexus greatly reduced in size.

DISTRIBUTION: *Rhipidomys caucensis* is known from middle to high elevations in the Western Cordillera of Colombia (departments of Antioquia, Valle and Cauca) and, disjunctly, on the eastern flanks of the Central Cordillera at the head of the Magdalena valley (department of Huila) and in the department of Caldas.

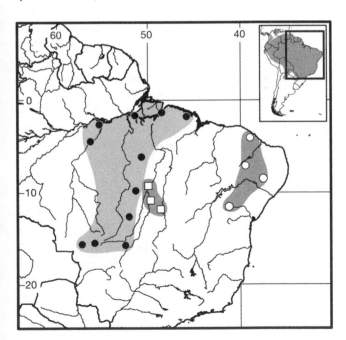

Map 309 Selected localities for *Rhipidomys cariri* (○), *Rhipidomys emiliae* (●), and *Rhipidomys ipukensis* (□).

SELECTED LOCALITIES (Map 310): COLOMBIA: Antioquia, Páramo Frontino (FMNH 71738); Caldas, Bosque de Florencia (Castaño et al. 2003); Cauca, Cerro Munchique (type locality of *Rhipidomys caucensis* J. A. Allen); Huila, San Antonio (FMNH 71736); Valle del Cauca, El Cairo, Cerro del Inglés (M. Alberico, in litt.).

SUBSPECIES: *Rhipidomys caucensis* is monotypic.

NATURAL HISTORY: Specimens of *R. caucensis* were taken at elevations of 2,800–3,500 m in Antioquia and at 2,540 m on the eastern and 1,830 m on the western slope of the Western Cordillera in Cauca; on the eastern slope of the Central Cordillera they were collected at 2,200–2,350 m in Huila and below 1,800 m in Caldas. These elevations generally correspond to cloud forest.

REMARKS: *Rhipidomys caucensis* is placed in the "fulviventer" section of the genus on the basis of cranial and pelage characters, despite its derived carotid circulation pattern. Apart from this and its usually shorter tail, it is morphologically similar to *R. wetzeli* to the extent that in a cladistic analysis of thomasomyine species based on morphological characters, which included seven *Rhipidomys* species, these were the only two of the seven that formed a clade to the exclusion of the others (Pacheco 2003). In both Cauca and Huila, *R. caucensis* occurs sympatrically with the larger *R. similis* at the same elevations but occupying higher elevations than *R. latimanus*.

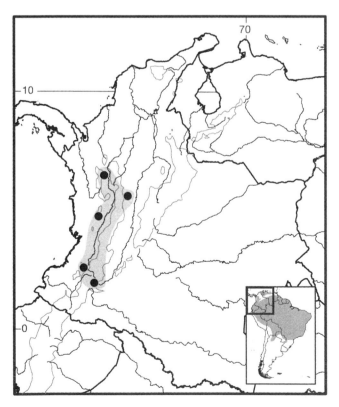

Map 310 Selected localities for *Rhipidomys caucensis* (●). Contour interval = 2,000 m.

Rhipidomys couesi (J. A. Allen and Chapman, 1893)
Coues's Rhipidomys
SYNONYMS:
Tylomys couesi J. A. Allen and Chapman, 1893:211; type locality "Princestown [i.e., Princes Town], Trinidad," further restricted to "a point twelve miles north of the southern coast and seven miles southeast of Princestown" (J. A. Allen and Chapman 1893:203).
Rhipidomys couesi: J. A. Allen and Chapman, 1897a:18; first use of current name combination.
Rhipidomys couesii Rendall, 1897:343; incorrect subsequent spelling of *couesi* J. A. Allen and Chapman.
Rhipidomys venezuelae cumananus Thomas, 1900c:271; type locality "Ipure, altitude 700 m," Sucre, Venezuela.
Rhipidomys sclateri couesi: Cabrera, 1961:42; name combination.
R[hipidomys]. sclateri cousei Furman and Tipton, 1961:192; incorrect subsequent spelling of *couesi* J. A. Allen and Chapman.

DESCRIPTION: Large to very large (head and body length 150–210 mm) with yellowish to reddish brown agouti dorsal pelage, conspicuously flecked with dark brown or black guard hairs; ventral pelage white or creamy yellow, sometimes with rarely noticeable pale to medium gray hair bases; pelage moderately coarse and short. Tail relatively short (100–120% of head and body length), unicolored, with rather large, dark scales and dark brown to reddish-brown hairs; well haired distally, terminating in pencil of hairs up to 20 mm in length. Hindfeet long (29–32 mm with claw) and broad, with distinctly or indistinctly outlined dark patch that extends at least to bases of digits II to IV. Skull large, robust, and rather narrow for genus, with flattened profile such that interparietal nearly horizontal, especially in older specimens; supraorbital ridges well developed, usually straight and weakly divergent posteriorly; braincase narrow, not inflated, with squamosals oriented dorsolaterally rather than laterally; auditory bullae small; carotid circulation pattern is derived (stapedial foramen small, no internal squamosal-alisphenoid groove, no sphenofrontal foramen). Maxillary toothrow long for genus, ranging typically from 5.6–6.1 mm, but relatively short in comparison to skull length.

DISTRIBUTION: *Rhipidomys couesi* is known from the island of Trinidad, the Venezuelan island of Margarita (Nueva Esparta state), the coastal region of Sucre, Anzoátegui, and Vargas states up to approximately 1,000 m elevation, the southeast-facing lower slopes of the Mérida Andes in Barinas state, and the foothills of the Eastern Cordillera near Villavicencio, Meta, Colombia. By extrapolation, it can be expected to occur throughout the narrow belt of lowland moist forest that borders the *llanos* along the southern and eastern foothills of the Venezuelan coastal range, Mérida Andes and Eastern Cordillera from Sucre in northeastern Venezuela to Meta in Colombia. From this point southward, where the Amazon-Orinoco forests reach the Andes, this species is replaced by the larger *R. leucodactylus.*

SELECTED LOCALITIES (Map 311): COLOMBIA: Meta, Villavicencio (AMNH 75239). TRINIDAD AND TOBAGO: Trinidad, Cumaca (AMNH 235071), 7 mi SE of Princes Town (AMNH 5956/4685). VENEZUELA: Barinas, Altamira (USNM 442302), Reserva Forestal de Ticoporo (Ochoa et al. 1988); Nueva Esparta, Isla Margarita, Cerro Matasiete, 3 km NE of La Asunción (USNM 406100); Sucre, Ensenada Cauranta, 7 km N and 5 km E of Guiria (USNM 409936), Ipure, near Cumaná (type locality of *Rhipidomys venezuelae cumananus* Thomas); Vargas: Los canales de Naiguatá, P. N. El Ávila (Rivas and Salcedo 2006).

SUBSPECIES: *Rhipidomys couesi* is monotypic.

NATURAL HISTORY: The monograph by Montserin (1937) includes data on reproduction, life history, nesting habits, and diet of *R. couesi* in Trinidad, as well as its status as a pest of cacao and the control measures taken against it. The species was found to be relatively common in plantations but difficult to trap. Trap-shyness may mean that abundance estimates based on trapping success are inadequate. Other reports (Urich 1911; Everard and Tikasingh 1973b) suggest it is a common arboreal species in Trinidad plantations and forests. In Barinas state, Venezuela, *R. couesi* was captured in dense rainforest (Ochoa et al. 1988), while in Sucre and Nueva Esparta it was found mainly in evergreen forest or plantations, and in one case inside a house (Handley 1976). For parasites of the species in Venezuela, see Brennan and Reed (1974a), Furman (1972a), Furman and Tipton (1961), P. T. Johnson (1972), and Saunders (1975).

REMARKS: *Rhipidomys couesi* is a large-bodied species in the "leucodactylus" section of the genus. Specimens from the islands of Trinidad and Margarita are larger on average than those from the Venezuelan mainland. If on the basis of new data the mainland form is demonstrated to be sufficiently distinct, the name *cumananus* Thomas is available. At the southwestern end of the range of the species, Colombian specimens have rather more slender skulls. This character, together with their shorter molar row, helps to distinguish skulls of *R. couesi* from the generally more robust skulls of *R. leucodactylus.* Aguilera et al. (1994) described the karyotype of *R. couesi* from Anzoátegui, Venezuela (under the name "*R. sclateri*"; note that *R. sclateri* is here considered a junior synonym of *R. leucodactylus*) as $2n = 44$, FN = 48.

Subfossil material from Tobago (off the northeast coast of Trinidad) identified by Eshelman and Morgan (1985)

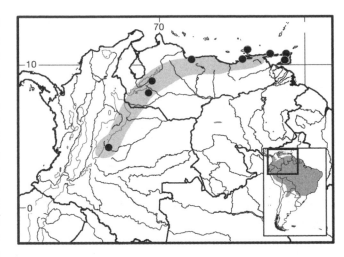

Map 311 Selected localities for *Rhipidomys couesi* (●). Contour line = 2,000 m.

as *Rhipidomys leucodactylus* likely represents *R. couesi* instead. Fossilized teeth and skull fragments from the island of Bonaire (Caribbean Netherlands), some 90 km from the Venezuelan coast, which were identified by Hooijer (1959) as "*Thomasomys spec.*" and compared by Tribe (1996) with *Rhipidomys couesi*, have recently been described as the oryzomyine species *Agathaeromys donovani* and *A. praeuniversitatis* (Zijlstra et al. 2010).

Rhipidomys emiliae (J. A. Allen, 1916)
Snethlage's Rhipidomys

SYNONYMS:

Oecomys emiliæ J. A. Allen, 1916c:525; type locality "Rio Mojú, Pará, Brazil."

Oryzomys [(*Oecomys*)] *emiliae*: Ellerman, 1941:358; name combination.

Rhipidomys emiliae: Goodwin, 1953:306; first use of current name combination.

DESCRIPTION: Medium sized (head and body length 120–150 mm) with dull grayish-brown to brighter orange-brown agouti dorsal pelage and cream or white ventral fur. Tail moderately long, medium to dark brown, moderately well haired, and ends in medium-length pencil (10–15 mm). Ears large, brown, and bare; some specimens have pale postauricular patch. Hindfeet moderate in length (25–28 mm), broad, with relatively long toes, and with narrow or broad dorsal dark patch, usually ill defined, leaving sides of feet and toes pale. Skull with moderately developed supraorbital ridges diverging posteriorly from point well forward, resulting in broad interorbital region; braincase narrow and somewhat angular, especially in older specimens, and moderately flattened; palatal bridge in adults often longer than molar row; auditory bullae small; carotid circulation pattern derived (small stapedial foramen, no internal groove on squamosal and alisphenoid, no sphenofrontal foramen); molars small, particularly in comparison with skull length, maxillary toothrow not exceeding 5 mm in length.

DISTRIBUTION: *Rhipidomys emiliae* occurs in the eastern Brazilian Amazon Basin, south of the Amazon River, from the west bank of the Rio Tapajós eastward to the state of Maranhão and southward to southern Mato Grosso state. It has not been found east of the Rio Araguaia in the states of Tocantins or Goiás.

SELECTED LOCALITIES (Map 309): BRAZIL: Maranhão, Alto da Alegria (BM 25.5.21.8); Mato Grosso, Chapada dos Guimarães, Casa de Pedra (MNRJ 24914), Estação Ecológica Serra das Araras (Santos-Filho et al. 2012), Fazenda Noirumbá, 34 km NW of Ribeirão Cascalheira (UFMG 2953), Fazenda São Luis, 30 km N of Barra do Garças (UFMG 2946), Vila Rica (MZUSP [APC 286]); Pará, Aramanaí, Rio Tapajós (AMNH 94810), Belém, Utinga (MNRJ 24195), Melgaço, Floresta Nacional de Caxiuañá (MPEG 34000), Parauapebas, Fazenda São Luiz (B. M. de A. Costa et al. 2011), Parque Nacional da Amazônia (T. K. George et al. 1988).

SUBSPECIES: *Rhipidomys emiliae* is monotypic.

NATURAL HISTORY: Emilie Snethlage, who collected the holotype of *Rhipidomys emiliae*, noted on the specimen tag that this rat was often found in houses and plantations. Pine (1973a) considered the species somewhat uncommon at Utinga, Belém. In the Serra do Roncador, Mato Grosso, one specimen was caught in a house, two in gallery forest, and one at the cerrado-dry forest ecotone; Fry (1970) provided details of the Roncador study area.

REMARKS: *Rhipidomys emiliae* is a member of the "leucodactylus" section of the genus. Specimens from south of the Amazon between the Tapajós and Tocantins rivers that were tentatively included by Tribe (1996) in *R. nitela* are here reallocated to *R. emiliae* on the basis of their karyotype (2n = 44, FN = 52: M. J. de J. Silva and Yonenaga-Yassuda 1999; Andrades-Miranda, Oliveira et al. 2002 [as "*R. leucodactylus* cytotype 1" from locality 5, Caxiuañá]; 2n = 44, FN = 46: M. S. Mattevi, pers. comm.) and molecular data (L. P. Costa 2003; R. G. Rocha, Ferreira, Costa et al. 2011; B. M. de A. Costa et al. 2011). They include the "*R. nitela*" sequenced by M. F. Smith and Patton (1999) and by L. P. Costa (2003). Specimens from this area are morphologically intermediate between *R. nitela* from north of the Amazon and *R. emiliae* from the Belém region. The size difference in specimens from Serra do Roncador considered by Tribe (1996) to represent two species (*R. nitela* and *R. emiliae*) is here attributed instead to individual variation within the latter species. Pine et al. (1970; also Pine 1973a) referred to specimens of this species from Serra do Roncador, Mato Grosso, and the Belém region of the eastern Amazon as *R. mastacalis*. Subfossil

Rhipidomys material from the Serra dos Carajás, southeastern Pará state, dated to between 2,000 and 8,000 ybp (Toledo et al. 1999), may be referable to *R. emiliae*.

In phylogeographic analyses based on the mtDNA cytochrome-*b* gene, *Rhipidomys emiliae* is most closely related to *R. ipukensis*, then *R. mastacalis*, followed by *R. gardneri*, then more distantly to a clade comprising *R. leucodactylus*, *R. itoan*, and *R. tribei* (= *Rhipidomys* "sp. nov." in L. P. Costa 2003), and more remotely still to *R. macrurus*, which is geographically its neighbor (M. F. Smith and Patton 1999 and L. P. Costa 2003 [both reported as "*R. nitela*"]; R. G. Rocha, Ferreira, Costa et al. 2011; B. M. de A. Costa et al. 2011). The observation by R. G. Rocha, Ferreira, Costa et al. (2011) that the Rio Araguaia appears to separate *R. emiliae* in the states of Pará and Mato Grosso (to the west of the river) from *R. ipukensis* in the state of Tocantins (to the east) suggests that the identification of all these specimens needs confirmation by a phylogenetic analysis that includes material from the type locality of *R. emiliae* at Rio Moju, just south of Belém, which lies to the east of the Rio Tocantins below its confluence with the Rio Araguaia. The situation is further complicated by reports that specimens collected even farther east in Pará (Ourém) and Maranhão (Bacabal), which are here provisionally referred to *R. emiliae*, have karyotypes with FN = 64 (C. R. Bonvicino, pers. comm.), substantially different from the FN = 46 and 52 mentioned above for this species.

Rhipidomys fulviventer Thomas, 1896
Tawny-bellied Rhipidomys

SYNONYMS:

Rhipidomys fulviventer Thomas, 1896:304; type locality "Agua Dulce, W. Cundinamarca, Colombia."

Rhipidomys fulviventer elatturus Osgood, 1914a:140; type locality "Paramo de Tama, head of Tachira River, Venezuela. Alt. 7,000 ft."

[*Rhipidomys*] *fulviventris* Tate, 1939:195; incorrect subsequent spelling of *fulviventer* Thomas.

Rhipidomys fulviventer fulviventer: Gyldenstolpe, 1932: 48; name combination.

Rhipidomys latimanus elatturus: Cabrera, 1961:419; name combination.

DESCRIPTION (amended from Tribe 1996, including his *R. f. fulviventer* and *R. f. elatturus* accounts): Small to medium sized with long, soft, and sleek fur; dorsal pelage mid-brown, often with olive tint, while gray-based ventral hairs have pale tips usually with orange suffusion throughout. Tail longer than head and body and generally medium to dark with slightly paler proximal area on ventral surface. Ears darker than the dorsal pelage. Hindfeet relatively narrow for genus, with dark metatarsal patch often indistinctly delineated. Skull with weakly developed supraorbital ridges and thus with rounded, hourglass-shaped interorbital region; braincase rounded, with poorly developed, curved parietal ridges and broad occiput; incisive foramina and mesopterygoid fossa typically narrow and almost parallel sided; posterior medial spine of palate small or absent, resulting in a posterior edge; parapterygoid fossae usually fenestrated; small sphenopalatine fissures usually present; carotid circulatory pattern primitive (large stapedial foramen, internal groove across squamosal and alisphenoid, and sphenofrontal foramen). Mandible with prominent incisor root capsule. Molars fairly large in comparison with skull length (maxillary toothrow length mostly 4.4–4.8 mm).

DISTRIBUTION: *Rhipidomys fulviventer* is known from elevations between 1,800 and 3,100 m in the Eastern Cordillera of the Colombian Andes, from Norte de Santander (and the Páramo de Tamá just across the Venezuelan border) southward to Cundinamarca and the Distrito Especial of Bogotá. In addition, a single specimen from the Serranía de la Macarena, an isolated tableland situated some 40 km east of the Eastern Cordillera in the department of Meta, is provisionally referred to this species.

SELECTED LOCALITIES (Map 312): COLOMBIA: Boyacá, Paipa, Parque Natural Ranchería (Vianchá-Sánchez et al. 2012); Cundinamarca, Guasca, Río Balcones (FMNH 71731), La Aguadita (BM 15.1.3.3); Meta, Serranía de la Macarena (AMNH 142140); Santander, Cachirí (CM 3192). VENEZUELA: Táchira, Páramo de Tamá (FMNH 18691).

SUBSPECIES: See comments under Remarks.

NATURAL HISTORY (information from Montenegro-Díaz et al. 1991, except where noted): *Rhipidomys fulviventer* (referred to as "*Rhipidomys latimanus*") inhabits relatively undisturbed cloud forest and the forest-páramo ecotone in the Eastern Cordillera of the Andes. The home range was found to be 0.2 ha in forest 25 m in height. Reproductive activity was fairly constant throughout the study period (February to August), two females having two embryos each (one Venezuelan specimen had three; USNM specimen tag); gestation lasted roughly one month. The content of five stomachs was 68% fruit and seeds, 26% leaves and other plant matter, and 6% insects. See also López-Arévalo et al. (1993). The individual from La Macarena was taken at 1,140 m (specimen tag). For parasites from the Venezuelan specimens, see Brennan and Reed (1974a), Furman (1972a), E. K. Jones et al. (1972), and Tipton and Machado Allison (1972).

REMARKS: This species belongs to the "fulviventer" section of the genus. Specimens representing the northern known population are smaller on average than those from Cundinamarca, although examined sample sizes are very

small. If the northern population is found to be genetically distinct and to warrant species status, the name *elatturus* Osgood is available. Its range is separated by the Táchira Depression (roughly 1,000 m elevation) from that of the next member of the "fulviventer" section, the moderately large *Rhipidomys venustus* in the Mérida Andes of Venezuela. Southward, lower elevations in the Eastern Cordillera separate *R. fulviventer* from *R. similis*, which also belongs to the same section. The single known specimen from the Serranía de la Macarena is provisionally referred to *Rhipidomys fulviventer* although it is a little larger and differs slightly in skull morphology. Tribe (1996) treated all these forms as subspecies of *Rhipidomys fulviventer*; here, however, *tenuicauda*, *venustus*, *fulviventer*, and *similis* are considered sufficiently isolated and morphologically distinct to warrant separate species status. See also the remarks under *Rhipidomys tenuicauda*. Cabrera (1961) regarded *Rhipidomys fulviventer* and *R. venustus* as subspecies of *R. latimanus*, but Handley (1976) recognized that they are clearly distinct from the latter species. Two specimens resembling *Rhipidomys fulviventer* have recently been collected in Bongará, Amazonas department, Peru, 1,000 km south of La Macarena (C. F. Jiménez and Pacheco 2012). They are not included in the species range given here pending their formal identification.

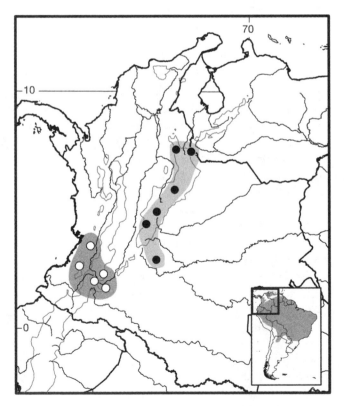

Map 312 Selected localities for *Rhipidomys fulviventer* (●) and *Rhipidomys similis* (○). Contour line = 2,000 m.

Rhipidomys gardneri Patton, da Silva,
and Malcolm, 2000
Gardner's Rhipidomys

SYNONYM:

Rhipidomys gardneri Patton, da Silva, and Malcolm, 2000:
165; type locality "Reserva Cusco Amazónico, left (=
north) bank of the Río Madre de Dios, 14 km east of
Puerto Maldonado, Departamento de Madre de Dios,
Perú, elevation about 200 m (12°33′S, 69°03′W)."

DESCRIPTION: Large-bodied (head and body length
150–190 mm) with relatively short and coarse dorsal fur,
gray to orange-brown agouti in color; ventral hairs fulvous to yellow, with gray-based hairs limited to midline
of throat and chest. Tail proportionately short to medium
in length (110–140% of head and body length), uniformly
brown in color, clothed with short hairs, and terminating
in short pencil that does not exceed 6 mm. Ears moderate
in size and dark brown in color. Hindfeet long (33–34 mm)
but broad, with narrow dark dorsal patch limited to metatarsals (not extending onto digits). Skull robust with short,
tapering rostrum, deep zygomatic notches when viewed
from above, and moderately developed supraorbital ridges
that extend imperceptibly onto elongated braincase; mesopterygoid fossa long and narrow, with scalloped anterior border with distinct, posteriorly projecting median
spine; auditory bullae very small; carotid circulatory pattern derived (small stapedial foramen, no internal groove
on squamosal or alisphenoid, no sphenofrontal foramen).
Maxillary toothrow long (>6 mm, with few exceptions),
but average in comparison to length of skull.

DISTRIBUTION: *Rhipidomys gardneri* occurs in the
upper reaches of the Juruá, Ucayali, Madre de Dios, and
Beni drainages (western Brazil, eastern Peru and northeastern Bolivia) from below 200 m to above 2,500 m elevation.

SELECTED LOCALITIES (Map 313): BOLIVIA: La
Paz, Chijchipani (AMNH 262991). BRAZIL: Acre, left
bank of the Rio Juruá opposite Igarapé Porongaba (Pat-
ton et al. 2000). PERU: Cusco, 2 km SW of Tangoshiari (B.
M. de A. Costa 2007), Machu Picchu, San Miguel Bridge
(USNM 194493), Centro de Investigación Wayqecha (C. E.
Medina et al. 2012), Hacienda Cadena (FMNH 68647); Lo-
reto, San Jerónimo, W bank Río Ucayali (BM 28.5.2.180);
Madre de Dios, Albergue Cusco Amazónico, Río Madre de
Dios, ca. 12 km E of Puerto Maldonado (type locality of
Rhipidomys gardneri Patton, da Silva, and Malcolm); Puno,
La Pampa (MCZ 39498).

SUBSPECIES: *Rhipidomys gardneri* is monotypic.

NATURAL HISTORY: Terborgh et al. (1984) and
Woodman et al. (1995) provided ecological information
on this species at Cocha Cashu and Cuzco Amazónico,
Peru, and Patton et al. (2000) described the capture site on
the upper Juruá. Specimens herein referred to *Rhipidomys*

gardneri from San Jerónimo, southern Loreto, Peru, were
captured in houses or thatched roofs. Stomach analysis of
a specimen from mature forest at the San Martín gas well
site in Ucayali, Peru, indicated that it had been feeding
on insect larvae living in bamboo stems. Females may become pregnant while still lactating: one individual (USNM
559408, Pakitza, Río Manu, Peru), captured when nursing
a litter of three young in a desk drawer, was pregnant with
four 20-mm embryos.

REMARKS: *Rhipidomys gardneri* belongs to the "leucodactylus" section of the genus. Tribe (1996) placed the
future holotype and a paratype of *R. gardneri* within *R.
leucodactylus* on the grounds that there were morphological intermediates between "typical" *R. leucodactylus* and
these slightly smaller specimens with less hairy tails from
southeastern Peru. Karyotype and molecular data confirmed the species status of this form (Patton et al. 2000).
Some morphologically intermediate specimens from the
Ucayali-Urubamba and Madre de Dios drainages (also
previously included in *R. leucodactylus* by Tribe, 1996)
are provisionally referred to *R. gardneri* in the absence
of diagnostic molecular or karyotype data, despite their
slightly larger molars and, in some cases, rather hairier
tails. Likewise, the *Rhipidomys* specimens from Chijchipani, at 850 m elevation in the upper Beni drainage of
La Paz, Bolivia ("*Rhipidomys* sp. 1" in Tribe 1996), are
here tentatively included in *R. gardneri* pending further
analysis. Molecular data link *R. gardneri* with the eastern

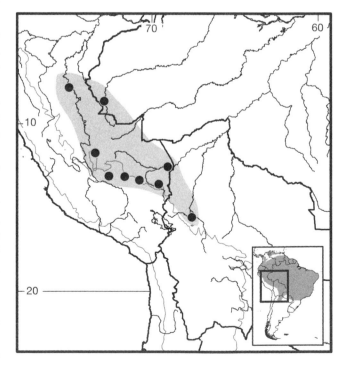

Map 313 Selected localities for *Rhipidomys gardneri* (●). Contour line =
2,000 m.

Brazilian species *R. mastacalis* and *R. emiliae* (reported as *R. nitela*), rather than with *R. leucodactylus* (L. P. Costa 2003). It should be noted that GenBank sequence U03550 (MVZ 196059), sometimes cited as *R. leucodactylus*, in fact corresponds to *R. gardneri*.

Rhipidomys ipukensis R. G. Rocha, Costa, and Costa, 2011

Ipuca Rhipidomys

SYNONYM:

Rhipidomys ipukensis R. G. Rocha, Costa, and Costa, 2011: 24; type locality "Fazenda Lago Verde, municipality of Lagoa da Confusão, state of Tocantins, Brazil, 10°52'09.1"S, 49°41'52.1"W, elevation 180 m."

DESCRIPTION: Small to medium sized with adult head and body length averaging 121 mm (observed range: 99–141 mm), tail longer than head and body (114–124%), and short and coarse dorsal pelage. Overall coloration yellowish gray brown to orangish brown, with dorsal hairs gray-based with yellowish to orangish tips mixed with completely black guard hairs; sides of body lighter than dorsum and with well-defined transition to the venter; ventral pelage woolly, varying from whitish to cream in color (some specimens have dark cream ventral midline). Juveniles completely light gray dorsally and white ventrally. Tail completely dark gray and covered with short hairs along anterior half, becoming more elongate along posterior half, ending in medium pencil 10–15 mm in length. Hindfeet broad, with dark dorsal metatarsal markings that vary from narrow and flanked with white fur to broad with poorly defined lateral limits. Ungual hairs white, completely covering but not surpassing claw tips. Ears large (20–23 mm), brown, often with small cream patch of fur at base.

Skull varies from small and more delicate in younger specimens to medium sized and more robust in older ones (occipitonasal length of adults averages 33.7 mm, range: 29.9–36.2 mm); rostrum moderately long (average 9.6 mm); zygomatic notches relatively shallow but visible in dorsal view; zygomatic plates broad with their posterior margins anterior to M1 alveoli; alisphenoid struts present and subsquamosal fenestrae small, almost occluded by hamular process of squamosal; carotid circulation pattern 3 (of Voss 1988), with small stapedial foramen, no squamosal-alisphenoid groove, and no sphenofrontal foramen; palatal bridge short and wide; incisive foramina bullet-shaped with parallel margins extending posteriorly to anterior edge of M1s; anterior margin of mesopterygoid fossa with median spine and extending to midline of M3s; posterior palatal pits simple and small on each side of mesopterygoid fossa; sphenopalatine vacuities either present as small slits or absent, such that mesopterygoid roof completely ossified; auditory bullae small.

Upper incisors opisthodont; molars rather small (maxillary toothrow length 4.5–5.0 mm); M1 with labial accessory root; m1 with both labial and lingual accessory roots; anterocones/conids of upper and lower molars divided into two conules/conulids by anteromedian flexus(id); relatively small, oblique paralophules present in all upper molars; posterior part of M3 reduced. Mental foramen opens laterally on mandibular body; capsular process well developed as conspicuous swelling.

DISTRIBUTION: This species is known from three localities in Tocantins state, Brazil. It is believed to be endemic to the Araguaia-Tocantins basin east of the Rio Araguaia in the Cerrado of central Brazil.

SELECTED LOCALITIES (Map 309): BRAZIL: Tocantins, Lagoa da Confusão, Fazenda Lago Verde (type locality of *Rhipidomys ipukensis* R. G. Rocha, Costa, and Costa), Pium, Parque Estadual do Cantão (MNRJ 73744), Rio Santa Teresa, 20 km NW of Peixe (UFMG 2952).

SUBSPECIES: *Rhipidomys ikpukensis* is monotypic.

NATURAL HISTORY: Little is known of the ecology or other aspects of the population biology of this species. Specimens were captured primarily in *ipucas*, seasonally flooded forest fragments, in traps placed both on the ground and in the understory, including pipe traps used as arboreal refuges for tree frogs. Adult males had scrotal testes, and adult females were lactating when captured in the month of September (R. G. Rocha, Ferreira, Costa et al. 2011).

REMARKS: *Rhipidomys ipukensis* is a member of the "leucodactylus" section of the genus. It has a close sister-relationship with *R. emiliae*, according to a phylogenetic analysis of mtDNA cytochrome-*b* sequences, with the two species differing by an average of 5.1%; together they form a clade with *R. mastacalis* from the Atlantic Forest (R. G. Rocha, Ferreira, Costa et al. 2011). The karyotype of *R. ipukensis* is as yet unknown. Should *Rhipidomys* specimens from northern Goiás state here provisionally referred to *R. mastacalis* because of their high FN=76 and 80 (Andrades-Miranda, Oliveira et al. 2002) prove instead to belong within *R. ipukensis*, it will not only extend the distribution of the species but also indicate that high fundamental numbers probably arose independently more than once in this group, given the low-FN karyotype of *R. emiliae*, the closest relation of *R. ipukensis*.

R. G. Rocha, Ferreira, Costa et al. (2011) noted that *R. ipukensis* was found to the east of the Rio Araguaia whereas *R. emiliae* was found to the west. Stability of the name *ipukensis* and the identity of the specimens found to the west of the river may depend on phylogenetic comparison with material from the type locality of *R. emiliae* at Rio Moju, Pará, which lies to the east of the Araguaia-Tocantins.

Rhipidomys itoan B. M. de A. Costa, Geise,
Pereira, and L. P. Costa, 2011
Sky Rhipidomys

SYNONYM:

Rhipidomys itoan B. M. de A. Costa, Geise, Pereira, and L. P. Costa, 2011:950; type locality "Garrafão (hillside of the Iconha River in Parque Nacional da Serra dos Órgãos), Guapimirim municipality, state of Rio de Janeiro, Brazil, 22°28′28″S, 42°59′86″W [*sic*], 700 m."

DESCRIPTION: Medium sized (mean head and body length = 136.4 mm) with tail varying from 107–148% of head and body length, mean 124% (*n* = 18). Dorsal pelage reddish or yellowish brown to dark reddish brown, with short (ca. 8 mm) body hairs and longer (ca. 16 mm) guard hairs; ventral pelage plain white or white with small areas of light gray at sides of abdomen, sometimes extending to forelimbs, with or without gray spots in pectoral region; dorsal pelage extends for short distance onto tail base; tail unicolored along length, covered by short, brown to black hairs that increase in length distally and terminate in short pencil extending up to 10 mm beyond its tip. Ears small to medium, ranging from 12–22 mm in length, externally brown in color with brown or whitish hairs internally. Hindfeet covered dorsolaterally with white hairs that extend onto toes; mid-dorsal surface of toes with patch of light brown or gray hairs.

Skull small to moderate in size (mean condyloincisive length 31.4 mm); in comparison with other species in eastern and southeastern Brazil, similar in size to *Rhipidomys macrurus* and *R. mastacalis*, and larger than *R. tribei*. Rostrum short and somewhat broad (mean length = 10.6 mm, mean width = 6.6 mm); lacrimal capsules slightly inflated; gnathic process present, nasal bones broad anteriorly but narrow posteriorly, and not contacting or extending posteriorly beyond maxillary-frontal sutures; anterior edges of zygomatic plates slightly rounded to rectangular but do not protrude anterior to maxillary roots of zygomatic arches; zygomatic notches narrow and quite deep; dorsal projections of lacrimal bones comparatively small, narrow, and slightly prominent; interorbital region broad with rounded, hourglass-shaped edges and slight supraorbital crests; braincase broader in relation to skull length than in other species occurring south of Amazon River; hamular process of squamosal divides postglenoid foramen and subsquamosal fenestra unequally; carotid circulation derived (pattern 3 of Voss 1988), with small stapedial foramen, no translucent internal groove across squamosal and alisphenoid, and no sphenofrontal foramen; thin alisphenoid strut present, as are oval and buccinator masticatory foramina; incisive foramina long, bullet shaped, with parallel margins that converge anteriorly and also slightly posteriorly, especially in specimens from state of São Paulo, ex-

tending to plane between M1s; mesopterygoid fossa long, with small postpalatal spine at anterior margin, extending anteriorly to posterior margins of M3 alveoli; sphenopalatine vacuities either absent or greatly reduced; triangular parapterygoid plates deeply excavated; auditory bullae small but inflated.

Upper incisors slightly opisthodont. Anteroflexus and anteromedian flexus in M1 usually long and deep, demarcating anterolabial and anterolingual conules; protoflexus and hypoflexus of M1 and M2 moderately long; paraflexus and metaflexus long and sharply angled, sometimes bifurcated at inner ends; both medial and labial fossettes frequently present in M2.

DISTRIBUTION: *Rhipidomys itoan* is found from the Serra da Mantiqueira southward to the coast in the Brazilian states of Rio de Janeiro and eastern São Paulo, including in the Serra dos Órgãos and Serra do Mar, and on inshore islands such as Ilha Grande and Ilha da Marambaia. Its range extends to the Serra de Paranapiacaba in southern São Paulo state.

SELECTED LOCALITIES (Map 314): BRAZIL: Rio de Janeiro, Garrafão (type locality of *Rhipidomys itoan* B. M. de A. Costa, Geise, Pereira, and L. P. Costa), Parque Estadual dos Três Picos (HGB 581), Praia Vermelha, Ilha Grande (MNRJ 24389), Parque Nacional do Itatiaia (HGB-CFVC 5); São Paulo, Estação Biológica de Boracéia (MZUSP 10816), Parque Estadual Intervales, Barra Grande (E. M. Vieira and Monteiro-Filho 2003 [EM 1130]), Pilar to Sul (B. M. de A. Costa et al. 2011 [CIT 1278]), São Sebastião (MZUSP 880).

SUBSPECIES: *Rhipidomys itoan* is monotypic.

NATURAL HISTORY: Specimens have been trapped in dense rain forest or coastal forests of the Serra do Mar range, where continuous cover may be punctuated with streams and rocky outcrops, occasionally with an open understory, including that caused by human disturbance (B. M. de A. Costa et al. 2011). In a capture-mark-recapture study at the type locality, 24 individuals of *Rhipidomys itoan* were captured a total of 61 times, always in trees, and remained on the grid for an average of 10 (± 6.73) months; the breeding period was August to December, in the early wet season (J. Macedo et al. 2007). Davis (1945a, 1947) provided life history notes on a sample from nearby Teresópolis (Rio de Janeiro state) and general ecological information on the habitat. Guitton et al. (1986), Martins-Hatano et al. (2002), and Neri-Bastos et al. (2004) investigated parasites on specimens from Rio de Janeiro and São Paulo states. A specimen from Parati in southwestern Rio de Janeiro state was infected with *Leishmania* (*Viannia*) (Soares et al. 1999).

REMARKS: With available taxon sampling, *R. itoan* and *R. tribei* form a sister-pair within the "leucodactylus"

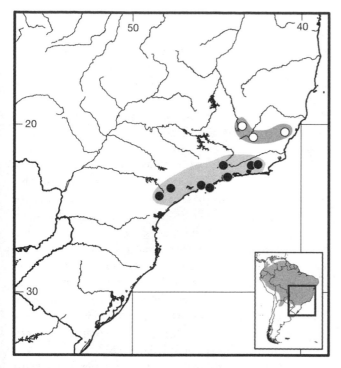

Map 314 Selected localities for *Rhipidomys itoan* (●) and *Rhipidomys tribei* (○).

section of the genus, based on mtDNA cytochrome-*b* sequences. The two species are, however, quite divergent, with the average sequence difference 8.3% (B. M. de A. Costa et al. 2011). *Rhipidomys itoan* is the correct name for those specimens identified by R. G. Rocha, Ferreira, Costa et al. (2011) as "*R.* sp. 2" in their phylogenetic analysis of mtDNA sequences. Previously, specimens of *R. itoan* had frequently been referred to as "*R. mastacalis*" (e.g. Davis 1945a, 1947), "*R.* cf./aff. *macrurus*" (e.g. Tribe 1996; E. M. Vieira and Monteiro-Filho 2003) or "*Rhipidomys* sp." (e.g. Zanchin, Langguth, and Mattevi 1992; Geise 1995; M. F. Smith and Patton 1999 [as "*Rhipidomys* sp. 2"]; L. P. Costa 2003).

The karyotype is $2n = 44$, with the number of autosomal arms varying from 48 to 50 due to polymorphism. The X chromosome is a large submetacentric, one of the largest of the complement; the Y chromosome is the smallest acrocentric. B. M. de A. Costa et al. (2011) also detailed the distribution of NORs within the complement.

Rhipidomys latimanus Tomes, 1860
Northwestern Rhipidomys

SYNONYMS:

H[esperomys]. latimanus Tomes, 1860a:213; type locality not given, but "the greater portion of these [species] are believed to have been collected at Pallatanga, on the western slope of the Cordillera" (Tomes 1860a:211), Chimborazo province, Ecuador.

[*Hesperomys*] (*M[yoxomys].*) *latimanus*: Tomes, 1861 [1862]:284; name combination.

Rhipidomys microtis Thomas, 1896:304; type locality "Saliña [for "Salina"] del Vatan [possibly near Quebrada Batán], Western Cundinamarca, Colombia."

Rhipidomys pictor Thomas, 1904e:193; type locality "Rio Verde, [Pichincha,] N.W. Ecuador. Alt. 1000 m."

Rhipidomys mollissimus J. A. Allen, 1912:78; type locality "Mira Flores [= Miraflores] (alt. 6200 ft.), west slope of Central Andes, near Palmira, Cauca [now Valle del Cauca], Colombia."

Rhipidomys cocalensis J. A. Allen, 1912:79; type locality "Cocal, Cauca, Colombia, Altitude, 4000 feet."

Rhipidomys quindianus J. A. Allen, 1913c:600; type locality "El Roble (altitude 7200 feet), Central Andes [Quindío department], Colombia."

Rhipidomys scandens Goldman, 1913:8; type locality "near head of Rio Limon (altitude 5,000 feet), Mount Pirri [= Cerro Pirre], Eastern Panama."

Rhipidomys latimanus latimanus: Cabrera, 1961:419; name combination.

Rhipidomys latimanus similis: Cabrera, 1961:419; part; not *Rhipidomys similis* J. A. Allen; attribution of *cocalensis*.

Rhipidomys latimanus microtis: Cabrera, 1961:419; name combination.

Rhipidomys latimanus mollissimus: Cabrera, 1961:419; name combination.

Rhipidomys latimanus pictor: Cabrera, 1961:419; name combination.

R[hipidomys]. scadens Cuervo Díaz, Hernández Camacho, and Cadena, 1986:498; incorrect subsequent spelling of *scandens* Goldman.

DESCRIPTION (amended from Tribe 1996, for the nominotypical subspecies he recognized): Medium sized (head and body length typically 120–140 mm) with intensely pigmented orange or reddish-brown dorsal pelage, finely flecked with dark brown or black guard hairs; ventral pelage creamy white, sometimes with inconspicuous gray bases to hairs. Hindfeet moderately large and often robust (26–28 mm in length) with rather narrow metatarsal dark patch, broadening at base of digits III to V but rarely extending onto digits. Tail long (120–150% of head and body length), medium to dark brown, with short hair on shaft, distally little longer, with terminal pencil 3–30 mm long but usually about 10 mm. Ears small to medium in size, and rather dark in color. Skull moderate in size and rather slender, with slender rostrum; supraorbital ledges or ridges pronounced, straight, and converge strongly to front of interorbital region; braincase oval, not angular, and not greatly flattened except in old specimens; parapterygoid fossae shallow and not fenestrated; auditory bullae small; carotid circulatory pattern derived (small stapedial foramen, no squamosal-alisphenoid groove, no sphenofrontal foramen). Incisor root capsule on mandible does not form prominent knob. Upper molar toothrow length 4.7–5.3 mm, average in relation to skull length in *Rhipidomys*.

DISTRIBUTION: At the southern limit of its range, *R. latimanus* occurs in the Chinchipe valley of northern Peru, close to the border with Ecuador, possibly linking northward along the eastern Andean slopes with the population in the upper Napo valley, Ecuador (700 to 1,100 m elevation). It is found along the western flanks and valleys of the Ecuadorian Andes (700 to 1,700 m) and also in the Colonche range of western Ecuador (about 450 m). In Colombia, it occurs from Nariño in the south, along both sides of the upper and middle Cauca valley (Cauca, Valle del Cauca, and Antioquia departments—700 to 2,000 m), spilling over in places to the opposite slopes of the Western and Central Cordilleras (2,200 m at El Roble), and extending eastward to the eastern side of the Magdalena valley in Cundinamarca and Boyacá (1,000 to 1,300 m); an apparently isolated population exists at the head of the Magdalena valley in Huila department (1,350 to 1,400 m). An isolated population in the Darién mountains of eastern Panama (1,250 to 1,550 m) probably extends into the hills across the Colombian border.

SELECTED LOCALITIES (Map 315, South American localities only): COLOMBIA: Antioquia, Valdivia, La Cabaña (FMNH 70255); Boyacá, Muzo (FMNH 71711); Cauca, Cocal (type locality of *Rhipidomys cocalensis* J. A. Allen); Huila, San Adolfo (FMNH 71704); Nariño, San Pablo (BM 99.9.3.1); Quindío, El Roble (type locality of *Rhipidomys quindianus* J. A. Allen); Valle del Cauca, Miraflores (type locality of *Rhipidomys mollissimus* J. A. Allen). ECUADOR: Chimborazo, Pallatanga (type locality of *Hesperomys latimanus* Tomes); Santa Elena, Cerro Manglar Alto (AMNH 66618); Napo, near Archidona (BM 34.9.10.112); Pichincha, Río Verde (type locality of *Rhipidomys pictor* Thomas). PERU: Cajamarca, Perico (MCZ 17043).

SUBSPECIES: As currently understood, *R. latimanus* is monotypic (but see Remarks).

NATURAL HISTORY: Little information has been published on the natural history of *R. latimanus*. Collection localities suggest that it occupies a range of habitats from moist to somewhat drier forests. Specimens from near Cali in the Cauca valley were trapped in coffee plantations and second-growth woodland and tested negative for *Leishmania* (Alexander et al. 1998). Specimen tags indicate that one individual from western Ecuador was taken in long grass, and one from eastern Panama was captured in cloud forest. Another Panamanian specimen was pregnant with three embryos in March.

REMARKS: *Rhipidomys latimanus* is a member of the "leucodactylus" section of the genus. Morphologically, it is closely allied to *R. venezuelae* (*q.v.*); both species encompass considerable morphological variation, with some populations of each taxon approaching the other in certain characters, especially in northern Colombia. Tribe (1996) therefore tentatively included *venezuelae* as a subspecies within *R. latimanus*, together with an unnamed subspecies corresponding to the specimens from Huila department, Colombia. As pointed out by Musser and Carleton (2005), however, the appreciable morphometric differentiation between the first two of these taxa suggests that they may be retained as separate species, and this arrangement is followed here. The Huila specimens are convergent in some characters with *R. fulviventer*, but a majority of features point to a position within *R. latimanus* (for details, see Tribe 1996). The specimens from the Darién mountains of extreme eastern Panama, considered by most authors a separate species (*R. scandens*), fall well within the range of variation found in Colombian *R. latimanus* and are therefore allocated to this species (Tribe 1996).

The karyotype of a *R. latimanus* specimen from Peñas Blancas, Valle del Cauca, Colombia ($2n=44$, FN$=48$) was

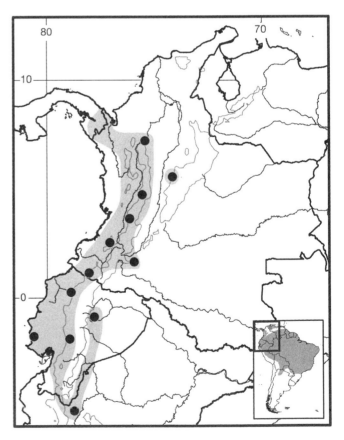

Map 315 Selected localities for *Rhipidomys latimanus* (●). Contour line = 2,000 m.

the first *Rhipidomys* karyotype to be published (A. L. Gardner and Patton 1976). It remains to be seen whether karyotypes and molecular data from other localities will help to clarify the large amount of morphological variation observed in the species as understood here.

Rhipidomys leucodactylus (Tschudi, 1845)
Great Rhipidomys
SYNONYMS:

H[*esperomys (Rhipidomys)*]. *leucodactylus* Tschudi, 1843: 252; *nomen nudum*.

H[*esperomys (Rhipidomys)*]. *leucodactylus* Tschudi, 1844: 16; *nomen nudum*.

H[*esperomys (Rhipidomys)*]. *leucodactylus* Tschudi, 1845: 183; type locality "im Oststriche," i.e., the eastern region of Peru visited by Tschudi, herein restricted to the Montaña de Vitoc area in Junín, Peru (see Remarks).

Nyctomys leucodactylus: Fitzinger, 1867b:89; name combination.

Hesperomys (Rhipidomys) sclateri Thomas, 1887:152 and plate XIX; type locality "Maccasseema" [= Makasima, on Pomeroon River, Pomeroon-Supinaam], Guyana.

Rhipidomys Goodfellowi Thomas, 1900c:270; type locality "upper Rio Napo, at mouth of Rio Coca," Napo, Ecuador.

Rhipidomys bovallii Thomas, 1911a:114; type locality "Potaro Highlands, towards Mt Roraima, British Guiana [Guyana]. Alt. 2000'."

Rhipidomys lucullus Thomas, 1911a:115; type locality "Garita del Sol, valley of Vitoc, Upper Peréné, [Junín], Central Peru. Alt. 5700'."

Rhipidomys equatoris Thomas, 1915:312; type locality "San Domingo, [Santo Domingo de los Tsáchilas,] W. Ecuador (79°6′W., 0°13′S.). Alt. 1600'."

Rhipidomys rex Thomas, 1927f:600; type locality "Chinchavita, 3000'," Huánuco, Peru.

Rhipidomys leucodactylus equatoris: Cabrera, 1961:421; name combination.

Rhipidomys leucodactylus goodfellowi: Cabrera, 1961:421; name combination.

Rhipidomys leucodactylus leucodactylus: Cabrera, 1961: 421; part, excluding *Rhipidomys ochrogaster*, as subspecies.

R[*hipidomys*]. *leucodactylus aratayae* Guillotin and Petter, 1986:541; type locality "saut Pararé, sur la rivière Arataye [Pararé falls, on the Arataye river] (4°3′N, 52°40′W)," French Guiana.

DESCRIPTION (amended from Tribe 1996, excluding specimens that are now recognized as *R. gardneri* [see Patton et al. 2000, and that account herein]): Largest species in genus *Rhipidomys*, with adult head and body length often exceeding 180 mm. Dorsal pelage typically medium-brown agouti, varying in tone from grayish to yellowish to reddish, with conspicuous dark guard hairs; flanks somewhat paler; ventral pelage white, cream, or yellowish, with gray-based hairs and often with orange suffusion in medial pectoral region; pelage texture somewhat coarse and often slightly woolly, especially on ventral surface. Tail proportionately short, only 95% to 130% of head and body length, reddish to very dark brown in color, unicolored, with shaft well clothed in hairs and tip terminating in long bushy pencil over 15 mm in length, sometimes reaching 40 mm. Ears moderately large. Hindfeet large (32–38 mm in length) and very broad, with dark patch covering most of dorsal surface and extending onto first or even second phalanges of digits II to V, sometimes also digit I; sides of foot and tips of toes silvery to golden. Skull large and robust (adult occipitonasal length usually exceeding 40 mm), relatively broad, with deep and broad rostrum; nasals expanded at anterior end and parallel-sided in posterior half; zygomatic notches small when viewed from above; supraorbital ridges well developed, sometimes converging anteriorly on top of interorbital region, with orbital wall ventrolateral to ridges forms narrowest part of interorbital constriction, providing general hourglass shape to region; braincase moderately broad and slightly inflated, with flatter profile in older specimens; interparietal nearly horizontal and occiput vertical; parietal ridges positioned somewhat dorsally on braincase such that squamosals are oriented dorsolaterally rather than laterally; mesopterygoid fossa broad, especially anteriorly; auditory bullae small and carotid circulation pattern derived (small stapedial foramen, no internal groove on squamosal or alisphenoid, and no sphenofrontal foramen). Molars large and maxillary toothrow long (6.0–7.0 mm), although slightly smaller on average than might be expected in relation to skull length.

DISTRIBUTION: *Rhipidomys leucodactylus* occurs in the lowland rainforests of the Amazon-Orinoco basin, from the Guiana coast and lower Amazon in the northeast to the Andean piedmont between Colombia (Serranía de la Macarena) and central Bolivia (Cochabamba). It penetrates the moist valleys of the Guiana highlands of Venezuela as well as submontane forests up to 1,750 m elevation in eastern Peru, but apparently not in the upper Urubamba and Madre de Dios drainages. The range of the species as understood here also extends through the Huancabamba Depression in northern Peru to the forests of the western Andean slopes and valleys in northwestern Peru and Ecuador, from where it may be expected to reach the Chocó region of western Colombia.

SELECTED LOCALITIES (Map 316): BOLIVIA: Cochabamba, Yungas (BM 34.9.2.172). BRAZIL: Amapá, Mazagão, Boa Fortuna, upper Igarapé Rio Branco (MPEG 2508); Amazonas, Altamira, right bank Rio Juruá (MVZ

190592); Mato Grosso, Aripuanã (MZUSP APC 260); Pará, Amorim, Rio Tapajós (AMNH 95508); Rondônia, Usina Hidrelétrica de Samuel, Rio Jamarí (UFPB 1259). COLOMBIA: Caquetá, Florencia (AMNH 33744); Meta, Serranía de la Macarena (AMNH 142141). ECUADOR: El Oro, Los Pozos (AMNH 67508); Orellana, San José Abajo (AMNH 68189); Pastaza, Mera (USNM 548392); Santo Domingo de los Tsáchilas, Santo Domingo (type locality of *Rhipidomys equatoris* Thomas). FRENCH GUIANA: Saut Pararé, Rivière Arataye (type locality of *Rhipidomys leucodactylus aratayae* Guillotin and Petter). GUYANA: Cuyuni-Mazaruni, Potaro Highlands, Venamo River (type locality of *Rhipidomys bovallii* Thomas); Pomeroon-Supenaam, Makasima, Pomeroon River (type locality of *Hesperomys (Rhipidomys) sclateri* Thomas). PERU: Ayacucho, Huanhuachayo (AMNH 241643); Huánuco, Chinchavito (type locality of *Rhipidomys rex* Thomas); Junín, La Garita del Sol (type locality of *Rhipidomys lucullus* Thomas); Loreto, Mouth of Río Curaray (AMNH 71906); Piura, Canchaque (FMNH 81293). VENEZUELA: Amazonas, Cerro Neblina, Camp VII, 5.1 km NE of Pico Phelps (USNM 560825), Pozón, 50 km NE of Puerto Ayacucho (T. E. Lee et al. 2000); Bolívar, Embalse Guri, Isla Panorama, Río Caroní (T. D. Lambert et al. 2003).

SUBSPECIES: As currently understood, *R. leucodactylus* is monotypic, but the considerable geographic variation in both morphology (Tribe 1996) and molecular sequences (B. M. de A. Costa et al. 2011) exhibited across its large range may underscore subspecific or even specific differentiation (see Remarks). As with many of the widespread species of this genus, larger series coupled with thorough analyses of this variation are needed to properly delimit geographic units.

NATURAL HISTORY: As a large, rat-sized, frugivorous rodent, *R. leucodactylus* is listed as a pest of agricultural crops (Aguilar et al. 1977); in fact, the holotype was captured while gnawing fruit in a pineapple field (Tschudi 1845), and other specimen tags mention sugar cane and yucca. Specimens have often been caught in rural dwellings (specimen tags; Handley 1976). The species generally occurs in moist, evergreen forest (Handley 1976; Ochoa et al. 1988), including floodable riparian habitats adjacent to the llanos (T. E. Lee et al. 2000). The holotype of *Rhipidomys leucodactylus aratayae* was observed to nest in a hole 15 m up in a tree, have a home range of 1.35 ha, and forage on the ground (Guillotin and Petter 1986). Three females from Huánuco, central Peru, were each pregnant with two embryos in August–September, and two from Napo, eastern Ecuador, had three in September and November (specimen tags). For parasites of Venezuelan specimens, see P. T. Johnson (1972) and Saunders (1975). A specimen of *Rhipidomys leucodactylus* captured in northern Brazil had uncharacterized *Leishmania* isolated from its skin (Lainson, Shaw, et al. 1981), and sylvatic plague (*Yersinia pestis*) was carried by *Polygenis litargus* fleas infesting individuals and nests of *R. leucodactylus* in Piura, northwestern Peru (Machiavello 1957 *apud* Pozo et al. 2005).

REMARKS: *Rhipidomys leucodactylus* is the type species of the genus. As explained in the Remarks on the genus, Tschudi's name *leucodactylus*, like *Rhipidomys*, becomes available from its use on page 183 of the *Untersuchungen über die Fauna peruana* (*Therologie*), published in the fourth installment of the work in 1845. Where it appears on page 16 of this work (published in 1844) and on page 252 of the "Mammalium conspectus" (Tschudi 1843) it is a *nomen nudum*. In his description of the material, Tschudi (1845:183–184) mentioned that he found this species "im Oststriche"—that is, the region east of the Andean Cordillera in central Peru—where it was associated with crop fields (his first specimen was shot while gnawing a pineapple) and human habitations. He also stated that Prof. Poeppig reported it from Tocache on the Río Huallaga. Cabrera's (1961:421) restriction of the type locality to the upper Huallaga "on the basis of Tschudi's text" is thus evidently based on a misinterpretation: Tschudi's own specimens came from the region he himself visited and not from the Huallaga, which lies much farther north. Tschudi (1846, 1847) describes his excursion from Lima over the Cordillera to the Montaña de Vitoc, the area between the rivers Anamayo and Tulumayo in Junín department. There he saw many plantations, especially of pineapples. Although he did not specifically mention *Rhipidomys* in this context, this is the only area east of the Andes that he described from firsthand experience. Given also the close similarity between the holotype of *leucodactylus* and other specimens collected later in the same area, it is justifiable to restrict the type locality of this taxon to the Montaña de Vitoc and to reject Cabrera's (1961) mistaken suggestion.

Wagner (1845a:147) also described a specimen of *Rhipidomys* under the name *Hesperomys leucodactylus*, but one now referable to a species different from Tschudi's material. I have been unable to resolve the question of which *leucodactylus*—Tschudi's or Wagner's—was published first in 1845: no month of publication appears in either work, and no clues have been found in the accession records of several libraries consulted. In the absence of any clear priority and in accordance with the principle of the first reviser (ICZN 1999:Art. 24.2), I consider that precedence should be given to *leucodactylus* Tschudi, 1845, over *leucodactylus* Wagner, 1845. This step best serves stability by retaining the name for the taxon with which it has almost always been associated. In contrast, to my knowledge Wagner's name has been cited only three times in the last 100 years (Tate 1932g:5; Moojen 1952b:75; and C. O. da C. Vieira 1955:419); of these authors, only Tate associated the name with *Rhipidomys*, the others retaining it with *Hesperomys*. Furthermore, it avoids the inconvenience of having to provide a replacement name for the type species of the genus. I additionally propose that *leucodactylus* Wagner be rejected both as a junior primary homonym (in *Hesperomys*) and as a secondary homonym (in *Rhipidomys*) (ICZN 1999:Art. 52.1–3, 53.3, 57.2). Because I attribute Wagner's type specimen to *Rhipidomys macrurus* (P. Gervais, 1955) (*q.v.*), the latter name replaces *leucodactylus* Wagner (ICZN 1999:Art. 60.2).

As treated herein, *Rhipidomys leucodactylus* displays substantial geographic and/or individual variation in characters such as overall size, relative tail length and hairiness, and ventral color. Patton et al. (2000) clearly distinguished between *R. leucodactylus* and *R. gardneri* on the basis of karyotype and mtDNA cytochrome-*b* sequence data as well as morphological characters. Unfortunately, some specimens do not fit neatly with Patton et al.'s morphological descriptions and, therefore, in the absence of karyotype and molecular data, have been allocated subjectively to one species or the other. For example, an individual from the Yungas (wet, forested Andean valleys) of Cochabamba, Bolivia, is here placed in *R. leucodactylus* because of its very large size, long molar row, and skull shape, although in its white ventral pelage and short tail pencil it resembles *R. gardneri*. Most specimens of *R. leucodactylus* from west of the Andes in northwestern Peru and western Ecuador closely resemble populations from the eastern valleys in central Peru and eastern Ecuador, suggesting contact at some time through the Huancabamba Depression.

Known karyotypes for *Rhipidomys leucodactylus* all show a diploid number of $2n = 44$ but vary in number of autosomal arms: FN = 46 (Patton et al. 2000), FN = 48 (Zanchin, Langguth, and Mattevi 1992), and FN = 52 (M. J. J. Silva and Yonenaga-Yassuda 1999, as "*R. cf. mastacalis*," from Aramanaí, Mato Grosso state). The karyotyped specimen identified by Aguilera et al. (1994) as "*R. sclateri*," from the coastal region of Venezuela, represents *Rhipidomys couesi* and not *R. leucodactylus*. It should also be noted that Andrades-Miranda, Oliveira et al. (2002) identified all their specimens with low-FN from Goiás and Pará, Brazil, as "*R. leucodactylus*" without reference to morphology; in fact, they represent *R. macrurus*, *R. emiliae*, and possibly *R. ipukensis*. A molecular phylogeny of 39 *Rhipidomys* specimens (B. M. de A. Costa et al. 2011) placed three *R. leucodactylus* in a sister-group relationship to a clade compris-

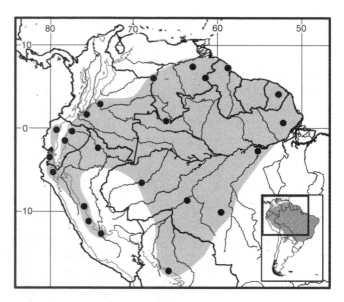

Map 316 Selected localities for *Rhipidomys leucodactylus* (●). Contour line = 2,000 m.

ing two subgroups corresponding to *R. itoan* and *R. tribei*. Within the *R. leucodactylus* clade a Kimura 2-parameter distance of 9.24% separated the above-mentioned specimen from Aramanaí from two specimens from different parts of the Rio Juruá, a distance greater than that separating several sister species within the genus. GenBank sequence U03550 (MVZ 196059), sometimes cited as *R. leucodactylus*, in fact corresponds to *R. gardneri*.

Rhipidomys leucodactylus has been collected at the same localities as *R. modicus* in central Peru, and as *R. nitela*, *R. wetzeli*, and *R. macconnelli* in Venezuela.

Rhipidomys macconnelli de Winton, 1900
Macconnell's Tepui Rhipidomys

SYNONYMS:

Rhipidomys Macconnelli de Winton, 1900:52; type locality "summit of Roraima, [Cuyuni-Mazaruni,] Demerara" [Guyana].

[*Rhipidomys*] *maccomelli* E.-L. Trouessart, 1904:409; incorrect subsequent spelling of *macconnelli* de Winton.

[*Thomasomys*] *macconnelli*: Thomas, 1917d:196; name combination.

Thomasomys macconnelli macconnelli: Tate, 1939:194; name combination.

Thomasomys macconnelli subnubis Tate, 1939:195; type locality "South slope Mt. Auyan-tepui, 3500 feet," Bolívar, Venezuela.

Rhipidomys macconelli Hershkovitz, 1959a:9; incorrect subsequent spelling of *macconnelli* de Winton.

DESCRIPTION: Small (head and body length 115–120 mm), resembling *Thomasomys* in many external characters but cranially closer to other *Rhipidomys*. Pelage sleep, long, and soft; colored dark chocolate brown dorsally and dark slate overlaid with cream or pale brown ventrally; juveniles darker and greyer than adults. Tail much longer than head and body length (>140%) but slenderer than in most other *Rhipidomys*; paler on its ventral surface and almost bare over much of its length but with pronounced terminal pencil (5–20 mm long). Hindfeet less specialized for climbing, being longer (relative to body length) than in other species of *Rhipidomys* due to longer, narrow metatarsals; plantar pads also smaller and separated, with gap between thenar and hypothenar; dark patch on dorsal surface poorly defined but mid-brown in color and covers most of surface; heels and ankles bare. Ears dark brown, matching dorsal pelage. Cranially, nasals narrow anteriorly and rostrum long and slender in comparison to rounded braincase; interorbital region bounded by distinct, slightly concave supraorbital ridges; the lateral process of the parietal usually triangular rather than rectangular in shape and contacts the occipital barely or not at all; occipital region broad; mesopterygoid fossa broadened anteriorly, parapterygoid fossae often slightly fenestrated; carotid circulation pattern derived (small stapedial foramen, no groove on the inner surface of the squamosal and alisphenoid, and no sphenofrontal foramen); upper molar row length (5.00 ± 0.157 mm, N = 214) large in relation to skull length for the genus.

DISTRIBUTION: *Rhipidomys macconnelli* lives on the slopes and summits of the Guiana Shield table mountains (*tepuis*) in Venezuela and adjacent areas of Brazil and Guyana, at elevations between 750 m and more than 2,600 m.

SELECTED LOCALITIES (Map 317): GUYANA: Cuyuni-Mazaruni, Mount Roraima, summit (type locality of *Rhipidomys Macconnelli* de Winton). VENEZUELA: Amazonas, Cerro Aracamuri, Cumbre Sur (MBUCV, in litt.), Cerro de Tamacuare (Ojasti, Guerrero, and Hernández 1992), Cerro Duida, Cabecera del Caño Culebra (USNM 406078), Cerro Neblina, Camp VII (USNM 560837); Bolívar, Acopán-tepuí (AMNH 176308), Auyán-tepuí (type locality of *Thomasomys macconnelli subnubis* Tate), km 125, 85 km SSE of El Dorado (USNM 387934).

SUBSPECIES: *Rhipidomys macconnelli* is monotypic (see Remarks).

NATURAL HISTORY: *Rhipidomys macconnelli* inhabits very moist cloud forest and dwarf vegetation on the *tepuis* and the evergreen forests surrounding them, where it occurs principally on the ground and on cliffs, making runways under rocks and trees (Handley 1976). Further habitat data may be found in Handley (1976), Guerrero et al. (1989), and Ojasti et al. (1992). From the large series collected, particularly on Cerro Duida, Auyán-tepui, and Roraima, *R. macconnelli* appears to be a common animal on the *tepuis*. The animals' molar teeth, even when relatively

unworn, acquire a thick tartar coating, more so than in most other *Rhipidomys* specimens. Museum specimens also show a considerable ectoparasite load of louse nits and carpet-beetle egg cases. For parasites, see Furman (1972a, 1972b), P. T. Johnson (1972), E. K. Jones et al. (1972), Machado-Allison and Barrera (1972), Tipton and Machado-Allison (1972), Brennan and Reed (1974a), Saunders (1975), and Brennan and Goff (1977) . Guerrero (1985) listed the ectoparasites recorded from all species of Venezuelan *Rhipidomys*.

REMARKS: *Rhipidomys macconnelli* is the sole member of the "macconnelli" section of the genus. Its generic allocation remained uncertain for many years because of the conflicting evidence of its *Thomasomys*-like external characters and *Rhipidomys*-like cranial morphology. Originally described as a *Rhipidomys* by de Winton in 1900, it was regarded as an aberrant form by Thomas (1906c), who later transferred the species to *Thomasomys* without explanation (Thomas 1917d). This arrangement was generally accepted for the next four decades until Hershkovitz (1959a) equally summarily restored *macconnelli* to *Rhipidomys*. His judgment was confirmed by the species' lack of a gall bladder (Voss 1991a) and by its karyotype (2*n* = 44, FN = 50; Aguilera et al. 1994), which ally it with other *Rhipidomys* species rather than with *Thomasomys*. Cladistic analyses based on molecular (M. F. Smith and Patton 1999) and morphological (Pacheco 2003) data corroborate this. Tate's (1939) description of a subspecies (*subnubis*) appears to have been founded on individual and clinal variation within the Auyán-tepui populations.

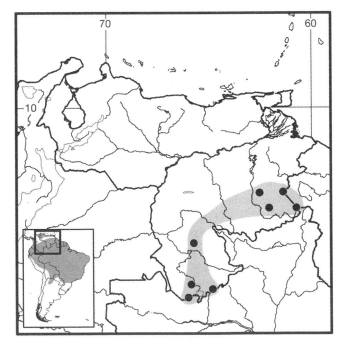

Map 317 Selected localities for *Rhipidomys macconnelli* (●). Contour line = 2,000 m.

Rhipidomys macrurus (P. Gervais, 1855)
Cerrado Rhipidomys
SYNONYMS:

Hesperomys leucodactylus Wagner, 1845a:147; not *Hesperomys (Rhipidomys) leucodactylus* Tschudi; type locality: "Rio Parana," Porto do Rio Paraná [= Igarapava], São Paulo, Brazil.

Mus (Hesperomys) macrurus P. Gervais, 1855:111, plate 16, Figs. 1, 1a; type locality "Trixas" [misspelling of Crixás], Goiás, Brazil.

Hesperomys macrourus P. Gervais, 1855:112; incorrect subsequent spelling of *macrurus* P. Gervais.

[*Hesperomys (Rhipidomys)*] *macrurus*: E.-L. Trouessart, 1880:136; name combination.

Rh[ipidomys]. macrurus: Thomas, 1896:303; first use of current name combination.

Rhipidomys mastacalis macrurus: Cabrera, 1961:423; name combination.

DESCRIPTION (amended from Tribe 1996, including his *Rhipidomys* sp. 2 and excluding specimens from the Serra da Ibiapaba reassigned herein to *R. mastacalis*): Medium sized (head and body length 125–145 mm, somewhat larger in southwestern part of range) with dull, reddish gray brown dorsal pelage; under parts white or pale cream, frequently with gray bases to hairs. Tail slightly longer than head and body length (100–120%, occasionally up to 140%) and terminates in pencil usually 5–15 mm long; tail medium to dark reddish brown, occasionally paler on proximal half. Ears large and medium brown in color. Hindfeet moderate in length (mostly 24–28 mm, but up to 33 mm in southwest), relatively broad and robust; dark dorsal patch either distinctly outlined or diffuse, either narrow or broad, occasionally extending onto digits; sides of feet and digits golden. Phallic morphology described and illustrated by Hooper and Musser (1964). Skull with short rostrum and small zygomatic notches in dorsal view; supraorbital edges squared or slightly ridged and, except in larger specimens, curved rather than straight, providing general rounded appearance to interorbital region, which is narrower than in neighboring species; braincase of average breadth and rounded, becoming more angular and flattened with age; incisive foramina roughly pear-shaped, broadening posteriorly, and palatal bridge short and broad; auditory bullae range from relatively small in southwest to moderately large elsewhere; carotid circulation pattern derived (small stapedial foramen, no internal groove on squamosal and alisphenoid, no sphenofrontal foramen). Maxillary toothrow length moderate to large in comparison with skull length and between 4.8–5.4 mm long in most of species' range; up to 5.8 mm in southwestern part.

DISTRIBUTION: *Rhipidomys macrurus* is found in gallery forests and patches of woodland throughout most of the Cerrado domain in Brazil, from southern Piauí state, where it has also been captured just within the Caatinga domain (M. A. N. Sousa 2005), south through eastern Tocantins state, western and central Bahia (including the semideciduous woods of the Rio São Francisco basin), Goiás, the Distrito Federal, northern and western Minas Gerais, northern and western São Paulo, and Mato Grosso do Sul. Specimens from eastern Paraguay were confirmed as *R. macrurus* by de la Sancha et al. (2011).

SELECTED LOCALITIES (Map 318): BRAZIL: Bahia, Andaraí, Fazenda Santa Rita (B. M. de A. Costa 2007), São Marcelo, Rio Preto (FMNH 21162); Goiás, Anápolis (MNRJ 4323), Crixás (type locality of *Mus (Hesperomys) macrurus* P. Gervais), Serra da Mesa, 20 km NW of Colinas do Sul (MNRJ 36449); Mato Grosso do Sul, Maracaju (MNRJ 4296), RPPN Cabeceira do Prata (MNRJ 69787); Minas Gerais, Caratinga, Fazenda Montes Claros (B. M. de A. Costa 2007), Coronel Murta, Ponte do Colatino, left bank Rio Jequitinhonha (B. M. de A. Costa 2007), Lavras (Mesquita 2009); Piauí, Caracol, Parque Nacional da Serra das Confusões (M. A. N. Sousa 2005), Estação Ecológica de Uruçuí-Una (MZUSP 30388); São Paulo, Itararé, Fazenda Ibiti (P. S. Martin et al. 2009), São José do Rio Preto (USNM 460533); Tocantins, Lajeado, Fazenda Santa Helena (MNRJ 76732). PARAGUAY: Amambay, Parque Nacional Cerro Corá (USNM 554544); Canindeyú, Sendero Morotí, Reserva Mbaracayú (de la Sancha et al. 2011).

SUBSPECIES: As currently understood, *R. macrurus* is monotypic (but see comments under Remarks).

NATURAL HISTORY: *Rhipidomys macrurus* typically inhabits the gallery forests lining rivers through the savanna-like cerrados of Brazil, as well as areas of woodland savanna (*cerradão*) and semideciduous forest. Mares, Braun, and Gettinger (1989) and Mares and Ernest (1995) provided ecological and population data for the species near Brasília; Gettinger (1992a, 1992b) reported on parasites from there. Mesquita (2009) gave details of the use of space by *R. macrurus* in a mosaic of cerrado, woodland, and farmland near Lavras, Minas Gerais state, where the species was relatively abundant. It was very abundant in semideciduous woodland in western Mato Grosso do Sul state, where Milano (2007) studied its population ecology and use of space. Bonvicino and Bezerra (2003) and Carmignotto and Aires (2011) reported specimens captured inside houses in southwestern Bahia and eastern Tocantins states, respectively. Dietz (1983) gave brief notes on the natural history of *R. macrurus* (as "*R. mastacalis*") in the Serra da Canastra National Park, Minas Gerais; parasites from there were described by Whitaker and Dietz (1987). M. F. S. Ribeiro et al. (2004) gave some details of water physiology in a specimen (reported as *Rhipidomys mastacalis*) from mesic forest at Morro do Chapéu (central Bahia; see Remarks).

REMARKS: *Rhipidomys macrurus* is a member of the "leucodactylus" section of the genus. Because the holotype of *R. macrurus*, collected by Castelnau and Deville in May 1844 (P. Gervais 1855), has not been located in the MNHN collections, the identity of this taxon is based on the hypothesis that the type locality (Crixás, Goiás), now deforested, falls within the range of this species of *Rhipidomys* as understood herein. A second specimen collected by Castelnau and marked "Co-type" [BM 49.10.15.32 (skin); 49.12.8.4 (skull)] has the provenance "Bahia"; this may mean that it was collected elsewhere but shipped from Bahia, where Castelnau was French consul from 1848, but equally there is no indication that it was collected at the type locality. Indeed, in his description of *Rhipidomys macrurus*, P. Gervais (1855) made no mention of a second specimen. Because eastern Brazilian species of *Rhipidomys* are morphologically very similar, there is no guarantee that the two specimens represent the same species, let alone the same population. No other specimens are known from that area. Work is in progress to select a suitable specimen to be designated as a neotype for *Rhipidomys macrurus* in order to clarify the identity of the taxon and ensure stability of usage (Tribe et al., in prep.). The specimen selected must not only be consistent with what is known of the geographical origin and morphology of the holotype, but also must agree in karyotype and DNA sequences with specimens assigned to *R. macrurus* (*sensu stricto*), given the importance of such evidence for species distinctions in eastern Brazil, as outlined herein.

Cabrera (1961) placed all *Rhipidomys* smaller than *R. leucodactylus* from Brazil, the Guianas, and Venezuela as subspecies of *R. mastacalis*, reflecting the difficulty in unambiguously distinguishing between them on morphological evidence alone. The advent of karyotyping revealed a dichotomy in fundamental number (number of autosomal arms) in eastern Brazilian *Rhipidomys*, with specimens with $2n=44$, FN ≥ 70, being associated with the name *R. mastacalis*, and those with the same diploid number but FN $= 48$–52, with *R. macrurus* (Zanchin, Langguth, and Mattevi 1992; Svartman and Almeida 1993a; Tribe 1996; M. A. N. Sousa 2005; L. G. Pereira and Geise 2007; Thomazini 2009; A. H. Carvalho et al. 2012). Accordingly, the population of *Rhipidomys* in the Serra da Ibiapaba, western Ceará state, tentatively included in *R. macrurus* by Tribe (1996, 2005), is here provisionally reassigned to *R. mastacalis* on the basis of its karyotype with FN $= 70$ (M. A. N. Sousa 2005). B. M. de A. Costa et al. (2011) also pointed out morphological differences between this population and *R. macrurus* from farther south in the cerrado domain, and

an mtDNA phylogeny (Thomazini 2009) suggests it belongs to a clade comprising *R. emiliae*, *R. ipukensis*, and *R. mastacalis*.

Specimens from Morro do Chapéu, central Bahia, have yielded fundamental numbers of 56 (A. L. G. Souza et al. 2007b) and 61 (M. J. de J. Silva and Yonenaga-Yassuda 1999); they may belong to the as-yet-unnamed clade mentioned below. They are sympatric with, and morphologically distinct from, a specimen referred to *R. cariri* (see that account). The fundamental number of 61 is intermediate between the numbers typical of *R. mastacalis* (FN $\geq$ 70) and of *R. cariri* and *R. macrurus* (FN $= 48$–52), suggesting that that particular specimen might be an interspecific hybrid (M. J. de J. Silva and Yonenaga-Yassuda 1999).

Recent work with molecular data has uncovered further diversity, separating additional species from the concept of *R. macrurus* employed by Tribe (1996), some of them overlapping geographically (L. P. Costa 2003; B. M. de A. Costa 2007; Thomazini 2009; B. M. de A. Costa et al. 2011; R. G. Rocha, Ferreira, Costa et al. 2011). For example, B. M. de A. Costa et al. (2011) have formalized the form found in Atlantic Forest of São Paulo and Rio de Janeiro states as *R. itoan*. Specimens from generally drier areas in central and western Bahia, northern and eastern Minas Gerais, northern Goiás, and southeastern Tocantins form a sister clade to *Rhipidomys cariri*, and together they are related to *Rhipidomys macrurus* (*sensu stricto*). This more inclusive clade is the sister clade to all other Brazilian *Rhipidomys* (L. P. Costa 2003; B. M. de A. Costa 2007; B. M. de A. Costa et al. 2011). The as-yet-undescribed form is retained here within *R. macrurus* pending formal description. Conversely, the same molecular studies and that of de la Sancha et al. (2011) demonstrate that the population from western Mato Grosso do Sul and eastern Paraguay belongs within *R. macrurus* (*s.s.*), although on morphological and morphometric grounds (not least its larger overall size and molar row length) Tribe (1996) considered it to represent a separate species. Clearly, considerable further research is needed to elucidate the status of all the forms included here in *R. macrurus*.

The specimen collected by Johann Natterer at Porto do Rio Paraná (now Igarapava, on the northern border of São Paulo state) and named *Hesperomys leucodactylus* by Wagner (1845a) was allocated to *R. mastacalis* by Tribe (1996) on biogeographical grounds. The distribution of the specimens used in the molecular studies mentioned herein suggests that Natterer's specimen should be reallocated to *R. macrurus*. Regarding the invalidity of the species name *leucodactylus* Wagner, see the Remarks section under *R. leucodactylus*.

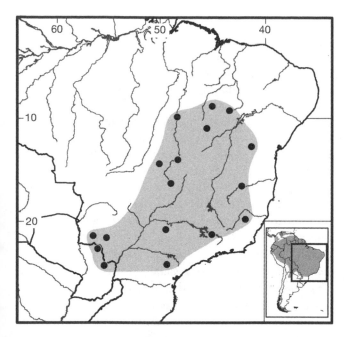

Map 318 Selected localities for *Rhipidomys macrurus* (●).

Rhipidomys mastacalis (Lund, 1840)
Northern Atlantic Forest Rhipidomys

SYNONYMS:

[*Mus*] *mastacalis* Lund, 1839a:233; *nomen nudum*.

Mus mastacalis Lund, 1840a:24 [1841b:240]; type locality "Rio das Velhas's Floddal" (Lund 1841b:264, 266), Lagoa Santa, Minas Gerais, Brazil.

Mus mustacalis Lund, 1840b:7 [1841c:279]; incorrect subsequent spelling of *mastacalis* Lund.

Mus mystacalis Wagner, 1843b:750; incorrect subsequent spelling of *mastacalis* Lund.

Mus maculipes Pictet and Pictet, 1844:67; type locality "Bahia," Brazil.

Mus masculipes Burmeister, 1854:185; incorrect subsequent spelling of *maculipes* Pictet and Pictet.

H[*esperomys* (*Calomys*)]. *mystacalis* Burmeister, 1855a:7; part; name combination and incorrect subsequent spelling of *mastacalis* Lund.

H[*esperomys* (*Calomys*)]. *maculipes*: Burmeister, 1855a:8; name combination.

[*Hesperomys* (*Rhipidomys*)] *mastacalis*: E.-L. Trouessart, 1880:136; name combination.

Rhipidomys mastacalis: Winge, 1887 [1888]:54; first use of current name combination.

[*Oryzomys*] *musculipes* E.-L. Trouessart, 1897:528; incorrect subsequent spelling of *maculipes* Pictet and Pictet and name combination.

Rhipidomys cearanus Thomas, 1910c:501; type locality "S[ão]. Paulo, on the top of the Serra [da Ibiapaba], at about 900 metres," Ceará, Brazil.

Rhipidomys venezuelae cearanus: Gyldenstolpe, 1932:47; name combination.

[*Oecomys*] *maculipes*: Tate, 1932g:21; name combination.

Rhipidomys maculipes: Laemmert, Ferreira, and Taylor, 1946:38; name combination.

Rhipidomys mastacalis cearanus: Cabrera, 1961:422; name combination.

Rhipidomys mastacalis mastacalis: Cabrera, 1961:423; name combination.

Rhipidomys masticalis A. L. Gardner, 1990:412; incorrect subsequent spelling of *mastacalis* Lund.

Rhipidomys masatacalis Voss and Myers, 1991:428; incorrect subsequent spelling of *mastacalis* Lund.

DESCRIPTION: Medium sized (head and body length 125–145 mm) with gray-brown to more intensely red-brown dorsal pelage and cream or white under parts. Tail moderate to long (120–140% of head and body length), light to dark brown, with short hairs proximally but longer ones on distal half, and with short to moderate-length pencil. Ears medium to large, usually medium brown. Hindfeet broad, medium in length (26–29 mm), with medium to dark brown dorsal patch, often broad and ill-defined, which does not extend onto toes; sides of foot and toes pale. Skull with moderately long, relatively slender rostrum; nasals taper regularly from anterior to posterior; supraorbital ridges straight, diverging from front of interorbital region; braincase not greatly inflated or rounded; incisive foramen narrow; palatal bridge long and narrow; carotid circulatory pattern derived (small stapedial foramen, no internal groove on squamosal and alisphenoid, no sphenofrontal foramen). Maxillary toothrow of moderate length for size of skull, averaging about 5.0 mm.

DISTRIBUTION: *Rhipidomys mastacalis* is found in the Atlantic rainforest of eastern Brazil, from the state of Paraíba in the north along the coastline to eastern Rio de Janeiro state in the south, and also inland on the Chapada Diamantina plateau of central Bahia and in eastern and central Minas Gerais. Apparently isolated populations in western Ceará and northern Goiás are also provisionally referred to this species.

SELECTED LOCALITIES (Map 319): BRAZIL: Alagoas, Anádia, Sítio Vale Verde (MNRJ 17450); Bahia, Fazenda Unacau, 8 km SE of São José (UFPB 425), Palmeiras, Pai Inácio (MNRJ 67751), São Felipe (MNRJ 22274); Ceará, Ibiapina, Sítio Pejuaba (MNRJ 17401); Espírito Santo, Linhares (MNRJ 34492); Goiás, Serra da Mesa, 40 km NE of Uruaçu (MNRJ 37350); Minas Gerais, Brumadinho, Rio Manso (UFMG 1649), Diamantina, Conselheiro Mata (B. M. de A. Costa 2007), Salinas (MNRJ 30015); Paraíba, Areia, Mata do Pau-Ferro (UFPB 3864); Pernambuco, Garanhuns, Sítio Cavaquinho (MNRJ 12500); Rio de Janeiro,

Fazenda União, Casimiro de Abreu (MNRJ 46803); Sergipe, south of Estância (Stevens and Husband 1998).

SUBSPECIES: *Rhipidomys mastacalis* is monotypic.

NATURAL HISTORY: *Rhipidomys mastacalis* occurs in remnants of Atlantic forest in eastern Brazil, including both highland enclaves (*brejos de altitude*) and lowland forest in the northeast (M. A. N. Sousa et al. 2004; Asfora and Pontes 2009), and less-moist forests in eastern Minas Gerais. It is also sometimes caught in the roofs of houses (data from specimen tags). Most captures of this species in forest fragments in Sergipe occurred well inside the fragments (Stevens and Husband 1998). About 59% of a large series of *R. mastacalis* collected for yellow fever research in the vicinity of Ilhéus, Bahia, was taken in young forest, 21% in old forest and 9% in swamp forest; *Rhipidomys* proved to be less important as a host for the virus than most rodents (Laemmert et al. 1946). Near Caruaru, Pernambuco, *R. mastacalis* was captured mainly in hillside forest both in trees and on the ground; only two females out of 17 collected were pregnant in October to December (M. A. N. Sousa et al. 2004). Cerqueira, Vieira, and Salles (1989) analyzed habitat and reproductive data for a population in the Serra da Ibiapaba (western Ceará state; see Remarks). *Rhipidomys mastacalis* is implicated in damage to cacao in southern Bahia, where it is relatively common in and around plantations (Cruz 1983; Cassano and Moura 2004; Pardini 2004). T. Araújo et al. (2006) described behavior in captivity. For parasites, see F. Fonseca (1959) and Hastriter and Peterson (1997).

REMARKS: *Rhipidomys mastacalis* is a member of the "leucodactylus" section of the genus. Having once been applied to *Rhipidomys* specimens from much of northern and eastern South America (Cabrera 1961), the name *R. mastacalis* is now restricted to a taxon occupying a moderately small range in eastern Brazil. This change was due not only to morphological reassessments (Musser and Carleton 1993; Tribe 1996), but especially to the results of karyotype analyses and molecular phylogeographic studies (Zanchin, Langguth, and Mattevi 1992; Svartman and Almeida 1993a; L. P. Costa 2003; B. M. de A. Costa 2007; Thomazini 2009; B. M. de A. Costa et al. 2011; R. G. Rocha, Ferreira, Costa et al. 2011). See also the Remarks under *Rhipidomys macrurus*.

The karyotype of *R. mastacalis* has the same diploid number ($2n=44$) as the majority of other *Rhipidomys* species, but it has a higher fundamental number (FN $\geq$ 70) than other eastern Brazilian forms. On the assumption that this combination of diploid and fundamental numbers characterizes a monophyletic group, certain specimens from Serra da Mesa in northern Goiás state with FN = 76 and 80 (Andrades-Miranda, Oliveira et al. 2002) are here provi-

sionally referred to *R. mastacalis*, although they may well represent a distinct species (see also the Remarks under *R. ipukensis*). They are sympatric with low-FN specimens referred here to *R. macrurus*.

Additionally, the population of *Rhipidomys* in the highland forest enclave in the Serra da Ibiapaba, western Ceará, which Tribe (1996, 2005) tentatively referred to *R. macrurus* on morphological grounds, is here provisionally reallocated to *R. mastacalis* because the first known karyotype from that population (with FN = 70) appears to be identical to that of *R. mastacalis* from eastern Pernambuco and Bahia (M. A. N. Sousa 2005). A phylogeny based on mitochondrial cytochrome oxidase I sequences places a specimen from Ibiapaba as a distant basal member of a clade allying *R. emiliae* and *R. ipukensis*, to which *R. mastacalis* forms the sister clade (Thomazini 2009). The *Rhipidomys* populations inhabiting isolated mesic forests (*brejos*) within the Caatinga domain of northeastern Brazil thus appear to be split between *R. mastacalis* (high FN) in the more peripheral enclaves and *R. cariri* (low FN) in the more central ones. Whereas specimens from the *brejos* in eastern Pernambuco are closely allied to *R mastacalis* from the coastal Atlantic Forest (Thomazini 2009), the Serra da Ibiapaba population on the far side of the Caatinga is more distantly related. Should further work demonstrate that the population in the Serra da Ibiapaba deserves species status, the name *cearanus* Thomas is available. The identity of specimens from Serra Negra, western Pernambuco, reported by Botêlho et al. (2001,

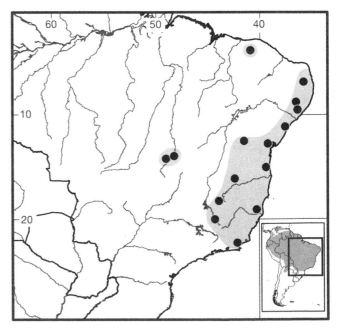

Map 319 Selected localities for *Rhipidomys mastacalis* (●); note isolated populations in central and northeastern Brazil.

2003) as "*Rhipidomys cearanus*" in a study of ectoparasites, has not been verified; they may represent either *R. mastacalis* or *R. cariri*.

Rhipidomys modicus Thomas, 1926
Lesser Peruvian Rhipidomys

SYNONYMS:

Rhipidomys modicus Thomas, 1926g:161; type locality Pucatambo ("Puca Tambo, 5100 [feet]"), San Martín, Peru.
Rhipidomys leucodactylus modicus: Cabrera, 1961:422; name combination as subspecies.

DESCRIPTION: Medium to large (head and body length 130–165 mm) with medium to dark yellowish to reddish brown agouti dorsal coloration, well streaked with dark guard hairs, and rather coarse and short; ventral pelage with white-tipped fur with very pale to dark gray bases; orange pectoral spot sometimes present. Tail long (up to 135% of head and body length) and medium to dark brown; shaft lightly haired and terminating in pencil that rarely exceeds 15 mm. Ears rather small. Hindfeet moderately large (28–30 mm) and broad, with broad dark patch on dorsal surfaces that sometimes extends onto digits. Skull moderate in size and robustness; rostrum short and somewhat pointed; nasals taper gradually along length; zygomatic plates not broad and notches small; supraorbital ridges squared rather than raised, diverging strongly posteriorly; interorbital region broad, with narrowest point forward; braincase oval, moderately broad, and not greatly flattened, although interparietal oriented more dorsally than posteriorly; mesopterygoid fossa broad and penetrating between hypocones of M3s; auditory bullae small; carotid circulation pattern derived (small stapedial foramen, no internal squamosal-alisphenoid groove, no sphenofrontal foramen). Maxillary toothrow 5.5–5.8 mm in length, comparatively long for size of skull.

DISTRIBUTION: *Rhipidomys modicus* is known from eastern Andean valleys in northern and central Peru, at elevations between 700 and 1,800 m. Collecting localities for *R. modicus* in central Peru are identical with or close to localities where specimens of *R. leucodactylus* have been taken.

SELECTED LOCALITIES (Map 320): PERU: Amazonas, 12 km E of La Peca Nueva (LSUMZ 21873); Huánuco, Chinchavito (BM 27.11.1.130); Junín, Chanchamayo (BM 7.6.15.13); San Martín, Moyobamba (FMNH 19363), Pucatambo (type locality of *Rhipidomys modicus* Thomas).

SUBSPECIES: *Rhipidomys modicus* is monotypic.

NATURAL HISTORY: The holotype was "trapped in brush near stream," and the specimen from Chinchavito was trapped in an abandoned shed (specimen tags).

REMARKS: *Rhipidomys modicus* is a member of the "leucodactylus" section of the genus. It is substantially

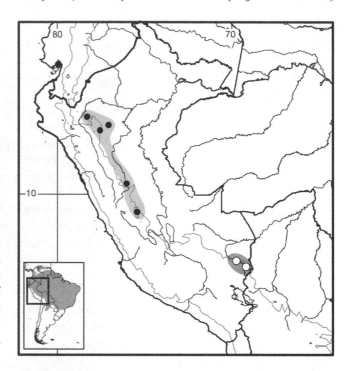

Map 320 Selected localities for *Rhipidomys modicus* (●) and *Rhipidomys ochrogaster* (○). Contour line = 2,000 m.

smaller than *R. leucodactylus* in external dimensions, skull length, and molar row length, and appears to occur sympatrically with the latter species in at least part of its range. Specimens exhibit considerable variation in the extent of gray bases to their ventral pelage, even at a single locality.

Rhipidomys nitela Thomas, 1901
Guianan Rhipidomys

SYNONYMS:

Rhipidomys nitela Thomas, 1901e:148; type locality Kwatata, Kanuku Mountains, Upper Takutu–Upper Essequibo, Guyana (see Remarks).
Rhipidomys venezuelæ fervidus Thomas, 1904d:34; type locality "La Unión, lower Orinoco," Caura valley, Bolívar, Venezuela.
Rhipidomys venezuelæ yuruanus J. A. Allen, 1913c:601; type locality "Río Yuruán, [Bolívar], Venezuela."
Rhipidomys milleri J. A. Allen, 1913c:602; type locality "Minehaha Creek (altitude 500 feet), Lower Essequibo River, British Guiana."
R[hipidomys]. v[enezuelae]. nitela: J. A. Allen, 1913c:602; name combination.
Rhipidomys venezuelae milleri: Gyldenstolpe, 1932:47; name combination.
Rhipidomys mastacalis nitela: Cabrera, 1961:423; name combination.
Rhipidomys mastacalis fervidus: Cabrera, 1961:423; name combination.

Rhipidomys nitela tobagi Goodwin, 1961:16; type locality
"Little Tobago, the West Indies," Little Tobago Island,
Trinidad and Tobago.

R[*hipidomys*]. n[*itela*]. *milleri*: Goodwin, 1961:16; name
combination.

DESCRIPTION: Small to medium sized (head and
body length typically 110–140 mm) with soft, short, dull
gray brown to richer reddish or orange brown dorsal pelage; ventral fur short and creamy white to roots or with
inconspicuous pale gray hair bases. Tail moderately long
(110–140% of head and body length) with short to medium terminal pencil 5–15 mm in length. Ears pale to dark
and moderately large in size, sometimes with pale postauricular patch. Hindfeet short (24–27 mm) and slender,
with small plantar pads and dorsal dark patch of varying
distinctness rarely extending onto digits. Skull small with
short, rather blunt rostrum; interorbit broad with squared
or ledged edges, sometimes raised; braincase oval, not inflated, and flattened in older individuals; incisive foramina
usually narrow and almost elliptical in shape; mesopterygoid fossa narrow and parapterygoid fossae may be fenestrated; auditory bullae small; carotid circulation pattern
derived (small stapedial foramen, no internal groove on
squamosal and alisphenoid, and no sphenofrontal foramen). Maxillary toothrow short, between 3.8–5.0 mm in
length, but average in size for genus in comparison with
skull length.

DISTRIBUTION: *Rhipidomys nitela* as understood
here occurs at lower elevations (below 1,400 m) throughout the Guiana Highlands and adjacent forested lowlands
in the region bounded by the Orinoco, Negro, and lower
Amazon rivers and the Atlantic coast. Specimens found on
the island of Little Tobago, northeast of Trinidad, are also
referred to this species.

SELECTED LOCALITIES (Map 321): BRAZIL:
Amapá, 50 km ESE of Porto Grande (MNRJ AN 188);
Amazonas, Km 50 Manaus-Itacoatiara road (MNRJ
19614); Pará, Oriximiná, Porto Trombetas, Rio Saraca-
zinho (MPEG 10113); Roraima, Ilha de Maracá (Barnett
and Cunha 1998). FRENCH GUIANA: Les Nouragues
(MNHN V.890), Paracou (AMNH 267021). GUYANA:
Cuyuni-Mazaruni, Kalakun (AMNH 207387); Upper
Takutu–Upper Essequibo, Kwatata, Kanuku Mountains
(type locality of *Rhipidomys nitela* Thomas). SURINAM:
Para, Matta, 15 km W of Zanderij airport (RMNH 21973).
TRINIDAD AND TOBAGO: Tobago, Little Tobago Is-
land (AMNH 184555). VENEZUELA: Amazonas, Cerro
Duida, Caño Culebra (USNM 406109); Bolívar, La Unión,
Río Caura (BM 4.5.7.34), Río Caroní, Embalse Guri, Isla
Panorama (T. D. Lambert et al. 2003), Río Yuruán (type lo-
cality of *Rhipidomys venezuelae yuruanus*), Serranía de los
Pijiguaos, 140 km SW of Caicara (Rivas and Linares 2006).

SUBSPECIES: As currently understood, *Rhipidomys
nitela* is monotypic, although the population from Little
Tobago Island has been regarded as a separate subspecies
(Goodwin 1961).

NATURAL HISTORY: *Rhipidomys nitela* occurs in
lowland rainforest, where it is primarily arboreal but also
comes down to the ground (Handley 1976, as "*R. mastacalis*"; Voss 1991a; Voss et al. 2001). Ecological data
for the species are provided by Malcolm (1988) near
Manaus in the central Amazon (see Remarks), and by
Mauffrey (1999) and Mauffrey and Catzeflis (2002) at Les
Nouragues, French Guiana. According to specimen tags
and published reports, *R. nitela* is often found in the rafters
or thatched roofs of buildings (J. A. Allen 1911; Husson
1978; Barnett and Cunha 1998; Lim, Engstrom, Genoways
et al. 2005). Specimens from Sinnamary (French Guiana)
and Bonda (northern Colombia; see Remarks) were shot
in bat roosts (Voss et al. 2001; L. Roguin, pers. comm.).
For parasites, see Brennan and Yunker (1969), Furman
and Tipton (1961), and Gettinger, Martins-Hatano et al.
(2005).

REMARKS: *Rhipidomys nitela* is a member of the "leucodactylus" section of the genus. The type locality given in
the published report (Thomas 1901e:149) and the locality
recorded in the accession register of the Natural History
Museum, London, for the whole type series is "Kwaimattat, Kanuku Mts."; the labels attached to all the skins,
however, are marked "Quatata [or "Quatatat"], Kanuku
Mts., B. Guiana, 240' [feet]." Kwaimattat and Quatatat,
now spelled Kwaimatta and Kwatata, are localities 27 km
apart in the savannas north of the Kanuku Mountains in
Guyana. Kwatata is here selected as the type locality, because the data accompanying the original material (in this
case the specimen tags) should be taken into account first
when a type locality is clarified (ICZN 1999:Recommenda-
tion 76A).

Specimens from northern Bolívar state, Venezuela,
including the holotypes of *Rhipidomys venezuelae fervidus* and *R. v. yuruanus*, tend to be larger and more
intensely pigmented than most other *Rhipidomys nitela*
specimens, and their skulls are more elongated with
strong supraorbital ridges. They are retained here within
R. nitela because the differences in character states do
not appear to be consistent enough for taxonomic separation. Specimens from south of the Amazon considered
by Tribe (1996) to be *Rhipidomys nitela* are here referred
to *R. emiliae* (see that account); these include the "*R.
nitela*" sequenced by M. F. Smith and Patton (1999) and
by L. P. Costa (2003). A single specimen from the isolated northern Colombian locality of Bonda, which Tribe
(1996) allocated to *R. nitela*, was considered by Voss
et al. (2001:127) to represent another taxon; in line with

their judgment, it is here regarded as belonging to an as yet undescribed species of *Rhipidomys*. Less convincing is their suggestion that the specimens from Little Tobago possibly "represent an insular form of the adjacent mainland species *R. venezuelae*" (Voss et al. 2001:127): the nearest part of the known range of *R. venezuelae* (in Estado Guárico, Venezuela) lies farther away from Little Tobago than that of *R. nitela* (in Estado Bolívar, Venezuela, and in Guyana), and the Little Tobago specimens cluster morphometrically with *R. nitela* and not with *R. venezuelae* (Tribe 1996). Additional material from the northern coast of South America and adjacent islands is clearly needed to resolve these relationships fully.

Rhipidomys nitela is the only species of the genus so far karyotyped to have a diploid number other than $2n = 44$, with reports of $2n = 48$ from southeast Venezuela (Voss et al. 2001); $2n = 48$, FN = 68, from Roraima state, Brazil (Andrades-Miranda, Oliveira et al. 2002); $2n = 48$, FN = 71, from French Guiana (Volobouev and Catzeflis 2000); and $2n = 50$, FN = 71 and 72, from near Manaus (M. J. de J. Silva and Yonenaga-Yassuda 1999, as *Rhipidomys* sp. B). Based on banding analysis, Volobouev and Catzeflis (2000) found considerable differences between the French Guiana and Manaus karyotypes, leading them to question the taxonomic status of the latter population. Pending further study, specimens from near Manaus are here provisionally retained in *R. nitela*.

Phylogenetic analyses based on mtDNA cytochrome-*b* sequences place specimens of *R. nitela* from French Guiana as the sister group to a clade consisting of *R. mastacalis*, *R. ipukensis*, and *R. emiliae* (R. G. Rocha, Ferreira, Costa et al. 2011). More extensive molecular analyses of *R. nitela* specimens are needed to elucidate the diversity contained within the species as understood here.

Rhipidomys ochrogaster J. A. Allen, 1901
Buff-bellied Rhipidomys

SYNONYM:

Rhipidomys ochrogaster J. A. Allen, 1901b:43.; type locality "Inca Mines, Peru (alt. 6000 feet)," Santo Domingo, Río Inambari, Puno, Peru.

DESCRIPTION: Large (head and body length 152–176 mm), long-tailed (length 198–228 mm) species with orange-brown dorsal fur and pale orange or melon-colored ventral fur. Tail dark, considerably longer than head and body (125–148%), and terminates in long pencil of hairs (20 mm long in holotype). Hindfeet moderately long (34–35 mm) and broad, with metatarsal patch extending onto first phalanges of toes. Skull moderately large, with rather narrow and pointed rostrum, broad interorbital region, and broad, rounded braincase; supraorbital shelf well developed and divergent posteriorly; postorbital ridges absent or weak; zygomatic plates narrow; carotid circulation pattern primitive (pattern 1, of Voss 1988), characterized by large stapedial foramen, squamosal-alisphenoid groove visible but not strong, and sphenofrontal foramen. Upper molar row long (6.1–6.3 mm), large in comparison to skull length for genus. Pacheco and Peralta (2011) provided photographs depicting external appearance (including dorsal, lateral, and ventral views of freshly dead specimen, dorsal and ventral views of hindfoot, and tail tip) as well as skull and both molar series in comparison with *R. gardneri* and *R. leucodactylus*.

DISTRIBUTION: Only five specimens of *Rhipidomys ochrogaster* are known from two localities in southeastern Peru at elevations between 1,200 and 1,940 m. Four of these are from the vicinity of the type locality in the upper Río Inambari drainage at the base of the Cordillera de Carabaya, and the fifth from the valley of the Río Tambopata.

SELECTED LOCALITIES (Map 320): PERU: Puno, Santo Domingo (type locality of *Rhipidomys ochrogaster* J. A. Allen), Yanacocha, valley of Río Tambopata (Pacheco and Peralta 2011).

SUBSPECIES: *Rhipidomys ochrogaster* is monotypic.

NATURAL HISTORY: J. A. Allen (1900a:219) described the type locality (erroneously referred to as Juliaca; see J. A. Allen 1901b:41), quoting the collector H. H. Keays: "The country is very broken, with deep

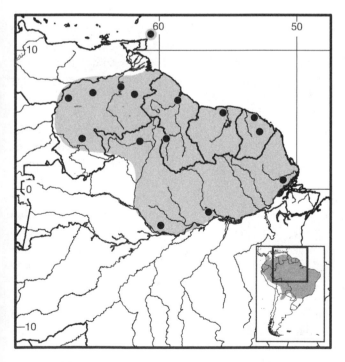

Map 321 Selected localities for *Rhipidomys nitela* (●); note insular population on Little Tobago Island northeast of Trinidad. Contour line = 2,000 m.

narrow cañons, and is covered with a dense undergrowth of shrubs and vines, with here or there a palmetto or a cedar rising above the surrounding vegetation." The recent specimen from the Río Tambopata drainage (Pacheco and Peralta 2011) was collected in a forest of slender trees of 15 m maximum height, dense understory, and forest floor covered with deep soil litter. Plant species included *Ilex* sp. (Aquifoliaceae), *Viburnum* sp. (Adoxaceae) *Miconia* sp. (Melastomataceae), *Myrsine* sp. (Myrsinaceae), *Myrcia* sp. (Myrtaceae), *Clusia* sp. (Clusiaceae), *Cecropia* sp. (Urticaceae), and *Clethra* sp. (Clethraceae). This habitat corresponds to the Yungas, the humid montane forest that flanks the eastern slope of the eastern cordillera of the Andes in southern Peru and Bolivia (Young and Leon 1999; Tovar Narváez et al. 2010). Pacheco and Peralta (2011) proposed that the species be classed as Endangered on account of the increasing deforestation of Yungas habitats in Puno department. Stomach contents of the Yanacocha specimen included both seeds and ants (Pacheco and Peralta 2011).

REMARKS: *Rhipidomys ochrogaster* is by far the largest member of the "fulviventer" section of the genus (not the "leucodactylus" section, as stated by Musser and Carleton 2005:1171), to which it clearly belongs on morphological grounds (see Key to the species of *Rhipidomys*). There is a gap of approximately 1,300 km along the Andean chain from the nearest other known members of the section, which were recently discovered in Amazonas department in northern Peru (C. F. Jiménez and Pacheco 2012; see Remarks under *Rhipidomys fulviventer*). Although the capture elevation of the holotype is given as "6800 ft" (2,075 m) on the specimen tag (AMNH 16481), the figure given in the published type description ("6000 feet," 1,830 m) is more plausible according to maps of the locality.

Rhipidomys similis J. A. Allen, 1912
Greater Colombian Rhipidomys

SYNONYMS:

Rhipidomys similis J. A. Allen, 1912:79; type locality "Cocal, Cauca, Colombia. Altitude, 6000 feet."

Rhipidomys latimanus similis: Cabrera, 1961:419; part; name combination.

DESCRIPTION: Moderately large with head and body length 125–155 mm and long tail measuring 125% to 150% of head and body length. Dorsum reddish brown to yellowish rufous lined with black, and somewhat darker medially than more orange yellow sides; ventral pelage cream or more usually pale orange hair tips over long, dark gray bases. Tail covered with black hairs, nearly concealing annular scales, and terminates in a pencil. Ears dull brown. Hindfeet moderately long (29–31 mm) and broad. Skull

with rounded interorbital region with slightly developed supraorbital ridges, and broad, rounded braincase; mesopterygoid fossa expanded at anterior end; sphenopalatine vacuities small to medium; parapterygoid fossae fenestrated; carotid circulation pattern primitive (large stapedial foramen, groove on inner surface of squamosal and alisphenoid, and sphenofrontal foramen). Maxillary toothrow larger than average for genus in comparison to skull length, ranging from 5.0–5.4 mm.

DISTRIBUTION: *Rhipidomys similis* is restricted to the Andes of southwestern Colombia. It is known from high elevations on both versants of the Western Cordillera in the departments of Cauca and Valle del Cauca, and from the mountains at the head of the Magdalena valley, department of Huila, including parts of the Central and Eastern Cordilleras. It is likely to occur in the knot of mountains connecting the three cordilleras in Cauca and Nariño departments and may also extend farther north along the Western and Central Cordilleras.

SELECTED LOCALITIES (Map 312): COLOMBIA: Cauca, Cocal (AMNH 32458); Huila, Belén (USNM 294984), San Adolfo, Río Aguas Claras (FMNH 71712), San Antonio (FMNH 71716); Valle del Cauca, Pichindé (Alberico 1983).

SUBSPECIES: *Rhipidomys similis* is monotypic.

NATURAL HISTORY: Specimens of *Rhipidomys similis* from the head of the Magdalena valley were found at between 1,600 m and 2,350 m. Specimens from the Western Cordillera were captured at elevations of 2,400 m on the eastern slope and 1,220 to 1,830 m on the cooler, moister western slope, corresponding to the differing elevations of cloud forest on the two versants. The remarkably low elevation of 1,220 m ("4000 ft") for a specimen from Cocal is possibly an error (see Chapman 1917:31, footnote 2, regarding bird specimens from the same locality).

REMARKS: *Rhipidomys similis* is the second largest species in the "fulviventer" section of the genus, after *R. ochrogaster*. It occurs sympatrically with *R. caucensis* in Cauca and Huila, and with *R. latimanus* in Valle, Cauca, and Huila, with its elevational range broadly overlapping that of the former and marginally that of the latter (if at all: see note on Cocal in Natural History section). Tribe (1996) treated the Western Cordillera and Huila units as separate subspecies of *Rhipidomys fulviventer*; see the Remarks under that species. Specimens allocated by Alberico (1983:65) to *R. latimanus* are of this species.

Rhipidomys tenuicauda (J. A. Allen, 1899)
Turimiquire Rhipidomys

SYNONYMS:

Oryzomys tenuicauda J. A. Allen, 1899b:211; type locality "Los Palmales, [Sucre,] Venezuela."

[*Rhipidomys*] *tenuicauda*: Tate, 1939:195; first use of current name combination.

Phipidomys mastacalis tenuicauda: Cabrera, 1961 423; name combination, and incorrect subsequent spelling of *Rhipidomys* Tschudi.

DESCRIPTION: Small species similar to *R. fulviventer*, with head and body length 109–128 mm (holotype exceptionally small at 88 mm). Dorsal pelage often bright orange brown agouti, sometimes more chestnut brown; ventral pelage with cream hair tips on middle to dark gray bases. Tail proportionately long (115–140% of head and body length, 145% in holotype), unicolored dark brown, varying from nearly naked at base but with increasingly long hairs distally, terminating in thick pencil about 10 mm long. Ears brown and lightly haired. Hindfeet small (23–27 mm in length) and narrow, with dorsal dark patch extending to base of toes and buff edges and digits. Skull with rounded interorbital region and slight supraorbital ridges; braincase broad, rounded; lateral processes of parietals narrow posteriorly; parapterygoid fossae sometimes fenestrated; sphenopalatine fissures variable, from long and moderately broad to short and narrow, or absent; carotid circulatory pattern primitive (large stapedial foramen, internal groove along squamosal and alisphenoid, and sphenofrontal foramen). Maxillary toothrow short, 4.2–4.6 mm, average for genus in comparison with skull length; molars proportionately somewhat narrower than average.

DISTRIBUTION: *Rhipidomys tenuicauda* appears to be restricted to the Serranía de Turimiquire of northeastern Venezuela, in the states of Sucre, Monagas, and Anzoátegui.

SELECTED LOCALITIES (Map 322): VENEZUELA: Anzoátegui, Cerro La Laguna, cumbre (F. J. García and Sánchez-González 2013); Monagas, San Agustín, 5 km NW of Caripe (USNM 409929).

SUBSPECIES: *Rhipidomys tenuicauda* is monotypic.

NATURAL HISTORY: *Rhipidomys tenuicauda* occurs in premontane humid forest (USNM specimen tags) as well as in coffee and cacao plantations (Tate 1931), at elevations between 945 m and 2,200 m. It may be found at lower elevations than other members of the "fulviventer" section (with the exception of *R. wetzeli*) because the vegetation zones in the Turimiquire Range lie at substantially lower levels than in the Andes and Venezuelan coastal range farther west (Chapman 1925). Two females captured by the Smithsonian Venezuelan Project had four embryos, and one had five in June–July (USNM specimen tags). For parasites, see Furman (1972a), P. T. Johnson (1972), and Tipton and Machado-Allison (1972). Guerrero (1985) listed the ectoparasites recorded from all species of Venezuelan *Rhipidomys*.

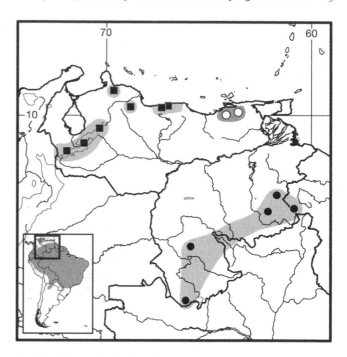

Map 322 Selected localities for *Rhipidomys tenuicauda* (○), *Rhipidomys venustus* (■), and *Rhipidomys wetzeli* (●). Contour line = 2,000 m.

REMARKS: This member of the "fulviventer" section of *Rhipidomys* resembles Colombian *R. fulviventer* in its moderately small size, although the holotype is an unusually small specimen (but with molars of normal size). The xeric lowlands of the Depresión de Unare in Anzoátegui state separate its range from that of the larger *R. venustus*, which inhabits the higher parts of the Venezuelan coastal range farther west, and by the llanos of the lower Orinoco drainage from the range of the smaller *R. wetzeli*, found in the region of the *tepui* inselbergs to the south. Musser and Carleton (2005) considered *tenuicauda* a subjective junior synonym of *R. fulviventer*. It is hypothesized here to be a species in its own right on biogeographical grounds, insofar as it is separated from the range of *R. fulviventer* in the Eastern Cordillera of the Andes by approximately 1,000 km of lowlands and mountains, which are partly occupied by the allied yet distinct form *R. venustus*. Future molecular data may help to demonstrate whether this chain of forms has resulted from a process of vicariant isolation on mountain blocks.

Rhipidomys tribei B. M. de A. Costa, Geise, Pereira, and L. P. Costa, 2011
Tribe's Rhipidomys

SYNONYM:

Rhipidomys tribei B. M. de A. Costa, Geise, Pereira, and L. P. Costa, 2011:947; type locality "Reserva Particular do Patrimônio Natural Santuário do Caraça, 25 km SW Santa Bárbara, Minas Gerais, Brazil, 20°05'S, 43°30'W, 1,300 m."

DESCRIPTION: Small, with average head and body length of 116.8 mm (observed range: 96–130 mm) and long tail 115–137% of body length. Dorsal pelage yellowish brown to reddish brown; coat comprises short (7–9 mm) tribanded hairs, black at tip, reddish brown or dark yellow in middle, with gray bases; longer guard hairs (11–15 mm) typically black but may bleach to reddish brown with translucent tips; ventral pelage white and distinct from dorsal color, sometimes with small areas of light gray hair at sides of abdomen; dorsal fur extends onto base of tail for maximum of 8 mm; tail unicolored; scales small (0.5–0.7 mm in length), three hairs emerging from under each one; tail hairs increase in length gradually from tail base to tip, forming short terminal brush or tuft up to 7 mm in length. Ears small to medium, between 16 and 21 mm in length, sparsely covered with short brown hairs on both surfaces. Hindfeet covered dorsolaterally with white hairs that extend onto toes; mid-dorsal surfaces with patch of light brown or gray hairs.

Skull with smallest average total length of all *Rhipidomys* species in southeastern Brazil, with mean condyloincisive length 28.5 mm. Rostrum short (less than one-third length of skull) and somewhat broad (more than one-half rostral length); lacrimal capsules slightly inflated; gnathic process present; nasal bones broad anteriorly, narrowing to posterior margin; nasals not contacting or extending posteriorly beyond maxillofrontal suture; anterior edges of zygomatic plates slightly rounded to rectangular and not protruding anterior to maxillary roots of zygomatic arches; zygomatic notches narrow and rather shallow; dorsal projections of lacrimal bones relatively large, elongated, and laterally expanded; interorbital convergent anteriorly, with slight medial constriction, strongly divergent posteriorly, with usually rounded borders and no supraorbital crest; hamular processes of squamosals separating enlarged subsquamosal fenestrae from smaller postglenoid foramina; slender alisphenoid struts present in all specimens; small stapedial foramen, no translucent groove across the inside of the braincase, and no sphenofrontal foramen signal derived carotid circulatory pattern 3 (of Voss 1988); margins of incisive foramina expanded medially but converging both anteriorly and posteriorly, giving overall diamond shape to opening; foramina extending posteriorly to line joining upper M1 alveoli; mesopterygoid fossa long, with small postpalatal projection, reaching line across posterior margins of M3 alveoli; sphenopalatine vacuities either absent or greatly reduced.

Upper incisors slightly opisthodont. Anteroflexus and anteromedian flexus of M1 small and separate anteroloph into anterolabial and anterolingual conules; protoflexus and hypoflexus of M1 typically small. M2 with Protoflexus absent or reduced and hypoflexus short; both paraflexus and metaflexus long and angled sharply, sometimes slightly bifurcated at their inner ends; medial fossette frequently present on M2, labial fossette usually absent.

DISTRIBUTION: *Rhipidomys tribei* is known from the type locality and two others in southeastern Minas Gerais state, Brazil, at middle elevations up to 1,300 m. Specimens with low-FN karyotypes from the hills of southern Espírito Santo state may be referred to this species.

SELECTED LOCALITIES (Map 314): BRAZIL: Espírito Santo, Vargem Alta, Hotel Monte Verde (UFPB 345); Minas Gerais, Santa Bárbara, Reserva Particular do Patrimônio Natural Santuário do Caraça (type locality of *Rhipidomys tribei* B. M. de A. Costa, Geise, Pereira, and L. P. Costa), Viçosa, Mata do Paraíso (MZUFV 386).

SUBSPECIES: *Rhipidomys tribei* is monotypic.

NATURAL HISTORY: Little is known about the ecology of this species. The few available specimens were collected in riparian semideciduous submontane forest patches surrounded by grasslands. The habitat at Serra do Brigadeiro was described by J. C. Moreira et al. (2009).

REMARKS: As mentioned in the account of *R. itoan*, this species and *R. tribei* form a sister-clade pair within the limited phylogeny of Brazilian species that have been compared using mtDNA cytochrome-*b* sequences. This pair of species, in turn, is sister to sequences of *R. leucodactylus* within the larger "leucodactylus" section of the genus (B. M. de A. Costa et al. 2011). This is the taxon referred to as "*R.* sp. 3" by R. G. Rocha, Ferreira, Costa et al. (2011) in their phylogenetic analysis of mtDNA cytochrome-*b* sequences. Specimens from southern Espírito Santo state herein referred to this species have karyotypes of $2n = 44$, $FN = 50$ (Zanchin, Langguth, and Mattevi 1992; Thomazini 2009).

Rhipidomys venezuelae Thomas, 1896
Venezuelan Rhipidomys

SYNONYMS:

Rhipidomys venezuelae Thomas, 1896:303; type locality "Merida, Venezuela, alt. 1630 metres."

R[*hipidomys*]. *v*[*enezuelae*]. *typicus* Thomas, 1900c:271; *nomen nudum*; referring to material from the type locality in comparison with his new subspecies *Rhipidomys venezuelae cumananus* (= *Rhipidomys couesi* herein).

Rhipidomys venezuelae venezuelae: Gyldenstolpe, 1932: 46; name combination.

[*Rhipidomys*] *venezuelanus* Tate, 1939:195; incorrect subsequent spelling of *venezuelae* Thomas.

Rhipidomys mastacalis venezuelae: Cabrera, 1961: 423; name combination.

DESCRIPTION: Medium to large (head and body length usually 130–150 mm) with long tail (110–135% of head and body length); dorsal pelage rich reddish to duller yellowish brown with slight to moderate agouti effect; sides of head and body generally paler; ventral pelage white or cream colored, sometimes with pale grey bases; pelage relatively soft and fine. Tail pale to medium brown, sometimes darker; shaft covered with short hairs, lengthening only close to tip, and terminal pencil short (rarely longer than 6 mm). Ears large relative to body size, pale to medium brown in color. Hindfoot length 26–29 mm; pes moderately broad with proportionally rather short and thick digits; colored pale brown with faint, darker patch over metatarsals. Skull elongated, becoming angular and flattened in older specimens; rostrum broad and nasals parallel sided along posterior half; supraorbital ridges not highly divergent, and interorbit and frontal bones both long; braincase not inflated; mesopterygoid fossa expanded anteriorly, sphenopalatine fissures small or absent; parapterygoid fossae not fenestrated and often recessed; auditory bullae moderate in size. Upper molar row generally 4.9–5.4 mm in length, rather larger in relation to skull length than average for genus.

DISTRIBUTION: *Rhipidomys venezuelae* occurs in northern Colombia and in northwestern and northern Venezuela. It is found at low to lower-middle elevations in the Sierra Nevada de Santa Marta, Serranía de Perijá, Mérida Andes (notably the Chama valley), Serranía de San Luis, and Cordillera de la Costa as far east as Guárico state as well as in the intervening and coastal lowlands.

SELECTED LOCALITIES (Map 323): COLOMBIA: César, El Orinoco, Río César (USNM 280447); La Guajira, Fonseca, Las Marimondas (USNM 280442); Magdalena, Hacienda Cincinati (AMNH 32667). VENEZUELA: Aragua, Estación Biológica Rancho Grande (USNM 517589); Falcón, Cerro Santa Ana, Península de Paranaguá (USNM 456363), La Pastora, 14 km ENE of Mirimire (USNM 442137); Guárico, Parque Nacional Guatopo, 40 km SSE of Caracas (O'Connell 1989); Lara, Caserío Boro, 13 km W of El Tocuyo (USNM 456374); Mérida, Hacienda Santa Catalina, Río Chama (FMNH 21826); Trujillo, Isnotú (USNM 371251); Zulia, Misión Tukuko (USNM 448629).

SUBSPECIES: As currently understood, *Rhipidomys venezuelae* is monotypic.

NATURAL HISTORY: *Rhipidomys venezuelae* inhabits both moist and dry forests in northern South America, its dorsal pelage generally being redder in the former and grayer and more agouti in the latter. Handley (1976) provided habitat data for Smithsonian Venezuelan Project specimens. O'Connell (1989) discussed the population dynamics of *R. venezuelae* (reported as "*R. mastacalis*") in premontane rainforest in Guatopo National Park, on the southern slopes of the Venezuelan coastal range. For parasites, see Ferris (1921), Hopkins (1949), Furman (1972a), P. T. Johnson (1972), E. K. Jones et al. (1972), Machado-

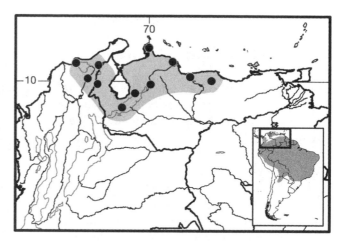

Map 323 Selected locality for *Rhipidomys venezuelae* (●). Contour line = 2,000 m.

Allison and Barrera (1972), Tipton and Machado-Allison (1972), Brennan and Reed (1974a), and Saunders (1975). Guerrero (1985) listed the ectoparasites recorded from all species of Venezuelan *Rhipidomys*.

REMARKS: *Rhipidomys venezuelae* is a member of the "leucodactylus" section of the genus. The taxon has been variously treated as a species in its own right or as a subspecies of *R. mastacalis* (e.g., Cabrera 1961). Tribe (1996) tentatively placed it as a subspecies of *R. latimanus*, to which it is closely allied morphologically (see Remarks under that species). López-Fuster et al. (2001) described cranial allometry in *R. venezuelae*.

A single small specimen from Bonda, on the seaward side of the Sierra Nevada de Santa Marta in northern Colombia, was found by Tribe (1996) to be morphometrically much closer to *Rhipidomys nitela* than to *R. venezuelae*. Voss et al. (2001) considered such an association implausible and suggested that this specimen represented another taxon, possibly allied to *R. venezuelae*. Further material will be needed to resolve this issue.

There have been several reports that a distinct species of *Rhipidomys* inhabits the llanos (Eisenberg et al. 1979; Poulton 1982; O'Connell 1982, 1989; August 1984). Because the only material I have seen from the llanos consists of a possibly mismatched juvenile skull and skin, it is not clear whether these reports involve a new taxon or a geographical variant of *R. venezuelae*, *R. couesi*, or *R. nitela*.

Rhipidomys venustus Thomas, 1900
Mérida Highland Rhipidomys

SYNONYMS:

Rhipidomys venustus Thomas, 1900a:152; type locality "Las Vegas del Chama, alt. 1400 m," Mérida, Venezuela.

Rhipidomys latimanus venustus: Cabrera, 1961:420; name combination.

DESCRIPTION: Medium to large (head and body length 120–145 mm) with moderately long tail (120–140% of head and body length). Pelage soft and dense, bright chestnut brown on dorsum and white-tipped over dark slate hair bases below. Tail sometimes reddish rather than dark, terminating in pencil of hairs 10–15 mm in length. Ears large, well haired, and dark. Hindfeet range from 26–30 mm in length and bear well-defined dark patches over metatarsals; sides of metatarsal portion and toes yellowish white. Skull has narrowed interorbital region with supraorbital ridges slightly more pronounced than in related species in "fulviventer" section; braincase rounded rather than angular; parapterygoid fossae often fenestrated; carotid circulation pattern primitive (large stapedial foramen, internal groove on squamosal and alisphenoid, and sphenofrontal foramen present). Maxillary toothrow averages 5 mm in length (range: 4.81–5.18 mm), average in comparison with skull length for genus.

DISTRIBUTION: *Rhipidomys venustus* is endemic to Venezuela, occurring in the Venezuelan Andes in the states of Mérida and Trujillo at elevations mostly above 2,000 m and potentially also in the adjoining, eastern part of Táchira state. Disjunct populations morphologically referable to this species have been found at elevations above 1,300 m in isolated mountain ranges farther northeast in Falcón and Yaracuy states, and in the Cordillera de la Costa in both Aragua and Vargas states and the Distrito Capital.

SELECTED LOCALITIES (Map 322): VENEZUELA: Aragua, Monumento Natural Pico Codazzi (F. J. García and Sánchez-González 2013); Distrito Capital, Pico Ávila, near Hotel Humboldt (USNM 371236); Falcón, Serranía de San Luis, Parque Nacional Juan Crisóstomo Falcón, Sector Cumbres de Uria, ca. 9 km N of Cabure, ca. 1,320–1,370 m (R. P. Anderson et al. 2012); Mérida, La Coromoto, 7 km SE of Tabay (USNM 387909), Páramo de Mariño, Mérida/Táchira border (Durant and Díaz 1995); Trujillo, Hacienda Misisí (USNM 374528); Yaracuy, Sierra de Aroa, Parque Nacional Yurubí, Sector El Silencio, La Trampa del Tigre (F. J. García and Sánchez-González 2013).

SUBSPECIES: *Rhipidomys venustus* is monotypic.

NATURAL HISTORY: Specimens of *R. venustus* and *R. tenuicauda* captured by the Smithsonian Venezuelan Project (N = 52) were collected in cloud forest (55%), evergreen forest (40%), and orchards (5%); most were found in trees and vines (68%), a few in houses (5%), and the remainder under bushes, on rock ledges, and in a log (Handley 1976; all identified as "*Rhipidomys venustus*"). Other specimens were taken in coffee plantations. Capture elevations were mostly between 2,000 and 3,350 m (specimen tag data), but the elevations of Salomón Briceño Gab-

aldón's capture localities near Mérida as given on specimen tags (from 1,400 to 4,000 m) are questionable. Near Mérida, Aagaard (1982) found the species to be predominantly arboreal, with 80% of captures in trees. Individuals were more likely to come down to the ground in the wetter forests than at higher elevations. Aagaard (1982) also provided data on population ecology and reported that one female had two embryos in the wet season. One USNM specimen had one embryo in September (specimen tag). For parasites, see Ferris (1921), Furman (1972a), P. T. Johnson (1972), E. K. Jones et al. (1972), Machado-Allison and Barrera (1972), Tipton and Machado-Allison (1972), and Brennan and Reed (1974a). Guerrero (1985) listed the ectoparasites recorded from all species of Venezuelan *Rhipidomys*.

REMARKS: *Rhipidomys venustus* is a moderately large species in the "fulviventer" section of the genus. The Táchira Depression (at roughly 1,000 m elevation) separates the range of this species from that of *R. fulviventer* to the southwest. The populations in the Cordillera de la Costa (Vargas to Aragua states) and in the Sierra de Aroá (Yaracuy state) and Sierra de San Luis (Falcón state) are isolated from the main species range in the Mérida-Trujillo Andes by dry lowlands. Further research, preferably using DNA sequences, is needed to determine how closely related these forms are to each other and to other members of the "fulviventer" group. Tribe (1996) treated the two forms in the Mérida-Trujillo Andes and in the Cordillera de la Costa as subspecies of *R. fulviventer*; see also the Remarks under both that species and *R. tenuicauda*.

Rhipidomys wetzeli A. L. Gardner, 1990
Wetzel's Rhipidomys

SYNONYMS:

Rhipidomys wetzeli A. L. Gardner, 1988:744; *nomen nudum.*
Rhipidomys wetzeli A. L. Gardner, 1990:417; type locality "Camp VII (00°50′40″N, 65°58′10″W), 1800 m, Cerro de la Neblina, Territorio Federal Amazonas, Venezuela."

DESCRIPTION: Very small (averaging <100 mm in head and body length) with bright orange dorsal pelage finely streaked with black, sides paler than dorsum, and ventral pelage with cream or pale orange hair tips over dark gray bases. Tail uniformly medium to dark brown, long (130–150% of head and body length), inconspicuously covered with short hair proximally but terminating in long pencil extending approximately 20 mm from terminal caudal vertebra. Ears conspicuously darker than back. Hindfeet short (21–25 mm), with dark patch over metatarsal region that may be restricted to small spot at base of toes. Skull small with rounded interorbit, very slight supraorbital ridges, and broad, globose braincase; carotid circulation pattern primitive (large stapedial foramen, a groove along inner surface of squamosal and alisphenoid, and sphenofrontal foramen); sphenofrontal foramina often contacting frontal bones, in contrast to their position in other "fulviventer" section members; sphenopalatine fissures usually present; paroccipital processes with bifurcated tips. Molar toothrows very short, 3.5–3.8 mm long in northern part of range to 3.8–3.9 mm in south.

DISTRIBUTION: *Rhipidomys wetzeli* is found at middle elevations on the *tepui* table mountains of the Guiana shield. The species is currently known from only a few localities within Parque Nacional Canaima (eastern Bolívar state) and in central and southern Amazonas state, Venezuela; it has been reported from the Guyanese side of Mount Roraima. It may also occur on the Brazilian side of the mountains along the Venezuelan-Brazilian border.

SELECTED LOCALITIES (Map 322): GUYANA: Cuyuni-Mazaruni, slopes of Mount Roraima (Lim, Engstrom, and Ochoa 2005). VENEZUELA: Amazonas, Cerro Duida, Cabecera del Caño Culebra (A. L. Gardner 1990), Cerro Neblina, Camp VII (type locality of *Rhipidomys wetzeli* A. L. Gardner); Bolívar, Churi-tepui, Camp 5 (A. L. Gardner 1990), km 125, 85 km SSE of El Dorado (A. L. Gardner 1990).

SUBSPECIES: *Rhipidomys wetzeli* is monotypic.

NATURAL HISTORY: Specimens of *Rhipidomys wetzeli* were captured at elevations between 1,032 m and 1,850 m. Those from Bolívar state were trapped in cloud forest, mostly on logs above the ground (data from specimen tags). The holotype was lactating when collected in December; its uterus had two embryo scars (A. L. Gardner 1990). For parasites, see Furman (1972a) and P. T. Johnson (1972).

REMARKS: *Rhipidomys wetzeli* is a very small member of the "fulviventer" section of the genus, closely resembling *R. caucensis* from the Colombian Andes except in its primitive carotid circulation pattern and slightly longer tail. The year of publication is often given as 1989 (e.g., Musser and Carleton 2005), but the book in which *wetzeli* Gardner is described was published on February 6, 1990.

Genus *Thomasomys* Coues, 1884

Víctor Pacheco

I recognize 44 species of *Thomasomys* (Pacheco 2003; Voss 2003; Salazar-Bravo and Yates 2007), all of which are endemic to South America in tropical and subtropical habitats. Members of the genus are distributed mainly along the Andean cordillera in premontane and montane forests and páramo, in shrubby and forested habitats, from about 09°N in the department of Trujillo, northern Venezuela to

nearly 18°S in Cochabamba and Santa Cruz, southern Bolivia, including isolated ranges such as the Serranía de la Macarena. The elevational range spans over 3,300 m, from about 1,200 m to above 4,500 m. *Thomasomys* is absent in the Guiana Shield, lowland tropical forests, desert regions, and oceanic islands.

Pacheco (2003) provided details of external and craniodental characters of all *Thomasomys*; the descriptions and terminology that follow stem from this treatise, unless otherwise noted. Species of *Thomasomys* span a considerable range in body size, with head and body length ranging 80–238 mm, and mass 14–335 g. Feet are usually narrow and long, but may be moderately broad and short; length from 14–59 mm. The ears are short or medium sized and hairless, with the length from 12 mm to 33 mm. The tail is slender, medium or long, 85–330 mm, representing from 76%–180% of head and body length. The tail is also naked or finely haired, occasionally with a short pencil of hair at the tip but never as developed as in *Rhipidomys*. The tail is also usually uniform in color but often has a terminal white tip. The hindfoot has relatively moderate or long metatarsals; digits I and V are usually longer than in nonthomasomyine rodents. Digit I of pes (with claw) extends close to or to the interphalangeal joint of dII; dV (with claw) of pes extends about half the length of the second phalanx or to the base of claw of dIV. Digit V is partly opposable in some species. The ventral base of claws on manus is usually closed, although is partially keeled or open in a few species. There are six large or medium plantar pads; the hypothenar and thenar are separated or not by a gap; and the hypothenar and fourth interdigital pad are separated by a gap. The dorsal surface of the metatarsus is uniformly colored, whitish or brownish, but a dark metatarsal patch is present in some species. The feet possess silvery ungual tufts. Female *Thomasomys* have six mammae, in inguinal, abdominal, and thoracic positions. Most *Thomasomys* have a conspicuously protuberant anus, usually approaching the size of the prepuce or larger (first noted by Stone 1914).

Thomasomys is characterized by a dull brown or bright reddish color. The dorsal coloration varies from olivaceous gray, dull olive fulvous, yellowish rufous, orange rufous, golden brown, reddish brown, and grayish brown to dark brown or almost black (Nowak 1999). Dorsal pelage is usually long, dense and soft in contrast to most *Rhipidomys* and *Chilomys*, and uniform except in a few species where a faint mid-dorsal band is conspicuous. The under parts are generally not much paler than the upper parts, with slate-based hairs uniformly extending from the gular to the inguinal region; pectoral spots are rare. Mystacial, superciliary, submental, and interramal vibrissae are present, but genal 1 may be present or absent, and both mystacial and superciliary vibrissae are long or very long, extending slightly or well beyond the pinnae when pressed back. The eyes are small to medium size, typically without an eye-ring. The muzzle is usually the same coloration as the dorsal pelage. A postauricular patch and a pectoral patch are occasionally present.

Adult skulls of *Thomasomys* range from about 24.9 to 50.7 mm in greatest length. In dorsal view, they have short and blunt to moderately long rostra, and a shallow zygomatic notch (depth less than half the width of notch), except in *T. niveipes* and *Thomasomys* sp. 2. Absence of a rostral tube is the common condition, but a moderate or distinct tube also occurs in some species in the genus. The zygomatic plates are usually moderately broad (i.e., the breadth is subequal to the length of M1), although it is narrow (i.e., the breadth is shorter than the length of M1) in a few species; the plate is either vertical or slopes backward conspicuously from its base. The zygomatic arches are variable, converging anteriorly, widely flaring, or almost parallel; squamosal roots of the zygomatic arches may or may not be laterally expanded. The interorbital region is usually narrow (i.e., the maxillary bridge of the molar rows and the base of the zygomatic plates are well to moderately exposed when the skull is seen in dorsal view), but it is moderately broad in some taxa with the interorbital region obscuring these structures (e.g., *T. erro* and few other species). It is usually hourglass in shape, with rounded or square supraorbital margins (distinct supraorbital crests are absent). Parietal ridges are usually absent, but postorbital ridges are conspicuous in some species, particularly those of the "aureus" group. The dorsal profile of the skull (in lateral view) is usually flattened from nasal tips to midfrontal region. The shape of the braincase varies, either globose or oblong, and is usually unmarked by prominent temporal scars or lambdoidal ridges (except in members of the "aureus" group). The interparietal bone is long (in the sagittal plane), usually exceeding 40% of the length of the interparietal suture, and wide with an oblong, half-oblong, or subrectangular shape. The posterodorsal extensions of the premaxillae (in dorsal view) are usually short, terminating anterior or subequal to the nasals (except in the "baeops" group). Ventrally, the skull has short or relatively long incisive foramina that may extend posteriorly between the M1 procingulum. The palatal bridge (*sensu* Hershkovitz 1962) is short, not extending posteriorly beyond the M3s, and usually broad (except in *T. macrotis* and the "aureus" group); its posterior margin is usually rounded, roughly squared, and with or without a small median process. Posterolateral palatal pits are small, simple inconspicuous perforations (without accessory pits and never recessed in shallow fossae), and placed level with or lateral to the anterior margin of the mesopterygoid fossa. The mesopterygoid fossa is usually broad,

with its breadth greater than the breadth of parapterygoid fossa determined at the level of presphenoid-basisphenoid suture; the anterior margin is usually U-shaped and more or less expanded; the sides are subparallel or converge posteriorly; and the bony roof is complete or pierced by short, slit-like to moderate sphenopalatine vacuities flanking the presphenoid-basisphenoid suture. The parapterygoid fossae are triangular, shallow, or moderately deep. Ethmoturbinals are usually large or moderate. An alisphenoid strut that separates the buccinator-masticatory and accessory oval foramina is usually present. The auditory bullae vary from small and uninflated to large and inflated, and are suspended anteriorly by a broad overlap between the tegmen tympani and the posterior suspensory process of the squamosal. The anterodorsal lamina of the ectotympanic does not contact the petrosal, thus producing a gap in most *Thomasomys* as the dorsal margin of the auditory meatus appears to be "open." The thin and long hamular process separates a distinct postglenoid foramen from the subsquamosal fenestra. The postglenoid foramen is comparatively small but may be large and elongated, whereas a conspicuous subsquamosal fenestra extends dorsally to the level of the posterior projection of the squamosal root of the zygomatic arch. The carotid circulation pattern is either primitive (pattern 1) or derived (pattern 3), as indicated by the presence or absence of a stapedial foramen, or an internal squamosal-alisphenoid groove and sphenofrontal foramen, respectively (see Voss 1988). A somewhat square-shaped malleus lamina and a slender processus brevis of incus are exhibited in most *Thomasomys*; the orbicular apophysis of the malleus is usually large.

The mandible ranges from slender to moderately robust; the coronoid process is slightly falciform, delimiting a small sigmoid notch, and is somewhat higher than or level to the condylar process. The angular process is usually shorter than the condylar process and delimits a shallow angular notch. A capsular process may be either present or absent. The anterior edge of the masseteric crest is placed distinctly posterior to the procingulum of m1 in most *Thomasomys*. The basihyal is somewhat straight (not strongly arched) and lacks an entoglossal process.

The upper incisors (*sensu* Hershkovitz 1962) are usually orthodont or slightly opisthodont (slightly proodont in a few species). The molars are pentalophodont, with distinct mesolophs in upper and lower molars, brachydont or moderately hypsodont, and lack well-developed cingula and stylar cusps. The labial and lingual folds of upper molars usually do not interpenetrate more than the toothrow axis, except in a few species that show a comparatively slight interpenetration (e.g., *T. kalinowskii*). The upper molars have three roots, and the lower have two, although M1 sometimes exhibits an accessory labial root. Typically, a deep anteromedian flexus divides the M1 anterocone in roughly two conules, although the flexus is indistinct in a few species; the conules are subequal in size except for the small anterolingual conule exhibited in *T. notatus*. The protostyle and enterostyle are usually absent in M1, although both are conspicuous in some species. M1 lacks additional accessory cusps, although the procingulum exhibits an accessory lophule in some members of the "aureus" group. The anteroloph in M1 is typically distinct and oriented perpendicularly to the labial margin, except in the "aureus" group where is labially and posteriorly oriented. The main cusps of M1 and M2 are arranged in slightly alternating pairs. The anteromedian flexid varies either present (e.g., *T. aureus*) or absent to indistinct, even in unworn dentitions (e.g., *T. ucucha*). A ridge on the anterolabial cingulum of m1 is present in most *Thomasomys* (A. L. Gardner and Romo 1993; Pacheco 2003). The anterolabial cingulid in m2 and m3 varies from distinctly present only in m2 to present and well developed in both molars.

The vertebral count is 7 cervical, 13 thoracic, 6 lumbar, 4 sacral, and 36 or more caudal; with 13 pairs of ribs (Steppan 1995; Pacheco 2003). The anconeal fossa of the distal humerus has a thin translucent lamina that may be pierced (most often) or not by a small perforation. Separate chevron bones are often seen in *Thomasomys*, although these elements may fuse, forming an arch without a spinous process in several species (Steppan 1995; Pacheco 2003). The trochlear process of the calcaneus is generally level to or overlaps with the posterior articular facet. The peroneal process of the fifth metatarsal is moderately long but does not quite extend to the proximal edge of the cuboid.

Most *Thomasomys* species possess a gall bladder (Voss 1991a; Pacheco 2003) and a unilocular-hemiglandular stomach (Carleton 1973; Pacheco 2003), but see exceptions in the species accounts. The soft palate has two complete diastemal and usually five incomplete intermolar ridges (Pacheco 2003). The glans penis is notably variable within the genus (Hooper and Musser 1964; Pacheco 2003). Its body may be short or long, always covered by spines, with shallow or moderate mid-dorsal and midventral grooves; the lateral bacular mounds are usually simple (except by a triple-curved or pointed shape) and extend to near the crater rim or just beyond it (Hooper and Musser 1964; Pacheco 2003). Lateral mounds, bifurcated urethral flaps, and dorsal papilla may or may not bear spines (Pacheco 2003).

Preputial glands vary in number; they may be absent, present as a single pair, or more numerous (Carleton 1980; Voss and Linzey 1981; Pacheco 2003). The medial and lateral ventral prostates are generally of equal size, although either the medial pair (e.g., *Thomasomys* sp. 5 and *T. emeri-*tus [reported as *laniger*]) or the lateral ventral pair (e.g., *T. cinereus*) may be distinctly larger (Carleton 1980; Voss and Linzey 1981; Pacheco 2003).

The known karyotypes of *Thomasomys* are conservative, typically with a diploid number of $2n = 40$, 42, or 44 (except *T. niveipes* with $2n = 24$). The number of autosomal arms varies slightly, with FN = 40, 42, or 44 (A. L. Gardner and Patton 1976; Gómez-Laverde et al. 1997; Aguilera et al. 2000; Salazar-Bravo and Yates 2007), with the exception of *Thomasomys* sp. 4 (FN = 47; Pacheco et al. 2012).

A phylogenetic analysis of thomasomyine rodents based on morphological characters, including all species of *Thomasomys* reported at that time, found a polyphyletic *Thomasomys* that included the species of *Aepeomys* (Pacheco 2003). After *Aepeomys* was removed as a separate genus from *Thomasomys* (see Remarks), Pacheco (2003) proposed that *Thomasomys* could be differentiated into six species groups ("cinereus," "aureus," "baeops," "gracilis," "macrotis," and "notatus") that he hypothesized constituted natural groups at the genus-group level. Further research has revealed a cluster of species allied to *T. incanus* (V. Pacheco, unpubl. data), which is included here as the "incanus" species group.

Pending further taxonomic and phylogenetic work, I tentatively follow my earlier arrangements as a working hypothesis. (1) The "cinereus" species group includes small and medium-sized species distributed throughout the montane forests from the cordilleras of Venezuela, Colombia, to southeastern Bolivia (although most of the species concentrate in the Northern Andes, from Colombia to northern Peru) with long and soft adult pelage, large protuberant anus, genal 1 vibrissae absent, rostral tube absent, auditory bullae inflated or uninflated, small anterior process of ectotympanic, maxillary palatal pits absent, and carotid circulation primitive (except *T. daphne*) with stapedial foramen, squamosal-alisphenoid groove, and sphenofrontal foramen present (pattern 1 of Voss 1988). M2 protoflexus is faint or absent, capsular process is absent or indistinct, glans penis usually without a notch on midventral crater margin, and preputial glands are usually absent.

(2) The "incanus" species group includes medium-sized rodents distributed throughout the montane forests of the Southern Andes (from northern Peru to northern Bolivia, south of the Huancabamba Deflection). This group is similar to the "cinereus" group except that included taxa exhibit a derived carotid circulation, pattern 3 (Voss 1988); and a rostral tube is usually present.

(3) The "aureus" species group includes mainly large-sized rodents distributed throughout the montane forests from the cordilleras of Venezuela and Colombia to southeastern Bolivia. The dorsal pelage is long and soft and generally brown reddish; hypothenar and thenar pad of pes

level or overlapping; digit I of pes (hallux) is long, extending to the first interphalangeal joint of digit II or beyond; digit V is very long, extending to the base of claw of digit IV or beyond; genal 1 vibrissae are present; mystacial vibrissae are very long, extending well beyond the posterior margin of the pinnae to a point greater than the length of one pinna; and the body pelage pattern is moderately countershaded. Their skulls are characterized by a narrow and convergent interorbital region, not beaded; a large postglenoid foramen, with the anterior margin at level with the tegmen tympani or extended forward; incisive foramina long extending between the M1 procingula; molar-bearing portion of the maxillae rectangular; carotid circulation pattern derived, pattern 3 (*sensu* Voss 1988); upper incisors opisthodont (*sensu* Hershkovitz 1962); the anteroloph labially and posteriorly oriented; M1 paraloph oriented backward to the mesoloph; the molar hypoflexus broad; and lower molars with a distinct anteromedian flexid and broad and curved posteroflexids. The stomach morphology has an intermediate incisura angularis; and preputial glands are numerous.

(4) The "baeops" species group includes small to medium-sized rodents distributed throughout the montane forests of the northern Andes (from Colombia to northern Peru, mostly north of the Huancabamba Deflection). The pelage is comparatively short, and generally grayish or brownish; genal 1 vibrissae are present; and the hindfeet are short and relatively broad. The skull is characterized by a short rostrum, narrow interorbital region, derived carotid circulation (pattern 3, *sensu* Voss 1988); mesopterygoid fossa breadth subequal; anterior process of the ectotympanic small; and distal portion of the lateral bacular mounds pointed.

(5) The "gracilis" species group includes small or medium-sized rodents distributed throughout the montane forests of the southern Andes (from northern Peru to Bolivia, south of the Huancabamba Deflection). The pelage is comparatively moderate or usually long, dense, and soft, and generally brownish; hindfeet are short and relatively broad. The skull has a short to moderate rostrum and narrow interorbital region. The carotid circulation is primitive (pattern 1, *sensu* Voss 1988); and bullae are either inflated with indistinct bony tubes or moderately inflated with short bonytubes.

(6) The "macrotis" species group includes only *T. macrotis*, a large mouse with comparatively large ears that is distributed only in the isolated mountains of the Cordillera Oriental of Peru, between the Río Marañón and Huallaga, south of the Huancabamba Deflection. This species group shares characteristics with both the "incanus" and "aureus" groups. The pelage is soft and long; genal 1 vibrissae are absent; the hindfoot is long and relatively broad;

the ventral base of manual claws are medially keeled; and the tail is moderately long with more than one third of its length white. The skull is characterized by a long rostrum, derived carotid circulation (pattern 3 of Voss 1988), a palate narrow, short premaxillae that terminate anterior to nasals and the zygomatic notch, large molars with the flexus(ids) with an incipient crenulation, and two lingual roots on M1.

(7) The "notatus" species group includes only *T. notatus*. These are medium-sized rodents distributed throughout the montane forests of the central Andes (north and central Peru, south of the Huancabamba Deflection). The dorsal pelage is comparatively short, generally orangish brown, and distinctly countershaded; hindfeet are short and relatively broad; and the tail is moderately long and with a terminal pencil. The skull has a short rostrum and narrow interorbital region, the primitive carotid circulation, a long postglenoid foramen that projects anterior to the tegmen tympani, sphenopalatine vacuities with moderately large perforations, uninflated bullae with distinct bony tubes, and a rectangular-shaped molar-bearing portion of the maxillae. The anterocone also presents a small anterolingual conule. Preputial glands are absent, and the glans penis has shallow middorsal and midventral grooves and large and spiny lateral bacular mounds that are triple curved (Hooper and Musser 1964; Pacheco 2003).

Apparently the Huancabamba Deflection has been a key geographic barrier that limits the distribution of several of these hypothetical species groups.

As mentioned previously, *Thomasomys* species exhibit considerable morphological variation that makes the diagnosis and determination of species and species groups difficult. Karyotype and molecular data remain limited despite the pivotal role these types of data have played in discriminating certain taxa and allying others (e.g., Gómez-Laverde et al. 1997; M. F. Smith and Patton 1999; Salazar-Bravo and Yates 2007). Much more work in the fields of genetic and molecular systematics, in conjunction with basic and detail morphological description, is needed to test the hypotheses of species and species groups presented here. The lack of complete skeletons and specimens preserved in fluid for many species have limited access to key morphological characters (e.g., glans penis, reproductive tract, stomach morphology); greater attention is needed to increase this type of preserved material.

SYNONYMS:

Hesperomys: Tomes, 1860a:219; part (description of *aureus*); not *Hesperomys* Waterhouse.

Hesperomys (*Rhipidomys*): Thomas, 1882:108; part (description of *cinereus* and *taczanowskii*); neither *Hesperomys* Waterhouse nor *Rhipidomys* Tschudi.

Hesperomys (*Vesperimus*): Thomas, 1884:449; part (listing of *cinereus* and *taczanowskii*); neither *Hesperomys* Waterhouse nor *Vesperimus* Coues.

Thomasomys Coues, 1884:1275; as subgenus of *Hesperomys*; type species *Hesperomys cinereus* Thomas, by original designation.

Oryzomys: Thomas, 1894b:349; part (description of *incanus* and *kalinowskii*); not *Oryzomys* Baird.

Peromyscus (*Thomasomys*): E.-L. Trouessart, 1897:512; part (listing of *cinereus* and *kalinowskii*); not *Peromyscus* Gloger.

Aepeomys: Thomas, 1898e:452; part (description of *vulcani*); not *Aepeomys* Thomas.

Thomasomys: Thomas, 1906c:443; raised to rank of genus.

Erioryzomys Bangs, 1900a:96; type species *Oryzomys monochromos* Bangs, by original designation.

Inomys Thomas, 1917d:197; type species *Oryzomys incanus* Thomas, by original designation.

REMARKS: *Thomasomys* is a speciose and taxonomically complex genus whose nomenclatural history is intertwined with *Aepeomys*, *Delomys*, and *Wilfredomys*, taxa that have been viewed as subgenera (e.g., Ellerman 1941; Cabrera 1961; Pine 1980) or as genera (see their accounts). *Thomasomys* has subsequently been restricted (Voss 1993; E. M. González 2000) to a smaller but still speciose group that is endemic to tropical Andean cloud forests. The center of diversity for the genus apparently includes eastern Ecuador (Voss 2003).

With very limited taxon sampling in molecular phylogenetic analyses, both M. F. Smith and Patton (1999) and D'Elía et al. (2003) found *Thomasomys* to the sister genus to *Chilomys*. Alternatively, Pacheco's (2003) cladistic analysis derived from morphological characters supported a polyphyletic *Thomasomys*, with species members of the "baeops," "gracilis," and "notatus" species groups connected to a monophyletic *Rhipidomys*. The phylogenetic analysis of Salazar-Bravo and Yates (2007), based on mtDNA cytochrome-*b* sequences from 15 taxa of *Thomasomys*, revealed some congruence with Pacheco (2003). One clade composed of *T. australis*, *T. daphne*, *T. caudivarius*, and *T. cinnameus* is consistent with my "cinereus" species group; a clade of *Thomasomys* sp. 9, *T. oreas*, and *T. andersoni* is consistent with my "gracilis" species group (although *T. gracilis* was found unrelated); the species pair of *T. aureus* and *T.* sp. 2 is consistent with my "aureus" species group; and a strongly supported clade of *T. baeops* and *T. taczanowskii* is consistent with my "baeops" species group. Disagreements between molecular versus the morphological hypotheses might be due to poor taxonomic coverage (e.g., 15 taxa versus 53 taxa, respectively) and/or misidentifications. Should future studies corroborate the monophyly of the species groups I recognize, it would be appropriate to grant generic status to each of these, an action that would parallel the recent division of *Oryzomys sensu lato* (e.g., Weksler et al. 2006; see also Weksler and Percequillo 2011). Except for the availability of *Thomasomys*, for which *cinereus* Thomas is the type species, and *Inomys* Thomas, for which *incanus* Thomas is the type species, no other genus-group names are currently available for the other species groups.

Species recognized herein basically follow Voss (2003), Pacheco (2003), Musser and Carleton (2005), and Salazar-Bravo and Yates (2007). Very little is known about fossils of *Thomasomys*. Ficcarelli et al. (1992) reported *Thomasomys* sp. as part of a micromammal assemblage of Lujanian Mammal Age in the upper horizon of the Cangahua Formation, Quebrada Cuesaca, Bolívar province, Ecuador.

Ectoparasites or endoparasites are reported for each species in the accounts that follow where these parasites can be unambiguously assigned to a host species. Citations listing only *Thomasomys* sp. are treated here. Jordan (1950) reported the flea *Plocopsylla hector* Jordan from *Thomasomys* sp. from Loja. Pozo et al. (2005) reported the fleas *Polygenis litargus* (Jordan and Rothschild) and *Craneopsylla minerva* (Rothschild) in *Thomasomys* sp. from Ayabaca, Piura. Hastriter and Schlatter (2006) reported *Dasypsyllus* (*Neornipsyllus*) *lewisi*, a new species of flea, on *Thomasomys* sp. from Huancabamba, Piura, Peru, collected by C. Kalinowski, but considered the host a likely accidental association with an undetermined species of ground dwelling/nesting bird. Based on species collected by C. Kalinowski from that area, *Thomasomys* sp. represents either *T. cinereus* or *T. taczanowskii*. Seevers (1955) reported the staphylinid beetle *Amblyopinus sanborni* Seevers off *Thomasomys* sp. from Marcapata, Limapunco, Cuzco (FMNH 75369); specimen currently determined as *Nephelomys levipes* (Thomas).

Durette-Desset (1970b) reported *Longistriata thomasomysi* as a new species of nematode from *Thomasomys* sp. from Pichiude, Valle del Cauca, Colombia. Lack of additional data impede the species determination of the host, although I think the author refers to the locality of Pichindé, a town above the Cauca Valley on lower eastern slope of western Andes near Cali (Paynter 1997). To my knowledge, the only specimens of *Thomasomys* captured in the Pichindé valley and vicinity are those of *T. popayanus* (reported as *T. aureus*) and *T. cinereiventer* (Trapido and Sanmartín 1971), which are proposed as hypothetical hosts. The current name of the nematode is *Hypocristata thomasomysi* (Durette-Desset and Guerrero 2006).

KEY TO THE SPECIES OF *THOMASOMYS*:

1. Molars comparatively large, palatal bridge narrow (except *T. nicefori*); genal 1 vibrissae present (except *T. macrotis*) 2

1'. Molars comparatively normal, palatal bridge broad; genal 1 present or absent 3

2. Genal 1 vibrissae absent, auditory bullae small and uninflated; postglenoid foramen comparatively small and not extended in front of the anterior margin of tegmen tympani; upper molars with normal anteromedian flexus, and narrow lingual and labial flexus; first lower molar with anteromedian flexid shallow; lower molars with normal posteroflexids .... *Thomasomys macrotis*

2'. Genal 1 vibrissae present; auditory bullae small and uninflated or large and inflated; postglenoid foramen large, long, and extended to level or in front of the anterior margin of tegmen tympani; upper molars with deep anteromedian flexus, unusually wide hypoflexus, broad mesoflexus and metaflexus; first lower molar with distinct anteromedian flexid; first and second lower molars with broad and curved posteroflexids ....
.... ("aureus" group) 7

3. Carotid circulation pattern derived (pattern 3, *sensu* Voss 1998) 4

3'. Carotid circulation pattern primitive (pattern 1, *sensu* Voss 1998) (except *Thomasomys daphne*) 5

4. Second and fifth digits of manus subequal (*sensu* Pacheco 2003); thenar and hypothenar pads of pes not separated by a gap; digit I of pes (with claw) long, extending to the interphalangeal joint of digit II, and digit V of pes (with claw) very long, extending to the base of claw of digit IV (*sensu* Pacheco 2003); genal 1 vibrissae present or absent; mystacial vibrissae very long, extending well beyond the posterior margin of pinnae to a point greater than the length of one pinna when bent; squamosal roots of zygomatic arch not expanded laterally, joining the braincase at oblique angle; mesopterygoid fossa subequal in breadth (not broad not narrow) with subparallel margins; wear surface of upper incisors more or less flat, facing backward (*sensu* Pacheco 2003); alveolus of first upper molar posterior to the posterior margin of the zygomatic plate; molars brachydont with weak labial cingula; m1 anteromedian flexi present; capsular process present; supratrochlear fenestra of humerus absent; distal portion of the lateral bacular mounds of phallus pointed (Pacheco 2003); preputial glands one pair. .....
............................("baeops" group) 15

4'. Second digit of manus distinctly longer than fifth (*sensu* Pacheco 2003); thenar and hypothenar pads of pes separated by a gap in most cases; digit I of pes (with claw) moderate, extending close to the interphalangeal joint of digit II, and digit V of pes (with claw) long, ex-tending about half the length of the second phalanx of digit IV (*sensu* Pacheco 2003); genal 1 vibrissae absent; mystacial vibrissae long, extending just beyond posterior margins of pinna when bent; squamosal roots of zygomatic arch laterally expanded, joining the braincase more perpendicularly; mesopterygoid fossa broad with subparallel or anteriorly expanded margins; wear surface of upper incisors oriented medially (*sensu* Pacheco 2003); alveolus of first upper molar anterior or subequal to the posterior margin of the zygomatic plate; molars moderately hypsodont; m1 anteromedian flexi absent; capsular process absent; supratrochlear fenestra of humerus present; lateral bacular mounds of phallus simple (Pacheco 2003); preputial glands absent ......
.... ("incanus" group) 16

5. Genal 1 present; dorsal pelage comparatively short; body pelage pattern distinctly countershaded; auditory bullae small and uninflated; capsular process distinct; molars brachydont even when unworn; no preputials; lateral bacular mounds of phallus "triple curved" (Pacheco 2003) .... *Thomasomys notatus*

5'. Genal 1 present or absent; dorsal pelage comparatively moderate or usually long; body pelage pattern moderately or not countershaded; auditory bullae large and inflated, or small and moderately inflated; capsular process present or absent; moderately hypsodont or brachydont even when unworn; preputials present or absent; lateral bacular mounds of phallus simple or triple curved (Pacheco 2003) 6

6. Genal 1 present; second and fifth digits of manus subequal; squamosal roots of zygomatic arch not flared laterally, joining the braincase at oblique angle; zygomatic arches project anteriorly producing a more parallel outline; capsular process present or absent; molars brachydont with weak labial cingula; wear surface of upper incisors more or less flat, facing backward; margins of lophs, lophids, flexus, and flexids crenulated; m3 mesolophid distinct; protostyle and/or enterostyle on M1 present; one pair of preputial glands; lateral bacular mounds of phallus simple or triple curved (Hooper and Musser 1964; Pacheco 2003) ....
...........................("gracilis" group) 20

6'. Genal 1 absent; second digit of manus distinctly longer than fifth; squamosal roots of zygomatic arch flared laterally, joining the braincase more perpendicularly; zygomatic arches convergent anteriorly producing a triangular outline in most cases; capsular process usually absent; molars moderately hypsodont (frequent condition) or brachydont with weak labial cingula; wear surface of upper incisors oriented medially; margins of lophs, lophids, flexus, and flexids usually not crenulated; m3 mesolophid coalesced to entoconid; protostyle

and/or enterostyle on M1 absent; preputial glands usually absent (when present are numerous); lateral bacular mounds of phallus simple (Pacheco 2003). . . . .
. . . . ("cinereus" group) 22

7. Body size very large (adult head and body length >235 mm); adult upper molar row >9.9 mm; zygomatic notch very shallow; about a third of terminal tail whitish. . . . .*Thomasomys apeco*

7′. Body size large or medium (adult head and body length <200 mm); adult upper molar row <8.5 mm; zygomatic notch shallow, tail usually unicolor. 8

8. Auditory bullae inflated, bony tube indistinct 9

8′. Auditory bullae small and not largely inflated, bony tube distinct 10

9. Auditory bullae comparatively smaller; sphenopalatine vacuities absent or slits; anterior edge on the procingulum of M1 usually present. . . . *Thomasomys auricularis*

9′. Auditory bullae largely swollen; sphenopalatine vacuities distinct and moderately large; anterior edge on the procingulum of M1 usually absent . . . .
. . . . *Thomasomys pyrrhonotus*

10. Dorsal pelage overall grayish; first and second lower molars with entolophids oriented medially joining the murid, first and second lower molars with mesolophids short or indistinct . . . .
. . . . *Thomasomys praetor*

10′. Dorsal pelage overall brownish; first and second lower molars with entolophids oriented forward joining the mesolophids, first and second lower molars with distinct mesolophids 11

11. Skull with moderately broad interorbital region, palatal bridge broad . . . . *Thomasomys nicefori*

11′. Skull with narrow interorbital region, palatal bridge narrow as usual for the group. 12

12. Lambdoid ridges present, well conspicuous. 13

12′. Lambdoid ridges indistinct. 14

13. Skull with posterior extension of nasals not extended beyond premaxillae; premaxillomaxillary-frontal intersection placed subequal to zygomatic notch; anterior process ectotympanic large; alveolus of first upper molar subequal to the posterior margin of the zygomatic plate; m1 with ectolophid absent; large specimens (head and body length 160–185 mm) with long tail (125–141% of head and body length) . . . .
. . . . *Thomasomys aureus*

13′. Skull with posterior extension of nasals usually longer than premaxillae; premaxillomaxillary-frontal intersection placed anterior to zygomatic notch; anterior process ectotympanic comparatively small; alveolus of first upper molar anterior to the posterior margin of the zygomatic plate; m1 with ectolophid present; comparatively larger specimens (head and body length 173–

198 mm) with shorter tail (113–131% of head and body length) . . . .*Thomasomys princeps*

14. Body pelage pattern no countershaded; premaxillae slightly longer than nasals in dorsal view; interorbital region narrow with square margins; postorbital ridge incipient; ethmoturbinals moderately developed; incisive foramina broad and long reaching level of M1 plane or slightly between; mesopterygoid fossa broad not greatly expanded anteriorly with closed roof; auditory bullae small, distinct bony tubes; anterior process of ectotympanic large; M1 without an additional anterior edge on the procingulum; M2 protoflexus faint or absent . . . . . .
. . . . *Thomasomys popayanus*

14′. Body pelage pattern moderately countershaded; premaxillae level nasals in dorsal view; interorbital region narrow with marked margins; postorbital ridge distinct; ethmoturbinals small; incisive foramina broader and longer extending between M1 almost to the posterior margin of anterocone; mesopterygoid fossa broad, distinctly expanded anteriorly and with moderately developed sphenopalatine vacuities; auditory bullae small but moderately inflated, bony tubes less distinct; anterior process of ectotympanic smaller; M1 with an additional anterior edge on the procingulum; M2 protoflexus distinct . . . . *Thomasomys rosalinda*

15. Interorbital region narrow (5.15–5.76 mm); auditory bullae small and uninflated; zygomatic arches narrow producing a somewhat parallel outline; incisive foramina narrow with extremes contracted and moderately long, extending to M1 anterior border, not between. . .
. . . .*Thomasomys baeops*

15′. Interorbital region moderately broad (5.38–6.51 mm); auditory bullae moderate size and inflated; zygomatic arches converging anteriorly; incisive foramina narrow extending to slightly between M1 anterocones . . . .
. . . . *Thomasomys taczanowskii*

16. Skull with distinct rostral tube; position of antorbital bridge low (ca. >25% of rostrum); interorbital region broad (most of molar row is concealed in dorsal view); optical foramen small (placed behind M3) . . .
. . . . 17

16′. Skull with rostrum not distinctly projected, rostral tube absent; position of antorbital bridge high (<25% of rostrum); interorbital region moderately broad (molar row is partially seen in dorsal view); optical foramen small (placed behind M3) or larger (level or partially overlap M3) 18

17. Parietosupraoccipital suture comparatively broad (suture is about one-third the posterior margin of parietal); body size small (adult head and body length usually <125 mm); upper molar row <5.1 mm. . . . .
. . . . *Thomasomys incanus*

17'. Parietosupraoccipital suture narrow (length of suture is less than one-fourth of the posterior margin of parietal); body size medium (adult head and body length usually >130 mm); upper molar row >5.1 mm.......
............................*Thomasomys ischyrus*

18. Body pelage moderately countershaded; optic foramen behind M3; the molar-bearing portion of the maxillae near triangular; molars normal, without interpenetrating flexi.................... *Thomasomys eleusis*

18'. Body pelage no countershaded; optic foramen level or overlap M3; the molar-bearing portion of the maxillae near rectangular; molars with interpenetrating flexi ...
....19

19. Body pelage brownish; mesopterygoid fossa broad and posteriorly convergent; upper incisors opisthodont....
........................*Thomasomys kalinowskii*

19'. Body pelage dark grayish; mesopterygoid fossa broad with subparallel margins; upper incisors orthodont ...
..........................*Thomasomys ladewi*

20. Auditory bullae small but moderately inflated; tail with small but conspicuous terminal tuff of long hairs; mesopterygoid fossa with closed roof; upper incisors orthodont.............. *Thomasomys andersoni*

20'. Auditory bullae large and inflated; tail lack pencil of long hairs; mesopterygoid fossa with moderately open sphenopalatine vacuities; upper incisors opisthodont ...
....21

21. Small size; body pelage pattern no countershaded; interfrontal fontanella absent; accessory labial root of M1 present; capsular process absent; glans penis body with shallow groove; lateral bacular mounds of phallus simple (Pacheco 2003), nonspined; spinous urethral flaps nonspined; gall bladder present ....
..........................*Thomasomys gracilis*

21'. Medium size; body pelage pattern moderately countershaded; interfrontal fontanella present; accessory labial root of M1 absent; capsular process present; glans penis body with deeper groove, extending to half of body; lateral bacular mounds of phallus "triple curved" (Pacheco 2003) and spined; spinous urethral flaps spined; gall bladder absent .... *Thomasomys oreas*

22. Angular process of mandible large, somewhat level condylar process in lateral view 23

22'. Angular process of mandible short, placed in front of posterior margin of condylar process in lateral view...
....27

23. Molar with interpenetrating flexi; incisive foramina moderate close or level M1 anterior plane; upper incisors opisthodont; molars with incipient crenulation 24

23'. Molars normal, without interpenetrating flexi; incisive foramina short, not close to the anterocone M1;

upper incisors usually orthodont; molars without trace of crenulation 25

24. Medium size (head and body length 112–130 mm, maxillary toothrow 4.7–5.5 mm); body pelage pattern no countershaded; molar bearing maxillae rectangular; interorbital region with rounded margins; upper incisors orthodont ....*Thomasomys bombycinus*

24'. Large size (head and body length 100–147 mm, maxillary toothrow 5.4–6.2 mm); body pelage pattern moderately countershaded; molar bearing maxillae triangular; interorbital region with incipient square margins; upper incisors opisthodont ...... *Thomasomys cinereiventer*

25. Feet with a gap between the thenar and hypothenar pads; m1 anteromedian flexi distinct 26

25'. Feet without a gap between the thenar and hypothenar pads; m1 anteromedian flexi absent ....
..........................*Thomasomys vestitus*

26. The fifth digit (dV) of pes moderately long; rostral tube present; zygomatic plate narrow, breadth shorter than the length of M1; optic foramen small and posterior to M3; palatal medial process present; auditory bullae small and noninflated .... *Thomasomys dispar*

26'. The fifth digit (dV) of pes long; premaxillae extends anteriorly beyond the incisors by not more than the depth of the incisors; breadth of zygomatic subequal to the length of M1; optic foramen level or overlap posterior border of M3; palatal medial process absent or weak; auditory bullae small but moderately inflated...
..........................*Thomasomys contradictus*

27. Zygomatic plate narrow, breadth shorter than the length of M1..................................28

27'. Zygomatic plate comparatively broad, breadth subequal to M1 or broader 32

28. Skull with distinct rostral tube 29

28'. Skull with rostrum not distinctly projected, rostral tube absent..................................30

29. Body size very small (adult head and body length <95 mm); upper molar row <3.6 mm; skull with peculiar rostrum, with a concave dorsal profile; auditory bullae inflated and bony tubes indistinct................
..........................*Thomasomys hudsoni*

29'. Body size medium (adult head and body length >95 mm); upper molar row >4.5 mm; skull with frontonasal profile somewhat straight, auditory bullae uninflated and bony tubes distinct or poorly distinct......
..........................*Thomasomys erro*

30. Body size very small (adult head and body length usually <95 mm); upper molar row <4.1 mm ....
..........................*Thomasomys cinnameus*

30'. Body size medium (adult head and body length >95 mm); upper molar row >4.4 mm.................31

31. Incisive foramen broad, with anterior and posterior margins contracted; M3 proportionally normal; short-tailed species (tail length usually <115% of head and body length) . . . . *Thomasomys fumeus*

31′. Incisive foramen oval and elongated; M3 comparatively smaller; long tail species (tail length usually >115% of head and body length) . . . . *Thomasomys silvestris*

32. Skull with auditory bullae large and inflated 33

32′. Skull with auditory bullae small or moderate, not inflated 36

33. Mystacial vibrissae very long, distinctly beyond posterior margins of pinnae when bent; rostral tube absent 34

33′. Mystacial vibrissae long, extending just beyond posterior margins of pinna when bent; premaxillae and nasals projected forward, incipient rostral tube present 35

34. Antorbital bridge low, usually 25% of rostrum; sphenopalatine vacuities moderately developed; upper incisors opisthodont . . . . *Thomasomys emeritus*

34′. Antorbital bridge higher, less than 25% of rostrum; sphenopalatine vacuities absent or present as slits; upper incisors more orthodont . . . . . *Thomasomys laniger*

35. Skull with zygomatic notch deeper; interorbital region narrow; interfrontal fontanella usually present; ethmoturbinals moderately developed; postglenoid foramen large and projected forward in front of tegmen tympani; palatal medial process absent; anterior process ectotympanic distinct . . . . *Thomasomys niveipes*

35′. Skull with zygomatic notch shallow; interorbital region moderately broad; interfrontal fontanella usually absent; ethmoturbinals large; postglenoid foramen smaller and shorter than level of tegmen tympani, palatal medial process present; anterior process ectotympanic small . . . .*Thomasomys paramorum*

36. Upper incisor proodont, procumbent 37

36′. Upper incisor usually orthodont, nonprocumbent 38

37. Tail usually lacking terminal white tip; palatal medial process absent or weak; primitive carotid circulatory pattern; capsular process absent or shelf . . . .*Thomasomys australis*

37′. Tail with a terminal white tip; palatal medial process present; derived carotid circulatory pattern; masseteric crest position subequal, capsular process distinct . . . . *Thomasomys daphne*

38. The m1 anteromedian flexi present 39

38′. The m1 anteromedian flexi absent 41

39. Body pelage pattern moderately countershaded; tail usually lack terminal white tip . . *Thomasomys cinereus*

39′. Body pelage pattern no countershaded; tail with a terminal white tip 40

40. Rostral tube absent; ethmoturbinals moderately developed . . . . *Thomasomys hylophilus*

40′. Premaxillae moderately projected, extending anteriorly beyond the incisors by not more than the depth of the incisors; ethmoturbinals larger . . . . *Thomasomys caudivarius*

41. Body pelage pattern moderately countershaded; ventral base of claw on manus partially keeled; position of antorbital bridge low (approx. >25% of rostrum); mesopterygoid fossa broad, more open anteriorly; zygomatic plate slope backward; optical foramen small (placed behind M3) . . . . *Thomasomys vulcani*

41′. Body pelage pattern no countershaded; claws on manus closed at base; position of antorbital bridge high (approx. <25% of rostrum); mesopterygoid fossa broad with subparallel margins; zygomatic plate vertical; optical foramen at level or partially overlap M3 42

42. Upper incisors proodont; molar-bearing maxillae rectangular; capsular process distinct . . . . *Thomasomys ucucha*

42′. Upper incisors orthodont; molar-bearing maxillae triangular; capsular process absent or shelf 43

43. Gap present between hypothenar and thenar pads of pes; interfrontal fontanella present; molars more brachydont; M1 anteroloph indistinct; hypoflexus M3 indistinct; M1 with accessorylabial root . . . . *Thomasomys monochromos*

43′. Hypothenar and thenar pads of pes closer, no gap; interfrontal fontanella absent; molars moderately hypsodont; M1 anteroloph distinct; hypoflexus M3 distinct; M1 lacks accessory labial root . . . . *Thomasomys onkiro*

The Key is based on original species descriptions, recent literature (e.g., L. Luna and Pacheco 2002; Voss 2003; Salazar-Bravo and Yates 2007), and my character descriptions and character matrix presented in the morphological phylogeny of thomasomyine rodents (Pacheco 2003). However, the hierarchical order in which characters are presented is not strictly phylogenetic. This key also includes my unpublished and ongoing research. Although the species of *Thomasomys* are treated alphabetically in the accounts that follow, the species I recognize can be grouped into seven species groups that hypothetically represent natural phylogenetic units; five of these are polytypic, the remaining two are monotypic: (1) "cinereus" species group, which includes *T. australis, T. bombycinus, T. caudivarius, T. cinereiventer, T. cinereus, T. cinnameus, T. contradictus, T. daphne, T. dispar, T. eleusis, T. erro, T. emeritus, T. fumeus, T. hudsoni, T. hylophilus, T. laniger, T. monochromos, T. niveipes, T. onkiro, T. paramorum, T. silvestris, T. ucucha, T. vestitus,* and *T. vulcani*; (2) "incanus" species group, including *T. eleusis*, *T. incanus*, *T. ischyrus*, *T. kalinowskii*, and *T. ladewi*; (3) "aureus" species group, including *T. apeco*, *T. aureus*, *T. auricularis*, *T. nicefori*, *T. popayanus*, *T. praetor*, *T. princeps*, *T. pyrrhonotus*, and *T. rosalinda*; (4) "baeops" species group, including only *T. baeops* and *T. taczanowskii*; (5) "gracilis" species group, including *T. andersoni*, *T. gracilis*, and *T. oreas*; (6) the monotypic "macrotis" species group, including only *T. macrotis*; and (7) the monotypic "notatus" species group, including only *T. notatus*.

Finally, the quantitative and qualitative descriptions of species that follow are in large part based on my personal examination of type material and additional series, original descriptions, and relevant literature (e.g., Voss 1996, 2003; Gómez-Laverde et al. 1997; Salazar-Bravo and Yates 2007). I also comment on population samples that appear to represent currently undescribed species in the Remarks section of the appropriate account. As a more detailed species description, including postcranial and soft anatomy will be presented elsewhere, this work does not constitute as yet a proper revision of the genus.

Thomasomys andersoni Salazar-Bravo and Yates, 2007
Anderson's Thomasomys

SYNONYM:

Thomasomys andersoni Salazar-Bravo and Yates, 2007: 752; type locality "an elfin forest near the headquarters of the Corani hydroelectric plant (17°12′43″S, 65°52′09″W, GPS coordinates, map datum WGS 84) at 2,630 m, Department of Cochabamba, Bolivia."

DESCRIPTION: Medium sized (mean head and body length 109 mm; condyloincisive length of skull 27.21 mm) with soft, dense, and comparatively long dorsal fur (mean length = 11 mm). Dorsal coloration dull brownish olive; ventral coloration olive buff with yellowish tinge and distinct yellow pectoral marking. Ears comparatively large (20–21 mm) and blackish brown. Mystacial vibrissae moderately long, extending slightly beyond posterior margin of pinnae when bent; genal 1 vibrissae present. Tail slightly longer than head and body (113–116%) with short terminal pencil (length 8–10 mm). Hindfoot relatively short (22–26 mm) and broad, with dorsal chocolate-brown metatarsal patch that contrasts with white digits. Hallux relatively short, not extending beyond metatarsal of dII, and dV long (claw of dV extending about half length of phalanx 2 of dIV). Skull of moderate proportions with narrow but not long rostrum, and oblong braincase and straight frontonasal profile; interorbital region narrow with rounded supraorbital margins, without ridges; zygomatic plates broad and vertically oriented; zygomatic arches converge anteriorly; incisive foramina long and relatively narrow, extending posteriorly slightly between M1s; auditory bullae small

but moderately inflated (ventral margins of bullae extend slightly ventral to hamular process of pterygoids when skull viewed in horizontal plane, bony tubes short (*sensu* Pacheco 2003), and nearly flask shaped; and orbicular apophysis of malleus comparatively small (not enlarged), not basally constricted. Carotid circulatory pattern primitive (large stapedial foramen, groove the inner surface of the squamosal and alisphenoid, and sphenofrontal foramen). Maxillary toothrow short (4.61 mm in average); molars bunodont and brachydont with weak labial cingula; first upper molar with accessory labial root present; upper incisors orthodont and yellowish; and capsular process comparatively small.

DISTRIBUTION: *Thomasomys andersoni* occurs on the eastern Andean slopes of the Bolivian Andes (La Paz and Cochabamba departments), between 2,000 and 2,630 m.

SELECTED LOCALITIES (Map 324): BOLIVIA: Cochabamba, near headquarters of the Corani hydroelectric plant (type locality of *Thomasomys andersoni* Salazar-Bravo and Yates); La Paz, 30 km by road N of Zongo, cement mine (UMMZ 156298).

SUBSPECIES: *Thomasomys andersoni* is monotypic.

NATURAL HISTORY: *Thomasomys andersoni* inhabits the upper montane rainforest and appears to be arboreal (Salazar-Bravo and Yates 2007). The two specimens from Corani were trapped at about 2 m above ground in a short tree at the bottom of a small ravine. The type locality has a flora dominated by the species *Ocotea jelskii* (Lauraceae) and *Podocarpus oleifolius* (Podocarpaceae). The climate of the region is very humid, with daily fog and mist, and low temperatures.

REMARKS: This species was originally reported only from the type locality (Salazar-Bravo and Yates 2007), but one specimen from 30 km north of Zongo in La Paz department (UMMZ 156298) extends the range to a second locality about 285 km to the northwest. S. Anderson (1997) incorrectly reported the Zongo specimen as *Rhipidomys nitela*; subsequently Voss et al. (2001) viewed this animal as resembling *T. oreas*. Based on the comparatively long dorsal pelage, lack of a dorsal band, small and little inflated auditory bullae, and lack of sphenopalatine vacuities, I regard this specimen as allied to *T. andersoni* and not to *T. oreas*, although other features (such as a midpalate process and wider incisive foramina) might suggest the Zongo specimen represents an unnamed taxon.

In a phylogenetic analysis based on the mtDNA cytochrome-*b* gene, *T. andersoni* was placed as sister to *T. oreas*, which supports its placement in the "gracilis" group. The species, however, does share morphological characters with members of both the "gracilis" and "no-

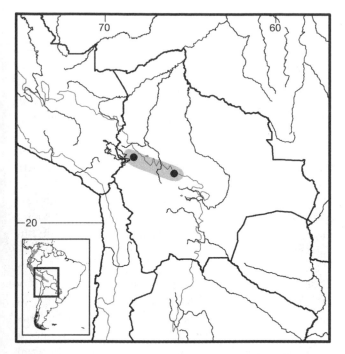

Map 324 Selected localities for *Thomasomys andersoni* (●). Contour line = 2,000 m.

tatus" groups, and its phylogenetic position may be annectant between these taxon groupings (Pacheco 2003; Salazar-Bravo and Yates 2007; V. Pacheco, unpubl. data).

The karyotype of *T. andersoni* has a diploid complement of 44 with all autosomes acrocentric (2*n* = 44, FN = 42); the X chromosomes also appear to be acrocentric (Salazar-Bravo and Yates 2007).

Thomasomys apeco Leo and Gardner, 1993
Apeco Thomasomys

SYNONYM:

Thomasomys apeco Leo and Gardner, 1993:417; type locality "Valle de Los Chochos, ca. 25 km NE Pataz, 3280 m, Parque Nacional Río Abiseo, San Martín, Perú."

DESCRIPTION: Large (head and body length 237–238 mm, condylobasal length of skull 47.7–47.8 mm, in two adult specimens) with long and dense dorsal fur (underfur as long as 22 mm mid-dorsally with guard hairs reaching 30 mm); dorsal coloration bright with color varying between individuals from raw sienna to ochraceous-tawny and tawny olive streaked with black guard hairs; ventral pelage ochraceous-buff although extension variable among individuals; very long mystacial vibrissae, extending well beyond pinnae; tail longer than head and body (135–137%), and without terminal pencil; interorbital region narrow, elevated but not beaded, with pronounced depression on frontals along midline of skull; zygomatic notches shallow; zygomatic

plates broad and moderately sloped backward from bottom to top; zygomatic arches convergent anteriorly; deep notch between lacrimal and zygomatic ramus of maxilla; thick jugal; auditory bullae small and uninflated; orbicular apophysis of malleus large and elongated; capsular process absent or represented by weak shelf; moderately hypsodont molars with weak labial cingula; accessory labial root of M1 present; orthodont upper incisors; upper molars robust, always longer (anteriorly-posteriorly) than wide, ranging from 10.0–10.2 mm in adults; long hallux and very long dV of pes as typical for "aureus" species group.

DISTRIBUTION: *Thomasomys apeco* is known only from the type locality and the immediately surrounding area in upper montane forests, from 3,200 to 3,380 m, in north central Peru.

SELECTED LOCALITIES (Map 325; from Leo and Gardner 1993): PERU: San Martín, Pampa del Cuy, ca. 24 km NE of Pataz, Puerta del Monte, ca. 26 km from Pataz, Valle de los Chochos, ca. 25 km NE of Pataz (type locality of *Thomasomys apeco* Leo and Gardner).

SUBSPECIES: *Thomasomys apeco* is monotypic.

NATURAL HISTORY: *Thomasomys apeco* inhabits upper montane forests and páramo-like habitats (locally called *jalca*) near forest borders. One specimen was caught at the end of a log and another on a bank above a small stream, on the forest floor in an isolated patch of elfin forest. Others were caught in a runway leading up from a rivulet of water in wet pampa habitat, in a small shallow stream bordered by a few small trees, or within continuous forest. Individuals may be terrestrial or arboreal (Leo and Gardner 1993). This species appears to breed in the dry season. One female was pregnant in August 1987 (dry season) with a single embryo; another female was captured in July with well-developed mammae. Subadult males had well-developed testes in July and August (Leo and Gardner 1993).

REMARKS: *Thomasomys apeco* is the largest living thomasomyine and among the largest Sigmodontinae known. It was differentially compared with species of the *Thomasomys aureus* complex group and *Megaoryzomys curioi*, an extinct Pleistocene form the Galapagos Islands (Leo and Gardner 1993), and with *T. macrotis* (A. L. Gardner and Romo 1993). Leo and Gardner (1993) suggested a closer affinity of *T. apeco* with the "aureus" species group, which was corroborated by Pacheco (2003) and followed here. *Thomasomys apeco* was compared also with *Thomasomys* sp., a specimen composed of pieces of maxillae (including the complete molar row), the right premaxilla (including the incisive teeth) and other braincase remains, obtained from Cueva Chaquil, Amazonas Department (O. Fabre et al. 2008). According to these authors, this specimen

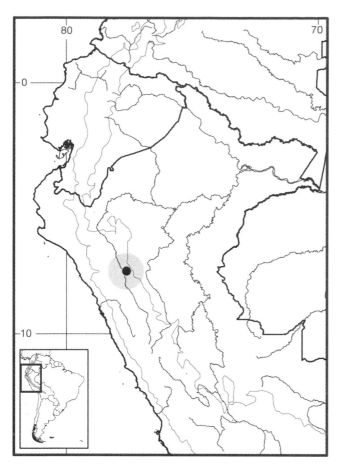

Map 325 Selected localities for *Thomasomys apeco* (●). Contour line = 2,000 m.

is slightly larger than *T. apeco*, based on comparisons of mean molar toothrow length (alveolar; 11.2 mm versus 9.9 mm); however, my own measurement of that specimen is 10.2 mm, the same as the larger specimen of *T. apeco* (Leo and Gardner 1993). Nonetheless, this similar size rat apparently corresponds to an unnamed taxon based on cranial characteristics that include the more pointed posterior ending of the incisive foramina, narrower palatal bridge, and slightly opisthodont upper incisors.

Thomasomys aureus (Tomes, 1860)
Golden Thomasomys

SYNONYMS:

H[*esperomys*]. *aureus* Tomes, 1860a:219; type locality undetermined.

O[*ryzomys*]. *aureus*: Thomas, 1900d:354; name combination.

Thomasomys aureus: Thomas, 1906c:443; first use of current name combination.

Thomasomys aureus altorum J. A. Allen, 1914b:200; type locality "Mt. Pichincha (altitude 11,000 ft.)," Pichincha, Ecuador.

Thomasomys pichinchius Lönnberg, 1921:38; *nomen nudum*.

Thomasomys aureus aureus: Gyldenstolpe, 1932a:51; name combination.

DESCRIPTION: Distinctively large species (head and body length 160–180 mm) with grizzled golden-brown dorsal fur indistinctly differentiated from yellow-washed, dark gray-based ventral fur. Dorsal pelage long, thick, and not very fine, with faint band of dark hairs along midline; long and blackish mystacial vibrissae extending distinctly posterior to pinnae; genal 1 vibrissae present. Hindfeet relatively long (36–40 mm) and moderately broad with dark patch distinctly lined with orange or whitish hairs extending to base of digits, becoming whitish on distal half; fifth digit semiopposable; gap between thenar and hypothenar absent. Hallux long, extending to interphalangeal joint of dII. Digit V of pes very long, claw extending to base of claw of dIV. Tail long (125–141% of head and body length), monocolored, finely annulated, covered with short hairs, and without terminal pencil. Protuberant anus prominent. Skull large, robust (condyloincisive length 35.4–39.5 mm), and zygomatic plates vertical and broad. Contact of premaxillomaxillary-frontal bones level with posterior margins of zygomatic notches. Interorbital region narrow and convergent anteriorly, with squared but not beaded margins, and with pronounced depression in frontals along midline of skull. Postorbital ridge distinctly present just behind, and parallel to, frontosquamosal suture. Braincase oblong, moderately broad, and not inflated; large interparietal subrectangular in shape. Incisive foramina long and moderately narrow, usually extending posteriorly between molar anterocones. Palatal bridge short and narrow with maxillary palatal pits present. Auditory bullae small, uninflated, and flask shaped; anterior process of ectotympanic is large. Ethmoturbinals moderately developed. Mesopterygoid fossa broad and somewhat parallel sided; sphenopalatine vacuities absent or appear only as slit-like vacuities flanking presphenoid-basisphenoid suture. Carotid circulatory pattern derived (stapedial foramen small, no internal squamosal-alisphenoid groove, no sphenofrontal foramen). Upper incisors opisthodont. Maxillary toothrow long, ranging typically from 6.8–7.8 mm.

DISTRIBUTION: *Thomasomys aureus* ranges broadly in Andean forests, from 1,460 to 3,850 m, from western Venezuela (see O. J. Linares 1998:Fig. 152), eastern and central Colombia, through Ecuador and Peru, to west-central Bolivia (Musser and Carleton 2005).

SELECTED LOCALITIES (Map 326): BOLIVIA: Cochabamba, 31 km by road W of Comarapa (AMNH 263183); La Paz, Murillo, Cuticucho (CBF 4463); Santa Cruz, approx. 25 km by road W of Comarapa, Siberia (UMMZ 156308). COLOMBIA: Huila, San Agustín, Río

Magdalena (FMNH 71290), San Agustín, Santa Marta (FMNH 71293); Risaralda, Río Termales (FMNH 71264). ECUADOR: Napo, near Papallacta (Voss 2003); Pichincha, Gualea (AMNH 36279). PERU: Cajamarca, San Ignacio, Tabaconas, Cerro La Viuda (Tabaconas-Namballe National Sanctuary Buffer Zone) (UMMZ 176556); Cusco, Cordillera de Vilcabamba (MUSM 13062), Machu Picchu (AMNH 91552), Marcapata, Limacpunco (FMNH 75228), 72 km NE (by road) Paucartambo; km 152 (MVZ 171496); Huánuco, Prov. 2 de Mayo, Cayna, Chiliatuna (MUSM 5252), Chinchuragra (MUSM 22892); Junín, Tarma, 22 mi E of Tarma (AMNH 231198); Piura, Batán on Zapalache-Carmen trail (LSUMZ 26945), Cerro Chinguela, ca. 5 km NE of Zapalache (LSUMZ 21896); Puno, 14 km W of Ya-nahuaya (MVZ 172597); San Martín, Puerta del Monte, ca. 30 km NE of Los Alisos (LSUMZ 27288), P.N. Río Abiseo, Puerta de Monte (MUSM 8051). VENEZUELA: Táchira, Buena Vista, 41 km SW of San Cristóbal, near Paramo de Tamá (USNM 442321).

SUBSPECIES: I consider *T. aureus* to be monotypic (but see Remarks).

NATURAL HISTORY: This species habits primary and secondary forest, preferring places with abundant vegetation such as dense grass and reeds, shrubs, and trees (Barnett 1999) in subalpine rainforest (Voss 2003), temperate forest, or páramo (Tirira 2007). Barnett (1999) pointed out that all known records of *T. aureus* come from forested areas, countering the argument of Reig (1986) and Tirira (2007) that the species may be found in páramo habitat as well. In northern and southern Peru it has been collected only in continuous and dense forest (Leo and Romo 1992; Patton 1987; V. Pacheco, unpubl. data). At Papallacta, Ecuador, Voss (2003) trapped specimens in well-worn paths through mats of moss and liverworts on horizontal tree limbs, and on the ground at the edge of a stream, among tall grass in a clearing, and in a runway through dense mats of moss. Elsewhere in Ecuador, Barnett (1999), Eisenberg and Redford (1999), Jarrín-V. (2001), and M. Brito et al. (2012) reported arboreality with nests built in trees at heights up to 6–7 m. In the Táchira Andes of Venezuela, this species was caught in a tree near a stream in cloud forest and in the lower montane very humid forest (Handley 1976). Individuals are known to feed on fruits, seeds, other plant material, and insects (Eisenberg and Redford 1999), with a preference for *Passiflora* fruits (Passifloraceae; Barnett 1999). Males with descended testes were found in September (Barnett 1999), and embryo counts of two to three have been recorded from Peruvian specimens (Eisenberg and Redford 1999). The species may be infrequently trapped, suggesting that it is either rare or difficult to capture (Barnett 1999).

Known ectoparasites include lice (*Hoplopleura* new sp. 2 on specimens from Cusco, Peru [V. S. Smith et al. 2008]), laelapid mites (*Hymenolaelaps princeps* on specimens from Táchira, Venezuela [Furman 1972]), fleas (*Polygenis impavidus* [Tipton and Machaco-Allison 1972], *Cleopsylla monticola*, *Polygenis thormani*, and *Neotyphloceras rosenbergi* [Maihuay et al. 1994], all on specimens from Cusco, Peru), and ticks (*Ixodes* sp., on specimens from Táchira, Venezuela [E. K. Jones et al. 1972]).

REMARKS: The type locality of *Thomasomys aureus* was unfortunately not determined in the original description but has been considered either to be Pallatanga, 4,950 ft (01°59′S, 78°57′W; 1,500 m), in the Ecuadorean province of Chimborazo (J. A. Allen 1914b; Ellerman 1941; Cabrera 1961), or Gualaquiza (03°24′S, 78°33′W; 750 m elevation) in the Ecuadorean province of Morona-Santiago (Thomas 1920h; A. L. Gardner 1983a). Musser and Carleton (2005) pointed out that a formal amendment of the type locality is needed, following arguments provided by A. L. Gardner (1983a) and Voss (2003). Voss (2003) stated that neither Pallatanga nor Gualaquiza is within the usual altitudinal range of the species (3,000–4,000 m), although, as I mentioned earlier, *T. aureus* has been reported well below this range. Pallatanga is thus in the range of known distribution of *T. aureus* but Gualaquiza is not (based on current records). In conclusion and based on this limited information, I consider Pallatanga the likely type locality. The Museum of Southwestern Biology holds two specimens (MSB 70707, 70708) collected on the Río Tatahuazo, near Cruz de Lizo, Bolivar province (01°43′S, 78°59′W), which is only 30 km from Pallatanga. To my knowledge these are the closest records to Pallatanga that could represent typical *T. aureus*, as compared with the Gualea series from Pichincha that J. A. Allen (1914b) suggested as typical of the species.

Thomas (1900d:355) was first to suggest that *Thomasomys praetor*, *T. aureus*, and *T. princeps* formed a small section of the genus. Later, based on *Thomasomys* material in the British Museum, Ellerman (1941) recognized as natural an *aureus* Group consisting of large bodied species in which he included *nicefori*, *popayanus*, *princeps*, and *praetor*. Cabrera (1961) considered each of these named taxa as subspecies within a highly variable *T. aureus*, without justification. Musser and Carleton (1993) followed this arrangement and also added *altorum* as a synonym or subspecies of *T. aureus*, supporting earlier doubts on the validity of *altorum* (Thomas 1920h:233). Because of the number of available names and the uncertain placement of taxon boundaries, A. L. Gardner and Romo (1993) coined the "*T. aureus* complex," which they suspected to consist of three or more valid species.

Voss (2003) postulated several attributes for separating *T. aureus* proper from *T. popayanus* and *T. praetor*, while maintaining *princeps* and *altorum* as synonyms of

T. aureus, and stating that *T. popayanus* probably included *nicefori*. Based on morphological traits, Pacheco (2003) also discriminated *popayanus* and *praetor* from *aureus*, but recognized *princeps* and *nicefori* as full species as well. Justifications for these decisions are provided in the separate species accounts that follow. Pacheco (2003) went further and recognized two additional unnamed species allied to *T. aureus* (e.g., *Thomasomys* sp. 1, a population from southern Peru and northern Bolivia, and *Thomasomys* sp. 2 from southern Bolivia). The first of these is large in size, similar to the Ecuadorean series, but the second is a small representative, probably the smallest in the group. Salazar Bravo and Yates (2007) sequenced a specimen of *Thomasomys* sp. 1 (reported as *T. aureus*; MVZ 170076, Cusco, 72 km NE [by road] Paucartambo, km 152), and a specimen of *Thomasomys* sp. 2 (AMNH 260422, Cochabamba, 28 km by road of Comarapa) reporting 12.3% of divergence between them, a large distance usually found between distinct species. Molecular analyses of *T. aureus* proper from Ecuador are still missing. S. Anderson (1997: 412) reported one undetermined *Thomasomys* (AMNH 263183) consisting of fragments of a skull from a fecal pellet of a carnivore from 30 km by road W of Comarapa, which I have identified as *Thomasomys* sp. 2 of Pacheco (2003).

My work in progress indicates that additional populations might warrant also species distinction. The population from the Mérida Andes in Venezuela departs in external and cranial characteristics from *T. aureus* proper and is somewhat isolated; further investigation may show that it merits recognition as a separate species. Similarly, other hypothetical new taxa are samples from Chinguela, Piura, Peru, that have characteristics of *T. pyrrhonotus* and *T. aureus*; the population from Río Abiseo, San Martín, Peru, with a whitish tail; and the population from Unduavi, Bolivia, where specimens are of moderate size, lack the genal 1 vibrissae, and exhibit a comparatively globose auditory bullae. Substantial work is still needed to resolve the taxonomy of this species complex.

Thomasomys aureus is highly differentiated in stomach, glans penis, and reproductive tract morphology relative to other species of *Thomasomys* (Carleton 1973; Hooper and Musser 1964; Voss and Linzey 1981; Pacheco 2003). Pacheco's (2003) phylogenetic analysis, which included these traits, supported a monophyletic "aureus" group.

A. L. Gardner and Patton (1976) reported a karyotype of 2n=44, FN=42, with X and Y acrocentric chromosomes, for specimens of *Thomasomys aureus* from Río Palca, Junín, and Yuraccyacu, Ayacucho, Peru; but this karyotype apparently belongs to *Thomasomys* sp. 1 of Pacheco (2003) (V. Pacheco, unpubl. data). Salazar-Bravo and Yates

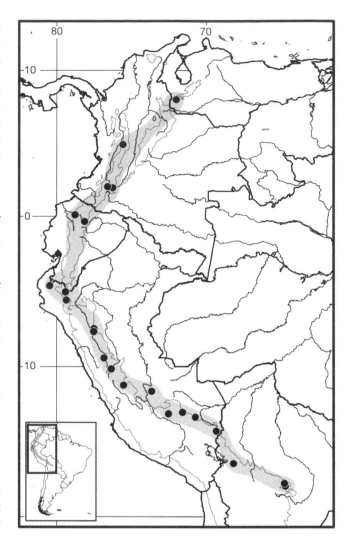

Map 326 Selected localities for *Thomasomys aureus* (●). Contour line = 2,000 m.

(2007:767) mistakenly listed the karyotype of *T. aureus* as 2n=42, FN=42, citing A. L. Gardner and Patton (1976). A karyotype of *T. aureus* proper from Ecuadorean populations remains unknown.

Thomasomys auricularis Anthony, 1923
Red Andean Thomasomys
SYNONYMS:

Thomasomys auricularis Anthony, 1923b:6; type locality "Taraguacocha, on trail from Zaruma to Zaraguro, altitude 10,250 feet, Cordillera de Chilla, Provincia del Oro, Ecuador."

Thomasomys pyrrhonotus: Cabrera, 1961:433, part (*auricularis* listed as a subspecies); not *pyrrhonotus* Thomas, 1886.

DESCRIPTION: Medium sized, slightly smaller than *T. aureus* (head and body length 138–155 mm), which it resembles superficially, but with light creamy-buff feet, an ochraceous postauricular patch, and larger auditory bullae. Dorsal pelage tawny olive, closely sprinkled with blackish hairs especially along midline of back, with flanks and sides more strongly tawny. Ventral pelage yellow-washed; color rather more intense over pectoral area; no line of demarcation in color between sides and under parts. Mystacial vibrissae moderately long, extending slightly beyond posterior margin of pinnae when bent; genal 1 vibrissae present. Tail relatively long (122–134% of head and body length), dark brown monocolored. Hindfoot moderately long (31–32 mm) with dI and dV long as in *T. aureus*; metatarsals usually covered with silvery white hairs; and no gap between thenar and hypothenar pads. Skull large and robust but smaller than *T. aureus* (condyloincisive length of skull 33.6–34.1 mm). Nasals slender and expanded anteriorly, and braincase not as inflated as in *aureus*. Incisive foramina long and extending backward between procingula of M1s; maxillary septum length long, about half foramina length. Interorbital region narrow, with parallel sides and squared and smooth, not beaded, edges. Interfrontal fontanella absent. Ethmoturbinals moderately developed. Mesopterygoid fossa extends to level of posterior border of last molars or between them; roof unfenestrated or with only reduced slits. Auditory bullae large and inflated (larger than *T. aureus* but comparatively smaller than *T. pyrrhonotus*). Postglenoid foramen and subsquamosal fenestra both long and narrow but subequal in size. Maxillary toothrow long, ranging typically from 6.6–7.4 mm. Additional anterior edge on procingulum of M1 usually present.

DISTRIBUTION: *Thomasomys auricularis* inhabits the western Andes of southwestern Ecuador in montane forests and isolated *Polylepis* (or *quenoa* [Rosaceae]) forests within the páramo (Barnett 1999), between 2,330 to 4,000 m.

SELECTED LOCALITIES (Map 327): ECUADOR: Azuay, The Cajas Plateau, Torreadora (Barnett 1999); Cañar, San Antonio (AMNH 63315), Chical (AMNH 63326); El Oro, Taraguacocha, small stream 500 ft above camp (type locality of *Thomasomys auricularis* Anthony); Loja, N side of Loja, Malacatos Divide (MCZ 58280).

SUBSPECIES: *Thomasomys auricularis* is monotypic.

NATURAL HISTORY: *Thomasomys auricularis* habits montane forest and páramo, degraded quenoa forests with open canopies and abundant shrub vegetation, and altered montane forest (Barnett 1999). The type specimen was taken on the bank of a small mountain stream in thick forest growth (Anthony 1923b). Individuals were also captured in traps placed on the lower limbs of *Polylepis* trees (Barnett 1999). Nothing about population biology, reproduction, or behavior is known.

REMARKS: This species has been treated as synonym or subspecies of *T. pyrrhonotus* (Cabrera 1961), and following synopses (e.g., Musser and Carleton 1993, 2005), although both taxa were never formally contrasted. It appears that Anthony (1923b) was not aware of *T. pyrrhonotus* because he compared *T. auricularis* only with *T. aureus* and commented on differences with *T. praetor*, *T. popayanus*, and *T. nicefori*, but not *T. pyrrhonotus*. Gyldenstolpe (1932) and later Pacheco (2003) recognized *T. auricularis* as a valid species, a decision followed by Tirira (1999, 2007).

Thomasomys auricularis is readily distinguished from *T. pyrrhonotus* by its slightly longer tail in relation to head and body length (130%, N = 1 adult and 122 to 134%, N = 5 subadults versus 120%, N = 2 adults, respectively); slightly larger molar toothrow (6.6 mm, N = 1 adult and 6.6 to 7.4 mm, N = 5 subadults versus 6.4 mm, N = 1 adult, respectively); inflated but comparatively smaller auditory bullae; the roof of the mesopterygoid fossa closed or with reduced slits versus moderately developed; subsquamosal fenestra less developed and more narrow; interfrontal fontanella absent versus present; ethmoturbinals medium in size versus small; and the M1 procingulum usually presenting an additional anterior edge versus almost absent (see also the account of *T. pyrrhonotus* for other morphological traits of the species). *Thomasomys auricularis* also ranges between medium and high elevations, *T. pyrrhonotus* occurs at lower and medium elevations.

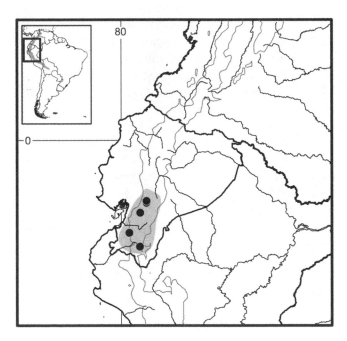

Map 327 Selected localities for *Thomasomys auricularis* (●). Contour line = 2,000 m.

The population from Cañar differs slightly from *T. auricularis* proper, and future studies might conclude this sample represents a separate taxon. These specimens have a dark and narrow metatarsal patch, the palatal bridge is narrower, the mesopterygoid fossa extends between M3, and their molar toothrow is slightly larger, among other features. This locality is also separated from the type locality by the deep Río Jubones.

Steppan (1995) reported the vertebral counts of this species (reported as *T. pyrrhonotus*). The karyotype is unknown.

Thomasomys australis Anthony, 1925
Austral Thomasomys

SYNONYMS:

Thomasomys daphne australis Anthony, 1925:4; type locality "Incachaca, 7700 feet, Prov. Cochabamba, Bolivia."

Thomasomys daphne: Voss, 1991a:112; part; not *daphne* Thomas, 1917.

Thomasomys australis: Voss, 2003:14; first use of current name combination.

DESCRIPTION: Small species (head and body length 89–91 mm; condyloincisive length of skull 24.1–25.5 mm) with cinnamon-brown to auburn dorsal pelage, and flanks almost as dark; ears small and covered with short brown hairs; and both hands and feet, wrists, and ankles washed with clove-brown. Ventral pelage cinnamon-buff. Pelage quite long, soft, and lax. Mystacial vibrissae moderately long, extending slightly beyond posterior margin of pinnae when bent; genal 1 vibrissae absent. Tail very long (144–155% of head and body length), covered by fine scales and unicolored. Hindfoot short (24–25 mm), slender, and dark brown. Skull with comparatively short rostrum and nasals; braincase moderately inflated; incisive foramina long and broad, reaching almost to anterior margin of M1s; and auditory bullae small. Carotid circulation with primitive condition (large stapedial foramen, groove on inner surface of squamosal and alisphenoid, and sphenofrontal foramen present). Capsular process absent or present as simple shelf. Upper incisors gracile, orthodont to barely proodont (*sensu* Hershkovitz 1962); maxillary toothrow short, ranging from 4.2–4.3 mm.

DISTRIBUTION: *Thomasomys australis* is known from the Bolivian Yungas, at elevations between 2,350 and 2,800 m.

SELECTED LOCALITIES (Map 328): BOLIVIA: Cochabamba, Corani (AMNH 268736), Incachaca (type locality of *Thomasomys daphne australis* Anthony); Santa Cruz, W of Comarapa, Siberia (UMMZ 156307).

SUBSPECIES: *Thomasomys australis* is monotypic.

NATURAL HISTORY: No detailed data are available on habits, behavior, or food. Four pregnant females taken in October had one embryo and three embryos. A lactating female was recorded in June (S. Anderson 1997). For central Bolivia, the dry season spans from May to September. The species was collected on the ground in upper montane elfin forest (Salazar-Bravo and Yates 2007).

REMARKS: Anthony (1925) described *T. australis* on p. 4, not p. 2 as stated by Cabrera (1961:428). Voss (2003:14) elevated *australis* to full species based on its opisthodont upper incisors (index value of 85°) versus the orthodont condition in *T. daphne* (index value 90°), an angle equivalent to the incisive index of Thomas (1919h). Salazar-Bravo and Yates (2007) followed Voss's decision in recognizing *T. australis* as a full species. Pacheco (2003) also recognized *T. australis* as a species different from *T. daphne* proper because of its primitive carotid circulation (pattern 1, *sensu* Voss 1988) versus the derived circulation (pattern 3, *sensu* Voss 1988), the unicolor tail versus the usually tip-white tail, and the capsular process indistinct versus distinct, respectively, among other traits. These character differences support the distinction provided by Anthony (1925), who stated that *T. australis* has a more rufous dorsal pelage, darker hindfeet, and longer incisive foramina than *T. daphne*. The Santa Cruz record constitutes the southernmost record for both this species and the genus *Thomasomys*.

Salazar-Bravo and Yates (2007), in a phylogenetic analysis based on the mtDNA cytochrome-*b* gene, found *T. australis* to be more closely related to *Thomasomys* sp. 3 (reported as *T. daphne*), an unnamed taxon (see *T. daphne*

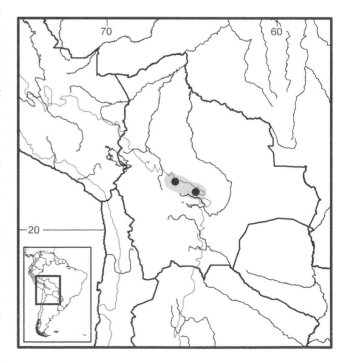

Map 328 Selected localities for *Thomasomys australis* (●). Contour line = 2,000 m.

account) proposed by Pacheco (2003). This result is consistent with Pacheco's (2003) morphological-based phylogeny that found *T. australis* related to both *T. daphne* and *Thomasomys* sp. 3.

Thomasomys baeops (Thomas, 1899)
Short-faced Thomasomys

SYNONYMS:

Oryzomys bæops Thomas, 1899b:152; type locality "Pita R[ío]., above the Chillo Valley, [Pichincha,] Ecuador. Alt. 3500 metres."

Thomasomys boeops: Thomas, 1906c:443; name combination.

Thomasomys baeops: Gyldenstolpe, 1932a:57; first use of corrected spelling.

DESCRIPTION: Medium sized (head and body length 106–128 mm; condyloincisive length of skull 25.2–27.5 mm) with dull grayish brown dorsal pelage, darker along middle line of back. Ventral pelage silvery gray with light brown tone extending anteriorly to base of mystacial vibrissae; base of hairs plumbeous. Hands whitish above; hindfeet pale brown above, but metatarsal region without distinct patch; digits whitish. Pelage quite long, soft, and woolly. Mystacial vibrissae very long and extend far behind ears when bent; genal 1 vibrissae present or absent. Tail moderately long (102–139% of head and body length), unicolored, with or without white tip. Hindfoot relatively short (22–27 mm) and broad. Digit I of pes (hallux) long, with its claw extending to interphalangeal joint of dII; and dV very long, extending to base of claw of dIV. Rostrum of skull short and delicate; nasals short and narrow; frontonasal profile quite flat, not convex; braincase moderately inflated; and interorbital region narrow, with edges faintly marked (neither rounded nor beaded). Zygomatic arches narrow, producing somewhat parallel outline. Incisive foramina narrow with extremes contracted and moderately long, extending to but not between M1s. Auditory bullae small, not inflated. Carotid circulatory pattern derived (stapedial foramen small, no internal squamosal-alisphenoid groove, no sphenofrontal foramen). Upper incisors opisthodont; capsular process distinct; and maxillary toothrow short, ranging from 4.2–4.9 mm.

DISTRIBUTION: *Thomasomys baeops* is endemic to the northern Andes in Ecuador and Colombia, north of the Huancabamba Deflection in northern Peru (Voss 2003; Pacheco 2003). It inhabits Andean forests and páramo habitats across an elevation range from 1,300 to 3,800 m.

SELECTED LOCALITIES (Map 329): COLOMBIA: Antioquia, Sonsón, Páramo, 7 km E (FMNH 70341); Cauca, 0.8 km E of Laguna San Rafael, Páramo de Puracé (KU 124030); Huila, San Agustín, Río Magdalena (FMNH 71472); Nariño, Pasto, faldas del Volcán Galeras (ICN

13269). ECUADOR: Azuay, Molleturo (AMNH 61949); Bolívar, Río Tatahuazo, 2.5 km E of Cruz de Lizo (MSB 70704); Loja, Las Chinchas, 1 km WSW (USNM 513595), San Bartolo (AMNH 61363); Morona Santiago, Tinajillas (QCAZ 8869); Napo, Antisana (EPN 944145); Pichincha, Peñas Blancas, comuna Itulcachi (EPN 954421), Volcán Pichincha, oriente (FMNH 91990); Tungurahua, Baños (AMNH 67501).

SUBSPECIES: I consider *T. baeops* to be monotypic (but see Remarks).

NATURAL HISTORY: This species lives in disturbed secondary forest sections of primary/secondary cloud forest mosaics, but prefers areas with abundant shrubby vegetation and little arborescent canopy development (Barnett 1999). A lactating female was obtained on September in the local dry season (Barnett 1999). In cloud forests, *T. baeops* coexists with *T. paramorum* and *T. aureus*. At Papallacta, it was found inside the shrubby páramo/forest ecotone, in subalpine rainforest, and in dense thickets of secondary growth at the bottom of a narrow ravine surrounded by pastures. It was found also in primarily alpine grassland dominated by *Stipa ichu* (Poaceae) with patches of *Polylepis* forests and secondary and riparian forests in Sangay National Park (T. E. Lee et al. 2011). Individuals have been captured most commonly on the ground, along the wet margins of small streams, in narrow trails through mossy debris and damp leaf litter, under mossy logs, or in holes under the roots of trees. Individuals have also been obtained on mossy branches of small trees (Voss 2003). Food includes seeds and insects (Jarrín-V. 2001). One female was captured with three embryos in Sangay National Park during the dry season (T. E. Lee et al. 2011). The species was considered common in Ecuador (Tirira 2007) but rare in Colombia (Gómez-Laverde et al. 2008).

REMARKS: Thomas (1899b) described and reported this taxon with the spelling *boeops*, which was followed by Thomas (1906c), and later by Cabrera (1961). Alternatively, Gyldenstolpe (1932), Tate (1932g), and Ellerman (1941) corrected the spelling to *baeops*. Gyldenstolpe (1932) and Tate (1932g) used the diphthong *æ* for both *T. baeops* and *T. praetor*. Although these spellings might be considered an unjustified emendation (ICZN 1999:Art. 33.2.3), as no explanation or justification was given by either author, it should be noted that the Greek root for "small " is "*baios*," which becomes "*baeo-*" when used within the English language (R. W. Brown 1956). Several mammalian genus-group names use the root *baeo-*, including *Baeodon*, *Baeus*, *Baeolophus*, *Baiostoma*, and *Ceratobaeus*. Recent authors (e.g., Voss 2003; Musser and Carleton 2005) have used the *baeops* spelling, although without an explicit justification. Both spellings (i.e., *boeops*, *baeops*) are homonyms (ICZN 1999:Art. 58.1), and according to the Principle of Homonymy the senior name should prevail (i.e., *boeops*). However, given the prevailing use of the spelling *baeops* and to maintain the stability, I consider *baeops* as a correct original spelling (ICZN 1999:Art. 33.3.1) because it is in prevailing usage and can be attributed to the original author and date.

Cabrera (1961) and Musser and Carleton (2005) erroneously gave the type locality as "El Oro Prov., Chilla Valley, Río Pita, 3500 m" although Thomas (1899b) clearly referred to the Río Pita, which passes through the Valle de los Chillos in the province of Pichincha (see also Voss 2003).

A phylogenetic analysis of discrete morphological characters for thomasomyine rodents found *T. baeops* more closely related to *T. taczanowskii*, and both members of the unnamed genus ("New Genus B," *sensu* Pacheco 2003). Salazar-Bravo and Yates (2007) confirmed this relationship in a molecular phylogenetic analysis based on mtDNA cytochrome-*b* sequences, who also reported a divergence of 6.4% between both taxa, the lowest between a species pairs within *Thomasomys*. T. E. Lee et al. (2011) reported *T. baeops* to be more closely related to the pair *T. ischyrus* and *T. silvestris*, also based on cytochrome-*b* sequences, but their sample of *T. ischyrus* (the GenBank sequence AF 108675 [MVZ 181999]) in fact corresponds to *T. taczanowskii* (see Pacheco 2003). Their result, therefore, is thus consistent with previous analyses, although the proposed relationship of *T. taczanowskii* with *T. silvestris* and then with *T. baeops* needs to be corroborated.

Although Pacheco's (2003) morphological analyses included only Ecuadorean specimens of *T. baeops*, several characters exhibited some degree of geographic differentiation. For example, genal 1 vibrissae were found to be present or absent in separate samples, the terminal tip of tail was white or not, and the premaxillae were positioned either posterior or subequal to nasals. The geographic concordance of these differences suggests that *T. baeops* may be composed of more than one taxon. This argument is in line with the analysis of cytochrome-*b* sequences that supported a specimen from Sangay National Park, identified morphologically as *T. baeops*, as an unknown taxon (T. E. Lee et al. 2011). Furthermore, populations from Caldas and Cauca in Colombia and Papallacta in Ecuador also appear to represent different taxa. The Caldas population exhibits a dorsal band faint and an auditory bullae small but comparatively more globose than typical *T. baeops*. The Cauca population is darker, genal 1 vibrissae are present, and the premaxillae are uniformly longer than nasal. The Papallacta sample lacks genal 1 vibrissae and exhibits both more robust upper incisors and wider incisive foramina and mesopterygoid fossa. These preliminary results still wait throughout analyses comparing side-by-side populations of these relatively common rodents.

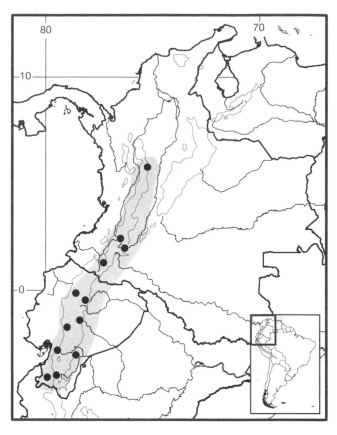

Map 329 Selected localities for *Thomasomys baeops* (●). Contour line = 2,000 m.

Thomasomys bombycinus Anthony, 1925
Silky Thomasomys

SYNONYM:

Thomasomys bombycinus Anthony, 1925:1; type locality "Paramillo, Western Andes, [Antioquia,] Colombia; altitude 12,500 ft."

DESCRIPTION: Medium bodied (head and body length 112–130 mm) with long, soft, and silky pelage. Dorsal color ranges between fuscous and clove-brown sprinkled with black-tipped hairs; sides paler; ears clove-brown; hands and feet are pale brown. Ventral pelage cinnamon-buff to tawny-olive, smoothly transitioned from darker color of flanks, not countershaded. Mystacial vibrissae moderately long, extending slightly beyond posterior margin of pinnae when bent; genal 1 vibrissae absent. Tail relatively short (tail length 99–124% of head and body length), drab colored above, somewhat lighter below, and thinly haired. Hindfoot moderately short (26–30 mm) and narrow, without metatarsal patch. First digit of pes (hallux) short, with claw not extending more than about half length of phalanx 1 of dII. Digit V long, claw extending about half length of phalanx 2 of dIV. Skull moderately long (condyloincisive length 26.8–29.2 mm) with relatively long and narrow rostrum. Interorbital region moderately broad with round margins. Incisive foramina short and open, barely reaching plane of M1s. Mesopterygoid fossa broad, especially anteriorly. Auditory bullae small and uninflated, with main axis less inclined to midline compared with congeners. Carotid circulatory pattern primitive (stapedial foramen, groove on inner surface of squamosal and alisphenoid, and sphenofrontal foramen present). Upper molars moderately hypsodont, with slight interpenetration mainly between hypoflexus and metaflexus of first and second upper molars. Maxillary toothrow short, ranging from 4.7–5.5 mm. Upper incisors orthodont.

DISTRIBUTION: *Thomasomys bombycinus* occurs in the northern part of the western Andes in Colombia.

SELECTED LOCALITIES (Map 330): COLOMBIA: Antioquia, Paramillo, W Andes (type locality of *Thomasomys bombycinus* Anthony), Urrao, Páramo Frontino (FMNH 71320), Urrao, Santa Bárbara (FMNH 71362); Risaralda, Finca Cañón, Papayal, Río San Rafael, Parque Nacional Tatamá (IAvH-M 5625).

SUBSPECIES: I consider *T. bombycinus* to be monotypic (but see Remarks).

NATURAL HISTORY: Nothing is known of the habitat and ecology of this species, but based on records it is likely found in moist montane habitats including forest and páramo (Delgado et al. 2008).

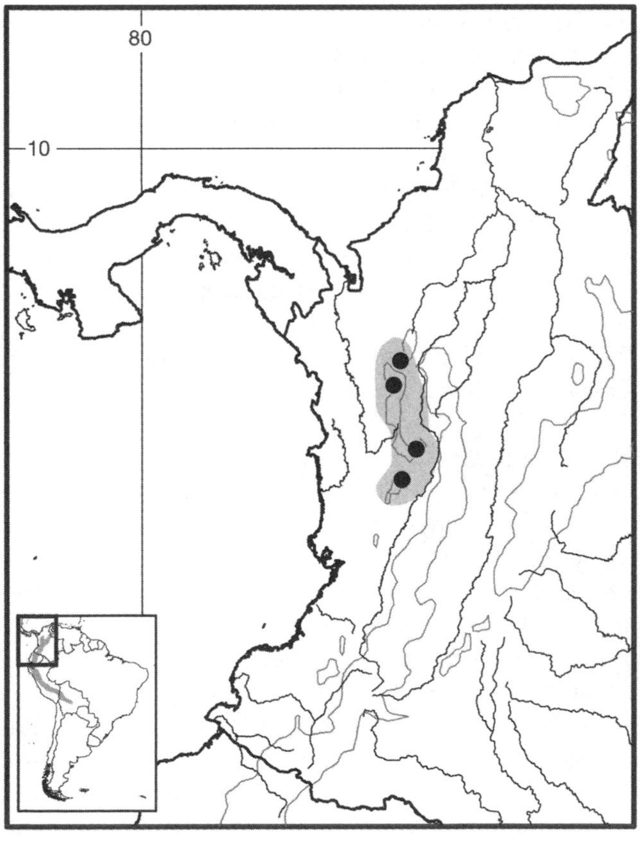

REMARKS: Anthony (1925) believed *T. bombycinus* to resemble *T. cinereus* more closely than *T. cinereiventer*, although Pacheco (2003) found *T. bombycinus* closer to *T. cinereiventer* in his cladistic analysis of morphological characters. Specimens from Páramo Frontino and Santa Bárbara exhibit some differences in contrast to the type series, such as the tail bicolored, a distinct gap between thenar and hypothenarpads of the pes, shorter and narrower incisive foramina with posterior margins extending close to the protocones of the M1s, not to the edge, and a medial process of palate small but distinct. These character differences suggest that *T. bombycinus* might be polytypic.

Thomasomys caudivarius Anthony, 1923
White-tipped Thomasomys

SYNONYMS:

Thomasomys caudivarius Anthony, 1923b:4; type locality "Taraguacocha, Cordillera de Chilla, 10,750 feet [3,277 m], Provincia del Oro, Ecuador."

Thomasomys cinereus caudivarius: Cabrera, 1961:427; name combination.

DESCRIPTION: Medium sized (head and body length, 116–125 mm) with soft, dense, and comparatively long dorsal fur. Dorsal coloration between mummy-brown and clove-brown with tip of the hairs colored; not countershaded with the ventral pelage, which is near chamois or buffy cream. Mystacial vibrissae moderately long, extending slightly beyond posterior margin of pinnae when bent; genal 1 vibrissae absent. Tail longer than head and body (116–134%) and terminates in white tip. Hindfoot moderately long (28.5–31 mm) and narrow; hands and feet approach color of upper parts but lighter; and distinct gap exists between thenar and hypothenar plantar pads. Digit I of pes (hallux) moderate, with its claw extending to about half length of phalanx 1 of dII. Digit V of pes long, with its claw extending about half length of phalanx 2 of dIV. Skull moderately long (condyloincisive length 27.5–29.7 mm) with relatively long and moderately broad rostrum, and anteriorly expanded nasals. Interorbital region moderately broad, hourglass in shape, and with round margins. Incisive foramina oval, more contracted anteriorly, and moderately long, reaching anterior margins of M1s or close. Mesopterygoid fossa broad, with subparallel margins and with weak medial palatal process usually present. Auditory bullae small and uninflated. Carotid circulatory pattern primitive (stapedial foramen, squamosal alisphenoid groove, and sphenofrontal foramen present). Upper molars moderately hypsodont, with interpenetration of flexi. Maxillary toothrow moderately long, ranging from 4.8–5.3 mm. Upper incisors orthodont. Capsular process absent.

Map 330 Selected localities for *Thomasomys bombycinus* (●). Contour line = 2,000 m.

DISTRIBUTION: *Thomasomys caudivarius* occurs from central Ecuador to northern Peru, in montane forest and páramo habitats of the western Andes, from 2,750 to 3,350 m (L. Luna and Pacheco 2002; Pacheco 2003).

SELECTED LOCALITIES (Map 331): ECUADOR: Bolívar, Río Tatahuazo, 4 km E of Cruz de Lizo (MSB 70712), Río Tatahuazo, 2.5 km E of Cruz de Lizo (MSB 70719); El Oro, Taraguacocha, Cordillera de Chilla, páramo 1000 ft above camp (type locality of *Thomasomys caudivarius* Anthony). PERU: Cajamarca, Las Ashitas, 4 km W of Pachapiriana (AMNH 268150), San Ignacio, Tabaconas, Piedra Cueva in Cerro Coyona (Tabaconas-Namballe National Sanctuary) (UMMZ 176719); Piura, Cerro Chinguela (MUSM 23743), Huancabamba, Tambo (FMNH 81331), Minera Majaz, Campamento Nueva York (MUSM 23565).

SUBSPECIES: I consider *T. caudivarius* to be monotypic (but see Remarks).

NATURAL HISTORY: This species is terrestrial and nocturnal, feeding on seeds and insects (Jarrín-V. 2001). Specimens have been captured along small mountain streams, in a valley with thick shrubbery and low stunted trees, and in open páramo with an abundance of low shrubbery and no trees (Anthony 1923b).

REMARKS: *Thomasomys caudivarius* was considered a large member of the "*cinereus* group" by Anthony (1923b), but was relegated to a subspecies of *T. cinereus* by Cabrera (1961) without justification. Tirira (1999), based on a personal communication from V. Pacheco, listed *caudivarius* as a valid species, a status fully supported by L. Luna and Pacheco (2002), who also reported the species from northern Peru. Steppan (1995) reported the vertebral counts of this species (reported as *T. cinereus*). Jarrín-V. (2001) reported *T. caudivarius* from Otonga, Cotopaxi, but his description and illustration of a species of dark general coloration appears to correspond better to *T. silvestris*. Furthermore, these specimens were taken at 2,000 m, an elevation well below the confirmed range of *T. caudivarius* but within that of *T. silvestris*. For these reasons, I have tentatively omitted this record pending further examination of his specimens; however, if valid, Otonga would represent the northernmost and the lowest record for *T. caudivarius*. Pozo-R. et al. (2007) reported another northern record of *T. caudivarius* from Finca Ana María, Pichincha, at 2,500 m; however, I have not included their record, as the authors provided no description to confirm their identification and because this locality is well outside the range I document herein.

L. Luna and Pacheco (2002) contrasted the species with *Thomasomys onkiro* and *T. silvestris*, but a phylogenetic analysis of morphological characters placed *T. caudivarius* within the "cinereus" species group, specifically most closely related to *T. hylophilus* (Pacheco 2003). Although *T. caudivarius* is considered monotypic, there is substantial morpho-

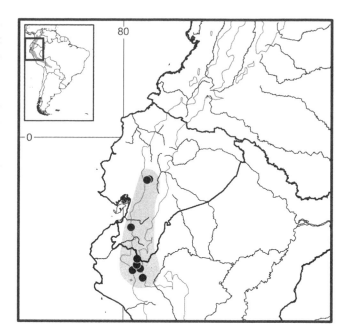

Map 331 Selected localities for *Thomasomys caudivarius* (●). Contour line = 2,000 m.

logical variation in the length of nasals, the breadth of incisive foramina and the degree of their posterior extension, breadth of mesopterygoid fossa, and size of molars among populations. Specimens from northern Peru, for example, exhibit a mesopterygoid fossa without a medial palatal process, and a roof perforated by narrow slits, among other features. These specimens appear more similar to two specimens P. Moreno and Albuja (2012) reported from Reserva de Tapichalaca (04°29′31.5″S, 79°07′47.2″W), Zamora-Chinchipe province, 2,500 m, Ecuador, as *T. onkiro* (see that account). They might represent another taxon allied to *T. caudivarius*. It is thus possible that this species is polytypic.

Thomasomys cinereiventer J. A. Allen, 1912
Ashy-bellied Thomasomys

SYNONYMS:

Thomasomys cinereiventer J. A. Allen, 1912:80; type locality "crest of Western Andes (altitude, 10,340 feet [3,152 m]), 40 miles [64 km] west of Popayan [= Popayán], Cauca, Colombia."

Thomasomys cinereiventer cinereiventer: Gyldenstolpe, 1932a:53; name combination.

DESCRIPTION: Large species (head and body length 100–147 mm) with very dark brown dorsal pelage, hairs of which tipped with lighter brown, and with flanks almost as dark as dorsal region. Ears rather small and brown in color. Ventral pelage ash gray with barely perceptible wash of grayish pale yellow, moderately countershaded with respect to dorsal pelage. Pelage quite long, soft, and lax. Mystacial vibrissae moderately long, extending slightly beyond posterior margin of pinnae when bent; genal 1 vibrissae absent. Tail variable in length, ranging from comparatively short or moderately long (107–128% of head and body length), uniformly pale brown, covered with short hairs, and without terminal pencil. Hindfoot long (32–36 mm), and pale brown. Digit I of pes (hallux) short, with claw not extending beyond half length of phalanx 1 of dII. Digit V of pes long, claw extending about half length of phalanx 2 of dIV. Skull large (condyloincisive length of skull 30.7–32.6 mm) with long and robust rostrum, tapering anteriorly; braincase oblong, comparatively more elongated than other species; zygomatic notches comparatively deeper; interorbital region intermediate with incipiently square margins. Zygomatic arches strongly convergent; zygomatic plates broad and vertical. Upper molars moderately hypsodont, with slight interpenetrating flexi mainly between hypoflexus and metaflexus of M1 and M2. Maxillary toothrow long, 5.4–6.2 mm. Upper incisors opisthodont and comparatively broad.

DISTRIBUTION: *Thomasomys cinereiventer* is known only from the western Andes in Colombia, from 1,828 to 3,152 m.

SELECTED LOCALITIES (Map 332): COLOMBIA: Antioquia, Mpio. Jardín, Vda. La Linda (ICN 16489); Cauca, Cocal (AMNH 32447), Munchique, E side (AMNH 181476), coast range W of Popayán (AMNH 32417); Chocó, con límite de dpto. Valle (carretera vereda pacifico-Mpio. San José del Palmar) (IAvH-M, 7340); Risaralda, Finca Cañón, Papayal, Río San Rafael, Parque Nacional Tatamá (IAvH-M 5622).

SUBSPECIES: *Thomasomys cinereiventer* is monotypic.

NATURAL HISTORY: Little is known about this species, but it is present in subtropical/tropical moist montane forests (Gómez-Laverde and Pacheco 2008). It was captured in Risaralda in a locality with steep terrain with patches of disturbed (selectively logged) primary cloud forest, agricultural fields, and pastures (Voss et al. 2002). Mendez (1971) described *Cummingsia inopinata* as a new species of mallophagan louse from specimens identified as *T. cinereiventer* from Laguna de la Cocha, Nariño, and Comis, Putumayo, Colombia. However, no voucher specimens are apparently available to confirm or check the identification of these host specimens. I suggest the alleged association of *C. inopinata* with *T. cinereiventer* be cautiously considered pending new evidence. Mendez (1971) also suggested that *T. cinereiventer* is partially arboreal based on the report of *Dasipsyllus gallinulae*, a true bird flea, on three different specimens of this species from Cerro Munchique. Individuals might temporarily occupy tree holes or other sites containing bird nests, from which the mentioned fleas were probably obtained. Brennan (1968) reported the chiggers *Crotiscus desdentatus*,

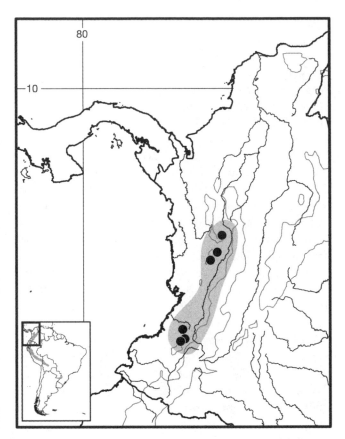

Map 332 Selected localities for *Thomasomys cinereiventer* (●). Contour line = 2,000 m.

Odontacarus munchiquensis, and *Euschoengastia pamelae* (the latter two as new species) from *T. cinereiventer*. Trapido and Sanmartín (1971) processed specimens of *T. cinereiventer* from the Pichindé valley, Munchique, and La Cocha for Pichindé virus, but failed to detect the virus.

REMARKS: Musser and Carleton (1993, 2005) treated *contradictus* Anthony, 1925 and *dispar* Anthony, 1925 as subspecies or simple synonyms of *T. cinereiventer*. Pacheco (2003) regarded these taxa as full species, a decision I follow herein (see these respective accounts for morphological support for this decision). Cabrera (1961) included *Thomasomys erro* as a subspecies of *T. cinereiventer*, but Voss (2003) recently elevated *erro* to full species by and convincingly contradicting Cabrera's hypothesis. A phylogenetic analysis based on morphological characters found instead found *T. cinereiventer* to be closely related to *T. bombycinus* (Pacheco 2003).

Thomasomys cinereus (Thomas, 1882)
Olive-gray Thomasomys
SYNONYMS:

H[esperomys]. (Rhipidomys) cinereus Thomas, 1882:108; type locality "Cutervo, 9,200′ [2,804 m]," Cajamarca, Peru.

Hesperomys (Vesperimus) cinereus: Thomas, 1884:449; name combination.

Hesperomys (Thomasomys) cinereus: Coues, 1884:1275; name combination.

Peromyscus (Thomasomys) cinereus: E.-L. Trouessart, 1897: 512; name combination.

Thomasomys cinereus: Thomas, 1906c:443; first use of current name combination.

DESCRIPTION: Medium sized (adult head and body length 107–146 mm) with long, soft, and dense dorsal fur. Dorsal coloration grizzled ashy gray, hairs being slate-colored at base and white at tips, sprinkled with longer and blackish hairs. Ventral pelage grayish white, hairs also slate-colored at base, and moderately countershaded with dorsal pelage. Mystacial vibrissae moderately long, extending slightly beyond posterior margin of pinnae when bent; genal 1 vibrissae absent. Tail comparatively thick, indistinctly bicolored, and lacks terminal white tip; it ranges from shorter to longer than head and body (96–145%). Hindfeet moderately long (28–32 mm) with metatarsals covered by pure white shining hairs; hands also whitish. Digit I of pes (hallux) moderately long, its claw extending close or to interphalangeal joint of dII. Digit V of pes long, with claw extending about half length of phalanx 2 of dIV. Skull moderately long (condyloincisive length 29.3–32.3 mm) with relatively long and broad rostrum, nasals well expanded anteriorly, and no rostral tube. Interorbital region narrow or moderate, hourglass in shape with rounded margins. Zygomatic plates moderately broad, subequal to length of M1, and vertical. Zygomatic arches converge anteriorly to moderate degree. Incisive foramina oval but elongated, and long, reaching anterior margins of M1s or slightly between them. Mesopterygoid fossa broad, wider anteriorly or with margins subparallel; medial process usually present. Auditory bullae small and uninflated. Carotid circulatory pattern primitive (stapedial foramen, groove on inner surface of squamosal and alisphenoid, and sphenofrontal foramen present). Upper molars moderately hypsodont, without interpenetration of flexi. Maxillary toothrow moderately long, ranging from 4.8–5.6 mm. Upper incisors orthodont. M1 lacks accessory labial root, and m1 with distinct anteromedian flexid. Capsular process absent.

DISTRIBUTION: *Thomasomys cinereus* is restricted to northwestern Peru, west of the Río Marañón, from 1,198 to 3,100 m.

SELECTED LOCALITIES (Map 333): PERU: Cajamarca, 35 mi WNW of Cajamarca (MVZ 137937), Cutervo, San Andrés de Cutervo, Cutervo National Park, 100 m over El Tragadero (UMMZ 176634), Pisit, T 25 (MUSM 25709); La Libertad, Cachicadán (MUSM 17285), south

of Huamachuco (BM 0.6.6.12); Lambayeque, Bosque de Chiñama (MUSM 5095), Seques (AMNH 73129); Piura Ayabaca, bosque de Huamba, 44 km E of Ayabaca (MUSM 511), 15 road km E of Canchaque (LSUMZ 19315), Cerro Chinguela, ca. 5 km NE of Zapalache (LSUMZ 27147), 33 road km SW of Huancabamba (LSUMZ 19301), Pariamarca Alto (MUSM 23758).

SUBSPECIES: *Thomasomys cinereus* is monotypic (but see Remarks).

NATURAL HISTORY: This species inhabits humid montane forests or Yungas of northern Peru where specimens have been caught on the ground among dense shrubs and on the forest floor. One female (LSUMZ 27081) from Piura was pregnant with three embryos in August (dry season). Several males with scrotal testes were caught at the same place from late July and early August. Known ectoparasites of *T. cinereus* include the flea *Plocopsylla kilya* (Schramm and Lewis 1987) and a staphylinid beetle, *Amblyopinus piurae* (Seevers 1955). P. T. Johnson (1972) and Durden and Musser (1994) reported *T. cinereus* as the host for the anopluran louse *Hoplopleura angulata*, but this allocation is corrected here to *T. ischyrus* (see that account).

REMARKS: Thomas (1882) described *T. cinereus* based mostly on external morphology and basic morphometric values, but it remained for Voss (1993) to detail descriptions of key external, cranial, dental, and soft characters. Pacheco (2003) confirmed these reported attributes and expanded on additional characters of the species.

Thomas (1884) identified a specimen, obtained by M. C. Jelski for the Warsaw Museum, from Maraynioc (near Chanchamayo), Junín department, as *Hesperomys (Vesperimus) cinereus*. However, this record is not included for the species because its description does not fit typical *T. cinereus* and I have not had the opportunity to examine the specimen. Osgood (1914b) also reported some specimens from near Uchco, Amazonas, as the subspecies *T. cinereus ischyrus*, but the taxon *ischyrus* is now regarded as a different species (Pacheco 2003; Musser and Carleton 2005). Previous records of the species from Ecuador (Musser and Carleton 1993) were based on the inclusion of *caudivarius* as a subspecies (following Cabrera 1961), but L. Luna and Pacheco (2002) treated this taxon as a valid species. Other Colombian records of the allegedly *T. cinereus* (from Antioquia [Hooper and Musser 1964] and Huila [Carleton 1973]) are likely based on misidentified specimens because these localities are far from the currently known distribution of the species. As a consequence, the descriptions of glans penis morphology (Hooper and Musser 1964) and stomach morphology (Carleton 1973) intended for *T. cinereus* do not correspond to this species. Rather, Pacheco (2003) reported on these morphological attributes based on correctly allocated specimens. The sample from La Libertad differs from

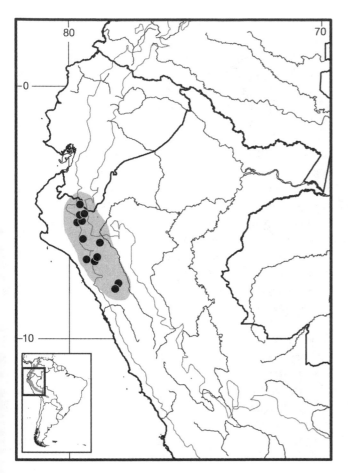

Map 333 Selected localities for *Thomasomys cinereus* (●). Contour line = 2,000 m.

specimens from the type locality in several discrete characters, such as a narrower and parallel mesopterygoid fossa. Detailed study might reveal that *T. cinereus* is also a complex of separate species. The karyotype is unknown.

Ellerman (1941) was the first to recognize a *cinereus* Group, but his concept was very broadly constructed, and it would include species allocated herein to the "baeops," "incanus," and "cinereus" groups of *Thomasomys* as well as some species of now assigned to *Rhipidomys*, *Wilfredomys*, and *Juliomys*. Therefore, Ellerman (1941)'s concept does not correspond to current taxonomy and is no longer useful. The "cinereus" group, in a more restrictive sense, was then first proposed by Pacheco (2003) and is elaborated here.

Thomasomys cinnameus Anthony, 1924
Cinnamon-colored Thomasomys

SYNONYMS:

Thomasomys cinnameus Anthony, 1924b:5; type locality "Hacienda San Francisco, east of Ambato, [Tungurahua,] central Ecuador, elevation 8000 feet [2,438 m]."

Thomasomys gracilis cinnameus: Cabrera, 1961:429; name combination.

DESCRIPTION: Small species (head and body length 84–90 mm) with long and soft dorsal fur. Dorsal coloration cinnamon-brown, gradually changing to clay color on sides, with ventral color cinnamon-buff. Mystacial vibrissae moderately long, extending slightly beyond posterior margin of pinnae when bent; genal 1 vibrissae absent. Tail long to very long (126–152% of head and body length), finely annulated and thinly haired, and uniformly clove-brown in color. Hindfoot short (22–24 mm) and slender, clove-brown color; hands of same color. Digit I of pes (hallux) moderately long, with claw not extending more than half length of phalanx 1 of dII. Digit V of pes very long, with its claw extending to base of claw of dIV. Skull small (condyloincisive length of skull 22–23.3 mm) with relatively long and slender rostrum and subglobose inflated braincase. Interorbital region narrow, hourglass in shape, and with round margins. Incisive foramina narrow, more contracted anteriorly, and long, reaching anterior margins of M1s or extending slightly between them. Zygomatic plates narrow, their breadth shorter than length of M1. Mesopterygoid fossa broad with subparallel margins and without medial process; its roof closed, without sphenopalatine vacuities. Auditory bullae small and uninflated, with distinct bony tubes. Carotid circulatory pattern primitive (stapedial foramen, groove on inner surface of squamosal and alisphenoid, and sphenofrontal foramen are present). Upper molars moderately hypsodont with weakly developed cingula and stylar cusps but no interpenetrating flexi. Maxillary toothrow short, ranging from 3.91–4.03 mm. Upper incisors opisthodont with incipient groove on anterior face.

DISTRIBUTION: *Thomasomys cinnameus* occurs in the Colombian and Ecuadorean Andes, in páramo and montane forests, from about 2,400 to 3,800 m.

SELECTED LOCALITIES (Map 334): COLOMBIA: Antioquia, Páramo, 7 km E of Sonsón (FMNH 70342); Cauca, 0.5 km E of Laguna San Rafael, Páramo de Puracé (KU 124058); Huila, San Agustín, Las Bardas (FMNH 71489). ECUADOR: Azuay, The Cajas Plateau (Barnett 1999); Carchi, Reserva Biológica Guandera (Tirira and Boada 2009); Chimborazo, Cochaseca (AMNH 63039); Morona-Santiago, Río Upano, Sangay National Park (T. E. Lee et al. 2011); Napo, 1.6 km W of Papallacta (UMMZ 155668); Tungurahua, Hacienda San Francisco, east of Ambato (type locality of *Thomasomys cinnameus* Anthony).

SUBSPECIES: I consider *T. cinnameus* to be monotypic (but see Remarks).

NATURAL HISTORY: Individuals of this species have been trapped among mossy boulders in an old lava flow that impounds the Río Tambo to form Laguna Papallacta, and on the ground in subalpine rainforest in the valley of the Río Papallacta (Voss 2003). Barnett (1999) collected the species only in páramo, in minimally disturbed quenoa forests, always on the ground in areas with dense canopy and little groundcover but deep moss, and in dense and isolated clumps of composite bushes on exposed slopes covered with tussock grass. T. E. Lee et al. (2011) collected the species in a variety of habitats, including elfin forest, bogs, and swampy cloud forest. One female was lactating, and males had scrotal testes in August, during the dry season (Barnett 1999).

REMARKS: Both Cabrera (1961) and Musser and Carleton (1993) treated *T. cinnameus* as a synonym of *T. gracilis*, but Voss (2003) amplified its unique traits in comparisons with both *T. gracilis* and *T. hudsoni* and recognized *T. cinnameus* as a valid species. Both Pacheco's (2003) cladistic assessment of morphological characters and Salazar-Bravo and Yates's (2007) phylogenetic analysis of mtDNA characters support this conclusion. Pacheco (2003) found *T cinnameus* closely related to *T. ucucha* while Salazar-Bravo and Yates (2007) placed it at the base of the group of species forming the *"cinereus"* group. Clearly, additional work is needed to resolve relationships, although the validity of *T. cinnameus* as a species seems without doubt.

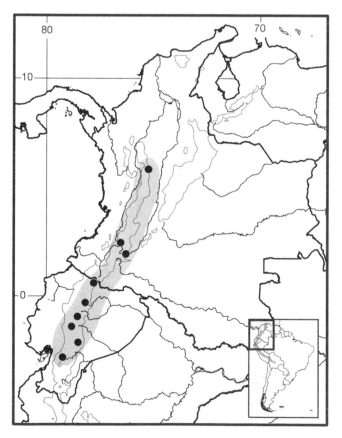

Map 334 Selected localities for *Thomasomys cinnameus* (●). Contour line = 2,000 m.

Voss (2003) considered *T. cinnameus* to be monotypic based on his review of the type specimen and Ecuadorean material, a decision I follow here. My assignment of some Colombian populations to this species results from their close morphological resemblance to the holotype (AMNH 67401) in most qualitative and quantitative characters, including the cinnamon-brown dorsal pelage and the incisors with incipient groove. However, several traits differ between the Ecuadorean and Colombian samples that suggest the presence of more than one taxon. For example, the sample from Huila is smaller (molar toothrow 3.59 mm on average) and with a narrow, slightly posteriorly inclined zygomatic plates. Furthermore, the Antioquia sample is probably the most distinct, as the genal 1 vibrissae are present, incisors are orthodont, and incisive foramina are shorter among other traits. Unfortunately, these samples are small, and more specimens would be necessary before a formal revision is possible.

Thomasomys contradictus Anthony, 1925
Central Andes Thomasomys

SYNONYMS:

Thomasomys cinereiventer: J. A. Allen, 1916b:211; part; not *Thomasomys cinereiventer* J. A. Allen.

Thomasomys cinereiventer contradictus Anthony, 1925:3; type locality "Santa Isabel, Quindio Andes, [Quindío,] Colombia; altitude, 12,700 feet [3,871 m]."

DESCRIPTION: Large species (head and body length 107–133 mm) with mummy brown coloration along sides ranging to fuscous on back, with some black-tipped hairs; ears brown; hands and feet drab. Ventral pelage between clay color and honey yellow, transition from darker color of sides very gradual. Pelage quite long, soft, and lax. Mystacial vibrissae moderately long, extending slightly beyond posterior margin of pinnae when bent; genal 1 vibrissae absent. Tail moderately long (111–138% of head and body length), brown above and below, coarsely annulated, sparsely haired, without pencil. Hindfoot long (30–34 mm), and drab in color above; gap between thenar and hypothenar pads present. Digit I of pes (hallux) short, its claw not extending more than about half length of phalanx 1 of dII. Digit V of pes long, its claw extending about half length of phalanx 2 of dIV. Skull large (condyloincisive length, 27.8–29.6 mm) with long and comparatively broad rostrum, and oblong braincase comparatively more rounded and less flat than that of *T. cinereiventer*. Zygomatic notches shallow, and nasals extend posteriorly behind premaxillae. Interorbital region moderately broad, hourglass in shape, with rounded margins. Incisive foramina short and broad, and do not reach plane of M1s. Jugal bones long and ethmoturbinals large. Optical foramen placed dorsal to M3. Zygomatic arches weakly convergent; zygomatic plates broad and slightly sloped backward from bottom to top. Mesopterygoid fossa broad with subparallel margins and without medial process; sphenopalatine vacuities present as narrow slits. Auditory bullae small to medium in size and moderately inflated. Carotid circulatory pattern primitive (stapedial foramen, groove on inner surface of squamosal and alisphenoid, and sphenofrontal foramen present). Upper molars moderately hypsodont, without interpenetrating flexi, and without crenulations. Anteroflexus of M1 distinct, and anteromedian flexid of m1 indistinct. Maxillary toothrow moderately long (4.8–6.6 mm), but shorter than *T. cinereiventer.* Upper incisors opisthodont.

DISTRIBUTION: *Thomasomys contradictus* is known only from the central Andes of Colombia, from ca. 2,400 to 3,871 m.

SELECTED LOCALITIES (Map 335): COLOMBIA: Antioquia, Mpio. La Unión, Vda. San Miguel de la Cruz, sitio Alto de San Miguel de la Cruz (ICN 17126), Medellín, Las Palmas (FMNH 70339), Sonsón, Páramo, 7 km E (FMNH 70273); Caldas, Manizales, Río Termales (FMNH 71319); Cauca, Almaguer (AMNH 37768), El Roble (AMNH 32960); Tolima, Hacienda Indostán (IAvH-M 5704).

SUBSPECIES: *Thomasomys contradictus* is monotypic.

NATURAL HISTORY: Little is known about any aspect of the ecology or population biology of this species, but it is present in subtropical/tropical moist montane forests (Gómez-Laverde and Pacheco 2008).

REMARKS: Cabrera (1961) and more recent taxonomic synopses (e.g., Musser and Carleton 2005) maintained *contradictus* as a subspecies of *T. cinereiventer.* Following my earlier analysis (Pacheco 2003) based on direct comparisons of available specimens, including the relevant type material, I regard *T. contradictus* as a species distinct from *T. cinereiventer. Thomasomys contradictus* is smaller, its zygomatic arches are not strongly convergent, the interorbital region lacks square margins, zygomatic notches are shallow, the rostrum and incisive foramina are broader, the upper molars lack interpenetrating flexi and crenulations, and the anteromedian flexid of m1 is indistinct. Nevertheless, some populations show a degree of variation that should be formally assessed with larger series. The samples from Sonsón and Las Palmas, Antioquia, have orthodont upper incisors and vertical zygomatic plates; these may prove to be a separate species. Anthony (1925) considered the specimens from Almaguer as intermediate between *T. cinereiventer* and *T. contradictus.* I consider this population tentatively as *T. contradictus* based on the ventral pelage with a tendency to yellowish and the lack of overlapping flexi and crenulation in the upper molars, although the rostrum is atypically quite robust. Anthony (1925) selected an adult male specimen as the holotype (AMNH 32955), but this specimen is a female, with a long prepuce covered by long hairs as typical of many *Thomasomys.*

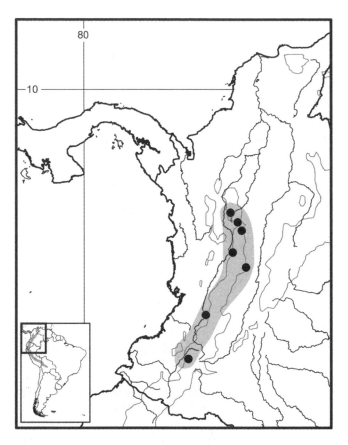

Map 335 Selected localities for *Thomasomys contradictus* (●). Contour line = 2,000 m.

Thomasomys daphne Thomas, 1917
Daphne's Thomasomys

SYNONYMS:

Thomasomys daphne Thomas, 1917g:2; type locality "Ocobamba Valley, 9,100 feet [2,774 m]," Cusco, Peru.

Thomasomys daphne daphne: Gyldenstolpe, 1932a:58; name combination.

DESCRIPTION: Small species (head and body length 88–109 mm; condyloincisive length of skull 24.5–26.3 mm) with dull brownish dorsal pelage and flanks almost as dark as dorsal region; ears small and brown, not contrasting markedly with head. Ventral pelage slaty gray but washed with buff, and not countershaded. Mystacial vibrissae moderately long, extending slightly beyond posterior margin of pinnae when bent; genal 1 vibrissae absent. Tail medium to long (106–166% of head and body length), covered by fine scales and unicolored brown, except for white tip in most specimens. Hindfoot short (23–27 mm), slender, and pale brown. Skull comparatively short and narrow rostra and nasals, moderately broad interorbital region with edges not ridged, and moderately inflated braincase. Incisive foramina narrow and relatively short, ending just in front of M1s, not extending to between them. Auditory bullae small. Carotid circulatory pattern derived (stapedial foramen small, no internal squamosal-alisphenoid groove, no sphenofrontal foramen). Capsular process distinct in adult specimens. Upper incisors gracile and proodont, and maxillary toothrow short, 3.8–4.2 mm.

DISTRIBUTION: *Thomasomys daphne* occurs along the eastern Andean slopes from southern Peru to northern Bolivia, in montane forests habitats at elevations from 2,000 to 3,540 m.

SELECTED LOCALITIES (Map 336): BOLIVIA: La Paz, Campamento Piara (CBF 6731), Murillo, Cuticucho (CBF 4450), Okara (AMNH 72131), ca. 15 km by road NE of Unduavi at old railroad crossing (UMMZ 155879), 30 km by road N of Zongo, cement mine (UMMZ 156190). PERU: Cusco, Chacarara, Hatunpampa (MUSM 17457), Occobamba Valley, Tocapoqueu (USNM 194902), 32 km NE (by road) of Paucartambo, km 112 (MVZ 166705), 90 km by road SE of Quillabamba (UMMZ 160749); Puno, 3 mi N of Limbani (MVZ 139525), 9 km N of Limbani (MVZ 171501).

SUBSPECIES: I consider *T. daphne* to be monotypic (but see Remarks).

NATURAL HISTORY: Little is known of the ecology or behavior of this species. Specimens in southern Peru were obtained on the ground in the dense understory of disturbed cloud forest and among clumps of bunch grasses in open areas within that forest formation. Known ectoparasites include the fleas *Plocopsylla viracocha* and *Neotyphloceras* sp. (Maihuay et al. 1994).

REMARKS: S. Anderson (1997) used *australis* as a subspecies of *T. daphne* for Bolivian populations, but Pacheco (2003), Voss (2003), and Salazar-Bravo and Yates (2007) each treated this taxon as a species. Although I regard *T. daphne* as monotypic, earlier I (Pacheco 2003) acknowledged this taxon was likely composed of at least two species, with the southern Peruvian population from the Limbani drainage in Puno department and northern Bolivian populations in La Paz department distinguished from topotypic samples from the Urubamba drainage in Cusco department. Compared with *T. daphne*, these populations exhibit the primitive condition of the carotid circulation (pattern 1, *sensu* Voss 1988), a broader interorbital region, and broader incisive foramina, among other features. For these reasons, Pacheco (2003) suggested these populations represent and unnamed taxa, informally named *Thomasomys* sp. 3. Salazar-Bravo and Yates (2007) included a Peruvian and a Bolivian sample of this taxon (reported as *T. daphne*) in a phylogenetic analysis based on mtDNA cytochrome-*b* gene, and found it closely related to *T. australis*, from which it was distinguished by 7.9% of uncorrected percent sequence divergence; a hypothesis also supported by Patton's (1986) species-limited allozyme analysis. The morphological phy-

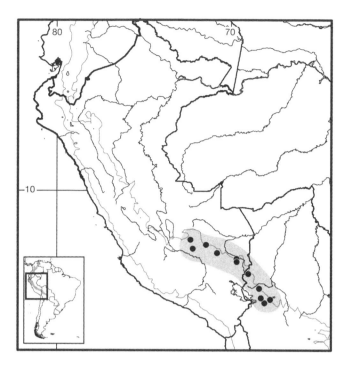

Map 336 Selected localities for *Thomasomys daphne* (●). Contour line = 2,000 m.

logeny of Pacheco (2003) placed *T. daphne* as closely related to *Thomasomys* sp. 3 followed by *T. australis*.

Thomasomys dispar Anthony, 1925
Colombian Thomasomys

SYNONYMS:

Thomasomys cinereiventer: J. A. Allen, 1916b:211; part; not *Thomasomys cinereiventer* J. A. Allen, 1912.

Thomasomys cinereiventer dispar Anthony, 1925:2; type locality "Andalucia, Eastern Andes, Huila, Colombia, elevation 7000 feet."

DESCRIPTION: Medium-sized (adult head and body length 98–119 mm) with soft, dense, and comparatively long dorsal fur. Pelage coloration mummy brown on back and sides, heavily washed with fuscous black in middorsal region. Ventral pelage ranges between ivory-yellow and cream-buff, gradually transitioning into darker color of flanks. Mystacial vibrissae long, extending well beyond posterior margin of pinnae when bent; genal 1 vibrissae absent. Tail very long (125–186% of head and body length), incipiently bicolored, with terminal white tip. Hindfoot moderately long (27.2–33 mm) and narrow; dorsal surface and that of digits light brown; gap present between hypothenar and thenar pads. Pinnae small (20.5–21.5 mm) and clothed with short, brown hairs. Digit I of pes (hallux) moderately long, with claw extending less than about half length of phalanx 1 of dII. Digit V of pes moderately long, with claw barely extending to interphalangeal joint of dIV. Skull moderately long (condyloincisive length 28.9–32.4 mm) with relatively long and moderately broad rostrum, anteriorly expanded nasals producing distinct rostral tube, and rather straight frontonasal profile. Braincase somewhat oblong and less flat than that of *T. cinereiventer*. Interorbital region moderately broad, hourglass in shape, and typically with smooth margins, rarely marked. Zygomatic arches converge anteriorly. Zygomatic plates narrow (breadth shorter than length of M1) and slightly sloped backward from bottom to top; anterior margins straight. Incisive foramina moderately broad, somewhat oval, but slightly broader posterior to premaxillomaxillary suture, and very short, not extending close to anterior margins of M1s. Mesopterygoid fossa comparatively broad with subparallel margins and with distinct medial process; sphenopalatine vacuities absent or occasionally present as slits between presphenoid-basisphenoid suture. Optic foramen small and placed posterodorsal to M3. Postglenoid foramen smaller than subsquamosal fenestra, and placed somewhat ventral to it. Auditory bullae small and uninflated; bony tubes distinct or poorly distinct. Carotid circulatory pattern primitive (stapedial foramen, groove on inner surface of squamosal and alisphenoid, and sphenofrontal foramen present). Alisphenoid strut present. Maxillary toothrow moderately long (5.2–5.5 mm); upper molars broad, comparatively brachydont, with no interpenetration and no crenulation, but with weak labial cingula. Anteromedian flexus of M1 distinct when unworn; anteromedian flexid of m1 indistinct, even in unworn condition. Upper incisors orthodont. On mandible, condylar and angular processes level, and capsular process absent.

DISTRIBUTION: *Thomasomys dispar* occurs in the central and eastern Andes of southern Colombia, in montane forests between 2,130 and 3,310 m.

SELECTED LOCALITIES (Map 337): COLOMBIA: Cauca, Almaguer (AMNH 33768), Coconuco, Parque Nacional Puracé, Río Bedón, Versalles, N° Calma 141-M (IAvH-M 266), Paletará, PNN Puracé (IAvH-M 1641), PNN Puracé, 0.5 km W of Laguna San Rafael (IAvH-M 3765), Huila, Andalucía (type locality of *Thomasomys cinereiventer dispar* Anthony), San Agustín, Las Bardas (FMNH 71377), San Agustín, Las Ovejeras (FMNH 71399), San Agustín, Santa Marta (FMNH 71405).

SUBSPECIES: I treat *T. dispar* as monotypic (but see Remarks).

NATURAL HISTORY: Little is known about this species, but it is present in subtropical/tropical moist montane forests (Gómez-Laverde and Pacheco 2008).

REMARKS: Cabrera (1961) maintained *dispar* Anthony as a subspecies of *T. cinereiventer*, as did Musser and Carleton (2005). Herein, I treat this taxon as a full spe-

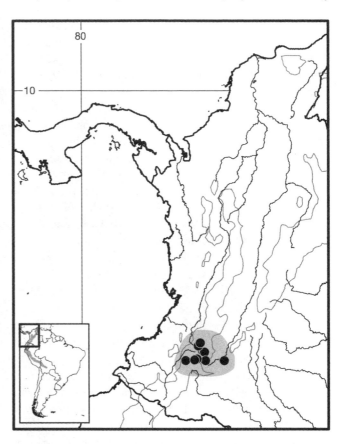

Map 337 Selected localities for *Thomasomys dispar* (●). Contour line = 2,000 m.

cies based on my earlier (Pacheco 2003) side-by-side comparisons of available specimens, including type material. *Thomasomys dispar* can be distinguished from *T. cinereiventer* by a combination of characters, including smaller size but relatively longer tail, moderately long digit V of the pes, rostral tube present, narrow zygomatic plates, small optic foramen posterior to M3, palatal medial process present, interorbital region with incipient smooth margins, and upper incisors orthodont, among other morphological features. Although I consider *T. dispar* as monotypic, a more thorough analysis of geographic trends may result in additional valid taxa. For example, the series from San Agustín, Huila, while generally similar to the type series, may prove to be an undescribed species based on the presence of a derived (pattern 3) carotid circulation, extremely large posterior alisphenoid channel, long jugal, and other morphological features (V. Pacheco, unpubl. data).

Thomasomys eleusis Thomas, 1926
Peruvian Thomasomys

SYNONYMS:

Thomasomys cinereus: Osgood, 1914b:161; part; not H[*esperomys*]. *Rhipidomys cinereus* Thomas.

Thomasomys ischyrus eleusis Thomas, 1926e:614; type locality "Tambo Jenes, 12,000′ [3,658 m]," Amazonas, Peru.

Thomasomys eleusis: Musser and Carleton, 1993:749; first use of current name combination.

DESCRIPTION: Medium size (head and body length 126–140 mm) with long, soft, and dense dorsal fur. Dorsal coloration dark gray with slight tinge of brownish on sides. Ventral pelage paler gray, with buffy overtone; hairs slate colored at base with tips dull whitish or slightly buffy; moderately countershaded relative to dorsal pelage. Mystacial vibrissae moderately long, extending slightly beyond posterior margin of pinnae when bent; genal 1 vibrissae absent. Tail approximates head and body length (93–112%), comparatively thick, indistinctly bicolored, and lacks terminal white tip. Hindfoot moderately long (28–31 mm); metacarpals and digits covered with pure white, shiny hairs; gap present between thenar and hypothenar pads; hands whitish. Digit I of pes (hallux) moderately long, with claw extending close or to interphalangeal joint of dII. Digit V of pes also long, with claw extending about half length of phalanx 2 of dIV. Skull moderate in size (condyloincisive length 28.6–30.3 mm) with moderately broad rostrum that tapers anteriorly. Nasals distinctly projected anteriorly but do not develop as rostral tube; long and narrow, moderately spatulate, with posterior margins distinctly longer than premaxillae. Antorbital bridge positioned high (<25% of rostrum). Zygomatic plates moderately broad, subequal in length relative to M1, and slightly sloped backward. Interorbital region moderately broad (molar row partially visible in dorsal view), hourglass in shape, and with incipiently square edges. Zygomatic arches converge slightly anteriorly. Incisive foramina long, with subparallel margins slightly broader at level of premaxillomaxillary suture, and extend to between M1 anterocones. Ethmoturbinals large; optical foramen small and placed posterodorsal to M3. Molar-bearing portion of maxillae nearly triangular. Mesopterygoid fossa broad, with subparallel margins and palatal medial process absent; roof closed, with sphenopalatine vacuities absent or represented only as slits. Auditory bullae small and uninflated. Carotid circulatory pattern derived (stapedial foramen small, no internal squamosal-alisphenoid groove, no sphenofrontal foramen). Molar rows moderately long (5.2–5.7 mm), individual teeth moderately hypsodont, with interpenetrating flexi and crenulation absent. M1 with accessory labial root; m1 anteromedian flexid weak, even in unworn teeth. Upper incisors orthodont. Capsular process absent.

DISTRIBUTION: *Thomasomys eleusis* is limited to northern Peru, east of the Río Marañón in the Cordillera Oriental, at elevations between 3,050 and 3,660 m.

SELECTED LOCALITIES (Map 338): PERU: Amazonas, Chachapoyas, Tambo Jenes, mountains E of Balsas (type locality of *Thomasomys ischyrus eleusis* Thomas), San Pedro (AMNH 72101); La Libertad, Piedra Negra, on trail from Los Alisos to Parque Nacional Río Abiseo (USNM 582212); San Martín, Parque Nacional Río Abiseo, Pampa de Cuy (Leo and Romo 1992).

SUBSPECIES: *Thomasomys eleusis* is monotypic.

NATURAL HISTORY: This species has been captured in dense humid forest (Osgood 1914b; Thomas 1926e). More recent specimens were found in habitats ranging from Puna grassland, bare or shrub rocky places, shrubby places bordering streams, or at the border of continuous humid montane forests, all within in the Subalpine rain páramo or Montane rainforest of the Holdridge system (Leo and Romo 1992).

REMARKS: Osgood (1914b) apparently had no true *Thomasomys cinereus* at hand to compare when he reported on the series of specimens from the mountain east of Balsas, Amazonas department, Peru, as true *T. cinereus*. He probably relied on Thomas's (1882) original descrip-

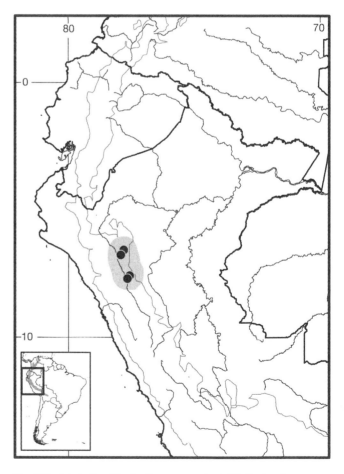

Map 338 Selected localities for *Thomasomys eleusis* (●). Contour line = 2,000 m.

tion of *T. cinereus* and on the geographical proximity of its type locality. Later, however, Thomas (1926e) compared the Balsas series with the holotype of *T. cinereus*, documented their distinction, and allied *T. eleusis* to *ischyrus* rather than to *cinereus* based on the general "champagne-bottle" outline of their skulls. Pacheco (2003) reported on external, cranial, dental, and soft morphological characters of this species and confirmed that *T. eleusis* is a species distinct from both *T. cinereus* and *T. ischyrus*. In addition, *T. eleusis* was found near the base of a clade that included *T. incanus* and allies in a phylogeny based on morphological attributes (Pacheco 2003). This clade is informally called here the "incanus" group. Leo and Romo (1992) reported this species from Parque Nacional Río Abiseo, San Martín department, under the name of *T. ischyrus eleusis*. The karyotype is unknown.

Thomasomys emeritus Thomas, 1916
Venezuelan Thomasomys

SYNONYMS:

Thomasomys laniger emeritus Thomas, 1916j:479; type locality "Montes de Escaguer, alt. 2500 m," Mérida, Venezuela.

Thomasomys laniger: P. T. Johnson, 1972:22; part; not *Oryzomys laniger* Thomas.

DESCRIPTION: Medium size (head and body length 104–120 mm) with long, soft, and fluffy dorsal fur. Dorsal coloration similar to that of *T. laniger*, but with more brownish tone, especially on head, and more rufous sides. Ventral pelage dull orangish, not contrasting with sides. Mystacial vibrissae very long, extending distinctly beyond posterior margin of pinnae when bent; genal 1 vibrissae absent. Postauricular patches whitish but not very conspicuous. Tail slightly shorter to slightly longer than head and body (92–120%), slender, finely haired, monocolored, and without terminal white tip. Hindfoot relatively short (23–25 mm), metacarpals and digits whitish, and gap present between thenar and hypothenar pads. Digit I of pes (hallux) of moderate length, with claw not extending more than about half length of phalanx 1 of dII. Digit V of pes long, with claw extending about half length of phalanx 2 of dIV. Skull relatively small (condyloincisive length 24.7–26.1 mm); braincase rounded; rostrum long and narrow; interorbital region moderately broad, hourglass in shape, with round margins. Zygomatic plates moderately broad, subequal to length of M1, and sloped backward from bottom to top. Zygomatic arches moderately convergent anteriorly. Incisive foramina of moderate length, extending to or only near to anterior margins of M1s; sides parallel to slightly contracted anteriorly. Mesopterygoid fossa more expanded anteriorly and contracted posteriorly; medial process small or absent; and roof usually perforated by moderate sized sphenopalatine vacuities. Parapterygoid vacuities present. Auditory bullae large and inflated. Carotid circulatory pattern primitive (stapedial foramen, groove on inner surface of squamosal and alisphenoid, and sphenofrontal foramen present). Upper molars moderately hypsodont, with interpenetration of flexi; anteroloph of M1 distinct; protoflexus of M2 inconspicuous; hypoflexus of M3 distinct; ectolophid of m1 present. Maxillary toothrow relatively short, 4.1–4.6 mm. Upper incisors opisthodont. Capsular process absent.

DISTRIBUTION: *Thomasomys emeritus* occurs only in the Venezuelan Andes, in the departments of Mérida and Trujillo, at elevations from 2,090 to 3,550 m.

SELECTED LOCALITIES (Map 339): VENEZUELA: Mérida, Bosque San Eusebio, 7 km SSE of La Azulita (CVULA 1234), Laguna Brava, Páramo de Mariño, CVULA, 2545), Laguna Negra, 5.75 km ESE of Apartaderos (USNM 579462), La Montana, 4.1 km SE of Mérida (USNM 579457), Montes de Escaguer (type locality of *Thomasomys laniger emeritus* Thomas), Monte Zerpa, 6 km N of Mérida (CVULA 4918), Paramito, 4 km W of Timotes (USNM 374554), Sierra de Mérida (FMNH 22121), Tabay, 9 km SE of Tabay, Laguna Verde (USNM 374591); Trujillo, Hacienda Misisi, 15 km E of Trujillo (USNM 495477), Macizo de Guaramacal, 9 km ESE of Boconó (CVULA 3342).

SUBSPECIES: *Thomasomys emeritus* is monotypic.

NATURAL HISTORY: Handley (1976) caught this species among lichen and moss-covered boulders and in rock slides, in thick, low shrubs on the ground, at the base of trees and logs, in trees, and in a house; most often in relatively dry habitats but frequently in damp places; in cloud forest and páramo. All localities are in the Lower Montane very humid forest, Montane very humid forest, Montane rainforest, or Subalpine páramo (in the vegetative life zones of Ewel and Madriz 1968).

An extensive list of ectoparasites is known from *T. emeritus*, though most with the host species incorrectly identified as *T. laniger*. These include the anopluran louse *Hoplopleura indiscreta* (P. T. Johnson 1972), the ascarid mites *Teratothrix truncata* and both *Eutrombicula cricetivora* and *E. webbi* (Brennan and Goff 1979) as well as *Laelaps paulistanensis*, *Gigantolaelaps canestrinii*, and *Androlaelaps fahrenholzi* (Furman 1972) and *Odontacarus dienteslargus*, *O. schoenesetosus*, *O. tuberculohirsutus*, and *O. vergrandi* (J. T. Reed and Brennan 1975), the chigger *Hoffmannina reedi* (Brennan 1972), the ticks *Ixodes jonesae* (Kohls et al. 1969) and *Ixodes* sp. (E. K. Jones et al. 1972), and the fleas *Cleopsylla monticola*, *Ctenidiosomus perplexus*, *Neotyphloceras rosenbergi*, *Pleochaetis smiti*, *Plocopsylla ulysses*, and *Sphinctopsilla tolmera* (Tipton and Machado-Allison 1972).

REMARKS: Thomas (1916j) thought that *emeritus* was larger than true *laniger*, but he had only the type specimens for comparisons. My unpublished comparisons (12 specimens of *T. laniger* versus five of *T. emeritus*, all the same "age" based on tooth wear) indicate that both species are rather similar in most external and cranial measurements. Although *T. laniger* is slightly larger than *T. emeritus* in most variables, the difference between the two species is more obvious in condyloincisive length (average of 26.1 mm versus 25.2 mm), diastema length (average of 7.96 mm versus 7.46 mm), and least interorbital breadth (average of 4.86 mm versus 4.55 mm). Furthermore, the dorsal color of *T. emeritus* is more brownish, its undersurface more orangish, the dorsal surface of hindfoot light brown, rostrum relatively longer and narrow, the mesopterygoid fossa wider anteriorly (lyre-shaped), the incisive foramina are less open, and upper incisors are opisthodont. Aguilera et al. (2000) reported the karyotype of *T. emeritus* (reported as *T. laniger*) as $2n = 42$, $FN = 40$, with an X metacentric and a Y acrocentric chromosomes. This complement is distinct from the karyotypic formula of *T. laniger* proper, reported as $2n = 40$, $FN = 40$, with an X chromosome metacentric or submetacentric and a Y acrocentric chromosome (Gómez-Laverde et al. 1997). These morphological and karyotypic differences support the hypothesis that *T. emeritus* is a valid species, distinct from *T. laniger*.

Pacheco (2003) presented a phylogeny based on morphological characters where *T. laniger* belonged to the clade "*cinereus*" group. Thomas (1916j) stated the type specimen of *emeritus* (BM 5.1.1.4) is a young adult male; however, my own examination of the type confirmed the specimen is a young adult female, with a long prepuce (typical of many *Thomasomys*) and six mammae.

Thomasomys erro Anthony, 1926
Wandering Thomasomys

SYNONYMS:

Thomasomys erro Anthony, 1926b:5; type locality "upper slopes of Mt. Sumaco, exact altitude unknown, but probably 8000–9000 feet [approx. 2,438–2,743 m] at head of the Rio Suno [= Río Suno], a tributary of the Rio Napo [= Río Napo]," Napo, Ecuador.

Thomasomys cinereiventer erro: Cabrera, 1961:427; name combination.

DESCRIPTION: Relatively large (adult head and body 120–130 mm) with long, soft, and lax dorsal fur. Dorsal pelage coloration grayish brown, paler on sides; ventral pelage washed with cream-buff, not countershaded relative to dorsal pelage. Mystacial vibrissae moderately long, extending slightly beyond posterior margin of pinnae when bent; genal 1 vibrissae absent. Tail long (130% to 139% of head and body length), thinly haired, uniformly brown in color, and terminates with 10–20 mm whitish tip extending as thin tuft of hairs. Hindfoot moderately long (30–32 mm), metacarpals similar in color to dorsum but paler, digits and ungual tuft of bristles whitish, and gap present between thenar and hypothenar pads. Ventral base of manual claw keeled. Digit I of pes (hallux) short, with first phalanx not extending beyond metatarsal of digit II and claw no more than about half the length of p1 of digit II. Digit V of pes moderately long, with claw extending to, but not beyond, first interphalangeal joint of digit IV. Skull moderately long (condyloincisive length 27.2–28.4 mm) and slender, with distinct rostral tube. Braincase inflated, rather globose in shape. Interorbital region broad, hourglass in shape, with rounded margins. Zygomatic plates narrow, shorter than length of M1, with anterior margin straight and vertical. Optical foramen small and placed posterior to M3. Incisive foramina oval, large, more contracted at anterior margins, but short, not reaching anterior margins of M1s. Mesopterygoid fossa broad, slightly expanded anteriorly, followed by subparallel margins posteriorly; medial process absent; and roof closed, or perforated by narrow slits on both sides of presphenoid-basisphenoid suture. Auditory bullae moderately inflated, bony tubes short. Postglenoid

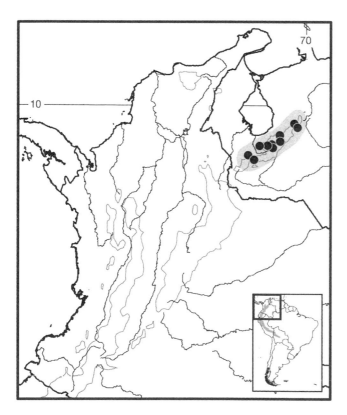

Map 339 Selected localities for *Thomasomys emeritus* (●). Contour line = 2,000 m.

foramen very small. Carotid circulatory pattern primitive (stapedial foramen, groove on inner surface of squamosal and alisphenoid, and sphenofrontal foramen are present). Upper molars moderately hypsodont, with no interpenetration of flexi. Maxillary toothrow moderately long, 4.5–4.8 mm. Upper incisors orthodont. M1 withaccessory labial root; M3 comparatively small; m1 with indistinct anteromedian flexid. Capsular process absent. Stomach unilocular with reduced glandular epithelium, and with lesser curvature usually lined with cornified squamous epithelium.

DISTRIBUTION: *Thomasomys erro* is known from the Cordillera Oriental of north-central Ecuador, across an elevational range from about 1,900 to 3,600 m.

SELECTED LOCALITIES (Map 340): ECUADOR: Napo, Cosanga, Río Aliso (EPN 902741), 9 km E of Papallacta (UMMZ 127133), 6.2 km W of Papallacta (UMMZ 155711), Volcán Sumaco (type locality of *Thomasomys erro* Anthony).

SUBSPECIES: *Thomasomys erro* is monotypic.

NATURAL HISTORY: Individuals have been captured in dense secondary vegetation, in subalpine rainforest and upper montane rainforest, on the ground, in runways through wet leaf litter and mossy debris, beneath mossy logs, branches, or roots, and inside the trunk of a hollow tree (Voss 2003). Voss (2003) only collected specimens at elevations ranging from 2,830 to 3,570 m, but T. E. Lee et al. (2006) reported the species from 1,900 m. Schramm and Lewis (1987) described *Plocopsylla nungui* as a new species of flea from specimens of *T. erro* from near Papallacta, Napo. Timm and Price (1985) reported the mallophagan *Cummingsia inopinata*, also on specimens from Papallacta, Napo.

REMARKS: *Thomasomys erro* was described on p. 5 in Anthony (1926b), not on p. 3 as given by Cabrera (1961) and Musser and Carleton (2005). Anthony (1926b) considered *T. erro* to be a highly distinctive species that was possibly derived from *T. cinereiventer*, a decision followed by Gyldenstolpe (1932) and Ellerman (1941) but not Cabrera (1961), who formalized the latter opinion as the trinomial *T. cinereiventer erro*. Musser and Carleton (1993) followed Cabrera's synonymy, but Tirira (1999) listed *erro* as a distinct species. Voss (2003) reported new records of *T. erro* from Papallacta, Napo province, enumerating morphological traits that clearly discriminated *T. cinereiventer* and *T. erro* as valid species. He also pointed out some differences between the Papallacta series and the type specimen from Volcán Sumaco, but with the exception of two dental measurements, judged these specimens to represent the same species. The phylogenetic position of *T. erro* is still unclear because the species shares characteristics with both *Thomasomys* and *Aepeomys* (Pacheco 2003).

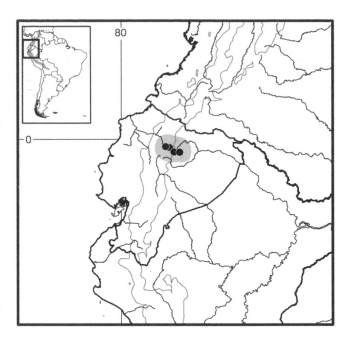

Map 340 Selected localities for *Thomasomys erro* (●). Contour line = 2,000 m.

Thomasomys fumeus Anthony, 1924
Smoky Thomasomys

SYNONYMS:

Thomasomys rhoadsi fumeus Anthony, 1924b:6; type locality "Hacienda San Francisco, east of Ambato, [Tungurahua,] central Ecuador, elevation about 8000 feet [2,438 m]."

T[*homasomys*]. *fumeus*: Tirira, 2007:196; first use of current name combination.

DESCRIPTION: Relatively small to medium sized (head and body length 101–125 mm) with long and soft dorsal fur. Dorsal coloration mixture of dusky neutral gray and pale neutral gray, darkest along back. Ventral color pale neutral gray, not countershaded with dorsal pelage. Mystacial vibrissae moderately long, extending slightly beyond posterior margin of pinnae when bent; genal 1 vibrissae absent. Tail short (76–116% of head and body length), comparatively thick and thinly haired, annular rings conspicuous, uniformly colored, and without terminal white tip. Hindfoot short (24–27 mm), with metatarsals light brown washed with white hairs. Digit I of pes (hallux) short, first phalanx does not extend beyond metatarsal of dII and its claw extends no more than about half length of phalanx 1 of dII. Digit V of pes long, with claw extending about half length of phalanx 2 of dIV. Skull comparatively short (condyloincisive length, 26.5–27.6 mm) with moderately broad rostrum, and anteriorly expanded nasals that taper posteriorly and extend beyond premaxillae. Interorbital region moderately broad, hourglass in shape, with rounded margins. Zygomatic plates narrow, breadth shorter than length of M1 and with anterior margins nearly straight and vertical. Zygomatic arches moderately convergent anteriorly. Incisive foramina widely open, with contracted anterior and posterior margins, and moderately long, reaching to anterior margins of M1s. Mesopterygoid fossa broad, wider anteriorly and convergent posteriorly; medial process absent. Auditory bullae small and uninflated. Postglenoid foramen and subsquamosal fenestra subequal in size. Carotid circulatory pattern primitive (stapedial foramen, groove on inner surface of squamosal and alisphenoid, and sphenofrontal foramen present). Upper incisors orthodont. Maxillary toothrow moderately long, 4.6–5.0 mm. Upper molars moderately hypsodont, with no interpenetration of flexi. M1 with distinct anteroloph; anteromedian flexid of m1 absent or weak; M3 relatively large and with distinct hypoflexus. Capsular process absent.

DISTRIBUTION: *Thomasomys fumeus* is endemic the Cordillera Oriental of Ecuador; it is known only from the type locality.

SELECTED LOCALITIES (Map 341): ECUADOR: Tungurahua, San Francisco, E of Ambato (type locality of *Thomasomys rhoadsi fumeus* Anthony).

SUBSPECIES: *Thomasomys fumeus* is monotypic.

NATURAL HISTORY: Nothing is known about the natural history of this species.

REMARKS: Anthony (1924b) distinguished the Tungurahua series from typical *rhoadsi* Stone (herein regarded as a synonym of *T. vulcani*; see that account), but used the trinomial combination to imply a relationship with *rhoadsi*. This decision was followed in recent synopses (e.g., Musser and Carleton 2005). At present, no revision or other studies of this taxon have been completed, although Voss (2003) recognized that *T. rhoadsi fumeus* was substantially smaller than typical *rhoadsi* and differed from it in qualitative external and cranial characters, as noted originally by Anthony (1924b). Pacheco (2003) was first to recognize *fumeus* as a valid species. Later, Tirira (2007), and based on a personal communication from me, also listed the species as valid, a decision followed by T. E. Lee et al. (2008). However, formal support for species distinction has yet to be established.

Herein, I argue that *T. fumeus* is a valid species, one distinct from *T. vulcani*, based on the following attributes: size generally smaller, hallux smaller, zygomatic plates narrower, anterior margin of zygomatic plates nearly straight, postglenoid foramen and subsquamosal fenestra subequal in size, M1 anteroloph distinct, and M3 hypoflexus distinct. In addition, several mensural attributes differ in size. For example, comparison of nine adult *T. fumeus* (Tungurahua, AMNH specimens) versus 21 adult *T. vulcani* (18 from Pichincha department [AMNH, FMNH, USNM, UMMZ specimens] and three from Napo department [FMNH

specimens]) reveals little or no overlaps in the following variables: head and body length (101–125 mm versus 118–153 mm), hindfoot (24–27 mm versus 27–31 mm), greatest length of skull (29.05–29.87 mm versus 30.85–33.69 mm), condyloincisive length (26.48–27.56 mm versus 28.83–31.05 mm), condylomolar length (17.22–18.10 mm versus 18.57–19.62 mm), length of nasals (11.25–12.87 mm versus 12.82–14.68 mm), length of maxillary toothrow (4.56–4.95 mm versus 5.07–5.59 mm), breadth of rostrum (4.65–5.06 mm versus 5.51–6.24 mm), breadth of palatal bridge (3.27–3.59 mm versus 3.62–4.40 mm), breadth of nasals (3.19–3.44 mm versus 3.46–3.98 mm), and breadth of zygomatic plate (1.83–2.22 mm versus 2.16–2.65 mm), respectively.

T. E. Lee et al. (2008) reported *T. fumeus* from Volcán Sumaco; however, I consider this record doubtful and suggest that their series more likely corresponds to *T. vulcani* proper. These authors reported that the external measurements are generally indicative of smaller size, but most of the variables either overlap the dimensions of *T. vulcani* or likely represent those of juvenile specimens that, by their inclusion, give the impression of overall small specimens. The Volcán Sumaco specimens should be reexamined and compared carefully with the type series of both *T. fumeus* and *T. vulcani*.

Pacheco (2003) found *T. fumeus* related to *Thomasomys* sp. 10, an undescribed species from Piura, Perú, followed by *T. vulcani*, in a cladistic analysis based on morphological characters, a general hypothesis in line with Anthony

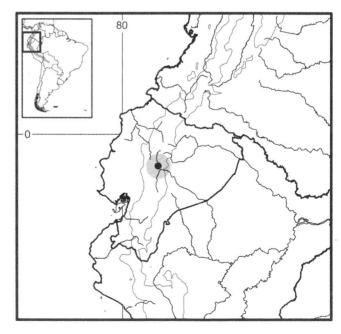

Map 341 Single known locality for *Thomasomys fumeus* (●). Contour line = 2,000 m.

(1924b). However, a phylogeny based on sequences of the mtDNA cytochrome-*b* gene disagreed with the morphological results, placing *T. fumeus* as allied to *T. aureus* and *Thomasomys* sp. (Salazar-Bravo and Yates 2007). Although it is very unlikely that *T. fumeus* is actually more closely related to *T. aureus* than to *T. vulcani*, given the number of trenchant morphological differences that differentiate members of the "cinereus" and "aureus" groups, more expansive studies are needed to resolve this issue.

Thomasomys gracilis Thomas, 1917
Gracile Thomasomys

SYNONYMS:

Thomasomys gracilis Thomas, 1917g:2; type locality "Matchu [*sic*] Picchu, 12,000 feet [3,658 m]," Cusco, Peru.

Thomasomys gracilis gracilis: Cabrera, 1961:429; name combination.

DESCRIPTION: Small (head and body length 83–101 mm) with soft, dense, and comparatively long dorsal fur (average length, 11 mm). Dorsal coloration dark grayish buff or clay, with undefined dark region along midline over rump. Ventral color soiled grayish buff with hairs slaty basally, and not sharply contrasting with dorsal color. Ears comparatively large (15–18 mm), blackish, and contrast with head color. Mystacial vibrissae moderately long, extending backward slightly beyond pinnae when bent; genal 1 vibrissae present. Tail long, moderately or distinctly longer than head and body (119–142%), incipiently bicolored but overall blackish brown and uniform, with tip sometimes white. Hindfoot relatively short (22–23 mm), with dark brown metatarsal patch, with edges and digits lighter. No gap between thenar and hypothenar pads. Hallux long, extending to interphalangeal joint of dII. Digit V of pes very long, with claw extending to base of claw of dIV. Skull short (condyloincisive length 22.4–23.6 mm), with narrow and relatively short rostrum, and rounded braincase with large parietals. Interorbital region narrow and with squared edges. Zygomatic plates narrow, subequal to M1, and vertical. Zygomatic arches very slightly flared laterally and parallel oriented forward. Incisive foramina well open and widest anteriorly, at or near premaxillomaxillary suture, and extend posteriorly to slightly between M1 anterocones. Mesopterygoid fossa comparatively broad, converging posteriorly and without medial process; roof perforated by medium-size sphenopalatine vacuities on each side of presphenoid-basisphenoid suture. Auditory bullae comparatively large and rather inflated. Subsquamosal fenestra widely open and larger than postglenoid foramen. Carotid circulatory pattern primitive (stapedial foramen, groove on inner surface of squamosal and alisphenoid, and sphenofrontal foramen present). Maxillary toothrow short (3.8–4.0 mm). Molars small and brachydont, without interpenetration of flexi but with distinct crenulation. M1 with distinct anteromedian flexus and accessory labial root; anteromedian flexid of m1 weak, even in unworn molars. Upper incisors slightly opisthodont. Capsular process absent.

DISTRIBUTION: *Thomasomys gracilis* is known only from the montane forest or Yungas of southeastern Peru, across an elevational range from 2,774 to 4,270 m.

SELECTED LOCALITIES (Map 342): PERU: Cusco, Cordillera del Vilcabamba (MUSM 13067), Idma road (AMNH 95207), Lucma, Cosireni pass, Chirapata (USNM 194813), Machu Picchu, above timberline, Runcaraccay ruins (Thomas 1917g), Occobamba valley, Tocapoqueu (USNM 194807), 90 km by road SE of Quillabamba (M. F. Smith and Patton 1999).

SUBSPECIES: I treat *T. gracilis* as monotypic (but see Remarks).

NATURAL HISTORY: Nothing has been published on the natural history of *T. gracilis*. The following comes from the field notes of J. L. Patton (MVZ archives): the habitat where the species was collected varied from very wet bunch grass on gently sloping ground to steep slopes dominated by melastome shrubs and ferns, with a dense understory of grasses, to elfin forest patches composed of bamboo thickets and large, moss and epiphyte covered trees. Runways were everywhere, particularly along the base of logs, through tree root systems, and even coursing through the thick moss growing on tree branches. All of the specimens collected (eight) were trapped on the ground, most in the patches of elfin forest, but two were taken on a steep slope covered by dense melastomaceous shrubs and ferns. Two males had scrotal testes (t = 7 or 8 mm, SV = 9–10 mm), two others had very small, nonscrotal testes (3 mm); two females were postlactating. All specimens were caught between 22 and 25 May, 1984, after the local rainy season.

REMARKS: *Thomasomys gracilis* was recognized as a monotypic species (Gyldenstolpe 1932; Ellerman 1941) until it was considered to include *cinnameus* and *hudsoni* as subspecies or referred synonyms (Cabrera 1961; Musser and Carleton 1993; respectively), hypotheses that extended the distribution of *T. gracilis* to Ecuador (Barnett 1999; Eisenberg and Redford 1999). However, these taxa were reinstated as species by Voss (2003) and independently by Pacheco (2003) on morphological grounds, a decision subsequently supported by Salazar-Bravo and Yates (2007) based on mtDNA sequences. Thus, the distribution of *T. gracilis* remains restricted to southern Peru. However, the proper identification of specimens assigned to *T. gracilis* from the Cajas Plateau, Ecuador by Barnett (1999), and allegedly sympatric with *T. cinnameus*, remains unclear. These may turn out to be *T. cinnameus* or an unknown taxon, but are not *T. gracilis* as understood herein (see the account of *T. hudsoni* for additional comments).

Reports of *T. gracilis* from Tres Cruces, Paucartambo, Cusco department, or Río Pilcopata, Cusco department (Patton et al. 1990; Pacheco et al. 1993), are erroneous (V. Pacheco, unpubl. data). Both reports were based on one specimen (MVZ 115645), determined herein as *T. oreas*. *Thomasomys gracilis* is thus absent from the upper parts of the Manu Biosphere Reserve (see also Solari et al. 2006). Furthermore, the single known specimen from the Cordillera Vilcabamba, Cusco department (Emmons et al. 2001), is provisionally referred to *T. gracilis*, although it is a little smaller and differs slightly in skull morphology than more typical specimens.

Thomas (1920h) postulated that *T. gracilis* was allied to *T. baeops* but also similar to a miniature *T. notatus*. To the contrary, Patton (1987), based on electrophoresis of 24 loci, found *T. gracilis* related to *T. oreas* as compared to other species of *Thomasomys* from southeastern Peru. Pacheco (2003) provided a phylogeny based on external, cranial, dental, and soft morphology that supported a relationship between *T. gracilis* and *T. oreas*. This hypothesis, however, was not supported by a molecular phylogeny based on mtDNA cytochrome-*b* gene sequences (Salazar-Bravo and Yates 2007), but resolution was likely limited by their small taxonomic sampling. The karyotype is unknown.

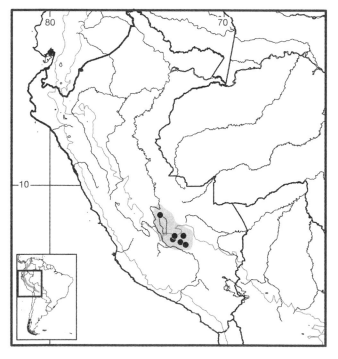

Map 342 Selected localities for *Thomasomys gracilis* (●). Contour line = 2,000 m.

Thomasomys hudsoni Anthony, 1923
Hudson's Thomasomys

SYNONYMS:

Thomasomys hudsoni Anthony, 1923b:3; type locality "Bestion [= Bestión], Provincia del Azuay, 10,100 feet [3,078 m], Ecuador."

Thomasomys gracilis hudsoni: Cabrera, 1961:429; name combination.

DESCRIPTION: Known only by holotype; apparently small species (head and body length 93 mm) with relatively long tail (129% of head and body length). Dorsal pelage grayish brown, between Dresden brown and mummy brown, darker along dorsal area. Fur long and soft. Undersurface grayish buffy, and not countershaded relative to dorsum. Mystacial vibrissae moderately long, extending slightly beyond posterior margin of pinnae when bent; genal 1 vibrissae absent. Hindfeet short (23 mm), metatarsals light brown, and gap present between thenar and hypothenar pads. Hallux moderate in length, with claw not extending more than about half length of phalanx 1 of dII. Digit V of pes is very long, with claw extending to base of claw of dIV. Tail long, slender, clothed in short hairs, uniformly colored brown, and without terminal white tip. Skull comparatively small (condyloincisive length 22 mm); braincase rounded and inflated; narrow rostrum peculiarly extended beyond incisors as flaring bony tube with concave (rather than convex) dorsal profile. Interorbital region moderately broad, hourglass in shape, with smooth edges. Zygomatic plates narrow and vertical, breadth shorter than length of M1. Incisive foramina narrow, although wider anteriorly (near premaxillomaxillary suture), and extending backward quite near anterior margins of M1s. Mesopterygoid fossa broad, anteriorly expanded, and posteriorly convergent; palatal medial process absent; roof closed, as sphenopalatine vacuities absent. Carotid circulatory pattern primitive (stapedial foramen, groove on inner surface of squamosal and alisphenoid, and sphenofrontal foramen present). Auditory bullae larger and more inflated than those of *T. cinnameus* but smaller and less inflated than those of *T. gracilis* (Voss 2003). Upper incisors orthodont. M1 with distinct anteroloph, anteromedian flexus, and anteroflexus distinct; M2 with indistinct protoflexus; M3 with indistinct mesoloph and hypoflexus. Anteromedian flexid of m1 distinct. Upper molar row small (3.5 mm). Capsular process absent.

DISTRIBUTION: *Thomasomys hudsoni* is known only from its type locality in southern Ecuador.

SELECTED LOCALITIES (Map 343): ECUADOR: Azuay, Bestión (type locality of *Thomasomys hudsoni* Anthony).

SUBSPECIES: *Thomasomys hudsoni* is monotypic.

NATURAL HISTORY: The type specimen was collected almost at the upper limit of forest where the habitat comprises rolling grassy meadows with scrub trees on some of the ridges and with thickets of brush on some of the slopes (Anthony 1923b).

REMARKS: *Thomasomys hudsoni* was recognized as a valid species (Gyldenstolpe 1932; Ellerman 1941), until Cabrera (1961) relegated it to a subspecies of *T. gracilis*, likely relying on Anthony's (1923b) argument that *T. hudsoni* and *T. gracilis* were closely related. Musser and Carleton (1993) followed Cabrera's decision. Recently, Voss (2003) contrasted *T. hudsoni* with *T. cinnameus* and *T. gracilis* and returned it to species rank. Independently, Pacheco (2003) also considered *T. hudsoni* a full species following comparisons with its congeners.

Although *T. hudsoni* is known only from the type specimen, it is possible that some specimens collected in Cajas National Park, Azuay (Barnett 1999), and reported as *T. gracilis* are indeed this species. Clearly, Barnett did not confuse his *T. gracilis* with the other small species from this region *T. cinnameus*, as he recorded the two in sympatry. Unfortunately, Barnett (1999) gave little information regarding his specimens beyond providing head and body lengths and mass for two adult female specimens (103 mm, 41 g [BM 82.818] and 86 mm, 16.25 g [BM 84.339]) and several grayish juveniles (head and body length 64 to 100 mm (mean 83.1 mm). These measurements and the grayish coloration are coincident with the head and body length 93 mm and grayish brown of the type specimen of *T. hudsoni*. This issue might be resolved after a close examination of Barnett's vouchers housed in the Natural History

Museum, London (BM). Barnett has reidentified some "*T. gracilis*" as either *T. cinnameus* or *T. baeops* (P. Jenkins, pers. comm.).

The phylogenetic relationships of this species remain unclear. Pacheco's (2003) cladistic analysis of morphological characters and my unpublished data refuted a relationship with *T. gracilis*, placing *T. hudsoni* with *T. niveipes*, *T. paramorum*, and other species allied to *T. cinereus*. The karyotype is unknown.

Thomasomys hylophilus Osgood, 1912
Woodland Thomasomys

SYNONYM:

Thomasomys hylophilus Osgood, 1912:50; type locality "Paramo de Tama [= Páramo de Tamá], head of Rio Tachira [= Río Táchira], Santander [= Norte de Santander], Colombia."

DESCRIPTION: Medium sized (head and body length 107–126 mm) with long tail (124–155 mm, 104% to 145% of head and body length) and relatively small ears (17–20 mm). Dorsal pelage colored tawny olive mixed with blackish, sometimes more blackish along middle of back; sides and face wood brown to tawny olive; under parts cinnamon, their hairs slate colored at base. Forefeet silvery gray, forelimbs blackish, and tail uniformly blackish except for white tip. Mystacial vibrissae moderately long, extending slightly beyond posterior margin of pinnae when bent; genal 1 vibrissae absent. Hindfoot small to medium in length (24.5–28 mm), whitish drab above with digits white. Skull of moderate size (condyloincisive length 26.4–28.4 mm); braincase large and oblong; rostrum comparatively long and narrow; frontonasal profile straight. Zygomatic plates moderately broad, subequal to length of M1, and vertical. Incisive foramina long and relatively narrow, extending to anterior margins of M1s or quite close to them. Interorbital region moderately broad, hourglass in shape, with slightly square posterior margins. Mesopterygoid fossa broad with sides subparallel; roof closed. Zygomatic arches converge anteriorly. Auditory bullae flask-shaped and moderately inflated, although not as much as in *T. laniger*; bony tubes short and broad; stapedial spine small. Postglenoid foramen and subsquamosal fenestra elongated and subequal in size, latter posteriorly placed. Alisphenoid strut separates buccinator-masticatory and accessory oval foramina present. Ethmoturbinals of moderate size. Posterior opening of alisphenoid canal of normal size. Carotid circulatory pattern primitive (stapedial foramen, squamosal alisphenoid groove, and sphenofrontal foramen present). Upper incisors orthodont, long, and moderately broad. Molars relatively hypsodont but lack well-developed cingula and stylar cusps. Anteroloph and mesolophs of M1 and M2 comparatively well developed;

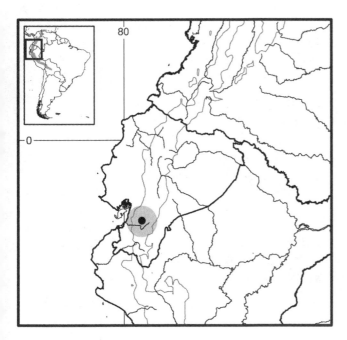

Map 343 Single known locality for *Thomasomys hudsoni* (●). Contour line = 2,000 m.

anteromedian flexus in M1 distinct; M3 with distinct hypoflexus; anteromedian flexid in m1 distinct. Maxillary toothrow moderately long, ranging from 4.6–5.0 mm. Capsular process either absent or present as an inconspicuous bony ridge.

DISTRIBUTION: *Thomasomys hylophilus* is known from the Cordillera Oriental, in eastern Colombia, and the Cordillera de Mérida, in western Venezuela.

SELECTED LOCALITIES (Map 344): COLOMBIA: Boyacá, East Andes, E side, Hacienda La Primavera (FMNH 92553); Santander, Mpio. Tona, Vda. Guarumales, Finca el Pajal (ICN 16680). VENEZUELA: Táchira, Buena Vista (UMMZ 156382), Buena Vista, 41 km SW of San Cristobal, near Páramo de Tamá (USNM 442306), Páramo de Tamá (CVULA 850).

SUBSPECIES: *Thomasomys hylophilus* is monotypic.

NATURAL HISTORY: The species was found in dense forests on the upper slopes of the páramo living among the innumerable galleries naturally formed under mossy roots, logs, and debris (Osgood 1912). Individuals are nocturnal, terrestrial to scansorial, solitary, and omnivorous. Handley (1976) described details where specimens were captured, including at the bases of trees and among tree roots, often in bamboo thickets, beside and under mossy rotting logs, in thick growths of shrubs and tree ferns, on mossy tree limbs, in litter on stream banks, and under tangled vines, near streams and in other damp places, most often in cloud forest, but also in clearings used for pasture and crops, in evergreen forest, all within the Lower Montane very humid forest (vegetative life zone of Ewel and Madriz 1968). Individuals feed on seeds and fruits (O. J. Linares 1998).

Known ectoparasites include the acarid mites *Crotiscus danae* (Brennan and Goff 1978) and *Eutrombicula goeldii*, *E. lukoschusi*, *E. tachirae*, and *E. webbi* (Brennan and Reed 1974a); the laelapid mite *Hymenolaelaps princeps* (Furman 1972); the chiggers *Peltoculus bobbiannae* (Brennan 1972) and *Aitkenius senticosus*, *Odontacarus comosus*, and *O. dienteslargus* (Brennan and Reed 1975); the tick *Ixodes* sp. (E. K. Jones et al. 1972); and the fleas *Cleopsylla monticola*, *Ctenidiosomus perplexus*, *Neotyphloceras rosenbergi*, *Pleochaetis apollinaris*, *Plocopsylla ulysses*, and *Sphinctopsilla tolmera* (Tipton and Machado-Allison 1972).

REMARKS: Little is known about intraspecific variation within *T. hylophilus*. Pacheco's (2003) morphological assessment revealed some degree of character polymorphism, in particular the Boyacá sample that differed in some attributes from the Páramo de Tamá samples. These Boyacá specimens are overall darker, with an undefined dorsal band, and are also darker ventrally, with less trace of brownish; the hindfoot lacks a gap between the thenar and hypothenar pads; nasals are narrower, their sides more parallel and less expanded anteriorly, and their posterior

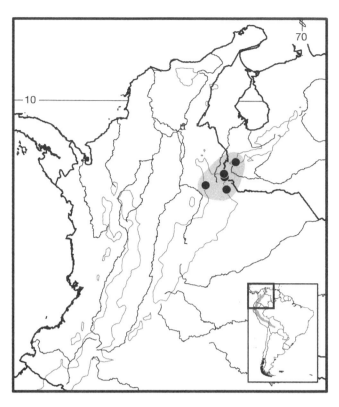

Map 344 Selected localities for *Thomasomys hylophilus* (●). Contour line = 2,000 m.

margin usually blunt; anterior margin of the nasals also exhibits a distinct notch at the midline. These differences might suggest the presence of an unnamed taxon that merits further investigation. Hooper and Musser (1964:Fig. 6) described and illustrated the glans penis of one specimen of *T. hylophilus* (FMNH 92555), but reported as *T. laniger*. Voss (2003) compared the morphology of this species with *T. ucucha*. The karyotype of *T. hylophilus* is unknown.

Thomasomys incanus (Thomas, 1894)

Inca Thomasomys

SYNONYMS:

Oryzomys incanus Thomas, 1894b:350; type locality "Valley of Vitoc [= Río Tulumayo]" Junín, Peru.

Thomasomys incanus: E.-L. Trouessart, 1897:409; first use of current name combination.

Inomys incanus: Thomas, 1917d:197; name combination.

Thomasomys fraternus Thomas, 1927f:602; type locality "Alcas, 25 miles N.E. of Cerro, Junin," Pasco, Peru.

Thomasomys incanus incanus: Ellerman, 1941:370; name combination.

Thomasomys incanus fraternus: Ellerman, 1941:370; name combination.

DESCRIPTION: Medium size (adult head and body length 111–134 mm) with relatively short tail (92–110%

of head and body length) and short hindfeet (22–28 mm). Dorsal pelage grizzled brownish gray in color; undersurface rather dirty fulvous and not countershaded. Mystacial vibrissae moderately long, extending slightly beyond posterior margin of pinnae when bent; genal 1 vibrissae absent. Ears small (18–21 mm). Gap present between thenar and hypothenar pads on plantar surface of hindfeet; distinct gap also present between hypothenar and digital pad 4; metatarsals and digits covered dorsally with silvery white hairs. Digit I of pes (hallux) moderate in length, with claw extending close or to interphalangeal joint of dII; and claw of digit V extends to interphalangeal joint of dIV. Tail finely annulated, clothed in short hairs, incipiently bicolored dark brownish above and paler below, and without whitish terminal tip. Skull of medium size (26.9–28.4 mm); rostrum long and narrow; braincase broad and rounded. Distinct rostral tube formed by premaxillae and nasals that project forward of upper incisors. Zygomatic arches slender, delicate, and converge anteriorly. Position of antorbital bridge low (about >25% of rostrum). Premaxillomaxillary-frontal intersection and posteromedian margin of premaxillae placed in front of zygomatic notch; posterior margin of nasals distinctly longer than premaxillae but does not surpass maxillary-frontal-lacrimal intersection. Anterior margin of M1 placed anterior to posterior border of zygomatic plate. Interorbital region broad (most of molar row concealed in dorsal view), hourglass in shape, with smooth margins. Optical foramen small (placed behind M3); ethmoturbinals large; alisphenoid strut present. Zygomatic plates moderately broad, breadth subequal to length of M1 and slightly sloped backward from bottom to top. Interparietal rather small; parietosupraoccipital suture moderately broad (suture about one third posterior margin of parietal). Incisive foramina long and moderately broad, extending to or quite close to anterior margins of M1s. Mesopterygoid fossa broad with either subparallel or posteriorly converging sides, and palatal medial process absent; roof closed, with sphenopalatine vacuities absent or represented as narrow slits. Auditory bullae small but moderately inflated, and flask shaped; bony tubes short and broad. Carotid circulatory pattern derived (stapedial foramen small, no internal squamosal-alisphenoid groove, no sphenofrontal foramen). Upper incisors small, delicate, and orthodont. Molars small and hypsodont without interpenetrating flexi. Upper molar row short, 4.3–5.0 mm in length. Mandible slender and delicate; capsular process absent.

DISTRIBUTION: *Thomasomys incanus* occurs along the eastern slope of the Andes in central and northern Peru, from Junín to San Martín departments, across an elevation range from 2,430 to 3,850 m.

SELECTED LOCALITIES (Map 345): PERU: Campamento Regional (MUSM 22951), Huánuco, mountains 15 mi NE of Huánuco (FMNH 23734); Junín, Maraynioc, 45 mi. NE of Tarma (MCZ 38537), Tarma, Tiambra (MUSM 22014), Vitoc valley (type locality of *Oryzomys incanus* Thomas); La Libertad, Mashua, E of Tayabamba, on trail to Ongón (LSUMZ 24809); Pasco, Estación Biológica de San Alberto (MUSM 11482), Millpo, E Tambo de Vacas on Pozuzo-Chaglla trail (LSUMZ 29169); San Martín, Las Papayas, ca. 5 km W of Pajaten ruins (LSUMZ 27300), Laurel (MUSM 24401), P.N. Río Abiseo, Pampa de Cuy (MUSM 8003).

SUBSPECIES: I consider *T. incanus* to be monotypic (but see Remarks).

NATURAL HISTORY: Thomas (1927f) mentioned that specimens were collected under rocks in scrub vegetation. Other individuals from Pasco and Huánuco were collected in dense humid forests, near streams; those from San Martín were collected in páramo vegetation, under rocks in grassland or scrub vegetation, in fragmented or continuous forest, and near streams either on riverine vegetation or on sandy banks (Leo and Romo 1992).

REMARKS: *Thomasomys incanus* is the type species of *Inomys* Thomas, a taxon recognized as valid by Gyldenstolpe (1932) but not Osgood (1933a), who found no diagnostic character to justify the generic-group name. *Inomys* has thus been listed as a synonym of *Thomasomys* in the more recent literature (Ellerman 1941; Cabrera 1961; Musser and Carleton 1993, 2005). However, Pacheco (2003) recovered a clade that included *T. incanus*, *T. ischyrus*, *T. kalinowskii*, *T. ladewi*, and other species with the derived carotid circulatory pattern in a phylogeny based on morphological characters. Although this clade included also *Aepeomys* and *Thomasomys erro*, probably caused by a long-branch effect (Bergsten 2005), the hypothesis of *Inomys* as a valid genus requires further examination and testing, particularly by molecular methods. Meanwhile, I recognize a group of species allied to *T. incanus* in an informal "incanus" group.

Osgood (1933a) doubted also the validity of *fraternus* Thomas, even as a subspecies, a decision followed by Cabrera (1961), Musser and Carleton (2005), and Pacheco (2003). Leo and Romo (1992) recorded the species from the Parque Nacional Río Abiseo in San Martín department, but this population is distinctly larger and with larger interparietals than the type series, and thus may prove to be a different species. Pacheco (2003) treated the sample from Las Papayas in San Martín department as the undescribed *Thomasomys* sp. 5. Other specimens, collected in Puerta del Monte, San Martín, are on the contrary quite small, with delicate and slender skulls, and small molar toothrow; Pacheco (2003) informally named them *Thomasomys* sp. 6. My more recent unpublished studies suggest that other populations might also constitute different species. The karyotype is unknown.

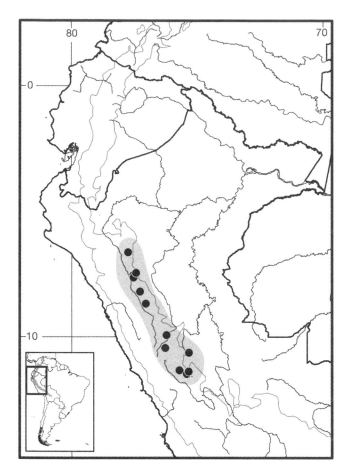

Map 345 Selected localities for *Thomasomys incanus* (●). Contour line = 2,000 m.

Thomasomys ischyrus Osgood, 1914
Long-tailed Thomasomys

SYNONYMS:

Thomasomys cinereus ischyrus Osgood, 1914b:162; type locality "Tambo Almirante, near Uchco"; defined as "Perú, Amazonas Dept., 65 km E Chachapoyas, near Uchco, Tambo Almirante, 5000 ft (1524 m)" by Musser and Carleton (2005).

Thomasomys ischyrus: Thomas, 1926e:613; first use of current name combination.

Thomasomys ischyrus ischyrus: Gyldenstolpe, 1932a:53; name combination.

Thomasomys ischyurus Cabrera, 1961:430; incorrect subsequent spelling of *ischyrus* Osgood.

Thomasomys ischyurus ischyurus: Cabrera, 1961:430; name combination, and incorrect subsequent spelling of *ischyrus* Osgood.

DESCRIPTION: Medium sized (adult head and body length 132–138 mm) with relatively short tail (100–121% of head and body length) and moderately long hindfoot (30–31 mm). Dorsal pelage colored rich burnt umber finely

sprinkled with sooty; ventral surface washed with fulvous brown and not countershaded. Mystacial vibrissae moderately long, extending slightly beyond posterior margin of pinnae when bent; genal 1 vibrissae absent. Ears medium in length (23 mm), blackish, hairs near inner bases with whitish roots. Hindfeet slender with dorsal surface of metatarsals brownish contrasting with whitish digits; gap present between thenar and hypothenar pads. Forefeet also brownish, ventral base of manual claw keeled along proximal half. Digit I of pes (hallux) moderately long, as claw does not extend more than about half length of phalanx 1 of dII. Digit V of pes long, with claw extending about half length of phalanx 2 of dIV. Tail dull brownish above, scarcely lighter below, and without terminal white tip. Skull medium in size (condyloincisive length 30.1–30.4 mm) with long and narrow rostrum and broad and rounded braincase, similar to *T. incanus* but larger; rostral tube formed by forward projecting premaxillae and nasals present. Zygomatic arches slender, delicate, and converge anteriorly. Position of antorbital bridge low (ca. >25% of rostrum). Premaxillomaxillary-frontal intersection lies in front of zygomatic notch; posterior margins of nasals distinctly longer than premaxillae, and level to or extend beyond lacrimal. Distal half of nasals expanded, but taper posteriorly. Interorbital region broad (most of molar row concealed in dorsal view), hourglass in shape, with smooth margins. Optical foramen small (placed behind M3); ethmoturbinals large; alisphenoid strut present or absent. Zygomatic plates moderately broad, breadth subequal to length of M1, with anterior margin convex and slightly sloped backward from bottom to top. Incisive foramina long and moderately broad, extending to anterior margins of M1s. Mesopterygoid fossa broad, anteriorly expanded with converging margins posteriorly; medial palatal process absent; roof closed, with sphenopalatine vacuities either absent or represented as narrow slits. A medial palatal process absent. Parietosupraoccipital suture narrow (length of suture less than one-fourth of posterior margin of parietal). Auditory bullae small but moderately inflated, and flask shaped; bony tubes short and broad. Carotid circulatory pattern derived (stapedial foramen small, no internal squamosal-alisphenoid groove, no sphenofrontal foramen). Upper incisors orthodont. Molars relatively large and hypsodont and upper molar row moderately long, 5.2–5.7 mm. Capsular process absent.

DISTRIBUTION: *Thomasomys ischyrus* is known only from the eastern slope of the Andes in northern and central Peru, in the departments of Amazonas, San Martín, and Huánuco, at an elevational range between 2,280 and 3,350 m.

SELECTED LOCALITIES (Map 346): PERU: Amazonas, ACP Abra Patricia–Alto Nieva (MUSM 25179), ACP

Huiquilla El Choctamal (MUSM 25758), Cordillera Colán, E of La Peca (LSUMZ 21857), Cordillera Colán, NE of La Peca (LSUMZ 21847), Goncha (USNM 297482), Rodríguez de Mendoza, Tambo Almirante, near Uchco (type locality of *Thomasomys cinereus ischyrus* Osgood); Huánuco, Caserío de San Pedro de Carpish (MUSM 19093); San Martín, Laurel (MUSM 24402).

SUBSPECIES: I treat *T. ischyrus* as monotypic (but see Remarks).

NATURAL HISTORY: Little is known about the ecology or population biology of this species. Specimens have been collected on the ground in pristine humid montane forests and in disturbed forests near potato crops in the Cordillera Carpish (V. Pacheco, unpubl. data), where trees are slender, reach heights of 10 to 15 m, and are densely covered by orchids, bromeliads, ferns, lichens, and mosses (Beltrán and Salinas 2010). Similarly, at Huiquilla in Amazonas department, individuals were obtained in montane forests with low or high canopies, where trees (*Morus insignis* [Moraceae] and some Lauraceae) reached 15 to 20 m and were covered with orchids and bromeliads as well as tree ferns of the genus *Cyathea* (C. Jiménez, pers. comm.). Ectoparasites include the anopluran louse *Hoplopleura angulata* (P. T. Johnson 1972; specimens reported as *T. cinereus*).

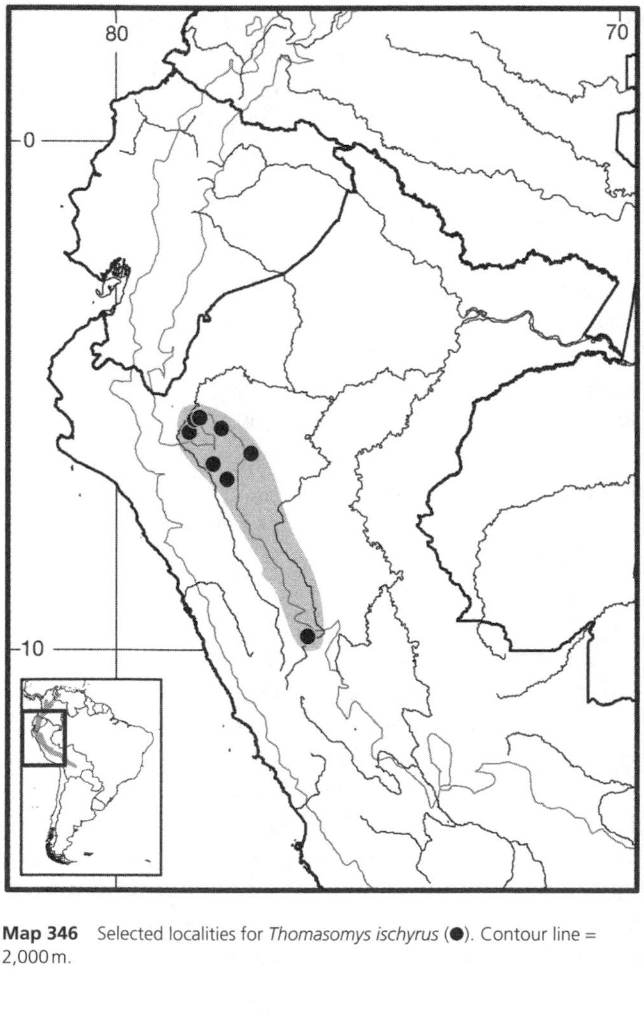

REMARKS: Leo and Romo (1992) reported specimens from Parque Nacional Río Abiseo, San Martín department as *T. ischyruseleusis*, but these belong to *T. eleusis* not to *T. ischyrus* as understood herein. The southern population from the Cordillera Carpish has a disjunct distribution and may represent an undescribed species (V. Pacheco, unpubl. data).

Osgood (1914b) considered *T. ischyrus* most similar to *T. cinereus*. However, Pacheco's (2003) phylogenetic assessment based on morphology included *T. ischyrus* in a small, resolved clade that contained *T. incanus*, *T. kalinowskii*, and their allies. The report by T. E. Lee et al. (2011), which placed *T. ischyrus* as related to *T. silvestris* based on mtDNA cytochrome-*b* sequences, needs reevaluation. The GenBank sequence (AF 108675; MVZ 181999) used by these authors is a specimen of *T. taczanowskii* (see the *T. baeops* account), not *T. ischyrus*. The karyotype is unknown.

Thomasomys kalinowskii (Thomas, 1894)
Kalinowski's Thomasomys

SYNONYMS:

Oryzomys kalinowskii Thomas, 1894b:349; type locality "Valley of Vitoc [= Río Tulumayo]" Junín, Peru.

Peromyscus (*Thomasomys*) *kalinowskii*: E.-L. Trouessart, 1897:513; name combination.

Thomasomys kalinowskii: E.-L. Trouessart, 1904:408; first use of current name combination.

DESCRIPTION: Moderately large (head and body length 136–143 mm) with relatively short tail (101–114% of head and body length), and moderately long hindfoot (28–35 mm). Dorsal pelage finely grizzled brownish gray, hairs dark slate at their bases with tips dull yellow; undersurface with dirty fulvous tone. Mystacial vibrissae moderately long, extending slightly beyond posterior margin of pinnae when bent; genal 1 vibrissae absent. Ears medium in length (21–26 mm). Plantar surface of hindfeet naked and pads prominent, with distinct gap between thenar and hypothenar pads; dorsal surface of metatarsals dark brown with sides and fingers silvery white. Forefeet also brownish; ventral base of manual claw keeled on proximal half. Digit I of pes (hallux) moderately long, with claw not extending more than about half length of phalanx 1 of dII. Digit V of pes long, with claw extending about half length of phalanx 2 of dIV. Tail uniformly black, finely haired, dull brownish above, scarcely lighter below, and without terminal white tip or pencil. Skull moderately long (condyloincisive length 30.1–31.9 mm)

Map 346 Selected localities for *Thomasomys ischyrus* (●). Contour line = 2,000 m.

with rostrum long and slender but not distinctly projected anteriorly (rostral tube absent) and braincase broad but not rounded. Zygomatic arches normally built, not delicate, and convergent anteriorly. Position of antorbital bridge high (<25% of rostrum). Premaxillomaxillary-frontal intersection lies in front of zygomatic notch, but posteromedian margin of premaxillae placed posterior to it. Posterior margin of nasals distinctly longer than premaxillae and level maxillary-frontal-lacrimal intersection or slightly longer. Interorbital region moderately broad (the molar row partially visible in dorsal view) with marked edges. Optical foramen partially dorsal to M3 and ethmoturbinals large. Zygomatic plates moderately broad, breadth subequal to length of M1, and slope slightly backward from bottom to top. Interparietal typically broad, parietosupraoccipital suture narrow (suture about one quarter or less posterior margin of parietal), and not distinctly large anteroposteriorly. Incisive foramina long and moderately broad, more contracted anteriorly, and barely extend between M1s. Mesopterygoid fossa broad with lateral margins converging backward; palatal medial process absent; roof closed, with sphenopalatine vacuities absent or represented as slits. Molar-bearing portion of maxillae somewhat triangular in shape. Auditory bullae small but moderately inflated and flask shaped; bony tubes short and broad. Carotid circulatory pattern derived (stapedial foramen small, no internal squamosal-alisphenoid groove, no sphenofrontal foramen). Upper incisors opisthodont. Molars broad, hypsodont, with slight interpenetration of flexi. Anteromedian flexus, anteroloph, and anteroflexus of M1 distinct; M2 protoflexus absent; M3 hypoflexus distinct even in worn molars. Molar toothrow moderately long (5.5–6.1 mm). Mandible elongated, moderately robust, not delicate as in *T. incanus*, and without capsular process.

DISTRIBUTION: *Thomasomys kalinowskii* occurs only in the Andes of central Peru from the departments of Huánuco south to Ayacucho, in montane forests from 2,050 to 3,673 m.

SELECTED LOCALITIES (Map 347): PERU: Ayacucho, Anchihuay (MUSM 26501), Yanamonte (MUSM 21476); Cusco, Cordillera del Vilcabamba (MUSM 13066); Huánuco, Bosque Zapatagocha above Acomayo (LSUMZ 18440), Dos de Mayo, Cayna, Chiliatuna (MUSM 1314), Iscarag (MUSM 22893), Kenwarajra (MUSM 23004), Pampa Hermoza (MUSM 23018), Shogos (MUSM 22679); Junín, Maraynioc, 45 mi NE of Tarma (FMNH 54202), 16 km NNE (by road) of Palca (MVZ 172599); Pasco, Cumbre de Ollón, ca. 12 km E of Oxapampa (LSUMZ 25899), Millpo, E of Tambo de Vacas on Pozuzo-Chaglla trail (LSUMZ 29165).

SUBSPECIES: I treat *T. kalinowskii* as monotypic (but see Remarks).

NATURAL HISTORY: *Thomasomys kalinowskii* was reported in Apurímac department in humid upper montane forest composed of tall shrubs, narrow trees of 5 to 10 m high, and stands of *Chusquea* bamboo, and near maize and squash crops at high relative abundance (Pacheco et al. 2007). It was collected also on the ground in humid montane pristine forests and in disturbed forests near potato crops in Carpish (V. Pacheco, unpubl. data), where trees are slender, reach heights of 10 to 15 m, and are densely covered by orchids, bromeliads, ferns, lichens, and mosses (Beltrán and Salinas 2010). Noblecilla and Pacheco (2012) found *T. kalinowskii* to feed mainly on plant material (67.7%, including mainly Piperaceae seeds) and arthropods (21.5%). Price and Emerson (1986) described *Cummingsia barkleyae* as a new species of Mallophaga from *Thomasomys* sp., from Unchog, Huánuco. The host indeed represents a new species allied to *T. kalinowskii*, named as *Thomasomys* sp. 7 by Pacheco (2003).

REMARKS: Thomas (1884: 455) reported *T. cinereus* from Maranyioc; his description, however, agrees more closely *T. kalinowskii* and *not T. cinereus*, which is restricted to northern Peru (see that account). Specimens from Panao, Huánuco, collected by J. T. Zimmer, and housed at the Field Museum were confirmed as *T. kalinowskii* (Pacheco 2002), although they differ slightly from the type series with paler metatarsals, alisphenoid strut present, and maxillary pits absent, among other features. Other populations that show some differences from the type series and may represent new taxa; these include a single specimen collected from the Cordillera Vilcabamba, Cusco department (Emmons et al. 2001), which exhibits an unusual straight ventral edge of the mandible, among other features. A sample from Unchog, Huánuco, named *Thomasomys* sp. 7 by Pacheco (2003), has darker dorsal pelage, a moderate rostral tube, and subparallel zygomatic arches. Finally, the sample from Yuraccyacu, Ayacucho has a white tipped tail, short and broad rostrum, and wide and squarish molars.

A. L. Gardner and Patton (1976) reported the karyotype ($2n = 44$, FN $= 44$), with X and Y acrocentric chromosomes, based on samples from Yuraccyacu, Ayacucho. M. F. Smith and Patton (1999) sequenced mtDNA cytochrome-*b* from two specimens of *T. kalinowskii* (reported as *Thomasomys* sp.) from Junín (MVZ 172598, 172599). A phylogenetic analysis based on morphological attributes suggested that *T. kalinowskii* is more closely related to *T. eleusis* in a clade here named the "incanus" group. However, Salazar-Bravo and Yates (2007), using mtDNA cytochrome-*b* sequences, placed *T. kalinowskii* as sister to a group that included *T. baeops*, *T. taczanowskii*, *T. notatus*, and *T. ladewi*, despite

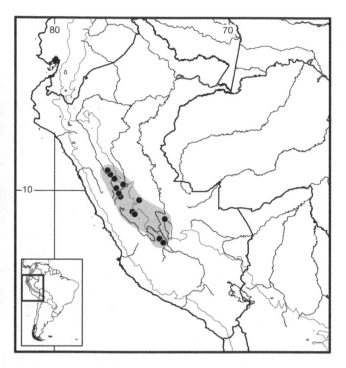

Map 347 Selected localities for *Thomasomys kalinowskii* (●). Contour line = 2,000 m.

the trenchant morphological differences that separates *T. kalinowskii* from the "baeops" and "notatus" groups. A similar phylogeny, also based on cytochrome-*b* gene sequences, found *T. kalinowskii* (reported as *T.* sp.) as a sister taxon to *T. notatus* and *T. ladewi* but not to members of the "baeops" group. These contrasting phylogenetic hypotheses may simply result from limited taxon sampling.

Thomasomys ladewi Anthony, 1926
Ladew's Thomasomys

SYNONYM:

Thomasomys ladewi Anthony, 1926a:1; type locality "Rio Aceramarca, northeast of La Paz [, La Paz], Bolivia, altitude 10,800 feet [3,292 m]."

DESCRIPTION: Medium sized (head and body length 120–138 mm) with soft, lax, and long fur (individual hairs up to 15–16 mm long). Pelage coloration above and below plumbeous, but tipped sooty black, fuscous, and brown on back, providing overall impression of very finely grizzled fuscous-black. Cheeks, sides, and lower rump darker than midback. Under parts washed with pale smoke gray to drab gray; pectoral area darker than rest of under parts; small patch of white hairs may be present on chest or throat; venter not countershaded relative to dorsal coloration. Hands, feet, and tail (above and below) grayish brown. Tail longer than head and body (107–132%), very sparsely haired, indistinctly bicolored, and terminates in white tip that extends 10–60 mm. Ears short

(20–23 mm). Mass 42–64 g. Mystacial vibrissae moderately long, extending slightly past posterior margins of the ears when bent; genal 1 vibrissae absent. Hindfoot moderately long (28–33 mm). Digit I of pes (hallux) moderately long, with claw not extending more than about half length of phalanx 1 of dII. Digit V of pes also long, with claw extending about half length of phalanx 2 of dIV. Skull relatively large, broad, and robust (condyloincisive length 30–31.7 mm), with rostrum broad and short and not distinctly projected; position of antorbital bridge high (<25% of rostrum). Zygomatic arches converge anteriorly; zygomatic plates broad, slightly sloped backward from bottom to top, anterior margin straight. Interorbital region moderately broad (molar row partially seen in dorsal view), hourglass in shape, with margins rounded to little edged; postorbital ridge present; and optic foramen level with or overlaps M3. Incisive foramina open, somewhat oval, with posterior margins reaching plane of M1s. Molar-bearing portion of maxillae near rectangular. Subsquamosal fenestra subequal and posterior to postglenoid foramen, which is elongated and level with anterior margin of tegmen tympani. Mesopterygoid fossa broad with subparallel margins. Carotid circulatory pattern derived (stapedial foramen small, no internal squamosalalisphenoid groove, no sphenofrontal foramen). Auditory bullae small and not inflated. Upper incisors orthodont. Molars moderately hypsodont, with interpenetrating flexi. Maxillary toothrow moderately long (5.5–5.8 mm). Capsular process absent.

DISTRIBUTION: *Thomasomys ladewi* inhabits montane forests from 2,360 to 3,300 m in a restricted region in the Andes in La Paz department of northwestern Bolivia.

SELECTED LOCALITIES (Map 348): BOLIVIA: La Paz, Río Aceramarca (type locality of *Thomasomys ladewi* Anthony), 2° Estación climatológica (CBF 7654), Franz Tamayo, Palcabamba (MNK 3597), Franz Tamayo, Pelechuco, Llamachaqui, on road to Apolo (CBF 4042).

SUBSPECIES: *Thomasomys ladewi* is monotypic.

NATURAL HISTORY: *Thomasomys ladewi* is terrestrial, with individuals captured on rocky and wooded hillsides. No other details about habits, behavior, diet, or reproduction are known (S. Anderson 1997).

REMARKS: Almost nothing is known about phylogenetic relationship of this species. Anthony (1926a) stated that *T. ladewi* was superficially more similar to *T. vulcani* (reported as *rhoadsi*), *T. cinereus*, or *T. cinereiventer*, although he thought the skull of *T. ladewi* matched more closely those of his *aureus* group. Pacheco (2003) placed *T. ladewi* in a clade that included *T. incanus* and allies in his phylogenetic hypothesis based on morphological characters. The karyotype of *T. ladewi* is unknown.

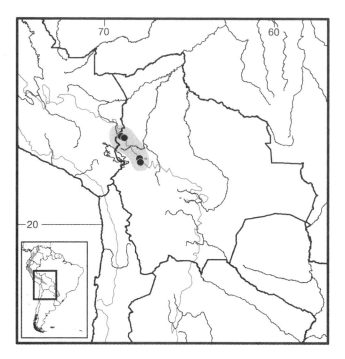

Map 348 Selected localities for *Thomasomys ladewi* (●). Contour line = 2,000 m.

Thomasomys laniger (Thomas, 1895)
Soft-furred Thomasomys

SYNONYMS:

Oryzomys laniger Thomas, 1895b:59; type locality "Bogota"; corrected to Bogotá Region, 8,750 ft (2,667 m), Cundinamarca, Colombia by Musser and Carleton (2005).

[*Oryzomys* (*Erioryzomys*)] *laniger*: E.-L. Trouessart, 1904: 423; name combination.

Thomasomys laniger: Thomas, 1906c:443; first use of current name combination.

Thomasomys laniger laniger: Gyldenstolpe, 1932a:56; name combination.

DESCRIPTION: Medium sized (head and body length 101–119 mm) with long, soft, and fluffy dorsal fur. Dorsal pelage colored dull olivaceous gray, with hairs dark slate-colored at base and olive at tips. Ventral pelage similarly grayish but paler, with hair tips dull yellowish; countershaded relative to dorsal pelage. Mystacial vibrissae very long, extending distinctly beyond posterior margin of pinnae when bent; genal 1 vibrissae absent. Ears relatively large and naked; postauricular patches conspicuous. Tail similar to or longer than head and body (103–123%), slender, finely haired, monocolored, and occasionally with terminal white tip. Hindfoot relatively short (24–26 mm), metatarsals dark brown mesially, edges and digits white, gap present between thenar and hypothenar pads. Digit I of pes (hallux) moderate, with its claw not extending more than about half length of phalanx 1 of dII. Digit V of pes long, with claw extending about half length of

phalanx 2 of dIV. Skull relatively small (condyloincisive length 25.6–27.3 mm), thin, and delicate, with large and rounded braincase, and relatively long and broad rostrum. Zygomatic plates moderately broad, subequal in length to M1, and slant backward from bottom to top. Interorbital region moderately broad, hourglass in shape, and with smooth to scarcely squared edges. Zygomatic arches moderately convergent anteriorly. Incisive foramina large and open, more contracted anteriorly, maximally extending to anterior margins of M1s. Mesopterygoid fossa broad with subparallel sides; sphenopalatine vacuities present as slits. Parapterygoid vacuities usually absent. Auditory bullae large and inflated. Subsquamosal fenestra smaller or subequal to postglenoid foramen. Carotid circulatory pattern primitive (stapedial foramen, groove on inner surface of squamosal and alisphenoid, and sphenofrontal foramen are present). Upper incisors orthodont. Molars broad; upper molars moderately hypsodont, with no interpenetration. Maxillary toothrow relatively small, 4.2–4.8 mm. M1 anteroloph indistinct; M2 protoflexus distinct; M3 hypoflexus indistinct; and m1 ectolophid absent. Capsular process absent.

DISTRIBUTION: *Thomasomys laniger* is known from the Cordillera Central and Oriental of Colombia, from 2,600 to 3,600 m.

SELECTED LOCALITIES (Map 349): COLOMBIA: Boyacá, Villa de Leyva, sector Chaina, vereda Río Abajo (IAvH-M 6922); Cundinamarca, Bogotá, San Francisco (FMNH 71459), Guasca, Río Balcones (FMNH 71408), Fusagasugá (AMNH 76717), Finca el Soche, 15 km W of Soacha (KU 124056); Huila, San Agustín, Páramo de las Papas (FMNH 71486).

SUBSPECIES: *Thomasomys laniger* is monotypic.

NATURAL HISTORY: This species is terrestrial and has been captured more often in Andean temperate forest and forest-páramo ecotone than in páramo proper (Lopez-Arévalo et al. 1993; Gómez-Laverde et al. 1997). Lopez-Arévalo et al. (1993) reported that density increased in the wet season (June to August), that sexual maturity was observed at approximately 3.5 months, and that the gestation period was about 24 days. The diet was composed of fruits, seeds, and insects, with an increase of insects in the wet season (Lopez-Arévalo et al. 1993). Density of *T. laniger* in Cerro Majuy, Cundinamarca was eight individuals/ha (Zuñiga et al. 1988). The latter authors provided also detailed information on soil, vegetation, and climate for the sample localities where *T. laniger* was found. Timm and Price (1985) reported the louse *Cummingsia inopinata* on specimens of *T. laniger* from Páramo, Antioquia, Colombia, which is likely a misidentification of the host species as *T. laniger* is unknown from this region. Rather, *Handleyomys intectus* (Thomas) (*sensu* Voss et al.

2002) is the likely host because several specimens were collected from Páramo and Río Negrito, near Sonsón, by Philip Hershkovitz, but mislabeled as *Thomasomys laniger*. This same interpretation applies to the report of Tipton and Machado-Allison (1972) of the flea *Pleochaetis smiti* from *T. laniger* from Páramo, a specimen also collected by P. Hershkovitz. Seevers (1955) described the staphylinid beetle *Amblyopinus emarginatus*, and Tipton and Machado-Allison (1972) reported the flea *Sphinctopsilla tolmera*.

REMARKS: Thomas (1916j) described *emeritus* as a subspecies of *T. laniger*, but this taxon was subsequently elevated to a valid species (see that account). Cabrera (1961) considered *niveipes* a synonym of *laniger* without justification; however, Gómez-Laverde et al. (1997) convincingly provided chromosomal, mensural, and ecological discrimination of *T. laniger* and *T. niveipes* in Colombia. The karyotypic formula revealed a diploid number of $2n=40$ and FN$=40$, with metacentric or submetacentric X chromosome and an acrocentric Y chromosome (Gómez-Laverde et al. 1997). The $2n=42$, FN$=40$ karyotype reported by Aguilera et al. (2000) for populations of *T. laniger* from Venezuela is based on misidentified samples of *T. emeritus* (see that account).

Ellerman (1941) included this species in his *cinereus* Group. Pacheco (2003) placed *T. laniger* in a clade with other species of the "cinereus" group in his cladistic analysis derived on morphological characters.

Thomasomys macrotis A. L. Gardner and Romo 1993
Large-eared Thomasomys

SYNONYM:

Thomasomys macrotis A. L. Gardner and Romo 1993:762; type locality "Puerta del Monte, ca. 30 km NE [of] Los Alisos, ca. 3250 m. [Parque Nacional Río Abiseo], San Martín, Perú."

DESCRIPTION: Large sized (head and body length 153–168 mm) with soft, dense, and comparatively long dorsal fur (mid-dorsal hairs 15 mm). Dorsal coloration bone brown to sepia, finely streaked with pale brown imparting an agouti pattern; paler brown tones cover sides and flanks; and venter, including inside of legs, pinkish cinnamon; ventral and dorsal pelage not countershaded. Mystacial vibrissae moderately long, extending slightly beyond posterior margin of pinnae when bent; genal 1 vibrissae absent. Ears large (28 to 33 mm). Hindfoot long (44–48 mm) and narrow, with metacarpal surface as dark as body dorsum, but with paler claws and digital bristles; metacarpals of forefeet dark with paler toes, claws, and digital bristles; ventral base of claws keeled; distinct gap present between thenar and hypothenar pads. Tail longer than head and body (126–143%), and uniformly dark brown except for terminal one-third to two-fifths where both scales and hairs are white. Digit I of pes (hallux) moderate, with claw not extending more than about half length of phalanx 1 of dII. Digit V of pes long, with claw extending about half length of phalanx 2 of dIV. Skull large and robust (condyloincisive length 34.3–41.0 mm) with comparatively broad and deep rostrum and spatulated nasals. Zygomatic plates broad and vertical. Incisive foramina comparatively broad, contracted at both anterior and posterior margins, and moderately long, reaching anterior margins of M1s. Interorbital region narrow, hourglass in shape, and with elevated and rounded margins. Palatal bridge narrow, and postorbital ridge present. Mesopterygoid fossa subequal in breadth to parapterygoid fossae; sides subparallel, not expanded anteriorly; lacks medial process; roof either closed or only slightly perforated by sphenopalatine vacuities. Auditory bullae relatively small and uninflated. Carotid circulatory pattern derived (stapedial foramen small, no internal squamosal-alisphenoid groove, no sphenofrontal foramen). Upper incisors orthodont. Molars large, moderately hypsodont, with no interpenetration, but flexi somewhat crenulated. M1 anteroflexus distinct; m1 anteromedian flexid is shallow and narrow. Accessory labial root of M1 present. Maxillary toothrow long, 8.0–8.3 mm. Capsular process absent.

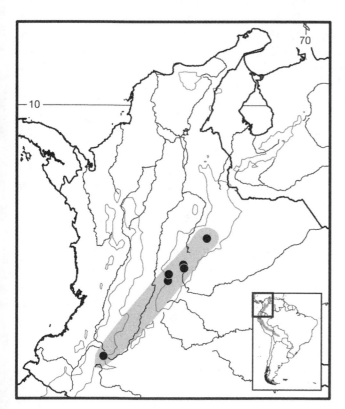

Map 349 Selected localities for *Thomasomys laniger* (●). Contour line = 2,000 m.

DISTRIBUTION: *Thomasomys macrotis* is known only from the type locality and vicinity, at elevations ranging from 3,250 to 3,380 m, in the north-central Peruvian Andes.

SELECTED LOCALITIES (Map 350): PERU: San Martín, 24 km NE of Pataz, Pampa del Cuy, Parque Nacional Río Abiseo (USNM 567243), Puerta del Monte, ca. 30 km NE de Los Alisos (type locality of *Thomasomys macrotis* A. L. Gardner and Romo).

SUBSPECIES: *Thomasomys macrotis* is monotypic.

NATURAL HISTORY: Individuals of this species occur in upper montane elfin forest, where they were captured in a rocky area, in grasslands near a fragmented forest in the Pampa del Cuy Valley (Leo and Romo 1992).

REMARKS: *Thomasomys macrotis* corresponds to *Thomasomys* sp. nov. 2 of Leo and Romo (1992). A. L. Gardner and Romo (1993) compared and differentiated *T. macrotis* from *T. aureus* and postulated that *T. macrotis* was more closely related to *T. ischyrus* based on dental and cranial features, although these were not explicitly mentioned. Pacheco (2003) suggested that *T. macrotis* might belong to its own group because the species exhibits characteristics of both the "aureus" and "incanus" groups. The karyotype is unknown.

Thomasomys monochromos Bangs, 1900
Unicolored Thomasomys

SYNONYMS:

Oryzomys (*Erioryzomys*) *monochromos* Bangs, 1900a:97; type locality "Paramo (= Páramo) de Macotama, Sierra Nevada de Santa Marta, 11,000 feet altitude," Magdalena, Colombia.

Thomasomys monochromos: Thomas, 1906c:443; first use of current name combination.

Thomasomys monochromus Gyldenstolpe, 1932a:57; incorrect subsequent spelling of *Oryzomys* (*Erioryzomys*) *monochromos* Bangs.

Thomasomys laniger monochromos: Cabrera, 1961:431; name combination.

DESCRIPTION: Relatively small (head and body length 105–120 mm) with rich brown dorsal pelage, more yellowish or raw umber on sides, and under parts dull yellowish wood-brown; venter not countershaded relative to dorsal pelage. Mystacial vibrissae moderately long, but not as long as in *T. laniger*, extending only slightly beyond posterior margin of pinnae when bent; genal 1 vibrissae absent. Hindfoot small to medium (24–27 mm), white above but irregularly marked with dusky. Tail short to medium for genus (96–124% of head and body length), uniformly dusky brown, and finely haired. Ears small (18–20 mm), nearly naked, and dusky in color. Skull small (condyloincisive length 25.9–27.4 mm) and delicate, with large and rounded braincase, comparatively broad rostrum, and nearly straight lateral profile. Zygomatic plates moderately broad, subequal to length of M1, and slightly slanting backward from their root. Incisive foramina wide and short, oval in shape, and extend to anterior margins of M1s or quite close. Interorbital region moderately broad, hourglass in shape, and with rounded margins. Interfrontal fontanella present. Mesopterygoid fossa broad with subparallel sides. Auditory bullae moderately inflated, but not as much as in *T. laniger*, with large stapedial spine. Postglenoid foramen and subsquamosal fenestra triangular shaped, but later larger and placed posteriorly. Carotid circulatory pattern primitive (stapedial foramen, squamosal alisphenoid groove, and sphenofrontal foramen present). Upper incisors orthodont. Maxillary toothrow length short, 4.3–4.6 mm. M1 with moderate accessory labial root.

DISTRIBUTION: *Thomasomys monochromos* occurs only in extreme northern Colombia, over an elevational range from 2,000 to 3,600 m in and around the Sierra Nevada de Santa Marta.

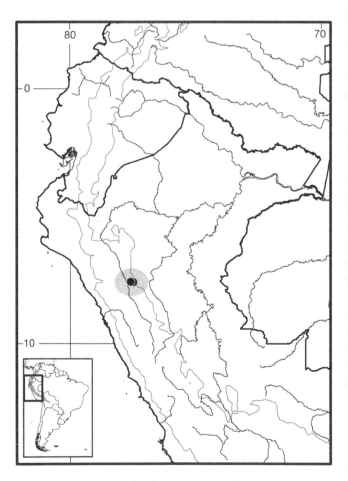

Map 350 Selected localities for *Thomasomys macrotis* (●). Contour line = 2,000 m.

SELECTED LOCALITIES (Map 351): COLOMBIA: Magdalena, Cerros San Lorenzo, San Lorenzo (USNM 507270), Mamancanaca, near (FMNH 69190), Páramo de Macotama, Sierra Nevada de Santa Marta (type locality of *Oryzomys* (*Erioryzomys*) *monochromos* Bangs), Santa Marta, Serranía San Lorenzo, Estación INDERENA (ICN 5415).

SUBSPECIES: *Thomasomys monochromos* is monotypic.

NATURAL HISTORY: The species is known from montane forest and páramos. Specimens were taken in National live traps on the ground in a tropical cloud (wet) forest remnant at the INDERENA (former Instituto Nacional de los Recursos Naturales Renovables y del Ambiente, Colombia) forestry station, where most of the adjacent land (steep hillsides on the slope of the Sierra de Santa Marta) had been converted to open pasture or planted to eucalyptus and Caribbean pine (A. L. Gardner, pers. comm.). No other information on ecology, population biology, or behavior has been reported.

REMARKS: Cabrera (1961) included *monochromos* Bangs as a subspecies of *T. laniger*, but A. L. Gardner and Patton (1976) considered it a valid species and reported the karyotype ($2n=42$, FN$=42$) with a submetacentric X and an acrocentric Y chromosomes. Bangs (1900a) compared

his species to *T. laniger*, with which it shares strong resemblance. Pacheco (2003) allied this species to *T. cinnameus* and *T. ucucha* in his phylogenetic assessment based on morphological features; however, the lack of skeletons and fluid specimens, and therefore access to a number of relevant characters, have probably limited a better resolution.

Thomasomys nicefori Thomas, 1921
Nicéforo María's Thomasomys

SYNONYMS:

Thomasomys nicefori Thomas, 1921j:355; type locality "San Pedro, north of the town [= Medellín]," Antioquia, Colombia.

Thomasomys aureus nicefori: Cabrera, 1961:425; name combination.

DESCRIPTION: Large sized (head and body length 154–172 mm) with long tail (114–136% of head and body length), and relatively small ears (21–23 mm). Dorsal pelage almost as buffy or ochraceous as that of *T. aureus*, although head and foreback rather less richly fulvous; under parts also buffy, and moderately countershaded. Mystacial vibrissae very long, extending distinctly beyond posterior margin of pinnae when bent; genal 1 vibrissae present. Hindfoot comparatively long (33–36 mm), buffy whitish above, with small, dark patches on ankles and metatarsals. Tail uniformly brown, clothed in short hairs, without terminal white tip. Skull large (condyloincisive length 32.9–35.2 mm), although smaller than *T. aureus*, with slender rostrum, and narrow, short nasals that taper posteriorly. Zygomatic plates moderately broad, breadth subequal to length of M1, and rather vertical. Interorbital region comparatively broad, margins marked or incipiently squared, with indistinct postorbital ridge. Incisive foramina long and moderately open, extend to level anterior margins of M1s or between anterocones. Palatal bridge short and comparatively broad; maxillary palatal pits present. Auditory bullae small, although moderately inflated, and bony tubes distinct. Anterior process of ectotympanic relatively small. Mesopterygoid fossa broad and somewhat parallel sided; sphenopalatine vacuities absent or present as slit-like vacuities flanking presphenoid-basisphenoid suture. Carotid circulatory pattern derived (stapedial foramen small, no internal squamosal-alisphenoid groove, no sphenofrontal foramen). Lambdoid ridge distinct. Upper incisors orthodont. Maxillary toothrow long (5.9–6.6 mm), but comparatively shorter than *T. aureus*. M1 with perceptible protostyle and enterostyle; m1 with ectolophid. Capsular process absent or represented as inconspicuous bony ridge.

DISTRIBUTION: *Thomasomys nicefori* occurs in Colombia, in the northern parts of both the Cordillera Central and Occidental of the Andes, from 1,900 to 3,810 m.

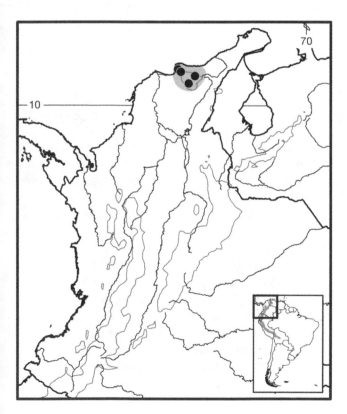

Map 351 Selected localities for *Thomasomys monochromos* (●). Contour line = 2,000 m.

SELECTED LOCALITIES (Map 352): COLOMBIA: Antioquia, Finca de Margarita Molina (ICN 16762), Medellín, Las Palmas (FMNH 70326), Paramillo (AMNH 37738), Páramo Frontino, Urrao (FMNH 71265), San Pedro (type locality of *Thomasomys nicefori* Thomas), SW of Santa Elena (AMNH 37736), Sonsón, Páramo, 7 km E (FMNH 70307), Valdivia, Las Ventanas (FMNH 70322); Caldas, Manizales, Río Termales (FMNH 71263); Quindío, Salento (AMNH 33941); Risaralda, La Pastora, PRN Ucumarí (ICN 12196).

SUBSPECIES: I treat *T. nicefori* as monotypic (but see Remarks).

NATURAL HISTORY: Little is known about the ecology of this species. Thomas (1921j) reported the holotype was "trapped in a wood." Kohls (1956) described *Ixodes tropicalis* as a new species of tick from *T. nicefori* (reported as *T. aureus*). Seevers (1955) reported *Amblyopinus colombiae* as a new species of staphylinid beetle from *T. nicefori* (reported as *T. aureus nicefori*). And Trapido and Sanmartín (1971) processed some specimens of *T. nicefori* (reported as *T. aureus*) for Pichindé virus, but none yielded the virus.

REMARKS: Thomas (1921j) defined *T. nicefori* based from a male specimen (BM 21.7.1.20); however, the holotype turns out to be a female specimen determined by myself by the presence of six teats and the long and narrow prepuce. Thomas compared his new species with both *T. aureus* and *T. princeps*. However, Voss (2003) suggested that *T. nicefori* might be included in *T. popayanus* based on similar size of feet and molar toothrow, a hypothesis not supported by Pacheco (2003), who accepted both *T. nicefori* and *T. popayanus* as valid species. *Thomasomys nicefori* can be readily separated from *T. popayanus* and *T. aureus* by the general smaller size and by the comparatively broader palatal bridge. In addition, *T. nicefori* compared with *T. popayanus* exhibits a light metatarsal patch, shorter nasals not distinctly longer than premaxillae, broader interorbital region, perceptible lambdoid ridge, auditory bullae moderately inflated, and smaller anterior process of ectotympanic. Additional comparisons are provided in the *T. popayanus* account.

The population on the western Cordillera, from Urrao, Páramo Frontino, is provisionally considered *T. nicefori*, although they have narrower zygomatic plates, broader incisive foramen, less convergent zygomatic arches among other features that suggest it might correspond to a different species (V. Pacheco, unpubl. data). This population is sympatric with *T. popayanus* (see that account). The specimen from Quindío, Salento (AMNH 33941), is considered conspecific, based on the relatively small size (33 mm hindfoot and 6.35 mm molar toothrow); it represents the southernmost record of the species along the Central Cordillera. The karyotype has not been described.

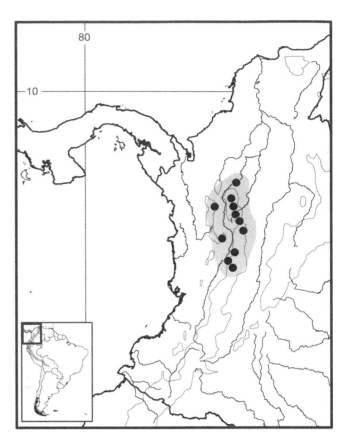

Map 352 Selected localities for *Thomasomys nicefori* (●). Contour line = 2,000 m.

Thomasomys niveipes (Thomas, 1896)
White-footed Thomasomys
SYNONYMS:

Oryzomys niveipes Thomas, 1896:305; type locality "La Oya del Barro, W. Cundinamarca, Colombia"; spelling corrected to La Hoya del Barro [= La Hoya del Barro Blanco], small basin situated at the head of the Río Teusacá and includes the Laguna El Verjón (Gómez-Laverde et al. 1997).

Thomasomys niveipes: Thomas, 1906c:443; first use of current name combination.

Thomasomys laniger laniger: Cabrera, 1961:431; part; not *laniger* Thomas.

DESCRIPTION: Medium sized (head and body length 116–130 mm) with comparatively short tail (96–122% of head and body length) and moderate-size ears (19–22 mm). Dorsal pelage dull olive-fulvous, darker mesially, paler on sides and below, with fulvous tone on ventral surface, and indistinctly countershaded. Forefeet silvery white. Tail brown, incipiently bicolored, thinly haired, and lacks whitish tip. Mystacial vibrissae moderately long, extending slightly beyond posterior margin of pinnae when bent; genal 1 vibrissae absent. Hindfoot of medium length (26–31 mm);

metatarsals silvery white. Skull of moderate size (condyloincisive length 27–29.4 mm); rostrum long and slender; braincase smoothly rounded; a moderately developed rostral tube present; frontonasal profile convex. Zygomatic notches comparatively deep for genus. Zygomatic plates moderately broad, subequal to length of M1, and vertical. Incisive foramina long and relatively narrow, and extend just past anterior margins of M1s. Interorbital region narrow with rounded edges. Mesopterygoid fossa somewhat narrow, subequal to parapterygoid fossae at level of presphenoid-basisphenoid suture, sides subparallel, and roof perforated by moderately open sphenopalatine vacuities; palatal medial process absent or weak. Zygomatic arches approach rectangular shape, converging only slightly anteriorly. Auditory bullae inflated, bony tubes indistinct, and anterior process of ectotympanic large. Postglenoid foramen smaller than large subsquamosal fenestra, and placed posterodorsally. Alisphenoid strut that separates buccinator-masticatory and accessory oval foramina absent or incomplete. Ethmoturbinals of moderate size. Posterior opening of the alisphenoid canal of normal size. Interfrontal fontanella present. Carotid circulatory pattern primitive (stapedial foramen, squamosal alisphenoid groove, and sphenofrontal foramen present). Upper incisors opisthodont. Molars relatively hypsodont and lack well-developed cingula and stylar cusps. Anterolophs and mesolophs of M1 and M2 comparatively well developed; M3 with distinct hypoflexus; anteromedian flexus in M1 distinct, parallel to distinct anteromedian flexid in m1; anterolabial cingulum distinct on m2, but not on m3. Maxillary toothrow length moderate, 4.7–5.2 mm. Capsular process absent.

DISTRIBUTION: *Thomasomys niveipes* occurs in the Cordillera Oriental of the Andes in central Colombia, in the departments of Boyacá and Cundinamarca, from 2,550 to 3,500 m.

SELECTED LOCALITIES (Map 353): COLOMBIA: Boyacá, Villa de Leyva, sector Chaina, vereda Río Abajo (IAvH-M 6921), Villa de Leyva, sector Chaina, vereda Río Abajo (IAvH-M 6915); Cundinamarca, Bogotá (MVZ 113965), Chipaque (AMNH 34589), Finca El Tabacal (IAvH-M 7088), Guasca, Río Balcones (FMNH 71465), La Regadera, Usme (IAvH-M 1169), Laguna Verde, Páramo de Guerrero (IAvH-M 7221), Páramo de Choachí (MCZ 27589), Reserva Forestal, Vda. Quebrada Honda (IAvH-M 7090), San Francisco, Bogotá (FMNH 71460), Usaquen (USNM 251973).

SUBSPECIES: *Thomasomys niveipes* is monotypic.

NATURAL HISTORY: This species was captured more often than expected in páramo habitat (Gómez-Laverde et al. 1997). Lopez-Arévalo et al. (1993) reported that *Thomasomys* sp. (here referred to *T. niveipes*) was terrestrial, that a pregnant specimen was captured in July (lo-

cal wet season), and that the diet was composed mainly of young leaves and other green part of plants. Schramm and Lewis (1987) described the flea *Plocopsylla nungui* from specimens of *Thomasomys* sp. collected by P. Hershkovitz from San Francisco, Bogotá. The actual host, however, may be attributed to either *T. niveipes* or *T. princeps* as Hershkovitz collected both species at this locality.

REMARKS: Thomas (1896) considered this species very similar externally to *T. laniger*, differing mainly in hindfoot color, notably the brownish feet in *T. laniger* versus whitish feet in *T. niveipes*. Cabrera (1961) included *niveipes* Thomas as a synonym of the nominate subspecies in his concept of a polytypic *T. laniger* (which also included *emeritus* Thomas and *monochromos* Bangs as subspecies, two taxa herein treated as species), without justification. Later, Musser and Carleton (1993) returned the taxon to species status based on their own specimen comparisons. Gómez-Laverde et al. (1997) described the karyotype of *T. niveipes* ($2n = 24$, FN = 42), and reinforced its separation from *T. laniger* ($2n = 40$, FN = 40), based on morphological, karyotypic, and ecological differences. Ellerman (1941) included this species in his *cinereus* Group, but Pacheco (2003) allied *T. niveipes* to *T. paramorum* and *T. hudsoni* in his phylogenetic assessment based on morphological characters, all members of his "cinereus" group.

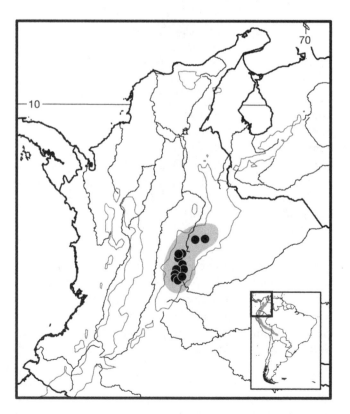

Map 353 Selected localities for *Thomasomys niveipes* (●). Contour line = 2,000 m.

Thomasomys notatus Thomas, 1917
Dusky-footed Thomasomys
SYNONYM:

Thomasomys notatus Thomas, 1917g:2; type locality "To-rontoy," Cusco, Peru.

DESCRIPTION: Medium sized (head and body length 110–128 mm) with close, thick, and rather woolly fur. Dorsal fur comparatively short (average length 6–9 mm, guard hairs 13 mm) and colored ochraceous tawny with perceptible dorsal stripe present, grayer on head, and more tawny posteriorly. Ventral coloration soiled whitish, hairs slaty basally but dull whitish terminally; lips and throat white, with hairs white basally; contrasts sharply with the dorsal coloration. Ears comparatively small (18–23 mm), blackish, contrasting with head coloration, and with whitish postauricular patch present. Dark orbital ring present. Mystacial vibrissae long, extending backward distinctly beyond pinnae when bent; genal 1 vibrissae present. Tail long (141–160% of head and body length), monocolored or incipiently bicolored, its ventral side indistinctly paler on proximal quarter, blackish brown, short haired, and with small tuft of long hairs at tip. Hindfoot relatively short (24–27 mm) and moderately broad, with dark brown metatarsal patch, sides and digits whitish; no gap between thenar and hypothenar pads. Hallux long, extending to interphalangeal joint of dII. Digit V of pes is very long, claw extending to base of claw of dIV. Protuberance anus raised, but not as prominent as in members of "cinereus" group. Skull moderate in size (condyloincisive length 26.1–29.3 mm) with narrow and relatively short rostrum, and rounded braincase with large parietals. Nasals narrow and long, tapering posteriorly beyond maxillary-frontal-lacrimal joint. Premaxillomaxillary-frontal joint lies posterior to zygomatic notch. Interorbital region narrow, hourglass in shape, and with squared and slightly raised edges. Zygomatic arches more parallel oriented not converging forward. Interfrontal fontanel present. Ethmoturbinals usually moderate in size. Molar bearing maxilla with rectangular shape. Zygomatic plates moderately broad, subequal to M1, and vertical. Incisive foramina long, narrow, with parallel sides, and extend posteriorly slightly between M1 anterocones. Mesopterygoid fossa somewhat narrow, breadth subequal to breadth of parapterygoid fossae at level of presphenoid-basisphenoid suture, and with lateral margins converging posteriorly or subparallel; no medial palatal process; roof perforated by slits on each side of presphenoid-basisphenoid suture. Auditory bullae small and uninflated, bony tubes distinct and narrow, and anterior process of ectotympanic comparatively large. Subsquamosal fenestra and postglenoid foramen narrow, elongated, and subequal; latter extends anteriorly even more than tegmen tympani. Carotid circulatory pattern primitive (stapedial foramen, groove on inner surface of squamosal and alisphenoid, and sphenofrontal foramen present). Upper incisors opisthodont. Maxillary toothrow short (4.1–4.5 mm); molars brachydont, without interpenetration but distinct crenulation. M1 exhibits distinct anteromedian flexus, anteroflexus, anteroloph, and mesoloph; hypoflexus narrow; protostyle and enterostyle present; and accessory labial root absent. Protoflexus on M2 and hypoflexus M3 distinct. The m1 anteromedian flexid absent or faint even in unworn molars. Mandible exhibits angular process shorter than condylar process and distinct capsular process present.

DISTRIBUTION: *Thomasomys notatus* occurs in montane forests on the eastern Andean slopes of Peru, at elevations between 1,400 to 3,400 m.

SELECTED LOCALITIES (Map 354): PERU: Amazonas, Río Utcubamba, Uchco (FMNH 19834); Ayacucho, Chiquintirca (MUSM 26308); Cusco, 72 km by road NE of Paucartambo (UMMZ 160588), Pillahuata (FMNH 172380), Torontoy (type locality of *Thomasomys notatus* Thomas); Huánuco, Campamento Provias (MUSM 22953), east slope Cordillera Carpish (LSUMZ 14368), Shogos (MUSM 22743); Junín, Cordillera del Vilcabamba (MUSM 13061), 22 mi E of Tarma (AMNH 231084); Pasco, Cumbre de Ollón, ca. 12 km E of Oxapampa (LSUMZ 25898), Santa Cruz, ca. 9 km SSE of Oxapampa (LSUMZ 25901); San Martín, Las Palmas, Parque Nacional Río Abiseo (MUSM 7947), San Martín, Pampa de Cuy, Parque Nacional Río Abiseo (MUSM 7956).

SUBSPECIES: I treat *T. notatus* as monotypic (but see Remarks).

NATURAL HISTORY: Few data on the natural history or ecology have been reported for this species. Individuals fed primarily on monocots in the families Cyperaceae and Poaceae (89.2% of diet) plus a few arthropods (< 5%; Noblecilla and Pacheco 2012). They are apparently in low abundance in montane forests (C. E. Medina et al. 2012).

REMARKS: Little is known on this species. After Thomas (1917g) described *T. notatus* from Cusco department, additional specimens were collected from Tambo Jenés, Amazonas (Thomas 1926e), and from Cordillera Carpish, Huánuco (A. L. Gardner 1976). Recent work reported the species (as *Thomasomys* sp. nov.) in Abiseo National Park, San Martín department (Leo and Romo 1992), in the Cordillera Vilcabamba, Junín (Emmons et al. 2001), and in Yanachaga National Park, Pasco (Vivar Pinares 2006). Based on these new records (25 in total), the potential distribution of the species was evaluated using the algorithm of Maxent, and estimated to be 182,420 km² (Quintana Navarrete 2011). Although, the species is currently considered endemic to the Peruvian Andes (Pacheco et al. 2009), the Maxent map predicted that the species could range to Ecuador and Bolivia along the Yungas habitats (Quintana Navarrete 2011).

Pacheco (2003) found that some characters varied regionally, which may suggest the presence of unknown taxa. In northern populations, from Amazonas and Pasco departments, the incisive foramina are long but do not penetrate between the M1s, the protostyle and enterostyle are more developed, and molar crenulation is weak or absent. Other variations include the absent of maxillary pits are in the sample from Huánuco, and the small ethmoturbinals in specimens from Pasco (V. Pacheco, unpubl. data). Additional studies are needed to determine the systematic status of these variable local populations.

Thomas (1920h) pointed out the difficulty in placing *T. notatus* within a systematic arrangement, stressing the broad hindfeet reminiscent of typical *Rhipidomys* rather than the long and narrow feet usually seen in *Thomasomys*, although the skull of this species does not have the broadly diverging interorbital region typical of *Rhipidomys*. Although Patton et al. (1990) noted that *T. notatus* exhibits considerable similarities in cranial structure and body proportions to *T. oreas*, these two species were not closer sister taxa in an analysis based on electrophoresis of 24 loci as compared to other species of *Thomasomys* from southeastern Perú (Patton 1987). A phylogenetic assessment based on morphological attributes found *T. notatus* more related to a monophyletic *Rhipidomys* and a cluster of species allied to *T. oreas* and *T. gracilis* than to

the remaining species of *Thomasomys* (Pacheco 2003). He proposed the species represents a distinct species group, probably at the genus-group level (his "New genus E"). However, Salazar-Bravo and Yates (2007), in a molecular phylogeny based on mitochondrial cytochrome-*b* gene sequence, found *T. notatus* closely related to *T. ladewi*, despite the trenchant morphological differences between these two taxa. This result needs to be verified with denser taxon sampling and a broader assortment of gene.

A. L. Gardner and Patton (1976) described the standard karyotype ($2n=44$, $FN=44$) with X and Y acrocentric chromosomes based on samples from the Cordillera Carpish, Huánuco department.

Thomasomys onkiro L. Luna and Pacheco, 2002
Ashaninka Thomasomys

SYNONYM:

Thomasomys onkiro L. Luna and Pacheco, 2002:835; type locality "PERU: Department of Cuzco, Province of La Convención, Vilcabamba Cordillera, eastern slope of the Andes, between the Ene and Urubamba rivers, 325 km northwest of city of Cuzco, 11°39′36″S, 73°40′02″W, at 3,350 m elevation in elfin forest patches of *Polylepis*, *Weinmannia*, and *Chusquea*," corrected to Province of Satipo, Junín department by Quintana Navarrete (2011).

DESCRIPTION: Medium sized (head and body length 94–114 mm) with soft, dense, and comparatively long dorsal fur. Dorsal pelage colored buffy brown and not countershaded with ventral pelage, which is buffy yellow. Mystacial vibrissae moderately long, extending slightly beyond posterior margin of pinnae when bent; genal 1 vibrissae absent. Tail long (144–159 mm), much longer than head and body (131–153%), and monocolored with terminal white tip. Hindfoot moderately long (26–28 mm) and narrow; dorsal surface mixture of dark brown and whitish hairs to base of digits, uniformly silvery white; proximal margin of hypothenar pad level with distal margin of thenar pad. Pinnae small (19–20 mm), outer and inner edges clothed with short, dark-brown hairs. Digit I of pes (hallux) moderate long, but claw does not extend more than about half length of phalanx 1 of dII. Digit V of pes long, with claw extending about half length of phalanx 2 of dIV. Skull moderately long (condyloincisive length 25.45–27.33 mm) with relatively long and slender rostrum and anteriorly expanded nasals. Interorbital region moderately broad, hourglass in shape, and with smooth margins. Zygomatic arches anteriorly convergent. Zygomatic plates moderately broad (subequal to length of M1), with anterior margins straight and vertical. Incisive foramina narrow, parallel sided but slightly broader at premaxillomaxillary suture, and moderately long, reaching close to anterior margins of M1s. Mesopterygoid fossa comparatively broad with subparallel

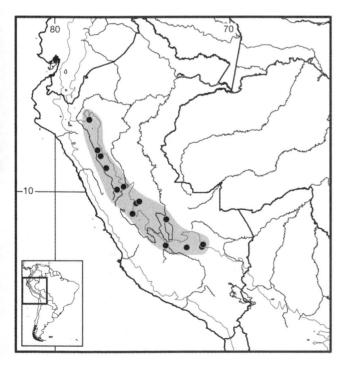

Map 354 Selected localities for *Thomasomys notatus* (●). Contour line = 2,000 m.

margins and without medial process; sphenopalatine vacuities absent or occasionally present as slits between the presphenoid-basisphenoid suture. Auditory bullae small and uninflated. Carotid circulatory pattern primitive (stapedial foramen, groove on inner surface of squamosal and alisphenoid, and sphenofrontal foramen are present). Alisphenoid strut present. Upper incisors slender and orthodont; lower incisors slender. Maxillary toothrow moderately long (4.5–4.65 mm), with upper molars moderately hypsodont, with no interpenetration of flexi, and weak labial cingula close to mesolophs of M1 and M2. M1 anteromedian flexus deep; m1 anteromedian flexid shallow; ectolophids present in m1 and m2. Capsular process absent; lower diastema slender.

DISTRIBUTION: *Thomasomys onkiro* is known only from elfin forest at the type locality in southern Peru.

SELECTED LOCALITIES (Map 355): PERU: Junín, Cordillera Vilcabamba (type locality of *Thomasomys onkiro* L. Luna and Pacheco).

SUBSPECIES: *Thomasomys onkiro* is monotypic.

NATURAL HISTORY: L. Luna and Pacheco (2002; also L. H. Emmons, pers. comm.) collected *T. onkiro* on the ground in elfin forest patches dominated by *Polylepis* (Rosaceae), *Weinmannia* (Cunoniaceae), and *Chusquea* (Poaceae) in places with deep moss on the ground. It was found in sympatry with *Mazama chunyi*, *Mustela frenata*, *Gracilinanus* cf. *aceramarcae*, *Cuscomys ashaninka*, *Cavia tschudii*, *Akodon torques*, *Microryzomys minutus*, *Thomasomys* cf. *aureus*, *T.* cf. *kalinowskii*, and *T.* cf. *gracilis* (Emmons et al. 2001).

REMARKS: L. Luna and Pacheco (2002) reported the type locality of *T. onkiro* to be in Cusco department, Province of La Convención, Cordillera de Vilcabamba, at 11° 39' 36" S, 73° 40' 02" W. However, mapping these coordinates in digital maps revealed this point is actually in Junín department, very near the border with Cusco (Quintana Navarrete 2011). The type locality is thus corrected accordingly. This change also applies to the type locality of the Ashaninka Arboreal Chinchilla Rat *Cuscomys ashaninka* Emmons, an abrocomid rodent reported as new species from the same locality of *T. onkiro* by Emmons (1999). L. Luna and Pacheco (2002) compared *T. onkiro* with both *T. caudivarius* and *T. silvestris*, finding it morphologically more similar to the latter. Phylogenetic relationships of this species with congeners are unresolved, although Pacheco (2003) argued that it was related to other species of the "cinereus" group. It is presumed to be a relict restricted to the Cordillera Vilcabamba (L. Luna and Pacheco 2002). Its karyotype is unknown.

P. Moreno and Albuja (2012) claimed to have the first record of *Thomasomys onkiro* from Ecuador, based on two specimens from Reserva de Tapichalaca (04°29'31.5"S,

Map 355 Selected localities for *Thomasomys onkiro* (●). Contour line = 2,000 m.

79°07'47.2"W), Zamora-Chinchipe province. However, these specimens do not have the slender upper and lower incisors, and the slender lower diastema observed in *T. onkiro* proper. Compared to *T. onkiro* proper, the Ecuadorean specimens have a mesopterygoid fossa with a medial process and roof perforated by narrow slits. I believe that these Ecuadorean specimens are best allocated to *T. caudivarius* (see *T. caudivarius* account), and this record is not included here.

Thomasomys oreas Anthony, 1926
Montane Fairy Thomasomys
SYNONYMS:

Thomasomys oreas Anthony, 1926a:2; type locality "Cocopunco [= Cocapunco], about 80 miles north of La Paz, Bolivia, altitude 10,000 feet."

Thomasomys taczanowskii: S. Anderson, 1993:37; part; not *taczanowskii* Thomas.

DESCRIPTION: Small sized (head and body length 90–108 mm) with long and soft dorsal fur (average length 15 mm, guard hairs 19 mm). Dorsal pelage colored Prout's brown, finely sprinkled on back with fuscous black; cheeks and sides brighter than back; dusky orbital ring present; ventral pelage pinkish buff, with hairs gray basally and moderately countershaded with dorsal coloration. Ears comparatively large (19 mm) and blackish brown. Mystacial vibrissae long, extending backward distinctly beyond pinnae when bent; genal 1 vibrissae present. Tail longer than head and body (126–139%), faintly bicolored, clothed with and short hairs, brown above but grayish below, and without terminal white tip or pencil tuft. Hindfoot relatively short (23–25 mm) with clove-brown metatarsal patch, edges and digits cartridge-buff; no gap between thenar and hypothenar pads. Hallux long, extending to interphalangeal joint of dII. Digit V of pes very long, its claw extending to base of claw of dIV. Protuberant anus raised but not as prominent as in *T. cinereus*. Skull short (condyloincisive length 25.3–25.8 mm) with long and slender rostrum and well-inflated braincase with rather anteroposteriorly broad interparietal. Interorbital region narrow, hourglass in shape, with rounded edges. Interfrontal fontanella present. Zygomatic arches with limited lateral flaring, either parallel or converging anteriorly. Zygomatic plates vertical and moderately broad, breadth subequal to length of M1. Molar bearing maxilla with rectangular shape. Distinct notch present between zygomatic maxillary root of zygomatic arch and lacrimal bone in dorsal view. Incisive foramina long and narrow, extending posteriorly between M1 anterocones. Mesopterygoid fossa moderately broad, slightly converging posteriorly, with weak or absent palatal medial process; roof perforated by medium-size sphenopalatine vacuities on each side of presphenoid-basisphenoid suture. Auditory bullae large and inflated. Subsquamosal fenestra well open and larger than postglenoid foramen. Carotid circulatory pattern primitive (stapedial foramen, groove on inner surface of squamosal and alisphenoid, and sphenofrontal foramen present). Upper incisors opisthodont. Maxillary toothrow short (4.1–4.3 mm); molars small and brachydont, without interpenetrating flexi but distinct crenulation. M1 exhibits distinct anteromedian flexus and anteroloph, but anteroflexus mostly coalesced; anteromedian flexid of m1 weak even in unworn molars. Capsular process present.

DISTRIBUTION: *Thomasomys oreas* occurs from northern Peru to central Bolivia, in eastern montane forests from 2,460 to 3,650 m.

SELECTED LOCALITIES (Map 356): BOLIVIA: Cochabamba, Ayopaya, El Choro (FMNH 74866), Totora, 2 km N of Cocapata (CBF 4954); La Paz, Chuncani (CBF 7427), Cocapunco (type locality of *Thomasomys oreas* Anthony), Franz Tamayo, Pelechuco, Llamachaqui, on road to Apolo (CBF 4041). PERU: Amazonas, mountains east of Balsas (FMNH 19830), 10 mi E of Molinopampa (FMNH 19832); Ayacucho, Comunidad Estera, Cerro Calle Nueva (MUSM 34718); Cusco, Ccachubamba (FMNH 75224), 32 km NE (by road) of Paucartambo; km 112 (MVZ 172344), Puesto Vigilancia Acjanaco (MUSM 6727), Tocopoqueu, Occobamba valley (USNM 194876).

SUBSPECIES: I provisionally consider *T. oreas* to be monotypic (but see Remarks).

NATURAL HISTORY: Bolivian records summarized by S. Anderson (1997) unfortunately lack data on habitat, habits, behavior, food, or reproduction. However, Salazar-Bravo and Yates (2007) reported that recent specimens were collected on the ground in upper montane elfin forest. One female specimen, with no embryos, was caught at Tres Cruces, Cusco department, Peru, on the edge of cloud forest (O. P. Pearson field notes, MVZ archives). Known ectoparasites include the lice *Hoplopleura tiptoni* and *Pterophthirus imitans* (although the latter may be a straggler; V. S. Smith et al. 2008) and the flea, *Sphintopsylla* sp. (Maihuay et al. 2004). Visceral parasites include the tapeworm *Taenia pisiformis* (Tantaleán and Chavez 2004).

REMARKS: S. Anderson (1997:91) corrected the spelling of the type locality of Cocopunco to Cocapunco. This species includes several records that appeared in the literature as *T. taczanowskii* (Thomas 1920h; A. L. Gardner 1976; A. L. Gardner and Patton 1976; S. Anderson 1993, 1997), based on my own examination of the reported specimens. Patton et al. (1990) tentatively referred *T. oreas* to one specimen from Cusco, Ocobamba valley as *T. taczanowskii* by Thomas (1920h). This locality belongs to the Río Urubamba, not to the Río Apurímac drainage as indicated in Patton et al. (1990). Pacheco (2003) observed considerable geographic and interspecific variation and suggested that *T. oreas* represents a species group that includes at least *T. oreas* s. s. and two other currently undescribed species, which he informally referred to as *Thomasomys* sp. 8 (from Yuraccyacu, Ayacucho department) and *Thomasomys* sp. 9 (from Paucartambo, Cusco department). The first exhibits larger and broader incisive foramina and a broader parapterygoid fossa; the second is distinguished by narrower incisive foramina, the lack of a palatal medial process, a larger postglenoid foramen, comparatively smaller auditory bullae, and subparallel zygomatic arches. Salazar-Bravo and Yates (2007) differentiated *T. oreas* from both *T. andersoni* and *T. notatus* by morphology as well as mtDNA cytochrome-*b* sequences. However, it should be noted that the genetic sequences came from a specimen from Cochabamba (MSB 87126) rather than from the type locality.

Cochabamba specimens are slightly smaller in body size and molar toothrow length than the type specimen from La Paz, and depart in other attributes, such as genal 1 absent, slightly broader rostrum, interparietal anteroposteriorly shorter, palatal medial process absent, and smaller squamosal fenestra, which might indicate the presence of another unknown taxon. S. Anderson (1997:411) was probably misled by these differences and misidentified two Cochabamba's specimens (FMNH 74866, 74867) as *T. taczanowskii* at the same time he recognized *T. oreas* for Bolivia. Pending further analyses, ideally based on additional specimens, I tentatively allocate the Cochabamba sample to *T. oreas*. A single specimen collected from the Cordillera Vilcabamba, Junín, might also represent an unknown taxon. This specimen has grayish tones, short dorsal pelage, and an undefined dorsal band, among other features.

Anthony (1926a) considered *T. oreas* as most closely related to *T. paramorum*. On the contrary, Patton (1987), based on electrophoresis of 24 loci, found *T. oreas* related to *T. gracilis* as compared with other species of *Thomasomys* from southeastern Perú. This result that was later supported by cladistic analysis of morphological characters where *T. oreas* appears related to a species cluster that included *Thomasomys* sp. 8, *Thomasomys* sp. 9, and *T. gracilis* (Pacheco 2003). The phylogenetic analysis based on mtDNA cytochrome-*b* sequences placed *T. oreas* as sister to *Thomasomys* sp. 9 and well removed from *T. gracilis* (Salazar-Bravo and Yates 2007).

The karyotype of *T. oreas* (reported as *T. taczanowskii*) with a diploid number $2n=44$ and, $FN=44$, X and Y autosomal chromosomes, from a specimen from Yuraccyacu, Ayacucho department, in the Río Apurímac drainage was reported by A. L. Gardner and Patton (1976), a sample that I earlier (Pacheco 2003) referred to as *Thomasomys* sp. 8. Given the extensive geographic variation in samples currently allocated to *T. oreas*, the karyotype of a topotype is critically lacking.

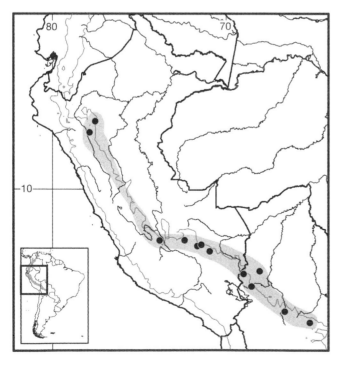

Map 356 Selected localities for *Thomasomys oreas* (●). Contour line = 2,000 m.

Thomasomys paramorum Thomas, 1898
Páramo Thomasomys

SYNONYM:

Thomasomys paramorum Thomas, 1898e:453; type locality "Paramo [= Páramo] south of Chimborazo [= Mt Chimborazo]," defined as "Urbina (1°30′S, 78°44′W) just a few kilometers SE of Chimborazo" by Voss (2007), Chimborazo, Ecuador.

DESCRIPTION: Small sized (head and body length 85–113 mm) with long, soft, and woolly dorsal fur. Dorsal color even brownish olive to reddish brown; ventral color pale gray to whitish; creamy line separating flank and abdomen; moderate countershading between dorsum and venter. Mystacial vibrissae moderately long, extending slightly beyond posterior margin of pinnae when bent; genal 1 vibrissae absent. Ears relatively large (15–22 mm), evenly rounded, well haired, and dark brown. Tail short to moderately long (101–147% of head and body length), well haired, incipiently bicolored brown above and paler below, and occasionally with terminal white tip. Hands and feet whitish above. Hindfoot relatively short (19–28 mm); gap present between thenar and hypothenar pads. Hallux moderately long, with claw not extending more than about half length of phalanx 1 of dII. Digit V of pes long, with claw extending about half length of phalanx 2 of dIV. Skull comparatively small (condyloincisive length 22.8–25.6 mm), slender and delicate, with narrow, long, and rounded braincase, and relatively narrow rostrum. Zygomatic plates narrow, breadth less than length of M1, or moderately broad, subequal to length of M1, and may slope backward or be vertically oriented. Interorbital region moderately broad, hourglass in shape, with smooth edges. Zygomatic arches slightly convergent anteriorly. Incisive foramina long, oval shaped, more contracted anteriorly, and extended backward between M1 procingula. Mesopterygoid fossa broad with subparallel sides; sphenopalatine vacuities present as slits or moderately open. Parapterygoid vacuities present or absent. Auditory bullae large and inflated. Subsquamosal fenestra slightly larger and more open than postglenoid foramen. Carotid circulatory pattern primitive (stapedial foramen, groove on inner surface of squamosal and alisphenoid, and sphenofrontal foramen present). Upper incisors narrow and opisthodont or orthodont. Maxillary toothrow relatively short, 3.8–4.5 mm; upper molars moderately hypsodont, with no interpenetration of flexi. Anteroloph of M1 indistinct; M3 with indistinct mesoloph and hypoflexus. Capsular process absent, or as only swollen ridge if present.

DISTRIBUTION: *Thomasomys paramorum* is an endemic species of the northern Andes, occurring north of the Huancabamba Deflection (Voss 2003) within montane forests and páramo habitats from southern Colombia to central Ecuador, between 2,000 and 4,300 m.

SELECTED LOCALITIES (Map 357): COLOMBIA: Nariño, Pasto, slopes of Volcán Galeras (ICN 13267). ECUADOR: Azuay, Cajas Plateau (Barnett 1999); Bolívar, Sinche (AMNH 66984); Carchi, Mantufar, Atal (UMMZ 77215); Chimborazo, Urbina (type locality of *Thomasomys paramorum* Thomas); Napo, Cerro Antisana (FMNH 53239); Napo, Cuyuco [probably Cuyuja], below Papallacta (AMNH 46629); Pichincha, E side of Mt. Pichincha (MVZ 137603), Guarumos West (FMNH 56238), Upper Río Pita (USNM 270151).

SUBSPECIES: *Thomasomys paramorum* is monotypic (but see Remarks).

NATURAL HISTORY: On the Cajas Plateau in Ecuador, this species occurs in primary, secondary, and recently disturbed forests, typically in habitats with moist mossy ground cover, fallen logs, and proximity to stands of *Chusquea* bamboo, but not in páramo (Barnett 1999). At Papallacta, also in northern Ecuador, individuals were captured in subalpine rainforest, in *Polylepis* (Rosaceae) thickets in the páramo, and in the shrubby páramo/forest ecotone (Voss 2003). It is mainly terrestrial, and uses runways through moss, along banks of small streams, in wet leaf litter under shrubs and branches, and under clumps of grass. It was also trapped in low, moss-covered trees (Voss 2003). M. Brito et al. (2012) described nets located under root masses and among dead leaves of *Espeletia* (Asteraceae) shrubs, at heights between 15 cm and 1.5 m. A female from the east side of Mt. Pichincha (MVZ) was collected with an embryo at the end of June.

REMARKS: The stomach morphology corresponds to the unilocular-hemiglandular model (Carleton 1973). The few karyotypes available suggest some geographic variation in Ecuador. Haynie et al. (2006) reported $2n=44$, FN= 42, for specimens from Tungurahua, similar to that reported previously by A. L. Gardner and Patton (1976) from Mount Pichincha (reported as *Thomasomys* sp.), except that X and Y chromosomes are acrocentric instead of submetacentric. A karyotype from the type locality is lacking.

Thomasomys paramorum exhibits some geographic polymorphism in several characters that might suggest a species complex (Pacheco 2003). Samples from southern Colombia in Nariño and northwestern Ecuador, Carchi, differ from typical *paramorum* by their reddish pelage, broader nasals, very inflated bullae that are strongly placed forward so that the middle lacerate foramen is closed, a broad mesopterygoid fossa with posteriorly convergent sides, the upper incisors are less opisthodont rather than strongly opisthodont, and the postglenoid foramen and subsquamosal fenestra are narrow and elongated, not rounded. Voss (2007) also noted than the topotypic series from Chimborazo average larger

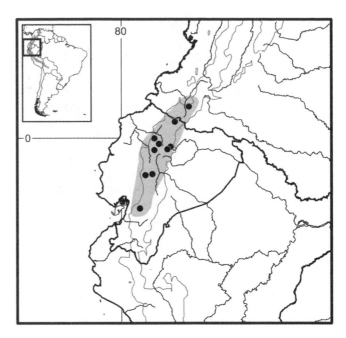

Map 357 Selected localities for *Thomasomys paramorum* (●). Contour line = 2,000 m.

than Papallacta specimens in most measurements and have proportionally narrower zygomatic plates, but he judged both to represent the same taxon. Pacheco (2003) also noted that the zygomatic plates are slightly sloped backward in these specimens, not vertical.

Little is known about the phylogenetic position of *T. paramorum*. Ellerman (1941) included this species in his *cinereus* Group. Pacheco (2003) found the species closely related to *T. niveipes* in his phylogenetic assessment based on morphology. And T. E. Lee et al. (2011) allied the species to an apparently unknown taxon, morphologically identified as *T. baeops*, from Sangay, based on an mtDNA cytochrome-*b* molecular phylogeny. Unfortunately, no additional data were provided on this taxon to understand this discrepancy. However, if this unknown taxon were certainly a type of *baeops*, the hypothesis of the "cinereus" and "baeops" groups presented here would be rejected despite the trenchant morphological differences between them. Because Robert S. Voss confirmed the identification of this taxon as *T. baeops* but the gene sequence suggests that the specimen is a member of the "cinereus" group, the possibility of a mismatch should be considered. Additional studies are needed to resolve this issue and the phylogenetic relationships of *T. paramorum*.

Thomasomys popayanus J. A. Allen, 1912
Popayán Thomasomys

SYNONYMS:

Thomasomys popayanus J. A. Allen, 1912:81; type locality "crest of Western Andes (alt. 10,340 ft [3,152 m]),

40 miles [64 km] west of Popayan [= Popayán], Cauca, Colombia."

Thomasomys aureus popayanus: Gyldenstolpe, 1932a:51; name combination.

DESCRIPTION: Large species (head and body length 145–164 mm) with general coloration similar to *T. aureus*. Dorsal pelage long, thick, and not very fine, yellowish rufous with faint but dark median band along midline; top of head darker than back and slightly grayish; ventral surface deep ochraceous buff. Hindfeet long (33–38 mm), moderately broad, and with upper metatarsal dark brown and digits whitish. Tail long (129–145% of head and body length), monocolored, finely annulated, covered with very short hairs, and without terminal pencil. Skull large (condyloincisive length 34.5–36 mm); nasals long, taper posteriorly, and longer than premaxillae. Zygomatic plates rather vertical and moderately broad. Interorbital region narrow with incipient square margins; postorbital ridge indistinct. Incisive foramina long and moderately narrow, extending posteriorly slightly between M1 anterocones. Palatal bridge short and narrow. Auditory bullae small, and not inflated; anterior process of ectotympanic is large. Mesopterygoid fossa broad and somewhat parallel sided; roof closed. Carotid circulatory pattern derived (stapedial foramen small, no internal squamosal-alisphenoid groove, no sphenofrontal foramen). Lambdoid ridge indistinct or absent. Upper incisors opisthodont. Maxillary toothrow long, 6.4–7.0 mm. M1 lacks perceptible protostyle and enterostyle; m1 with ectolophid. Capsular process absent.

DISTRIBUTION: *Thomasomys popayanus* occurs in the Cordillera Occidental of Colombia, between 1,830 m and 3,300 m.

SELECTED LOCALITIES (Map 358): COLOMBIA: Antioquia, Urrao, Santa Bárbara (FMNH 71287), Urrao, Páramo Frontino (FMNH 71269); Cauca, coast range W of Popayán (MCZ 17317), Cocal (AMNH 32460), Munchique (FMNH 89273); Valle del Cauca, San Antonio (AMNH 32187).

SUBSPECIES: I consider *T. popayanus* to be monotypic.

NATURAL HISTORY: Little is known about this species. Delgado-V. (2009) reported it from the Reserva San Sebastián-La Castellana, Valle de Aburrá, Antioquia, Colombia, by both captures and from remains in scat of the Crab-eating Fox *Cerdocyon thous*. Trapido and Sanmartín (1971) screened specimens of *T. popayanus* (reported as *T. aureus*) for Pichindé virus, but failed to detect the virus.

REMARKS: *Thomasomys popayanus* was described only on the basis of coloration and standard measurements (J. A. Allen 1912). Without substantive arguments, Thomas (1921j) doubted that *popayanus* was separable from *T. aureus*, and Ellerman (1941) formally reduced it to a subspecies of *T. aureus*, an action followed by Ca-

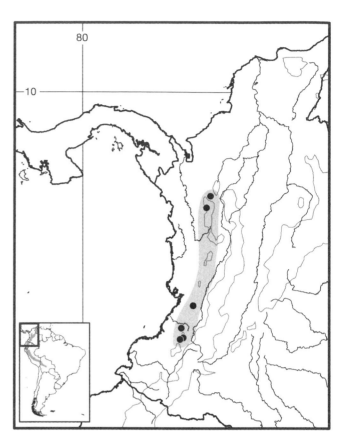

Map 358 Selected localities for *Thomasomys popayanus* (●). Contour line = 2,000 m.

brera (1961) and Musser and Carleton (1993) but not by Voss (2003) or Pacheco (2003). Voss (2003) recognized *T. popayanus* (in which he provisionally included *nicefori*) based on shorter (33–34 mm) hindfeet and smaller (6.0–6.6 mm) molar toothrows than *T. aureus*. However, for unknown reasons, Voss's (2003) measurements do not coincide with my own data of hindfeet (33–38 mm, N = 4, AMNH adult specimens) or molar toothrow (6.44–6.96 mm, N = 4, AMNH adult specimens), or J. A. Allen's (1912) data (35–38 mm, 7.0 mm, respectively), both sets of which are slightly larger than hindfoot (33–36 mm), and molar toothrow (5.9–6.6 mm) of *T. nicefori* (N = 10). In addition, *T. popayanus* has the tail usually longer than *T. nicefori* (129–145% versus 114–136% of head and body length). Pacheco (2003) presented additional morphological traits to separate both species, and which are detailed above (see *T. nicefori* account). The karyotype is unknown.

Thomasomys praetor (Thomas, 1900)

Cajamarca Thomasomys

SYNONYMS:

Oryzomys praetor Thomas, 1900d:354; type locality "Eastern slope of Paramo [= Páramo] between San Pablo and Cajamarca, [Cajamarca,] Peru," 4,000 m elevation.

Thomasomys praetor: Thomas, 1906c:443; first use of current name combination.

Thomasomys aureus praetor: Cabrera, 1961:426; name combination.

DESCRIPTION: Large sized (head and body length 161–187 mm) with comparatively short tail (107–116% of head and body length). Fur long, soft, close, and thick, with hairs on midback 13–15 mm long. Dorsal color olivaceous grayish, darkened on middle and posterior back; head, cheeks, shoulders, and sides clearer yellowish gray, undersurface buffy white with slate-based hairs. Mystacial vibrissae moderately long, extending slightly beyond posterior margin of pinnae when bent; genal 1 vibrissae present. Hindfoot long (35–38 mm); metatarsals silvery white. Hallux long, extending to interphalangeal joint of dII, or moderate, not reaching that joint. Digit V of pes moderately long, claw extending close to base of claw of dIV. Tail incipiently bicolored, uniformly brown above, rather paler below, short haired, and usually with terminal white tip. Ears of medium size (23–25 mm), blackish brown, and conspicuous white postauricular patch present. Skull large (condyloincisive length 36.9–37 mm), within range of *T. aureus*; braincase rather broad and flat; rostrum short and comparatively broad; nasals expanded anteriorly and taper posteriorly, with posterior margins level with or longer than premaxillae. Zygomatic plates moderately broad, breadth subequal to length of M1, rather vertical, and more projecting such that zygomatic notches well defined. Interorbital region narrow, sides parallel or slightly convergent, and flat above; edges squared and traceable across parietals to outer margins of interparietal; postorbital ridge distinct. Molar bearing maxillae with triangular shape. Parietals with large lateral projections. Interparietal broad, strap-like, and shorter anteroposteriorly. Incisive foramina large and open, extending backward usually between M1 anterocones. Palatal bridge short and narrow, with maxillary palatal pits present. Auditory bullae small but moderately inflated, with bony tubes short but distinct; anterior process of ectotympanic relatively small; processus brevis of incus unusually robust. The mesopterygoid fossa is broad and somewhat parallel sided; its roof perforated by moderate sphenopalatine vacuities flanking presphenoid-basisphenoid suture. Carotid circulatory pattern derived (stapedial foramen small, no internal squamosal-alisphenoid groove, no sphenofrontal foramen). Lambdoid ridge distinct. Upper incisors orthodont, molars large. Maxillary toothrow length long, ranging from 7.1–7.5 mm, similar to *T. aureus*. The m1 entolophid oriented to murid, and coalesced with minute mesolophid. Capsular process absent.

DISTRIBUTION: *Thomasomys praetor* is known from the western versant and inter-Andean valleys of the northern Andes in Peru, between the Huancabamba Deflection and the Río Marañón, from 2,050 to 4,000 m.

SELECTED LOCALITIES (Map 359): PERU: Ancash, Carhuaz, Jangas, Antauran, Cordillera Negra, Quebrada Lancash (MUSM 11644), Cuenca del Río Pampas, below Conzuso (MUSM 23326). Quebrada Chalhuacocha, Acrana (MUSM 23332); Cajamarca, Celendín, Hacienda Limón (FMNH 19256), San Pablo (type locality of *Oryzomys praetor* Thomas); La Libertad, 5 mi SW of Otuzco (MVZ 137941), Sanagorán (MUSM 17301).

SUBSPECIES: *Thomasomys praetor* is monotypic.

NATURAL HISTORY: This species has been collected in páramo thickets (Thomas 1900d) and at the edge of the timberline in the dry forests of the Río Marañón valley (Osgood 1914b). I have obtained additional specimens in dense and tall shrub habitats on rocky substrates, near rivers or streams.

REMARKS: *Thomasomys praetor* is another of the large *Thomasomys* relegated to a subspecies of *T. aureus* by Cabrera (1961). However, Voss (2003) and Pacheco (2003) independently recognized *T. praetor* as a valid species. Voss (2003) followed and highlighted Thomas's (1900d) main characteristics provided in the original diagnosis: grayish dorsal fur, pale-silvery ventral fur, pale hindfeet, shorter tail, narrower interorbital, and a broad and distinctively flattened braincase. Pacheco (2003) scored new and numerous external, hard, and soft morphological characters in his phylogenetic analysis of the genus. In addition to Thomas's description, *T. praetor* can be readily separated from *T. aureus* by the smaller hallux and digit V, shorter mystacial vibrissae, deeper and better defined zygomatic notch, and entolophids oriented to the murid and coalesced with a minute mesolophid.

T. E. Lee et al. (2011) claimed to have the first record of *T. praetor* from Ecuador, based on four specimens collected in Atillos lagoons, Chimborazo, Sangay National Park, that have the following external range measurements: total length 376–432 mm, tail length 211–249 mm, hindfoot 38–42 mm, and ear 24–25 mm. However, these specimens are much larger in the first three dimensions than the holotype of *T. praetor* (total length 348 mm, tail length 180 mm, hindfoot 35 mm, and ear 25 mm), an adult specimen of age 3 (*sensu* Voss 1991a), and the average of 13 other specimens (total length 337.15 mm, tail length 180.69 mm, hindfoot 34.92 mm, and ear 23.54 mm). Although the fur color of their specimens did not match that described for *T. praetor* proper (Thomas 1900d; Voss 2003), T. E. Lee et al. (2011) were convinced of their determination and further claimed that *T. praetor* and *T. aureus* were not sister taxa. Although I have not examined the specimens in question, I doubt that the Atillos' sample corresponds to *T. praetor*

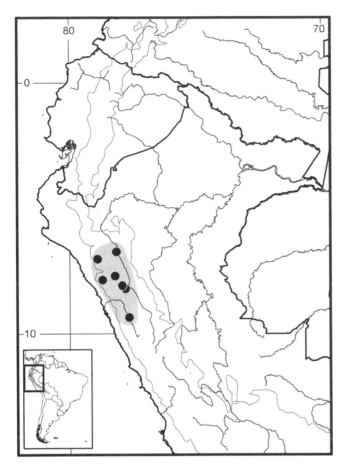

Map 359 Selected localities for *Thomasomys praetor* (●). Contour line = 2,000 m.

because Peruvian specimens (including the holotype) are uniformly smaller and all specimens I have examined exhibit the grayish dorsal fur, pale-silvery ventral fur, and pale hindfeet as stated by Thomas (1900d), Voss (2003), and herein.

Thomas (1900d) allied *T. praetor* to *T. aureus* and *T. princeps*, and suggested they comprised a section of the genus. Pacheco (2003)'s phylogenetic assessment based on morphological attributes found *T. praetor* allied to other species of his "aureus" group, but unexpectedly in a basal position. The karyotype is unknown.

Thomasomys princeps (Thomas, 1895)
Principal Thomasomys

SYNONYMS:

Oryzomys princeps Thomas, 1895b:58; type locality "Bogota," corrected to Bogotá Region, 8,750 ft (2,667 m), Cundinamarca, Colombia by Musser and Carleton (2005).

Thomasomys princeps: Thomas, 1906c:443; first use of current name combination.

Thomasomys aureus princeps: Cabrera, 1961:426; name combination.

DESCRIPTION: Distinctively large species (head and body length 173 198 mm), slightly larger than Ecuadorean *T. aureus*. Dorsal pelage long, thick, and not very fluffy, rich orange-rufous in color with faint dark band along midline, richer along sides; undersurface clear orange-buff, not countershaded. Mystacial vibrissae long, extending distinctly posterior to pinnae; genal 1 vibrissae present. Hindfeet relatively long (35–39 mm) and moderately broad, with dark brown patch on metatarsals and silvery whitish on sides and digits. Tail long (113–131% of head and body length), monocolored blackish brown, finely annulated, covered with short hairs, and without terminal pencil. Hallux long, extending to interphalangeal joint of dII. Digit V of pes very long, claw extending to base of claw of dIV. Protuberant anus prominent. Skull large and robust (condyloincisive length 36.7–39.4 mm); rostrum relatively short and broad; braincase oblong, moderately broad, and not inflated. Zygomatic plates broad and vertical. Anterior borders of premaxillae do not project much beyond upper incisors. Premaxillomaxillary-frontal intersection anterior to zygomatic notches. Anterior margin of M1 placed anterior to posterior border of zygomatic plates. Interorbital region narrow, slightly convergent anteriorly, with square (no beaded) margins; postorbital ridge distinct; parietals large with comparatively small and less rectangular lateral processes; interparietal large and half-oblong in shape; lambdoid ridge well developed. Incisive foramina large and comparatively broad, and extend posteriorly level to or between anterocones of M1; maxillary septum usually long, approaching half length of incisive foramen. Palatal bridge short and narrow with maxillary palatal pits present. Auditory bullae small, uninflated, and flask-shaped; anterior process ectotympanic comparatively small. Mesopterygoid fossa broad and somewhat parallel-sided; roof closed, with sphenopalatine vacuities absent or present only as slit-like vacuities flanking presphenoid-basisphenoid suture. Parapterygoid fossae possess large posterior opening of alisphenoid canal. Carotid circulatory pattern derived (stapedial foramen small, no internal squamosal-alisphenoid groove, no sphenofrontal foramen). Upper incisors opisthodont. Maxillary toothrow long, 7.2–7.6 mm. M1 with anteroflexus coalesced with anterocone, showing indistinct anteroloph; protoflexus of M2 indistinct; ectolophid of m1 present. Capsular process absent but area strongly swollen producing distinct shelf.

DISTRIBUTION: *Thomasomys princeps* is known only from the Cordillera Oriental of Colombia, from about 2,650 to 3,060 m.

SELECTED LOCALITIES (Map 360): COLOMBIA: Cundinamarca, Choachí (USNM 251958), Bogotá region (BM 95.8.1.37, holotype), Laguna Vergon (USNM 251976), Bogotá, San Cristobal (FMNH 71307), Bogotá,

San Francisco (FMNH 71305), Guasca, Río Balcones (FMNH 71304).

SUBSPECIES: *Thomasomys princeps* is monotypic.

NATURAL HISTORY: Individuals have been collected in montane forests. Lopez-Arévalo et al. (1993) collected the species (reported as *T. aureus*) at the forest of the Reserva Biológica Carpanta, Cundinamarca, Colombia, where trees reach heights of 25 m and are covered with epiphytes. Schramm and Lewis (1987) described a new species of flea, *Plocopsylla nungui*, from specimens they only identified as *Thomasomys* sp., which were collected by P. Hershkovitz from San Francisco, Bogotá. However, Hershkovitz collected both *T. niveipes* and *T. princeps* from this locality, so the type host of this flea could be either of these two species.

REMARKS: Thomas (1900d) and Ellerman (1945) recognized *T. princeps* as a valid species, but Cabrera (1961) relegated is as a subspecies of *T. aureus* without justification. Musser and Carleton (1993, 2005) and Voss (2003) followed this arrangement. Alternatively, Pacheco (2003) regarded Thomas's *princeps* as a valid species, one that can be differentiated from *T. aureus* by a number of characters: the short anterior border of the premaxillae not projecting beyond upper incisors; short posterior margin of premaxillae producing a premaxillomaxillary-frontal intersection that lies in front of the zygomatic notch rather than level to it; longer nasals that extend posteriorly usually beyond the lacrimals; broader and oval-shaped incisive foramina rather than narrow and elongated, with maxillary septum about half the length of foramina opening rather than shorter; the comparatively small anterior process of ectotympanic; comparatively larger posterior opening of the alisphenoid canal; anterior margin of M1 placed anterior to the posterior border of zygomatic plates rather than ventral to them; a strongly swollen capsular process producing a distinct shelf; an indistinct anteroflexus of M1; an indistinct protoflexus of M2; and an ectolophid on m1.

Thomasomys princeps was found to cluster unresolved with other species of the "aureus" group in a phylogenetic analysis based on morphological characters (Pacheco 2003). However, resolution was limited by the lack of skeletons and specimens in alcohol. The karyotype is unknown.

Thomasomys pyrrhonotus Thomas, 1886
Reddish-backed Thomasomys

SYNONYMS:

Hesperomys [*Rhipidomys*] *pyrrhorhinus*: Thomas, 1882: 107; part; not *Mus pyrrhorhinos* Wied-Neuwied.

H[*esperomys*]. *pyrrhonotus* Thomas, 1886b:422; type locality "Tambillo, [Río Malleta,] 5800 ft [1,768 m]," Cajamarca, Peru.

Thomasomys pyrrhonotus: Thomas 1906c:443; first use of current name combination.

DESCRIPTION: Medium sized comparable to *T. auricularis* (head and body length 137–144 mm), which it resembles superficially, but with slightly shorter tail and slightly shorter toothrow length. Dorsal pelage rich rufous, head similar but lighter; undersurface whitish with small orangish pectoral streak. Bases of dorsal and ventral hairs slated-color. Ears of moderate length (19.5–22 mm), and whitish postauricular patch present. Blackish mystacial vibrissae moderately long, extending slightly or distinctly beyond posterior margin of pinnae when bent; genal 1 vibrissae present. Tail relatively long (120% of head and body length), uniformly dark brown except for basal half-inch portion covered with red-tipped body hairs. Hindfoot moderately long (30–31 mm) with metatarsals pale orange and digits brown. Digit I of pes (hallux) long, extending to interphalangeal joint of dII; digit V moderately long or long, extending to base of claw of dIV or close. Skull large and robust but similar to that of *T. auricularis* (condyloincisive length of skull, 33.9); rostrum long and slender; nasals slender and expanded anteriorly; the braincase oblong and somewhat inflated. Incisive foramina long, open, and extend backward between M1 procingula; maxillary septum length short, less than half incisive foramina length. Interorbital region narrow, sides parallel, edges squared but smooth and not beaded, and interfrontal fontanella present. Ethmoturbinals small and poorly

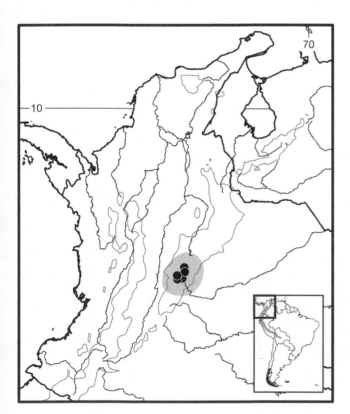

Map 360 Selected localities for *Thomasomys princeps* (●). Contour line = 2,000 m.

developed. Mesopterygoid fossa broad, sides subparallel, and roof moderately fenestrated. Auditory bullae unusually large and inflated (larger than *T. auricularis*). Postglenoid foramen conspicuous and narrow but smaller than more open subsquamosal fenestra; hamular process narrow and very long. Lamina of malleus with well-defined edge on ventral and posterior margin. Incisors very broad and strong; anterior surface dark orange. Maxillary toothrow long (6.4 mm); molars proportionately large; additional anterior edge on the procingulum of M1 usually absent.

DISTRIBUTION: *Thomasomys pyrrhonotus* occurs only in montane forests in northwestern Peru, from 1,200 to 3,100 m.

SELECTED LOCALITIES (Map 361): PERU: Cajamarca, Bosque Cachil, between Cascas and Contumaza (MUSM [VPT 1670]), 7 km N and 3 km E of Chota (LSUMZ 22317), 1 km NW of Cutervo (LSUMZ 21874), Santa Cruz, Catache, 3.81 km NE of Monteseco (UMMZ, uncatalogued), Tambillo, Río Malleta (type locality of *Hesperomys pyrrhonotus* Thomas); Piura, Ayabaca (FMNH 81296), Cerro Chinguela, ca. 5 km NE of Zapalache (LSUMZ 26947), Huancabamba, Canchaque (FMNH 81295).

SUBSPECIES: *Thomasomys pyrrhonotus* is monotypic.

NATURAL HISTORY: Thomas (1882) stated that this species lived in trees, but few other data on ecology or natural history are known. The species has been collected in tall shrub habitat in Cachil, Cajamarca (V. Pacheco, unpubl. data), in montane secondary forests and montane bamboo forests on the ground and near a stream in Monteseco, Cajamarca (L. Luna, pers. comm.), and in montane cloud forest at Cutervo, Cajamarca (LSUMZ 21874). Short observations on one captured specimen revealed this species is an excellent climber, making good use of the digits of the hindfoot to grasp branches and its long tail for balance (V. Pacheco, unpubl. obs.).

REMARKS: This species has been treated as polytypic after including *auricularis* Anthony as a subspecies (Cabrera 1961), or either as synonym or subspecies in following synopses (e.g., Musser and Carleton 1993, 2005). Pacheco (2003) recognized *T. pyrrhonotus* and *T. auricularis* as species; arguments supporting these hypotheses are provided in the account of *T. auricularis*.

Page 421 of Thomas (1886b) has been commonly cited for the description of this species (Gyldenstolpe 1932; Cabrera 1961; Musser and Carleton 1993, 2005), but the name first appears on p. 422.

Ellerman (1941:368) recognized a *pyrrhonotus* Group, which included *T. pyrrhonotus* and *T. auricularis* based on their enlarged auditory bullae, separate from his *aureus* Group. Pacheco's (2003) phylogenetic assessment found *T. pyrrhonotus* allied to *T. rosalinda*, followed by *T. auricula-*

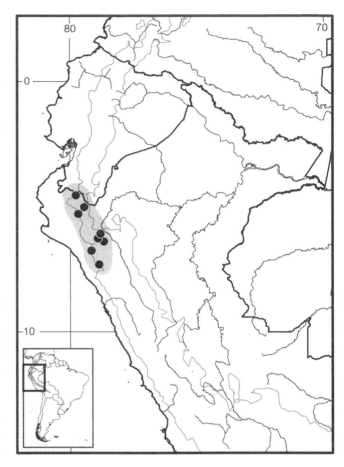

Map 361 Selected localities for *Thomasomys pyrrhonotus* (●). Contour line = 2,000 m.

ris, all within a partially resolved monophyletic clade comprising his "aureus" group, and thus partially supporting Ellerman's hypothesis. The karyotype is unknown.

Thomasomys rosalinda Thomas and St. Leger, 1926
Rosalinda's Thomasomys

SYNONYM:

Thomasomys rosalinda Thomas and St. Leger, 1926a:347; type locality "Goncha, Amazonas, 8500′ [2,591 m]," Peru.

DESCRIPTION. Medium size (head and body length 135 mm), comparatively smallest reported species of "aureus" group. Dorsal color reddish gray overall, more grayish on head, foreback, and cheeks, and smoothly more rufous on posterior back, rump, and hindlimbs. Undersurface tawny cinnamon, paler on mental, throat, and inner sides of forelimbs, with indistinct rufous patch on chest, and more rufous on inner sides of hind limbs, inguinal region, and base of tail. Bases of hairs plumbeous all over body. Hairs of moderate length (12 mm), with guard hairs up to 15 mm long. Nose reddish gray contrasting with dark spot at base of mystacial vibrissae. Mystacial vibrissae moderately long, extending slightly beyond posterior margin of pinnae when bent; genal 1 vibrissae present. Ears of moderate length (20.5 mm), brown, and covered by long hairs for two-thirds of length, surrounded by inconspicuous tawny area. Carpal region tawny silvery, not countershaded with inner sides of forelimbs; metacarpals and fore digits dark. Tail moderately long (126% of head and body length), unicolored brown, and covered by short hairs. Hindfoot of medium length (27.5 mm); metatarsals silvery light brown proximally, becoming slightly darker on distal surface and digits, with ungual tuft silvery white. Hallux long, extending to interphalangeal joint of dII. Digit V of pes very long, claw extending to base of claw of dIV. Skull also of medium length (condyloincisive length 31 mm), with relatively long and narrow rostrum and oblong braincase. Interorbital region narrow with sides angular and divergent. Zygomatic arches converge slightly anteriorly, although maxillary roots exhibit subrectangular outline. Interparietal unusually large, rectangular shaped, with outer corners preventing parietals from contacting occipital bone. Lambdoid ridge weakly developed. Moderately broad and vertical zygomatic plates project slightly forward, producing well-defined zygomatic notches. Incisive foramina large and open, capsular-shaped, and extend backward deeply to posterior margins of M1 anterocones. Mesopterygoid fossa broad, expanded anteriorly and narrowing behind; roof moderately perforated by sphenopalatine vacuities on both sides of presphenoid-basisphenoid suture; palatal medial process absent. Auditory bullae small, although moderately inflated; bony tubes short and broad. Postglenoid foramen and subsquamosal fenestra very large, latter more open. Orbicular apophysis of malleus small. Upper incisors opisthodont. Maxillary toothrow moderate in length (6.4 mm). Procingulum of M1 exhibits distinct additional border; anteroloph and anteroflexus of this tooth distinct, whereas protoflexus of M2 distinct. The m1 exhibits both distinct anteromedian flexid and ectolophids.

DISTRIBUTION: *Thomasomys rosalinda* is known only from its type locality in northern Peru.

SELECTED LOCALITIES (Map 362): PERU: Amazonas, Goncha (type locality of *Thomasomys rosalinda* Thomas and St. Leger).

SUBSPECIES: *Thomasomys rosalinda* is monotypic.

NATURAL HISTORY: This species apparently occurs in montane forests, with the type specimen trapped in wet brush (Thomas and St. Leger 1926a). No other information on ecology or behavior is known.

REMARKS: This species, still known only by the holotype, is one of the most poorly known in the genus. Ellerman (1941:371) unexpectedly placed *T. rosalinda* in his widely constructed *cinereus* Group, but Pacheco (2003) confirmed the species as a member of the "aureus" group in his phylogenetic assessment, apparently related to both

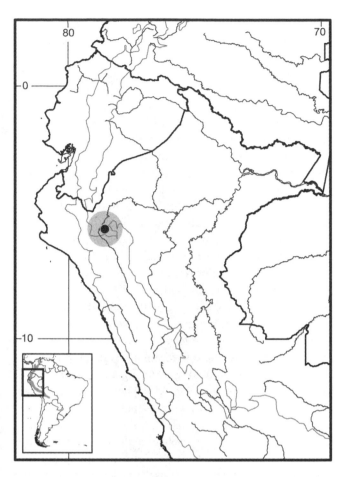

Map 362 Single known locality for *Thomasomys rosalinda* (●). Contour line = 2,000 m.

T. pyrrhonotus and both with *T. auricularis*. Both the large interparietal, precluding a parietal-occipital contact, and the deep incisive foramina readily differentiate this species from other members of the "aureus" group. The karyotype is unknown.

Thomasomys silvestris Anthony, 1924
Sylvan Thomasomys

SYNONYMS:

Thomasomys silvestris Anthony, 1924a:2; type locality "Las Maquinas [= Máquinas], on Santo Domingo trail, west of Corazon [= Corazón], Western Andes, [Pichincha,] Ecuador; altitude 7,000 feet [2,134 m]."

Thomasomys ischyurus silvestris: Cabrera, 1961:430; name combination.

DESCRIPTION: Medium sized (head and body length 95–136 mm) with soft, dense, and comparatively long dorsal fur. Dorsal color varies from brownish cast (hairs slate-colored basally and tipped with raw umber or mummy-brown) to much darker, more sooty tone; ventral pelage uniformly gray plumbeous, washed with light brown, and not countershaded with dorsal pelage. Mystacial vibrissae moderately long (ratio of longest mystacial vibrissae to length of head and body 0.44), and extend slightly beyond posterior margin of pinnae when bent; genal 1 vibrissae absent. Ear moderately long (15–19 mm), clove-brown in color. Tail distinctly longer than head and body (112–150%), slender, almost naked, almost uniformly clove-brown above and slightly lighter below, and with terminal white tip. Hindfoot moderately long (25–30 mm), narrow, and clove-brown above, and with distinct gap between thenar and hypothenar pads. Digit I of pes (hallux) moderate, with claw not extending more than about half length of phalanx 1 of dII. Digit V of pes long, with claw extending about half length of phalanx 2 of dIV. Skull moderately long (condyloincisive length 24.7–28.1 mm) with relatively long and moderately narrow rostrum, rounded braincase, and nasals expanded anteriorly but tapering posteriorly. Interorbital region moderately broad, hourglass in shape, and with rounded margins. Zygomatic plates narrow with straight, vertical anterior margins; breadth shorter than length of M1. Zygomatic arches converge anteriorly. Incisive foramina oval, contracted anteriorly, and moderately long, reaching anterior margins of M1s or quite close to them. Mesopterygoid fossa broad, parallel sided, and usually without palatal medial process; roof either closed or with narrow, slit-like sphenopalatine vacuities present. Auditory bullae small and uninflated. Alisphenoid strut present. Carotid circulatory pattern primitive (stapedial foramen, squamosal alisphenoid groove, and sphenofrontal foramen are present). Postglenoid foramen small. Lamina of malleus exhibits unusually well-developed cephalic peduncle and cephalic process. Upper molars moderately hypsodont, with no interpenetration. Upper incisors long and orthodont. Maxillary toothrow moderately long, 4.5–5.1 mm. M1 with anteromedian flexus either absent or indistinct, and anteroloph indistinct; M3 comparatively small; m1 with anteromedian flexid weak or absent; ectolophids absent in m1 and m2. Capsular process absent.

DISTRIBUTION: *Thomasomys silvestris* occurs in Ecuador in montane forest and páramo habitats from Bolívar province in the south to Napo and Pichincha provinces in the north, at elevations from 1,800 to 4,500 m.

SELECTION LOCALITIES (Map 363): ECUADOR: Bolívar, Carmen, near Sinche (AMNH 66886); Napo, Cerro Antisana, oriente (FMNH 43251); Pichincha, Cotopaxi, Atacazo (EPN 225), Gualea, Ilambo Valley, 1,800 m (FMNH 94993), Hacienda Monjas, 4,500 m (EPN 190), Las Máquinas, Santo Domingo trail, W of Corazón (type locality of *Thomasomys silvestris* Anthony), San José, occidente (FMNH 53247), Volcán Pichincha, oriente (FMNH 92000), West Mindo, W Andes (USNM 304850), Zapadores, Río Saloya (USNM 513590).

SUBSPECIES: *Thomasomys silvestris* is monotypic.

NATURAL HISTORY: Anthony (1924a) stated that this species appeared to be an exclusive forest-dwelling form, not exploring the thickets of the páramo region. However, Tirira (2007) stated that the species occurred in subtropical, temperate, and adjacent páramo vegetation.

REMARKS: Cabrera (1961) relegated *silvestris* to a subspecies of *T. ischyrus*, although Anthony (1924a) explicitly distinguished these as separate species. Tirira (1999) accepted *T. silvestris* as a valid species following a personal communication from me, a position later supported by L. Luna and Pacheco (2002) and Voss (2003). L. Luna and Pacheco (2002) provided detail comparisons differentiating this species from *T. onkiro* and *T. caudivarius*. T. E. Lee et al. (2011) reported *T. silvestris* from Sangay National Park, in the provinces of Chimborazo and Morona Santiago, claiming they represented the southernmost record of the species. However, the Bolívar record is further south. The specimens collected by T. E. Lee et al. (2011) are also slightly larger than my measurements or those of Voss (2003) in hindfoot length (reaching to 34 versus 30 mm) and ear length (reaching to 25 versus 19 mm), suggesting that additional examination of their specimens is warranted.

Thomasomys silvestris exhibits little variation across its range. Voss (2003) stated that one specimen from Antisana (FMNH 43251), in the eastern Andes, was similar in qualitative characters to the type series collected on the western Andes, although he suggested that the specimen was collected on the western side of that volcano, as was true for specimens of *T. vulcani* from the same area (see that account). The dorsal pelage of specimens from Cotopaxi is

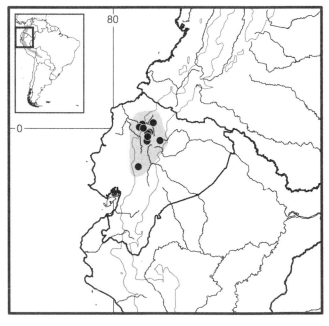

Map 363 Selected localities for *Thomasomys silvestris* (●). Contour line = 2,000 m.

brownish, resembling *T. caudivarius*, but the grayish ventral pelage and skull features correspond to *T. silvestris*.

Little is known on the phylogenetic relationships of *T. silvestris*. Anthony (1924a) compared the species with *T. cinereus*, *T. rhoadsi* (a synonym of *T. vulcani*; see that account), and *T. ischyrus*, and suggested that *T. silvestris* was more nearly related to *T. ischyrus*. A phylogenetic assessment based on morphological characters places *T. silvestris* more related to *T. caudivarius*, both members of the "cinereus" group (Pacheco 2003). This result, however, is not supported by a molecular phylogeny based on mtDNA cytochrome-*b*, where *T. silvestris* was found related to *T. taczanowskii* (reported as *T. ischyrus*, see both species accounts), a member of the "baeops" group, despite the extensive morphological differences between them. Additional studies are necessary to resolve this discrepancy.

Thomasomys taczanowskii (Thomas, 1882)
Taczanowski's Thomasomys

SYNONYMS:

Hesperomys (*Rhipidomys*) *taczanowskii* Thomas, 1882:109; type locality "Tambillo, [Río Malleta,] 5800′ [1,768 m]," Cajamarca, Peru.

Hesperomys (*Vesperimus*) *taczanowskii*: Thomas, 1884: 449; name combination.

Hesperomys (*Thomasomys*) *taczanowskii*: Coues, 1884: 1275; name combination.

Thomasomys taczanowskii: Gyldenstolpe, 1932a:56; first use of current name combination.

DESCRIPTION: Medium sized (head and body length 97–120 mm; condyloincisive length 25.1–27.7 mm) with soft and woolly fur of medium length. Dorsal color grayish yellow with hairs tipped rufous yellow; head grayer and less yellow. Ventral pelage grayish white, moderately countershaded. Mystacial vibrissae very long, extending far behind ears when bent; genal 1 vibrissae present. Tail moderately long (98–132% of head and body length), unicolored pale brown, and with terminal white tip usually present. Hindfoot relatively short (24–28 mm). Digit 1 of pes (hallux) long and dV very long, as in *T. baeops*. Skull with short and delicate rostrum; short nasals; frontonasal profile quite flat, not convex; and braincase moderately inflated. Interorbital region narrow with faintly marked (not rounded nor beaded) edges, but slightly broader than *T. baeops*. Zygomatic arches narrow and converging anteriorly. Incisive foramina narrow and long, extending backward between anterior borders of M1s. Auditory bullae small but comparatively more inflated than in *T. baeops*. Carotid circulatory pattern derived (stapedial foramen small, no internal squamosal-alisphenoid groove, no sphenofrontal foramen). Upper incisors opisthodont. Maxillary toothrow short, 4.1–4.6 mm. Capsular process distinct.

DISTRIBUTION: *Thomasomys taczanowskii* occurs in montane forests in the Andes of Ecuador and northern Peru, at elevations between approximately 1,150 and 3,350 m.

SELECTED LOCALITIES (Map 364): ECUADOR: Azuay, Girón (FMNH 53242); Carchi, Montufar, Atal (UMMZ 77160); El Oro (above) Salvias, Cordillera de Chilla (AMNH 47671); Loja, Celica (AMNH 47672). PERU: Cajamarca, 35 mi WNW of Cajamarca (MVZ 137928), Las Juntas, 3 km W of Pachapiriana (AMNH 268147), San Ignacio, Tabaconas, Piedra Cueva in Cerro Coyona, Tabaconas-Namballe National Sanctuary (UMMZ 176751), Tambillo, Río Malleta (type locality of *Hesperomys* (*Rhipidomys*) *taczanowskii* Thomas); Lambayeque, Seques (AMNH 73124), Uyurpampa (MUSM 21800); Piura, Huancabamba, Canchaque (FMNH 81313), Laguna (USNM 304539), Pariamarca Alto (MUSM 23782), 2 km W of Porculla Pass (MVZ 137942), Portachuelo (MUSM 21741), Quebrada El Gallo, Campamento Bomba Quemada 1 (MUSM 23592); San Martín, Mariscal Caceres Huicungo, P.N. Río Abiseo, La Playa (MUSM 7983).

SUBSPECIES: I consider *T. taczanowskii* to be monotypic (but see Remarks).

NATURAL HISTORY: Little is known about the ecology, population biology, or other natural history aspects of this species. Individuals have been captured in tall shrub mixed with short and slender trees in upper montane forests at Las Ashitas, Cajamarca (V. Pacheco, pers. obs.). In Pariamarca Alto, the species was taken in Queñua Andean forests (*Polylepis weberbaueri*), among small trees (about 5 m high), densely covered by mosses and lichens, and the ground by short grasses browsed by cattle (V. Pacheco, pers. obs.). Specimens taken at Cerro Chinguela and Quebrada El Gallo, in Piura (about 3,000 m elevation) came from pristine upper montane forest with trees reaching heights of 10 to 25 m high, and with numerous species of shrubs, epiphytes, tree ferns, vines, and epiphytes present (V. Pacheco, pers. obs.), and in montane secondary forests and montane bamboo forests near a stream in Monteseco, Cajamarca (L. Luna, pers. comm.). My brief observations in the field suggest that individuals are excellent and agile climbers. Schramm and Lewis (1987) described a new species of flea, *Plocopsylla kilya*, as from Peruvian specimens of *T. taczanowskii*.

REMARKS: *Thomasomys taczanowskii* has been confused with *T. oreas* in the literature. This includes A. L. Gardner (1976:13) report of *T. taczanowskii* from Yuraccyacu, Ayacucho, based on two specimens (LSUMZ 16735, 16736), and the karyotype of these specimens reported by A. L. Gardner and Patton (1976). Pacheco (2003) re-identified these specimens as *Thomasomys* sp. 8, an undescribed species allied to *T. oreas* (see *T. oreas* account). Thomas (1920h:236) provisionally identified a specimen

(USNM 194876) from the Occobamba Valley, Cusco, as *T. taczanowskii*, but Pacheco (2003) corrected this assignment as well to *Thomasomys* sp. 9, another undescribed form allied also to *T. oreas* (see that account). S. Anderson (1993:37) recorded *T. taczanowskii* from El Choro, Cochabamba, Bolivia, based on one specimen (FMNH 74866), which also has been tentatively reidentified as *T. oreas* (V. Pacheco, unpubl. obs.; see the account of that species).

A phylogenetic analysis of discrete morphological characters for *Thomasomys* recovered *T. taczanowskii* as most closely related to *T. baeops* (Pacheco 2003), a hypothesis corroborated by mtDNA cytochrome-*b* sequences (Salazar-Bravo and Yates 2007). As mentioned earlier (see *T. baeops* account), the GenBank sequence AF 108675 (MVZ 181999) corresponds to *T. taczanowskii* and not to *T. ischyrus* (see Pacheco 2003).

Two specimens from Carchi, Ecuador, representing the northernmost record of *T. taczanowskii*, are unfortunately both fragmented and juveniles (age class 1, *sensu* Voss 1991a). Cranial and external characteristics of these specimens agree with the typical representatives of the

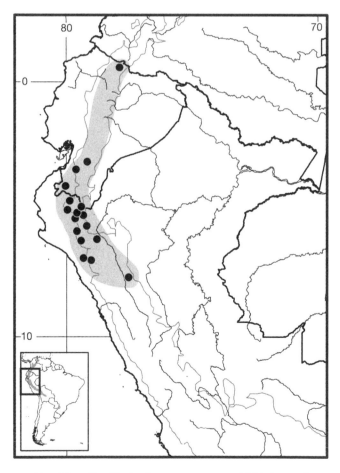

Map 364 Selected localities for *Thomasomys taczanowskii* (●). Contour line = 2,000 m.

species, but a revision based on a larger series is required to confirm this hypothesis. A single specimen from Río Abiseo, San Martín, tentatively determined as *T. taczanowskii*, represents the southernmost record of the species, although several features (including absent genal 1 vibrissae, longer nasals, narrow zygomatic plates, and large ethmoturbinals) suggest it might represent an undescribed taxon.

Thomasomys ucucha Voss, 2003
Ucucha Thomasomys

SYNONYM:

Thomasomys ucucha Voss, 2003:10; type locality "11,100 ft (3384 m) in the valley of the Río Papallacta (ca. 3–5 km by trail NNW Papallacta), Provincia Napo, Ecuador."

DESCRIPTION: Medium sized (head and body length, 94–119 mm) with dense, fine, and soft fur. Dorsal coloration near brownish olive along flanks, shading to dark grayish brown middorsally. Ventral color dark neutral gray basally, with superficial wash of light neutral gray or glaucous, not sharply contrasting with dorsal coloration. Mystacial vibrissae moderately long, extending just behind pinnae when laid back alongside head; genal 1 vibrissae absent. Ears small (17–20 mm) and sparsely covered with short, blackish hairs, not contrasting conspicuously with color of the head. Tail long (122–151 mm), and substantially longer than combined length of head and body (122–125%). Hindfoot of moderate size (26–30 mm) and neither very narrow nor conspicuously broad, without gap between thenar and hypothenar pads. Digit I of pes (hallux) moderately long, with claw not extending more than about half length of phalanx 1 of dII. Digit V of pes long, with claw extending about half length of phalanx 2 of dIV. Skull moderately long (condyloincisive length 25.8–29.1 mm) with relatively short and blunt rostrum, and straight frontonasal profile. Interorbital region narrow with rounded supraorbital margins. Zygomatic arches rounded and flare widely. Zygomatic plates moderately broad (subequal to length of M1) and vertically oriented. Incisive foramina widest just behind premaxillomaxillary suture and short, not approaching M1 anterocones. Mesopterygoid fossa broad and straight sided; bony roof complete or perforated by narrow, slit-like sphenopalatine openings flanking presphenoid-basisphenoid suture. Alisphenoid strut present, separating buccinatormasticatory from accessory oval foramina. Posterior opening of alisphenoid canal comparatively reduced. Carotid circulatory pattern primitive (stapedial foramen, squamosal alisphenoid groove, and sphenofrontal foramen present). Auditory bullae small and uninflated. Upper incisors large, broad, and conspicuously procumbent. Maxillary toothrow relatively short (4.2–4.6 mm), primarily due to conspicuously smaller M3. Molars hypsodont and lack well-developed cingula and stylar cusps. In upper molars, anterolophs and mesolophs of M1 and M2 comparatively less developed, and M3 usually lacks distinct hypoflexus. In M1, anteromedian flexus distinct. The m1 anteromedian flexid indistinct and ectolophids absent. Mandible with distinct capsular process and short angular process relative to condylar process. Stomach unilocular-hemiglandular (Carleton 1973); gall bladder present; macroscopic preputial glands absent.

DISTRIBUTION: *Thomasomys ucucha* is known only from the crest of the Cordillera Oriental (between ca. 3,380 and 3,800 m) just south of the equator in the Ecuadorean provinces of Carchi, Pichincha, and Napo (Voss 2003; Arcos et al. 2007).

SELECTED LOCALITIES (Map 365): ECUADOR: Carchi, Los Encinos (Arcos et al. 2007), Potrerillos, Ipuerán (Arcos et al. 2007); Napo, Río Papallacta valley (type locality of *Thomasomys ucucha* Voss); Pichincha, Tablón, road to Papallacta (AMNH 46621).

SUBSPECIES: *Thomasomys ucucha* is monotypic.

NATURAL HISTORY: This species has been taken in grassy páramo, in the shrubby páramo/forest ecotone, in grassy glades surrounded by forest, and deep inside subalpine rainforest. Most recorded captures were on the ground, along rabbit trails or runways through dense grass or low herbs, in runways through moss or damp litter, under mossy debris, at the bases of mossy trees, or along the wet margins of small streams. The species is sympatric and syntopic (in the same habitats) with *Akodon mollis, Anotomys leander, Chilomys instans, Microryzomys altissimus, M. minutus, Neusticomys monticolus, Neomicroxus latebricola, T. aureus, T. baeops, T. cinnameus, T. erro,* and *T. paramorum* (Voss 2003).

REMARKS: Records of this species include those of *Thomasomys* sp., as reported by Voss (1988:Table 43) in a previous list of Papallacta mammals, and *Thomasomys* sp. 11 in Pacheco (2003). A phylogenetic assessment based on morphological characters finds *T. ucucha* in a cluster with other species of the "cinereus" group, specifically related to *T. cinnameus,* although with weak support (Pacheco 2003). Voss (2003) contrasted *T. ucucha* principally with *T. hylophilus,* another member of the "cinereus" group.

Arcos et al. (2007) reported additional specimens of *Thomasomys ucucha* from Carchi province in northern Ecuador, close to the Colombian border. They reported a hindfoot length of 17.3 mm for one specimen (MECN 5918), which is shorter than the range (26 to 30 mm) given by Voss (2003). The correct measurement of this specimen is, however, 27.5 mm (L. Albuja, pers. comm.). This species is then very likely present in Colombia.

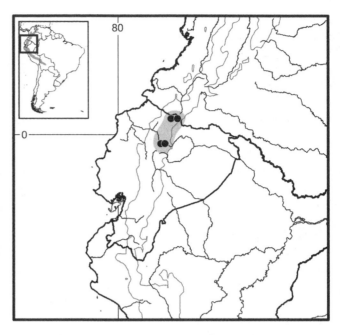

Map 365 Selected localities for *Thomasomys ucucha* (●). Contour line = 2,000 m.

Thomasomys vestitus (Thomas, 1898)
Mérida Thomasomys

SYNONYMS:

Oryzomys vestitus Thomas, 1898e:454; type locality "Rio Milla [= Río Milla], Merida [= Mérida], 1630 m," Venezuela.

Thomasomys vestitus: Thomas, 1906c:443; first use of current name combination.

DESCRIPTION: Large size (head and body length 138–141 mm) with long, soft, and woolly dorsal fur, long tail (138–141 mm, 120% of head and body length), and medium-sized ears (17–21 mm). Dorsal pelage colored grizzled fuscous gray, more rufous posteriorly, with grayish head. Under parts uniformly grayish, washed with dull buff; not countershaded relative to the dorsum. Ears thinly haired and pale brown. Hands dull whitish above. Tail monocolored pale brown, and comparatively well haired; terminal white tip absent. Mystacial vibrissae moderately long, extending slightly beyond posterior margin of pinnae when bent; genal 1 vibrissae absent. Hindfoot of medium length (33–33.3 mm) and dull whitish above. Digit I of pes (hallux) moderately long, with claw not extending more than about half length of phalanx 1 of dII. Digit V of pes long, with claw extending about half length of phalanx 2 of dIV. Skull large (condyloincisive length 32.5–33.9 mm), broad and depressed in cranial region; braincase large and oblong; rostrum comparatively short and narrow; and frontonasal profile straight. Nasals short, narrow, and taper posteriorly. Zygomatic plates vertical and moderately broad, subequal to length of M1. Incisive foramina moderate length, relatively narrow, and unusually separated by broad maxillary septum; they extend to anterior margins of M1s, distance equal to length of M1 anterocone. Interorbital region narrow (molar-bearing maxilla rectangular in shape and visible from dorsal view), hourglass in shape, with rounded or faintly squared margins. Mesopterygoid fossa broad, with lateral margins posteriorly convergent; roof closed. Zygomatic arches converge anteriorly. Auditory bullae small and flask shaped, bony tubes short and broad, and stapedial spine small. Postglenoid foramen small, and the subsquamosal fenestra slightly larger and rounded. Alisphenoid strut that separates buccinator-masticatory and accessory oval foramina present. Ethmoturbinals of moderate size. Posterior opening of alisphenoid canal comparatively reduced. Carotid circulatory pattern primitive (stapedial foramen, squamosal alisphenoid groove, and sphenofrontal foramen present). Upper incisors orthodont, long, broad, and robust. Maxillary toothrow length moderate, 5.5–6.6 mm. Molars relatively hypsodont but lack well-developed cingula and stylar cusps. In M1, anteromedian flexus indistinct, but anteroloph and anteroflexus distinct; hypoflexus unusually wide. M2 with distinct protoflexus, and M3 with distinct hypoflexus. Anteromedian flexid in m1 present although weak. Condylar and angular processes subequal, and capsular process absent.

DISTRIBUTION: *Thomasomys vestitus* in known only from the Mérida Andes in western Venezuela, at elevations between 1,600 and 2,500 m.

SELECTED LOCALITIES (Map 366): VENEZUELA: Mérida, El Baho, 5 km SW of Santo Domingo (CVULA 2537), Río Milla (type locality of *Oryzomys vestitus* Thomas); Trujillo, Hacienda Misisi, 15 km E of Trujillo (USNM 374544).

SUBSPECIES: *Thomasomys vestitus* is monotypic.

NATURAL HISTORY: These are nocturnal, terrestrial to arboreal, solitary, and omnivorous mice (O. J. Linares 1998). As with other *Thomasomys, T. vestitus* is associated with moist habitats and cloud forest habitats (Eisenberg and Redford 1999). Handley (1976) reported one specimen trapped under a mossy log on damp ground and another on a log over a small stream, both in cloud forest (the Lower Montane very humid forest in the Vegetative Life Zone system of Ewel and Madriz 1968). Individuals feed on small fruits, plant material, and insects (O. J. Linares 1998). J. T. Reed and Brennan (1975) described the chiggers *Odontacarus schoenesetosus* and *O. tiptoni* as new species, with *T. vestitus* as the type host. Tipton and Machado-Allison (1972) reported the fleas *Neotyphloceras rosenbergi* (Rothschild) and *Cleopsylla monticola* Smit.

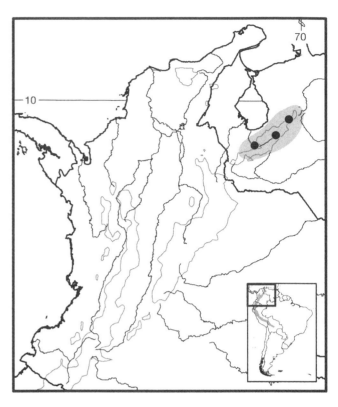

Map 366 Selected localities for *Thomasomys vestitus* (●). Contour line = 2,000 m.

REMARKS: Aguilera et al. (2000) described a karyotype with $2n = 44$, $FN = 42$, including acrocentric X and Y chromosomes, a complement noted for its general conformity to the hypothesized primitive condition for the genus.

Almost nothing is known on the phylogenetic relationships of *T. vestitus*. My phylogenetic assessment based on morphological characters (Pacheco 2003) placed *T. vestitus* as related to a cluster of species of the "cinereus" group, which included forms allied to *T. cinereiventer* and *T. vulcani*.

Thomasomys vulcani (Thomas, 1898)

Pichincha Thomasomys

SYNONYMS:

Aepeomys vulcani Thomas, 1898e:452; type locality "Mount [Volcán] Pichincha, [Pichincha,] Ecuador, at about 12,000 feet [= 3,658 m]."

Thomasomys rhoadsi Stone, 1914:12; type locality "Hacienda Garzon, Mt. [Volcán] Pichincha, 10,500 feet," Pichincha, Ecuador.

Thomasomys rhoadsi rhoadsi: Gyldenstolpe, 1932a:54; name combination.

Thomasomys vulcani: Ellerman, 1941:369; first use of current name combination.

Thomasomys lugens vulcani: Cabrera, 1961:432; name combination.

Aepeomys lugens: Musser and Carleton, 1993:688, part; not *lugens* Thomas.

DESCRIPTION: Medium sized (adult head and body length 118–153 mm) with long, soft, and woolly dorsal fur. Underfur about 12 mm long, guard hairs about 15 mm in length. Dorsal color dark buffy gray, slightly darker along top of loins, providing overall very dark appearance, nearly black on back. Ventral pelage moderately countershaded, hairs slate-colored at their bases with tips buffy. Ventral base of manual claws keeled. Mystacial vibrissae moderately long, extending slightly beyond posterior margin of pinnae when bent; genal 1 vibrissae absent. Ears short (18–25 mm) and well haired. Tail comparatively short (78–101% of head and body length), fine scaled, clothed in short hairs, and uniformly brown. Hindfoot moderately long (27–31.5 mm), with metatarsals light brown above, darker mesially; sides and digits whitish; gap present between thenar and hypothenar pads. Digit I of pes (hallux) moderately long, with claw not extending more than about half length of phalanx 1 of dII. Digit V of pes long, with claw extending about half length of phalanx 2 of dIV. Skull moderately long (condyloincisive length 28.8–31.1 mm), with relatively long and broad rostrum but no rostral tube. Nasals expanded anteriorly, taper posteriorly, with posterior margins extending posteriorly beyond premaxillae. Premaxillomaxillary-frontal joint placed anterior to zygomatic notches. Interorbital region moderately broad, hourglass in shape, and with rounded margins. Ethmoturbinals large. Zygomatic plates moderately broad, subequal to length of M1, and vertical or only very slightly sloped backward from bottom to top. Zygomatic arches converge very slightly anteriorly. Incisive foramina large, open, and rhomboid-shaped, with anterior half more contracted, and long, extending to anterior margins of M1s or close. Mesopterygoid fossa broad, wider anteriorly, converging posteriorly, lacks palatal medial process, but with closed roof. Optic foramen small and posterodorsal to M3. Alisphenoid strut present. Auditory bullae small and uninflated. Carotid circulatory pattern primitive (stapedial foramen, groove on inner surface of squamosal and alisphenoid, and sphenofrontal foramen present). Upper incisors orthodont. Maxillary toothrow moderately long, ranging from 5.1 to 5.6 mm, and upper molars moderately hypsodont, with no interpenetration of flexi. M1 with distinct anteromedian flexus and lacks accessory labial root; anteromedian flexid of m1 shallow. Capsular process absent.

DISTRIBUTION: *Thomasomys vulcani* occurs only in páramo and montane forests of the north-central Andes of Ecuador, in Napo and Pichincha provinces, from 1,400 to 4,500 m.

SELECTED LOCALITIES (Map 367): ECUADOR: Napo, Cerro Antisana, oriente (FMNH 43246); Pichincha, Gualea, Ilambo Valley (FMNH 94992), Hacienda Monjas (EPN 323), Huila North, oriente (FMNH 53406), E side of Mt. Pichincha (MVZ 137610), Mount Pichincha (type locality of *Aepeomys vulcani* Thomas), Pinantura (AMNH 46718), 15 mi S of Quito (AMNH 213548), San José, occidente (FMNH 53210).

SUBSPECIES: I treat *T. vulcani* as monotypic.

NATURAL HISTORY: This species is terrestrial and lives in páramo and both subtropical and temperate forest (I. Castro and Román 2000; Tirira 2007). Specimens have been taken on rocky wooded slopes, where they have burrows from 1 to 3 inches below the surface of the soil and debris (Stone 1914). Pozo-R. et al. (2006) captured individuals in disturbed plant fences that separate cattle ranches, although in lower abundance than in undisturbed forests.

REMARKS: *Thomasomys vulcani* was described as a species of *Aepeomys* and thereafter classified as a subspecies or synonym of *lugens* Thomas, whether this taxon was treated as an *Aepeomys* (Musser and Carleton 1993) or *Thomasomys* (Cabrera 1961). However, Pacheco (in Voss et al. 2002) considered *vulcani* to be a form of *Thomasomys* proper based on his own examination of type material. I elaborate on this view here.

The type specimen of *vulcani* (BM 98.5.1.10) is a female specimen, consisting of a skin in good condition but badly damaged skull. In dorsal view, the occipital and parietal bones are squashed, and zygomatic arches are broken and incomplete. In ventral view, the anterior half of the skull is somewhat complete, with the rostrum and the first molars practically complete; however, the entire posterior part of the skull is destroyed except for the right bulla, which is still in place. To complicate the determination, the type specimen is a juvenile, considered TWC 1 of Voss (1991a) because the m3 is essentially unworn (M3 is lost). Nonetheless, and in spite of the incomplete material, several attributes of the type specimen of *vulcani* are not typical of *Aepeomys* but are of *Thomasomys* (see also *Aepeomys* account, this volume). These include the short tail (although apparently incomplete), lack of a rostral tube, nondeveloped masseteric tubercle, large incisive foramina that extend to the level M1, and paraflexus and metaflexus not oriented anteroposteriorly along the midline of M1–M2.

Based on my own examination of the type specimen of *vulcani* Thomas, paratypes of *rhoadsi* Stone, and available specimens from Pichincha, Ecuador, I hypothesize that *rhoadsi* is a junior synonym of *T. vulcani*. In spite of the incompleteness of the type skull of *T. vulcani*, several attributes can be observed that match those of Stone's *rhoadsi*. These include (1) the nasals, which are both expanded anteriorly and posteriorly are longer than premaxillae; (2) the premaxillomaxillary-frontal joint placed anterior to the zygomatic notch; (3) the moderately broad interorbital region with rounded margins; (4) the open, large, and rhomboid-shaped incisive foramina that extend posteriorly to be level to the anterior plane of M1; (5) the moderately broad and slightly sloped zygomatic plates; (6) the primitive circulatory pattern 1 of Voss (1998); (7) the well-developed anteroloph and mesoloph of M1 with a weak cingulum; (8) the large and rounded orbicular apophysis; and (9) the malleus lamina with a well-differentiated cephalic process. Some of these morphologies are observable in photos of an additional specimen from Mount Pichincha (BM 98.8.1.10; illustrated by Gyldenstolpe 1932: plate III, Figs. 2, 2a, 2b), which agree with the type specimen. Mensural comparisons between *vulcani* and *rhoadsi* are not very appropriate in this case because of the incomplete specimen and the juvenile condition of the type. Nonetheless, 10 measurements obtained from the holotype of *vulcani* are each smaller than a series of *rhoadsi* (10 adult AMNH specimens from Pichincha), except for nasal breadth (3.65 mm versus an average of 3.66 mm), M1 breadth (1.6 mm versus an average of 1.56 mm), and zygomatic plate breadth (1.9 mm versus an average of 2.35 mm, range: 2.16 mm to 2.49 mm), for which both taxa are quite similar. In addition, the holotype skin of *vulcani* has long and soft hair, a short tail (although apparently incomplete), and is similar in color to specimens of *rhoadsi*. Based on this evidence, I regard *rhoadsi* Stone as a synonym of *T. vulcani*.

Voss (2003) compared specimens from Volcán Antisana that included a series collected by R. Olalla in 1934 from "Cerro Antisana, Andes Orientales" and another collected by C. S. Webb in 1937 from "Mt. Antisana, E. Andes 12,500–13,000 ft" with topotypes of *T. rhoadsi*. Despite slightly larger averages of several external and craniodental dimensions, Voss concluded these samples represented the same taxon. He also remarked that these Antisana specimens were likely collected on the western side of Antisana. The specimens of *T. vulcani* reported herein from Antisana are from the eastern slopes of the mountain. Pozo-R. et al. (2006) reported *Thomasomys* cf. *vulcani* (as *T.* cf. *rhoadsi*) from Hacienda El Prado, Parroquia San Fernando, Cantón Rumiñahui, Provincia de Pichincha (78°24′44″W, 0°23′20″N), arguing that they were smaller than the topotypes examined by Voss (2003). A morphometric analysis is needed to assess intraspecific variability.

Stone (1914) thought that *T. vulcani* was apparently allied to *T. cinereus*. Pacheco's (2003) phylogenetic assessment based on morphological characters suggested that *T. vulcani* (reported as *T. rhoadsi*) was related to *T. fumeus* and *Thomasomys* sp. 10, both members of his "cinereus" group. *Thomasomys* sp. 10 is an undescribed species collected from the montane forests of Piura and Cajamarca, Peru (Pacheco 2003).

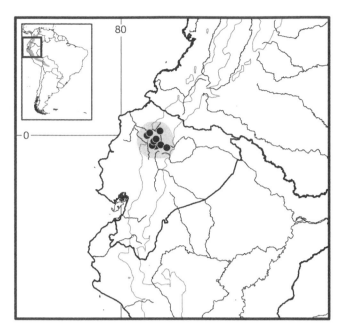

Map 367 Selected localities for *Thomasomys vulcani* (●). Contour line = 2,000 m.

The proposal that *T. vulcani* should be regarded by the IUCN as Critically Endangered (Moreno-Cárdenas and Tirira 2011) needs to be revised because the authors considered *T. vulcani* and *T. rhoadsi* as separate species.

Tribe Wiedomyini Reig, 1980
Cibele R. Bonvicino

Reig (1980) erected the tribe Wiedomyini to include the fossil genus *Cholomys* Reig (from eastern Argentina) and the Recent genus *Wiedomys* Hershkovitz (from northeastern Brazil). The phylogenetic position of the tribe is problematic. Historically (see review by Osgood [1933b]), *Wiedomys* has been variously classified as an *Oryzomys* (Thomas 1886b, 1928a; Gyldenstolpe 1932; Moojen 1943a), *Rhipidomys* (Thomas 1882, 1884; Tate 1932g; Osgood 1933b), *Oecomys* (Osgood 1933b), or *Thomasomys* (Moojen 1952b). Moreover, phallic similarities (Langguth and Silva Neto 1993) suggested a linkage with the phyllotines, but primitive karyological traits (Maia and Langguth 1987) and craniodental attributes (Reig 1980; Steppan 1995) suggested that *Wiedomys* was an early offshoot of the South American sigmodontine radiation, possibly derived from a basal thomasomyine grade. Weksler's (2006) cladistic analysis of morphological traits placed *Wiedomys* as basal to the oryzomyines. Early phylogenetic analyses of DNA sequences linked the genus either to the abrotrichines (mtDNA sequences; M. F. Smith and Patton 1999) or to the oryzomyines (nucDNA; D'Elía 2003; Weksler 2003, 2006). More recent taxon-dense analyses utilizing a larger number of genes and state-of-the-art analytical methods confirmed a sister relationship between *Wiedomys* and the abrotrichine genera (Parada et al. 2013; Schenk et al. 2013). While M. F. Smith and Patton (1999) suggested that *Wiedomys* was one of the most ancient lineages of the Sigmodontinae, placing the basal divergence between 14 and 10 mya, Parada et al. (2013) and Schenk et al. (2013) estimated a more realistic divergence time of 7 to 5 mya.

Genus *Wiedomys* Hershkovitz, 1959

Wiedomys, restricted to the semiarid zone of northeastern Brazil, is the only extant representative of the tribe Wiedomyini (Reig 1980; J. A. Oliveira et al. 2003). Three species are currently recognized, including the fossil *W. marplatensis* from the late Pliocene (Sanandresian) of Argentina (C. A. Quintana 2002) and two extant: *W. pyrrhorhinos* (Wied-Neuwied) and the recently recognized *W. cerradensis* P. R. Gonçalves, Almeida, and Bonvicino.

Externally, *Wiedomys* is characterized by a conspicuous coloration, with an overall grizzled-brown dorsal pelage but with the rump, ears, and muzzle bright orange. The sides of the body are slightly grayer than the dorsum, and the ventral pelage is well delineated with entirely white hairs extending from the gular to the inguinal region, including the ventral region of forelimbs and medial region of the hind limbs. The eyes are surrounded by a conspicuous orange eye-ring due to the presence of short orange hairs. The muzzle is well furred by small orange hairs. The feet, furred dorsally by entirely orange hairs, possess silvery ungual tufts. The tail, entirely dark brown with its proximal ventral surface slightly paler, is larger than the combined head and body length. Adult females possess eight mammae: two pectoral, two thoracic, two abdominal, and two inguinal.

The skull has a relatively short rostrum, laterally expanded braincase, and sharply defined supraorbital edges but lacks distinct supraorbital crests. The interparietal is broad and wide, and does not contact the squamosal laterally. The zygomatic plates are narrow, do not project anteriorly, and have moderately deep and broad zygomatic notches. The incisive foramina are long, extending posteriorly to the protoflexi of the M1s. The palate is short, not extending posteriorly beyond the hypoflexi of the M3s. The mesopterygoid fossa is narrow and U-shaped. The sphenopalatine vacuities are present and well developed. An alisphenoid strut may be present or absent . Parapterygoid fossae are shallow but large. The ectotympanic portion of the auditory bullae is notably inflated and with a globular shape on the ventral surface. The mandible has a

small coronoid process delimiting a small sigmoid notch. The condylar process is well developed, extends posterodorsally to the level of the angular process, and possesses a deep angular notch.

The upper incisors are opisthodont with rounded, yellow enameled bands on the anterior face. The molars are brachydont and crested with alternating cusps. The anteromedian flexus of M1 is well defined, unequally dividing the anterocone in a small anterolingual conule and larger anterolabial conule. The mesoloph is distinct in M1 and M2. The procingulum of m1 is narrower than the breadth of the m2, and the anteroconid of m1 is equally divided by the anteromedian flexid, but the mesolophid is absent.

The phallus of *Wiedomys* is similar to that of some phyllotines and different from those of oryzomyines. The medial bacular mound is longer than the lateral mounds; a small dorsal papilla is present; the urethral flap is divided into two lobes; the base of baculum is convex with a triangular shape; and the cartilage of the lateral digits is thinner and shorter than the medial (Langguth and Silva Neto 1993).

The chromosome complement of this genus varies considerably, with diploid numbers ranging from 60 to 62 and autosome fundamental numbers ranging from 86 to 104 (Maia and Langguth 1987; P. R. Gonçalves, Almeida, and Bonvicino 2005; A. L. G. Souza, Corrêa et al. 2011).

SYNONYMS:

Mus: Wied-Neuwied, 1821:177; part (description of *pyrrhorhinos*); not *Mus* Linnaeus.

Hesperomys (*Rhipidomys*): Thomas, 1882:107; part (allocation of *pyrrhorhinus* [sic]; neither *Hesperomys* Waterhouse nor *Rhipidomys* Tschudi.

[*Hesperomys*] (*Oryzomys*): Thomas, 1886b:422; part (allocation of *pyrrhorhinus* [sic]); neither *Hesperomys* Waterhouse nor *Oryzomys* Baird.

Oryzomys pyrrhorhinus: Thomas, 1928b:154; name combination.

Wiedomys Hershkovitz, 1959a:5; type species *Mus pyrrhorhinos* Wied-Neuwied, by original designation.

REMARKS: The generic placement of *Mus pyrrhorhinos* has been plagued with uncertainty since the taxon was originally described. As noted by Osgood (1933b), Thomas (1882, 1884) initially referred *Mus pyrrhorhinus* [sic] to *Rhipidomys*, previous authors having placed it in the very inclusive *Hesperomys*. Two years later, however, Thomas (1886b:422) erroneously stated that it was a member of the subgenus *Oryzomys*, based on specimens he thought were of this species but were actually those of *Thomasomys* [= *Wilfredomys*] *oenax*. Subsequently, Thomas (1928a), with actual specimens of *pyrrhorhinos* in hand, placed it in *Oryzomys*, which had been elevated to generic rank in 1890 by Elliot Coues. This placement was continued until Moojen (1952b) regarded *pyrrho-rhinos* a species of *Thomasomys*. Hershkovitz (1959a) finally recognized the generic distinction of this species, listing it as the type (and only known) species of his newly erected *Wiedomys*.

KEY TO THE SPECIES OF *WIEDOMYS*:

1. Maxillary toothrow short (<4.4 mm); incisive foramina narrow (<1.8 mm); alisphenoid strut present in majority of specimens . . . . *Wiedomys cerradensis*
1'. Maxillary toothrow long (>4.6 mm, on average); incisive foramina broad (>2.22 mm, on average); alisphenoid strut absent in majority of specimens . . . .
. . . . *Wiedomys pyrrhorhinos*

Wiedomys cerradensis P. R. Gonçalves, Almeida, and Bonvicino, 2005

Cerrado Wiedomys, Rato-de-Focinho-Vermelho

SYNONYM:

Wiedomys cerradensis P. R. Gonçalves, Almeida, and Bonvicino, 2005:51; type locality "Fazenda Sertão do Formoso (14°37′888″S and 45°51′293″W), Jaborandi, state of Bahia," Brazil.

DESCRIPTION: Small mouse (head and body length 107 mm) with long tail (tail length 142 mm) and pelage color of muzzle, ears, eye-ring, and rump bright orange. Skull small, like that of *W. pyrrhorhinos*, but with narrow incisive foramina (<1.8 mm), short maxillary toothrow (<4.35 mm), and buccinator-masticatory and accessory oval foramina separated by distinct alisphenoid strut; otherwise, nearly identical to *W. pyrrhorhinos*.

DISTRIBUTION: *Wiedomys cerradensis* is endemic to Brazil, ranging from southeastern Bahia to Goiás, Tocantins, and Ceará states.

SELECTED LOCALITIES (Map 368): BRAZIL: Bahia, Fazenda Sertão do Formoso Jaborandi (type locality of *Wiedomys cerradensis* P. R. Gonçalves, Almeida, and Bonvicino); Goiás, São Domingo, Fazenda Cruzeiro do Sul (UNB 2593, Bezerra et al. 2013); Tocantins, Paranã, Fazenda São João (UNB 2594, Bezerra et al. 2013); Ceará, Jaguaruana (LBCE 5205).

SUBSPECIES: *Wiedomys cerradensis* is monotypic.

NATURAL HISTORY: Specimens of *W. cerradensis* have been collected in semideciduous forest (P. R. Gonçalves, Almeida, and Bonvicino 2005), cerrado *sensu stricto* (N. F. Camargo et al. 2012), and in seasonally dry tropical forest and gallery forest (Bezerra et al. 2013). Individuals were active during the night but also early in the morning; one young specimen was observed foraging on the ground after sunrise.

REMARKS: *Wiedomys cerradensis* differs from *W. pyrrhorhinos* in karyotype, smaller cranial and body di-

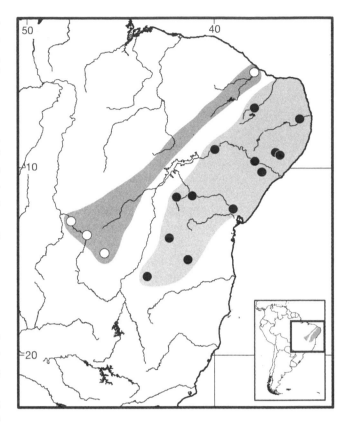

Map 368 Selected localities for *Wiedomys cerradensis* (○) and *Wiedomys pyrrhorhinos* (●).

mensions, presence of an alisphenoid strut separating the buccinator-masticatory foramen from the accessory oval foramen in the majority of specimens (see Bezerra et al. 2013), and deeply divergent and reciprocally monophyletic mtDNA sequences (P. R. Gonçalves, Almeida, and Bonvicino 2005). Both species display the same color pattern, with rump, ears, and eye-ring bright orange and strongly contrasting with the grizzled-brown color of the head and flanks. The karyotype of topotypes of *Wiedomys cerradensis* is $2n = 60$, FN = 88. The specimen used by Weksler (2003, 2006) in his phylogenetic analyses of DNA sequences, which he referred to *W. pyrrhorhinos*, is instead *W. cerradensis*.

Wiedomys pyrrhorhinos (Wied-Neuwied, 1821)

Red-nosed Wiedomys, Rato-de-palmatória

SYNONYMS:

Mus pyrrhorhinos Wied-Neuwied, 1821:177; type locality "Riacho da Ressaque [= Ressaca; see Moojen 1952b]," a small tributary of the Rio Gavião (which flows into the Rio Contas), Bahia, Brazil (Hershkovitz 1959a:6).

Mus pyrrhorhinus Wied-Neuwied, 1823:plate 2; unjustified Latinization of *Mus pyrrhorhinos* Wied-Neuwied.

Hesperomys pyrrhorhinus: Burmeister, 1854:172; name combination and unjustified Latinization of *Mus pyrrhorhinos* Wied-Neuwied.

Hesperomys [Rhipidomys] pyrrhorhinus: Thomas, 1882: 107; name combination and unjustified Latinization of *Mus pyrrhorhinos* Wied-Neuwied.

H[esperomys]. (Rhipidomys) pyrrhorhinus: Thomas, 1884: 448; name combination and unjustified Latinization of *Mus pyrrhorhinos* Wied-Neuwied.

Oryzomys pyrrhorhinus: Thomas, 1886b:422; name combination (misidentified specimens of *Wilfredomys oenax*); unjustified Latinization of *Mus pyrrhorhinos* Wied-Neuwied

Orizomys [sic] pyrrhorhinus: Moojen, 1943a:11; name combination and unjustified Latinization of *Mus pyrrhorhinos* Wied-Neuwied.

Thomasomys pyrrhorhinos: Moojen, 1952b:61; name combination.

Wiedomys pyrrhorhinos Hershkovitz, 1959a:5; first use of current name combination.

DESCRIPTION: As for genus, except that *W. pyrrhorhinos* has longer maxillary toothrow (mean=4.75 mm, range: 4.62–4.90 mm) and broader incisive foramina (mean 2.22 mm, range: 2.02–2.45 mm); majority of specimens lack alisphenoid strut.

DISTRIBUTION: *Wiedomys pyrrhorhinos* is endemic to the Brazilian Caatinga, occurring in the states of Ceará, Paraíba, Pernambuco, Alagoas, Sergipe, Bahia, and Minas Gerais.

SELECTED LOCALITIES (Map 368): BRAZIL: Alagoas, Quebrângulo (MNRJ 18549); Bahia, Caetité (MNRJ 63357), Curaça (MNRJ [LBCE 5202]), Fazenda Salinas, Morro do Chapéu (MNRJ 71607), Feira de Santana (MNRJ 18715), Ibipeba (UNB 1379), Tremendal, Rio Riacho da Ressaca (type locality of *Mus pyrrhorhinos* Wied-Neuwied); Minas Gerais, Juramento (UFRJ [FC 50]); Paraíba, Pombal (UFPB 243), Salgado São Felix (UFPB 254); Pernambuco, Bom Conselho (Maia and Langguth 1987); Sergipe, Canindé do São Francisco (A. L. G. Souza, Pessôa et al. 2011), Nossa Senhora da Glória (A. L. G. Souza, Pessôa et al. 2011).

SUBSPECIES: *Wiedomys pyrrhorhinos* is monotypic.

NATURAL HISTORY: *Wiedomys pyrrhorhinos* breed at high rates throughout the year, even during periods of intense, prolonged water stress, but very few young reach adulthood, resulting in typical low populations levels (Streilen 1982c). They are deft climbers and make extensive use of their long tail for balance. Lactating females readily adopt other young; one pregnant female still had a subadult pelage (Streilen 1982a). The preference of *W. pyrrhorhinos* for thorn scrub habitats in the Caatinga morphoclimatic domain suggests a specialist habit

(Streilen 1982d). This species, despite being endemic to semiarid Caatinga, has a very weakly developed urine-concentrating capability (Streilen 1982b). It uses abandoned bird nests, termite nests previously occupied by parrots, and ground bromeliads. In one termite nest, Moojen (1952b) found 21 individuals, including eight adults and 13 young of different size classes. Nests are constructed with leaves and grasses and litter size varies from one to six, usually with five young (Moojen 1952b). There is a report of natural infection of *W. pyrrhorhinos* by *Trypanosoma cruzi* (Pereira Barreiro and Domingos Ribeiro 1972). In Bahia state, population numbers can dramatically increase, equivalent to *ratada* events known in other sigmodontines (A. M. P. de Almeida, pers. comm.).

REMARKS: The type specimen is a skin from a female (AMNH 574). In the original description, Hershkovitz (1959a) erroneously placed the southern limit of his new genus in the state of Rio Grande do Sul, following earlier authors (Ihering 1892; Thomas 1886b). However, Thomas (1928a) had correctly restricted the species distribution to northeastern Brazil, from the states of Minas Gerais to Ceará. Three karyotypes have been reported: $2n=62$, $FN=86$ or 90, with the variation in autosomal arm number due to a pericentric inversion affecting one member of a medium-size acrocentric pair (P. R. Gonçalves, Almeida, and Bonvicino 2005; Maia and Langguth 1987) and $2n=62$, $FN=104$, a nearly completely biarmed complement (A. L. G. Souza, Pessôa et al. 2011). The limits of geographic distribution of this species remain controversial, as preliminary phylogenetic studies suggest that the species is restricted to the right bank of Rio São Francisco (A. L. G. Souza, Pessôa et al. 2011).

Subfamily Tylomyinae Reig, 1984
Sergio Ticul Álvarez-Castañeda

The subfamily Tylomyinae comprises four genera (*Nyctomys*, *Otonyctomys*, *Ototylomys*, and *Tylomys*) of largely Middle American mice and rats specialized for an arboreal habitus (Reig 1984; Steppan 1995). Only members of the genus *Tylomys*, which together with *Ototylomys* are in the tribe Tylomyini (Carleton 1980; Bradley et al. 2004; Musser and Carleton 2005), extend into South America. Musser and Carleton (2005:1186) provided an emended diagnosis. Early molecular phylogenetic analyses were unable resolve relationships of the Tylomyinae relative to the Sigmodontinae and/or Neotominae among the New World Cricetidae (D'Elía et al. 2003, 2005; Jansa and Weksler 2004; Steppan, Adkins, and Anderson 2004), but the most recent studies (Parada et al. 2013; Schenk et al. 2013) suggested a deep sister relationship with the Sigmodontinae.

Genus *Tylomys* Peters, 1866

Members of this genus are large (over 380 mm in total length, and 45 mm in skull length). Externally, they are recognizable by their tail, which is longer than the head and body, naked, and shiny, with large scales forming wide annular rings. The tail is often held loosely coiled. The hindfeet are short and broad; digit V is nearly equal to digits II–IV; plantar pads are large and closely proximal; ungual tufts are present. The color of the tops of the feet and toes vary among species from white to brown. The eyes are rather small, shine bright, with a reddish reflection. The ears are blackish, large, and naked. Mystacial vibrissae are thick and long. The fur is long, dense, and woolly. Individual hairs are black or brown at the base for about half their length; the tip varies from white to yellowish, depending on the species. Dorsal color ranges from orange tawny, gray, gray-brown to blackish, with an indistinct broad dorsal band of glossy dark hair that extends from just before the ear to the rump. The frontal and the supraorbital areas have a different color than the top of the head; vibrissae arise from dark patches on the muzzle, and dark orbital rings are strikingly defined; the anterior muzzle varies in color from cinnamon-brown to dark gray. The venter varies from creamy white to fulvous, depending on the species.

The skull has pronounced and dorsally reflected supraorbital shelves that extend as a well-defined ridge across the temporal region; the interparietal bone is conspicuous, long, and wide, laterally contacting the squamosal fenestra; the zygomatic plates are relatively narrow; an alisphenoid strut is present; the postglenoid foramen is tiny; the subsquamosal fenestra is absent; the hamular process is undefined; the parapterygoid fossa is shallow and relatively narrow; and the roof of the mesopterygoid fossa is fully ossified (Goodwin 1955; Schaldach 1966; Carleton 1980; F. A. Reid 1997; Musser and Carleton 2005).

The three cheek teeth above and below are pentalophodont, brachydont, and strongly cuspidate with principal cones opposite one another. The mesoloph(id) and other accessory enamel ridges are well developed; M2 has four roots; and m3 is large in size and with coronal topography resembling m2 (Carleton 1980; Musser and Carleton 2005). *Tylomys* shares with *Ototylomys* and *Nyctomys* an entepicondylar foramen located on the medial epicondyle of the humerus, regarded as a primitive condition in the neotomine-peromyscine group (Carleton 1980). It is possible that the three genera (and presumably *Otonyctomys*) represent survivors of a phyletically ancient group (Carleton 1980).

Limited soft anatomical data are available for the Mexican and Middle American *T. nudicaudus* and *T. fulviventer*, but few data of any kind are known from *T. mirae*. For example, Hartung and Dewsbury (1978) described both the lack of a copulatory plug and a locking copulatory system with a mechanical tie between the penis and vagina in *T. nudicaudus*, and Hooper (1960) described and figured the gland penis and baculum of *T. nudicaudus* and *T. fulviventer* (Hooper 1960). The glans is broad and relatively short (about one-third the length of the hind foot, densely armed with unusually long spines over its exterior; the urinary meatus is unusually wide, extending across the distal glans face and guarded by a pair of small, soft, spine-tipped papillae; there is no terminal or internal crater (but termed "incipient" by Carleton 1980). The baculum is completely imbedded in dense tissue; it is short and broad, and tipped with a single cone of cartilage. Females of *T. nudicaudus* have a flattened, nonossified penis-like structure placed immediately anterior to the vagina with two pairs of inguinal mammae positioned close to the urogenital openings (Hall and Dalquest 1963). Carleton (1973) detailed the structures of a unilocular-hemiglandular stomach.

In captivity *T. nudicaudus* were active during light hours but accepted food whenever it was offered (Helm 1975). R. H. Baker and Petersen (1965) characterized vocalizations as "a catlike, spitting sound" and "piglike squeals." Pathak et al. 1973) described the karyotypes of *T. nudicaudus* (including from the subspecies *T. n. gymnurus* and *T. n. villai*) and *T. panamensis*, but the karyotype of the South American *T. mirae* remains unknown.

SYNONYMS:

Tylomys Peters, 1866:404; as subgenus; type species *Hesperomys (Tylomys) nudicaudus*, by original designation.

Neomys Gray, 1873:417, Fig. 1; type species *Neomys panamensis*, by original designation (= *Tylomys panamensis*; see Alston 1880a:150); preoccupied by *Neomys* Kaup, 1829 (Mammalia, Soricidae).

Tylomys: Thomas, 1884:450; elevation to generic rank.

Tylomis Villa, 1941:763; incorrect subsequent spelling of *Tylomys* Peters.

REMARKS: Seven species are currently recognized for the genus *Tylomys* (Musser and Carleton 2005), three of which comprise the *nudicaudus* complex (*tumbalensis*, *bullaris*, and *nudicaudus*) and four the *panamensis* complex (*fulviventer*, *panamensis*, *mirae*, and *watsoni*). However, as noted by Musser and Carleton (2005:1188), a taxonomic review appropriately targeting species boundaries is long overdue.

Hall (1981:626–627) mapped the ranges of the six North and Central American species that range from southern

Mexico through Panama (see also Handley 1966). Cabrera (1961:435–436) listed a single species, *T. mirae* Thomas, in South America, with two subspecies, *T. m. mirae* (from southern Colombia and northern Ecuador) and *T. m. bogotensis* (from the eastern Andes of Colombia). Possibly the Panamanian species *T. fulviventer*, *T. panamensis*, and/or *T. watsoni*, all of which occur in extreme eastern Panamá (as mapped by Hall 1981), extend into adjacent Colombia. Fieldwork in the general region of the Serranía de los Altos is critically needed.

Tylomys mirae Thomas, 1899
Southern Climbing Rat

SYNONYMS:

Tylomys mirae Thomas, 1899c:278; type locality "Paramba, River Mira, [Imbabura,] N. Ecuador, altitude, 1100 meters."

Tylomys mirae bogotensis Goodwin, 1955:4; type locality "Volcán Caparrapi on the western slopes of the Cordillera Oriental, elevation about 1270 meters, 90 kilometers north-northwest of Bogotá and 10 kilometers west of La Palma," Cundinamarca, Colombia.

Tylomys mirae mirae: Cabrera, 1961:435; name combination.

DESCRIPTION. Among largest species in genus (head and body length averages about 470 mm) exceeded only by some specimens of *T. nudicaudus*. Fur dense and thick, with individual hairs on back reaching lengths of 20–23 mm. Tail longer than head and body, naked on base with black hairs extending from scales; tail tip colored white to yellowish white. Dorsal body dark gray to brownish in color, with slight tinge of fawn; sides grayer and less fawn (or snuff brown), somewhat buffy. Head similar in color to dorsal body, or grayer; no eye-ring. Under parts white but not very sharply defined; on belly, white area narrow, like broad line down midline, with completely self-colored white fur in pectoral and inguinal regions and rest of under parts with hairs tinged slightly grayish at their bases with white tips. Forelimbs gray laterally, white medially; hindlimbs entirely gray except for narrow white line extending to ankles. Upper surfaces of both forefeet and hindfeet chocolate brown, terminal phalanges and hairs at base of claws white. Ears rather small, naked, and gray.

Skull large and angular without obvious zygomatic notches; nasals narrow, with posterior part truncate in *T. m. bogotensis* and pointed in *T. m. mirae*, ending on line or scarcely surpassed by premaxillaries. Supraorbital ridges well defined with prominent postorbital angle from which they run backward, and curved from little to strongly outward in parietal region. Molariform teeth relatively small and narrow; V-shaped mesopterygoid fossa extends forward to near posterior margin of M2 (Goodwin 1955; Thomas 1899c).

DISTRIBUTION: The few available specimens of *T. mirae* are known from tropical forest on the lower slopes of the Andean cordilleras of Colombia and Ecuador south along through the Pacific lowlands to northwestern Ecuador, over an elevational range from 200 to 1,300 m (Cabrera 1961; Eisenberg 1989).

SELECTED LOCALITIES (Map 369): COLOMBIA: Antioquía, 4 km NE of Bellavista above Río Porce (FMNH 79553), 25 km S and 22 km W of La Tirana (USNM 499617), Puerto Berrío "Las Virginias" (IAvH), Valdivia, Quebrada Valdivia (FMNH 70554); Boyacá, Muzo (FMNH 71216); Caldas, Samana, Río Hondo (FMNH 71215); Cundinamarca, Nariño, near Río Mira (ICN, Bogotá); Nariño, Buenavista (AMNH 34204), Candelilla (FMNH 89564); Valle de Cauca, 6 km N of Buenaventura (USNM 554237). ECUADOR: Carchi, El Pailón (Diego Tirira, pers. comm.); Esmeraldas, El Salta, estero (Diego Tirira, pers. comm.), Pamibilar (USNM 113318); Imbabura, Paramba, Río Mira (type locality of *Tylomys mirae* Thomas).

SUBSPECIES: Goodwin (1955) distinguished *T. m bogotensis*, in comparison with the nominotypical form *T. m. mirae*, on the basis of paler dorsal color, dusky instead of white under parts, and smaller teeth. Cabrera (1961) continued to list these two subspecies as valid, but Musser and Carleton (2005:1189) simply list *bogotensis* Goodwin as a synonym. Validation of subspecies boundaries requires a more thorough evaluation of geographic variation than currently available.

NATURAL HISTORY: Current data indicate no strong difference between the habitats of different species of the genus (F. A. Reid 1997). The climbing rats are found in cloud forest and tropical evergreen forest, in dense forest with tall trees (up to 30 m) and a dense shrub layer, where the crowns harbor great masses of vines and creepers with much parasitic and epiphytic growth. Most records state specimens were taken either on the ground, in shrubs, or in trees, in areas often associated with rock outcrops or cliffs. Climbing rats move very quickly through the branches, stems, and vines.

REMARKS: Musser and Carleton (2005) suggested that both *fulviventer* and *watsoni* might be junior synonyms of *T. panamensis*. Goodwin (1955:5), in his description of *T. mirae bogotensis*, remarked that his new subspecies "is not very unlike the type of *T. fulviventer*." *Tylomys mirae* can be distinguished generally from this complex by gray-based under parts with pectoral and inguinal white spots as opposed to entirely white venters, and skulls with V-shaped versus U-shaped mesopterygoid fossae, narrower rostra and smaller bullae, and heavier supraorbital shelves and their temporal extensions.

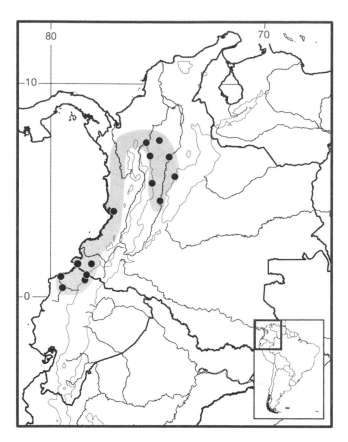

Map 369 Selected localities for *Tylomys mirae* (●). Contour line = 2,000 m.

Suborder Hystricomorpha Brandt, 1855
Infraorder Hystricognathi Tullberg, 1899
James L. Patton

The New World hystricognathous rodents comprise 10 to 13 families, depending on the authority, typically arranged into four superfamilies that are, in turn, often united into the Caviomorpha (at the categorical level of suborder [Wood 1965] or infraorder [B. Patterson and Wood 1982]). Woods (1993) questioned whether all caviomorphs were part of a single radiation, and McKenna and Bell (1997) excluded the erethizontids from their parvorder Caviida that contained all other New World hystricognaths. However, all available molecular data consistently support the monophyly of the caviomorphs relative to Old World hystricognaths. These same analyses also suggest a single ancestral entry to the New World (Huchon and Douzery 2001; Houchon et al. 2007; Opazo 2005; Poux et al. 2006; Rowe et al. 2010), as does the microstructure of incisor enamel (T. Martin 2004, 2005), cladistic analyses of fossil hystricognathous forms (Sallam et al. 2009; Antoine et al. 2012), and a very stable, but unique, placentation pattern (A. M. Carter and Mess 2013).

The group has a long and very diverse fossil history in the New World, beginning with certainty in the Oligocene (B. Patterson and Wood 1982; Vucetich et al. 1999; Wyss et al. 1993; Pascual 2006; R. do V. Vilela et al. 2009) and possibly as early as the late Eocene (Frailey and Campbell 2004). Huchon and Douzery (2001) suggested that the early caviomorph radiation corresponded to two ecological events, an arrival that coincided with cooling between 25 and 36 mya in which euhypsodont teeth evolved as a key adaptation and subsequent diversification during the climatic perturbations of the mid-Miocene (Vucetich et al. 1999). The basal date of the Caviomorpha radiation estimated from molecular studies ranges from 36.7 ± 3.7 mya (Poux et al. 2006; Huchon et al. 2007), 34.1 mya (31.6–37.4 confidence interval; Upham and Patterson 2012), 33.8 ± 1.6 mya (Opazo 2005), to 30.7 mya (95% credibility interval 29.1–33.4 mya; Sallam et al. 2009), estimates that are consistent with the 31–37 mya age span suggested by Wyss et al. (1993) and Frailey and Campbell (2004). The newly discovered diverse caviomorph assemblage from the late Middle Eocene Yahuarango Formation in Peru (approximately 41 mya) supports an earlier entry of stem caviomorphs from Africa into South America, one that corresponds with the warmer and wetter conditions of the Mid Eocene Climatic Optimum (Antoine et al. 2012). This date is consistent with the estimate of the 40.4 mya (36.1–45.6 confidence interval) for the divergence of the Caviomorpha and African Phiomorpha split (Upham and Patterson 2012).

By the early Miocene (Santacruzian Land Mammal Age), caviomorphs had reached the diverse array of body sizes, locomotory modes, and presumed feeding behaviors observed among extant taxa, with considerable commonality in morphofunctional groupings across these disparate temporal assemblages (Candela et al. 2012). Several early members of extant lineages appeared to eat harder and more abrasive foods than those consumed by their living relatives, and the earliest caviomorphs likely lived in an environment similar to the Brazilian Cerrado today.

The arrangement of genera into family and subfamily groups has been relatively stable for many decades, with a few notable exceptions (e.g., *Chaetomys* as an Erethizontidae or Echimyidae; Woods 1982 versus T. Martin 1994 or R. do V. Vilela et al. 2009) as has been the division of caviomorphs into four superfamilies. However, there has been considerable flux in family and subfamily membership in each superfamily (Simpson 1945; Landry 1957; Wood 1965; Woods 1982, McKenna and Bell 1997). The categorical rank accorded to some groups traditionally given family status has also been debated (e.g., placement of Hydrochoeridae and Myocastoridae as subfamilies within the Caviidae and Echimyidae, respectively; Woods and Kilpatrick 2005).

The families and superfamilies recognized herein generally follow Woods and Kilpatrick (2005), modified by molecular phylogenetic analyses (Opazo 2005; Rowe et al. 2010; Upham and Patterson 2012). These studies, combining both mitochondrial and nuclear gene sequences and broad taxon sampling, posit the same hypotheses of relationships among families, with two exceptions. First, the Erethizontidae are placed as part of a clade that otherwise comprised the Chinchilloidea and Octodontoidea (Opazo 2005; Rowe et al. 2010) or as the sister to the Cavioidea (Upham and Patterson 2012). Second, these analyses support different topologies of family relationships within the Cavioidea (Dasyproctidae [Caviidae + Cuniculidae], Opazo 2005; Rowe et al. 2010; versus Cuniculidae [Caviidae + Dasyproctidae], Upham and Patterson 2012). Herein, I retain the Erethizontidae within its own superfamily (Erethizontoidea), place the Dinomyidae + Chinchillidae within the Chinchilloidea, group the Abrocomidae + Echimyidae + Ctenomyidae + Octodontidae within the Octodonoidea, and include the Dasyproctidae + Caviidae + Cuniculidae within the Cavioidea. Following the well-supported cladistic relationships established by each of the molecular analyses, the capybaras are regarded as a subfamily (Hydrochoerinae) within the Caviidae and the coypus as a subfamily (Myocastorinae) within the Echimyidae. Resolution of the differences in phyletic placement of the Erethizontidae, and thus potential changes in the higher categorical groups of Caviomorpha, awaits further analyses. Because no molecular analysis can (at least, as yet) encompass the rich diversity of extinct forms, their inclusion requires cladistic analyses of morphological characters of both fossil and extant taxa (e.g., G. A. S. Carvalho and Salles 2004; Candela and Rasia 2012; Olivares et al. 2012). The Capromyidae are not covered, as this group resides completely outside the boundaries of South America (Hall 1981; Woods and Kilpatrick 2005).

Despite highly varied body form and size, all caviomorphs are characterized by an hystricognathous lower jaw, in which the root of the angular process is deflected lateral to the sheath of the lower incisor, an hystricomorphous jaw musculature, with a large masseter medialis penetrating a greatly enlarged infraorbital foramen to insert on the side of the rostrum, hypsodont, often ever-growing cheek teeth with planned occlusal surfaces, and a dental formula of I 1/1, C 0/0, PM 1/1, M 3/3 = 20. The deciduous fourth premolars (dP4/dp4) are apparently replaced in all families with the exception of the Echimyidae (e.g., G. A. S. Carvalho 2000).

KEY TO THE FAMILIES OF SOUTH AMERICAN HYSTRICOGNATHI:

1. Dentary with prominent ridge and groove on lateral side of dentary below and parallel to cheek teeth. . . . . Caviidae
1′. No prominent ridge and groove on lateral side of dentary. 2
2. Zygomatic arch greatly expanded, forming thickened and rugose cheekplate; body size large, mass >5 kg, skull length >100 mm . . . . .Cuniculidae
2′. Zygomatic arch normally developed, without thickened cheekplate; body size and skull length variable. 3
3. Cheek teeth composed of two to four simplified transverse plates. 4
3′. Cheek teeth with more complex occlusal surfaces . . . 5
4. Body bauplan stocky, limbs short, hindfeet with four well-developed digits; pelage short, hairs coarse, color black with rows of white and spots on back; toothrows diverge slightly posteriorly; paroccipital processes short, blunt, extending to ventral plane of bullae; bullae uninflated . . . . Dinomyidae
4′. Body bauplan more slender, limbs elongated, hindfeet with three well-developed digits; pelage long, thick, hairs fine, no rows of white spots on back; toothrows diverge strongly posteriorly; paroccipital processes elongated, extending well below ventral plane of bullae when bullae uninflated, or curved over bullae when latter inflated . . . . Chinchillidae
5. Fur conspicuously spiny, some hairs modified as rounded, barbed quills; hindfeet greatly modified for arboreal life, with hallux replaced by broad, movable pad; cheek teeth either with very wide reentrant folds or nearly laminate, rooted; paroccipital process not greatly lengthened . . . . Erethizontidae
5′. Spines may be present in fur, but if so always flattened, without barbs; hindfeet usually not modified for arboreal life (except some Echimyidae, with broad soles and short digits, or elongated digits with nail-like claws), hallux never modified into broad, movable pad 6
6. Body bauplan modified for cursorial locomotion, with elongated limbs and short tail; three digits on hindfeet; claws thickened, almost hoof-like; no spines in fur; cheek teeth semirooted, reentrant folds isolating as narrow islands on crown surfaces in adults . . . . . .Dasyproctidae
6′. Body bauplan not modified for cursorial life; digits of hindfeet four or five; claws not hoof-like; spines may be present in fur; cheek teeth variable when ever-growing occlusal surfaces rather simplified, when rooted occlusal surface pattern of isolated islands. 7
7. Cheek teeth greatly simplified, labial side of upper series with only one reentrant fold. 8

7'. Cheek teeth more complex, labial side of upper series always with at least two, typically three, sometimes more reentrant folds . . . . Echimyidae

8. Lower cheek teeth prismatic and angular; occlusal surfaces of upper series figure-8 shaped; part of lacrimal canal open on side of rostrum. . . . . Abrocomidae

8'. Lower cheek teeth not prismatic; occlusal surfaces of both upper and lower cheek teeth with same figure-8 shape or with kidney shape pattern; no part of lacrimal canal opens on side of rostrum 9

9. Occlusal surfaces of cheek teeth weakly kidney shaped, size decreasing notably from front to back; skull heavily ridged; elongated rostral tubes extend dorsolaterally from bullae; foreclaws greatly enlarged . . Ctenomyidae

9'. Occlusal surfaces of cheek teeth figure-8 shaped, similar in size from front to back; skull not heavily ridged; no elongated auditory canal extending from bullae; foreclaws not greatly enlarged . . . . Octodontidae

Superfamily *Cavioidea G. Fischer, 1817*

The Cavioidea comprises three families, Caviidae (which includes the subfamilies Caviinae, Hydrochoerinae, and Dolichotinae), Cuniculidae, and Dasyproctidae. Exclusive of the Cuniculidae and Dasyproctidae, fossils assignable to Cavioidea *sensu stricto* date from as early as the early Oligocene (Deseadan; M. E. Pérez and Vucetich 2011). As noted above, molecular estimates of 26.5 mya for the basal diversification of extant genera, and thus families, are fully consistent with the fossil record (Opazo 2005; Upham and Patterson 2012). Also as noted above, however, there remains ambiguity among family-group relationships, as separate molecular analyses support either the Dasyproctidae (Rowe and Honeycutt 2003; Opazo 2005; Rowe et al. 2010; M. E. Pérez and Pol 2012) or Cuniculidae (Upham and Patterson 2012) as the basal member of the superfamily.

Family Caviidae G. Fischer, 1817
Jonathan L. Dunnum

Six Recent genera (*Cavia*, *Galea*, *Microcavia*, *Kerodon*, *Dolichotis*, and *Hydrochoerus*) in three subfamilies (Caviinae, Dolichotinae, and Hydrochoerinae) were recognized (Woods and Kilpatrick 2005). The hydrochoerines were included as a subfamily by Tate (1935) and Ellerman (1940), but others (Pocock 1922; Simpson 1945; Woods 1993; McKenna and Bell 1997) have recognized it as a distinct family. Molecular phylogenetic analyses (Rowe and Honeycutt 2002; Opazo 2005; Rowe et al. 2010; Upham and Patterson 2012) provide strong support for the three subfamilies and the position of the Hydrochoerinae as sister to the Dolichotinae. M. E. Pérez and Pol (2012) corroborated these relationships using both molecular and fossil data. Their topology is recognized here.

The family is well represented in the fossil record from the middle to late Miocene (B. Patterson and Pascual 1972; C. A. Quintana 1998, 2001; Ubilla and Rinderknecht 2003; M. E. Pérez and Vucetich 2011). All current genera are present since the Pleistocene (McKenna and Bell 1997). Recent molecular work has placed the divergence date for the family in the early (18.5 ± 2.5 mya; Opazo 2005) to middle Miocene (14–17 mya; Huchon and Douzery 2001), dates that are consistent with the limited fossil record (M. E. Pérez and Vucetich 2011). In the most recent treatment, M. E. Pérez and Pol (2012) dated the Caviidae split at about 13.5 mya based on combined fossil and molecular data. Molecular data resulted in a mean date of about 16.5 mya (95% HPD 11–24 mya).

The cheek teeth are ever growing, unilaterally hypsodont, simplified in structure, but with sharp folds and projections exhibiting a more or less prismatic effect. The toothrows tend to converge anteriorly. The tibia and fibula are not fully fused (Ellerman 1940). The lower mandible exhibits the least hystricognathous condition of the extant hystricognath rodents. The lateral process of the supraoccipital is lacking (Eisenberg 1989). Body form is robust, the head is large, and the ears and limbs are short in the Caviinae and Hydrochoerinae, while the Dolichotinae exhibit the hare-like traits of long ears and limbs. Head and body length ranges from ca. 200 mm in *Galea* and *Microcavia* to over 1,300 mm in *Hydrochoerus*. The tail is absent in all taxa except *Dolichotis*. The manus is pentadactyl, and the pes is tridactyl, with nails blunt and either claw-like or hoof-like. Members of the family are generally long-lived, and young are well developed at birth and reach maturity quickly. The family exhibits a wide degree of variation in social and mating systems, from highly social and monogamous in the Dolichotinae, to polygynous and highly social in *Hydrochoerus* and *Kerodon*, to promiscuous with low sociality in *Cavia*, *Galea*, and *Microcavia* (Novak 1999; Rowe and Honeycutt 2002; Adrian and Sachser 2011). Diploid number ranges from 52 to 68 (W. George et al. 1972; Maia 1984; Dunnum and Salazar-Bravo 2006). M. H. Gallardo et al. (2002) described sperm morphology. The family occupies a wide range of habitats, including mesic and xeric grasslands, forest edges, and marsh from sea level to 5,000 m over much of South America and extending north into Panama.

KEY TO THE SUBFAMILIES AND GENERA
OF RECENT CAVIIDAE:
1. Head and body length <450 mm 2
1'. Head and body length >450 mm 5

2. Toes with nails; sternum narrow and rounded. . . . .
. . . . .Hydrochoerinae, *Kerodon*
2′. Toes with claws; sternum broad and flat. . . Caviinae, 3
3. Incisors pigmented yellow; orbital branch of maxillary completely interrupted by lacrimal . . . . *Galea*
3′ Incisors white; orbital branch of maxillary continuous as a narrow strip in front of lacrimals. 4
4. Head and body length <225 mm; prominent eye-ring present; ears with well-developed tragus and antitragus; teeth without cement interleaves between prisms; PM4 with an anterior accessory prolongation in the first prism . . . . *Microcavia*
4′. Head and body length >225 mm; eye-ring not prominent or absent; ears with poorly developed tragus and antitragus; teeth with cement interleaves between prisms; PM4 without an anterior accessory prolongation in the first prism . . . . *Cavia*
5. Size large, head and body length ca. 1,000 mm; feet semiwebbed; M3 longer than combined length of other three molariform teeth . . . .
. . . . Hydrochoerinae, *Hydrochoerus*
5′. Size medium, head and body length 450–800 mm; feet not webbed; hind toes hoof-like; teeth not as above . . .
. . . . .Dolichotinae, *Dolichotis*

Subfamily Caviinae G. Fischer, 1817

The Caviinae include the three guinea pig-like genera (*Cavia*, *Galea*, and *Microcavia*). Fossil taxa date from the mid-Miocene (C. A. Quintana 1998, 2001; Ubilla and Rinderknecht 2003; M. E. Pérez and Vucetich 2011), and divergence of the subfamily ranges from 16.2 ± 2.5 mya (Opazo 2005). Diploid numbers span 62 to 68 (W. George et al. 1972; Maia 1984; Gava et al. 1998; Dunnum and Salazar-Bravo 2006), but see Lobato and Araujo (1980) for a possible exception in the Ecuadorean Cavy.

Genus *Cavia* Pallas, 1766

Medium-sized herbivores, the wild forms of *Cavia* are endemic to South America, occurring over most of the continent, with the exception of portions of Amazonia and the austral parts of Chile and Argentina (Ximénez 1980; Eisenberg 1989; Redford and Eisenberg 1992; Eisenberg and Redford 1999). They occur from mesic lowlands (10 m) to arid highlands (4,200 m) and while primarily associated with grasslands, they also occur in forest edges and swampland. Due to its utility to *Homo sapiens*, the domesticated form is now found nearly worldwide in conjunction with human habitation. Woods and Kilpatrick (2005) recognized six species, but Dunnum and Salazar-Bravo (2010a) elevated the Ecuadorean *C. patzelti* to specific status.

The fossil record for *Cavia* dates from the late Pliocene (Verzi and Quintana 2005) to the middle Pleistocene (Ubilla and Alberdi 1990; McKenna and Bell 1997) and Holocene (Hadler et al. 2008). However, molecular data suggest the genus has been present since the Miocene/Pliocene boundary (Opazo 2005; Dunnum and Salazar-Bravo 2010a). *Cavia* specimens have been taken from archeological excavations in Peru and Colombia dating to at least 9,000 years BP and they have been domesticated for the last 4,500–7,000 years for food and spiritual uses (Wing 1986). The human facilitated expansion of the domesticated form and a lack of broad systematic work on the wild forms have complicated our understanding of the evolutionary history, systematics, taxonomy, and biogeography of the genus, although recent work (Dunnum 2003; Spotorno, Valladares, Marín, and Zeballos 2004; Spotorno et al. 2006; Dunnum and Salazar-Bravo 2010a) has provided much resolution.

The skull has a limited interorbital constriction with a sagittal crest developed in adults; the infraorbital foramen is broader below than above; auditory bullae are relatively large; the paroccipital processes are elongated; the hard palate is short, extending to the front of M3; palatal foramina are short and narrow; and the jugal is of medium length and does not approach the lacrimal. Upper cheek teeth are about equal in size and divided into two lobes by an inner reentrant fold in the upper series, the posterior lobe being larger than the anterior one, and with a deeply indenting fold in its outer border. M3 has a posterior projection. Lower cheek teeth have one deep outer fold dividing the tooth into two lobes with an inner fold in the posterior lobe. Incisors are not pigmented (Ellerman 1940). Mean condyloincisive length is 53.8 mm (range: 44.5–63.5 mm; Hückinghaus 1961a). Upper cheek teeth are divided into two lobes by inner reentrant fold, the posterior lobe being larger than the anterior and with a deeply indented fold on its labial border. The third upper molar has a posterior projection. Lower cheek teeth have one deep outer fold dividing each tooth into two lobes and with an inner fold on the posterior lobe. The hindfeet have three digits, each terminated by a sharp claw, with the central digit longest. The forefeet have four digits. The ears are relatively short and rounded (Ellerman 1940). Females have a single pair of inguinal mammae.

SYNONYMS:

Mus: Linnaeus, 1758:59; part (description of *porcellus*); not *Mus* Linnaeus.

Cavia Pallas, 1766:30; type species *Cavia cobaya* by subsequent designation (Palmer 1904) [= *C. cobaya* Marcgraf, = *Mus porcellus* Linnaeus].

Cauia Storr, 1780:39; incorrect subsequent spelling of *Cavia* Pallas.

Lepus: G. I. Molina, 1782:306; part; not *Lepus* Linnaeus; either a *Cavia* or *Galea*, see Tate (1935:344).

Calva Gmelin, 1788:120; incorrect subsequent spelling of *Cavia* Pallas.

Scavia Blumenbach, 1797:80; unjustified emendation of *Cavia* Pallas.

Agouti: Daudin, 1802:166; part (listing of *aperea*); not *Agouti* Lacépède.

Savia Treviranus, 1802:211; unjustified emendation of *Cavia aperea* Erxleben.

Hydrochoerus: É. Geoffroy St.-Hilaire, 1803:163; part (listing of *cobaya*); not *Hydrochoerus* Brisson.

Hydrochaerus G. Fischer, 1814:593; part (listing of *cobaya*); incorrect subsequent spelling of *Hydrochoerus* Brisson.

Anoëma F. Cuvier, 1809:394; type species *Cavia cobaya* Pallas.

Anoema F. Cuvier, 1812:292; justified emendation of *Anoëma* F. Cuvier; type species *Cavia cobaya* Pallas.

Cobaya F. Cuvier, 1817b:481; type species *Cavia cobaya* Pallas.

Anoemas F. Cuvier, 1829a:493; incorrect subsequent spelling of *Anoema* F. Cuvier.

Coiza Billberg, 1827:45; new name for *Cavia* Pallas.

Prea Lund, 1839b:206; included *Cavia aperea* ("Aperea de Marcgraf" = *C. cobaya* Margrave = *Mus porcellus* Linnaeus) and *Cavia rufescens* Lund (= *Cavia aperea* Erxleben).

Kerodon: Wagner, 1844:64; listing of *obscurus*; not *Kerodon* F. Cuvier.

Prea Liais, 1872:541; part (listing of *cobaya*); new name for *Cavia* Pallas.

Mamcaviaus A. L. Herrera, 1899:13; unavailable name because it is based on zoological formulae (ICZN 1987:318).

REMARKS: *Mus porcellus* Linnaeus and *Cavia cobaya* Pallas were both based on the *Cavia cobaya* of Marcgraf (1648), a domesticated form sent to Europe from an unknown place of origin in South America. The name *porcellus* Linnaeus, indicated by Thomas (1916f) as the type species of *Cavia*, is excluded from consideration because it was not included under the generic name at the time of its original publication (see Tate 1935). The generic name *Calva* appears (p. 120) in some but not all versions of Gmelin, 1788; however, *Cavia* is correctly listed in the running header in each. G. I. Molina (1782:306) used *Lepus minimus* in describing a "Cuy," which Tate (1935:344, 347) thought was likely a *Galea* and thus included as a subspecies of *G. musteloides*. Hückinghaus (1961a) followed Tate and assigned *minimus* G. I. Molina to *Galea*, whereas Cabrera (1961) listed it in the synonymy of *Cavia porcellus*.

Tate (1935) summarized the taxonomy of the genus up to 1930, recognizing 11 species. The number of species delimited in subsequent works has varied from three to eight species (Cabrera 1961; Hückinghaus 1961a; Corbet and Hill 1991; Novak 1999; Woods 1993). Woods (1993) recognized five species (*aperea*, *fulgida*, *magna*, *porcellus*, and *tschudii*), following Hückinghaus (1961a) by including *C. nana* and *C. guianae* in the synonymy of *C. aperea* and *C. anolaimae* in the synonymy of *C. porcellus*. Woods and Kilpatrick (2005) followed Woods (1993) but added the Intermediate Cavy, *C. intermedia*, to the list of species they recognized. In this account I recognize seven species within *Cavia*, including the Ecuadorean cavy *C. patzelti*, and despite its clear derivation from *C. tschudii*, I also treat the domesticated form *C. porcellus* as a species to avoid further complicating the taxonomic understanding of the group.

KEY TO THE SPECIES OF *CAVIA*:

1. Dorsal pelage multicolored white, black, or tan and occurring in human-occupied areas throughout South America and worldwide as a pet or laboratory animal . . . . *Cavia porcellus*

1′. Dorsal pelage uniformly brown, tan, or red-brown, not multicolored; occurring in the wild in South America 2

2. M3 without deep notch on the outer side of the anterior end of the posterior lobe 3

2′. M3 with deep notch on the outer side of the anterior end of the posterior lobe, found in coastal regions of eastern Brazil . . . . *Cavia fulgida*

3. Found in highlands above 2,500 m or west of the Andes, in Peru, Bolivia, Ecuador, and Argentina 4

3′. Found at elevations below 2,500 m east of the Andes or in the highlands of Colombia 5

4. Occurring in the highlands of Ecuador . . . *Cavia patzelti*

4′. Occurring in highlands above 2,500 m or west of the Andes, in Peru, Bolivia, and Argentina . . . . . *Cavia tschudii*

5. Interdigital membranes enlarged 6

5′. Interdigital membranes not enlarged; occurs on eastern versant of the Andes (Bolivia through Colombia) and the non-Amazonian lowlands east of the Andes . . . . *Cavia aperea*

6. Coastal marshlands of extreme southeastern Brazil and Uruguay; $2n = 64$, except $2n = 62$ population on Ilha Marinheiros . . . . *Cavia magna*

6′. Islands of Moleques do Sul, Brazil; $2n = 62$ . . . . *Cavia intermedia*

Cavia aperea Erxleben, 1777
Brazilian Guinea Pig

SYNONYMS: See under Subspecies.

DESCRIPTION: Varies among subspecies. Nominotypical subspecies of eastern Brazil among largest, with mean greatest length of skulls from Minas Geraes and São Paulo states 68.7 mm, with largest 73 mm. Overall coloration grizzled, brownish gray; venter dull whitish with clear white spot usually present at midline of chest below brown collar (Thomas 1917a).

More southerly Brazilian form, *C. a. rosida*, somewhat smaller (greatest length of skull about 63 mm), with reduction in skull length due primarily to reduced rostrum. Dorsal color heavily grizzled mummy brown; median part heavily blackened, with lumbar region sometimes almost completely black; under surfaces and insides of limbs dull cinnamon buff, with hairs pale gray basally; throat markings not easily distinguished, usual dark collar overlaid with dull buffy and white chest patch absent or reduced to small spot. Dimensions of holotype of *rosida* head and body length 395 mm, hindfoot length 46 mm, ear length 20 mm, and greatest skull length 62 mm (Thomas 1917a).

The Argentinean *C. a. pamparum* similar to Brazilian forms but more grayish or olivaceous, less brown; venter whitish or slightly drab, chest pattern well marked. Greatest length of skull 62–63 mm. Dimensions of holotype: head and body length 251 mm, hindfoot length 40 mm, ear length 23 mm, and greatest skull length 62 mm (Thomas 1901e, 1917a).

Colombian *C. a. anolaimae* uniformly dark yellowish gray, with exposed portion of hairs black with narrow subapical band of olivaceous buff; ventral surfaces dingy brownish gray, hairs uniform to base; no white around eyes or ears (J. A. Allen 1916a).

Specimens of *C. a. guianae* from Guyana similar in general characteristics to Brazilian forms but paler and grayer above and less buffy below; general coloration olive gray; dorsally heavily grizzled and lined with black; head, neck, shoulders, flanks, and rump more grayish; under parts whitish or slightly buffy, bases of hair slaty gray; eyes circumscribed by indistinct whitish rings; ears thinly haired in white. Dimensions of holotype head and body length 275 mm, hindfoot length 46 mm, ear length 19 mm, and greatest skull length 59 mm (Thomas 1901c). Husson (1978) reported that a Surinam specimen differed from holotype of *C. a. guianae* only in that type was much whiter ventrally; remainder type series indistinguishable from Surinam specimen. Dimensions of other Surinam specimens, head and body length 160–265 mm, hindfoot length 36–45 mm, and ear length 22 mm.

The Pygmy Cavy of the Bolivian Andean versant, *C. a. nana*, is smallest. Dorsal coloration dark olive brown, with light rings on hairs being buffy or cinnamon; no darkening along median area of back; ventral surface creamy whitish and gray collar well marked; hands and feet pale brown. Dimensions of holotype head and body length 215 mm, hindfoot length 38 mm, ear length 23 mm, and greatest

skull length 52 mm (Thomas 1917a). Specimens from Bolivian lowlands similar in appearance to Pygmy Cavy but overall darker olivaceous and exhibit medial dorsal darkening; ventrally hair longer and tips much more yellow, giving more striking color than muted creamy gray of *C. a. nana* (J. Dunnum, pers. obs.).

DISTRIBUTION: *Cavia aperea* has a disjunct range, with southern populations occurring south of the Amazon Basin from the eastern slopes of the Andes in southeastern Peru and Bolivia, through Paraguay, northeastern Argentina, Uruguay, and to the eastern coast of Brazil, and northern populations from central Colombia east through Venezuela, Guyana, Surinam, and northern Brazil.

SELECTED LOCALITIES (Map 370): ARGENTINA: Buenos Aires, Arroyo Sauce Chico (Massoia 1973b), Lobería (Massoia 1973b), Zelaya (Cabrera 1953); Chaco, El Colorado (Massoia 1973b); Córdoba, Noetinger (Massoia 1973b); Corrientes, Goya (type locality of *pamparum* Thomas); San Luis, Pedernera (J. R. Contreras 1980); Santa Fé, Tostado (Massoia 1973b). BOLIVIA: Beni, Río Itenez, 20 km above mouth (S. Anderson 1997), Santa Rosa (S. Anderson 1997); Chuquisaca, Chuquisaca (S. Anderson 1997); Cochabamba, Incachaca (S. Anderson 1997); La Paz, Chulumani (type locality of *nana* Thomas); Santa Cruz, Parque Nacional Noel Kempff Mercado, 6 km S of Los Fierros (S. Anderson 1997). BRAZIL: Distrito Federal, Brasília (Ximénez 1980); Goiás, Anápolis (Ximénez 1980); Minas Gerais, Serra Caparaó, Fazenda Cardozo (AMNH 61846); Pernambuco, Saltinho, Rio Formoso (Ximénez 1980); Rio de Janiero, Itatiaia (Avila-Pires 1977); Rio Grande do Sul, Tramandaí (Cherem et al. 1999); Roraima, Ilha Maracá (Barnett and Cunha 1994); São Paulo, Ribeirão Preto (Asher et al. 2004); Sergipe, 1 km W of Pacatuba (Ximénez 1980). COLOMBIA: Cundinamarca, Anolaima (type locality of *anolaimae* J. A. Allen); Meta, Condominio Camino Real (Zúñiga 2000). GUYANA: Berbice, Berbice (Thomas 1917a); Upper Takutu–Upper Essequibo, Kanuku Mountains (type locality of *guianae* Thomas). PARAGUAY: Concepción, 7 km NE of Concepción (MSB 82571); Presidente Hayes, Río Negro (Riacho Negro) (AMNH 36517). PERU: Madre de Dios, Pampas de Heath, ca. 50 km S of Puerto Pardo (LSUMZ 22683). SURINAM: Sipaliwini, Sipaliwini airstrip (S. L. Williams et al. 1983). URUGUAY: Canalones, Bañado de Tropa Vieja (Barlow 1969), Treinta y Tres, Boca del Río Tacuari (Barlow 1969). VENEZUELA: Bolívar, Altagracia (type locality of *Cavia porcellus venezuelae* J. A. Allen), San Ignacio Yuraní (Voss 1991a); Carabobo, Montalbán (Handley 1976); Monagas, San Agustín (Handley 1976).

SUBSPECIES: I recognize seven subspecies of *C. aperea*.

C. a. anolaimae J. A. Allen, 1916

SYNONYMS:

Cavia anolaimae J. A. Allen, 1916a:85; type locality "Anolaima, on a branch of the Rio Bogotá, west of Bogotá, Colombia," Cundinamarca.

Cavia porcellus anolaimae: Cabrera, 1961:578; name combination.

Cavia aperea guianae: Hückinghaus, 1961a:58; part (not *guianae* Thomas).

Cavia aperea anolaimae: O. J. Linares, 1998:209; first use of current name combination.

This subspecies is restricted to high elevations (2,400 to 4,000 m) of the Cordillera Oriental in the departments of Boyacá, Cundinamarca, and Santander, Colombia, and the Mérida Andes and Cordillera Central (600 to 1,200 m) in western Venezuela (O. J. Linares 1998; Zúñiga 2000).

Hückinghaus (1961a) stated that *anolaimae* J. A. Allen was likely identical to *guianae* Thomas, and Cabrera (1961) placed it in the species *C. porcellus*. However, some authors (Zúñiga 2000; Zúñiga et al. 2002) suggested that *anolaimae* J. A. Allen, *guianae* Thomas (= *aperea* Erxleben), and *porcellus* Linnaeus were each distinct species based on karyotypic and morphologic characters. Allen's *anolaimae* is included here as a subspecies of *C. aperea* as molecular analyses placed it solidly within the *aperea* complex; however, its distinction from *C. a. guianae* was not supported and merits further investigation (Dunnum and Salazar-Bravo 2010a).

It is possible that the holotype was collected in the Florida swamps in the Engativá region of the Bogotá savanna and that J. A. Allen (1916a) mistakenly deduced that the holotype came from La Florida, a train station in the town of Anolaima (Zúñiga et al. 2002:118). This possibility is supported by the fact that no records other than the type series are known from the temperate region on the eastern flank of the Cordillera.

C. a. aperea Erxleben, 1777

SYNONYMS:

[*Cavia*] *aperea* Erxleben, 1777:348; type locality "Brasiliae"; restricted to Pernambuco, Brazil (Cabrera 1961).

Calva aperea: Gmelin, 1788:122; name combination.

Agouti aperea: Daudin, 1802:166; name combination.

Hydrochaerus aperea: F. Cuvier, 1817a:18; name combination.

S[*avia*]. *aperea*: Olfers, 1818:214; name combination.

Anoéma hilaria É. Geoffroy St.-Hilaire, 1820:2, text to plate 282; *nomen nudum*.

Cavia leucopyga Brandt, 1835:436; type locality "Patria. Brasilia."

Anoema Aperea: Lund, 1838:48; name combination.

Cavia gracilis Lund, 1839b:208; *nomen nudum* (a fossil).

Cavia apereoides Lund, 1840b (1841c:294, plate 25, Fig. 16): type locality "Rio das Velhas's Floddal" (Lund 1841c: 292, 294), Lagoa Santa, Minas Gerais, Brazil (a fossil).

Cavia robusta Lund, 1842:14; *nomen nudum*.

Prea obscura: Liais, 1872:541; part; name combination.

[*Cavia*] *porcellus*: E.-L. Trouessart, 1897:637; part; name combination.

[*Cavia*] *aperea aperea*: Tate, 1935:343; first use of current name combination.

The nominotypical subspecies occurs in eastern Brazil; its western limits are unclear.

Tate (1935:342) stated that true *Cavia azarae* Wagner (1844:63, footnote, based upon *Dasyprocta azarae* Lichtenstein) is likely a synonym of *C. a. aperea*. However, Ximénez (1980:152–153) argued that Lichtenstein's specimen was referable to *Dasyprocta*, not *Cavia*. Massoia (1973b) supported the subspecific status of *C. a. aperea* in comparisons with *C. a. pamparum* using external and cranial measurements. Maia (1984) reported a $2n = 64$, FN = 116, karyotype from Pernambuco composed of 27 pairs of biarmed and four pairs of acrocentric chromosomes; the X chromosome was submetacentric.

C. a. guianae Thomas, 1901

SYNONYMS:

[*Cavia*] *porcellus*: E.-L. Trouessart, 1897:637; part; name combination.

Cavia porcellus guianae Thomas, 1901e:152; type locality "Four skins of different ages from the Kanuku Mountains, 7th and 8th December, 600 feet, and one from Berbice, on the coast." Restricted to "Kanuku Mountains, 600 feet" by Husson (1978) as the holotype originates from that locality; Upper Takutu–Upper Essequibo, Guyana.

C[*avia*]. *leucopyga*: Cabanis, 1848:780; not *Cavia leucopyga* Brandt; name combination.

[*Cavia rufescens*] *guianae*: E.-L. Trouessart, 1904:525; name combination.

Cavia porcellus venezuelae J. A. Allen, 1911:250; type locality "Altagracia, Immataca district, Venezuela," Bolívar.

C[*avia*]. *rufescens venezuelae*: Osgood, 1915:195; name combination.

Cavia guianae: Thomas, 1917a:153; name combination.

Cavia aperea guianae: Hückinghaus, 1961a:58; first use of current name combination.

Cavia guianae caripensis Ojasti, 1964:148; type locality "Sabana de Piedras, Caripe, Estado Monagas, Venezuela; 1100 m alt."

Cavia aperea caripensis: O. J. Linares, 1998:209; name combination.

This subspecies occurs north of the Amazon Basin in Surinam, the Guianas, Venezuela, and lowland Colombia.

In his original description, Thomas (1901e:152) considered *guianae* a subspecies of *C. porcellus*, but he elevated it to full species in his 1917 revision of *Cavia*. Cabrera (1961) included it in *C. porcellus*. Husson (1978:451) stated "since the relationship between wild and domestic forms is not quite clear, I prefer to reserve the name *C. porcellus* (Linnaeus, 1758) for the domestic form and to treat the wild forms as distinct species." He followed Hückinghaus (1961a), who considered the Guianan form to be a subspecies of *C. aperea* Erxleben. Woods (1993) also followed Hückinghaus (1961a:58); however, Corbet and Hill (1991:201) listed *guianae* Thomas as a distinct species. Nowak (1999) wrote "information presented by Woods (1993) and Eisenberg (1989) suggests that the populations of *C. aperea* in Ecuador and Surinam, as well as the entire species *C. anolaimae* and *C. guianae*, may represent feral domestic guinea pigs." Zúñiga (2000) and Zúñiga et al. (2002) compared populations of Colombian *Cavia* and recognized *guianae* as distinct from *anolaimae* J. A. Allen and *porcellus* Linnaeus, based on karyological and morphologic data. Phylogenetic analyses based on molecular data (Dunnum 2003; Dunnum and Salazar-Bravo 2010a) supported placement within the *aperea* complex.

C. a. hypoleuca Cabrera, 1953
SYNONYMS:
Cavia aperea hypoleuca Cabrera, 1953:58; type locality "Paso de la Patria, Corrientes," Argentina.
Cavia cobaya: Moreau de Saint Mery, in Azara, 1801:65; name combination, but not *cobaya* Pallas.
Cavia aperea: Rengger, 1830:274; name combination, but not *aperea* Erxleben.
[*Cavia*] *leucopyga*: E.-L. Trouessart, 1880b:194; part; name combination.
[*Cavia*] *porcellus*: E.-L. Trouessart, 1897:637; part; name combination.
[*Cavia*] *rufescens*: E.-L. Trouessart, 1904:525; part; name combination.
Cavia azarae: Thomas, 1901g:534; name combination, but not *azarae* Lichtenstein [Dasyproctidae].
Cavia porcellus aperea: Bertoni, 1914a:74; name combination.
Cavia rufescens pamparum: J. A. Allen, 1916d:567; name combination, but not *pamparum* Thomas.
Cavia aperea azarae: Thomas, 1917a:154; name combination.

This subspecies occurs in Paraguay and northern Argentina.

Thomas (1901g, 1917a), in referring to Paraguayan specimens, erroneously assumed that *C. azarae* Wagner was based upon the Paraguayan "*Aperea*" of Azara (1801). Cabrera (1953) thus placed *C. azarae*, in the sense of Thomas (1901g), in synonymy of *C. a. hypoleuca*. See Ximénez (1980:152–153) and Tate (1935:343) for more discussion on *C. azarae*. Ximénez (1980:152) could find no differences between *C. a. hypoleuca* and *C. a. pamparum*. *Cavia a. hypoleuca* forms a clade with the lowland Bolivian *C. a. aperea*, which in turn is sister to *C. a. pamparum* (Dunnum and Salazar-Bravo 2010a).

C. a. nana (Thomas, 1917)
SYNONYMS:
Cavia nana Thomas, 1917a:158; type locality "Chulumani, Yungas, 2000 m," La Paz, Bolivia.
Cavia aperea sodalis: Hückinghaus, 1961a:57; part; name combination.
Cavia tschudii sodalis: Cabrera, 1961:579; part; name combination.
Cavia tschudii nana: S. Anderson, 1997:480; part; name combination.
Cavia tschudi nana: Salazar-Bravo, Tarifa, Aguirre, Yensen, and Yates, 2003:10; part.
C[*avia*]. *a*[*perea*]. *nana*: Dunnum and Salazar-Bravo, 2010a: 5; first use of current name combination.

This subspecies occurs in the Yungas of central Bolivia along the eastern slope of the Andes. The northern and southern extent of its distribution is unresolved.

Cabrera (1961) recognized *nana* as a valid species, as did Corbet and Hill (1991) and Novak (1999). Hückinghaus (1961a) believed *nana* and *C. t. sodalis* to be synonymous (as did Cabrera 1953) and included both under the name *C. aperea sodalis*. Woods (1993) followed Hückinghaus (1961a) and included *nana* in *C. aperea*. Most recently, Woods and Kilpatrick (2005) included *nana* in the synonymy of *tschudii*. However, molecular (Dunnum 2003; Dunnum and Salazar-Bravo 2010a) and karyological ($2n = 64$, FN = 114; Dunnum and Salazar-Bravo 2006) data supported inclusion within *aperea*. Dunnum and Salazar-Bravo (2010a) found this Bolivian Andean subspecies to be sister to the *aperea* clade north of the Amazon as opposed to the lowland Bolivian clade, suggesting a possible prior Andean corridor linking the disjunct populations of *aperea* north and south of the Amazon.

C. a. pamparum (Thomas, 1901)
SYNONYMS:
Cavia cobaia Waterhouse, 1839:89; part; not *C. cobaya* Pallas, 1766; new name for *aperea* Erxleben.
Cavia aperea: Waterhouse, 1848:185; part; name combination.
Cavia leucopyga: Burmeister, 1861:424; name combination, but not *Cavia leucopyga* Brandt.
Cavia australis: Arechavaleta, 1882; name combination, but not *Cavia australis* I. Geoffroy St.-Hilaire and d'Orbigny.

[*Cavia*] *porcellus*: E.-L. Trouessart, 1897:637; part; name combination; not *Mus porcellus* Linnaeus.

Cavia porcellla: Ameghino, 1898:183; name combination and incorrect gender agreement of *porcellus* Linnaeus; not *Mus porcellus* Linnaeus.

Cavia rufescens pamparum Thomas, 1901h:538; type locality "Goyas [= Goya], Corrientes," Argentina.

Cavia pamparum: Thomas, 1917a:155; name combination.

Cavia aperea pamparum: Hückinghaus, 1961a:57; first use of current name combination.

This subspecies occurs in eastern Argentina, Uruguay, and southern Brazil.

Cabrera (1961) recognized *C. pamparum* as a valid species while Hückinghaus (1961b) included it as a subspecies of *C. aperea*. Massoia and Fornes (1967a) revalidated the synonymy of *C. pamparum* Thomas and *C. aperea* Erxleben. Molecular sequence data support its subspecies status within *aperea* and a sister relationship with the lowland Bolivian *aperea* and *C. a. hypoleuca* of Paraguay (Dunnum and Salazar-Bravo 2010a). Various authors (e.g., Massoia 1973b; Ximénez 1980) supported the subspecific designation of *C. a. pamparum* and provided distribution maps. J. R. Contreras (1972, 1980) described distributional range extensions and new records. W. George et al. (1972) and Pantaleão (1978) reported identical karyotypes ($2n = 64$, FN = 124) from Entre Rios, Argentina, and Paraná and Rio de Janeiro, Brazil.

C. a. rosida Thomas, 1917

SYNONYMS:

Cavia rosida Thomas, 1917a:154; type locality "Roca Nova. Alt. 1000 m," Serra do Mar, Paraná, Brazil.

Cavia aperea rosida: Cabrera, 1961:577; first use of current name combination.

This subspecies is known only from the Serra do Mar in eastern Paraná state, Brazil.

NATURAL HISTORY: *Cavia aperea* are primarily grazers but will eat grass inflorescences as well. Individuals typically forage in open areas close to shelter. They do not excavate burrows but make interconnecting surface tunnels and runs in the vegetation (Eisenberg and Redford 1999). In Parque Nacional Noel Kempff Mercado [NKM, Santa Cruz, Bolivia], cavys were collected in both Sherman and Tomahawk traps along clear runs and tunnels in the grasses of the wet savannas (grasses about 2 feet high) near "Los Fierros" (J. Dunnum, pers. obs.). Emmons et al. (2006) reported that individuals are found in the upland savannas and grasslands of NKM, which experience brief, shallow flooding, and that they constitute a main prey item of the Maned wolf (*Chrysocyon brachyurus)*. In the cerrado of Brazil, *C. aperea* occupied the gallery forest and flooded areas near forest and foraged in the grassy fields in the early evening (Alho et al. 1987).

Individuals of *Cavia a. anolaimae* are diurnal, active primarily in the mornings and evenings. It apparently reproduces year-round as pregnant females and reproductive males have been collected through the year. Females averaged one to three embryos, but all animals close to term except one had a single embryo, the other embryos likely having been reabsorbed. Two pairs of inguinal mammae are present. Avian predators include Black-crested Buzzard Eagle (*Geranoaetus melanoleucus*) and hawks of the genus *Buteo*; known mammalian predators are *Cerdocyon thous*, *Mustela frenata*, *Nasua nasua*, domestic dogs, and man.

At the type locality of *C. a. nana*, I observed Pygmy cavies in the early mornings and evenings grazing in grassy areas. Well-developed runs were evident going from grassy areas into dense shrubby vegetation. I never observed cavies venturing far from protected areas, and they retreated quickly when disturbed.

Cavia aperea has a polygynous social and mating system, with linear dominance hierarchies within each sex (Rood 1972; Sachser et al. 1999). Asher et al. (2004) investigated a wild population exhibiting a system of single males with one to two females occupying stable home ranges (880 ± 217 m^2 for males and 549 ± 218 m^2 for females) based on resource availability. In the moist savannas of the Atlantic coast in Rocha department, Uruguay, densities ranged from 27–32 adults/ha. Male home ranges overlapped multiple female ranges. Strong sexual dimorphism existed in the investigated cavy population. Adult males were 17% heavier overall than adult, nonpregnant females. Resident males were 32% heavier than females, 18% heavier than roaming males, and 56% heavier than satellite males. Roaming and satellite males did not differ in weight from females (Asher et al. 2008). Asher et al. (2004) and Rood (1972) reported population densities that ranged from 12.5 individuals/ha in Brazil to 38 individuals/ha in Argentina. Activity periods were highest between 07:00–11:00 and 17:30–20:00 hours. This species exhibits strong sexual dimorphism with males 11–14% heavier than females (Sachser 1998; Asher et al. 2004). Rood (1972) reported individual swimming ability of several kilometers during floods. Captive populations derived from Argentinean stock exhibited a wide acoustic repertoire of ten different call types (Monticelli and Ades 2013).

REMARKS: The specific name *aperea* Erxleben is conserved for this wild form although it is predated by *porcellus* Linnaeus, which is based on the domestic form (ICZN 2003). Furthermore, molecular sequence data do not support *C. aperea* as the source of domestication (Dunnum 2003; Spotorno, Valladares, Marín, and Zeballos 2004; Dunnum and Salazar-Bravo 2010a). Hadler et al. (2008) recorded Mid to Late Holocene fossils from Rio Grande do Sul state, Brazil.

Karyotypes consist of $2n = 64$, with reported FN = 96 (Zúñiga 2000, for individuals of the nominotypical subspe-

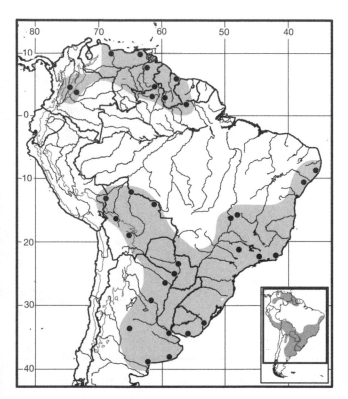

Map 370 Selected localities for *Cavia aperea* (●). Contour line = 2,000 m.

cies), 114 (Dunnum and Salazar-Bravo 2006), 116 (Maia 1984), and 124 (W. George et al. 1972; Gava et al. 1998).

Cavia fulgida Wagler, 1831
Shiny Guinea Pig

SYNONYMS:

Cavia fulgida Wagler, 1831a:511 footnote; type locality "Herr von Spix von seiner Reise am Amazonenstrom zurückbrachte," (likely an error as the species does not occur in the Amazon Basin; see Thomas 1917a:159; Cabrera 1961:577).

Cavia obscura Lichtenstein, 1823:3; *nomen nudum.*

Cavia rufescens Lund, 1839b:206; type locality "bassin du Rio das Velhas," Lagoa Santa, Minas Gerais, Brazil.

C[*avia*]. *nigricans* Wagner, 1844:64; type locality "Als Heimath ist Brasilien angeben"; a redescription of *C. obscura* Lichtenstein.

Kerodon obscurus: Wagner, 1844:64; name combination.

C[*avia*]. *aperea*: Giebel, 1855:459; name combination.

Prea obscura: Lias, 1872:541; part; name combination.

Prea rufescens: Liais, 1872:541; name combination.

Prea fulgida: Liais, 1872:541; name combination.

[*Cavia fulgida*] *rufescens*: E.-L. Trouessart, 1880b:194; name combination.

DESCRIPTION: Dorsal color rich grizzled reddish-brown, ventral color deep buffy or ochraceous. Relatively

small cavy, with greatest length of skull rarely exceeding 60 mm. Only species in genus diagnosable by consistent dental character: deep notch on outer (labial) side of anterior end of posterior lobe of M3 (Thomas 1917a).

DISTRIBUTION: *Cavia fulgida* is known from the eastern coast of Brazil from Lagoa Santa, Minas Gerais, to Santa Catarina (Thomas 1917a; Ximénez 1980; Woods 1993).

SELECTED LOCALITIES (Map 371): BRAZIL: Bahia, Caravelas (Ximénez 1980), Vitoria da Conquista (Ximénez 1980); Espírito Santo, Santa Teresa (Ximénez 1980); Minas Gerais, Alem Paraíba (Ximénez 1980), Parque Estadual do Rio Doce (Stallings 1989), Poços de Caldas (Cherem et al. 1999); Paraná, Estação de Roça Nova (Cherem et al. 1999); Santa Catarina, Rancho Queimado (Cherem et al. 1999); São Paulo, Ubatuba (Ximénez 1980).

SUBSPECIES: *Cavia fulgida* is monotypic.

NATURAL HISTORY: Eisenberg and Redford (1999) stated that this species does not overlap in range with *C. magna* but is sympatric with *C. aperea.* Myers and Hanson (1980) reported their presence along with *Oxymycterus roberti* and *Necromys lasiurus* in the Parque Nacional de Brasília in flooded (3–4 cm deep) marshland dominated by perennial bunch grass of 53 cm in height. This species is the prey of Crab-eating raccoons, *Procyon cancrivorus*, in the coastal plains of Espírito Santo state (Gatti et al. 2006).

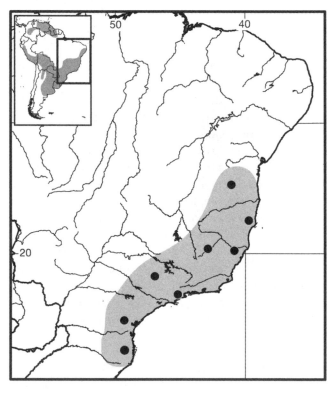

Map 371 Selected localities for *Cavia fulgida* (●).

REMARKS: Lund (1839b) described *Cavia rufescens* among the extant ("*vivans*") mammals of the Lagoa Santa region as a small and reddish ("*plus petit, rougeâtre*") species; he subsequently (1841c, plate 25, Fig. 15) illustrated the last upper right molar. These characteristics match those of Wagler's *C. fulgida*, described eight years earlier.

Pantaleão (1978) reported a karyotype of $2n = 64$, FN = 124.

Cavia intermedia Cherem, Olimpio, and Ximénez, 1999
Intermediate Guinea Pig

SYNONYMS:

Cavia aff. *magna*: Gava, Freitas, and Olimpio, 1998:77; not *magna* Ximénez.

Cavia intermedia Cherem, Olimpio, and Ximénez, 1999: 100; type locality "A major das três ilhas que formam o Arquipélago de Moleques do Sul, a 27°51'S, e 48°26'W," Santa Catarina, Brazil.

DESCRIPTION: *Cavia intermedia* intermediate color and size between *C. aperea* and *C. magna*. Dorsal pelage gray basally and tipped with bands of black, yellow, or both; darker medial line runs from head to rump; ventral pelage gray basally with dull yellow tips; white patch present in throat area. Enlarged interdigital membranes present on hindfeet, as in *C. magna*. Cranially *C. intermedia* differs from other Brazilian *Cavia* by flat sagittal crest, large occipital condyles, broad foramen magnum, shallow labial notch on second prism of upper molars, prisms of lower premolar of similar width, and first prism of last mandibular molar wider than second with small lingual notch on latter prism (Cherem et al. 1999).

DISTRIBUTION: *Cavia intermedia* is known only from the type locality.

SELECTED LOCALITIES (Map 372): BRAZIL: Santa Catarina, Ilhas Moleques do Sul (type locality of *Cavia intermedia* Cherem, Olimpio, and Ximénez).

SUBSPECIES: *Cavia intermedia* is monotypic.

NATURAL HISTORY: This is a predominantly nocturnal herbivore, a major portion of its diet consists of the grasses *Stenotaphrum secundatum* and *Paspalum vaginatum*. Four females collected in May were all lactating (Cherem et al. 1999). Salvador and Fernandez (2008a) examined population dynamics and conservation status, and suggested that the species may be one of the rarest species on the planet and should be listed as critically endangered. Females were reproductively active throughout the year. Litter size was small (one or two young per litter), with well-developed offspring that weighed approximately 19% of the mass of an adult female. Sexual maturity was reached later than in other species of this genus, at around 59 days of age or 70% of adult body size (Salvador and Fernandez 2008b).

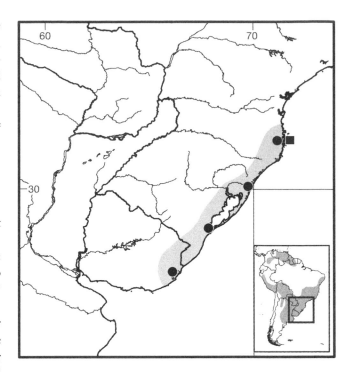

Map 372 Selected localities for *Cavia intermedia* (■) and *Cavia magna* (●).

REMARKS: Cherem et al. (1999) suggested that the species may have originated as an insular population derived from the mainland *C. magna* as little as 8,000 years ago. *Cavia intermedia* is one of two species in the genus with a diploid number other than 64 ($2n = 62$, FN = 108; see account for *C. patzelti*). The autosomal complement has 24 biarmed pairs and six acrocentric pairs. The sex chromosomes include a large metacentric X and a large acrocentric Y (Gava et al. 1998).

Cavia magna Ximénez, 1980
Greater Guinea Pig

SYNONYMS:

Cavia aperea rosida: Ximénez, 1967:1; name combination, but not *rosida* Thomas.

Cavia pamparum: Barlow, 1969:43; part; name combination.

Cavia magna Ximénez, 1980:148; type locality "En las orillas del arroyo Imbé, municipio de Tramandaí, estado de Rio Grande del Sur, Brasil."

DESCRIPTION: Largest species in genus (head and body length 310 mm, hindfoot length 52.6 mm, ear length 26.4 mm, and mass 635 g; Ximénez 1980). Dark agouti brown dorsally and reddish ventrally. Semiaquatic with enlarged interdigital membranes.

DISTRIBUTION: *Cavia magna* occurs in the coastal marshes of extreme southeastern Brazil and eastern Uruguay, from Rocha department (34°S, 54°W) in the south extending north along the littoral zone through Rio Grande do Sul and Santa Catarina states as far as the city of Circiuma (29°S, 49°W; Ximénez 1980; Eisenberg and Redford 1999; E. M. González 2001).

SELECTED LOCALITIES (Map 372): BRAZIL: Rio Grande do Sul, Ilha Marinheiros (Gava et al. 2012), Tramandaí (Cherem et al. 1999); Santa Catarina, Palhoça (Cherem et al. 1999). URUGUAY: Rocha, Parque Nacional Refugio de Fauna Laguna de Castillos (Trillmich et al. 2004).

SUBSPECIES: *Cavia magna* is monotypic.

NATURAL HISTORY: This species occupies mesic grasslands and marshes where individuals use complex networks of tunnels and runs. *Cavia magna* has a solitary social system and overlap promiscuity is the likely mating system (Kraus et al. 2003). Home ranges showed very little stability, and range size was correlated with sex, body size, and water level. The gestation period is 64 days. Reproduction occurred year-round but was concentrated in the spring and early summer (Kraus et al. 2003). Among the Caviinae, *C. magna* is particularly precocial, with individual neonates weighing on average 18% of maternal mass. Individual females produced on average three litters per year. Some females born in early spring conceived successfully between the age of 30 and 45 days (Kraus et al. 2005).

REMARKS: Pantaleão (1978) reported a karyotype from mainland Brazil ($2n = 64$, FN = 124). Gava et al. (2012) found a $2n = 62$, FN = 102, karyotype in an isolated island population. This represents the first case of intraspecific diploid number variation within the Caviidae and provides further support for the sister relationship between *C. magna* and the insular *C. intermedia*, which also has $2n = 62$. Dunnum and Salazar-Bravo (2010a) found that the *C. magna* lineage represented the first split among the living *Cavia* species, diverging around 6.7 mya. This species is known from Holocene deposits in Rio Grande do Sul state, Brazil (Hadler et al. 2008).

Cavia patzelti Schliemann, 1982
Sacha Guinea Pig
SYNONYMS:

Cavia aperea patzelti Schliemann, 1982:81; type locality "Hochkordillere von Alao, Provinz Chimborazo, Equador," Alao, Chimborazo, Ecuador.

Cavia patzelti: Dunnum and Salazar-Bravo, 2010a:5; first use of current name combination.

DESCRIPTION: General dorsal coloration of body and head dark reddish brown; hairs with pale apical bands with only short dark tips. Skull very long and wide, wider in all respects than specimens from Colombia and Peru, particularly so for rostrum and maxilla at M3. Maxillary toothrow especially long, between 16.8–18.7 mm. Head and body length 280–290 mm, hindfoot length from 46–

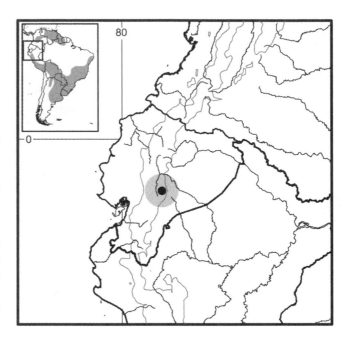

Map 373 Selected localities for *Cavia patzelti* (●). Contour line = 2,000 m.

50 mm, mean ear length 50 mm, greatest length of skull 63–69 mm, and mass 700–725 g (Schliemann 1982).

DISTRIBUTION: This species is restricted to the highlands of Chimborazo, Ecuador, where it is known only from the type locality; specimens exist only from the type locality but observational records suggest a wider distribution.

SELECTED LOCALITIES (Map 373): ECUADOR: Chimborazo, Alao (type locality of *Cavia aperea patzelti* Schliemann; see Paynter 1993).

NATURAL HISTORY: The type locality is in the páramo between 3,000–3,800 m in an area composed of wide marshy areas and steep grassy slopes; individuals were taken in the moist hollows. The vegetation is dominated by grass of the genus *Stipa*, various shrubs in the families Asteraceae (*Chuquiragua* sp.), Ericaceae, and Hypericaceae (*Hypericum* sp.) as well as *Polylepis* (Rosaceae) forest (Schliemann 1982).

REMARKS: In his description, Schliemann (1982) reported large morphological differences in comparisons with other *Cavia* taxa. However, he assigned the specific epithet *aperea* to this cavy, following Hückinghaus (1961a) who included all highland *Cavia* in *aperea*. Molecular (Dunnum and Salazar-Bravo 2010a) and karyological ($2n = 56$; Lobato and Araujo 1980) data support the species status.

Cavia porcellus (Linnaeus, 1758)
Domestic Guinea Pig
SYNONYMS:

Mus porcellus Linnaeus, 1758:59; type locality "Brasilia," based on the *Cavia cobaya* of Marcgraf (1648:224).

Cavia cobaya Pallas, 1766:31; type locality "Brazil."

[*Cavia*] *porcellus*: Erxleben, 1777:349; first use of current name combination.

Prea cobaya: Liais, 1872:541; name combination.

Lepus Minimus G. I. Molina, 1782:306; but see Tate (1935: 344), who believed this taxon to possibly be a *Galea*, and Ellerman (1940:243), who listed it as a valid species of *Galea*.

Calva cobaya: Gmelin, 1788:122; name combination.

[*Scavia*] *porcellus*: Blumenbach, 1797:80; name combination.

S[*avia*]. *cobaya*: Treviranus, 1802:211; name combination.

Agouti cobaya: Daudin, 1802:166; name combination.

Hydrochoerus cobaya: É. Geoffroy St.-Hilaire, 1803:163; name combination.

Hydrochaerus cobaya: G. Fischer, 1814:593; name combination.

Savia Cavia porcellus: Oken, 1816:824; name combination, but unavailable (ICZN 1956).

Coiza porcellus: Billberg, 1827:45; name combination.

[*Cavia aperea*] *porcellus*: J. B. Fischer, 1829:383; name combination.

Anoema Cobaya: Poeppig, in Froriep, 1829:279, footnote; name combination.

Cavia cutleri Bennett, 1836:191; type locality "Chile."

Cavia cobaia: Poeppig, 1836:45; name combination.

[*Cavia cobaya*] *alba* Fitzinger, 1867b:155; type locality "Eur," *Cavia porcellus* var. *alba* Erxleben, introduced.

Cavia cobaya helvola Fitzinger, 1867b:155; type locality "Eur," introduced.

[*Cavia cobaya*] *nigra* Fitzinger, 1867b:155; type locality "Eur," introduced.

Cavia cobaya longipilis Fitzinger, 1880:431; type locality "Japan," introduced.

Prea cobaya: Liais, 1872:541; name combination.

[*Cavia cutleri*] *cobaya*: E.-L. Trouessart, 1897:638; name combination, but not *cutleri* Tschudi.

[*Cavia*] *brasiliensis*: E.-L. Trouessart, 1897:637; name combination, republication of invalid pre-Linnaean name (ICZN 1910)

Mamcaviaus cobaya: A. L. Herrera, 1899:13; name combination.

Cavia porcellus porcellus: Cabrera, 1961:578; name combination.

DESCRIPTION: General characteristics as for genus but wide range of colors, hair lengths, and sizes due to breeding for specific traits. Pelage ranges from brown, tan, black, reddish, white, or combinations of those colors. Mass reaches 1,200 g, and overall length ranges 200–500 mm.

DISTRIBUTION: Worldwide as a domestic animal (not mapped).

NATURAL HISTORY: Domesticated in pre-Columbian times by Amerindian peoples (Wing 1986), *Cavia porcel-lus* now occupies human inhabited areas throughout South America and still serves an important role in native religion, folk medicine, and as a food source in Andean countries. This species is domesticated worldwide as a pet and laboratory animal. The domestication process was likely the result of a single ancient event in the western Andes, followed by multiple modern events from the sixteenth century on leading to the various strains seen today (Spotorno, Valladares, Marín, and Zeballos 2004; Spotorno et al. 2006, 2007). Pigière et al. (2012) reported on archaeological finds and radiocarbon dating of guinea pigs from Europe, suggesting multiple importations by Spanish and Dutch during the 1500s. Three currently recognized wild forms (*C. aperea*, *C. tschudii*, and *C. fulgida*) have at one time been proposed as the possible ancestral form. The characteristics of the M3 (Hückinghaus 1961b), production of infertile males from *porcellus* × *fulgida* crosses (Detlefson 1914), and molecular data (Dunnum 2003; Dunnum and Salazar-Bravo 2010a) are sufficient to eliminate *C. fulgida* as a likely ancestor. Fertile hybrids have been produced in crosses of *porcellus* to both *tschudii* (Weir 1974c) and *aperea* (Trillmich et al. 2004), but molecular data place *porcellus* solidly within *C. tschudii* (Dunnum 2003; Spotorno, Valladares, Marín, and Zeballos 2004; Dunnum and Salazar-Bravo 2010a). Spotorno et al. (2007) suggested two potential points of domestication: southern Peru and the highlands of Colombia. They also noted that examination of archaeological material from Tequendama, near Bogota in Colombia, was warranted as another potential region for initial domestication. Analyses including all potential progenitors from a broad geographic distribution supported an initial derivation from *C. tschudii tschudii* populations in the region of Ica, Peru (Dunnum and Salazar-Bravo 2010a), an area where the oldest archaeological site containing guinea pig remains (Ayamachay) is located (Sandweiss and Wing 1997).

Sachser (1986) found long-lasting relationships between individual males and females and a polygynous mating system, with resource availability critical in defining social organization.

REMARKS: Thomas (1901g:533) restricted *porcellus* Linnaeus to include only the domestic form. Subsequently, the International Commission (ICZN 2003) conserved the name *C. porcellus* for the domestic form, although incorrectly assuming *C. aperea* to be the wild progenitor. The domesticated form is included here as a species to maintain taxonomic stability, although it is clear that it is derived from *C. tschudii*. The type locality "Brasilia" for the domestic form is certainly in error (Cabrera 1961), and no type specimen exists. Ellerman (1940) included the fossil taxa *C. gracilis* (Lund, 1839b), *C. apereoides* (Lund, 1840b [1841c]), and *C. robusta* (Lund, 1842), all from Lagoa

Santa, Brazil (the first and third being *nomina nuda*), in the synonymy of *porcellus* Linnaeus. Cabrera (1961:578) suggested that the evidence for this placement was tentative at best, and our current knowledge of the group would suggest these are not referable to the domestic *C. porcellus* or its progenitor *C. tschudii* but likely to *C. aperea*. The correct assignment of Molina's Cuy, *Lepus minimus*, is in doubt. No type specimen exists, and the description is inadequate for an unambiguous decision. Tate (1935:344) regarded it as "apparently a cavy, either of the genus *Cavia* or *Galea*, and more probably the latter." Ellerman (1940) listed it as a valid species of *Galea*. But Cabrera (1961) listed it as a synonym of *Cavia p. porcellus*. I follow Cabrera.

The karyotype consists of $2n = 64$, FN $= 90, 94$ (Awa et al. 1959; Cohen and Pinsky 1966).

Cavia tschudii Fitzinger, 1867
Montane Guinea Pig

SYNONYMS: See under Subspecies.

DESCRIPTION: Body size and dorsal pelage color vary greatly geographically. The Peruvian forms vary from grizzled cinnamon in *C. t. tschudii* and *C. t. festina*, to dark reddish brown in *C. t. osgoodi*, dark olive brown in *C. t. stolida*, and grizzled gray to blackish in *C. t. umbrata*. Subspecies *C. t. atahualpae* large and dark, with upper parts grizzled cinnamon and blackish, bases of hair broadly dark drab, followed by two or more cinnamon and blackish annulations, sides and lateral parts only slightly paler than back, throat mixed cinnamon and blackish, not different from dorsal area, chin and submaxillary region buffy, forefeet and hindfeet grizzled pale drab, ears thinly haired blackish and not contrasting with surrounding parts, and without a definite eye-ring (Osgood 1913). Specimens assigned to *C. t. sodalis* from Argentina and Bolivia much paler brown; ventral coloration v whitish to gray to buffy. Skull length 59–69 mm (Thomas 1917a, 1926d, 1926g, 1927f; Sanborn 1949a). In *C. t. festina* chin, chest, and venter ochraceous buffy, and collar below neck brown and well marked. Dimensions of holotype head and body length 272 mm, hindfoot length 47 mm, ear length 26 mm, and greatest skull length 69 mm (Thomas 1927f). Venter of *C. t. sodalis* decidedly paler than other subspecies, being almost white. Dimensions of holotype head and body length 218 m, hindfoot length 36.5 mm, ear length 35 mm, and greatest skull length 55 mm (Thomas 1926d). In *C. t. stolida*, venter dull soiled buffy color; feet grizzled brown with lighter digits. Skull broad and heavy, nasals particularly broad, exceeding that of all other forms. Dimensions of holotype head and body length 313 mm, hindfoot length 49 mm, ear length 29 m, greatest skull length 63 m, zygomatic breadth 39.3 mm, interorbital breadth 17 mm, and

breadth of braincase 25.8 mm (Thomas 1926g). Individuals of *C. t. umbrata* differ from other subspecies by grayer color throughout with rings on hairs whitish instead of cinnamon or buffy; median area of back blackish due to reduction of light-colored rings to about 1 mm in length as opposed to 3–4 mm in other subspecies; venter soiled drab, collar and middle line of chin grayish brown; feet pale brown. Dimensions of holotype hindfoot length 42 mm, ear length 20 mm, and greatest skull length 60 mm (Thomas 1917a). Nominotypical subspecies fairly dark, strongly grizzled dorsally with light rings on hairs being buffy or cinnamon; median area of back not darkened; venter strongly buffy. Skull length 59–61 mm (Thomas 1917a). Subspecies *pallidior* similar to nominotypical subspecies but color much lighter, pale rings on hairs paler buffy, and venter pale creamy buff approaching whitish; collar gray and hands and feet buffy whitish and slightly browner proximally. Dimensions of holotype head and body length 242 mm, hindfoot length 24 mm, ear length 29 mm, and greatest skull length 59.5 mm (Thomas 1917a).

DISTRIBUTION: *Cavia tschudii* is known from the highlands and Pacific coast of Peru, south through the Altiplano of Bolivia, northern Chile, and the Andes of northern Argentina, across an elevational range from sea level to 4,500 m.

SELECTED LOCALITIES (Map 374): ARGENTINA: Catamarca, Highway 4, 10 km S of El Rodeo (P. E. Ortiz and Jayat 2012a); Tucumán, Concepción (type locality of *Cavia tschudii sodalis* Thomas). BOLIVIA: La Paz, Río Desaguadero (S. Anderson 1997); Santa Cruz, 3 km N by road of Torrecillas (S. Anderson 1997); Tarija, Entre Rios (S. Anderson 1997). CHILE: Arica y Parinacota, Valle de Azapa. PERU: Amazonas, Río Utcubamba, about 15 mi S of Chachapoyas (type locality of *Cavia tschudii stolida* Thomas); Arequipa, Tambo (Thomas 1917a); Cajamarca, Cajamarca (type locality of *Cavia tschudii atahualpae* Osgood); Cusco, Urubamba (Thomas 1917a); Ica, 70 mi E of Pisco (type locality of *Cavia tschudii* Fitzinger); Junín, Incapirca, Zezioro (type locality of *Cavia tschudii umbrata* Thomas); La Libertad, Mashua, east of Tayabamba on trail to Ongón (LSUMZ 24845); Puno, Limbani (type locality of *Cavia tschudii osgoodi* Sanborn).

SUBSPECIES: I tentatively recognize eight subspecies within *C. tschudii* pending further revision.

C. t. atahualpae Osgood, 1913
SYNONYMS:

Cavia atahualpae Osgood, 1913:98; type locality "Cajamarca, [Cajamarca,] Peru. Alt. 9100 ft."

Cavia tschudii atahualpae: Thomas, 1917a:156; first use of current name combination.

Cavia aperea tschudii: Hückinghaus, 1961a; part; name combination.

This subspecies occurs in the interandean valleys of northern Peru.

Cabrera (1961) included *stolida* Thomas in the synonymy of *C. t. atahualpae*, a taxon I regard as a valid subspecies herein.

C. t. festina Thomas, 1927

SYNONYMS:

Cavia tschudii festina Thomas, 1927f:604; type locality "Huariaca, Junín, 9000'," Peru.

Cavia tschudii umbrata: Cabrera, 1961:580; name combination.

Cavia aperea festina: Hückinghaus, 1961a:57; name combination.

This subspecies occurs in the Andes of central Peru, in Junín department.

Woods and Kilpatrick (2005) included *umbrata* in the nominotypical subspecies. Cabrera (1961) considered *festina* Thomas a synonym of *C. t. umbrata* Thomas.

C. t. osgoodi Sanborn, 1949

SYNONYMS:

Cavia tschudii osgoodi Sanborn, 1949a:133; type locality "Limbani, 12,000 feet, District Limbani, Province Sandia, Department Puno, Peru."

C. aperea osgoodi: Hückinghaus, 1961a:57; name combination.

This subspecies is known from the Altiplano of southern Peru, north of Lake Titicaca.

C. t. pallidior Thomas, 1917

SYNONYMS:

Cavia tschudii pallidior Thomas, 1917a:158; type locality "Arequipa, 2500 m," Peru; not *Kerodon niata pallidior* Thomas [= *Cavia (Monticavia) niata pallidior*, = *Microcavia niata*; see Osgood (1919:34) and *Microcavia* account].

Cavia tschudii arequipae Osgood, 1919:34; new name for *C. t. pallidior* Thomas.

Cavia tschudii tschudii: Sanborn, 1949a:133; name combination.

Cavia tschudii nana: S. Anderson, 1997:479; part; name combination.

This subspecies occurs from lowland northern Chile and southern Peru into the highlands of Bolivia.

Osgood (1919:34) proposed *arequipae* as a replacement name for *pallidior* Thomas on the grounds that "if *Monticavia* be regarded as no more than a subgenus of *Cavia*, then *C. t. pallidior* Thomas, 1917 is preoccupied

by *K[erodon]. N[iata]. pallidior* Thomas, 1902, and requires a new name. Mr. Thomas, who proposed both names, treats *Monticavia* as a full genus and therefore would make no change." *Monticavia* is now regarded as a synonym of *Microcavia* (C. A. Quintana 1996; Woods and Kilpatrick 2005), thus validating Thomas's original proposal and making Osgood's *arequipae* an objective synonym of *pallidior* Thomas. Sanborn (1949a:133) synonymized *pallidior* Thomas (renamed *arequipae* by Osgood) with *C. t. tschudii* while Woods and Kilpatrick (2005) regarded *arequipae*, but not *pallidior*, as a valid subspecies. S. Anderson (1997) incorrectly referred all Bolivian specimens to *C. tschudii nana* (= *Cavia nana* Thomas, which herein is regarded as a valid subspecies of *C. aperea*; see account of that species). The specimens that Anderson examined, therefore, comprise two separate species. Pine et al. (1979) referred Chilean specimens to the nominotypical subspecies. However, Chilean specimens from Valle de Lluta formed a clade with those from Arequipa, Peru, and highland Santa Cruz, Bolivia (Dunnum and Salazar-Bravo 2010a).

C. t. sodalis Thomas, 1926

SYNONYMS:

Cavia tschudii sodalis Thomas, 1926d:607; type locality "Ñorco, Vipos, 2500 m," Tucumán, Argentina.

Cavia tschudii pallidior: Thomas, 1925b:580; name combination, but not *pallidior* Thomas.

Cavia aperea sodalis: Hückinghaus, 1961a:57; name combination.

This subspecies minimally occurs in the higher elevations of Tucumán and Jujuy provinces in northern Argentina, and historically occurred in southwest Salta, Argentina (Tonni 1984). The taxonomic status of *Cavia* populations in Tarija department in extreme southern Bolivia is in need of further investigation.

C. t. stolida Thomas, 1926

SYNONYMS:

Cavia tschudii stolida Thomas, 1926g:166; type locality "Rio Utcubamba, 8500'," about 15 miles south of Chachapoyas (Thomas 1926g:157), Amazonas, Peru.

Cavia stolida: Hückinghaus, 1961a:58; name combination.

Cavia tschudii atahualpae: Cabrera, 1961:579; part (*stolida* Thomas listed as synonym of *atahualpae* Osgood).

This subspecies is known from the Utcubamba Valley in northwestern Peru.

Cabrera (1961) synonymized *stolida* Thomas with *atahualpae* Osgood. Hückinghaus (1961a) considered *stolida* a valid species. Woods (1993) included it in the synonymy of *C. tschudii*.

C. t. tschudii Fitzinger, 1867

SYNONYMS:

Cavia cutleri Tschudi, 1845:195; type locality "in der Umgegend der Stadt Yca," Ica, Peru. [Tschudi incorrectly assumed this form to be the *C. cutleri* that King collected (Bennett 1836), which was actually the domestic form, *C. porcellus* (Thomas 1917a)].

[*Cavia*] *Tschudii* Fitzinger, 1867b:154 (pg. 98 in separates); new name for *C. cutleri* Tschudi, 1845, type locality "Ica, 70 miles E of Pisco, Peru."

C[*avia*]. *leucopyga*: Giebel, 1855:461; part; name combination.

[*Cavia leucopyga*] *Tschudii*: E.-L. Trouessart, 1880b:195; name combination.

Cavia tschudii tschudii: Thomas, 1917a:157; first use of current name combination.

Cavia aperea tschudii: Hückinghaus, 1961a:57; name combination.

Dunnum and Salazar-Bravo (2010a) restricted the nominotypical subspecies to populations in the Ica region of Peru. This leaves populations from the highlands of Cusco department without an available name and removes the northwestern Peruvian *C. t. umbrata* Thomas (type locality "Incapirca, Zezioro," Junín, Peru) and *C. t. atahualpae* Osgood (type locality "Cajamarca, Peru. Alt. 9100 ft.") from the synonymy of the nominotypical species as included by Woods and Kilpatrick (2005). Cabrera (1961) recognized both as valid subspecies, including *festina* in the synonymy of *umbrata*. I refrain from formally recognizing these subspecies pending resolution of the taxonomy of cavys in northwestern Peru.

Thomas (1917a:156) discussed *C. cutleri* Bennett, originally referred to as the "Peruvian cavy," and concluded that it represented the domestic form, *C. porcellus*. Bennett's name, therefore, is removed from the synonymy of *C. tschudii*.

C. t. umbrata Thomas, 1917

SYNONYMS:

Cavia tschudii umbrata Thomas, 1917a:157; type locality "Incapirca, Zezioro," Junín, Peru.

Cavia aperea tschudii: Hückinghaus, 1961a:57; part; name combination.

This subspecies is restricted to the Andes of Junín department in central Peru.

Cabrera (1961) included *festina* Thomas in his synonymy of *C. t. umbrata*; I regard both as valid subspecies, pending further review.

NATURAL HISTORY: In northern Argentina and the Altiplano of Peru, this species occupies brush or grasslands and utilizes well-defined runways (Thomas 1926d;

Pearson 1957). In northern Chile, it occurs in the desert regions in the vicinity of rivers, the humid pampas, and from the coast up to 4,200 m, where it inhabits brush edges, pastures and alfalfa cultivation, as in the Lluta river valley (Muñoz-Pedrero 2000). In captivity, the gestation period is 63 days, with a mean litter size of 1.9 (range: 1–4); the average age of first reproduction is two months (Weir 1974b).

REMARKS: Woods (1993) incorrectly cited the date of Fitzinger's (1867b) description as 1857, and many subsequent authors have thus followed suit (S. Anderson 1997; Woods and Kilpatrick 2005; Spotorno et al. 2006). Hückinghaus (1961a) included the *C. tschudii* group (except *C. t. stolida*) in *C. aperea*. Woods (1993) followed Cabrera (1961) and recognized *C. tschudii* as a valid species. Woods and Kilpatrick (2005) recognized six subspecies (*C. t. arequipae*, *C. t. festina*, *C. t. osgoodi*, *C. t. sodalis*, *C. t. stolida*, and *C. t. tschudii*). Recent revisionary work (Dunnum and Salazar-Bravo 2010a) did not include material of *festina*

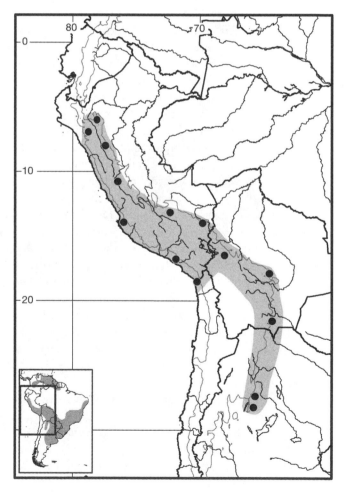

Map 374 Selected localities for *Cavia tschudii* (●). Contour line = 2,000 m.

or *stolida* but supported the status of the other four subspecies and restricted the distribution of the nominotypical subspecies. A thorough understanding of the *tschudii* group is still lacking and further work is needed, primarily on the populations in northern and western Peru (*festina* Thomas, *stolida* Thomas, with its synonym *atahualpae* Osgood and *umbrata* Thomas). *Cavia t. festina* is recognized here following Woods and Kilpatrick (2005); however, if future work shows *C. t. festina* to be a synonym of *C. t. umbrata* as suggested by Cabrera (1961), *umbrata* Thomas is the older name. Dunnum and Salazar-Bravo (2010a) followed Woods and Kilpatrick (2005) in using the name *C. t. arequipae*; however, see comments in the account of *C. t. pallidior* for reasons why *pallidior* should be applied to this taxon.

Dunnum and Salazar-Bravo (2006) reported the karyotype of a specimen from the Bolivian highlands as $2n = 64$, FN = 104–108.

Genus *Galea* Meyen, 1833

Members of the genus *Galea* are terrestrial, diurnal herbivores that inhabit grasslands and rocky scrub areas at both high and low elevations. They are largely colonial, preferring open habitat with good visibility. Females have one pair of inguinal and one pair of lateral thoracic mammae. There is no external tail. Dorsal coloration varies from olivaceous to agouti; the venter is uniformly grayish white. A prominent submandibular gland is present. The cheek teeth are less complex than in *Cavia*, with the upper teeth separated into two lobes by a single lingual fold. M3 has a weak, backwardly projecting heel. Lower cheek teeth have two lobes, and pm4 has a short anterior prolongation. The incisors are yellow rather than white, as in both *Cavia* and *Microcavia*. The orbital branch of maxilla is completely interrupted by the lacrimal (Ellerman 1940). The mean condyloincisive length of the skull is 46.5 mm (range: 41–56 mm; Hückinghaus 1961a). *Galea* is known in the fossil record only from the Pleistocene (Ubilla and Rinderknecht 2001; McKenna and Bell 1997). Opazo (2005) estimated the base of the radiation of the extant species of *Galea*, from molecular sequence data, at 5.9 ± 1.6 mya.

SYNONYMS:
Cavia: Wagler, 1831a:512; part (description of *spixii*); not *Cavia* Pallas.
Galea Meyen, 1833:597; type species *Galea musteloides* Meyen, by monotypy.
Cerodon: Lund, 1840c:313; listing of *bilobidens*; not *Cerodon* Wagler.
Kerodon: Lesson, 1842:102; part (listing of *rupestris*); not *Kerodon* F. Cuvier.

Anoema: Burmeister, 1861:424 (description of *leucoblephara*); not *Anoema* F. Cuvier.
Prea Liais, 1872:541; part (listing of *saxatilis*); replacement name for *Cavia saxatilis* Lund (= *Galea spixii* Wagler).
Gelea Woods, 1982:386; incorrect subsequent spelling of *Galea* Meyen.

REMARKS: Woods (1993) and S. Anderson (1997) incorrectly listed the date of Meyen's description, which was for the year 1832, but published in 1833. Woods and Kilpatrick (2005) listed three species in their concept of this genus (*flavidens* Brandt, *musteloides* Meyen, and *spixii* Wagler). Bezerra (2008) revised the genus based on morphological characters, and analysis largely congruent with the taxa delimited by molecular phylogenetics. Dunnum and Salazar-Bravo (2010b) revised the genus and recognized four species (*G. musteloides*, *G. leucoblephara*, *G. comes*, and an undescribed form from mid elevations [2,000-4,500 m] in Tarija department, Bolivia) within the *musteloides* group. Aditionally, they suggested that *spixii* Wagler might warrant generic status based on molecular (K2p distances for mtDNA cytochrome-*b* sequences of 19–21% between *spixii* and the *musteloides* group species) and karyological data ($2n = 68$ in *musteloides* group members versus $2n = 64$ in *spixii*). Herein, I follow the revised taxonomy of Dunnum and Salazar-Bravo (2010b) for the *musteloides* group but include *spixii* within *Galea* and provisionally regard *flavidens* Brandt as a valid species distinct from *C. spixii* pending a thorough evaluation of these forms. Dating estimates suggest a late Miocene divergence between the *musteloides* group and *spixii*, followed by species level divergence within the *musteloides* group in the pre-Puna of the Andes during the Pliocene (Dunnum and Salazar-Bravo 2010b). Tate (1935), Hückinghaus (1961a), Cabrera (1961), Solmsdorff et al. (2004), and Bezerra (2008) have reviewed the genus.

KEY TO THE SPECIES OF *GALEA*:
1. Processus alaris superior absent; second prism of the PM3, M1, and M2 extended lateroaborally; nasals relatively short; parietals relatively long; in most cases mesopterygoid fossa relatively narrow and V-shaped anteriorly; white spot behind ear absent; occurs from the Bolivia-Brazil border west to southern Peru and south to southern Argentina. 2
1′. Processus alaris superior present; second prism of the PM3, M1, and M2 not extended lateroaborally; nasals relatively long; parietals relatively short; in most cases mesopterygoid fossa broad and U-shaped anteriorly; white spot behind ear present; occurs in eastern Brazil 4
2. Distributed in the Andes of southern Peru, Bolivia, Argentina, and northern Chile 3

2′. Distributed in the lowlands of Bolivia, Paraguay, and Argentina . . . . *Galea leucoblephara*

3. Range limited to southern Bolivia and northern Argentina . . . . *Galea comes*

3′. Range limited to northern and central Bolivia, Peru, and northern Chile . . . . *Galea musteloides*

4. Dorsal pelage blackish from the eyes to the nape . . . . *Galea flavidens*

4′. Dorsal pelage not as above . . . .*Galea spixii*

Galea comes Thomas, 1919
Southern Highland Yellow-toothed Cavy

SYNONYMS:

Kerodon boliviensis: Thomas, 1913a:143; name combination, but not *Cavia boliviensis* Waterhouse, 1848.

Galea comes Thomas, 1919f:134; type locality "Maimara, 2230 m," Jujuy, Argentina.

Galea musteloides: Thomas, 1926c:327; part; not *musteloides* Meyen.

DESCRIPTION: Dorsal coloration more mottled than *G. musteloides*, blend of olive, brown and tan; venter mix of yellow and white, with line between dorsum and venter not as distinct as in *G. musteloides*. Dorsal hairs gray at base, tan or brown at midshaft, and black at tip; ventral hairs gray at base and either buffy yellow or white at midshaft and tip. Ears covered with orange hairs; yellow ring surrounds the eyes (J. Dunnum, pers. obs.). Dimensions of holotype include head and body length 243 mm, hindfoot length 39 mm, ear length 22 mm, and greatest skull length 54 mm (Thomas 1919f).

DISTRIBUTION: *Galea comes* is known only from the Andes of southern Bolivia and northern Argentina.

SELECTED LOCALITIES (Map 375): ARGENTINA: Jujuy, Abrapampa (Thomas 1919f), Maimara (type locality of *Galea comes* Thomas), Sierra de Zenta (Thomas 1921k), 12.3 km N and 11.5 km W of San Antonio de los Cobres (Dunnum and Salazar-Bravo 2010b); La Rioja, La Invernada, ca. 35 km N of Nevada de Famatina (Thomas 1920g; not located). BOLIVIA: Tarija, Iscayachi (Dunnum and Salazar-Bravo 2010b).

SUBSPECIES: *Galea comes* is monotypic.

NATURAL HISTORY: The collector, Emilio Budin, reported this species as common in rough and brambly ground [at Maimara] and that it was harmful to cultivation (Thomas 1913a:143). In the Rio Tomayapo river valley (Tarija department, Bolivia), colonies composed of many individuals were interspersed among *Ctenomys lewisi* burrows and were observed foraging around burrows during the day (J. Dunnum, pers. obs.).

REMARKS: Thomas' (1919f) description includes specimens from Abrapampa but he designated a specimen collected from Maimara (Thomas 1913a) as the type.

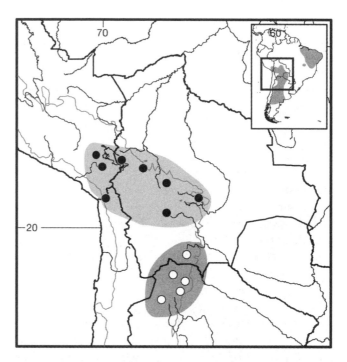

Map 375 Selected localities for *Galea comes* (○) and *Galea musteloides* (●). Contour line = 2,000 m.

Thomas later (1926c:327) synonymized *comes* Thomas with *musteloides* Meyen, a decision followed by most subsequent authors (Tate 1935; Cabrera 1961, Woods and Kilpatrick 2005). Dunnum and Salazar-Bravo (2010b) considered *comes* Thomas a distinct species, a decision I follow here. The karyotype is $2n = 128$, FN = 136 (Dunnum and Salazar-Bravo 2010b).

Galea flavidens (Brandt, 1835)
Eastern Yellow-toothed Cavy

SYNONYMS:

Cavia flavidens Brandt, 1835:439; type locality "Patria. Brasilia."

[*Cavia*] *bilobidens* Lund, 1839b:208; *nomen nudum.*

Cavia bilobidens Lund, 1840a:32 [1841b:248, plate 21, Fig. 6]; type locality "Rio das Velhas's Floddal" (see Lund 1841b:264–266), Lagoa Santa, Minas Gerais, Brazil (a fossil).

[*Cerodon*] *bilobidens*: Lund, 1840c:313; name combination.

Cavia (*Cerodon*) *flavidens*: Waterhouse, 1848:168; name combination.

K[*erodon*]. *flavidens*: Giebel, 1855:462; name combination.

[*Cavia* (*Galea*)] *flavidens*: E.-L. Trouessart, 1880b:195; name combination.

[*Cavia* (*Kerodon*)] *flavidens*: E.-L. Trouessart, 1897:639; name combination.

[*Galea*] *flavidens*: Thomas, 1916f:303; first use of current name combination.

DESCRIPTION: Similar in appearance to *G. spixii*, but with blackish dorsal areas stretching from eyes to nape (Brandt 1835). Incisors opisthodont; lingual prisms of PM4 and M1 narrower than labial ones, and labial prism of M2 not elongated posteriorly. Skull with broad zygomatic arches; lateral profile from orbits to rostrum deeply concave; posterior palatal emargination V-shaped; and angular process of mandible short (Solmsdorff et al. 2004). Bonvicino, Lima, and Weksler (2005) provided measurements for specimens she attributed to *C. flavidens*: head and body length 205–231 mm, hindfoot 43–47 mm, ear 24–25 mm, and mass 150–330 g.

DISTRIBUTION: *Galea flavidens* is restricted to the cerrado of central Brazil (Redford and Fonseca 1986).

SELECTED LOCALITIES (Map 376): BRAZIL: Goiás, Parque Nacional da Chapada dos Veadeiros (Bonvicino, Lima, and Weksler 2005).

SUBSPECIES: *Galea flavidens* is monotypic.

NATURAL HISTORY: Little information is available for this species. Bonvicino, Lima, and Weksler (2005) reported a single pregnant female (with one embryo) collected in November in *cerrado rupestre* between 1,300 and 1,500 m in the Chapada dos Veadeiros National Park. Other rodents taken at the same time included species of *Thrichomys* and *Oligoryzomys*.

REMARKS: Solmsdorff et al. (2004) described the two historically known specimens and commented on cranial similarities with *G. spixii*. Bonvicino, Lima, and Weksler (2005) tentatively recognized the validity of this species and commented on three newly collected specimens. Bezerra

(2008) discussed the difficulty in morphologically assessing *G. flavidens* due to the lack of good diagnostic characters in the original description, lack of a holotype, and ambiguous type locality; she suggested that *flavidens* was likely a synonym of *G. spixii*. Paula Couto (1950:232) considered *bilobidens* Lund, a fossil species from Lagoa Santa, Brazil, a synonym of *spixii* Wagler. Cabrera (1961:573), on the other hand, listed *bilobidens* Lund in the synonymy of *G. flavidens*, noting that, if he was correct, then the type locality of *flavidens* Brandt could be in Minas Gerais, Brazil). Cabrera also recognized that Paula Couto could have been correct in his assignment of *bilobidens* Lund to *G. spixii*. In either case, both the taxonomic and geographic status of *G. flavidens* remains uncertain, as does the allocation of Lund's *bilobidens*.

Galea leucoblephara Burmeister (1861)
Lowland Yellow-toothed Cavy

SYNONYMS: See under Subspecies.

DESCRIPTION: General characters as for genus but more lightly colored than *M. musteloides*, in part due to paler hue of subterminal area of hair. Overall color ranges from olive to reddish dorsally and yellow on flanks; ventral areas dirty yellowish white (Cabrera 1953). Burmeister (1861) remarked on pure white eye-rings, from which his name derives.

Subspecies *G. l. littoralis* similar to nominotypical form but more finely speckled, with less yellow; hairs on back indistinctly annulated to their bases with alternating light and dark rings of gray; subterminal ring buffy; tips black. Rings around eyes lighter than head but not conspicuously so. Venter dull buffy white and not sharply defined; hairs gray at base. Upper surfaces of feet buffy. Dimensions of holotype head and body length 199 mm, hindfoot length 39 mm, ear length 18 mm, greatest skull length 47 mm (Thomas 1901c).

Specimens of *G. l. demissa* from Chuquisaca, Bolivia mottled tan dorsally and buffy cream ventrally; dorsal hair light gray at base, gradually darkening to near black, followed by subterminal tan ring, and dark tip; both gray band around neck and eye-ring present; feet light tan; ears well furred and covered with fine orange hair. Pelage of animals from Salta province, Argentina, with speckled appearance, with grayish to brownish dorsum, whitish venter, and denuded area on neck. White eyering not readily visible and some individuals have dark band around neck. Head and body length 198–235 mm, mass from 180–280 g (Mares, Ojeda, and Barquez 1989).

DISTRIBUTION: *Galea leucoblephara* occurs in the lowlands from central Bolivia through Paraguay to southern Argentina.

SELECTED LOCALITIES (Map 377): ARGENTINA: Buenos Aires, Bahía Blanca (type locality of *Cavia boliviensis littoralis* Thomas), Laguna Chasticó (J. R. Contreras 1972); La Rioja, La Rioja (Cabrera 1953); Neuquén,

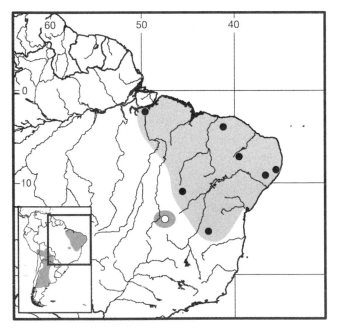

Map 376 Selected localities for *Galea flavidens* (○) and *Galea spixii* (●).

Collón Curá (Thomas 1927e), Zapala (Thomas 1927d); Río Negro, Pilcaniyeu (type locality of *Galea negrensis* Thomas); Salta, La Represa, Metán (Cabrera 1953); San Juan, Angaco Sud (AMNH 41894); San Luis, 9 km N of Paso del Rey (Dunnum and Salazar-Bravo 2010b); Santiago del Estero, Suncho Corral (AMNH 41589). BOLIVIA: Chuquisaca, 12 km N and 11 km E of Tarabuco (Dunnum and Salazar-Bravo 2010b); Santa Cruz, Ascención de Guarayos (S. Anderson 1997), Campo de Guanacos (type locality of *Galea spixi campicola* Doutt), 3.5 km W of Estación El Pailón (Dunnum and Salazar-Bravo 2010b); Tarija; 2 km S and 5 km E of Palos Blancos (Dunnum and Salazar-Bravo 2010b). PARAGUAY: Alto Paraguay, Palmar de Las Islas (Dunnum and Salazar-Bravo 2010b), Estancia Tres Marias (Dunnum and Salazar-Bravo 2010b).

SUBSPECIES: I follow Dunnum and Salazar-Bravo (2010b) and recognize three subspecies.

G. l. demissa Thomas, 1921

SYNONYMS:

Galea boliviensis demissa Thomas, 1921l:623; type locality "San Antonio, Parapiti, [Santa Cruz,] Bolivia. Alt. 600 m."

Galea musteloides demissa: Tate, 1935:347; name combination.

Galea spixi campicola Doutt, 1938:100; type locality "Campo de Guanacas, eastern Bolivia; altitude 450 meters," western part of Cordillera Province, about 90 miles south of Santa Cruz de la Sierra, Santa Cruz, Bolivia.

[*Galea spixi*] *spixi*: Doutt, 1938:101; part, not *Cavia spixii* Wagler.

Galea spixii campicola: Cabrera, 1961:575; name combination.

G[alea]. l[eucoblephara]. demissa: Dunnum and Salazar-Bravo, 2010b:251; first use of current name combination.

This subspecies occurs mainly in the Chaco ecoregion of lowland central and southern Bolivia, western Paraguay, and as far south as Santiago del Estero and Catamarca provinces in Argentina. This subspecies may also extend into the eastern slopes and valleys of the pre-Andean chains in Tucumán province, Argentina (Dunnum and Salazar-Bravo 2010b).

S. Anderson (1997) regarded *campicola* as a subspecies of *Galea spixii*, but it is included here based on morphological (Bezerra 2008) and molecular analyses (Dunnum and Salazar-Bravo 2010b).

G. l. leucoblephara (Burmeister, 1861)

SYNONYMS:

Anoema leucoblephara Burmeister, 1861:425; type locality "Die Art war haüfig bei Mendoza wie bei Tucuman; sie bewohnt ohne Zweifel den ganzen Westen der La Plata-Staaten, lebt in Erdlöchern an Wegen, wo Büsche

stehen und wurde in allen Altern und Geschlechtern von mir gesammelt." Type locality restricted to Mendoza, Argentina by Yepes (1936:40); lectotype designated by Solmsdorff et al. (2004).

[*Cavia (Galea)*] *leucoblephara*: E.-L. Trouessart, 1880b:195; name combination.

[*Cavia (Kerodon)*] *leucoblephara*: E.-L. Trouessart, 1897:640; name combination.

[*Cavia (Cerodon) boliviensis*] *leucoblephara*: E.-L. Trouessart, 1904:526; name combination.

Kerodon boliviensis: Thomas, 1902c:229; part; name combination.

Kerodon leucoblepharus: Thomas, 1911c:254; name combination but incorrect gender concordance.

G[alea]. leucoblephara: Thomas, 1919b:212; name combination.

C[avia]. (Galea) boliviensis leucoblephara: Osgood, 1915:195; name combination.

[*Galea*] *musteloides leucoblephara*: Tate, 1935:347; name combination.

Galea l[eucoblephara]. leucoblephara: Dunnum and Salazar-Bravo, 2010b:252; first use of current name combination.

This subspecies occurs in western Argentina from southern Catamarca province in the north through La Rioja, San Juan, San Luis, Mendoza, and Córdoba provinces in the south (Cabrera 1953; Dunnum and Salazar-Bravo 2010b).

G. l. littoralis (Thomas, 1901)

SYNONYMS:

Cavia boliviensis littoralis Thomas, 1901c:195; type locality "Bahia Blanca, [Buenos Aires.] Argentina."

[*Cavia (Cerodon) boliviensis*] *littoralis*: E.-L. Trouessart, 1904:526; name combination.

G[alea]. b[oliviensis]. littoralis: Thomas, 1919b:212; name combination.

Galea negrensis Thomas, 1919b:211; type locality "Pilcañeu, Upper Rio Negro. 1400 m [41°S., 71°W.]," Río Negro, Argentina.

Galea littoralis: Thomas, 1929:44; name combination.

[*Galea*] *musteloides littoralis*: Tate, 1935:347; name combination.

[*Galea*] *musteloides negrensis*: Ellerman, 1940:243; name combination.

G[alea]. leucoblephara littoralis: Dunnum and Salazar-Bravo, 2010b:251; first use of current name combination.

This subspecies occurs in southern Argentina, extending from northern Chubut in the south to lower Mendoza and La Pampa in the north and from the Atlantic coast of Buenos Aires province to the base of the Andes (Cabrera 1961). Bezerra (2008) recognized *littoralis* as a distinct species based on cranial characters.

NATURAL HISTORY: This species is primarily a diurnal, terrestrial herbivore, inhabiting both mesic and xeric grasslands of lowland central and southern South America. Original studies on population biology, behavior, and reproduction were reported as *G. musteloides*. Males maintain a linear dominance hierarchy through aggression; the female hierarchy is less stable. Breeding occurs year-round and females can produce up to seven litters per year. The gestation period is about 53 days and average litter size is three (Rood 1972). Females have one inguinal and one lateral abdominal pair of mammae. The average age of first reproduction is about three months for males and two months for females, with a minimum of 11 days (Weir 1974b). Substantial work has been done on their promiscuous mating behavior, mate choice, and reproductive success (Keil et al. 1999; Sachser et al. 1999; Hohoff et al. 2003). F. Fonseca (1960) and Hopkins and Rothschild (1966) reported ectoparasites for Bolivian specimens.

REMARKS: Dunnum and Salazar-Bravo (2010b) elevated *G. leucoblephara* from within *G. musteloides* based on molecular phylogenetic analyses. These authors also

suggested that the current distribution of the *leucoblephara* group is the result of two dispersal events into the Argentinian lowlands from the pre-Puna during the late Pliocene, followed by northward expansion during the Pleistocene into the Chaco regions of Bolivia and Paraguay. The karyotype consists of $2n = 68$ and FN = 136 (W. George et al. 1972; Dunnum and Salazar-Bravo 2010b).

Galea musteloides Meyen, 1833
Highland Yellow-toothed Cavy

SYNONYMS: See under Subspecies.

DESCRIPTION: General hue of fur brownish with yellowish or olivaceous tinge dorsally and white ventrally. Dorsal hairs pale gray at base, with pale brownish yellow median ring, and dark tip; venter and throat whitish with distinct line demarking venter from dorsal pelage; gray band present around throat area; eyelids whitish yellow and ears clothed in yellow brown hairs. Subspecies *G. m. auceps* differs greatly from *G. m. musteloides* and *G. m. boliviensis* in pelage coloration: paler yellow, tan dorsum and sides, with hairs gray at base and tipped black, but with wide middle section of yellow tan; venter buffy white and eyelids and lashes black. Cranial and tooth characters of *G. musteloides* as for genus. This species differs from *G. spixii* in that upper surface of skull less arched when viewed from side; outline nearly straight from tip of nasals back to point just anterior to frontoparietal suture, but descends slightly between orbits and toward apex of nasals; palate between the molars nearly flat, not concave as in *G. spixii* (Waterhouse 1848).

DISTRIBUTION: *Galea musteloides* is found in the Andes of central Bolivia, southeastern Peru, and extreme northeastern Chile.

SELECTED LOCALITIES (Map 375): BOLIVIA: Cochabamba, Valle Hermoso (type locality of *Galea monasteriensis* Solmsdorff, Kock, Hohoff, and Sachser); La Paz, Achacachi (S. Anderson 1997), Irupana (S. Anderson 1997); Santa Cruz, Estancia Laja (Dunnum and Salazar-Bravo 2010b); Oruro, 7 km S and 4 km E of Cruce Ventilla (Dunnum and Salazar-Bravo 2010b). CHILE: Arica y Parinacota, Putre (Mann 1950). PERU: Puno, Pairumani, 22 mi SW of Llave on the Río Huanque (Pearson 1951), Sillustani (Dunnum and Salazar-Bravo 2010b).

SUBSPECIES: I recognize three subspecies of *Galea musteloides*.

G. m. auceps (Thomas, 1911)
SYNONYMS:

Kerodon auceps Thomas, 1911c:255; type locality "Guarina (alt. 4000 m.), near the south-east, the Bolivian, end of the lake [Titicaca]," La Paz, Bolivia.

C[avia]. (Galea) auceps: Osgood, 1915:195; name combination.

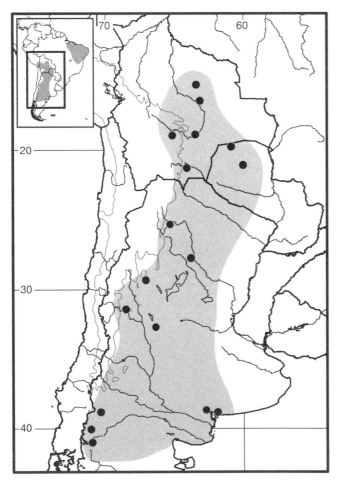

Map 377 Selected localities for *Galea leucoblephara* (●). Contour line = 2,000 m.

[*Galea*] *auceps*: Thomas, 1916f:303; name combination.

G[*alea*]. *anceps* Thomas, 1919f:134; incorrect subsequent spelling of *Kerodon auceps* Thomas.

[*Galea*] *musteloides auceps*: Tate, 1935:347; first use of current name combination.

This subspecies occurs in the highland areas around the southern edge of Lake Titicaca in western La Paz department, Bolivia, and adjacent areas of Puno department, Peru.

Cabrera (1961) included *auceps* Thomas in the synonymy of *G. musteloides musteloides*. S. Anderson (1997:481) disagreed, stating "the pale and yellowish subspecies *G. m. auceps* is rather distinct in comparison to the darker nominotypical subspecies," a decision followed by Woods and Kilpatrick (2005).

G. m. boliviensis (Waterhouse, 1848)

SYNONYMS:

Cavia boliviensis Waterhouse, 1848:175; type locality "high tableland between Cochabamba and La Paz [Bolivia]". Thomas (1911c:255) designated a lectotype and restricted the type locality to "Paratani [= Parotani], close to Cochabamba," Cochabamba, Bolivia.

K[*erodon*]. *boliviensis*: Giebel, 1855:463; part; name combination.

[*Cerodon*] *boliviensis*: Fitzinger, 1867b:155; part; name combination.

[*Cavia (Galea)*] *boliviensis*: E.-L. Trouessart, 1880b:195; part; name combination.

[*Galea*] *boliviensis*: Thomas, 1916f:302; name combination.

Cavia musteloides boliviensis: Osgood, 1916:210; name combination.

Galea musteloides musteloides: Tate, 1935:347; part; name combination.

Galea monasteriensis Trillmich, Kraus, Künkele, Asher, Clara, Dekomien, Epplen, Saralegui, and Sachser, 2004: footnote 517; *nomen nudum*.

Galea monasteriensis Solmsdorff, Kock, Hohoff, and Sachser, 2004:137–156; type locality "Valle Hermoso, 2557 m a.s.l., 66°09′W, 17°23′S, S of Cochabamba, Cordillera Oriental, Cochabamba Province, Bolivia."

Galea musteloides boliviensis: Solmsdorff, Kock, Hohoff, and Sachser, 2004:147; first use of current name combination.

This subspecies occurs in the eastern region of the Bolivian Altiplano.

Osgood (1916:211) erroneously stated that Thomas (1911c:255) had designated Sajama, Bolivia, as the type locality of Waterhouse's *boliviensis*, based on Thomas's (1911b:255) referral of a specimen from this locality to the nominotypical subspecies. Subsequently, Thomas (1926c:327) reduced *boliviensis* Waterhouse to a synonym of *musteloides* Meyen and most subsequent authors have

followed this taxonomy (Tate 1935; Cabrera 1961; Woods 1993; Woods and Kilpatrick 2005), but recent reviews have recognized the subspecies (Solmsdorff et al. 2004; Dunnum and Salazar-Bravo 2010b). Bezerra (2008) found insufficient morphological differences to warrant separation of *monasteriensis* and *musteloides*; Dunnum and Salazar-Bravo (2010b) treated *G. monasteriensis* as a synonym of *G. m. boliviensis* based on molecular data.

G. m. musteloides Meyen, 1833

SYNONYMS:

Galea musteloides Meyen, 1833:598; type locality "Sie bewohnen die Hochebenen der westlichen Cordillerenkette, auf dem Passe von Tacna nach dem Alpen-see von Titicaca. Das Plateau von Tacora, besonders der Kamm desselben, zwischen dem Bolivianischen Indianerdorfe Morocallo, und dem Peruanischen Kirchdorfe Pisacoma [Pisacoma, 16°54′S, 69°23′W, Puno, Peru], gänzlich bestend aus trachytischer Lava und Conglomeraten, wird von ihnen bewohnt," = Paso de Tacna, on road to Lake Titicaca, Peru.

Kerodon rupestris: Lesson, 1842:102; name combination.

K[*erodon*]. *boliviensis*: Giebel, 1855:463; part; name combination.

[*Cerodon*] *boliviensis*: Fitzinger, 1867b:155; part; name combination.

Galea musteloides musteloides: Tate, 1935:347; first use of current name combination.

Cavia musteloides musteloides: Pearson, 1951:153; name combination.

This subspecies occurs in the Andean Altiplano of southwestern Peru, western Bolivia, and extreme northeastern Chile (Dunnum and Salazar-Bravo 2010b).

NATURAL HISTORY: *Galea musteloides* is primarily a diurnal, terrestrial herbivore, inhabiting the grasslands of the central Andes. Individuals excavate their own burrow systems but also utilize abandoned burrows of other mammals, such as *Lagostomus*, *Ctenomys*, and *Chaetophractus*. They appear to respond to the alarm calls of *Ctenomys* by seeking cover (Sanborn and Pearson 1947). Muñoz-Pedreros (2000) reported their burrow systems to be shallow with grass-lined alcoves and that they cohabitate with *Ctenomys opimus*. In northern Chile, they consume the bark and leaves of *Lepidophyllum* and *Baccharis* shrubs (Asteraceae) and the grasses *Stipa* and *Festuca* (Muñoz-Pedreros 2000). Females reach sexual maturity at one month of age, males at two to three months. The gestation period averages 55.3 days (range: 50–58 days; Solmsdorff et al. 2004). Young weigh around 40 g at birth, and litter size ranges from two to seven (average of three) individuals that are suckled for three to five weeks (Muñoz-Pedreros 2000). This species exhibits low compatibility among unfamiliar adults as well as differences in courtship and copulatory posture in comparison to *G. leucoblephara* (Trillmich et al. 2004). This is the only species of *Galea* to exhibit a monogamous mating system (Hohoff et al. 2002). F. Fonseca (1960), Hopkins and Rothschild (1966), and Smit (1987) report ectoparasites from Bolivian specimens.

REMARKS: Woods and Kilpatrick (2005) gave the date of Meyen's description as 1832; the correct publication date of Meyen's work, however, is 1833.

Dunnum and Salazar-Bravo (2010b) restricted *G. musteloides* to populations from the Altiplano of southern Peru, northern Chile, and west-central Bolivia, identifying two molecular clades within this range. One of these comprised individuals from the Lake Titicaca area in Peru and Bolivia assignable to the subspecies *G. m. musteloides* and *G. m. auceps*. The second included samples from Oruro, Cochabamba, and Santa Cruz departments, Bolivia, which they assigned to *G. m. boliviensis*. These authors suggested that diversification of the subspecies occurred about the Pliocene-Pleistocene boundary. G. I. Molina (1782:306) used *Lepus minimus* in describing a "Cuy" from Chile, which Tate (1935:344) thought likely to be *Galea* and thus included it as a subspecies of *G. musteloides*. Ellerman (1940:243) regarded Molina's *minimus* to be a valid species of *Galea*, but also noted that, if Tate were correct in his assignment to *musteloides*, Molina's name predated *musteloides* Meyen and as such Meyen's name, and all others associated with it, would become synonyms of *G. minimus* G. I. Molina. Hückinghaus (1961a) agreed with Ellerman and referred to this species as *Galea minimus*. Alternatively, Cabrera (1961) listed Molina's name in the synonymy of *Cavia porcellus*. Because the correct assignment of Molina's name remains unresolved, I follow Cabrera (1961) and regard it as a synonym of *Cavia porcellus* (Linnaeus); see Remarks under that account.

The karyotype of *G. musteloides* consists of a $2n = 68$ and FN = 136 (Dunnum and Salazar-Bravo 2010b).

Galea spixii (Wagler, 1831)
Spix's Yellow-toothed Cavy

SYNONYMS: See under Subspecies.

DESCRIPTION: General coloration grayish, with somewhat indistinct brownish tint on back; dorsal hair pale gray at base and dusky at tip with broad subterminal ring of pale brownish yellow; hairs on sides of body lighter at tip with subterminal ring almost whitish; throat, abdomen, and inner surfaces of limbs white; small white mark present above eye and more distinct white patch behind the ear. *Galea s. palustris* lacks postauricular spot seen in *G. s. spixii* and *G. s. wellsi* (Osgood 1915); feet nearly white. Lateral profile of skull forms gentle convex curve from base of nasals to posterior part of the parietals, where it descends quickly to the occiput, not straight or concave between orbits as in *G. musteloides* (Waterhouse 1848). Measurements of holotype *G. s. palustris* head and body length 225 mm, hindfoot length 46 mm, ear length 25.5 mm, and greatest skull length 55 mm (Thomas 1911c), and those of *G. s. wellsi* head and body length 234 mm, hindfoot length 51 mm, and greatest skull length 57.5 m (Osgood 1915).

DISTRIBUTION: *Galea spixii* is known from eastern Brazil (Moojen 1952b).

SELECTED LOCALITIES (Map 376): BRAZIL: Alagoas, Quebrangulo (Mares, Willig et al. 1981); Bahia, São Marcello, junction of Rio Preto and Rio Sapão (type locality of *Cavia* [*Galea*] *wellsi* Osgood); Ceará, Crato (Moojen 1943a), São Paulo (Thomas 1910c); Minas Gerais, Campos Geraes de San Felipe (Osgood 1915); Pará, Cametá, lower Rio Tocantins (type locality of *Kerodon palustris* Thomas); Pernambuco, Saltinho, Rio Formoso (Ximénez 1980).

SUBSPECIES: I recognize three subspecies.

G. s. palustris (Thomas, 1911)
SYNONYMS:

Kerodon palustris Thomas, 1911b:608; type locality "Cametá, Lower Tocantins," Pará, Brazil.

C[*avia*]. (*Galea*) *palustris*: Osgood, 1915:195; as subgenus, name combination.

[*Galea*] *palustris*: Thomas, 1916f:303; name combination.

[*Galea spixi*] *palustris*: Doutt, 1938:101; first use of current name combination and incorrect subsequent spelling of *Cavia spixii* Wagler.

This subspecies occurs in northeastern Brazil, south of the Rio Amazonas.

G. s. spixii (Wagler, 1831)
SYNONYMS:

Cavia spixii Wagler, 1831a:512; type locality "Die Ordung der Nager bin ich durch zwey neue Gattungen Ferkelmäuse (Cavia), welche Herr von Spix von seiner Reise am Amazonianstrome, zurückbrachte, und die sich daher in unserem Museum befinden, zu bereichern im Stande," = Rio Amazonas, Brazil. Osgood (1915:197) discussed and restricted the type locality to Campos Geraes de San Felipe, east of Januária, Bahia [a locality actually in Minas Gerais state], Brazil. Cabrera (1961:575) stated that the type locality was unknown but, by his inclusion of *saxatilis* Lund in the synonymy of this subspecies, he restricted the type locality to Lagoa Santa, Minas Gerais, Brazil.

[*Cavia*] *sexatilis* Lund, 1839b:208; *nomen nudum*.

C[*avia*]. (*Cerodon*) *saxatilis* Lund, 1840b:14 [1841c:287, plate 25, Figs. 14, 18]; type locality "Rio das Velhas's Floddal" (see Lund 1841b:292, 294), Lagoa Santa, Minas Gerais, Brazil.

Cerodon saxatilis: Lund, 1840c:313; name combination.

Cavia (Cerodon) spixii: Waterhouse, 1848:173; name combination.

K[erodon]. spixi Giebel, 1855:463; name combination and incorrect subsequent spelling of *Cavia spixii* Wagler.

Prea saxatilis: Liais, 1872:541; name combination.

[*Cavia (Galea)*] *Spixii*: E.-L. Trouessart, 1880b:195; part; name combination.

[*Cavia (Kerodon)*] *spixii*: E.-L. Trouessart, 1897:639; name combination.

[*Cavia (Cerodon) boliviensis*] *saxatilis*: E.-L. Trouessart, 1904:526; name combination.

C[avia]. (Galea) spixi: Osgood, 1915:195; incorrect subsequent spelling of *Cavia spixii* Wagler.

[*Galea*] *spixi*: Thomas, 1916f:303; name combination and incorrect subsequent spelling of *Cavia spixii* Wagler.

[*Galea*] *spixii*: Tate, 1935:347; name combination.

This subspecies occurs in eastern Brazil, but its western limits remain unclear.

G. s. wellsi (Osgood, 1915)

SYNONYMS:

[*Cavia (Galea)*] *Spixii*: E.-L. Trouessart, 1880b:195; part; name combination.

[*Cavia (Kerodon)*] *spixii*: E.-L. Trouessart, 1897:639; part; name combination.

Cavia spixi: Goeldi and Hagmann, 1906:74; name combination and incorrect subsequent spelling of *spixii* Wagler.

Cavia (Galea) wellsi Osgood, 1915:196; type locality "São Marcello [= São Marcelo], junction of Rio Preto and Rio Sapaõ [= Sapão], Bahia, Brazil."

[*Galea*] *wellsi*: Thomas, 1916f:303; name combination.

[*Galea spixi*] *wellsi* Doutt, 1938:101; first use of current name combination and incorrect subsequent spelling of *Cavia spixii* Wagler.

This subspecies is limited to northeastern Brazil (Cabrera 1961).

NATURAL HISTORY: *Galea spixii* occupies only low-lying areas within the Caatinga ecoregion, where individuals use a network of well-worn runways. Individuals are mostly crepuscular but activity may occur at any time (Mares, Willig et al. 1981). Nest sites appeared to be temporary and were typically found under rocks or low, overhanging vegetation. Areas frequented by *G. spixii* were readily distinguished by their runways and cleared patches used for sand bathing (Streilein 1982a). *Galea spixii* exhibited a wide range of vocalizations in response to varying situations. The mating system approximated a male dominance polygyny (Lacher 1981). The gestation period ranged from 49 to 52 days and litter size averaged 2.2 (range: 1–5). Lacher (1981) and Mares, Willig et al. (1981) reported re-

production to occur in both captivity and the wild. Mares et al. (1982) described molt. This species is difficult to trap except when traps were placed along the runways. Trapped individuals frequently displayed a death-feigning behavior when handled (Streilein 1982a).

REMARKS: Emmons et al. (2006) reported *G. spixii* from the Parque Nacional Noel Kempff Mercado, but identifications need to be verified as these may be referable to *G. l. demissa*. Dunnum and Salazar-Bravo (2010b) placed *G. spixii campicola* in synonymy with *G. leucoblephara demissa*. The concept of *G. spixii* I employ here includes *wellsi* Osgood and *saxatilis* Lund (Cabrera 1961:575; Corbet and Hill 1991:201), but Hückinghaus (1961a:71) considered *wellsi* a distinct species and both Avila-Pires (1982) and Solmsdorff et al. (2004) recognized *saxatilis* as a valid species. Karyotype consists of a $2n = 64$ and FN = 118 (Maia 1984).

Genus *Microcavia* H. Gervais and Ameghino, 1880

The genus as currently recognized is composed of three species and occurs from the highlands of southwestern Bolivia and northern Chile south to Santa Cruz province in southern Argentina and likely southern Chile (Osgood 1943b; C. A. Quintana 1996; S. Anderson 1997). Fossil materials from four species are known from mid-Pliocene to Recent deposits in Argentina and Uruguay (C. A. Quintana 1996; Ubilla et al. 1999). Mountain cavies occupy a wide variety of arid and semiarid habitats from near sea level to above 4,000 m. They are medium-sized caviids, slightly smaller than *Cavia* (head and body length ranges from 170 to 250 mm, mass from 200 to 500 g, and mean condyloincisive length is 42.2 mm [range: 36–46 mm; Hückinghaus 1961a]). Individuals possess no external tail; females have four mammae; eyes are large and surrounded by a conspicuous white ring; dorsal coloration is olive gray agouti and under parts are pale gray; and the submandibular gland present in *Galea* is lacking. The skull is shortened and bowed compared to other Caviinae. The highest part of the skull occurs over the posterior zygomatic root and the orbital branch of the maxillary is continuous as a narrow strip in front of the lacrimal. The incisive foramen is larger than in *Cavia* and *Galea*, triangular in shape, and placed more closely to the toothrows. The bullae are larger and the orbit is more rounded than in either *Cavia* or *Galea*. Cheek teeth are like those of *Galea* with each tooth divided into two lobes by one inner fold. M3 has a deeper posterior fold than in *Galea*. The lower teeth are bilobed. Incisors are nonpigmented. Descriptive information is from Osgood (1915), Thomas (1916f), and Ellerman (1940).

The genus *Microcavia* is known from late Pliocene through mid Pleistocene deposits (Verzi and Quintana 2005). Cabrera (1961) recognized two subgenera, *Microcavia* containing *M. australis* and *M. shiptoni*, and *Monticavia* containing *M. niata*. C. A. Quintana (1996), however, found no support for subgeneric divisions.

SYNONYMS:

Cavia: I. Geoffroy St.-Hilaire and d'Orbigny, 1833:plate 12; not *Cavia* Pallas.

Kerodon: Bennett, 1836:190; part (description of *kingii*); not *Kerodon* F. Cuvier.

Cerodon Waterhouse, 1848:180; as a subgenus of *Cavia* (allocation of *australis*); unjustified emendation of *Kerodon*, but not *Kerodon* F. Cuvier.

Anaema Blainville, 1855:26; incorrect subsequent spelling of *Anoema*, but not *Anoema* F. Cuvier; referred to *C. aperea* but a *Microcavia* skull is figured.

Anoema: Burmeister, 1879:271; part (listing of *australis*); as a subgenus of *Cavia*, but not *Anoema* F. Cuvier.

Microcavia H. Gervais and Ameghino, 1880:50; type species *Cavia australis* I. Geoffroy St.-Hilaire and d'Orbigny, by subsequent designation (Kraglievich 1930a).

Caviella Osgood, 1915:194; as subgenus; type species *Cavia australis* I. Geoffroy St.-Hilaire and d'Orbigny, by original designation.

Monticavia Thomas, 1916f:303; type species *Monticavia niata* (= *Cavia niata* Thomas), by original designation.

Nanocavia Thomas, 1925a:419; type species *Nanocavia shiptoni* Thomas, by original designation.

Microavia Cabrera, 1953:20; incorrect subsequent spelling of *Microcavia* H. Gervais and Ameghino.

REMARKS: In the original description of the genus, H. Gervais and Ameghino (1880:50–55) did not designate a type species among the four they listed under *Microcavia*. Subsequently, Kraglievich (1930a) selected C[*avia*]. *australis* I. Geoffroy St.-Hilaire and d'Orbigny (1833) as the type species, a choice followed by C. A. Quintana (1996) in his review of the genus. Cabrera (1953), however, argued that the type species should be *Microcavia typus*, one of the fossil taxa described by H. Gervais and Ameghino (1880:52) under their description of their new genus and now listed as a synonym of *M. australis* (e.g., C. A. Quintana 1996). Cabrera (1961) and Tognelli et al. (2001) accepted this argument as each gave *Microcavia typus* H. Gervais and Ameghino as the type species of *Microcavia*. Both erred, however, as H. Gervais and Ameghino (1880) clearly neither explicitly nor implicitly selected *M. typus* among the four species they listed, as required by genus-level names authored prior to 1930 (ICZN 1999:Art 67.4.1). Thus, Kraglievich's (1930a) selection of *australis* I. Geoffroy St.-Hilaire and d'Orbigny is correct. Thomas (1921e), Hück-

inghaus (1961), and C. A. Quintana (1996) reviewed the genus.

KEY TO THE SPECIES OF *MICROCAVIA* (FROM TOGNELLI ET AL. 2001):

1. Incisors orthodont, lateral mandibular fossa deepened anteriorly . . . . *Microcavia australis*
1'. Incisors proodont, lateral mandibular fossa not deepened anteriorly 2
2. Cranial profile strongly convex, length of auditory bullae >10 mm . . . . *Microcavia niata*
2'. Cranial profile not strongly convex, length of auditory bullae <10 mm . . . . *Microcavia shiptoni*

Microcavia australis (I. Geoffroy St.-Hilaire and d'Orbigny, 1833)
Southern Mountain Cavy

SYNONYMS: See under Subspecies.

DESCRIPTION: Head and body length 170–245 mm, hindfoot 35–50 mm, ear 14–20 mm (Cabrera 1953), and mass 141–340 g (Mares, Ojeda, and Barquez 1989). General coloration olive-gray agouti with speckled appearance, ventrum paler gray. Eyes large and encircled by white ring. Teeth described for genus but incisors orthodont. Lateral mandibular fossa not deepened anteriorly as in *M. niata* and *M. shiptoni* (Tognelli et al. 2001). Tognelli et al. (2001) provided external and cranial measurements for each subspecies.

DISTRIBUTION: *Microcavia australis* occurs in Argentina from Jujuy province in the northwest, east to Buenos Aires province, and south to Santa Cruz province, and to southern Chile. Woods (1993) included extreme southern Bolivia in the distribution, but S. Anderson (1997) reported no known specimens from that country. Pine et al. (1979) discussed its presence in Aysén, Chile.

SELECTED LOCALITIES (Map 378; from C. A. Quintana 1996, except as noted): ARGENTINA: Buenos Aires, Copetonas, San Blas (Cabrera 1953); Catamarca, Recreo (type locality of *Caviella australis salinia* Thomas), Tinogasta, Santa María; Chubut, Lago Colhue-Huapi, Rawson; Jujuy, Yavi; La Rioja, Guandacol; Neuquén, Zapala (Thomas 1927d), Collón Curá (Thomas 1927e); San Juan, Cañada Honda (type locality of *Caviella australis joannia* Thomas); Santa Cruz, Lago San Martín (Cabrera 1953); Santa Fe, Tostado. CHILE: Aysén, Chacabuco Valley (Pine et al. 1979).

SUBSPECIES: Thomas (1921e) recognized five subspecies within *M. australis* while Cabrera (1953) recognized only three: *M. a. australis* I. Geoffroy St.-Hilaire and d'Orbigny (including *kingii* Bennett, *nigriana* Thomas, and *joannia* Thomas, as synonyms); *M. a. maenas* Thomas; and *M. a. salinia* Thomas. I follow Cabrera's (1953) treatment but note that if southern populations prove recognizable, then *kingii* Bennett is the available name. Sassi et al.

(2011) found high levels of molecular divergence between highland and lowland populations, suggesting possible cryptic taxa within *M. australis*.

M. a. australis (I. Geoffroy St.-Hilaire and d'Orbigny, 1833)

SYNONYMS:

C[avia]. australis I. Geoffroy St.-Hilaire and d'Orbigny, 1833:1 [unnumbered text for plate 12]; type locality "sur les bords du Rio Negro, vers le Quarante-unième degré," emended to "on the Lower Rio Negro," Río Negro, Argentina, by Thomas (1929:44).

Kerodon Kingii Bennett, 1836:190; type locality "apud Portum Desire dictum, ad Patagoniæ littus orientale," Puerto Deseado, Santa Cruz, Argentina.

[Cerodon] Kingii: Wagner, 1844:70; name combination.

Cavia (Cerodon) australis: Waterhouse, 1848:180; name combination.

Cavia Kingii: Waterhouse, 1848:plate 3, Fig. 2; name combination.

Anaema aperea: Blainville, 1855:26; name combination, referring to plate, but not *aperea* Erxleben.

C[avia]. (Anaema) Aperea: Blainville, 1855:5 of Atlas; name combination, referring to plate, but not *aperea* Erxleben.

C[avia]. aperea: Blainville, 1855:plate 2 in Atlas; name combination, but not *aperea* Erxleben; caption reads *C. aperea* but the skull figured is of *M. australis*.

K[erodon]. australis: Giebel, 1855:463; name combination.

Cavia (Anoema) australis: Burmeister, 1879:272; name combination, but not *Anoema* F. Cuvier.

[Cavia (Galea)] australis: E.-L. Trouessart, 1880b:195; name combination.

[Cavia (Kerodon)] australis: E.-L. Trouessart, 1897:639; name combination.

C[avia]. (Caviella) australis: Osgood, 1915:195; name combination.

[Caviella] australis: Thomas, 1916f:302; name combination.

C[aviella]. australis: Thomas, 1916f:302; name combination.

Caviella australis australis: Thomas, 1921e:445; name combination.

Caviella australis nigriana Thomas, 1921a:446; type locality "Neuquen, R. Negro," Neuquén, Argentina.

Caviella australis joannia Thomas, 1921e:446; type locality "Cañada Honda, San Juan. Alt. 500 m," Argentina.

[Microcavia] australis: Kraglievich, 1927:579; name combination.

C[aviella]. a[ustralis]. kingi: Thomas, 1929:44; name combination and incorrect subsequent spelling of *kingii* Bennett.

Microcavia australis australis: Yepes, 1935a:242; first use of current name combination.

M[icrocavia]. australis kingii: Yepes, 1935a:242; name combination.

M[icrocavia]. australis joannia: Yepes, 1935a:242; name combination.

Cavia (Microcavia) australis: Osgood, 1943b:142; name combination.

The nominotypical subspecies is found in Argentina from San Juan province in the west and southern Buenos Aires province in the east, south to Santa Cruz province.

M. a. maenas (Thomas, 1898)

SYNONYMS:

Cavia maenas Thomas, 1898d:284; type locality "Chilecito, Rioja, 1200 metres," La Rioja, Argentina.

[Cavia (Cerodon)] moenas E.-L. Trouessart, 1904:527; name combination and incorrect subsequent spelling of *maenas* Thomas.

C[avia]. (Caviella) maenas: Osgood, 1915:195; name combination.

[Caviella] maenas: Thomas, 1916f:302; name combination.

Caviella australis maenas: Thomas, 1921e:447; name combination.

[Microcavia] moenas: Kraglievich, 1927a:579; name combination and incorrect subsequent spelling of *maenas* Thomas.

M[icrocavia]. australis maenas: Yepes, 1935a:242; first use of current name combination.

Microcavia australis moenas: Yepes, 1936:40; incorrect subsequent spelling of *maenas* Thomas.

Microcavia australis joannia: Yepes, 1936:39; part, but not *joannia* Thomas.

This subspecies is found in the montane zones of La Rioja, Catamarca, Salta, and Jujuy provinces of northwestern Argentina at elevations up to 2,500 m.

M. a. salinia (Thomas, 1921)

SYNONYMS:

Caviella australis salinia Thomas, 1921e:447; type locality "Recreo," Catamarca, Argentina.

M[icrocavia]. australis salinia: Yepes, 1935a:242; first use of current name combination.

This subspecies occupies the saline areas of eastern Catamarca and La Rioja, the southwestern part of Santiago del Estero, and the northwestern part of Córdoba provinces, south to northern San Luis province, Argentina.

NATURAL HISTORY: Rood (1970, 1972) provided extensive information on the behavior and natural history of the species. Mountain cavies are diurnal and active all year. They are herbivores, feeding mostly on leaves, shoots, flowers, and fruits, but may gnaw on the bark of shrubs and trees when necessary (Rood 1970; Tognelli et al. 2001). They exhibit a male dominance hierarchy based upon

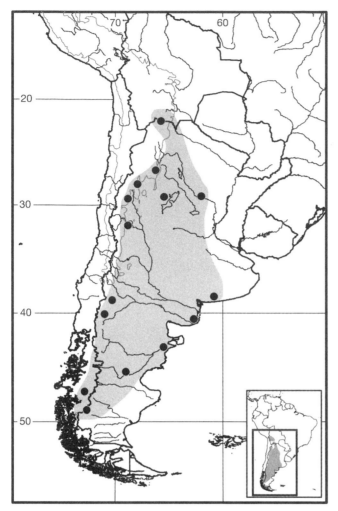

Map 378 Selected localities for *Microcavia australis* (●). Contour line = 2,000 m.

agonistic interactions. However, cooperation among individuals in the form of group huddling, mutual grooming, and indiscriminant nursing is common. Colony size ranged from 4 to 38 individuals in the Monte Desert (J. R. Contreras and Roig 1978). Rood (1970) also reported densities of 24 individuals/ha from Buenos Aires Province. Home ranges were approximately 0.75 ha for males and half that for females. Reproduction occurred from August to April, with most litters born between September and October. The gestation period in captivity was 54 days (Rood 1970). Mean litter size was 2.8 (range: 1–5; Rood 1972). De la Barrera (1940) provided disease and parasite information.

Microcavia niata (Thomas, 1898)
Northern Mountain Cavy

SYNONYMS: See under Subspecies.

DESCRIPTION: Size small, similar to that of *M. australis*. Dorsal hair length 15–18 mm. General coloration pale yellowish buff; face, cheeks, hairs on and around base of ears, and upper surface of feet whitish buff; venter and anal region white with slight buffy tinge, not sharply defined on sides (Thomas 1898c). General cranial characters as for genus but varies as follows: incisors more proodont than in *M. australis*, with angle of line of toothrow about 115°; M3 less complicated, with heel a short projection without internal notch; lateral mandibular fossa not deepened anteriorly. Differs from *M. shiptoni* in that skull bowed more anteriorly (Ellerman 1940), providing strongly convex profile, and length of auditory bullae >10 mm (Tognelli et al. 2001). Auditory bullae large, with maximum diameter greater than one-third zygomatic width (Kraglievich 1930). Measurements of holotype of *M. n. niata* include head and body 190 mm, hindfoot 34 mm, and ear 13 mm (Thomas 1898c), and that of *M. n. pallidior* head and body 200 mm, hindfoot 41 mm, ear 22 mm, and greatest skull length 46.5 mm (Thomas 1902c).

DISTRIBUTION: *Microcavia niata* occupies the Altiplano of southwestern Bolivia and adjacent northern Chile.

SELECTED LOCALITIES (Map 379; from S. Anderson 1997, except as noted): BOLIVIA: La Paz, Huaraco-Antipampa, Esperanza, near Mount Sajama (type locality of *Cavia niata* Thomas); Oruro, Condo, 30 km S and 25 km E of Sajama; Potosí, Pampa de Talapaca. CHILE: Tarapacá, vicinity of Enquela (Marquet et al. 1993).

SUBSPECIES: I recognize two subspecies, following S. Anderson (1997) and Woods and Kilpatrick (2005).

M. n. niata (Thomas, 1898)
SYNONYMS:

Cavia niata Thomas, 1898c:282; type locality "Esperanza, a 'tambo' in the neighbourhood of Mount Sahama, Bolivia . . . at an altitude of 4000 metres in the 'Puna' region," La Paz, Bolivia [see S. Anderson 1997].

[*Cavia* (*Cerodon*)] *niata*: E.-L. Trouessart, 1904:527; name combination.

Kerodon niata: Neveu-Lemaire and Grandidier, 1911:16; name combination.

C[*avia*]. (*Caviella*) *niata*: Osgood, 1915:195; name combination.

M[*onticavia*]. *niata*: Thomas, 1916f:303; name combination.

[*Monticavia*] *niata niata*: Tate, 1935:349; name combination.

Caviella [(*Monticavia*)] *niata niata*: Ellerman, 1940:246; name combination.

Cavia (*Microcavia*) *niata*: Pearson, 1951:170; name combination.

Microcavia (*Monticavia*) *niata*: Hückinghaus, 1961a:83; name combination.

Microcavia niata: Cabrera, 1961:572; name combination.

Microcavia niata niata: Cabrera, 1961:572; first use of current name combination.

The nominotypical subspecies is found in the border region of Bolivia and Chile, in extreme southwestern La Paz department, northwestern Oruro department, and the area around Colchane, Chile.

M. n. pallidior (Thomas, 1902)

SYNONYMS:

Kerodon niata pallidior Thomas, 1902c:229; type locality "Pampa Aullaga. Alt. 3700 m," west of Lago Poopó, Oruro, Bolivia (see S. Anderson 1997).

[*Cavia* (*Cerodon*) *niata*] *pallidior*: E.-L. Trouessart, 1904: 527; name combination.

C[*avia*]. (*Caviella*) *niata pallidior*: Osgood, 1915:195; name combination.

M[*onticavia*]. *niata pallidior*: Thomas, 1916f:303; name combination.

Caviella [(*Monticavia*)] *niata pallidior*: Ellerman, 1940:246; name combination.

Microcavia (*Monticavia*) *niata*: Hückinghaus, 1961a:83; part; name combination, but did not recognize subspecies.

Microcavia niata palldiior Hückinghaus, 1961a:84; incorrect subsequent spelling of *pallidior* Thomas.

Microcavia niata pallidior: Cabrera, 1961:572; first use of current name combination.

This subspecies is found in the departments of Oruro and Potosí, Bolivia.

NATURAL HISTORY: Marquet et al. (1993) reported on populations occurring in bog areas in the Chilean highlands (3,700–4,000 m). They observed colonies in shrubby areas and in close proximity to *Ctenomys* burrows, possibly occupying the abandoned burrows as reported for *Galea musteloides* by Mann (1978). Here, the diet of *M. niata* consisted of the herbaceous genera *Eleocharis* (Cyperaceae) and *Werneria* (Asrteraceae), the grasses *Distichlis* and *Deyeuxia* (Poaceae), and aquatic plants such as *Liolaeopsis* (Apiaceae). A single colony was composed of 5 males and 10 females, 8 were juveniles and 7 were adults, ranging in size from 81–380 g. No sexual or agonistic behavior was observed among members of the same colony. However, individuals of a colony were highly territorial and aggressive towards those of other colonies and elicited alarm calls in response to the approach of potential predators.

In the Sajama area of Bolivia (3,800 m), *M. niata* occupied small isolated salt flats located within very sandy Puna habitat dominated by Peruvian feather grass (*Stipa ichu*) and woody shrubs. Their burrows were interspersed throughout an an area also occupied by *Ctenomys*; no *Microcavia* were found in the surrounding areas. Individuals were active from dawn to dusk. Most active burrows appeared to contain a

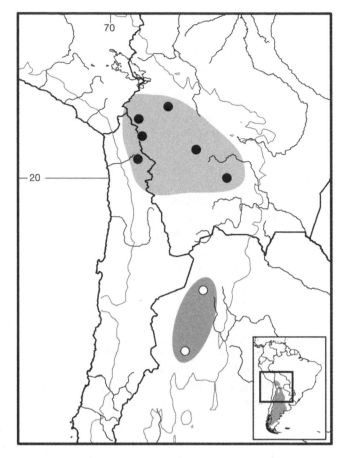

Map 379 Selected localities for *Microcavia niata* (●) and *Microcavia shiptoni* (○). Contour line = 2,000 m.

single animal, however a pregnant female with two embryos and two subsadults were observed foraging together an were eventually taken from a single tunnel system no longer than 5 m in length (J. Dunnum, pers. obs.).

Microcavia shiptoni (Thomas, 1925)
Shipton's Mountain Cavy

SYNONYMS:

Nanocavia shiptoni Thomas, 1925a:419; type locality "Laguna Blanca, Catamarca Altitude 3400 m," Argentina.

Caviella [(*Nanocavia*)] *shiptoni*: Ellerman, 1940:246; name combination.

Monticavia (*Nanocavia*) *shiptoni*: Kraglievich, 1930a:94; name combination.

Microcavia shiptoni: Cabrera, 1953:34; first use of current name combination.

DESCRIPTION: General characteristics as for genus. Dorsal color dull brownish buffy to tawny, venter buffy white or whitish with base of hairs brownish slaty (Mares, Ojeda, and Barquez 1989). Ears short and covered with fine buffy hairs. Similar in appearance to *M. australis*, but somewhat smaller. Head and body length 181–220 mm, hindfoot

36–39 mm, and mass 150–220 g. Dimensions of holotype include head and body 220 mm, hindfoot 37 mm, and greatest skull length 45 mm (Thomas 1925a). Bullae considerably smaller than in other two species, with portion appearing on occipital surface nearly uninflated (Ellerman 1940).

DISTRIBUTION: *Microcavia shiptoni* occurs in the highlands of northwestern Argentina between the elevations of 3,000 and 4,000 m in the provinces of Catamarca, Tucumán, and Salta.

SELECTED LOCALITIES (Map 379): ARGENTINA: Catamarca, Laguna Blanca (type locality of *Nanocavia shiptoni* Thomas); Salta, Chorrillos (C. A. Quintana 1996).

SUBSPECIES: *Microcavia shiptoni* is monotypic.

NATURAL HISTORY: Habits and behavior are unknown (Mares, Ojeda, and Barquez 1989).

REMARKS: Mares, Ojeda, and Barquez (1989) suggested *M. shiptoni* may be a race of *M. australis*, however, recent work utilizing molecular data supports its recognition as a distinct species (Dunnum 2009). The phylogenetic placement of *M. shiptoni* in relation to the other *Microcavia*, however, is not fully resolved. Quintana (1996) suggested that *M. shiptoni* was most similar to *M australis* and the extinct *M. robusta*. Based on molecular data, the best supported hypothesis is a *shiptoni*–*niata* sister relationship (Dunnum 2009; Mahadeshwar 2010).

Subfamily Dolichotinae Pocock, 1922

The Dolichotinae comprises the single Recent genus *Dolichotis*. This subfamily appears as sister to the Hydrochoerinae in molecular phylogenetic analyses (Opazo 2005; Upham and Patterson 2012) as well as combined fossil and molecular work (M. E. Pérez and Pol 2012), with the divergence estimates for these subfamilies ranging from the early Middle Miocene (Opazo 2005; Rowe et al. 2010) to the late Miocene (M. E. Pérez and Pol 2012).

Genus *Dolichotis* Desmarest, 1819

I include both species of Mara within the genus *Dolichotis*, although many authors (Cabrera 1961; Mares and Ojeda 1982; Mares, Ojeda, and Barquez 1989; Rowe and Honeycutt 2002; Barquez, Díaz, and Ojeda 2006) have placed *D. salinicola* in the genus *Pediolagus*. The Maras are currently distributed throughout the arid and semi-arid lowlands of southern Bolivia, Paraguay, and Argentina, and are present in the fossil record from early and mid-Pleistocene deposits (McKenna and Bell 1997). Opazo (2005) estimated divergence time of the two extant species at 7.5 ± 4.8 mya, Dunnum (2009) at 5.9 mya (95% HPD

1.8–9.9 mya), both based on molecular data but utilizing differing methods. These dates are similar to those within other caviid genera and, I argue, provide further support for the inclusion of both species within the single genus *Dolichotis*.

Maras are hare-like in appearance and are adapted for a cursorial life. The limbs are long in comparison to other caviids, with the hind limbs much longer and more muscular than the forelimbs. The radius is longer than the humerus. The hind limbs have three hoof-like digits with the central digit elongated. The forefeet have four sharply clawed digits with the middle two equally enlarged. The maras are much larger than other Caviinae, ranging from 450 mm head and body length in *D. salinicola* to 800 mm in *D. patagonum*. The hairy ears are long, with a simple supertragus. The short tail is nearly hairless. The dorsal pelage is agouti gray, and a distinct white rump band is present across the tail. Under parts are white; the flanks and chin are orange. A single pair of anal glands is present above the anus, in contrast to a position anterior to the anus in all other caviids (C. M. Campos, Tognelli, and Ojeda 2001).

Cranially, the frontals are broadened, with the orbits roofed in by expansion of the frontal bone, which is deeply notched anteriorly. The nasals are specialized and do not extend as far forward as the premaxillae. The paroccipital processes are elongated more so than in genera of Caviinae or *Kerodon*, although less than in *Hydrochoerus*. The palate is short, extending posteriorly only to about the level of M2. The jugal is broad and short. Auditory bullae are moderately large. The cheek teeth are ever-growing and unilaterally hypsodont. The upper cheek teeth are bilobed, except for M3, which is divided into three lobes by two reentrant folds. Lower cheek teeth have one labial fold dividing each tooth into two lobes (Ellerman 1940). The diastema is longer than the length of the maxillary toothrow, the nasolacrimal foramen is absent in the lateral view, and the mesopterygoid fossa is positioned at the level of the anterior and posterior prisms of the M2 (Ubilla and Rinderknecht 2003).

SYNONYMS:

Cavia: Zimmermann, 1780:328; description of *patagonum*; not *Cavia* Pallas.

Lepus: Oken, 1816:825; inclusion of *patagonicus*; unavailable name (ICZN 1956).

Dolichotis Desmarest, 1819g:211; type species *Cavia patachonica* Shaw, by original designation (= *Cavia patagonum* Zimmermann).

Dasyprocta: Desmarest, 1822:358; part (listing of *patachonica*); not *Dasyprocta* Illiger.

Chloromys: Desmoulins, 1823:47; part (listing of *patagonicus*); not *Chloromys* F. Cuvier.

Mara d'Orbigny, 1829:220; based on "La Lièvre de Patagonie."

Pediolagus Marelli, 1927:5; type species *Dolichotis centralis* Weyenbergh, by original designation (= *Dolichotis salinicola* Burmeister, 1876).

Weyenberghia Kraglievich, 1927a:578; type species *Dolichotis salinicola* Burmeister; proposed as subgenus; objective synonym of *Pediolagus* Marelli.

Paradolichotis Kraglievich, 1927b:594; replacement name for *Weyenberghia* Kraglievich, type species *Dolichotis salinicola* Burmeister (= *Dolichotis centralis* Weyenbergh).

Lagospedius Marelli, 1928:103; type species *Lagospedius centralis* (= *Dolichotis centralis* Weyenbergh).

KEY TO THE SPECIES OF *DOLICHOTIS*:

1. Size smaller, head and body length <500; legs relatively short; ulna approximately the same length as the skull or the tibia; nasal bones with the edges not projected forward; two pairs of mammae; penis with two spikes at the base of the glandular sac. . . *Dolichotis salinicola*

1′. Size larger, head and body length >500; legs relatively long; ulna substantially longer than the skull or the tibia; nasal bones with the edges projected forward; four pairs of mammae; penis with no spikes at base of glandular sac . . . .*Dolichotis patagonum*

Dolichotis patagonum (Zimmermann, 1780)
Patagonian Mara

SYNONYMS: See under Subspecies.

DESCRIPTION: Second largest caviid, after capybaras total length 600–800 mm, greatest length of skull 125–133 mm, and mass 7–8 kg (C. M. Campos, Tognelli, and Ojeda 2001). Dorsal coloration agouti gray; lower portions of rump with sharply defined white patch separated from dorsal area by contrasting darker area; venter white and flanks, cheeks, chin and chest orange brown. Cranial and tooth characters as for genus.

DISTRIBUTION: *Dolichotis patagonum* occurs in lowland forest, shrub, and grasslands of Argentina, between approximately 28°S to 50°S (C. M. Campos, Tognelli, and Ojeda 2001) but locally extirpated from some regions such as Buenos Aires province (Cabrera 1953).

SELECTED LOCALITIES (Map 380): ARGENTINA: Buenos Aires, D'Orbigny (Cabrera 1953); Catamarca, Catamarca (Cabrera 1953); Córdoba, Cruz del Eje (Thomas 1902d), near border of San Luis on road between Río Cuarto and Mercedes (Cabrera 1953); La Rioja, La Rioja (Cabrera 1953); San Luis, San Luis (Cabrera 1953); Santa Cruz, Bahía del Fundo (Krumbiegel 1941); Río Negro, east of Nahuel Huapi (Krumbiegel 1941).

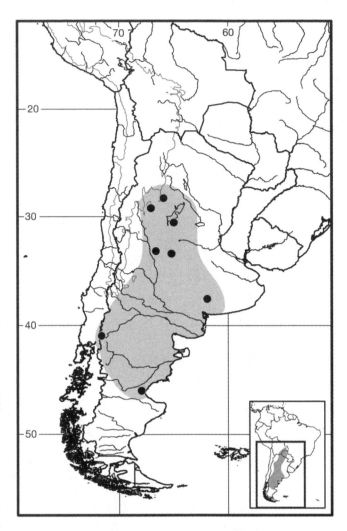

Map 380 Selected localities for *Dolichotis patagonum* (●). Contour line = 2,000 m.

SUBSPECIES: I follow Cabrera (1953) and C. M. Campos, Tognelli, and Ojeda (2001) and recognize two subspecies.

D. p. patagonum (Zimmermann, 1780)
SYNONYMS:

Cavia patagonum Zimmermann, 1780:328; type locality unknown. Tate (1935:355) stated that Zimmermann "alluded to Pennant's knowledge of the animal, but no mention of it by Pennant appears until his second edition of 'History of Quadrepeds'." Tate (1935) listed the type locality as "Patagonia."

Cavia magellanica Kerr, 1792:220; type locality "inhabits the country about Port Desire in Patagonia," Argentina, based on the Patagonian Cavy of Pennant (1781).

Cavia Patachonica Shaw, 1801:226; type locality "Patagonia," Argentina, based on the Patagonian Cavy of Pennant (1781).

Lepus patagonicus Oken, 1816:825; name combination and incorrect subsequent spelling of *patagonum* Zimmermann, 1780; unavailable name (ICZN 1956).

Dasyprocta patagonum: Illiger, 1815:108; name combination.

[*Dolichotis*] *patachonica*: Desmarest, 1819g:210; name combination.

[*Dolichotis*] *patagonicha* Desmarest, 1819h:40; incorrect subsequent spelling of *patachonica* Shaw, 1801.

Dolichotis [*magellanica*]: Say, 1821:331, footnote for *Cavia magellanica*; name combination.

Dasyprocta patachonica: Desmarest, 1822:358; name combination; placed *Cavia patachonica* Shaw, 1801 in *Dasyprocta* Illiger [Dasyproctidae] but called attention to the genus *Dolichotis* in a footnote.

Chloromys patagonicus: Desmoulins, 1823:47; name combination, and incorrect subsequent spelling of *Cavia patachonica* Shaw.

D[*asyprocta*]. *patagonica* Gray, 1827:272; incorrect subsequent spelling of *patachonica* Shaw.

Mara magellanica: Lesson, 1831:113; name combination.

Mara patagonica Lesson, 1831:plate xlii; name combination, and incorrect subsequent spelling of *patachonica* Shaw.

Chloromys patachonicus Waterhouse, 1839:89; incorrect subsequent spelling of *patachonica* Shaw.

D[*olichotis*]. *patagonica*: Wagner, 1844:iv:66; name combination and incorrect subsequent spelling of *patachonica* Shaw.

Dolichotis patichonica Waterhouse, 1848:plate 3, Fig. 1; incorrect subsequent spelling of *patachonica* Shaw.

Dolichotis patagona: J. A. Allen, 1902a:22; argued that *patagonum* Zimmermann took precedence over *patachonica* Shaw.

Dolichotis magellanicus: Thomas, 1929:44; justified emendation of *magellanica* Kerr; considered *Dolichotis* a masculine noun.

D[*olichotis*]. *magellanica magellanica*: Yepes, 1935a:245; name combination and incorrect gender agreement.

Dolichotis australis: Cabrera and Yepes, 1940:233; name combination.

Dolichotis [(*Dolichotis*)] *patagona patagona*: Ellerman, 1940:248; name combination and incorrect gender agreement.

Dolichotis patagonum: Krumbiegel 1941:21; name combination.

Dolichotis patagonum patagonum: Krumbiegel, 1941b:24; first use of current name combination.

The nominotypical subspecies is known from central Buenos Aires province west to Córdoba, San Luis and Mendoza provinces, and south to Santa Cruz province in Argentina (Cabrera 1953).

D. p. centricola (Thomas, 1902)

SYNONYMS:

Dolichotis magellanicus centricola Thomas, 1902d:242; type locality "Cruz del Eje, Central Cordova," Córdoba Province, Argentina.

D[*olichotis*]. *salinicola*: Thomas, 1902d:243; part; not *salinicola* Burmeister.

Dolichotis [(*Dolichotis*)] *patagona centricola*: Ellerman, 1940; name combination and incorrect gender agreement of *patagonum* Zimmermann.

Dolichotis australis centricola: Cabrera and Yepes, 1940:233; name combination.

Dolichotis patagonum centricola: Krumbiegel, 1941:24; first use of current name combination.

This subspecies occurs in Catamarca, the eastern part of La Rioja, northwestern Córdoba, and southwestern Santiago del Estero provinces, Argentina (Cabrera 1953).

Cabrera (1953) remarked on the misleading nature of the type locality "Cruz del Eje, Central Cordova," which refers to the name of a train station and not actually to central Córdoba province. The subspecies only reaches the northern portion of Córdoba province.

NATURAL HISTORY: C. M. Campos, Tognelli, and Ojeda (2001) reviewed the ecology and life history of the Patagonian Mara. This species is an herbivore, feeding predominantly on fruits and green vegetation (Bonino et al. 1997; C. M. Campos 1997; C. M. Campos and Ojeda 1997) with a preference for grasses over shrubs and forbs (Sombra and Mangione 2005). Home range size varied from 33.25 to 197.5 ha, with a mean of 97.87 ha (Taber 1987). They are strongly monogamous, with pairs remaining together until death (Genest and Dubost 1975). Breeding is either communal or solitary, with increased pup survival in dens with higher numbers. Selection of open habitats for breeding warrens resulted in significantly higher pup survival in Mara populations in Península Valdés (Baldi 2007). Breeding in southern Argentina occurred from August to December or January (Baldi 2007; Taber and MacDonald 1992). Gestation was 100 days in the wild and most females produced one litter per year (Taber, 1987). Four pairs of ventral mammae are present in females (Weir 1974b). Eisenberg (1974) described vocalizations. Rues et al. (2013) delineated trophic interactions between maras and the introduced European hare (*Lepus europaeus*) in San Juan province, Argentina. Sutton and Hugot (1987), Sutton and Durette-Desset (1995), and Porteous and Pankhurst (1998) provided information on parasites.

REMARKS: C. M. Campos, Tognelli, and Ojeda (2001) summarized the nomenclatural history and recognized two subspecies, following Cabrera (1953). Tate (1935:355) noted that Lesson (1827) employed the combination *Lepus magellanicus*. Cabrera (1953), however, argued that Lesson's name actually referred to the European hare, *Oryctolagus cuniculus*, not *Dolichotis*. Wurster et al. (1971) described a karyotype with $2n = 64$, FN = 124, and metacentric sex chromosomes.

Dolichotis salinicola Burmeister, 1876
Chacoan Mara
SYNONYMS:

Dolichotis salinicola Burmeister, 1876a:635; type locality "Central Argentine Railway . . . near the stations Totoralejo and Recreo, about lat. 29° S and long. 65° W," Catamarca, Argentina.

Dolichotis centralis Weyenbergh, 1877:247; type locality "der Sierra de Cerdoba (*sic*) . . . die tusschen de dorpen Perchal en Quilpolos," Sierra de Córdoba, presumably between the pueblos of Perchal and Quilpo, Córdoba province, Argentina.

Dolichotis salinica Weyenbergh, 1877:247; incorrect subsequent spelling of *salinicola* Burmeister.

Dolichotis patagonica: Burmeister, 1879:260; part (see Berg 1898:24); not *patachonica* Shaw [= *patagonum* Zimmerman].

[*Dolichotis patagonica*] *salinicola*: E.-L. Trouessart, 1880b: 196–197; name combination, and incorrect subsequent spelling of *patachonica* Shaw.

[*Dolichotis*] *centralis*: E.-L. Trouessart, 1904:528; name combination.

Pediolagus centralis: Marelli, 1927:plates 1, 2, and 5; name combination.

Weyenberghia salinicola: Kraglievich, 1927a:578; name combination as subgenus.

[*Dolichotis*] (*Paradolichotis*) *salinicola*: Kraglievich, 1927b: 594; name combination.

Pediolagus salinicola: Yepes, 1935a:245; name combination.

Dolichotis [(*Paradolichotis*)] *salinicola*: Ellerman, 1940: 248; name combination.

Dolichotis salinicola ballivianensis Krumbiegel, 1941:22; type locality "Fortin Ballivian am Pilcomayo," Río Pilcomayo, Salta, Argentina.

Dolichotis salinicola salinicola: Krumbiegel, 1941:23; first use of current name combination.

Pediolagus salinicola cyniclus Cabrera, 1953:75; type locality "La Florencia, Formosa," Argentina.

DESCRIPTION: General appearance similar to *D. patagonum* but overall size smaller and pelage coloration duller with less orange. Total length 439–515 mm, hindfoot length 91–101 mm, ear length 58–64 mm, and mass 1,800–2,300 g (Mares, Ojeda, and Barquez 1989). Dorsum agouti brown, extending onto sides and throat; sides somewhat lighter and venter white and tan; cheeks and chin washed with tan; white patches occur in front of and behind eyes. Limbs shorter than in *D. patagonum*; ulna approximately same length as skull or tibia. Two pairs of mammae present and penis contains two spikes at base of glandular sac. Cranial characters as for genus but, in contrast to *D. patagonum*, lower anterior prolongation of nasals rudimentary or absent (Ellerman 1940).

DISTRIBUTION: *Dolichotis salinicola* occurs in the Chaco ecoregions of extreme southern Bolivia, Paraguay and northwestern Argentina as far south as Córdoba province.

SELECTED LOCALITIES (Map 381): ARGENTINA: Córdoba, Cruz del Eje (Thomas 1902d); Formosa, La Florencia (Cabrera 1953); La Rioja, La Rioja (Cabrera 1953); Salta, Fortín Ballivian (*Dolichotis salinicola ballivianensis*), Macapillo (Cabrera 1953). BOLIVIA: Santa Cruz, San José Chiquitos (S. Anderson 1997); Tarija, Pilcomayo (S. Anderson 1997). PARAGUAY: Alto Paraguay, Estancia 3 Marias (TTU 79838); Presidente Hayes, Estancia Samaklay (TTU 80407).

SUBSPECIES: *Dolichotis salinicola* is monotypic.

NATURAL HISTORY: The ecology of *D. salinicola* is similar to that of *D. patagonum*. It occupies the dry, low, thorny scrub of the arid Chaco. The species is diurnal and social, often found in pairs or groups, living in self-constructed burrows or those abandoned by viscachas. Females have two pairs of mammae. Gestation period is about 77 days, with one to two young per litter (Weir 1974b), although, Mares, Ojeda, and Barquez (1989) reported a gestation of two months and litter sizes ranging from two to five.

REMARKS: Burmeister's (1876a) original description of *D. salinicola* was based on young individuals about half adult size; he subsequently (1876b) elaborated on the external characteristics of this species based on his observations of two living adults. Cabrera (1961) included *D. salinicola* in the genus *Pediolagus* and recognized three subspecies, *ballivianensis* Krumbiegel, *cyniclus* Cabrera, and *salinicola* Burmeister. Woods (1984, 1993) followed Ellerman (1940) and regarded *Pediolagus* as a subgenus of *Dolichotis*. Verzi and Quintana (2005) recorded this species from the San Andrés Formation of the late Pliocene of coastal Buenos Aires province, Argentina.

Kraglievich (1927a:578) proposed *Weyenberghia* as a subgenus; however, later in the same volume of *Physis* (Kraglievich 1927b:594), he proposed the replacement name *Paradolichotis*, stating that *Weyenberghia* had been previously applied to an insect and was thus unavailable

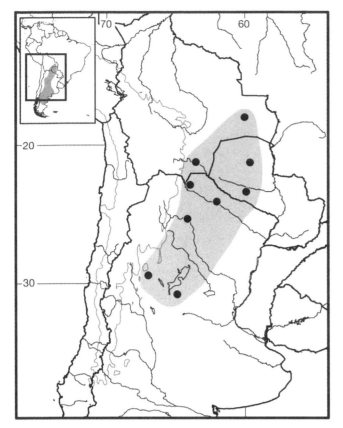

Map 381 Selected localities for *Dolichotis salinicola* (●). Contour line = 2,000 m.

as a homonym. I am unable to find generic use of *Weyenberghia* predating Kraglievich's publication. Kraglievich may have been mistaken in thinking the name was preoccupied by *Lasiodora weijenberghii* Thorell, 1894 (an arachnid). This tarantula species was subsequently included within *Weyenberghia* Mello-Leitao 1941. Cabrera (1961) was not sure if *Weyenberghia* was indeed preoccupied and followed it with a question mark ("Preocupado?").

Subfamily Hydrochoerinae Gray, 1825

Kerodon is included within this subfamily due to the consistently observed sister taxon relationship between *Hydrochoerus* and *Kerodon* in both morphological and molecular sequence analyses (Woods 1984; Rowe and Honeycutt 2002; Opazo 2005; Upham and Patterson 2012). Divergence between the two genera occurred in the mid-Miocene, from between 8 mya (M. E. Pérez and Pol 2012) and 12.28 ± 2.3 mya (Opazo 2005; see also Upham and Patterson 2012). Several distinct lineages of the Hydrochoerinae are known in the fossil record and some fossil forms extended into North America during the late Pliocene and Pleistocene (Mones and Ojasti 1986; Woods 1984; Mones 1991; Walton 1997; Vucetich et al. 2005, 2013; Deschamps et al. 2012). A complete

review of virtually all aspects of the evolution and ecology of capybara is available in J. R. Moreira et al. (2013).

Genus *Hydrochoerus* Brisson, 1762

Two species are recognized, *H. hydrochaeris* and *H. isthmius*. The genus is found in the fossil record from the Pleistocene in North America and the early or mid-Pleistocene in South America (McKenna and Bell 1997).

Capybaras are the largest living rodent, similar in appearance to *Cavia* only much larger with a head and body length varying from 100 to 130 cm and mass from 27 to 79 kg. The tail is vestigial. The pelage is sparse, long, and coarse, reddish brown to grayish dorsally and yellow brown ventrally. Mature males possess enlarged sebaceous glands on the snout. The forefoot has four digits and hindfoot has three, as in the other Caviids. The digits are arranged in a radial pattern and are partially webbed. Females have six pairs of ventral mammae. The limbs are short, the head is broad and large, and the muzzle is heavy and truncate. The eyes are located dorsally and far back on the head. Toothrows converge anteriorly. The incisors are white, with shallow grooves on their anterior surfaces. Cheek teeth are evergrowing and more complex than the other caviids, composed of multiple plates or lophs. M3 is enormously enlarged, exceeding the length of the other three molariform teeth combined (Ellerman 1940; Woods 1984).

SYNONYMS:

Mus: Linnaeus, 1758:59; part (description of *porcellus*); not *Mus porcellus* Linnaeus.

Hydrochoerus Brisson, 1762:80; type species *Sus hydrochaeris* Linnaeus, by subsequent designation (ICZN 1998).

Cavia: Pallas, 1766:31; part (description of *capybara*); not *Cavia* Pallas.

Sus: Linnaeus, 1766:102; part (description of *hydrochaeris*); not *Sus* Linnaeus.

Hydrochaeris Brünnich, 1771:44; incorrect subsequent spelling of *Hydrochoerus* Brisson.

Hydrochaerus Erxleben, 1777:191; part (description of *capybara*); incorrect subsequent spelling of *Hydrochoerus* Brisson.

Calva Gmelin, 1788:120; part (listing of *capybara*); incorrect subsequent spelling of *Cavia* Pallas.

Savia Treviranus, 1802:211; part (listing of *capybara*); unjustified emendation of *Cavia* Pallas.

Hydrochoenus Gray, 1821:304; incorrect subsequent spelling of *Hydrochoerus* Brisson.

Hydrocharus Gray, 1825; incorrect subsequent spelling of *Hydrochoerus* Brisson.

Hydrocherus F. Cuvier, 1829a:492; incorrect subsequent spelling of *Hydrochoerus* Brisson.

Capibara Moussy, 1860:13; *nomen nudum.*

Capiguara Liais, 1872:545; new name for *Hydrochoerus* Brisson.

Hydrochoeris J. A. Allen, 1916d:568; incorrect subsequent spelling of *Hydrochoerus* Brisson.

Xenohydrochoerus Rusconi, 1934a:21; type species *Xenohydrochoerus ballesterensis* Rusconi, by original designation.

Hydrocheirus Hollande and Batisse, 1959:1; incorrect subsequent spelling of *Hydrochoerus* Brisson.

REMARKS: Husson (1978:456) discussed the spelling of the generic name and concluded that *Hydrochoerus* Brisson was not valid because Brisson's (1762) work was not consistently binominal. The International Commission on Zoological Nomenclature (ICZN 1998) subsequently ruled that Brisson's names, including *Hydrochoerus* Brisson, were valid. Moussy (1860) is a summary of the wild and domestic animals occurring in Argentina and includes no descriptions. His *Capibara* (based on Marcgraf 1648) is listed in the synonymy for the capybara and should be regarded as a *nomen nudum.* Mones (1991) reviewed the genus.

KEY TO THE SPECIES OF *HYDROCHOERUS*:
1. Size large, head and body length >1 m; premaxillae with posteriorly extended portions ending well in front of posterior plane of incisive foramina; $2n = 66$, FN $= 102$ . . . . *Hydrochoerus hydrochaeris*
1'. Size small, head and body length <1 m; premaxillae with posteriorly extended portions reaching to near posterior plane of incisive foramina; $2n = 64$, FN $= 104$ . . . . *Hydrochoerus isthmius*

Hydrochoerus hydrochaeris (Linnaeus, 1766)

Capybara

SYNONYMS:

Mus porcellus Linnaeus, 1758:59; part; type locality "Brasilia," Brazil.

[*Sus*] *hydrochaeris* Linnaeus, 1766:103; type locality "Habitat in Surinamo" (= Surinam); restricted by Feijó and Langguth (2013:100) to "Rio São Francisco, 2 km suboeste da cidade de Penedo, estado de Sergipe, Brasil" by selection of neotype (see pp. 103–107).

[*Cavia*] *capybara* Pallas, 1766:31; type locality unknown.

Mus cobiai P. L. S. Müller, 1776:40; in reference to "*Le Cobiai*," the common name given to the capybara by Buffon (1776:470, plate 28).

Sus hydrochoerus Zimmermann, 1777:552; incorrect subsequent spelling of *Sus hydrochaeris* Linnaeus.

Hydrochaerus capybara Erxleben, 1777:193; type locality "America australi ad fluvios maiores."

[*Calva*] *capybara*: Gmelin, 1788:123; name combination.

Cavia cobaya: Lacépède, 1799:9; name combination.

S[*avia*]. *capybara*: Treviranus, 1802:211 ; name combination.

Hydrochoerus capybara: É. Geoffroy St.-Hilaire, 1803:163; name combination.

Sus hydrochaerus Bewick, 1804:381; incorrect subsequent spelling of *Sus hydrochaeris* Linnaeus.

H[*ydrochoerus*]. *capibara* Wied-Neuwied, 1826:475; incorrect subsequent spelling of *capybara* Erxleben.

Capibara brasiliensis Moussy, 1860:13; based on "Capybara brasiliensibus" of Marcgraf (1648); *nomen nudum.*

Capiguara americana Liais, 1872:545; new name for *Hydrochaerus capybara* Erxleben.

Hydrochoerus capivara Winge, 1887:69; incorrect subsequent spelling of *capybara* Erxleben.

Hydrochoerus irroratus Ameghino, 1889:911 and plate 79; type locality "Las barrancas de los alrededores de la ciudad de Paraná," Entre Ríos, Argentina.

Hydrochoerus hydrochaeris: Merriam, 1895b:376; first use of current name combination.

Hydrochoerus hydrochoerus Berg, 1900:221; name combination, but incorrect subsequent spelling of *hydrochaeris* Linnaeus.

Hydrochoerus uruguayensis C. Ameghino and Rovereto, in Rovereto, 1914:144; type locality unknown, restricted to "Castillos, Depto. de Rocha, Uruguay" by Ximénez et al. (1972:22).

Hydrochoerus capybara dabbenei Rovereto, 1914:144; type locality "del territorio de Misiones," Argentina.

Hydrochoerus hydrochaeris notialis Hollister, 1914a:58; type locality "Paraguay."

Hydrochoeris hydrochoeris: J. A. Allen, 1916d:568; name combination and incorrect subsequent spelling of *hydrochaeris* Linnaeus.

Hydrochoerus hydrochaeris hydrochoeris: F. W. Miller, 1930:17; name combination and incorrect subsequent spelling of *hydrochaeris* Linnaeus..

H[*ydrochoerus*]. *hydrochoeris uruguayensis*: Kraglievich, 1930b:243; name combination.

Hydrochoerus hidrochoerus Pittier and Tate, 1932:265; incorrect subsequent spelling of *hydrochaeris* Linnaeus.

Hydrochoerus hydrochoeris notialis: Devincenzi, 1935:78; name combination.

Hydrocheirus capybara: Hollande and Batisse, 1959:1; name combination.

Hydrochaeris hydrochaeris: Cabrera, 1961:583; name combination.

Hydrochaeris hydrochaeris dabbenei: Cabrera, 1961:583; name combination.

Hydrochaeris hydrochaeris uruguayensis: Cabrera, 1961:584; name combination.

Hydrochoerus cololoi Berro, in Francis and Mones, 1968:46; *nomen nudum.*

Hydrochoerus hydrochaeris uruguayensis: Barlow, 1969: 44; name combination.

DESCRIPTION: General characteristics as for genus; body large and heavy, limbs short, and forefeet and hindfeet mesaxonic and webbed. Larger of two species, with length over 1,300 mm and mass over 75 kg. Hair course, from dark brown and reddish to light brown and yellowish. Hair length 30–120 mm. Skull with strong zygomatic, broad jugal, and extremely elongated paroccipital process. Palate long, reaching posterior part of M3. Bullae proportionally smaller than in other Caviidae. Pterygoid fossa deep but not perforated. Angular process of mandible not everted as in typical Hystricognath rodents. First three upper cheek teeth composed of two Y-shaped prisms; M3 11 to 14 prisms, first Y-shaped, others comprised of single plates, except posterior ones, which may vary. Lower cheek teeth composed of three prisms, which may be separated into up to six plates in m3. Prisms on upper and lower teeth separated by thick cement lamina. Description based on Mones and Ojasti (1986).

DISTRIBUTION: *Hydrochoerus hydrochaeris* occurs east of the Andes in eastern Colombia through Venezuela, the Guyanas, Ecuador, and Peru, south through Brazil, Bolivia, Paraguay, northwestern and eastern Argentina, and Uruguay.

SELECTED LOCALITIES (Map 382): ARGENTINA: Entre Ríos, islands of the lower Río Paraná delta (R. D. Quintana 2002), Jujuy, Agua Salada (M. M. Díaz and Barquez 2007), Paraná (type locality of *Hydrochoerus irroratus* Ameghino). BOLIVIA: Beni, camino San Borja to Trinidad (S. Anderson 1997); Santa Cruz, Buena Vista (S. Anderson 1997). BRAZIL: Ceará, Lagoa do Catu (Feijó and Langguth 2013); Maranhao, Miritiba (Mones 1991); Pará, 52 km SSW of Altamira, right bank Rio Xingu, (Voss and Emmons 1996), Ilha de Marajó, Fazenda Eco-Búfalos (J. R. Moreira et al. 1997); Paraíba, Rio Cuiá, bairro Valentina (Feijó and Langguth 2013); Rio de Janiero, Vale do Paraíba, Itatiaia (Avila-Pires 1977); Sergipe, Rio São Francisco, 2 km SW of Penedo (Feijó and Langguth 2013). COLOMBIA: Arauca, Campo Petrolero de Caño Limón (Forero-Montana et al. 2003). ECUADOR: Pastaza, Mera (Rageot and Albuja 1994). FRENCH GUIANA: Cayenne, Les Nouragues (Voss and Emmons 1996). GUYANA: Mazaruni-Potaro, Kartabo (Voss and Emmons 1996). PARAGUAY: Boquerón, Fortín Juan de Zalazar (Wetzel and Lovett 1974). PERU: Amazonas, Huampami, Río Cenepa (Patton et al. 1982); Madre de Dios, Cocha Cashu Biological Station (Voss and Emmons 1996). SURINAM: Wanica, Kwatta (Husson 1978). URUGUAY: Canelones, Arroyo Tropa Vieja (Barlow 1969); Trienta Tres, 16 km SSW of Boca del Río Tacuari (Barlow 1969). VENEZUELA: Apure, Río Cinaruco, 65 km NW of Pto. Páez (Hand-

ley 1976); Aragua, Lake Valencia (Osgood 1910); Bolívar, Río Supamo, 50 km SE of El Manteco (Handley 1976); Monagas, Hato Mata de Bejuco (Handley 1976).

SUBSPECIES: Cabrera (1961) recognized three subspecies (*dabbenei* Rovereto from Paraguay and northeastern Argentina, *uruguayensis* Ameghino and Rovereto from Uruguay and eastern Argentina, and the nominotypical form from the remainder of the range). Mones and Ojasti (1986:1), however, noted that populations exhibited a latitudinal cline in body size and mass, and that subspecies recognition could only be "based on extreme populations and arbitrary limits." They chose to view *H. hydrochaeris* as monotypic although highly variable geographically.

NATURAL HISTORY: Mones and Ojasti (1986) reviewed the ecology, behavior, reproduction, and other aspects of population biology, including parasites. Capybaras occupy a wide variety of mesic lowland habitats including forested riverbanks, mangrove swamps, and brackish wetlands. Regarded as a semiamphibious mammal, it requires water for drinking, wallowing, and protection. They are diurnal herbivores and feed on grasses, sedges, and semi-aquatic plants. They live in herds typically ranging in size from two to 30 animals controlled by a dominant male. During drought periods, temporary herds of up to a hundred animals may aggregate around remaining waterholes. Home range size varies from 10 ha in high-density populations to over 200 ha. Capybara breed throughout the year but mating frequency appears to increase at the onset of the rainy season. The gestation period is 150 days and females normally breed once per year, twice if conditions remain favorable. Litter size ranges from one to seven, and averaged four in Venezuelan populations. Over 80 parasites are listed in the specialized literature. Borges-Landáez et al. (2012) examined potential gene flow effects generated by hunting pressure, and thus the preservation of genetic diversity, in populations of capybara in the llanos of western Venezuela. Moriera et al. (2013) reviewed capybara biology, evolution, and ecology.

REMARKS: Marcgraf (1648:230) was the first to describe and illustrate a capybara in Willem Piso's *Historia naturalis Brasiliae*, a description upon which Linnaeus (1766:103) subsequently based his *Sus hydrochaeris*. Hollister (1914) mentioned Surinam as the type locality of the capybara, following the statement by Linnaeus (1766) of "habitat in Surinamo." Tate (1935:355), however, gave the type locality as "Brazil," acknowledging that, while Linnaeus wrote Surinam, he had based his *Sus hydrochaeris* wholly upon Marcgarv's description and his own examination of a young animal, apparently without a stated locality of origin. Marcgraf's travels and observations were confined to northeastern Brazil. In concert with Tate (1935), Cabrera (1961:583) gave the type locality as "Pernambuco, Brazil,"

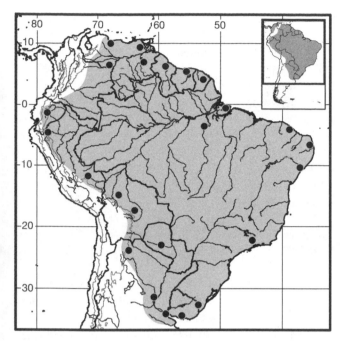

Map 382 Selected localities for *Hydrochoerus hydrochaeris* (●). Contour line = 2,000 m.

Carvalho (1965) cited it as "banks of the São Francisco river, in the State of Alagoas," and Mones (1975) listed it as "Río São Francisco en la frontera entre los estados de Alagoas y Sergipe, Brasil." Husson (1978:451) selected the juvenile specimen apparently seen by Linnaeus as the lectotype and restricted the type locality to Surinam. Most recently, Feijó and Langguth (2013:103–107) designated a neotype for *Sus hydrochaeris* Linnaeus, with its precise locality of "Rio São Francisco, 2 km suboeste da cidade de Penedo, estado de Sergipe, Brasil," because (1) the type locality of *Sus hydrochaeris* Linnaeus remained uncertain, (2) Husson's choice of a lectotype did not by itself define Surinam as the type locality since the origin of Linnaeus's specimen was uncertain, (3) there was no specimen corresponding to the lectotype, and (4) *Sus hydrochaeris* Linneaus was polytypic, with multiple subspecies subsequently named.

The karyotype consists of a $2n = 66$ and $FN = 102$ (Wurster et al. 1971).

Hydrochoerus isthmius Goldman, 1912
Lesser Capybara

SYNONYMS:

Hydrochaerus hydrochaeris: J. A. Allen, 1904c:444; listing of a specimen from Mamatoca, Magdalena, Colombia.

Hydrochoerus hydrochaeris: Osgood, 1912:56; listing of specimens from El Panorama, Río Aurare, Zulia, Venezuela.

Hydrochoerus isthmius Goldman, 1912b:11; type locality "Marragantí, near the head of tide-water on the Río Tuyra, eastern Panama," Darién.

Hydrochaeris isthmius: Hall and Kelson, 1959:785; name combination.

Hydrochaeris hydrochaeris isthmius: Cabrera, 1961:584; name combination.

Hydrochaeris hydrochaeris: Handley, 1966:785; name combination.

Hydrochoerus hydrochaeris isthmius: D. Heinemann, 1975:446; name combination.

DESCRIPTION: Smaller than *H. hydrochaeris*, with total length approximating 100 cm and greatest skull length of holotype 200 mm. Pelage color ranges from dark reddish to dull clay color, usually darker above than below, and blackish in some specimens on middle of face, cheeks, lower part of rump, and outer sides of hind legs. Eyerings, sides of muzzle, and spots at base of ears paler. The ears and feet brownish (Goldman 1912b). Measurements of holotype include head and body length 1,025 mm and hindfoot length 200 mm (Goldman 1912b). Skull with pterygoid processes shorter, thicker, and more rounded, less produced posteriorly.

DISTRIBUTION: *Hydrochoerus isthmius* occurs in Panama, western Colombia, and northwestern Venezuela.

SELECTED LOCALITIES (Map 383; South America only): COLOMBIA: Magdalena, Santa Marta, Mamatoco

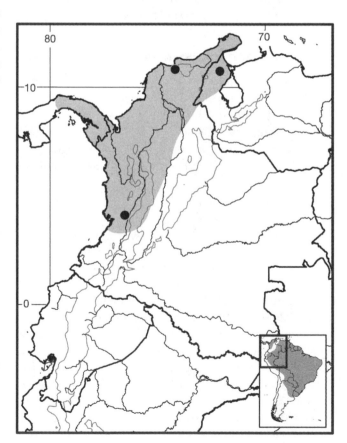

Map 383 Selected localities for *Hydrochoerus isthmius* (●). Range in Panama from Hall (1981). Contour line = 2,000 m.

(J. A. Allen 1904c); Valle del Cauca, Río Frio (J. A. Allen 1916b). VENEZUELA: Zulia, El Panorama (Osgood 1912).

SUBSPECIES: *Hydrochoerus isthmius* is monotypic.

NATURAL HISTORY: Little is known about the population biology and ecology of this species. Populations in Panama are scarce and shy and although small groups can be seen at dusk or dawn around Gamboa pond and on the shores of islands in the Panama Canal, in other areas they are mostly nocturnal (F. A. Reid 1997). The gestation period ranges from 104 to 111 days (Weir 1974b). In northwestern Colombia in Córdoba department, Lesser Capybara's were abundant where they used stream edges, swamps, and permanent lagoons (Ballesteros Correa and Jorgenson 2009). Herd sizes averaged 4.9 individuals, with the threat of local populations resulting from hunting and habitat destruction. Principal predators included domestic dogs, *Boa constrictor*, the crocodilians *Caiman crocodiles* and *Crocodilus acutus*, and the Crested Eagle (*Morphnus guianensis*).

REMARKS: Handley (1966:785) and Mendez (1993) regarded *H. isthmius* as a subspecies of *H. hydrochaeris*, but Mones (1991) in his systematic and phylogenetic review recognized it as a distinct species. The karyotype is $2n = 64$ and FN = 104 (Peceño 1983).

Genus *Kerodon* F. Cuvier, 1823

The genus *Kerodon* contains two species, *K. acrobata* and *K. rupestris*, distributed in eastern and central Brazil, from the state of Piauí to the northern part of Minas Gerais (Cabrera 1961; Moojen et al. 1997). Both are rock specialists that inhabit the semiarid Caatinga region. The genus dates from the late Pleistocene (McKenna and Bell 1997).

Mocos, or rock cavies, are gray-agouti dorsally and light brown on the venter. Typical adult weight can reach 1 kg and body size is equal to or larger than that of *Cavia*. The genus is similar in body form to members of the Caviinae but differs in a number of unique characteristics, including clawless toes of the manus and pes covered with a thick leather-like epidermis, and all digits with subcutaneous nails except the innermost digit of the pes, which has a small grooming claw. A tail is absent or only a vestigial projection. The sternum is narrow and rounded, not broad and flat as in the Caviinae. The skull, especially the rostrum, is much narrower and longer than that of the Caviinae and the length of the diastema is proportionately larger as well (Lacher 1981; Moojen 1952b). The infraorbital foramen lacks a canal for nerve transmission. The incisive foramen is excessively narrowed (Ellerman 1940). G. Lessa et al. (2005) reported on geographic variation in cranial characters among populations of *K. rupestris* and between this species and *K. acrobata*.

Traditionally the genus *Kerodon* has been placed within the Caviinae (Ellerman 1940; Cabrera 1961; Woods 1993; McKenna and Bell 1997) although behavioral and reproductive data suggested the genus was distinct from the other caviines (Lacher 1981). Woods (1984) placed *Kerodon* within the Dolichotinae based on morphological characters but molecular phylogenetic analyses strongly support a sister relationship between *Kerodon* and *Hydrochoerus* (Rowe and Honeycutt 2002; Opazo 2005; M. E. Pérez and Pol 2013).

SYNONYMS:

Cavia Wied-Neuwied, 1820:43; type species *Cavia rupestris* Wied-Neuwied, by original designation; not *Cavia* Pallas.

Kerodon F. Cuvier, 1823b:151; type species the "Moco" of É. Geoffroy St.-Hilaire (= *Kerodon moco* Lesson, = *Cavia rupestris* Wied-Neuwied), by monotypy.

Kerodons F. Cuvier, 1829a:493; incorrect subsequent spelling of *Kerodon* F. Cuvier.

Cerodon Wagler, 1830:18, footnote; unjustified emendation of *Kerodon* F. Cuvier.

Ceratodon Wagler, 1830:18 footnote; preoccupied by *Ceratodon* Brisson [= *Monodon* Linnaeus; Cetacea: Monodontidae].

Prea Liais, 1872:541; part (listing of *rupestris*); new name for *Kerodon* F. Cuvier.

Ceredon Parsons, 1894:252; incorrect subsequent spelling of *Kerodon* F. Cuvier.

REMARKS: Many authors have used the year 1825 for F. Cuvier's description of *Kerodon*, however, his *Des Dents des Mamifères* was published from 1821 to 1825 in parts, with the rodent section appearing in 1823. Isidore Geoffroy St.-Hilaire and d'Orbigny (1833:plate 1 of unnumbered text for *Cavia australis*) designated the type species *Cavia rupestris* "elle est devenue le type d'un petit groupe a part, le genre *Kerodon*." Osgood (1915) restricted *Kerodon* to include only *K. rupestris*, to which Thomas (1916f) concurred, indicating the genus was monotypic. Moojen et al. (1997) more recently described the second species currently recognized.

KEY TO THE SPECIES OF *KERODON*:

1. Size large, mean adult weight 1,040 g, mean head and body length 384 mm, mean hindfoot length 72 mm, mean greatest length of skull 87.6 mm mean incisive foramina length 9.6 mm; rostrum long, ratio of diastema to width of braincase 91.8%; dorsal pelage darker gray, dorsal coloration of feet dark-brown ferruginous, hip pelage not conspicuously colored. Known only from NE Goiás, Brazil . . . . *Kerodon acrobata*

1'. Size small, mean weight 612 g, mean head and body length 297 mm, mean hindfoot length 62 mm, mean greatest length of skull 70.6 mm; mean incisive foramina length 4.3 mm; rostrum short, ratio of diastema to width of braincase 68.8%; dorsal pelage lighter gray, dorsal coloration of feet light-brown with a white band in the internal half, hip pelage brilliant ferruginous color........................ *Kerodon rupestris*

Kerodon acrobata Moojen, Locks, and Langguth, 1997
Acrobatic Moco

SYNONYMS:

Kerodon acrobata Mares and Ojeda, 1982:394; *nomen nudum.*

Kerodon acrobata Moojen, Locks, and Langguth, 1997:1; type locality "Fazenda Santa Helena, at Rio São Mateus, about 72 km from São Domingos and 60 km from Posse (by road), 13°50'S, 46°50'W, Goiás, Brazil."

DESCRIPTION: General characteristics as for genus. Overall color of upperparts dark gray to light brown agouti. Dorsal pelage formed by long, dark guard hairs (30–35 mm), with no light bands, and shorter overhairs (22–25 mm), with subterminal buffy ring and dark tip; all hairs gray at base; sides same color as dorsum although paler; hairs on lower rump and outer sides of legs have light ferruginous colored ring; venter gray with buffy wash, mainly at midline hairs; feet brown-orange dorsally with darker band up middle, with forefeet more lightly colored than hindfeet. Mystacial vibrissae long (up to 105 mm).

Skull with flat profile in middorsal region; rostrum is conspicuously stretched forward; posterior half of frontal not inflated and supraorbital ridges slightly beaded; palatal bridge short and mesopterygoid fossa V-shaped, reaching second prism of M2; pterygoid bones well developed; occiput concave with medial posterior crest; orbital branch of zygomatic process of maxilla not interrupted by lacrimal; occipital foramen as tall as wide, with short, blunt median process projecting ventrally. Mandible with long symphysis, and the angular process is wide at the base; crista masseterica well developed on anterior part and fossa masseterica shallow on posterior part. Incisors orthodont or opisthodont, with white enamel. Upper molars with two cordiform prisms except in PM4, with three that diminish in size posteriorly, and M3, with additional distal extension of posterior prism. Labial face of upper molars with obsolete sulcus, more developed in anterior prism of PM4; sulcus deeper in lower molars. Anterior prism of PM4 smaller than second; second largest in molar series. Moojen et al. (1997) provided descriptive characters of pelage and skull, and Bezerra et al. (2010) compared cranial variation for both sexes and described and figured the glans penis and baculum.

DISTRIBUTION: *Kerodon acrobata* is known from a small region in central Brazil in northeastern Goiás state and adjacent Tocantins state, west of the Espigão Mestre, Serra Geral de Goiás (Moojen et al. 1997; Bezerra et al. 2010).

SELECTED LOCALITIES (Map 384): BRAZIL: Goiás, Fazenda Cana Brava, Nova Roma (Moojen et al. 1997), Fazenda Canadá (Bezerra et al. 2010), Fazenda Santa Helena, Rio São Mateo (type locality of *Kerodon acrobata* Moojen, Locks, and Langguth); Tocantins, Dianópolis (Bezerra et al. 2010).

SUBSPECIES: *Kerodon acrobata* is monotypic.

NATURAL HISTORY: The Acrobatic Moco inhabits seasonally dry tropical forest west of the Serra Geral de Goiás and appears to be restricted to Cerrado habitat. This species is similar in ecology to *K. rupestris*. It is a terrestrial herbivore, well adapted for climbing on rocks and nesting in crevices within the rocks and feeding on Cactaceae, manioc and leaves of the local vegetation. It is extremely agile and able to spring from branch to branch within the tallest bushes (Moojen et al. 1997; Bezerra et al. 2010).

REMARKS: Mares and Ojeda (1982:394) used the name *Kerodon acrobata* for a second species from Brazil but never provided a formal description (Woods 1993:780). Moojen et al. (1997:4) thus correctly regarded *acrobata* Mares and Ojeda a *nomen nudum* and provided the necessary description to establish the validity of the name.

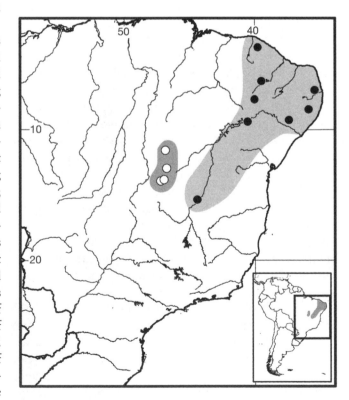

Map 384 Selected localities for *Kerodon acrobata* (○) and *Kerodon rupestris* (●).

Kerodon rupestris (Wied-Neuwied, 1820)
Moco, Rock Cavy
SYNONYMS:
Cavia rupestris Wied-Neuwied, 1820:43; type locality "Rio Grande de Belmonte, am Rio Pardo, am S. Francisco," Bahia, Brazil.
[*Kerodon*] *moco* F. Cuvier, 1823b:151, plate 48; "Nous ne connaissons encore dans ce genre que l'espèce qui nous a donné ce système de dentition et qui recoit au Brésil, suivant M. de Saint Hilaire, le nom de moco."
Kerodon sciureus I. Geoffroy St.-Hilaire, 1826:120; type locality "L'Amérique meridionale est la patrie de cette espèce."
Kerodon rupestris: Rengger, 1830:274, footnote; first use of current name combination.
Kerodon rupestre Lund, 1838:50, footnote; incorrect subsequent spelling of *rupestris* Wied-Neuwied.
Cerodon rupestris: Lund, 1840b:13 [1841c:285]; name combination.
Cavia (Cerodon) rupestris: Waterhouse, 1848:164; name combination.
Prea rupestris: Liais, 1872:541; name combination.
Cavia (Kerodon) ruspestris E.-L. Trouessart, 1880b:196; name combination, and incorrect subsequent spelling of *rupestris* Wied-Neuwied.
Ceredon rupestris: Parsons, 1894:252; name combination.
Cavia sciurus Tate, 1935:337; name combination, and incorrect subsequent spelling of *sciureus* I. Geoffroy St.-Hilaire.
Kerodon aciureus Cabrera, 1961:580; incorrect subsequent spelling of *Kerodon sciureus* I. Geoffroy St.-Hilaire.
DESCRIPTION: General characteristics as for genus. Size large (mass 700–1,000 g), although smaller than *K. acrobata*. Dorsal coloration gray-agouti with black and white mottling; dorsal hairs gray-brown at roots, annulated with grayish white near tip, and black at tip; black prevails onto crown of head but on back dark and pale colors nearly equal; hind part of back assumes brownish color due to pale subterminal ring of each hair having rufous white hue; rump and hinder part of hind legs red chestnut color; throat white, suffused with pale ochraceous yellow; abdomen sometimes white, and chest tinted with rufous in some individuals and yellowish in others; dorsal portions of tarsi whitish and suffused with rufous or pale rust color; ears clothed in pale yellowish hairs. Mystacial hairs black and very long (Waterhouse 1848).

Skull, especially rostrum, narrow and long, with length of diastema also proportionally larger than in caviine taxa (Lacher 1981; Moojen 1952b). Infraorbital foramen lacks canal for nerve transmission; incisive foramen excessively narrowed (Ellerman 1940); occipital surface with round convex area as opposed to concave with median posterior crest as in *K. acrobata*. Masseteric fossa reaches anteriorly only to end of m2. External crest of angle of mandible well developed (Moojen et al. 1997). Upper cheek teeth bilobed with M3 possessing weak, backwardly projecting heel. Lower pm4 with well-marked extra anterior projection and talonid of lower m3 poorly defined (Ellerman 1940). Mean condyloincisive length of skull 73 mm (range: 59–65.9 mm; Hückinghaus 1961a).

DISTRIBUTION: Northeastern Brazil from the state of Minas Gerais in the southwest to Ceará in the northeast.

SELECTED LOCALITIES (Map 384): BRAZIL: Alagoas, Santana do Ipanema, Sítio Goiabeira (Mares, Willig et al. 1981); Bahia, Juàzeiro (Mares, Willig et al. 1981); Ceará, Itapajé (Mares 1981), Juá (Moojen et al. 1997); Minas Gerais, Riacho da Cruz (Mares, Willig et al. 1981); Paraiba, Mulungú (Moojen 1943a); Pernambuco, Bodocó (Moojen et al. 1997), Pernambuco, Bonito (USNM 14761).

SUBSPECIES: *Kerodon rupestris* is monotypic.

NATURAL HISTORY: *Kerodon rupestris* is a terrestrial herbivore and an extreme habitat specialist, restricted to rocky outcrops that occur throughout the Caatinga bioregion of northeastern Brazil. Sexual and agonistic data suggested that they exhibit resource defense polygyny in relation to these outcrops (Lacher 1981). This species exhibited a longer gestation period (75 days) and smaller litter size (mean = 1.4) than any species of the Caviinae. Reproduction occurred year round in the laboratory (Streilen 1982c). As is typical of caviids, juveniles are born precocious and the growth rate is very rapid. Animals can weigh 500 g and are reproductively mature by three months (Lacher 1979). Despite the lack of a tail and claws, individuals are very adept climbers, venturing out of rock outcrops both day and night to forage in adjacent trees and shrubs (Lacher 1981). Streilein (1982a) reported that they gain the majority of their sustenance from treetop foraging. Mares et al. (1982) described molting patterns in relation to two other genera of caviids in the Caatinga. Lacher (1979) suggested that *Kerodon* could be domesticated for use as a supplemental protein source due to their rapid growth rate, early maturation, and ease of reproduction in captivity.

REMARKS: Moojen (1952b:126) listed the type locality as "Rio Grande de Belmonte, Rio Pardo, rio São Francisco," following Wied-Neuwied's (1820) original designation. Cabrera (1961:580) simply listed the type locality as "Río Belmonte, estado de Bahía." The karyotype is $2n = 52$, with variation in FN from 92 to 94 (Maia 1984; G. Lessa et al. 2013).

Family Cuniculidae G. S. Miller and Gidley, 1918
James L. Patton

The pacas include two living species within a single genus that together are among the largest of the extant caviomorph rodents (mass reaching 12–13 kg; Collett 1981; E. M. Pérez 1992; Emmons and Feer 1997). Males are typically larger than females. Both species have moderately robust and stout bodies, a large head, swollen cheeks, thick and fleshy lips, large nares, prominent and stiff vibrissae, short and rounded ears, a large rump, and a tail reduced to a small stump. The eyes are large, dorsally positioned, and widely spaced. The limbs are short; forefeet have four sub-equal toes with a small pollex bearing a claw, hindfeet have three large toes with both a short hallux and fifth digit, neither of which typically contact the ground. The claws are extremely thick, almost hoof-like. The pelage is coarse, without distinct underfur, varying from short and shiny in appearance in *C. paca* to thick, long, and relatively soft in *C. taczanowskii*. In both species, lines of large white spots that sometimes coalesce into stripes mark the dorsal and lateral pelage. Females have one inguinal and one axillary pair of mammae. External genitalia are hidden within an anal sac in both sexes; there is no scrotum in males.

The skull is massive and broad, with short nasals but long and broad frontals, and uniquely notable for a greatly inflated zygomatic region modified by outgrowths of the maxilla and jugal to form large, internally hollowed cheek-plates, the surface of which are highly rugose in males but less so in females. The cheekplates develop quickly from the more typical slender and smooth zygoma of post-natal young (T. W. Nelson and Shump 1978). Both external and internal cheek pouches are associated with the expanded zygoma (A. B. Howell 1940) and, along with the cheek-plate, may serve for amplification of vocalizations and tooth-grinding sounds (Hershkovitz 1955a). The jugal does not approach the lacrimal and lacks a jugal fossa. The lacrimal canal is large and positioned on the side of the rostrum. The infraorbital foramen is rather small, in part because of the anterior expansion of the zygoma, and with a distinct canal for nerve transmission on its floor. A sagittal crest is present in adults and postorbital processes are present but at the very rear of the orbit. The auditory bullae are small and the paroccipital processes are narrow, recurved in distal aspect, and extend vertically below the occiput. The lower jaw is hypsignathous with a rounded angular process only moderately displaced laterally.

Upper incisors are moderately small and narrow, especially in comparison to the size of the skull, and have an orange enamel layer. Cheek teeth are hypsodont, semi-rooted, flat-crowned, and have a complex re-entrant fold pattern. Three labial and two lingual folds characterize PM4 through M2 of the maxillary toothrow; M3 typically has three labial and three lingual folds. All mandibular cheek teeth usually have one labial and three lingual folds. Shortly after eruption in all teeth, most folds isolate into narrow, transverse islands with wear (Friant 1968).

F. S. Oliveira et al. (2006) detailed dental characteristics and tooth dimensions during ontogeny using radiographic methods.

The genus is known only from the Recent from Mexico, Central America, and South America (McKenna and Bell 1997).

Phylogenetic affinities of the pacas have been somewhat problematic historically, but an understanding of their position within the New World caviomorphs has stabilized markedly in the past decade. Landry (1957) placed the Cuniculidae within the superfamily Octodontoidea and Wood (1965) listed it within the superfamily Chinchilloidea. Virtually all other authors, however, have consistently allied the pacas with the cavies, capybara, and, especially, the agoutis in the superfamily Cavioidea, a placement solidified by molecular phylogenetic analyses (Rowe and Honeycutt 2003; Opazo 2005; Rowe et al. 2010; Upham and Patterson 2012; P.-H. Fabre et al. 2012). Classifications have also varyingly treated the pacas as a distinct family (Cabrera 1961; Wood 1965; Husson 1978; Woods 1982, 1984a, 1993) or as a subfamily within a group that also included the dasyproctids (Simpson 1945 [Dasyproctidae, Cuniculinae]; McKenna and Bell 1997 [Agoutidae, Agoutinae]). Both mitochondrial and nuclear gene sequences support the Cuniculidae as sister to the Caviidae (Rowe and Honeycutt 2003; Opazo 2005) or to a clade comprising both the Dasyproctidae + Caviidae (Upham and Patterson 2012; P.-H. Fabre et al. 2012). While the two extant species of *Cuniculus* share a relatively recent divergence (within the Plio-Pleistocene, estimated at 3.5 ± 1.2 mya), the family links only deeply to other members of the Cavioidea (the late Oligocene, estimated at 26.5 ± 2.5 mya by Opazo 2005).

REMARKS: Earlier use of the family name Agoutidae Gray (e.g., Cabrera 1961; McKenna and Bell 1997) instead of Cuniculidae stemmed from the unacceptability of Brisson's name *Cuniculus*, a position made moot by the explicit acceptance of Brisson's mammal names (ICZN 1998:Opinion 1894).

Genus *Cuniculus* Brisson, 1762

SYNONYMS:

Cuniculus Brisson, 1762:98; type species by subsequent designation "*Paca*, Brisson, p. 99, based on *Cuniculus major palustris, fasciis albis notatus* Barrère, 1741" = *Mus paca* Linnaeus (Hollister 1913:79); not *Cuniculus* Meyer, 1790, a synonym of *Oryctolagus* (Mammalia: Leporidae), nor *Cuniculus* Wagler, 1830, a synonym of *Dicrostonyx* (Mammalia: Cricetidae, Arvicolinae).

Mus: Linnaeus, 1766: 79; part (description of *paca*); not *Mus* Linnaeus.

Cavia: Erxleben, 1777:348; part (description of *paca*); not *Cavia* Pallas.

Agouti Lacépède, 1799:9; type species *Agouti paca* Lacépède (= *Mus paca* Linnaeus), by monotypy.

Coelogenus F. Cuvier, 1807:203; type species *Mus paca* Linnaeus, by subsequent designation (Thomas 1924g).

Coelogenys Illiger, 1811:92; emendation of *Coelogenus* F. Cuvier.

Paca G. Fischer, 1814:85; type species *C[avia]. paca* G. Fischer (= *Mus paca* Linnaeus), by original designation.

Caelogenus J. Fleming, 1822:192: incorrect subsequent spelling of *Coelogenus* F. Cuvier.

Osteopera Harlan, 1825:126; type species *Osteopera platycephala* Harlan, by monotypy.

Caelogenys Agassiz, 1842:5; incorrect subsequent spelling of *Coelogenys* Illiger.

Genyscoelus Liais, 1872:537; substitute name for *Coelogenus* F. Cuvier.

Mamcoelogenysus A. L. Herrera, 1899:26; unjustified renaming of *Coelogenys* Illiger and unavailable name because it is based on zoological formulae (ICZN 1987:318).

Stictomys Thomas, 1924b:238; type species *Stictomys taczanowskii* (Stolzman, 1885), by original designation.

Agouti (*Coelogenys*): Krumbiegel, 1940: 236; name combination.

REMARKS: Tate (1935) and E. M. Pérez (1992) reviewed the complex nomenclatural history of this genus. Brisson (1762:98) erected the genus *Cuniculus* but did not designate a type species. The fourth species in his list (pp. 99) was "Paca," based directly on his own "Le Pak," initially published in the first edition of his *Regnum Animale* (1756). This description was accompanied by **, which, on the third page of the preface (1762), was explained as indicating that he had personally examined a specimen. Thus, both editions of Brisson's *Regnum Animale* alluded to a paca that Brisson had personally inspected. Since Brisson's generic bird names were accepted (ICZN 1911:Opinion 37), Tate (1935:310) argued that Brisson's generic mammal names, which were strictly analogous to his bird names, should also be accepted. However, J. A. Allen (1910a:322) had previously argued that Brisson's specific names, both bird and mammal, were unacceptable because they were not binomial, although they were binary. Consequently, Brisson's "Paca" has been interpreted as an invalid technical name. Linnaeus (1776), however, formally described *Mus paca*, listing "Paca" Brisson, 1756, as the author, and thus validated the specific epithet for this taxon.

Merriam (1895b:375–376) pointed out that *Cuniculus* Brisson, 1762, was a composite genus comprised of *Cavia*, *Lemmus*, *Coelogenus*, *Dasyprocta*, *Anisonyx* [= *Spermophilus*], and *Allactaga*. He attempted to fix the type as "*alactaga* (Olivier) 1800" by the process of elimination.

Hollister (1913:79) noted that Merriam's (1895b designation by elimination of *alactaga* Olivier as the type of *Cuniculus* Brisson (1762) was an unacceptable method to determine a genotype and designated "Paca, Brisson, p. 99, based on *Cuniculus major palustris, fasciis albis notatus* Barrère, 1741" the type of the genus. As Tate (1935:310) documented, Brisson actually examined the animal he named "Paca" and thus it was upon that specimen that *paca* Linnaeus (1766) was based.

Thomas (1914g:285) remarked upon "fiat" versus "priority" names and advocated the retention of "*Coelogenys*" [*sic*] with type *Mus paca* Linnaeus instead of *Agouti* Lacépède or *Cuniculus* Brisson, both of which antedated *Coelogenus* F. Cuvier.

Lönnberg (1921:43) remarked that the "*Paca*" of Marcgraf (1648:224), Ray (1691:226), and others be regarded to constitute the basis of *Mus paca* Linnaeus, and suggested that Pernambuco be designated the type locality. This argument had been, however, superseded by Hollister's (1913:79) prior designation of "Paca, Brisson" as the type of the genus *Cuniculus*.

The ICZN (1925:Opinion 90) ruled against the suspension of the Rule of Priority in the case of *Coelogenys* (Illiger's [1811] emendation of *Coelogenus* F. Cuvier, 1807).

Cabrera (1961:595) listed J. Stolzmann as the author of *Coelogenys*, which he regarded as different from *Coelogenys* Illiger (an emendation of *Coelogenus* F. Cuvier). The basis for Cabrera's decision, however, is unclear as Stolzmann's description (1885:161) clearly mentioned "*Coelogenys paca*," giving the name of the paca in common use in the early and mid-1800s (e.g., Wied-Neuwied 1821; Rengger 1830), before he went on to name *Coelogenys taczanowskii* as a new taxon.

Woods (1982, 1984a, 1993) and McKenna and Bell (1997), among others, accepted *Agouti* Lacépède, 1799, as the valid generic name for Linnaeus's *Mus paca* on the argument that *Cuniculus* Brisson, 1762, was unavailable because his work was not consistently binomial, citing Hopwood (1947) and the International Commission (ICZN 1955a:197). However, and finally, the International Commission (ICZN 1998:Opinion 1894) removed the nomenclatural instability of the generic name for the pacas by explicitly ruling to conserve *Cuniculus* Brisson.

Thomas (1905a:590) noted that the pacas fell into two groups, one characteristic of lowland forests and the other with a montane distribution, groups that are now regarded as the two species recognized in the current literature.

KEY TO THE SPECIES OF *CUNICULUS*:

1. Pelage short and sparse; spots aligned in rows largely on sides; plantar surfaces smooth; zygoma more greatly expanded laterally, compressing the infraorbital foramen

as viewed from the front; nasals short; incisive foramina shallow and narrow; foramen ovale large. . . . .
. . . . *Cuniculus paca*
1'. Pelage long and dense; spots aligned in rows on sides, neck and mid-back; plantar surfaces granulated; zygoma less expanded laterally so that infraorbital foramen as viewed from the front is broader; nasals longer and more robust; postorbital process more conspicuous; foramen ovale small . . . .*Cuniculus taczanowskii*

Cuniculus paca (Linnaeus, 1766)
Paca

SYNONYMS:

[*Mus*] *paca* Linnaeus, 1766:81; type locality "Brasilia [= Brazil], Guiana"; restricted to "French Guiana" by Hollister (1913:79; see Tate 1935:315) and to Pernambuco, Brazil, by Lönnberg (1921:43).

[*Cavia*] *paca* Erxleben, 1777:356; = *Mus paca* Linnaeus.

Agouti paca Lacépède, 1799:9; = *Mus paca* Linnaeus.

C[*avia*] *paca alba* Kerr, 1792:216; type locality "environs of the river St. Francis in South America," Rio São Francisco, Minas Gerais, Brazil.

Coelogenus subniger F. Cuvier, 1807:206; type locality "Tobago," Trinidad and Tobago.

Coelogenus fulvus F. Cuvier, 1807:207; based on Brisson (1762), Buffon (1763), and É. Geoffroy St.-Hilaire (1803) and, therefore, with type locality "Cayenne, French Guiana."

[*Paca*] *maculata* G. Fischer, 1814:87; type locality "Guiana, Brasilia [= Brazil]."

[*Coelogenys*] *rufa* Olfers, 1818:213; type locality "Sudamerica."

Coelogenys paca: Wied-Neuwied, 1821:254; name combination.

Osteopera platycephala Harlan, 1825:126; based on a skull of unknown origin found on "the shore of the river Delaware," United States.

C[*oelogenys*]. *fulvus*: Wied-Neuwied, 1926:247; name combination.

Coelogenys subniger: Wied-Neuwied, 1826:457, name combination.

Coelogenys paca: Rengger, 1830: 251; name combination.

Coelogenys laticeps Lund, 1839c [Lund, 1841a:102]; type locality "Rio das Velhas Floddal" (see Lund 1841a: 133–134), Lagoa Santa, Minas Gerais, Brazil; a fossil taxon considered a synonym of *C. paca* by Paula Couto (1950:169).

Coelogenys rugiceps Lund, 1839c [Lund, 1841a:102]; type locality "Rio das Velhas Floddal" (see Lund 1841a: 133–134), Lagoa Santa, Minas Gerais, Brazil; a fossil taxon considered a synonym of *C. paca* by Paula Couto (1950:169), following Winge (1887).

Coelogenys fuscus Lesson, 1842:103; type locality "Guyane; Brésil."

Coelogenys sublaevis P. Gervais, 1854:326; type locality "Colombie."

Paca americana Liais, 1872:538; type locality "le Brésil, la Guyane et le Paraguay."

Agouti paca virgatus Bangs, 1902:47; type locality "Divala," Chiriquí, Panama.

Coelogenys paca mexianae Hagmann, 1908:25; type locality "Insel Mexiana [= Ilha Mexiana]," Ilha Mexiana, Pará, Brazil.

Agouti paca nelsoni Goldman, 1913:9; type locality "Catemaco, southern Vera Cruz, Mexico."

Cuniculus paca nelsoni: Hollister, 1913:79; name combination.

Cuniculus paca paca: Hollister, 1913:79; name combination.

Cuniculus paca virgatus: Hollister, 1913:79; name combination.

Cuniculus paca sublaevis: Hollister, 1913:79; name combination.

C[*oelogenys*]. *p*[*aca*]. *virgata*: Lönnberg, 1921:45; name combination.

Coelogenys paca guanta Lönnberg, 1921:45; type locality "Gualea, 5,000 feet," Pichincha, Ecuador.

[*Coelogenys*] *guanta*: Thomas, 1924b:239; name combination.

[*Cuniculus*] *paca alba*: Tate, 1935:316; name combination.

[*Cuniculus*] *subniger*: Tate, 1935:316: name combination.

[*Cuniculus*] *fulvus*: Tate, 1935:316; name combination.

[*Cuniculus*] *sublaevis*: Tate, 1935:316; name combination.

[*Cuniculus*] *paca mexianae*: Tate, 1935:316; name combination.

[*Cuniculus*] *paca guanta*: Tate, 1935:316; name combination.

A[*gouti*]. *p*[*aca*]. *mexiannae*: Krumbiegel, 1940:233; name combination.

Agouti paca venezuelica Krumbiegel, 1940:233; type locality "Maracay," Aragua, Venezuela.

[*Agouti paca*] *guanta*: Krumbiegel, 1940:237; name combination.

Cuniculus paca laticeps: Paula Couto, 1950:169; name combination.

[*Agouti paca*] *mexianae* Woods, 1993:783; incorrect subsequent spelling of *mexianae* Hagmann.

[*Cuniculus paca*] *mexianae*: Woods and Kilpatrick, 2005: 1559; incorrect subsequent spelling of *mexianae* Hagmann.

DESCRIPTION: Chestnut red to dark brown upperparts with three or four lines of large white spots, often coalescing into stripes, from sides of neck to rump; lower cheeks, throat, chest, and venter white. Hair coarse, sparse, and short over entire body (Emmons and Feer 1997). Soles of feet generally smooth, not granular (Thomas 1905a).

Cranially, cheekplates more greatly expanded, constricting infraorbital foreman to greater degree than in Mountain Paca, shorter nasal bones, less conspicuous postorbital processes, narrower and more shallow incisive foramina, and larger foramina ovale (Ríos-Uzeda et al. 2004). Otherwise, two species of paca similar in all other attributes that diagnose genus *Cuniculus*.

DISTRIBUTION: *Cuniculus paca* occurs throughout the Neotropical lowland forests from eastern and southern Mexico south through Central America and in the South American countries of Colombia, Venezuela, the Guianas, Ecuador, Peru, Bolivia, Paraguay, and most of Brazil, and has been introduced into Cuba and the Lesser Antilles. The elevational range is from sea level to 1,600 m (Hall 1981; E. M. Pérez 1992) or to 2,000 m (Cuervo Díaz et al. 1986).

SELECTED LOCALITIES (Map 385): ARGENTINA: Misiones, Parque Nacional Iguazú (Crespo 1982b). BOLIVIA: Cochabamba, San Antonio (S. Anderson 1997); La Paz, 8 km SE of Caranavi (S. Anderson 1997); Santa Cruz, 10 km N of San Ramón (S. Anderson 1997), Santa Cruz, Buena Vista (S. Anderson 1997). BRAZIL: Bahia, Santo Amaro (FMNH 20435); Ceará, Serra do Covão (Feijó and Langguth 2013); Goais, Anápolis (AMNH 134086); Pará, Belem (MCZ 31700), Ilha Mexiana (type locality of *Coelogenys paca mexianae* Hagmann); Pernambuco, Pernambuco (type locality of [*Mus*] *paca* Linnaeus, as restricted by Lönnberg 1921:43); Rio de Janeiro, Bemposta (AMNH 134098); Santa Catarina, Ilha de Santa Catarina (Graipel et al. 2001). COLOMBIA: Antioquia, Río Currulao, 20 km SE of Turbo (FMNH 69635); Cauca, El Payapo (FMNH 90067); La Guajira, Villanueva, Sierra Negra (USNM 281824); Magdalena, Buritaca (AMNH 23470); Nariño, near Guayabetal (MSU 29484); Sucre, Las Campanas (FMNH 72413). ECUADOR: El Oro, Portovelo (AMNH 46547); Esmeraldas, La Bocana (MSU 9513); Santa Elena, near Manglaralto, Cordillera de Colconche (UMMZ 80186). FRENCH GUIANA: Paracou (Voss et al. 2001). GUYANA: Barima-Waini, Baramita (ROM 100935); Demerara-Mahaica, 1 mi E of Buxton (FMNH 46223). PARAGUAY: Canindeyú, Mbaracayu Reserve (UAM 46601); Misiones, 5 km ENE of Ayolas (UMMZ 125719). PERU: Amazonas, Huampami, Río Cenepa (Patton et al. 1982); Cusco, Kiteni (LSUMZ 19344); Huánuco, Hacienda Exito (FMNH 41144); Junín, Chanchamayo (FMNH 65791); Puno, San Juan (Sanborn 1953). SURINAM: Paramaribo, Paramaribo (USNM 13311). TRINIDAD AND TOBAGO: Princestown (AMNH 6041). VENEZUELA: Falcón, near La Pastora, 11 km ENE of Mirimiri (Handley 1976); Mérida, Mérida (AMNII 21645); Miranda, 1 km S of Río Chico (Handley 1976); Monagas, Hato Mata de Bejuco (Handley 1976); Zulia, Kasmera (Handley 1976).

SUBSPECIES: Ellerman (1940) recognized seven subspecies, including four from South America in addition to the nominotypical form (*alba* Kerr, *mexianae* Hagmann, *guanta* Lönnberg, and *sublaevis* P. Gervais). Cabrera (1961) recognized only three, including *guanta* Lönnberg and *mexianae* Hagmann in addition to the nominotypical form. Both E. M. Pérez (1992) and Woods and Kilpatrick (2005) recognized four subspecies in addition to the nominotypical form, only two with South American distributions (*guanta* Lönnberg, and *mexianae* Hagmann [incorrectly spelled *mexicanae* by Woods and Kilpatrick, 2005]). However, no review of geographic variation that would adequately diagnose and delineate the geographic range of any of these taxa has yet been accomplished.

NATURAL HISTORY: E. M. Pérez (1992) reviewed published data on reproduction, population ecology, and behavior. These large rodents inhabit primarily tropical rainforest, including a wide range of forest types, from mangrove swamps, riparian communities, and dense upland scrub. They prefer areas near water (Collett 1981). Mostly frugivorous, pacas are opportunistic feeders that also feed on leaves, buds, and flowers, and are considered to play a major role in seed predation and dispersal (Beck-King et al. 1999; Dubost and Henry 2006). Population densities, measured in quite different habitats in Venezuela, Colombia, Panama, and Costa Rica ranged from 25 to >70 adults per square kilometer (Eisenberg et al. 1979; Collett 1981; Smythe et al. 1983; Glanz 1983; Beck-King et al. 1999). These are important members of tropical forest communities, representing up to 16% of the biomass of nonvolant mammals (Eisenberg and Thorington 1973; Eisenberg et al. 1979). Collett (1981) reported on age structure in natural populations in Colombia, noting that animals can live up to 13 years. Although fecundity is low, survivorship of adults is high. Primary mammalian predators are all species of cats (Felidae) and both coyotes and bush dogs; non-mammalian predators include crocodiles and the boa constrictor (Ramdial 1972; Deutsch 1983). Pacas are hunted extensively for food, with the meat preferred by many local people, including indigenous communities where it may constitute as much as 8% of wild meat consumed (Ojasti 1983). Smythe (1987, 1991) advocated captive breeding for the potential of paca as a domestic protein source in lowland forest communities; Emmons (1987a) presented counter-arguments.

Pacas are a nocturnal, solitary animal that seeks refuge in burrows, hollows at the base of trees, or among rocks; their burrows are modified from those originally dug by armadillos (Lander 1974). Extensive camera-trap surveys in northern Mato Grosso state, Brazil, documented activity from dusk to dawn, with activity levels unaffected by the lunar cycle, except for the onset of dusk, season, temperature, or rainfall (Michalski and Norris 2011). A. Link et al. (2012) also used camera traps to establish an important year-round use of mineral licks by paca in eastern Ecuador. Individuals use fixed trails during nightly foraging or other travel purposes. They breed year-round, with no indication of seasonal clustering of births or significant age-specific variation in fecundity (Collett 1981; Dubost et al. 2005). Females reach reproductive maturity at about nine months of age, males at one year. Litter size is a single young with individual females breeding from one to three times per year (Mondolfi 1972; Matamoros 1981). Gestation ranges from 114 to 119 days in the laboratory, with an interbirth interval of 191 days (Lander 1974; Collett 1981; Matamoros and Pashov 1984). Neonates are precocial, with eyes open, well-developed pelage, and erupted incisors and cheek teeth; weaning occurs at about 12 weeks (Matamoros 1981).

An extensive array of organisms parasitize pacas, including nematode, cestode, and protozoan endoparasites (Gonzáles and Espino 1974; Guerrero 1985; Naiff et al. 1985; S. L. Gardner et al. 1988; Sato et al. 1988; A. Q. Gonçalves et al. 2006; S. L. Gardner et al. 2013) and ectoparasitic ticks and fleas (Darskaya and Malygin 1996; Barros-Battesti and Knysak 1999). Pacas also serve as a reservoir for important tropical diseases, such as leishmania (F. T. Silveira et al. 1991) and trypanosomiasis (Loyola et al. 1987).

REMARKS: As Tate (1935:315) noted in his treatment of the taxonomy of caviomorph rodents, there has been some debate as to what *Mus paca* Linnaeus was, including its type locality. Linnaeus (1766) cited under *paca* the descriptions of Marcgraf (1648), Ray (1691), and Brisson (1762), among others, and gave the habitat as "Brasilia, Guiana." Brisson cited Marcgraf and Ray, among others, but his description was marked by a double **, indicating that he had examined the specimen himself. Except for Brisson, all citations ultimately focused on the work of Marcgraf. Hence, as Tate stated (p. 315) "the question then remains whether *paca* Linnaeus must be held founded upon Marcgraf's paca of Brazil or upon Brisson's paca from 'Guiana and Brazil.'" Tate used the Determination of the First Reviser (ICZN 1999:Art. 24.2) to decide between the two alternatives, noting that Hollister (1913), who had selected *paca* Linnaeus as the type of *Cuniculus*, selected as type for the species "*Paca*, Brisson, p. 99, based on *Cuniculus major palustris, fasciis albis notatus* Barrère, 1741." Tate then argued that Barrère's paca "was undoubtedly from French Guiana" and thus that "Hollister may be considered as having fixed the type locality of *paca* as French Guiana." Lönnberg (1921:43) subsequently, and incorrectly, restricted the type locality to Pernambuco, Brazil. Moojen (1952b:119) also cited Pernambuco for

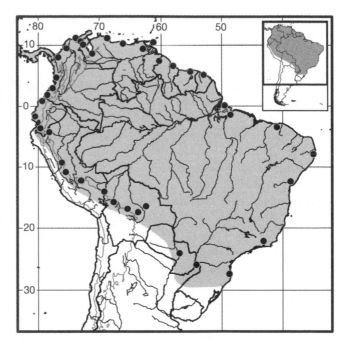

Map 385 Selected locality for *Cuniculus paca* (●). Contour line = 2,000 m.

the type locality, but other reviewers (Cabrera 1961:595; Woods and Kilpatrick 2005:1559) have followed Tate's argument by listing the type locality of *paca* Linnaeus as Cayenne, French Guiana. Thomas (1905a) initially thought that *sublaevis* P. Gervais could be either *C. paca* or *C. taczanowskii*, but later both he (Thomas 1924b) and Lönnberg (1921) treated it as a female *Cuniculus paca*.

The karyotype consists of a $2n = 74$ and FN of 56 (Fredga 1966). Van Vuuren et al. (2004) examined population genetic structure in French Guiana using molecular sequence markers, with a signature characteristic of recently expanded populations.

Cuniculus taczanowskii (Stolzmann, 1885)
Mountain Paca
SYNONYMS:

Coelogenys taczanowskii Stolzmann, 1885:161; type locality "habite les montagnes de l'Equadeur entre 6000 et 10,000 pieds au dessus de niveau de la mer. Elle n'est pas très rare dans les fôrets des deux versants des Andes," Pichincha, Ecuador.

A[gouti]. taczanowskii: Thomas, 1905a:589; name combination.

Agouti sierrae Thomas, 1905a:589; type locality "Pedregosa Montañas, Sierra de Merida, Alt. 2000m," Mérida, Venezuela.

Cuniculus taczanowskii: Hollister, 1913:79; first use of current name combination.

Cuniculus sierrae: Hollister, 1913:79; name combination.

Agouti [*Coelogenys*] *sierrae andina* Lönnberg, 1913:28; type locality "rocks of Pichincha . . . altitude of 9 to 12,000 feet." Pichincha, Ecuador.

Agouti paca taczanowskii: J. A. Allen, 1916b:205; name combination.

Agouti thomasi Eaton, 1916:89; type locality "small kitchen-midden near eastern limit of the city," Machu Picchu, Cusco, Peru.

Coelogenys taczanowskii andina: Lönnberg, 1921:48; name combination.

S[*tictomys*]. *taczanowskii*: Thomas, 1924b:238; name combination.

[*Stictomys*] *andina*: Thomas, 1924b:238; name combination.

[*Stictomys*] *sierrae*: Thomas, 1924b:238; name combination.

Coelogenys sierrae: Pittier and Tate, 1932:265; name combination.

[*Stictomys*] *taczanowskii taczanowskii*: Tate, 1935:317; name combination.

[*Stictomys*] *taczanowskii andina*: Tate, 1935:317; name combination.

Stictomys taczanowskii: Tate, 1939:184: name combination.

Cuniculus taczanowskii taczanowskii: Ellerman, 1940:225; name combination.

Cuniculus taczanowskii andina: Ellerman, 1940:225; name combination.

Cuniculus taczanowskii sierrae: Ellerman, 1940:225; name combination.

Stictomys taczanowskii andina: Cabrera, 1961: 595; name combination.

Stictomys taczanowskii sierrae: Cabrera, 1961:596; name combination.

Stictomys taczanowskii taczanowskii: Cabrera, 1961:596; name combination.

Cuniculus hernandezi Castro, López, and Becerra, 2010: 130; *nomen nudum*; no type locality designated, but inferred from text to be "Cordillera Central, Colombia."

DESCRIPTION: Same general body size and conformation as Common Paca, but differs externally by having much longer, thicker, and softer dorsal fur, spots on top of back as well as on sides, and distinctly granulated soles of feet. General color also darker, with blackish brown tones rather than reddish brown, and venter pale but washed terminally with pale brown. Cranial character differences documented by Ríos-Uzeda et al. (2004) include longer and more robust nasals, less expanded cheekplates resulting in broader and more open infraorbital foramen, more conspicuous postorbital processes, broader and deeper incisive foramina, and much smaller foramen ovale.

DISTRIBUTION: *Cuniculus taczanowskii* is known from forested slopes of the Andes of northwestern Venezu-

ela, Colombia, Ecuador, Peru, and Bolivia, at an elevational range from approximately 2,000 to 4,000 m (Cuervo Díaz et al. 1986; Ríos-Uzeda et al. 2004; J. Castro et al. 2010).

SELECTED LOCALITIES (Map 386): COLOMBIA: Antioquia, El Cedro, Caicedo (FMNH 70801), Paramillo (AMNH 37795); Cauca, La Florida (J. A. Allen 1916b); Cundinamarca, Río Balcones (FMNH 70803); Huila, Río Ovejeras (FMNH 70802); Nariño, Barbacoas (J. A. Allen 1916b); Quindío, La Guneta (J. A. Allen 1916b). ECUADOR: Napo, Río Ansu (YPM, no number); Azuay, Molleturo (AMNH 61877); Pichincha, Pichincha (type locality of *Agouti* [*Coelogenys*] *sierrae andina* Lönnberg) San Ignacio (AMNH 66727). PERU: Ayacucho, Yuraccyacu (LSUMZ 14436); Cusco, Machu Picchu (type locality of *Agouti thomasi* Eaton), Marcapata (FMNH 75125), Puesto de Vigilancia Acjanaco (B. D. Patterson et al. 2006); Huánuco, Cordillera Carpish (A. L. Gardner 1971); Piura, Tambo (FMNH 80974); Puno, Sandia (FMNH 79917); San Martín, Puerta del Monde (LSUMZ 27383). BOLIVIA: La Paz, San Juanito (Ríos-Uzeda et al. 2004). VENEZUELA: Lara, Paramo de los Rosas (J. A. Allen 1911); Mérida, Paramito (Handley 1976); Tachira, Buena Vista (Handley 1976).

SUBSPECIES: Cabrera (1961) recognized three subspecies (*andina* Lönnberg, from the central Andes of Ecuador; *sierrae* Thomas, from the northern Andes of Colombia and the Sierra de Mérida of northwestern Venezuela; and the nominotypical form from the Andes of Ecuador and southern Colombia). Woods and Kilpatrick (2005:1560) regarded *C. taczanowskii* to be monotypic. No adequate review of the few available specimens has been attempted. See Remarks for comments about the uniqueness of southern Peruvian specimens.

NATURAL HISTORY: The mountain paca has not been studied in the field, and little information on life history, ecology, or behavior is known. The single known specimen from Bolivia (Ríos-Uzeda et al. 2004) was collected in montane forest dominated by bamboo (*Chusquea* and *Neurolepis* [Poaceae]) and upland trees (*Clusia* [Clusiaceae] and *Podocarpus* [Podocarpaceae]). In Ecuador, the species was taken in subtropical forest on the edge of páramo (I. Castro and Román 2000).

REMARKS: Both Cabrera (1961:596, but not on p. 595) and Woods and Kilpatrick (2005:2560) incorrectly gave the citation for Stolzmann's description of *Coelogenys taczanowskii* as 1865, not 1885.

Tate (1935:314) suggested that *Agouti thomasi* Eaton was a *nomen nudum*, as he had not discovered a published description of this taxon. The name, however, stems from Eaton (1916:89, plates 38, Figs. 6–8, and 39, Figs. 1–3) who described *thomasi* based on partial skulls from the archaeological site at Machu Picchu in southern Peru. While Edmund Heller, the collector of modern mammals during the 1915 Yale Peruvian Expedition of which Eaton was a part, failed to obtain representatives of this species, Thomas (1920h:220) concluded that "closely allied to living Andean forms, [*thomasi*] will perhaps be found still to exist in the neighborhood [of Machu Picchu]." Sanborn (1953:7) wrote "a subspecies, *thomasi*, was described from Machu Picchu by Eaton (1916) from skulls found in burial caves. The validity of this race can not be established however until complete specimens are collected." Thomas (1924b:239) linked the Peruvian *thomasi* to the Ecuador *guanta* Lönnberg and *sublaevis* P. Gervais, two names currently listed as synonyms of the lowland *C. paca*, not to the montane *C. taczanowskii* despite the high elevation of and cloud forest habitat at the type locality of Machu Picchu. The expanded nasals, less protrusive cheekplates, prominent postorbital processes, and small foramina ovale depicted in the drawings of the holotype (Eaton 1916:plate 39) are all characteristics of *C. taczanowskii* (Ríos-Uzeda et al. 2004). While Ellerman (1940) simply repeated Tate's (1935) suggestion that *Agouti thomasi* Eaton was a *nomen nudum*, neither Cabrera (1961) nor Woods and Kilpatrick (2005) listed the name in their respective taxonomic compilations of *Cuniculus*.

The karyotype is variable, ranging from $2n = 78$, FN = 82 for individuals from the Cordillera Oriental and $2n = 72$, FN = 82 for those from the Cordillera Central, both in Colombia (J. Castro et al. 2010) to $2n = 42$, FN = 80 (A. L. Gardner 1971), based on a specimen from Junín department in Peru. J. Castro et al. (2010) designated the $2n = 72$ karyotypic form as a distinct species, *Cuniculus hernandezi*, an action supported by limited molecular and morphometric data. However, *hernandezi* is a *nomen nudum*; the proposed name does not conform to the requirements of the Code of Nomenclature (ICZN 1999), neither Article 13.1.1 (the need for an explicit description or definition that clearly differentiates a new species from other taxa) nor Article 16.4 (the need to designate a name-bearing type specimen). The authors of *hernandezi* also did not follow either Recommendation 16C (to deposit a type specimen in a research collection that meets the requirements of Recommendation 72F) or Recommendation 16F (that the type be illustrated). Furthermore, since the type locality of *taczanowskii* Stolzmann is from Ecuador, and no comparisons were made to Stolzmann's description or to specimens from at or near to the type locality, it is unclear why J. Castro et al. (2010) allocated specimens with $2n = 78$ from the Cordillera Oriental to *taczanowskii* and not those they called *hernandezi*. While *Cuniculus taczanowskii* appears to be polytypic as currently understood, adequate geographic analyses of multiple character sets are required to delineate taxon boundaries and apply appropriate names to each.

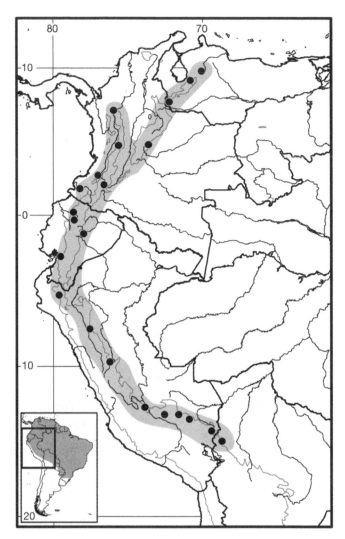

Map 386   Selected localities for *Cuniculus taczanowskii* (●). Contour line = 2,000 m.

Family Dasyproctidae Bonaparte, 1838
James L. Patton and Louise H. Emmons

Agoutis (*Dasyprocta*) and acouchys (*Myoprocta*) are largely diurnal, rabbit-like denizens of lowland to midelevation tropical forests, dry forests, savanna edge, and human-disturbed habitats including cultivated land. *Dasyprocta* ranges from southern Mexico to western Ecuador on the west side of the Andes and east of the Andes throughout the Guianas to the mouth of the Amazon and south to northern Argentina and southern Brazil, as well as on several islands of the Greater and Lesser Antilles. *Myoprocta* is limited to the Guianan region and Amazon Basin. The bodies of members of both genera are relatively slender; the forelimbs and hindlimbs are elongated and thin, with a decreased femoral robustness indicative of speed specialization (Elissamburu and Vizcaíno 2004). The eyes are large, ears are short but conspicuous and sparsely haired, and the tail is either short (*Myoprocta*) or almost vestigial (*Dasyprocta*). The forefeet have five digits, the middle three well developed with digit III elongated; the pollex is reduced to a knob-like structure, and digit V is short but still obvious to the eye. The hindfeet have only three digits, again the digit III elongated. All digits have sharp, hoof-like claws. The soles for forefeet and hindfeet are naked and smooth; large plantar pads are present. The mesaxonic and digitigrade foot structure coupled with elongated limbs and a vestigial clavicle (Rocha-Barbosa et al. 2002) signal cursorial locomotion with a quadrupedal, saltational gait; individuals can run and jump well. Body size ranges from 6 kg in the largest species of *Dasyprocta* to a maximum of 1.5 kg in species of *Myoprocta*.

The pelage is thick and coarse, with hairs appearing glossy. Long and coarse mystacial, supercilliary, and genal vibrissae are present, with the mystacial reaching past the ear when laid back against the side of the head. Overall color varies greatly from black, brown, or reddish to yellowish with individual hairs having multiple, alternating eumelanic and pheomelanic color bands. The rump may have elongated hair often contrasting in color with midback and thighs; these hairs are erected to flash contrasting pale bases when individuals are alarmed. The under parts are generally paler, with sparse hair. Forefeet and hindfeet may be colored as the legs or have a contrastingly darker, typically black, tone.

The skull is elongate and smooth, lacking heavy ridging, although a slight sagittal crest over the posterior portion of the braincase is present in adults. The premaxillae and nasals, of approximately equal length, project beyond the upper incisors. The rostrum is deep and heavy. The zygomatic arches are short and weak in appearance, the jugal being especially thin and not in contact with an expanded lacrimal. There is no jugal fossa present. The lacrimal canal opens on the side of the rostrum above and slightly forward of the root of PM4. The infraorbital foramen is moderate in size and lacks a distinct grove or canal on its floor for nerve passage. The incisive foramina are small with the opening nearly occluded by an expanded medial septum. The anterior border of the mesopterygoid fossa is rounded and extends at least to the posterior margin of M2. Elongated but narrow sphenopalatine foramina are present on the bony roof of the fossa. The auditory bullae are moderately inflated; the external auditory meatus is noticeably small; the paroccipital process is short but distinct in *Dasyprocta*, nearly vestigial in *Myoprocta*. The supraoccipital lacks a lateral process but a distinct vertical median ridge is present. Blood supply to the brain is via the vertebral and external carotid arteries; both the internal carotid artery

and stapedial artery are missing (Bugge 1971, 1974). The dentary lacks a ridge or groove on its lateral surface; the angular process is extremely deflected; the condyloid process has a prominent postcondyloid process; and the triangular-shaped coronoid process is only weakly developed.

The upper toothrows are approximately parallel. The molariform teeth are strongly hypsodont, semirooted, and have flat occlusal surfaces. The single premolar (PM4) is preceded by a deciduous precursor (dPM4). Each tooth has a single lingual and four labial reentrant folds forming five crests, the pentalophodont pattern regarded as primitive for caviomorphs (Reig 1986; Vucetich and Verzi 1993). The folds isolate quickly into lakes with wear; these lakes may subdivide into as many as eight elements in transverse lines. Upper and lower incisors are thin and laterally compressed, with the enamel pigmented yellow or orange. Both extant genera share a combination of dental traits that include the lingual separation of the metaloph from the posteroloph in early stages of wear and the presence of a spur on the posteroloph; the posterior projection of the paracone; the early union of the lingual ends of the mesolophs, metalophs, and posterolophs; bulbous enamel on the occlusal surface; separation of the mesoloph from the mure, at least in unworn teeth; and a posterloph that is separated from the hypocone in early stages of wear (Vucetich and Verzi 2002). F. S. Oliveira et al. (2012) detailed the macrostructure of the cheek teeth of *Dasyprocta azarae* using radiography and histological sections.

Pocock (1922), Hooper (1961), Iack-Ximenes (1999), and Menezes et al. (2001, 2003) describe morphology of the phallus, which is similar to that of the paca (*Cuniculus paca*) with the characteristic intromittent sac, epidermis on both inside and out covered by dense comb-like spines, and a pair of elongated spikes on the floor that extend distally beyond the remainder of the glans when the intromittent sac is everted. The baculum is a subcylindrical bony rod expanded laterally at the base, deeply concave ventrally and convex dorsally, with a truncate and angular tip capped by a mass of cartilage. In the female, the genital and urinary openings are visible externally as distinct apertures, with the prepuce, containing the urinary orifice, and clitoris separate from the sphincter surrounding the anus (Pocock 1922). A pair of anal glands is present, represented externally by small pouches opening on each side of and within the sphincter musculature of the anus. M. A. M. Carvalho et al. (2008) described the arterial supply to the penis of *Dasyprocta prymnolopha*; Mollineau et al. (2006, 2009) described the male reproductive system, including the position and both gross and microanatomy of the accessory glands.

Females have four pairs of mammae, all in a ventral position (Weir 1974b). Ovulation is generally spontaneous, with a mean estrous cycle length of 34 to 42 days, and seasonally structured. A postpartum estrous occurs within 50 days if young are present, and within 16 days if lost (in *Myoprocta*; Weir 1974b). As with other caviomorphs, dasyproctids have small litters (one to three offspring), with young precocial at birth following a long gestation period (99 to 120 days). The lactation period is short (about 20 days), and puberty is reached at an average age of 6 months (*Dasyprocta*) or up to 12 months (*Myoprocta*; Weir 1974b). The precocial nature of neonates is reflected by a large ratio of newborn to adult brain weight (Pilleri 1959). Miglino et al. (2002, 2004) and R. F. Rodrigues et al. (2006) provided details of placentation in *Dasyprocta*. The eyes of agoutis are well adapted for their diurnal lifestyle, with a high proportion of cones relative to rods in the retina (Farias Rocha et al. 2009).

Agoutis and acouchys have been traditionally placed within the Cavioidea, either as a family unto themselves (Winge 1887; Tullberg 1899; Wood 1965) or one that also included the pacas (Pocock 1922; Simpson 1945; Landry 1957; and McKenna and Bell 1997; see comment in Woods 1993). Molecular analyses are completely concordant in documenting the Dasyproctidae as part of the Cavioidea, but support a basal position of either the Dasyproctidae (Rowe and Honeycutt 2002; Opazo 2005; Rowe et al. 2010) or the Cuniculidae (Upham and Patterson 2012). Divergence dates for the base of the family are estimated to be as recent as 27.9 ± 2.4 mya in the Late Oligocene (Opazo 2005) or as old as 31 mya (95% posterior probability range: 28–47 mya; Rowe et al. 2010), in the Late Eocene. The earliest fossils are recorded from the early Oligocene (B. Patterson and Wood 1982; McKenna and Bell 1996). However, as posited by B. Patterson and Wood (1982), Vucetich (1984), and Walton (1997), these early, now extinct forms are a clade apart from the two extant genera. Divergence estimates based on molecular sequences (Opazo 2005) suggest a split between *Dasyprocta* and *Myoprocta* in the early part of the Late Miocene, at about 10.7 ± 2.3 mya. Both extant genera have been restricted to tropical and subtropical forested regions throughout their histories, although at least one extinct member of their lineage (*Plesiaguti totoi*) extended to coastal Buenos Aires province in Argentina during a warm period of the lower-middle Pleistocene (Ensendadan) (Vucetich and Verzi 2002).

KEY TO THE GENERA OF DASYPROCTIDAE:

1. Size large (2–6 kg); tail minute, inconspicuous, and naked; teeth large with toothrow >17 mm in length . . . . .
. . . . *Dasyprocta*

1'. Size smaller (<1.5 kg); tail short but conspicuous, and hairy; teeth small, with toothrow <14 mm in length . . .
. . . . *Myoprocta*

REMARKS: The skulls of both *Dasyprocta* and *Myoprocta* share the same cranial conformation and general set of qualitative features, with most obvious differences between them associated with overall size. Ellerman (1940) believed the two genera sufficiently similar to be considered as subgenera of a single genus-group taxon.

Genus *Dasyprocta* Illiger, 1811

Members of the genus *Dasyprocta*, or agoutis, are among the most recognizable rodents of the Neotropical forests, ranging from southern Mexico through Central America to Ecuador on the west side of the Andes and from the north coast of South America, including Trinidad and Tobago and several islands of the Lesser Antilles, to northern Argentina and coastal Brazil east of the Andes. The genus can be generally characterized by the set of external and cranial characters described above for the family.

The ubiquity and easy recognition of an agouti, however, belie the rather chaotic current taxonomy of the genus. The number of species, their respective morphological boundaries, and thus their geographic ranges are poorly understood, with substantive conflicts in both names applied and mapped or described ranges among modern syntheses (Eisenberg 1989; Eisenberg and Redford 1999; Emmons and Feer 1990, 1997; Woods and Kilpatrick 2005; Bonvicino, Oliveira, and D'Andrea 2008). As nearly every modern author has lamented, the genus is in critical need of revision.

The 10 species we recognize, their distributions, and their morphological characters are based primarily on four sources: (1) the descriptions of external characters and mapped ranges of species recognized by Emmons and Feer (1990, 1997), which, while not documented by referenced specimens, were based on the senior author's examination of specimens in U.S., European, and Brazilian collections, her access to the notes of Charles O. Handley Jr. at the National Museum of Natural History, and her personal field experiences; (2) the evaluation of Venezuelan species by Ojasti (1972), a study that detailed character differences of morphological and taxonomic groups from throughout that country; (3) the reconstruction of part of the convoluted nomenclatural history of the genus by Husson (1978) and, especially, Voss et al. (2001) in their respective reviews of agoutis of the Guianan region; and (4) the unpublished masters thesis of Iack-Ximenes (1999), who reviewed Brazilian species based primarily on specimens housed in collections of that country. Iack-Ximenes detailed trends in color and color pattern and provided a hypothesis of species relationships derived from cladistic analysis of those characters. He also emphasized the general allopatric nature of populations with different color patterns, using this criterion as a major basis for the species he delimited. Our review here is clearly provisional and will undoubtedly be modified substantively as even regional units are studied in appropriate detail. Indeed, as this book was in the final stages of production, Feijó and Langguth (2013) described *Dasyprocta iacki* from northeastern Brazil and reviewed the taxonomy of the red-rumped agoutis from south of the Amazon herein considered as *D. leporina*. Readers should examine Feijó and Langguth (2013) for the details of their arguments.

SYNONYMS:

Mus: Linnaeus, 1758:58; part (description of *leporinus*, based on *Lepus javensis* of Catesby); not *Mus* Linnaeus.

Calva Gmelin, 1788:120; part; incorrect subsequent spelling of *Cavia* Pallas [*Calva* written in text although *Cavia* is on page headers].

Cavia: Erxleben, 1777:353; part (listing of *aguti* and *leporina*); not *Cavia* Pallas.

Agouti: Daudin, 1802:166; part (description of *cayanus*); not *Agouti* Lacépède.

Dasyprocta Illiger, 1811:93; part; included species *Cavia aguti* (= *Mus aguti* Linnaeus) and *Cavia acouschy* [*sic*] Erxleben; type species *Mus aguti* Linnaeus, by subsequent designation (Thomas 1903d:464, who erected *Myoprocta* to contain *Dasyprocta acouchy* [Linnaeus]).

Platypyga Illiger, 1811:93; included in the synonymy of *Dasyprocta*; *nomen nudum*.

Cloromis F. Cuvier, 1812:290, plate 15, Fig. 10; no type species given.

Chloromys Rafinesque, 1815:56; incorrect subsequent spelling of *Cloromis* F. Cuvier.

Platipyga Desmarest, 1816:210; incorrect subsequent spelling of *Platypyga* Illiger.

Chloromys G. Cuvier, 1817:214; incorrect subsequent spelling of *Cloromis* F. Cuvier; included species *Cavia acuti* (= *Mus aguti* Linnaeus) and *Cavia acuchi* (= *Myoprocta acouchy* [Erxleben]).

Chloromys Lesson, 1827:300; inclusion of *acuti* (= *Mus aguti* Linnaeus) and *cristatus*; not *Chloromys* Schlosser, 1884, a subgenus of *Steneofiber* (see McKenna and Bell 1997:197); year incorrectly given as 1927 in Woods and Kilpatrick (2005).

Cutia Liais, 1872:534; part; type species *Mus aguti* Linnaeus; *nomen oblitum*.

Mamdasyproctaus A. L. Herrera, 1899:29; unavailable name because it is based on zoological formulae (ICZN 1987:318).

REMARKS: Illiger (1811), when he erected *Dasyprocta*, included both *Cavia aguti* [= *Mus aguti* Linnaeus] and

Cavia acouschy (*sic* = *Cavia acouchy* Erxleben) without selecting either as the type of his new genus. Thomas's (1903d:464) erection of *Myoprocta* to contain "the long-tailed Agouti, '*Dasyprocta*' *acouchy*, Linn,'" resulted in the de facto selection of *Mus aguti* Linnaeus as the type species of Illiger's *Dasyprocta*.

Emmons and Feer (1990, 1997) recognized seven species in the genus, six with ranges at least in part of South America and five within Brazil alone. They only treated, however, those South American species that occurred in lowland rain or dry forest areas. Iack-Ximenes (1999) delineated 12 species in Brazil, including three that as yet have no available name; many of these have been included in the recent guides of Brazilian mammals (N. R. Reis et al. 2006; Bonvicino, Oliveira, and D'Andrea 2008; Paglia et al. 2011). Herein, we take a somewhat intermediate approach, recognizing all of the taxa shared in common by Emmons and Feer and Iack-Ximenes plus additional taxa not considered by either of these authors. Not all of those taxa recognized by Iack-Ximenes seem worthy of recognition in our opinion, given geographic variation in color and color pattern in the widespread species, although we comment on each in the various accounts that follow. Although Iack-Ximenes (1999) and Voss et al. (2001) provided linear measurements of cranial dimensions for samples of Brazilian and Guianan taxa, as yet no treatise has adequately reviewed geographic variation in these characters within species or delineated the differences among them. Nor has anyone yet described the qualitative cranial characters of use in defining species or any geographic units. Rather, our current understanding of species boundaries stems almost exclusively from pelage color and color pattern, supplemented limitedly by overall body size and, to a lesser extent, the apparent lack of true sympatry among species. What we provide here is best viewed as a simple set of hypotheses waiting to be tested by an adequate integration of external and craniodental morphology, especially with molecular phylogeographic and phylogenetic data and methods.

A word of further caution: the poorly delineated ranges and different taxonomies employed by authors over the past decades make the allocation of ecological or life history data to the species recognized herein difficult. Thus, for the information summarized for each species in the natural history notes, we have excluded references citing a particular species name unless the locality of field studies is identified and clearly falls within that part of a species' range where no other agouti might be present.

Contrary to so many Neotropical rodent genera, there is substantial karyotypic stability among the species of *Dasyprocta*. All literature reports to date give a diploid number of 64 and FN = 122 or 124, regardless of the species named or the geographic locality identified.

KEY TO THE SOUTH AMERICAN SPECIES OF *DASYPROCTA* (BECAUSE OF EXTENSIVE GEOGRAPHIC VARIATION IN COLOR, SOME INDIVIDUALS MIGHT FAIL TO BE DISTINGUISHED BY THIS KEY; CONSULT BOTH THE DETAILED DESCRIPTIONS AND RANGE MAPS IN THE ACCOUNTS THAT FOLLOW):

1. Body size smallish, typically <3 kg in mass 2
1'. Body size larger, typically >3 kg in mass 3
2. Overall color grayish with yellow to orange tones, hairs finely banded by five to seven phaeomelanin alternating with an equal number of eumelanin bands; mid-back and rump darker than sides but of same tone . . . . *Dasyprocta azarae*
2'. Forequarters gray, rump and flanks bright, pale orange; flanks with few dark bands on hairs; distribution limited to eastern Amazonia south of the Rio Amazonas. . . . . *Dasyprocta croconota*
3. Dorsal and lateral color a mixture of black and white, without reddish or yellow pigments 4
3'. Dorsal and lateral color contains reddish or yellowish pigments. 6
4. Head, back, and rump uniformly blackish flecked with white tipped hairs; fore and hind limbs and feet black . . . .*Dasyprocta fuliginosa*
4'. Head, back, and rump not blackish flecked with white tipped hairs; forelimbs and hindlimbs not black but forefeet and hindfeet either brown or black 5
5. Head and back blackish flecked with yellow to rufous tipped hairs; long hairs overhanging rump entirely white basally, with black tips; feet blackish. . . . . *Dasyprocta kalinowskii*
5'. Head, back, and rump uniform reddish brown or blackish with yellowish hairs; mid-body and rump with long, black; pale-tipped hairs overhanging rump; feet brown or blackish . . . .*Dasyprocta punctata*
6. Rump orange to red with a wide black dorsal stripe from mid-back across rump to tail; flanks of thighs orange. . . . . *Dasyprocta prymnolopha*
6'. No blackish dorsal stripe present over mid-back or rump; flanks of thighs vary in color expression. 7
7. Head and forequarters grizzled olivaceous; rump and flanks orange; fore and hind limbs blackish 8
7'. Head and upper back grizzled olivaceous, black flecked with reddish, yellowish, brownish, whitish; rump and flanks not yellow-orange. 9
8. Broadly distributed from the Guianan region in the north to the south-central coast of Brazil, except for the northeastern region. . . . . *Dasyprocta leporina*
8'. Restricted to the coastal margin of northeastern Brazil . . . .*Dasyprocta iacki*
9. Head, mid-back, and rump uniformly finely grizzled black and tawny yellow, brown and yellowish, or black and orange; midline of back usually darker than sides but of same color tone; feet brown, with hindfeet relatively short (in comparison to skull length). . . . . *Dasyprocta variegata*
9'. Head, mid-back, and rump uniformly blackish with or without yellowish or reddish bands dominating the black tones; hindfoot relatively long (in comparison to length of skull) . . . .*Dasyprocta guamara*

Dasyprocta azarae Lichtenstein, 1823
Azara's Agouti

SYNONYMS:

Dasyprocta Azarae Lichtenstein, 1823:3; based on Felix d'Azara's (1801:26) "*l'acuti*," but type locality given as "E provincia San. Paulo [= São Paulo] Brasil."

Chloromys Acuti: Rengger,1830:259; name combination.

Dasyprocta caudata Lund, 1840b:12 [1841c:287]; type locality "Rio das Velhas's Floddal" (see Lund 1841c:292, 294), Lagoa Santa, Minas Gerais, Brazil.

Cutia azarae: Liais, 1872:536; part; name combination.

Cutia azarae, var. *brasiliensis* Liais, 1872:536; part; no type locality mentioned; implied type locality Minas Gerais, Brazil.

[*Cutia azarae*] var. *paraguayensis* Liais, 1872:536; based on Felix d'Azara's (1801:26) "*à tort l'acuti*"; therefore, the type locality is near Asunción, Paraguay.

Dasyprocta aurea Cope, 1889:138; type locality "Chapada" [= Chapada dos Guimarães], Mato Grosso, Brazil.

D[asyprocta]. rarea: Cope, 1989:139; typographical error.

Dasyprocta variegata yungarum Thomas, 1910d:505; type locality "Chimosi [= Chimasi], alt. 1700 m. Yungas, [La Paz,] Bolivia."

Dasyprocta azarai Miranda-Ribeiro, 1914:45, table; name combination and incorrect subsequent spelling of *azarae* Lichtenstein.

Dasyprocta variegata urucuma J. A. Allen, 1915b:634; type locality "Urucum (near Curumbá [= Corumbá]) Matto Grosso [= Mato Grosso do Sul], Brazil."

Dasyprocta felicia Thomas, 1917f:310; type locality "near Concepcion, [Concepción department,] Paraguay."

Dasyprocta variegata boliviae Thomas, 1917f:312; type locality "Yacuiba, [Tarija,] southern Bolivia, on Argentine boundary south of Caiza."

Dasyprocta azarae catrinae Thomas, 1917f:311; type locality "Santa Catharina, [Santa Catarina,] S. Brazil."

Dasyprocta paraguayensis: Osgood, 1921:39; name combination.

Dasyprocta azarae azarae: Ellerman, 1940:194; name combination.

Dasyprocta azarae aurea: C. O. da C. Vieira, 1945:396; name combination.

[*Dasyprocta azarae*] *azarae*: C. O. da C. Vieira, 1945:425; name combination.

Dasyprocta azarae paraguayensis: Cabrera, 1961:587; name combination.

Dasyprocta punctata urucuma: Cabrera, 1961:590; name combination.

DESCRIPTION: One of smallest members of genus, head and body length 430–575 mm, tail length 10–35 mm, hindfoot length 100–120 mm, ear length 25–45 mm, and mass 2.4–3.4 kg (Emmons and Feer 1997; Iack-Ximenes 1999). Dorsal color overall gray washed with dull tawny yellow or olivaceous to bright orange, with hairs finely banded by multiple (5–7) white, yellowish, or orange bands interspersed with similar number of blackish ones. In sunlight, living individuals appear golden, often with black rump. Mid-body broadly washed with tawny yellow orange, rump hairs black contrasting with mid-body and sparsely speckled with white or pale orange (or gray, from frosted with white tips on hairs, and area around tail (as viewed from rear) often orange. Sides of head from nose to ear all, or partly, washed with orange or tawny; base of vibrissae pink and naked, with white patch around corner of mouth. Legs orange or tawny, extremities of feet have blackish "socks" that may extend up to heels and elbows or only to toes; posterior thighs and hind legs black. Under parts, flanks, and anterior thighs bright to pale orange; inguinal region often white. Young with same color and color pattern as adults, but with shorter rump hairs. Animals from drier regions, particularly in Argentina, Paraguay, and southwestern Brazil, tend to be mostly gray or olivaceous. *Dasyprocta azarae* easily distinguished from other species by overall homogeneous yellow-olivaceous color coupled with grayish to blackish hindquarters.

DISTRIBUTION: Azara's Agouti ranges throughout southwestern and south-central Brazil, from Mato Grosso and Goiás states in the north to the coast in Rio Grande do Sul and São Paulo states, Bolivia south of the Río Beni, eastern Paraguay, and northeastern Argentina, at elevations ranging from near sea level to 700 m. Iack-Ximenes (1999) included a specimen (INPA 99) from a locality within a patch of cerrado vegetation in Rondônia state, Brazil, an area that otherwise has only *Dasyprocta fuliginosa*. S. Anderson (1997) did not recognize more than one Bolivian species (which he listed as *D. punctata*), but we provisionally assign all Bolivian *Dasyprocta* from Chuquisaca, La Paz, Santa Cruz, and Tarija departments to *D. azarae*. The species probably ranges north as far as the Río Beni, but northern limits require verification. L. H. Emmons (unpubl. data) has both collected a specimen (from El Encanto, Santa Cruz; LHE 781) and obtained many photographs of *D. azarae* from the cerrado of Santa Cruz de-

partment. Iack-Ximenes (1999) included another (MNRJ 5618) of unspecified locality from this same region in Bolivia. The type locality of *D. variegata boliviae* Thomas is on the Argentina border at the western edge of the Bolivian Chaco. The species also occurs in Kaa-Iya National Park, in the Bolivian Chaco (20°03′03″S, 62°26′04″W), based on photo documentation (E. Cuellar and A. Noss, pers. comm. of photos to L. H. Emmons). Because the *Dasyprocta* of southern Bolivia and Paraguay all conform to *D. azarae*, we have provisionally grouped the neighboring populations in Argentina in this species; these samples have been assigned to *D. punctata* in previous synopses of Argentinian species (e.g., Barquez, Díaz, and Ojeda 2006; M. M. Díaz and Barquez 2007).

SELECTED LOCALITIES (Map 387): ARGENTINA: Jujuy, Mesada de la Colmenas (Heinonen and Bosso 1994), Villa Carolina (Thomas 1920a); Misiones, Parque Nacional Iguazú (Crespo 1982b); Salta, Aguaray (R. A. Ojeda and Mares 1989), Vado de Arrayazal, 20 km NW of Aguas Blancas (R. A. Ojeda and Mares 1989). BOLIVIA: Chuquisaca, Porvenir (S. Anderson 1997); La Paz, Chimosi [= Chimasi] (type locality of *Dasyprocta variegata yungarum* Thomas), Puente de Choculo (S. Anderson 1997); Santa Cruz, El Encanto, Cascades (LHE 781), Floripondio (S. Anderson 1997); Tarija, Yacuiba (type locality of *Dasyprocta variegata boliviae* Thomas), Parque Nacional Noel Kempff Mercado, El Encanto (LHE 781). BRAZIL: Bahia, Fazenda Jatobá-Floryl (UNB 1560); Goiás, Pousso Alto [= Pirancanjuba] (USNM 283473), Rio São Miguel (MNRJ 2305); Mato Grosso, Chapada [= Chapada dos Guimarães] (USNM 113431), Jacaré, alto Rio Xingú (MNRJ 11974), São Domingos (MZUSP 6986), São Luiz de Cáceres (MNRJ 5833), Tapirapoan (MNRJ 904); Mato Grosso do Sul, Urucum, near Corumbá (type locality of *Dasyprocta variegata urucuma* J. A. Allen), Maracaju (MNRJ 4612); Minas Gerais, Lagoa Santa, Rio das Velhas (type locality of *Dasyprocta caudata* Lund), Lassance (MNRJ 2304); Paraná, Monte Alegre (MHNCI 485); Rio Grande do Sul, São Lorenço do Sul (MZUSP 249); Santa Catarina, Corupá (MSCJ [no number]); São Paulo, Cotia (MZUSP 25531), Serra da Bocaina (MZUSP 25792). PARAGUAY: Caaguazú, Sommerfield Colony #11 (MNRJ 43170); Concepción, near Concepción (type locality of *Dasyprocta felicia* Thomas).

SUBSPECIES: Thomas (1904a:241) remarked that he could find little tangible difference between specimens collected at localities as wide ranging as the Chapada dos Guimarães in Mato Grosso state, Brazil, eastern Paraguay, or at the type locality of São Paulo, Brazil. He commented, however, on the extensive variation among individuals taken at any one of these localities. Cabrera (1961:587) listed three valid subspecies: the nominotypical form (with *caudata* Lund and *aurea* Cope as synonyms) from east-central Brazil in Minas Gerais, São Paulo, Paraná, and Mato Grosso states; *D. a. catrinae* Thomas from the southwestern Brazilian states of Santa Catarina and Rio Grande do Sul; and *D. a. paraguayensis* Liais (with *felicia* Thomas a synonym) from eastern Paraguay and northeastern Argentina. S. Anderson (1997), while including all Bolivian *Dasyprocta* in his concept of *D. punctata*, listed two subspecies, *boliviae* Thomas and *yungarum* Thomas, both of which we assign herein to *D. azarae*. Iack-Ximenes (1999) regarded both *aurea* Cope and *catrinae* Thomas to be valid species (see Remarks). Given the number of available names but complete lack of any comprehensive review of character variation over the species range, it is premature to validate any of these taxa as recognizable geographic variants.

NATURAL HISTORY: Azara's Agouti occurs in forest patches within savannas (*floresta estacional semidecidual*) and lowland Atlantic Forest (*floresta ombrófila densa*). It is common in cerrado habitats in eastern Bolivia and central Brazil, especially dense in groves of *Attalea* palms that invade burned deforested areas. It feeds on seeds of these (L. B. de Almeida and Galetti 2007) and other palms and is an important disperser of palm seeds (F. R. da R. Silva et al. 2011). Individuals are active during daylight and cool temperatures just after dawn and before nightfall. This agouti adapts to human disturbance, where it is fond of mangos, avocados, cashews, and other domestic crops (L. H. Emmons, pers. comm.). It was the most abundant medium-sized mammal recorded from camera-trap surveys in Rio Grande do Sul (Kasper et al. 2007), and the major prey item of bush dogs, *Speothos venaticus*, in Paraguay (Zuecher et al. 2005). Basic physiological parameters, such a basal metabolic rate, have been measured in captive populations (H. F. Brito et al. 2010). Known internal parasites include Trichostrongyloid nematodes (Pudicinae; Cassone and Durett-Desset 1991), ciliated protozoa (*Amylophorus*; C. Pereira and Almeida 1942), and the flagellate protozoan, *Trypanosoma cruzi* (J. C. Oliveira and Pattolli 1964). Ticks of the genus *Amblyomma* were recovered from agoutis inhabiting Parque Nacionaldas Emas in Goias state, Brazil (T. F. Martins et al. 2011).

REMARKS: Felix d'Azara (1801) recorded "*l'acouti*" from near Asunción, Paraguay, and subsequently Lichtenstein (1823:3) briefly described *Dasyprocta azarae*, basing it on Azara's animal, but curiously giving São Paulo, Brazil, as the locality. Liais (1872:536) described *Cutia azarae* var. *paraguayensis*, also based on Azara's "*l'acuti*." Thomas (1917f:310) argued that Lichtenstein's *azarae* from São Paulo was a different species from Azara's Paraguay ani-

mal and, apparently unaware of Liais' *paraguayensis*, proposed *Dasyprocta felicia* for Azara's animal. Subsequently, Osgood (1921:39) stated that *paraguayensis* Liais "although provisionally and somewhat irregularly proposed, this name is clearly based on Azara's '*l'acuti*.' This species, therefore, should take the name *paraguayensis* and *Dasyprocta felicia* Thomas becomes a synonym." Tate (1935), in his taxonomic history of *Dasyprocta*, neither listed Liais's *paraguayensis* nor referred to Osgood's 1921 paper discussing the topic. Ellerman (1940), however, agreed with Osgood and listed *felicia* Thomas as a synonym of *paraguayensis* Liais. Cabrera (1961) also considered *felicia* Thomas a synonym of *paraguayensis* Liais while listing the latter as a valid subspecies of *D. azarae*. The most recent published compilation of taxa in the genus, that of Woods and Kilpatrick (2005), simply regarded both *paraguayensis* Liais and *felicia* Thomas as synonyms of *azarae* Lichtenstein. Iack-Ximenes (1999), in his unpublished review of Brazilian species, expressed the same view.

Lund (1840b [1841c]) described *D. caudata* from Lagoa Santa, Minas Gerais, Brazil, recognizing his new taxon by its larger size and longer tail than the Paraguayan agouti, which he called *D. azarae*. Thomas (1917f:310) believed that *caudata* was the same as Lichtenstein's *D. azarae* from São Paulo, and Lund's name has been listed as a synonym of this species by essentially all subsequent authors, regardless of varying views as to whether Lichtenstein's *azarae* and Liais's *paraguayensis* (including *felicia* Thomas as a synonym) were the same or different forms.

Cope (1889:138) described *Dasyprocta aurea* based on a single, well-preserved skin entirely yellow-orange in color and supposedly collected at Chapada (= Chapada dos Guimarães), Mato Grosso by H. H. Smith. In comparisons to *croconota* Wagler, *prymnolopha* Wagler, and *azarae* Lichtenstein, Cope believed the relationship of his new species lay with *D. azarae*. J. A. Allen (1915b:633–634), upon examining Cope's type, suggested that the animal was a yellow albino, although "the long rump hairs indicate that *aurea* is not an albinism of *D. azarae*." On the other hand, Thomas (1917f:311), based on apparently normal-colored specimens from Chapada in the British Museum, stated that *aurea* Cope was barely distinguishable subspecifically from *D. azarae*. Two other specimens from Chapada, collected by H. H. Smith in 1883 (USNM 113430–113431), are gray-frosted dorsally and on the rump, and slightly paler but otherwise similar to *D. azarae* from Mato Grosso (Caceres, USNM 390350) and Goiás (Pousso Alto [= Piracanjuba], USNM 283473). Most subsequent authors (C. O. da C. Vieira 1945, 1951, 1953, 1955; Cabrera 1961; Woods 1993; Woods and Kilpatrick 2005) followed Thomas in listing *aurea* Cope as a synonym of *azarae* Lichtenstein. Moojen (1952b), Bishop (1974), and Iack-Ximenes (1999), however, each sided with J. A. Allen (1915b) in asserting that *aurea* was a valid species separate from *azarae*. Iack-Ximenes (1999) questioned whether Chapada dos Guimarães was, in fact, the place where Smith obtained the specimen Cope described, in part because Smith's travels took him widely through central and coastal Brazil and also because none of the many other agoutis from this area in collections exhibited the unique color characters of Cope's animal. Iack-Ximenes further argued that the rarity of true sympatry among agoutis supported his belief that Cope's type was collected elsewhere. He cited a second specimen (MNRJ 2286) that matched the characters of *aurea* but that, unfortunately, lacked associated provenance or other information. Smith's locality may or not be questionable, but it is not possible at present to refute either Allen's suggestion that the animal is a yellow albino of a unique species or Thomas's, and others, conclusion that *aurea* Cope simply represents a mutant *D. azarae*.

Thomas (1910d, 1920a) concept of *Dasyprocta variegata* was rather expansive, and he described both *yungarum* from the Yungas of La Paz department, Bolivia, in 1910 and later *boliviae* from Tarija department in 1917 in this species. However, both *D. azarae* and *D. variegata* share some color characters, such as black or brown feet and finely annulated black and colored hairs, although these traits are geographically and individual variable and descriptions are ambiguous. S. Anderson (1997) recognized *yungarum* Thomas as one of two subspecies of agoutis in Bolivia, both of which he incorrectly assigned to *D. punctata*. He erred, however, on two accounts: first, by mapping *yungarum* to southern Bolivia (Santa Cruz, Chuquisaca, and Tarija departments) and, second, by listing its type locality of Chimosi (= Chimasi) in Cochabama rather than La Paz department. We have not seen the holotype of *yungarum*, but the description could fit that of *D. azarae* as well as *D. variegata*. Because *azarae*-like agoutis extend to near the base of the Andes in the vicinity of the savannas of Ixiamas in La Paz department (photo documentation by R. Wallace, pers. comm. to L. H. Emmons), to the east of Chimasi but at lower elevation, we provisionally assign *yunganus* Thomas to *D. azarae*.

J. A. Allen (1915b) described *Dasyprocta variegata urucuma* based on single specimen collected by Leo E. Miller during the American Museum's Roosevelt Brazilian expedition. He explicitly stated that *urucuma* belonged to the "*D. variegata* group" and needed no comparison to *D. azarae* (J. A. Allen 1915b:634), but his reasoning was unstated, and his concept of *variegata* at the time included three species we recognize herein (*D. punctata* from northwestern South America, *D. azarae*, and true *D. variegata*).

On both geographic and habitat grounds, we are inclined to associate *urucuma* J. A. Allen with *D. azarae*. However, Robert S. Voss (pers. comm.), who examined Allen's holotype, noted that "all of these open-country agoutis [= *D. azarae* and *D. variegata*] look more or less the same," a comment that only highlights the need for careful revision of existing specimens of both species. This need is exemplified by Iack-Ximenes's (1999) assignment of a specimen (INPA 99) from Jaciparaná in Rondônia state, Brazil, to *D. azarae*, a locality just a few kilometers south of Porto Velho from where Osgood (1916:209) assigned specimens to *D. variegata*.

Thomas (1917f:311) described *Dasyprocta azarae catrinae* from an unspecified locality in Santa Catarina state, Brazil, "a locality further south than any place from which a *Dasyprocta* has been recorded." He regarded this animal to represent a new subspecies of *D. azarae*, distinguishing it from Lichtenstein's "true" *azarae* from São Paulo by its paler and grayer general color, with the foreback and sides "scarcely tinged at all with yellow; the rump also yellowish, not hoary grayas in the usual *azarae*." Iack-Ximenes (1999:134) believed *catrinae* to be specifically distinct from *D. azarae*, distinguishing the single specimen he examined (MZUSP 1631, from Joinville, Santa Catarina) by the same color differences noted by Thomas. Other recent authors had followed Thomas and regarded *catrinae* either as a valid subspecies of *D. azarae* (Cabrera 1961) or placed the name in a list of synonyms of that species (Woods and Kilpatrick 2005). A specimen purportedly from Santa Catarina (USNM 19691) has long, bright red rump hairs annulated with black, and is indistinguishable from an agouti from Rio de Janeiro (USNM 283472), both identifiable as *D. leporina*, and unlike *D. azarae*. However, the USNM specimen was acquired in 1891 from a Berlin dealer (August Müller), so its provenance is questionable. Given that apparently only three specimens are available from Santa Catarina state (the holotype of *catrinae* in the British Museum and those in the MZUSP and USNM), only one with provenance, we take the conservative approach and list *catrinae* Thomas as a synonym of *D. azarae* Lichtenstein.

While we believe Woods and Kilpatrick (2005) appropriately included *aurea* Cope and *catrinae* Thomas as synonyms of *D. azarae*, they also listed *acuti* F. Cuvier, 1812 as a synonym. This must be an inadvertent error because Cuvier's name is a misspelling of *aguti* Linnaeus, 1766, and is thus correctly placed in the synonymy of *Dasyprocta leporina* (see the account of that species).

Finally, Thomas (1917f:312) also described *D. variegata boliviae*, from Yacuiba, in Tarija department at the edge of the Bolivian Chaco. Because his description matches specimens we have examined from Paraguay and both the

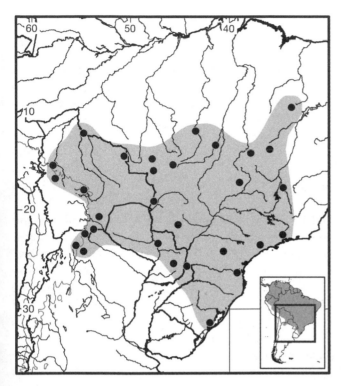

Map 387 Selected localities for *Dasyprocta azarae* (●). Contour line = 2,000 m.

Cerrado and Chaco of Bolivia, we assign *boliviae* Thomas to *D. azarae*. As noted here, S. Anderson (1997) assigned all Bolivian agoutis to *D. punctata*, but probably all records he listed from the dry forests of eastern Santa Cruz department are *D. azarae*, although a reexamination of each of these specimens is required to confirm identifications (see Remarks for *D. variegata*). In general, the distinction between *D. azarae* and *D. variegata* is not absolute, and it will likely take an appropriate analysis using DNA methodologies to determine whether these are, in fact, separate biological entities or whether *variegata* might simply be a northern extension of *D. azarae*.

Specimens from the northern Pantanal in Mato Grosso state, Brazil, have a karyotype with $2n = 64$, FN = 122, one that differs in individual chromosome morphology from the complement of other species (A. L. G. Souza et al. 2007a).

Dasyprocta croconota Wagler, 1831
Orange Agouti
SYNONYMS:

Dasyprocta croconota Wagler, 1831c:618; type locality "Brasilia ad flumen Amazonum," incorrectly restricted by J. A. Allen (1915b:628) to "the immediate vicinity of the mouth of the Rio Madeira"; emended by Iack-Ximenes (1999:142) to Santarém, Rio Tapajós, Pará, Brazil; see Remarks).

D[asyprocta]. croconota: Wagner, 1844:44, Fig. 172b; name combination.

[Dasyprocta] croconota: E.-L. Trouessart, 1897:634; name combination.

Dasyprocta croconota croconota: J. A. Allen, 1916d:568; name combination.

Dasyprocta aguti maraxica Thomas, 1923a:341; type locality "Caldeirão, Marajó Island [= Ilha de Marajó]," Pará, Brazil.

[Dasyprocta] aguti maraxica: Tate, 1935:330; name combination.

[Dasyprocta] a[guti]. mar[axica].: Krumbiegel, 1941:104, Fig. 8; name combination.

[Dasyprocta] crocon[ota].: Krumbiegel, 1941:104, Fig. 8; name combination.

Dasyprocta aguti croconota: Cabrera, 1961:586; name combination.

DESCRIPTION: Another relatively small species, head and body length 465–560 mm, tail length 10–25 mm, hindfoot length 105–110 mm, ear length 40–45 mm, and mass 2.2–2.8 kg. Distinguished from other species, particularly from *D. leporina*, by bright orange coloring over hindquarters and rump and general absence of black or brown pigment on hindquarters. Rump hairs orange to ruby red with strongly faded, or even whitish, basal portion; some specimens may have rump sprinkled with flecks of black, but not to degree that these affect general monochromatic orange coloring. Dorsal streak on mid-back well demarcated and body, other than rump, sprinkled with black to clear brown and yellow to reddish hairs.

DISTRIBUTION: This species occurs in the eastern Amazon of Brazil from the lower Rio Tapajós to the left margin of the lower Rio Tocantins and on Ilha de Marajó and Ilha Mexiana at the mouth of the Rio Amazon. All known localities are between 0 and 200 m in elevation.

SELECTED LOCALITIES (Map 388): BRAZIL: Pará, Cametá, west bank of Rio Tocantins (MZUSP 5363), Curralinho, Ilha de Marajó (MNRJ 4960), Fazenda Gavinha, Ilha de Marajó (MPEG 982), Foz do Rio Curuá (MZUSP 5353), Ilha Mexiana (Krumbiegel 1941a), Monte Cristo, Rio Tapajós (MZUSP 3791), Rio Arapiuns, Santarém (MPEG 704), Rio Curuá-Tinga, affluent of Rio Curuá-Una (MNRJ 11602), Santarém, Rio Tapajós (MZUSP 3790), Sítio Calandrinho, Rio Tocantins (MPEG 8871), Urucurituba, Rio Tapajós (MZUSP 2347).

SUBSPECIES: We treat *Dasyprocta croconota* as monotypic, although Thomas (1923a) recognized specimens from Ilha de Marajó as a subspecies of *D. aguti* [= *D. leporina* herein]; see comment under Remarks.

NATURAL HISTORY: This species has apparently not been studied in the wild, as no data on ecology, population biology, life history, or behavior have been published (at least under this species name). It inhabits lowland rainforest, including second growth and forest edge, within the *Floresta Ombrófilo Densa* vegetative cover type (Iack-Ximenes 1999).

REMARKS: *Dasyprocta croconota* has usually been listed as a subspecies (E.-L. Trouessart 1880:192–193; Cabrera 1961:586) or as a simple synonym (Woods and Kilpatrick 2005) of *D. aguti* [= *D. leporina*]. Thomas (1923a:342) believed that *croconota* "may be an Amazon representative" of *D. aguti* [= *D. leporina*]. Moojen (1952b), however, regarded *D. croconota* as a species distinct from *D. aguti*, and Iack-Ximenes (1999) presented evidence of character differences between and broadly overlapping distributions of *D. croconota* and *D. aguti* south of the Rio Amazon in eastern Brazil. We follow Iack-Ximenes and recognize two species of red-rumped agoutis in east-central Brazil: *D. croconota*, characterized by its bright orange to red flanks and rump and a range limited to Pará state, and *D. leporina* with yellow-orange hindquarters sprinkled with black and a range extending from the Guianan region through eastern Amazonia and the gallery forests of the Cerrado to the forested central coast of Brazil. Appropriate future molecular analyses will determine whether Iack-Ximenes's hypothesis is correct, or whether specimens he allocated to *D. croconota* are only part of a color polymorphism present in populations from Pará state within a single species, *D. leporina*.

Wagler gave only "Brasilia ad flumen Amazonum" as the locality for his *croconota*, which was based on a specimen collected by Johann von Spix during his travels in Amazonia (see Hershkovitz 1987b:28, Fig. 5 for a map of Spix's route). J. A. Allen (1915b:628–629) stated "specimens in the American Museum collected by Leo E. Miller (Roosevelt Brazilian Expedition) at Calama, on the lower Rio Madeira [actually upper Rio Madeira in northernmost Rondônia state; see Paynter and Traylor 1991], agree satisfactorily with Wagler's description," with the footnoted comment "except that the incisors are not white." Allen thus posited, because Spix had spent considerable time on the Amazon River between the Rio Madeira and Rio Negro, ascending the lower parts of both, that "it seems not unreasonable to indicate the immediate vicinity of the mouth of the Rio Madeira as the type locality of *Dasyprocta croconota*." Most subsequent authors (e.g., Tate 1935; Ellerman 1940; Moojen 1952b; C. O. da C. Vieira 1955; Cabrera 1961) followed Allen's argument. Iack-Ximenes (1999:149–152), however, questioned Allen's supposition on multiple grounds. First, only Black Agoutis are known from the lower Rio Madeira and adjacent areas of Amazonia, near where Wagner's (1842c) holotype of *Dasyprocta nigricans* had originated, and where Allen himself had restricted the type locality of Wagler's

Dasyprocta fuliginosa (see Remarks in the account of that species). Second, even Miller's specimens of an obvious Red-rumped Agouti from Chalama are of questionable provenance because, once again, only Black Agoutis are known from the middle Rio Madeira in the vicinity of Chalama today. Third, although C. O. da C. Vieira (1955) followed Allen's opinion regarding the type locality of *croconota*, the distribution of this species that he articulated was primarily that area between the lower Rio Tapajós and Rio Tocantins in Pará state, well to the east of the Rio Madeira. Finally, because Spix also traveled extensively along the lower Amazon between the Tapajós and Tocantins, Iack-Ximenes argued that a locality in this region from which a specimen of the same color characteristics as the one Spix collected, and that Wagler described, should be chosen as the type locality. He suggested Santarém, on the lower Rio Tapajós, as most appropriate. We agree with this assessment, regard Allen's restriction of *croconota* Wagler to the lower Rio Madeira as incorrect, and accept Iack-Ximenes's emendation to Santarém, Rio Tapajós, Pará, Brazil, as a logical alternative.

Thomas (1923a) described *Dasyprocta aguti maraxica* from Caldeirão on Ilha de Marajó at the mouth of the Amazon, discriminating it from Wagler's *D. croconota* with its "brownish fore-back and light centre of belly" while emphasizing the "more strongly tawny colour of the anterior half of the body" of *maraxica* as a characteristic more typical of *D. aguti* (p. 342). Krumbiegel (1941) identified populations from Ilha Mexiana, on the north side of Ilha de Marajó as *Dasyprocta aguti* and *Dasyprocta*

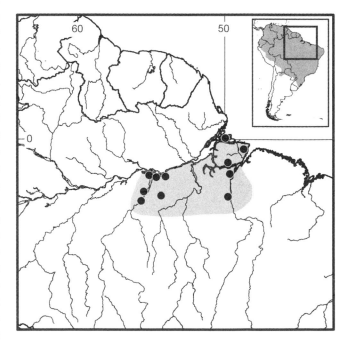

Map 388 Selected localities for *Dasyprocta croconota* (●).

rubrata, both now synonyms of *D. leporina*. However, based on his examination of specimens from these islands in Brazilian collections, Iack-Ximenes (1999) placed *maraxica* Thomas in the synonymy of *croconota* Wagler and regarded Krumbiegel's specimens from Ilha Mexiana to be *D. croconota* as well.

The karyotype of *D. croconota* is apparently unknown.

Dasyprocta fuliginosa Wagler, 1832
Black Agouti

SYNONYMS:

D[asyprocta]. fuliginosa Wagler, 1832:1220; type locality "habitat in Brasilia versus flumen Amazonum"; restricted by J. A. Allen (1915b:626) to "Villa de Borba," Rio Madeira, Amazonas, Brazil.

Dasyprocta nigra Gray, 1842:264; type locality "South America."

Dasyprocta nigricans Wagner, 1842c:362; type locality "Von Borba, am Madeiro und vom Rio Negro oberhalf des Cocuy," based on specimens collected by Johann Natterer in part at Villa de Borba, on the lower Rio Madeira, with which Wagner (1842c:362, 1844:48) compared Wagler's type and considered it identical with Natterer's specimens.

Dasyprocta pallida Waterhouse, 1848:389; *nomen nudum* (see Husson 1978:468).

Dasyprocta variegata zamorae J. A. Allen, 1915b:627; type locality "Zamora (altitude 2000 feet), [Zamora-Chinchipe], Ecuador."

Dasyprocta fuliginosa fuliginosa: J. A. Allen, 1916d:568; name combination.

Dasyprocta fuliginosa mesatia Cabrera, 1917:53; type locality "Tarapoto, en el bajo Napo, Ecuador," about 2°S latitude, Oriente, Ecuador.

Dasyprocta apurensis Delacour, 1922:198; type locality "San Fernando-de-Apure," Apure, Venezuela.

Dasyprocta fuliginosa apurensis: Ojasti, 1972:184; name combination.

D[asyprocta]. fuliginosa nigricans: Ojasti, 1972:188; name combination.

DESCRIPTION: Among largest of agoutis, with head and body length 540–760 mm, tail length 20–40 mm, hindfoot length 120–145 mm, ear length 35–50 mm, and mass 3.5–6 kg. Distinguished by entirely black upper parts, finely grizzled with white (individual hairs are black with short, white tips). Hairs over rump black, sometimes with inconspicuous white tips, and usually do not overhang the rump as distinct fringe; nape with slight crest of longer, black hairs; tail black, as are feet; throat white or strongly grizzled white; venter dark brown and more or less grizzled with white, often with belly midline white. Animals from Venezuela (Amazonas state) with long, overhanging rump

hair, with long white tips, similar to condition characteristic of *D. punctata isthmica*, perhaps indicating taxonomic distinctiveness or hybridization with contiguous populations of *D. punctata*.

Readily separable from *Dasyprocta leporina*, which ranges to immediate east, by overall dark coloring, especially black hindquarters, sprinkled with white or pale yellow; from *Dasyprocta azarae* to southeast, which possesses hairs with multiple, alternating reddish-yellow and blackish-brown bands, producing golden-hued forequarters and midbody, rather than uniform black hairs with pale tips, and is much smaller; and differs from *D. variegata* and *D. azarae*, which replaces it from central Peru south to Argentina, by larger size and overall black color rather than grizzled black with yellow, brown and yellow, or black and orange overall appearance.

DISTRIBUTION: Black Agoutis occur almost exclusively in the western Amazon Basin of Brazil west of the Rio Negro and Rio Madeira, in southern Venezuela, southeastern Colombia, eastern Ecuador, and eastern Peru north of Junín department. However, Ojasti's (1972) assignment of *apurensis* Delacour from Apure state, Venezuela, to *D. fuliginosa* extends the species' distribution into the western tributaries of the Río Orinoco (see map in O. J. Linares 1998). A specimen in the USNM (443450) from Apure is dark brown with white annulations on hairs rather than black, resembling animals from Zulia that are cataloged as *D. punctata* but is also quite like some *D. fuliginosa* from Amazonas. Its smaller size (head and body 522 mm) and browner hue cause us to provisionally retain it under *D. punctata*, but with some doubts. Elevational range extends to 1,000 m on the eastern flank of the Andes.

SELECTED LOCALITIES (Map 389): BRAZIL: Acre, Seringal do Oriente, Rio Juruá (MPEG 835); Amazonas, Borba, Rio Madeira (type locality for *Dasyprocta fuliginosa* Wagler and *Dasyprocta nigricans* Wagner), km 62 on BR 319 (INPA 94), Cucuy [= Cucuí], Rio Negro (Wagner 1842c), Codajeas (MPEG 8913), João Pessôa [= Eirunepé], Rio Juruá (MNRJ 5987); Mato Grosso, Aripuanan [= Aripuanã] (MNRJ 2307), Porto do Braço das Lontras, Rio Comemoração Floriano, alto Rio Jamarí (MNRJ 2292); Rondônia, Alvorada d'Oeste, km 87 on BR 429 (MPEG 19708), UHE Samuel, Rio Jamarí (MPEG 22576). COLOMBIA: Meta, Restrepo (AMNH 136304), San Juan de Arama, Plaza Bonita (FMNH 87891); Putumayo, Río Mecaya (FMNH 70785). ECUADOR: Chimborazo, Pallatanga (Tomes 1860a); Orellana, San José de Payamino (MVZ 170936); Zamora-Chinchipe, Zamora (type locality of *Dasyprocta variegata zamorae* J. A. Allen). PERU: Amazonas, vicinity of Huampani, Río Cenepa (MVZ 155205), Chachapoyas (Thomas 1927a); Loreto, Balsapuerto (Osgood 1914b), Iquitos (Thomas 1928c), near Río Curaray

(Lönnberg 1921); Ucayali, 59 km SW of Pucallpa (USNM 499236). VENEZUELA: Amazonas, Belén, Río Cunucunuma, 56 km NNW of Esmeralda (Handley 1976), Boca Mavaca, 84 km SSE of Esmeralda (Handley 1976), Cerro Yapacana (USNM 256432); Apure, Hato El Frío, 30 km W of El Samán (Ojasti 1972), San Fernando de Apure (type locality of *Dasyprocta fuliginosa apurensis* Delacour).

SUBSPECIES: Some authors (e.g., Cabrera 1961; Ojasti 1972) have recognized multiple subspecies within *D. fuliginosa*, but we allocate some of those names to other species (e.g., *candelensis* J. A. Allen and *colombiana* Bangs, both synonyms of *D. punctata*). J. A. Allen (1915b) noted the strong similarity of the large, black *D. colombiana* from Santa Marta, to those he called *D. variegata* (regarded herein as *D. punctata*) and to *D. fuliginosa*, while Ojasti (1972) remarked that Venezuelan specimens that he referred to *D. f. fuliginosa* exceeded the size of those from the type locality in Brazil, and could be separated as a valid subspecies, to which he applied the name *D. f. nigricans*. However, Wagler's *fuliginosa* and Wagner's *nigricans* both share the same type locality (Borba on the Rio Madeira). Thus, should appropriate geographic analyses eventually conclude that the southern Venezuela samples deserve recognition, another name must be coined. Ojasti also noted that *apurensis* Delacour, a name that had been ignored by previous workers, was applicable to samples of *D. fuliginosa* from Apure state in Venezuela. These populations share the same chromatic pattern as typical *D. fuliginosa* but are much smaller in all dimensions. Although the color and color pattern of *D. fuliginosa* do vary geographically, the lack of any review of this variation prevents one from delineating to any realistic degree appropriate geographic units that might be formally recognized, much less deciding on which, if any, available name might apply.

NATURAL HISTORY: Little is known about the population biology, behavior, or natural history of this species. Gorchov et al. (2004) documented *D. fuliginosa*, along with the acouchys (probably *Myoprocta pratti*), as a major disperser of the seeds of *Hymenaea courbaril* (Fabaceae) in logged rainforest in eastern Peru. Similarly, Tuck Haugaasen et al. (2012) found Black Agoutis to be both major predators and dispersers of Brazil nut (*Bertholletia excelsa* [Lecythidaceae]) seeds on the lower Rio Purus in central Amazonian Brazil. Munari et al. (2011), using both line transects and camera traps, determined that Black Agoutis were primarily diurnal, a characteristic of other agouti species where studied, and present in both *várzea* and *terra firme* forests along the Rio Juruá in western Amazonian Brazil. Reproduction and pregnancy was homogenously distributed in every month of the year in the Peruvian Amazon, with the same pregnancy rate in both wet and dry seasons (Mayor et al. 2011). Mean ovu-

lation rate was 2.5 follicles, and litter sizes were 2.1 per pregnancy. Reproductive potential in this area is believed adequate to sustain density despite intense hunting pressure of the local indigenous people (Mayor et al. 2011). Goff and Brennan (1978) described ectoparasitic mites (family Trombiculidae), and both Durette-Desset et al. (2006) and A. Q. Gonçalves et al. (2006) have identified internal nematode (*Trichostrongylina* and *Helminthoxys*) and cestode (*Raillietina*) parasites in animals they attribute to this species.

REMARKS: The type locality of Wagler's *fuliginosa* was not explicitly specified in the original publication. The type specimen is another one of those that had been obtained by Johann von Spix during his travels along the central and lower Rio Amazon in the early 1800s (see Hershkovitz 1987b:28, Fig. 5). Wagner (1842c:362; see also 1844:48) described *Dasyprocta nigricans* based on specimens collected by Johann Natterer, also from the central Amazon of Brazil, stating that Spix's type had come from the same region as his *nigricans*, for which he gave the type locality as Villa de Borba, on the lower Rio Madeira. Wagner (1842c, 1844) compared Wagler's type and considered it identical with Natterer's specimens, but he chose not to use Wagler's earlier name. Because *fuliginosa* and *nigricans* are the same taxon, and because both are based on specimens collected in the same general region, J. A. Allen (1915b:626) restricted the type locality of Wagler's *fuliginosa* to Borba, lower Rio Madeira, the same locality as that of Wagner's *nigricans*.

Iack-Ximenes (1999) argued that the Black Agouti should properly bear the name *Dasyprocta cristata* (É. Geoffroy St.-Hilaire, 1803), a concept he also extended to include Tschudi's *Dasyprocta variegata*, or the Brown Agouti, as envisioned by Emmons and Feer (1990, 1997) and herein. His reasoning was based primarily on the supposition that the specimen examined by Geoffroy St.-Hilaire was one of those collected by the Brazilian naturalist Alexandre Rodrigues Ferreira but subsequently stolen from the Museu Real d'Ajuda in Lisbon by Napoleon's army during its invasion of Portugal in 1808, and then sent to the Muséum National d'Histoire Naturelle in Paris. Because Ferreira collected through central Amazonia, Iack-Ximenes (1999:68) corrected the type locality of *cristata* from "découvert a la Guyane hollandaise" [= Surinam], as originally given, to Igarapé Grande on the Rio Juruá, Amazonas, Brazil. Although É. Geoffory St.-Hilaire as well as A.-G. Desmarest did base descriptions of some South American mammals on the purloined Ferreira collection (e.g., *Mesomys hispidus*, see that account and Orlando et al. 2003; see also Hershkovitz 1987b), Geoffory St.-Hilaire's *cristata* cannot be one of those as he named the taxon five years before Napoleon's army plundered

Lisbon. For further details regarding *cristata* É. Geoffory St.-Hilaire, see Remarks in the account for *D. leporina*.

Cabrera (1961) listed *nigra* Gray (1842) as a synonym of *D. fuliginosa* Wagler, but Woods and Kilpatrick (2005) ignored Gray's name. Despite Gray's (p. 264) unhelpful type locality of "South America," the description he gave ("black, grizzled with white; shoulder and haunches blacker; legs black") associates *nigra* with Wagler's *fuliginosa* (but see Remarks under *D. kalinowskii*).

Contrary to Cabrera (1961), who regarded *zamorae* J. A. Allen, 1915, as a valid subspecies of *Dasyprocta punctata*, or Woods and Kilpatrick (2005), who followed Cabrera but simply included Allen's name as a synonym of *D. punctata*, we regard *zamorae* J. A. Allen a synonym of *D. fuliginosa* Wagler. Our belief is based on the hypothesis that the agoutis on both sides of the northern Andes (e.g., *punctata* on the western and *fuliginosa* on the eastern versant) represent separate species, despite similarities that some samples may share in color and color pattern. Allen's description (p. 627) highlighted a reddish ("deep ochraceous tawny") overall tone of *zamorae* rather than the blackish tones typical of *fuliginosa*, although he did note the rump hairs had rather long, white tips, a characteristic of some specimens of *fuliginosa*. He believed, however, that *zamorae* was closely related to *chocoensis* (which Allen described in the same paper from the Chocó of western Colombia) and *colombiana* Bangs (from the Santa Marta region, both of which are now regarded as representatives of *D. punctata* (see their account).

Husson's (1978:466–468) inclusion of *D. fuliginosa* in the fauna of Surinam is based on a single specimen labeled as from that country but without additional details as to locality, collector, or even date of collection (see Voss et al. 2001). Husson assigned the specimen to this species, and not to the widespread and common agouti of the country, *D. leporina*, because of its dark blackish brown dorsal surface and the lack of even a trace of rufous on the rump. Whether this specimen (RMNH 23963) is a mutant, or local color variant of *D. leporina*, or something truly different, without additional specimens from anywhere in the Guianan region of unquestionable provenance, we do not include Surinam within the currently understood geographic range of *D. fuliginosa*.

The specimens that Tomes (1860a:216) referred to *Dasyprocta caudata* Lund from Pallatanga, Ecuador, are listed here as *D. fuliginosa*, but those that Lönnberg (1921) called *D. fuliginosa* from Gualea (on the western slope of the Andes in Pichincha department), Ecuador, are either best attributed to *D. punctata* or the locality may be erroneous (as is the case for Lönnberg's specimen of *Pattonomys occasius* [Thomas 1921f], also said to have come from Gualea, although this species is restricted to the Amazon

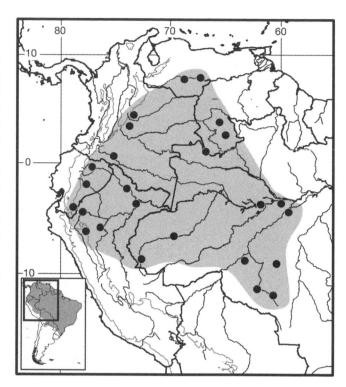

Map 389 Selected localities for *Dasyprocta fuliginosa* (●). Contour line = 2,000 m.

lowlands [Emmons 2005]). We have not examined any of these specimens.

The karyotype of *D. fuliginosa* is $2n = 64$, FM = 122 (J. F. S. Lima and Langguth 1998; R. S. L. Ramos et al. 2003).

Dasyprocta guamara Ojasti, 1972
Guamara Agouti
SYNONYM:

Dasyprocta guamara Ojasti, 1972:176; type locality "Caño Araguabisi, Territorio Delta Amacuro, Venezuela."

DESCRIPTION: Medium-sized agouti, equivalent in total length (513–604 mm) to some of larger species in genus but with smaller mass, 3–4.4 kg. Remarkable characteristic are proportionally elongated hindfeet, which average about 130% of greatest skull length in comparison, for example, to <115% among geographically proximate samples of *D. leporina*. Dorsal coloration brown to black finely mixed with ochraceous and without any chromatic separation between rump and rest of dorsal surface. Skull notable for its relatively wide and short proportions (condylobasilar length <2 times zygomatic width), conical contour of zygomatic arches when viewed from above, and well-developed lateral crests on jugal.

DISTRIBUTION: *Dasyprocta guamara* is known only from two localities in the Orinoco delta of northeastern Venezuela.

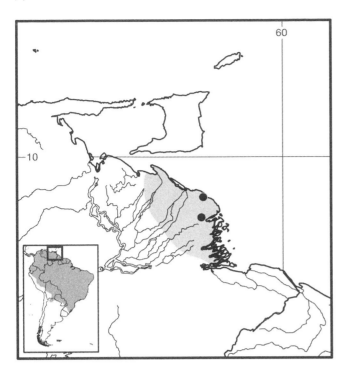

Map 390 Selected localities for *Dasyprocta guamara* (●).

SELECTED LOCALITIES (Map 390): VENEZUELA: Delta Amacuro, Cañõ Araguabisi (type locality of *Dasyprocta guamara* Ojasti), Caño Guiniquina (Ojasti 1972).

SUBSPECIES: *Dasyprocta guamara* is monotypic.

NATURAL HISTORY: Little is known of the natural history and ecology of this species. It inhabits the dense and marshy forests of the Orinoco delta, habitat that is always wet and partly inundated during high tide. Ojasti (1972) conjectured that the exceptionally elongated feet facilitate locomotion through the muddy substrate. Individuals seek refuge in hollow trunks of fallen trees. This is one of the most abundant mammals in the region and forms an important protein source for local peoples. Two of five adult females collected in the month of November were pregnant, each with two embryos.

REMARKS: The karyotype of *D. guamara* is unknown.

Dasyprocta iacki Feijó and Langguth, 2013
Iack's Red-rumped Agouti

SYNONYM:

Dasyprocta iacki Feijó and Langguth, 2013:112 [English version], 115 [Porguguese version]; type locality "Biological Reserve Guaribas (60°[sic, should be 6°] 44′S, 35°9′W), municipality of Mamanguape, state of Paraíba, Brazil."

DESCRIPTION: Medium-sized species (head and body 479 mm); dorsal part of head brown speckled with yellow-orange; cheeks more yellowish in color than crown; mystacial vibrissae long and black, some reading posteriorly to cars; chin, circumbucal, and intermandibular regions with white hairs; ventral region of neck yellow with basal white bands; ears average 43 mm in length, covered with sparse hairs; innerside of pinna with yellow-orange hairs with brown base; hairs on back of body gradually increase in length caudally; general color of body agouti, mid-dorsal region with wide but subtle stripe of darker brown speckled with orange that becomes gradually paler on the sides; general color of rump dark brown speckled with orange; sides orange; venter with yellow patch in thoracic region that narrows caudally, becoming grayish agouti over belly; limbs brown speckled with orange on outer side and yellow on inner side; forefeet and hindfeet covered dorsally with short brown hairs speckled with orange.

DISTRIBUTION: *Dasyprocta iacki* is known for certain from only the littoral zone in Paraíaba and Pernambuco states, Brazil. As noted by Fejió and Langguth (2013), however, this is probably the same species as that identified by Iack-Ximenes (1999) as *Dasyprocta aguti* from south of the lower Rio Amazonas between the Madeira and Tocantins rivers (and different from *D. leporina*, which Iack-Ximenes restricted to the red-rumped agoutis from the left bank of the lower Rio Amazonas north to the Guianan region; see account for *Dasyprocta leporina*, particularly the Remarks section).

SELECTED LOCALITIES (Map 391): BRAZIL: Paraíba, Reserva Biológica Guaribas, Mamanguape (type

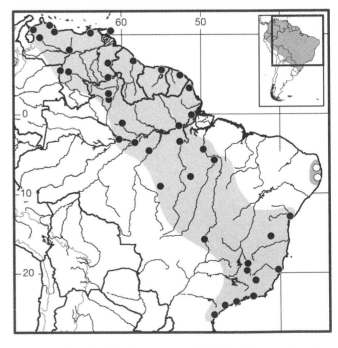

Map 391 Selected localities for *Dasyprocta iacki* (○) and *Dasyprocta leporina* (●). Contour line = 2,000 m.

locality of *Dasyprocta iacki* Feijó and Langguth); Pernambuco, Mata de Dois Irmáos (Feijó and Langguth 2013).

SUBSPECIES: *Dasyprocta iacki* is monotypic.

NATURAL HISTORY: No information on ecology is available; the two known localities are within the Atlantic Forest domain.

REMARKS: Feijó and Langguth (2013) suggested that *D. iacki* would likely apply to other red-rumped agoutis from the right (south) bank of the Rio Amazonas south to the Rio São Franciaco, should they be recognized as a separate species, as posited by Iack-Ximenes (1999); see account for *D. leporina*.

Specimens referred to *D. aguti* by J. F. S. Lima and Langguth (1998) with $2n = 64$, FN = 122 or 124 represent this species (see Feijó and Langguth 2013:114).

Dasyprocta kalinowskii Thomas, 1897
Kalinowski's Agouti

SYNONYM:

Dasyprocta Kalinowskii Thomas, 1897e:219; type locality "Idma, Valley of Santa Ana, Cuzco, Peru. Alt. 4600 feet."

DESCRIPTION: Thomas (1897e:219) wrote that this species was of large size, giving head and body length of holotype as 630 mm, hindfoot length as 125 mm, and tail length about 24 mm. Upper parts of body colored with hairs banded in black and yellowish rufous, with elongated hairs overhanging rump uniquely white with black tips. Area at base of tail black and abruptly defined from elongated and white rump hairs. Under parts yellowish, grizzled with brown, and feet blackish. White rump hairs tipped black readily distinguish this species from *D. fuliginosa*, with its all black rump and hairs tipped in white, as well as from *D. variegata*, which occurs lower on same slopes in southern Peru, and which is colored grizzled black and tawny yellow, brown and yellowish, or black and orange. Black feet of *D. kalinowskii* also differ from the brown feet and legs of *D. variegata*.

DISTRIBUTION: *Dasyprocta kalinowskii* is an endemic species of southern Peru known only from the eastern slopes of the Andes in the Urubamba and Marcapata drainages, at an elevational range between 1,000 and 2,000 m.

SELECTED LOCALITIES (Map 392): PERU: Cusco, Aymaibanba [= Amaybamba], Urubamba Valley (MCZ 58244), Hacienda Cadena, Río Marcapata (FMNH 78664), Idma, Santa Ana Valley (type locality of *Dasyprocta kalinowskii* Thomas).

SUBSPECIES: *Dasyprocta kalinowskii* is monotypic.

NATURAL HISTORY: No information is available on local ecology, behavior, or life history traits of this species. The species occupies upper tropical forest (*yungas* ecoregion; Brack-Egg 1986; Pacheco et al. 2009) on the steep slopes of the eastern Andes.

REMARKS: In keeping with most previous authors (e.g., Cabrera 1961; Pacheco 2002; Woods and Kilpatrick 2005; Pacheco et al. 2009), we list *Dasyprocta kalinowskii* as a species distinct from the neighboring *D. variegata*, which occurs at lower elevations in the eastern Andes and lowlands throughout southern Peru. Emmons and Feer (1990, 1997) made no mention of this species, although their mapped range of *D. variegata* encompassed the limited range of *D. kalinowskii*. Eisenberg and Redford (1999) also regarded *D. kalinowskii* as a species, but noted that its status was uncertain and that careful comparisons to both *D. variegata* and *D. fuliginosa* were warranted. Gray (1842:264) described *D. nigra* as "black, grisled with white; shoulder and haunches blacker; legs black; throat gray; belly rather grayer; hair of the back elongated, flattened, white at the base. *Hab.* South America." This description is suspiciously like *D. kalinowskii*, but Thomas did not mention Gray's earlier species, although Alston (1876) placed *nigra* in synonymy with *D. fuliginosa*, and redescribed it as with "rump hoary, the very long and soft hairs beige broadly tipped and often ringed at the base with white." Specimens of *D. fuliginosa* do not usually have rump hairs annulated with white to the base, although a specimen from Venezuela (USNM 388224) has very narrow white bands alternating with brown along the entire length of its rump hairs.

Thomas (1897e:219), in his description of *D. kalinowskii*, stated "this handsome Aguti may be readily distinguished from any species hitherto described by the peculiar

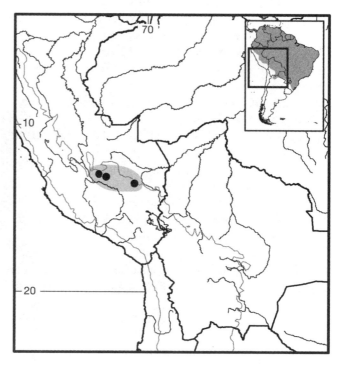

Map 392 Selected localities for *Dasyprocta kalinowskii* (●). Contour line = 2,000 m.

coloration of its long rump-hairs, which, white with black tips, are quite unlike those of any other species." This color pattern of the rump hairs is opposite that typical for *D. variegata*, as Thomas himself noted (1910d:506) when he remarked that Peruvian specimens had yellowish light tips and that Tschudi, in his description of *variegata*, had noted that the tips were whitish.

The karyotype of *D. kalinowskii* is unknown.

Dasyprocta leporina (Linnaeus, 1758)
Red-rumped Agouti

SYNONYMS:

[*Mus*] *leporinus* Linnaeus, 1758:80; based on *Lepus javensis* of Catesby (1754:18, plate 18); type locality fixed by neotype selection to "forest near the boarding-school Peninika, near the confluence of the Peninika Creek and the upper Commewijnw River," Commewijnw District, Surinam (Husson 1978:457, 463).

[*Mus*] *aguti* Linnaeus, 1766:80; type locality "Brasilia, Surinamo, Guiania"; based on "*Aguti vel acuti Brasilensibus*" of Marcgraf (1648:224); restricted to Brazil by Thomas (1898f:272) and given incorrectly as "Pernambuco" by Cabrera (1961).

Cavia aguti: Erxleben, 1777:353; name combination.

Cavia leporina: Erxleben, 1777:355; name combination.

[*Cavia*] *bicolor* Boddaert, 1784:103; type locality "Habitat in Americae meridionalis Calidiora," listed as equivalent to *leporinus* Linnaeus.

Calva Aguti: Gmelin, 1788:121; name combination.

Cavia Agouti Brongniart, 1792:115; name combination and incorrect subsequent spelling of *aguti* Linnaeus.

C[*avia*]. *aguti leporina*: Kerr, 1792:217; name combination.

C[*avia*]. *aguti cunicularis* Kerr, 1792:217; part; type locality "Brasil, Giana, Cayenne e Índias Ocidentalis."

Agouti Cayanus Daudin, 1802:166; type locality "Cayenne," French Guiana.

Cavia cristata É. Geoffroy St.-Hilaire, 1803:catalog; type locality "Patrie. Surinam.

[*Dasyprocta*] *aguti*: Illiger, 1811:93; name combination.

Dasyprocta acuti Illiger, 1815:114; name combination and incorrect subsequent spelling of *aguti* Linnaeus, 1766.

[*Dasyprocta*] *S. acuty* Oken, 1816:823; [*S. = Savia?*]; incorrect subsequent spelling of *aguti* Linnaeus.

Chloromys acuti G. Cuvier, 1817:214; name combination and incorrect subsequent spelling of *aguti* Linnaeus, 1766.

[*Dasyprocta aguti*] *leporina* B: Olfers, 1818:214; part; name combination.

D[*asyprocta*]. *aguti*: Olfers, 1818:214; part; name combination.

Dasyprocta acuti Desmarest, 1822:357; part; name combination and incorrect subsequent spelling of *aguti* Linnaeus, 1766.

Dasyprocta cristata: Desmarest, 1822:358; name combination.

Dasyprocta aguti: Lichtenstein, 1823:5; name combination.

[*Dasyprocta*] *Agouti*: Minding, 1829:91; name combination, and incorrect subsequent spelling of *aguti* Linnaeus, 1766.

D[*asyprocta*]. *agouti*: Waterhouse, 1848:388; name combination; part; *lapsus calami* of *aguti* Linnaeus.

Dasyprocta leporina: Waterhouse, 1848:391; first use of current name combination.

Cutia cristata: Liais, 1872:536; name combination; "découvert a la Guyane hollandaise," with reference to É. Geoffroy St.-Hilaire's *Cavia cristata*, which Liais stated was the same as "l'Akouchy" [= *Cavia acouchy* Erxleben].

Cutia azarae var. *brasiliensis* Liais, 1872:536; part; type locality "Minas-Geraes [= Minas Gerais], Brazil"; based on "l'Agouti de Buffon & Marcgraaf."

Dasyprocta rubrata Thomas, 1898f:273; type locality "Savanna Grande, Trinidad."

Dasyprocta rubrata flavescens Thomas, 1898f:274; type locality "Caripé [= Caripe], Cumana [= Cumaná], [Monagas,] Venezuela."

Dasyprocta lucifer Thomas, 1903b:491; type locality "Caicara, River Orinoco," Bolívar, Venezuela.

Dasyprocta lucifer cayennae Thomas, 1903b:492; type locality "Approuague, Cayenne," French Guiana.

Dasyprocta croconota lucifer: J. A. Allen, 1915b:629; name combination.

Dasyprocta cayennae: Thomas, 1917e:259; name combination.

D[*asyprocta*]. *a*[*guti*]. *flavescens*: Thomas, 1917e:259; name combination.

Dasyprocta aguti lunaris Thomas, 1917e:259; type locality "Moon Mountains, Southern British Guiana [Guyana], about 1°N., 59°W" [presumed by Voss (1991a:8) to be synonymous with the Serra da Lua, about 60 mi SE Boa Vista, Roraima, Brazil, labeled as the "Kai-Irite" mountains on older maps].

D[*asyprocta*]. *a*[*guti*]. *aguti*: Thomas, 1923a:342; name combination.

[*Dasyprocta*] *cayanus*: Tate, 1935:330; name combination, and incorrect gender concordance.

[*Dasyprocta*]. *aguti aguti*: Tate, 1935:330; name combination.

[*Dasyprocta*]. *lucifer lucifer*: Tate, 1935:330; name combination.

[*Dasyprocta*] *rubrata rubrata*: Tate, 1935:330; name combination.

[*Dasyprocta*] *flavescens*: Tate, 1939:183; name combination.

Dasyprocta aguti aguti: Ellerman, 1940:193; name combination.

Dasyprocta cayanus: Simpson, 1941:7; name combination, and incorrect gender concordance.

D[*asyprocta*]. *aguti aguti*: Krumbiegel, 1941:98; part; name combination.

Dasyprocta lunaris: Krumbiegel, 1941:101; name combination.

[*Dasyprocta aguti*] *aguti*: Moojen, 1952b:115; name combination.

Dasyprocta agutti C. O. da C. Vieira, 1953:158; incorrect subsequent spelling of *aguti* Linnaeus, 1766.

Dasyprocta leporina leporina: Husson, 1978:373; name combination.

Dasyprocta a[*guti*]. *aguti*: Tribe, Leher, Doglio, Rebello, Mello, and Guimaráes, 1985:296; name combination.

Dasyprocta leporina aguti: Hershkovitz, 1987b:37; name combination.

[*Dasyprocta leporina*] (*aguti*): Corbet and Hill, 1991:202; name combination.

DESCRIPTION: One of more variable species in body size although typically in moderate to large size range for genus; head and body length 470–650 mm, tail length 10–30 mm, hindfoot length 118–148 mm, ear length 40–50 mm, and mass 2.1–5.9 kg (Emmons and Feer 1997; Iack-Ximenes 1999). Color variable geographically but generally characterized by finely grizzled olivaceous head and forequarters; rump dark red to yellowish-orange, covered by long, straight hairs, which overhang rump in fringe, usually paler yellow or orange at base (color visible when hairs are erect). Individual rump hairs with yellow-orange bands, beginning with base and ending at tips, alternating with two to five brown or black bands. Top of head, neck, and mid-back between shoulders sometimes blackish or with crest of longer, pure blackish hairs. Vibrissae stiff, black, and reach to base of ear. Chin naked. Ears short, rounded, and mostly naked. Eyes large. Tail tiny, inconspicuous, and naked stub. Both forefeet and hindfeet dark brown. Under parts grizzled brownish orange or orange with white midline. Individuals range in color from dull brown-olivaceous to bright orange-olivaceous forequarters; specimens from Trinidad and Tobago smaller than mainland representatives of species and generally colored dark brown. Populations north of Rio Amazonas larger in size and have narrower skull (Iack-Ximenes 1999). Distinguished from *D. croconota* by bright orange rump and flanks of latter, from *D. fuliginosa* by yellowish-red instead of black rump, from *D. prymnolopha* by lacking black mid-dorsal stripe over rump, and from *D. azarae* by larger size and brownish overall color with red to yellow-orange rump contrasting with tawny gray washed with orange tones of latter.

DISTRIBUTION: *Dasyprocta leporina* occurs throughout the Guianan region, from northern Venezuela (east of Lake Maracaibo) to French Guiana and Brazil north of the Amazon and east of the Rio Negro, then south of the Amazon east of the Rio Madeira to central coastal Brazil.

SELECTED LOCALITIES (Map 391): BRAZIL: Amapá, Macapá (MPEG 589), Vila Velha do Cassiporé (MPEG 6608); Amazonas, Fortaleza, Paraná do Urariá (MZUSP 25519), Manaus (MNRJ 43131), right bank Rio Uatumã, 5 km S mouth of Rio Pitinga (INPA 629); Bahia, Ilha de Madre de Deus (MZUSP 3827), Itambé (MNRJ 43140); Espirítu Santo, Linhares (MNRJ 43139); Goiás, Fazenda Formiga (MZUSP 3944); Minas Gerais, Fazenda Jaguara (MNRJ 43138), 104 km N of Lagoa Santa (MNRJ 13521), Viçosa (MNRJ 1369); Pará, Agrovila União, 18 km S and 19 km W of Altamira, Rio Tocantins (MPEG 20160), Gorotire, Rio Xingu (MZUSP 25515), Ilha Tocantins, Rio Tocantins (MPEG 12310), Km 19, Transamazônica Pará-Itaitube, Jacareacange, Rio Tapajós (MPEG 8480), Mariussú (MNRJ 11981), Serra do Cachinbo (MNRJ 43137); Rio de Janeiro, Caxias (MNRJ 7310); Roraima, Colônia do Apiaú, margen direita do Igarapé Serrinha, Rio Mucajaí (MPEG 21880), Taperinha [= Lago do Cobra, Caracaraí] (MPEG 22375); São Paulo, Ariri, Cananéia (MHNCI 3021), Fazenda Banaura, left bank Rio Branco (Iack-Ximenes 1999), São Sebastião (MZUSP 6783); Tocantins, Araguatins (MZUSP 25533). FRENCH GUIANA: Cayenne (type locality of *Agouti cayanus* Daudin). GUYANA: Demerara-Mahaica, Demerara (Thomas 1917e). SURINAM: Commewijnw, Peninika boarding school (type localities of *Mus leporinus* Linnaeus and *Mus aguti* Linnaeus). TRINIDAD AND TOBAGO: Trinidad, Savanna Grande (type locality of *Dasyprocta rubrata* Thomas). VENEZUELA: Amazonas, 32 km S of Puerto Ayacucho (Handley 1976), San Juan, Río Manapiare, 163 km ESE of Puerto Ayacucho (Handley 1976); Bolívar, 43 to 45 km NE of Icabarú (Handley 1976), Caicara, Río Orinoco (type locality of *Dasyprocta lucifer* Thomas), El Manaco, 56 to 59 km SE of El Dorado (Handley 1976); Carabobo, San Esteban (J. A. Allen 1911); Falcón, near Mirimire (Handley 1976); Lara, Antoátequi, W of Guarico (J. A. Allen 1911); Monagas, Caripe, Cumaná (*Dasyprocta rubrata flavescens* Thomas); Zulia, El Panorama, Río Aurare (Osgood 1912).

SUBSPECIES: There is considerable geographic variation in color over the back and flanks of *D. leporina*, and multiple subspecies have been recognized in limited geographic parts of the species range. For example, Cabrera (1961) and Ojasti (1972) each recognized several subspecies, and O. J. Linares (1998) mapped the ranges of two subspecies in Venezuela (*cayana* Daudin to the south of the Río Orinoco, at elevations ranging between 10 and 860 m, and *flavescens* Thomas from north of the Río Orinoco in the Cordillera Central, Cordillera Oriental, east side of Lago de Maracaibo, and throughout the llanos, at an elevation range from 10 to 1,450 m). We agree, however, with Voss et al. (2001:142), who argued "the necessity for a trinomial classification remains to be convincingly established by a comprehensive study of geographic variation." These authors also noted (p. 142) that "for future revisionary work, a large series of specimens collected by H. A. Beatty in the Wilhelmina Mountains of Surinam . . . provides a useful sample of individual variation from a single local population of this nomenclaturally important species."

NATURAL HISTORY: *Dasyprocta leporina* has been studied extensively, mostly in the Guianan region but also in parts of both eastern Amazonian and Atlantic forests of Brazil. In French Guiana (e.g., Dubost 1988), the species can occur in all available vegetation types but primarily in open forest, usually distant from both water and dense vegetation. Density may vary with forest type, with greater numbers in fragmented patches than in continuous forest (Jorge 2008). The basic social unit of this agouti is the family, comprising a breeding pair with their young of the year. Subadult and juvenile males are either solitary or may group into small units of two to three individuals. Each group occupies areas about 200 m in diameter with groups separated from one another by 50 m or less, although there is intergroup territorial defense. This distributional pattern is clumped, as a result of the aggregation of individual home ranges. Home range, or site, fidelity is nearly permanent over time. Individuals frequently use well-worn trails and often travel in pairs or as the entire family group. Consistent with the longevity of family groups and a clumped distribution, several separate mitochondrial maternal lineages may overlap within an area, even over relatively small geographic regions such as French Guiana (van Vurren et al. 2004).

Both diurnal and nocturnal resting places are used by single individuals, with the number varying by individual not by sex or age (Dubost 1988). As is characteristic of all others in the genus, this species is primarily diurnal but with a bimodal activity pattern concentrated in the early morning and late evening when a second phase of foraging occurs (Norris et al. 2010), with foraging activity varying in response to available resources and temperature (Jorge and Peres 2005). Foraging time depends upon forest age and tree structure, both of which affect resource availability. Adult sex ratio is skewed in favor of females because of high juvenile male mortality. Females typically reproduce once per year, and while births may occur year-round most are concentrated in the dry season (Henry 1994). In French Guianan forests, pregnancy peaks between November and April, beginning with the early rainy season and coincidental with an increase in the number of fruiting trees (Dubost et al. 2005). The gestation period is 110–112 days, and litter size ranges from one to three (Weir 1971b; Henry 1997;

Dubost et al. 2005). Neonates are precocial, well furred, and with eyes open at birth (Henry 1994).

Considerable research has been directed to the role this species plays in the removal, dispersal, and survival of the seeds of many rainforest trees (Forget 1990; Forget and Cuijpers 2008), particularly those large-seeded species (Galetti et al. 2010), including important cash crops such as Brazil nuts (Peres and Baider 1997; Scoles and Gribel 2012) and palms (Pires and Galetti 2012). Seed dispersal is over short distances, given the average size of home ranges (3 to 8.5 ha in eastern Brazil; Silvius and Fagoso 2003, Jorge and Peres 2005). Henry (1999), Dubost and Henry (2006), and Silvius and Fragoso (2003) documented the importance of frugivory, including both pulp and seeds, in a diet that was broadened with the use of other plant parts (e.g., cotyledons, leaves, flowers) and even animal materials under varying rainfall and fruit production seasonal schedules. Henry (1997) detailed the interplay between diet and seasonal reproduction in French Guiana, also including the seasonal use of invertebrates by pregnant females. Ghilardi and Alho (1991) described seasonal dietary trends in eastern Amazonian Brazil. Individuals scatter hoard seeds that they later retrieve and consume (Jorge and Howe 2009) and also the pilfered caches of others (P. R. Guimarães et al. 2005). They avoid plant species whose seeds accumulate toxic secondary compounds, such as quinolizidine alkaloids (P. R. Guimarães et al. 2003).

A wide range of endoparasites may infect individual agoutis, including nematodes (Cameron and Reesal 1951; Hugot 1986; A. Q. Gonçalves et al. 2006), apicomplexans of the genus *Eimeria* (Lainson et al. 2007), tapeworms (Cameron and Reesal 1951; Zimmerman et al. 2009), and round worms (Cameron and Reesal 1951). This species is also an important reservoir for *Trypanosoma cruzi*, the parasitic euglenoid trypanosome associated with Chagas disease (Deane 1967; R. D. Ribeiro and Barretto 1977).

REMARKS: The technical name applied to the Red-rumped Agouti has varied among authors for more than a century, with *Mus aguti* Linnaeus (1766) initially almost universally applied (Tate 1935; Ellerman 1940; Cabrera 1961; Handley 1976; Corbet and Hill 1980) but *Mus leporinus* Linnaeus (1758) used in more recent synopses (Woods 1993; Woods and Kilpatrick 2005). Part of the variance in name application has been taxonomic (the number of species recognized) and part as been nomenclatural (the correct name applied to those species recognized, and why). What follows is a temporal summary of the major players in this debate.

Thomas (1898f:272–273) believed that the red- and yellow-rumped agoutis of the Guianas and Brazil (presumably including that portion of Brazil south of the Amazon River), while superficially similar, were different from one another. Hence, he argued that fixing Linnaeus's "*Mus aguti*" to either of these geographic groups was difficult because, while Linnaeus had included Surinam and Guiana along with Brazil in the locality of his *Mus aguti* (Linnaeus 1766:80, geographic range given as "Brasilia, Surinamo, Guiania"), he had based his name on the description provided by Marcgraf (1648) of a Brazilian animal, specifically that part of Brazil where Marcgraf's geographical work was centered. Consequently, Thomas (p. 272) assigned *Mus aguti* Linnaeus to "the true Brazilian species" (i.e., the red- and yellow-rumped form that occurs south of the Amazon River). Thomas (p. 273) then proposed the name *Dasyprocta rubrata* to represent the red- and yellow-rumped agoutis inhabiting Trinidad and the Guianas.

Husson (1978:462–463) reviewed the nomenclatural history of the Guianan agouti and, contrary to Thomas (1898f), stated that *Dasyprocta aguti* (Linnaeus, 1766) has been generally used for that species. He noted, however, that *Mus leporinus* Linnaeus, 1758 (based on the description and color plate of "*Lepus javensis*" of Catesby [1754:18, plate 18]), was older and thus had priority. Husson also noted that, although some authors had accepted this synonymy, others had not because Linnaeus had stated that Catesby's animal had originated from "Java and Sumatra." However, Pennant (1781:366) had recognized that this geographic provenance was in error and remarked that the "Javan Cavy" inhabited "Surinam and the hotter parts of South America, where it is a common food." Shaw (1801:26) also stated that Catesby's animal was a "native of Surinam, and other parts of South America. It is altogether an American animal . . . notwithstanding its common title of the Java Hare." He correctly regarded *Mus leporinus* as an agouti and treated it as a variety of *Mus aguti*. Gray (1843b:124) incorrectly identified *Mus leporinus* as an acouchy (now *Myoprocta acouchi* [Erxleben, 1777]). Waterhouse (1848:391–392, footnote 1) corrected Gray's error, but, although he agreed that *Mus leporinus* was undoubtedly an agouti, he did not use the name for any species in the genus *Dasyprocta*. Thomas (1911d:146) and Tate (1935:329, 331) had also assigned *Mus leporinus* to *Dasyprocta*, but while neither of these authors believed the species to be identifiable, Tate did suggest that it might be referable to one of the Central American agoutis. Finally, Husson (1978) argued that Surinam was the probable origin of Catesby's animal, and, to stabilize nomenclature, he selected a neotype from that country (RMNH 20752) for *Mus leporinus* Linnaeus, as Catesby's specimen was no longer extant. This action thus made *Dasyprocta leporinus* (Linnaeus) the oldest available name for the Guianan red- and yellow-rumped species.

Hershkovitz (1969:23) used the name *Dasyprocta leporina* in discussing South American mammals that had ex-

panded into Middle America after the closure of the Panamanian isthmus but did not identify the geographic scope of his concept. Eisenberg (1989) did use *Dasyprocta leporina* to apply to the Guianan agouti, basing his decision on Husson (1978). Later (in Eisenberg and Redford 1999), he extended his concept of *Dasyprocta leporina* to include Brazil on both sides of the lower Amazon ranging south to the Atlantic coast, and specifically mentioned that his *D. leporina* included *D. agouti* [*sic*]. Emmons and Feer (1990, 1997) continued in the same vein, mapping a single species of red-rumped agouti from the Guianas south to coastal Brazil, but used the name *D. agouti* [*sic*] rather than *D. leporina*, commenting only that "some recent works use the species name *D. leporina*" (Emmons and Feer 1997:226). Iack-Ximenes (1999) followed Thomas (1898f) and regarded red- and yellow-rumped agoutis in the Guianan region as a species distinct from those in Brazil south of the Amazon. He applied *Dasyprocta leporina* (Linnaeus) to the former and *Dasyprocta aguti* (Linnaeus) to the latter. He also argued that samples south of the Rio São Francisco in southern Brazil were distinct from those north of that river to the right (= south) bank of the lower Amazon.

We follow Eisenberg (1989), Emmons and Feer (1990, 1997), Eisenberg and Redford 1999), and Voss et al. (2001) and recognize a single, broadly distributed species of red-rumped agouti, *Dasyprocta leporina*. Should future analyses of morphological and molecular characters support Iack-Ximenes's (1999) hypothesis that this single species is actually divisible into three, one north of the Rio Amazonas and two south of that river and divided by the Rio São Francisco, the name *D. aguti* (Linnaeus) can no longer apply to either, as argued by Thomas (1898f) and used by Iack-Ximenes (1999). Voss et al. (2001) selected as neotype of Linnaeus's *Mus aguti* the same specimen that Husson (1978) had selected as neotype of *Mus leporinus* Linneaus, thus unequivocally making *Mus aguti* a junior subjective synonym of *Mus leporinus* with both having the same type locality in Surinam (see Voss et al. 2001:143–144 for details). Feijó and Langguth (2013) suggested that their newly described *D. iacki* might apply to those samples distributed between the right (= south) bank of the lower Rio Amazonas and the left (= north) bank of the Rio São Francisco, and it is also conceivable that Wagler's earlier *croconota* might apply, as Iack-Ximenex's (1999) separation of this taxon as a species has yet to be tested (see Remarks under *D. croconota*). There is no currently available name for the red-rumped agouti distributed between the Rio São Francisco and the Atlantic coast of Brazil should those samples eventually be recognized as specifically distinct.

The name *Agouti cayanus* is usually credited to Lacépède, 1802 (e.g., Tate 1935; Cabrera 1961; Woods and Kilpatrick 2005), but, as noted by Sherborn (1899:409 and

1924b:1154; see also Husson 1978:483), F. M. Daudin is the author of those species names that first appeared in the 1802 (*Didot*) edition of Lacépède's *Tableau des divisions*.

The status of *Cavia cristata* É. Geoffroy St.-Hilaire, 1803, with its type locality given as "Surinam," is uncertain. Tate (1935:331) listed it questionably in his "Dark gray agouti" group; Ellerman (1940), Cabrera (1961), and Husson (1978) each regarded *cristata* as a species distinct from the widespread *D. leporina* in the Guianan region, although Husson (p. 466) acknowledged its "uncertain status." Hershkovitz (1972:329), minimizing the value of color differences alone in such variable organisms as agoutis, regarded those variants as belonging to a single species. Voss et al. (2001:142) suggested, because neither the holotype nor a second specimen in the Leiden Museum mentioned by Husson (1978) were "accompanied by names of collectors, dates of collection, or any additional geographic information," that "we assume that these old and poorly documented Surinamse records are erroneous." Iack-Ximenes (1999) argued that *Cavia cristata* É. Geoffroy St.-Hilaire was the correct name for the Black Agouti, to which herein we refer as *D. fuliginosa* (see that account). Because of the uncertainty in the provenance of Geoffroy St.-Hilaire's *cristata* (Voss et al. 2001) and the inability, at present, to document levels of both within and among population variation, even within the limited region of the Guianas, we provisionally list *cristata* É. Geoffroy St.-Hilaire as a synonym of *D. leporina*.

Fredga (1966), Hsu and Benirschke (1968a) and R. S. L. Ramos et al. (2003) described the karyotype of *D. leporina* (as *D. aguti*) with $2n = 64$, FN = 122 or 124.

Dasyprocta prymnolopha Wagler, 1831
Black-rumped Agouti

SYNONYMS:

Dasyprocta prymnolopha Wagler, 1831c:619; type locality "Habitat in Guiana," but corrected to "foz do Amazonas, Pará, Brazil" by Goeldi and Hagmann (1904:73) and further restricted to "Belém, Pará, Brazil" by Iack-Ximenes (1999).

Dasyprocta croconota prymnolopha: J. A. Allen, 1915b: 629; name combination.

Dasyprocta nigriclunis Osgood, 1915:192; type locality "São Marcello [= São Marcelo], upper Rio Preto, Bahia, Brazil."

DESCRIPTION: Another relatively small species, head and body length 450–525 mm, tail length 18–30 mm, hindfoot length 95–115 mm, and ear length 36–43 mm. Emmons and Feer (1990, 1997) characterized general color and color pattern as yellow-orange grizzled with black foreparts that become red-orange from midback over outer thighs, with top of rump covered by contrasting wedge of long, completely black hairs that are often pale yellow at base and may be visible when rump hairs are erect. Crown of head blackish, and neck with crest of longer, black hairs. Both forefeet and hindfeet dark brown. Under parts pale orange sprinkled with white hairs down midline. Young share same color and color pattern as adults but are generally duller overall. Combination of black stripe from midback over rump and sharply demarcated rusty-red upper thighs make this species one of more recognizable in genus.

DISTRIBUTION: This species is limited to northeastern Brazil from eastern Pará state south of the Rio Amazonas along the coast to Alagoas and Bahia states, and inland into Minas Gerais and Tocantins states. The elevational range is from sea level to 900 m (Iack-Ximenes 1999).

SELECTED LOCALITIES (Map 393): BRAZIL: Alagoas, Fazenda Canoas (MZUSP 7371); Bahia, Fazenda Jatobá-Floryl (UNB 1599), Lamarão (Thomas 1917e), São Marcelo, upper Rio Prêto (type locality of *Dasyprocta nigriclunis* Osgood); Ceará, Sítio Camará, Milarges (MNRJ 43146), Sítio Piraguara, São Benedito (MNRJ 43144); Maranhão, Fazenda Pé de Coco, Estreito (MPEG 22683), Ilha de São Luis do Maranhão (MZUSP 7984), Palmeiral, Cantanhede (MPEG 23175), Pedra Chata, Rio Gurupi, Carutapera (MPEG 10947); Minas Gerais, Mocambinho, Manga (MNRJ 29059); Pará, Paraná do Sumaúma, mouth of Rio Tocantins (MZUSP 25516), Vila Brabo, right bank of Rio Tocantins (MPEG 11874), Vista Alegre, Praia do Rio Camará (MPEG 10202); Pernambuco, Recife (MNRH 7637); Tocantins, Cana Brava (MZUSP 3963).

SUBSPECIES: As currently understood, *D. prymnolopha* is monotypic, although the inclusion of *nigriclunis* Osgood within this species may signal a regional color variant in the southwestern portion of the range worthy of recognition (see Remarks).

NATURAL HISTORY: This species can be found within a wide diversity of phytogeographic assemblages in eastern Brazil, from *Floresta Ombrófila Densa* to *Savana* and *Savana Estépica*. It is the only agouti of the Caatinga biome. As with other species in the genus, *D. prymnolopha* is a seed predator and plays a central role in the dispersal and survival of rainforest trees (Melo and Tabarelli 2003). No other aspects of the natural history of this species have been published based on field studies of natural populations, but components of the reproductive biology are known from captive animals. For example, D. A. Guimarães et al. (1997) documented a 30-day estrus period and an average gestation length of 104 days, D. A. Guimarães et al. (2009) reported on the importance of the presence of a male to stimulate the onset of puberty in females, and D. A. Guimarães et al. (2011) described sterol concentrations during the estrous cycle.

REMARKS: *Dasyprocta prymnolopha* has been rather uniformly regarded as a valid species of agouti throughout the past century (E.-L. Trouessart 1904; Tate 1935; Moojen 1952b; Cabrera 1961; Emmons and Feer 1990, 1997; Woods 1993; Woods and Kilpatrick 2005), although Ojasti (1972) suggested that its status was uncertain. J. A. Allen (1915b:628–629) applied *prymnolopha* Wagler to the red-rumped agoutis of the Guianan region, based on the errant locality "Guianas" given by Wagler (1831c) in his description, listing *prymnolopha* as the Guianan subspecies of the red or orange-rumped agoutis, to which he applied the name *D. croconota* (see Remarks for that species). Thomas (1917e:260–261) corrected Allen's use of both *croconota* Wagler and *prymnolopha* Wagler for the red-rumped agoutis because *prymnolopha* was so easily distinguished from all other agoutis by its distinctly black rump.

Osgood (1915), in his description of *D. nigriclunis*, believed it to be distinct from but with a general relationship to *D. prymnolopha*. Both forms were acknowledged to have the characteristic black nape and rump, but Osgood differentiated his new species by the banded (grizzled) rather than monochromatic (clear) orange hairs on the thighs and sides of the rump. Thomas (1917e:259), however, doubted whether *nigriclunis* Osgood was separable from *prymnolopha*, citing variability among specimens from Bahia in the grizzled versus clear color of the thighs and rump sides in support of his argument. These opinions were followed by virtually all subsequent authors (Cabrera 1961; Emmons and Feer 1990, 1997; Woods 1993; Woods and Kilpatrick 2005). Iack-Ximenes (1999), however, regarded *nigriclunis* a valid species closer in relationship to *D. fuliginosa*, with its range limited to southern Tocantins and northern Bahia states.

The type locality of "Guianas" given by Wagler (1831c) is clearly in error, as noted by Goeldi and Hagmann (1904:73), who believed the type must have been collected near the mouth of the Amazon River in Pará state, Brazil (see also Voss et al. 2001:142). J. A. Allen (1915b:629), in accepting Wagler's "Guiana" locality, suggested that "it would be necessary to designate a definite type locality . . . , which may or may not have come from Cayenne." Thomas (1917e:260), however, agreed with Goeldi and Hagmann's assessment, as have all subsequent authors (Cabrera 1961; Woods 1993; Woods and Kilpatrick 2005). Iack-Ximenes (1999:182) proposed to restrict the type locality of *D. prymnolopha* Wagler to Belém, Pará state, Brazil. A specimen of *prymnolopha* in the USNM (no. 19645) has as its locality "Guyana." It was acquired in 1891 from the dealer E. Gerrard Jr. in a mixed lot including specimens from Rio Grande do Sul (Brazil), Argentina, Costa Rica, Ecuador, Medellín (Colombia), and British Guiana (Guyana). The clearly erroneous locality can be attributed

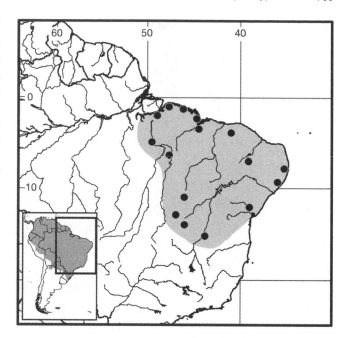

Map 393 Selected localities for *Dasyprocta prymnolopha* (●). Contour line = 2,000 m.

to poor record keeping by the dealer. A similar occurrence may have involved Wagler's type, as traders, both then and now, travel up the coast from Brazil to the Guianas.

The karyotype of *D. prymnolopha* is $2n = 64$, $FN = 122$ (J. F. S. Lima and Langguth 1998; R. S. L. Ramos et al. 2003).

Dasyprocta punctata Gray, 1842
Central American Agouti

SYNONYMS (South American only; see Hall 1981 for names applied to Mexican and Central American forms):

Dasyprocta punctata Gray, 1842:264: type locality "South America," restricted to Realejo, Nicaragua by Goodwin (1946:417).

Dasyprocta colombiana Bangs, 1898a:163; type locality "Santa Marta, Colombia."

Dasyprocta variegata colombiana: Osgood, 1910:28; name combination.

Dasyprocta fuliginosa candelensis J. A. Allen, 1915b:625; type locality "La Candela (altitude 6500 feet), Huila, Colombia."

Dasyprocta variegata chocoensis J. A. Allen, 1915b:627; type locality "Los Cisneros (altitude 600 feet), Chocó district, Colombia."

Dasyprocta fuliginosa colombiana: J. A. Allen, 1916b:205; name combination.

Dasyprocta pandora Thomas, 1917f:313; type locality "Gorgona Island, off west coast of Colombia."

Dasyprocta punctata colombiana: Cabrera, 1961:589; name combination.

Dasyprocta punctata chocoensis: Cabrera, 1961:589; name combination.

Dasyprocta punctata pandora: Cabrera, 1961:590; name combination.

Dasyprocta punctata zuliae Ojasti, 1972:190; type locality "Matera Cusare, Cuenca del río Negro en la sierra de Perijá, Edo. Zulia, Venezuela."

DESCRIPTION: Somewhat intermediate in size relative to small species, such as *D. azarae*, and larger ones, such as *D. fuliginosa*, head and body length 480–600 mm, tail length 20–55 mm, hindfoot length 120–156 mm, ear length 36–37 mm, and mass 3.2–4.2 kg (Emmons and Feer 1997). Two basic color patterns largely, but not completely, separable on geographic grounds: Populations from Pacific slope from southern Mexico to Panama, Colombia, and Ecuador with upper parts uniformly warm red-brown, yellow-brown, or gray-yellow; fur banded with fine black striations throughout when viewed closely; nape and rump hairs not different from remainder of upper parts; chin and inguinal region either clear orange or white in gray-yellow individuals; chest grizzled like back. Samples from along Atlantic slope from Costa Rica, Panama, and northern Colombia, foreparts brown to blackish, finely grizzled with tawny or olivaceous; crown and nape often blackish, midbody forward of rump with band of brighter orange-banded hairs, and rump hairs long, black, and with long yellow to white tips; long hairs overhang rump in fringe; chin and inguinal region whitish, belly brown. Some animals from western Colombia and Ecuador generally exhibit second pattern but have pale yellow frosting on rump, tawny foreparts, and necks with grayish sides. Extreme individuals with brown to black color phase may resemble overall color of *D. fuliginosa* while less extreme specimens are similar to *D. variegata*, both of which occur east of Andes (see Remarks for both *D. fuliginosa* and *D. variegata*).

DISTRIBUTION (South America only): This species is largely cis-Andean, ranging from northern Colombia east to northwestern Venezuela in the Sierra de Perijá and western slopes of the Sierra de Mérida (Ojasti 1972; O. J. Linares 1998; Prieto-Torres et al. 2011) and south along the Pacific coast of Colombia (including Isla Gorgona) and Ecuador. It apparently extends onto the eastern slope of the eastern Andes in Colombia and into the headwaters of the Río Sarare, a tributary of the Río Orinoco in Venezuela. Elevation in Colombia extends from sea level to 1,600 m (Alberico et al. 2000; Castaño et al. 2003).

SELECTED LOCALITIES (Map 394): COLOMBIA: Antioquia, Río Currulao, 20 km SE of Turbo (FMNH 69556); Arauca, Fátima, Río Cobaria (FMNH 92628); Bolívar, San Juan Nepomuceno (FMNH 68907); Cauca, Gorgona Island (type locality of *Dasyprocta pandora*

Thomas), Río Mechengue (FMNH 86764); Chocó, Baudo (J. A. Allen 1912), Unguia (FMNH 69553); Cordoba, Catival, upper Río San Jorge (FMNH 68913); Huila, La Candela (type locality of *Dasyprocta fuliginosa candelensis* J. A. Allen), Río Aguas Claras, Río Suaza (FMNH 70794); Magdalena, Santa Marta (type locality of *Dasyprocta colombiana* Bangs); Nariño, Guayacana (FMNH 89512); Tolima, Quebrada Aico between Coyaima and Chaparral (MVZ 104956); Valle de Cauca, Los Cisneros (type locality of *Dasyprocta variegata chocoensis* J. A. Allen). ECUADOR: Bolívar, San José, 12 mi SW of Huigra (MVZ 184026); Esmeraldas, Rosa Zarate, near Río Quininde (FMNH 44326); Guayas, Naranjo [= Naranjito] (J. A. Allen 1915b); Pichincha, Gualea (Lönnberg 1921). VENEZUELA: Apure, Nuitla, Selvas de San Camilo, 29 km SSW of Santo Domingo (USNM 443450); Táchira, Orope (Osgood 1910); Zulia, El Rosario, 45 to 51 km WNW of Encontrados (Handley 1976), Río Limón (FMNH 18767).

SUBSPECIES: Hall (1981) recognized 11 subspecies in Central America and Cabrera (1961) eight for South America, although only three of these (*colombiana* Bangs, *chocoensis* J. A. Allen, and *pandora* Thomas) are included herein in the synonymy of *D. punctata*. Cabrera (1961:587) regarded *candelensis* J. A. Allen to be a valid subspecies of *D. fuliginosa* rather than of *D. punctata*, as we do herein. Agoutis from the type locality in Nicaragua are uniform reddish in color, with short, undifferentiated rump hairs banded throughout their length. The earliest name for the form with long, pale-tipped rump hairs is *D. isthmica* Alston, 1876, from Colón in Panama. The reddish forms in coastal Ecuador and Colombia seem to resemble *isthmica*, with variations in color (e.g., *chocoensis* J. A. Allen) whereas those extending east into central and northern Colombia and northwestern Venezuela are darker, brown or blackish, but also with long, pale-tipped rump hairs (e.g., *colombiana* Bangs, 1898; *dariensis* Goldman, 1913). J. A. Allen (1915b) discussed these forms at length. See Remarks for *D. variegata*.

Pending an adequate review of geographic trends in color and cranial characters, it is premature to validate any of forms as subspecies, although the regional differences in color may well argue for such. If so, Bang's name *colombiana* would apply to the form along the Caribbean coast, with *zuliae* Ojasti either a synonym or a localized valid taxon in the Sierra de Perijá and adjacent areas of northwestern Venezuela (but see comment by Ojasti in the addendum to his 1972 paper). Allen's *candelensis* would be the earliest name applicable to a Pacific coast animal (by page priority over *chocoensis* J. A. Allen), unless *isthmica* Alston, 1876, proves applicable to the agoutis of the Chocó. We regard the remainder of the South American forms listed by Cabrera (1961) as subspecies of his *D. punctata* to be more appropriately linked to *D. azarae* (*boliviae* Thomas and *urucuma* J. A. Allen), *D. fuliginosa* (*zamorae* J. A. Allen), or *D. variegata* (*yungarum* Thomas).

NATURAL HISTORY: Populations of *D. punctata* in Costa Rica and Panama have been well studied, with numerous publications on topics ranging from general natural history to details on density, social organization, group membership, spacing behavior, and foraging ecology (Smythe 1978). Additional studies include those on scatter hoarding and cache-finding behavior (J. O. Murie 1977; Galves et al. 2009), home range (Aliaga-Rossel et al. 2008), predators (Aliaga-Rossel et al. 2006), reproductive biology (Merritt 1983), diel activity pattern (T. D. Lambert et al. 2009), and role in seed predation and dispersal (Smythe 1970; Hallwachs 1986; Forget 1992; Forget et al. 1994). Few data are available from populations of this species within its South American range, as mapped here, but presumably these would display the same kinds of characteristics detailed for those in Central America as well as the basic ones typical of all agouti species.

REMARKS: Following Emmons and Feer (1990, 1997), we restrict *D. punctata* to those agoutis with a cis-Andean

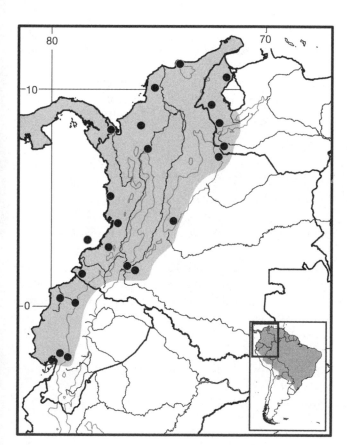

Map 394 Selected localities for *Dasyprocta punctata* (●). Distribution in Panama from Hall (1981). Contour line = 2,000 m.

distribution on the Pacific coast of Ecuador and Colombia, Caribbean coast of Colombia and extreme northwestern Venezuela, and inter-Andean valleys west of the Cordillera Oriental. This species name has been commonly applied, including recently, to agoutis on the eastern side of the Andes, especially in Peru (Pacheco et al. 1995), Bolivia (S. Anderson 1997), and northwestern Argentina (R. A. Ojeda and Mares 1989; Barquez, Díaz, and Ojeda 2006; M. M. Díaz and Barquez 2007), all records that we consider to represent *D. azarae* or *D. variegata* (see those accounts).

Matson and Shump (1981), in one of the few such papers available for any species of agouti, described sex and age-related variation in a large sample of skulls of *D. punctata* from a single locality in northwestern Ecuador.

The karyotype is apparently unknown.

Dasyprocta variegata Tschudi, 1845
Brown Agouti

SYNONYMS:

D[*asyprocta*]. *variegata* Tschudi, 1845:plate XVI; type locality "hingegen höher hinauf bis an die Gränze der obern Waldund Ceja-region dis zu 6000′ ü. M.," [= "the boundary of the upper forest- and Ceja region, 6000 feet,"] Chanchamayo region, Junín, Peru.

Dasyprocta variegata variegata: J. A. Allen, 1912:76; name combination.

DESCRIPTION: Somewhat smallish but heavy-bodied species, with head and body length 445–540 mm, tail length 11–38 mm, hindfoot length 94–120 mm, ear length 41–45 mm, and mass 3.0–5.2 kg (Emmons and Feer 1990, 1997). Upper parts finely grizzled black and tawny yellow, brown and yellowish, or black and orange, giving overall appearance blackish washed with tawny, plain brown, or orange from distance. Hairs of rump not elongated into overhanging fringe, and finely banded like hairs on rest of upper parts, from which they little differentiated. Head often blackish, and midline of back usually darker than sides. Feet dark brown, except in orange animals, colored like back. Chin and throat, and often midline of venter, white. Rest of under parts brownish yellow in brown individuals to bright pale orange in orange animals. The Brown Agouti becomes progressively paler and more orange from north to south and east to west. Those from northwestern part of range in Peru (Junín department, near likely type locality) blackish speckled with white and yellow, and, successively to south, those from Río Urubamba plain brown, and those from Madre de Dios yellowish to reddish brown.

DISTRIBUTION: This species occurs in the lowland rainforest in the western Amazon Basin from central Peru (Junín department) and western Brazil (upper Rio Purus and Rio Madeira drainages) to extreme northern Bolivia.

The southern limits of *D. variegata* and the northern boundary *of D. azarae* are unclear; we map it at about the Río Beni in northern Bolivia.

SELECTED LOCALITIES (Map 395): BOLIVIA: Beni, Riberalta (S. Anderson 1997); Pando, Independencia (S. Anderson 1997). BRAZIL: Rondônia, Porto Velho, Rio Madeira (Osgood 1916). PERU: Cusco, Río Urubamba, 50 mi NE of Cusco (MVZ 116712); Junín, Chanchamayo (type locality of *Dasyprocta variegata* Tschudi); Puno, Bella Pampa, 16 mi N and 2 mi E of Limbani (MVZ 116711); Ucayali, Balta, Río Curanja (MVZ 136642).

SUBSPECIES: We regard *D. variegata* as monotypic, pending an adequate review of geographic variation.

NATURAL HISTORY: The ecology of this species is poorly known throughout its range, although Enrique Ortiz (pers. comm. to L. H. Emmons) has studied its role as a disperser of Brazil nuts (*Bertholletia excelsa*[Lecythidaceae]) in Peru. The species apparently maintains dense populations wherever there are Brazil nuts, as do other taxa of agoutis (e.g., *D. leporina*, Peres and Baider 1997) as they can open the giant, hard exterior shell. *Dasyprocta variegata* also eats and caches nuts of the palm genera *Attelya* and *Astrocaryum*, with their idiosyncratic behavior as a seed disperser an important contributor to spatial association and genetic structure in the tree species (Choo et al. 2012). They sleep at night under fallen brush and in hollow logs (L. H. Emmons, pers. comm.). H. Gómez et al. (2005), using camera traps, determined a primarily diurnal activity pattern but one extending into the twilight hours at localities in Madidi National Park in Amazonian Bolivia.

REMARKS: Tschudi's description and illustration of the type of *variegata* indicate a blackish animal with entirely black neck and rump hairs (Tschudi 1845; Alston 1876), speckled with some yellow-banded and white-tipped hairs. Although this color combination is reminiscent of *D. fuliginosa*, a specimen from Chanchamayo, Peru (USNM 255135), nearly matches the description, but has dark-brown overtones caused by narrow ochraceous bands on most hairs, which produce speckles throughout the midbody and sides. This animal likewise lacks the long, pale-tipped rump hairs characteristic of South American *D. punctata*.

Thomas (1899a:40) used the name *Dasyprocta isthmica* Alston to refer to specimens collected by Otto Garlepp in the "district of Cuzco," southern Peru. This is clearly an error, as *isthmica* was the name Alston (1876b, 1880a) gave to agoutis from Central America (see *D. punctata* account). Specimens from Beni and Pando departments, Bolivia, that S. Anderson (1997) included under his concept of *D. punctata* we refer herein to *D. variegata*, pending a thorough review of the boundaries between *D. varie-*

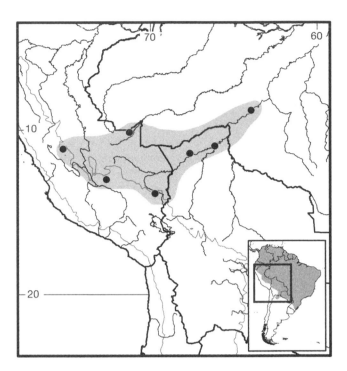

Map 395 Selected localities for *Dasyprocta variegata* (●). Contour lines = 2,000 m.

gata and *D. azarae* (as noted in the account of the latter species).

Hungerford and Snyder (1964) provided the karyotype of a live animal they obtained from the Philadelphia Zoological Garden and identified as *Dasyprocta variegata*. The diploid number was 64; the entire complement, including sex chromosomes, was composed of entirely biarmed elements (FN = 124).

Genus *Myoprocta* Thomas, 1903

Acouchys are a smaller version of the agouti, about one-third the mass on average and with a conspicuous, but still short, tail. Other than overall size and length and hairiness of the tail, the two genera share the same body form with large hindquarters and humped back, elongated, thin limbs, digitigrade foot structure, cursorial gait, general conformation of the skull, and even details of tooth and cranial structures.

SYNONYMS:

Cavia: Erxleben, 1977:353; part (description of *acouchy*); not *Cavia* Pallas.

Agouti: Daudin, 1802:166; part (listing of *acuschy*); not *Agouti* Lacépède.

Dasyprocta: Illiger, 1811:93; part; included *Cavia acouschy* [sic] Erxleben.

Chloromys G. Cuvier, 1817:214; part; included *Cavia acouschy* [sic] Erxleben; incorrect subsequent spelling of *Cloromis* F. Cuvier.

Cutia Liais, 1872:534; part (listing of *acuchy*); *nomen oblitum.*

Myoprocta Thomas, 1903d:464; type species "*Dasyprocta*" *acouchy* Linnaeus (= *Cavia acouchy* Erxleben 1777), by original designation.

REMARKS: Tate (1939), Cabrera (1961), Woods (1993), Voss et al. (2001), and Woods and Kilpatrick (2005) each recognize two species of acouchys, generally the "red" (or "reddish") eastern Amazonian and "green" (or "greenish") western Amazonian forms, although the names applied to each differ among these authors. Iack-Ximenes (1999) recognized three species, which do not conform exactly to the "red" and "green" types. Herein, I follow Voss et al. (2001), who provided characters that, in combination, unambiguously distinguished specimens of "red" (*M. acouchy*) and "green" (*M. pratti*) acouchys they examined in the major European and U.S. museums.

Voss et al. (2001) stated that all *M. acouchy* specimens they examined closely resembled one another and found no compelling evidence that more than a single species was represented. Alternatively, these authors noted but did not document significant geographic variation among populations of *M. pratti* that may suggest this taxon is a complex of closely related species. Both species are in need of appropriate revisionary analyses.

KEY TO THE SPECIES OF *MYOPROCTA*:
1. Color reddish-brown dorsally, with uniformly orange or reddish under parts; distinct rump patch of very long (60–80 m) and highly polished hairs more heavily pigmented than those of sides, middle back, or forequarters, and that form a blackish to deep mahogany brown, glossy fringe extending to the base of the tail; cranial dimensions average larger (see Voss et al. 2001:Table 41); sphenopalatine vacuities narrow slits . . . .
. . . . *Myoprocta acouchy*
1'. Color drab yellowish- or grayish-brown dorsally with yellowish under parts (usually with a distinct white midventral streak); distinct rump patch never present, with fur over hindquarters similar in length, color, and texture to the rest of the dorsal pelage; cranial dimensions average smaller; sphenopalatine vacuities wide and teardrop in shape. . . . .*Myoprocta pratti*

Myoprocta acouchy (Erxleben, 1777)
Red Acouchy
SYNONYMS:
[*Cavia*] *acouchy* Erxleben, 1777:354; type locality fixed by neotype selection to "Paracou, French Guiana" (Voss et al. 2001:148).
Cavia akouchi Zimmermann, 1777:508; different spelling for the same animal upon which Erxleben's name was

based (see J. A. Allen [1902a] for the preferential use of Erxleben's name).
Calva Acuschy Gmelin, 1788:121: name combination and incorrect subsequent spelling of *acouchy* Erxleben.
Agouti acuschy Daudin, 1802:166; name combination and incorrect subsequent spelling of *acouchy* Erxleben.
Cavia acuschi É. Geoffroy St.-Hilaire, 1803:166; type locality "l'Amérique méridionale."
Dasyprocta acuchi Illiger, 1811:93; name combination and incorrect subsequent spelling of *acouchy* Erxleben.
Dasyprocta acuchy Illiger, 1815:108; name combination and incorrect subsequent spelling of *acouchy* Erxleben.
[*Dasyprocta*] *S. Acuchy* Oken, 1816:823; [*S.* = *Savia?*]; name combination and incorrect subsequent spelling of *acouchy* Erxleben.
Cavia Akuchi Desmarest, 1816:212; name combination and incorrect subsequent spelling of *acouchy* Erxleben.
Cavia acuchi G. Cuvier, 1817:214; name combination and incorrect subsequent spelling of *acouchy* Erxleben.
Dasyprocta acuschy Desmarest, 1822:358; name combination and incorrect subsequent spelling of *acouchy* Erxleben.
[*Dasyprocta*] *Acouchy*: Minding, 1829:91; name combination.
Cavia exilis Wagler, 1831c:620; type locality "Habitat in Brasilia ad flumen Amazonum," restricted to "near the mouth of Rio Negro," Amazonas, Brazil (J. A. Allen 1916b:205).
D[asyprocta]. leptura Wagner, 1844:49; type locality "Die Heimath ist am Rio negro, einem der nördlichen haupzuflüsse des Amazonenstroms," possibly at Barra do Rio Negro [= Manaus], Amazonas, Brazil (see Pelzeln 1883; Voss et al. 2001).
Cutia acuchy: Liais, 1872:536; based on "*l'Akouchy de Buffon*" from "la Guyane et le nord de l'Amérique du Sud" [= *Cavia acouchy* Erxleben].
Myoprocta acouchy: Thomas, 1903d:464; first use of current name combination.
Dasyprocta acouchy: Goeldi and Hagmann, 1904:73; name combination.
Dasyprocta (Myoprocta) exilis exilis: J. A. Allen, 1916d: 569; name combination.
Dasyprocta exilis: Thomas, 1920f:278; name combination.
D[asyprocta]. exilis: Thomas, 1920f:280; name combination.
[*Myoprocta*] *exilis exilis*: Tate, 1935:335; name combination.
[*Myoprocta*] *leptura*: Tate, 1935:335; name combination.
Myoprocta exilis demararae Tate, 1939:182; type locality "Bonasica, Essequibo River, British Guiana."
M[yoprocta]. e[xilis]. lepture Tate, 1939:182; incorrect subsequent spelling of *leptura* Wagner.

DESCRIPTION: Small bodied, head and body length 335–390 mm, tail length 51–78 mm, hindfoot length 90–104 mm, ear length 25–40 mm, mass 1.05–1.45 kg. Upper parts colored dark chestnut red or orange on sides and legs, grizzled with black and some yellow on crown and neck. Mid-back and rump glossy black or very dark red. Rump hairs long, straight, and overhang rear of body in straight fringe; hairs entirely dark, not banded. Eyes large; ears quite large, naked, with tips well above crown; areas behind ears, around mouth and eyes, and under chin almost naked. Vibrissae well developed, black, and reach to behind ear. Tail short and slender, white below and with small white tuft at tip. Behaviorally, tail often wagged or tipped straight up, exposing white underside. Legs long and thin; sole of feet black. Under parts thinly haired, pale to dark rust-red or orange. Some individuals grizzled with yellow and appear olivaceous, but these retain long, dark rump hairs otherwise characteristic of *M. acouchy* (Emmons and Feer 1997).

Skull averages larger that that of *M. pratti* (Voss et al. 2001), with noticeably longer toothrows (mean 13.8 mm, range: 12.4–15.5 mm) and bullae (mean 16.1 mm, range: 14.5–18.5 mm), broader interorbit (mean 22.5 mm, range: 20.9–24.2 mm), and longer nasals (mean 27.3 mm, range: 25.0–31.8 mm). Of four qualitative traits identified by Tate (1939) in his distinction between red and green acouchys, Voss et al. (2001) confirmed that size of sphenopalatine vacuities that perforate the bony roof of the mesopterygoid fossa ("palatal foramina" of Tate) were most useful; very narrow slits (<1 mm wide) in *M. acouchy* while in most *M. pratti* both larger (>1 mm wide) and openings teardrop shaped.

DISTRIBUTION: Lowland rainforest in Guyana, Surinam, and French Guiana, and that part of northeastern Brazil east of the Rio Branco and north of the Rio Amazonas. Voss et al. (2001:145) noted two records outside of this domain, both substantiated by vouchered specimens: one in the AMNH from the "Lower Solimões" believed to be Manacaparu on the north (= left) bank of the Solimões just west of the mouth of the Rio Negro; and two from the south (= right) bank of the Amazon between the Rio Madeira and Rio Tapajós. Emmons and Feer (1997) extended the mapped range of their Red Acouchi to the region between the lower Rio Xingu and Rio Tocantins, presumably based on unvouchered reports (e.g., T. K. George et al. 1988; Voss and Emmons 1996). Apparently no Brazilian collection contains specimens of any acouchy from the Rio Xingu or Rio Tocantins basins (Iack-Ximenes 1999).

SELECTED LOCALITIES (Map 396): BRAZIL: Amapá, Rio Vila Nova, Mazagão (MPEG 411), Terezinha, Rio Amaparí, Serra do Navio (MNRJ 43207), Vila Velha do Cassiporé (MPEG 6601); Amazonas, Lago do Baptista (FMNH 50895), Lago do Serpa (FMNH 50897), Manaus (type locality of *Dasyprocta leptura* Wagner; Voss et al. 2001), Rio Apuaú, left bank Rio Negro (MPEG 7148), Rio Pitinga, left bank Rio Uatumã (MVZ 194289), Villa Bella da Imperatriz, south bank Amazon River (AMNH 93044); Pará, Monte Alegre (MPEG 2285), Lago Cuiteua, Rio Amazonas (MCZ 32356); Roraima, Conceição, Rio Branco (FMNH 20007). FRENCH GUIANA: Cayenne, Paracou (type locality of *Cavia acouchy* Erxleben). GUYANA: Essequibo Islands–West Demerara, Bonasika River (type locality of *Myoprocta exilis demararae* Tate); Upper Demerara–Berbice, Potaro River (ROM 29120); Upper Takutu–Upper Essequibo, Nappi Creek, foot of Kanuku Mountain (ROM 31539). SURINAM: Paramaribo, Paramaribo (USNM 13309).

SUBSPECIES: Cabrera (1961) listed three subspecies, the nominotypical form (with *demararae* Tate as a synonym) from the Guianan region and eastern Amazonian Brazil north of the Rio Amazonas, *exilis* Wagler (with *leptura* Wagner as a synonym) from the lower Rio Negro to the lower Rio Tapajós in the central Amazonian Brazil, and *parva* Lönnberg, from east of the Andes in southern Colombia, Ecuador, and northern Peru. This latter name has been applied to the Green Acouchy, *Myoprocta pratti*, by more recent authors (see Remarks under that species). The validity of *M. a. acouchy* versus *M. a. exilis* has yet to be tested by rigorous analyses of any type of character set.

NATURAL HISTORY: Acouchys of both species share similar life history, ecology, and behavior to those attributes of agoutis. In French Guiana (Dubost 1988), Red acouchys frequented all forest habitats except disturbed areas. Individuals were strictly diurnal, with most activity confined to morning hours. The family, with an adult male and female plus their immediate offspring, comprised the basic social unit, although each member lived alone within individual home ranges that, while clumped, exhibited little overlap. These small home ranges had multiple heavily frequented areas and many resting sites, often located in hollow logs, that consisted of a single nest in current use and several older leaf-heaps. Males preferred open forests, females more closed habitats. The size of home ranges and the level of individual activity decreased in the dry season when food resources are most scarce. Familial units were widely separated, so intergroup interactions were rare. Food was not stored in or adjacent to nests but buried in small caches through the entire home range. Reproduction is strongly seasonal, with more than half of all births concentrated between November and January during the period when the number of fruiting tree species and individuals of each begins their annual increase; no births occurred from August to October, when fruit availability is at its lowest (Dubost et al. 2005). Litter size ranged from one to three, with a median of two young per pregnancy. The juvenile sex ratio was skewed in favor of males, but the adult sex ratio was near one, resulting from increased risks taken by juvenile males when dispersing. Red acouchys eat primarily fruits, predominantly pulp followed by seeds and exocarp; other food types, including insects, comprised less than 1% of the diet (Dubost and Henry 2006). Diet varies seasonally, depending upon both the number of fruiting plant species and overall fruit production and nutritional needs associated with reproduction. Seeds were collected and scatter-hoarded by burying, with caches often more than 100 m distant from the tree that dropped them (Jansen et al. 2004), both important components of tree dispersal and regeneration (see also Forget 1990, 1991). Red acouchys will hoard rapidly germinating seeds, such as from the tree *Carapa procera* (in the mahogany family Meliaceae), but manipulate the seed by removing the protruding radicle and epicotyl to stop seed development and thus to prolong storage life in the cache (Jansen et al. 2006). In central Amazonian Brazil where the forest has been modified into patches of varying size, scatter hoarding and seed removal was inversely related to patch size (Jorge and Howe 2009), and population density dropped nearly an order of magnitude from continuous forest to patches of about 1 ha in size (Jorge 2008).

REMARKS: Thomas (1926f:639), Cabrera (1961:591), and Emmons and Feer (1990, 1997) applied Erxleben's *acouchy* to the Red Agouti while Tate (1939), Husson (1978), Woods (1993), Alberico et al. (2000), and Iack-Ximenes (1999) used this name for the Green Acouchy. Voss et al. (2001), however, noted that this disagreement in usage stemmed from whether geography or color was viewed as the determining factor in how Erxleben's name should be applied. These authors favored geography, as the type locality for Erxleben's animal could be unambiguously established as French Guiana ("Cayenne"). They then (p. 148) stabilized the nomenclature of species-name assignments by selecting a neotype for *Cavia acouchy* Erxleben from Paracou, French Guiana. For further details, see Voss et al. (2001:144–151).

Wagler's (1831c) description of *exilis* clearly identifies a reddish animal ("*notaeo toto castaneo-fuscescente*") and thus aligns Wagler's name with *Myoprocta acouchy*. Similarly, the "*ferrugineo-rufa*" color given by Wagner (1844) to his *leptura* solidifies this name as a synonym of *M. acouchy* as well. Both original specimens came from the lower Rio Negro, as specified by Wagner in his paper describing *leptura* and by J. A. Allen (1916b:205) in his restriction of the locality of Wagler's *exilis*. Voss et al. (2001:151, footnote *d*) noted that Wagner's type series was collected by Johann Natterer and possibly at "Barra do Rio Negro" (= Manaus), which was the only definite local-

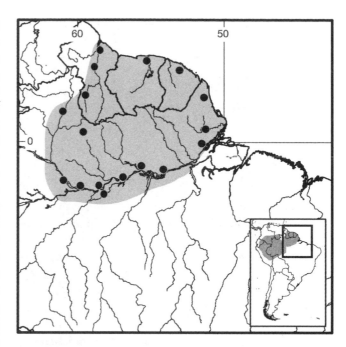

Map 396 Selected localities for *Myoprocta acouchy* (●).

ity where Natterer apparently obtained acouchys (Pelzeln 1883).

Emmons and Feer (1997) suggested that *milleri* J. A. Allen represented an isolated population of *M. acouchy* in the upper Río Vaupés of Colombia, where some hybridization with *M. pratti* might occur. These authors also noted that both red and green acouchys may co-occur in the Serranía de la Macarena in eastern Colombia, presumably based on the list of Colombian mammals assembled by Cuervo Díaz et al. (1986). However, Voss et al. (2001:151), who actually examined Allen's holotype, unambiguously regarded *milleri* as a green acouchy.

Fredga (1966) and Hsu and Benirschke (1968b) described a karyotype with $2n = 62$, FN = 118 for specimens they called *Myoprocta acouchy* but that lacked provenance data.

Myoprocta pratti Pocock, 1913
Green Acouchy
SYNONYMS:

Myoprocta pratti Pocock, 1913:110; type locality "Amazons," Peru; restricted to Pongo de Rentema, Río Marañón, Amazonas, about 78°20′ by Thomas (1920f:279).

Myoprocta milleri J. A. Allen, 1913a:477; type locality "La Murelia, Caquetá, altitude 600 feet," Colombia.

Dasyprocta (Myoprocta) exilis milleri: J. A. Allen, 1916b: 205; name combination.

Myoprocta pratti limanus Thomas, 1920f:279; type locality "Acajutuba, near mouth of Rio Negro, above Manaos [= Manaus]," Amazonas, Brazil.

M[*yoprocta*]. p[*ratti*]. *milleri*: Thomas, 1920f:279; name combination.

Myoprocta exilis parva Lönnberg, 1921:41; type locality "near Rio Curaray, El Oriente, Ecuador."

Myoprocta pratti archidonae Lönnberg, 1925:274; type locality "Archidona on the eastern side of the Andes [Napo province] in Ecuador, 2400 feet above sea level."

Myoprocta pratti limana: Thomas, 1926f:638; corrected gender concordance.

Myoprocta pratti caymanum Thomas, 1926f:638; type locality "Canabouca, Parana do Jacaré, south side of the Solimoens [= Rio Solimões], about 120 km. to the southwest of Manaos [= Manaus]"; = Paraná do Jacaré, Lago Canabouca, Amazonas, Brazil (Vanzolini 1992).

Myoprocta pratti puralis Thomas, 1926f:639; type locality "Ayapua [Aiapuá], about 300 km. S.W. of Manaos [= Manaus]"; an island on west side of the lower Rio Purus below Arumã, Amazonas, Brazil (Paynter and Traylor 1991).

[*Myoprocta*] *pratti pratti*: Tate, 1935:335; name combination.

M[*yoprocta*]. a[*couchy*]. *milleri*: Tate, 1939:182; name combination.

M[*yoprocta*]. a[*couchy*]. *pratti*: Tate, 1939:182; name combination.

M[*yoprocta*]. a[*couchy*]. *limanus*: Tate, 1939:182; name combination.

M[*yoprocta*]. a[*couchy*]. *archidonae*: Tate, 1939:182; name combination.

M[*yoprocta*]. a[*couchy*]. *caymanum*: Tate, 1939:182; name combination.

M[*yoprocta*]. a[*couchy*]. *puralis*: Tate, 1939:182; name combination.

DESCRIPTION: Slightly smaller in external dimensions than *M. acouchy*, head and body length 298–383 mm; tail length 40–58 mm, hindfoot length 74–98 mm, ear length 31–37 mm, and mass 0.8–1.2 kg (Emmons and Feer 1997). Color on upper parts and legs finely grizzled olivaceous, with each hair possessing multiple alternating narrow bands of black and yellow; back and sides usually uniformly colored, sometimes washed with reddish tone, and rump may occasionally be darker, almost blackish, but with hairs always banded. Rump hairs long but usually do not overhang as fringe. Sides of muzzle, cheeks, and postauricular patch rusty orange; chin and area around eyes almost naked. Under parts pale orange, thinly haired over yellowish skin; throat, chest, and midline of belly often white. Young animals colored as adults but often with stronger orange tints. Overall body shape and proportions, including tail length and color, like those of *M. acouchy*. Green acouchys from Venezuelan Amazon more typically gray-brown on sides, thighs grizzled with white hairs.

Cranially the two species share most features, although skull of *M. pratti* somewhat smaller in most dimensions, notably so in maxillary toothrow length (mean 13.2 mm, range: 12.4–14.2), interorbital breadth (mean 20.0 mm, range: 18.4–22.0), and bullar length (mean 14.7 mm, range: 13.7–15.6; Voss et al. 2001:Table 41). In contrast to condition in *M. acouchy*, sphenopalatine vacuities in bony roof of mesopterygoid fossa of *M. pratti* are teardrop-shaped apertures, rather than narrow slits, >1 mm in width (Tate 1939; Voss et al. 2001).

DISTRIBUTION: *Myoprocta pratti* ranges from the eastern foothills of the Andes to west of the lower Rio Negro, north of the Amazon in southern Venezuela, Brazil, Colombia, Ecuador, and Peru, and south of the Amazon River in Peru, Bolivia, and Brazil to at least the Rio Madeira. S. Anderson (1997) mapped two additional localities in Pando department, Bolivia, citing an unpublished report by L. H. Emmons. Elevational range extends from the floor of the Amazon Basin at about 50 to 1,200 m on the Andean slopes.

SELECTED LOCALITIES (Map 397): BOLIVIA: Pando, Luz de América, Reserva Nacional de Vida Silvestre Amazónica Manuripi (Miserendino and Azurduy 2005). BRAZIL: Acre, Seringal Santo Antônio (MZUSP 11234); Amazonas, Acajutuba, lower Rio Negro (type locality of *Myoprocta pratti limanus* Thomas), Ayapua [= Aiapuá] (type locality of *Myoprocta pratti puralis* Thomas), Canabouca, Paraná do Jacaré (type locality of *Myoprocta pratti caymanum* Thomas), Carvoeiro (MZUSP 20187), Foz do Rio Castanhas (MNRJ 2316), São João, Aripuanan (MNRJ 2313), Tauá, Rio Uaupés (AMNH 78599), Umarituba, Rio Negro (AMNH 78598); Rondônia, Calama (AMNH 37122). COLOMBIA: Meta, La Macarena Parque, Río Guapaya (FMNH 88023); Putumayo, Río Mecaya (FMNH 71133). ECUADOR: Napo, 12 km NE of Lago Agrio (FMNH 125085); Pastaza, Río Capahuari (FMNH 43191), Río Pindo Yacu (FMNH 43186). PERU: Amazonas, Pongo de Rentema, Río Marañón (type locality of *Myoprocta pratti* Pocock), Tseasim, Río Huampami of Río Cenepa (MVZ 153579); Cusco, Quincemil (FMNH 75195); Loreto, Sarayacu, Ucayali River (AMNH 76277); Madre de Dios, Cocha Cashu Biological Station (B. D. Patterson et al. 2006); Pasco, Cumbre de Ollon, ca. 12 km E of Oxapampa (LSUMZ 25891); Ucayali, Lagarto, upper Ucayali River (AMNH 76621), 59 km SW of Pucallpa (USNM 499237). VENEZUELA: Amazonas, Belen, 56 km NNW of Esmeralda, Río Cunucunuma (USNM 388222), Boca Mavaca, 84 km SSE of Esmeralda (USNM 406522).

SUBSPECIES: Cabrera (1961) recognized six subspecies: *parva* Lönnberg (which he considered a form of *M. acouchy*) from eastern Ecuador and probably southern Colombia and northern Peru; *caymanum* Thomas from south of the Amazon River in the lower Rio Purus drainage in Brazil; *limana* Thomas from the Rio Negro drainage in central Brazil and southern Venezuela; *milleri* J. A. Allen from south Colombia and northeastern Ecuador; *pratti* Pocock from south of the Río Marañón to at least Cusco department, Peru; and *puralis* Thomas in the region south of the Rio Solimões and west of the Rio Purus. As noted by Voss et al. (2001), there is considerable variation in color and mensural characteristics over the range of the species, but the degree to which the available names map on to coherent geographic units must wait for appropriate character analyses. As these authors also noted, *M. pratti* as currently envisioned may be a complex of closely related taxa worthy of recognition at the species level.

NATURAL HISTORY: The Green Acouchy has not been extensively studied in the field, with most information available anecdotal or based on captive observations. In general, these animals are diurnal and feed on fallen seeds (that they scatter hoard, as do other dasyproctids; Morris 1967) and fruits, and can den in abandoned armadillo burrows (L. H. Emmons, pers. comm.). J. A. Allen (1916d:595) noted, "In the Caquetá I found them [*milleri*] apparently living in colonies, in wet forests. They were out and about in the mornings, and took refuge in holes in the riverbank. The flesh is white and possesses a delicious flavor." Using both line transect and camera traps, Munari et al. (2011) demonstrated that Green Acouchys along the Rio Juruá in western Amazonian Brazil were active only during daylight hours and specialized in upland, *terra firme* forest without venturing into flooded, or *várzea* forest. Kleiman (1970) and Weir (1971a) described an estrous cycle lasting about 40 days and a gestation length of 99 days in captive females. Under these conditions, sexual maturity was reached within 8 to 12 months. Breeding success was greatest when the female was housed with a single male, which suggests the importance of paired adult family groups similar to the basic social unit described for the Red Acouchy in French Guiana (e.g., Dubost 1988). Kleiman (1971, 1972) summarized courtship, copulatory, and maternal behaviors in her captive population.

REMARKS: Both *pratti* Pocock and *milleri* J. A. Allen were named in 1913; Pocock's name has date priority, however, as his paper was published in July of that year while Allen's appeared in September.

J. A. Allen (1916d:569) believed that Pocock's *Myoprocta pratti* agreed well with Wagler's description of *exilis* and with a specimen he had from Calama (from the middle Rio Madeira) in central Brazil. He also regarded the type locality of *pratti* as "indefinite as the type locality of *exilis*" and continued with "it seems perfectly safe to assume that both came from that portion of the Amazon drainage which includes the lower Rio Negro and the lower Rio Madeira." Allen thus

assigned *pratti* as a synonym of *Dasyprocta* (*Myoprocta*) *exilis exilis*. Thomas (1920f:279), however, found it "impossible to accept Dr. Allen's identification of [*Dasyprocta exilis* Wagler] with one of the greenish-coloured *pratti* group." Thomas also reported that, although Pocock did not record the exact locality of his *pratti*, Thomas had learned from Mr. A. E. Pratt that the specimen had been obtained at the Pongo de Rentema on the Río Marañón, which is just east of the boundary between the regions of Cajamarca and Amazonas on that river, at about 78°20′W latitude.

Lönnberg (1921) assigned his *parva* to *Myoprocta exilis* and subsequently in his description of *archidonae* (Lönnberg 1925) contrasted the two as representing "red" (*parva*) and "green" (*archidonae*) acouchys, respectively. Both Tate (1939) and Cabrera (1961) followed Lönnberg, listing *parva* as a subspecies of the Red Acouchy (Tate under the name *M. exilis*, Cabrera as *M. acouchy*). Lönnberg (1921:41–42) contrasted his *parva* with J. A. Allen's *milleri* from nearby in southern Colombia as not having the pale olivaceous yellow washed with black characteristic of *milleri*. Rather, he described *parva* as dark grizzled brownish overall color, the back "with only a little sprinkling of pale yellow," the top of the head blackish brown sprinkled with orange yellow, and the under parts "yellowish buff." He made no mention of the distinctive long fringe of hairs on the rump that characterize *M. acouchy* (as envisioned by Voss et al. 2001 and herein) but did note the small dimen-

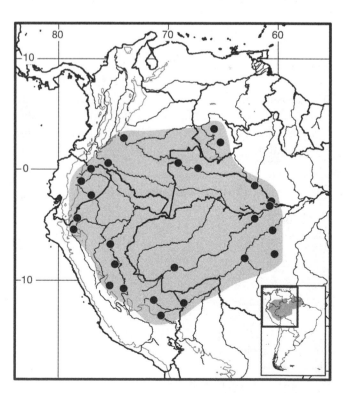

Map 397 Selected localities for *Myoprocta pratti* (●). Contour line = 2,000 m.

sions of the skull, giving a measurement of 11 mm for the length of the upper toothrow for both a male and female specimen. Although the specimens Lönnberg had may be darker than other *M. pratti* from the western Amazon, the characteristics given are clearly within the range expressed among populations and even individuals from single localities of that species. There seems every reason, therefore, to associate Lönnberg's *parva* with the Green Acouchy *M. pratti*, a conclusion also reached by Voss et al. (2001) and Woods and Kilpatrick (2005).

As currently understood, Green and Red acouchys, like so many other species in the west-central Amazon, are apparently allopatric and separated by the lower Rio Negro (*M. pratti* to the west and *M. acouchy* to the east). Their ranges closely approach near the mouth of the Rio Negro in the vicinity of Manaus, a limited area with the respective type localities of Thomas's *limanus* (a Green Acouchy; see Voss et al. 2001:151, Table 41) and Wagler's *exilis* and Wagner's *leptura*, both Red acouchys.

The karyotype of *Myoprocta pratti* is apparently unknown.

Superfamily Chinchilloidea Bennett, 1833
Angel E. Spotorno and James L. Patton

This superfamily includes only the Chinchillidae (which, in turn is composed of the subfamilies Chinchillinae and Lagostominae) plus Dinomyidae. The sister relationship between these two family-group taxa has become well established in all molecular phylogenetic analyses (Opazo 2005; Rowe et al. 2010; Upham and Patterson 2012; P.-H. Fabre et al. 2012), with the divergence between the two lineages beginning in the Early Miocene, at about 10.1 ± 2.7 mya (Opazo 2005). Among others, Woods (1982) and McKenna and Bell (1997) placed the Dinomyidae within the Cavioidea.

Family Chinchillidae Bennett, 1833

This family comprises the chinchillas (*Chinchilla*), mountain viscachas (*Lagidium*), and plains viscachas (*Lagostomus*), mostly limited to the drier regions of western and southern South America from the highlands of Ecuador through the Andes of Peru and Bolivia to the coastal mountains of Chile and Patagonian steppe of Argentina.

These are rabbit-like rodents with short forelimbs and elongated hindlimbs, large eyes and ears, and with size varying from a head and body length of nearly 400 mm and average mass of 500 g (*Chinchilla*) to 700 mm body length and mass of 5 kg or greater (*Lagostomus*). Sexual dimorphism can be extreme, with females nearly twice the mass of males in *Chinchilla* but with males larger than females in *Lagostomus*. The tail varies in length, but is always less than half of the length of the total length, proportionally shortest in *Lagostomus* (about 33% of head and body length) and longest in *Lagidium* (about 75% of head and body length). It is always well furred, with a distinctly brushlike appearance in species of both *Chinchilla* and *Lagidium*. The tail of *Lagostomus* breaks easily at the 4th caudal vertebrae and is thus often lost. Both forefeet and hindfeet have fleshy plantar pads; the distal toe pads are enlarged and obscure, with small, curved claws in *Chinchilla* and *Lagidium*. Elongated and nearly straight claws on both forefeet and hindfeet dwarf small toe pads in *Lagostomus*. The pollex (digit I) is lacking (*Lagidium* and *Lagostomus*) or severely reduced in size (*Chinchilla*), so the forefeet have only four functional digits in all three genera. Both *Chinchilla* and *Lagidium* have four digits on the hindfeet (lacking the hallux, digit I); *Lagostomus* has only three (lacking digits I and V). Digit II of the hindfeet in both *Chinchilla* and *Lagidium* has a cluster of stiffened hairs on one side of the claw; a similar cluster is present on the central digit (III) of *Lagostomus* (drawings in Pocock 1922:Figs. 14, 15). Mystacial vibrissae in all three genera are coarse, stiff, and long, extending to the tip of the ear when laid back. Superciliary and genal vibrissae are soft and fine in *Chinchilla* and *Lagidium*, but longer and coarser in *Lagostomus*. This latter genus also has an extensive, thick mat of short bristles on the cheek below the eye. Overall dorsal coloration varies from gray to bluish black with yellowish white under parts in *Chinchilla* and *Lagidium* and brownish or grayish black with strikingly bold black and white facial bands in *Lagostomus*. The pelage is especially soft and dense, notably so in both *Chinchilla* and *Lagidium*. *Chinchilla* has been nearly exterminated in the wild because of the value of their fur (Spotorno, Zuleta et al. 2004).

The skull is elongated and relatively narrow, with a broad interorbital region, long and narrowed rostrum, and a braincase varying in shape from slightly rounded (in *Lagidium*) to triangular (in *Chinchilla*, due to the enlarged mastoid bullae), to squared (in *Lagostomus*), which is especially robust in appearance. A postorbital process on the frontal is weakly developed in *Chinchilla* and *Lagidium* but expanded, flat, and triangular in shape in *Lagostomus*. The lacrimal canal is positioned on the side of the rostrum just in front of the infraorbital foramen, which itself is large and with a deep canal for nerve passage in *Lagostomus* but not in *Chinchilla* or *Lagidium*. The zygomatic arch is horizontally thin but deep, with a well-developed jugal contacting the lacrimal in *Lagidium* and *Lagostomus* but not in *Chinchilla*, and a moderate postorbital spine comprised solely by the jugal. The upper zygomatic root is positioned far behind the lower root, over the posterior molars. Both the tympanic and mastoid portions of the auditory bullae are greatly inflated in *Chinchilla*, less so in *Lagidium*, and only weakly inflated in *Lagostomus*, with the mastoid portion visible on the dorsal surface of the cranium in *Chinchilla* and *Lagidium*. The auditory tube is directed dorsally rather than laterally. The incisive foramina are long and narrow, encompassing more than half the length of the diastema in *Chinchilla* and *Lagidium* but only about one-third that distance in *Lagostomus*. The maxillary toothrows of all three genera diverge posteriorly, so the minimum distance between the fourth premolars is only about 20% to 30% of the maximum distance between the third molars. The palate is short, with the anterior border of the mesopterygoid fossa reaching to M2 in all three genera. Sphenopalatine vacuities are well developed in *Chinchilla*, small and contained completely within the pterygoid in *Lagidium*, and reduced to small slits at the palatine-pterygoid junction in *Lagostomus*. The basisphenoid varies considerably in width in conjunction with the degree of hypertrophy of the tympanic bullae among the three genera. The paroccipital processes, both in size and relationship to the bullae, vary as well among the genera, being small and closely appressed to the bullae in *Chinchilla*, longer but free from and curved over the distal portion of the bullae in *Lagidium*, and elongated, vertical, and completely free of the bullae in *Lagostomus*.

Upper incisors are narrow with little expressed pigment in the anterior enamel, orthodont to opisthodont in *Chinchilla* and *Lagidium* but proodont in *Lagostomus*. Cheek teeth are flat crowned and hypsodont. All teeth have three transverse plates, or laminae, in *Chinchilla* and *Lagidium*; all teeth in also have three plates in *Lagostomus* except M3, which possesses only two plates. The deciduous premolars (dPM4 and dpm4) are replaced.

The lower jaw lacks a masseteric crest on the dentary in all three genera. The coronoid process is small in both *Chinchilla* and *Lagidium* but elongate in *Lagostomus*, set off from the ramus by a moderately deep depression lateral to the mandibular incisure for the insertion of the posterior masseter. The angular process is thin and elongate in both *Chinchilla* and *Lagidium*, much more robust in *Lagostomus*; it is only slightly deflected laterally to the sheath of the incisor, again except in *Lagostomus*.

Bugge (1971) described the cephalic arterial system for *Chinchilla*; that for the other two genera is unknown but presumably is similar. As is characteristic of most other hystricognaths, both the internal carotid artery and the tympanic part of the stapedial artery are lacking. The vertebral artery supplies the brain whereas the stapedial area of supply is taken over by the external carotid. The persistent parts of the stapedial artery include the distal part of the stem, the proximal part of the supraorbital branch with the middle meningeal artery, the orbital part of the supraorbital branch, and the entire infraorbital and mandibular branches. This combination of features sets chinchillids apart from other caviomorph lineages.

Lagostomus is apparently unique among hystricognath rodents in having a narrow, cylindrical penis that lacks the intromittent sac (= *sacculus urethralis*) that is characteristic of all other members of this larger group (Pocock 1922; Hooper 1961), including *Chinchilla* (Tullberg 1899; Spotorno 1979). The phallus of *Lagidium* has not been described.

Chinchillids can be distinguished from other South American caviomorph rodents by the combination of a large lacrimal bone and lacrimal canal opening on the side of the rostrum; by the dorsally deflected auditory canal with an accessory foramina at the base; by an angular process that is delicate and not strongly hystricognathous; by the lack of a masseteric crest on the dentary; and by an occlusal pattern of the cheek teeth composed of closely packed but distinctly transverse lamellar plates, numbering three or fewer.

Chinchillids extend from the Deseadan (Oligocene) to Recent in South America. The family is commonly divided into two subfamilies, the Chinchillinae, which contains only the extant genera *Chinchilla* and *Lagidium*, and the Lagostominae, which contains at least two extinct genera that date to the Early and Middle Miocene as well as the extant *Lagostomus* (McKenna and Bell 1997). Neither *Chinchilla* nor *Lagidium* have a fossil record, but *Lagostomus* is known from the early Pleistocene, and possibly late Pliocene (Kraglievich 1926).

Wood and Patterson (1959) and Wood (1965) grouped the chinchillids with dasyproctids, dinomyids, and cuniculids into the superfamily Chinchilloidea based on similarities of dental morphology. Vucetich (1975) and B. Patterson and Wood (1982) proposed a more restricted Chinchilloidea, containing among the extant South American caviomorphs only the Chinchillidae. However, available molecular data support a strong sister relationship between the Chinchillidae and Dinomyidae (Huchon and Douzery 2001; Opazo 2005; Sallam et al. 2009; Rowe et al. 2010), with these two families comprising the Chinchilloidea as the sister group to the Octodontoidea (families Abrocomidae, Echimyidae, Ctenomyidae, Octodontidae). The stem split within a Chinchilloidea that encompasses both the Chinchillidae and Dinomyidae is placed in the Early Miocene, at an estimated age 19.1 ± 2.7 mya (Opazo 2005), at 19.2 mya (95% credibility interval 16.5–23.3; Sallam et al. 2009), or within the range of 17 to 21 mya (Huchon and Douzery 2001), which spans the Oligocene-Miocene boundary.

KEY TO THE SUBFAMILIES AND GENERA
OF CHINCHILLIDAE:

1. Paroccipital processes long, separated from bulla; bullae not inflated and not visible on dorsal surface of cranium; occipital region of skull strong, prominent with enlarged lambdoidal ridge; skull more prominently ridged; postorbital processes of frontal well developed; incisors proodont; cheek teeth (except M3) bilaminate; three digits on hindfeet with heavy, prominent, and sharp claws. . . . . .Lagostominae, *Lagostomus*
1'. Paroccipital processes small and closely aligned to bulla; bullae inflated, visible on dorsal surface of cranium; occipital region smooth, without sharply delineated lambdoidal ridge; skull without prominent ridges for muscle attachment; postorbital process of frontals weakly developed; incisors orthodont or slightly opisthodont; cheek teeth trilaminate; four digits on hindfeet with weak, blunt claws. . . . . .Chinchillinae, 2
2. Bullae greatly inflated, with large exposure of mastoid portion on dorsal surface of cranium; jugal usually not in contact with lacrimal; laminae of cheek teeth straight, not curved . . . . *Chinchilla*
2'. Bullae less inflated, with only small exposure of mastoid portion on dorsal surface of cranium; jugal always in contact with lacrimal; laminae of cheek teeth strongly curved, not straight. . . . . *Lagidium*

Subfamily Chinchillinae Bennett, 1833

The Chinchillinae comprises two genera, *Chinchilla* and *Lagidium*, characterized by small to medium size (head and body length ranging from 300 to 650 mm; mass 400 to 1700 g), very soft and lax fur, four toes on both forefeet and hindfeet with enlarged toe pads obscuring small, curved claws, a skull with small paroccipital processes either appressed to or conforming to the curvature of inflated auditory bullae with the mastoid portion visible on the dorsal surface, and three laminae on each cheek tooth. Limited molecular sequence data confirm a sister relationship between *Chinchilla* and *Lagidium* with respect to *Lagostomus* (Spotorno, Valladares, Marín et al. 2004; Ledesma et al. 2009).

Genus *Chinchilla* Bennett, 1829

As currently understood, there are two species within the genus *Chinchilla*, the Chilean Chinchilla (*Chinchilla lanigera*) and the Short-tailed Chinchilla (*Chinchilla chinchilla* [= *C. brevicaudata*]). *Chinchilla lanigera* is smaller in body size (less than 260 mm in body length) than *C. chinchilla* (>320 mm long), with shorter and more rounded ears (<32 mm versus >45 mm), and with a longer tail (>130 mm

versus <100 mm). External characters in common to both species include very soft fur; long and narrow hindfeet with three main digits and digit V placed high on the foot and not reaching the base of digit IV, and an ungual bundle of stiff bristle hairs on digit II; short forefeet with four main digits, pollex represented by a small, clawless tubercle; and rudimentary cheek pouches (Pocock 1922).

Cranially, the rostrum is long and narrow, the interorbital region broad, but with a well-developed constriction. The postorbital processes of the frontal are very small, and the skull lacks prominent ridges of muscle attachment. The most conspicuous aspect of the skull are the greatly inflated tympanic and mastoid portions of the auditory bullae, with an expanded dorsal mastoid portion clearly visible such that the interparietal is narrow and long and the lateral portion of the supraoccipital is restricted to narrow bands of bone. The mastoid bullae extend posteriorly from the occipital region. Paroccipital processes are small and closely appressed to and dwarfed by the bullae. The floor of the infraorbital foramen is smooth, lacking a canal for nerve transmission. The jugal is especially deep but usually does not extend to reach the lacrimal. Palatal foramina are very long and narrow, occupying more than 75% of the length of the diastema. The mandible has narrow and elongated angular processes and the ridge lateral to the condylar process for the attachment of the posterior masseter is weakly developed.

The cheek teeth diverge strongly, with the distance between the anterior PM4s about 30% of the distance between the posterior M3. Each cheek tooth has three laminae, or plates, with the posterior one of M3 directed backward rather than transversely. The anterior lobe of each lower cheek tooth is reduced relative to the medial and posterior lobes. The laminae of all teeth are straighter than are those of *Lagidium*.

Delimitation of the number of valid species is constrained by a generally inadequate understanding of geographic continuity in all types of characters. Osgood (1943b) recognized a single species (*Chinchilla chinchilla*) with three subspecies: the Peruvian *C. c. chinchilla* Lichtenstein, *C. c. velligera* Prell [with *lanigera* Bennett a synonym] from Chile, and *C. c. boliviana* Brass from Bolivia and Argentina. Cabrera (1961), Woods (1993), and Woods and Kilpatrick (2005) listed two (*Chinchilla chinchilla* [= *C. brevicaudata* of other authors] and *C. lanigera*), and Bidlingmaier (1937) recognized three (*chinchilla* Lichtenstein, *boliviana* Bass, and *velligera* Prell). Waterhouse (1848) commented on differences in overall size and tail length between the Chilean and Peruvian chinchillas, describing the latter as *C. brevicaudata* in reference to its short tail. Pine et al. (1979) included *brevicaudata* in *lanigera* without comment. S. Anderson (1997:473) stated "no convincing case, based on adequately documented specimens of known provenance, has been made for more than one species." He went on to write (p. 474) that geographic variation "was known in a general way to the fur trade. Pelts from higher and colder regions had longer, finer, warmer fur, probably were larger, and the animals may have had shorter tails and ears. However, variation is so poorly known that the use of [even] subspecies seems more confusing than clarifying." Alternatively, Spotorno, Valladares, Marín et al. (2004) found that samples from northern Chile (Antofagasta; which they referred to *brevicaudata*) differed from those from central Chile (Coquimbo; referred to *lanigera*) by 5.9% in mitochondrial sequence divergence, and argued that these represented separate species. However, given that only a single wild population of one (*lanigera*) and two populations of the second (*brevicaudata*) were sampled for each taxon, that these localities are nearly 1,000 km distant, and that none of the range in Peru, Bolivia, or Argentina was sampled, it is perhaps premature to use these limited data as definitive evidence for species boundaries. Unfortunately, it is unclear whether we will ever be able to delineate the number of species within the genus, as Osgood (1943b:135) noted more than 60 years ago. There are few specimens of known provenance in collections, and the local extermination of wild populations over much of the historic range due to overexploitation for their valuable fur greatly restricts the level of geographic sampling that would be necessary to adequately delimit both species and geographic ranges. Nevertheless, there is a recent record at a geographically intermediate locality (F. Valladares et al. 2012), so there is hope that additional populations will be uncovered that can help resolve taxonomic boundaries. Herein, we follow the most recent reviews (Spotorno, Valladares, Marín et al. 2004; Spotorno, Zuleta et al. 2004; Woods and Kilpatrick 2005) and provisionally recognize two species.

SYNONYMS:

Mus: G. I. Molina, 1782:301; part (description of *laniger*); not *Mus* Linnaeus.

Cricetus: É. Geoffroy St.-Hilaire 1803:197; part (listing of *laniger*); not *Cricetus* Leske.

Lemmus: Tiedemann, 1808:476; part (listing of *laniger*); not *Lemmus* Link.

Chinchilla Bennett, 1829:1; type species *Chinchilla lanigera* Bennett, derived from *Mus laniger* G. I. Molina, by monotypy.

Eriomys Lichtenstein, 1830:plate 28, and unnumbered text; type species *Eriomys chinchilla* Lichtenstein, by monotypy.

Callomys d'Orbigny and Geoffroy St.-Hilaire, 1830:282; part; a composite genus with no type species selected.

Lagostomus: F. Cuvier, 1830:plate 64; part (listing of *laniger*); not *Lagostomus* Brookes.

Aulacodus: Kaup, 1832:column 1211; not *Aulacodus* Temminck.

REMARKS: Bennett's *Chinchilla* has been varyingly dated to 1829, 1830, and 1831, resulting in differing opinions as to the proper genus-group name applied to the chinchilla. Wiegmann (1835:205), for example, argued that *Eriomys* Lichtenstein had precedence over *Chinchilla* Bennett, although he wrote "in den *Gardens and Menagerie of the Zoological Society. 1831. Vol. 1, p. 1.,* bereits [already] 1829 gedruckt [printed]."

Alternatively, as noted by Osgood (1941:410), *Chinchilla* had been given precedence historically because Bennett's description was more complete and accurate.

The confusion over date priority likely stems from the fact that *The Gardens and Menagerie of the Zoological Society of London*, which was edited by Bennett, was published in three editions, one each in the years 1829, 1830, and 1831. Tate (1935:357) apparently had in hand the 1831 edition when he noted that Bennett's own preface was dated June 30, 1830. Sherborn (1922) initially gave 1831 for the date of Bennett's publication but subsequently (Sherborn 1925a:1233, 1927b:1829) listed 1829 for Bennett's *Chinchilla* and *lanigera*, respectively. Most subsequent authors have cited 1829 as the publication date for Bennett's name (E.-L. Trouessart 1897; Ellerman 1940; Osgood 1941; Cabrera 1961 [p. 567; on p. 566 Cabrera gave 1820, which is likely a printer's error]). None of these authors, however, referenced the words of Bennett himself who wrote (Bennett 1833:58), "To complete the history of Chinchilla he [= Bennett] also gave an account of the various notices regarding it, which have appeared since September 1829, the date of his account of it in the 'Gardens and Menagerie of the Zoological Society'" and (Bennett 1835c:45) "Dr. Rousseau translated into French my account of the *Chinchilla*, from the 'Gardens and Menagerie of the Zoological Society', attributing its date to 1831, which some of the later published copies of the volume bear upon the title-page, instead of 1829, when the number containing the *Chinchilla* was published." There seems little doubt that the initial publication of Bennett's *Chinchilla* stems from 1829, removing any question of the correct use of his name for the chinchilla, and with precedence over those proposed subsequently.

D'Orbigny and Geoffroy St.-Hilaire's (1830) *Callomys* is a composite genus that included *Callomys viscaccia* (assigned to *Viscaccia* Schinz [1825a], a synonym of *Lagostomus* Brookes), *Callomys laniger* (= *Chinchilla lanigera*), and *Callomys aureus* (an indeterminate species based on a furrier's skin that lacked the feet, ears, and tail). As none of these three names were specifically chosen a type species, *Callomys* d'Orbigny and Geoffroy St.-Hilaire is indeterminate as well (see J. A. Allen 1901d).

There has also been considerable confusion and debate over the type species of *Chinchilla* Bennett, originally described in a semipopular publication (Bennett 1829) but subsequently in a more technical and fully illustrated treatment (Bennett 1835c). G. I. Molina (1782:301) described "La Chinchilla, *Mus Laniger* (I. Mus cauda mediocri, palmis tetradactylis, plantis pentadactilis, corpore cinereo lanato)" from Chile. Bennett (1829:1) gave the type of his new genus as *Chinchilla lanigera*, using the feminine ending to Molina's name but clearly describing an animal that he had in hand at the time; his specimen also came originally from Chile. Lichtenstein (1830:plate 28 and accompanying unnumbered text) described *Eriomys* with the type species *Eriomys chinchilla*, from Peru. Finally, Prell (1934a), who recognized that there was confusion over Molina's *laniger*, named the Chilean chinchilla *Chinchilla velligera*, stating clearly that his name was intended to apply to the animal described by Bennett.

Osgood (1941) argued that the correct type species for Bennett's *Chinchilla* was *Chinchilla velligera* Prell, 1934a, rather than *Mus laniger* G. I. Molina, 1782 (as used by Tate 1935:359; Ellerman 1940:227; and Cabrera 1961:566, for example) or *Chinchilla lanigera* Bennett, 1829 (as used by Woods and Kilpatrick 2005). Osgood's reasons were fivefold: (1) *Mus laniger* G. I. Molina was clearly based on a composite because Molina "confounded the true chinchilla, the mountain viscacha, the chinchilla rat [= *Abrocoma*], and perhaps the degu (*Octodon*)" (p. 409), and thus its designation as the type of the genus "is unjustified not only because it is unidentifiable but also because it was a 'species inquirendae'" (p. 410); (2) that *Eriomys chinchilla* Lichtenstein, 1830, while valid for the Peruvian form, was "not available as type of the genus *Chinchilla*, in spite of the tautonomy, and in spite of its priority" (p. 410) because "the type *must have a name based on the animal described by Bennett*, that is, the Chilean form as at present understood" [italic in the original] (p. 410); (3) that *Cricetus chincilla* [sic] G. Fischer, 1814, cannot be used as it was proposed only as a substitute name for *Mus laniger* G. I. Molina, which itself is unidentifiable; (4) that "*Chinchilla lanigera* as used by Bennett is, at best, also to be regarded as a substitute for *Mus laniger*, and, although it is not an exact homonym, *any attempt to give it status with Bennett as author is likely to be futile*" [italic ours] (p. 410); and (5) that the Chilean chinchilla had no other name subsequently applied to it until Prell (1934a:100) used *Chinchilla velligera* and stated plainly that the name was intended to apply to the animal described by Bennett. Osgood thus argued that Prell's name is the only one that provided what he considered the requirements for a type for the genus if *Mus laniger* G. I. Molina was removed from consideration.

S. Anderson (1997:474) acknowledged that Molina's *laniger* could be rejected as a dubious name of uncertain synonymy, and used *Eriomys chinchilla* Lichtenstein as the type for Bennett's *Chinchilla*. He stated that the type locality of Lichtenstein's taxon was actually uncertain, said to be Peru by Prell (1934a) and Osgood (1941) or to be uncertain but perhaps Chile by G. M. Allen (1942). Thus, without explicitly stating so, Anderson's action indicated that the second objection raised by Osgood (1941; point 2, above) was invalid.

Finally, and most recently, Spotorno, Zuleta et al. (2004) gave the type species as *Mus laniger* G. I. Molina, arguing that Molina's *laniger* was neither ambiguous nor unidentifiable and that Bennett's emendation of *laniger* to the feminine *lanigera* was incorrect, as *laniger* "is perfectly acceptable as a noun in apposition" (p. 6). These authors then cited J. P. Valladares and Spotorno (2003:Case 3278) as a petition submitted to the International Commission on Zoological Nomenclature to conserve Molina's name, although they stated that "the petition is not yet published, and the Commission has not rendered an opinion." In response to this petition, however, the Secretariat answered by letter to Spotorno that it "is not necessary to refer to the Commission to resolve these problems, because the genus-group name *Chinchilla* Bennett (1829) appears to be the oldest available name for the taxon and as such is the valid name (Article 23 of the Code). Also, the different genus-group referable to it does not need to be suppressed for absolute tautonomy (Article 68.4 of the Code) or for junior subjective synonym of *Chinchilla* (Article 23 of the Code)." The logical extension of the Secretariat's response is that, if Bennett's *Chinchilla* is valid, then his species-group name *C. lanigera* is also valid.

KEY TO THE SPECIES OF *CHINCHILLA*:
1. Body size small (head and body length <260 mm); ears short (<32 mm) and more rounded; and tail long (>130 mm) . . . . *Chinchilla lanigera*
1'. Body size larger (head and body length >320 mm); ears longer (>45 mm) and less rounded; and tail short (<100 mm) . . . . *Chinchilla chinchilla*

Chinchilla chinchilla (Lichtenstein, 1830)
Short-tailed Chinchilla
SYNONYMS:

Eriomys chinchilla Lichtenstein, 1830:plate 28 and unnumbered text; type locality "Peru."

Lagostomus laniger: Wagler, 1831b:614; part; not *laniger* G. I. Molina.

Lagostomus chinchilla: F. J. F. Meyen, 1833:586; name combination; = *Eriomys chinchilla* Lichtenstein.

Chinchilla brevicaudata Waterhouse, 1848:241; type locality "Peru."

[*Chinchilla brevicaudata*] var. *major*: E.-L. Trouessart, 1897:628; name combination, attributed to Burmeister, 1879.

Chinchilla boliviana Brass, 1911:613; type locality "aus Peruist sehr selten geworden. –Die fiensten Felle kommen aus der Gegend von Tacna und Arica."

Chinchilla laniger boliviana: G. M. Allen, 1942:389; name combination.

Chinchilla laniger brevicaudata: G. M. Allen, 1942:389; name combination.

[*Chinchilla*] *chinchilla*: Osgood, 1941:410; first use of current name combination.

Chinchilla chinchilla chinchilla: Osgood, 1943b:136; name combination.

Chinchilla chinchilla boliviana: Osgood, 1943b:136; name combination.

Chinchilla brevicaudata boliviana: Cabrera, 1961:567; name combination.

Chinchilla brevicaudata brevicaudata: Cabrera, 1961:567; name combination.

[*Chinchilla chinchilla*] *brevidauda* Woods and Kilpatrick, 2005:1550; incorrect subsequent spelling of *Chinchilla brevicaudata* Waterhouse.

DESCRIPTION: Distinguishable from *C. lanigera* by larger size (head and body length >320 mm) and both shorter ears (<32 mm) and tail (<100 mm). Modal number of caudal vertebrate 20 (Grau 1986). Otherwise, two species apparently share same overall external and craniodental traits. Prell (1934a) stated that Peruvian (*chinchilla* or *brevicaudata*) and Bolivian (*boliviana*) forms differed only slightly in color.

DISTRIBUTION: *Chinchilla chinchilla* was historically known from the coast and Andes of central Peru (Tschudi 1845) south through the highlands of west-central Bolivia (S. Anderson 1997), northern Chile (Spotorno, Valladares, Marín et al. 2004), and adjacent northwestern Argentina (Osgood 1943b; Barquez, Díaz, and Ojeda 2006). Based on information gleaned from local informants, Cajal and Bonaventura (1998) believed that relictual populations remain in the northwestern Argentinean provinces of Catamarca, Jujuy, and Salta. M. M. Díaz and Barquez (2007) recorded specimens from the highlands of Jujuy province, but failed to find any extant populations during their five years of fieldwork, although Chebez (1994) stated that a Provincial Reserve had been established for the species. Chilean populations were thought to be extinct due to overexploitation by 1953 (J. E. Jiménez 1996), but have since been rediscovered in the highlands of Antofagasta (Spotorno, Zuleta et al. 1998; Cortes et al. 2003; Spotorno, Valladares, Marín et al. 2004). Bolivian populations may be extinct, as the most recent specimen was collected in 1939 (S. Anderson 1997). Eisenberg and Redford (1999) stated

that the last Peruvian specimens had been taken more than 50 years ago, and Pacheco (2002) considered the species to be extinct in the wild in Peru. However, the Royal Ontario Museum has a single specimen collected in 1984 at a restaurant in Cerro de Pasco east of Lima (ROM 91301; identification verified by Judith Eger, pers. comm.). It is unclear if this animal came from a nearby natural population or might have been from a captive colony, but the possibility that an extant population remains in the remote highlands of central Peru is noteworthy nevertheless. J. E. Jiménez (1996) provided a generalized map of the presumed historical distribution.

SELECTED LOCALITIES (Map 398): ARGENTINA: Jujuy, Abra Pampa (M. M. Díaz and Barquez 2007), Susques (Cajal and Bonaventura 1998); Salta, Los Andes (R. A. Ojeda and Mares 1989), Socompa (R. A. Ojeda and Mares 1989). BOLIVIA: Cochabamba, mountains near Chaquecamata (S. Anderson 1997); La Paz, Frankfurt, at border with Chile, Bolivia, and Peru (S. Anderson 1997). CHILE: Antofagasta, Calama (AMNH 137276), El Laco, 56 km SW of Socaire (Spotorno, Valladares, Marín et al. 2004), Llullaillaco (Spotorno, Valladares, Marín et al. 2004); Arica y Parinacota, Parinacota (USNM 391825); Atacama, Laguna Santa Rosa (F. Valladares et al. 2012), Quebrada Piedras Lindas, Parque Nacional Nevado Tres Cruces (F. Valladares et al. 2012). PERU: Pasco, Cerro de Pasco (ROM 91301).

SUBSPECIES: Brass (1911) distinguished populations from Bolivia and Chile as a subspecies (*boliviana*) separate from the nominal form from Peru. However, given the difficulty in establishing species boundaries, we agree with S. Anderson (1997) that delineating subspecies is probably more confusing than it is useful. Thus, we regard *C. chinchilla* to be monotypic until adequate studies of geographic variation are undertaken and published.

NATURAL HISTORY: Little information is known about the ecology or population biology of this species, with none known for Peruvian or Bolivian localities. In Peru, before most populations were exterminated by over exploitation for fur or meat, *C. chinchilla* was reported as abundant, ranging from the coast near Lima to elevations as high as 11,000 feet (Osgood 1943b:135; quoting from Tschudi 1845) and Walle (1914) made similar comments about Bolivian populations in La Paz, Oruro, and Potosí departments. Walle (1914:370), however, noted "the extinction of the species will only be a matter of time if measures are not taken to preserve it." In Argentina, García Fernández et al. (1997) and Barquez, Díaz, and Ojeda (2006) listed conservation status as "in critical danger" due to overexploitation. In Chile as well, *C. chinchilla* is regarded as in risk of extinction (Cofré and Marquet 1999; SAG 2000), although a few populations have been rediscovered in Antofagasta (II Región) where it lives at elevations between 3,400 and 5,000 m (Muñoz-Pedreros and Yáñez 2000; Tirado et al. 2012). Tirado et al. (2012) reported that individuals in Parque Nacional Llullaillaco at 3,400 m were herbivorous, consuming primarily bunch grass (*Stipa chrysophylla*) and secondarily shrubs of the genus *Adesmia* (Papilionaceae). As an adaptation to high elevations, this species has a higher hemoglobin oxygen affinity (P50 at 23.3 ± 0.06 mm Hg) than that of the coastal *C. lanigera* (27.7 ± 0.03 mm Hg; Ostojic et al. 2002). It is also well adapted to the low water and food availability at these extreme elevations, as basal metabolic rate, thermal conductance, and the energetic cost for maintenance of water balance were each lower than predicted from body size (Cortés et al. 2003). Spotorno, Zuleta et al. (1998) argued that the short ears and tail of *C. chinchilla* are external adaptations to the cold environments in which these animals occur.

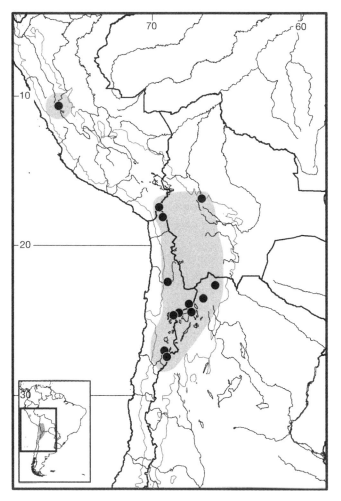

Map 398 Selected localities for *Chinchilla chinchilla* (●); questionable locality in central Peru shown as disjunct from contiguous range. Contour line = 2,000 m.

REMARKS: Many authors who recognized more than a single species of *Chinchilla* have used and continue to use *brevicaudata* Waterhouse, 1848 as the name for the Short-tailed Chinchilla (e.g., E.-L. Trouessart 1897, 1904; Cabrera 1961; Woods 1993; Barquez et al. 2006; M. M. Díaz and Barquez 2007). It seems clear, however, that Lichtenstein's earlier name *chinchilla*, with its type locality in Peru, has priority (see Osgood 1943b; S. Anderson 1997; Woods and Kilpatrick 2005). S. Anderson (1997:475) listed *Chinchilla intermedia* Dennler, 1939, as a synonym of *C. chinchilla*, citing Cabrera (1961) and without having seen Dennler's paper. The reference that Anderson cited (p. 529), and stated to come from Prell (1934a:103), was a paper by Dennler published in 1934 regarding the European fur trade and farm breeding of chinchillas, with no mention of the name *C. intermedia*. Rather, as noted by Prell (1934a:102–103), Dennler referred only to the common names given in the fur trade, "chinchilla cordillerana" in South America and "chinchilla Indiana" in Europe. Cabrera (1961:567) cited a second Dennler paper published in another source as "*Chinchilla intermedia* Dennler, Rev. Anim. Pelif., 2, 1939:95." We have been unable to locate either of these papers, and have not listed Dennler's apparent name in the synonymy above as a result.

The karyotype is completely biarmed, with $2n = 64$ and FN = 128 (Vidal-Rioja et al. 1973a, as *C. brevicaudata*).

Chinchilla lanigera Bennett, 1829
Chilean Chinchilla

SYNONYMS:

Mus Laniger G. I. Molina, 1782:301; type locality "Provincie Boreali del Chile" [= northern provinces of Chile]; given as "Coquimbo" by Osgood (1943b:134).

Cricetus laniger: É. Geoffroy St.-Hilaire, 1803:197; name combination.

L[*emmus*]. *laniger*: Tiedemann, 1808:476; name combination.

C[*ricetus*]. *chincilla* G. Fischer, 1814:55; type locality "Chile borealibus"; substitute name for *Mus laniger* G. I. Molina.

[*Hypudaeus*] ? *laniger* Illiger, 1815:108; *nomen nudum*; name given in a list of South American species and without a description.

Chinchilla lanigera Bennett, 1829:1; type locality "Coquimbo, Chile"; based on *Mus laniger* G. I. Molina.

Callomys laniger: d'Orbigny and Geoffroy St.-Hilaire, 1830:289; name combination.

Lagostomus laniger: F. Cuvier, 1830:plate 64; name combination.

Aulacodus laniger: Kaup, 1832:column 1211; name combination.

Chinchilla velligera Prell, 1934a:100; type locality "Chile-vom Rio Chuapa"; replacement name for *Mus lanigera* Bennett.

DESCRIPTION: As previously noted, *Chinchilla lanigera* differs from *C. chinchilla* by smaller body size (<260 mm), more rounded and longer ears (>45 mm), and longer tail (>130 mm; Spotorno, Zuleta et al. 2004). Modal number of caudal vertebrate 23 (Grau 1986). Otherwise, the two species share same external and craniodental traits that characterize genus. This species exhibits reverse sexual dimorphism, with females substantially larger than males; wild animals, but smaller and less sexually dimorphic than domestic individuals, with mass averaging 412 g in wild-caught males, 422 g in wild-caught females (J. E. Jiménez 1990a) in contrast to weights up to 600 and 800 g, respectively, for two sexes in domestic stock (Neira et al. 1989).

DISTRIBUTION: The historic distribution included the hills and low mountains of coastal Chile, between 26° and 35°S latitude (Bennett 1835c; Spotorno, Zuleta et al. 2004). By the mid-nineteenth century, however, *C. lanigera* had been exterminated from latitudes south of 32° (Río Choapa, the type locality of *Chinchilla velligera* Prell) and north of 29°S latitude (Gay 1847; J. E. Jiménez 1990a,b). The two currently known natural populations are restricted to the region between 31° and 29°S latitude (Spotorno, Zuleta et al. 2004). An earlier report (E. P. Walker 1968) of populations as far south as 52° is apparently in error (Spotorno, Zuleta et al. 2004). Both J. E. Jiménez (1996) and Spotorno, Zuleta et al. (2004) map the presumed historical distribution.

SELECTED LOCALITIES (Map 399; known extant populations only, from Spotorno, Valladares, Marín et al. 2004): CHILE: Coquimbo, Illapel, 15 mi N of Aucó, La Higuera.

SUBSPECIES: *Chinchilla lanigera* is monotypic.

NATURAL HISTORY: Spotorno, Zuleta et al. (2004) reviewed the ecology, reproductive biology, physiology, and behavior of both wild and captive *C. lanigera*. The literature on captive animals is voluminous, covering virtually all aspects of form and function but especially reproductive anatomy and physiology as well as basic husbandry. Data from natural populations are very limited. The native range is within the barren, arid, and rugged areas of the transverse mountain chains in north-central Chile that connect the coastal ranges with the Andes, at elevations from 400 to 1,650 m. Here, rainfall is seasonal and limited, and the typical habitat is rocky or sandy substrates covered by plant communities that are a matrix of typical xeric species from the northern desert and mesic southern Mediterranean zones. Individuals shelter in crevices between rocks, in holes dug below rocks, or within large bromeliads (*Puya*

beteroniana; J. E. Jiménez 1990a). This species is social, living in colonies of up to 100 individuals (Albert 1900) or greater (J. E. Jiménez 1990a) that range in extent from 1.5 to >100 ha, usually on northfacing slopes. The natural population structure is viewed as a classic metapopulation, with frequent local extinction followed by recolonization of suitable habitat patches. Population density averaged 4.37 individuals per hectare and did not fluctuate between years of low and high precipitation (J. E. Jiménez et al. 1992). Natural predators include the Culpeo (*Lycalopex culpaeus*), which take all age classes, and the Magellanic Horned Owl (*Bubo magellanicus*), which preys mostly on the young (J. E. Jiménez 1993). Chinchillas are herbivores that feed on a wide range of plant species, mostly grasses and forbs, with diet shifting both seasonally and annually (Serra 1979; Cortés et al. 2002).

Chinchillas are noted for being quiet and relatively shy but nevertheless quite active animals that ricochet across rocks with remarkable agility by use of their long hindlimbs. Individuals are highly philopatric, using conspicuous rocks within their home range for resting, as observation posts, and as latrines that advertise their presence. They keep their fur clean by sand or soil bathing (Stern and Merari 1969), repeatedly using the same small patch and thus leaving a barren circle of fine dust up to 40 cm in diameter near the entrance to their burrow.

The gestation period is 111 days (range: 105–118), based on captive animals, with litter size ranging from one to six but usually two or three (Neira et al. 1989). Lactation normally lasts six to eight weeks, and sexual maturity of both sexes is reached by eight months of age, on average (Weir 1974a). Reproduction in the wild apparently occurs during the winter (between May and November; Spotorno, Zuleta et al. 2004). The maximum recorded life span in the wild has been six years (J. E. Jiménez 1990a), but captive animals may live more than twenty years. Lammers et al. (2001) examined the ontogeny leading to sexual dimorphism in adults and Cortés, Rosenmann, and Bozinovic (2000a,b) examined the efficiency of water regulation by minimizing evaporative water loss and maximizing metabolic water production.

This species is highly endangered in the wild (Thornback and Jenkins 1982; Glade 1988), and has the second highest conservation priority among Chilean mammals (Cofré and Marquet 1999). Due to severe overharvesting for their valuable fur (Iriarte and Jaksic 1986), the species was nearly exterminated from the wild by the early 1900s (J. E. Jiménez 1996). J. E. Jiménez (1993) estimated the total wild population at only 2,500 to 11,700 individuals, with numbers apparently declining.

Chilean Chinchillas were initially domesticated for their valuable pelts but now are raised both for the pet trade

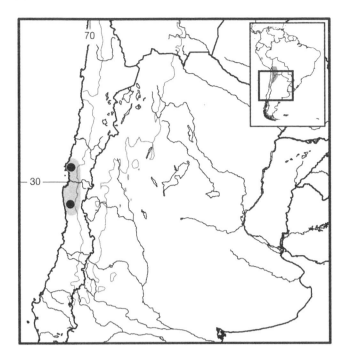

Map 399 Selected localities for *Chinchilla lanigera* (●). Contour line = 2,000 m.

(range: 322 to 405 mm) and mass 1.24 kg (range: 1.2 to 1.6 kg; Pearson 1948). Unlike *Chinchilla*, however, while females are slightly larger in average measurements than males, they are not significantly so. The tail is shorter than the head and body, although longer absolutely and proportionally than in *Chinchilla*, averaging about 78% of the body length. Coarse, pale hairs form a crest along the top of the tail and are sharply demarcated from the shorter, finer dark hairs along the bottom. The tip of the tail varies from black to red-brown. As with *Chinchilla*, the fur is thick and soft, comprised of dense tufts that contain a dozen or so individual hairs; the sparse guard hairs have slender shafts and stout tips. Pearson (1948) noted that adult coloration of populations from southern Peru he called *L. peruanum* was unusually variable, with specimens in the same local population ranging from creamy buff over the entire back to a dark grizzled gray. The venter is always paler than the back, gray washed with buffy yellow. Buff or orange fur usually covers the rump and upper thighs, and there may be axillary and inguinal patches of white. The palms and soles are naked and black, but the backs of both forefeet and hindfeet are furry, usually quite pale or even white. The ears are gray, or gray-brown, fringed with pale gray or white hairs. The darkest region is over the rump, and a narrow, dark, mid-dorsal line running up to the shoulder region is frequently present. The ears are large. The hindfeet are narrow with four digits, with digit V extremely short. The forefeet have four digits. All digits terminate in short, curved claws dwarfed by enlarged toe pads. The tail is longer absolutely and proportionally than in *Chinchilla*, though typically less than 50% of the head and body length, well furred, and terminates in a black tip that is characteristically curled dorsally. Females of *L. peruanum* have a single pair of mammae, located on the side of the thorax (Pearson 1949).

The skull is narrow, with a long rostrum. The interorbital region is broad, with weakly developed postorbital processes of the frontals that tend to be depressed between the orbits. The braincase is flat and rounded, with no sagittal ridge and only weakly developed parietal ridging. The bullae are enlarged, with the mastoid portion visible on the dorsal surface of the skull, but not to the degree as in *Chinchilla*, and not extending posterior to the occiput. Thus, the interparietal is wider than it is long, in contrast to the condition in *Chinchilla*. The paroccipital processes are longer than in *Chinchilla*, not in intimate contact with the bullae but curved to conform to the shape of the tympanic portion. The jugal is broad and in contact with the lacrimal anteriorly. The floor of the infraorbital foramen is smooth, lacking a canal indicative of nerve passage. Incisive foramina are long and narrow, but occupy only bout 60% of the diastemal length. The anterior palate is greatly

(R. A. Webb 1991) and as laboratory animals, where they are used in auditory research (Heffner and Heffner 1991; L. Robles and Ruggero 2001).

REMARKS: For a discussion of whether *laniger* Molina or *lanigera* Bennett is the correct specific epithet to be applied to this species, see Remarks for the genus.

The karyotype is completely biarmed, with 2n = 64 and FN = 128 (Makino 1953; Nes 1963; Hsu and Benirschke 1967).

Genus *Lagidium* Meyen, 1833

We recognize three species of *Lagidium*, one with an extensive distribution down the spine of the Andes, from southern Peru through Bolivia and northern Chile to central Argentina (*L. viscacia*, including *L. peruanum* which has been recognized by many authors as a valid species, both in the past and presently [e.g., Woods and Kilpatrick 2005; Ledesma et al. 2009]), a second from a small area in the mountains of west-central Argentina and adjacent Chile (*L. wolffsohni*), and a third, recently described from southern Ecuador (*L. ahuacaense*). While multiple species names are available along the extended distribution of *L. viscacia*, hypotheses of species boundaries or geographic limits are lacking at the present time (see Remarks).

Externally, *Lagidium* is larger in body size than *Chinchilla*, with head and body length averaging 344 mm restricted between the premolars, with the horizontal distance less than 25% of that between the last molars. The mandible is similar to that of *Chinchilla*, with elongated and narrowed angular processes and a weakly developed ridge lateral to the reduced coronoid process for muscle attachment.

The cheek teeth each have three, distinctly curved rather than straight, laminae, particularly so in M3. In the maxillary series the third plate of each tooth is shorter than the other two, with that of M3 forming a posteriorly pointed heel. In the mandibular series, the first plate of each tooth is reduced relative to the median and posterior plates.

SYNONYMS:

Lepus: G. I. Molina, 1782:307; part (description of *viscacia*); not *Lepus* Linnaeus.

Viscaccia Oken, 1816:835; part (a composite based on *Lepus chilensis* Oken, the "*le vizcache*" of Azara [1801], and *Mus laniger* G. I. Molina [see J. A. Allen 1902c:196]); ruled a synonym of *Lagidium* Meyen (ICZN 1925) and an unavailable name (ICZN 1956).

Callomys d'Orbigny and Geoffroy St.-Hilaire, 1830:282; part (description of *aureus*); a composite genus with no type species selected.

Lagidium Meyen, 1833:576; type species *Lagidium peruanum* Meyen, by monotypy (recommended for legalization by fiat, instead of *Viscaccia* Oken, by Thomas [1914g:285; see also Thomas 1924g:347], with *Lagidium* Meyen officially adopted in preference to *Viscaccia* Oken [ICZN 1925:Opinion 90; ICZN 1929:Opinion 110]).

Lagotis Bennett, 1833:58; type species *Lagotis Cuvieri* Bennett, by monotypy; not *Lagotis* Desmarest, a junior synonym of *Pedetes* Illiger (Palmer 1904).

REMARKS: Oken (1816:835) erected *Viscaccia*, based both on *Lepus chilensis* (a replacement name for Molina's *Lepus viscacia*) and *Mus laniger* G. I. Molina, a composite but a name that is associated with *Chinchilla* (see Remarks for that genus). *Viscaccia* Oken has often been associated with the mountain viscachas (e.g., E.-L. Trouessart 1897; Thomas 1907a). J. A. Allen (1902c), however, argued that, because Molina's *Lepus chilensis* was based wholly on "*le vizcache*" of Azara (1801), which is the Plains Viscacha, *Lagostomus* (see that account), Oken's *Viscaccia* applied to *Lagostomus* rather than *Lagidium*. Initially the International Commission (ICZN 1925:Opinion 90) ruled "*Lagidium* Meyen, 1833, type species *Lagidium peruanum* Meyen, is adopted in preference to *Viscaccia* Oken, 1816, genotype '*Lepus chilensis* G. I. Molina.'" Subseqeuntly, the ICZN (1929:Opinion 110) suppressed *Viscaccia* for the purpose of Priority but not of Homonomy and then, in Opinion 417 (ICZN 1956), rejected Oken's entire *Lehrbuch* (Vol.

3) for nomenclatural purposes, and thus making the generic association of his *Viscaccia* moot.

Lagidium Meyen and *Lagotis* Bennett were both proposed in 1833, the former in March of that year and the latter initially read before the Zoological Society of London in May. Bennett, however, asserted (1835a:492; 1835c:492) that his *Lagotis* should have priority as it was "characterized by me . . . at a Meeting of the Committee of Science and Correspondence [of the Zoological Society of London] in June 1832, and the name then given was affixed, through the life of the individual, to the cage in which the Society's animal was kept." Despite Bennett's protestations to the contrary, however, *Lagotis* as a valid name stems from its first appearance in the published literature, two months after Meyen's *Lagidium* (see Wiegmann 1835:204). Furthermore, *Lagotis* Bennett is preoccupied by *Lagotis* Desmarest (based on a manuscript name of M. de Blainville), a junior synonym of *Pedetes* Illiger (Tate 1935:361).

Callomys d'Orbigny and Geoffroy St.-Hilaire is a composite genus clearly based on species now placed in both *Lagostomus* (*Callomys viscacia*) and *Chinchilla* (*Callomys laniger*), but also a third species, *Callomys aureus*, based on furrier's skins without feet, ears, or tail that Bennett (1833:59) considered a questionable synonym of *Chinchilla lanigera* and Tate (1935:360) thought might well represent a *Lagidium*. E.-L. Trouessart (1897:627) listed *aureus* as a synonym of *Lagidium peruanum* Meyen. Even so, *Callomys* as a genus is indeterminate and *C. aureus* as a species is as well.

The genus *Lagidium* is in need of a revision that incorporates analyses of adequate geographic samples for both morphological and molecular character variation. The poorly delineated species boundaries are highlighted by the various published opinions, all of which lack substantive character analyses.

Thomas (1907a:441) reviewed the characters and distribution of the species named during the nineteenth century, referring all to the genus *Viscaccia*, as follows: *viscacia* G. I. Molina, a "large deep gray animal from the Chilean Andes," *cuvieri* Bennett and *pallipes* Bennett, both "referable to a strongly yellowish form found in Northern Chile," and *peruana* Meyen, which "I cannot certainly identify." He then described 10 additional species as new. In 1914, Thomas (p. 285) reversed his earlier opinion and advocated use of *Lagidium* over *Viscaccia*. He subsequently described seven additional species between 1919 and 1926 (*lockwoodi* [1919d], *vulcani* [1919f], *famatinae* [1920g], *boxi* [1921c], *tontalis* and *viatorum* [1921i], and *sarae* [Thomas and St. Leger 1926b]), each largely known only from one or a few specimens and from their respective type localities. Ellerman (1940:230–231) cited R. W. Hayman, who examined all specimens in the British Museum and believed that the

21 named forms could be grouped readily into four species: *L. viscacia* (with *lutea, cuscus, perlutea, vulcani, tucumana, lockwoodi, famatinae, tontalis, viatorum,* and *moreni* synonyms), *L. peruanum* (including *pallipes, inca, subrosea, saturata, punensis,* and *arequipae*); *L. boxi* (including *sarae*); and *L. wolffsohni,* which he stated (p. 231) "differs clearly in colour pattern from all the remainder." Osgood (1943b) placed all Chilean and Argentinean taxa within a polytypic *L. viscacia,* but did not comment on those taxa described from Bolivia or Peru. Cabrera (1961) recognized three species, the polytypic and broadly ranging *L. peruanum* and *L. viscacia* and the monotypic *L. wolffsohni* with a limited range in Santa Cruz, Argentina. S. Anderson (1997) recognized a single species in Bolivia (*L. viscacia*) even though he mapped localities only minimally distant from those of *L. peruanum* from the adjacent Altiplano of Peru. Similarly, Pacheco et al. (2009) applied *L. viscacia* to Peruvian records, noting the imprecise definition of this species and difficulty in delimiting *L. peruanum.* Woods and Kilpatrick (2005) followed Cabrera in their recent taxonomic compilation of the genus, and recognized three species (*L. peruanum, L. viscacia,* and *L. wolffsohni*). Finally, Ledesma et al. (2009) described a new species, unexpectedly encountered in southern Ecuador (see Werner et al. 2006).

It is impossible at the present time to either effectively diagnose species or to map their ranges with any certainty, much less to sort the plethora of available names to one "species" or another, except by the geographic proximity of their respective type localities. Only Thomas (1907a), when he described 10 new "species," and R. W. Hayman, when he "looked through the considerable British Museum material with a view of getting the twenty-one 'distinct species' in this genus into some sort of revision" (in Ellerman 1940:230), had examined a broad spectrum of specimens and named forms at one time. Hayman placed considerable weight on overall size and coloration, particularly the distinctness of the dorsal stripe, in establishing his groups. However, as Pearson (1948:346) noted, "the color of specimens from a single locality is extremely variable [including the dorsal stripe] and changes considerably with age—facts not fully realized by Thomas when he described five full species from Peru (see especially Thomas 1907a) and three from adjacent Bolivia." Moreover, in some cases, Hayman's allocation of names to his "species" severely conflicted with geography (i.e., his linkage of *pallipes* to *L. peruanum,* even as a possible synonym of *L. p. peruanum* from southern Peru [Ellerman 1940:231], when the type of Bennett's *pallipes* came from far to the south, in Mendoza province, Argentina [see Osgood 1943b:138–139]). These conflicts only highlight Pearson's (1948) caution about intralocality variation not to mention the complete lack of any substantive geographic analyses of any kind,

for any character. Even the limited mitochondrial sequence data available (Spotorno, Valladares, Marín et al. 2004; Ledesma et al. 2009) has so far failed to resolve species boundaries although it offers hope as more samples are added. However, if the general clade structure obtained by these authors is upheld by additional work, then both the number of species and the names applied to each will substantially change, as will the geographic limits of each species so defined. For example, the recent analysis of Ledesma et al. (2009) recovered six clades that together formed a basal polytomy with nearly equivalent levels of divergence (maximum 9%). These clades included (1) the single specimen of *L. ahuacaense* from Ecuador, (2) two specimens assigned to *L. viscacia perlutea* from Bolivia and northern Argentina, (3) *L. viscacia boxi* (from Neuquén, Argentina) plus *L. wolffsohni,* (4) *L. viscacia viscacia* from north-central Chile, (5) *L. peruanum* from northern Chile, and (6) *L. peruanum* from southern Peru. Only clades 5 and 6 share limited sister-group support.

Herein, we take a conservative position and recognize only a single, widely ranging and likely polytypic species, *L. viscacia* (including *L. peruanum* and *L. boxi*), and two highly localized, peripheral isolates, the southern *L. wolffsohni,* despite the fact that it is apparently nested phyletically within *L. viscacia,* and the northern *L. ahuacaense.* As we noted earlier, with additional molecular sampling targeted both to type localities and to infill geographic lacunae, the resolution of species limits and the applicability of available names to each will certainly redefine taxonomic diversity in the genus. Fortunately, *Lagidium* has not been exploited to the degree as has *Chinchilla,* and sufficient natural populations remain that can be sampled.

KEY TO THE SPECIES OF *LAGIDIUM*:

1. Body size larger (head and body length >470 mm); dorsal coloration rich brownish slate; mid-dorsal dark stripe weakly developed; ears short with black outer and lighter inner surfaces; distribution limited to southern Argentina and adjacent Chile . . . . *Lagidium wolffsohni*

1'. Body size smaller (head and body length ranges from 360 to 425 mm); dorsal coloration highly variable, from creamy buff to dark gray; mid-dorsal dark stripe usually well developed; distribution from southern Ecuador to northern Argentina and Chile 2

2. Dorsal color dark gray; dorsal surface of feet blackish; ears short (ca. 60 mm) and black on both surfaces; least interorbital breadth of skull narrow (ca. 16.5 mm); rostrum broad (ca. 16.2 mm); braincase flat; known only from southern Ecuador . . . . *Lagidium ahuacaense*

2'. Dorsal color highly variable, but typically buff to dark gray; dorsal surface of feet pale, not black; ears longer

(>65 mm, on average); broader interorbital breadth (>16.8 mm); narrower rostrum (<14 mm); braincase more elevated; broad range from central Peru to central Chile and Argentina . . . . *Lagidium viscacia*

Lagidium ahuacaense Ledesma, Werner, Spotorno, and Albuja, 2009
Ecuadorean Viscacha
SYNONYM:

Lagidium ahuacaense Ledesma, Werner, Spotorno, and Albuja, 2009:47; type locality "Cerro El Ahuaca, Parroquia Cariamanga, Canton Calvas, Loja Province, Ecuador (04°18′2″S, 79°32′47″W)."

DESCRIPTION: Medium sized for genus, head and body length of holotype 403 mm. Dorsal color brownish-gray dorsal; ventral color yellowish gray; ears blackish with cream-colored fringes; forefeet and hindfeet bear black hairs dorsally; palms and soles naked and blackish in color. Dorsally, tail with long and coarse hairs, maroon with some cream in color; ventral surface of tail clothed in short, blackish-brown hairs; tail tip clothed in long hairs. Vibrissae mostly black, with mystacial vibrissae thick and long (up to nearly 150 mm), superciliary vibrissae scarce but thick, and single genal vibrissa present.

In comparison with other species, *L. ahuacaense* darker than average specimen of *L. viscacia* and of *L. wolffsohni*. Skull, in dorsal view, with more pronounced convex premaxillofrontal sutures, wider rostrum and narrower interorbital constriction, less arched braincase, more robust jugal; in ventral view, mesopterygoid fossa with more parallel sides and somewhat narrower basioccipital, constricting space between bullae.

DISTRIBUTION: *Lagidium ahuacaense* is known only from the type locality in the Cerro El Ahuaca, a steep granite peak about 1 km from the town of Cariamanga. Here, the species inhabits the full elevational extent of the peak (1,950–2,480 m).

SELECTED LOCALITIES (Map 400): ECUADOR: Loja, Cerro El Ahuaca (type locality of *Lagidium ahuacaense* Ledesma, Werner, Spotorno, and Albuja).

SUBSPECIES: *Lagidium ahuacaense* is monotypic.

NATURAL HISTORY (available data from Werner et al. 2006): Individuals at the type locality of Cerro Ahuaca are restricted to the immediate vicinity of extensive rocky surfaces, on a steep cliff face (varying from a 40° to 80° incline) at 2,450 m, where a small colony of individuals (at least two adults and one juvenile) was observed and photographed. The vegetation is classified as dry montane scrub (Sierra 1999), with the surrounding area heavily deforested in recent decades. Observations suggested that individuals are timid and stay close to their den entrance, which was placed in a deep rock crevice on

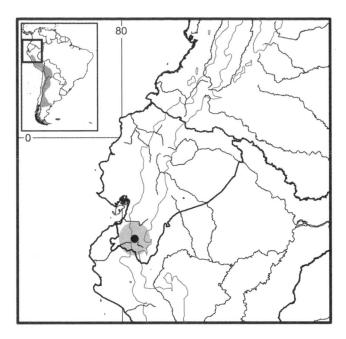

Map 400 Single known locality for *Lagidium ahuacaense* (●). Contour line = 2,000 m.

an 80° slope. The viscacha at Cerro Ahuaca was unknown to the local people.

REMARKS: This species is known only from the holotype as a preserved specimen. Werner et al. (2006) published a color photograph of a living individual.

Lagidium viscacia (G. I. Molina, 1782)
Mountain Viscacha
SYNONYMS:

Lepus viscacia G. I. Molina, 1782:307; no type locality given; restricted to "Chilean Andes; cordillera of Santiago" (Osgood 1943b:137).

Lepus viscaccica Brandis, 1786:272; unjustified emendation of *viscacia* G. I. Molina.

Callomys aureus d'Orbigny and Geoffroy St.-Hilaire, 1830:291; no type locality given; *nomen dubium* (Bennett 1833; Tate 1935).

Lagidium peruanum Meyen, 1833:578; type locality "Es lebt auf den Hoch-ebenen Perus, stets über die Höhe von 12 bis 13000 Fuss hinaus" (= high plateau, Peru, altitude, 12,000 to 13,000 feet), probably Pisacoma, Puno, Peru (Pearson 1948:346, citing pers. comm. from C. C. Sanborn).

Lagotis Cuvieri Bennett, 1833:59; no type locality given; "purchased, in June 1832, from a dealer, who was completely ignorant of the locality from which it was originally obtained" (Bennett 1833:59).

Lagotis pallipes Bennett, 1835b:67–68; type locality "in Chiliae montosis," actually from Mendoza province,

Argentina, "at an elevation of 4000 to 5000 feet, between Villavicencia [= Villavicencio] and Uspallata " (Bridges 1843:132; see also Osgood 1943b:138–139).

Lagidium Cuvieri: Wagner, 1843a:306; name combination.

Lagidium pallipes: Wagner, 1843a:308; name combination.

Lagidium lutescens Philippi, 1896:8, plate II, Fig. 2, plate III, Fig. 5; type locality "provincia Tarapacá inter Copacoya et Inocaliri, nec non ad Socaire prope oppidum Atacama," Chile.

Lagidium crassidens Philippi, 1896:10; no type locality given.

Lagidium crinigerum Philippi, 1896:10, plate III, Fig. 2; no type locality given.

[*Lagidium*] *pallipes*: E.-L. Trouessart, 1897:627; name combination.

Lagidium Moreni Thomas, 1897b:467; type locality "Chubut, Patagonia," (stated [p. 466] to have been obtained "in the hills near Chubut Eastern Patagonia," but subsequently considered as "unknown, as 'Chubut' is a province of considerable size, and there is no evidence as to where in it the specimen was obtained" [Thomas 1921c:181]), Argentina.

Viscaccia viscaccia Thomas, 1907a:441; name combination and incorrect subsequent spelling of *viscacia* G. I. Molina.

V[*iscaccia*]. *Cuvieri*: Thomas, 1907a:441; name combination.

V[*iscaccia*]. *pallipes*: Thomas, 1907a:441; name combination.

V[*iscaccia*]. *peruana*: Thomas, 1907a:441; name combination.

Viscaccia inca Thomas, 1907a:442; type locality "Incapirca, Zezioro," Junín, Peru.

Viscaccia arequipae Thomas, 1907a:442; type locality "Sumbay, near Arequipa. Alt. 4000 m," Arequipa, Peru.

Viscaccia subrosea Thomas, 1907a:442; type locality "Galera, W. of Oroya, Dept. Lima. Alt. 4800 m," Peru.

Viscaccia saturata Thomas, 1907a:442; type locality "Limbane [= Limbani], Inambari, Dept. Puno. Alt. 3500 m," Peru.

Viscaccia punensis Thomas, 1907a:443; type locality "Puno, Lake Titicaca. Alt. 3800 m," Puno, Peru.

Viscaccia cuscus Thomas, 1907a:443; type locality "Paratani, [Cochabamba,] Bolivia (about 66° W., 17° 5′ S.). Alt. 2600 m."

Viscaccia lutea Thomas, 1907a:443; type locality "Esperanza, [near Mt.] Sahama, [La Paz,] Bolivia. Alt. 4000 m."

Viscaccia perlutea Thomas, 1907a:443; type locality "Pampa Aulliaga, [Oruro,] Bolivia (67°W., 19° 30′ S.). Alt. 3800 m."

Viscaccia tucumana Thomas, 1907a:444; type locality "Cumbre de Mala-Mala, Sierra de Tucuman. Alt. 3000 m," Tucumán, Argentina.

Lagidium lockwoodi Thomas, 1919d:499; type locality "Otro Cerro, North-eastern Rioja," a ranch at the south-ern end of the Sierra de Ambato, Catamarca, Argentina (Pardiñas et al. 2007).

Lagidium tucumanum: Thomas, 1919f:133; name combination.

Lagidium vulcani Thomas, 1919f:133; type locality "Cerro Casabindo, 4800 m," Jujuy, Argentina.

Lagidium famatinae Thomas, 1920g:421; type locality "La Invernada, in the northern half of the chain, about 35 km. north of the mountain 'Nevada de Famatina,' and situated at an altitude of about 3800 m," La Rioja, Argentina.

Lagidium boxi Thomas, 1921c:180; type locality "Pilcañeu [= Pilcaniyeu], Upper Rio Negro, N.W. Patagonia. 1200 m," Río Negro, Argentina.

Lagidium tontalis Thomas, 1921i:219; type locality "Sierra Tontal, a north and south range of mountains some 60 km. W. of San Juan. Collection made at Los Sombreros, an estancia at about 2700 m. altitude, and 35 km. N.W. of Pedernal" (p. 214), San Juan, Argentina.

Lagidium viatorum Thomas, 1921i:220; Punta de Vacas, N.W. Mendoza. Alt. 2300 m," Argentina.

Lagidium sarae Thomas and St. Leger, 1926b:639; type locality "Pino Hachado, Neuquen, 1500 m," Argentina.

Lagidium viscaccia: Tate, 1935:363; name combination and incorrect subsequent spelling of *viscacia* G. I. Molina.

[*Lagidium viscaccia*] *cuvieri*: Tate, 1935:363; name combination and incorrect subsequent spelling of *viscacia* G. I. Molina.

[*Lagidium viscaccia*] *lutescens*: Tate, 1935:363; name combination and incorrect subsequent spelling of *viscacia* G. I. Molina.

[*Lagidium viscaccia*] *crassidens*: Tate, 1935:363; name combination and incorrect subsequent spelling of *viscacia* G. I. Molina.

[*Lagidium viscaccia*] *chilensis*: Tate, 1935:363; name combination and incorrect subsequent spelling of *viscacia* G. I. Molina.

[*Lagidium viscaccia*] *crinigerum*: Tate, 1935:363; name combination and incorrect subsequent spelling of *viscacia* G. I. Molina.

[*Lagidium viscaccia*] *viscaccia*: Tate, 1935:363; name combination and incorrect subsequent spelling of *viscacia* G. I. Molina.

[*Lagidium*] *inca*: Tate, 1935:364; name combination.

[*Lagidium*] *arequipae*: Tate, 1935:364; name combination.

[*Lagidium*] *subrosea*: Tate, 1935:364; name combination.

[*Lagidium*] *saturata*: Tate, 1935:364; name combination.

[*Lagidium*] *punensis*: Tate, 1935:364; name combination.

[*Lagidium*] *cuscus*: Tate, 1935:364; name combination.

[*Lagidium*] *lutea*: Tate, 1935:364; name combination.

[*Lagidium*] *sublutea*: Tate, 1935:364; name combination.

Lagidium viscaccia lutea: Ellerman, 1940:231; name combination and incorrect subsequent spelling of *viscacia* G. I. Molina.

Lagidium viscaccia cuscus: Ellerman, 1940:231; name combination and incorrect subsequent spelling of *viscacia* G. I. Molina.

Lagidium viscaccia perlutea: Ellerman, 1940:231; name combination and incorrect subsequent spelling of *viscacia* G. I. Molina.

Lagidium viscaccia vulcani: Ellerman, 1940:231; name combination and incorrect subsequent spelling of *viscacia* G. I. Molina.

Lagidium viscaccia tucumana: Ellerman, 1940:231; name combination and incorrect subsequent spelling of *viscacia* G. I. Molina.

Lagidium viscaccia lockwoodi: Ellerman, 1940:231; name combination and incorrect subsequent spelling of *viscacia* G. I. Molina.

Lagidium viscaccia famatinae: Ellerman, 1940:232; name combination and incorrect subsequent spelling of *viscacia* G. I. Molina.

Lagidium viscaccia tontalis: Ellerman, 1940:232; name combination and incorrect subsequent spelling of *viscacia* G. I. Molina.

Lagidium viscaccia viatorum: Ellerman, 1940:232; name combination and incorrect subsequent spelling of *viscacia* G. I. Molina.

Lagidium viscaccia moreni: Ellerman, 1940:232; name combination and incorrect subsequent spelling of *viscacia* G. I. Molina.

Lagidium peruanum peruanum: Ellerman, 1940:232; name combination.

Lagidium peruanum pallipes: Ellerman, 1940:232; name combination.

Lagidium peruanum inca: Ellerman, 1940:232; name combination.

Lagidium peruanum subrosea: Ellerman, 1940:232; name combination.

Lagidium peruanum saturata: Ellerman, 1940:232; name combination.

Lagidium peruanum punensis: Ellerman, 1940:232; name combination.

Lagidium peruanum arequipae: Ellerman, 1940:232; name combination.

Lagidium boxi boxi: Ellerman, 1940:232; name combination.

Lagidium boxi sarae: Ellerman, 1940:232; name combination.

Lagidium viscacia boxi: Osgood, 1943b:141; name combination.

Lagidium viscacia sumuncurensis Crespo, 1963:61; type locality "Campana Mahuida, depot. Valcheta, Río Negro, Argentina, 700 m."

DESCRIPTION: As for genus, and as distinguished from *L. ahuacaense* and *L. wolffsohni* in those accounts.

DISTRIBUTION: *Lagidium viscacia* occurs along the spine of the Andes from central and southern Peru, western Bolivia, northern to central Chile, and northwestern and western Argentina as far south as Chubut province (42°S latitude). Osgood (1943b) considered the Mountain Viscacha to be the only Chilean mammal definitively characteristic of his Puna Zone, that region above timberline and below snowline. The species ranges from 700 m in Río Negro province, Argentina, to above 4,800 m in the mountains from central Peru through Bolivia, Chile, and northwestern Argentina. Pearson (1951) recorded the species as confined almost entirely to the Altiplano in Peru, but subsequently (Pearson 1957) described a population in the fog-shrouded hills (*lomas*) near Lima in central Peru at an elevation of 2,200 ft elevation (ca. 670 m). In northwestern Argentina, the species is restricted to Puna habitat, but at middle and southern latitudes within the country it is found in pre-Puna and Patagonian steppe (R. A. Ojeda and Mares 1989; Barquez, Díaz, and Ojeda 2006).

SELECTED LOCALITIES (Map 401): ARGENTINA: Catamarca, Otro Cerro (type locality of *Lagidium lockwoodi* Thomas); Chubut, El Hoyo (LACM 59595); Jujuy, Cerro Casabindo (type locality of *Lagidium vulcani* Thomas); La Rioja, La Invernada (type locality of *Lagidium famatinae* Thomas); Mendoza, between Villavicencio and Uspallata (type locality of *Lagotis pallipes* Bennett), Puesto El Peralito, Malargüe (R. A. Ojeda and Mares 1989); Neuquén, Piño Hachado (type locality of *Lagidium sarae* Thomas and St. Leger); Río Negro, Campana Mahuida (type locality of *Lagidium viscacia sumuncurensis* Crespo), 40 mi S of Maquinchao (MCZ 19222), Pilcaniyeu (type locality of *Lagidium boxi* Thomas); Salta, Chorrillos (R. A. Ojeda and Mares 1989); San Juan, Los Sombreros, Sierra Tontal (type locality of *Lagidium tontalis* Thomas); Tucumán, Cumbre de Mala-Mala, Sierra de Tucumán (type locality of *Viscaccia tucumana* Thomas). BOLIVIA: Chuquisaca, 6 mi NE of Tarabuquillo (CAS 14848); Cochabamba, Altamachi (USNM 271426), 9.5 km by road SE of Rodeo (AMNH 260946); La Paz, 10–15 km N of Achacachi (USNM 271392), 1 km W of pass between Antequilla and Pelechuco (MVZ 164778), border of Chile, Bolivia, and Peru (S. Anderson 1997), Esperanza (type locality of *Viscaccia lutea* Thomas), La Paz (S. Anderson 1997); Oruro, Pampa Auliaga (type locality of *Viscaccia perlutea* Thomas); Potosí, 9 km W of Villa Alota (S. Anderson 1997), Llica (S. Anderson 1997); Tarija, Tambo

(S. Anderson 1997). CHILE: Antofagasta, near Copacoya (type locality of *Lagidium lutescens* Philippi), 20 mi E of San Pedro (Osgood 1943b), Talabre (Spotorno, Valladares, Marín et al. 2004); Arica y Parinacota, Timar (USNM 391824); Atacama, Santa Rosa (Spotorno, Valladares, Marín et al. 2004); Coquimbo, Vicuña (Spotorno, Valladares, Marín et al. 2004); O'Higgins, Sewell (Osgood 1943b). PERU: Apurímac, Grau, Virundo (MUSM 2212); Arequipa, Sumbay (type locality of *Viscaccia arequipae* Thomas); Ayacucho, 15 mi WNE of Puquio (MVZ 138147); Cusco, Laguna Sibinacocha (MUSM 17553), Occobamba Pass (USNM 194467); Junín, 20 km E of Lurin (MVZ 137762); Lima, Naña (Pearson 1957); Pasco, Chipa (AMNH 60576); Puno, Limbani (type locality of *Viscaccia saturata* Thomas), Pisacoma (type locality of *Lagidium peruanum* Meyen); San Martín, Pampa de Cuy, Parque Nacional Río Abiseo (MUSM 7889).

SUBSPECIES: Osgood (1943b) recognized seven subspecies that occurred, or might occur, in Chile; Crespo (1963) recognized five subspecies (including *wolffsohni*) in central and southern Argentina; and S. Anderson (1997) mapped the ranges of three subspecies in Bolivia. *Lagidium viscacia* is highly variable in color, body size, and tail length over its large range, as exemplified by the large number of "species" that were described in the early 1900s. Although variation can be extensive within populations and no assessment of geographic patterns has been made, it is highly likely that geographic units can and should be formally recognized as subspecies. At the moment, however, it is not possible to do so, as even the question as to whether the single species recognized here is composite or not remains an open issue. As a consequence, we make no effort to delineate infraspecific taxa within *L. viscacia* at the present time.

NATURAL HISTORY: This species has been studied extensively in the field in the Altiplano of southern Peru (as *L. peruanum*; Pearson 1948, 1949) and in the Patagonian step of Argentina (as *L. viscacia*; R. S. Walker et al. 2000, 2003, 2007, 2008; Galende and Raffaele 2012). In southern Peru the habitat is dry, mountainous country above 3,800 m, with sparse vegetative cover composed mostly of bunch grasses (*Stipa* and *Festuca*) and short *tola* shrubs (predominantly *Lepidophyllum*, *Baccharis*, and *Senecio* [Asteraceae]), and where viscachas "may make his home any place . . . where there are rocks, water, and food—no rocks, no viscachas" (Pearson 1948:346). Preferred habitat was vast tumbles of boulders where crevices provide ready-made burrows, but animals were also found on steeply vertical cliffs or other rock faces as long as there are adequate crevices or narrow, stony tunnels to provide nesting sites. Animals are highly gregarious, living in colonies that may range widely in size (from 4 to 75 individuals), at an overall density of 0.4 individuals per acre (0.162 ha) because of the highly clumped distribution of occupied boulder fields. Each colony is composed of smaller groups consisting usually of two to five individuals; these groups appear to be families, although their composition does not remain constant throughout the year. Each "family" contains a mature male and female (presumably the parents) and one, two, or occasionally three young of different ages. The oldest male offspring may be sexually mature yet still living with the family. There is no defense of individual or family territories, and the entrances of different "family" dens may be only a few feet apart within the same colony. Individual viscachas occupy the same burrow for long periods. Wariness is a group as well as individual characteristic, and all individuals pay attention to the sharp alarm whistles of a single individual. They appear to rely most heavily on vision as the dominant sense. Lucherini et al. (2009) estimated relative abundance of Mountain Vizcachas at multiple sites in northwestern Argentina using direct observations, pellet counts, and photographic camera "traps."

Mountain viscachas are poor diggers with only small claws, so dens are within natural crevices or under boulder piles. The rare self-dug burrows, typically found under boulders in loose gravel or soil, probably require multiple individuals and perhaps multiple generations to construct. The presence of viscachas is readily detected by distinctive and copious fecal pellets that can be found throughout the colony, on top of rocks or in sheltered crannies. There is a tendency to use certain latrine areas where multiple generations may "build" fecal piles that occupy several square feet. Presence is also evident by occasional tufts of fur lodged in plants or among rocks, by tola bushes stripped of bark, and by the presence of beaten pathways between outcrops within the larger home range of a colony.

Unlike either Chinchillas or Plains Viscachas, *Lagidium* are diurnal, up by sunrise, often spending the initial hours of the day perched on sentinel boulders where they dress their fur and soak up heat from the morning sun. In the evening, they typically feed until dark or shortly thereafter, before returning to their shared burrows. They are herbivorous, specializing on green leaves, flowers, fruits, stems, and bark rather than dried plant materials, but eat a wide variety of plants in the local environment, so the range of plants available will differ considerably from one geographic region to another. They do not cache green vegetation for future use, but actively forage daily. Individuals descend from their bounder field or rocky cliff colonies to feed on the green vegetation where pools of melt water accumulate.

Predators detected by Pearson (1948) in Peru included the Culpeo (*Lycalopex culpaeus*), Lesser Grison (*Galictis cuja*), Puma (*Puma concolor*), and Andean Mountain Cat

(*Leopardus jacobitus*) as well as various avian raptors. Napolitano et al. (2008) found the remains of *L. viscacia* in the feces of both the Pampas Cat (*Leopardus colocolo*) and Andean Mountain Cat (*L. jacobitus*) in northern Chile. Mountain Viscachas are agile, quadrupedal bounders, like rabbits, with propulsion provided by their long hindfeet and limbs. Their "reckless, carefree bounds from rock to rock, or ledge to ledge" (Pearson 1948:360) and long stride of up to six feet makes them difficult to capture. Individuals are very vigilant and vocalize warning with a series of very high-pitched whistles. Both sexes whistle, and all individuals within the group will respond to the calls of a single animal. Different whistles may be given for aerial and terrestrial potential predators.

In southern Peru, the reproductive season begins in late October and extends through the "summer" months, with pregnant females taken in October, November, and December as well as July and August, but not in September or most of October. Females breed after they reach approximately 1 kg in mass and thus may have their first estrous the year that they are born. Males reach breeding condition at the same mass, or at about seven months of age; they are capable of breeding at all seasons thereafter and for the remainder of their life. A large vaginal plug is formed in the female tract after copulation. Ovulation occurs near the time of copulation, is almost always from the right ovary, with implantation thus nearly always in the right uterine horn. Gestation is approximately three months with a single, precocial young born at a birth weight of about 225 g. A postpartum estrus and pregnancy is possible, and a single female may have two to three litters per year. Lactation is limited in temporal duration to about one month, as the young are born alert, fully furred, and with open eyes at birth, and are capable of feeding on green vegetation immediately. Pearson (1948) described the ontogenetic molt and color change with age; there is no season molting pattern in adults, so that molting is likely continuous. Life expectancy of both sexes is equal, maximally two to three years.

Mountain Viscachas overlap with grazing/browsing artiodactyls, such as vicuñas, guanacos, and huemuls as well as with a number of small rodents, predominantly the sigmodontine genera *Abrothrix*, *Auliscomys*, *Chinchillula*, *Phyllotis*, and *Punomys*. Individuals typically host roundworms and tapeworms of the genus *Cittotaenia*, and ectoparasites such as lice of the genera *Gyropus* and *Philandesia*. Their use of dust-baths and the bristle-like combs on their hind toes help dislodge troublesome "boarders."

Indigenous peoples of the Andes mixed viscacha hair with that of alpaca, vicuña, or other fibers to spin yarn for making exceedingly fine cloth, with viscacha hair providing color variety to the finished product. The use of viscacha fur or pelts for clothing, however, has been largely discontinued throughout their range. This species is, however, still used for food. Historically, however, Mountain Viscachas have not been persecuted by humans with the same fervor as have chinchillas, and as a result natural populations are still reasonably abundant throughout the long latitudinal range of the species. Current IUCN (2011) conservation status is in the "Least Concern" category.

Pearson (1957) recorded a colony at Naña, 2,200 ft [= 670 m] in the fog belt, or *lomas*, of the coastal desert in Lima department, Peru. These *lomas* are hilltops that support only clumps of fog-nourished *Tillandsia* (Bromeliaceae), where no vegetation occurs on the lower slopes but where irrigated fields are present on valley floors. The single individual Pearson observed "appeared to be similar to the Mountain viscachas of southern Peru but with more than average buffy color" (p. 5).

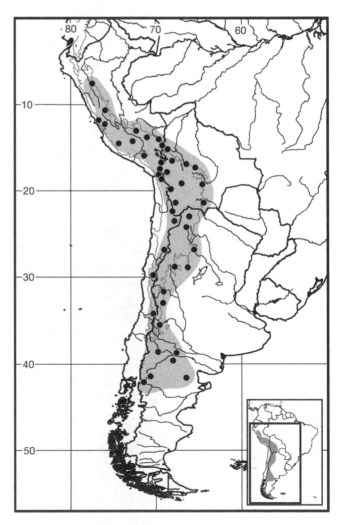

Map 401 Selected localities for *Lagidium viscacia* (●). Contour line = 2,000 m.

REMARKS: The specific epithet *viscacia* has been spelled with either one second *c* (*viscacia*: Tate 1935; Osgood 1943b; Cabrera 1961; Woods and Kilpatrick 2005) or two (*viscaccia*: Thomas 1907a; Ellerman 1940). Because the name stems from G. I. Molina (1782), who spelled it with a single second *c*, *viscacia* is the correct spelling (ICZN 1999:Art. 32.3). Molina described *Lepus viscacia* in 1782. As pointed out by Lahille (1906), there is little doubt that Molina had reference to a *Lagidium*, although Bennett (1835a) showed that Molina confused the habits of the lowland (*Lagostomus*) and mountain viscachas (*Lagidium*). Bennett (1835d) elaborated on his earlier (1835b) description of *Lagotis pallipes*, providing a beautiful color illustration of this taxon.

The karyotype of specimens referred to both *peruanum* Meyen and to *boxi* Thomas is $2n = 64$, FN = 126 (W. George and Weir 1974).

Lagidium wolffsohni (Thomas, 1907)
Wolffsohn's Mountain Viscacha
SYNONYMS:

Viscaccia Wolffsohni Thomas, 1907a:440; type locality "Sierra de los Baguales y de las Vizcachas, lat. 50°50′S., long. 72°20′ W., [Santa Cruz] on the boundary between Chili [= Chile] and Argentina."

Lagidium peruanum: Burmeister, 1879:253; part (*wolffsohni* as a synonym of *peruanum* Meyen).

[*Lagidium*] *wolffsohni*: Tate, 1935:363; name combination.

Lagidium viscacia wolffsohni: Osgood, 1943b:142; name combination.

L[*agidium*]. v[*iscacia*]. *wolffsohni*: Crespo, 1963:63; name combination.

DESCRIPTION: Large species with long fur strongly suffused with orange and very bushy tail. Head and body length of holotype 470 mm, tail length 305 mm, hindfoot length 107 mm, and ear length 65 mm. Size appears even larger because of very long and rich fur, with wool hairs exceeding 35 mm in length. General color of head and body clay gray, but brownish slate wool hairs give overall darker tone to fur; only a hint of dark dorsal stripe characteristic of other species; lower cheeks, throat, chest, and belly color richer and redder, almost tawny; distinctive white spots present on each axilla and on each side of inguinal region; ears comparatively short, thickly and closely haired, outside surface black but inner surface clothed with whitish hairs, and with marked line of creamy-tipped hairs along base; forelimbs tawny or yellowish to tips of toes; hind limbs duller, more brownish clay colored. Hindfeet large and heavy. Tail stated as "finer" than any other taxon, immensely bushy with hairs forming dorsal crest over 150 mm long; color of upper, crested side mixed black and buff or ochraceous buff; underside of tail black, finely grizzled with glossy ochraceous buff. Skull comparatively large and heavy, with expanded nasals, short incisive foramina, and less swollen bullae than other forms.

DISTRIBUTION: *Lagidium wolffsohni* is known from the Sierra de los Baguales on both sides of the border between Argentina (Santa Cruz province; Barquez, Díaz, and Ojeda 2006) and Chile (Magallanes y Antártica Chilena region; Osgood 1943b; Mann 1978) and from a single locality in Aysén province, Chile (Ledesma et al. 2009).

SELECTED LOCALITIES (Map 402): ARGENTINA: Santa Cruz, Sierra de los Baguales y de las Vizcachas (type locality of *Viscaccia Wolffsohni* Thomas). CHILE: Aysén, Estancia Chacabuco, Cochrane (Ledesma et al. 2009); Magallanes y Antártica Chilena, Sierra Baguales, Última Esperanza (Spotorno, Valladares, Marín et al. 2004).

SUBSPECIES: *Lagidium wolffsohni* is monotypic.

NATURAL HISTORY: Nothing is known about the natural history or ecology of this species, although its conservation status in Argentina is considered "En Peligro," in part because of its extremely restricted range (Barquez, Díaz, and Ojeda 2006).

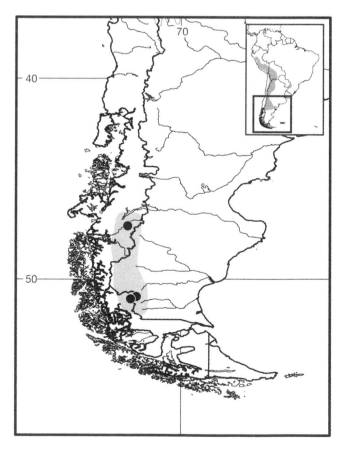

Map 402 Selected localities for *Lagidium wolffsohni* (●). Contour line = 2,000 m.

REMARKS: Thomas (1907a:441) stated that *L. wolffsohni* "is readily distinguishable from all other members of the genus by its large size, rich colour, long fur, immensely bushy tail, and short black ears." Alternatively, Osgood (1943b:142), in listing this taxon as a valid subspecies of *L. viscacia* noted that "its published measurements indicate the size to be about the same as in *moreni* and *boxi*, and the description of the skull offers no unusual characters, but the color is distinctive." Mann (1978:244), however, followed Thomas and differentiated *wolffsohni* at the species level from *viscacia* by virtue of its larger size and short, black ears.

The karyotype of *L. wolffsohni* is unknown. As noted above, the specific status of this taxon is questionable, especially given the molecular evidence of its nested phyletic position within other samples of *L. viscacia* (Ledesma et al. 2009). Osgood (1943b) and Crespo (1963) regarded *wolffsohni* as a subspecies of *L. viscacia*.

Subfamily Lagostominae Wiegmann, 1835

This subfamily comprises the single Recent genus *Lagostomus*, with two described species. One of these, *L. crassus*, is known only from a single skull found buried in the sand at a locality in southern Peru (Thomas 1910a). The second, *L. maximus* or the Plains Viscacha, occurs from the dry lowlands of south-central Bolivia, western Paraguay, south through chaco, pampas, and steppe habitats of central Argentina as far south as the provinces of Neuquén and Río Negro. McKenna and Bell (1997) regarded *Pliolagostomus* Ameghino, 1887 and *Prolagostomus* Ameghino, 1887, both from the Early to Middle Miocene, to be extinct representatives of the subfamily. *Lagostomopsis*, described by Kraglievich (1926) as a subgenus of *Lagostomus*, is known from the Pliocene and Pleistocene of Argentina.

REMARKS: McKenna and Bell (1997:210) dated the subfamily name Lagostominae to Wiegmann, 1832 (Lagostomi, used as a family name), which is in error. Wiegmann's paper was published in 1835, in the inaugural year of his new journal *Archiv für Naturgeschichte* (see Bauer and Adler 2001).

Genus *Lagostomus* Brookes, 1829

Lagostomus differs chiefly from members of the Chinchillinae externally by much larger size, with total length averaging 641 mm for females (range: 530 to 735 mm) and 753 mm for males (range: 685 to 820); tail length 160 mm (range: 135 to 173 mm) and 181 mm (range: 154 to 205 mm), respectively; and mass 4.0 kg (range: 3.5 to 5.0) and 6.3 (range: 5.0 to 8.8), respectively. The dorsal and lateral coloration is gray-brown with bold black and white

markings on the face, and fur that is less soft to the touch. A patch of short and stiff bristles is positioned below the eyes. The tail is short, proportionately much more so than in other genera (about 33% of head and body length). The forefeet possess four digits, with digit I missing entirely. The hindfeet are notably long and narrow, with only three digits due to complete loss of digits I and V. Each digit possesses only a very small terminal toe pad and stout and distinctly thick and sharp claws. Females have two pairs of mammae located laterally on the thorax.

The skull is flat, with broad frontals that bear well-marked and triangular-shaped postorbital processes, and heavy parasagittal ridging for muscle attachment. In comparison with the other genera in the family, the rostrum appears robust, but the nasals remain relatively long and narrow. The jugal is deep and thick, extending upward to meet an expanded lacrimal, and has a moderately developed postorbital process. The floor of the infraorbital foramen has a deep groove bordered by a lateral flange for nerve passage. The tympanic portion of each bulla is expanded ventrally, but not to the degree in the other genera. The mastoid portion is not inflated and, while visible on the occiput when viewed from behind, is not visible on the dorsal surface of the cranium. Paroccipital processes are lengthened, vertical, and both extend well below and are free from contact with the bullae. The incisive foramina are short, occupying less than half the length of the diastema. The palate is strongly constricted anteriorly, with the distance between the upper premolars only about 20% of that between the last molars.

The mandible is thick and stout, with a reduced angular process that flares laterally more so than in the other genera, a low but conspicuous coronoid process, and a well-developed ridge lateral to the condyle for the attachment of the posterior masseter muscle.

The upper incisors are distinctly proodont, with weakly pigmented enamel (if at all), and their surfaces covered with faint longitudinal grooves. Each maxillary cheek tooth has only two straight and rather thick laminae, except M3 that has three, with the last plate essentially lacking a posteriorly directed heel. All mandibular teeth have only two laminae.

The male phallus is elongated and thin, and is unique among all other hystricognath genera, including *Chinchilla*, in lacking the intromittent sac (Pocock 1922).

Fossils of *Lagostomus* are known from the Late Miocene (Salicas Formation; Tauber 2005), Pliocene (Chapadmalal Formation; Kraglievich 1926; J. L. Prado et al. 1998), and late Pleistocene-Holocene (G. Gómez et al. 1999) in Argentina.

SYNONYMS:

Viscacia Rafinesque, 1815:56; *nomen nudum* (Tate 1935: 365).

Viscaccia Oken, 1816:835; part (a composite based on *Lepus chilensis* Oken, the "*le vizcache*" of Azara [1801], and *Mus laniger* G. I. Molina [see J. A. Allen 1902c:196]); unavailable name (ICZN 1956).

Dipus: Desmarest, 1817d:117; part (description of *maximus*); not *Dipus* Zimmermann, 1780.

Viscaccia Schinz, 1825a:429; type species *Viscaccia Americana* Schinz (= *Dipus maximus* Desmarest, 1817); a homonym of *Viscaccia* Oken.

Vizcacia Schinz, 1825b [or 1826; see Palmer 1897a]:117; variant spelling of *Viscaccia* Schinz.

Lagostomus Brookes, 1828:134; *nomen nudum*.

Lagostomus Brookes, 1829:96; type species *Lagostomus trichodactylus* Brookes (= *Dipus maximus* Desmarest), by monotypy.

Chinchilla C. H. Smith, 1842:309, plate 26; no type locality given; based on an engraving (facing p. 171) in Griffith et al. (1827) *The Animal Kingdom*, the English translation of G. Cuvier's *Le règne animal*.

Callomys d'Orbigny and Geoffroy St.-Hilaire, 1830:282; part (inclusion of *viscacia* [= *Dipus maximus* Desmarest]); a composite genus with no type species selected.

Viscacia Rengger, 1830:272, footnote; incorrect subsequent spelling of *Viscaccia* Schinz (a homonym of *Viscaccia* Oken, which is unavailable [ICZN 1956]).

Lagostomopsis Kraglievich, 1926:45; an extinct taxon proposed as a subgenus of *Lagostomus* Brookes.

REMARKS: Félix de Azara (1801) observed and described "*le vizcache*," an animal that could only have been the Plains Viscacha based on the geographic area of his travels (the La Plata river basin of present day Paraguay, Argentina, Brazil, and Uruguay). The nomenclatural history of the Plains Viscacha is complicated, but not to the same degree as that of either *Chinchilla* or *Lagidium* (see Palmer 1897a; J. A. Allen 1900b, 1905; Tate 1935). As with the latter two genera, much of this contorted history revolves around the common name "viscacha" (in its various spellings) and the formalization of that common name as *Viscacia*, *Vizcacia*, or *Viscaccia* by early nineteenth century authors, compounded by the often unclear application of these names to one or more of the three genera of chinchillids currently recognized. Fortunately, both the nonoverlapping geographic ranges of each of these genera and the specimen(s) or authority upon which each early name was based help unravel the historical confusion in the applicability of names to these three taxa.

Joshua Brookes (1828) first used the name *Lagostomus* in a paper read before the Zoological Club of the Linnaean Society on May 27 of that year, with a two-sentence synopsis published in the *Zoological Journal* in July 1828 (Sherborn 1927a). Here he proposed to designate *Lagostomus trichodactylus* for Desmarest's *Dipus maxi-mus* (= *Lagostomus maximus*). Brookes subsequently read a considerably more detailed paper before the Zoological Club on both June 3 and 17, 1828, although the issue of the *Transactions of the Linnaean Society* in which his paper was published did not appear until 1829. In this second paper, Brookes (1829) detailed characteristics of the skin and skeleton of his *Lagostomus*, again proposed *trichodactylus* for Desmarest's *maximus*, and documented similarities and differences of the specimen he had at his disposal with the descriptions given of a living specimen by F. Cuvier and M. de Blainville that had been summarized by Desmarest (1817d).

Brookes's (1829) *Lagostomus* thus became the name applied to the Plain's Viscacha for most of the nineteenth century (Bennett 1833, 1835c; Waterhouse 1848; Burmeister 1879; E.-L. Trouessart 1897). However, J. A. Allen (1902b:375; 1905:31) argued that Azara's "*le vizcache*" formed the basis of Oken's (1816) *Viscaccia*, because Oken included within his *Viscaccia* both *Lepus chilensis* Oken, which he wrote was based wholly on "*le vizcache*," and *Mus laniger* G. I. Molina, a name that is now associated with *Chinchilla*. Furthermore, because Oken diagnosed his *Viscaccia* by the possession of four toes on the fore and three toes on the hindfeet ("Zehen vorn 4, hinten 3"), *Mus laniger*, with four toes on both forefeet and hindfeet, was excluded as the type for his genus. Thus, J. A. Allen (1905) used *Viscaccia* Oken for the Plains Viscacha, with *Lagostomus* Brookes considered a junior synonym. The year previously E.-L. Trouessart (1904) employed *Viscacia* [*sic*] Schinz (1825a) for the Plains Viscacha. Oken's *Viscaccia* is unavailable, first because his name was initially ruled a synonym of Bennett's *Lagidium* (ICZN 1925:Opinion 90), then suppressed under the Plenary Powers of the International Commission in Opinion 110 (ICZN 1929; see also ICZN 1955b:Direction 24), and finally because his entire *Lehrbuch* (Vol. 3) was rejected for nomenclatural purposes (ICZN 1956:Opinion 417). A year before Oken, Rafinesque (1815:56) had used the name *Viscacia* in a list, but as he did not provide a description his name is a *nomen nudum* (Palmer 1897a). Both *Viscaccia* Schinz, 1825a (with type species *Viscaccia Americana* Schinz) and *Vizcacia* Schinz, 1825b (variant spelling of *Viscaccia* Schinz; year of publication uncertain, see Palmer 1897a:22; unfortunately, C. D. Sherborn, in his multivolume *Index Animalium*, made no reference to either *Vizcacia* Schinz or *pamparum* Schinz) have been historically considered homonyms of Oken's earlier *Viscaccia*, and remain so even after Oken's names were declared unavailable (as per ICZN 1929:Opinion 110). Thus, *Lagostomus* Brookes is the earliest valid name for the Plains Viscacha (see also Cabrera 1961:559). D'Orbigny and Geoffroy St.-Hilaire (1830) erected *Callomys*, a composite genus that probably included species in each of the three chinchillid genera recognized today, but certainly those of *Lagostomus* and *Chinchilla* (see Tate 1935:358, 360, and 365). Tate (1935:364–368) provided a detailed summary of this complicated history.

KEY TO THE SPECIES OF *LAGOSTOMUS*:

1. Distribution southern Peru; skull large (condylobasilar length = 122 mm) . . . . *Lagostomus crassus*
1.′. . . . . Distribution southern Bolivia, Paraguay, and Argentina; skull smaller (condylobasilar length <110 mm, on average). . . . .*Lagostomus maximus*

Lagostomus crassus Thomas, 1910
Peruvian Plains Viscacha

SYNONYM:

Lagostomus crassus Thomas, 1910a:246; type locality "Santa Ana, District of Cuzco, Peru."

DESCRIPTION: Larger in size and stouter in form than *L. maximus*, with stout and heavy skull larger in all di-

mensions than largest specimen of *L. maximus* in the British Museum collections (Thomas 1910a). Anterior frontal region particularly broad and convex upward; breadth across antorbital foramina greatly exceeds that of *L. maximus*; incisive foramina minute in size and palatal foramina large.

DISTRIBUTION: *Lagostomus crassus* is known only from the type locality.

SELECTED LOCALITIES (Map 403): PERU: Cusco; Santa Ana (type locality of *Lagostomus crassus* Thomas).

SUBSPECIES: *Lagostomus crassus* is monotypic.

NATURAL HISTORY: This species is known from only a single skull found by J. Kalinowski "buried in sand" (Thomas 1910a:245). Thomas stated further "no viscachas of this genus are known to live in Peru, and the animal is probably now extinct, but the skull is in no way fossilized, and indicates that these animals lived in Peru at a very recent date." No specimens of *Lagostomus* have been collected in the adjoining Altiplano regions of Bolivia (S. Anderson 1997). Given the lack of any additional specimens in the modern or archaeological record of such an obvious rodent in the Altiplano of Peru or Bolivia, we agree with Jackson et al. (1996) that the single skull of *L. crassus*, which is at the upper end of the size range of *L. maximus*, probably represents a Plains Viscacha that was taken to Peru.

Lagostomus maximus (Desmarest, 1817)
Plains Viscacha

SYNONYMS:

Lepus chilensis Oken, 1816:835; based on Azara's (1801) "*le vizcache*" (J. A. Allen 1902c:196); unavailable name (ICZN 1956).

Dipus maximus Desmarest, 1817d:117; no type locality given; based on a live animal observed and described in 1814 by M. de Blainville in England, stated to have originated in "Nouvelle-Hollande" [= Java, Indonesia].

V[iscaccia]. Americana Schinz, 1825a:429; no type locality designated.

Vizcacia pamparum Schinz, 1825b [circa 1825 or 1826; see Palmer 1897a]:244; no type locality designated.

Lagostomus trichodactylus Brookes, 1828:134; *nomen nudum.*

Lagostomus trichodactylus Brookes, 1829:102; no type locality designated.

Callomys viscacia: d'Orbigny and Geoffroy St.-Hilaire, 1830:291; = *Dipus maximus* Desmarest, not *Lepus viscacia* G. I. Molina; name combination.

Lagostomus viscacha: Meyen, 1833:584; name combination.

Lagotis criniger Lesson, 1842:105; *nomen nudum*; stated to be from "Plata" and thus likely referable to *Lagostomus* rather than *Lagotis* Bennett [= *Lagidium* Meyen].

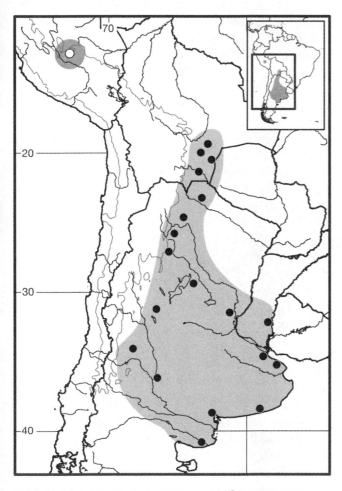

Map 403 Selected localities for *Lagostomus crassus* (○) and *Lagostomus maximus* (●). Contour line = 2,000 m.

Chinchilla Diana C. H. Smith, 1842:309; no type locality designated.

Lagostomus tricodactylus Waterhouse, 1848:448; incorrect subsequent spelling of *trichodactylus* Brookes.

Vizcacia maxima: Palmer, 1897a:21; name combination.

Viscacia maxima: Berg, 1900:221; name combination.

Viscaccia chilensis: J. A. Allen, 1902c:196; name combination.

Viscacia viscacia: E.-L. Trouessart, 1904:517; name combination.

Viscaccia chilensis: J. A. Allen, 1905: 32; name combination.

Lagostomus maximus: Thomas, 1910a:245; first use of current name combination.

Lagostomus maximus immollis Thomas, 1910a:245; type locality "Tapia, Tucuman. Alt. 700 m," Tucumán, Argentina.

Lagostomus maximus petilidens Hollister, 1914a:58; type locality "8 miles north of Carmen de Patagones, [Buenos Aires,] Argentina" (not "8 miles south of Carmen de Patagones" as given by Tate [1935:368] and Ellerman [1940:236]).

[*Lagostomus*] *maximus maximus*: Tate, 1935:368; name combination.

[*Lagostomus*] *maximus immollis*: Tate, 1935:368; name combination.

[*Lagostomus*] *maximus petilidens*: Tate, 1935:368; name combination.

Lagidium chilensis: Devincenzi, 1935:75; name combination.

Lagostomus maximus maximus: Ellerman, 1940:236; name combination.

Viscaccia americana: Osgood, 1943b:238; name combination.

DESCRIPTION: As for genus, except as distinguishable from *L. crassus*.

DISTRIBUTION: *Lagostomus maximus* is distributed in the pampas and adjoining semiarid Monte and Chaquenean regions of South America, from southeastern Bolivia, western Paraguay, and northern, eastern, and central Argentina. As an agricultural pest, the Plains Viscacha has been extirpated from a substantial portion of its native range, but it has also expanded this historic range in areas due to anthropogenic modification of natural habitats. Detailed data on current distribution are lacking (Jackson et al. 1996).

SELECTED LOCALITIES (Map 403): ARGENTINA: Buenos Aires, Bahia Blanca (KU 3023), Buenos Aires (MCZ 5040), 8 mi N of Carmen de Patagones (type locality of *Lagostomus maximus petilidens* Hollister), Punta Blanca (UWBM 73509), Necochea (USNM 172825); Entre Ríos, Arroyo Perucho Verne (Udrizar Sauthier, Abba et al. 2008); La Pampa, Estancia Los Ranqueles (MVZ 177626) Tello (USNM 172849); Mendoza, Ñacuñán (OMNH 15158); Salta, El Breal (R. A. Ojeda and Mares

1989), Las Lajitas (R. A. Ojeda and Mares 1989), Rearte Norte, Rosario de la Frontera (R. A. Ojeda and Mares 1989); Santa Fé, Esperanza (USNM 118480); Santiago del Estero, Buena Vista (OMNH 23723); Tucumán, Río Colorado (USNM 172822). BOLIVIA: Chuquisaca, 1 km W of Sargento Rodriguez [Paraguay] (S. Anderson 1997); Santa Cruz, Com. Guirapembi (S. Anderson 1997), Parapeti (S. Anderson 1997); Tarija, ca. 35 km by road SE of Villa Montes (UMMZ 155896).

SUBSPECIES: Three subspecies are recognized in the literature (Llanos and Crespo 1952; Jackson et al. 1996) based on coat color, skull morphology, and incisor width: *L. m. immollis* Thomas from Bolivia, Paraguay, and north-central Argentina; *L. m. maximus* Desmarest from central Argentina; and *L. m. petilidens* Hollister from southern Buenos Aires, La Pampa, and Río Negro provinces, Argentina. However, as noted by Jackson et al. (1996), the ranges of the three subspecies are uncertain, and, in the absence of an adequate review of morphological and/or molecular variation over this range, the validity of each remains to be determined.

NATURAL HISTORY: Jackson et al. (1996) reviewed aspects of the natural history, ecology, reproduction, and conservation of this species. This species lives in lowland habitats that include subtropical, humid grasslands in northeastern Argentina, dry thorn scrub in Bolivia, Paraguay, and north-central Argentina, and desert scrub in the southwestern part of its range. These are colonial animals, with one to three adult males, two to four times as many females, and immatures living in a communal burrow system called a "vizcachera" (Llanos and Crespo 1952). The burrow system is complex, with multiple entrances and branching underground tunnels. Sticks, bones, dry cow dung, and other materials may be placed around the entrances; males use these items for scent marking (Branch 1993b).

In the semiarid scrub of La Pampa in central Argentina, groups of Plains Viscachas live in vizcacheras that range in number of burrows from 18 to 93, with satellite burrows scattered among them (Branch, Villarreal, Sosa et al. 1994). These satellites are used for escaping high-risk situations during the nonbreeding season (winter months) but serve as residences and display sites for adult males during the breeding season (spring to fall months). All viscachas from a single vizcachera share a common home range, which averages 1.29 ha (Branch 1993a), with minimal overlap between the home ranges of adjacent vizcacheras. Home range boundaries are not defended, but all members of a group will defend the vizcachera from intruders of either sex or age class. Mutual grooming by members of a vizcachera is common, and all individuals used a communal dust bath. All individuals of the vizcachera also have access to all burrows in the system (Branch 1993b). Abandoned burrows serve as important refuges for burrowing owls (*Athene cunicularia*; Machicote et al. 2004) and as breeding sites for various reptiles and amphibian species (J. M. Gallardo 1970).

Plains Viscachas are nocturnal and active year-round, emerging from their burrows shortly before dusk (Branch 1993c). In winter, adult males forage with both females and immatures, but during the remainder of the year they forage alone. Groups of up to 30 individuals may be observed foraging together (Llanos and Crespo 1952; Branch 1993b), and individuals may forage at distances considerably beyond the boundaries of the home range.

Breeding is seasonal in the wild, with most females reproductively active in the autumn and young born in the spring (Llanos and Crespo 1952; Jackson 1989; Weir 1971c; Branch et al. 1993). The estrous cycle is forty days long. Females hyperovulate, producing 200 to 800 ova per estrous event (Jensen et al. 2006, 2008); following a delayed implantation, up to five blastocysts implant in each uterine horn (Weir 1971c,d). The gestation day period is long: 154 days, including the delay of implantation of 18 days. Young are precocial at birth. Lactation lasts two to three months (Jackson 1989). Average litter size in natural populations is about 1.9 young (Llanos and Crespo 1952; Jackson 1989). Females reach sexual maturity earlier than do males, with a mean time to first conception of 214 days and first birth of 368 days (Jackson 1989). Males reach sexual maturity at 12 to 16 months (Llanos and Crespo 1952; Jackson 1989). The testicular cycle in males is highly seasonal, with peak conception rates and male fertility in March and April, at which time 91% of males in the population were fertile.

The sex ratio at birth in the wild is approximately 1:1 (Jackson 1990a), or slightly skewed in favor of females (Branch et al. 1993). Females are recruited into their natal vizcacheras and may breed at eight months of age. Males, on the other hand, typically disperse although they may continue to reside in their natal vizcachera when environmental conditions are poor. There is usually a complete turnover of adult males each year, with resident males replaced by immigrants from other vizcacheras. Individuals may live seven to eight years in the wild (Llanos and Crespo 1952). Branch, Villarreal, and Fowler (1994) summarized factors affecting population dynamics.

Plains Viscachas are herbivorous and coprophagous (Jackson 1985a; Kufner et al. 1992; Puig et al. 1998; Kufner and Monge 1998), foraging on a wide range of plant species. In the grasslands of central Argentina, not surprisingly the majority of forage comprises species of grasses, but in dry scrub habitats, viscachas eat the leaves of a variety of forbs and shrubs as well as grasses (Branch, Villarreal, Sbriller, and Sosa 1994). By selective grazing, viscachas al-

ter the species composition, cover, and vegetative structure around their burrow systems (Branch, Villarreal et al. 1996; Branch et al. 1999; Arias et al. 2003; Villareal et al. 2008).

Predators comprise a variety of mammalian carnivores, including pumas (*Puma concolor*), Geoffroy's Cat (*Leopardus geoffroyi*), Pampas Fox (*Lycalopex gymnocercus*), Crab-eating Fox (*Cerdocyon thous*), and Lesser Grison (*Galictis cuja*; Branch 1995; Branch, Pessino, and Villarreal 1996; Jackson et al. 1996; Pessino et al. 2001). The boa (*Boa constrictor*) may have been an important predator in the northern part of the range in Argentina in earlier periods (Llanos and Crespo 1952). Individuals give a loud, two-syllable antipredator call (Branch 1993a). The call is given by adult males most frequently and is similar to the call these individual use in response to male intruders.

A variety of nematode and cestode helminthes (Sutton and Durette-Desset 1995; G. W. Foster et al. 2002) and protozoans of the family Eimeriidae parasitize Plains Viscachas (Couch et al. 2001). Ectoparasites include ticks of the genus *Amblyomma* (Guglielmona et al. 1990).

This species is of considerable economic interest in Argentina both because of its status as an agricultural pest and commercial value of its meat and pelt. Substantial numbers of skins are exported annually.

REMARKS: Both *Dipus maximus* Desmarest and *Lagostomus trichodactylus* Brookes are based on the same animal. Desmarest's name came directly from a manuscript description, provided to him by M. de Blainville, of a live animal seen in England (Waterhouse 1848:213). And, as Brookes had detailed (1829:95), his own description was based on an individual obtained "from Mr. Cross, in whose Vivarium at Exeter Change it had been seen while living, and especially noticed, both by M. de Blainville and by M. F. Cuvier. Each of these distinguished naturalists has described its general characters and habits."

The karyotype is $2n = 56$, FN $= 110$ (Hsu and Benirschke 1971; Wurster et al. 1971; Vidal-Rioja et al. 1973b; W. George and Weir 1974).

Family Dinomyidae Alston, 1876
James L. Patton

The single living species is among the largest of extant caviomorph rodents, with a head and body length up to 800 mm and a mass reaching 15 kg. The ears are short and rounded; the upper lip has a deep cleft; the vibrissae are long; the tail is short, thick, and well furred; the limbs are short and stout; the feet are plantigrade, relatively short and broad, with four digits, each with a long claw. The palmar and plantar surfaces are naked with pads containing the remaining skeletal elements of the pollex and hallux. The animal walks with only the heel of the hindfoot raised above the ground, which results in a waddling gait (Goeldi 1904; Pocock 1926). Webbing extends from the plantar pad to about half way to the tips of the small digital pads (Grand and Eisenberg 1982; Pocock 1926). Overall color is black or dark brown (Lönnberg [1921] suggested that the sexes are dichromatic, with adult males black and adult females brown), typically with two discontinuous white stripes down the back beginning from the shoulders and two or more shorter rows of white spots down each side. The under parts are paler than the dorsum and lack any marks. The pelage itself is coarse, with the hairs of varied length. Females have two lateral-thoracic and two lateral-abdominal pairs of mammae (Weir 1974b).

The skull is massive and blocky, but only slightly ridged, and nearly flat in dorsal profile. The zygoma is deep and heavy and lacks a jugal fossa; the jugal does not touch the lacrimal. The lacrimal canal does not open on the side of the rostrum. The infraorbital foramen is large but does not exhibit a ventral groove for nerve passage. The lacrimal bone lacks a descending flange, and there are no lateral processes of the supraoccipital. The paroccipital processes are short and vertically directed; the bullae are of medium size. The angular process of the dentary is strongly deflected, and the coronoid process is vestigial.

The upper incisors are broad and thick, their width greater than that of the cheek teeth. The cheek teeth are tetralophodont, have flat occlusal surfaces and high crowns, are rootless, and grow continuously throughout life (or euhypsodont, following the terminology of Mones [1968]). The two anterior laminae of the maxillary teeth are isolated and the posterior two laminae are united; the mandibular teeth exhibit a reversed pattern, with the anterior two laminae connected and the posterior two isolated. Deciduous premolars are replaced by permanent teeth, as in most mammals but not all hystrocimorph rodents. The maxillary toothrows converge strongly anteriorly.

The second and third cervical vertebrae are fused (Ray 1958). Mones (1997) provided a thorough description of the postcranial skeleton of *Dinomys*.

A monotypic genus, *Dinomys*, is present in the Recent fauna of South America, although the family has a rich and deep historical record in the Cenozoic (McKenna and Bell 1997; Woods 1984a). Either living populations of *D. branickii* or fossil taxa are known from every country except Chile, Paraguay, and the Guianas. Earliest fossils are recorded from the Deseadan Oligocene of Bolivia (B. Patterson and Wood 1982). Mones (1981), in his revision of the group, recognized four subfamilies, three of which are now extinct. The Dinomyinae is composed of two tribes, one now extinct and the Dinomyini containing the single Recent genus and species. The two largest known rodents (*Phoberomys* from the Late Miocene of Venezuela, estimated to be 700 kg in mass [Sánchez-Villagra et al. 2003; Horovitz et al. 2006], and *Josephoartigasia* from Plio-Pleistocene of Uruguay, estimated at nearly 1000 kg [Mones 2007; Rinderknecht and Blanco 2008]) are members of this family.

At one time or another the Dinomyidae have been placed within each of the three caviomorph superfamilies recognized herein: the Erethizontoidea (Grand and Eisenberg 1982), the Cavioidea (Simpson 1945; Fields 1957; Woods 1982; B. Patterson and Wood 1982; McKenna and Bell 1997), and Octodontoidea (Landry 1957), or even within a superfamily (Dinomyoidea) of its own (see White and Alberico 1992). Tate (1935) went so far as to include *Dinomys* as a genus within the Caviidae. The limited molecular sequence data available (Spotorno, Valladares, Marín et al. 2004; Opazo 2005; Rowe et al. 2010; Upham and Patterson 2012) clearly support *Dinomys* (and thus the Dinomyidae) as the sister to a clade containing the chinchillid genera *Chinchilla*, *Lagidium*, and *Lagostomus* and well separated from erethizontids, caviids, or octodontids. Alternatively, cladistics analyses of craniodental, chromosomal, and reproductive and gestation patterns of caviomorphs supported a sister-group relationship between *Dinomys* and the prehensile-tailed porcupine *Coendou* (Lidia Nasef 2010).

REMARKS: Simpson (1945), Starrett (1967), and Woods (1984a), among others, attributed the family name Dinomyidae to Alston (1876a) whereas S. Anderson (1997) and Woods (1993) attributed authorship to Troschel (1874). However, Mones (1995) clearly documented that the name properly stems from Peters (1873a:552; see also Peters 1873b:233), with Alston's name only an emendation and Troschel's an incorrect emendation. McKenna and Bell (1997) and Woods and Kilpatrick (2005) followed Mones's position.

Genus *Dinomys* Peters, 1873

As the single Recent genus in a monotypic family, *Dinomys* is characterized by the same set of features that diagnose the family.

SYNONYM:

Dinomys Peters, 1873a:551; type species *Dinomys branickii* Peters, 1873, by original designation.

Dinomys branickii Peters, 1873
Pacarana
SYNONYMS:

Dinomys branickii Peters, 1873a:551; type locality "der Colonia Amable Maria, in der Montaña de Vitoc, in den Hochgebrigen Perus erlegt," Junín, Peru.

Dinomys pacarana Miranda-Ribeiro, 1919:13; type locality "procedente do Amazonas," Brazil, "probably came from Río Purús region" (Sanborn 1931a:151).

Dinomys branickii occidentalis Lönnberg, 1921:49; type locality "road to Gualea, about 6,000 feet; . . . llambo near Gualea, about 5,000 feet," ca. 1,500 to 1,800 m, Pichincha, Ecuador (White and Alberico 1992).

Dinomys gigas Anthony, 1921a:6; type locality "La Candela, Huila, Colombia, altitude, 6,500 ft."

DESCRIPTION: As for the family.

DISTRIBUTION: *Dinomys branickii* is known from the Mérida Andes of northwestern Venezuela, all three Andean cordilleras in Colombia, south along both Andean slopes of Ecuador, the eastern Andean slope and western Amazon Basin of Peru and Bolivia, and western Brazil in the upper Purus and Juruá drainages. Gottdenker et al. (2001) mapped the known localities in Bolivia. The elevational range is approximately 250 to 3,200 m, spanning the western margins of lowland Amazonian rainforest to upper montane tropical forest on the Andean slopes.

SELECTED LOCALITIES (Map 404): BOLIVIA: Cochabamba, El Palmar (S. Anderson 1997); La Paz, Yolocito (S. Anderson 1997); Santa Cruz, San Rafael de Amboró (Rumiz and Eulert 1996). BRAZIL: Acre, Iquiri [= Rio Ituxi, near Rio Branco; Paynter and Traylor 1991] (C. O. da C. Vieira 1952), Paraná do Natal, Rio Juruá (C. O. da C. Vieira 1955), Parque Nacional da Serra do Divisor (Calouro 1999). COLOMBIA: Antioquia, Sierra Santa Elena (Nicéforo-María 1923), 9 km S of Valdivia (FMNH 69593); Boyacá, Páramo de Toquilla (White and Alberico 1992); Caldas, Bosque de Florencia (Castaño et al. 2003); Cauca, Inza (AMNH 149283); Huila, La Candela (type locality of *Dinomys gigas* Anthony). ECUADOR: Pichincha, Gualea (type locality of *Dinomys branickii occidentalis* Lönnberg), Santo Domingo de los Colorados (Lönnberg 1921). PERU: Amazonas, Kagka, Río Kagka (MVZ 153246); Cusco, Consuelo, 15.9 km SW of Pilcopata (Solari et al. 2006), Marcapata (FMNH 57186); Huánuco, Hacienda Buena Vista, Río Chinchao (FMNH 34408); Junín, Colonia Amable Maria (type locality of *Dinomys branickii* Peters); Loreto, Iquitos (AMNH 98576); Madre de Dios, Cocha Cashu (Pacheco et al. 1993); Pasco, Pozuzo (FMNH 34702); Puno, Sandia (FMNH 78430); Ucayali, Balta, Río Curanja (LSUMZ 16768). VENEZUELA: Táchira, El Caimito (Boher et al. 1988).

SUBSPECIES: In the absence of any review of geographic trends in any character set, it is premature to judge the taxonomic value of the names now available. I thus treat *Dinomys branickii* as monotypic.

NATURAL HISTORY: White and Alberico (1992) reviewed the ecology, behavior, and reproductive biology of *Dinomys branickii*. Most of what is known comes from captive animals (Mohr 1937; Collins and Eisenberg 1972; Merritt 1984; Iwata 1989), with few studies available on natural populations. Their calm, gentle nature in captivity (Goeldi 1904; Edmund Heller, in Sanborn 1931a) belies their generic name *Dinomys*, which means "terror mouse."

Alho (1982:Table 1) gave the preferred habitat as seasonally flooded tropical evergreen rainforest (or *várzea*), but this habitat occurs only limitedly along the Amazonian margins of the species range. Nearly every specimen obtained to date has come from upland rainforest, typically on the middle to lower slopes of the Andes or outlier ridges (Grimwood 1969). These animals are herbivorous, eating palm and other fruits, tender stems, and leaves. They apparently nest in rock crevices, hollow logs, or self-dug burrows, either as solitary individuals or in small groups ranging from two to five individuals (Boher and Marín 1988). The most extensive field study on pacaranas is that of Saavedra-Rodríguez et al. (2012), who examined habitat and space use in the central Colombian Andes. Here, animals occupied patches of primary and secondary forest as well as disturbed areas near streams, but where tree cover was always >20%. Only the largest patches were occupied. Individuals lived in groups of four to five composed of two adults (presumably a male and female) and two to three young. Each group occupied a den in deep caves found on sloping rocky outcrops, and had an average home range of 2.45 ha around the den. Population density varied between 5.5 and 9.9 groups per km^2, with those estimates based on sign from footprints and group latrines (see also Osbahr 2010). Animals fed on a wide variety of plants, mostly consuming leaves and stems but also fruits, nuts, and roots. These authors concluded that while pacaranas can survive in forest fragments, the main factor limiting distribution and abundance was the availability of adequate dens. Individuals are parasitized by ascarid and nematod worms (Merritt 1984) as well as cestodes and protozoans (Boher and Marín 1988).

Osbahr et al. (2009) and Osbahr and Azumendi (2009) examined the relationship between limb structure and movement and the functional ecology of habitat use, specifically comparing the slow walking cursorial gait characteristic of this species with the more fossorial activity of the sympatric mountain paca *Cuniculus taczanowskii*.

REMARKS: This rare, large, and extremely unique rodent was completely unknown to science until Peters's 1873 brief description, and few specimens have been collected since. Goeldi (1904) gave the first description of living animals, and Preller (1907) provided the first complete description of the skull, based on Goeldi's specimens. Miranda-Ribeiro (1919), Anthony (1921a), and Lönnberg (1921) in sequence described three additional taxa based on differences in dorsal coloration (black versus brown) and stripe-spot patterns. Nicéforo-María (1923) described a specimen from the Sierra Santa Elena, near Medellín, Colombia, which he thought might represent a distinct species based on its yellow instead of white lines and spots. Sanborn (1931a) reviewed available specimens and concluded that all named forms were only geographic or age variants of the single species *D. branickii*. All subsequent authors (Cabrera 1961; Ellerman 1940; Tate 1935; Woods 1993; Woods and Kilpatrick 2005) have followed this taxonomy.

Verified records of this species in Brazil are limited to the headwaters of the Rio Juruá and Rio Purus in Acre state (Miranda-Ribeiro 1919; C. O. da C. Vieira 1952, 1955; Whitney et al. 1996 [cited in Vriesendorp et al. 2006]; Calouro 1999), although the precise locality is unknown for several of these. The locality in Amazonas state cited by C. O. da C. Vieira (1955; Paraná do Natal, Rio Juruá) is most likely in Acre, as the Rio Paraná do Natal is in the municipality of Porto Walter, with the community of Natal approximately 7 km up river from that town. The locality of a single specimen of *D. branickii* in the mammal collection of the Funcação Universidade de Rondônia (UFROM 038)

is unknown, but certainly from Brazil (Y. L. R. Leite, pers. comm.). As that collection primarily houses voucher specimens from the hydroelectric plants at Santo Antônio and Jirau along the Rio Madeira, this species likely extends into the extreme northern part of Rondônia state. Eisenberg and Redford (1999:map 16.151) plotted two Brazilian localities, one in the mid-Rio Purús (Amazonas) and the second in the mid-Rio Madeira (Rondônia), but did not identify vouchers or provide references. Alho (1982:Table 2) suggested that the range extended to Pará state in eastern Amazonia, likely an error based on Goeldi's (1904) report.

Superfamily Erethizontoidea Bonaparte, 1845

Family Erethizontidae Bonaparte, 1845

Robert S. Voss

Members of the family Erethizontidae, commonly known as New World porcupines, comprise three genera in two subfamilies that collectively range from boreal North America (Alaska and northern Canada) to subtropical South America (Uruguay and northern Argentina). Whereas the subfamily Chaetomyinae contains a single species (*Chaetomys subspinosus*) that is endemic to the Atlantic Forest of southeastern Brazil, the subfamily Erethizontinae contains the North American porcupine (*Erethizon dorsatum*) and at least 13 Neotropical species, all of which are here referred to the genus *Coendou* (including *Echinoprocta* and *Sphiggurus*; Voss 2011, Voss et al. 2013). Although several fossil genera have been referred to the family Erethizontidae (McKenna and Bell 1997), few exhibit diagnostic familial synapomorphies, so their relationships to Recent taxa are correspondingly ambiguous.

Insofar as known, Neotropical erethizontids are nocturnal, arboreal, and herbivorous, consuming leaves, bark, fruit, and immature seeds (Gaumer 1917; Goldman 1920; Enders 1935; Montgomery and Lubin 1978; Charles-Dominique et al. 1981; Janzen 1983; Armond and Chagas 1990). At least two species routinely den in hollow trees or tree cavities, which accumulate piles of feces and have a strong odor (Miles et al. 1981; Janzen, 1983); other recorded diurnal retreats are in vine tangles, tree forks, and palm crowns. Most reports suggest that erethizontids normally give birth to a single largeprecocial offspring after a long gestation (Shadle 1948; Roberts et al. 1985; Roze 1989; Armond and Chagas 1990; Pasamanni 2009; but see Gaumer, 1917). Evidence of predation on Neotropical porcupines is scarce, but they are known to be eaten, at least occasionally, by harpy eagles (Izor 1985; Piana 2007), ocelots (Emmons 1987b; R. S. Moreno et al. 2006;

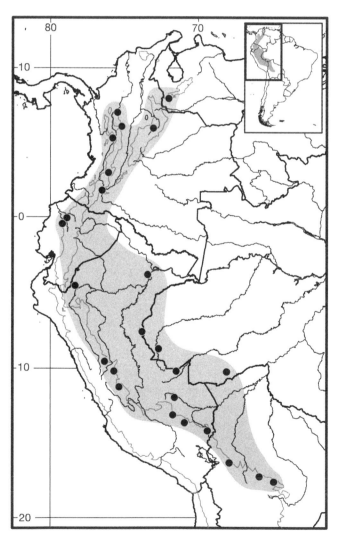

Map 404 Selected localities for *Dinomys branickii* (●). Contour line = 2,000 m.

R. C. Bianchi and Mendes 2007), pumas (Chinchilla 1997; Novack et al. 2005; R. J. Foster et al. 2010), and large snakes (M. R. Duarte 2003). Although the ectoparasites of erethizontids have not been reviewed in the literature, chewing lice of the trichodectid genus *Eutrichophilus* and the ixodid tick *Amblyomma longirostre* seem to be consistently associated with these rodents (Timm and Price 1994; Nava et al. 2010). South American erethizontids inhabit a wide variety of forested landscapes from sea level to about 3,500 m, including lowland and montane rain forests, deciduous (dry) forests, and gallery forests surrounded by savannas.

Erethizontids are relatively large rodents, but they nevertheless exhibit substantial taxonomic variation in size. The adult weight of the smallest known species (*Coendou roosmalenorum*) is probably in the range of 500–1,000 g (Voss and da Silva 2001), whereas the largest living species (the North American porcupine, *Erethizon dorsatum*) is said to exceed 10 kg in some populations (Roze 1989). In general aspect, porcupines are stoutly built, with short limbs, blunt heads, small eyes and pinnae, spiny pelage, and prehensile feet. In the absence of any adequate published account of erethizontid comparative morphology, the following paragraphs provide a brief summary of essential characters and descriptive terminology.

Porcupine vibrissae include most of the usual mammalian complement of tactile hairs (Pocock 1914; J. C. Brown 1971; J. C. Brown and Yalden 1973) and others of more restricted taxonomic occurrence. All porcupines have mystacial, superciliary, and genal vibrissae on the face, and a small tuft of short submental vibrissae on the chin, but interramal vibrissae are apparently absent (Pocock 1922). The mystacial hairs are the most abundant and obvious of the facial vibrissae; in *Erethizon* these are short, scarcely reaching the base of the ear when laid back against the face, but the mystacial vibrissae of *Chaetomys* and *Coendou* are long and extend well behind the ear. Postcranial vibrissae are apparently absent in *Erethizon*, but are prominent in the other two genera. In *Chaetomys* and *Coendou*, the forelimb is provided with vibrissae between the elbow and wrist, and vibrissae also occur on the hindlimb from the upper leg to the ankle. Bilaterally paired vibrissae distributed along the ventral surface of the thorax and abdomen in *Chaetomys* and *Coendou* appear to be unique among hystricognath rodents, although similar tactile hairs occur in tree squirrels (Hyvärinen et al. 1977).

Soft fur that is long enough to completely or partially conceal the quills is always part of the pelage of juvenile porcupines. A sparse covering of short hairs hidden among the bases of the quills is invariably retained in species that appear to lack any soft pelage as adults (e.g., *Coendou pre-*

hensilis). Long or short, most porcupine fur is composed of wavy wool hairs, but the emergent fur of some species also contains distinctively stiffer and coarser guard hairs. Unlike wool hairs, guard hairs are straight (not wavy), but they are flexible enough to be tied in loose knots without kinking or breaking (unlike spines or bristles). In *Erethizon*, wool hairs are shed seasonally en masse, whereas guard hairs are lost and replaced individually throughout the year (Po-Chedley and Shadle 1955). Unfortunately, molting phenomena have not been studied in any Neotropical erethizontid species.

Erethizontid quills are stiff spines that are normally round in cross-section; cut transversely and viewed under low magnification, quills can be seen to consist of a thin but tough outer cortex surrounding a spongy pith (Chernova 2002:Fig. 1a); they cannot be bent double without kinking. Two types of quills are usefully distinguished. Defensive quills are effective weapons capable of inflicting painful, sometimes crippling and even fatal wounds; easily detached from the owner's skin, they have sharp, hard tips with microscopic barbs (Vincent and Owers 1986:plate VIIc; Chernova 2002:Fig. 1b). Bristle-quills, by contrast, are nondefensive spines; firmly rooted in the skin, they have soft, flexible tips (Voss and da Silva 2001:Fig. 1C). Both *Erethizon* and *Coendou* have ordinary (defensive) quills, but bristle-quills are found only in a few species of the latter genus (comprising the "*Coendou vestitus* group" of Voss and da Silva 2001). Erethizontid quills come in two basic color patterns: bicolored quills are whitish or yellowish basally and blackish distally, whereas tricolored quills have whitish or yellowish bases, blackish middle bands, and paler (whitish, yellowish, reddish, or pale-brown) tips. The black-and-white (or black-and-yellow) motif of porcupine quills is an often-cited example of aposematic coloration, which is also associated with dangerous spines in other animal groups (Inbar and Lev-Yadun 2005; Speed and Ruxton 2005; Caro 2009).

Most of the dorsal body pelage of *Chaetomys* consists of another distinctive pelage structure that could be called broomstraw-spines. Firmly rooted in the skin (like bristle-quills), they are distinctively kinked or wavy, somewhat resemble broomstraws or phocid vibrissae (Voss and Angermann 1997:Fig. 8). Nothing like these unusual hairs is known from the pelage of any other rodent.

The underside of the basal part of the tail in all erethizontids is densely covered with stiff bristles that point toward the tail tip. These distinctive hairs are round in cross-section and have no hollow or spongy pith; they lack barbed points and are very difficult to detach from the skin. In some taxa, caudal bristles are restricted to the underside of the tail, but in others they extend onto the lateral and dorsal surfaces as well. These bristles probably function to brace the animal from slipping backward on inclined surfaces when the tail is pressed against the substrate.

The ventral pelage of some species of *Coendou* includes another distinctive hair type in the form of soft spines that are flattened in cross-section and are much thicker, stiffer, and straighter than the wool hairs with which they are usually mixed. Although spinous in form, they have soft, flexible tips. They can be bent double without kinking or breaking in some species, but not in others.

Erethizontids are distinctive in several other external features. Unlike most rodents, they do not have a naked glandular rhinarium surrounding the nostrils, and they also lack a naked philtrum dividing the upper lip into left and right halves (see J. C. Brown 1971, for facial anatomical terminology). Instead, the entire muzzle, including the nostril margins, is densely covered with short, velvety hairs, and the upper lip is undivided.

The small but distinct pinna of *Erethizon* (illustrated by Pocock 1922:Fig. 6C) is densely furred and has an evenly convex free margin; the simple fleshy structure of the external ear in this genus is unremarkable, broadly resembling the morphology seen in other hystricognaths (op. cit.). By contrast, the pinna in *Chaetomys* and *Coendou* is reduced to an angular ridge of almost naked skin that can apparently be closed by folds of the internal cartilaginous structure to protect the external auditory canal (Pocock 1922:Fig. 6D); the antitragus is much enlarged and bears a small but distinct tuft of soft spines. The essential characters of the pinna can usually be determined from properly prepared dry skins.

Erethizontids are plantigrade climbers whose feet are conspicuously modified for prehension. The manus and pes of all species have naked soles densely covered with small undifferentiated tubercles, none of which can be homologized with the large plantar pads (thenar, hypothenar, and interdigitals; J. C. Brown and Yalden 1973) that are present in most rodents and in all other nonaquatic plantigrade hystricognaths (Pocock 1922). All erethizontids likewise have strong, curved claws that can be powerfully closed over the plantar surface. Unlike most rodents, in which the pollex is small but distinct and provided with a nail, no external trace of a thumb is normally present in any erethizontid. In other respects, however, the external morphology of the feet differs among erethizontid genera.

The manus of *Erethizon* is about as broad across the metacarpal-phalangeal articulations as it is at the wrist, and it has short subequal digits; the stout claws of this genus are laterally compressed by comparison with those of many other hystricognaths, but the sides of the claws are not joined ventrally at the base, so the subunguis is exposed from the digital pad to the tip. By contrast, the manus of *Chaetomys* and *Coendou* is conspicuously broadened proximally by a fleshy pad projecting from the inner (medial) side, and the forefoot is narrowed distally by elongation of the digits, of which the middle two (III and IV) much exceed the outer pair (II and IV); the claws are much elongated and more laterally compressed than in *Erethizon*, with the result that the sides are often joined at the base, concealing the subunguis.

The short pes of *Erethizon* has five clawed digits, of which the hallux is smallest and the rest are subequal in length. The digits are set side by side, without special separations among them, and together oppose a fleshy expansion of the inner margin of the sole, but the shape of the foot is not otherwise unusual. In *Chaetomys* and *Coendou*, however, the hallux lacks any external vestige and the remaining pedal digits are elongated. These digits oppose a large prehensile pad (supported internally by metatarsal I and a greatly expanded medial sesamoid; Candela and Picasso 2008:Fig. 22) on the medial side of the foot; this grasping structure, easily recognizable even on dried skins, is unique in Mammalia.

The tail of *Erethizon* is a stubby organ, somewhat less than half the length of the head and body, covered with defensive quills above and with stiff bristles below; there is no naked prehensile tip. The tail of the North American porcupine functions importantly in defense (tail-slaps drive quills deep into the skin of would-be attackers) and, pressed against tree bark, as a brace that prevents climbing animals from slipping backward (Roze 1989).

By contrast, the tail of most *Coendou* species is substantially longer than the head and body and has a naked prehensile surface near the tip. Whereas most other prehensile-tailed mammals grasp arboreal supports using the ventral caudal surface, porcupine tails are dorsally prehensile. The very short tail (somewhat less than half the length of head and body) of *C. rufescens* has been described as nonprehensile by authors who probably never saw a living example (e.g., Ellerman 1940), but most specimens have sparsely-haired tail-tips that are worn and calloused dorsally, suggesting that this organ retains some prehensile function (Alberico et al. 1999). The basal part of the tail in *Coendou* is always covered dorsally with body pelage (easily detached defensive quills with or without emergent fur, depending on the species), but the ventral surface, sides, and at least some of the caudal dorsum proximal to the naked prehensile surface is densely covered with stiff backward-pointing bristles that probably have a mechanical bracing function like that described above for *Erethizon*. The tail of *Chaetomys* has stiff, backward-pointing bristles on the ventral surface near the base, and the dorsal surface near the tip is more sparsely haired than the ventral surface, suggesting that it is dorsally prehensile.

Female erethizontids have four mammae in two thoracic-abdominal pairs on the ventral surface (well away from the quills), each surrounded by a small hairless areola. Males sometimes also have well-developed mammae, however, so skins cannot be confidently sexed by the presence of teats.

In general aspect, erethizontid skulls have a short, blunt rostrum; broad interorbital region; strong, laterally projecting zygomatic arches; and prominent auditory bullae. As in other caviomorph rodents, the infraorbital foramen is very large (transmitting the anterior part of the medial masseter; Woods 1972) and the pterygoid fossae are perforated (communicating with the orbital fossae). The lacrimal bone is usually absent (or very small and indistinguishably fused to the frontal or to the maxilla); it is seldom visible as a separate ossification even in juvenile skulls. The paroccipital process, highly developed in many other caviomorphs, is comparatively small in erethizontids and never projects ventrally below the auditory bulla. Well-developed postorbital processes and bony auditory tubes (surrounding the external auditory meatus on the lateral aspect of each bulla) are present in *Chaetomys* but not in any other Recent erethizontid. Three species of *Coendou* have highly inflated nasofrontal sinuses, possibly a defensive adaptation that provides increased cranial surface area for quill deployment and allows erected quills to point forward and sideways, thus protecting otherwise vulnerable soft tissues from attacking predators (Voss et al. 2013).

The upper cheek tooth rows are parallel or weakly convergent anteriorly in Neotropical porcupines, never strongly convergent as in some large sympatric caviomorphs (e.g., caviids and *Dinomys*). The coronoid process of the mandible is small but distinct in most species (*Coendou vestitus* is an exception; Voss and da Silva 2001), and the lower incisor root does not extend posteriorly beyond m3. The cheek teeth are rooted and have either four or five transverse crests. The deciduous fourth premolars (dP4/dp4) are replaced in all erethizontids (including *Chaetomys*; G. A. S. Carvalho 2000).

Gupta (1966) described the axial postcranial skeleton of erethizontids, and Candela and Picasso (2008) have described the appendicular skeleton.

KEY TO THE NEOTROPICAL GENERA OF
ERETHIZONTIDAE:
1. Quills wavy, without hard, sharp points (except on head); tail without a well-developed hairless prehensile surface; skull with well-developed postorbital processes, bony auditory tubes, and ontogenetically persistent sutures; upper cheek teeth longer than wide, high-crowned, and essentially laminar in occlusal design, without connecting median crests . . . . *Chaetomys*

1'. Quills straight or gently curved but never wavy, with hard, sharp points on head, rump, and usually over the rest of the dorsal surface from nose to tail base; tail with a well-developed hairless dorsal prehensile surface (except in a few specimens of *Coendou rufescens*); skull without well-developed postorbital processes or bony auditory tubes; cranial sutures tending to disappear with age; upper cheek teeth wider than long, low-crowned, always with a median crest linking second and third transverse lophs. . . . . *Coendou*

Genus *Chaetomys* Gray, 1843

The monotypic genus *Chaetomys* is a small (ca. 1 kg) erethizontid with a tail that is shorter than the combined length of its head and body (the ratio of tail to head and body length is about 70% in the only examined adult specimen with external measurements). Its visible dorsal pelage consists of thin, wavy, dull-brown spines; except on the head, most of the spines are nondefensive because they lack hard, sharp points. The tail is slender and lacks a well-developed prehensile surface, although the dorsal surface is hairless near the tip in some specimens. The skull is very heavily built, with well-developed postorbital processes of the frontal bones that almost contact correspondingly well-developed postorbital processes of the jugals, and each auditory bulla is provided with a bony tube that surrounds the external auditory meatus. Cranial sutures are ontogenetically persistent, remaining visible even in adult specimens. The frontal sinuses are never inflated. The upper cheek teeth are longer than they are wide, high-crowned, and laminar in occlusal design, consisting of transverse plates without connecting median crests. Illustrations of external and/or craniodental characters are in Ellerman (1940), Oliver and Santos (1991), and Voss and Angermann (1997).

SYNONYMS:
Chaetomys Gray, 1843b:123; type species *Hystrix subspinosa* Olfers, by monotypy.
Plectrochoerus Pictet, 1843b:227; type species *Plectrochoerus moricandi* Pictet, by monotypy.

REMARKS: Although G. S. Miller and Gidley (1918), Stehlin and Schaub (1951), and B. Patterson and Wood (1982) all referred *Chaetomys* to the family Echimyidae, there is no compelling evidence that it is anything other than a porcupine. The only morphological trait cited as evidence for a close phylogenetic relationship between *Chaetomys* and echimyids was the alleged retention of the deciduous premolars (dP4/dp4) throughout adult life (B. Patterson and Wood 1982: 394). However, the milk dentition of *Chaetomys* is, in fact, shed and replaced by P4/p4 as in all other nonechimyid hystricognaths (G. A. S. Carvalho 2000).

Furthermore, *Chaetomys* lacks the derived enamel ultrastructural traits of Octodontoidea (the higher-level clade to which Echimyidae belongs; T. Martin 1994), and it shares with other erethizontids the derived presence of a posterior carotid foramen (T. Martin 1994; Bryant and McKenna 1995; G. A. S. Carvalho 2000). These arguments, together with the identical prehensile structure of the manus and pes that *Chaetomys* uniquely shares with *Coendou* (see above) strongly support the traditional placement of *Chaetomys* with other New World porcupines (Thomas 1897a; Ellerman 1941; Landry 1957). Recent molecular analyses (R. do V. Vilela et al. 2009; Voss et al. 2013) provide additional compelling evidence that Erethizontidae, in the sense herein understood, is monophyletic.

Chaetomys subspinosus Olfers, 1818
Broomstraw-spined Porcupine

SYNONYMS:

H[ystrix]. subspinosa Olfers, 1818:211; type locality Salvador, Bahia, Brazil (Voss and Angermann 1997).

H[ystrix]. tortilis Olfers, 1818:211; type locality Salvador, Bahia, Brazil (Voss and Angermann 1997).

H[ystrix]. Subspinosa Kuhl, 1820; a junior synonym based on the same ZMB material in Berlin previously named by Olfers (Voss and Angermann 1997).

Chaetomys subspinosus: Gray, 1843b:21; first use of current name combination.

Plectrochoerus moricandi Pictet, 1843b:227; type locality "Bahia" (= Salvador), Bahia, Brazil (Voss and Angermann 1997).

C[ercolabes]. subspinosus: Wagner, 1844:35; name combination.

Chaetomys tortilis: Moojen, 1952b:100; name combination.

DESCRIPTION: As for the genus.

DISTRIBUTION: *Chaetomys subspinosus* occurs in the humid Atlantic coastal lowlands and mountains (to at least 650 m) of southeastern Brazil, formerly from Sergipe to northern Rio de Janeiro (roughly between 11° and 22° S latitude; including easternmost Minas Gerais), but many local populations are now extinct due to widespread deforestation (I. B. Santos et al. 1987; Oliver and Santos 1991).

SELECTED LOCALITIES (Map 405; from Oliver and Santos 1991): BRAZIL: Bahia, Araçás, Cravolândia, Firmino Alves; Espírito Santo, Jirau, Nova Lombardia Biological Reserve; Minas Gerais, Bandeira; Rio de Janeiro, Fazenda São Pedro; Sergipe, Arauá.

SUBSPECIES: I treat *Chaetomys subspinosus* as monotypic.

NATURAL HISTORY: This species is said to occur in primary and secondary rainforest, gallery forest, *restinga* forest (a low-canopy coastal formation including palms and mangroves), macega scrub (severely degraded thickets without large trees), and cabruca plantations (cacao orchards with an intact canopy of native trees). In all of these habitats of the Atlantic coastal region, *Chaetomys subspinosus* co-occurs with, but is less common than, another small erethizontid, *Coendou insidiosus* (see I. B. Santos et al. 1987; Oliver and Santos 1991).

A radio-collared adult male observed by Chiarello et al. (1997) was active only at night and moved through the forest at an average height of about 10 m above the ground, descending only to defecate; collected feces appeared to be composed of finely ground leaves, and tree leaves were the only items that the animal was seen eating. Observed movements were sluggish, and the animal spent most of each night resting. Most observed locomotion was on small-diameter (<2 cm) substrates, and the animal's tail was usually seen coiled around branches and twigs. Subsequent field studies, also based on radio-collared animals (Souto Lima et al. 2010; Giné et al. 2010, 2011; Zortéa and de Brito 2010; Giné et al. 2011; P. A. Oliveira et al. 2011), have confirmed that this species is nocturnal, arboreal, and solitary. Several unusual behaviors, such as using concealed latrines and leaving offspring alone most of the time, suggest that *Chaetomys* is highly vulnerable to predation and survives by being highly cryptic (Giné et al. 2011). *Chaetomys* is a specialized herbivore, perhaps more so than other erethizontids (Souto Lima et al. 2010). The diet of four radio-collared individuals studied by Giné et al. (2010) included leaves of 17 species of trees, but only three species of Fabacea with protein-rich leaves accounted for 78% of recorded feeding observations. The anatomically simple stomach of this species, its long large intestine, and its sacculated caecum suggest that *Chaetomys* is a hindgut fermenter with typical erethizontid adaptations for a high-fiber diet and water conservation (no individuals were observed to drink water in >900 hours of observation; Giné et al. 2010). The screwworm, *Chochliomyia hominivorax*, may infest the skin of *C. subspinosus* (Kuniy and Santos 2003), and helminth endoparasites of the genera *Hymenolepis* and *Trichuris* are known endoparasites (Kuniy and Brasileiros 2006). Molecular genetic diversity is mostly contained within local populations (C. G. Oliveira et al. 2011).

REMARKS: Voss and Angermann (1997) discussed the complex taxonomic history of this species (the original authorship of which is often misattributed) and redescribed Olfers' (1818) type material. Other useful descriptions of the external and/or craniodental characters of *Chaetomys semispinosus* are in Waterhouse (1848), Ellerman (1940), Emmons and Feer (1997), T. Martin (1994), and G. A. S. Carvalho (2000). The karyotype is $2n = 52$, FN = 76 (R. do V. Vilela et al. 2009).

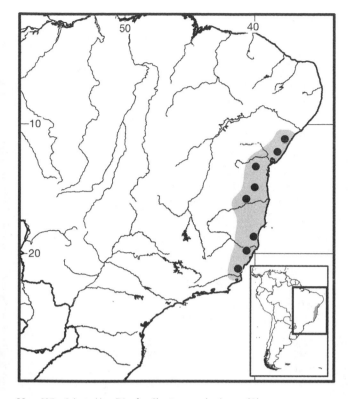

Map 405 Selected localities for *Chaetomys subspinosus* (●).

Genus *Coendou* Lacépède, 1799

The 13 living species of *Coendou* (including *Echino-procta* and *Sphiggurus*; see Voss 2011; Voss et al. 2013) collectively range from Mexico to Uruguay and northern Argentina. Twelve species occur in South America (only *C. mexicanus* does not), where they inhabit tropical and subtropical moist and dry forests from sea level to about 3,500 m. Although many fossils have been referred to *Coendou*, assignments of extinct species to this genus (versus *Erethizon*) are sometimes controversial (Sussman 2011).

Recent species of *Coendou* are small to large (ca. 500 g to 5 kg) erethizontids with tails that vary in relative length from about 40% to >100% of the combined length of the head and body (Voss 2011). The visible dorsal pelage is completely spiny in some species, but in other species the quills may be partially or completely concealed by long fur. Whether concealed by fur or not, the quills of *Coendou* are straight or gently curving, never wavy. At least on the rump, the quills of all species are defensive spines with hard, sharp, microscopically barbed points. The tail of most species is a robust, muscular organ with a well-developed dorsal prehensile surface. The skull lacks well-developed postorbital processes and bony auditory tubes. The usual sutures among adjacent cranial bones are visible in juvenile and in many subadult skulls, but most cranial sutures disappear with age (only the sutures surrounding

the premaxillae and the auditory bullae persist in old adult specimens). The frontal sinuses are conspicuously inflated in some species, but not in others. The roof of the external auditory meatus is smooth and featureless in some species, but in others it is provided with a transverse bony ridge (Voss et al. 2001:Fig. 70A) and is said to be keeled. The upper cheek teeth are low-crowned and wider than they are long; a median crest connecting the second and third transverse lophs is always present. Citations to published illustrations of external and craniodental morphology are provided in the species accounts that follow.

Several long-furred species of *Coendou* were previously referred to the genus *Sphiggurus* (e.g., by Husson 1978; Woods and Kilpatrick 2005), and the short-tailed species *C. rufescens* was previously referred to the monotypic genus *Echinoprocta*. As traditionally recognized, however, these taxa are not reciprocally monophyletic, and allegedly diagnostic "generic" characters (e.g., size, long versus short fur, cranial sinus inflation) appear to have evolved homoplastically in the course of erethizontid phylogeny (Voss et al. 2013). Contradictory results (mtDNA sequence analyses interpreted as supporting the reciprocal monophyly of *Sphiggurus* and *Coendou*; Bonvicino, Penna-Firme, and Braggio 2002) were based on misidentified specimens (Voss 2011:9, footnote 2).

SYNONYMS:

Coendou Lacépède, 1799:11; type species *Hystrix prehensilis* Linnaeus, by monotypy.

Coendus E. Geoffroy, 1803:157; incorrect subsequent spelling of *Coendou* Lacépède.

Coandu G. Fischer, 1814:102; incorrect subsequent spelling of *Coendou* Lacépède.

Sinethere F. Cuvier, 1823a:427; type species *S. prehensilis* (Linnaeus), by original designation.

Sphiggure F. Cuvier, 1823a:427; type species *S. spinosa* F. Cuvier, by original designation.

Sinoetherus F. Cuvier, 1823c:256; incorrect subsequent spelling of *Sinethere* F. Cuvier.

Sphiggurus F. Cuvier, 1823c:256; incorrect subsequent spelling of *Sphiggure* F. Cuvier (but to be maintained due to prevailing subsequent usage; Voss 2011:6).

Coendu Lesson, 1827:290; incorrect subsequent spelling of *Coendou* Lacépède.

Synethere Lesson, 1827:291; incorrect subsequent spelling of *Sinethere* F. Cuvier.

Sinetheres J. B. Fischer, 1829:369; incorrect subsequent spelling of *Sinethere* F. Cuvier.

Cercolabes Brandt, 1835:391; type species *Hystrix prehensilis* Linnaeus (proposed as a replacement name for *Coendou* Lacépède).

Synoetheres Lund, 1839a:227; incorrect subsequent spelling of *Sinethere* F. Cuvier.

Sphingurus Tschudi, 1844:185; incorrect subsequent spelling of *Sphiggure* F. Cuvier.

Echinoprocta Gray, 1865: 321; type species *Erethizon* (*Echinoprocta*) *rufescens* Gray, by monotypy.

Cryptosphingurus Miranda-Ribeiro, 1936:975; type species *C. villosus* (F. Cuvier), by original designation.

KEY TO THE SOUTH AMERICAN SPECIES
OF *COENDOU*:

1. Size small (adult head and body length <400 mm; adult condyloincisive length of skull <70 mm); dorsal pelage includes defensive quills (with hard, sharp, microscopically barbed points) intermingled with bristle-quills (with long, flexible, wiry tips; see Voss and da Silva 2001: Fig. 1) except on head, rump, and base of tail (which have only defensive quills); tail always shorter than the combined length of head and body; frontal sinuses not inflated 2

1'. Size small or large; dorsal pelage does not include bristle-quills intermingled with defensive quills; tail longer or shorter than head and body; frontal sinuses inflated or not 5

2. Tail very short (about 50% of head and body length); dorsal pelage includes long blackish fur that partially or completely conceals defensive quills; bristle-quills bicolored (whitish or yellowish basally and dark brown or blackish distally); buccinator-masticatory foramen and foramen ovale confluent (Voss and da Silva 2001: Fig. 6A); mandible lacks a distinct coronoid process (op. cit.:Fig. 4) . . . .*Coendou vestitus*

2'. Tail >50% of head and body length (except, perhaps, in high-montane populations of *C. pruinosus*); dorsal pelage with or without long fur; bristle-quills tricolored (with pale tips); buccinator-masticatory foramen and foramen ovale separate; mandible with distinct coronoid process 3

3. Tail usually <70% of head and body length (never much longer); dorsal pelage always includes long fur that partially or completely conceals defensive quills; auditory bullae small, uninflated, not contacting paraoccipital processes. . . . .*Coendou pruinosus*

3'. Tail usually >75% of head and body length (never much shorter); dorsal pelage with or without long fur; auditory bullae large and inflated, contacting paraoccipital processes. 4

4. Dorsal pelage appears completely spiny (long fur absent); at least some quills on head tricolored (with pale tips); postcranial defensive quills one-third to one-half black; ventral pelage includes clusters of soft spines mixed with sparse wool hairs . . . .*Coendou ichillus*

4'. Dorsal pelage includes long fur that partially or completely conceals defensive quills; no pale-tipped quills

on head; postcranial defensive quills mostly pale (only tips dark brown or blackish); ventral pelage consists of guard hairs and sparse wool . . . .
. . . . *Coendou roosmalenorum*

5. Tail very short (usually <50% of head and body length); dorsal body pelage without long fur; postcranial quills with hard, sharp, microscopically barbed points only on rump (the remainder nondefensive, with soft points) . .
. . . . *Coendou rufescens*

5'. Tail consistently longer (>50% of head and body except in highland populations of *C. quichua*); dorsal body pelage with or without long fur; all postcranial quills with hard, sharp, microscopically barbed points 6

6. Large (combined length of adult head and body >400 mm; condylo incisive length of adult skull >77 mm; adult maxillary toothrow >18 mm); dorsal body pelage appears completely spiny (long fur absent); frontal sinuses inflated in most fully adult specimens 7

6'. Usually smaller (combined length of adult head and body usually <400 mm; condyloincisive length of adult skull usually <77 mm; adult maxillary toothrow usually <18 mm); dorsal body pelage with or without long fur; frontal sinuses seldom inflated 8

7. Quills uniformly bicolored and abruptly shorter on lower back and rump than on shoulders and upper back (resulting in a visually conspicuous cape or mantle of long quills); mesopterygoid fossa usually extends anteriorly between second upper molars . . . .*Coendou bicolor*

7'. Long tricolored quills (with white tips) intermingled with shorter bicolored quills over most of the dorsal surface, including the rump (except in specimens from the southern Maracaibo Basin, which have uniformly bicolored quills); dorsal quills decrease gradually in length from upper back to rump, without abrupt transition; mesopterygoid fossa usually extends anteriorly between third molars . . . . *Coendou prehensilis*

8. Dorsal pelage does not include long fur (appearing completely spiny) 9

8'. Dorsal pelage usually includes long fur that partially or completely conceals the quills. 11

9. Occurs in northern and western Colombia and western Ecuador; roof of external auditory meatus usually strongly keeled . . . .*Coendou quichua*

9'. Occurs in Brazil; roof of external auditory meatus smooth or weakly keeled 10

10. Occurs along south bank of the Amazon east of the Rio Madeira to Maranhão (on the Atlantic coast); dorsal body quills mostly bicolored (whitish or yellowish basally with black tips), but at least a few lateral quills usually with pale (usually whitish but sometimes brownish) tips; ventral pelage consists of soft spines;

dorsal surface of tail covered with blackish bristles ...
.... *Coendou nycthemera*
10′. Occurs in extreme E Brazil (Pernambuco and Alagoas); most dorsal body quills tricolored, with long reddish tips; ventral pelage consists of soft fur; dorsal surface of tail covered with brownish or reddish bristles ..
.... *Coendou speratus*
11. Occurs north of Amazon and east of the Rio Negro-Casiquiare-Orinoco; tail about as long as combined length of head and body; dorsal fur blackish streaked with long yellowish guard hairs; quills all bicolored; roof of external auditory meatus smooth ....
.... *Coendou melanurus*
11′. Occurs in southeastern Brazil; tail substantially shorter than combined length of head and body; dorsal fur dark or pale but not streaked with long yellowish guard hairs; quills all bicolored or some quills tricolored; roof of external auditory meatus smooth 12
12. Long dorsal fur (when present) dark basally with pale hair tips (except in some northern specimens which have pale-gray fur); at least some of the longer quills tricolored (with yellowish, orange, or reddish tips); vibrissae and caudal bristles bicolored (dark basally with yellowish or orange tips); adult maxillary toothrow >15 mm
.... *Coendou spinosus*
12′. Long dorsal fur uniformly pale (white, yellowish, or grayish), or pale basally with darker (brownish) hair tips; vibrissae and caudal bristles all-dark (without pale tips); adult maxillary toothrow <15 mm ....
.... *Coendou insidiosus*

Coendou bicolor (Tschudi, 1844)
Bicolor-spined Porcupine
SYNONYMS:
Sphingurus bicolor Tschudi, 1844:186; type locality "in den Urwäldern zwischen den Flüssen Tullumayo und Chanchamayo," Junín, Peru.
Coendou simonsi Thomas, 1902b:141; type locality "Charuplaya, 1400 m," Cochabamba, Bolivia.
Coendou (*Coendou*) *bicolor*: Tate, 1935:306; name combination.
Coendou (*Coendou*) *simonsi*: Tate, 1935:306; name combination.
Coendou [(*Coendou*)] *bicolor bicolor*: Ellerman, 1940:187; name combination.
Coendou [(*Coendou*)] *bicolor simonsi*: Ellerman, 1940: 187; name combination.
DESCRIPTION: Length of adult head and body 405–500 mm; ratio of tail to head and body 90% to 105% (varies geographically); condyloincisive length of adult skull 85–95 mm; length of adult maxillary toothrow 19–22 mm; long dorsal fur absent (appears completely spiny); defen-

sive quills bicolored, very long over shoulders and upper back but abruptly shorter on lower back and rump; bristle-quills absent; frontal sinuses usually inflated; roof of external auditory meatus smooth.

DISTRIBUTION: *Coendou bicolor* occurs along the eastern Andean foothills and in adjacent Amazonian lowland forest from San Martín in northeastern Peru southward to northeastern Bolivia (Beni). An apparently isolated population occurs in relictual montane forest on the west side of the Andes in northern Peru (Cajamarca), and the species is also credibly reported from northwestern Argentina (M. M. Lucero 1987). By contrast, published reports of *C. bicolor* from Amazonas in northern Peru (Patton et al. 1982), Brazil (Lara et al. 1996), Ecuador (Tirira 1999), Colombia (Alberico et al. 1999), and Venezuela (O. J. Linares 1998) appear to be based on misidentifications (Voss 2011).

SELECTED LOCALITIES (Map 406): ARGENTINA: Jujuy, Yuto (M. M. Lucero 1987). BOLIVIA: Beni, Puerto Caballo (AMNH 214615), Yucumo (AMNH 262271); Cochabamba, Charuplaya (BM 2.1.1.103). PERU: Cajamarca, 2.5 km N of Monte Seco (MUSM 9398); Huánuco, Tingo María (FMNH 91303); Junín, Chanchamayo (FMNH 65799); Madre de Dios, Reserva Cuzco Amazónico (KU 144560); San Martín, Área de Conservación Municipal Mishquiyacu-Rumiyacu y Almendra (FMNH 203679); Ucayali, Río Alto Ucayali (AMNH 147500).

SUBSPECIES: No subspecies of *Coendou bicolor* are recognized (see Remarks).

NATURAL HISTORY: Nothing has been recorded in the literature about the natural history of this species, but specimens have been collected in both montane and lowland rainforest. The relictual montane rainforest in which *Coendou bicolor* occurs the west side of the Andes, for example, was described by Koepcke and Koepcke (1958), and the lowland habitat of the species east of the Andes was described by Duellman and Koechlin (1991).

REMARKS: Cabrera (1961) treated several nominal taxa as subspecies of *Coendou bicolor*, including *quichua* (herein recognized as a valid species), *richardsoni* (a synonym of *quichua*), and *sanctaemartae* (a synonym of *prehensilis*). Cabrera's concept of *C. bicolor* and other classifications based on it (e.g., Woods 1993; Woods and Kilpatrick 2005) were discussed by Voss (2003, 2011) and Voss et al. (2013), whose recommended usage is followed here. Recent phylogenetic analyses of cytochrome-*b* sequence data suggest that *C. bicolor* is the sister species of *C. nycthemera*, a much smaller congener from eastern Amazonia (Voss et al. 2013). As discussed by Voss (2011), mtDNA cytochrome-*b* sequences previously attributed to *C. bicolor* by Bonvicino, Penna-Firme, and Braggio (2002), R. do V. Vilela et al. (2009), and Y. L. R. Leite et al. (2011) were obtained by Lara et al. (1996) from misidentified specimens of *C. prehensilis*.

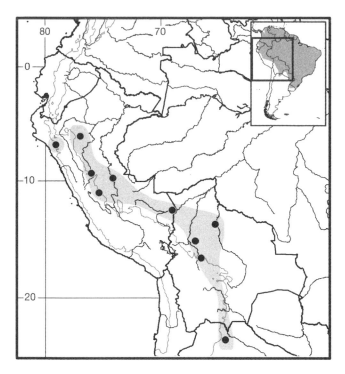

Map 406 Selected localities for *Coendou bicolor* (●). Contour line = 2,000 m.

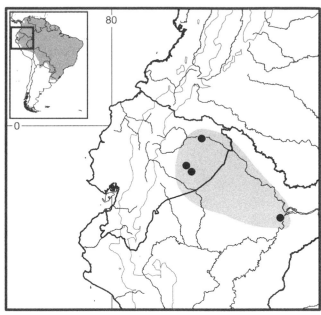

Map 407 Selected localities for *Coendou ichillus* (●). Contour line = 2,000 m.

Coendou ichillus Voss and da Silva, 2001
Western Amazonian Dwarf Porcupine

SYNONYMS:

Coendou ichillus Voss and da Silva, 2001:17; type locality "Río Pastaza," Pastaza, Ecuador (exact locality unknown; see Voss and da Silva 2001:footnote, p. 7).

Sphiggurus ichillus: Woods and Kilpatrick, 2005:1548; name combination.

DESCRIPTION: Length of adult head and body ca. 260–290 mm (estimated values from only two specimens); average ratio of tail to head and body ca. 85%; condyloincisive length of adult skull 58–65 mm; length of adult maxillary toothrow 14–16 mm; long dorsal fur absent (appears completely spiny); defensive quills bicolored (except on head where some quills are tricolored); bristle-quills present (tricolored); frontal sinuses not inflated; roof of external auditory meatus smooth.

DISTRIBUTION: All known specimens and reliable sight records of *Coendou ichillus* are from the rainforested lowlands of eastern Ecuador and northeastern Peru (Voss and da Silva 2001), but it would be reasonable to expect this species to occur throughout most of northwestern Amazonia (north of the Amazon and west of the Rio Negro).

SELECTED LOCALITIES (Map 407): ECUADOR: Napo, La Selva Jungle Lodge (Voss and da Silva 2001); Pastaza, Río Conambo (EPN 807), Río Yana Rumi (FMNH 43289). PERU, Loreto, Iquitos (FMNH 112565).

SUBSPECIES: No subspecies of *Coendou ichillus* are recognized.

NATURAL HISTORY: No museum specimen of *Coendou ichillus* is accompanied by habitat information, but all of them come from landscapes dominated by lowland rainforest. Two sight records, however, one of which is vouchered by a photograph (Voss and da Silva 2001:Fig. 12), are from lowland rainforest in a white-water river floodplain; both observations suggest that, like other erethizontids, *C. ichillus* is nocturnal and arboreal.

REMARKS: This species appears to be closely related to *Coendou pruinosus*, *C. roosmalenorum*, and *C. vestitus*, all of which also have bristle-quills in addition to ordinary defensive quills and soft fur in their dorsal pelage (Voss et al. 2013). Voss and da Silva (2001) provided a detailed morphological description of *C. ichillus*, photographs, measurements, and relevant comparisons with congeneric taxa. A recently published photograph of the skin of an Ecuadorean porcupine in the Gothenburg Museum of Natural History (Johansson and Högström 2008) is captioned as "*Sphiggurus ichillus*," but the specimen in question is unambiguously identifiable by its very short tail and reddish quills as *Coendou rufescens*.

Coendou insidiosus (Olfers, 1818)
Bahian Hairy Dwarf Porcupine

SYNONYMS:

H[ystrix]. insidiosa Olfers, 1818:211; type locality "Brazil," probably in or near the city of Salvador in northern Bahia (Voss and Angermann 1997).

Cercolabes pallidus Waterhouse, 1848:434; type locality unknown (originally said to be from the West Indies, but presumably from southeastern Brazil; Voss and Angermann 1997).

Synetheres (Sphiggurus) pallidus: E.-L. Trouessart, 1880b: 184; name combination.

Coendu pallidus: E.-L. Trouessart, 1897:622; name combination.

Coendu insidiosus: E.-L. Trouessart, 1897:623; name combination.

Coendou (Sphiggurus) insidiosus: Tate, 1935:307; name combination.

Coendou (Sphiggurus) pallidus: Tate, 1935:307; name combination.

Coendou [(Sphiggurus)] pallidum Ellerman, 1940:187; incorrect subsequent spelling of *pallidus* Waterhouse, 1848).

Coendou (Sphiggurus) insidiosus insidiosus: Cabrera, 1961: 600; name combination.

Sphiggurus insidiosus: Husson, 1978:484; part (not including referred Surinam specimens); name combination.

Sphiggurus pallidus: Woods, 1993:776; name combination.

DESCRIPTION: Length of adult head and body ca. 310–350 mm; average ratio of tail to head and body approximately 65%; condyloincisive length of adult skull 61–71 mm; length of adult maxillary toothrow 13–15 mm; long dorsal fur present (uniformly pale or pale basally with darker hair tips); defensive quills bicolored; bristle-quills absent; frontal sinuses not inflated; roof of external auditory meatus smooth.

DISTRIBUTION: I have only examined specimens of *Coendou insidiosus* from the southeastern Brazilian state of Bahia, but the species is said to occur in moist forests from southern Sergipe to northern Rio de Janeiro (roughly between 11° and 22° S latitude), including easternmost Minas Gerais (I. B. Santos et al. 1987; Oliver and Santos 1991). Pale-gray-furred individuals of another southeastern Brazilian species, *C. spinosus*, however, are superficially similar to *C. insidiosus*, so unvouchered sightings are perhaps unreliable where both species could be expected (e.g., in Espírito Santo, northeastern Minas Gerais, and northern Rio de Janeiro states).

SELECTED LOCALITIES (Map 408; from Oliver and Santos 1991, except as noted): BRAZIL: Bahia, Itanhem, Itirussu (AMNH 76838), Ponto do Pitu, Salvador (ZMB 1298), Trancoso); Espírito Santo, Anchieta, Ecoporanga, 1 km S of São Mateus; Minas Gerais, Jordania; Sergipe, Arauá.

SUBSPECIES: No subspecies of *Coendou insidiosus* are recognized.

NATURAL HISTORY: This species is said to be nocturnal, arboreal, and to occur in primary and mature-secondary rainforest, gallery forest, *restinga* forest (a low-canopy coastal formation including palms and mangroves), macega scrub (severely degraded thickets without large trees), cabruca plantations (cacao orchards with an intact canopy of native trees), and capoeira (dense young secondary growth); in all but the last of these habitats *Coendou insidiosus* co-occurs with and appears to be more numerous than *Chaetomys subspinosus* (see Oliver and Santos 1991). The species is hunted for food and for its quills, which are thought to have medicinal properties (Oliver and Santos 1991).

REMARKS: The name *insidiosus* has often been used for several other valid species of *Coendou*, including *C. melanurus* and *C. spinosus*. For a review of the complex taxonomic history of *C. insidiosus* (the authorship of which is often misattributed), see Voss and Angermann (1997), who also provided measurements, photographs, and a detailed description of Olfers' (1818) type material. Diagnostic comparisons with *C. melanurus* are in Voss et al. (2001), and color photographs of living animals are in Oliver and Santos (1991). *Coendou insidiosus* can be distinguished from superficially similar pale-gray-furred specimens of *C. spinosus* by their unicolored-dark vibrissae and caudal bristles (the vibrissae and caudal bristles of *C. spinosus* are dark basally and orange or yellowish

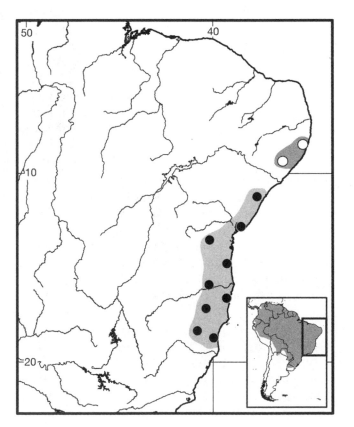

Map 408 Selected localities for *Coendou insidiosus* (●) and *Coendou speratus* (○). Contour line = 2,000 m.

distally), consistently bicolored quills (at least some cranial and postcranial quills are always tricolored in *C. spinosus*), and shorter toothrows (maxillary toothrow length is >15 mm in *C. spinosus*). Although this species was not included in their sequencing study, Voss et al. (2013) hypothesized that *Coendou insidiosus* is most closely related to *C. spinosus*, another southeastern Brazilian species.

F. S. Lima (1994) reported the karyotype of a specimen identified as *Coendou insidiosus* to be $2n = 62$, FN = 76).

Coendou melanurus (Wagner, 1842)
Black-tailed Porcupine

SYNONYMS:

Cercolabes melanurus Wagner, 1842c:360; type locality "Rio Negro [Barra]" (= Manaus), Amazonas, Brazil.

Sphiggurus melanurus Gray, 1842:262; type locality "Brazil."

Synetheres (Sphiggurus) melanurus: E.-L. Trouessart, 1880b: 184; name combination.

Coendu melanurus: E.-L. Trouessart, 1897:622; name combination.

Coendou (Sphiggurus) melanurus: Tate, 1935:307; name combination.

Coendou (Sphiggurus) insidiosus melanurus: Cabrera, 1961: 601; name combination.

Sphiggurus melanura: Bonvicino, Penna-Firme, and Braggio, 2002:1071; epithet spelled with incorrect gender.

DESCRIPTION: Length of adult head and body 330–435 mm; average ratio of tail to head and body approximately 95%; condyloincisive length of adult skull 69–80 mm; length of adult maxillary toothrow 15–20 mm; long dorsal fur present (blackish streaked with long yellow-tipped guard hairs); defensive quills bicolored; bristle-quills absent; frontal sinuses seldom inflated; roof of external auditory meatus with well-developed transverse bony ridge.

DISTRIBUTION: *Coendou melanurus* apparently occurs throughout the northeastern Amazonian lowlands (north of the Amazon and east of the Orinoco–Rio Negro), including eastern Venezuela, Guyana, Surinam, French Guiana, and northern Brazil.

SELECTED LOCALITIES (Map 409): BRAZIL: Amapá, Serra do Navio (USNM 394732); Amazonas, Manaus (NMW 42010); Pará, Faro (AMNH 94174). FRENCH GUIANA: Paracou (AMNH 266565), St.-Laurent du Maroni (MNHN 1909–242). GUYANA: Cuyuni-Mazaruni, Kartabo (AMNH 70120); Demerara-Mahaica, Georgetown (FMNH 17762); Upper Takutu–Upper Essequibo, Nappi Creek near Letham (ROM 31683). SURINAM: Brokopondo, Afobaka (Husson 1978); Paramaribo, Paramaribo (RMNH 23982). VENEZUELA: Amazonas, Sierra Parima (O. J. Linares 1998); Bolívar, Imataca (Ochoa 1995).

SUBSPECIES: No subspecies of *Coendou melanurus* are recognized.

NATURAL HISTORY: Most specimens of *Coendou melanurus* are unaccompanied by habitat information, but the known range of this species is mostly covered by lowland rainforest, in which some individuals are definitely known to have been collected. One specimen was shot at night as it fed on the leaves of a subcanopy tree about 15–20 m above the ground in well-drained primary forest (Voss et al. 2001). Although this species is rarely seen, large numbers recovered by faunal rescue teams during hydroelectric dam construction (e.g., at Petit Saut, French Guiana; Vié 1999) suggest that it can be locally abundant. Throughout its range, *C. melanurus* is sympatric with the much larger and more commonly collected *C. prehensilis*.

REMARKS: Photographs, measurements, detailed descriptions of external and craniodental morphology, and diagnostic comparisons with other long-furred species of *Coendou* are in Voss et al. (2001). Husson (1978) misidentified this species as *C. insidiosus*, as did others who based their porcupine classifications on his work (Honacki et al. 1982; Woods 1993). A recent maximum-likelihood analysis of cytochrome-*b* sequence data (Voss et al. 2013) recovered *C. melanurus* as the sister taxon of a clade that included *C. vestitus*, *C. pruinosus*, and *C. ichillus*.

Bonvicino, Penna-Firme, and Braggio (2002) described the karyotype of a specimen (UFPB 3001) identified as *Coendou melanurus* from Roraima state, Brazil, with $2n = 72$, FN = 76.

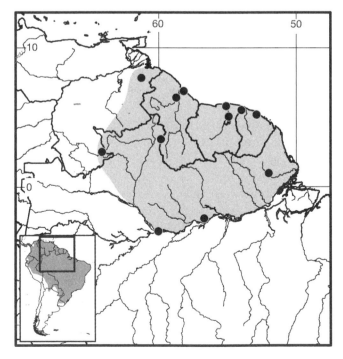

Map 409 Selected localities for *Coendou melanurus* (●). Contour line = 2,000 m.

Coendou nycthemera (Olfers, 1818)
Eastern Amazonian Dwarf Porcupine

SYNONYMS:

H[ystrix]. nycthemera Olfers, 1818:211; type locality "Brazil," probably somewhere on the south (right) bank of the Amazon below Óbidos (Voss and Angermann 1997).

Cercolabes nycthemera: Waterhouse, 1848:417; name combination.

Synetheres (Synetheres) nycthemera: E.-L. Trouessart, 1880b: 183; name combination.

Coendu nycthemera: E.-L. Trouessart, 1897:622; name combination.

Coendou (Sphiggurus) nycthemera: Tate, 1935:307; name combination.

Coendou [(Coendou)] nycthemera: Ellerman, 1940:187; name combination.

Coendou koopmani Handley and Pine, 1992:238; type locality "Belém, Pará, Brazil."

Coendou nycthemerae Eisenberg and Redford, 1999:450; incorrect subsequent spelling of nycthemera Olfers.

DESCRIPTION: Length of adult head and body 290–380 mm; average ratio of tail to head and body about 90%; condyloincisive length of adult skull 61–78 mm; length of adult maxillary toothrow 14–17 mm; long dorsal fur absent (appears completely spiny); defensive quills mostly bicolored (skins appear predominantly blackish), but some tricolored quills present on many specimens; bristle-quills absent; frontal sinuses not inflated; roof of external auditory meatus smooth.

DISTRIBUTION: Specimens of Coendou nycthemera have been collected from the Rio Madeira eastward along the right (south) bank of the Amazon to Ilha de Marajó in the Brazilian states of Amazonas and Pará (Handley and Pine 1992). Interviews with indigenous hunters suggest that the range of this species also extends into the adjacent Atlantic coastal watershed of Maranhão (T. G. Oliveira et al. 2007).

SELECTED LOCALITIES (Map 410): BRAZIL: Amazonas, Auará Igarapé (AMNH 91726), Villa Bella da Imperatriz (AMNH 92898); Pará, Belém (USNM 394733), Cametá (AMNH 96316), Curralinho (AMNH 134074), Mocajuba (AMNH 96330), Muaná (USNM 105527).

SUBSPECIES: No subspecies of Coendou nycthemera are recognized.

NATURAL HISTORY: Nothing has been recorded in the literature about the natural history of this species, but all known specimens are from landscapes dominated by lowland rainforest. Throughout its geographic range Coendou nycthemera probably occurs sympatrically with the much larger species C. prehensilis (see Handley and Pine 1992). Coendou nycthemera is one of two small porcupines (the other is C. roosmalenorum) that inhabits the right (eastern) bank of the Rio Madeira, but known collecting localities

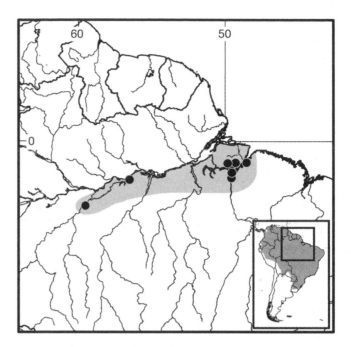

Map 410 Selected localities for Coendou nycthemera (●).

for these species are widely separated, and it is not known whether their ranges overlap (Voss and da Silva 2001).

REMARKS: Detailed descriptions of external and craniodental morphology are in Handley and Pine (1992) and Voss and Angermann (1997); the latter authors additionally provided black-and-white photographs of the holotype skin and skull. Recent analyses of cytochrome-b sequence data recovered Coendou nycthemera as the sister taxon of C. bicolor, a much larger western Amazonian species (Voss et al. 2013).

Coendou prehensilis (Linnaeus, 1758)
Brazilian Porcupine

SYNONYMS:

Hystrix prehensilis Linnaeus, 1758:57; type locality "Mata Xanguá, Usina Trapiche, municipality of Sirihaém, state of Pernambuco, Brazil, 8°38′50″S, 35°10′15″W, elevation 100 m" (fixed by neotype selection; Y. L. R. Leite et al. 2011).

Coendou prehensilis: Lacépède, 1799:11; first use of current name combination.

Coendou longicaudatus Daudin, 1802:172; type locality Cayenne, French Guiana (based on the "Coendou à longue queue" of Buffon).

Hystrix cuandu Desmarest, 1822:346; type locality "Le Brésil, la Guyane, l'île de la Trinité."

S[inethere]. prehensilis: F. Cuvier, 1823a:433; name combination.

Cercolabes (Synetheres) prehensilis: Brandt, 1835:396; name combination.

Cercolabes (*Synetheres*) *platycentrotus* Brandt, 1835:399; type locality "America australis?" (see Remarks).

Cercolabes prehensilis: Wagner, 1844:29; name combination.

Cercolabes boliviensis Gray, 1850:380; type locality "Bolivia," apparently near Santa Cruz de la Sierra (Voss 2011).

Cercolabes tricolor Gray, 1850:381; type locality unknown (Voss 2011).

Hystrix brandtii Jentink, 1879a:96; type locality "Surinam" (fixed by lectotype selection; Husson 1978).

Synetheres (*Synetheres*) *boliviensis*: E.-L. Trouessart, 1880b: 183; name combination.

Synetheres (*Synetheres*) *prehensilis*: E.-L. Trouessart, 1880b: 183; name combination.

Synetheres (*Synetheres*) *Brandtii*: E.-L. Trouessart, 1880b: 183; name combination.

Coendu brandtii: E.-L. Trouessart, 1897:621; name combination.

Coendu tricolor: E.-L. Trouessart, 1897:622; name combination.

Coendu boliviensis: E.-L. Trouessart, 1897:622; name combination.

Coendou centralis Thomas, 1904a:240; type locality "Chapada [= Chapada dos Guimarães]," Mato Grosso, Brazil.

Coendou sanctaemartae J. A. Allen, 1904c:411; type locality "Bonda, Santa Marta district, Colombia."

Coendou (*Coendou*) *prehensilis*: Tate, 1935:306; name combination.

Coendou (*Coendou*) *boliviensis*: Tate, 1935:306; name combination.

Coendou (*Coendou*) *brandtii*: Tate, 1935:306; name combination.

Coendou (*Coendou*) *sanctaemartae*: Tate, 1935:306; name combination.

Coendou (*Coendou*) *tricolor*: Tate, 1935:306; name combination.

Coendou (*Coendou*) *centralis*: Tate, 1935:306; name combination.

Coendou (*Coendou*) *platycentrotus*: Tate, 1935:306; name combination.

Coendou [(*Coendou*)] *prehensilis prehensilis*: Ellerman, 1940:187; name combination.

Coendou [(*Coendou*)] *prehensilis boliviensis*: Ellerman, 1940: 187; name combination.

Coendou prehensilis platycentrotus: Miranda-Ribeiro, 1936: 971; name combination.

Coendou (*Coendou*) *baturitensis* Feijó and Langguth, 2013: 124; type locality "Community Sitio Barreiros, municipality of Aratuba, Baturite Range, Ceará, Brazil."

DESCRIPTION: Length of adult head and body 400–530 mm; average ratio of tail to head and body about 100%; condyloincisive length of adult skull 77–103 mm; length of adult maxillary toothrow 18–24 mm; long dorsal fur absent (appears completely spiny); defensive quills bicolored and tricolored (most skins appear heavily speckled or streaked with white); bristle-quills absent; frontal sinuses highly inflated; roof of external auditory meatus weakly keeled.

DISTRIBUTION: *Coendou prehensilis* occurs from northern Colombia (Magdalena, Cesar, Bolívar departments) eastward across northern South America (including Venezuela and the Guianas) and southward throughout most of the forested cis-Andean lowlands of Ecuador, Peru, and Brazil to eastern Bolivia (Beni, Cochabamba, La Paz, Santa Cruz departments), eastern Paraguay, and northwestern Argentina (Salta province).

SELECTED LOCALITIES (Map 411): ARGENTINA: Salta, Tartagal (Olrog 1976). BOLIVIA: Beni, 8 km N of Exaltación (AMNH 214613); Cochabamba, Campamento Yuquí (Anderson 1997); La Paz, Río Madidi (MSB 56078); Santa Cruz, Buena Vista (FMNH 21396). BRAZIL: Bahia, Barra (FMNH 21709); Goiás, Anápolis (AMNH 134062); Mato Grosso, Acurizal (Schaller 1983); Mato Grosso do Sul, Maracaju (AMNH 134073); Minas Gerais, Lagoa Santa (CN 520); Pará, Igarapé Assu (BM 4.7.4.82), Marajó (USNM 519688); Pernambuco, Mata Xanguá (Y. L. R. Leite et al. 2011); Rio Grande do Sul, Venâncio Aires (Scheibler and Christoff 2007). COLOMBIA: Magdalena, Bonda (AMNH 15459); Meta, Villavicencio (AMNH 73680). ECUADOR: Sucumbíos, Limoncocha (USNM 528360). FRENCH GUIANA: Paracou (Voss et al. 2001). GUYANA: Upper Takutu–Upper Essequibo, 20 mi E Dadanawa (USNM 362242). PARAGUAY: Concepción, Yby-Yaú (UMMZ 146507); Paraguarí, Parque Nacional Ybycuí (Yahnke et al. 1998). PERU: Amazonas, Huampami (MVZ 153571); Pasco, Pozuzo (ZMB, uncataloged). SURINAM: Paramaribo, Paramaribo (FMNH 95783). TRINIDAD AND TOBAGO: Trinidad, Caparo (AMNH 24202). VENEZUELA: Apure, Nulita (USNM 442610); Falcón, Mirimire (USNM 406750); Sucre, Manacal (USNM 406748); Táchira, Las Mesas (USNM 443409); Vargas, Los Venados (USNM 371277).

SUBSPECIES: I do not recognize any subspecies of *Coendou prehensilis*.

NATURAL HISTORY: *Coendou prehensilis* occurs in lowland rainforest, dry (deciduous) forests, and riparian woodlands in savanna landscapes. Arboreal, nocturnal, and solitary, these large porcupines seem to be less folivorous than smaller erethizontids (e.g., *C. spinosus*; see below), apparently feeding mostly on fruit, immature seeds, and sometimes bark (Charles-Dominique et al. 1981; Emmons and Feer 1997). Diurnal retreats are in hollow trees (Miles et al. 1981) or in shady places in the subcanopy (Montgomery and Lubin 1978). Captive females give birth aseasonally to a single precocial offspring after a gestation of about 200 days; despite their advanced state of development at birth, young *C. prehensilis* are not nutritionally independent until about 15 weeks after birth (Roberts et al. 1985). Three animals radio-collared in the Venezuelan llanos moved 50–700 m each night and used a different sleeping site each day; at least some nocturnal travel must have been on the ground because trees were widely spaced in this habitat; the largest home range recorded over the course of a two-month study period was 38 hectares (Montgomery and Lubin 1978). Large numbers of this species have been recovered by faunal rescue projects after hydroelectric dam construction, suggesting that it can be locally abundant in some situations.

REMARKS: Illustrations and descriptions of the external and craniodental morphology of *Coendou prehensilis* are in Husson (1978) and Leite et al. (2011); summary tabulations of measurement data are in Voss (2011). Specimens of *C. prehensilis* misidentified as *C. bicolor* were sequenced by Lara et al. (1996), whose molecular data were subsequently analyzed by Bonvicino et al. (2002), Vilela et al. (2009), and Leite et al. (2011). Husson (1978), Leite et al. (2011), and Voss (2011) discussed the nomenclature of this very widespread species. Mitochondrial DNA sequence data suggest that the nominotypical Pernambuco population is the most divergent among the 19 samples analyzed, whereas other specimens (from northern South America, Amazonia, and the Cerrado) exhibit very little genetic variation (Voss et al. 2013). The oldest available name that applies to the latter clade is *longicaudatus* Daudin, 1802, from French Guiana. Sequence data are,

unfortunately, unavailable from the Ceará population that Feijó and Langguth (2013) named as *C. baturitensis*, the phenotypic characteristics of which do not seem to fall outside the range of variation among material that I refer to *C. prehensilis*. Surprisingly, *C. prehensilis* appears to be most closely related to a clade of trans-Andean and Andean species that includes *C. mexicanus*, *C. quichua*, and *C. rufescens* (see Voss et al. 2013).

F. S. Lima (1994) described the karyotype of this species as $2n = 74$, FN $= 82$.

Coendou pruinosus Thomas, 1905
Frosted Porcupine

SYNONYMS:

Coendou pruinosus Thomas, 1905b:310; type locality "Montañas de la Pedregosa," Mérida, Venezuela.

Coendou (Sphiggurus) pruinosus: Tate, 1935:307; name combination.

Coendou (Sphiggurus) vestitus pruinosus: Cabrera, 1961: 602; name combination.

Sphiggurus vestitus: Honacki et al., 1982:572; name combination, part.

Sphiggurus pruinosus: Woods and Kilpatrick, 2005:1549; name combination.

DESCRIPTION: Length of adult head and body 320–380 mm; ratio of tail to head and body about 50–70% (varies geographically); condyloincisive length of adult skull 61–71 mm; length of adult maxillary toothrow 14–16 mm; long dorsal fur present (usually frosted-blackish in fresh highland specimens); defensive quills bicolored (except on head, which has some tricolored quills); bristle-quills present (tricolored); frontal sinuses not inflated; roof of external auditory meatus weakly keeled.

DISTRIBUTION: This species is known from mountains, foothills, and adjacent lowlands (from 54 to 2,600 m) in northern Colombia (Meta, Norte de Santander departments) and northern Venezuela (Aragua, Vargas, Mérida, and Zulia states).

SELECTED LOCALITIES (Map 412): COLOMBIA: Meta, Villavicencio (AMNH 136312); Norte de Santander, Alturas de Pamplona (FMNH 140260). VENEZUELA: Aragua, Colonia Tovar (EBRG 20243), Rancho Grande (EBRG 126); Mérida, Mérida (AMNH 21350); Zulia, El Rosario (USNM 496172); Vargas, Caracas (BM 26.11.4.11); Zulia, Misión Tukuko (MHNLS 7692).

SUBSPECIES: No subspecies of *Coendou pruinosus* are recognized.

NATURAL HISTORY: Most specimens are unaccompanied by habitat information, but two are known to have been shot from trees at night; recorded habitats include lowland rainforest and cloud forest (Handley 1976; Concepción and Molinari 1991; Voss and da Silva 2001).

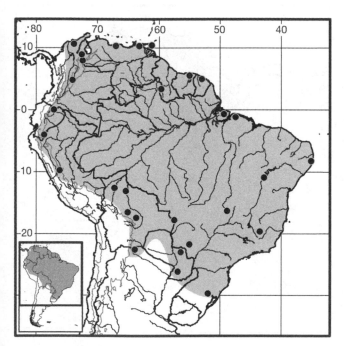

Map 411 Selected localities for *Coendou prehensilis* (●). Contour line = 2,000 m.

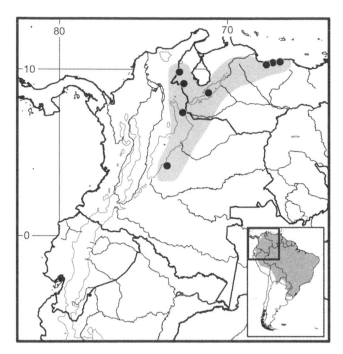

Map 412 Selected localities for *Coendou pruinosus* (●). Contour line = 2,000 m.

REMARKS: Voss and da Silva (2001) provided a detailed morphological description of this species and compared it with other congeneric taxa. Recent phylogenetic analyses of cytochrome-*b* sequence data (Voss et al. 2013) suggest that *Coendou pruinosus* may be closely related to *C. ichillus*, *C. roosmalenorum*, and *C. vestitus*, all of which also have bristle-quills in addition to ordinary defensive quills and soft fur in their dorsal pelage.

Concepción and Molinari (1991) reported the karyotype of this species as $2n = 42$, FN = 76, based on a specimen from Mérida state, Venezuela (CVULA I-3030).

Coendou quichua Thomas, 1899
Quichua Porcupine

SYNONYMS:

Coendou quichua Thomas, 1899c:283; type locality "Puembo, Upper Guallabamba River, Province of Pichincha, Ecuador; altitude about 2500 metres."

Coendou rothschildi Thomas, 1902e:169; type locality allegedly "Sevilla Island, off Chiriqui, Panama," but this information is fraudulent according to Olson (2008), who claimed that the type was probably collected near Boquerón on the adjacent mainland of Chiriquí province, Panama.

Coendou quichua richardsoni J. A. Allen, 1913a:478; type locality "Esmeraldas (near sea level), Ecuador."

Coendou (Coendou) quichua quichua: Tate, 1935:306; name combination.

Coendou (Coendou) quichua richardsoni: Tate, 1935:306; name combination.

Coendou (Coendou) rothschildi: Tate, 1935:306; name combination.

Coendou (Coendou) bicolor quichua: Cabrera, 1961:598; name combination.

Coendou (Coendou) bicolor richardsoni: Cabrera, 1961:598; name combination.

DESCRIPTION: Length of adult head and body 330–440 mm; ratio of tail to head and body about 55–90% (varies geographically); condyloincisive length of adult skull 69–87 mm; length of adult maxillary toothrow 16–19 mm; long dorsal fur absent (appears completely spiny); defensive quills bicolored and tricolored; bristle-quills absent; frontal sinuses seldom inflated; roof of external auditory meatus strongly keeled in most specimens.

DISTRIBUTION: *Coendou quichua* is known from Panama, trans-Andean Colombia, and western Ecuador. Elevations recorded on specimen labels suggest that the species ranges from sea level to about 3,300 m.

SELECTED LOCALITIES (Map 413): COLOMBIA: Cesar, San Alberto (LACM 27376); Chocó, Acandí (Alberico et al. 1999); Cundinamarca, San Juan de Río Seco (AMNH 73679); Santander, San Vicente de Chucurí (Alberico et al. 1999). ECUADOR: Cotopaxi, Otonga (Jarrín-V. 2001); Esmeraldas, Esmeraldas (AMNH 33242); Pichincha, Mindo (BM 34.9.10.186), Puembo (BM 99.2.18.17), Tablón (NRM A58/2822).

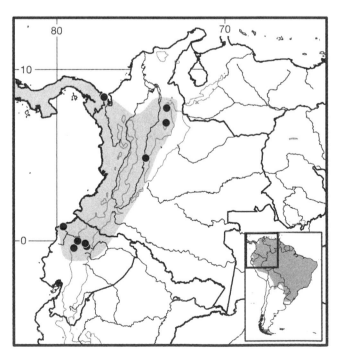

Map 413 Selected localities for *Coendou quichua* (●). Contour line = 2,000 m.

SUBSPECIES: I do not recognize any subspecies of *Coendou quichua*.

NATURAL HISTORY: Panamanian specimens have been observed in trees at night and are said to conceal themselves among matted vines in the forest canopy; anecdotes suggest that they feed on leaves and fruits, and that they are attacked by ocelots (*Leopardus pardalis*) on occasion (Goldman 1920; Enders 1935).

REMARKS: For a discussion of the taxonomic status of this species, formerly regarded as a subspecies of *Coendou bicolor*, see Voss (2003, 2011). Panamanian populations of *C. quichua* were traditionally identified as *C. rothschildi* (e.g., by; Emmons and Feer 1997; F. A. Reid 1997; Woods and Kilpatrick 2005), but these nominal taxa seem to be part of a single widespread trans-Andean species (Voss 2011). Phylogenetic analyses of cytochrome-*b* sequence data (Voss et al. 2013) recover *C. quichua* (represented by Panamanian, Colombian, and Ecuadorean samples) as a member of a clade that also includes *C. mexicanus* and *C. rufescens*.

W. George and Weir (1974) described a karyotype with $2n = 74$, $FN = 82$, from a Panamanian specimen of *Coendou rothschildi*.

Coendou roosmalenorum Voss and da Silva, 2001
Roosmalens's Porcupine
SYNONYMS:

Coendou roosmalenorum Voss and da Silva, 2001:24; type locality "Novo Jerusalem [= Nova Jerusalem] near the left bank of the middle Rio Madeira in the Brazilian state of Amazonas."

Sphiggurus roosmalenorum: Woods and Kilpatrick, 2005: 1549; name combination.

DESCRIPTION: Length of adult head and body 290 mm (the only two specimens accompanied by measurements have identical values for this dimension); average ratio of tail to head and body about 85%; condyloincisive length of adult skull 59–65 mm; length of adult maxillary toothrow 12–14 mm; long dorsal fur present (brownish or grayish); defensive quills bicolored; bristle-quills present (tricolored); frontal sinuses not inflated; roof of external auditory meatus weakly keeled.

DISTRIBUTION: *Coendou roosmalenorum* is currently known from just three localities on the Rio Madeira, but it is likely to be widely distributed south of the Amazon in western Amazonia.

SELECTED LOCALITIES (Map 414): BRAZIL: Amazonas, Nova Jerusalem (INPA 2586); Rondônia, 49 km E of Porto Velho (INPA 677).

SUBSPECIES: No subspecies of *Coendou roosmalenorum* are recognized.

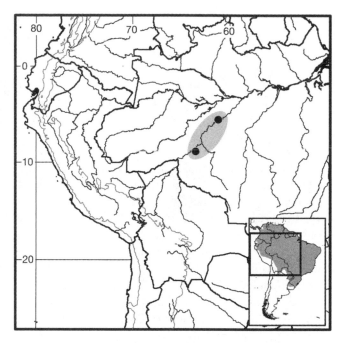

Map 414 Selected localities for *Coendou roosmalenorum* (●). Contour line = 2,000 m.

NATURAL HISTORY: Of the three known specimens, the holotype was captured as it emerged from a felled tree, one paratype was shot on the ground at night, and another paratype was taken during faunal rescue operations at a hydroelectric dam site; all three capture sites were in lowland rainforest. Kept alive as a pet for several months, one individual was only active at night and ate a variety of seeds and fruit (Voss and da Silva 2001).

REMARKS: This species may be closely related to *Coendou ichillus*, *C. pruinosus*, and *C. vestitus*, all of which also have bristle-quills in addition to ordinary defensive quills and soft fur in their dorsal pelage (Voss et al. 2013). Voss and da Silva (2001) provided a detailed morphological description of *C. roosmalenorum*, photographs, measurements, and relevant comparisons with congeneric taxa.

Coendou rufescens (Gray, 1865)
Stump-tailed Porcupine
SYNONYMS:

Erethizon (Echinoprocta) rufescens Gray, 1865:321; type locality "Columbia" [= Colombia].

Coendou prehensilis rufescens: E.-L. Trouessart, 1897:621; name combination.

Echinoprocta rufescens: Cabrera, 1901:159; name combination.

Coendou sneiderni Lönnberg, 1937:17; type locality "Munchique, Cauca" in the Cordillera Occidental (western Andes) of Colombia.

Coendou rufescens: Alberico et al., 1999:606; first use of current name combination.

DESCRIPTION: Length of adult head and body 340–410 mm; average ratio of tail to head and body about 40%; condyloincisive length of adult skull 65–78 mm; length of adult maxillary toothrow 16–19 mm; long dorsal fur absent (appears completely spiny); most quills tricolored (with brownish or reddish tips), but defensive quills on rump and base of tail bicolored; bristle-quills absent; frontal sinuses not inflated; roof of external auditory meatus smooth.

DISTRIBUTION: *Coendou rufescens* occurs in all three Andean cordilleras of Colombia, on both slopes of the Ecuadorean Andes, in northern Peru (Lambayeque), and in northern Bolivia (Cochabamba). Elevations recorded on specimen tags and in the literature document an elevational range from about 800 to 3,500 m, but most records are clustered in the interval from 1,500 to 3,000 m (Voss 2011).

SELECTED LOCALITIES (Map 415): BOLIVIA: Cochabamba, Incachaca (CM 5255). COLOMBIA: Antioquia, Concordia (Alberico 1999), Reserva San Sebastián-La Castellana (Delgado-V. 2009); Caldas, Reserva Río Blanco (F. Sánchez et al. 2004); Cauca, Munchique (FMNH 88524), Paispamba (ROM 57254); Cundinamarca, Fómeque (AMNH 150028), La Aguadita (AMNH 73678); Huila, Meremberg (Alberico et al. 1999); Nariño, Reserva Natural La Planada (Alberico et al. 1999); Valle del Cauca, 4 km NW of San Antonio (MVZ 124088). ECUADOR: Azuay, Valle de Yunguilla (QCAZ 7591); Carchi, Reserva Biológica Guandera (C. Hice, pers. comm. 2005); Napo, Río Napo (Alberico et al. 1999); Tungurahua, Baños (MCZ 36327). PERU: Lambayeque, Reserva Ecológica Chaparri (Pacheco et al. 2009).

SUBSPECIES: *Coendou rufescens* is monotypic.

NATURAL HISTORY: Although the natural vegetation throughout most of the known range of this species is wet montane ("cloud") forest, specimens have also been collected in seasonally dry forest in northwestern Peru (Linares-Palomino and Ponce-Alvarez 2009; Pacheco et al. 2009). Virtually nothing is known about the behavior or natural diet of this species, but I have seen a photograph of an individual eating leaves in the wild, and Orcés and Albuja (2004) reported a captive animal to have eaten bananas.

REMARKS: Morphological descriptions of *Coendou rufescens* are in Cabrera (1901), M. E. Trouessart (1920), Lönnberg (1937), Emmons and Feer (1997), and Alberico et al. (1999). See Voss (2011) for measurement data and

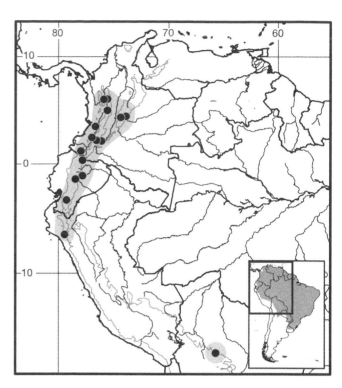

Map 415 Selected localities for *Coendou rufescens* (●); disjunct population in Bolivia mapped separately from continuous range. Contour line = 2,000 m.

comments on the taxonomy and geographic distribution of this species. Phylogenetic analyses of cytochrome-*b* sequence data (Voss et al. 2013) suggest that this species belongs to a clade that also includes *C. quichua* and *C. mexicanus*. Voss (2011) speculated that the Bolivian population might have descended from animals transported by the Inca.

Coendou speratus Mendes Pontes, Gadelha, Melo, de Sá, Loss, Caldara Junior, Costa, and Leite, 2013
Pernambuco Dwarf Porcupine

SYNONYM:

Coendou speratus Mendes Pontes et al., 2013:424; type locality "Mata Tauá, Usina Trapiche, municipality of Sirinhaém, state of Pernambuco, Brazil."

Coendou (Sphiggurus) speratus: Feijó and Langguth, 2013:131; name combination.

DESCRIPTION: Length of adult head and body 330–440 mm; average ratio of tail to head and body about 85%; condyloincisive length of adult skull 62–76 mm; length of adult maxillary toothrow 14–16 mm; long dorsal fur absent (appears completely spiny); defensive quills mostly tricolored (skins appear heavily streaked with red); bristle-quills absent; frontal sinuses not inflated; roof of external auditory meatus smooth or weakly keeled.

DISTRIBUTION: Specimens of *Coendou speratus* are only known from the eastern Brazilian states of Pernambuco and Alagoas (Mendes Pontes et al. 2013).

SELECTED LOCALITIES (Map 408): BRAZIL: Alagoas, near Viçosa (Mendes Pontes et al. 2013); Pernambuco, Mata Tauá (type locality of *Coendou speratus* Mendes Pontes et al.).

SUBSPECIES: *Coendou speratus* is monotypic.

NATURAL HISTORY: This recently described species is only known to occur in remnant patches of submontane Atlantic Forest, where it occurs sympatrically with the much larger *Coendou prehensilis*. All recorded observations of *C. speratus* were of solitary individuals or pairs encountered at night while perched 10–20 m above the ground in trees. Dens are said to be in hollow trees (all natural history information from Mendes Pontes et al. 2013).

REMARKS: Phylogenetic analyses of mtDNA cytochrome-*b* sequence data (in Mendes Pontes et al. 2013) suggest that *Coendou speratus* is closely related to other Brazilian dwarf porcupines with geographically adjacent ranges (*C. insidiosus*, *C. nycthemera*, and *C. spinosus*). Unfortunately, these analyses did not include sequence data from *C. bicolor*, a much larger western Amazonian species that was found to be closely related to *C. nycthemera* in another phylogenetic study (Voss et al. 2013).

Coendou spinosus (F. Cuvier, 1823)
Paraguayan Hairy Dwarf Porcupine

SYNONYMS:

Hystrix couiy Desmarest, 1822:345; part; type locality "Le Mexique et sans doute le Brésil; le Paraguay."

Sphiggure spinosa F. Cuvier, 1823a:433; type locality Sapucaí, Paraguarí, Paraguay (fixed by neotype selection; Voss 2011).

Sphiggure villosa F. Cuvier, 1823a:434; type locality unknown (Voss 2011).

Cercolabes (Sphiggurus) nigricans Brandt, 1835:403; type locality "Brasilia" (= Brazil).

Cercolabes (Sphiggurus) affinis Brandt, 1835:412; type locality "Brasilia" (= Brazil).

Sphingurus sericeus Cope, 1889:136; type locality São João do Monte Negro, Rio Grande do Sul, Brazil (Koopman 1976).

Synetheres (Synetheres) spinosus: E.-L. Trouessart, 1880b:183; name combination.

Synetheres (Sphiggurus) villosus: E.-L. Trouessart, 1880b:184; name combination.

Coendu affinis: E.-L. Trouessart, 1897:623; name combination.

Coendu spinosus: E.-L. Trouessart, 1897:622; name combination.

Coendu sericeus: E.-L. Trouessart, 1897:623; name combination.

Coendu villosus: E.-L. Trouessart, 1897:622; name combination.

Coendou Roberti Thomas, 1902a:63; type locality "Roça Nova, Serra do Mar of the Province of Paraná," Brazil.

Coendou (Sphiggurus) spinosus: Tate, 1935: 307; name combination.

Coendou (Sphiggurus) villosus: Tate, 1935: 307; name combination.

Coendou (Sphiggurus) nigricans: Tate, 1935: 307; name combination.

Coendou (Sphiggurus) affinis: Tate, 1935: 307; name combination.

Coendou (Sphiggurus) sericeus: Tate, 1935: 307; name combination.

Coendou (Sphiggurus) roberti: Tate, 1935: 307; name combination.

Coendou (Sphiggurus) spinosus nigricans: Cabrera, 1961: 601; name combination.

Coendou (Sphiggurus) spinosus roberti: Cabrera, 1961: 602; name combination.

Coendou (Sphiggurus) spinosus spinosus: Cabrera, 1961: 602; name combination.

DESCRIPTION: Length of adult head and body 285–470 mm; average ratio of tail to head and body about 75%; condyloincisive length of adult skull 59–76 mm; length of adult maxillary toothrow 15–17 mm); long dorsal fur usually present (geographically variable in color); defensive quills bicolored and tricolored; bristle-quills absent; frontal sinuses not inflated; roof of external auditory meatus smooth.

DISTRIBUTION: *Coendou spinosus* occurs in humid tropical and subtropical forests of the Mata Atlântica of southeastern Brazil (from Espírito Santo southward to Rio Grande do Sul) and in similar contiguous habitats of northeastern Argentina (Misiones), eastern Paraguay (Caazapá, Concepción, Guaira, Itapúa, Paraguarí, San Pedro), and northern Uruguay (Artigas, Rivera, Salto, Tacuarembó). Examined specimens document an elevational range from near sea level to at least 900 m.

SELECTED LOCALITIES (Map 416): ARGENTINA: Misiones, Profundidad (Crespo 1974), Tobunas (Crespo 1974). BRAZIL: Espírito Santo, Engenheiro Reeve (BM 3.9.4.87); Minas Gerais, Lagoa Santa (UZM 501); Paraná, Roça Nova (BM 3.7.1.97); Rio de Janeiro, Nova Friburgo (USNM 259793); Rio Grande do Sul, São João do Monte Negro (ANSP 4804); Santa Catarina, Colônia Hansa (BM 28.10.11.48); São Paulo, Butantan (ZMB 35529), São Sebastião (BM 2.11.8.14). PARAGUAY: Caazapá, Estancia Dos Marías (UMMZ 174975); Concepción, Horqueta

(UMMZ 68132); Itapúa, El Tirol (MSB 54078); Paraguarí, Sapucay (USNM 115122). URUGUAY: Artigas, Paso del Campamento (Ximénez et al. 1972); Tacuarembó, near Pueblo Ansina (Ximénez et al. 1972).

SUBSPECIES: I do not recognize any subspecies of *Coendou spinosus*.

NATURAL HISTORY: The only published information about the natural history of this species was obtained by following a radio-collared adult female for one year in an Atlantic Forest fragment in Espírito Santo, Brazil (Passamani 2010). The animal was only active at night, but it spent about 50% of its time resting between episodes of moving and feeding from dusk to dawn. Diurnal retreats were in dense lianas or tree forks, at an average height of 4.5 m above the ground. The average height at which the animal was active at night was 8 m, and the only time it was observed to descend to the ground was when it defecated. The average distance travelled per night was 280 m, and the total home range was 6.3 ha. The only observed food eaten by this animal was new leaves of six tree species. A single young was born in December.

REMARKS: The external morphology of Brazilian and Paraguayan specimens (referred to "*Sphiggurus villosus*" and "*S. paragayensis*," respectively) was described by Emmons and Feer (1997). See Voss (2011) for measurement data and a discussion of the complex taxonomic history of this species. Recent analyses of cytochrome-*b* sequences (Voss et al. 2013) recovered *Coendou spinosus* as the sister taxon of a clade that included *C. bicolor* and *C. nycthem-*

era, but *C. insidiosus* (not sequenced by Voss et al. 2013) is probably a close relative.

Bonvicino et al. (2000) described a karyotype with $2n = 42$, FN = 76, from two male specimens (MNRJ 46937 and 46938) collected in Rio de Janeiro state, Brazil.

Coendou vestitus Thomas, 1899
Blackish Hairy Dwarf Porcupine

SYNONYMS:

Coendou vestitus Thomas, 1899c:284; type locality "Colombia."

Coendou (Sphiggurus) vestitus: Tate, 1935:307; name combination.

Coendou (Sphiggurus) vestitus vestitus: Cabrera, 1961:603; name combination.

Sphiggurus vestitus: Honacki et al., 1982:572; name combination.

DESCRIPTION: Length of adult head and body about 330–370 mm; average ratio of tail to head and body about 50%; condyloincisive length of adult skull 65–69 mm; length of adult maxillary toothrow 14–17 mm; long dorsal fur present (blackish); defensive quills bicolored; bristlequills present (bicolored); frontal sinuses not inflated; roof of external auditory meatus weakly keeled.

DISTRIBUTION: *Coendou vestitus* is currently known from only two definitely identifiable localities, both of which are in the western foothills of the eastern Andean cordillera in the Colombian department of Cundinamarca (but see Remarks).

SELECTED LOCALITIES (Map 417): COLOMBIA: Cundinamarca, Quipile (AMNH 70529), San Juan de Río Seco (AMNH 70596).

SUBSPECIES: No subspecies of *Coendou vestitus* are recognized.

NATURAL HISTORY: Nothing has been recorded about the natural history of this species.

REMARKS: Voss and da Silva (2001) provided a detailed morphological description of this species, together with photographs, measurements, and comparisons with congeneric taxa. Recent analyses of cytochrome-*b* sequence data (Voss et al. 2013) recovered *Coendou vestitus* as part of a clade that also included *C. ichillus* and *C. pruinosus*; together with *C. roosmalenorum* (not sequenced by Voss et al. 2013), all of these species have bristle-quills in addition to ordinary defensive quills and soft fur in their dorsal pelage.

Alberico et al. (1999) reported a skin that they identified as *Coendou vestitus* in the Instituto de Ciencias Naturales (Universidad Nacional, Bogotá) from Villavicencio, on the eastern side of the Andes in the Colombian department of Meta, but the only specimen of *Coendou* that I have seen from that locality (AMNH 136312) is referable to *C. pruinosus*. Although it is possible that two closely

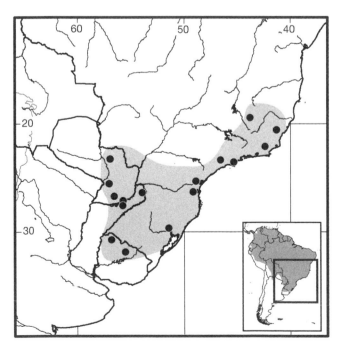

Map 416 Selected localities for *Coendou spinosus* (●). Contour line = 2,000 m.

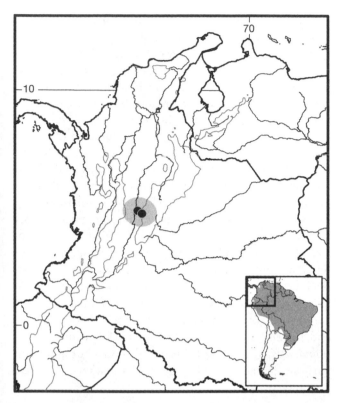

Map 417 Selected localities for *Coendou vestitus* (●). Contour line = 2,000 m.

related species of small porcupines are sympatric near Villavicencio, the identification of the ICN specimen remains to be confirmed by cranial characters (illustrated by Voss and da Silva 2001).

Superfamily Octodontoidea Waterhouse, 1839

James L. Patton

The most taxonomically diverse group of hystricognaths, the Octodontoidea comprise 70% of extant genera and more than 75% of extant species of the infraorder, with species density concentrated within the tropical rainforests of Amazonia and Atlantic Forest (Upham and Patterson 2012). As recognized herein, the superfamily within South America comprises four family-group taxa, the Abrocomidae, Echimyidae, Ctenomyidae, and Octodontidae. Most molecular phylogenetic analyses support the coupling of these families as a monophyletic assemblage, and with the same pattern of relationships (Abrocomidae (Echimyidae (Ctenomyidae + Octodontidae))); Rowe and Honeycutt 2003; Opazo 2005; Rowe et al. 2010; Upham and Patterson 2012; P.-H. Fabre et al. 2012). These analyses place the divergence of the extant lineages as early as the Late Oligocene (26.8 mya, confidence interval 24.8–28.9; Up-

ham and Patterson 2012) and as late as the Early Miocene (20.6 ±2.4 mya), with divergence between Echimyidae versus Ctenomyidae + Octodontidae at 17.5 ± 2.2 mya, and that between the Ctenomyidae and Octodontidae at 15.0 ± 2.1 mya (Opazo 2005). Estimates derived by Upham and Patterson (2012) from a larger dataset of taxa and gene sequences are somewhat older: Echimyidae vs. Ctenomyidae + Octodontidae at 25.3 mya (24.6–26.7 mya confidence interval) and Ctenomyidae vs. Octodontidae, 19.1, 14.3–23.5 mya.

Family Abrocomidae G. S. Miller and Gidley, 1918

James L. Patton and Louise H. Emmons

Members of this family are called "chinchilla rats" because their soft and luxuriant fur resembles that of true chinchillas (Chinchillidae). They are similar in cranial morphology to the Octodontidae but differ in a number of readily observable characters, namely, having the lacrimal canal opening on the rostrum just above the root of the zygoma; nearly parallel maxillary toothrows that curve slightly outward posteriorly; hypselodont and flat-crowned cheek teeth in a general figure-8 occlusal pattern, the maxillary teeth with single lingual and labial folds (except for M3, which has an extra labial fold) and the mandibular teeth with two lingual and one labial fold; narrowed lower incisors and anterior part of dentaries; narrow and elongate incisive foramina; no groove on the medioventral floor of the infraorbital canal indicative of nerve passage; a delicate zygoma with the anterior ascending process of the maxilla reduced to a slender strut, the jugal reduced to the ventral base of the bridge and thus well separated from the lacrimal, and no posterior jugal extension beyond the curvature at the base of the arch; enlarged bullae with mastoids that appear dorsally on the skull between the parietal, squamosal, and occipital bones (mastoid "islands"), and thus with correspondingly reduced supraoccipital processes; a narrowed, elongated, and delicate angular process of the dentary turned medially close to the axis of the alveolus of the incisor (referred to as subhystricognathus); a wide semilunar notch between the condyloid and angular processes; and a reduced coronoid process.

In general, these are rat-like animals of moderate (*Abrocoma*, head and body length 200 to 250 mm; skull length <55 mm) to large (*Cuscomys*, head and body length >300 mm; skull length >63 mm) size with short forelimbs and hindlimbs; a large head with blunt snout; large eyes; large and rounded ear pinnae; long and soft dorsal pelage and dense underfur, with the feet and under parts paler than the typically gray or brown dorsum; a tail of moderate to short length and well furred; forefeet with four digits and hindfeet with five; both forefeet and hindfeet short and broad with short toes; claws weakly developed and hollow beneath, with stiff hairs projecting over claws of the middle three digits of hindfeet (*Abrocoma*) or strong and curved and with less developed comb-like hairs over digits II–IV (*Cuscomys*); and soles of feet naked but covered with small tubercles except for larger terminal toe pads beneath each claw.

Glanz and Anderson (1990) and McKenna and Bell (1997) considered abrocomids members of the Chinchilloidea. Virtually all other authors have, however, consistently placed them within the superfamily Octodontoidea, either as a distinct family (e.g., Simpson 1945; Wood 1955, 1965; Cabrera 1961; S. Anderson and Jones 1967; B. Patterson and Wood 1982; Woods 1982; Reig 1986) or subfamily of the Echimyidae (Ellerman 1940) or Octodontidae (Landry 1957). Molecular analyses using allozymes (Köhler et al. 2000), DNA-DNA hybridization (M. H. Gallardo and Kirsch 2001), and both mitochondrial and nuclear DNA sequences (Huchon and Douzery 2001; Opazo 2005; Opazo et al. 2005) have confirmed a phyletic link of abrocomids with the echimyids and octodontids, as opposed to chinchillids, and as the basal lineage within the Octodontoidea. Opazo (2005) estimated the basal split between the Abrocomidae and Echimyidae + Octodontidae to be 20.6 ± 2.4 mya, during the end of the Early Miocene, slightly older than the estimate given by Huchon and Douzery (2001; 15–18 mya) but younger that that of M. H. Gallardo and Kirsch (2001; 25–30 mya).

Until the last decade, the family contained the single genus *Abrocoma* with just two recognized species. Active systematic analyses of existing museum specimens as well as sustained fieldwork in critically important geographic areas has doubled the number of genera and increased the number of recognized species fivefold (Braun and Mares 2002; Emmons 1999a). Given that members of the family range primarily along the high-elevation spine of the central Andes, much of which is difficult to access and remains biologically unexplored, the discovery of additional diversity is likely. As five taxa of Abrocomidae are known only from their type localities, mostly from only their holotypes, the family is among the most poorly known and potentially at risk in the Neotropics.

The family Abrocomidae includes the extinct genus *Protabrocoma* from the Late Miocene of Argentina and Bolivia (Cione et al. 2000; V. H. Contreras 1991; Kraglievich 1927b; Rovereto 1914; Verzi and Quintana 2005) and the extant genera *Abrocoma* and *Cuscomys*. Pliocene fossils of *Abrocoma* are known from the coastal plain of Argentina, well east of the Andes where all extant taxa range today (Verzi and Quintana 2005).

Among the Recent genera, *Abrocoma* comprises *A. bennettii*, *A. boliviensis*, and the six closely related species within the *A. cinerea* complex (Braun and Mares 2002); *Cuscomys* contains the two species, *C. ashaninka* and *C. oblativus* (Emmons 1999a), the latter originally known from the archaeological record but rediscovered in 2009 with a recent sighting and photos (Julio G. Ochoa, pers. comm. to L. H. Emmons). Verzi and Quintana (2005) assigned *Abrocoma boliviensis* Glanz and Anderson to *Cuscomys*, based on similar features of the cheek teeth (broad anteroposterior lophs with the anterior loph nearly triangular in shape [*Cuscomys*] versus narrow anteroposterior lophs with the anterior loph both subrectangular in shape and slightly concave on its anterior face [*Abrocoma*]) and size of the coronoid process (moderate in *Abrocoma*, reduced in *Cuscomys*). However, *A. boliviensis* does not share the other dramatic characters of hindfoot and cranium detailed by Emmons (1999a:2) as diagnostic for her new genus *Cuscomys*. For the moment, therefore, we follow Glanz and Anderson (1990) and Emmons (1999a) in retaining *boliviensis* within *Abrocoma* until more thorough comparisons, perhaps including molecular markers, have been undertaken.

KEY TO THE GENERA OF ABROCOMIDAE:
1. Body size medium (condylobasilar length of skull <55 mm); tail short, less than 75% of head and body length (except *A. boliviensis*); hallux short, not reaching beyond base of second toe, and stout; claws short and nail-like; frontals behind postorbital process broad; mastoid islands medium to large and inflated; face uniformly colored; tail dorsoventrally bicolored . . . .*Abrocoma*
1'. Body size larger (condylobasilar length of skull >60 mm); tail long, greater than 75% of head and body length; hallux stout but longer, reaching beyond base of 2nd toe; claws strong and curved; frontals behind postorbital process noticeably constricted; mastoid island small and not inflated; face with pure white blaze down center; tail anteroposteriorly bicolored dark basally, white distally . . . . *Cuscomys*

Genus *Abrocoma* Waterhouse, 1837

These are moderately sized rats with long, soft, and dense fur, large and readily visible ear pinnae, moderate to short tails (typically <75% of the head and body length, with *A. boliviensis* an exception), and short but broad feet. The toes are short; the claws weak and small, almost nail-like. Four digits are present on the forefoot, five on the hindfoot. The hallux is reduced, barely reaching the base of the digit II, and the digit IV is shortened. Stiff bristles of hair extend distally above the claws of the hindfoot, except the hallux, and hide the claws when viewed from above. The claw of digit I of the hindfoot is wide and hollow ventrally. Palmar and plantar surfaces of the feet, including the bottoms of the digits, are unhaired and covered with well-developed tubercles; large pads are present at the base of the digits, and a terminal pad is present beneath each claw. Overall coloration ranges from grayish to brownish-gray, the venter is typically paler than the dorsum, and the well-furred tail is weakly bicolored above and below.

The skull is moderate in length (condylobasilar length <55 mm), somewhat delicate in appearance, with a narrowed and elongated rostrum and an arched dorsal profile in lateral view. The nasals are evenly narrow for most of their length, tapering to a point posteriorly. The frontals are not constricted posterior to the postorbital process so that the frontal width at this point is greater than that of the interorbital region. The braincase is rounded, and both parietal ridges and the supraoccipital crest are only weakly developed, even in old specimens. Mastoid islands that are sandwiched between the occipitals and parietals are medium to large in size. The lateral processes of the supraoccipital are narrow and elongated. Paroccipital processes are short, broad, and closely adhered to the posterior surface of the bullae.

SYNONYMS:

Abrocoma Waterhouse, 1837:30; type species *Abrocoma bennettii* Waterhouse, fixed by Tate (1935:384).

Habrocoma Wagner, 1842a:5; unjustified emendation of *Abrocoma* Waterhouse.

REMARKS: Waterhouse (1837) described two species in his new genus *Abrocoma*, *A. bennettii* and *A. cuvieri*. Tate (1935:384) selected the former as the genotype, stating (footnote on p. 384) "I cannot find that the type of *Abrocoma* has previously been designated." Prell (1934b) regarded G. I. Molina's (1782) *Mus laniger* to represent a chinchilla rat (*Abrocoma*) and not a chinchilla (*Chinchilla*, Chinchillidae) where Molina's name has most commonly been associated (see Osgood 1941:409).

The seven species of *Abrocoma* recognized herein fall into three groups. Two contain single taxa: *A. bennettii* from the western Andean slope and coastal region of north-central Chile, and *A. boliviensis* from the east-central Andes of Bolivia. The third group includes the five remaining species that together comprise the *cinerea* complex (*A. cinerea*, *A. budini*, *A. famatina*, *A. schistacea*, *A. uspallata*, and *A. vaccarum*). These are distributed along the spine of the Andes from southern Peru through Bolivia, northeastern Chile, and northwestern Argentina, and extend onto isolated ranges to the immediate east of the Andes in Argentina (Braun and Mares 2002). All species of the *cinerea* complex are considered specialized for life in rocky cliff

faces, with many saxicolous adaptations (Braun and Mares 1996; Mares, 1997). The rock fissures these species typically inhabit accumulate crystalline fecal and urine piles (middens), which as with the North American woodrats of the genus *Neotoma* offer opportunities to examine temporal changes in paleoenvironments through macrofossil analysis and dating (Betancourt and Saavedra 2002; Holmgren et al. 2001; Kuch et al. 2002; Latorre et al. 2002; F. P. Díaz et al. 2012). Braun and Mares (2002) argued that speciation in the genus has been allopatric due to low dispersal abilities coupled with a very patchy habitat and the dietary specialization necessary to cope with unique vegetation components encountered. These authors also suggest that the most basal species in the *cinerea* complex is that which has the least affinity for saxicolous habitats, namely, the more broadly distributed *A. cinerea*. Hypotheses of both allopatric speciation and the basal position of *A. cinerea* require testing with molecular phylogenetic methods.

Verzi and Quintana (2005) assigned fossils from Santa Isabel of the San Andrés Formation (along coastal Buenos Aires Province, Argentina, south of Punta San Andrés), dated at between 2.58 and 2.14 mya, to *Abrocoma*. This locality is substantially distant both geographically and ecologically from the Chilean coastal and Andean region where species in this genus are currently distributed (Osgood 1943b; Braun and Mares 2002:Fig. 2).

Tarifa et al. (2009) detailed the habitat and behavior of a putative undescribed species from the *Polylepis* forests in Cochabamba department, Bolivia, which shares an external morphology similar to *A. boliviensis* and may represent geographic variants of this species. Tarifa et al. watched individuals use their long tail for balance, climb well, and move easily among branches, and suggested that both this animal and *A. boliviensis* are suited for arboreal activity in contrast the more terrestrial oriented *A. bennettii*, *A. cinerea*, and *A. vaccarum* (Glanz and Anderson 1990; Braun and Mares 1996). These observations highlight the potential increased diversity in this genus, both taxonomic and biological, awaiting detailed field studies in remote parts of the range.

KEY TO THE SPECIES OF *ABROCOMA*:

1. Body size medium (head and body length ca. 175 mm); tail proportionately long (approximately 81% of head and body length); dorsal color dark (brownish); ventral color pale (whitish); mastoid islands distinctly small (<16% of mastoid breadth) . . . . . *Abrocoma boliviensis*

1'. Body size medium to large; tail not long (<75 head and body length); dorsal color dark or pale (grayish); ventral color drab (dark grayish) to pale; mastoid islands medium to large (>20% of mastoid breadth) 2

2. Body size large (head and body length >200 mm); tail moderately long (ca. 71% head and body length); hindfoot long (>32 mm); dorsal color dark (brownish); ventral color dark (brownish); interorbital region broad (breadth >8 mm); bullae small (width <23% condylobasilar length)............... *Abrocoma bennettii*

2′. Body large to medium (head and body length <200 m); tail moderate in length (60–70% of head and body length) or short (<55%); hindfoot length shorter (<31 mm); dorsal color paler brownish gray to gray; ventral color dark to pale gray; interorbital region narrow to medium (breadth <7.8 mm); bullae medium to large (width >27% of condylobasilar length) 3

3. Dorsal color pale gray; tail very short (35–42% of head and body length); tail weakly bicolored above and below; interorbital region broad (breadth >7.5 mm); mastoid bullae inflated (width >27% of condylobasilar length) .... *Abrocoma cinerea*

3′. Dorsal color gray with brownish overtones; tail of medium proportions (53–68% head and body length); tail moderately to strongly bicolored; interorbital region narrow (breath <7.0 mm); mastoid bullae moderately to well inflated (width 25–29% of condylobasilar length) 4

4. Hindfeet long (>30 mm); mastoid islands large (>23% of mastoid breadth) 5

4′. Hindfeet smaller (<30 mm, on average); mastoid islands medium (<23% of mastoid breadth 6

5. Body size large (head and body length ca. 200 mm); dorsal color darker (brownish gray); tail moderately bicolored; maxillary toothrow proportionately short (<22% condylobasilar length); posterior border of palate pointed; upper incisors large and broad .... *Abrocoma budini*

5′. Body size medium (head and body length <180 m); dorsal color paler (grayish brown); tail strongly bicolored; maxillary toothrow proportionately long (24% of condylobasilar length); posterior border of palate rounded; upper incisors small and narrow................. .... *Abrocoma uspallata*

6. Tail relatively short (<53% of head and body length, on average); upper incisors small and narrow; bullar inflation great (width >28% condylobasilar length .... *Abrocoma famatina*

6′. Tail relatively longer (>58% of head and body length); upper incisors large and broad; bullar inflation medium (width <27% condylobasilar length)............... 7

7. Dorsal color grayish drab; mesopterygoid fossa wide; upper incisors orthodont .... *Abrocoma schistacea*

7′. Dorsal color gray; mesopterygoid fossa narrow; upper incisors opisthodont .... *Abrocoma vaccarum*

Abrocoma bennettii Waterhouse, 1837
Bennett's Chinchilla Rat

SYNONYMS:

Abrocoma Bennettii Waterhouse, 1837:31; type locality "Chile," refined to "flanks of the Cordillera, near Aconcagua" [near old village of Aconcagua, = Plaza Vieja, ca. 5 km W Los Andes] by C. Darwin (in Waterhouse 1839:85), Valparaíso, Chile.

Abrocoma Cuvieri Waterhouse, 1837:32; type locality "Valparaiso," refined to "dry hills near Valparaiso, Chile" by C. Darwin (in Waterhouse 1839:86).

Habrocoma helvina Wagner, 1842a:7; type locality "Die Heimath des von mir beschriebenen Exemplars ist, wie schon erwähnt, Chile."

H[abrocoma]. Bennettii: Wagner, 1842a:7; name combination.

Habrocoma Bennettii: Wagner, 1842b:288; name combination.

[Habrocoma] Cuvieri: Wagner, 1842b:288; name combination.

Habrocoma cuvieri: Schinz, 1845:100; name combination.

Abrocoma Murrayi Wolffsohn, 1916:6; type locality "Vallenar, Alt. 600 m," Atacama, Chile.

Abrocoma laniger: Prell, 1934b:208; considered to be *Mus laniger* G. I. Molina, 1782 (a *nomen dubium*, based on a composite [Osgood 1941; see also G. M. Allen 1942:390]).

Abrocoma bennetti bennetti Ellerman, 1940:154, Figs. 32–34; name combination, and incorrect subsequent spelling of *bennettii* Waterhouse.

Abrocoma bennetti murrayi: Ellerman, 1940:154; name combination.

DESCRIPTION: Largest species in genus (mean head and body length 206 mm); tail absolutely and proportionately long (length averaging 81% of head and body length, range: 74–88%), long hindfeet (length ≥ 35 mm), and large ears (pinna length >28 mm). Dorsal fur dark brownish, with gray overtones, becoming slightly paler on sides; ventral fur very dark gray basally tipped with pale gray and thus dark, lightly frosted appearance that contrasts with pale venters of other species (Glanz and Anderson 1990:Fig. 19). Tail uniformly colored around, or only slightly paler below; well haired, but not as densely so as either *A. boliviensis* or members of *A. cinerea* complex. Dorsal surfaces of both forefeet and hindfeet clothed in whitish hairs with palmar and plantar surfaces naked but darkly pigmented. Sternal gland present, demarcated by line of whitish hairs over chest in some specimens.

Intermediate between *A. boliviensis* and members of *A. cinerea* complex in most cranial dimensions (Glanz and Anderson 1990:Table 2), although closer and typically overlapping with former species. Skull averages shorter overall (mean condylobasal length 48–49 mm) than that of *A. cinerea* (ca. 50 mm), but is considerably longer than that of *A. boliviensis* (ca. 40 mm). Tympanic and posterior mastoid bullae also intermediate in size between those of *A. boliviensis* and the *A. cinerea* complex, even including variation present in comparing near sea level to high-elevation populations. Interorbital region of *A. bennettii* averages broader (>8 mm) than in other species; equivalent to *A. boliviensis* in nasal length (19–20 mm), but considerably longer than nasals of *A. cinerea* complex (13–14 mm).

DISTRIBUTION: *Abrocoma bennettii* is restricted to the western Andean slopes of north-central Chile, from Antofagastain the north to O'Higgins in the south, and from sea level to 4,000 m (Osgood 1943b; Mann 1978; Guzmán and Siefeld 2011).

SELECTED LOCALITIES (Map 418): CHILE: Antofagasta, Paposo (Guzmán and Siefeld 2011); Atacama, Altamira, coastal mountains of Vallenar (USNM 391845), Domeyko (Osgood 1943b), Piedra Colgada (P. Valladares and Campos 2012), Ramadilla (Osgood 1943b), Vallenar (type locality of *Abrocoma Murrayi* Wolffsohn); Coquimbo, Paihuano (Osgood 1943b), Parque Nacional Fray Jorge (MVZ 118668), Romero (Osgood 1943b); O'Higgins, Banos de Cauquenes (Osgood 1943b); Santiago, Fundo Santa Laura (MVZ 150113); Valparaíso, Aconcagua (type locality of *Abrocoma Bennettii* Waterhouse), La Lingua, Papudo (Osgood 1943b), Limache (Osgood 1943b).

SUBSPECIES: *Abrocoma bennettii* is polytypic, with two subspecies recognized by recent authors (Ellerman 1940; Cabrera 1960; Woods and Kilpatrick 2005): The darker and larger nominotypical subspecies distributed in the central Chilean regions of Aconcagua, Valparaíso, Santiago, and O'Higgins, and the paler and smaller *A. b. murrayi* Wolffsohn known from the northern Chilean regions of Antofagasta, Atacama, and Coquimbo. Osgood (1943b:107) regarded these two taxa to intergrade based on the specimens available to him, thus confirming Ellerman's (1940) earlier decision to list both as valid subspecies of a polytypic *A. bennettii*.

NATURAL HISTORY: Charles Darwin (in Waterhouse 1839:86) noted that his specimen "was caught amongst some thickets in a valley on the flanks of the Cordillera, near Aconcagua. On the elevated plain, near the town of Santa Rosa, in front of the same part of the Andes, I saw two others, which were crawling up an acacia tree, with so much facility, that this practice must be, I should think, habitual with them." Subsequent field observers have confirmed the partial arboreal habitus of *A. bennettii* (E. C. Reed 1877). Darwin (in Waterhouse 1839:87) also stated

that "the species is abundant on the dry hills, partly covered with bushes, near Valparaiso," a reference to the characteristic Mediterranean habitat of Chile (see also Cofré et al. 2007). Fulk (1976) observed *A. bennettii* using the burrows of the degu (*Octodon degus*).

This species is part of a small mammal community that has been intensively studied in the Mediterranean habitat of central Chile. It inhabits rocky matorral and shrublands from sea level and coastal canyon to 2,000 m in the Andes. In the coastal desert zones in northern Chile, it occurs in local habitats influenced by fog and dominated by cacti and large, succulent shrubs (Guzmán and Siefeld 2011), but habitats include agricultural, wetland, and riparian shrublands (P. Valladares and Campos 2012). Data are available on population ecology, including the interplay between biotic and abiotic factors (Meserve et al. 1999), species assemblage (Iriarte, Contreras, and Jaksic 1989), patterns of temporal abundance and persistence (Jaksic et al. 1981; J. E. Jiménez et al. 1992), demographic responses to an El Niño Southern Oscillation (ENSO) event (Meserve et al. 1995), trophic relationships (Meserve 1981a), feeding ecology (Meserve 1983), and resource partitioning (Meserve 1981b), primarily but not exclusively on populations within Parque Nacional Fray Jorge. This species is also a prey item of the grison (*Galictis cuja*) and two species of local foxes (*Lycalopex culpaeus* and *L. griseus*) as well as the Barn Owl (*Tyto alba*) (Durán et al. 1987; Ebensberger et al. 1991; Iriarte, Jimenez et al. 1989; Jaksic et al. 1980; Meserve et al. 1987). Beaucournu et al. (2011) described a species of flea in the genus *Delostichus* and M. S. Gómez (1998) described the louse *Eulinognathus chilensis*, both taken from Chilean specimens.

Durden and Webb (1999), Gomez (1998), Moreno Salas et al. (2005), described ectoparasites.

REMARKS: Wolffsohn (1916:6) noted that skins of this species were "frequently offered for sale as true chinchilla [= *Chinchilla*] skins." Thomas (1927b:553) regarded *A. cuvieri* Waterhouse to be "unquestionably the young of *A. bennettii*," certainly true since the third molar has yet to erupt in the maxilla and mandible figured by Waterhouse (1839:plate 33, Fig. 1e,f). Osgood (1943b:107) noted that the old village of Aconcagua (the type locality) "has now disappeared and is represented only by a 'fundo' called Plaza Vieja. This is some five kilometers west of the present town of Los Andes." L. C. Contreras et al. (1993) described the soft anatomy of the glans penis and baculum. M. H. Gallardo et al. (2002) detailed sperm morphology. Spotorno et al. (1988) described the karyotype as $2n = 64$, FN = 110; M. H. Gallardo (1992) illustrated the C-banded karyotype; and Spotorno et al. (1995) discussed chromosomal divergence within the Octodontoidea. The species

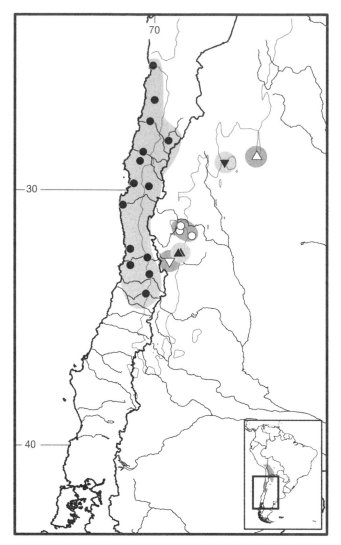

Map 418 Selected localities for *Abrocoma bennettii* (●), *Abrocoma budini* (△), *Abrocoma famatina* (▼), *Abrocoma schistacea* (○), *Abrocoma uspallata* (▲), and *Abrocoma vaccarum* (▽). Contour line = 2,000 m.

has been recovered from Holocene deposits spanning the last 8,000 years in central Chile (Simonetti 1994).

Abrocoma boliviensis Glanz and Anderson, 1990
Bolivian Chinchilla Rat

SYNONYMS:

Abrocoma boliviensis Glanz and Anderson, 1990:23; type locality "Comarapa, 2500 m, province of Manual M. Caballero, department of Santa Cruz, Bolivia; 17°54′S and 64°29′W."

Cuscomys boliviensis: Verzi and Quintana, 2005:310; name combination.

DESCRIPTION: Medium sized, head and body length 170–178 mm, with relatively small hindfeet (length 30 mm), moderately long tail (132–150 mm), and short ears (22–

25 mm). Tail well furred, more so than in either *A. bennettii* or members of *A. cinerea* complex. Dorsal fur short and not as soft as in other species. Overall dorsal coloration dark gray-brown, contrasting strongly with gray-based but white-tipped hairs of venter; patches of hair on chest and inguinal areas white to their base; dark throat patch present in both known specimens, more distinctly obvious in holotype; narrow midline white streak on chest may indicate presence of a sternal gland. Tail indistinctly bicolored, paler below. Dorsal surfaces of forefeet and hindfeet clothed with whitish hairs, palmar and plantar surfaces naked and less pigmented than other species.

Skull smaller than in other species, with condylobasal length averaging 40.1 mm (range: 39.0–41.1 mm) as opposed to >48 mm. Nasals relatively shorter and do not extend posterior to anterior edge of the orbit. Tympanic and mastoid portions of bullae notably less inflated, although more similar to bullae of low elevation populations of *A. bennettii* in coastal Chile.

DISTRIBUTION: *Abrocoma boliviensis* is known only from the vicinity of the type locality.

SELECTED LOCALITIES (Map 419): BOLIVIA: Santa Cruz, Comarapa (type locality of *Abrocoma boliviensis* Glanz and Anderson).

SUBSPECIES: *Abrocoma boliviensis* is monotypic.

NATURAL HISTORY: Nothing is known about the behavior, ecology, or other aspects of the population biology of this species. O. P. Pearson (field notes, MVZ archives) collected the second known specimen in a trap set "along a ridge with rock outcrops plus bushes, grass, and many succulents, plus a few orchids . . . and ferns" in an area otherwise characterized as brushy vegetation with patches of cloud forest.

REMARKS: As noted by Glanz and Anderson (1990:24) there is some uncertainty about the place of origin of the holotype, and thus details about the exact type locality. The town of Comarapa is at 1,815 m (Atlas Censal de Bolivia), but the terrain around the town reaches 2,500 m within a few kilometers to the north, west, and south. The specimen thus likely came not from the town of Comarapa itself, but from the surrounding high country. The second, and only other specimen known of this species, was collected 8 km west of Comarapa, by road, at an elevation of 7,500 ft (ca. 2,270 m; O. P. Pearson field notes, MVZ archives). The area where the MVZ specimen was collected in 1955 had been greatly altered by the mid-1980s when S. Anderson (AMNH) visited the region, by increased intensity of human use, notably heavy grazing and cultivation (Glanz and Anderson 1990:26). Possibly this presumably rare species has either gone locally extinct or its abundance has been severely impacted. Additional fieldwork is critically needed in the vicinity of the type locality and other areas in the

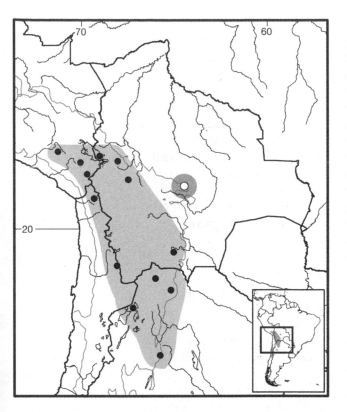

Map 419 Selected localities for *Abrocoma boliviensis* (○) and *Abrocoma cinerea* (●). Contour line = 2,000 m.

diverse "valles" region of central Bolivia (Glanz and Anderson 1990:26). Valim and Linardi (2008) described the host association of ectoparasitic lice in the genus *Gyropus*.

Verzi and Quintana (2005) referred *boliviensis* to *Cuscomys*, based on shared features of the molar teeth and similar small coronoid processes. Pending future phylogenetic analyses that should include both morphological and molecular traits, we retain *boliviensis* within *Abrocoma* because the species possesses none of the diagnostic characters of *Cuscomys* presented by Emmons (1999a) in her description of this genus. Interestingly, Emmons (1999a) considered *A. boliviensis* along with *Cuscomys* to be the morphologically least specialized Abrocomidae.

Abrocoma budini Thomas, 1920
Budin's Chinchilla Rat

SYNONYMS:

Abrocoma budini Thomas, 1920b:476; type locality "Otro Cerro, Catamarca. Alt. 3000 m" (about 18 km NNW Chumbicha), now considered an abandoned ranch at the southern end of Sierra de Ambato, NE Chumbicha, Capayán, Catamarca, Argentina, 28°45′S, 66°17′W, 2,023 m (Pardiñas et al. 2007).

Abrocoma cinerea budini: Ellerman, 1940:155; name combination.

DESCRIPTION: Largest species in *A. cinerea* complex, total length >320 mm, head and body length >130 mm, hindfoot length >30 mm; darker both dorsally and ventrally than other species in complex. Coloration uniformly brownish gray, back slightly darker than sides, venter grayish drab, with hairs dark grayish or slate for about three-fourths of their length and with pale drab tips. Hairs of sternal gland white to base, thus forming white patch contrasting strongly with surrounding ventral coloration; chin grayish white, underside of neck darker than other ventral areas; hairs surrounding perianal region with buffy or whitish tips. Like *A. famatina*, drab throat patch present. Forefeet and hindfeet covered dorsally with whitish hairs. Tail dark drab to grayish above contrasting with grayish white to whitish below. Cranially, nasals evenly narrow for about half to two-thirds of length, becoming markedly convergent posteriorly, ending in a point. Dorsal edge of zygoma slanted outward such that medial edge is visible from above; anterior ascending maxillary portion of zygoma thin. Zygomatic arches expanded laterally; frontals noticeably constricted. Ventrally, posterior border of palate pointed, mesopterygoid fossa wide, mastoid islands large, and bullae slightly inflated. Upper incisors broad, large, and opisthodont. Hypoflexus and metaflexus of upper molars about equal in size and positioned more or less horizontally. Major cusps of upper cheek teeth more rounded and less anteroposteriorly compressed then in other species; all teeth subequal in size and shape. The overall length of upper toothrow short in relation to length of skull, similar in proportional length to that of *A. famatina* and *A. vaccarum*, smaller than that of *A. cinerea* and *A. schistacea*, and much smaller than that of *A. uspallata*. In comparison to other species in complex, *A. budini* significantly larger in cranial length, diastema length, nasal length (except *A. schistacea*), and mandible length (except *A. schistacea*). Toothrows and palate significantly longer, and wider zygomatic arches, than *A. famatina*, but interorbital breadth significantly narrower than *A. cinerea*.

DISTRIBUTION: *Abrocoma budini* is known only from the Sierra de Ambato in Catamarca province, Argentina.

SELECTED LOCALITIES (Map 418): ARGENTINA: Catamarca, Cerro Ambato (Glanz and Anderson 1990), Otro Cerro (type locality of *budini* Thomas).

SUBSPECIES: *Abrocoma budini* is monotypic.

NATURAL HISTORY: No data on ecology, behavior, or any aspect of life history are known for this species, although Barquez et al. (1991) stated that the species (regarded as a subspecies of *A. cinerea*) lives in rocks and is apparently colonial. Thomas (1920b:476), in his description of this taxon, noted that the collector, Emilio Budin, found it "among rocks, in the clefts of which it lives."

REMARKS: As with other species in the *Abrocoma cinerea* complex, *A. budini* is a geographic outlier to the broad range of *A. cinerea* and, as such, was considered only subspecifically distinct from that species by Ellerman (1940) and Cabrera (1961). Braun and Mares (2002) elevated *A. budini* to species status. The karyotype is unknown.

Abrocoma cinerea Thomas, 1919
Ashy Chinchilla Rat
SYNONYMS:

Abrocoma cinerea Thomas, 1919f:132; type locality "Cerro Casabindo, 4800 m," Jujuy, Argentina.

Abrocoma cinerea cinerea: Ellerman, 1940:154; name combination.

DESCRIPTION: Medium sized, total length <286 mm, head and body length <196 mm, tail length <96 m, hindfoot length typically ≤ 27 mm). Tail notable absolutely shorter than any other species, total length and hindfoot length less than those of *A. budini*, *A. famatina*, and *A. schistacea* (note, however, that the measurements given by Braun and Mares [2002:Table 1] include very small dimensions [lower range: 15.6 mm], likely from either juvenile specimens or those measured without the claw; Glanz and Anderson [1990:Table 1] give the range for nine specimens as 23–30 mm, mean 27.8 mm). Mean ratio of tail length to body length less than that for other species. Overall dorsal coloration pale gray, with ventral hairs gray basally with whitish tips, giving under parts whitish appearance. Sternal gland covered with white hairs, but does not contrast with very pale venter; hairs of perianal region white to the base. Both forefeet and hindfeet covered with whitish hairs. Tail indistinctly bicolored, pale gray above and whitish or grayish-white below. Two pairs of mammae in adult females, one axillary, other placed laterally on abdomen.

Skull smaller than that of *A. budini* but larger than *A. famatina*. Interorbital width greater than that of other species in group; nasals broad at midlength, tapering gradually posteriorly; zygomata not slanted outward but oriented more or less vertically, in contrast to all other species except *A. bennetti*; zygomatic arches moderately expanded; frontals not constricted; mastoid islands only of medium size; posterior border of palate terminates in sharp point; mesopterygoid fossa wide, paroccipital processes narrow and not flattened; bullae inflated. Upper incisors broad, large, and opisthodont. Hypoflexi appear slightly narrower than in other species. Length of maxillary toothrow medium in relation to skull length, about equal to that of *A. schistacea*, larger than that of *A. budini*, *A. famatina*, and *A. vaccarum*, but much smaller than that of *A. uspallata*.

DISTRIBUTION: *Abrocoma cinerea* is distributed throughout the high Andes (above 3,700 m) from Arequipa in southern Peru, south through Bolivia, northwestern Argentina (Catamarca), and northeastern Chile (Antofagasta and Tarapacá). Late Pleistocene fossils from 1,562 m in Catamarca, Argentina, support a lower elevational and likely broader range, at least at the end of the last glacial maxima (P. E. Ortiz et al. 2011a).

SELECTED LOCALITIES (Map 418): ARGENTINA: Jujuy, Cerro Casabindo (type locality of *Abrocoma cinerea* Thomas), Sierra de Zenta (Thomas 1919f); Catamarca, Pasto alto (Braun and Mares 2002); Salta, Los Andes (Braun and Mares 2002); Tucumán, Concepción (Braun and Mares 2002). BOLIVIA: La Paz, Copacabana (S. Anderson 1997), Tembladerani (S. Anderson 1997); Oruro, Huancaroma (S. Anderson 1997); Tarija, Sama (S. Anderson 1997). CHILE: Antofagasta, Tocone (MVZ 116801); Arica y Parinacota, Arica (FMNH 134301). PERU: Arequipa, 2 km W of Sumbay (MVZ 174291); Puno, Caccachara (Pearson 1951); Tacna, 2 km N of Nevado Livine (Pearson 1951).

SUBSPECIES: As currently understood (Braun and Mares 2002), *A. cinerea* is monotypic.

NATURAL HISTORY: Emilio Budin collected the type specimen "among the rocks on the Volcano of Casabindo" (Thomas, 1919f:133) and specimens from the Sierra de Zenta "among the rocky volcanic mountains" (Thomas, 1921i:216). In northern Chile, Koford (1955) collected specimens "in their track ways or at the mouths of their open burrows [2 inches in diameter] at the bases of boulders or low resinous *tola* shrubs. On dry sandy slopes between low bushes and tufts of fescue, the round footprints, usually placed alternately as if the animal most often moved by running rather than by hopping, were followed for a distance as great as 200 yards." He also noted that captive individuals readily ate the flowers and branch tips of the three most common shrubs (*Lepidophyllum rigidum*, *L. quadrangulare*, and *Baccharis microphylla* [Asteraceae]). The shrub *Baccharis tola* was the main plant consumed in the Parque Nacional Llullaillaco in northern Chile, with the bunch grass *Stipa chrysophylla* of secondary importance (Tirado et al. 2012). Mann (1978), also in northern Chile, found this species inhabiting large rocks and carpets of yareta (*Laretia* [Apiaceae]). Pearson (1951) collected the species in the Puno department, southern Peru, at burrows dug at the edge of shale outcroppings or at the base of shrubs, and J. L. Patton (field notes, MVZ archives) obtained a single specimen in Arequipa department, Peru, among weathered granite boulders in the Altiplano. S. Anderson (1997) stated that this species was widely distributed in the Altiplano of Bolivia, where often it could be found in stonewalls. Both Pearson (1951) and Mann (1978) suggested that *A. cinerea* lives in colonies, where they feed primarily on shrubs in the family Asteraceae (*Senecio* and *Parastrephia* [Chile] and *Senecio* or *Lepidophyllum* [Peru]). Gut morphology, in particular a long, convoluted caecum and a thick intestine with tight flexures, suggests a diet high in cellulose. Reproductively active females (pregnant or lactating) have been taken in the months of August, November, and December in both Chile and Peru (Pearson 1951; Mann 1978). Litter size in the few pregnant females collected was two or three. For populations in Salta Province, Argentina, Mares, Ojeda, and Barquez (1989) stated that the species was diurnal, herbivorous, and an excellent climber that lived in rock fissures or in burrows under rocks. These authors also suggested that the species is colonial. Cortes et al. (2002) described food habits of populations in northern Chile. Hugot and Gardner (2000) describe endoparasitic nematodes, based on specimens from Bolivia.

REMARKS: The locality record for the Santiago region in Chile (10 km W Tiltil) in Braun and Mares (2002:9 and map, pg. 7) continues an error in identification originally given by Glanz and Anderson (1990:9, caption to Fig. 3). The specimen in question (MVZ 150117) is an *Abrocoma bennettii*, not *A. cinerea* as stated. L. C. Contreras et al. (1993) described and figured the morphology of the glans penis and baculum, based on specimens from southern Bolivia. M. H. Gallardo et al. (2007) gave the diploid number as 64.

Abrocoma famatina Thomas, 1920
Famatina Chinchilla Rat

SYNONYMS:

Abrocoma famatina Thomas, 1920g:419; type locality "La Invernada, Famatina Range [= Sierra de Famatina], Rioja, 3800 m," Argentina.

Abrocoma cinerea famatina: Ellerman, 1940:155; name combination.

DESCRIPTION: Medium sized for genus, head and body length 164–182 mm, tail length 108–117 mm, hindfoot length 28–30 mm. Ratio of tail length body length similar to that of *A. budini*, but *budini* considerably larger in overall size. Dorsal color overall grayish with brownish tints along midline and across rump. Hairs of venter gray basally with whitish or pale drab tips; hairs of throat dark drab; hairs of sternal gland white to base, forming distinct white patch on mid-venter; hairs of perianal region whitish. Forefeet and hindfeet covered with white or whitish hairs. Tail bicolored, grayish above and whitish below. In color, *A. famatina* somewhat darker than *A. cinerea*, similar to *A. vaccarum*, but paler than both *A. budini* and *A. schistacea*. Presence of drab throat patch in common only with *A. budini*. Cranially, nasals slightly wider at midlength and taper posteriorly; dorsal edges of zygoma slanted outward such that medial edges easily visible in dorsal view; ascending maxillary portions of zygoma thin; zygomatic arches expanded, as in other species in this group; frontals noticeably constricted and mastoid islands medium in size. Ventrally, posterior border of palate rounded; mesopterygoid fossa wide; bullae inflated. Upper incisors narrow, small, and opisthodont to orthodont. Length of maxillary toothrow medium in relation to skull length, about equal to that of *A. budini* and *A. vaccarum*, smaller than for *A. cinerea* and *A. schistacea*, and much smaller than for *A. uspallata*.

DISTRIBUTION: *Abrocoma famatina* is known only from the Sierra de Famatina in Rioja, Argentina. Cabrera (1961) suggested that the species extends to the Sierra del Valle Fertil and Sierra Pie de Palo in eastern San Juan province, but this is unlikely due to geographic barriers separating these ranges from the Sierra de Famatina (Braun and Mares 2002).

SELECTED LOCALITIES (Map 418): ARGENTINA: Rioja, Famatina (Braun and Mares 2002), La Invernada (type locality of *Abrocoma famatina* Thomas).

SUBSPECIES: *Abrocoma famatina* is monotypic.

NATURAL HISTORY: No data are available on the behavior or ecology of this species.

REMARKS: As with other members of the *A. cinerea* complex, *A. famatina* is limited to an isolated extension of the Andes in northwestern Argentina, and is thus an geographic outlier to the broad range of *A. cinerea*. Thomas (1920g) regarded this species as closest to *A. budini*. Considered a subspecies of *A. cinerea* by Ellerman (1940) and Cabrera (1961), *A. famatina* was raised to species status by Braun and Mares (2002). The karyotype is unknown.

Abrocoma schistacea Thomas, 1921
Sierra Tontal Chinchilla Rat

SYNONYMS:

Abrocoma schistacea Thomas, 1921i:216; type locality "Los Sombreros, Sierra Tontal. Alt. 2700 m," San Juan, Argentina.

Abrocoma cinerea schistacea: Ellerman, 1940:155; name combination.

DESCRIPTION: Medium-sized rat, head and body length 160–196 mm, tail length 96–120 mm, hindfoot length 25–30 mm. Overall dorsal color grayish drab, slightly darker along midline, with under parts similar but slightly paler to dorsum. Drab throat patch lacking; sternal gland present, covered by hairs white to their base; whitish hairs surround perianal region. Tail bicolored, similar to dorsal body color above but whitish yellow below. Forefeet and hindfeet covered dorsally with whitish hairs. All other species in genus, except *A. budini*, paler in both dorsal and ventral coloration. External measurements significantly smaller than those of *A. budini*, significantly larger than *A. cinerea*, but similar to *A. famatina* except for its smaller hindfoot. Mean ratios of tail to body length similar to values for *A. uspallata* and *A. vaccarum*. Cranially, characterized by nasals wider at midlength and taper posteriorly; dorsal edge of zygoma slanted laterally such that medial edge easily visible in ventral view; anterior ascending maxillary portion of zygoma thin; zygomatic arches expanded; frontals noticeably constricted; mastoid islands of medium size. Ventrally, posterior border of palate rounded; mesopterygoid fossa wide; bullae only slightly inflated. Upper incisors broad, large, and more orthodont than in other species. Molariform teeth appear broad, but length of maxillary toothrow medium in relation to length of skull about equal to that of *A. cinerea*, longer than *A. budini*, *A. famatina*, and *A. vaccarum*, but much shorter than that of *A. uspallata*. Hypoflexus narrower in breadth than in other species. Taraborelli et al. (2011) noted, in particular, presence of buccal epidermal denticles (or "false teeth") on midline of palate and tongue with horny pad.

DISTRIBUTION: *Abrocoma schistacea* is known from three localities in southern San Juan province, Argentina (Thomas 1921i; Braun and Mares 2002; Taraborelli et al. 2011), all within the Monte Desert biome at elevations between 1,100 and 2,900 m.

SELECTED LOCALITIES (Map 418): ARGENTINA: San Juan, Los Sombreros (type locality of *Abrocoma schistacea* Thomas), Parque Nacional El Leoncito (Taraborelli et al. 2011), Pedernal (Thomas 1921i).

SUBSPECIES: *Abrocoma schistacea* is monotypic.

NATURAL HISTORY: Few data on behavior, ecology, or reproduction are available. Taraborelli et al. (2011) found the species in shrub habitats dominated by *Larrea nitida* and *L. divaricata* (Zygophyllaceae) in the Parque Nacional El Leoncito, San Juan province, Argentina. Individuals occupied burrow systems within crevices on hillsides with rocky flagstone walls. Animals lived in groups of three to four, consisting of a single male and two females or two of each sex; density averaged 0.15 individuals per ha^{-1}. They used readily identifiable latrine sites composed of piles of fecal pellets. Animals were active primarily in the early morning. Feeding was selective, rather than in relation to plant availability, with *Larrea* being the dominant food plant, especially in summer and autumn.

REMARKS: A geographic outlier to the broadly distributed *A. cinerea*, this taxon was regarded as a subspecies of *A. cinerea* by Ellerman (1940) and Cabrera (1961) but raised to species status by Braun and Mares (2002) based on comparative morphometrics analyses. The karyotype is unknown.

Abrocoma uspallata Braun and Mares, 2002
Uspallata Chinchilla Rat

SYNONYM:

Abrocoma uspallata Braun and Mares, 2002:9; type locality "Argentina: Mendoza Province: Quebrada de la Vena,

ca. 7 km SSE of the village of Uspallata, 32°39.405′S, 69°20.970′W, 1,880 ± 150 m."

DESCRIPTION: Medium sized, mean head and body length 175.3 mm, tail length 118.7 mm, hindfoot length 29.2 mm, and mass 157.3 g; hindfeet and ears large relative to body size. The dorsal fur grayish brown with lighter tones at tips on some hairs but black on others; under parts gray basally with creamish tips, and thus an overall appearance of lightly colored venter; sternal gland and area surrounding the perianal region covered with hairs white to base and not contrasting with overall ventral color. Forefeet and hindfeet covered with white hairs. Skull with nasals evenly narrow along entire length, tapering only at posterior end; frontals constricted; mastoid islands large; zygomatic arches moderately expanded but do not extend beyond greatest width of skull; zygomata delicate, with anterior ascending processes of maxillae reduced to slender struts; jugal parts of zygomatic bridges reduced to ventral bases of bridges and, thus jugals well separated from lacrimals; incisive foramina elongated and narrow, broadening slightly posteriorly; palate short and narrow, extending posteriorly to about posterior margins of M2, with rounded posterior border; posterior palatal foramina present as two small, separate entities; pterygoids delicate and adjacent foramina large; mesopterygoid fossa ; narrow; bullae inflated, covering middle lacerate foramina, and large relative to cranial length; paroccipital processes broad, flattened, and fused to bullae posteroventrally. Upper incisors small, narrow, short, opisthodont, with orange enamel. Upper toothrows converge only slightly anteriorly. Lophs of cheek teeth transverse in orientation with cusps of upper molars opposite and compressed anteroposteriorly.

Distinguished from other species in genus by grayer dorsal and whiter ventral coloration (except *A. cinerea*), longer tail (except *A. budini*, which is longer, and *A. famatina*, about equal), larger hindfoot, larger ears, narrower nasals, moderately expanded zygomatic arches (except *A. cinerea*), larger mastoid islands (except *A. budini*), narrower mesopterygoid fossa (except *A. vaccarum*), relatively larger bullae, wide and flattened paroccipital processes, smaller incisors (except *A. famatina*), and relatively longer toothrows. Braun and Mares (2002) provided additional characters explicitly separating this species from other members of the *A. cinerea* complex.

DISTRIBUTION: *Abrocoma uspallata* is known from two localities in the Sierra de Uspallata, northwestern Mendoza Province, Argentina (Braun and Mares 2002; Taraborelli et al. 2011), within the Monte Desert biome at elevations between 1,850 and 2,150 m.

SELECTED LOCALITIES (Map 418): ARGENTINA: Mendoza, Quebrada de la Vena (type locality of *Abrocoma uspallata* Braun and Mares), Uspallata (Taraborelli et al. 2011).

SUBSPECIES: *Abrocoma uspallata* is monotypic.

NATURAL HISTORY: This species occurs at the lowest elevation of any in the genus, with the holotype collected in a rock outcrop within the precordilleran Monte Desert characterized by creosote bush, palo verde, and various cacti (Braun and Mares 2002; Taraborelli et al. 2011). Burrows were found in crevices among rock blocks or flagstones on steep slopes, where vegetation was dominated by bunch grasses (*Stipa* sp.) and shrubs (*Larrea divaricata* and *Atriplex lampa* [Zygophyllaceae]; Taraborelli et al. 2011). Braun and Mares (2002) argued that the species was a specialist on creosote bush, which suggested a long association with the Monte Desert because of the need to evolve necessary adaptations to cope with the toxic compounds characterizing this plant. Taraborelli et al. (2011), however, found *Larrea* to comprise only 30% of the diet, with shrubs of the genera *Lycium* (Solanaceae) and *Schinus* (Anacardiaceae) taken in nearly equal proportions. Individuals are diurnal, being most active in the morning hours, and use common latrines near their burrow systems.

REMARKS: This species is known from a single specimen, the holotype, so aspects of individual variation are unknown. The karyotype has a diploid number of 66; the fundamental number could not be determined, but the majority of the chromosomes are apparently biarmed (Braun and Mares 2002).

Abrocoma vaccarum Thomas, 1921
Mendozan Chinchilla Rat

SYNONYMS:

Abrocoma vaccarum Thomas, 1921i:217; type locality "Punta de Vacas, Altitude 3000 m, North-western Mendoza," Argentina.

Abrocoma cinerea vaccarum: Ellerman, 1940:155; name combination.

DESCRIPTION: Medium sized, head and body length 165–191 mm, tail length 108–117 mm, hindfoot length 25–27 mm). Overall dorsal tones grayish, with some brownish tints, especially along midline and over rump. Hairs of venter gray basally for most of length, with whitish tips; sternal gland and perianal area covered with hairs white to base, contrasting sharply with generally gray under parts. The forefeet and hindfeet are covered with whitish hairs. Tail bicolored, grayish above and whitish below. In other species, except *A. cinerea*, tail darker dorsally and somewhat darker below, although all species share generally bicolored tail. Drab throat patch present, as in *A. budini* and *A. famatina*. Skull with nasals slightly wider at midlength and tapering posteriorly; dorsal edge of zygoma slanted outward such that medial edge easily visible in dorsal view;

anterior ascending maxillary portion of zygoma thin; frontals well constricted, as in other species of group; mastoid islands medium in size; posterior border of palate rounded; mesopterygoid fossa narrow; bullae only slightly inflated. Upper incisors broad, large, opisthodont to orthodont, with white to very pale orange enamel. Molariform teeth appear slightly broader than those of other species, except *A. schistacea*. Length of maxillary toothrow in relation to cranial length about equal to that of *A. budini* and *A. famatina*, shorter than that of *A. cinerea* and *A. schistacea*, and much shorter than that of *A. uspallata*.

DISTRIBUTION: *Abrocoma vaccarum* is known only from the type locality in Mendoza Province, Argentina. Cabrera (1961) extended the range to include southwestern San Juan, but no specimens are known from that region.

SELECTED LOCALITIES (Map 418): ARGENTINA: Mendoza, Punta de Vacas (type locality of *Abrocoma vaccarum* Thomas).

SUBSPECIES: *Abrocoma vaccarum* is monotypic.

NATURAL HISTORY: No data on are available on ecology, behavior, or other aspects of the biology of *Abrocoma vaccarum*. The habitat at the type locality consists of the short grass and low shrub vegetation characteristic of the Puna-like high Andean vegetation (Braun and Mares 2002). Beaucournu and Gallardo (2005) described ectoparasitic fleas.

REMARKS: *Abrocoma vaccarum* is another geographic outlier from the range of the broadly distributed and presumably closely related species, *A. cinerea* (Braun and Mares 2002:7). Considered a subspecies of *A. cinerea* by Ellerman (1940) and Cabrera (1961), *A. vaccarum* was raised to species status by Braun and Mares (2002). The karyotype is unknown.

Genus *Cuscomys* Emmons, 1999

The genus contains the largest species in the family (*C. ashaninka*), with the second known species (*C. oblativus*) only slightly smaller. Because the latter is only known from osteological remains, little external morphology was included in Emmons's (1999a) diagnosis of the genus. However, many of the very distinctive external morphological features of the genotype, detailed below, may also be diagnostic of the genus. In 2009 a living *Cuscomys* specimen was captured and later released about 3 km from the type locality of *C. oblativus*. Based on a series of photos (Julio G. Ochoa, pers. comm., with photographs, to L. H. Emmons), this animal showed few differences from *C. ashaninka*. We presume this individual to have been *C. oblativus*, and describe its external morphology as such, pending collection of a complete specimen for definitive comparison.

Emmons (1999a:2) diagnosed *Cuscomys* as the largest known living abrocomid rodent and the only genus morphologically adapted for an arboreal habitus. It has a long tail, greater than 75% of the head and body length; broad feet, with strong curved claws; and a stout and long hallux, reaching beyond the base of the second toe. The skull is large (condylobasal length >60 mm), with a robust and long rostrum nearly as broad as the interorbital region. It may be either flat or arched in dorsal profile when viewed laterally. The nasals are inflated distally, and are either blunt posteriorly (*C. ashaninka*) or separated (*C. oblativus*). The frontals are constricted behind the postorbital process such that the width at this point is equal to or less than the interorbital breadth. The palate between the upper toothrows is only slightly concave, with a prominent ridge along the median suture of the palatine bones. A temporal foramen is present posterior to the postglenoid vacuity. The dorsal extension of the mastoid between the occipitals and parietals is small and not inflated (it does not bulge dorsally). The sigmoid notch of the mandible is long and the medial condyloid ridge does not approach or meet the posterior edge of the mandible. The head has a pure white blaze from crown to chin, and the tail is sharply bicolored dark for the basal half, white for the distal half.

SYNONYM:

Cuscomys Emmons, 1999a:2; type species, *Cuscomys ashaninka* Emmons, by original designation.

REMARKS: Woods and Kilpatrick (2005:1575) incorrectly cited *A. oblativus* Eaton as the type species. These authors also used the feminine ending for the species epithet despite the fact that Emmons (1999a:2) clearly stated "the generic name is masculine." Verzi and Quintana (2005) included *Abrocoma boliviensis* within this genus, based on shared features of the maxillary teeth and small coronoid process. But, as noted in that account, *A. boliviensis* shares none of the diagnostic characters of *Cuscomys* detailed by Emmons (1999a) in her description of the genus. We thus leave *boliviensis* within *Abrocoma*, pending the necessary phylogenetic analyses of either, or preferably both, morphological and molecular characters.

KEY TO THE SPECIES OF *CUSCOMYS* (SKULL CHARACTERS ONLY, BECAUSE EXTERNAL FEATURES ARE KNOWN ONLY FOR ONE OF THE TWO KNOWN SPECIES):

1. Larger skull (greatest skull length >66 mm); skull flat in dorsal profile; foramen magnum much wider than high; premaxillae narrow posteriorly; posterior loph of M3 directed labially, posterior border straight . . . . *Cuscomys ashaninka*

1′. Smaller skull (greatest skull length >64 mm); skull arched in dorsal profile; foramen magnum subcircular,

slightly higher than wide; premaxillae widening distinctly posteriorly; posterior loph of M3 directed posteriorly, posterior border strongly concave . . . . *Cuscomys oblativus*

Cuscomys ashaninka Emmons, 1999
Ashaninka Arboreal Chinchilla Rat

SYNONYM:

Cuscomys ashaninka Emmons, 1999a:2; type locality "Peru: Departamento de Cusco, northern Cordillera de Vilcabamba, 11°39′36″S; 73°40′02″W (by GPS, map datum WGS-84); elevation 3370 m," corrected to Junín department by Quintana Navarrete (2011).

DESCRIPTION: Large (mass >900 g; head and body length >300 mm), with long tail (tail length >200 mm), and long but wide hindfeet (length >60 mm). Fur dense and long, pale gray dorsally with abundant black guard hairs 40 mm long extending 20–25 mm beyond underfur in mid-dorsum, and giving blackish coloration along midback; sides paler, sprinkled with mixture of white and dusky overhairs; dorsal underfur pale gray, soft, and somewhat wavy. Head with narrow white blaze from nose to crown; lips and chin white. Venter gray, with lightly frosted hairs, and not contrasting with sides; midline of chest between forelegs with streak of short hairs, probably associated with a sternal gland. Mystacial vibrissae dense, stiff, and either white or dusky, and extend behind shoulder when flattened against body. Three supercilliary vibrissae present but no genal vibrissae evident; single vibrissa present on mid-forelimb of holotype, and only known specimen. Ear pinnae nearly naked, with fur at base sufficiently long to reach ear tips. Tail robust and long (76% of head and body length), clothed with stiff hairs that nearly hide underlying scales, and sharply bicolored dark for basal half and white for distal half. Forefeet have four toes, without evidence of a pollex; hindfeet with hallux without ungual tuft, other four digits with long, dusky tufts reaching beyond tips of stout, strongly curved claws. Feet thinly clothed with dusky hairs; broad with soles pigmented proximally but not distally (white in fresh specimen), covered with tiny tubercles; palmar and plantar pads soft and poorly defined. Only single pair of inguinal mammae was discovered, but specimen was primaparous, and its nipples small.

Skull large and robust (greatest skull length >65 mm), dorsal profile flat, and rostrum deep and broad. Nasal bones broad and slightly inflated distally, two posterior tips meet in straight line with no intercalation of the frontals; premaxillary bones of rostrum do not widen posteriorly; frontals widest anterior to postorbital process and constricted posteriorly; dorsal projections of mastoid bones small and uninflated; parietal ridges and supraoccipital crest strongly developed. Incisors large and robust; upper incisors proodont. Maxillary toothrows slightly divergent posteriorly. Mandible robust in comparison to species of *Abrocoma*, with angular and condyloid processes dorsoventrally deep; medial condyloid ridge passing midway up center of process, not approaching its posterior edge; external tip of condyloid process with well-defined, short medial crest; and low coronoid process. Occlusal pattern of cheek teeth similar to other species in family, but with lingual flexi at shallow angle and nearly equal in length with labial flexi such that mid-tooth mure short and nearly parallel to tooth axis.

DISTRIBUTION: *Cuscomys ashaninka* is known only from the type locality.

SELECTED LOCALITIES (Map 420): PERU: Junín, Cordillera de Vilcabamba (type locality of *Cuscomys ashaninka* Emmons).

SUBSPECIES: *Cuscomys ashaninka* is monotypic.

NATURAL HISTORY: The holotype, and single known specimen, was found dead, probably killed by a long-tailed weasel (*Mustela frenata*). The specimen was found on a steep slope in tall, wet, mossy cloud forest dominated by *Weinmannia fagaroides/microphylla* (Cunoniaceae) and *Polylepis* cf. *pauta* (Rosaceae) trees with abundant *Chusquea* sp. scandent bamboo. Alonso et al. (1999) provided details of the vegetation and other fauna at the type locality. The stomach of the holotype was greatly expanded with 140 cc of finely triturated plant material, including fruit and unidentified plant tissue. When collected 15 June, it contained a single 8 × 5 mm embryo.

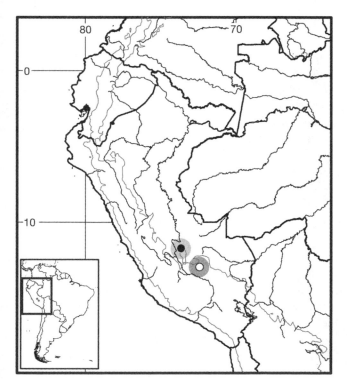

Map 420 Selected localities for species of *Cuscomys ashaninka* (●) and *Cuscomys oblativus* (○). Contour line = 2,000 m.

REMARKS: The type locality is the northernmost recorded for the family, about 200 km north of the Inca burials at Machu Picchu, where *C. oblativus* was found. Between these two localities are the deep canyons of the Urubamba and Apurimac rivers and the 5,000 m snow-capped peaks of the Nevado de Sacsarayoc. Upper elevations of the northern Vilcabamba ridge are essentially a habitat island. Other mammals collected at the type locality show relationships to those to the south from the Machu Picchu area, but the area is likely to have a high local mammalian endemism because of its size and degree of isolation (Emmons 1999a).

Price and Timm (2002) described two genera of chewing lice (Phthriraptera, Gryopidae), one new, from the holotype of *C. ashaninka*.

Cuscomys oblativus (Eaton, 1916)
Machu Picchu Arboreal Chinchilla Rat

SYNONYMS:

Abrocoma oblativa Eaton, 1916:87; type locality "Machu Picchu," Cusco, Peru.

Cuscomys oblativus: Emmons, 1999a:2; first use of current name combination.

DESCRIPTION: See Emmons (1999a:6–8, Figs. 7–9, Table 1) for detailed descriptions and figures of two nearly complete skulls of *C. oblativus* recovered at Machu Picchu. Specimens slightly smaller than *C. ashaninka* in both cranial length and breadth, with much longer tympanic bullae; incisors slightly larger in width; toothrow length larger; interorbital constriction is broader. One specimens with M3 unerupted and is immature. Skull more strongly arched in dorsal profile, much like that of *Abrocoma*; upper incisors opisthodont rather than proodont; tympanic bullae much less inflated; bony auditory tubes are shorter, with their openings more laterally, as opposed to upwardly, directed. Maxillary process of superior zygomatic root broader anteroposteriorly, and canted such that superior root is anterior to inferior root. Narrower zygomatic process of *C. ashaninka* slants in opposite direction. Nasals also separated posteriorly by projection from frontal bones, and premaxillae of rostrum widen posteriorly when viewed dorsally. Foramen magnum subcircular and slightly higher than wide, contrasting with much wider than high condition of *C. ashaninka*. Finally, maxillary toothrows more nearly parallel for entire length, with less posterior divergence; posterior loph of M3 directed posteriorly rather than laterally, and with strongly concave posterior border in contrast to nearly flat border. Medial loph of M3 broad and square, not tapered as in *C. ashaninka*.

A living specimen captured near Machu Picchu (J. G. Ochoa E., pers. comm. to L. E. Emmons, with photographs) very like *C. ashaninka*, but with ears relatively longer, paler, and more pointed than round ears of *C. ashaninka*; smaller area of pure white on upper lips and base of vibrissae, and much paler head, apparently lacking black guard hairs on crown that border white blaze of *C. ashaninka*. Sex and age of this individual undocumented, but it shows a molt-line on rump, suggesting that it is subadult.

DISTRIBUTION: *Cuscomys oblativus* is known only from the type locality. A living rat presumably of this species was recently found and photographed on the Inca Trail near Wiñay Huayna about 3 km (airline) south of Machu Picchu (J. G. Ochoa E., pers. comm. to L. H. Emmons).

SELECTED LOCALITIES (Map 420): PERU: Cusco, Machu Picchu (type locality of *Abrocoma oblativus* Eaton).

SUBSPECIES: *Cuscomys oblativus* is monotypic.

NATURAL HISTORY: No data on behavior, ecology, or reproductive biology are known for *Cuscomys oblativus*.

REMARKS: For nearly a century this species was known only from the archaeological record at Machu Picchu, the famous Incan redoubt on the alto Río Urubamba. The tombs from which the original specimens were obtained date from between 1450 and 1532 AD (R. Burger, cited as pers. comm. in Emmons 1999a). Thomas (1920h:220) had concluded that the species was probably extinct "judging by the fact that . . . relatives are no longer found in this faunal area, but only far south in Chile and Argentina." With the discovery of *C. ashaninka*, however, to the north in the Cordillera de Vilcabamba, Thomas's statement regarding geographically distant relatives is no longer valid. A living individual was found and photographed near the type locality at Wiñay Huayna (J. G. Ochoa E., pers. comm. to L. H. Emmons), but no recent specimen exists for further comparison with *C. ashaninka*. Emmons (1999a:13) surmised that individuals of this species might have been kept as pets, or even domesticated as food or amusement because parts of individuals were found alongside four human burials, including a complete animal in a ceramic pot. One had elongated incisors, as if it has been maintained in captivity.

Family Ctenomyidae Lesson, 1842
Claudio J. Bidau

The family Ctenomyidae consists of the single Recent genus *Ctenomys* and four to six extinct genera (McKenna and Bell 1997), including *Actenomys* Burmeister (Pliocene), *Eucelophorus* Ameghino (Pliocene- Pleistocene), *Megactenomys* Rusconi (Pliocene), *Praectenomys* Villarroel (Pliocene), *Paractenomys* Ameghino (Pliocene), and *Xenodontomys* Kraglievich (Late Miocene). Both *Megactenomys* and *Paractenomys* are now viewed as a either subgenera or synonyms of *Ctenomys* (Verzi and Lezcano

1996; Verzi 2002, 2008). Molecular phylogenetic analyses, both DNA sequences (e.g., A. H. Castillo et al. 2005; Cook and Lessa 1998; D'Elía et al. 1999; E. P. Lessa and Cook 1998; Opazo et al. 2005; Parada et al. 2011) and DNA-DNA hybridization (M. H. Gallardo and Kirsch 2001), strongly support a sister group relationship between the ctenomyids and the family Octodontidae. Authors, however, have differed as to whether these two taxa should be recognized at the level of the family (e.g., Simpson 1945; Woods and Kilpatrick 2005) or subfamily (e.g., McKenna and Bell 1997). The family is known only from southern South America, where much of its history has been arguably tied directly to the profound cooling and drying that began in the Miocene, especially with the Late Pliocene pulse detected worldwide at about 2.5 mya (Verzi 2002; Verzi and Quintana 2005). Earliest known fossils of *Ctenomys* date from the Late Miocene (Verzi 2002, 2008; Verzi and Quintana 2005; Reguero et al. 2007; Vieytes et al. 2007; Verzi et al. 2010).

The differentiation of the ctenomyids involved diverse adaptations to open habitats, especially those related to digging and life underground (Reig and Quintana 1992; Verzi 2002; Verzi and Olivares 2006). These adaptations included (1) acquisition of molar euhypsodonty and associated changes (occlusal simplification, secondary acquisition of radial enamel on the molar leading edge), (2) acquisition of a crescent-shaped occlusal morphology adapted to an oblique mastication, and (3) subterranean specializations of the external body and limbs, skull, and skeleton (Verzi 2002). Verzi (2008) hypothesized phyletic relationships among fossil and extant ctenomyid genera based on cladistic analysis of craniodental characters. Unambiguous synapomorphies for a monophyletic Ctenomyidae include an anterior extension of the maxillary (maxillary extended anterodorsally with respect to premaxillary septum, constraining the incisive foramen posteriad), a dorsal position of the rostral masseteric fossa (which provides attachment for the origin of the *M. masseter medialis anterior* muscle; its depth indicates strong development of this muscle), a protruding external auditory meatus with the anterodorsal and anterior margin oriented anteriad, molars with flexi(id)s vestigial or absent, and a reduced M3, with the posterior or postcrolingual face flat and anterior lobe protruding. Synapomorphies of the genus *Ctenomys* (including *Paractenomys*) include a rostral masseteric fossa positioned dorsal to the alveolar sheath of the upper incisor, deep, and ending in a curved crest slightly anterior to or level with the premaxillomaxillary suture; the bottom of the alveolar sheath of the upper incisor lodged in a cavity of the maxillary, positioned anterior to the alveolar sheath of M1; an external auditory meatus forming a protruding tube with its anterior wall moderately to very concave, and with the ectotympanic recess forming a horizontal flat surface; a well-developed postcondyloid process, with a strong lateral apophysis on its ventral margin, and a crescent-shaped occlusal design of dpm4–m2 with a weak labial fold and a wider lingual concavity, with the posterior lobe of each molar more lingually oriented (Verzi 2008; Verzi et al. 2010).

Genus *Ctenomys* Blainville, 1826

The genus *Ctenomys* is the most speciose of any South American rodent with 63 species recognized herein (J. R. Contreras et al. 2000; Azurduy 2005a). However, as noted by Woods and Kilpatrick (2005), some 85 names have been applied to this genus, and many taxa remain to be adequately delimited geographically or systematically. It is thus clear that the number of species will increase with the combination of much needed additional field collections and revisionary work. In fact, within the past 20 years alone, eight new Recent species have been described (summary in Reeder et al. 2007; T. R. O. Freitas et al. 2012) along with several known only from the fossil record (Verzi et al. 2004; Azurduy 2005a; S. O. Lucero et al. 2008; see also Mones 1986). Additional species await description (Montes et al. 2001; Parada et al. 2011).

The earliest naturalists who explored southern South America made the initial observations and collections of *Ctenomys*. Rusconi (1931b) stated that Alcide d'Orbigny was the first one to collect a *Ctenomys* fossil, which formed the basis of *C. bonariensis* (= *C. bonaerenses*). Charles Darwin (1839, 1845) made observations on living tuco-tucos and also collected fossil mandibular fragments at Monte Hermoso in Buenos Aires province, Argentina, that Sir Richard Owen used to describe *Ctenomys priscus*, later transferred to the genus *Actenomys* (Villarroel 1975).

Living *Ctenomys* have been divided into three subgenera, based on morphological criteria: *Chacomys*, to include the single species *C. conoveri* (Osgood 1946); *Haptomys* to include *C. leucodon* (Thomas 1916g); and the nominotypical *Ctenomys* for all other species (Rusconi 1928). However, molecular evidence fails to support the validity of these subgenera, as both *C. conoveri* and *C. leucodon* are either deeply nested with the complex of species comprising the nominotypical subgenus (Cook and Yates 1994) or form part of a large basal polytomy (E. P. Lessa and Cook 1998; A. H. Castillo et al. 2005; Parada et al. 2011).

The genus is distributed from the Peruvian highlands, at approximately 15° south latitude, to Tierra del Fuego, with representative species in Bolivia, Chile, Argentina, Paraguay, Uruguay, and southwestern Brazil. The majority of species occur in Argentina where they are found in all provinces except Misiones (Bidau 2006), and where they occupy

a great diversity of habitats from sea level to an elevation of almost 5,000 m. Tuco-tucos are also one of the mammal groups with highest karyotypic diversification, including species with diploid numbers ranging from $2n = 10$ to $2n = 70$ (Bidau 2006). The oldest fossils of *Ctenomys* are from the Uquía Formation in northwestern Argentina, providing a minimum age for the genus at about 3.5 mya (Reguero et al. 2007; Verzi et al. 2010), with the central Andes suggested as the possible area of origin. The molecular estimate of the time to the most recent common ancestor of extant *Ctenomys* species is considerably older, estimated at 9.2 mya (range: 6.4–12.6 mya), although most species groups stem from within the past 3 mya (Parada et al. 2011).

All species of tuco-tucos are of similar external morphology, each adapted for subterranean existence. However, there is considerable diversity in body size, ranging from the very small species such as *C. pundti*, *C. talarum*, or *C. sericeus* at less than 100 g and 140 mm head and body length (A. I. Medina et al. 2007) to very large ones, like the Chacoan species *C. conoveri*, where males can reach 1,200 g and 558 mm in head and body length (S. Anderson 1997). All species are sexually size dimorphic, with males the larger of the two sexes, but the degree of dimorphism increases with general body size (A. I. Medina and Bidau 2007; Bidau and Medina 2013; Steiner-Souza et al. 2010). The pelage ranges from short to long, is typically very thick, and general color varies extensively from gray or creamy buff to black on the upper surface (Nowak 1999). As in the North American pocket gophers (Geomyidae), which tuco-tucos resemble in external appearance and habits, fringes of hair are present on both forefeet and hindfeet. In tuco-tucos, however, the hairs on the hindfeet are stiffened and form comb like bristles that are the origin of the generic (and familial) names *Ctenomys* and Ctenomyidae (Greek root *ktenos* = comb; Blainville 1826).

The body is stout and cylindrical, the tail is short, and the limbs and neck are both short and very stout. All digits on both forefeet and hindfeet have very strong claws. The ears are very small, and the eyes are small but not as reduced as in other subterranean mammals. The incisors usually have bright orange enamel (except for *C. leucodon*, as per the name, where is it white or pale yellow). The molars are kidney-shaped, and the last one is reduced in size.

Data on reproduction are available only for a few species. Females typically produce one litter per year but some species can produce two (i.e., *C. talarum*; Malizia and Busch 1991). Species may reproduce in the dry (*C. opimus* in Peru) or the wet season (*C. torquatus* in Uruguay; Nowak 1999). Average gestation is 120 days, and litter size varies between one and three young in *C. opimus* to one to seven in *C. talarum*. Bidau and Medina (2013) investigated testis size allometry and its relationship to sperm competition.

Males (and to a lesser degree, females) vocalize underground to mark territory, maintain social groups, or express fear. Burrow systems vary between species and type of soil but usually consist of a main tunnel up to 30 cm underground and of varying diameter, with several lateral ones with either blind ends or connecting to food sources on the surface, and a nest chamber. External openings are usually plugged when the burrow is occupied. Apparently, only *C. sociabilis* maintains burrows that are permanently open. Internal microclimatic conditions (temperature and humidity) are remarkably constant (20–22° C) in *C. torquatus* and *C. talarum* (A. I. Medina et al. 2007). Digging is typically a daylight activity. To dig, the substrate is loosened with the forefeet and removed with the hindfeet while incisors are used for cutting through roots (Weir 1974a). Tuco-tucos are completely herbivorous, eating roots and aerial parts of available plants, especially grasses (Barlow 1969).

Vieytes et al. (2007; see also Justo et al. 1995) investigated the ultrastructure of the incisor enamel and delineated morphofunctional traits among octodontoid genera with disparate digging adaptations. Those taxa that use their teeth in digging have a higher inclination of the Hunter-Schreger bands, higher relative thickness of the external index, and a higher enamel zone, traits which uniquely or in combination provide for enamel reinforcement to withstand the higher tension forces encountered while digging in hard soils. Vassallo and Mora (2007) examined scaling and growth trajectories in craniomandibular traits relevant to tooth digging, including the development of the mandibular angle and masseteric crest and the robustness of the incisors, and documented significant shape differences between ctenomyid and octodontid genera. And Steiner-Souza et al. (2010) documented shape diversity in the humerus with respect to digging adaptations. Within *Ctenomys*, these traits scale allometrically, such that large and small species do not differ in their respective ontogenetic trajectories. As a consequence, morphological diversification within this genus follows simple allometric scaling laws but changes in development resulting in departures from ontogenetic scaling underlie differentiation among genera within the superfamily. Similarly, Mora et al. (2003) examined cranial morphological attributes of tooth digging within *Ctenomys* and noted that incisor procumbency (a generalized character of tooth-digging) scales with rostral length but not overall size, but that incisor width and thickness do exhibit positive allometry such that larger species have proportionally more powerful incisors to resist greater bending forces.

The phylogenetic relationships within *Ctenomys* using molecular markers have only recently begun (E. P. Lessa and

Cook 1998; J. R. Contreras and Bidau 1999; D'Elía et al. 1999; Mascheretti et al. 2000; Montes et al. 2001; Slamovits et al. 2001; Giménez et al. 2002; A. H. Castillo et al. 2005; Parada et al. 2011). Interestingly, the substitution rate in the mtDNA cytochrome-*b* gene, which is the most common gene examined today, is elevated in comparison to above-ground octodontid relatives (C. C. Da Silva et al. 2009). Although results of these studies tend to be concordant, especially when considered jointly with morphological, chromosomal, parasitological, and biogeographic data (J. R. Contreras and Bidau 1999), species boundaries and phylogenetic relationships within the genus remain complex and incompletely understood. In this account, I recognize 64 species, 56 of which are monotypic and eight polytypic (with a combined 24 subspecies, including the nominotypical races of each). Two of these species (*C.* "yolandae" and *C.* "mariafarelli") have yet to be properly described, but an account is provided for both because these names have been employed in the literature as though they were valid (e.g., Mascheretti et al. 2000; Woods and Kilpatrick 2005). Multiple other taxa defined by chromosomal or molecular sequences, each probably valid species, await formal description (see, for example, Parada et al. 2011). I include data on diploid number (2*n*), autosomal fundamental number (FN), presence of chromosomal polymorphism, and sperm morphology (J. R. Contreras 1996; Bidau 2006) in each species account for which such data are available.

SYNONYMS:

Orycteromys Blainville, 1826:63; used in the French form "Oryctéromes" for *Ctenomys* Blainville (see Palmer 1897b).

Ctenomys Blainville, 1826:64; type species *Ctenomys brasiliensis* Blainville, by monotypy.

Ratton Brants, 1827:187; part; inclusion of *Ratton tucotuco* Brants (1827:187), based on Azara's (1802:69) "*el tucotuco*"; suppressed for the purpose of Principle of Priority (but not Homonomy) by the International Commission (ICZN 1982:Opinion 1232).

Spalax: J. B. Fischer, 1829:300; part (listing of *brasiliensis*); not *Spalax* Gueldenstaedt, 1770.

Georychus: Lichtenstein, 1830:plate 31; part; not *Georychus* Illiger, 1811.

Haptomys Thomas, 1916g:305; as a subgenus of *Ctenomys*; type species *Ctenomys leucodon* Waterhouse, by original designation.

Chacomys Osgood, 1946:47; as a subgenus of *Ctenomys*; type species *Ctenomys conoveri* Osgood, by original designation.

Stenomys Avila-Pires, 1968:162; incorrect subsequent spelling of *Ctenomys* Blainville.

Both Cook et al. (1990:4–5) and S. Anderson (1997:58–59) provided illustrated keys to Bolivian species known to them at that time, but, as documented in the accounts herein, I do not follow all of their taxonomic conclusions. Unfortunately, the production of a key for all 64 species herein covered is impossible at present. Species boundaries are both poorly defined morphologically and delimited geographically for too many taxa, 11 of which are known only from their type localities. The few specimens available preclude an adequate assessment of morphological variation and thus delineation of geographic trends for all but very few species. Moreover, while each species is described based on the available literature in the accounts here, these descriptions are uneven, incomplete, and all too often superficial in their coverage of the morphological characters provided. A thorough review of virtually every species, necessarily accompanied by widespread additional collecting, is critical to advance the current state of species-level systematics in the genus.

Recently, Parada et al. (2011) recognized eight polytypic species groups of *Ctenomys* plus seven additional apparently monophyletic species lineages using mtDNA sequences. These groups are almost coincident with those proposed by J. R. Contreras and Bidau (1999) on the basis of morphological, biogeographic, chromosomal, and incipient molecular data. The polytypic groups include (from Parada et al. 2011, followed by that by J. R. Contreras and Bidau 1999): (1) "boliviensis" or Boliviano-Matogrossense, including *C. boliviensis*, *C. goodfellowi*, *C. nattereri*, *C. robo*, and *C. steinbachi*; (2) "frater" or Bolivian-Paraguayan, including *C. conoveri*, *C. frater*, *C. lewisi*, and an undescribed species from Llathu, Cochabamba, Bolivia; (3) "magellanicus" or Patagonian (in part), including *C. colburni*, *C. coyhaiquensis*, *C. fodax*, *C. haigi*, *C. magellanicus*, and *C. sericeus*; (4) "mendocinus" or *C. mendocinus* complex plus Eastern Lineage, including *C. australis*, *C. flamarioni*, *C. mendocinus*, *C. porteousi*, and *C. rionegrensis*; (5) "opimus" or Chacoan, including *C. fulvus*, *C. opimus*, *C. saltarius*, and *C. scagliai*; (6) "talarum" or Ancestral, including *C. pundti* and *C. talarum*; (7) "torquatus" or Corrientes and Eastern, including *C. lami*, *C. minutus*, *C. pearsoni*, *C. perrensi*, *C. roigi*, and *C. torquatus*; and (8) "tucumanus" or Chacoan, including *C. argentinus*, *C. latro*, *C. occultus*, and *C. tucumanus*. Single species lineages, those that lacked supported relationships to any other taxa, included *C. maulinus*, *C. leucodon*, *C. tuconax*, and *C. sociabilis*. J. R. Contreras and Bidau (1999) either regarded these unique, single species lineages as of uncertain position (*C. leucodon* and *C. tuconax*) or did not include them in their study (*C. sociabilis*). T. R. O. Freitas et al. (2012) placed their newly described species *C. ibicuiensis* within the "torquatus" clade, also based on mtDNA sequences.

The phylogenetic hypothesis posited by Parada et al. (2011) and T. R. O. Freitas et al. (2012) is a remarkable development, linking species that share biogeographic as well as biological characteristics. Unfortunately, however, these studies include only 40 of the 64 species recognized in the accounts that follow. Thus, the degree to which a relationship between clade membership and both distributional as well as other characters will remain robust must await the future inclusion of all known species into an enlarged analysis.

Note added in proof: After this book went into production, S. L. Gardner et al. (2014) characterized each of the 12 Bolivian species they recognized, describing four as new (*C. andersoni*, *C. erikacuellarae*, *C. lessai*, and *C. yatesi*).

Ctenomys argentinus J. R. Contreras and Berry, 1982
Argentine Tuco-tuco

SYNONYM:

Ctenomys argentinus J. R. Contreras and Berry, 1982b:166; type locality "Establecimiento Invernizzi, Campo Aráos, legua 2, 27 km norte de General San Martín, Departamento Libertador General San Martín, Provincia del Chaco, a 26°36' de latitud sur y 59°15' de longitud oeste," Argentina.

DESCRIPTION: Medium sized (mean total length 257.5 mm in females and 260.5 mm in males; mean tail length 78.0 mm in females and 84.3 mm in males). Color very characteristic, as all specimens exhibit black dorsal band that starts on muzzle, passes between eyes to become enlarged on crown, and then narrower again to extend along dorsal surface onto tail; remainder dorsal region brown, becoming paler toward flanks where hair becomes yellowish brown and then pale grayish-brown ventrally. Light-colored collar between ears and throat another conspicuous character (J. R. Contreras and Berry 1982b). Skull robust with strong crests in adult specimens and notably allometric in growth, resulting in "exaggerated" rostrum described by Thomas (1921a) for *C. boliviensis*. Tympanic bullae small and not very inflated; zygomatic arches strong with emergence of anterior apophysis of jugal, with greatest transverse width at junction of third and four sections; visible end of vomer well developed; interpremaxillary foramen, conspicuously separate in many species (Reig et al. 1966), not separable from incisive foramina; dorsal surface of nasals anteriorly flattened, much more than in *C. validus*; orbit relatively small but with postorbital apophysis of frontal and paraorbital apophysis of jugal, well developed; frontosquamosal suture coincides with internal margin of temporal crest; frontals become narrow behind postorbital apophyses; in old individuals, temporal crests increasingly closer to midline and may even make contact. Upper molariform teeth moderate to large. Upper incisors moderately proodont with anterior surface at about 105° relative to horizontal plane. Baculum laminar and well ossified with enlarged proximal end and convex dorsal surface; length about 7 mm.

DISTRIBUTION: This species is known only from northern Argentina (Chaco, Formosa, Santiago del Estero, and possibly Santa Fe provinces; J. R. Contreras and Berry 1982b, 1985; Bidau et al. 2005; Bidau 2006).

SELECTED LOCALITIES (Map 421; from J. R. Contreras and Berry 1982b, except as noted): ARGENTINA: Chaco, Campo Aráos (type locality of *Ctenomys argentinus* J. R. Contreras and Berry), Campo Bermejo, Colonia Benítez, General Pinedo, Las Breñas Pinedo, Pampa Chica, Presidencia Roca, Tres Isletas Pinedo; Formosa, Pozo del Tigre (Massoia 1970), San Martín Número Uno; Santa Fe, Tostado; Santiago del Estero, Bandera.

SUBSPECIES: *Ctenomys argentinus* is monotypic.

NATURAL HISTORY: *Ctenomys argentinus* is widely but sparsely distributed within its range, probably because it inhabits sandy soil patches separated by ample extensions of noncolonizable lands subjected to periodic floods that cause temporal cycles of contraction and expansion of tuco-tuco populations. Populations extend across the *Chaco Húmedo* and *Chaco Seco* ecoregions (Burkart et al. 1999).

REMARKS: Previous to its formal description by J. R. Contreras and Berry (1982b) specimens of Formosa province had been assigned to *C. boliviensis* and *C. conoveri*

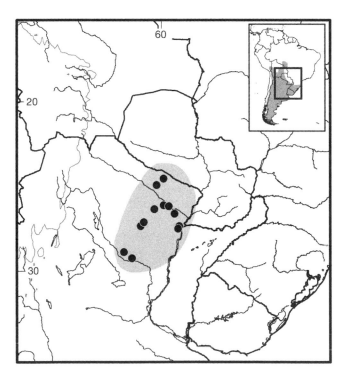

Map 421   Selected localities for *Ctenomys argentinus* (●). Contour line = 2,000 m.

(Massoia 1970, 1977). *Ctenomys argentinus* is closely allied to the species of the "Chacoan lineage" of J. R. Contreras and Bidau (1999) such as *C. occultus, C. tucumanus, C. latro,* and *C. pilarensis* of Paraguay. The karyotype is $2n = 44$ with the FN = 50, 51, or 52, due to pericentric inversion polymorphism (Ortells et al. 1990; Ortells 1995; Bidau et al. 2005). Morphology of the sperm is simple symmetric (Bidau et al. 2005).

Ctenomys australis Rusconi, 1934
Dune Tuco-tuco

SYNONYMS:

Ctenomys porteousi australis Rusconi, 1934b:108; type locality "Necochea (F.C.S.), Provincia de Buenos Aires," Partido de Necochea, Buenos Aires, Argentina, 38°33'S, 58°45'W, 10 m.

Ctenomys australis: J. R. Contreras and Reig, 1965:165; first use of current name combination.

DESCRIPTION: Medium-sized to moderately large species with total length approximately 300 mm, mean range in head and body length 212–215 mm, tail length 96–103 mm, hind foot length 41–43 mm, and mass 349–366 g (J. R. Contreras and Reig 1965:175, 180). Skull longer and broader than that of *C. brasiliensis* but zygomatic and interorbital widths smaller; rostrum robust, especially transversally, and mandible very robust with very narrow incisors.

DISTRIBUTION: *Ctenomys australis* is distributed along the Atlantic coast of southern Buenos Aires province, Argentina, within the *Pampa* ecoregion. J. R. Contreras and Reig (1965:171) provided a detailed map of localities showing the juxtaposition of the range of this species with those of the adjacent *C. talarum* and *C. porteousi*.

SELECTED LOCALITIES (Map 422): ARGENTINA: Buenos Aires, Claromecó (Slamovitz et al. 2001), Cristiano Muerto (J. R. Contreras 1972), Monte Hermosa (A. I. Medina et al. 2007), Necochea (type locality of *Ctenomys porteousi australis* Rusconi), Punta Alta (J. R. Contreras and Reig 1965), Río Quequén Salado (J. R. Contreras and Reig 1965).

SUBSPECIES: *Ctenomys australis* is monotypic.

NATURAL HISTORY: *Ctenomys australis* is an exclusive inhabitant of the first line of vegetated dunes immediately behind the beach. In some parts of its range, this species is parapatric or sympatric with *Ctenomys talarum recessus*, which occupies the second line of dunes (see that account, and Comparatore et al. 1992). This is one of the more thoroughly studied species with regard to behavior, life history, and ecology of natural populations (Zenuto and Busch 1995; C. Busch et al. 2000). Mora, Kittlein et al. (2013) documented craniodental variation within and among populations, attributing interpopulation differences to demo-

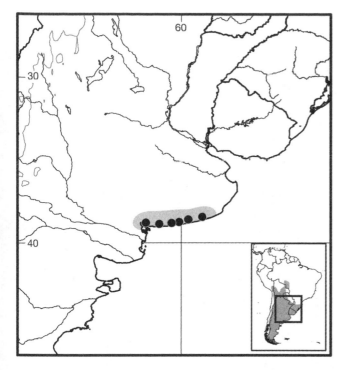

Map 422 Selected localities for *Ctenomys australis* (●). Contour line = 2,000 m.

graphic characteristics delineated by phylogeographic analyses (see Mora et al. 2010) and local adaptation.

REMARKS: Although it was initially described as a subspecies of *C. porteousi*, J. R. Contreras and Reig (1965) and Roig and Reig (1969) established the specific status of this taxon. The species is closely related to the Argentinean species of the "mendocinus" group, including *C. azarae, C. mendocinus,* and *C. porteousi* (Mascheretti et al. 2000), as well as to *C. flamarioni,* the most divergent "mendocinus" species from southern Brazil (Massarini, Barros, Ortells, and Reig 1991; D'Elía et al. 1999; J. R. Contreras et al. 2000; Parada et al. 2011). The karyotype is $2n = 48$, FN = 80 (Massarini, Barros, Ortells, and Reig 1991), similar to that of other "mendocinus" group species, but especially to *C. flamarioni* (see Massarini and Freitas 2005). Sperm morphology is simple asymmetric (Vitullo et al. 1988). Mora et al. (2006) examined phylogeographic structure using mtDNA sequences and Apfelbaum et al. (1991) compared levels of electromorphic variation within and between *C. australis* and *C. porteousi.*

Ctenomys azarae Thomas, 1903
Azara's Tuco-tuco

SYNONYM:

Ctenomys azarae Thomas, 1903a:228; type locality "Sapucay, Paraguay"; corrected to "Province of Buenos Ayres, on the central pampas, lat. 37°45', long. 65°W, 780 kilometers south-west of the Capital" by Thomas

(1903c), which now equals General Acha, Departamento Utracán, La Pampa, Argentina, 216 m.

DESCRIPTION: Relatively small species (external dimensions of holotype head and body length, 158 mm, tail length, 77 mm, hindfoot length (with claw), 35 mm). Dorsal color uniformly brown; venter pale buff, with darker markings on upper surface or white patches below; top of muzzle slightly darker than rest of body. Skull with same general shape as that of *C. mendocinus,* comparatively narrow and slender, not flattened and squared; nasals short and narrow; interorbital region ridged, with rudimentary postorbital processes; parietal ridges more marked than in similar species; zygomatic arches slope backward gradually to broadest point instead of being evenly rounded or square shouldered; mesopterygoid fossa penetrates to posterior edge of m2 instead of center of this tooth, as is more common in other species; bullae more swollen than those of *C. mendocinus.* Cheek teeth comparatively small, broad, and rounded in section, almost completely circumscribed by enamel and with only very small gap, at best, at their enteroexternal and posterointernal corners.

DISTRIBUTION: *Ctenomys azarae* inhabits central Argentinean provinces of La Pampa, Mendoza, and San Luis, within the *Espinal, Monte de Llanuras, Mesetas,* and *Pampa* ecoregions.

SELECTED LOCALITIES (Map 423; from J. R. Contreras and Reig 1965, except as noted): ARGENTINA: La Pampa, Algarrobo del Aguila, Colonia 25 de Mayo, General Acha (type locality of *Ctenomys azarae* Thomas), 5 km S of Santa Rosa, Naicó, Utracán, Victorica; Mendoza, General Alvear (Braggio et al. 1999); San Luis, Arizona.

SUBSPECIES: *Ctenomys azarae* is monotypic.

NATURAL HISTORY: This species of tuco-tuco constructs burrows in sandy soils, in zones of low and open forest, and also on hills with psammophylic vegetation. Kin and Justo (1995) provided data on diet in natural and laboratory settings.

REMARKS: Thomas (1903a:229) erroneously gave the type locality as "Sapucai, Paraguay" in his original description of this species, but corrected his mistake in a second publication that same year (Thomas 1903c:243) to "Province of Buenos Ayres, on the central pampas, lat. 37°45'S, long. 65°W, 780 kilometers south-west of the Capital, a region from which no examples of *Ctenomys* have been recorded." This region is now recognized as General Acha in La Pampa Province. *Ctenomys azarae* is related to the "*mendocinus-porteousi-australis*" assemblage, and both Cabrera (1961) and J. R. Contreras and Reig (1965) regarded *azarae* as a subspecies of *C. mendocinus.* The karyotype is very similar to that of *C. mendocinus* ($2n = 46, 47, 48$), with the two species share the same chromosomal polymorphisms (Massarini, Barros, Ortells, and

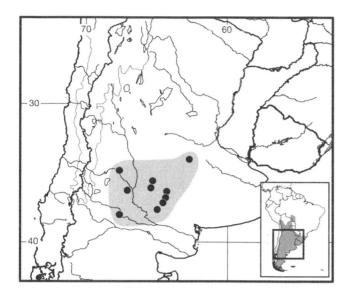

Map 423 Selected localities for *Ctenomys azarae* (●). Contour line = 2,000 m.

Reig 1991; Massarini et al.1998; Braggio et al. 1999). The sperm is of the simple asymmetric type (M. H. Gallardo 1979; Vitullo et al. 1988). De Santis et al. (2001) described enamel microstructure.

Ctenomys bergi Thomas, 1902
Córdoba Tuco-tuco

SYNONYMS:

Ctenomys bergi Thomas, 1902d:241; type locality "Cruz del Eje," Departamento Cruz del Eje, Córdoba, Argentina, 30°44′S, 64°48′W, 449 m.

Ctenomys mendocinus bergi: Cabrera, 1961:551; name combination.

DESCRIPTION: Small to medium-sized species; external dimensions of holotype, a male, head and body length 190 mm, tail length 74 mm; head and body ranges (from A. I. Medina et al. 2007) in males 157–163 mm and females 132–144 mm. General color above uniform sandy fawn while sides and ventral surfaces pale fawn; center of face from muzzle to between ears dark brown and conspicuously different from rest of body. Skull small, narrow, and not flattened in lateral profile but rather slightly and evenly convex; nasals parallel-sided or evenly narrowing posteriorly; posterior width across auditory meatus broader than zygomatic breadth; bullae large and inflated; mesopterygoid fossa extends to level of M2.

DISTRIBUTION: This species is restricted to the *Chaco Seco* ecoregion of Córdoba province, Argentina, extending northwest from the type locality along the course of the Río Cruz del Eje. Populations at Salinas Grandes reach almost 1,000 m. Contrary to Cabrera (1961), there are no confirmed records of this species from La Rioja Province.

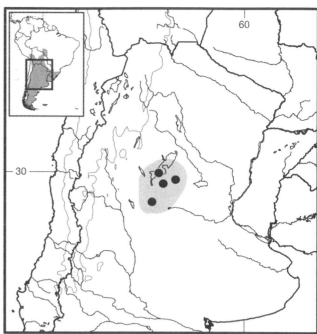

Map 424 Selected localities for *Ctenomys bergi* (●). Contour line = 2,000 m.

SELECTED LOCALITIES (Map 424; from A. I. Medina et al. 2007, except as noted): ARGENTINA: Córdoba, Cruz del Eje (type locality of *Ctenomys bergi* Thomas, 1902), Guanaco Muerto, Las Toscas, Salinas Grandes.

SUBSPECIES: *Ctenomys bergi* is monotypic.

NATURAL HISTORY: This is a poorly known species sparsely distributed in low-density populations.

REMARKS: Although Cabrera (1961:552) considered *Ctenomys bergi* a subspecies of *C. mendocinus*, the species is well differentiated from the latter by the possession of a unique karyotype ($2n = 50$, FN = 90; Giménez et al. 1999) and through mtDNA phylogenetic analyses (Mascheretti et al. 2000). The sperm has the simple asymmetric morphology (Giménez et al. 1999).

Ctenomys bicolor Miranda-Ribeiro, 1914
Bicolored Tuco-tuco

SYNONYMS:

Ctenomys bicolor Miranda-Ribeiro, 1914:41, plates 20 and 23, Figs. 2–2a′, 4, and 6; type locality "Mato Grosso" (Avila-Pires 1968:182); restricted by Bidau and Avila-Pires (2009) to Rôndonia (Brazil) between 11°50′10″S and 12°00′00″S, and 60°51″35″W and 61°19′29″W.

Ctenomys minutus bicolor: Cabrera, 1961:553; name combination.

DESCRIPTION: The holotype and only known specimen (sex unknown) large; head and body length 230 mm, tail length 95 mm. Skull relatively short (length 46 mm) but wide (zygomatic breadth 31 mm). Miranda-Ribeiro (1914) compared his single specimen to *C. minutus* and recorded a number of cranial differences, including zygomatic breadth and curvature, which are larger and more pronounced, respectively; presence of postocular process of frontals in *C. bicolor* but not in *C. minutus*; and general shape of cranium different. Dorsal color blackish gray flanks similar but hairs with ochraceous tips, limbs whitish-ochraceous, and tail whitish (Miranda-Ribeiro 1914; Avila-Pires 1968).

DISTRIBUTION: This species is known only from its rather ambiguous type locality.

SELECTED LOCALITIES (Map 425): BRAZIL: Rondônia, no more precise locality is known (see above; type locality of *Ctenomys bicolor* Miranda-Ribeiro).

SUBSPECIES: *Ctenomys bicolor* is monotypic.

NATURAL HISTORY: No ecological or other biological data are known for *Ctenomys bicolor*.

REMARKS: The only known specimen of *C. bicolor* was collected on October 9, 1912, by the Comissão Rondon although this information was not cited in the original description (Miranda-Ribeiro 1914). The specimen is deposited in the collection of the Museu Nacional (Rio de Janeiro) as MNRJ 2052 (Langguth et al. 1997), erroneously cited as MNRJ 2025 by João Moojen (in Miranda-Ribeiro 1955). The only information about the locality is "Matto Grosso," which at the time of collection comprised vast territories in the present-day state of Rondônia, Brazil. Avila-Pires (1968: 182), who redescribed the holotype,

gave the same information. However, Bidau and Avila-Pires (2009), based on the memoirs of Colonel Cândido Rondon who led the expedition during which the single specimen was collected (Viveiros 1958), restricted the type locality to a 1,000 km² area of Rôndonia state between Primavera de Rondônia and the Rio Barão de Melgaçõ (possibly on the banks of Rio Pimenta Bueno) between the coordinates 11°50′10″S and 12°00′00″S, and 60°51′35″W and 61°19′29″W.

Cabrera (1961) listed *C. bicolor* as a valid subspecies of *C. minutus*, without justification.

Ctenomys boliviensis Waterhouse, 1848
Bolivian Tuco-tuco

SYNONYMS:

Ctenomys boliviensis Waterhouse, 1848:278; type locality "Plains of Santa Cruz de la Sierra"; Thomas (1921a:136) selected as lectotype (BM 46.7.28.57, an adult male) the cotype of which the skull was figured by Waterhouse, and refined the type locality to "Santa Cruz de la Sierra," Santa Cruz, Bolivia, 480 m, 17°48′S 63°10′W.

Ctenomys boliviensis boliviensis: Cabrera, 1961:546; name combination.

DESCRIPTION: Large species, with total length of males reaching 356 mm, tail length 80 mm, and mass up to 650 g; females smaller, total length 316 mm, tail length 95 mm, and mass 420 g (S. Anderson et al. 1987; S. Anderson 1997). Fur soft, very glossy, and rather short; general hue bright, rufous brown; upper surface of head and muzzle blackish brown, with same color continuing as broad but poorly defined band along back of neck to fore part of back; under parts bright rusty yellow color, with exception of space between hind legs, and large patch covering fore part of abdomen, where hairs are entirely white; tail dark brown above, pale brown beneath; fur tinged gray next skin on under parts, but dark slate gray at base on upper parts; sides of body rufous brown; back darker shade of same color, becoming gradually deeper toward head, where dark color contracted into band which runs along neck, and joins black-brown color of upper surface of head; broad mark of somewhat paler hue descends from behind ears to sides of body. Vibrissae dirty white. Feet sparsely clothed with pale brown hairs, with whitish hairs on sides. Tail clothed with dirty white color, very short hairs, except at base where they may be 10 mm in length. Upper incisors notably very broad, proportionally more so than in *C. brasiliensis*. Skull remarkable for its narrowed middle part of rostrum; nasal bones broader at their extremities than in *C. brasiliensis*; auditory bullae narrow with portion that extends ontooccipital plane considerably less extended in vertical direction; ridge traversing outer surface of malar bone very well developed. Cheek teeth broader relative to

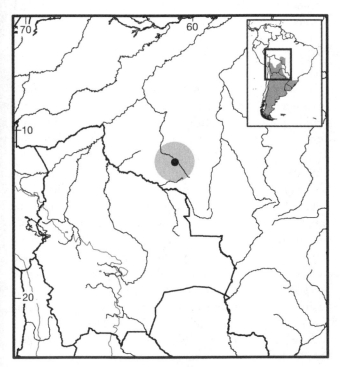

Map 425 Approximate position of the single known locality for *Ctenomys bicolor* (●) (see Remarks). Contour line = 2,000 m.

length, with length of first molar shorter than in *C. brasiliensis*. About 20% of adult specimens show an open frontoparietal fenestrae (S. L. Gardner and Anderson 2001).

DISTRIBUTION: *Ctenomys boliviensis* is distributed in south-central Santa Cruz, Bolivia. Massoia's (1970) report of *C. boliviensis* for Argentina is in error; these specimens are *C. argentinus* (J. R. Contreras, pers. comm.).

SELECTED LOCALITIES (Map 426; from S. Anderson 1997, except as noted): BOLIVIA: Santa Cruz, 2 km SE of Cotoca, Escatia Cachuela Esperanza, 3.5 km W of Estación Pailón, 4 km SW of La Bélgica, 5 km S of Mineros, Palmar de Osorio, 1 km SE of Puerto Pacay, San Miguel Rincón, Santa Cruz de la Sierra (type locality of *Ctenomys boliviensis* Waterhouse), ca. 55 km SE of Santa Cruz at Brecha.

SUBSPECIES: Cabrera (1961) recognized two subspecies, *C. b. boliviensis* and *C. b. goodfellowi*. The latter was originally described as a full species (Thomas 1921d) based on his comparisons with *C. boliviensis* that established many clear differences. S. Anderson et al. (1987) maintained the subspecific status of *C. goodfellowi*, but also cited a number of characters to suggest its specific condition, a position later taken formally by Cook and Yates (1994) and S. Anderson (1997). Anderson et al. (1987) also considered *Ctenomys nattereri* (from Mato Grosso, Brasil) to be a subspecies of *C. boliviensis*, although knowledge of this taxon is extremely limited. Thus, in this account, I treat *C. boliviensis* as monotypic and regard both *C. goodfellowi* and *C. nattereri* as full species.

NATURAL HISTORY: *Ctenomys boliviensis* inhabits friable soils in areas not frequently flooded where food consists mainly of underground tubers and roots (S. Anderson 1997). As is characteristic of other species of tucotucos that have been studied in the field, this one eats primarily underground tubers and roots. Mean litter size is 1.7 (range: 1–5; N = 55; S. Anderson 1997). S. L. Gardner and Duszynski (1990) described eimerian parasites.

REMARKS: *Ctenomys boliviensis* belongs to the Chacoan lineage of *Ctenomys* (J. R. Contreras and Bidau 1999) along with *C. argentinus*, *C. conoveri*, *C. latro*, *C. goodfellowi*, *C. steinbachi*, *C. pilarensis*, and possibly *C. nattereri* (J. R. Contreras and Berry 1982a; S. Anderson et al. 1987). This species exhibits substantial karyotypic diversity over its limited geographic range, with 2*n* ranging from 36 to 46 (S. Anderson et al. 1987; Cook et al. 1990). Vitullo and Cook (1991) described sperm morphology with a symmetric head; Biknevicius (1993) provided data on functional anatomy; Ruedas et al. (1993) gave data on genome size estimates; and Cook and Yates (1994) evaluated species boundaries and relationships based on electromorphic characters.

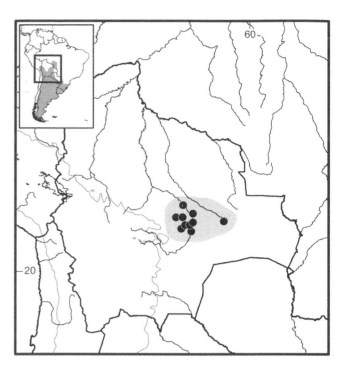

Map 426 Selected localities for *Ctenomys boliviensis* (●). Contour line = 2,000 m.

Ctenomys bonettoi J. R. Contreras and Berry, 1982
Bonetto's Tuco-tuco

SYNONYM:

Ctenomys bonettoi J. R. Contreras and Berry, 1982a: 123; type locality "7.5 kilómetros al sudeste de Capitán Solari, Departamento Sargento Cabral, Provincia del Chaco, a 26°48′ de latitud sur y a 59°33′ de longitud oeste," = Colonia Elisa, Argentina.

DESCRIPTION: Moderate-sized tuco-tuco, with males reaching 183 mm in head and body length, 77 mm in tail length, 37 mm in hindfoot length, and 220 g in mass; equivalent female dimensions 171 mm, 61 mm, 33 mm, and 185 g, respectively. Dorsal color uniformly brown marbled with dark hairs with median black band, more noticeable on head and in specimens with worn pelage; flanks yellowish brown, becoming yellowish on venter; hands and hindfeet covered dorsally with silvery gray hairs; palmar and plantar surfaces pink in life; tail yellowish with slightly more darkened dorsal stripe. Skull moderately robust, but with gracile rostrum; broadest part of zygomatic arch at level of posterior part of orbital fossa; bullae large and easily visible from above; anterior notch of mesopterygoid fossa V-shaped and penetrates palate to level of middle of M2; orbit relatively well developed, with postorbital process of frontal blunt and poorly defined. Upper incisors nearly opisthodont, with anterior face at about a 98° angle relative to horizontal plane.

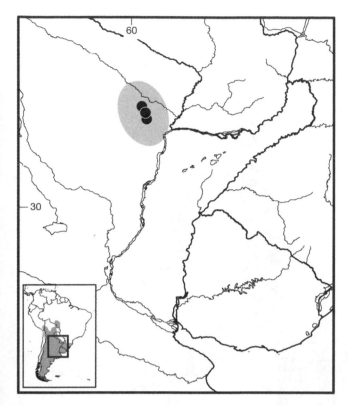

Map 427 Selected localities for *Ctenomys bonettoi* (●).

DISTRIBUTION: *Ctenomys bonettoi* occurs only from a few populations around the type locality in Chaco province, Argentina, within the *Chaco Húmedo* ecoregion. The entire distribution of this species is contained within the range of *C. argentinus*.

SELECTED LOCALITIES (Map 427; from J. R. Contreras and Berry 1982a): ARGENTINA: Chaco, Ciervo Petiso, Colonia Elisa (type locality of *Ctenomys bonettoi* J. R. Contreras and Berry), Laguna Limpia.

SUBSPECIES: *Ctenomys bonettoi* is monotypic.

NATURAL HISTORY: This is a poorly known species inhabiting sandy patches within periodically flooded humid Chaco areas.

REMARKS: The karyotype of *C. bonettoi* is $2n = 50$, FN = 70 (Bidau et al. 2005), one that shares many affinities to that of *C.* "yolandae" (Bidau et al. 2005; see that account, below), with which it appears closely related based on molecular sequences (Mascheretti et al. 2000). *Ctenomys bonettoi* has a simple asymmetric sperm (Bidau et al. 2005).

Ctenomys brasiliensis Blainville, 1826
Brazilian Tuco-tuco

SYNONYMS:

Ctenomys Brasiliensis Blainville, 1826:64; type locality "des parties intérieures du Brésil, de la Province de Las Minas," Minas Gerais, Brazil; redefined herein as Minas, Lavalleja, Uruguay (see Remarks).

Spalax Brasiliensis: J. B. Fischer, 1829:304; name combination.

DESCRIPTION: Blainville's original description and measurements indicate relatively large body size. No other information on external characters, including color or craniodental attributes, have been published. Fernandes et al. (2012) compared skull of Blainville's holotype with representatives of *C. pearsoni* and *C. torquatus* using geometric morphometrics but did not provide actual linear dimensions for any craniodental attribute.

DISTRIBUTION: The range of *C. brasiliensis* is uncertain; see Remarks.

SELECTED LOCALITIES (Map 428): URUGUAY: Lavalleja, Minas (type locality of *Ctenomys Brasiliensis* Blainville).

SUBSPECIES: *Ctenomys brasiliensis* is monotypic.

NATURAL HISTORY: No data are available.

REMARKS: The name *C. brasiliensis* cannot be applied with certainty to any geographic group of tuco-tucos because Blainville's citation of Minas Gerais as the type locality is likely in error. Despite the frequent listing of this species as part of the fauna of Minas Gerais (e.g., J. A. Oliveira and Bonvicino 2006), there are no known specimens of any tuco-tuco collected from this part of Brazil, nor are subfossil specimens of the genus known from the famous Lagoa Santa caves studied by the Danish naturalist Peter Wilhelm Lund and others. It is possible that Blainville's type locality actually corresponds to the locality of Minas in present-day Uruguay, which in the early 1800s was part of the Brazilian Empire (Reig et al. 1966). In this sense, *C. minutus* (Reig et al. 1966) and *C. torquatus* (C. O. da C. Vieira 1955) have been suggested as synonyms of *C. brasiliensis*. A second possibility is that *C. brasiliensis* is one of the forms of the "*pearsoni*" complex of Uruguay (including *C. dorbignyi* and *C. pearsoni*), the distribution of which includes the town of Minas. However, no detailed comparisons have been made to support either of these hypotheses.

Recently, Fernandes et al. (2012) compared the skull of the holotype of Blainville's *C. brasiliensis* (collected by M. Florent Prévost and currently in the Museum National d'Histore Naturelle, Paris; NMHNP 397) to several other species, especially those from Uruguay, with geometric morphometric methods. These data and rediscovered information on the accompanying label written as "St. Paul, prov. las Minas" support Minas, in the department of Lavalleja, Uruguay, as the type locality of *C. brasiliensis*. At the time the specimen was collected, the entire region of what is now Uruguay was dominated by the Portuguese Empire ("Provincia Cisplatina do Reino Unido de Portugal, Brasil e Algarves"), and after Brazilian independence

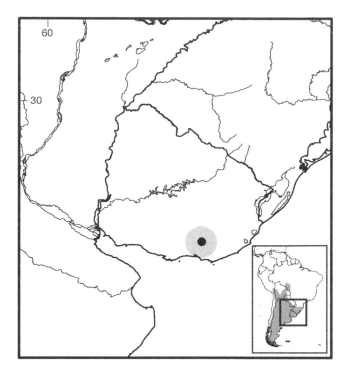

Map 428 Single known locality for *Ctenomys brasiliensis* (●).

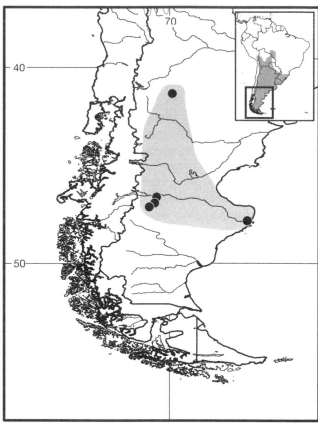

Map 429 Selected localities for *Ctenomys colburni* (●).

in 1822, the Uruguayan territory was considered part of a new country, Brazil. It was not until after 1825 that the province that later became Uruguay gained independence. Today, the area along the southern coast of Uruguay contains only populations of *C. pearsoni* E. P. Lessa and Langguth (see that account). Although Blainville's holotype falls within the morphospace of a large sample of *C. pearsoni* from southern Uruguay but not that of *C. torquatus* from northern Uruguay, Fernandes et al. (2012) argued that additional studies were required before the taxonomic relationship of *C. brasiliensis*—and thus the possible application of Blainville's name—to the *C. pearsoni* complex of tuco-tucos can be established. Available evidence, however, supports the hypothesis that *C. brasiliensis*, the type species of *Ctenomys*, was collected in what is now Uruguay and never ranged into southeastern Brazil.

Karyotype and sperm morphology are unknown. Gerard (1862) listed two Bolivian specimens at the British Museum without exact localities as *C. brasiliensis*, but these probably belong to *C. boliviensis* (S. Anderson 1997).

Ctenomys colburni J. A. Allen, 1903
White-bellied Tuco-tuco

SYNONYM:

Ctenomys colburni J. A. Allen, 1903b:188; type locality "Arroyo Ayke, in the basalt canyons, 50 miles southeast of Lake Buenos Ayres, Patagonia," = Cañones basálticos, Arroyo Aikén, 50 miles SE Lago Buenos Aires, De-

partamento Lago Buenos Aires, Santa Cruz, Argentina, 46°40′S, 70°30′W, ca. 500 m.

DESCRIPTION: Small tuco-tuco, with mean head and body length 161 mm and tail length 71 mm for males, and 146 mm and 63 mm, respectively, for females (J. A. Allen 1903b; Kelt and Gallardo 1994). Dorsal color yellowish gray strongly suffused with fulvous as opposed to black; color varies among individuals, especially with respect to distinctness of dorsal tail stripe that may be absent, present only as faint trace, or present with strongly developed black stripe.

DISTRIBUTION: This species is known only from a few localities in Santa Cruz and Río Negro provinces of Argentina.

SELECTED LOCALITIES (Map 429): ARGENTINA: Río Negro, Estancia Huanuluan (Kelt and Gallardo 1994); Santa Cruz, Arroyo Aikén (type locality of *Ctenomys colburni* J. A. Allen), Casa de Piedra, Río Ecker (A. I. Medina et al. 2007), Estancia La Cantera (Kelt and Gallardo 1994), Río Deseado (Kelt and Gallardo 1994).

SUBSPECIES: *Ctenomys colburni* is monotypic.

NATURAL HISTORY: No data are available.

REMARKS: *Ctenomys colburni* is morphologically similar to the neighboring Chilean species *C. coyhaiquensis*, although the latter appears chromosomally more similar to the *C. maulinus* group (Kelt and Gallardo 1994). J. A. Allen (1903b) also pointed to similarity with *C. sericeus* from the Argentine Patagonia. The karyotype is $2n = 34$ (M. H. Gallardo 1991). Morphology of the sperm is simple asymmetric (Kelt and Gallardo 1994).

Ctenomys coludo Thomas, 1920
Puntilla Tuco-tuco

SYNONYMS:

Ctenomys coludo Thomas, 1920d:119; type locality "La Puntilla, near Tinogasta, Catamarca," Argentina, 28°04′S, 67°34′W, 1,000 m.

Ctenomys coludo coludo: Tate, 1935:390; name combination.

Ctenomys fulvus coludo: Cabrera, 1961:548; name combination.

DESCRIPTION: Relatively large species with total length of 302 mm and tail length of 97 mm in type specimen (a male), readily recognizable by uniformly pale color, long tail (>45% of head and body length), narrow skull, and large bullae; easily differentiated from two geographically nearest species: *C. fochi*, which is more than 20% smaller, and *C. knighti*, which is much darker, has shorter tail, larger teeth, and broader frontal region.

DISTRIBUTION: *Ctenomys coludo* is known only from the type locality, which is "a few miles out from Tinogasta toward Copacabana, at an altitude of about 1000 m above sea level" (Thomas 1920d:116).

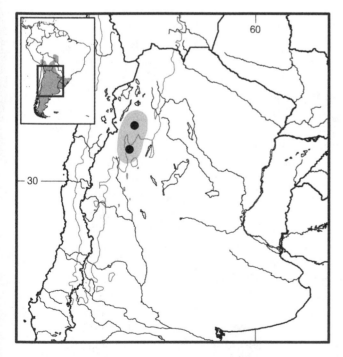

Map 430 Selected localities for *Ctenomys coludo* (●). Contour line = 2,000 m.

SELECTED LOCALITIES (Map 430): ARGENTINA: Catamarca, La Puntilla (type locality of *Ctenomys coludo* Thomas), Portezuelo de Pasto Ventura (FMNH 46137).

SUBSPECIES: *Ctenomys coludo* is monotypic.

NATURAL HISTORY: No ecological or other biological data are available for this species.

REMARKS: *Ctenomys coludo* is apparently related to *C. fulvus* from Chile. Karyotype and type of sperm are unknown.

Ctenomys conoveri Osgood, 1946
Chacoan Tuco-tuco

SYNONYM:

Ctenomys (Chacomys) conoveri Osgood, 1946:47; type locality "Colonia Fernheim, 16 km. west of Filadelfia, Paraguayan Chaco. Approximately Long. 60°10′W., Lat. 22°15′S.," Boquerón, Paraguay.

DESCRIPTION: Largest species in genus, with males reaching 558 mm in head and body length, 122 mm in tail length, 60 mm in hindfoot length, and mass of 1,200 g (S. Anderson 1997). Pelage long and rather coarse, color nearly uniform cinnamon rufous with light mixture of dusky and scattered white hairs above and clear below. Tail heavily haired, rufous above and pale below, with median line of white toward tip. Skull massive and angular; jugal possesses unique cresentic excavation in front of high and broad dorsal process. Upper incisors very broad and heavy, slightly proodont, and grooved on anterior surface. Some adult specimens exhibit open frontoparietal fenestrae (S. L. Gardner and Anderson 2001).

DISTRIBUTION: This species occurs through the *Chaco Boreal* ecoregion of northwestern Paraguay and southern Bolivia.

SELECTED LOCALITIES (Map 431; from S. Anderson 1997, except as noted): BOLIVIA: Chuquisaca, 9 km E of Carandaytí, 1.5 km NW of Povenir; Santa Cruz, 26 km E of Boyuibe; Tarija, 10 km S of Capirenda, Palo Marcado, Villa Montes. PARAGUAY: Boquerón, Colonia Fernheim (type locality of *Ctenomys conoveri* Osgood).

SUBSPECIES: *Ctenomys conoveri* is monotypic.

NATURAL HISTORY: No data are available on ecology, food habits, or reproduction. S. L. Gardner and Duszynski (1990) described eimerian oocysts in one specimen.

REMARKS: Because Osgood (1946:47) believed that *C. conoveri* differed "widely from any previously described species," he placed it as the sole member of his subgenus *Chacomys*. However, electrophoretic analyses do not support subgeneric rank but rather indicate a sister relationship between *C. conoveri* and both *C. frater* and *C. lewisi*, two species of the nominotypical subgenus, among those species examined (Cook and Yates 1994). A relationship to other species of a Chacoan group such as *C. argentinus*,

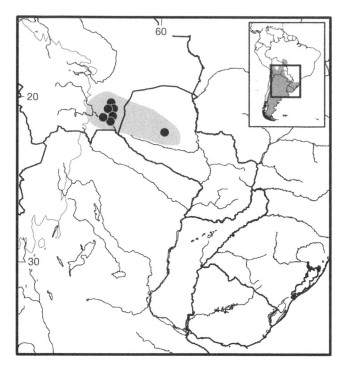

Map 431 Selected localities for *Ctenomys conoveri* (●). Contour line = 2,000 m.

C. boliviensis, and *C. goodfellowi* had been suggested earlier on morphological grounds (J. R. Contreras and Berry 1982b). Osgood (1946) compared the skull of *C. conoveri* to that of *C. robustus* (now = *C. fulvus* Philippi), another large species, in his description. Eisentraut (1933) misidentified specimens of this species from Villa Montes, Palo Marcado, and near Samahuate in Bolivia as *C. brasiliensis*. Later, he (Eisentraut 1983) also referred specimens from the Chaco region of Bolivia to *C. brasiliensis*, considered by S. Anderson (1997:493) as "perhaps *C. conoveri*." Krumbiegel (1941a) also misidentified specimens of *C. conoveri* from Villa Montes, Bolivia, as *C. leucodon* (see S. Anderson 1997). Reports of the presence of *C. conoveri* in Argentina (Massoia 1977) are based on misidentified specimens of *C. argentinus* (J. R. Contreras and Berry 1982a). The karyotype for Bolivian specimens is $2n = 48$ (S. Anderson et al. 1987:15), and for Paraguayan individuals is $2n = 50$ (Ortells 1995). Sperm morphology is simple asymmetric (J. R. Contreras 1996).

Ctenomys coyhaiquensis Kelt and Gallardo, 1994
Coyahique Tuco-tuco

SYNONYM:

Ctenomys coyhaiquensis Kelt and Gallardo, 1994:344; type locality "2 km S Chile Chico and 1 km W Chile Chico aeródromo, Provincia General Carrera, XI Región de Aisén [= Aysén], Chile. 46°33'S, 71°46'W, 330 m."

DESCRIPTION: Small species of tuco-tuco, head and body length between 195–264 mm, tail length 60–81 mm, hindfoot length 28–34 mm, and mass 72–182 g. Pelage tawny brown washed with yellow and black. Dorsal pelage consists of two kinds of hairs, one long and black, scattered throughout, and other tricolored with dark gray base, buff median band, and black tips. Middorsal region darker, as typical of many other species of tuco-tucos; dorsal coloration fades laterally toward ventral region, where hairs are bicolored dark gray proximally and pale buff or ochraceous buff distally. Feet of adults covered dorsally with silvery-white hairs. Tail bicolored, with darker dorsal color extending ventrally along terminal one-third of tail. Skull with large orange and mildly proodont incisors; sturdy zygomatic arches; large auditory bullae; auditory meatus extending laterally just beyond zygomatic arches; nasals slightly flared anteriorly; bony capsule of incisor root protrudes from side of rostrum; pronounced lateral ridge separates dorsal and ventral zygomatic fossae; sphenoid bones large and extend posteriorly to occipital crest; infraorbital canal large and ovoid, with small dorsoposterior projection where zygomatic plate meets frontal bone; parietal bones possess small postorbital process, complementing similar process on jugal; incisive foramina never more than one-half length of diastema; alisphenoid-presphenoid bridge straight and runs perpendicular to sagittal plane; postalisphenoid canal oval; and generally single lateral foramen present in alisphenoid. Both M3 and m3 reduced. Combination of traits diagnostic for species.

DISTRIBUTION: *Ctenomys coyhaiquensis* is known only from two localities in Chile between 330 and 730 m. Kelt and Gallardo (1994) suggested that the species might extend into Argentina, because both collecting sites are within 1 km of the border between the two countries.

SELECTED LOCALITIES (Map 432): CHILE: Aysén, 2 km S Chile Chico and 1 km W of Chile Chico Aeródromo (type locality of *Ctenomys coyhaiquensis* Kelt and Gallardo), Fundo Los Flamencos, 4.5 km SE of Coyhaique Alto (Kelt and Gallardo 1994).

SUBSPECIES: *Ctenomys coyhaiquensis* is monotypic.

NATURAL HISTORY: At the type locality (Kelt and Gallardo 1994), *C. coyhhaiquensis* occupied sandy or rocky soils that supported a sparse shrub and herbaceous community. Principal shrubs included *Mullinum spinosum* (Apiaceae), *Colliguaya saliscifolia* (Euphorbiaceae), *Escallonia rubra* (Escalloniaceae), and *Berberis buxifolia* (Berberidaceae). At the higher Coyhaique Alto site, these tuco-tucos inhabited a gravelly soil with bunchgrasses and a few shrubs (*B. buxifolia* and *E. rubra*; Kelt and Gallardo 1994).

REMARKS: M. H. Gallardo et al. (1996) documented the loss of genetic (electromorphic protein) variation in a population following an eruption of Volcán Hudson in

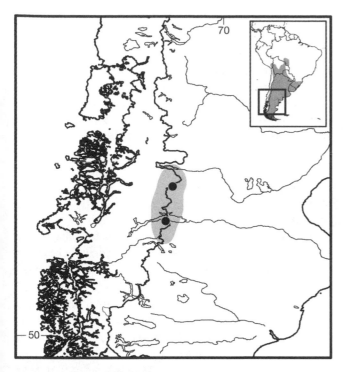

Map 432 Selected localities for *Ctenomys coyhaiquensis* (●). Contour line = 2,000 m.

southern Chile in 1991. The karyotype consists of $2n = 28$, $FN = 44$ (Kelt and Gallardo 1994), and the sperm is of the simple asymmetric type (Kelt and Gallardo 1994).

Ctenomys dorbignyi J. R. Contreras and Contreras, 1984
D'Orbigny's Tuco-tuco

SYNONYMS:

Ctenomys d'orbignyi J. R. Contreras and Contreras, 1984: 131; type locality "Paraje Mbarigüí, Departamento Berón de Astrada, Provincia de Corrientes, a 27°33′ de latitud Sur y a 57°31′ de longitud oeste," Argentina, 54 m; note incorrect original spelling with apostrophe (ICZN 1999:Art. 32.5.2).

Ctenomys dorbignyi: J. R. Contreras and Scolaro, 1986:21; corrected spelling of *d'orbignyi* J. R. Contreras and Contreras.

DESCRIPTION: Relatively large tuco-tuco, with average head and body length 206 mm (range: 190–224 mm), tail length 91 mm (range: 86–99 mm), hindfoot length (including claws) 40.4 mm (range: 38.7–43.0 mm), and mass 331 g (range: 269–376 g). Dorsal color fairly uniform brown tone overall; venter pale; there no collar of differentiated color; white postauricular spots as well as white areas in axilla and inguinal regions present, although extent of latter variable. Skull robust and relatively long, with broad and slightly expanded rostrum, and moderately proodont upper incisors; width of interorbital constriction

and zygomatic breadth proportionately smaller than species of Chaco region, and tympanic bullae more developed than in that species group; frontoparietal sutures gently curved forward; line through apex of spinous processes and postorbital process of jugal slants slightly forward; free end of vomer between premaxillaries well developed and clearly visible; no visible interpremaxillary foramen; orbital expansion of maxilla extends above jugal with crescent shaped external depression extending onto adjacent bones, which not as deep as in *C. argentinus* or *C. conoveri*, but more marked than in *C. perrensi*, with vertical, thin but well-developed sickle-shaped blade.

DISTRIBUTION: *Ctenomys dorbignyi* occurs in a series of isolated areas in Corrientes and Entre Ríos provinces within the *Esteros del Iberá* and *Pampa* ecoregions of Argentina (J. R. Contreras and Contreras 1984; J. R. Contreras and Scolaro 1986; see Remarks).

SELECTED LOCALITIES (Map 433; from J. R. Contreras and Contreras 1984, except as noted): ARGENTINA: Corrientes, Costa Grande, Mbarigüí (type locality of *Ctenomys dorbignyi* J. R. Contreras and Contreras), Tres Bocas, Esquina, Los Angeles, Laguna Itá, Arerunguá, La Tilita, Ibahay, Paraje Yabyrahá, Caa Catí, Palmar Grande, Colonia Romero, Tacuaracuarendy Sarandicito (Argüelles et al. 2001); Entre Ríos, Paso Vera (Giménez et al. 2002), San Joaquín de Miraflores (Giménez et al. 2002), Tiro Federal (Giménez et al. 2002).

SUBSPECIES: *Ctenomys dorbignyi* is monotypic.

NATURAL HISTORY: *Ctenomys dorbignyi* inhabits sandy patches far from the forested islets that are frequent within the area of its distribution. Population size is strongly affected by the cycles of annual rainfall. Litter size averages 1.3 ± 0.58, $N = 18$ (J. R. Contreras, pers. comm.). Population density is low in northern populations and much higher in the southern ones. J. R. Contreras and Contreras (1984) suggested that the species is at least semisocial, although this claim needs confirmation.

REMARKS: *Ctenomys dorbignyi* is readily distinguished from neighboring species (J. R. Contreras and Contreras 1984). From *C. perrensi*, *C. dorbignyi* differs in its color and color pattern, in less pronounced procumbency, by the reduction or obliteration of the interpremaxillary foramen, larger tympanic bullae, higher and wider rostrum, and a more convex cranium. From *C. argentinus*, *C. dorbignyi* differs in all diagnostic cranial characters. This species does show some superficial resemblance to *C. minutus* but differs in the morphology of the occipital foramen and relative zygomatic breadth. The karyotype of *C. dorbignyi* is $2n = 70$ in all the populations with minor variation in FN (80 to 84). The karyotypic similarity to *C. pearsoni* ($2n = 68-70$) reinforces other commonalities between the two taxa (Argüelles et al. 2001).

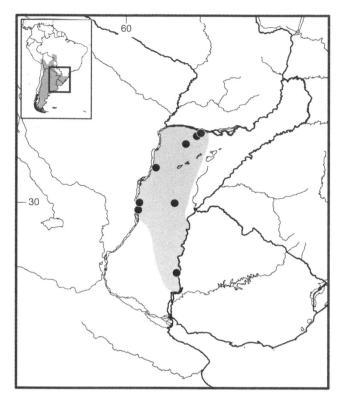

Map 433 Selected localities for *Ctenomys dorbignyi* (●).

J. R. Contreras and Scolaro (1986) compared cranial measurements among four geographically partitioned samples covering the species range in Corrientes province, noting that the eastern sample (*oriental nucleo*), centered at the locality of Contreras Cué, was substantially divergent in multivariate morphospace relative to the other three samples. Interestingly, this same population does not share the typical $2n = 70$ karyotype characteristic of *C. dorbignyi* but has diploid numbers ranging between $2n = 41$ and $2n = 47$ (Ortells et al. 1990; Giménez et al. 2002). Analyses of mtDNA suggest that this latter group is likely an undescribed species (informally called *Ctenomys* sp. by Giménez et al. 2002; see also Remarks for *C. perrensi*). Pending further field and laboratory analyses, the Contreras Cué sample is best considered as a taxon distinct from *C. dorbignyi*; it is not included in the mapped range. Sperm morphology is simple symmetric (Vitullo et al. 1988).

Ctenomys dorsalis Thomas, 1900
Black-backed Tuco-tuco
SYNONYM:
Ctenomys dorsalis Thomas, 1900g:385; type locality "Northern Chaco of Paraguay"; a more specific locality is in-

determinable at present, but presumably either in Alto Paraguay or Boquerón departments.

DESCRIPTION: Small tuco-tuco, holotype (female) 156 mm in head and body length, 46 mm in tail length, and 30 mm in hindfoot length. Fur soft and fine. General dorsal color buffy fawn with marbled black dorsal line extending from nose to rump; line sharply defined on head, about 10 mm broad, but widens to more diffuse band on back; dark lateral face markings around eye or ear lacking, but well-marked light collar present behind cheeks and chin extending to each side of ear. Throat, chest, outer edges of belly, and narrow midline stripe pale buffy with slate-colored basal hairs; remaining ventral hairs self-colored white. Dorsal surfaces of both hands and feet dirty white. Tail hairs mixed black and white. Craniodental characters have not been described, except that incisors orange above and below.

DISTRIBUTION: The geographic distribution of *C. dorsalis* is unknown, but presumably encompasses the *Chaco Boreal* ecoregion of Paraguay and perhaps adjacent Bolivia and/or Brazil (see Remarks).

SELECTED LOCALITIES (Map 434): PARAGUAY: Alto Paraguay or Boquerón, precise locality unknown.

SUBSPECIES: *Ctenomys dorsalis* is monotypic.

NATURAL HISTORY: No data are available.

REMARKS: The type, and only known specimen, was collected in 1900 and presented to the Natural History Museum (London) by J. G. Kerr; it consists of a skin and upper incisors. Because of the uncertainty of its type locality, *C. dorsalis* is another of the problematic species of tuco-tucos, although various authors have provided arguments to restrict the type locality. For example, Rusconi (1928) proposed Resistencia (in Chaco province, Argentina) as the type locality. Moojen (1952b) suggested a distribution in present-day Chaco Boreal (which was part of Bolivia in 1900, the year the holotype was collected), perhaps extending into adjacent Brazil. Cabrera (1961), Mares and Ojeda (1982), Redford and Eisenberg (1992), and Woods and Kilpatrick (2005) give a variation on "northern Paraguayan Chaco." Redford and Eisenberg (1992:map 11.129) map two localities without documentation. However, until additional specimens matching the details of the holotype are collected, the type locality of *C. dorsalis* will remain ambiguous. A useful hint for identifying at least the general Chacoan area where the type specimen was obtained might be in the local vernacular name *Sumkum* (Thomas 1900g), possibly *sunkum*. Careful examination of the vocabularies of the various indigenous tribes in the general Chaco region may help to pinpoint a more limited area of occurrence (J. R. Contreras and Roig 1992). Karyology and sperm morphology are unknown.

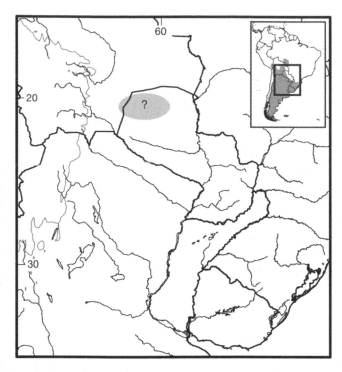

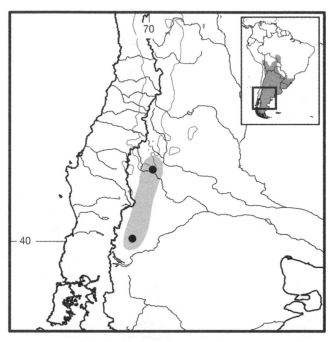

Map 434 Possible range of *Ctenomys dorsalis* (?) (see Remarks).

Map 435 Selected localities for *Ctenomys emilianus* (●). Contour line = 2,000 m.

Ctenomys emilianus Thomas and St. Leger, 1926
Emily's Tuco-tuco

SYNONYM:

Ctenomys emilianus Thomas and St. Leger, 1926b:637; type locality "Chos Malal, Prov. Neuquén. Altitude 805 m," Departamento Chos Malal, Neuquén, Argentina, 37°23′S, 70°16′W.

DESCRIPTION: Large, pale tuco-tuco, "readily distinguishable by its clear fawn color, contrasting with the whitish belly and sides, and its low flattish skull" (Thomas and St. Leger 1926b:638). External dimensions of holotype head and body length 211 mm, tail length 91 mm, and hindfoot length 39 mm. General color above pale glossy fawn, uniform on head and back, without frontal blackening; sides and venter, including chin and throat, buffy whitish, with this color extending further up sides and onto hips and thighs more than in most other species, and thus contrasting with brown color of rump. Hairs of forefeet and hindfeet and tail buffy whitish; tail lacks any blackening of its terminal tuft. Skull slender in build, flat, with zygomatic arches expanded but not convex; nasals narrow; interorbital region flattened; braincase low with masseteric ridge well defined; lambdoidal ridges well defined; vertical occipital central ridge absent; and bullae large and well inflated.

DISTRIBUTION: *Ctenomys emilianus* is only known from the type locality and neighboring areas in the Patagonian steppe of Neuquén province, Argentina.

SELECTED LOCALITIES (Map 435): ARGENTINA: Neuquén, Chos Malal (holotype of *Ctenomys emilianus* Thomas and St. Leger), Río Quilquihue (Massoia 1988a).

SUBSPECIES: *Ctenomys emilianus* is monotypic.

NATURAL HISTORY: *Ctenomys emilianus* lives in sandy soils and dunes at elevations up to 800 m. It occurs parapatrically with a small and more darkly colored species (probably *C. haigi*) in the vicinity of the type locality of Chos Malal (Thomas and St. Leger 1926b; Pearson 1984).

REMARKS: Neither karyotype nor sperm morphology have been described.

Ctenomys famosus Thomas, 1920
Famatina Tuco-tuco

SYNONYMS:

Ctenomys famosus Thomas, 1920g:420; type locality "Potrerillo, at about 2600 m," Sierra de Famatina, Departamento Famatina, La Rioja, Argentina, 25°50′S, 67°27′W, 2,600 m.

Ctenomys fulvus famosus: Cabrera, 1961:549; name combination.

DESCRIPTION: Relatively small tuco-tuco, with holotype (female) dimensions head and body length 160 mm, tail length 74 mm, and hindfoot length 31.5 mm. Color above and below uniformly pale, similar to that of *C. coludo*, but tail considerably shorter. Skull also similar to that of *C. coludo* but overall smaller, with smaller bullae,

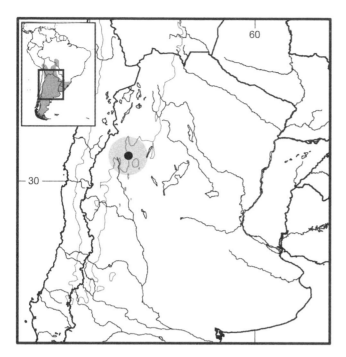

Map 436 Single known locality for *Ctenomys famosus* (●). Contour line = 2,000 m.

with less widely expanded zygomatic arches, and less heavy cheek teeth; supraorbital region with small ledges.

DISTRIBUTION: *Ctenomys famosus* is known from the type locality and probably nearby areas within the Sierra de Famatina in La Rioja province, Argentina, at elevations to 3,800 m.

SELECTED LOCALITIES (Map 436): ARGENTINA: La Rioja, Potrerillo (type locality of *Ctenomys famosus* Thomas).

SUBSPECIES: *Ctenomys famosus* is monotypic.

NATURAL HISTORY: No ecological data are available.

REMARKS: Thomas (1920g) regarded *C. famosus* as morphologically similar to the geographically adjacent *C. coludo*, which he had named earlier in the same year, differing from the latter primarily by a relatively shorter tail and smaller bullae. A year later Thomas (1921g) suggested that *famosus* was only a subspecies of *coludo*. Cabrera (1961) treated *famosus* as a subspecies of *C. fulvus*. The karyotype and sperm morphology are unknown.

Ctenomys flamarioni Travi, 1981
Flamarion's Tuco-tuco

SYNONYM:

Ctenomys flamarioni Travi, 1981:123; type locality "Fazenda Caçapava, Estação Ecológica do Taim, Rio Grande, Rio Grande do Sul, Brasil," 32°52′S 52°32′W, 5–10 m.

DESCRIPTION: Medium to large-sized species, with holotype (female) measuring 289 mm in total length and

74 mm of tail length. Upper incisors orthodont (Travi 1981). Overall color sandy yellow, making *C. flamarioni* palest species of those inhabiting southern Brazil and adjacent Uruguay (Rebelato 2006). Baculum elongated (mean length 10.4 mm) and thin, with rounded tip. Glans penis with spines of various shapes (from sharp to rounded); spine density 12/mm^2, mean spine length 226.26 ± 32.42 μm s.d. (Rocha-Barbosa et al. 2013).

DISTRIBUTION: *Ctenomys flamarioni* occurs within the coastal dunes in Santa Catarina and Rio Grande do Sul states, Brazil.

SELECTED LOCALITIES (Map 437; from Travi 1981): BRAZIL: Rio Grande do Sul, Cidreira, Fazenda Caçapava (type locality of *Ctenomys flamarioni* Travi), Hermenegildo, Lagoa do Peixe, Tramandaí; Santa Catarina, Morro dos Conventos.

SUBSPECIES: *Ctenomys flamarioni* is monotypic.

NATURAL HISTORY: This is one of the best-known species of tuco-tucos with respect to ecology and behavior. Males occupy territories almost five times larger than females (G. P. Fernandez 2002). Tunnel systems are very long, especially in males, and constructed at a depth of about 30 cm (Bretschneider 1987). Food items are mainly species of Poaceae and Cyperaceae (Bretschneider 1987). The overall sex ratio does not differ from 1:1 in the gen-

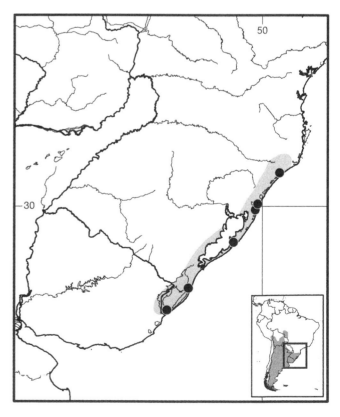

Map 437 Selected localities for *Ctenomys flamarioni* (●).

eral population, but in adults the ratio is biased toward females (2:1–2.7:1), and the species is polygynous in mating behavior (G. P. Fernandez 2002). Gestation is about 120 days long (Bretschneider 1987), and litter size ranges between two and three (Travi 1983). Fernandez-Stolz et al. (2007) described the role of dispersal in local population bottlenecks.

REMARKS: *Ctenomys flamarioni* is distinguished from the neighboring species *C. torquatus*, *C. minutus*, and *C. lami* by its whitish-yellow color and zygomatic region. The karyotype of *C. flamarioni* is $2n = 48$, $FN = 50–78$, partly associated with geographic variation in whole-arm constitutive heterochromatin. G-banding suggests a close cytogenetic relationship between *C. flamarioni* and species of the Argentinian "*mendocinus*-group," such as *C. australis*, *C. mendocinus*, and *C. porteousi* (T. R. O. Freitas 1994; Massarini and Freitas 2005), a hypothesis strongly supported by mtDNA sequence analysis (Parada et al. 2011). Sperm morphology is the simple asymmetric type (T. R. O. Freitas 1995a).

Ctenomys fochi Thomas, 1919
Foch's Tuco-tuco

SYNONYMS:

Ctenomys fochi Thomas, 1919a:117; type locality "Chumbicha, Catamarca," Departamento Capayán, Catamarca, Argentina, 28°52′S, 66°14′W, 415 m.

Ctenomys mendocinus fochi: Cabrera, 1961:552; name combination; considered a valid subspecies of a polytypic *C. mendocinus* Philippi.

C[tenomys]. fachi Redford and Eisenberg, 1992:374; incorrect subsequent spelling of *Ctenomys fochi* Thomas.

DESCRIPTION: Small species, with external dimensions of holotype (young adult male) head and body length 162 mm, tail length 76 mm, and hindfoot length 30 mm. General body color drab buffy brown with hairs of venter washed with paler tones; hairs throughout slaty at base; muzzle and crown nearly black. Skull similar to that of *C. bergi*, but with bullae uniformly more inflated and line connecting anterointernal angle with meatal tube more distinctly convex forward.

DISTRIBUTION: *Ctenomys fochi* is known only from the vicinity of the type locality in Catamarca province, Argentina.

SELECTED LOCALITIES (Map 438): ARGENTINA: Catamarca, Chumbicha (type locality of *Ctenomys fochi* Thomas).

SUBSPECIES: *Ctenomys fochi* is monotypic.

NATURAL HISTORY: No ecological data are available. According to the collector, Emilio Budin, the species "lives in very dry red earth" (Thomas 1919a:119).

REMARKS: *Ctenomys fochi* is apparently closely related to *C. bergi* from Córdoba province (Thomas 1919a),

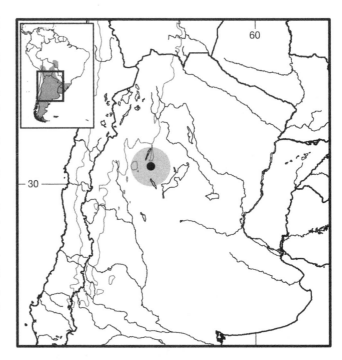

Map 438 Single known locality *Ctenomys fochi* (●). Contour line = 2,000 m.

but is distinguishable in aspects of color, particularly the nearly black muzzle and crown, and by the more greatly inflated bullae. The karyotype and sperm morphology are unknown.

Ctenomys fodax Thomas, 1910
Lago Blanco Tuco-tuco

SYNONYMS:

Ctenomys fodax Thomas, 1910a:243; type locality "Valle del Lago Blanco, Cordillera region of Southern Chubut, Patagonia (about 46°S., 71°W.)"; restricted to Estancia Valle Huemules, Departamento Río Senguerr, Chubut, Argentina (45°57′S, 71°31′W, 575 m) by Pardiñas et al. (2007).

Ctenomys talarum fodax: Rusconi, 1928:243; name combination.

DESCRIPTION: One of larger species in genus, with dimensions of holotype (adult male) head and body length 260 mm, tail length 98 mm, and hindfoot length (without claw) 40.4 mm. Fur soft; dorsal color generally cinnamon. Skull large and bowed, with nasals broad in front giving entire rostrum broad appearance; nasals extend posteriorly to frontopremaxillary processes; auditory bullae sufficiently expanded to be visible when braincase is viewed from above; interparietal longer than broad, commonly divided by median suture.

DISTRIBUTION: *Ctenomys fodax* is restricted to the vicinity of the type locality in the Patagonian steppe of Chubut province, Argentina.

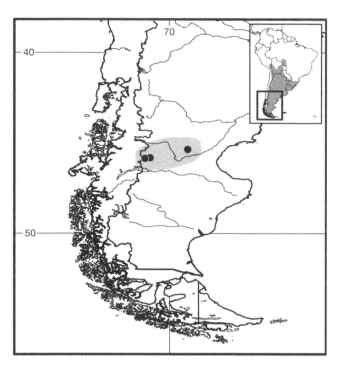

Map 439 Selected localities for *Ctenomys fodax* (●).

SELECTED LOCALITIES (Map 439): ARGENTINA: Chubut, Estancia Valle Huemules, Lago Blanco (type locality of *Ctenomys fodax* Thomas, as restricted by Pardiñas et al. 2007), Lago Musters (A. I. Medina et al. 2007), Valle del Lago Blanco (FMNH 18191).

SUBSPECIES: *Ctenomys fodax* is monotypic.

NATURAL HISTORY: No ecological data are available.

REMARKS: Considered by Thomas (1910a) to be most closely related to *C. osgoodi* J. A. Allen (now *C. magellanicus osgoodi*; see that account), but larger in body size and with distinctive cranial differences. Listed by Osgood (1943b) and Cabrera (1961) as a synonym of *C. magellanicus osgoodi* J. A. Allen. The karyotype is $2n = 28$, FN = 38, and sperm morphology is of the simple asymmetric type (Montes et al. 2001).

Ctenomys frater Thomas, 1902
Forest Tuco-tuco, Red Tuco-tuco

SYNONYMS:

Ctenomys frater Thomas, 1902c:228; type locality "Potosí, 4300 m," Potosí, Bolivia, 19°35′S, 65°45′W, ca. 4,300 m.

Ctenomys budini Thomas, 1913a:141; type locality "Cerro de Lagunita, E. of Maimara. 4500m," Jujuy, Argentina.

Ctenomys sylvanus Thomas, 1919g:155; type locality "Tartagal, 600 m"; Salta, Argentina.

Ctenomys sylvanus utibilis Thomas, 1920a:193: type locality "Yuto, Rio San Francisco. Alt. 500 m."; about 70 km N Villa Carolina, Jujuy, Argentina.

Ctenomys budini barbarus Thomas, 1921d:185; type locality "Sierra Santa Barbara, S. E. Jujuy. Type from Sunchal, alt. 1200 m"; Jujuy, Argentina.

Ctenomys budini budini: Thomas, 1921d:186; name combination.

Ctenomys budini sylvanus: Thomas, 1921d:186; name combination.

Ctenomys budini utibilis: Thomas, 1921d:186; name combination.

Ctenomys sylvanus mordosus Thomas, 1926c:325; type locality "Tambo, 2200 m," Tarija, Bolivia; considered by Cabrera (1961:548) to be a valid subspecies of a polytypic *C. frater* Thomas.

[*Ctenomys*] *sylvanus sylvanus*: Tate, 1935:390; name combination.

Ctenomys frater barbarus: Cabrera, 1961:547; name combination.

Ctenomys frater budini: Cabrera, 1961:548; name combination.

Ctenomys frater frater: Cabrera, 1961:548; name combination.

Ctenomys frater mordosus: Cabrera, 1961:548; name combination.

Ctenomys frater sylvanus: Cabrera, 1961:548; name combination.

DESCRIPTION: Small to medium-sized species; specimens of *C. f. mordosus* from Bolivia with total length 229–272 mm, tail length 55–76 mm (S. Anderson 1997); those from Jujuy, Argentina, larger, with total length 230–300 mm, tail length 50–85 mm (M. M. Díaz and Barquez 2002). General color of nominotypical subspecies and *C. f. mordosus* brownish fawn with dull buffy undersurface without inguinal white patches (Thomas 1902c; S. Anderson 1997). Specimens from Salta province, Argentina, dark russet to black (M. M. Díaz et al. 1997), and those from Jujuy Province reddish or coppery (hence, common name "*tuco colorado*" [red tuco-tuco]; M. M. Díaz and Barquez 2002). Skull greatly vaulted. Thomas (1926c:325–326) characterized *C. s. mordosus* as having "size and general external appearance quite as in true *sylvanus*," but with much broader and heavier incisors, "their combined breadth attaining 7 mm in old specimens as compared with 5.5 mm. in equally aged specimens of *sylvanus*." Upper incisors nearly orthodont, with angle of anterior face 98° of horizontal plane. Rostrum of *mordosus* thickened, in a manner seen "to a still greater extent in *C. boliviensis*" and in *C. steinbachi*. "An allied species, with equally broad incisors, is the *Ctenomys frater* of Potosi, but that has peculiarly narrowed bullae and is of a decidedly lighter colour."

DISTRIBUTION: *Ctenomys frater* occurs in southern Bolivia (Potosí and Tarija departments) and northwestern Argentina (Salta and Jujuy provinces).

SELECTED LOCALITIES (Map 440): ARGENTINA: Jujuy, Caimancito (FMNH 29048), Calilegua (FMNH 23235), Cerro de la Lagunita (type locality of *Ctenomys budini* Thomas), mountains W of Yala (FMNH 23241), San Rafael (FMNH 29053), Sunchal (type locality of *Ctenomys budini barbarus* Thomas), Yuto, Río San Francisco (type locality of *Ctenomys sylvanus utibilis* Thomas); Salta, Aguaray (CML 80), Tartagal (type locality of *Ctenomys sylvanus* Thomas). BOLIVIA: Potosí, Potosí (type locality of *Ctenomys frater* Thomas); Tarija, Caraparí (Anderson 1997), 3 km SE of Cuyambuyo (S. Anderson 1997), Tambo (type locality of *Ctenomys sylvanus mordosus* Thomas), Yacuiba (S. Anderson 1997).

SUBSPECIES: *Ctenomys frater* currently includes five subspecies (synonymies listed here under the species): *Ctenomys f. frater* Thomas, an isolated population in eastern Potosí, Bolivia; *Ctenomys f. barbarus* Thomas, from the mountains in southern Jujuy and adjacent Salta, Argentina; *Ctenomys f. budini* Thomas, from the high elevations in Jujuy and adjacent west-central Salta, Argentina; *Ctenomys f. mordosus* Thomas, from southern Tarija, Bolivia, and perhaps extreme northern Jujuy, Argentina; and *Ctenomys f. sylvanus* Thomas, from the base of the Andes in eastern Jujuy and western Salta, Argentina.

NATURAL HISTORY: In Argentina, *Ctenomys frater* has been captured in moist forests and grassy meadows between 2,000 and 4,500 m in Jujuy Province, and in flat areas with deep soils near creeks in Salta Province (Mares, Ojeda, and Kosco 1981; see also Thomas 1919g). In Bolivia, the subspecies *C. f. mordosus* occurs at elevations from 600 to 2,700 m, and *C. f. frater* at 4,300 m. No detailed data on habitat, habits, behavior, food, or reproduction is known for Bolivian populations (S. Anderson 1997). S. L. Gardner and Duszynski (1990) describe eimerian oocysts in the feces of three specimens of *C. f. mordosus*.

REMARKS: Woods (1993) erroneously gave the type locality of *C. frater* Thomas as in Argentina. In 1921, Thomas (p. 186) argued that *sylvanus* Thomas and *sylvanus utibilis* Thomas, along with *budini* Thomas and a new form he described in that paper (*budini barbarus*) "should be united specifically, and be considered as four subspecies of one species" (i.e., *C. b. budini*, *C. b. sylvanus*, *C. b. utibilis*, and *C. b. barbarus*). In 1926, ignoring his earlier view that *sylvanus* was only a subspecies of *budini*, Thomas (p. 325) described *Ctenomys sylvanus mordosus* but noted that it was allied to *Ctenomys frater*. Cabrera (1961) grouped each of these names, in various combinations, into five subspecies of *C. frater*: *C. f. barbarus*, *C. f. budini*, *C. f. frater*, *C. f. mordosus*, and *C. f. sylvanus* (with *utibilis* Thomas as a synonym). Woods and Kilpatrick (2005:1562) listed *budini* Thomas as a

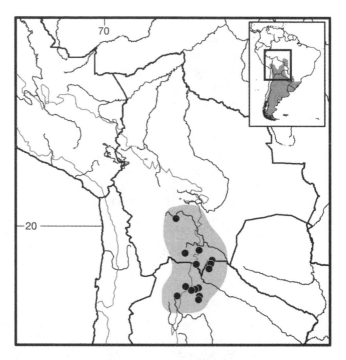

Map 440 Selected localities for *Ctenomys frater* (●). Contour line = 2,000 m.

species distinct from *C. frater*, with *barbarus* Thomas as a valid subspecies. These authors listed *sylvanus* Thomas both in the synonymy of *C. frater* (p. 1564, with an incorrect year of publication of 1925) and as a separate species, *C. sylvanus* (p. 1569, citing the correct publication year of 1919).

Cook et al. (1990) compared *C. f. frater* and *C. f. mordosus* to other Bolivian species in external and craniodental characters. Cook and Yates (1994) regarded *C. frater* (as represented by *C. f. mordosus*) and *C. lewisi* to be more closely related to each other, based on electrophoretic analyses, than to any other Bolivian species these authors examined. Cook et al. (1990) described a karyotype consisting of $2n = 52$, $FN = 78$ for *C. f. mordosus*, and Ruedas et al. (1993) reported on genome size and other karyological details of this same subspecies. The sperm is the simple symmetric morphological type (Vitullo and Cook 1991; Bidau 2006).

Ctenomys fulvus Philippi, 1860
Long-tailed Tuco-tuco
SYNONYMS:

Ctenomys fulvus Philippi, 1860:157, Zool. tab. L; type locality "Reise durch die Wüste Atacama," or vicinity of Pingo-Pingo, Atacama desert, Chile, 2,700 to 3,300 m, 24°00′S 69°00′W (Osgood 1943b:127).

Ctenomys atacamensis Philippi, 1860:158, Zool. Table II, Fig. 1; type locality "Reise durch die Wüste Atacama," or Tilopozo, about lat. 23°20′S, Atacama Desert, Chile (Osgood 1943b:127).

Ctenomys robustus Philippi, 1896:11, plate 4, Fig. 2, and plate 5, Fig. 1a–d; type locality "la provincia de Tarapacá, cerca de Pica, en los lugares llamados 'canchones' en una elevacion de cerda de 12200 m," Chile; Canchones (20°15′S 69°20′W) is on the open plain between Pica and Noria (Osgood 1943).

Ctenomys pallidus Philippi, 1896:13:plate 4, Fig. 1 and plate 5, Fig. 3a–c; type locality "Breas, en el desierto de Atacama," southwest of Antofagasta de la Sierra, about 26°3′S and 67°56′W, alt. 9,000–10,000 ft (Osgood 1943b:127; the coordinates, however, place the locality in Catamarca, Argentina).

Ctenomys pernix Philippi, 1896:15, plate 5, Fig. 5 and plate 6, Fig. 2; type locality "Aguas Calientes," near Socaire, east of Salar de Atacama, about lat. 23°S., long. 68°16′W., Chile (Osgood 1943b:127).

Ctenomys chilensis Philippi, 1896:16, plate 6, Fig. 1; type locality "la cordillera de Linares" but probably from Atacama Desert (Osgood 1943b:127).

Ctenomys fulvus robustus: Mann, 1978:292; name combination.

DESCRIPTION: Very large species, with total length 280–350 mm; samples assignable to *robustus* may exceed 350 mm (Mann 1978). Upper parts slightly grizzled clay color with sides considerably paler than back; forehead and narrow line around mouth darkened or even slightly blackish; under parts uniform, clear cinnamon buff; tail blackish brown above and terminates in light pencil; feet whitish buff with some darkening medially and proximally. Skull large and heavily ridged with swollen auditory bullae and broad, wedge-shaped nasals.

DISTRIBUTION: *Ctenomys fulvus* occurs in northern Chile at elevations above 2,700 m.

SELECTED LOCALITIES (Map 441): CHILE: Antofagasta, Aguas Calientes (type locality of *Ctenomys pernix* Philippi), Pingo-Pingo; (type locality of *Ctenomys fulvus* Philippi, Salar de Atacama (Slamovitz et al. 2001), San Pedro de Atacama (Feito and Gallardo 1982), 20 mi E of San Pedro de Atacama (Osgood 1943b), Tilopozo (type locality of *Ctenomys atacamensis* Philippi); Tarapacá, Canchones, La Huayca (Feito and Gallardo 1982), near Pica (type locality of *Ctenomys robustus* Philippi).

SUBSPECIES: Two subspecies of *Ctenomys fulvus* are currently recognized: *C. f. fulvus* (with *atacamensis* Philippi, *pallidus* Philippi, *pernix* Philippi, and *chilensis* Philippi as synonyms) and *C. f. robustus* Philippi (see Mann 1978; Woods and Kilpatrick 2005). The nominotypical form occurs in western Antofagasta and *C. f. robustus* is restricted to the Oasis de Pica in the desert of Tarapacá. The two ranges are separated by four degrees of latitude.

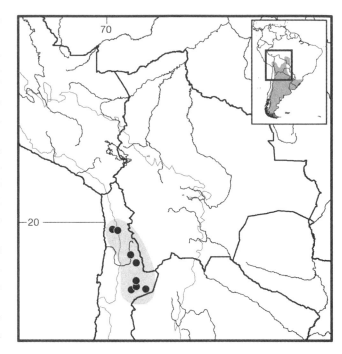

Map 441 Selected localities for *Ctenomys fulvus* (●). Contour line = 2,000 m.

NATURAL HISTORY: *Ctenomys fulvus* inhabits sandy soils in desert flats with creosote bush (*Larrea*, Zygophyllaceae) or other desert shrubs, and riparian forests along dry gullies. Burrow systems are deep (more than 25 cm below the surface), and burrow temperature is maintained between 19° and 25°C while the outside temperature may vary between 6° and 62°C. Individuals are active early in the morning (5:00 to 7:00 AM) and especially in the dry season. They are known to eat *Larrea* leaves (Rosenmann 1959; Mann 1978). Hickman (1985, 1988) described mound-building behavior and swimming ability, and Cortés, Miranda et al. (2000) described thermal biology.

REMARKS: Cabrera (1961) included both *C. coludo* Thomas and *C. famosus* Thomas as valid subspecies of *C. fulvus*. Both of these taxa are considered separate species herein. The two subspecies of *C. fulvus* share the same 2*n* = 26 karyotype (M. H. Gallardo 1979). Sperm morphology is symmetric in both subspecies (Feito and Gallardo 1982).

Ctenomys goodfellowi Thomas, 1921
Goodfellows's Tuco-tuco
SYNONYMS:

Ctenomys goodfellowi Thomas, 1921a:136; type locality "Esperanza, near Conception, Prov. Nuflo de Chaves, E. Bolivia," Santa Cruz, Bolivia, 16°15′S 62°04′W, 400 m.

Ctenomys boliviensis goodfellowi: Cabrera, 1961:546; name combination.

DESCRIPTION: Large tuco-tuco, slightly smaller than related *C. boliviensis*; total length of females 257–288 mm, tail length 71–79 mm; for males, total length >330 mm and tail length >93 mm (S. Anderson 1997). Colors also very similar to those of *C. boliviensis*, although "the dark dorsal line is heavier and the white of the undersurface is reduced to inconspicuous axillary and inguinal patches" (Thomas 1921a:136). Broad, heavy orange incisors and very large premolars characteristics of both species. In general, cranium of *C. goodfellowi* smaller, less ridged, with less thickened zygomata, and smaller and less inflated bullae than *C. boliviensis* (Thomas 1921a).

DISTRIBUTION: *Ctenomys goodfellowi* is known only from two localities in north-central Santa Cruz department, Bolivia (see S. Anderson 1997).

SELECTED LOCALITIES (Map 442): BOLIVIA: Santa Cruz, Esperanza (type locality of *Ctenomys goodfellowi* Thomas), 10 km N of San Ramón, La Laguna (S. Anderson 1997).

SUBSPECIES: *Ctenomys goodfellowi* is monotypic.

NATURAL HISTORY: No data are available on ecology, habitat, behavior, or food. A single pregnant female was trapped in June (S. Anderson 1997). S. L. Gardner and Duszynski (1990) report eimerian oocysts in the feces.

REMARKS: Listed as a subspecies of *C. boliviensis* by Cabrera (1961) and S. Anderson et al. (1987), but electromorphic data (Cook and Yates 1994) support *C. good-*

fellowi as a distinct species. Mitochondrial DNA sequence data (E. P. Lessa and Cook 1998), however indicate that *C. boliviensis* and *C. goodfellowi* are sister species, at least among the limited set of taxa these authors examined. The karyotype is 2n = 46, FN= 68 (S. Anderson et al. 1987), and the sperm is symmetric in its morphology.

Ctenomys haigi Thomas, 1919
Patagonian Tuco-tuco

SYNONYMS:

Ctenomys haigi Thomas, 1919b:210; type locality "Maitén, W. Chubut. 700 m," El Maitén, Departamento Cushamen, Chubut, Argentina, 42°03'S, 71°10'W, 700 m.

Ctenomys haigi lentulus Thomas, 1919b:211; type locality "Pilcañeu, Upper Rio Negro, 1400 m," Pilcaniyeu, Departamento Pilcaniyeu, Provincia de Rio Negro, Argentina, 41°13'S, 70°61'W.

Ctenomys haigi haigi: Thomas, 1919b:211; name combination.

Ctenomys mendocinus haigi: Thomas, 1927e:201; name combination.

Ctenomys haigi luteolus Rusconi, 1928:245; incorrect subsequent spelling of *lentulus* Thomas; not *luteolus* Thomas.

Ctenomys haigii: Pearson, 1995:116; inadvertent spelling error.

DESCRIPTION: Medium-sized and short-tailed species (mean body measurements from five specimens of females and males, respectively, in MVZ collection: head and body length, 172 and 194 mm; tail length, 81 and 88 mm; hindfoot length, 34 and 36 mm; mass, 152 and 229 g). Dorsal color grayish brown, without median line on crown or rump; top of nose dark brown; sides clearer gray and flanks and venter light to washed buffy. Mandible lacks deep channel. About 56% of adult specimens show open frontoparietal fenestrae (S. L. Gardner and Anderson 2001).

DISTRIBUTION: *Ctenomys haigi* occurs in the Neuquén, Rio Negro and Chubut provinces, Argentina in Patagonian steppe and the *Monte de Llanuras y Mesetas* ecoregions.

SELECTED LOCALITIES (Map 443; from Pearson and Christie 1985, except as noted): ARGENTINA: Chubut, El Maitén (type locality of *Ctenomys haigi* Thomas); Neuquén, 37 km SW of Chos Malal, Collón Cura (Thomas 1927e), Collón Cura, 16 SE of La Rinconada, 1 km W of Confluencia, Estancia Alicurá, Parque Nacional Laguna Blanca, Paso del Cordoba (MVZ 183291); Río Negro, 10 km S of Comallo (MVZ 185053), 19 km SE of Bariloche, 43 km SSW of Bariloche, Campo Viejo, Estancia San Ramón (MVZ 186062), 7.5 km N and 8 km E of Cerro Ñireco (MVZ 183295), 5 km W of Comico, Lago Steffen (MVZ 162284), Laguna Tromen (MVZ 193693), Pilcaniyeu (type

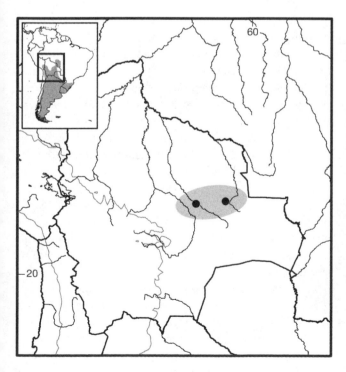

Map 442 Selected localities for *Ctenomys goodfellowi* (●). Contour line = 2,000 m.

locality of *Ctenomys haigi lentulus* Thomas), 5 km W of Sierra Grande.

SUBSPECIES: *Ctenomys haigi* is monotypic.

NATURAL HISTORY (from Pearson 1984, except as noted): As is characteristic of all tuco-tucos, *C. haigi* is restricted to self-dug underground burrows where it may be active day or night. It is rarely seen above ground, but occasional emergence is indicated by its frequent occurrence in owl pellets as well as the toothmarks on the small-sized, above-ground shrubs upon which it feeds. Individuals are vocal, uttering a single, short syllable and with as many as 30 such "calls" in a single bout. The species inhabits treeless steppe vegetation. Litter size ranges between two and four (mean 2.6). Lacey et al. (1998) described solitary burrow use by adults of both sexes. Hambuch and Lacey (2002) and Lacey (2001) contrasted patterns of variation at both MHC and microsatellite loci, respectively, between the asocial *C. haigi* and the geographically adjacent but social *C. sociabilis*.

REMARKS: Thomas (1919b) considered *C. haigi* to be allied to *C. colburni*, although smaller and with less pronounced bullae. Subsequently, Thomas and St. Leger (1926b:639) thought that *haigi* "will be found to grade into" *C. mendocinus*. And, finally, Thomas (1927e) formalized this earlier suggestion by stating that his subspecies *lentulus* was indistinguishable from typical *haigi*, which he then placed as a subspecies of *C. mendocinus*.

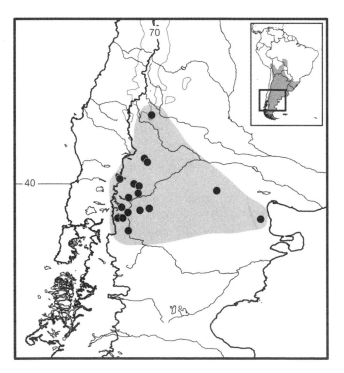

Map 443 Selected localities for *Ctenomys haigi* (●). Contour line = 2,000 m.

Pearson (1984) argued that *C. haigi* was a species distinct from *C. mendocinus* on the basis of their different karyotypes, a position supported by substantial divergence in electromorphic and mtDNA sequence data (Sage et al. 1986; E. P. Lessa and Cook 1998). Pearson (1984) also regarded specimens from Chos Malal, Neuquén Province, Argentina, in the Natural History Museum, London, that Thomas identified as *C. mendocinus* were best assigned to *C. haigi* on the basis of size and coloration. The karyotype of *C. haigi* is 2n = 50, FN = 66 (Pearson 1984; M. H. Gallardo 1991). Sperm morphology is of the simple asymmetric type (M. H. Gallardo et al. 2002; Bidau 2006).

Ctenomys ibicuiensis T. R. O. Freitas, Fernandes, Fornel, and Roratto, 2012
Ibicuí Tuco-tuco

SYNONYMS:

Ctenomys ibicuiensis T. R. O. Freitas, Fernandes, Fornel, and Roratto, 2012:1358: type locality "Manoel Viana, in central-western Rio Grande do Sul, southern Brazil, 29°23′37″S, 55°25′43″W."

DESCRIPTION: Species defined primarily by chromosomal and mtDNA cytochrome-*b* sequence differences relative to other species of "*torquatus*-group" from southwestern Brazil and adjacent parts of Argentina and Uruguay; few data available for external characteristics, including pelage color and color pattern, although species moderate in body size (head and body length 159 mm, tail length 75 mm, hindfoot length 37 mm, and mass 200 g; T. R. O. Freitas et al. 2012). Cranially, *C. ibicuiensis* most similar to geographically adjacent populations of *C. torquatus*, but differs with shorter skull, less robust zygomatic arches, and more compact cranium. Multivariate analyses of dorsal, ventral, and lateral landmarks of skull separate *C. ibicuiensis* from six other species, including *C. lami*, *C. minutus*, *C. pearsoni*, *C. perrensi*, *C. roigi*, and *C. torquatus*.

DISTRIBUTION: *Ctenomys ibicuiensis* is known from only six closely adjacent localities in the municipalities of Manoel Viana and Macambará in Rio Grande do Sul state, Brazil, at an elevation approximately 200 m.

SELECTED LOCALITIES (Map 444; from T. R. O. Freitas et al. 2012): BRAZIL: Rio Grande do Sul, BR 176, Manoel Viana, Piraju, Manoel Viana (type locality of *Ctenomys ibicuiensis* T. R. O. Freitas et al.), Passo do Narciso, Maçambará.

SUBSPECIES: *Ctenomys ibicuiensis* is monotypic.

NATURAL HISTORY: No data ecology, reproduction, behavior, or other aspects of the population biology of this species have been reported. T. R. O. Freitas et al. (2012:1359) recorded the habitat as "sand dunes and grasslands, disturbed by agricultural activities and desertification."

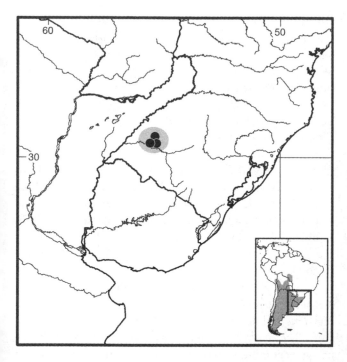

Map 444   Selected localities for *Ctenomys ibicuiensis* (●).

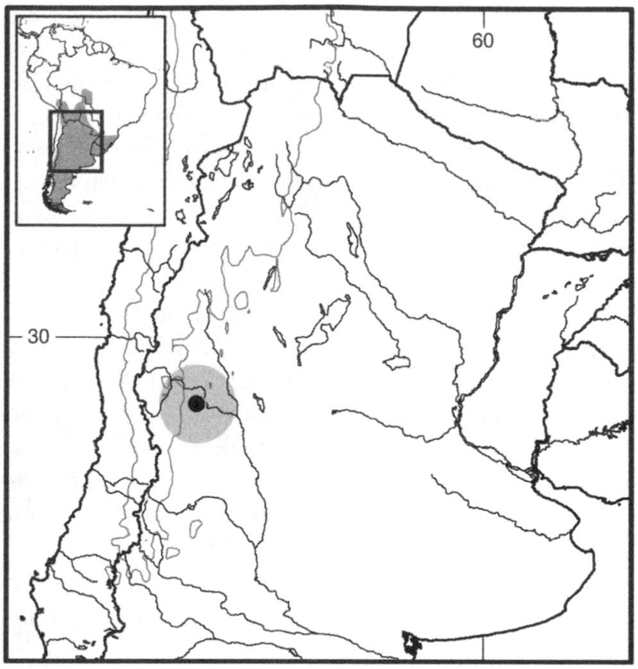

Map 445   Single known locality for *Ctenomys johannis* (●). Contour line = 2,000 m.

REMARKS: The karyotype consists of $2n = 50$, FN = 68, with 10 biarmed and 14 acrocentric autosomal pairs and a biarmed X chromosome (T. R. O. Freitas et al. 2012). A distinctly large pair of biarmed autosomes is diagnostic. In their phylogenetic analysis of mtDNA cytochrome-*b* sequences (T. R. O. Freitas et al. 2012), *C. ibicuiensis* was sister to a clade comprised of *C. lami*, *C. minutus*, and *C. torquatus*.

Ctenomys johannis Thomas, 1921
San Juan Tuco-tuco

SYNONYMS:

Ctenomys coludo johannis Thomas, 1921g:523; type locality "Department of San Juan. Type from Cañada Honda. Alt. 500 m," Departamento Sarmiento, San Juan, Argentina, 31°59′S 68°33′W, 600 m.

Ctenomys fulvus johannis: Cabrera, 1961:549; name combination.

Ctenomys johanis: J. R. Contreras, Roig, and Suzarte, 1977: 162; first use of current name combination but inadvertent misspelling of *johannis* Thomas.

DESCRIPTION: Similar in body size and other characters to *C. coludo* from Catamarca province, Argentina, but color of *C. johannis* less strongly buffy (termed "drab-gray"), with nape, foreback, and sides of head gray; nasal patch present, more strongly marked than in *C. coludo*. Ventral hairs with whitish tips, faintly tinged buffy. Caudal crest blackened terminally, darkening essentially absent in

C. coludo. Skull similar to that of *C. coludo*, but zygomatic arches less widely spread so that breadth across arches distinctly less than across opening of auditory tubes; interparietal sutures barely perceptible, or not at all. External dimension of holotype, head and body length 199 mm, tail length 97 mm, and hindfoot length 36 mm.

DISTRIBUTION: *Ctenomys johannis* is only known from the type locality in the *Monte de Llanuras y Mesetas* ecoregion of San Juan province, Argentina.

SELECTED LOCALITIES (Map 445): ARGENTINA: San Juan, Cañada Honda (type locality of *Ctenomys coludo johannis* Thomas).

SUBSPECIES: *Ctenomys johannis* is monotypic.

NATURAL HISTORY: According to Emilio Budin, the collector, this species was found "under trees; holes made in naked soil" (Thomas 1921g:524).

REMARKS: The karyotype and sperm morphology are unknown.

Ctenomys juris Thomas, 1920
Jujuy Tuco-tuco

SYNONYMS:

Ctenomys juris Thomas, 1920a:194; type locality "El Chaguaral, between San Pedro and Villa Carolina. Alt. 500 m," Departamento San Pedro, Jujuy, Argentina, 24°16′S, 64°48′W, 500 m (see J. R. Contreras 1984b:239).

Ctenomys mendocinus juris: Cabrera, 1961:552; name combination.

DESCRIPTION: Small species, resembling *C. fochi* but with much smaller bullae (Thomas 1920a; A. I. Medina et al. 2007). General color uniformly pale brown with buffy or whitish half-collar extending to ear on side of neck; as with *C. fochi*, marked darkening on top of muzzle; inner sides of forelimbs whitish, paler than venter; hands and feet white above; tail dull buffy whitish, with dark brown terminal crest. Skull with broad nasals, slightly narrowed posteriorly; zygomatic arches widely expanded, anterior breadth often greater than posterior; mesopterygoid fossa penetrates palate to level of middle of M2; bullae small and narrow but smoothly contoured, not compressed, and markedly smaller than those of *C. fochi*. Upper incisors particularly proodont, reaching angle of 108°, more so than either *C. bergi* (100°) or *C. fochi* (94°). External measurements of holotype, head and body length 177 mm, tail length 72 mm, and hindfoot length 29 mm.

DISTRIBUTION: *Ctenomys juris* is known only from the type locality in the *Chaco Seco* ecoregion of Jujuy province, Argentina.

SELECTED LOCALITIES (Map 446): ARGENTINA: Jujuy, El Chaguaral (type locality of *Ctenomys juris* Thomas).

SUBSPECIES: *Ctenomys juris* is monotypic.

NATURAL HISTORY: According to Emilio Budin (Thomas 1920a:194), this tuco-tuco was found "in stony ground in ravines running down to the river."

REMARKS: Cabrera (1961) treated this species as a subspecies of *C. mendocinus* Philippi, a position main-

tained by Redford and Eisenberg (1992) and Woods (1993). Galliari et al. (1996) listed *C. juris* as a species, without comment. The diploid number of *C. juris* is 2*n* = 26 (C. J. Bidau, unpubl. data). Sperm morphology is the simple asymmetric type (Bidau 2006).

Ctenomys knighti Thomas, 1919
Catamarca Tuco-tuco

SYNONYMS:

Ctenomys knighti Thomas, 1919d:498; type locality "Otro Cerro, North-eastern Rioja"; an abandoned ranch at the southern end of Sierra de Ambato, NE Chumbicha, Departamento Capayán, Catamarca, Argentina, 28°45′S, 66°17′W, 2,023 m (see Pardiñas et al. 2007).

Ctenomys knighti knighti: Cabrera, 1961:549; name combination.

DESCRIPTION: Small to medium-sized species; dimensions of holotype (adult male) head and body length 203 mm, tail length 82 m, and hindfoot length 36 mm. Dorsal color dark brown; flanks, especially over thighs, buffy; venter uniformly ochraceous with cinnamon-buff hair tips; no collar present around the throat; muzzle, both on top and sides of mouth and tip of chin, distinctly blackish. Skull elongated with narrow nasals; parietal region lacks any trace of separate interparietal, even in the youngest specimens; incisive foramina elongated; bullae large and, when viewed from behind, have conspicuously smooth posterosuperior portion not covered by plastering bones of occipital series.

DISTRIBUTION: *Ctenomys knighti* is found in Catamarca and La Rioja provinces of Argentina within the *Chaco Seco* and *Monte de Sierras y Bolsones* ecoregions, at elevations from 770 to 2,025 m.

SELECTED LOCALITIES (Map 447): ARGENTINA: Catamarca, Otro Cerro (type locality of *Ctenomys knighti* Thomas); La Rioja, Machigasta (Llanos 1947).

SUBSPECIES: *Ctenomys knighti* is monotypic.

NATURAL HISTORY: No ecological data are available. Emilio Budin, the collector, found the species "on stony ground" (Thomas 1919d: 499).

REMARKS: The location of Otro Cerro has been debatable since Thomas (1919d) first described specimens from this locality (see Thomas 1920b; Pearson 1958; Cabrera 1961; and Myers et al. 1990 for varying opinions). Pardiñas et al. (2007:399–400 and Fig. 4, p. 403) reviewed this history and provided detailed information that placed this site at the southern end of the Sierra de Ambato in Catamarca, not Rioja, province. Fortunately, the resolution of this issue by Pardiñas et al. (2007) fixes the type locality for *C. knighti* and several other mammal species described by Thomas.

Thomas (1919d) believed *C. knighti* allied to *C. budini* (= *C. frater budini* herein) from Jujuy Province, but distin-

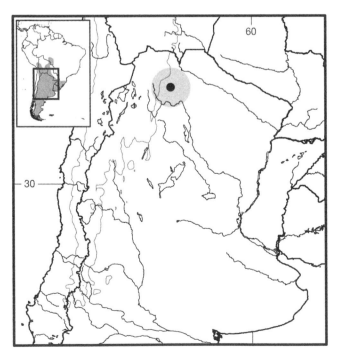

Map 446 Single known locality for *Ctenomys juris* (●). Contour line = 2,000 m.

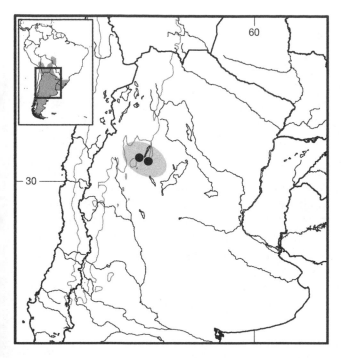

Map 447 Selected localities for *Ctenomys knighti* (●). Contour line = 2,000 m.

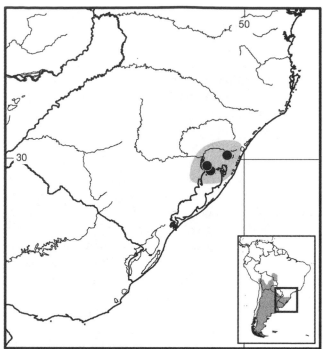

Map 448 Selected localities for *Ctenomys lami* (●).

guishable from that species by its dark muzzle, more buffy sides, absence of a collar, and cranially notably by the absence of a separate interparietal. Sperm morphology and karyotype are unknown.

Ctenomys lami T. R. O. Freitas, 2001
Lami Tuco-tuco
SYNONYM:

Ctenomys lami T. R. O. Freitas, 2001:2; type locality "Beco dos Cegos, Lami Beach in the Guaiba river, near Porto Alegre, at 30°51′S, 51°10′W," Rio Grande do Sul, Brazil.

DESCRIPTION: Quite variable in size, total length 231–310 mm, tail length 67–92 mm, hindfoot length 31–42 mm, and mass 170–307 g. Dorsal color dark brownish, with each hair dark grayish basally and brownish near tip; ventral pelage paler. Cranially, nasal bones wide anteriorly and taper posteriorly; frontal region flat from interorbital region posteriorly; frontoparietal suture round; braincase oval-shaped, with length about one-third length of skull; zygomatic arches squared and robust; incisive foramina small; mesopterygoid fossa wide posteriorly and flared anteriorly; auditory bullae large and ventrally inflated. Mandible with long and narrow coronoid process, heavy condyloid process, and large angular process. All molariform teeth kidney-shaped, with M3 and m3 much reduced in size. Upper incisors large, orthodont, and faced with orange enamel.

DISTRIBUTION: *Ctenomys lami* is known from multiple localities in Rio Grande do Sul state, Brazil, in the small area south and east of Porto Alegre between the Rio Guaiba and Lago Barros.

SELECTED LOCALITIES (Map 448; from T. R. O. Freitas 2001): BRAZIL: Rio Grande do Sul, Beco dos Cegos (type locality of *Ctenomys lami* T. R. O. Freitas), Chico Lomã, Parque de Itapuã.

SUBSPECIES: *Ctenomys lami* is monotypic.

NATURAL HISTORY: T. R. O. Freitas (2001) recorded pregnant females of *C. lami* in the months of June and December. Litter size was one to three embryos (mean = 2.04 ± 0.73; N = 25).

REMARKS: T. R. O. Freitas (2001) distinguished this species from the parapatric *C. minutus* primarily by its unique but highly variable karyotype. At least seven karyotypes are known, with diploid number ranging from $2n = 54$ to 58 with fusion/fission and pericentric inversion systems described in detail (T. R. O. Freitas 2007). El Jundi and Freitas (2004) provided details on genetic and demographic structure.

Ctenomys latro Thomas, 1918
Mottled Tuco-tuco
SYNONYM:

Ctenomys latro Thomas, 1918a:38; type locality "Tapia, about 20 miles north of Tucuman City. Alt. 600 m," Departamento Trancas, Tucumán, Argentina, 26°36′ S, 65°18′ W, 689 m.

DESCRIPTION: Small to medium-sized species, head and body length 161–172 mm (A. I. Medina et al. 2007).

Overall color buffy fawn on sides of head, back, and flanks; crown and middle of face dark brown; venter pale buff with throat white and darker chest patch present; colors of the upper and under surfaces sharply differentiated; light buffy patch behind ear that runs downward and backward onto side of neck; tail dark brown along entire length of dorsal surface but pale buffy on sides and below. Skull low and flattened, with narrow braincase; rostrum unusually broadened, due to thickening of bone outside anterior half of incisor alveolus; nasals broad, abruptly and squarely truncated behind, considerably surpassed by premaxillary processes, and with sides forming straight converging lines; zygomatic arches widely expanded, but distinctly short anteroposteriorly; temporal ridges unite to form low median sagittal crest rather than remaining separated; supraoccipital smooth, without median ridge; bullae small but smooth and well inflated. Cheek teeth small and appear rather delicate; PM4 exceeds molars in diagonal diameter.

DISTRIBUTION: *Ctenomys latro* is known only from Tucumán province, Argentina, in the *Chaco Seco* ecoregion.

SELECTED LOCALITIES (Map 449): ARGENTINA: Tucumán, Burruyacu, E Cerro Campo (Thomas 1926d), Río Choromoro (A. I. Medina et al. 2007), Tapia (type locality of *Ctenomys latro* Thomas), Ticucho (Ipucha et al. 2008).

SUBSPECIES: *Ctenomys latro* is monotypic.

NATURAL HISTORY: No data on ecology are available. Specimens have been collected in areas of open Chacoan forest with extremely stony soils.

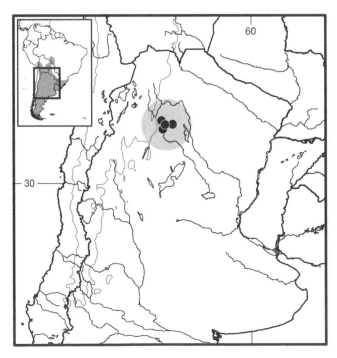

Map 449 Selected localities for *Ctenomys latro* (●). Contour line = 2,000 m.

REMARKS: Cabrera (1961:553) synonymized *C. latro* with *C. mendocinus*, listing the name as a synonym under the subspecies *C. m. tucumanus* Thomas, part of his concept of a polytypic *C. mendocinus*. Reig and Kiblisky (1969) and Roig and Reig (1969) considered both *C. latro* and *C. tucumanus* to be distinct species based on karyotypic differences, a conclusion followed both by subsequent authors and as viewed herein. *Ctenomys latro* is closely allied to *C. pilarensis*, *C. occultus*, and *C. argentinus* in mtDNA sequence analyses, with this clade apparently sister to *C. tucumanus* (Mascheretti et al. 2000; Parada et al. 2011; see also J. R. Contreras and Bidau 1999). Contrary to Cabrera's (1961) hypothesis, *C. latro* shares no molecular relationship with *C. mendocinus* (Parada et al. 2011). Several related karyotypes have been reported for the species, including $2n = 42$, FN = 50 (Reig and Kiblisky 1968a), $2n = 40$, FN = 44 (Ortells 1995) and $2n = 40–42$, FN = 48 (Ipucha 2002; Ipucha et al. 2008).

Ctenomys leucodon Waterhouse, 1848
White-toothed Tuco-tuco

SYNONYMS:

Ctenomys leucodon Waterhouse, 1848:281; type locality "San Andrés de Machaca, . . . south of the Lake Titicaca, Department of La Paz, Bolivia"; ca. 4,000 m, 16°44'S 69°01'W. Thomas, 1927b:552.

Ctenomys (Haptomys) leucodon Thomas, 1916g:305; designation as type species of new subgenus *Haptomys*.

DESCRIPTION: Medium sized with total length of adults 200–278 mm, tail length 79–85 mm, and hindfoot length averaging 34 mm (Sanborn and Pearson 1947; S. Anderson 1997). Back clay colored with base of hairs slate gray followed by either broad or narrow clay band and tipped with black; head darker and sides of muzzle and cheeks buckthorn brown; undersurface near tawny-olive overall but with reddish tinge on chest; tail dark brown above and faintly lighter beneath, not dirty white as given in Waterhouse's description, and without crest of long whitish hairs (Sanborn and Pearson 1947). Upper incisors strongly procumbent and almost unique within genus with white or pale yellowish white enamel (Waterhouse 1848; Pearson 1959).

DISTRIBUTION: *Ctenomys leucodon* occurs in the Altiplano highlands surrounding Lake Titicaca, extending from La Paz, Bolivia, to southern Puno, Peru (Pearson 1959).

SELECTED LOCALITIES (Map 450): BOLIVIA: La Paz, Chilalaya (Cook et al. 1990), Comauchi (S. Anderson 1997), San Andrés de Machaca (type locality of *Ctenomys leucodon* Waterhouse). PERU: 10 km SW of Huacullani (Pearson 1959), Río Ccallacami, near Huacullani (Sanborn and Pearson 1947).

SUBSPECIES: *Ctenomys leucodon* is monotypic.

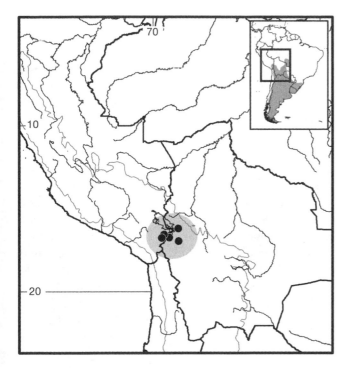

Map 450 Selected localities for *Ctenomys leucodon* (●). Contour line = 2,000 m.

NATURAL HISTORY: Few data on ecology are available. A single pregnant female has been recorded in the month of August in Bolivia (S. Anderson 1997). In one local Peruvian population, Pearson (1959) noted that individuals were nonvocal, made large earthen mounds in sloping grassland, and occurred at slightly lower elevation than adjacent populations of *C. opimus*. S. L. Gardner and Duszynski (1990) examined feces of this species but found no eimerian oocysts.

REMARKS: Thomas (1916g) placed *C. leucodon* into the monotypic subgenus *Haptomys*, a position that has mixed support from molecular markers. Cook et al. (1990) compared this species to others they recognized as occurring in Bolivia. Phylogenetic analyses of electromorphic allozymes (Cook and Yates 1994) placed *C. leucodon* in a near basal-polytomy with species of the nominotypical subgenus, whereas mtDNA sequences positioned *C. leucodon* in a near-basal polytomy, excluding *C. sociabilis* (Parada et al. 2011). Cook et al. (1990) described a karyotype with $2n = 36$, FN = 68. Sperm morphology is unknown.

Ctenomys lewisi Thomas, 1926
Lewis's Tuco-tuco

SYNONYM:

Ctenomys lewisi Thomas, 1926c: 323; type locality "Sama, 4000 m," 50 km W Tarija, Tarija, Bolivia, 21°3′S 65°10′W.

DESCRIPTION: Relatively large, reddish-brown tuco-tuco with unusually procumbent incisors, similar to those of *C. leucodon* but more robust. Head and body length 204–219 mm, tail length 68–71 mm, and hindfoot length averages 37 mm. General color of upper parts uniform dark cinnamon brown without special markings; muzzle only slightly washed with black; under surfaces brighter cinnamon, often very bright, sometimes throughout but always in inguinal region; throat with slight, darker collar; forefeet and hindfeet dull whitish above; tail thinly haired, blackish above at base but whitish otherwise. Skull large, long, but neither broad nor robust; nasals rather short and broad; interorbital region broad, without projecting supraorbital ledges; lambdoidal crest strongly developed, with well-marked median forwardly directed angle; zygoma with simple angular projection above, not broadened at its tip; concave groove on outer upper side of malar absent; bullae appear somewhat compressed. Upper incisors very large and heavy, flattened and strongly pigmented orange in front, very proodont (incisive angle of holotype 117°), and readily visible in front of premaxilla when viewed from above. About 33% of adult specimens show open frontoparietal fenestrae (S. L. Gardner and Anderson 2001).

DISTRIBUTION: *Ctenomys lewisi* is known from two closely adjacent localities in northwestern Tarija, Bolivia (S. Anderson 1997:492, Fig. 770).

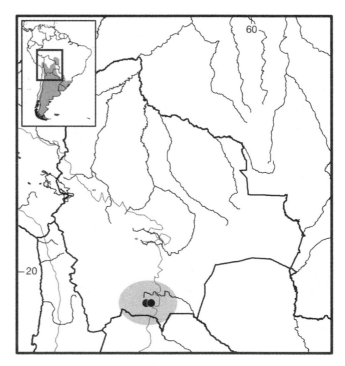

Map 451 Selected localities for *Ctenomys lewisi* (●). Contour line = 2,000 m.

SELECTED LOCALITIES (Map 451): BOLIVIA: Tarija: 1 km E of Iscayachi, Río Tomayapo (S. Anderson 1997), Sama (type locality of *Ctenomys lewisi* Thomas).

SUBSPECIES: *Ctenomys lewisi* is monotypic.

NATURAL HISTORY: *Ctenomys lewisi* inhabits deep soils of the Tomayapo Valley and Sama between 3,450 and 4,000 m. Thomas (1926c) inferred a semiaquatic habit based on captures in holes in wet stream banks, but more recent captures have not confirmed this suggestion (S. Anderson 1997). Pregnant females have been recorded in the month of July. S. L. Gardner and Duszynski (1990) reported eimerian oocysts from feces.

REMARKS: Both electromorphic (Cook and Yates 1994) and mtDNA sequence (E. P. Lessa and Cook 1998; Parada et al. 2011) analyses suggest a close sister relationship between *C. lewisi* and *C. frater*, and then with *C. conoveri* and an apparently undescribed species from Llathu, Cochabamba, Bolivia (Parada et al. 2011). Cook et al. (1990) described a karyotype with 2*n* = 56 and FN = 74; Ruedas et al. (1993) reported on genome size and other karyological attributes. Sperm morphology is the simple symmetric type (Vitullo and Cook 1991).

Ctenomys magellanicus Bennett, 1836
Magellanic Tuco-tuco

SYNONYMS:

Ctenomys magellanicus Bennett, 1836:190, type locality "Port Gregory [= Bahía San Gregorio], near eastern end of north side of Straits of Magellan, Chile," Aysén , 52°25′S, 69°45′W, 0 m (see J. A. Allen 1905:35; Rusconi 1928:238; Osgood 1943b).

Ctenomys fueginus Philippi, 1880:276, type locality "östlichen Insel," eastern island or Isla Grande, Tierra del Fuego, Magallanes y Antártica Chilena, Chile (Osgood 1943b).

Ctenomys neglectus Nehring, 1900b:535; type locality "Patagonien."

Ctenomys robustus J. A. Allen, 1903b:185; type locality "Rio Chico de Santa Cruz, near the Cordilleras," Santa Cruz, Argentina; restricted by Pardiñas (2013) as "río Tucu Tucu, ca. 8 km aguas abajo desde su nacimiento (48.47°S, 71.87°W, departamento Río Chico, Santa Cruz, Argentina)." Preoccupied by *Ctenomys robustus* Philippi (1896), although not the same species (see account for *C. fulvus*).

Ctenomys osgoodi J. A. Allen, 1905:191; substitute name for *C. robustus* J. A. Allen.

Ctenomys magellanicus fueginus: Osgood, 1943b:119; name combination.

Ctenomys magellanicus osgoodi: Osgood, 1943b:120; name combination.

Ctenomys magellanicus dicki Osgood, 1943b:123; type locality "Estancia Ponsonby, east end of Riesco Island, Ma-

gallanes, Chile," Isla de Riesco, Provincia de Magallanes, Magallanes y Antártica Chilena region, ca. 52°50′S 71°45′W.

Ctenomys magellanicus obscurus Texera, 1975: 163; type locality "Estancia Lago Escondido, 20 km S de la Sección Rio Grande, cerca de Lago Blanco, Tierra Del Fuego, Magallanes y Antártica Chilena, Chile, ca. 500 m," 53°53′S 68°52′W.

DESCRIPTION: Relatively large species with total length from 267 mm in *C. m. magellanicus* to 304 mm in *C. m. fueginus*. Generally pale-colored, with pale grizzled grayish buff upper parts and clear cinnamon buff under parts in *C. m. magellanicus* and *C. m. fueginus*; prevailing color of populations allocated to *C. m. osgoodi*; much darker, brownish ochraceous rather than grayish buff; color of *C. m. dicki* wholly mixed blackish and buffy smoke gray both above and below; that of *C. m. obscurus* homogeneously darker than other subspecies (Texera 1975). Skull notably angular with many sharp ridges and pointed processes, particularly so in specimens of nominotypical race. Skull of *C. m. fueginus* similar to that of *C. m. magellanicus* but with slightly more swollen and bulbous auditory bullae, broader interorbital region, and broader rostral or antemolar part. Skull of *C. m. osgoodi* also similar but with narrower and more laterally compressed auditory bullae. Skull of *C. m. dicki* essentially as that of *C. m. fueginus*, except that auditory bullae slightly smaller and shorter and interorbital region wide, as in *C. m. magellanicus*.

DISTRIBUTION: *Ctenomys magellanicus* is distributed in extreme southern Chile and southern Argentina, including Tierra del Fuego.

SELECTED LOCALITIES (Map 452): ARGENTINA: Santa Cruz, Cordilleras, west of upper Río Chico (type locality of *Ctenomys osgoodi* J. A. Allen), 30 mi S of Santa Cruz (Osgood 1943b); Tierra del Fuego, Estancia San Martín, Bahía San Sebastián (MVZ 164145), Estancia Viamonte (Osgood 1943b), Lago Fagnano (Osgood 1943b), near east end of Lake Fagnano (Osgood 1943b). CHILE: Aysén, Río Ñirehuao (Osgood 1943b); Magallanes y Antártica Chilena, Bahía Felipe (USNM 398541), Estancia Lago Escondido (*Ctenomys magellanicus obscurus* Texera), Estancia Ponsonby, east end of Riesco Island (type locality of *Ctenomys magellanicus dicki* Osgood), La Cumbre, Baguales (Feito and Gallardo 1982), Port Gregory (type locality of *Ctenomys magellanicus* Bennett), Punta Arenas (AMNH 17444).

SUBSPECIES: Both Osgood (1943b) and Cabrera (1961) considered *C. magellanicus* to be a polytypic species with four subspecies (*dicki* Osgood, *magellanicus* Bennett, *fueginus* Philippi, and *osgoodi* Allen). Texera (1975) later described a fifth subspecies (*obscurus*).

NATURAL HISTORY: An inhabitant of the Patagonian steppe, *C. magellanicus* lives usually in open meadows with dense grass cover. Burrow systems are constructed 30 cm or more below the surface. Individual feed on the roots of grasses and shrubs (Mann 1978). Native Americans from Tierra del Fuego called these tuco-tucos *cururu*, a name also applied to other medium-sized rodents in the area, and at least the indigenous Selk'nam included them in their diet (Pepa 2001; Pardo 2007).

REMARKS: Thomas (1898h) recorded a specimen of this species from Tombo Point (Punta Tombo), Chubut, Argentina, without justification. This tuco-tuco is very likely not *C. magellanicus*, given the large expanse of Patagonian steppe separating this locality from the main distribution of the species. Osgood (1943b:122) referred a topotype of *C. fodax* in the Field Museum collection to *C. m. osgoodi*; *C. fodax* is herein considered a species distinct from *C. magellanicus* (see that account).

The type locality of J. A. Allen's *C. robustus* (subsequently renamed *C. osgoodi*) was given in the original description as "Rio Chico de Santa Cruz, near the Cordilleras (J. A. Allen 1903b:185). Subsequently, Allen (1905:40) gave additional information to narrow the exact locality, stating first "in the alluvial river valley of the upper Rio Chico, at the eastern base of the Andes" and then a few sentences later "most specimens of this species appear to have been taken in Mayer Basin, at the edge of the Cordilleras." Earlier in the same report, Allen (1905:2) had written about the route taken by the field party who "then followed a continuous journey of some 259 miles up the Rio Chico de Santa Cruz, to the mouth of the Rio Belgrano" with mammal specimens taken "north from the Rio Belgrado to the vicinity of Lake Buenos Aires." Allen (1905:footnote 2) further wrote "most specimens are labeled 'Arroyo Eche' (= Aike)." Lago Buenos Aires (= Lago General Carrera) is about 180 km north of the mouth of the Río Belgrano where it joins the Río Chico (at about 48.24°S, 71.24°W). The type locality of *osgoodi* J. A. Allen is thus likely at a point between these two places. I have been unable to locate either "Mayer Basin" or "Arroyo Eche," although present-day RN 40 crosses the Río Ecker about 60 km south of Perito Moreno, which is just east of Lago Buenos Aires.

Ctenomys magellanicus is chromosomally polytypic. Reig and Kiblisky (1969) recorded diploid numbers of 36 and 38 for *C. m. fueginus*, and M. H. Gallardo (1979) described a diploid number of 34 for *C. m. magellanicus*. Lizarralde et al. (2003) examined telomeric sequence localization in the $2n = 34$ and $2n = 36$ cytotypes. Sperm morphology is the simple asymmetric type (Feito and Gallardo 1982). Fasanella et al. (2013) reported on historical demography and spatial structure in mtDNA sequences

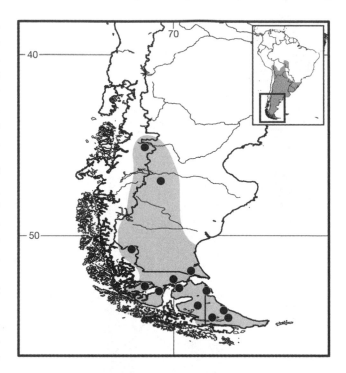

Map 452 Selected localities for *Ctenomys magellanicus* (●).

and nuclear microsatellite loci for population samples on Tierra del Fuego that encompassed the two cytotypes.

Ctenomys "mariafarelli" Azurduy, 2005
Maria Farell's Tuco-tuco

SYNONYM:

Ctenomys mariafarelli Azurduy, 2005b:70; type locality "5.5 Km al noreste de la localidad de Vallegrande (18°28′S, 64°08′O; 1800 m), provincia Florida, del Departamento de Santa Cruz, Bolivia"; *nomen nudum* (see Remarks).

DESCRIPTION: Relatively large tuco-tuco, with total length of males (N = 2) 291–295 mm, tail length 74–85 mm, hindfoot length 39–41 mm, and mass 307–364 g; females (N = 3) smaller, total length 254–273 mm; tail length 73–81 mm; hindfoot length 33–38 mm, and mass 173–227 g. Dorsal color brownish with poorly defined median blackish stripe, lacking half-collar typical of *C. boliviensis*, purportedly its closest relative; flanks paler brown or ochraceous; venter yellowish with white inguinal and axillary spots. Skull elongated and narrow with respect to expansion of zygomatic arches; nasal bones wide posteriorly and taper anteriorly; frontal region narrow; zygomatic arches slightly square; tympanic bullae moderately large and inflated. Upper incisors conspicuously proodont, large, and faced with orange enamel.

DISTRIBUTION: *Ctenomys "mariafarelli"* is known only from the type locality in Santa Cruz department, Bolivia.

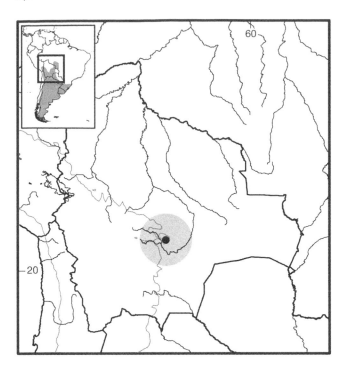

Map 453 Single known locality for Ctenomys "mariafarelli" (●). Contour line = 2,000 m.

SELECTED LOCALITIES (Map 453): BOLIVIA: Santa Cruz: Vallegrande (type locality of *Ctenomys mariafarelli* Azurduy).

SUBSPECIES: *Ctenomys* "mariafarelli" is monotypic.

NATURAL HISTORY: No data are available on ecology, behavior, reproduction, or food preferences.

REMARKS: According to Azurduy (2005b), the species is closely allied to *C. boliviensis*. Neither the karyotype nor sperm morphology is known. The name *Ctenomys mariafarelli* was published in the electronic journal *Kempffiana*, and thus at the time of publication did not meet the requirements of Article 8.6 of the International Code for Zoological Nomenclature (ICZN 1999). The recently amended Code (International Commission on Zoological Nomenclature 2012) accepts electronic publication of new names, but only if (a) the date of issue is subsequent to 2011, (b) the publication date is included in the work itself, and (c) the publication is registered in the *Official Register of Zoological Nomenclature* (ZooBank). Since none of these requirements apply to Azurduy's name, *mariafarelli* Azurduy is a *nomen nudum*.

Ctenomys maulinus Philippi, 1872

Maule Tuco-tuco

SYNONYMS:

Ctenomys maulinus Philippi, 1872: 442; type locality "D. Toribio Medina, hat diese Art im Januar und Februar d.J. in der hohen Cordillere der Prov. Maule gefunden und zwar in der Hähe des Sees laguna de Maule, aus

welchem der bie Talca vorbeifliessende und bei Constitucion Mündende Fluss gleichen Namens entspringt," given as Laguna de Maule, Talca province, Chile, ca. 36°00′S, 70°30′W" by Osgood (1943b:125).

Ctenomys mendocinus maulinus: Thomas, 1927d:657; name combination.

Ctenomys maulinus maulinus: Osgood, 1943b:24; name combination.

Ctenomys maulinus brunneus Osgood, 1943b:125; type locality "Rio Colorado, Province of Malleco, Chile. Alt. 3,000 ft," 38°27′S, 71°22′W, 900 m.

DESCRIPTION: Medium sized, total length varying from 275 mm in nominotypical subspecies to 305 mm in *C. m. brunneus*. Dorsal color of *C. m. maulinus* uniformly light brown and tail typically with short, white pencil at tip. Color of *C. m. brunneus* richer and darker brown (Mann 1978); feet dull buffy white; tail brown above, pale buffy below with buffy white pencil at the tip. Skull with persistent frontoparietal fontanelle; wide flat interorbital space; imperceptible postorbital processes; auditory bullae relatively short and swollen in nominotypical subspecies but narrower and more elongate in *C. m. brunneus*.

DISTRIBUTION: The nominotypical subspecies *C. m. maulinus* occurs only in Talca province, Chile, whereas *C. m. brunneus* is distributed in the provinces of Cautín and Malleco between 1,000 and 2,000 m (Osgood 1943b; Greer 1965). Yepes (1935a) and Pearson (1984, 1995) extended the range of this species into Neuquén and Río Negro provinces of Argentina, but without comments on subspecies allocation.

SELECTED LOCALITIES (Map 454): ARGENTINA: Neuquén, Aeropuerto Caviahue (MVZ 183299), 2 km E and 1 km N of Caviahue (MVZ 183298), Chapelco (MVZ 183304), 1 km E of Copahue, Las Maquinitas (MVZ 183310), Lago Epulafquen (MVZ 185055), Lago Lolog, NE coast, 5 km W of Rio Quilquihue (MVZ 183312), near Estancia Los Helechos, 2.5 km W and 5 km S of Cerro Colorado (MVZ 163462); Río Negro, ridge above Refugio Neumeyer (MVZ 163823). CHILE: Araucanía, Río Colorado (type locality of *C. maulinus brunneus* Osgood), Lonquimay, Río Colorado (MVZ 154132); Maule, Arroyo del Valle, Río Maule (FMNH 50731), Río Maule, 14 km above Curillanque (FMNH 50733), Laguna del Maule (type locality of *Ctenomys maulinus* Philippi).

SUBSPECIES: Both Osgood (1943b) and Cabrera (1961) recognized two subspecies, the paler nominotypical form in Talca province, Chile, and the darker *C. m. brunneus* from Malleco and Cautín provinces, Chile. Specimens from Neuquén and Río Negro provinces in Argentina are, as yet, unassigned to subspecies (see Pearson and Christie 1985).

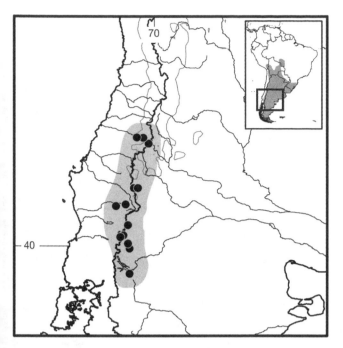

Map 454 Selected localities for *Ctenomys maulinus* (●). Contour line = 2,000 m.

NATURAL HISTORY: *Ctenomys maulinus* inhabits widely divergent habitats such as *Nothophagus* (Nothofagaceae) – *Araucaria* (Araucariaceae) woodlands and open volcanic sands from 900 to 2,000 m elevation (Redford and Eisenberg 1992). M. H. Gallardo and Anrique (1991) describe population parameters and burrow systems, and M. H. Gallardo and Kohler (1994) document demographic changes and loss of genetic variation in a local population following a volcanic eruption.

REMARKS: Skins and skulls of Argentinean specimens are indistinguishable from *C. m. maulinus* from Chile (Pearson 1995). Numerous large tuco-tucos, probably *C. maulinus*, have been recovered from owl pellets more than 5,000 years old at Cueva Traful, near Confluencia, Neuquén province, Argentina (Pearson 1984). Both subspecies of *C. maulinus* share the same $2n = 26$, $FN = 48$, karyotype (M. H. Gallardo 1979) and an asymmetrical simple sperm morphology (Feito and Gallardo 1976, 1982).

Ctenomys mendocinus Philippi, 1869
Mendoza Tuco-tuco

SYNONYMS:

Ctenomys mendocina Philippi, 1869:38; type locality "Mendoza," = Mendoza Province, Argentina, Departamento Capital, 32°53′S, 68°49′W, 980 m.

Ctenomys mendocinus: Philippi, 1896:17; correction of gender agreement.

Ctenomys eremophilus J. R. Contreras and Roig, 1975:19; unavailable (ICZN 1999:Art. 13; see Parada et al. 2012).

Ctenomys eremicus J. R. Contreras, 1979a:45; *nomen nudum*.

Ctenomys eremophilus J. R. Contreras, 1981:22; *nomen nudum*.

DESCRIPTION: Medium sized (mean total body length 247.3 mm in females and 262.12 mm in males; tail length 77.4 mm females and 82.2 mm males; Rosi et al. 2002, 2005; A. I. Medina et al. 2007). Dorsal pelage light brown to grayish red with transverse waves of blackish hair that disappear in paler ventral region; tail whitish with median dorsal stripe of longer hairs; feet covered with sparse white hair. Cranial features include broad rostrum, prominently ridged parietals without sagittal crest, well-developed lambdoid crest, jugals with prominent dorsally projected processes, enlarged infraorbital foramina with no canal for nerve transmission, and large bullae with flat and closely appressedparoccipital processes. Mandible with moderately developed coronoid process and wide angular process that flares outward. Upper and lower third molars vestigial. Upper incisors nearly orthodont, with roots extending to anterior edge of cheek teeth; enamel colored dark orange.

DISTRIBUTION: *Ctenomys mendocinus* is distributed along the eastern base of the Andes of west-central Argentinean provinces of Mendoza, San Luis, San Juan, and La Pampa provinces (Braggio et al. 1999; Rosi et al. 2002). Most populations occupy the *Monte de Llanuras y Mesetas* ecoregion between 174 and 3,400 m (Rosi et al. 2002; Bidau 2006).

SELECTED LOCALITIES (Map 455): ARGENTINA: La Plata, Estancia la Pastoril (Braggio et al. 1999), Lihuel-Calel (Wilkins and Cunningham 1993), Naicó (Braggio et al. 1999), Río Salado (Braggio et al. 1999), Santa Rosa (Braggio et al. 1999); Mendoza, Desaguadero (Wilkins and Cunningham 1993), 3 km S of Laguna Diamante (Sage et al. 1986), La Paz (Wilkins and Cunningham 1993), Las Higueras (Rosi, Cona et al. 1996), Las lajas (Rosi, Cona et al. 1996), Lavalle (Wilkins and Cunningham 1993), Mendoza (type locality of *Ctenomys mendocinus* Philippi), Papagallos (Roig and Reig 1969), Tambillos (Rosi et al. 2002), Tupungato (Slamovitz et al. 2001), 25 de Mayo (Pearson and Lagiglia 1992); San Luís, Villa Mercedes (Rosi et al. 1992).

SUBSPECIES: *Ctenomys mendocinus* today is considered monotypic.

NATURAL HISTORY: *Ctenomys mendocinus* is one of the best-studied tuco-tuco species and one with the largest known geographic distribution (see Rosi et al. 2005). Albanese et al. (2010) described resource use and distribution and Camin (2010) reviewed gestation, maternal behavior, and growth and development. This species constructs burrows of the linear type with those of males significantly longer (mean 51 m) than those of females (22 m; Camin et al. 1995; Rosi, Cona et al. 1996). As is true for most tuco-tucos, *C. mendocinus* has solitary habits even in the reproductive season and is strongly territorial (Puig et al. 1992; Rosi et al. 2000). Reproduction occurs from July to March; gonadal activity of females begins in the winter, but male testes seem to be active throughout the year (Dacar et al. 1998). Gestation lasts three months, and the annual mean litter size is 2.8 (range: 1–4; Rosi, Puig et al. 1996). Camin (1999) described mating behavior. Dietary habitats of *C. mendocinus* are also well known, with specialization primarily on the above ground parts of grasses and shrubs (Camin and Madoery 1994; Madoery et al. 1997; Puig et al. 1999; Tort et al. 2004).

REMARKS: Cabrera (1961:551) included *C. bergi*, *C. fochi*, *C. haigi*, *C. juris*, *C. occultus*, *C. recessus* (= *C. talarum recessus*), and *C. tucumanus* as subspecies of *C. mendocinus*. He also included *C. azarae* and *C. latro* in the synonymy of *C. m. mendocinus* and *C. m. tucumanus*, respectively. Herein, each of these taxa is considered good species (Bidau 2006). M. M. Díaz et al. (2000) cited *C. mendocinus* for Salta province, Argentina, but those specimens are more appropriately assigned to *C. fochi*. *Ctenomys mendocinus* has its closest affinities (based on morphological, chromosomal, and molecular data) with other species in central Argentina that collectively comprise a "mendocinus" group, namely *C. azarae*, *C. porteousi*, and *C. australis* (J. R. Contreras and Bidau 1999; Mascheretti et al. 2000), which has been expanded more recently to include *C. flamarioni* and *C. rionegrensis* (Parada et al. 2011) based on mtDNA sequences.

J. R. Contreras and Roig (1975) coined the name *Ctenomys eremophilus* for a new species they proposed from the Ñacuñán Biosphere Reserve in Mendoza province, Argentina, in the abstracts of a scientific meeting. J. R. Contreras (1979a, 1981) subsequently referred to the same sample as *C. eremicus* or *C. eremophilus*, respectively. Parada et al. (2012), however, documented that *eremophilus* J. R. Contreras and Roig was unavailable, as it did not meet the requirements for publication (ICZN 1999:Art. 13). And, Galliari et al. (1996) concluded that both *eremicus* J. R. Contreras and *eremophilus* J. R. Contreras were *nomina nuda* as neither name was accompanied by a description, illustration, or reference to such (ICZN 1999:Art.12). Unfortunately, *eremophilus* has been used in the subsequent literature as though it were a valid species (Giannoni et al. 1996; Borruel et al. 1998; Rosi et al. 2009). Bidau (2006) regarded *eremophilus* as a synonym of *C. pontifex* Thomas, without comment (see Parada et al. 2012).

The karyotype of *C. mendocinus* is polymorphic, with diploid numbers of 48 or 50 and number of autosomal arms varying between 68 and 80, mostly due to variation in short-arm heterochromatin (Massarini, Barros, Ortells,

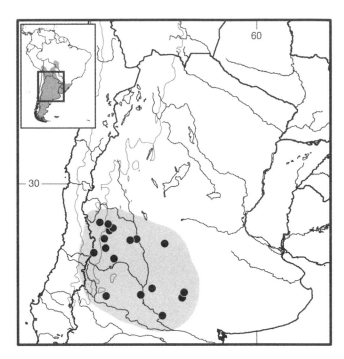

Map 455 Selected localities for *Ctenomys mendocinus* (●). Contour line = 2,000 m.

and Reig 1991; Massarini, Barros, Roig, and Reig 1991; Braggio et al. 1999; Parada et al. 2012). Individuals with both $2n = 48$ and 50 were present in the Ñacuñán Biosphere Reserve, with these samples sharing high similarity to other samples of *C. mendocinus* in mtDNA cytochrome-*b* sequences (Parada et al. 2012). It has a simple asymmetrical type of sperm (Vitullo et al. 1988).

Ctenomys minutus Nehring, 1887
Minute Tuco-tuco

SYNONYMS:

Ctenomys minutus Nehring, 1887:47; type locality "stammen aus den 'Campos', welche sich östlich von Mundo Novo (also ziemlich weit nördlich in der Provinz Rio Grando do Sul)," Brazil (= Taquara do Mundo Novo, Reig. et al. 1965; or = Taquara, Belton 1985), ca. 10 m, 29°39′S 50°47′W; restricted by Langguth and Abella (1970:18), following Nehring (1900c:206), to the oceanic coast at a resort near the mouth of the Rio Tramandahy, 98 km E of Porto Alegre, Rio Grande do Sul, Brazil.

Ctenomys minimus S. Anderson, 1985:14; incorrect subsequent spelling of *Ctenomys minutus* Nehring.

DESCRIPTION: Described from three skulls sent to Alfred Nehring by Th. Bischoff. Nehring's original description included only a few dimensions of each skull (approximate total lengths 41.5, 39, and 35.5 mm, respectively; basilar lengths 34, 32.3, and 28.5 mm; widths of upper incisors, 4.6, 4.0, and 3.8 mm; and lengths of upper toothrow, 31.5, 27.0 [?], and 27.6 mm). Nehring noted that all three skulls were of adult animals, because all teeth had erupted and were worn, although smallest of three was younger than largest. Other features apparent from illustrations provided in Nehring (1887, 1900c) include slightly proodont upper incisors, enlarged but anterior-posteriorly constricted auditory bullae, proportionally broad interorbital region, braincase with weakly developed parasagittal ridged, approximately equal distances across zygomatic arches and auditory tubes, and vertically directed postorbital spine of the jugal. Overall dimensions, contrary to implication of species epithet, place *C. minutus* among medium-sized species of tuco-tucos, with total length averaging 250 mm (Reig et al. 1966). Overall color medium to dark brown dorsally and light brown ventrally; young individuals usually paler than adults (Rebelato 2006). Baculum paddle shaped, mean length 9.34 mm, with V-shaped proximal tip. Glans penis with dense (12/mm²), single-tipped epidermal spines averaging 245.14 ± 49.81 μm s.d. in length (Rocha-Barbosa et al. 2013).

DISTRIBUTION: *Ctenomys minutus* is known from the southern Brazilian states of Santa Catarina and Rio Grande do Sul.

SELECTED LOCALITIES (Map 456; from T. R. O. Freitas 1997, except as noted): BRAZIL: Rio Grande do Sul, Capão Novo, Cidreira, Mostardas, Osório, Palmares do Sul, Palmital, Passinhos, Taquara (type locality of *Ctenomys minutus* Nehring), Tavares, Tôrres; Santa Catarina, Jaguaruna, Morro dos Conventos.

SUBSPECIES: Cabrera (1961:553) recognized two subspecies, *C. m. minutus* and *C. m. bicolor*, the latter corresponding to *C. bicolor* (Miranda-Ribeiro 1914), herein regarded as a separate species. Consequently, *C. minutus* is monotypic.

NATURAL HISTORY: This species usually occupies the highest parts of sandy and dry fields and pastures not far from water bodies along the coastal plain of southern Brazil. This is a solitary and territorial species, with low vagility and sedentary behavior and sexual size dimorphism in favor of males (Marinho and Freitas 2006). Population densities may vary from 7 to 42 individuals per hectare (Gastal 1994; M. B. Fonseca 2003). Sex ratio is 1:1 in juveniles and subadults but biased toward females in adults, with 1.3 times more females in general and 1.26:1 females in the adult cohort (M. B. Fonseca 2003; Marinho and Freitas 2006). Gestation time is about 90 days, and sexual maturity is reached within 6–7 months (M. B. Fonseca 2003).

REMARKS: Thomas (1898d:285) believed that the original series of three skulls upon which Nehring (1887)

based his *C. minutus* were immatures, "so much so, indeed, that I doubt if the full-grown animal is any smaller than the Uruguayan *Ct. torquatus*, Licht." Nehring (1900a,c), however, rebutted Thomas's assertion in great detail and provided line-drawings as well as a photograph of the largest of three skulls he used in his original description to emphasize its "adult" features, including well-developed ridging, broad zygomatic arches, and almost complete fusion of the sphenobasilar suture.

S. Anderson (1997:497; see also S. Anderson et al. 1987) tentatively assigned Bolivian specimens from Santa Cruz department to *C. minutus* "on the basis of size and geographic occurrence." This assignment, however, is probably in error. Anderson's "geographic occurrence" rationale was based on Cabrera's (1961) unjustified consideration of *C. bicolor* from Mato Grosso (or Rôndonia; see account of that species), Brazil, as a valid subspecies of *C. minutus*. Moreover, the holotype, and only known specimen of *C. bicolor* is large, not small in size, making the linkage of this taxon to either the Bolivian specimens or to true *C. minutus* from southeastern Brazil untenable, and thereby negating a geographic connection between the small Bolivian specimens and *C. minutus*. Indeed, E. P. Lessa and Cook (1998) referred to these Bolivian specimens as "*Ctenomys* sp. 'minut,'" thus suggesting that

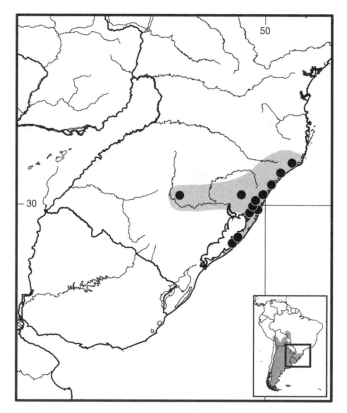

Map 456 Selected localities for *Ctenomys minutus* (●).

they had been incorrectly assigned to *C. minutus*. Whether or not these Bolivian specimens belong to another of those species that have been described or represent an as yet unnamed taxon must await further revisionary work on the genus.

Populations of *C. minutus* from Brazil are very polymorphic for Robertsonian and tandem fusions, resulting in a variation of diploid number from 42 to 50 (T. R. O. Freitas 1997, 2005; Freygang et al. 2004). Marinho and Freitas (2000) and Gava and Freitas (2002, 2004) characterized contact zones between different cytotypes. The sperm has the symmetric morphology (T. R. O. Freitas 1995).

Ctenomys nattereri Wagner, 1848
Natterer's Tuco-tuco

SYNONYMS:

Ctenomys Nattereri Wagner, 1848:72; type locality "Caissora (Provinz Matogrosso)" (= Caiçara, Nehring 1990b: 537 and Cabrera 1961:553, or = Cáceres, S. Anderson et al. 1987:1), Mato Grosso, Brazil, ca. 200 m, 16°00′S 57°45′W.

Ctenomys boliviensis nattereri: S. Anderson et al., 1987:11; name combination.

DESCRIPTION: Moderately large, head and body length about 245 mm, tail length 71 mm, ear length 11 mm, and hindfoot length (including claw) 54 mm. Head and particularly muzzle very broad and flat, rhinarium nearly naked, eyes small, and incisors very broad and ungrooved; ear opening very small, within small pinna. Claws on forefeet especially long (medial claw about 16 mm), and typical tuco-tuco comb-like hairs extend from sides of hindfeet. Overall dorsal color uniform brownish and shiny, flecked with black that vaguely defines a dorsal stripe passing from middle of head along back, gradually disappearing toward rump. Under surface all black but with light shading, interrupted by white spotting, particularly in axillary and inguinal areas. Upper incisors faced with "vivid saffron-red" enamel. A third known specimen, from José Bonifácio, Rondônia state, Brazil, smaller, with head and body 218 mm and tail 72 mm (J. A. Allen 1916d).

DISTRIBUTION: *Ctenomys nattereri* is known only from three localities, one each in Santa Cruz department, Bolivia, and both Mato Grosso and Rôndonia states in west-central Brazil.

SELECTED LOCALITIES (Map 457): BOLIVIA: Santa Cruz, Santa Cruz de la Sierra (Parada et al. 2011). BRAZIL: Rondônia, José Bonifácio (J. A. Allen 1916d); Mato Grosso, Cáceres (probable type locality of *Ctenomys nattereri* Wagner).

SUBSPECIES: *Ctenomys nattereri* is monotypic.

NATURAL HISTORY: No data are available ecological characteristics of *C. nattereri* except for the account of the specimen that Leo E. Miller collected on February 24, 1914, during the Roosevelt-Rondon Expedition (J. A. Allen 1916d; L. E. Miller 1918). According to L. E. Miller (1918:239), the single excavated burrow "measured fifteen feet [4.60 m] long, eight feet [2.45 m] deep and three feet [0.92 m] wide, and it required half a day for the Indians to complete the work." This is a fairly deep system for tuco-tucos. The burrow, however, was simple because "at the end [of the 15 feet long tunnel] was a small cavity, but no nest. Small bunches of grass were found in the gallery which had been pulled down by the roots." The local Nhambiquara, who helped Miller collect the specimen, were very fond of the animal's flesh "and often dig them out to eat" (L. E. Miller 1918:238). Although "the animal seemed bewildered above ground and could not run fast" (L. E. Miller 1918:239), Theodore Roosevelt, who was watching the whole procedure, wrote that "the animal dug hard to escape, but when taken and put on the surface of the ground it moved slowly and awkwardly," but "it showed vicious courage" (Roosevelt 1914:246). In his field notes, L. E. Miller (in J. A. Allen, 1916d:595) calls this tuco-tuco "cururu," probably mistaking this with "curu-curu," a common name given to tuco-tucos in parts of Brazil (A. B. de H. Ferreira 1986:513). Interestingly, *cururu* or *cururú* was the name given by indigenous Fueguians to *C. magellanicus* (see that account).

REMARKS: The true geographic distribution of *C. nattereri* is almost completely unknown. Nehring (1900b) described and figured the skull upon which Wagner based his name *C. nattereri*, giving the locality as "Caiçara in Matto Grosso." Cabrera (1961:553) also restated Wagner's (1848) type locality of "Caissora" as "Caicara (= Caiçara)." S. Anderson et al. (1987:11), however, equated "Caissora" with "Cáceres" in the Brazilian state of Mato Grosso (16°00′S 57°45′W). No locality named "Caissora" occurs in Brazil, and of the six Caiçara localities found in maps and gazetteers, none relates geographically to the probable type locality of *C. nattereri*. Thus, "Cáceres" is the most probable geographic location of the type of *C. nattereri*.

J. A. Allen (1916d:569) reported finding one specimen of *C. nattereri* at 12°19′S 60°30′W near the community of José Bonifácio (misspelled as José Bonefacio [pp. 560, 595] and José Bonesfascio [p. 569]) in the Brazilian state of Rondônia, although he gave no reason for this assignment. This locality is, however, quite distant from Cáceres, and separated from it by the high Chapada dos Parecís, suggesting an atypically (for a tuco-tuco) large distribution. Moreover, L. E. Miller (in J. A. Allen 1916d:595) detected the sporadic presence of *Ctenomys* mounds from "Tapiropoan" (= Tapirapoãn or Tapirapuã) at 14°51′S and 57°45′W, up to José Bonifácio. S. Anderson et al. (1987) compared the skull of this specimen with those of *C. boliviensis* and found a number of small but insignificant differences. These authors thus concluded that *nattereri* Wagner was a subspecies of *C. boliviensis* Waterhouse, which they (as first reviewers) selected as having date priority. Their action thus expanded the range of this tuco-tuco species even more. A further complication is the description of *C. rondoni* by Miranda-Ribeiro (1914), also from Mato Grosso but with the imprecise locality of Maria de Molina and Rio Juruena (see the account of *C. rondoni*). Again without justification, Cabrera (1961:553) synonymized *C. rondoni* with *C. nattereri*, a decision followed most recently by Woods and Kilpatrick (2005:1561). Herein, these two are treated as separate species. Clearly, detailed surveys for tuco-tucos in the Brazilian states of Mato Grosso and Rôndonia and the adjacent Bolivian province of Santa Cruz are required to determine the current biological boundaries of these various "species" and map their distributions correctly.

Both karyotype and sperm morphology is unknown. MtDNA cytochrome-*b* sequence analysis (Parada et al. 2011) places *C. nattereri* as a member of the "*boliviensis*" group, along with *C. boliviensis* and *C. goodfellowi*, a well-supported monophyletic clade identified as the

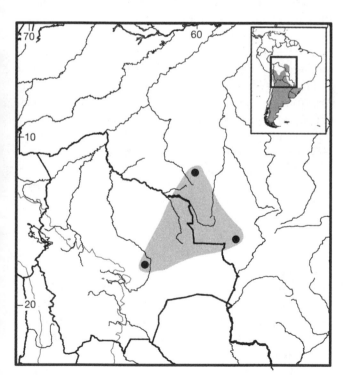

Map 457 Selected localities for *Ctenomys nattereri* (●). Contour line = 2,000 m.

"Boliviano-Matogrossense" group by J. R. Contreras and Bidau (1999).

Ctenomys occultus Thomas, 1920
Furtive Tuco-tuco, Oculto

SYNONYMS:

Ctenomys occultus Thomas, 1920e:243; type locality "Monteagudo, about 80 km. S.E. of Tucuman City," San Miguel de Tucumán, Departamento Simoca, Tucumán, Argentina, 27°31′S, 65°17′W, 296 m.

Ctenomys mendocinus occultus: Cabrera, 1961:552; name combination.

DESCRIPTION: Small species, head and body length 1380–151 mm (A. I. Medina et al. 2007). Overall color warm brown, with upper surfaces glossy cinnamon brown with some vague darkening on crown, but without blackish forehead; under surfaces washed with pale drabby, chin and throat more sharply drab. Specimens range from whitish to drab below. Skull similar to that of *C. juris* with small bullae, proodont incisors, and presence of small sharp-edged ledges projecting over orbital fossae, which are short and more abruptly cut; zygomata widely spaced with middle region markedly convex outward; often small medial additional foramen present anterior to incisive foramina; mesopterygoid fossa extends forward to middle of M2. Anterior face of the proodont incisors angled at 102° to 104° of horizontal plane.

DISTRIBUTION: *Ctenomys occultus* is known from only a few localities in Tucumán province in the *Chaco Seco* ecoregion.

SELECTED LOCALITIES (Map 458): ARGENTINA: Tucumán, Alberdi (J. R. Contreras, pers. comm.), La Cocha (J. R. Contreras, pers. comm.), Lamadrid (Thomas 1920e), Monteagudo (type locality of *Ctenomys occultus* Thomas), Simoca (Parada et al. 2011).

SUBSPECIES: *Ctenomys occultus* is monotypic.

NATURAL HISTORY: No data on ecology are available; Redford and Eisenberg (1992) stated that individuals live in areas of xeric vegetation, but without documentation.

REMARKS: Thomas (1920e:244) regarded *C. occultus* to be closely allied to *C. juris* from Jujuy province, differing from it "by so many little characters that it seems to deserve a special name, and also to *C. latro* from Tucumán province, despite their differences in size." Reig and Kiblisky (1968a) used the unique karyotype of $2n = 20$, FN = 40, of *C. occultus* to validate it as a species distinct from *C. mendocinus*. This species belongs to the "*tucumanus* group," along with *C. argentinus*, *C. latro*, and *C. tucumanus*, based on mtDNA sequence analyses (Parada et al. 2011). The sperm is the simple asymmetric type (Bidau 2006).

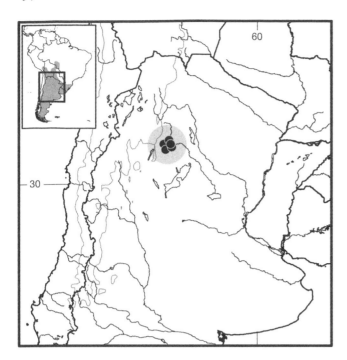

Map 458 Selected localities for *Ctenomys occultus* (●). Contour line = 2,000 m.

Ctenomys opimus Wagner, 1848
Andean Tuco-tuco

SYNONYMS:

Ctenomys opimus Wagner, 1848:75; type locality: "Bolivia," restricted by Thomas (1900g:383) to "Sahama," Mount Sahama, Oruro, Bolivia, ca. 17°49′S, 67°28′W, 3,000–4,000 m.

Ctenomys opimus nigriceps Thomas, 1900g:383; type locality "Tetiri, about 40 miles [64 km] W. of Puno, on the Puno-Moquegua road. Altitude 16,000 feet [4,880 m]," Pampa de Titiré, Puno, Peru.

Ctenomys opimus luteolus Thomas, 1900g:384; type locality "Cordilleras of Jujuy, [Jujuy,] Argentina Republic."

Ctenomys opimus opimus: Yepes, 1930:325; name combination.

Ctenomys luteolus: Yepes, 1935a:252; name combination.

Ctenomys optimus S. L. Gardner and Anderson, 2001:12; incorrect subsequent spelling of *Ctenomys opimus* Wagner.

DESCRIPTION: Large, total length of Bolivian and Peruvian specimens 270–340 mm and those from Argentina (Salta and Jujuy provinces) 275–300 mm (Pearson 1959; Cook et al. 1990; S. Anderson 1997; M. M. Díaz and Barquez 2002; A. I. Medina et al. 2007). Pelage long, lax, and soft in texture, with dorsal color uniformly pale yellowish, tending to become darker as tips of hairs wear, sometimes in patches, as on top of head. Majority of adult specimens show open frontoparietal fenestrae (S. L. Gardner and Anderson 2001).

DISTRIBUTION: *Ctenomys opimus* occurs widely across the Altiplano from southern Peru to northwestern Argentina, northern Chile, and southwestern Bolivia, all at elevations above 3,200 m.

SELECTED LOCALITIES (Map 459): ARGENTINA: Catamarca, Antofagasta de la Sierra (R. A. Ojeda 1985); Jujuy, Tres Cruces (A. I. Medina et al. 2007); Salta, Chorrillos (Yepes 1930), La Poma (Mares, Ojeda, and Barquez 1989), Los Cardones (Ipucha et al. 2008), Salar Pastos Grandes (M. M. Díaz et al. 2000), Tolar Grande (M. M. Díaz et al. 2000), Vega Cortadera (M. M. Díaz et al. 2000). BOLIVIA: La Paz, Esperanza, near Mount Sahama (Cook et al. 1990), Huaraco-Antipampa (Cook et al. 1990); Oruro, Challapata (Thomas 1902c), Cruce Ventilla (Cook et al. 1990), Huancaroma (Cook et al. 1990), Mount Sajama (type locality of *Ctenomys opimus* Wagner), Oruro (Cook et al. 1990); Potosí, Livichuco (Thomas 1902c), Potosí (Thomas 1902c), Uyuni (Cook et al. 1990). CHILE: Arica y Parinacota, Choquelimpie (Sanborn and Pearson 1947), Lagho Chungara (Feito and Gallardo 1982). PERU: Puno, Caccachara (Sanborn and Pearson 1947; Pearson 1951), Lago Loriscota (MVZ 139609), Pampa de Ancomarca (MVZ 115976), Pampa de Tetirí (type locality of *Ctenomys opimus nigriceps* Thomas), 25 km SW of Pisacoma (MVZ 115981), Río Santa Rosa, 8 mi W of Mazocruz (MVZ 115982); Tacna, Lago Suche (Pearson 1959).

SUBSPECIES: Three subspecies are recognized in the current literature (e.g., Cabrera 1961; Woods and Kilpatrick 2005): *nigriceps* Thomas, from Moquegua, Tacna, and Puno departments in southern Peru; *opimus* Wagner, from northern Chile and throughout the highlands of Bolivia; and *luteolus* Thomas, from the mountains of northwestern Argentina in Jujuy, Catamarca, and Salta provinces. These taxa differ primarily by differences in color pattern.

NATURAL HISTORY: This species occurs at elevations up to 5,000 m, where it can be very abundant in areas of sparse vegetation and loose sandy, gravely, or cindery soils. Pearson (1951, 1959) described burrow systems, food habits, digging behavior, parasites and associates, reproductive biology, sexual dimorphism, and population densities at several study sites in southern Peru. Bennett (1846:8) found animals on sandy slopes and valleys "at no great distance from water" where "large patches of land were completely undermined by its workings." He believed that animals ate below ground bulbs and grass roots, rarely leaving their burrows to feet on grass stems. Their burrowing activity was heightened in the morning. S. Anderson (1997) recorded reproduction season in Bolivia. C. R. Lambert et al. (1988) and S. L. Gardner and Duszynski (1990) reported on eimerian oocysts in the feces of Bolivian specimens. This species is

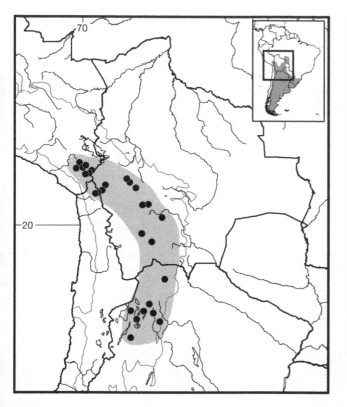

Map 459 Selected localities for *Ctenomys opimus* (●). Contour line = 2,000 m.

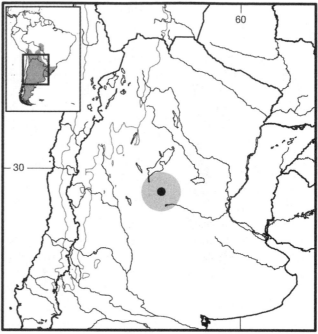

Map 460 Single known locality for *Ctenomys osvaldoreigi* (●). Contour line = 2,000 m.

the type host of a species of sucking louse (Durden and Musser 1994).

REMARKS: George R. Waterhouse (in Bennett 1846:8) referred to specimens from Potosí, Bolivia recorded by Bennett (1946:8–9) as *C. brasiliensis*; these, however, most likely represent *C. opimus*. Feito and Gallardo 1982) regarded *C. opimus* to be closely allied to *C. fulvus* based on sperm morphology; molecular analyses either place *C. opimus* in a basal position (electrophoretic characters; Cook and Yates 1994), within a group of Bolivian species with respect to those from geopolitical areas outside of Bolivia (mtDNA sequences; E. P. Lessa and Cook 1998), or within a "*opimus* group" that also included *C. fulvus*, *C. scagliai*, and *C. saltarius* (mtDNA sequences; Parada et al. 2011). All analyzed subspecies have the same karyotype of $2n = 26$ metacentric chromosomes, or FN= 48 (M. H. Gallardo 1979; Cook et al. 1990; Toloza et al. 2004; Ipucha et al. 2008). Ruedas et al. (1993) provided data on genome size. The sperm type is simple symmetric (Feito and Gallardo 1982; Vitullo and Cook 1991).

Ctenomys osvaldoreigi J. R. Contreras, 1995
Osvaldo Reig's Tuco-tuco

SYNONYM:

Ctenomys osvaldoreigi J. R. Contreras, 1995a:1; type locality "Estancia San Luis, Ruta Provincial 20, 10 kiló-

metros al oeste del Río Yuspe, en el extremo sudeste del Departamento Cruz del Eje, Provincia de Córdoba, 31°24′S, 64°48′W, aproximadamente a 2000 metros," Argentina.

DESCRIPTION: Medium sized, total body length of male holotype 275 mm with mass, 244 g; equivalent data for female paratype 254 mm and 203.2 g, respectively; and for female topotypic series, body length 235–254 mm, mass 176–279 g (J. R. Contreras 1995a; A. I. Medina et al. 2007). General color uniformly ochraceous brown dorsally (corresponding to "antique brown" in Smithe's [1975] atlas) with tawny tips to hairs; venter ochraceous ("cinnamon") and lacking inguinal and axillary white spots common in many species in genus; white collar that characterizes most Bolivian and Chacoan species is lacking; tail bicolored, brown above and cream below; dorsal surfaces of feet same general color as body. Skull gracile with elongated rostrum that exhibits little evidence of lateral widening; weakly developed temporal crests; slender zygomatic arches, with breath across them significantly greater than that across external auditory tubes; interparietalscarcely visible or fused with parietal and squamosal bones; bregmatic fontanelles lacking in adults; auditory bullae small and uninflated. Small fossa that characterizes most Chacoan species absent.

DISTRIBUTION: *Ctenomys osvaldoreigi* is only known from its type locality in Córdoba province, Argentina.

SELECTED LOCALITIES (Map 460): ARGENTINA: Córdoba, Estancia San Luis (type locality of *Ctenomys osvaldoreigi* J. R. Contreras).

SUBSPECIES: *Ctenomys osvaldoreigi* is monotypic.

NATURAL HISTORY: No data on behavior, ecology, reproduction, or food preferences are available.

REMARKS: Giménez et al. (1999) described a karyotype with $2n = 52$, $FN = 56$ (see also Ipucha et al. 2008), and simple, asymmetric sperm morphology.

Ctenomys paraguayensis J. R. Contreras, 2000
Paraguayan Tuco-tuco

SYNONYM:

Ctenomys paraguayensis J. R. Contreras, 2000:62; type locality "Corate-í, 12 kilómetros al oeste de la ciudad de Ayolas, departamento Misiones, República del Paraguay, aproximadamente en las coordenadas: 27°24′S–57°01′W."

DESCRIPTION: Medium sized with head and body length of three known specimens 166–174 mm, tail length 73–80 mm, hindfoot (with claw) 34.8–37 mm, and mass 146–187 g. General color very similar to that of *C. pilarensis*, the other tuco-tuco species from eastern Paraguay, including presence of prominent, pale semicollar that passes obliquely from below ear around back of head, reaching base of neck; cheeks of *C. paraguayensis*, however, paler with face lacking any blackish coloration. Ventral coloration clear creamy gray without white spots in inguinal or axillary regions; tail only slightly bicolored. Skull graceful in overall appearance; width across zygomatic arches always greater than width across auditory tubes; postorbital spine of jugal rises almost perpendicular to occlusal plane of cheek teeth rather than obliquely; rim of zygomatic arch lacks foramen that may be present in other species; superior placement of auditory meatus relative to cheek teeth, a characteristic shared with species in northeastern Argentina and adjacent Uruguay, one of most notable features; two lateral openings through which alisphenoid-presphenoid bridge is visible present both sides of mesopterygoid fossa; this bridge is thin, with medial posterolateral expansion shaped like sharp stylus about 4 mm in length.

J. R. Contreras (2000) provided detailed comparisons, particularly of cranial features, between *C. paraguayensis*, *C. dorbignyi*, and *C. pilarensis*, and somewhat less so to *C. argentinus*.

DISTRIBUTION: *Ctenomys paraguayensis* is known only from the type locality in Misiones department, Paraguay.

SELECTED LOCALITIES (Map 461): PARAGUAY: Misiones: Corate-í (type locality of *Ctenomys paraguayensis* J. R. Contreras).

SUBSPECIES: *Ctenomys paraguayensis* is monotypic.

NATURAL HISTORY: No ecological data are available.

REMARKS: *Ctenomys paraguayensis* was described based on three specimens secured at the type locality. The species is distinct in both cytogenetic and morphological characters relative to all other known species in the genus. The karyotype is $2n = 52$ (J. R. Contreras 2000), more similar to species occurring in northern Corrientes province, Argentina (see accounts of *C. perrensi* and allies), and not to the neighboring *C. pilarensis* from Ñeembucú department, Paraguay. Comparative analysis of morphometric, osteological, and general somatic characteristics indicated a clear differentiation of this species with *C. pilarensis* as well as no close affinity with those species from the northeastern Argentinean provinces of Corrientes and Entre Ríos, such as *C. dorbignyi*. *Ctenomys paraguayensis* may represent Brants (1827) *Ratton tucotuco* Brants, a bibliographical species based on Azara's (1802) "*el tucutuco*" from Paraguay, but definitive evidence for this hypothesis is lacking (see Langguth 1966, 1978). Brant's *tucotuco*, however, was ruled (ICZN 1982:Opinion 1232) to be a vernacular name and unavailable for use in zoological nomenclature. Because of the great environmental change resulting from the construction of the Yacyretá dam, *C. paraguayensis* is very likely reduced in numbers if it has not already vanished from its very restricted geographic range in extreme eastern Paraguay. Sperm morphology is unknown.

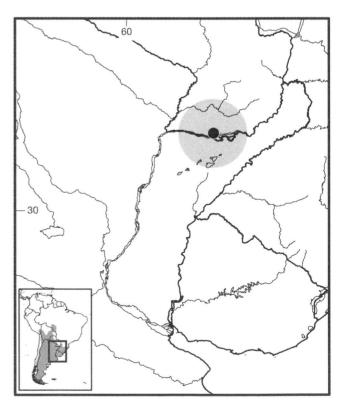

Map 461 Single known locality for *Ctenomys paraguayensis* (●).

Ctenomys pearsoni E. P. Lessa and Langguth, 1983
Pearson's Tuco-tuco
SYNONYM:
Ctenomys pearsoni E. P. Lessa and Langguth, 1983:86; type locality "Arroyo Limetas, 25 km SE de Carmelo, Dep. Colonia, Uruguay."
DESCRIPTION: Medium sized, total length 245–277 mm, tail length 72–82 mm, and mass 165–300 g (E. P. Lessa and Langguth 1983; E. M. González 2001). Body color similar to that of *C. torquatus*, and both species possess same clear half-collar. Skull depressed and elongated; dorsal surface of frontals almost flat from level of interorbital constriction toward rear; frontoparietal sutures straight and almost transverse; parietals show anterior medial elongated zone limited by temporal crests, in same plane as frontals; rostrum narrow and elongated, with markedly proodont incisors; frontals narrowed because of very limited interorbital constriction; braincase relatively narrow; interpremaxillary and incisive foramina enlarged and highly visible; palate narrow and excavated between molariform teeth, with thin bony septum present in sagittal plane; auditory bullae elongated and project from behind occipital condyles; paroccipital processes well developed; preorbital foramina high and elongated; mandibular apophyses of jugals triangular and moderately robust. Coronoid process of mandible subtriangular, with wide base; posterior border elevated toward front relative to plane of molariform teeth. E. P. Lessa and Langguth (1983) detailed number of small but clear cranial differences that distinguish this taxon from both geographically neighboring (such as *C. torquatus*, which it strongly resembles) and distant species. Penile morphology distinctive feature of *C. pearsoni*, with paddle-shaped, single-tipped baculum wider, shorter, and better defined than in other tuco-tucos (Altuna and Lessa 1985).

DISTRIBUTION: *Ctenomys pearsoni* occurs along the coast of Uruguay from Rocha department in the east to Colonia in the west, and in adjacent Entre Ríos province in Argentina.

SELECTED LOCALITIES (Map 462): ARGENTINA: Entre Ríos, Médanos (Reig et al. 1966). URUGUAY: Canelones, Cuchilla Alta (Tomasco and Lessa 2007); Colonia Arroyo Limetas (type locality of *Ctenomys pearsoni* E. P. Lessa and Langguth), Estanzuela (E. P. Lessa and Langguth 1983); Maldonado, Chihuahua (Tomasco and Lessa 2007), José Ignacio (Tomasco and Lessa 2007), Solís (Tomasco and Lessa 2007); Montevideo, Carrasco (Tomasco and Lessa 2007); Rocha, Laguna de Rocha (Tomasco and Lessa 2007), Laguna Negra (Tomasco and Lessa 2007), Valizas (Tomasco and Lessa 2007); San José, Arazatí (E. P. Lessa and Langguth 1983), Río Santa Lucía (E. P. Lessa and Langguth 1983); Soriano, 1 km N of Nueva Palmira (E. P. Lessa and Langguth 1983).

SUBSPECIES: *Ctenomys pearsoni* is monotypic.

NATURAL HISTORY: *Ctenomys pearsoni* inhabits coastal sandy soils along the edge or near river mouths (Altuna et al. 1999). Altuna (1985) described details of the burrow microclimate. This is a solitary, highly territorial, and aggressive species, typical of most tuco-tucos. Females are sexually mature and mate within their first year, but males do not. The baculum and other reproductive structures of males are not fully developed until an age of one year (Altuna and Lessa 1985). The mating system is polygynous. Reproduction occurs during the winter months, and litter sizes range between two and four. As in other tuco-tucos, young are weaned after 30–40 days and are fully independent at about 2 months' age. Maternal behavior is highly developed in this species, as is probably true for most other tuco-tucos (Altuna et al. 1999). Vocal signals in pups and geographic variation in vocalizations have been studied by Francescoli (2001, 2002). As for other well-studied tuco-tucos, *C. pearsoni* is an herbivorous generalist preferring grasses such as *Cynodon dactylon* and *Panicum racemosum*, the more abundant species in its habitat (Altuna et al. 1999). Feeding behavior of these tuco-tucos has been well studied; individuals are autocoprophagic (Altuna et al. 1998, 1999). The caecum is unique, weighing 30% of total mass, thus making it the (relatively) largest among all hystricognaths for which this feature has been measured (Altuna et al. 1998).

REMARKS: *Ctenomys pearsoni* shares a pair of spiny bulbs on the male phallus with *C. dorbignyi*, *C. perrensi*, and *C. roigi* from Argentina (Altuna and Lessa 1985). The karyotype of *C. pearsoni* is distinctive but highly variable, with diploid numbers ranging from 56 and 70 in Uruguayan populations (A. Novello and Altuna 2002; A. F. Novello and Lessa 1986) with the chromosome morphology of the $2n = 70$ form almost the same as the karyotype of *C. dorbignyi* (Argüelles et al. 2001). In Uruguay, a series of chromosomally differentiated populations extending from Montevideo to Punta del Este (east of the Santa Lucia river in the departments of Carrasco, Salinas, and Maldonado) are included in the "*pearsoni* complex" (E. P. Lessa and Langguth 1983; E. M. González 2001; A. Novello and Altuna 2002). The Argentinean Médanos population of *C. pearsoni* has $2n = 68$ (Reig et al. 1966; originally assigned to *C. torquatus*), but differences from the $2n = 70$ *pearsoni-dorbignyi* pattern are minimal. The morphologically very similar *C. torquatus* has a completely different karyotype (see below). However, populations of the "*pearsoni*-complex," that is, those populations morphologically close to typical *C. pearsoni*, are karyotypically distinct ($2n = 56$–64; A. F. Novello and Lessa 1986). Sperm type is symmetrical. Tomasco and Lessa (2007) described the molecular phylogeography of the chromosomally variable

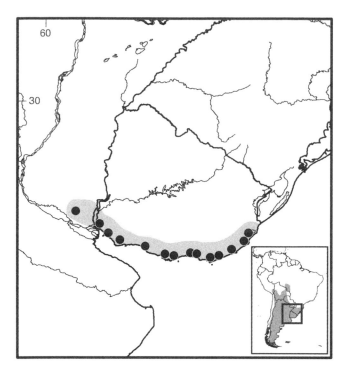

Map 462 Selected localities for *Ctenomys pearsoni* (●).

populations (2*n* = 56, 58, 64, 66, and 70) along the coast of Uruguay and discussed the role of chromosomal diversification in speciation within with species and tuco-tucos more generally. D'Anatro and D'Elía (2011) detailed morphometric variation among populations representing all karyotypic forms. With a single exception, all populations were indistinguishable, supporting their view that not all fixations of chromosomal rearrangements define independent evolutionary units that would warrant species recognition.

Ctenomys perrensi Thomas, 1896
Perrens's Tuco-tuco
SYNONYMS:

Ctenomys Perrensi Thomas, 1896:311; type locality "Goya, Corrientes, Argentina," Departamento Goya, Corrientes, Argentina, 29°08′S, 59°16′W, 37 m.

Ctenomys perrensis Cabrera, 1961:554; incorrect subsequent spelling of *Ctenomys perrensi* Thomas repeated by several subsequent authors (e.g., Redford and Eisenberg 1992; Woods 1993; Nowak 1999).

DESCRIPTION: Medium sized, total length 230–270 mm (A. I. Medina et al. 2007). General color dark buff or clay, heavily mixed with black along median line from face extending down back; eye to ear and below is lighter patch; cheeks brownish. Under surfaces from throat to belly rich buff, but patches of pure and sharply

contrasting white occur in axillary and inguinal regions. Upper surfaces of hands and feet thinly haired white. Skull short, broad, and rounded, not heavily ridged; zygomatic arches broader than breadth across posterior skull as measured at auditory meatus; nasals short, tapering backward, unusually broad and truncated behind; interorbital region short and very broad, postorbital process and ledges of limited development; frontoparietal suture almost directly transverse, little bowed backward; bullae small and little inflated, especially anteriorly. Upper incisors distinctly opisthodont, forming arc of smaller circle than in most tuco-tucos.

DISTRIBUTION: *Ctenomys perrensi* occurs in west-central Corrientes province, Argentina (see detailed map in J. R. Contreras et al. 1985, and Remarks).

SELECTED LOCALITIES (Map 463; from Giménez et al. 2002, except as noted): *Ctenomys perrensi* (*sensu stricto*): ARGENTINA: Corrientes, Colonia 3 de Abril, Goya (type locality of *Ctenomys perrensi* Thomas). *Ctenomys perrensi* complex: ARGENTINA: Corrientes, Arroyo Peguajó (Mirol et al. 2010), Chavarría, Contreras Cué, Curuzú Laurel, Estancia Tacuarita (Mirol et al. 2010), Loma Alta, Loreto, Manantiales, Pago Alegre, Paraje Angostura (Mirol et al. 2010), Paraje Caiman, Saladas, San Miguel (Parada et al. 2011).

SUBSPECIES: *Ctenomys perrensi* is monotypic.

NATURAL HISTORY: Very little ecological data are available. Mean litter size is 2.45 ± 0.89 s.d. (N = 11; J. R. Contreras, pers. comm.).

REMARKS: Both Cabrera (1961:554) and Woods and Kilpatrick (2005:1567) incorrectly cited the year of publication of Thomas's description as 1898. However, the issue of the *Annals and Magazine of Natural History* in which Thomas proposed *Ctenomys perrensi* (series 6, volume 18) appeared in 1896.

Ctenomys perrensi is closely allied at the molecular level to all other *Ctenomys* inhabiting Corrientes Province, including *C. roigi*, *C. dorbignyi*, and unnamed chromosomally distinct taxa inhabiting central and northwestern Corrientes (referred to as *Ctenomys* sp. and *Ctenomys* sp. in Giménez et al. 2002), and also to *C. pearsoni* from Entre Ríos province, Argentina, and Uruguay (J. R. Contreras and Bidau 1999; Mascheretti et al. 2000; Giménez et al. 2002). Parada et al. (2011) have named this assemblage of species the "*torquatus*" group (equivalent to the "Corrientes" group of J. R. Contreras and Bidau, 1999) on molecular evidence. However, none of the five taxa from Corrientes form reciprocally monophyletic groups in either allozyme or mtDNA or microsatellite analyses (Ortells and Barrantes 1994; Giménez et al. 2002; Mirol et al. 2010), al-

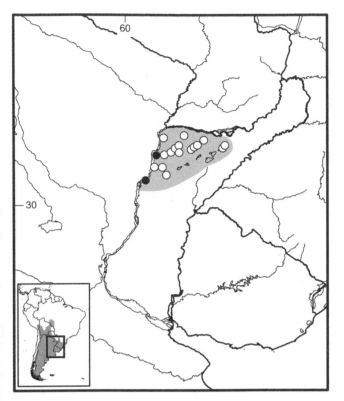

Map 463 Selected localities for *Ctenomys perrensi* (●) and those of the *Ctenomys "perrensi*-complex" (○; see Remarks).

though each is well differentiated chromosomally, suggesting that the origin of the whole assemblage has been recent and that collectively they form a "superspecies" complex (Giménez et al. 2002; Mirol et al. 2010; Gómez Fernández et al. 2012). The karyotype of *C. perrensi* (*sensu stricto*) is 2n = 50; however, the "*perrensi*-complex" (*C. perrensi* + *Ctenomys* sp. α + *Ctenomys* sp. β) members range in diploid number from 2n = 41 to 2n = 66 (Ortells 1995; Giménez et al. 2002). This species has a simple symmetric type of sperm (Vitullo et al. 1988).

The map of localities (Map 463) separates those records for *C. perrensi sensu stricto* from the karyotypically differentiated forms referred to as the "*perrensi*-complex," which extend toward central Corrientes province. Additional karyotypically distinct populations occur in northeastern Corrientes province, but these seem more related at the molecular level to *C. dorbignyi*, despite their low chromosome numbers. All references to the presence of *C. perrensi* in Misiones province are erroneous (Yepes 1935a; Honacki et al. 1982; Redford and Eisenberg 1992; Woods 1993). Misiones is the only Argentinean province where no *Ctenomys* populations are known to exist. There are also references to the presence

of *C. perrensi* in Entre Ríos province (Woods 1993; Redford and Eisenberg 1992), but these records actually correspond to *C. pearsoni*.

Ctenomys peruanus Sanborn and Pearson, 1947
Peruvian Tuco-tuco

SYNONYM:

Ctenomys peruanus Sanborn and Pearson, 1947: 135; type locality "Pisacoma, alt. 14,000 ft., Department of Puno, southern Peru," 16°55′S 69°22′W, 3,965 m.

DESCRIPTION: Large, pale, dark-footed species, with broadly expanded zygomatic arches, sagittal crest, and reduced last molar. Mean head and body length 220 mm, mean tail length 88 mm, and mean hindfoot length 41 mm (A. I. Medina et al. 2007). Back, sides, and belly colored creamy-buff, heavily lined in black to give overall brown tone; nose, lips, ears, and surrounding fur dark brown; hindfeet brown above, forefeet the same color as body; tail tawny. Skull with broadened zygoma expanded anteriorly so much so that zygomatic width greater than distance between outer edges of bony auditory meatus; sagittal crest present, even in smaller females; top of skull moderately arched. Last molar above and below about one quarter size of other molars.

DISTRIBUTION: *Ctenomys peruanus* is limited to Puno department in southern Peru.

SELECTED LOCALITIES (Map 464): PERU: Puno, Pampa de Quellecota (Pearson 1959), Hacienda Pichupichuni, Río Callacame, 8 km NW of Huacullani (Pearson 1959), Quellecota (MVZ 114767), Mazocruz (Sanborn and Pearson 1947), Pisacoma (type locality of *Ctenomys peruanus* Sanborn and Pearson).

SUBSPECIES: *Ctenomys peruanus* is monotypic.

NATURAL HISTORY (from Pearson 1951, 1959): This species inhabits open, often heavily grazed pampa shrub steppe composed of thorn bushes, closely cropped grass, dwarf flowers, and occasional tufts of taller grasses in southern Peru, where its range meets that of *C. opimus*. In the sandy and gravelly soils these tuco-tucos likely compete seriously with domesticated livestock (alpaca, llama, and sheep). This is the only Peruvian species that vocalizes conspicuously, with an alarm call described as "a musical bubbling" (p. 157). They are active during the day, will sit motionless with their heads at the entrance to their tunnels, and forage above ground to a meter or more from their burrow opening. Stomach contents contained coarse vegetable matter. The breeding season begins in November and lasts at least through April, with a gestation period approximating 4 months. There is likely only one litter produced by each female in a year. Litter size varied from 1 to 5, with a mean of 3.5, although resorption rates reduce the mean litter to about three young. Young

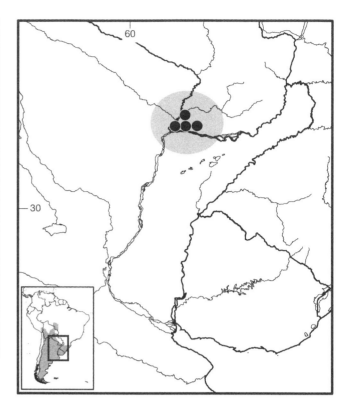

Map 464 Selected localities for *Ctenomys peruanus* (●). Contour line = 2,000 m.

Map 465 Selected localities for *Ctenomys pilarensis* (●).

are precocial, with late-term embryos well furred and neonates well developed at birth. Sex ratio may slightly favor females (Pearson 1951) or be strongly skewed to females (Pearson 1959). Several other vertebrates (mammals, birds, reptiles, and amphibians) use abandoned burrow systems. Pearson (1959) estimated density by counting active burrow opening and observing, with binoculars, the heads of active tuco-tucos at their entrances. He also suggested that *C. peruanus* was social, living in colonies, a behavior that contrasts with other tuco-tucos except *C. sociabilis*.

REMARKS: Sanborn and Pearson (1947) believed that *C. peruanus* was closely related to *C. fulvus* from northern Chile. The karyotype and sperm type of *C. peruanus* have not been reported.

Ctenomys pilarensis J. R. Contreras, 1993
Pilar Tuco-tuco

SYNONYM:

Ctenomys pilarensis J. R. Contreras, 1993:44; type locality "Pilar, [Ñeembucú,] Paraguay."

DESCRIPTION: Medium sized, head and body length of males 189–209 mm and that of females 169–177 mm (A. I. Medina et al. 2007). No other details published.

DISTRIBUTION: *Ctenomys pilarensis* occurs in extreme southern Paraguay.

SELECTED LOCALITIES (Map 465; from Giménez et al. 1997, except as noted): PARAGUAY: Ñeembucú, Des-

mochado, Mayor Martínez, Paso Pucú, Pilar (type locality of *Ctenomys pilarensis* J. R. Contreras).

SUBSPECIES: *Ctenomys pilarensis* is monotypic.

NATURAL HISTORY: *Ctenomys pilarensis* inhabits fine sandy soils and is often found in manioc plantations. No specific data on behavior, reproduction, or ecology are known.

REMARKS: I have been unable to locate the original description of this species, which was apparently published in the abstracts and proceedings of a scholarly meeting. If the rules of the Code were followed in this brief presentation, then the name *pilarensis* J. R. Contreras is valid. However, if eventual access to these proceedings shows that the Code requirements were not followed, then *pilarensis* J. R. Contreras will become a *nomen nudum*, and this taxon will require a new name.

Mascheretti et al. (2000) included *C. pilarensis* in a group of Chacoan species that otherwise included *C. argentines*, *C. latro*, *C. occultus*, and *C. tucumanus*, which Parada et al. (2011) called the "tucumanus group." Both studies were based on mtDNA sequences.

This species is chromosomally polytypic with the few populations sampled either $2n = 48$ or $2n = 50$. Each karyotype, however, has an FN = 50, supporting differentiation through Robertsonian changes (Giménez et al.

1997). The sperm is the simple asymmetric morphology (Giménez et al. 1997).

Ctenomys pontifex Thomas, 1918
Brown Tuco-tuco

SYNONYM:

Ctenomys pontifex Thomas, 1918a:39; type locality "East side of the Andes near Fort San Rafael, Province of Mendoza," Argentina; restricted to Volcán Peteroa, Malargüe department, ca. 35°26′S, 70°20′W by Pearson and Lagiglia (1992).

DESCRIPTION: Medium sized; external dimensions of holotype (adult female) head and body length 183 mm, tail length 77 mm, and hindfoot length 34 mm. Color above uniform drab-brown without darker markings; under parts paler and buffier; tail brown above, whitish below. Skull rather narrow; zygomatic arches not expanded widely and with median ascending process placed posteriorly, making orbit proportionally large compared with temporal fossa; nasals long and nearly parallel sided, slightly surpassed behind by premaxillary processes; braincase lacks strong ridges; mesopterygoid fossa narrow; bullae long and narrow, in marked contrast with those of nearby *C. mendocinus.*

DISTRIBUTION: *Ctenomys pontifex* is known only from Mendoza province, Argentina.

SELECTED LOCALITIES (Map 466): ARGENTINA: Mendoza, Ñacuñán (J. R. Contreras 1979a), Volcán Peteroa (type locality of *Ctenomys pontifex* Thomas).

SUBSPECIES: *Ctenomys pontifex* is monotypic.

NATURAL HISTORY: Giannoni et al. (1996) described the burrowing behavior of *C. pontifex* (under *the nomen nudum C. eremophilus*; see Remarks). No other ecological data are available.

REMARKS: *Ctenomys pontifex* is another tuco-tuco species for which the type locality is uncertain and (as in the case of *C. knighti* discussed previously) where other rodent species share the same type locality. Thomas (1918a:40) gave this locality as "East side of the Andes near Fort San Rafael, Province of Mendoza"; Cabrera (1961:555) listed it simply as "San Rafael, Mendoza" and not, as Rusconi (1928:244) had thought, to "Fuerte de San Rafael, Provincia de Mendoza." This locality is within the Department of 25 de Mayo of Mendoza province. Pearson and Lagiglia (1992:35) assigned to *C. mendocinus* tuco-tucos collected in the vicinity of the presumed type locality of *C. pontifex*, and suggested that Thomas Bridges had likely collected the type specimen of *pontifex* in the neighbourhood of the Peteroa volcano, near the border between Argentina and Chile. This locality is in the Department of Malargüe, Mendoza, about 200 km WSW of San Rafael

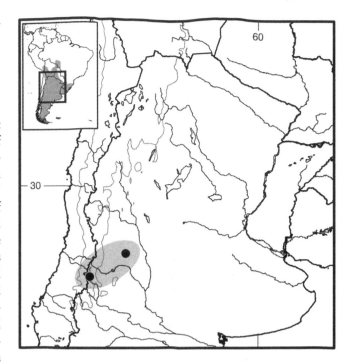

Map 466   Selected localities for *Ctenomys pontifex* (●). Contour line = 2,000 m.

between 2,400 and 3,000 m elevation (Pearson and Lagiglia 1992).

Bidau (2006) regarded *eremophilus* J. R. Contreras and Roig as a synonym of *C. pontifex*, but Parada et al. (2012) clearly linked this unavailable name to *C. mendocinus* (see that account).

The diploid number is 50 (M. O. Ortells, pers. comm.), but the sperm type is unknown.

Ctenomys porteousi Thomas, 1916
Cinnamon Tuco-tuco

SYNONYMS:

Ctenomys porteousi Thomas, 1916g:304; type locality "Bonifacio, S.W. Buenos Ayres Province, about 36°40′S, 62°W," Departamento Guaminí, Buenos Aires, Argentina, 36°49′S, 62°13′W, 100 m.

Ctenomys porteousi porteousi: Yepes, 1935a:252; name combination.

DESCRIPTION: Medium sized with mean head and body length 186 mm (range: 174–205), tail length 77 mm (range: 68–89), hindfoot 30 mm (range: 28–33), and mass 192 g (range: 115–240; J. R. Contreras and Reig 1965). General color above cinnamon-brown lined with black; mid-dorsal area darkened, sometimes black, but dark area not sharply defined; top of muzzle and crown also blackish; under parts near vinaceous buff or drabby, bases of hairs dark slaty; upper surfaces of hands and feet buffy whitish; tail dull whitish or pale brown, with darker, often blackish, terminal crest. Skull large but with smooth crown, parietal ridges imperceptible even in oldest individuals; incisive foramina medium in size; foramen magnum placed unusually high, with either one or two projections downward on its upper edge; bullae large. Upper incisors proodont, angled in comparison to toothrows between 97° and 100°; their enameled surfaces dark orange.

DISTRIBUTION: *Ctenomys porteousi* is found in the *Pampas* ecoregion of southeastern Buenos Aires province, Argentina.

SELECTED LOCALITIES (Map 467; J. R. Contreras and Reig 1965, except as noted): ARGENTINA: Buenos Aires, Álamos, Bonifacio (type locality of *Ctenomys porteousi* Thomas), Daireaux, Papín (Thomas 1916g), Pirovano.

SUBSPECIES: *Ctenomys porteousi* is monotypic.

NATURAL HISTORY: No information on ecology is available.

REMARK: *Ctenomys porteousi* is a member of the "mendocinus" group based on mtDNA sequences, along with *C. australis*, *C. flamarioni*, *C. mendocinus*, and *C. rionegrensis* (Parada et al. 2011). It has the typical "mendocinus" karyotype of $2n=46$, 47, or 48 (Massarini, Barros, Ortells, and Reig 1991). Massarini et al. (1992) described the genetic structure of populations of this species based on electromorphic markers. The sperm is simple asymmetric in morphology (Vitullo et al. 1988). Apfelbaum et al. (1991) compared levels of electromorphic variation within and between *C. porteousi* and *C. australis*, and Mapelli et al. (2012) detailed phylogeography and posited Late and post-Pleistocene population history based on mtDNA sequences.

Ctenomys pundti Nehring, 1900
Pundt's Tuco-tuco, Small Tuco-tuco

SYNONYMS:

Ctenomys Pundti Nehring, 1900a:420; type locality "Alejo Ledesma im Süden der Provinz Cordoba," Departamento Marcos Juárez, Córdoba, Argentina, 33°38′S, 62°37′W, 113 m.

Ctenomys mendocinus pundti: Cabrera, 1961:553; name combination.

DESCRIPTION: One of smallest species; overall head and body length of specimen used by Nehring in his description 170 mm, tail 43 mm, and hindfoot (including claws) only 20 mm. Color of upper parts yellowish-brown; under parts whitish-gray; both forefeet and hindfeet whitish above; tail bicolored, blackish above and whitish below. Dorsal hairs with yellow-brown tips and dark blue-gray bases. A. I. Medina et al. (2007) collected fully mature specimens with head and body length ranging between 133 and 138 mm in females, and 144 and 157 mm in males; some specimens weighed less than 100 g. Skull likewise small, with dimensions given by Nehring of total length 31.3 mm, basilar length 25.5 mm, zygomatic breadth 19.5 mm, width across auditory meatus 19.6 mm, least interorbital constriction 6.6 mm, nasal length 10.2 mm, diastema length 8.0 mm, upper toothrow length 6.9 mm, and width across both lower incisors 3.2 mm. Nehring noted, in particular, that occiput was not sharp-edged but founded, reflecting weakly developed lambdoidal crest. Other cranial features apparent in illustrations provided by Nehring (1900a,c) include proodont upper incisors, greatly expanded and ventrally rounded auditory bullae, auditory meatus set on same plane as upper cheek teeth, and weakly developed but verticalpostorbital process of jugal. Many of these characters are attributes of small size.

DISTRIBUTION: *Ctenomys pundti* occurs in the provinces of Córdoba and San Luis in north-central Argentina within the *Pampa* and *Espinal* ecoregions.

SELECTED LOCALITIES (Map 468, from A. I. Medina et al. 2007, except as noted): ARGENTINA: Córdoba, Alejo Ledesma (type locality of *Ctenomys pundti* Nehring), La Carlota, Manantiales, Puente Olmos; San Luis, Eleodoro Lobos (J. R. Contreras, pers. comm.)

SUBSPECIES: *Ctenomys pundti* is monotypic.

NATURAL HISTORY: No ecological data are available.

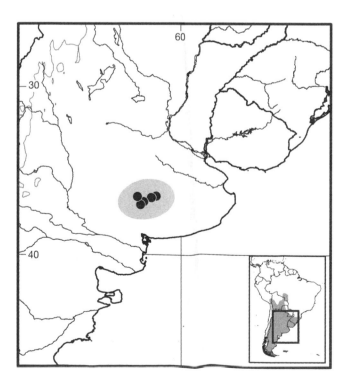

Map 467 Selected localities for *Ctenomys porteousi* (●). Contour line = 2,000 m.

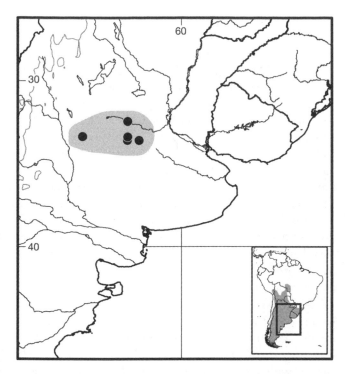

Map 468 Selected localities for *Ctenomys pundti* (●). Contour line = 2,000 m.

REMARKS: The karyotype of *C. pundti* includes a 2*n* = 50 and FN = 84 (Ipucha et al. 2008). Tiranti et al. (2005) discussed the morphological and karyological relationships of this species to other tuco-tucos, concluding that *C. pundit* is closest to *C. talarum*. The sperm morphology is the simple symmetric type (Bidau 2006).

Ctenomys rionegrensis Langguth and Abella, 1970
Rio Negro Tuco-tuco

SYNONYMS:

Ctenomys minutus rionegrensis Langguth and Abella, 1970: 13; type locality "Balneario Las Cañas, boca del Arroyo Las Cañas, 7 km al Sudoeste de Fray Bentos, Río Negro, Uruguay," ca. 100 m, 33°06′S 58°20′W.

Ctenomys rionegrensis: Altuna and Lessa, 1985:483; first use of current name combination.

DESCRIPTION: Relatively small tuco-tuco; external and cranial measurements for both sexes separately: males, head and body length 178–190 mm, tail length 71–83 mm, hindfoot length 30–33 mm, greatest skull length 45.9–47.9 mm, condylobasilar length 41–42.8 mm; females, head and body length 162–190 mm, tail length 65–83 mm, hindfoot length 29–33 mm, greatest skull length 41–44.6 mm, condylobasilar length 37.1–40.0 mm. Dorsal color varies extensively, with individuals either clear brownish-orange, dark-backed, or melanic (D'Elía et al. 1998; Wlasiuk et al. 2003); "agouti" individuals with small, yellow parauricular spots and yellowish venters; no inguinal or axillary

spots present; base of all body hairs, above and below, dark gray; tail bicolored with dorsal dark brown line along length. Skull with sides of rostrum parallel when viewed from above; temporal crests of parietal run almost parallel getting more near posteriorly; dorsal face of frontals and parietals flat. In comparison with *C. torquatus*, skull more rectilinear in shape; nasals flat or slightly convex longitudinally; dorsal border of foramen magnum with wide medial indentation; lateral processes seen in *C. torquatus* barely conspicuous; dorsal radius of upper incisors shorter such that incisors more opisthodont; most external point of zygomatic arch near its posterior end; zygoma dorsoventrally narrow with respect to *C. torquatus*; vertex of mandibular process of zygoma more rostral with respect to vertex of frontal process; longitudinal crest running along the external surface of the zygomatic bone is almost straight. The ventral root of zygomatic process of maxillary, at its origin, projected backward forming open angle; rostral edge of dorsal root of zygomatic arch projected forward and upward with respect to plane that passes to masticatory surface; Hill's (1935) interpremaxillary foramen very small.

DISTRIBUTION: *Ctenomys rionegrensis* inhabits the southeastern most part of the Río Negro department in Uruguay, between the Río Negro and Río Uruguay, and the coast of the Río Uruguay in Entre Ríos province, Argentina.

SELECTED LOCALITIES (Map 469): ARGENTINA: Entre Ríos, Concordia (Langguth and Abella 1970), Colonia Yerua (Langguth and Abella 1970), Paranacito (Reig et al. 1966), Parque Nacional El Palmar (Reig et al. 1966), Ubajay (Reig et al. 1966), Victoria (Reig et al. 1966). URUGUAY: Río Negro, Arrayanes (Wlasiuk et al. 2003), Balneario Las Cañas (type locality of *Ctenomys minutus rionegrensis* Langguth and Abella), La Guarida (Wlasiuk et al. 2003), La Tabaré (Wlasiuk et al. 2003), Mafalda (Wlasiuk et al. 2003).

SUBSPECIES: *Ctenomys rionegrensis* is monotypic.

NATURAL HISTORY: *Ctenomys rionegrensis* inhabits sand dunes between the Río Negro and Río Uruguay in Uruguay and in adjacent Argentina, where it has undergone recent demographic expansion (E. P. Lessa et al. 2005). Population densities are high, up to 40 adults per hectare (Tassino 2006), with multiple animals, including multiple adults, sometimes captured at the same burrow entrance (E. P. Lessa et al. 2005). These data suggested that *C. rionegrensis* might exhibit a level of sociality beyond that of the typical tuco-tuco solitary habits. However, ratio telemetric studies (Tassino et al. 2011) documented that *C. rionegrensis* is not social, as individuals neither exhibited the extensive spatial overlap nor sharing of nest sites characteristic of the group-living *C. sociabilis* (Lacey et al. 1997; Lacey 2000). Nevertheless, overlap in areas occupied

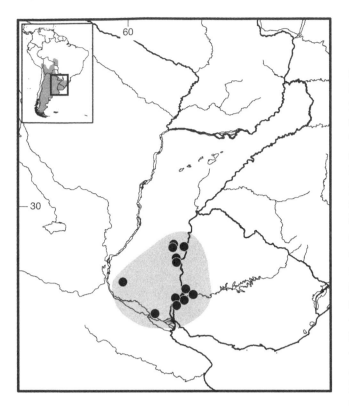

Map 469 Selected localities for *Ctenomys rionegrensis* (●).

by adults was detected, suggesting an intermediate pattern of space use between the extremes of complete solitariness and group living. Breeding is highly seasonal, with activity beginning in the late austral autumn with initial births occurring in the early spring (Tassino and Passos 2010).

REMARKS: *Ctenomys rionegrensis* shows a remarkable variation in pelage color with three primary morphs: melanic, agouti, and dark-backed (D'Elía et al. 1998; Wlasiuk et al. 2003). These authors also described population genetic structure and gene flow rates using molecular markers (see also E. P. Lessa et al. 2005). Altuna and Lessa (1985) established that *C. rionegrensis* is a species distinct from *C. minutus* from nearby Rio Grande do Sul in Brazil based on characters of the male phallus. The karyotype of this species is both polymorphic with populations and polytypic among them. At the type locality, specimens show $2n = 50$ (Kiblisky et al. 1977), but diploid numbers ranging from 48 to 56 have been reported for other Argentinean and Uruguayan populations (Bidau 2006). Sperm morphology is the simple asymmetric type (Altuna et al. 1985).

Ctenomys roigi J. R. Contreras, 1988
Roig's Tuco-tuco

SYNONYM:

Ctenomys roigi J. R. Contreras, 1988:53; type locality "Procedente de Costa Mansión, 10 km. al sur de Em-

pedrado, Departamento Empedrado, Provincia de Corrientes, a 28°02′S, 58°49′W, 60 m," Argentina.

DESCRIPTION: Relatively large tuco-tuco, mean total length 277.8 mm (range: 260–299 mm), head and body length 198.1 mm (176–200 mm), tail length 88.7 mm (76–99 mm), hindfoot (with claw) length 38.6 mm (34.5–43.8), and mass 231.1 g (175.5–278.8 g). Sexually dimorphic, with males averaging 105% larger in head and body length and 131% heavier than females. Dorsal color brownish hues, more tawny on back, clay on flanks, and lightening progressively from head to rump; head in particular with dark area extending from nose to neck, more accentuated in worn than newly molted pelage; many specimens with small tufts of white hairs on head dispersed within normal coat; light collar typical of other species distributed nearby (*C. argentinus*, *C. pearsoni*, *C. torquatus*, or *C. validus*) not present; pale zone surrounds pinna and area of white rostral vibrissae. Ventral color cinnamon with pale creamy wash; ventral white spots present in majority of specimens in both axillary and inguinal regions. Legs and feet pale, with hair almost white, even ungual combs. Tail sparsely covered with hair, moderately bicolored in fresh specimens, but blackens evenly after skin dries in preserved material; tip with short pencil of hairs.

Skull robust, solid, and strong, endowed with only moderately expanded tympanic bullae not visible in dorsal view; zygomatic arch strong with greatest breadth across arches wider than that across external auditory canals; rostrum short, robust, and with very broad nasal bones anteriorly that taper and truncate before leading edge of orbit; premaxillary bones extend to posterior limit of nasals or only slightly beyond; frontal bones slightly concave above; interorbital constriction broad, terminating in well-marked postorbital process; bregmatic fenestra, conspicuous in *C. dorbignyi* and *C. perrensi*, is rarely present; temporal crests well developed, so squamosal does not extend onto dorsal surface of skull, although degree of development is both age and sex dependent; well-developed lambdoidal crests present, with interparietal barely developed or not apparent at all. Ventrally, paroccipital processes broad, covering nearly half of posterolateral surface of auditory bullae, greater extent then in other species (such as *C. argentinus* and *C. bonettoi*); fossa for insertion of masseter superficialis muscle shallow, appearing in some older specimens simply as an ossification of terminal tendon of insertion; mesopterygoid fossa V-shaped notch. Upper incisors distinctly proodont, with their anterior face at angle of 109° relative to horizontal plane formed by upper cheek teeth, much more so than other closely adjacent species, such as *C. dorbignyi*, *C. bonettoi*, *C. perrensi*, or *C. argentinus*. Upper molariform teeth medium in size; the anterior prism of PM4 very enlarged and with visible inner groove; M3 small and either circular or sub-circular in shape.

Baculum small, spatulate-shaped bone, with ventral face flattened and more rugose and slightly convex dorsal face, usually with slight depression of varying length along median axis. Average bacular length 6.74 mm (range: 5.7–7.6 mm), average maximum width 1.85 mm (range: 1.4–2.4 mm).

DISTRIBUTION: *Ctenomys roigi* inhabits a very limited area no more than 12 km long and 3 km wide along the coast of the Río Paraná in Corrientes province, Argentina.

SELECTED LOCALITIES (Map 470): ARGENTINA: Corrientes, Colonia Brougnes (Giménez et al. 2002), Costa Mansión (type locality of *Ctenomys roigi* J. R. Contreras), Empedrado (Parada et al. 2011), Estancia Yacyretá (A. I. Medina et al. 2007).

SUBSPECIES: *Ctenomys roigi* is monotypic.

NATURAL HISTORY: Little is known about the population biology or ecology of this species. As summarized by J. R. Contreras (1988), individuals are solitary living in individually exclusive but closely adjacent burrow systems. Breeding takes place during the spring-summer. Fully 50% of females were either obviously pregnant or lactating in October, with 11% young of the year, while two months later, in December, 40% of the population were young of the year with pregnant and lactating individuals comprising only

10% and 30%, respectively. Litter size is low, with seven of eight females with a single fetus, and one female with two, for an overall average of 1.124 young per pregnancy.

This species occupies sandy loam soils in prairie habitats covered by herbaceous plants, particularly invasive Solanaceae, in areas strongly subject to overgrazing by cattle. These areas are not seasonally inundated. Individuals feed from within their burrows, either harvesting tubers and roots of plants such as *Rhunchosia pallida*, *Ipomoea bonariensis*, *Senecio grisebachi*, or *Byptia mutabilis*, or the basal stems of the dominant grasses. Individuals also access above-ground plant materials by pulling whole plant into their tunnels from below rather than by venturing onto the ground surface.

REMARKS: The karyotype of *C. roigi* is $2n = 48$, FN $= 76$ (Giménez et al. 2001, 2002), and the sperm is the simple symmetric type (Bidau 2006). Ortells et al. (1990) reported the FN as 80 (and cited as such by Woods and Kilpatrick 2005:1568), but their calculation included the sex chromosomes. In *C. roigi*, both X and Y chromosomes are biarmed, thus increasing the autosomal FN by four.

Ctenomys rondoni Miranda-Ribeiro, 1914
Rondon's Tuco-tuco

SYNONYMS:

Ctenomys rondoni Miranda-Ribeiro, 1914:39, plates 20 and 23, Figs. 1–1a′, 3, and 5; type locality (based on the lectotype chosen by Moojen, in Miranda-Ribeiro 1955:415) "Maria de Molina," = Campos dos Palmares de Maria de Molina, Vilhena, Rondônia, Brazil, 12°07′12″ S 60°28′56″ W.

Stenomys rondoni: Avila-Pires, 1968:162; inadvertent misspelling of genus name.

DESCRIPTION: Medium sized with head and body length about 230 mm and tail length 80 mm. Dorsal pelage comprises pale hairs at base but turning to sepia at tips; head and especially venter slightly rufous; tail uniformly brownish. Skull robust and depressed, approximately 54 mm in total length and 34 mm across zygomatic arches (Miranda-Ribeiro 1914). Intermaxillaries robust with lateral protruding expansion that decreases toward front and projects past incisors; supraorbital process protrudes; interparietal contacts frontals along narrow zone; transverse occipital-temporal crest straight; tympanic bullae inflated; occipital foramen round anteriorly and square posteriorly; maxillaries narrow; body of ethmoid bones very narrow, pterygoids small but thick. Mandible particularly strong and wide.

DISTRIBUTION: *Ctenomys rondoni* is known only from two specimens, one each from localities in Mato Grosso and Rondônia states in western Brazil.

SELECTED LOCALITIES (Map 471): BRAZIL: Mato Grosso, Juruena (Miranda-Ribeiro 1914); Rondônia,

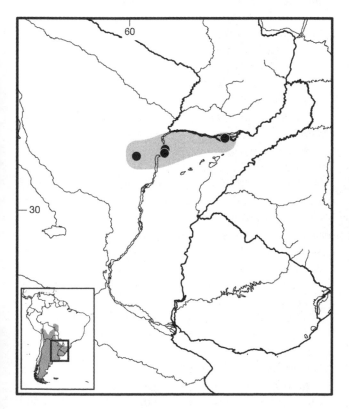

Map 470 Selected localities for *Ctenomys roigi* (●).

Campos dos Palmares de Maria de Molina (type locality of *Ctenomys rondoni* Miranda-Ribeiro).

SUBSPECIES: *Ctenomys rondoni* is monotypic.

NATURAL HISTORY: No ecological data are available.

REMARKS: The type locality of *C. rondoni* has been varyingly given as Rio Juruena (S. Anderson et al. 1987) or Juruena (Avila-Pires 1968). Miranda-Ribeiro (1914) did not select a holotype and based his description on both the skin and partial cranium of the single specimen from Juruena as well as on the more complete cranium of the specimen from Maria de Molina. João Moojen (in Miranda-Ribeiro 1955:415) designated the Maria de Molina specimen as lectotype (MNRJ 2051) while MNRJ 2050 from Juruena remained as paralectotype. However, in a catalog of type specimens at the Museo Nacional, Avila-Pires (1968:183) incorrectly switched the specimen numbers. Thus, Maria de Molina and not Juruena should be considered the type locality of *Ctenomys rondoni*. Subsequently, MNRJ 2051 was corrected in the museum records, and the cranium received the same number as the skin, MNRJ 2048 (see Langguth et al. 1997).

Despite clarification of the locality and catalog number of the lectotype, both known localities remain troublesome. Maria de Molina does not exist on maps or in gazetteers of Brazil, either old or new. In fact, the 1909 expedition led by Coronel Cándido Rondon (when the specimens of *C. rondoni* [as well as those of *C. bicolor*] were collected) established camp in an unpopulated site near José Bonifácio. Rondon (in Viveiros 1958:298) named the place "Campos dos Palmares de Maria de Molina," which establishes the coordinates given herein.

Although all recent authors have regarded *C. rondoni* as a synonym of *C. nattereri* (e.g., Cabrera 1961:553) or as a valid subspecies of *C. boliviensis* (e.g., Woods and Kilpatrick 2005:1561), I believe these decisions are unjustified given theambiguous geographic status of both *boliviensis* and *nattereri*.

No data are available on either karyotype or sperm morphology.

Ctenomys rosendopascuali J. R. Contreras, 1995
Rosendo Pascual's Tuco-tuco

SYNONYM:

Ctenomys rosendopascuali J. R. Contreras, 1995b:1; type locality "Mar Chiquita, Departamento San Justo, provincia de Córdoba, Argentina, 30°48′S, 62°53′W, 65 m"; *nomen nudum?* (see Remarks).

DESCRIPTION: Relatively small with head and body length of males ranging from 160–177 mm, and 159–166 mm for females (A. I. Medina et al. 2007). Few other character data published.

DISTRIBUTION: *Ctenomys rosendopascuali* is only known from a few localities in Córdoba province, Argentina, in the *Chaco Seco* ecoregion.

SELECTED LOCALITIES (Map 472): ARGENTINA: Córdoba, Candelaria (Giménez et al. 1999), Los Mistoles (Mascheretti et al. 2000), Mar Chiquita (type locality of *Ctenomys rosendopascuali* J. R. Contreras).

SUBSPECIES: *Ctenomys rosendopascuali* is monotypic.

NATURAL HISTORY: No data on ecology, behavior, or reproduction are available.

REMARKS: Giménez et al. (1999:91) described a polytypic chromosome complement of $2n = 52$ with a variable FN of 62, 64, and 66 (see also Ipucha et al. 2008). The sperm is the simple asymmetric type (Giménez et al. 1999). Mascheretti et al. (2000) concluded that *C. rosendopascuali* was part of a lineage that also included *C. azarae*, *C. bergi*, *C. bonettoi*, *C. mendocinus*, and *C. "yolandae"* based on mtDNA sequences.

It is unclear whether the issue of *Nótulas Faunísticas* (volume 86, pp. 1–6) in which J. R. Contreras (1995b) described *C. rosendopascuali* was actually published (Yolanda Davies, daughter of Julio R. Contreras, pers. comm. to U. F. J. Pardiñas, Oct. 25, 2012). If not, then the taxon to which subsequent authors have referred under this name will require a formal description.

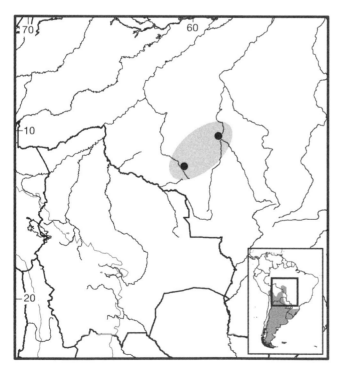

Map 471 Selected localities for *Ctenomys rondoni* (●). Contour line = 2,000 m.

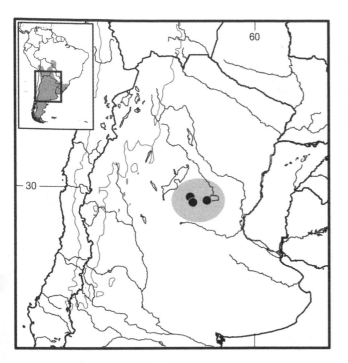

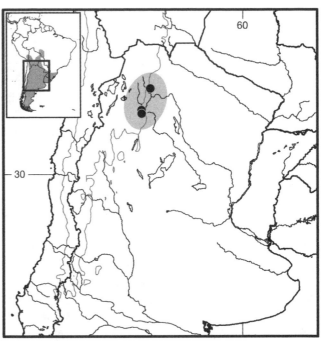

Map 472 Selected localities for *Ctenomys rosendopascuali* (●). Contour line = 2,000 m.

Map 473 Selected localities for *Ctenomys saltarius* (●). Contour line = 2,000 m.

Ctenomys saltarius Thomas, 1912
Salta Tuco-tuco

SYNONYMS:

Ctenomys saltarius Thomas, 1912c:639; type locality "Salta, [Salta,] N Argentina."

Ctenomys talarum saltarius: Yepes, 1935a:252; name combination.

DESCRIPTION: Fairly large with head and body length (of holotype, adult female) 200 mm, tail length 90 mm, and hindfoot length 33 mm. Fur short and overall dull in tone, having neither length nor glossiness of neighboring species (e.g., *C. opimus luteolus* Thomas and *C. tucumanus* Thomas). Upper parts dull raw umber, browner along dorsal midline, paler on sides; under parts dull buffy whitish; area around snout whitish with inconspicuous dark collar further behind; feet dull white above; tail markedly bicolored, blackish above and dull white below. Skull very narrow, with zygomatic arches less spread than distance from front of incisors to back of toothrow, a condition that contrasts with other species; frontal region also very narrow, with interorbital breadth less than length of molar series; postorbital processes practically absent.

DISTRIBUTION: *Ctenomys saltarius* is known only from a few localities in Salta province, Argentina, all within either the *Yungas* or *Chaco Seco* ecoregions (J. R. Contreras 1984a).

SELECTED LOCALITIES (Map 473): ARGENTINA: Salta, Cafayate (CML 854), Salta (holotype of *Ctenomys saltarius* Thomas), Tolombón (A. I. Medina et al. 2007).

SUBSPECIES: *Ctenomys saltarius* is monotypic.

NATURAL HISTORY: *Ctenomys saltarius* inhabits hillsides and valleys of the Monte Desert, and is especially common in *Larrea* (Zygophyllaceae) flats and areas with *Prosopis* (Fabaceae) where it feeds on creosote and other shrubs (Mares, Ojeda, and Kosco 1981).

REMARKS: Both karyotype and sperm are unknown.

Ctenomys scagliai J. R. Contreras, 1999
Scaglia's Tuco-tuco

SYNONYMS:

C[*tenomys*]. *scagliai* J. R. Contreras and Bidau, 1999:3; *nomen nudum*.

C[*tenomys*]. *scagliai* J. R. Contreras, Castro, and Cicchino, 1999:5; *nomen nudum*.

Ctenomys scagliai J. R. Contreras, 1999:10; type locality: "Los Cardones, sobre la Ruta Nacional N° 307, Kilómetro 101, entre las localidades tucumanas de El Infiernillo y Amaicha del Valley, aproximadamente a 2.500 metros sobre el nivel del mar (26°38′S–65°49′W)," Tucumán, Argentina.

DESCRIPTION: Medium sized with external measurements of holotype including head and body length 195 mm, tail length 85 mm, hindfoot length (with claw) 37.0 mm, ear length 10 mm, and mass 270 g. General color clear ochraceous brown on back to yellow-brown (straw) on flanks and venter; hairs long, silky, and strongly bicolored, with black basal part for two-fifths length of dorsal hairs and about half the length of ventral hairs; top of muzzle and head black, extending back as dorsal stripe; both vibrissae and supra-ungual combs white; area surrounding vibrissae same color as rest of face; tail covered by short and sparsely distributed hairs, bicolored, with proximal part dirty white becoming progressively darker until distal third, which is black; ears very small, with creamy white subauricular areas that suggest pale collar; no white spots in either axillary or inguinal regions; legs covered with sparse coat of fur colored similar to that of belly.

Skull robust but looks more delicate than those of many other species; nasal bones short and relatively narrow, tapering only slightly posteriorly, with premaxilla and incisors in clear view from above and broadened premaxillary and maxillary wings on either side; posterior border of frontal bones convex; interorbital region relatively broad with weakly developed postorbital processes; frontals extend laterally to form part of overhanging roof that forms part of orbit; squamosal forms part of dorsal roof of skull, with well-developed post-tympanic projections; pair of sesamoid bones in front of lambdoidal crest between parietals and post-tympanic process of squamosal gives latter spatulate shape; interparietal enlarged and V-shaped, both uncommon features for genus; distance across external meatus on both sides typically greater than greatest width across zygomatic arches; laterally, alveoli of upper incisors positioned in common depression located posterior to proximal edge of vomer; no interpremaxillary opening, which is apomorphic condition for species; external auditory meatus positioned high on side of skull, with its lower border above plane formed by occlusal surface of upper toothrow; zygomatic arch relatively delicate, with zygomatic process of maxilla forming moderately obtuse angle; bullae appear inflated and evenly rounded in lateral view, but transversely compressed in ventral view, with expanded paroccipital process appressed only to very posterior margins; palate relatively short, with acutely V-shaped mesopterygoid fossa extending anteriorly to middle of M2; openings in alisphenoid asymmetrical, with one on right side and two on left; foramen ovale wide and within it alisphenoid-presphenoid bridge runs anteromedially from alisphenoid to become trapezoidal plate linked by short isthmus to presphenoid. Upper incisors orthodont.

DISTRIBUTION: *Ctenomys scagliai* is only known from the vicinity of the type locality in Tucumán province, Argentina.

SELECTED LOCALITIES (Map 474): ARGENTINA: Tucumán, Los Cardones (type locality of *Ctenomys scagliai* J. R. Contreras), Km 95 of Hwy 307 near Tafí del Valle (A. I. Medina et al. 2007).

SUBSPECIES: *Ctenomys scagliai* is monotypic.

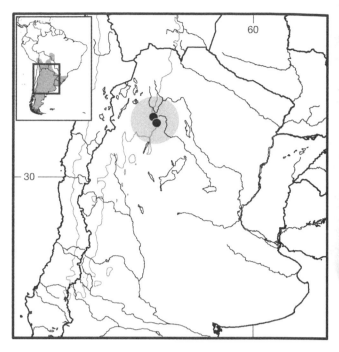

Map 474 Selected localities for *Ctenomys scagliai* (●). Contour line = 2,000 m.

NATURAL HISTORY: Ecology, reproduction, behavior, or other aspects of the population biology are unknown.

REMARKS: The karyotype is 2n = 36, FN = 64 (Ortells 1995, as *C. knighti*; see J. R. Contreras 1999). Sperm morphology is symmetrical, the plesiomorphic condition for the genus (J. R. Contreras 1999).

Ctenomys sericeus J. A. Allen, 1903
Silky Tuco-tuco

SYNONYMS:

Ctenomys sericeus J. A. Allen, 1903b:187; type locality "Cordilleras, upper Rio Chico de Santa Cruz, Patagonia"; restricted to "confluencia de los ríos Belgrano y Chico (~48.26°S, 71.20°W, departamento Río Chico, Santa Cruz, Argentina" (Pardiñas 2013).

Ctenomys seriseus Rusconi, 1928:243; incorrect subsequent spelling of *sericeus* J. A. Allen.

DESCRIPTION: Small with total length of males averaging 200 mm (range: 195–208 mm), tail length 56.6 mm (51–62 mm), and hindfoot length 26.2 mm (25–28 mm; A. I. Medina et al. 2007). Pelage short, soft, and glossy. General color above yellowish-gray strongly varied with black, hairs being plumbeous on basal three-fourths, then banded narrowly with pale yellowish brown, and tipped with black; flanks and ventral surface buff; sides of nose yellowish-brown; top of nose and top of head like midback; ears small and blackish, upper surfaces of feet dingy gray with slight yellowish cast; tail pale yellowish, with

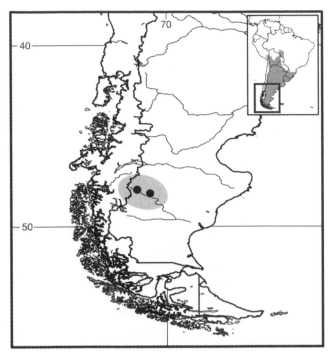

Map 475 Selected localities for *Ctenomys sericeus* (●).

median dusky stripe along apical half of upper surface. Skull with interparietal either lacking or greatly reduced in size.

DISTRIBUTION: *Ctenomys sericeus* is only known from two localities in Santa Cruz province, Argentina.

SELECTED LOCALITIES (Map 475): ARGENTINA: Santa Cruz, La Porteña, Río Vista (Parada et al. 2011), confluence of Río Chico and Río Belgrano (J. A. Allen 1905).

SUBSPECIES: *Ctenomys sericeus* is monotypic.

NATURAL HISTORY: J. A. Allen (1905:41) stated that *Ctenomys sericeus* was parapatric to *C. osgoodi*. No other data are available.

REMARKS: J. A. Allen (1903b) compared *C. sericeus* to both *C. pundti* Nehring and *C. bergi* Thomas, two species from nearby Córdoba province and from both of which it differs in coloration. Subsequently, J. A. Allen (1095:40) narrowed the type locality of the confluence of the Chico and Belgrano rivers in Santa Cruz province. This species has a karyotype of $2n = 28$–30 chromosomes and a simple asymmetric sperm type (Montes et al. 2001; Bidau 2006).

Ctenomys sociabilis Pearson and Christie, 1985
Colonial Tuco-tuco

SYNONYM:

Ctenomys sociabilis Pearson and Christie, 1985:338; type locality "Estancia Fortín Chacabuco, 1075 m, 3 km S y 2 km W Cerro Puntudo, 71°11′40″W, 40°58′00″S," Departamento Los Lagos, Neuquén, Argentina.

DESCRIPTION: Medium sized with head and body length averaging 192 mm (range: 168–247 mm), tail 68 mm (range: 67–80) mm, hindfoot length 34 mm (range: 32–36 mm), and mass 182 g (range: 180–234 g). Color overall ochraceous-tawny above mottled with black hairs, especially in frontal region of head; ventral color similar, but lacks black hairs; diagnostic ochraceous-orange spot on both sides of nose and conspicuous black and white moustache uniformly present. Claws large, relatively narrow, and only slightly curved. Cranially, rostrum elongated, arc of upper incisors wide, and auditory bullae uninflated and exceptionally narrow.

DISTRIBUTION: *Ctenomys sociabilis* has a narrow range between the Río Traful and Lago Nahuel Huapi, west of the Río Limay in the Reserva Nacional Nahuel Huapi in Neuquén province, Argentina (Pearson 1995).

SELECTED LOCALITIES (Map 476): ARGENTINA: Neuquén, Estancia Fortín Chacabuco (type locality of *Ctenomys sociabilis* Pearson and Christie); Estancia La Primavera, 11 km NW of Confluencia (MVZ 179310), 4 km N and 4 km E of Estancia Paso Coihue (MVZ 184869), Estancia Rincon Grande (MVZ 186111).

SUBSPECIES: *Ctenomys sociabilis* is monotypic.

NATURAL HISTORY: *Ctenomys sociabilis* is one of the more thoroughly studied tuco-tuco species in the field. It is the only clearly documented social tuco-tuco (although *C. peruanus* is suspected to be social; see Pearson 1959). Their populations consist of aggregates of collective burrows at the margins of *mallines*, the local name for the moist meadows in the Andean-Patagonian steppe, where somewhat lush vegetation grows (Pearson and Christie 1985). The majority of active burrow systems surveyed across the species range, however, were in non-*mallín* habitats (Tammone et al. 2012). It inhabits black, wet, and fine soils. It is also the only tuco-tuco that does not maintain a burrow with the mouth permanently closed and exhibits a substantial level of above-ground activity. The grass *Poa* is a main food item. Reproduction occurs in the winter, with young typically born in November. Gestation is about three months, and litter size varies from two to six young. Populations of this species may contact those of *C. haigi* and *C. maulinus* (Pearson and Christie 1985). Lacey et al. (1997) studied burrow sharing by multiple individuals; Lacey and Wieczorek (2004) described kinship patterns and group composition; Lacey (2004) documented that sociality reduces direct individual fitness; Woodruff et al. (2010) examined stress physiology and social biology in wild and captive populations; and Schwanz and Lacey (2003) showed that individuals have the capacity to discriminate gender by olfactory cues. Hambuch and Lacey (2002) and Lacey (2001) contrasted patterns of variation at both MHC and microsatellite loci, respectively, between the colonial *C. sociabilis*

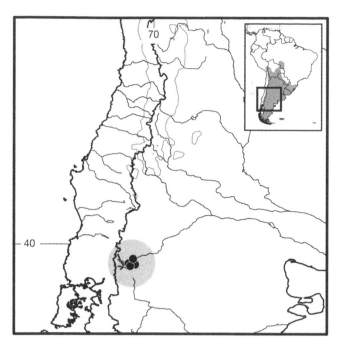

Map 476 Selected localities for *Ctenomys sociabilis* (●). Contour line = 2,000 m.

and the geographically adjacent but asocial *C. haigi*. Chan et al. (2005) and Chan and Hadly (2011) documented a substantial loss of genetic diversity in modern population samples of *C. sociabilis* over the past 10,000 years.

REMARKS: The chromosome complement of *C. sociabilis* has $2n = 56$, FN = 72 (M. H. Gallardo 1991); its sperm is the simple asymmetric type (Pearson and Christie 1985; M. H. Gallardo et al. 2002). Mason (2004) described the middle ear apparatus.

Ctenomys steinbachi Thomas, 1907
Steinbach's Tuco-tuco

SYNONYM:

Ctenomys Steinbachi Thomas, 1907b:164; type locality "Campo of Province Sara, near Santa Cruz de la Sierra, Bolivia"; restricted to 6 km N of Buen Retiro, 17°13′S 63°38′W by S. Anderson et al. (1987:13).

DESCRIPTION: Very large with males reaching at least 350 mm in body length, and with males and females of equivalent size (S. Anderson et al. 1987). Measurements of holotype (an adult male) head and body length 245 mm, tail length 86 mm; hindfoot length (with claw) 45 mm. Fur straight, fine, and glossy, with hairs of back about 13 mm in length. General color unusual dark drabby-gray-brown or coppery, quite unlike that of any other *Ctenomys* known to Thomas when he described this species. Color uniform over entire head, upper surface, and sides; under surface creamy white, hairs dull slaty for their basal two-thirds; line of demarcation

on sides sharply defined; vibrissae white; chin and band across lower neck in front of arms brown, separated by broad whitish patch running across interramal region and narrowing on sides to point below ear; arms and legs pale-colored, except for narrow band running down front of forelimbs; hands and feet almost naked above, pale brown, with lateral fringes whitish; tail very thinly clothed, with dull white, sparse hairs.

DISTRIBUTION: *Ctenomys steinbachi* occurs in west-central Santa Cruz province, Bolivia.

SELECTED LOCALITIES (Map 477, from S. Anderson 1997, except as noted): BOLIVIA: Santa Cruz, 6 km N of Buen Retiro (type locality of *Ctenomys steinbachi* Thomas), 2 mi S of Caranda, Río Surutó, San Rafael de Amboró, 10 km S of Zanja Honda.

SUBSPECIES: *Ctenomys steinbachi* is monotypic.

NATURAL HISTORY: No published data on the autecology available. A pregnant female was collected in August with a single embryo (S. Anderson 1997).

REMARKS: Molecular analyses support a close phylogenetic relationship with the geographically adjacent *C. boliviensis* and *C. goodfellowi* (Cook and Yates 1994; E. P. Lessa and Cook 1998). Parada et al. (2011) placed this species in their "*boliviensis*" group, along with *C. boliviensis*, *C. goodfellowi*, *C. nattereri*, and *C. robo*, based on mtDNA sequences. *Ctenomys steinbachi* has a karyotype with $2n = 10$ and FN = 18 (S. Anderson et al. 1987), the lowest chromosome number known for a rodent, shared

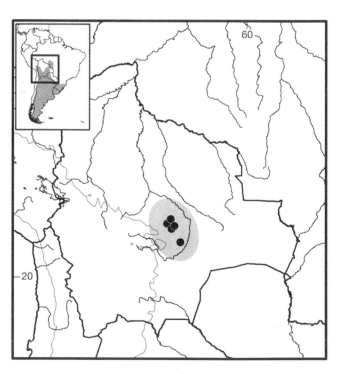

Map 477 Selected localities for *Ctenomys steinbachi* (●). Contour line = 2,000 m.

only with the grass mouse *Akodon* sp. 2n = 10. Vitullo and Cook (1991) described a simple symmetric type of sperm morphology.

Ctenomys talarum Thomas, 1898
Talas Tuco-tuco

SYNONYMS:

Ctenomys talarum Thomas, 1898d:285; type locality "'Las Talas,' Ensenada, La Plata," on the Río de la Plata, east of La Plata, Partido de Berisso, Buenos Aires, Argentina, 34°52'S, 57°53'W, 10 m.

Ctenomys talarum antonii Thomas, 1910a:240; type locality "Los Yngleses, in the eastern part of the Province of Buenos Ayres," Argentina.

Ctenomys talarum recessus Thomas, 1912a:241; type locality "Bahía Blanca," Buenos Aires, Argentina.

Ctenomys talarum anthonii: Rusconi, 1928:243; inadvertent misspelling of *antonii* Thomas.

Ctenomys mendocinus talarum: Thomas, 1929:44; name combination.

Ctenomys mendocinus recessus: Thomas, 1929:44; name combination.

Ctenomys talarum talarum: Tate, 1935:391; name combination.

Ctenomys talarum occidentalis Justo, 1992:35; type locality "Luan Toro [= Luán Toro]," La Pampa, Argentina.

DESCRIPTION: Small with three subspecies currently recognized similar in size and degree of sexual dimorphism (A. I. Medina et al. 2007). Mean body length (both sexes combined) of nominotypical subspecies 233.4 mm, range: 212–254 mm, and tail 66.7 mm, range: 56.0–75.0 mm (Reig et al. 1966). Tail short (<39% of head and body length). Overall color dark hazel grayish red with prominent axillary white patches; conspicuous white patch occurs at lower edge of ears (Justo et al., 2003). Distinctive cranial characteristics include mastoid breadth less than zygomatic breadth, and conspicuous interpremaxillary foramen (Justo et al. 2003). Baculum short, narrow, and not very expanded at either end; mean length 6.4 mm, proximal width 1.32 mm, distal width 0.9 mm (N = 12; Justo et al. 2003).

DISTRIBUTION: The nominotypical subspecies, now extirpated from the type locality, extends along coastal Buenos Aires province, Argentina, from Magdalena to Santa Clara del Mar; *C. t. recessus* occupies the second or third line of sand dunes from Necochea to Punta Alta, also in Buenos Aires province; and *C. t. occidentalis* is limited to two localities in La Pampa province. Sparse populations of unknown subspecific status occur in northwestern Buenos Aires province, possibly entering La Pampa province close to the range of *occidentalis*.

SELECTED LOCALITIES (Map 478): ARGENTINA: Buenos Aires, Bahía Blanca (type locality of *Ctenomys ta-* *larum recessus* Thomas), Cerro de la Gloria (Bidau et al. 2000), Ciudad de Lincoln (J. R. Contreras 1972), Ciudad de Pehuajó (J. R. Contreras 1972), Fortín Olavarría (J. R. Contreras 1972), Laguna Las Encadenadas (J. R. Contreras 1972), Las Talas (type locality of *Ctenomys talarum* Thomas), Mar Chiquita (Bó et al. 2002), Monte Hermosa (A. I. Medina et al. 2007), Necochea (Vasallo 1998), Pechuén-có (J. R. Contreras 1972), Punta Médanos (Bidau et al. 2000), Quiroga (Massoia 1988a), Saladillo (Parada et al. 2011), San Clemente del Tuyú (Bidau et al. 2000), Sierra de la Ventana (J. R. Contreras 1972); La Pampa, El Guanaco (A. I. Medina et al. 2007), La Florida (A. I. Medina et al. 2007), Luán Toro (type locality of *Ctenomys talarum occidentalis* Justo).

SUBSPECIES: *Ctenomys talarum* includes three recognized subspecies (Justo et al. 2003): *C. t. talarum* Thomas, *C. t. antonii* Thomas, and *C. t. occidentalis* Justo. Reig et al. (1966:299) incorrectly considered *antonii* Thomas to be a synonym of *recessus* Thomas. However, *antonii* Thomas has date priority, so the southern subspecies of *C. talarum* is properly named *C. talarum antonii*, with *recessus* Thomas a subjective junior synonym (ICZN 1999:Art. 23.1).

NATURAL HISTORY: *Ctenomys t. antonii* occurs in partial sympatry with the larger *C. australis* in Buenos Aires province (J. R. Contreras and Reig 1965) and *C. t. occidentalis* overlaps with *C. azarae* in La Pampa province (Justo et al. 2003). Comparatore et al. (1992) examined the juxtaposition of *C. talarum* and *C. australis* in relation to edaphic and plant composition in the Necochea district, also in Buenos Aires province. Justo et al. (2003) provided a thorough summary of life history, ecology, behavior, and other attributes of this species. These data are perhaps more extensive than for any other species of tucotuco, and include information on social structure, spatial distribution, and age composition (Pearson et al. 1968); burrow structure (Antinuchi and Busch 1992); home range and activity patterns (A. P. Cutrera, Antinuchi et al. 2006; A. P. Cutrera, Mora et al. 2010); chemical communication (Zenuto et al. 2004); vocal communication (Schleich and Busch 2002a,b); social transmission (Vassallo 2006); energetics (Schleich and Busch 2004; Antinuchi et al. 2007; Perissinotti et al. 2009), thermoregulation (C. Busch 1989; F. Luna and Antinuchi 2007a,b), including its ontogeny (A. P. Cutrera et al. 2003); foraging costs (F. Luna and Antinuchi 2006, 2007a, b), food preferences, feeding selectivity, and digestive strategies (del Valle et al. 2001; del Valley and Lopez Mananes 2008, 2011); predation (Baladron et al. 2009); parasitism (Rossin 2004); reproductive biology, including seasonality and litter size (Malizia and Busch 1991; Fanjul et al. 2006); population ecology and behavior (Vasallo 1998; C. Busch et al. 2000); polygynous mating system (Zenuto et al. 1999), with expected low sperm production (Zenuto et al. 2003); and various aspects of population genetic structure based on molecular markers, including effective population size (A. P. Cutrera, Lacey, and Busch 2006), kinship and dispersal (A. P. Cutrera et al. 2005), and demographic influences on selection (A. P. Cutrera and Lacey 2006, 2007; A. P. Cutrera, Lacey et al. 2010). Incisor procumbency is not the determinant factor during excavation, because the *masseter superficialis*, *masseter lateralis profundus*, *masseter medialis*, *temporalis*, *pterygoideus internus*, *pterygoideus externus*, *digastricus*, *acromiotrapezius*, *cleidomastoideus*, and *sternomastoideus* muscles assist the forelimbs in construction and compaction of tunnels (De Santis et al. 1998). Microstructure of incisor enamel indicates that incisors may assist in excavation and transportation and movement of obstacles in tunnels (Justo et al. 1995).

REMARKS: The nominotypical subspecies exhibits a polymorphic karyotype with diploid number varying from 44 to 48 (Kiblisky and Reig 1966; Bidau et al. 2000; Massarini et al. 2002). Populations of *C. t. antonii* have karyotypes of $2n = 48$ to 50 (Massarini et al. 1995; under the name *recessus*), and those of *C. t. occidentalis* are uniform at $2n = 48$ (Braggio et al. 1999). Sperm morphology is the simple symmetric type (Vitullo et al. 1988; Zenuto et al. 2003). De Santis et al. (2001) described enamel microstructure; Garcia Esponda et al. (2009) examined craniometric variation; and Mora, Cutrera et al. (2013) delineated historical demography through phylogeographic and population genetic analyses of mtDNA gene sequences.

It is possible that *C. talarum* is present in northeastern Buenos Aires province at Lima farther north than the type locality (Agnolín et al. 2010), but this claim needs confirmation.

Ctenomys torquatus Lichtenstein, 1830
Collared Tuco-tuco

SYNONYM:

Ctenomys torquatus Lichtenstein, 1830:plate 31, Fig. 1; type locality "Das Vaterland dieses Thiers sind die Südlichen Provinzen Brasiliens und die Ufer des Uruguay, wo es malwurfartig unter der Erde lebt" (see Remarks).

DESCRIPTION: Small; external measurements for males, head and body 167–230 mm, tail length 55–85 mm, hindfoot length 28–35 mm; for females, head and body 152–200 mm, tail length 60–83 mm, hindfoot length 29–33 mm (data for two Uruguayan samples; Langguth and Abella 1970). Dorsal color brownish-orange becoming clearer toward sides; whitish spots always present below and behind ears; ventral color pale brownish-orange, with whitish spots present in both inguinal and axillary regions; collar of pale brownish-orange present; basal portion of brown hairs dark gray, and that of whitish ones, white. Melanistic individuals reported for some populations (Fernandes et al. 2007). Skull much more curvilinear than that of *C. minutus*, with dorsal surface of frontal and parietal bones with small bulges and with temporal crests that converge near midline posteriorly without forming sagittal crest; nasals dorsally and longitudinally convex; dorsal border of foramen magnum with two processes, occasionally fused into single one; zygomatic arch dorsoventrally wider than in *C. minutus*; longitudinal crest on exterior face of zygomatic arch curvilinear with inferior concavity; ventral root of zygomatic process of maxilla almost perpendicular to major axis of skull; dorsal root approximately vertical; interpremaxillary foramen (J. E[ric]. Hill 1935) inconspicuous. Radius of arch described by incisors wider than in *C. minutus* such that incisors more proodont. Baculum relatively short (mean length 6.4 mm), broad, with proximal and distal tips rounded. Glans penis with sharp and convex dorsal spines, rounded and concave ventral spines; spine density 17.5/mm²; mean spine length 189.0 ± 23.9 μm s.d. (Rocha-Barbosa et al. 2013).

DISTRIBUTION: *Ctenomys torquatus* is distributed from southwestern Brazil in the state Rio Grande do Sul into neighboring Uruguay, in lowlands with fields and gallery forests termed *campos sulinos* (T. R. O. Freitas 1995b, 2006, Fernandes et al. 2007, 2009).

SELECTED LOCALITIES (Map 479): BRAZIL: Rio Grande do Sul, Alegrete (Fernandes et al. 2009), Butiá

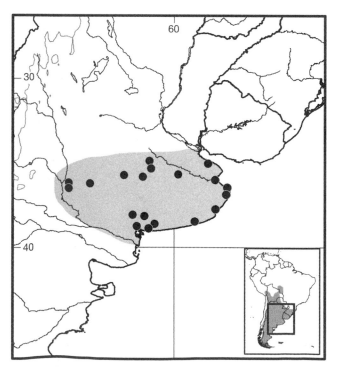

Map 478 Selected localities for *Ctenomys talarum* (●). Contour line = 2,000 m.

(J. da Silva, Freitas, Heuser et al. 2000), Cachoeira do Sul (Fernandes et al. 2007), Candiota (J. da Silva, Freitas, Marinho et al. 2000), Dom Pedrito (Fernandes et al. 2009), General Câmara (Fernandes et al. 2009), Itaqui (Fernandes et al. 2009), Pelotas (J. da Silva, Freitas, Marinho et al. 2000), Quaraí (Fernandes et al. 2009), Rio Grande (Fernandes et al. 2009), Santa Maria (Fernandes et al. 2009), Taim (T. R. O. Freitas and Lessa 1984), Uruguaiana (Fernandes et al. 2009). URUGUAY: Paysandú, Guabiyú (T. R. O. Freitas and Lessa 1984); Rivera, Cuñapirú (T. R. O. Freitas and Lessa 1984); Taquarembó, Ansina (T. R. O. Freitas and Lessa 1984).

SUBSPECIES: *Ctenomys torquatus* is monotypic, although it is chromosomally polytypic (see Remarks).

NATURAL HISTORY: Individuals of this species always construct burrows in sandy soils in open fields (Travi 1983), with the species found in cultivated fields, pastures, and vacant lots within the city of Montevideo (Barlow 1969). Burrow temperature is maintained constantly at 20–22°C (Weir 1974a). Gestation lasts 107 days on average (Weir 1974b), and litter size is 2–3 young (Nowak 1999). This species has been used as a native rodent monitor for environmental toxicity (J. da Silva, Freitas, Marinho et al. 2000; J. da Silva, Freitas, Heuser et al. 2000).

REMARKS: The type locality of *C. torquatus* is uncertain. Lichtenstein (1830) wrote that the animal came from "the southern provinces of Brazil on the shores of the Uruguay [River]." Moojen (1952b:188) repeated this same generalized locality as "margens do rio Uruguai, e estados do sul do Brasil." Reig and Kiblisky (1969) regarded Maldonado, Uruguay, as the type locality, presumably based on the skull of "*Ctenomys torquatus*" from this locality illustrated by Nehring (1900c:203). However, only *C. pearsoni*, and not *C. torquatus*, occurs along the coast of Uruguay. And, finally, Langguth and Abella (1970) suggested that the type specimen came from southern Brazil near Lages, on the Alto Rio Pelotas in Santa Catarina state, but *C. torquatus*, as defined herein, does not occur in that area (see T. R. O. Freitas 1995, 2006).

T. R. O. Freitas and colleagues (T. R. O. Freitas and Lessa 1984; T. R. O. Freitas 1995, 2006; Fernandes et al. 2007, 2009) restricted *C. torquatus* to those populations in northern Uruguay and Rio Grande do Sul state, Brazil, with $2n = 40$–46, $FN = 72$, karyotypes. Samples allocated to this species from Entre Rios province, Argentina, and southern Uruguay with $2n = 56, 64, 68$, and 70 (Reig et al. 1966; Reig and Kiblisky 1969; Kiblisky et al. 1977) are cranially distinct from the $2n = 40$–46 forms and were incorrectly assigned to *C. torquatus* (T. R. O. Freitas and Lessa 1984). A. F. Novello and Lessa (1986) subsequently identified the $2n = 56$ and $2n = 70$ individuals as *C. pearsoni* (see that account).

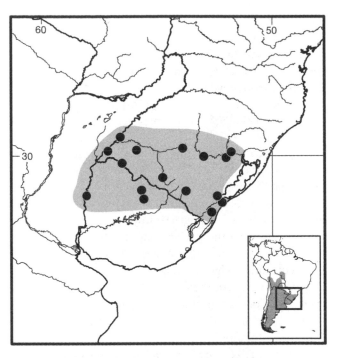

Map 479 Selected localities for *Ctenomys torquatus* (●).

Ctenomys tuconax Thomas, 1925
Robust Tuco-tuco

SYNONYM:

Ctenomys tuconax Thomas, 1925c:583; type locality "Concepción, 500 m," Departamento Concepción, Tucumán, Argentina, 27°20′S, 63°35′W.

DESCRIPTION: Among largest and most robust species in genus; total length can reach 330 mm and mass nearly 600 g (A. I. Medina et al. 2007). Fur of medium length and rather thin. General color above uniformly dark chestnut, and back lacks darkened median line; under parts similar in color to back but paler, in some specimens approaching cinnamon; axillary white patches rarely present; head like body in color, without special markings, although top of head may be slightly darker and chin paler; hands and feet thinly haired and dull whitish above; tail pale brown. Skull very large, about equal in size to specimens of *C. opimus*, which it also resembles in shape; upper surface convex; supraorbital edges and processes, although well developed, not raised to level higher than midline of skull; nasals of medium size, not markedly narrowed posteriorly; zygomatic arches robust but not especially expanded laterally, nearly same degree as auditory meatal breadth; postorbital process of jugal of average size with small and shallow but fairly well-defined fossa on surface of bone in front of process; lambdoidal crest well developed, sharply defined, and projects forward mesially, then backward and then again forward laterally to form distinct W across skull; mesopterygoid fossa narrow and extends forward to level of posterior third of M2; bullae not especially developed, less inflated than in *C. opimus*. Upper incisors broad, heavy, and slightly proodont, their angle about 107°. Cheek teeth strongly developed with anterior teeth proportionally larger.

DISTRIBUTION: *Ctenomys tuconax* occurs in the *Yungas* ecoregion of Tucumán province, Argentina. The species apparently inhabits two disjunct population nuclei in the Nevados de Aconquija, the southern extension of the Calchaquies Valleys, and the easternmost mountain range before the Chaco-Pampean flats, one at about 3,000 m elevation, and the other (including the type locality where the species may be extinct) in the lower Aconquija, between 262 and 469 m. Although Thomas (1925c) gives the elevation as 500 m, this is either an error or the species no longer occurs at low elevations.

SELECTED LOCALITIES (Map 480): ARGENTINA: Tucumán, Concepción (type locality of *Ctenomys tuconax* Thomas), El Infiernillo (MSUMMR 19212), La Calera (A. I. Medina et al. 2007).

SUBSPECIES: *Ctenomys tuconax* is monotypic.

NATURAL HISTORY: No ecological data are available.

REMARKS: Thomas (1925c:584) recognized this species on the basis of its large size and color, "being far larger than *C. tucumanus* and others of this neighbourhood, and is, indeed, but little surpassed by *C. boliviensis*. From *C. opimus* of the Bolivian plateau, which it most resembles in skull characters, it is distinguished by its uniform chestnut-

colour, widely different from that of the pale Bolivia animal." Diploid chromosome number varies between 58 and 61, FN = 80 (Reig and Kiblisky 1968a, 1969; Slamovits et al. 2001). Sperm morphology is the simple asymmetric type (Bidau 2006).

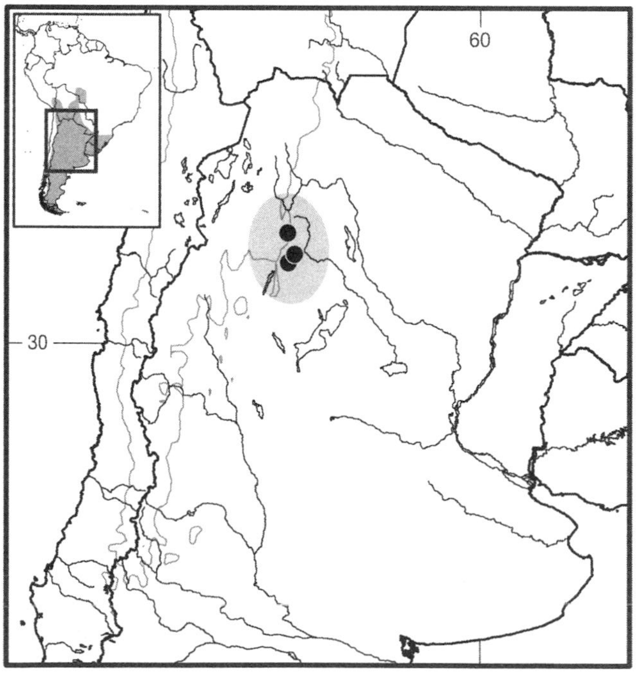

Ctenomys tucumanus Thomas, 1900
Tucumán Tuco-tuco

SYNONYMS:

Ctenomys tucumanus Thomas, 1900f:301; type locality "Tucumán [San Miguel de Tucumán]. Altitude 450 m," Departamento Capital, Tucumán, Argentina, 26°49′S, 65°13′W.

Ctenomys mendocinus tucumanus: Cabrera, 1961:553; name combination.

DESCRIPTION: Small to medium; external dimensions of holotype (old adult male) head and body length 172 mm, tail length 71 mm, and hindfoot length (with claw) 30.5 mm. General color above brownish fawn, with faint reddish suffusion; midline of face blackish; cheeks brownish fawn like back, with faint and lighter patch below the ear; under parts pale buffy, hairs gray basally; large white axillary and inguinal patches present, former almost extending across chest; upper surfaces of hands well haired and whitish, those of hindfeet nearly naked with few hairs white; tail practically naked, with few whitish hairs forming slight terminal crest. Near nakedness of hindfeet and tail possibly due to age and wear. Skull broad and flattened, much more so than in *C. mendocinus*; nasals evenly tapering backward, and terminating behind level of antorbital bridge; interorbital region and braincase flat, short, and broad no discernible interparietal; mesopterygoid fossa broad and open, more so than in other species.

DISTRIBUTION: *Ctenomys tucumanus* is known only from the type locality and immediate environs in Tucumán province, Argentina.

SELECTED LOCALITIES (Map 481): ARGENTINA: Tucumán, Camino Ticucho (Ortells 1995), San Miguel de Tucumán (type locality of *Ctenomys tucumanus* Thomas), Yerba Buena (J. R. Contreras 1999).

SUBSPECIES: *Ctenomys tucumanus* is monotypic.

NATURAL HISTORY: Populations occupy deep, humid soils in the piedmont and forest borders of Tucumán city (J. R. Contreras, pers. comm.). No other ecological, behavior, or other population data are known for this species.

REMARKS: Thomas (1900f) believed that *C. tucumanus* was most closely related to *C. talarum* from the coastal plain of Buenos Aires province, although it is geographically closest to *C. mendocinus*. Cabrera (1961), however, listed *tucumanus* as a valid subspecies in his concept of a polytypic *C. mendocinus*. Alternatively, Parada et al. (2011) placed this species in their "*tucuma-*

Map 480 Selected localities for *Ctenomys tuconax* (●). Contour line = 2,000 m.

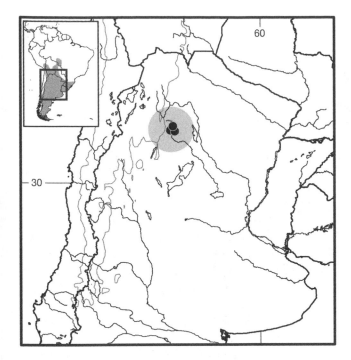

Map 481 Selected localities for *Ctenomys tucumanus* (●). Contour line = 2,000 m.

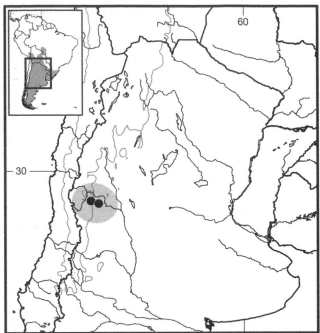

Map 482 Selected localities for *Ctenomys tulduco* (●). Contour line = 2,000 m.

nus" group, along with *C. argentinus*, *C. lato*, and *C. occultus*, based on mtDNA sequences. The karyotype of *C. tucumanus* is $2n = 28$ and FN = 52 (Reig and Kiblisky 1968a), and its sperm is of the simple symmetric type (Bidau 2006).

Ctenomys tulduco Thomas, 1921
Sierra Tontal Tuco-tuco, Tulduco

SYNONYMS:

Ctenomys tulduco Thomas, 1921i:218; type locality "Los Sombreros, Sierra Tontal, Alt. 2700 m," Departamento Calingasta, San Juan, Argentina, 31°33′S, 69°11′W.

Ctenomys fulvus tulduco: Cabrera, 1961:549; name combination.

DESCRIPTION: Moderate sized with head and body length of holotype (adult male) 190 mm, tail length 69 mm, and hindfoot length 32.6 mm. General color above drabby gray, similar to that of *C. johannis*, and not as warm as that of *C. coludo*; ventral color strongly drabby as in *C. coludo* but unlike light undersurface of *C. johannis*; inconspicuous dull nasal patch present; tail short, with line along upper side black or blackish, varying in definition but always more pronounced that in allied species. Skull similar to that of *C. johannis*, but smaller and with rather smaller bullae, which are still far larger than those of *C. mendocinus*.

DISTRIBUTION: *Ctenomys tulduco* is known only from the vicinity of its type locality.

SELECTED LOCALITIES (Map 482): ARGENTINA: San Juan, Los Sombreros (type locality of *Ctenomys tulduco* Thomas), Pedernal (Thomas 1921i).

SUBSPECIES: *Ctenomys tulduco* is monotypic.

NATURAL HISTORY: No biological or ecological data are available for *C. tulduco*.

REMARKS: This species is similar, and possibly allied, to *C. coludo* although it is smaller (Thomas 1921i). Neither the karyotype nor sperm morphology is known.

Ctenomys validus J. R. Contreras, Roig, and Suzarte, 1977
Guaymallén Tuco-tuco

SYNONYM:

Ctenomys validus J. R. Contreras, Roig, and Suzarte, 1977:160; type locality "El Algarrobal, Médanos del Borbollón, Departamento Guaymallén, Provincia de Mendoza," Argentina, 32°49′S, 68°46′W, 775 m.

DESCRIPTION: Relatively large with total length and mass ranging from 281–314 mm and 187–341 g in males to 260–278 mm, 176–218 g in females (measurements of male holotype total length 296 mm, mass 312.4 g). General color grayish brown to yellowish brown, ventrally yellowish; conspicuous clear band between ears and through throat; hairs of feet and hands white, as in *C. coludo* and *C. johannis*; tail bicolored, yellowish ventrally, and with black longitudinal stripe dorsally. Skull large and robust with well-developed crests similar to those of *C. johan-*

nis; bullae large and moderately inflated; zygomatic arches strong; auditory meatus conspicuous; mesopterygoid fossa V-shaped; premaxillary bones extend posteriorly beyond nasals much more so than in *C. johannis*; interparietal sutures not visible in adult individuals; nasals widest at tips, narrowing posteriorly; skull achieves greatest depth at level of PM4; interpremaxillary foramen always visible; postorbital processes well developed, with frontal bones narrowing markedly behind. Upper cheek teeth moderately sized; occlusal surface of M3 cylindrical to subcylindrical in shape. In lower cheek teeth, pm4 similar to or smaller than m1. Baculum relatively large (length ca. 9 mm) and well ossified, ventrally flat, and slightly concave dorsally.

DISTRIBUTION: *Ctenomys validus* is only known from the type locality.

SELECTED LOCALITIES (Map 483): ARGENTINA: Mendoza, El Algarrobal (type locality of *Ctenomys validus* J. R. Contreras, Roig, and Suzarte).

SUBSPECIES: *Ctenomys validus* is monotypic.

NATURAL HISTORY: No biological or ecological data are available for *C. validus*.

REMARKS: J. R. Contreras et al. (1977) believed that *C. validus* was related to *C. johannis* Thomas, and with that species plus *C. coludo* Thomas, *C. famosus* Thomas, *C. tulduco* Thomas, and possibly *C. fulvus* Philippi formed an arc of precordilleran, inter-sierran, and Andean species that extended from Mendoza province north and west to the Atacama of Chile. This hypothesis has yet to be tested by any character set. Karyotype and sperm type are unknown.

Ctenomys viperinus Thomas, 1926
Monte Tuco-tuco, Vipos Tuco-tuco

SYNONYMS:
Ctenomys viperinus Thomas, 1926d:605; type locality "Ñorco, near Vipos, 2500 m," Departamento Trancas, Tucumán, Argentina, 26°29′S, 65°22′W, 2,500 m.
Ctenomys knighti viperinus: Cabrera, 1961:550; name combination.

DESCRIPTION: Medium size with head and body length of holotype (adult male) 213 mm, tail length 76 mm, and hindfoot length 36 mm. General color warm brown above, paler and more drab below; many specimens with blackened muzzles and crowns, others lack these color features; majority of specimens with white axillary patches, and some have white inguinal spots. Skull somewhat small, with zygomata decidedly expanded and antorbital foramen widely open; bullae rather narrow; paroccipital processes less forwardly expanded on their lower surfaces. Upper incisors broad, enamel surfaces rather dark orange; incisive index of type 105°. PM4 considerably wider than M1.

DISTRIBUTION: *Ctenomys viperinus* is known from the type locality and neighboring areas between 320 and 2,500 m altitude in the *Yungas* ecoregion of Tucumán province, Argentina.

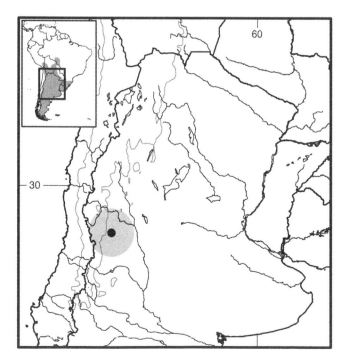

Map 483 Single known locality for *Ctenomys validus* (●). Contour line = 2,000 m.

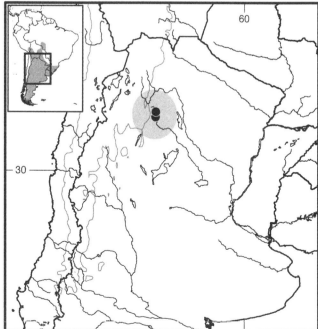

Map 484 Selected localities for *Ctenomys viperinus* (●). Contour line = 2,000 m.

SELECTED LOCALITIES (Map 484): ARGENTINA: Tucumán, Ñorco (type locality of *Ctenomys viperinus* Thomas), Villa San Javier (A. I. Medina et al. 2007).

SUBSPECIES: *Ctenomys viperinus* is monotypic.

NATURAL HISTORY: No biological or ecological information is available for *C. viperinus*.

REMARKS: No data on karyotype or sperm morphology are available.

Ctenomys "yolandae" J. R. Contreras and Berry, 1984
Santa Fe Tuco-tuco, Yolanda's Tuco-tuco

SYNONYM:

Ctenomys yolandae J. R. Contreras and Berry, 1984:75; type locality, "Las Palmas, Departamento General Obligado, Provincia de Santa Fe," Argentina, 29°25′S, 59°40′W, 50 m; *nomen nudum* (see Remarks).

DESCRIPTION: Relatively small species (total length, ca. 230 mm) that has yet to be diagnosed or adequately described.

DISTRIBUTION: According to J. R. Contreras and Berry (1984), the species occurs in eastern Santa Fe province along the Paraná and San Javier rivers. This distribution is disjunct, separated by a low, floodable area unsuitable for colonization by tuco-tucos.

SELECTED LOCALITIES (Map 485; from J. R. Contreras, pers. comm., except at noted): ARGENTINA: Santa Fe, Cacique Ariacaiquín, Coronda, Helvecia, Las Palmas (J. R. Contreras and Berry 1984), Reconquista, San José del Rincón.

SUBSPECIES: *Ctenomys "yolandae"* is monotypic.

NATURAL HISTORY: No ecological data are available.

REMARKS: Woods and Kilpatrick (2005:1570) listed this taxon as a valid species. However, the name *Ctenomys* "yolandae" appeared originally only in a short abstract of presentations of a scientific meeting, and a formal designation of a holotype, diagnosis, and description has yet to be published. The name "yolandae" thus does not meet the criteria established by the International Code of Zoological Nomenclature (ICZN 1999:Art. 8) and is appropriately regarded as *nomen nudum*. Nevertheless, I list this taxon here as *C.* "yolandae" because this tuco-tuco is a distinct entity at both karyotypic (Ortells et al. 1990; Bidau et al. 2005) and molecular (Mascheretti et al. 2000) levels. In the latter case, mtDNA studies determined the phyletic proximity of *C.* "yolandae" with *C. bergi* and *C. bonettoi*, as might be expected from their morphological similarities. Furthermore, *C.* "yolandae" possesses a unique type of sperm, the complex asymmetric morphology, not found in any other tuco-tuco (Vitullo et al. 1988). The diploid number is $2n = 50$, FN = 78, in specimens from the type locality (Ortells et al. 1990), and $2n = 50$, FN = 67, 70 (Bidau et al. 2005) elsewhere. A proper description of this species is needed to validate the name.

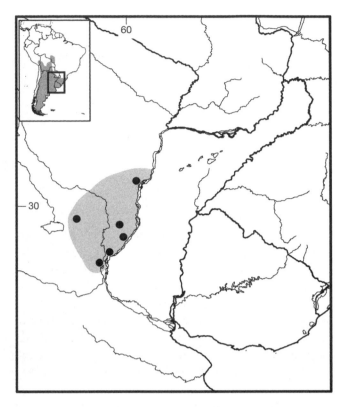

Map 485 Selected localities for *Ctenomys "yolandae"* (●). Contour line = 2,000 m.

Family Echimyidae Gray, 1825
Louise H. Emmons, Yuri L. R. Leite, and James L. Patton

The family Echimyidae is the most diverse of the South American hystricognath rodent families, both with regard to numbers of taxa and variety of body plans (see, for example, S. I. Perez et al. 2009; Upham and Patterson 2012; P.-H. Fabre et al. 2012). The taxonomic history has been chaotic, with a number of generic names proposed, several abandoned, and the contents of both the family and genera highly unstable. Recent reviews and phylogenetic analyses (Lara et al. 1996; Y. L. R. Leite and Patton 2002; G. A. S. Carvalho and Salles 2004; Emmons 2005; Galewski et al. 2005; Iack-Ximenes et al. 2005a,b; Upham and Patterson 2012; P.-H. Fabre et al. 2012; Upham et al. 2013) have clarified relationships, and thus membership within the family, as well as both genus and species boundaries. Although great clarity has been achieved by these studies, much remains to be understood about the composition and evolutionary history of the family at all taxonomic levels.

The family Echimyidae is of ancient origin in South America, found among the oldest fossil rodents, with the earliest records from the mid-Eocene (Antoine et al. 2012) and Oligocene (B. Patterson and Wood 1982; Vucetich and Verzi 1991; Deschamps et al. 2012). Recent work documents a diverse fauna of largely grassland (pampa) echimyids in the Miocene (Vucetich et al. 1993) as well as a rich assemblage of taxa from presumptively tropical forest habitats at the eastern base of the Andes (Frailey and Campbell 2004). Molecular-based estimates place the diversification of extant lineages in the Miocene, between 18.8 mya (95% credibility interval 20.2–17.5 mya; P.-H. Fabre et al. 2012) and 16.0 mya (21.7–11.3 mya; Upham et al. 2013).

Unfortunately, most extant taxa lack a fossil record altogether, and few of these genera can be tied phyletically to any of the more than 20 fossil genera currently recognized (G. A. S. Carvalho and Salles 2004; but see Candela and Rasia 2012; Deschamps et al. 2012; and Olivares et al. 2012 for current efforts to connect fossil and extant genera within a phylogenetic framework). The family is currently divided into three or four subfamilies. The living Eumysopinae (or Heteropsomyinae, see Patton and Reig 1990) includes six extant genera of terrestrial species and two of arboreal ones, with greatest generic diversity among grassland or shrubland taxa, but greatest species number in the rainforest genus *Proechimys*. There are three genera of specialized arboreal folivores in the Dactylomyinae, all denizens of lowland and montane forests. The Echimyinae contains four to nine genera of arboreal rats of forested or wooded habitats; we recognize nine genera here. Following recent molecular studies (Galewski et al. 2005; Opazo 2005; Rowe et al. 2010; Upham and Patterson 2012; P.-H. Fabre et al. 2012; Upham et al. 2013), we regard the Myocastorinae as a fourth subfamily of the Echimyidae.

Phylogenetic support for the traditional subfamily structure as well as relationships among genera within each remains uncertain. Part of this uncertainty results from the lack of inclusion of all extant genera in any character-based analysis, be it morphological or molecular, and also likely from the apparent very rapid diversification of the Recent lineages (Lara et al. 1996; Y. L. R. Leite and Patton 2002). For example, analyses of both morphological (G. A. S. Carvalho and Salles 2004; Emmons 2005; Candela and Raisa 2012) and molecular traits (Lara et al. 1996; Y. L. R. Leite and Patton 2002; Galewski et al. 2005; P.-H. Fabre et al. 2012; Upham et al. 2013) have so far failed to recover either the Eumysopinae or Echimyinae as monophyletic units. Although each of these analyses supports the monophyly of the Dactylomyinae, this group of three genera appears nested among the remaining extant echimyine genera. Furthermore, the most recent molecular analyses support the phylogenetic of the spiny tree rats *Mesomys* and *Lonchothrix*, two genera that combine echimyine and eumysopine characters (see those accounts) with the echimyine tree rats (P.-H. Fabre et al. 2012; Upham et al. 2013). Thus, the allocation of genera into phylogenetically well-supported suprageneric taxa remains a challenge for

students of echimyid rodents. Given the important representation of teeth in the fossil record, the posited occlusal character homologies proposed by Candela and Rasia (2012) are noteworthy, especially for the integration of the diverse fossil and extant taxa into a coherent set of hypothesis of relationships as well as rates and timing of lineage diversification.

The following synopsis of the nomenclatural history of the family Echimyidae is based largely on G. A. S. Carvalho and Salles (2004) and Emmons (2005). Zimmermann (1780) described the first echimyid, *Myoxus chrysurus*, placing it within a European rodent genus as was typical of that period. *Echimys* was the first echimyid genus, defined by F. Cuvier in 1809 and in which he placed the "*lérot à queue dorée*" of Allamand (= *Echimys chrysurus*) and the "*rat épineux*" of Azara (= *Euryzygomatomys spinosus*). Desmarest (1817b), crediting names to É. Geoffroy St.-Hilaire, described six new echimyids, placing these along with *Echimys chrysurus* (which he called *E. cristatus*) in the genus *Echimys*. Desmarest's concept of *Echimys* thus included both the type species of *Echimys* and six additional species now allocated to other genera: *E. dactylinus* (a *Dactylomys*), *E. spinosus* (a *Euryzygomatomys*), *E. hispidus* (a *Mesomys*), *E. didelphoides* (a *Makalata*), *E. cayennensis* (a *Proechimys*), and *E. setosus* (a *Trinomys*). Isidore Geoffroy St.-Hilaire (1838, 1840) divided the genus *Echimys*, as viewed by Desmarest, into three genera: *Dactylomys* (including only *D. dactylinus*), *Echimys* (including *E. setosus*, *E. albispinus*, *E. myosurus* [all three *Trinomys*], *E. cayennensis* [a *Proechimys*], *E. hispidus* [a *Mesomys*], and *E. spinosus* [a *Euryzygomatomys*]), and *Nelomys* (including *N. cristatus* [= *Echimys chrysurus*] and *N. paleaceus* [an *Echimys*], *N. semivillosus* [a *Pattonomys*], *N. blainvillii* [a *Phyllomys*], and *N. armatus* and *N. didelphoides* [both *Makalata*]). In these two works, I. Geoffroy St.-Hilaire established the basis for the subsequent division of extant echimyids into the three subfamilies (Dactylomyinae, Echimyinae, and Eumysopinae) recognized in most modern classifications.

Tate (1935) sifted though 150 years of publications in a heroic review of the taxonomic history of all caviomorph rodents. He reconciled many nomenclatural discrepancies and proposed a revised generic classification, but he did not provide an extensive review of characters. His interpretation of the Echimyidae, in which he segregated the species of *Echimys* largely on the degree of hairiness of the tail, left many unresolved questions. With the major exception of Ellerman (1940), who placed nine subfamilies within his concept of the Echimyidae (including both New World [Abrocomidae, Capromyidae, Octodontidae] and Old World [Petromuridae, Thryonomyidae] groups now widely recognized as separate families), Tate's classification was adopted by most systematists until Cabrera (1961) again reviewed the family along with all other South American mammals and synonymized many names with little or no explanation. More recently, Emmons and Feer (1990, 1997) proposed some revisions of genera in field guide format, without explanation, including segregating *Nelomys* (now *Phyllomys*) from *Echimys*, and allocated additional species to *Makalata*. Patton and Emmons (1985) reviewed the genus *Isothrix*, from which Emmons and Vucetich (1998) subsequently segregated a new genus, *Callistomys*. Emmons, Leite et al. (2002) and Y. L. R. Leite (2003) revised the Brazilian Atlantic tree rats *Phyllomys* and documented that the name *Nelomys* had been incorrectly applied to this taxon. Emmons (1997a) presented an outline of many of the characters that diagnose the arboreal echimyines, and she later thoroughly documented both systematic and nomenclatural conclusions of that initial report (Emmons 2005), describing two new genera (*Pattonomys* and *Santamartamys*). Iack-Ximenes et al. (2005a) placed *Loncheres grandis* Wagner into the new genus *Toromys*, while commenting on characters and other taxonomic issues particularly within *Echimys*.

G. A. S. Carvalho and Salles (2004) examined all extant as well as a rich assemblage of fossil genera and delineated 50 morphological traits that they then used to provide a phylogenetic hypothesis of generic relationships. They defined and diagnosed the crown-group Echimyidae and presented a classification encompassing both Recent and extinct genera and subfamilies based on their phylogenetic hypothesis. Independently, Emmons (2005) also undertook the characterization and phylogenetic analysis of morphological characters, and reached the same set of conclusions. She also thoroughly diagnosed the subfamilies and all genera she included within the Echimyinae, which was the emphasis of her work. Lara et al. (1996), Y. L. R. Leite and Patton (2002), Galewski et al. (2005), Upham and Patterson (2012), P.-H. Fabre et al. (2012), and Upham et al. (2013) have each examined mitochondrial and nuclear gene sequence variation among 14 to 16 extant genera. A common conclusion of both morphological and most molecular analyses has been the inclusion of *Myocastor* within the Echimyidae as the monotypic subfamily Myocastorinae. Woods and Kilpatrick (2005), in their more recent classification of the Hystricomorpha, did not follow this position. Another consistent component of both morphological and molecular analyses has been a difficulty in corroborating reciprocally monophyletic assemblages of genera within both the subfamilies Echimyinae and Eumysopinae. The three genera that comprise the Dactylomyinae (*Dactylomys*, *Kannabateomys*, and *Olallamys*) do form a monophyletic unit, but this group is otherwise nested among genera of the Echimyinae. Among the extant Eumysopinae, clades comprising *Clyomys* + *Euryzygomatomys* + *Carterodon* are consistently recovered (but see Vucetich and Verzi 1991, who do not include *Carterodon* in a sister relationship with both *Clyomys* and *Euryzygomatomys*), whereas *Proechimys* + *Hoplomys* + *Thrichomys* + *Trinomys* are linked with morphological but not always with molecular support. These two clades together lack any support as the monophyletic group that would be hypothesized by their inclusion in the same subfamily. Furthermore, while *Mesomys* + *Lonchothrix*, commonly placed within the Eumysopinae because of shared dental characters, appear to be phyletically related by molecular analyses to various echimyine and dactylomyine genera (Upham and Patterson 2012; P.-H. Fabre et al. 2012; Upham et al. 2013). Most phyletic relationships based either on morphology or gene sequences are, however, for the most part weakly supported, and a fully resolved phylogenetic hypothesis and resulting classification must await further data and analyses.

Herein, we follow G. A. S. Carvalho and Salles (2004) in their concept of the crown-group Echimyidae to include four subfamilies, two monophyletic (Dactylomyinae and Myocastorinae), one paraphyletic (Echimyinae), and the fourth apparently either paraphyletic or polyphyletic (Eumysopinae) (see also Olivares et al. 2012). We keep these apparently nonphylogenetic groupings for convenience until a more completely resolved phylogeny is forthcoming. We exclude the Antillean Capromyidae (contra Y. L. R. Leite and Patton 2002; but see Galewski et al. 2005; Kilpatrick et al. 2012; P.-H. Fabre et al. 2012; Upham et al. 2013), which appears to be the sister to the Echimyidae (Huchon and Douzery 2001; Galewski et al. 2005; Rowe et al. 2010) or part of an unresolved polytomy at the base of the echimyid radiation (P.-H. Fabre et al. 2012). We also exclude the bristle-spined porcupine *Chaetomys* (contra B. Patterson and Wood 1982; Woods 1993) because this genus unambiguously belongs to the family Erethizontidae (T. Martin 1994; G. A. S. Carvalho 2000; R. do V. Vilela et al. 2009; Voss et al. 2013). Emmons (2005) placed the three subfossorial genera (*Euryzygomatomys*, *Clyomys*, and *Carterodon*) in the tribe Euryzygomatomini within the subfamily Eumysopinae. Given the lack of documented equivalency in rank for other generic assemblages in this subfamily or others, we follow G. A. S. Carvalho and Salles (2004) and do not use tribal groupings.

The crown-group Echimyidae is diagnosed by six transformations (G. A. S. Carvalho and Salles 2004): presence of the central portion of the neolophid in dpm4; absence of the metalophid in m1–3: shallow or absent lingual opening of the mesoflexid in little-worn m1–3; shallow or absent labial opening of the mesoflexus in little-worn upper teeth;

retention of the deciduous premolar throughout life; and presence of well-developed sphenopalatine vacuities.

KEY TO THE SUBFAMILIES OF ECHIMYIDAE:

1. Large (mass >5 kg); hindfeet with interdigital webbing; pelage with long, coarse guard hairs over thick, soft underfur; maxillary teeth with two to three labial and three lingual flexi; teeth increasing in size from front to back; paroccipital process greatly elongated, vertical, and completely free from bulla . . . . Myocastorinae
1′. Smaller (mass <1 kg); hindfeet lack interdigital webbing; pelage soft or spiny and not divided into obvious elongated guard hairs covering dense underfur; maxillary teeth with typically only one lingual flexus with labial flexi varying from two to four; teeth either uniform or decreasing in size from front to back; paroccipital process short and curved to follow contour of bulla 2
2. Digits of forefeet and hindfeet elongate and slender; plantar surfaces without distinct pads; maxillary toothrows diverge posteriorly; maxillary teeth with three labial and one lingual flexi . . . . Dactylomyinae
2′. Digits of forefeet and hindfeet not elongate and slender; plantar surfaces with distinct pads; maxillary toothrows parallel; maxillary cheek teeth with variable number of labial and lingual flexi or with transverse laminar plates 3
3. Maxillary cheek teeth rounded in occlusal view; labial and lingual flexi do not coalesce to form transverse laminar plates . . . . .Eumysopinae
3′. Maxillary cheek teeth rectangular in occlusal view; labial and lingual flexi combining to form variable number of transverse laminar plates . . . . Echimyinae

Subfamily Dactylomyinae Tate, 1935
Louise H. Emmons, James L. Patton, and Yuri L. R. Leite

The Dactylomyinae comprise three genera, each confined to the major tropical forest regions of South America: *Dactylomys* is distributed in elfin forests of the eastern central Andean slopes, throughout lowland rainforest of the Amazon Basin, and gallery forests in the eastern Cerrado of central Brazil; *Kannabateomys* occurs in coastal and montane Atlantic Forest of Brazil south to Argentina; and *Olallamys* has a limited range in montane forests of the northern Andes in Colombia and Venezuela. Bamboo patches form a dominant component of the habitat and, where diet is known, all taxa feed chiefly on bamboo leaves and shoots (but see Emmons 1981).

These are moderate-sized, folivorous rodents adapted to an arboreal habitus. The pelage lacks spines or bristles. The tail is typically much longer than head and body

length, and may be naked or covered with sparse fur. Both forefeet and hindfeet lack raised, smooth plantar pads; rather, the plantar skin of both manus and pes is evenly and densely covered with tiny tubercles. The toes are elongate and slender, with the middle two digits especially elongated relative to lateral ones in *Dactylomys* and *Kannabateomys*.

The skulls of all three genera share the same essential cranial and dental characters, except as noted. Rostra are short and broad; the interorbital region is broad, approximately equal to the braincase breadth, and edged with overhanging ledges that possess weak postorbital processes. Incisive foramina are small to nearly obsolete. The palate is narrow and maxillary toothrows diverge posteriorly, more strongly so in *Dactylomys* and *Olallamys* than in *Kannabateomys*. Cheek teeth are brachydont, with four roots and with occlusal surfaces in approximately the same plane as the palate. Maxillary cheek teeth are very large, wider than the greatest palatal width, and split by three deep labial flexi and a single lingual flexus. In *Dactylomys* and *Olallamys*, the hypoflexi and mesoflexi coalesce so that the four lophs form two independent V- or Y-shaped pairs. In *Kannabateomys*, a mure connects the protoloph and metaloph so that the hypoflexi and mesoflexi do not unite, although the same general V- or Y-shaped pairs of lophs are apparent. The lower molars have two lingual flexids and one labial flexid. In *Kannabateomys*, these remain separate, but in *Dactylomys* and *Olallamys* the hypoflexids and metaflexids unite to form an anterior V-shaped loph and a single, posterior laminar loph. The alisphenoid bone has a reduced or no bony bridge extending from the foramen ovale to the basisphenoid posterior to base of parapterygoid processes. Paroccipital processes are free from bullar contact but curved over the bullae; in occipital view, these processes are angled laterally rather than vertically.

The first molecular phylogenetic study to include all three genera (Upham et al. 2013) suggests either that the Colombian *Olallamys* and the Atlantic Forest *Kannabateomys* form a sister group relative to *Dactylomys* or that the three genera are part of an unresolved trichotomy. The basal age of the radiation has been posited to be as old as 10.2 mya (Upham and Patterson 2012) or as recent as 3.5 mya (Leite and Patton 2002).

KEY TO THE GENERA OF DACTYLOMYINAE:

1. Maxillary toothrows nearly parallel; zygomatic arches widest anteriorly; confined to Atlantic Forest . . . . *Kannabateomys*
1′. Maxillary toothrows converge sharply anteriorly; zygomatic arches widest posteriorly; do not occur in Atlantic Forest 2

2. Digits 3 and 4 of forefeet and hindfeet elongated, with nails on all digits; greatest breadth across zygomatic arches >30 mm; minimum interorbital breadth >16 mm; found on east slope of central Andes, Amazon Basin, and gallery forests of Cerrado. . . . .*Dactylomys*

2'. Digits 3 and 4 of forefeet and hindfeet not especially elongated, with claws on all digits; greatest breadth across zygomatic arches <30 mm; minimum interorbital width <15.5 mm; range in northern Andes . . *Olallamys*

Genus *Dactylomys* I. Geoffroy St.-Hilaire, 1838

Dactylomys are moderate-sized (mass up to 700 g) arboreal rats with nonspiny fur, long tails (typically >120% head and body length), feet with elongated central toes and pointed nails on all digits, and large, partially laminated cheek teeth that greatly restrict the anterior palate and diverge sharply posteriorly. There is a gap between the fourth and fifth toes of the hindfoot that permits the fifth toe to be used in opposition to the others to grasp thin stems, which may also be held between the middle toes. Despite their conspicuous size and morphology, their remarkable and loud vocalizations (Emmons 1981), and the strong, musky odor that permeates their living areas, *Dactylomys* are among the more poorly known and only rarely collected members of the family. They are folivores that do not readily enter traps, because of the baits commonly used or for other, unknown reasons. Their weak eye-shine and slow, methodical movements make them difficult to see at night (Emmons 1981; Emmons and Feer 1997). The vernacular name bamboo rat stems from their characteristic association with patches of bamboo, apparently a major food item, but they also occur in riverine vegetation, especially in canebrakes, and in upland forest on the rich soils of western Amazonia as well as on the intermediate slopes of the eastern Peruvian Andes. They are typically absent from blackwater riverine communities.

Two or three species are recognized in the recent literature (Woods 1993; Emmons and Feer 1990, 1997; Patton et al. 2000). One of these (*D. peruanus*) occurs in the upper elevation elfin forests on the eastern slope of the Andes in Peru and Bolivia, and is unique in its fully furred tail. Two other species occur in the lowland rainforest of the Amazon Basin and gallery forests south of the Amazon, but character variation and geographic ranges of both are uncertain. Patton et al. (2000) reviewed the characters and geographic distribution of populations in western Amazonian Brazil, and recognized two species (*D. dactylinus* and *D. boliviensis*) based on concordant craniodental characters, vocal call-note, and mtDNA sequences. We follow this assessment in the accounts that follow, but the degree

of concordant color and vocal characters appears to vary geographically. The selected localities and maps we provide are based on the assignment of museum specimens by color pattern because this character can be assessed for all such specimens. It remains to be determined whether color alone is an adequate diagnosis for these two species; a complete geographic review of all characters, but especially the relationship between call structure and morphological and genetic attributes, remains an important area of future research.

SYNONYMS:

Echimys: Desmarest, 1817b:54; part (description of *dactylinus*); not *Echimys* F. Cuvier.

Dactylomys I. Geoffroy St.-Hilaire, 1838a:888; type species *Dactylomys typus* I. Geoffroy St.-Hilaire (= *Echimys dactylinus* Desmarest, 1817b:54), by original designation.

Lachnomys Thomas, 1916e:299; defined as a subgenus of *Dactylomys*; type species *Dactylomys peruanus* J. A. Allen, by original designation.

Dactylinus Anthony, 1920:83; incorrect subsequent emendation of *Dactylomys* I. Geoffroy St.-Hilaire.

REMARKS: As noted by Emmons, Leite et al. (2002), the original generic description of *Dactylomys* by Geoffroy St.-Hilaire was published in four different French journals in 1838. Following Article 24.2.1 of the International Code (ICZN 1999), Emmons, Leite et al. (2002) fixed the account in the *Comptes Rendus Hebdomadaires des Sèances de L'Academie des Sciences* 6(26):884–888 as the first publication.

KEY TO THE SPECIES OF *DACTYLOMYS*:

1. Tail naked; pelage coarse. 2

1'. Tail furred to tip; pelage long, soft, and dense . . . . *Dactylomys peruanus*

2. Upper parts fairly uniform dark olivaceous, thighs and tail base little, if at all, tinged rusty, top of head with white banded hairs forming pale streaks above and below the eyes, and usually with an olivaceous-black stripe darkening from crown to nape; undercolor of back blackish . . . . *Dactylomys boliviensis*

2'. Top of head and muzzle pale to medium brown with unbanded hairs above eyes and to well down neck, gradually darkening to the nape; rear of thighs and tail-base rusty orange; undercolor of back black to chestnut . . . . *Dactylomys dactylinus*

Dactylomys boliviensis Anthony, 1920
Bolivian Bamboo Rat

SYNONYM:

Dactylomys boliviensis Anthony, 1920:82; type locality "Mission San Antonio, Rio Chmore [*sic*, Chimoré], Prov. Cochabamba, Bolivia; altitude 1300 feet."

DESCRIPTION: Large olivaceous rat (head and body length about 270 mm), with very long tail (about 150% of head and body length), long hindfeet (>60 mm), and short ears (about 20 mm). Dorsal color tones more muted than in *D. dactylinus*; grizzled grayish-olivaceous streaked with black, becoming paler on sides, and gradually merging with sparsely furred, pure white venter; face distinctly grizzled gray with white banded hairs above and below eyes and with olivaceous black stripe bordered by pale-tipped hairs extends on midline from nose posteriorly between ears to nape, darkening progressively to blackish brown from above eyes to neck, but in some samples black not pronounced. Mystacial vibrissae in some samples short, extending only to one-half length of superciliary vibrissae, in contrast to much longer mystacial vibrissae of *D. dactylinus*. Tail with well-furred base that extends for about 65 mm; it is heavily scaled and naked in appearance from furred base to tip, but scales appear finer, averaging six annuli per cm near tail base and 7.5 per cm near tip, and less distinctly pentagonal than those of *D. dactylinus*; each caudal scale-hair distinctly dark brown or black over basal half, becoming completely colorless along terminal half to third of tail; median hair extends 1.5–2.0 scale rows. Tail bicolored, especially over anterior two-thirds of its length, but with broad, dark dorsal surface becoming gradually paler toward tip.

Skull large, rostrum short and broad, and supraorbital ledges proportionally broader than in other species but, with these others, form subtriangular postorbital processes. Skull differs from that of *D. dactylinus* in that paroccipital processes more anteriorly directed, following curvature of tympanic bullae, and postorbital process of zygomatic arch composed primarily of jugal. Cheek teeth of both upper and lower jaws are as in the other species.

DISTRIBUTION: Southwestern margins of the Amazon Basin, from central Peru south to central Bolivia, and east as far as the central Rio Juruá in western Brazil. This range is largely, but not exclusively, coincidental with the large area of bamboo (primarily *Guadua* spp.) dominated forest in the western Amazon (see J. M. Jacobs and von May 2012). In general, this species seems to occur at higher elevations than *D. dactylinus* or is restricted to the headwaters where the two species co-occur in the same river basin.

SELECTED LOCALITIES (Map 486): BOLIVIA: Cochabamba, Mission San Antonio, Rio Chimoré; La Paz, La Reserva (AMNH 264884). BRAZIL: Acre, Fazenda Santa Fé (= Flora), left bank Rio Juruá (Patton et al. 2000). PERU: Ayacucho, Hda. Luisiana (LSUMZ 16761); Cusco, Consuelo, 15.9 km SW of Pilcopata (FMNH 175250), Quincemil (FMNH 65680), 2 km SW of Tangoshiari (USNM 588070); Madre de Dios, Tambopata (Emmons

1981); Ucayali, Balta, Río Curanja (LSUMZ 12422), Sepahua (MVZ 173099).

SUBSPECIES: *Dactylomys boliviensis* is monotypic.

NATURAL HISTORY: Summaries of reproduction, ecology, and behavior of *D. boliviensis* are given in Emmons (1981; Peruvian animals mentioned are now assigned to this species) and Dunnum and Salazar-Bravo (2004). Habitat includes bamboo and cane thickets typically along river margins (Salazar-Bravo et al. 1994; Patton et al. 2000). Adaptations for arboreal browsing are detailed by Emmons (1981). High pulse number that ranges from a mean of 20 to 44 characterize staccato vocalizations (Emmons 1981; M. N. F. da Silva and Patton 1993). Pregnant females were recorded in July at La Reserva in Bolivia (Salazar-Bravo et al. 1994).

REMARKS: Calouro (1999) reported bamboo rats from four localities on the Brazilian side of the Serra do Divisor in the upper Rio Juruá basin of Acre state, assigning these sightings to *D. dactylinus*. These records most likely represent *D. boliviensis*, given the analyses presented by Patton et al. (2000) which were also based on specimens from the upper Rio Juruá.

Dunnum et al. (2001) described a chromosome complement with $2n = 118$ and $FN = 168$, the highest diploid number known for any mammal. Didier (1962) figured and described the baculum of a specimen from southeastern Peru, which averaged 3 mm longer than that of a specimen of *D. dactylinus* from northern Peru.

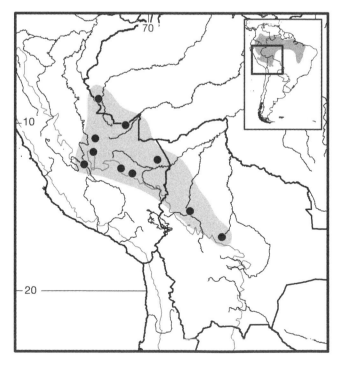

Map 486 Selected localities for *Dactylomys boliviensis* (●). Contour line = 2,000 m.

Dactylomys dactylinus (Desmarest, 1817)
Amazon Bamboo Rat

SYNONYMS:

Echimys dactylinus Desmarest, 1817b:54; type locality not given; restricted to "Upper Amazon area" (Thomas 1912d:88); see Remarks.

Dactylomys dactylinus: I. Geoffroy St.-Hilaire, 1838a:888; first use of current name combination.

Dactylomys typus I. Geoffroy St.-Hilaire, 1838a:888; type locality "l'Amérique méridionale, probablement le Brésil"; a renaming of *Echimys dactylinus* Desmarest, 1817.

D[*actylomys*]. *dactylinus*: J. A. Allen, 1900a:221; name combination.

Dactylomys dactylinus canescens Thomas, 1912d:87; type locality "Itacoatiara, below Manaos [=Manaus], Middle Amazons," Amazonas, Brazil.

D[*actylomys*]. *dactylomys* J. A. Allen, 1914c:389; incorrect subsequent spelling of *dactylinus* Desmarest.

Dactylomys dactylinus dactylinus: J. A. Allen, 1916b:208; name combination.

Dactylomys dactylinus modestus Lönnberg, 1921:38; type locality "on the banks of the river Curaray, El Oriente, Ecuador, 1,000 ft," presumably Pastaza.

DESCRIPTION: Large olivaceous to yellowish rat; head and body length about 315 mm, with long tail (about 125% head and body length), long hindfeet (>60 mm), and short ears (about 20 mm). Face pale tan-cinnamon or umber-brown, without white bands on hairs above eyes; self-colored hairs extend above and below eyes, nose, and over rostrum, gradually darkening to blackish, red, or chestnut between ears and posteriorly onto neck. This stripe contrasts sharply with grizzled yellowish to blackish dorsum, which is streaked with black hairs (see Patton et al. 2000:177, Fig. 116). Dorsal hairs chestnut or black at their bases, with subterminal black band and yellow or pale yellow tips. Sides become progressively more fulvous, with posterior thighs and sides of tail base bright burnt orange. Mid-dorsal hairs are of two types: Heavier and longer hairs bicolored, black basally with short pale yellow tip; more abundant thinner hairs blackish or reddish basally with short dark tips. Venter sparsely covered with completely white hairs. Tail base fully haired for about 60 mm, then appears naked to tip; not bicolored at base. Scales large, prominent, pentagonal, and coarse in appearance; average about five annuli per cm near base to six per cm at tip; scale hairs appear correspondingly shorter, extending less than 1.5 scale rows. All scale hairs colorless along entire length of tail distal to furred base.

Skull large, with short rostrum and broad, well-developed supraorbital ledges that form subtriangular postorbital processes; paroccipital processes more vertically oriented than in *D. boliviensis*, and thus not following contour of bullae as closely.

DISTRIBUTION: *Dactylomys dactylinus* is distributed in the Amazon Basin from eastern Colombia south through eastern Ecuador, northern and central Peru to northeastern Bolivia and east along the Amazon River to its mouth in Brazil, extending north into the headwaters of the Río Orinoco in Venezuela and south through gallery forests along Amazonian tributaries into the eastern Cerrado of Brazil.

SELECTED LOCALITIES (Map 487): BOLIVIA: El Beni, 4 km S of Guayaramerin (AMNH 210356); Pando, Ingavi, N bank of Río Orton (USNM 579620). BRAZIL: Amazonas, Borba (AMNH 91865); Colocação Vira-Volta, left bank Rio Juruá (Patton et al. 2000), Ilha das Onças (Patton et al.2000), right bank Rio Purus (Patton et al. 2000); Pará, 52 km SSW of Altamira, right bank Rio Xingu (USNM 549596), Fazenda Santana, Ilha Mexiana (Silva-Júnior and Nuñes 2000), Fordlândia (FMNH 92916), Reserva Biológica de Trombetas (H. Sick 31573), Serra dos Carajás (Moraes-Santos et al. 1999); Goias, Usina Hidroelétrica Serra da Mesa (Bezerra Silva, and Marinho-Filho 2007); Maranhão, Palmeiral, Matöes (Silva-Júnior and Nunes 2000); Roraima, Caracaraí (MPEG 6766). COLOMBIA: Meta, Serranía de La Macarena (ICN 2013), Villavicencio (J. A. Allen 1916b). ECUADOR: Pastaza, Montalvo (FMNH 41471); Sucumbios, Limoncocha (USNM 528370). PERU: Loreto, Orosa (AMNH 73771), Sarayacu (AMNH 76427); Ucayali, Yarinacocha (FMNH 55501). VENEZUELA: Amazonas, Coyowateri (C. Molina et al. 1995), Río Siapa (Ojasti et al. 1992; C. Molina et al. 1995).

SUBSPECIES: Whether both *modestus* Lönnberg and *canescens* Thomas represent valid geographic variants of *D. dactylinus* or whether other species should be segregated from the genus must await confirmation of the true locality of Desmarest's holotype and an appropriate analysis of geographic variation in both morphology, call structure, and molecules over the known range of this species. We, therefore, treat *D. dactylinus* as monotypic and list both *canescens* Thomas and *modestus* Lönnberg as synonyms.

NATURAL HISTORY: This species typically inhabits bamboo and cane thickets on the margins of seasonally inundated and upland rainforest, or in Ecuador only, the canopy of tall forests (LaVal 1976; Emmons 1981 [Ecuador material only]). Few data are available on reproductive biology, although pregnant females have been reported at the end of the rainy season in June, in the central Amazon (Patton et al. 2000). These are vocal rats, with calls composed of staccato pulses in series of 5 to 10 notes in the southwestern and central Amazon (mean = 7; Patton et al. 2000), but in Ecuador of a mean of 42 pulses was emitted during 14 second calls (Emmons 1981). Calls of *D. boliviensis* from Peru (Manu) had longer individual pulses with longer interpulse intervals than those of *D. dactylinus* from Ecuador, but without much difference in pulse numbers (Emmons 1981). An adequate understanding of geographic variation in call structure is needed for both species.

REMARKS: Thomas's (1912d) restriction of the type locality to the upper Amazon is probably in error. Lönnberg (1921) suggested that Desmarest's holotype in the Muséum National d'Histoire Naturelle in Paris came from somewhere downriver from Iquitos in northwestern Peru. When Deville (1852) named the genus, he used both Desmarest's holotype in Paris and three additional specimens (male, female, and young) collected by the Castelnau expedition, of which Deville was a part, at "La Mission de Sarayacu, rivière de l'Ucayale, Pampa del Sacramento." He noted that the holotype was in very poor condition, and he described it as having fur varying red-brown, black, and fawn (fauve) above, with a small crest of reddish brown on the head. He described the Ucayali specimens as red, brown, and black above, tawny (fauve) red on the flanks, clear red on the outer hind thighs, and with a small crest of stiff reddish hairs on the head and nose. These specimens clearly differ in coloration from both *modestus* Lönnberg and *boliviensis* Anthony as detailed in the original descriptions of each of these taxa. We assign specimens from Limoncocha on the Río Napo in Ecuador to *D. dactylinus*, as these correspond to the description of *D. d. modestus* (Lönnberg 1921). Specimens from the Napo have black underfur and a short self-colored brown crown patch that ends at the level of the ears, with some black hairs above the ears. However, other specimens from further east along the Rio Xingu in Brazil also have

similar coloration. Aniskin (1993) described a chromosome complement of $2n = 94$, FN = 144, for specimens assigned to this species from the Loreto in northern Peru. As we have not seen these specimens, nor are they described in Aniskin's report, we cannot verify their identification. Didier (1962) described and figured the baculum based on a specimen from northern Peru. O. J. Linares (1998) tentatively assigned specimens from southern Venezuela to the subspecies *canescens* Thomas.

Dactylomys peruanus J. A. Allen, 1900
Montane Bamboo Rat

SYNONYMS:

Dactylomys peruanus J. A. Allen, 1900a:220; type locality "Juliaca, Peru, altitude 6000 feet," corrected to "Inca Mines [= Santo Domingo Mine], about 200 miles northeast of Juliaca, on the east side of the Andes, on the Inambary [Inambarí] River," Puno, Peru by J. A. Allen (1901b:41).

Dactylomys [*Lachnomys*] *peruanus*: Thomas, 1927f:604; name combination.

DESCRIPTION: Smallest species ingenus; head and body length of holotype 240 mm, tail length 320 mm, hindfoot length 51 mm, and ear length 14 mm. Fur distinctly soft to touch, lacking the strong bristles characteristic of both *D. boliviensis* and *D. dactylinus*. Tail longer than head and body length (about 130%); unique in being heavily furred along basal one-fifth of its length and lightly furred along distal four-fifths, instead of naked; tail hairs long, brownish-black in color; tail terminates in dark tufted tip; scale annuli remain visible even though tail is hairy. Dorsal color uniformly warm yellowish brown (olivaceous), slightly more gray-brown to clear gray on muzzle and forehead, but without stripe on top of head. Dorsal hairs bicolored, grayish black beneath and buffy at tip, with distal light band increasing in length over rump and flanks. Ventral hairs white to base, with some tipped in gray; overall ventral color whitish, or midventer may be dusky washed with some buff. Forelegs and hindlegs thickly furred with long hair to wrists and heels. Color of thighs and adjacent lateral tail base ranges from indistinct to bright orange. Vibrissae long, when appressed extending back to shoulders, predominantly black in color with pale tips. Ears small and do not protrude above hair on top of head.

Skull smaller than that of other species in genus, but conforms to them in most details of cranium and teeth; paroccipital processes curve forward following bullar contour, as in *D. boliviensis*. Species distinguishable primarily by its uniform olivaceous coloration, dense and softer pelage, and furry tail, rather than by cranial characters.

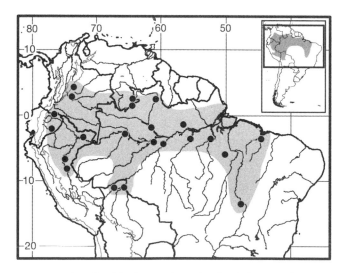

Map 487 Selected localities for *Dactylomys dactylinus* (●). Contour line = 2,000 m.

DISTRIBUTION: *Dactylomys peruanus* occurs in the middle elevations of the eastern Andean slopes of southern Peru and northern Bolivia.

SELECTED LOCALITIES (Map 488): BOLIVIA: La Paz, Cotapata (Salazar-Bravo et al. 2003), La Paz, Astillero (BM 1.6.7.57). PERU: Cusco, Amaybamba (MVZ 173100); Junín, Acobamba (BM 27.11.1.221), Cordillera Vilcabamba (USNM 582148); Puno, Inca Mines (type locality of *Dactylomys peruanus* J. A. Allen), San Juan (FMNH 71131).

SUBSPECIES: *Dactylomys peruanus* is monotypic.

NATURAL HISTORY: Little is known about the natural history of this rarely encountered species. The holotype was collected on the ground by a stream bank; the specimen from Amaybamba, Peru, was taken at night in a clump of bamboo about 2 m above ground; and the Bolivian specimens were taken in a bamboo thicket in wet mossy cloud forest at 2,000 m. Emmons found them to be common in the canopy of bamboo thickets at 2,000 m in wet cloud forest of the Vilcabamba region in Peru, where they vocalized with distinctive birdlike calls, unlike those known for any of its congeners. The species was also photo-documented in wet *Chusquea* bamboo forest at 2,600 m in the Machu Picchu Sanctuary, Peru (R. Williams, pers. comm. to L. H. Emmons), where *Dactylomys* sp. remains were found with a pre-Colombian Inca burial (Yale Peabody Mus. osteo. 3322, Ant. 196396).

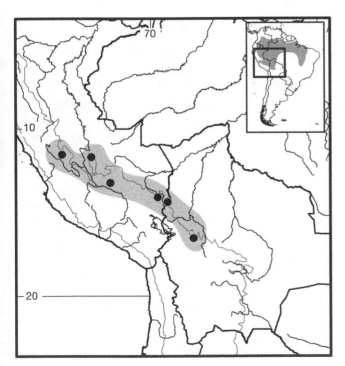

Map 488 Selected localities for *Dactylomys peruanus* (●). Contour line = 2,000 m.

REMARKS: This species differs from both *D. dactylinus* and *D. boliviensis* in its softer and longer dorsal pelage and in its furred tail, the basis for Thomas's (1916e) assignment to the monotypic subgenus *Lachnomys*. Tate (1935) gave generic rank to *Lachnomys*, but Ellerman (1940:136) wrote "it scarcely seems even a valid subgenus." Upham et al. (2013) found *D. peruanus* basal to the sister-pair of *D. boliviensis* and *D. dactylinus* in DNA sequence analyses. The karyotype is unknown. Alberico et al. (2000) record the species in the Cundinamarca, Colombia, at an elevation of 1,700 m in the Andes Biogeographical Region, but without reference to a museum voucher. Given the documented distribution, this assignment seems likely in error, perhaps due to confusion with the externally somewhat similar *Olallamys*.

Genus *Kannabateomys* Jentink, 1891

Kannabateomys is a monotypic genus closely allied to, and strongly resembling, *Dactylomys* (Y. L. R. Leite and Patton 2002; Emmons 2005), but restricted in distribution to the Atlantic rainforest of Brazil, Uruguay, Argentina, and Paraguay (Emmons and Feer 1997). It is smaller than *Dactylomys* (head and body length 230–347 mm; mass up to 570 g) with a similarly long tail (tail length 300–420 mm; >120% of head and body length), moderately long hindfeet (45–57 mm), and relatively short ears (16–27 mm). The upper parts are uniform glossy olivaceous to fulvous, finely streaked with black, and the sides of the body and tail are sometimes fulvous. The fur is long and soft, the hairs blackish with buffy tips. The head is large with a square muzzle; vibrissae are long, dense, and coarse, with the mystacial vibrissae reaching the shoulder; the cheeks are grayish. The tail is much longer than the head and body, robust, furred to the tip, and tapering from thick body hair on the base to sparse hair distally, bicolored pale below, dark brown above at the base paling through dusky to white or buff near the tip. The tip sometimes has a tuft of dark brown hairs. The hands and feet have long, grasping toes; the hands have four long fingers, the center two of which are separated by a wide gap. Nails, not claws, are present on all digits. The under parts are orange with variable amounts of white on the throat and chest. The lips around the mouth are white.

The skull resembles that of other genera in the subfamily, except that the maxillary toothrows are nearly parallel and labial and lingual flexi of the cheek teeth remain separate, with the mesoflexus and hypoflexus not coalescing to divide the tooth into two Y- or V-shaped lophs.

SYNONYMS:

Dactylomys: Wagner, 1845a:146; part (description of *amblyonyx*); not *Dactylomys* I. Geoffroy St.-Hilaire.

Kannabateomys Jentink, 1891:109; type species *Dactylomys amblyonyx* Wagner, by original designation.

Cannabateomys Lydekker, 1896:91; incorrect subsequent spelling of *Kannabateomys* Jentink.

Kannabateomys amblyonyx (Wagner, 1845)
Atlantic Bamboo Rat

SYNONYMS:

Dactylomys amblyonyx Wagner, 1845a:146; type locality "Ypanema" (= Floresta Nacional de Ipanema, 20 km NW Sorocaba, São Paulo, Brazil, 23°26′7″S 47°37′41″W, 701 m; L. P. Costa et al. 2003).

Kannabateomys amblyonyx: Jentink, 1891:109; first use of current name combination.

Kannabateomys amblyonyx pallidior Thomas, 1903b:489; type locality "Sapucay," Paraguarí, Paraguay.

Kannabateomys amblyonyx amblyonyx: Tate, 1935 435; name combination.

DESCRIPTION: As for the genus.

DISTRIBUTION: *Kannabateomys amblyonyx* occurs along the Atlantic coast of Brazil south to Uruguay and inland to northeastern Argentina and eastern Paraguay (Emmons and Feer 1997).

SELECTED LOCALITIES (Map 489): ARGENTINA: Misiones, Parque Nacional Iguazú (Crespo 1982b). BRAZIL: Paraná, Usina Hidroelétrica Salta Caixas (Y. L. R. Leite and Patton 2002); Rio de Janeiro, Reserva Biológica de Poço das Antas (Y. L. R. Leite and Patton 2002); Rio Grande do Sul, Parque Estadual do Itapuã (R. B. Silva et al. 2008); Santa Catarina, Theresopolis (NMW 1691); São Paulo, Ipanema (type locality of *Dactylomys amblyonyx* Wagner). PARAGUAY: Paraguaí, Sapucay (type locality of *Kannabateomys amblyonyx pallidior* Thomas).

SUBSPECIES: Specimens from the southern part of the range in Argentina and Paraguay are notably dull, buff yellowish with the venter and tail almost white (Emmons and Feer 1997). Thomas (1903b) gave these southern populations the formal subspecies name *pallidior*, but an appropriate analysis of geographic variation and thus the validity of subspecies designations remains to be accomplished.

NATURAL HISTORY: This is an arboreal rat typically living singly or in pairs. The combination of paternal care of young, delayed juvenile dispersal, a reduced degree of sexual dimorphism, and exclusive home ranges for members of the same sex but with male and female ranges overlapping are attributes that suggest social monogamy as the basic mating system (R. B. Silva et al. 2008). Individuals feed on the young shoots and leaves of bamboo and other plants, and inhabit bamboo thickets, especially at watersides, and dense thickets without bamboo in swamps, in Atlantic coastal

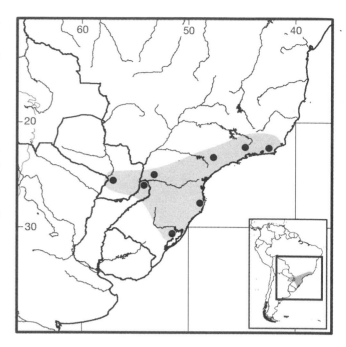

Map 489   Selected localities for *Kannabateomys amblyonyx* (●). Contour line = 2,000 m.

forests, inland rainforest, wet gallery forest, and bamboo patches such as planted bamboo hedgerows between agricultural fields (Hensel 1872a; Kierulff et al. 1992; Olmos et al. 1993; Stallings et al. 1994). Paresque et al. (2004) described a karyotype with $2n = 98$ and FN = 126.

Genus *Olallamys* Emmons, 1988

This is a poorly known genus whose distribution is limited to the Andes of northern Colombia and northwestern Venezuela, at elevations generally above 2,000 m and in association with moist bamboo thickets. The smallest genus among the dactylomyines, with head and body varying between 150–225 mm, also with a long tail (>140% of head and body length). The tail is sparsely haired, faintly to strongly bicolored, and with the distal half paler than the proximal part. The fur is soft, without spines or bristles. Upper parts are a shade of reddish brown and the under parts are white to yellow. The genus is characterized by claws on all digits of forefeet and hindfeet, except d2 of the hindfoot that bears an oblique, asymmetrical nail. Two species are usually recognized, although the differences between them, beyond size, are slight, and it is possible that these represent a single, geographically differentiated species with subpopulations in the separate Andean cordilleras of northwestern South America.

SYNONYMS:

Thrinacodus Günther, 1879; type species *Thrinacodus albicauda* Günther, 1879:144, by original designation; preoccupied by *Thrinacodus* St. John and Worthen, 1875 (Chondrichthyes: Elasmobranchii).

Olallamys Emmons, 1988:241; replacement name for *Thrinacodus* Günther, with the same type species, *Thrinacodus albicauda* Günther.

KEY TO THE SPECIES OF *OLALLAMYS*:

1. Maximum zygomatic breadth <27 mm; maxillary toothrow <15 mm . . . . *Olallamys albicaudus*
1′. Maximum zygomatic breadth >28 mm; maxillary toothrow >15 mm . . . . *Olallamys edax*

Olallamys albicaudus (Günther, 1879)
White-tailed Olalla Rat

SYNONYMS:

Thrinacodus albicauda Günther, 1879:145; type locality "the vicinity of Medellin, Columbian Confederation," Antioquia, Colombia.

Thrinacodus apolinari J. A. Allen, 1914c:387; type locality "Tomeque [= Fómeque], (altitude 6500 feet), Bogotá district," Cundinamarca, Colombia.

Thrinacodus albicauda apolinari: Cabrera, 1961:545; name combination.

Olallamys albicauda: Woods, 1993:790; first use of current name combination, but incorrect gender of specific epithet.

DESCRIPTION: As for the genus; distinguishable from *O. edax* (below) primarily by tail color, for which proximal half is reddish-brown and terminal half white, with grayish tinge toward tip. Smaller than *O. edax*; head and body length 150 and 180 mm, respectively for holotypes of *albicaudus* Günther and *apolinari* J. A. Allen, respective tail lengths 255 and 260 mm.

DISTRIBUTION: Known only from the Andes of Colombia. Alberico et al. (2000) indicated that this species ranged in elevation from 2,000–3,200 m in the Andes Biogeographical Region in the departments of Antioquia, Cundinamarca, Nariño, and Quindío. These authors listed specimens from three Colombian collections (IAvH, ICN, and UV) without reference to the vouchers or specific localities.

SELECTED LOCALITIES (Map 490): COLOMBIA: Antioquia, Páramo de Belmira (Delgado and Zurc 2005), Reserva San Sebastián-La Castellana (Delgado and Zurc 2005); Cundinamarca, Reserve Biológica Carpanta (ICN 17047), San Cristobal (FMNH 71128); Nariño, Laguna de La Cocha (UV [RBM 4143]).

SUBSPECIES: Cabrera (1961) recognized two subspecies, the nominotypical form from the Cordillera Oriental near Medellín and *apolinari* J. A. Allen, from the Cordillera Central near Bogotá. No analysis of geographic variation in character traits has been done to validate this hypothesis.

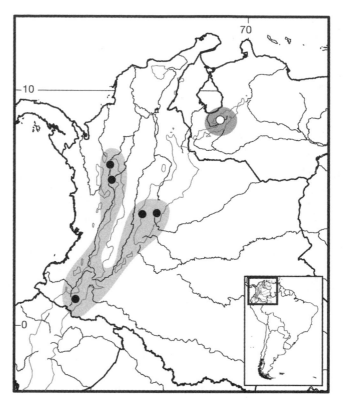

Map 490 Selected localities for *Olallamys albicaudus* (●) and *Olallamys edax* (○). Contour line = 2,000 m.

NATURAL HISTORY: Little is known about the ecology of this species. Recent records (Delgado and Zurc 2005) are from disturbed primary forest containing homogeneous patches of dense bamboo at or near tree line.

REMARKS: Didier (1962) described and figured the baculum. J. A. Allen (1914c:387) defined *apolinari* as a species separate from Günther's *albicauda* primarily on the basis of uniform pale brown color along the length of the tail as opposed to the terminal white present in the latter. Cabrera (1961) assigned *apolinari* J. A. Allen as a subspecies of *Thrinacodus* (= *Olallamys*) *albicauda* without comment.

Olallamys edax (Thomas, 1916)
Greedy Olalla Rat

SYNONYMS:

Thrinacodus edax Thomas, 1916e:299; type locality "Sierra de Mérida. Alt. 2800 m.," Mérida, Venezuela.

Olallamys edax: Woods, 1993:791; first use of current name combination.

DESCRIPTION: Similar in color and other external characters to *O. albicaudus*, but tail completely white along its ventral surface and distal half white above. Larger than *O. albicaudus*, with head and body length of holotype 225 mm, tail length 345 mm, and skull length 57 mm.

DISTRIBUTION: Known only from the vicinity of the type locality in the Mérida Andes (O. J. Linares 1998).

SELECTED LOCALITIES (Map 490): VENEZUELA: Mérida, Sierra de Mérida (type locality of *Thrinacodus edax* Thomas).

SUBSPECIES: *Olallamys edax* is monotypic.

NATURAL HISTORY: This species is known from high-elevation bamboo thickets; its ecology and reproduction are undescribed.

REMARKS: *Olallamys edax* was separated from *O. albicauda* on the basis of tail color, skull size, and tooth size, characters that vary among specimens of the latter species from the Central Cordillera in Colombia. For example, the tail of the holotype of *apolinari* J. A. Allen is pale brown above and paler below (J. A. 1914c:388), and the terminal half of the tail of the holotype of *albicauda* Günther is white. Future studies may well show that *O. edax* is not a valid species, but too few specimens are currently available in museum collections for a proper analysis.

Subfamily Echimyinae Gray, 1825
Louise H. Emmons, Yuri L. R. Leite, and James L. Patton

All genera in this subfamily are arboreal rodents with a pelage that may be spiny, bristly, or soft, and a tail that is typically as long as or longer than the head and body length. The feet are short and broad, and possess raised, smooth plantar pads developed under the joints. Molars are brachydont, with four roots (except in *Callistomys*), but crowns are often high (hypselodont or coronal hypsodont). The occlusal plane is angled away from plane of the palate, achieved by tipping otherwise straight-sided teeth so that the labial gumline is higher than the lingual (except in *Callistomys*). The cheek teeth have deep reentrant flexi(id)s that may split teeth into separate, parallel laminae. One to three deep flexi are present on the labial side of the maxillary cheek teeth. The posterior tip of the lower incisor root terminates anterior to m3. Both the maxillary zygomatic process and the jugal are usually slender. Toothrows are long, over 25% of basilar length of the skull.

G. A. S. Carvalho and Salles (2004) assigned six genera to this subfamily, and Emmons (2005:299; Table 3) included eight. Five genera (*Callistomys*, *Diplomys*, *Echimys*, *Makalata*, and *Phyllomys*) are in common between these two recent reviews, but Emmons (2005) described two as new (*Pattonomys* and *Santamartomys*). She placed three genera as *incertae sedis* within the subfamily because of their individually unique sets of attributes. One of these, *Isothrix*, has been consistently included within the subfamily by earlier authors (e.g., McKenna and Bell 1997; Woods and Kilpatrick 2005; G. A. S. Carvalho and Salles

2004)), and two, *Lonchothrix* and *Mesomys*, have usually been included within the Eumosopinae (McKenna and Bell 1997; Woods and Kilpatrick 2005), despite sharing some characters with echimyine genera (see discussion in Patton and Reig 1990). Iack-Ximenes et al. (2005a) have recently added a ninth genus (*Toromys*).

Phylogenetic analyses of taxon-limited mitochondrial and nuclear DNA sequence data (Y. L. R. Leite and Patton 2002; Galewski et al. 2005; B. D. Patterson and Velazco 2008; Upham and Patterson 2012; P.-H. Fabre et al. 2012; Upham et al. 2013) consistently recover a well-supported clade consisting of (*Makalata* (*Echimys* + *Phyllomys*)), (*Makalata*, (*Echimys* + *Toromys*)), or (*Makalata*, *Echimys* (*Phyllomys* + *Toromys*)), depending upon the genera included in the respective analyses. All molecular studies indicate that *Isothrix* groups at the base of this clade of three or four genera. Furthermore, the recent analysis by P.-H. Fabre et al. (2012) and Upham et al. (2013) supported Emmon's contention that *Lonchothrix* and *Mesomys*, long recognized as sister taxa, also group with these echimyine genera. In the expanded morphological analysis of G. A. S. Carvalho and Salles (2004), *Isothrix* is contained within a clade otherwise composed of (*Pattonomys* (*Echimys* (*Isothrix*, *Makalata*, *Toromys*, (*Phyllomys* + *Diplomys*)))). *Callistomys* was placed outside this clade in their analysis, as were both *Lonchothrix* and *Mesomys*. The rarely collected *Callistomys*, *Diplomys*, *Pattonomys*, and *Santamartamys*, have not been included as yet in molecular phylogenetic studies.

SYNONYMS:

Echimyinae Gill, 1872:22; a subset of the genera contained within the Echimyina of Waterhouse (1848:286–359).

Echinomyinae Alston, 1876a:92; name attributed to Waterhouse (1848) on p. 72.

Loncherinae Thomas, 1897a:1024.

KEY TO THE GENERA OF ECHIMYINAE (SEE EMMONS 2005 FOR DESCRIPTIONS AND ILLUSTRATIONS OF CHARACTERS):

1. Occlusal surface of maxillary cheek teeth with one lingual and three labial flexi/fossettes 2
1'. Occlusal surface of maxillary cheek teeth with pattern other than one lingual and three labial flexi/fossettes. . . 4
2. Pelage of lower back soft. 3
2'. Pelage of lower back spiny . . . . *Echimys*
3. Maxillary cheek teeth hypselodont with three roots; occlusal plane strongly angled labially . . . . . . *Callistomys*
3'. Maxillary cheek teeth brachydont with four roots; occlusal plane weakly angled labially . . . . *Isothrix*
4. Occlusal surface of maxillary cheek teeth with four separate and parallel laminae. 5
4'. Occlusal surface of maxillary cheek teeth with two labial and two lingual flexi (or fossettes) 6

5. Lower molars with three separate lophs; auditory meatus close to squamosal suture; guard hairs unbanded . . . . *Diplomys*

5′. Lower molars with anteroloph separate; auditory meatus well separated from squamosal suture; guard hairs banded . . . . *Phyllomys*

6. Anteroloph of lower premolar lacks fossetid; upper incisor root at level of or outside zygoma; maxillary and premaxillary portions of incisive foramina septum separate 7

6′. Anteroloph of lower premolar with central fossetid or with flexid opening lingually; upper incisor root within maxillary root of zygoma; maxillary and premaxillary portions of incisive foramina septum broadly fused . . 8

7. Dorsal pelage stiff; mesoloph of third upper molar approximate in size to protoloph; metalophid of lower premolar and anterolophid of lower molars separate; auditory tube directed forward; inferior jugal process inconspicuous . . . . *Santamartamys*

7′. Dorsal pelage spiny; mesoloph of third upper molar much smaller than protoloph; metalophid of lower premolar and anterolophid of lower molars not separate; auditory tube directed laterally; inferior jugal process elongate . . . . *Pattonomys*

8. Dorsal pelage spiny; plantar surface of hindfeet without tubercles; third upper molar with four or more well-developed lophs . . . . *Makalata*

8′. Dorsal pelage bristly; plantar surface of hindfeet with tubercles between pads; posteroloph of third upper molar reduced, or M3 with 3 lophs only . . . . *Toromys*

Genus *Callistomys* Emmons and Vucetich, 1998

Louise H. Emmons and Yuri L. R. Leite

Callistomys can be diagnosed by the following set of characters (taken from Emmons and Vucetich 1998:3; Emmons 2005): cheek teeth have three roots; both upper and lower cheek teeth are high crowned; PM4 and M1 are unilaterally hypsodont (lingual side of crown higher than labial side, crown curved outward); PM4–M3 are tetralophodont, with three labial flexi and one lingual flexus; hypoflexi and mesoflexi are deep, with PM4 completely divided by the joined hypoflexus-metaflexus into two U-shaped lophs with no mure; M1–3 have a narrow mure connecting protocone and hypocone. The hypoflexids of pm4–m3 are set at a strong oblique angle; the medial end of the flexid extends farther anterior then the labial end. Lower premolars are tetralophodont; the anteroconid and protoconid are united, enclosing the anteroexternal flexid as a slit-like fossetid; the anterior half

of the tooth approximates a triangle with its axis slightly tipped anterolabially; and the hypoflexids and metaflexids do not join (pm4 is not divided by a continuous flexid). The lower incisors are robust and strongly curved. Cranially, the jugals are expanded dorsoventrally; the lateral jugal fossa is wide and diffuse anteriorly, not coming to a sharp point, with the anterior edge of the fossa positioned above PM4, anterior to a line extended from the posterior border of the ascending maxillary process of the zygomatic arch. The superior zygomatic root of the maxillary is expanded posteriorly. The tympanic auditory bullae are inflated, with a large auditory meatus at the end of a strongly developed auditory tube. The angular process of the mandible is strongly projected ventrally with respect to the inferior projection of the symphysis such that an angle drawn between the ventral posterior tip of the angular process and the occlusal plane of the toothrow, with the apex at the anterior edge of the occlusal surface of pm4, is greater than 30°.

The single living species is a moderate-sized (head and body length 250–295 mm, tail length 273–325 mm, hindfoot length 43–47 mm, and ear length 16 mm) and arboreally adapted rat with a striking black (or brown) and white patterned pelage. The fur is long and dense, with hairs of uneven lengths; overhairs are slender and soft, lacking spines or bristles, and the underfur is dense, long, and wavy. Dorsal hairs are brown at their base with tips black or white such that color is entirely whitish except for the sharply defined glossy black saddle (fading to brown in old museum specimens) covering the dorsum from shoulders to tail, and across upper forelegs. A wide, black diamond-shaped patch extends over the nape and crown to between the eyes. The tail is black at its base, with the distal part silky white above, golden yellow below; it is thickly covered with short, dense, flat-lying hairs. Ear pinnae are small and inconspicuous. Vibrissae are slender; longest mystacial vibrissae reaches to shoulder when flattened back; superciliary vibrissae reach to ear tip. Lower legs and feet are stout and broad, toes are long. There are two pairs of lateral mammae.

SYNONYMS:

Nelomys: Pictet, 1843a:205; part (description of *pictus*).

Loncheres (*Isothrix*): Waterhouse, 1848:327; part (listing of *picta*); neither *Loncheres* Illiger nor *Isothrix* Wagner.

Echimys: Tate, 1935:414; part (listing of *pictus*); not *Echimys* F. Cuvier.

Callistomys Emmons and Vucetich, 1998:3; type species *Nelomys pictus* Pictet, by original designation.

REMARKS: This genus was erected to include two species from coastal Brazil. One of these is known only from a mandibular fragment collected by Lund at Lagoa Santa, Minas Gerais, and described by Winge (1887:71), who referred it to

"*Lasiuromys villosus* Dev. ? 1852)," a subjective synonym of *Isothrix bistriata*. This species is apparently without a name, and is listed by Emmons and Vucetich (1998:7) as *Callistomys* sp. The second species is *Callistomys pictus* (Pictet 1843a), which was originally placed in the genus *Nelomys* (= *Phyllomys*; see that account), an action continued by Emmons and Feer (1990 1997), but which has been varyingly assigned also to *Isothrix* (E.-L. Trouessart 1880; Ellerman 1940; Cabrera 1961; Patton and Emmons 1985; Eisenberg and Redford 1999) or *Echimys* (Tate 1935; Moojen 1952b; Woods 1993). These different assignments highlight the overall uniqueness of this taxon.

Indeed, several features distinguish *Callistomys* from other genera in the subfamily Echimyinae and raise questions about whether it belongs in the subfamily: The possession of the eumysopine cheek tooth root pattern and hypselodonty, and the dorsoventrally expanded jugals and anterior zygomatic arch, likewise features common with the Eumysopinae (especially *Euryzygomatomys* and *Clyomys*) and not shared by any other living echimyine. The phylogenetic analysis of G. A. S. Carvalho and Salles (2004) place the genus as sister to a clade otherwise containing the dactylomyines and all other genera of echimyines. Leite's preliminary molecular sequence data suggest that *Callistomys* has no close relative among all living echimyid genera.

Callistomys pictus (Pictet, 1843)

Painted Tree Rat

SYNONYMS:

Nelomys pictus Pictet, 1841a:29; *nomen nudum*.

Nelomys pictus Pictet, 1843a:205; type locality "Bahia"; probably from Fazenda Almada in the region north of the city of Ilhéus, Bahia, Brazil (Moojen 1952b).

Loncheres (*Isothrix*) *picta*: Waterhouse, 1848:327; name combination.

Loncheres (*Isothrix*) *pictus*: E.-L. Trouessart, 1897:606; name combination.

[*Isothrix*] *picta*: Thomas, 1916e:295; name combination.

[*Echimys*] *pictus*: Tate, 1935:414; name combination.

Echimys pictus: Moojen, 1952b:133; name combination.

Isothrix picta: Cabrera, 1961:537; name combination.

Callistomys pictus: Emmons and Vucetich, 1998:3; first use of current name combination.

DESCRIPTION: As for the genus.

DISTRIBUTION: *Callistomys pictus* is endemic to the Atlantic Forest with known localities limited to three adjacent municipalities near Ilhéus and Itabuna, in coastal Bahia, Brazil (Vaz 2002). Eisenberg and Redford (1999) list a specimen from the Rio de Birasa without documentation. Peter W. Lund collected subfossil remains from Lagoa Santa, Minas Gerais, but the species is unknown from the modern fauna of this region (Emmons and Vucetich 1998).

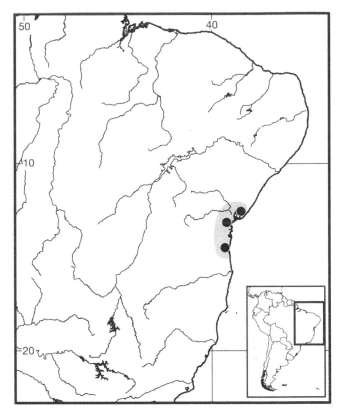

Map 491: Selected localities for *Callistomys pictus* (●).

SELECTED LOCALITIES (Map 491): BRAZIL: Bahia, Fazenda Almada (probable type locality of *Nelomys pictus* Pictet; Moojen 1952b), Reserva Particular do Patrimônio Natural Jequitibá (Vaz 2002), Rio do Braço (MNRJ 11207).

SUBSPECIES: *Callistomys pictus* is monotypic.

NATURAL HISTORY: An apparent inhabitant of evergreen lowland rainforest, but only 13 specimens are known. Recent fieldwork (Vaz 2002) found this species in cocoa plantations, but substantial habitat loss and modification has likely led to severe restriction in the number of extant populations. The species is locally hunted for food (Vaz 2002). Valim and Linardi (2008) described the host association of ectoparasitic lice in the genus *Gyropus*.

REMARKS: The first use of the name stems from Pictet (1841a:29), in a list of new species obtained by the Musé de Genève. Pictet (1843a:203) later provided the description of his new taxon. K. Ventura et al. (2008) described a karyotype with $2n = 42$, FN = 76.

Genus *Diplomys* Thomas, 1916

Louise H. Emmons and James L. Patton

These are moderate-sized arboreal rats without spines but with guard hairs flattened and slightly stiff, each with a deep dorsal sulcus. The dorsal guard hairs are 1.8–2.0 cm long; underhairs are few, fine, and inconspicuous. The dorsal body fur extends from 1.5–5.0 cm onto base of the tail, and the distal part is fully but moderately clothed with stiff, bristle-like, brown hairs, and with scales visible beneath. The feet are without tiny tubercles between pads.

The skull is long and narrow, with the dorsal surface is curved throughout its length from nasals to occiput when viewed laterally. The jugal is narrow below the postorbital process, with an inconspicuous ventral process positioned well forward of the dorsal process. The tympanic bullae are not inflated, the auditory meatus is positioned close to the squamosal, and auditory tubes are well developed and directed strongly forward such that the middle ear bones are not visible in lateral view. The distance between the foramen ovale and the masticatory foramen of the alisphenoid is typically very short, and the ventral roof of the canal formed by these foramina is often nearly obsolete. The hamular process of the pterygoid has a prominent, anteriorly directed spur. A large, oval sphenopalatine vacuity is present anterior to the presphenoid-basisphenoid suture. The squamosotympanic fenestra is tube-like, opens forward, and is without a depression on squamosal below. The angle of the sigmoid notch, between angular and condyloid processes of the mandible, is shallow, and the mandibular foramen on the side of ramus is positioned well anterior to the condyloid ridge, not in a fossa. The masseteric crest is large.

Maxillary cheek teeth are large and the toothrow is relatively long. Morphology of the upper teeth resembles that of *Phyllomys* spp. (see below), with the occlusal surface divided by three flexi into four laminae. These laminae are parallel in M1–3, but in PM4 the labial ends of the two anterior lophs are bent forward. The flexi are of even depth, such that with wear, three straight fossettes are centered on the molars. In the mandibular cheek teeth, all flexids may traverse all lophs in unworn teeth, but with wear the metaflexid divides m1–3 into a free posterolophid. The middle loph (entoconid + hypoconid) is usually connected to the anterolophid at near its midpoint by a short mure, forming a highly distinctive pattern of two crescent-shaped lophids connected by a central stem. The hypoflexids are positioned at an oblique angle. The lower premolar is split into three or four parts by two or three flexids; the anterior loph forms a rounded triangle enclosing a flattened oval transverse fossette, metalophid is a separate lamina, and the entoconid and posterolophid are either joined labially or divided into separate parts by the metaflexid. The lower incisors are strongly curved.

There are two pairs of lateral mammae. The phallus lacks deep lateral ridges, but the ventral surface has a long, longitudinal slit distal to a short transverse fold. The dorsal tip possesses a long, slender, bacular papilla, and an inconspicuous urethral lappet is present.

Two species are now recognized within the genus, subsequent to Emmons's (2005) removal of *rufodorsalis* J. A. Allen to her new genus *Santamartamys*: *D. caniceps* (Günther) from Colombia and *D. labilis* (Bangs, 1901, with *darlingi* Goldman, 1912b, a synonym) from Panama south along the Pacific coast of Colombia to Ecuador. Hall (1981:874–875) provided a taxonomic summary for and mapped the distribution of *D. labilis* in Panama. Few specimens are available from either species, and given the lack of more than cursory comparisons in the literature, the number of species and their range(s) remain questions for the future to address. Largely by virtue of its relatively soft pelage and well-haired tail, *Diplomys* has often been associated with Amazonian and Guianan genus *Isothrix* (e.g., E.-L. Trouessart 1880; Goldman 1912b) or the Atlantic Forest *Phyllomys* (Goldman 1916). The phylogenetic analyses of G. A. S. Carvalho and Salles (2004) supported a sister relationship to the Atlantic Forest *Phyllomys*. Molecular data are as yet unavailable to test any phyletic hypothesis.

SYNONYMS:

Loncheres: Günther, 1876:745; part (description of *caniceps*); not *Loncheres* Illiger.

Loncheres (*Isothrix*): E.-L. Trouessart, 1880b:178; part (listing of *caniceps*); neither *Loncheres* Illiger nor *Isothrix* Wagner.

Echimys (*Isothrix*): E.-L. Trouessart, 1904:504; part (listing of *caniceps*); neither *Echimys* F. Cuvier nor *Isothrix* Wagner.

Phyllomys: Goldman, 1916a:126; part (listing of *caniceps*); not *Phyllomys* Lund.

Diplomys Thomas, 1916d:240; type species *Loncheres caniceps* Günther, by original designation.

KEY TO THE SPECIES OF *DIPLOMYS*:

1. Range restricted to Cauca Valley, Colombia; maxillary and mandibular molar toothrows divergent, with ends curved outward such that a straight edge held against the first and last tooth shows a gap between them and the middle teeth . . . . *Diplomys caniceps*

1'. Range (in South America) restricted to western Colombia and northwestern Ecuador; both maxillary and mandibular toothrows nearly straight . . . . . . *Diplomys labilis*

Diplomys caniceps (Günther, 1876)
Colombian Rufous Tree Rat
SYNONYMS:

Loncheres caniceps Günther, 1876:745; type locality "Medellin," Antioquia, Colombia.

Loncheres (Isothrix) caniceps: E.-L. Trouessart, 1880b:178; name combination.

Echimys (Isothrix) caniceps: E.-L. Trouessart, 1904:504; name combination.

Phyllomys caniceps: Goldman, 1916a:126; name combination.

Diplomys caniceps: Thomas, 1916d:240; first use of current name combination.

DESCRIPTION: Larger of two species; head and body length 212–320 mm, tail length 178–267 mm; hindfoot length 40–48 mm, ear length 12–20 mm, and mass 360–430 g. Upper parts dull or bright rusty red-brown faintly streaked with dark brown hairs, reddish brightest on rear half of body, and forequarters sometimes fulvous. Fur stiff and dense but lacks spines. Nose brown; head forward of the ears is gray streaked with black, face marked with small whitish spots over the eye and at base of vibrissae, and pale yellow spot present behind and below the ears. Ears short and generally do not protrude above crown. Vibrissae coarse and long, reaching to shoulder. Chin often white, throat whitish or orange, and remainder of the under parts range from pale to rich orange, often dark rust-orange along the midline of the abdomen and underside and sides of the tail base. Tail robust, subequal in length to body length (90–105%), and moderately but completely covered with dark brown hairs that stand outward to give bristly appearance; scales of tail just visible beneath the hair. Hindfeet broad, yellow-brown to gray or brown above.

Cranial and dental characters as for the genus.

DISTRIBUTION: *Diplomys caniceps* is apparently limited to the Caribe Biogeographical Region (Alberico et al. 2000) in northwestern Colombia, in the Cauca Valley in the vicinity of Medellín (Emmons and Feer 1997). Cuervo Díaz et al. (1986) extended the range to the Serranía de San Lucas in the state of Bolívar, but without documentation. Eisenberg (1989: map 13.73) reversed the ranges of *D. caniceps* and *D. rufodorsalis* (= *Santamartamys rufodorsalis*), incorrectly mapping the former to northeastern Colombia.

SELECTED LOCALITIES (Map 492): COLOMBIA: Antioquia, Finca Katiri (CTUA, C. Delgado-V., pers. comm.), vicinity of Medellín (type locality of *Loncheres caniceps* Günther); Bolívar, Corregimiento de Ventura (INDERENA 3856).

SUBSPECIES: *Diplomys caniceps* is monotypic.

NATURAL HISTORY: This species has not been studied in its natural habitat.

REMARKS: *Diplomys caniceps* is superficially similar to *D. labilis* (below) in external characters, but the two species seem to differ in cranial characters. The exact type locality is unknown. The collector, T. K. Salmon, shipped

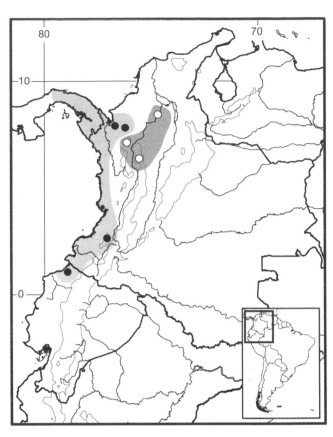

Map 492 Selected localities for *Diplomys caniceps* (○) and *Diplomys labilis* (●). Range of *D. labilis* in Panama from Hall (1981). Contour line = 2,000 m.

specimens from Medellín to England, but worked in the Magdalena, Neche, Pocuné, Río Sucio, and Cauca valleys in all directions from that city (Sclater and Salvin 1879). It is unclear if *D. caniceps* is geographically isolated from coastal *D. labilis*. This species is represented only by specimens in the Natural History Museum in London, the Instituto de Investigación de Recursos Biológicos Alexander von Humboldt in Bogotá, Colombia (Alberico et al. 2000), and Colección Teriológica Universidad de Antioquia (CTUA), Medellín, Colombia (C. Delgado-V., pers. comm.).

Diplomys labilis (Bangs, 1901)

Central American Rufous Tree Rat

SYNONYMS:

Loncheres labilis Bangs, 1901:638; type locality "San Miguel Island, [Archipiélago de las Perlas], Panama."

Isothrix darlingi Goldman, 1912b:12; type locality "Marrangantí (near Real de Santa María), on the Río Tuyra, eastern Panama."

Echimys (Isothrix) labilis: E.-L. Trouessart, 1904:504; name combination.

Phyllomys labilis: Goldman, 1916:126; name combination.

Phyllomys darlingi: Goldman, 1916:126; name combination.

Diplomys labilis: Thomas, 1916d:240; first use of current name combination.

Diplomys darlingi: Thomas, 1916d:240; name combination.

D[iplomys]. l[abilis]. darlingi: Handley, 1966:787; name combination.

DESCRIPTION: Externally similar to *D. caniceps* but somewhat smaller (head and body length about 190 mm, tail length 267 mm, hindfoot 40 mm, and ear length 12 mm). Upper parts ochraceous buff to bright rusty, slightly darkened by admixture of black-tipped hairs, becoming grayish brown on middle of face and cheeks; nose, lips, and muzzle gray, with white vertical streak at base of vibrissae (see photograph in Tesh 1970a); wrists, ankles, and chin gray; feet silvery white above. A small pale spot present above eyes, contrasting with a black eye-ring; ears dark, with marginal fringe of longer hairs; narrow strip of white encircles the pinnae. Tail generally proportionally shorter than that of *D. caniceps*, 70–100% of the head and body length; unicolored brownish, tapering evenly to tip; thinly clothed with short hairs, which become somewhat longer although not denser or forming a distinct brush near tip. Dentition is similar to that of *D. caniceps*. Cranially, palate narrower and premaxillae sometimes extend posterior to nasals in *D. labilis* in comparison with conditions in *D. caniceps*. There is considerable variation of ventral color from dark to paler rufous, of the rostrum from broad to quite narrow, and of the premaxillary bones either even with, or extending posterior to nasals. Verification of current species boundaries, and therefore their geographic ranges, awaits analyses of existing specimens and further collecting.

DISTRIBUTION: *Diplomys labilis* occurs from central Panama, including Isla San Miguel in the Gulf of Panama, south along the Pacific coast of Colombia (Pacífico Biogeographic Region; Alberico et al. 2000) and extends into northwestern Ecuador. Specimens and localities from western Colombia were listed as "probably *D. labilis*" by Emmons and Feer (1997), and those in the collections of Museo de Historia Natural de la Universidad del Cauca and Universidad del Valle in Colombia from Cauca, Chocó, and Nariño departments were assigned to *D. labilis* by Alberico et al. (2000). Specific localities for these latter specimens, however, were not provided. We assign specimens in the Field Museum from western Colombia to *D. labilis*. Tirira (2007:223) mapped the range of this species as extending into northwestern Ecuador but records no specific localities. A subadult specimen collected in 1901 from "Bayone, N. Ecuador" exists in the collections of the Natural History Museum, London (BM). We have been unable to locate this locality, although it may be on the Río Borbón in Esmeraldas province as specimens of other species in the BM were collected there on the same date (Paynter 1993).

SELECTED LOCALITIES (Map 492; South America only): COLOMBIA: Antioquia, Turbo (FMNH 70101); Cauca, Río Saija, La Boca (FMNH 90109); Córdoba, Socorre, upper Río Sinu (FMNH 69119). ECUADOR: Esmeraldas, Bayone (BM 2.7.26.2).

SUBSPECIES: Two subspecies occur in Panama, the insular nominotypical form and *darlingi* Thomas on the mainland (Hall 1981). There has been no evaluation of variation in populations from Colombia or Ecuador.

NATURAL HISTORY (from Tesh 1970a, based on studies in Panama): This is a nocturnal, solitary, arboreal species that feeds on both fruits and leaves. Individuals nest in tree holes, usually in trees near the water, and can be seen sticking their heads out of the entrance if disturbed. They are agile climbers and aggressive when captured. Their habitat includes both mature and secondary rainforest and deciduous forest, including lowland and montane, mangroves, and plantations with trees. Litter size is one or two.

REMARKS: Both Thomas (1916d, e) and Goldman (1916) suggested that *labilis* Bangs and *darlingi* Goldman were allied, but retained each as separate species. Handley (1966:787) listed the mainland *darlingi* as a subspecies of the insular *labilis*, a conclusion followed by Hall (1981:874). Valim and Linardi (2008) described the host association of ectoparasitic lice in the genus *Gyropus*.

Genus *Echimys* F. Cuvier, 1809

Louise H. Emmons, Yuri L. R. Leite, and James L. Patton

Members of this genus are moderate-sized arboreal rats with the pelage including stout spines. The tail is longer than the head and body, completely covered with dense hair that forms a pencil at tip. Four pairs of mammae are present, three laterals and one inguinal. The phallus of *E. chrysurus* (Emmons 2005) is stout, strongly ridged on its sides, with a short, triangular bacular papilla at the dorsal tip; the ventral lip of urethral crater forms a short, U-shaped fold above a straight-line lip.

The skull has small to medium-sized auditory bullae with a small meatus on a short auditory tube directed laterally at nearly a right angle to the cranial axis; the meatus is positioned near (less than half its diameter) to the squamosal suture. The lateral process of the supraoccipital (erroneously termed mastoid process by Emmons et al. 2002 and Y. L. R. Leite 2003; see Verzi et al. 2013 for anatomical details) is short. The squamosotympanic fenestra is small and subcircular, with no raised ventral lip, and is entirely anterior to the auditory meatus. The premaxillary part of the septum of the incisive foramina is free posteriorly. The jugal is deep below the postorbital process, and the jugal fossa is very broad, forming an angle of at least 40°. The postorbital process of the zygoma is formed only by the jugal. The mandibular foramen is anterior to the condyloid ridge, not in a fossa with base in ridge; the masseteric crest is weakly developed.

Maxillary toothrows are parallel, not converging anteriorly. The cheek teeth are rectangular, longer than wide, and with tall crowns. Only one lingual flexus is present (the hypoflexus), which usually joins with mesoflexus to completely divide tooth into two U-shaped lophs (that is, there is no mure connecting the protoloph and mesoloph); three labial flexi are present. The four lophs of PM4–M2 are subequal in length and width and nearly parallel. The paraflexus and metaflexus are long, about two-thirds or more of the width of the tooth, and parallel; the hypoflexus or its trace is short, about one quarter to one-third of the width of the tooth. Hypoflexids of the lower molars reach less than half way across tooth, open at a wide angle, and are scarcely, if at all angled forward, such that lophids adjoining the tip of the hypoflexid are about equal in width. The metaflexids of m1 and more prominently, m2, have an angled internal tip with two points (perhaps outlining a mesoconid) that bends slightly posteriorly; the mesoflexid has an internal tip that bends slightly anteriorly. The lower premolar is pentalophodont, completely divided by two flexids into three parts: a symmetrical triangular anterior loph, rounded at its anterior tip and enclosing a subcircular fossetid; a single, laminar, central metalophid bar; and posterolophid and entoconids of posterior third of tooth united labially into a single, V-shaped loph that opens lingually. The lower incisors are not strongly curved.

SYNONYMS:

Myoxus: Zimmermann, 1780:352; part (description of *chrysurus*); not *Myoxus* Zimmermann.

Hystrix: Schreiber, 1792: table 170B; part (description of *chrysuros*); not *Hystrix* Linnaeus.

Glis: Treviranus, 1803:182: part (listing of *chrysurus*); not *Glis* Brisson.

Echimys F. Cuvier, 1809:394; based on Allamand's (1778) "*lérot a queue dorée*" and Azara's (1801:73) "*rat épineux*"; type species *Echimys cristatus* Desmarest (= *Myoxus chrysurus* Zimmermann) by subsequent designation (Tate 1935:430).

Loncheres Illiger, 1811:90; type species *Myoxus chrysurus* Zimmermann, by original designation.

Echimys Desmarest, 1817b:54; type species *Echimys cristatus* Desmarest, by original designation (see Remarks).

Loncheris Olfers, 1818; incorrect subsequent spelling of *Loncheres* Illiger.

Nelomys Jourdan, 1837:522; type species *Echimys cristatus* Desmarest (see Remarks).

Echinomys Wagner, 1840:199; unjustified emendation of *Echimys* F. Cuvier (same unjustified emendation cited by I. Geoffroy St.-Hilaire 1840:1); see Remarks.

Echimys I. Geoffroy St.-Hilaire, 1840:30; not *Echimys* F. Cuvier (see Remarks).

Enchomys Gloger, 1841 (1842):100; replacement name for *Echimys* Desmarest, and thus an objective synonym (see Thomas 1895a:190).

Echmiys Miranda-Ribeiro, 1914:48; incorrect subsequent spelling of *Echimys* F. Cuvier.

REMARKS: J. A. Allen (1899c), Thomas (1916e), and Tate (1935) outlined the convoluted and intertwined taxonomic history of *Echimys* and *Loncheres*, two generic names that have been commonly applied to the arboreal spiny rats, including *Echimys* as we envision this genus herein. *Echimys* was first used by É. Geoffroy St.-Hilaire in a manuscript apparently written in either 1808 or 1809 but never published (I. Geoffroy St-Hilaire 1838, 1840). F. Cuvier (1809) first published the name basing it on the "*lérot à queue dorée*" (from Allamand's [1778] supplements to Buffon's *Histoire Naturelle*, which was also the basis of *Myoxus chrysurus* by Zimmermann [1780]), and the "*rat épineux*" of Azara (1801). F. Cuvier (1812) described and figured the dentition of the only two species mentioned by him as belonging to this genus, namely the same "*le lérot à queue dorée*" (plate xv, Fig. 15) and "*rat épineux*" (plate xv, Fig. 14). Illiger (1811) described *Loncheres*, referring to it two species: *Loncheres paleacea*, a *nomen nudum* first made available by Olfers (1818:212; also see Hershkovitz 1959b) and *Myoxus chrysurus* Zimmermann. Because an undescribed species cannot be taken as the type of a genus, *Myoxus chrysurus* Zimmermann becomes the type of Illiger's *Loncheres*. The seven species of spiny rats that É. Geoffroy St.-Hilaire referred to the genus *Echimys* in his unpublished manuscript were then described by Desmarest (1817b), who specifically equated the "*rat épineux de d'Azara*" to *Echimys spinosus* (a *Euryzygomatomys*) and É. Geoffroy St.-Hilaire's "*lérot à queue dorée*" to *Echimys cristatus*, a renaming of *Myoxus* [= *Loncheres*] *chrysurus*. Desmarest wrote explicitly of *Echimys* "Le lérot à queue dorée d'Allamand et de Buffon en est le type" (1817b:54) and thus designated *Echimys cristatus*, which he equated with "*lérot à queue dorée*," as the type species of the genus *Echimys*. Tate (1935:418) considered this "clearly a designation of the type of *Echimys*," as do we. Many authors have overlooked Desmarest's designation, however, and since Cuvier's description and figures clearly indicated that his two species are not congeneric, J. A. Allen (1899c:262–263) argued that one of these needed to be selected as the type of his *Echimys* and further argued that the erection of *Loncheres* Illiger, with its type *M. chrysurus* (= *E. cristatus* Desmarest), automatically made *Echimys spinosus* Desmarest (or *"rat épineux de d'Azara"*) the type of *Echimys* F. Cuvier. Thomas (1916c:71) disagreed, pointing out that J. Fleming (1822:191) had much earlier selected *"Hystrix chrysurus* Lin Gmel" (= *Myoxus chrysurus* Zimmermann) as the type of *Echimys*. Finally, Tate (1935:430) stated that J. Fleming had merely listed the name *Hystrix chrysurus* without clearly and unambiguously selecting it as the type to F. Cuvier's *Echimys*. He then (p. 430) designated "as the type species of *Echimys* F. Cuvier, 1809, the *"lérot à queue dorée"* (= *Myoxus chrysurus* Zimmermann, 1780, = *Echimys cristatus* Desmarest, 1817)." É. Geoffroy St.-Hilaire, followed by F. Cuvier and Desmarest, may have originally named the taxon because the Paris Museum had earlier acquired a specimen from French Guiana that is noted in É. Geoffroy St.-Hilaire's 1803 catalogue as the basis for the description of the *"lérot à queue dorée"* (MHN A-7694 old no. 403), although Allamand stated that his (1778) description is of a Surinam specimen, which was sent to him by Klockner. It is not clear what became of the holotype from Surinam or if É. Geoffroy St.-Hilaire ever saw it. The lovely figure in Allamand (1778) was clearly drawn from the specimen he described in great detail, but subsequent figures in later editions of Buffon were progressively poorer copies. Zimmermann (1780) condensed his description of *Myoxus chrysurus* from Allamand's more detailed one, and translated its name directly into Latin, without evidence that he ever saw a specimen.

In his 1899 paper, J. A. Allen also removed the terrestrial spiny rats from *Echimys* (and *Loncheres*) to his new genus *Proechimys* (p. 264), a broad concept in which he included species now grouped into four genera within the subfamily Eumysopinae (*Proechimys* proper, *Trinomys*, *Hoplomys*, and *Mesomys*). Many subsequent researchers have shared Allen's narrowed view of the genera of arboreal spiny rats (Thomas 1928c, d; Tate 1935; Ellerman 1940; Cabrera 1961). Moojen (1952b), however, was the first to recognize further groups in the traditional *Echimys*, elevating the Atlantic Forest arboreal spiny rats to the genus *Phyllomys* Lund. Husson (1978) erected the genus *Makalata* to include *Nelomys armatus* I. Geoffroy St.-Hilaire, and more recently Emmons and Vucetich (1998) created *Callistomys* to include *Nelomys pictus* Pictet. Emmons, Leite et al. (2002; see also Y. L. R. Leite 2003) confirmed Moojen's separation of *Phyllomys*. Most recently, Iack-Ximenes et al. (2005a) placed *Echimys grandis* Wagner in the new genus *Toromys*, and Emmons (2005) removed *Echimys semivillosus* I. Geoffroy St.-Hilaire to her new genus *Pattonomys*. What remains of *Echimys* F. Cuvier, therefore, is only a remnant of the large number of species once, and varyingly, included within the genus.

Echinomys has been commonly listed as a synonym of *Echimys* F. Cuvier, with the author given as either I. Geof-

froy St.-Hilaire, 1840 (e.g., Tate 1935) or Wagner, 1840 (Simpson 1945; Cabrera 1961; Woods 1993; McKenna and Bell 1997; Woods and Kilpatrick 2005). It is unclear which of these two publications has date priority. *Echinomys* is, however, an unjustified emendation of *Echimys* F. Cuvier (rationale for name modification given by Wagner [1840:199; 1843a:339]). Moreover, all species listed by Wagner (1843a:339) under his *Echinomys* belong not to *Echimys* but to other genera of echimyids, notably *Euryzygomatomys*, *Mesomys*, *Proechimys*, or *Trinomys* (see those accounts).

Cabrera (1961) included 19 valid species or subspecies in his concept of the genus *Echimys*. Most of these are now placed in other genera, three of which have been described subsequent to his work. Herein, we follow the recent revisionary analyses noted immediately above in the allocation of the names Cabrera listed within *Echimys* as follows: *Makalata* Husson (including *armatus* I. Geoffroy St.-Hilaire and *macrura* Wagner), *Pattonomys* Emmons (*carrikeri* J. A. Allen, *occasius* Thomas, *punctatus* Thomas, and *semivillosus* I. Geoffroy St.-Hilaire), *Phyllomys* Lund (*blainvillei* Jourdan, *braziliensis* Lund, *dasythrix* Hensel, *lamarum* Thomas, *medius* Thomas, *nigrispinus* Wagner, *thomasi* Ihering, and *unicolor* Wagner), and *Toromys* Iack-Ximenes, de Vivo, and Percequillo (*grandis* Wagner and *rhipidurus* Thomas). With the removal of these names to other genera, *Echimys* as currently understood comprises three species, one of which was only recently described.

KEY TO THE SPECIES OF *ECHIMYS*:

1. Distributed in eastern Amazonia and Guianan region 2
1'. Distributed in western Amazonia; head glossy black . . . . *Echimys saturnus*
2. Wide to narrow white blaze from nose to crown, may be reduced to white tuft on top of head . . . . *Echimys chrysurus*
2'. Top of head dark maroon. . . . . *Echimys vieirai*

Echimys chrysurus (Zimmermann, 1780)
White-faced Tree Rat
SYNONYMS:

Myoxus chrysurus Zimmermann, 1780:352; type locality "Surinam," based on Allamand's (1778) *"lérot à queue dorée"* with a nearly verbatim translation of Allamand's description.

Hystrix chrysuros Schreiber, 1792:Table 170B; see also Schreiber, 1843:333; incorrect subsequent spelling of *chrysurus* Zimmermann.

M[yoxus]. chrysuros: H. F. Link, 1795:77; name combination.

Glis chrysurus: Treviranus, 1803:182; name combination.
Loncheres chrysurus: Illiger, 1811:90; name combination.
Loncheres paleacea Illiger, 1811:90; *nomen nudum*.
[*Loncheres*] *chrysura*: Illiger, 1815:108, *lapsus* to be ignored.
Echimys cristatus Desmarest, 1817b:54; a renaming of Allamand's "*lérot à queue dorée*"; attributed the name to É. Geoffroy St.-Hilaire, not to Zimmermann.
Loncheres paleacea Olfers, 1818:212; type locality "Brazilien"; renaming of *Loncheres paleacea* Illiger, a *nomen nudum*; antedates Lichtenstein (1820) and Kuhl (1820), who both redescribed the same Berlin specimen seen by Olfers and Illiger.
Nelomys paleaceus: I. Geoffroy St.-Hilaire, 1838a:887; name combination.
Nelomys cristatus: I. Geoffroy St.-Hilaire, 1838a:887; name combination.
Loncheres cristata: Wagner, 1843a:332; name combination.
Echimys chrysurus: E.-L. Trouessart, 1904:503; first use of current name combination.
E[*chimys*]. *paleaceus*: Thomas, 1916e:295; reference to Lichtenstein's name.
[*Echimys*] *paleacea*: Tate, 1935: 431; name combination and incorrect gender concordance.
Echimys paleacea: Ellerman, 1940:113; incorrect gender concordance, following Tate (1935).

DESCRIPTION: A relatively large species, mean head and body length 277.3 mm (range: 242–300 mm, N = 11), tail length 336.8.0 mm (range: 305–370 mm), hindfoot length 50.8 mm (range: 42–54 mm), and ear length 19.5 mm (range: 19–20 mm). Upper parts colored dark to pale gray-brown, pelage of upper parts densely covered with wide, flat spines from neck to rump with remainder of pelage composed of stiff bristles. A pale-colored (whitish to yellowish) blaze occurs down center of head, from crown to nose, either completely covering face between the eyes or only as narrow stripe or small white tuft on crown; remainder of crown usually dark brown; cheeks and sides of neck sometimes reddish. Ears small, thinly haired, and brown. Eyes dark brown, with weak, dull red eyeshine. Tail longer than the head and body (mean 120%), fully haired, dusky basally with distal half to two-thirds white, yellow, or chestnut. Chin, throat, inguinal region, and a band across chest behind the elbows white; remainder of under parts gray-brown, not sharply differentiated from sides. Considerable individual variation present in color, particularly in extent of white on head, under parts, and tail. Feet large, broad, with strong claws on all toes; colored gray-brown above.

DISTRIBUTION: *Echimys chrysurus* is known from the eastern Amazon Basin and Guianan region of Brazil, Guyana, Surinam, and French Guiana, east of the Rio Negro north of the Rio Solimões and east of the Rio Xingu south of the Rio Solimões.

SELECTED LOCALITIES (Map 493): BRAZIL: Amapá, Serra do Navio (Iack-Ximenes et al. 2005b); Amazonas, Rio Pitinga (M. N. F. da Silva and Patton 1993); Maranhão, Vargem Grande (T. G. Oliveira and Mesquita 1998); Pará, 52 km SSW of Altamira, right bank Rio Xingu (Y. L. R. Leite and Patton 2002), Baião (Iack-Ximenes et al. 2005b), Peixe-Boi (Iack-Ximenes et al. 2005b), São Domingo do Capim (Iack-Ximenes et al. 2005b). FRENCH GUIANA: Cayenne (Iack-Ximenes et al. 2005b). GUYANA: East Berbice-Corentyne, Potaro Highlands (Iack-Ximenes et al. 2005b), Essequibo Islands–West Demerara, Supinaam River (Iack-Ximenes et al. 2005b); Upper Takutu–Upper Essequibo, Tamton (ROM 36840). SURINAM: Nickerie, Polder, near Nieuw Nickerie (Husson 1978), Paramaribo, near Paramaribo (Husson 1978).

SUBSPECIES: *Echimys chrysurus* is monotypic.

NATURAL HISTORY: Nocturnal, arboreal, and usually solitary, *E. chrysurus* occurs in mature lowland rainforest where it feeds on fruit and probably leaves. Individuals use the middle and upper levels of the forest, especially in viney areas, but are also present in open forest (Miles et al. 1981). They may run quickly through the branches but are more typically slow and methodical in their movements, making them difficult to spot. When disturbed they creep quietly into thick vegetation and sit motionless for many minutes, then sneak away silently (Emmons and Feer 1997). They make nests of leaves in tree holes. Known predators include the Spectacled Owl (Haverschmidt 1968).

REMARKS: Cuvier's (1809) *Echimys* included several species now classified in different genera, but Desmarest (1817b) clearly fixed the type species of *Echimys* by writing "Le lérot à queue dorée d'Allamand et de Buffon en est le type" (p. 54), which is *Myoxus chrysurus* Zimmermann. Iack-Ximenes et al. (2005b:57) followed Tate (1935:430) in considering *E. cristatus* Desmarest as a junior synonym of *Echimys chrysurus* Zimmermann. Lichtenstein (1820) has been generally regarded as the author of *Loncheres paleacea* (e.g., Thomas 1916e; Ellerman 1940; Cabrera 1961; Woods 1993; Woods and Kilpatrick 2005), with Illiger's (1811) first use of the name considered a *nomen nudum*. Iack-Ximenes et al. (2005b) suggested that *E. paleaceus* is a valid species, a conclusion that followed Moojen's (1952b) earlier opinion. Cabrera (1961:541) listed *paleaceus* as a valid subspecies of *E. chrysurus*, but cited Kuhl (1820) as the author of the name because his work apparently appeared earlier in 1820 than did Lichtenstein's description. However, both Iack-Ximenes et al. (2005b) and Cabrera (1961) overlooked Olfers' 1818 description of *Loncheris* [*sic*] *paleacea* (see Hershkovitz 1959h, for discussion), which was based on the same specimen in the Berlin Museum described later by both Kuhl and Lichtenstein. Thus, Kuhl and Lichtenstein only redescribed Olfer's *paleacea*.

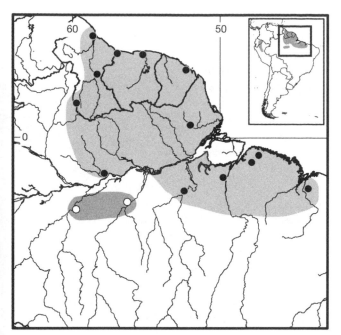

Map 493 Selected localities for *Echimys chrysurus* (●) and *Echimys vieirai* (○).

DESCRIPTION: Also moderately large arboreal rat, head and body length 272–335 mm (N = 2), tail length 295–383 mm, hindfoot length 50–51, ear length 17–18 mm. Upper parts generally glossy brown and head, mid-back, and tail base deep glossy black; sides chestnut or coppery in color. Dorsal pelage composed of stiff bristles and mid-back has wide, flat spines that do not extend onto rump; spines slaty with black tips. Ears short, buried in hair of head. Under parts white, white-spotted, or whitish with buff tinge; chin dark. Tail much longer than the head and body (140% in one individual); densely haired; black at base with variable length of white on or near tip. Hindfeet broad and strong, with strong claws on each toe. Some melanistic individuals known.

DISTRIBUTION: *Echimys saturnus* is known from the eastern Andean foothills and adjacent lowlands of Ecuador and northern Peru; all records are below 1,000 m.

SELECTED LOCALITIES (Map 494): ECUADOR: Napo, Cerro Galeras (AMNH 71903), Río Pucuno (MCZ 41569); Orellana, Río Cotapino (KU 68093), Tiputini Biodiversity Station (Blake et al. 2010). PERU: Amazonas,

Hershkovitz (1959b) argued that Illiger's (1811:90) *L. paleacea* is actually valid, as it is contained within his diagnosis of the genus *Loncheres*. However, because Illiger (1811) listed two species under his new genus, one of which was a valid taxon at the time (*chrysurus*), the nonspecific description of the genus cannot be said to diagnose his new species (*paleacea*). Thus, *paleacea* Illiger remains as a *nomen nudum*, as thought by J. A. Allen (1899c). Lichtenstein (1820) redescribed the same Berlin specimen that formed the basis for Illiger's *paleacea* as like a porcupine with a white stripe on the head and a tail with a scaly base and tufted tip, features that clearly link *paleaceus* to *E. chrysurus*. Finally, Iack-Ximenes et al. (2005b:58) assigned a specimen from Peixe-Boi, Pará state, Brazil in the Natural History Museum of London (BM 14.6.10.1) to *E. chrysurus*, the same specimen which Thomas (1916e:295) remarked was "the closely allied but smaller Amazonian species *E. paleaceus*, Licht., [which] has been received from the Goeldi Museum, Para (locality, Peixe-Boi)." We have examined this specimen and agree that it represents *E. chrysurus*. Therefore, pending further field and museum work, we regard *paleaceus* as a synonym of *E. chrysurus* and accept Olfers as the valid author of the name.

Echimys saturnus Thomas, 1928
Dark Tree Rat
SYNONYM:

Echimys saturnus Thomas, 1928d:409, type locality "Rio Napo, Oriente of Ecuador. Alt. 3300'," probably Napo province.

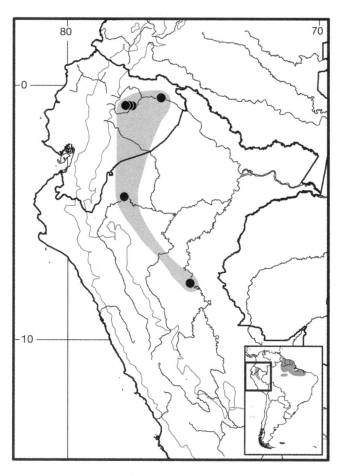

Map 494 Selected localities for *Echimys saturnus* (●). Contour line = 2,000 m.

mouth of Río Santiago (AMNH 98262); Ucayali, Río Pisqui (AMNH 98261).

SUBSPECIES: *Echimys saturnus* is monotypic.

NATURAL HISTORY: This species has not been studied in the field. Blake et al. (2010) obtained nine cameratrap photographs of this rare species on the ground at night beside a stream by a mineral lick. One photo included an apparent adult pair and one young traveling together in a tight group (color photo in Blake et al. 2010). Repeated presence at a lick suggests an herbivorous diet (Emmons and Stark 1979).

REMARKS: Virtually nothing is known about this very infrequently seen or collected species. Thomas (1928d:409) considered *E. saturnus* closely related to *E. chrysurus*, but "distinguishable by the absence of a white crown-patch, by its deep black head and centre area of the back, its coppercoloured sides, white middle line below, and nearly wholly black tail." Iack-Ximenes et al. (2005b: table 1) provided comparative measurements for a single specimen of *E. saturnus* and a sample of similar aged (by tooth wear categories) *E. chrysurus*. The former species is slightly larger in all dimensions of the skull with the exception that it has smaller bullae. Emmons (2005:286, Fig. 14) provided dorsal, ventral, and lateral photographic views of the cranium and a lateral view of the left mandible of the holotype (BM 34.9.10.182).

Echimys vieirai Iack-Ximenes, de Vivo, and Percequillo, 2005

Vieira's Tree Rat

SYNONYM:

Echimys vieirai Iack-Ximenes, de Vivo, and Percequillo, 2005b:52; type locality "Barreirinha, right bank of the Tapajós River, near São Luis do Tapajós, Pará State, Brazil; geographic coordinates 04°25′S, 56°13′W."

DESCRIPTION: Smallest species in genus, but only known from young individuals; head and body length of one specimen 245 mm, tail length 340 mm, hindfoot length 50 mm, and ear length 15 mm. Upper parts generally dark brown and head from muzzle to crown, midback, and tail base black; head with dark maroon medial stripe from rostrum to crown; sides dull brown only slightly distinct from back. Dorsal pelage composed of stiff bristles and mid-back has wide, flat spines that do not extend onto rump. Ears short, buried in hair of head. Under parts light brown. Tail long (139% of head and body length), densely haired, brown at base with terminal third white. Hindfeet broad and strong, with strong claws on each toe.

DISTRIBUTION: *Echimys vieirai* is known only from two localities south of the Rio Amazon in Amazonas and Pará states, Brazil (Iack-Ximenes et al. 2005b).

SELECTED LOCALITIES (Map 493): BRAZIL: Amazonas, Virgen Guajará (Iack-Ximenes et al. 2005b); Pará, Barreirinha (type locality of *Echimys vieirai* Iack-Ximenes, de Vivo, and Percequillo).

SUBSPECIES: *Echimys vieirai* is monotypic.

NATURAL HISTORY: This species has not been studied in the field.

REMARKS: Known from only two juvenile specimens, this species is distinguished from *E. chrysurus* primarily by the lack of the white blaze on the top of the head that is so characteristic of the latter throughout its broad range. In this respect, *E. vieirai* is similar to *E. saturnus* from the northwestern Amazon, but other features of the pelage and cranium, as well as the large hiatus between the known ranges, make it unlikely that they represent only geographic variants of the same taxon.

Genus *Isothrix* Wagner, 1845

Louise H. Emmons and James L. Patton

These are medium to relatively large arboreal rats with soft fur, and a tail covered with long, unbanded hair that curls outward like a bottlebrush but may not completely hide the underlying scales. The tail of both *I. pagurus* and *I. sinnamariensis* has the tendency to coil at the tip into a tight downward spiral of loops (Patton and Emmons 1985; Vié et al. 1996).

The cranium is broad, with a short and broad rostrum; zygomatic arches are strongly bowed outward anterior to squamosal. The tympanic bullae are moderately inflated, and the auditory meatus is medium-sized, positioned near the squamosal, and with a short auditory tube strongly slanted forward. The squamosotympanic fenestra is a long, open slit that reaches posteriorly to the base of the lateral process of the supraoccipital. The postpalatal notch is deep, extending to the middle of M2; the incisive foramina are large and wide. In the pterygoid region, there is no shelf of bone behind the base of parapterygoid (hamular) processes on the wall of the buccinator foramen. Anterior to the presphenoid-basisphenoid suture is a round vacuity connected to a slit-like vacuity along the side of the presphenoid that is shaped like a written musical note. The base of the mandible has strongly developed masseteric and pterygoid crests; the mandibular foramen is anterior to the condyloid ridge and not in a fossa.

The maxillary toothrows are short, ≤26.8% of basilar length of Hensel. Maxillary cheek teeth are small and short, with a somewhat subcircular shape in occlusal view. The plane of the maxillary teeth is not strongly tipped laterally as in other echimyines; rather, the occlusal surface is nearly on a plane parallel to the palate (see Emmons 2005:258, Fig. 2). All hypoflexi(id)s of both upper and lower premolars and molars are oval to subcircular with a mure; the adjacent lophs curve around the hypoflexi to nearly close at the rim of the tooth. Hypoflexids of the mandibular cheek teeth slant slightly backward, with the internal end of the flexid slightly posterior to the external end. The lower premolar is pentalophodont, hypoflexid is short, with a mure, and other flexids are close to parallel, and all are open lingually when unworn. This tooth quickly wears to a pointed oval, with three parallel lingual fossettes and a round labial fossette derived from the hypoflexid.

Mammae range in number from three to five lateral pairs with usually one inguinal pair. The phallus of *I. bistriata* is short and blunt, with deep longitudinal grooves on sides, the lappet above urethra is not salient from the crater (figured in Emmons 2005), and the baculum is long and narrow (Patton and Emmons 1985).

We recognize six species, which fall into two readily separable groups based on body size and pelage color and color pattern. The species within each group, however, lack easily observable morphological differences. The three species of the "*bistriata*" group (*I. bistriata, I. negrensis,* and *I orinoci*) are all large, but separable from one another only by multivariate analyses of cranial mensural variables and, in the case of two of them, by molecular data (data for *I. orinoci* are lacking). The other three species (*I. barbarabrownae, I. pagurus,* and *I. sinnamariensis*) are all smaller; the latter two of this triad are largely distinguishable by karyotype as they share similar but reciprocally monophyletic mitochondrial sequences. Adult specimens documenting overall size and color pattern are unknown for the recently described *I. barbarabrownae*. Each of the six species has allopatric ranges, so species assignments of museum specimens can be confidently made by a combination of morphology coupled with distributional data.

SYNONYMS:

Isothrix Wagner, 1845a:145; type species *Isothrix bistriata* Wagner, by subsequent designation (Goldman, 1916:125.)

Loncheres (Isothrix): Waterhouse, 1848:327; part, as subgenus (listing of *bistriata* and *pagurus*).

Lasiuromys Deville, 1852:557; type species *Lasiuromys villosus* Deville, by monotypy.

Echimys (Isothrix): E.-L. Trouessart, 1904:504; part, as subgenus (listing of *bistriata, pagurus,* and *villosus*); not *Echimys* F. Cuvier.

REMARKS: Members of this genus, like *Callistomys* spp., have some characters shared with eumysopines and not with other echimyines. These include small and somewhat rounded cheek teeth with only the hypoflexi(id)s open (other flexi(id)s appearing as closed enamel folds), and an open squamosotympanic fenestra. Some molecular sequence analyses (Lara et al. 1996; Y. L. R. Leite and Patton 2002; Galewski et al. 2005; B. D. Patterson and Velazco 2008; Upham and Patterson 2012) failed to firmly support inclusion of *Isothrix* within the Echimyinae, although other studies did (P.-H. Fabre et al. 2012). Phylogenetic analyses of morphological characters (G. A. S. Carvalho and Salles 2004; B. D. Patterson and Velazco 2006) do place *Isothrix* within a group of echimyine genera. As mentioned previously, Emmons (2005) regarded the phylogenetic position to be sufficiently equivocal to list *Isothrix* as Echimyinae *incertae sedis*. The two most recent and comprehensive molecular analyses both place *Isothrix* as basal to other echimyine plus the dactylomyine genera (P.-H. Fabre et al. 2012; Upham et al. 2013).

The genus was reviewed by Patton and Emmons (1985), who included *Callistomys pictus*, following previous authors and without any character evaluation or other justification. Exclusive of *pictus*, these authors recognized two species: *Isothrix bistriata*, a wide-ranging western Amazonian taxon (in which they included the named forms *pachyura* Wagner, *villosa* Deville, *orinoci* Thomas, *negrensis* Thomas, *molliae* Thomas, and *boliviensis* Petter and Cuenca Aguirre) varyingly characterized by black or dark-brown supraorbital stripes extending onto the nape of the neck; and *Isothrix pagurus*, a smaller species with a reddish nose that lacks stripes on the head and that is limited to both sides of the lower Amazon River east of the Rio Negro-Madeira axis. Patton and Emmons (1985), however, documented that populations of *I. bistriata* from the Venezuelan Amazon (including the holotype of *orinoci* Thomas 1899) were separable from those of the western Amazon of Peru and Bolivia in multivariate morphometric characters as well as in color pattern. They lacked samples from the central Amazon of Brazil, except the holotype of *negrensis* Thomas, which by itself fell within a western Amazon morphometric group. Based on similar color and color patterns, Patton and Emmons (2005:12) treated *orinoci* Thomas as a valid subspecies of a polytypic *Isothrix bistriata*, and listed *negrensis* Thomas as a junior synonym of the nominotypical subspecies. Subsequently, based on newly collected materials and expanded analyses, Patton et al. (2000; see also M. N. F. da Silva and Patton 1993) showed that samples from the central Amazon of Brazil, including the geographic area from which the holotype of *negrensis* was obtained, were phyletically separable from those of western Amazonia in mtDNA sequences, and that these former samples were also different in a multivariate morphometric analysis from both western Amazonian samples and those from southern Venezuela. Bonvicino, Menezes, and Oliveira (2003) documented chromosomal differences between populations from the western and central Amazon and expanded on the molecular divergence analysis of Patton et al. (2000). These authors concluded that *negrensis* Thomas was a valid species, diagnosable by morphometric, chromosomal, and molecular characters, a conclusion followed by Emmons (2005), Woods and Kilpatrick (2005), B. D. Patterson and Velazco (2006, 2008), and Lim et al. (2006). Emmons (2005) listed *orinoci* Thomas as a valid species but without comment, and B. D. Patterson and Velazco (2006) provided a thorough rationale for the species status of this taxon. Herein, we follow these recent trends and divide *Isothrix bistriata* (*sensu* Patton and Emmons 1985) into three species: *bistriata* Wagner, *negrensis* Thomas, and *orinoci* Thomas.

Additionally, two newly recognized species have been discovered in the past two decades. Vié et al. (1996) described *I. sinnamariensis* from French Guiana, closely related to *I. pagurus* but with a radically different chromosome complement. And, B. D. Patterson and Velazco (2006) described *I. barbarabrownae* from the eastern Andean slope in southern Peru.

Finally, *pachyura* Wagner, 1845 has been commonly referred to *Isothrix* as a valid species or subspecies (Palmer 1904; Tate 1935; Ellerman 1940; Cabrera 1961) or simply arranged as a junior synonym of *I. bistriata* (Patton and Emmons 1985). However, Wagner (1848) himself, while renaming *I. pachyura* as *I. crassicaudus*, stated that it was probably the same as Lund's *Nelomys antricola*, to which E.-L. Trouessart (1880b:179, as *Echimys* [*Thrichomys*] *antricola*) agreed. Thomas (1916e) also correctly allocated *pachyura* Wagner to *Cercomys* (= *Thrichomys*, see that account), and Moojen (1952b:172) regarded it as a valid subspecies of *C. cunicularis* (a composite taxon [see Remarks in the *Thrichomys* account] but as understood by Moojen would equal *Thrichomys apereoides* of today).

KEY TO THE SPECIES OF *ISOTHRIX*:

1. Size large (mean head and body length >235 mm; mean condyloincisive length of skull >49 mm); black stripes on top of head 2
1'. Size smaller (mean head and body length <225 mm; mean condyloincisive length of skull <44 mm); head uniformly colored without black stripes 4
2. Black head stripes remain separate behind ears; distribution in western Amazon . . . . *Isothrix bistriata*
2'. Black head stripes fuse behind ears into a black band on the neck 3
3. Distribution in central Amazonian Brazil . . . . *Isothrix negrensis*
3'. Distribution in upper Orinoco drainage of Venezuela . . . .*Isothrix orinoci*
4. Tail color monochromatic; distribution in eastern Amazon and Guianan Shield 5

4'. Tail with terminal one-third white; distribution limited to eastern Andean slope of southern Peru . . . . *Isothrix barbarabrownae*
5 Postauricular dark patch absent; 2*n* = 22; distributed in east-central Amazonian Brazil . . . . *Isothrix pagurus*
5'. Postauricular dark patch present; 2*n* = 28; distributed in Guianan region . . . . *Isothrix sinnamariensis*

Isothrix barbarabrownae B. D. Patterson and Velazco, 2005

Barbara Brown's Brush-tailed Rat

SYNONYM:

Isothrix barbarabrownae B. D. Patterson and Velazco, 2006:179; type locality "Km 138.5 on the Carretera Paucartambo-Shintuya, near 'Suecia' (a roadside restaurant), 1900 m, Provincia de Paucartambo, Departamento de Cusco, Peru," in the Cultural Zone of the Manu Biosphere Reserve at 13°6.032'S, 71°34.124'W.

DESCRIPTION: Size small for genus (condyloincisive length of skill of subadult, 40.36 mm). Fur long and lax, hairs reach 20 mm on back, 15 mm on flanks, and 10 mm on belly. Head lacks conspicuous supraorbital stripes although blackish dorsal crest or mane is present. Tail uniformly haired and tricolored, with terminal 5 cm white, middle part black, and base brown. Skull compact, with long axis bowed; braincase not swollen, and temporal fossa not deep; infraorbital foramen narrow and triangular in shape; septum of incisive foramina with well-developed maxillary component; palate narrow; foramen magnum especially broad and with distinct dorsal median notch; auditory bullae small.

The holotype, and only known specimen, is a subadult without fully erupted cheek teeth. B. D. Patterson and Velasco (2006) provided craniodental measurements in comparison with other species in genus.

DISTRIBUTION: *Isothrix barbarabrownae* is known only from the type locality in southeastern Peru.

SELECTED LOCALITIES (Map 495): PERU: Cusco, near "Suecia" (type locality of *Isothrix barbarabrownae* B. D. Patterson and Velazco).

SUBSPECIES: *Isothrix barbarabrownae* is monotypic.

NATURAL HISTORY: This is the only species in the genus distributed in upper tropical forest rather than lowland rainforest. The holotype was collected on a steep, rocky slope covered by ferns, mosses, and bryophytes in elfin forest; a second individual was seen in similar circumstances but was not secured.

REMARKS: This species can be distinguished from all congeners by its long, lax, and reddish pelage. No other species have a dorsal crest or mane of blackish hairs. Cranially, *I. barbarabrownae* resembles *I. bistriata* but comparisons are limited because the holotype and only known specimen

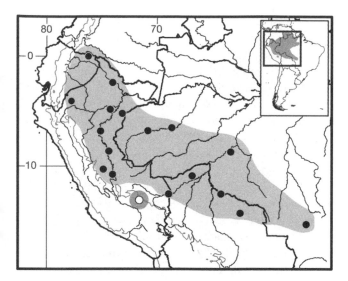

Map 495 Selected localities for *Isothrix barbarabrownae* (○) and *Isothrix bistriata* (●). Contour line = 2,000 m.

is a subadult where PM4 has not yet erupted. Phylogenetic analysis of mitochondrial and nuclear DNA sequences (B. D. Patterson and Velazco 2008) placed *I. barbarabrownae* at the base of a clade comprising the remaining species in the genus (e.g., [*barbarabrownae* [[[*pagurus, sinnamariensis*]] [[*negrensis, orinoci*] *bistriata*]]]).

Isothrix bistriata Wagner, 1845
Yellow-crowned Brush-tailed Rat

SYNONYMS:

Isothrix bistriata Wagner, 1845a:146; type locality "Rio Guapore," fixed herein to Isla Leomil, Río Guaporé, Beni, Bolivia (see Remarks).

Lasiuromys villosus Deville, 1852:560, plate 15, plate 16, Fig. 5, 5a; type locality "la Mission de Sarayacu (Pérou, Pampa del Sacramento, et le village de Saint-Paul, Haut-Amazone, partie brésilienne)," Río Ucayali, Loreto, Peru.

Loncheres (*Isothrix*) *bistriatus*: Thomas, 1899d:383; name combination.

Loncheres (*Isothrix*) *villosus*: E.-L. Trouessart, 1897:606; name combination.

Echimys (*Isothrix*) *villosus*: E.-L. Trouessart, 1904:504; name combination.

[*Isothrix*] *villosa*: Thomas, 1916e:295; name combination.

Isothrix villosa molliae Thomas, 1924c:534; type locality "Tushemo, near Masisea, R. Ucayali. Alt. 1000′," Loreto, Peru.

Isothrix villosa villosa: Thomas, 1928c:291); name combination.

[*Isothrix*] *bistriatus bistriatus*: Tate, 1935:414; name combination but incorrect gender concordance.

[*Isothrix*] *villosus villosus*: Tate, 1935:416; name combination but incorrect gender concordance.

Isothrix bistriata boliviensis Petter and Cuenca Aguirre, 1982:191; type locality "une région de forêt subtropicale humide (350 m d'altitude) au bord de la rivière Saint-Louis [= Río San Luis], sud-est du Béni (Bolivie)"; subsequently stated to be (p. 199) "Barranca Colorada, Beni, Bolivie [= Bolivia]."

Isothrix bistriata bistriata: Patton and Emmons, 1985:12; name combination.

DESCRIPTION: One of larger species in genus, mean head and body length 249.2 mm (range: 214–275 mm; N = 18), tail length 242.8 mm (range: 215–271 mm; N = 18), hindfoot length 48.3 mm (range: 45–54 mm; N = 18), ear length 18.5 mm (range: 15–20 mm; N = 18), and mean condyloincisive length 50.7 mm (range: 44.5–55.0 mm; N = 69). Dorsal color grizzled yellow-brown to olive mixed with black hairs, and well-defined black or dark-brown supraorbital stripes extending over forehead to nape and bordering a median pale creamy patch on crown. Brightness of crown patch and black supraorbital stripes varies, due to age, individual, or interpopulation variation. Unlike two similar species, *I. orinoci* and *I. negrensis*, black supraorbital stripes do not coalesce behind ears but remain separate onto neck. Venter pale yellow to buff. Tail fully haired, typically rust or golden colored over basal third to half with terminal portion usually black, but tail color varies considerably, with pale yellow to almost white tip in some animals but black in others. Tail about same length as body (mean tail: body ratio 97%, range: 87.6–108.9%). Baculum long and broad, more so than that of *I. orinoci* (comparison of Peru and Venezuela samples in Patton and Emmons 1985). Skull as for the genus.

DISTRIBUTION: *Isothrix bistriata* is known from the western Amazon Basin of Peru, Bolivia (S. Anderson 1997), and Brazil (Patton et al. 2000), and probably extends north to eastern Ecuador and possibly to southeastern Colombia. Tirira (1999:141) recorded a visual record from the Reserva de Producción Faunística Cuyabeno, Sucumbiós, and subsequently (Tirira 2007:225–226) he listed two localities from which photographic documentation was available and mapped a distribution in eastern Sucumbiós and Orellana provinces. L. H. Emmons (unpubl. obs.) likewise observed an *Isothrix* sp. in riverside forest on the Ecuador border near the mouth of the Río Napo, close to one of Tirira's localities. We include the observational and photographic records in the map (Map 492), with the reservation that we cannot confirm identifications of either. However, there is certainly no obvious reason why *Isothrix bistriata* would not be part of the fauna of eastern Ecuador or even southern Colombia.

SELECTED LOCALITIES (Map 495): BOLIVIA: Beni, Barranca Colorada, Río San Luis (type locality of *Isothrix bistriata boliviensis* Petter and Cuenca Aguirre), Isla Leomil (type locality of *Isothrix bistriata* Wagner; see Remarks);

Pando, Ingavi (Patton et al. 2000). BRAZIL: Amazonas, Jainu, left bank Rio Juruá (Patton et al. 2000), Sacado (Patton et al.2000); Mato Grosso, Cuiabá (Moojen 1952b; Bonvicino, Menezes, and Oliveira 2003); Rondônia, Samuel (Bonvicino, Menezes, and Oliveira 2003). ECUADOR: Sucumbíos, Reserva de Producción Faunística Cuyabeno (Tirira 1999). PERU: Amazonas, La Poza (Patton et al. 2000); Loreto, Boca Río Cururay (Patton and Emmons 1985), Nuevo San Juan (AMNH 272808), Río Tigre, 1 km below Río Tigrillo (B. D. Patterson and Velazco 2006), Sarayacu (Patton and Emmons 1985); Madre de Dios, Reserva Cusco Amazonico (B. D. Patterson and Velazco 2006); Pasco, Nevati Mission (Patton and Emmons 1985); Ucayali, Lagarto (Patton and Emmons 1985), Tushemo (type locality of *Isothrix villosa molliae* Thomas).

SUBSPECIES: As currently construed, *Isothrix bistriata* is monotypic. Thomas (1928c:291) placed his *molliae* (Thomas 1924c) as a synonym of *villosa* Deville and suggested that the latter might prove to grade into *bistriata* from Brazil, of which *villosa* would then "form an Upper Amazonian subspecies." Cabrera (1961) listed *negrensis* Thomas, *orinoci* Thomas, *pagurus* Wagner, and *pachyura* Wagner as valid subspecies in his concept of *I. bistriata*, the first three of which are herein elevated to species status and the latter referred to the genus *Thrichomys* (see above, and account of *Thrichomys*). Contrary to Thomas's (1924c) suggestion, Cabrera (1961) maintained *villosa* Deville as a species separate from *I. bistriata*, and included *molliae* Thomas as a valid subspecies. However, Patton and Emmons (1985) were unable to distinguish samples from Peru and Bolivia (including the holotype of *molliae* Thomas and topotypes of *villosa* Deville) in a multivariate analysis of craniodental mensural characters, and listed both names as synonyms of *I. bistriata bistriata* Wagner. This is likewise the case with *I. b. boliviensis* (Petter and Cuenca Aguirre 1982), also from Beni, Bolivia. The type locality of *boliviensis* (Río San Luis, Barranca Colorada), although not precisely located, is <300 km from the type locality of *I. bistriata* (the maximum length of the Río San Luis)

NATURAL HISTORY: Nearly all specimens of *I. bistriata* have been collected in seasonally inundated or floodplain forest (*várzea*) along whitewater rivers. These are arboreal animals that nest in tree holes, and where male and female pairs have been observed on several occasions (Patton et al. 2000). Litter size averages two, based on embryo scar counts, and breeding may be continuous throughout the year (Patton et al. 2000).

REMARKS: Wagner (1845a:145) gave the type locality of *I. bistriata* as "Rio Guapore," which has been uniformly considered to be in Mato Grosso state, Brazil, by subsequent authors (Ellerman 1940; Cabrera 1961; Patton and Emmons 1985; Woods and Kilpatrick 2005). However, the

catalogue sheet for Wagner's holotype in the Naturhistorisches Museum Wien, Vienna (NMW B914), reads "nahe Mündung des Rio Itonamas in den Guaporé 'gegenuber Liomil am Guaporé' (von Pelzeln 1883) Rondonia, Brasil oder Bolivia." The Río Itonamas is in Bolivia, its mouth at 12.750°S, 64.633°W on the Río Guaporé. There is an Isla Leomil or Leamel in the Río Guaporé, Beni, Bolivia (64.1667°S, 12.50°W; Alexandria Digital Library). We thus fix the type locality of *I. bistriata* Wagner to Isla Leomil, Río Guaporé, Beni, Bolivia.

Patton et al. (2000) assigned Wagner's *bistriata* to the western Amazon clade they defined on morphological and molecular grounds. Although they did not include specimens from the type locality in their analyses, one sample from Pando ("c," p. 179; actually from Ingavi and not San Juan de Nuevo Mundo) is 350 km NW of the type locality. This sample clusters with those from Madre de Dios (Peru) and the upper Rio Juruá (Brazil) (Patton et al. 2000). If subsequent studies indicate that populations from southwestern Amazonia are separable, then this western Amazon clade would be appropriately named *I. villosa* Deville, with *I. bistriata* restricted in distribution to the area south of the Rio Juruá to southeastern Peru and Bolivia south and east to Cuiabá, Brazil. The karyotype (based on specimens from the Rio Jamarí, Rondônia, and the Rio Juruá, Acre and Amazonas, Brazil) consists of $2n = 60$, $FN = 116$, with a large acrocentric X chromosome and a very small, metacentric Y chromosome (Leal-Mesquita 1991, who illustrated both the nondifferentially stained and C-banded complement; Patton et al. 2000). Valim and Linardi (2008) described the host association of ectoparasitic lice in the genus *Gyropus*.

Isothrix negrensis Thomas, 1920
Rio Negro Brush-tailed Rat
SYNONYMS:

Isothrix bistriata negrensis Thomas, 1920f:277, type locality "Acajutuba, lower Rio Negro, near its mouth," a little above Manaus on the Rio Negro, Amazonas, Brazil.

[*Isothrix*] *bistriatus negrensis*: Tate, 1935:416; name combination but incorrect gender concordance.

Isothrix bistriata negrensis: Patton and Emmons, 1985:12; name combination.

Isothrix negrensis: Bonvicino, Menezes, and Oliveira, 2003: 206; first use of current name combination.

DESCRIPTION: Size large for genus, with mean head and body length 254.1 mm (range: 203–292 mm; N = 3), tail length 226.5 mm (range: 182–271 mm; N = 3), hindfoot length 49.3 mm (range: 43–45 mm; N = 3), ear length 18.7 mm (range: 17–20 mm; N = 18), and mean condyloincisive length 53.7 mm (range: 50.9–55.9 mm; N = 3). Similar in color and color pattern to *I. orinoci*, and thus differing from *I. bistriata* by having marked, light postauricular patches and a shorter median light crown-patch, which ends opposite the middle of the ears instead of continuing down the nape so that the nape is completely blackish instead of divided down the center (Thomas 1920f); differs from *I. orinoci* in more ochraceous color, especially on rump, buffy rather than whitish ear patches, median crown patch distinct and buffy instead of whitish, and strongly ochraceous buffy venter. Tail proportionally longer than head and body length (mean 106.7%). Cranial vault deeper, in lateral view, and diastema shorter in this species than in either *I. bistriata* or *I. orinoci*; otherwise all three species are similar in cranial morphometric characters (Patton et al. 2000; B. D. Patterson and Velazco 2006).

DISTRIBUTION: *Isothrix negrensis* occurs in the central Amazon Basin of eastern Colombia and Brazil, from the lower Rio Juruá east to the Rio Madeira on the south side of the Rio Solimões, and on the north side throughout the Rio Negro drainage.

SELECTED LOCALITIES (Map 496): BRAZIL: Amazonas, Acajutuba (type locality of *Isothrix bistriata negrensis* Thomas), Colocação Três Barracas (Bonvicino, Menezes, and Oliveira 2003), Ilha das Onças (Patton et al. 2000), Lago Três Unidos (Patton et al. 2000), alto Rio Urucu (Patton et al. 2000). COLOMBIA: Vaupés, Yay Gojes (FMNH 88236).

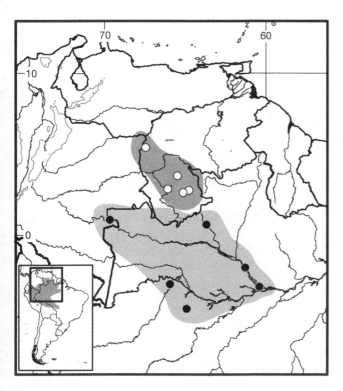

Map 496 Selected localities for *Isothrix negrensis* (●) and *Isothrix orinoci* (○). Contour line = 2,000 m.

SUBSPECIES: *Isothrix negrensis* is monotypic.

NATURAL HISTORY: This species has been taken in seasonally inundated, lowland rainforest, along both whitewater and blackwater rivers.

REMARKS: *Isothrix negrensis* corresponds to the central Amazon clade of "*I. bistriata*" defined on mtDNA sequences by Patton et al. (2000:179, Fig. 117). Emmons (2005:278, Fig. 11) provided dorsal, ventral, and lateral photographs of the skull and lateral view of the mandible of the holotype (BM 20.7.1.20). Treated as a synonym of *I. bistriata bistriata* by Patton and Emmons (1985) but as a valid subspecies by Patton et al. (2000). Elevated to species status by Bonvicino, Menezes, and Oliveira (2003), who described a karyotype with 2*n*=60, FN=112, and provided additional molecular sequence to support its separation from *I. bistriata*.

Isothrix orinoci (Thomas, 1899)
Orinoco Brush-tailed Rat

SYNONYMS:

Loncheres (*Isothrix*) *bistriatus orinoci* Thomas, 1899d: 382; type locality "Maipures, Upper Orinoco," Vichada, Colombia.

Echimys (*Isothrix*) *bistriatus orinoci*: E.-L. Trouessart, 1904: 504; name combination.

Isothrix bistriatus orinoci: Pittier and Tate, 1932:264; name combination but incorrect gender concordance.

[*Isothrix*] *bistriatus orinoci*: Tate, 1935:416; name combination but incorrect gender concordance.

Isothrix bistriata orinoci: Patton and Emmons, 1985:12; name combination.

Isothrix orinoci: Emmons, 2005:299; first use of current name combination.

DESCRIPTION: Similar in size, color, and color pattern to *I. bistriata* and *I. negrensis*. Mean head and body length 236.0 mm (range: 206–257 mm; N=19), tail length 242.9 mm (range: 168–278 mm; N=19), hindfoot length 48.4 mm (range: 45–50 mm; N=19), and mean condyloincisive length 49.9 mm (range: 47.0–52.2 mm; N=19). but with a bright yellow stripe over the head between the eyes and ears dulled to grizzled gray, passing on each side without sharp contrast into the dark supraorbital lines, which coalesce on the neck immediately behind the head to form a broad, black band; prominent whitish postauricular patch contrasting sharply with the blackish nuchal part of the coalesced supraorbital lines. The skull has a shorter rostrum but proportionally longer nasals than either *I. bistriata* or *I. negrensis* (Patton et al. 2000). The baculum is long and narrow, as in *I. bistriata*, but slightly smaller in all dimensions (Patton and Emmons 1985).

DISTRIBUTION: *Isothrix orinoci* is known from the Amazon Basin of the upper Río Orinoco in eastern Colombia and Venezuela as well as the Casiquiare of southern

Venezuela; it possibly extends into adjacent northern Brazil (Patton and Emmons 1985; B. D. Patterson and Velazco 2006).

SELECTED LOCALITIES (Map 496): COLOMBIA: Maipures, upper Orinoco (type locality of *Loncheres bistriatus orinoci* Thomas). VENEZUELA: Amazonas, Boca Mavaca (Patton and Emmons 1985), Amazonas, Capibara (Handley 1976), Cerro Duida (Patton and Emmons 1985).

SUBSPECIES: *Isothrix orinoci* is monotypic.

NATURAL HISTORY: In Venezuela, *Isothrix orinoci* was found mostly in trees but rarely on the ground, always near streams, and always in lowland evergreen forest (Handley 1976).

REMARKS: In his original description, Thomas (1899d:383) remarked that he would have considered *orinoci* a species distinct from *bistriata* (from the Rio Madura [= Madeira]) "were it not that Natterer's second specimen, from the Rio Negro, is, as I am kindly informed by Dr. Lorenz, more or less intermediate in character, as in locality, between the two." These specimens of "*bistriata*" from the Rio Madeira and Natterer's specimen from the Rio Negro, however, are actually *I. negrensis* as defined herein. It was treated as a valid subspecies of *I. bistriata* by Patton and Emmons (1985) and Patton et al. (2000), but elevated to species status by Emmons (2005) and B. D. Patterson and Velazco (2006). O. J. Linares (1998) suggested that the holotype, and thus the type locality, probably came from the right bank of the Río Orinoco at the Raudal de Maipures in Venezuela rather than on the left bank in Colombia, because the Colombian side is dominated by open savanna while that in Venezuela is evergreen forest, a more likely habitat for this arboreal rodent.

Isothrix pagurus Wagner, 1845
Plain Brush-tailed Rat

SYNONYMS:

Isothrix pagurus Wagner, 1845a:146, type locality "Borba," Rio Madeira, Amazonas, Brazil.

[*Loncheres*] *pagurus*: Thomas, 1888a:326; name combination.

Loncheres (*Isothrix*) *pagurus*: E.-L. Trouessart, 1897:606; name combination.

Echimys (*Isothrix*) *pagurus*: E.-L. Trouessart, 1904:504; name combination.

DESCRIPTION: One of three smaller species in genus, with mean head and body length 221.8 mm (range: 170–234 mm; N = 5), tail length 216.8 mm (range: 170–233 mm; N = 5), hindfoot length 41.5 mm (range: 39–45 mm; N = 6), and condyloincisive length 43.5 mm (range: 41.5–45.9 mm; N = 6). Dorsal color from grizzled yellow or gray brown from forehead to midback, with flanks and rump distinctly russet; snout reddish, with that color circumscribing eyes;

no black supraorbital stripes over crown; tail grades uniformly from gray brown at base to yellow-brown toward tip; venter pale yellowish or buffy. Tail not only lacks black hairs characteristic of *bistriata*-complex species but typically exhibits opposite color pattern, being paler at the tip; it is subequal to the body length (mean ratio of tail: body 98%, range: 95.7–113.7%). It is smaller in each cranial dimension measured than any of the three *bistriata*-complex species (Patton and Emmons 1985). The baculum is similar in overall shape but substantially shorter in length in comparison to those of *I. bistriata* and *I. orinoci* (Patton and Emmons 1985).

DISTRIBUTION: *Isothrix pagurus* is known from the central Amazon Basin of Brazil, from the left (= east) bank of the lower Rio Negro east to the Rio Pitinga, north of the Rio Solimões, and from the lower Rio Madeira east to the lower Rio Tapajós, south of the Rio Solimões, in the states of Amazonas and Pará (Patton et al. 2000; B. D. Patterson and Velazco 2006).

SELECTED LOCALITIES (Map 497): BRAZIL: Amazonas, Borba (type locality of *Isothrix pagurus* Wagner), Lago Meduinim (Patton et al. 2000), 80 km N (by road) of Manaus (Patton and Emmons 1985); UHE Pitinga, Rio Pitinga (Patton et al. 2000); Pará, Boim (Thomas 1912d), Casa Nova (B. D. Patterson 1992b).

SUBSPECIES: *Isothrix pagurus* is monotypic.

NATURAL HISTORY: On the left (= east) bank of the lower Rio Negro in central Brazil, this species is found in seasonally inundated black water (*igapó*) forest while *I.*

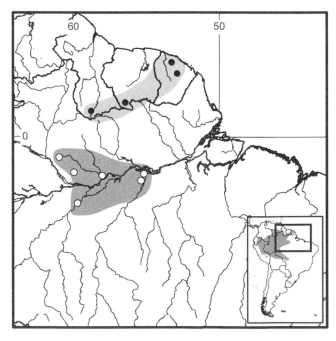

Map 497 Selected localities for *Isothrix pagurus* (○) and *Isothrix sinnamariensis* (●).

bistriata occurs in *várzea*, or the floodplain forest of white-water rivers and streams. B. D. Patterson (1992b) noted that specimens have been taken in mature rainforest from tree holes in both living and dead trees.

REMARKS: Cabrera (1961) listed *I. pagurus* as a subspecies of *I. bistriata*, but Patton and Emmons (1985) and Patton et al. (2000) established its structural, karyotypic, and molecular distinctness. Patton and Emmons (1995) illustrated the karyotype ($2n=22$, FN$=38$), including both nondifferentially stained and C-band patterns.

Isothrix sinnamariensis Vié, Volobouev, Patton, and Granjon, 1996
Sinnamary Brush-tailed Rat

SYNONYM:

Isothrix sinnamariensis Vié, Volobouev, Patton, and Granjon, 1996:395; type locality "Sinnamary River right bank, 21 km upstream from the Petit Saut Dam, French Guiana (4°56.80′N–53°01.90′W—altitude below 20 m)."

Isothrix sinnamariensis Lim and Joemratie, 2011:144; incorrect subsequent spelling of *sinnamariensis* Vié, Volobouev, Patton, and Granjon.

DESCRIPTION: Small, with head and body length 212–215 mm, tail length 262 mm, hindfoot length 39–42 mm, and ear length 14–16 mm. Coloration generally dark brown, becoming reddish brown on hindparts and, less markedly, on nose; small but very distinct black spot apparent behind each eye, insertion point of a group of long genal vibrissae. Tail with short muff of dense fur at base, otherwise covered with curled hairs along entire length, terminating in tuft of long hairs (ca. 35 mm) coiled downward. Dorsal hairs bicolored, with gray base and reddish brown tip; ventral hairs shorter and also bicolored, with gray base and pale orange to reddish-brown tip. Zone of white hairs present in anogenital region. Hindfeet broad, with strong claws, and covered dorsally with short, pale hairs. Ears short and rounded, with small tufts of hairs on both inner and outer surfaces. Females have four pairs of lateral and one pair of inguinal mammae. Most similar to *I. pagurus* but tail longer (124% of head and body length) and less densely haired, and maxillary toothrow shorter (mean length 9.81 mm versus 10.22 mm).

DISTRIBUTION: *Isothrix sinnamariensis* is restricted to the Guiana Shield, where it is known only from four localities, two in French Guiana and one each in Surinam and Guyana.

SELECTED LOCALITIES (Map 497): FRENCH GUIANA: Sinnamary River (type locality of *Isothrix sinnamariensis* Vié, Volobouev, Patton, and Granjon), Les Nouragues (B. D. Patterson and Velazco 2006). GUYANA: Upper Takutu–Upper Essequibo, Chodikar River (Lim et al.

2006). SURINAM: Sipaliwini, camp 1, Kutari River (Lim and Joemratie 2011).

SUBSPECIES: *Isothrix sinnamariensis* is monotypic.

NATURAL HISTORY: The two known specimens from French Guiana were both collected by hand during the daytime in trees, during flooding subsequent to dam construction, of undisturbed lowland rainforest within 100 m of the bank of the Sinnamary River (Vié et al. 1996). The single specimen from Guyana was caught in a live-trap placed on a vine about 15 cm above the ground in nonflooded, tall, evergreen, hill-land forest (Lim et al. 2006).

REMARKS: Vié et al. (1996) described and figured the chromosome complement ($2n=28$, FN$=42$), including G- and C-banding patterns, and document phylogenetic relationships to other species in the genus based on mtDNA sequences.

Genus *Makalata* Husson, 1978
Louise H. Emmons and James L. Patton

These are arboreal rats with short legs, long backs, and coarse pelage. The tail is about as long as the head and body, naked or sparsely clothed with stiff hair. Females typically have two pairs of functional lateral mammae, with a third, apparently obsolete, pair occasionally present. The phallus of *M. didelphoides* is slender, heavily ridged on the sides, with a long pointed dorsal bacular papilla positioned posterior to an extruded, large, and pointed urethral lappet (Emmons 2005:276, Fig. 10). The ventral lip of crater forms a deep V-shaped notch proximally abutting a straight, horizontal fold. The maxillary toothrows are parallel or slightly divergent at either end; the cheek teeth are square to rectangular, longer than wide, with crowns of medium height, and usually two lingually opening flexi (hypoflexus, metaflexus) and two labially opening flexi (paraflexus and mesoflexus) such that M1 and M2 have two, U-shaped lophs opening in opposite directions. Several species, however, have one lingual and three labial flexi in some or all molars. The para-, meso-, and metaflexi extend two-thirds or more across PM4 as well as M1–M3, but hypoflexi are short, crossing about one quarter of the width of the tooth. Hypoflexids of the lower molars are positioned at an oblique angle such that the ectolophids at anterior tip of flexids are narrower than the lophid posterior to tip. The lower premolar is pentalophodont, with the anteroconid pointed; when unworn usually the flexid is not enclosed by lophids (no central fossette), but opens posteriorly such that anterior loph is an inverted V. The metaflexids of m1–m2 bend posteriorly at the internal tip, and mesoflexids are bent forward; the hypoflexid is often curved forward. The lower incisors are not strongly curved. The auditory tympanic bullae are moderately inflated; the auditory meatus is medium sized to small; the auditory tube is short, outwardly or slightly forwardly directed; and the meatus is close to the squamosal suture. The squamoso-tympanic fenestra is slit-like but short, with its ventral lip raised in a ridge with a depression below. The maxillary septum of the incisive foramina is broadly fused to the premaxillary portion. The mandibular foramen is positioned in a fossa with its base on the condyloid ridge. The condyloid process of the mandible is short and dorsoventrally wide; the masseteric crest is strongly developed.

SYNONYMS:

Echimys: Desmarest, 1817b:54; part (description of *didelphoides*); not *Echimys* F. Cuvier.

Nelomys: I. Geoffroy St.-Hilaire 1838:887; part (listing of *didelphoides* and *armatus*); not *Nelomys* Jourdan.

Loncheres: Wagner, 1840:196; part (description of *obscura*); not *Loncheres* Illiger.

Makalata Husson, 1978:445; type species *Nelomys armatus* I. Geoffroy St.-Hilaire, a junior synonym of *Echimys didelphoides* Desmarest (Emmons 1993), by original designation; name specifically noted to be treated as gender feminine.

REMARKS: Emmons (2005) divided the genus into two groups, the "*didelphoides* group" in which she placed seven taxa (*didelphoides* Desmarest, *guianae* Thomas, *macrura* Wagner, *castaneus* Allen and Chapman, *longirostris* Anthony, *handleyi* Goodwin, and *obscura* Wagner) and the "*grandis* group" with *grandis* Wagner and *rhipidurus* Thomas. That same year Iack-Ximenes et al. (2005a) erected the genus *Toromys* with Wagner's *Loncheres grandis* as its type species. Herein we restrict the genus *Makalata* to encompass only those taxa in Emmons's "*didelphoides* group."

The number of species present in the concept of *Makalata* that we invoke here remains uncertain. The description and map given in Emmons and Feer (1997) encompass the ranges of both *M. macrura* and *M. didelphoides*, as herein recognized. However, these authors (p. 238) commented that large-bodied animals from the central Amazon with heavily black-streaked backs and entire gray-brown under parts might represent a species (*M. macrura*) distinct from *M. didelphoides* (see also Emmons 1993; and the following accounts). B. D. Patterson (1992b) had previously noted that specimens from western Amazonian Brazil contrasted with those to the east in both coloration and stiffness of the spines. Patton et al. (2000) documented strongly differentiated mtDNA clades that also segregated by these same morphological and distributional attributes, and recognized *M. didelphoides* and *M. macrura* as separate species. These authors, however, noted that each species included substantial geographic structure. Finally, Emmons (2005) listed all seven taxa in her "*didelphoides* group" as species but without comment or documentation. Pending an adequate review of available museum materials, we recognize only three species. One of these (*obscura* Wagner) is a *nomen dubium* (see Emmons 1993), a second (*macrura* Wagner) is reasonably well defined morphologically and geographically, and a third (*didelphoides* Desmarest) is highly variable and most likely composite. As detailed below, the genus also contains several formally named taxa that have been varyingly applied at the species or subspecies levels. Several of these, as well as some other unnamed but distinctive forms, are likely to be recognized at the species level once new collections fill geographic gaps and both more expansive morphological and molecular analyses are accomplished.

KEY TO THE SPECIES OF *MAKALATA* (EXCLUDING *M. OBSCURA* WAGNER, WHICH CANNOT BE IDENTIFIED NOR BOUNDED WITH ANY CERTAINTY AT PRESENT):

1. Distribution in Guianan region, eastern Amazonia from Pará southeast to Ceará, and southwestern Amazonia; throat, inner thigh, and sometimes venter pale cream or buff; size smaller, mass generally <220 g . . . . *Makalata didelphoides*

1' Distribution west-central Amazonia; venter gray-brown not sharply differentiated from sides; size larger, mass often >300 g . . . .*Makalata macrura*

Makalata didelphoides (Desmarest, 1817)
Red-nosed Armored Tree Rat
SYNONYMS:

Echimys didelphoides Desmarest, 1817b:58; type locality "Amérique méridionale"without further restriction (see Remarks).

L[oncheres]. didelphoides: J. B. Fischer, 1829:307; name combination.

Mus hispidus Lichtenstein, 1830:plate 35, and unnumbered text; not *hispidus* Desmarest.

Nelomys armata I. Geoffroy St.-Hilaire, 1838a:887; type locality "Cayenne" (= French Guiana); a renaming of *Mus hispidus* Lichtenstein.

Nelomys armata: I. Geoffroy St.-Hilaire, 1840:plate 24; illustration of specimen upon which the manuscript name *didelphoides* of É. Geoffroy St.-Hilaire was formally described by Desmarest (1817b).

Loncheres armata: Wagner, 1840:335; name combination.

Loncheres guianae Thomas, 1888a:326; type locality "British Guiana," Guyana, possibly Demerara region.

Echimys castaneus J. A. Allen and Chapman, 1893:222; type locality "Princestown [= Princes Town], Trinidad," Trinidad and Tobago.

Loncheres (Loncheres) armatus: E.-L. Trouessart, 1897:605; name combination.

Echimys macrourus Jentink, 1879b:97; type locality "Dieperinck, Surinam"; not *macrura* Wagner; the difference in spelling is acceptable (ICZN 1999:Art. 58), as noted by Tate (1939:181).

Loncheres (*Loncheres*) *guianae*: E.-L. Trouessart, 1897: 605; name combination.

Loncheres (*Loncheres*) *castaneus*: E.-L. Trouessart, 1897: 605; name combination.

N[elomys]. armatus: J. A. Allen, 1899c:261; name combination.

Echimys (*Echimys*) *armatus*: E.-L. Trouessart, 1904:503; name combination.

Echimys (*Echimys*) *guianae*: E.-L. Trouessart, 1904:503; name combination.

Echimys (*Echimys*) *castaneus*: E.-L. Trouessart, 1904:503; name combination.

Loncheres armatus: Thomas, 1912d:88; name combination.

Phyllomys armatus: Goldman, 1916:126; name combination, but in which the author placed *Mus hispidus* Lichtenstein, *Nelomys armatus* I. Geoffroy St.-Hilaire (as a new name for Lichtenstein's *M. hispidus*), *Phyllomys brasiliensis* Lund, and *Loncheres armatus* Winge as synonyms. Note that only the first two names apply to *Makalata didelphoides* as we define this taxon here.

Echimys longirostris Anthony, 1921a:5; type locality "Kartabo, British Guiana [= Guyana]."

E[chimys]. longirostris: Thomas, 1928c:292; name combination.

Echimys guianae: Pittier and Tate, 1932:264; name combination.

Mesomys didelphoides: Tate, 1935:413; incorrect assignment of *didelphoides* Desmarest, 1817 to *Mesomys* Wagner (Emmons, 1993).

Echimys armatus: Tate, 1935:432; name combination.

[Echimys] guianae: Tate, 1935:432; name combination.

[Echimys] longirostris: Tate, 1935:432; name combination.

Echimys armatus armatus: Tate, 1939:181; name combination.

Echimys armatus macroura: Tate, 1939:181; considered *macroura* Jentink different from (and thus not a homonym of) *macrura* Wagner.

Echimys armatus castaneus: Hershkovitz, 1948a:127; name combination.

Echimys armatus handleyi Goodwin, 1962:1; type locality "Speyside, St. John Parish, eastern Tobago, the West Indies, at an altitude of 300 feet," Trinidad and Tobago.

Makalata armata armata: Husson, 1978:445; name combination.

Makalata didelphoides: Emmons, 1993; first use of current name combination.

DESCRIPTION: This is a medium-sized arboreal rat, with mean head and body length 217.1 mm (range: 180–245 mm; N = 21), tail length 207.1 mm (range: 180–234 mm; N = 18), hind foot length 39.1 mm (range: 35–43 mm; N = 22), ear length 15.8 mm (range: 14–17 mm), and mass 340–405 g. Dorsal color drab reddish brown to yellowish brown heavily streaked with black, becoming more rusty on the rump and tail base; midback and rump finely speckled by paler, reddish brown tips on blackish bristles. Dorsal fur mixture of thin bristles and flat, flexible, and sharp spines, sometimes with thin rusty hairs between them. Muzzle colored faint to bright rust-red, from top of nose to between eyes. Vibrissae coarse and long, reaching to shoulder, and brown in color. Ears small and thinly haired. Eyes dark brown and eyeshine is faint dull red. Tail usually shorter than head and body, thickly haired at base for 3–4 cm, with remainder nearly naked, scaly, and thinly haired with inconspicuous brown hairs that do not extend beyond tip as tiny tuft. Throat and inner thighs or entire venter pale cream or buff, sometimes with white areas; central part of abdomen may be gray-based or brownish; one specimen from Bolivia with entirely dark under parts. Rear of thighs and haired tail base below faint to bright rust. Hindfeet short and broad; dorsal color pale reddish or yellowish.

DISTRIBUTION: *Makalata didelphoides* occurs throughout the Guianan region, including the islands of Trinidad and Tobago, from northern Venezuela, Guyana, Surinam, French Guiana, to eastern Amazonia generally east of the Rio Negro and Rio Madeira, to Maranhão and Ceará, and southwest to northeastern Bolivia.

SELECTED LOCALITIES (Map 498): BOLIVIA: Beni, opposite Costa Marques (AMNH 210355); Santa Cruz, El Refugio (USNM 584590), Flor de Oro (Patton et al. 2000). BRAZIL: Amazonas, Itacoatiara (B. D. Patterson 1992b); Ceará, São Benedito (MNRJ 21490), Trairuçu Aguiraz (MNRJ 21493); Maranhão, Miritiba (Thomas 1916e); Pará, Belem (AMNH 203047), Ilha de Marajo (FMNH 19502), Iroçanga (B. D. Patterson 1992b); Tocantíns, Rio Santa Teresa (MVZ 197569). FRENCH GUIANA: Cayenne (type locality of *Nelomys armata* I. Geoffroy St.-Hilaire). GUYANA: Mazaruni-Potaro, Kartabo (type locality of *Echimys longirostris* Anthony). SURINAM: Saramacca, La Poule (FMNH 95696). TRINIDAD AND TOBAGO: Tobago, Speyside (type locality of *Echimys armatus handleyi* Goodwin); Trinidad, Princes Town (type locality of *Echimys castaneus* J. A. Allen and Chapman). VENEZUELA: Apuré, Nulita (Handley 1976); Bolivar, Río Supamo (Handley 1976); Monagas, Hato Mata de Bejuco (Handley 1976); Sucre, Cumanacoa (AMNH 69898).

SUBSPECIES: Some recent authors have recognized varying numbers of subspecies, but there is no consistency among these treatments. For example, Ellerman listed *guianae* Thomas and *castaneus* Allen and Chapman as simple synonyms of the nominotypical subspecies that he contrasted with *armatus occasius* (now a species of *Pattonomys*). Goodwin (1962), in his description of *handleyi* from Tobago and citing a personal communication from C. O. Handley Jr. also stated that *castaneus* was indistinguishable from the nominotypical subspecies. However, Cabrera (1961) listed *castaneus* as a valid subspecies distinct from the nominotypical form, in which he placed *hispidus* Lichtenstein, *guianae* Thomas, and *longirostris* Anthony as synonyms. Given that there is clearly no consensus among authors and that there has never been even a cursory, much less reasonable, review of geographic variation, we hesitate to list subspecies except in the synonymy of this species.

NATURAL HISTORY: These are nocturnal, arboreal, and solitary rats that feed primarily on fruit, unripe seeds, and some leaves. Allen and Chapman (1893) noted that the collectors found this species (which they listed as *Loncheres guianae*) residing in mangroves at the mouth of the Río Caroní in Venezuela where it fed on the fruits of those trees. Individuals were commonly seen in the middle to late afternoon, suggesting a crepuscular activity period, but we have seen them active only at night. This species is generally restricted to riverine forests (see, for example, R. G. Rocha, Ferreira, Costa et al. 2011; R. G. Rocha, Ferreira, Leite et al. 2011), but has been trapped in gardens surrounding Amerindian houses (Catzeflis 2012).

REMARKS: The assignment of Desmarest's *didelphoides* to those populations of *Makalata* with a pale venter and distributed in the Guianas and eastern Amazonia, in place of the more commonly applied name *armatus* I. Geoffroy St.-Hilaire, is now well-established (Emmons 1993). However, given that this species is likely polytypic, based on the limited geographic sampling of molecularly quite distinct clades (Patton et al. 2000), further refinement of the very imprecise type locality of Desmarest's *didelphoides* would be welcome. Without further research, we hesitate to make such a refinement but note that the history of Desmarest's holotype mirrors that of the holotype of *Mesomys hispidus*, another taxon named by Desmarest (1817b) in his paper describing É. Geoffroy St.-Hilaire's manuscript names. For this latter taxon, Orlando et al. (2003) reviewed the available historical documents in the Museum National d'Histoire Naturelle in Paris and restricted the type locality of *M. hispidus* to the state of Amapá in eastern Amazonian Brazil. It is thus likely that this same region also represents that from which the holotype of *M. didelphoides* originally came.

Tate (1935:432) listed *guianae* Thomas and *longirostris* Anthony as valid species along with *armatus* I. Geoffroy St.-Hilaire in his "naked-tailed group" from Venezuela,

Colombia, Guiana, and Trinidad. Thomas (1916e:295) had earlier stated that his *guianae* "is unquestionably *E. armatus*, Geoff.," and later (Thomas, 1928c:292) doubted also that *longirostris* was distinct from *armatus*. Anthony (1921c) placed *castaneus* J. A. Allen and Chapman in synonymy with *armatus* I. Geoffroy St.-Hilaire. Cabrera (1961:539) listed both *guianae* Thomas and *longirostris* Anthony in the synonymy of the nominotypical subspecies *E. armatus armatus*. As above, whether or not any, or all, of these named populations warrant formal recognition as either species or subspecies must await adequate analyses of existing museum specimens. Patton et al. (2000) documented extensive mitochondrial sequence divergence between samples they assigned to this species from western Bolivia and eastern Amazonia, variation that likely signals separate species status. However, as molecular data are not as yet available from within the geographic boundaries of any of the available named taxa, it is premature both to further divide *didelphoides* much less to assign possible names to any subdivision.

This species is not known from Colombia, based on documented specimens, despite literature references to the contrary. Cuervo Díaz et al. (1986:500) verbalized the range of *Makalata armata* as a broad region across northern Colombia encompassing the western slope of the Cordillera Oriental, Pacific coast, lower Cauca region, Magdalena valley, and llanos piedmont. However, in the more recent synopsis of Colombian mammals, Alberico et al. (2000:67) limited the range of this species in Colombia to the Orinocan biogeographic region or upper llanos, citing the map in Emmons and Feer (1997) and the American Museum of Natural History as the "collection for reference." The American Museum contains no specimens of

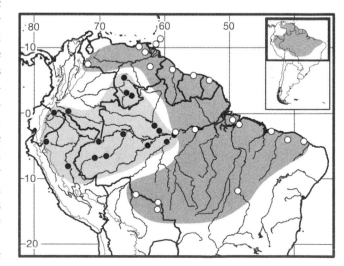

Map 498 Selected localities for *Makalata didelphoides* (○) and *Makalata macrura* (●). Contour line = 2,000 m.

this species from Colombia. The confusion with regard to the presence of this species in Colombia likely stems from a specimen of "*Echimys armatus*" cited by Hershkovitz (1948a:127). The specimen in question, however, is apparently USNM 282119, a *Mesomys* collected from the same locality in northern Colombia cited by Hershkovitz. The original label identified this specimen as "*Echimys armatus* near *occasius.*"

J. F. S. Lima et al. (1998) reported a karyotype with $2n=66$ and $FN=106$ for specimens from the Balbina hydroelectric dam on the Rio Uatumã, Amazonas, Brazil.

Makalata macrura (Wagner, 1842)
Long-tailed Armored Tree Rat
SYNONYMS:

Loncheres macrura Wagner, 1842c:360; type locality "Borba," mouth of Rio Madeira, Amazonas, Brazil.
[*Echimys*] *macrura*: Tate, 1935:432; name combination.
Echimys macrurus: Cabrera, 1961:542; name combination.
E[*chimys*]. *macrurus*: Emmons, 1993:3; name combination.
Makalata macrura: Patton, da Silva, and Malcolm, 2000: 188; first use of current name combination.

DESCRIPTION: Similar to *M. didelphoides* in general external and cranial features, but larger in size, with mean head and body length 236.2 mm (range: 207–280 mm; N=38), tail length 203.2 mm (range: 170–242 mm), hind-foot length 41.8 mm (range: 38–46 mm), ear length 16.3 mm (range: 15–17 mm), and with dull gray-brown rather than pale cream or buff venter (B. D. Patterson 1992b; Patton et al. 2000). Both species with reddish muzzle and dorsal fur composed of a mixture of bristles, flexible spines, and fine hairs.

DISTRIBUTION: *Makalata macrura* occurs throughout the western Amazon Basin, including southern Colombia and Venezuela, eastern Ecuador, and north-central Peru east through western Brazil to the left bank of the lower Rio Negro and at least to the lower Rio Madeira. The geographic boundaries are poorly understood in the southern and eastern parts of its range, where it might contact or even overlap with *M. didelphoides*.

SELECTED LOCALITIES (Map 498): BRAZIL: Amazonas, Arquipélago da Anavilhanas (Patton et al. 2000), Borba (type locality of *Loncheres macrura* Wagner), Colocação Vira-Volta (Patton et al. 2000), Itacoatiara (B. D. Patterson 1992b), Nova Empresa (Patton et al. 2000), Redenção (B. D. Patterson 1992b), Tambor (Patton et al. 2000). COLOMBIA: Caquetá, Tres Troncos (FMNH 71123). ECUADOR: Sucumbios, Santa Cecilia (KU 114007). VENEZUELA: Amazonas, Boca Mavaca (Handley 1976), San Juan (Handley 1976), Tamatama (Handley 1976). PERU: Amazonas, Huampami, Río Cenepa (Patton et al.2000); Loreto, Nazareth (= Amelia) (FMNH 19854); Ucayali, Yarinacocha (LSUMZ 14409).

SUBSPECIES: *Makalata macrura* is monotypic (but see Remarks).

NATURAL HISTORY: This species is typically found in lowland rainforest, in the seasonally inundated floodplain of whitewater (*várzea*) or blackwater (*igapó*) rivers (e.g., Hice and Velazco 2012). Nearly all specimens have been obtained in traps placed in trees or were shot from perches in the canopy. On the Rio Juruá in western Brazil, a pregnant female was taken in the dry season in September (Patton et al. 2000). In northeastern Peru, a lactating female was obtained in May, and a pregnant female, which gave birth to two young, was collected in June (Hice and Velazco 2012).

REMARKS: Some geographic variation in mtDNA cytochrome-*b* sequences is apparent with the limited samples available to date (Patton et al. 2000), indicating a deep phylogeographic history for this species and perhaps future recognition of further formal taxa. B. D. Patterson (1992b) allocated one specimen each from Itacoatiara, on the north bank of the Rio Amazonas below its confluence with the Rio Madeira, to his two morphological groups, which generally map to *M. didelphoides* and *M. macrura* as we construe these two species herein. If correct, this would represent the only known locality of overlap between them, and would further support their recognition as distinct species.

Makalata obscura (Wagner, 1840)
Dark Armored Tree Rat
SYNONYMS:

Loncheres obscura Wagner, 1840:196; type locality "Brasilien."
[*Echimys*] *obscura*: Tate, 1935: 432; name combination.
M[*akalata*]. *obscura*: Emmons, 2005:299; first use of current name combination.

DESCRIPTION: Waterhouse (1848:323), based on the two specimens J. B. von Spix brought from Brazil, stated that this species has a robust body with the short feet and thick tail that characterizes other species of his *Loncheres*. The pelage color is brown, flecked with yellow, and the venter is yellowish. The tail is equal to head and body length, covered with very short, scattered hairs. The upper parts of the body, including the shoulders and thighs, are clothed in spines mixed with scattered coarse hairs, while elsewhere only coarse hairs are found. Hence, in all respects this species shares the major characteristics of other members of the genus and, with respect to the yellowish venter, is most similar to (and perhaps the same as) *M. didelphoides*.

DISTRIBUTION: Known only from the imprecise type locality, which cannot be placed geographically. J. A. Oliveira and Bonvicino (2006) provisionally gave the range as Pará and Maranhão states, in eastern Amazonian Brazil, following the suggestion of Moojen (1952b). Bonvicino, Oliveira, and D'Andrea (2008) mapped this same range.

SUBSPECIES: *Makalata obscura* is monotypic.

NATURAL HISTORY: Nothing is known about this species.

REMARKS: The holotype of *Loncheres obscura* Wagner, 1840, collected by Spix in Brazil, has apparently been lost (Dr. Richard Kraft, pers. comm. to L. H. Emmons; see Emmons 2005). This taxon appears from the original description and figures to be a *Makalata* (Wagner 1840), but the name currently cannot be assigned to a specific population. Because of its yellowish ventral color ("*gastraeo flavicante*"), Emmons (2005:299) provisionally placed *obscura* Wagner within her *didelphoides*-group of *Makalata* and listed it as a species of questionable application (a *nomen dubium*) in her Table 3. We follow her decisions herein.

Genus *Pattonomys* Emmons, 2005

Louise H. Emmons and James L. Patton

Species in this genus are medium to large in body size and arboreally adapted rats. Their pelage color is generally gray on the head and sides, often with a yellowish cast; the back is tinged brown; and the tail is uniform reddish brown, lightly clothed throughout its length with fine hairs. The dorsal pelage includes strong spines, many of which on the rump are tipped whitish, imparting an overall speckled appearance. There are two pairs of lateral mammae. The phallus is long and slender, with a long and pointed bacular papilla, a small urethral lappet, and with the border of the ventral crater wall a straight line, without a V- or U-shaped ventral fold (phallus of *P. carrikeri* figured by Emmons [2005:276, Fig. 10]). The maxillary cheek teeth are square to rectangular and longer than wide. The posterior lophs are rounded. All cheek teeth have an uneven occlusal appearance due to unequal loph sizes. Two labial and two lingual flexi are always present on PM4–M3. The paraflexi and mesoflexi are short and usually reach only to midtooth; the hypoflexi are long and likewise reach midway across tooth, opening at a wide angle. PM4 and M1 always have a mure in the center of the tooth; a mure may or may not be present in M2–M3. The chief diagnostic features of the upper teeth include widely open flexi, some nearly as wide as the lophs themselves, and an anteroposteriorly expanded protocone and posteroloph (probably metaloph and posteroloph combined). These expanded lophs become accentuated with wear. The four lophs of PM4–M2 are markedly unequal in length and width. The anteroloph is broad and squared lingually, tapering labially; and the posteroloph is pointed labially and strongly curved along the posterior margin of the tooth. The paraf-

lexus and metaflexus of PM4–M2 slant in markedly opposite directions, with the metaflexi slanting posteriorly from the labial to medial edges of the tooth. The lower premolar is tetralophodont, with two labial and two lingual flexids. This tooth is usually divided by one central flexid into two V-shaped lophs, but it may have a central mure. The flexids of m1–m2 form nearly straight-sided Vs. The lower incisors are not strongly curved. The cranium is short and broad, with an expanded, winglike supraorbital shelf that curves upward from the frontals. The supraorbital region is broad, especially well developed in *P. carrikeri* but much narrower in *P. semivillosus*. The auditory bullae are moderately to considerably inflated; the auditory meatus is high, positioned near the squamosal, and directed outward; the auditory tubes short, their role taken by the formation of two overlapping bony rings, especially well developed in old individuals. The lateral process of the supraoccipital is short, usually positioned only to the middle of the auditory meatus and does not extend ventrally beyond the lower edge of the meatus. The angular process of the mandible is slender; the condyloid process is narrow; the mandibular foramen is usually placed in a fossa on the condyloid ridge; and the masseteric crest is strongly developed anteriorly. The teeth of members of this genus are so distinctive that any single tooth is diagnostic except m1–m3, and often these are as well.

Cabrera (1961) and subsequent authors synonymized most of the named forms we refer here to *Pattonomys* under the species *Echimys semivillosus* I. Geoffroy St.-Hilaire. Cabrera (1961) considered *carrikeri* J. A. Allen (with *flavidus* Hollister a synonym) and *punctatus* Thomas as valid subspecies of this species, as did O. J. Linares (1998), but who listed *flavidus* as valid subspecies. However, all but *P. flavidus* are cranially distinctive and readily diagnosable. Emmons (2005) considered them all to be valid species, but each merits further study. The species *P. flavidus* is cranially similar to coastal populations of *P. carrikeri* and may be synonymous, but as it also has distinctive features, such as an extremely deep mandible, we provisionally recognize it pending additional specimens and molecular genetic analysis (fide Emmons 2005). *Echimys occasius* Thomas was formerly placed in *Makalata* (Emmons and Feer 1997), which it resembles in pelage color, smaller size, and Amazonian geographic distribution. It does not possess, however, several of the more diagnostic characters *Makalata*, and groups in parsimony analysis with *Pattonomys* (Emmons 2005). We follow this placement, pending additional analyses, perhaps including molecular studies, which hopefully will better clarify its relationships. All five species replace one another geographically, with four serially positioned across northern Colombia and Venezuela and one in the western Amazon. To date, no species in the genus *Pattonomys* has been included in any molecular phylogenetic analysis.

O. J. Linares's (1998:245) statement that *P. semivillosus* in Venezuela (which herein equals the three species: *P. carrikeri, P. flavidus,* and *P. punctatus*) is very common belies the few specimens known for all species of this genus in museum collections.

SYNONYM:

Nelomys: I. Geoffroy St.-Hilaire, 1838a:887; part (description of *semivillosus*); not *Nelomys* Jordan.

Loncheres: Wagner, 1843a:330; part (listing of *semivillosus*); not *Loncheres* Illiger.

Echimys: E.-L. Trouessart, 1904:503; part (listing of *punctatus*); not *Echimys* F. Cuvier.

Pattonomys Emmons, 2005:282; type species *Nelomys semivillosus* I. Geoffroy St.-Hilaire, 1838, by original designation.

KEY TO THE SPECIES OF *PATTONOMYS*:

1. Spines over mid-back and rump with rusty tips; muzzle reddish; greatest length of skull (in adults) <51 mm . . . .*Pattonomys occasius*
1′. Spines over mid-back and rump with white tips; muzzle not reddish; greatest length of skull (in adults) >51 mm. 2
2. Nonspinous dorsal hairs; color yellow-buff; range restricted to Isla Margarita, Venezuela. . . . . *Pattonomys flavidus*
2′. Dorsal hairs spiny, with spines white-tipped; color varying shades of gray; distributed on mainland of Colombia or Venezuela 3
3. Dorsal spines extend from shoulder to rump and down hind leg to heel . . . . *Pattonomys semivillosus*
3′. Dorsal spines do not extend much forward of rump and onto hind leg 4
4. Throat in front of forelegs with dusky-based band, midabdomen dusky based; least interorbital constriction <14 mm; range north of the Río Orinoco in llanos and Cordillera Central of Venezuela . . . .*Pattonomys carrikeri*
4′. Throat and chest, and usually abdomen, with hairs white to base; least interorbital constriction >14 mm; distributed south of the Río Orinoco and its delta . . . .*Pattonomys punctatus*

Pattonomys carrikeri (J. A. Allen, 1911)

Carriker's Speckled Tree Rat

SYNONYMS:

Loncheres carrikeri J. A. Allen, 1911:251; type locality "San Esteban," Carabobo, Venezuela.

[*Echimys*] *carrikeri*: Thomas, 1916e:295; name combination.

[*Echimys*] *carrikeri*: Tate, 1935:432; name combination.

E[*chimys*]. *semivillosus carrikeri*: Tate, 1939:180; name combination.

Pattonomys carrikeri: Emmons, 2005:282; first use of current name combination.

DESCRIPTION: One of three relatively smaller *Pattonomys*, with mean head and body length 222.8 mm (range: 200–250 mm; N = 12), tail length 218.2 mm (138–242 mm), hindfoot length 36.3 mm (range: 35–40 mm), and ear length 11.9 mm (range: 11–18 mm). Similar in color to *P. punctatus*, with head and flanks overall grayish; mid-dorsal region grayish mixed with black and rufous; long stiff hairs and bristles black tipped; broad spines black with conspicuous white tips, most abundant on lower back; underfur finer, rufous in color on surface between bristles and spines. Venter white, with fairly distinct pectoral band of dusky-based hairs; center of abdomen usually same color as pectoral band. Outer surfaces of forefeet and hindfeet, limbs, and ventral surface at the base of the tail gray. Ears brownish black. Tail well covered with short stiff hairs, although scales remain visible along length. Skull robust, but rostral portion narrow and somewhat tapering, with anterior nares narrow and shallow, bullae swollen and spherical, and basioccipital short and narrow, all features that contrast with skull of *P. punctatus*.

DISTRIBUTION: *Pattonomys carrikeri* is known from the coastal cordillera and central llanos of Venezuela. O. J. Linares (1998) mapped localities from Caracas, Falcón, Lara, Yaracuy, Carabobo, Aragua, and Guárico states, without reference to museum vouchers or identification of specific localities.

SELECTED LOCALITIES (Map 499): VENEZUELA: Carabobo, San Esteban (type locality of *Loncheres carrikeri* J. A. Allen); Caracas, Caracas (AMNH 130815); Falcón, Coro (Aguilera et al. 1998), 19 km NW of Urama (Handley 1976); Guárico, Zaraza (AMNH 135442), Guárico, Río Orituco (ROM 107955); Lara, Caserio Boro (Handley 1976).

SUBSPECIES: *Pattonomys carrikeri* is monotypic.

NATURAL HISTORY: At the time the holotype was collected, the habitat in the vicinity of San Esteban was "heavy forest" (J. A. Allen 1911:239). Field notes of the collector (M. A. Carriker, in Allen 1911:252) state that the type specimen was shot from the lower branches of a large tree, "likely disturbed from the trunk where it ran out on a long limb." Carriker further remarked that local people had never seen such an animal before and that, therefore, the species was probably rare. Tate (1939:180) noted that this species lives in the hollow trunks of dwarfed trees in the arid or semiarid cactus belt of northern Venezuela, and thus are chiefly inhabitants of xerophytic vegetation or

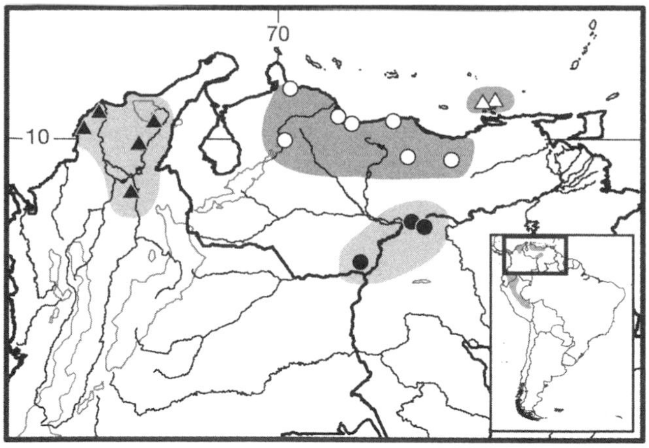

Map 499 Selected localities for *Pattonomys semivillosus* (▲), *Pattonomys carrikeri* (○), *Pattonomys punctatus* (●), and *Pattonomys flavidus* (△). Contour line = 2,000 m.

savanna-like areas with sparse groves of trees and gallery forests. Of the large series (52) of *P. carrikeri* and the few *P. punctatus* collected by the Smithsonian Venezuelan Project (1965–1968) and grouped under the name *Echimys semivillosus*, nearly all (98%) were taken in trees near streams or other moist places, usually in thorn forest but occasionally in scattered trees in the savanna and rarely in evergreen forest (Handley 1976).

REMARKS: Considered a valid subspecies of *P. semivillosus* by some authorities (Cabrera 1961; O. J. Linares 1998), but Hershkovitz (1948a:127) regarded *carrikeri* to be a synonym of *flavidus* Hollister without comment or documentation. Specimens assigned to *P. semivillosus* by Handley (1976:58) from Carabobo, Falcón, and Lara states are here assigned to *P. carrikeri*. Aguilera et al. (1998) described and figured a karyotype of $2n=94$, $FN=114$, from a single specimen of "*P. semivillosus*" from Coro, Falcón. Although we have not seen the voucher specimen, by geography it would fall at the western margins of the range of *P. carrikeri*, as we currently understand this species.

Pattonomys flavidus (Hollister, 1914)
Yellow Speckled Tree Rat

SYNONYMS:

Loncheres flavidus Hollister, 1914c:143, type locality "El Valle [= El Valle del Espíritu Santo], Margarita Island," Nueva Esparta, Venezuela.

[*Echimys*] *flavidus*: Tate, 1935:432; name combination.

Echimys flavidus: Pittier and Tate, 1932:264; name combination.

E[*chimys*]. *armatus castaneus*: Tate, 1939:180; considered *flavidus* Hollister a synonym of *castaneus* J. A. Allen and Chapman, 1893, a species of *Makalata* Husson.

Echimys semivillosus flavidus: Hershkovitz, 1948a:127; name combination.

Pattonomys flavidus: Emmons, 2005:282; first use of current name combination.

DESCRIPTION: Slightly larger in external and cranial dimensions than either *P. carrikeri* or *P. punctatus*, with head and body length 252 mm, tail length 248 mm, hind foot length 40.8 mm (measurements of holotype). In comparison with *P. carrikeri* and *P. punctatus*, hairs between spines yellowish-buff or ochraceous rather than gray, and flanks more buffy, less streaked with darker colors. Under parts self-colored cream, without dusky pectoral or abdominal bands. Spines on lower back and rump strong and tipped white. Pelage long. Skull shares narrower rostrum, more spatulate nasals, and more inflated bullae of *P. carrikeri* from the adjacent mainland coastal mountains. It differs from all other taxa in having zygomatic arches, viewed dorsally, that taper anteriorly from jugal-squamosal suture, and strikingly deep mandible from below toothrow to base of angular process.

DISTRIBUTION: *Pattonomys flavidus* is known only from two localities on Isla Margarita, off the north-central coast of Venezuela.

SELECTED LOCALITIES (Map 499): VENEZUELA: Nueva Esparta, El Valle, Isla Margarita (type locality of *Loncheres flavidus* Hollister), Nueva Esparta, Península Macanao, Isla Margarita (Aguilera et al. 1998).

SUBSPECIES: *Pattonomys flavidus* is monotypic.

NATURAL HISTORY: This species has not been studied in the field, and no data on habitat or life history traits are available. However, Isla Margarita is arid with chaparral and cactus dominated flora. It has some gallery forests along watercourses, but these are vanishing. If *P. flavidus* occupies such forests, it could be endangered.

REMARKS: Hollister (1914c), in his description of *flavidus*, considered it "an insular form of *Loncheres punctatus*." Emmons (2005) stated that *P. flavidus* was cranially similar to coastal specimens of *P. carrikeri*, with which it may be synonymous, but that it possessed uniquely distinctive features, such as an extremely deep mandible. Aguilera et al. (1998) described a karyotype of $2n=94$, $FN=114$, for three specimens from the Península Macanao, the western portion of Isla Margarita. These authors assigned their specimens to *P. semivillosus*, but as construed herein, they should be identified as *P. flavidus*, the only species of *Pattonomys* known from that island.

Pattonomys occasius (Thomas, 1921)
Bare-tailed Tree Rat

SYNONYMS:

Echimys occasius Thomas, 1921f:450, type locality "Gualea, west of Pichincha. Alt. 4000'," Pichincha, Ecuador; probably in error (see Emmons 2005, and Remarks).

E[*chimys*]. *occasius:* Thomas, 1928c:292.

[*Echimys*] *occasius:* Tate, 1935:432.

Echimys armatus occasius: Ellerman, 1940:112; name combination but incorrect assignment as subspecies of what is now known as *Makalata didelphoides.*

Makalata occasius: Emmons and Feer, 1997:237; name combination.

Pattonomys occasius: Emmons, 2005:282; first use of current name combination.

DESCRIPTION: Smallest species in genus, with head and body length 218 mm, tail length 225 mm, and hindfoot length 37 mm (measurements of holotype). Upper parts colored dull buff-brown streaked with dark brown. Body from shoulder to rump on both back and sides thickly covered with wide, flat, sharp, flexible spines that lie flat and may not be apparent from a distance; spines brown or black on back, many with pale rusty tips, especially on rump, and those flanks gray-brown; tips of spines blunt, not drawn out into hair-like processes. Fine, soft, and sparse orange hairs occur between spines over body. Head clothed with stiff, thin bristles. Ears short and naked. Tip of muzzle and cheeks reddish. Vibrissae fine, thin, but long, with mystacial vibrissae reaching shoulder. Tail robust and long, equal to or longer than the head and body. Body hair ends abruptly 1–2 cm from tail base; remainder of tail conspicuously naked and scaly. Under parts clear orange, often with white thoracic and inguinal regions. Undersurface of tail base and thighs dull brown. Hindfeet short and broad, brownish in color with whitish toes. Skull smaller than those of other species; auditory bullae not inflated, and supraorbital ridges only very weakly developed.

DISTRIBUTION: Except for the type locality (which may be in error; see Tate 1935:428; Emmons 2005), all referred specimens of *Pattonomys occasius* are known only from the western Amazon in Ecuador and Peru.

SELECTED LOCALITIES (Map 500): ECUADOR: Napo, Río Jatun Yacu (= Río Rutuno) (MCZ 37964); Orellana, Río Suno (AMNH 68177); Pichincha, Gualea (type locality of *Echimys occasius* Thomas; probably an error, see Remarks). PERU: Loreto, boca Río Curaray (AMNH 71897), mouth of Rio Santiago (AMNH 98262); Madre de Dios, Cocha Cashu Biological Station (B. D. Patterson et al. 2006), Río Colorado (FMNH 84259).

SUBSPECIES: *Pattonomys occasius* is monotypic.

NATURAL HISTORY: The habitat of the few known specimens is lowland rainforest, but the species may extend upward in elevation into more montane forests (Emmons and Feer 1997). There are no published reports of any aspect of the ecology, behavior, or reproductive biology of this species.

REMARKS: Thomas (1921f) did not give the etymology of his name *occasius.* Emmons (2005) surmised, from its

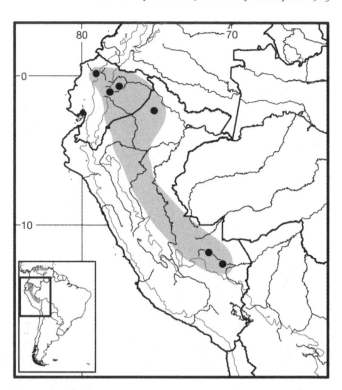

Map 500 Selected localities for *Pattonomys occasius* (●). Contour line = 2,000 m.

purported type locality west of the Andes, that Thomas derived the name from the Latin *occasus,* meaning western (to set as the sun), with the comparative superlative ending *-ius,* or westernmost. The suffix therefore does not change with the gender of the generic epithet. The type locality on the western slope of the Andes in northern Ecuador is certainly erroneous as is the accompanying natural history note that it was captured between rocks in a stream (Thomas 1921f) because all other known specimens are from the Amazonian lowlands on the eastern base of the Andes (see Tate 1935:428). Unfortunately, Thomas (1921f:450) in his description compared this species mostly with *E. armatus* (= *Makalata didelphoides*), although he did note that in coloration and white incisors "it shows an approach to the *E. punctatus* group." Emmons and Feer (1997) placed *P. occasius* in the genus *Makalata,* comparing it to *M. didelphoides* and *M. rhipidurus* in external features.

Pattonomys punctatus (Thomas, 1899)
Orinocan Speckled Tree Rat

SYNONYMS:

Loncheres punctatus Thomas, 1899b:153; type locality "Caicara, Orinoco," Bolívar, Venezuela.

Echimys (*Echimys*) *punctatus:* E.-L. Trouessart, 1904:503; name combination.

[*Echimys*] *punctatus:* Thomas, 1916e:295; name combination.

[*Echimys*] *punctatus*: Tate, 1935:432; name combination.

E[*chimys*]. *semivillosus punctatus*: Tate, 1939:180; name combination.

Pattonomys punctatus: Emmons, 2005:282; first use of current name combination.

DESCRIPTION: One of larger rats in genus, with head and body length 236 mm, tail length 233 mm, hindfoot length 34 mm, and ear length 20 mm (measurements of holotype). Fur spiny, with spines on back reaching 20 mm in length and 2.3 mm in width. Overall color pale gray with whitish speckles and rufous tint to back due to hairs with reddish tips and gray bases. Spines of rump grayish, white basally, and tipped white, most numerous posteriorly and sharply contrasting with general grayish color. Head above and laterally coarsely mixed black and white. Ears clothed with fine black hairs on edges, with small tuft of whitish hairs on antitragus and distinct whitish postauricular patch. Flanks coarsely mixed whitish gray. Venter white, sharply demarcated from sides. Inner surfaces of limbs white, outer surfaces gray. Upper surfaces of forefeet gray; hindfeet gray above, while along medial margin, with digits whitish. Tail of medium length, approximately equal to head and body, thinly haired with scales visible, and uniformly brown in color, only slightly paler below if at all. Skull with nasals equal in length to premaxillary processes, flattened frontal region, broadly expanded supraoccipital ridges, and narrow, not spatulate pterygoid processes. Few available specimens indicate that cranial dimensions are largest for genus.

DISTRIBUTION: *Pattonomys punctatus* is restricted to Venezuela south of the Río Orinoco and its delta and west of the Río Orinoco in Apure state. Tate (1939:180) stated that this species is known from the "Caura district" near the mouth of the Río Caura. O. J. Linares (1998) mapped two localities in the Orinoco delta (Delta Amacuro state) that presumably are of this species, but he did not provide locality details or refer to voucher specimens.

SELECTED LOCALITIES (Map 499): VENEZUELA: Apure, Río Cinaruco (Handley 1976); Bolívar, Caicara (type locality of *Loncheres punctatus* Thomas), Hato La Florida, 45 km SE of Caicara (Handley 1976).

SUBSPECIES: *Pattonomys punctatus* is monotypic.

NATURAL HISTORY: Nothing is known about the ecological range or life history of this species. The habitat at Hato La Florida, on the southern margins of the llanos, is tropical dry forest (Handley 1976).

REMARKS: Thomas (1899b) noted that *P. punctatus* was readily distinguished by the prominent white speckling on its lower back, which results from broad white tips of the spines in that region. This pattern, however, is also characteristic of both *P. semivillosus* and especially the geographically adjacent *P. carrikeri*, which was described

12 years later (J. A. Allen 1911). He considered it closely allied to *P. semivillosus*. He also (p. 154) provisionally referred a specimen "said to have come from Caracas" in the Natural History Museum, London, to this species, a locality that is well west of the Orinoco delta and thus probably a record of *P. semivillosus* itself, not *P. punctatus* as herein construed. Tate (1939), Cabrera (1961), and O. J. Linares (1998) regarded *P. punctatus* a subspecies of *P. semivillosus*.

Pattonomys semivillosus (I. Geoffroy St.-Hilaire, 1838a)
Colombian Speckled Tree Rat

SYNONYMS:

Nelomys semivillosus I. Geoffroy St.-Hilaire, 1838a:887; type locality "Carthagène, (Nouvelle-Grenade)," Cartagena, Bolívar, Colombia.

Loncheres semivillosa: Wagner, 1843a:330; name combination.

Loncheres (*Loncheres*) *semivillosa*: E.-L. Trouessart, 1880b: 177; name combination.

Echimys (*Echimys*) *semivillosus*: E.-L. Trouessart, 1904:503; name combination.

[*Echimys*] *semivillosus*: Thomas, 1916e:295; name combination.

E[*chimys*]. *semivillosus semivillosus*: Tate, 1939:180; name combination.

Pattonomys semivillosus: Emmons, 2005:282; first use of current name combination.

DESCRIPTION: Moderately large species in genus, with head and body length ranging from 200–268 mm, tail length 210–261 mm, hindfoot length 36–43 mm, ear length 17–22 mm, and mass 194–407 mm. Distinctly white muzzle contrasts sharply with almost black crown of head; neck and shoulders often gray streaked with black; midback gray washed with fulvous; rump or mid-back, rump, and hind legs speckled with spines tipped white. Flanks either colored like back or are gray. Shoulder to rump heavily clothed with flat but flexible spines. Ears small, often with pale postauricular and subauricular patch. Vibrissae numerous, black, and long, extending to behind ear. Tail robust, 80–120% of the head and body in length, and appears naked and scaly, thinly covered with stiff, pale brown hairs. Hindfeet broad, gray or gray yellow above. Under parts white or pale orange, sometimes with gray across abdomen. Cranium more slender and lightly built than those of Venezuelan species, with relatively narrow supraorbital ledges, auditory bullae little inflated, and narrow incisive foramina. Incisors distinctly white. Emmons (2005:284) published photographs of skull and mandible.

DISTRIBUTION: *Pattonomys semivillosus* is known only from the lower Río Magdalena drainage in northeastern Colombia at elevations to at least 600 m. The species may extend into northwestern Venezuela (O. J. Linares 1998).

SELECTED LOCALITIES (Map 499): COLOMBIA: Atlántico, lower Río Magdalena opposite Barranquilla (FMNH 69117); Bolívar, Cartagena (type locality of *Nelomys semivillosus* I. Geoffroy St.-Hilaire), Morales (ICN 513); Cesar, El Orinoco (Hershkovitz 1948a), Río Guaimaral (Hershkovitz 1948a).

SUBSPECIES: *Pattonomys semivillosus* is monotypic.

NATURAL HISTORY: These rats are presumably nocturnal and den in tree holes. The few available specimens were all collected from swampy lowland *cienagas* of the lower Magdalena Valley. This area appears more heavily forested and evergreen then the habitats of *P. carrikeri* and *P. punctatus*, a difference that may be associated with the small bullae of *P. semivillosus*.

REMARKS: The designation of Cartagena as the type locality may be in error, as this was the port from which the specimens that formed the basis of I. Geoffroy St.-Hilaire's description were shipped (Hershkovitz 1948a:127). However, recent specimens have been collected in swamps near to Cartagena. Thomas (1916e) considered *P. semivillosus*, *P. punctatus*, and *P. carrikeri* to be closely related, but noted that insufficient material was available for him to make a proper comparison among them. He did not comment on the fourth species, *P. flavidus*, which had been described two years earlier. O. J. Linares (1998) noted that *P. semivillosus* is not presently known from Venezuela, but suggested that it might occur in the Lago de Maracaibo region.

Genus *Phyllomys* Lund, 1839
Yuri L. R. Leite and Ana Carolina Loss

Y. L. R. Leite (2003) and Emmons (2005) provided detailed diagnoses of the genus, summarized as follows. These are medium-sized to large arboreal rats with the pelage spiny to soft fur, large eyes, small and rounded ears, and long vibrissae reaching the shoulders. The limbs are short, and the hindfeet are broad and short, with strong claws on all digits except the thumb, which bears a nail. Dorsal color varies from brown to reddish golden brown, ventral color from pure white to light gray-brown. Spines are conspicuous in most species, especially on the rump, where they can reach 1.5 mm in width. Mammae occur in four pairs, three lateral and one inguinal. A sternal gland is well developed in both sexes, but is larger in males. The tail is typically slightly shorter or longer than head and body; its base is covered with body-like hair for approximately 20 mm, with the remaining length varying from almost naked to thickly haired with a bushy tuft at the tip. The skull is strong, with a relatively short rostrum, usually inflated bullae with the meatus low and directly slightly forward, and with a space as wide as the meatus between it and the squamosal, wide interorbital region with well-developed supraorbital ledges; the palate is narrow with nearly parallel toothrows; and the incisive foramina vary from small to relatively large. The upper cheek teeth are rectangular and are composed of four transverse laminae separated by three labial flexi (para-, meso-, and metaflexus). In unworn teeth, the enamel of one lamina does not contact the others, but in older individuals two or more laminae coalesce at the labial and/or lingual side, forming a transverse U-shaped loph, similar to that in unworn teeth of *Echimys* and *Makalata*. The pattern of coalescence varies (Winge 1887; Moojen 1952b), but usually the two posterior lophs unite labially with wear and the anterior lophs unite lingually (Emmons 2005). The hypoflexid is deep and angled forward in the lower teeth, connecting to the mesoflexid on M1. A slender mure is typically present on M2 and M3, separating the hypoflexid and mesoflexid, and forming the shape of the Arabic numeral 3. The lower premolar is pentalophodont, the first two lophids coalesce at an early age forming a triangle with a circular island of enamel inside. The lack of complete transverse laminae on unworn lower cheek teeth, but their presence on the uppers distinguishes *Phyllomys* from *Diplomys*.

SYNONYMS:

Phyllomys Lund, 1839a:226; type species *Phyllomys brasiliensis* Lund, by subsequent designation (Emmons, Leite et al. 2002:5).

Lonchophorus Lund, 1839b:206; proposed as a subgenus; type species *Lonchophorus fossilis* Lund (1839b:208), by monotypy (see also Lund 1840b:plate 1, Fig. 9; 1841c: plate 25, Fig. 9).

Nelomys: Jordan, 1837:522; part (description of *blainvilii*); see Remarks.

Loncheres: Wagner, 1840:203; part (listing of *blainvillei*); not *Loncheres* Illiger.

Mesomys: Ihering, 1897:171; part (description of *thomasi*); not *Mesomys* Wagner.

Echimys: E.-L. Trouessart, 1904:503; part (listing of *blainvillei, armatus, dasythrix, nigrispina, unicolor*); not *Echimys* F. Cuvier.

Euryzygomatomys: E.-L. Trouessart, 1904:506; part (listing of *thomasi*); not *Euryzygomatomys* Goeldi.

REMARKS: The taxonomy, systematics, and natural history of the Atlantic Forest tree rats have been recently reviewed and updated by Emmons, Leite et al. (2002) and Y. L. R. Leite (2003). The accounts that follow stem directly from these treatments, the first complete revisions of the genus and its member species. Until these publications appeared, the taxonomic and nomenclatural history of the genus was complex and exceedingly confusing, and most recent authors either placed this group of species under the broadly encompassing genus *Echimys* F. Cuvier (Ellerman 1940; C. O. da C. Vieira 1955; Cabrera 1961; Woods 1993; McKenna and Bell 1997; Eisenberg and Redford 1999) or under *Nelomys* Jourdan (Emmons and Feer 1990, 1997). Although Moojen (1952b), in his compilation of Brazilian rodents, correctly allocated these species to *Phyllomys* Lund, it was not until the publication of Emmons, Leite et al. (2002) that the validity of this application was documented. Y. L. R. Leite (2003:2–3) succinctly documented the correct use of the name *Phyllomys* Lund for this group of rats, and the convoluted history of this and other names commonly applied to them, as follows.

"The name *Nelomys* Jourdan, 1837, although older than *Phyllomys* Lund, 1839, is a junior synonym of *Echimys* F. Cuvier, 1809, and, therefore, it is not available (Emmons, Leite et al. 2002). Most recent authors . . . erroneously attribute the name *Nelomys* to G. Cuvier. Cuvier's report, based on a memoir written by Jourdan, was presented to the French Academy of Sciences on 2 January 1838. Although the paper came out in the 1837 volume of the *Annales des Sciences Naturelles*, the date on the paper itself takes precedence and the correct citation should therefore be G. Cuvier (1838, not 1837). As members of the Academy, G. Cuvier and Deméril were in charge of introducing Jourdan and presenting his research, since he was not a member. Part of Jourdan's memoir, however, had already been published on 9 October 1837 in the *Comptes Rendus Hebdomadaires de Séances de L'Academie des Sciences* (vol. 15, date of publication printed on page 69), predating G. Cuvier's report . . .

"Jourdan begins his description of *Nelomys* by stating: 'Ce genre formé aux dépens du genre Échymys des auteurs, a pour type l'Échimys cristatus [*sic*]' (Jourdan, 1837:522). This statement makes *Echimys cristatus* Desmarest, 1817 the type of his new genus *Nelomys*. However, *Echimys cristatus* is a junior synonym of *Myoxus chrysurus* Zimmermann, 1780, which is the genotype of *Echimys* F. Cuvier, 1809, and therefore *Nelomys* becomes a junior synonym of *Echimys* . . .

"The next available name for the Atlantic tree rats is *Phyllomys* Lund, first published in 1839 with a sentence describing the dentition in which the upper molars comprise four simple transverse laminae (Lund 1839a:226). This article in French is an extract based on three of Lund's monumental memoirs, which were published a year later in Danish (Lund, 1840a; for the explanation as to why the date 1840 is correct, see Musser et al. 1998:330–331). *Phyllomys* was treated on page 243 and illustrated in plate 21, Figures 12 and 13. The figure captions reading 'Phyl-

lostomus' are obviously wrong as pointed out by Paula Couto (1950:559)."

Lund (1839b:206) coined the name *Lonchophorus* as a subgenus to include the fossil species *L. fossilis* (Lund 1839b:208) from "du bassin de Rio das Velhas" (Lagoa Santa, Minas Gerais, Brazil), and for which he subsequently illustrated the first and second lower right molars (Lund 1840b:plate 1 [1841c:294, plate 25, Fig. 9]). Emmons, Leite et al. (2002) regarded *L. fossilis* as a *nomen dubium* but perhaps a synonym of *P. brasiliensis* Lund.

Y. L. R. Leite (2003:45) incorrectly listed the type species of *Phyllomys* as *Nelomys blainvilii* Jourdan, the first species recognized that is now allocated to this genus. Emmons, Leite et al. (2002:5) had one year earlier clearly documented that the type species of this genus is *Phyllomys brasiliensis* Lund.

Oldfield Thomas (1916d,e) recognized that *Phyllomys* was generically distinct from *Echimys*, based on the laminar, transverse plates of the occlusal surface of the upper teeth, and from *Diplomys*, in which both upper and lower teeth have transverse plates. He erroneously believed, however, that *Phyllomys* was a junior synonym of *Nelomys*. Tate (1935), however, considered both *Nelomys* and *Phyllomys* as synonyms of "*Echimys*," and most authors (as above) have followed the same path.

Lara et al. (1996), Y. L. R. Leite and Patton (2002), Galewski et al. (2005), and P.-H. Fabre et al. (2012) provided hypotheses of phylogenetic relationship among most extant genera of echimyids and documented a well-supported sister relationship between *Phyllomys* and *Echimys* based on DNA sequences. G. A. S. Carvalho and Salles (2004), Emmons (2005), and Candela and Rasia (2012), using a cladistic assessment of qualitative morphological characters, showed that member species of *Phyllomys* form a monophyletic assemblage, but with closer relationship to *Diplomys* (G. A. S. Carvalho and Salles 2004; Candela and Rasia 2012) or part of an unresolved polytomy (see comments above under the family). Y. L. R. Leite (2003) provided a phylogenetic hypothesis for 8 of the 13 species currently recognized, based on mtDNA sequence data. Y. L. R. Leite et al. (2007) added *P. unicolor* to the molecular assessment, and most recently Loss and Leite (2011) expanded the phylogenetic hypothesis of species relationships with both mitochondrial and nuclear gene sequences. In this latter study, Loss and Leite (2011) uncovered three monophyletic lineages, each with strong support, that identify apparently undescribed taxa. Loss and Leite (2011) obtained a different mtDNA sequence from the same specimen of *P. unicolor* originally examined by Y. L. R. Leite et al. (2007). This new sequence, when combined with additional nucDNA sequences, placed *P. unicolor* with *P. pattoni*. Therefore, the sequence of *P.*

unicolor reported by Y. L. R. Leite et al. (2007) should be considered erroneous.

Y. L. R. Leite (2003) reviewed species limits, defined species by a combination of a hypothesis of phyletic relationships based on molecular sequence data and qualitative characters, diagnosed each taxon morphologically (and chromosomally, where such data exist), reviewed all available specimens in museum collections in the United States and Brazil, and mapped their respective distributions. He summarized what few data are available on natural history, habitat range, and reproductive biology, and commented on conservation issues. These taxa, like others inhabiting the Atlantic Forest of coastal Brazil, have been severely and negatively impacted by habitat loss and modification in recent decades, and are thus of considerable conservation concern. A new species of *Phyllomys* has been more recently described (Y. L. R. Leite et al. 2008), bringing the total of extant species now known to 13. Four additional lineages remain to be described (Loss and Leite 2011; N. P. Araújo et al. 2014).

Lund (1839b) described a single fossil species, *Lonchophorus fossilis*, based on mandibular fragments from late Pleistocene cave deposits at Lapa dos Ossinhos, near Serra das Abelhas and Poções, Lagoa Santa, Minas Gerais, Brazil. Hadler et al. (2008) described abundant fossil material from Mid to Late Holocene (9,400–3,730 ybp) deposits in Rio Grande do Sul state, southern Brazil, but were unable to assign these to an extant species. Specimens allocated only to *Phyllomys* sp. have been reported from Quaternary deposits in São Paulo (M. C. Castro and Langer 2011) and Pernambuco (D. Ferreira et al. 2012) states, in southeastern and northern Brazil, respectively.

KEY TO THE SPECIES OF *PHYLLOMYS*:

1. Head and body length >280 mm; greatest skull length >56 mm. 2
1'. Head and body length <280 mm; greatest skull length <56 mm. 3
2. Fur long and spiny, aristiforms on rump >30 mm in length; incisive foramina teardrop in shape; rostral premaxillary foramen absent; endemic to Ilha São Sebastião, off the coast of São Paulo, Brazil . . . . *Phyllomys thomasi*
2'. Fur short and stiff, aristiforms on rump <25 mm in length; incisive foramina ovate; rostral premaxillary foramen present; known only from the vicinity of Helvécia, southern Bahia, Brazil . . . . *Phyllomys unicolor*
3. Aristiforms on rump with black tips 4
3'. Aristiforms on rump with orange tips 9
4. Spines conspicuous, aristiforms on rump >0.7 mm in width 5

4'. Spines inconspicuous, aristiforms on rump <0.7 mm in width 6
5. Maxillary toothrow length >11.5 mm; rostral breadth <8 mm; interorbital constriction <11.5 mm . . . . *Phyllomys nigrispinus*
5'. Maxillary toothrow length <11.5 mm; rostral breadth >8 mm; interorbital constriction >11.5 mm . . . . *Phyllomys kerri*
6. Aristiforms on rump >26 mm in length; body size larger, head and body length >215 mm; hindfoot length >40 mm. . . . . *Phyllomys medius*
6'. Aristiforms on rump <26 mm in length; body size smaller, head and body length <215 mm; hindfoot length <40 mm. 7
7. Pelage coarse, aristiforms on rump reach 0.6 mm in width . . . . *Phyllomys sulinus*
7'. Pelage soft, aristiforms on rump <0.2 mm in width . . 8
8. Tail slightly longer than head and body, without terminal tuft; coronoid process shorter than condyloid process . . . . *Phyllomys dasythrix*
8'. Tail as long as head and body, with terminal tuft; coronoid process stands higher than condyloid process . . . . *Phyllomys mantiqueirensis*
9. Aristiforms on rump ending with a blunt tip; incisive foramina bullet-shaped . . . . *Phyllomys pattoni*
9'. Aristiforms on rump ending with a whip-like tip; incisive foramina ovate. 10
10. Aristiforms on rump >25 mm in length; mesopterygoid fossa >4 mm in width; Interorbital constriction >12 mm . . . . *Phyllomys brasiliensis*
10'. Aristiforms on rump <25 mm in length; mesopterygoid fossa <4 mm in width; Interorbital constriction <12 mm 11
11. Tail thickly haired with terminal tuft; aristiforms on rump <0.9 mm in width . . . . *Phyllomys blainvilii*
11'. Tail nearly naked, with visible scales and no terminal tuft; aristiforms on rump >0.9 mm in width 12
12. Aristiforms on rump reach 1.3 mm in width; supraorbital ridges well developed; infraorbital foramen length <4.5 mm; ventral mandibular spine present. . . . . *Phyllomys lamarum*
12'. Aristiforms on rump <0.9 mm in width; supraorbital ridges weakly developed; infraorbital foramen length >4.5 mm; ventral mandibular spine absent . . . . *Phyllomys lundi*

Phyllomys blainvilii (Jourdan, 1837)
Golden Atlantic Tree Rat
SYNONYMS:

Nelomys blainvilii Jourdan, 1837:522; type locality "un petite île sur les côtes du Brésil, près du Bahia," erroneously given by most subsequent authors as "Ilha de

Deos" based on I. Geoffroy St.-Hilaire's (1840) listing as "Brazil, near Bahia; the small island Deos, off the coast of Brazil, likewise near Bahia" (translation in Emmons, Leite et al. 2002:15); restricted to "Seabra, Bahia, Brazil, ca. 12°25′S 41°46′W" based on itinerary of original collector (Emmons, Leite et al. 2002:15); lectotype chosen by Emmons, Leite et al. (2002).

Loncheres (Nelomys) blainvillei Wagner, 1840:203; part; name combination but subsequence incorrect spelling of *blainvilii* Jourdan.

[*Loncheres (Loncheres)*] *Blainvillii*: E.-L. Trouessart, 1880b: 177; part; name combination but subsequent incorrect spelling of *blainvilii* Jourdan.

[*Echimys (Echimys)*] *blainvillei*: E.-L. Trouessart, 1904:503; name combination but subsequent incorrect spelling of *blainvilii* Jourdan.

Nelomys blainvillei: Thomas, 1916d:240; name combination but subsequent incorrect spelling of *blainvilii* Jourdan.

Phyllomys lamarum: Moojen, 1952b:140, plate 20; not *Nelomys lamarum* Thomas.

Echimys blainvillei blainvillei: Cabrera, 1961:539; name combination but subsequent incorrect spelling of *blainvilii* Jourdan.

Echimys (Nelomys) blainvillei: Eisenberg and Redford, 1999: 487; name combination but subsequent incorrect spelling of *blainvilii* Jourdan.

Phyllomys blainvilii: Emmons, Leite, Kock, and Costa 2002: 9; first use of current name combination.

DESCRIPTION: Medium sized for genus, with head and body length 192–250 mm, tail length 155–220 mm, hindfoot length 33–40 mm, and ear length 14–20 mm. Dorsal surface spiny, ochraceous-brown streaked with black, paler on sides. Aristiforms average 24 mm long and 1 mm wide, are pale at base, darker in middle, and orange near thin, whip-like tip. Venter pale cream with hint of yellowish overwash. Tail robust, usually slightly longer than head and body, thickly covered with long blackish-brown or gold hair forming a 15 mm, slightly wavy tuft at tip; it darkens from base to tip, usually strongly contrasting with pale color of the body. Skull with well-developed supraorbital ridges; interorbital region with nearly straight edges that diverge posteriorly; zygomatic arch slender, with maximum height less than or equal to one-third of the length of jugal; postorbital process of zygoma usually spinose and formed mainly by jugal; lateral process of supraoccipital short, extending to horizontal midline of external auditory meatus; narrow, forming an angle of about 45°, mesopterygoid fossa reaches anterior lamina of M3; incisive foramina oval. Cheek teeth narrow, with palatine width greater than tooth width at M1; toothrows slightly divergent at either end. Mandible with

ventral root of angular process deflected laterally. Upper incisors orthodont.

DISTRIBUTION: *Phyllomys blainvilii* occurs in the inland region of northeastern Brazil, from the southern portion of the state of Ceará (ca. 7° S) to the northernmost part of Minas Gerais (ca. 15° S), perhaps extending to the coast in Alagoas and Pernambuco states.

SELECTED LOCALITIES (Map 501; from Y. L. R. Leite 2003, except as noted): BRAZIL: Alagoas, Sítio Angelim; Bahia, Bom Jesus da Lapa, Fazenda Santa Rita, Seabra (type locality of *Nelomys blainvilii* Jourdan, as restricted by Emmons, Leite et al. 2002); Ceará, Chapada do Araripe; Minas Gerais, Mocambinho; Pernambuco, Igarassú (but see Remarks), Sítio Cavaquinho.

SUBSPECIES: *Phyllomys blainvilii* is monotypic.

NATURAL HISTORY: This species occupies fragmented habitat throughout a large area of northeastern Brazil, primarily in isolated areas of semideciduous forest, especially forest islands within the Caatinga termed *brejos* or *brejos de altitude*, and forest patches along the Rio São Francisco.

REMARKS: Identification of specimens from the coastal area of northeastern Brazil (Alagoas and Pernambuco states) should be reviewed, as these might represent an undescribed species. This is certainly the case for the specimen from Paraíba state reported by Campos and Per-

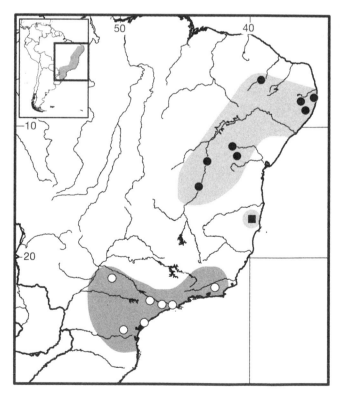

Map 501 Selected localities for *Phyllomys blainvilii* (●), *Phyllomys nigrispinus* (○), and *Phyllomys unicolor* (■).

cequillo (2007) as *P. blainvilii* (see Loss and Leite 2011). M. J. Souza (1981) and Y. L. R. Leite (2003) described a karyotype with $2n = 50$, FN = 94–96. This species has been often confused in the literature with *P. lamarum*. Moojen (1952b), for example, had them reversed; his plate 19 actually portrays an adult *P. lamarum*, identifiable by the speckled fur on the back and nearly naked tail, while plate 20 pictures a young *P. blainvilii* with a more uniform body color and a tufted tail.

Phyllomys brasiliensis Lund, 1840
Brazilian Atlantic Tree Rat

SYNONYMS:

Phyllomys brasiliensis Lund, 1839b:208; *nomen nudum.*

Phyllomys brasiliensis Lund, 1840b [1841c:294]; type locality fixed to "Lapa das Quatro Bocas," Lagoa Santa, Minas Gerais, Brazil, by selection of a lectotype (Emmons, Leite et al. 2002:7).

[*Loncheres*] *braziliensis* Waterhouse, 1848:330; name combination but subsequent incorrect spelling of *brasiliensis* Lund.

[*Loncheres* (*Loncheres*) *armatus*] *brasiliensis*: E.-L. Trouessart, 1880b:177; name combination.

Loncheres armatus: Winge, 1887:71; name combination; not *armatus* I. Geoffroy St.-Hilaire.

Echimys (*Echimys*) *armatus*: E.-L. Trouessart, 1904:503; part; name combination; not *armatus* I. Geoffroy St.-Hilaire.

Phyllomys armatus: Goldman, 1916:126; part.

[*Nelomys*] *brasiliensis*: Thomas, 1916d:240; name combination.

[*Echimys*] *braziliensis*: Tate, 1935:431; name combination but subsequence incorrect spelling of *brasiliensis* Lund.

Echimys (*Nelomys*) *brasiliensis*: Eisenberg and Redford, 1999:487; name combination.

DESCRIPTION: Medium sized for genus, head and body length 210 mm, hindfoot length 36 mm, and ear length 18 mm (Reinhardt 1849). Dorsal pelage spiny, colored orange-brown sprinkled with black. Aristiforms tipped with orange and are relatively long (27 mm), with whip-like tip, and wide (1.3 mm) on rump. Venter cream-yellow with cream-white inguinal and axillary regions with white-based hair along midline. Tail about equal to the head and body length; covered with short brown hair such that scales remain visible except near tip, where longer hairs reach 20 mm. Skull with well-developed and beaded supraorbital ridges; interorbital region slightly divergent posteriorly, with moderately developed postorbital processes; zygomatic arch slender to moderately robust, maximum height less than or equal to one-third the length of the jugal; spinose postorbital process of zygoma without squamosal contribution; lateral process of supraoccipital

long, extending below horizontal midline of external auditory meatus; mesopterygoid fossa sharply pointed anteriorly, wide posteriorly, forming angle >60°, with anterior point reaching posterior lamina of M2. Incisive foramina oval. Cheek teeth narrow, palatine width equal to tooth width at M1; toothrows nearly parallel, diverging posteriorly in older individuals. Mandible with ventral root of angular process deflected laterally and ventral spine present posterior to junction of mandibular rami. Upper incisors are orthodont.

DISTRIBUTION: *Phyllomys brasiliensis* is known only from the Lagoa Santa region and Fazenda Santa Cruz, 150 km to the northwest, all in Minas Gerais state, Brazil, in the valleys of the Paraopeba and das Velhas rivers.

SELECTED LOCALITIES (Map 502): BRAZIL: Minas Gerais, Fazenda Santa Cruz (Y. L. R. Leite 2003), Sumidouro (Emmons, Leite et al. 2002).

SUBSPECIES: *Phyllomys brasiliensis* is monotypic.

NATURAL HISTORY: *Phyllomys brasiliensis* inhabits the area between semideciduous forest and Cerrado, between 500 and 800 m.

REMARKS: *Phyllomys brasiliensis* has been frequently confused with the recently named *P. pattoni* (Emmons, Leite et al. 2002), but the description and measurements of *P. brasiliensis* given by Reinhardt (1849) and Winge (1887)

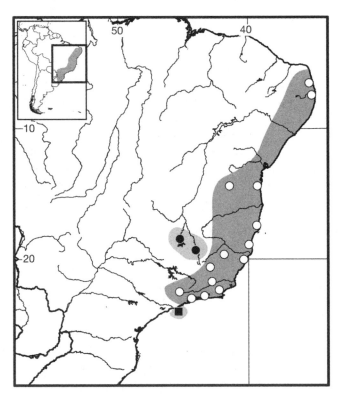

Map 502 Selected localities for *Phyllomys brasiliensis* (●), *Phyllomys pattoni* (○), and *Phyllomys thomasi* (■). Contour line = 1,000 m.

are rich in detail and provide unambiguous distinctions between these two species: "aristiforms grayish at the base, orange distally and grayish-brown at the tip, ending in a very fine, curved hairlike point" (Reinhardt 1849); "tail hairs get longer distally, forming a thick tuft at the tip" (Winge 1887). Known primarily from fragmentary skulls from owl pellets collected by Lund (1840a,b [1841b,c]) and described in more detail by Winge (1887), there is only one record of *P. brasiliensis* from the twentieth century.

Phyllomys dasythrix Hensel, 1872
Drab Atlantic Tree Rat

SYNONYMS:

Phyllomys dasythrix Hensel, 1872b:49, plate 1, Figs. 11 and 12; type locality "Rio Grande do Sul, Süd-Brasiliens"; restricted to "Porto Alegre, Rio Grande do Sul, Brazil, 30°04′S 51°07′W" (Emmons, Leite et al. 2002:21).

[*Loncheres (Loncheres)*] *dasythrix*: E.-L. Trouessart, 1880b: 177; name combination.

[*Echimys (Echimys)*] *dasythrix*: E.-L. Trouessart, 1904:503; name combination.

[*Nelomys*] *dasythrix*: Thomas, 1916d:240; name combination.

Echimys dasythrix dasythrix: Cabrera, 1961:540; name combination.

Echimys (Nelomys) dasythrix: Eisenberg and Redford, 1999: 487; name combination.

DESCRIPTION: Medium-sized for genus, head and body length 190–195 mm, tail length 205–225 mm, hindfoot length 32–39 mm, and ear length 13–19 mm (Y. L. R. Leite et al. 2008). Pelage soft, with very fine and long (0.2 × 26 mm) aristiforms on rump, paler at base and blackish distally, with thin whip-like tips. Tail moderately covered with brown hairs to tip, which lacks tuft. Skull with weakly developed supraorbital ridges; interorbital region slightly divergent posteriorly, sometimes with small postorbital process present; zygomatic arches robust with maximum height approximately one-third length of jugal; lateral process of supraoccipital short, extending to horizontal midline of external auditory meatus; mesopterygoid fossa narrow, forming angle of 45–60°, reaching to last lamina of M2. Incisive foramina oval. Cheek teeth large with palatine width less than tooth width at M1, and toothrows parallel. Mandible with ventral root of angular process not deflected laterally. Upper incisors orthodont.

DISTRIBUTION: The known range of *Phyllomys dasythrix* is along the south coast of Brazil; from the state of Paraná (ca. 26°S) to Rio Grande do Sul (ca. 30°S).

SELECTED LOCALITIES (Map 503; from Y. L. R. Leite 2003:67): BRAZIL: Paraná, Palmira (BM 0.6.29.20);

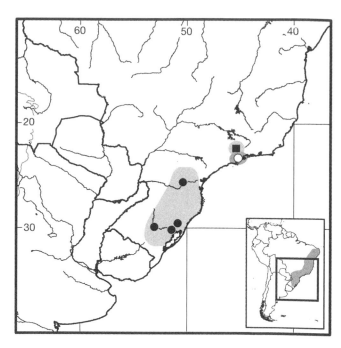

Map 503 Selected localities for *Phyllomys dasythrix* (●), *Phyllomys lundi* (○), and *Phyllomys mantiqueirensis* (■).

Rio Grande do Sul, Pinheiros (Y. L. R. Leite 2003), Porto Alegre (type locality of *Phyllomys dasythrix* Hensel, as restricted by Emmons, Leite et al. 2002), São Francisco de Paula (Y. L. R. Leite 2003).

SUBSPECIES: *Phyllomys dasythrix* is monotypic.

NATURAL HISTORY: The habitat range extends from semideciduous, and perhaps *Araucaria*, forests inland, usually below 800 m elevation.

REMARKS: While *P. dasythrix* has distinctly soft fur, Y. L. R. Leite (2003) recognized a second and sympatric animal with spiny fur but the same qualitative cranial characters, which he referred to as *P.* aff. *dasythrix*. Oldfield Thomas (1909) apparently mixed these two forms in his comparison of "*P. dasythrix*" with *P. medius*, which he described in that paper. Y. L. R. Leite et al. (2008) have now described this second taxon; see account for *P. sulinus*, below. Y. L. R. Leite (2003) gave the diploid number as 72, but details of the karyotype are unknown.

Phyllomys kerri (Moojen, 1950)
Kerr's Atlantic Tree Rat

SYNONYMS:

Echimys (Phyllomys) kerri Moojen, 1950:489, Figs. 1–5; type locality "Ubatuba, S[ão]. Paulo, Brasil," restricted to "Estação Experimental de Ubatuba, Ubatuba, São Paulo, Brazil, 23°25′S 45°07′W" (Emmons, Leite et al. 2002:28).

Phyllomys kerri: Moojen, 1952b:142; first use of current name combination.

Echimys kerri: C. O. da C. Vieira, 1955:437; name combination.

Nelomys kerrei Emmons and Feer, 1990:274; name combination, and incorrect subsequent spelling of *Echimys (Phyllomys) kerri* Moojen.

DESCRIPTION: Medium-sized for genus, head and body length 212 mm, tail length 223 mm, and hindfoot length 38 mm, ear length 17 mm (measurements of holotype; Emmons, Leite et al. 2002). Dorsal pelage spiny, colored reddish brown streaked with black. Aristiforms hairs on rump black, averaging 27 mm in length and 1 mm in width, and terminated by thin, whip-like tips. Ventral color yellowish orange with hairs usually gray at base. Skull long and narrow; supraorbital ridges well developed; interorbital region diverges posteriorly, with inconspicuous postorbital process; zygomatic arch robust with maximum height approximately one-third the length of jugal; postorbital process of zygoma spinose and formed mainly by jugal; lateral process of supraoccipital short, reaching horizontal midline of external auditory meatus; palatine width equal to or greater than tooth width at M1; toothrows nearly parallel; mesopterygoid fossa wide, forming an angle of nearly 60°, reaching last laminae in M2; incisive foramina oval. Mandible with ventral root of angular process not deflected laterally. Upper incisors slightly opisthodont.

DISTRIBUTION: *Phyllomys kerri* is known only from the its type locality.

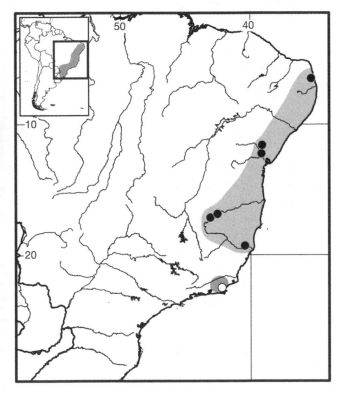

SELECTED LOCALITIES (Map 504): BRAZIL: São Paulo, Estação Experimental de Ubatuba (type locality of *Echimys [Phyllomys] kerri* Moojen, 1950, as restricted by Emmons, Leite et al. 2002).

SUBSPECIES: *Phyllomys kerri* is monotypic.

NATURAL HISTORY: The type locality of *Phyllomys kerri* is within broadleaf evergreen rainforest close to sea level. No other information is known.

REMARKS: This species is known from only three specimens, all collected at the type locality. Y. L. R. Leite (2003:73) stated that the morphological differences between *P. kerri* and *P. nigrispinis* identified by Moojen (1950) in his description of the former fall within the natural range of variation of the latter species. Nevertheless, he provided additional diagnostic characters to separate these two taxa (shorter maxillary toothrow, broader rostrum, wider interorbital region, and reduced bullae) and thus provisionally maintained *P. kerri* as a valid species, at least until sufficient specimens of both taxa become available for adequate comparisons.

Phyllomys lamarum (Thomas, 1916)
Pallid Atlantic Tree Rat
SYNONYMS:

Nelomys lamarum Thomas, 1916e:297; type locality "Lamaraõ [sic], Bahia, about 70 miles N.W. of Bahia city [now Salvador]. Alt. 300 mm [sic]," amended to "Lamarão, about 70 miles NW Salvador, Bahia, Brazil, 11°47′S 38°53′W, elev. 300 m" (Emmons, Leite et al. 2002:26).

[*Echimys*] *lamarum*: Tate, 1935:431; name combination.

Phyllomys blainvilii: Moojen, 1952b:139, plate 19; not *blainvilii* Jourdan.

Echimys dasythrix lamarum: Cabrera, 1961:541; name combination.

Echimys (Nelomys) lamarum: Eisenberg and Redford, 1999:487; name combination.

Phyllomys lamarum: Emmons, Leite, Kock, and Costa, 2002:26; first use of current name combination.

DESCRIPTION: Medium sized for genus, with head and body length 195–230 mm, tail length 210–233 mm, hindfoot length 32–35 mm, and ear length 13–16 mm (Emmons and Feer 1997). Dorsum yellow-brown with speckled pattern given by short (24 mm), broad (1.3 mm) aristiforms. Spines pale at base and darken toward tip where they become orange, ending in dark, thin, whip-like tips. Venter pale brown with white patches, varying to pure white; fulvous lateral line typically present. Tail relatively slender, slightly shorter than or equal to head and body length, and thinly covered with pale brown hairs such that scales remain visible to the eye along entire length. Skull with posteriorly diverging interorbital region, almost straight-edged due to inconspicuous or absent postorbital processes, with well-developed supraorbital ledges; zygomatic arch relatively robust with maximum height less than or equal to one-third the length of jugal; postorbital process of zygoma spinose, usually without contribution from squamosal; lateral process of supraoccipital short, extending to midline of external auditory meatus; wide, forming an angle of more than 60°, mesopterygoid fossa reaches last lamina of M2; incisive foramina oval. Mandible with ventral root of angular process deflected laterally; ventral spine present near posterior junction of mandibular rami. Upper incisors orthodont.

DISTRIBUTION: *Phyllomys lamarum* extends along coastal Brazil from the state of Paraíba in the northeast (ca. 7°S) to Espírito Santo in the southeast (ca. 19°S), and inland to northern Minas Gerais. The three geographic sets of known localities are each about 700 km apart, and thus may represent truly disjunct populations.

SELECTED LOCALITIES (Map 504): BRAZIL: Bahia, Lamarão (type locality of *Nelomys lamarum* Thomas, as amended by Emmons, Leite et al. 2002), São Gonçalo (Y. L. R. Leite 2003); Espírito Santo, Pancas (Loss and Leite 2011); Minas Gerais, Estação Ecológica de Acauã (Y. L. R. Leite 2003), Fazenda Paiol (UMFG 1491); Paraíba, Mamanguape (Y. L. R. Leite 2003).

SUBSPECIES: *Phyllomys lamarum* is monotypic.

NATURAL HISTORY: *Phyllomys lamarum* is found mainly in semideciduous forests but extends to the coastal forests in Paraíba state.

REMARKS: Identification of specimens assigned to this species from the coastal area of northeastern Brazil (states of Alagoas, Pernambuco, and Paraíba) should be reviewed; these likely represent an undescribed species of *Phyllomys* (see Loss and Leite 2011). As mentioned in the account of *P. blainvilii*, Moojen (1952b) had these two species reversed. Specimens of *P. lamarum* from the arid parts of the range, such as the type locality of Lamarão, are paler than specimens from moister areas, for example, in Minas Gerais (Y. L. R. Leite 2003).

N. P. Araújo et al. (2014) described the karyotype of a male with $2n = 56$, $FN = 102$ collected at Araçuaí, Minas Grais (16.85°S, 42.07°W).

Phyllomys lundi Y. L. R. Leite, 2003
Lund's Atlantic Tree Rat

SYNONYM:

Phyllomys lundi Y. L. R. Leite, 2003:19; type locality "Fazenda do Bené, 4 km SE Passa Vinte, Minas Gerais, Brazil, 22°14′S 44°12′W 680 m."

DESCRIPTION: Among smallest species in genus, with head and body length 209 mm, tail length 204 mm, hindfoot length 36 mm, ear length 16 mm, and mass 145 g (measurements of holotype; Y. L. R. Leite 2003). Dorsal pelage colored predominantly orange intermixed with black; neck and thighs markedly orange. Spines conspicuous from neck to tail. Ventral hairs cream with white base, providing overall washed aspect. Tail brown in color and hairy from base to tip. The forefeet are covered with brown-yellow hairs, except for gray-white fingers. Dorsal surfaces of hindfeet covered with golden-creamy hairs and toes with silver hair. Skull delicate with relatively narrow and long rostrum, and wide and convex interorbital region. M3 with only three transverse plates. Mandible with short coronoid process and shallow sigmoid notch.

DISTRIBUTION: *Phyllomys lundi* is known only from two localities separated by nearly 200 km, one each in the states of Minas Gerais (the type locality) and Rio de Janeiro, southeastern Brazil (Y. L. R. Leite 2003).

SELECTED LOCALITIES (Map 503): BRAZIL: Minas Gerais, Fazenda do Bené (type locality of *Phyllomys lundi* Y. L. R. Leite); Rio de Janeiro, Reserva Biológica de Poço das Antas (Y. L. R. Leite 2003).

SUBSPECIES: *Phyllomys lundi* is monotypic.

NATURAL HISTORY: *Phyllomys lundi* occurs in second growth, broadleaf evergreen rainforest with a dense understory, where most trees are at least 20 cm in diameter and crowns 20 m in height, and also in selectively logged mature forest on dry ridge-top soils, discontinuous overstory with emergents reaching 32 m, sparse understory, and lack of discernible boundaries between forest strata. Both specimens were taken in traps placed on tree limbs or lianas within 1–1.5 m above the ground.

REMARKS: The phylogenetic position of *P. lundi* remains uncertain. Analyses of mtDNA and nucDNA sequences suggest that this species is closely related to either a southern clade formed by *P. dasythrix*, *P. sulinus*, and *P. nigrispinus* or a northeastern clade comprised by *P. blainvilii*, *P. lamarum*, and *P. brasiliensis* (Loss and Leite 2011).

Phyllomys mantiqueirensis Y. L. R. Leite, 2003
Serra da Mantiqueira Atlantic Tree Rat

SYNONYM:

Phyllomys mantiqueirensis Y. L. R. Leite, 2003:25; type locality "Fazenda da Onça, 13 km SW Delfim Moreira, Minas Gerais, Brazil, 22°36′S 45°20′W 1850 m."

DESCRIPTION: Medium sized for genus, head and body length 217 mm, tail length 216 mm, hindfoot length 41 mm, ear length 18 mm, mass 207 g (measurements of holotype; Y. L. R. Leite 2003). Soft, densely furred arboreal rat with hairy tail tufted at tip. Dorsal body color brownish gray. Tail equal to head and body length and darker in color than body; covered with brown hairs that become denser and longer toward tip, forming a long (30 mm) terminal tuft. Skull with short and broad rostrum, narrow interorbital region, well-developed lacrimal processes, small and oval incisive foramina. Mandible with slim and long coronoid process and deep sigmoid notch.

DISTRIBUTION: *Phyllomys mantiqueirensis* is known only from the type locality.

SELECTED LOCALITIES (Map 503): BRAZIL: Minas Gerais, Fazenda da Onça (type locality of *Phyllomys mantiqueirensis* Y. L. R. Leite).

SUBSPECIES: *Phyllomys mantiqueirensis* is monotypic.

NATURAL HISTORY: The type locality is in cool, wet montane forest with an open overstory composed of a 10-m canopy and emergent araucaria trees, on a steep slope with a moderately dense understory composed of shrubs, lianas, vines, abundant bamboos and lichens, and some ferns. The holotype, and only known specimen, was collected in a trap placed 2 m high on a liana connecting two trees.

REMARKS: This species is very distinctive, readily diagnosed by both morphological and molecular sequence characters. Phylogenetic analysis based on mtDNA and nucDNA sequences suggest a close relationship to *P. pattoni* and that, together, this pair of species may form the oldest lineage within the genus, one sister to all other extant forms (Loss and Leite 2011).

Phyllomys medius (Thomas, 1909)
Long-furred Atlantic Tree Rat
SYNONYMS:

Loncheres medius Thomas, 1909:239; type locality "Roça Nova Serro [*sic*] do Mar, Parana, S[outh]. Brazil. Alt. 1000 m," amended to "Roça Nova, Serra do Mar, Paraná, Brazil, 25°28′S 49°01′W, elevation 1000 m" (Emmons, Leite et al. 2002:24).

[*Nelomys*] *medius*: Thomas, 1916d:240; name combination.

[*Echimys*] *medius*: Tate, 1935:432; name combination.

Phyllomys medius: Moojen, 1952b:138; name combination.

Echimys blainvillei medius: Cabrera, 1961:540; name combination.

Echimys blainvillei: Woods, 1993:791; part; not *blainvilii* Jourdan.

Phyllomys medius: Emmons, Leite, Kock, and Costa, 2002: 23; first use of current name combination.

DESCRIPTION: Medium sized for genus, head and body length 230 mm, tail length 240 mm, hindfoot length 40 mm, and ear length 17 mm (measurements of holotype; Emmons, Leite et al. 2002). Fur stiff but relatively soft. Dorsum dark brown sprinkled with black, darkest on mid-dorsum. Aristiforms on rump very long (36 mm), thin (0.4 mm), black distally, with thin whip-like tip. Ventral hairs distinctly gray-based, with fulvous tips. Skull robust

and long; supraorbital ledges well developed and interorbital region parallel sided or divergent posteriorly with inconspicuous, or absent, postorbital processes; zygomatic arch robust with its maximum height more than one-third the length of jugal; postorbital process of zygoma spinose and formed mainly by jugal; lateral process of supraoccipital short, extending to horizontal midline of external auditory meatus; incisive foramina are small and teardrop in shape; maxillary teeth narrow, with palatine width equal to tooth width at M1; toothrows parallel; mesopterygoid fossa narrow, forming angle of about 45°, reaching last lamina of M2; tympanic bullae small. Mandible with ventral root of angular process not deflected laterally. Upper incisors opisthodont.

DISTRIBUTION: *Phyllomys medius* is known in southern Brazil from the states of Minas Gerais and Rio de Janeiro in the north to Rio Grande do Sul in the south.

SELECTED LOCALITIES (Map 505; from Y. L. R. Leite 2003, except as noted): BRAZIL: Minas Gerais, Alto da Consulta; Paraná, Rio Paracaí, Roça Nova (type locality of *Loncheres medius* Thomas, as amended by Emmons, Leite et al. 2002); Rio Grande do Sul, Banhado do Pontal; Rio de Janeiro, Fazenda Comari; São Paulo, Barra do Ribeirão Onça Parda; Santa Catarina, Florianópolis.

SUBSPECIES: *Phyllomys medius* in monotypic.

NATURAL HISTORY: This species is found mainly along the coast in areas of broadleaf evergreen rainforest but extends into *Araucaria* forests in Paraná state (ca. 52°W). Northern records are from elevations as high

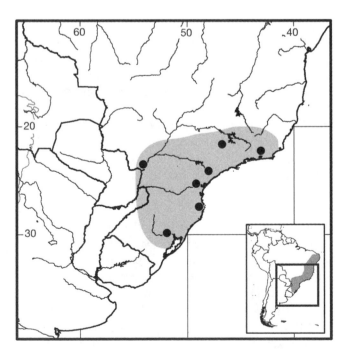

Map 505 Selected localities for *Phyllomys medius* (●).

as 1,000 m; those in the south are at lower elevations, suggesting a preference for cooler climates. Davis (1945a) provided detailed descriptions of the habitat of this species at Terezópolis, in Rio de Janeiro state.

REMARKS: Sbalqueiro et al. (1988) reported a diploid number of 96 for two specimens from Santa Catarina (as *Echimys dasythrix*, but one specimen identified by Y. L. R. Leite 2003:69 as *P. medius*).

Phyllomys nigrispinus (Wagner, 1842)
Black-spined Atlantic Tree Rat
SYNONYMS:

Loncheres nigrispina Wagner, 1842c:361; type locality "Ypanema," amended to "Floresta Nacional de Ipanema, 20 km NW Sorocaba, São Paulo, Brazil, 23°26'S 47°37'W, elev. 550–970 m" (Emmons, Leite et al. 2002:17).

[*Loncheres* (*Loncheres*) *Blainvillii*] *nigrispina*: E.-L. Trouessart, 1880b:177; name combination.

[*Loncheres* (*Loncheres*)] *nigrispina*: E.-L. Trouessart, 1897: 604; name combination.

[*Echimys* (*Echimys*)] *nigrispina*: E.-L. Trouessart, 1904: 503; name combination.

[*Nelomys*] *nigrispina*: Thomas, 1916e:297; name combination.

Echimys nigrispina: Tate, 1935:432; name combination.

Phyllomys nigrispina: Moojen, 1952b:141; name combination but incorrect gender concordance.

Echimys (*Nelomys*) *nigrispinus*: Eisenberg and Redford, 1999:488; name combination.

Phyllomys nigrispinus: Emmons, Leite, Kock, and Costa, 2002:17; first use of current name combination, and correct gender ending.

DESCRIPTION: Medium sized for genus, head and body length 215 mm, tail length 240 mm, hindfoot length 40 mm, and ear length 18 mm (Emmons and Feer 1997). Substantial variation within and among populations, but generally dorsum spiny with reddish brown hairs streaked with black, clothed with relatively narrow (1 mm), medium length (27 mm), and black aristiform hairs that terminate with thin, whip-like tips. Ventral color extremely variable, ranging from buffy-white to yellowish gray; hairs are usually white at base. Skull with supraorbital ridges well developed; interorbital region diverging posteriorly with absent or inconspicuous postorbital processes; zygomatic arch slender to moderately robust, maximum height between one-fourth and one-third length of jugal; postorbital process of zygoma rounded or spinose, formed mainly by jugal; lateral process of supraoccipital long, extending below horizontal midline of external auditory meatus; incisive foramina oval; palatine width equal to or broader than tooth width at M1; toothrows parallel; wide mesoptery-

goid fossa, forming an angle of about 60° and reaching last laminae of M2 or first of M3. Mandible with ventral root of angular process not deflected laterally. Upper incisors slightly opisthodont.

DISTRIBUTION: *Phyllomys nigrispinus* occurs in southeastern Brazil from the state of Rio de Janeiro in the north (ca. 22°S) to Paraná in the south (ca. 26°S), mainly along the coastal zone although extending inland to western São Paulo state (ca. 50°W).

SELECTED LOCALITIES (Map 501; from Y. L. R. Leite 2003, except as noted): BRAZIL: Paraná, Guajuvira; Rio de Janeiro, Fazenda Alpina; São Paulo, Floresta Nacional de Ipanema (type locality of *Loncheres nigrispina* Wagner, as amended by Emmons, Leite et al. 2002), Ilha do Cardoso, Interlagos, Santos, Vanuire.

SUBSPECIES: *Phyllomys nigrispinus* is monotypic.

NATURAL HISTORY: *Phyllomys nigrispinus* occurs in both broadleaf evergreen rainforest and semideciduous forest.

REMARKS: Y. L. R. Leite (2003) noted that the extensive variation in morphological characters among the limited series of specimens available might signal multiple species within this taxon as he defined it. He reported a diploid number of 52 for specimens from Cachoeiras de Macacu, Rio de Janeiro state, but molecular data suggest these specimens represent an undescribed species (Loss and Leite 2011).

Phyllomys pattoni Emmons, Leite, Kock, and Costa, 2002
Patton's Atlantic Tree Rat
SYNONYMS:

Loncheres armatus: Burmeister, 1854:196; part; not *armatus* I. Geoffroy St.-Hilaire.

Nelomys brasiliensis: Thomas, 1916e:297; not *Phyllomys brasiliensis* Lund.

Phyllomys brasiliensis: Moojen, 1952b:141; not *Phyllomys brasiliensis* Lund.

Nelomys sp.: Emmons and Feer, 1990:220; under the common name "rusty-sided tree rat."

Phyllomys pattoni Emmons, Leite, Kock, and Costa, 2002: 30; type locality "Mangue do Caritoti, Caravelas, Bahia, Brazil, 17°43'30"S 39°15'35"W; at sea level."

DESCRIPTION: Medium-sized for genus, head and body length 209–241 mm, tail length 190–223 mm, hind foot length 40–42 mm, and ear length 16–18 mm. Most heavily spined species in genus; aristiforms on rump average 23 mm in length and 1.0 in width. Dorsal spines light gray at base, gradually darkening toward black medially, and terminate in blunt orange tip. Most aristiforms on rump lack characteristic whip-like tip present in all other species (Emmons, Leite et al. 2002). Overall, dorsum dark brown, with speckled aspect given by rusty-tipped spines. Tail slightly shorter to slightly longer than head and body length and covered with fine hair to tip, although scales remain visible to eye. Skull robust and relatively broad; supraorbital ledges well developed and beaded; interorbital region divergent posteriorly and lacks postorbital process; zygomatic arch robust with maximum height approximately one-third the length of jugal; postorbital process of zygoma spinose, with contributions of both jugal and squamosal; lateral process of supraoccipital long, extending below horizontal midline of external auditory meatus; incisive foramina bullet shaped; maxillary toothrows relatively short and narrow, with palatine width equal to tooth width at M1. Mandible with ventral root of angular process deflected laterally. Upper incisors orthodont.

DISTRIBUTION: *Phyllomys pattoni* is a wide-ranging species of coastal Brazil, extending from the state of Paraíba (ca. 7°S) in the northeast to São Paulo (ca. 23°S) in the southeast.

SELECTED LOCALITIES (Map 502; see Emmons, Leite et al. 2002, except as noted): BRAZIL: Bahia, Fazenda Pirataquicê, Mangue do Caritoti (type locality of *Phyllomys pattoni* Emmons, Leite, Kock, and Costa, 2002), São Felipe; Espírito Santo, Fazenda Santa Terezinha, Parque Estadual da Fonte Grande; Minas Gerais, Além Paraíba, Fazenda Monte Claros, Mata Paraíso; Paraíba, João Pessoa; Pernambuco, Dois Irmãos; Rio de Janeiro, Fazenda União, Ilha Grande, Tijuca; São Paulo, Piquete.

SUBSPECIES: *Phyllomys pattoni* is monotypic.

NATURAL HISTORY: *Phyllomys pattoni* is found chiefly in broadleaf evergreen rainforest and associated habitats, such as mangroves, from sea level to 1,000 m.

REMARKS: Zanchin (1988) gave a diploid number of 80 for a specimen he identified as "*Echimys* sp." and Severo described a karyotype of $2n=80$, $FN=112$, under the name "*Echimys thomasi*." Both are reports of the same specimen from Espírito Santo state. Paresque et al. (2004) describe a karyotype of $2n=80$, $FN=100$, for *P. pattoni*, also a specimen from Espírito Santo. Whether the reported differences in the number of autosomal arms reflect real morphological distinctions or are methodological artifacts remains to be established. Specimens from Rio de Janeiro are $2n=72$, $FN=114$ (Y. L. R. Leite 2003). As documented by Emmons, Leite et al. (2002), this species was incorrectly associated with Lund's *P. brasiliensis* for more than a century (Burmeister 1854; Thomas 1916e; Moojen 1952b). Emmons and Feer (1990, 1997) referred to this species as the "Rusty-Sided Atlantic Tree Rat, *Nelomys* sp."

Phyllomys sulinus Y. L. R. Leite, Christoff, and Fagundes, 2008
Southern Atlantic Tree Rat

SYNONYM:

Phyllomys sulinus Y. L. R. Leite, Christoff, and Fagundes, 2008: 846; type locality "South bank of the Uruguay River, municipality of Aratiba, Rio Grande do Sul, Brazil, 27°23′39″S 52°18′01″W, 420 m elevation."

DESCRIPTION: Medium sized for genus for genus, head and body length 200–210 mm, tail length 160–248 mm, hindfoot length 35–40 mm, and ear length 11–18 mm (Y. L. R. Leite et al. 2008). Pelage coarse, with both aristiform and setiform hairs; aristiforms on rump medium in length (26 mm) and width (0.6 mm), wider and paler at base, thinning gradually into black and with thin whip-like tip; setiforms on rump (24 mm) slightly shorter than aristiforms, with basal portion gray, gradually turning to dark brown, with 4 mm orange band near tip. Overall dorsal color orange brown. Ventral region covered with short (10 mm), cream-colored hairs with gray bases; white patches of fur occur on chin, throat, axillary, and inguinal regions. Tail dark brown above and light brown below; hairy along entire length, usually with scales hidden; tail hairs longer toward the tip, forming conspicuous (20 mm) terminal tuft. Head dark brown, contrasting with lighter body; nose blunt and gray brown; cheeks pale cream. Mystacial vibrissae dark brown and long (65 mm); longest supercilliary and genal vibrissa reach 45 mm. Ears dark brown and nearly naked, sparsely covered with long (15 mm) and very thin hairs, especially along edges. Tail long (about 120% of head and body length); hindfeet short and broad (with claws, about 21% of head and body length); ears round and small (about 8% of head and body length). Dorsal surfaces of forefeet and hindfeet silver gray; ungual tufts grayish white and long, extending slightly beyond claws. Ventral surface of forefoot with thenar, hypothenar, and three equidistant interdigital pads; ventral surface of hindfoot with thenar, hypothenar, and four equidistant interdigital pads. Glans penis elongate, with slight medial swelling; terminal portion cone shaped, ending in long pointed bacular papilla; urethral lappets present near proximal end intromittent sac.

Skull with well-developed supraorbital ridged; interorbital region slightly divergent posteriorly, sometimes with small postorbital process; zygomatic arch robust, with maximum approximately one-third the length of jugal; postorbital process of zygoma spinose, formed mainly by jugal; lateral process of supraoccipital short, extending to horizontal midline of the external auditory meatus; incisive foramina oval; mesopterygoid fossa narrow, forming an angle of 45–60°, and reach last lamina of M2. Cheek teeth large; palatine width usually less than tooth width at M1; upper toothrows parallel. Mandible with ventral root of angular process not deflected laterally. Upper incisors orthodont.

DISTRIBUTION: *Phyllomys sulinus* occurs in the subtropical region of southern Brazil, from the states of São Paulo to Rio Grande do Sul. Y. L. R. Leite (2003) mentioned a possible occurrence in the state of Minas Gerais, just north of São Paulo, based on an unlabeled voucher (UFMG 3015) found in a freezer by E. L. Sábato and M. T. da Fonseca. Y. L. R. Leite's (2003) suggestion that *P. sulinus* is a highland species is incorrect because specimens have been collected below 500 meters at the type locality as well as on the island of Santa Catarina.

SELECTED LOCALITIES (Map 506; from Y. L. R. Leite et al. 2008, except as noted): BRAZIL: Paraná, Parque Barigüi; Rio Grande do Sul, Aratiba (type locality of *Phyllomys sulinus* Leite, Christoff, and Fagundes, 2008), Parque Nacional dos Aparados da Serra; Santa Catarina, Ilha de Santa Catarina; São Paulo, Teodoro Sampaio, Estação Biológica de Boracéia, Parque Estadual da Serra do Mar.

SUBSPECIES: *Phyllomys sulinus* is monotypic.

NATURAL HISTORY: Similar to the other species of *Phyllomys*, *P. sulinus* is a nocturnal, arboreal folivore, nesting above ground, usually in tree hollows (Y. L. R. Leite 2003; Y. L. R. Leite et al. 2008). Specimens of *P. sulinus* have been recorded in coastal Atlantic rainforest, as well as *Araucaria* and semideciduous forests inland. The Atlantic rainforest experiences warm and wet climate without a dry season, whereas a seasonal climate with a relatively severe dry season (generally from April to September) predominates in the Atlantic semideciduous and *Araucaria* forests (Morellato and Haddad 2000).

REMARKS: Y. L. R. Leite (2003) described and mapped this species as *P. aff. dasythrix*. Y. L. R. Leite et al. (2008) described and figured a karyotype with $2n = 92$, $FN = 102$ of a paratype from Rio Grande do Sul. Yonenaga (1975) reported a diploid number of 90 for a specimen from São Paulo state that likely is this species (see Y. L. R. Leite 2003:77 for discussion).

Phyllomys thomasi (Ihering, 1897)

Thomas's Atlantic Tree Rat

SYNONYMS:

Mesomys thomasi Ihering, 1897:171; type locality "Ilha de São Sebastião," amended to "Ilha da São Sebastião, São Paulo, Brazil, 23°46'S 45°21'W (Emmons, Leite et al. 2002: 23).

Loncheres nigrispina: Ihering, 1898:506; part; not *Loncheres nigrispina* Wagner.

[*Euryzygomatomys*] *thomasi*: E.-L. Trouessart, 1904:506; name combination.

L[*oncheres*]. *thomasi*: Thomas, 1909:239; name combination.

[*Nelomys*] *thomasi*: Thomas, 1916d:240; name combination.

[*Echimys*] *thomasi*: Tate, 1935:432; name combination.

Phyllomys thomasi: Moojen, 1952b:138; first use of current name combination.

Echimys blainvillei thomasi: Cabrera, 1961:540; name combination.

Echimys (Nelomys) thomasi: Eisenberg and Redford, 1999: 488; name combination.

DESCRIPTION: Largest species in genus: measurements of holotype include head and body length 287 mm, hindfoot length 47 mm (Emmons, Leite et al. 2002); unvouchered specimen with head and body length 270 mm, tail length 270 mm, mass 432 g (Olmos 1997). Dorsal pelage spiny; overall color reddish brown streaked with black, darker on mid-dorsum. Venter cream to light gray, grading gradually from sides. Aristiforms on rump long (33 mm), relatively narrow (0.7 mm), and gray-brown proximally and black distally with thin whip-like tip. Tail robust and covered with dark brown hair from base to tip. Skull with well-developed supraorbital ledges; interorbital region nearly parallel, with inconspicuous postorbital processes; zygomatic arch robust, reaching maximum height more than one-third the length of jugal; postorbital process of zygoma usually rounded and formed mainly by jugal; lateral process of supraoccipital

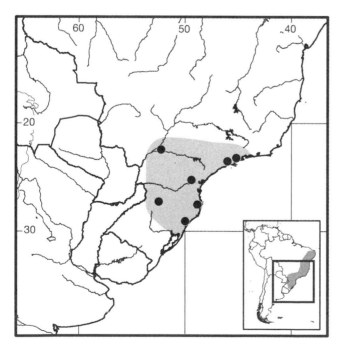

Map 506 Selected localities for *Phyllomys sulinus* (●).

short, extending to midline of external auditory meatus; incisive foramina teardrop in shape; mesopterygoid fossa wide, forming an angle of about 60°, reaching posterior lamina of M2. Cheek teeth wide; palatine width equal to or greater than tooth width at M1; toothrows parallel to slightly divergent posteriorly. Mandible with ventral root of angular process not deflected laterally. Upper incisors slightly opisthodont.

DISTRIBUTION: *Phyllomys thomasi* is endemic to the island of São Sebastião, off the coast of São Paulo.

SELECTED LOCALITIES (Map 502): BRAZIL: São Paulo, Ilha da São Sebastião (type locality of *Mesomys thomasi* Ihering, 1897, as amended by Emmons, Leite et al. 2002).

SUBSPECIES: *Phyllomys thomasi* is monotypic.

NATURAL HISTORY: The habitat of this species is broadleaf evergreen rainforest over rugged topography, from sea level to 1,379 m. Olmos (1997) found a nest 13 m up in a fork of a tree, it was constructed of "interwoven dried and twisted leaves of *Terminalia* . . . forming a roughly oval dome with a base" and measured 380 × 300 × 200 mm. The female he caught in this nest and kept for three days ate fruit but not green fodder. Olmos (1997) noted that the rat was then (in 1994) uncommon. It is classified as Endangered (IUCN 2011) because of its small total range size endemic to a single island and declining population.

REMARKS: Ihering based his original description of *P. thomasi* on the fact that the only specimen he had examined had no tail (Ihering 1897). After he had inspected the remaining specimens collected in the original series, some with tails and others without, he concluded that it was conspecific with *P. nigrispinus* (Ihering 1898). *Phyllomys thomasi* is, however, distinguished from *P. nigrispinus* by its much larger size although it otherwise shares a similar general morphology. It is also very similar to *P. kerri* in cranial shape but is proportionally larger in all skull measurements (Y. L. R. Leite 2003:68). A photograph of a living specimen taken by F. Olmos (pers. comm. to L. H. Emmons) shows a handsome rat with a striking pitch black and thickly furred tail, blackish back, and warm brown sides. As all museum specimens are many decades to over a century old, possibly their original black colors have bleached to red-brown.

Phyllomys unicolor (Wagner, 1842)
Unicolored Atlantic Tree Rat
SYNONYMS:

Loncheres unicolor Wagner, 1842c:361; type locality "Brasilia," restricted to "Colônia Leopoldina [now Helvécia], 50 km SW Caravelas, Bahia, Brazil, 17°48′S 39°39′ W, elev. 59 m" (Emmons, Leite et al. 2002:19).

[*Loncheres* (*Loncheres*) *Blainvillii*] *unicolor*: E.-L. Trouessart, 1880b:177; name combination, but incorrect subsequent spelling of *blainvilii* Jourdan.

[*Loncheres* (*Loncheres*)] *unicolor*: E.-L. Trouessart, 1897: 604; name combination.

[*Echimys* (*Echimys*)] *unicolor*: E.-L. Trouessart, 1904:503; name combination.

Nelomys unicolor: Thomas, 1916e:297; name combination.

[*Echimys* (*Phyllomys*)] *unicolor*: Moojen, 1950:491; name combination.

Phyllomys unicolor: Moojen, 1952b:142; first use of current name combination.

DESCRIPTION: Second of two large species in genus, with estimated measurements of holotype including head and body length 280 mm, tail length 202 mm, hindfoot length 37–40.4 mm, and ear length 16.4 mm (Emmons, Leite et al. 2002). Pelage with short (<2.0 cm on dorsal mid-rump) and stiff narrow hairs of uniform color throughout length. Overall dorsal color uniformly pale rusty red-brown grading to buff on venter. Tail rust colored, completely covered with long hairs (ca. 5 mm) that hide scales and that become longer toward tip. Hindfeet broad but not long (average hindfoot length 40.5 mm), with stout claws and yellowish above. Ears nearly naked, short (average ear length 16.2 mm), with tuft on anterior rim. Skull massively built, dorsally flat in lateral profile, and with short and robust rostrum; jugals broad dorsoventrally; jugal fossa deeply concave and a strong, beaded ventral lip, with tip of fossa reaching anteriorly to ventral maxillojugal suture; postorbital process of zygoma rounded and formed mainly by jugal; lateral process of supraoccipital extends to lower edge of external auditory meatus; incisive foramina oval; mesopterygoid fossa wide, angle of fossa averaging 60°, reaching level of posterior edge of first lamina of M3; tympanic bullae conspicuously inflated. Cheek teeth especially large (maxillary toothrow length averaging 13.85 mm); toothrows parallel. Mandible with ventral root of angular process deflected laterally. Upper incisors broad, orange, and orthodont.

DISTRIBUTION: *Phyllomys unicolor* is known only from its type locality in the southernmost part of Bahia state, Brazil (Emmons, Leite et al. 2002).

SELECTED LOCALITIES (Map 501): BRAZIL: Bahia, Colônia Leopoldina [now Helvécia] (type locality of *Loncheres unicolor* Wagner, as restricted by Emmons, Leite et al. 2002).

SUBSPECIES: *Phyllomys unicolor* is monotypic.

NATURAL HISTORY: The only known locality is within broadleaf evergreen rainforest close to the sea. This species appears rare and restricted to the region surrounding the type locality in southern Bahia, where it is sympatric with the more abundant and widespread *P. pattoni*.

REMARKS: The only known specimen is the holotype, collected before 1824, and redescribed fully by Emmons, Leite et al. (2002). Moojen (1952b:491) provided a brief description of a specimen he allocated to this species, and compared it to *P. kerri*, but this reference is likely a misidentification of some other species (Y. L. R. Leite 2003:66). After intensive field surveys at the type locality, Y. L. R. Leite et al. (2007) reported the rediscovery of *P. unicolor*. Their report was based on a young specimen identified as *P. unicolor* by a combination of morphological characters and a high level of mtDNA cytochrome-*b* sequence divergence from syntopic *P. pattoni*. However, Loss and Leite (2011) resequenced this same specimen and obtained a different cytochrome-*b* sequence. These new data, combined with additional nuclear sequences, unambiguous place the specimen within the *P. pattoni* clade. Thus, the original identification of the specimen obtained by Y. L. R. Leite et al. (2007) is erroneous, leaving *P. unicolor* still known solely by the holotype (Loss and Leite 2011).

Genus *Santamartamys* Emmons, 2005
Louise H. Emmons and James L. Patton

Santamartamys are medium-sized (head and body length 190 mm, tail length 267 mm, hind foot length 40 mm, and ear length 12 mm), bright rust-red arboreal rats, with sides and legs paling to yellowish orange and gray underfur (Emmons and Feer 1997:243). The throat and chest are pale orange, the tip of the chin is white, and the venter is washed with orange. The coat is not stiff or bristly but instead has long, lax overhairs, 2.0–3.8 cm in length on the dorsum. Overhairs are so slender that they are difficult to distinguish from other pelage. Dense woolly pelage covers the legs to the ankles and wrists. A crest of long hair, bright red in color like the back, is present on the crown between the ears. The muzzle and sides of the head are tawny yellow-gray streaked with black. The vibrissae are fine and relatively short, extending to behind the ears. The longest genal and mystacial vibrissae are 5 cm; vibrissae are present on the wrist. The ears are short, almost naked, with thin tufts of long brown hair sprouting from the inner rim. Two pairs of lateral mammae are present on the abdominal edge of lateral pelage, in the ventral pelage field. The tail is longer than the head and body, robust, thickly haired with rusty fur for 2 cm at the base but abruptly changing to fine brown or black hair that covers the scales for three-fifths the length but with a terminal two-fifths pure white. Forefeet and hindfeet are brown above, washed with silver; both palmar and plantar surfaces lack tiny tubercles between pads; and the pollux has a nail. The unusual and

spectacular color and color pattern of a living animal were published in a press release announcing the rediscovery of this species after a 113-year lapsus (Conservation International 2011).

Maxillary cheek teeth are rectangular in shape, longer than wide; the teeth and toothrow are relatively long and bowed inward. Each upper tooth (PM4–M3) is split by a deep flexus into two parts; the anterior lophs have flexi opening labially and posterior lophs have flexi opening lingually. Protocones are broad, conferring a wishbone shape to joined anteroloph/protoloph pairs, especially on PM4. Lower molars are split by mesoflexid and hypoflexid into two parts, a curved and laminar anterolophid and a somewhat wishbone-shaped entoconid/posterlophid. The lower premolar is split by two flexids into a small, closed triangular anterior loph apparently lacking a fossette; a laminar metalophid, and a wishbone-shaped posterior loph opens lingually. Hypoflexids are strongly oblique; protoconids large and squarish. The lower incisors are strongly curved. The cranium is conspicuously curved in dorsal profile; the rostrum short and broad. Auditory bullae are small and flattened, merging with the alisphenoid at a shallow angle. The auditory meatus is small and placed high near the squamosal; the auditory tube is short and strongly directed anteriorly; the lateral process of the supraoccipital is extremely short. A bony bridge present between the foramen ovale and the masticatory foramen is exceptionally long; there seems to be a small oval vacuity below the presphenoid-basisphenoid suture. The condyloid process of mandible is deep; the angle of the sigmoid notch between angular and condyloid processes of mandible are shallow; and the mandibular foramen is set in a fossa beside the condyloid ridge. The masseteric crest of lower edge of mandible is poorly developed and shallow; the pterygoid shelf is small.

SYNONYM:

Isothrix: J. A. Allen, 1899b:197; description of *rufodorsalis*; not *Isothrix* Wagner.

Echimys: E.-L. Trouessart, 1904:504; part (listing of *rufodorsalis*); not *Echimys* F. Cuvier.

Diplomys: Tate, 1935:415; part (listing of *rufodorsalis*); not *Diplomys* Thomas.

Santamartamys Emmons, 2005:282; type species *Isothrix rufodorsalis* J. A. Allen, by original designation.

REMARKS: *Santamartamys* seems closely allied to members of the genus *Diplomys*, where until recently it had been placed. It shares with *Diplomys* a number of probable apomorphies as well as close geographic affinity. Emmons (2005) segregated it as a genus primarily because of its distinctive cheek tooth occlusal morphology, which is different from all other living echimyids. This taxon likewise has unique pelage, mammae placement, bullar, and alisphenoid configurations. Emmons based her description above largely on AMNH 34392, a young female with little-worn dentition. The teeth of the holotype, AMNH 14606, are extremely worn. The genus is monotypic.

Santamartamys rufodorsalis (J. A. Allen, 1899)
Red-Crested Tree Rat

SYNONYMS:

Isothrix rufodorsalis J. A. Allen, 1899b:197; type locality "Onaca, Santa Marta District," Magdalena, Colombia.

Echimys (Isothrix) rufodorsalis: E.-L. Trouessart, 1904: 504; name combination.

Diplomys rufodorsalis: Tate, 1935:415; name combination.

Santamartamys rufodorsalis: Emmons, 2005; first use of current name combination.

DESCRIPTION: As for the genus.

DISTRIBUTION (from Emmons and Feer 1997:243): *Santamartamys rufodorsalis* is known only from the Sierra Nevada de Santa Marta, between 700 and 2,000 m. Eisenberg (1989: map 13.73) mapped a locality (incorrectly as *Diplomys caniceps*) in the Sierra de Parijá without documentation, and O. J. Linares (1998) suggested that the species might range to this mountain area in Venezuela.

SELECTED LOCALITIES (Map 507): COLOMBIA: Magdalena, El Dorado Nature Reserve, Sierra Nevada de

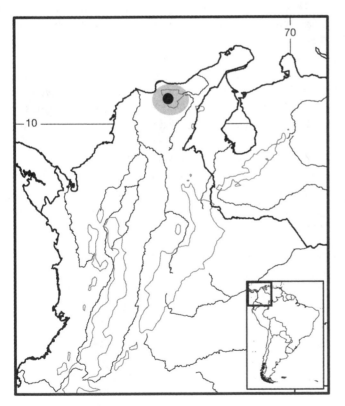

Map 507 Selected localities for *Santamartamys rufodorsalis* (●). Contour line = 2,000 m.

Santa Marta (Conservation International 2011); Onaca (type locality of *Isothrix rufodorsalis* J. A. Allen).

SUBSPECIES: *Santamartamys rufodorsalis* is monotypic.

NATURAL HISTORY: Presumably a montane forest inhabitant, *S. rufodorsalis* is among the most rare and poorly known species of South American mammals (Emmons and Feer 1997).

REMARKS: The single species is apparently known from only three specimens in the American Museum of Natural History in New York (AMNH 14605, 14606, and 34392) and an unrecorded number of specimens in the collection of the Instituto de Investigación Alexander von Humboldt in Colombia (Alberico et al. 2000). A living individual found at the El Dorado Nature Reserve Eco-lodge in the Sierra Nevada de Santa Marta was recently observed and photographed (Conservation International 2011). Rodríguez-Mahecha et al. (1995) incorrectly identified O. Bangs as the author of the name.

Genus *Toromys* Iack-Ximenes, de Vivo, and Percequillo, 2005
Louise H. Emmons, Yuri L. R. Leite, and James L. Patton

This is a genus of large arboreal rats with an average head and body length ranging from 237 (*T. rhipidurus*) to 307 (*T. grandis*) mm, tail length averaging 80–90% of the body, and short, rounded ears. The genus is diagnosed primarily by its well-haired tail, either fully furred (*T. grandis*), or with sparser hairs that decrease in density from base to tip (*T. rhipidurus*). The body is covered with stiff setiform and aristiform hairs that terminate in hair-like processes instead of sharp points (termed "stiff bristles" as opposed to spines by Emmons 2005). Forefeet and hindfeet have small tubercular rugosities between the plantar and palmar pads. The skull shares the same basic morphological plan of *Echimys*, *Phyllomys*, and *Makalata*, although the nasals are constricted medially, the squamosotympanic fenestra is a narrow horizontal slit, the supraorbital ridge is expanded into a shelf, and the protoloph and metaloph of PM4–M3 are often connected by a slender mure. The mean maxillary toothrow length (13–15 mm) is longer than those of any *Makalata* (modified to include *T. rhipidurus* from Iack-Ximenes et al. 2005).

SYNONYMS:

Loncheres: Wagner, 1845a:145; part (description of *grandis*); not *Loncheres* Illiger.

Echimys: E.-L. Trouessart, 1904:504; part (listing of *grandis*); not *Echimys* F. Cuvier.

Makalata: Emmons and Feer, 1997:236; part (listing of *grandis* and *rhipidurus*); not *Makalata* Husson.

Toromys Iack-Ximenes, de Vivo, and Percequillo, 2005a:91; type species *Loncheres grandis* Wagner, by original designation.

REMARKS: The genus was considered monotypic when described, but Emmons (2005) included *Echimys rhipidurus* Thomas within a "*grandis* group" in her concept of *Makalata*, in which she also placed *Loncheres grandis* Wagner, the type species of *Toromys*. Provisionally, therefore, we include these two species within the genus *Toromys*, pending a more thorough review. *Toromys* appears as the sister to a clade comprising *Phyllomys* + *Diplomys* in the phylogenetic analysis of G. A. S. Carvalho and Salles (2004), although in the same larger clade as *Echimys chrysurus* (the type species of *Echimys*), *Makalata* spp., and *Isothrix*. Molecular analyses place *Toromys* (as represented by *T. grandis*) in a clade with *Phyllomys*, *Echimys*, and *Makalata* (Upham and Patterson 2012; Upham et al. 2013).

KEY TO THE SPECIES OF *TOROMYS*:

1. Overall color golden and black; tail black and well furred; size larger, head and body length >270 mm, skull length >60 mm; distributed in lower Amazon of Brazil . . . . *Toromys grandis*

1′ Overall color blackish to reddish brown; tail dark brown or grayish, lightly furred; size smaller, head and body length <260 mm, skull length <59 mm; distributed in upper Amazon of Peru . . . .*Toromys rhipidurus*

Toromys grandis (Wagner, 1845)
Black Toro

SYNONYMS:

Loncheres grandis Wagner, 1845a:146; type locality "vom Amazonenstrom," but according to the catalogue "Manaqueri im Mündungsbereich des Rio Solimões" (Pelzeln 1883; now Manaquiri, lower Rio Solimões, Amazonas, Brazil [Moojen 1952b; Paynter and Traylor 1991]).

Loncheres (*Loncheres*) *grandis*: E.-L. Trouessart, 1897:604; name combination.

Echimys (*Echimys*) *grandis*: E.-L. Trouessart, 1904:503; name combination.

Echimys grandis: Thomas, 1916e:295; name combination.

E[*chimys*]. *grandis*: Thomas, 1928d:410; name combination.

[*Echimys*] *grandis*: Tate, 1935:431; name combination.

Makalata grandis: Emmons and Feer, 1997:236; name combination.

Toromys grandis: Iack-Ximenes, de Vivo, and Percequillo, 2005a:91; first use of current name combination.

DESCRIPTION: Larger species in genus, mean head and body length 303 mm (range: 275–354 mm), tail length 285.1 (range: 244–361 mm), hind foot 52.8 mm (range: 40–65 mm), and ear length 19.5 (range: 15–25 mm; Iack Ximenes et al. 2005a). Pelage coarse, harsh, with stiff bristles. Upper parts mixture of golden and black, head black sprinkled with gold; tail completely furred, terminating in

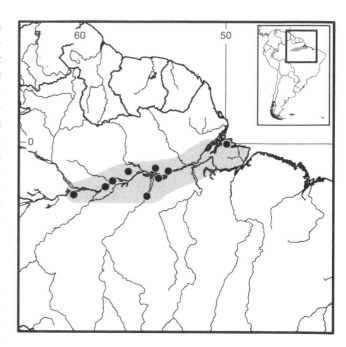

Map 508 Selected localities for *Toromys grandis* (●).

very short pencil 2–3 mm long. Craniodental features as for genus, except anterior part of jugal greatly expanded dorsoventrally (Iack-Ximenes et al. 2005a:Fig. 8).

DISTRIBUTION: Iack-Ximenes et al. (2005a) mapped the 16 known localities along both sides of the lower Rio Amazonas, from near it confluence with the Rio Negro to Ilha Caviana at its mouth.

SELECTED LOCALITIES (Map 508; from Iack-Ximenes et al. 2005a, except as noted): BRAZIL: Amazonas, Lago do Batista, Manaqueri (type locality of *Loncheres grandis* Wagner), Urucurituba; Pará, Fazenda Paraíso, Fazenda Recreio, Fazenda São Pedro, Igarapé Açú, Lago Cuiteua, Santarém.

SUBSPECIES: *Toromys grandis* is monotypic.

NATURAL HISTORY: These are nocturnal, arboreal rats that nest in tree holes, typically near water, and only in lowland rainforest. They are likely restricted to floodplain forest (Emmons and Feer 1997; Iack-Ximenes et al. 2005a).

Toromys rhipidurus (Thomas, 1928)
Peruvian Toro

SYNONYMS:

Echimys rhipidurus Thomas, 1928c:291; type locality "Pebas, 300′," Loreto, Peru.

Echimys armatus rhipidurus: Sanborn, 1949b:286; name combination.

Makalata rhipidurus: Emmons and Feer, 1997:239; name combination.

M[*akalata*]. *rhipidura*: Emmons, 2005:299; name combination.

DESCRIPTION: Smaller species in genus, mean head and body length 229.8mm (range: 210–260; N=16), tail length 196.6mm (range: 180–215mm; N=14), hindfoot length 38.3 (range: 31–49mm; N=15), ear length 14.1mm (range: 11–17mm; N=16), and mass 315g (N=1). Upper parts reddish to yellowish brown, finely to heavily streaked with black hairs. Head and forequarters more grayish or yellowish; rump and tail base more rusty or reddish. Pelage slightly glossy, a combination of stiff, narrow bristles and flexible spines with tips drawn out into hair-like processes; sharp and flat spines are lacking on forequarters or sides of body. Rump sometimes lightly sprinkled with few white speckles from white-tipped spines, but generally spines lack pale tips seen in *Makalata* and *Pattonomys* spp. Base of whiskers dull rust red, but this color does not extend onto muzzle or between the eyes (as it does in *Makalata* spp.). Vibrissae coarse, numerous, black in color, and long, reaching to shoulder. Ears small, almost naked. Tail robust, about 70–95% of head and body length, furred like the back over the initial 3.5cm with remainder covered to tip with short, glossy dark brown or grayish bristly hairs that do not quite hide scales, and extending as very short, 2–3mm tuft beyond tip. From a distance, however, tail appears naked. Base of tail and rear of thighs dull to bright rust-red. Hindfeet short and broad, reddish above; forefeet small and gray above. Hairiness of tail decreases in specimens from north to south.

Skull differs from that of *T. grandis* in that wedge of petrosal separating auditory meatus from squamosal approximately twice as deep (character illustrated by Iack-Ximenes et al. 2005a:Fig. 7). Otherwise, craniodental features as described for genus.

DISTRIBUTION: *Toromys rhipidurus* is known only from northern Peru in the western Amazon Basin. Tirira (2007:227) extended the range of this species throughout eastern Ecuador, citing a personal communication with L. H. Emmons regarding a specimen seen years earlier in a collection in Quito (presumably the Museo de Ciencias Naturales de la Escuela Politécnica Nacional), a specimen now apparently lost.

SELECTED LOCALITIES (Map 509): PERU: Loreto, Nazareth FMNH 19854), Pebas (type locality of *Echimys rhipidurus* Thomas), Puerto Indiana (AMNH 73277), Río Yavarí (FMNH 19854); Ucayali, Suayo (AMNH 98667), Yarinacocha (Sanborn 1949b).

SUBSPECIES: *Toromys rhipidurus* is monotypic.

NATURAL HISTORY: This species occurs in lowland rainforest, with most specimens apparently coming from riverine forests. Other aspects of its biology are unknown.

REMARKS: Emmons (2005) placed *T. rhipidurus* in her "*grandis* species-group" of *Makalata*, which is herein considered equivalent to *Toromys* Iack-Ximenes, de Vivo, and Percequillo. A sister relationship between *T. rhipidu-*

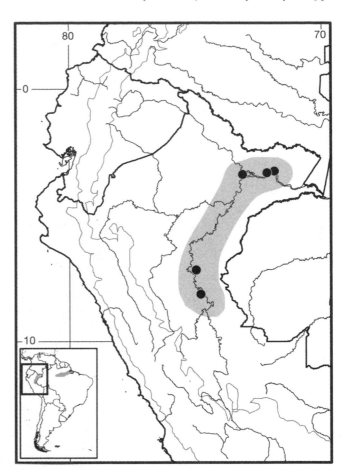

Map 509 Selected localities for *Toromys rhipidurus* (●). Contour line = 2,000m.

rus and *T. grandis*, however, has yet to be established unambiguously. Emmons (2005:281, Fig. 12) also provided photographs of the dorsal, ventral, and lateral views of the skull and both medial and lateral view of the left mandible of the holotype (BM 20.7.1.20).

Subfamily Eumysopinae Rusconi, 1935
James L. Patton and Louise H. Emmons

The nine genera we place in this subfamily include four that are generalized terrestrial rats (elongated limbs, narrow and long hindfeet, short recurved claws on forefeet and hindfeet, tail nearly as long as head and body length; *Hoplomys*, *Proechimys*, *Thrichomys*, and *Trinomys*), three that exhibit semifossorial adaptations (short limbs, well-developed claws on the forefeet, short tail; *Carterodon*, *Clyomys*, and *Euryzygomatomys*), and two that are adapted for climbing (short and broad hindfeet with enlarged plantar pads and well-developed digits, elongated tail as long as or longer than head and body length; *Lonchothrix* and *Mesomys*). Despite the range of body form resulting for these differences in lifestyle, all nine genera share a common tooth structure, with both upper and lower cheek teeth rounded in occlusal outline, as opposed to square or rectangular, and with narrow folds (flexi, flexids) that become isolated as enamel islands in adults but for which the lophs never develop into transverse plates. Two to four, rarely five, flexi are present on the labial side of the maxillary cheek teeth. The maxillary zygomatic process and the jugal are typically robust. The toothrows are short, usually less than 20% of the basilar length of the skull.

The extant eumysopines have long been considered as a natural unit (Ellerman 1940) and thus comprise a formal group in all recent classifications (the Eumysopinae [Woods 1993; Woods and Kilpatrick 2005] or Heteropsomyinae [McKenna and Bell 1997]; see Patton and Reig 1990 for reasons to regard Eumysopinae Rusconi as the appropriate subfamily name). Vucetich and Verzi (1991), however, questioned the implicit monophyly of the included genera, a view supported by molecular sequence analyses (Lara et al. 1996; Y. L. R. Leite and Patton 2002; Galewski et al. 2006; P.-H. Fabre et al. 2012) and the cladistic analysis of morphological characters (G. A. S. Carvalho and Salles 2004). Most of these studies found the various eumysopine genera to form a basal polytomy with the dactylomyine and echimyine genera varyingly nested within. Within this basal group the generic pairs of *Hoplomys*+*Proechimys*, *Mesomys*+*Lonchothrix*, and *Clyomys*+*Euryzygomatomys* are consistently recovered as sisters, even if little other phylogenetic structure is apparent. The most recent molecular analysis (P.-H. Fabre et al. 2012), however, placed the *Mesomys*+*Lonchothrix* clade among the other echimyine genera. As noted in the introduction to the family, full resolution of clades, their generic membership, and thus subfamilial grouping within the Echimyidae remain vexing problems.

KEY TO THE GENERA OF EUMYSOPINAE (CRANIODENTAL CHARACTERS LARGELY FROM G. A. S. CARVALHO AND SALLES 2004:463, TABLE 2):

1. Hindfeet short and broad (width at base of digits >30% of length); pelage consisting of dense, flat, broad, and very stiff spines; skull with short, wide rostrum; interorbital region with parallel, broadly overhanging supraoccipital ledges. 2
1'. Hindfeet narrow and elongated (width <25% of length); pelage varying in expression of aristiform spines; skull with either short and broad or long and narrow rostrum; interorbital region concave with varyingly developed supraorbital ledges. 3
2. Dorsal spines longer (>25 mm) and wider (>1.8 mm); spines tipped pale yellow; tail with long terminal tuft, up to 45 mm in length; skull with more parallel-sided

nasals and narrower mesopterygoid fossa that reaches to the middle of M2 . . . . *Lonchothrix*
2'. Dorsal spines shorter (<25 mm) and narrower (<1.5 mm); spines tipped orange; tail with short terminal tuft, <30 mm in length; skull with convergent-sided nasals and wider mesopterygoid fossa that reaches to M3 . . . .*Mesomys*
3. Forefeet and hindfeet with elongated, straight, and strong claws (modified for digging); tail short (<50% of head and body length); dorsal pelage stiff but with nonspinous aristiform hairs; rostrum short (<32% of skull length) and broad; premaxilla projects anterior to tip of nasals; interorbital region broad (>80% of rostral length); partial contact between ectotympanic and squamosal resulting in cleft (see G. A. S. Carvalho and Salles 2004:462, Fig. 17); sphenopalatine vacuities greatly enlarged 4
3'. Forefeet and hindfeet with short, slightly recurved claws (not modified for digging); dorsal pelage varying from soft to heavily spinose; rostrum elongated (>39% of skull length) and narrow; premaxilla and nasals subequal in anterior length; interorbital region narrower (<60% rostral length); mastoid portion of bullae uninflated; ectotympanic and squamosal in contact along entire margin; sphenopalatine vacuities present or absent 6
4. Anterior face of upper incisors smooth; protoloph of dPM4 to M3 restricted to labial margin of tooth. 5
4'. Anterior face of upper incisors grooved; protoloph of dPM4 to M3 absent . . . . *Carterodon*
5. Dorsal color yellowish to reddish; inferior zygomatic root not exposed ventrally; incisive foramina narrow and elongate; mesopterygoid fossa wide, extending to M3; both mastoid and tympanic portions of bullae inflated . . . . *Clyomys*
5'. Dorsal color yellowish brown to dark brown; inferior zygomatic root exposed ventrally; incisive foramina short and oval; both mastoid and tympanic portions of bullae not inflated. . . . . *Euryzygomatomys*
6. Fur soft, without developed aristiform hairs modified into bristles or spines; tail well haired and completely covering scales; anterior projection of premaxilla well developed; sphenopalatine vacuities greatly enlarged . . . .*Thrichomys*
6'. Fur coarse, with aristiform hairs modified into bristles or spines; tail typically sparsely haired with scales visible; anterior projection of premaxilla very well developed with dorsal portion a well-marked anteromesial expansion in the region of the nasal openings; sphenopalatine vacuities small or absent 7
7. Cheek teeth typically simplified with no more than two folds, with the main fold almost completely separating

the protocone and protoloph and between hypolophid and hypoconid; labial portion of neolophid absent; no contact between metalophid and ectolophid . . . .*Trinomys*

7′. Cheek teeth more complex, typically with a minimum of three folds and sometimes with four or five, with main fold not deeply dividing protocone and protoloph and between hypolophid and hypoconid; labial portion of neolophid present or absent; contact between metalophid and ectolophid present 8

8. Dorsal pelage comprised predominantly of aristiform hairs modified into long, broad, flat, and stiff spines that extend onto the flanks; maxillary cheek teeth complex with four folds on each tooth; only lingual portion of neolophid present so that neolophid of mandibular teeth does not contact metalophid and thus does not form crest C (of G. A. S. Carvalho and Salles 2004) . . . .*Hoplomys*

8′. Dorsal pelage composed of both setiform and aristiform hairs, with the latter ranging in development as spines from weak bristles to stiff and broad, but when the latter, never extending onto the flanks; maxillary cheek teeth complex but varying greatly in number of folds, typically with three but occasionally four or only two; neolophid of mandibular cheek teeth uniformly in contact with metalophid forming crest C (see G. A. S. Carvalho and Salles 2004) . . . .*Proechimys*

Genus *Carterodon* Waterhouse, 1848

Alexandra M. R. Bezerra and Cibele R. Bonvicino

The genus *Carterodon* contains the single species *C. sulcidens* (Lund, 1838), which occurs only in the tropical savanna and grassland habitats of the Cerrado biome in central Brazil, at elevations between 250 and 1,100 m (Eisenberg and Redford 1999; Carmignotto 2005; Bezerra, Marinho-Filho, and Carmignotto 2011). This is a rare, medium-sized, semifossorial rodent (head and body length 135–250 mm and mass 92–195 g; Bezerra, Marinho-Filho, and Carmignotto 2011), with dense and soft hairs (setiforms) intermixed with soft spines (aristiforms), short limbs with long and powerful claws, short ears, and a short tail (less than half of head and body length) densely covered with short and unbanded hair partially covering the scales. The overall pelage color is predominantly brownish and grizzled, with black aristiforms on the dorsal surface, pale brown hairs on the gular and abdominal surfaces, and without a defined lateral line (Moojen 1952b; Bishop 1974).

The phallus of *Carterodon* has not been described.

The following craniodental descriptions are largely from G. A. S. Carvalho and Salles (2004). *Carterodon* has conspicuous, grooved upper incisors (Bishop 1974), unique among the extant echimyids. The occlusal surface of the upper molariform teeth is nearly on a plane parallel to palate. Maxillary cheek teeth are unilateral hypsodont with three roots, one lingual root and two labial roots (G. A. S. Carvalho 1999). Both upper and lower premolar and molar surfaces are semicircular. On the upper toothrow, M2 is the largest tooth, M1 of intermediate size, and both dPM4 and M3 smallest and approximately equal in size. Molariform teeth have three transversal lophs, an anteroloph, a metaloph, and a posteroloph. A protoloph is lacking on all four maxillary teeth, and the metaloph is only partially inclined as a mure with the metaflexus more persistent with tooth wear. The hypoflexus is oriented anterolabially. Mandibular molars also have three distinctive transversal lophs, an anterolophid, hypolophid, and posterolophid. A fourth lophid continuous to lingual end of anterolophid may be visible on unworn lower teeth. The central portion of neolophid does not contact other crown structures in the dPM4. In the few available specimens, G. A. S. Carvalho (1999) observed that the hypolophid was not in contact with hypoconid. Two flexids, the mesoflexid and metaflexid, open lingually in the unworn condition.

The cranium is broad and the rostrum short and wide. An anterior projection of the premaxillary bone is well developed. The inferior zygomatic root is short and ventrally exposed in relation to the ventral surface of rostrum, and positioned at the same level as the palatal region. The auditory meatus is medium sized, with partial contact between the ectotympanic and squamosal, restricted to posterior portion of the dorsal margin of ectotympanic, and forming a cleft between these two bones. Incisive foramina are short and broadly oval in shape; palatine foramina are reduced in size or absent. Both the sphenopalatine foramen and sphenopalatine vacuities are well developed.

SYNONYMS:

Echimys: Lund, 1838:49; part (description of *sulcidens*); not *Echimys* F. Cuvier.

Nelomys: Lund, 1839a:227; part (listing of *sulcidens*); not *Nelomys* Jourdan.

Aulacodus: Lund, 1840c:315; part (*Aulacodus temminckii* Lund as a renaming of *Echimys sulcidens* Lund).

Carterodon Waterhouse, 1848:351; part; type species *Echimys sulcidens* Lund, by original designation and monotypy.

REMARKS: Several authors (Woods 1993; G. A. S. Carvalho and Salles 2004; Galewski et al. 2005) argued that the three genera *Carterodon*, *Clyomys*, and *Euryzygomatomys* formed a monophyletic group. However, Vucetich and Verzi (1991) suggested that *Carterodon* belonged to a different group, a hypothesis supported by unpublished molecular sequence data (C. R. Bonvicino, A. M. R. Bezerra and colleagues).

Carterodon sulcidens (Lund, 1838)
Groove-toothed Spiny Rat

SYNONYMS:

Echimys sulcidens Lund, 1838:49: type locality initially given as "bassin du Rio das Velhas" (Lund 1839a:231–234) and subsequently "Rio das Velhas Floddal" (Lund 1841a: 133), Lagoa Santa, Minas Gerais, Brazil.

N[*elomys*]. *sulcidens*: Lund, 1839a:227; name combination.

Aulacodus Temminckii Lund, 1840:315; renaming of *sulcidens* Lund.

Carterodon sulcidens: Waterhouse, 1848:351; first use of current name combination.

DESCRIPTION: As for the genus.

DISTRIBUTION: *Carterodon sulcidens* is known only for the Cerrado domain of central and western Brazil (Eisenberg and Redford 1999; Carmignotto 2005; Woods and Kilpatrick 2005).

SELECTED LOCALITIES (Map 510): BRAZIL: Distrito Federal, Brasília (G. A. S. Carvalho 1999); Goiás, UHE Serra da Mesa, Minaçu (G. A. S. Carvalho 1999), Parque Estadual da Serra de Caldas Novas, Caldas Novas (Bezerra, Marinho-Filho, and Carmignotto 2011); Mato Grosso, APH Manso, Chapada dos Guimarães (Bezerra, Marinho-Filho, and Carmignotto 2011), Caiê-Malu, Chapada do Parecis (Carmignotto 2005), Campos Novos, Serra do Norte, Chapada do Parecis (Miranda-Ribeiro 1914), Estação Ecológica Serra das Araras, Porto Estrela (Carmignotto 2005), Serra do Roncador, 264 km north of Xavantina, Ribeirão Cascalheira (Bishop 1974; municipality inferred from geographical coordinates in Bishop 1974); Mato Grosso do Sul, "Pedra am Rio Pardo," Pedras Bataguassu (Bezerra, Marinho-Filho, and Carmignotto 2011; locality and municipality inferred from map and text in Krieg 1948); Minas Gerais, Fazenda Lapa Vermelha, Pedro Leopoldo (Carmignotto 2005), Rio das Velhas Floddal, Lagoa Santa (type locality of *Echimys sulcidens* Lund; see also Reinhardt 1851; Winge 1887; Lund 1950), Parque Nacional Grande Sertão Veredas (Carmignotto 2005).

SUBSPECIES: *Carterodon sulcidens* is monotypic.

NATURAL HISTORY: Few ecological, behavioral, or life history data are available for *Carterodon*. Lund's original specimens were subfossil fragments obtained from cave deposits. The first recent specimens were collected by Reinhardt (1851) and came from "the open Pampas," and all subsequent specimens have been collected in open areas of the Cerrado biome. This savanna environment encompasses a wide structural complexity of habitats (Eiten 1972). Bishop (1974) reported specimens captured in "campo" but did not specify the type. Moojen (1952b) reported that the German naturalist H. Burmeister found *Carterodon* specimens at Lagoa Santa in grass nests 30 cm deep within burrows. This species is preyed upon by the

Barn Owl (*Tyto alba*; Moojen 1952b; data on label of museum specimens). Based on surveys in southeastern Mato Grosso do Sul, Krieg (1948) reported *Carterodon* within Cerrado vegetation types such as grass, shrubs, and *cerradão* (a forest savanna dominated by trees above ca. 8 m in height), although he did not record actual localities where the species was taken. Bishop (1974) captured two pregnant females in the Serra do Roncador, one in August with a single fetus weighing 16 g and another in December with two embryos weighing 12 g each. Bishop (1974) also collected juveniles in both the dry (August and October) and wet seasons (November and December).

REMARKS: The publication date of Lund's *Echimys sulcidens* is usually given as 1841 (e.g., Cabrera 1961, as p. 99; Woods and Kilpatrick 2005, as p. 49). However, while the name was discussed by Lund in several papers dealing with his Lagoa Santa collection, he described this species in 1838, on page 49, where he wrote "ved en Fure paa den forreste Flade af dens Skaeretaender, hvorfor jeg foreslaaer for den Navnet *Echimys sulcidens*" (approximate translation, "by a furrow on the anterior surface of the incisor, which is why I suggested the name *Echimys sulcidens*") (see also Paula Couto 1950:123, 549). Lund's description of the grooved incisors, unique among echimyid genera, clearly both diagnosed his new species and gave the foundation for Waterhouse's (1848) later establishment of the genus-group name *Carterodon*, with *Echimys sulcidens* Lund as the type species. Subsequent to 1838, Lund (1841a) continued to regard *sulcidens* as a species of *Echimys* F. Cuvier, transferred it to *Nelomys* Jourdan (Lund

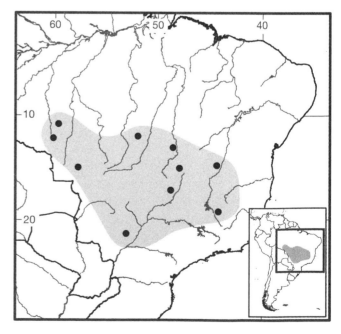

Map 510 Selected localities for *Carterodon sulcidens* (●).

1839a, 1841b), or renamed it *Aulacodus temminckii* (Lund 1840c, 1842).

This species has been encountered rarely, with only 32 completely intact specimens known in museum collections. One of these is in the FMNH; three are at the UNB; four are in the MZUSP; seven are in the MNRJ; five are at the ZMUC; and 12 in the BM. Salles et al.(1999) recorded additional subfossil remains from Quaternary deposits in the Igrejinha cavern in the Serra do Mesa region of Goiás state. The karyotype of *C. sulcidens* is unknown.

Genus *Clyomys* Thomas, 1916

Alexandra M. R. Bezerra and Cibele R. Bonvicino

The genus *Clyomys* contains a single species, *C. laticeps* (Thomas 1909), which occurs in tropical savanna and grassland habitat in Brazil and Paraguay, over an elevational range from about 100–1,100 m. This is a medium-sized, semifossorial rodent (head and body length 150–295 mm and mass 158–334 g; Bezerra 2002), with coarse and spinous fur, short limbs, long and powerful claws, and a short tail (less than the half of head-body size) sparsely covered with short and unbanded hair, partially covering scales, and short ears (Thomas 1909; Bishop 1974). General pelage color varies from grizzled rufous and black to grizzled yellow and black on the dorsal surface, and whitish or buffy, with or without the presence of grayish or pale rufous patches, on the gular and ventral surface; the sides are colored as the back (Thomas 1909; Moojen 1952a; Bishop 1974; Bezerra 2002).

Bezerra (2002) described the phallus of *Clyomys* as subcylindrical, elongated, and straight. Some specimens show a slight median constriction on the sides and/or an extension in the region of the intromittent sac. There are longitudinal grooves on sides of the glans, with simple, small spines uniformly covering the epidermis and extending into the intromittent sac. The dorsal side of the glans is longer than ventral side and distally is convergent on the apical mount, the lappet of which does extend only slightly from the crater. The baculum is simple, long, and narrow; in lateral view it is slightly concave at its proximal end.

G. A. S. Carvalho and Salles (2004) provided a thorough description of craniodental characters. *Clyomys* has robust incisors, unilateral hypsodont teeth with three roots, with the occlusal surface of the maxillary teeth nearly on a plane parallel to the palate, very similar to *Euryzygomatomys* (G. A. S. Carvalho 1999). Upper molarifom teeth have one large lingual root and two almost inconspicuous labial roots. The surface of both upper and lower premolars and molars is semicircular in shape. M3 of the upper toothrow is reduced in size, with dPM4 and M1 only slightly smaller

than M2; with enamel wear, M1 becomes the largest tooth in toothrow. Maxillary cheek teeth have three transverse lophs, the anteroloph, metaloph, and posteroloph. Unworn teeth may exhibit a small protoloph restricted to the labial margin. The metaloph is oriented posterolabially from the mure, narrowing and enclosing the metaflexus. A mesoflexus becomes more persistent with tooth wear. The hypoflexus is anterolabially oriented. Worn dPM4–M2 teeth show two round labial fossettes derived from the metaflexus and mesoflexus and a round lingual fossette derived from the hypoflexus (see also Bezerra 2002). Mandibular molars have three distinctive transverse lophs, the anterolophid, hypolophid, and posterolophid. Unworn teeth may show another lophid continuous to the lingual end of the anterolophid. The hypolophid is always in contact with the hypoconid and ectolophid, and is directed posterolingually as is the hypoflexid. The mesoflexid and metaflexid open lingually when unworn. The lower m1 may show a round labial fossette derived from the hypoflexid with increasing wear.

The cranium is broad, the rostrum short and wide. The mesopterygoid fossa extends to the hypoflexus of M2 but not beyond (Bezerra 2002). The anterior projection of the premaxillary bone is short. Similarly, a short inferior zygomatic root is present at the same level in relation to the ventral surface of the rostrum and palatal region. Tympanic bullae are hypertrophied, being highly visible beyond the paraoccipital process when viewed from behind (Thomas 1916e). The auditory meatus is medium sized, with partial contact between the ectotympanic and squamosal restricted to the posterior portion of the dorsal margin of ectotympanic, forming a cleft between these two bones. The incisive foramina are narrow and elongate (Thomas 1909). Both the sphenopalatine foramen and sphenopalatine vacuities are well developed. A well-developed canal of the infraorbital nerve is present on the floor of the infraorbital foramen.

SYNONYMS:

Loncheres: Lund, 1839a:233; part (description of *laticeps* Lund, a *nomen nudum*); not *Loncheres* Illiger.

Mesomys: Winge, 1887:923; part (listing of *laticeps* Lund, a *nomen nudum*, as a synonym of *Mesomys spinosus* Desmarest).

Echimys: Thomas, 1909:240; part (description of *laticeps*); not *Echimys* F. Cuvier.

Clyomys Thomas, 1916e:300; type species *Echimys laticeps* (Thomas), by original designation.

REMARKS: The initial use of the name *laticeps* was by Lund (1839a), who simply included it in a list of species of *Loncheres* without a formal description. Winge (1887) wrongly placed Lund's *Loncheres laticeps* as a junior synonym of *Mesomys spinosus* Desmarest (= *Euryzygomatomys spinosus*). Thomas (1909) recognized that Lund's *laticeps* was a *nomen nudum* and formally described *laticeps* under *Echimys*. Later, Thomas (1916e) described *Clyomys* with *Echimys laticeps* as the type species. Reig (1986:409, footnote *f*) regarded *Clyomys* to be inseparable from, and thus should be placed within, *Euryzygomatomys*. Morphological and molecular phylogenetic analyses indicate a very close sister relationship between these two taxa (Y. L. R. Leite and Patton, 2002; G. A. S. Carvalho and Salles 2004; Galewski et al. 2005; P.-H. Fabre et al. 2012). Fossils of the genus, assigned to the extinct species *C. riograndensis*, are known from Mid to Late Holocene (9,400–3,730 ybp) deposits in Rio Grande do Sul state, Brazil (Hadler et al. 2008).

Clyomys laticeps (Thomas, 1909)
Broad-headed Spiny Rat
SYNONYMS:

L[*oncheres*]. *laticeps* Lund, 1839a:226, 233; *nomen nudum* (Thomas 1909:240).

Mesomys spinosus: Winge: 1887:92; name combination; not *spinosus* Desmarest.

Echimys laticeps Thomas, 1909:240; redescription of *Loncheres laticeps* Lund; type locality given initially on page 241 as "Lagoa Santa, Minas Geraes" and subsequently, on the same page, as "Lagoa Santa, on the Rio São Francisco, Minas Gerais," Brazil.

Clyomys laticeps: Thomas, 1916e:300; name combination.

Clyomys laticeps whartoni Moojen, 1952a:102; type locality "1 km north of Aca-Poi, long. 56°7'W., lat. 23°5'S., Department of San Pedro, Partido de Tacuatí, Paraguay; approximately 60 km east-northeast of Puerto Ybapobo and 10 km south of the Rio Ypané."

Clyomys bishopi Avila-Pires and Wutke, 1981:529; type locality "Itapetininga, São Paulo, Brasil."

DESCRIPTION: As for the genus.

DISTRIBUTION: *Clyomys laticeps* is known from the Cerrado domain of central Brazil, extending south to São Paulo state (Bishop 1974; Avila-Pires and Wutke 1981), and the Chaco of Paraguay (Moojen 1952a).

SELECTED LOCALITIES (Map 511): BRAZIL: Bahia, Fazenda Jatobá (Bezerra 2002); Distrito Federal, Parque Nacional de Brasília (Svartman 1989), Reserva Biológica de Águas Emendadas (Bezerra 2002); Goiás, Santa Rita do Araguaia, Assentamento Babilônia (Bezerra 2002); Mato Grosso, Ribeirão Cascalheira (Bishop 1974); Mato Grosso do Sul, Corumbá, Fazenda Alegria, Corumbá (Bezerra 2002), Bonito, Fazenda Bonito (Bezerra 2002); Minas Gerais, São Roque de Minas, Fazenda Chico Cera (Bezerra 2002), Lagoa Santa (type locality of *Echimys laticeps* Thomas), São Paulo, Campininha (Bezerra 2002); São Paulo, Itapetininga (type locality of *Clyomys bishopi* Avila-Pires and Wutke). PARAGUAY: San Pedro, 1 km N of Aca-Poi (type local-

ity of *Clyomys laticeps whartoni* Moojen); Canindeyú, Reserva Natural del Bosque Mbaracayú (Bezerra 2002); Concepción, 28 km S junction Rte 3 and Rte 5 on Rte 3 (MVZ 145318).

SUBSPECIES: We treat *Clyomys laticeps* as monotypic. Moojen (1952a) described *C. laticeps whartoni* based on two specimens collected in the Paraguayan Chaco, which he distinguished by the presence of grayish patches on the ventral surface, larger tympanic bullae, and a shorter palate. Bishop (1974) observed that only the shorter palate was a good character separating *whartoni* from the nominate subspecies. A multivariate analysis of craniodental measurements and qualitative data from both baculum and skull, however, did not corroborate the Paraguayan population as a different taxonomic unit (Bezerra, 2002; Bezerra and Oliveira 2010).

NATURAL HISTORY: All specimens have been collected in open areas of the Cerrado biome, a savanna environment that encompasses a wide structural complexity of habitats, including unflooded grassland and shrubby vegetation with ground cover of grasses, and areas of the Paraguayan Chaco, a somewhat drier western extension of the Cerrado (Moojen 1952a; Bishop 1974; Avila-Pires and Wutke 1981; Lacher and Alho 1989; M. V. Vieira 1989, 1997). Burrows can often be easily located due to semifossorial habits of this species (Lacher and Alho 1989; Amante 1975; C. T. Carvalho and Bueno 1975; A. M. R. Bezerra, unpubl. data). A. Bueno et al. (2004) verified the strong relationship between the burrows of armadillos and the presence of *Clyomys* in São Paulo state. Individuals are also dominant predators and dispersers of seeds of the palms *Attalea geraensis* and *Syagrus petraea* (L. B. de Almeida and Galetti 2007). Breeding may be seasonal. A pregnant female was captured at Itirapina, São Paulo state, at end of the dry period, and weaning and growth was completed during wet season (M. V. Vieira 1997). Bishop (1974) captured pregnant females in Mato Grosso state June, September (both dry season), and December (wet season), respectively, all with a single embryo. Juveniles were taken in December (two specimens) and April (one).

REMARKS: For the taxonomic history of this taxon, see Tate (1935:406) and the comments under Remarks for the genus. Bezerra (2002) revised the genus, placing *Clyomys bishopi* Avila-Pires and Wutke (1981) in the synonymy of *C. laticeps*, a decision supported by morphometric and qualitative analyses of cranial variables (Bezerra and Oliveira 2010). Woods and Kilpatrick (2005), however, maintained a concept of the genus *Clyomys* with two species, *C. laticeps* and *C. bishopi*. The comprehensive reviews of Cabrera (1961) and Woods and Kilpatrick (2005) continued Tate's (1935:406) incorrect listing of the type locality as "Joinville, Santa Catherina, Brazil" while other authors (Bishop 1974; Avila-Pires and Wukte 1981) called attention to this mistake. Thomas (1909:241), in his original description of *Echimys laticeps*, was unambiguous in listing the type locality as "Lagoa Santa, Minas Geraes." We suspect that the recent literature confusion stemming from Tate (1935) arose because Thomas (1909) compared the holotype of *C. laticeps* with specimens of *Echimys spinosus* (= *Euryzygomatomys spinosus*) from Paraguay and Joinville (Bezerra and Oliveira 2010).

The karyotype of *C. laticeps* is variable, with five separate chromosomal complements described to date. Yonenaga (1975) and M. J. Souza and Yonenaga-Yassuda (1984) described two karyotypes in specimens from São Paulo state: $2n=34$, $FN=58$ with 13 pairs of biarmed pairs and three pairs of acrocentric autosomes (Pagnozzi, Fagundes et al. 2000), and $2n=34$, $FN=60$, with 14 pairs of biarmed pairs and two pairs of acrocentric autosomes. Svartman (1989) recorded a karyotype of $2n=34$ and $FN=62$ with 14 metacentric/submetacentric pairs, one subtelocentric pair and one pair of acrocentric autosomes in specimens from the Distrito Federal. Specimens from Corumbá, Mato Grosso do Sul state, have a $2n=34$ and $FN=58$ karyotype with 13 biarmed pairs and three pairs of acrocentric autosomes (C. R. Bonvicino, unpubl. data). Finally, specimens from the Parque Nacional das Emas, Goiás state, had $2n=32$ and $FN=54$ with one submetacentric, 11 metacentric, and three acrocentric autosome pairs (Pagnozzi et al. 2000). It remains to be determined if these karyological differences signal species boundaries or merely represent intraspecific polymorphism.

Genus *Euryzygomatomys* Goeldi, 1901

Cibele R. Bonvicino and Alexandra M. R. Bezerra

The genus *Euryzygomatomys* contains the single species *E. spinosus* (G. Fischer, 1814), occurring in forested habitats of southwestern Brazil, northeastern Argentina, and southern Paraguay. This is a medium sized, semifossorial rodent (head and body length 163–205 mm and mass 180–210 g), characterized by a short tail (less than half of head and body length) sparsely covered with short and unbanded hair, coarse and spinous fur, short limbs, long and powerful claws, and short ears. The dorsal pelage coloration varies from yellow-brown to dark brown, with black guard hairs; the sides are lighter in color; and the venter varies from white to slightly yellowish, except for a more ferruginous throat. The dorsal surfaces of both forefeet and hindfeet are dark brown, and the digits are paler. Manual claws are more developed than those of pes (Thomas 1909; 1916e; Moojen 1952b). Three pairs of mammae are present, one pectoral and two inguinal (J. A. Oliveira and Bonvicino 2006). The phallus of *Euryzygomatomys* has not been described.

Craniodental characters described here are taken from G. A. S. Carvalho and Salles (2004), unless otherwise indicated. *Euryzygomatomys* has large and heavy incisors (Thomas 1909), unilateral hypsodont teeth with three roots, with the occlusal surface of the maxillary teeth nearly on a plane parallel to that of the palate, very similar to the condition in *Clyomys* (G. A. S. Carvalho 1999). Upper cheek teeth have one large lingual root and two almost inconspicuous labial ones. All upper and lower cheek teeth are semicircular in outline. In the upper toothrow, M3 is slightly reduced in size relative to the other teeth, with dPM4–M1 almost the same size and M2 only slightly smaller. With increasing wear M1 becomes the largest tooth in toothrow. All upper cheek teeth possess three transverse lophs, an anteroloph, metaloph, and posteroloph. A small protoloph restricted to the labial tooth margin may be present in unworn teeth. The metaloph is oriented posterolabially from the mure, narrowing and closing the metaflexus. The mesoflexus persists even with increasing tooth wear. The hypoflexus is oriented anterolabially. Worn dPM4–M2 typically have two round labial fossettes derived from metaflexus and mesoflexus and a round lingual fossette derived from the hypoflexus. The mandibular cheek teeth also have three distinctive transversal lophs, an anterolophid, hypolophid, and posterolophid. Unworn teeth may exhibit another lophid continuous with lingual end of anterolophid. The

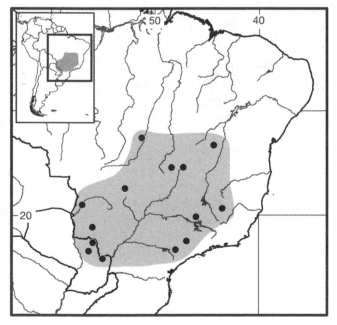

Map 511: Selected localities for *Clyomys laticeps* (●).

hypolophid is always in contact with the hypoconid and ectolophid, and both it and the hypoflexid are posterolingually oriented. The mesoflexid and metaflexid are deep and open lingually in unworn teeth. The hypoflexid is always internally elongated in the lower m1 when worn.

The cranium is broad and the rostrum is short and wide. The mesopterygoid fossa extends to the hypoflexus of M2 (Thomas 1909, referred to as posterior palatal notch). The anterior projection of premaxillary bone is truncated. A short inferior zygomatic root is exposed ventrally in relation to ventral surface of rostrum, and at same level of palatal region. The auditory meatus is medium sized, with partial contact between the ectotympanic and squamosal restricted to the posterior portion of dorsal margin of the ectotympanic and thus forming a cleft between these two bones. The incisive foramina are short and oval (Thomas 1909). Sphenopalatine vacuities with well-developed openings are present as is a groove on the floor of the infraorbital foramen that houses the infraorbital nerve.

SYNONYMS:

Rattus: G. Fischer, 1814:105; part (description of *spinosus*, based on Azara's [1801:73] "*rat épineux*"); not *Rattus* Fischer.

Loncheres: Illiger, 1815:108; part (listing of *?brachyura*, a *nomen nudum*); not *Loncheres* Illiger.

Echimys: Desmarest, 1817b:57; part (description of *spinosus*, based on Azara's [1801:73] "*rat épineux*"); not *Echimys* F. Cuvier.

Loncheris: Olfers, 1818:212; part (description of *brachyura*, based on Azara's [1801:73] "*rat épineux*"); incorrect subsequent spelling of, but not *Loncheres* Illiger.

Ratton Brants, 1827:186; part (description of *espinoso*, based on Azara's [1802:76] "*el espinoso*"; suppressed for the purpose of Principle of Priority [but not Homonomy] by the International Commission [ICZN 1982:Opinion 1232]).

Hypudaeus: Brandt, 1835:432; part (description of *guiara*); not *Hypudaeus* Illiger.

Echinomys Wagner, 1840:199; part (listing of *spinosus*); unjustified emendation of *Echimys* F. Cuvier.

Mesomys: Burmeister, 1854:205; part (listing of *spinosus*); not *Mesomys* Wagner.

Euryzygomatomys Goeldi, 1901:179; type species *Echimys spinosus* Rengger (= *Rattus spinosus* G. Fischer), by original designation.

REMARKS: Tate (1935:402–405) thoroughly summarized the nomenclatural history of this genus, but he was wrong in stating that *spinosus* was based on Desmarest (1817b) rather than G. Fischer (1814). Both authors provided a name for Azara's (1801:73) "*rat épineux*," although because Fischer did so three years earlier his name has priority. Tate, however, failed to list *Ratton* Brants,

and *Ratton espinoso*, one of the species Brants included within his genus and which he based on "*el espinoso*" from the Spanish edition of Azara's book (1802:76), and the same animal as the "*rat épineux*" from the French version (Azara 1801). The International Commission (ICZN 1982:Opinion 1232) suppressed *Ratton* Brants for the purpose of the Principle of Priority, following petitions by Langguth (1966, 1978). Goeldi (1901) regarded *Echimys spinosus* Renggeras the type species of his new genus *Euryzygomatomys*. Likewise, however, Rengger's name is a junior synonym of *Rattus spinosus* G. Fischer. Reig (1986:409, footnote *f*) argued that *Clyomys* and *Euryzygomatomys* were inseparable and should be placed in the same genus. Morphologic and molecular phylogenetic studies based on nuclear and mitochondrial strongly support a sister relationship between *Euryzygomatomys* and *Clyomys* (Y. L. R. Leite and Patton 2002; G. A. S. Carvalho and Salles 2004; Galewski et al. 2005). Fossils of the genus (assigned to an extinct species, *E. mordax*) are known from Mid to Late Holocene deposits in Rio Grande do Sul state, Brazil (Hadler et al. 2008).

Euryzygomatomys spinosus (G. Fischer, 1814)

Guiara (Brazil), Rata guira or Ratón espinoso (Paraguay)

SYNONYMS:

Rattus spinosus G. Fischer, 1814:105; type locality "peuplade d'Atira, un peu plus de huit lieues (44 kilomètres), a l'Orient de l'Assomption," = Atyrá, 44 km E of Asunción, Cordillera, Paraguay; based on Azara's (1801:73) "*rat épineux*."

Loncheres ?brachyura Illiger, 1815:108; *nomen nudum* (Tate 1935:403).

Echimys spinosus Dsemarest, 1817b:57; also based on Azara's "*rat épineux*."

L[oncheris]. (sic) brachyura Olfers, 1818:212; type locality "Paraguay"; also based on Azara's "*rat épineux*."

L[oncheres]. rufa Lichtenstein, 1820:192; also based on Azara's "*rat épineux*."

Ratton espinoso Brants, 1827:186: based on of Azara's (1802:76) "*el espinoso*"; ruled by the International Commission (ICZN 1982:Opinion 1232) to be a vernacular name and unavailable for use in zoological nomenclature.

Hypudaeus Guiara Brandt, 1835:432; type locality "Ypanema Provinciae St. Pauli Brasiliae" [= Floresta Nacional de Ipanema, 20 km NW Sorocaba, São Paulo, Brazil, 23°26′7″S 47°37′41″W, 701 m; L. P. Costa et al. 2003].

[Echinomys] spinosus: Wagner, 1840:199; name combination.

Echinomys brachyurus Wagner, 1843a:346; not *brachyura* Illiger nor *brachyura* Olfers; also based on Azara's [1802:76] "*el espinoso*."

Mesomys spinosus: Burmeister, 1854:205; part (listing of *spinosus* Desmarest and *brachyurus* Wagner).

Euryzygomatomys spinosus: Thomas, 1916e:301; first use of current name combination.

Euryzygomatomys catellus Thomas, 1916e:301; type locality "Joinville, Santa Catarina, Brazil."

Euryzygomatomys guiara: Tate, 1935:405; name combination.

Euryzygomatomys spinosus spinosus: Ellerman, 1940:125; name combination.

Euryzygomatomys spinosus catellus: Ellerman, 1940:125; name combination.

Euryzygomatomys spinosus guiara: Cabrera, 1961:532; name combination.

SUBSPECIES: As currently construed, *Euryzygomatomys spinosus* is monotypic.

DISTRIBUTION: *Euryzygomatomys spinosus* occurs in southern Brazil, from the states of Espírito Santo, Rio de Janeiro, and Minas Gerais to Rio Grande do Sul, southern Paraguay, in La Cordillera and Guaira departments, and northeastern Argentina, in Misiones province. Cabrera (1961) also reported this species to northern Corrientes, Argentina, but without documentation. Santos-Filho, Frieiro-Costa et al. (2012) reported this species in an area within the Cerrado at the Serra das Araras Ecological Station, Porto Estrela, Mato Grosso state, Brazil (specimens misidentified as *Carterodon sulcidens*, with correction made by A. M. R. Bezerra from photos of an adult skull sent by M. Santos-Filho and confirmed by M. N. F. da Silva by examination of the skullsof vouchers at INPA).

SELECTED LOCALITIES (Map 512): ARGENTINA: Misiones, Posadas, Estancia Santa Inés (Massoia 1990; Pardiñas et al. 2005). BRAZIL: Espírito Santo, Viana, Coacas (I. S. Pinto, Loss et al. 2009); Minas Gerais, Caratinga (G. A. S. Carvalho 1999), Conceição do Mato Dentro (G. A. S. Carvalho 1999), Passos (G. A. S. Carvalho 1999); Rio Grande do Sul, Cambará do Sul (G. L. Gonçalves et al. 2007), Tapes (M. A. F. Andrade and Christoff 2001); Rio de Janeiro, Caxias (Moojen 1952b), Nova Friburgo (MNRJ [LBCE 5195]), Sumidouro (MNRJ [LBCE 73]); Santa Catarina, Itapiranga (Cherem, Simões-Lopes et al. 2004), Joinville (type locality of *Euryzygomatomys catellus* Thomas), Santo Amaro da Imperatriz (Cherem, Simões-Lopes et al. 2004); São Paulo, Casa Grande (Yonenaga 1975), Ipanema (type locality of *Hypudaeus guiara* Brandt). PARAGUAY: Guaira, Villarica (AMNH 66785); Cordillera, Atyrá (type locality of *Rattusspinosus* G. Fischer).

NATURAL HISTORY: *Euryzygomatomys spinosus* is a semifossorial species with features apparently related to burrowing habits, like short ears, short tail, and long foreclaws (Eisenberg and Redford 1999). This species is an herbivore that is active at night (Mares and Ojeda 1982), and

harmful to *Pinus* plantations (G. L. Gonçalves et al. 2007). Although primarily found under the riparian vegetation in dense forests, it is apparently a habitat generalist (Mares and Ojeda 1982). In Brazil, individuals occur in both altered and conserved vegetation communities (Bonvicino, Lindbergh, and Maroja 2002). Stallings (1988, 1989) found this species in wet meadows in Minas Gerais state. Pardiñas, Teta, and Heinonen Fortabat (2005) recorded the species in natural grasslands in Misiones, Argentina. Captured females contained one or two embryos (Davis 1947); a female had three embryos in November (Moojen 1952b). This rodent can be naturally infected by the tick *Amblyomma* (Barros-Battesti and Knysak 1999) and lice of the genera *Gliricola* and *Gyropus* (Cardosode Almeida et al. 2003; Valin and Linardi 2008), as well the Boraceia virus, a member of the *Anopheles* B arbovirus group (O. S. Lopes and Sacchetta 1974). M. Robles, Galliari, and Navone (2012) described nematode parasites from specimens collected in Misiones province, Argentina. Also in Argentina, this species is prey of the Barn Owl (*Tyto alba*; Massoia 1990).

REMARKS: Woods and Kilpatrick (2005:1582) suggested that, should G. Fischer (1814) become unavailable by future action of the International Commission on Zoological Nomenclature, *rufa* Lichtenstein (1820) would be the next available name for this species. However, these authors overlooked Desmarest's (1817b) *Echimys spinosus* and Olfer's (1818) *Loncheris* [sic] *brachyura*, both of which are valid and predate Lichtenstein. Consequently, if G. Fischer's *Rattus spinosus* does become unavailable,

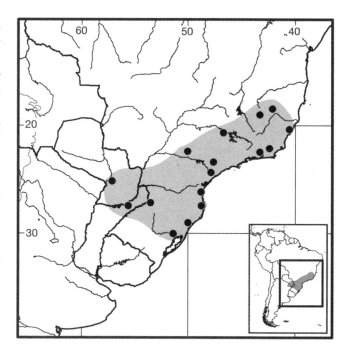

Map 512: Selected localities for *Euryzygomatomys spinosus* (●).

then the correct name would be *Echimys spinosus* Desmarest, 1817. Tate (1935:405) provisionally listed *guiara* as a species of *Euryzygomatomys*, primarily based on Brandt's description of a short tail. However, Cabrera (1961:532) noted that the taxonomic position of *guiara* was dubious, but that if Tate's assignment were correct, *guiara* would only be a subspecies of *spinosus*.

The karyotype of *E. spinosus* is $2n=46$, $FN=82$, with 19 pairs of metacentric or submetacentric and three pairs of acrocentric autosomes (Yonenaga 1975, specimens from São Paulo state; D. I. Lima et al. 2006, a specimen from Santa Catarina state, without data on fundamental autosome number).

This species is considered critically endangered in Paraguay (Resolution No. 524/06 March 17 2006, Secretaria del Ambiente de Paraguay Lista Roja de Fauna y Flora de Paraguay); the IUCN (2011) lists *E. spinosus* in the Least Concern category.

Genus *Hoplomys* J. A. Allen, 1908
James L. Patton and Louise H. Emmons

Hoplomys is a terrestrial, large-bodied rat (mass up to 800 g; Adler et al. 1998) with a stout body, short and thick legs, relatively narrow and elongated hindfeet, a nearly naked and proportionately short tail (60–70% of head and body length), and short, rounded ears. Head and body length ranges from 212–300 mm, tail length from 114–240 mm, hind foot length from 47–61 mm, and ear length from 20–29 mm (Emmons and Feer 1997). The most characteristic external features are the broad, flat, and stiff spines that cover the dorsum from behind the shoulders to the flanks and rump. These range from 26–33 mm in length and 1.7–2.2 mm in width, and tend to obscure the soft underfur (Handley 1959). The spines are tipped black along the mid-dorsum but may have pale tips on the sides, which gives the lateral aspect of the skin a distinctive speckled pattern. As with other echimyids with well-developed spines (e.g., *Lonchothrix* and *Mesomys*), those of *Hoplomys* extend down the sides to the junction with the ventral fur. Coloration varies geographically, in South America ranging from reddish orange dorsally in the southern part of the range to more yellowish in the north (Handley 1959). Young are more typically colored dull brown dorsally. Individuals vary with regard to the presence or absence of black ocular and crown areas. The venter is predominantly white, although there may be a partial or complete dark brown collar across the throat, and buff lateral bands or a buff venter. Some extralimital individuals from Panama are melanistic, almost black, with dark brown venters variably spotted with white. Such coloration could occur in bordering areas of Colombia.

The skull is conformed as in most nonfossorial eumysopine genera, relatively narrow in shape with an elongate rostrum, concave interorbital region, and round cheek teeth with deeply oblique lateral folds that do not transect the crown and that become isolated as fossettes with wear. Prominent temporal ridges are present, the supraorbital shelf is beaded, the rostrum narrows toward the tip, the auditory bullae are very small, and the floor of the infraorbital foramen typically flat, although a slight groove that marks the passage of the infraorbital nerve is present in some individuals (Patton 1987). The size, flatness, and ridging of the skull of *Hoplomys* all increase with age (as judged by tooth eruption and wear), with growth apparently continuing well after all teeth are erupted and fully functional. As a consequence, skulls of "adult" individuals will exhibit a considerable range in overall size (Handley 1959), a characteristic of other eumysopine genera, most notably *Proechimys* (Patton and Rogers 1983; Lara et al. 1992) and *Trinomys* (Pessôa and Reis 1991b).

In his description of the genus, J. A. Allen (1908:650) diagnosed *Hoplomys* by the combination of the impressively broad and stiff dorsal spines and an enamel pattern with all maxillary and mandibular cheek teeth possessing four well developed and obliquely positioned folds, a character that he contrasted with the three folds typical of *Proechimys*. However, in nearly all characters, craniodental and well as pelage, *Hoplomys* is either indistinguishable from or only at the extreme end of a continuum represented by the diverse species of *Proechimys*. For example, the counterfold formula of *Hoplomys* does vary, particularly in the lower molars (from 4/4, 4/4, 4/4, 4/4 to, rarely, 4/4, 4/3, 4/3, 4/3), and some species of *Proechimys* also exhibit four counterfolds on most, or all cheek teeth (most notably *Proechimys semispinosus* Tomes and *Proechimys quadruplicatus* Hershkovitz; see Patton 1987, and below). Moreover, some species of *Proechimys* approach *Hoplomys* in aristiform spine development, such as *Proechimys hoplomyoides* Tate, which has been allocated to *Hoplomys* by some authors (Moojen 1948; Cabrera 1961) and *Proechimys echinothrix* Silva, which has aristiform spines nearly as long and broad (22 mm × 1.4 mm, on average; M. N. F. da Silva 1998). The range of spine depth, robustness, and particularly length of *Hoplomys* reported by Hoey et al. (2004) is well beyond that observed for the eight species of *Proechimys* these authors examined, but their complement of species did not include either *P. hoplomyoides* or *P. echinothrix*.

Both A. L. Gardner and Emmons (1984) and Patton and Reig (1990) questioned the generic distinction of *Hoplomys* relative to *Proechimys* based on morphological characters of the bulla and analysis of protein electrophoretic characters, which nested *Hoplomys* within a group of

10 species of spiny rats. MtDNA sequence data (Lara et al. 1996; Y. L. R. Leite and Patton 2002) also indicate a close phyletic relationship between *Hoplomys* and *Proechimys*, but too few species of the latter have been examined as yet to determine whether *H. gymnurus* is nested among the species of this large genus, as suggested by the allozyme analyses, or is sister to a clade otherwise comprising all species of *Proechimys*. Alternatively, G. A. S. Carvalho and Salles (2004) showed that *Hoplomys* has a uniquely complete neolophid in the mandibular molars among a phyletic triad of genera that includes *Proechimys* and *Trinomys*. All three of these genera are otherwise defined by a long and narrow rostrum relative to other echimyids and the presence of a very well-developed anterior projection of the premaxilla. Clearly, the generic status of *Hoplomys* remains an open question for future research to resolve.

Hoplomys is monotypic, with a single species *H. gymnurus* distributed from Honduras in Central America south to northwestern Ecuador along the Pacific coast of South America.

SYNONYMS:

Echimys: Thomas, 1897f:550; description of *gymnurus*; not *Echimys* F. Cuvier.

Hoplomys J. A. Allen, 1908:649; type species *Hoplomys truei* J. A. Allen, by original designation.

Hoplomys gymnurus (Thomas, 1897)
Armored Rat

SYNONYMS:

Echimys semispinosus: Alfaro, 1896:41; not *Echimys semispinosus* Tomes (= *Proechimys semispinosus* [Tomes])

Echimys gymnurus Thomas, 1897f:550; type locality "Cachavi [= Cachabí], [Esmeraldas,] N. Ecuador, altitude 170 metres," approximately 00°58′N 78°48′W (Paynter 1993).

[*Proechimys*] *gymnurus*: J. A. Allen, 1899c:264; name combination.

Hoplomys truei J. A. Allen, 1908:650; type locality "Lavata [= Savala], Matagalpa Province, Nicaragua."

H[*oplomys*]. *gymnurus*: J. A. Allen, 1908:651; first use of current name combination.

Hoplomys goethalsi Goldman, 1912a:10; type locality "Rio Indio, near Gatun, Canal Zone, Panama."

Hoplomys gymnurus goethalsi: Goldman, 1920:123; name combination.

Hoplomys gymnurus truei: Goldman, 1920:124; name combination.

Hoplomys gymnurus gymnurus: Ellerman, 1940:123; name combination.

Hoplomys gymnurus wetmorei Handley, 1959:9, type locality "Isla Escudo de Veraguas, Prov. Bocas del Toro, Panama."

DESCRIPTION: As for the genus.

DISTRIBUTION: In Central America, from southeastern Honduras (Pine and Carter 1970) south through Panama (Handley 1959; Hall 1981); in South America, from northwestern Colombia south along the Pacific coast and west of the Cordillera Occidental to northwestern Ecuador, from elevations ranging between 180–1,250 m in northwestern Colombia to sea level in southwestern Colombia and adjacent northwestern Ecuador.

SELECTED LOCALITIES (Map 513; South America only): COLOMBIA: Antioquia, Alto Bonito (Handley 1959), Villa Arteaga (FMNH 70102), Zaragoza (USNM 499753); Cauca, Río Saija (FMNH 90110); Chocó, Bagadó (Handley, 1959), Río Docampado (FMNH 90115); Nariño, Buenavista (Handley 1959); Valle del Cauca, 6 km N of Buenaventura (MVZ 162309), 1 km from San Isidro above Río Calima (ICN 8313). ECUADOR: Esmeraldas, Cachabí (type locality of *Echimys gymnurus* Thomas); Imbabura, 10 km E of Santa Rosa (T. E. Lee et al. 2010); Pichincha, Mindo, Río Blanco (Handley 1959).

SUBSPECIES: In his revision of the genus, Handley (1959) recognized a single species with four subspecies. Two of these have exclusively Central American distributions (*truei* J. A. Allen from mainland Nicaragua and Costa Rica and the Panamanian insular *wetmorei* Handley). One subspecies (*goethalsi* Goldman) was mapped from Panama to the central Pacific coast of Colombia in the provinces of Antioquia and Chocó, and the nominotypical subspecies was recorded from northwestern Ecuador to southwestern Colombia in the state of Nariño. However, given that many of the geographic gaps in the map illustrated by Handley (1959:3, Fig. 1) are now represented by specimens in museum collections, and that no adequate review of these has as yet taken place, we hesitate to recognize the subspecies boundaries depicted by Handley herein.

NATURAL HISTORY: Nocturnal and terrestrial, these rats feed primarily on fruit and seeds. They are the most numerous rodents in the pluvial rainforests of western Colombia, the wettest forest habitat in the New World. They frequent fallen logs, brush piles, rocks, and stream edges in evergreen rainforest, and in adjacent grassy clearings (see Adler et al. 1998; Goldman 1920; Handley 1959; Pine and Carter 1970). Tomblin and Adler (1998) described details of microhabitat structure and use by *Hoplomys* in Panama. Alberico and González (1993) recorded higher densities in primary as opposed to secondary forest, and measured seasonal shifts in population density, home range size, and movement distances. Alberico and González (1993) and Tesh (1970b) provided limited details on reproductive biology, including litter size and seasonality, for western Colombia and Panama, respectively. The species is sexually dimorphic, with males in Panama averaging 38% larger than females (Adler et al. 1998). Tail autotomy is also common, a trait *Hoplomys* shares with other echimyids, especially terrestrial species (Adler et al. 1998). Aboriginal groups in Panama utilized *Hoplomys* as a protein food resource (O. F. Linares 1976). Brennan (1969), Brennan and Reed (1974b), Fain et al. (1982), Mendez (1969), and Tipton et al. (1967) described external parasites.

REMARKS: J. A. Allen (1908:650) suggested that *Echimys semispinosus* Tomes (1860b) might be referable to his new genus *Hoplomys*, along with *Echimys gymnurus* Thomas (1897f). On the same page, Allen referred also to *Echimys subspinosus* Tomes, which is clearly an inadvertent misspelling of *semispinosus*. However, Tomes' *E. semispinosus* has been consistently placed within *Proechimys* in all subsequent literature (Tate 1935; Ellerman 1940; Cabrera 1961; Patton 1987; Woods 1993; Woods and Kilpatrick 2005). Valim and Linardi (2008) described the host association of ectoparasitic lice in the genus *Gyropus*.

Honeycutt et al. (2003) reported mitochondrial and nuclear DNA sequences for a specimen of "*Hoplomys gymnurus*" from 52 km SSW of Altamira, Brazil (GenBank accession numbers AF520668 and AF520661). Either this specimen (Texas A&M University number AK 9671) has been misiden-

tified or the locality given is in error. There are no records of this genus from Brazil or from anywhere east of the Andes.

Genus *Lonchothrix* Thomas, 1920
James L. Patton

This monotypic genus is similar in all morphological and ecological features to the spiny tree rats, *Mesomys*, with which it also forms a close sister relationship (Lara et al. 1996; Y. L. R. Leite and Patton 2002). Head and body length varies from 155–220 mm, tail length from 150–230 mm, hind foot length from 30–37 mm, and ear length is 13 mm (Emmons and Feer 1997). As with *Mesomys*, this is a relatively small rat with the body covered with dense, wide (1.8–2.0 mm), long (26–28 mm), flat, and flexible spines mixed with fine hairs. The spines, in general, are wider and stiffer than those of *Mesomys*. The upper parts are gray-brown, the sides and rump speckled yellowish white, and the wide spines on the sides and rump have pale yellowish tips, in contrast with a more orange or buff tips on the spines in *Mesomys*. The ears are short, naked, brown in color, and buried among the spines of the head. The tail is robust, scaly, speckled with short, dark, evenly spaced but flat, scale-like hairs, and terminates in a tuft of long, coarse, red-brown hairs that may reach 45 mm in length. *Lonchothrix* departs from *Mesomys* primarily in these accentuated characteristics of the tail, although the recently described *Mesomys occultus* (Patton et al. 2000) has a terminal tail tuft averaging 30 mm in length. The under parts are white, grayish, or buff, often tending to buff along the midline. The chest is often pale gray-brown and the throat has white-tipped hairs. The feet are dirty white. The hindfeet are broad and short, with pink soles; the forefeet are small and also with a pink palmar surface. The skull and cheek teeth are generally as in *Mesomys*, although as Thomas (1920c) noted in his description, *Lonchothrix* differs slightly as follows: muzzle broader with nasals more parallel sided and less tapered posteriorly, the malar projects anteriorly, the incisive foramina are longer and more open, the mesopterygoid fossa is narrower (reaching the middle of M2), and the hamular processes of the pterygoids is narrower, not spatulate. Upper cheek teeth have one short lingual fold and have four labial folds, or flexi, that penetrate deep into the tooth and become isolated as fossettes with wear. All lower cheek teeth have a single, short labial fold but varying numbers of lingual folds: pm4 with four folds, m1 with three, with the most posterior quite small, and both m2 and m3 with just two.

SYNONYM:

Lonchothrix Thomas, 1920c:113; type species *Lonchothrix emiliae* Thomas, 1920, by original designation.

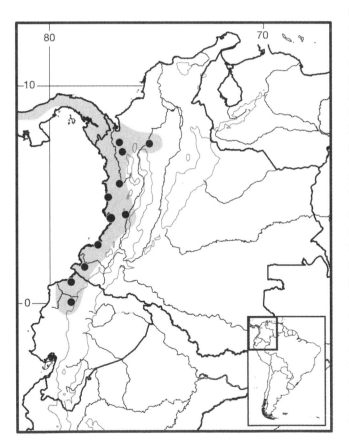

Map 513 Selected localities for *Hoplomys gymnurus* (●). See Hall (1981) for distributional records in Central America. Contour line = 2,000 m.

REMARKS: A strict phylogenetic definition could argue for the placement of *Lonchothrix* and *Mesomys* into a single genus in recognition of their strongly supported sister relationship (Lara et al. 1996; Y. L. R. Leite and Patton 2002). Indeed, as noted by Thomas (1920c:114), the skull of *Lonchothrix* is "so like that of *Mesomys hispidus* that it is difficult to find any character of more than specific value." Despite this statement, however, Thomas was unsure of the closest relative, remarking that its short, climbing feet and strongly spiny pelage resembled that of *Echimys*, although its brachydont and nearly rounded molars set it well apart from this genus or from either *Dactylomys* and *Kannabateomys*. He noted that the teeth resemble those of *Cercomys* (= *Thrichomys*). Although he compared his new genus to *Mesomys*, for some reason he made no special connection between these two genera as the close relatives they clearly are. However, I note here that the number of cheek tooth laminae differ between the genera, *Lonchothrix* with three reentrant folds and *Mesomys* spp. generally with four.

Lonchothrix emiliae Thomas 1920
Tuft-tailed Spiny Tree Rat

SYNONYM:

Lonchothrix emiliae Thomas 1920c:114; type locality "Villa Braga, on the left bank of the Rio Tapajós, just above the first rapids," Pará, Brazil.

DESCRIPTION: As for the genus.

DISTRIBUTION: Known only from the lower reaches of the Rio Madeira, Rio Tapajós, and Rio Xingu; all south of the Amazon River in eastern Brazil.

SELECTED LOCALITIES (Map 514): BRAZIL: Amazonas, Igarapé Auara (AMNH 91868); Pará, Alter do Chão (Lara et al. 1996), Igarapé Amorim (AMMH 95549), Jacareacanga (Auricchio 2001), Tauari (AMNH 95776), Villa Braga (type locality of *Lonchothrix emiliae* Thomas).

SUBSPECIES: *Lonchothrix emiliae* is monotypic.

NATURAL HISTORY (from Auricchio 2001): This is a poorly known and rarely seen arboreal rat that has yet to be studied in the field. Collectors' field notes suggest that it is nocturnal, lives in lowland mature and secondary (capoeira) rainforest, nests in tree holes, readily enters and uses the rafters of wooden structures, and is locally common despite its limited geographic range. Stomach contents of a few specimens included only the pulp of unidentified plant materials, most likely fruit of *Brysonima* sp., a common small tree in the family Malpighiaceae. A pregnant female with two embryos was taken in March.

REMARKS: Auricchio (2001) questioned the identification of the specimen from the Rio Madeira (AMNH 91868). The specimen in question lacks the terminal portion of its tail, but in its overall size, broader, stiffer, and

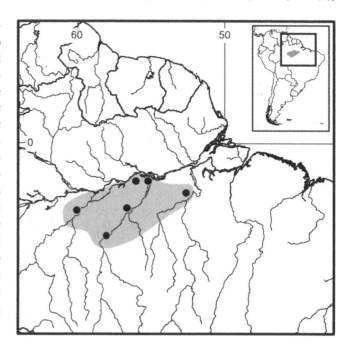

Map 514 Selected localities for *Lonchothrix emiliae* (●).

pale-tipped spines, and more naked appearing proximal portion of the tail, it matches *Lonchothrix* and not *Mesomys*. Barring additional character discovery, we regard AMNH 91868 as a *Lonchothrix emiliae*, and thus extend the known range of the species west to the lower Rio Madeira.

The report of the karyotype of *L. emiliae* by Aniskin (1993) based on specimens from Peru must be in error. His specimens most likely represent *Mesomys hispidus*, both because of the unlikely 3,000-km range extension and the fact that the karyotype reported is identical to that illustrated for this species by Patton et al. (2000:197, Fig. 129B) from multiple localities in western Amazonian Brazil.

Genus *Mesomys* Wagner, 1845
James L. Patton and Louise H. Emmons

These are also relatively small, heavily spined rats (head and body length averaging about 180 mm) with a tail as long or slightly longer than the body, thinly haired, and terminating in a tuft of hairs of varying length. Hindfeet are short and broad. The pelage possesses heavily developed and stiffened aristiform spines throughout the back, sides, and rump. The spines are shorter (24–25 mm) and somewhat more narrowed (1.5 mm) than those of the closely related *Lonchothrix*. Individuals are common in the canopy and lower strata of lowland rainforest throughout the Amazon Basin and the Guianan region, but extend to an elevation of 2,000 m in the lower montane forest along the eastern slope of the Andes. They may be especially common in tree falls, in dense liana tangles, or in disturbed areas, as around villages where they can often be found on the rafters of thatched houses.

Within the family, *Mesomys* and it sister genus *Lonchothrix* are unique in their combination of a dorsal pelage composed of broad and stiff spines, a sparely haired tail that terminates in a distinct brush or pencil, and small, rounded cheek teeth with flexi wearing to isolated oval fossettes or pits rather than forming transverse laminae. These tooth characters are shared with other members of the subfamily Eumysopinae, while the foot structure and elongated tail are more characteristic of the Echimyinae, which are similarly arboreal in their habits. This combination of characters of taxa otherwise belonging to separate subfamilies was noted by Wagner (1845a) in his description of the genus and reemphasized by Thomas (1905a). As a consequence, the subfamilial placement of both *Mesomys* and *Lonchothrix* has been uncertain (see account above, and discussion in Patton and Reig 1990; Emmons 2005). These two genera do, however, form a well-supported sister pair in molecular sequence analyses (Lara et al. 1996; Y. L. R. Leite and Patton 2002; Galewski et al. 2005; P.-H. Fabre et al. 2012).

Thomas (1905a:591) provided an expanded description of *Mesomys* that we augment as follows: external form similar to that of *Proechimys*, but with an arboreal bauplan of long back, short legs, short and broad pink-soled feet, strong, curved claws, and long whiskers. The ears are shorter and the fifth hind toe is longer, reaching to the distal end of the first phalanx of the fourth toe. Aristiform hairs are in the form of stiff, broadened, prominent spines distributed over the mid-dorsum and rump. The tail is long, haired lightly to varying degrees, and sometimes terminated by a tuft or pencil. The cheek teeth are rounded, as in *Proechimys*, with four or five labial folds on each upper tooth, forming five or six lophs, which become isolated fossettes with wear. The lower premolar has four oblique lingual folds, and each lower molar only three. The general conformation of the skull is as in genera of Echimyinae, with a short, parallel-sided rostrum and a broadened, square interorbital region. Along with *Lonchothrix*, these features contrast markedly with the elongated rostrum and generally concave interorbital region characteristic of other genera in the subfamily Eumysopinae. Tate (1935:413) noted that "*Mesomys* comprises small, densely spinous, buff-brown-bellied rats with short, rather broad feet, slightly tufted tails and *Proechimys*-like, rounded molars." The tail is particularly susceptible to autotomy and is often truncated, as was that of one of the first individuals to be described, unfortunately, as *M. ecaudatus* (= tailless *Mesomys*, Wagner 1845a). Apart from *Lonchothrix*, the

sympatric taxon most easily confused with a *Mesomys* spp. is *Pattonomys occasius*, which has a conspicuously naked tail (Emmons and Feer 1997).

As emphasized by Voss et al. (2001:152), the genus *Mesomys* has never been revised, and recognition of its constituent species has been in flux essentially since Desmarest (1817b) described the first species (see Tate [1935] for a general review of the nomenclatural history of the genus). Tate (1939:179) considered *Mesomys* a monotypic genus "with possibly a few weakly differentiated races" because of the "obvious morphological stability throughout the known range." Alternatively, Ellerman (1940), Moojen (1948), and Cabrera (1961) listed two or more species within the genus, although not the same ones in all cases. E.-L. Trouessart (1897) placed *ecaudatus* Wagner (= *hispidus* Desmarest) in the genus *Mesomys* (p. 608), but listed *hispidus* Desmarest and *ferrugineus* Günther (a synonym of *hispidus*) as species of *Echimys* Desmarest (p. 610). Emmons (1993) reassigned two of the five species listed by Woods (1993) in *Mesomys* to the *Makalata* (*didelphoides* Desmarest and *obscurus* Wagner) and Patton et al. (2000) described an additional species (*occultus*). Voss et al. (2001) documented two morphological groups, a small-toothed form from the Guianan region and eastern Amazonia, and a large-toothed form from the western and central Amazon, and suggested that if these toothrow groups represented species, then *M. hispidus* (with *stimulax* Thomas as a synonym) would apply to the small-toothed form and *M. ferrugineus* Günther to the large-toothed form. However, they also noted that toothrow length was insufficient as a basis for taxonomic inference because that character was discordant with the molecular clade structure identified by Patton et al. (2000). They also suggested that Tate's (1939) restriction of the type locality of Desmarest's *M. hispidus* to Borba, Rio Madeira, in the central Amazon was likely in error. Orlando et al. (2003) expanded on the molecular analysis of Patton et al. (2000), and importantly included sequence data from the holotype of Desmarest's *M. hispidus* housed in the Paris Museum. Their analyses confirmed the discordance between geographic clades and the toothrow classes recognized by Voss et al. (2001). Desmarest's holotype was molecularly nearly identical to samples from French Guiana, and, after a review of the most likely collection history of this specimen (see also discussion in Voss et al. 2001:153, footnote 22), Orlando et al. (2003:114) corrected the type locality from Tate's Borba, Rio Madeira to Amapá state in eastern Brazil. The molecular data presented by Orlando et al. (2003) support the recognition of three species: a widely distributed and both morphologically and molecularly polytypic *M. hispidus*, a highly geographically localized *M. occultus* that is sympatric with

M. *hispidus*, and a southeastern Amazonian and small-toothed *M. stimulax*. The molecular data of both Patton et al. (2000) and Orlando et al. (2003), however, identify a number of divergent and geographically structured clades within *M. hispidus* that may warrant species status when additional sampling and analyses are forthcoming. Given this possibility, we comment on the application of available names to these clades in the account for *M. hispidus*, below. Finally, until *M. leniceps*, which is restricted to montane forests in northern Peru, is examined molecularly, we follow previous authors by treating it as a distinct species, pending further analyses.

SYNONYMS:

Echimys: Desmarest, 1817b:58; part (description of *hispidus* Desmarest); not *Echimys* F. Cuvier.

Loncheres: J. B. Fischer, 1829:307; part (listing of *hispida* Desmarest); not *Loncheres* Illiger.

Nelomys: I. Geoffroy St.-Hilaire, 1838b:122; part (listing of *hispidus* Desmarest); not *Nelomys* Jourdan.

Echinomys: Wagner, 1843a:345; part (listing of *hispidus* Desmarest); unjustified emendation of *Echimys* F. Cuvier.

Mesomys Wagner, 1845a:145; type species, *Mesomys ecaudatus* Wagner, by original designation.

Proechimys: J. A. Allen, 1899c:262; part (listing of *hispidus* Desmarest); not *Proechimys* J. A. Allen.

KEY TO THE SPECIES OF *MESOMYS*:

1. Head, sides, mid-back, and rump covered with broad, long, and stiff spines with interspersed setiform hairs sparse and generally not visible externally 2
1'. Head covered with nonstiff hairs; sides, mid back, and rump covered with narrowed spines interspersed with long and softer hairs . . . . *Mesomys leniceps*
2 Body size small (head and body length averaging <170 mm; upper toothrow averaging <6.3 mm); distributed south of the Amazon and east of the Rio Tapajós in Brazil . . . . *Mesomys stimulax*
2'. Body size usually larger (head and body length averaging >178 mm, upper toothrow averaging >7.14 mm); distributed elsewhere throughout the Amazon Basin; or body size small (head and body length averaging 169 mm) and distributed in the Guianan region 3
3 Dorsal spines without orange mid-band; tail with short terminal tuft of hair (<25 mm) . . . . . *Mesomys hispidus*
3'. Dorsal spines with orange mid-band; tail with long terminal tuft of hair (>30 mm) . . . . *Mesomys occultus*

Mesomys hispidus (Desmarest, 1817)

Ferreira's Spiny Tree Rat

SYNONYMS:

Echimys hispidus Desmarest, 1817b:58; type locality "Amérique meridionale"; restricted to "Borba, Rio Madeira, Brazil" (Tate 1939:179) but more recently to the state of Amapá in eastern Brazil (Orlando et al. 2003:119), based on the known travel route of the original collector.

L[oncheres]. hispida: J. B. Fischer, 1829:307; name combination.

E[chinomys]. hispidus: Wagner, 1843a:345; name combination.

Mesomys ecaudatus Wagner, 1845a:145; type locality "Borba in Brasilien," Borba, Rio Madeira, Amazonas, Brazil.

Echimys ferrugineus Günther, 1876:750; type locality "Chamicuros, Huallaga River"; upper Río Samíria [Stephens and Traylor 1983], Loreto, Peru.

Proechimys hispidus: J. A. Allen, 1899c:262; name combination.

Proechimys ferrugineus: J. A. Allen, 1899c:264; name combination.

M[esomys]. ferrugineus: Thomas, 1924c:535; name combination.

Mesomys ferrugineus spicatus Thomas, 1924c:535; type locality "Tushemo, near Masisea, R. Ucayali. Alt. 1000'," Ucayali, Peru.

Mesomys hispidus: Tate, 1935:413; first use of current name combination.

Mesomys hispidus ferrugineus: Cabrera, 1961:535; name combination.

Mesomys hispidus hispidus: Cabrera, 1961:535; name combination.

Mesomys hispidus spicatus: Cabrera, 1961:536; name combination.

DESCRIPTION: Orlando et al. (2003) defined this species as an assemblage of six monophyletic mtDNA clades distributed broadly across Amazonia from Peru to the Guianan region, exclusive of the partially sympatric *M. occultus* from west-central Brazil and *M. stimulax* from eastern Brazil south of the Amazon River. As acknowledged by these authors, *M. hispidus* thus encompasses a taxon that varies substantially in size, being smaller in the Guianan part of its range (mean head and body length 161 mm) and larger in western Amazonia (mean head and body length 178–186 mm, depending upon clade), but is otherwise characterized by well-developed, pale buffy to orangish-tipped spines covering the dorsum and a long, moderately hairy tail terminating with a short tuft of hairs. In most essential external and cranial characters, *M. hispidus* cannot be distinguished from sympatric *M. occultus*, except in its tendency to lack a median orange band on spines, have petiolate and black hairs among tail scales, and shorter tuft of tail tip. Cranially, these two species also are similar, but *M. hispidus* is larger with proportionately shorter rostrum and palate, longer and narrower incisive foramina that have absolutely longer premaxillary septum, and PM4–M3 with more strongly developed 4th fold.

The following external description is from Emmons and Feer (1997): Upper parts uniform pale to medium brown, with mid-back often heavily streaked with black. Pelage consists of conspicuous short, wide, flat, flexible spines from shoulder to rump, each spine brown with pale tip. Tail often broken short or completely missing. Spines on head and neck softer and narrower than those on back. Ears brown, short and rounded, thinly haired, and with small tuft of longer hairs at base. Eyes large, and eyeshine moderately bright yellow. Vibrissae fine and long, reaching shoulder. Tail robust, brown, and thinly covered with long, red-brown hairs that do not hide scales, and with short and sparse tuft at tip, 5–21 mm in length (Patton et al. 2000); six rows of tail scales per 5 mm (holotype of *M. ferrugineus*), with central, darkly pigmented hair of each scalar triad at base of tail sharp and spine-like. Under parts uniformly pinkish orange, which contrasts sharply with color of sides. Hindfeet short and broad, claws sharp and strongly curved, and soles pink and have large plantar pads.

DISTRIBUTION: *Mesomys hispidus* occurs in the Amazon Basin from northern and eastern Bolivia, north through eastern Peru, eastern Ecuador, southeastern Colombia, southern Venezuela, the Guianas, and all of Brazilian Amazon drainage except that part east of the Rio Tapajós and south of the Rio Amazon.

SELECTED LOCALITIES (Map 515): BOLIVIA: La Paz, 1 mi W Puerto Linares (TTU 34988), right bank Río Alto Madidi (Emmons 1993); Santa Cruz, El Refugio (USNM 584592). BRAZIL: Amapá, Amapá (Orlando et al. 2003); Amazonas, Cucui (USNM 255162), Tunui (AMNH 77034); Mato Grosso, left bank Rio Cristalino (Patton et al. 2000); Pará, Igarapé Amorin (Voss et al. 2001). COLOMBIA: Amazonas, Caserio "Kuiru" (Lemke et al. 1982); Putumayo, Río Mecaya (FMNH 71125); Vaupes, Río Apaporis, Iino Goje (FMNH 57242). ECUADOR: Orellana, Río Suno (AMNH 64015); Pastaza, Montalvo (FMNH 41461); Sucumbíos, Santa Cecilia (MSU 11780). FRENCH GUIANA: Cayenne, Paracou (Voss et al. 2001). GUYANA: Upper Takutu–Upper Essequibo, 5 km SE of Surama (Voss et al. 2001). PERU: Amazonas, headwaters Río Kagka (of Río Cenepa) (MVZ 155158); Cusco, Quincemil (FMNH 75197), 2 km SW of Tangoshiari (Patton et al. 2000); Loreto, Yurimaguas (FMNH 19631); Pasco, Delfin, Pozuzo (MUSM 12135), Nevati Mission (AMNH 230928). VENEZUELA: Amazonas, Acanana (Handley 1976), Capibara (Handley 1976).

SUBSPECIES: Cabrera (1961) recognized five subspecies, including *ferrugineus* Günther (northern Peru), *leniceps* Thomas (Andes of northern Peru), *spicatus* Thomas (upper Amazon of Ecuador and northeastern Peru), and *stimulax* Thomas (lower Amazon of eastern Brazil). He

gave the range of the nominotypical form as the central Amazon of Brazil, including southern Venezuela, based on Tate's (1939) restriction of the type locality to Borba, on the lower Rio Madeira. We treat both *stimulax* Thomas and *leniceps* Thomas as species, although the latter with reservation. The validity of the other taxa at the species or subspecies level must await adequate comparisons of museum samples available, coupled with increased molecular sampling.

NATURAL HISTORY: Of the series of specimens taken along the Rio Juruá in western Brazil by Patton et al. (2000), 97% were taken in canopy traps with the remainder on the ground. *Mesomys hispidus* was more common in upland, nonflooded forest than in the seasonally flooded *várzea*, but was found in a variety of forest types, including naturally and human-disturbed habitats as long as some trees remained. Pregnant females were found throughout the year, suggesting continuous breeding; litter sizes ranged from one to three, with the modal number of 1. In northeastern Peru, 60% of captures were in canopy traps in mature forest, or the oldest disturbed forest (Hice and Velasco 2012). Litter size of a single pregnant female was two. In southern Venezuela 94% were taken in trees or houses, mostly near streams or other moist places, and all in lowland evergreen rainforest (Handley 1976). This species has also been captured in tropical semideciduous forest in Bolivia (L. H. Emmons, unpubl. data).

REMARKS: Thomas (1905a:591; and 1911b:607, in his description of *M. stimulax*) equated *Echimys ferrugineus* Günther with *M. ecaudatus* Wagner. Five years later Thomas (1916e:298) argued that the type skull of Desmarest's *Echimys hispidus* "about which so much confusion has arisen from time to time, belongs to none of the genera to which it has been hitherto referred, but is a *Mesomys*, apparently quite similar to *M. ecaudatus* Wagner. As a result, the early and suitable name *hispidus* will happily replace the unfortunate term *ecaudatus*, given to a specimen which had lost its tail." Goeldi (1897, 1901) discussed the history of *ecaudatus* but provided no resolution as to the validity of Wagner's name. Consequently, most recent literature has listed both *ecaudatus* Wagner and *ferrugineus* Günther as synonyms of *hispidus* Desmarest. The type locality of Wagner's *ecaudatus* (Borba, on the lower Rio Madeira, in central Amazonian Brazil) is within the known range of *M. hispidus* but also just east of the known range of *M. occultus*. Wagner's holotype of *M. ecaudatus* is extant (NMW 915, examined by L. H. Emmons). Although the skin is in poor condition, characters of the skull place it near *M. hispidus* and do not conform to those of either *M. occultus* or *M. stimulax*; so we follow previous authors and maintain it in the synonymy of *M. hispidus*.

Thomas (1924c) described *spicatus* as a subspecies of *ferrugineus* Günther on the basis of browner, less ferruginous color, and the spines having a dark brown subterminal and pale buffy terminal band, imparting "a conspicuously speckled appearance to the animal." The nominotypical subspecies was noted to range in the Huallaga drainage to the west of the central Ucayali, the vicinity of the type locality of *spicatus*. Thomas recognized *ferrugineus* Günther as a species separate from *hispidus* Desmarest primarily on the basis of a distinct, longer tail tuft (about "an inch long" [=24 mm]), while in typical *hispidus* these hairs do not exceed 10–11 mm. However, within the large sample of *hispidus* from the Rio Juruá in western Brazil (Patton et al. 2000:195), the terminal tuft was extremely variable among individual specimens, ranging from 5–21 mm. Thus, while there might be geographic variation in the length of the tail tuft, the specimens available to us do not segregate geographically into short- and long-tufted samples that might correspond with a lower Amazonian *M. hispidus* and an upper Amazonian *M. ferrugineus*. However, because there is extensive geographically ordered clade structure within what is now considered to be the single species *M. hispidus* (Patton et al. 2000; Orlando et al. 2003), additional analyses and inclusions of specimens from type localities may well delineate those clades as species, with one corresponding to Günther's *ferrugineus*.

Voss et al. (2001) recognized a small-toothed form of *Mesomys* in the Guianan region and eastern Amazonia (maxillary toothrow length ranging from 5.9–6.6 mm) and a large-toothed form from western Amazonia (with maxillary toothrow length >7 mm; see Patton et al. 2000: tables 56, 58), and suggested that each represented a separate species. Because Desmarest's type of *hispidus* is small toothed (maxillary toothrow length = 6.4 mm), these authors suggested that this name would apply to the eastern, small-toothed form and argued that Tate's (1939) restriction of the type locality to Borba on the Rio Madeira was incorrect. Orlando et al. (2003:114) provided an expanded history of Desmarest's holotype and further restricted the type locality of *hispidus* to Amapá state, north of the Rio Amazon delta in eastern Brazil. As currently understood, *M. hispidus* is sympatric with *M. occultus* at both localities where *M. occultus* is known to occur (Patton et al. 2000) and maybe sympatric with *M. stimulax* at one locality on the lower Rio Tapajós (Voss et al. 2001).

Future studies may divide the morphologically variable *M. hispidus*, as we conceive it herein, into separate species coincident with those reciprocally monophyletic molecular clades defined by Orlando et al. (2003). If so, then *hispi-*

dus Desmarest would apply at least to that part of clade F from the eastern Guianan region defined by Orlando et al. (2003), the source of Desmarest's holotype. This would make Wagner's (1845a) *M. ecaudatus* from Borba available for the large-toothed central Amazonian animals. We note that there seems no reason to doubt the provenance of Borba for Natterer's specimen: his holotype of *I. pagurus* (Wagner 1845a), also from Borba, and also extant, is consistent with the geographic range of that species. However, Wagner's *ecaudatus* could apply to either clade C from the central Amazon or to the southwestern portion of clade F, as its type locality lies between the mapped ranges of both. Because the type localities of both *ferrugineus* Günther and *spicatus* Thomas are probably contained within the range of clade A from the western Amazon of northern Bolivia, eastern Peru, and western Brazil, Günther's name would apply to this clade because of priority. The remaining molecular clades, each currently delineated only by single localities in northeastern Bolivia, southern Venezuela, or Guyana, lack available names, at least on geographic grounds, should there be a desire to recognize each formally.

Leal-Mesquita (1991) and Patton et al. (2000) described and figured a chromosome complement of $2n=60$, FN=116. Aniskin (1993) reported the same karyotype from specimens he identified as *Lonchothrix emiliae* from northern Peru, several thousand kilometers west of the known range of this species (see *Lonchothrix* account). Aniskin's specimens are mostly likely *M. hispidus* (Patton et al. 2000). Patton et al. (1994) documented two mitochondrial clades along the Rio Juruá and discussed the pattern of differentiation with reference to riverine barriers. Patton et al. (2000) showed that these two clades partially overlap in cranial multivariate morphometric analyses.

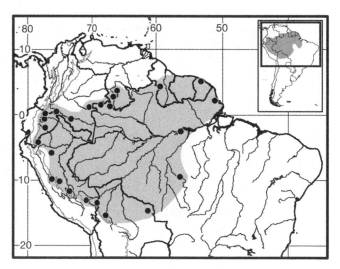

Map 515 Selected localities for *Mesomys hispidus* (●). Contour line = 2,000 m.

Mesomys leniceps Thomas, 1926
Long-haired Spiny Tree Rat

SYNONYMS:

Mesomys leniceps Thomas, 1926i:348; type locality "Yambrasbamba, Amazonas, 6500'," Peru.
Mesomys hispidus leniceps: Cabrera, 1961:536; name combination.

DESCRIPTION: Size and general characters similar to those of the larger, western samples of *M. hispidus*, but fur much less strongly spinose throughout, and spines longer, thinner, and more liberally mixed with fur. Individual spines average about 20 mm in length (Thomas 1926i). Ventral fur also more hairy than spiny, as characteristic of other species in genus. Head distinctly hairier than back and rump, noted to be "scarcely hispid to the touch" (Thomas 1926i:348), with long hair of crown covering tops of ears. Both dorsal and ventral colors are similar to those of *M. hispidus*, in particular rich buffy ventral tones and large white patches on chin and in axillary and inguinal regions. Dorsal surfaces of both forefeet and hindfeet pale buffy; digits white. Tail long for a *Mesomys* (about 120% of head and body length), brown throughout, with body hair extending farther onto tail base than in other taxa (ca. 2.5 cm); tail hairs more appressed to body of tail than usual in genus; tail terminates in long, well-developed pencil. Tail scales smaller than those of other Peruvian *Mesomys*, with eight scale rows per 5 mm (data from the holotype). Skull resembles that of *M. hispidus* but is somewhat smaller, with more slender rostrum and nasals that do not extend behind premaxillary processes. Jugal and bones surrounding antorbital foramina slender such that vertical height of zygoma distinctly narrowed. Incisive foramina constricted posteriorly. Only type and one juvenile specimen known, so extent of variation in any feature of species completely unknown.

DISTRIBUTION: Known from only two localities in upper montane forest of northern Peru.

SELECTED LOCALITIES (Map 516): PERU: Amazonas, Yambrasbamba (type locality of *Mesomys leniceps* Thomas); San Martín, Puca Tambo (Thomas 1926i).

SUBSPECIES: *Mesomys leniceps* is monotypic.

NATURAL HISTORY: Nothing is known about the biology of this species, other than the comment in the original description that the holotype was "trapped under roots of tree in heavy forest, at same burrow as *Proechimys simonsi*" (Thomas 1926i:349).

REMARKS: Thomas (1926i:348) described *M. leniceps* as an "upland representative of the Amazonian *M. ferrugineus*," gave the elevation as 5,100 feet, and characterized it mostly by its less spiny pelage, with a more liberal mix of hairs resulting in a comparatively long coat. These pelage features are consistent with a distributional occurrence at a

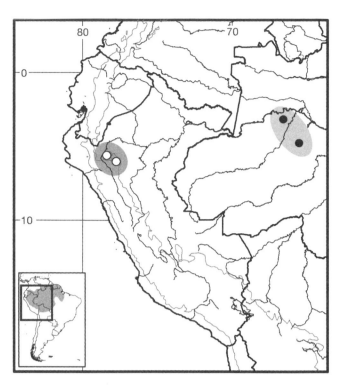

Map 516 Selected localities for *Mesomys leniceps* (○) and *Mesomys occultus* (●). Contour line = 2,000 m.

higher elevation (see, for example, the discussion by Patterson and Velazco [2006] in their description of a montane species of *Isothrix*). In his description, Thomas allocated a second specimen to *leniceps* from Puca Tambo, San Martín, that he had previously (1926g:162) referred to *ferrugineus* Günther. Most reviewers (e.g., Tate 1935; Ellerman 1940; Woods 1993; Woods and Kilpatrick 2005) since Thomas's (1926i) original description have consistently listed *leniceps* as a distinct species. The sole exception is Cabrera (1961:536), who regarded *leniceps* as a subspecies of *M. hispidus*. No justification for these various decisions, however, has been provided by any set of authors subsequent to Thomas's original description. As with so many cases for the Echimyidae, a thorough review of character variation among available museum specimens of *Mesomys* is badly needed to establish the status of *M. leniceps* at any level.

Mesomys occultus Patton, da Silva, and Malcolm, 2000
Hidden Spiny Tree Rat

SYNONYM:

Mesomys occultus Patton, da Silva, and Malcolm, 2000: 194; type locality "Colocação Vira-Volta, left bank Rio Juruá on Igarapé Arabidi, affluent of Paraná Breu, Amazonas, Brazil (3°17'S, 66°14'W)."

DESCRIPTION: Equivalent in size and similar in color, color pattern, and degree of aristiform spine development to *M. hispidus* (see above). In contrast to sympatric *M. hispidus*, spines on neck and shoulders possess orange midband. Hairs along tail thin and colorless, and tuft at tip of tail contains longer hairs that average more than 30 mm in length. At base of tail, central pigmented hair of each scale elongated and not spinelike as in *M. hispidus*. Cranially, compared with sympatric *M. hispidus*, incisive foramina shorter and wider, with shorter premaxillary septum; grooves on anterior palate less conspicuous; and paroccipital processes flatter and broader. Four labial flexi present on P2, M1, and M2, with the fourth very small and usually lost with wear. M3 usually with only three labial flexi, or, if four, the two middle ones may coalesce into Y-branched structure.

DISTRIBUTION: *Mesomys occultus* is known only from two localities, one each on the lower Rio Juruá and on the Rio Urucu, to the south of the Rio Solimões in the central Amazon Basin of the state of Amazonas in Brazil.

SELECTED LOCALITIES (Map 516): BRAZIL: Amazonas, Colocação Vira-Volta, left bank Rio Juruá, Igarapé Arabidi (type locality of *Mesomys occultus* Patton, da Silva, and Malcolm), alto Rio Urucu (Patton et al. 2000).

SUBSPECIES: *Mesomys occultus* is monotypic.

NATURAL HISTORY: This species is known from only four specimens, all caught in upland, nonflooded (= *terra firme*) rainforest in traps placed at least 1.5 m above the ground. A pregnant female was captured in early June (Patton et al. 2000).

REMARKS: Externally, *M. occultus* is so similar to sympatric *M. hispidus* that it was not recognized as distinct until molecular sequence and chromosomal data were obtained. Patton et al. (2000) described and figured the karyotype as $2n = 42$, $FN = 54$, and compared it to the $2n = 60$, $FN = 116$, complement of *M. hispidus*.

Mesomys stimulax Thomas, 1911

Pará Spiny Tree Rat

SYNONYMS:

Mesomys stimulax Thomas, 1911b:607; type locality "Cametá, Lower Tocantins," Pará, Brazil.

Mesomys hispidus stimulax: Cabrera, 1961:536; name combination.

DESCRIPTION: Relatively small (head and body length averaging 169 mm, range: 154–196 mm; Voss et al. 2001:154, table 44), with a short toothrow (mean 6.3 mm, range: 5.8–6.8 mm; Voss et al. 2001:154, table 44). Maxillary toothrows slightly converge anteriorly, and M1 with three laminae rather than four, which is characteristic of *M. hispidus*. In size, *M. stimulax* similar to samples of *M. hispidus* from Guianan region (Voss et al. 2001; Orlando et al. 2003). Color above paler and buffier than in western samples of *M. hispidus* but difference minimal (Thomas 1911b). In contrast to uniformly buffy venter of *M. his-*

pidus, however, throat, axillae, chest, and inguinal region white, and belly more fawn-colored. Vibrissae and long hairs around ears black. Forefeet and hindfeet are whitish above and lack any buffy suffusion. Tail uniformly brown and terminates in thin tuft. Overall, skull smaller than western samples of *M. hispidus*, including among specimens that are apparently sympatric along lower Rio Tapajós (see Voss et al. 2001:153). Thomas (1911b:607) stated that difference in overall size is "almost entirely due to the considerable shortening of the muzzle." He noted that nasals are shorter than those of *ecaudatus* Wagner (which to him was equivalent to western samples of *M. hispidus*), parallel sided mesially, and narrowed in front rather than decidedly broader anteriorly than posteriorly. Tympanic bullae also smaller than in *M. hispidus*. The holotype (BM 11.4.28.29) has distinctive incisive foramina, with premaxillary portion converging anteriorly to narrow, straight-sided point (arrowhead shape); posterior portion with sides slightly pinching in to form shallow palatal grooves. Mesopterygoid fossa straight sided, rather than slightly bowed outward, and nasals not expanded anteriorly as in holotypes of *ferrugineus*, *hispidus*, *leniceps*, *occultus*, and *ecaudatus*.

DISTRIBUTION: *Mesomys stimulax* is known from the eastern Amazon Basin of Brazil, south of the Rio Amazonas from the Rio Tapajós east to at least the Rio Tocantins.

SELECTED LOCALITIES (Map 517): BRAZIL: Pará, 52 km SSW of Altamira, right bank Rio Xingu (Patton et al. 2000), Cametá (type locality of *Mesomys stimulax* Thomas), Igarapé Amorin (Voss et al. 2001), opposite Iroçanga (B. D. Patterson 1992b), Mata do Sequeirinho (CJ 7), Tauari (AMNH 94862), Vilarinho do Monte (AMNH 95971).

SUBSPECIES: *Mesomys stimulax* is monotypic.

NATURAL HISTORY: Along the Rio Xingu in eastern Amazonian Brazil, two individuals of this species were collected (by L. H. Emmons) at heights of 4 and 10 m, respectively, in vines on trees of tall evergreen forest.

REMARKS: Included in *M. hispidus* as a valid subspecies by Cabrera (1961). However, available samples from the southeastern Amazon of Brazil are strongly separated from *M. hispidus* by molecular sequence analyses (M. N. F. da Silva and Patton 1993; Patton et al. 1994, 2000; Orlando et al. 2003). Voss et al. (2001) examined the holotype and grouped it with the "small-toothed" forms from the Guianas, the basis for Husson's (1978) belief that *stimulax* was a species distinct from *M. hispidus*. However, Orlando et al. (2003) documented the molecular distinction between small-toothed populations from the Guianan region and the small-toothed *M. stimulax* from southeastern Amazonia (see the account of *M. hispidus*). On the lower Rio Tapajós, mixtures of the two toothrow-size classes are present in

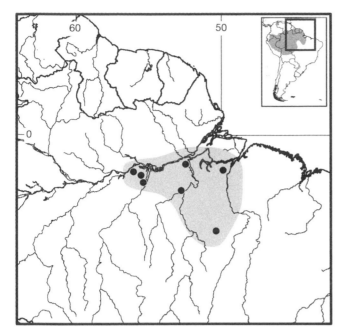

Map 517 Selected localities for *Mesomys stimulax* (●).

specimens in museum collections, and Voss et al. (2001:53) identified the small-toothed form as *M. stimulax*. However, the geography of the molecular clades hypothesized by Orlando et al. (2003) suggests that small-toothed animals from the Tapajós are more likely to represent true *M. hispidus* (Guyana clade) than *M. stimulax*, which may be restricted to east of the Rio Xingu. Both hypotheses can be tested by expanded collections, with appropriate molecular data, from the lower Rio Tapajós. The chromosomal complement is $2n = 60$, $FN = 116$, the same as that for *M. hispidus* (Patton et al. 2000).

Genus *Proechimys* J. A. Allen, 1899
James L. Patton and Rafael N. Leite

Proechimys is the most speciose and geographically most widely distributed genus of the family Echimyidae. All species are essentially limited to lowland rainforest habitats in Central America and both cis- and trans-Amazonian South America, although a few reach elevations of 2,000 m on the lower slopes of the eastern Andes or extend into dry forests in southeastern Bolivia and northern Paraguay and the Cerrado of central Brazil. These terrestrial spiny rats are often the most abundant nonvolant mammal of lowland Neotropical forests, and perhaps the most easily recognizable. Individuals can be easily heard scurrying through the leaf litter and are readily seen at night, by virtue of their bright red or yellow eyeshine, freezing when caught in a spotlight, or bounding off with a distinctive loping gait. They are important components of the terrestrial forest community, serving as seed predators and dispersal agents and as reservoirs of zoonoses, such as leishmaniasis (Telford et al. 1972; Pajot et al. 1982; Dedet et al. 1984, 1989), encephalitis (Carrara et al. 2005), trypanosomiasis (Morales and Carreno 1976), and arboviruses (Travassos da Rosa et al. 1983).

In contrast to many other echimyids, all species of this diverse genus are terrestrial, with elongated heads and long rostra, large and erect ears, and narrow and long hindfeet. Body size ranges extensively among species, with the smallest (e.g., *P. kulinae* or *P. pattoni*) averaging about 180 mm in head and body length and the largest (e.g., *P. semispinosus*, *P. steerei*, and *P. quadruplicatus*) nearly 300 mm, and most are larger than sympatric sigmodontines rodents. The tail is always shorter than the head and body length (typically 65–70% of head and body length, but up to 85% in somespecies [e.g., *P. simonsi*]), and the dorsal pelage is a mixture of soft setiform hairs and expanded aristiform spines, varyingly stiffened among species and/or populations (see M. N. F. da Silva 1998 and Hoey et al. 2004 for descriptions of aristiform development among species). The dorsal and lateral color is generally reddish brown to gray-brown, often streaked with black particularly along the midline, and the venter is white, although it can be tinged ochraceous or grayish, especially in the throat, thorax, and inner thighs. The narrow and elongate feet have slender toes and small plantar tubercles, the number of which may help define species (Patton and Gardner 1972). The color pattern on the dorsal surface of the foot is also often species specific (Patton and Gardner 1972; A. L. Gardner and Emmons 1984; Patton et al. 2000). The ears are distinctly larger than those of all similar-sized arboreal echimyid genera but are similar in length to those of the terrestrial spiny rats of the Atlantic Forest of Brazil, *Trinomys*.

The skull retains the same relatively narrow and elongated shape in all species currently recognized and despite substantial differences in overall size. A long rostrum, relatively narrow interorbital region with concave sides, and rounded cheek teeth with varying number of folds characterize the skull of all species. Characters such as the shape and structure of the incisive foramina, the degree of expression in temporal ridging, the degree of development of the groove on the floor of the infraorbital foramen produced by the passage of the maxillary nerve, the width and posterior palatal penetration of the mesopterygoid fossa, the size and construction of the postorbital process of the zygoma, the number and pattern of bullar septae, and the number of folds on both maxillary and mandibular cheek teeth (Patton and Gardner 1972; A. L. Gardner and Emmons 1984; Patton 1987; M. N. F. da Silva 1998; Patton et al. 2000) have proven useful in diagnosing species, as have bacular size and shape or the soft anatomy of the male glans penis (Patton and Gardner 1972; Patton 1987; M. N. F. da Silva 1998; Patton et al. 2000).

Moojen (1948) revised those species of the genus distributed in Brazil, following earlier authors in distinguishing two subgenera: *Proechimys*, which contains those species from Central America south through greater Amazonia, and *Trinomys*, which contains those species limited to the coastal Atlantic Forest of Brazil. Cladistic analysis of morphological characters supports a sister relationship between *Proechimys* and *Trinomys*, along with *Hoplomys* (G. A. S. Carvalho and Salles 2004), but molecular sequence analyses do not support the sister status of *Proechimys* and *Trinomys* (Lara et al. 1996; Y. L. R. Leite and Patton 2002; Galewski et al. 2005). As a result, all recent literature has separated *Trinomys* from *Proechimys* (G. A. S. Carvalho and Salles 2004; Woods and Kilpatrick 2005) at the generic level. We agree with and follow these recent trends.

The extent of our knowledge of the systematics of this large and complex genus stand in stark contrast to their ubiquitous presence in all forest types, disturbed or pristine. The historically poor state of our understanding of the taxonomy of *Proechimys* is underscored by Oldfield Thomas's (1928b:262) famous statement: "The bewildering instability of the characters of these spiny rats makes it at present impossible to sort them according to locality into separate species, subspecies, or local races . . . I confess myself defeated in any attempt at present to distinguish the local races." A half-century later, Pine et al. (1981:267) echoed this sentiment, writing "among the rodents, *Proechimys* remains what may be the most problematical genus taxonomically in all mammaldom."

Moojen (1948) recognized only five species within *Proechimys* (*sensu stricto*) while Woods and Kilpatrick (2005) listed 25. Three to five species might be present at a single locality (Patton et al. 2000); these may be truly syntopic, including sharing the same den sites on successive nights (Emmons 1982; Malcolm 1992), but they are often segregated by microhabitat (Patton et al. 2000; Voss et al. 2001). Although it is sometimes relatively easy to distinguish species when sympatric, identifying living animals to species takes a skilled eye, and defining species boundaries over larger segments of geography has proven very difficult. An extreme level of character variability, both within and among populations, hampered earlier studies attempting to diagnose species (Thomas 1928b; Moojen 1948; Hershkovitz 1948a). Even karyotypes, which have proven useful in differentiating sympatric taxa (Patton and Gardner 1972; M. N. F. da Silva 1998), may be quite variable geographically (Reig, Aguilera et al. 1980; A. L. Gardner and Emmons 1984; Patton et al. 2000; T. Machado et al. 2005). Patton and Rogers (1983) documented that individual spiny rats essentially grow continuously throughout

life, so mensural variables may increase substantially with advancing age. As a consequence, one must be exceedingly careful in morphometric comparisons between taxa and not be confused by age-related variation.

Despite these complexities and problems, considerable advances have been made in our understanding of species boundaries, definable geographic ranges, and morphological variation in the past two decades. Much of this progress has been fueled by the application of new methodologies (chromosomes and molecular genetic characters) and detailed examination of qualitative morphological characters. For example, Patton and Gardner (1972) sorted co-occurring species in eastern Peru using concordant karyotypic, soft anatomical, and phallic (including bacular) characters. Patton and Rogers (1983) documented that samples needed to age-grouped for effective morphometric comparisons of different species because of almost indeterminate growth. A. L. Gardner and Emmons (1984) used the septal patterns of the tympanic bullae to define species and group these into coherent units. Patton (1987) documented the utility of several qualitative cranial variables and bacular size and shape in allocating the 59 available names to one of nine species groups that he recognized. For each of these, he provided hypotheses of species units and remarked on the probable geographic range of each. M. N. F. da Silva (1998) described four new species from the western Amazon of Brazil and presented the initial mtDNA sequence data to help define these and other sympatric species. Subsequently, in the most thorough analysis of the genus to date Patton et al. (2000) defined, described, compared, and mapped the eight species that are now recognized in western Amazonia. These authors also made recommendations regarding species units elsewhere in Amazonia, based on patterns of morphological, karyotypic, and molecular phylogeographic variation. Reig, Aguilera et al. (1980) and Corti and Aguilera (1995) documented the distribution and limits of many species in Venezuela. Petter (1978), Steiner et al. (2000), and Voss et al. (2001) delineated sympatric species in French Guiana, describing morphological, molecular, and/or ecological differences. Weksler et al. (2001) revised one of the complexes of species occurring in eastern Amazonia and central Brazil. And, finally, Bonvicino, Otazu, and Vilela (2005) delineated karyotypic and molecular diversification among populations in the Rio Negro drainage of Amazonian Brazil, and suggested limited taxonomic changes.

J. A. Allen (1899c) proposed *Proechimys* to replace *Echimys* I. Geoffroy St.-Hilaire (1840), which is not the same as *Echimys* F. Cuvier (1809). Fortunately, Allen designated *Echimys trinitatis* Allen and Chapman as the type species of his new genus, as *E. setosus*, the type of Geoffroy St.-Hilaire's *Echimys* is a species in the genus

Trinomys Thomas. Voss and Abramson (1999) stated that the type species of the sigmodontine genus *Holochilus* Brandt at that time was *Mus leucogaster* Brandt, 1835, which was based on an example of the echimyid *Trinomys*, an oversight not previously recognized and which would have necessitated the application of *Holochilus* Brandt as the senior synonym of *Trinomys* Thomas. In Opinion 1894 the International Commission (ICZN 2001) asserted its plenary powers and fixed the type species of *Holochilus* Brandt as *H. sciureus* Wagner, thus keeping both *Proechimys* J. A. Allen and *Trinomys* Thomas on the Official List of Generic Names in Zoology.

SYNONYMS:

Mus: É. Geoffroy St.-Hilaire, 1803:194; part (description of *guyannensis*); not *Mus* Linnaeus.

Echimys: Desmarest, 1817b:58; part (listing of *cayennensis* Desmarest); not *Echimys* F. Cuvier.

Loncheres: Lichtenstein, 1830:plate 36, Fig. 2; figure of *longicaudatus*, not *myosurus* Lichtenstein (= *Trinomys setosus* [Desmarest]) as indicated.

Echinomys: Wagner, 1843a:341; part (listing of *cayennensis* Desmarest and *longicaudatus* Rengger); unjustified emendation of *Echimys* F. Cuvier.

Thricomys: E.-L. Trouessart, 1897:607; part (listing of *brevicauda* Günther); incorrect subsequent spelling of, but not *Thrichomys* E.-L. Trouessart.

Proechimys J. A. Allen, 1899c:264; type species, *Echimys trinitatis* J. A. Allen and Chapman, by original designation.

Tricomys: E.-L. Trouessart, 1904:504; part (listing of *brevicauda* Günther); incorrect subsequent spelling of, but not *Thrichomys* E.-L. Trouessart.

Proechinomys Elliott, 1904b:385; renaming of *Proechimys* J. A. Allen, based on the belief that it was misspelled.

Hoplomys: Moojen, 1948:315; listing of *hoplomyoides* Tate; not *Hoplomys* J. A. Allen.

We organize the species accounts of *Proechimys* around the species groups defined by Patton (1987), although these were established based simply on character similarity rather than on an explicit phylogenetic hypothesis. One group recognized includes species described subsequent to Patton's analysis (M. N. F. da Silva 1998), and other species that do not fit comfortably into these groups are treated separately. Phylogenetic data based on limited mtDNA sequences generally support the groups recognized (M. N. F. da Silva 1998; Patton et al. 2000; Weksler et al. 2001; Bonvicino, Lemos, and Weksler 2005; Bonvicino, Otazu, and Vilela 2005). However, while a reasonable understanding of species limits and distribution is now available for some geographic regions (e.g., western Amazonia, Guianan region, northern Venezuela), our understanding of the number of taxa for other large areas

(particularly trans-Amazonian Colombia and Ecuador) has progressed little since the initial taxon descriptions a century ago.

In the accounts that follow, 22 species of *Proechimys* are recognized. This list, however, clearly underestimates the actual number of species in the genus as even the limited molecular sequence data now available delineate deeply divergent phylogenetic clades within several of these taxa (Patton et al. 2000; Bonvicino, Otazu, and Vilela 2005) and substantial karyotypic differences characterize others (Reig, Aguilera et al. 1980; A. L. Gardner and Emmons 1984; Patton et al. 2000; Bonvicino, Otazu, and Vilela 2005; T. Machado et al. 2005; N. A. B. Ribeiro 2006). These common and easily trapped rats are ripe for both field and museum studies that must necessarily associate karyotypes and molecular sequences to defined morphological entities, especially with reference to holotypes and type series, so that available names can be properly assigned and truly new forms recognized and properly described.

REMARKS: G. A. S. Carvalho and Salles (2004) questioned the monophyly of all species that have been traditionally included within the genus *Proechimys* (Moojen 1948 [subgenus *Proechimys*]; Patton 1987; Woods and Kilpatrick 2005), suggesting that some are more closely related to *Hoplomys* while others are aligned to *Trinomys*. Because most of the characters that support their phylogenetic hypothesis are details of the cheek teeth, especially construction of folds and contact between loph(id)s, it is perhaps not surprising that highly varying fold number encountered among species of *Proechimys* partitions this genus into groups that align with the high fold count of *Hoplomys* and the low fold count of *Trinomys*. An expanded molecular analysis that includes the full range of taxa of all three genera is needed to resolve both generic limits and species assignments.

The holotype of Brandt's *Mus* (*Holochilus*) *langsdorffii* is now recognized to represent a species of *Proechimys* (Voss and Abramson 1999) rather than a sigmodontine rodent. This name, however, cannot be connected to any extant member of this large genus until adequate comparisons are made between the currently recognized species and the type specimen (a skin with skull, lacking mandibles, cataloged as ZINRAS 218). Once accomplished, *langsdorffii* will likely become a senior synonym of one of the species we recognize below.

KEY TO THE SPECIES GROUPS OF *PROECHIMYS*:

1. Head and body length short (<185 mm); pelage harsh to touch, aristiform spines with blunt tips but moderate to broad in width; distribution in western Amazonia......
.... *gardneri* species group (*Proechimys gardneri*, *Proechimys kulinae*, and *Proechimys pattoni*)

1'. Head and body length larger (>200 mm); pelage varying from very harsh to soft, aristiform spines usually with whip-like tip, rarely blunt; distribution including full range of genus 2

2. Head and body length very large, up to 300 mm 3

2'. Head and body length moderate to large, usually <250 mm. 4

3. Dorsal color sandy fawn; pelage soft to touch, aristiform spines narrow; rostrum short and broad; floor of infraorbital foramen smooth; mesopterygoid fossa narrow and deep, reaching to M2; never four folds on upper cheek teeth, always two folds on lower m2 and m3. . . . .
. . . *decumanus* species group (*Proechimys decumanus*)

3'. Dorsal color reddish brown; pelage stiff, aristiform spines moderate in width; rostrum long and narrow; floor of infraorbital with groove; mesopterygoid fossa moderately wide, reaching only to M3; four folds commonly on upper cheek teeth, rarely two folds on m2 and m3. . . . . . .
. . . . *goeldii* species group (*Proechimys goeldii*, *Proechimys quadruplicatus*, *Proechimys steerei*)

4. Distribution non-Amazonian (western Peru, Ecuador, Colombia; northern Colombia and Venezuela [except *Proechimys hoplomyoides*]) 5

4'. Distribution Amazonian, Guianan, or central Brazil . . . 7

5. Dorsal color grayish; pelage very soft, aristiform spines narrow with long whip-like tip; tail with very narrow scale annuli (13–16 per cm); rostrum short and broad; postorbital process of zygoma well developed; upper and lower molars always with two folds. . . . .
. . . . . . *canicollis* species group (*Proechimys canicollis*)

5'. Dorsal color reddish yellow to reddish brown; pelage stiff, aristiform spines of moderate width or broad; tail with moderate to large scales (7–12 per cm); rostrum long and narrow; postorbital process of zygoma obsolete; always three (or more) folds on upper cheek teeth, usually three on lower cheek teeth 6

6. Dorsal color reddish yellow; tail indistinctly bicolored; incisive foramina oval in shape, anterior palate typically smooth; baculum long but stout . . . .
. . . . *trinitatis* species group (*Proechimys chrysaeolus*, *Proechimys guairae*, *Proechimys hoplomyoides*, *Proechimys mincae*, *Proechimys trinitatis*)

6'. Dorsal color reddish brown; tail sharply bicolored; incisive foramina lyrate in shape, anterior palate with median ridge; baculum short and broad with well-developed apical wings . . . . *semispinosus* species group (*Proechimys semispinosus*, *Proechimys oconnelli*)

7. Body size moderate (head and body length <220 mm); pelage very stiff, aristiform spines very broad and blunt; floor of infraorbital foramen with groove and well-developed lateral flange . . . . *echinothrix* species group (*Proechimys echinothrix*)

7'. Body size moderate to large (head and body length >220 mm); pelage soft to moderately stiff, aristiform spines narrow to moderate in width, with whip-like tip; floor of infraorbital foramen smooth or with weakly developed groove 8

8. Tail proportionately short (60–70% head and body length) and indistinctly bicolored; tail scales moderate, with annuli 9–10 per cm; pelage stiff, aristiform spines moderately wide; incisive foramina strongly lyrate in shape, anterior palate with strongly developed grooves and median ridge; mesopterygoid fossa broad and shallow; baculum massive and broad, with apical wings or extensions. . . . .
. . . . . *longicaudatus* species group (*Proechimys brevicauda*, *Proechimys cuvieri*, *Proechimys longicaudatus*)

8'. Tail proportionately long (typically >80% head and body length) and sharply bicolored; tail scales small, with annuli 10–13 per cm; pelage may be stiff, with aristiforms moderately broad, or soft, with narrow aristiforms; incisive foramina oval in shape, anterior palate smooth or with only weak grooves and no median ridge; mesopterygoid fossa narrow and deep; baculum long and narrow, without apical extensions or wings. 9

9. Tail proportionately very long (>85% head and body length); plantar surface of hindfeet with five pads, lacking hypothenar; pelage soft to touch, aristiform spines narrow, terminated with whip-like tip; range in western Amazonia . . . . *simonsi* species group (*Proechimys simonsi*)

9'. Tail shorter (<85% head and body length); plantar surface of hindfeet with six pads, including hypothenar; pelage stiff to touch, aristiforms of moderate width, but with whip-like tip; range in Guianan region, eastern Amazonia, and central Brazil . . . .*guyannensis* species group (*Proechimys guyannensis*, *Proechimys roberti*).

Proechimys canicollis species group

According to Patton (1987), this group is monotypic, containing the single species *P. canicollis* (J. A. Allen).

Proechimys canicollis (J. A. Allen, 1899)
Colombian Spiny Rat
SYNONYMS:

Echimys canicollis J. A. Allen, 1899b:200; type locality "Bonda, Santa Marta District," on Río Manzanares, where joined by Quebrada Matogiro, 9 mi E of Santa Marta (Paynter 1997), Magdalena, Colombia.

[*Proechimys*] *canicollis*: J. A. Allen, 1899c:264; first use of current name combination.

Proechimys canicollis: J. A. Allen, 1904c:440; name combination.

DESCRIPTION: Moderate sized for genus, with largest individuals reaching head and body length approximately 225 mm, and with proportionately short tail (75% of head and body length). Pelage relatively soft, sparsely intermixed with weakly developed aristiform spines. Aristiform hairs long (22–25 mm) and quite narrow (0.02–0.03 mm), terminating in long whip-like tip. Dorsal color pale yellowish brown or pale golden brown sprinkled with black-tipped spines on back and paler and more gray on sides. Top of head and nape grayish, varied with black; sides of head and neck clear gray, extending onto sides of throat but becoming paler on cheeks. Ventral color white along midline from throat to inguinal region, but gray lateral bands encroach on midline, completely coalescing across belly in some specimens; sides and chin and jowls gray. Insides of limbs white; white of hind limbs may be continuous across ankles to meet whitish-gray dorsal surface of hindfeet. Ears broad but short, brown, and appear naked. Tail indistinctly bicolored, blackish above and dull-flesh colored below, and moderately haired, with hairs partially concealing narrow scale annuli (13–16 per cm at midpoint).

Skull moderate in size and, with exception of distinctively short and broad rostrum, conforms to general shape of other species in genus. Temporal ridges absent, or extend only weakly from supraorbital ledge onto anterior parietals. Incisive foramina broad and oval in shape, with weakly developed posterolateral flanges and thus an anterior palate that exhibits only faint grooves; premaxillary portion of septum broad and long, occupying at least half of opening; maxillary portion moderately to weakly developed, always in contact with premaxilla, and sometimes with a keel that extends limitedly onto anterior palate; vomerine portion of septum usually hidden from view. Floor of infraorbital foramen smooth; few specimens with barely perceptible groove. Mesopterygoid fossa relatively deep, extending to anterior half of M3s, and forming acute angle, averaging 54°. Postorbital process of zygoma moderately well developed, and formed by squamosal or of squamosal and jugal in equal frequency. Cheek teeth with fewest counterfolds of any species; counterfold formula uniform 3–2–2–2 for upper series and 2–2–2–2 for lower; only species of *Proechimys* with only two folds on 4th lower premolar.

Baculum relatively short (sample length 7.74–8.11 mm) and stout (mean distal width 2.64–22.80 mm; mean proximal width 2.55–2.78 mm) with rounded base, weakly concave sides, and rather flat distal tip with only weakly

developed apical wings (Patton 1987); general size and characters similar to members of *goeldii*-species group.

DISTRIBUTION: *Proechimys canicollis* inhabits the foothills and flatlands around the Sierra de Perijá and Sierra de Santa Marta in northeastern Colombia and northwestern Venezuela.

SELECTED LOCALITIES (Map 518): COLOMBIA: Atlántico, Ciénaga de Guájaro (Hershkovitz 1948a); Bolívar, San Juan de Nepumoceno (Patton 1987); Cesar, Río Guaimaral (= Río Garupal) (Patton 1987); Magdalena, Bonda (type locality of *Echimys canicollis* J. A. Allen), El Orinodo (Hershkovitz 1948a); La Guajira, Villanueva (USNM 280145). VENEZUELA: Zulia, Perijá (Patton 1987), Río Cachiri (Corti and Aguilera 1995).

SUBSPECIES: *Proechimys canicollis* is monotypic.

NATURAL HISTORY: This species occurs in dry tropical forest where it may be sympatric with both *P. mincae* and *P. chrysaeolus*. Handley (1976) reported that all captures in northwestern Venezuela were on the ground and equally in moist evergreen and dry deciduous forest, cropland, or orchards.

REMARKS: A. L. Gardner and Emmons (1984) placed *P. canicollis* in their *brevicauda*-group based on bullar

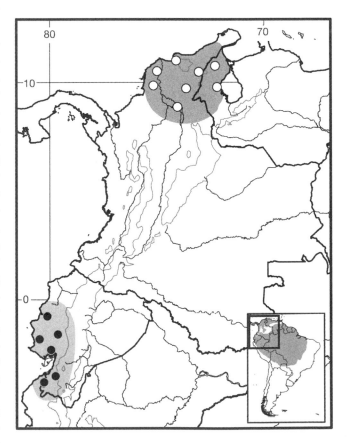

Map 518 Selected localities for *Proechimys canicollis* (○) and *Proechimys decumanus* (●). Contour line = 2,000 m.

septal patterns, but Patton (1987) considered it the sole member of its own group, not closely related to any other species in the genus, and "one of the more readily recognizable in the entire genus" by virtue of its reduced counterfold pattern and distinctive, almost bicolored, dorsal color pattern with the gray head and yellow brown back and rump. The karyotype consists of a $2n = 24$ and FN = 44, with minor differences in the number of arms in an autosomal pair between samples from northwestern Venezuela (Aguilera et al. 1979) and topotypes from Bonda, northeastern Colombia (A. L. Gardner and Emmons 1984). This species is positioned at the base of a group of nine other species of spiny rats in a tree based on allozyme electrophoretic characters (Patton and Reig 1990).

Proechimys decumanus species group
As currently understood, the *decumanus*-group is monotypic, and with no apparent close relatives among other spiny rats (Patton 1987).

Proechimys decumanus (Thomas, 1899)
Pacific Spiny Rat
SYNONYMS:
Echimys decumanus Thomas, 1899c:282; type locality "Chongon, Guayas Province, west of Guayaquil, Ecuador," ca. 100 m, 25 km W of Guayaquil (Paynter 1993).
[*Proechimys*] *decumanus*: J. A. Allen, 1899c:264; first use of current name combination.
Proechimys [(*Proechimys*)] *cayennensis decumanus*: Ellerman, 1940:120; name combination.
[*Proechimys semispinosus*] *decumanus*: Moojen, 1948:316; name combination.

DESCRIPTION: Large spiny rat, head and body length of adults 260–300 mm, but with proportionately short tail (about 65% of head and body length). Dorsal color coarsely grizzled sandy fawn with intermixed with darker hairs. Sides paler and grayer; face grizzled gray. Inner sides of forelimbs and hips pure white, with hairs white to base. Upper surfaces of forefeet and hindfeet white or slightly washed yellow, and white on inner thighs continuous across ankle onto foot. Aristiform spines long (length 25–27 mm) and thin (width 0.5 mm) and tipped with long, flexible filament; pelage thus inconspicuously spiny to eye and touch. Plantar pads of hindfoot large, particularly both thenar and hypothenar pads. Tail bicolored, dark above and pale below, and uniformly but thinly haired with scale annuli narrow (13 per cm at midpoint) and visible to eye.

Skull large and elongated, with short and broad rostrum. Temporal ridges moderately developed and either continuous or interrupted across parietals from posterior end of supraorbital flange. Incisive foramina oval to slightly lyrate

in shape and large, with weakly developed posterolateral flanges and weak grooves extending onto anterior palate; premaxillary portion of septum long, but tapers posteriorly and in direct contact with maxillary portion, which is varyingly developed as either thin, spiculate bone or broad shelf, often perforated by small foramen, and with either no or only limited keel that extends onto anterior palate; vomer not visible along septum. Floor of infraorbital foramen flat, rarely with limited evidence of lateral flange indicating passage of maxillary nerve. Mesopterygoid fossa moderately deep, reaching to anterior half of M3s, and rather narrow, with its angle averaging 53°. Postorbital process of zygoma obsolete. Cheek teeth simple, typically with three counterfolds on all upper teeth (although rarely only two on M3) and with three folds on pm4 and m1 but only two on m2 and m3; counterfold formula 3-3-3-(2)3/3-3-2-2.

Baculum long but stout, among longest of any *Proechimys*, mean length range: 9.2–11.8 mm, distal width 2.9–3.8 mm, proximal width 2.8–3.4 mm. In both size and shape, baculum most similar to those *trinitatis*-group members; sides nearly parallel, base rounded, and distal tip slightly expanded with weak median depression (Patton 1987).

DISTRIBUTION: *Proechimys decumanus* is known only from the tropical dry forests of southwestern Ecuador and adjacent northwestern Peru.

SELECTED LOCALITIES (Map 518): ECUADOR: El Oro, Santa Rosa (Patton 1987); Guayas, Chongón (type locality of *Echimys decumanus* Thomas); Los Rios, Hacienda Pijigual (Patton 1987); Manabí, Bahía de Caraques (Patton 1987); Santa Elena, Manglar Alto (UMMZ 80040). PERU: Piura, Quebrada Bandarrango (Patton 1987; not located); Tumbes, Matapalo (FMNH 81197).

SUBSPECIES: *Proechimys decumanus* is monotypic.

NATURAL HISTORY: An inhabitant of tropical semi-deciduous forest, this large spiny rat has not been studied for any aspect of its ecology, reproduction, or behavior. It may be sympatric, or nearly so, with *Proechimys semispinosus* in SW Ecuador, with the latter presumably inhabiting patches of more mesic, evergreen forest.

REMARKS: Thomas (1899c) considered this species closely allied to *P. semispinosus* (Thomas), and A. L. Gardner and Emmons (1984) placed it with in their broadly encompassing *brevicauda*-group. In bacular characters, *P. decumanus* appears most similar to members of the *trinitatis*-group, and not at all close to any species within the *semispinosus* or *longicaudatus* groups (Patton 1987). A. L. Gardner and Emmons (1984) described a chromosomal complement of $2n = 30$ and FN = 54 for topotypes, one indistinguishable from that of specimens of *P. semispinosus* from Costa Rica, but different from those of sympatric, or near-sympatric *P. semispinosus* from Ecuador.

Proechimys echinothrix species group

The single species within this group is the recently described *Proechimys echinothrix* Silva, which apparently has no close relatives within the genus (M. N. F. da Silva 1998). For the present, therefore, we list this as the sole member of another monotypic species group.

Proechimys echinothrix M. N. F. da Silva, 1998
Stiff-spined Spiny Rat

SYNONYM:

Proechimys echinothrix M. N. F. da Silva, 1998:441; type locality "Colocação Vira-volta, left bank Rio Juruá on Igarapé Arabidi, affluent of Paraná Breu, 66°14′W, 3°17′S, Amazonas, Brazil."

DESCRIPTION: Moderately large for genus, length of head and body 141–245 mm, with robust body, long ears (length 19–28 mm), moderately and proportionately long tail (length 106–209 mm, 77% of head and body length), and large hindfeet (length 41–54 mm). Dorsal color uniformly reddish-brown, coarsely streaked on back and sides with varying amounts of black; interspersed with heavy, dark brown guardhairs that make mid-dorsum darker, which grades evenly onto brighter and paler sides. Aristiform hairs long (length 19.3–24.1 mm) and much broader (width 1.2–1.6 mm) than other sympatric species, with strong and blunt tips conspicuous to eye and touch, especially in mid-dorsal region. Venter and inner surfaces of limbs pure white. Tail indistinctly bicolored, dark above and white below; well haired, with scales nearly completely obscured from view; scales small, 12 annuli per cm at midpoint. Hindfeet nearly unicolored white on dorsal surfaces; all six pads present on plantar surfaces, but hypothenar weakly developed in relation to thenar pad.

Skull moderately large (condyloincisive length 39.8–49.0 mm), with long and narrow rostrum and well-developed supraorbital ledge extending over orbits but discontinuous across parietals as weakly developed temporal ridge. Zygoma usually lacks postorbital process or, if present, it is low and rounded with equal contributions by jugal and squamosal. Well-developed groove with lateral flange present on floor of infraorbital foramen. Incisive foramina ovate to lyrate in shape, with posterolateral margins mostly flat, or only weakly flanged with very shallow grooves extending onto anterior palate that lacks median ridge; premaxillary portion of septum long and narrow, extending between one-half and two-thirds of length of opening; maxillary portion typically attenuate, with weak to no contact with premaxillary portion; vomer visible in most specimens. Mesopterygoid fossa moderately deep, extending to front of M3s, but broad, with sides defining angle about 70°. Median number of folds on upper cheek

teeth three; m3 occasionally with two; counterfold formula 3–3–3–3/3–3–3– (2)3.

Baculum massive and relatively short (length 8.3–8.6 mm); shaft broad with thick and expanded base (proximal width 4.8–5.4 mm), distal end with pair of divergent apical extensions separated by shallow median depression (distal width 4.2–5.1 mm; M. N. F. da Silva 1998; Patton et al. 2000). M. N. F. da Silva (1998) described the soft anatomy of male phallus.

DISTRIBUTION: *Proechimys echinothrix* is known only from the western Amazon Basin of Brazil, west of the Rio Madeira and on both sides of the Rio Solimões (Patton et al. 2000:230, Fig. 145; Schetino 2008:12, Fig. 2), possibly extending into Colombia in the upper Rio Negro drainage.

SELECTED LOCALITIES (Map 519): BRAZIL: Amazonas, Barro Vermelho (M. N. F. da Silva 1998), Colocação Vira-Volta, left bank Rio Juruá (type locality of *Proechimys echinothrix* Silva), Comunidade Colina, Rio Tiquié (Patton et al. 2000), Lago Vai-Quem-Quer, right bank Rio Juruá (M. N. F. da Silva 1998), Macaco, Rio Jaú (Patton et al. 2000), alto Rio Urucu (M. N. F. da Silva 1998), Tambor, Rio Jaú (Patton et al. 2000), Comunidade Bela Vista (Schetino 2008).

SUBSPECIES: *Proechimys echinothrix* is monotypic.

NATURAL HISTORY (from Patton et al. 2000): This species typically inhabits upland, nonseasonally inundated (*terra firme*) forest, although it may be found along the margins of flooded *várzea* or *igapó* forest. Pregnant fe-

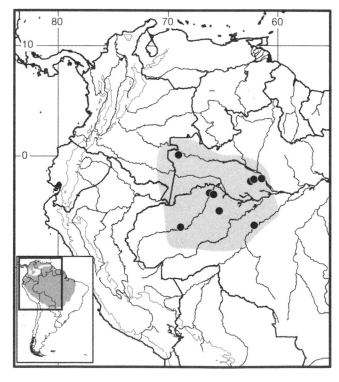

Map 519 Selected localities for *Proechimys echinothrix* (●). Contour line = 2,000 m.

males have been taken in both dry and wet seasons in western Brazil, where modal litter size was two (range: 1–3).

REMARKS: *Proechimys echinothrix* is one of the most readily recognizable among the sympatric assemblage of spiny rats of western Brazil, primarily due to the dense mid-dorsal cover of broad, stout, and blunt aristiform spines. Populations on opposite sides of the Rio Solimões are markedly distinct in mtDNA sequences (Patton et al. 2000), which might signal species-level differences. Both groups of populations, however, share the same karyotype and set of morphological attributes.

Proechimys gardneri species group

This complex includes three recently described species from the western Amazon of Brazil, Peru, and northern Bolivia (M. N. F. da Silva 1998). These are all of similar body size, with small ears, short hindfeet, and a similarly proportioned tail. Species in this group differ markedly in characters of the cranium, phallus (including baculum), and karyotype, yet appear to form a monophyletic clade based on mtDNA sequence data (M. N. F. da Silva 1998; Patton et al. 2000). The ranges of the three species are nonoverlapping, as all three replace one another along the length or on opposite banks of the Rio Juruá in western Amazonian Brazil.

KEY TO THE SPECIES OF THE *PROECHIMYS GARDNERI* SPECIES GROUP:

1. Aristiform spines on mid-dorsum thin so that pelage is relatively soft to the touch; tail not sharply bicolored; baculum stout and broad, with well-developed apical wings . . . . *Proechimys pattoni*
1'. Aristiform spines on mid-dorsum thick so that pelage is distinctly coarse to the touch; tail sharply bicolored; baculum without well-developed apical wings 2
2. Six tubercles on plantar surface of hindfeet; scales of tail annuli small (11 per cm); postorbital process of zygoma obsolete; mesopterygoid fossa long and narrow, reaching to M2s; baculum broad with weak apical wings . . . . *Proechimys gardneri*
2'. Five tubercles on plantar surface of hindfeet (missing hypothenar); scales of tail annuli large (nine per cm); postorbital process of zygoma well-developed and formed by squamosal; mesopterygoid fossa short but narrow, reaching only M3s; baculum short and narrow, without apical wings. . . . .*Proechimys kulinae*

Proechimys gardneri M. N. F. da Silva, 1998
Gardner's Spiny Rat

SYNONYM:

Proechimys gardneri M. N. F. da Silva, 1998:460; type locality "Altamira, right bank Rio Juruá, 68°54'W, 6°35'S, Amazonas, Brazil."

DESCRIPTION: One of three small species in western Amazonia, head and body length 154–209 mm, tail length 88–152 mm, hindfoot length 32–45 mm, and ear length 18–24 mm. Tail relatively short, about 70% of head and body length. Body color reddish brown or auburn, coarsely streaked with varying amounts of black both on dorsum and sides; midback appears darker, especially over rump, due to presence of heavy, dark brown aristiform spines. Aristiforms short (length 15.4–19.9 mm) and moderately wide (0.7–1.0 mm), with blunt tip, providing distinctly coarse or stiff texture to dorsal pelage. Venter and chin pure white; most specimens with white lips. Tail sharply bicolored, dark brown above and cream to white below; scales relatively small (11 annuli per cm at midpoint), but not completely hidden by hair. Inner surfaces of hindlimbs pure white, extending across ankle onto foot so that ankle lacks complete circular dark band. Dorsal surfaces of hindfoot yellowish white, not pure white, often with distal parts of toes brownish. Plantar surface of hindfoot with six tubercles.

Skull small (condyloincisive length 34.5–46.1 mm) and delicate, with relatively long and narrow rostrum, and beaded supraorbital ledges above orbits, which extend posteriorly as weakly developed ridges on anterior parietals. Postorbital process of zygoma obsolete. Floor of infraorbital foramen smooth, lacking ventral groove. Incisive foramina ovate to slightly lyrate in shape, with posterolateral margins lying flat or weakly flanged, and outlining shallow groove on anterior palate; maxillary portion of septum dorsoventrally compressed posteriorly and narrow anteriorly, visible over almost half length of foraminal opening, and fully connected to broad premaxillary portion about half foramina length; vomer not visible on ventral margin of septum. Palate smooth, without median ridge. Mesopterygoid fossa long, extending to middle of M2s, and narrow, with angle of indentation averaging 61°. Cheek teeth very small, maxillary toothrow 6.9–8.2 mm. Upper teeth typically with three folds, M3 occasionally with two; pm4 with four (occasionally three) folds, m1 and m2 with three, and m3 with either two or three; counterfold formula 3–3–3– (2)3/(3)4–3–3–2(3).

Baculum massive and relatively long (length 6.5–8.6 mm), especially relative to body size, with short, broad, and distolaterally directed apical extensions separated by shallow median depression; midshaft relatively broad and base thick and expanded (distal width 3.1–4.2 mm, proximal width 4.7–5.1 mm; M. N. F. da Silva 1998; Patton et al. 2000).

DISTRIBUTION: *Proechimys gardneri* is known only from only nine localities in western Amazonian Brazil and northern Bolivia, between the Rio Juruá and Rio Madeira (M. N. F. da Silva 1998; Schetino 2008).

SELECTED LOCALITIES (Map 520): BRAZIL: Amazonas, Altamira, right bank Rio Juruá (type locality of *Proechimys gardneri* M. N. F. da Silva), alto Rio Urucu

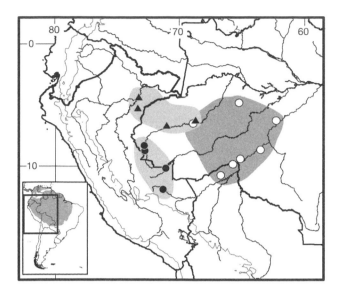

Map 520 Selected localities for *Proechimys gardneri* (○), *Proechimys kulinae* (▲), and *Proechimys pattoni* (●). Contour line = 2,000 m.

(M. N. F. da Silva 1998), Comunidade Bela Vista (Eler et al. 2012); Rondônia, Abunã, left bank Rio Madeira (Schetino 2008), Jirau, right bank Rio Madeira (Schetino 2008). BOLIVIA: Pando, San Juan de Nuevo Mundo (USNM 579615), Río Negro (USNM 579617).

SUBSPECIES: *Proechimys gardneri* is monotypic.

NATURAL HISTORY: This species has only been observed in upland, nonseasonally flooded (*terra firme*) forest. Pregnant females are known from the rainy season in western Brazil, although age structure suggests that breeding extends into the dry season. Modal litter size is two, with a range of one to three young (Patton et al. 2000). Subadults and juveniles were collected in Bolivia in July (L. H. Emmons, pers. comm.). Mendonça et al. (2011) reported this species in the stomach contents of the colubrid snake *Spilotes pullatus*.

REMARKS: This species is similar in most characters to *P. pattoni*, with which it is parapatric in distribution in western Brazil. On the Rio Juruá the two species share the same $2n = 40$, FN = 56, karyotype but differ in details of the baculum (M. N. F. da Silva 1998) and are strongly differentiated in mtDNA sequences (M. N. F. da Silva 1998; Patton et al. 2000). The karyotype of *P. gardneri* from the mid Rio Madeira is slightly different, with $2n = 40$, FN = 54 (Eler et al. 2012).

Proechimys kulinae M. N. F. da Silva, 1998
Kulina Spiny Rat

SYNONYM:

Proechimys kulinae M. N. F. da Silva, 1998:451; type locality "Seringal Condor, left bank Rio Juruá, 70°51′W, 6°45′S, Amazonas, Brazil."

DESCRIPTION: Another of smallest species in genus, head and body length 144–190 mm, tail length 95–140 mm, hindfoot length 38–44 mm, and ear length 17–23 mm. General appearance of slight build, small ears, short hindfeet, and short tail (about 70% of head and body length). Dorsal color uniform reddish brown, coarsely streaked with varying amounts of black on back and sides. Dorsal pelage interspersed with moderately short (16.4–20.0 mm) and thick (0.8–1.0 mm), dark brown aristiform hairs forming darker medial band contrasting with sides of body. Tip of each aristiform blunt. Venter, chin, and undersurfaces of forelimbs and hindlimbs pure white; upper lips dark, generally without patches of white hair; tarsal joint either ringed by dark and rusty-colored hair or interrupted by white hair confluent with that of undersurface of hind limbs. Dorsal surfaces of hindfeet, including digits, white, tinged golden in some individuals. Plantar surfaces typically with five tubercles, lacking hypothenar pad in most specimens. Tail appears almost naked, distinctly bicolored with dark brown above and white below; scales larger than in other species of group (nine annuli per cm at midpoint).

Skull relatively small (condyloincisive length 33.9–40.8 mm), with short and narrow rostrum, and well-developed supraorbital ledge extending onto anterior portion of parietals. Postorbital process of zygoma well developed and formed mostly by squamosal. Floor of infraorbital foramen generally smooth, a demonstrable groove for maxillary nerve. Incisive foramina mostly square to oval in shape, with nearly flat posterolateral margins; anterior palate smooth, lacking both grooves extending posteriorly from the incisive foramina and median ridge; premaxillary portion of septum short, extending for less than half length of foramen; maxillary portion variable, attenuate to expanded anteriorly, and usually in contact with premaxillary part; vomer either completely enclosed or only barely visible. Mesopterygoid fossa narrow, with angle of indentation averaging 57°, moderately deep, usually extending well into M3s. Upper cheek teeth with three folds, M3 sometimes with two; four folds on pm4, three on m1 and m2, and two on m3; counterfold formula 3-3-3– (2)3/4-3-3-2.

Baculum elongate (length 5.4–8.2 mm) and relatively narrow (proximal width 1.6–2.1 mm, distal width 1.7–2.6 mm), with stout and short apical extensions (M. N. F. da Silva 1998; Patton et al. 2000).

DISTRIBUTION: *Proechimys kulinae* is known only from northeastern Peru, south of the Río Amazonas, and western Brazil along both sides of the central Rio Juruá.

SELECTED LOCALITIES (Map 520): BRAZIL: Amazonas, Barro Vermelho (M. N. F. da Silva 1998), Seringal Condor (type locality of *Proechimys kulinae* M. N. F. da Silva). PERU: Loreto, Jenaro Herrera (AMNH 276708), Nuevo San Juan (MUSM 13340), San Pedro (MVZ 198489).

SUBSPECIES: *Proechimys kulinae* is monotypic.

NATURAL HISTORY (from Patton et al. 2000): This species is only known from primary or second growth upland, seasonally noninundated (*terra firme*) forest. Pregnant females were taken only in the dry season in the Rio Juruá basin; the modal litter size was one young, maximum number two.

REMARKS: Substantial differences in mtDNA sequences are present between sampled localities, with samples from the central Rio Juruá in western Brazil and those from northeastern Peru strongly separable from those toward the mouth of the Rio Juruá. Despite these molecular differences, all known specimens share the same general morphology, including bacular type, and karyotype (Patton et al. 2000). The chromosomal complement is $2n = 34$, $FN = 52$ (M. N. F. da Silva 1998). Aniskin (1993) described the same karyotype for specimens identified as *Proechimys* sp. from northeastern Peru; as noted by M. N. F. da Silva (1998), these likely represent *P. kulinae* as she allocated specimens from nearby localities to this species.

Proechimys pattoni M. N. F. da Silva, 1998
Patton's Spiny Rat

SYNONYM:

Proechimys pattoni M. N. F. da Silva, 1998:454; type locality "Igarapé Porongaba, right bank Rio Juruá, 72°47'W, 8°40'S, Acre, Brazil."

DESCRIPTION: Third small species in western Amazonia, equal in size to *P. gardneri* and *P. kulinae* (head and body length 146–193 mm); individuals slender with short hindfeet (37–43 mm), short ears (20–22 mm), and proportionally medium length tail (106–141 mm; 70% of head and body length). General body colorfrom reddish brown and auburn, coarsely streaked with varying amounts of black both on dorsum and sides. Interspersed dark brown aristiform spines give dorsum dark aspect, but contrast between color of back and sides not sharp. Dorsal pelage stiff to touch, withblunt tipped and shorter (13.2–15.9 mm), narrower (0.5–0.8 mm) aristiform spines compared with other species in group. Color of venter, chin, and upper lips pure white. Dark rings circumscribe tarsal joints, interrupting white of inner hind limbs and dorsal surfaces of hindfeet. Plantar surface of hindfeet with six pads. Dark brown dorsal surface of tail grades evenly, rather than sharply, with paler brown to cream color of its ventral surface. Tail scales small, 11 annuli per cm at midpoint.

Skull small (condyloincisive length 36.1–39.5 mm) and delicate, with overhanging supraorbital ledges but with only weakly developed beading extending onto temporal regions. Low but distinct postorbital process of zygoma present, usually formed solely by squamosal. Floor of infraorbital foramen smooth, lacking even hint of groove.

Incisive foramina ovate to nearly square, with flat posterolateral margins, an attenuate or dorsoventrally compressed maxillary portion of septum, and broad and short premaxillary portion usually not in contact with maxillary portion. Palate smooth, without median ridge. Mesopterygoid fossa long and narrow, angle of indentation acute (50° to 60°), and may penetrate to M2s. Cheek teeth markedly small, with entire toothrow <7.5 mm in length. Upper teeth with three lateral folds; lower cheek teeth with three folds, but pm4 sometimes with four and m3 occasionally with two; counterfold formula 3–3–3–3/3(4)–3–3–(2)3.

Baculum massive (length 7.4–9.1 mm, proximal width 4.2–5.1 mm, distal width 3.1–4.2 mm), especially in proportion to body size; shaft broad, base thick and expanded, divergent apical extensions long and separated by wide and deep median depression (Patton and Gardner 1972; M. N. F. da Silva 1998; Patton et al. 2000).

DISTRIBUTION: *Proechimys pattoni* is known only from two localities in the headwaters of the Rio Juruá in western Amazonian Brazil and adjacent parts of eastern and southern Peru (M. N. F. da Silva 1998).

SELECTED LOCALITIES (Map 520): BRAZIL: Acre, Sobral (M. N. F. da Silva 1998), Acre, Igarapé Porongaba (type locality of *Proechimys pattoni* M. N. F. da Silva). PERU: Madre de Dios, Pakitza (M. N. F. da Silva 1998); Puno, Putina Punco (M. N. F. da Silva 1998); Ucayali, Balta (Patton and Gardner 1972, as *P. guyannensis*).

SUBSPECIES: *Proechimys pattoni* is monotypic.

NATURAL HISTORY (from Patton et al. 2000): This species is known only from upland, nonseasonally flooded (*terra firme*) forest, where it might be found in undisturbed forest or in disturbed areas dominated by bamboo. Pregnant females were taken in the rainy season in western Brazil; modal litter size was two, with a range from one to two young.

REMARKS: This species was described and figured by Patton and Gardner (1972) based on specimens from eastern Peru, but under the name *P. guyannensis*. The karyotype is $2n = 40$, $FN = 56$ (Patton and Gardner, 1972), similar to that of *P. gardneri* (M. N. F. da Silva 1998).

Proechimys goeldii species group

Patton (1987) included 12 named taxa in this complex, but was unsure of how many of these might represent valid species. Considerable karyotypic and molecular data are now available to help define species limits and geographic ranges. Some of these data have been published (M. N. F. da Silva 1998; Patton et al. 2000), but much remains unpublished. Because of the relatively thorough geographic sampling, especially in regions that encompass most of the type localities of the names Patton (1987) placed in the group, we recognize three species and assign with confidence the remainder of the available names as synonyms to one or the other of these three.

Members of this group are restricted to the lowland rainforest of the Amazon Basin where they most commonly inhabit the seasonally inundated *várzea* or *igapó* forests, in contrast to their sympatric congeners that live in nonseasonally flooded upland, or *terra firme*, forests (Patton et al. 2000). All three species have parapatric ranges, with rivers forming their common boundaries. These species are united by a uniformly large body size; by a common short, stout, but relatively narrow baculum (Patton 1987:316, Fig. 6); by skulls with weakly lyre-shaped or parallel-sided incisive foramina with a short premaxillary portion of the septum, typically attenuate maxillary portion sometimes not in contact with the premaxillary portion, and with the vomer only rarely exposed ventrally (Patton 1987:324, Fig. 15); a moderately narrow mesopterygoid fossa (angle 60° to 68°) that penetrates to the anterior part of M3s (Patton 1987:331, Table 4); and by upper cheek teeth typically characterized by four counterfolds (Patton 1987:332, Fig. 25; 334–335, Table 5).

KEY TO THE SPECIES OF THE *PROECHIMYS GOELDII* SPECIES GROUP:

1. Distributed north of the Marañón-Solimões-Negro rivers in northern Peru, Ecuador, Colombia, north-central Brazil, and southern Venezuela; counterfold pattern of maxillary cheek teeth most commonly 4–4–4–4 . . . . *Proechimys quadruplicatus*
1′. Distributed south or east of the Marañón-Solimões-Negro rivers in Peru, Bolivia, and Brazil; counterfold pattern of maxillary cheek teeth usually 3–3–3–3, but occasionally M3 and M4 with four folds 2
2. Distributed in western Amazonia, west of the Rio Madeira; M3 and M4 commonly with four folds; dpm4 always with four folds; m1, m2 with three folds . . . . *Proechimys steerei*
2′. Distributed in eastern Amazonia, east of the Rio Madeira and mostly south of the Rio Amazonas; M3 and M4 with three folds, only rarely with four; dpm4 often with three folds; m1, m2, and m3 sometimes with two folds . . . . *Proechimys goeldii*

Proechimys goeldii Thomas, 1905
Goeldi's Spiny Rat
SYNONYMS:

Proechimys goeldii Thomas, 1905a:587; type locality "Santarem [= Santarém], Lower Amazon," Rio Tapajós, Pará, Brazil (locality of holotype given as "Santarem, Barras de Tapajoz" in catalog of the Naturhistorisches Museum der Stadt Bern, Switzerland [Güntert et al. 1993]).

Proechimys [(Proechimys)] cayennensis goeldii: Ellerman, 1940:121; name combination.
Proechimys goeldii goeldii: Moojen, 1948:340; name combination.
Proechimys guyannensis hyleae Moojen, 1948:361; type locality "Tauarí, Rio Tapajós, Porto de Moz, Pará, Brazil; approximately 87 kilometers south of Santarem."
Proechimys guyannensis nesiotes Moojen, 1948:363; type locality "Ilha de Manapirí, Rio Tocantins, Pará, Brazil."
Proechimys guyannensis leioprimna Moojen, 1948:364; type locality "Cametá, left bank of Tocantins River, near its mouth, Cametá, Pará, Brazil."

DESCRIPTION: Size moderately large (head and body length 188–271 mm), hindfeet long (44–61 mm), and tail proportionately short (length 100–171 mm; about 65% of head and body length). Dorsal color dark reddish brown strongly mixed with black, especially over mid-back and rump, but specimens from more southern localities distinctly paler and more orangish-red in color. Venter clear white from chin to inguinal region; inner thighs white but separated from dorsal foot color by dark band. Dorsal surface of hindfeet bicolored, outer side dark, inner side pale, a pattern also characteristic of *P. quadruplicatus* and *P. steerei*. Six well-developed plantar pads, thenar and hypothenar large and subequal in size. Tail clothed with short, sparsely distributed hairs, so that it appears naked to eye; scales small, 12 annuli per cm at midpoint. Aristiform spines stout and stiff, length 18–20 mm, width about 1.1 mm in width, tip blunt.

Skull large (greatest skull length 49–68 mm), broad across zygomatic arches (zygomatic breadth 22.4–30.9 mm), with relatively long but broad rostrum (mean length 22.7 mm, mean width 8.3 mm). Older specimens exhibit weakly continuous temporal ridge extending from supraorbital ledge across parietals. Incisive foramina broadly open, weakly lyre-shaped or parallel sided, with posterior margins slightly flanged and extending onto anterior palate forming grooves; premaxillary portion of septum short, usually less than one-half length of opening; maxillary portion varies greatly, from weak and attenuate, perhaps not in contact with premaxillary portion, tospatulated and filling much of posterior opening; maxillary portion with slight keel that continues onto palate as median ridge; vomerine portion typically enclosed in premaxillary sheath and thus not visible. Floor of infraorbital foramen may be smooth, lacking any evidence of groove, or with groove defined by moderately developed. Mesopterygoid fossa moderately broad, defining angle of 65°, penetrating posterior palate to about midpoint of M3s. Postorbital process of zygoma moderately developed and comprised mostly by squamosal. Counterfold pattern of cheek teeth varies, with number of folds decreasing from west to east and from north to south: Western samples with upper cheek teeth 3–(3)4–(3)4–3 and lower cheek teeth 4(rarely 3)-3-3-3 folds; eastern and southern samples with reduced fold number of 3–3-3-3 above and 3(rarely 4)-3-3-3 or 2 below.

Baculum same in general size and shape as those of other members of group (sample mean length 6.58–8.96 mm, proximal width 2.06–2.93 mm, distal width 1.88–2.75 mm); sides straight and parallel, base slightly expanded, and tip with faint development of apical wings and median depression (Patton 1987).

DISTRIBUTION: *Proechimys goeldii* occurs in eastern Amazonian Brazil along both banks of the Rio Amazonas largely east of the Rio Madeira, in the states of Amazonas and Pará, and south along the Tapajós, Xingu, and Tocantins-Araguaia rivers and their tributaries, reaching the Cerrado biome in Mato Grosso state.

SELECTED LOCALITIES (Map 521; from Patton 1987, except as noted): BRAZIL: Amazonas, Cachoeirinha (Schetino 2008), Itacoatiara; Mato Grosso, Reserva Ecológica Cristalino (MVZ 197575), Serra da Chapada, Utiariti; Pará, Baião, Cametá (type locality of *Proechimys guyannensis leioprimna* Moojen), Fazenda Paraíso (Moojen 1948), Itaituba, Óbidos (Moojen 1948), Project Pinkiati Research Station, Kayapo Indigenous Reserve (MVZ 199560), Vilarinho do Monte.

SUBSPECIES: Both Moojen (1948) and Cabrera (1961) listed *steerei* Goldman as a valid subspecies, along with the nominotypical form. We treat *steerei* as a separate species, following Patton et al. (2000). However, the several species-level names we list above as synonyms of *P. goeldii* may warrant formal recognition when appropriate analyses of geographic variation in morphological as well as molecular characters are undertaken.

NATURAL HISTORY: This species has not been studied in the field. Moojen (1948) stated that the holotype of his *leioprimna*, from the Rio Tocantins, was collected in seasonally flooded forest, which is consistent with the general habitat range of the two other species in the group. Moojen (1948) also recorded pregnant females in January for samples collected on the Rio Tapajós (*hyleae* Moojen) and Tocantins (*leioprimna* Moojen). *Proechimys goeldii* may be sympatric with two other species of spiny rats, commonly with *P. roberti* (of the *guyannensis*-group) and less commonly with *P. cuvieri* (of the *longicaudatus*-group).

REMARKS: Thomas (1920f:277) referred specimens from Manacaparu, west of Manaus, and Acajutuba, near Manaus, to *P. goeldii*. These are, however, most likely *P. steerei*, as both localities are within the known distribution of this species, which extends north of the Rio Solimões into the Imerí region between that river and the Rio Negro (see Patton et al. 2000, and the account for *P. steerei*). Moreover, the specimens to which Thomas referred (in

the British Museum and American Museum collections) have the four folds that characterize the cheek teeth of *P. steerei*. Moojen (1948:341–342) also listed specimens from Manaus as *P. goeldii*, but Patton et al. (2000) allocated all specimens from the left bank of the lower Rio Negro to *P. quadruplicatus* based on DNA sequence analyses.

Osgood (1944:199) suggested that *P. goeldii* was a synonym of *P. oris* (= *P. roberti* herein), which he also thought indistinguishable from *P. cayennensis* (= *P. guyannensis*). We agree that *P. oris* shares characters with *P. guyannensis*, and for that reason Patton (1987) placed it within his *guyannensis*-group. However, both *P. oris* (= *P. roberti* herein) and *P. goeldii* are broadly sympatric throughout eastern Amazonia, are readily separable by a number of external and craniodental traits (Patton 1987), and share no close phyletic relationship (R. N. Leite and J. L. Patton, unpubl. DNA sequence data).

Moojen (1948:340) considered *P. goeldii* and *P. steerei* "clearly related," so much so that he included the latter as a subspecies of the former, an opinion followed by Cabrera (1961). While these two species are related, phylogenetic relationships based on mtDNA place *P. goeldii* as a basal node to the sister-species pair of *P. steerei* and *P. quadruplicatus* (Patton et al. 2000).

T. Machado et al. (2005) described and illustrated a karyotype with $2n = 15$, FN = 16, for a specimen from Juruena, Mato Grosso, Brazil (MZUSP 31924), which they identified as "*Proechimys* gr. *goeldii*." From their description of craniodental morphology, this specimen clearly belongs to the *goeldii*-group and is most likely assignable

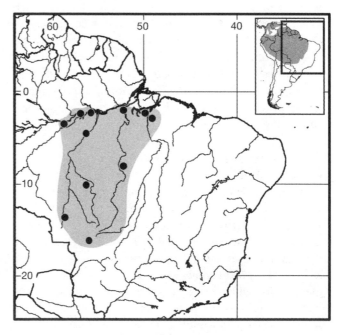

Map 521 Selected localities for *Proechimys goeldii* (●).

to *P. goeldii*, as currently understood, based on specimens that J. L. Patton examined from other localities in Mato Grosso. However, specimens from the Rio Xingu have a quite different karyotype of $2n = 24$, $FN = 44$ (L. H. Emmons, unpubl. data). Available data from mtDNA sequences (R. N. Leite and J. L. Patton, unpubl. data; R. do V. Vilela 2005; Schetino 2008) also suggest two well-defined clades within *P. goeldii*, one of which occurs through eastern Amazonia in an area encompassing the type localities of *goeldii* Thomas, *hyleae* Moojen, *nesiotes* Moojen, and *leioprimna* Moojen. The second clade is known from Mato Grosso state and two additional localities in the middle to upper Rio Madeira in Amazonas and Rondônia states. The likelihood, therefore, is that each clade maps to the different karyotypes described herein. Should these two clades be recognized as separate species based on future studies, there is no name currently available for the Mato Grosso clade.

Proechimys quadruplicatus Hershkovitz, 1948
Napo Spiny Rat

SYNONYMS:

Proechimys quadruplicatus Hershkovitz, 1948a:138; type locality "Llunchi, an island in the Río Napo, about 18 kilometers below the mouth of the Río Coca, [Orellano,] eastern Ecuador."

Proechimys semispinosus amphichoricus Moojen, 1948:344; type locality "Mount Duida, Esmeralda, Amazonas, Venezuela; altitude 325 m."

DESCRIPTION: So similar in external and cranial characters to *P. steerei* that, at least until adequate analyses of morphological characters are made, identification may rest solely on either chromosomes or molecules or on general geographic ranges (Patton 1987; Patton et al. 2000).

Size among largest in genus, with adult weights averaging 450–500 g, head and body length about 211–295 mm, hindfoot length 48–64 mm, ear length 20–28. Tail proportionately short (length 125–202 mm), approximately 70% of head and body length. Overall color ochraceous-orange, darkening along midline from head to rump. Venter may be pure white or lightly washed pale buff; pale thigh stripe extends across ankle onto hindfoot; dorsal surfaces of hindfeet characteristically bicolored, with outside dark longitudinal band encompassing digits IV and V contrasted with pale inner band encompassing at least digits I and II, and usually III. Tail weakly bicolored, appearing nearly naked as scales large and conspicuous, eight annuli per cm at midpoint. Dorsal pelage stiff to touch, with well-developed aristiform spines about 20 mm long and 1.0 mm wide, terminated by weakly developed whip-like tip; distal one-third black, which gives the overall darkened tone to midline of back and rump; spines most well developed on midback, less so on shoulder or over rump.

Skull large (greatest skull length 51.7–70.4 mm) and elongate with long rostrum and heavy supraorbital ridges that extend posteriorly onto parietals as distinct ridges, especially in older individuals. Incisive foramina vary from weakly lyre-shaped to oval, with lateral margins tapering slightly posteriorly or parallel sided; premaxillary portion of septum short, usually one-half or less length of foramen; maxillary portion varies greatly, typically weak and attenuate, often not in contact with premaxillary portion but sometimes broadly spatulated and filling much of opening; maxillary portion may be slightly ridged but never keeled; only rarely does ridge extend posteriorly to form moderately developed grooves on anterior palate. Posterolateral margins of foramen moderately flanged. Specimens for which Patton (1987:324, Fig. 15) illustrated incisive foramina and called *P. steerei* are actually *P. quadruplicatus* as now understood. Moderately developed groove present on floor of infraorbital foramen, although development varies among individuals. Mesopterygoid fossa narrower than either *P. goeldii* or *P. steerei*, with angle averaging 61°, and extends anteriorly to middle of M3s. Counterfold number varies slightly, but characteristic four folds upon which Hershkovitz based his name *quadruplicatus* is most common condition; pattern is 4(3)–4(3)–4(3)–4(3)/4–3(4)–3(4)–3 for samples from Colombia, Ecuador, and northern, but 3–3–3–3/3–3–3–3 in half of specimens examined from Venezuela and northern Brazil (including holotype of *amphichoricus* Moojen).

Baculum comparatively short and stout (sample mean length 7.11–8.28 mm, proximal width 2.45–2.87 mm, distal width 2.46–3.02 mm), with nearly straight sides and only slightly flared apical wings and expanded base, similar to those of *P. goeldii* and *P. steerei*. Of the nine specimens of the "*goeldii*-group" figured by Patton (1987), those of Figure 6a–d and f–g are *P. quadruplicatus* as this species is currently understood; others in figure either of *P. goeldii* or *P. steerei* (see those accounts). As with *P. steerei*, phallus remarkably small for body size, especially in comparison to those of other, smaller species (M. N. F. da Silva 1998).

DISTRIBUTION: *Proechimys quadruplicatus* occurs throughout the northwestern Amazon Basin essentially north of the Río Marañón–Rio Solimões axis, in northern Peru, eastern Ecuador, southern Colombia, southern Venezuela, and northern Brazil on the left (= north and east) bank of the Rio Negro to its mouth near Manaus. This species is replaced to the south by the morphologically and ecologically similar *P. steerei* (see account of that species).

SELECTED LOCALITIES (Map 522): BRAZIL: Amazonas, Arquipélago da Anavilhanas (Patton et al. 2000), Lago Meduinim (Patton et al. 2000), Manaus (Moojen 1948), Tahuapunta (= Tausa) (Patton 1987). COLOMBIA: Caquetá, Manganite (Patton 1987). ECUADOR: Napo,

Llunchi (type locality, Hershkovitz 1948a); Orellana, San José de Pay amino (MVZ 170291). PERU: Amazonas, La Poza (Patton et al. 2000); Loreto, El Chino (MVZ 198518), Peas (Patton 1987), Río Tigre (FMNH 123013). VENEZUELA: Amazonas, Boca del Río Orinoco (AMNH 78031), Neblina base camp (USNM 560667), 18 km SSE of Puerto Ayacucho (Patton 1987), San Carlos de Río Negro (Patton et al. 2000).

SUBSPECIES: We treat *Proechimys quadruplicatus* as monotypic, but adequate comparisons between western samples of the nominotypical form and *amphichoricus* Moojen have not as yet been made.

NATURAL HISTORY: As is true of *P. steerei*, *P. quadruplicatus* is most commonly found in seasonally flooded forest during the dry season, or along the margins of seasonally wet habitats during the flood season, in both blackwater and whitewater river systems, although this species may also be locally common in *terra firme* forest. Hice and Velazco (2012) recorded pregnant or lactating females in most months from January through at least September, and suggested that breeding was year-round. Litter size averaged 2.5 (range: 1–4). In northern Peru and eastern Ecuador, it may be found sympatric with *P. brevicauda*, *P. cuvieri*, and *P. simonsi* (Hershkovitz 1948a; Patton et al. 1982; Hice and Velazco 2012).

REMARKS: The combination of large size, stiff pelage, bicolored hindfeet, and four folds on most cheek teeth

make this one of the most readily recognizable species of spiny rats in the northwestern Amazon Basin. In comparison to other species, *P. quadruplicatus* is remarkable for uniformity in, and with low levels of, molecular diversification across the known range (Patton et al. 2000; Matocq et al. 2000). Both *quadruplicatus* Hershkovitz and *amphichoricus* Moojen were published in the same year, but as noted by Patton et al. (2000), Hershkovitz's paper appeared earlier and thus has date of publication priority. Minor geographic variation in karyotype has been recorded, with 2*n* varying from 26 (southern Venezuela) to 28 (Ecuador, Peru, and Brazil) and FN ranging from 42–44 (Reig and Useche 1976; A. L. Gardner and Emmons 1984; Patton et al. 2000; Bonvicino, Otazu, and Vilela 2005).

Specimens in the Field Museum of Natural History from Gualaquiza, Río Santiago, Ecuador, referred to as *P. semispinosus* (Tomes) by Osgood (1944:200–201), are *P. quadruplicatus*. Osgood's misidentification probably stemmed from the confusion regarding the type locality of Tomes's rat (see A. L. Gardner 1983a), as both species share similarly complex occlusal patterns.

Proechimys steerei Goldman, 1911
Steere's Spiny Rat

SYNONYMS:

Proechimys steerei Goldman, 1911a:238; type locality "Rio Purus, a southern tributary of the Amazon, in northwestern [Amazonas] Brazil," amended to "Hyutánahan [= Huitanaã], a small village of rubber gatherers, on the north side of the Rio Purus, in the upper part of its course" (Goldman 1912d).

Proechimys kermiti J. A. Allen, 1915b:629; type locality "Lower Rio Solimoens," amended to "Lower Rio Solimões (up the Solimões 50 to 60 miles on the north bank of the river), Manacaparú, Amazonas, Brazil" (Moojen 1948:345).

Proechimys pachita Thomas, 1923b:694; type locality "Puerto Leguia, 2000', Rio Pachitea," Huánuco, Peru.

Proechimys hilda Thomas, 1924c:534; type locality "San Lorenzo. Alt. 500'. Marañon, just above the mouth of the Huallaga," Loreto, Peru.

Proechimys rattinus Thomas, 1926g:164; type locality "Ucayale, Tushemo, Masisea, 1000'," Ucayali, Peru.

Proechimys [(Proechimys)] cayennensis pachita: Ellerman, 1940:121; name combination.

Proechimys [(Proechimys)] cayennensis hilda: Ellerman, 1940:121; name combination.

[Proechimys guyannensis] steerei: Hershkovitz, 1948a:133; name combination.

[Proechimys guyannensis] hilda: Hershkovitz, 1948a:133; name combination.

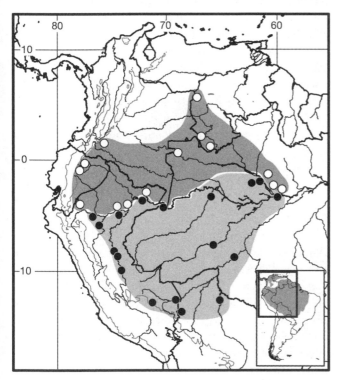

Map 522 Selected localities for *Proechimys quadruplicatus* (○) and *Proechimys steerei* (●). Contour line = 2,000 m.

[*Proechimys guyannensis*] *kermiti*: Hershkovitz, 1948a: 133; name combination.

[*Proechimys guyannensis*] *pachita*: Hershkovitz, 1948a: 138; name combination.

Proechimys goeldii steerei: Moojen, 1948:338; name combination.

[*Proechimys longicaudatus*] *pachita*: Moojen, 1948:316; name combination.

[*Proechimys longicaudatus*] *rattinus*: Moojen, 1948:316; name combination.

[*Proechimys semispinosus*] *hilda*: Moojen, 1948:316; name combination.

Proechimys semispinosus liminalis Moojen, 1948:343; type locality "Rio Quichito, affluent from the south of the Javarí River, near Benjamin Constant, Benjamin Constant, Amazonas, Brazil."

Proechimys semispinosus kermiti: Moojen, 1948:345; name combination.

DESCRIPTION: Large, average mass 450 g, with some individuals exceeding 800 g. Head and body length 215–493 mm; hindfeet (length 49–63 mm) and ears large (length 20–26 mm); tail proportionately short (120–207 mm; 70% of head and body length). Dorsal color light reddish brown, only faintly streaked with darker hairs; most individuals lack dark midback characteristic of *P. quadruplicatus*. Venter pure white, and texture of ventral fur thicker and more velvety both to eye and touch than in other species. Aristiforms of dorsum distinctly long (about 15 mm), narrow (about 0.5 mm), delicate, soft to touch, with whip-like tip; pelage markedly softer that other sympatric species in the genus, including other *goeldii*-group members. Tail bicolored with dark above and pale below, and clothed in fine hairs but with scales conspicuous to eye; 7–8 scale annuli per cm. The color of dorsal surface of the hindfoot distinctly bicolored, with pale to dark brown outer band and whitish inner band along length, from tarsal joint to end of toes.

Skull large (condyloincisive length 41.0–55.4 mm), but smaller than other members of group; rostrum long and narrow (mean length 24.03 mm, mean width 8.73 mm); supraorbital ledge that extends onto parietals as weakly developed ridge well developed. Incisive foramina lyrate to oval in outline, with slightly to strongly flanged posterolateral margins that form grooves extending onto anterior palate; premaxillary portion of septum short, less than half length of opening, maxillary portion distinctly narrow, and both in contact in most specimens; vomer usually not visible. Groove present on floor of infraorbital foramen, but lateral flange is only weakly developed. Mesopterygoid fossa relatively broad (angle averaging 67°), but penetrates to midpoint of M3s. Counterfold number varies in upper teeth, but is constant in lowers; pattern 3(4)–3(4)–3(4)–4(3)/4–3–3–3.

Baculum comparatively short and moderately wide (sample mean length 7.11–8.67 mm, proximal width 2.45–2.93 mm, distal width 2.46–3.11 mm), with parallel or slightly concave sides, similar in size and shape to those of *P. quadruplicatus* and *P. goeldii*. Male phallus remarkably small, particularly in comparison to body size, with prepuce extended as narrow tube and terminated by characteristic tuft of hairs, rather than rounded and blunt as in most *Proechimys* phalli.

DISTRIBUTION: *Proechimys steerei* is known from eastern and southern Peru south of the Río Marañón-Río Amazonas, northwestern Bolivia, and western Brazil south of the Rio Solimões to at least the Rio Purus, but extending north of the Rio Solimões into the Imerí region south and west of the lower Rio Negro (Patton et al. 2000:250, Fig. 153). It is unclear if specimens from the lower Rio Madeira are of this species or represent *P. goeldii*.

SELECTED LOCALITIES (Map 522): BOLIVIA: Beni, Río Mamoré, 17 km NNW Nuevo Berlín (S. Anderson 1997, as *Proechimys hilda*); La Paz, Río Madidi, Moira camp (Patton et al. 2000). BRAZIL: Amazonas, Huitanaã (type locality of *Proechimys steerei* Goldman), Ilha Paxiuba (Patton et al. 2000), right bank Rio Jaú above mouth (Patton et al. 2000), Rio Quichito (type locality of *Proechimys semispinosus liminalis* Moojen), mouth of Rio Solimões, near Manacaparú (type locality of *Proechimys kermiti* J. A. Allen), Tambor (Patton et al. 2000); Rondônia, Pista Nova (Patton 1987). PERU: Loreto, Orosa (Patton 1987), San Lorenzo (type locality *Proechimys hilda* Thomas), Loreto, Santa Elena (Patton 1987), Yurimaguas (Patton 1987); Madre de Dios, Albergue Cusco Amazónico (Patton et al. 2000), Itahuanía (Patton 1987); Ucayali, Cumaría (Patton 1987), Tushemo (type locality of *Proechimys rattinus* Thomas), Yarinacocha (Patton and Gardner 1972).

SUBSPECIES: By current understanding, *Proechimys steerei* is monotypic. The various synonyms listed above generally belie a rather fairly uniform character variation among populations of this species, although samples from that portion of the central Brazilian Amazon between the lower Rio Negro and Rio Solimões are sharply distinct from the remainder of the species range south of the Marañón-Solimões axis in western Brazil and northeastern Peru in mtDNA sequences (see Remarks).

NATURAL HISTORY: This species prefers the seasonally flooded white-water (*várzea*) or black-water (*igapó*) forests, where it is typically the only spiny rat present. It also occurs in secondary and disturbed *terra firme* forests where water or moist areas are nearby, active and abandoned gardens, and riverine or margins of flooded grasslands within the forest. Breeding apparently takes place throughout the year in western Brazil, females begin to breed before they have molted into their adult pelage, and modal litter size is three, with a range from one to seven young (Patton et al. 2000). These reproductive characteristics suggest a more r-selected life history, one adapted for living in seasonally available habitats. Emmons (1982) examined the population ecology of *P. steerei* (she used the name *P. brevicauda* for this species, following Patton and Gardner 1972) and sympatric *P. simonsi* in southern Peru, including nightly movement pattern, seasonal home range, diet, and density. A notable finding was that both species feed on sporocarps of mycorrhizal fungi, potentially a pivotal ecological service (Emmons 1982; Janos et al. 1995).

Matocq et al. (2000) described molecular population genetic structure in relation to riverine barriers and habitat range in western Brazil, contrasting those patterns with the codistributed *P. simonsi*. And, E. P. Lessa et al. (2003) provided estimates of late and post-Pleistocene population expansion and stability based on coalescence analysis of molecular sequence data.

REMARKS: Thomas (1926g:164) considered *rattinus* close to *P. hendeei* (= *P. simonsi*) in both pelage and cranial characters. Osgood (1944:202) also thought that *rattinus* Thomas might be a "lowland race or phase of *hendeei*." The holotype of *rattinus* Thomas is an adult, represented by a skull but not a skin; its cranial features are clearly those of *P. steerei*, not *P. simonsi* (= *P. hendeei*); see Patton (1987). Hershkovitz (1948a) noted that Thomas's description of *rattinus* was based on the skull of an adult female (designated as the type, BM 24.2.22.16) and the skin of an immature female (BM 24.2.22.18). He argued that the skull was referable to *guyannensis*, which is clearly wrong, and that the skin "agrees closely with that of the type of *P. hendeei*" (= *P. simonsi*). However, the skin to which Hershkovitz referred has a gray throat patch, which is not typical of *simonsi*, although a note on the label reads: "the only skin of a dozen with belly other than wholly white." One of us (JLP) has examined all specimens of spiny rats in the BM collection; each of those from the type locality of *rattinus* (including the holotype skull) are *P. steerei*.

Thomas (1927f) assigned specimens from Tingo Maria and Chinchavita, Huánuco, Peru to his *Proechimys pachita* (which we treat herein as a synonym of *P. steerei*); the entire series in the British Museum from both localities, however, are specimens of *P. brevicauda* Günther (see account of that species). Osgood (1944:202) also considered *pachita* Thomas a synonym of *P. simonsi*, noting that it differed primarily by "having a longer palatal foramina." Thomas's holotype, however, clearly displays the incisive foramina and other characters of *P. steerei* and not *P. simonsi*, as described and defined by Patton (1987).

Moojen's *liminalis* can be assigned as a junior synonym to either *P. steerei* Goldman or *P. quadruplicatus* Hershkovitz based on Moojen's (1948:343) description of the structure and shape of the incisive foramina, the vomerine sheath, as well as the four counterfolds on both M2 and M3. However, because these two species appear separated south and north, respectively, of the Marañón-Amazonas-Solimões axis, and the type locality of *liminalis* is on the south side, we assign the name to *P. steerei*. Should this prove incorrect, *liminalis* Moojen would become a junior synonym of *quadruplicatus* Hershkovitz, which has date of publication priority in 1948.

Proechimys steerei comprises two sharply divergent mtDNA clades (Patton et al. 2000), one limited to the Imerí region sandwiched between the Rio Solimões on the south and Rio Negro on the north and east, and the other south of the Rio Marañón–Rio Solimões axis in western Brazil, Peru, and Bolivia. Should these two clades be recognized in the future as separate taxa, then J. A. Allen's *kermiti*, with its type locality near Manacaparú on the north bank of the Rio Solimões, would be an available name for the clade in the Imerí region.

Karyotypic variation is minimal geographically, with multiple samples available to represent both mtDNA clades. All individuals for which voucher specimens have been examined by J. L. Patton as assignable to *P. steerei* have a karyotype with $2n=24$ and FN$=40$–42 (Patton and Gardner 1972; A. L. Gardner and Emmons 1984; Patton et al. 2000; M. N. F. da Silva and J. L. Patton, unpubl. data). N. A. B. Ribeiro (2006) also described a karyotype of $2n=24$, FN$=42$, from specimens from Pauiní, Amazonas, Brazil, which he ascribed to *P. steerei*. We have not examined these. Patton et al. (2000) described a non-Robertsonian polymorphism in one pair of biarmed autosomes in samples from the Rio Juruá in western Brazil.

Proechimys guyannensis species group

We list two species in this complex, which is distributed mostly in eastern Amazonia, from east of the Rio Negro, including the Guianan region, and extending south into central Brazil. One species, *P. guyanensis*, occurs north of the Amazon, the other, *P. roberti*, occurs only to the south of this river. Both species, however, exhibit phylogeographic structure in mtDNA sequences (Weksler et al. 2001; Steiner et al. 2000; L. P. Costa, Y. L. R. Leite, and R. N. Leite, unpubl. data). Moreover, what we refer to here as *P. guyannensis* is also composed of multiple karyotypic forms (Reig and Useche 1976; Petter 1978; Weksler et al. 2001; Bonvicino, Otazu, and Vilela 2005; T. Machado et al. 2005). Both of these "species" thus may prove to be composite. Because there are multiple names already available but currently listed only as synonyms, the challenge for future investigators will be to tie molecular clades and/ or karyotypes directly to these names. Given that multiple species of spiny rats can co-occur at single sites, it is not sufficient to simply obtain "topotypes" for name application, as acknowledged by Hershkovitz (1948a) nearly 60 years ago; care must be taken to compare karyotyped or sequenced voucher specimens with holotypes.

KEY TO THE SPECIES OF THE *PROECHIMYS GUYANNENSIS* SPECIES GROUP:

1. Distributed north of the Rio Amazonas largely in the Guianan subregion; tail long, ≥77% of head and body length; tail scales small with 11–14 annuli per cm; middorsal aristiforms stout (0.9–1.0 mm in width); mesopterygoid fossa narrow (angle <58°), penetrating to the level of M2s . . . . *Proechimys guyannensis*
1′. Distributed south of the Rio Amazonas in eastern and central Brazil; tail short, ≤70% of head and body length; tail scales large with 9–10 annuli per cm; middorsal aristiforms narrow (0.6–0.8 mm); mesopterygoid fossa moderately wide (angle >64°), penetrating only to level of M3s . . . . *Proechimys roberti*

Proechimys guyannensis (É. Geoffroy St.-Hilaire, 1803)
Guyenne Spiny Rat

SYNONYMS:

Mus guyannensis É. Geoffroy St.-Hilaire, 1803:194; type locality "Cayenne," French Guiana.

Echimys cayennensis Desmarest, 1817b:58; a redescription of *Mus guyannensis* É. Geoffroy St.-Hilaire.

echymis cayennensis: Desmarest 1822:292; inadvertent *lapsus*.

E[chinomys]. cayennensis: Wagner, 1843a:341; name combination.

Proechimys cayennensis: J. A. Allen, 1899c:264; name combination.

Echimys cherriei Thomas, 1899d:382; type locality "Munduapo [= Monduapo, Paynter 1982], Upper Orinoco," on the right bank of the upper Río Orinoco at 4°45′N, 67°48′W, Amazonas, Venezuela.

[Proechimys] cherriei: J. A. Allen, 1899c:264; name combination.

Proechimys vacillator Thomas, 1903b:490; type locality "Kanuku Mountains, British Guiana. Altitude 600 feet," Upper Takutu–Upper Essequibo, Guyana.

Proechimys warreni Thomas, 1905b:312; type locality "Comackka, 80 miles up the Demerara River, British Guiana. Alt. 50 feet," Upper Demerara-Berbice, Guyana.

Proechimys [(Proechimys)] cayennensis cherriei: Ellerman, 1940:121; name combination.

Proechimys [(Proechimys)] cayennensis warreni: Ellerman, 1940:121; name combination.

[Proechimys guyannensis] cherriei: Moojen, 1948:316; name combination.

[Proechimys guyannensis] vacillator: Moojen, 1948:316; name combination.

[Proechimys guyannensis] warreni: Moojen, 1948:316; name combination.

Proechimys guyannensis: Moojen, 1948:355; first use of current name combination.

Proechimys guyannensis riparum Moojen, 1948:367; type locality "Manaus, Manaus, Amazonas, Brazil."

Proechimys guyannensis arabupu Moojen, 1948:369; type locality "Arabupu, Mount Roraima, Boa Vista, Territ. Rio Branco; about 1540 meters altitude," Roraima, Brazil. [Moojen's designation of the type locality in Brazil is in error; Arabupu [=Arabopó] is on the Río Arabopó, Bolívar, Venezuela; Paynter 1982.]

Proechimys canicollis vacillator: Cabrera, 1961:518; name combination.

DESCRIPTION: Moderate-sized spiny rat, with head and body length 180–230 mm (Moojen 1948; Catzeflis and Steiner 2000; Voss et al. 2001). Tail proportionately long (length 110–186 mm; 77–87% of head and body length (Catzeflis and Steiner 2000; Voss et al. 2001). Samples across northern Guianan region light reddish to yellowish brown lined with black along midback, distinctly paler on lower sides, but abruptly meeting pure white venter from chin to inguinal region. White inner thigh stripes typically continuous across ankle to dorsal surface of hindfoot, which is typically light colored with only slight brown patch on tarsus below digit I; all digits generally white. Plantar pads on hindfeet only moderately developed, but both thenar and hypothenar present and subequal in size. Tail sharply bicolored, light brown above and cream below; hairs on tail sparsely distributed and very short such that tail appears completely naked from distance; scales small and annuli consequently narrow (11–14 per cm at midpoint). Pelage stiff to touch, particular along midback, with aristiform spines rather short (length 16–19 mm) and stout (with 0.9–1.0 mm); tip either blunt or terminates with very short filament.

Skull like other species in general shape, but because of moderate body size, appears small (condyloincisive length 37.3–47.6 mm; Voss et al. 2001; Weksler et al. 2001) and rather delicate, lacking heavy ridging that may be present in skulls of larger species. Consequently, temporal ridges generally poorly developed, if at all, and maximally just a short and weak posterior extension from supraorbital ledges. Incisive foramina oval or teardrop in shape, with either no or only weakly developed posterolateral flanges so that anterior palate typically flat or only very slightly grooved; premaxillary portion of septum short, occupying less than half opening, and usually not in contact with very attenuate maxillary portion; maxillary part short and unkeeled; anterior palate lacks medial ridging; vomerine portion to septum typically not visible in ventral view. Foraminal shape and structure similar to that of *P. roberti*, described below, and also similar to *P. simonsi*. Floor of infraorbital foramen either lack evidence of groove or have moderately developed lateral flange that forms groove for passage of the maxillary nerve. Width of mesopterygoid fossa ranges widely among geographic samples, but generally narrow with angle <58°. Nevertheless, fossa penetrates to level of posterior half of M2s in nearly all specimens. Postorbital process of zygoma obsolete or only weakly developed, formed entirely by squamosal. Number of folds on upper cheek teeth relatively constant, with three characterizing PM4, M1, and M2, and either three or, more commonly, two present on M3; lower cheek teeth more variable, pm4 with three folds and m1–m3 typically only two, but occasionally three; counterfold pattern 3–3–3–(2)3/3–2(3)–2(3)–2(3). In comparison with sympatric *P. cuvieri*, the cheek teeth of *P. guyannensis* also notably small in size, with toothrow ≤8.0 mm in length (Voss et al. 2001).

Baculum relatively long and narrow (mean sample length 6.38–8.87 mm, proximal width 1.87–2.29 mm, distal width 1.99–2.38 mm); shaft straight with little dorsoventral curvature and only slightly tapered lateral indentations near midshaft, proximal end usually evenly rounded and paddle shaped, and distal tip with only slight development of apical wings and moderate median depression (Patton 1987).

DISTRIBUTION: *Proechimys guyannensis* is an Amazonian endemic largely confined in the Guiana subregion of eastern and southern Venezuela, Guyana, Surinam, French Guiana, and Brazil north and east of the Rio Negro and north of the Rio Amazonas (Patton 1987; Weksler et al. 2001; Voss et al. 2001:163).

SELECTED LOCALITIES (Map 523): BRAZIL: Amapá, Serra do Navio (Patton 1987); Amazonas, Faro (Patton 1987), Igarapé Araújo (Bonvicino, Otazu, and Vilela 2005, as "sp. A"), Manaus (type locality of *Proechimys guyannensis riparum* Moojen), São Gabriel da Cachoeira (INPA [JPB 0]). FRENCH GUIANA: Cayenne (type locality of *Mus guyannensis* É. Geoffroy St.-Hilaire). GUYANA: Demerara-Mahaica, Loo Creek (ROM 98216). SURINAM: Surinam, Santa Boma (Husson, 1978). VENEZUELA: Amazonas, Capibara (Patton 1987), Monduapo (type locality of *Echimys cherriei* Thomas), Neblina base camp (USNM 560675), Pozon (ROM 107892); Bolívar, Arabupu [= Arabopó] (type locality of *Proechimys guyannensis arabupu* Moojen), 3 km E of Puerto Cabello del Caura (ROM 107944).

SUBSPECIES: We treat *P. guyannensis* as monotypic. The lack of an adequate review of any character set for available specimens precludes the delineation of geographic units that might warrant recognition, despite the plethora of names listed in the synonymy. This is a highly variable species, however, and likely polytypic (see comments above and Remarks).

NATURAL HISTORY: At Paracou in French Guiana, Voss et al. (2001) caught this species mostly in well-drained forest but also occasionally in creekside forest, and significantly more commonly in primary forest than its sympatric congener *P. cuvieri*. All but two specimens were collected on the ground; those that were not came from traps placed 0.5–1.5 m above ground in liana tangles. Spiny rats are important seed predators and dispersers of tropical forest trees in French Guiana (Forget 1991, 1996), although field ecological studies have been hampered by the difficulty in distinguishing sympatric species (Malcolm 1992). Consequently, the several published studies in French Guiana, or elsewhere in the Guianan region and northeastern Amazonian Brazil, probably include data on *P. guyannensis*, but were unable to differentiate this species from sympatric *P. cuvieri*.

REMARKS: Woods (1993:795) used *cayennensis* Desmarest as the name for this species, following the argument of Wilson and Reeder (1993:831) that É. Geoffroy St.-Hilaire's 1803 publication did not meet the requirements of the International Code of Zoological Nomenclature and thus that his names were unavailable. However, the International Commission (ICZN 2002:Opinion 2005) subsequently validated É. Geoffroy St.-Hilaire, 1803, an action that makes *cayennensis* Desmarest a junior synonym of *guyannensis* É. Geoffroy St.-Hilaire.

As noted by Voss et al. (2001), Husson (1978) used the name *P. warreni* to refer to this species in his book on Surinam mammals, and applied *P. guyannensis* incorrectly to the larger-bodied species that has been identified by subsequent authors as *P. cuvieri*.

Patton (1987) incorrectly mapped the type locality of *cherriei* Thomas in the lower, instead of upper, Orinoco basin (see Voss et al. 2001).

Proechimys guyannensis, as defined here, is most likely composite given the extensive karyotype diversity described in the literature and the limited DNA sequence data available (Weksler et al. 2001; Bonvicino, Otazu, and Vilela 2005). Reig, Tranier, and Barros (1980) described a karyotype of $2n=40$, FN=54, from the type locality (Cayenne, French Guiana). T. Machado et al. (2005) reported a karyotype of $2n=44$, FN=52, from a specimen from Manaus, in Amazonas state, Brazil, which probably represents *P. guyannensis*, as L. H. Emmons (unpubl. data) obtained the same karyotype from a specimen assignable to this species on morphological grounds and taken at 80 km N of Manaus (USNM 555638). Furthermore, Bonvicino, Otazu, and Vilela (2005) reported two different karyotypes for specimens that belong to *P. guyannensis* (*sensu lato*), what they called "*Proechimys* sp. A," with $2n = 38$ and FN = 52 (in the upper Rio Negro–Rio Aracá in Amazonas state, Brazil), and "*Proechimys* sp. B," with $2n = 46$, FN = 50 (from the Rio Anauá, a tributary of the Rio Branco in Roraima state; the same karyotype obtained from the Rio Uatumã, in Amazonas state by C. E. F. Silva et al. 2012). They suggested that each karyotypic form represented a different species, and further suggested that *arabupu* Moojen was the correct name to apply to their "sp. B" because its type locality is close to the locality of their specimen. However, the holotype of *arabupu* (AMNH 75816) came from Arabopó in southeastern Venezuela at the base of Mt. Roraima (not in Brazil as Moojen 1948 stated), some 450 km to the north of the locality for "sp. B." Moreover, there is a karyotype of $2n = 40$, FN = 50, for a specimen (MVZ 160094) collected near Icabarú, Bolívar, Venezuela, only 130 km west of the type locality of *arabupu* Moojen and with which it shares the same craniodental morphology. It is thus not possible to assign the name *arabupu* Moojen to the karyomorph Bonvicino, Otazu, and Vilela (2005) termed "sp. B" with any certainty. Eler et al. (2012) has also extended the distribution of the $2n = 38$, FN = 52, karyotype to the Rio Jari basin on the Amapá-Pará state boundary in eastern Amazonia, giving this chromosomal form an extensive range.

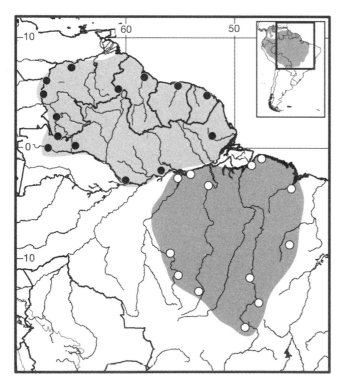

Map 523 Selected localities for *Proechimys guyannensis* (●) and *Proechimys roberti* (○). Contour line = 2,000 m.

Clearly, care must be taken before applying an available name to a karyotypic form because different karyotypes do not always signal reproductive incompatibility within the genus (Aguilera, Reig, and Pérez-Zapata 1995; M. N. F. da Silva 1998). Moreover, an effort must be made to obtain karyotypes and molecular sequences from topotypes that have been compared directly to relevant holotypes or type series, as two or more species of spiny rats may co-occur at single sites throughout the range of the genus.

Van Vuuren et al. (2004) described molecular sequence variation among populations in French Guiana, noted low molecular diversity and the lack of haplotype structure within this limited geographic region, and suggested that the sampled region had undergone a recent population expansion.

Proechimys roberti Thomas, 1901
Robert's Spiny Rat
SYNONYMS:

Proechimys Roberti Thomas, 1901g:531; type locality "Rio Jordão, S. W. Minas Geraes [*sic*], Alt. 960 meters," Araguarí, Minas Gerais, Brazil.

Proechimys oris Thomas, 1904e:195; type locality "Igarapé-Assu, near Pará. Alt. 50 m," near Belém, Pará, Brazil (Moojen 1948:365).

Proechimys boimensis J. A. Allen, 1916c:523; type locality "Boim, Rio Tapajos," Pará, Brazil."

Proechimys [(Proechimys)] boimensis: Ellerman, 1940:119; name combination.

Proechimys [(Proechimys)] cayennensis roberti: Ellerman, 1940:121; name combination.

Proechimys [(Proechimys)] cayennensis oris: Ellerman, 1940: 121; name combination.

Proechimys cayennensis arescens Osgood, 1944:198; type locality "Roca, near Fazenda Inhuma, below Santa Philomena [= Filomena], upper Rio Parnahyba [= Vitoria do Alto Parnaiba], Maranhão, Brazil" (Moojen 1948:366).

[*Proechimys guyannensis*] *boimensis*: Hershkovitz, 1948: 128; name combination.

Proechimys longicaudatus boimensis: Moojen, 1948:350; name combination.

Proechimys longicaudatus roberti: Moojen, 1948:353; name combination.

Proechimys guyannensis oris: Moojen, 1948:365; name combination.

Proechimys guyannensis arescens: Moojen, 1948:366; name combination.

DESCRIPTION: Similar in size to *P. guyannensis*, with head and body length 200–230 mm, but tail distinctly shorter, both absolutely and proportionally (70% of head and body length). Dorsal color ranges from reddish brown in northern part of range in Amazonia proper to pale buff in samples at southern terminus of range in dry and gallery forests of Cerrado, a difference in color that reinforced species status of *P. roberti* when Thomas (1904e) described *P. oris* a few years later. Venter white from chin to inguinal region, but white of inner thigh discontinuous, broken by dark ankle band, with pale surface of hindfeet; brownish patch extending from tarsus to the lateral toes in many specimens. Plantar pads of hindfoot moderately developed, both thenar and hypothenar present and subequal in size. As in *P. guyannensis*, tail very sparsely covered with short hairs so that it appears completely naked to the eye from a distance; scales large (9–10 annular rings per cm at midpoint). Pelage stiff to touch, but softer in southern samples; aristiform spines relatively short (length 20 mm) and narrow (width 0.6–0.8 mm), each terminating in long whiplike tip. Difference in softness of pelage between Amazonian and Cerrado specimens mostly in aristiform density, not to appreciable differences in width or other features.

Skull similar in all respects to that of *P. guyannensis*; medium-sized (mean condyloincisive length in five samples 40.8–43.6 mm; Weksler et al. 2001) but rather narrow and with narrowed, tapering rostrum. Slight temporal ridges extend posteriorly onto parietals from supraorbital ledge in older individuals, but most specimens lack ridges altogether. Incisive foramina cannot be distinguished from those of *P. guyannensis* in any feature; they are relatively wide, oval to teardrop in shape, with little or no posterolateral flange so that anterior palate is either flat of only weakly grooved; premaxillary portion of septum short, less than half opening, and either connected to very attenuate maxillary portion or separated entirely; maxillary portion unkeeled so that anterior palate lacks medial ridge; vomerine portion inconsistently visible ventrally. Floor of infraorbital foramen either flat or with only hint of groove formed by weakly developed lateral flange. Mesopterygoid fossa intermediate in width, with angle 64°–67°, and penetrating posterior palate to anterior half of M3s. Postorbital process of zygoma weakly to moderately developed, and formed completely by squamosal. Number of folds on upper cheek teeth three except M3, occasionally with two. Number of folds on lower teeth more variable, with three (or occasionally four) on pm4, and either two (usually) or three on each lower molar; counterfold count 3–3–3–(2)3/3(4)–2(3)–2(3)–2(3). As with other species of spiny rats whose populations span a range of forest types, southernmost samples of *P. roberti* in the drier Cerrado forests typically have simpler teeth, with higher proportion of individuals with only two folds on lower molars.

Baculum relatively short and narrow (sample mean length 5.43–6.08 mm, proximal width 1.74–2.27 mm, distal width 1.28–1.79 mm), with same general features as *P. guyannensis*, except specimens from southern part of range tend to be smaller in length and width (Patton 1987).

DISTRIBUTION: *Proechimys roberti* occurs throughout the rainforest of Amazonian Brazil south of the Rio Amazonas and extends south to the Cerrado biome in east-central Brazil, in the states of Pará, Maranhão, Tocantins, Mato Grosso, Goiás, and Minas Gerais (Weksler et al. 2001).

SELECTED LOCALITIES (Map 523): BRAZIL: Goiás, Anápolis (Patton 1987), Fazenda Fiandeira (Weksler et al. 2001); Maranhão, Fazenda Lagoa Nova (Weksler et al. 2001); Mato Grosso, Gaúcho do Norte (T. Machado et al. 2005), Cláudia (T. Machado et al. 2005), Reserva Ecológica Cristalino (MVZ 197576); Minas Gerais, Rio Jordão (type locality of *Proechimys roberti* Thomas); Pará, 52 km SSW of Altamira, right bank Rio Xingu (Weksler et al. 2003), Belém (MVZ 196096), Boim, Rio Tapajós (type locality of *Proechimys boimensis* J. A. Allen), Curuá-Una (Weksler et al. 2001), Igarapé-Assú (type locality of *Proechimys oris* Thomas); Piauí, Estação Ecológica de Uruçuí-Una (T. Machado et al. 2005); Tocantins, Rio Santa Teresa (MVZ 197585).

SUBSPECIES: Populations in lowland Amazonian rainforest could be considered a subspecies (*P. r. oris*) separate from those in the semideciduous forests of the Cerrado (*P. r. roberti*), although the limited analyses available suggest clinal variation between these geographic and ecological extremes (Weksler et al. 2001).

NATURAL HISTORY: Neither the ecology nor life history of *P. roberti* has been studied extensively in the field, although Alho (1981) examined homing ability using radio telemetry in the Brazilian Cerrado. This species is sympatric with *P. goeldii* along the southern margins of the Rio Amazonas in eastern Brazil, and with *P. longicaudatus* in the dry forests of the Cerrado in central Brazil.

REMARKS: Thomas (1904e:196) suggested that his new species agreed "very closely with *P. oris* in its cranial characters, but differs by its paler and more uniformly buffy color, its fully haired under surface, and its much longer and softer fur, of which the spines form a less considerable proportion than usual." The pelage characters, both color and spine development, are typical of populations of spiny rats that live in more open, drier forests in comparison with conspecific populations in denser, more humid forests. Weksler et al. (2001) reviewed geographic variation in both *P. roberti* and *P. oris* and concluded that the two belonged to the same taxon on morphological, karyotypic, and molecular grounds, a conclusion contrary to the more geographically limited morphometric study of Pessôa et al. (1990). Weksler et al. (2001) examined specimens that filled the large geographic hiatus between the ranges of the Amazonian lowland *P. oris* Thomas and the Cerrado isolate *P. roberti* Thomas, samples that documented clinal rather than discordant variation in morphological characters across the north-to-south range encompassing both taxa, and thus reinforcing the conclusion of conspecificity. Molecular sequence data reported by Weksler et al. (2001) for specimens spanning this geographic range, from the mouth of the Amazon to south of Brasília, indicate only minor interpopulation differences (maximum sequence divergence 2.4%). However, specimens from Mato Grosso and Pará states, in the Tapajós-Xingu interfluvial region, not included in the Weksler et al. analyses, are somewhat different morphologically and twice as divergent in mtDNA sequence (R. do V. Vilela 2005; Schetino 2008; Y. L. R. Leite, L. P. Costa, and R. N. Leite, unpubl. data). Further analyses may warrant segregation of more than one species unit within what is considered as a single species (this account; Weksler et al. 2001).

A. L. Gardner and Emmons (1984), Leal-Mesquita (1991), Weksler et al. (2001), T. Machado et al. (2005), and N. A. B. Ribeiro (2006) described and figured the karyotype, as $2n = 30$ and FN = 54–56, noting limited regional differences in the number of small biarmed versus uniarmed autosomal pairs. Valim and Linardi (2008) described the host association of ectoparasitic lice in the genus *Gyropus*.

Proechimys longicaudatus species group

Patton (1987) included nine taxa within his concept of the *longicaudatus*-group, and suggested that at least two species were recognizable among them: *P. longicaudatus* Rengger, from the dry forests of southeastern Bolivia east across northern Paraguay to the Cerrado of central Brazil, and *P. brevicauda* Günther, from the lowland rainforest in northern Bolivia, eastern Peru, eastern Ecuador, southeastern Colombia, and western Amazonian Brazil. A. L. Gardner and Emmons (1984) had previously suggested that samples from northern Peru and eastern Ecuador might be separable from *P. brevicauda* as a third species, *P. gularis* Thomas, based on chromosomal differences. Although Patton (1987) did not include *P. cuvieri* Petter within his *longicaudatus*-group, available DNA sequence data suggest a relationship between this species and those members of the *longicaudatus*-group that he recognized (R. do V. Vilela 2005; Schetino 2008; R. N. Leite and J. L. Patton, unpubl. data). Because of this apparent phyletic coupling and the fact that each of these species shares a similar suite of craniodental features, we include *P. cuvieri* along with *P. longicaudatus* and *P. brevicauda* within the *longicaudatus*-group here. Common characters that uniquely unite this group of species include lyrate and strongly flanged incisive foramina, broad and long maxillary portion to the septum with the vomerine portion exposed, and deep groves extending onto the anterior palate; a broad and relatively shallow mesopterygoid fossa;

a smooth floor to the infraorbital foramen; and a generally uniform three counterfolds on each cheektooth, with virtually never four folds on any upper tooth, but with two folds characterizing the lower posterior molars of a substantial portion of individual samples. All three share a wide baculum.

Two of these species (*P. brevicauda* and *P. cuvieri*) have been found in sympatry at several localities in the western Amazon (Patton et al. 2000), but the "test of sympatry" has not been established for *P. brevicauda* and *P. longicaudatus*. These two seem to grade from one to the other through the transition between the lowland rainforest of western Amazonia (*brevicauda*) to the dry forests of eastern Bolivia and central Brazil (*longicaudatus*). Whether these two are best treated as a single species that varies substantially over this transition area or as the two that we recognize herein must await further critical sampling. Regardless of the status of *brevicauda* and *longicaudatus*, however, the number of species in the group is likely much greater. The limited mtDNA sequence data synthesized in Patton et al. (2000) indicate well-defined geographic clades that differ by considerable levels of divergence within both *P. brevicauda* and *P. cuvieri*, and unpublished morphological and sequence data as well as published karyotypes (T. Machado et al. 2005; Eler et al. 2012) suggest the presence of an undescribed species in the middle and upper part of the Rio Madeira in western Brazil. Clearly, this is a group ripe for additional research.

KEY TO THE SPECIES OF THE *PROECHIMYS LONGICAUDATUS* SPECIES GROUP:

1. Dorsal color dark, rich reddish brown; distributed in lowland rainforest of the Guianan region and Amazon Basin. 2
1′. Dorsal color pale, yellowish brown; distributed in the dry forests of eastern Bolivia, northern Paraguay, and central Brazil . . . . *Proechimys longicaudatus*
2. Ventral color always white; aristiform spines long, stout, and with a blunt tip; well-developed postorbital process of zygoma comprised mostly of squamosal; mesopterygoid fossa moderately wide (angle <73°), penetrating to middle of M3s; baculum nearly as wide as long, with deep median depression and elongated apical extensions . . . . *Proechimys cuvieri*
2′. Ventral color often brownish gray or reddish-buff; aristiform spines short and narrow with elongated tip; postorbital process of zygoma obsolete; mesopterygoid fossa broad (angle >73°) and shallow, penetrating barely to posterior edge of M3s; baculum nearly twice as long as wide, with a shallow, or no, median depression and without apical extensions . . . . . *Proechimys brevicauda*

Proechimys brevicauda (Günther, 1876)
Short-tailed Spiny Rat

SYNONYMS:

Echimys brevicauda Günther, 1876:748; type locality "Chamicuros, Huallaga river" (lectotype selected by Thomas, 1900f:301), Loreto, Peru.

Thricomys brevicauda: E.-L. Trouessart, 1897:607; name combination.

Tricomys brevicauda: E.-L. Trouessart, 1904:504; name combination.

Proechimys bolivianus Thomas, 1901h:537; type locality "Mapiri, Upper Rio Beni, N.W. [La Paz,] Bolivia. Altitude 1000m."

Proechimys securus Thomas, 1902b:140; type locality "Charuplaya, 1350–1400 m," Río Sécure, Cochabamba, Bolivia.

Proechimys gularis Thomas, 1911c:253; type locality "Canelos, Rio Bobonaza, Oriente of [= Pastaza] Ecuador. Alt. 2100'."

Proechimys brevicauda: Ihering, 1904:422; first use of current name combination.

Proechimys brevicaudus securus: Osgood, 1916,209; name combination.

Proechimys [(*Proechimys*)] *cayennensis brevicauda*: Ellerman, 1940:120; name combination.

Proechimys [(*Proechimys*)] *cayennensis gularis*: Ellerman, 1940:120; name combination.

Proechimys [(*Proechimys*)] *cayennensis bolivianus*: Ellerman, 1940:121; name combination.

Proechimys [(*Proechimys*)] *cayennensis securus*: Ellerman, 1940:121; name combination.

Proechimys hendeei elassopus Osgood, 1944:203; type locality "Santo Domingo, Rio Inambari, Puno, Peru. Altitude 6,000 ft."

Proechimys "hendeei" elassops [sic]: Hershkovitz, 1948a:138; incorrect spelling of *elassopus* Osgood.

P[*roechimys*]. *guyannensis gularis*: Hershkovitz, 1948a:138; name combination.

[*Proechimys guyannensis*] *brevicauda*: Hershkovitz, 1948a:138; name combination.

[*Proechimys guyannensis*] *bolivianus*: Moojen, 1948:316; name combination.

[*Proechimys longicaudatus*] *elassopus*: Moojen, 1948:316; name combination.

Proechimys longicaudatus brevicauda: Moojen, 1948:349; name combination.

[*Proechimys longicaudatus*] *securus*: Moojen, 1948:316; name combination.

[*Proechimys semispinosus*] *gularis*: Moojen, 1948:316; name combination.

Proechimys guyannensis bolivianus: Cabrera, 1961:519; name combination.

Proechimys guyannensis gularis: Cabrera, 1961:520; name combination.

DESCRIPTION: Moderate sized, mean head and body length of samples from northern Peru and western Brazil 189–253 mm, tail length 137–167 mm, hind foot length 46–53 mm, and ear length 21–23 mm (Patton and Rogers 1983; Patton et al. 2000). Color of head, back, and rump reddish brown, but not as dark along the midline as in *P. cuvieri*, although overall tones lighten in samples in central Bolivia where range of *P. brevicauda* approaches that of *P. longicaudatus*. Fulvous lateral stripe characteristically separates darker dorsal pelage from paler venter; venter varyingly colored in different parts of geographic range, brownish or grayish in parts of eastern Ecuador (as in the named form *gularis* Thomas) or reddish-buff throughout most of northern and central Peru and western Brazil, and clear white in most individuals from southern Peru and Bolivia (Patton and Gardner 1972 [as *P. longicaudatus*]; Patton et al. 2000). Osgood (1914b:168) noted extensive variation in color and color pattern on venter in specimens from Yurimaguas in northern Peru, near type locality of Günther's *brevicauda*, "which can scarcely be said to be exactly alike in any two individuals. Fulvous and white are distributed in varying proportions, in general occupying about equal areas of the under parts. The chin and throat with scarcely any exception are fulvous and likewise the sides of the belly. Sometimes the white is reduced to a small pectoral and an inguinal patch or it may cover practically the entire belly and run forward to the middle of the throat." Fulvous lateral stripe continues across ankle to small paler patch at proximal base of metatarsal area on dorsal surface of hindfeet. Otherwise, upper surfaces of hindfeet, including all five toes, uniformly dark. Tail sparsely haired, much less so and with distinctly shorter hairs than that of *P. cuvieri*; scale annuli 9–10 per cm at midpoint. Dorsal pelage stiff to touch, but aristiform hairs less well developed than those of *P. cuvieri*; length 18–20 mm, width 0.6–0.8 mm; distinctly tapering tip present on all aristiforms.

Skull of *P. brevicauda* similar to other species in *longicaudatus*-group, moderate size (condyloincisive length 39.1–48.8 mm) with elongated but relatively broad rostrum. Temporal ridge moderately to weakly developed, often with anterior parietal portion separated from posterior lambdoidal portion. Incisive foramina typically strongly lyrate in shape, distally flanged so that anterior palate is deeply grooved with median ridge, and with complete and keeled septum. Postorbital process of zygoma nearly obsolete and fashioned predominantly by jugal. Mesopterygoid fossa shallow, generally only barely reaching posterior margins of M3s, and wide (angle 73°–80° among geographic samples). Floor of infraorbital foramen smooth, lacking grove indicative of infraorbital nerve. Counterfold pattern of cheek teeth uniformly 3–3–3–3 above and 3(4)–(2)3–(2)3–(2)3 below. In rare individuals, upper M2 and M3 may have remnant fourth fold, and southern samples exhibit higher frequency of only two folds on m3.

Baculum massive, long and wide (sample mean length 7.79–11.67 mm, proximal width 3.72–5.11 mm, distal width 4.61–5.87 mm), with slight but broad apical wings and expanded base (Patton and Gardner 1972; Patton 1987; Patton et al. 2000). Massive size and length of baculum make phallus long and distinctly heavy or broad in appearance; palpating phallus in live males provides easy way to distinguish this species from sympatric congeners in the field.

DISTRIBUTION: *Proechimys brevicauda* occurs throughout the western Amazon Basin, from southern Colombia and eastern Ecuador south throughout eastern Peru, northwestern Bolivia, and east into Acre state in western Brazil.

SELECTED LOCALITIES (Map 524): BOLIVIA: Beni, Río Mamoré, 5 km NE of Río Grande mouth (Patton 1987), 10 km E of San Antonio de Lora (S. Anderson 1997); Cochabamba, Charuplaya (type locality of *Proechimys securus* Thomas), El Palmar (Patton 1987), Mission San Antonio (Patton 1987); La Paz, Mapirí (type locality of *Proechimys bolivianus* Thomas). BRAZIL: Acre, Sobral, Rio Juruá (Patton et al. 2000). COLOMBIA: Caquetá, La Murelia (Patton 1987); Putumayo, Río Mecaya (Patton 1987). ECUADOR: Orellana, San José Abajo (Patton 1987); Pastaza, Canelos, Río Bobonaza (type locality of *Proechimys gularis* Thomas). PERU: Amazonas, Huampami (Patton et al. 2000); Huánuco, Chinchavito (Thomas 1927f); Loreto, Pebas (Thomas 1928c), San Fernando, Río Yavari (Patton 1987); Pasco, Montsinery (Patton 1987); Puno, Santo Domingo (type locality of *Proechimys hendeei elassopus* Osgood); Ucayali, Balta, Río Curanja (Patton and Gardner 1972, as *P. longicaudatus*).

SUBSPECIES: The number of taxa assigned to this species, as well as both variation in karyotype and the limited molecular sequence data (see Remarks), could signal valid geographic units, but the analyses needed to determine such have as yet to be done. For the present we regard *P. brevicauda* as monotypic.

NATURAL HISTORY: This species typically occupies upland, nonseasonally inundated (*terra firme*) lowland rainforest, both undisturbed and disturbed forest and second growth, where it is likely to be the most common spiny rat present. It is often found in garden plots (*chacras*), where individuals feed on yucca and plantains (Osgood 1914b). The few reproductive data, from northern Peru and western Brazil, suggest that breeding commences by the end of the dry season (Patton et al. 2000). Emmons (1982) recorded low densities, with males occupying larger

home ranges than females, and females with exclusive-use home ranges, suggesting a possible polygynous mating system (Adler 2011). *Proechimys brevicauda* may be sympatric with up to four other species of spiny rats throughout much of its range, typically with *P. simonsi*, *P. quadruplicatus*, and more rarely *P. cuvieri* in northern Peru, Ecuador, and Colombia; with *P. simonsi*, *P. steerei*, *P. cuvieri*, and *P. pattoni* in eastern Peru and western Brazil (Patton and Gardner 1972; Patton et al. 2000); and with *P. steerei* and occasionally *P. simonsi* in northern Bolivia (S. Anderson 1997, who listed *P. steerei* as *P. hilda*). Valim and Linardi (2008) described the host association of ectoparasitic lice in the genus *Gyropus*.

REMARKS: Günther (1876) based his description of *brevicauda* on two specimens, one from Chamicuros, Río Huallaga, and the other from a locality recorded only as "Upper Amazons." Thomas (1900f:301) selected the specimen (a skin with a distinctly rufous throat and belly) from Chamicuros as the lectotype. Stephens and Traylor (1983) stated that Chamicuros is on the upper Río Samiria, about 35 miles east of Santa Cruz, which is on the Río Huallaga.

Specimens in the American Museum from Inca Mines, on the Río Inambari, Puno, Peru, and referred by J. A. Allen (1900a, 1901b) to *P. simonsi*, are *P. brevicauda*. Patton and Gardner (1972) applied the name *longicaudatus* Rengger to specimens from eastern Peru that A. L. Gardner and Emmons (1984) and Patton (1987) subsequently, and correctly, referred to *P. brevicauda* (Günther).

Thomas (1901h) regarded his *bolivianus* to be "most nearly allied to *P. simonsi*, but larger and different in cranial details." The skull of the holotype, however, possesses all of the cranial attributes of *P. brevicauda* although the under parts of the skin are pure white, not with the varying degrees of fulvous in typical *brevicauda*, but characteristic of the white venter of all specimens we assign to this species from the southern part of its range in southern Peru and Bolivia. If this assignment of *bolivianus* Thomas to *brevicauda* Günther, following Patton (1987), is in error, the only other species we recognize to which this name could apply based on its characters is *P. steerei* Goldman. If this were true, then *bolivianus* Thomas would be senior to Goldman's name.

Thomas (1911c) considered *P. brevicauda* to be the closest ally to his *gularis*, differing primarily by the dark-colored throat of the latter in comparison to the buffy venter of *brevicauda*. A. L. Gardner and Emmons (1984) suggested that *gularis* Thomas was a valid species, based on karyotypic and color differences. However, Osgood (1944:201) noted that variation in specimens of *Proechimys gularis* Thomas from the vicinity of the type locality "are practically identical with some of the variations of typical *brevicauda*. It is, therefore, doubtful that the name should stand, even for a subspecies."

Patton et al. (2000) noted the molecular uniqueness of *elassopus* Osgood from southern Peru but nevertheless included it within their concept of *P. brevicauda*. A. L. Gardner and Emmons (1984) described the karyotype as $2n=28$, $FN=48$, based on specimens from nearby localities in southern Peru. This karyotype differs from that of *P. brevicauda* from northern Peru with the same $2n$ and FN in the number of different classes of biarmed autosomes (Patton et al. 2000). The type series of *elassopus* Osgood has cranial features generally consistent with those of typical *longicaudatus*-group animals, yet differs in a more attenuate and weak maxillary portion to the septum of the incisive foramina with the vomer not visible and in a wide but much deeper anterior margin of the mesopterygoid fossa that penetrates the posterior palate to the level of M2s. In many respects, these are features also shared with *bolivianus* Thomas and *securus* Thomas, both from Bolivia and also from the lower Andean slopes. Future studies may well conclude that these three taxa deserve species status. If so, and if all three names belong to the same entity, then *bolivianus* Thomas would be the earliest name available.

The karyotype is geographically variable, ranging from $2n=30$, $FN=48$, in northern Peru and Ecuador (A. L. Gardner and Emmons 1984) to $2n=28$, $FN=48-50$, in central Peru and western Brazil (Patton and Gardner 1972; A. L. Gardner and Emmons 1984; M. N. F. da Silva 1998; Patton et al. 2000). Aniskin et al. (1991) described a similar karyotype, presumably of this species, from Peru south of the Río Marañon-Río Amazonas axis. The specimen referred to *P. longicaudatus* from Usina Hidroelétrica Samuel, in Rondônia state, Brazil, by T. Machado et al. (2005) is probably *P. brevicauda*. It has a karyotype with $2n=28$ and $FN=48$ and an acrocentric, not biarmed, Y chromosome typical of other samples of *P. brevicauda* from eastern Peru and western Brazil, and the locality is within lowland Amazonian rainforest rather than the tropical dry forest habitat characterizing *P. longicaudatus*. To whichever species these specimens can be assigned with confidence, it is apparently sympatric with an undescribed member of the *longicaudatus*-group that has a different karyotype ($2n=30$, $FN=52$), and that T. Machado et al. (2005) assign to the *longicaudatus*-group (MZUSP 27396). Specimens with this karyotype have both the enlarged baculum similar to other members of this group (R. E. Martin 1970; Patton 1987) and their mtDNA sequences are positioned with other group members. This undescribed entity is known from Amazonas, Rondônia, and Mato Grosso states, where it can be sympatric with *P. roberti* and *P. goeldii* in the middle to upper Rio Madeira and

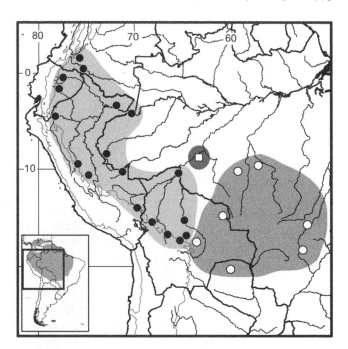

Map 524 Selected localities for *Proechimys brevicauda* (●), *Proechimys longicaudatus* (○), and *Proechimys* sp. from the central Rio Madeira (□). Contour line = 2,000 m.

headwaters of the Rio Tapajós, but likely extends to near the mouth of these rivers as well (R. do V. Vilela 2005; Schetino 2008; Eler et al. 2012; M. N. F. da Silva, J. L. Patton, and R. N. Leite, unpubl. data). Additional sampling for DNA sequences and careful morphological comparisons of sequenced specimens to those already present in museum collections will be required to accurately map the ranges of *P. brevicauda*, *P. longicaudus*, and this undescribed taxon in the southwestern Amazon Basin of Bolivia and Brazil.

Proechimys cuvieri Petter, 1978
Cuvier's Spiny Rat

SYNONYM:

Proechimys cuvieri Petter, 1978:263; type locality "Saül (S 21), Guyane française," French Guiana.

DESCRIPTION: Moderate-sized species closely similar in external and craniodental characters to sympatric samples of *P. brevicauda* in western part of its range (Patton et al. 2000), but markedly distinct from sympatric *P. guyannensis* (Catzeflis and Steiner 2000; Voss et al. 2001) in Guianan region. Mean head and body length for western Brazilian and French Guiana samples 211 and 234 mm, respectively, tail length 144 and 168 mm, hindfoot length 48 and 51 mm, and ear length 22 and 24 mm. Tail proportionately short (approximately 70% of head and body length). Overall color is dark reddish orange, with midline of back darker than sides, which contrast sharply with white venter.

Specimens from western Brazil may have slight fulvous edge to ventral fur, but bright white contrasts sharply with typically buffy venter of sympatric *P. brevicauda*. Dorsal surfaces of hindfeet dark on toes and lateral margin, but short hairs above the metatarsals are silverish in color, or at least distinctly paler than toes, a pattern also contrasting with that of *P. brevicauda*, where dorsal surfaces dark and dull. Tail sharply bicolored and clothed in long, slightly curved and dark hairs, which convey distinctly "shaggy" appearance, rather remarkable for a *Proechimys* (Malcolm 1992; Voss et al. 2001). Nevertheless, tail scales visible to eye; annuli from 9–12 per cm at midpoint. Dorsal pelage stiff to touch, with well-developed aristiform spines averaging 0.9 mm in width and 20–21 mm in length. Some geographic variation in spine development apparent in samples J. L. Patton has examined, with narrower spines in those from Guianan region and tips with whip-like extension in those from south of Amazon in eastern Brazil; otherwise, samples characterized by rather blunt-tipped spines (Patton et al. 2000).

Skull relatively moderate-sized (mean condyloincisive length 39.7–47.1 mm), with long but relatively broad rostrum and well-developed supraorbital ridges, but temporal ridges weakly developed. Incisive foramina lyrate in shape with only moderate posterior constrictions; posterolateral margins flanged, but not as strongly so as in *P. brevicauda*, and extend onto palate forming only weak. Premaxillary portion of septum long and typically in contact with maxillary portion, which may be either keeled or smooth; vomer slightly to well exposed ventrally. Groove on floor of infraorbital foramen absent or only weakly developed, as is lateral flange of this groove. Postorbital process of zygoma well developed and formed completely by squamosal, or with only minimal jugal contribution. Mesopterygoid fossa relatively broad but penetrates posterior palate into M3s, with mean angle 66°–73° among on geographic samples. Cheek teeth large (mean length of maxillary toothrow 8.6–8.7. Three folds typically present on all four upper and lower cheek teeth, with limited variation among geographic samples). In particular, dpm4 may have either three or four folds, and m3 may likewise have either two or three folds; counterfold formula 3–3–3–3/3(4)–3–3–(2)3.

Baculum short and massive (sample mean width 6.13–8.61 mm, proximal width 5.30–6.70 mm, distal width 5.18–7.40 mm), with broad but short shaft, expanded base, and deep notch in distal portion resulting in distinct apical extensions (Patton 1987; M. N. F. da Silva 1998; Patton et al. 2000).

DISTRIBUTION: *Proechimys cuvieri* is widely distributed throughout the Amazon Basin, from eastern Ecuador and Peru to eastern Brazil, Venezuela, and the Guianas.

SELECTED LOCALITIES (Map 525): BRAZIL: Acre, Igarapé Porongaba (Patton et al. 2000); Amapá, 4 km N of Amapá (Patton 1987); Amazonas, Barro Vermelho (Patton et al. 2000), Comunidade Colina, Rio Tiquié (Patton et al. 2000), Lago Meduinim, left bank Rio Negro (Patton et al. 2000); Pará, Floresta Nacional Tapirapé-Aquiri (Patton et al. 2000), Ilha do Taiuno (Patton 1987). ECUADOR: Sucumbios, Laguna Grande, Río Cuyabeno (FMNH 125088). FRENCH GUIANA: vicinity of Cayenne (Catzeflis and Steiner 2000). GUYANA: Barima-Waini, Baramita (ROM 100890). PERU: Amazonas, La Poza (Patton et al.2000); Loreto, Santa Luisa, Río Nanay (Patton 1967), Sarayacu, Río Ucayali (Patton 1987). SURINAM: Surinam, Lelydorpplan (Patton 1987). VENEZUELA: Bolívar, 69 km SE of Río Cuyuni (Patton et al. 2000).

SUBSPECIES: *Proechimys cuvieri* is monotypic.

NATURAL HISTORY: Habitat associations of *P. cuvieri* have been studied limitedly both in western (Patton et al. 2000) and central (Malcolm 1992) Amazonian Brazil, and French Guiana (Guillotin 1982, 1983; Voss et al. 2001; Adler et al. 2012). In all three areas, the species inhabits upland, or *terra firme* rainforest, but may be found equally in locally inundated forest or secondary upland forest and abandoned gardens. Malcolm (1992), for example, found *P. cuvieri* proportionally more abundant than sympatric *P. guyannensis* in early-successional and edge-dominated habitats. B. D. Patterson (1992b) also reported on specimens from the central Rio Juruá in western Brazil that were taken from dense virgin forest in hilly terrain, particularly within palm stands along an igarapé and on the margins of *igapó* or *várzea* seasonally inundated forest. At Paracou in French Guiana, both Voss et al. (2001) and Adler et al. (2012) took most specimens in traps placed on the ground, but Voss et al. captured two individuals in liana tangles up to 1 m above the ground. The species was taken in well-drained primary forest, creek-side primary forest, and secondary vegetation beside logs, at the bases of trees, among stilt roots, on top of logs and under masses of fallen branches.

Proechimys cuvieri is broadly sympatric with *P. guyannensis* throughout the Guianan region (Voss et al. 2001) and northern Amazonian Brazil (Malcolm 1992), and may co-occur with up to four other species along the Rio Juruá in western Brazil, being absolutely syntopic with as many as three others on the same traplines (Patton et al. 2000).

In western Brazil pregnant females were only observed in the months of February and March, during the wet season, but data are too limited to define the actual length of the breeding season. Only adult females reproduce, and litter size was always composed of two young (Patton et al. 2000). In French Guiana, juveniles were taken in every month of the year, although at a substantially higher proportion in March through May, suggesting that reproduction is continuous but with a peak in births also coinciding with the rainy season (Guillotin 1983). Males had larger home ranges than females, which were exclusive (Guillotin 1982), suggesting a polygynous mating system (Adler 2011).

Proechimys cuvieri has also been implicated as a vector for leishmania in French Guiana (Dedet et al. 1984).

REMARKS: Husson (1978) incorrectly identified specimens of this species from Surinam as *P. guyannensis guyannensis* (see Voss et al. 2001). Four strongly divergent geographic clades are defined by the limited mtDNA sequence data available (up to 10% divergence; Patton et al. 2000). One of these is distributed across the Guianan region and eastern Amazonian Brazil, east of the lower Rio Negro along both sides of the Rio Amazonas. A second is known from the Rio Juruá basin in western Brazil and northeastern Peru south of the Río Marañón. A third is known only from northern Peru, north of the Río Marañón. And a fourth is known only from the upper Rio Negro in Brazil, near the Colombian border. Each of these varied geographic clusters share the same basic set of morphological attributes, including the same bacular and glans characters and karyotype, with $2n = 28$ and $FN = 46–48$ (Reig, Tranier, and Barros 1980; Maia and Langguth 1993; M. N. F. da Silva 1998; Patton et al. 2000; N. A. B. Ribeiro 2006; Eler et al. 2012; C. E. F. Silva et al. 2012). However, as noted by Voss et al. (2001), differences in the relative proportions of the qualitative cranial features described by Patton (1987) are apparent in separate geographic regions such that more rigorous analyses of larger samples may well distinguish morphological groupings coincidental with the markedly distinct mtDNA clades. Should more than a single species be recognized for this group of geographic units, only that from the Guianan region, which contains the type locality of *P. cuvieri* Petter, would have a name to which it can be reliably referred. The remaining clades would lack a name. As with so many other "species" of *Proechimys*, here is another rich opportunity for further research to elucidate species boundaries and their geographic ranges. Steiner et al. (2000) and Van Vuuren et al. (2004) provided more detailed analyses of molecular genetic population structure for samples in the Guianan region, where there is higher sequence diversity and greater geographic structure than in the co-distributed populations of *P. guyannensis*.

Guillotin and Ponge (1984) doubted that *P. cuvieri* could be distinguished from sympatric *P. guyannensis* in French Guiana by standard craniodental measurements, but Catzeflis and Steiner (2000) countered this conclusion by a thorough morphometric analysis of appropriately aged samples. These latter authors also provided a detailed distribution map of known localities in French Guiana. Voss et al. (2001)

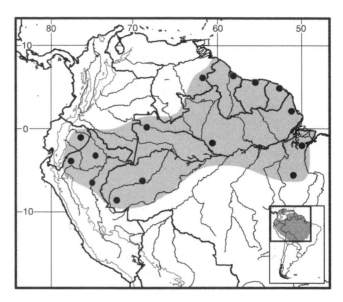

Map 525 Selected localities for *Proechimys cuvieri* (●). Contour line = 2,000 m.

extended the Catzeflis and Steiner (2000) analysis and evaluated a number of qualitative characters that distinguish *P. cuvieri* from *P. guyannensis*. For example, the single, and easily measured, metric of maxillary toothrow length is non-overlapping between their large series of both species (7.0–8.0 mm in *guyannensis* and 8.2–9.3 in *cuvieri*; table 45, p. 157). Malcolm (1992) documented that these two species were also readily separable by toothrow length as well as hindfoot length for populations in the central Brazilian Amazon near Manaus. In males, the difference in baculum width and shape is easily palpated in living specimens, even in juveniles (L. H. Emmons, pers. comm.). Lara et al. (1992) examined age and sex components of craniodental metrical dimensions in a large sample from Amapá state, eastern Brazil.

Proechimys longicaudatus (Rengger, 1830)
Long-tailed Spiny Rat
SYNONYMS:

Echimys longicaudatus Rengger, 1830 236, type locality "Northern Paraguay."

Loncheres myosurus: Lichtenstein, 1830:plate 36, Fig. 2, and unnumbered text; not *myosurus* Lichtenstein.

Echimys myosurus: I. Geoffroy St.-Hilaire, 1840:15, 17; name combination; not *myosurus* Lichtenstein.

Echimys cayennensis: Pictet, 1841b,145; name combination; not *Echimys cayennensis* Desmarest.

E[chinomys]. longicaudatus: Wagner, 1843a:341; name combination.

P[roechimys]. longicaudatus: Thomas, 1901g:532; first use of current name combination.

Proechimys [(Proechimys)] cayennensis longicaudatus: Ellerman, 1940:121; name combination.

Proechimys leucomystax Miranda Ribeiro, 1914:42, type locality "Utiarití, Rio Papagaio," Mato Grosso, Brazil.

Proechimys longicaudatus: Moojen, 1948:346; name combination.

Proechimys longicaudatus longicaudatus: Moojen, 1948: 351; name combination.

Proechimys longicaudatus leucomystax: Moojen, 1948: 352; name combination.

Proechimys guyannensis villacauda Moojen, 1948:355; type locality "Tapirapoã, Rio Sepotuba, Cáceres, Mato Grosso, Brazil."

Proechimys guyannensis ribeiroi Moojen, 1948:361; type locality "Rio 12 de Outubro, affluent of the Camararé, Mato Grosso, Mato Grosso, Brazil; about 190 kilometers west of Utiarití; altitude 414 meters."

DESCRIPTION: Medium sized for genus, head and body length averaging about 220–250 mm, and with proportionately short tail, averaging about 60–63% of head and body length. Dorsal color distinctly pale reddish or yellowish brown, streaked with dark brown, contrasting sharply with dark reddish brown mixed with black of both *P. brevicauda* and *P. cuvieri*. Venter pure white, without hint of buffy overtones. Dorsal surfaces of hindfeet highly variable, from uniformly white, including toes (most specimens), to mostly dusky, or dusky on outside and white on inside. Dark ankle band may or may not separate white inner thigh from white foot. Tail bicolored, thinly haired with scales obvious; scale annuli 9–10 per cm at midpoint. Fur relatively soft to touch, with aristiforms narrow (0.6–0.7 mm), rather short (17–18 mm), and terminating with long whip-like tip.

Skull similar in nearly all features to other members of group, but the skull generally smaller than that of either *P. brevicauda* or *P. cuvieri*. Temporal ridge varies from weakly continuous across parietals or limited to simple posterior extension of supraorbital ledge. Incisive foramina lyrate in shape, as in *P. brevicauda* and *P. cuvieri*, with typically well-developed posterolateral flanges that extend onto anterior palate forming deep groves, and an expanded and long premaxillary portion of septum with short, often thin, but typically keeled maxillary portion that continues as median ridge onto anterior palate; vomerine portion of septum short but exposed. Floor of infraorbital foramen smooth, without groove for infraorbital nerve. Postorbital process of zygoma well developed and comprised equally by both jugal and squamosal. Mesopterygoid fossa broad (angle averages 78°), but penetrates posterior palate to middle of M3s. All maxillary cheek teeth have three counterfolds; mandibular cheek teeth vary from three (rarely four) on pm4 to two or three folds on each molar. Counterfold formula 3-3-3-3/3(4)-2(3)-2(3)-2(3). Population samples of *P. longicaudatus* have higher percentage of individuals with two folds on lower molars than do other two species. Pattern fits general trend of decreasing counterfold number in species, and/or their populations, along an environmental gradient from extremely wet forests (trans-Amazonian Cocó and western Amazon) to dry forests (eastern Amazonia and central Brazil; Patton 1987). Other examples of this trend are *P. quadruplicatus* in the western Amazon versus *P. goeldii* in east, or samples of *P. semispinosus* from Chocó in western Colombia versus those of isolated population of this species (*rosa* Thomas) in dry forests of southern Ecuador.

Baculum virtually identical to that of *P. brevicauda*; robust, long, and wide (sample mean width 10.17–11.08 mm, proximal width 3.20–5.02 mm, distal width 5.10–5.22 mm), with short, stout apical wings (Patton 1987).

DISTRIBUTION: *Proechimys longicaudatus* occurs in the dry tropical forests of eastern Bolivia, northern Paraguay, and central Brazil.

SELECTED LOCALITIES (Map 524): BOLIVIA: Santa Cruz, Buenavista (S. Anderson 1997), Santa Cruz, El Refugio (USNM 588195). BRAZIL: Goiás, Parque Nacional das Emas (T. Machado et al. 2005); Mato Grosso, Apiacás (T. Machado et al. 2005), Aripuanã (T. Machado et al. 2005), Fazenda São Luís (MVZ 197661); Mato Grosso do Sul, Urucum (Patton 1987). PARAGUAY: Alto Paraguay, 54 km E of Agua Dulce (Patton 1987).

SUBSPECIES: The number of taxa assigned to this species could signal valid geographic units, but the analyses needed to determine such have not as yet been done. Consequently, *P. longicaudatus* is treated as monotypic.

NATURAL HISTORY: Little has been published about the natural history of this species. S. Anderson (1997) recorded pregnant females in March and August in Bolivia. Emmons (2009) found that reproduction seems to begin after a hiatus about the first week in September, with appearance of lactating and late-gestation females, shortly after which young began to appear in the traps. This species was the most common rodent in some dry forests, where its populations remained stable over 10 years. The species also bred in mid-savanna grassland habitats shaded by trees and shrubs, where it denned in holes in termite mounds and in armadillo burrows (Emmons 2009, and pers. comm.) Three species of fleas have been reported, also from Bolivian specimens (Smit 1987). Valim and Linardi (2008) described the host association of ectoparasitic lice in the genus *Gyropus*.

REMARKS: The exact locality from which the type specimen was collected has not been determined, although Thomas (1903c:240) wrote "Rengger's type was obtained on the 21st parallel of latitude, therefore not far south of Corumbá" (state of Mato Grosso do Sul, Brazil). He considered it "nearly allied to my *P. bolivianus* [herein = *P. brevicauda* Günther], which differs from it by the cranial characteristics given in the description of the latter." Moojen (1948) assigned specimens from Utiarití on the Rio Papagaio in Mato Grosso, the type locality of Miranda-Ribeiro's *leucomystax*, to both *leucomystax* and *villacauda* Moojen, considering these to belong to separate species based on differences in counterfold counts of the lower pm4 and m1. Our view is that these are the same taxon with a varying expression of counterfold number, which is common in samples examined by one of us (JLP) from eastern Bolivia and western Brazil (Patton 1987:334–335, Table 5). However, as noted in the account of *P. brevicauda*, a clear understanding of both geographic and character range must wait additional sampling that explicitly focuses on the undescribed species we mention above. It is likely that northern localities in Brazil herein ascribed to *P. longicaudus* may actually represent this new taxon.

T. Machado et al. (2005) report karyotypic variation of $2n = 28$ with the FN varying from 48–50 for samples from Mato Grosso and Goiás states in west-central Brazil. These karyotypes are the same as that described for *P. brevicauda* from Amazonian Brazil and Peru (Patton et al. 2000), except the Y chromosome is biarmed rather than uniarmed.

Proechimys semispinosus group

Patton (1987) included 13 named taxa in his concept of the *semispinosus*-group, most of which had been regarded in earlier literature as subspecies of *P. semispinosus* or as synonyms. He further suggested that only two valid species were contained within the group, *P. semispinosus* and *P. oconnelli*. Among South American taxa, the insular *P. gorgonae* Bangs and *P. rosa* Thomas from southern Ecuador had been recognized by some authors as a species, but listed as subspecies by others. Here, we follow A. L. Gardner and Emmons (1984) and Patton (1987) in recognizing two species, the widespread *P. semispinosus* and the limited range *P. oconnelli*.

KEY TO THE SPECIES OF THE *PROECHIMYS SEMISPINOSUS* SPECIES GROUP:

1. Distributed in Central American south from Honduras along the west coast of Colombia and Ecuador; temporal ridge well developed; cheek teeth complex, typically with four counterfolds commonly present on both upper and lower molar series and especially on lower pm4 . . . . *Proechimys semispinosus*
1'. Distributed east of the Cordillera Oriental in the northwestern Amazon; temporal ridge nonexistent or only weakly developed; cheek teeth simplified, with three counterfolds on upper cheek teeth and lower pm4, and two to three counterfolds on lower molar series . . . . *Proechimys oconnelli*

Proechimys oconnelli J. A. Allen, 1913
O'Connell's Spiny Rat
SYNONYMS:

Proechimys o'connelli J. A. Allen, 1913a:479; type locality "Villavicencio (alt. 1600 ft.), Colombia," Meta.

Proechimys cayennensis o'connelli: Tate, 1939:178; name combination but incorrect use of an apostrophe (ICZN 1999:Art. 32.5.2.3).

[*Proechimys guyannensis*] *oconnelli*: Hershkovitz, 1948a: 133; name combination.

DESCRIPTION: Moderately large, head and body length up to 250 mm in adult individuals, and with a medium-length tail (about 70% of head and body length). Dorsal color orange rufous finely lined with black, paler on sides than mid-back and rump. Venter pure white, sharply defined against color of sides. Pale inner thigh stripes continuous across ankles onto dorsal surfaces of hindfeet, which are two-toned, pale cream on inner half and light brown on outer half, with dark color typically extending to digits IV and V; plantar pads moderate in size, with thenar and hypothenar pads subequal. Tail sharply bicolored, dark brown above and creamy-white below, and thinly clothed with short, fine hairs; visible scale annuli are relatively wide, averaging nine per cm at midpoint. Pelage neither distinctly nor heavily spinous, as aristiforms are weakly developed, long (18–21 mm) and thin (0.8–0.9 mm), and tipped with long whip-like filament.

Skull unremarkable, with elongated and tapering rostrum. Temporal ridge either weakly developed or nonexistent, extending posteriorly from supraorbital ledge onto parietals, contrasting sharply with condition in *P. semispinosus*. Incisive foramina angular or lyrate in shape, with moderately developed posterolateral flanges that extend onto anterior palate forming grooves on either side of midline, and despite only moderate development of maxillary keel and median palatal ridge. Premaxillary portion of septum well developed and elongated, encompassing more than half opening; premaxillary portion well developed and always in contact with premaxillary portion; vomer completely encased and not visible in ventral aspect. Floor of infraorbital foramen has obvious groove supporting passage of maxillary nerve, and formed by distinct lateral flange. Mesopterygoid fossa moderate in width, with angle averaging 63°, and penetrating to middle of M3s. Postorbital process of zygoma obsolete but formed completely by jugal. Cheek teeth both above and below simplified, with three counterfolds on each upper tooth and lower pm4 and two to three counterfolds on lower molars; counterfold pattern 3–3–3–3/3–(2)3–(2)3–2(3).

Baculum of medium length but broad (sample mean length 8.16–8.43 mm, proximal width 4.93–4.94 mm, distal width 4.63–5.26 mm), with blunt and thickened base and

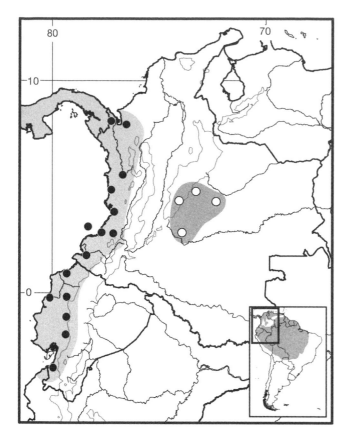

Map 526 Selected localities for *Proechimys oconnelli* (○) and *Proechimys semispinosus* (●). Contour line = 2,000 m.

indented sides typical of *P. semispinosus*, but without distal apical extensions characteristic of that species (Patton 1987).

DISTRIBUTION: *Proechimys oconnelli* is known only from east central Colombia, east of the Cordillera Oriental in the headwaters of the Río Meta and Río Guaviare.

SELECTED LOCALITIES (Map 526): COLOMBIA: Cundinamarca, Mámbita (USNM 240036); Meta, Barrigona (Patton 1987), Meta, La Macarena Parque (FMNH 88051); Cundinamarca, La Aguadita (Patton 1987).

SUBSPECIES: *Proechimys oconnelli* is monotypic.

NATURAL HISTORY: There have been no published studies on the ecology or population biology of this species.

REMARKS: Both A. L. Gardner and Emmons (1984), using bullar septal patterns and karyotypes, and Patton (1987), using bacular and other craniodental characters, suggested that *P. oconnelli* was the only Amazon-drainage spiny rat to have a close affinity with the trans-Andean *P. semispinosus*. This hypothesis, however, has yet to be tested by a cladistic analysis of any character set, morphological or molecular. A. L. Gardner and Emmons (1984) described and illustrated a karyotype of $2n=32$, FN=52, from a topotype, which differs from that of *P. semispinosus* mainly in the presence of two medium acrocentrics instead of a single large submetacentric.

Proechimys semispinosus (Tomes, 1860)
Tomes's Spiny Rat
SYNONYMS:

Echimys semispinosus Tomes, 1860b:265; type locality "Gualaquiza, southeastern Ecuador," amended to "Esmeraldas, Prov. Esmeraldas, on the Pacific coast of Ecuador" (A. L. Gardner 1983a).

Echinomys semispinosus: Thomas, 1882:101; name combination.

Echinomys semispinosus True, 1889:467; type locality "San Emilio, Lake Nicaragua, Nicaragua"; not *semispinosus* Tomes.

Echinomys centralis Thomas, 1896:312; renaming of *semispinosus* True.

[*Proechimys*] *semispinosus*: J. A. Allen, 1899c:264; first use of current name combination.

Proechimys rosa Thomas, 1900b:219; type locality "Santa Rosa, S. W. [= El Oro] Ecuador. Alt. 10 meters."

Proechimys centralis panamensis Thomas, 1900b:220; type locality "Savanna near Panama," Panama, Panama.

Proechimys centralis chiriquinus Thomas, 1900b:220; type locality "Bugava [=Bugaba], Chiriqui, N. W. Panama," Chiriqui, Panama.

Proechimys burrus Bangs, 1901:640; type locality "Isla San Miguel, Archipelago de las Perlas" Golfo de Panama, Panama.

Proechimys gorgonae Bangs, 1905:89; type locality "Gorgona Island," about 50 km W Punta las Reys, Cauca, Colombia.

Proechimys semispinosus calidior Thomas, 1911c:254; type locality "San Javier, Lower Cachavi R., N.W. [Esmeraldas] Ecuador. Alt. 60 feet."

Proechimys rubellus Hollister, 1914a:57; type locality "Angostura Valley, Costa Rica," but holotype actually came from "Pacuare, Costa Rica" (True, 1889:467; A. L. Gardner, 1983a:136).

Proechimys centralis colombianus Thomas, 1914c:60; type locality "Condoto, Choco, W. Colombia. Alt. 300'."

Proechimys semispinosus panamensis: Goldman, 1920:120; name combination.

Proechimys semispinosus burrus: Goldman, 1920:122; name combination.

Proechimys semispinosus goldmani Bole, 1937:178; type locality "altos Cacao, Prov. Veraguas, Panama."

Proechimys [(*Proechimys*)] *cayennensis semispinosus*: Ellerman, 1940:120; name combination.

Proechimys [(*Proechimys*)] *cayennensis burrus*: Ellerman, 1940:120; name combination.

Proechimys [(*Proechimys*)] *cayennensis centralis*: Ellerman, 1940:120; name combination.

Proechimys [(*Proechimys*)] *cayennensis panamensis*: Ellerman, 1940:120; name combination.

Proechimys [(*Proechimys*)] *cayennensis rubellus*: Ellerman, 1940:120; name combination.

Proechimys [(*Proechimys*)] *cayennensis colombianus*: Ellerman, 1940:120; name combination.

Proechimys [(*Proechimys*)] *cayennensis calidior*: Ellerman, 140:120; name combination.

Proechimys [(*Proechimys*)] *cayennensis gorgonae*: Ellerman, 140:120; name combination.

Proechimys [(*Proechimys*)] *cayennensis rosa*: Ellerman, 1940: 120; name combination.

Proechimys semispinosus ignotus Kellogg, 1946:61, type locality "Isla San José, Archipelago de las Perlas, Golfo de Panama, Panama."

[*Proechimys guyannensis*] *calidior*: Hershkovitz, 1948a:133; name combination.

[*Proechimys guyannensis*] *semispinosus*: Hershkovitz, 1948a: 133; name combination.

Proechimys semispinosus gorgonae: Moojen, 1948:316; name combination.

Proechimys semispinosus rosa: Moojen, 1948:316; name combination.

Proechimys guyannensis gorgonae: Cabrera, 1960:520; name combination.

DESCRIPTION: Moderately variable species geographically, larger in body size in Central America and northern Colombia than in southern part of range in Ecuador. Panamanian specimens may range to nearly 1 kg in mass (Adler 1996). Head and body length varies from about 290 mm in northern Colombia to 240 mm in northwestern Ecuador, with tail length varying proportionally from about 63% of head and body length in the north to 70% in the south; hindfoot length 45–60 mm; ear length 18–30 mm. Dorsal color in South American samples consistently dark reddish brown liberally speckled with black, with sides only slightly paler to contrast sharply with uniformly white venter. Pale inner thigh stripes do not continue across ankle onto uniformly dark dorsal surfaces of hindfeet. Plantar pads of hindfeet well developed; thenar and hypothenar pads enlarged and subequal in size. Tail sharply bicolored, dark brown above and pale below, particularly in northern samples, but less bicolored in southern samples from northwestern Ecuador. Hairiness of tail varies among individuals, with some moderately clothed in elongated hairs such that tail scales nearly hidden from view while others, from same population sample, with more sparsely haired tails; scales large, annuli obvious to eye, 7–8 per cm at midpoint. Dorsal pelage stiff to touch, but aristiform development varies from north to south. Spines consistently long (19–21 mm) in all populations, but vary in width, and hence stiffness (width 0.9–1.1 mm in northern samples and 0.6–0.8 in southern ones). Each aristiform terminates in elongated, filament-like, not blunt, tip.

Skull large (greatest skull length 51.2–67.0 mm) and broad across the zygomatic arches, but with an elongated and narrowed rostrum. Species uniquely characterized among spiny rats by its well-developed temporal ridges extending from supraorbital ledge across length of parietals, only rarely interrupted into anterior and posterior segments. Incisive foramina of specimens from Colombia and Ecuador narrow, with almost parallel sides, or weakly lyre-shaped; posterolateral margins usually strongly flanged, creating deep grooves extending onto anterior palate despite only moderate development of medial ridge; premaxillary portion of septum long, encompassing nearly entire length of opening; maxillary portion varies from well developed to attenuate, only weakly keeled at best, and nearly always in contact with premaxillary portion; vomer completely hidden from ventral view. Floor of infraorbital foramen with groove supporting maxillary nerve, formed by well-developed lateral flange. Mesopterygoid fossa of moderate width, but angle becomes broader from north to south (57°–62°). In contrast to *P. oconnelli*, postorbital process of zygoma moderately well developed and more commonly formed (especially in northern samples, less so in southern ones) by jugal. Counterfold count of cheek teeth similar to that of species of *goeldii*-group, with four folds commonly present on all upper teeth and on pm4, less commonly on lower m2 and even m3. However, fold number decreases in samples from northern Colombia to southern Ecuador (samples of *rosa* Thomas), where four folds are rare on all teeth and two folds may be found on M3 and all lower molars (Patton 1987:334–335, Table 5).

Baculum of massive, long, and broad (sample mean length 8.54–9.66 mm, proximal width 4.21–4.94 mm, distal width 5.11–6.19 mm), with deeply concave margins, broadly expanded and thickened base, and wide distal portion with well-developed apical wings separated by median depression (Patton 1987).

DISTRIBUTION (South America only): *Proechimys semispinosus* is found along the coastal lowlands of western Colombia and Ecuador, from the Panama border in the north to near the Peruvian border in the south.

SELECTED LOCALITIES (Map 526; from Patton 1987, except as noted): COLOMBIA: Cauca, Isla Gorgona (type locality of *Proechimys gorgonae* Bangs), Cauca, Río Saija; Nariño, Barbacoas (AMNH 34172); Chocó, , Río Docompado, Unguía; Cordoba, Socorré; Valle de Cauda, Zabaletas (FMNH 86766). ECUADOR: Chimborazo, Puente de Chimbo; El Oro, Santa Rosa (type locality of *Proechimys rosa* Thomas); Esmeraldas, Esmeraldas (type locality of *Echimys semispinosus* Tomes); Guayas, Bucay; Manabí, Cuaque; Pichincha, Santo Domingo.

SUBSPECIES: Hall (1981) recognized six subspecies in Central America (*burrus* Bangs, *centralis* Thomas, *goldmani* Bole, *ignotus* Kellogg, *panamensis* Thomas [with *chiriquinus* Thomas as a synonym], and *rubellus* Hollister. The number and range of subspecies in the South American part of the distribution has not been assessed. The insular *gorgonae* Bangs averages smaller than samples from the Colombian mainland, and is unique in its very dark brown, almost black, dorsal color, and gray-brown infusion across the chin, throat, upper chest, and encroachment along the sides to the inguinal region. This taxon might justify status as a subspecies on the basis of its unique coloration alone.

NATURAL HISTORY: Extensive ecological and life history studies of this species have been published for populations in Panama, including microhabitat use and spacing patterns (Adler et al. 1997; Tomblin and Adler 1998; T. D. Lambert and Adler 2000; Endries and Adler 2005); density, home range overlap, and mating system (Adler et al. 1997; Seamon and Adler 1999; Endries and Adler 2005); seed predation and food habits (Bonaccorso et al. 1980; Adler 1995; Hoch and Adler 1997; Adler and Kestell 1998; Mangan and Adler 1999; Adler 2000); growth and reproduction (Tesh 1970b; Gliwicz 1984; Adler and Beatty 1997); life span (Oaks et al. 2008); and comparative life history (T. H. Fleming 1971). Adler et al. (2003) described infestation by bot flies, Durette-Desset (1970a) described nematode parasites, and McKee and Adler (2002) documented tail autotomy. Little apparent ecological or life history data are available for populations in South America. Bangs (1905) did note, however, that spiny rats on Isla Gorgona were so common that not all specimens trapped could be preserved before the crabs ate them. Valim and Linardi (2008) described the host association of ectoparasitic lice in the genus *Gyropus*.

REMARKS: Thomas (1900b:219), in describing *P. rosa*, considered it "most nearly allied to *P. chrysaeolus*" and not belonging to *P. semispinosus* Tomes, although he also thought *P. rosa* was "allied to the Central American species *P. centralis*," which now is considered a synonym (and valid subspecies) of *P. semispinosus* (Hall 1981).

Bangs (1905), in his description of *P. gorgonae* from Isla Gorgona off the Pacific coast of southwestern Colombia, stated that this "species" was marked only by its unique color, as its skull was indistinguishable from mainland samples of *P. centralis* (= *P. semispinosus*).

Specimens in the Field Museum from Lagunas, near the junction of the Marañón and Ucayali rivers, in northern Peru that Osgood (1944:200–2001) referred to *P. semispinosus* are *P. steerei* (see that account).

Patton and Gardner (1972) described and figured a karyotype of $2n = 30$, FN = 54, from Costa Rica, and A. L. Gardner and Emmons (1984) reported intraspecific variation in the morphology of two pairs of small autosomes,

with $2n = 30$ but FN = 50–54, in samples from Panama, Colombia, and Ecuador, including near-topotypes of *rosa* Thomas. Gómez-Laverde et al. (1990) described and figured the karyotype of *gorgonae* Bangs as $2n = 30$, FN = 56, and M. L. Bueno and Gómez-Laverde (1993) recorded two karyotypes for the Pacific lowlands of Colombia, both $2n = 30$, FN = 56, that differ in the amount and pattern of constitutive heterochromatin. Based on similarities in C-banding patterns, M. L. Bueno and Gómez-Laverde (1993) suggested a close relationship between *gorgonae* Bangs and mainland populations of *P. semispinosus* in southern Colombia and western Ecuador.

Proechimys simonsi species group

This group is defined on the basis of its relatively large size; long and slim body form with elongated and narrow head and skull; absolutely and proportionately long tail, bright white venter and white hindfeet; distinctive, rather oval shaped incisive foramina with a noticeably weak maxillary part to the septum often not in contact with the premaxillary portion; flat and smooth anterior palate without a median ridge; narrow and deeply penetrating mesopterygoid fossa; and long and narrow baculum (Patton 1987). Only a single species is currently recognized, but deeply divergent and geographically structured DNA sequence clades may underlie greater levels of species diversity than presently understood (Patton et al. 2000; R. N. Leite and J. L. Patton, unpubl. data).

Proechimys simonsi Thomas, 1900
Simons's Spiny Rat

SYNONYMS:

Proechimys simonsi Thomas, 1900f:300; type locality "Perené River, Junin Province, Peru. Altitude 800 m."

Proechimys hendeei Thomas, 1926g:162; type locality "Puca Tambo, 5100′," 1,480 m, on trail from Chachapoyas to Moyobamba, Río Huallaga drainage (Stephens and Traylor 1983), San Martín, Peru.

Proechimys [(*Proechimys*)] *cayennensis simonsi*: Ellerman, 1940:121; name combination.

Proechimys hendeei nigrofulvus Osgood, 1944:199; type locality "Montalvo, Rio Bobonaza, southeast of Sarayacu," Pastaza, Ecuador.

[*Proechimys longicaudatus*] *hendeei*: Moojen, 1948:316; name combination.

[*Proechimys longicaudatus*] *nigrofulvus*: Moojen, 1948:316; name combination.

[*Proechimys longicaudatus*] *simonsi*: Moojen, 1948:316; name combination.

Proechimys rattinus: Cabrera, 1961:524; referral of paratype (BM 24.2.22.18), but not holotype, to *P. hendeei* (= *P. simonsi* Thomas).

DESCRIPTION: One of largest species in genus, equaled or exceeded in body size only by *P. semispinosus*, *P. quadruplicatus*, and *P. steerei* (head and body length 165–275 mm). Body of living individuals distinctly elongated, face long and narrow, ears long (length 21–28 mm), tail absolutely (length 118–231 mm) and proportionately long (mean 85% of head and body length), hindfeet long (length 45–57 mm). Tail sharply bicolored, dark above and white below; covered by sparse, fine hair, although small scales remain conspicuous to eye (9–13 annuli per cm at midpoint). Mid-dorsal color darker than sides, reddish brown, coarsely streaked with black hairs and interspersed by dark brown aristiforms. Aristiform hairs long (22–24 mm) and thin (0.2–0.4 mm) with distinctly whip-like tip. Venter, chin, sides of upper lips, undersurfaces of forelimbs, and hind limbs pure white. White of inner legs extends across tarsal joint onto white dorsal surfaces of hindfeet. As noted by Patton and Gardner (1972), *P. simonsi* apparently unique among all species of genus in lacking hypothenar pad on plantar surface of hindfeet.

Skull large (greatest skull length 44.1–63.7 mm), rostrum distinctly long and narrow (mean length 21.9 mm, mean width 8.2 mm). Supraorbital ridges well developed, but do not extend onto parietals as temporal ridge. Incisive foramina diagnostically ovoid in shape, sometimes slightly elongated but never with strongly constricted posterior margins, flat posterolateral margins lacking grooves extending onto anterior palate, short and rounded premaxillary portion of septum usually no more than half length of opening, and attenuate maxillary portion usually not in contact with premaxillary part. Floor of infraorbital foramen usually grooved, with moderately developed lateral flange. Anterior border of mesopterygoid fossa acutely angled (49°–53°), penetratingly deeply into palate, reaching anterior half of M3s or middle of M2s. PM4 and M1 typically with three folds while M2–M3 have three or four; number of folds, particularly on lower pm4, varies from north to south from four to three. Counterfold formula 3-3-3(4)–3(4)/(3)4-3-3-(2)3.

Baculum is long and narrow (sample mean length 8.42–10.09 mm, proximal width 1.93–2.50 mm, distal width 1.51–1.89 mm), with rounded and slightly broadened base (Patton and Gardner 1972 [as *P. hendeei*]; see Patton 1987).

DISTRIBUTION: *Proechimys simonsi* occurs in the western Amazon Basin, including the eastern Andean slopes, from southern Colombia through eastern Ecuador, eastern Peru, northern Bolivia, and into western Brazil.

SELECTED LOCALITIES (Map 527): BOLIVIA: Beni, Río Mamoré (S. Anderson 1997); La Paz, Alto Río Madidi, Moira Camp (USNM 579259); Pando, 18 km N of San Juan de Nuevo Mundo (USNM 579616). BRAZIL: Amazonas, Altamira, Rio Juruá (Patton et al. 2000), Colo-cação Vira-Volta, Rio Juruá (Patton et al. 2000), alto Rio Urucu (Patton et al. 2000). COLOMBIA: Caquetá, La Morelia (Patton 1987); Putumayo, Río Mecaya (Patton 1987). ECUADOR: Morona-Santiago, Gualaquiza (Patton 1987); Napo, San Francisco (UMMZ 80046); Orellana, San José Abajo (Patton 1987); Pastaza, Río Pindo Yacu (Patton 1987). PERU: Amazonas, Yambrasbamba (Patton 1987); Cajamarca, Huarandosa (Patton 1987); Cusco, Consuelo, 15.9 km SW of Pilcopata (Solari et al. 2006), 2 km SW of Tangoshiari (USNM 588058); Huánuco, Tingo Maria (FMNH 24800); Loreto, Boca Río Curaray (Patton 1987), Orosa (Patton 1987); Madre de Dios, Río Tavara, Fila Boca Guacamayo (USNM 579696); Pasco, San Pablo (Patton 1987); San Martín, Puca Tambo (type locality of *Proechimys hendeei* Thomas).

SUBSPECIES: As currently understood, *P. simonsi* is monotypic (but see Remarks).

NATURAL HISTORY: This species primarily inhabits upland, nonseasonally flooded or *terra firme* forests, including undisturbed rainforest and secondary or disturbed forests and garden plots in the western Amazon Basin. However, it extends upward to above 2,000 m in montane forest on the eastern slope of the Andes, and thus has the broadest elevational range of any species of *Proechimys*. In western Brazil, reproductive females (those pregnant or lactating) were taken in all seasons, suggesting that breeding may take place throughout the year. The modal litter size was two, with the range in embryo number from one to three. Both males and females apparently do not reach reproductive maturity until they are fully grown and have molted into their adult pelage (Patton et al. 2000).

Emmons (1982) examined the population ecology, including habitat association, home range size, nightly movement patterns, and density in sympatric populations of *P. simonsi*, *P. steerei*, and *P. brevicauda* in southeastern Peru (for which she used the names *hendeei*, *brevicauda*, and *longicaudatus*, respectively, following Patton and Gardner 1972). Females appear to be territorial, which is likely the reason they reproduce only as adults (they may be unable to acquire territories on which to breed as subadults; Emmons 1982, and pers. comm.).

REMARKS: In his description of *P. simonsi*, Thomas (1900f) considered it externally similar to *P. chrysaeolus* and *P. rosa* (= *P. semispinosus* herein), with the skull "scarcely distinguishable from that of the outwardly very different *P. brevicauda*" (p. 300–301). Osgood (1944:202) referred specimens from the Río Perené at an elevation of 800 ft in central Peru to *P. simonsi*, stating that the species is "probably allied to *brevicauda*, from which it differs at least in immaculate white under parts." It is, however, difficult to reconcile either of these comparisons, as *P. simonsi* is perhaps the most easily recognizable spiny rat, readily separable from all others externally by virtue of its elongated body, pure white venter, elongated and sharply bicolored tail, and five hindfoot pads; and cranially by its uniquely shaped and constructed incisive foramina and narrow, deeply penetrating mesopterygoid fossa (Patton 1987; Patton et al. 2000). Neither of these conditions even approach those of the other species to which both Thomas and Osgood referred.

Cabrera (1961:524) referred the paratype of *Proechimys rattinus* used by Thomas (1926g) in his description of this species to *P. hendeei* (= *P. simonsi* herein). This specimen (BM 24.2.22.18) is a skin only of a young animal, with juvenile pelage still on the lower back and rump. It is dark colored with a white venter and gray throat patch but a proportionately long tail (79% of head and body length). Without a skull, identification is difficult, but Cabrera's allocation appears to be correct.

Two markedly divergent geographic clades are apparent in the limited mtDNA sequence data available (Patton et al. 2000). One of these is in northern Peru, north of the Río Marañón–Río Amazonas axis, and the second covers the remainder of the range in eastern and southern Peru, western Brazil, and northern Bolivia. If subsequent studies document that these clades correspond to diagnosable taxa, then *nigrofulvus* Osgood would be an available name for the northern clade. The karyotype varies only minimally throughout the sampled range, from eastern Ecuador to southern Peru and western Brazil, with $2n = 32$ and $FN = 56$–58 (Patton and Gardner 1972; A. L. Gardner and Emmons 1984; Patton et al. 2000; N. A. B. Ribeiro 2006). Reig and Useche (1976) described an identical karyotype for specimens from southern Colombia, but we have not examined the vouchers and thus cannot unequivocally allocate them to *P. simonsi*. Similarly, Aniskin et al. (1991) described a $2n = 32$, $FN = 58$, karyotype from northern Peru but without a morphological description to assign those specimens to *P. simonsi* with confidence.

Matocq et al. (2000) described molecular population genetic structure in relation to riverine barriers and habitat range in western Brazil, contrasting those patterns with the codistributed *P. steerei*. E. P. Lessa et al. (2003) provided estimates of late and post-Pleistocene population growth and stability based on coalescence analysis of these same molecular sequence data. Milishnikov (2006) contrasted patterns of variation in allozyme loci between *P. simonsi* and other co-occurring species of spiny rats in northern Peru.

Proechimys trinitatis species group

Patton (1987) assigned nine nominal taxa to this group, but made no recommendations as to the actual number of species. Curiously, most members of this group have been better studied with a broader range of methods and characters than any other group of spiny rats, yet species boundaries remain uncertain and phylogenetic relationships among species are unclear. Karyotypic variation is extensive (reviewed by Reig and Useche 1976; Reig et al. 1979; Reig, Aguilera et al. 1980; Reig 1981), and both karyotypic and electromorphic data (Benado et al. 1979; Reig, Aguilera et al. 1980) differentiate a *guairae* superspecies (including *guairae* Thomas and *poliopus* Osgood) and a *trinitatis* superspecies (composed of *trinitatis* Allen and Chapman and *urichi* Allen). The *guairae* superspecies exhibits a sequence of karyotypic diversity from $2n = 42$–62, $FN = 72$–76, that form a ring of ill-defined species or subspecies around the eastern-most branch of the northern Andes (Benado et al. 1979; Reig, Aguilera et al. 1980; Corti et al. 2001). Although we acknowledge the extensive karyotypic diversity, for the present we consider these forms to represent a single species, *P. guairae*. Karyotypic intermediates are known from points of geographic contact between some of these chromosomal forms (Aguilera, Reig, and Pérez-Zapata 1995), all karyotypic forms are very closely related, based on protein allozyme comparisons (Benado et al. 1979; Pérez-Zapata et al. 1992), and all share a similar craniometric morphology (Aguilera and Corti 1994; Corti and Aguilera 1995; Corti et al. 2001). Following the arguments of Reig, Aguilera et al. (1980)

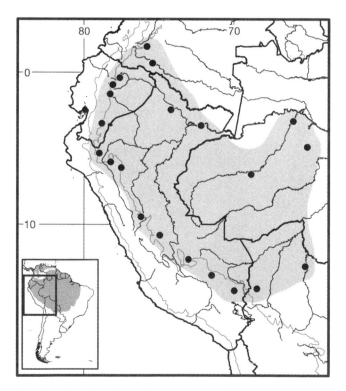

Map 527 Selected localities for *Proechimys simonsi* (●). Contour line = 2,000 m.

in their delineation of the *trinitatis* superspecies, we place both *trinitatis* J. A. Allen and Chapman and *urichi* J. A. Allen into the single species *P. trinitatis*. O. J. Linares (1998), however, included all members of both the *P. guairae* and *P. trinitatis* superspecies (*sensu* Reig, Aguilera et al. 1980) in the single species *P. trinitatis*, recognizing three subspecies (*P. t. trinitatis* from the Cordillera Oriental and Orinoco delta; *P. t. guairae* from the Cordillera Central, Llanos, and flanks of the Mérida Andes; and *P. t. ochraceus* from the Maracaibo Basin). We retain the usual consideration of *P. mincae* as a valid species, although A. L. Gardner and Emmons (1984) suggested that it is also a member of the *guairae* superspecies. Finally, we recognize that *P. chrysaeolus* Thomas (including *magdalenae* Hershkovitz) and *P. hoplomyoides* Tate, although they are members of Patton's (1987) *trinitatis* group, are outside of either the *P. guairae* or *P. trinitatis* superspecies. We thus retain each as a separate species, at least until additional data can verify both the phylogenetic cohesion of this group of taxa as well as help resolve species boundaries within them.

The *trinitatis*-group of *Proechimys* is defined by a moderate to large body, relatively soft fur with narrow and elongated aristiform spines (except in *P. chrysaeolus* or *P. hoplomyoides*), a long but stout baculum in males, weakly developed or nonexistent temporal ridges, large and open incisive foramina with an attenuated maxillary portion to the septum that is seldom connected to the premaxillary portion, a groove present on the floor of the infraorbital foramen, a narrow and deeply penetrating mesopterygoid fossa, and simplified cheek teeth, with many individuals expressing only two folds on the lower molars.

KEY TO THE SPECIES OF THE *PROECHIMYS TRINITATIS* SPECIES GROUP:

1. Body size large (average head and body length >260 mm); pelage relatively soft to the touch, as aristiform spines are thin (0.4–0.5 mm in width) and terminate with a long whip-like tip . . . . *Proechimys trinitatis*
1'. Body size moderate (average head and body length <240 mm); pelage stiff to touch, with of aristiforms moderate (0.6–0.8 mm) to wide (0.9–1.1) and often with blunt tip 2
2. Tail proportionately long (85–90% of head and body length); aristiform spines of moderate width; lower molars with only two counterfolds 3
2'. Tail proportionately short (70–75% of head and body length); aristiform spines well developed and wide; all lower molars usually with three counterfolds 4
3. Floor of infraorbital foramen with groove and lateral flange to accommodate maxillary nerve; range limited to lower Magdalena Valley in northern Colombia . . . .
. . . . *Proechimys mincae*

3'. Floor of infraorbital foramen smooth, or with only weakly developed groove; range in Colombia east of Sierra de Perijá and Cordillera Oriental, otherwise range in Venezuela . . . . *Proechimys guairae*
4. Narrowed mesopterygoid fossa, penetrating to M2s; baculum long and narrow; range Guianan and Amazonian Venezuela . . . . *Proechimys hoplomyoides*
4'. Mesopterygoid fossa wide, penetrated only to M3s; baculum long but broad; range restricted to northern Colombia . . . . *Proechimys chrysaeolus*

Proechimys chrysaeolus (Thomas, 1898)
Boyacá Spiny Rat
SYNONYMS:

Echimys chrysaeolus Thomas, 1898b:244; type locality "Muzo, N. of Bogota," Cundinamarca, Colombia (but see Hershkovitz 1948a, who questioned the designation of Muzo as the type locality).

Proechimys chrysaeolus: J. A. Allen, 1899c:264; first use of current name combination.

P[roechimys]. xanthaeolus Thomas, 1914c:61; incorrect subsequent spelling of *chrysaeolus* Thomas.

Proechimys [(Proechimys)] cayennensis chrysaeolus: Ellerman, 1940:120; name combination.

Proechimys guyannensis chrysaeolus: Hershkovitz, 1948a:136; name combination.

Proechimys guyannensis magdalenae Hershkovitz, 1948a:136; type locality "Río San Pedro, a small stream in the northern foothills of the Cordillera Central, above the village of Norosí altitude 178 meters, department of Bolívar, Colombia."

[Proechimys] magdalenae: A. L. Gardner and Emmons, 1984:14; name combination.

DESCRIPTION: Moderate-sized spiny rat, head and body length averaging 210–220 mm, with medium length tail averaging 70–75% of head and body length. Overall dorsal color dark yellowish to reddish brown speckled with black, with sides only marginally paler than midback and abruptly contrasting with white venter from chin to inguinal region. White inner thigh stripe discontinuous with few white hairs on dorsal surface of hindfoot by broad, dark ankle band. Hindfoot varies in color above, but typically dark or may have silvery hairs sprinkled across metatarsals; toes always dark brown. Plantar pads well developed, with thenar and hypothenar subequal in size. Tail dark brown above and pale below, sparsely haired, appears naked to eye, with scale annuli forming distinct rings, more so than in most other species (10–13 per cm at midpoint). Pelage stiff and bristly to touch, with well-developed aristiform hairs (length 20–22 mm, width 0.9–1.1 mm); tip either blunt or terminating in very short whip.

Skull robust but long and narrow, with particularly elongated rostrum. Temporal ridge absent or only weakly developed, extending onto anterior parietals from supra-orbital ledge if present. Incisive foramina oval to teardrop in shape, tapering slightly with weak posterolateral flanges extending onto anterior palate forming shallow groove. Premaxillary portion of septum well developed, broad, and extends at least half length of opening; maxillary portion may be broad or narrow, but always contacts premaxillary portion so that vomerine portion not visible; maxillary part of septum unkeeled, so palate lacks median ridge. Floor of infraorbital foramen may be entirely smooth or with only moderately developed groove resulting from slight lateral flange. Mesopterygoid fossa moderately wide, with an angle averaging 59°, and penetrating to middle of M3s. Counterfold pattern of all maxillary cheek teeth uniformly 3–3–3–3; that of lower cheek teeth varies slightly, pm4 typically with three folds but rarely four, m1 always with three folds, and m2 and m3 with either two or three folds, and in about equal frequencies. Counterfold formula 3–3–3–3/3(4)–3–2(3)–2(3).

Baculum long but relatively broad (sample mean length 9.01–10.97 mm, proximal width 2.74–3.11 mm, distal width 2.34–2.62 mm), base bulbous with median depression, and apical wings slightly developed with median notch at distal end; similar in size and shape to other members of *trinitatis*-group, and to baculum of *P. decumanus* (Patton 1987).

DISTRIBUTION: *Proechimys chrysaeolus* occurs in northern Colombia from the Caribbean coast into the lower Cauca and Magdalena valleys west of the Cordillera Oriental.

SELECTED LOCALITIES (Map 528): COLOMBIA: Antioquia, Pure (Patton 1987); Bolívar, above Norosí (type locality of *Proechimys guyannensis magdalenae* Hershkovitz), San Juan Nepomuceno (Patton 1987); Boyacá, Muzo (type locality of *Echimys chrysaeolus* Thomas), Puerto Boyacá (M. L. Bueno et al. 1989); Córdoba, Datival (Patton 1987), Socorré (Patton 1987); Norte de Santander, Guamalito (Hershkovitz 1948a); Santander, Finca San Miguel (MVZ 196095); Sucre, Las Campanas (Patton 1987).

SUBSPECIES: *Proechimys chrysaeolus* is monotypic, although the relationships between typical *P. chrysaeolus* and *magdalenae* Hershkovitz are unresolved and may signal racial, or even species-level, differences.

NATURAL HISTORY: This species has not been studied in the field. *Proechimys chrysaeolus* may be found in sympatry with *P. canicollis*. It is apparently an important zoonotic vector of equine encephalitis in Colombia, where a laboratory colony has been established in the Instituto Nacional de Salud in Bogotá (M. L. Bueno et al. 1989).

REMARKS: In his original description (Thomas 1898b:244) stated that *P. chrysaeolus* was similar in pelage and cranial characters to *P. trinitatis*. Hershkovitz (1948a:136) suggested that the holotype must have come from farther down the Magdalena Valley and not from Muzo, which he argued was at an elevation (1,240 m) too high for this species. We provisionally include *magdalenae* Hershkovitz in our concept of *P. chrysaeolus*, although Hershkovitz (1948a:136–137) specifically compared his *magdalenae* to the geographically adjacent *P. chrysaeolus*, both of which he assigned as valid subspecies of *P. guyannensis* É. Geoffroy St.-Hilaire. Hershkovitz believed that *magdalenae* was more similar to what he termed the "western" forms of the races of *P. guyannensis* (in which he included *decumanus* Thomas, *panamensis* Thomas, and *calidior* Thomas; the former now a species unto itself and the latter two are synonyms of *P. semispinosus*) than to the geographically adjacent *mincae* J. A. Allen or *chrysaeolus* Thomas by virtue of a consistent three folds on each lower molar. However, *P. decumanus* has only two folds on each lower molar, samples of *P. semispinosus* from western Colombia and Ecuador may have either two or four folds, in addition to the more typical three, and most individuals of *P. chrysaeolus* have three folds on their lower molars (Patton 1987:335, Table 5). Consequently, *magdalenae*

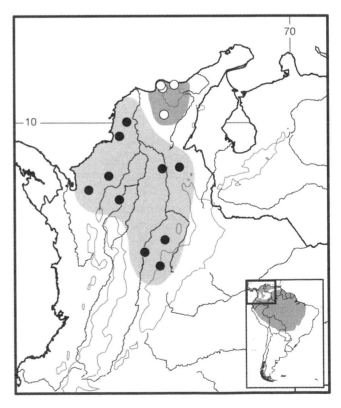

Map 528 Selected localities for *Proechimys chrysaeolus* (●) and *Proechimys mincae* (○). Contour line = 2,000 m.

Hershkovitz falls within the range of counterfold number found in *P. chrysaeolus*. A. L. Gardner and Emmons (1984) placed *magdalenae* Hershkovitz in their *brevicauda*-group, not in their *guairae*-group, on the basis of bullar septal pattern. These authors did not examine specimens of *P. chrysaeolus*. Because one of us (JLP) has seen specimens from the INS laboratory colony in Bogotá which he regards as *P. chrysaeolus*, we believe specimens from the same colony for which M. L. Bueno et al. (1989) described and figured a karyotype of $2n = 32$, FN = 54, are this species.

Proechimys guairae Thomas, 1901
La Guaira Spiny Rat
SYNONYMS:

Proechimys guairae Thomas, 1901a:27; type locality "La Guaira," Vargas, Venezuela.

Proechimys ochraceus Osgood, 1912:56; type locality "El Panorama, Rio Aurare, Zulia, Venezuela."

Proechimys poliopus Osgood, 1914a:141; type locality "San Juan de Colon, State of Tachira, Venezuela. Altitude, 2500 ft."

Proechimys [(Proechimys)] cayennensis guairae: Ellerman, 1940:121; name combination.

[Proechimys guyannensis] ochraceus: Hershkovitz, 1948a: 133; name combination.

Proechimys guyannensis poliopus: Hershkovitz, 1948a:133; name combination.

[Proechimys guyannensis] guairae: Moojen, 1948:316; name combination.

DESCRIPTION: Moderate to large, with head and body length of adults 210–240 mm. Tail absolutely and proportionately long, averaging nearly 85% of head and body length. Dorsal coloration light reddish brown lined with black, along southern slopes of Merida Andes, but distinctly paler and more yellowish brown in drier forests around Lake Maracaibo. Ventral color white from chin to inguinal region, including white inner thighs where stripe may be continuous across ankles onto dorsal surface of hindfeet. Hindfeet pale above, but often with outer light brown stripe extending from ankle to cover digits IV and V. Plantar pads of hindfeet well developed, with thenar and hypothenar pads large and subequal in size. Tail bicolored, light brown above and pale below, lightly haired so that large scales obvious to eye; scale annuli 7–8 per cm at midpoint. Pelage coarse but relatively soft to touch, with long (length 20–22 mm) and narrow (width 0.5–0.7 mm) aristiform spines tipped with whip-like extension.

Skull unremarkable, sharing conformational shape of most spiny rats. Temporal ridges undeveloped, or present only as weak and short posterior extension from supraorbital ledge. Incisive foramina broad and long, oval to slightly teardrop in shape, and with weakly developed

posterolateral flanges that extend onto the anterior palate forming slight moderate grooves. Premaxillary portion to septum narrow, extends at least to midpoint of septal opening, but may be only weakly connected to maxillary portion or not at all; maxillary portion of septum thin and attenuated, slightly keeled such that medial ridge may be present on anterior palate; vomer visible between premaxillary and maxillary septal elements in ventral view. Floor of infraorbital foramen may be either smooth or with slight groove developed by short lateral flange. Mesopterygoid fossa varies from narrow to moderate in width, with angle 47°–55°, penetrating palate to at least posterior margins of M2s and commonly even deeper. Postorbital process of zygoma moderately developed and composed of jugal alone or by equal contributions of jugal and squamosal. Three counterfolds uniformly present on PM4, M1, and M2, with M3 having either two or three folds in about equal proportions within samples. Lower cheek teeth uniform with three folds on pm4 and two folds on molar series. Counterfold formula 3-3-3-(2-3)/3-2-2-2.

Baculum with same shape and general size as other species in *trinitatis*-group (sample mean length 8.62–10.01 mm, proximal width 2.93–3.57 mm, distal width 2.78–3.34 mm), with bulbous base notched at midline and weakly developed apical wings (Patton 1987).

DISTRIBUTION: *Proechimys guairae* occurs in northern and western Venezuela in the foothills of both sides of the Andes and upper Llanos north of the upper Río Orinoco, including around Lake Maracaibo and the eastern slopes of the Sierra de Perijá, and into adjacent northeastern Colombia.

SELECTED LOCALITIES (Map 529): COLOMBIA: Arauca, Fatima (FMNH 92608), Río Arauca (FMNH 92576); Norte de Sandander, San Calixto (USNM 280182), Tarrá (Reig, Aguilera et al. 1980). VENEZUELA: Anzoátequi, Cueva de Agua (Corti and Aguilera 1995); Apure, Guasdualito (Reig, Aguilera et al. 1980); Aragua, Ocumare de la Costa (Aguilera, Reig, Pérez-Zapata 1995); Barinas, Buena Vista (Reig, Aguilera et al. 1980); Cojedes, El Baul (Reig, Aguilera et al. 1980); Falcón, Carrizalito (Aguilera, Reig, Pérez-Zapata 1995), Sanare (Reig, Aguilera et al. 1980); Monagas, San Juan (Pérez-Zapata et al. 1992); Guárico, Dos Caminos (Reig, Aguilera et al. 1980); Táchira, San Juan de Colón (Reig, Aguilera et al. 1980); Vargas, La Guaira (type locality of *Proechimys guairae* Thomas); Zulia, El Panorama (type locality of *Proechimys ochraceus* Osgood), Los Angeles del Tucuco (Corti and Aguilera 1995).

SUBSPECIES: Reig, Aguilera et al. (1980) delineated the extensive karyotypic variation they described in *P. guairae* and regarded each chromosomal "segment" in the circular distribution of forms as either species (*poliopus*

Osgood plus the unnamed "Barinas" species) or subspecies (*guairae* Thomas and *ochraceus* Osgood, plus two to four unnamed races). Counterclockwise around the ring, beginning in Apure and Barinas states in southwestern Venezuela, these included: "Barinas" ($2n=62$, FN=72; Reig and Useche 1976; Reig, Aguilera et al. 1980), "ssp." ($2n=48$–50, FN=72; Reig and Useche 1976; Reig, Aguilera et al. 1980), *guairae guairae* ($2n=46$, FN=72; Reig and Useche 1976), *guairae ochraceous* ($2n=44$, FN=76; Reig, Aguilera et al. 1980), and, finally, *guairae poliopus* ($2n=42$, FN=76; Reig and Useche 1976; Reig, Aguilera et al.1980). A geographically disjunct $2n=52$, FN=74, karyomorph is also known from Anzoátegui and Monagas states. Samples of each of these races can be delineated by both standard multivariate morphometrics (Aguilera and Corti 1994; Corti and Aguilera 1995) and geometric morphometrics (Corti et al. 2001), but a formal taxonomy defining and diagnosing each subspecies has yet to be produced.

NATURAL HISTORY: Aguilera (1999) examined the population ecology of *P. guairae* at a single site in coastal Venezuela, including home range size, density, reproductive schedule, and spatial organization of the sexes, and estimated the census effective population size. Males had larger home ranges than females, but those of females were exclusive, which suggested a polygynous or promiscuous mating system to Adler (2011). Handley (1976) included both *P. guairae* and *P. quadruplicatus* within a taxon he recognized as "*P. semispinosus*" from Venezuela. Although

it is possible to allocate the specimens in his list of localities (p. 57) to either of these two species, it is unfortunately not possible to parse the habitat and trap data he provides in the same manner. Osgood (1912) collected specimens from the roots of the wild pineapple in the arid parts of the northeastern slope of Lake Maracaibo, where he found spiny rats to be common. Valim and Linardi (2008) described the host association of ectoparasitic lice in the genus *Gyropus*.

REMARKS: Thomas (1901a:28), in his description of *P. guairae*, recognized that it was "evidently closely allied to *P. trinitatis* and its continental representatives of *P. urichi* and *P. mincae*" but differed from them primarily by a much paler color. Osgood (1912) believed that *P. ochraceus* was most similar to *P. guairae* but smaller and paler. Two years later when he described *P. poliopus*, Osgood (1914a:141) regarded this new species as being most similar to *ochraceus*. He also made comparisons to *P. guairae* as well as with *P. mincae* and *P. canicollis*, stating that *poliopus* was "doubtless related" to *P. guairae*, which is distinguishable by its larger size and white feet.

Proechimys hoplomyoides Tate, 1939
Guianan Spiny Rat
SYNONYMS:

Proechimys cayennensis hoplomyoides Tate, 1939:179; type locality "Rondon Camp, Mt. Roraima, 6800 feet," Bolívar, Venezuela.

Hoplomys hoplomyoides: Moojen, 1948:315; name combination.

Hoplomys gymnurus hoplomyoides: Cabrera, 1961:532; name combination.

Proechimys hoplomyoides: Handley, 1976:57; first use of current name combination.

Proechimys guyannensis hoplomyoides: O. J. Linares, 1998: 238; name combination.

DESCRIPTION: Moderate-sized spiny rat (head and body length of holotype 213 mm) with moderately long tail (72% of head and body length). Tate diagnosed species by its blackish-brown and heavily spined dorsal pelage and slightly larger cheek teeth in comparison to other spiny rats nearby (which would have been samples of *P. guyannensis*). Broad and very stiff aristiform spines so reminiscent of those of *Hoplomys* that Moojen (1948) and subsequently Cabrera (1961) placed *P. hoplomyoides* in that genus. (No data are available on aristiform dimensions, characters of tail scales and scale annuli, or number and size of plantar tubercles; indeed, with only six specimens known, an adequate understanding of character variation is not possible at present.) As detailed by Patton (1987), skull relatively long and narrow, with elongated rostrum characteristic of all spiny rats. Temporal ridges extend pos-

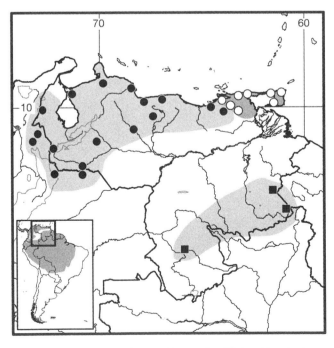

Map 529 Selected localities for *Proechimys guairae* (●), *Proechimys hoplomyoides* (■), and *Proechimys trinitatis* (○). Contour line = 2,000 m.

teriorly from supraorbital ledges absent; floor of infraorbital foramen smooth, without groove indicative of passage of maxillary nerve; incisive foramina somewhat lyre shaped with weakly flanged posterolateral margins that define grooves extending on anterior palate; premaxillary portion of septum of foramen enlarged, extending at least one-half length of opening, but maxillary portion attenuate and not in contact with premaxilla. Mesopterygoid fossa narrow and deeply penetrating, to at least level of posterior half of M2s. Counterfold pattern simple, with three folds on each upper and lower tooth; counterfold formula 3–3–3–3/3–3–3–3.

Baculum long and rather narrow (length of single specimen 11.64 mm, proximal width 2.47 mm, distal width 1.79 mm), more so than any other species of spiny rat, including other members of *trinitatis*-group (Patton 1987).

DISTRIBUTION: The known range of *Proechimys hoplomyoides* is limited to the tepui area of southeastern and southern Venezuela. This the only species in the *trinitatis*-group with a distribution south of the Río Orinoco and within Amazonia and the Guianan region.

SELECTED LOCALITIES (Map 529): VENEZUELA: Amazonas, Tamatama (Handley 1976); Bolívar, km 125, 85 km SSE of El Dorado (Handley 1976), Rondon Camp (type locality of *Proechimys cayennensis hoplomyoides* Tate).

SUBSPECIES: *Proechimys hoplomyoides* is monotypic.

NATURAL HISTORY: *Proechimys hoplomyoides* is known from evergreen rainforest and gardens, near streams or other moist areas (Handley 1976). The holotype was taken at the highest elevation (6,800 ft, or 2,000 m) recorded for any spiny rat in the Guianan region (Tate 1939).

REMARKS: Moojen (1948) placed *P. hoplomyoides* in the genus *Hoplomys* because of the heavy and dense spines, an action followed by Cabrera (1961). The bullar septal pattern, however, is close to other species in the *trinitatis*-group (A. L. Gardner and Emmons 1984) although the long and narrow baculum is more similar to that of *P. simonsi* than to those of other species in the *trinitatis*-group (Patton 1987:Fig. 12, p. 321). O. J. Linares (1998) assigned the species as a valid subspecies of *Proechimys guyannensis*, likely following Tate (1939) in his original description, but without comment. The species is known from only six specimens, including the holotype, and four localities in southern Venezuela.

Proechimys mincae (J. A. Allen, 1899)
Minca Spiny Rat

SYNONYMS:

Echimys mincae J. A. Allen, 1899b:198; type locality "Minca, Santa Marta District," Magdalena, Colombia.

[*Proechimys*] *mincae*: J. A. Allen, 1899c:264; first use of current name combination.

Proechimys [(*Proechimys*)] *cayennensis mincae*: Ellerman, 1940:120; name combination.

Proechimys guyannensis mincae: Hershkovitz, 1948a:134; name combination.

DESCRIPTION: Moderately large spiny rat (head and body length 220–230 mm) with a proportionately long tail (90% of head and body length). Coloration across dorsum reddish brown speckled with black, becoming slightly paler on sides, and with white venter from chin to inguinal region but varyingly bordered by light buff margins. White inner thigh stripe passes weakly across ankle to be confluent with basically dirty white dorsal surface of hindfoot, with toes only slightly darker. Plantar pads of hindfeet enlarged and well developed, with thenar and hypothenar large and subequal in size. Tail bicolored, brownish gray above and pale cream below. Tail thinly haired with scales large, irregular in shape, and readily visible to eye; scale annuli average 8–9 per cm at midpoint. Aristiform spines long (length 20–22 mm) and thin (width 0.6–0.8 mm) but stiff, giving pelage raspy texture when brushed. Tip of each aristiform may terminate in short, whip-like extension or be blunt.

Skull similar to that of most species of *Proechimys*, elongated, relatively narrow, and with tapering rostrum. Temporal region of braincase smooth, lacking virtually any evidence of ridges extending posterior to supraorbital ledge). Incisive foramina wide and with somewhat rounded sides conferring oval shape to opening; posterolateral flanges either nonexistent or so weakly developed that anterior palate lacks grooves. Premaxillary portion of septum well developed and long, filling more than half distance of opening, but either does not or only rarely contacts greatly attenuated maxillary portion; midpalate may have small medial ridge, but small maxillary portion of septum lacks any hint of keel; vomer can often be seen in ventral view. Floor of infraorbital foramen typically grooved, resulting from moderate development of lateral flange. Mesopterygoid fossa moderately broad, opening at angle averaging 57°, and penetrating posterior palate to level of M2s. All cheek teeth relatively simple, with never more than three folds in upper series, but often only two folds on M2 and especially M3; lower cheek teeth typically with three folds on pm4 but only two on each molar. Counterfold formula 3–3–2(3)–2(3)/3–2–2–2.

Baculum of *P. mincae* long and stout, similar in size and shape to that of other species in *trinitatis*-group, except that of *P. hoplomyoides* (mean length 7.29–9.98 mm, proximal width 2.04–2.17 mm, distal width 2.42–2.94 mm; Patton 1987).

DISTRIBUTION: *Proechimys mincae* is known only from the lower Magdalena valley in northern Colombia to the west and south of the Sierra Nevada de Santa Marta.

SELECTED LOCALITIES (Map 528): COLOMBIA: Cesar, Colonia Agricola de Caracolicito (USNM 280198); Magdalena, Bonda (Patton 1987), Don Dago (Patton 1987), Minca (type locality of *Echimys mincae* J. A. Allen).

SUBSPECIES: *Proechimys mincae* is monotypic.

NATURAL HISTORY: *Proechimys mincae* has not been studied in the field. It is sympatric with *P. canicollis* at Bonda, the type locality of the latter species (J. A. Allen 1899b).

REMARKS: A. L. Gardner and Emmons (1984) described and figured the karyotype of *P. mincae* as $2n = 48$ and FN = 68, similar to that of some karyotypes of *P. guairae*. In his description of this species, J. A. Allen (1899b:199) noted that it belonged to the same group as *P. trinitatis*, with the "same white belly, the same general proportions, and practically the same dentition." He considered it separate from *P. trinitatis* by virtue of smaller size and golden brown rather than dark chestnut brown dorsal color. However, the skull of *P. mincae* is more different in geometric shape than any members of Reig, Aguilera et al.'s (1980) *P. guairae* or *P. trinitatis* superspecies groups, and even the unrelated *P. canicollis* (Corti et al. 2001).

Proechimys trinitatis (J. A. Allen and Chapman, 1893)
Trinidad Spiny Rat

SYNONYMS:

Echimys trinitatis J. A. Allen and Chapman, 1893:223; type locality "Princestown [= Princes Town], Trinidad, " Trinidad and Tobago.

E[*chinomys*]. *trinitatis*: Thomas, 1896:313; name combination.

Echimys urichi J. A. Allen, 1899b:199; type locality 'Quebrada Secca," Villarroel, Sucre, Venezuela.

[*Proechimys*] *trinitatis*: J. A. Allen, 1899c:264; first use of current name combination.

[*Proechimys*] *urichi*: J. A. Allen, 1899c:264; name combination.

Proechimys urichi: Pittier and Tate, 1932:264; name combination.

Proechimys [(*Proechimys*)] *cayennensis urichi*: Ellerman, 1940:121; name combination.

Proechimys [(*Proechimys*)] *cayennensis trinitatis*: Ellerman, 1940:121; name combination.

[*Proechimys guyannensis*] *trinitatis*: Moojen, 1948:316; name combination.

[*Proechimys guyannensis*] *urichi*: Moojen, 1948:316; name combination.

Proechimys guyannensis urichi: Cabrera, 1961:523; name combination.

DESCRIPTION: Moderately large (head and body length 265–270 mm in mature adult individuals), with a proportionately long tail (75–80% of head and body length). Dorsal color brownish orange heavily mixed with black lines, only slightly paler on sides, and contrasting with pure white venter. White inner thigh stripe may continue across ankle to inner side of hindfoot surface, which is otherwise brownish in color. Plantar pads of hindfeet well developed with both thenar and hypothenar pads enlarged and subequal in size. Tail covered very sparsely with short hairs such that it appears quite naked to eye; scales large, almost oval in shape, with 6–7 annuli per cm at midpoint. Pelage relatively soft to touch, with aristiform spines long (length 20–22 mm), thin (width 0.3–0.5 mm), and terminating in extended whip-like tip.

Overall shape as in other species of spiny rats, relatively narrow and elongated, with long and tapering rostrum. Weak temporal ridges extend back to midparietal from supraorbital ledges present in older individuals less obvious in younger specimens. Incisive foramina wide and oval, with only weakly posterolateral flanges, if at all, so that anterior palate either flat or weakly grooved. Premaxillary portion of septum variable in size laterally and may extend virtually entire length of opening or only to midpoint; maxillary portion narrow and weakly developed, barely in contact with premaxilla, and only weakly keeled, typically without extending onto anterior palate as continuing medial ridge; vomerine portion of septum usually not visible ventral view. Floor of infraorbital foramen grooved, bordered laterally by typically moderately developed flange. Mesopterygoid fossa relatively narrow, forming angle of about 53°, and deeply penetrating to anterior edges of M3s or posterior edges of M2s. Paroccipital processes distinctly broad, flattened, and tightly appressed to each bulla, more so than in any other species of *Proechimys*. Postorbital process of zygoma weakly to moderately developed, and composed entirely of squamosal. Upper cheek teeth uniform with three folds on each, except that M3 occasionally has only two. Lower cheek teeth all have three folds, with each molar often with only two. Counterfold formula 3–3–3–(2)3/3–(2)3–(2)3–(2)3.

Baculum elongated but moderately wide (sample mean length 9.97–10.46, proximal width 3.10–3.15, distal width 3.14–3.32), with slightly expanded base with median notch, straight sides, and slight apical wings, similar to that of other species within the *trinitatis*-group (Patton 1987).

DISTRIBUTION: *Proechimys trinitatis* occurs on the island of Trinidad and the adjacent coastal lowlands of northern Venezuela (see maps in Pérez-Zapata et al. 1992; Corti and Aguilera 1995).

SELECTED LOCALITIES (Map 529): TRINIDAD AND TOBAGO: Trinidad, Chaguaramas (MVZ 168949), Oropuche Heights (FMNH 20960), Princes Town (type locality of *Echimys trinitatis* J. A. Allen and Chapman). VENEZUELA: Monagas, 6.5 km of NE Cachipo (Pérez-Zapata et al. 1992), Guanaguana (Aguilera, Reig, Pérez-Zapata

1995); Sucre, El Argarrobo (Aguilera, Reig, Pérez-Zapata 1995), Guaraunos (Aguilera, Reig, Pérez-Zapata 1995), Quebrada Seca (type locality of *Echimys urichi* J. A. Allen).

SUBSPECIES: Spiny rats from the Venezuelan mainland are somewhat smaller than those on the island of Trinidad. Ellerman (1940), Moojen (1948), and Cabrera (1961) regarded these mainland populations as a distinct subspecies, but listed it as *P. guyannensis urichi* (J. A. Allen).

NATURAL HISTORY: Everard and Tikasingh (1973a) examined the population ecology of *P. trinitatis* on Trinidad, including recapture rates, home range, and movement patterns, longevity, reproduction, sex ratio, and parasites. Here, the species is common in secondary rainforest, reaching monthly densities from 3.7–5.3 individuals per acre, with mean distances moved by individuals between successive captures up to 300 ft, average home range size of 0.425 acres, and extensive overlap of male and female home ranges. Individuals lived at least 20 months under natural conditions in the field, and a maximum of 41 months in captivity. Captive breeding indicated a gestation period of 62–64 days, a mean litter size of 2.4, mean birth weight of 24.7 g, and reproductive maturity about five months following birth. Individuals were docile in captivity, but were aggressive when kept together, often inflicting wounds on each other. Everard and Tikasingh (1973a) and Valim and Linardi (2008) described the host association of ectoparasitic lice, ticks, and fleas. One of 285 individuals examined had *Cuterebra* larvae in the skin (Everard and Tikasingh 1973a). Everard et al. (1974) examined filarial nematodes, and Stunkard (1953) described cestodes.

REMARKS: Reig and colleagues (Reig et al. 1979; Reig, Aguilera et al. 1980; Pérez-Zapata et al. 1992) report a karyotype of 2n = 62, FN = 80, for both Trinidad and Venezuelan populations of *P. trinitatis*. Benado et al. (1979) examined allozyme relationships and show that *P. trinitatis* is the sister to a group of four karyomorphs of the *P. guairae* superspecies. However, Patton and Reig (1990) showed that *urichi* Allen (= *P. trinitatis*) is nested phylogenetically within three karyomorphs of *P. guairae*, including two examined by Benado et al. (1979; 2n = 46 and 62 forms) plus 2n = 42 *poliopus* Osgood. If more sophisticated molecular analyses support the phylogenetic position of *P. trinitatis* within the *P. guairae* superspecies, then O. J. Linares's (1998) conspecific hypothesis should be reevaluated.

Genus *Thrichomys* E.-L. Trouessart, 1880
Leila M. Pessôa, William C. Tavares, Antonio C. A. Neves, and André L. G. da Silva

Species of *Thrichomys* comprise a morphologically and ecologically distinct group within the family Echimyidae. The genus is distributed widely across different biomes from Caatinga, Cerrado, and Pantanal in Brazil to the Chaco of Bolivia and Paraguay. In Brazil, this distribution is closely associated with the diagonal belt of open vegetation that stretches in a northeast to southwest direction sandwiched between the coastal Atlantic Forest and Amazon Basin (Alho 1982; Mares and Ojeda 1982). In Bolivia, *Thrichomys* inhabits the northern edge of the Chaco, which is characterized by scattered shrubs without large rock outcrops (S. Anderson 1997), but it is also associated with rocky outcrops of the Brazilian Shield in Chiquitania dry forest and Cerrado (Emmons et al. 2006, and unpubl. data). Species are terrestrial but may climb into the lower strata of trees and shrubs.

The oldest fossil unambiguously assignable to *Thrichomys* dates from Pleistocene deposits at Lagoa Santa, Minas Gerais (Lund 1839, 1840a, 1841b; Paula Couto 1950). McKenna and Bell (1997) suggested that the oldest record of the genus was from the late Miocene, but Reig (1989) argued that the identity of this material was uncertain because the specimen is very fragmented (Verzi et al. 1995). Olivares et al. (2012) phyletically linked the late Miocene *Pampamys emmonsae* (Huayquerian South American Land Mammal Age, from central Argentina) to *Thrichomys*, which both supports 6 mya as a minimum age for the genus and relates the simplified-molar echimyids to expansion of open environments during the late Miocene. Nascimento et al. (2013), based on mtDNA and nucDNA sequences, estimated the basal divergence of living member of the genus at 8.46 mya, and posited an ancestral origin in the central Caatinga and/or northern Cerrado of Brazil.

The generic name *Thrichomys* is etymologically associated with the presence of soft fur and a thickly haired tail, characteristics not shared with any other terrestrial echimyid (Ellerman 1940). The tail is moderate in length, approximately 86% of the head and body length. It is bicolored, uniformly covered by blackish hairs dorsally and whitish or grayish hairs ventrally. The dorsum of the body is generally ash gray in color in all species; setiform hairs are gray basally, black near the tip, and have a yellowish subapical zone. The venter is whitish, contrasts with dorsal coloration, but is not sharply demarcated from the lateral coloration. A grayish stripe is typically present in the gular region. The eye is outlined by contrasting white lines above and below. Long vibrissae extend posteriorly to the shoulder. A spot of white hairs is present at the base of each ear. White, sparse hairs cover the ears (Ellerman 1940; Moojen 1952b). Females have four abdominal mammae. In adult specimens, the head and body length varies from 170–310 mm and tail length from 175–226 mm (Moojen 1952b; data from museum specimens MNRJ 22066,

MNRJ 67899). Mass varies from 150–700 g (MNRJ 67760, MNRJ 64460).

As in other terrestrial echimyids, the feet are elongated and narrow, ranging from 39–42 mm in adults. The hindfeet bear six well-developed plantar pads. The thenar and hypothenar pads are very elongated, extending toward the heel. The hallux is very short and does not reach the base of the second toe. The second toe is long, almost reaching the tip of the third and fourth toes, which are the longest and subequal in length. The fifth toe reaches the level of the proximal third of the fourth toe. Each toe bears a claw covered by long white ungual hairs that do not extend more than 1 mm beyond the claws. The short forefeet bear five plantar pads, three interdigitals, hypothenar, and thenar. Only the pollex does not possess a claw. This tiny digit bears a thin nail not covered by hairs. The second and fifth digits are short, subequal in length, and extend to the proximal third of the third and fourth digits. Dense, short hairs that are either white or gray at the base and white at the tip cover the dorsal surfaces of both forefeet and hindfeet. All plantar pads are surrounded by smaller tubercles that are most evident in fresh specimens. The surface between pads is rough.

Neither the external morphology of the phallus nor baculum of *Thrichomys* has been described previously. The following description is based on MNRJ 64623 from Barão de Melgaço in Mato Grosso. The glans is an elongated and cylindrical structure with a slight constriction apparent on the lateral and dorsal general outlines of the body wall. The main body is uniformly covered with spinous epidermis and lacks evidence of any groove along the length of its ventral or dorsal surfaces. A dorsal constriction that delimits the intromittent sac is present midway along the length of the shaft. The apical tip shows three lobes, a median one that extends beyond the lateral two. This glans measured 17 mm in total length (from the base of the prepuce to the tip) with a greatest diameter of 4.3 mm.

The baculum of adult males is elongate and narrow with a straight shaft with a slight dorsoventral curvature and broadened proximal end. Both the proximal and distal ends vary in shape. The proximal end is typically paddle-shaped but the distal end can be straight or possess a median depression with slightly developed apical wings. This description is based on one individual from Matozinho (MNRJ 68158) and three from Santana do Riacho (MNRJ 68602–68604), all from in Minas Gerais; eight individuals from four localities from Chapada Diamantina (MNRJ 67504, MNRJ 67571, MNRJ 67597, MNRJ 67626, MNRJ 67698, MNRJ 67771, MNRJ 67772, MNRJ 67783) in Bahia; and four individuals from Barão de Melgaço (MNRJ 64029, MNRJ 64460, MNRJ 64478,

MNRJ 64623) in the northern part of the Pantanal in Mato Grosso.

The adult skull is broad and robust with conspicuous supraorbital ridges, extending almost in parallel along the anterior half of the frontals and diverging toward the squamosoparietal suture. The rostrum is short and wide. The nasal bones do not extend forward of the posterior limit of the premaxillae. An anterior projection of the premaxilla is well developed. The inferior zygomatic root is short and, in lateral view, located almost at the same level of the ventral surface of the rostrum and the palate. The interpremaxillary foramen is very marked. The incisive foramina are wide and oval or lyre shaped; the septum is discontinuous, with the premaxillary portion robust and either and elongated or barely reaching the level of the premaxillo-maxillary suture; the maxillary part is short. The vomer is visible in ventral view. Palatal grooves are well developed and house a pair of palatine foramina placed anterior to or at the same level as PM4; more posterior foramina, when present, vary in number, position, and size. A groove with a well-developed lateral flange is present on the floor of the infraorbital foramen. The mesopterygoid fossa is V- or U-shaped, and its extension into the posterior palate varies with age. Among most adults, it typically extends to the middle of M3. The sphenopalatine foramen and vacuities are well developed. The masticatory foramen and foramen ovale accessorius are separated either by a narrow strut or a broader plate. The bullae vary in size from large to small and thus is well or only slightly inflated. The lateral tube of the auditory meatus either extends at right angles to the long axis of the skull or is slanted slightly forward. The contact between the ectotympanic and squamosal is often restricted to the central portion of the dorsal margin of the ectotympanic; this condition is associated with the formation of anterior and posterior clefts between these bones. The jugal may be broad or narrow but always contains a deep fossa. The inferior jugal process is moderately developed, often pointed; it is projected as far as the level of postorbital process of zygoma, which is pointed and formed only by squamosal or by squamosal and jugal. The mandibular foramen is positioned between the posterior limit of the toothrow and the tip of condylar process. The angular process of the dentary extends to or beyond the posterior edge of the condylar process. The coronoid process of the dentary is lower than condylar process.

The dental formula is I 1/1, C 0/0, PM 1/1, M 3/3, total 20. Three roots are present in each maxillary tooth, one lingual and two labial. The occlusal surfaces of the cheek teeth are rounded in the upper toothrow but more prismatic in the lower, where dpm4 is elongated. The first and second molars are larger than the premolar and third molar, with the difference more pronounced in the upper toothrow. The molar series are parallel or converge slightly anteriorly.

The homology of dental structures remains debatable among living and fossil echimyid genera; here we follow the nomenclature of G. A. S. Carvalho and Salles (2004). Occlusal surfaces of unworn upper teeth have three or four major lophs, including an anteroloph, protoloph, metaloph, and posteroloph. Two or three moderately short and shallow transverse flexi open at the labial surface and extend half way across the tooth width; these become closed islands with wear. Continuous connections are present between the protoloph and protocone and between posteroloph and hypocone; rarely in young individuals both connections may be interrupted by a shallow sulcus. The hypoflexus is short, extends only to the middle of each tooth, and opens primarily to lingual surface; it is the last flexus to close as a dentine island with wear.

The lower dpm4 has four major lophids, including the anterolophid, metalophid, hypolophid, and posterolophid. The metalophid-protoconid and hypolophid-hypoconid pairs are each connected by an ectolophid without interruption by a sulcus. Continuous connections are present between the hypolophid and hypoconid and the anterolophid and protoconid; rarely both connections may be interrupted by a shallow sulcus in young individuals. Three shallow flexids (anteroflexid, mesoflexid, and metaflexid) open from the labial surface, but each becomes fully closed, forming dentine islands, by the time all teeth are fully erupted. A hypoflexid opens primarily from the labial surface but quickly becomes isolated with wear.

Three major lophids are present on m1 to m3, including the anterolophid, hypolophid, and posterolophid. The ectolophid is orientated anteroposteriorly in unworn teeth. The ectolophid of m3 in subadults is sometimes interrupted by a connection between the metaflexid and hypoflexid. There is a continuous connection between the hypolophid and hypoconid and another between the anterolophid and protoconid; rarely both connections are interrupted by a shallow sulcus in young individuals. Two short flexids open primarily from the lingual surface but become closed with wear; these extend about halfway across the occlusal surface. The hypoflexid is short, extending toward the lingual surface maximally to midwidth of the tooth; it opens primarily from the labial surface but also becomes isolated with wear.

Five nominal taxa are identified in the taxonomic literature of *Thrichomys*, including *Echimys apereoides* Lund, *Echimys inermis* Pictet, *Isothrix pachyura* Wagner, *Thrichomys fosteri* Thomas, and *Thrichomys laurentius* Thomas. For much of the twentieth century, these names were placed either in *Cercomys* F. Cuvier (*apereoides*, *inermis*, *fosteri*, and *laurentius*) or *Isothrix* (*pachyura*; see Tate 1935). Moreover, until recently most authorities recognized only a single species, *Thrichomys apereoides*, with all other taxa treated as subspecies of as synonyms (Cabrera 1961; Woods 1993; S. F. Reis and Pessôa 2004).

An initial set of morphometric analyses of *Thrichomys* attempted to determine the degree and pattern of geographic structure in the genus, including studies that examined patterns of age and sex variation within specific localities (Moojen et al. 1988; Monteiro et al. 1999), and those that utilized standard (Bandouk and dos Reis 1995; Bandouk et al. 1996) and geometric morphometrics (L. C. Duarte et al. 2000; S. F. Reis et al. 2002a,b; Monteiro et al. 2003). These analyses restricted their geographic scope to samples from central and northeastern Brazil. The resulting geographic patterns were complex, with single population samples often sharply segregated from all others (Bandouk and Reis 1995; L. C. Duarte et al. 2000) or with geometric shape descriptors of the skull and mandible significantly associated with a latitudinal environmental gradient (Monteiro et al. 2003). Nevertheless, these analyses recovered two geographic groupings that differentiated samples from central and northeastern Brazil. Interlocality distances within both the central and northeastern groups exhibited isolation by distance, suggestive of recurrent gene flow, but no such pattern was observed in the among-group comparisons (S. F. Reis et al. 2002b). This pattern was hypothesized as evidence that each group comprised an independent evolutionary unit, without taxonomic implications on the monotypy of the genus (S. F. Reis et al. 2002b). Subsequently, Pessôa et al. (2004) employed standard morphometrics and karyological data to samples from southeastern Brazil (including the material collected by Lund used to describe *apereoides* and more recently collected topotypes) and southwestern Brazil. This study recognized marked divergences between the two geographic samples and referred the latter to *pachyurus* Wagner. Additional analyses in unpublished graduate theses (A. L. G. da Silva 2005; Teixeira 2005; Neves 2008) recognized subtle qualitative discontinuities among some Brazilian populations and have helped to clarify the limits among the described nominal taxa, including different ontogenetic growth patterns among samples from northeastern, central, and southeastern Brazil (Neves 2008; Neves and Pessôa 2011).

Especially for samples from northeastern and central Brazil, results from morphometric studies contrast with published cytogenetic data. The latter have found a wide range of karyological variation, with diploid numbers from 26–34. At least five karyotypes have been described, each with a relatively defined geographic range within Brazil (M. J. Souza and Yonenaga-Yassuda 1982; Svartman 1989; Leal-Mesquita 1991; Leal-Mesquita et al. 1993; J. F. S. Lima 2000; Bonvicino, Otazu, and D'Andrea

2002; Pessôa et al. 2004; Bonvicino, Lemos, and Weksler 2005; A. H. Carvalho and Fagundes 2005; A. L. G. da Silva 2005; L. G. Pereira and Giese 2007; A. H. Carvalho et al. 2008). These samples include one topotype (*T. apereoides* Lund) and three samples collected close to type localities of three other nominal taxa (*T. pachyurus* Wagner, *T. inermis* Pictet, and *T. laurentius* Thomas). One karyotype, however, including samples from central Brazil and northern Tocantins, does not correspond to any available name (A. H. Carvalho and Fagundes 2005; A. H. Carvalho et al. 2008).

Molecular analyses of the mtDNA cytochrome-*b* gene that included most of the known karyomorphs of *Thrichomys* have helped to elucidate intrageneric diversity. These data documented a general correspondence between clade structure and karyotypic units. Braggio and Bonvicino (2004) found high levels of molecular divergence among samples and the recognition of four monophyletic lineages that they named *T. inermis*, *T. pachyurus*, an undescribed species from Tocantins state, and a clade formed by *T. a. apereoides* and *T. a. laurentius*, this last clade composed of two distinct karyomorphs. Most subsequent workers have treated *Thrichomys* as polytypic, with three valid species (Woods and Kilpatrick 2005; J. A. Oliveira and Bonvicino 2006). However, tests of postzygotic reproductive isolation based on crossbreeding experiments indicated that *T. apereoides*, *T. laurentius*, and *T. pachyurus* are best considered as separate biological species (Borodin et al. 2006). In each set of crosses, F1 males were sterile.

Despite the complexity in establishing species boundaries within the genus, including some incongruence among morphometric, cytogenetic, and molecular analyses, recent advances in each have independently documented that at least five species can be recognized, some yet to be described (S. Anderson 1997; Bonvicino, Otazu, and D'Andrea 2002; S. F. Reis et al. 2002a,b; Braggio and Bonvicino 2004; Pessôa et al. 2004; A. L. G. da Silva 2005; Bonvicino, Oliveira, and D'Andrea 2008). After examining type specimens of all nominal forms associated with the genus *Thrichomys* and all available sources of information, especially new data from ontogenetic series (Neves 2008; Neves and Pessôa 2011), we here recognize four named species: *T. apereoides*, *T. inermis*, *T. pachyurus*, and *T. laurentius*. We regard *fosteri* Thomas as a synonym of *T. pachyurus* based on our examination of the qualitative morphology of type material and other specimens from Sapucay and Concepción, Paraguay. We base the morphological characterization of each species on specimens from localities as near as possible of their type localities. However, we recognize that the number of valid species will likely increase in the near future.

We limit our accounts to the four taxa for which available names can be applied with some confidence, but also map the known range of the karyotypic morph that may represent an undescribed species without providing additional details. Further studies that combine data sets and more expansive geographic sampling are required to fully delineate species boundaries and their geographic ranges.

SYNONYMS:

Echimys: Lund, 1839c:98; part (description of *apereoides*); not *Echimys* F. Cuvier.

Nelomys: Lund, 1841:242; part (listing of *antricola* as replacement name for *apereoides*); not *Nelomys* Jourdan.

Isothrix: Wagner, 1845a:146; part (description of *pachyura*); not *Isothrix* Wagner.

Thrichomys E.-L. Trouessart, 1880b:179; no type species designated; type species *Echimys apereoides* Lund, by subsequent designation (S. F. Reis and Pessôa 2004).

Loncheres: Pelzeln, 1883:61; part (listing of *pachyura* in synonymy, accorded to Waterhouse 1848); not *Loncheres* Illiger.

Thricomys E.-L. Trouessart, 1897:606; incorrect subsequent spelling of *Thrichomys* E.-L. Trouessart.

Tricomys E.-L. Trouessart, 1904:504; incorrect subsequent spelling of *Thrichomys* E.-L. Trouessart.

Trichomys Miranda-Ribeiro, 1914:42; incorrect subsequent spelling of *Thrichomys* E.-L. Trouessart.

Cercomys: Thomas, 1912e:116; part (listing of *fosteri* and *laurentius*); incorrect generic usage, see Remarks.

REMARKS: In 1880, E.-L. Trouessart proposed the generic name *Thrichomys* to include the names *antricola* Lund (= *Echimys apereoides* Lund), *inermis* Pictet, and *brevicauda* Günther (now a *Proechimys*) but did not designate a type species. By 1912, however, F. Cuvier's much earlier name *Cercomys* became associated with E.-L. Trouessart's *Thrichomys*. Goldman (1912c:94) examined the plates of the skull and dentition of *Cercomys cunicularis* in F. Cuvier (1829b) and concluded that the skull was similar to that of a *Proechimys* J. A. Allen. Thomas (1912e:115–116) analyzed the same plates and agreed that the skull and dentition were similar to *Proechimys*, but argued that adult specimens of *Proechimys* always had spiny pelage, a feature clearly not evident in the specimen figured in Cuvier's plate. To solve this problem, Thomas used a specimen identified as *Cercomys* that he borrowed from Paris Museum, and concluded that this skull was the same as *Thrichomys*, not *Proechimys*, an action that for the next 60 years relegated E.-L. Trouessart's name as a synonym of *Cercomys*. However, the tail figured in Cuvier's plate of *Cercomys cunicularis* was naked, not fully furred as in *Thrichomys*. The incorrect assignment of *Thrichomys* to *Cercomys* was retained in the literature (Ellerman 1940; Moojen 1952b; Cabrera 1961) until Petter (1973) examined the holotype of *Cercomys cunicularius* F. Cuvier in the Paris Museum and concluded that it was a composite consisting of a skin referable to *Nectomys squamipes* (Brants), a skull referable to *Proechimys* J. A. Allen, and a mandible referable to *Echimys* F. Cuvier. Consequently, since no elements of F. Cuvier's type of *C. cunicularis* represented any taxon connected to E.-L. Trouessart's *Thrichomys*, *Cercomys* F. Cuvier must be removed from its association with that genus. Thus, *Thrichomys* is restored as the correct name applied to the soft-haired, furry-tailed rodents from northeastern Brazil southwest to Paraguay and Bolivia that Thomas (1912e), Tate (1935), and others had placed within *Cercomys*. S. F. Reis and Pessôa (2004) designated *Echimys apereoides* Lund, 1839 as the type species of E.-L. Trouessart's *Thrichomys* (but see Remarks under this species).

KEY TO THE SPECIES OF *THRICHOMYS*:

1. Incisive foramina wide, with a slight constriction due ossification in premaxillomaxillary suture; distal end of hamular processes of pterygoid bone broadening to the tip and often rectangular 2
1'. Incisive foramina elongated without constriction between premaxillomaxillary suture; distal end of hamular processes of pterygoid bone becoming thinner and pointed to the tip 3
2. Skull robust; bullae large and well inflated (bullar length 12–15 mm); jugal wide dorsoventrally; paroccipital process not appressed to each bulla; wide plate (ca. 2 mm) between oval and masticatory foramina . . . .
. . . . *Thrichomys pachyurus*
2'. Skull slender; bullae small and uninflated (bullar length 10–13 mm); jugal narrow dorsoventrally; paroccipital process appressed to each bulla; narrow strut (ca. 0.7 mm) between ovale and masticatory foramina . . . .
. . . . *Thrichomys inermis*
3. Lateral flange present on posterior margins of incisive foramina; mesopterygoid fossa U-shaped; palate keeled with well-developed and deep palatal grooves . . . .
. . . . *Thrichomys apereoides*
3'. Lateral flange on posterior margins of incisive foramina absent; mesopterygoid fossa V-shaped; palate almost flat, with shallow palatal grooves . . . .
. . . . *Thrichomys laurentius*

Thrichomys apereoides (Lund, 1839)

Lagoa Santa Punaré

SYNONYMS:

N[*elomys*]. *antricola* Lund, 1839a:227, 233; *nomen nudum*.
E[*chimys*]. *apereoides* Lund, 1839c [1841a:98]; type locality "Rio das Velhas Floddal" (see Lund 1841a:133–134), Lagoa Santa, Minas Gerais, Brazil.

E[*chimys*]. *antricola* Lund, 1840a [1841b:242, footnote; plates 22 and 23]; renaming of E[*chimys*]. *apereoides* Lund.
Nelomys antricola Lund, 1841:242; name combination.
[*Thrichomys*] *antricola*: E.-L. Trouessart, 1880b:179; name combination.
[*Thrichomys*] *apereoides*: E.-L. Trouessart, 1880b:179; first use of current name combination.
[*Thrichomys*] *antricola*: E.-L. Trouessart, 1897:606; name combination.
[*Tricomys*] *apereoides*: E.-L. Trouessart, 1904:504; name combination.
Trichomys apereoides: Miranda-Ribeiro, 1914:42; name combination.
[*Cercomys*] *apereoides*: Tate, 1935:409; name combination.
Cercomys cunicularius apereoides: Moojen, 1952b: 173; name combination.
Nelomys apereoides: Petter, 1973:422; incorrect name combination justified as a synonym of *antricola* Lund.
Nelomys aperoides Woods, 1993:798; incorrect subsequent designation of type species and incorrect subsequent spelling of *apereoides* Lund.
T[*hrichomys*]. a[*pereoides*]. *apereoides*: S. F. Reis and Pessôa, 2004:1; name combination.

DESCRIPTION: Larger in body size than *T. laurentius* and *T. inermis*. Dorsal coloration brownish gray, ventral hairs completely white or with light gray base and contrast with dorsal coloration. Dorsal tail hairs often dark brown, whereas ventral tail hairs whitish gray. The greatest skull length in adults averages 55.61 ± 3.18 mm (N = 9). Skull broad with supraorbital ridges extending almost in parallel to middle of frontal bones, and then diverging toward squamosoparietal suture; rostrum short and wide; nasals relatively short, not extending to posterior limit of premaxillae. Anterior projection of premaxilla well developed. Inferior zygomatic root short and, in lateral view, located almost at same level as ventral surface of rostrum and palatal. Interpremaxillary foramen elongated and incisive foramina slender and elongated, lyre shaped, and with lateral flange. Septum of incisive foramina incomplete, its premaxillary portion elongated and extending posterior to premaxillomaxillary contact; maxillary part reduced to small process, and portion of vomer visible in ventral view. Palatal grooves well developed and contain pair of palatal foramina placed at same level as anterior portion PM4. More than two pairs of palatal foramina present, one in median region of palate and other located near level of third molar. Infraorbital foramen large, and floor houses canal with well-developed lateral flanges. Mesopterygoid fossa U-shaped and among most adults extends to middle of M3s. Hamular processes of pterygoid pointed. Sphenopalatine foramen and vacuities well developed. Narrow strut separates masticatory foramen and foramen ovale accessorius. Bullae large but barely inflated. Lateral tube of auditory meatus perpendicular to longitudinal axis of skull. Jugal moderately wide; jugal fossa very deep. Inferior jugal process conspicuous, often pointed, and projecting at least to level of postorbital process of zygoma. Postorbital process of zygoma pointed and formed only by squamosal or by squamosal and jugal. Mandibular foramen midway between posterior limit of toothrow and tip of condylar process. Angular process of dentary extends as far as condylar process or exceeds its posterior limit; coronoid process lower than condylar process.

DISTRIBUTION: *Thrichomys apereoides* is known from the type locality and a few other localities in Minas Gerais state, Brazil.

SELECTED LOCALITIES (Map 530): BRAZIL: Minas Gerais, Fazenda Canoas, Juramento (Bonvicino, Otazu, and D'Andrea 2002), Lagoa Santa (type locality of E[*chimys*]. *apereoides* Lund).

SUBSPECIES: *Thrichomys apereoides* is monotypic.

NATURAL HISTORY: Lund (1839c, 1841a) reported that *T. apereoides* often inhabited caves and could be found in human dwellings, stating that the species was an important agricultural pest. He also noted a large number of insect remains at the entrance of burrows and believed that arthropods were important dietary items. The recent dietary study by L. G. Lessa and Costa (2009) confirmed Lund's observation of the importance of insectsin the diet, especially during the dry season when fruit availability is low. These authors further documented that seeds comprise the most frequent component of the diet, after arthropods, with large seeds severely damaged after chewing while small seeds of pioneer plants retained a high capacity for germination after passage through the gut. Their data suggested that *T. apereoides* is an important potential disperser of small seed (<1 mm) plants. Arends and McNab (2001) provided physiological attributes for what they referred to as *T. apereoides*, but we cannot confirm the taxon because these authors did not provide data on their voucher specimens.

REMARKS: *Thrichomys apereoides* was one of several species of hystricognath rodents that P. W. Lund described during his work at Lagoa Santa, Minas Gerais, Brazil. He initially (Lund 1839c) described *Echimys apereoides* based on fossil and living material found in and near the caves at Lagoa Santa. Nevertheless, he did not specifically designate a holotype. Later, Lund (1840a, preprint of Lund 1841b:243) divided *Echimys* into four different genera based on the occlusal structure of their cheek teeth. Within one of these, *Nelomys* Jourdan, Lund placed *N. sulcidens* (now *Carterodon sulcidens*) and *N. antricola*, both larger species with short ears and densely haired tails. Lund

(1840a,1841b:241–242, footnote) coined *antricola* as a replacement name of his earlier name *apereoides*, an allusion to this rodent's habit of living in holes. He (1840a:plate 19; 1841b:plate 22) illustrated the skull and mandibles of his *Nelomys antricola*, drawings that correspond to specimen L. 117 of the Zoologisk Museum, Copenhagen (UZM), and listed there as the "syntypus of *Echimys apereoides* Lund, also syntypus of *Nelomys antricola* Lund." As this is apparently the only specimen of *apereoides* or *antricola* in the Lund collection at UZM (http://www.zmuc.dk/VerWeb /lund/lund_mammals.html), this specimen can be considered the holotype of both *Echimys apereoides* Lund and *Nelomys antricola* Lund.

In 1880, E.-L. Trouessart proposed the name *Thrichomys* as a substitute for *Nelomys* Lund, which was preoccupied by *Nelomys* Jourdan. E.-L. Trouessart did not give priority to *apereoides*, the earliest name used by Lund (1839c), but instead listed *Thrichomys antricola*. Only in 1904 did E.-L. Trouessart allocate *apereoides* to *Tricomys* [*sic*], when he again misspelled the generic name.

Thrichomys apereoides is known from Pleistocene deposits at Lagoa Santa (Paula Couto 1950).

The chromosomal complement from specimens from Lagoa Santa today is $2n = 28$ and $FN = 50$ (Pessôa et al. 2004). Bonvicino, Otazu, and D'Andrea (2002) described additional fundamental numbers of 50 and 52 from Juramento, Minas Gerais, and Jaborandi, Bahia, respectively,

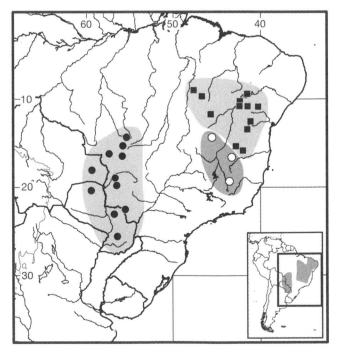

Map 530 Selected localities for *Thrichomys apereoides* (○), *Thrichomys inermis* (■), and *Thrichomys pachyurus* (●). Contour line = 2,000 m.

referring both to *T. apereoides*. Herein, we associate specimens with the 2*n* = 28, FN = 50, karyotype with Lund's *T. apereoides*, because this karyotype is present in individuals from the type locality. It is unclear at present whether the 2*n* = 28, FN = 52, karyotype belongs to the same species or might represent one that is separate and as yet undescribed. Laboratory crosses imply reproductive isolation of topotypical *T. apereoides* from both *T. laurentius* and *T. pachyurus* due to postzygotic barriers (Borodin et al. 2006).

Thrichomys inermis (Pictet, 1843)
Jacobina Punaré, Rabudo

SYNONYMS:

Echimys inermis Pictet, 1843a:207; type locality "la Jacobine (dans les montagnes de las Almas, province de Bahia," Jacobina, Bahia, Brazil.

[*Thrichomys*] *inermis*: E.-L. Trouessart, 1880b:179; first use of current name combination.

[*Thrichomys*] *inermis*: E.-L. Trouessart, 1897:607; name combination.

[*Tricomys*] *inermis*: E.-L. Trouessart, 1904:504; name combination.

Echimys inermis: Miranda-Ribeiro, 1914:48; name combination.

[*Cercomys*] *inermis*: Tate, 1935:409; name combination.

Cercomys cunicularius inermis: Moojen, 1952b:173; name combination.

T[*hrichomys*]. *apereoides inermis*: Braggio and Bonvicino, 2004:316; name combination.

DESCRIPTION: Smallest species of punaré. Dorsal color ash gray, but slightly more yellow than that of *T. pachyurus* and less yellowish than other two species. Ventral hairs usually completely white, rarely with light-gray base; dorsal tail hairs darker, blackish brown, contrasting sharply with mid-dorsal body color. Ventral tail hairs whitish gray. Skull small and less robust, with greatest length in nonsenile adults averaging 50.58 ± 2.3 mm (N = 16). Supraorbital ridges extend only to contact between frontal and squamosal bones. Inferior zygomatic root short and, in lateral view, displaced slightly below level of rostrum and palate. Interpremaxillary foramen broad and elongated. Incisive foramina broad, ovate, and sometimes slightly elongated, constricted in region of premaxillomaxillary suture, but not to degree as found in *T. pachyurus*. Septum of incisive foramina discontinuous; its premaxillary portion robust, extending posterior to premaxillomaxillary suture. Vomerine sheath visible in ventral view, at level of premaxillary portion of septum. Maxillary part reduced to small process continuous with well-developed palatal crest, more prominent than in *T. laurentius*. Groove with well-developed lateral flange for passage of infraorbital nerve

present on floor of infraorbital foramen. Mesopterygoid fossa V-shaped and extends anteriorly the middle of M3s in adults. Hamular processes of pterygoid almost always rectangular at tip. Bullae small to medium; lateral tube of auditory meatus directed perpendicular to long axis of skull. Jugal thinner than in other species, but jugal fossa remains very deep. Postorbital process of zygoma pointed and often formed mainly by squamosal in adults. Thin strut separates masticatory foramen, and foramen ovale accessorius separated. Mandibular foramen with middle position between posterior limit of toothrow and tip of condylar process. Angular process of dentary slender and extends to posterior limit of coronoid process; coronoid process of dentary spiny and lower than condylar process.

DISTRIBUTION: *Thrichomys inermis* is known from the type locality and few other localities in the central region of Bahia state north to central Tocantins state in central Brazil.

SELECTED LOCALITIES (Map 530): BRAZIL: Bahia, Barreiras (L. C. Duarte et al. 2000), Chapada Diamantina (L. G. Pereira and Geise 2007), Jacobina (type locality of *Echimys inermis* Pictet), Lençóis (J. A. Oliveira and Pessôa 2005), Mucujê (Leal-Mesquita et al. 1993), Santo Inácio (Leal-Mesquita et al. 1993), Sento Sé (Bonvicino, Otazu, and D'Andrea 2002), Vacaria (Leal-Mesquita et al. 1993); Tocantins, Jalapão (A. H. Carvalho and Fagundes 2005), Rio Sono (A. H. Carvalho et al. 2008).

SUBSPECIES: *Thrichomys inermis* is monotypic.

NATURAL HISTORY: The Caatinga, the semiarid biome where both *T. inermis* and *T. laurentius* occur, is characterized by an unpredictable rainfall regime (Streilein 1982a). Compared to areas where species of echimyid rodents occur, the Caatinga also records the highest insolation, lowest degree of cloudiness, highest mean temperature, and lowest relative humidity (Nimer 1979). Some physiological and behavioral characteristics of species of *Thrichomys* may have evolved as a response to this extreme environment (Streilein 1982a, 1982b, 1982c; Roberts et al. 1988; Meyerson-McCormick et al. 1990; Mendes et al. 2004). Individuals frequent granite or sandstone formations, locally called *lajeiros* or *lajedos*, which provide crevices for temporary refuges and secure shelter (Streilein 1982d). The nests constructed of cotton, straw, or dry leaves are built in rocky crevices or in hollow trees (Moojen 1943a). For a small mammal that lives in this extreme biome, the heat-retaining and insulating properties of rock help in thermoregulation, providing a cool refuge during warm hours and a warm home during the cold night (Nutt 2007).

In common with Lund's (1839c, 1841a) observations on *T. apereoides*, we captured individuals of *T. inermis* inside the entrance of the Lapa de Brejões, a large cave in the northern Chapada Diamantina, Bahia. Moojen (1943a) also reported that northeastern *rabudos* might live in tree cavities or more rarely among stems of the spiny macambira bromeliad (*Bromelia lacinosa*), likely a characteristic of both *T. inermis* and *T. laurentius*. Although *T. inermis* appears primarily terrestrial, we captured it at a height of 2 m on a tree branch at Morro do Chapéu, thus indicating at least a limited climbing ability.

Few data are available on reproductive biology of *T. inermis*. Moojen (1952b) recorded females with two, three, or four embryos or young at two localities in February and March in Bahia state.

Mendes et al. (2004) reported on physiological responses to stressful environmental conditions in a comparative study on populations from Curaçá and Ibiraba, Bahia state. Although these authors treated both populations they studies as *T. apereoides*, we believe the sample from Curaçá is more appropriately referred to *T. inermis* and the population from Ibiraba to *T. laurentius* based on karyological data of individuals from Ibiraba (Leal-Mesquita et al. 1993) and on the small mass of the specimens form Curaçá (Mendes et al. 2004). These authors argued that some feeding parameters were partially driven by precipitation conditions. Individuals from Ibiraba, where mean annual precipitation is 700 mm, decrease their relative food intake and relative urine volume while increasing the urine osmolality when exposed to water scarcity. On the other hand, individuals from Curaçá, where annual precipitation is 50% less, maintained stability in these parameters under the same stimulus.

Although *T. laurentius* has been given as the type host of the lice *Gyropus cercomydis* and *G. scalaris*, the type locality of these two parasites is Xique-Xique, Rio São Francisco, Bahia, Brazil. Only *T. inermis*, not *T. laurentius*, is present at this locality (Valim and Linardi, 2008).

REMARKS: Woods and Kilpatrick (2005:1589) erred in giving the publication date for Pictet's *Echimys inermis* as 1841 rather than 1843. The chromosomal complement is $2n = 26$, FN $= 48$ (Pessôa et al. 2004). The species status of *T. inermis* is supported by both chromosomal data (Bonvicino, Otazu, and D'Andrea 2002) and molecular phylogenetics (Braggio and Bonvicino 2004).

Thrichomys laurentius Thomas, 1904
São Lourenço Punaré

SYNONYMS:

Thrichomys laurentius Thomas, 1904c:254; type locality "São Lourenço [=São Lourenço da Mata], near Pernambuco, Alt. 50 m," Pernambuco, Brazil.

C[ercomys]. laurentius: Thomas, 1912e:116; name combination.

Cercomys cunicularius laurentius: Ellerman, 1940:124; name combination.

T[hrichomys]. apereoideslaurentius: Braggio and Bonvicino, 2004:316; name combination.

[*Thrichomys apereoides*] *laurenteus* Woods and Kilpatrick, 2005:1589; incorrect subsequent spelling of *laurentius* Thomas.

DESCRIPTION: Medium-sized punaré when compared with other species. Dorsum generally yellowish gray and ventral hairs completely white or with gray base, contrasting with dorsal coloration. Dorsal tail hairs yellowish gray and markedly longer on dorsal surface; ventral tail hairs whitish. Cranium of *T. laurentius* small, but larger than that of *T. inermis*. Average length of adult skulls 53.75 $\pm$ 3.28 mm (N = 15). Supraorbital ridges extend posteriorly to squamosoparietal suture. Rostrum short and wide; nasal bones do not reach posterior limit of premaxillary bones. Anterior projection of premaxilla well developed. Inferior zygomatic root short and, in lateral view, placed slightly below level of rostrum and palate. Interpremaxillary foramen narrow with constrictions on both ends. Incisive foramina wide and oval shaped; septum of incisive foramina occupies more than half foraminal opening. Palatal grooves little developed and contain single pair of foramina placed near front of PM4. Palate almost flat. Floor of infraorbital foramen with groove for passage of infraorbital nerve, with well-developed lateral flange. Mesopterygoid fossa V-shaped, extending in adults to middle of M3s. Hamular processes of pterygoid bones pointed at tip. Sphenopalatine foramen and vacuities well developed. Narrow strut separates masticatory foramen and foramen ovale accessorius. Bullae medium in size and only slightly inflated; lateral tube of auditory meatus short. Jugal narrow but contains very deep fossa. Inferior jugal process pointed, projecting to level of postorbital process of zygoma, which is pointed and formed mainly by squamosal. Mandibular foramen located closer to tip of condylar process when compared with other species. Angular process of dentary exceeds posterior edge of condylar process; coronoid process of dentary is lower than condylar process.

DISTRIBUTION: *Thrichomys laurentius* occurs in northeastern Brazil.

SELECTED LOCALITIES (Map 531): BRAZIL: Alagoas, Delmiro Gouveia (Braggio and Bonvicino 2004), Palmeira dos Índios (L. C. Duarte et al. 2000); Bahia, Cactité (Bonvicino, Otazu, and D'Andrea 2002), Queimada (Leal-Mesquita et al. 1993); Ceará, Chapada Ibiapaba (Bonvicino, Otazu, and D'Andrea 2002), Itapagé (Bandouk and Reis 1995), Jaguaruana (Bonvicino, Otazu, and D'Andrea 2002); Pernambuco, São Lourenço da Mata (type locality of *Thrichomys laurentius* Thomas); Piauí, Fazenda Felicidade (Bonvicino, Otazu, and D'Andrea 2002), João Costa (Braggio and Bonvicino 2004).

SUBSPECIES: *Thrichomys laurentius* is monotypic.

NATURAL HISTORY: Moojen (1952b) reported that in dry season *T. laurentius* ate e*macambira* buds (*Bromelia lacinosa*), leaves of the cactus locally named *mandacaru* (*Cereus jamacaru*), and the *catolé* coconut (*Attalea* sp.). In Caatinga habitat, *T. laurentius* also forages on the pads of the cactus *Opuntia palmadora*, using them as a source of water (Streilein 1982a). At Exu in Pernambuco state, this species has an activity peak at dusk but may also be active during short periods of both day and night (Streilein 1982a). In the same population, female home ranges rarely overlapped, but those of males overlapped with multiple females but were relatively exclusive to other males. Streilein (1982e) showed that reproductive activity occurred in every month of the year, with a peak from September to December. Adult females produced two or three litters per year, with an interbirth interval of four to six months. Litter size averaged 3.1 but varied from one to six (Streilein 1982c). Pregnant females in Ceará state were captured in July with one to four embryos, with two most common (Moojen 1952b). Roberts et al. (1988) found litter size in captive individuals to vary from one to seven, with an average of 3.2 (1.42 s.d.), a positive correlation between the mother's age and litter size, and a gestation period of 95 to 98 days. The milk of captive females has high fat content, a possible adaptation to minimize water loss in an arid environment (Meyerson-McCormick et al. 1990). Neonates are precocial, born at an average weight of 2.1 g, with sufficient motor coordination to eat solid food a few hours after birth, with eyes and ears open, a fully furred body, and upper and lower incisors erupted (Roberts et al. 1988).

Records of predators of *T. laurentius* are rare, but include the Barn Owl (*Tyto alba*; Roda 2006), and the Maned wolf (*Chrysocyon brachyurus*; F. H. G. Rodrigues et al. 2007). In northeastern Brazil, punaré may be included in diet of local human populations (C. T. Carvalho 1979).

Thrichomys laurentius is the type host of the louse *Gyropus lenti lenti*, whose type locality is Ceará, Brazil. The same louse also parasitizes *T. apereoides* and *T. inermis* from localities in the states of Bahia, Ceará, Goiás, Minas Gerais, and Pernambuco (Valim and Linardi 2008). While *T. inermis* is given as the type host of another louse, *G. freitasi*, with type locality Serra do Pacoti, Ceará, only *T. laurentius* occupies this area (Valim and Linardi 2008).

REMARKS: M. J. Souza and Yonenaga-Yassuda (1982) recorded a karyotype of $2n = 30$ and $FN = 54$ from several localities in Pernambuco state. Later, Bonvicino, Otazu, and D'Andrea (2002) found this karyotype in other localities from northeastern Brazil, which they allocated to *T. laurentius*, although the karyotype of specimens from the type locality is unknown. Other specimens collected in central Brazil (Brasília, Distrito Federal; Fazenda Osara, Tocantins state; and Cavalcante, Mimoso de Goiás,

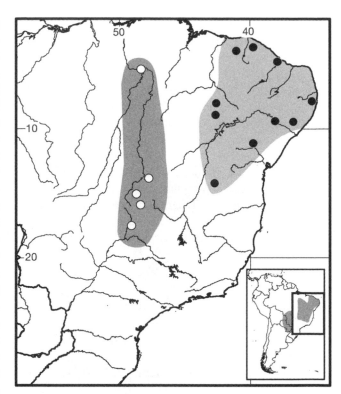

Map 531 Selected localities for *Thrichomys laurentius* (●) and *Thrichomys* sp. [$2n = 30$, FN = 56 karyomorph] (○).

and Teresina de Goiás, Goiás state) have the same diploid number but a different fundamental number ($FN = 56$; Svartman 1988; J. F. S. Lima 2000; Bonvicino, Otazu, and D'Andrea 2002). These specimens cannot currently be assigned to any described taxon, but Bonvicino, Otazu, and D'Andrea (2002), Braggio and Bonvicino (2004), and Bonvicino, Oliveira, and D'Andrea (2008) argued they represented an undescribed species. We map these localities separately from the range of *T. laurentius* in Map 531.

Specimens collected at Jaíba, in the northern part of Minas Gerais, were also $2n = 30$, but their fundamental number was undetermined (Pessôa et al. 2004), and thus their allocation to karyomorph, and perhaps to species, cannot be determined as yet. It is of note that Moojen (1952b), based on morphological data, recognized two taxa from Minas Gerais state, one from Lagoa Santa and another from further to the north.

Thrichomys pachyurus Wagner, 1845
Paraguayan Punaré; Rato-boiadeiro
SYNONYMS:

Isothrix pachyura Wagner, 1845a:146; type locality "Cuyaba [= Cuiabá]," Mato Grosso, Brazil, by subsequent designation (Pelzeln 1883:61).

Isothrix pachyurus: Wagner, 1848:291; name combination, but incorrect gender concordance.

Isothrix crassicaudus Wagner, 1848:291; replacement name for *Isothrix pachyurus* Wagner; incorrect gender concordance.

Loncheres (*Isothrix*) *pachyura*: Pelzeln, 1883:61; name combination.

Thrichomys pachyurus: E.-L. Trouessart, 1880b:179; first use of current name combination.

[*Thricomys*] *pachyurus*: E.-L. Trouessart, 1897:607; name combination.

Thrichomys Fosteri Thomas, 1903a:227; type locality "Sapucay," Paraguarí, Paraguay.

[*Tricomys*] *fosteri*: E.-L. Trouessart, 1904:504; name combination.

[*Tricomys*] *pachyurus*: E.-L. Trouessart, 1904:504; name combination.

C[*ercomys*]. *fosteri*: Thomas, 1912e:116; name combination.

Cercomys cunicularius fosteri: Ellerman, 1940: 124; name combination.

Cercomys cunicularius pachyurus: Moojen, 1952b:172; name combination.

Isothrix bistriata pachyura: Cabrera, 1961:537; name combination.

Thrichomys apereoides fosteri: S. Anderson, 1997:512; name combination.

T[*hrichomys*]. *apereoides pachyurus*: Braggio and Bonvicino, 2004:316; name combination.

DESCRIPTION: Largest species in genus. Dorsum generally ash gray, and ventral hairs either completely white or a light gray base, thus contrasting with dorsal coloration. Dorsal tail hairs blackish brown and contrast strongly with color of mid-dorsum; underside of tail whitish gray. Skull broadest and most robust of any species in genus; greatest skull length averages 55.43 ± 2.06 mm ($N = 10$). Supraorbital ridges extend at least to frontosquamosal suture. Rostrum short and wide, and nasal bones do not extend posterior to premaxillae. Anterior projection of premaxilla well developed. Inferior zygomatic root short and, at lateral view, placed at slightly lower level then rostrum and palate. Interpremaxillary foramen rounded and intermediate in size when compared with other species. Incisive foramina oval, and widest of all congeners, with indentation caused by ossification of premaxillomaxillary suture; septum of incisive foramina incomplete, its premaxillary portion elongated in younger individuals but shorter in adults; its posterior margin does not extend to premaxillomaxillary suture; maxillary part of septum reduced to small process, and significant portion of vomer visible ventrally. Palatal grooves and median ridges more strongly developed than in other species; palatal foramina of varying number and size found within these grooves, although first is always placed at front of palate, often inside groove formed by incisive foramina. Groove for passage of infraorbital nerve present on floor of infraorbital foramen; it possesses well-developed lateral flange. Mesopterygoid fossa U-shaped, its extension varying ontogenetically but among adults often reaching to middle of M3s. Hamular process of pterygoid rectangular in shape and broad at tip. Sphenopalatine foramen and vacuities well developed. Wide and complex plate separates masticatory foramen and foramen ovale accessorius. Bullae largest and most inflated among all species, and lateral tube of auditory meatus not directed outward as in other species but curves up in both juveniles and adults. Jugal robust and the major contributor to zygomatic arch, which possesses very deep fossa. Inferior jugal process often pointed, projecting as far as level of postorbital process of zygoma. Postorbital process of zygoma pointed and formed mainly by squamosal. Mandibular foramen positioned half way between posterior limit of toothrow and tip of condylar process. Angular process of dentary slender and extends to posterior limit of coronoid process, which is spiny and lower than condylar process.

DISTRIBUTION: *Thrichomys pachyurus* has the westernmost range of the genus, occurring from western Brazil, in states of Mato Grosso and Mato Grosso do Sul, through Paraguay and Bolivia.

SELECTED LOCALITIES (Map 530): BOLIVIA: Santa Cruz, Roboré (S. Anderson 1997). BRAZIL: Mato Grosso, Cuiabá (type locality of *Isothrix pachyura* Wagner), Reserva Particular do Patrimônio Natural do Serviço Social do Comércio (Pessôa et al. 2004), Tucum (MNRJ 1947), Usina Hidroelétrica Manso (Lacher and Alho 2001); Mato Grosso do Sul, Corumbá, Fazenda Alegria, Corumbá (Bonvicino, Otazu, and D'Andrea 2002); Fazenda Princesinha (Cáceres et al. 2007). PARAGUAY: Alto Paraguay, Fortín Madrejón (UMMZ 125552); Amambay, Pedro Juan Caballero (UMMZ 125301); Concepción, 8 km E of Concepción (MVZ 145320); Paraguarí, Sapucay (type locality of *Thrichomys fosteri* Thomas).

SUBSPECIES: *Thrichomys pachyurus* is monotypic.

NATURAL HISTORY: Pelzeln (1883) reported that individuals from Cuiabá were collected inside human dwellings. Moojen (1952b) reported a lactating female in July taken near Cuiabá. During our own fieldwork at Barão de Melgaço (from 1999–2004) in the northern Pantanal region, this species was collected in traps set in trees about 1.5 m above the ground during the rainy season in February. We also observed a female with two young in a nest constructed inside an abandoned house.

In contrast to the Caatinga, the Pantanal is a very humid area with predictable rainfall characterized by distinct dry and rainy seasons. In this area, rocky outcrops do not occur, and *T. pachyurus* are commonly found among aggregations of the *acuri* palm, *Attalea phalerata* (Z. Campos and Mourão 1988; Camilo-Alves and Mourão 2010). These authors also reported that animals took palm fruits to their borrows or nearby, where they ate both mesocarp flesh and seed. *Thrichomys pachyurus* is primarily terrestrial but may climb trees. Püttker et al. (2012; see also Püttker, Prado et al. 2013) and Oliveira-Santos et al. (2013) estimated home range by telemetry and distances between successive captures.

Thrichomys pachyurus and *T. laurentius* are natural hosts of *Trypanosoma cruzi*, the vector agent of Chagas disease (Rodrigues Roque et al. 2005; Xavier et al. 2007). As a consequence, breeding colonies of both species have been established to serve as experimental models in various parasitological studies. Both taxa exhibit attributes that make them excellent laboratory models: they can be easily maintained in standard laboratory conditions, breed throughout the year, do not have any special dietary demands, and can be sustained on food pellets designed for laboratory mice. In captivity, *T. pachyurus* produces precocial offspring that have both eyes and ears open, teeth erupted, fur well developed, and that are capable of eating solid food in the first week of life. *Thrichomys pachyurus* has smaller litter sizes with larger masses at birth in comparison to *T. laurentius*. Teixeira et al. (2005) compared reproductive traits of *T. pachyurus* and *T. laurentius*, noting that females of the former became reproductively active later and at a larger mass than those of the latter.

Thrichomys pachyurus is the type host of the louse *Gyropus lenti distinctus*, with type locality of Salobra, Mato Grosso do Sul, Brazil (Valim and Linardi 2008). Gentili et al. (2012) recorded 12 species of helminths in specimens from two localities in the Pantanal of Brazil. Species richness of worms was higher during the wet season, with only a few very common species present in the dry season. These authors related this seasonal pattern with increased habitat availability and host population increases during the dry season.

REMARKS: Cabrera (1961) followed Wagner (1845a) and allocated *pachyura* (including *crassicaudus* Wagner) to the genus *Isothrix* as a valid subspecies of *I. bistriata*. Patton and Emmons (1985) maintained this error in their review of the species of *Isothrix*. Our comparisons of the type specimens of *pachyura* Wagner and *bistriata* Wagner in the Natural History Museum in Vienna, Austria, confirmed the hypothesis posed by Pessôa et al. (2004) that Wagner's *pachyura* is indeed a *Thrichomys* and not an *Isothrix*, as E.-L. Trouessart established in 1880. Tate (1935:414) incorrectly dated *crassicaudus* Wagner to 1850 instead of 1848 (see Sherborn 1925b).

S. F. Reis and Pessôa (2004) followed S. Anderson (1997) and regarded *fosteri* Thomas as a subspecies of *T. apereoides* (*sensu lato*). Qualitative characters used by Thomas in the original description of *fosteri* were applied in comparisons to the type specimen of *pachyurus* Wagner, and no differences were found. Thus, we regard *fosteri* as a junior synonym of *pachyurus*. Future analyses of geographic variation in morphology, karyotype, and molecular sequences that include samples from Paraguay and the Pantanal from Mato Grosso are needed to confirm this hypothesis.

Salles et al. (2005) recovered subfossil remains of *Thrichomys* from cave deposits in the Serra da Bodoquena, Mato Grosso do Sul. These presumably represent *T. pachyurus* because this species in present in the area today (Cáceres et al. 2007).

The chromosomal complement in *T. pachyurus* is $2n = 34$, $FN = 54$ (Pessôa et al. 2004).

Genus *Trinomys* Thomas, 1921
Leila M. Pessôa, William C. Tavares, João A. de Oliveira, and James L. Patton.

Species within *Trinomys* are terrestrial, cursorial rats that share a similar gross craniodental and body morphology to those of *Proechimys*. Primarily as a result of these shared similarities, species of both taxa, popularly called spiny rats, were united within a single genus (*Proechimys*) for most of the past century, divisible only at the subgeneric level. However, the combination of biogeographic data, details of dental morphology, and molecular phylogenetic analyses have progressively characterized *Trinomys* as a singular entity in the echimyid radiation, one that does not share a sister relationship with *Proechimys*. Today, *Trinomys* is treated as a genus that comprises one of the most diversified and complex hystricognathous groups in eastern Brazil, with 10 described species currently recognized. Species in the genus are distributed over a wide area in the Atlantic Forest and Caatinga of eastern and southern Brazil, within a myriad of local environments ranging from very humid pluvial forests to semiarid sand dunes (Moojen 1948; Pessôa and Reis 1992a; P. L. B. Rocha 1996; Lara and Patton 2000; Attias et al. 2009; Tavares and Pessôa 2010).

The oldest fossil unambiguously assignable to *Trinomys* dates from the Pleistocene deposits at Lagoa Santa, Minas Gerais (Lund 1839c, 1841a; Paula Couto 1950; McKenna and Bell 1997). However, molecular clock studies suggest that genus began diversification by the Late Miocene (Y. L. R. Leite and Patton 2002; Galewski et al. 2005; Upham and Patterson 2012).

These are moderately large rats with the head and body length of adults varying from 152 mm in *T. albispinus* to 246 mm in *T. paratus* (with respective tail length 148–246 mm), and mass from 120–350 g (Pessôa and Reis 2002). As in other cursorial echimyids, the tail is moderate in length (ranging from 80–120% of head and body length) and the hindfeet are elongated and narrow (ranging from 39 mm in *T. albispinus* to 55 mm in *T. eliasi*; Pessôa and Reis 2002; Tavares and Pessôa 2010). The tail is markedly bicolored in all species, although there are rare localized exceptions. Hairs along tail are sparse and very short, but the distal tip of the tail may or may not present a hairy bush, which is variable in color and size. General dorsal coloration ranges from light cinnamon to dark brownish, gradually becoming lighter on sides. The venter is always uniformly white, contrasting sharply with the darker dorsolateral coloration. Dorsal aristiforms extend from head to the base of tail. There is a broad range of variation in the general appearance of aristiform spines, with their mean width varying greatly among species. The vibrissae are elongated and extend posteriorly as far as the shoulder. Females have two abdominal mammae.

The adult skull is slender with a moderately long rostrum and supraorbital ridges that diverge posteriorly along the frontals toward the frontoparietal suture. The nasal bones extend forward of the posterior limit of the premaxillae. An anterior projection of the premaxilla is well developed. The inferior zygomatic root is short and, in lateral view, located almost at the same level of the ventral surface of the rostrum and the palate. The interpremaxillary foramen is absent or dot-like. Incisive foramina are moderately wide and oval or lyre shaped; the septum may be either continuous or discontinuous; and the vomer is either visible or not in ventral view. Palatal grooves are not well developed and house a variable number of palatine pits. The floor of the infraorbital foramen lacks a groove for the passage of the infraorbital nerve. The mesopterygoid fossa is V- or U-shaped, and its extension into the posterior palate varies with age. The sphenopalatine foramina and vacuities are moderately developed. The bullae vary in size from large to small, ranging from well to slightly inflated. The contact between the ectotympanic and squamosal is often restricted to the central portion of the dorsal margin of the ectotympanic; this condition is associated with the formation of anterior and posterior clefts between these bones. The jugal may be broad or narrow and may present a moderately deep fossa. The inferior jugal process is moderately developed, often pointed. The postorbital process of the zygoma is pointed or rounded and formed by squamosal or by the jugal alone, or by both elements. The mandibular foramen is positioned between the posterior limit of the toothrow and the tip of condylar process. The angular process of the dentary extends to or beyond the posterior edge of the condylar process. The coronoid process of the dentary is lower than condylar process.

The first and second molars are larger than the premolar and third molar, with the difference more pronounced in the upper toothrow. The molar series are parallel or converge slightly anteriorly. Homology of dental structures among living and fossil echimyid genera remains debatable (G. A. S. Carvalho and Salles 2004; Candela and Rasia 2012). Occlusal surfaces of unworn upper teeth have three to five major lophs, including an anteroloph, protoloph, metaloph, and posteroloph. Herein, we follow Moojen's (1948) nomenclature of mainfolds and counterfolds in the diagnosis of species.

The baculum of adult males in *Trinomys* species is elongate and narrow, with a shaft that is straight or with a slight dorsoventral curvature and a broadened proximal end. Both the proximal and distal ends vary in shape. The proximal end is typically paddle shaped, but the distal end can be straight or possess a median depression with slightly or developed distal wings (Pessôa and Reis 1992a; P. L. B. Rocha 1996; Pessôa et al. 1996; Lara et al. 2002; Tavares and Pessôa 2010).

Moojen subsumed 14 nominal taxa then known into five species, three polytypic and one included in *Trinomys* only as *incertae sedis*. Subsequent researchers have invested considerable effort to understand the variation of taxa allocated to the genus *Trinomys*. These have resulted in the reordering of existing taxa as well as the description of new ones, at both the species (Pessôa et al. 1992; P. L. B. Rocha 1996; Lara et al. 2002) and subspecies levels (Pessôa and Reis 1993a; S. F. Reis and Pessôa 1995). Each of these studies used morphological markers (as, for example, unique qualitative aspects of the bacular morphology) to define limits of variation at the species level (Pessôa and Reis 1992a, 1994; P. L. B. Rocha 1996; Pessôa et al. 1996; Lara et al. 2002; Iack-Ximenes 2005). Lara and Patton (2000) also developed the initial hypothesis of specific and infraspecific structure within the genus based on mtDNA cytochrome-*b* sequences. Their analyses recovered three clades, each of which had some morphological cohesion (Lara and Patton 2000), including general concordance between the ordination of taxa based on cranial shape, using geometric morphometrics, and the molecular phylogeny (Nicola et al. 2003). Two nominal taxa (*panema* Moojen, 1948 and *moojeni* Pessôa, Oliveira, and Reis, 1992) have yet to be included in any molecular phylogenetic analysis. Karyotypic data are now available for nine of the nominal taxa in the genus (Yonenaga-Yassuda et al. 1985; Leal-Mesquita et al. 1992; Corrêa et al. 2005; Pessôa et al. 2005; A. L. G. Souza et al. 2006), with A. L. G. Souza et al. (2006) documenting congruence between cytogenetic and molecular data. Thus, despite the complexity in establishing morphological boundaries that map to species units within the genus, further integration of diverse data sets offers the hope that more clearly defined limits will be forthcoming in the near future.

Of the 19 nominal taxa now allocated to the genus (e.g., Woods and Kilpatrick 2005), we limit our accounts to 10 species for which available names can be applied with some security. As noted above, further studies that combine data sets and more expansive geographic sampling are required to delineate fully species boundaries and their geographic ranges. The species we regard as unequivocally valid include: *Trinomys albispinus* (I. Geoffroy St.-Hilaire), *Trinomys dimidiatus* (Günther), *Trinomys eliasi* (Pessôa and Reis), *Trinomys gratiosus* (Moojen), *Trinomys iheringi* (Thomas), *Trinomys mirapitanga* Lara, Patton, and Hingst-Zaher, *Trinomys moojeni* (Pessôa, Oliveira, and Reis), *Trinomys paratus* (Moojen), *Trinomys setosus* (Desmarest), and *Trinomys yonenagae* (Rocha). We comment on hypothesized relationships, based on molecular or other character traits, in the accounts that follow.

SYNONYMS:

Echimys: Desmarest, 1817b:59; part (description of *setosus*); not *Echimys* F. Cuvier.

Loncheres: Lichtenstein, 1820:192; part (description of *myosurus*); not *Loncheres* Illiger.

Mus.: Brants, 1827:150; part (description of *leptosoma*); not *Mus* Linnaeus.

Echinomys: Wagner, 1843a:339; part (listing of *albispinus*); unjustified emendation of *Echimys* F. Cuvier.

Proechimys: J. A. Allen, 1899c:264; part (listing of *albispinus*, *dimidiatus*, and *setosus*); not *Proechimys* J. A. Allen.

Trinomys Thomas, 1921h:140; part; proposed as a subgenus of *Proechimys* J. A. Allen, 1899; type species *Proechimys albispinus* I. Geoffroy St.-Hilaire, by original designation.

Prochimys: Cerqueira, Gentile, Fernandez, and D'Andrea, 1994:512, Fig. 2; listing of *iheringi*; *lapsus calami*.

Trinomys: Lara and Patton, 2000:661; first regarded as full genus.

REMARKS: Historically, the generic name *Trinomys* was etymologically based on of the number of counterfolds in the molariform teeth. Thomas (1921h) recognized two subgenera in the genus *Proechimys* (*Proechimys* and *Trinomys)*, defining *Proechimys* as those terrestrial spiny rats with four counterfolds and *Trinomys* as those with three. According to this definition, the subgenus *Proechimys* encompassed all species occurring in Central America, the Amazon Basin, and central Brazil plus one species, *P. iheringi*, from the Atlantic Forest in southeastern Brazil. Moojen (1948:312) observed, however, that the main fold in molariform teeth was short in all forms allocated by Thomas to the subgenus *Proechimys*, except *P. iheringi*, while in all Atlantic Forest taxa, including *P. iheringi*, this fold extended entirely across the occlusal surface. Moojen (1948) then used this character difference to diagnose the two subgenera, rather than the number of counterfolds,

which he documented as highly variable at the subspecific level. The subgeneric division between *Trinomys* and *Proechimys* espoused by both Thomas and Moojen assumed a close, sister relationship between the two taxon groups, a hypothesis supported by cladistic analysis of craniodental traits (G. A. S. Carvalho and Salles 2004) and dental characters alone (Candela and Rasia 2012) but refuted by all molecular phylogenetic studies using either, or both, mitochondrial and nuclear gene sequences (Lara et al. 1996; Y. L. R. Leite and Patton 2002; Galewski et al. 2005; Upham and Patterson 2012). All molecular studies document a strongly supported relationship between *Trinomys* species and *Clyomys* + *Euryzygomatomys* to the exclusion of *Proechimys* species, which in turn are sisters successively to *Hoplomys* and then *Thrichomys* within the Eumysopinae.

Voss and Abramson (1999) pointed out that *Mus leucogaster* Brandt, then regarded as the type species of the sigmodontine rodent *Holochilus*, was in fact based on a specimen of *Trinomys*. If allowed to stand, *Holochilus* Brandt would become the senior synonym for the spiny rats grouped by Thomas into *Trinomys* 90 years ago, an action that would contravene name associations that have stood in the literature for more than a century. Hence, Voss and Abramson (1999) petitioned the International Commission on Zoological Nomenclature to conserve *Holochilus* Brandt for the myomorphous marsh rats and both *Proechimys* and *Trinomys* for the hyscricomorphous spiny rats, to which the International Commission (ICZN 2001:Opinion 1984) agreed, using its plenary power to designate *Holochilus sciureus* Wagner as the type species of *Holochilus* and preserve *Trinomys* Thomas for the southern Brazilian spiny rats. Because the holotype of Brandt's *leucogaster* is a specimen of *Trinomys*, this name is appropriately connected to the genus. Comarisons of the holotype (a skin with skull and mandibles, cataloged as ZINRAS 219; see Voss and Abramson 1999:257) are needed to link Brandt's name to one of the species we recognize herein, and once done, *leucogaster* may well become a senior synonym of one of them.

KEY TO THE SPECIES OF *TRINOMYS*:

1. Tail shorter than head and body length (about 87–96%), without a conspicuous brush of elongated hairs in the tip. 2

1′. Tail length similar or longer than head and body length (from 100–130%), with conspicuous brush of elongated hairs in the tip. 7

2. White aristiforms on back sides, rump and at base of tail; septum of incisive foramen often incomplete, with premaxillary portion at a level lower than maxillary portion; molars with only one counterfold; baculum longer than 11 mm in adult specimens, with large distal wings . . . . *Trinomys albispinus*

2'. Without marked white aristiforms on back sides, rump and at base of tail; septum of incisive foramina variable; molariform teeth with more than one counterfold; baculum less than 10 mm in length in adult specimens, without conspicuous large distal wings. 3

3. Incisive foramina medium sized to large (3.3–5.9 mm), premaxillary portion of septum broad and maxillary portion very constricted, septum often incomplete; baculum with a broad basal paddle-shaped expansion, lateral indentation near distal end. 4

3'. Incisive foramina small to medium sized (2.8–4.8 mm); septum always complete; baculum often with basal expansion rounded, with lateral indentation near midshaft. 5

4. Incisive foramina long and constricted posteriorly; 2*n* = 56; NA = 108. . . . . *Trinomys gratiosus*

4'. Incisive foramina large and wide posteriorly; 2*n* = 56; NA = 106 . . . . *Trinomys moojeni*

5. General mid-dorsal coloration very dark tending to blackish; mid-dorsal aristiform hairs very wide (mean width 1.49 mm) and . . . . *Trinomys mirapitanga*

5'. General mid-dorsal coloration dark tending to brownish; very narrow mid-dorsal aristiforms (mean width 0.6 mm) 6

6. Differentiated light-colored pelage on outer thighs; incisive foramen constricted on posterior end; postorbital process of zygoma contributed by jugal and squamosal; presence of dot-like supernumerary chromosomes, 2*n* = 61–65, NA = 116. . . . . *Trinomys iheringi*

6'. No differentiated light-colored pelage on outer thighs; incisive foramen wide posteriorly; postorbital process of zygoma contributed primarily by the squamosal; absence of dot-like supernumerary chromosomes, 2*n* = 60; NA = 116 . . . . *Trinomys dimidiatus*

7. Molariform teeth with only one counter-fold; very inflated auditory bullae (mean bullar length 11.70 mm); skull small to medium sized (mean basilar length 32.79 mm) and slender; dorsal pelage pale brownish; very elongated hindfeet (about 27% the head and body length in adults); tail length about 130% of head and body length; terminal tail brush highly developed. . . . .
. . . . *Trinomys yonenagae*

7'. Molariform teeth almost always with two or more counterfolds; auditory bullae small to moderately inflated; skull usually large and stout; dorsal pelage variable from dark brownish to cinnamon 8

8. Incisive foramina oval and long, with posterolateral margins of maxillary flat or slightly flanged; septum of incisive foramina complete and columnar; vomer not visible under ventral view; mid-dorsal aristiforms very wide (mean width 1.58 mm) and dark. . . . *Trinomys paratus*

8'. Incisive foramen often lyre shaped and long, with posterolateral margins of maxillary markedly flanged; septum of incisive foramina comprised mainly by premaxillary portion, often becoming gradually narrower toward maxillary portion, and may be complete or incomplete; mid-dorsal aristiforms width variable. 9

9. Conspicuously spiniform and long postorbital process of zygoma formed primarily by jugal; auditory bullae small and least inflated (mean 10.08 mm); 2*n* = 56, NA = 104, uniarmed X and Y chromosomes . . . . . *Trinomys setosus*

9'. Moderately spiniform and short postorbital process of zygoma formed primarily by jugal, but sometimes with important contribution of squamosal; auditory bullae moderately to well inflated (mean bullar length 10.67 mm); 2*n* = 58, NA = 112, biarmed X and Y chromosomes. . . . . *Trinomys eliasi*

Trinomys albispinus (I. Geoffroy St.-Hilaire, 1838a)
White-spined Spiny Rat

SYNONYMS: See under Subspecies.

DESCRIPTION: Small spiny rat with wide and stiff guard hairs, distinguishable from other species smaller body and skull size, by presence of lanceolate and clavate guard hairs, and by whitish aristiforms on back, sides, rump, and base of tail (Moojen 1948; Pessôa and Strauss 1999; Pessôa and Reis 2002). Mean external measurements of each subspecies (Pessôa and Reis 2002): *P. a. albispinus* from Jaguaquara, Bahia (N = 10): head and body length 177 mm (153–190 mm), tail length 166 mm (148–175 mm), hindfoot length 40 mm (37–42 mm), ear length 26 mm (22–30 mm), and mass 173 g (120–230 g); *T. a. minor* from Morro do Chapéu, Bahia: (N = 6) head and body length 163 mm (152–165 mm); tail length 141 mm (126–160 mm); hindfoot length 36 mm (34–38 mm); ear length 25 mm (23–26); and mass 134 g (N = 1).

Dorsal pelage ochraceous-tawny gradually changing to ochraceous-buff on sides. Differentiated light-colored guard hairs occur on back, sides, rump, and at base of tail. Aristiform guard hairs in mid-dorsal region with whitish base that gradually blackens toward tip, interrupted by ochraceous-tawny subdistal zone. Lanceolate guard hairs also with whitish base that gradually blackens toward tip, but lack visible subdistal zone; total length 25–28 mm, maximum width 1.2 mm. Guard hairs on outer thighs show two color patterns, one with whitish base that gradually blackens toward tip and another that also gradually blackens distally but with distal fifth, including tip, colored ochraceous-tawny. Ventral pelage of body and inner sides of legs totally white, and well delimited from the sides. Forefeet and hindfeet are whitish on dorsal parts, with some specimens darker outer margins of hindfeet. Tail blackish above, white below, without hairy pencil at tip (Moojen 1948; Pessôa and Reis 1995; Pessôa and Strauss 1999).

Skull short and smooth in comparison with other species, average cranial measurements of each subspecies (Pessôa and Reis 1991b; Pessôa and Reis 2002; Pessôa and Strauss 1999) are: *T. a. albispinus* (N = 18; Moojen 1948): skull length 45.0 mm (37.6–48.9 mm), condyloincisive length 40.1 mm (38.3–41.4 mm), zygomatic breadth 23.9 mm (22.8–24.5 mm), nasal length 15.9 mm (13.5–16.8 mm), interorbital constriction width 10.8 mm (9.7–11.6 mm), palatal length 15.9 mm (15.2–16.4 mm), toothrow length 7.7 mm (7.3–8.3 mm); *T. a. minor* (N = 4): skull length 40.8 mm (39.0–42.5 mm), condyloincisive length 34.7 mm (32.2–36.9 mm), zygomatic breadth 22.6 mm (21.5–23.2 mm), nasal length 13.9 mm (13.4–14.8 mm), interorbital constriction 9.5 mm (9.1–9.8 mm), palatal length 14.1 mm (13.5–14.7 mm), toothrow length 7.0 mm (6.9–7.1 mm). Sexual dimorphism in cranial measurements negligible (Pessôa and Reis 1991b). Incisive foramina oval and short, about 33% of diastema length; septum often incomplete, with premaxillary portion at level lower than maxillary portion, maxillary part small, and vomer not visible ventrally. Posterior palatine foramina at plane of first molars; postorbital process of zygoma well developed and involves both jugal and squamosal. Auditory bullae small, smooth, elongated, and least inflated. Molar teeth with only one counterfold; PM4 rarely with two counterfolds in *T. a. albispinus* (Moojen 1948; Pessôa and Strauss 1999).

Baculum larger than those of other species (mean length 11.4 mm, N = 10 adults); shaft with dorsoventral curvature and slightly tapered lateral indentation near midshaft; proximal end paddle shaped, and distal end with well-developed distal wings with pronounced median depression (Pessôa and Reis 1992a; Pessôa et al. 1996).

DISTRIBUTION: *Trinomys albispinus* is known from Sergipe, Bahia, and Minas Gerais states in coastal Brazil.

SELECTED LOCALITIES (Map 532): *T. albispinus albispinus*—BRAZIL: Bahia, Cachoeira do Ferro Doido (MNRJ 67903), Feira de Santana (MNRJ 14019), Ibiraba (MZUSP 26772), Jequié (MNRJ 13820), Lamarão (Thomas 1921h), Seabra (MNRJ 13790), Senhor do Bonfim (MNRJ 6454), Vitória da Conquista (MNRJ 13778); Minas Gerais, Bocaiúva (MNRJ 73436), Irapé (MNRJ 73429); Sergipe, Canindé de São Francisco (Iack-Ximenes 2005]), Cristinápolis (MNRJ 30537). *T. albispinus minor*—BRAZIL: Bahia, Caetité (MNRJ 63343), Catolés de Cima, Abaíra (MNRJ 67814), Fazenda Juramento, Morro do Chapéu (S. F. Reis and Pessôa 1995).

SUBSPECIES: We recognize two subspecies of *T. albispinus*.

T. a. albispinus (I. Geoffroy St.-Hilaire, 1838a)
SYNONYMS:
Echimys albispinus I. Geoffroy St.-Hilaire, 1838a:886; type locality "Ile Deos, sur la cotê du Brésil," emended by Moojen (1948) to Itaparica, near Salvador, Bahia, Brazil.

E[chinomys]. albispinus: Wagner, 1843a:344; name combination.

Loncheres albispinis Schinz, 1845:111; incorrect subsequent spelling of *albispinus* I. Geoffroy St.-Hilaire.

Echimys albispinosus Waterhouse, 1848:341; incorrect subsequent spelling of *albispinus* I. Geoffroy St.-Hilaire.

[Echimys (Echimys) cayennensis] albispinus: E.-L. Trouessart, 1880b:180; name combination.

[Proechimys] albispinus: J. A. Allen, 1899c:264; name combination.

[Proechimys (Trinomys)] albispinus albispinus: Thomas 1921h:141; name combination.

Proechimys (Trinomys) albispinus sertonius Thomas, 1921h: 142; type locality "Lamarão, Bahia, about 70 miles north of Bahia City. Alt. 300 m," Brazil.

[Proechimys] serotinus Woods, 1993:795; incorrect subsequent spelling of *sertonius* Thomas.

Trinomys albispinus: Lara, Patton, and da Silva, 1996; name combination.

[Trinomys albispinus] serotinus: Woods and Kilpatrick, 2005:1590; name combination and incorrect spelling of *sertonius* (Thomas).

The nominotypical subspecies occurs in coastal and interior parts of Bahia, Minas Gerais, and Sergipe states, Brazil (combined distributions of *albispinus* and *sertonius* in Pessôa and Strauss 1999; Pessôa and Reis 2002).

T. a. minor (S. F. Reis and Pessôa, 1995)
SYNONYMS:
Proechimys albispinus minor S. F. Reis and Pessôa, 1995: 239; type locality "Morro do Chapéu, 800 m above sea level, state of Bahia, Brazil."

Trinomys albispinus minor: Manaf, Morato, and Oliveira, 2003:129; first use of current name combination.

Trinomys minor: J. A. Oliveira and Bonvicino, 2006:398; name combination.

This subspecies is known from upland areas in the interior of Bahia state in southeastern Brazil (Pessôa and Strauss 1999; Pessôa and Reis 2002).

NATURAL HISTORY: *Trinomys albispinus* is one of the most specialized forms in the genus for drier habitats (Moojen 1948). This habitat is part of the Caatinga biome, which is characterized by unpredictable rainfall and semiarid conditions, with vegetation including plants of the families Cactaceae and Bromeliaceae (Ab'Saber 1974). The subspecies *T. a. minor* occurs in northern edge of the

Espinhaço range in the state of Bahia, an area characterized by rock formations produced by a geosincline of pre-Cambrian age, which is 800–2,000 m. From 800–1,000 m, the vegetation is characteristically savanna, grading into grassland between 1,000 and 1,100 m (S. F. Reis and Pessôa 1995; Pessôa and Strauss 1999). Museum labels indicate that females are pregnant between January and June, with numbers of embryos varying from two to four.

REMARKS: Thomas (1921h) described *Proechimys albispinus sertonius* from Lamarão, Bahia, an inland form that he distinguished from *P. a. albispinus*, which he believed to be a coastal form. S. F. Reis and Pessôa (1995) described *P. a. minor*, a smaller form from Morrão, Morro do Chapéu, Bahia. Pessôa and Strauss (1999) subsequently showed that craniometric discriminant scores of *P. a. albispinus* and *P. a. sertonius* overlapped widely, although samples of the two taxa differed in average greatest skull length, with these two together well differentiated from *P. a. minor*, which presented smaller skulls and body size. Iack-Ximenes (2005) regarded *minor* as a distinct species, and documented cranial qualitative and quantitative variation along geographic transects in *T. albispinus*, although he did not regard this variation worthy of subspecific distinction. He also examined the type specimens of *T. a. albispinus* and *T. a. sertonius* and considered them indistinguishable.

A. L. G. Souza et al. (2006) and A. L. G. Souza (2011) examined population samples identified as *T. a. sertonius* and *T. a. minor* from two localities near Morro do Chapéu, Bahia, and showed that they have the same diploid and fundamental numbers and shared mtDNA cytochrome-*b* haplotypes. These samples also have cytochrome-*b* haplotypes similar to a specimen from Sergipe identified as *T. a. sertonius*, differing from it by only 2.2%. Two other mtDNA sequences belonging to specimens from Abaíra and Caetité, Bahia, both identified as *T. a. minor*, formed a sister clade with respect to the Morro do Chapéu–Sergipe clade. The authors argued that the region near Morro do Chapéu was a contact zone between *T. a. minor* and *T. a. sertonius* (= *T. a. albispinus*). A. L. G. Souza (2011) also postulated that the two subspecies had diverged recently or were in a process of divergence, a condition that would justify their taxonomic distinction at subspecific level.

We reexamined the specimens deposited in the Museu Nacional in Rio de Janeiro, including the extensive series obtained by the Serviço Nacional de Peste, the type and topotypical series of *T. a. minor* from Morro do Chapéu, as well as more recently collected samples, including a series from northern Minas Gerais and the Chapada Diamantina, in Bahia, in the light of recent unpublished analyses of Iack-Ximenes (2005) and A. L. G. Souza (2011). We here regard specimens from Morrão and Fazenda Jaboticaba

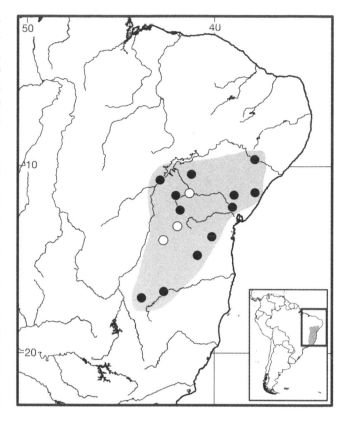

Map 532 Selected localities for *Trinomys a. albispinus* (●) and *Trinomys a. minor* (○).

(Morro do Chapéu), and from Catolés de Cima (Abaíra) and Caetité, as the museum representatives of *T. a. minor*. We also examined specimens from north of Minas Gerais that had previously been allocated to *T. a. minor*, which we reidentified as *T. a. albispinus*.

The karyotype is $2n = 60$ and NF = 116, with 29 pairs of biarmed autosomes, 17 of which are metacentric, 7 submetacentric, and 5 subtelocentric. The X chromosome is a large submetacentric intermediate in size between autosomal pairs 1 and 2. The Y chromosome is a small acrocentric. Two NORs, based on Ag-NOR staining, are located in the interstitial region of the long arm of chromosome pair 10 (A. L. G. Souza et al. 2006).

Trinomys dimidiatus (Günther, 1876)
Rio de Janeiro Spiny Rat
SYNONYMS:

Echimys dimidiatus Günther, 1876:747; type locality unknown; regarded as "probably southwestern Rio de Janeiro, Brazil," by Moojen (1948:371); herein fixed to "Parque Nacional da Tijuca, Rio de Janeiro, Estado do Rio de Janeiro, Brasil" (see Remarks).

[*Proechimys*] *dimidiatus*: J. A. Allen, 1899c:264; name combination.

P[*roechimys*]. [*Proechimys*] *dimidiatus*: Thomas, 1921h: 141; name combination.

[*Proechimys* (*Proechimys*)] *dimidiatus*: Ellerman, 1940: 122; name combination.

Proechimys [(*Trinomys*)] *dimidiatus*: Moojen, 1948:371; name combination.

Trinomys dimidiatus: Lara, Patton, and da Silva, 1996:405; first use of current name combination.

DESCRIPTION: Large species, with head and body length 203 mm (170–244 mm; s.d. 14.5; N=1 32), tail length 185 mm (150–220 mm; s.d. 14.0; N=90), hindfoot length 48 mm (40–55 mm; s.d. 2.7; N=137), ear length 25 mm (19–30 mm; s.d. 2.0; N=136), and mass 204 g (100–290 g; s.d. 33.5; Iack-Ximenes 2005). General color of upper parts ochraceous-buff, finely lined with blackish brown, gradually becoming lighter on sides, and contrasting with white ventral coloration (Moojen 1948). Tail bicolor, brownish above and white below, averaging 90% of head and body length, without conspicuous hairy pencil at tip. Aristiforms narrow and soft, length from 16–19 mm and average maximum width 0.5 mm, imparting nonspiny appearance to pelage (Pessôa and Reis 1993b). Setiforms on mid-dorsal region whitish basally, gradually blackening toward tip but interrupted by ochraceous-buff subdistal zone; setiforms on outer thighs whitish on their basal half then gradually become grayish at middle and finally light ochraceous-buff on distal third, or have blackish tip and ochraceous-buff subdistal zone (Moojen 1948).

Skull elongated and slender, with no conspicuous ridges, with average length of 51.7 mm (s.d. 1.9; N=217), basilar length 36.2 mm (s.d. 1.7, N=227), and zygomatic width 25.6 mm (s.d. 0.9; N=237) (data from Moojen 1948; Pessôa and Reis 1993b; Iack-Ximenes 2005). Incisive foramina oval, short (averaging about 32% of diastema length) and wide posteriorly; septum complete and columnar with thick and wide maxillary part, and vomer not visible ventrally. Posterior palatine foramina situated at plane of first molars or slightly anterior to them. Postorbital process of zygoma rounded and involves only squamosal. Auditory bullae small, smooth, elongated, and uninflated.

Baculum of medium size, with average length 7.64 mm (s.d. 0.57; N=7; elongate, with straight shaft without dorsoventral curvature but with lateral indentation near midshaft; proximal and distal ends evenly rounded, without median depressions or wings (Pessôa and Reis 1992a, 1993b).

DISTRIBUTION: *Trinomys dimidiatus* is known from Rio de Janeiro state and along the northern coastal border of São Paulo state (Moojen 1948; Pessôa and Reis 1993b; Pessôa et al. 2005; Attias et al. 2009).

SELECTED LOCALITIES (Map 533): BRAZIL: Rio de Janeiro, Barro Branco, Duque de Caxias (Iack-Ximenes 2005), Fazenda Carlos Guinle, Estrada Rio-Teresópolis

(MNRJ 33725), Fazenda Nova Freiburgo (Lara and Patton 2000), Fazenda do Tenente, São João de Marcos (MNRJ 4019), Gávea (MNRJ 10358), Mambucaba, Angra dos Reis (Lara and Patton 2000), Mata da Rifa, Parque Estadual do Desengano (Lara and Patton 2000), Rio Bonito (Pessôa et al. 2005); São Paulo, Picinguaba, Ubatuba (Pessôa et al. 2005).

SUBSPECIES: *Trinomys dimidiatus* is monotypic.

NATURAL HISTORY: *Trinomys dimidiatus* is regularly found in climax forests, especially in moist places near windfalls (Davis 1945a). The typical habitat is the relatively open interior of these forests, characterized by sparse vegetation near the ground and a marked middle layer of vegetation under large trees, up to 30 m in height and belonging predominantly to families Lauraceae and Melinaceae, plus smaller trees of the families Anonaceae, Melastomataceae, Myrsinaceae, and Rubiaceae. Lianas and epiphytes are also common (Davis 1947). The best shelter and nesting ground is usually under boulders, commonly not farther than 10 m from water sources; nesting places were also located at the base of trees and near fallen logs, where litter accumulates. No more than two adult animals (male and female) were captured in the same shelter, but occasionally young were captured in the same place with adults. At Teresópolis, the species was found breeding from March to May and from September to November. The number of juveniles varied from one to five but was commonly two. Specimens in captivity survived more than two years, and one specimen, already adult when captured, lived an additional 1,139 days (Moojen 1948, 1952).

REMARKS: Günther (1876) described *Trinomys dimidiatus* on the basis of a juvenile specimen of unknown provenance. Miranda-Ribeiro (1905), commenting on a specimen from Mont-Serrat, Itatiaia, identified by O. Thomas as *P. dimidiatus*, suggested Itatiaia as the *patria* of the species. Thomas (1921h:141) assigned two specimens from Itatiaia (BM 14.2.23.19 and BM 14.2.23.19) to *P. dimidiatus*, prompting Moojen (1948:373) to suggest "southwestern Rio de Janeiro" as the probable type locality for the species. Recent samples of spiny rats from Itatiaia, however, were allocated to *T. gratiosus* (Geise, Pereira et al. 2004). Iack-Ximenes (2005) examined museum specimens from Itatiaia, including those analyzed by Thomas, and assigned them to either *T. gratiosus* or *T. panema*. Furthermore, there are no known localities documenting the occurrence of *T. dimidiatus* from the Itatiaia region (Attias et al. 2009). Therefore, Moojen's (1948) suggestion that southwestern Rio de Janeiro is the probable type locality for this species is untenable. Here, we fix the type locality of *T. dimidiatus* to "Parque Nacional da Tijuca, Rio de Janeiro, Estado do Rio de Janeiro, Brasil." This locality, a national park within the city limits of Rio de Janeiro, is in the central portion of the known distribution of this species and from which *T. dimidiatus* is both known and the only species of *Trinomys* present. As documented by Attias et al. (2009), other regions of the state of Rio de Janeiro where *T. dimidiatus* is known may also house other species.

Analysis of 393 bp of the mtDNA cytochrome-*b* (Lara and Patton 2000) revealed relatively high levels of DNA sequence divergence (4.0–4.8%) between sequences from populations that are 150–200 km apart (Parque Nacional do Desengano and Serra dos Órgãos), but high levels of similarity between populations separated by 180 km (Serra dos Órgãos and Mambucaba, 0.81%). Despite this variability, all samples of *T. dimidiatus* formed a monophyletic assemblage, consistent with its current species status. *Trinomys dimidiatus* is a member mtDNA Clade 1 along with *T. iheringi* as its sister species, with these two species differing by 15.6–17.1% (Lara and Patton 2000). Moojen (1948:321) had previously called attention to the morphological similarity between *dimidiatus* and *iheringi*. Lara et al. (2002:240) observed extreme morphological bacular variation in population samples of *T. dimidiatus* when compared with other species such as *T. albispinus*. This fact prompted these authors to suggest that this variation may not reflect solely ontogenetic differences.

Cytogenetic analyses of one female and one male collected in Rio Bonito (about 100 km east of Rio de Janeiro city) and two females and one male collected at Ubatuba (São Paulo state, near the western limit of Rio de Janeiro state and 300 km west of Rio de Janeiro city) revealed a diploid number of $2n = 60$ and $FN = 116$ (Pessôa et al. 2005). This karyotype comprises 29 pairs of metacentric and submetacentric autosomes, of which pair 10 has a secondary constriction on

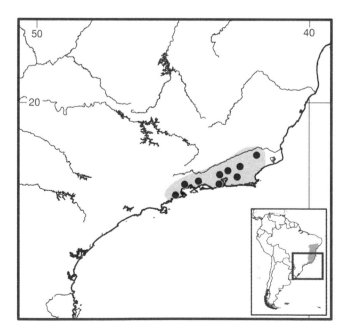

Map 533 Selected localities for *Trinomys dimidiatus* (●).

the long arm, probably coincident with a nucleolar organizer regions. The X chromosome is a large submetacentric (corresponding to the second pair of the complement), and the Y is a metacentric intermediary to pairs 20 and 21.

Trinomys eliasi (Pessôa and Reis, 1993)
Elias's Spiny Rat
SYNONYMS:

Proechimys iheringi eliasi Pessôa and Reis, 1993a:182; type locality "Restinga da Barra de Maricá, Maricá, Rio de Janeiro."

[*Proechimys* (*T*[*rinomys*].) *iheringi*] *eliasi*: Pessôa, Reis, and Pessôa, 1996:130; name combination.

Trinomys eliasi: Lara, Patton, and da Silva, 1996:405; first use of current name combination.

Trinomys iheringi eliasi: Lara and Patton, 2000:666; name combination.

P[*roechimys*]. *eliasi*: Lara and Patton, 2000:681; name combination.

DESCRIPTION: Large member of genus, with mean head and body length 197 mm (range: 186–215 mm; s.d. 7.8; N = 15), tail length 206 mm (181–231 mm; s.d. 13.1; N = 11), hindfoot length 52 mm (47–55 mm; s.d. 2.2; N = 15), ear length 28 mm (25–31 mm; s.d. 2.4; N = 15), and mass 221 g (124–262 g; s.d. 43.8; N = 13; Iack-Ximenes 2005). General color of upper parts varies from olivaceous-brown to cinnamon-brown, becoming paler on sides, and contrasting with white ventral coloration. Tail long, averaging 110% of head and body length, markedly bicolored (dark above, white below) in samples from some localities (e.g., Maricá and Morro do Itaoca), but in central part of geographic distribution bicolor pattern not very marked (e.g., in sample from Silva Jardim), and terminating in conspicuous hairy pencil. Aristiforms stiffest and longest for genus, similar to those of *T. paratus* and *T. mirapitanga*. Mid-dorsal aristiforms very long and wide, averaging 22.6 mm by 1.5 mm, colored white to pale buff basally, gradually becoming sepia in apical region and black at tip; on rump aristiforms white basally transitioning to ochraceous-tawny toward tip but sometimes interrupted by subapical zone pale brownish-olive band. Setiforms in mid-dorsal region whitish or pale-buff basally, becoming sepia toward tip but interrupted by cinnamon-brown subapical zone; on rump, setiforms white basally becoming sepia toward tip, but interrupted by cinnamon or pale pinkish-cinnamon subapical zone (Pessôa and Reis 1993a; Tavares 2007).

Skull elongated and stout, with marked ridges in most specimens, an average length of 51.5 mm (s.d. 2.0, N = 16), basilar length of 35.2 mm (s.d. 1.7; N = 16), and zygomatic width of 25.8 mm (s.d. 0.8). Incisive foramina lyre shaped, of medium size (mean length 4.8 mm; occupying about 45%

of diastema length), and broader at premaxillomaxillary suture level but becoming narrower and constricted toward posterior end; septum almost always complete in samples from type locality, in which it tends to become narrower toward posterior end, but often incomplete at other localities; premaxillary portion always more developed than maxillary portion, comprising about 70% of septum length; the vomer often visible in ventral view as tiny structure in septum. Two pairs of palatine foramina often present, first at level of PM4, and second at level of M1. Postorbital processes of zygoma well developed and often slightly pointed, often formed from both jugal and squamosal (Pessôa and Reis 1993a; Tavares and Pessôa 2010). Auditory bullae large, smooth, and well inflated (mean bullar length 10.7; s.d. 0.2). PM4–M2 typically with single major fold, often two counterfolds in samples from type locality, rarely one or three counterfolds in other samples.

Bacular shaft of medium length and thin, without marked lateral indentation, with slight dorsoventral curvature proximally; proximal end paddle shaped, large and wide, corresponding to about 20% of total length of bone, with marked edges in specimens from Barra de Maricá, while those from Silva Jardim general form more oval and smooth; distal end with slight median depression (Pessôa and Reis 1993a; Tavares and Pessôa 2010).

DISTRIBUTION: *Trinomys eliasi* occurs in the coastal sandy plains and contiguous forested lowland habitats of eastern and northeastern Rio de Janeiro state, Brazil.

SELECTED LOCALITIES (Map 534; from Tavares and Pessôa 2010, except as noted): BRAZIL: Rio de Janeiro, Cabiúnas, Macaé (MNJR [PRG 1649]), Mata do Carvão, São Francisco do Itabapoana, Morro do Itaoca, Campos dos Goytacazes, Quissamã, Reserva Biológica Poço das Antas, Silva Jardim, Reserva Biológica União, Rio das Ostras, Restinga da Barra e Maricá, Maricá (type locality of *Proechimys iheringi eliasi* Pessôa and Reis).

SUBSPECIES: *Trinomys eliasi* is monotypic.

NATURAL HISTORY: *Trinomys eliasi* preferentially inhabits forested formations, both semideciduous and evergreen moist forests (Tavares and Pessôa 2010). Even in coastal sandy lowland ecosystems (*restingas*), where open vegetation predominates, this spiny rat occurs exclusively in forest patches (F. A. S. Fernandez 1989; Cerqueira et al. 1990; Gentile and Cerqueira, 1995; Bergallo et al. 2004).

Trinomys eliasi was the third most frequently captured terrestrial small mammal in a long-term study at the type locality, with an average population density of 1.6 individuals per hectare (F. A. S. Fernandez 1989). However, recent field surveys indicated that this species is relatively rare in other areas (Bergallo et al. 2004; our unpubl. field data). Population size remained relatively constant throughout the year at the type locality, with limited dispersal, and an average residence time by individuals of 9.7 months per year (Cerqueira et al. 1994; Gentile and Cerqueira 1995). Longevity varied from 6.1 to 20.6 months, with an average of 11.5 months (s.d. = 7.3; F. A. S. Fernandez 1989), and sex ratio did not differ significantly from 1:1. Reproductive activity was found throughout the year, except in April, with the largest proportion of sexually active individuals in January (F. A. S. Fernandez 1989). At Campos dos Goytacazes a female and a juvenile male were captured together in the same trap, indicating that pups may follow their mother during foraging (our unpubl. data).

Trinomys eliasi is cursorial and has ricocheteal-like movements when liberated from live-traps (Cerqueira et al. 1990). Temporal capture pattern indicates a crepuscular peak of activity (Gentile and Cerqueira 1995).

Until recently, only specimens from the type locality were known (Tavares and Pessôa 2010). *Trinomys eliasi* has been classified as Endangered in a list of endangered species of Rio de Janeiro state because of endemism and human population pressure in the area immediately around its small range (Bergallo et al. 2000; see also D. Brito 2004). D. Brito and Figueiredo (2003) estimated that the Maricá population was not viable in the medium to longer time frame based on a population viability analysis. The IUCN (2011) red list classifies *T. eliasi* as Endangered (category B1ab.iii).

REMARKS: *Trinomys eliasi* was described as a subspecies of *Proechimys (Trinomys) iheringi* but later recognized as a species based on phylogenetic analyses of mtDNA cytochrome-*b* sequence data (Lara and Patton 2000), which indicated the polyphyly of *Proechimys iheringi* (*sensu* Moojen, 1948). *Trinomys eliasi* was placed in clade 2 (*sensu* Lara and Patton 2000) as sister to *T. paratus* with a relatively low (for the genus) molecular distance of about 7%.

Tavares and Pessôa (2010) identified remarkable morphological qualitative and quantitative variation along the geographic range of *T. eliasi*, including both bacular and craniodental traits usually recognized as good species-specific markers. Specimens collected at the type locality were statistically distinct from those from more eastern localities within the range, and pointed out that such a degree of intraspecific variation had not been recorded in other congeners. Tavares and Pessôa (2010) hypothesized that environmental heterogeneity along the present geographic distribution or possible regional environmental rearrangements in the Quaternary might be associated with the origin of the remarkable morphological variation recorded in the species.

In contrast to morphological variation, the karyotype is apparently conservative in this species, with $2n = 38$, $FN = 112$, for specimens from the type locality (Pessôa et al. 2005) and also from Silva Jardim (Tavares 2007) as well as four other localities in the eastern part of the species range (our unpubl. data).

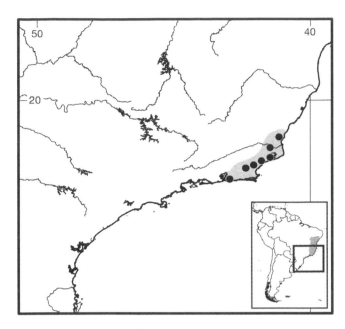

Map 534 Selected localities for *Trinomys eliasi* (●).

Following Galewski et al. (2005), the divergence between *T. eliasi* and *T. paratus* occurred after the Middle Pliocene, a relatively recent split considering that major speciation events in the genus *Trinomys* were apparently centered in the Miocene (see also Upham and Patterson 2012).

Trinomys gratiosus (Moojen, 1948)
Gracile Atlantic Spiny Rat
SYNONYMS: See under Subspecies.

DESCRIPTION: Medium to large member of genus, with head and body length 200 mm (163–232 mm; s.d. = 14.5; N = 93), tail length 178 mm (150–220 mm; s.d. = 12.7; N = 82), hindfoot length 47 mm (41–57 mm; s.d. = 2.8 N = 82), ear length 25 mm (21–36 mm; s.d. = 2.1; N = 92), and mass 209 g (162–270 g; s.d. = 26.6; N = 68; Iack-Ximenes 2005). General color of upper parts ochraceous-buff contrasting with white venter (Moojen 1948). Tail bicolored, brownish above and white below, averaging 90% of head and body length, lacking conspicuous hairy pencil at tip. Aristiform hairs darker along mid-dorsal region, forming back that contrasts with paler setiform hairs on sides of body. Aristiform hairs long, varying from 21–27 mm in length, and relatively narrow, maximally 0.6 mm wide, but become shorter and more narrowed laterally (18–21 mm and 0.5 mm, respectively); gray at base, blackening toward tip. Setiforms on mid-dorsal region gray basally, blackening toward tip, but interrupted by ochraceous-buff subdistal zone; on outer thighs setiforms gray basally, gradually becoming black toward middle and entirely ochraceous-buff distally, occasionally with blackish tip. Hairs on both forefeet and hindfeet dorsally white (Moojen 1948).

Skull slender, with no conspicuous ridges, with mean greatest length 50.6 mm (s.d. 1.7; N = 92), basilar length 35.6 mm (s.d. 1.4; N= 79), and zygomatic width 25.9 mm (s.d. 0.97; N = 81) (Moojen 1948; Iack-Ximenes 2005). Incisive foramina lyre shaped, relatively long (mean length 4.8 mm; averaging about 46% of diastema length), and constricted posteriorly; septum always incomplete, with maxillary part lacking or, if present, slender; premaxillary portion extends well over half foramina length; vomer does not contribute to ventral aspect of septum. Posterior palatine foramina positioned at plane of first molars or slightly anterior to them. Postorbital process of zygoma rounded and involves only squamosal. Auditory bullae small, smooth, elongated, and inflated. Upper molariform teeth with two primary folds (Moojen 1948).

Baculum with accentuated dorsoventral curvature over proximal third and lateral indentation near distal end; it broadens near proximal end and tapers at tip; distal end with weak medial depression and without distal wings (Pessôa and Reis 1992a; Pessôa et al. 1996).

DISTRIBUTION: This species is known from several localities along the Serra do Mar and Serra da Mantiqueira mountains of Espírito Santo, Minas Gerais, and Rio de Janeiro states, Brazil (Moojen 1948; Iack-Ximenes 2005).

SELECTED LOCALITIES (Map 535): *T. gratiosus bonafidei*—BRAZIL: Rio de Janeiro, Fazenda Boa Fé, Teresópolis (type locality of *Proechimys iheringi bonafidei* Moojen), Fazenda São José da Serra Bonsucesso, Sumidouro (Iack-Ximenes 2005) Fazenda Santo Antônio da Aliança, Valença (Attias et al. 2009). *T. gratiosus gratiosus*—BRAZIL: Espírito Santo, Campinho, Colatina (type locality of *Proechimys iheringi panema* Moojen), Castelinho, Cachoeiro do Itapemirim, Parque Nacional do Caparaó (Lara and Patton 2000), Hidroelétrica São Pedro, Domingos Martins (Iack-Ximenes 2005), Floresta da Caixa d'Água, Santa Teresa (type locality of *Proechimys iheringi gratiosus* Moojen); Minas Gerais, Estação Biológica Mata do Sossego, Simonésia (UFMG [YL 42]), Fazenda do Bené, 4 km SE of Passa Vinte (Lara and Patton 2000), Parque Estadual da Serra do Brigadeiro (J. C. Moreira et al. 2009); Rio de Janeiro: Parque Nacional do Itatiaia, Itatiaia (Geise, Pereira et al. 2004).

SUBSPECIES: We recognize two subspecies of *T. gratiosus*.

T. g. gratiosus (Moojen, 1948)
SYNONYMS:

Proechimys iheringi gratiosus Moojen, 1948:379; type locality "Floresta da Caixa D'água, Santa Teresa, Espírito Santo, Brazil; altitude 750 meters."

Proechimys iheringi panema Moojen, 1948:380; type locality "Campinho, Colatina, Espírito Santo, Brazil; altitude 500 meters."

[*Proechimys (Trinomys) iheringi*] *gratiosus*: Pessôa, Reis, and Pessôa, 1996:131; name combination.

[*Proechimys (Trinomys) iheringi*] *panema*: Pessôa, Reis, and Pessôa, 1996:131; name combination;

Trinomys iheringi gratiosus: Lara and Patton, 2000:665; name combination.

T[rinomys]. i[heringi]. panema: Lara and Patton, 2000:681; name combination.

Trinomys gratiosus: Lara and Patton, 2000:682; name combination.

Trinomys g[ratiosus]. gratiosus: Lara and Patton, 2000:683; first use of current name combination.

Trinomys panema: J. A. Oliveira and Bonvicino, 2006:397; name combination.

The nominotypical subspecies is known from coastal forests in Espírito Santo and northeastern Rio de Janeiro, inland to southern Minas Gerais and the Parque Nacional do Itataia in northwestern Rio de Janeiro, Brazil.

T . g. bonafidei (Moojen, 1948)

SYNONYMS:

Proechimys iheringi bonafidei Moojen, 1948:378; type locality "Fazenda Boa Fé, Teresópolis, Rio de Janeiro, Brazil; alt. 850 meters."

[*Proechimys (Trinomys) iheringi*] *bonafidei*: Pessôa, Reis, and Pessôa, 1996:130; name combination.

Trinomys iheringi bonafidei: Lara and Patton, 2000:665; name combination.

Trinomys gratiosus bonafidei: Lara and Patton, 2000:682; first use of current name combination.

Trinomys bonafidei: G. A. S. Carvalho and Salles, 2004:452; name combination.

This subspecies occurs in forested habitats in the coastal ranges of Rio de Janeiro state, Brazil.

NATURAL HISTORY: The holotype of *T. g. gratiosus* and four paratypes were collected in climax forest in Floresta da Caixa D'água, altitude 750 m. Females of *T. g. bonafidei* were captured with one or two embryos in April and September, with juveniles most common in April. Paratypes were collected in secondary forest with elevated humidly (Moojen 1952). A specimen assigned to *T. gratiosus panema* was recorded at Fazenda do Bené, Minas Gerais, in a forest patch running along the Rio Preto at an elevation of 680 m, in a habitat described as old second-growth rainforest with a dense overstory, where most trees are 20 cm in diameter with crowns 20 m in height. The terrain here was moderately steep (30°–60°), with rocky outcrops and relatively few fallen logs on the ground, and sparse understory usually reaching 1.5 m height, few lianas and vines, some arboreal bromeliads, abundant arboreal lichens, and many palm and fruit trees (Y. L. R. Leite 2003).

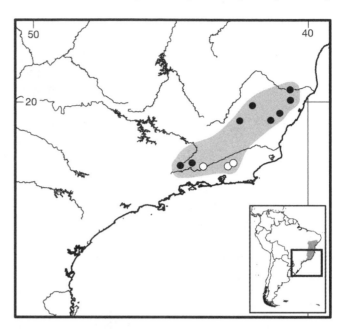

Map 535 Selected localities for *Trinomys g. gratiosus* (●) and *T. g. bonafidei* (○).

REMARKS: Moojen (1948:380) distinguished *P. i. gratiosus* from *P. i. panema* by the presence of only one counterfold in lower molariform teeth in a small percentage of specimens and by an ochraceous-buff rather than cinnamon subdistal zone of aristiforms. Lara and Patton (2000:181) did not consider *P. i. panema* to be morphologically distinct from *Trinomys gratiosus* (= *P. i. gratiosus* in Moojen's sense) and included *P. i. bonafidei* Moojen as a subspecies of *T. gratiosus*, based on the results of mtDNA analysis. Here, we follow Lara and Patton (2000) in considering *bonafidei* as subspecies of *T. gratiosus gratiosus*, and *panema* as a junior synonym of *T. gratiosus*.

Cytogenetic analyses specimens of *T. gratiosus gratiosus* collected in Monte Verde, Espírito Santo, revealed a diploid number of $2n = 56$ and FN = 108. This karyotype consists of 26 pairs of metacentric and submetacentric autosomes (Zankin 1988). The karyotype of *T. g. bonafidei* from type locality showed the same diploid and fundamental number; the X chromosome was a large submetacentric and Y chromosome a small metacentric (Pessôa et al. 2005). See Remarks for *T. paratus* for further comments.

Trinomys iheringi (Thomas, 1911)
São Paulo Spiny Rat
SYNONYMS:

Proechimys iheringi Thomas, 1911c:252: type locality "Island of São Sebastião, off São Paulo, Brazil."

P[roechimys]. [Proechimys] iheringi: Thomas, 1921h:141; name combination.

Proechimys [(*Proechimys*)] *iheringi*: Ellerman, 1940:122; name combination.

Proechimys [(*Trinomys*)] *iheringi iheringi*: Moojen, 1948: 378; name combination.

Trinomys iheringi: Lara, Patton, and da Silva, 1996:405; first use of current name combination.

Trinomys iheringi iheringi: Lara and Patton, 2000:666; name combination.

DESCRIPTION: Medium-sized member of genus the tail shorter than head and body; average head and body length 191 mm (s.d. 16.1; N = 82), tail length 171 mm (s.d. 12.4; N = 68), hindfoot length 44 mm (s.d. 2.5; N = 81), ear length 25 mm (s.d. 2.26; N = 79), and mass 209 g (s.d. 33.5; Iack-Ximenes 2005). General color of upper parts cinnamon-buff, gradually becoming lighter on sides, and contrasting with white venter (Moojen 1948). Tail bicolored, brownish above and white below, averaging 87% of head and body length, and lacking conspicuous hairy pencil at tip (Pessôa and Reis 1996). Aristiforms hairs narrow and soft, 18–23 mm in length and with maximum width 0.6 mm (Pessôa and Reis 1996). Setiforms on mid-dorsal region whitish basally, gradually blackening toward tip but interrupted by cinnamon-buff subdistal zone; setiforms on outer thighs whitish on basal half but gradually turning to grayish at midshaft and finally pale cinnamon-buff along distal third. Hairs on both forefeet and hindfeet white dorsally (Moojen 1948).

Skull elongated and slender, with conspicuous ridges, average length 51.0 mm (s.d. 2.2; N = 119), basilar length 35.6 (s.d. 1.7, N = 125), and zygomatic width 25.6 mm (s.d. 0.8; N = 125) (Moojen 1948; Iack-Ximenes 2005). Incisive foramina oval, short (length 3.7 mm, averaging about 48% of diastema), and constricted posteriorly; septum always complete and columnar, with maxillary part short and vomer not visible ventrally. Posterior palatine foramina positioned at plane of premolars. Postorbital process of zygoma rounded and formed by both jugal and squamosal. Auditory bullae small, smooth, and uninflated (Moojen 1948; Pessôa and Reis 1996). Upper molariform teeth with three primary folds, with rudimentary fourth fold on posterior of M3; lower teeth also with three folds, first smaller than others (Moojen 1948; Pessôa and Reis 1996).

Baculum shorter than those of most other species (length 6.7 mm); dorsoventral curvature and lateral indentations lacking (Pessôa and Reis 1992a, 1996; Pessôa et al. 1996); distal end evenly rounded, lacking distal wings or median depression; proximal end with overall rounded shape.

DISTRIBUTION: This species is known from several localities along the littoral and plateau of São Paulo state and from Ilha Grande on the southern coast of Rio de Janeiro state, Brazil (Moojen 1948, Pessôa and Reis 1996; Iack-Ximenes 2005).

SELECTED LOCALITIES (Map 536): BRAZIL: Rio de Janeiro: Ilha Grande, Abraão (MNRJ 62285); São Paulo, Alto da Serra (Iack-Ximenes 2005), Barra do Icapara (MZUSP 23857), Caraguatatuba (MZUSP 26646), Cotia (MZUSP 10212), Fazenda Intervales, Capão Bonito (MZUSP 23397), Ilha de São Sebastião (type locality of *Proechimys iheringi* Thomas), Ilha do Cardoso (MZUSP 27756), Iporanga (Yonenaga-Yassuda et al. 1985), Juréia (Iack-Ximenes 2005), Ubatuba (Lara and Patton 2000), Varjão do Guaratuba, Bertioga (MZUSP 25904).

SUBSPECIES: *Trinomys iheringi* is monotypic.

NATURAL HISTORY: Bergallo (1994) examined reproductive seasonality and population ecology at Ilha do Cardoso, São Paulo state over a two-year period. Both reproduction and survival rate in females was directly related to food availability and rainfall. Individuals reproduced throughout the year, but with peaks during the rainy season (Bergallo and Margnusson 1999). Home range size was larger than those of other rodent species of comparable body size inhabiting rainforests (Manaf, Morato, and Oliveira 2003). Bergallo (1995) also reported on home range size, with males using greater area than females and female home ranges exclusive, both elements suggesting a polygynous mating system (see also Adler 2011).

REMARKS: *Trinomys iheringi* was reviewed by Pessôa and Reis (1991a, 1992b, 1993a, 1996) and S. F. Reis et al. (1992), who recognized the six subspecies listed by Moojen (1948)—*bonafidei*, *denigratus*, *gratiosus*, *iheringi*, *panema*, *paratus*—and described a seventh, *eliasi*. These authors concluded that *bonafidei*, *denigratus*, *eliasi*, *gratiosus*, and *panema* might be specifically distinct from *iheringi*. Lara and Patton (2000) based on molecular analysis concluded that *T. iheringi* was a monotypic taxon, to the exclusion of the remaining forms listed as subspecies by Moojen (1948) and Pessôa and Reis (1996). *Trinomys iheringi* is a member of the mtDNA clade formed by *T. dimidiatus*, *T. mirapitanga*, *T. g. gratiosus*, and *T. g. bonafidei* (Lara and Patton 2000; Lara et al. 2002).

The karyotype of *T. iheringi* has a variable diploid number ($2n = 60$–65) due to the presence of B chromosomes; excluding these elements, the FN = 116 (Yonenaga-Yassuda et al. 1985; Fagundes et al. 2004). The basic autosomal complement consists of 25 biarmed and four uniarmed pairs; the X chromosome is a large submetacentric, and the Y a minute metacentric. Because all populations of *T. iheringi* sampled do date show variable numbers (from 1–5) of supernumerary chromosomes (Yonenaga-Yassuda et al. 1985), both the morphology of the Y chromosome and the presence of supernumerary chromosomes are useful markers to differentially diagnose *T. iheringi* from *T. dimidiatus*. Typically these chromosomes are heteroge-

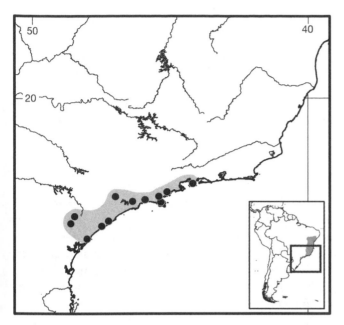

Map 536 Selected localities for *Trinomys iheringi* (●).

neous relative to size, morphology, banding patterns, presence/absence of NORs, and presence/absence of interstitial telomeric signals after fluorescence in situ hybridization (Pessôa et al. 2005).

Trinomys mirapitanga Lara, Patton, and Hingst-Zaher, 2002

Pau Brasil Spiny Rat

SYNONYM:

Trinomys mirapitanga Lara, Patton, and Hingst, 2002:237; type locality "Estação Ecológica do Pau Brasil, 16 km W Porto Seguro, Bahia, Brazil, 40 m alt."

DESCRIPTION: Medium sized in comparison to congeners (Iack-Ximenes 2005), with average head and body length 216 mm (s.d. 15.3; N=9), tail length 201 mm (s.d. 13.3; N=7), hindfoot length 50 mm (s.d. 2.8; N=8), ear length 28 mm (s.d. 1.9; N=9), and mass 259 g (s.d. 33.8). Overall coloration much darker along mid-dorsal region than in other species, forming dark cape on middle of back that contrasts with paler setiform hairs and much paler sides of body. Aristiforms in midback smoky-gray at their base with ochraceous-buff midsection and dark brown tip. Lateral setiform hairs typically with four color bands, drabgray basally followed by smoky-gray, ochraceous-buff, and a black tip. Tail less than head and body in length (ca. 80%) and covered with short hairs. Aristiforms very long (mean 24.9 mm) and wide (1.5 mm) at mid-dorsum, but shorter and more narrowed laterally (mean 19.5 mm and 1.3 mm, respectively). Hairs on both forefeet and hindfeet white dorsally.

Skull large, with average length 52.4 mm (s.d. 1.5; N=4), basilar length 36.8 (s.d. 1.2, N=4), zygomatic width 25.8 mm (s.d. 1.0; N=4) (Lara et al. 2002; Iack-Ximenes 2005). Incisive foramina oval, short (averaging about 32% of diastema length), and constrict posteriorly; septum complete and columnar; vomer does not contribute to ventral aspect of septum. Posterior palatine foramina located at plane of first molars. Postorbital process of zygoma rounded and comprising both squamosal and jugal; it never forms sharp spine. Auditory bullae smooth and of moderate size (11.3 mm, on average). Upper molariform teeth with three primary folds, with rudimentary fourth fold on posterior end of M3; lower teeth also have three folds, first smaller than rest.

Baculum shorter than most other species (length 7.9 mm in only known male), with weakly developed distal wings and broad basal expansion that is strongly flexed ventrally. Median trough evident along dorsal surface of basal portion.

DISTRIBUTION: This species is known from only two localities along the southern coast of Bahia state, Brazil (Lara et al. 2002). Iack-Ximenes (2005) recorded specimens he identified as *T. mirapitanga* from Estação Ecológica Nova Esperança-Wenceslau Guimarães (UFMG 2155) and Fazenda São João (UFMG 2157) on the central coast of Bahia; however, these specimens are *T. setosus*.

SELECTED LOCALITIES (Map 537): BRAZIL: Bahia, Cumuruxatiba Prado (MNRJ 48012), Estação Ecológica do Pau Brasil (type locality of *Trinomys mirapitanga* Lara, Patton, and Hingst).

SUBSPECIES: *Trinomys mirapitanga* is monotypic.

NATURAL HISTORY: The type specimen and two paratypes were collected at the Estação Ecológica do Pau Brasil, a biological reserve established to protect one of the largest remaining stands of *Caesalpinia echinata* (Fabaceae). No data are otherwise available on ecology, behavior, or any other aspect of the population biology of *T. mirapitanga*.

REMARKS: *Trinomys mirapitanga* is a member of mtDNA Clade 1 (*sensu* Lara and Patton 2000; see also Lara et al. 2002), sister to the *T. dimidiatus* and *T. iheringi* species pair. It differs from *T. eliasi*, *T. paratus*, *T. setosus*, and *T. yonenagae* in having a tail less than the head and body length and no brush of elongated hairs at its tip, a round rather than speculate postorbital process of the zygoma, oval incisive foramina, and a baculum with both distal and basal expansions. In comparison to *T. dimidiatus*, *T. iheringi*, and *T. gratiosus*, *T. mirapitanga* differs by having a well-developed maxillary portion of the septum of the incisive foramina, a postorbital process of the zygoma formed by both squamosal and jugal elements, a more well-developed basal expansion of the baculum relative to the distal, strong ventral flexion to the bacular base, and a median trough on its dorsal surface.

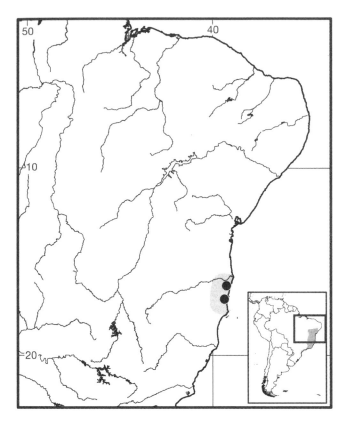

Map 537 Selected localities for *Trinomys mirapitanga* (●).

Trinomys moojeni (Pessôa, Oliveira, and Reis, 1992)
Moojen's Spiny Rat

SYNONYMS:

Proechimys moojeni Pessôa, Oliveira, and Reis, 1992:40; type locality "Mata do Dr. Daniel, Conceição do Mato Dentro, Minas Gerais, Brazil."

T[rinomys]. moojeni: Lara, Patton, and Hingst, 2002: 235; first use of current name combination.

DESCRIPTION: Medium-sized species in comparison to congeners, average head and body length 187 mm (s.d. 18.2; N = 7), tail length 179 mm (s.d. 15.7; N = 6), hindfoot length 44 mm (s.d. 2.6; N = 7), and ear length 24 mm (s.d. 1.5; N = 7) (Pessôa et al. 1992; Iack-Ximenes 2005). Dorsal and lateral color orange-brown, with orangish areas over neck and shoulders due to predominance of orange-banded setiform hairs. Dorsal pelage not particularly stiff, with relatively long (mean 20.4 mm) and narrow (mean 0.6 mm) blackish tipped aristiform hairs that become less common, smaller (mean 15.9 mm), and narrower (mean 0.5 mm) on sides and lateral thighs, where some aristiform hairs have distal half ochraceous rather than gradually darkening toward tip. Ventral pelage completely white, well defined from sides and dorsum. Tail bicolored on its anterior two-thirds, brownish dorsally and whitish below, with distal third whitish. Tail covered by short hairs along most of length, colored as skin to which they are attached. Hairs of tail tip much longer but do not form notable pencil. Hairs on both forefeet and hindfeet dorsally white, with whitish ungual tufts barely covering claws feet (Pessôa et al. 1992).

Skull slender, of moderate size, and slightly convex with average length 50.2 mm (s.d. 2.1; N = 7), basilar length 35.5 mm (s.d. 1.9, N = 7), and zygomatic width 25.5 mm (s.d. 0.7; N = 7) (Pessôa et al. 1992; Iack-Ximenes 2005). Incisive foramina oval in shape, long (averaging about 40% of diastema length), and wide posteriorly; septum incomplete and formed almost exclusively by premaxillae; maxillary part short. Posterior palatine foramina present at plane of first molars. Postorbital process of zygoma well developed and formed by both jugal and squamosal bones. Auditory bullae small and smooth. Two counterfolds on both upper and lower molariform teeth; upper incisors orthodont (Pessôa et al. 1992).

Baculum short (6.6 mm in length the only specimen examined), with slight dorsoventral curvature and tapered lateral indentation near midshaft; proximal end straight and paddle shaped, and distal end has weakly developed distal wings (Pessôa et al. 1992, 1996).

DISTRIBUTION: *Trinomys moojeni* is apparently restricted to the southern Espinhaço massif in east-central Minas Gerais state, Brazil. The original series composed of seven individuals was obtained in August 1954 by Cory T. Carvalho and João Moojen at two sites about 1,200 m elevation along the road that connects the cities of Lagoa Santa and Conceição do Mato Dentro, at the section in which it crosses the upper parts of Serra do Cipó, Espinhaço Massif (Avila-Pires 1960b). More recently, *T. moojeni* was reported from Parque Nacional da Serra do Cipó (A. M. O. Paschoal et al. 2004) and Serra do Caraça Private Reserve (Cordeiro and Talamoni 2006).

SELECTED LOCALITIES (Map 538): BRAZIL: Minas Gerais, Mata do Dr. Daniel, Conceição do Mato Dentro (type locality of *Proechimys moojeni* Pessôa, Oliveira, and Reis), Serra do Caraça Private Reserve, Tanque Grande (MNRJ 73489).

SUBSPECIES: *Trinomys moojeni* is monotypic.

NATURAL HISTORY: Data recorded on specimen labels of the type series suggest that *T. moojeni* is restricted to the upper forest formations. A total of 29 specimens of *T. moojeni* were captured, marked, and recaptured 49 times in the Serra do Caraça between January 2001 and January 2002 (Cordeiro and Talamoni 2006). The sex ratio of adults was skewed in favor of females (1:2.3), with males predominantly captured in the dry season and females in equal numbers in both dry and rainy seasons. Mean dispersal distance of males (174.4 ± 40.0 m, N = 5) was greater than that of females (88.5 ± 24.1 m, N = 6).

Sexual dimorphism in body measurements was not significant, and reproductive animals were observed throughout the year. Likewise, young animals as well as pregnant and lactating females were captured during both dry and rainy seasons, indicating that there is no seasonal pattern to the reproduction in this species; recruitment probably occurs shortly after weaning.

Trinomys moojeni is classified as Endangered (B1ab.iii) by the IUCN (2011).

REMARKS: The phylogenetic relationship of *T. moojeni* and other species in the genus has been uncertain. Lara and Patton (2000; see also Lara et al. 2002) hypothesized a link between this species and the molecular clade including *T. setosus*, *T. paratus*, *T. eliasi*, and *T. yonenagae* based on morphological similarity. However, Lara and Patton (2000) also called attention to the fact that *T. moojeni* was the only taxon in the genus, apart from *T. gratiosus gratiosus* and *T. g. panema*, with an incomplete vomerine septum of the incisive foramina. Our close inspection of other usually relevant taxonomic characters also suggested a closer morphological affinity between *T. gratiosus* and *T. moojeni*, including a tail of similar length as the head and body and without a pencil, a postorbital process of the zygoma well developed and formed by jugal and squamosal bones, upper and lower molariform teeth with two counterfolds, and a short baculum with weakly developed distal wings. Karyotypes (2n=56, FN=106; autosomal complement with 26 pairs of biarmed and one pair of uniarmed elements, with the X a large submetacentric and the Y a medium-sized metacentric; Corrêa et al. 2005) from Morro do Pilar, Minas Gerais, near the type locality of *T. moojeni*,

are also similar to those of *T. gratiosus* from Monte Verde, Espírito Santo (2n=56; FN=108). Finally, our unpublished mtDNA cytochrome-*b* sequence analyses support a sister relationship between *T. moojeni* and *T. gratiosus*, with the divergence level between the two species is about 10%.

Trinomys paratus (Moojen, 1948)
Rigid-spined Atlantic Spiny Rat

SYNONYMS:

Proechimys [(*Trinomys*)] *iheringi paratus* Moojen, 1948:382: type locality "Floresta da Capela de São Braz, Santa Teresa, Espírito Santo, Brasil; altitude 630 meters."

Trinomys paratus: Lara, Patton, and da Silva, 1996:405; first use of current name combination.

Trinomys iheringi paratus: Lara and Patton, 2000:670; name combination.

DESCRIPTION: Large for genus, with mean head and body length 211 mm (187–246 mm; s.d. 14.1; N=23), tail length 212 mm (171–235 mm; s.d. 15.2; N=21), hindfoot length 498 mm (25–54 mm; s.d. 5.8; N=28), ear length 28 mm (18–33 mm; s.d. 3.4; N=27), and mass 259 g (195–350 g; s.d. 40.2; N=21). General color of upper parts varies from cinnamon to blackish-brown, becoming paler on sides, and contrasting with white venter. Tail approximately same length as head and body, and markedly bicolored (dark above, white below), with conspicuous hairy pencil at tip. Aristiform spines most rigid and large in genus; very long and wide at mid-dorsum, with average length 24.2 mm and mean width 1.6 mm. Aristiform color at mid-dorsum whitish or pale buff basally, becoming gradually sepia toward apical region and black at tip; on rump aristiforms white basally becoming ochraceous toward to tip but sometimes interrupted by subapical zone light brownish-olive color. Setiforms on mid-dorsal region white or pale buff basally, becoming sepia toward tip but interrupted by tawny or orange-cinnamon subapical zone; setiforms on rump white basally, becoming sepia toward tip but also interrupted by cinnamon or pale pinkish-cinnamon subapical zone (Moojen 1948; Tavares 2007).

Skull large, elongated, and stout, with well-marked ridges in most specimens; average length of 52.4 mm (s.d. 1.4; N=10), basilar length 37.0 mm (s.d. 1.3 mm; N=10), and zygomatic width 25.9 mm (s.d. 0.7; N=10). Incisive foramina fusiform or oval in shape, and medium sized (mean length 4.4 mm, occupying about 40% of diastema length); septum complete and columnar, its premaxillary portion long, constituting about 75% of septum length; vomer not visible in ventral view and maxillary portion small and wide. Posterior palatine foramina present at plane of premolars. Postorbital processes of zygoma well developed and often slightly pointed, often composed of both jugal

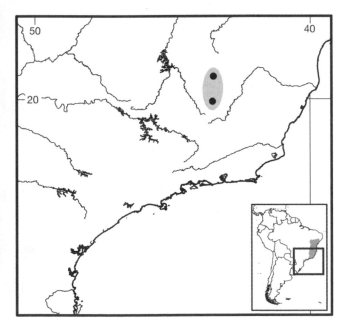

Map 538 Selected localities for *Trinomys moojeni* (●).

and squamosal. Auditory bullae large, smooth, and slightly inflated (mean bullar length 10.3; s.d. 0.5; N = 10; Tavares 2007). PM4–M1 in specimens from type locality with one major fold and often two counterfolds.

Baculum elongated and moderate to thin (average length 9.8 mm), with slight dorsoventral curvature and without expansions or apical wings; proximal end slightly broader than shaft, elongated and almost rectangular; distal end concave (Lara and Patton 2000; Lara et al. 2002; Iack-Ximenes 2005; see Remarks).

DISTRIBUTION: *Trinomys paratus* is known from few localities in the southern half of Espírito Santo and east-central Minas Gerais states (Iack-Ximenes 2005).

SELECTED LOCALITIES (Map 539): BRAZIL: Espírito Santo, Aracruz Florestal, Aracruz (Lara and Patton 2000), Barra do Jucu, Vila Velha (MNRJ 33776), Cariacica (Tavares and Pessôa 2010), Floresta da Capela de São Bráz, Santa Teresa (type locality of *Proechimys* [(*Trinomys*)] *iheringi paratus* Moojen), Lagoa de Sete Pontas, Itapemirim (Tavares and Pessôa, 2010); Minas Gerais, Estação Biológica de Caratinga, Caratinga (Tavares and Pessôa 2010).

SUBSPECIES: *Trinomys paratus* is monotypic.

NATURAL HISTORY: Information about natural history of *Trinomys paratus* is very limited. The geographic range extends along Atlantic Forest formations in relatively low elevations (below 400 m). According to label data, specimens of the type series were captured in *mata virgem* (undisturbed original forest). The type locality is in a submontane region of moist dense forest, but the species also inhabits lowland dense moist forests (as at Viana, Espírito Santo), coastal sand vegetation *restinga* close to the littoral (Lagoa de Sete Pontas, Itapemirim), and inland semideciduous forests (as in Caratinga). Compared to other terrestrial small mammals, *Trinomys paratus* was very abundant in a parasitological survey carried out in Reserva Biológica Duas Bocas (I. S. Pinto, Botelho et al. 2009; I. S. Pinto, Loss et al. 2009; I. S. Pinto, Santos et al. 2009; T. M. Pereira et al. 2010), with a relatively high prevalence of siphonaptera infestation (I. S. Pinto, Botelho et al. 2009). At Viana, Espírito Santo, individuals of *T. paratus* collected in a secondary forest were infected with *Leishmania forattini*, with the parasite isolated from skin lesions (Falqueto et al. 1985, 1998).

REMARKS: Lara and Patton (2000) obtained partial mtDNA cytochrome-*b* sequences from specimens collected at Caratinga, Minas Gerais state, and Aracruz, Espírito Santo state, which showed to be very similar (0.5% of divergence), despite the geographic distance of about 250 km between these localities. In their analyses, *T. paratus* formed a clade with *T. eliasi*, the two lineages having probably diverged in a relatively recent cladogenic event (see Remarks under *T. eliasi*).

Based on cranial qualitative and quantitative data, Tavares and Pessôa (2010) delineated a notable cohesion along all known samples of *T. paratus*, confirming Iack-Ximenes (2005) assertion that samples from Minas Gerais and Espírito Santo are indistinguishable based in cranial and external morphology. Contrasting to this uniformity, different bacular morphologies have been described for *T. paratus* from Santa Teresa, Espírito Santo (Pessôa et al. 1996) and those from Caratinga, Minas Gerais (Lara and Patton 2000; Lara et al. 2002). Whereas the proximal end of the baculum from Caratinga is rectangular, in the specimen from Santa Teresa it is pointed. We reanalyzed the baculum described by Pessôa et al. (1996) and noted that it is damaged at the proximal end. If this broken piece is replaced to its original position, the general morphology will have the same general appearance of the baculum described from Caratinga. More specimens from Espírito Santo state should be analyzed to allow a more reliable assessment of bacular variation in the species.

The karyotype *T. paratus* is unknown. Paresque et al. (2004) published a karyotype attributed to *Trinomys iheringi* from Espírito Santo, with specimens collected from Estação Biológica de Santa Lúcia, Santa Teresa (where *T. gratiosus gratiosus* is known to occur), and in Reserva Biológica Duas Bocas, Cariacica (a locality of *T. paratus*). Unfortunately, the authors did not specify the provenance of the published karyotype, which had the same 2*n* and FN described and attributed to *T. gratiosus* by Zanchin (1988).

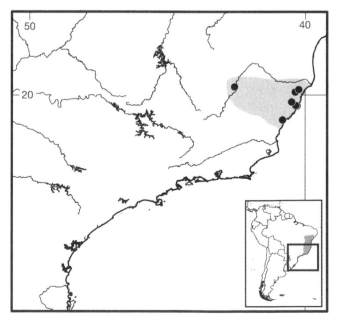

Map 539　Selected localities for *Trinomys paratus* (●).

Trinomys setosus (Desmarest, 1817)
Elegant-spined Atlantic Spiny rat

SYNONYMS: See under Subspecies.

DESCRIPTION: Medium to large spiny rat, with mean head and body length 204 mm (range: 180–226 mm, s.d. 13.9; N=28), tail length 209 mm (168–230 mm, s.d. 16.1; N=27), hindfoot length 49 mm (43–53 mm, s.d. 2.7; N=31), ear length 27 (20–33 mm, s.d. 2.0; N=31), and mass 203.6 g (s.d. 33.5; N=22) (Iack-Ximenes 2005). General color of upper parts varies from brownish-olive to cinnamon, gradually changing to olive or pale cinnamon on sides and outer thighs, contrasting with white venter; marked cinnamon color at nape present in most individuals. Tail bicolored, brownish above and white below, with conspicuous whitish or cinnamon hairy pencil at tip (Moojen 1952), and slightly longer than head and body (about 103%). In northernmost part of geographic range, specimens tend to have very spiny appearance, whereas those from southernmost part have softer pelage. Aristiforms of mid-dorsal region long (ranging from 17–22 mm), wide in T. s. setosus (mean 1.1 mm; N=10) but narrow in T. s. elegans (mean 0.7 mm; N=10). Setiforms on mid-dorsal region gray basally, gradually blackening toward tip but interrupted by cinnamon subdistal zone. Setiforms on outer thighs gray on basal half then gradually become cinnamon at tip (Moojen 1948).

Skull large, without conspicuous ridges, and with length averaging 51.3 mm (s.d. 2.4; N=42), basilar length 36.7 mm (s.d. 1.8, N=47), and zygomatic width 25.2 mm (s.d. 1.0; N=53) (Iack-Ximenes 2005). Incisive foramina lyre shaped, long, and narrow (4.7 × 2.2 mm, averaging about 44% of diastema length), usually without conspicuous constriction at posterior end; foramina slightly shorter in specimens from Sergipe state; septum complete, with premaxillary portion elongated and vomer visible, extending along most of length of septum with its posterior part touching maxillary part, which is reduced to a small process. Posterior palatine foramina obsolete but located at level of M1. Postorbital process of zygoma well developed and markedly spiniform, slender, and involves only jugal. Auditory bullae large and inflated. Upper premolars with two counterfolds; upper molars have one to three counterfolds.

Bacular morphology varies both among and within subspecies (Pessôa and Reis 1992a; Pessôa et al. 1996; Lara et al. 2002; Iack-Ximenes 2005). Two bacular types present in specimens of T. s. setosus. Specimens from Itabuna, Bahia, with dorsoventrally curved shaft on proximal third and lateral indentation near middle, broadened near proximal end, and round at tip; distal end concave and without median depressions or distal wings. In contrast, specimens from Conceição do Mato Dentro, Minas Gerais, shaft narrow and lacks dorsoventral curvature and lateral indenta-tions; proximal end round and distal end thin and concave. In T. s. elegans, bacular shaft straight without dorsoventral curvature or lateral indentations; distal end lacks both median depression and distal wings; and proximal end straight, without lateral expansions.

DISTRIBUTION: Trinomys setosus has the broadest geographic range of any species in the genus, occurring throughout the states of Sergipe, Bahia, Minas Gerais, Espírito Santo, and Rio de Janeiro, Brazil.

SELECTED LOCALITIES (Map 540): T. setosus setosus—BRAZIL: Bahia, Fazenda Aldeia, 7 km NNW of Valença (Lara and Patton 2000), Itambé (MNRJ 8323), Mata do Nono, São Felipe (Iack-Ximenes 2005), Mata do Ribeirão da Fortuna, 40 km W of Ilhéus, Itabuna (Proechimys [(Trinomys)] iheringi denigratus Moojen); Espírito Santo, Linhares (Pessôa et al. 1996), São Mateus (Iack-Ximenes 2005); Minas Gerais, Almenara (MNRJ 73409), Mata da Praúna, 5 km N (by road) of Conceição do Mato Dentro (Lara and Patton 2000); Sergipe, Fazenda Cedro, near Estância, outside Crasto (MNRJ 29370), Fazenda Cruzeiro, 13km SSE of Cristinápolis (Lara and Patton 2000). T. setosus elegans—BRAZIL: Minas Gerais, Fazenda Cauaia, Matozinhos (Iack-Ximenes 2005), Fazenda Esmeralda, 30 km E and 4 km N (by road) of Rio Casca (Lara and Patton 2000), Juiz de Fora (MNRJ 34127), Santa Bárbara (Iack-Ximenes 2005); Rio de Janeiro, Cambuci (Attias et al. 2009).

SUBSPECIES: We recognize two subspecies, following Moojen (1948) and Lara and Patton (2000).

T. s. setosus (Desmarest, 1817)
SYNONYMS:

Echimys setosus Desmarest, 1817b:59; type locality "Amérique"; Moojen (1948) suggested that the type came from "(southern?) Bahia," Brazil.

L[oncheres]. myosurus Lichtenstein, 1820:192, Fig. 2; type locality "Provinz Bahia," Brazil.

M[us]. leptosoma Brants, 1827:150; type locality "Bahia," Brazil.

[Mus] cinnamomeus Lichtenstein, 1830:plate 36 Fig. 2; no type locality designated.

Echimys cayennensis: Pictet, 1841:146; not E. cayennensis Desmarest [= Proechimys guyannensis (É. Geoffroy St.-Hilaire)].

Echinomys leptosoma: Tschudi, 1845:174; name combination.

L[oncheres]. cayennensis: Schinz, 1845:109; name combination; not cayennensis Desmarest [= Proechimys guyannensis (É. Geoffroy St.-Hilaire)].

[Echimys] cinnamomeus: E.-L. Trouessart, 1880b:180; name combination.

[Echimys] [cayennensis] leptosomus: E.-L. Trouessart, 1880b:180; name combination.

[*Echimys*] [*cayennensis*] *myosurus*: E.-L. Trouessart, 1880b: 180; name combination.

[*Proechimys*] *setosus*: J. A. Allen, 1899c:264; name combination.

[*Proechimys*] *cinnamomeus*: E.-L. Trouessart, 1904:609; name combination;

[*Echimys*] *leptosoma*: Thomas, 1921h:41; name combination.

Proechimys [(*Proechimys*)] *leptosoma*: Ellerman, 1940:119; name combination.

Proechimys [(*Proechimys*)] *myosurus*: Ellerman, 1940:119; name combination.

Proechimys [(*Trinomys*)] *setosus*: Ellerman, 1940:122; name combination.

Proechimys [(*Trinomys*)] *setosus setosus*: Moojen, 1948: 369, 385; name combination.

Proechimys [(*Trinomys*)] *iheringi denigratus* Moojen, 1948: 375; type locality "Mata do Ribeirão da Fortuna, 40 kilometers west of Ilhéus, Itabuna, Bahia, Brazil."

Trinomys setosus: Lara, Patton, and da Silva, 1996:405; name combination.

Trinomys myosurus: Eisenberg and Redford, 1999:499; name combination.

Trinomys setosus setosus: Lara and Patton, 2000:665; first use of current name combination.

Trinomys iheringi denigratus: Lara and Patton, 2000:665; name combination.

T[*rinomys*]. *s*[*etosus*]. *denigratus*: Lara, Patton, and Hingst-Zaher, 2002:235; name combination.

The nominotypical subspecies is distributed in the northeastern part of the species range, from the coastal forests of Bahia and Espírito Santo, extending to the interior of Minas Gerais, Brazil.

T. s. elegans (Lund, 1839)

SYNONYMS:

Loncheres elegans Lund, 1839a:233; *nomen nudum*.

E[*chimys*]. *elegans* Lund, 1839c:59 [1841a:99]; type locality "Rio das Velhas Floddal" (Lund 1841a:134), Lagoa Santa, Minas Gerais, Brazil.

E[*chinomys*]. *fuliginosus* Wagner, 1843a:343; type locality "Brasilien."

Lonch[*eres*]. *fuliginosa*: Schinz, 1845:115; name combination.

Echimys fuliginosus: Waterhouse 1848:341; name combination.

Echinomys cajennensis: Winge, 1887:71: name combination; misapplication and misspelling of *cayennensis* Desmarest to Lund's specimens from Lagoa Santa.

[*Proechimys*] *fuliginosus*: E.-L. Trouessart, 1904:609; name combination.

Proechimys setosus: Thomas, 1921h:141; part.

[*Proechimys*] *fuliginosa*: Hershkovitz, 1948a:128; name combination but incorrect gender concordance.

[*Proechimys*] *elegans*: Tate, 1935:400; name combination.

Proechimys [(*Proechimys*)] *elegans*: Ellerman, 1940:119; name combination.

Proechimys [(*Trinomys*)] *setosus elegans*: Moojen, 1948: 387; name combination.

Proechimys (*Trinomys*) *setosus elegans*: Paula Couto, 1950: 166; name combination.

Trinomys setosus elegans: Lara and Patton, 2000:665; first use of current name combination.

This subspecies occurs in the southwestern part of the species range, in interior Minas Gerais and northern Rio de Janeiro, Brazil.

NATURAL HISTORY: A single female captured at Itabuna, Bahia, on 9 January 1944 gave birth to two young on 26 January 1945 (Moojen 1948). Each measured 177 mm in total length and weighed 27.8 g shortly after birth. Moojen (1948) gave external and craniodental measurements at the age of two months, by which time head and body length had increased to 120 mm and skull length to 35–36 mm, for example. Young specimens were captured from January to May. Despite the high levels of rainfall in the area of the Itabuna, Moojen (1948) mentioned that the type series of *T. denigratus* came from a forest with deciduous trees, and that all animals were trapped near water. On the basis of observations reported by Lund (1841b), Moojen (1952) reported that *elegans* is found near water bodies, being a good swimmer and building nests on the grass at the margin of lakes.

G. A. B. Fonseca and Kierulff (1988) found that *T. setosus* was the most abundant rodent in a 17-month study in semideciduous forest in Minas Gerais state. Individuals were captured in all months of the study, but with a peak in the late dry season. Pregnant and lactating females were present year-round, although breeding individuals were more frequent in mid-dry and midrainy seasons. They classified *T. setosus* as entirely terrestrial, lacking any scansorial ability. Analysis of stomach contents indicated that this spiny rat is primarily frugivorous, but also opportunistically feeds on insects and seeds.

Individuals from Três Braços, Bahia state, served as a natural reservoir of *Leishmania forattini* (Barretto et al. 1985; Yoshida et al. 1993; reported as *Proechimys iheringi denigratus*).

REMARKS: The precise type locality of *Echimys setosus* Desmarest is unknown. Thomas (1921h) reviewed the type specimens of *setosus* and *elegans* and treated them as synonyms. Moojen (1948) identified specimens collected by Wied-Neuwied as *setosus*, and only examined one specimen of *elegans*, recognizing two subspecies, *P. setosus setosus* and *P. s. elegans*. Moojen (1948) also described *P. iheringi denigratus* based on a series from Itabuna, southern Bahia.

In their mtDNA cytochrome-*b* phylogenetic study, Lara and Patton (2000) analyzed specimens from Cristinápolis (Sergipe) and Valença (Bahia), regarding them as *T. s. setosus*, a sample from Rio Casca (Minas Gerais), which they identified as *T. s. elegans*, and specimens from Conceição do Mato Dentro (Minas Gerais), which they referred to as *T. i. denigratus*. These samples were grouped in a monophyletic lineage belonging to their "Clade 2," leading these authors to reallocate the three subspecies to *T. setosus*. Iack-Ximenes (2005) examined holotypes of these three taxa and, based on a lack of definable morphological differences between *setosus* and *denigratus*, proposed that the latter be regarded as a junior synonym of *T. setosus*. He also suggested that *elegans* be elevated to species category, recognizing both *T. setosus* and *T. elegans* in his "*setosus* group." The topology of Clade 2 (Lara and Patton, 2000), however, does not corroborate the monophyly of *T. setosus sensu* Iack-Ximenes (2005) because a haplotype of an individual identified by him as *T. elegans* emerges as a lineage inside *T. setosus* in the Lara and Patton (2000) analysis. Despite the absence of monophyly, a relatively high divergence in mtDNA sequences (7.88%) was found between samples from Sergipe and Minas Gerais (Lara and Patton 2000), indicating some degree of differentiation between the *setosus* from Sergipe and *elegans* from Minas Gerais.

We analyzed recently collected samples from central and northern Minas Gerais state, and observed that combinations of characters usually considered diagnostic between the two forms in the extremes of the distribution (e.g., constricted incisive foramina characteristic of *elegans* and the basally expanded baculum of *setosus*) could be found in a same individual. Likewise, aristiform width and coloration, which are quite distinct at the extremes of the distribution, showed intermediate conditions in this part of the geographic range. Therefore, the molecular and morphological data currently available do not support a clear distinction between *setosus* and *elegans* and suggest a contact zone between these two forms. We thus recognize *denigratus* and *setosus* as synonyms, following Iack-Ximenes (2005), but regard *Trinomys setosus* as a single taxon with two subspecies: *T. s. setosus* (occurring in the northernmost part of range) and *T. s. elegans* (in the southernmost part of range).

The karyotype of *T. s. setosus* from Almenara, Minas Gerais state, is $2n=56$ and FN=108 (unpublished data). *Trinomys setosus elegans* from Santa Bárbara, Minas Gerais has $2n=56$ and FN=104. This karyotype consists

Map 540 Selected localities for *Trinomys s. setosus* (●) and *Trinomys s. elegans* (○).

of 25 pairs of metacentric, submetacentric, and subtelocentric autosomes and two pairs of acrocentric autosomes (pairs 26 and 27). The X chromosome is a large acrocentric corresponding in size to the third autosomal pair; the Y chromosome is acrocentric and one of the smallest chromosomes in the set (Corrêa et al. 2005).

Trinomys yonenagae (P. L. B. Rocha, 1996)
Torch-tail Spiny Rat, Yonenaga's Spiny Rat
SYNONYMS:

Proechimys yonenagae P. L. B. Rocha 1996:540; type locality "Ibiraba, Bahia, Brazil (10°48′S, 42°50′W)."

Trinomys yonenagae: Lara and Patton, 2000:665; first use of current name combination.

Trinomys yonenga: Lara and Patton, 2000:668; *lapsus calami* in the spelling of *yonenagae* (Rocha).

DESCRIPTION: Small to medium-sized member of genus, with mean head and body length 164 mm (141–195 mm; s.d. 14.2; N=23), tail length 191 mm (169–225 mm; s.d. 10.1; N=23); hindfoot length 44 mm (41–51 mm; s.d. 2.3; N=23), and ear length 24 (20–27 mm; s.d. 1.7; N=23) (Iack-Ximenes 2005). General color of upper parts paler than any other congeneric species, generally pale cinnamon mixed with darker hairs on mid-dorsum, gradually becoming lighter on sides, and only slightly contrasting with white ventral coloration. Pelage relatively soft for genus; aristiforms on mid-dorsum of medium length and soft, with average total length 16–19 mm and maximum width about 0.6 mm. Tail markedly bicolored (brownish above and white below) and long, averaging about 120% of head and body length, with very long hairy pencil at tip (hairs extend more than 30 mm beyond tail tip). Hindfeet especially long for genus (P. L. B. Rocha 1996).

Skull medium in length and relatively narrow, being more robust in parietal and occipital region, and with marked ridges, but becoming progressively more slender and delicate toward rostrum. Average skull length 44.7 mm (s.d. 4.4; N = 31), basilar length 30.8 mm (s.d. 3.5; N = 31), and zygomatic width 21.9 mm (s.d. 1.8; N = 31). Incisive foramina often oval and elongated (mean length × length, 4.3 × 2.4 mm), occupying about 45% of diastema; septum complete, with long premaxillary portion, occupying about 75% of septum, short maxillary portion, and vomer not visible in ventral view. Postorbital process of zygoma well developed and typically formed only by jugal; unique among other species in having laminar inflection of jugal. Auditory bullae large, smooth, and distinctively very inflated (mean bullar length 10.3 mm; s.d. 0.5). PM4–M1 with one major fold and one counterfold.

Baculum elongated (mean length 8.1 mm; s.d. 0.7; N = 7) and slender, with slight dorsoventral curvature; distal end median depression or apical wings; shaft with slight lateral indentations near distal end and broadening toward proximal end, but interrupted by terminal constriction at about its fifth part; proximal end pointed (P. L. B. Rocha 1996).

DISTRIBUTION: *Trinomys yonenagae* is known from sandy dunes on the left bank of middle Rio São Francisco, in northwestern Bahia state, Brazil.

SELECTED LOCALITIES (Map 541): BRAZIL: Bahia, Ibiraba (type locality of *Proechimys yonenagae* Rocha), Queimadas (P. L. B. Rocha 1996).

SUBSPECIES: *Trinomys yonenagae* is monotypic.

NATURAL HISTORY: This species is restricted to a small area in a semiarid sandy dune habitat characterized by a sparse vegetative covering and subject to the macroclimatic conditions of Caatinga biome (P. L. B. Rocha 1996). Local herbaceous vegetation is composed mainly of spiny bromeliads and cacti, but with the genus *Eugenia* (Myrtaceae) comprising 35% of shrub and tree species (J. W. Santos and Lacey 2011). Some ecological, behavioral, and morphological traits of this species are very distinctive relative to its forest-dwelling congeners and apparently represent generalized convergent traits of desert rodents broadly recognized as adaptations to semiarid conditions. Among these traits are the medium to small sized body, very inflated tympanic bullae, long hindfeet and tail, strongly penciled tail, asymmetrical pattern of limb coordination, colonial behavior, and burrow-dwelling habits (P. L. B. Rocha 1996; P. L. B. Rocha et al. 2007; J. N. S. Freitas et al. 2008, 2010; J. W. Santos and Lacey 2011).

Trinomys yonenagae is the only small mammal found on the sand dunes near Ibiraba, and should be the prey of multiple predators (P. L. B. Rocha 1996; J. W. Santos and Lacey 2011). Most food and water consumed by this spiny rat is obtained from seeds of *Eugenia* (Myrtaceae). Individuals live in self-dug burrow systems, the majority of which are located in valleys among sand dunes, where vegetation is sparse, resulting in high soil temperature (60°C) during daylight hours (J. W. Santos and Lacey 2011). Mean density of occupied burrow systems is 9.9 ±1.6 per hectare, with each system having 1 to 13 entrances (mean 5.4 ± 4.2) per burrow system. Field data suggested that adults of both sexes share burrow systems and, in some cases, nest sites (J. W. Santos and Lacey 2011). *Trinomys yonenagae* meets the spatial criteria used to diagnose sociality in fossorial rodents (J. W. Santos and Lacey 2011). In accordance with sociality indices found in the field, captive experiments showed that the intensity of affiliative behavior in *T. yonenagae* is high, both in intersexual and intrasexual encounters (J. N. S. Freitas et al. 2003, 2008, 2010). This species was thus markedly more affiliative than other *Trinomys* or *Thrichomys* species, lacking the types of anxiety signals often associated with social interactions (J. N. S. Freitas et al. 2008). Affiliation was mediated by acoustic communication by for males and females, and does not differ between animals from the same social group or from different social groups. However, females tended to spend more time near familiar males than unfamiliar males, a pattern interpreted as an evidence of social monogamy (Manaf and Oliveira 2009). Alternatively, J. W. Santos and Lacey (2011) provided data on extensive overlap in home ranges of males and females and suggested a promiscuous mating system. *Trinomys yonenagae* may use secretions from an anal scent gland as communication signals in social interaction (Manaf, Brito-Gitirana, and Oliveira 2003).

In captive experiments, *T. yonenagae* employed stop-and-go search (in accordance with field data; P. L. B. Rocha 1991), a pattern of locomotion characterized by a notable component of vigilance (Manaf, Brito-Gitirana, and Oliveira 2003). Individuals use saltatorial locomotion and leave burrows at night (P. L. B. Rocha 1991); captive studies indicated a regular pattern of circadian activity restricted to nocturnal hours (Marcomini and Oliveira 2003).

Trinomys yonenagae is classified as Endangered (B1ab. iii) by the IUCN (2011).

REMARKS: P. L. B. Rocha (1996) hypothesized that *T. yonenagae* shared a closer relationship with *T. albispinus*, forming a natural clade in the genus, based in tooth morphology and geography. Alternatively, Pessôa et al. (1998) suggested a close affinity with a polytypic *Trinomys iheringi* (which then included *eliasi* and *paratus* as valid

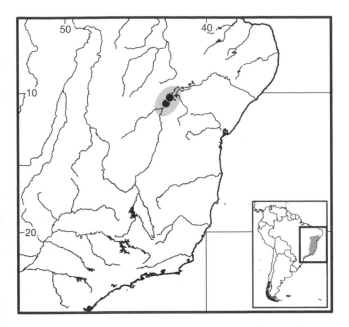

Map 541 Selected localities for *Trinomys yonenagae* (●).

species and *bonafidei, denigratus, gratiosus,* and *panema* as subspecies of *iheringi*) based on bacular structure and quantitative cranial variation. Finally, in their mtDNA phylogenetic analysis, Lara and Patton (2000) placed *T. yonenagae* within a clade comprising *T. paratus, T. setosus,* and *T. eliasi,* and to the exclusion of both *T. albispinus* and *T. iheringi.* Although this analysis supported *T. yonenagae* as a very distinct taxon, its relationship to the remaining species in its clade remains unclear.

The karyotype of *T. yonenagae* is $2n = 54$ and $FN = 104$, including an autosomal complement of 26 pairs of metacentrics or submetacentric chromosomes; the X chromosome is a large-sized acrocentric; the Y is a small metacentric (Leal-Mesquita et al. 1992).

Subfamily Myocastorinae Ameghino 1902
James L. Patton

These are large-bodied, robust, and heavy rodents adapted for an aquatic lifestyle. Head and body length ranges to 600 mm, tail length to 450 mm, hindfoot length to 120 mm, and mass averages more than 1 kg. The ears are small (maximally 35 mm in height), nearly hidden in the thick fur; the mouth is closable behind the incisors; the tail is long, round, thick, scaly, and sparsely haired; the limbs are short; the hindfeet have four toes, each with a long claw, and webbed between; the pollex is reduced in size and the fifth toe of the forefoot is unwebbed; and the pelage is soft and thick with dense underfur and long guard hairs. Sebaceous glands are located near the corner of the mouth

and near the anus. There are four or five pairs of mammae, each lateral in position and located high on the side.

As with other echimyids, the fourth upper and lower premolars are not replaced and are thus generally considered deciduous. All cheek teeth, above and below, are flat-crowned and extremely hypsodont but not hypselodont. The upper toothrows converge anteriorly, but not to the degree as in *Dactylomys,* and are inclined outward. Upper cheek teeth have two labial flexi and three lingual flexi; lower cheek teeth have one labial and three lingual flexids, although pm4 has an additional anterior lake. The reentrant folds of both upper and lower cheek teeth become lakes with wear. The incisors are broad and deep, with the enamel strongly pigmented orange. The skull is long and broad, flat in lateral profile, and with a broad, deep, and elongated rostrum. A sagittal crest is well developed. The zygomatic arch is broad and flaring, the jugal is thick but does not approach the lacrimal, and the jugal fossa is reduced. The infraorbital foramen is large without a distinct groove on the ventral surface for nerve passage. The angular process of the mandible is strongly deflected and the coronoid process is vestigial. Stems of the internal carotid artery and stapedial artery are both missing (description from Woods 1984). The subfamily contains the single, monotypic extant genus *Myocastor,* but the fossil record is rich, with nine genera recognized in deposits from the Early Miocene to Late Pliocene (McKenna and Bell 1997).

As noted earlier, myocastorines have been frequently treated as a subfamily of the Capromyidae (Hall 1981; Simpson 1945). However, Woods and Howland (1979) documented numerous important morphological differences between these two groups that supported their phyletic distinctness. Landry (1957) placed the myocastorines as a separate family in his superfamily Octodontoidea. B. Patterson and Pascual (1968a), B. Patterson and Wood (1982), and McKenna and Bell (1997) placed *Myocastor* as a subfamily of the Echimyidae. Woods (1984:445) considered it "slightly more closely related to echimyids than to capromyids or octodontids" and treated it as a separate family in the same superfamily as the echimyids and octodontids. He concluded, however, that the myocastorids had their origin from echimyids of the subfamily Adelphomyinae in the Oligocene, which would make the Echimyidae paraphyletic to his Myocastoridae. Both molecular sequence analyses (Y. L. R. Leite and Patton 2002; Galewski et al. 2005; Upham and Patterson 2012) as well as a cladistic analysis of morphological traits (G. A. S. Carvalho and Salles 2004) solidly place *Myocastor* as nested within the Echimyidae. These studies form the basis for the inclusion of myocastorines as a subfamily of the Echimyidae herein.

Genus *Myocastor* Kerr, 1792

SYNONYMS:

Mus: G. I. Molina, 1782:287; part (listing of *huidobrius*, a *nomen dubium*, and description of *coypus*); not *Mus* Linnaeus.

Myocastor Kerr, 1792:225; type species *Mus coypus* G. I. Molina by subsequent designation (Palmer 1904:438).

Hydromys É. Geoffroy St.-Hilaire, 1804:253; partly based on *Mus coypus* G. I. Molina; not *Hydromys* É. Geoffroy St.-Hilaire in current usage (Muridae; see Tate 1936:642; Ellerman 1941:298; Musser and Carleton 2005:1333).

Myopotamus É. Geoffroy St.-Hilaire, 1805:82; based on manuscript name of the French naturalist Philibert Commerson (see Palmer 1904:440).

Hydromis É. Geoffroy St.-Hilaire, 1805:86; part (description of *coypou*); incorrect subsequent spelling of *Hydromys* É. Geoffroy St.-Hilaire.

potamys F. Cuvier, 1823b:184; name given under *Myopotamus* but stated as different from Commerson's animal ("qui est plus régulièrement formé que celui de Commerson"); see also Desmarest (1826:491).

Potamys Larranhaga, 1823:83; based on the "*quouiya*" of Azara (1801:5), which equals *Mus coypus* G. I. Molina.

Castor J. B. Fischer, 1829:286; part (listing of *huidobrius* and *coypus*); not *Castor* Linnaeus.

Mastonotus Wesmael, 1841:61; type species *Mastonotus popelairi* Wesmael (= *Mus coypus* G. I. Molina).

Guillinomys Lesson, 1842:126; type species *Guillinomys Chilensis* Lesson, which was based on *Castor Huidobrius* G. I. Molina.

Tramyocastor Rusconi, 1936:1; based on two fossil species, *T. andiai* Rusconi and *T. majus* Rusconi.

REMARKS: J. A. Allen (1895a), Thomas (1920h), Tate (1935), and Osgood (1941, 1943b) have discussed the taxonomic history of *Myocastor* Kerr and other names associated with it. G. I. Molina (1776), under the vernacular names *guillin* (or *guillino*) and *coipu*, was the first to write about the Coypu. He subsequently (G. I. Molina 1782) described *Castor Huidobrius* (p. 285) for the *guillino* and *Mus coypus* (p. 287) for the *coipu*, the former "with the dimensions and habits of an otter, and the dentition of a rodent" (Osgood 1941:408–409). Thus, both Thomas (1920h:225) and Osgood (1941:408–409, 1943b:132) argued that Molina's *huidobrius* and subsequent names derived from it (such as *Guillinomys chilensis* Lesson 1842) should be regarded as composite and unidentifiable. This leaves Molina's second species, *Mus coypus*, which Molina thoroughly described, as the type for *Myocastor*.

Étienne Geoffroy St.-Hilaire's name *Myopotamus* was universally used for the coypu throughout the nineteenth century (e.g., Waterhouse 1848; E.-L. Trouessart 1880) until J. A. Allen (1895a:182–183) pointed out that *Myocastor* Kerr was a valid, earlier name. Kerr's *Myocastor* was based on both *Mus coypus* G. I. Molina and *Castor zibethicus* Linnaeus [= *Ondatra zibethicus*, the muskrat], and Allen selected the coypu as the type species by elimination. This action was solidified in Opinion 55 of the ICZN (1913), which fixed *zibethicus* Linnaeus as the type of *Ondatra* Link, leaving *Mus coypus* G. I. Molina as the sole member of Kerr's *Myocastor*. Other nineteenth century names (e.g., *Potamys* Larranhaga or *Mastonotus* Wesmael) were either based directly or indirectly, through Azara's (1801) "*guouiya*," on Molina's *Mus coypus*, and *Guillinomys* Lesson was based on Molina's indeterminable *Castor Huidobrius*. Cabrera (1961:568) and Woods and Kilpatrick (2005:1593) gave Desmarest, 1825 as the author and date for *Potamys*, but both F. Cuvier (1823b) and Larranhaga (1823) had proposed the name two years earlier. Furthermore, Desmarest's paper appeared in 1826, not 1825.

B. Patterson and Pascual (1968a:7) regarded *Tramyocastor* Rusconi, proposed for two fragmentary fossil species, as indistinguishable from *Myocastor*, and Verzi, Deschamps, and Vucetich (2002) reviewed the status of *Paramyocastor* and other generic names assigned to it from the Late Miocene to Late Pliocene. Earliest records of *Myocastor* are from the Late Miocene Mesopotamian deposits in Argentina (Candela and Noriega 2003; Candela 2004).

Myocastor coypus (G. I. Molina, 1782)
Coypu, Nutria

SYNONYMS:

Castor Huidobrius G. I. Molina, 1782:287; Latinized name for the 'guillino' of G. I. Molina (1776); see also Alsop's English translation of Molina (1808:240); regarded as a *nomen dubium* by Thomas (1920g:225) and Osgood (1941:408–409).

Mus coypus G. I. Molina, 1782:287; type locality "Chili"; given as "Rio Maipo, Santiago Province, Chile" by Woods et al. (1992:1).

Myocastor coypus: Kerr, 1792; first use of current name combination.

Ondatra coypus: H. F. Link, 1795:76; name combination.

Hydromys coipou É. Geoffroy St.-Hilaire, 1804:254; name combination and incorrect subsequent spelling of *coypus* G. I. Molina.

Myopotamus Bonariensis É. Geoffroy St.-Hilaire, 1805:82; type locality given as "Rio Parana, Paraguay" by Woods et al. (1992).

Hydromis coypou É. Geoffroy St.-Hilaire, 1805:86; name combination and incorrect subsequent spelling of *coypus* G. I. Molina.

Mus castorides Burrow, 1815:168; type locality "the Brazils."

Castor ? huidobrius: Illiger, 1815:108; name combination.

Hydromys coypus: Illiger, 1815:108; name combination.

Castor Chilensis Oken, 1816:886; unavailable name (ICZN 1956).

Potamys coypou: Desmarest, 1826:491; name combination and incorrect subsequent spelling of *coypus* G. I. Molina.

Myopotamus canariensis Gray, 1827:223; probably a spelling lapsus for *bonariensis* É. Geoffroy St.-Hilaire (Cabrera 1961:568).

Myopotamus coypus: I. Geoffroy St.-Hilaire, 1827:374; name combination.

Castor coypus: J. B. Fischer, 1829:288; name combination.

Myopotamus antiquus Lund, 1840b [1841c:289, plate 21, Figs. 1–5]; type locality "Rio das Velhas's Floddal" (Lund 1841c:292, 294), Lagoa Santa, Minas Gerais, Brazil; described from fossil remains.

Mastonotus Popelairi Wesmael, 1841:61; type locality "Ces animaux, qui habitent les bords des eaux douces du Chile, ont les plus grands rapports avec les Couia [*Myopotamus* Comm.], etc," given as "Bobica, Bolivia" (S. Anderson 1997) but type specimen stated as coming from Chile, not Bolivia (Woods et al. 1992).

Guillinomys Chilensis Lesson 1842:126; renaming of *Castor huidobrius* (Osgood 1943b: 90).

Myopotamus Coypu rufus Fitzinger, 1867b:134, type locality "Am. Paraguay"; based on J. B. Fischer's (1829:288) C[astor]. Coypus β.

Myopotamus Coypu dorsalis Fitzinger, 1867b:134, type locality "Am. Paraguay"; based on J. B. Fischer's (1829:288) C[astor]. Coypus γ.

Myopotamus Coypu albomaculatus Fitzinger, 1867b:134, type locality "Am. Chili"; based on J. B. Fischer's (1829:288) C[astor]. Coypus δ.

Myopotamus coypu Thomas, 1881b:6; incorrect subsequent spelling of *coypus* G. I. Molina.

Myocastor coypus santacruzae Hollister, 1914a:57; type locality "north bank Rio Salado, near Los Palmares, Santa Cruz [corrected to Santa Fe, see Poole and Schantz 1942:490], Argentina."

Myocastor coypus coypus: Hollister, 1914a:57; name combination.

[*Myocastor coypus*] *bonariensis*: Hollister, 1914a:58; name combination.

Myocastor coypus brasiliensis Marelli, 1931:45; *nomen nudum*; contained in a list of synonyms.

Myocastor coypus melanops Osgood, 1943b:132; type locality "Quellon, Chiloe Island," Los Lagos, Chile.

Myocastor coypus popelairi: Osgood, 1943b:132; name combination.

Myocastor coypus castoroides: Paula Couto, 1950:291, footnote 323; name combination.

[*Myocastor coypus*] *sanctaecruzae*: Woods and Kilpatrick, 2005:1593; incorrect subsequent spelling of *santacruzae* Hollister.

DESCRIPTION: As for the subfamily.

DISTRIBUTION: *Myocastor coypus* is native to Chile and Argentina, northward to Bolivia, Paraguay, Uruguay, and southern Brazil, in wetland habitats within a broad range of biomes, including humid and dry Chaco, Patagonian steppe, pampas grassland, and coastal margins. The native range in Brazil is limited to the southern states of Rio Grande do Sul and Santa Catarina; feral populations are now established in São Paulo state following introduction by commercial breeders (Bonvicino, Oliveira, and D'Andrea 2008). Lund (1840a, 1841b:266, plate 21, Figs. 1–5) recorded this species in the fossil record from the caves near Lagoa Santa, Minas Gerais (as *Myopotamus antiquus*, which was equated to the extant *M. coypus* by Paula Couto 1959:291, footnote 323). The species is now introduced widely into northern South America, North America, Europe, central Asia, and east Africa (Woods et al. 1992; J. Carter and Leonard 2002; Bertolino et al. 2012).

SELECTED LOCALITIES (Map 542; native range only): ARGENTINA: Buenos Aires, Delta del Paraná (FMNH 21694), Partido Balcarce (MSU 20469); Jujuy, Aguas Negras (Heinonen and Bosso 1994); La Pampa, Laguna Guatrache (TTU 66631); Neuquén, 11 km NNE of Nahuel Huapi (MVZ 179309); Santa Fe, Los Palomares (type locality of *Myocastor coypus santacruzae* Hollister); Tucumán, Los Sarmientos (AMNH 41543). BOLIVIA: Beni, Piedras Blancas (S. Anderson 1997); Cochabamba, Todos Santos (S. Anderson 1997); Santa Cruz, Buena Vista (S. Anderson 1997). BRAZIL: Rio Grande do Sul, Itaqui (MZUSP 3190), São Lourenço do Sul (AMNH 134145); Santa Catarina, Paulo Lopes (Cherem et al. 2004); São Paulo, Pedreira (Bonvicino, Lindbergh, and Maroja 2002), Guarapiranga Reservoir, Interlagos (MZUSP 25808). CHILE: Araucanía, Temuco (MSU 9652); Biobío, Concepción (Wolffsohn 1923); Los Lagos, Chiloé Island, Quellon (type locality of *Myocastor coypus melanops* Osgood); O'Higgins, Cachapoal (Wolffsohn 1923). PARAGUAY: Alto Paraguay, Estancia Tres Marias (UMMZ 79836); Neembucú, Estancia Yacaré (UMMZ 80583); Paraguarí, Ybycui (UMMZ 124717), Presidente Hayes, Estancia Juan de Zalazar, 2 km S of headquarters (UMMZ 57389). URUGUAY: Río Negro, 15 km S of Paysandu (AMNH 206455); Soriano, 15 km SW of Dolores (Sanborn 1929); Treinta y Tres, 16 km SSW mouth of of Río Tacuari (AMNH 206459).

SUBSPECIES: Hollister (1914a) recognized three races within the native range: *coypus* G. I. Molina from Chile (with *popelairi* Wesmael a synonym), *bonariensis* É. Geoffroy St.-Hillaire from northern Argentina and Paraguay, and *santacruzae* Hollister from southern Argentina. Osgood (1943b) described *melanops*, from Chiloé Island in southern Chile, and recognized four other subspecies: the nominotypical form from central Chile, *bonariensis* from northeastern Argentina and Uruguay, *sanctacruzae* from southwestern Argentina, and *popelairi* from southeastern Bolivia. Cabrera (1961) extended the range of *bonariensis* to include the subtropical pampas of northeastern Argentina, Uruguay, Paraguay, and southwestern Brazil. No modern analysis of geographic variation in any character set, morphological or molecular, has been undertaken to test the validity of these geographic units.

NATURAL HISTORY: Willner et al. (1979), Willner (1982), Woods et al. (1992), and Bounds et al. (2003) reviewed the life history, ecology, reproductive biology, and behavior of *Myocastor coypus*. Most research on natural populations has centered on those introduced into the United States or Europe. However, there has been substantial recent fieldwork on populations within the native range in southern South America, particularly in northern and eastern Argentina, covering nearly all aspects of the population biology, in part because of the importance of nutria in the fur market trade. Bó et al. (2006a,b) summarized their "Proyecto Nutria" program to evaluate population density and other parameters, habitat suitability, and hunting pressure, among those aspects necessary for the sustainable management of this species within Argentinean wetlands. Individually, there have been recent studies covering a wide range of natural history topics, from diet (Borgnia et al. 2000; Colares et al. 2010), habitat use (D'Adamo et al. 2000; Corriale et al. 2006), population structure, social behavior, and group formation (Guichón, Borgnia et al. 2003; Guichón, Doncastor, and Cassini 2003), foraging behavior (Guichón, Benítez et al. 2003), and genetic variability and relatedness with respect to social organization (Túnez et al. 2005, 2009). In Chile, the predators of nutria include the puma, *Puma concolor* (Rau et al. 2002), and in Brazil Geoffroy's cat, *Leopardus geoffroyi* (K. S. Sousa and Bager 2008). A variety of cestode, trematode, and nematode internal parasites of copyus have been detailed by F. A. Martínez (1988), Paulsen and Brum (1999), Paulsen et al. 1999), Flores et al. (2007), Gayo et al. (2011), and P. E. Martino et al. (2012).

Coypus are aquatic, inhabiting slow-moving streams, lakes, freshwater marshes, and brackish water. Their density in the Argentinean Pampas was positively associated with the width of the alluvial plain of watercourses and agricultural land uses, negatively with urbanization and semiurban landscapes (Leggieri et al. 2011). They feed on grains and the roots, rhizomes, stems, and leaves of a wide range of aquatic plants, as well as mussels and snails, using logs or other floating debris as feeding stations. They are

excellent swimmers, can remain submerged for up to five minutes, and nest in self-dug burrows in banks. Burrows are usually short with no branches and a single nest chamber, although more complex burrows have been described in parts of their native range. The gestation period is long, up to 132 days, and litter size varies from three to five, on average. Coypus are capable of breeding throughout the year in all areas of their range, both native and introduced. Sexual maturity may be reached within six months, but more typically between 12 and 15 months of age. Their chief predators are crocodilians, fish, large snakes, turtles, raptors, and canids. Where they have been introduced, coypus have become a major agricultural pest. The fur can be a major income source in some areas.

REMARKS: Both *Castor chilensis* Oken and *Guillinomys chilensis* Lesson are renamings of *Castor huidobrius* G. I. Molina, an indeterminate name based on characters of both an otter and coypu (Thomas 1920g; Osgood 1941, 1943b). Furthermore, all names Oken proposed in his *Lehrbuch* (Vol. 3) are unavailable (ICZN 1956). Woods et al. (1992) stated that the type specimen of *popelairi* Wesmael, 1841 came from Chile, not Bobica, Bolivia. These authors considered the single distinguishing character of this taxon (the nipples located high on the back) to be an individual anomaly. As a result, they concluded that *popelairi* Wesmael was a junior synonym of the nominotypical subspecies, *M. c. coypus*. Tsigalidou et al. (1966), W. George and Weir (1974), and S. González and Brum-Zorrilla (1995) described a karyotype of $2n=42$, $FN=76$.

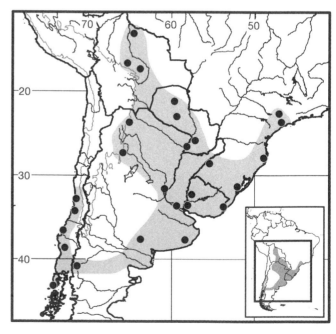

Map 542 Selected localities for *Myocastor coypus* (●). Contour line = 2,000 m.

Family Octodontidae Waterhouse, 1839

Diego H. Verzi, M. Mónica Díaz, and Rubén M. Barquez

The family Octodontidae comprises an assemblage of medium- to small-sized caviomorph rodents endemic of southern South America. Previous to Tate (1935), *Ctenomys* and the extinct allied genera were usually included among octodontids without a category to separate them from the other genera (Ameghino 1889; Rovereto 1914; G. S. Miller and Gidley 1918; Rusconi 1931a; Kraglievich 1934). Indeed, Bennett (1841) showed that his genus *Octodon* was intermediate in molar occlusal structure between F. Cuvier's *Poephagomys* (= *Spalacopus*) and *Ctenomys*. Tate (1935) and later Reig (1958; see justification in Pascual et al. 1965) segregated *Ctenomys* and related extinct genera from the other octodontids. This proposal is currently adopted in the literature but without agreement in relation to the taxonomic rank assigned to both lineages (as families Octodontidae and Ctenomyidae [Simpson 1945; Wood 1955; Cabrera 1961; Woods 1993; M. H. Gallardo and Kirsch 2001; Honeycutt et al. 2003], as subfamilies of Octodontidae [Pascual et al. 1965; Reig 1989], or as tribes of Octodontinae [McKenna and Bell 1997]). Here we follow the strictest definition of the Octodontidae (after Honeycutt et al. 2003), which excludes the Ctenomyidae. According to this criterion, Octodontidae comprises 16 extant species grouped in six living genera with surface-dwelling (*Octomys*), fossorial (*Octodontomys, Octodon, Tympanoctomys* [including *Pipanacoctomys* and *Salinoctomys*], semisubterranean (*Aconaemys*), and completely subterranean (*Spalacopus*) habits (Reig 1986, 1989; R. A. Ojeda et al. 1996; M. H. Gallardo et al. 2007; E. P. Lessa, Vassallo et al. 2008).

Although the Octodontidae probably originated in the Late Oligocene (23.4–25.5 mya; Voloch et al. 2013; Verzi et al. 2013), most analyses of molecular evidence (Köhler et al. 2000; M. H. Gallardo and Kirsch 2001; Honeycutt et al. 2003; Opazo 2005) suggest that the six extant genera diversified into two major clades in the Late Miocene, a time period consistent with the fossil record indicating a major radiation of octodontids in response to climatic change (Vucetich et al. 1999; Verzi et al. 2013). The combined mtDNA/nucDNA gene analyses by Voloch et al. (2013) provided an older, Early Miocene age for this divergence. The two clades comprise the genera (*Octomys + Tympanoctomys*), estimated to have diverged at 4.28 ± 1.08 mya, and (*Octodontomys* (*Octodon* (*Aconaemys, Spalacopus*))), with a basal divergence date of 6.07 ± 1.34 mya (Opazo 2005). Protein electrophoresis (Köhler et al. 2000), DNA/DNA hybridization (M. H. Gallardo and Kirsch 2001), and both mitochondrial and nuclear DNA sequences (Honeycutt et al. 2003; Opazo 2005) have each

recovered the same basic topology. The position of *Octodontomys* is somewhat ambiguous, however, as Upham and Patterson (2012) place this genus as sister to *Octomys + Tympanoctomys* in their Bayesian analysis while this genus is grouped with (*Octodon + (Aconaemys, Spalacomys*)) in a chronogram using fossil calibrations. Most studies also have not recovered the monophyly of *Aconaemys* species (but see Gallardo et al. 2007), with *A. porteri* forming a sister-relationship with *Spalacopus* and not to *A. fuscus + A. sagei*. The association of these taxa is, however, weak, and their correct phyletic relationships remain to be resolved (Upham and Patterson 2012). Diversification of the extant species within each genus apparently stems from the late Pliocene or mid-Pleistocene, with dates ranging from 2.57 ± 0.76 mya for three species of *Octodon* to 1.74 ± 0.59 mya for two species of *Tympanoctomys* (Opazo 2005).

The only living Octodontidae genus recognized in the fossil record is *Tympanoctomys* (Verzi, Tonni et al. 2002; but see Reig 1986). Modern (i.e., hypsodont) members of the family are known in the fossil record of Argentina since the Late Miocene, but their phylogenetic relationships to living species remain unknown (Reig and Quintana 1991; Verzi 2001; Deschamps et al. 2012).

Species of octodontids are distributed in mesic to arid open land biomes, in the Andean region or adjacent lowlands, between 16°S latitude in southwestern Bolivia, to 43°S latitude in southern Argentina and Chile (L. C. Contreras et al. 1987; Woods 1993; Hutterer 1994; Mares et al. 2000; M. H. Gallardo and Kirsch 2001; Verzi, 2001; M. H. Gallardo et al. 2007, 2009). These modern representatives originated in the Patagonian region (including the Monte desert and southern Andes), the northern portion of the Monte desert being the most probable ancestral area (Upham and Patterson 2012; A. A. Ojeda et al. 2013). The genera *Octomys* and especially *Tympanoctomys* are the South American rodents most highly adapted to desert environments (Mares 1975, 1993, 1997; Bozinovic and Contreras 1990; R. A. Ojeda et al. 1996, 1999; M. M. Díaz and Ojeda 1999; Ojeda and Tabeni 2009). *Spalacopus* (and *Aconaemys* to a lesser degree) is one of the more highly specialized subterranean rodents in the South American fauna, with independent developmentof a number of highly derived morphological attributes also found in the tuco-tucos (*Ctenomys*; see Reig 1970; E. P. Lessa, Vassallo et al. 2008; Morgan and Verzi 2011; Morgan and Alvarez 2013).

Octodontids are medium-size species, ranging in head and body length from 150 to almost 400mm and body mass from 100 to nearly 300g. The pelage is usually long, dense, and soft. The hind feet have a comb of stiff hairs extending beyond the middle digits. Most species are grayish or drab in dorsal color with white or cream venters, except *Spalacopus*, which is uniformly darkish brown to black. In general, apart from the semisubterranean and subterranean genera, most species have a long tail with a terminal tuft. The feet are short, with four well-developed digits ending in sharp and curved claws, whereas the first digit (especially the pollex) is strongly reduced and usually bears only a small nail. Asin other South American hystricognath rodents, the glans penis has an intromittent sac below the urethra that is everted during sexual activity. Spines, highly variable in number and morphology, are present in the floor of the sac (Spotorno 1979; L. C. Contreras et al. 1993).

Cranial morphology is diverse, corresponding to different types of mastication, from oblique to propalinal, and terrestrial to subterranean habits (Olivares et al. 2004; E. P. Lessa, Vassallo et al. 2008). Endocranial volume (a proxy of the brain size) of octodontids is relatively larger than those expected for a caviomorph, except for the fossorial and subterranean *Aconaemys* and *Spalacopus* (Vasallo and Echeverría 2009). The tympanic bullae are hypertrophied in desert-adapted genera *Octodontomys*, *Octomys*, and especially *Tympanoctomys* (Ojeda et al. 1999), but general characteristics of the middle ear are shared between subterranean specialists (e.g., *Spalacopus*) and nonsubterranean taxa (e.g., *Octodon*). Argyle and Mason (2008) presented evidence that low-frequency hearing, usually considered a specialization of subterranean taxa, is plesiomorphic in the Octodontoidea. The paroccipital process is short, ventrolaterally oriented, and completely joined to the bulla (Verzi 2001). The mandible is strongly hystricognathous, with markedly expanded masseteric crests, in the fossorial to subterranean *Aconaemys* and *Spalacopus*. The origin of the masseteric crest, posterior to the notch for the tendon of the medial masseter muscle, is ventrally deflected (Verzi et al. 2013).

The dental formula is I 1/1, C 0/0, PM 1/1, M 3/3 = 20. Incisors are long and deeply implanted in the skull and mandible, especially in the subterranean *Spalacopus*. Molariform teeth are euhypsodont, with a typical to modified figure-8 occlusal design; the first molariform teeth are the deciduous fourth premolars (DP4/dp4), which are retained throughout life. Adult DP4/dp4-M3/m3 lack some folds (flexi): paraflexus, metaflexus(id), and posteroflexus, or the corresponding fossettes(id)s. The alveolar sheaths of the molars protrude into the orbital region. The humerus of the fossorial to subterranean *Aconaemys* and *Spalacopus* is robust, with wide epicondyles (Lessa et al. 2008; Morgan and Alvarez 2013), the hand of these genera showing relatively short carpal bones, robustmetacarpals, and incipient mesaxony (metacarpal III > metacarpal IV > metacarpal II); the metacarpal V is remarkably broad and shortened in *Spalacopus* (Morgan and Verzi 2011).

Octodontids are mainly herbivorous (M. H. Gallardo et al. 2007). The gestation period is long, and young are precocial at birth, typically well haired and, in some species,

with eyes and ears open. Although the deserticolous *Octomys* and *Tympanoctomys* are solitary, sociality is common in the clade of mesic-adapted *Octomys-Aconaemys-Spalacopus* (Lacey and Ebensperger 2007; Ebensperger et al. 2008), and an evolutionary trend from solitary to social habits has been proposed for the family (Ebensperger et al. 2008).

Octodontids show extensive chromosomal repatterning. Members of this family exhibit one of the largest ranges in chromosomal number of any mammal, with diploid numbers from $2n = 38$ in *Octodontomys* to $2n = 102$ in *Tympanoctomys* (L. C. Contreras et al. 1990; M. H. Gallardo 1992, 1997; M. H. Gallardo et al. 2007). M. H. Gallardo et al. (1999; see also M. H. Gallardo, Garrido et al. 2004; M. H. Gallardo, Kausel et al. 2004; M. H. Gallardo et al. 2006) proposed by both *T. barrerrae* and *T. aureus* are the only known cases of mammalian poloploidy (but see Svartman et al. 2005), their putative tetraploid condition acquired via hybridization (M. H. Gallardo et al. 2007; Suárez-Villota et al. 2012).

KEY TO THE GENERA OF OCTODONTIDAE:
1. Tail short, up to 50% of head and body length, and without terminal tuft of hairs 2
1'. Tail long, more than 70% of head and body length, and with a terminal tuft of hairs 3
2. Pelage uniformly darkish brown to black; upper incisors highly procumbent, with the incisor alveolus extended up to M1–2; molariforms with lingual and labial folds not connected at the middle . . . . *Spalacopus*
2'. Pelage dark drab; upper incisors orthodont and normally implanted; molariforms with lingual and labial folds connected at the middle . . . . *Aconaemys*
3. Upper and lower molariforms with figure-8 occlusal pattern; dorsal wall of the epitympanic recess of the bulla forming a large bony "island" on the braincase roof. 4
3'. Upper and lower molariforms with asymmetrical occlusal pattern; the upper ones with the posterior lobe not labially extended; dorsal wall of the epitympanic recess as a small dorsal ossification on the braincase roof or absent 5
4. Auditory bullae large but not hypertrophied; length of the braincase less than 3.5 times the maxillary toothrow length . . . . *Octomys*
4'. Auditory bullae hypertrophied; length of the braincase nearly four times the maxillary toothrow length. . . . . *Tympanoctomys*
5. Upper and lower molariforms with folds (flexi/flexids) shallow or absent; the upper ones without a penetrating lingual fold (hypoflexus) . . . . *Octodontomys*
5'. Upper and lower molariforms with folds (flexi/flexids) well developed; the upper ones with a deep lingual fold (hypoflexus) . . . . *Octodon*

Genus *Aconaemys* Ameghino, 1891

Diego H. Verzi, M. Mónica Díaz, and Rubén M. Barquez

Aconaemys is a semisubterranean octodontid that inhabits open areas and forests, on both eastern and western slopes of the Andes (Cabrera 1961; Pine et al. 1979; Pearson 1984; M. H. Gallardo and Mondaca 2002), where it builds complex systems of burrows (Greer 1965; Pearson 1983; Nowak 1999). Species in this genus are presumed to be social (Lacey and Ebensperger 2007). Three species are currently included within the genus: *A. fuscus*, *A. porteri*, and *A. sagei*.

Aconaemys is a medium-sized octodontid (mass of males averages 118.9 g; length of head and body length ranges from 135–190 mm, tail length from 57–100 mm, and ear length from 15–21 mm; Greer 1965; Pearson 1984; M. H. Gallardo and Reise 1992; specimens in the MLP). As in the other subterranean octodontid *Spalacopus*, the body is short and robust, and the tail is short, representing up to 50% of the head and body length, and has no terminal tuft. The forefeet have long claws, except for the first digit, which is reduced. The pelage is dark drab, but varies regionally and locally related to soil variation. Specimens from more xeric areas have a paler coloration than those from humid regions, revealing a north-to-south gradient (M. H. Gallardo and Reise 1992). More northern specimens have a brown dorsum that changes gradually to the ventral region, while the specimens from the south have a more cinnamon venter. The coloration of the tail is highly variable. The dorsal surface of the feet is washed with white or cream (Pine et al. 1979).

Brain size is smaller than expected in comparison to those of caviomorph rodents (Vassallo and Echeverría 2009). The glans penis has two spines on each side within the intromittent sac, although some specimens of both *A. fuscus* and *A. sagei* exhibit individual variation, with patterns of 2–3 and/or 3–3 (Spotorno 1979; L. C. Contreras et al. 1993). The bacula of both *A. fuscus* and *A. sagei* have a slightly expanded base that gradually tapers to the tip; that of *A. fuscus* is longer (average length 7.6 mm), wider (average width 1.7 mm), and thus more massive (index of robustness = 0.23) than that of *A. sagei* (comparable values are length 5.8 mm, width 0.9, and robustness 0.14; L. C. Contreras et al. 1993).

The skull has a flattened interorbital region with divergent edges. As in *Spalacopus*, but unlike the remaining genera, the sphenopalatine foramen in the orbital region is small, vertical, and extends in front of the alveolar sheath of the M1. The zygomatic arches are parallel. As in *Spalacopus*, they are expanded toward outside, and the ventral zygomatic root (area for the origin of the masseter lateralis muscle) is wide and thin. The lateral flange of the canal for the infraorbital nerve is joined to the alveolar sheath of the incisor. The auditory bullae are small and oval as in *Octodon*, unlike the much larger condition in desert-adapted octodontids. The epitympanic recess is only slightly developed, as in *Octodon* and *Spalacopus*. The incus and malleus are unfused, also as in *Spalacopus* and *Octodon degus* (Wood and Patterson 1959:292–293).

The mandible is strongly built and deep. The angle is expanded laterally, and the coronoid and condyloid processes are vertical as in *Spalacopus*. In rodents, this morphology is related to subterranean habits (Laville et al. 1989; Olivares et al. 2004). Vassallo and Mora (2007), in examining the scaling and ontogenetic growth of the skull and mandible of ctenomyid and octodontid genera, documented an intensification of the lateral expansion of the mandibular angle and masseteric crest in *Aconaemys* (as well as in *Spalacopus* and ctenomyids; E. P. Lessa, Vassallo et al. 2008), features indicative of the evolution of convergent morphologies in these subterranean or partially subterranean genera.

The upper incisors are broad, orthodont, and with orange enamel face. Molariform teeth are similar to those of *Octomys*, with their occlusal surfaces shaped as a figure-8, with the labial and lingual extremes of each lobe more acute than in *Octomys*. The two lobes are separated by lingual and labial folds (mesoflexus/flexid, and hypoflexus/flexid) that are transverse and meet in the middle; each fold is occupied by cement. The first lobe of the lower molariform teeth is triangular in shape in occlusal view. The last molars (M3/m3) are moderately reduced, and especially the lower is highly variable in morphology. Chewing is propalinal (Koenisgwald et al. 1994; Olivares et al. 2004). Microstructure of the molar enamel has been studied in *A. fuscus*; two layers of radial enamel enclose a layer of multiserial Hunter-Schreger bands (Koenigswald et al. 1994).

Ameghino (1891) reported on mandibular remains and isolated teeth of *Aconaemys* for the "Pampeano inferior" of Córdoba (Pleistocene). Unfortunately, we have been unable to locate these specimens in the MACN collections. Simonetti (1989) and Saavedra and Simonetti (2003) recorded *A. fuscus* in Holocene archaeological sites in Andean areas close to Santiago through coastal sectors of the Río Maule, up to 120 km north of its present distribution.

Molecular data suggest that *Aconaemys* is very closely related to *Spalacopus* (M. H. Gallardo and Kirsch 2001; Honeycutt et al. 2003; Opazo 2005). Most trees constructed from either DNA/DNA hybridization data (M. H. Gallardo and Kirsch 2001) or both mitochondrial and nuclear gene sequences (Honeycutt et al. 2003; Opazo 2005) support a sister relationship between *A. porteri* and *Spalacopus* with respect to the sister-pair of *A. fuscus* and *A. sagei*. This paraphyletic condition of *Aconaemys*, however, weak, and their correct phylogenetic relationships remain to be resolved (see Upham and Patterson 2012). The estimated date of divergence between these two pairs of species is 2.77 ± 0.75 mya, in the late Pliocene; divergence between *A. fuscus* and *A. sagei* is estimated at 7 kya (Opazo 2005). Paleontological data suggest, alternatively, the early Pliocene (nearly 5.0 mya) as a minimum age for the divergence between the lineages of *Spalacopus* and *Aconaemys* (Reig 1986; Verzi et al. 2013).

The number of species within *Aconaemys* has been uncertain (Pearson 1984; L. C. Contreras et al. 1987; M. H. Gallardo and Reise 1992; M. H. Gallardo and Mondaca 2002). Several authors have recognized only two species, *A. fuscus* and *A. sagei*, with *A. porteri* included as a subspecies of *A. fuscus* (Osgood 1943b; Cabrera 1961; Mann 1978; Pine et al. 1979; Pearson 1984). M. H. Gallardo (1992), M. H. Gallardo and Reise (1992), and M. H. Gallardo and Mondaca (2002) described different karyotypes for each of the nominal taxa, and M. H. Gallardo and Reise (1992) documented concordant morphometric and color differences between them. These authors have also reassigned specimens from localities in Chile previously referred to *A. fuscus* to *A. sagei* based on karyotype, and thus expanded greatly the known range of this recently described species. Thus, the three species we recognize herein (*A. fuscus*, *A. porteri*, and *A. sagei*) are based the analyses and conclusions of M. H. Gallardo and his colleagues, an action also accepted by other authors, such as Pearson (1995) and Woods and Kilpatrick (2005). M. H. Gallardo and Mondaca (2002) map contiguously allopatric ranges from that of *A. fuscus* in the north, *A. sagei* in the middle, and *A. porteri* in the south. In general, *A. porteri* may be more easily distinguished from the other two species by the pelage coloration and cranial morphology (Pearson 1984; M. H. Gallardo and Reise 1992). We caution, however, that morphological variation among populations is extensive along with both clinal or mosaic distribution patterns (M. H. Gallardo and Mondaca 2002), and that there are large collecting gaps within the known ranges of each species such that their distributional limits remain poorly known.

SYNONYMS:

Schizodon Waterhouse, 1842:91; type species *Schizodon fuscus* Waterhouse, by original designation; preoccupied by *Schizodon* Agassiz, in Spix and Agassiz (1829; Characiformes: Anostomidae).

Aconaemys Ameghino, 1891:245; replacement name for *Schizodon* Waterhouse, with the same type species, *Schizodon fuscus* Waterhouse.

REMARKS: Tate (1935) reviewed the rather uncomplicated taxonomic history of the genus *Aconaemys*. Waterhouse (1842) erected the genus *Schizodon*, with *S. fuscus* as the type species. But Ameghino (1891) showed that *Schizodon* Waterhouse was preoccupied by *Schizodon*

Agassiz, and proposed *Aconaemys* as a replacement name. Reig (1986:408) regarded *Aconaemys* as inseparable at the generic level from the Pliocene genus *Pithanotomys* Ameghino, 1887, a position accepted by McKenna and Bell (1997). Although the sister-group relationship between *Pithanotomys* and *Aconaemys* suggested by Verzi et al. (2013) is consistent with this proposal, the generic status of *Aconaemys* remains to be tested; if Reig's statement is correct, *Pithanotomys* Ameghino would have priority over *Aconaemys* Ameghino.

KEY TO THE SPECIES OF *ACONAEMYS*:

1. Dorsal color rich brown, venter bright rufous; inferior jugal process of zygoma more posterior than paraorbital process; lobes of the molariform teeth lateromedially elongated; last molars (M3/m3) with deep flexi/flexids . . . . *Aconaemys porteri*
1'. Dorsal color between Mummy Brown and Proust's Brown, venter tawny; inferior jugal process of zygoma more anterior or level with paraorbital process; lobes of the molariform teeth not lateromedially elongated; last molars (M3/m3) with reduced flexi/flexids   2
2. $2n = 56$, $FN = 108$ . . . . *Aconaemys fuscus*
2'. $2n = 54$, $FN = 104$ . . . . *Aconaemys sagei*

Aconaemys fuscus (Waterhouse, 1842)

Chilean Rock Rat

SYNONYMS:

Schizodon fuscus Waterhouse, 1842:91; type locality "from Chile," = "the Valle de las Cuevas, on the eastern side of the Andes, about six leagues from the Volcano of Peteroa" (Waterhouse 1848:265; Thomas 1917b:282; Pearson 1984:228); lectotype chosen by Thomas (1927b:553).

Aconaemys fuscus: Ameghino, 1891:245; first use of current name combination.

Aconaemys fuscus fuscus: Osgood, 1943b:112; name combination.

DESCRIPTION: Somewhat intermediate in size between small *A. sagei* and larger *A. porteri*. Two samples assignable species by karyotype (Siete Tazas and Chillán; M. H. Gallardo and Reise 1992: table 1) range in average head and body length from 156.1–156.9 mm, tail length 65.1–74.3 mm, hindfoot length 29.2–29.4 mm, and ear length 18.1–19.1 mm. Dorsal color paler brown than that of other species, with slightly tawny venter, presumably as adaptation to more xeric environments; tail uniformly bicolored (M. H. Gallardo and Reise 1992). Skull illustrated in Osgood (1943b:Fig. 14) shows interorbital region diverging gradually from anterior part of orbit, contrasting with pattern particularly of *A. sagei* (see Pearson 1984). Length of upper toothrow equivalent to that of *A. porteri* (9.6–9.9 mm) and much larger than that of *A. sagei* (8.6 mm in type series; Pearson 1984). Breadth of incisors at their tip also intermediate between other two species with mean of 4.2–4.3 mm (versus 3.9 for *A. sagei*; Pearson 1984). Only 17% of specimens examined in Siete Tazas sample of *A. fuscus* had persistent fontanelle at frontoparietal juncture in fully grown specimens (M. H. Gallardo and Reise 1992).

DISTRIBUTION: *Aconaemys fuscus* is found in Andean areas of central-west Argentina and in central Chile from Curicó province to Ñuble province (M. H. Gallardo and Reise 1992; M. H. Gallardo and Mondaca 2002).

SELECTED LOCALITIES (Map 543): ARGENTINA: Mendoza, Valle de Las Cuevas (type locality of *Schizodon fuscus* Waterhouse). CHILE: Biobío, Las Quilas, Termas de Chillán (Pine et al. 1979); Maule, Lircay (M. H. Gallardo and Mondaca 2002), Radal (M. H. Gallardo and Mondaca 2002), Valle de Río Teño (M. H. Gallardo and Reise 1992).

SUBESPECIES: *Aconaemys fuscus* is currently considered monotypic.

NATURAL HISTORY: This species inhabits highland forests of *Araucaria araucana* (Araucariaceae) as well as flat and sandy areas covered with xerophytic vegetation and dunes (Muñoz-Pedreros 2000). It feeds on pinenuts (*A. aracucana*) and roots as well as subterranean bulbs of Amaryllidacea and Alstromeriaceae (Muñoz-Pedreros 2000). This species forms colonies up to seven individuals, with its burrows forming mounds 0.5 m high and as much as 2 m in diameter. They build nests within the tunnel system where vestiges of vegetable and insect (primarily beetles) are found (Muñoz-Pedreros 2000). Newly born young and pregnant females of *A. fuscus* have been recorded in the spring, between October and November (Nowak 1999; Muñoz-Pedreros 2000).

Aconaemys fuscus is host to a rich ectoparasitic fauna that includes several species of fleas (Lewis 1976; Beaucournu and Torres-Mura 1986; Hastriter 2001), ticks (Keirans et al. 1976), and lice (Moreno Salas et al. 2005).

Aconaemys fuscus has suffered significant reduction in population distribution at its northern limits within the Holocene (Saavedra and Simonetti 2003). The cause of this historical retraction is unknown, but it may result from the complex changes in vegetative cover due both to long-term climate shifts and more recent human-induced landscape disturbances. At the local level, extinction risk exists because of the habitat alteration for livestock grazing and the conversion of native vegetation communities to plantations of *Pinus radiata* (Muñoz-Pedreros 2000).

REMARKS: Earlier descriptions purported of this species (e.g., Osgood 1943b) are compromised because they were based on composite samples that included what are now regarded as multiple species. Pearson (1984) summarized the taxonomic history of *A. fuscus*. Waterhouse (1842) described *A. fuscus* based on two specimens col-

lected by Thomas Bridges but failed to designate a type or provide a more precise locality than "from Chile." Waterhouse (1848:265) subsequently stated, based on communication with Bridges, that the specimens came from "the Valle de las Cuevas, on the eastern side of the Andes, about six leagues from the Volcano of Peteroa (S. lat. about 75° [clearly an error, should be 35°]), at an elevation from five to seven thousand feet." Volcán Peteroa is in Mendoza Province, Argentina. Thomas (1927b) designated BM 55.12.24.195 as the lectotype, the second of two specimens from the same locality originally listed by Waterhouse. The publication year for Waterhouse's *Schizodon fuscus* has been given as either 1841 (Thomas 1927b; Tate 1935; Osgood 1943b) or 1842 (Cabrera 1960; Woods and Kilpatrick 2005). Waterhouse presented his description to the Zoological Society of London on November 9, 1841, but the *Proceedings* for that year were not published until March 1842 (Sherborn 1930).

The karyotype of *A. fuscus* is $2n = 56$, $FN = 108$ (M. H. Gallardo and Reise 1992; M. H. Gallardo and Mondaca 2002), although Venegas (1976) reported $FN = 112$ for animals with the same diploid number. Specimens from Tolhuaca and Nahuelbuta (Malleco Province, Chile) that were previously assigned to this species (Venegas 1976;

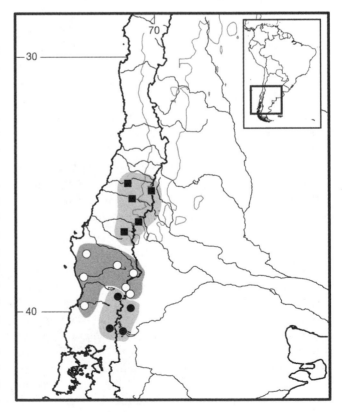

Map 543 Selected localities for *Aconaemys fuscus* (■), *Aconaemys sagei* (○), and *Aconaemys porteri* (●). Contour line = 2,000 m.

L. C. Contreras et al. 1987; M. II. Gallardo and Reise 1992) belong to *A. sagei* based on their karyotypes (M. H. Gallardo and Mondaca 2002).

Aconaemys porteri Thomas, 1917
Porter's Rock Rat

SYNONYMS:

Aconaemys porteri Thomas, 1917b:281; type locality "Osorno, S. Chile," slopes of Volcán Osorno, Los Lagos Region.

Aconaemys fuscus porteri: Osgood, 1943b:114; name combination.

DESCRIPTION: Large for *Aconaemys*, 40% larger in all body and cranial measurements than, and nonoverlapping with, *A. sagei* (Pearson 1984). Average head and body length 176.4 mm (148–192 mm), tail length 77.8 mm (68–85 mm), hindfoot length 34.7 mm (30–37 mm), ear length 21.0 mm (20–22 mm), and mass 131.5 g (105–160 g; N = 10, specimens in MVZ, see also Pearson 1984:228–229). Mean measurements slightly smaller for sample from Chile reported by M. H. Gallardo and Reise (1992: table 1, Huerquehue sample). As with *A. fuscus*, dorsal hairs long (averaging 13.5 mm), with dark, rich brown dorsal color and bright rufous venter, and strikingly bicolored tail (Pearson 1984; M. H. Gallardo and Reise 1992). Molariform teeth with lobes more lateromedially elongated, and last molars (M3/m3) with deeper flexi/flexids; 33% of sample from Huerquehue reported by M. H. Gallardo and Reise (1992; presumptively as *A. porteri*), with conspicuous fontanelle at frontoparietal junction.

DISTRIBUTION: *Aconaemys porteri* extends from Volcán Villarica to Puyehue in Chile, and from both Parque Nacional Lanín and Parque Nacional Nahuel Huapi in Argentina (Pearson 1984, 1995; M. H. Gallardo and Reise 1992; M. H. Gallardo and Mondaca 2002).

SELECTED LOCALITIES (Map 543): ARGENTINA: Neuquén, Lago Curruhué, Parque Nacional Lanín (Verzi 2001), Ruca Malén, Parque Nacional Nahuel Huapi (Pearson 1984). CHILE: Araucanía, Quetropillán, Parque Nacional Villarrica (M. H. Gallardo and Mondaca 2002); Los Lagos, Volcán Osorno (type locality of *Aconaemys* M. H. Gallardo Thomas), Puyehue, Parque Nacional Puyehue (M. H. Gallardo and Mondaca 2002).

SUBSPECIES: *Aconaemys porteri* is monotypic.

NATURAL HISTORY: This species inhabits dense bamboo and southern beech forests (*Nothofagus dombeyi*; Pearson 1983, 1984; M. H. Gallardo and Reise 1992) where it lives in small groups in communal burrow systems, which are large and conspicuous. Some burrows contained small piles of thin bamboo stems and large aggregations of fecal pellets. In captivity, individuals displayed a high tolerance for conspecifics. They make a high-pitched squeak that is audible only to humans with good hearing. Both diurnal and nocturnal activity has been recorded. They feed on leaves, sprouts, little branches of bamboo, and probably other type of vegetation. One female caught at the end of October had three embryos. The sexes are monomorphic (Pearson 1983).

REMARKS: Several authors considered *A. porteri* as a subspecies of *A. fuscus* (Osgood 1943b; Cabrera 1961; Mann 1978; Pine et al. 1979; Pearson 1983, 1984), but others have supported Thomas's (1917b) original proposal as a valid species (M. H. Gallardo and Reise 1992; Pearson 1995; M. H. Gallardo and Mondaca, 2002). This species is clearly distinguishable from the other two by its morphology, and molecular sequence analyses indicate that *A. porteri* is phylogenetically basal to the sister pair of *A. fuscus* and *A. sagei* (Honeycutt et al. 2003; Opazo 2005). *Aconaemys porteri* also has a unique karyotype of 2n = 58, FN = 112, which represents an ancestral condition for the genus and one shared with other octodontids (M. H. Gallardo and Reise 1992; M. H. Gallardo and Mondaca 2002). Specimens of *Aconaemys* sp. reported by Verzi and Alcover (1990) from Parque Nacional Lanín should be assigned to *A. porteri* based on morphology.

Aconaemys sagei Pearson, 1984
Sage's Rock Rat

SYNONYM:

Aconaemys sagei Pearson, 1984:229; type locality "Pampa de Hui Hui, 4 km west and 2 km south of Cerro Quillén, 1050 m, Province of Neuquén, Argentina."

DESCRIPTION: Smallest species in genus *Aconaemys*; head and body length of type series averaging 150.5 mm (140–159 mm), tail length 61.3 mm (58–68 mm), hindfoot length 28.3 mm (26–29 mm), ear length 18.3 (17–19 mm), and mass 95.7 g (range: 83–110 g; N = 6, MVZ). Specimens from Chilean populations somewhat larger in most dimensions (M. H. Gallardo and Reise 1992: table 1, Nahuelbuta and Tolhuaca samples). All dimensions of skin and skull at least 40% smaller relative to samples of *A. porteri* from nearby localities in Argentina (Pearson 1984). Pearson (1984) characterized species by slightly darker dorsal coloration ranging between mummy brown and Proust's brown and grading indistinctly to slightly more tawny venter. Tail lightly furred and slightly bicolored, but much thinner than series of *A. porteri* from nearby in Argentina. Also in comparison to *A. porteri*, dorsal fur flatter and shorter in length, averaging only 9 mm; claws distinctly smaller than those of *A. porteri*. Skull smaller in all dimensions, anterior border of mesopterygoid fossa is more pointed, and interorbital region flat with straight edges that diverge to side of cranium at their posterior edge rather than continually diverging from their anterior border. Upper incisors especially narrow at tips, width averaging 3.9 mm, and maxillary toothrow short, averaging 8.6 mm in length. All individuals in type series (Pearson 1984) and 90% of those from Chilean localities assigned to *A. sagei* based on karyotype (M. H. Gallardo and Reise 1992; M. H. Gallardo and Mondaca 2002), with conspicuous frontoparietal fontanelle.

DISTRIBUTION: *Aconaemys sagei* is known from Lago Quillén and Lago Hui Hui in Neuquén Province in Argentina and Malleco and Cautín provinces in Chile (Pearson 1984, 1995; M. H. Gallardo and Mondaca 2002).

SELECTED LOCALITIES (Map 543): ARGENTINA: Neuquén, Pampa de Hui Hui (type locality of *Aconaemys sagei* Pearson). CHILE: Araucanía, 27 km WNW of Angol (Greer 1965), Parque Nacional Nahuelbuta (M. H. Gallardo and Mondaca 2002), Parque Nacional Tolhuaca (M. H. Gallardo and Mondaca 2002), Pedregosa (M. H. Gallardo and Mondaca 2002), Reigolil Pass (M. H. Gallardo and Mondaca 2002); Los Lagos, vicinity of Valdivia (MVZ 142176).

SUBESPECIES: *Aconaemys sagei* is monotypic.

NATURAL HISTORY: This is a diurnal herbivore that Pearson (1984) found to be especially abundant in ungrazed bunchgrass but that he also encountered in second growth *Nothofagus* forest with a bamboo understory. Stomachs of individuals contained only green vegetable matter, and the caecum was enormous, indicative of herbivory. Areas occupied by individuals had numerous open burrows with small piles of excavated earth in from; these were frequently trampled to form runways, a characteristic described by Osgood (1943b) and Greer (1965) for Chilean populations likely assignable to all three species. Pearson (1984) argued that *A. sagei* in particular, and *Aconaemys* in general, is moderately specialized for a subterranean lifestyle but suggested that runway construction indicated that individuals spent considerable time above ground, with behavioral characteristics more like those of Holarctic microtine rodents or ground squirrels than highly specialized subterranean genera such as the closely related *Ctenomys*.

REMARKS: This species is apparently the sister to *A. fuscus*, based on molecular phylogenetic analyses (M. H. Gallardo and Kirsch 2001; Honeycutt et al. 2003; Opazo 2005). The karyotype is $2n = 54$, FN = 104 (M. H. Gallardo and Mondaca 2002). M. H. Gallardo and Mondaca (2002) assigned specimens from localities in Malleco (Tolhuaca National Park, Río Colorado, Pedregoso, and Nahuelbuta National Park) and Cautín (Reigolil) provinces to *A. sagei* based on karyotype. Specimens from these localities had been previously allocated to *A. fuscus* (Osgood 1943b:113; Greer 1965:133; Pine et al. 1979:365). M. H. Gallardo and Mondaca (2002) also recorded specimens

from Huerquehue National Park as *A. sagei* in their Table 1 (p. 107) but included this locality within the mapped range of *A. porteri* in Fig. 1 (p. 106). M. H. Gallardo and Reise (1992) had earlier stated that specimens from this locality had both the karyotype of *A. porteri* ($2n = 58$) and the morphological features (large size, long and dense fur) of this species. Because of this apparent confusion in the identity of these specimens, we do not include this locality in the list for either *A. porteri* or *A. sagei*.

Genus *Octodon* Bennett, 1832
M. Mónica Díaz, Rubén M. Barquez, and Diego H. Verzi

This is the most speciose octodontid genus, comprising the four species *O. bridgesii*, *O. degus*, *O. lunatus*, and *O. pacificus*. Three of these are endemic to Chile, occurring between approximately 28°S and 38°S latitude, and the fourth is distributed primarily in Chile but extends to Neuquén Province, Argentina. *Octodon* has also been recorded from Peru (Tschudi 1845; Yepes 1930; Cabrera and Yepes 1960), but this highly disjunct record is believed to correspond to an escaped pet (Woods and Boraker 1975). The genus spans an elevational range from sea level to 1,800 m. All species are primarily herbivorous (A. L. G. da Silva 2005), may be either diurnal or nocturnal in activity pattern, and are typically social (or presumably so; Lacey and Ebensperger 2007), with several individuals of both sexes occupying what may be an extensive, common burrow system.

Octodon is a medium-sized octodontid, with the total length ranging from 200–390 mm and tail length from 81–170 mm. The ears are large (23–30 mm), except in *O. pacificus* (20 mm); the hindfeet are adapted for jumping; the tail is as long as 70–80% of the head and body length, and it terminates in a small tuft developed to different degrees depending upon the species. Forefeet have four well-developed digits, with sharp claws on each; the fifth toe is poorly developed and bears a nail. Hindfeet have granular plantar pads. The dorsal coloration is grayish or drab with orange shades, and the venter is yellowish cream; tail color is similar to that of the dorsum, with the tuft black.

The glans penis is characterized by a variable number of medium sized spikes on each side of the intromittent sac (about five on each side, but more frequently a pattern of two on each side; Spotorno 1979; L. C. Contreras et al. 1993).

The skull has a slender rostrum, narrower than the interorbital region. The zygomatic arches are parallel sided. The jugal fossa for the origin of the masseter posterior muscle is nearly horizontal and continuous with the area for the origin of the anterior part of the masseter lateralis muscle on the zygomatic root. The lateral flange separating the canal for the infraorbital nerve is large. In the orbital region, the dorsal portion of the alveolar sheath of M1 is hidden inside of an elongated fissure-like sphenopalatine foramen. The auditory bullae are piriform and medium to small sized. Morphological details of the middle ear are shared with subterranean octodontoids (e.g., *Ctenomys* and *Spalacopus*; Argyle and Mason 2008).

The incisors may be orthodont to slightly opisthodontin different species, but the enameled face is uniformly pale orange. Molariform cheek teeth have a peculiar and asymmetrical morphological pattern. The upper molariform teeth (dPM4 through M3) have the posterior lobe not extended labially, similar to those of Ctenomyidae, but with a moderate to deep lingual fold (hypoflexus) filled with cement. The three anterior lower molarifom teeth (dpm4 through m2) have an oblique figure-8 occlusal design, with two oblique lobes separated by a labial (hypoflexid) and a lingual (mesoflexid) fold filled with cement. The last lower molariform (m3) almost lacks a labial fold (hypoflexid). Chewing is oblique (Verzi 2001; Olivares et al. 2004). Microstructure of the molar enamel in *O. bridgesii* is three-layered, as in *Aconaemys*, with two layers of radial enamel enclosing one of Hunter-Schreger bands (Koenigswald et al. 1994).

Octodon is the sister genus to a clade comprising *Aconaemys* and *Spalacopus* (M. H. Gallardo and Kirsch 2001; Honeycutt et al. 2003; Opazo 2005), with an estimated time of divergence of 3.69 ± 0.93 mya, in the Pliocene (Opazo 2005). Of the living species, both *O. bridgesii* and *O. lunatus* form a sister-species pair relative to *O. degus*, with the basal radiation of these lineages estimated at 2.57 ± 0.76 mya, also in the Pliocene (Opazo 2005). The phylogenetic position of the insular *O. pacificus* is not known. The genus has no fossil record, although specimens assignable to the extant species have been recorded from archaeological sites (Massoia 1979a; Massoia et al. 1980; Massoia and Silveira 1996; Pearson 1987, for Neuquén Province, Argentina; Saavedra and Simonetti 2003, for central Chile). Saavedra and Simonetti (2003) reported coexistence of the three continental *Octodon* species at the Holocene archaeological site of Quivolgo, Chile.

SYNONYMS:

Sciurus: G. I. Molina, 1782:303; part (description of *degus*); not *Sciurus* Linnaeus.

Myoxus: Bechstein, 1800:179, part (listing of *degus*); not *Myoxus* Zimmermannn.

Octodon Bennett, 1832b:46; type species *Octodon Cumingii* Bennett (− *Sciurus degus* G. I. Molina), by monotypy.

Dendroleius Meyen, 1833: table xliv; type species *Dendrobius degus* G. I. Molina (= *Sciurus degus* G. I. Molina), by monotypy.

Dendrobius Meyen, 1833:601; incorrect subsequent spelling of *Dendroleius* Meyen.

REMARKS: The taxonomic history of the genus *Octodon* has been relatively uncomplicated (Tate 1935:378–379). G. I. Molina (1782) described *Sciurus degus* and Bennett (1832b) erected *Octodon* with the single species, *O. cumingii*. Meyen (1833) wrote of *Dendroleius*, with *Dendrobius* (sic) *degus* as its single species. Waterhouse (1839) detailed *Octodon*, stating that *cumingii* Bennett was the same as the "degu" of Molina, and thus fixed *degus* G. I. Molina as the senior synonym of *cumingii* Bennett. Bennett (1841) provided additional details of his *cumingii*, compared his genus *Octodon* to *Poephagomys* F. Cuvier (= *Spalacopus*) and *Ctenomys*, and illustrated the skull, mandibles, and colored plate of a living animal. Bridges (1844) also equated *cumingii* Bennett with both *Sciurus degus* G. I. Molina and *Dendrobius degus* Meyen.

KEY TO THE SPECIES OF *OCTODON*:

1. Size small (total length maximally to 300 mm); tail terminates in well-developed tuft; anterior molariform teeth with shallow lingual fold (hypoflexus) filled with little cement . . . . *Octodon degus*
1'. Size medium to large (average total length >300 mm); tail terminates in short or inconspicuous tuft; anterior molariform teeth deep lingual fold nearly touching labial enamel and filled with abundant cement 2
2. Size medium (average total length 350 mm); tail terminated with moderate tuft; M3 without lingual fold. . . . . *Octodon lunatus*
2'. Size medium to large; tail with inconspicuous terminal tuft; M3 with moderate to deep hypoflexus 3
3. Size medium (total length<370 mm;) dorsal color blackish brown; lower molariform teeth with moderately transverse lobes . . . . *Octodon bridgesii*
3'. Size large (total length >385 mm); dorsal color orange drab; lower molariform teeth with strongly transverse lobes . . . . *Octodon pacificus*

Octodon bridgesii Waterhouse, 1845
Bridges's Degu
SYNONYMS:

Octodon Bridgesii Waterhouse, 1845:155; type locality "Chile," restricted to "Río Teno, Colchagua," O'Higgins, Chile (Thomas 1927b:553).

Octodon bridgesi Tate, 1935:378; incorrect subsequent spelling of *bridgesii* Waterhouse.

DESCRIPTION: Medium-sized species, smaller than *O. lunatus* but larger than *O. degus* (total length 250–370 mm, tail length 102–167 mm). Tail straight and shorter than head and body, with terminal tuft only slightly developed. Eyes larger, ears smaller, and pelage softer and more grayish in color in comparison to *O. degus*. Ventral color uniformly ochraceous coffee, with white axillary and inguinal spots. Molariform teeth similar to those of *O. lunatus*, but M3 possesses moderate fold on lingual side (Hutterer 1994:Fig. 6).

DISTRIBUTION: In Chile, *Octodon bridgesii* occurs in the mountains of the coast of Cauquenes (Maule Region), Tomé (Biobío Region) and Nahuelbuta (Araucanía Region) west of the Central Valley, and through the Andes from Baños de Cauquenes (O'Higgins Region) to Baños del Río Blanco (Araucanía Region). In Argentina, it is found in the Parque Nacional Lanín in Neuquén Province (Verzi and Alcover 1990; Pearson 1995; Podestá et al. 2000). Elevational range is from sea level up to 1,200 m (Muñoz-Pedreros 2000).

SELECTED LOCALITIES (Map 544): ARGENTINA: Neuquén, Estancia La Querencia (Podestá et al. 2000), Refugio (MVZ 184958), 0.8 km E and 3.3 km N junction Rte 234 and Río Pichi Traful (MVZ 199699). CHILE: Araucanía, 4 km W of Baños Río Blanco (Greer 1965), Nahuelbuta, Parque Nacional Nahuelbuta (Muñoz-Pedreros 2000); Biobío, Quirihue, Los Remates (Verzi 2001), Tomé (Muñoz-Pedreros 2000); Maule, Constitución (Muñoz-Pedreros et al. 1988), Parque Inglés–Siete Tazas (Reise and Venegas 1987); O'Higgins, Baños de Cauquenes (Osgood 1943b).

SUBSPECIES: *Octodon bridgesii* is monotypic.

NATURAL HISTORY: This species inhabits rocky areas with dense shrubs (primarily *Aristotelia* [Elaeocarpaceae], *Lithraea* [Anacardiaceae], *Peumus* [Monimiaceae], and *Ugni* [Myrtaceae], among others), prefers a low density of trees, bare ground, and dead branches (Muñoz-Pedreros 2000). It is a skilled climber and builds nests within vegetation; some nests have been found among bamboo (*Chusquea*; Redford and Eisenberg 1992; Muñoz-Pedreros 2000). Verzi and Alcover (1990) found individuals in burrows in Argentina, and Pearson (1995) collected specimens in both semiopen forests and jumbled lava blocks at the upper edge of the forest, also in Argentina.

Muñoz-Pedreros (2000) recorded reproductive activity in the months of April and December, with a litter size of two to three offspring. *Octodon bridgesii* is primarily nocturnal and rarely seem, but its presence can be detected because of the noisy alarm screams produced when it perceives the presence of strangers (Mann 1978). It feeds on herbaceous plants (A. L. G. da Silva 2005), of which the most important item may be the introduced *Pinus radiata* (Muñoz-Pedreros 2000). Major predators include the Barn Owl (*Tyto alba*), the Black-chested Eagle Buzzard (*Geranoaetus melanoleucus*), and the Culpeo (*Lycalopex culpaeus*; Muñoz-Pedreros 2000; Podestá et al. 2000). Maintenance of water balance is more difficult in *O. bridgesii* than *O.*

degus, a fact likely restricting the distribution of the species to more mesic habitats (L. C. Contreras et al. 1987). Massoia (1979a) recovered skulls that may belong to this species (see Pearson 1995) from owl pellets at an estancia in west-central Neuquén Province, Argentina. Muñoz-Pedreros et al. (1988) also collected skulls of *O. bridgesii* from pellets of Barn Owls in Chile.

The local extinction of this species in the Andes and in the coastal range close to Santiago (Chile) in the Holocene has resulted mainly from human activities modifying natural habitats (Simonetti 1989; Simonetti and Saavedra 1998). *Octodon bridgesii* is a dense vegetation specialist, and the development of agriculture and intense woodcutting triggered the local extinction of the species (Simmonetti and Saavedra 1998). This species is considered vulnerable in Chile as its distribution range has

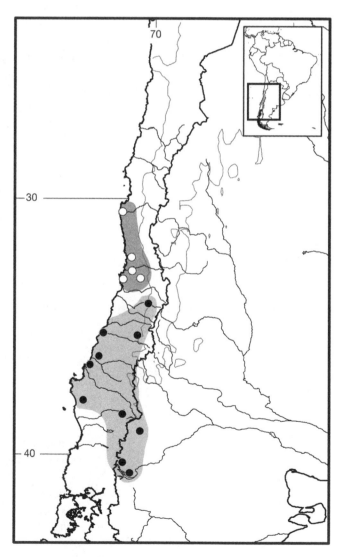

Map 544 Selected localities for *Octodon bridgesii* (●) and *Octodon lunatus* (○). Contour line = 2,000 m.

diminished in recent years and currently only in three discrete populations are known. Habitat fragmentation is a major influence of local population persistence (Saavedra and Simmonetti 2005). However, *O. bridgesii* is also considered a plague species because it can severely damage plantations of *Pinus radiata* (Redford and Eisenberg 1992; Muñoz-Pedreros 2000; Mella et al. 2002). In Argentina, the species is also considered vulnerable (G. B. Diaz and Ojeda 2000).

REMARKS: Thomas (1927b:553) selected one of the two co-types collected by Bridges and used by Waterhouse (1845) in his description of *Octodon bridgesii*. This specimen (BM 55.12.24.196) is from the Río Teno, Colchagua province; he thus emended the type locality to this site. The Argentinean populations are provisionally regarded as *O. bridgesii*, but the affinities of these specimens are in need of careful review (Pearson 1995). The year of publication of Waterhouse's *Octodon bridgesii* is varyingly given as 1844 (Thomas 1927b; Tate 1935; Osgood 1943b; Cabrera 1960) or 1845 (Woods and Kilpatrick 2005). Waterhouse read his paper before the Zoological Society of London on October 8, 1844, but that issue of the *Proceedings* was not published until February 1845 (Sherborn 1924a; Duncan 1937).

The karyotype is $2n = 58$, FN = 116 (Venegas 1976).

Octodon degus (G. I. Molina, 1782)
Common Degu
SYNONYMS:

Sciurus Degus G. I. Molina, 1782:303; type locality "St. Jago," Santiago, Chile (see Osgood 1943b:108).

Myoxus degus: Bechstein, 1800:479; name combination.

Myoxus ? degus: Illiger, 1815:107; *nomen nudum*; in a list of South American species.

Myoxus getulinus Poeppig, in Froriep, 1829:278; type locality "Santiago, Chile."

Octodon Cumingii Bennett, 1832b:47; type locality "in the neighbourhood of Valparaiso, . . . in the high-road between that place and St. Jago," Chile.

Dendroleius degus: Meyen, 1833:601; name combination.

Dendrobius degus: Meyen, 1833:610; inadvertent subsequent spelling of *Dendroleius* Meyen.

Octodon kummingii Schinz, 1845:101; incorrect subsequent spelling of *cumingii* Bennett.

Octodon pallidus Wagner, 1845b:33; type locality "Chili."

O[*ctodon*]. *cummingii* Tschudi, 1845:171; incorrect subsequent spelling of *cumingii* Bennett.

Octodon cummingii var. *peruana* Tschudi, 1845:172; type locality "Quebrada von San Mateo, in der nähe des dorfes San Juan de Matucana, ctwa 9000' ü. m," Lima, Peru" (probably based on escaped pet; Thomas, 1927c:557).

Octodon degus: Waterhouse, 1848:253; first use of current name combination.

[*Octodon Cumingii*] var. *albus* Fitzinger, 1867b:131; *nomen nudum* (see Osgood 1943b:108).

Octodon degus clivorum Thomas, 1927c:556; type locality "Puente Alto, E. of Santiago. Altitude 800 m," Santiago, Chile.

DESCRIPTION: Smallest species in genus, with total length 200–307 mm, tail length 81–138 mm, and mass 200–300 g, Ears and eyes large, and tail dorsally curved and terminates in well-developed tuft; individuals can autotomize tail as mechanism to escape predators, but without regenerative capability. Pelage soft, yellowish brown dorsally, and yellowish cream ventrally; albino specimens occasionally observed. First three upper molarifom teeth (dPM4–M2) have short fold (hypoflexus) on lingual margin but lack lingual fold on M3 (Hutterer 1994:Fig. 6; Verzi 2001).

DISTRIBUTION: *Octodon degus* is endemic to Chile (Mella et al. 2002), where it occurs between 28°S and 35°S latitude and from sea level to 1,800 m (Woods and Boraker 1975; Muñoz-Pedreros 2000). As noted by Woods and Boraker (1975), Waterhouse's (1848) report of this species from Peru is in error (see also Thomas 1927c).

SELECTED LOCALITIES (Map 545): CHILE: Atacama, Vallenar (Osgood 1943b), 40 km E of Vallenar (TTU 28636); Coquimbo, 10.5 km E of Illapel (USNM 541813), Parque Nacional Fray Jorge (Verzi 2001), 10 km N of Pte. Los Molles (FMNH 119658), Romero (Osgood 1943b); Maule, Curicó (Woods and Boraker 1975; Muñoz-Pedreros 2000); O'Higgins, Rapel (Muñoz-Pedreros 2000); Santiago, Las Condes (USNM 391827); Santiago, Puente Alto (type locality of *Octodon degus clivorum* Thomas), Rinconada de Maipú (Osgood 1943); Valparaíso, La Calera (Osgood 1943b).

SUBSPECIES: *Octodon degus* is currently viewed as monotypic (Osgood 1943b; Cabrera 1961; Woods and Boraker 1975; Woods and Kilpatrick 2005), although Ellerman (1940) listed five subspecies, while Thomas (1927c) and Yepes (1930) distinguished lowland (*O. d. degus*) and highland (*O. d. clivorum*) races.

NATURAL HISTORY: *Octodon degus* is one of the most well-known small mammals of central Chile (see L. C. Contreras et al. 1987 and literature therein). It has also been studied extensively in the laboratory, including determinants of circadian rhythm (Ocampo-Garces 2005; Vivanco et al. 2007), reproductive features such as placentation (Bosco et al. 2007), behavior, including vocalizations (Long 2007), communal versus solitary breeding (Ebensperger et al. 2007), and mother-offspring relationships (Jesseau et al. 2008). In captivity, degus reproduce and live between five and seven years and show hallmarks of neurodegenerative diseases (including Alzheimer's disease), diabetes, and cancer (Ardiles et al. 2012, 2013; see also husbandry and breeding guidelines in Palacios and Lee 2013).

This species inhabits semiarid environments on the western slopes of the Andes, in open areas near rocks, stonewalls, and groves, associated with steppes dominated by *Acacia caven* (Mann 1978; L. C. Contreras et al. 1987; Kenagy et al. 2002b). It digs tunnels among rocks and brushes; dens have several tunnels and openings where a few individuals or even large groups may live (Mann 1978; Muñoz-Pedreros 2000). In general, the social group includes a male, two or three females with young, and four or six juvenile females (Redford and Eisenberg 1992; Muñoz-Pedreros 2000); these groups occupy communal burrow systems (Ebensperger and Bozinovic 2000a; Ebensperger, Chech et al. 2011) and engage in collective detection of predators (Ebensperger, Hurdado, and Ramos-Jiliberto 2006). These communally nesting groups are composed of female kin (Ebensperger et al. 2004) and a breeding female does not discriminate between her own offspring and those of other females during lactation (Ebensperger, Hurtado, and Valdivia 2006). Per capita fitness decreases as a function of group size (Ebensperger, Ramírez-Estrada et al. 2011). Degus spend significantly less time being vigilant as group size increases ("group size effect"; Vásquez 1997). Dust bathing behavior plays an important role in social communication among unfamiliar, same-sex conspecifics (Ebensperger 2000; Ebensperger and Caiozzi 2002). Ebensperger and Bozinovic (2000b) examined the energetic costs of burrowing. Burrowing costs and predation risk are interpreted as drivers of total group size and group composition (Ebensperger et al. 2012), while social competition explains juvenile dispersal (Quirici et al. 2011). Also, ecologically based explanations of sociality in degus are controversial, but it could represent a case of phylogenetic inertia given that group living is common within hystricognath families basal to Octodontidae (Ebensperger et al. 2012).

In a study of fluctuating asymmetry, Saavedra (2003) demonstrated that males of *O. degus* preferred asymmetric females during nonreproductive periods but symmetric females during the reproductive season. Both *O. degus* and *O. bridgesii* showed preferences based on visual cues, despite their contrasting diurnal and nocturnal lifestyles (Saavedra 2003).

The four to six offspring are born mainly in the spring and summer (September–October and December–January); they are precocial at birth with open eyes, fur, and the capability to eat vegetation within a few days (Woods and Boraker 1975; Muñoz-Pedreros 2000).

This species is primarily diurnal although its activity is greatest at twilight, with activity peaks both in the morning and in the last hours of the night (Mann 1978; Redford and Eisenberg 1992; Muñoz-Pedreros 2000; Bacigalupe

et al. 2003; Kenagy et al. 2004). In laboratory conditions, an individual's temporal feeding strategy seems mainly arrhythmic, whereas its wheel-running pattern is driven by a strongly circadian bimodal rhythm (García-Allegue et al. 2006). In the wild, individuals feed outside and can be observed climbing in brushes while searching for seeds and sheaths (Woods and Boraker 1975; Mann 1978; Muñoz-Pedreros 2000). The species is thus primarily herbivorous-folivorous, feeding on grasses, seeds, fruits, roots, and bark of a variety of plants, such as *Stipa plumosa*, *Cestrum palqui*, *Mimosa cavenia*, *Erodium cicutarum*, *Chenopodium petiolare*, *Proustia cuneifolia*, *Atriplex repunda*, and *Acacia caven* (Redford and Eisenberg 1992; Muñoz-Pedreros 2000). Mass-independent basal metabolic rate is strongly influenced by habitat productivity, with populations from more productive habitats having "faster" metabolic rates (Bozinovic et al. 2009). In the field, degus are constrained to specific foraging areas, mainly by their limited thermal tolerance and by environmental food quality, with food availability strongly influencing home range sizes (Quirici et al. 2010). Individuals thus balance their diet selection by maximizing nutrients relative to digestible energy intake, but food selection is compromised by seasonal and spatial changes in food quality and environmental temperatures, coupled with feeding time and digestive and thermoregulatory constraints (Torres-Contreras and Bozinovic 1997). Naya et al. (2008) examined the complex interplay among energy acquisition, storage, and expenditure during lactation. Veloso and Bozinovic (2000) examined the effect of food quality on ingestion, digestion, and metabolic rate during pregnancy and lactation. Sabat and Veloso (2003) analyzed ontogenetic development of intestinal disaccharides. Vega-Zuniga et al. (2013) compared vision characteristics between two species (*O. degus* and *O. lunatus*), relating the degree of frontal binocular overlap to differences in diel activity pattern.

The limited capacity for thermoregulation also likely prevents individuals from leaving their tunnels in unfavorable periods of the year (Bozinovic and Vásquez 1999; Kenagy et al. 2002a). Thermal limitation is also a factor determining space use, with individuals restricted above ground to covered as opposed to open areas (Lagos et al. 1995). However, animals do not drink and lose little water through respiration (Muñoz-Pedreros 2000). They also have a high capacity of renal concentration, which implies a strong physiological specialization to relatively xeric conditions (Woods and Boraker 1975; Mann 1978; Muñoz-Pedreros 2000).

Primary predators include mammals such as foxes and the grison (*Pseudalopex*, *Galictis*), and both diurnal and nocturnal birds of prey (*Buteo*, *Parabuteo*, *Athene*, and *Tyto*; Woods and Boraker 1975; Jaksic 1986; Meserve et al. 1993; Torres-Contreras et al. 1994).

Among ectoparasites, fleas of the genus *Ectinorus* have been recorded (also present in other rodents) as well as the lice *Abrocomphaga hellenthali*, *Ferrisella chilensis*, and *F. disgrega* (the latter also present in *Octodontomys gliroides*; Torres-Mura and Contreras 1998; Muñoz-Pedreros 2000; D. Castro and Verzi 2002). *Octodon degus* is also parasitized by a number of endoparasites, including a pinworm (*Octodonthoxys gigantea*), a whipworm (*Trichuris bradleyi*), the cestodes (*Aprostatandrya octodonensis* and *Echinococcus granulosus*), and several nematodes (Woods and Boraker 1975; Muñoz-Pedreros 2000), and serves as a host for the protozoan *Trypanosoma cruzi* (Rozas et al. 2005), which causes Chagas disease.

Muñoz-Pedreros (2000) suggested that this species is of conservation concern, but Mella et al. (2002) regarded it to be excessively abundant. *Octodon degus* was used as

food by the Araucanian Indians as well as by European colonists. Currently, it may infrequently cause damage in vineyards as a result of local population irruption (Mann 1978). Unlike *O. bridgesii* and *O. lunatus*, this species prefers habitats with limited vegetation cover, typical of human-modified habitats. Based on this, *O. degus* may have been favored by human-induced changes in habitat cover (Mann 1978; Simonetti 1989). However, these same human-modified areas have been invaded by European rabbits, which have resulted in a reduced food supply and probably have had a negative impact on native species such as *O. degus* (Simonetti 1983).

REMARKS: Bridges (1844) equated *Octodon cumingii* Bennett with *Sciurus degus* G. I. Molina and *Dendrobius degus* Meyen. The highly disjunct Peruvian record of *O. degus*, the basis for Tschudi's *peruana* and which he stated was from San Juan de Matucana, east of Lima, is likely an error. Thomas (1927c:556–557) examined the holotype of Tschudi's *peruana* and stated "the specimen proves in every detail of size and colour to be so exactly like ordinary Valparaiso specimens of *O. degus* that I must frankly express my disbelief that it ever lived in Peru, unless as a pet." Furthermore, Thomas noted that the presumptive type locality is quite close to areas where both Perry O. Simons and Russell Hendee made substantial collections, yet failed to secure any belonging to *Octodon*. No member of the family Octodontidae is known from Peru. The karyotype of *O. degus* is $2n = 58$, FN $= 116$ (R. Fernández 1968).

Octodon lunatus Osgood, 1943
Coastal Degu
SYNONYM:
Octodon lunatus Osgood, 1943b:110; type locality "Olmue, Province of Valparaiso, Chile."

DESCRIPTION: Similar in size to *O. bridgesii*, with total length 328–382 mm, tail length 152–161 mm, and mass averaging 233 g. Tail straight as in *O. bridgesii*. Pelage denser than in *O. degus* but similar in coloration to *O. bridgesii*, only slightly more chestnut, and tail ventrally blackish for least half of its length. Also like *O. bridgesii*, *O. lunatus* with white axillary and inguinal spots. Skull similar to *O. bridgesii* with interorbital region slightly wider (Osgood 1943b). Molariform teeth with deep folds, as in *O. bridgesii*, but last upper molar (M3) lacks fold on lingual side (Hutterer 1994:Fig. 6).

DISTRIBUTION: *Octodon lunatus* is restricted to the coastal hills on the west side of the central valley of Chile (Osgood 1943b; L. C. Contreras et al. 1987; Redford and Eisenberg 1992).

SELECTED LOCALITIES (see Fig. 544): CHILE: Coquimbo, Parque Nacional Fray Jorge (Verzi 2001); Santi-

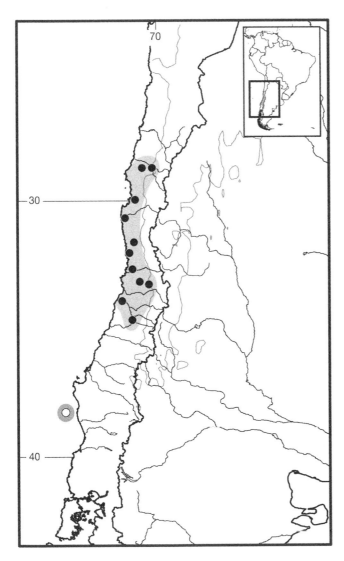

Map 545 Selected localities for *Octodon degus* (●) and *Octodon pacificus* (○). Contour line = 2,000 m.

ago, Fundo Santa Laura (MVZ 150088); Valparaíso, Los Molles (MVZ 150087), Olmué (type locality of *Octodon lunatus* Osgood), Papudo (Osgood 1943b).

SUBSPECIES: *Octodon lunatus* is a monotypic species.

NATURAL HISTORY: This species lives along the coast, in rocky areas with dense heaths and groves, from sea level to 1,200 m. Unlike *O. degus*, this species has been little studied in its natural environment. Apparently individuals have a variable activity rhythm, with authors reporting diurnal, crepuscular, or nocturnal activity patterns (Muñoz-Pedreros 2000). Cortés, Rosenmann, and Bozinovic (2000a) reported low respiratory water loss, indicating an excellent ability to conserve water. Lice are a known ectoparasite of *O. lunatus*, including one species shared with *O. degus* (D. Castro and Cicchino 2002). Only the Burrowing (*Athene cunicularia*) and Barn Owls (*Tyto alba*) are known as predators (Muñoz-Pedreros 2000). Like *O. bridgesii*, this species prefers areas with dense vegetation and has suffered a reduction in its range during the Holocene, based on the zooarchaeological record, likely due to human-induced habitat modifications (Saavedra and Simonetti 2003). Currently, this species is considered vulnerable (Muñoz-Pedreros 2000; Mella et al. 2002). Vásquez and Simonetti (1999) assessed life history traits in relation to vulnerability to habitat change, concluding that *O. lunatus* was among a group of birds and mammals that exhibited a high sensitivity index to change, primarily due to the combination of low reproductive effort and narrow habitat requirements.

REMARKS: Some authors have suggested that *O. lunatus* is conspecific with, and thus a synonym of, *O. bridgesii*, because the two are very similar and hard to distinguish (Tamayo and Frassinetti 1980; L. C. Contreras et al. 1987). However, *O. lunatus* has a distinct karyotype of $2n=78$, $FN=128$ (Spotorno et al. 1995).

Octodon pacificus Hutterer, 1994
Mocha Degu

SYNONYM:

Octodon pacificus Hutterer, 1994:28; type locality "Isla Mocha (38°22′ S, 73°55′ W), Arauca Province," Biobío, Chile.

DESCRIPTION: Larger and heavier than other species in the genus (two available specimens have total length of 389 and 390 mm, tail length of 165 and 170 mm, and a mass of 290 g, respectively). Pelage soft and long; ears relatively short; and tail long, 74–77% of head and body length. Tail ends in inconspicuous tuft. Dorsal coloration orange drab, venter paler. Skull long, especially diastema. Zygomatic arch diagnostic, with insertion higher than in other species, leaving considerable space between inferior jugal process of zygoma and upper toothrow. Mandible

larger and heavier than in other species. Incisors opisthodont with deep notch at tips. Both upper and lower molariform teeth wider than in other species, with deeper folds that filled with abundant cement, especially third molars (Hutterer 1994:Fig. 6).

DISTRIBUTION: *Octodon pacificus* is known only from the type locality.

SELECTED LOCALITIES (Map 545): Chile, Biobío, Isla Mocha (type locality of *Octodon pacificus* Hutterer).

SUBSPECIES: *Octodon pacificus* is monotypic.

NATURAL HISTORY: The only known locality of *Octodon pacificus*, Isla Mocha, is near the edge of the Valdivian rainforest zone (Hutterer 1994). There is little information available for this species. The only known specimens were collected in 1959 during an expedition made by Dr. Francisco Behn, who was primarily interested in birds. The mammals collected were held in a private collection for more than 30 years (Hutterer 1994). Reproductive activity apparently begins in September, and offspring are born in December (Hutterer 1994). Saavedra et al. (2003) recorded this species from the Holocene zooarchaeological record of Isla Moche.

REMARKS: The karyotype is unknown. Hutterer (1994:39), in his commentary about insular species, including *O. pacificus*, stated that "what seems certain is that the natural range of the species covers only a few square kilometers and for this reason alone it must be regarded as vulnerable." Muñoz-Pedreros (2000) also considered this species as vulnerable, but Mella et al. (2002) regarded the conservation status as undefined. E. P. Lessa, Ojeda, and Bidau (2008a) considered *O. pacificus* as Critically Endangered, with deforestation the major threat. The lowlands of Isla Moche are now dominated by grasslands, and the original Valdivian forest is restricted to the highest elevations on the island.

Genus *Octodontomys* Palmer, 1903
Diego H. Verzi, M. Mónica Díaz, and Rubén M. Barquez

This genus includes only the single species *Octodontomys gliroides*. It is a mainly a fossorial octodontid that inhabits the Andean and sub-Andean zones of southwestern Bolivia, northern Chile, and northwestern Argentina (Cabrera 1961; Pine et at. 1979; L. C. Contreras et al. 1987; S. Anderson 1997).

Octodontomys is a medium to large-sized octodontid, with a body mass ranging from 100–200 g, total length from 200–380 mm, and tail length from 100–190 mm. The pelage is relatively long and silky; dorsal coloration is grayish drab; and the venter has white hairs with gray bases, except on the chin and throat where they are pure white, and contrasts strongly with the dorsum. The ears are large, finely covered by short grayish hairs, with a white tuft on their anterior end (Ipinza et al. 1971). The tail is nearly 80% of the head and body length; it is bicolored, with an elongated, brown or ochraceous brush extending from the tip. Plantar surfaces have fine granulations (L. C. Contreras et al. 1987). Juveniles are darker with grayer venters, and have an essentially unicolored tail that terminates in a short and black or dark drab brush.

The glans penis of *Octodontomys* has two long spines on each side of the intromittent sac, although some specimens show a two–three pattern (Spotorno 1979; L. C. Contreras et al. 1993). These spines are the longest among octodontids (average length 5.6 mm; L. C. Contreras et al. 1993). As in other genera of the family, the baculum gradually tapers from its base to the tip.

The skull is roughly similar to that of *Octomys*. The rostrum is relatively long and narrow. Incisive foramina are large and with their posteriormost ends behind the premaxillary-maxillary suture. Unlike other genera, the intra-alveolar portion of the first molariform tooth (dP4) is slightly inclined forwardly, medial to the incisor. Consequently, an anterior extension of the maxillaries into the posterior margins of the incisive foramina, where the dP4 are housed, is scarcely developed to nearly absent. The lateral flange of the canal for the infraorbital nerve is well developed and has the dorsal end free, not joined to the maxillary. The zygomatic arches are little expanded, and with a low and slightly deep jugal fossa for the masseter posterior muscle. The paroccipital processes are completely joined to the bullae, with each root is medially oriented by the development of the mastoid bullae. As is characteristic for the other desert-adapted octodontids, the bullae are large. The mandible is slender, low, and moderately hystricognathous.

The incisors are orthodont, narrow, and short. Upper molariform teeth are roughly similar to those of primitive, extinct members of the Ctenomyidae. They lack folds (flexi) and have the posterior lobe not extended labially. Lower molariform teeth have a nearly figure-8 shaped occlusal morphology, with slightly oblique and rounded lobes and shallow traces of folds (mesoflexids and hypoflexids). The last upper and lower molars (M3/m3) are reduced in size. Chewing is oblique (Olivares et al. 2004). T. Martin (1992) and Vieytes et al. (2007) described incisor enamel microstructure.

Historically *Octodontomys* was a genus with uncertain affinities (W. George and Weir 1972; Rossi et al. 1990; M. H. Gallardo 1992, 1997; Verzi 2001). Although most molecular data (M. H. Gallardo and Kirsch 2001; Honeycutt et al. 2003; Opazo 2005; Rowe et al. 2010) support a sister relationship to a clade comprising (*Octodon* [*Acon-*

aemys + *Spalacopus*]), Upham and Patterson (2012) place it basal to the desert-adapted *Octomys* + *Tympanoctomys*. Although no fossil record is known, *Octodontomys* exhibits similarities with some species from the Mio-Pliocene of Argentina (Verzi and Iglesias 1999; Verzi et al. 1999; Verzi et al. 2013), suggesting that it may belong to an early octodontid offshoot. Opazo (2005) estimated the divergence of *Octodontomys* from its sister clade to be 6.07 ± 1.34 mya, near the end of the Miocene, a value in agreement with M. H. Gallardo and Kirsch's (2001) estimate based on DNA hybridization but older than the date estimated by Honeycutt et al. (2003).

SYNONYMS:

Octodon: P. Gervais and d'Orbigny, 1844:22; part (description of *gliroides*); not *Octodon* Bennett.

Neoctodon Thomas, 1902g:114; type species *Neoctodon simonsi* Thomas (= *Octodon gliroides* P. Gervais and d'Orbigny) by original designation; preoccupied by *Neoctodon* Bedel, 1892 (Insecta: Coleoptera).

Octodontomys Palmer, 1903:873; replacement name for *Neoctodon* Thomas.

REMARKS: Originally regarded as a species of *Octodon*, this monotypic taxon was raised to generic status by Thomas (1902g) under the name *Neoctodon*. Palmer (1903), however, pointed out that *Neoctodon* Thomas was preoccupied by *Neoctodon* Bedel, 1892, a genus of beetle, and provided *Octodontomys* as a replacement name. The taxonomy of this species has been stable since Palmer's action.

Octodontomys gliroides (P. Gervais and d'Orbigny, 1844)
Mountain Degu, Chozchori

SYNONYMS:

Octodon gliroides P. Gervais and d'Orbigny, 1844:22; type locality "des Andes boliviennes, à Lapaz"; La Paz, Bolivia.

Neoctodon simonsi Thomas, 1902g:115; type locality "Mountainous region south and south-east of the Titicaca-Poopo basin. Potosí, 4400 metres," Bolivia.

Octodontomys gliroides: Thomas, 1913a:143; first use of current name combination.

DESCRIPTION: As for the genus.

DISTRIBUTION: *Octodontomys gliroides* has a disjunct distribution in the Andean and sub-Andean areas of southwestern Bolivia (from La Paz to Potosí departments), northern Chile (Tarapacá Province), and northwestern Argentina (from Jujuy to La Rioja provinces; Cabrera 1961; Mann 1978; Pine et al. 1979; L. C. Contreras et al. 1987). This species has a wide elevational range, from 200–300 m at Saladillo in Jujuy, Argentina to 4,400 m in Potosí, Bolivia.

SELECTED LOCALITIES (Map 546): ARGENTINA: Jujuy, Saladillo (G. B. Diaz and Ojeda 1999), 8.2 km S of Sey (OMNH 34965), Sierra de Zenta (M. M. Díaz and Barquez 2007); La Rioja, Villa Unión (Yepes 1936); Salta, 16 km S and 1.8 km W of Barrancas (OMNH 34964), 17 km NW of Cachi (ARG 6396): BOLIVIA: La Paz, Andes de Bolivia, vicinity of La Paz (type locality of *Octodon gliroides* P. Gervais and d'Orbigny); Oruro, Oruro (S. Anderson 1997); Potosí, Lipez (S. Anderson 1997), Potosí, Livichuco (S. Anderson 1997), Potosí (type locality of *Neoctodon simonsi* Thomas), Yuruma (S. Anderson 1997); Tarija, 2 km SE of Cieneguillas (FMNH 162891). CHILE: Arica y Parinacota, Cuesta Zapahuira (USNM 541823); Tarapacá, Miñita (Mann 1978).

SUBSPECIES: *Octodontomys gliroides* is monotypic.

NATURAL HISTORY: This species occupies dry Andean and sub-Andean zones (L. C. Contreras et al. 1987) and is the only octodontid present in the high areas of the Puna (Reig 1986). It inhabits rocky areas with vegetation dominated by columniform cacti (*Browningia*, *Cereus*, *Polycererus*, and *Opuntia*), shrubs, and herbaceous vegetation (Mann 1978). It possesses a tail with well-developed terminal brush, which is capable of autotomy similar to that of *Octodon degus*. When climbing, the tail is used for substrate support (Muñoz-Pedreros 2000). Although *O. gliroides* has no obvious morphological adaptations of the skull or limbs for a fossorial lifestyle (E. P. Lessa, Vassallo et al. 2008), it digs short burrows at the base of cacti and acacias and inhabits rocky caves, including Indian tombs (Mann 1978; L. C. Contreras et al. 1987; Nowak 1997). In northwestern Argentina, it can be seen during the day and in the first hours at night (M. M. M. Díaz 1999). Individuals feed on acacia sheaths in winter and cactus fruits during summer, and eat leaves and bark of resinous shrubs (Mann 1978). Similar to *Octodon*, it can use cactus tissues to satisfy water requirements (Mann 1978). Eisenberg (1974) identified the menace vocalizations of *Octodontomys* as a gurgle or twitter similar to those produced by some birds. *Octodontomys gliroides* and *Octodon degus* emit not only low and middle frequency calls, but also several high frequency or ultrasonic vocalizations (Eisenberg 1974), which is in contrast with the predominant low-frequency range of vocalizations of subterranean rodents (Schleich et al. 2007). Arends and McNab (2001) reported physiological parameters of basal metabolic rate, conductance, and body temperature. G. B. Diaz and Ojeda (1999) suggested poor physiological adaptation to desert conditions and thus survival on a water-rich diet of cacti because the kidney has a small renal papilla, low renal indices, and low urine concentration.

Pine et al. (1979) reported a lactating female and a juvenile in November, in the spring in Chile. In Argentina, in Jujuy Province, M. M. Díaz (1999) recorded juveniles of different ages in the same month, and S. Anderson (1997) found both pregnant females and juveniles in January and May. Weir (1974b) reported a gestation period of 100–109 days, with a litter of one to three offspring. The young are precocial, born completely furred and with open eyes.

Individuals of *O. gliroides* are hosts for *Trypanosoma cruzi*, the hemoflagellate responsible for Chagas disease, and the species may have been involved with the domestic cycle of transmission with humans from prehistoric times (Schweigmann et al. 1992). Other ectoparasites include fleas of the genus *Ectinorus* (Hastriter and Sage 2009), and lice of the genera *Ferrisella* (D. Castro, Gonzalez, and Cicchino 1998; D. Castro and Verzi 2002), *Amblycera*, and *Anoplura* (Moreno Salas et al. 2005).

The species is known from archaeological sites in Jujuy Province, Argentina, from 2,000 years BP (Teta and Ortiz 2002). It has no apparent conservation problems (G. B. Diaz and Ojeda 2000; Muñoz-Pedreros 2000), and the IUCN (2011) considers *O. gliroides* a species of Least Concern.

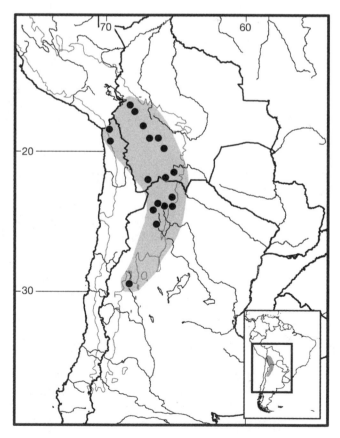

Map 546 Selected localities for *Octodontomys gliroides* (●). Contour line = 2,000 m.

REMARKS: D'Orbigny and Gervais (1847:plate 16) illustrated the living animal and line drawings of the skull and teeth, presumably of the type specimen. The chromosome complement is $2n = 38$, FN $= 64$ (W. George and Weir 1972; M. H. Gallardo 1992; L. C. Contreras et al. 1994). This diploid number is interpreted to result from a reduction from the family modal $2n = 56$–58 (M. H. Gallardo 1992, 1997, 2007).

Genus *Octomys* Thomas, 1920
Diego H. Verzi, M. Mónica Díaz, and Rubén M. Barquez

This genus includes the single species *Octomys mimax*, which is primarily a surface-dwelling, desert-adapted octodontid limited to the sub-Andean areas of the Monte biome in northwestern Argentina (Cabrera 1961; L. C. Contreras et al. 1987; R. A. Ojeda and Tabeni 2009).

Octomys is a medium-sized octodontid, with recorded mean body mass 95.8–31 g, mean total length 279.3–314.3 mm, and mean tail length 145.4–162.5 mm (Redford and Eisenberg 1992; Sobrero et al. 2010). The dorsal coloration is pale tan, with the venter and feet whitish. The tail is longer than the head and body, bicolored, well haired, and has a well-developed tuft at the tip. The ears are large, appearing hairless but covered with fine white hairs and elongated guard hairs in front of the meatus (Thomas 1920d; Redford and Eisenberg 1992; Sobrero et al. 2010). Berman (2003) described the superficial facial musculature and oral cavity morphology. The dorsal oral cavity of *Octomys* has dense concentrations of vibrissae, caudal to the incisors on either side of a palatal median cleft, similar to those in *Tympanoctomys*. These vibrissae have follicles less robust than those of *Tympanoctomys*, operated by the intermaxillaris muscle as in the latter genus (Berman 2003).

The glans penis has one spine on each side of the intromittent sac. This pattern is regarded as ancestral for the family (Spotorno 1979; L. C. Contreras, Torres-Mura et al. 1993).

The skull has a slender rostrum and weak zygomatic arches, similar to that of *Octodontomys*. The supraorbital borders of the frontals are straight and divergent; the distance between the frontal bone and the zygomatic arch is shorter than in *Octodon*, *Aconaemys*, *Spalacopus*, and *Octodontomys*, but similar to that in *Tympanoctomys*. The lateral flange of the canal for the infraorbital nerve is well developed, and its dorsal border is free, not attached to the maxillary. As in *Tympanoctomys*, only the squamosal bone forms the postorbital process of the zygomatic arch, as is visible in lateral view. The roots of the paroccipital processes in the occiput are medially oriented by the development of the mastoid bullae, similar to the condition in *Octodontomys*. The bullae are large but smaller than those of *Tympanoctomys* (Ojeda et al. 1999); the dorsal wall of the epitympanic recess forms a large bony "island" on the roof of the braincase, also as in *Tympanoctomys*.

Upper incisors are orthodont to opisthodont, short, and narrow. Molariform teeth have a figure-8 occlusal design, with two transverse lobes separated by folds (mesoflexus/flexid and hypoflexus/flexid) on both the labial and lingual faces. There is little cement in the folds. The last upper and lower molars (M3/m3) are reduced in size. Chewing is propalinal (Verzi 2001; Olivares et al. 2004).

The general external aspect of *Octomys* is similar to that of *Octodontomys* and *Tympanoctomys*, but with a shorter tail and a less developed terminal tuft. The skull is long and slender, and with large bullae, although these are less developed than in *Tympanoctomys*.

Thomas (1920d:118) pointed out that his new genus *Octomys* was related to *Spalacopus* and *Aconaemys*. However, cladistic analyses of morphological traits (Verzi 2001) as well as all molecular data (Köhler et al. 2000; M. H. Gallardo and Kirsch 2001; Honeycutt et al. 2003; Opazo 2005; Upham and Patterson 2012) strongly support a sister relationship between *Octomys* and *Tympanoctomys*. The time of divergence between these generic lineages has been estimated at 6.5 mya (based on DNA hybridization data; M. H. Gallardo and Kirsch 2001), 4.28 ± 1.08 mya (DNA sequences; Opazo 2005), or 3.57 mya (95% confidence interval 2.77–4.37 mya; DNA sequences; M. H. Gallardo et al. 2013). The genus is monotypic and has no fossil record. Two hypothetical *Octomys* lineages and the living species *Octomys mimax* have been proposed as ancestors of tetraploid *Tympanoctomys aureus* and *Tympanoctomys barrerae*, the implied evolutionary processes involving production of unreduced gametes and introgressive hybridization (M. H. Gallardo et al. 2004, 2007).

SYNONYM:

Octomys Thomas, 1920d:117; type species *Octomys mimax* Thomas, by original designation.

Octomys mimax Thomas, 1920
Viscacha Rat
SYNONYMS:

Octomys mimax Thomas, 1920d:118; type locality "La Puntilla, near Tinogasta . . . a few miles out from Tinogasta towards Copacabana, at an altitude of about 1000 metres" (Thomas 1920d:116), Catamarca, Argentina.
Octomys joannius Thomas, 1921i:217; type locality "Pedernal, about 60 km. S.W. of San Juan, and 30 W. of Cañada Honda. Altitude, 1200 m." (Thomas 1921i:214), San Juan, Argentina.

Octomys mimax mimax: Ellerman, 1940:157; name combination.

Octomys mimax joannius: Ellerman, 1940:157; name combination.

DESCRIPTION: As for the genus.

DISTRIBUTION: An endemic species of Argentina (Redford and Eisenberg 1992; R. A. Ojeda et al. 2002), *Octomys mimax* is found only in La Rioja, Catamarca, Mendoza, San Juan, and San Luis provinces in northwestern Argentina.

SELECTED LOCALITIES (Map 547): ARGENTINA: Catamarca, La Puntilla (type locality of *Octomys mimax* Thomas); La Rioja, Cañón de la Reserva, Talampaya (IADIZA 03785), Chilecito (Yepes 1930); San Juan, Parque Provincial Ischigualasto, Valle Fértil (IADIZA 03067), Pedernal (type locality of *Octomys joannius* Thomas); San Luis, 6 km W of Hualtaran, Parque Provincial Sierra de las Quijadas (ARG 3262).

SUBSPECIES: Currently *O. mimax* is considered monotypic (Cabrera 1961; Woods and Kilpatrick 2005). However, Yepes (1930, 1942) regarded *Octomys joannius* Thomas as a valid species while Ellerman (1940) and Lawrence (1941a) listed it as a subspecies of *O. mimax*.

NATURAL HISTORY: Sobrero et al. (2010) summarized available information on ecology, including habitat range, space use, behavior, food habits, and other populational attributes. Ebensperger et al. (2008) reported general behavior, including diel activity pattern, nesting, resting patterns, habitat use, and home range derived from radio telemetry. Individuals live among rocks in the Monte biome of northwestern Argentina, at elevations ≤800 m. In San Luis province, *O. mimax* occurs in a transitional area between the Monte and a semiarid partition of the Chaco biome (Sobrero et al. 2010). Individuals use rock crevices for refuges, and there is no evidence that they modify daytime resting sites through digging (Ebensperger et al. 2008). Rocky habitat provides a thermally stable environment, and nocturnal activity could allow *O. mimax* to avoid the high temperatures and low relative humidity of the day (Sobrero et al. 2010). This nocturnal activity is in accord with the relatively good thermoregulatory capabilities of these rodents (Bozinovic and Contreras 1990). However, although *O. mimax* has specialized oral vibrissae like those used by *T. barrerae* to strip salt-laden epidermis from leaves of halophytes, this species does not eat halophytes. At Ischigualasto (San Juan province), its diet is mainly leaves and fruits of *Prosopis*, (Fabaceae), but also includes *Larrea* (Zygophyllaceae), cacti, seeds, and arthropods (Oyarce et al. 2003). Water conservation capability of the kidney of *O. mimax* is less developed than in *Tympanoctomys*, and more like that of *Octodontomys* in aspects of the renal papilla, re-

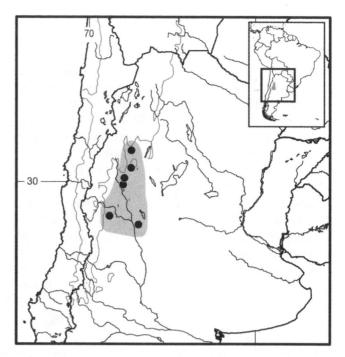

Map 547 Selected localities for *Octomys mimax* (●). Contour line = 2,000 m.

nal indices, and low urine concentration (G. B. Díaz and Ojeda 1999).

At Valle de la Luna (also San Juan province), one of us (RMB) captured young individuals in the same trap in November, suggesting that breeding occurs in the late winter or early spring. Another individual was also observed during the day at this same locality. *Octomys mimax* is solitary, a social system shared with its sister genus *Tympanoctomys*, and different from most octodontid genera, which are group living (Lacey and Ebensperger 2007).

Both fleas of the genus *Ectinorus* (Beaucournu et al. 2004) and lice of the genus *Ferrisella* (D. Castro, Gonzales, and Cicchino 1998) are part of the ectoparasitic fauna of *O. mimax*. G. B. Diaz and Ojeda (2000) regarded *O. mimax* as vulnerable; E. P. Lessa, Ojeda, and Bidau (2008b) and the IUCN (2011) placed this species in the Least Concern category while stating that if current agricultural trends persist it may qualify for "Near Threatened" status in the future.

REMARKS: The chromosome complement is $2n = 56$, $FN = 108$ (M. H. Gallardo 1992; L. C. Contreras et al. 1994).

Genus *Spalacopus* Wagler, 1832

Diego H. Verzi, Mónica G. Díaz, and Rubén M. Barquez

The genus *Spalacopus* has only one species, *S. cyanus*, distributed in central Chile from the coast to 3,400 m in the

Andes, and from Copiapó to Ñuble province (Pine et al. 1979; Nowak 1999; Muñoz-Pedreros 2000). The genus has no fossil record deeper than the Holocene. L. C. Contreras et al. (1987) suggested that *Spalacopus* became specialized in situ during the Pleistocene. Biogeographic (Reig 1970), molecular (M. H. Gallardo and Kirsch 2001; Honeycutt et al. 2003; Opazo 2005), and morphological evidence (Verzi 2001), however, support an older origin, dating at least to the Pliocene. Opazo (2005) date the split of *Spalacopus* and its sister genus *Aconaemys* at 2.77 ± 0.75 mya. *Spalacopus cyanus* has not modified its distribution during the Holocene, although it became sympatric with the other subterranean octodontid *Aconaemys fuscus* in central Chile during this time period (Saavedra and Simonetti 2003).

Spalacopus is a medium to small-sized octodontid, with mass ranging from 80–120 g, total length 140–225 mm, and tail length 28–57 mm (Torres-Mura and Contreras 1998). Specimens from lower elevations are smaller and have more striking sexual dimorphism than those from the Andes (J. R. Contreras 1986). Like other subterranean rodents, the body is short and robust, the eyes are small and dorsally located, and the ears are short (8–12 mm). The eyes, however, are not especially small, in contrast to those of many other subterranean rodents (Stein 2000). The forefeet are strong and with long claws, except for the first digit which is reduced, and its claw does not extend beyond the digital pad (Nowak 1999; Torres-Mura and Contreras 1998). The tail is shorter than those of other octodontid genera, approximately 25% of the total length; it is cylindrical, scaly, lacks a terminal tuft, and has hairs longer than those of either *Aconaemys* or *Octodon*. The pelage is soft, lustrous, and short. Color overall the entire body is uniformly darkish brown or black, but some specimens have pectoral and pelvic spots colored white or coffee brown. Complete or partial albinism has also been recorded (Nowak 1999; Torres-Mura and Contreras 1998; Muñoz-Pedreros 2000).

As in the sister genus *Aconaemys*, *Spalacopus* has smaller brain size compared with those expected for a caviomorph rodent (Vassallo and Echeverría 2009). Eyes show lower rod density than that of nocturnal surface-dwelling rodents, suggesting adaptation to visual demands of diurnal surface activity (Peichel et al. 2005). Ear morphology includes an enlarged eardrum and mobile-type ossicles that are believed to improve hearing sensitivity at lower frequencies (Begall and Burda 2006). The glans penis is characterized by a variable number of spines on each side of the intromittent sac (up to five on each side, but more frequently in a 3–3 or 2–2 pattern; Spotorno 1979; L. C. Contreras 1993). The spikes are smaller than those of other octodontids. Females have two pairs of abdominal

and one pair of inguinal mammae (Torres-Mura and Contreras 1998).

The skull is robust, short, broad, and flat. A vertical sphenopalatine foramen, similar to that of *Aconaemys*, is present in the orbitotemporal region in front of the alveolar sheath of M1, partially covered by the alveolar sheath of the incisor. The zygomatic arches are expanded, with the jugal fossa lower, shallow, and placed more posteriorly than in other octodontids. The lateral flange of the canal for the infraorbital nerve is reduced or absent. The bullae are piriform, smaller than in other octodontids, and nearly filled with small, bony septae. The incus and malleus are not fused as in *Aconaemys* and *Octodon*. The mandible is moderately hystricognathous, with the condyloid and coronoid processes nearly vertical. The angular (lunar) notch is shallow.

The upper incisors are long, narrow, and with the enameled face white or pale yellow. They are strongly procumbent (Thomas's angle about 117° [Thomas 1919h; E. P. Lessa and Thaeler 1989; Reig and Quintana 1992]), contrasting sharply with the orthodont or opisthodont condition of all other octodontids. The incisor alveolar sheath is extended posteriorly through the orbital region and ends at the level of the two first upper molars (E. P. Lessa, Vassallo et al. 2008). The base of the lower incisor extends to the level of the mandibular condyle. The first to third upper and lower molariform teeth (DP4/dp4-M2/m2) have a figure-8 occlusal shape with two transverse lobes separated by dual-faced folds (mesoflexus/flexid and hypoflexus/flexid). Neither pair of folds contact in the middle of the tooth, unlike *Aconaemys*, and little cement is present. The last molars (M3/m3) are slightly reduced in size. Chewing is propalinal with a slight oblique component (Koenisgwald et al. 1994; Olivares et al. 2004). Microstructure of the molar enamel is one layered and consists entirely of multiserial Hunter-Schreger bands, except for the trailing edge of the lower molars, which also has an inner layer of radial enamel (Koenigswald et al. 1994).

The clavicle is wide distally and both shallow and slender proximally. The deltoid crest of the humerus is well developed. The femur has the third trochanter enlarged, as in *Octodon degus*. The tibia and fibula are fused proximally (Torres-Mura and Contreras 1998).

SYNONYMS:

Mus: G. I. Molina, 1782:300; part (description of *cyanus*); not *Mus* Linnaeus.

Bathyergus Froriep, 1829:279; not *Bathyergus* Illiger, 1811; *nomen nudum*; based on *Bathyergus maritimus* Poeppig, a manuscript name quoted from a letter written by E. Poeppig to L. F. Froriep.

Spalacopus Wagler, 1832:1219; type species *Spalacopus poeppigii* Wagler (= *Mus cyanus* G. I. Molina), by monotypy.

Poephagomys F. Cuvier, 1834:323; type species *Poephagomys ater* F. Cuvier, by original designation.

Psammomys Poeppig, 1835a:166; type species *Psammomys noctivagus* Poeppig, by original designation; preoccupied by *Psammomys* Cretzschmar, 1828 (Rodentia: Gerbillinae).

Psammoryctes Poeppig, 1835b:252; replacement name for *Psammomys* Poeppig, with the same type species, *Psammomys noctivagus* Poeppig.

REMARKS: The taxonomic history of the genus *Spalacopus* has been relatively uncomplicated (Tate 1935:381–382). Wagler (1832) erected the genus *Spalacopus* with the single species *S. poeppigii*, which he based on a description by R. F. Froreip contained in a letter received by him from E. L. Poeppig. Cuvier (1834) described *Peophagomys ater* and Poeppig (1935a) described *Psammomys noctivagus*. Poeppig (1835b) subsequently erected *Psammoryctes* as a replacement for his earlier *Psammomys*, which he realized was preoccupied by *Psammomys* Cretzschmar, 1828, a gerbilline rodent from North Africa and the Middle East. Wagner (1843a) recognized *noctivagus* with *poeppigii* and *ater* in its synonymy, but under the Poeppig's *Psammoryctes*. Bridges (1844) then argued that *Poephagomys ater* F. Cuvier was the same as *Mus cyanus* G. I. Molina. E.-L. Trouessart (1880b:174–175) placed both *ater* and *noctivagus* in the synonymy of *poeppigii*, with *cyanus* listed as a questionable subspecies. Thomas (1925e) finalized the current species-level taxonomy by arguing that *Mus cyanus* G. I. Molina was the same species as that represented by Wagler's *poeppigii*, Cuvier's *ater*, and Poeppig's *noctivagus*.

Spalacopus cyanus (G. I. Molina, 1782)
Cururo, Coruro, Guanque

SYNONYMS: See under Subspecies.

DESCRIPTION: As for the genus.

DISTRIBUTION: *Spalacopus cyanus* is endemic to central Chile, between 27°S to 36°S latitude, where it ranges from the coast to about 3,400 m in the Andes (Reig et al. 1972; L. C. Contreras et al. 1987; Nowak 1999; Torres-Mura and Contreras 1998; Muñoz-Pedreros 2000).

SELECTED LOCALITIES (Map 548): CHILE: Atacama, Quebrada de Pajonales, 22 km N of Caldera (L. C. Contreras et al. 1987); Biobío, Quirihue (type locality of *Spalacopus cyanus maulinus* Osgood); Coquimbo, Huetelauquén (Opazo et al. 2008), La Serena (Opazo et al. 2008), Parque Nacional Fray Jorge (Urrejola et al. 2005); O'Higgins, Los Cipreses, Rancagua (L. C. Contreras et al. 1987; Torres-Mura and Contreras 1998; Bacigalupe et al. 2002); Santiago, Alicahue (Torres-Mura and Contreras 1998), Lagunillas (Opazo et al. 2008), Santuario de la naturaleza Yerba Loca (Urrejola et al. 2005), Santiago, Villa

Alhué (USNM 541832); Valparaíso, 3 km N of Los Molles (Mares et al. 2000), 1 km N of Pte. Los Molles (FMNH 119604), Quilpué (MSU 8720).

SUBSPECIES: Osgood (1943b) and Woods and Kilpatrick (2005) recognized two subspecies that correspond to coastal and Andean populations. L. C. Contreras et al. (1987) and Torres-Mura and Contreras (1998) recognized three subspecies, a view that we follow here.

Spalacopus cyanus cyanus (G. I. Molina, 1782)
SYNONYMS:

Mus Cyanus G. I. Molina, 1782:300; no type locality identified, but restricted by Osgood (1943b:114) to "Chile; Province of Valparaiso."

Hypudaeus ? cyanus: Illiger,1815:108; name combination; in a list of South American species.

Mus cyaneus Lesson, 1827:271; incorrect subsequent spelling of *Mus cyanus* G. I. Molina.

Bathyergus maritimus Poeppig, in Froriep, 1829:279; *nomen nudum*.

Poephagomys ater F. Cuvier, 1834:323; type locality "Coquimbo, Chile."

Psammomys noctivagus Poeppig, 1835a:166; type locality "sand dunes in the coastal region of northern Chile."

Psammoryctes noctivagus Poeppig, 1835b:252; renaming of *Psammomys noctivagus* Poeppig.

Peophagomys cyanus: Bridges, 1844:55; name combination, equated *ater* F. Cuvier with *cyanus* G. I. Molina.

Psammoryctes ater: Schinz, 1845:103; name combination.

Spalacopus cyaneus Wolffsohn and Porter, 1908:80; incorrect subsequent spelling of *cyanus* G. I. Molina.

Spalacopus cyaneus: Cabrera, 1917:52; incorrect subsequent spelling of *cyanus* G. I. Molina.

Spalacopus cyanus cyanus: Osgood, 1943b:114; first use of current name combination.

This subspecies occurs in central Chile from Caldera (Copiapó province, Atacama Region) to Curicó (Curicó province, Maule Region) from sea level to 1,000 m (Torres-Mura and Contreras 1998).

Spalacopus cyanus maulinus Osgood, 1943
SYNONYM:

Spalacopus cyanus maulinus Osgood, 1943b:115; type locality "Quirihue, Province of Maule," now Ñuble Province, Biobío, Chile.

This subspecies is found in the coastal area of Ñuble province, Biobío Region, south of the range of *S. c. cyanus*.

Similar in size to *S. c. cyanus* (mass 64–130 g, total length 140–195 mm, tail length 30–43 mm; Torres-Mura and Contreras 1998), but Osgood (1943b:116) noted cranial and dental differences with *S. c. cyanus*. L. C. Contreras et al. (1987) retained this subspecies but questioned its validity. *Spalacopus c. maulinus* is considered at risk (Muñoz-Pedreros 2000).

Spalacopus cyanus poeppigii Wagler, 1832

SYNONYMS:

Spalacopus Poeppigii Wagler, 1832:1219; type locality "vom Fuss der Anden Chile's."

Spalacopus tabanus Thomas, 1925c:585; type locality "South Chili"; exact type locality unknown (Osgood 1943b:116).

Spalacopus cyanus tabanus: Osgood, 1943b:116; name combination.

S[*palacopus*]. c[*yanus*]. *poeppigii*: L. C. Contreras et al. 1987: 405; first use of current name combination (see also Reig et al. 1972).

This subspecies is found on the western slope of the Andes, from Cordillera (Santiago Region) south to Curicó (Maule Region) provinces and at an elevational range from 1,500 to 3,400 m (Torres-Mura and Contreras 1998). It averages larger than *S. c. cyanus*, with total length 162–225 mm, tail length 39–57 mm, and mass 68–151 g (Torres-Mura and Contreras 1998). Both Osgood (1943b) and Mann (1978) regarded *poeppigii* Wagler as a synonym of *S. c. cyanus*, and Mann (1978) also considered *tabanus* Thomas to be a junior synonym of the nominotypical form. However, Reig et al. (1972) pointed out that Andean populations of *Spalacopus*, assignable to *S. c. poeppigii* (including *S. tabanus* Thomas), are phenotypically distinct from the coastal ones, a separation verified by Yáñez and Zülch (1981) through morphometric analyses. L. C. Contreras et al. (1987) concluded that *S. c. poeppigii* Wagler was a valid subspecies distinct from the nominotypical form.

NATURAL HISTORY: Torres-Mura and Contreras (1998) reviewed the literature covering aspects of the ecology, reproduction, physiology, and behavior of *S. cyanus*. This species is social, an unusual attribute for a strictly subterranean animal. Populations occur in discrete patches (M. H. Gallardo et al. 1992) from alpine grasslands in the Andes to sand dunes on the coast (L. C. Contreras et al. 1987; Torres-Mura and Contreras 1998). Local populations comprise many small colonies, each with a relatively stable home range (L. C. Contreras and Gutierrez 1991 vs. Reig 1970) of about 16 (up to 26) animals (Begall et al. 1999; Begall and Gallardo 2000). Burrow systems consist of a complex network of interconnected tunnels (Begall and Gallardo 2000; E. P. Lessa, Vassallo et al. 2008). A burrow system excavated at El Alamo (Ñuble province) was estimated to be 600 m long (Begall and Gallardo 2000). Incisors are used in digging, and individuals inhabiting harder soils in the Andes have longer incisors, in relation to a more frequent use with respect to those of the populations of the coastal dunes (Bacigalupe et al. 2002).

Cururos exhibit a lower basal metabolic rate relative to similar-sized above-ground species, a physiological feature in common with other subterranean rodents, but there is no difference in either basal or digging metabolic rates among populations inhabiting contrasting climatic and soil conditions (Bozinovic et al. 2005). They have lower ventilation in normoxia than expected for their body mass and a blunted hyperventilatory response, characteristic of fossorial mammals, when exposed to hypercapnia (Tomasco et al. 2010). The suggestion that ambient temperature limits above-ground activity (Rezende et al. 2003) may be an artifact of captivity because field studies document diurnality with substantial time spent on the surface (Urrejola et al. 2005).

Activity in free-living populations in both the coast and Andes is largely diurnal, and significant time is spent on the ground surface away from the burrow (Urrejola et al. 2005), a pattern contrasting sharply with the pattern observed in captive populations (Begall et al. 2002). Diurnal, above-ground activity is in accord with eyes of typical size for a nonspecialized surface-dwelling rodent and dichromatic color vision (Peichel et al. 2005). Cururos are also highly vocal, with a complex acoustic repertoire composed of a number of different true vocalizations and at least one mechanical sound (Veitl et al. 2000). The most distinctive and unique vocalization is a long musical trill that may last up to two minutes. Vocalizations in the wild are largely restricted to daylight hours, in accord with the diurnal activity pattern (Urrejola et al. 2005). The middle and inner ear morphology is typical of the pattern generally seen in subterranean rodents, which are generally characterized by specializations for low frequency and low-sensitivity hearing (Begall and Burda 2006). Nevertheless, cururos have a broader hearing range (covering frequencies at least between 0.25 and 20 kHz) coupled with an unusually high sensitivity, both atypical features in audiograms of subterranean rodents (Begall et al. 2004). Odor receptors are well developed, and cururos can discriminate among individuals within their colony or adjacent ones (Hagemeyer and Begall 2006).

Cururos are strictly herbivorous, foraging on tubers (*Libertia* [Iridaceae], *Leucocoryne* [Amaryllidaceae]) and bulbs (*Rodophiala*,[Amaryllidacea], *Dioscorea* [Discoreaceae]) (Ipinza et al. 1971; Reig 1970; Torres-Mura and Contreras 1998; Begall and Gallardo 2000). At Los Maitenes, individuals were found foraging on the ground on leaves of *Convolvulus* (Convolulaceae) (Begall and Gallardo 2000). *Spalacopus* damage cultivated crops both by their construction of tunnel systems and because they forage on food plants, but at the same time they can increase the biomass of herbaceous plants in tunnel areas (L. C. Contreras and Gutiérrez 1991) and on above-ground establishment through soil disturbance (Torres-Díaz et al. 2012).

Reproductive activity has been recorded for the months of March through January (Begall et al. 1999). The sex ratio of reproductive adults in skewed in favor of females (about 2:1), the gestation period is 77 days, and the mean litter size 3.5 (range: 2–5). Young are born altricial, but the eyes open at two to eight days, pups are weaned after two months, and adult body weight is attained by six to seven months of age.

This species is able to swim short distances. Interestingly, northern populations regard water as a novel element, but more southern ones will enter water without hesitation (Hickman 1988; Reise and Gallardo 1989).

REMARKS: *Spalacopus* shares the same karyotype with *Octodon degus*, *Octodon bridgesii*, and *Aconaemys porteri* (Reig et al. 1972; M. H. Gallardo 1992), supporting the close phyletic relationship among these three genera derived from molecular analyses (Opazo 2005). Zülch

et al. (1982) compared G- and C-bands of montane and coastal samples of *Spalacopus* and of *Octodon*.

Population genetic attributes have been examined by both allozyme (M. H. Gallardo et al. 1992) and mtDNA sequence methodologies (Opazo et al. 2008). Both independent data sets indicate that diversity is medium to low, mostly contained within populations, and with little genetic contact among regions. Phylogeography of mtDNA documented asymmetrical migration from the Andes to the coast, and suggested a montane origin of the species (Opazo et al. 2008). Both Bacigalupe et al. (2000) and Bozinovic et al. (2002) had previously proposed this biogeographic scenario.

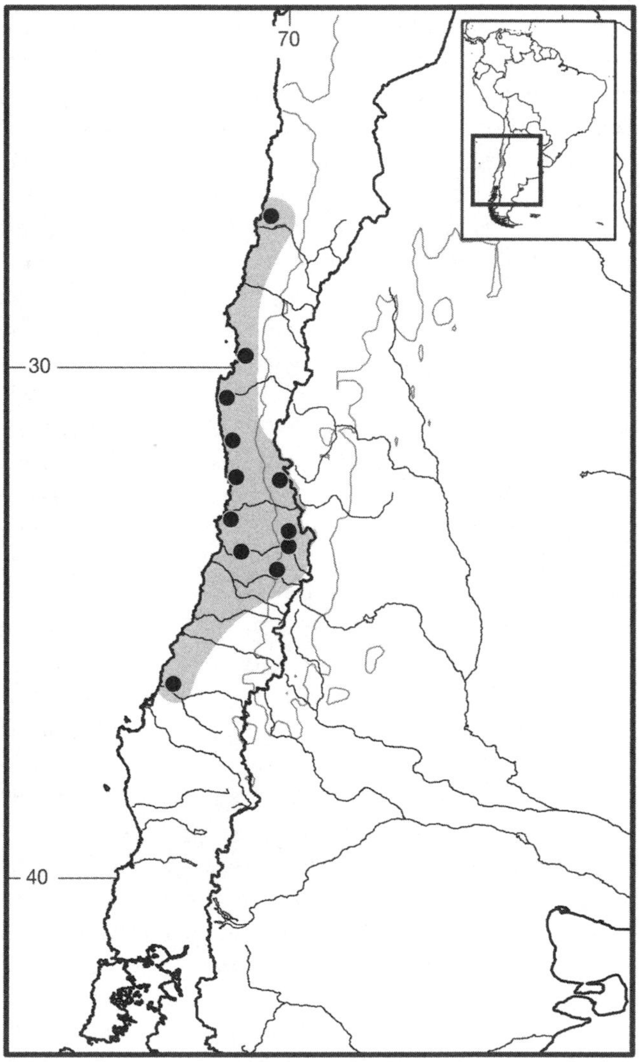

Genus *Tympanoctomys* Yepes, 1942
M. Mónica Díaz, Rubén M. Barquez, and Diego H. Verzi

Four living and one extinct species comprise the genus *Tympanoctomys*. One extant species was recently described from Patagonia (Teta et al. 2014) and two were originally placed in different genera, *Salinoctomys* and *Pipanacoctomys* (Mares et al. 2000). Barquez et al. (2002) subsequently treated both as *Tympanoctomys* based on morphological characters. Alternatively, Woods and Kilpatrick (2005) and M. H. Gallardo et al. (2007) have continued to recognize all three as valid genera. We concur with the arguments of Barquez et al. (2002) and recognize four species herein.

Species of *Tympanoctomys* are endemic to the arid regions of central and western Argentina, within the Monte, Chaco, and Patagonian desert biomes. Within these habitats they are known from salt basins (*salares*), sand dunes, and open scrubland habitats in Catamarca, Chubut, La Rioja, Mendoza, San Juan, La Pampa, and Neuquén provinces (Justo et al. 1985; R. A. Ojeda et al. 1996; Mares et al. 2000; A. A. Ojeda et al. 2007; Udrizar Sauthier, Teta et al. 2011, 2014). We regard this patchy distribution as relictual, and the particular habitats occupied as probable interglacial refuges. Verzi, Tonni et al. (2002) recorded the extinct *Tympanoctomys cordubensis* in the Early-Middle Pleistocene of central Argentina, in sediments representing a glacial event.

Tympanoctomys is a small-sized octodontid, with mean mass ranging from 67–104 g, mean head and body length from 130–170 mm, and mean length of tail from 114–135 mm (Mares et al. 2000; Teta et al. 2014). Overall coloration is different in each species, but ranges from dark brown to golden yellow. The tail is longer than or subequal to head and body length and terminates in a distinct tuft. Both the length and color of this tuft vary with age, as in *Octodontomys*, but in general young individuals have a short and black tuft while in adults it is longer and reddish. Claws are not well developed on the forefeet.

Map 548 Selected localities for *Spalacopus cyanus* (●). Contour line = 2,000 m.

Tympanoctomys is unique because it possesses a pair of bundles of stiffened hairs located on either side of the palate, just posterior to the upper incisors, and two oral glands not seen in other rodents (Mares, Ojeda, Borghi et al. 1997; Berman 2003). The bristle bundles vibrate against the lower incisors and remove the hypersaline epidermis from their major food plants (Berman 2003).

The glans penis of *T. barrerae* has one spike on each side of the intromittent sac, a condition similar to that in *Octomys* and likely the ancestral pattern for the family (Spotorno 1979; L. C. Contreras et al. 1993).

The skull has short nasals. The frontal width increases markedly posteriorly and the borders are straight and divergent. Consequently, the distance between the frontal bone and the zygomatic arch is shorter than in other octodontids, except *Octomys*. The zygomatic arches are short, anteriorly deep, and with their ventral border descending posteriorly. The jugal fossa (for the origin of the masseter posterior muscle) is deep, level with the paraorbital apophysis, and is directed posteriorly. The paraorbital process, in lateral view, is formed only by the squamosal, as in *Octomys*. The lateral flange of the canal for the infraorbital nerve is small or absent. In the orbital region, between the alveolus for the incisor and the sphenopalatine foramen, the nasolacrimal canal is exposed through a hole larger than in other genera. The auditory bullae are the largest for any taxon in the family, with the mastoid portion extending posteriorly well past the plane of the occiput, and the dorsal wall of the epitympanic recess forms a large bony "island" on the roof of the braincase. The root of the paroccipital process is strongly medial and close to the occipital condyle, due to enlargement of the mastoid. The mandible is short and wide, with the angular process thin and not flattened because the pterygoid shelf is very narrow. The angular (lunar) notch is deep and the postcondyloid process is laterally oriented.

The incisors are orthodont to opisthodont, and deeply implanted. The base of the lower incisor is posterodorsal to the last molar (m3). The occlusal pattern of the molariform teeth is figure-8 shaped as in *Octomys*, *Aconaemys*, and *Spalacopus*, but with lobes compressed anteroposteriorly. The mesoflexus/flexid and hypoflexus/flexid are in contact at the midline; each is wide and with abundant cement. The first lower molariform, dpm4, usually has a secondary lingual fold in the anterior lobe. The last lower molar, m3, is very reduced, variable in morphology, and usually has only an oblique labial fold (hypoflexid). The lower m3 of the newly described *T. kirchnerorum* is uniquely figure-8 shaped, similar in morphology to m2 (Teta et al. 2014). Chewing is propalinal (Verzi 2001; Olivares et al. 2004).

Ameghino (1889) was the first to describe species of *Tympanoctomys*, an extinct species that he included it in

the genus *Pithanotomys* as *P. cordubensis*. Thus, a fossil of *Tympanoctomys* was known for more than 50 years before the genus was described as part of the living fauna. Yepes (1942) named *Tympanoctomys* based on the extant species, *Octomys barrerae* Lawrence, and recognized the supraspecific level of differences between Lawrence's *O. barrerae* and *O. mimax*, the type species of *Octomys* Thomas. Subsequent analyses of morphology (De Santis et al. 1991; Verzi 2001), chromosomes (L. C. Contreras et al. 1990; M. H. Gallardo 1992), and molecular data (M. H. Gallardo and Kirsch 2001; Honeycutt et al. 2003; Opazo 2005) have confirmed Yepes's decision. Although cladistic analyses of both craniodental characters (Verzi 2001; Verzi et al. 2013) and molecular data based on DNA hybridization (M. H. Gallardo and Kirsch 2001) and sequences (Honeycutt et al. 2003; Opazo 2005) support a close phylogenetic relationship with *Octomys*, the extinct Pliocene genus *Abalosia* is the apparent sister to *Tympanoctomys* (Verzi 2001). Verzi (2001) suggested that this triad of genera was part of a clade of desert specialists that differentiated east of the Andes during the Late Pliocene, a time of expansion of arid environments. The living representatives are thus likely remnants of an earlier much wider radiation. Opazo (2005) estimated the mean divergence between *Tympanoctomys* and *Octomys* at 4.28 mya, in the Pliocene, and between two living species *T. barrerae* and *T. aureus* at 1.74 mya, at the Plio-Pleistocene boundary. M. H. Gallardo et al. (2013) estimated that *T. kirchnerorum* diverged from *T. barrerae* during the Daniglacial phase of the Middle Pleistocene, at approximately 1.47 mya, a period following the Great Patagonian Glaciation.

Milton Gallardo and colleagues (M. H. Gallardo 1997; M. H. Gallardo et al. 1999, 2006; M. H. Gallardo, Garrido et al. 2004; M. H. Gallardo, Kausel et al. 2004; Teta et al. 2014) presented evidence that three species of this genus (the karyotype of *T. loschalchalerosorum* is unknown) are the first known tetraploid mammals, although Svartman et al. (2005) argued against this conclusion.

SYNOMYMS:

Pithanotomys: Ameghino, 1889:165; part; not *Pithanotomys* Ameghino, 1887.

Tympanoctomys Yepes, in de la Barrera, 1940:569; *nomen nudum.*

Octomys: Lawrence, 1941a:43; part (description of *barrerae*); not *Octomys* Thomas.

Tympanoctomys Yepes, 1942:75; type species *Octomys barrerae* Lawrence, by original designation.

Pipanacoctomys Mares, Braun, Barquez, and Díaz, 2000:3; type species *Pipanacoctomys aureus* Mares, Braun, Barquez, and Díaz, by original designation.

Salinoctomys Mares, Braun, Barquez, and Díaz, 2000:6; type species *Salinoctomys loschalchalerosorum* Mares, Braun, Barquez, and Díaz, by original designation.

REMARKS: Woods and Kilpatrick (2005:1573) incorrectly cited "Yepes, 1941. Rev. Inst. Bact., 9, 1940 [1941]:569" as the basis for the genus *Tympanoctomys*. This citation, however, is a *nomen nudum* since the taxon was neither formally described nor was a holotype designated in that paper. Rather, *Tympanoctomys* as a valid name stems from Yepes (1942:75).

KEY TO THE SPECIES OF *TYMPANOCTOMYS*:

1. Lateral flange of the canal for the infraorbital nerve reduced to absent; bristle bundle with stiff hairs well developed; dorsal color pale brown to tan washed with black. 2
1'. Lateral flange of the canal for the infraorbital nerve present; bristle bundle absent or not well developed 3
2. Tail proportionally longer (>80% head and body length); lower m3 distinctly unilobed, "comma"-shaped. . . . . . .
. . . . *Tympanoctomys barrerae*
2'. Tail proportionally shorter (50% or less of head and body length); lower m3 distinctly bilobed, figure-8 shaped . . . . *Tympanoctomys kirchnerorum*
3. Posterolateral border of incisive foramina raised; posterior lobe of M3 anteroposteriorly not shortened, longer than the anterior lobe; dorsal color golden blond. . . . .
. . . .*Tympanoctomys aureus*
3'. Posterolateral border of incisive foramina not raised; posterior lobe of M3 anteroposteriorly shortened as the anterior one; dorsal color brownish-black . . . .
. . . .*Tympanoctomys loschalchalerosorum*

Tympanoctomys aureus (Mares, Braun, Barquez, and Díaz, 2000)
Golden Viscacha Rat

SYNONYM:

Pipanacoctomys aureus Mares, Braun, Barquez, and Díaz, 2000:3; type locality "28 km S, 9.3 km W Andalgalá, Departamento Pomán, Catamarca Province, Argentina, 27°50'03"S 66°15'59"W; elevation 680 m," Salar de Pipanaco, Saujil department.
Tympanoctomys aureus: Barquez, Díaz, and Ojeda, 2002:38; first use of current name combination.

DESCRIPTION: Largest species in genus, with mean head and body length 170 mm (166–178 mm). Tail long, averaging 135 mm (127–145), 76–85% of head and body length. Ears short and rounded, averaging 20 mm in length; hindfeet long for genus, averaging 37 mm. Dorsal coloration uniquely golden blond and venter white to pale cream. Tail with well-developed rufous terminal tuft in adults. Juveniles gray with black, short-tufted tail, color pattern contrasting sharply with that of adult specimens. Bristle bundles present, but not as well developed as in *T. barrerae*. Diagnostic characters of cranium include distinct

groove and flange for infraorbital nerve, postorbital process or protuberance located on frontal, long palate that extends beyond posterior borders of M2s, wide and flat paroccipital processes completely appressed to bullae, raised posterolateral border of incisive foramina that appears as protuberance anterior to DPM4 in lateral view, sharply acute angle formed by jugal and zygomatic process of maxilla, small interpremaxillary foramen, large and inflated auditory bullae, fused masticatory and buccinator foramina, light orange upper incisors with narrow white lateral edge, long upper and lower toothrows, and well-developed metacone and hypocone forming triangular-shaped posterior loph on M3.

DISTRIBUTION: *Tympanoctomys aureus* is known only from the type locality in the Salar de Pipanaco, within the Monte Phytogeographic Province, of Catamarca Province, Argentina.

SELECTED LOCALITIES (Map 549): ARGENTINA: Catamarca, 28 km S and 9.3 km W of Andalgalá (type locality of *Pipanacoctomys aureus* Mares, Braun, Barquez, and Díaz).

SUBSPECIES: *Tympanoctomys aureus* is monotypic.

NATURAL HISTORY (data from Mares et al. 2000, and M. M. Díaz and R. M. Barquez, pers. obs.): This species lives on the sandy substrate of the narrow strip of halophytic vegetation between the central, bare salt flat ("*salar*") and the more typical surrounding Monte Desert vegetation composed of *Prosopis*,(Fabaceae), *Larrea* (Zygophyllaceae) and other perennial trees and shrubs. Nocturnal, individuals feed exclusively on chenopodiaceous shrubs of the Amaranthaceae genera *Heterostachys*, *Atriplex*, *and Suaeda* that dominate the perisaline shrubland. They construct burrows in the earthen mounds that accumulate at the base of these same shrubs, with apparently more than one individual inhabiting a single mound, different from *T. barrerae*. Burrows are shallow and have several openings; in the active burrows, food and feces eliminated by individuals cleaning the burrows can be observed at the entrances. A birth was observed in October; the newborn was well haired, and the eyes and ears were open within a few hours after birth. Ectoparasites are unknown.

This species is at risk of extinction because the single known locality is experiencing expansion of olive tree cultivation that causes the loss of habitat, reducing both food and living space. Strong conservation measures are needed to avoid the extinction of this recently described species.

REMARKS: The karyotype of $2n=92$ is characterized by 44–45 pairs of biarmed chromosomes that mostly group into quartets, and a biarmed X chromosome and uniarmed Y chromosome. M. H. Gallardo, Kausel et al.

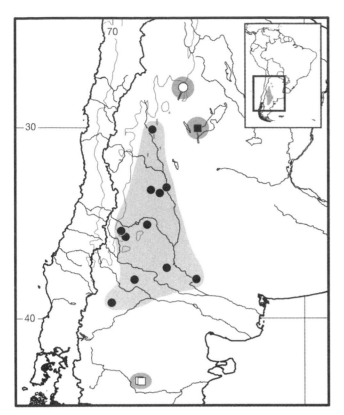

Map 549 Selected localities for *Tympanoctomys aureus* (○), *Tympanoctomys barrerae* (●), *Tympanoctomys kirchnerorum* (□), and *Tympanoctomys loschalchalerosorum* (■). Contour line = 2,000 m.

(2004) argued that the karyotype of *T. aureus* is a tetraploid derivative, like that of *T. barrerae*. According to the model proposed by M. H. Gallardo (2007), *T. aureus* originated through a reticulated speciation event.

Tympanoctomys barrerae (Lawrence, 1941)

Red Viscacha Rat

SYNONYMS:

Tympanoctomys barrerai Yepes, in de la Barrera, 1940:69; *nomen nudum.*

Octomys barrerae Lawrence, 1941a:43; type locality "La Paz, Mendoza Province, Argentina."

Tympanoctomys barrerai: Yepes, 1942:69; name combination and incorrect gender concordance.

Tympanoctomys barrerae: Cabrera, 1961:516; first use of current name combination.

Tympanoctomys barrerai: M. M. Díaz and Verzi, 2006:235; name combination and incorrect gender concordance.

DESCRIPTION: Middle sized of three species in genus, only slightly larger than *T. loschalchalerosorum* but considerably smaller than *T. aureus*. Mean head and body length 145 mm (125–160 mm), tail length 118 mm (93–147 mm), hindfoot 34 mm (30–36 mm), and ear length 16.3 mm (13–19 mm; Mares et al. 2000). Dorsal color-

ation tan washed with black; venter white. Tail proportionally longer (82% of head and body length) than in other species, and terminates in black brush. Forefeet and hindfeet small with claws covered by white hairs. Oral bristle bundles well developed and composed of stiff hairs, absent or not well developed in other species. Skull with flange for infraorbital nerve absent; small flange in other two species.

DISTRIBUTION: *Tympanoctomys barrerae* is endemic to the arid regions of central and western Argentina, within the Monte and Patagonian desert biomes. The geographic range is restricted and the preferred habitat is patchy (G. B. Diaz et al. 2000). The species is known only from scattered localities of salt flats, sand dunes, and open scrubland habitats in the provinces of Neuquén, La Pampa, Mendoza, and San Juan (Justo et al. 1985; R. A. Ojeda et al. 1989, 1996; G. B. Diaz et al. 2000; A. A. Ojeda et al. 2007). Populations range in elevation from 300–1,400 m.

SELECTED LOCALITIES (Map 549): ARGENTINA: La Pampa, Gran Salitral (A. A. Ojeda et al. 2007), Lihuel Calel (De Santis et al. 1991); Mendoza, 27 km N of Desaguadero (Mares et al. 2000), La Batra, SW of Laguna Llancanelo (Torres-Mura et al. 2003), La Paz (type locality of *Octomys barrerae* Lawrence), Las Catitas (G. B. Dias et al. 2000), Malargüe (G. B. Diaz et al. 2000), Nihuil Dam (Mares, Braun, and Channell 1997); Neuquén, Salar de Añelo (A. A. Ojeda et al. 2007); San Juan, Ischigualasto (A. A. Ojeda et al. 2007).

SUBSPECIES: *Tympanoctomys barrerae* is monotypic.

NATURAL HISTORY: G. B. Diaz et al. (2000) summarized the literature on the ecology, behavior, reproductive biology, and physiology of *T. barrerae* prior to 2000. This species inhabits salt basins and sand dunes in lowland habitats. Individuals are nocturnal and solitary (Mares, Braun, and Channell 1997), living in complex burrow systems built in soft soil mounds (R. A. Ojeda et al. 1996; Mares, Braun, and Channell 1997; G. B. Diaz 2003; Torres et al. 2003). Litters have been born in captivity from pregnant females captured in the wild (from Nihuil, Mendoza Province) in February. Neonates were altricial, with eyes closed and only 4 g in weight at birth. This species is strictly herbivorous, specializing on halophytic vegetation of the Amaranthaceae shrub genera such as *Atriplex*, *Allenrolfea*, *Heterostachys*, and *Suaeda* (Torres-Mura et al. 1989; R. A. Ojeda et al. 1996). They scrape off the surface of *Atriplex* leaves with their lower incisors and bristle brushes of stiff hairs before consumption, greatly reducing their electrolyte intake (Mares, Ojeda, Borghi et al. 1997; G. B. Diaz et al. 2000; Giannoni et al. 2000; Berman 2003). They are able to concentrate urine, to a maximum of 7,080 mosm/l, similar to other desert rodents that forage on halophytes (such as the North American kangaroo rat, *Dipodomys microps* [Heteromyidae] and the African and Middle Eastern sand rat, *Psammomys obesus* [Gerbillinae]; Bozinovic and Contreras 1990; G. B. Diaz and Ojeda 1999; R. A. Ojeda et al. 1999; G. B. Diaz 2003; G. B. Diaz et al. 2006).

Justo et al. (1985) recovered skulls of *T. barrerae* from pellets of the Barn Owl (*Tyto alba*). De la Barrera (1940) recorded ectoparasites, such as fleas of the genera *Hectopsylla* and *Parapsyllus*, on both individuals and in nests. Two species of *Parapsyllus* (De la Barrera 1940) and one of the louse genus *Ferrisella* (D. Castro and Verzi 2002) are exclusive parasites of *T. barrerae*. The Red Viscacha Rat is considered a Near Threated species (E. P. Lessa, Ojeda, and Bidau 2008c), although it was considered as "vulnerable" under prior criteria (Reca et al. 1996; R. A. Ojeda and Díaz 1997).

REMARKS: Lawrence (1941a:43) described *Octomys barrerae*, naming her new taxon for Sr. J. M. de la Barrera "through whose kind offices the specimen was obtained." Although Sr. Barrera was a male, Lawrence was correct to use the Latinized feminine genitive ending (ICZN 1999:Art. 31.1.1) rather than the masculine genitive ending *barrerai* employed by Yepes (1940, 1942) and M. M. Díaz and Verzi (2006).

A. A. Ojeda (2010) described molecular phylogeographic structure based on mtDNA sequences that uncovered two geographically overlapping clades with a modest level of differentiation, low levels of gene flow, and evidence for post-Pleistocene expansion in the central part of the current species range.

The completely biarmed karyotype of *T. barrerae* is among the largest in both diploid and fundamental number ($2n = 102$, FN $= 200$) for any mammal, and includes 36 pairs of metacentric to submetacentric chromosomes, 14 pairs of subtelocentric autosomes, and an XY sex-chromosome system (L. C. Contreras et al. 1990; M. H. Gallardo 1992). All chromosomes have discrete blocks of centromeric heterochromatin (M. H. Gallardo 1992). The cellular DNA content is also larger than that of any other mammal, with an average 2C-value of 16.8 pg per nucleus (M. H. Gallardo et al. 1999, 2003). This species was proposed as the first tetraploid mammal known (M. H. Gallardo et al. 1999, 2006; M. H. Gallardo, Garrido et al. 2004), and Bacquet et al. (2008) suggest that both genetic and epigenetic mechanisms were important in the functional diploidization of a tetraploid genome. Svartman et al. (2005), however, presented evidence that the genome of *T. barrerae*, while large, is diploid rather than polyploid. Finally, Suárez-Villota et al. (2012) provided detailed analyses of repetitive DNA sequence composition and suggested that the unique karyotype of *T. barrerae* originated by hybridization.

The macrocephalic and distinctive paddle-like sperm head of *T. barrerae* is the largest reported for rodents, differing from the common oval-shaped ones found in other Octodontidae (M. H. Gallardo et al. 2002).

Tympanoctomys kirchnerorum Teta, Pardiñas, Udrizar Sauthier, and Gallardo, 2014
Kirchners' Viscacha Rat

SYNONYM:

Tympanoctomys kirchnerorum Teta, Pardiñas, Udrizar Sauthier, and Gallardo, 2014:62; type locality "Argentina, Province of Chubut, Department of Gastre, 18 km NNE by road of Los Adobes, conjunction between Ruta Provincial 58 and Zanjón El Colorado (43°13′51.6″, 68°38′49.1″, 517 m)."

DESCRIPTION: Slightly smaller than *T. barrerae*; mean head and body length 129.4 mm (range: 111–136 mm), tail length 114 mm (111–114 mm), hindfoot 31.7 mm (29–35 mm), and ear length 14.5 mm (14–15 mm). Dorsal coloration pale brown; venter whitish or pure white; forefeet and hindlegs whitish. Tail bicolored, sparsely haired, proportionally short (ca. 43–50% of head and body length), with medium terminal pencil. Forelegs and hindlegs whitish. Oral bristle bundles well developed, composed mostly of thick white hairs. Skull more bowed than that of *T. barrerae* in dorsal profile, with posteriorly cuneate nasals; supraoccipital without lateral wings; dpm4 with very shallow lingual secondary fold and conspicuous labial secondary fold; m3 distinctly figure-8 shaped, similar to morphology of m2.

DISTRIBUTION: Known range of extant populations are limited to the type locality and vicinity; fossil remains (Late Holocene) are known from the middle and lower parts of the Río Chubut well to the east of the type locality (see Udrizar Sauthier et al. 2009; Pardiñas, Udrizar Sauthier, and Teta 2012).

SELECTED LOCALITIES (Map 549): ARGENTINA: Chubut, 18 km NNE by road of Los Adobes (type locality of *Tympanoctomys kirchnerorum* Teta, Pardiñas, Udrizar Sauthier, and Gallardo), 3 km W of Los Adobes (Teta et al. 2014).

SUBSPECIES: *Tympanoctomys kirchnerorum* is monotypic.

NATURAL HISTORY: *Tympanoctomys kirchnerorum* inhabits sandy terrain within a general badlands landscape (M. H. Gallardo et al. 2009), with the type locality located within the *Chuquiraga avellanedae* (Asteraceae) steppe of the Central District of the Patagonian Phytogeographic Region (León et al. 1998). Here, plant colver is sparse, consisting mostly of low shrubs in the families Amaranthaceae, Anacardiaceae, Asteracea, and Fabaceae, such as *Atriplex*, *Chuquiraga*, *Prosopis*, *Prosopidastrum*, and *Schinus*. Burrows are readily recognized by their typical semicircular entrances about 10 cm high, well curved dorsally and flat below, and usually placed under *Atriplex* bushes. Other small mammals found together with *T. kirchnerorum* in owl pellets include the didelphid marsupial *Lestodelphys halli*, cricetid rodents *Akodon iniscatus*, *Eligmodontia* sp, *Euneomys chinchilloides*, *Graomys griseoflavus*, and *Reithrodon auritus*, and the caviomorph rodents *Microcavia australis* and *Ctenomys* sp. (Teta et al. 2014).

REMARKS: M. H. Gallardo et al. (2009:158) recorded the type locality as "Estancia La Porfía (43°13′51.6″, 68°38′49.1″, 517 masl . . .) approximately 20 km NE of Los Adobes, on provincial route 58." Phylogenetic analyses of mtDNA indicate that *T. kirchnerorum* is sister to the two clades identified within *T. barrerae* among members of the genus (A. A. Ojeda 2010; M. H. Gallardo et al. 2013). However, phylogenetic analyses including larger data sets of all species in the genus remain necessary. The karyotype is similar to that of *T. barrerae* with $2n = 102$, with approximately 32 pairs of metacentric and submetacentreic chromosomes and 18 pairs of subtelocentric elements (Teta et al. 2014). M. H. Gallardo et al. (2009) described sperm morphology as smaller in size but similar in shape to that of *T. barrerae*.

Tympanoctomys loschalchalerosorum (Mares, Braun, Barquez, and Díaz, 2000)
Chalchalero Viscacha Rat
SYNONYM:

Salinoctomys loschalchalerosorum Mares, Braun, Barquez, and Díaz, 2000:6; type locality "26 km SW Quimilo, Departamento Chamical, La Rioja Province, Argentina (30°02′43.4″S 65°31′13.4″W; elevation 581 m."

Tympanoctomys loschalchalerosorum: Barquez, Díaz, and Ojeda, 2002:38; first use of current name combination.

DESCRIPTION: Medium-sized species, with total length 255–275 mm. Tail long (111–119 mm), >75% of head and body length (144–156 mm); ears short and rounded (16–17 mm); hindfeet short (28–30 mm). Dor-sal color brown-black dorsally and white ventrally; hairs with pale gray basal band. Proximal one-third to half of tail bicolored; terminal tuft black. Bristle bundles present but not well developed, and composed of soft hairs. Cranially, characterized by distinct groove and small flange for infraorbital nerve, large interpremaxillary foramen, supraoccipital with distinct medial crest, short palate that extends only to middle of M2s, wide and flattened paroccipital processes completely appressed to bulla, sharply acute angle of suture between jugal and zygomatic process of maxilla, distinct "knob" posterior to suture between jugal and zygomatic process of maxilla, large and inflated auditory bullae, posterior opening of alisphenoid canal and foramen ovale present with latter displaced toward bulla, foramen magnum directed distinctly downward, light orange upper incisors with narrow white edge, and M3 with well-developed metacone and hypocone.

DISTRIBUTION: *Tympanoctomys loschalchalerosorum* is only known from its type locality, in the Chaco Phytogeographic Province of La Rioja province, Argentina.

SELECTED LOCALITIES (Map 549): ARGENTINA: La Rioja, 26 km SW of Quimilo (type locality of *Salinoctomys loschalchalerosorum* Mares, Braun, Barquez, and Díaz).

SUBSPECIES: *Tympanoctomys loschalchalerosorum* is monotypic.

NATURAL HISTORY: *Tympanoctomys loschalchalerosorum* inhabits perisaline shrublands in a salt basin associated with the Salinas Grandes. The area is in to the Chaco Phytogeographic Province, in a thorn scrub island dominated by cactus, shrubs, and grasses (Amarahthaceae, Scrophulariaceae, Gramineae, Asteraceae, Solanaceae, Fabaceae, Leguminosae) that arise within a salt flat. Like the other species of *Tympanoctomys*, it is associated with halophytic plants. The mounds of this species are not as obvious as are those of the other two, but large nests with accumulated grasses occur in the center of a burrow system. A pregnant female with two embryos was collected in July (Mares et al. 2000). Ectoparasites are unknown.

REMARKS: The karyotype is unknown. No information on conservation status is available.

Literature Cited

Aagaard, E. M. J. 1982. "Ecological distribution of mammals in the cloud forests and páramos of the Andes, Merida, Venezuela." Ph.D. diss., Colorado State University, Fort Collins, 277 pp.

Abrahamovich, A. H., M. Lucia, N. B. Díaz, M. F. R. Batiz, and D. D. Castro. 2006. Types of lice (Insecta, Phthiraptera) housed in the Museo de La Plata, Argentina. *Zootaxa* 1344:43–58.

Abravaya, J. P., and J. O. Matson. 1975. Notes on a Brazilian mouse *Blarinomys breviceps* (Winge). *Contr. Sci. Nat. Hist. Mus. Los Angeles Co.* 270:1–8.

Ab'Sáber, A. N. 1974. O domínio morfoclimático semi árido das Caatingas Brasileiras. *Geomorfologia* 43:1–37.

Abud, C. 2011. "Variación genética y estructura filogeográfica de *Abrothrix olivaceus* en la platagonia argentina y el sur chileno." Master's thesis, Universidad de la República, Montevideo, Uruguay, 56 pp.

Adler, G. H. 1995. Fruit and seed exploitation by Central American spiny rats, *Proechimys semispinosus*. *Stud. Neotrop. Fauna Environ.* 30:237–44.

———. 1996. The island syndrome in isolated populations in a tropical forest rodent. *Oecologia* 108:694–700.

———. 2000. Tropical tree diversity, forest structure and the demography of a frugivorous rodent, the spiny rat (*Proechimys semispinosus*). *J. Zool. Lond.* 250:57–74.

———. 2011. Spacing patterns and social mating systems of echimyid rodents. *J. Mamm.* 92:31–38.

Adler, G. H., A. Carvajal, S. L. Davis-Foust, and J. W. Dittel. 2012. Habitat associations of opossums and rodents in a lowland forest in French Guiana. *J. Mamm. Biol.* 77:84–89.

Adler, G. H., and R. P. Beatty. 1997. Changing reproductive rates in a neotropical forest rodent, *Proechimys semispinosus*. *J. Anim. Ecol.* 66:472–80.

Adler, G. H., S. L. Davis, and J. Carvajal. 2003. Bots (Diptera: Oestridae) infesting a neotropical forest rodent, *Proechimys semispinosus* (Rodentia: Echimyidae), in Panama. *J. Parasitol.* 89:693–97.

Adler, G. H., M. Endries, and S. Poitter. 1997. Spacing patterns within populations of a tropical forest rodent, *Proechimys semispinosus*, on five Panamanian islands. *J. Zool. Lond.* 241:43–53.

Adler, G. H., and D. W. Kestell. 1998. Fates of neotropical tree seeds influenced by spiny rats (*Proechimys semispinosus*). *Biotropica* 30:677–81.

Adler, G. H., D. C. Tomblin, and T. D. Lambert. 1998. Ecology of two species of echimyid rodents (*Hoplomys gymnurus* and *Proechimys semispinosus*) in central Panamá. *J. Trop. Ecol.* 14:711–17.

Adrian, O., and N. Sachser. 2011. Diversity of social and mating systems in cavies; a review. *J. Mamm.* 92:39–53.

Aellen, V. 1970. Catalogue raisonné des chiroptères de la Colombie. *Rev. Suisse Zool.* 77:1–37.

Agassiz, L. 1842. *Nomenclator zoologicus, continens nomina systematica generum animalium tam viventium quam fossilium, secundum ordinem alphabeticum deposita, adjectis auctoribus, libris in quibus reperiuntur, anno editionis, etymologia et families, ad quas pertinent, in variis classibus.* Fasciculus I. Continens

Mammalia, Echinodermata et Acalephas. Soloduri: Jent et Gassmann, xii + 38 pp., iv + 14 pp., iv + 7 pp.

———. 1846. *Nomenclator zoologicus, continens nomina systematica generum classium, ordinum, familiarum et generum animalium ómnium, tam viventium quam fossilium, secundum ordinem alphabeticum unicum disposita, adjectis homonymiis planarum, nec non variis adnotationibus et emendationibus. Index universalis.* Soloduri: Jent et Gassmann, ix + 393 pp.

Agnolín, F. L., N. R. Chimento, E. Guerrero, and S. O. Lucero. 2010. Presencia del género *Ctenomys* (Rodentia, Ctenomyidae) en el noreste de la provincia de Buenos Aires, Argentina. *Rev. Mus. Argentino Cienc. Nat.,* n.s., 12:17–22.

Agrellos, R., C. R. Bonvicino, E. S. T. Rosa, A. A. R. Marques, P. S. D'Andrea, and M. Weksler. 2012. The taxonomic status of the Castelo dos Sonhos Hantavirus reservoir, *Oligoryzomys utiaritensis* Allen 1916 (Rodentia, Cricetidae, Sigmodontinae). *Zootaxa* 3220:1–28.

Agüero, D. 1978. Analisis reproductivo de una población de *Holochilus brasiliensis* (Rodentia-Cricetidae) en cultivos de arroz del estado Portuguesa. *Agron. Trop.* 28:101–16.

Aguilar F., P. G., O. D. Beingolea G., A. J. Brack E., and I. Ceballos B. 1977. Vertebrados importantes en la agricultura peruana. *Rev. Peru. Entomol.* 20:25–32.

Aguilera, M. 1985. Growth and reproduction in *Zygodontomys microtinus* (Rodentia, Cricetidae) from Venezuela in a laboratory colony. *Mammalia* 40:75–83.

———. 1987. "Ciclo de vida, morfometría craneana y cariología de *Holochilus venezuelae* Allen 1904 (Rodentia, Cricetidae)." Ph.D. diss., Universidad Simón Bolívar, Caracas, Venezuela.

———. 1999. Population ecology of *Proechimys guairae* (Rodentia: Echimyidae). *J. Mamm.* 80:487–98.

Aguilera, M., and M. Corti. 1994. Craniometric differentiation and chromosomal speciation of the genus *Proechimys* (Rodentia: Echimyidae). *Z. Säugetierk.* 59:366–77.

Aguilera, M., A. Pérez-Zapata, and A. Martino. 1995. Cytogenetics and karyosystematics of *Oryzomys albigularis* (Rodentia, Cricetidae) from Venezuela. *Cytogen. Cell Genet.* 69:44–49.

Aguilera M., M. A. Pérez-Zapata, A. Martino, M. A. Barros, and J. L. Patton. 1994. Karyosystematics of *Aepeomys* and *Rhipidomys* (Rodentia: Cricetidae). *Acta Cient. Venez.* 48:247–48.

Aguilera M., M. A. Pérez-Zapata, J. Ochoa G., and P. Soriano. 2000. Karyology of *Aepeomys* and *Thomasomys* (Rodentia: Muridae) from the Venezuelan Andes. *J. Mamm.* 81:52–28.

Aguilera, M., A. Pérez-Zapata, N. Sanginés, and A. Martino. 1993. Citogenética evolutiva en dos géneros de roedores suramericanos: *Holochilus* y *Proechimys*. *Bol. Soc. Zool. Urug.* 8:49–61.

Aguilera, M., O. A. Reig, M. A. Barros, and M. G. Besáñez. 1979. Sistematica, citogenetica y datos reproductivos de una poblacion de *Proechimys canicollis* del Noroeste de Venezuela. *Acta Cient. Venez.* 30:408–17.

Aguilera, M., O. A. Reig, and A. Pérez-Zapata. 1995. G- and C-banding karyotypes of spiny rats (*Proechimys*) of Venezuela. *Rev. Chilena Hist. Nat.* 68:185–96.

Aguilera, M., N. Sanginés, and A. Pérez-Zapata. 1998. *Echimys semivillosus*, a rodent species with a very high chromosomal number. *Caryologia* 51:181–87.

Aguilera, M., and A. Zapata. 1989. Karyotyping of *Holochilus venezuelae* (Rodentia, Cricetidae). *Acta Cient. Venez.* 40:198–207.

Aguirre, L. F., R. Aguayo, J. Balderrama, C. Cortez, and T. Tarifa, eds. 2009. *Libro rojo de la fauna silvestre de vertebrados de Bolivia.* La Paz, Bolivia: Ministerio de Medio Ambiente y Agua.

Alarcón, O., G. D'Elía, E. P. Lessa, and U. F. J. Pardiñas. 2011. Phylogeographic structure of the fossorial long-clawed mouse *Chelemys macronyx* (Cricetidae: Sigmodontinae). *Zool. Stud.* 50:682–88.

Albanese, S., D. Rodriguez, M. A. Dacar, and R. A. Ojeda. 2010. Use of resources by the subterranean rodent *Ctenomys mendocinus* (Rodentia, Ctenomyidae), in the lowland Monte desert, Argentina. *J. Arid Environ.* 74:458–63.

Albanese, S., D. Rodriguez, and R. A. Ojeda. 2011. Differential use of vertical space by small mammals in the Monte Desert, Argentina. *J. Mamm.* 92:1270–77.

Alberico, M. 1983. Lista anotada de los mamíferos del Valle. *Cespedesia* 45/46:51–72.

———. 1990. A new species of pocket gopher (Rodentia: Geomyidae) from South America and its biogeographic significance. In *Vertebrates in the Tropics: Proceedings of the International Symposium on Vertebrate Biogeography and Systematics in the Tropics*, ed. G. Peters and R. Hutterer, 103–11. Bonn: Alexander Koenig Zoological Research Institute and Zoological Museum, 424 pp.

Alberico, M., A. Cadena, J. I. Hernández-Camacho, and Y. Muñoz-Saba. 2000. Mamíferos (Synapsida: Theria) de Colombia. *Biota Colomb.* 1:43–75.

Alberico, M., and A. González-M. 1993. Relaciones competitivas entre *Proechimys semispinosus* y *Hoplomys gymnurus* (Rodentia: Echimyidae) en el occidente Colombiano. *Caldasia* 17:325–32.

Alberico, M., V. Rojas-Díaz, and J. G. Moreno. 1999. Aporte sobre la taxonomía y distribución de los puercoespines (Rodentia: Erethizontidae) en Colombia. *Rev. Acad. Colomb. Cienc. Exact. Fís. Natur.* 23(Suppl.):595–612.

Albert, F. 1900. La chinchilla. *Actes Soc. Sci. Chili* 10:379–407.

Albuja V., L. 1991. Mamíferos. *Politécnica.* 16:163–203.

———. 1992. Mammal list; July trip. In *Status of forest remnants in the Cordillera de la Costa and adjacent areas of southwestern Ecuador*, ed. T. A. Parker, III, and J. L. Carr, 124–27. RAP Working Papers 2. Washington, DC: Conservation International.

Aldrich, J. W., and B. P. Bole Jr. 1937. The birds and mammals of the western slope of the Azuero Peninsula (Republic of Panama). *Sci. Publ. Cleve. Mus. Nat. Hist.* 7:1–196.

Alexander, B., C. Lozano, D. C. Barker, S. H. E. Mc-Cann, and G. H. Adler. 1998. Detection of *Leishmania* (*Viannia*) *braziliensis* complex in wild mammals from Colombian coffee plantations by PCR and DNA hybridization. *Acta Trop.* 69:41–50.

Alfaro, A. 1896. *Primera exposición Centroamericana de Guatemala.* San José, Costa Rica: Museo Nacional.

Alho, C. J. R. 1981. Small mammal populations of Brazilian Cerrado: the dependence of abundance and diversity on habitat complexity. *Rev. Bras. Biol.* 41:223–30.

———. 1982. Brazilian rodents: their habitats and habits. In *Mammalian biology in South America*, ed. M. A. Mares and H. H. Genoways, 377–92. Pittsburgh: Pymatuning Laboratory of Ecology, University of Pittsburgh, xii + 1–539.

Alho, C. J. R., and L. A. Pereira. 1985. Population biology of a Cerrado rodent community in central Brazil. *Rev. Bras. Biol.* 45:597–607.

Alho, C. J. R., L. A. Pereira, and A. C. Paula. 1987. Patterns of habitat utilization by small mammal populations in cerrado biome of central Brazil. *Mammalia* 50:447–60. [1986 vol. 50, no. 4: pp. 425ff. "Achevé d'imprimer de 15 avril 1987."]

Alho, C. J. R., and O. M. M. Villela. 1984. Scansorial ability in *Oryzomys eliurus* and *Oryzomys subflavus* (Rodentia: Cricetidae) from the Cerrado. *Rev. Bras. Biol.* 44:403–8.

Aliaga-Rossel, E., R. W. Kays, and J. M. V. Fragoso. 2008. Home-range use by the Central American agouti (*Dasyprocta punctata*) on Barro Colorado Island, Panama. *J. Trop. Ecol.* 24:367–74.

Aliaga-Rossel, E., R. S. Moreno, R. W. Kays, and J. Giacaione. 2006. Ocelot (*Leopardus pardalis*) predation on agouti (*Dasyprocta punctata*). *Biotropica* 38:691–94.

Aliaga-Rossel, E., R. Ríos-Uzeda, and J. Salazar-Bravo. 2009. Observations on the growth of *Chinchillula*

sahamae (Rodentia, Sigmodontinae) in captivity. *Mammalia* 73:261–63.

Al-kahtani, M. A., C. Zuleta, E. Caviedes-Vidal, and T. Garland Jr. 2004. Kidney mass and relative medullary thickness of rodents in relation to habitat, body size, and phylogeny. *Physiol. Biochem. Zool.* 77:346–65.

Allamand, J. N. S. 1778. Le Lérot à queue dorée. In *Histoire naturelle, générale et particulière, servant de suite à l'histoire des animaux quadrupèdes par feu M. Le Compte de Buffon*, ed. G. L. L. Buffon, 164–66, plate 72. Amsterdam: J. H. Schneider.

Allen, G. M. 1902. The mammals of Margarita Island, Venezuela. *Proc. Biol. Soc. Washington* 15:91–97.

———. 1942. *Extinct and vanishing mammals of the Western Hemisphere with marine species of all cceans.* American Committee for International Wild Life Protection, Special Publication 11. Lancaster, PA: Intelligencer Printing, xv + 1–620.

Allen, J. A. 1877. Monograph XI. Sciuridae. In *Monographs of North American Rodentia*, ed. E. Coues and J. A. Allen, 631–940. Washington, DC: U.S. Geological Survey 11:i–xii, i–x, 1087 pp.

———. 1891a. Descriptions of two supposed new species of mice from Costa Rica and Mexico, with remarks on *Hesperomys melanophrys* of Coues. *Proc. U.S. Natl. Mus.* 14:193–95.

———. 1891b. Notes on a collection of mammals from Costa Rica. *Bull. Am. Mus. Nat. Hist.* 3:203–18.

———. 1891c. Notes on new or little-known North American mammals, based on recent additions to the collection of mammals in the American Museum of Natural History. *Bull. Am. Mus. Nat. Hist.* 3:263–310.

———. 1892. On a small collection of mammals from the Galapagos Islands, collected by Dr. G. Baur. *Bull. Am. Mus. Nat. Hist.* 4:47–50.

———. 1893. Further notes on Costa Rica mammals, with description of a new species of *Oryzomys*. *Bull. Am. Mus. Nat. Hist.* 5:237–40.

———. 1894a. Descriptions of ten new North American mammals, and remarks on others. *Bull. Am. Mus. Nat. Hist.* 13:317–32.

———. 1894b. Recent progress in the study of North American mammals. *Proc. Linn. Soc. N.Y.* 1894:17–45.

———. 1895a. On the names of mammals given by Kerr in his "Animal Kingdom." published in 1792. *Bull. Am. Mus. Nat. Hist.* 7:179–92.

———. 1895b. Descriptions of new American mammals. *Bull. Am. Mus. Nat. Hist.* 7:327–40.

———. 1897a. Additional notes on Costa Rican mammals, with descriptions of new species. *Bull. Am. Mus. Nat. Hist.* 9:31–44.

———. 1897b. Further notes on mammals collected in Mexico by Dr. Audley C. Buller, with descriptions of new species. *Bull. Am. Mus. Nat. Hist.* 9:47–58.

———. 1897c. On a small collection of mammals from Peru, with descriptions of new species. *Bull. Am. Mus. Nat. Hist.* 9:115–19.

———. 1897d. Description of a new species of *Sigmodon* from Bogota, Colombia. *Bull. Am. Mus. Nat. Hist.* 9:121–22.

———. 1899a. Descriptions of five new American rodents. *Bull. Am. Mus. Nat. Hist.* 12:11–17.

———. 1899b. New rodents from Colombia and Venezuela. *Bull. Am. Mus. Nat. Hist.* 12:195–218.

———. 1899c. The generic names *Echimys* and *Loncheres*. *Bull. Am. Mus. Nat. Hist.* 12:257–64.

———. 1900a. On mammals collected in southeastern Peru, by H. H. Keays, with descriptions of new species. *Bull. Am. Mus. Nat. Hist.* 13:219–27.

———. 1900b. The proper name of the viscacha. *Proc. Biol. Soc. Washington* 13:183–84.

———. 1901a. Descriptions of two new species of South American Muridae. *Bull. Am. Mus. Nat. Hist.* 14:39–40.

———. 1901b. On a further collection of mammals from southeastern Peru, collected by Mr. H. H. Keays, with descriptions of new species. *Bull. Am. Mus. Nat. Hist.* 14:41–46.

———. 1901c. New South American Muridae and a new *Metachirus*. *Bull. Am. Mus. Nat. Hist.* 14:405–12.

———. 1901d. The proper generic names of the viscacha, chinchillas, and their allies. *Proc. Biol. Soc. Washington* 19:181–82.

———. 1902a. Zimmerman's "Zoologie Geographicae" and "Geographische Geschichte" considered in their relation to mammalian nomenclature. *Bull. Am. Mus. Nat. Hist.* 16:13–22.

———. 1902b. Mammal names proposed by Oken in his "Lehrbuch der Zoologie." *Bull. Am. Mus. Nat. Hist.* 16:373–79.

———. 1902c. A further note on the name of the Argentine viscacha. *Proc. Biol. Soc. Washington.* 15:196.

———. 1903a. Description of a new species of *Sigmodon* from Ecuador. *Bull. Am. Mus. Nat. Hist.* 19:99–100.

———. 1903b. Descriptions of new rodents from southern Patagonia, with a note on the genus *Euneomys* Coues, and an addendum to article IV, on Siberian mammals. *Bull. Am. Mus. Nat. Hist.* 19:185–96.

———. 1904a. Mammals from southern Mexico and central South America. *Bull. Am. Mus. Nat. Hist.* 20:29–80.

———. 1904b. New mammals from Venezuela and Colombia. *Bull. Am. Mus. Nat. Hist.* 20:327–35.

———. 1904c. Report on mammals from the District of Santa Marta, Colombia, collected by Mr. Herbert H. Smith, with field notes by Mr. Smith. *Bull. Am. Mus. Nat. Hist.* 20:407–68.

———. 1905. The mammalia of southern Patagonia. In *Reports on the Princeton University Expeditions to Patagonia, 1896–1899*. Vol. 3, *Zoölogy*, ed. W. B. Scott, Part 1, 1–210, 29 pls. Princeton, NJ: Princeton University/Stuttgart: E. Schweizwerbart'sche Werlagshandlung (E. Nägele).

———. 1908. Mammals from Nicaragua. *Bull. Am. Mus. Nat. Hist.* 24:647–70.

———. 1910a. Collation of Brisson's genera of birds with those of Linnaeus. *Bull. Am. Mus. Nat. Hist.* 27:317–35.

———. 1910b. Additional mammals from Nicaragua. *Bull. Am. Mus. Nat. Hist.* 28:87–115.

———. 1910c. Mammals from the Caura District of Venezuela, with description of a new species of *Chrotopterus*. *Bull. Am. Mus. Nat. Hist.* 28:145–49.

———. 1911. Mammals from Venezuela collected by Mr. M. A. Carriker Jr., 1909–1911. *Bull. Am. Mus. Nat. Hist.* 30:239–73.

———. 1912. Mammals from western Colombia. *Bull. Am. Mus. Nat. Hist.* 31:71–95.

———. 1913a. New mammals from Colombia and Ecuador. *Bull. Am. Mus. Nat. Hist.* 32:469–84.

———. 1913b. Revision of the *Melanomys* group of American Muridae. *Bull. Am. Mus. Nat. Hist.* 32:533–54.

———. 1913c. New South American Muridae. Bull. Am. Mus. Nat. Hist. 32:597–604.

———. 1914a. Review of the genus *Microsciurus*. *Bull. Am. Mus. Nat. Hist.* 33:145–65.

———. 1914b. Two new mammals from Ecuador. *Bull. Am. Mus. Nat. Hist.* 33:199–200.

———. 1914c. New South American bats and a new octodont. *Bull. Am. Mus. Nat. Hist.* 33:381–89.

———. 1914d. New South American Sciuridae. *Bull. Am. Mus. Nat. Hist.* 33:585–97.

———. 1915a. Review of the South American Sciuridae. *Bull. Am. Mus. Nat. Hist.* 34:147–309.

———. 1915b. New South American mammals. *Bull. Am. Mus. Nat. Hist.* 34:625–34.

———. 1916a. New South American mammals. *Bull. Am. Mus. Nat. Hist.* 35:83–87.

———. 1916b. List of mammals collected in Colombia by the American Museum of Natural History Expeditions, 1910–1915. *Bull. Am. Mus. Nat. Hist.* 35:191–238.

———. 1916c. New mammals collected on the Roosevelt Brazilian Expedition. *Bull. Am. Mus. Nat. Hist.* 35:523–30.

———. 1916d. Mammals collected on the Roosevelt Brazillian Expedition, with field notes by Leo E. Miller. *Bull. Am. Mus. Nat. Hist.* 35:559–610.

Allen, J. A., and F. M. Chapman. 1893. On a collection of mammals from the island of Trinidad, with descriptions of new species. *Bull. Am. Mus. Nat. Hist.* 5:203–34.

———. 1897a. On a second collection of mammals from the island of Trinidad, with descriptions of new species, and a note on some mammals from the island of Dominica, W. I. *Bull. Am. Mus. Nat. Hist.* 9:13–30.

———. 1897b. On a collection of mammals from Jalapa and Las Vigas, state of Veracruz, Mexico. *Bull. Am. Mus. Nat. Hist.* 9:197–208.

Almeida, C. R. de, A. M. Paiva de Almeida, and D. P. Brasil. 1982. Observations sur lê comportement de fouissement de *Zygodontomys lasiurus pixuna* Moojen, 1943. Reproduction au laboratoire (Rongeurs, Cricetidés). *Mammalia* 45:415–21. [1981 Volume 45, no. 4:403ff) "Achevé d'imprimer le 15 février 1982."]

Almeida, E. J. C. 1980. "Variabilidade citogenética nos gêneros *Oryzomys* e *Thomasomys* (Cricetidae, Rodentia)." Ph.D. diss., Universidade de São Paulo, Brazil.

Almeida, E. J. C., and Y. Yonenaga-Yassuda. 1974. Estudo citogenético em roedores do gênero *Oryzomys*. *Ciênc. Cult.* 26:243.

Almeida, F. C., C. R. Bonvicino, and P. Cordeiro-Estrela. 2007. Phylogeny and temporal diversification of *Calomys* (Rodentia, Sigmodontinae): implications for the biogeography of an endemic genus of the open/dry biomes of South America. *Mol. Phylogenet. Evol.* 42:449–66.

Almeida, L. B. de, and M. Galetti. 2007. Seed dispersal and spatial distribution of *Attalea geraensis* (Arecaceae) in two remnants of Cerrado. *Acta Oecol.* 32:180–87.

Almendra, A. L., D. S. Rogers, and F. X. González-Cózatl. 2014. Molecular phylogenetics of the *Handleyomys chapmani* complex in Mesoamerica. *J. Mamm.* 95:26–40.

Alonso, L. E., A. Alonso, T. S. Schulenberg, and F. Dallmeier, eds. 1999. *Biological assessment of the Cordillera Vilcabamba, Perú.* RAP Working Papers 12. Washington, DC: Conservation International., 296 pp.

Alston, E. R. 1876a. On the classification of the order Glires. *Proc. Zool. Soc. Lond.* 1876, part 1:61–98.

———. 1876b. On the genus *Dasyprocta*; with description of a new species. *Proc. Zool. Soc. Lond.* 1876, part 2:347–53.

———. 1877. On two new species of *Hesperomys*. *Proc. Zool. Soc. Lond.* 1876, part 4:755–57. [Part 4 of the 1876 volume of the *Proceedings* was published in April 1877; see Duncan 1937.]

———. 1878. On the squirrels of the Neotropical region. *Proc. Zool. Soc. Lond.* 1878, part 3:656–70.

———. 1880a. Mammalia. In *Biologia Centrali-Americana. Zoology, Botany, and Archaeology*, ed. F. D. Godman and O. Salvin, part 8, 177–200. London: Taylor and Francis, 2:xx+220 pp.+22 pls. [Printed in nine parts between September 1879 and October 1881; prepage materials dated December 1882.]

———. 1880b. Note on the *Perognathus bicolor* of Gray. *Ann. Mag. Nat. Hist.*, ser. 5, 6:118–19.

Altuna, C. A. 1985. Microclima de cuevas de *Ctenomys pearsoni* (Rodentia, Octodontidae) en arroyo Carrasco (Montevideo). *Actas J. Zool. Urug.* 1985:59–60.

Altuna, C. A., L. Bacigalupe, and S. Corte. 1998. Food-handling and feces reingestion in *Ctenomys pearsoni* (Rodentia, Ctenomyidae) from Uruguay. *Acta Theriol.* 43:433–37.

Altuna, C. A., G. Francescoli, G. Izquierdo, and B. Tassino. 1999. Ecoetología y conservación de mamíferos subterráneos de distribución restringida: el caso de *Ctenomys pearsoni* (Rodentia, Octodontidae) en el Uruguay. *Etologia* 7:47–54.

Altuna, C. A., and E. P. Lessa. 1985. Penial morphology in Uruguayan species of *Ctenomys* (Rodentia: Octodontidae). *J. Mamm.* 66:483–88.

Altuna, C. A., M. Ubilla, and E. P. Lessa. 1985. Estado actual del conocimiento de *Ctenomys rionegrensis* Langguth y Abella, 1970 (Rodentia: Octodontidae). *Actas J. Zool. Urug.* 1:8–9.

Alvarado-Serrano, D. F. 2005. "Caracterización morfométrica y distribución del género Akodon (Muridae: Sigmodontinae) en Ecuador." Undergraduate thesis, Pontificia Universidad Católica del Ecuador, Quito, Ecuador, 175 pp.

Alvarado-Serrano, D. F., and G. D'Elía. 2013. A new genus for the Andean mice *Akodon latebricola* and *A. bogotensis* (Rodentia: Sigmodontinae). *J. Mamm.* 94:995–1015.

Alvarado-Serrano, D. F., L. Luna, and L. L. Knowles. 2013. Localized versus generalist phenotypes in a broadly distributed tropical mammal: how is intraspecific variation distributed across disparate environments? *BMC Evol. Biol.* 13:160.

Alvarenga, C. A., and S. A. Talamoni. 2005. Nests of the Brazilian squirrel *Sciurus ingrami* Thomas (Rodentia Sciuridae). *Rev. Bras. Zool.* 22:816–18.

———. 2006. Foraging behavior of the Brazilian squirrel *Sciurus aestuans* Linnaeus, 1766 (Rodentia, Sciuridae). *Acta Theriol.* 51:69–74.

Álvarez-Castañeda, S. T., and P. Cortés-Calva. 2003. Peromyscus pembertoni. *Mamm. Species* 734:1–2.

Amante, E. 1975. Prejuízos causados pelo roedor *Clyomys laticeps* (Echimydae) em *Pinus* spp., em áreas reflorestadas de Sacramento, Minas Gerais. In *Resumos 27a: Reunião Anual da Sociedade Brasileira para o Progresso da Ciência*, 373–74. São Paulo: Sociedade Brasileira para o Progresso da Ciência.

Ameghino, F. 1887. Enumeración sistemática de las especies de mamíferos fósiles coleccionados por Cárlos Ameghino en las terrenos eocenos de la Patagonia austral y depositados en el Museo de La Plata. *Bol. Mus. La Plata* l:1–26.

———. 1889. Contribución al conocimiento de los mamiferos fósiles de la República Argentina. *Actas Acad. Nac. Cienc. Repub. Argent. Córdoba, Buenos Aires* 6:xxxii+11027 pp.; atlas, 98 pls., each with text.

———. 1891. Mamíferos y aves fósiles Argentinas: especies nuevas, adiciones y correcciones. *Rev. Argent. Hist. Nat.* 1:240–59.

———. 1898. Sinopsis geológico-paleontológica de la Argentina. In *Segundo censo Nacional de la República Argentina*, ed. D. G. de la Fuente for the Comisión Directiva del Censo, Tomo 1 territorio, capitulo 1, tecera parte, pp. 111–225. Buenos Aires: Taller tip. de la Penitenciaria nacional.

———. 1902. Première contribution a la connaissance de la faune mammalogique des couches a Colpodon. *Bol. Acad. Nac. Cienc. Repub. Argent. Córdoba, Buenos Aires* 17:71–138.

———. 1908. Las formaciones sedimentarias de la región litoral de Mar del Plata y Chapalmalán. *An. Mus. Nac. Buenos Aires* 3:343–428.

Amico, G. C., and M. A. Aizen. 2000. Mistletoe seed dispersal by a marsupial. *Nature* 408:929–30.

Amman, B. R., and R. D. Bradley. 2004. Molecular evolution in *Baiomys* (Rodentia: Sigmodontinae): evidence for a genetic subdivision in *B. musculus*. *J. Mamm.* 85:162–66.

Anderson, R. P. 1999 [2000]. Preliminary review of the systematics and biogeography of the spiny pocket mice (*Heteromys*) of Colombia. *Rev. Acad. Colomb. Cienc. Exact. Fis. Nat.* 23(Suppl.):613–30.

———. 2003. Taxonomy, distribution, and natural history of the genus *Heteromys* (Rodentia: Heteromyidae) in western Venezuela, with the description of a dwarf species from the Península de Paraguaná. *Am. Mus. Novit.* 3396:1–43.

Anderson, R. P., and E. E. Gutiérrez. 2009. Taxonomy, distribution, and natural history of the genus *Heteromys* (Rodentia: Heteromyidae) in central and eastern Venezuela, with the description of a new species from

the Cordillera de la Costa. In *Systematic Mammalogy: Contributions in Honor of Guy G. Musser*, ed. R. S. Voss and M. D. Carleton. *Bull. Am. Mus. Nat. Hist.* 331:33–93.

Anderson, R. P., E. E. Gutiérrez, J. Ochoa-G., F. J. García, and M. Aguilera. 2012. Faunal nestedness and species–area relationship for small non-volant mammals in "sky islands" of northern Venezuela. *Stud. Neotrop. Fauna Environ.* 47:157–70.

Anderson, R. P., and P. Jarrín-V. 2002. A new species of spiny pocket mouse (Heteromyidae: *Heteromys*) endemic to western Ecuador. *Am. Mus. Novit.* 3382:1–26.

Anderson, R. P., and E. Martínez-Meyer. 2004. Modeling species' geographic distributions for preliminary conservation assessments: an implementation with the spiny pocket mice (*Heteromys*) of Ecuador. *Biol. Conserv.* 116:167–79.

Anderson, R. P., and A. Raza. 2010. The effect of the extent of the study region on GIS models of species geogrpahic distributions and estimates of niche evolution: preliminary tests with montane rodents (genus *Nephelomys*) in Venezuela. *J. Biogeogr.* 37:1378–93.

Anderson, R. P., and P. J. Soriano. 1999. The occurrence and biogeographic significance of the southern spiny pocket mouse *Heteromys australis* in Venezuela. *Z. Säugetierk.* 64:121–25.

Anderson, R. P., and R. M. Timm. 2006. A new montane species of spiny pocket mouse (Rodentia: Heteromyidae: *Heteromys*) from northwestern Costa Rica. *Am. Mus. Novit.* 3509:1–38.

Anderson, R. P., M. Weksler, and D. S. Rogers. 2006. Phylogenetic analyses of spiny pocket mice (Heteromyidae: Heteromyinae) based on allozymic and morphological data. *J. Mamm.* 87:1218–33.

Anderson, S. 1985. Lista preliminar de mamíferos bolivianos. *Cuad. Acad. Nac. Cienc. Boliv.* 65:5–16.

———. 1993. *Los mamíferos Bolivianos: notas de distribución y claves de identificación*. La Paz: Publicación Especial del Instituto de Ecología, Colección Boliviana de Fauna, 159 pp.

———. 1997. Mammals of Bolivia, taxonomy and distribution. *Bull. Am. Mus. Nat. Hist.* 231:1–652.

Anderson, S., and J. K. Jones Jr., eds. 1967. *Recent Mammals of the World: A Synopsis of Families*. New York: Ronald Press, 453 pp.

Anderson, S., K. F. Koopman, and G. K. Creighton. 1982. Bats of Bolivia: An annotated checklist. *Am. Mus. Novit.* 2750:1–24.

Anderson, S., and N. Olds. 1989. Notes on Bolivian mammals. 5. Taxonomy and distribution of *Bolomys* (Muridae, Rodentia). *Am. Mus. Novit.* 2935:1–22.

Anderson, S., and T. L. Yates. 2000. A new genus and species of phyllotine rodent from Bolivia. *J. Mamm.* 81:18–36.

Anderson, S., T. L. Yates and J. A. Cook. 1987. Notes on Bolivian mammals 4: the genus *Ctenomys* (Rodentia, Ctenomyidae) in the eastern lowlands. *Am. Mus. Novit.* 2891:1–20.

Andrade, A. 2008. Mammalia, Rodentia, Cricetidae, *Notiomys edwardsii* (Thomas, 1890): distribution extension and geographic distribution map. *Check List* 4:33–36.

Andrade, A., and P. Teta. 2003. Micromamíferos (Rodentia y Didelphimorphia) del Holoceno Tardío del sitio arqueológico alero Santo Rosario (provincia de Río Negro, Argentina). *Atek Na* 1:273–87.

Andrade, A., P. Teta, and C. Panti. 2002. Oferta de presas y composición de la dieta de *Tyto alba* (Aves: Tytonidae) en el sudoeste de la provincia de Río Negro, Argentina. *Hist. Nat. (Corrientes)* 1:9–15.

Andrade, A., D. E. Udrizar Sauthier, and U. F. J. Pardiñas. 2004. Vertebrados depredados por la Lechucita Vizcachera (*Athene cunicularia*) en la meseta de Somuncurá (Río Negro, Argentina). *Hornero* 19:91–93.

Andrade, A. F. B. de, and C. R. Bonvicino. 2003. A new karyological variant of *Oecomys* (Rodentia: Sigmodontinae) and its phylogenetic relationship based on molecular data. *Genome* 46:195–203.

Andrade, A. F. B. de, C. R. Bonvicino, D. C. Briani, and S. Kasahara. 2004. Karyologic diversification and phylogenetics relationships of the genus *Thalpomys* (Rodentia, Sigmodontinae). *Acta Theriol.* 49:181–90.

Andrade, M. A. F. de, and A. U. Christoff. 2001. Roedores equimídeos (Rodentia: Echimyidae) do Rio Grande do Sul. In *Livro de Resumos.13th Salão de Iniciação Científica (12.2000, Porto Alegre)*, Resumo 053, p. 315. Porto Alegre: UFRGS, http://hdl.handle.net/10183/80397.

Andrade-Lima, D. de. 1982. Present-day forest refuges in northeastern Brazil. In *Biological Diversification in the Tropics*, ed. G. T. Prance, 245–51. New York: Columbia University Press, 741 pp.

Andrades-Miranda, J., L. F. B. Oliveira, C. A. V. Lima-Rosa, D. A. Sana, A. P. Nunes, and M. S. Mattevi. 2002. Genetic studies in representatives of genus *Rhipidomys* (Rodentia, Sigmodontinae) from Brazil. *Acta Theriol.* 47:125–35.

Andrades-Miranda, J., L. F. B. Oliveira, C. A. V. Lima-Rosa, A. P. Nunes, N. I. T. Zanchin, and M. S. Mattevi. 2001. Chromosome studies of seven species of *Oligoryzomys* (Rodentia: Sigmodontinae) from Brazil. *J. Mamm.* 82:1080–91.

Andrades-Miranda, J., L. F. B. de Oliveira, N. I. T. Zanchin, and M. S. Mattevi. 2001. Chromosomal description of the rodent genera *Oecomys* and *Nectomys* from Brazil. *Acta Theriol.* 46:269–78.

Andrades-Miranda, J., N. I. T. Zanchin, L. F. B. de Oliveira, A. R. Langguth, and M. S. Mattevi. 2002. Cytogenetic studies in nine taxa of the genus *Oryzomys* (Rodentia, Sigmodontinae) from Brazil. *Mammalia* 65:461–72. DOI: 10.1515/mamm.2001.65.4.461. [2001 volume 65, no. 4:417ff. "Achevé d'imprimer lell 01 (=janvier) 2002."]

Aniskin, V. M. 1993. Three new karyotypes of prickly chinchillas of the family Echimyidae (Rodentia). *Genetika* 29:1500–7. Article in Russian.

———. 1994. Karyological characterization of mammals from three region of the Republic of Peru. In *Mammals of Peruvian Amazonia*, ed. M. Hayka, 33–47. Moscow: Nauka, 301 pp. Text in Russian.

Aniskin, V. M., V. M. Malygin, A. A. Warshavski, and S. I. Isyev. 1991. Karyological interrelations of three chromosomes forms of spiny rats of the genus *Proechimys* (Rodentia, Echimyidae). *Genetika* 27:1066–75. Article in Russian.

Aniskin, V. M., and V. T. Volobouev. 1999. Comparative chromosomal banding of two South American species of rice rats of the genus *Oligoryzomys* (Rodentia, Sigmodontinae). *Chromosome Res.* 7:557–62.

Anthony, H. E. 1916. Panama mammals collected in 1914–1915. *Bull. Am. Mus. Nat. Hist.* 35:357–75.

———. 1920. New rodents and new bats from neotropical regions. *J. Mamm.* 1:81–86.

———. 1921a. New mammals from British Guiana and Colombia. *Am. Mus. Novit.* 19:1–7.

———. 1921b. Preliminary report on Ecuadorean mammals. No. 1. *Am. Mus. Novit.* 20:1–6.

———. 1921c. Mammals collected by William Beebe at the British Guiana Tropical Research Station. *Zoologica* 3:265–85.

———. 1922. Preliminary report on Ecuadorean mammals. No. 2. *Am. Mus. Novit.* 32:1–7.

———. 1923a. Mammals from Mexico and South America. *Am. Mus. Novit.* 54:1–10.

———. 1923b. Preliminary report on Ecuadorean mammals. No. 3. *Am. Mus. Novit.* 55:1–14.

———. 1924a. Preliminary report on Ecuadorean mammals. No. 4. *Am. Mus. Novit.* 114:1–6.

———. 1924b. Preliminary report on Ecuadorean mammals. No. 6. *Am. Mus. Novit.* 139:1–9.

———. 1925. New species and subspecies of *Thomasomys*. *Am. Mus. Novit.* 178:1–4.

———. 1926a. Two new rodents from Bolivia. *Am. Mus. Novit.* 178:1–4.

————. 1926b. Preliminary report on Ecuadorean mammals. No. 7. *Am. Mus. Novit.* 240:1–6.

————. 1929. Two new genera of rodents from South America. *Am. Mus. Novit.* 383:1–6.

————. 1932. A new species of *Thomasomys* from Venezuela. *Am. Mus. Novit.* 548:1–2.

Anthony, H. E., and G. H. H. Tate. 1935. Notes on South American Mammalia. No. 1. *Sciurillus*. *Am. Mus. Novit.* 780:1–13.

Antinuchi, C. D., and C. Busch. 1992. Burrow structure in the subterranean rodent *Ctenomys talarum*. *Z. Säugetierk.* 57:163–68.

————. 1999. Intrageneric comparisons in water deprivation responses, urine concentrating capacity and renal morphology among three species of *Akodon* from different geographic rainfall regimens. *Z. Säurgetierkd.* 64:277–84.

————. 2000. Tasas metabólicas y características termorregulatorias de *Akodon azarae* (Rodentia: Sigmodontinae). *Rev. Chilena Hist. Nat.* 73:131–38.

Antinuchi, C. D., R. R. Zenuto, F. Luna, A. P. A. P. Cutrera, P. P. Perissinotti, and C. Busch. 2007. Energy budget in subterranean rodents: insights from the tuco-tuco *Ctenomys talarum* (Rodentia: Ctenomyidae). In *The quintessential naturalist: Honoring the life and legacy of Oliver P. Pearson*, ed. D. A. Kelt, E. P. Lessa, J. Salazar-Bravo, and J. L. Patton, 111–39. *Univ. California Publ. Zool.* 134:v-xii + 1–981.

Antoine, P.-O., L. Marivaux, D. A. Croft, G. Billet, M. Ganerød, C. Jaramillo, T. Martin, M. J. Orliac, J. Tejada, A. J. Altamirano, F. Duranthon, G. Fanjat, S. Rousse, and R. S. Gismondi. 2012. Middle Eocene rodents from Peruvian Amazonia reveal the pattern and timing of caviomorphs origns and biogeography. *Proc. R. Soc. Lond. B Biol. Sci.* 279:1319–26.

Antunes, P. C., M. A. A. Campos, L. G. R. Oliveira-Santos, and M. E. Graipel. 2009. Population dynamics of *Euryoryzomys russatus* and *Oligoryzomys nigripes* (Rodentia, Cricetidae) in an Atlantic Forest area, Santa Catarina Island, southern Brazil. *Biotemas* 22:143–51.

Apfelbaum, L. I., and A. Blanco. 1985. Temporal variation of allele frequencies in populations of *Akodon dolores* (Rodentia, Cricetidae). *Theoret. Appl. Genet.* 70:569–72.

Apfelbaum, L. I., R. C. Liascovich, and O. A. Reig. 1993. Relaciones citogenéticas y genético-alozímicas entre roedores akodontinos (Cricetidae: Sigmodontinae). *Bol. Soc. Zool. Urug.*, 2nd ser., 8:62–78.

Apfelbaum, L. I., A. I. Massarini, L. E. Daleffe, and O. A. Reig. 1991. Genetic variability in the subterranean rodents *Ctenomys australis* and *Ctenomys porteousi* (Rodentia: Octodontidae). *Biochem. System. Ecol.* 19:467–76.

Apfelbaum, L. I., and O. A. Reig. 1989. Allozyme genetic distances and evolutionary relationships in species of akodontine rodents (Cricetidae: Sigmodontinae). *Biol. J. Linn. Soc.* 38:257–80.

Arana, M., O. Ramirez, S. Santa Maria, C. Kunimoto, R. Velarde, C. De La Cruz, and M. L. Ruiz. 2002. Population density and reproduction of two Peruvian leaf-eared mice (*Phyllotis* spp.). *Rev. Chilena Hist. Nat.* 75:751–56.

Arana-Cardó, R., and C. Ascorra. 1994. Observaciones sobre la distribución de algunos sigmodontinos (Rodentia, Muridae) altoandinos del departamento de Lima, Perú. *Publ. Mus. Hist. Nat. Univ. Nac. Mayor de San Marcos A Zool.* 47:1–7.

Araripe, L. O., P. Aprigliano, N. Olifierso, P. Borodin, and R. Cerqueira. 2006. Comparative analysis of life-history traits in two species of *Calomys* (Rodentia: Sigmodontinae) in captivity. *Mammalia* 70:2–8.

Arata, A. 1964. The anatomy and taxonomic significance of the male accessory reproductive glands of muroid rodents. *Bull. Fla. State Mus. Biol. Sci.* 9:1–42.

Araújo, D. S. D. 1992. Vegetation types of sandy coastal plains of tropical Brazil: a first approximation. In *Coastal Plant Communities of Latin America*, ed. U. Seeliger, 337–67. New York: Academic Press, 392 pp.

Araújo, N. P., A. C. Loss, D. A. Cordeiro-Junior, K. R. da Silva, Y. L. R. Leite, and M. Svartman. 2014. New karyotypes of Atlantic tree rats, genus *Phyllomys* (Rodentia: Echimyidae). *Genome* 57:1–8.

Araújo, T. A. Z. V., R. J. Young, and M. O. G. Lopes. 2006. Descrição do comportamento em cativeiro de *Rhipidomys mastacalis* (Lund, 1840). 26th *Congresso Brasileiro de Zoologia;* Londrina, Brazil; February 12–17, 2006, abstract 1175.

Arcay, L. 1982. Nuevos Coccidia de roedores silvestres de Venezuela: *Eimeria guerlingueti* sp. nov. y *Wenyonella maligna* sp. nov. de *Sciurus* (*Guerlinguetus*) *granatensis* y *Eimeria akodoni* sp. nov., de *Akodon urichi venezuelensis* (Rodentia, Myomorpha, Cricetidae, Cricetinae). *Acta Biol. Venezuel.* 11:125–48.

Arcos D., R., L. Abuja, and P. Moreno. 2007. Nuevos registros y ampliacíon del rango de distribución de algunos mamíferos del Ecuador. *Rev. Politéc. Ser. Biol.* 27:126–32.

Ardiles, A. O., J. Ewer, M. L. Acosta, A. Kirkwood, A. D. Martinez, L. A. Ebensperger, F. Bozinovic, T. M. Lee, and A. G. Palacios. 2013. *Octodon degus* (Molina 1782): a model in comparative biology and biomedicine. *Cold Spring* Harb. Protoc. 4:312–18.

Ardiles A. O., C. C. Tapia-Rojas, M. Mandal, F. Alexandre, A. Kirkwood, N. C. Inestrosa, and A. G. Palacios. 2012. Post-synaptic dysfunction is associated with spatial and object recognition memory loss in a natural model of Alzheimer's disease. *Proc. Natl. Acad. Sci. USA* 109:13835–40.

Arechavaleta, J. 1882. Reino animal. In *Album de la República Oriental del Uruguay presentado en la Exposición Continental de Buenos Aires*, ed. F. A. Berra, A. de Vedia, and C. M. de Pena, 41–54. Montevideo: Rius y Becchi, 341 pp + 20 maps.

Arellano, E., F. X. González-Cózatl, and D. S. Rogers. 2005. Molecular systematics of Middle American harvest mice *Reithrodontomys* (Muridae), estimated from mitochondrial cytochrome-*b* gene sequences. *Mol. Phylogenet. Evol.* 37:529–40.

Arellano, E., D. S. Rogers, and F. A. Cervantes. 2003. Genic differentiation and phylogenetic relationships among tropical harvest mice (*Reithrodontomys:* subgenus *Aporodon*). *J. Mamm.* 84:129–43.

Arends, A., and B. K. McNab. 2001. The comparative energetics of 'caviomorph' rodents. *Comp. Biochem. Physiol. A. Mol. Int. Physiol.* 130:105–22.

Argüelles, C. F., P. Suárez, M. D. Giménez, and C. J. Bidau. 2001. Intraspecific chromosome variation between different populations of *Ctenomys dorbignyi* (Rodentia, Ctenomyidae) from Argentina. *Acta Theriol.* 46:363–73.

Argyle, E. C., and M. J. Mason. 2008. Middle ear structures of *Octodon degus* (Rodentia: Octodontidae), in comparison with those of subterranean caviomorphs. *J. Mamm.* 89:1447–55.

Arias, S. M., N. Madanes, and R. D. Quintana. 2003. Vegetation structure and composition in active and inactive vizcacheras in the Parana River Delta region. *Mastozool. Neotrop.* 10:9–20.

Armada, J., L. A. Pereira, P. G. S. Filho, and H. N. Seuánez. 1984. Número cromosómico de *Akodon reinhardti* Langguth, 1975 (= *Thalpomys lasiotus* Thomas, 1916) (Rodentia, Cricetinae). *Ciênc. Cult.* 36:765.

Armond, A. D., and E. P. Chagas. 1990. Alguns aspectos da biologia do ouriço-cacheiro (*Sphiggurus villosus*). In *Resumos: 27th Congresso Brasileiro de Zoologia*, 224. Londrina: Sociedade Brasileira de Zoologia.

Arvy, L. 1974. Contribution à la connaissance de l'appareil mammaire chez les rongeurs. *Mammalia* 38:108–38.

Asfora, P. H., and A. R. M. Pontes. 2009. The small mammals of the highly impacted North-eastern Atlantic Forest of Brazil, Pernambuco Endemism Center. *Biota Neotrop.* 9:31–35.

Asfora, P. H., A. R. Torre Palma, D. Astúa, and L. Geise. 2011. Distribution of *Oecomys catherinae* Thomas, 1909 (Rodentia: Cricetidae) in northeastern Brazil with karyotypical and morphometrical notes. *Biota Neotrop.* 11:415–24.

Asher, M., T. Lippmann, J. T. Epplen, C. Kraus, F. Trillmich and N. Sachser. 2008. Large males dominate: ecology, social organization, and mating system of wild cavies, the ancestors of the guinea pig. *Behav. Ecol. Sociobiol.* 62:1509–21.

Asher, M., E. Spinelli de Oliveira, and N. Sachser. 2004. Social system and spatial organization of wild guinea pigs (*Cavia aperea*) in a natural population. *J. Mamm.* 85:788–96.

Attias, N., D. S. L. Raíces, F. S. Pessôa, H. G. Albuquerque, T. Jordão-Nogueira, T. C. Modesto, and H. G. Bergallo. 2009. Potential distribution and new records of *Trinomys* species (Rodentia: Echimyidae) in the state of Rio de Janeiro. *Zoologia* 26:305–15.

August, P. V. 1981. "Population and community ecology of small mammals in northern Venezuela." Ph.D. diss., Boston University, Massachusetts.

———. 1984. *Population ecology of small mammals in the llanos of Venezuela*. Special Publications of the Museum 22. Lubbock: Texas Tech University Press, 71–104.

Auricchio, P. 2001. Notes on the tuft-tailed-spine-tree-rat, *Lonchothrix emiliae* Thomas, 1920 (Rodentia, Echimyidae). *Publ. Avulsas Inst. Pau Bras.* 4:21–26.

Autino, A., and M. Lareschi. 1998. Capítulo 27: Orden Siphonaptera. In *Biodiversidad de artrópodos argentinos. Una perspectiva biotaxonómica*, ed. J. Morrone and S. Coscarón, 279–90. La Plata: Ediciones Científicas Americanas, 599 pp.

Avila-Pires, F. D. de. 1959a. Nota prévia sobre uma nova espécie de *Oryzomys* do Brasil (Rodentia, Cricetidae). *Atas Soc. Biol. Rio J.* 3:2.

———. 1959b. Nota prévia sobre nova subespécie de *Oryzomys ratticeps* (Hensel, 1872) (Rodentia, Cricetidae). *Atas Soc. Biol. Rio J.* 3:3.

———. 1960a. Sobre *Oryzomys* do grupo *ratticeps*. In *Congreso Sudamericano de Zoologia*, 1°. Actas y Trabajos, Tomo IV, Sectión V:3–7.

———. 1960b. Roedores colecionados na região de Lagoa Santa, Minas Gerais, Brasil. *Arq. Mus. Nac., Rio de Janeiro* 50:25–45.

———. 1960c. Um novo gênero de roedor sulamericano. *Bol. Mus. Nac., Rio de Janeiro*, nova sér., zool., 220:1–6.

———. 1964. Mamíferos colecionados na região do rio Negro (Amazonas, Brasil). *Bol. Mus. Paraense Emilio Goeldi*, sér. zool., n.s., 42:1–23.

————. 1965. The type specimens of Brazilian mammals collected by Prince Maximilian zu Wied. *Am. Mus. Novit.* 2209:1–21.

————. 1968. Tipos de mamíferos recentes no Museu Nacional, Rio de Janeiro. *Arq. Mus. Nac., Rio de Janeiro* 53:161–91.

————. 1972. A new subspecies of *Kunsia fronto* (Winge, 1888) from Brazil (Rodentia, Cricetidae). *Rev. Bras. Biol.* 32:419–22.

————. 1977. Mamíferos do Parque Nacional do Itatiaia. *Bol. Mus. Nac., Rio de Janeiro*, nova sér., zool., 291:1–29.

————. 1982. Notas taxonômicas sobre Caviidae Brasileiros (Mammalia, Rodentia). *Rev. Nordest. Biol.* 5:259–68.

————. 1994. Mamíferos desritos do Estado do Rio Grande do Sul, Brdasil. *Rev. Bras. Biol.* 54:367–84.

Avila-Pires, F. D. de, and E. Gouveia. 1977. Mamíferos do Parque Nacional do Itatiaia. *Bol. Mus. Nac., Rio de Janeiro*, nova sér., zool., 291:1–29.

Avila-Pires, F. D. de, and M. R. C. Wutke. 1981. Taxonomia e evolução de *Clyomys* Thomas, 1916 (Rodentia, Echimyidae). *Rev. Bras. Biol.* 4:529–34.

Awa, A., M. Sasaki, and S. Takayama. 1959. An in vitro study of the somatic chromosomes in several mammals. *Jpn. J. Zool.* 12:257–65.

Axenovich, T. I., P. S. D'Andrea, F. Fernandes, C. R. Bonvicino, I V. Zorkoltseva, and P. M. Borodin. 2004. Inheritance of white head spotting in natural populations of South American water rat (*Nectomys squamipes* Rodentia: Sigmodontinae). *J. Hered.* 95:76–80.

Azara, F. d.' 1801. *Essais sur l'histoire naturelle des quadrupeds de la province du Paraguay.* Translated by Pra. M. L. E. Moreau-Saint-Mércy. Paris: Charles Pougens, 2:1–499.

————. 1802. Apuntamientos para la historia natural de los quadrúpedos del Paraguay y Río de la Plata. Madrid: La Imprenta de la Viuda de Ibarra, 2:2 (unnumbered), x + 1–328.

Azurduy, H. 2005a. Una nueva especie fósil de *Ctenomys* (Rodentia) y breve panorama paleontológico del género en Bolivia. *Kempffiana* 1:29–39.

————. 2005b. Descripción de una nueva especie de *Ctenomys* (Rodentia, Ctenomyidae) de los valles interandinos de Bolivia. *Kempffiana* 1:70–74.

Bacigalupe, L., D. Bustamante, F. Bozinovic, and R. Nespolo. 2010. Phenotypic integration of morphology and energetic performance under routine capacities: a study in the leaf-eared mouse *Phyllotis darwini*. *J. Comp. Physiol. B.* 180:293–99.

Bacigalupe, L. D., J. Iriarte-Díaz, and F. Bozinovic. 2002. Functional morphology and geographic variation in the digging apparatus of cururos (Octodontidae: *Spalacopus cyanus*). *J. Mamm.* 83:145–52.

Bacigalupe, L. D., E. L. Rezende, G. J. Kenagy, and F. Bozinovic. 2003. Activity and space use by degus: a trade-off between thermal conditions and food availability? *J. Mamm.* 84:311–18.

Bacquet, C., T. Imamura, C. A. Gonzalez, I. Conejeros, G. Kausel, T. M. A. Neildez-Nguyen, A. Paldi, and M. G. Gallardo. 2008. Epigenetic processes in a tetraploid mammal. *Mamm. Genome* 19:439–47.

Badzinski, C., D. Galiano, and J. Marinho. 2012. Mammalia, Rodentia, Cricetidae, *Calomys laucha* (Fischer, 1814): distribution extension in southern Brazil. *Check List* 8:264–66.

Baird, S. F. 1857 [1858]. Mammals. In Reports of Explorations and Surveys to Ascertain the Most Practicable and Economical Route for a Railroad from the Mississippi River to the Pacific Ocean. Part 1. General Report upon the Zoology of the Several Pacific Railroad Routes. Washington, DC: Beverly Tucker, 8(1):xix–xlviii, 1–757, 60 plates..

Baker, R. H., and M. K. Petersen. 1965. Notes on a climbing rat, *Tylomys*, from Oaxaca. *J. Mamm.* 46:694–95.

Baker, R. J., B. F. Koop, and M. W. Haiduk. 1983. Resolving systematic relationship with G-bands: a study of five genera of South American cricetine rodents. *Syst. Zool.* 32:403–16.

Baladron, A. V., A. I. Malizia, and M. S. Bó. 2009. Predation upon the subterranean rodent *Ctenomys talarum* (tuco-tucos) by *Buteo polyosoma* (red-backed hawks) in coastal grasslands of Argentina. *Stud. Neotrop. Fauna Environ.* 44:61–65.

Baldi, R. 2007. Breeding success of the endemic mara *Dolichotis patagonum* in relation to habitat selection: conservation implications. *J. Arid Environ.* 68:9–19.

Ball, S. S., and V. L. Roth. 1995. Jaw muscles of New World squirrels. *J. Morphol.* 224:265–91.

Ballesteros Correa, J., and J. P. Jorgenson. 2009. Aspectos poblacionales del Cacó (*Hydrochoerus hydrochaeris isthmius*) y amenazas para su conservación en el nor-occidente de Colombia. *Mastozool. Neotrop.* 16:27–38.

Balzan, L. 1893. *De Irupana a Covendo, de Covendo a Reyes.* Informes presentados a la Sociedad Geografica Italiana, translated by R. P. Fray Nicolas Armentia. La Paz: Imprenta de "La Revolucion," 58 pp.

————. 2008. *A carretón y canoa. La obra del naturalists Luigi Balzan en Bolivia y Paraguay (1885–1893).* Edición, estudio, notas y traducción de Clara López Beltrán. La Paz: Plural editores, 419 pp.

Bancroft, E. 1769. *An essay on the natural history of Guiana, in South America : containing a description of many curious productions in the animal and vegetable systems of that country. Together with an account of the religion, manners, and customs of several tribes of its Indian from a gentleman of the medical faculty during his residence in that country.* London: T. Becket and P. A. De Hondt, 1 (unnumbered) + i–iv + 402 + 2 (unnumbered).

Bandouk, A. C., and S. F. dos Reis. 1995. Craniometric variation and subspecific differentiation in *Thrichomys apereoides* in northeastern Brazil (Rodentia: Echimyidae). *Z. Säugetierk.* 60:176–85.

Bandouk, A. C., S. F. dos Reis, and B. Bordin. 1996. Cranial phenotypic evolution in *Thrichomys apereoides* (Rodentia: Echimyidae). *J. Zool.* 239:65–71.

Bangs, O. 1898a. Descriptions of some new mammals from Sierra Nevada de Santa Marta, Colombia. *Proc. Biol. Soc. Washington* 12:161–65.

———. 1898b. On *Sciurus variabilis* from the Santa Marta region of Colombia. *Proc. Biol. Soc. Washington* 12:183–86.

———. 1898c. A new *Sigmodon* from the Santa Marta region of Colombia. *Proc. Biol. Soc. Washington* 12:189–90.

———. 1899. A new pygmy *Oryzomys* from the Santa Marta region of Colombia. *Proc. Biol. Soc. Washington* 13:9–10.

———. 1900a. List of the mammals collected in the Santa Marta region of Colombia by W. W. Brown Jr. *Proc. New England Zool. Club* 1:87–102.

———. 1900b. Description of a new squirrel from Panama. *Proc. New England Zool. Club* 2:43–44.

———. 1901. The mammals collected in San Miguel Island, Panama, by W. W. Brown Jr. *Am. Nat.* 35:631–44.

———. 1902. Chiriqui Mammalia. *Bull. Mus. Comp. Zool.* 39:17–51.

———. 1905. The vertebrata of Gorgona Island, Colombia. Introduction and Mammalia. *Bull. Mus. Comp. Zool.* 45:87–102.

Barlow, J. C. 1969. Observations on the biology of rodents in Uruguay. *Life Sci. Contrib. R. Ont. Mus.* 75:1–59.

Barnett, A. A. 1999. Small mammals of the Cajas plateau, southern Ecuador: ecology and natural history. *Bull. Fla. State Mus. Nat. Hist.* 42:161–217.

Barnett, A. A., and A. C. da Cuhna. 1994. Notes on the small mammals of Ilha de Maracá, Roraima State, Brazil. *Mammalia* 58:131–37.

———. 1998. Small mammals of the Ilha de Maracá. In *Maracá: The biodiversity and environment of an Amazonian rainforest*, ed. W. Milliken and J. A. Ratter, 189–210. New York: John Wiley and Sons, xxii + 1–508.

Barquez, R. M. 1983. La distribución de *Neotomys ebriosus* Thomas en la Argentina y su presencia en la provincia de San Juan (Mammalia, Rodentia, Cricetidae). *Hist. Nat.* 3:189–91.

Barquez, R. M., M. M. Díaz, and M. Goytia. 1994. Una nueva especie de *Akodon* (Rodentia: Cricetidae: Sigmodontinae) de las Cumbre del Taficillo, las Aguitas, Tucumán. *Novenas Jornadas Argentinas de Mastozoología (Vaquerías, Córdoba, Argentina): Libro de Resúmenes*, p. 23.

Barquez, R. M., M. M. Díaz, and R. A. Ojeda. 2006. *Mamíferos de Argentina: Sistemática y distribución.* Tucumán: Sociedad Argentina para el Estudio de los Mamíferos, 359 pp.

Barquez, R. M., L. I. Ferro, and M. S. Sanchez. 2006. *Tapecomys primus* (Rodentia: Cricetidae), new genus and species for Argentina. *Mastozool. Neotrop.* 13:117–21.

Barquez, R. M., D. A. Flores, M. M. Díaz, N. P. Giannini, and D. H. Verzi. 2002. Análisis filogenético preliminar de los octodóntidos vivientes basado en caracteres morfológicos. In *Libro de resúmenes de las XVII Jornadas Argentinas de Mastozoología*, p. 38. Mar del Plata: Sociedad para el Estudio de Los Mamíferos.

Barquez, R. M., M. A. Mares, and R. A. Ojeda. 1991. *Mamíferos de Tucumán—Mammals of Tucumán.* Norman: Oklahoma Museum of Natural History.

Barquez, R. M., D. F. Williams, M. A. Mares, and H. H. Genoways. 1980. Karyology and morphometrics of three species of *Akodon* (Mammalia: Muridae) from northwestern Argentina. *Ann. Carnegie Mus.* 49:379–403.

Barrantes, G. E., M. O. Ortells, and O. A. Reig. 1993. New studies on allozyme genetic distance and variability in akodontine rodents (Cricetidae) and their systematic implications. *Biol. J. Linn. Soc.* 48:283–98.

Barrera, J. M. de la. 1940. Estudios sobre le peste selvática en Mendoza. *Rev. Inst. Bacteriol.* 5:565–86.

Barrère, P. 1741. *Essai sur l'histoire naturelle de la France équinoxiale, ou Dénombrement des plantes, des animaux et des minéraux qui se trouvent dans l'isle de Cayenne, les isles de Remire, sur les clotes de la mer et dans le continent de la Guyane.* Paris: Chez Piget, i–xxiv + 1–24 (unnumbered) + 1–215 + 1–7 (unnumbered).

Barreto, G. R., and S. García-Rangel. 2005. Holochilus sciureus. *Mamm. Species* 780:1–5.

Barretto, A. C., N. E. Peterson, E. Lago, A. C. Rosa, R. S. M. Braga, C. A. Cuba, J. A. Vexenat, and P. D. Marsden. 1985. *Leishmania mexicana* in *Proechimys iheringi denigratus* Moojen (Rodentia, Echimyidae) in a region endemic for American Cutancous Leishmaniasis. *Rev. Bras. Med. Trop.* 18:243–46.

Barriga-Bonilla, E. 1966. Estudios mastozoologicos colombianos II. *Caldasia* 45:491–502.

Barros, M. A., R. C. Liascovich, L. Gonaález, M. Lizarralde, and O. A. Reig. 1990. Banding pattern comparisons between *Akodon iniscatus* and *Akodon puer* (Rodentia, Cricetidae). *Z. Säugetierk.* 55:115–27.

Barros, M. A., and O. A. Reig. 1979. Doble polimorfismo robertsoniano en *Microxus bogotensis* (Rodentia, Cricetidae) del Páramo de Mucujají (Mérida). *Acta Cient. Venez.* 30:106.

Barros, M. A., O. A. Reig, and A. Pérez-Zapata. 1992. Cytogenetics and karyosystematics of South American oryzomyine rodents (Cricetidae: Sigmodontinae). *Cytogenet. Cell Genet.* 59:34–38.

Barros, M. C., I. Sampaio, H. Schneider, and A. Langguth. 2009. Molecular phylogenies, chromosomes and dispersion in Brazilian akodontines (Rodentia, Sigmodontinae). *Iheringia Zool.* 99:373–80.

Barros-Battesti, D. M., M. Arzua, P. M. Linardi, J. R. Botelho, and I. V. Sbalqueiro. 1998. Interrelationship between ectoparasites and wild rodents from Tijucas do Sul, State of Paraná, Brazil. *Mem. Inst. Oswaldo Cruz Rio J.* 93:719–25.

Barros-Battesti, D. M., and I. Knysak. 1999. Catalogue of the Brazilian *Ixodes* (Acari: Ixodidae) material in the Mite Collection of Instituto Butantan, São Paulo, Brazil. *Pap. Avulsos Zool.* (São Paulo) 41:49–57.

Bartoli, G., M. Sarnthein, M. Weinelt, H. Erlenkeuser, D. Garbe-Schönberg, and D. W. Lea. 2005. Final closure of Panama and the onset of northern hemisphere glaciation. *Earth Planet. Sci. Lett.* 237:33–44.

Baskin, J. A. 1978. *Bensonomys, Calomys*, and the origin of the phyllotine group of neotropical cricetines (Rodentia, Cricetidae). *J. Mamm.* 59:125–35.

———. 1986. *The Late Miocene radiation of neotropical sigmodontine rodents in North America.* Contributions to Geology, University of Wyoming, Special Paper 3:287–303.

Basso, B., A. J. Eraso, E. R. Moretti, I. Albesa, and F. O. Kravetz. 1977. Natural infection of *Calomys musculinus* (Rodentia, Cricetidae) by *Trypanosoma cruzi. Rev. Asoc. Argent. Microbiol.* 9:11–16.

Baud, F. 1977. Catalogue des types de mammifères et d'oiseaux du Muséum d'Histoire naturelle de Genéve. *Rev. Suisse Zool.* 84:201–20.

Bayard, V., P. T. Kitsutani, E. O. Barria, L. A. Ruedas, D. S. Tinnin, C. Muñoz, I. B. de Mosca, G. Guerrero, R. Kant, A. Garcia, L. Caceres, F. G. Gracia, E. Quiroz, Z. de Castillo, B. Armien, M. Libel, J. N. Mills, A. S. Khan, S. T. Nichol, P. E. Rollin, T. G. Ksiazek, and

C. J. Peters. 2004. Outbreak of hantavirus pulmonary syndrome, Los Santos, Panama, 1999–2000. *Emerg. Infect. Dis.* 10:1635–42.

Beaucournu, J.-C., and M. N. Gallardo. 1988. Puces nouvelles d'Argentine (Insecta, Siphonaptera). *Rev. Suisse Zool.* 95:99–112.

Beaucournu, J.-C., and M. N. Gallardo. 1991. Siphonapteres du Chili; description de quatre especes nouvelles (Siphonaptera). *Bull. Soc. Entomol. Fr.* 96:185–203.

———. 2005. Deux puces nouvelles (Siphonaptera: Rhopalopsyllidae: Parapsyllinae) d'Argentine, parasites d'*Abrocoma uspallata* Braun et Mares, 2002 (Rodentia: Abrocomidae). *Parasite* 12:39–43.

Beaucournu, J.-C., L. Moreno, and D. González-Acuña. 2011. Deux espèces de puces (Siphonaptera: Ctenophthalmidae & Rhapalopsyllidae) du Chili. *Parasite* 18:241–46.

Beaucournu, J.-C., and J. C. Torres-Mura. 1986. A new *Tetrasyllus* from Chile (Siphonaptera, Rhopalosyllidae). *Rev. Fr. Entomol.* (n.s.) 8:9–12.

Becker, R. G., G. Paise, L. C. Baumgarten, and E. M. Vieira. 2007. Estrutura de comunidades de pequenos mamíferos e densidade de *Necromys lasiurus* (Rodentia, Sigmodontinae) em áreas abertas de Cerrado no Brasil Central. *Mastozool. Neotrop.* 14:157–68.

Beck-King, H., O. von Helversen, and R. Beck-King. 1999. Home range, population density, and food resources of *Agouti paca* (Rodentia: Agoutidae) in Costa Rica: a study using alternative methods. *Biotropica* 31:675–85.

Bechstein, J. M. 1800. *Thomas Pennant's Allgemeine uebersicht der vierfüssigen thiere. Aus dem Englischen übersetzt und mit anmerkungen und zusätzen versehen, von Johann Matthäus Bechstein.* Weimar: Im verlage des Industrie-Comptoir's, 2:x + 2 unnumbered, 3236–66, plates 35–54.

Bedel, L. 1892. Révision des *Scarabaeus* palearctiques. *L'Abeile J. Entomol.* 27:281–88.

Bee de Speroni, N., and A. M. Pellegrini de Gstaldo. 1988. Encephalization and brain composition in three South American rodents *Dolichotis patagonum, Lagostomus maximus* and *Calomys musculinus. Phys. A Ocean. Organis.* 46:31–39.

Begall, S., and H. Burda. 2006. Acoustic communication and burrow acoustics are reflected in the ear morphology of the coruro (*Spalacopus cyanus*, Octodontidae), a social fossorial rodent. *J. Morphol.* 267:382–90.

Begall, S., H. Burda, and M. H. Gallardo. 1999. Reproduction, postnatal development and growth of social coruros, *Spalacopus cyanus* (Octodontidae, Rodentia) from Chile. *J. Mamm.* 80:210–17.

Begall, S., H. Burda, and B. Schneider. 2004. Hearing in coruros (*Spalacopus cyanus*): special audiogram features of a subterranean rodent. *J. Comp. Physiol. A Neuroethol. Sens. Neural. Behav. Physiol.* 190:963–69.

Begall, S., S. Daan, H. Burda, and G. J. F. Overkamp. 2002. Activity patterns in a subterranean social rodent, *Spalacopus cyanus* (Octodontidae). *J. Mamm.* 83:153–58.

Begall, S., and M. H. Gallardo. 2000. *Spalacopus cyanus* (Octodontidae, Rodentia): an extremist in tunnel constructing and food storing among subterranean mammals. *J. Zool, Lond.* 251:53–60.

Bekker, J. P. 1999. *The Mammals of Aruba: Mammalia: Chiroptera, Rodentia, Lagomorpha.* Utrecht: Vereniging voor Zoogdierkunde en Zoogdierbescherming, 78 pp.

Bell, D. M., M. J. Hamilton, C. W. Edwards, L. E. Wiggins, R. Muniz Martinez, R. E. Strauss, R. D. Bradley, and R. J. Baker. 2001. Patterns of karyotypic megaevolution in *Reithrodontomys*: evidence from a cytochrome-*b* phylogenetic hypothesis. *J. Mamm.* 82:81–91.

Bellocq, M. I., and F. O. Kravetz. 1994. Feeding strategy and predation of the barn owl (*Tyto alba*) and the burrowing owl (*Speotyto cunicularia*) on rodent species, sex, and size, in agrosystems of central Argentina. *Ecol. Austral* 4:29–34.

Belmar-Lucero, S., P. Godoy, M. Ferrés, P. Vial, and R. E. Palma. 2009. Range expansion of *Oligoryzomys longicaudatus* (Rodentia, Sigmodontinae) in Patagonian Chile, and first record of hantavirus in the region. *Rev. Chilena Hist. Nat.* 82:265–75.

Belton, W. 1985. Birds of Rio Grande do Sul, Brazil. Part 2. Formicariidae through Corvidae. *Bull. Am. Mus. Nat. Hist.* 180:1–242 (95 maps, 1 table).

Beltrán, H., and I. Salinas. 2010. Flora vascular y vegetación de los Bosques Montanas Húmedos de Carpish (Huánuco–Perú). *Arnaldoa* 17:107–30.

Benado, M., M. Aguilera, O. A. Reig, and F. J. Ayala. 1979. Biochemical genetics of chromosome forms of Venezuelan spiny rats of the *Proechimys guairae* and *Proechimys trinitatis* superspecies. *Genetica* 50:89–97.

Bennett, E. T. 1829. The chinchilla. In *Gardens and Menagerie of the Zoological Society of London,* vol. 1, *Quadrupeds,* ed. E. T. Bennett, 1–12. London: Charles Tilt, xii+308 pp.

———. 1832a. Characters of a new specis of otter (*Lutra,* Erxl.) and of a new species of mouse (*Mus,* L.) collected in Chili by Mr. Cuming. *Proc. Comm. Sci. Correspond.. Zool. Soc. Lond.* 1832, part 2:1–4.

———. 1832b. Characters of a new genus of rodent Mammalia, from Chile, presented by Mr. Cuming.

Proc. Comm. Sci. Correspond.. Zool. Soc. Lond. 1832, part 2:46–48.

———. 1833. On the family of Chinchillidae, and on a new genus referrible [*sic*] to it. *Proc. Zool. Soc. Lond.* 1833, part I:57–60.

———. 1835a. Note of some recent publications on the Chinchillidae. *Zool. J.* 5 (for 1832–1834):491–95.

———. 1835b. On a second species of *Lagotis* (*Lag. pallipes*). *Proc. Zool. Soc. Lond.* 1835 (part III):67–68.

———. 1835c. On the Chinchillidae, a family of herbivorous Rodentia, and on a new genus referrible [*sic*] to it. *Trans. Zool. Soc. Lond.* 1:35–64 (+ 4 plates).

———. 1935d. Additional remarks on the genus *Lagotis*, with some account of a second species referrible [*sic*] to it. *Trans. Zool. Soc. Lond.* 1:331–34 (+ 1 plate).

———. 1836. On a new species of *Ctenomys* and other rodents collected near the Straits of Magellan by Capt. P. P. King, R. N. *Proc. Zool. Soc. Lond.* 1835 (part 3, no. 36):189–91. [Part III, no. XXXVI of the 1835 volume of the *Proceedings* was published April 8, 1836; see Duncan 1937.]

———. 1841. On the genus *Octodon*, and on its relations with *Ctenomys*, Blainv., and *Poephagomys*, F. Cur.; including a description of a new species of *Ctenomys*. *Trans. Zool. Soc. Lond.* 2:75–86, 2 plates.

———. 1846: Letter addressed to G. R. Waterhouse, dated February 10, 1846. *Proc. Zool. Soc. Lond.* 1846, part XIV:7–10.

Berg, C. 1898. *Dolichotis salinicola* Burm. est bona species. *Comm. Mus. Nac. Buenos Aires* 1:23–24.

———. 1900. Notas sobre los nombres de algunos mamiferos sudamericanos. *Comm. Mus. Nac. Buenos Aires.* 1:219–22.

Bergallo, H. G. 1994. Ecology of a small mammal community in an Atlantic Forest area in southeastern Brazil. *Stud. Neotrop. Fauna Environ.* 29:197–217.

———. 1995. Comparative life-history characteristics of two species of rats, *Proechimys iheringi* and *Oryzomys intermedius*, in an Atlantic Forest of Brazil. *Mammalia* 59:51–64.

Bergallo, H. G., L. Geise, C. R. Bonvicino, R. Cerqueira, P. S. D'Andrea, C. E. Esberárd, F. A. S. Fernandez, C. E. Grelle, A. Peracchi, S. Siciliano, and S. M. Vaz. 2000. Mamíferos. In *A Fauna ameaçada de extinção do Estado do Rio de Janeiro*, eds. H. G. Bergallo, C. F. D. Rocha, M. A. S. Alves, and M. Van Sluys, 125–35. Rio de Janeiro: Editora da Universidade do Estado do Rio de Janeiro, 166 pp.

Bergallo, H. G., F. M. Hatano, D. S. Raíces, T. T. L. Ribeiro, A. G. Alves, J. L. Luz, R. Mangolin, and M. A. R. Mello. 2004. Os mamíferos da Restinga de Jurubatiba. In *Pesquisasecológicas de longa duração na Restinga de Jurubatiba: ecologia, história natural e conservação*, ed. C. F. D. Rocha, F. A. Esteves, and F. R. Scarano, 215–30. Rio de Janeiro, Brazil: RIMA Editora, 376 pp.

Bergallo, H. G., J. L. Luz, D. S. Raíces, F. H. Hatano, and F. Martins-Hatano. 2005. Habitat use by *Oryzomys subflavus* (Rodentia) in an open shrubland formation in Restinga de Jurubatiba National Park, RJ, Brazil. *Brazil. J. Biol.* 65:583–88.

Bergallo, H. G., and W. E. Magnusson. 1999. Effects of climate and food availability on four rodent species in southeastern Brazil. *J. Mamm.* 80:472–86.

Bergallo, H. G., F. Martins-Hatano, D. S. Raíces, T. T. L. Ribeiro, A. G. Alves, J. L. Luz, R. Mangolin, and M. A. R. Mello. 2004. Os Mamíferos da Restinga de Jurubatiba. In *Pesquisas de longa duração na Restinga de Jurubatiba—Ecologia, história natural e conservação*, eds. C. F. D. Rocha, F. A. Esteves, and F. R. Scarano, 215–30. São Carlos: Editora Rima, 376 pp.

Bergmans, W. 2011. An annotated list of mammal type specimens in the collections of the former Zoological Museum of the University of Amsterdam (1890–2010). *Zool. Mededelingen* 85: pages not numbered.

Bergsten, J. 2005. A review of long-branch attraction. *Cladistics* 21:163–93.

Berman, S. L. 2003. A desert octodontid rodent, *Tympanoctomys barrerae*, uses modified hairs for stripping epidermal tissue from leaves of halophytic plants. *J. Morphol.* 257:53–61.

Bertolino, S., M. L. Guichon, and J. Carter. 2012. *Myocastor coypus* Molina (coypu). In *A Handbook of Global Freshwater Invasive Species*, ed. R. A. Francis, 357–68. London: Routledge, 456 pp.

Bertoni, A. de W. 1914a. Fauna Paraguaya. Catálogos sistemáticos de los vertebrados del Paraguay, peces, batracios, reptiles, aves, y mamíferos conocidos hasta 1914. In *Descripción física y económica del Paraguay*, ed. M. S. Bertoni. Asunción: M. Brossa, 86 pp.

———. 1914b. Fauna Paraguaya. Catálogos sistemáticos de los vertebrados del Paraguay. Classe Primera: Mamíferos. *Rev. Soc. Cient. Paraguay* 73:5–14.

Betancourt, J. L., and B. Saavedra. 2002. Paleomadrigueras de roedores, un Nuevo método para el estudio del Cuaternario en zonas áridas de Subamérica. *Rev. Chilena Hist. Nat.* 75:527–46.

Bewick, T. 1804. *A general history of quadrupeds. The figures engraved on wood chiefly copied from the original of T. Bewick, by A. Anderson. First American edition, with an appendix, containing some American animals not hitherto described.* New York: G. and R. Waite, New York, x + 531 pp.

Bezerra, A. M. R. 2002. "Variabilidade morfologia e status taxonômico das amostras populacionais do gênero *Clyomys* (Rodentia: Echimyidae)." Master's thesis, Museu Nacional and Universidade Federal do Rio de Janeiro, Brazil.

———. 2005. Phallic morphology of *Kunsia tomentosus* (Rodentia: Sigmodontinae). *Mastozool. Neotrop.* 12:227–32.

———. 2008. "Revisão taxonômica do gênero *Galea* Meyen, 1832 (Rodentia, Caviidae, Caviinae)." Ph.D. diss., Universidade de Brasília, Brazil.

———. 2011. Collection records of *Gyldenstolpia planaltensis* (Avila-Pires, 1972)(Rodentia, Cricetidae) suggest the local extinction of the species. *Mastozool. Neotrop.* 18:119–23.

Bezerra, A. M. R., C. R. Bonvicino, A. A. N. Menezes, and J. Marinho-Filho. 2010. Endemic climbing cavy *Kerodon acrobata* (Rodentia: Caviidae: Hydrochoerinae) from dry forest patches in the Cerrado domain: new data on distribution, natural history, and morphology. *Zootaxa* 2724:29–36.

Bezerra, A. M. R., A. P. Carmignotto, A. P. Nunes, and F. H. G. Rodrigues.2007. New data on the distribution, natural history and morphology of *Kunsia tomentosus* (Lichtenstein, 1830) (Rodentia: Cricetidae: Sigmodontinae). *Zootaxa* 1505:1–18.

Bezerra, A. M. R., A. P. Carmignotto, and F. H. G. Rodrigues. 2009. Small non-volant mammals of an ecotone region between the Cerrado hotspot and the Amazonian rainforest, with comments on their taxonomy and distribution. *Zool. Stud.* 48:861–74.

Bezerra, A. M. R., A. Lazar, C. R. Bonvicino, and J. Marinho Filho. 2013. *Wiedomys cerradensis* Gonçalves, Almeida and Bonvicino, 2005 (Mammalia: Rodentia: Cricetidae) in Tocantins and Goiás states, central-northern Brazil. *Check List* 9:680–83.

Bezerra, A. M. R., J. Marinho-Filho, and A. P. Carmignotto. 2011. A review of the distribution, morphology, and habitof the Owl's spiny rat *Carterodon sulcidens* (Lund, 1841) (Rodentia: Echimyidae). *Zool. Stud.* 50:566–76.

Bezerra, A. M. R., and J. A. de Oliveira. 2010. Taxonomic implications of cranial morphometric variation in the genus *Clyomys* Thomas, 1916 (Rodentia: Echimyidae). *J. Mamm.* 91:260–72.

Bezerra, A. M. R., N. J. da Silva Jr., and J. Marinho-Filho. 2007. The Amazon bamboo rat *Dactylomys dactylinus* (Rodentia: Echimyidae: Dactylomyinae) in the Cerrado of central Brazil, and the role of gallery forests in its distribution. *Biota Neotrop.* 7:235–37.

Bharadwaj, M., J. Botten, N. Torrez-Martínez, and B. Hjelle. 1997. Rio Mamore virus: genetic characterization of a newly recognized hantavirus of the pygmy rice rat, *Oligoryzomys microtis*, from Bolivia. *Am. J. Trop. Med. Hyg.* 57:368–74.

Bi, Z., P. B. H. Formenty, and C. E. Roth. 2008. Hantavirus infection: a review and global update. *J. Infect. Dis. Dev. Ctries.* 2:3–23.

Bianchi, N. O. 2002. *Akodon* sex reversed females: the never-ending story. *Cytogenet. Genome Res.* 96:60–65.

Bianchi, N. O., and J. R. Contreras. 1967. The chromosomes of the field mouse *Akodon azarae* (Cricetidae, Rodentia) with special reference to sex chromosome anomalies. *Cytogenetics* 6:306–13.

Bianchi, N. O., and S. Merani. 1984. Cytogenetics of South American akodont rodents (Cricetidae). X. Karyological distances at generic and intergeneric levels. *J. Mamm.* 65:206–19.

Bianchi, N. O., S. Merani, and M. Lizarralde. 1979. Cytogenetics of the South American akodont rodents (Cricetidae). VI. Polymorphism in Akodon Dolores Thomas. *Genetica* 50:99–104.

Bianchi, N. O., O. A. Reig, O. J. Molina, and F. N. Dulout. 1971. Cytogenetics of the South American akodont rodents (Cricetidae). I. A progress report of Argentinian and Venezuelan forms. *Evolution* 25:724–36.

Bianchi, N. O., L. Vidal Rioja, and M. S. A. Bianchi. 1973. Constitutive heterochromatin, G-bands and Robertsonian rearrangements in the chromosomes of *Akodon molinae* (Rodentia, Cricetidae). *Can. J. Genet. Cytol.* 15:855–61.

———. 1976. Cytogenetics of the South American akodont rodents (Cricetidae): II. Interspecific homology of G-banding patterns. *Cytologia* 41:139–44.

Bianchi, R. C., and S. L. Mendes. 2007. Ocelot (*Leopardus pardalis*) predation on primates in Caratinga Biological Station, southeast Brazil. *Am. J. Primatol.* 69:1173–78.

Bianchini, J., and L. Delupi. 1994. Consideraciones sobre el estado sistemático de *Deltamys kempi* Thomas, 1917 (Cricetidae, Sigmodontinae). *Physis Sec. C* 49:27–35.

Bidau, C. J. 2006. Familia Ctenomyidae. In *Mamíferos de Argentina: sistemática y distribución*, eds. R. M. Barquez, M. M. Díaz, and R. A. Ojeda, 212–31. Tucumán: Sociedad Argentina para el Estudio de los Mamíferos, 359 pp.

Bidau, C. J., and F. D. de Avila-Pires. 2009. On the type locality of *Ctenomys bicolor* Miranda Ribeiro, 1914 (Rodentia: Ctenomyidae). *Mastozool. Neotrop.* 16:445–47.

Bidau, C. J., M. D. Giménez, Y. E. Davies, and J. R. Contreras. 2005. New *Ctenomys* karyotypes from lower Chaco, Argentina (Rodentia, Ctenomyidae). *Nucleus* 48:135–42.

Bidau, C. J., M. D. Giménez, J. R. Contreras, C. F. Argüelles, E. Braggio, R. D'Errico, C. Ipucha, C. Lanzone, M. Montes, and P. Suárez. 2000. Variabilidad cromosómica y molecular inter- e intraespecíficas en *Ctenomys* (Rodentia, Ctenomyidae, Octodontoidea): múltiples patrones evolutivos. *Actas IX Congreso Iberoamericano de Biodiversidad y Zoología de Vertebrados, Buenos Aires, Argentina*, pp. 127–30.

Bidau, C. J., and A. I. Medina. 2013. Sexual size dimorphism and testis size allometry in tuco-tucos (Rodentia: Ctenomyidae). *Mammalia* 77:81–93.

Bidlingmaier, T. C. 1937. Notes on the genus *Chinchilla*. *J. Mamm.* 18:159–63.

Biknevicius, A. R. 1993. Biomechanical scaling of limb bones and differential limb use in caviomorph rodents. *J. Mamm.* 74:95–107.

Bilenca, D. N., and F. O. Kravetz. 1998. Seasonal variations in microhabitat use and feeding habits of the pampas mouse, *Akodon azarae*, in agroecosystems of central Argentina. *Acta Theriol.* 43:195–203.

———. 1999. Seasonal changes in microhabitat use and niche overlap between *Akodon azarae* and *Calomys laucha* (Rodentia, Muridae) in agroecosystems of central Argentina. *Stud. Neotrop. Fauna Environ.* 34:129–36.

Billberg, G. J. 1827. *Synopsis Faunae Scandinaviae, complectens Animalia Sveciae Norvegiae. Mammalia.* Holimiae: Ordinum Equestrium, 1(1):viii + 55.

Bingham, H. 1916. Further exploration in the land of the Incas. *Nat. Geogr. Mag.* 26(5):431–473.

Bisbal-E., F. J. 1990. Inventario preliminar de la fauna del Cerro Santa Ana, Península de Paraguaná-Estado Falcón, Venezuela. *Acta Cient. Venez.* 41:177–85.

Bisceglia, S. B. C., J. A. Pereira, P. Teta, and R. D. Quintana. 2011. Rodent selection by Geoffroy's cats in a semi-arid scrubland of central Argentina. *J. Arid Environ.* 75:1024–28.

Bishop, I. R. 1974 [1975]. An annotated list of caviomorph rodents collected in north-eastern Mato Grosso Brazil. *Mammalia* 38:489–502. [1974 vol. 38, no. 3: pp. 355–66. "Achevé d'imprimer le 31 mars 1975."]

Blainville, H. M. D. de. 1826. Sur une nouvelle espéce de rongeur fouisseur du Brésil. *N. Bull. Sci. Soc. Phil. Paris* 2:62–64.

Blainville, H. M. D. de. 1855. *Ostéographie ou description iconographique comparée du squelette et du système dentaire des cinq classes d'animaux vertébrés récents et fossiles pour servir de base a la zoologie et a la géologie: Mammifères*, Part 4, fascicule 25, 63 pp. + Atlas, 41 plates. Paris: Arthus Bertrand.

Blake, J., D. Mosquera, J. Guerra, and D. Romo. 2010. New locality records and the first photographs of living *Echimys saturnus* (dark tree rat, Echimyidae) from eastern Ecuador. *Ecotropica* 16:141–44.

Blaustein, S. A., R. C. Liascovich, L. I. Apfelbaum, L. Daleffe, R. M. Barquez, and O. A. Reig. 1992. Correlates of systematic differentiation between two closely related allopatric populations of the *Akodon boliviensis* group from NW Argentina (Rodentia, Cricetidae). *Z. Säugetierk.* 57:1–13.

Blumenbach, J. F. 1797. *Handbuch der Naturgeschichte.* Fünfte Auglange. Göttingen: Johann Christian Dieterich, xviii + 714 pp., 32 [unnumbered], 11 folded leaves of plates.

Bó, M. S., J. P. Isacch, A. I. Malizia, and M. M. Martínez. 2002. Lista comentada de los mamíferos de la Reserva de la Biósfera Mar Chiquita, Provincia de Buenos Aires, Argentina. *Mastozool. Neotrop.* 9:5–11.

Bó, R. F., G. M. Porini, S. M. Arias, and M. J. Corriale. 2006a. Estudios ecologicos basicos para el manejo sustenable del coipo (*Myocastor coypus*) en los grandes sistemas de humedales de Argentina. In *Humedales fluviales de América del Sur. Hacia un manejo sustentable,* ed. J. Peteán and J. Cappato, 111–27. Buenos Aires: Ediciones Proteger, 570 pp.

Bó, R. F., G. M. Porini, M. J. Corriale, and S. M. Arias. 2006b. Proyecto Nutria. Estudios ecológicos básicos para el manejo sustentable de *Myocastor coypus* en la Argentina. In *Manejo de fauna Silvestre en la Argentina. Programas de uso sustentable,* ed. M. L. Bolkovic and D. Ramadori, 93–104. Buenos Aires: Dirección de Fauna Silvestre, Secretaria de Ambiente y Desarrollo Sustentable, 168 pp. + 8 illust.

Boada, C., M. Gómez-Laverde, and R. Anderson. 2008. *Akodon latebricola. IUCN Red List of Threatened Species,* version 2013.1, accessed July 30, 2013, http://www.iucnredlist.org.

Boddaert, P. 1784. *Elenchus animalium.* Roterdami: C. R. Hake, 1:xxxviii + 2 unnumbered + 174 pp.

Boher, S., and B. Marín. 1988. El pacarana (*Dinomys branickii*) en Venezuela. *Natura* 84:14–18.

Boher, S., J. Naveua, and L. Escobar. 1988. First record of *Dinomys branickii* for Venezuela. *J. Mamm.* 69:433.

Bole, B. P., Jr. 1937. Annotated list of mammals of the Mariato River district of Azuero Peninsula. *Sci. Publ. Cleve. Mus. Nat. Hist.* 7:140–88.

Bonaccorso, F. J., W. E. Glanz, and C. M. Sanford. 1980. Feeding assemblages of mammals at fruiting *Dipteryx panamensis* (Papilionaceae) trees in Panama: seed predation, dispersal, and parasitism. *Rev. Biol. Trop.* 28:61–72.

Bonaparte, C. L. J. L. 1838. Synopsis Vertebratorum Systematis. *N. Ann. Sci. Nat. Bologna* 2:105–33.

———. 1845. *Catalogo methodico dei mammiferi Europei.* Milano: L. de Giacomo Pirola.

Bonaventura, S. M., A. M. Balabusic, M. C. Sabatini, A. M. Miranda, F. Marcelino, F. Ferrero, and D. M. Conrado. 1998. Diversidad y biomasa de pequeños roedores en el desierto del Monte, Argentina. *Bol. Soc. Biol. Concepc. Chile* 69:39–45.

Bonaventura, S. M., M. J. Piantanida, L. Gurini, L., and M. I. S. Lopez. 1991. Habitat selection in population of cricetine rodents in the region Delta (Argentina). *Mammalia* 55:339–54.

Bonecker, S. T., L. G. Portugal, S. F. Costa-Neto, and R. Gentile. 2009. A long term study of small mammal populations in a Brazilian agricultural landscape. *Mamm. Biol.* 74:469–79.

Bonhote J. L. 1901. On the squirrels of the *Sciurus erythraeus* group. *Ann. Mag. Nat. Hist.,* ser. 7, 7:160–67.

Bonino, N., A. Sbriller, M. M. Manacorda, and F. Larosa. 1997. Food partitioning between the mara (*Dolichotis patagonum*) and the introduced hare (*Lepus europaeus*) in the Monte Desert, Argentina. *Stud. Neotrop. Fauna Environ.* 32:129–34.

Bonvicino, C. R. 1994. "Especiação do rato d'água Nectomys. Abordagem cariológica, morfológica e geográfica." Ph.D. diss., Universidade Federal do Rio de Janeiro, Brazil.

———. 1999. New taxonomic status to the French Guianan *Nectomys parvipes* Petter (Sigmodontinae, Rodentia). *Rev. Bras. Zool.* 16:253–55.

———. 2003. A new species of *Oryzomys* (Rodentia, Sigmodontinae) of the *subflavus* group from the Cerrado of Central Brazil. *Mamm. Biol.* 68:78–90.

Bonvicino, C. R., and F. C. Almeida. 2000. Karyotype, morphology and taxonomic status of *Calomys expulsus* (Rodentia : Sigmodontinae). *Mammalia* 64:339–51.

Bonvicino, C. R., F. C. Almeida, and R. Cerqueira. 2000. The karyotype of *Sphiggurus villosus* (Rodentia: Erethizontidae) from Brazil. *Stud. Neotrop. Fauna Eviron.* 35:81–83.

Bonvicino, C. R., P. S. D'Andrea, and P. Borodin. 2001. Pericentric inversions: a study in natural populations of *Oligoryzomys nigripes* (Rodentia: Sigmodontinae). *Genome* 44:791–96.

Bonvicino, C. R., P. S. D'Andrea, R. Cerqueira, and H. N. Seuánez. 1996. The chromosome of *Nectomys* (Rodentia, Cricetidae) with 2n = 52, 2n = 56 and interspecific hybrids (2n = 54). *Cytogenet. Cell Genet.* 73:190–93.

Bonvicino, C. R., and A. M. R. Bezerra. 2003. Use of regurgitated pellets of Barn Owl (*Tyto alba*) for inventorying small mammals in the Cerrado of central Brazil. *Stud. Neotrop. Fauna Environ.* 38:1–5.

Bonvicino, C. R., R. Cerqueira, and V. A. Soares. 1996. Habitat use by small mammals of upper Araguaia river. *Rev. Bras. Biol.* 56:761–67.

Bonvicino, C. R., F. A. Fernandes, M. C. Viana, B. R. Teixeira, and P. S. D'Andrea. 2013. *Scapteromys aquaticus* (Rodentia: Sigmodontinae) in Brazil with comments on karyotype and phylogenetics relationships. *Zoologia* 30:242–47.

Bonvicino, C. R., and A. L. Gardner. 2001. A new karyotype in *Nectomys squamipes* complex (Rodentia, Sigmodontinae). *Rev. Nordest. Biol.* 15:69–72.

Bonvicino, C. R., and L. Geise. 1995. Taxonomic status of *Delomys dorsalis collinus* Thomas, 1917 (Rodentia, Cricetidae) and description of a new karyotype. *Z. Säugetierk.* 60:124–27.

Bonvicino, C. R., P. R. Gonçalves, J. A. de Oliveira, L. F. B. de Oliveira, and M. S. Mattevi. 2008. Divergence in *Zygodontomys* (Rodentia: Sigmodontinae) and distribution of Amazonian savannas. *J. Hered.* 100:322–28.

Bonvicino, C. R., A. Langguth, S. M. Lindbergh, and A. C. de Paula. 1997 [1998]. An elevational gradient study of small mammals at Caparaó National Park, southeastern Brazil. *Mammalia* 61:547–60. [1997 vol. 61, no. 4: pp. 467ff. "Achevé d'imprimer le 23 fevier 1998."]

Bonvicino, C. R., B. Lemos, and M. Weksler. 2005. Small mammals of Chapada dos Veadeiros National Park (Cerrado of central Brazil): ecologic, karyologic, and taxonomic considerations. *Brazil. J. Biol.* 65: 395–406.

Bonvicino, C. R., J. F. S. Lima, and F. C. Almeida. 2003. A new species of *Calomys* Waterhouse (Rodentia, Sigmodontinae) from the Cerrado of central Brazil. *Rev. Bras. Zool.* 20:301–7.

Bonvicino, C. R., S. M. Lindbergh, and L. S. Maroja. 2002. Small non-flying mammals in altered and conserved areas of Atlantic Forest and Cerrado: comments on their potential use for monitoring environment. *Brazil. J. Biol.* 62:1–12.

Bonvicino, C. R., L. S. Maroja, J. A. de Oliveira, and J. R. Coura. 2003. Karyology and morphology of *Zygodontomys* (Rodentia, Sigmodontinae) from the Brazilian Amazon, with a molecular appraisal of phylogenetic relationships within this genus. *Mammalia* 67:119–31.

Bonvicino, C. R., A. R. E. A. N. de Menezes, and J. A. Oliveira. 2003. Molecular and karyologic variation in the genus *Isothrix* (Rodentia, Echimyidae). *Hereditas* 139:206–11.

Bonvicino, C. R., and M. A. M. Moreira. 2001. Molecular phylogeny of the genus *Oryzomys* (Rodentia: Sig-

modontinae) based on cytochrome-b DNA sequences. *Mol. Phylogenet. Evol.* 18:282–92.

Bonvicino, C. R., J. A. de Oliveira, and P. S. D'Andrea. 2008. *Guia dos Roedores do Brasil, com chaves para gêneros baseadas em caracteres externos*. Rio de Janeiro: Centro Pan-Americano de Febre Aftosa–OPAS/OMS, Série de Manuais Técnicos, 11:1–120.

Bonvicino, C. R., J. A. de Oliveira, P. S. D'Andrea, and R. W. Carvalho. 2001. The endemic Atlantic Forest rodent *Phaenomys ferrugineus* (Thomas, 1894) (Sigmodontinae): new data on its morphology and karyology. *Bol. Mus. Nac., Rio de Janeiro*, nova sér., zool., 467:1–12.

Bonvicino, C. R., J. A. de Oliveira, and R. Gentile. 2010. A new species of *Calomys* (Rodentia: Sigmodontinae) from eastern Brazil. *Zootaxa* 2336:19–25.

Bonvicino, C. R., and I. B. Otazu. 1999. The *Wilfredomys pictipes* (Rodentia: Sigmodontinae) karyotype with comments on the karyosystematics of Brazilian Thomasomyini. *Acta Theriol.* 44:329–32.

Bonvicino, C. R., I. B. Otazu, and P. M. Borodin. 1999. Chromosome variation in *Oryzomys subflavus* species group (Sigmodontinae, Rodentia) and its taxonomic implication. *Cytologia* 64:327–32.

Bonvicino, C. R., I. B. Otazu, and P. S. D'Andrea. 2002. Karyologic evidence of diversification of the genus *Thrichomys* (Rodentia, Echimyidae). *Anim. Cytogenet. Genome Res.* 97:200–4.

Bonvicino, C. R., I. B. Otazu, and J. F. Vilela. 2005. Karyologic and molecular analsysis of *Proechimys* Allen, 1899 (Rodentia, Echimyidae) from the Amazon Region. *Arq. Mus. Nac., Rio de Janeiro* 63:191–200.

Bonvicino, C. R., V. Penna-Firme, and E. Braggio. 2002. Molecular and karyologic evidence of the taxonomic status of *Coendou* and *Sphiggurus* (Rodentia: Hystricognathi). *J. Mamm.* 83:1071–76.

Bonvicino, C. R., V. Penna-Firme, H. N. Seuánez. 1998. The karyotype of *Brucepattersonius griserufescens* Hershkovitz. 1998 (Rodentia, Sigmodontinae) with comments on distribution and taxonomy. *Z. Säugetierk.* 63:329–35.

Bonvicino, C. R. and M. Weksler. 1998. A new species of *Oligoryzomys* (Rodentia, Sigmodontinae) from central Brazil. *Z. Säugetierk.* 63:90–103.

Boom, B. M. 1981. The Ladew expedition to Bolivia and Peru: George Tate's botanical collections. *Brittonia* 33:482–89.

Borchert, M., and R. L. Hansen. 1983. Effects of flooding and wildfire on valley side wet campo rodents in central Brazil. *Rev. Bras. Biol.* 43:229–40.

Bordignon, M., and E. L. A. Monteiro-Filho. 1999. Seasonal food resources of the squirrel *Sciurus in-*

grami (Thomas, 1901) in a secondary araucaria forest in southern Brazil. *Stud. Neotrop. Fauna Environ.* 34:137–40.

———. 2000. Behaviour and daily activity of the squirrel *Sciurus ingrami* in a secondary araucaria forest in southern Brazil. *J. Can. Zool.* 78:1732–39.

Borges-Landáez, P. A., G. Perdomo, and E. A. Herrera. 2012. Estructura y diversidad genética en poblaciones manejadas de chigüire in los llanos Venezolanos. *Interciencia* 37:227–33.

Borghi, E., and M. Eguaras. 2000. Primera cita de *Androlaelaps dasymys* (Acari: Laelapidae) para la region del Monte. *Rev. Soc. Entomol. Argent.* 59:59–60.

Borgnia, M., M. L. Galante, and M. H. Cassini. 2000. Diet of the coypu (nutria, *Myocastor coypus*) in agrosystems of Argentinean pampas. *J. Wildl. Manag.* 64:354–61.

Borodin, P. M., S. C. Barreiros-Gomez, A. I. Zhelezova, C. R. Bonvicino, and P. S. D'Andrea. 2006. Reproductive isolation due to genetic incompatibilities between *Thrichomys pachyurus* and two subspecies of *Thrichomys apereoides*. *Genome* 49:159–67.

Borruel, N., C. M. Campos, S. M. Giannoni and C. E. Borghi. 1998. Effect of herbivorous rodents (cavies and tuco-tucos) on a shrub community in the Monte Desert, Argentina. *J. Arid Environ.* 39:33–37.

Bosco, C., C. Buffet, M. A. Bello, R. Rodrigo, M. Gutierrez, and G. Garcia. 2007. Placentation in the degu (*Octodon degus*): analogies with extra subplacental trophoblast and human extravillous trophoblast. *Comp. Biochem. Physiol. A Mol. Integr. Physiol.* 146:475–85.

Bossle, R. C., C. R. Alzão, A. D. Beck, I. J. Sbalqueiro, and B. R. Lange. 1988. Estudos cromossômicos em dois gêneros de roedores da família Cricetidae coletados na serra do mar do Paraná. *Resumos do XV Congresso Brasileiro de Zoolologia*, pp. 521.

Botelho, J. R., and P. M. Linardi. 1980. Alguns ectoparasitas de roedores silvestres do município de Caratinga, Minas Gerais, Brasil. I. Relações pulga/hospedeiro. *Rev. Bras. Entomol.* 24:127–30.

Botelho, J. R., P. M. Linardi, P. Williams, and L. Nagem, 1981. Alguns hospedeiros reais de ectoparasitas do município de Caratinga, Minas Gerais. *Mem. Inst. Oswaldo Cruz* 76:57–59.

Botêlho, M. C. N., J. B. de Oliveira, M. D. B. Cavalcanti, L. M. R. M. Leite, I. P. Bastos Neto, L. A. M. da Silva, M. L. C. B. Campello, et al. 2001. Acarofauna de marsupiais e roedores silvestres da reserva biológica de Serra Negra, Pernambuco, Brasil. *Entomol. Vectores* 8:193–204.

Botêlho, M. C. N., J. B. de Oliveira, L. M. R. M. Leite, I. P. Bastos Neto, L. A. M. da Silva, L. M. C. B.

Campello, M. C. de A. Aguiar, et al. 2003. Sifonápteros parasitos de marsupiais e pequenos roedores silvestres da Reserva Biológica de Serra Negra, Pernambuco, Brasil—Registro de novos hospedeiros. *Rev. Univ. Rural Ser. Cien. Vida* 22:71–74.

Botto-Mahan, C., A. Bacigalupo, J. P. Correa, E. Oda, and A. Solari. 2012. Field assessment of *Trypanosomacruzi* infection and host survival in the native rodent *Octodondegus*. *Acta Trop.* 122:164–67.

Bounds, D. L., M. H. Sherfy, and T. A. Mollett. 2003. Nutria (*Myocastor coypus*). In *Wild mammals of North America*, 2nd ed., ed. G. A. Feldhamer, B. C. Thompson, and J. A. Chapman, 1119–47. Baltimore: Johns Hopkins University Press, xiv + 1126 pp.

Boyce, M. S., M. D. Carleton, A. L. Gardner, R. Guenzel, A. Langguth, R. A. Ojeda, J. Ramirez-Pulido, and O. A. Reig. 1982. *Oryzomys*. In *Mammal species of the world: A taxonomic and geographic reference*, ed. J. H. Honacki, K. E. Kinman, and J. W. Koeppl, 435–44. Lawrence, KS: Allen Press and Association of Systematic Collections, ix + 694 pp.

Bozinovic, F., M. J. Carter, and L. A. Ebensperger. 2005. A test of the thermal-stress and the cost-of-burrowing hypotheses among populations of the subterranean rodent *Spalacopus cyanus*. *Comp. Biochem. Physiol. A Mol. Integr. Physiol.* 140:329–36.

Bozinovic, F., and L. C. Contreras. 1990. Basal rate of metabolish and temperature regulation of two desert herbivorous octodontid rodents: *Octomys mimax* and *Tympanoctomys barrerae*. *Oecologia* 84:567–70.

Bozinovic, F., A. P. Cruz-Neto, A. Cortés, G. B. Diaz, R. A. Ojeda, S. M. Giannoni. 2007. Physiological diversity in tolerance to water deprivation among species of South American desert rodents. *J. Arid Environ.* 70:427–42.

Bozinovic, F., and P. A. Marquet. 1991. Energetics and torpor in the Atacama desert-dwelling rodent *Phyllotis darwini rupestris*. *J. Mamm.* 72:734–38.

Bozinovic, F., and R. F. Nespolo. 1997. Effect of ambient temperature and energy demands on digestive functions in leaf-eared mice (*Phyllotis darwini*) from central Chile. *Int. J. Biometeorol.* 41:23–25.

Bozinovic, F., and M. Rosenmann. 1988a. Comparative energetics of South American cricetid rodents. *Comp. Biochem. Physiol. A Comp. Physiol.* 91:195–202.

———. 1988b. Daily torpor in *Calomys musculinus*, a South American Rodent. *J. Mamm.* 69:150–52.

Bozinovic, F., and R. A. Vásquez. 1999. Patch use in a diurnal rodent: handling and searching under thermoregulatory costs. *Funct. Ecol.* 13:602–10.

Brach-Egg, E. 1986. Las ecorregiones del Perú. *Bol. Lima* 44:57–70.

Bradley, R. D., N. D. Durish, D. S. Rogers, J. R. Miller, M. D. Engstrom, and C. W. Kilpatrick. 2007. Toward a molecular phylogeny for *Peromyscus*: evidence from mitochondrial cytochrome-*b* sequences. *J. Mamm.* 88:1146–59.

Bradley, R. D., C. W. Edwards, D. S. Carroll, and C. W. Kilpatrick. 2004. Phylogenetic relationships of Neotomine-Peromyscine rodents: based on DNA sequences from the mitochondrial cytocrome-*b* gene. *J. Mamm.* 85:389–95.

Bradley, R. D., D. D. Henson, and N. D. Durish. 2008. Re-evaluation of the geographic distribution and phylogeography of the *Sigmodon hispidus* complex based on mtDNA sequences. *Southwest. Nat.* 53:301–10.

Bradley, R. D., F. Mendez-Harclerode, M. J. Hamilton, and G. Ceballos. 2004. A new species of *Reithrodontomys* from Guerrero, Mexico. *Occas. Papers Mus. Texas Tech Univ.* 231:1–12.

Braggio, E., and C. R. Bonvicino. 2004. Molecular divergence in the genus *Thrichomys* (Rodentia: Echimyidae). *J. Mamm.* 85:316–20.

Braggio, E., M. D. Giménez, J. R. Contreras, E. Justo, and C. J. Bidau. 1999. Karyotypic variation in populations of *Ctenomys* (Rodentia, Ctenomytidae) from La Pampa Province, Argentina. *Caryologia* 52:131–40.

Branch, L. C. 1993a. Social organization and mating system of the plains vizcacha (*Lagostomusmaximus*). *J. Zool.*, London 229:473–91.

———. 1993b. Intergroup and intragroup spacing in the plains vizcacha, *Lagostomus maximus*. *J. Mamm.* 74:890–900.

———. 1993c. Seasonal patterns of activity and body mass in plains vizcacha, *Lagostomus maximus* (family Chinchillidae). *Can. J. Zool.* 59:152–56.

———. 1995. Observations of predation by pumas and Geoffroy's cats on the plains vizcacha in semi-arid scrub of central Argentina. *Mammalia* 59:152–56.

Branch, L. C., J. L. Hierro, and D. Villarreal. 1999. Patterns of plant species diversity following local extinction of the plains vizcacha in semi-arid scrub. *J. Arid Environ.* 41:173–82.

Branch, L. C., M. Pessino, and D. Villarreal. 1996. Response of mountain lions to a population decline of the plains vizcacha. *J. Mamm.* 77:1132–40.

Branch, L. C., D. Villarreal, and G. S. Fowler. 1993. Recruitment, dispersal, and group fusion in a declining population of the plains vizcacha (*Lagostomus maximus*; Chinchillidae). *J. Mamm.* 74:9–20.

———. 1994. Factors influencing population dynamics of the plains vizcacha (*Lagostomus maximus*, Mammalia, Chinchillidae) in scrub habitat of central Argentina. *J. Zool.*, London 232:383–95.

Branch, L. C., D. Villarreal, J. L. Hierro, and K. M. Portier. 1996. Effects of local extirpation of the plains vizcacha (*Lagostomus maximus*) on vegetation patterns in seni-arid scrub. *Oecologia* 106:389–99.

Branch, L. C., D. Villarreal, A. P. Sbriller, and R. A. Sosa. 1994. Diet selection of the plains vizcacha (*Lagostomus maximus*, family Chinchillidae) in relation to resource abundance in semi-arid scrub. *Can. J. Zool.* 72:2210–16.

Branch, L. C., D. Villarreal, A. Sosa, M. Pessino, M. Machicote, P. Lerner, P. Borraz, M. Urioste, and J. L. Hierro. 1994. Estructura de las colonias de vizcacha y problemas asociados con la estimación de la densidad poblacional en base a la actividad de las vizcacheras. *Mastozool. Neotrop.* 1:1–7.

Brandão-Filho, S. P., M. E. Brito, F. G. Carvalho, E. A. Ishokawa, E. Cupolillo, L. Floeter-Wintes, and J. J Saw. 2003. Wild and synanthropic hosts of *Leishmania* (*Viannia*) *brasiliensis* in the endemic cutaneous leishmaniasis locality of Amaraji, Pernambuco state, Brazil. *Trans. Roy. Soc. Trop. Med. Hyg.* 97:291–96.

Brandis, J. D. 1786. *Versuch einer Naturgeschichte von Chili*. Leipzig: Friedrich Gotthold Jacobäer, 2 [unnumbered] + 1–328. [German edition of Molina, G. I. 1782. *Saggio sulla storia naturale del Chili*.]

Brandt, J. F. 1835. Mammalium rodentium exoticorum novorum vel minus rite cognitorum Musei Academici Zoologici, descriptiones et icones. Sectio I. Hystricum, quae in museo academico Servantur Generum atque specierum illustrationes, pp. 357–425, pls. 1–10. Sectio II. Sciuri langsdorffii, Muris leucogastri, Muris anguyae, Hypudaei guiara et Criceti fuscati illustrationes, pp. 425–36, pls. 11–15. Sectio III. Caviae leucopygae et Caviae flavidentis descriptio. pp. 436–42, pls. 16–17. *Mémoires de l'Académie Impériale des Sciences de SaintPétersbourg, série 6, Sciences Mathématiques, Physiques et Naturelles, Tome 3* (pt. 2):357–442 + 17 plates. [Also published separately as Brandt, J. F. 1835. *Mammalium exoticorum novorum vel minus rite cognitorum Museu Academici Zoologici descriptions et icons, ex Academiae Imperalis Scientiarum CommentariorumVI. Series Tomo II. et III. separatism impressae*. Petropoli: Lipsiae apud L. Voss, pp. 1–106 + 19 pls.]

———. 1855. *Beiträge zur nähern Kenntniss der Säugethiere Russland's*. *Mem. Acad. Imp. Sci. St.-Petersbourg.*, series 6, 7:1–365.

Brants, A. 1827. *Het Geslacht der Muizen door Linnaeus opgesteld, Volgens de Tegenswoordige toestand der Wetenschap in Familien, Geslachten en Soorten verdeeld*. Berlin: Gedrukt ter Akademische Boekdrukkery, xii + 190 pp, 1 plate.

Brass, E. 1911. *Aus dem Reiche der Pelze*, 2 vols. Berlin: Neue Pelzwaren-Zeitung, xxl + 709 pp.

Braun, J. K. 1993. *Systematic relationships of the tribe Phyllotini (Muridae: Sigmodontinae) of South America*. Norman: Special Publication Oklahoma Museum of Natural History, University of Oklahoma, 50 pp.

Braun, J. K., B. S. Coyner, M. A. Mares, and R. A. Van Den Bussche. 2008. Phylogenetic relationships of South American grass mice of the *Akodon varius* group (Rodentia, Cricetidae, Sigmodontinae) in South America. *J. Mamm.* 89:768–77.

Braun, J. K., and M. A. Mares. 1995. A new genus and species of phyllotine rodent (Rodentia: Muridae: Sigmodontinae: Phyllotini) from South America. *J. Mamm.* 76:504–21.

———. 1996. Unusual morphologic and behavioral traits in *Abrocoma* (Rodentia: Abrocomidae) from Argentina. *J. Mamm.* 77:891–97.

———. 2002. Systematics of the *Abrocomacinerea* species complex (Rodentia: Abrocomidae), with a description of a new species of *Abrocoma. J. Mamm.* 83:1–19.

Braun, J. K., M. A. Mares, B. S. Coyner, and R. A. Van Den Bussche. 2010. New species of *Akodon* (Rodentia: Cricetidae: Sigmodontinae) from central Argentina. *J. Mamm.* 91:387–400.

Braun, J. K., M. A. Mares, and R. A. Ojeda. 2000. A new specif of grass mouse, genus *Akodon* (Muridae: Sigmodontinae), from Mendoza Province, Argentina. *Z. Säugetierk.* 65:216–25.

Brennan, J. M. 1957. Revival of *Crotiscus* Ewing, 1944, and descriptions of four new species of the genus from Peruvian rodents (Acarina: Trombiculidae). *J. Parasitol.* 43:673–80.

———. 1968. A collection of chiggers (Acarina: Trombiculidae) from rodents in southwestern Colombia. *J. Parasitol.* 54:679–85.

———. 1969. Five new species of the unique qenus *Polylopadium* (Acarina: Trombiculidae). *J. Parasitol.* 55:866–71.

———. 1972. A new genus and two new species of Venezuelan chiggers with eyes in the scutum (Acarina: Trombiculidae). *J. Med. Entomol.* 9:16–18.

Brennan, J. M., and M. L. Goff. 1977. The Neotropical genus *Hoffmannina*: four new species and other records from Mexico, Panama and Venezuela (Acarina, Trombiculidae). *J. Parasitol.* 63:908–14.

———. 1978. Two new species of *Crotiscus* (Acari: Trombiculidae) from northern South America. *J. Med. Entomol.* 14:569–72.

———. 1979. A new species of *Teratothrix* (Acari: Trombiculidae) from Venezuela. *J. Med. Entomol.* 15:272–73.

Brennan, J. M., and E. K. Jones. 1961. Chiggers of Peru (Acarina: Trombiculidae). *Acarologia* 3:172–205.

Brennan, J. M., and J. T. Reed. 1974a. The genus *Eutrombicula* in Venezuela (Acarina: Trombiculidae). *J. Parasitol.* 60:699–711.

———. 1974b. Endoparasitic chiggers: 8. The intradermal genus *Intercutestrix* Brennan and Yunker 1966, and three new species (Acarina: Trombiculidae). *J. Parasitol.* 60:185–87.

———. 1975. A list of Venezuela chiggers, particularly of small mammalian hosts (Acarina: Trombiculidae). *Brigham Young Univ. Sci. Bull.*, biol. ser., 20:45–75.

Brennan, J. M., and C. E. Yunker. 1969. Endoparasitic chiggers. V. New genera, species and records from Venezuela and Brazil (Acarina: Trombiculidae). *J. Med. Entomol.* 6:299–304.

Brennand, P. G. G., A. Langguth, and A. R. Percequillo. 2013. The genus *Hylaeamys* Weksler, Percequillo, and Voss 2006 (Rodentia: Cricetidae: Sigmodontinae) in the Brazilian Atlantic Forest: geographic variation and species definition. *J. Mamm.* 94:1346–1363.

Bretschneider, D. S. 1987. "Algunos aspectos da biologia e ecologia de *Ctenomys flamarioni* Travi, 1981 (Rodentia, Ctenomytidae)." Master's thesis, Universidade Federal de Rio Grande do Sul, Brazil.

Briani, D. C., E. M. Vieira, and M. V. Vieira. 2001. Nests and nesting sites of Brazilian forest rodents (*Nectomys squamipes* and *Oryzomys intermedius*) as revealed by a spool-and-line device. *Acta Theriol.* 46:331–34.

Bridges, T. 1844. On the habits, &c. of some of the smaller species of Chilian rodents. *Ann. Mag. Nat. Hist.* 14:53–57.

Brisson, M. J. 1756. *Le règne animal, divisé en IX classes, ou méthode contenant la division generale des animaus en IX classes, & la division particulière des deux première classes, sçavoir de celle des Quadrupèdes & de celle des Cetacées, en orders, sections, genres & espéces* [etc.]. Paris: Cl. Jean-Baptiste Bauche, i-x + 1–382 + 2 (unnumbered).

———. 1762. *Regnum animale in classes IX*. Lugduni Batavorum: Theordorum Haak, Leiden, 1–7 + 296 pp.

Brito, D. 2004. Lack of adequate taxonomic knowledge may hinder endemic mammal conservation in the Brazilian Atlantic Forest. *Biodiver. Conserv.* 13:2135–44.

Brito, D., and M. S. L. Figueiredo. 2003. Minimum viable population and conservation status of the Atlantic Forest spiny rat *Trinomys eliasi. Biol. Conserv.* 112:153–58.

Brito, H. F. V., R. R. Lange, J. R. Pachaly, and I. Deconto. 2010. Determination of the basal metabolic rate in agoutis, *Dasyprocta azarae*, by indirect calorimetry. *Pesqui. Vet. Bras.* 30:471–78.

Brito M. J., and A. Arguero. 2012. Nuevos datos sobre la distribución de *Scolomys ucayalensis* (Rodentia: Cricetidae) y *Phylloderma stenops* (Chiroptera: Phtyllostomidae) en Ecuador. *Mastozool. Neotrop.* 19:293–98.

Brito M. J., W. R. Teska, and R. Ojala-Barbour. 2012. Descripción del nido de dos especies de *Thomasomys* (Cricetidae) en un bosque alto-andino en Ecuador. *Therya* 3:263–68.

Brongniart, A. 1792. Catalogue de mammifères envoyés de Cayenne par M. le Blond. *Act. Soc. Hist. Nat. Paris* 1:115.

Brookes, J. 1828. On a new genus of Rodentia. *Zool. J.* 4(13):133–34.

———. 1829. On a new new genus of the order Rodentia. *Trans. Linn. Soc. Lond.* 16, pt. 1:95–104 + 1 plate. [This paper was read before the Linnean Society on both June 3 and 17, 1828, but the volume of the *Transactions* in which it appeared is dated 1829.]

Brooks, D. M., R. J. Baker, R. J. Vargas M., T. Tarifa, H. Aranifar, and J. M. Rojas. 2004. A new species of *Oryzomys* (Rodentia: Muridae) from an isolated pocket of Cerrado in eastern Bolivia. *Occas. Papers Mus. Texas Tech Univ.* 241:1–11.

Brown, A. D., H. R. Grau, L. R. Malizia, and A. Grau. 2001. Argentina. In *Bosques Nublados del Neotrópico*, ed. M. Kappelle and A. D. Brown, 623–59. Santo Domingo de Heredia, Costa Rica: Editorial INBIO, 698 pp.

Brown, J. C. 1971. The description of mammals. 1. The external characters of the head. *Mamm. Rev.* 1:151–68.

Brown, J. C., and D. W. Yalden. 1973. The description of mammals. 2. Limbs and locomotion of terrestrial mammals. *Mamm. Rev.* 3:107–34.

Brown, R. W. 1956. *Composition of scientific words.* Washington, DC: Smithsonian Institution Press, 882 pp.

Brum-Zorrilla, N. 1965. Investigaciones citogeneticas sobre algunas especies de Cricetidae (Rodentia) del Uruguay. *An. Segundo Congr. Lat-Am. Zool.* 2:315–20.

Brum-Zorrilla, N., G. H. De Catalfo, C. Degiovanangelo, R. L. Wainberg, and T. Gentile de Fronza. 1990. *Calomys laucha* (Rodentia, Cricetidae) chromosomes from Uruguay and Argentina. *Caryologia* 43:65–77.

Brum-Zorrilla, N., N. Lafuente, and P. Kiblisky. 1972. Cytogenetic studies in the cricetid rodent *Scapteromys tumidus* (Rodentia-Cricetidae). *Specialia* 28:1373.

Brum-Zorrilla, N., G. Oliver, T. Gentile de Fronza, and R. Wainberg. 1986. Karyological studies of South American rodents (Rodentia: Cricetidae). I. Comparative chromosomic analysis in *Scapteromys* taxa. *Caryologia* 39:131–42.

Brünnich, M. T. 1771. *Zoologiae fundamenta praelectionibus academicis accomodata.* Grunde I Dyrelaeren. Hafniae et Lipsiae: F. C. Pelt, 253 pp.

Bryant, J. D., and M. C. McKenna. 1995. Cranial anatomy and phylogenetic position of *Tsaganomys altaicus* (Mammalia: Rodentia) from the Hsanda Gol Formation (Oligocene), Mongolia. *Am. Mus. Novit.* 3156:1–42.

Bryant, M. C. 1945. Phylogeny of Nearctic Sciuridae. *Am. Midl. Nat.* 33:257–390.

Bueno, A. de A., M. J. Lapenta, F. Oliveira, and J. C. Motta-Junior. 2004. Association of the "IUCN vulnerable" spiny rat *Clyomys bishopi* (Rodentia: Echimyidae) with palm trees and armadillo burrows in southeastern Brazil. *Rev. Biol. Trop.* 52:1009–11.

Bueno, A. de A., and J. C. Motta-Junior. 2008. Small mammal prey selection by two owl species in southeastern Brazil. *J. Raptor Res.* 42:248–55.

Bueno, M. L., and M. Gómez-Laverde. 1993. Variación heterochromatica en *Proechimys semispinosus* (Rodentia: Echimyidae) de la region Pacífica Colombiana. *Caldasia* 17:333–40.

Bueno, M. L., M. Gómez-Laverde, and A. Morales. 1989. Caracterization cariologica de *Proechimys sp.* (Rodentia: Echimyidae) de una colonia experimental. *Biomedica* 9:13–22.

Buffon, G. L. le Clerc. 1763. *Histoire naturelle, générale et particulière, avec la description du cabinet du Roi.* Paris: L'Imprimerie Royale, 10:1–6 (unnumbered) + 368 pp., 57 plates.

———. 1767. *Histoire naturelle, générale et particulière, avec la description du cabinet du Roi.* Paris: L'Imprimerie Royale, 15:5 (unnumbered) + 1–207, 17 plates.

———.1776. Histoire natuirelle générale et particuliére, servant de suite à l'histoire des animaux quadrupeds. In *Histoire naturelle, avec la description du cabinet du Roi*, ed. G. L. L. de Buffon. Paris: L'Imperimerie Royale, Supplément, 3:11 (unnumbered) + 330 + xxi pp., 65 plates.

———. 1778. Histoire naturelle générale et particuliére, servant de suite à l'histoire des animaux quadrupèdes. In *Histoire naturelle, avec la description du cabinet du Roi*, G. L. L. de Buffon. Paris: L'Imprimerie Royale, Supplement, 4:1–601, 84 plates.

———. 1789. Histoire natuirelle générale et particuliére, servant de suite à l'histoire des animaux quadrupeds. In *Histoire naturelle, avec la description du cabinet du Roi*, ed. G. L. L. de Buffon. Paris: L'Imperimerie Royale, Supplément, 7:1–8, ix-xx, + 364 pp., 81 plates.

Bugge, J. 1970. The contribution of the stapedial artery to the cephalic arterial supply in muroid rodents. *Acta Anat.* 76:313–36.

———. 1971. The cephalic arterial system in New and Old World hystricomorphs, and in bathyergoids, with special reference to the systematic classification of rodents. *Acta Anat.* 80:516–36.

———. 1974. The cephalic arteries of hystricogmorph rodents. In *The Biology of Hystricomorph Rodents*, ed. I. W. Rowlands and B. J. Weir, 61–78. *Symp. Zool. Soc. Lond.*, 34:xix + 482.

Buiton-Jurado, G., and M. Tobar. 2007. Posible asociación de la ardilla enana *Microsciurus flaviventer* (Rodentia: Sciuridae) y bandadas mixtas de aves en la Amazonia ecuatoriana. *Mastozool. Neotrop.* 14:235–40.

Burger, J. R., A. S. Chesh, P. Muñoz, F. Fredes, L. A. Ebensperger, and L. D. Hayes. 2012. Sociality, exotic ectoparasites, and fitness in the plural breeding rodent *Octodon degus. Behav. Ecol. Sociobiol.* 66:57–66.

Burkart, R., N. O. Bárbaro, R. O. Sánchez, and D. A. Gómez (eds). 1999. *Eco-regiones de la Argentina.* Buenos Aires: Administración de Parques Nacionales, Programa de Desarrollo Institucional Ambiental, 42 pp.

Burmeister, H. 1854. *Systematische Uebersicht der Thiere Brasiliens: welche während einer Reise durch die Provinzen von Rio de Janeiro und Minas geraës gesammelt oder beobachtet wurden von Dr. Hermann Burmeister. Säugethiere (Mammalia).* Berlin: George Reimer, 1:x + 342 pp.

———. 1855a. [Prof. Burmeister über südamerikanische Murinen.] *Abhandl. Naturfors. Gesell. Halle* 2, Sitzungsberichte, 1, Quartal:3–10.

———. 1855b. Über eine neue Ratte, *Lasiomys hirsuta* von Maracaibo. *Abhandl. Naturfors. Gesell. Halle* 2, Sitzungsberichte, 1, Quartal:15–17.

———. 1861. *Reise durch die La Plata-Staaten, mit besonderer Rücksicht auf die physische Beschaffenheit und den Culturzusttand der argentinischen Republick ausgefürt in den Jahren 1857, 1858, 1859 und 1860.* Halle: H. M. Schmidt, 2:vi + 538 pp., 1 map.

———. 1876a. Description of a new species of *Dolichotis. Proc. Zool. Soc. Lond.* 1875 (part IV):634–37 + 1 plate. [Part IV of the 1875 volume of the *Proceedings* was published in April 1876; see Duncan 1937.]

———. 1876b. Additional notes on *Dolichotis salinicola. Proc. Zool. Soc. Lond.* 1876 (part III):461–62.

———. 1879. Description physique de la République Argentine d'après des observations personnelles et étrangères. Buenos Aires: Paul-Emile Coni, 3(1):vi + 556 pp.

Burrow, E. J. 1815. Description of *Mus Castorides*, a new species. *Trans. Linn. Soc. Lond.* 11:167–69.

Burt, W. H. 1934. A new rice rat (*Oryzomys*) from Sonora, Mexico. *Proc. Biol. Soc. Washington.* 47:107–8.

Busch, C. 1989. Metabolic-rate and thermoregulation in two species of tuco-tuco, *Ctenomys talarum* and *Ctenomys australis* (Caviomorpha, Octodontidae). *Comp. Biochem. Physiol. A Physiol.* 93:345–47.

Busch, C., C. D. Antinuchi, J. C. del Valle, M. J. Kittlein, A. I. Malizia, A. I. Vasallo, and R. R. Zenuto 2000. Population ecology of subterranean rodents. *Life underground: The biology of subterranean rodents*, eds. E. A. Lacey, J. L. Patton, and G. N. Cameron, 183–226. Chicago: University of Chicago Press, xi + 449 pp.

Busch, C., and J. Del Valle. 2003. Body composition and gut length of *Akodon azarae* (Muridae: Sigmodontinae): relationship with energetic requirements. *Acta Theriol.* 48:347–57.

Busch, M. 1986. Identificación de algunas especies de pequeños mamíferos de la Provincia de Buenos Aires mediante características de sus pelos. *Physis* 44(107):113–18.

Busch, M., M. R. Alvarez, E. A. Cittadino, and F. O. Kravetz. 1997. Habitat selection and interspecific competition in rodents in pampean agroecosystems. *Mammalia* 61:167–84.

Busch, M., and F. O. Kravetz. 1992. Competitive interactions among rodents (*Akodon azarae, Calomys laucha, C. musculinus* and *Oligoryzomys flavescens*) in a two-habitat system. I. Spatial and numerical relationships. *Mammalia* 56:45–56.

Busch, M., M. Miño, J. Dadon, and K. Hodara. 2001. Habitat selection by *Akodon azarae* and *Calomys laucha* (Rodentia, Muridae) in pampean agroecosystems. *Mammalia* 65:29–48.

Buzzio, O. L., and A. Castro-Vázquez. 2002. Reproductive biology of the corn mouse, *Calomys musculinus*, a Neotropical sigmodontine. *Mastozool. Neotrop.* 9:135–58.

Cabanis, J. 1848. Saeugethiere. In *Versuch einer Fauna und Flora von Britisch-Guiana.* In Reisen in Britisch-Guiana in den Jahren 1840–1884, ed. R. Schomburgk, 766–86. Leipzig: J. J. Weber, 3:vii + 531–1260 pp.

Cabrera, A. 1901. Sobre las caracteres y la clasificación del puerco espín pequeño de Colombia. *Bol. Soc. Esp. Hist. Nat.* 1901:158–62.

———. 1905. Notas sobre algunos mamíferos chilenos. *Rev. Chilena Hist. Nat.* 9:15–16.

———. 1912. Catálogo metódico de las colleciones de mamíferos del Museo de Ciencias Naturales de Madrid. *Trab. Mus. Cien. Nat. Madr.* 7:1–147.

———. 1917. Mamíferos del viaje al Pacifico. *Trab. Mus. Cien. Nat. Madr. Ser. Zool.* 31:1–62.

———. 1926. Dos roedores nuevos de las montañas de Catamarca. *Rev. Chilena Hist. Nat.* 30:319–21.

————. 1934. Dos nuevos micromamíferos del Norte argentino. *Notas Preliminares del Museo de la Plata* 3:123–28.

————. 1953. Los roedores Argentinos de la familia Caviidae. *Facultad de Agronomía y Veterinaria, Universidad de Buenos Aires, Escuela de Veterinaria Publicación* 6:1–93.

————. 1960. Acerca de las chinchillas. *Actas y trabajos del Primer Congreso Sudamericano de Zoología*, Universidad Nacional de La Plata, Argentina 4:195–202.

————. 1961. Catálogo de los mamíferos de America del Sur. *Rev. Mus. Argentino Cienc. Nat. "Bernardino Rivadavia," Cien. Zool.* 4(2):xxii + 309–732.

Cabrera, A., and J. Yepes. 1940. *Mamíferos Sud-Americanos (vida, costumbres y descripción)*. Historia natural Ediar. Buenos Aires: Compañía Argentina de Editores, 370 pp, 1 map, 78 pls.

————. 1960. *Mamíferos Sud-Americanos*, Vol. 2, 2nd ed. Buenos Aires: Ediar, 160 pp.

Cáceres, N. C., M. C. Bornschein, W. H. Lopes, and A. R. Percequillo. 2007. Mammals of Bodoquena Mountains, southwestern Brazil: an ecological and conservation analysis. *Rev. Bras. Zool.* 24: 426–35.

Cáceres, N. C., A. P. Carmignotto, E. Fischer, and C. Ferreira Santos. 2008. Mammals from Mato Grosso do Sul, Brazil. *Check List* 4:321–35.

Cáceres, N. C., R. P. Napoli, J. Casella, and W. Hannibal. 2010. Mammals in a fragmented savanna landscape in south-western Brazil. *J. Nat. Hist.* 44:491–512.

Cáceres, N. C., R. P. Napoli, and W. Hannibal. 2011. Differential trapping success for small mammals using pitfall and standard cage traps in a woodland savanna region of southwestern Brazil. *Mammalia* 75:45–52.

Cademartori, C. V., M. E. Fabián, and J. O. Menegheti. 2004. Variações na abundância de roedores (Rodentia, Sigmodontinae) em duas áreas de floresta ombrófila mista, Rio Grande do Sul, Brasil. *Rev. Bras. Zoociên.* 6:147–67.

————. 2005. Reproductive biology of *Delomys dorsalis* (Hensel, 1872)—Rodentia, Sigmodontinae—in an area of mixed forest with conifers, at Rio Grande do Sul, Brasil. *Mastozool. Neotrop.* 12:133–44.

Cademartori, C. V., R. V. Marques, and S. M. Pacheco. 2008. Estratificação vertical no uso do espaço por pequenos mamíferos (Rodentia, Sigmodontinae) em área de floresta ombrófila mista, RS, Brasil. *Rev. Bras. Zoocien.* 10:189–96.

Cadena, A., R. P. Anderson, and P. Rivas-Pava. 1998. Colombian mammals from the Chocoan slopes of Narino. *Occas. Papers Mus. Texas Tech Univ.* 180:1–15.

Cajal, J. L., and S. M. Bonaventura. 1998. Densidad, biomasa y diversidad de mamíferos en la Puna y Cordillera Frontal. In *Bases para la conservación y manejo de la Puna y Cordillera Frontal de Argentina. El rol de las Reservas de la Biósfera*, ed. J. L. Cajal, R. Tecchi, and J. G. Fernández, Capitulo 15, 191–212. Buenos Aires: FUCEMA, 336 pp.

Cajimat, M. N. B., M. L. Milazzo, B. D. Hess, M. P. Rood, and C. F. Fulhorst. 2007. Principal host relationships and evolutionary history of the North American arenaviruses. *Virology* 367:235–43.

Calderon, G., N. Pini, J. Bolpe, S. Levis, J. Mills, E. Segura, N. Guthmann, G. Cantoni, J. Becker, A. Fonollat, C. Ripoll, M. Bortman, R. Benedetti, and D. Enria. 1999. Hantavirus reservoir hosts associated with peridomestic habitats in Argentina. *Emerg. Infect. Dis.* 5:792–97.

Calouro, A. M. 1999. Riqueza de mamíferos de grande e médio porte do Parque Nacional da Serra do Divisor (Acre, Brazil). *Rev. Bras. Zool.* 16(suppl. 2):195–213.

Camargo, C. R., E. Colares, and A. M. L. Castrucci. 2006. Seasonal pelage color change: news based on a South American rodent. *An. Acad. Brasil. Ciênc.* 78:77–86.

Camargo, N. F., J. F. Ribeiro, R. Gurgel-Gonçalves, A. R. T. Palma, A. F. Mendonça, and E. M. Vieira. 2012. Is footprint shape a good predictor of arboreality in Sigmodontinae rodents from a Neotropical savanna? *Acta Theriol.* 57:1–7.

Cameron, T. W. M., and M. R. Reesal. 1951. Studies on the endoparasitic fauna of Trinidat mammals. VII. Parasites of hystricomorph rodents. *Can. J. Zool.* 29:276–89.

Camilo-Alves, C. S. P., and G. M. Mourão. 2010. Palms use a bluffing strategy to avoid seed predation by rats in Brazil. *Biotropica* 42:167–73.

Camin, S. 1999. Mating behaviour of *Ctenomys mendocinus* (Rodentia, Ctenomyidae). *Z. Säugetierk.* 64:230–38.

————. 2010. Gestation, maternal behaviour, growth and development in the subterranean caviomorph rodent *Ctenomys mendocinus* (Rodentia, Hystricognathi, Ctenomyidae). *Anim. Biol.* 60:79–95.

Camin, S., and L. Madoery. 1994. Feeding behavior of the tuco-tuco (*Ctenomys mendocinus*): its modifications according to food availability and the changes in the harvest pattern and consumption. *Rev. Chilena Hist. Nat.* 67:257–63.

Camin, S., L. Madoery, and V. Roig. 1995. The burrowing behavior of *Ctenomys mendocinus* (Rodentia). *Mammalia* 59:9–17.

Campos, B. A. T. P., and A. R. Percequillo. 2007. Mammalia, Rodentia, Echimyidae, *Phyllomys blainvilii*

(Jourdan, 1837): range extensión and new geographic distribution map. *Check List* 3:18–20.

Campos, C. M. 1997. "Utilización de recursos alimentarios por mamíferos medianos y pequeños del desierto del Monte." Ph.D. diss., Universidad de Córdoba, Argentina).

Campos, C. M., S. M. Giannoni, P. Taraborelli, and C. E. Borghi. 2007. Removal of mesquite seeds by small rodents in the Monte desert, Argentina. *J. Arid Environ.* 13:259–62.

Campos, C. M. and R. A. Ojeda. 1997. Dispersal and germination of *Prosopis flexuosa* (Fabaceae) seeds by desert mammals in Argentina. *J. Arid Environ.* 35:707–14.

Campos, C. M., R. A. Ojeda, S. Monge, and M. Dacar. 2001. Utilization of food resources by small and medium-sized mammals in the Monte desert biome, Argentina. *Austral Ecol.* 26:142–49.

Campos, C. M., M. F. Tognelli, and R. A. Ojeda. 2001. Dolichotis patagonum. *Mamm. Species* 652:1–5.

Campos, Z. M. da S., and G. de M. Mourão. 1988. Predação e dispersão de sementes por dois pequenos roedores no Pantanal. In *Congresso Brasileiro de Zoologia, 15, 1988, Curitiba. Resumos*, p. 509. Curitiba: Sociedade Brasileira de Zoologia/Universidade Federal do Paraná.

Candela, A. M. 2004. Los coipos (Rodentia, Caviomorpha, Myocastoridae) del "Mesopotamiense" (Mioceno tardío; Formación Ituzaingó) de la provincia de Entre Ríos, Argentina. *Historia* 12:77–82.

Candela, A. M., and J. I. Noriega. 2003. Los coipos (Rodentia, Caviomorpha, Myocastoridae) del "Mesopotamiense" (Mioceno tardío; Formación Ituzaingó) de la provincia de Entre Ríos, Argentina. *INSUGEO. Misc.* (Tucumán):12:5–12.

Candela, A. M., and M. B. J. Picasso. 2008. Functional anatomy of the limbs of Erethizontidae (Rodentia: Caviomorpha): indicators of locomotor behavior in Miocene porcupines. *J. Morphol.* 269:552–93.

Candela, A. M., and L. L. Rasia. 2012. Tooth morphology of Echimyidae (Rodentia, Caviomorpha): homology assessments, fossils, and evolution. *Zool. J. Linn. Soc.* 164:451–80.

Candela, A. M., L. L. Rasia, and M. E. Perez. 2012. Paleobiology of Santacrucian caviomorph rodents: a morphofunctional approach. In *Early Miocene Paleobiology in Patagonia: High-Latitude Paleocommunities of the Santa Cruz Formation*, ed. S. F. Vizcaino, R. F. Kay, and M. S. Bargo, 287–305. Cambridge: Cambridge University Press, 378 pp.

Cangussu, S. D., F. G. Vieira, and R. B. Rossoni. 2002. Sexual dimorphism and seasonal variation in submandibular gland histology of *Bolomys lasiurus* (Rodentia, Muridae). *J. Morphol.* 254:320–27.

Cañón, C., G. D'Elía, U. F. J. Pardiñas, and E. P. Lessa. 2010. Phylogeography of *Loxodontomys micropus* with comment on the alpha taxonomy of *Loxodontomys* (Cricetidae: Sigmodontinae). *J. Mamm.* 91:1449–58.

Cañón Valenzuela, C. 2012. "Sistemática molecular de los roedores de la tribu Abrotrichini (Cricetidae: Sigmodontinae)." Master's thesis, Facultad de Ciencias Naturales y Oceanográficas, Universidad de Concepción, Concepción, Chile.

Cantoni, G., P. Padula, G. Calderón, J. N. Mills, E. Herrero, P. P. Sandoval, V. Martínez, N. Pini, and E. Larrieu. 2001. Seasonal variation in prevalence of antibody to hantaviruses in rodents from southern Argentina. *Trop. Med. Int. Health* 6:811–16.

Capllonch, P., A. Autino, M. M. Díaz, R. M. Barquez, and M. Goytia. 1997. Los mamíferos del Parque Biológico Sierra de San Javier, Tucumán, Argentina: observaciones sobre su sistemática y distribución. *Mastozool. Neotrop.* 4:49–71.

Cardoso, E. M. 2011. "Inferências filogenéticas e diversidade molecular no gênero *Neacomys* Thomas 1900 (Rodentia: Cricetidae)." Master's thesis, Universidade Federal do Pará, Belém, Brazil, 59 pp.

Cardozode Almeida, M., P. M. Linardi, and J. Costa. 2003. The type specimens of chewing lice (Insecta: Mallophaga) deposited in the entomological collection of Instituto Oswaldo Cruz, Rio de Janeiro, RJ, Brazil. *Mem. Inst. Oswaldo Cruz* 98:233–40.

Carleton, M. D. 1973. *A Survey of Gross Stomach Morphology in New World Cricetinae (Rodentia, Muroidea), with Comments on Functional Interpretations.* Miscellaneous Publications of the Museum of Zoology University of Michigan 146. Ann Arbor: Museum of Zoology, University of Michigan.

———. 1980. *Phylogenetic Relationships in Neotomine-Peromyscine Rodents (Muroidea) and a Reappraisal of the Dichotomy within New World Cricetinae.* Miscellaneous Publications of the Museum of Zoology University of Michigan 157. Ann Arbor: Museum of Zoology, University of Michigan.

———. 1989. Systematics and evolution. In *Advances in the study of Peromyscus (Rodentia)*, eds. G. L. Kirkland and J. N. Layne, 7–141. Lubbock: Texas Tech University Press, vi + 367 pp.

Carleton, M. D., and J. Arroyo-Cabrales. 2009. Review of the Oryzomys couesi complex (Rodentia: Crietidae: Sigmodontinae) in western Mexico. In *Systematic mammalogy: Contributions in honor of Guy G. Musser*, ed. R. S. Voss and M. D. Carleton, 94–127. *Bull. Am. Mus. Nat. Hist.* 331:1–450.

Carleton, M. D., and G. G. Musser. 1984. Muroid rodents. In *Orders and families of Recent mammals of the world*, ed. S. Anderson and J. K. Jones Jr., 289–379. New York: John Wiley and Sons, 686 pp.

Carleton, M. D., and G. G. Musser. 1989. Systematic studies of oryzomine rodents (Muridae, Sigmodontinae): a synopsis of *Microryzomys*. *Bull. Am. Mus. Nat. Hist.* 191:1–83.

Carleton, M. D., and S. L. Olson. 1999. Amerigo Vespucci and the rat of Fernando de Noronha: a new genus and species of Rodentia (Muridae: Sigmodontinae) from a volcanic island off Brazil's continental shelf. *Am. Mus. Novit.* 3256:1–59.

Carmignotto, A. P. 2005. "Pequenos mamíferos terrestres do bioma Cerrado: padrões faunísticos locais e regionais." PhD. diss., Universidade de São Paulo, Brazil.

Carmignotto, A. P., and C. C. Aires. 2011. Mamíferos não voadores (Mammalia) da Estação Ecológica Serra Geral do Tocantins. *Biota Neotrop.* 11:307–22.

Carmignotto, A. P., M. de Vivo, and A. Langguth. 2012. Mammals of the Cerrado and Caatinga. Distribution patterns of the tropical open biomes of central South America. In *Bones, clones, and biomes: The history and geography of Recent Neotropical mammals*, ed. B. D. Patterson and L. P. Costa, 307–50. Chicago: Chicago University Press, 419 pp.

Caro, T. 2009. Contrasting coloration in terrestrial mammals. *Philos. Trans. R. Soc. Lond. B Biol. Sci.* 364:537–48.

Carrara, A.-S., M. Gonzales, C. Ferro, M. Tamayo, J. Aronson, S. Paessler, M. Anishchenko, J. Boshell, and S. C. Weaver. 2005. Venezuelan equine encephalitis virus infection of spiny rats. *Emerg. Infect. Dis.* 11:663–69.

Carrizo, L. V., and M. M. Díaz. 2011. Postcranial description of *Rhipidomys austrinus* and *Graomys griseoflavus* (Rodentia, Cricetidae, Sigmodontinae). *Iheringia Ser. Zool.* 101:207–19.

Carroll, D. S., and R. D. Bradley. 2005. Systematics of the genus *Sigmodon*: DNA sequences from beta-fibrinogen and cytochrome-*b*. *Southwest. Nat.* 50:342–49.

Carroll, D. S., J. N. Mills, J. M. Montgomery, D. G. Bausch, P. J. Blair, J. P. Burans, V. Felices, A. Gianella, N. Tisoshi, S. T. Nichol, S. T. Olson, D. S. Rogers, M. Salazar and T. G. Ksiazek. 2005. Hantavirus pulmonary syndrome in central Bolivia: relationships between reservoir hosts, habitats, and viral genotypes. *Am. J. Trop. Med. Hyg.* 72:42–46.

Cartaya, E. 1983. "Estudio de la comunidad de roedores plaga asociada a un cultivo de arroz (*Oryzasativa* L.) a lo largo de su ciclo de vida, en el Estado Portuguesa." Trabajo Especial de Grado, Universidad Simón Bolívar, Sartenejas, Venezuela, 89 pp.

Cartaya, E., and M. Aguilera. 1984. Area de acción de *Holochilus venezuelae* Allen, 1904 (Rodentia, Cricetidae) en un cultivo de arroz. *Acta Cien. Venezol.* 35:162–63.

———. 1985. Estudio de la comunidad de roedores plaga en un cultivo de arroz. *Acta Cien. Venezol.* 36:250–57.

Carter, A. M., and A. M. Mess. 2013. Conservation of placentation during the Tertiary radiation of mammals in South America. *J. Morphol.* 274:557–69.

Carter, J., and B. P. Leonard. 2002. A review of the literature on the worldwide distribution, spread of, and efforts to eradicate the coypu (*Myocastor coypus*). *Wildl. Soc. Bull.* 30:162–75.

Carvalho, A. H., and V. Fagundes. 2005. Área de ocorrência de três táxons do gênero *Thrichomys* (Echimyidae, Rodentia) baseados em identificação cariotípica. In *Livros de Resumos, III Congresso Brasileiro de Mastozoologia*. ed. V. Fagundes, L. P. Costa, Y. L. R. Leite, and S. L. Mendes, p. 102. Espírito Santo: Aracruz, 150 pp.

Carvalho, A. H., M. O. G. Lopes, and M. Svartman. 2008. Cariótipo de *Thrichomys inermis* (Rodentia, Echimyidae) do Tocantins. In *Congresso Brasileiro de Zoologia, Curitiba*, 2008.

———. 2012. A new karyotype for *Rhipidomys* (Rodentia, Cricetidae) from southeastern Brazil. *Comp. Cytogenet.* 6:227–37.

Carvalho, C. T. de. 1959. Lectótipos e localidades das espécies de Goeldi (Primates, Carnivora e Rodentia). *Rev. Bras. Biol.* 19:459–61.

———. 1962. Lista preliminar dos mamíferos do Amapá. *Pap. Avulsos Zool.* (São Paulo) 15:283–97.

———. 1965. Comentários sobre os mamíferos descritos e figurados por Alexandre Rodrigues Ferreira em 1790. *Arq. Zool., São Paulo* 12:7–70.

———. 1979. *Dicionários dos Mamíferos do Brasil*. 2nd ed. São Paulo: Livraria Nobel.

———. 1983. Lista nominal dos mamíferos brasileiros . *Bol. Téc. Inst. Florestal Sao Paulo* 37:31–115.

Carvalho, C. T. de, and Bueno, R. A. 1975. Animais causando danos em plantios (Mammalia, Rodentia). *Silvicultura* 9:39–46.

Carvalho, C. T. de, and A. J. Toccheton. 1969. Mamíferos do nordeste do Pará, Brasil. *Rev. Biol. Trop.* 15:215–26.

Carvalho, G. A. S. 1999. "Relações filogenéticas entre formas recentes e fósseis de Echimyidae (Rodentia: Hystricognathi) e aspectos da evolução da morfologia dentária." Master's thesis, Museu Nacional and Universidade Federal do Rio de Janeiro, Brazil.

———. 2000. Substitution of the deciduous premolar in *Chaetomys subspinosus* (Olfers, 1818) (Hystricognathi,

Rodentia) and its taxonomic implications. *Z. Säuget-ierk.* 65:187–90.

Carvalho, G. A. S., and L. O. Salles. 2004. Relationships among extant and fossil echimyids (Rodentia: Hystricognathi). *Zool. J. Linn. Soc.* 142:445–77.

Carvalho, M. A. M. de, A. A. N. Machado Junior, R. A. Silva, E. Bezerra, D. J. A. Menezes, A. M. Conde Junior, and D. R. Righi. 2008. Arterial supply of the penis in agoutis (*Dasyprocta prymnolopha*, Wagler, 1831). *Anatom. Hist. Embryol.* 37:60–62.

Casado, F., J. Vilela, P. Gonçalves, R. Pardini, and C. Bonvicino. 2006. Novos registros de *Bibimys labiosus* (Rodentia: Sigmodontinae) no leste do Brasil. I. *Congreso Sul-Americano de Mastozoologia, Gramado, Brasil, Resumos*, p. 100.

Casella, J., and N. C. Cáceres. 2006. Diet of four small mammal species from Atlantic Forest patches in South Brazil. *Neotrop. Biol. Conserv.* 1:5–11.

Cassone, J., and M.-C. Durette-Desset. 1991. Five species of three new trichostrongyloid nematodes coparasites of *Dasyprocta azarae* from Paraguay. *Rev. Suisse Zool.* 98:229–42.

Cassano, C., and R. Moura. 2004. Mamíferos em sistemas produtivos de cultura permanente no entorno da Reserva Biológica de Una, Bahia. *XXV Congresso Brasileiro de Zoologia: Resumos.* Brasília: Sociedade Brasileira de Zoologia, 271.

Castaño, J. H., Y. Muñoz-Saba, J. E. Botero, and J. H. Vélez. 2003. Mamíferos del Departamento de Caldas–Colombia. *Biota Colomb.* 4:247–59.

Castellarini, F., H. Agnelli, and J. Polop. 1998. Study on the diet and feeding preferences of *Calomys venustus* (Rodentia, Muridae). *Mastozool. Neotrop.* 5:5–11.

Castellarini, F., C. Dellafiore, and J. Polop. 2003. Feeding habits of small mammals in agroecosystems of central Argentina. *Mamm. Biol.* 68:91–101.

Castillo, A. H., M. N. Cortinas, and E. P. Lessa. 2005. Rapid diversification of South American tuco-tucos (*Ctenomys*; Rodentia, Ctenomyidae): contrasting mitochondrial and nuclear intron sequences. *J. Mamm.* 86:170–79.

Castillo, E., J. Priotto, A. M. Ambrosio, M. C. Provensal, N. Pini, M. Morales, A. Steinmann, and J. J. Polop. 2003. Comensal and wild rodents in an urban area of Argentina. *Int. Biodeterior. Biodegrad.* 52:135–41.

Castro, D. del C., and A. Cicchino. 1987. Lista referencial de los Anoplura y Mallophaga (Insecta) conocidos como parásitos de mamíferos en la Argentina. *Rev. Soc. Entomol. Argent.* 44:357–70.

———. 1998. Capitulo 10. Anoplura. In *Biodiversidad de artrópodos argentinos. Una perspectiva biotax-onómica*, ed. J. J. Morrone and S. Coscarón, 125–39. La Plata, Argentina: Ediciones Sur, vii + 599 pp.

———. 2002. The species of the genus *Gyropus* Nitzch, 1818 (Phthiraptera: Gyropidae) parasitic on the Octodontidae (Mammalia: Rodentia). *Rev. Chilena Hist. Nat.* 75:293–98.

Castro, D. del C., and A. González. 1997. A new species of the genus *Hoplopleura* Enderlein, 1904 (Anoplura, Hoplopleuridae) parasitic on *Andinomys edax* (Rodentia-Cricetidae). *Graellsia* 52:31–35.

———. 2003. Una nueva especie de *Hoplopleura* (Phthiraptera, Anoplura) parásita de tres especies de *Bibimys* (Muridae, Sigmodontinae, Rodentia). *Iheringia Ser. Zool.* 93:183–88.

Castro, D. del C., A. Gonzalez, and A. Cicchino. 1998. Contribution to the knowledge of the genus *Hoplopleura* Enderlein, 1904 (Anoplura, Hoplopleuridae): a new species parasitic on Octodontidae (Rodentia). *Rev. Bras. Entomol.* 41:245–48.

Castro, D. del C., A. Gonzalez, and M. Dreon. 2001. A new species of sucking louse (Insecta, Anoplura) parasitic on *Auliscomys* (Mammalia, Rodentia) in Argentina. *J. Parasitol.* 87:263–66.

Castro, D. del C., A. Gonzalez, and E. C. Vidal. 1998. *Hoplopleura nicolai* sp. nov. (Anoplura, Hoplopleuridae), parasita de *Andalgalomys* sp. (Muridae, Rodentia). *Rev. Acad. Cien. Exact. Fisic. Quim. Nat. Zaragoza* 53:297–306.

Castro, D. del C., and D. H. Verzi. 2002. A new species of *Ferrisella* (Phthiraptera, Anoplura, Hoplopleuridae) parasitic on the desert-adapted rodent *Tympanoctomys barrerae* (Rodentia, Octodontidae). *Rudolstadter Nat. Hist. Schr.*, Suppl. 4:113–23.

Castro, E. C., M. S. Mattevi, S. W. Maluf, and L. F. B. Oliveira. 1991. Distinct centric fusions in different populations of *Deltamys kempi* (Rodentia, Cricetidae) from South America. *Cytobios* 68:153–59.

Castro, I., and H. Román. 2000. Evaluación Ecológica rápida de la mastofauna en el Parque Nacional Llanganates. In *Biodiversidad en el Parque Nacional Llanganates: Un Reporte de las Evaluaciones Ecológicas y Socioeconómicas Rápidas*, eds. M. Vásquez, M. Larrea, and L. Suárez, 129–47. Quito, Ecuador: Ministerio del Ambiente, 203 pp. + 1 map.

Castro, M. C., anbd M. C. Langer. 2011. The mammalian fauna of Abismo Iguatemi, southeastern Brazil. *J. Cave Karst Stud.* 73:83–92.

Catanesi, C. I., L. Vidal-Rioja, J. V. Crisci, and A. Zambelli. 2002. Phylogenetic relationships among Robertsonian karyomorphs of *Graomys griseoflavus* (Rodentia, Muridae) by mitochondrial cytochrome-*b* DNA sequencing. *Hereditas* 136:130–36.

Catanesi, C. I., L. Vidal-Rioja, and A. Zambelli. 2006. Molecular and phylogenetic analysis of mitochondrial control region in Robertsonian karyomorphs of *Graomys griseoflavus* (Rodentia, Sigmodontinae). *Mastozool. Neotrop.* 13:21–30.

Catesby, M. 1754. *The natural history of Carolina, Florida and the Bahama Islands: containing the figures of birds, beasts, fishes, serpents, insects and plants: particularly the forest-trees, shrubs, and other plants, not hitherto described, or very incorrectly figues by authors, together with their descriptions in English and French, to which are added, observations on the air, soil, and waters: with remarks upon agriculture, grain, pulse, roots, etc. to the whole is prefixed a new and correct map of the countries treted of.* London: C. Marsh and T. Wilcox, 2:1–100 + xliv + 6 (unnumbered).

Catzeflis, F. 2012. A survey of small non-volant mammals inhabiting Wayampi Amerindian houses in French Guiana. *Mammalia* 76:327–30.

Catzeflis, F., A. W. Dickerman, J. Michaux, and J. A. Kirsch. 1993. DNA hybridization and rodent phylogeny. In *Mammalian phylogeny: Placentals*, eds. F. S. Szalay, M. J. Novacek, and M. C. McKenna, 159–72. New York: Springer-Verlag, 321 pp.

Catzeflis, F., and C. Steiner. 2000. Nouvelles données sur la morphologie comparée et la distribution des rats épineux *Proechimys cuvieri* et *P. cayennensis* (Echimyidae: Mammalia) en Guyane française. *Mammalia* 64:209–20.

Catzeflis, F., and M.-K. Tilak. 2009. Molecular systematics of Neotropical spiny mice (*Neacomys*: Sigmodontinae, Rodentia) from the Guianan Region. *Mammalia* 73:239–347.

Cavia, R., A. Andrade, M. E. Zamero, M. S. Fernandez, E. Muschetto, G. Cueto, and O. V. Suárez. 2008. Hair structure of small rodents from central Argentina: a tool for species identification. *Mammalia* 72:35–43.

Caviedes-Vidal, E., E. Caviedes-Codelia, V. Roig, V., and R. Dona 1990. Facultative torpor in the South-American rodent *Calomys venustus* (Rodentia, Cricetidae). *J. Mamm.* 71:72–75.

Cendrero, L. 1972. *Zoología Hispanoamericana. Vertebrados.* México: Editorial Porrúa, xii + 1–1160.

Cerda, J., and R. Sandoval. 2004. Wild rodents population fluctuations in the region de Aysen and their relationship to the hantavirus cadiopulmonary syndrome (hcps) epidemiology. *Boletín Veterinario Oficial,* accessed April 15, 2011, http://www.bvo.sag.gob.cl.

Cerqueira, R. 1975. A localidade tipo de *Holochilus brasiliensis vulpinus* Brants, 1835 (Rodentia, Cricetidae). *Rev. Bras. Biol.* 35:31–34.

———. South American landscapes and their mammals. In *Mammalian biology in South America*, ed. M. A. Mares and H. H. Genoways, 53–75. The Pymatuning Symposia in Ecology 6, Special Publications Series. Pittsburgh: Pymatuning Laboratory of Ecology, University of Pittsburgh, xii + 539.

Cerqueira, R., F. A. S. Fernandez, and M. F. S. Quintela. 1990. Mamíferos da Restinga de Barra de Maricá, Rio de Janeiro. *Pap. Avulsos Zool.* (São Paulo) 37:141–57.

Cerqueira, R., R. Gentile, F. A. S. Fernandez, and P. S. D'Andrea. 1994. A five-year population study of an assemblage of small mammals in southeastern Brazil. *Mammalia* 57:507–17. [1993 vol. 57, no. 4: pp 482ff. "Achevé d'imprimer le 20 janvier 1994."]

Cerqueira, R., and L. B. Klakczo. 1975. Biometric studies on *Holochilus brasiliensis* (Rodentia, Cricetidae) I. Ontogenetic variation of a population at Crato, northeastern Brazil. *Rev. Bras. Biol.* 35:35–38.

Cerqueira, R., R. T. Santori, R. Gentile, and S. M. S. Guapiassi. 2003. Micrographical ecological differences between two populations of *Akodon cursor* (Rodentia, Sigmodontinae) in a Brazilian restinga. *J. Adv. Zool. Gorakhpur* 24:46–52.

Cerqueira, R., M. V. Vieira, and L. O. Salles. 1989. Habitat and reproduction of *Rhipidomys cearanus* at São Benedito, Ceará (Rodentia, Cricetidae). *Ciênc. Cult.* 41:1009–13.

Cestari, A. N., and J. Imada. 1968. Os cromossomos de roedor *Akodon arviculoides cursor* Winge 1888 (Cricetidae–Rodentia). *Ciênc. Cult.* 20:758–62.

Chan, Y. L., and E. A. Hadly. 2 011. Genetic variation over 10 000 years in *Ctenomys*: comparative phylochronology provides a temporal perspective on rarity, environmental change and demography. *Mol. Ecol.* 20:4592–605.

Chan, Y. L., E. A. Lacey, O. P. Pearson, and E. A. Hadly. 2005. Ancient DNA reveals Holocene loss of genetic diversity in a South American rodent. *Biol. Lett.* 1:423–26.

Chapman, F. M. 1917. The distribution of bird life in Colombia; a contribution to a biological survey of South America. *Bull. Am. Mus. Nat. Hist.* 36:1–729.

———. 1925. Remarks on the life zones of northeastern Venezuela with descriptions of new species of birds. *Am. Mus. Novit.* 191:1–15.

———. 1926. The distribution of bird-life in Ecuador; a contribution to a study of the origin of Andean bird-life. *Bull. Am. Mus. Nat. Hist.* 55:1–738.

Charles-Dominique, P., M. Atramentowicz, M. Charles-Dominique, H. Gérard, A. Hladik, C. M. Hladik, and M. F. Prévost. 1981. Les mammifères frugivores arboricoles nocturnes d'une forêt guyanaise: inter-relations plantes-animaux. *Rev. Ecol. (Terre Vie)* 35:341–435.

Charrel, R.N., and X. De Lamballerie. 2010. Zoonotic aspects of arenavirus infections. *Vet. Microbiol.* 140:213–20.

Chebez, J. C. 1994. *Los Que se Van. Especies Argentinas en Peligro*. Buenos Aires: Editorial Albatros, 604 pp.

Chehébar, C., and S. Martín. 1989. Guía para el reconocimiento microscópico de los pelos de los mamíferos de la Patagonia. *Donana Acta Vert.* 16:247–91.

Chenu, J. C. 1850. *Encyclopédie d'histoire naturelle / ou / Traité complet di cette science / d'après / les travaux des naturalists les plus éminents de tous les pays et de toutes les époques/Buffon, Daugenton, Lacépède, G. Cuvier, F. Cuvier, Geoffroy Saint-Hilaire, Latreille, de Jussieu, Brongniart, etc. etc. / Ouvrage résumant les observations des auteurs anciens et comprenant toutes les découvertes modernes jusqu'à nos jours. Rongeurs et pachyderms*. Paris: Chez Marescq et Compagnie, Chez Gustave Havard, 2:6 (unnumbered) + 312.

Cherem, J. J. 2005. Registros de mamíferos não voadores em estudos de availação ambiental no sul do Brasil. *Biotemas* 18:169–202.

Cherem, J. J., M. E. Graipel, K. Abati, M. Moraes, and T. Moreira. 2005. Registro de *Abrawayaomys ruschii* Cunha e Cruz, 1979 (Rodentia, Sigmodontinae) para o Estado de Santa Catarina, Sul do Brasil. *3 Congresso Brasileiro de Mastozoologia, Resumos*, p. 110.

Cherem, J. J., J. Olimpio, and A. Ximénez. 1999. Descrição de uma nova espécie do gênero *Cavia* Pallas, 1766 (Mammalia–Caviidae) das Ilhas dos Moleques do Sul, Santa Catarina, Sul do Brasil. *Biotemas* 12:95–117.

Cherem, J. J., and D. M. Perez. 1996. Mamíferos terrestres de floresta de araucaria no município de Três Barras, Santa Catarina, Brasil. *Biotemas* 9:29–46.

Cherem, J. J., P. C. Simões-Lopes, S. L. Althoff, M. E. Graipel. 2004. Lista dos mamíferos do estado de Santa Catarina, sul do Brasil. *Mastozool. Neotrop.* 11:151–84.

Chernova, O. F. 2002. New findings of a specialized spine cuticle in porcupines (Rodentia: Hystricomorpha) and tenrecs (Insectivora: Tenrecidae). *Dokl. Biol. Sci.* 384:267–70.

Chiaperro, M. B., G. M. Panzetta-Dutari, D. Gómez, D. Castillo, J. J. Polop, and C. N. Gardenal. 2011. Contrasting genetic structure of urban and rural populations of the wild rodent *Calomys musculinus* (Cricetidae, Sigmodontinae). *Mamm. Biol.* 76:41–50.

Chiarello, A. G., M. Passamani, and M. Zortéa. 1997. Field observations on the thin-spined porcupine, *Chae-*

tomys subspinosus (Rodentia: Echimyidae). *Mammalia* 61:29–36.

Chinchilla, F. A. 1997. La dieta del jaguar (*Panthera onca*), el puma (*Felis concolor*), y el manigordo (*Felis pardalis*) (Carnivora: Felidae) en el Parque Nacional Corcovado, Costa Rica. *Rev. Biol. Trop.* 45:1223–29.

Choo, J., T. E. Juenger, and B. B. Simpson. 2012. Consequences of frugivore-mediated seed dispersal for the spatial and genetic structures of a neotropical palm. *Mol. Ecol.* 21:1019–31.

Christoff, A. U., V. Fagundes, I. J. Sbalqueiro, M. S. Mattevi, and Y. Yonenaga-Yassuda. 2000. Description of a new species of *Akodon* (Rodentia: Sigmodontinae) from southern Brazil. *J. Mamm.* 81:838–51.

Chu, Y.-K., R. D. Owen, L. M. Gonzalez, and C. B. Johnson. 2003. The complex ecology of hantavirus in Paraguay. *Am. J. Trop. Med. Hyg.* 69:263–68.

Ciccioli, M. A. 1991. Classical, C-banding and Cd-banding karyotypes in mitotic and meiotic chromosomes of *Calomys musculinus* (Rodentia, Cricetidae). *Caryologia* 44:177–86.

Ciccioli, M. A., and L. Poggio. 1993. Genome size in *Calomys laucha* and *Calomys musculinus* (Rodentia, Cricetidae). *Genet. Select. Evol.* 25:109–19.

Cione, A. L., M. M. Azpelicueta, M. Bond., A. A. Carlini, J. R. Casciotta, M. A. Cozzuol, M. De La Fuente, Z. Gasparini, F. J. Goin, J. Noriega, G. J. Scillato-Yane, L. Soibelzon, E. P. Tonni, D. H. Verzi, and M. G. Vucetich. 2000. Miocene vertebrates from Entre Rios, eastern Argentina. In *El Neógeno de Argentina*, ed. F. G. Aceñolaza and R. Herbst, 191–237. UNSUGEO, *Ser. Corr. Geol.*, vol. 14.

Cirignoli, S., C. A. Galliari, U. F. J. Pardiñas, D. H. Podestá, and R. Abramson. 2011. Mamíferos de la Reserva Valley del Cuña Pirú, Misiones, Argentina. *Mastozool. Neotrop.* 18:25–43.

Clark, D. B. 1980. Population ecology of an endemic neotropical island rodent: *Oryzomys bauri* of Santa Fe Island, Galapagos, Ecuador. *J. Anim. Ecol.* 49:185–98.

———. 1984. Native land mammals. In *Key Environments: Galápagos*, ed. R. Perry, 225–31. Oxford: Pergamon Press, x + 321 pp.

Cockerell, T. D. A., L. I. Miller, and M. Printz. 1914. The auditory ossicles of American rodents. *Bull. Am. Mus. Nat. Hist.* 33:347–80.

Cofré, H. L., and P. A. Marquet. 1999. Conservation status, rarity, and geographic priorities for conservation of Chilean mammals: an assessment. *Biol. Conserv.* 88:53–68.

Cofré, H. L., H. Samaniego, and P. A. Marquet. 2007. Patterns of small mammal species richness in Mediterranean and temperate Chile. In *The quintessential* naturalist: Honoring the life and legacy of Oliver P. Pearson, ed. D. A. Kelt, E. P. Lessa, J. Salazar-Bravo, and J. L. Patton, 275–302. *Univ. California Publ. Zool.* 134:v-xii + 1–981.

Cohen, M. M., and L. Pinsky. 1966. Autosomal polymorphism via a translocation in the guinea pig, *Cavia porcellus* L. *Cytogenetics* 5:120–32.

Colares, I. G., R. N. V. Oliveira, R. M. Oliveira, and E. P. Colares. 2010. Feedings habits of coypu (*Myocastor coypus* Molina 1782) in the wetlands of the southern region of Brazil. *An. Acad. Brasil. Ciênc.* 82:671–78.

Collett, S. F. 1981. Population characteristics of *Agouti paca* (Rodentia) in Colombia. *Publ. Mus. Mich. State Univ.* 5:485–602.

Collins, L. R., and J. F. Eisenberg. 1972. Notes on the behaviour and breeding of pacaranas *Dinomys branickii* in captivity. *Int. Zoo Yearbk.* 12:108–14.

Colombetti, P. L., A. G Autino, G. L. Claps, M. I. Carma, and M. Lareschi. 2008. Primera cita de *Cleopsylla townsendi* (Siphonaptera: Stephanocircidae: Craneopsyllinae) en la Argentina. *Rev. Soc. Entomol. Argent.* 67:179–82.

Colombi, V. H., A. B. Martinelli, and V. Fagundes. 2011. Low molecular divergence and distinct karyotypes in *Calomys cerqueirai*: one or two species? *2a Reunião Brasileira de Citogenética*, Águas de Lindóia, São Paulo, Brazil, p. 30.

Comparatore, V. M., M. Agnusdel, and C. Busch. 1992. Habitat relations in sympatric populations of *Ctenomys australis* and *Ctenomys talarum* (Rodentia, Octodontidae) in a natural grassland. *Z. Säugetierk.* 57:47–55.

Concepción, J. L., and J. Molinari. 1991. *Sphiggurus vestitus pruinosus* (Mammalia, Rodentia, Erethizontidae): the karyotype and its phylogenetic implications, descriptive notes. *Stud. Neotrop. Fauna Environ.* 26:237–41.

Conservation International. 2011. Spectacular mammal rediscovered after 113 years—first ever photographs taken. Conservation International Press Release, May 18, 2011, www.conservation.org/newsroom/pressreleases/Pages/Rodent-Rediscovered-Colombia.aspx. (See also Long-lost tree rat reappears. 2011. *Science* 332:1017–18.)

Contreras, J. R. 1966. Una nueva localidad para el género *Scapteromys* (*Mammalia, Rodentia*). *Physis* 26:65–66.

———. 1968. *Akodon molinae* una nueva especie de ratón de campo del sur de la provincial de Buenos Aires. *Zool. Plat. Invest. Zool. Paleontol.* 1(2):9–12.

———. 1972. Nuevos datos acerca de la distribucion de algunos roedores en las provincias de Buenos Aires,

La Pampa, Entre Ríos, Santa Fe y Chaco. *Neotropica* 18:27–30.

———. 1979a. Los vertebrados de la Reserva de Ñacuñán. I. Lista Faunística preliminar. *Cuaderno Técnico* (IADIZA) 1:39–47.

———. 1979b. Ecología del ratón de campo *Akodon azarae* en la región semiárida del sudoeste bonaerense, Laguna Chasiceo, partido de Villarino. *Deserta* (Mendoza) 4:15–24.

———. 1980. Sobre el limite occidental de la distribución geográfica del cuis grande *Cavia aperea pamparum* en la Argentina. *Hist. Nat.* 11:73–74.

———. 1981. El tunduque, un modelo de ajuste adaptativo. *Ser. Cien. (Mendoza)* 21:22–25.

———. 1982a. Nota acerca de *Bolomys temchuki* (Massoia, 1982) en el noreste argentino con la descripción de dos nuevas subespecies (Rodentia, Cricetidae). *Hist. Nat.* 2:174–76.

———. 1982b. *Graomys griseoflavus* (Waterhouse, 1837) en la provincial del Chaco, República Argentina (Rodentia, Cricetidae). *Hist. Nat.* 2:252.

———. 1984a. Notas para servir de base a una revisión del género *Ctenomys* (Mammalia: Rodentia). II. *Ctenomys saltarius*. *Hist. Nat.* 3:249–52.

———. 1984b. Notas para servir de base a una revisión del género *Ctenomys* (Mammalia: Rodentia). IV. *Ctenomys juris*. *Hist. Nat.* 4:239–40.

———. 1984c. Nota sobre *Bibimys chacoensis* (Shamel, 1931) (Rodentia, Cricetidae, Scapteromyini). *Hist. Nat.* 4:280.

———. 1986. Bioenergetics and distribution of fossorial *Spalacopus cyanus* (Rodentia): thermal stress, or cost of burrowing? *Physiol. Zool.* 59:20–28.

———. 1988. *Ctenomys roigi*, una nueva especie de "änguyá-tutú" de la provincia de Corrientes, Argentina (Rodentia, Ctenomyidae). In *Homenaje a Antonio Cano Gea*, ed. F. Cano Gea, A. Duque, J. Martínez Oña, et alia, 51–67. *Bol. Inst. Estud. Almerien.* 6:437 pp.

———. 1992. Acerca de la localidad tipica de *Calomys callosus* (Rengger, 1830) (Mammalia, Rodentia, Cricetidae). *Nótulas Faun.* 35:1–5.

———. 1993. Una nueva especie de roedor excavador del género *Ctenomys* procedente del Paraguay Oriental (Rodentia, Octodontinae, Ctenomyinae). Descripción preliminar. *Resúmenes VI Congreso Iberoamericano de Conservación y Zoología de Vertebrados* (Santa Cruz, Bolivia): 44–46.

———. 1995a. *Ctenomys osvaldoreigi*, una nueva especie de tucu-tuco procedente de la llanura cordobesa nororiental, República Argentina. *Nótulas Faun.* 84:1–4.

———. 1995b. Una nueva especie de tuco-tuco procedente de la llanura cordobesa nororiental, República Argentina (Rodentia, Ctenomyidae). *Nótulas Faun.* 86:1–6.

———. 1996. Acerca de la distribución geográfica de la morfología espermática en el género *Ctenomys* (Rodentia, Ctenomyidae). *Nótulas Faun.* 88:1–5.

———. 1999. El género *Ctenomys* en la Provincia de Tucumán, República Argentina, con la descripción de una nueva especie (Rodentia, Ctenomyidae). *Ciencia Siglo XXI, Fundación Bartolomé Hidalgo, Buenos Aires, Argentina* 3:1–31.

———. 2000. *Ctenomys paraguayensis*, una nueva especie de roedor excavador procedente del paraguay Oriental (Mammalia, Rodentia, Ctenomyidae). *Rev. Mus. Argent. Cien. Nat.*, n.s., 2(1):61–68.

Contreras, J. R., and L. M. Berry. 1982a. *Ctenomys bonettoi*, una nueva especie de tucu-tucu procedente de la provincia del Chaco (Rodentia, Octodontidae). Diagnosis preliminar. *Hist. Nat.* 2:123–24.

———. 1982b. *Ctenomys argentinus*, una nueva especie de tucu-tucu procedente de la provincia del Chaco, República Argentina (Rodentia, Octodontidae). *Hist. Nat.* 2:165–73.

———. 1984. Una nueva especie del género *Ctenomys* procedente de la Provincia de Santa Fe (Rodentia, Ctenomyidae). *Resúmenes de las VII Jornadas Argentinas de Zoología*, Mar del Plata, Argentina, p. 75.

———. 1985. Acerca de la distribución de *Ctenomys argentinus* (Rodentia: Ctenomyidae). *Hist. Nat.* 5:104.

Contreras, J. R., and C. J. Bidau. 1999. Líneas generales del panorama evolutivo de los roedores excavadores del género *Ctenomys* (Rodentia: Ctenomyidae). *Ciencia Siglo XXI, Fundación Bartolomé Hidalgo, Buenos Aires, Argentina* 1:1–22.

Contreras, J. R., C. J. Bidau, M. D. Giménez, and A. S. Di Giacomo. 2000. Nuevas consideraciones acerca de la taxonomía y la historia paleobiogeográfica y evolutiva del género *Ctenomys* (Rodentia: Caviomorpha: Ctenomyidae). *Actas IX Congreso Iberoamericano de Biodiversidad y Zoología de Vertebrados*, Buenos Aires, Argentina, pp. 139–41.

Contreras, J. R., D. C. Castro, and A. C. Cicchino. 1999. Relaciones del los Phthiraptera (Insecta, Amblycera, Gyropidae) con la evolución taxonómia del genero Ctenomys (Mammalia: Rodentia: Ctenomyidae). *Siglo XXI, Fundación Bartolomé Hidalgo, Buenos Aires, Argentina* 2:1–32.

Contreras, J. R., and A. N. Ch. de Contreras. 1984. Diagnosis preliminar de una nueva especie de "Anguyá-tutú" (género *Ctenomys*) para la provincia de Corrientes, Argentina (Mammalia, Rodentia). *Hist. Nat.* 4:131–32.

Contreras J. R., Y. E. Davies, A. O. Contreras, and M. Alvarez M. 1985. Acerca de la distribución de *Ctenomys perrensi* Thomas 1896 y sus relaciones geográficas con las demás especies del género (Rodentia, Ctenomyidae). *Hist. Nat.* 5:173–78.

Contreras, J. R., and E. R. Justo. 1974. Aportes a lo mastozoología pampeana. I. Nuevas localidades para roedores Cricetidae (Mammalia, Rodentia). *Neotrópica* 20:91–96.

Contreras, J. R., and O. A. Reig. 1965. Datos sobre la distribución del género *Ctenomys* (Rodentia, Octodontidae) en la zona costera de la provincia de Buenos Aires comprendida entre Necochea y Bahía Blanca. *Physis* 25:169–86.

Contreras, J. R., and V. G. Roig. 1975. *Ctenomys eremophilus*, una nueva especie de tucu-tuco de la región de Ñacuñán, provincia de Mendoza (Rodentia, Octodontidae). *Resúmenes IV Jornadas Argentinas de Zoología*, Corrientes, Argentina, p. 19.

———. 1978. Observaciones sobre la organización social, la ecología y la estructura de los habitáculos de *Microcavia australis* en Ñacuñán, provincia de Mendoza. *Ecosur* 5:191–99.

———. 1992. Las especies del género *Ctenomys* (Rodentia: Octodontidae). I. *Ctenomys dorsalis* Thomas, 1900. *Nótulas Faun.* 34:1–4.

Contreras, J. R., V. G. Roig, and C. M. Suzarte. 1977. *Ctenomys validus*, una nueva especie de "tunduque" de la provincia de Mendoza (Rodentia, Octodontidae). *Physis* 36:159–62.

Contreras, J. R. and M. L. Rosi. 1980a. Una nueva subespecie del ratón colilargo para la Provincia de Mendoza: *Oligoryzomys flavescens occidentalis* (Mammalia, Rodentia, Cricetidae). *Hist. Nat.* 1:157–60.

———. 1980b. Acerca de la presencia en la Provincia de Mendoza del ratón de campo *Akodon molinae* Contreras, 1968 (Rodentia: Cricetidae). *Hist. Nat.* 1:181–84.

———. 1981. Notas sobre los Akodontini argentinos (Rodentia, Cricetidae). II. *Akodon andinus andinus* (Philippi, 1868) en la provincia de Mendoza. *Hist. Nat. Corrientes Argent.* 1:233–36.

Contreras J. R., and J. A. Scolaro. 1986. Distribución y relaciones taxonómicas entre los cuatro núcleos geográficos disjuntos de *Ctenomys dorbignyi* en la Provincia de Corrientes, Argentina (Rodentia, Ctenomyidae). *Hist. Nat.* 6:21–30.

Contreras, J. R., and P. Teta. 2003. Acerca del estatus taxonômico y de la localidad típica de *Oxymycterus rufus* (Fischer, 1814) (Rodentia: Muridae: Sigmodontinae). *Nótulas Faun.* (n.s.): 14:1–5.

Contreras, J. R., P. Teta, and A. Andrade. 2003. Comentarios sobre el estatus de *Calomys callosus* (Rengger) y nuevos datos sobre la distribucion de micromamiferos en el noroeste de la provincia de Corrientes (Argentina). *Rev. Mus. Argent. Cienc. Nat.,* n.s., 5:73–78.

Contreras, L. C., and J. R. Gutiérrez. 1991. Effects of the subterranean herbivorous rodent *Spalacopus cyanus* on herbaceous vegetation in arid coastal Chile. *Oecologia* 87:106–9.

Contreras, L. C., J. Torres-Mura, and A. Spotorno. 1990. The largest known chromosome number for a mammal in a South American desert rodent. *Experientia* 46:506–8.

Contreras, L. C., J. C. Torres-Mura, A. E. Spotorno, and F. M. Catzeflis. 1993. Morphological variation of the glans penis of South American octodontid and abrocomid rodents. *J. Mamm.* 74:926–35.

Contreras, L. C., J. C. Torres-Mura, A. E. Spotorno, and L. I. Walker. 1994. Chromosomes of *Octomys mimax* and *Octodontomys gliroides* and relationships of octodontid rodents. *J. Mamm.* 75:768–74.

Contreras, L. C., J. C. Torres-Mura, and J. L. Yáñez. 1987. Biogeography of octodotid rodents: an eco-evolutionary hypothesis. *Fieldiana Zool.* 39:401–11.

Contreras, V. H. 1991. Abrocomidae (Rodentia: Caviomorpha) in the Huayquerian (Upper Miocene) of San Juan, Argentina. *Ameghiniana* 28:405.

Cook, J. A., S. Anderson, and T. L. Yates. 1990. Notes on Bolivian mammals. 6. The genus *Ctenomys* in the highlands. *Am. Mus. Novit.* 2980:1–27.

Cook, J. A., and E. P. Lessa. 1998. Are rates of diversification in subterranean South American tuco-tucos (genus *Ctenomys*, Rodentia: Octodontidae) unusually high? *Evolution* 52:1521–27.

Cook, J. A., and T. L. Yates. 1994. Systematic relationships of the Bolivian tuco-tucos, genus *Ctenomys* (Rodentia: Ctenomyidae). *J. Mamm.* 75:583–99.

Cope, E. D. 1889. On the Mammalia obtained by the Naturalist Exploring Expedition to southern Brazil. *Am. Nat.* 23:128–50.

Corbalán, V. E. 2006. Microhabitat selection by murid rodents in the Monte desert of Argentina. *J. Arid Environ.* 65:102–10.

Corbalán, V. E., and G. Debandi. 2006. Microhabitat use by *Eligmodontia typus* (Rodentia: Muridae) in the Monte Desert (Argentina). *Mamm. Biol.* 71:124–27.

Corbalán, V. E., and R. A. Ojeda. 2004. Splatial and temporal organization of small mammal communities in the Monte desert, Argentina. *Mammalia* 68:5–14.

———. 2005. Áreas de acción en un ensamble de roedores del Desierto del Monte (Mendoza, Argentina). *Mastozool. Neotrop.* 12:145–52.

Corbalán, V. E., S. Tabeni, and R. A. Ojeda. 2006. Assessment of habitat quality for four small mammal

species of the Monte Desert, Argentina. *Mamm. Biol.* 71:227–37.

Corbet, G. B., and J. E. Hill. 1980. *A world list of mammalian species.* London: British Museum (Natural History).

———. 1991. *A world list of mammalian species,* 3rd ed. Oxford: Oxford University Press.

Cordeiro-Estrela, P., M. Baylac, C. Denys, and J. Marinho-Filho. 2006. Interspecific patterns of skull variation between sympatric Brazilian vesper mice: geometric morphometrics assessment. *J. Mamm.* 87:1270–79.

Cordeiro Jr., D. A., and S. A. Talamoni. 2006. New data on the life history and occurrence of spiny rats *Trinomys moojeni* (Rodentia: Echimyidae), in southeastern Brazil. *Acta Theriol.* 51:163–68.

Corley, J. C., G. J. Fernandez, A. F. Capurro, A. J. Novaro, M. C. Funes, and A. Travaini. 1995. Selection of cricetine prey by the culpeo fox in Patagonia: a differential prey vulnerability hypothesis. *Mammalia* 59:315–25.

Corrêa, M. M. de O., M. O. G. Lopes, E. V. C. Câmara, L. C. Oliveira, and L. M. Pessôa. 2005. The karyotypes of *Trinomys moojeni* (Pessôa, Oliveira and Reis, 1992) and *Trinomys setosus elegans* (Lund, 1841) (Rodentia, Echimyidae) from Minas Gerais, eastern Brazil. *Arq. Mus. Nac., Rio de Janeiro* 63:169–74.

Corriale, M. J., S. M. Arias, R. F. Bó, and G. Porini. 2006. Habitat-use patterns of the coypu *Myocastor coypus* in an urban wetland of its original distribution. *Acta Theriol.* 51:295–302.

Cortés, A., E. Miranda, M. Rosenmann, and J. R. Rau. 2000. Thermal biology of the fossorial rodent *Ctenomys fulvus* from the Atacama desert, northern Chile. *J. Therm. Biol.* 25:425–30.

Cortés, A., J. R. Rau, E. Miranda, and J. E. Jimenéz. 2002. Habitos alimenticios de *Lagidium viscacia* y *Abrocoma cinerea*: roedores sintopicos en ambientes altoandinos del norte de Chile. *Rev. Chilena Hist. Nat.* 75:583–93.

Cortés, A., M. Rosenmann, and F. Bozinovic. 2000a. Water economy in rodents: Evaporative water loss and metabolic water production. *Rev. Chilena Hist. Nat.* 73:311–21.

———. 2000b. Relación costo beneficio en la termoregulación en *Chinchilla lanigera*. *Rev. Chilena Hist. Nat.* 73:351–57.

Cortés, A., C. Tirado, and M. Rosenmann. 2003. Energy metabolish and thermoregulation in *Chinchilla brevicaudata*. *J. Therm. Biol.* 28:489–95.

Corti, M., and M. Aguilera. 1995. Allometry and chromosomal speciation of the casiraguas *Proechimys* (Mammalia, Rodentia). *J. Zool. Syst. Evol. Res.* 33:109–15.

Corti, M., M. Aguilera, and E. Capanna. 2001. Size and shape changes in the skull accompanying speciation of South American spiny rats (Rodentia: *Proechimys* spp.). *J. Zool. Lond.* 253:537–47.

Costa, B. M. de A. 2007. "Sistemática de *Rhipidomys* (Mammalia: Sigmodontinae) do leste do Brasil." Master's thesis, Universidade Federal do Espírito Santo, Vitória, Brazil.

Costa, B. M. de A., L. Geise, L. G. Pereira, and L. P. Costa. 2011. Phylogeography of *Rhipidomys* (Rodentia: Cricetidae: Sigmodontinae) and description of two new species from southeastern Brazil. *J. Mamm.* 92:945–62.

Costa, L. P. 2003. The historical bridge between the Amazon and Atlantic Forest of Brazil: a study of molecular phylogeography with small mammals. *J. Biogeogr.* 30:71–86.

Costa, L. P., Y. L. R. Leite, and J. L. Patton. 2003. Phylogeography and systematic notes on two species of gracile mouse opossums, genus *Gracilinanus* (Marsupialia: Didelphidae) from Brazil. *Proc. Biol. Soc. Washington* 116:275–92.

Costa, L. P., S. E. Pavan, Y. L. R. Leite, and V. Fagundes. 2007. A new species of *Juliomys* (Mammalia: Rodentia: Cricetidae) from the Atlantic Forest of southeastern Brazil. *Zootaxa* 1463:21–37.

Couch, L., G. W. Foster, M. Machicote, and L. C. Branch. 2001. Descriptions of two new species of *Eimeria* (Apicomplexa: Eimeriidae) and of *Eimeria chinchillae*-like oocysts from the plains vizcacha *Lagostomus maximus* (Desmarest, 1817) (Rodentia: Chinchillidae) from Argentina. *J. Parasitol.* 87:144–47.

Coues, E. 1874. Synopsis of the Muridae of North America. *Proc. Acad. Nat. Sci. Philadelphia* 26:173–96.

———. 1884. *Thomasomys*, a new subgeneric type of *Hesperomys. Am. Nat.* 18:1275.

———. 1890. *Oryzomys*. In *The century dictionary and cyclopedia*, vol. V, ed. W. D. Whitney, 4164. New York: The Century Company, 3557–4436.

Courtalon, P., and M. Busch. 2010. Community structure and diversity of sigmodontine rodents (Muridae: Sigmodontinae) inhabiting maize and soybean fields in pampean agroecosystems, Argentina. *Interciencia* 35:812–17.

Coutinho, L. C., J. A. Oliveira, and L. M. Pessôa. 2013. Morphological variation in the appendicular skeleton of Atlantic Forest sigmodontine rodents. *J. Morphol.* 274, acccessed April 2013, doi:10.1002/jmor.20134.

Couto, D., and S. A. Talamoni. 2005. Reproductive condition of *Akodon montensis* Thomas and *Bolomys lasiurus* (Lund) (Rodentia, Muridae) based on histological and histometric analyses of testes and external characteristics of gonads. *Acta Zool.* 86:111–18.

Couve Montane, E. 1975. Citogenetica de *Phyllotis* (*Auliscomys*) *boliviensis* Waterhouse, con referencia especial al subgenero *Auliscomys* (Rodentia, Cricetidae). *An. Mus. Hist. Nat. Valparaiso* 8:114–23.

Coyner, B. S. 2010. "Phylogenetic relationships and historical biogeography of the genus *Akodon* (Rodentia: Cricetidae)." Ph.D. diss., Oklahoma State University, Stillwater.

Coyner, B. S., J. K. Braun, M. A. Mares, and R. A. van den Bussche. 2013. Taxonomic validity of species groups in the genus *Akodon* (Rodentia, Cricetidae). *Zool. Scr.* 42(4):335–50. doi:10.1111/zsc.12014.

Cracraft, J. A. 1985. Historical biogeography and patterns of differentiation within the South American avifauna: areas of endemism. In *Neotropical Ornithology*, ed. P. A. Buckley, M. S. Foster, E. S. Morton, R. S. Ridgely, and F. G. Buckley, 49–84. *Ornithol. Monogr.* 36:1–1,041, 8 plates.

Crespo, J. A. 1963. Dispersión del chinchillón, *Lagidium viscacia* (Molina) en el noreste de Patagonia y descripción de una nueva subespecie (Mammalia; Rodentia). *Neotrópica* 9:61–63.

———. 1964. Descripción de una nueva subespecie de roedor filotino. *Neotrópica* 10:99–101.

———. 1966. Ecología de una comunidad de roedores silvestres en el Partido de Rojas, Provincia de Buenos Aires. *Rev. Mus. Argentino Cien. Nat. "Bernardino Rivadavia," Ecol.* 1:79–134.

———. 1974. Comentarios sobre nuevas localidades para mamíferos de Argentina y de Bolivia. *Rev. Mus. Argentino Cien. Nat. "Bernardino Rivadavia," Cien. Zool.* 11:1–31.

———. 1982a. Introducción a la ecología de los mamíferos del Parque Nacional El Palmar. *An. Parq. Nac.* 15:1–13.

———. 1982b. Ecología de la comunidad de mamíferos del parque Nacional Iguazú, Missiones. *Rev. Mus. Argentino Cien. Nat. "Bernardino Rivadavia," Cien. Zool.* 3:45–162.

Cretzschmar, P. J. 1826–1830. Säugethiere. In *Atlas zu der Reise im nördlichen Afrikavon Eduard Rüppel, Erste Abtheilung: Zoologie*, 1–78 + 30 plates. Senckenbergische naturforschende Gesellschaft, Frankfurt am Main: Heinrich Ludwig Brönner, not continuously paginated.

Cruz, P. F. N. 1983. Occurrence and evaluation of damages caused by rodents, cacao pests in Bahia, Brazil. *Rev. Theobroma* 13:59–61.

Cuartas-Calle, C. A., and J. Muñoz-Arango. 2003. Lista de los mamíferos (Mammalia: Theria) del departmento de Antioquia, Colombia. *Biota Colomb.* 4:65–78.

Cuervo Díaz, A., J. Hernández Camacho, and A. Cadena G. 1986. Lista actualizada de los mamíferos de

Colombia anotaciones sobre su distribución. *Caldasia* 15:471–501.

Carus, J. V. 1868. *Handbuch der Zoologie*. Leipzig: Verlag von Wilhelm Englemann. ii + 1–432.

Cueto, G. R., P. Teta, and P. De Carli. 2008. Rodents from southern Patagonian semi-arid steppes (Santa Cruz Province, Argentina). *J. Arid Environ.* 72:56–61.

Cueto, V. R., M. Cagnoni, and M. J. Piantanida. 1995a. Habitat use of *Scapteromys tumidus* (Rodentia: Cricetidae) in the delta of the Paraná River, Argentina. *Mammalia* 59:25–34.

Cueto, V. R., M. I. S. Lopez, and M. J. Piantanida. 1995b. Variación estacional del area de campeo de *Oxymycterus rufus* (Rodentia: Cricetidae), en el Delta del Río Paraná, Argentina. *Doñana Acta Vertebr.* 22:87–95.

Cueto, V. R., M. J. Piantanida, and M. Cagnoni. 1995c. Population demography of *Oxymycterus rufus* (Rodentia: Cricetidae) inhabiting a patchy environment of the delta of the Paraná river, Argentina. *Acta Theriol.* 40:123–30.

Cunha, A. B. A. 2005. "Estrutura genética populacional de *Akodon cursor* (Rodentia: Sigmodontinae) em fragmentos florestais de Mata Atlântica." Master's thesis, Universidade Federal do Rio de Janeiro, Rio de Janeiro, 50 pp.

Cunha, F. L. S., and J. F. Cruz. 1979. Novo gênerode Cricetidae (Rodentia) de Castelo, Espírito Santo, Brasil. *Bol. Mus. Biol. Prof. Mello Leitão Ser. Zool.* 96:1–5.

Cutrera, A. P., C. D. Antinuchi, and C. Busch. 2003. Thermoregulatory development in pups in the subterranean rodent *Ctenomys talarum*. *Physiol. Behav.* 79:321–30.

Cutrera, A. P., C. D. Antinuchi, M. S. Mora, and A. I. Vassallo. 2006. Home-range and activity patterns of the south American subterranean rodent *Ctenomys talarum*. *J. Mamm.* 87:1183–91.

Cutrera, A. P., and E. A. Lacey. 2006. Major histocompatibility complex variation in talas tuco-tucos: the influence of demography on selection. *J. Mamm.* 87:706–16.

———. 2007. Trans-species polymorphism and evidence of selection on class II MHC loci in tuco-tucos (Rodentia: Ctenomyidae). *Immunogenetics* 59:937–48.

Cutrera, A. P., E. A. Lacey, and C. Busch. 2005. Genetic structure in a solitary rodent (*Ctenomys talarum*): implications for kinship and dispersal. *Mol. Ecol.* 14:2511–23.

———. 2006. Interspecific variation in effective population size in talar tuco-tucos (*Ctenomys talarum*): the role of demography. *J. Mamm.* 87:108–16.

Cutrera, A. P., E. A. Lacey, M. S. Mora, and E. P. Lessa. 2010. Effects of contrasting demographic histories on selection at major histocompatibility complex loci in two sympatric species of tuco-tucos (Rodentia: Ctenomyidae). *Biol. J. Linn. Soc.* 99:260–77.

Cutrera, A. P., M. S. Mora, C. D. Antenucci, and A. I. Vassallo. 2010. Intra-and interspecific variation in home-range size in sympatric tuco-tucos, *Ctenomys australis* and *C. talarum*. *J. Mamm.* 91:1425–34.

Cutrera, R. A., N. B. Carreno, and A. Castro-Vázquez. 1992. Correlative genital tract morphology and plasma progesterone levels during the ovarian cycle in corn mice *Calomys musculinus*. *Z. Säugetierk.* 57:14–22.

Cuvier, F. 1807. Du genre paca. Coelogenus (*Cavia paca*, Lin.). *Ann. Mus. Nat. Hist. Nat. Paris* 10:203–8.

———. 1809. Extrait des premiers Mémoires de M. F. Cuvier, sur les Dents des mammiferes considérées comme caractères génerériques. *N. Bull. Sci. Soc. Philomath. Paris* 1(24):393–95.

———. 1812. Du mémoire intitulé: Essai sur de nouveaux caractères pour les genres des Mammifères (1). Des rongeurs. *Ann. Mus. Natl. Hist. Nat.* 19:268–95, 1 plate.

———. 1817a. *Dictionnaire des sciences naturelles, dans lequel on traite méthodiquement des différens êtres de la nature, considérés soit en eux-mêmes, d'après l'état actuel de nos connoissances, soit relativement à l'utilité qu'en peuvent retirer la médecine, l'agriculture, le commerce et les artes.* Strasbourg: F. G. Levrault; Paris: Le Normant, 6:1–108.

———. 1817b. *Dictionnaire des sciences naturelles, dans lequel on traite méthodiquement des différens êtres de la nature, considérés soit en eux-mêmes, d'après l'état actuel de nos connoissances, soit relativement à l'utilité qu'en peuvent retirer la médecine, l'agriculture, le commerce et les artes.* Strasbourg: F. G. Levrault; Paris: Le Normant, 9:1–560.

———. 1819. Ecureuil (Mamm.). In *Dictionnaire des sciences naturelles*, Vol. 14: 240–49. Paris: F.G. Levrault et Le Normant, 1–558.

———. 1823a. Examen des espèces du genre porc-épic, et formation des genres ou sous-genres *Acanthion*, *Eréthizon*, *Sinéthère* et *Sphiggure*. *Mem. Mus. Hist. Nat. Paris* 9:413–37. [See Sherborn (1914) for date of publication.]

———. 1823b. Sur les rapports qui existent entre les animaux de la famille des Écureuils; c'est-à-dire, les Tamias, les Macroxus, les Ecureuils, les Sciuroptères et les Ptéromys. *Mem. Mus. Hist. Nat. Paris* 10:116–28 + 1 plate

———. 1823c. *Des dents des mammifères, considérées comme caractères zoologiques.* Strasbourg and Paris: F. G. Levrault; Paris: Le Normant, lvi + 258 pp., 103 pls. [Commonly cited with a publication date of 1825, but *Des dents* . . . was issued in parts from 1821 to

1825 (see Sherborn 1922) with the rodent section first published in 1823.]

——. 1823d. Description du *Saccomys anthophile*. *Mem. Mus. Hist. Nat. Paris* 10:419–28.

——. 1829a. Zoologie = Mammalogie. In *Dictionnaire des sciences naturelles, dans lequel on traite méthodoquement des différens êtres de la nature, considérés soit en eux-mêmes, d'après l'état actuel de nos connoissances, soit relativement a l'utilité qu'en peuvent retirer la médecine, l'agriculture, le commerce et les arts*, 357–519. Strasbourg and Paris: F. G. Levrault; Paris: Le Normant, 59:1–520.

——. 1829b. *Cercomys du Bresil*. In *Histoire naturelle des mammifères, avec des figures originales, coloriées, dessinées, d'apres des animaux vivans*, ed. É. Geoffroy St.-Hilaire and F. Cuvier, Series 3, 6:1–2, pl. 276. Paris: Chez A. Berlin, Libraire-Editeur.

——. 1829c. Écureuil de la Californie. In *Histoire naturelle des mammifères, avec des figures originales, coloriées, dessinées, d'apres des animaux vivans*, ed. É. Geoffroy St.-Hilaire and F. Cuvier, Series 3, 60:2. Paris: Chez A. Berlin, Libraire-Editeur.

——. 1830. *Histoire naturelle des mammifères, avec des figures originales, coloriées, dessinées, d'apres des animaux vivans*, ed. É. Geoffroy St.-Hilaire and F. Cuvier, vol. 2:21–40. Paris: Museum d'Histoire Naturelle.

——. 1831. Les écureuils. In *Supplément a l'Histoire Naturelle Générale et Particuliere de Buffon*. Tome 1. Mammifères, pp. 271–312. Paris: Chez F. D. Pillot, 470 pp.

——. 1832. Description des characterres propres aux genres *Graphiure* et *Cercomys* de l'ordre des rongeurs. *N. Ann. Mus. Hist. Nat. Paris* 1:449–52, plate 18, figure 2.

——. 1834. Description d'une nouvelle espèce de rongeur et établissement du genre *Poephagomys*. *Ann. Sci. Nat., Zoologie*, ser. 2, 1:321–26 + plate 13.

——. 1837. Du genre Eligmodonte et de l'Eligmodonte de Buenos-Ayres, *Eligmodontiatypus*. *Ann. Sci. Nat., Zoologie*, ser. 2, 7:168–71 + p. 5.

Cuvier, G. 1817. *Le règne animal distribué d'après son organization, pour servir de base a l'histoire naturelle des animaux et d'introduction a l'anatomie comparée. Tome I. L'introduction, les mammifères et les oiseaux*. Paris: Deterville, 1:xxxviii + 540 pp.

——. 1838. Rapport sur un mémoire de M. Jourdan, de Lyon, concernant quelques mammifères nouveaux. *Ann. Sci. Nat., Zool.*, ser. 2, 8:367–74.

Dacar, M., S. Monge, S. Brengio, and V. Roig. 1998. Estudio histológico en gónadas de *Ctenomys mendocinus* en la localidad de Cacheuta (Mendoza, Argentina). *Mastozool. Neotrop.* 5:13–19.

D'Adamo, P., M. L. Guichon, Roberto F. Bó, and M. H. Cassini. 2000. Habitat use by coypu *Myocastor coypus* in agro-systems of the Argentinean pampas. *Acta Theriol.* 45:25–33.

Darskaya, N. F., and V. M. Malygin. 1996. On the fleas of mammals from the Ucayali River basin, Peruvian Amazonia. *Parazitologiya* 30:187–90.

Dalby, P. L. 1975. Biology of pampa rodents (Balcarce Area, Argentina). *Publ. Mus. Mich. State Univ.* 5:153–271.

Dalby, P. L., and M. A. Mares. 1974. Notes on the distribution of the coney rat, *Reithrodon auritus*, in northwestern Argentina. *Am. Midl. Nat.* 92:205–6.

Dalecky, A., S. Chauvet, S. Ringuet, O. Claessens, J. Judas, M. Larue, and J.-F. Cosson. 2002. Large mammals on small islands: short term effects of forest fragmentation on the large mammal fauna in French Guiana. *Rev. Ecol. Terre Vie* 8(Suppl.):145–64.

Dalmagro, A. D., and E. M. Vieira. 2005. Patterns of habitat utilization of small rodents in an area of Araucaria forest in southern Brazil. *Austral Ecol.* 30:353–62.

D'Anatro, A., and G. D'Elía. 2011. Incongruent patterns of morphological, molecular, and karyotypic vartiation among populations of *Ctenomys pearsoni* Lessa and Langguth, 1983. *Mamm. Biol.* 76:36–40.

D'Andrea, P. S., C. Horta, R. Cerqueira, and L. Rey. 1996. Breding of the water rat (*Nectomys squamipes*) in the laboratory. *Lab. Anim.* 30:369–76.

D'Andrea P. S., L. Maroja, R. Gentile, R. Cerqueira, A. Maldonado Junior, and L. Rey. 2000. The parasitism of *Schistosoma mansoni* (Digenea-Trematoda) in a naturally infected population of water rats, *Nectomys squamipes* (Rodentia-Sigmodontinae) in Brazil. *Parasitology* 120:573–82.

Darwin, C. 1838. "A notice of their habits and ranges," in Mammalia, described by George R. Waterhouse, Esq. In *The zoology of the voyage of H.M.S. Beagle under the command of Captain Firzroy, R.N., during the years 1832–1836*, ed. C. Darwin, Fascicles 2, 4, and 5 (pages i–vi + 1–48, pls. 1–24, 33). London: Smith, Elder and Co., 2:xii + 97 pp., 35 pls. 1838–1839. [See Sherborn 1897 for dates of publication of parts.]

——. 1839. *Narrative of the surveying voyages of His Majesty's ships Adventure and Beagle, between the years 1826 and 1836, describing their examination of the southern shores of South America, and the Beagle's circumnavgation of the globe*. Vol. 3, Journal and remarks. London: Henry Colburn, xiv + 615 pp.

——. 1845. *Journal of researches into the natural history and geology of the countries visited during the voyage of H.M.S. Beagle round the world, under the*

Command of Capt. FitzRoy, R.N. 2nd ed. London: John Murray, xiii + 519 pp.

Da Silva, C. C., I. H. Tomasco, F. G. Hoffmann, and E. P. Lessa. 2009. Genes and ecology: accelerated rates of replacement substitutions in the cytochrome *b* gene of subterranean rodents. *Open Evol. J.* 3:17–30.

Da Silva, M. C. 2011. "Filogeografía del género *Eligmodontia* (Rodentia: Cricetidae) en la Patagonia Argentina." Master's thesis, Universidad de la Republica, Mondevideo, Uruguay.

Daudin, F. M. 1802. Tableau des divisions, sous-divisions, ordres et genres des mammifères, Par le C^{en} Lacépède; Avec l'indication de toutes les espèces décrites par Buffon, et leur distribution dans chacun des genres. In *Histoire naturelle par Buffon*, ed. B. G. E. de la V. Lacépède, 143–96. Paris: P. Didot et Firmin Didot, 14:1–346. [See Richmond (1899) for date of publication. See Sherborn (1899:409) and Husson (1978:483 and references cited therein) for the attribution of names dating from this work.]

Davis, D. E. 1944. The capture of the Brazilian mouse *Blarinomys breviceps. J. Mamm.* 25:367–69.

———. 1945a. The annual cycle of plants, mosquitos, birds, and mammals in two Brazilian forests. *Ecol. Monogr.* 15:243–95.

———. 1945b. The home range of some Brazilian mammals. *J. Mamm.* 26:116–27.

———. 1947. Notes on the life histories of some Brazilian mammals. *Bol. Mus. Nac., Rio de Janeiro*, nova sér., zool.,, 76:1–8.

Deane, L. M. 1967. Mammalian trypanosomes from the Amazon region of Brazil: direct blood examination and xenodiagnosis of wild animals from the forest along the Belén-Brasilia highway. *Rev. Inst. Med. Trop. Sao Paulo* 9:143–48.

Debrot, A. O., J. A. de Freitas, A. Brouwer, and M. V. M. Kooy. 2001. The Curacao Barn Owl: status and diet, 1987–1989. *Caribb. J. Sci.* 37:185–93.

De Catalfo, G. E., and R. L. Wainberg. 1974. Citogenetica de *Calomys callosus callosus* (Rengger 1830) (Rodentia, Cricetinae). Analisis metrico del cariotipo somatico. *Physis* 33:215–19.

De Conto, V. D. 2007. "Estratégias bionômicas e padrões de variação e covariação do crânio de *Akodon cursor* (Rodentia: Sigmodontinae." Ph.D. diss., Universidade Federal do Rio de Janeiro, Brazil, 241 pp.

De Conto, V. D., and R. Cerqueira. 2007. Reproduction, development and growth of *Akodon lindberghi* (Hershkovitz, 1990) (Rodentia, Muridae, Sigmodontinae). *Brazil. J. Biol.* 67:707–13.

Dedet, J.-P., P. Desjeux, P. Goyot, and H. Gasselin. 1984. Infestation naturelle de *Proechimys cuvieri* Petter,

1978 (rongeurs, echimyides) par *Leishmania mexicana amazonensis* Lainson et Shaw, 1972 (kinetoplastide, trypanosomatide) en Guyane francaise. *Compt. Rend. Acad. Sci. Ser. III Sci. Vie* 298:85–87.

Dedet, J.-P., F. Gay, G. Chatenay. 1989. Isolation of *Leishmania* species from wild mammals in French Guiana. *Trans. Roy. Soc. Trop. Med. Hyg.* 83:613–15.

De la Barrera, J. M. 1940. Estudios sobre peste selvática en Mendoza. *Rev. Inst. Bacteriol.* 9:565–86.

Delacour, J. 1922. Notes sur quelques mammifères rapportés vivant de la Guyane Française et du Vénézuéla. *Rev. Hist. Nat. Appl. (Mamm.)* 3:197–99.

de la Sancha, N. U. 2010. "Effects of habitat fragmentation on non-volant small mammals of the interior Atlantic Forest of eastern Paraguay." Ph.D. diss., Texas Tech University, Lubbock, xv + 183 pp.

de la Sancha, N. U., G. D'Elía, F. Netto, P. Pérez, and J. Salazar-Bravo. 2009. Discovery of *Juliomys* (Rodentia, Sigmodontinae) in Paraguay, a new genus of Sigmodontinae for the country's Atlantic Forest. *Mammalia* 73:162–67.

de la Sancha, N. U., G. D'Elía, C. J. Tribe, P. E. Perez, L. Valdez, and R. H. Pine. 2011. *Rhipidomys* (Rodentia, Cricetidae) from Paraguay: noteworthy new records and identity of the Paraguayan species. *Mammalia* 75:269–76.

Delgado, C., M. Gómez-Laverde, and V. Pacheco. 2008. *Thomasomys bombycinus*. IUCN Red List of Threatened Species, version 2011.2, accessed February 23, 2012, http://www.iucnredlist.org.

Delgado-V., C. A. 2009. Non-volant mammals, Reserva San Sebastián-La Castellana, Valle de Aburrá, Antioquia, Colombia. *Check List* 5:1–4.

Delgado-V., C. A., and D. Zurc. 2005. New records of *Olallamys albicauda* (Rodentia: Echimyidae) in Antioquia, Colombia. *Brenesia* 63–64:131–32.

D'Elía, G. 2000. Comments on recent advances in understanding sigmodontine phylogeny and evolution. *Mastozool. Neotrop.* 7:47–55.

———. 2003a. Phylogenetics of Sigmodontinae (Rodentia, Muroidea, Cricetidae), with special reference to the akodont group, and with additional comments on historical biogeography. *Cladistics* 19: 307–23.

———. 2003b. Rats, mice, and relatives IV: Sigmodontinae. In *Grzimek's Animal Life Encyclopedia, Mammals V.*,Vol. 16, eds. D. G. Kleiman, V. Geist, M. Hutchins, and M. C. McDade, 263–79. Michigan: Thomson-Gale, 704 pp.

D'Elía, G, E. M. González, and U. F. J. Pardiñas. 2003. Phylogenetic analysis of sigmodontine rodents (Muroidea), with special reference to the akodont genus *Deltamys. Mamm. Biol.* 68:351–64.

D'Elía, G., J. P. Jayat, P. E. Ortiz, J. Salzar-Bravo, and U. F. J. Pardiñas. 2011. *Akodon polopi* Jayat et al., 2010 is a senior subjective synonym of *Akodon viridescens* Braun et al., 2010. *Zootaxa* 2744:62–64.

D'Elía, G., E. P. Lessa, and J. A. Cook. 1998. Geographic structure, gene flow, and maintenance of melanism in *Ctenomys rionegrensis* (Rodentia: Octontidae). *Z. Säugetierk.* 63:285–96.

———. 1999. Molecular phylogeny of tuco-tucos, genus *Ctenomys* (Rodentia, Octodontidae): evaluation of the *mendocinus* species group and the evolution of asymmetric sperm. *J. Mamm. Evol.* 6:19–38.

D'Elía, G., L. Luna, E. M. González, and B. D. Patterson. 2006. On the Sigmodontinae radiation (Rodentia, Cricetidae): an appraisal of the phylogenetic position of *Rhagomys*. *Mol. Phylogenet. Evol.* 38:558–64.

D'Elía, G., I. Mora, P. Myers, and R. D. Owen. 2008. New and noteworthy records of Rodentia (Erethizontidae, Sciuridae, and Cricetidae) from Paraguay. *Zootaxa* 1784:39–57.

D'Elía, G., A. A. Ojeda, F. Mondaca, and M. H. Gallardo. 2006. New data of the long-clawed mouse *Pearsonomys annectens* (Cricetidae, Sigmodontinae) and additional comments on the distinctiveness of *Pearsonomys*. *Mamm. Biol.* 71:39–51.

D'Elía, G., and U.F.J. Pardiñas. 2004. Systematics of Argentinean, Paraguayan, and Uruguayan swamp rats of the genus *Scapteromys* (Rodentia, Cricetidae, Sigmodontinae). *J. Mamm.* 85:897–910.

D'Elía, G., U. F. J. Pardiñas, J. P. Jayat, and J. Salazar-Bravo. 2008. Systematics of *Necromys* (Rodentia, Cricetidae, Sigmodontinae): species limits and groups, with comments on historical biogeography. *J. Mamm.* 89:778–90.

D'Elía, G., U. F. J. Pardiñas, and P. Myers. 2005. An introduction to the genus *Bibimys* (Rodentia: Sigmodontinae): phylogenetic position and alpha taxonomy. In *Mammalian diversification: from chromosomes to phylogeraplhy*, eds. E. A. Lacey and P. Myers, 211–46. *Univ. California Publ. Zool.* 133:v-viii + 1–383.

D'Elía, G., U. F. J. Pardiñas, P. Teta, and J. L. Patton. 2007. Definition and diagnosis of a new tribe of sigmodontine rodents (Cricetidae: Sigmodontinae), and a revised classification of the subfamily. *Gayana* 71:187–94.

D'Elía, G., P. Teta, and U. F. J. Pardiñas. 2006. Incertae sedis. In *Mamíferos de Argentina: sistemática y distribución*, eds. R. M. Barquez, M. M. Díaz, and R. A. Ojeda, 197–202. Tucumán: Sociedad Argentina para el Estudio de los Mamíferos, 359 pp.

Della-Flora, F., G. Leal Melo, J. Sponchiado, and N. C. Cáceres. 2013. Association of the southern Amazon red squirrel *Urosciurus spadiceus* Olfers, 1818 with mixed-species bird flocks. *Mammalia* 77:113–17.

del Valle, J. C., M. I. Lohfelt, V. M. Comparatore, M. S. Cid, and C. Busch. 2001. Feeding selectivity and food preference of *Ctenomys talarum* (tuco-tuco). *Mamm. Biol.* 66:165–73.

del Valle, J. C. and A. A. Lopez Mananes. 2008. Digestive strategies in the South American subterranean rodent *Ctenomys talarum*. *Comp. Biochem. Physiol. A. Mol. Int. Physiol.* 150:387–94.

———. 2011. Digestive flexibility in females of the subterranean rodent *Ctenomys talarum* in their natural habitat. *J. Exp. Zool. A Ecol. Genet. Physiol.* 315A:141–48.

Dennler, G. 1934. Die Chinchilla in Zoologie und Pelzhandel und ihre Farmzucht. *Landwirtschaftliche Pelztierzucht*, Jahrg. 5:2–9, 34–37, 66–70, 1934. [Not seen, citation from Prell 1934:103 and Anderson 1997:529.]

Debrot, A. O., J. A. de Freitas, A. Brouwer, and M. V. M. Kooy. 2001. The Curacao Barn Owl: status and diet, 1987–1989. *Caribb. J. Sci.* 37:185–93.

De Santis, L. J. M., N. Basso, I. Noriega, and M. F. Grossman. 1994. Explotación del recurso trófico por la lechuza de los campanarios (*Tyto alba*) en el oeste de Chubut, Argentina. *Stud. Neotrop. Fauna Environ.* 29:43–47.

De Santis, L. J. M., C. M. García Esponda, and G. J. Moreira. 1996. Mamíferos integrantes de la dieta de *Athene cunicularia* (Aves: Strigidae) en la región costera de la provincia del Chubut (Argentina). *Neotrópica* 43:125–26.

De Santis, L. J. M., and E. R. Justo. 1980. *Akodon (Abrothrix) mansoensis*, sp. nov., un nuevo ratón lanoso de la Provincia de Río Negro, Argentina (Rodentia, Cricetidae). *Neotropica* 26:121–27.

De Santis, L. J. M., C. I. Montalvo, and E. R. Justo. 1983. Mamíferos integrantes de la dieta de *Tyto alba* (Aves, Strigiformes, Tytonidae) en la provincia de La Pampa, Argentina. *Hist. Nat. Corrientes Argent.* 3:187–88.

De Santis, L. J. M., G. J. Moreira, and C. M. Garcia Esponda. 2001. Enamel microstrucure of upper incisors of *Ctenomys azarae* and *C. talarum* (Rodentia, Ctenomyidac). *Mastozool. Neotrop.* 8:5–14.

De Santis, L. J. M., G. J. Moreira, and E. R. Justo. 1998. Anatomy of branchiomeric muscles in some species of Ctenomys Blainville, 1826 (Rodentia, Ctenomyidae): Adaptive characters. *Bol. Soc. Biol. Concepc.* 69:89–107.

De Santis, L. J. M., G. J. Moreira, and G. O. Pagnoni. 1997. Mammals integral in the diet of *Athene cunicularia* (Aves: Strigidae) in the coastal region of the province of Chubut (Argentina). *Neotrópica* 43:109–10.

De Santis, L. J. M., and G. O. Pagnoni. 1989. Alimentación de *Tyto alba* (Aves: Tytonidae) en localidades costeras de la provincial del Chubut (República Argentina). *Neotrópica* 35:43–49.

De Santis, L., V. Roig, and E. Justo. 1991. La anatomía craneo-dentaria de *Tympanoctomys barrerae* (Lawrence). Comparación con *Octomys mimax* y consideraciones acerca de su Estado taxonómico (Rodentia: Octodontidae). *Neotropica* 37:113–22.

Deschamps, C. M., M. G. Vucetich, D. H. Verzi, and A. I. Olivares. 2012. Biostratigraphy and correlation of the Monte Hermoso Formation (early Pliocene, Argentina): the evidence from caviomorph rodents. *J. South Am. Earth Sci.* 35:1–9.

Desmarest, A. G. 1804. Tableau méthodique des mammifères. In *Tableau méthoidiques d'histoire naturelle*, 5–38. In *Nouveau dictionnaire d'histoire naturelle, appliquée aux arts, principalement à l'argriculture, à l'économie rurale et domestique: Par une sociétee de naturalists et d'agriculteurs: Avec des figures tirees des trios règnes de la nature.* Paris: Deterville, vol. 24.

———. 1816. *L'Acouchi ou Akouchi* (*Cavia Akuchi*, Lin.). In *Nouvelle disctionaire d'historie naturalle, appliquée aux arts, à l'agriculture, à l'économie rurale et domestique, à la médecine, etc. Par une société de naturalists et d'agriclturs*, 212–13. Nouv. éd. Paris: Deterville, 1:1–3, i-lxxix, 1–552.

———. 1817a. Genre Gerboise du Cap (*Lagotis*). In *Nouveau dictionaire d'histoire naturelle, appliquée aus arts, à l'agrruiculture, à l'économie rurale et domestique, à la medicine, etc. Par une société de naturalists et d'agriculteurs*, 284. Nouv. éd. Paris: Deterville, 9:1–3 (unnumbered) + 1–624.

———. 1817b. Echimys, *Echimys*. In *Nouveau dictionaire d'histoire naturelle, appliquée aus arts, à l'agrruiculture, à l'économie rurale et domestique, à la medicine, etc. Par une société de naturalists et d'agriculteurs*, 54–59. Nouv. éd. Paris: Deterville, 10:1–6 (unnumbered) + 1–591.

———. 1817c. Ecureuils guerlinguets. In *Nouveau dictionaire d'histoire naturelle, appliquée aus arts, à l'agrruiculture, à l'économie rurale et domestique, à la medicine, etc. Par une société de naturalists et d'agriculteurs*, 109–10. Nouv. éd. Paris: Deterville, 10:1–6 (unnumbered) + 1–591.

———. 1817d. Le grande Gerboise, *Dipus maximus*, Blainv. In *Nouveau dictionaire d'histoire naturelle, appliquée aus arts, à l'agrruiculture, à l'économie rurale et domestique, à la medicine, etc. Par une société de naturalists et d'agriculteurs*, 117–19. Nouv. éd. Paris: Deterville, 13:3 (unnumbered) + 1–586 + 7 pls.

———. 1817e. Neuvième Espèce.—Le Hamster anomal, *Cricetus anomalus*, Nob.; *Mus anomalus*, J.-V. Thompson, Trans. Soc. Linn. In *Nouveau dictionaire d'histoire naturelle, appliquée aus arts, à l'agrruiculture, à l'économie rurale et domestique, à la medicine, etc. Par une société de naturalists et d'agriculteurs*, 180–81. Nouv. éd. Paris: Deterville, 14:1 (unnumbered) + 1–627.

———. 1819a. Le Rat du Brésil, *Mus brasiliensis*, Geoffr. In *Nouveau dictionnaire d'histoire naturelle, appliquèe aux art, principalement à l'agriculture et à l'économie rurale et domestique, à la medicine, etc. Par une société de naturalists et d'agriculteurs*, 62. Nouv. éd. Paris: Deterville, 29:1–576 + 7 plates.

———. 1819b. Le Rat angouya, Mus angouya, Nob.—Rat troisième ou Rat angouya, d'Azara. In *Nouveau dictionnaire d'histoire naturelle, appliquèe aux art, principalement à l'agriculture et à l'economie rurale et domestique, à la medicine, etc. Par une société de naturalists et d'agriculteurs*, 62. Nouv. éd. Paris: Deterville, 29:1–576 + 7 plates.

———. 1819c. Le Rat roux du Paraguay, Mus rufus, Nob.—Rat cinquième ou Rat roux, d'Azara. In *Nouveau dictionnaire d'histoire naturelle, appliquèe aux art, principalement à l'agriculture et à l'economie rurale et domestique, à la medicine, etc. Par une société de naturalists et d'agriculteurs*, 62–63. Nouv. éd. Paris: Deterville, 29:1–576 + 7 plates.

———. 1819d. Le Rat a grosse tête, Mus cephalotes, Nob.—Rat second ou rat a grosse tête, d'Azara. In *Nouveau dictionnaire d'histoire naturelle, appliquèe aux art, principalement à l'agriculture et à l'economie rurale et domestique, à la medicine, etc. Par une société de naturalists et d'agriculteurs*, 63–64. Nouv. éd. Paris: Deterville, 29:1–576 + 7 plates.

———. 1819e. Le Rat Oreillard, Mus auritus, Nob.—Le Rat quatrième ou Rat oreillard d'Azara. In *Nouveau dictionnaire d'histoire naturelle, appliquèe aux art, principalement à l'agriculture et à l'economie rurale et domestique, à la medicine, etc. Par une société de naturalists et d'agriculteurs*, 64. Nouv. éd. Paris: Deterville, 29:1–576 + 7 plates.

———. 1819f. Le Rat a tarse noir, Mus nigripes, Nob.—Le Rat sixième a tarse noir d'Azara. In *Nouveau dictionnaire d'histoire naturelle, appliquèe aux art, principalement à l'agriculture et à l'economie rurale et domestique, à la medicine, etc. Par une société de naturalists et d'agriculteurs*, 64. Nouv. éd. Paris: Deterville, 29:1–576 + 7 plates.

———. 1819g. Note sur un mammifère de l'ordre Rongeurs, mentionné par quelques auteurs, mais dont l'existence n'est pas encore génèralment admise par les naturalistes nomenclateurs. *J. Phys. Chim. Hist. Nat. Arts* 88:205–11.

———. 1819h. Note sur un mammifére de l'ordre des rongeurs, mentionné par quelques auteurs, mais dont l'existence n'est par encore généralement admise par les naturalistes. *N. Bull. Sci. Soc. Phil. Paris* (ser. 3) 6:40.

———. 1820. *Mammalogie ou description des espèces de mammifères. Premiere partie, contenant les ordres des Bimans, des Quadrumanes et des Carnassiers.* Paris: Veuve Agasse, viii + 1–555.

———. 1822. *Mammalogie ou description des espèces de mammifères.* Seconde partie, contenant les orders de Rongeurs, des Édentés, des Pachydermes, des Ruminans et des Cetacés, viii, 277–555. In *Encyclopédie méthodidque . . .* , Paris: Veuve Agasse, 196 vols.

———. 1823. "Chloromys ou Agouti." In *Dictionnaire classique d'histoire naturelle*, tome 4, ed. J. B. G. M. Bory de Saint-Vincent, 45–47. Paris: Rey et Gravier, 628 pp.

———. 1826. "Rat coypu ou Quouya." *Dictionaire des sciences naturelles* 44:490–92. Strasbourg: F. G. Levrault; Paris: Le Normant, 526 pp.

Detlefsen, J. A. 1914. Genetic studies on a cavy species cross. *Publ. Carnegie Inst.* 205:1–134.

Deutsch, L. A. 1983. An encounter between bush dog (*Speothos venaticus*) and paca (*Agouti paca*). *J. Mamm.* 64:532–33.

De Villafañe, G. 1981. Reproducción y crecimiento de *Akodon azarae azarae* (Fischer, 1829). *Hist. Nat.* 1:193–204.

De Villafañe, G., F. O. Kravetz, M. J. Piantanida, and J. A. Crespo. 1973. Dominancia, densidad e invasión en una comunidad de roedores de la localidad de Pergamino (Provincia de Buenos Aires). *Physis Sec. C* 32:47–59.

Deville, M. E. 1852. Notice sur le genre *Dactylomys*, et établissement d'un genre nouveau, le *Lasiuromys*, provenant de l'expédition de M. de Castelnau. *Rev. Mag. Zool. Pur Appl.* 1852, ser. 4:553–61 + 3 plates.

Devincenzi, G. J. 1935. Mamíferos del Uruguay. *Anal. Mus. Hist. Nat. Montevideo*, sér. 2, 4:1–96, 12 plates.

de Winton, W. E. 1896. On some mammals from Ecuador. *Proc. Zool. Soc. Lond.* 1896 (part II):507–13 + 1 plate.

———. 1900. Report on a collection made by Messrs F. V. McConnell and J. J. Quelch at Mount Roraima in British Guiana - Mammalia. *Trans. Linn. Soc. Lond. Zool.* 8:52.

Dexter, N, R. C. Dowler, J. P. Flanagan, S. Hart, M. A. Revelez, and T. E. Lee Jr. 2004. The influence of Feral Cats (*Felis catus*) on the distribution and abundance of introduced and endemic Galápagos rodents. *Pac. Conserv. Biol.* 10:210–15.

Dias, L .C., F. D. Avila-Pires, and A. C. Pinto. 1978. Parasitological and ecological aspects of *Schistosomiasis mansoni* in the valley of the Paraíba do Sul river (São Paulo state, Brazil) I. Natural infection of small mammals with *Schistosoma mansoni. Trans. Roy. Soc. Trop. Med. Hyg.* 72:496–500.

Diaw, O. T. 1976. Contribution a l'etude denematodes Trichostrongyloidea parasites de xenarthre, marsupiaux et rongeurs neotropicau. *Bull. Mus. Hist. Nat. Zool. Paris*, no. 282:1065–89.

Díaz, F. P., C. Latorre, A. Maldonado, J. Quade, and J. L. Betancourt. 2012. Rodent middens reveal espisodic, long-distance plant colonizations across the hyperarid Atacama Desert over the last 34,000 years. *J. Biogeogr.* 39:510–25.

Diaz, G. B. 2003. La rata vizcacha colorada, una especialista en el desierto. *Cien. Hoy* 13:22–23.

Diaz, G. B., and R. A. Ojeda. 1999. Kidney structure and allometry of Argentina desert rodents. *J. Arid Environ.* 41:453–61.

———. 2000. *Libro rojo de mamíferos amenazados de la Argentina.* Mendoza: Sociedad Argentina para el Estudio de los Mamíferos, 106 pp.

Díaz, G. B., R. A. Ojeda, M. H. Gallardo, and S. M. Giannoni. 2000. Tympanoctomys barrerae. *Mamm. Species* 646:1–4.

Diaz, G. B., R. A. Ojeda, and E. L. Rezende. 2006. Renal morphology, phylogenetic history, and desert adaptation of South American hystricognath rodents. *Funct. Ecol.* 20:609–20.

Díaz, I. 1999. Food habits of the Rufous-legged Owl (*Strix rufipes*) in the Mediterranean sclerophyllous forest of central Chile. *J. Raptor Res.* 33:260–63.

Díaz, M. M. 1999. "Mamíferos de la Provincia de Jujuy: Sistemática, distribución y ecología." Ph.D. diss., Universidad Nacional de Tucumán, Argentina.

———. 2000. Key to the native mammals of Jujuy Province, Argentina. *Occas. Papers, Sam Noble Oklahoma Mus. Nat. Hist.* 7:1–29.

Díaz, M. M. and R. M. Barquez. 1999. Contributions to the knowledge of the mammals of Jujuy Province, Argentina. *Southwest. Nat.* 44: 324–33.

———. 2002. *Los mamíferos de Jujuy, Argentina.* Buenos Aires, Argentina: L.O.L.A. (Literature of Latin America).

———. 2007. The wild mammals of Jujuy Province, Argentina: systematics and Distribution. In *The quintessential naturalist: Honoring the life and legacy of Oliver P. Pearson*, ed. D. A. Kelt, E. P. Lessa, J. Salazar-Bravo, and J. L. Patton, 417–578. *Univ. California Publ. Zool.* 134:v-xii + 1–981.

Díaz, M. M., R. M. Barquez, J. K. Braun, and M. A. Mares. 1999. A new species of *Akodon* (Muridae: Sigmodontinae) from northwestern Argentina. *J. Mamm.* 80:786–98.

Diaz, M. M., R. M. Barquez, and L. V. Carrizo. 2009. Mammalia, Rodentia, Cricetidae, *Tapecomys primus* Anderson and Yates, 2000: new locality record. *Check List* 5:439–41.

Díaz, M. M., J. K. Braun, M. A. Mares, and R. M. Barquez. 1997. Key to mammals of Salta Province, Argentina. *Occas. Pap. Sam Nobel Okla. Mus. Nat. Hist.* 2:1–10.

———. 2000. An update of the taxonomy, systematics, and distribution of the mammals of Salta Province, Argentina. *Occas. Pap. Sam Nobel Okla. Mus. Nat. Hist.* 10:1–52.

Díaz, M. M., P. Teta, U. F. J. Pardiñas, and R. Barquez. 2006. Tribu Phyllotini. In *Mamíferos de Argentina: Sistemática y distribución*, ed. R. M. Barquez, M. M. Díaz, and R. A. Ojeda, 175–89. Tucumán: Sociedad Argentina para el Estudio de los Mamíferos, 359 pp.

Díaz, M. M., and D. H. Verzi. 2006. Familia Octodontidae Waterhouse, 1839. In *Mamíferos de Argentina: Sistemática y distribución*, ed. R. M. Barquez, M. M. Díaz, and R. A. Ojeda, 231–37. Tucumán: Sociedad Argentina para el Estudios de los Mamíferos, 359 pp.

Díaz de Pascual, A. 1993. Caracterización del habitat de algunas especies de pequeños mamíferos de la selva nublada de Monte Zerpa, Mérida. *Ecotropicos* 6:1–9.

Díaz de Pascual, A., and J. Pefáur. 1982. Morfología del baculum de algunos roedores cricétidos venezolanos. In *Zoología Neotropical*, ed. P. Salinas, 665–80. Mérida, Venezuela: Actas del VIII Congreso Latinoamericano de Zoología, 1,531 pp.

Dickerman, A. W., and T. E. Yates. 1995. Systematics of *Oligoryzomys*: protein-electrophoretic analyses. *J. Mamm.* 76:172–88.

Didier, R. 1955: L'os pénien des ecureuils de l'Amérique du Sud. *Mammalia* 21:416–26.

———. 1959. Note sur les os péniens des rongeurs collectés au Pérou par J. Dorst. *Mammalia* 23:172–79.

———. 1962. Note sur l'os penien de quelques rongeurs de l'Amérique du Sud. *Mammalia* 26:408–30.

Dietz, J. M. 1983. Notes on the natural history of some small mammals in central Brazil. *J. Mamm.* 64:521–23.

Digiani, M. C., and M. C. Durette-Desset. 2003. Two new species of Nippostrongylinae (Nematoda: Trichostrongylina: Heligmonellidae) from the grey leaf-eared mouse *Graomys griseoflavus* (Sigmodontinae) in Argentina. *Parasite* 10:21–29.

———. 2007. Trichostrongylina (Nematoda) parasitic in *Phyllotis* sp. (Rodentia: Sigmodontinae) from Argentina, with description of three new species. *Parasitol. Int.* 56:9–18.

Digiani M. C., C. A. Sutton, and M. C. Durette-Desset. 2003. A new genus of Nippostrongylinae (Nematoda: Heligmonellidae) from the water rat *Scapteromys aquaticus* (Sigmodontinae) in Argentina. *J. Parasitol.* 89:124–32.

Domínguez-Bello, M. G., and M. D. Robinson. 1991. A comparison of digestive adaptations in two Neotropical cricetid rodents (*Holochilus venezuelae* and *Zygodontomys microtinus*). *Physiol. Zool.* 64:1542–51.

Donatti, C. I., M. Galetti, and P.R. Guimarães. 2009. Seed dispersal and predation of an endemic Atlantic Forest palm in a gradient of seed disperser's abundance. *Ecol. Res.* 24:1187–95.

d'Orbigny, A. 1829. Voyage de M. Alcide d'Orbigny dans le sud de l'Amérique Méridionale. In *Histoire naturelle génerale*, 124. In *Bulletin des sciences naturelles et de géologie*, tome 19, no. 11:212–22.

d'Orbigny, A., and I. Geoffroy St.-Hilaire. 1830. Notice sur la Viscache et le Chinchilla, considérés comme les types d'un genre particulier, nommé *Callomys*, et description d'une espèce nouvelle. *Ann. Sci. Nat.* 21:282–97.

d'Orbigny, A., and P. Gervais. 1847. Mammifères. In *Voyage dans l'Amérique méridionale (le Brésil, la République orientale de l'Uruguay, la République Argentine, la Patagonie, la République du Chili, la République de Bolivia, la République du Pérou), exécuté pendant les années 1826, 1827, 1828, 1829, 1830, 1831, 1832 et 1833*, ed. A. d'Orbigny. Paris: P. Bertrand; Strasbourg: V. Levrault, 4(2):1–32 + 32 pls. [See Sherborn and Griffin 1934 for dates of publication.]

Dorst, J. 1971. Nouvelles recherches sur l'ecologie des rongeurs des haut plateaux Peruviens. *Mammalia* 35:515–47.

———. 1973a. Poids relatif du cœur de quelques rongeurs des hautes andes Peruviennes. *Mammalia* 36:389–94. [1972 vol. 36, no. 3:315–542, "Dépôt lépal [sic], 1er trimestre 1973."]

———. 1973b. Morphologie de l'estomac et régime alimentaire de quelques rongeurs des hautes Andes de Pérou. *Mammalia* 36:647–56. [1972 vol. 36, no. 4:543ff. "Dépôt légal, 1er trimestre 1973."]

Doutt, J. K. 1938. Two new mammals from South America. *J. Mamm.* 19:100–1.

Dowler, R. C., and D. S. Carroll. 1996. The endemic rodents of Isla Fernandina: population status and conservation issues. *Noticias Galápagos* 57:8–13.

Dowler, R. C., D. S. Carroll, and C. W. Edwards. 2000. Rediscovery of rodents (Genus *Nesoryzomys*) considered extinct in the Galápagos Islands. *Oryx* 34:109–17.

Dragoo, J. W., J. Salazar-Bravo, L. J. Layne, and T. L. Yates. 2003. Relationships within the *Calomys callosus* species group based on amplified fragment length polymorphisms. *Biochem. Syst. Ecol.* 31:703–13.

Duarte, L. C., L. R. Monteiro, F. J. Von Zuben, and S. F. dos Reis. 2000. Variation in mandible shape in *Thrichomys apereoides* (Rodentia: Echimyidae): geometric analysis of a complex morphological structure. *Syst. Biol.* 49:563–78.

Duarte, M. R. 2003. Prickly food: snakes preying upon porcupines. *Phyllomedusa* 2:109–12.

Dubois, J. Y., F. M. Catzeflis, and J. J. Beintema. 1999. The phylogenetic position of "Acomyinae" (Rodentia, Mammalia) as sister group of a Murinae plus Gerbillinae clade: evidence from the nuclear ribonuclease gene. *Mol. Phylogenet. Evol.* 13:181–92.

Dubost, G. 1988. Ecology and social life of the red acouchy, *Myporocta exilis*; comparison with the orange-rumped agouti, *Dasyprocta leporina*. *J. Zool. Lond.* 214:107–23.

Dubost, G. and O. Henry. 2006. Comparison of diets of the acouchy, agouti and paca, the three largest rodents of the French Guianan forests. *J. Trop. Ecol.* 22:641–51.

Dubost, G., O. Henry, and P. Comizzoli. 2005. Seasonality of reproduction in the three largest terrestrial rodents of French Guiana forest. *Mamm. Biol.* 70:93–109.

Dubost, G., and F. Petter. 1979. Une espèce nouvelle de "rat-pêcheur" de Guyane française: *Daptomys oyapocki* sp. nov. (Rongeurs, Cricetidae). *Mammalia* 42:435–39. [1978 vol. 42, no. 4:391ff. "Achevé d'imprimer le 15 janvier 1979."]

Duckworth, J. W., and R. H. Pine. 2003. English names for a world list of mammals, exemplified by species of Indochina. *Mamm. Rev.* 33:151–73.

Duellman, W. E., and J. E. Koechlin. 1991. The Reserva Cuzco Amazónico, Peru: biological investigations, conservation, and ecotourism. *Occas. Papers, Univ. Kansas Mus. Nat. Hist.* 142:1–38.

Dulout, F. N., R. M. Ponssa, A. E. Schugurensky de Adris, and L. L. Dulout. 1976. Multiformidad y polimorfismo cromosómico en *Akodon* (*Abrothrix*) *illutea* Thomas (Rodentia, Cricetidae). *Mendeliana* 1:49–52.

Duncan, F. M. 1937. On the dates of publication of the Society's 'Proceedings,' 1859–1926. With an Appendix containing the dates of publication of 'Proceedings,' 1830–1858, compiled by the late F. H. Waterhouse, and of the 'Transactions,' 1833–1869, by the late Henry Peavot, originally published in P.Z.S. 1893, 1913. *Proc. Zool. Soc. Lond. ser. A* 107:71–81.

Dunnum, J. L. 2003. "Systematics of Bolivian *Cavia*, with biogeographic notes on the genus." Master's thesis, University of New Mexico, Albuquerque.

———. 2009. "Phylogeny, evolution and systematics within the family Caviidae (Mammalia: Rodentia)." Ph.D. diss., Texas Tech University, Lubbock.

Dunnum, J. L., and J. Salazar-Bravo. 2004. Dactylomys boliviensis. *Mamm. Species* 745:1–4.

———. 2006. Karyotypes of some members of the genus *Cavia* (Rodentia: Caviidae) from Bolivia. *Mamm. Biol.* 71:366–70.

———. 2010a. Molecular systematics, taxonomy and biogeography of the genus *Cavia* (Rodentia: Caviidae). *J. Zool. Syst. Evol. Res.* 48:376–88.

———. 2010b. Phylogeny, evolution, and systematics of the *Galeamusteloides* complex (Rodentia: Caviidae). *J. Mamm.* 91:243–59.

Dunnum, J. L., J. Salazar-Brazo, and T. L. Yates. 2001. The Bolivian bamboo rat, *Dactylomys boliviensis* (Rodentia: Echimyidae), a new record for chromosome number in a mammal. *Mamm. Biol.* 66:121–26.

Dúran, J. C., P. E. Cattan, and J. L. Yáñez. 1987. Food habits of foxes (*Canis* sp.) in the Chilean National Chinchilla Reserve. *J. Mamm.* 68:179–81.

Durant, P., and A. Díaz. 1995. Aspectos de la ecologia [sic] de roedores y musarañas de las cuencas hidrograficas [sic] andino-venezolanas. *Caribb. J. Sci.* 31:83–94.

Durden, L. A., and G. G. Musser. 1994. The sucking lice (Insecta, Anoplura) of the world: a taxonomic checklist with records of mammalian hosts and geographical distributions. *Bull. Am. Mus. Nat. Hist.* 218:1–90.

Durden, L. A., and J. P. Webb Jr. 1999. *Abrocomaphthirus hoplai*, a new genus and species of sucking louse from Chile and its relevance to zoogeography. *Med. Vet. Entomol.* 13:447–52.

Durette-Desset, M. C. 1970a. Nematodes héligmosomes d'Amérique du Sud. 7. Etude de trois especes nouvelles, parasites de *Proechimys semispinosus* (rongeurs echimyides). *Bull. Mus. Natl. Hist. Nat., Paris* 42:601–8.

———. Nématodes Héligmosomes d'Amérique du Sud. 8. Description de six nouvelles espèces, parasites de Cricétidés. *Bull. Mus. Natl. Hist. Nat., Paris* 42:730–44.

Durette-Desset, M. C., A.G. Chabaud, and C.A. Sutton. 1997. *Tapironema coronatum* n. gen., n. sp. (Trichostrongyloidea-Cooperiidae-Obeliscoidinae), a parasite of *Holochilus brasiliensis* and *Tapirus terrestris*. *Parasite* 4:227–32.

Durette-Desset, M. C., and R. A. Guerrero. 2006. A new species of *Hypocristata* (Nematoda, Trichostrongylina, Heligmosomoidea) a parasite of *Sigmodon hispidus* (Cricetidae, Sigmodontinae) from Venezuela. *Parasite* 13:201–4.

Durette-Desset, M. C., A. Q. Gonçalves, and R. M. Pinto. 2006. *Trichostrongylina* (Nematoda, Heligmosomoidea) coparasides in *Dasyprocta fuliginosa* Wagler

(Rodentia, Dasyproctidae) from Brazil, with the re-establishment of the genus *Avellaria* Freitas & Lent and the description of two new species. *Rev. Bras. Zool.* 23:509–19.

Durette-Desset, M. C., and C. Sutton. 1985. Contribución al conocimiento de la fauna parasitológica Argentina. X. Nematodes (Trichostrongyloidea) en *Akodon azarae azarae* (Fischer) y *Reithrodon auritus* Fischer. *Rev. Mus. La Plata Zool.*, n.s., 14:21–33.

Durnford, H. 1878. Notes on the birds of central Patagonia. *Ibis*, ser. 4, 2:389–406.

Dyzenchauz, F.J., and A. I. Massarini. 1999. First cytogenetic analysis of the genus *Bibimys* (Rodentia, Cricetidae). *Z. Säugetierk.* 64:59–62.

Eaton, G. F. 1916. The collection of osteological material from Machu Picchu. *Mem. Conn. Acad. Arts Sci.* 5:1–96 + 39 plate.

Ebensperger, L. A., and F. Bozinovic. 2000a. Communal burrowing in the hystricognath rodent, *Octodon degus*: a benefit of sociality? *Behav. Ecol. Sociobiol.* 47:365–69.

———. Energetics and burrowing behaviour in the semi-fossorial degu *Octodon degus* (Rodentia: Octodontidae). *J. Zool. Lond.* 252:179–86.

Ebensperger, L. A., A. S. Chesh, A. Rodrigo, L. Ortiz Tolhuysen, V. Quirici, J.R. Berger, R. Sobrero, L. D. Hayes. 2011. Burrow limitations and group living in the communally rearing rodent, *Octodon degus*. *J. Mamm.* 92:21–30.

Ebensperger, L. A., M. J. Hurtado, and C. Leon. 2007. An experimental examination of the consequences of communal versus solitary breeding on maternal condition and the early postnatal growth and survival of degu, *Octodon degus*, pups. *Anim. Behav.* 73:185–94.

Ebensperger, L. A., M. A. J. Hurtado, and R. Ramos-Jiliberto. 2006. Vigilance and collective detection of predators in degus (*Octodon degus*). *Ethology* 112:879–87.

Ebensperger, L. A., M. J. Hurtado, M. Soto-Gamboa, E. A. Lacey, and A. T. Chang. 2004. Communal nesting and kinship in degus (*Octodon degus*). *Naturwissenchaften* 91:391–95.

Ebensperger, L. A., M. J. Hurtado, and I. Valdivia. 2006. Lactating females do not discriminate between their own young and unrelated pups in the communally breeding rodent, *Octodon degus*. *Ethology* 112:921–29.

Ebensberger, L. A., J. E. Mella, and J. A. Simonetti. 1991. Trophic-niche relationships among *Galictis cuja*, *Dusicyon culpaeus*, and *Tyto alba* in central Chile. *J. Mamm.* 72:820–23.

Ebensperger, L. A., J. Ramírez-Estrada, C. Leon, R. A. Castro, L. Ortiz Tolhuysen, R. Sobrero, V. Quirici,

et al. 2011. Sociality, glucocorticoids and direct fitness in the communally rearing rodent, *Octodon degus*. *Horm. Behav.* 60:346–52.

Ebensperger, L. A., R. Sobrero, V. Campos, and S. M. Giannoni. 2008. Activity, range areas, and nesting patterns in the viscacha rat, *Octomys mimax*. *J. Arid Environ.* 72:1174–83.

Ebensperger, L. A., R. Sobrero, V. Quirici, R. A. Castro, L. Ortiz Tolhuysen, F. Vargas, J. R. Burger, et al. 2012. Ecological drivers of group living in two populations of the communally rearing rodent, *Octodon degus*. *Behav. Ecol. Sociobiol.* 66:261–74.

Echave Llanos, J. M., and C. A. Vilchez. 1964. Anatomía microscópia del estómago del ratón hocicudo (*Oxymycterus rutilans*). *Rev. Soc. Argent. Biol.* 40:187–92.

Eiris, G. C., and G. R. Barreto. 2008. Home range of marsh rats, *Holochilus sciureus*, a rodent pest in rice fields of Venezuela. *Interciencia* 34:400–5.

Eisenberg, J. F. 1963. The behavior of heteromyid rodents. *Univ. California Publ. Zool.* 69:1–100.

———. 1974. The function and motivational basis of hystricomorph vocalizations. *Symp. Zool. Soc. Lond.* 34:211–47.

———. 1989. *Mammals of the Neotropics: The northern Neotropics*. Vol. 1: *Panama, Colombia, Venezuela, Guyana, Surinam, French Guiana*. Chicago: University of Chicago Press, x + 449 pp. + 21 plates.

———. 1999. Biodiversity reconsidered. In *Mammals of the Neotropics*, Vol. 3: *The central Neotropics, Ecuador, Peru, Bolivia, Brazil*, J. F. Eisenberg and K. H. Redford, 527–48. Chicago: University of Chicago Press, x + 609 pp. + 19 plates.

Eisenberg, J. F., M. A. O'Connell, and P. V. August. 1979. Density, productivity and distribution of mammals in two Venezuelan habitats. In *Vertebrate ecology in the northern Neotropics*, ed. J. F. Eisenberg, 187–207. Washington, DC: Smithsonian Institution Press, 271 pp.

Eisenberg, J. F., and K. H. Redford. 1999. *Mammals of the Neotropics: The central Neotropics*. Vol. 3: *Ecuador, Peru, Bolivia, Brazil*. Chicago: University of Chicago Press, x + 609 pp. + 19 plates.

Eisenberg, J. F., and R. W. Thorington Jr. 1973. A preliminary analysis of a Neotropical mammal fauna. *Biotropica* 5:150–61.

Eisentraut, M. 1933. Biologischer Studien im bolivianischen Chaco. Beitrag zur biologie der säugetierfauna. *Z. Sügetierk.* 8:47–69.

———. 1983. Im Land der Chaco-Indianen, Begegnungen mit Tieren und Menschen in Sudost Bolivien. Baden-Baden: Biotropic-Verlag, 108 pp.

Eiten, G. 1972. The Cerrado vegetation of Brazil. *Bot. Rev.* 38:205–341.

Eler, E. S., M. N. F. da Silva, C. E. F. Silva, and E. Feldberg. 2012. Comparative cytogenetics of spiny rats of the genus *Proechimys* (Rodentia, Echimyidae) from the Amazon region. *Genet. Mol. Res.* 11:830–46.

Elissamburu, A., and S. F. Vizcaíno. 2004. Limb proportions and adaptations in caviomorph rodents (Rodentia: Caviomorpha). *J. Zool. Lond.* 262:145–59.

El Jundi, T. A. R. J., and T. R. O. de Freitas. 2004. Genetic and demographic structure in a population of *Ctenomys lami* (Rodentia-Ctenomyidae). *Hereditas* 140:18–23.

Ellenberg, H. 1959. Über den Wasserhaushalt tropischer Nebel-Oasen in der Kusten-wuste Perus. *Bericht über das Geobotanisches Forschungsinstitut Rubel in Zürich* 1958:47–74.

Ellerman, J. R. 1940. *The families and genera of living rodents*. Vol. 1. *Rodents other than Muridae*. London: Trustees of the British Museum (Natural History), xxvi + 689 pp.

———. 1941. *The families and genera of living rodents*. Vol. 2. *Family Muridae*. London: Trustees of the British Museum (Natural History), xii + 690 pp.

Elliot, D. G. 1903. A list of a collection of Mexican mammals with descriptions of some apparently new forms. *Field Columbian Museum, Zool. Ser.*, Publ. 71, 3:141–49.

———. 1904a. Descriptions of apparently new species and subspecies of mammals and a new generic name proposed. *Field Columbian Museum, Zool. Ser.*, Publ. 90, 3:263–70.

———. 1904b. The land and sea mammals of Middle America and the West Indies. *Field Columbian Museum, Zool. Ser.* 4(1):xxi + 489 + i-xiix.

———. 1917. *A check-list of mammals of the North American continent, the West Indies, and the neighboring seas*. Supplement. New York: American Museum of Natural History, iii-iv + 1–192 pp.

Ellis, B. A., J. N. Mills, J. E. Childs, M. C. Muzzini, K. T. McKee Jr., D. A. Enria, and G. E. Glass. 1997. Structure and floristics of habitat associated with five rodent species in an agroecosystem in central Argentina. *J. Zool.* 243:437–60.

Ellis, B. A., J. N. Mills, G. E. Glass, K. T. McKee Jr., D. A. Enria, and J. E. Childs. 1998. Dietary habits of the common rodents in an agroecosystem in Argentina. *J. Mamm.* 79:1203–20.

Ellis, B. A., J. N. Mills, E. J. T. Kennedy, J. I. Maiztegui, and J. E. Childs. 1994. The relationship among diet, alimentary tract morphology, and life history for five species of rodents from the central Argentine pampa. *Acta Theriol.* 39:345–55.

Emmons, L. H. 1981. Morphological, ecological, and behavioral adaptation for arboreal browing in *Dactylomys dactylinus* (Rodentia, Echimyidae). *J. Mamm.* 62:183–89.

———. 1982. Ecology of *Proechimys* (Rodentia, Echimyidae) in southeastern Peru. *Trop. Ecol.* 23:280–90.

———. 1987a. Ecological considerations on the farming of game animals: capybaras yes, pacas no. *Vida Silvestre Neotrop.* 1:54–55.

———. 1987b. Comparative feeding ecology of felids in a neotropical rainforest. *Behav. Ecol. Sociobiol.* 20:271–83.

———. 1988. Replacement name for a genus of South American rodent (Echimyidae). *J. Mamm.* 69:421.

———. 1991. Mammals of Alto Madidi andmammal list. In *A biological assessment of the Alto Madidi region and adjacent areas of northwest Bolivia, May 18–June 15, 1990*, ed. T. Parker and B. Bailey, 23–25, 72–73. RAP Working Papers 1. Washington, DC: Conservation International, 108 pp.

———. 1993. On the identity of *Echimys didelphoides* Desmarest, 1817 (Mammalia: Rodentia: Echimyidae). *Proc. Biol. Soc. Washington* 106:1–4.

———. 1994. New locality records of *Mesomys* (Rodentia: Echimyidae). *Mammalia* 58:148–49.

———. 1997a. A revision of the genera of arboreal echimyid rodents (Echimyidae, Echimyinae). *Seventh International Theriological Congress*, Acapulco, Mexico, Abstracts, p. 96.

———. 1997b. Mammals of the Río Urucuti basin, south central Chuquisaca, Bolivia. In *A rapid assessment of the humid forests of south central Chuquisaca, Bolivia*, ed. T. S. Schulenberg and K. Awbrey, 81–82 (Appendix 5). RAP Working Papers 8, Conservation International, Washington, DC, 84 pp.

———. 1998. Mammal fauna of Parque Nacional Noel Kempff Mercado. In *A biological assessment of Parque Nacional Noel Kempff Mercado, Bolivia*, ed. T. J. Killeen and T. S. Schulenberg, 129–43 + Appendix 3 (341–47). RAP Working Papers 10. Washington, DC: Conservation International, 372 pp.

———. 1999a. A new genus and species of Abrocomid rodent from Peru (Rodentia: Abrocomidae). *Am. Mus. Novit.* 3279:1–14.

———. 1999b. Two new species of *Juscelinomys* (Rodentia: Muridae) from Bolivia. *Amer. Mus. Novitates* 3280:1–15.

———. 2005. A revision of the genera of arboreal Echimyidae (Rodentia: Echimyidae, Echimyinae), with descriptions of two new genera. In *Mammalian diversification: from chromosomes to phylogeography*, ed. E. A. Lacey and P. Myers, 247–334. *Univ. California Publ. Zool.* 133:v-viii + 1–383 pp.

———. 2009. Long-term variation in small mammal abundances in forest and savanna of Bolivian Cerrado. *Biotropica* 41:493–502.

Emmons, L. H., V. Chávez, N. Rocha, B. Phillips, I. Phillips, L. F. del Aguila, and M. Swarner. 2006. The non-flying mammals of Noel Kempff Mercado National Park (Bolivia). *Rev. Bol. Ecol. Conserv. Ambiental* 15:23–46.

Emmons, L. H., and F. Feer. 1990. *Neotropical rainforest mammals, A field guide.* Chicago: University of Chicago Press, xiv + 1–281.

———. 1997. *Neotropical rainforest mammals, A field guide.* 2nd ed. Chicago: University of Chicago Press, xvi + 307 pp., 29 + A_G pls..

Emmons, L. H., Y. L. R. Leite, D. Kock, and L. P. Costa. 2002. A review of the named forms of *Phyllomys* (Rodentia: Echimyidae) with the description of a new species from coastal Brazil. *Am. Mus. Novit.* 3380:1–40.

Emmons, L. H., L. Luna W., and M. Romo R. 2001. Mammals of the northern Vilcabamba mountain range, Peru. In *Biological and social assessments of the Cordillera de Vilcabamba, Peru,* ed. L. E. Alonso, A. Alonso, T. S. Schulenberg, and F. Dallmeier, 105–9, 255–61. RAP Working Papers 12. Washington, DC: Conservation International, 296 pp.

Emmons, L. H., and J. L. Patton. 2005. A new species of *Oryzomys* (Rodentia: Muridae) from eastern Bolivia. *Am. Mus. Novit.* 3478:1–14.

———. 2012. Taxonomic revision of Bolivian *Juscelinomys* (Rodentia, Cricetidae) with notes on morphology and ecology. *Mammalia* 76:285–94.

Emmons, L. H., M. Romo, L. Luna, A. C. Farfán, and C. Kopper. 2002. Anexo 4. Comparación de ocurrencia de especies de mamíferos de SNPH (RAP 1992 y 1996) con otras localidades de Madre de Dios. In *Informe de las evaluaciones biológicas Pampas del Heath, Perú, Alto Madidi, Bolivia y Pando, Bolivia,* ed. J. R. Montambault, 106–10. RAP Bulletin of Biological Assessment 24. Washington, DC: Conservation International, 125 pp.

Emmons, L. H., and N. Stark. 1979. Elemental composition of a natural mineral lick in Amazonia. *Biotropica* 11:311–13.

Emmons, L. H., and M. G. Vucetich. 1998. The identity of Winge's *Lasiuromys villosus* and the description of a new genus of echimyid rodent (Rodentia: Echimyidae). *Am. Mus. Novit.* 3223:1–12.

Enders, R. K. 1935. Mammalian life histories from Barro Colorado Island. *Bull. Mus. Comp. Zool.* 78:383–502 + 5 plates.

Endries, M. J., and G. H. Adler. 2005. Spacing patterns of a tropical forest rodent, the spiny rat (*Proechimys semispinosus*), in Panama. *J. Zool.* 265:147–55.

Engle, S. R., K. M. Hogan, J. F. Taylor, and S. K. Davis. 1998. Molecular systematics and paleobiogeography of the South American sigmodontine rodents. *Mol. Biol. Evol.* 15:35–49.

Englis, A., Jr., R. E. Cole, and T. Caro. 2012. Small mammal survey of Chiquibul Forest Reserve, Maya Mountains, Beliz, 2001. *Occas. Papers Mus. Texas Tech Univ.* 308:1–23.

Engstrom, M. D., H. H. Genoways, and P. K. Tucker. 1987. Morphological variation, karyology, and systematic relationships of *Heteromys gaumeri* (Rodentia: Heteromyidae). In *Studies in neotropical mammalogy, essays in honor of Philip Hershkovitz,* ed. B. D. Patterson and R. M. Timm, 289–303. *Fieldiana Zool.* 39:frontispiece, viii + 1–506.

Enría D. A. M., and S. C. Levis. 2004. Zoonosis virales emergentes: las infecciones por Hantavirus. *Rev. Sci. Tech. Office Int. Epizooties* 23:595–611.

Enría, D. A., J. N. Mills, D. Bausch, W.-J. Shieh, and C. J. Peters. 2011. Arenavirus infections. In *Tropical infectious diseases: Principles, pathogens and practice,* 3rd ed., ed. R. L. Guerrant, D. H. Walker, and P. F. Weller, 449–61. Edinburgh, Saunders Elsevier, v–xxiv + 1–1130 pp.

Ernest, K. A. 1986. Nectomys squamipes. *Mamm. Species* 265:1–5.

Ernest, K. A., and M. A. Mares. 1986. Ecology of *Nectomys squamipes*, the neotropical water rat, in central Brazil: home range, habitat selection, reproduction and behavior. *J. Zool. Lond.* 210:599–612.

Erxleben, J. C. P. 1777. *Systema regni animalis per classes, ordines, genera, species, varietates cum synonymia et historia animalium. Classis 1. Mammalia.* Lipsiae: Impensis Weygandianis, xlvii + 636 pp. + 64 (unnumbered).

Eshelman, R. E., and G. S. Morgan. 1985. Tobagan Recent mammals, fossil vertebrates, and their zoogeographic implications. *Nat. Geogr. Soc. Res. Rep.* 21:137–43.

Espinal T., L. S., and E. Montenegro M. 1963. *Formaciones vegetales de Colombia: memoria explicativa sobre el mapa ecológico.* Bogotá: Instituto Geografica Militar "Agustín Codazzi," [booklet + 4 folding maps].

Espinosa, M. B., A. Lasserre, M. Piantanida, and A. D. Vitullo. 1997. Cytogenetics of vesper mice, *Calomys* (Sigmodontinae): a new karyotype from the Puna region and its implication for chromosomal phylogeny. *Cell. Mol. Life Sci.* 53:583–86.

Espinosa, M. B. and O. A. Reig. 1991. Cytogenetics and karyosystematics of South American oryzomyine rodents (Cricetidae, Sigmodontinae). III. Banding karyotypes of Argentinean *Oligoryzomys*. *Z. Säugetierk.* 56:306–17.

Espinosa, M. B., A. D. Vitullo, and M. S. Merani. 1991. Chromosomes of the Argentine Andean mouse, *Akodon andinus* (Cricetidae: Sigmodontinae). *Z. Säugetierk.* 56:124–25.

Estrada-Franco, J. G., R. Navarro-Lopez, J. E. Freier, D. Cordova, T. Clements, et al. 2004. Venezuelan equine encephalitis virus, southern Mexico. *Emerg. Infect. Dis.* 10:2113–21.

Everard, C. O. R., and E. S. Tikasingh. 1973a. Ecology of the rodents, *Proechimys guyannensis trinitatis* and *Oryzomys capito velutinus*, on Trinidad. *J. Mamm.* 54:875–86.

———. 1973b. The abundance of small mammals in Turure Forest, Trinidad. *J. Trinidad Field Nat. Club* (1973):85–90.

Everard, C.O.R., E. S. Tikasingh, E.S. and J. B. Davies. 1974. The biology of *Dipetalonema proechimyis* Esslinger, 1974 (Nematoda: Filarioidea) in Trinidad. *J. Parasitol.* 60:556–58.

Ewel, J. J., and A. Madriz. 1968. *Zonas de vida de Venezuela.* Caracas: Ministerio de Agricultura y Cria, 265 pp, 119 figs., map.

Eydoux, J. F. T., and L. F. A. Souleyet. 1841. *Voyage autour du Monde exécuté pendant les années 1836 et 1837 sur la corvette La Bonite commandée par M. Vaillant.* Paris: Arthus Bertrand, 1:xl + 132 pp.; Atlas: plates 1–12 (Mammifères).

Fabre, P.-H., T. Galewski, M. Tilak, and E. J. P. Douzery. 2012 [2013]. Diversification of South American spiny rats (Echimyidae): a multigene phylogenetic approach. *Zool. Scr.* 42:117–34. [Published on-line 15 September 2012, with print copy in March 2013.]

Fabre, O., J. L. Guyot, R. Salas-Gismondi, M. Malaver-Pizarro, and E. Maniero. 2008. Los chachapoya de la región de Soloco: Chaquil, del sitio de hábitat a la cueva funeraria. *Bull. Inst. Fr. Etud. Andines* 37:271–92.

Fagundes, V., J. P. M. Camancho, and Y. Yonenaga-Yassuda. 2004. Are dot-like chromosomes in *Trinomys iheringi* (Rodentia, Echimyidae) B chromosomes? *Cytogenet. Genome Res.* 106:159–64.

Fagundes, V., A. U. Christoff, R. C. Amaro-Ghilard, D. R. Scheibler, and Y. Yonenaga-Yassuda. 2003. Multiple interstitial ribosomal sites (NORs) in the Brazilian squirrel *Sciurus aestuans ingrami* (Rodentia, Sciuridae) with 2n = 40. *Genet. Mol. Biol.* 26:253–57.

Fagundes, V., A. U. Christoff, and Y. Yonenaga-Yassuda. 1998. Extraordinary chromosomal polymorphism with 28 different karyotypes in the Neotropical species *Akodon cursor* (Muridae, Sigmodontinae), one of the smallest diploid number in rodents (2n = 16, 15 and 14). *Hereditas* 129:263–74.

Fagundes, V., and L. P. Costa. 2009. Um cariótipo novo para *Blarinomys breviceps* (Winge, 1887): um caso de erro de identificação ou uma nova espécie no gênero?. *Resumos do 54° Congresso Brasileiro de Genética*, p. 198.

Fagundes, V., and C. D. de A. Nogueira. 2007. The use of PCR-RFLP as an identification tool for three closely related species of rodents in the genus *Akodon* (Sigmodontinae, Akodontini). *Genet. Mol. Biol.* 30:698–701.

Fagundes, V., Y. Sato, M. J. de J. Silva, F. Rodrigues, and Y. Yonenaga-Yassuda. 2000. A new species of *Calomys* (Rodentia, Sigmodontinae) from central Brazil identified by its karyotype. *Hereditas* 133:195–200.

Fagundes, V., A. M. Vianna-Morgante, and Y. Yonenaga-Yassuda. 1997. Telomeric sequence localization and G-banding patterns in the identification of a polymorphic chromosomal rearrangement in the rodent *Akodon cursor* (2n = 14, 15 and 16). *Cytogenet. Cell Genet.* 78:224–28.

Fagundes, V., and Y. Yonenaga-Yassuda. 1998. Evolutionary conservation of whole homeologous chromosome arms in the Akodont rodents *Bolomys* and *Akodon* (Muridae, Sigmodontinae): maintenance of interstitial telomeric segments (ITBs) in recent event of centric fusion. *Chromosome Res.* 6:643–48.

Fain, A. 1976. Nouveaux acariens parasites de la superfamille listrophoroidea (astigmates). *Acta Zool. Pathol. Antverpiensia* 64:37–67.

Fain, A., and F. S. Lukoschus. 1982. Diagnoses de nouveaux Listrophoridae neotropicaux. *Bull. Ann. Soc. R. Belge. Entomol.* 118:100–1.

Fain, A., F. S. Lukoschus, and E. Mendez. 1982. Two new species of the genus *Echimytricalges* Fain, 1970 (Acari, Astigmata, Lobagidea) from American spiny rats (Echimyidae). *Bull. Ann. Soc. R. Belge. Entomol.* 118:121–30.

Faissol, S. 1952. *O Mato Grosso de Goiás.* Rio de Janeiro: Biblioteca Geográfica Brasileira (Série A, publicação 9, IBGE), 140 pp.

Falqueto, A., E. Cupolillo, G. M. C. Machado, L. E. Carvalho-Paes, and G. Grimaldi Jr. 1998. A new enzymatic variant of *Leishmania* (*Leishmania*) *forattinii* isolated from *Proechimys iheringi* (Rodentia, Echimyidae) in Espírito Santo, Brazil. *Mem. Inst. Oswaldo Cruz* 93:795–98.

Falqueto, A., G. Grimaldi Jr., P. A. Sessa, J. B. M. Varejão, and L. M. Daene. 1985. *Lutzomyia gasparviannai* Martins, Godoy & Silva, 1962, probable vector of *Leishmania mexicana* ssp. in Viana municipality, Espírito Santo State, Brazil. *Mem. Inst. Oswaldo Cruz* 80:497.

Fanjul, M. S., R. R. Zenuto, and C. Busch. 2006. Seasonality of breeding in wild tuco-tucos *Ctenomys talarum* in relation to climate and food availability. *Acta Theriol.* 51:283–93.

Farias Rocha, F. A. de, P. K. Ahnelt, L. Peichi, C. A. Saito, and L. C. L. Silveira. 2009. The topography of cone photoreceptors in the retina of a diurnal rodent, the agouti (*Dasyprocta aguti*). *Vis. Neurosci.* 26:167–75.

Fasanella, M., C. Bruno, YT. Cardoso, and M. Lizarralde. 2013. Historical demography and spatial genetic structure of the subterranean rodent *Ctenomys magellanicus* in Tierra del Fuego (Argentina). *Zool. J. Linn. Soc.* 169:697–710.

Feijó, A., and A. Langguth. 2013. Mamíferos de médio e grande porte do nordeste do Brasil: distribuição e taxonomia, com descrição de novas espécies. *Rev. Nordest. Biol.* 22(1/2):3–225.

Feijoo, M., G. D'Elía, U. F. J. Pardiñas, and E. P. Lessa. 2010. Systematics of the southern Patagonian-Fueguian endemic *Abrothrix lanosus* (Rodentia: Sigmodontinae): Phylogenetic position, karyotypic and morphological data. *Mamm. Biol.* 75:122–37.

Feito, R., and M. H. Gallardo. 1976. Notes on the sperm morphology of *Ctenomys maulinus* (Rodentia, Octodontidae). *Experientia* 32:48–53.

———. 1982. Sperm morphology of the Chilean species of *Ctenomys* (Octodontidae). *J. Mamm.* 63:658–61.

Fejfar, O., A. Blasetti, G. Calderoni, M. Coltorti, G. Ficcarelli, F. Masini, L. Rook, and D. Torre. 1993. New finds of cricetids (Mammalia, Rodentia) from the Late Pleistocene-Holocene of northern Ecuador. *Doc. Lab. Geol. Lyon* 125:151–67.

Fernandes, F. A., G. P. Fernández-Stolz, C. M. Lopes, and T. R. O. de Freitas. 2007. The conservation status of the tuco-tucos, genus *Ctenomys* (Rodentia: Ctenomyidae), in southern Brazil. *Brazil. J. Biol.* 67(4, suppl.):839–47.

Fernandes, F. A., R. Fornel, and T. R. O. Freitas. 2012. *Ctenomys brasiliensis* Blainville (Rodentia: Ctenomyidae): clarifying the geographic placement of the type species of the genus *Ctenomys*. *Zootaxa* 3272:57–68.

Fernandes, F. A., G. L. Gonçalves, S. S. F. Ximenes, and T. R. O. de Freitas. 2009. Karyotypic and molecular polymorphisms in *Ctenomys torquatus* (Rodentia: Ctenomyidae): taxonomic considerations. *Genetica* 136:449–59.

Fernandez, F. A. S. 1989. "Dinâmica de populações e uso do espaço e do tempo por uma comunidade de pequenos mamíferos na Restinga de Barra de Maricá, Rio de Janeiro." Master's thesis, Universidade Estadual de Campinas, São Paulo, Brazil.

Fernández, F., F. Idoeta, C. GarcíaEsponda, J. Carrera, G. Moreira, F. Ballejo, and L. De Santis. 2012. Small mammals (Didelphimorphia, Rodentia and Chiroptera) from Pampean Region, Argentina. *Check List* 8:130–34.

Fernández, F. J., P. Teta, R. Barberena, and U. F. J. Pardiñas. 2012. Small mammal remains from Cueva Huenul 1 (Late Pleistocene-Holocene), Patagonia, Argentina. Taphonomy and paleoenvironmental significance. *Quarternary Int.* 278:22–31.

Fernandez, G. P. 2002. "Analise da etrutura populacional e da variabilidade genética em três populações de *Ctenomys flamarioni* (Rodentia: Ctenomyidae) a través de loci de microsatélites." Master's thesis, Universidade Federal do Rio Grande do Sul, Brazil.

Fernández, R. 1968. The karyotype of *Octodon degus* (Rodentia: Octodontidae). *Arch. Biol. Med. Exp.* 5:33–37.

Fernandez-Stolz, G. P., J. F. B. Stolz, and T. R. O. de Freitas. 2007. Bottlenecks and dispersal in the tuco-tuco das dunas, *Ctenomys flamarioni* (Rodentia: Ctenomyidae), in southern Brazil. *J. Mamm.* 88:935–45.

Ferreira, A. B. de H. 1986. *Novo dicionário aurélio da língua Portuguesa.* 2nd ed. Rio de Janeiro: Editora Nova Frontera, xxiii+ 1838 pp.

Ferreira, D., A. K. B. Silva, C. H. C. Matos, P. Hadler, and A. S. Hsiou. 2012. Assembleia holocênica de vertebrados de pequeno porte do Sítio Alcobaça, estado de Pernambuco, Brasil. *Rev. Bras. Paleontol.* 15:359–70.

Ferris, G. F. 1921. Contributions toward a monograph of the sucking lice; part II. *Stanf. Univ. Publ. Univ. Ser. Biol. Sci.* 2:53–133.

Ferro, L. I., and R. M. Barquez. 2008. Comments on the distribution of *Abrothrix andinus* and *Calomys lepidus* (Rodentia: Cricetidae) in Tucumán province, Argentina. *Mastozool. Neotrop.* 15:197–201.

Ferro, L. I., and J. J. Martínez. 2009. Molecular and morphometric evidence validates a Chacoan species of the grey leaf-eared mice genus *Graomys* (Rodentia: Cricetidae: Sigmodontinae). *Mammalia* 73:265–71.

Ficcarelli, G., A. Azzaroli, V. Borselli, M. Coltorti, F. Dramis, O. Fejfar, A. Hirtz, and D. Torre. 1992. Stratigraphy and paleontology of upper Pleistocene deposits in the Interandean Depression, northern Ecuador. *J. South Am. Earth Sci.* 6:145–50.

Fields, R. W. 1957. Hystricomorph rodents from the late Miocene of Colombia, South America. *Univ. Calif. Publ. Geol. Sci.* 32:273–309.

Figueiredo, G. G. de, A. A. Borges, G. M. Campos, A. M. Machaco, F. P. Saggioro, G. dos S. Sabino Júnior, S. J. Badra, et al. 2010. Diagnosis of hantavirus infection in humans and rodents in Ribeirão Preto, State of São Paulo, Brazil. *Rev. Soc. Bras. Med. Trop.* 43:348–54.

Figueiredo, L. T. M., M. L. Moreli, R. L. M. Souza, A. A. Borges, G. G. Figueiredo, A. M. Machado, I. Bisordi, et al. 2009. Distinct hantaviruses causing pulmonary syndrome in central plateau, southeastern and southern Brazil. *Emerg. Infect. Dis.* 15:561–67.

Figueroa, R. A., R. Cádiz, R. Figueroa, E. S. Corales, and R. Murúa. 2012. Abundance, habitat and body measurements ofc the rare Long-chawed Mouse (*Pearsonomys annectens*) in the coastal temperate rainforest of southern Chile. *Stud. Neotrop. Fauna Environ.* 47:1–10.

Figueroa, R. A., E. S. Corales, and R. López. 2001. Record of the White-throated Hawk (*Buteo albigula*), and notes on its hunting methods and movements, in the Andes of central-southern Chile. *Int. Hawkwatch* 4:3–9.

Finotti, R., A. Cunha, and R. Cerqueira. 2003. Alimentação macroscópica do trato digestivo de *Abrawayaomys ruschii.2 Congresso Brasileiro de Mastozoologia, Resumos*, p. 265.

Fischer, G. 1814. *Zoognosia tabulis synopticis illustrata.* Volumen tertium. Quadrupedum reliquorum, cetorum et montrymatum descriptionem continens. Mosquae: Nicolai Sergeidis Vsevolozsky Moscow, 3:xxiv + 1–732.

———. 1817. Adversaria zoologica. Fasciculus primus. *Quaedam ad Mammalium systema et genera illustranda. Mem. Soc. Imp. Nat. Mosc.* 5:357–446 [error for 428], 2 pls.

Fischer, J. B. 1829. *Synopsis mammalium.* Stuttgardtiae: J. G. Cottae, xlii + 752 pp.

Fitzinger, L. J. 1867a. Versuch einer natürlichen Anordnung der Nagethiere (Rodentia). *Sitzungsber. Math. Naturwiss. Cl. Kl. Akad. Wiss. Wien.* 55:453–514 [separately numbered from pp. 1–63].

———. 1867b. Versuch einer natürlichen Anordnung de Nagethiere (Rodentia). *Sitzungsber. Math. Naturwiss. Cl. Kl. Akad. Wiss. Wien.* 56:57–168 [separately numbered from pp. 1–112].

———. 1880. Der langhaarische gemeine Ferkelhase (*Cavia cobaya, longipilis.*) Eine bisher noch nich beschriebenc Form. *Sitzungsber. Math. Naturwiss. Cl. Kl. Akad. Wiss. Wien.* 80(1):431–438.

FitzRoy, R. 1839. *Narrative of the surveying voyages of His Majesty's Ships Adventure and Beagle between the years 1826 and 1836, describing their examination of the southern shores of South America, and the Beagle's circumnavigation of the globe.* Vol. 2. London: Henry Colburn, vii-xiv + 1–695 pp.

Fleming, J. 1822. *The philosophy of zoology; or a general view of the structure, functions, and classifications of animals.* Edinburgh: Archibald Constable and Co.; London: Hurst, Robinson, 2:1–618.

Fleming, T. H. 1970. Notes on the rodent fauna of two Panamanian forests. *J. Mamm.* 51:473–90.

———. 1971. Population ecology of three species of Neotropical rodents. *Misc. Publ. Mus. Zool., Univ. Michigan* 143:1–77.

Flores, V. R., N. Brugni, and C. A. Rauque. 2007. Redescription of *Hippocrepis myocastoris* Babero, Cabello et Kinoed, 1979 (Digenea: Notocotylidae) in the coypu *Myocastor coypus* (Rodentia: Myocastoridae) from Patagonia, Argentina. *Comp. Parasitol.* 74:233–36.

Flower, W. H., and R. Lydekker. 1891. *An introduction to the study of mammals living and extinct.* London: Adam and Charles Black, xvi + 763 pp.

Fonollat, A. M. P. d. 1984. Cricétidos de la provincia de Tucumán (Argentina). *Acta Zool. Lilloana* 37:219–25.

Fonseca, F. da. 1959. Notas de acarologia. XLIV. Inquérito sôbre a fauna acarológica de parasitas no nordeste do Brasil. *Mem. Inst. Butantan* 28:99–186.

———. 1960. Notas de Acarologia XLVI–Acarofauna zooparasite na Bolivia. *Mem. Inst. Butantan* 29:89–141.

Fonseca, G. A. B. da, G. Herrmann, Y. L. R. Leite, R. A. Mittermeier, A. B. Rylands, and J. L. Patton. 1996. Lista Anotada dos Mamíferos do Brasil. *Occas. Papers Conserv. Biol.* 4:ii + 1–38.

Fonseca, G. A. B. da, and M. C. M. Kierulff. 1989. Biology and natural history of Brazilian Atlantic Forest small mammals. *Bull. Fla. State Mus. Biol. Sci.* 34:99–152.

Fonseca, M. B. da. 2003. "Biologia populacional e classificaçao etaria do roedor subterrâneo tuco-tuco *Ctenomys minutus* Nehring 1887 (Rodentia: Ctenomyidae) na planicie costeira do Rio Grande do Sul." Master's thesis, Universidade Federal do Rio Grande do Sul, Brazil.

Fonseca, R., H. G. Bergallo, A. C. Delciellos, O. Rocha-Barbosa, and L. Geise. 2013. *Juliomys rimofrons* Oliveira and Bonvicino, 2002 (Rodentia: Cricetidae): distribution extension. *Check List* 9:684–85.

Fontes, S. V., M. Passamani, C. H. Jacinto, M. S. Pereira, and A. P. P. Sant'Ana. 2007. Área de vida e deslocamento de *Akodon montensis* e *Gracilinanus microtarsus* em um fragmento no sul de Minas Gerais. *Anais do VIII Congresso de Ecologia do Brasil, 23 a 28 Setembro de 2007*, Caxambu, MG.

Forcone, A., M. V. Luna, F. O. Kravetz, and J. A. Lisanti. 1981. C bands and G bands of *Calomys musculinus* (Rodentia, Cricetidae). *Mendeliana* 5:57–66.

Forero-Montana, J., J. Betancur, and J. Cavelier. 2003. Dieta del capibara *Hydrochaeris hydrochaeris* (Rodentia: Hydrochaeridae) in Caño Limón, Arauca, Colombia. *Rev. Biol. Trop.* 51:579–90.

Forget, P.-M. 1990. Seed-dispersal of *Vouacapoua americana* (Caesalpiniaceae) by caviormorph rodents in French Guiana. *J. Trop. Ecol.* 6:459–68.

———. 1991. Scatterhoarding of *Astrocaryum paramaca* by *Proechimys* in French Guiana: comparison with *Myoprocta exilis*. *Trop. Ecol.* 32:155–67.

———. 1992. Seed removal and seed fate in *Gustavia superba* (Lecythidaceae). *Biotropica* 24:408–14.

———. 1996. Removal of seeds of *Carapa procera* (Meliaceae) by rodents and their fate in rainforest in French Guiana. *J. Trop. Ecol.* 12:751–61.

Forget, P.-M., and L. Cuiljpers. 2008. Survival and scatterhoarding of frugivores-dispersed seeds as a function of forest disturbance. *Biotropica* 40:380–85.

Forget, P.-M., E. Muñoz, and E. G. Leigh Jr. 1994. Predation by rodents and bruchid beetles on seeds of *Scheelea zonensis* on Barro Colorado Island, Panama. *Biotropica* 26: 420–26.

Formoso, A. E., D. E. U Sauthier, and U. F. J. Pardiñas. 2010. Mammalia, Rodentia, Sigmodontinae, *Holochilus brasiliensis* (Desmarest, 1819): distribution extention. *Check List* 6:195–97.

Fornes, A., and E. Massoia. 1965. Micromamíferos (Marsupialia y Rodentia) recolectados en la localidad bonaerense de Miramar. *Physis Sec. C* 25:99–108.

Foster, G. W., L. C. Branch, M. Machicote, J. M. Kinsella, D. Villarreal, and D. J. Forrester. 2002. Gastrointestinal helminths of the plains vizcacha *(Lagostomus maximus)* from Argentina, with observations on interspecific interactions between nematodes and cestodes. *Comp. Parasitol.* 69:26–32.

Foster, R. J., B. J. Harmsen, B. Valdes, C. Pomilla, and C .P. Doncaster. 2010. Food habits of sympatric jaguars and pumas across a gradient of human disturbance. *J. Zool.* 280:309–18.

Frailey, C. D., and K. E. Campbell Jr. 2004. Paleogene rodents from Amazonian Peru: the Santa Rosa local fauna. In *The Paleogene mammalian fauna of Santa Rosa, Amazonian Peru*, K. E. Campbell Jr., ed., 71–130. *Sci. Ser. Nat. Hist. Mus. Los Angeles Co.* 40:1–163.

Francés, J., and G. D'Elía. 2006. *Oligoryzomys delticola* es sinónimo de *O. nigripes* (Rodentia, Cricetidae, Sigmodontinae). *Mastozool. Neotrop.* 13:123–31.

Francescoli, G. 2001. Vocal signals from *Ctenomys pearsoni* pups. *Acta Theriol.* 46:327–30.

Francescoli, G. 2002. Geographic variation in vocal signals in *Ctenomys pearsoni*. *Acta Theriol.* 47:35–44.

Francis, J. C., and A. Mones. 1968. Los roedores fósiles del Uruguay. *Bol. Lab. Paleontol. Vert.* 1:35–55.

Fredga, G. 1966. Chromosome studies in five species of South American rodents (suborder Hystricomorpha). *Mamm. Chromo. Newslett.* 20:45–46.

Freitas, C. A. de. 1957. Notícia sôbre a peste no Nordeste. *Rev. Bras. Malariol. Doenças Trop.* 9:123–33.

Freitas, J. N. S., L. A. S. Carvalho, C. N. El-Hani, and P. L. B Rocha. 2010. Affiliation in the social interactions in captivity of the torch tail rat, *Trinomys yonenagae* (Rodentia: Echimyidae). *J. Ethol.* 28:105–12.

Freitas, J. N. S., C. N. El-Hani, and P. L. B. Rocha. 2003. Affiliation in the torch tail rat, *Trinomys yonenagae* (Rodentia: Echimyidae), a sand-dwelling rodent from Brazilian semiarid Caatinga: evolutionary implications. *Rev. Etol.* 5:61–73.

Freitas, J. N. S., C. N. El-Hani, and P. L. B. Rocha. 2008. Affiliation in four echimyid rodent species based on intrasexual dyadic encounters: evolutionary implications. *Ethology* 114:389–97.

Freitas, T. R. O. de. 1980. "Estudos citogenéticos em roedores do sul do Brasil." Master's thesis, Universidade Federal do Rio Grande do Sul, Porto Alegre, Brazil.

———. 1994. Geographic variation of heterochromatin in *Ctenomys flamarioni* (Rodentia-Octodontidae) and its cytogenetic relationships with other species of the genus. *Cytogenet. Cell Genet.* 67:193–98.

———. 1995a. Geographical distribution of patterns of sperms in the genus *Ctenomys* (Rodentia-Octodontidae). *Braz. J. Genet.* 18:43–46.

———. 1995b. Geographic distribution and conservation of four species of the genus *Ctenomys* in southern Brazil. *Stud. Neotrop. Fauna Environ.* 30:53–59.

———. 1997. Chromosome polymorphism in *Ctenomys minutus* (Rodentia-Octodontidae). *Braz. J. Genet.* 20:1–7.

———. 2001. Tuco-tucos (Rodentia, Octodontidae) in southern Brazil: *Ctenomys lami* spec. nov. separated from *C. minutus* Nehring 1887. *Stud. Neotrop. Fauna Environ.* 36:1–8.

———. 2005. Analisis of skull morphology in 15 species of the genus *Ctenomys* including seven karyological distinct forms of *Ctenomys minutus*. In *Mammalian diversification: from chromosomes to phylogeography*, ed. E. A. Lacey and P. Myers, 131–54. *Univ. California Publ. Zool.* 133:v-viii + 1–383.

———. 2006. Cytogenetic status of four *Ctenomys* species in the south of Brazil. *Genetica* 126:227–35.

———. 2007. *Ctenomys lami*: the highest chromosome variability in *Ctenomys* (Rodentia, Ctenomyidae) due to a centric fusion/fission and pericentric inversion system. *Acta Theriol.* 52:171–80.

Freitas, T. R. O. de, F. A. Fernandes, R. Fornel, and P. A. Roratto. 2012. An endemic new species of tuco-tuco, genus *Ctenomys* (Rodentia: Ctenomyidae), with a restricted geographic distribution in southern Brazil. *J. Mamm.* 93:1355–67.

Freitas, T. R. O. de, and E. P. Lessa. 1984. Cytogenetics and morphology of *Ctenomys torquatus* (Rodentia-Octodontidae). *J. Mamm.* 65:637–42.

Freitas, T. R. O., M. S. Mattevi, and L. F. B. Oliveira. 1983. G- and C-banded karyotype of *Reithrodon auritus* from Brazil. *J. Mamm.* 64:318–21.

———. 1984. Unusual C-band in three karyotypically rearranged forms of *Scapteromys* (Rodentia, Cricetidae) from Brazil. *Cytogenet. Cell Genet.* 38:39–44.

Freitas, T. R. O., M. S. Mattevi, L. F. B. Oliveira, M. J. Souza, Y. Yonenagba-Yassuda, and F. M. Salzano. 1983. Chromosome relationships in three representatives of the genus *Holochilus* (Rodentia, Cricetidae) from Brazil. *Genetica* 61:13–20.

Freygang, C. C., J. R. Marinho, and T. R. O. de Freitas. 2004. New karyotypes and some considerations about the chromosomal diversification of *Ctenomys minutus* (Rodentia: Ctenomyidae) on the coastal plain of the Brazilian state of Rio Grande do Sul. *Genetica* 121:125–32.

Friant, M. 1968. La morphologie des dents jugales du paca. *Ann. Soc. R. Zool. Belge.* 98:139–46.

Fronza, T., R. L. Wainberg, and B. E. Llorente. 1976. Polimorfismo del cromosoma X y significacion filogenética del cariotipo de la "Rata aquatica" *Scapteromys aquaticus* (Rodentia, Cricetidae) de la ribera de Punta Lara (Argentina). *Mendeliana* 1:41–48.

Froriep, L. F. 1829. Schreiben des jesst in Chile reifenden hrn. Dr. Pöppig. In *Notizen aus dem Gebiete der Natur - und Heilkunde*, L. F. v. Froriep, no. 502, 23(18):273–88.

Fry, C. H. 1970. Ecological distribution of birds in northeastern Mato Grosso state, Brazil. *An. Acad. Brasil. Ciênc.* 42:275–318.

Fuentes, L., and C. Poleo. 2005. Bioecología de las principales especies de roedores cricétidos de Venezuela. *Revista Digital CENIAP HOY* Número 8 mayo-agosto 2005. Maracay, Aragua, Venezuela.

Fulhorst, C. F., M. D. Bowen, R. A. Salas, G. Duno, A. Utrera, T. G. Ksiazek, N. M. De Man zione, et al. 1999. Natural rodent host associations of Guanarito and Pirital viruses (family Arenaviridae) in central Venezuela. *Am. J. Trop. Med. Hyg.* 61:325–30.

Fulhorst, C. F., M. N. B. Cajimat, M. L. Milazzo, H. Paredes, N. M. C. de Manzione, R. A. Salas, P. E. Rollin, and T. G. Ksiazek. 2008. Genetic diversity between and with the arenavirus species indigenous to western Venezuela. *Virology* 378:205–13.

Fulhorst, C .F., M. N. B. Cajimat, A. Utrera, M. L. Milazzo, and G .M. Duno. 2004. Maporal virus, a hantavirus associated with the fulvous pigmy rice rat (*Oligoryzomys fulvescens*) in western Venezuela. *Virus Res.* 104:139–44.

Fulhorst, C. F., T.G. Ksiazek, C.J. Peters, and R.B. Tesh. 1999. Experimental infection of the cane mouse *Zygodontomys brevicauda* (family Muridae) with Guanarito virus (Arenaviridae), the etiologic agent of Venezuelan Hemorrhagic Fever. *J. Infect. Dis.* 180:966–69.

Fulhorst, C. F., M. C. Monroe, R. A. Salas, G. Duno, A. Utrera, T. G. Ksiazek, S. T. Nichol, N. M. de Manzione, D. Tovar, and R. B. Tesh. 1997. Isolation, characterization and geographic distribution of Cano Delgadito virus, a newly discovered South American hantavirus (family Bunyaviridae). *Virus Res.* 51:159–71.

Fulk, G. W. 1975. Population ecology of rodents in the semiarid shrublands of Chile. *Occas. Papers Mus. Texas Tech Univ.* 33:1–40.

———. 1976. Notes on the activity, reproduction, and social behavior of *Octodon degus*. *J. Mamm.* 57:495–505.

———. 1977. Owl predation and rodent mortality: a case study. *Mammalia* 40:423–28. [1976 vol. 40, no. 3: pp. 355–530. "Achevé d'imprimer le 20 mars 1977."]

Funari, N. 2013. New findings on the origin of *Cavia intermedia*, one of the world's rarest mammals. *Mamm. Rev.* 43:323–326.

Furman, D. P. 1972a. Laelapid mites (Laelapidae: Laelapinae) of Venezuela. *Brigham Young Univ. Sci. Bull.*, biol. ser., 17:1–58.

———. 1972b. New species of *Laelaps* (Acarina: Laelapidae) from Venezuela. *J. Med. Entomol.* 9:35–46.

Furman, D. P., and V. J. Tipton. 1961. Acaros parasitos laelaptine (Acarina: Laelaptidae) de Venezuela. *Mem. Soc. Cien. Nat. La Salle* 21:166–212.

Furtado, V. de V. 1981. "Diversidade cromossômica em roedores das famílias Cricetidae e Caviidae em Pernambuco, Brasil." Ph.D. diss., Universidade do Rio Grande do Sul, Porto Alegre, Brazil.

Galende, G. I., and E. Raffaele. 2012. Diet selection of the southern vizcacha (*Lagidium viscacia*): a rock specialist in north western Patagonian steppe, Argentina. *Acta Theriol.* 57:333–41.

Galetti, M. 1990. Predation on the squirrel *Sciurus aestuans* by capuchin monkeys, *Cebus apella*. *Mammalia* 54:152–54.

Galetti, M., C. I. Donatti, C. Steffler, J. Genini, R. S. Bovendrop, and M. Fleury. 2010. The role of seed mass on the caching decision by agoutis, *Dasyprocta leporina* (Rodentia: Agoutidae). *Zoologia* 27:472–76.

Galewski, T., J.-F. Mauffrey, Y. L. R. Leite, J. L. Patton, and E. J. P. Douzery. 2005. Ecomorphological diversification among South American spiny rats (Rodentia: Echimyidae): a phylogenetic and chronological approach. *Mol. Phylogenet. Evol.* 34:601–15.

Galiano, D., B. B. Kubiak, C. Estevan, and J. R. Marinho. 2007. A floração da taquara-lixa e a explosão populacional de roedores silvestres. Ratada? *Anais do VIII Congresso de Ecologia do Brasil*, pp. 1–2.

Gallardo, J. M. 1970. Ecological study of amphibians and reptiles of the southwest of Buenos-Aires Province, Argentina. *Rev. Mus. Argentino Cien. Nat. "Bernardino Rivadavia," Cien. Zool.* 10:27–63.

Gallardo, M. H. 1979. Las especies chilenas de *Ctenomys* (Rodentia, Octodontidae). I. Estabilidad cariotípica. *Arch. Biol. Med. Exp.* 12:71–82.

———. 1982. Chromosomal homology in southern *Akodon*. *Experientia* 38:1485–87.

———. 1991. Karyotypic evolution in *Ctenomys* (Rodentia: Ctenomyidae). *J. Mamm.* 72:11–21.

———. 1992. Karyotypic evolution in octodontid rodents based on C-band analysis. *J. Mamm.* 73:89–98.

———. 1997. A saltation model of karyotypic evolution in the Octodontoidea (Mammalia, Rodentia). In *Chromosomes today*, vol. 12, eds. N. Henríque-Gill, J. S. Parker, and M. J. Puertas, 347–65. London: Chapman and Hall, 379 pp.

Gallardo, M. H., G. Aguilar, and O. Goicoechea. 1988. Systematics of sympatic cricetid *Akodon* (*Abrothrix*) rodents and their taxonomic implications. *Medio Ambiente* 9:65–74.

Gallardo, M. H., and J. A. Anrique. 1991. Population parameters and burrow systems in *Ctenomys maulinus brunneus* (Rodentia, Ctenomyidae). *Medio Ambiente* (Valdivia) 11:48–53.

Gallardo, M. H., C. Araneda, and N. Köhler. 1992. Genetic divergence in *Spalacopus cyanus* (Rodentia, Octodontidae). *Z. Säugetierk.* 57:231–37.

Gallardo, M. H., J. W. Bickham, R. L. Honeycutt, R. A. Ojeda, and N. Köhler. 1999. Discovery of tetraploidy in a mammal. *Nature* 401:341.

Gallardo, M. H., J. W. Bickham, G. Kausel, N. Köhler, and R. L. Honeycutt. 2003. Gradual and quantum genome size shifts in the hystricognath rodents. *J. Evol. Biol.* 16:163–69.

Gallardo, M. H., O. Garrido, R. Bahamonde, and M. González. 2004. Gametogenesis and nucleotypic effects in the tetraploid red vizcacha rat, *Tympanoctomys barrerae* (Rodentia, Octodontidae). *Biol. Res.* 37:767–75.

Gallardo, M. H., C. A. González, and I. Cerbrián. 2006. Molecular cytogenetics and allotetraploidy in the red vizcacha rat, *Tympanoctomys barrerae* (Rodentia, Octodontidae). *Genomics* 88:214–21.

Gallardo, M. H., G. Kausel, A. Jiménez, C. Bacquet, C. González, J. Figueroa, N. Köhler, and R. Ojeda. 2004. Whole-genome duplications in South American desert rodents (Octodontidae). *Biol. J. Linn. Soc.* 82:443–51.

Gallardo, M. H., and J. A. W. Kirsch. 2001. Molecular relationships among Octodontidae (Mammalia: Rodentia: Caviomorpha). *J. Mamm. Evol.* 8:73–89.

Gallardo, M. H., and N. Köhler. 1994. Demographic changes and genetic losses in populations of a subterranean rodent (*Ctenomys maulinus brunneus*) affected by a natural catastrophe. *Z. Säugetierk.* 59:358–65.

Gallardo, M. H., N. Köhler, and C. Araneda. 1996. Loss of genetic variation in *Ctenomys coyhaiquensis* (Rodentia, Ctenomyidae) affected by vulcanism. *Mastozool. Neotrop.* 3:7–13.

Gallardo, M. H., and F. Mondaca. 2002. The systematics of *Aconaemys* (Rodentia, Octodontidae) and the distribution of *A. sagei* in Chile. *Mamm. Biol.* 67:105–12.

Gallardo, M. H., F. C. Mondaca, R. J. Ojeda, N. Köhler, and O. Garrido. 2002. Morphological diversity in the sperms of caviomorph rodents. *Mastozool. Neotrop.* 9:159–70.

Gallardo, M. H., R. A. Ojeda, C. A. González, and C. A. Ríos. 2007. The Octodontidae revisited. In *The quintessential naturalist: Honoring the life and legacy of Oliver P. Pearson*, eds. D. A. Kelt, E. P. Lessa, J. Salazar-Bravo, and J. L. Patton, 695–719. *Univ. California Publ. Zool.* 134:v-xii+1–981.

Gallardo, M. H., and E. Palma. 1990. Systematics of *Oligoryzomys longicaudatus* (Rodentia: Muridae) in Chile. *J. Mamm.* 71:333–42.

Gallardo, M. H., and B. D. Patterson. 1985. Chromosomal differences between two nominal subspecies of *Oryzomys longicaudatus* Bennett. *Mamm. Chromo. Newslett.* 25:49–53.

Gallardo, M. H., and D. Reise. 1992. Systematics of *Aconaemys* (Rodentia, Octodontidae). *J. Mamm.* 73:779–88.

Gallardo, M. H., E. Y. Suárez-Villota, J. J. Nuñez, R. A. Vargas, R. Haro, and N. Köhler. 2013. Phylogenetic analysis and phylogeography of the tetraploid rodent *Tympanoctomys barrerae* (Octodontidae): insights on its origin and the impact of Quaternary climate changes on population dynamics. *Biol. J. Linn. Soc.* 108:453–69.

Gallardo, M. H., D. E. Udrizar Sauthier, A. A. Ojeda, and U. F. J. Pardiñas. 2009. Discovery of desert-adapted *Tympanoctomys barrerae* in central Patagonia, Argentina. *Mammalia* 73:158–61.

Galliari, C. A., and U. F. J. Pardiñas. 1999. *Abrothrix lanosus* (Rodentia, Muridae) en la Patagonia continental, Argentina. *Neotropica* 45:119–20.

———. 2000. Taxonomy and distribution of the sigmodontine rodents of genus *Necromys* in central Argentina and Uruguay. *Acta Theriol.* 45:211–32.

Galliari, C. A., U. F. G. Pardiñas, and F. J. Goin. 1996. Lista comentada de los mamíferos argentinos. *Mastozool. Neotrop.* 3:39–61.

Galvez, D., B. Kranstauber, R. W. Kays, and P. A. Jansen. 2009. Scatter hoarding by the Central American agouti: a test of optimal cache spacing theory. *Anim. Behav.* 78:1327–33.

Gamarra de Fox, I., and A. J. Martín. 1996. Lista de Mamíferos del Paraguay. In *Colecciones de fauna y flora del Museo Nacional de Historia Natural del Paraguay*, ed. O. Romero, 469–573. Asunción, Paraguay: Dirección de Parques Nacionales y Vida Silvestre, viii + 573 pp.

Gambarotta, J. C., A. Saralegui and E. M. González. 1999. Vertebrados tetrápodos del Refugio de Fauna Laguna de Castillos, departamento de Rocha. *Relev. Biodivers.* 3:1–31.

Ganzorig, S., Y. Oku, M. Okamoto, R. Malgor, and M. Kamiya. 1999. A new nematode, *Ansiruptodera scapteromi* sp nov (Nematoda : Aspidoderidae), recovered from the Argentinean water rat *Scapteromys tumidus* (Waterhouse, 1837) in Uruguay. *Parasitol. Res.* 85:597–600.

Garagna, S., A. Pérez-Zapata, M. Zuccotti, S. Mascheretti, N. Marziliano, C. A. Redi, M. Aguilera, and E. Capanna. 1997. Genome composition in Venezuelan spiny-rats of the genus *Proechimys* (Rodentia, Echimyidae). I. Genome size, C-heterochromatin and repetitive DNAs in situ hybridization patterns. *Cytogenet. Cell Genet.* 78:36–43.

García, F. J., M. Delgardo-Jaramillo, M. Machado, and L. Aular. 2012. Preliminary inventory of mammals from Yurubí National Park, Yaracuy, Venezuela with some comments on their natural history. *Rev. Biol. Trop.* 60:459–72.

García, F. J., M. Machado, M. I. Delgado-Jaramillo, L. Aular, and Y. Mújica. 2012. Nuevo registro de *Ichthyomys pittieri* (Rodentia: Cricctidae) para la Cordillera de la Costa Central de Venezuela, con notas sobre su historia natural y distribución. *Mastozool. Neotrop.* 19:303–9.

García, F. J., and E. Sánchez-González. 2013. Morfometría geométrica craneal en tres especies de roedores arborícolas neotropicales (Rodentia: Cricetidae: *Rhipidomys*) en Venezuela. *Therya* 4:157–78.

García, K., J. C. Ortiz, M. Aguayo, and G. D'Elía. 2011. Mammalia, Rodentia, Sigmodontinae, *Irenomys tarsalis* (Philippi, 1900) and *Geoxus valdivianus* (Philippi, 1858): significant ecological range extension. *Check List* 7:276–78.

García, L. F. 1999. "Molecular phylogenetics of Neotropical oryzomyine rodents." Ph.D. diss., University of California, Berkeley.

García-Allegue, R., P. Lax, A. M. Madariaga, and J. A. Madrid. 2006. Locomotor and feeding activity rhythms in a light-entrained diurnal rodent, *Octodon degus*. *Am. J. Physio. Reg. Integr. Comp. Physiol.* 277:523–31.

García Esponda, C. M., L. J. M. De Santis, J. I. Noriega, G. O. Pagnoni, G. J. Moreira, and M. N. Bertellotti. 1998. The diet of *Tyto alba* (Strigiformes: Tytonidae) in the lower Chubut valley river (Argentina). *Neotrópica* 44:57–63.

García Esponda, C. M., G. J. Moreira, E. R. Justo, and L. J. M. De Santis. 2009. Analysis of craniometric variation in *Ctenomys talarum* (Rodentia, Ctenomidae). *Mastozool. Neotrop.* 16:69–81.

García Fernández, J. J., R. A. Ojeda, R. M. Fraga, G. B. Diaz, and R. J. Baigún, eds. 1997. *Libro rojo de mamíferos y aves amenazados de la Argentina*. Fundación para la Conservación de las Especies y el Medio Ambiente, Sociedad Argentina para el Estudio de los Mamíferos, Asociación Ornitológica del Plata, and Administración de Parques Nacionales, 221 pp.

García-Olaso, F. 2008. Evaluación de los caracteres diagnósticos de *Oxymycterus josei* Hoffmann, Lessa y Smith, 2002 (Rodentia: Cricetidae) con comentarios sobre la diferenciación de las espécies uruguayas del gênero. *Mastozool. Neotrop.* 15:117–23.

Gardner, A. L. 1971. Karyotypes of two rodents from Perú, with a description of the highest diploid number recorded for a mammal. *Experientia* 26:1088–89.

———. 1976. The distributional status of some Peruvian mammals. *Occas. Papers Mus. Zool. Louisiana State Univ.* 48:1–18.

———. 1983a. *Proechimys semispinosus* (Rodentia: Echimyidae): distribution, type locality, and taxonomic history. *Proc. Biol. Soc. Washington* 96:134–44.

———. 1983b. *Oryzomys caliginosus* (raton pardo, raton arrocero pardo, Costa Rican dusky rice rat). In *Costa Rican natural history*, ed. D. H. Janzen, 483–85. Chicago: University of Chicago Press, xi + 816 pp.

———. 1988. The mammals of Parque Nacional Serranía de la Neblina, Territorio Federal Amazonas, Venezuela. In *Cerro de la Neblina. Resultados de la Expedición 1983–1987*, ed. C. Brewer-Carias, 695–765. Caracas: Editorial Sucre, viii + 922 pp.

———. 1990. Two new mammals from southern Venezuela and comments on the affinities of the highland fauna of Cerro de la Neblina. In *Advances in Neotropical mammalogy*, eds. K. H. Redford and J. F. Eisenberg, 411–24. Gainesville, Florida: Sandhill Crane Press, x+614 pp. [Dated 1989; published 6 February 1990.]

———., ed. 2008. *Mammals of South America. Volume 1, Marsupials, Xenarthrans, Shrews, and Bats.* Chicago: University of Chicago Press, i–xx+1–669 pp. [Copyright date is 2007, but book was published in 2008.]

Gardner, A. L., and L. H. Emmons. 1984. Species groups in *Proechimys* (Rodentia, Echimyidae) as indicated by karyology and bullar morphology. *J. Mamm.* 65:10–25.

Gardner, A. L., and J. L. Patton. 1976. Karyotypic variation in oryzomyine rodents (Cricetinae) with comments on chromosomal evolution in the Neotropical cricetine complex. *Occas. Papers Mus. Zool. Louisiana State Univ.* 49:1–48.

Gardner, A. L., and M. Romo R. 1993. A new *Thomasomys* (Mammalia: Rodentia) from the Peruvian Andes. *Proc. Biol. Soc. Washington* 106:762–74.

Gardner, S. L., and S. Anderson. 2001. Persistent fontanelles in rodent skulls. *Am. Mus. Novit.* 3327:1–15.

Gardner, S. L., A. T. Dursahinhan, G. R. Rácz, N. Batsaikhan, S. Ganzorig, D. S. Tinnin, D. Damdinbazar, et al. 2013. Sylvatic species of *Echinococcus* from rodent intermediate hosts in Asia and South America. *Occas. Papers Mus. Texas Tech Univ.* 318:1–13.

Gardner, S. L., and D. W. Duszynski. 1990. Polymorphism of eimerian oocysts can be a problem in naturally infected hosts: an example from subterranean rodents in Bolivia. *J. Parasitol.* 76:805–11.

Gardner, S. L., R. L. Rausch, and O. C. J. Camacho. 1988. *Echinococcus vogeli* Rausch and Bernstein, 1972, from the paca, *Cuniculus paca* L. (Rodentia: Dasyproctidae), in the Departamento de Santa Cruz, Bolivia. *J. Parasitol.* 74:399–402.

Gardner, S. L., J. Salazar-Bravo, and J. A. Cook. 2014. *New species of* Ctenomys *Blainville 1826 (Rodentia: Ctenomyidae) from the lowlands and central valleys of Bolivia.* Special Publications of the Museum 62. Lubbock: Texas Tech University Press, 1–34.

Gasco, A., M. I. Rosi, and V. Durán. 2006. Análisis arqueofaunístico de microvertebrados en "Caverna de las Brujas" (Malargüe-Mendoza-Argentina). *An. Arqueol. Etnol.*, special issue, 61:135–62.

Gastal, M. L. A. 1994. Densidade, razao sexual e dados biométricos de uma população de *Ctenomys minutus* Nehring, 1887 (Rodentia: Caviomorpha: Ctenomyidae). *Iheringia Ser. Zool.* 77:35–43.

Gatti, A., R. Bianchi, C. Regina, X. Rosa, and S.L. Mendes. 2006. Diet of two sympatric carnivores, *Cerdocyon thous* and *Procyon cancrivorus*, in a restinga area of Espírito Santo State, Brazil. *J. Trop. Ecol.* 22:227–30.

Gaumer, G. F. 1917. *Monografía de los mamíferos de Yucatán.* Mexico City: Secretaria de Fomento., i–xli, 331 pp.

Gava, A., and T. R. O. de Freitas. 2002. Characterization of a hybrid zone between chromosomally divergent populations of *Ctenomys minutus* (Rodentia: Ctenomyidae). *J. Mamm.* 83:843–51.

———. 2004. Microsatellite analysis of a hybrid zone between chromosomally divergent populations of *Ctenomys minutus* from southern Brazil (Rodentia: Ctenomyidae). *J. Mamm.* 85:1201–6.

Gava, A., T. R. O. de Freitas, and J. Olimpio. 1998. A new karyotype for the genus *Cavia* from a southern island of Brazil (Rodentia–Caviidae). *Genet. Mol. Biol.* 21:77–80.

Gava, A., M. B. dos Santos, and F. M. Quintgela. 2012. A new karyotype for *Caviamagna* (Rodentia: Caviidae) from an estuarine island and *C. aperea* from adjacent mainland. *Acta Theriol.* 57:9–14.

Gay, C. 1847. *Historia física y política de Chile: segun documentos adquiridos en esta república durante doce años de residencia en ella y publicada bajo los auspicios del supremo gobierno.* Zoologia. 1. Paris: the author; Chile: Museo de Historial Natural de Santiago, 1–496 pp.

Gayo, V., P. Cuervo, D. Rosadilla, S. Birriel, L. Dell'Oca, A. Trelles, U. Cuore, and R. Mera y Sierra. 2011. Natural *Fasciola hepatica* infection in nutria (*Myocastor coypus*) in Uruguay. *J. Zoo Wildl. Med.* 42:354–56.

Geise, L. 1995. "Os roedores sigmodontíneos (Rodentia, Muridae) do estado do Rio de Janeiro. Sistemática, citogenética, distribuição e variação geográfica." Ph.D. diss., Universidade Federal do Rio de Janeiro, Brazil.

———. 2012. Akodon cursor (Rodentia: Cricetidae). *Mamm. Species* 44(893):33–43.

Geise, L., C. Aires, A. Freires, and A. Percequillo. 2006. Novos dados sobre morfologia, cariótipo e diversidade molecular de *Bibimys labiosus* (Winge, 1887). I. Congresso Sul-Americano de Mastozoologia, Gramado, Brasil, Resumos, 99.

Geise, L., H. G. Bergallo, C. E. L. Esbérard, C. F. D. Rocha, and M. Van Sluys. 2008. The karyotype of *Blarinomys breviceps* (Mammalia: Rodentia: Cricetidae) with comments on its morphology and some ecological notes. *Zootaxa* 1907:47–60.

Geise, L., F. C. Canavez, and H. N. Seuánez. 1998. Comparative karyology in *Akodon* (Rodentia, Sigmodontinae) from southeastern Brazil. *J. Hered.* 89:158–63.

Geise, L., R. Cerqueira, and H. N. Seuánez. 1996. Karyological characterization of a new population of *Akodon lindberghi* in Minas Gerais state (Brazil). *Caryologia* 49:59–63.

Geise, L., G. Marroig, and L. G. Pereira. 2007. Comparative craniofacial morphometry, karyotypic and mtDNA in the *Akodon cursor* (Rodentia, Muridae) from the South American Atlantic Forest: integrative approaches. *J. Morphol.* 268:1076.

Geise, L., D. A. Moraes, and H. S. Silva. 2005. Morphometric differentiation and distributional notes of three species of *Akodon* (Muridae, Sigmodontinae, Akodontini) in the Atlantic coastal area of Brazil. *Arq. Mus. Nac., Rio de Janeiro* 63:63–74.

Geise, L., R. Paresque, H. Sebastão, L. T. Shirai, D. Astúa, and G. Marroig. 2010. Non-volant mammals, Parque Nacional do Catimbau, Vale do Catimbau, Buíque, Pernambuco State, Brazil, with karyologic data. *Check List* 6:180–86.

Geise, L., and L. G. Pereira. 2008. Rodents (Rodentia) and marsupials (Didelphimorphia) in the municipalities of Ilhéus and Pau Brasil, state of Bahia, Brazil. *Check List* 4:174–77.

Geise, L., L. G. Pereira, D. E. P. Bossi, and H. G. Bergallo. 2004. Pattern of elevational distribution and richness of nonvolant mammals in Itatiaia National Park and surroundings, in southeastern Brazil. *Brazil. J. Biol.* 82:92–101.

Geise, L., M. F. Smith, and J. L. Patton. 2001. Diversification in the genus *Akodon* (Rodentia, Sigmodontinae) in southeastern South America: mtDNA sequence analysis. *J. Mamm.* 82:92–101.

Geise, L., M. Weksler, and C. R. Bonvicino. 2004. Presence or absence of gall bladder in some Akodontini rodents (Muridae, Sigmodontinae). *Mamm. Biol.* 69:210–14.

Genest, H., and G. Dubost. 1975. Pair living in the mara (*Dolichotis patagonum* Z.). *Mammalia* 38:155–62. [1974 vol. 38, no. 3: pp. 355–566. "Achevé d'imprimer le 31 mars 1975."]

Genoways, H. H. 1973. *Systematics and evolutionary relationships of spiny pocket mice, genus Liomys.* Special Publications of the Museum 5. Lubbock: Texas Tech University Press, 368 pp.

Genoways, H. H., S. L. Williams, and J. A. Groen. 1981. Results of the Alcoa Foundation-Suriname Expeditions. V. Noteworthy records of Surinamese mammals. *Ann. Carnegie Mus.* 50:319–32.

Gentile, R., and R. Cerqueira. 1995. Movement patterns of five species of small mammals in a Brazilian Restinga. *J. Trop. Ecol.* 11:671–77.

Gentile, R., S. F. Costa-Neto, M. M. L. Goncalves, S. T. Bonecker, F. A. Fernandes, J. S. Garcia, M. G. M. Barreto, et al. 2006. An ecological field study of the water-rat *Nectomys squamipes* as a wild reservoir indicator of *Schistosoma mansoni* transmission in an endemic área. *Mem. Inst. Oswaldo Cruz* 101:111–17.

Gentile, R., P. S. D'Andrea, and R. Cerqueira. 1997. Home ranges of *Philander frenata* and *Akodon cursor*

in a Brazilian Restinga (Coastal shrubland). *Mastozool. Neotrop.* 4:105–12.

Gentile de Fronza, T. 1970. Cariotipo de *Oxymycterus rutilans platensis* (Rodentia: Cricetidae). *Physis* 30:343.

Gentile de Fronza, T., R. Wainberg, and G. Hurtado de Catalfo. 1979. Cromosomas sexuales múltiples en *Deltamys kempi* (Rodentia, Cricetidae): estadio preliminar a la instauración del sistema XY1Y2/XX. *Mendeliana* 4:28–35.

Gentile de Fronza, T., R. Wainberg, and G. Hurtado de Catalfo. 1981. Multiple sex chromosomes in *Deltamys kempi* (Rodentia, Cricetidae): preliminary steps towards the stablishment of the XY1Y2/XX system. *Caryologia* 34:457–66.

Geoffroy St.-Hilaire, É. 1803. *Catalogue des mammifères du Muséum National d'Histoire Naturelle.* Paris: publisher unknown, 272 pp.

———. 1804. Note sure un nouveau genre de mammifères, de l'ordre des rongeurs, sous le nom d'hydromys. *Bull. Sci. Soc. Philomat. Paris* 3(93):253–54.

———. 1805. Mémoire sur un nouveau genre de mammifères nommé *Hydromys. Ann. Mus. Nat. Hist. Nat. Paris* 6:81–90, 2 plates.

———. 1820. *Histoire Naturelle des Mammiferes, avec des figures originales, colorées, dessinées d'après des animaux vivans; publié sous l'autorité de l'administation de Musée d'histoire naturelle,* vol. 3, p. 2 (text to pl. 282). Paris: C. de Lasteyrie.

Geoffroy St.-Hilaire, I. 1826. Le Moco, *Kerodon Sciureus.* In *Dictionnaire classique d'histoire naturelle,* 120–21. Paris: Rey et Gravier, Baudouin Frères, 9(Io-Macis):1 (unnumbered) + 1–596.

———. 1827. Myopotame. *Myopotamus.* In *Dictionnaire classique d'histoire naturelle,* 372–74. Paris: Rey et Gravier, Baudouin Frères, 11(Mo-Nso):1 (unnumbered) + 1–615.

———. 1828. Porc-Épic. *Hystrix.* In *Dictionnaire classique d'histoire naturelle,* 212–17. Paris: Rey et Gravier, Baudouin Frères, 14(Pla–Roy):1 (unnumbered) + 1–710.

———. 1832 [1833]. Essai sur le genre *Sciurus*, et description de six nouvelles espèces. Écureuil. *Sciurus.* Linn. *Mag. Zool.* 2, Classe 1:8 pg. [unnumbered], plates 4–6.

———. 1838a. Notice sur les rongeurs épineux désignés par les auteurs sous les noms d'*Echimys, Loncheres, Heteromys* et *Nelomys. C. R. Hebd. Seances Acad. Sci.* 6(2):884–88.

———. 1838b. Notice sur les rongeurs épineux désignés par les auteurs sous les noms d'*Echimys, Loncheres, Heteromys* et *Nelomys. Ann. Sci. Nat.* ser. 2, 10(August):122–27.

———. 1840. Notice dur les rongeurs épineux, désignés par les auteurs sous les noms d'*Echimys, Loncheres, Heteromys* et *Nelomys*. *Mag. Zool. Anat. Comp. Palaeontol.* 1840 (12):1–57, 24 plates.

———. 1855. Mammifères. In *Voyage autour du monde sur La Frégate la Vénus, commandee par Abel du Petit-Thouars. Zoologie, mammifères, oiseaus, reptiles et poissons*, 1–176 + 13 pls. Paris: Gide et J. Baudry, 1–351 + i–iii + 23 plates.

Geoffroy St.-Hilaire, I., and A. D. d'Orbigny. 1833. *Cobaye. Cavia.* Linn. *Rev. Mag. Zool. Pure Appl.* Paris: Chez Lequien Fils Libraire, 3 (Classe 1):1–4 (unnumbered), plate 12.

George, T. K., S. A. Marques, M. de Vivo, L. C. Branch, N. Gomes, and R. Rodrigues. 1988. Levantamento de mamíferos do Pará—Tapajós. *Bras. Florest.* 63:33–41.

George, W., and B. J. Weir. 1974. Hystricomorph chromosomes. In *The biology of hystricomorph rodents*, ed. I. W. Rowlands and B. J. Weir, 79–108. *Symp. Zool. Soc. Lond.*, 34:xix + 1–482.

George W., B. J. Weir, and J. Bedford. 1972. Chromosome studies in some members of the family Caviidae (Mammalia: Rodentia). *J. Zool. Soc. Lond.* 168:81–89.

Gerrard, E. 1862. Catalogue of the bones of Mammalia in the collection of the British Museum. London: British Museum (Natural History), 296 pp.

Gervais, H., and F. Ameghino. 1880. *Los mamíferos fósiles de la América del Sud.* Paris: F. Savy; Buenos Aires: Igon Hermanos, xi + 1–225 pp.

Gervais, P. 1841. Mammifères. In *Zoologie*, ed. Eydoux, J. F. T., and L. F. A. Souleyet, 1–68. In *Voyage autour du Monde exécuté pendant les années 1836 et 1837 sur la corvette La Bonite commandée par M. Vaillant.* Paris: Arthus Bertrand, 1:xl + 132 pp.; Atlas: plates 1–12 (Mammifères).

———. 1854. *Histoire naturelle de mammifères avec l'indication de leurs moeurs, et de leurs rapports avec les arts, le commerce et l'agriculture. Primates, cheiroptères, insectivores et rongeurs.* Paris: L. Curmer, 1:xxiv + 1–418, 19 plates.

———. 1855 [1856]. Cinquième mémoire. Énumération des principals espèces de mammifères. In *Mammifères*, ed. P. Gervais, 107–16, 4 pls. In *Animaux nouveaux ou rares recueillis pendant l'expédition dans les parties centrales de l'Amérique du Sud, de Rio de Janeiro à Lima, et de Lima au Pará; exécutée par ordre du gouvernement français pendant les années 1843 à 1847, sous la direction du comte Francis de Castelnau*, ed. F. de Castelnau. Paris: P. Bertrand, 7ᵉ. Partie, Zoologie, Tome 1:1–116, 20 pls., 1855. [Received by Académie Français on 30 June 1856; see Sherborn and Woodward 1901.]

Gervais, P., and A. d'Orbigny. 1844. Mammalogie.—La description d'une espèce nouvelle de Mammifères rongeurs, du genre *Octodon* de Bennett. *Soc. Phil. Paris* 1844:22.

Gettinger, D. 1992a. Three new species of *Laelaps* (Acari: Laelapidae) associated with small mammals in central Brazil. *J. Med. Entomol.* 29:66–70.

———. 1992b. Host specificity of *Laelaps* (Acari: Laelapidae) in central Brazil. *J. Med. Entomol.* 29:71–77.

Gettinger, D., and K. Ernest. 1995. Small-mammal community structure and the specificity of ectoparasite associations in central Brazil. *Rev. Bras. Biol.* 55:331–41.

Gettinger, D., and S. L. Gardner. 2005. Bolivian ectoparasites: a new species of laelapine mite (Acari: Parasitiformes, Laelapidae) from the rodent *Neacomys spinosus*. *J. Parasitol.* 91:49–52.

Gettinger, D., F. Martins-Hatano, M. Lareschi, and J. R. Malcolm. 2005. Laelapine mites (Acari: Laelapidae) associated with small mammals from Amazonas, Brazil, including a new species from marsupials. *J. Parasitol.* 91:45–48.

Gettinger, D., and R. D. Owen. 2000. *Androlaelaps rotundus* Fonseca (Acari: Laelapidae) associated with akodontine rodents in Paraguay: a morphometric examination of a pleioxenous ectoparasite. *Rev. Bras. Biol.* 60:425–34.

Ghilardi, R., Jr., and C. J. R. Alho. 1990 (1991). Forest seasonal productivity and animal forage activity in terra firme habitat in Amazonia. *Acta Amaz.* 20:61–76.

Giannoni, S. M., C. E. Borghi, M. Dacar, and C. M. Campos. 2005. Main food categories in diets of sigmodontine rodents in the Monte (Argentina). *Mastozool. Neotrop.* 12:181–87.

Giannoni, S. M., C. E. Borghi, and V. G. Roig. 1996. The burrowing behavior of *Ctenomys eremophilus* (Rodentia, Ctenomyidae) in relation with substrate hardness. *Mastozool. Neotrop.* 3:161–70.

Giannoni, S. M., C. E. Borghi, and R. A. Ojeda. 2000. Feeding behaviour of *Tympanoctomys barrerae*, a rodent specialized in consuming Atriplex leaves. *J. Arid Environ.* 46:117–21.

Giannoni, S. M., M. Dacar, P. Taraborelli, and C. E. Borghi. 2001. Seed hoarding by rodents of the Monte Desert, Argentina. *Austral Ecol.* 26:259–63.

Giebel, C. G. 1855. *Die säugethiere in zoologischer, anatomischer und palaeontologischer beziehung umfassend dargestellt.* Leipzig: A. Abel, xii + 1,108 pp.

———. 1857. Beiträge zur Osteologie der Nagetiere. *Abhandlungen Naturwissenschaften VereinSachsen Thüringen Halle* 1: 91–261, 5 plates.

———. 1859. Die Säugethiere in zoologischer, anatomischer und palaeontologischer Beziehung, 2nd ed. Leipzig: Ambrosius Abel, xvi + 1108 pp.

Giglioli, E. H. 1874. Ricerche intorno alla distribuzione geografica generale o corologie degli animali vertebrati. III. Regione Boreo-Americana. *Boll. Soc. Geogr. Ital.* 2:321–66.

Gil, G. 1989. *Informe del estudio florofaunístico del área de Monte León–Mamíferos terrestres.* Informe interno inédito, Fundación Vida Silvestre Argentina, 33 pp.

Gill, T. 1872. Arrangement of the families of mammals with analytical tables. *Smithsonian Misc. Coll.* 11(1):vi + 98 pp.

Giménez, M. D., C. J. Bidau, C. F. Argüelles, and J. R. Contreras. 1999. Chromosomal characterization and relationship between two new species of *Ctenomys* (Rodentia, Ctenomyidae) from northern Córdoba Province, Argentina. *Z. Säugetierk.* 64:91–106.

Giménez, M. D., C. J. Bidau, and J. B. Searle. 2001. Chromosomal and molecular delimitation of *Ctenomys* (Rodentia, Ctenomyidae) from Argentine Mesopotamia. *Chromosome Res.* 9:109–10.

Giménez, M. D., J. R. Contreras, and C. J. Bidau. 1997. Chromosomal variation in *Ctenomys pilarensis*, a recently described species from eastern Paraguay (Rodentia, Ctenomyidae). *Mammalia* 61:385–98.

Giménez, M. D., P. M. Mirol, C. J. Bidau, and J. B. Searle. 2002. Molecular analysis of populations of *Ctenomys* (Caviomorpha, Rodentia) with high karyotypic variability. *Cytogenet. Genome Res.* 96:130–36.

Giné, G. A. F., J. M. B. Duarte, and D. Faria. 2010. Feeding ecology of a selective folivore, the thin-spined porcupine (*Chaetomys subspinosus*) in the Atlantic Forest. *J. Mamm.* 91:931–41.

Giné, G. A. F., J. M. B. Duiarte, T. C. S. Motta, and D. Baria. 2011. Activity, movement and secretive behavior of a threatened arboreal folivore, the thin-spined porcupine, in the Atlantic Forest of southern Bahia, Brazil. *J. Zool.* 286 (2012):131–39.

Glade, A. (editor) 1988. *Libro rojo de los vertebrados terrestres chilenos.* Corporación Nacional Forestal, Ministerio de Agricultura, Santiago, Chile, 65 pp.

Glanz, W. E. 1983. The terrestrial mammal fauna of Barro Colorado Island: censuses and long-term changes. In *The ecology of a tropical forest: Seasonal rhythms and long-term changes*, ed. E. Leigh Jr., A. Rand, and D. Windsor, 455–68. Washington, DC: Smithsonian Institution Press, 468 pp.

———. 1984. Food and habitat use by two sympatric *Sciurus* species in central Panama. *J. Mamm.* 65:342–47.

Glanz, W. E., and S. Anderson. 1990. Notes on Bolivian mammals. 7. A new species of *Abrocoma* (Rodentia) and relationships of the Abrocomidae. *Am. Mus. Novit.* no. 2991:1–32.

Glanz, W. E., R. W. Thorington Jr., J. Giacalone Madden, and L. R. Heaney. 1982. Seasonal food use and demographic trends in *Sciurus granatensis*. In *The ecology of a tropical forest: Seasonal rhythms and long-lerm changes*, ed. E. G. Leigh, A. S. Rand, and D. M. Windsor, 239–52. Washington, DC: Smithsonian Institution Press, 468 pp.

Gliwicz, J. 1984. Age indicators in the spiny rat *Proechimys semispinosus*. *Trop. Ecol.* 24:299–304.

Gloger, C. W. L. 1841–1842. *Gemeinnutziges Hand- und Hilfsbuch der Naturgeschichte. Für gebildete Leser aller Stände, besonders für die reifere Jugend und ihre Lehrer.* Breslau: U. Schulz und Co., 1:xxxxiv + 496. [Most of the section on mammals is included in the first 160 pages, which were published in 1841; see Thomas 1895a.]

Gmelin, J. F. 1788. *Caroli a Lineeé . . . Systema naturae per regna tria naturae secundum classes ordines, genera, species com characteribus, differentiis, synonymis, locis.* Editio decima tertia, aucta, reformata. Lipsiae: Georg. Emanuel. Beer, 1–2 + 10 (unnumbered), 3–500 pp.

Godoi, M. N., N. L. da Cunha, and N. C. Cáceres. 2010. Efeito do gradiente floresta-cerrado-campo sobre a comunidade de pequenos mamíferos do Alto do Maciço do Urucum, oeste do Brasil. *Mastozool. Neotrop.* 17:263–77.

Goeldi, E. A. 1893. *Os mammiferos do Brasil.* Rio de Janeiro: Livraría Classica de Alves, iv + 182 pp.

———. 1897. *Mesomys edaudatus.* Um roedor esquecido durante meio seculo. *Bol. Mus. Paraense* 2(2):253–55.

———. 1901. Dois roedores notáveis da família dos ratos do Brasil. *Bol. Mus. Paraense* 3(2):166–79.

———. 1904a. On the rare rodent *Dinomys branickii* Peters. *Proc. Zool. Soc. Lond.* 1904(2):158–65.

Goeldi, E. A., and G. Hagmann. 1904. Prodromo de um catalogo critico, comentado de collecção de mamíferos no museo do Pará. *Bol. Mus. Paraense* 4:38–122.

———. 1906. Catálogo de Mammíferos. *Bol. Mus. Paraense Emilio Goeldi Hist. Nat. Ethnograph.*, 1904–1906, 4, fasc. 4.

Goff, M. L. 1981. The genus *Teratothrix* (Acari: Trombiculidae), with descriptions of three new species and a key to the species. *J. Med. Entomol.* 18:244–48.

Goff, M. L., and J. M. Brennan. 1977. The genus *Speleocola* Lipovsky (Acari: Trombiculidae) with descriptions of two new species from Venezuela. *J. Parasitol.* 63:1089–91.

———. 1978. Three new species of *Colicus acari* (Trombiculidae) from Venezuela. *J. Med. Entomol.* 14:565–69.

Goff, M. L., and D. Gettinger. 1995. New genus and six new species of chiggers (Acari: Trombiculidae and Leeuwenhoekiidae) collected from small mammals in Argentina. *J. Med. Entomol.* 32:439–48.

Goff, M. L., and J. O. Whitaker Jr. 1984. A small collection of chiggers (Acari: Trombiculidae) from mammals collected in Paraguay. *J. Med. Entomol.* 21:327–35.

Goff, M. L., J. O. Whitaker Jr., and J. M. Dietz. 1983. Three new species of chiggers (Acari: Trombiculidae) from Brazil. *J. Med. Entomol.* 20:183–88.

Goldfuss, G. A. 1809. *Vergleichende Naturbeschreibung der Säugethiere.* Erlangen: Verlage der Taltherschen Kunst-und Buchgandlung, xix + 314 pp., 36 plates pp. [Not seen; cited by Hershkovitz 1959.]

Goldman, E. A. 1911a. Three new mammals from Central and South America. *Proc. Biol. Soc. Washington* 24:238–39.

———. 1911b. Revision of the spiny pocket mice (genera *Heteromys* and *Liomys*). *N. Amer. Fauna* 34:1–70.

———. 1912a. Descriptions of twelve new species and subspecies of mammals from Panama. *Smithsonian Misc. Coll.* 56:1–11.

———. 1912b. New mammals from eastern Panama. *Smithsonian Misc. Coll.* 60:1–18.

———. 1912c. The generic name *Cercomys* and *Proechimys*. *Proc. Biol. Soc. Washington* 25:94.

———. 1912d. The type locality of *Proechimys steerei* Goldman. *Proc. Biol. Soc. Washington* 25:186.

———. 1913. Descriptions of new mammals from Panama and Mexico. *Smithsonian Misc. Coll.* 60:1–20.

———. 1915. Five new rice rats of the genus *Oryzomys* from Middle America. *Proc. Biol. Soc. Washington* 28:127–30.

———. 1916. Notes on the genera *Isothrix* Wagner and *Phyllomys* Lund. *Proc. Biol. Soc. Washington* 29:125–26.

———. 1918. The rice rats of North America (Genus *Oryzomys*). *N. Amer. Fauna* 43:1–100.

———. 1920. Mammals of Panama. *Smithsonian Misc. Coll.* 69:1–309.

———. 1933. A new climbing mouse from Panama. *J. Washington Acad. Sci.* 23:525–26.

Gómez, G., J. L. Prado., and M. T. Alberdi. 1999. Small mammals from Arroyo Seco 2 (Buenos Aires province, Argentina). Their implications for taphonomy and paleoenvironment. *Estud. Geol.* (Madrid) 55:273–81.

Gómez, H., G. Ayala, R. B. Wallace, and F. Espinoza. 2003. Densidad de la ardilla roja amazónica (Família Sciuridae, *Sciurus spadiceus*) en el valle del río Tuichi (Parque Nacional y Area Natural de Manejo Integrado Madidi, La Paz, Bolivia). *Ecol. Bolivia* 38:79–88.

Gómez, H., R. B. Wallace, G. Ayala, and R. Tejada. 2005. Dry season activity periods of some Amazonian mammals. *Stud. Neotrop. Fauna Environ.* 40:91–95.

Gómez, M. S. 1998. Two Anoplura species from rodents in Chile: *Hoplopleura andina* Castro, 1981 (Hoplopleuridae) from *Geoxus valdivianus* (Cricetidae) and *Eulinognathus chilensis* n. sp. (Polyplacidae) from *Abrocoma bennetti* (Abrocomidae). *Res. Rev. Parasitol.* 58:49–54.

Gómez Fernández, M. J., O. E. Gaggiotti, and P. Mirol. 2012. The evolution of a highly speciose group in a changing environment: are we witnessing speciation in the Iberá wetlands? *Mol. Ecol.* 21:3266–82.

Gómez-Laverde, M. 1994. Los pequeños mamíferos no voladores del Parque Regional Natural Ucumarí. In *Ucumarí, un caso típico de la diversidad biótica andina,* ed. J. O. Rangel-Ch., 377–96. Pereira: Corporación Autónoma Regional de Risaralda, 451 pp.

Gómez-Laverde, M., R. P. Anderson, and L. F. García. 2004. Integrated systematic reevaluation of the Amazonian genus *Scolomys* (Rodentia: Sigmodontinae). *Mamm. Biol.* 69:119–40.

Gómez-Laverde, M., M. L. Bueno, and A. Cadena. 1990. Poblaciones de ratas (*Proechimys semispinosus*) (Rodentia: Echimyidae). In *Biota y ecosistemas de Gorgona,* ed. J. Aguirre and O. Rangel, 244–51. Bogotá: Fondo FEN Colombia, Editorial Presencia, 303 pp.

Gómez-Laverde, M., M. L. Bueno, and H. Lopéz-Arevalo. 1999. Descripción cariológica y morfológica de *Nectomys magdalenae* (Rodentia: Muridae: Sigmodontinae). *Rev. Acad. Colomb. Cien.* (suplemento especial) 23:631–40.

Gómez-Laverde, M., O. Montenegro-Díaz, H. López-Arévalo, A. Cadena, and M. L. Bueno. 1997. Karyology, morphology, and ecology of *Thomasomys laniger* and *T. niveipes* in Colombia. *J. Mamm.* 78:1282–89.

Gómez-Laverde, M., and V. Pacheco. 2008. *Thomasomys cinereiventer.* IUCN Red List of Threatened Species, version 2011.2, accessed February 23, 2012, http://www.iucnredlist.org.

Gómez-Laverde, M., V. Pacheco, and D. Tirira. 2008. *Thomasomys baeops.*IUCN 2011. IUCN Red List of Threatened Species. Version 2011.2, accessed February 23, 2012, http://www.iucnredlist.org.

Gómez-Laverde, M., and B. Rivas. 2008. *Akodon bogotensis.* IUCN Red List of Threatened Species, version 2013.1, accessed July 30, 2013, http://www.iucnredlist.org.

Gonçalves, A. Q., M. N. Boia, J. R. Coura, and R. M. Pinto. 2006. New records for helminths of hystricognath rodents from the middle and high Rio Negro micro-region, State of Amazonas, Brazil. *Rev. Bras. Zool.* 23:716–26.

Gonçalves, G. L., M. A. Faria-Correa, A. S. Cunha, and T. R. O. de Freitas. 2007. Bark consumption by the spiny rat *Euryzygomatomys spinosus* (G. Fischer) (Echimyidae) on a *Pinus taeda* Linnaeus (Pinaceae) plantation in South Brazil. *Rev. Bras. Zool.* 24:260–63.

Gonçalves, P. R. 2006. "Diversificação dos roedores sigmodotíneos em formações alto-montanas da Mata Atlântica." Ph.D. diss., Universidade Federal do Rio de Janeiro, Brazil.

Gonçalves P. R., F. C. Almeida, and C. R. Bonvicino. 2005. A new species of *Wiedomys* (Rodentia: Sigmodontinae) from Brazilian Cerrado. *Mamm. Biol.* 70:46–60.

Gonçalves, P. R., P. Myers, J. Vilela, J., and J. A. Oliveira. 2007. Systematics of species of the genus *Akodon* (Rodentia: Sigmodontinae) in southeastern Brazil and implications for the biogeography of the *campos de altitude. Misc. Publ. Mus. Zool., Univ. Michigan* 197:1–24.

Gonçalves, P. R., and J. A. Oliveira. 2004. Morphological and genetic variation between two sympatric forms of *Oxymycterus* (Rodentia: Sigmodontinae): an evaluaton of hypotheses of differentiation within the genus. *J. Mamm.* 85:148–61.

Gonçalves, P. R., and J. A. Oliveira. 2014. An integrative appraisal of the diversification in the Atlantic forest genus *Delomys* (Rodentia: Cricetidae: Sigmodontinae) with the description of a new species. *Zootaxa* 3760:1–38.

Gonçalves, P. R., J. A. Oliveira, M. Oliveira Corrêa, and L. M. Pessoa. 2005. Morphological and cytogenetic analyses of *Bibimys labiosus* (Winge, 1887) (Rodentia, Sigmodontinae): implications for its affinities with the scapteromyine group. In *Mammalian diversification: from chromosomes to phylogeography,* ed. E. A. Lacey and P. Myers, 175–210. *Univ. California Publ. Zool.* 133:v-viii + 1–383.

Gonnet, J. M., and R. A. Ojeda. 1998. Habitat use by small mammals in the arid Andean foothills of the Monte Desert of Mendoza, Argentina. *J. Arid Environ.* 38:349–57.

Gonzáles, A., and M. Espino. 1974. Un caso de salmonellosis en majaz (*Cuniculus paca*). *Rev. Invest. Pecurias* 3:93.

González, E. M. 1996. Mamíferos silvestres del Parque Lecocq y adyacencias. Lista preliminar y comentarios sobre su abundancia relativa y distribución en la zona. CIPFE CLAES, *Contribuciones en Biología* 16:5–6.

———. 1997. Aplicación de un método de evaluación rural rápida para el relevamiento de mamíferos terrestres en una isla del Delta del Paraná (Pcia. de Buenos Aires, Argentina). *XII Jornadas Argentinas de Mastozoología, Resúmenes,* p. 62.

———. 2000. Un nuevo genero de roedor sigmodontino de Argentina y Brasil (Mammalia: Rodentia: Sigmodoninae). *Comun. Zool. Mus. Hist. Nat. Montevideo* 12(195):1–12.

———. 2001. *Guía de campo de los mamíferos de Uruguay. Vida silvestre.* Montevideo: Sociedad Uruguaya para la Conservación de la Naturaleza, 339 pp.

González, E. M. and G. Fregueiro. 1999. Mamíferos no voladores de Laguna del Cisne, Departamento de Canelones, Uruguay (Mammalia). *Relevamientos de Biodiversidad, Publicación de Vida Silvestre, Sociedad Uruguay a para la Conservación de la Naturaleza* 2:1–7.

González, E. M., A. Langguth, and L. F. Oliveira. 1998. A new species of *Akodon* from Uruguay and southern Brazil (Mammalia: Rodentia: Sigmodontinae). *Comun. Zool. Mus. Hist. Nat. Montevideo* 191:1–8.

González, E. M. and J. A. Martínez Lanfranco. 2010. *Mamíferos de Uruguay. Guía de campo e introducción a su estudio y conservación.* Montevideo: ONG Vida Silvestre Uruguay, Museo Nacional de Hisoria Natural, Editoria Banda Oriental, 464 pp.

González, E. M., and E. Massoia. 1995. Revalidación del género *Deltamys* Thomas, 1917, con la descripción de una nueva subespecie de Uruguay y Sur del Brasil (Mammalia, Rodentia: Cricetidae). *Comun. Zool. Mus. Hist. Nat. Montevideo* 12:1–8.

González, E. M., and J. A. de Oliveira. 1997. La distribución geográfica de *Wiedomys pyrrhorhinos* (Wied, 1821) y *Wilfredomys oenax* (Thomas, 1928) (Rodentia: Muroidea). *12th Jornadas Argentinas de Mastozooología Resúmenes.*

González, E. M., and U. F. J. Pardiñas. 2002. Deltamys kempi. *Mamm. Species* 711:1–4.

González, E. M., and A. Saralegui. 1996. Análisis de componentes mastozoológicos en regurgitados de *Athene cunicularia* (Aves, Strigiformes) del Parque Santa Teresa, Rocha, Uruguay. *Contribuciones en Biología,* Uruguay 16:4.

González, J. 1994. Análisis biostadístico del género *Scapteromys* en Uruguay (Mammalia: Rodentia: Cricetidae). *Comun. Zool. Mus. Hist. Nat. Montevideo* 12:1–6.

González, L. A., and R. Murúa. 1985. Caracteristicas del periodo reproductivo de tres especies de roedores cricetidos del bosque higrofilo templado. *Anales del Museo de Historia Natural, Valparaiso* 16:87–99.

González, L. A., R. Murúa, and R. Feitl. 1982. Densidad poblacvional y padrones de actividad espacial de *Akodon olivaceus* (Rodentia: Cricetidae) en habitats diferentes. In *Zoología Neotropical*, ed. P. Salinas, 935–47. Mérida, Venezuela: Actas del VIII Congreso Latinoamericano de Zoología, 1531 pp.

González, L. A., R. Murúa, and C. Jofré. 2000. Habitat utilization of two muroid species in relation to population outbreaks in southern temperate forests of Chile. *Rev. Chilena Hist. Nat.* 73:489–95.

González, P., Y. E. Sawyer, M. Avila, A. G. Armién, B. Armién, and J. A. Cook. 2010. Variation in cytochrome-*b* haplotypes suggests a new species of *Zygodontomys* (Rodentia: Cricetidae) endemic to Isla Coiba, Panama. *Zoología* 27:660–65.

González, S., and N. Brum-Zorrilla. 1995. Karyological studies of the South American rodent *Myocastor coypus* Molina 1782 (Rodentia: Myocastoridae). *Rev. Chilena Hist. Nat.* 68:215–26.

González Dela Valle, M., A. Edelstein, S. Miguel, V. Martínez, J. Cortez, M. L. Cacace, G. Jurgelenas, S. S. Estani, and P. Padula. 2002. Andes virus associated with hantavirus pulmonary syndrome in northern Argentina and determination of the precise site of infection. *Am. J. Trop. Med. Hyg.* 66:713–20.

González-Ittig, R. E., J. L. Patton, and C. N. Gardenal. 2007. Analysis of cytochrome-*b* nucleotide diversity confirms a recent range expansion in *Calomys musculinus* (Rodentia, Muridae). *J. Mamm.* 88:777–83.

Gonzáles-Acuna, D., E. Briones, K. Ardiles, G. Valenzuela-Dellarossa, S. Corales, and R. A. Figueroa. 2009. Seasonal variation in the diet of the white-tailed kite (*Elanus leucurus*) in a surburban area of southern Chile. *J. Raptor Res.* 43:134–41.

González-Ittig, R. E., J. Salazar-Bravo, R. M. Barquez, and C. N. Gardenal. 2010. Phylogenetic relationships among species of the genus *Oligoryzomys* (Rodentia, Cricetidae) from Central and South America. *Zool. Scr.* 39:511–26.

González-M., A., and M. Alberico. 1993. Seleccion de habitat en una comunidad de mamiferos pequenos en la costa Pacifica de Colombia. *Caldasia* 17:313–24.

Goode, M., C. W. Radcliffe, K. Estep, A. Odum, and D. Chiszar. 1990. Field observations on feeding behavior in an Aruba island (West Indies) rattlesnake *Crotalus durissus unicolor* strike-induced chemosensory searching and trail following. *Bull. Psychonomic Soc.* 28:312–14.

Goodin, D. G., R. Paige, R. D. Owen, K. Ghimire, D. E. Koch, Y.-K. Chu, and C. B. Jonsson. 2009. Microhabitat characteristics of *Akodon montensis*, a reservoir for hantavirus, and hantaviral seroprevalence in an Atlantic Forest site in eastern Paraguay. *J. Vector Ecol.* 34:104–13.

Goodwin, G. G. 1943. Two new squirrels from Costa Rica. *Am. Mus. Novit.* 1218:1–2.

———. 1946. Mammals of Costa Rica. *Bull. Am. Mus. Nat. Hist.* 87:271–473.

———. 1953. Catalogue of the type specimens of recent mammals in the American Museum of Natural History. *Bull. Am. Mus. Nat. Hist.* 102:207–412.

———. 1955. New tree-climbing rats from Mexico and Colombia. *Am. Mus. Novit.* 1738:1–5.

———. 1956. A preliminary report on the mammals collected by Thomas MacDougall in southeastern Oaxaca, Mexico. *Am. Mus. Novit.* 1757:1–15.

———. 1959. Descriptions of some new mammals. *Am. Mus. Novit.* 1967:1–8.

———. 1961. The murine opossums (genus *Marmosa*) of the West Indies, and the description of a new subspecies of *Rhipidomys* from Little Tobago. *Am. Mus. Novit.* 2070:1–20.

———. 1962. Descriptions of two new rodents from Tobago, the west Indies, and of *Zygodontomys brevicauda tobagi* Thomas. *Am. Mus. Novit.* 2096:1–9.

———. 1965. A new subspecies of *Zygodontomys brevicauda* from Soldado Rock, Trinidad, the West Indies (Rodentia, Cricetidae). *Am. Mus. Novit.* 2238:1–10.

Gorchov, D. L., J. M. Palmeirim, M. Jaramillo, and C. F. Ascorra. 2004. Dispersal of seeds of *Hymenaea courbaril* (Fabaceae) in a logged rain forest in the Peruvian Amazonian. *Acta Amaz.* 34:251–59.

Gorostiague, M., and H. A. Regidor. 1993. La captura comercial del coypo *Myocastor coypus* (Mammalia: Myocastoridae) en Laguna Adela, Argentina. *Stud. Neotrop. Fauna Environ.* 28:57–63.

Gottdenker, N., R. B. Wallace, and H. Gómez. 2001. La importancia de los atropellos para la ecología y conservación: *Dinomys branickii* un ejemplo de Bolivia. *Ecol. Boliv.* 35:61–67.

Gouveia, A. L. 2009. "Pequenos mamíferos não voadores no sul de Minas Gerais com caracterização cromossômica das espécies de roedores da sub-família Sigmodontinae." Undergraduate thesis, Universidade Federal de Lavras, Minas Gerais, Brazil.

Graipel, M. E. 2003. A simple ground-based method for trapping small mammals in the forest canopy. *Mastozool. Neotrop.* 10:177–81.

Graipel, M. E., J. J. Cherem, E. L. A. Monteiro-Filho, and L. Glock. 2006. Dinâmica populacional de marsupiais e roedores no Parque Municipal da Lagoa do Peri, Ilha de Santa Catarina, sul do Brasil. *Mastozool. Neotrop.* 13(1):31–49.

Graipel, M. E., J. J. Cherem, and A. Ximénez. 2001. Mamíferos errestres não voadores da Ilha de Santa Catarina, sul do Brasil. *Biotemas* 14:109–40.

Graipel, M. E., P. R. M. Miller, and L. Glock. 2003. Padrão de atividade de *Akodon montensis* e *Oryzomys russatus* na Reserva Volta Velha, Santa Catarina, sul do Brasil. *Mastozool. Neotrop.* 10:255–60.

Grand, T. I., and J. F. Eisenberg. 1982. On the affinities of the Dinomyidae. *Saugetierkd. Mitt.* 30:151–57.

Grant, T. K., and G. B. Estes. 2009. *Darwin in Galapagos: Footsteps to a New World.* Princeton, NJ: Princeton University Press, xi + 362 pp.

Granzinolli, M.A., and J. C. Motta-Junior. 2006. Small mammal selection by the white-tailed hawk in southeastern Brazil. *Wilson J. Ornithol.* 118:91–98.

Grassé, P. P., and P. L. Dekeyser. 1955. Ordre des rongeurs. In *Traite de zoologie, Tome 17(2), Mammifères, les orders: anatomie, ethologie, systematique,* ed. P. P. Grassé, 1321–525. Paris: Masson, 1173–2300 pp.

Grau, J. 1986. *La chinchilla, su crianza en todos los climas,* 3rd ed. Buenos Aires: El Ateneo, 270 pp.

Gray, J. E. 1821. On the natural arrangement of vertebrose animals. *Lond. Med. Reposit.* 15:297–311.

———. 1825. An outline of an attempt at the disposition of Mammalia into tribes and families, with a list of the genera apparently appertaining to each tribe. *Ann. Phil.,* n.s., ser. 2, 10:337–44.

———. 1827. Synopsis of the species of the class Mammalia, as arranged with reference to their organization, by Cuvier, and other naturalists, with specific characters, synonyma. In *The animal kingdom arranged in conformity with its organization, by the Baron Cuvier, with additional descriptions of all the species hitherto named, and of many not before noticed,* vol. 5, ed. E. Griffith and others, 1–296. London: printed for Geo. B. Whittaker, 1–392.

———. 1842. Descriptions of some new genera and fifty unrecorded species of Mammalia. *Ann. Mag. Nat. Hist.,* ser. 1,10:255–67.

———. 1843a. [Published Excerpts of letter from J. Gray to the Curator of the Zoological Society of London.] *Proc. Zool. Soc. Lond.* 1843 (part XI):79.

———. 1843b. *List of the specimens of Mammalia in the collection of the British Museum.* London: British Museum (Natural History), xxviii + 216 pp.

———. 1850. On the species of *Cercolabes* confounded under the name of *C. prehensilis. Ann. Mag. Nat. Hist.,* ser. 2, 5:380–81.

———. 1861. Description of a new squirrel, in the British Museum, from New Grenada. *Proc. Zool. Soc. Lond.* 1861 (part I):92 + 1 plate.

———. 1865. Notice of an apparently undescribed species of American porcupine. *Proc. Zool. Soc. Lond.* 1865 (part I):321–22 + 1 plate.

———. 1867. Synopsis of the species of American squirrels in the collection of the British Museum. *Ann. Mag. Nat. Hist.,* ser. 3, 20:415–34.

———. 1868. Synopsis of the species of Saccomyinae, or pouched mice, in the collection of the British Museum. *Proc. Zool. Soc. Lond.* 1868 (part I):199–206.

———. 1972. On *Macroxus tephrogaster. Ann. Mag. Nat. Hist.,* ser. 4, 10:408.

———. 1873. Notes on the rats; with the description of some new species from Panama and the Aru Islands. *Ann. Mag. Nat. Hist.,* ser. 4, 12:416–19.

Greer, J. K. 1965. Mammals of Malleco Province Chile. *Publ. Mus. Mich. State Univ.* 3(2):49–152 + 8 plates.

Grelle, C. E. V. 2003. Forest structure and vertical stratification of small mammals in a secondary Atlantic Forest, southeastern Brazil. *Stud. Neotrop. Fauna Environ.* 38:81–85.

Grenha, V., M. V. Macedo, A. S. Pires, and R. F. Monteiro. 2010. The role of *Cerradomys subflavus* (Rodentia, Cricetidae) as seed predator and disperser of the palm *Allagoptera arenaria. Mastozool. Neotrop.* 17:61–68.

Griffith, E., C. Hamilton-Smith, and E. Pidgeon. 1827. The class Mammalia arranged by the Baron Cuvier, with specific descriptions. In *The Animal Kingdom arranged in conformity with its organization, by the Baron Cuvier, . . . with additional descriptions of all the species hitherto names, and of many not before noticed, by Edward Griffith . . . and others.* London: Geo. B. Whittaker, 3:1–468.

Grimwood, I. R. 1969. Notes on the distribution and status of some Peruvian mammals. *Amer. Comm. Internat. Wild Life Protection and New York Zool. Soc.,* Special Publ., no. 21:vi + 86 pp., 4 pls.

Gudinho, F. S., and G. E. Iack-Ximenes. 2010. Estudo de variação geográfica em *Blarinomys breviceps* (Winge, 1888). *V Congresso Brasileiro de Mastozoologia. Posters, ST12* (available on CD).

Gueldenstaedt, A.I. 1770. *Spalax,* novvm glirivm genus. *N. Comment. Acad. Sci. Imp. Petropolitanae* 14:409–40, 2 plates.

Guerrero, R. 1985. Parasitología. In *El estudio de los mamíferos en Venezuela: evaluación y perspective,* ed. M. Aguilera, 35–91. Caracas, Venezuela: Fondo Editorial del Acta Científica Venezolana, 256 pp.

Guerrero, R., O. Bain, C. Martin, and M. Barbuto. 2011. A new species of *Litomosoides* (Nematoda: Onchocercidae), parasite of *Nectomys palmipes* (Rodentia: Cricetidae: Sigmodontinae) from Venezuela: description, molecular evidence, *Wolbachia pipientis* screening. *Folia Parasitol.* 58:149–56.

Guerrero, R., R. Hoogesteijn, and P. Soriano. 1989. Lista preliminar de los mamíferos del Cerro Marahuaca, T. F. Amazonas, Venezuela. *Acta Terramaris* 1:71–77.

Guglielmone, A. A., A. J. Mangold, and J. E. Keirans. 1990. Redescription of the male and female of *Amblyomma parvum* Aragao 1908 and description of the nymph and larva of all stages of *Amblyomma pseudoparvum* new-species (Acari: Ixodida: Ixodidae). *Acarología* 31:143–60.

Guglielmone, A. A., and S. Nava. 2011. Rodents of the subfamily Sigmodontinae (Myomorpha: Cricetidae) as hosts for South American hard ticks (Acari: Ixodidae) with hypotheses on life history. *Zootaxa* 2904:45–65.

Guichón, M. L., V. B. Benítez, A. Abba, M. Borgnia, and M. H. Cassini. 2003. Foraging behavior of coypus *Myocastor coypus*: why do coypus consume aquatic plants? *Acta Oecol.* 24:241–46.

Guichón, M. L., M. Borgnia, C. F. Righi, G. H. Cassini, and M. H. Cassini. 2003. Social behavior and group formation in the coypu (*Myocastor coypus*) in the Argentinean pampas. *J. Mamm.* 84:254–62.

Guichón, M. L. C. P. Doncastor, and M. H. Cassini. 2003. Population structure of coypus (*Myocastor coypus*) in their region of origin and comparison with introduced populations. *J. Zool.* 261:265–72.

Guillotin, M. 1982. Rhymes d'activite et remimes alimentaires de *Proechimys cuvieri* et d'*Oryzomys capito velutinus* (Rodentia) en foret Guyanaise. *Rev. Ecol. (Terre Vie)* 36:337–81.

———. 1983. Place de *Proechimys cuvieri* (Rodentia, Echimyidae) dans les peuplements micromammaliens terrestres de la forêt guyanaise. *Mammalia* 46:299–318. [1982 vol. 46, no. 4: pp. 419ff. "Achevé d'imprimer le 10 fevrier 1983."]

Guillotin, M., and F. Petter. 1986. Un *Rhipidomys* nouveau de Guyane française, *R. leucodactylus aratayae* ssp. nov. (Rongeurs, Cricétidés). *Mammalia* 48:541–44. [1985 vol. 49, no. 4: pp. 445ff. "Achevé d'imprimer le 20 mars 1986."]

Guillotin, M., and J.-F. Ponge. 1984. Identification de deux espèces de rongeurs de Guyane français, *Proechimys cuvieri* et *Proechimys guyannensis* (Echimyidae) par l'analyse des correspondences. *Mammalia* 48:287–91.

Guimarães, D. A., R. S. Luz Ramos, G. W. Garcia, and O. M. Ohashi. 2009. The stimulatory effect of male agouti (*Dasyprocta prymnolopha*) on the onset of female puberty. *Acta Amaz.* 39:759–62.

Guimarães, D. A., D. Moreira, and W. G. Vale. 1997. Determination of agouti (*Dasyprocta prymnolopha*)

reproductive cycle by colpocytologyc diagnostic. *Acta Amaz.* 27:55–63.

Guimarães, D. A., R. L. Ramos, O. M. Ohashi, G. W. Garcia, and W. G. Vale. 2011. Plasma concentration of progesterone and 17β-estradiol of black-rumped agouti (*Dasyprocta prymnolopha*) during the estrous cycle. *Rev. Biol. Trop.* 59:29–35.

Guimarães, L. R. 1972. Contribuição a epidemiologia da peste endêmica no nordeste do Brasil e estado da Bahia: estudo das pulgas encontradas nessa região. *Rev. Bras. Malariol. Doenças Trop.* 24:95–163.

Guimarães, P. R., B. Z. Gomes, Y. J. Ahn, and M. Galetti. 2005. Cache pilferage in red-rumped agoutis (*Dasyprocta leporina*) (Rodentia). *Mammalia* 69:431–34.

Guimarães, P. R., Jr., J. Jose, M. Galetti, and J. R. Trigo. 2003. Quinolizidine alkaloids in *Ormosia arborea* seeds inhibit predation but not hoarding by agoutis (*Dasyprocta leporina*). *J. Chem. Ecol.* 29:1065–72.

Guitton, N., N. A. Araújo Filho, and I. A. Sherlock. 1986. Ectoparasitas de roedores e marsupiais no ambiente silvestre de Ilha Grande, estado do Rio de Janeiro, Brasil. *Mem. Inst. Oswaldo Cruz* 81:233–34.

Güntert, M., K. Grossenbacher, C. Huber, A. Aerni, and H. U. Morgenthaler. 1993. The E. A. Goeldi zoological collection in the Natural History Museum Bern: comments on an inventory. *Jahrbuch Nat. Mus. Stadt Bern* 11:147–61.

Günther, A. 1876. Report on some of the additions to the collection of Mammalia in the British Museum. 4. on some new mammals from tropical America. *Proc. Sci. Meet. Zool. Soc. Lond.* 1876:743–51.

———. 1879. On a new rodent from Medellin. *Proc. Sci. Meet. Zool. Soc. Lond.* 1879:144–45 + 1 plate.

Gupta, B. B. 1966. Skeleton of *Erethizon* and *Coendou*. *Mammalia* 30:495–97.

Guthmann, N., M. Lozada, J. A. Monjeau, and K. M. Heinemann. 1997. Population dynamics of five sigmodontine rodents of northwestern Patagonia. *Acta Theriol.* 42:143–52.

Guzmán, J. A., and W. Sielfeld. 2011. A new northern distribution limit of *Abrocoma bennettii* (Rodentia, Abrocomidae) in the coastal Atacama Desert, Paposo, north of Chile. *Mastozool. Neotrop.* 18:131–34.

Gwinn, R. N., J. R. Koprowski, R. R. Jessen, and M. J. Merrick. 2012. Sciurus spadiceus (Rodentia: Sciuridae). *Mamm. Species* 44:59–63.

Gyldenstolpe, N. 1932a. A manual of Neotropical sigmodont rodents. *Kungliga Svenska Vetenskapsakademiens Handlingar*, ser. 3, band 11(3):1–164, 18 plates.

———. 1932b. A new *Scapteromys* from Chaco Austral, Argentina. *Ark. Zool.* 24B(1):1–2.

Haag, T., V. C. Muschner, L. B. Freitas, L. F. B. Oliveira, A. R. Langguth, and M. S. Mattevi. 2007. Phylogenetic relationships among species of the genus *Calomys* with emphasis on South American lowland taxa. *J. Mamm.* 88:769–76.

Hadler, P., and J. Ferigolo. 2004. Roedores pleistocênicos da planície costeira do Estado do Rio Grande do Sul, Brasil. *Rev. Bras. Paleontol.* 7:231–38.

Hadler, P., D. H. Verzi, M. G. Vucetich, J. Ferigolo, and A. M. Ribeiro. 2008. Caviomorphs (Mammalia, Rodentia) from the Holocene of Rio Grande do Sul State, Brazil: systemtics and paleonenvironmental context. *Rev. Bras. Paleontol.* 11:97–116.

Hafner, D. J., M. S. Hafner, G. L. Hasty, T. A. Spradling, and J. W. Demastes. 2008. Evolutionary relationships of pocket gophers (*Cratogeomys castanops* species group) of the Mexican altiplano. *J. Mamm.* 89:190–8.

Hafner, J. C., and M. S. Hafner. 1983. Evolutionary relationships of heteromyid rodents. *Great Basin Nat. Mem.* 7:3–29.

Hafner, J. C., J. E. Light, D. J. Hafner, M. S. Hafner, E. Reddington, D. S. Rogers, and B. R. Riddle. 2007. Basal clades and molecular systematics of heteromyid rodents. *J. Mamm.* 88:1129–45.

Hafner, M. S. 1981. A biochemical investigation of geomyoid systematics. *Z. Zool. Syst. Evol.* 20:118–30.

Hafner, M. S., L. J. Barkley, and J. M. Chupasko. 1994: Evolutionary genetics of New World tree squirrels (Tribe Sciurini). *J. Mamm.* 75:102–9.

Hafner, M. S., A. R. Gates, V. L. Mathis, J. W. Demastes, and D. J. Hafner. 2011. Redescription of the pocket gopher *Thomomys atrovarius* from the Pacific coast of mainland Mexico. *J. Mamm.* 92:1367–82.

Hafner, M. S., and D. J. Hafner. 1987. Geographic distribution of two Costa Rican species of *Orthogeomys,* with comments on dorsal pelage markings in the Geomyidae. *Southwest. Nat.* 32:5–11.

———. 2009. Systematic and conservation status of the pocket gophers of Mexico. In *Sixty years of the National Mammal Collection of the Institute of Biology, National University of Mexico (UNAM): Contributions to knowledge and conservation of Mexican mammals,* ed. F. A. Cervantes, Y. Hortelano, and J. Vargas C., 301–8. Mexico City: Instituto de Biología, Universidad Nacional Autónoma de México, 317 pp.

Hafner, M. S., D. J. Hafner, J. W. Demastes, G. L. Hasty, J. E. Light, and T. A. Spradling. 2009. Evolutionary relationships of pocket gophers of the genus *Pappogeomys* (Rodentia: Geomyidae). *J. Mamm.* 90:47–56.

Hafner, M. S., J. E. Light, D. J. Hafner, S. V. Brant, T. A. Spradling, and J. W. Demastes. 2005. Cryptic species in the Mexican pocket gopher, *Cratogeomy merriami. J. Mamm.* 86:1095–8.

Hafner, M. S., T. A. Spradling, J. E. Light, D. J. Hafner, and J. R. Demboski. 2004. Systematic revision of pocket gophers of the *Cratogeomys gymnurus* species group. *J. Mamm.* 85:1170–83.

Hagemeyer, P., and S. Begall. 2006. Individual odour similarity and discrimination in the coruro (*Spalacopus cyanus,* Octodontidae). *Ethology* 112:529–36.

Hagmann, H. G. 1908. Die Landsäugetiere del Insel Mexiana. Als Beispiel der Einwirkung der Isolation auf die Umbildung der Arten. *Arch. Rasse Gessell. Biol.* 5:1–35, 2 plates.

Haiduk, M. W., J. W. Bickham, and D. J. Schmidly. 1979. Karyotype of six species of *Oryzomys* from Mexico and Central America. *J. Mamm.* 60:610–15.

Hajduk, A., A. Albornoz, and M. J. Lezcano. 2004. El "Mylodon" en el Patio de Atrás. Informe preliminar sobre los trabajos en el Sitio El Trébol, Ejido Urbano de San Carlos de Bariloche, Provincia de Río Negro. In *Contra viento y marea. Arqueología de Patagonia,* ed. M. T. Civalero, P. M. Fernández, and A. G. Guráieb, 715–31. Buenos Aires: Sociedad Argentina de Antropología.

Hall, E. R. 1960. *Oryzomys couesi* only subspecifically different from the marsh rice rat, *Oryzomys palustris. Southwest. Nat.* 5:171–73.

———. 1981. *The mammals of North America.* 2nd ed. New York: John Wiley and Sons, 1:xviii + 600 + 90 pp.

Hall, E. R., and K. R. Kelson. 1959. *The mammals of North America.* New York: Ronald Press, 2:viii + 2 (unnumbered) + 547–1083 + 79 pp.

Hallwachs, W. 1986. Agoutis (*Dasyprocta punctata*): the inheritors of guapinol (*Hymenaea courbaril:* Leguninosae). *Tasks Veg. Sci.* 15:285–304.

Hambuch, T. M., and E. A. Lacey. 2002. Enhanced selection for MHC diversity in social tuco-tucos. *Evolution* 56:841–45.

Hamilton, M.J., R. L. Honeycutt, and R. J. Baker. 1990. Intragenomic movement, sequence amplification and concerted evolution in satellite DNA in harvest mice, *Reithrodontomys:* evidence from in situ hybridization. *Chromosoma* 99:321–29.

Handley, C. O., Jr. 1959. A review of the genus *Hoplomys* (thick-spined rats), with description of a new form from Isla Escudo de Veraguas, Panamá. *Smithson. Misc. Collect.* 139 (4):1–10.

———. 1966. Checklist of the mammals of Panama. In *Ectoparasites of Panama,* ed. R. I. Wenzel and V. J. Tipton, 753–95, map. Chicago: Field Museum of Natural History, xii + 861 pp.

———. 1976. Mammals of the Smithsonian Venezuelan project. *Brigham Young Univ. Sci. Bull.*, biol. ser. 20(5):1–89.

Handley, C. O., Jr., and E. Mondolfi. 1963. A new species of fish-eating rat, *Ichthyomys*, from Venezuela (Rodentia, Cricetidae). *Acta Biol. Venezuel.* 3:417–19.

Handley, C. O., Jr., and R. H. Pine. 1992. A new species of prehensile-tailed porcupine, genus *Coendou* Lacépède, from Brazil. *Mammalia* 56:237–44.

Hannibal, W., and N. C. Cáceres. 2010. Use of vertical space by small mammals in gallery forest and woodland savanna in south-western Brazil. *Mammalia* 74:247–55.

Hansen, K. L. 2012. *E Museo Lundii: Addendum.* Copenhagen: Statens Naturhistoriske Museum, 104 pp.

Hanson, J. D. and R. D. Bradley. 2008. Molecular diversity within *Melanomys caliginosus* (Rodentia: Oryzomyini): evidence for multiple species. *Occas. Papers Mus. Texas Tech Univ.* 275:1–11.

Hanson, J. D., J. L. Indorf, V. J. Swier, and R. D. Bradley. 2010. Molecular divergence within the *Oryzomys palustris* complex: evidence for multiple species. *J. Mamm.* 91:236–47.

Hanson, J. D., A. Utrera, and C. F. Fulhorst. 2011. The delicate pygmy rice rat (*Oligoryzomys delicatus*) is the principal host of Maporal virus (Family *Bunyaviridae*, Genus *Hantavirus*). *Vector-Borne Zoonotic Dis* 11:691–96.

Harlan, R. 1825. *Fauna Americana: Being a description of the mammiferous animals inhabiting North America.* Philadelphia: Anthony Finley, 318 pp.

———. 1837. Description of a new species of Quadruped, of the order Rodentia, inhabiting the United States. *Am. J. Sci. Art.* 31:385–86.

Harris, D. B., S. B. Gregory, and D. W. Macdonald. 2006. Space invaders? A search for patterns underlying the coexistence of alien black rats and Galápagos rice rats. *Oecologia* 149:276–88.

Harris, D. B., and D. W. MacDonald. 2007. Population ecology of the endemic rodent *Nesoryzomys swarthi* in the tropical desert of the Galápagos Islands. *J. Mamm.* 88:208–19.

Harris, W. P., Jr. 1932. Four new mammals from Costa Rica. *Occas. Papers Mus. Zool., Univ. Michigan* 248:1–6.

———. 1944. Additions and corrections to the section of Sciuridae in Ellerman's *Families and Genera of Living Rodents. Occas. Papers Mus. Zool., Univ. Michigan* 484:1–21.

Harris, W. P., Jr., and P. Hershkovitz. 1938. Two new squirrels from Ecuador. *Occas. Papers Mus. Zool. Univ., Michigan* 391:1–6.

Harttwig, W. C., and C. Cartelle. 1996. A complete skeleton of the giant South American *Protopithecus. Nature* 6580:307–11.

Hartung, T. G., and D. A. Dewsbury. 1978. A comparative analysis of copulatory plugs in muroid rodents and their relationship to copulatory behavior. *J. Mamm.* 59:717–23.

Hass, I. 2001. "Polimorfismo anônimo de DNA em seis espécies de *Akodon* (Rodentia, Muridae) da região sul do Brasil." Master's thesis, Universidade Federal do Paraná, Curitiba, Brazil.

———. 2006. "Análise filogenética por pintura cromossômica multicolor, em roedores da tribo Akodontini (Rodentia, Cricetidae), ocorrentes na região sul do Brasil." Ph.D. diss., Universidade Federal do Paraná, Curitiba, Brazil.

Hass, I., I. Sbalqueiro, and S. Müller. 2008. Chromosomal phylogeny of four Akodontini species (Rodentia, Cricetidae) from southern Brazil established by Zoo-FISH using *Mus musculus* (Muridae) painting probes. *Chromosome Res.* 16:75–88.

Hastriter, M. W. 2001. Fleas (Siphonaptera: Ctenophthalmidae and Rhopalopsyllidae) from Argentina and Chile with two new species from the Rock Rat, *Aconaemys fuscus*, in Chile. *Ann. Carnegie Mus.* 70:169–78.

Hastriter, M. W., and N. E. Peterson. 1997. Notes on some fleas (Siphonaptera) from Amazonas and Bahia states, Brazil. *Entomological News* 108:290–96.

Hastriter, M. W. and R. D. Sage. 2009. A description of two new species of *Ectinorus* (Siphonaptera: Rhopalopsyllidae) from Laguna Blanca National Park, Neuquén Province, Argentina. *Proc. Entomol. Soc. Wash.* 111:581–97.

Hastriter, M. W., and R. P. Schlatter. 2006. Revision of the fleas in the subgenus *Dasypsyllus* (*Neornipsyllus*) (Siphonaptera: Ceratophyllidae). *Ann. Carnegie Mus.* 75:247–57.

Hatcher, J. B. 1903. Narrative of the expeditions. Geography of southern Patagonia. In *Reports of the Princeton University Expeditions to Patagonia, 1896–1899,* vol. 1. Princeton, NJ: Princeton University, and Stuttgart: E. Schweizwerbart'sche Werlagshandlung (E. Nägele), xvi+314, 26 pls, 1 map.

Haverschmidt, F. 1968. *Birds of Surinam.* Oliver & Boyd, Edinburgh and London, 445 pp.

Haynie, M. L., J. G. Brant, L. R. McAliley, J. P. Carrera, M. A. Revelez, D. A. Parish, X. Viteri, C. Jones, and C. J. Phillips. 2006. Investigations in a natural corridor between two national parks in central Ecuador: Results from the Sowell Expedition, 2001. *Occas. Papers Mus. Texas Tech Univ.* 263:1–16.

Heaney, L. R. 1978. Ecology of Neotropical red-tailed squirrels, *Sciurus granatensis*, in the Panama Canal Zone. *J. Mamm.* 59:846–51.

Heaney, L. R., and R. S. Hoffmann. 1978. A second specimen of the Neotropical montane squirrel, *Syntheosciurus poasensis*. *J. Mamm.* 59:854–55.

Heffner, R. S., and H. E. Heffner. 1991. Behavioral hearing range of the chinchilla. *Hearing Research* 52:13–16.

Heinemann, D. 1975. Superfamily: cavies. In *Animal Life Encyclopedia 11 (Mammals II)*, ed. B. Grizmek, 441–53. New York: Van Nostrand Reinhold, 635 pp.

Heinemann, K. M., N. Guthmann, M. Lozada, and J. A. Monjeau. 1995. Area de actividad de *Abrothrix xanthorhinus* (Muridae, Sigmodontinae) e implicaciones para su estrategia reproductiva. *Mastozool. Neotrop.* 2:15–21.

Heinonen, S., and A. Bosso. 1994. Nuevos aportes para el conocimiento de las mastofauna del Parque Nacional Calilegua (Provincia de Jujuy, Argentina). *Mastozool. Neotrop.* 1:51–60.

Heinonen Fortabat, S. 2001. Los mamíferos del Parque Nacional Río Pilcomayo, Provincia de Formosa, Argentina. *Facena* 17:15–34.

Heinonen Fortabat, S., and E. H. Haene. 1994. Primeros aportes al conocimiento de los micromamíferos del Monumento Natural de los Bosque Petrificado (Provincia de Santa Cruz, República Argentina), con algunos comentarios biogeográficos. *Nótulas Faun.* 58:1–4.

Heller, E. 1904. Mammals of the Galapagos Archipelago, exclusive of the Cetacea. *Proc. Calif. Acad. Sci.* 3:233–49.

Helm, J. D., III. 1975. Reproductive biology of *Ototylomys* (Cricetidae). *J. Mamm.* 56:575–90.

Henriques, R. P. B., D. C. Briani, A. R. T. Palma, and E. M. Vieira. 2006. A simple graphical model of small mammal succession after fire in the Brazilian cerrado. *Mammalia* 70:226–30.

Henry, O. 1994. Saisons de reproduction chez trois rongeurs et un artiodactyle en Guyane française, en fonction des facteurs du milieu et de l'alimentation. *Mammalia* 58:183–200.

———. 1996. The influence of sex and reproductive state on diet preference in four terrestrial mammals of French Guianan rainforest. *Can. J. Zool.* 75:929–35.

———. 1999. Frugivory and the importance of seeds in the diet of the orange-rumped agouti (*Dasyprocta leporina*) in French Guiana. *J. Trop. Ecol.* 15:291–300.

Hensel, R. 1872a. Beiträge zur Kentniss der Thierwelt brasiliens. *Zoologische Garten* 13:76–87.

———. 1872b. Beiträge zur Kentniss der Säugethiere Süd-Brasiliens. *Abhand. König. Akad. Wiss. Berlin* 1872:

1–130, 3 pls. [Hensel's paper is dated to 1872, but this issue of *Abh. Konigl. Preuss. Akad. Wiss. Berlin* was published in both 1872 and 1873.]

Henson, D. D., and R. D. Bradley. 2009. Molecular systematics of the genus *Sigmodon*: results from mitochondrial and nuclear gene sequences. *Can. J. Zool.* 87:211–20.

Hercolini, C. 2007. "Efectos de la urbanización sobre las comunidades de pequeños roedores del Área Metropolitana de Buenos Aires, Argentina." Master's thesis, Universidad de Buenos Aires, Argentina, 52 pp.

Hernández-Camacho, J. 1956. Una subespecie nueva de *Heteromys anomalus* (Mammalia: Rodentia). *Lozania* 10:1–15.

———. 1957. Mammalia. InInforme preliminar sobre aves y mamíferos de Santander, Colombia, ed. J.I. Borrero-H. and J. Hernández-C., 213–30. *An. Soc. Biol. Bogotá* 7:197–230.

———. 1960. Primitiae Mastozoologicae Colombianae. I. Status taxonomico de *Sciurus pucheranii santanderensis*. *Caldasia* 8:359–68.

Herrera, A. L. 1899. *Sinonimía vulgar y científica de los principales vertebrados Mexicanos*. Mexico: Officina Tipográfica de la Secretería de Formento, 31 pp.

Herrera, E. J. R., M. H. Mino, J. Notarnicola, and M. D. Robles. 2011. A new species of Syphacia (Nematoda: Oxyuridae) from *Calomys laucha* (Rodentia: Cricetidae) in an agroecosystem of central Argentina. *J. Parasitol.* 97:676–81.

Herron, M. D., T. A. Castoe, and C. L. Parkinson. 2004. Sciurid phylogeny and the paraphyly of Holarctic ground squirrels (*Spermophilus*). *Mol. Phylogenet. Evol.* 31:1015–30.

Hershkovitz, P. 1940a. Four new oryzomyine rodents from Ecuador. *J. Mamm.* 21:78–84.

———. 1940b. Notes of the distribution of the akodont rodent, *Akodon mollis*, in Ecuador with a description of a new race. *Occas. Papers Mus. Zool., Univ. Michigan* 418:1–3.

———. 1940c. A new spiny rat of the genus Neacomys from Eastern Ecuador. *Occas. Papers Mus. Zool., Univ. Michigan* 419:1–4.

———. 1941. The South American harvest mice of the genus *Reithrodontomys*. *Occas. Papers Mus. Zool., Univ. Michigan* 441:1–7.

———. 1944. Systematic review of the Neotropical water rats of the genus *Nectomys* (Cricetinae*). Misc. Publ. Mus. Zool., Univ. Michigan* 58:1–101.

———. 1947. Mammals of northern Colombia. Preliminary report no. 1: squirrels (Sciuridae). *Proc. U.S. Natl. Mus.* 97:1–46.

———. 1948a. Mammals of northern Colombia. Preliminary report no. 2: spiny rats (Echimyidae), with supplemental notes on related forms. *Proc. U.S. Natl. Mus.* 97:125–40.

———. 1948b. Mammals of northern Colombia. Preliminary report no. 3: water rats (genus *Nectomys*), with supplemental notes on related forms. *Proc. U.S. Natl. Mus.* 98: 49–56.

———. 1955a. On the cheek pouches of the tropical American paca, *Agouti paca* (Linnaeus, 1766). *Saugetierkd. Mitt.* 3:67–70.

———. 1955b. South American marsh rats, genus *Holochilus*, with a summary of sigmodont rodents. *Fieldiana Zool.* 37:639–73 + 13 plates.

———. 1959a. Two new genera of South American rodents (Cricetinae). *Proc. Biol. Soc. Washington* 72:5–9.

———. 1959b. Nomenclature and taxonomy of the Neotropical mammals described by Olfers, 1818. *J. Mamm.* 40:337–53.

———. 1960. Mammals of northern Colombia. Preliminary report n° 8: arboreal rice rats, a systematic revision of the subgenus *Oecomys*, genus *Oryzomys*. *Proc. U.S. Natl. Mus.* 110:513–68.

———. 1962. Evolution of Neotropical cricetine rodents (Muridae) with special reference to the phyllotine group. *Fieldiana: Zoology* 46:1–524.

———. 1966a. South American swamp and fossorial rats of the scapteromyine group (Cricetinae, Muridae), with comments on the glans penis in murid taxonomy. *Z. Säugetierk.* 31:81–149.

———. 1966b. Mice, land bridges and Latin American faunal interchange. In *Ectoparasites of Panama*, ed. R. L. Wenzel and V. J. Tipton, 725–51. Chicago: Field Museum of Natural History, xii + 861 pp.

———. 1968. Metachromism or the principle of evolutionary change in mammalian tegumentary colors. *Evolution* 22:556–75.

———. 1969. The evolution of mammals on southern continents. VI. The Recent mammals of the neotropical region: a zoogeographical and ecological review. *Q. Rev. Biol.* 44:1–70.

———. 1970a. Metachromism like it is. *Evolution* 24:644–48.

———. 1970b. Supplementary notes on neotropical *Oryzomys dimidiatus* and *Oryzomys hammondi*. *J. Mamm.* 51:789–94.

———. 1971. A new rice rat of the *Oryzomys palustris* group (Cricetinae, Muridae) from northwestern Colombia, with remarks on distribution. *J. Mamm.* 52:700–9.

———. 1972. The recent mammals of the neotropical region: a zoogeographic and ecologic review. In *Evolution, mammals, and southern continents*, ed. A. Keast, F. C. Erk, and B. Glass, 311–431. Albany: State University of New York Press, 543 pp.

———. 1987a. First South American record of Coues' marsh rice rat, *Oryzomys couesi*. *J. Mamm.* 68:152–54.

———. 1987b. A history of the Recent mammalogy of the Neotropical Region from 1492 to 1850. In *Studies in Neotropical mammalogy, essays in honor of Philip Hershkovitz*, ed. B. D. Patterson and R. M. Timm, 11–98. *Fieldiana Zool.* 39:frontispiece, viii + 1–506.

———. 1990a. The Brazilian rodent genus *Thalpomys* (Sigmodontinae, Cricetidae) with a description of a new species. *J. Nat. Hist.* 24:763–83.

———. 1990b. Mice of the *Akodon boliviensis* size class (Sigmodontinae, Cricetidae), with the description of two species from Brazil. *Fieldiana Zool.*, n.s., 57:1–35.

———. 1993. A new central brazilian genus and species of sigmodontine rodent (Sigmodontinae) transitional between akodonts and oryzomyines, with a discussion of muroid molar morphology and evolution. *Fieldiana Zool.*, n.s., 75:1–18.

———. 1994. The description of a new species of South American Hocicudo, or long-nose mouse genus *Oxymycterus* (Sigmodontinae, Muroidea), with a critical review of the generic content. *Fieldiana Zool.*, n.s., 79:1–43.

———. 1998. Report on some sigmodontine rodents collected in southeastern Brazil with descriptions of a new genus and six new species. *Bonner Zoologisch. Beitr.* 47:193–256.

Heusser, C. 2003. *Ice Age southern Andes: A chronicle of paleoecological events*. Amsterdam: Elsevier Science, xvi + 256 pp.

Heymann, E. W., and C. Knogge. 1997. Field observations on the Neotropical pygmy squirrel, *Sciurillus pusillus* (Rodentia: Sciuridae) in Peruvian Amazonia. *Ecotropica* 3:67–69.

Hibbard, C. W. 1944. Stratigraphy and vertebrate paleontology of Pleistocene deposits of southwestern Kansas. *Bull. Geol. Soc. Am.* 55:707–54, 3 plates.

Hice, C. L. 2001. Records of a few rare mammals from northeastern Peru. *Mamm. Biol.* 66:317–19.

Hice, C. L., and D. J. Schmidly. 2002. The effectiveness of pitfall traps for sampling small mammals in the Amazon Basin. *Mastozool. Neotrop.* 9:85–89.

Hice, C. L., and P. M. Velazco. 2012. *The non-volant mammals of the Reserve Nacional Allpahuayo-Mishana, Loreto, Peru*. Special Publications of the Museum 60. Lubbock: Texas Tech University Press, 135 pp.

Hickman, G. C. 1985. Surface-mound formation by the tuco-tuco, *Ctenomys fulvus* (Rodentia: Ctenomyidae), with comments on earth-pushing in other fossorial mammals. *J. Zool.* 205:385–90.

———. 1988. The swimming ability of *Ctenomys fulvus* (Ctenomyidae) and *Spalacopus cyanus* (Octodontidae), with reference to swimming in other subterranean mammals. *Z. Säugetierk.* 53:11–21.

Hill, J. Edwards. 1990. A memoir and bibliography of Michael Rogers Oldfield Thomas, F.R.S. *Bull. Br. Mus. Nat. Hist.* (Hist. Ser.) 18:25–113.

Hill, J. Eric. 1935. The cranial foramina in rodents. *J. Mamm.* 18:121–29.

Hillyard, J. R., C. J. Phillips, E. C. Birney, J. A. Monjeau, and R. S. Sikes. 1997. MtDNA analysis and zoogeography of two species of silky desert mice, *Eligmodontia*, in Patagonia. *Z. Säugetierk.* 62:281–92.

Hingst, E., D. Astúa de Moraes, F. S. Rocha, L. C. Araripe, M. Weksler, and R. Cerqueira. 1997. Diversidade em uma cumunidate de pequenos mamíferos de uma região de contato Caatinga-Cerrado. In *Contribuição ao conhecimento ecológico do Cerrado*, ed. L. L. Leite and C. H. Saito, 157–63. Brasília: Universidade de Brasília, v + 325 pp.

Hingst-Zaher, E., L. F. Marcus, and R. Cerqueira. 2000. Application of geometric morphometrics to the study of postnatal size and shape changes in the skull of *Calomys expulsus*. *Hystrix* 11:99–113.

Hinojosa P., F., S. Anderson, and J. L. Patton. 1987. Two new species of *Oxymycterus* (Rodentia) from Peru and Bolivia. *Am. Mus. Novit.* 2898:1–17.

Hock, G. A., and G. H. Adler. 1997. Removal of black palm (*Astrocaryum standleyanum*) seeds by spiny rats (*Proechimys semispinosus*). *J. Trop. Ecol.* 13:51–58.

Hodara, K., M. Busch, and F. Kravetz. 2000. Effects of shelter addition on *Akodon azarae* and *Calomys laucha* (Rodentia, Muridae) in agroecosystems of central Argentina during winter. *Mammalia* 64:295–306.

Hodara, K., O. Suárez, and F. Kravetz. 1997. Nesting and digging behavior in two rodent species (*Akodon azarae* and *Calomys laucha*) under laboratory and field conditions. *Z. Säugetierk.* 62:23–29.

Hodara, V. L., C. J. Quintans, M. B. Espinosa, and M. S. Merani. 1994. Hematology, serum chemistry and urinalyses values of two *Calomys* species (Rodentia, Cricetidae). *Rev. Bras. Biol.* 54:385–89.

Hoekstra, H. E., and S. V. Edwards. 2000. Multiple origins of XY female mice (genus *Akodon*): phylogenetic and chromosomal evidence. *Proc. R. Soc. Lond. B Biol. Sci.* 267:1825–31.

Hoey, K. A., R. R. Wise, and G. H. Adler. 2004. Ultrastructure of echimyid and murid rodent spines. *J. Zool. Lond.* 263:307–15.

Hoffmann, F. G., E. P. Lessa, and M. F. Smith. 2002. Systematics of *Oxymycterus* with the description of a new species from Uruguay. *J. Mamm.* 83:408–20.

Hoffmann, R. S., C. G. Anderson, R. W. Thorington Jr., and L. R. Heaney. 1993. Family Sciuridae. In *Mammal species of the world*, 2nd ed., ed. D. E. Wilson and D. M. Reeder, 419–65. Washington, DC: Smithsonian Institution Press, xviii + 1,206 pp.

Hoffstetter, R. 1968. Ñuapua, un gisement de vertébrés pléistocènes dans le Chaco Bolivien. *Bull. Mus. Natl. Hist. Nat.*, 2nd. ser., 40:823–36.

———. 1986. High Andean mammalian faunas during the Plio-Pleistocene. In *High Altitude Tropical Biogeography*, ed. F. Vuilleumier and M. Monasterio, 218–45. New York: Oxford University Press, xi + 671 pp.

Hohoff, C., K. Franzen, and N. Sachser. 2003. Female choice in a promiscuous wild guinea pig, the yellow-toothed cavy (*Galea musteloides*). *Behav. Ecol. Sociobiol.* 53:341–49.

Hohoff, C., K. Solmsdorff, P. Löttker, K. Kemme, J. T. Epplen, T. G. Cooper, and N. Sachser. 2002. Monogamy in a new species of wild guinea pig (*Galea* sp.). *Naturwissenshaften* 89:462–65.

Hollande, A., and A. Batisse. 1959. Contribution à l'étude des infusories parasites du coecum de l'hydrocheire (*Hydrocheirus capybara* L.). I. La famille des Cycloposthiilidae. *Mem. Inst. Oswaldo Cruz* 57:1–16, plates 1–13.

Hollister, N. 1913. The type species of *Cuniculus* Brisson. *Proc. Biol. Soc. Washington* 26:79.

———. 1914a. Four new Neotropical rodents. *Proc. Biol. Soc. Washington* 27:57–60.

———. 1914b. Four new mammals from tropical America. *Proc. Biol. Soc. Washington* 27:103–6.

———. 1914c. Descriptions of four new mammals from tropical America. *Proc. Biol. Soc. Washington* 27:141–44.

Holmgren, C., J. L. Betancourt, K. A. Rylander, J. Roque, O. Tovar, H. Zeballos, E. Linares, and J. Quade. 2001. Holocene vegetation history from fossil rodent middens near Arequipa, Peru. *Q. Res.* 56:242–51.

Honacki, J. H., K. E. Kinman, and J. W. Koeppl. 1982. *Mammal species of the world: A taxonomic and geographic reference.* Lawrence, KS: Allen Press and Association of Systematic Collections, ix + 694 pp.

Honeycutt, R. L., D. L. Rowe, and M. H. Gallardo. 2003. Molecular systematics of the South American caviomorph rodents: relationships among species and genera in the family Octodontidae. *Mol. Phylogenet. Evol.* 26:476–89.

Hooijer, D. A. 1959. Fossil rodents from Curaçao and Bonaire. *Studies on the Fauna of Curaçao and other Caribbean Islands* 9:1–27, pl. I-III.

———. 1967. Pleistocene vertebrates of the Netherlands Antilles. In *Pleistocene extinctions: The search for a cause*, ed. P. S. Martin and H. E. Wright, 399–406. New Haven: Yale Univ. Press, x+453 pp.

Hooper, E. T. 1952. A systematic review of harvest mice (genus *Reithrodontomys*) of Latin America. *Misc. Publ. Mus. Zool., Univ. Michigan.*77:1–255.

———. 1959. The glans penis in five genera of cricetid rodents. *Occas. Papers Mus. Zool., Univ. Michigan* 613:1–11.

———. 1960. The glans penis in *Neotoma* (Rodentia) and allied genera. *Occas. Papers Mus. Zool., Univ. Michigan* 618:1–21.

———. 1961. The glans penis in *Proechimys* and other caviomorph rodents. *Occas. Papers Mus. Zool., Univ. Michigan* 623:1–18.

———. 1962. The glans penis in Sigmodon, Sigmomys, and Reithrodon (Rodentia, Cricetidae). *Occas. Papers Mus. Zool., Univ. Michigan* 625:1–11.

———. 1968. Classification. In *Biology of* Peromyscus *(Rodentia)*, ed. J. A. King, 27–74. Special Publication 2. [Provo, UT]: American Society of Mammalogists.

Hooper, E. T., and G. G. Musser. 1964. The glans penis in Neotropical cricetines (Family Muridae), with comments on classification of muroid rodents. *Occas. Papers Mus. Zool., Univ. Michigan* 123:1–57.

Hopkins, G. H. E. 1949. The host-associations of the lice of mammals. *Proc. Zool. Soc. Lond.* 119:387–604.

Hopkins, G. H. E., and M. Rothschild. 1953. *An illustrated catalogue of the Rothschild Collection of fleas (Siphonaptera) in the British Museum (Natural History)*. London: British Museum, 1:1–361 pp.

Hopwood, A. T. 1947. The generic names of the mandrill and baboons, with notes on some of the genera of Brisson, 1872. *Proc. Zool. Soc. Lond.* 117:533–36.

Horn, G. B. 2005. "A assembléia de pequenos mamíferos da floresta paludosa do Faxinal, Torres-RS; sua relação com a borda e o roedor *Akodon montensis* (Rodentia, Muridae)." Master's thesis, Universidade Federal do Rio Grande do Sul, Porto Alegre, Brazil.

Horovitz, I., M. R. Sánchez-Villagra, T. Martin, and O. Aguilera. 2006. The fossil record of *Phoberomys pattersoni* Mones 1980 (Mammalia, Rodentia) from Urumaco (Lake Miocene, Venezuela), with an analysis of its phylogenetic relationships. *J. Syst. Paleontol.* 4:293–306.

Howell, A. B. 1940. Cheek pouches of the paca. *J. Mamm.* 21:361.

Howell, A. H. 1914. Revision of the American harvest mice (genus *Reithrodontomys*). *N. Amer. Fauna* 36:1–97.

Hsu, T. C., and K. Benirschke. 1967. *Chinchilla lanigera* (Chinchilla). In *An Atlas of Mammalian Chromosomes*, 1: Folio 19. New York: Springer-Verlag.

———. 1968a. *Dasyprocta aguti* (Orange-rumped agouti). In *An Atlas of Mammalian Chromosomes*, 2: Folio 74. New York: Springer-Verlag.

———. 1968b. *Myoprocta acouchy* (Red acouchy). In *An Atlas of Mammalian Chromosomes*, 2: Folio 75. New York: Springer-Verlag.

———. 1971. *Lagostomus maximus* (Viscacha). In *An Atlas of Mammalian Chromosomes*, 6: Folio 281. New York: Springer-Verlag.

———. 1973. *Akodon orophilus.* In *An Atlas of Mammalian Chromosomes*, 7: Folio 314.

Huchon, D., P. Chevret, U. Jordan, C. W. Kilpatrick, V. Ranwez, P. D. Jenkins, J. Brosius, and J. Schnitz. 2007. Multiple molecular evidence for a living mammalian fossil. *Proc. Natl. Acad. Sci. USA* 104:7495–99.

Huchon, D., and E. J. P. Douzery. 2001. From the Old World to the New World: A molecular chronicle of the phylogeny and biogeography of hystricognath rodents. *Mol. Phylogenet. Evol.* 20:238–51.

Hückinghaus, F. 1961a. Vergleichende Untersuchungen über die Formenmannig-faltigkeit der Unterfamilie Caviinae Murray 1886. (Ergebnisse der Südamerika-expedition Herre/Röhrs 1956–1957). *Z. Wissenschaf. Zool.* 166:1–98, 62 plates.

———. 1961a. 1961b. Zur Nomenklatur und Abstammung des Hausmeerschweinchens. *Z. Säugetierk.* 26:109–11.

Hueck, K. 1972. *As florestas da América do Sul: Ecologia, composição e importância econômica*. São Paulo: Poligono, Editora da Universidade de Brasília , xxvii+466 pp.

Hugot, J.-P. 1986. Morphological study of *Helminthoxys urichi* (Oxyurata, Nematoda) parasitic of *Dasyprocta aguti* (Caviomorpha, Rodentia). *Bull. Mus. Natl. Hist. Nat. A Zool. Biol. Ecol. Anim.* 8:133–38.

Hugot, J.-P., and S. L. Gardner. 2002. *Helminthoxys abrocomae* n. sp. (Nematoda: Oxyurida) from *Abrocoma cinerea* in Bolivia. *Syst. Parasitol.* 47:223–30.

Humboldt, A. de. 1811. Mémoire sur l'os hyoïde et le larynx des oiseau, des singes et du crocodile. In *Recueil d'observations de Zoologie et d'anatomie comparée*, vol. 1 (1805), A. de Humboldt and A. Bonpland, 1–13. Paris: F. Schoell and G. Dufour, i–viii+368 pp.+34 plates.

Hungerford, D. A., and R. L. Snyder. 1964. Karyotypes of two more mammals. *Am. Nat.* 98:125–27.

Hunt, J. L., J. E. Morris, and T. L. Best. 2004. Nyctomys sumichrasti. *Mamm. Species* 754:1–6.

Husson, A. M. 1978. *The mammals of Surinam.* Zoölogische Monographieën van het Rijksmuseum van Natuurlijke Historie No. 2. Leiden: E. J. Brill, xxxiv + 569 pp. + 160 plates.

Hutterer, R. 1994. Island rodents: a new species of *Octodon* from Isla Mocha, Chile (Mammalia: Octodontidae). *Z. Säugetierk.* 59:27–41.

Hutterer, R., and U. Hirsch. 1979. Ein neuer *Nesoryzomys* von der Insel Fernandina, Galápagos. *Bonner Zool. Beitr.* 30:276–83.

Hutterer, R., and P. Oromi. 1993. La rata gigante de la Isla Santa Cruz, Galapagos: algunos datos y problemas. *Result. Cien. Proyecto Galapagos Patr. Hum. Mus. Cien. Nat. Tenerife* 4:63–76.

Hutterer, R., and G. Peters. 2010. Type specimens of mammals (Mammalia) in the collections of the Zoologisches Forschungsmuseum Alexander Koenig, Bonn. *Bonn Zool. Bull.* 59:3–27.

Hyvärinen, H., H. Kangasperko, and R. Peura. 1977. Functional structure of the carpal and ventral vibrissae of the squirrel (*Sciurus vulgaris*). *J. Zool. Lond.* 182:457–66.

Iack-Ximenes, G. E. 1999. "Sistemática de família Dasyproctidae Bonaparte, 1838 (Rodentia, Histrocognathi) no Brasil." Master's thesis, Universidade de São Paulo, São Paulo, Brazil.

———. 2005. "Sistemática de Trinomys Thomas, 1921 (Rodentia, Hystricognathi, Echimyidae)." Ph.D. diss., Universidade de São Paulo, Departamento de Zoologia, São Paulo, Brazil.

Iack-Ximenes, G. E., M. de Vivo, and A. R. Percequillo. 2005a. A new genus for *Loncheres grandis* Wagner, 1845, with taxonomic comments on other arboreal echimyids (Rodentia, Echimyidae). *Arq. Mus. Nac., Rio de Janeiro* 63:89–112.

———. 2005b. A new species of *Echimys* Cuvier, 1809 (Rodentia, Echimyidae) from Brazil. *Pap. Avulsos Zool.* (São Paulo) 45:51–60.

Ibàñez, C. 1980. Ritmo de actividad de algunos ratones de los Llanos de Apure. *Doñana Acta Vert.* 7:117–20.

Ibañez, C., J. Cabot, and S. Anderson. 1994. New records of Bolivian mammals in the collection of the Estación Biológica de Doñana. *Doñana Acta Vert.* 21:79–83.

Ibáñez, C., and S. Moreno. 1982. Ciclo reproductor de algunos cricétidos (Rodentia, Mammalia) de los Llanos de Apure (Venezuela). *Actas VIII Congresso de Latinoamericana de Zoológia* Tomo 1:471–80.

IBGE [Instituto Brasileiro de Geografia e Estatística]. 2004. Mapa de Vetação do Brasil. Map 1:500,000. 3rd ed. Rio de Janeiro, Instituto Brasileiro de Geografia e Estatística.

ICZN [International Commission on Zoological Nomenclature]. 1910. Opinion 5. Status of certain pre-Linnean names reprinted subsequent to 1757. In Opinions rendered by the Intrnational Commission on Zoological Nomenclature. Opinions 1 to 25, p. 6. *Publ. Smithsonian Inst.* 1938:1–68.

———. 1911. Opinion 37. Shall the genera of Brisson's "Ornithologia," 1760, be accepted. In Opinions rendered by the International Commission on Zoological Nomenclature. Opinions 30 to 37, pp. 87–88. *Publ. Smithsonian Inst.* 2013:69–88.

———. 1913. Opinion 55. The type of the genus *Ondatra* Link. In Opinions rendered by the Intrnational Commission on Zoological Nomenclature. Opinions 52 to 56, pp. 126–27. *Publ. Smithsonian Inst.* 2169:119–30.

———. 1925. Opinion 90. Report on sixteen generic names of mammals for which suspension of rules was requested. In Opinions rendered by the International Commission on Zoological Nomenclature. Opinions 82–90, pp. 34–39. *Smithsonian Misc. Coll.* 73(3):1–39.

———. 1929. Opinion 110. Suspension of rules for *Lagidium* 1833. Opinions rendered by the International Comission on Zoologial Nomenclature. *Smithsonian Misc. Coll.* 73(6):17.

———. 1955a. Second report on the status of the generic names "*Odobenus*" Brisson, 1762, and "*rosmarus*" Brünnich, 1771 (Class Mammalia) (A report prepared at the request of the Thirteenth International Congress of Zoology, Paris, 1948). *Bull. Zool. Nomencl.* 11 (part 6):196–98.

———. 1955b. Direction 24. Completion of the entries relating to names of certain genera in the class Mammalia made on the *Official List of Generic Names in Zoology* in the period up to the end of 1936. *Opinions and Declarations Rendered by the International Comission on Zoologial Nomenclature, London* 1:219–46.

———. 1956. Opinion 417. Rejection for nomenclatural purposes of volume 3 (Zoologie) of the work by Lorenz Oken entitled Okens Lehrbuch der Naturgeschischte published in 1815–1816. *Opinions and Declarations Rendered by the International Commission on Zoological Nomenclature, London* 14:1–42.

———. 1982. Opinion 1232. Supression of names for South American rodent published by Brants, 1827. *Bull. Zool. Nomencl.* 39:247–49.

———. 1985. *International Code of Zoological Nomenclature, Third Edition, adopted by the XX General Assembly of the International Union of Biological*

Sciences. London: International Trust for Zoological Nomemclature, xx + 388 pp.

———. 1987. *Official lists and indexes of names and works in zoology.* London: International Trust for Zoological Nomenclature, 4 (unnumbered) + 365 pp.

———. 1998. Opinion 1894. *Regnum Animale . . .* , ed. 2 (M. J. Brisson, 1762): Rejected for nomenclatural purposes, with the conservation of the mammalian generic names *Philander* (Marsupialia), *Pteropus* (Chiroptera), *Glis, Cuniculus,* and *Hydrochoerus* (Rodentia), *Meles, Lutra,* and *Hyaena* (Carnivora), *Tapirus* (Perissodactyla), *Tragulus* and *Giraffa* (Artiodactyla). *Bull. Zool. Nomencl.* 55:64–71.

———. 1999. *International Code of Zoological Nomenclature.* 4th ed. London: The International Trust for Zoological Nomenclature, xxix + 306 pp.

———. 2001. Opinion 1984. *Holochilus* Brandt, 1835, *Proechimys* J.A. Allen, 1899 and *Trinomys* Thomas, 1921 (Mammalia, Rodentia): conserved by the designation of *H. sciureus* Wagner, 1842 as the type species of *Holochilus. Bull. Zool. Nomencl.* 58:245–46.

———. 2002. Opinion 2005 (Case 3022). *Catalogue des mammifères du Muséum National d'Histoire Naturelle* by Étienne Geoffroy Saint-Hilaire (1803): Placed on the Official List of Works Approved as Available for Zoological Nomenclature. *Bull. Zool. Nomencl.* 59: 153–54.

———. 2003. Opinion 2027. Usage of 17 specific names based on wild species which are pre-dated by or contemporary with those based on domestic animals (Lepidoptera, Osteichthyes, Mammalia): conserved. *Bull. Zool. Nomencl.* 60:81–84.

———. 2012. Amendment of Articles 8, 9, 10, 21, and 78 of the International Code of Zoological Nomenclature to expand and refine methods of publication. *Zoo-Keys* 219:1010. [Also published in *Zootaxa* 3450:1–7 (2012).]

Ihering, H. von. 1892. Os mammiferos do Rio Grande do Sul. *Annuário do Estado do Rio Grande do Sul* 1893:96–123.

———. 1894. *Os mamíferos de São Paulo: catálogo.* São Paulo, Brazil: Diario Official, 3–30 pp.

———. 1897. A Ilha de São Sebastião. *Rev. Mus. Paulista, São Paulo* 2:9–171.

———. 1898. Bibliographia (História natural e Anthropologia). *Rev. Mus. Paulista, São Paulo* 3:505–7.

———. 1904. O Rio Juruá. *Rev. Mus. Paulista, São Paulo* 6:385–460.

———. 1927. Die Geschichte des Atlantischen Ozeans. Jena: G. Fischer, vii + 237 pp, 9 maps.

Illiger, J. K. W. 1811. *Prodromus systematis mammalium et avium additis terminis zoographicis utriusque classis,* *eorumque versione germanica.* Berolini: C. Salfield, xviii + 302 pp.

———. 1815. Üeberblick der Säugthiere nach ihrer Vertheilung über die Welttheile. *Abh. Konigl. Preuss. Akad. Wiss. Berlin* 1804–1811:39–159.

Inbar, M., and S. Lev-Yadun. 2005. Conspicuous and aposematic spines in the animal kingdom. *Naturwissenschaften* 92:170–72.

Ipinza, J., M. Tamazo, and J. Torrmann. 1971. Octodontidae in Chile. *Notic. Mens. Mus. Nac. Hist. Nat. Chile* 16:3–10.

Ipucha, M. C. 2002. "Caracterización de linajes del género *Ctenomys* (Rodentia, Ctenomyidae) en base de patrones de bandeo cromosómico con endonucleasas de restricción." Undergraduate thesis, Universidad Nacional de Misiones, Argentina), 85 pp.

Ipucha, M. C., M. D. Giménez, and C. J. Bidau. 2008. Heterogeneity of heterochromatin in the genus *Ctenomys* revealed by a combined analysis of C- and RE-banding. *Acta Theriol.* 53:57–71.

Iriarte, J. A., L. C. Contreras, and F. M. Jaksic. 1989. A long-term study of a small-mammal assemblage in the central Chilean matorral. *J. Mamm.* 70:79–87.

Iriarte, J. A., and F. M. Jaksic. 1986. The fur trade in Chile: an overview of seventy-five years of export data (1910–1984). *Biol. Conserv.* 38:243–53.

Iriarte, J. A., J. E. Jinenez, L. C. Contreras, and F. M. Jaksic. 1989. Small-mammal availability and consumption by the fox, *Dusicyon culpaeus,* in central Chilean scrublands. *J. Mamm.* 70:641–45.

Iriarte, J. A., and J. A. Simonetti. 1986. *Akodon andinus* (Philippi, 1858): visitante ocasional del matorral esclerófilo centrochileno. *Notic. Mens. Mus. Nac. Hist. Nat. Chile* 311:607.

IUCN [International Union for Conservation of Nature]. 2011. *IUCN Red List of Threatened Species.* Version 2011.1, accessed January 1, 2013, http://www.iucnredlist.org.

Iwata, K. 1989. Breeding physiology of pacarana, *Dinomys branickii. J Jpn. Assoc. Zool. Gard. Aquar.* 31:47–50.

Izor, R.J. 1985. Sloths and other mammalian prey of the harpy eagle. In *The evolution and ecology of armadillos, sloths, and vermilinguas,* ed. G.G. Montgomery, 343–46. Washington, DC: Smithsonian Institution Press, 10 (unnumbered) + 451 pp.

Jacobs, J. M., and R. von May. 2012. Forest of grass. *Nat. Hist.* Dec. 2011/Jan. 2012:22–29.

Jacobs, L. L., and E. H. Lindsay. 1984. Holarctic radiation of Neogene muroid rodents and the origin of South American cricetids. *J. Vert. Paleontol.* 4:265–72.

Jaksic, F. M. 1986. Predation upon small mammals in shrublands and grasslands of southern South America: ecological correlates and presumable consequences. *Rev. Chilena Hist. Nat.* 59:209–11.

Jaksic, F. M., and M. Lima. 2003. Myths and facts on ratadas: Bamboo blooms, rainfall peaks and rodent outbreaks in South America. *Austral Ecol.* 28:237–51.

Jaksic, F. M., J. Rau, and J. Yáñez. 1978. Oferta de presas y predación por Bubo virginianus (Strigidae) en el Parque Nacional "Torres del Paine." *An. Inst. Patagonia Ser. Cien. Nat.* 9:199–202.

Jaksic, F. M., R. P. Schlatter, and J. L. Yáñez. 1980. Feeding ecology of central Chilean foxes, *Dusicyon culpaeus* and *Dusicyon griseus*. *J. Mamm.* 61:254–60.

Jaksíc, F. M., J. C. Torres-Mura, C. Cornelius, and P. A. Marquet. 1999. Small mammals of the Atacama Desert (Chile). *J. Arid Environ.* 42:129–35.

Jaksic, F. M., J. L. Yáñez, and E. R. Fuentes. 1981. Assessing a small mammal community in central Chile. *J. Mamm.* 62:391–96.

Jackson, J. E. 1985. Ingestión voluntaria y digestibilidad en la vizcacha (*Lagostomys maximus*). *Rev. Argent. Prod. Anim.* 5:113–19.

———. 1989. Reproductive parameters of the plains vizcacha in San Luis Province, Argentina. *Vida Silvestre Neotrop.* 2:57–62.

———. 1990. Growth rates in vizcacha (*Lagostomus maximus*) in San Luis, Argentina. *Vida Silvestre Neotrop.* 2:52–55.

Jackson, J. E., L. C. Branch, and D. Villarreal. 1996. Lagostomus maximus. *Mamm. Species* 543:1–6.

Janos, D. P., C. T. Sahley, and L. H. Emmons. 1995. Rodent dispersal of vesicular-arbuscular mycorrhizal fungi in Amazonian Peru. *Ecology* 76:1852–58.

Jansa, S. A., and M. Weksler. 2004. Phylogeny of muroid rodents: relationships within and among major lineages as determined by IRBP gene sequences. *Mol. Phylogenet. Evol.* 31:256–76.

Jansen, P. A., F. Bongers, and L. Hemerik. 2004. Seed mass and mast seeding enhance dispersal by a neotropical scatter-hoarding rodent. *Ecol. Monogr.* 74:569–89.

Jansen, P. A., F. Bongers, H. H. T. Prins. 2006. Tropical rodents change rapidly germinating seeds into long-term food supplies. *Oikos* 113:449–58.

Janzen, D. H. 1983. *Coendou mexicanum* [sic]. In *Costa Rican natural history*, ed. D. H. Janzen, 460–61. Chicago: University of Chicago Press, 816 pp.

Jarrín-V., P. 2001. *Mamíferos en la neblina Otonga, un bosque nublado del Ecuador*. (Publicación Especial 5 del Museo de Zoología). Quito: Centro de Biodiversidad y Ambiente, Pontificia Universidad Católica de Ecuador, 244 pp.

Jayat, J. P., G. D'Elía, U. F. J. Pardiñas, M. D. Miotti, and P. E. Ortiz. 2008. A new species of the genus *Oxymycterus* (Mammalia: Rodentia: Cricetidae) from the vanishing Yungas of Argentina. *Zootaxa* 1911:31–51.

Jayat, J. P., G. D'Elía, U. F. J. Pardiñas, and J. G. Namen. 2007. A new species of *Phyllotis* (Rodentia, Cricetidae, Sigmodontinae) from the upper montane forest of the yungas of northwestern Argentina. In *The quintessential naturalist: Honoring the life and legacy of Oliver P. Pearson*, ed. D. A. Kelt, E. P. Lessa, J. Salazar-Bravo, and J. L. Patton, 775–98. *Univ. California Publ. Zool.* 134:v-xii + 1–981.

Jayat, J. P., and P. E. Ortiz. 2010. Mamíferos del pedemonte de Yungas de la Alta Cuenca del Río Bermejo en Argentina: una línea de base de diversidad. *Mastozool. Neotrop.* 17:69–86.

Jayat, J. P., P. E. Ortiz, and F. R. González. 2013. First record of *Abrothrix jelskii* (Thomas, 1894) (Mammalia: Rodentia: Cricetidae) in Salta province, northwestern Argentina: filling gaps and distribution map. *Check List* 9:902–5.

Jayat, J. P., P. E. Ortiz, R. González, R. L. Allende, and M. C. Madozzo Jaén. 2011. Mammalia, Rodentia, Sigmodontinae Wagner, 1843: new locality records, filling gaps and geographic distribution maps from La Rioja province, northwestern Argentina. *Check List* 7:614–18.

Jayat, J. P., P. E. Ortiz, and M. D. Miotti. 2008. Distribución de roedores sigmodontinos (Rodentia: Cricetidae) en pastizales de neblina del noroeste de Argentina. *Acta Zool. Mex.*, n.s., 24:137–78.

———. 2009. Mamíferos de la selva pedemontana del noroeste argentino. In *Ecología, historia natural y conservación de la selva pedemontana de las yungas australes*, ed. A. D. Brown, P. G. Blendinger, T. Lomáscolo, and P. García Bes, 273–316. Tucumán, Argentina: Ediciones del Subtrópico.

Jayat, J. P., P. E. Ortiz, S. Pacheco, and R. González. 2011. Distribution of sigmodontine rodents in northwestern Argentina: main gaps in information and new records. *Mammalia* 75:53–68.

Jayat, J. P., P. E. Ortiz, U. F. J. Pardiñas, and G. D'Elía. 2007. Redescripción y posición filogenética del ratón selvático *Akodon sylvanus* (Rodentia: Cricetidae: Sigmodontinae). *Mastozool. Neotrop.* 14:201–25.

Jayat, J. P., P. E. Ortiz, J. Salazar-Bravo, U. F. J. Pardiñas, and G. D'Elía. 2010. The *Akodon boliviensis* species group (Rodentia: Cricetidae: Sigmodontinae) in Argentina: species limits and distribution, with the description of a new entity. *Zootaxa* 2409:1–61.

Jayat, J. P., P. E. Ortiz, P. Teta, U. F. J. Pardiñas, and G. D'Elía. 2006. Nuevas localidades argentinas para algunos roedores sigmodontinos (Rodentia: Cricetidae). *Mastozool. Neotrop.* 13:51–67.

Jayat, J. P., and S. E. Pacheco. 2006. Distribución de *Necromys lactens* y *Phyllotis osilae* (Rodentia: Cricetidae: Sigmodontinae) en el noroeste argentino: modelos predictivos basados en el concepto de nicho ecológico. *Mastozool. Neotrop.* 13:69–88.

Jayat, J. P., S. E. Pacheco, and P. E. Ortiz. 2009. A predictive distribution model for Andinomys edax (Rodentia: Cricetidae) in Argentina. *Mastozool. Neotrop.*16:321–32.

Jenkins, P. D., and A. A. Barnett. 1997. A new species of water mouse, of the genus *Chibchanomys* (Rodentia, Muridae, Sigmodontinae) from Ecuador. *Bull. Nat. Hist. Mus. Lond. Zool.* 63:123–28.

Jensen, F., M. A. Willis, M. S. Albamonte, M. B. Espinosa, and A. D. Vitullo. 2006. Naturally suppressed apoptosis prevents follicular atresia and oocyte reserve decline in the adult ovary of *Lagostomus maximus* (Rodentia, Caviomorpha). *Reproduction* 132:301–8.

Jensen, F., M. A. Willis, N. P. Leopardo, M. B. Espinosa, and A. D. Vitullo. 2008. The ovary of the gestating South American plains vizcacha (*Lagostomus maximus*): Suppressed apoptosis and corpora lutea persistence. *Biol. Reprod.* 79:240–46.

Jentink, F. A. 1879a. On a new porcupine from South-America. *Notes Leyden Mus.* 1:93–96.

———. 1879b. On a new species of *Echimys*. *Notes Leyden Mus.* 1:97–98.

———. 1891. Note VII. On *Dactylomys dactylinus* and *Kannabateomys amlyonyx*. *Notes Leyden Mus.* 8:105–10 + 1 plate.

Jesseau, S. A., W. G. Holmes, and T. M. Lee. 2008. Mother-offspring recognition in communally nesting degus, *Octodon degus*. *Anim. Behav.* 75:573–82.

Jessen, R. R., M. J. Merrick, J. L. Koprowski, and O. Ramirez. 2010. Presence of Guayaquil squirrels on the central coast of Peru: an apparent introduction. *Mammalia* 74:443–44.

Jessen, R. R., G. H. Palmer, and J. L. Koprowski. 2013. Maternity nest of an Amazon red squirrel in a bromeliad. *Mastozool. Neotrop.* 20:159–61.

Jiménez, C. F., and V. Pacheco. 2012. Nuevo registro de *Rhipidomys* (Rodentia: Sigmodontinae) del grupo *fulviventer* de los bosques montanos de Amazonas, Perú. Paper presented at *III Congreso Peruano de Mastozoología*, 14–18 October 2012.

Jiménez, C. F., V. Pacheco, and D. Vivas. 2013. An introduction to the systematics of *Akodon orophilus* Osgood, 1913 (Rodentia: Cricetidae) with the description of a new species. *Zootaxa* 3669:223–42.

Jiménez, J. E. 1990a. Bases biológicas para la conservación y manejo de la chinchilla chilena Silvestre: proyecto conservación de la chinchilla chilena (*Chinchilla lanigera*). Final report. Corporación Nacional Forestal-World Wildlife Fund, Santiago, Chile.

———. 1990b. Proyecto conservación de la chinchilla chilena (*Chinchilla lanigera*). Final report. Corporación Nacional Forestal, Minisgterio de Agricultura, Santiago, Chile.

———. 1993. "Comparative ecology of *Dusicyon* foxes at the Chinchilla National Reserve in northcentral Chile." Master's thesis, University of Florida, Gainesville, 100 pp.

———. 1996. The extirpation and current status of wild chinchillas (*Chinchilla lanigera* and C. *brevicaudata*). *Biol. Conserv.* 77:1–6.

Jiménez, J. E., P. Feinsinger, and F. M. Jaksic. 1992. Spatiotemporal patterns of an irruption and decline of small mammals in northcentral Chile. *J. Mamm.* 73:356–64.

Jiménez-Ruiz, F. A., and S. L. Gardner. 2003. The nematode fauna of long-nosed mice Oxymycterus spp. from the Bolivian Yungas. *J. Parasitol.* 89:299–308.

Johnson, A. M., M. D. Bowen, T. G. Ksiazek R. J. Williams, R. T. Bryan, J. N. Mills, C. J. Peters, and S. T. Nichol. 1997. Laguna Negra virus associated with HPS in western Paraguay and Bolivia. *Virology* 238:115–27.

Johnson, P. T. 1972. Sucking lice of Venezuelan rodents, with remarks on related species (Anoplura). *Brigham Young Univ. Sci. Bull.*, biol. ser. 17:1–62.

Johnson, W. E., W. F. Franklin, and J. A. Iriarte. 1990. The mammalian fauna of the northern Chilean Patagonia: a biogeographical dilemma. *Mammalia* 54:457–69.

Jones, E. K., C. M. Clifford, J. E. Keirans, and G. M. Kohls. 1972. The ticks of Venezuela (Acarina: Ixodoidea) with a key to the species of Amblyomma in the western hemisphere. *Brigham Young Univ. Sci. Bull.*, biol. ser. 17:1–40.

Jones, J. K., Jr., and M. D. Engstrom. 1986. Synopsis of the rice rats (genus *Oryzomys*) of Nicaragua. *Occas. Papers Mus. Texas Tech Univ.* 103:1–23.

Jonkers, A. H., L. Spence, W. G. Downs, T. H. G. Aitken, and E. S. Tikasingh. 1968. Arbovirus studies in Bush bush forest, Trinidad, W.I., September 1959—December 1964. VI. Rodent-associated viruses (VEE and agents of groups C and Guamá): isolations and further studies. *Am. J. Trop. Med. Hyg.* 17:285–98.

Jordan, K. 1950. Notes on a collection of fleas from Peru. *Bull. WHO* 2:597–609.

Jorge, M. L. S. P. 2008. Effects of forest fragmentation on two sister genera of Amazonian rodents (*Myoprocta acouchy* and *Dasyprocta leporina*). *Biol. Conserv.* 141:617–23.

Jorge, M. L. S. P., and H. F. Howe. 2009. Can forest fragmentation disrupt a conditional mutualism? A case from central Amazon. *Oecologia* 161:709–18.

Jorge, M. L., and C. A. Peres. 2005. Population density and home range size of red-rumped agoutis (*Dasyprocta leporina*) within and outside a natural Brazil nut stand in southeastern Amazonia. *Biotropica* 37:317–21.

Jourdan, C. 1837. Mémoire sur quelques mammifères nouveaux. *C. R. Hebd. Sèances Acad Sci.* 15:521–24.

Justo, E. R. 1992. *Ctenomys talarum occidentalis*, una nueva subespecie de tuco-tuco (Rodentia, Octodontidae) en La Pampa, Argentina. *Neotropica* 38:35–40.

Justo, E. R., L. E. Bozzolo, and L. J. M. De Santis. 1995. Microstructure of the enamel of the incisors of some ctenomyid and octodontid rodents (Rodentia, Caviomorpha). *Mastozool. Neotrop.* 2:43–51.

Justo, E. R., and L. J. M. De Santis. 1977. *Akodon serrensis serrensis* Thomas en la Argentina (Rodentia Cricetidae). *Neotrópica* 23:47–48.

Justo, E. R., L. J. M. De Santis, and M. S. Kin. 2003. Ctenomys talarum. *Mamm. Species* 730:1–5.

Justo, E. R., C. I. Montalvo, and J. M. De Santis. 1985. Nota sobre la presencia de *Tympanoctomys barrerae* (Lawrence, 1941) en La Pampa (Rodentia: Octodontidae). *Hist. Nat. Argent.* 28:243–44.

Kajon, A. E., O. A. Scaglia, C. Horgan, C. Velazquez, M. S. Merani, and O. A. Reig. 1984. Tres nuevos cariotipos de la tribu Akodontini (Rodentia, Cricetidae). *Rev. Mus. Argentino Cien. Nat. "Bernardino Rivadavia," Cien. Zool.* 13:461–69.

Karimi, Y., C. R. de Almeida, and F. Petter. 1976. Note sur les rongeurs du nord-est du Brésil. *Mammalia* 40:257–66.

Kasahara, S., and Y. Yonenaga-Yassuda. 1984. A progress report of cytogenetic data on Brazilian rodents. *Rev. Bras. Genet.* 7:509–33.

Kasper, C. B., F. D. Mazim, J. B. G. Soares, T. G. de Oliveira, and M. E. Fabián. 2007. Composição e abundância relativa dos mamíferos de médio e grande porte no Parque Estadual to Durvo, Rio Grande do Sul, Brasil. *Rev. Bras. Zool.* 24:1087–100.

Kaup, J. 1832. Berichtigung, bie Gattung *Callomys* d'Orb. betressend. *Isis von Oken* 25(11):columns 1208–11.

Keil, A., J. T. Epplen, and N. Sachser. 1999. Reproductive success of males in the promiscuous-mating yellow-toothed cavy (*G. musteloides*). *J. Mamm.* 80:1257–63.

Keirans, J. E., C. M. Clifford, and D. Corwin. 1976. *Ixodes sigelos*, n. sp. (Acarina: Ixodidae), a parasite of rodents in Chile, with a method for preparing ticks for examination by scanning electron microscopy. *Acarologia* 18:217–25.

Kellogg, R 1946. Three new mammals from the Pearl Islands, Panama. *Proc. Biol. Soc. Washington* 59:57–62.

Kelt, D. A. 1993. Irenomys tarsalis. *Mamm. Species* 447:1–3.

———. 1994. The natural history of small mammals from Aisén Region, southern Chile. *Rev. Chilena Hist. Nat.* 67:183–207.

———. 1996. Ecology of small mammals across a strong environmental gradient in southern South America. *J. Mamm.* 77:205–19.

Kelt, D. A., A. Engilis Jr., I. E. Torres, and A. T. Hitch. 2008. Ecologically significant range extension for the Chilean tree mouse, *Irenomys tarsalis. Mastozool. Neotrop.* 15:125–28.

Kelt, D. A., and M. H. Gallardo. 1994. A new species of tuco-tuco, genus *Ctenomys* (Rodentia: Ctenomyidae) from Patagonian Chile. *J. Mamm.* 75:338–48.

Kelt, D. A., R. E. Palma, M. H. Gallardo, and J. A. Cook. 1991. Chromosomal multiformity in *Eligmodontia* (Muridae, Sigmodontinae), and verification of the status of *E. morgani. Z. Säugetierk.* 56:352–58.

Kenagy, G. J., R. A. Vásquez, R. F. Nespolo, and F. Bozinovic. 2002a. A time-energy analysis of day-time surface activity in degus, *Octodon degus. Rev. Chilena Hist. Nat.* 75:149–56.

———. 2002b. Daily and seasonal limits of time and temperature to activity of degus. *Rev. Chilena Hist. Nat.* 75:567–81.

Kenagy, G. J., R. A. Vásquez, B. M. Barnes, and F. Bozinovic. 2004. Microstructure of summer activity bouts of degus in a thermally heterogeneous habitat. *J. Mamm.* 85:260–67.

Kerr, R. 1792. *The animal kingdom or zoological system, of the celebrated Sr. Charles Linnaeus. Class I. Mammalia: Containing a complete systematic description, arrangement, and nomenclature, of all known species and varieties of the Mammalia, or animals which give suck to their young; being a translation of that part of the Syswtema Naturae, as lately published, with great improvements, by Professor Gmelin of Goettingen. Together with numerous additions from more recent zoological writers, and illustrated with copperplates.* Edinburgh:, A. Strahan, T. Cadell, and W. Creech, xii + 1–32 + 30 (unnumbered) + 33–400 pp., 7 plates.

Key, G., and E. Munoz Heredia. 1994. Distribution and current status of rodents in the Galápagos. *Noticias Galápagos* 53:21–25.

Kiblisky, P. 1969. Chromosomes of two species of the genus *Oryzomys* (Rodentia: Cricetidae). *Experientia* 25:1338–39.

Kiblisky, P., N. Brum-Zorrilla, G. Pérez, and F. A. Sáez. 1977. Variabilidad cromosómica entre diversas poblaciones uruguayas del roedor cavador del género *Ctenomys* (Rodentia-Octodontidae. *Mendeliana* 2:85–93.

Kiblisky, P., and O. A. Reig. 1966. Variation in chromosome number within the genus *Ctenomys* and description of the male karyotype of *Ctenomys talarum talarum* Thomas. *Nature* 212:436–38.

Kierulff, M. C., J. R. Stallings, and E. L. Sabato. 1992. A method of capture of the bamboo rat (*Kannabateomys amblyonyx*) in bamboo forests. *Mammalia* 55:633–36. [1991 vol. 55, no. 4: pp. 489ff. "Achevé d'imprimer le 20 mars 1992."]

Kilpatrick, C. W., R. Borroto-Paez, and C. A. Woods. 2012. Phylogenetic relationships of recent capromyid rodents: A review and analyses of karyological, biochemical and molecular data. In *Terrestrial mammals of the West Indies: Contributions*, ed. R. Borroto-Páez, C. A. Woods, and F. E. Sergil, 51–69. Gainesville: Florida Museum of Natural History and Wacahooto Press, 482 pp.

Kin, M. S., and E. R. Justo. 1995. Observaciones sobre la dieta de *Ctenomys azarae* Thomas (Rodentia: Ctenomyidae) en el campo y en cautividad. *Nótulas Faun.* 77:1–3.

Kittlein, M. 1997. Assessing the impact of owl predation on the growth rate of a rodent prey population. *Ecol. Model.* 103:123–34.

Kleiman, D. G. 1970. Reproduction in the female green acouchy, *Myoprocta pratti*. *J. Reprod. Fertil.* 23:55–65.

———. 1971. The courtship and copulatory behavior of the green acouchy, *Myoprocta pratti*. *Z. Tierpsychol.* 29:259–78.

———. 1972. Maternal behavior of the green acouchy, *Myoprocta pratti*, a South American caviomorph rodent. *Behaviour* 43–48–84.

Klompen, J. S. H., and M. W. Nachman. 1990. Occurrence and treatment of the mange mite *Notoedresmuris* in marsh rats from South America. *J. Wildl. Dis.* 26:135–36.

Koenigswald, W. von, P. M. Sander, M. B. Leite, B. Mörs, and W. Santel. 1994. Functional symmetries in the schmelzmuster and morphology of rootless rodent molars. *Zool. J. Linn. Soc.* 110:141–79.

Koepcke, H. W., and M. Koepcke. 1958. Los restos del bosques en las vertientes occidentales de los Andes peruanos. *Bol. Com. Nac. Protec. Nat. Lima* 16:22–30.

Koford, C. B. 1955. New rodent records for Chile and for two Chilean Provinces. *J. Mamm.* 36:465–66.

———. 1968. Peruvian desert mice: water independence competition and breeding cycle near equator. *Science* 160:552–53.

Köhler, N., M. H. Gallardo, L. C. Contreras, and J. C. Torres-Mura. 2000. Allozymic variation and systematic relationships of the Octodontidae and allied taxa (Mammalia, Rodentia). *J. Zool.* 252:243–50.

Kohls, G. M. 1956. Eight new species of *Ixodes* from Central and South America (Acarina: Ixodidae). *J. Parasitol.* 42:636–49.

Kohls, G. M, D. E. Sonenshine, and C. M. Clifford. 1969. *Ixodes* (*Exopalpiger*) *jonesae* sp. n. (Acarina, Ixodidae), a parasite of rodents in Venezuela. *J. Parasitol.* 55:447–52.

Koop, B. F., R. J. Baker, M. W. Haiduk, and M. D. Engstrom. 1984. Cladistical analysis of primitive G-band sequences for the karyotype of the ancestor of the Cricetidae complex of rodents. *Genetica* 64:199–208.

Koopman, K. F. 1976. Catalog of type specimens of Recent mammals in the Academy of Natural Sciences at Philadelphia. *Proc. Acad. Nat. Sci. Philadelphia* 128:1–24.

Korth, W. W. 1994. *The Tertiary record of rodents in North America*. New York: Plenum, 319 pp.

Kosloski, M. A. 1997. Morfometria craniana de uma populacao de *Akodon serrensis* Thomas, 1902 (Rodentia-Cricetidae) do Municipio de Araucaria, Parana, Brasil. *Estudos de Biologia PUC-PR, Curitiba* 41:61–88.

Kraglievich, L. 1926. Humeral aperture in the viscachas and in *Pachyrucos* of the Chapadmalal formation, with a description of "*P. imperforatum*."—I. Presence of the humeral "canalis supracondyloideus (entepicondyloideus)" permits the separation of the Chapadmalal viscachas into a distinct subgenus: "*Lagostomopsis*," n. subg. II. Absence of the humeral "canalis supra condyloideus" in some Chapadmalal species of *Pachyrucos*, with a description of "*P. imperforatum*" (F. Amegh.). *An. Mus. Nac. His. Nat. Buenos Aires* 34:45–88.

———. 1927a. Los géneros vivientes de la subfamilia Caviinae con descripción de *Weyenberghia salinicola*, nuevo subgénero. *Physis* 8:578–79.

———. 1927b. Nota preliminary sobre nuevos generos y especies de roedores de la fauna argentina. *Physis* 8:591–98.

———. 1930a. Diagnosis osteológico-dentaria de los géneros vivientes de la subfamilia Caviinae. *An. Mus. Nac. Hist. Nat. Buenos Aires* 36:59–96.

———. 1930b. Los más grandes carpinchos actuales y fósiles de la subfamilia Hydrochoerinae. *An. Soc. Cien. Argen.* 110:234–50, 340–58, pls. 1–9.

———. 1934. *La antigüedad pliocena de las faunas de Monte Hermoso y Chapadmalal, deducidas de su comparación con las ques le precedieron*. Montevideo: Imprenta "El Siglo Ilustrado," 133 pp.

Kramer, K. M., J. A. Monjeau, E. C. Birney, and R. S. Sikes. 1999. Phyllotis xanthopygus. *Mamm. Species* 617:1–7.

Kraus, C., J. Künklee, and F. Trillmich. 2003. Spacing behavior and its implications for the mating system of a precocial small mammal: an almost asocial cavy *Cavia magna*? *Anim. Behav.* 66:225–38.

Kraus, C., F. Trillmich, and J. Künklee. 2005. Reproduction and growth in a precocial small mammal, *Cavia magna*. *J. Mamm.* 86:763–72.

Kravetz, F. O. 1972. Estudio del regimen alimentario, períodos de actividad y otros rasgos ecológicos en una población de "Ratón Hocicudo" (*Oxymycterus rufus platensis* Thomas) de Punta Lara. *Acta Zool. Lilloana* 29:201–12.

Kravetz, F. O., M. Busch, R. E. Percich, M. C. Manjon, and P. N. Marconi. 1981. Population dynamics of *Calomys laucha* (Rodentia, Cricetidae) in Rio Cuarto, Córdoba, Argentina. 2. Determination of age morphometric characteristics and their seasonal variations. *Ecología (Buenos Aires)* 6:35–44.

Kravetz, F. O., M. C. Manjon, M. Busch, R. E. Percich, P. Marconi, and M. P. Torres. 1981. Population dynamics of *Calomys laucha* in Rio Cuarto, Argentina. 1. Population density structure and reproductive study. *Ecología (Buenos Aires)* 6:15–22.

Krieg, H. 1948. *Zwischen Anden und Atlantik. Reisen eines biologen in Südamerika.* C. Hansen, München, 491 pp.

Krumbiegel, I. 1940. Die Säugetiere der Sudamerika-Expeditionen Prof. Dr. Kriegs. 7. Pakas. *Zool. Anzeiger* 132:223–38.

———. 1941a. Die Säugethiere der Südamerika-Expecitionen Prof. Dr. Kriegs. 8. Agutis. *Zool. Anzeiger* 133:97–113.

———. 1941b. Die Säugetiere der Südamerika-Expeditionen Prof. Dr. Kriegs 9, Maras. *Zool. Anzeiger* 134:18–26.

Kuch, M., N.l Rohland, J. L. Betahcourt, C. Latorre, S. Steppan, and H. N. Poinar. 2002. Molecular analysis of an 11700-year-old rodent midden from the Atacama Desert, Chile. *Mol. Ecol.* 11:913–24.

Kufner, M. B., G. Gavier, and D. Tamburini. 2004. Comunidades de roedores de pampas de altura en las Sierras Grandes en Córdoba, Argentina. *Ecol. Apl.* 3:118–21.

Kufner, M. B., and S. Monge. 1998. *Lagostomus maximus* (Rodentia, Chinchillidae) diet in areas under human intervention in the Monte Desert, Argentina. *Iheringia Ser. Zool.* 10:175–84.

Kufner, M. B., A. P. Sbriller, and S. Monge. 1992. Relaciones tróficas de una comunidad de herbívoros del desierto del Monte (Argentina) durante la sequía invernal. *Iheringia* 72:113–19.

Kuhl, H. 1820. Beiträge zur Zoologie. In *Beiträge zur Zoologie und vergleichenden Anatomie*, 1–152. Frankfurt am Main: Verlag der Hermannschen Buchhandlung, 6 unnumbered, 1:1–152; 2:(2 unnumbered), 1–213 + 11 plates.

Kühlhorn, F. 1954. Säugetierkundliche Studien aus Süd-Mattogrosso. 2. Teil: Edentata; Rodentia. *Säugetierkunde Mitt.* 2:66–72.

Kuniy, A. A., and M. T. R. Brasileiro. 2006. Occurrence of helminths in bristle-spined porcupin (*Chaetomys subspinosus*) (Olfers, 1818), Salvador, Brazil. *Braz. J. Biol.* 66:379–80.

Kuniy, A. A., and C. N. Santos. 2003. Myiasis in bristle-spined porcupine, *Chaetomys subspinosus* (Olfers, 1818), in Bahia, Brazil. *Entomol. News* 114:291–92.

Lacépède, B. G. E. de la V. 1799. Tableau des divisions, sous-divivisions, orders et genres des Mammifères. Supplement to *Discours d'ouverture et de clôture du cours d'histoire naturelle donné dans le Muséum national d'Histoire naturelle, l'an VII de la République, et tableau méthodiques des mammifères et de oiseaux.* Paris: Plassan, 18 pp. [Originally issued as a pamphlet dated "L'An VII de la République" (1799), but subsequently reprinted in other works; see Sherborn (1899).]

Lacey, E. A. 2000. Spatial and social systems of subterranean rodents. In *Life underground: The biology of subterranean rodents*, ed. E. A. Lacey, J. L. Patton, and G. N. Cameron, 255–96. Chicago: University of Chicago Press, i-xi + 449 pp.

———. 2001. Microsatellite variation in solitary and social tuco-tucos: molecular properties and population dynamics. *Heredity* 86:628–37.

———. 2004. Sociality reduces individual direct fitness in a communally breeding rodent, the colonial tuco-tuco (*Ctenomys sociabilis*). *Behav. Ecol. Sociobiol.* 56:449–57.

Lacey, E. A., S. H. Braude, and J. R. Wieczorek. 1997. Burrow sharing by colonial tuco-tucos (*Ctenomys sociabilis*). *J. Mamm.* 78:556–62.

———. 1998. Solitary burrow use by adult Patagonian tuco-tucos (*Ctenomys haigi*). *J. Mamm.* 79:986–91.

Lacey, E. A., and L. A. Ebensperger. 2007. Social structure in Octodontid and Ctenomyid rodents. In *Rodent societies: An ecological and evolutionary perspective*, ed. J. O. Wolff and P. W. Sherman, 257–96. Chicago: University of Chicago Press, 610 pp.

Lacey, E. A., and J. C. Wieczorek. 2004. Kinship in colonial tuco-tucos: evidence from group composition and population structure. *Behav. Ecol.* 15:988–96.

Lacher, T. E. 1979. Rates of growth in *Kerodon rupestris* and an assessment of its potential as a domesticated food source. *Pap. Avulsos Zool.* (São Paulo) 33:67–76.

———. 1981. The comparative social behavior of *Kerodon rupestris* and *Galea spixii* and the evolution of behavior in the Caviidae. *Bull. Carnegie Mus.* 17:1–71.

Lacher, T. E., Jr., and C. J. R. Alho. 1989. Microhabitat use among small mammals in the Brazilian Pantanal. *J. Mamm.* 70:396–401.

———. 2001. Terrestrial small mammal richness and habitat associations in an Amazon Forest-Cerrado contact zone. *Biotropica* 33:171–81.

Lacher, T. E., M. A. Mares, and C. J. R. Alho. 1990. The structure of a small mammal community in a central Brazilian savanna. In *Advances in Neotropical mammalogy*, ed. J. Eisenberg and K. Redford, 137–62. Gainesville, Florida: The Sandhill Crane Press, Inc., x+614 pp. [Dated 1989; published 6 February 1990.]

Laconi, M. R., and A. Castro-Vázquez. 1999. Nest building and parental behaviour in two species of *Calomys* (Muridae, Sigmodontinae): a laboratory study. *Mammalia* 63:11–20.

Laemmert, H. W. Jr., L. C. Ferreira, and R. M. Taylor. 1946. An epidemiological study of jungle yellow fever in an endemic area in Brazil. Part II. Investigations of vertebrate hosts and arthropod vectors. *Am. J. Trop. Med. Hyg.* 26 (Suppl), 6:23–69.

Lagos, V. O., F. Bozinovic, and L. C. Contreras. 1995. Microhabitat use by a small diurnal rodent (*Octodon degus*) in a semiarid environment: thermoregulatory constraints or predation risk? *J. Mamm.* 76:900–5.

Lahille, F. 1906. El nombre científico de las vizcachas. *An. Soc. Cien. Argen.* 62:39–44.

Lainson, R., L. A. Carneiro, and F. T. Silveira. 2007. Observations on *Eimeria* species of *Dasyprocta leporina* (Linnaeus, 1758) (Rodentia : Dasyproctidae) from the state of Para, North Brazil. *Mem. Inst. Oswaldo Cruz* 102:183–89.

Lainson, R., and J. J. Shaw. 1969. Some reservoir-hosts of *Leishmania* in wild animals of Mato Grosso state two distinct strains of parasites isolated from man and rodents. *Trans. Roy. Soc. Trop. Med. Hyg.* 63:408–9.

Lainson, R., J. J. Shaw, P. D. Ready, M. A. Miles, and M. Povoa. 1981. Leishmaniasis in Brazil: XVI. Isolation and identification of *Leishmania* species from sandflies, wild mammals and man in north Para State, with particular reference to *L. braziliensis guyanensis* causative agent of "pian-bois." *Trans. Roy. Soc. Trop. Med. Hyg.* 75:530–36.

Lambert, C. R., S. L. Gardner, and D. W. Duszynski. 1988. Coccidia (Apicoplexa: Eimeriidae) from the subterranean rodent *Ctenomys opimus* Wagner (Ctenomidae) from Bolivia, South America. *J. Parasitol.* 74:1018–22.

Lambert, T. D., and G. H. Adler. 2000. Microhabitat use by a tropical forest rodent, *Proechimys semispinosus*, in central Panama. *J. Mamm.* 81:70–76.

Lambert, T. D., G. H. Adler, C. Mailen Riveros, L. Lopez, R. Ascanio, and J. Terborgh. 2003. Rodents on tropical land-bridge islands. *J. Zool.* 260:179–87.

Lambert, T. D., R. W. Kays, P. A. Jansen, E. Aliaga-Rossel, and M. Wikelski. 2009. Nocturnal activity by the primarily diurnal central American agouti (*Dasyprocta punctata*) in relation to environmental conditions, resource abundance and predation risk. *J. Trop. Ecol.* 25:211–15.

Lambert, T. D., J. R. Malcom, and B. L. Zimmerman. 2005. Variation in small mammal species richness by trap height and trap type in southeastern Amazonia. *J. Mamm.* 86:982–90.

Lammers, A. R., H. A. Dziech, and R. Z. German. 2001. Ontogeny of sexual dimorphism in *Chinchilla lanigera* (Rodentia: Chinchillidae). *J. Mamm.* 82:179–89.

Lander, A. 1974. Observationes preliminaries sobre lapas *Agouti paca* (Linn. 1766) (Rodentia, Agoutidae) en Venezuela. Venezuela: Trabajo de Ascenso, Universidad Central de Venezuela, 104 pp.

Landry, S. O., Jr. 1957. The interrelationships of the New and Old World hystricomorph rodents. *Univ. California Publ. Zool.* 56:1–118, plates 1–5.

Langguth, A. 1965. Contribución al conocimiento de los Crcetinae del Uruguay (especies halladas en las regurgitados de Búho). *Anais 2do Congresso Latinoamericano de Zoología* (São Paulo, Brasil) 2:327–35.

———. 1966a. Application to place on the Official Index of Rejected Names in Zoology, the generic name *Ratton* and the specific names *R. agreste*, *R. blancodebaxo*, *R. colibreve*, *R. espinoso*, and *R. tucotuco*, dated from Brants 1827. Z.N.(S.) 1775. *Bull. Zool. Nomencl.* 23:243–44.

———. 1966b. Application to place on the appropriate official list the names given by G. Fischer 1814 to the cricetids described by Felix de Azara in the French translation of "Essais sur l'histoire naturelle des quadrupedes du Paraguay" 1801. *Bull. Zool. Nomencl.* 23:285–88.

———. 1975. La identidad de *Mus lasiotis* Lund y el status del género *Thalpomys* Thomas (Mammalia, Cricetidae). *Pap. Avulsos Zool.* (São Paulo) 29:45–53.

———. 1978. Revived application in the case of the names for South American rodents published by Brants (1927). Z. N.(S.) 1775. *Bull. Zool. Nomencl.* 35:115–20.

Langguth, A., and A. Abella. 1970. Las especies uruguayas del género *Ctenomys* (Rodentia Octodontidae). *Comun. Zool. Mus. Hist. Nat. Montevideo* 10:1–20.

Langguth, A., and C. R. Bonvicino. 2002. The *Oryzomys subflavus* group, with description of two new species (Rodentia, Muridae, Sigmodontinae). *Arq. Mus. Nac., Rio de Janeiro* 60:285–94.

Langguth, A., V. L. A. G. Limeira, and S. Franco. 1997. Nuevo catálogo do material-tipo da coleção de mamíferos do Museu Nacional. *Publ. Avulsos Mus. Nac., Rio de Janeiro* 70:1–29.

Langguth, A., V. Maia, and M. S. Mattevi. 2005. Karyology of large size Brazilian species of the genus *Oecomys* Thomas, 1906 (Rodentia, Muridae, Sigmodontinae). *Arq. Mus. Nac,. Rio de Janeiro* 63:183–90.

Langguth, A., and E. J. Silva Neto. 1993. Morfologia do pênis em *Pseudoryzomys wavrini* e *Wiedomys pyrrhorhinos* (Rodentia–Cricetidae). *Rev. Nordest. Biol.* 8:55–59.

Lanzone, C. 2009. "Sistemática y evolución del género *Eligmodontia* (Rodentia, Muridae, Sigmodontinae)." Ph.D. diss., Universidad de Córdoba, Argentina.

Lanzone, C., V. Chillo, D. Rodríguez, M. A. Dacar, and C. M. Campos. 2012. Dry season diet composition of *Eligmodontia moreni* (Rodentia, Cricetidae, Sigmodontinae) in a hyper-arid region of the Monte desert (Mendoza, Argentina). *Multequina* 21:25–30.

Lanzone, C., A. Novillo, N. S. Suárez, and R. A. Ojeda. 2007. Cytogenetics and redescription of *Graomys* (Rodentia, Sigmodontinae) from Chumbicha, Catamarca, Argentina. *Mastozool. Neotrop.* 14:249–55.

Lanzone, C., R. A. Ojeda, S. Albanese, D. Rodríguez, and M. Dacar. 2005. Karyotypic characterization and new geographical record of *Salinomys delicatus* (Rodentia, Cricetidae, Sigmodontinae). *Mastozool. Neotrop.* 12:257–60.

Lanzone, C., A. A. Ojeda, R. A. Ojeda, S. Albanese, D. Rodríguez, and M. A. Dacar. 2011. Integrated analyses of chromosome, molecular and morphological variability in the Andean mice *Eligmodontia puerulus* and *E. moreni* (Rodentia, Cricetidae, Sigmodontinae). *Mamm. Biol.* 76:555–62.

Lanzone, C., and R. A. Ojeda. 2005. Citotxonomía y distribución del género *Eligmodontia* (Rodentia, Cricetidae, Sigmodontinae). *Mastozool. Neotrop.* 12:73–77.

Lanzone, C., R. A. Ojeda, and M. H. Gallardo. 2007. Integrative taxonomy, systematics, and distribution of the genus *Eligmodontia* (Rodentia, Cricetidae, Sigmodontinae) in the temperate Monte Desert of Argentina. *Mamm. Biol.* 5:299–312.

Lanzone, C., D. Rodríguez, P. Cuello, S. Albanese, A. A. Ojeda, V. Chillo, and D. A. Martí. 2011. XY$_1$Y$_2$ chromosome system in *Salinomys delicatus* (Rodentia, Cricetidae). *Genetica* 139:1143–47.

Lara, M., M. A. Bogan, and R. Cerqueira. 1992. Sex and age components of variation in *Proechimys cuvieri* (Rodentia: Echimyidae) from northern Brazil. *Proc. Biol. Soc. Washington* 105:882–93.

Lara, M. C., and J. L. Patton. 2000. Evolutionary diversification of spiny rats (genus Trinomys, Rodentia: Echimyidae) in the Atlantic Forest of Brazil. *Zool. J. Linn. Soc.* 130:661–86.

Lara, M. C., J. L. Patton, and E. Hingst-Zaher. 2002. *Trinomys mirapitanga*, a new species of spiny rat (Rodentia: Echimyidae) from the Brazilian Atlantic Forest. *Mamm. Biol.* 67:233–42.

Lara, M. C., J. L. Patton, and M. N. F. da Silva. 1996. The simultaneous diversification of South American echimyid rodents (Hystrigonathi) based on complete cytochrome-b sequences. *Mol. Phylogenet. Evol.* 5:403–13.

Lareschi, M. 2000. "Estudio de la fauna ectoparásita (Acari, Phthiraptera y Siphonaptera) de roedores sigmodontinos (Rodentia: Muridae) de Punta Lara, provincia de Buenos Aires." Ph.D. diss., Universidad Nacional de La Plata, Argentina.

———. 2006. The relationship of sex and ectoparasite infestation in the water rat *Scapteromys aquaticus* (Rodentia: Cricetidae) in La Plata, Argentina. *Rev. Biol. Trop.* 54:673–79.

———. 2010. Ectoparasite occurrence associated with males and females of wild rodents *Oligoryzomys flavescens* (Waterhouse) and *Akodon azarae* (Fischer) (Rodentia: Cricetidae: Sigmodontinae) in the Punta Lara wetlands, Argentina. *Neotrop. Entomol.* 39:818–22.

———. 2011. Laelapid mites (Parasitiformes, Gamasida) parasites of *Akodon philipmyersi* (Rodentia, Cricetidae) in the northern Campos grasslands, Argentina, with the description of a new species. *J. Parasitol.* 97:795–99.

Lareschi, M., A. G. Autino, M. M. Díaz, and R. M. Barquez. 2003. New host and locality records for mites and fleas associated with wild rodents from northwestern Argentina. *Rev. Soc. Entomol. Argent.* 62:60–64.

Lareschi, M., and D. Gettinger. 2009. A new species of *Androlaelaps* (Acari: Parasitiformes) from the akodontine rodent *Deltamys kempi* Thomas, 1919, in La Plata River Basin, Argentina. *J. Parasitol.* 95:1352–55.

Lareschi, M., D. Gettinger, S. Nava, A. Abba, and M. Merino. 2006. First report of mites and fleas associated with sigmodontine rodents from Corrientes Province, Argentina. *Mastozool. Neotrop.* 13: 251–54.

Lareschi, M., and P. M. Linardi. 2009. Morphological variability in *Polygenis* (*Polygenis*) *platensis* (Jordan and Rothschild) (Siphonaptera: Rhopalopsyllidae: Rhopalopsyllinae) and taxonomic consequences. *Zootaxa* 2310:35–42.

Lareschi, M., and R. Mauri. 1998. Capítulo 58: Acari: Dermanyssoidea (ácaros ectoparásitos de vertebrados). In *Biodiversidad de artrópodos argentinos. Una perspectiva biotaxonómica*, ed. J. Morrone and S. Coscarón, 581–90. La Plata: Ediciones Científicas Americanas, 599 pp.

Lareschi, M., R. Ojeda, and P. M. Linardi. 2004. Flea parasites of small mammals in the Monte Desert biome in Argentina with new host and locality records. *Acta Parasitol.* 49:63–66.

Larranhaga, D. 1823. Note sur le *Megaterium* de Cuvier, l'*Hydromis*, et une variété nouvelle de *Maïs*. *Bul. Sci. Soc. Philo.*, 10:83.

Latorre, C. L. 1998. Paleontologia de mamiferos del alero Tres Arroyos I, Tierra del Fuego, XII Region, Chile. *An. Inst. Patagon. Ser. Cien. Nat.* 26:77–90.

Latorre, C. L., J. L. Betancourt, K. A. Rylander, and J. A. Quade. 2002. Vegetation invasions into absolute desert: a 45,000-year rodent midden record from the Calama-Salar de Atacama Basins, Chile. *Geol. Soc. Am. Bull.* 114:349–66.

La Val, R. K. 1976. Voice and habitat of *Dactylomys dactylinus* (Rodentia, Echimyidae) in Ecuador. *J. Mamm.* 57:402–4.

Laville, P. E., A. Casinos, J.-P. Gasc, S. Renous, and J. Bou. 1989. Les mécanismes du fouissage chez *Arvicola terrestris* et *Spalax ehrenbergi*: Étude fonctionelle et Évolutive. *Anat. Anzeiger.* 169:131–44.

Lawrence, B. 1933. New name for *Sciurus milleri* J. A. Allen. *J. Mamm.* 14:369–70.

———. 1941a. A new species of *Octomys* from Argentina. *Proc. New England Zool. Club* 18:43–46.

———. 1941b. *Neacomys* from northwestern South America. *J. Mamm.* 22:418–27.

Layne, J. N. 1960. The glans penis and baculum of the rodent *Dactylomys dactylinus* Desmarest. *Mammalia* 24:87–92.

Leal-Mesquita, E. R. 1991. "Estudos citogenéticos em dez espécies de roedores brasileiros da família Echimyidae." Master's thesis, Universidade de São Paulo, Brazil.

Leal-Mesquita, E. R., V. Fagundes, Y. Yonenaga-Yassuda, and P. L. B. Rocha. 1993. Comparative cytogenetic studies of two karyomorphs of *Thrichomys apereoides* (Rodentia, Echimyidae). *Genet. Mol. Biol.* 16:639–51.

Leal-Mesquita, E. R., Y. Yonenaga-Yassuda, T. H. Chu, P. L. B. and Rocha. 1992. Chromosomal characterization and comparative cytogenetic analysis of two species of *Proechimys* (Echimyidae, Rodentia) from the

Caatinga domain of state of Bahia, Brazil. *Caryologia* 45:197–212.

Leche, W. 1886. Ueber einige südbrasilianische *Hesperomys*-Arten. *Zoolgischer Jahrbücher* 1(1886):687–702 + 1 plate.

Le Conte, J. L. 1852. An attempt at a synopsis of the genus *Geomys* Raf. *Proc. Acad. Nat. Sci. Philadelphia* 6:157–63.

———. 1853. Descriptions of three new species of American Arvicolae, with remarks upon some other American rodents. *Proc. Acad. Nat. Sci. Philadelphia* 6:404–20.

Lecroy, M., and R. Sloss. 2000. Type specimens of birds in the American Museum of Natural History. Part 3. Passeriformes: Eurylaimidae, Dendrocolaptidae, Furnariidae, Formicariidae, Conopophagidae, and Rhinocryptidae. *Bull. Am. Mus. Nat. Hist.* 257:1–88.

Ledesma, K. J., F. A. Werner, A. E. Spotorno, and L. H. Albuja V. 2009. A new species of Mountain Viscacha (Chinchillidae: *Lagidium* Meyen) from the Ecuadorean Andes. *Zootaxa* 2126:41–57.

Lee, D., and D. P. Furman. 1970. *Gigantolaelaps trapidoi*, a new Neotropical mite (Acarina: Laelapidae). *J. Med. Entomol.* 7:497–99.

Lee, T. E., Jr, D. F. Alvarado-Serrano, R. N. Platt, and G. G. Goodwiler. 2006. Report on mammal survey of the Cosanga river drainage, Ecuador. *Occas. Papers Mus. Texas Tech Univ.* 260:1–10.

Lee, T. E. Jr., C. Boada-Terán, A. M. Scott, S. F. Burneo, and J. D. Hanson. 2011. Small mammals of Sangay National Park, Chimborazo Province and Morona Santiago Province, Ecuador. *Occas. Papers. Mus. Texas Tech Univ.* 305:1–14.

Lee, T. E., Jr., S. F. Burneo, T. J. Cochran, and D. Chávez. 2010. Small mammals of Santa Rosa, southwestern Imbabura Province, Ecuador. *Occas. Papers Mus. Texas Tech Univ.* 290: 1–14.

Lee, T. E. Jr., S. F. Burneo, M. R Marchán, S. A. Roussos, and R. S. Vizcarra-Váscomez. 2008. The mammals of the temperate forests of Volcán Sumaco, Ecuador. *Occas. Papers Mus. Texas Tech Univ.* 276:1–12.

Lee, T. E., Jr., B. K. Lim, and J. D. Hanson. 2000. Noteworthy records of mammals from the Orinoco river drainage of Venezuela. *Texas J. Sci.* 52:264–66.

Leggieri, L. R., M. L. Guichon, and M. H. Cassini. 2011. Landscape correlates of the distribution of coypu *Myocastor coypus* (Rodentia, Mammalia) in Argentinean Pampas. *Ital. J. Zool.* 78:124–29.

Lehmann, E. von, and H.-E. Schaefer. 1979. Cytologisch-taxonomische Studien an einer Kleinsäugeraufsammlung aus Honduras (Spermienmorphologie und vergleichende Cytochemie). *Z. Zool. Syst. Evol.* 17:226–36.

Leite, R. do N. 2006. "Comunidade de pequenos mamíferos em um mosaico de plantações de eucalipto, florestas primárias e secundárias da Amazônia Ocidental." Master's thesis, Instituto Brasileiro de Pesquisa da Amazônia, Manaus, Brazil.

Leite, R. do. N., M. N. F. da Silva, and T. A. Gardner. 2007. New records of *Neusticomys oyapocki* (Rodentia, Sigmodontinae) from a human-dominated forest landscape in northeastern Brazilian Amazonia. *Mastozool. Neotrop.* 14:257–61.

Leite, Y. L. R. 2003. Evolution and systematics of the Atlantic tree rats, genus *Phyllomys* (Rodentia, Echimyidae), with descripton of two new species. *Univ. California Publ. Zool.* 132:1–118.

Leite, Y. L. R., A. U. Christoff, and V. Fagundes. 2008. A new species of Atlantic Forest tree rat, genus *Phyllomys* (Rodentia, Echimyidae) from southern Brazil. *J. Mamm.* 89:845–51.

Leite, Y. L. R., V. Caldera Jr., A. C. C. Loss, L. P. Costa, E. R. A. Melo, J. R. Gadelha, and A. R. M. Pontes. 2011. Designation of a neotype for the Brazilian porcupine, *Coendou prehensilis* (Linnaeus, 1758). *Zootaxa* 2791:30–40.

Leite, Y. L. R., S. Lóss, R. P. Rego, L. P. Costa, and C. R. Bonvicino. 2007. The rediscoverey and conservation status of the Bahian giant tree rat *Phyllomys unicolor* (Mammalia: Rodentia: Echimyidae) in the Atlantic Forest of Brazil. *Zootaxa* 1638:51–57.

Leite, Y. L. R., and J. L. Patton. 2002. Evolution of South American spiny rats (Rodentia, Echimyidae): the star-phylogeny hypothesis revisited. *Mol. Phylogenet. Evol.* 25:55–464.

Lemke, T. O., A. Cadena, R. H. Pine, and J. Hernandez-Camacho. 1982. Notes on opossums, bats, and rodents new to the fauna of Colombia. *Mammalia* 46:225–34.

Leo, L. M., and A. L. Gardner. 1993. A new species of a giant *Thomasomys* (Mammalia: Muridae: Sigmodontinae) from the Andes of Northcentral Peru. *Proc. Biol. Soc. Washington* 106:417–28.

Leo L., M., and M. Romo. 1992. Distribución altitudinal de roedores sigmodontines (Cricetidae) en el Parque Nacional Río Abiseo, San Martín, Perú. *Mem. Mus. Hist. Nat., UNMSM (Lima)* 21:105–118.

León, R. J. C., D. Bran, M. Collantes, J. M. Paruelo, and A. Soriano. 1998. Grandes unidades de vegetación de la Patagonia extra andina. In *Ecosistemas patagónicas*, ed. M. Oesterheld, M. R. Aguiar, and J. M. Paruelo, 125–44. *Ecol. Austral* 8:75–308.

Leonard, K. M., B. Pasch, and J. L. Koprowski. 2009. *Sciurus pucheranii* (Rodentia: Sciuridae). *Mamm. Species* 841:1–4.

Leske, N. G. 1779. *Anfangsgründe der Naturgeschichte.* Leipsig: S. L. Crusius, 1:xliv+681 pp..

Lessa, E. P., and J. A. Cook. 1998. The molecular phylogenetics of tuco-tucos (genus *Ctenomys*, Rodentia: Octodontidae) suggests an early burst of speciation. *Mol. Phylogenet. Evol.* 9:88–99.

Lessa, E. P., J. A. Cook, and J. L. Patton. 2003. Genetic footprints of demographic expansion in North America, but not Amazonia, following the Late Pleistocene. *Proc. Natl. Acad. Sci. USA* 100:10331–34.

Lessa, E. P., G. D'Elía, and U. F. J. Pardiñas. 2010. Genetic footprints of late Quaternary climate change in the diversity of Patagonian-Fueguian rodents. *Mol. Ecol.* 19:3031–37.

Lessa, E. P., and A. Langguth. 1983. *Ctenomys pearsoni*, n. sp. (Rodentia, Octodontidae), del Uruguay. *Res. Comun. Jorn. Cien. Nat.* (Montevideo) 3:86–88.

Lessa, E. P., and C. S. Thaeler Jr. 1989. A reassessment of morphological specializations for digging in pocket gophers. *J. Mamm.* 70:689–700.

Lessa, E. P., R. Ojeda, and C. Bidau. 2008a. *Octodon pacificus. 2010 IUCN Red List of Threatened Species,* accessed January 1, 2011, http://www.iucnredlist.org.

———. 2008b.*Octomys mimax. 2010 IUCN Red List of Threatened Species,* accessed January 1, 2011, http://www.iucnredlist.org.

———. 2008c. *Tympanoctomys barrerae.* In *2010 IUCN Red List of Threatened Species,* accessed January 1, 2011, http://www.iucnredlist.org.

Lessa, E. P., A. I. Vassallo, D. H. Verzi, and M. S. Mora. 2008. Evolution of morphological adaptations for digging in living and extinct ctenomyid and octodontid rodents. *Biol. J. Linn. Soc.* 95:267–83.

Lessa, E. P., G. Wlasiuk, and J. C. Garza. 2005. Dynamics of genetic differentiation in the Rio Negro Tuco-tuco (*Ctenomys rionegrensis*) at the local and geographical scales. In *Mammalian diversification: from chromosomes to phylogeography*, ed. E. A. Lacey and P. Myers, 153–73. *Univ. California Publ. Zool.* 133:v-viii+1–383.

Lessa, G., C. A. C. Braga, B. C. Resende, and M. R. S. Pires. 2009. The characterization of small mammal fauna in the Ouro Branco mountain range (Minas Gerais, Brazil). Mendoza, Argentina: 10th International Mammalogical Congress, 9–14 August 2006, Abstract 54, p. 204.

Lessa, G., M. M. O. Corrêa, L. M. Pessôa, and I. A. Zappes. 2013. Chromosomal differentiation in *Kerodon rupestris* (Rodentia: Caviidae) from the Brazilian semi-arid region. *Mastozool. Neotrop.* 20:399–405.

Lessa, G., P. R. Gonçalves, M. M. Morais Jr., F. M. Costa, R. F. Pereira, A. P. Paglia. 1999. Caracterização e monitoramento da fauna de pequenos mamíferos terrestres em um fragmento de mata secundária em Viçosa, Minas Gerais. *Bios* 7:41–49.

Lessa, G., P. B. Gonçalves, and L. M. Pessôa. 2005. Variação geográfica em caracteres cranianos quantitativos de *Kerodon rupestris* (Wied, 1820) (Rodentia, Caviidae). *Arq. Mus. Nac., Rio de Janeiro* 63:75–88.

Lessa, L. G., and F. N. Costa. 2009. Food habits and seed dispersal by *Thrichomys apereoides* (Rodentia: Echimyidae) in a Brazilian Cerrado reserve. *Mastozool. Neotrop.* 16:459–63.

Lessa, L. G., and S. A. Talamoni. 2000. New information on range and habitat characteristics of *Pseudoryzomys simplex* (Rodentia: Muridae) for southeastern Brazil. *Bios* 8:19–23.

Lesson, R.-P. 1827. *Manuel de mammalogie ou histoire naturelle des mammifères.* Paris: Roret, xv + 1–441 pp.

———. 1831. *Centurie zoologique, ou choix d'animaux rares, nouveaux ou imparfaitement connus.* Paris: F. G. Levrault, x + 1–244 pp, 80 leaves of plates.

———. 1835. *Illustrations de zoologie, ou recueil de figures d'animaux peintes d'après nature*, Plate 43 Paris: Arthus Bertrand.

———. 1838. Le rongeurs. In *Compléments de Buffon*, 463–505. Paris: P. Pourrat Freres, 1:1–660 + 2 [unnumbered] + 72 plates.

———. 1842. *Nouveau tableau du Règne Animal. Mammifères.* Paris: Arthus-Bertrand, 204 pp.

Leveau, L. M., P. Teta, R. Bogdaschewsky, and U. F. J. Pardiñas. 2006. Feeding habits of the barn owl (*Tyto alba*) along a longitudinal-latitudinal gradient in central Artgentina. *Ornitol. Neotrop.* 17:353–62.

Levis, S., J. Garcia, N. Pini, G,. Calderón, J. Ramírez, D. Bravo, S. St. Jeor, et al. 2004. Hantavirus Pulmonary Syndrome in northwestern Argentina: circulation of Laguna Negra virus associated with *Calomys callosus*. *Am. J. Trop. Med. Hyg.* 71:658–63.

Lew, D., B. Rivas, H. Rojas, and A. Ferrer. 2009. Mamíferos del Parque Nacional Canaima. In *Biodiversidad del Parque Nacional Canaima: Bases técnicas para la conservación de la Guayana venezolana*, ed. J. C. Señaris, D. Lew, and C. Lasso, 153–79. Caracas: Fundación La Salle de Ciencias Naturales and Nature Conservancy, 256 pp.

Lewis, R. E. 1976. A review of the South American flea subgenus *Ectinorus* Jordan 1942, with descriptions of two new species and a key (Siphonaptera: Rhopalopsyllidae). *J. Parasitol.* 62:1003–9.

Lewis, R. E., and A. E. Spotorno. 1984. A new subspecies of *Agastopsylla nylota* (Siphonaptera: Hystrichopsyllidae) from Chile, with a key to the known taxa. *J. Med. Entomol.* 21:392–94.

Lezcano, M. J., C. Reboledo, and C. E. Schreiber. 1992. Bioestratigrafía de los sedimentos de la Cuenca alta del río de La Reconquista (Pleistocene tardio, noreste del la provincial de Buenos Aires, Argentina). *Ameghiniana* 29:387.

Liais, E. 1872. *Climats, géologie, faune et géographie botanique du Brésil.* Paris: Garnier Frères, viii + 1–640 pp., 1 map.

Liascovich, R. C., R. M. Barquez, and O. A. Reig. 1989. A karyological and morphological reassessment of *Akodon (Abrothrix) illuteus* Thomas. *J. Mamm.* 70:386–91.

———. 1990. Multiple autosomal polymorphism in populations of *Akodon simulator simulator* Thomas, 1916 from Tucumán, Argentina (Rodentia, Cricetidae). *Genetica* 82:165–75.

Liascovich, R. C., and O. A. Reig. 1989. Low chromosomal number in *Akodon cursor montensis* Thomas, and karyologic confirmation of *Akodon serrensis* Thomas in Misiones, Argentina. *J. Mamm.* 70:391–95.

Libardi, G. S. 2013. "Variação não-geográfica em *Necromys lasiurus* (Lund, 1840) (Cricetidae: Sigmodontinae) no Brasil." Master's thesis, Escola Superior de Agricultura "Luiz de Queiroz," Universidade de São Paulo, Piracicaba, Brazil, 137 pp.

Lichtenstein, H. 1818. Die werke von Marcgrave und Piso über dir naturgeschichte Brasiliens, erläuter aus den wieder aufgefundenen Originalzeichnungen. *Abhand. König. Akad. Wiss. Berlin* 1814–1815:201–22.

———. 1820. Ueber die Ratten mit platten Stacheln. *Abhand. König. Akad. Wiss. Berlin* 1818–1819:187–96. [Often citred as 1818, but the volume was published in 1820.]

———. 1823. *Verzeichniss der Doubletten des Zoologischen Museums der Königl. Universität zu Berlin nebst Beschreibung vieler bisher unbekannter Arten von Säugethieren, Vögeln, Amhibien und Fischen.* Berlin: Commission bie T. Trautwein, x + 1–118 pp.

———. 1830. Darstellungen neuer oder wenig bekannte SäugethiereAbbildungen und Beschreibungen von fünf und sechzig Arten und füntzig colorirten Steindrucktafeln nach den Originalen des Zoologischen Museum der Universität zu Berlin. Berlin: C. G. Luderitz, unpaginated text belonging to 50 plates. [Lichtenstein's Darstellungen was published in parts from 1827 to 1834; see Sherborn 1922:lxxxi for publication dates of plate sets.]

Lidia Nasef, N. 2010. Phylogenetic position of *Dinomys branickii* Peters (Dinomyidae) in the context of the Caviomorpha (Hystricognathi, Rodentia). *Cladistics* 26:219.

Lim, B. K., and M. D. Engstrom. 2005. Mammals of Iwokrama Forest. *Proc. Acad. Nat. Sci. Philadelphia* 154: 71–108.

Lim, B. K., M. D. Engstrom, H. H. Genoways, F. M. Catzeflis, K. A. Fitzgerald, S. L. Peters, M. Djosetro, et al. 2005. Results of the Alcoa Foundation-Surinam expeditions. XIV. Mammals of Brownsberg Nature Park, Surinam. *Ann. Carnegie Mus.* 74:225–74.

Lim, B. K., M. D. Engstrom, and J. Ochoa G. 2005. Mammals. In *Checklist of terrestrial vertebrates of the Guiana Shield*, ed. T. Hollowell and R. P. Reynolds, 77–92, pl. 6. *Bull. Biol. Soc. Washington* 13:x + 98 pp.

Lim, B. K., M. D. Engstrom, J. C. Patton, and J. W. Bickham. 2006. Systematic relationships of the Guianan brush-tailed rat (*Isothrix sinnamariensis*) and its first occurrence in Guyana. *Mammalia* 70:120–25.

Lim, B. K., and S. Joemartie. 2011. Rapid Assessment Program (RAP) survey of small mammals in the Kwamalasamutu reigon of Surinam. *RAP Bulletin of Biological Assessment* 63:144–49.

Lima, D. I. F., P. S. D'Andrea, R. C. Oliveira, A. C. S. Caldas, E. R. S. Lemos, and C. R. Bonvicino. 2006. Contribuição ao conhecimento da fauna de pequenos mamíferos no estado de Santa Catarina. *Livro de resumos do I Congresso Sul-Americano de Mastozoologia*, Gramado, RS, 5 a 8 de outubro, Resumo 391:147.

Lima, D. O., B. O. Azambuja, V. L. Camilotti, and N. C. Cáceres. 2010. Small mammal community structure and microhabitat use in the austral boundary of the Atlantic Forest, Brazil. *Zoologia* 27:99–105.

Lima, F. S. 1994. Cariótipos em espécies de Dasyproctidae e Erethizontidae, com discussão da evolução cromossômico (Rodentia, Caviomorpha). *Braz. J. Genet.* 17(Suppl.):135.

Lima, J. F. de Sousa. 2000. "Diversidade cariológica de roedores de pequeno porte do estado do Tocantins, Brasil." Master's thesis, Universidade Estadual Paulista Júlio de Mesquita Filho, Rio Claro, Brazil.

Lima J. F. de Sousa, C. R. Bonvicino, and S. Kasahara. 2003. A new karyotype of *Oligoryzomys* (Sigmodontinae, Rodentia) from central Brazil. *Hereditas* 139:1–6.

Lima J. F. de Sousa, and S. Kasahara. 2001. A new karyotype of *Calomys* (Rodentia, Sigmodontinae). *Iheringia Ser. Zool.* 91:133–36.

Lima, J. F. de Sousa, and A. Langguth. 1998. The karyotypes of three Brazilian species of the genus *Dasyprocta* (Rodentia, Dasyproctidae). *Iheringia Ser. Zool.* 85:141–45.

———. 2002. Karyotypes of Brazilian squirrels: *Sciurus spadiceus* and *Sciurus alphonsei* (Rodentia, Sciuridae). *Folia Zoologica* 51:201–4.

Lima, J. F. de Sousa, A. Langguth, and L. C. de Sousa. 1998. The karyotype of *Makalata didelphoides* (Rodentia, Echimyidae). *Z. Säugetierk.* 63: 315–18.

Lima, M., and F. M. Jaksic. 1999. Survival, recruitment and immigration processes in four subpopulations of the leaf-eared mouse in semiarid Chile. *Oikos* 85:343–55.

Lima, M., R. Julliard, N. C. Stenseth, and F. M. Jaksic. 2001. Demographic dynamics of a Neotropical small rodent (*Phyllotis darwini*): feedback structure, predation and climatic factors. *J. Anim. Ecol.* 70:761–75.

Lima, M., J. E. Keymer, and F. M. Jaksic. 1999. El Nino-southern oscillation-driven rainfall variability and delayed density dependence cause rodent outbreaks in western South America: linking demography and population dynamics. *Am. Nat.* 153:476–91.

Lima Francisco, A., W. Magnusson, and T. Sanaiotti. 1995. Variation in growth and reproduction of *Bolomys lasiurus* (Rodentia: Muridae) in an Amazonian savanna. *J. Trop. Ecol.* 11:419–28.

Lima-Rosa, C. A. V., M. H. Hutz, L. F. B. de Oliveira, J. Andrades-Miranda, and M. S. Mattevi. 2000. Heterologous amplification of microsatellite loci from mouse and rat in oryzomyine and thomasomyine South American rodents. *Biochem. Genet.* 38:97–108.

Linardi, P. M., and L. R. Guimarães. 2000. *Sifonápteros do Brasil*. São Paulo, Brazil: MZUSP, FAPESP, 291 pp.

Linares, O. F. 1976. "Garden hunting" in the American tropics. *Hum. Ecol.* 4:331–49.

———. 1998. *Mamíferos de Venezuela*. Caracas: Sociedad Conservationista Audubon de Venezuela, 691 pp.

Linares-Palomino, R., and S. I. Ponce-Alvarez. 2009. Structural patterns and floristics of a seasonally dry forest in Reserva Ecológica Chaparri, Lambayeque. *Trop. Ecol.* 50:305–14.

Lindsay, E. H. 2008. Cricetidae. In *Evolution of Tertiary mammals of North America. Volume 2, Small mammals, xenarthrans, and marine mammals*, ed. C. M. Janis, G. E. Gunnell, and M. D. Uhen, 456–79. Cambridge, UK: Cambridge University Press, 802 pp.

Lindsay, E. H., and N. J. Cazplewski. 2011. New rodents (Mammalia, Rodentia, Cricetidae) from the Verde Fauna of Arizona and the Maxum Fauna of California, USA, early Blancan Land Mammal Age. *Palaeontol. Electron.* 14:29A:16p; palaeo-electronica.org/2011_3/5_lindsay/index.html.

Link, A., A. Di Fiore, N. Galvis, and E. Fleming. 2012. Patterns of mineral lick visitation by lowland tapir (*Tapirus terrestris*) and lowland paca (*Cuniculus paca*) in a western Amazonian rainforest in Ecuador. *Mastozool. Neotrop.* 19:63–70.

Link, H. F. 1795. *Beyträge zur Naturgeschichte.* Rostock und Leipzig: Karl Christoph Stiller, 1:1–8 (unnumbered) + 1–126.

Linnaeus, C. 1758. *Systema naturae per regnum tria naturae, secundum classes, ordines, genera, species, cum characteribus, differentiis, synonymis, locis.* Editio decima, reformata. Holmiae: Laurentii Salvii, 1:1–824.

———. 1766. *Systema naturae per regna tria naturae, secundum classes, ordines, genera, species, cum characteribus, differentiis, synonymis, locis.* Editio duodecima, reformata. Holmiae: Laurentii Salvii, 1:1–532.

Lisanti, J. A., F. O. Kravetz and C. L. del V. Ramírez. 1976. Los cromosomas de *Calomys callosus* (Rengger) (Rodentia Cricetidae) de la provincia de Córdoba. *Physis* 35:221–30.

Lizzaralde, M., A. Bolzan, and M. Bianchi. 2003. Karyotype evolution in South American subterranean rodents *Ctenomys magellanicus* (Rodentia: Octodontidae): Chromosome rearrangements and (TTAGGG)n telomeric sequence localization in 2n = 34 and 2n = 36 chromosomal forms. *Hereditas* 139:13–17.

Lizzaralde, M., J. Escobar, S. Alvarez, and G. Deferrari. 1994. Un nuevo registro de *Euneomys* en Tierra del Fuego. *9th Jornadas Argentinas de Mastozoología,* Resume 76.

Llanos, A. C. 1944. Apreciaciones de campo con motivo de una concentración de roedores en las provincias de Salta y Jujuy. *Rev. Argent. Zoogeograf.* 4:51–59.

———. 1947. Informe sobre la ecologia de lós roedores indígenas de Chilecito. *Instituto de Sanidad Vegetal, Ministerio de Agricultura y Ganadería de la Nación,* serie A, año III (27):1–55.

Llanos, A. C., and J. A. Crespo. 1952. Ecología de la vizcacha (*Lagostomus maximus maximus* Blainv.) en el nordeste de la Provincia de Entre Ríos. *Rev. Invest. Agricol.* 6:289–378.

Lobato, L., and B. Araujo. 1980. Estudio chromosomico de Sacha cuy. *Rev. Pontif. Univ. Catolic. Ecuad.* 27:171–72.

Lobato, L., G. Cantos, B. Araujo, N. O. Bianchi, and S. Merani. 1982. Cytogenetics of the South American akodont rodents (Cricetidae) X. *Akodon mollis:* a species with XY females and B chromosomes. *Genetica* 57:199–205.

Long, C. V. 2007. Vocalisations of the degus, *Octodon degus,* a social caviomorph rodent. *Bioacoustics* 16:223–44.

Lönnberg, E. 1913. Mammals from Ecuador and related forms. *Ark. Zool. (Stockh.)* 8(16):1–36, 1 plate.

———. 1921. A second contribution to the mammalogy of Ecuador, with some remarks on *Caenolestes. Arkiv Zool.,* Stockholm 14(4):1–104.

———. 1925. Notes on some mammals from Ecuador. *J. Mamm.* 6:271–75.

———. 1937. Notes on some South-American mammals. *Arkiv Zool.,* Stockholm 29A(19):1–29.

Lopes, O. de Sousa, and L. de Abreu Sacchetta. 1974. Epidemiology of Boraceia virus in a forested area in São Paulo, Brazil. *Am. J. Epidemiol.* 100:410–13.

Lopes, R. P. and F. S. Buchmann. 2011. Pleistocene mammals from the southern Brazilian continental shelf. *J. South Am. Earth Sci.* 31:17–27.

López-Arévalo, H., O. Montenegro-Díaz, and A. Cadena. 1993. Ecología de los pequeños mamíferos de la Reserva Biológica Carpanta, en la Cordillera Oriental colombiana. *Stud. Neotrop. Fauna Environ.* 28:193–210.

López-Fuster, M. J., R. Pérez-Hernández, and J. Ventura. 2001. Variación craneométrica de *Rhipidomys latimanus venezuelae* (Muridae, Sigmodontinae). *Orsis* 16:111–20.

Lord, R. 1999. *Wild mammals of Venezuela.* Caracas, Venezuela: Armitano Editores, 344 pp.

Lorvelec, O., M. Pascal, and C. Pavis. 2001. Inventaire et statut des Mammifères des Antilles françaises (hors Chiroptères et Cétacés). *Rep. AEVA* 27:1–22.

Loss, A. C., and Y. L. R. Leite. 2011. Evolutionary diversification of *Phyllomys* (Rodentia: Echimyidae) in the Brazilian Atlantic Forest. *J. Mamm.* 92:1352–66.

Loyola, E. G., J. L. Freyre, A. F. Holquin, A. Sanchez, A. Gonzales, and M. Barreto. 1987. *Trypanosoma cruzi* infections in sylvatic hosts on the Pacific coast of Colombia. *Trans. Roy. Soc. Trop. Med. Hyg.* 81:760.

Lozada, M., and N. Guthmann. 1998. Microhabitat selection under experimental conditions of three sigmodontine rodents. *Ecoscience* 5:51–55.

Lozada, M., J. A. Monjeau, K. M. Heinemann, N. Guthmann, and E. C. Birney. 1996. Abrothrix xanthorhinus. *Mamm. Species* 540:1–6.

Lozada, M., M. de Torres Curth, K. M. Heinemann, and N. Guthmann. 2001. Space use in two rodent species (*Abrothrix xanthorhinus* and *Eligmodontia morgani*) in north-western Patagonia. *Int. J. Ecol. Environ. Sci.* 27:39–43.

Lucero, M. M. 1987. Sobre la presencia de dos taxa de *Coendou* en el Norte Argentino (Mammalia, Erethizontidae). *Acta Zool. Lilloana* 39:37–41.

Lucero, S. O., F. L. Agnolín, R. E. Obredor, R. F. Lucero, M. M. Cenizo, and M. L. de los Reyes. 2008. Una nueva especie del género *Ctenomys* (Mammalia; Rodentia) del Plioceno Tardío-Pleistoceno Medio del sudeste de la provincia de Buenos Aires. *Stud. Geol. Salman.* 44:163–75.

Lucherini, M., J. I. Reppucci, and E. Luegnos Vidal. 2009. A comparison of three methods to estimate variations in the relative abundance of mountain vizcachas (*Lagidium viscacia*) in the high Andes ecosystems. *Mastozool. Neotrop.* 16:223–28.

Luna, F., and C. D. Antinuchi. 2006. Cost of foraging in the subterranean rodent *Ctenomys talarum*: effect of soil hardness. *Can. J. Zool.* 84: 661–67.

———. 2007a. Energetics and thermoregulation during digging in the rodent tuco-tuco (Ctenomys talarum). *Comp. Biochem. Physiol. A Mol. Integr. Physiol.* 146: 559–64.

———. 2007b. Effect of tunnel inclination on digging energetics in the tuco-tuco, *Ctenomys talarum* (Rodentia: Ctenomyidae). *Naturwissenschaften* 94:100–6.

Luna, L., and V. Pacheco. 2002. A new species of *Thomasomys* (Muridae: Sigmodontinae) from southeast Andes of Perú. *J. Mamm.* 83:834–42.

Luna, L., and B. D. Patterson. 2003. A remarkable new mouse (Muridae: Sigmodontinae) from southeastern Peru, with comments on the affinities of *Rhagomys rufescens* (Thomas, 1886). *Fieldiana Zool.*, n.s., 101:1–24.

Luna Wong, L. A. 2002. "A new genus and species of rodent from Peru (Muridae: Sigmodontinae) and its phylogenetic relationships." Master's thesis, University of Illinois at Chicago.

Lunaschi, L. I., and F. B. Drago. 2007. Checklist of digenean parasites of wild mammals from Argentina. *Zootaxa* 1580:35–50.

Lund, P. W. 1838. Blik paa Brasiliens Dyreverden för sidste Jordomvaeltning. Förste Afhandling: Indledning. Lagoa Santa den 14de Febr. 1837. *K. Danske Vidensk. Selskabs Naturv. Math. Afhandl.* 8:27–60.

———. 1839a. Coup-d'oeil sur le espèces éteintes de mammifères du Brésil; extrait de quelques mémoires présentés à l'Academie royale des Sciences de Copenhague. *Ann. Sci. Nat. (Paris)*, ser. 2, 11(Zoologie):214–34.

———. 1839b. Nouvelles observations sur la faune fossile du Brésil; extraits d'une lettre adressée aux rédacteurs par M. Lund. *Ann. Sci. Nat. (Paris)*, ser. 2, 12(Zoologie):205–8.

———. 1839c. Pattedyrene. Lagoa Santa de 16de Novbr. 1837. *K. Danske Vidensk. Selskabs Naturv. Math. Afhandl.* 2:1–84, 13 pls. [Preprint of Lund, P. W. 1841a.]

———. 1840a. Fortsaettelse af Pattedyrene. Lagoa Santa den 12te Septbr. 1838. *K. Danske Vidensk. Selskabs Naturv. Math. Afhandl.* 3:1–56, 11 pls. [Preprint of Lund, P. W. 1841b.]

———. 1840b. Tillaeg til de to sidste Afhandlinger over Brasiliens Dyreverden för sidste Jorgomvaeltning. Lagoa Santa, den 4de April 1839. *K. Danske Vidensk. Selskabs Naturv. Math. Afhandl.* 3:1–24, 3 pls. [Preprint of Lund, P. W. 1841c.]

———. 1840c. Nouvelles recheres sur la Fauna fossile du Brésil. (Extraites d'une letter adressée aux Rédacteurs, et date de Lagos-Santa, 1er avril 1840). *Ann. Sci. Nat. (Paris)*, ser. 2, 13(Zoologie):310–19.

———. 1841a. Blik paa Brasiliens Dyreverden för sidste Jordomvaeltning. Anden Afhandling: Pattedyrene. Lagoa Santa de 16de Novbr. 1837. *K. Danske Vidensk. Selskabs Naturv. Math. Afhandl.* 8:61–144, pls. 1–13.

———. 1841b. Blik paa Brasiliens Dyreverden för sidste Jordomvaeltning. Tredie Afhandling: Fortsaettelse af Pattedyrene. Lagoa Santa. d. 12te Septbr. 1838. *K. Danske Vidensk. Selskabs Naturv. Math. Afhandl.* 8:217–72, pls. 14–24.

———. 1841c. Tillaeg til de to sidste afhandlinger over Brasiliens Dyreverden för sidste Jordomvaeltning. Lagoa Santa den 4de April 1839. *K. Danske Vidensk. Selskabs Naturv. Math. Afhandl.* 8:273–96, pls. 25–27.

———. 1842. Fortsatte Bemaerkninger over Brasiliens uddöne Dryskabning. Lagoa Santa D. 27de Marts 1840. *K. Danske Vidensk. Selskabs Naturv. Math. Afhandl.* 9:1–16.

———. 1950 Segunda memória sobre a fauna das cavernas. In *Memórias sobre a Paleontologia Brasileira*, ed. C. de Paula Couto, 131–203. Rio de Janeiro, Brazil: Instituto Nacional do Livro.

Lydekker, R. 1896. *A geographical history of mammals.* Cambridge: Cambridge University Press, xii + 1–400.

Macchiavello, A. 1957. Estudios sobre peste selvática en América del Sur: II. Peste selvática en la región fronteriza de Perú y Ecuador. 2. El foco de peste selvática del distrito de Lancones, departamento de Piura, Perú. *Bol. Oficina. Sanit. Panam.* 43:225–50.

Macedo, J., D. Loretto, M. C. S. Mello, A. R. Freitas, M. V. Vieira, and R. Cerqueira. 2007. História natural dos mamíferos de uma área perturbada do Parque Nacional da Serra dos Órgãos. In *Ciência e Conservação na Serra dos Órgãos*, ed. C. Cronemberger and E. B. Viveiros de Castro, 165–81. Brasília: Instituto Chico Mendes de Conservação da Biodiversidade, 296 pp.

Macêdo, R., and M. A. Mares. 1987. Geographic variation in the South American cricetine rodent *Bolomys lasiurus*. *J. Mamm.* 68: 578–94.

Machado, A. B. M., G. M. Drummond, and A. P. Paglia (eds.). 2008. *Livro vermelho da fauna brasileira ameaçada de extinção*, 2 vols. Brasília, D.F.: Ministério do Meio Ambiente, Biodiversidade 19.

Machado, L. F., Y. L. R. Leite, A. U. Christoff, and L. G. Giugliano. 2013. Phylogeny and biogeography of tetralophodont rodents of the tribe Oryzomyini (Cricetidae: Sigmodontinae). *Zool. Scr.*, doi:10.1111/zsc.12041.

Machado, L. F., R. Paresque, and A. U. Christoff. 2011. Anatomia comparada e morfometria de *Oligoryzomys nigripes* e *O. flavescens* (Rodentia, Sigmodontinae) no Rio Grande do Sul, Brasil. *Pap. Avulsos Zool.* (São Paulo)51:29–47.

Machado, T., M. J. de J. Silva, E. R. Leal-Mesquite, A. P. Carmignotto, and Y. Yonenage-Yassuda. 2005. Nine karyomorphs for spiny rats of the genus *Proechimys* (Echimyidae, Rodentia) from north and central Brazil. *Genet. Mol. Biol.* 28:682–92.

Machado-Allison, C. E., and A. Barrera. 1972. Venezuelan Amblyopinini (Insecta: Coleoptera; Staphylinidae). *Brigham Young Univ. Sci. Bull.*, biol. ser. 17:1–14.

Madoery, L., M. Kufner, and S. Monge. 1997. Comparación de la dieta obtenida a partir de muestras estomacales y fecales del tuco-tuco, *Ctenomys mendocinus*, en dos poblaciones de la Precordillera de los Andes, Argentina. *Doñana Acta Vertebr.* 24:217–25.

Madrigal, J., D. A. Kelt, P. L. Meserve, J. R. Gutierrez, and F. A. Squeo. 2012. Bottom-up control of consumers leads to top-down indirct facilitation of invasive annual herbs in semiarid Chile. *Ecology* 92:282–88.

Magrini, L., and K. G. Facure. 2008. Barn owl (*Tytoalba*) predation on small mammals and its role in the control of hantavirus natural reservoirs in a periurban area in southeastern Brazil. *Brazil. J. Biol.* 68:733–40.

Mahadeshwar, H. 2010. "Phylogenetics and systematics of the South American genus *Microcavia*, Rodentia." Master's thesis, Texas Tech University, Lubbock, TX.

Mahnert, V. 1982. Two new flea species in the genera *Plocopsylla* Jordan and *Hectopsylla* Frauenfeld (Insecta, Siphonaptera) from Argentina. *Rev. Suisse Zool.* 89:567–72.

Mahoney, M. M., B. V. Rossi, M. H. Hagenauer, and T. M. Lee. 2011. Characterization of the estrous cycle in *Octodon degus. Biol. Reprod.* 84:664–71.

Maia, V. 1984. Karyotypes of three species of Caviinae (Rodentia, Caviidae). *Experentia* 40:564–66.

———. 1990. Karyotype of *Oryzomys capito oniscus* (Rodentia), from northeastern Brazil. *Rev. Bras. Genet.* 13:377–82.

Maia, V., and A. Hulak. 1981. Robertsonian polymorphism in chromosomes of *Oryzomys subflavus*(Rodentia, Cricetidae). *Cytogenet. Cell Genet.* 31:33–39.

Maia, V., J. M. Lafayette, and M. W. Pinto. 1983. Dados cromossômicos (bandas C, G e RONs) em *Oryzomys* aff. *eliurus* (Cricetidae, Rodentia) de Pernambuco. *Ciên. Cult.* 35 (Suppl.):713.

Maia, V., and A. Langguth. 1981. New karyotypes of Brazilian akodont rodents with notes on taxonomy. *Z. Säugetierk.* 46:241–49.

———. 1987. Chromosomes of the Brazilian rodent *Wiedomys pyrrhorhinus* (Wied, 1821). *Rev. Bras. Genet.* 10:229–33.

———. 1993. Constitutive heterochromatin polymorphism and NORs in *Proechimys cuvieri* Petter, 1978 (Rodentia, Echimyidae. *Rev. Bras. Genet.* 16:145–54.

Maia, V., Y. Yonenaga-Yassuda, J. R. O. Freitas, S. Kasahara, M. Suñé-Mattevi, L. F. Oliveira, M. A. Galindo, and I. J. Sbalqueiro, 1984. Supernumerary chromosomes in *Nectomys squamipes* (Cricetidae- Rodentia). *Genética* 63:121–28.

Maihuay, C., V. Pacheco, and S. Solari. 1994. Shiphonaptera (Insecta) en roedores silvestres del Cusco. *Rev. Peru. Entomol.* 36:27–28.

Makino, S. 1953. Notes on the chromosomes of the porcupine and the chinchilla. *Experientia* 9:213–14.

Malcolm, J. R. 1988. Small mammal abundances in isolated and non-isolated primary forest reserves near Manaus, Brazil. *Acta Amaz.* 18:67–83.

———. 1990. Estimation of mammalian densities in continuous forest north of Manaus. In *Four Neotropical rainforests*, ed. A. H. Gentry, 339–57. New Haven, CT: Yale University Press, iv–xvi + 620 pp.

———. 1991. Comparative abundances of neotropical small mammals by trap height. *J. Mamm.* 72:188–92.

———. 1992. Use of tooth impressions to identify and age live *Proechimys guyannensis* and *P. cuvieri* (Rodentia: Echimyidae). *J. Zool. Lond.* 22:537–46.

Malcolm, J. R., C. Liu, R. P. Neilson, L. Hansen, and L. Hannah. 2006. Global warming and extinctions of endemic species from biodiversity hotspots. *Conserv. Biol.* 20:538–48.

Maldonado Junior, A., J. R. Machado e Silva, R. Rodrigues e Silva, H. L. Lenzi, and L. Rey. 1994. Evaluation of the resistance to *Schistosoma mansoni* infection in *N. squamipes* (Rodentia: Cricetidae), a natural host of infection in Brazil. *Rev. Inst. Med. Trop. São Paulo* 36:193–98.

Maldonado Junior, A., J. Pinheiro, R. D. O. Simoes, and R. M. Lanfredi. 2010. *Canaania obesa* (Platyhelminthes: Dicrocoeliidae): redescription and new hosts records. *Zoologia* 27:789–94.

Malizia, A. M., and C. Busch. 1991. Reproductive parameters and growth in the fossorial rodent *Ctenomys talarum* (Rodentia). *Mammalia* 55:293–305.

Mallmann, A. S., M. Finokiet, A. C. Dalmaso, G. L. Melo, V. L. Ferreira, and N. C. Cáceres. 2011. Population dynamics and reproduction of cricetid rodents in a deciduous forest of the Urucum mountains, western Pantanal, Brazil. *Neotrop. Biol. Conserv.* 6:94–102.

Malygin, V. M., V. M. Aniskin, S. I. Isaev, and A. N. Milishnikov. 1994. *Amphinectomys savamis* Malygin gen.

et. sp. n., a new genus and a new species of water rat (Cricetidae, Rodentia) from Peruvian Amazonia. *Zool. Zh.* 73:195–208.

Malygin, V. M., and M. Rosmiarek. 2005. Notes on ecology of cricetinae from Peruvian amazonia and southern Bolivia [in Russian]. *Zool. Zh.* 84:492–506.

Manaf, P., L. Brito-Gitirana, and E. S. Oliveira. 2003. Evidence of chemical communication in the spiny rat *Trinomys yonenagae* (Echimyidae): anal scent gland and social interactions. *Can. J. Zool.* 81:1138–43.

Manaf, P., S. Morato, and E. S. Oliveira. 2003. Profile of wild Neotropical spiny rats (*Trinomys*, Echimyidae) in two behavioral tests. *Physiol. Behav.* 79:129–33.

Manaf, P., and E. S. Oliveira. 2009.Female choice in *Trinomys yonenagae*, a spiny rat from the Brazilian Caatinga. *Rev. Bras. Zoociên.* 11:201–8.

Mangan, S. A., and G. H. Adler. 1999. Consumption of arbuscular mycorrhizal fungi by spiny rats (*Proechimys semispinosus*) in eight isolated populations. *J. Trop. Ecol.* 15:779–90.

Mann F., G. 1944. Dos nuevas especies de roedores. *Biológica (Santiago)* 1:95–113.

———. 1945. Mamífeos de Tarapacá. Observaciones realizadas durante una expedición al alto Norte de Chile. *Biológica (Santiago)* 2:23–98.

———. 1950. Nuevos mamíferos de Tarapaca. *Invest. Zool. Chil.* 1:4–6.

———. 1978. *Los pequeños mamíferos de Chile.* Zoología (Gayana) 40:1–342.

Manzione, N. de, R.A. Salas, H. Paredes, O. Godoy, L. Rojas, F. Araoz, C. F. Fulhorst, et al. 1998. Venezuelan Hemorrhagic Fever: clinical and epidemiological studies of 165 cases. *Clin. Infect. Dis.* 26:308–13.

Mapelli, F. J., M. S. Mora, P. M. Mirol, and M. J. Kittlein. 2012. Effects of Quaternary climatic changes on the phylogeography and historical demography of the subterranean rodent *Ctenomys porteousi. J. Zool.* 286:48–57.

Marcgraf de Liebstad, G. 1648. Historiae rerum naturalium Brasiliae, libri octo . . . Cum appendice de Tapuyis, et Chilensibus. Ioannes de Laet, Antvverpianus, In ordinem digessit & annotationes addidit, & varia ab auctore omissa supplevit & illustravit. *In* W. Piso, *Historia naturalis Brasiliae, in qua non tantum plantae et animalia, sed et indigenarum morbi, ingenia et mores discribuntur et iconibus supra quingentas illustrantur.* Lugdun. Batavorum: apud Fransiscum Hackium, et Amstelodami: apud Lud. Elzevirium, 293 pp. + 13 (unnumbered).

Marchais, Chevalier des. 1731. *Voyage du Chevalier des Marchais en Guinée, isles voisines, et à Cayenne, faid en 1725, 1726 & 1727* (4 vols.). Paris: Pierre Prault, 345–681 + 36 unnumbered.

Marcomini, M., and E. S. Oliveira. 2003. Activity pattern of echimyid rodent species from the Brazilian caatinga in captivity. *Biol. Rhythm Res.* 34:157–66.

Marconi, P. N. 1988. "Efecto de las perturbaciones intensas sobre la estructura de las comunidades de roedores." Master's thesis, Universidad Nacional de Buenos Aires, Argentina.

Marelli, C. A. 1927. Notas anatómicas que fundamentan el género *Pediolagus* de roedores hystricomorfos. *Mem. Jard. Zool. La Plata Argent.* 3:1–11, 5 plates.

———. 1928. Notas anatomicas que fundamentan el género *Lagospedius* de roedores historicomorfos. *Physis* 9:102–4.

———. 1931. Los vertebrados exhibidos en los zoológicos del Plata. *Mem. Jard. Zool. La Plata Argent.* 4:1–275, 84 plates.

Mares, M. A. 1975. South American mammal zoogeography: evidence from convergent evolution in desert rodents. *Proc. Natl. Acad. Sci. USA* 72:1702–6.

———. 1977a. Water independence in a South American non-desert rodent, *J. Mamm.* 58:653–56.

———. 1977b. Water economy and salt balance in a South American desert rodent, *Eligmodontia typus. Comp. Biochem. Physiol.* 56:325–32.

———. 1993. Heteromyids and their ecological counterparts: a pandesertic view of rodent ecology and evolution. In *Biology of the Heteromyidae,* ed. H. H. Genoways and J. H. Brown, 652–714. Special Publication 10. [Provo, UT]: American Society of Mammalogists, v–xii + 1–719 pp.

———. 1997. The geobiological interface: granitic outcrops as a selective force in mammalian evolution. *J. Roy. Soc. West. Aust.* 80:131–39.]

Mares, M. A., and J. K. Braun. 1996. A new species of phyllotine rodent, genus *Andalgalomys* (Muridae: Sigmodontinae), from Argentina. *J. Mamm.* 77:928–41.

———. 2000a. Three new species of *Brucepattersonius* (Rodentia: Sigmodontinae) from Missiones province, Argentina. *Occas. Papers, Sam Noble Oklahoma Mus. Nat. Hist.* 9:1–13.

———. 2000b. *Graomys*, the genus that ate South America: a reply to Steppan and Sullivan. *J. Mamm.* 81:271–76.

Mares, M. A., J. K. Braun, R. M. Barquez, and M. M. Díaz. 2000. Two new genera and species of halophytic desert mammals from isolated salt flats in Argentina. *Occas. Papers Mus. Texas Tech Univ.* 203:1–27.

Mares, M. A., J. K. Braun, and R. Channell. 1997. Ecological observations on the octodontid rodent, *Tym-*

panoctomys barrerae, in Argentina. *Southwest. Nat.* 42:488–504.

Mares, M. A., J. K. Braun, B. S. Coyner, and R. A. van den Bussche. 2008. Phylogenetic and biogeographic relationships of gerbil mice *Eligmodontia* (Rodentia, Cricetidae) in South America, with a description of a new species. *Zootaxa* 1753:1–33.

Mares, M. A., J. K. Braun, and D. Gettinger. 1989. Observations on the distribution and ecology of the mammals of the Cerrado grasslands of central Brazil. *Ann. Carnegie Mus.* 58:1–60.

Mares, M. A., and K. A. Ernest. 1995. Population and community ecology of small mammals in a gallery forest of central Brazil. *J. Mamm.* 76:750–68.

Mares, M. A., K. A. Ernest, and D. A. Gettinger. 1986. Small mammal community structure and composition in the Cerrado province of central Brazil. *J. Trop. Ecol.* 2:289–300.

Mares, M. A., and R. A. Ojeda. 1982. Patterns of diversity and adaptation in South American hystricognath rodents. In *Mammalian biology in South America*, ed. M. A. Mares and H. H. Genoways, 393–432. The Pymatuning Symposia in Ecology 6, Special Publications Series. Pittsburgh: Pymatuning Laboratory of Ecology, University of Pittsburg, xii + 539.

Mares, M. A., R. A. Ojeda, and R. M. Barquez. 1989. *Guide to the mammals of Salta Province, Argentina.* Norman: University of Oklahoma Press, 303 pp.

Mares, M. A., R. A. Ojeda, C. E. Borghi, S. M. Giannoni, G. B. Diaz, and J. Braun. 1997. How desert rodents overcome halophytic plant defenses. *BioScience* 47:699–704.

Mares, M.A., R. A. Ojeda, J. K. Braun, and R. M. Barquez. 1997. Systematics, distribution, and ecology of the mammals of Catamarca Province, Argentina. In *Life among the muses: Papers in honor of James S. Findley*, ed. T. L. Yates, W. L. Gannon, and D.E. Wilson, 89–141. Special Publication 3:6. Albuquerque: Museum of Southwestern Biology, 6 (unnumbered) + 290 pp.

Mares, M. A., R. A. Ojeda, and M. P. Kosco. 1981. Observations on the distribution and ecology of the mammals of Salta province, Argentina. *Ann. Carnegie Mus.* 50:151–206.

Mares, M. A., K. E. Streilen, and M. P. de La Rosa. 1982. Nonsynchronous molting in three genera of tropical rodents from the Brazilian Caatinga (*Trichomys*, *Galea*, and *Kerodon*). *J. Mamm.* 63:484–88.

Mares, M. A., M. R. Willig, K. Streilen, and T. Lacher. 1981. The mammals of northeastern Brazil: a preliminary assessment. *Ann. Carnegie Mus.* 50:81–137.

Marinho, J. R., and T. R. O. de Freitas. 2000. Intraspecific craniometric variation in a chromosome hybrid zone of *Ctenomys minutus* (Rodentia, Hystricognathi). *Z. Säugetierk.* 65:226–31.

Marinho-Filho, J., M. L. Reis, P. S. Oliveira, E. M. Vieira, and M. N. Paes. 1994. Diversity standards and small mammal numbers: conservation of the Cerrado biodiversity. *An. Acad. Brasil. Ciênc.* 66 (supl.1):149–57.

Marinho-Filho, J., F. H. G. Rodrigues, M. M. Guimarães, M. M., and M. L. Reis. 1998. Os mamíferos da Estação Ecológica de Águas Emendadas, Planaltina, DF. In *Vertebrados da Estação Ecológica de Águas Emendadas, história natural e ecologia de um fragmento de Cerrado do Brasil Central*, ed. J. Marinho-Filho, F. H. G. Rodrigues, and M. M. Guimarães, 34–63. Brasília: Governo do Distrito Federal, Instituto de Ecologia e Meio Ambiente do Distrito Federal, 94 pp.

Markham, B. J. 1970. Reconocimiento faunístico del área de los fiordos Toro y Cóndor, Isla Riesco, Magallanes. *An. Inst. Patagonia*, Punta Arenas, 1:41–59.

Marques, R. V. 1988. O gênero *Holochilus* (Mammalia: Cricetidae) no Rio Grande do Sul: taxonomia e distribuição. *Rev. Bras. Zool.* 4:347–60.

Marquet, P., L. Contreras, S. Silva, J. Torres-Mura, and F. Bozinovic. 1993. Natural history of *Microcavia niata* in the high Andean zone of northern Chile. *J. Mamm.* 74:136–40.

Márquez, E. J., M. Aguilera, M. Corti. 2000. Morphometric and chromosomal variation in populations of *Oryzomys albigularis* (Muridae: Sigmodontinae) from Venezuela: multivariate aspects. *Z. Säugetierk.* 65:84–99.

Marshall, L. G. 1979. A model of paleobiogeography of South American cricetine rodents. *Paleobiology* 5:126–32.

Marshall, L. G., A. Berta, R. Hoffstetter, R. Pascual, O. A. Reig, M. Bombin, and A. Mones. 1984. Mammals and stratigraphy: geochronology of the continental mammal-bearing Quaternary of South America. *Paleovertebrata, Mémoire Extraordinaire*, pp. 1–76.

Marshall, L. G., and T. Sempere. 1991. The Eocene to Pleistocene vertebrates of Bolivia and their stratigraphic context: a review. In *Fósiles y facies de Bolivia, Vol. I–Vertebrados*, ed. R. Suarez-Soruco, 631–52. *Revista Técnica de Yacimientos Petrolíferos Fiscales Bolivianos* 12(3–4).

———. 1993. Evolution of the Neotropical Cenozoic land-mammal fauna in its geochronologic, stratigraphic and tectonic contest. In *Biological relationships between Africa and South America*, ed. P. Goldblatt, 329–92. Newhaven, CT: Yale University Press, viii + 630 pp.

Martin, G. M. 2010. Mammalia, Rodentia, Criceti-dae, *Irenomys tarsalis* (Philippi, 1900). *Check List* 6:561–63.

Martin, G. M., and M. Archangelsky. 2004. Aportes al conocimiento de *Notiomys edwardsii* (Thomas, 1890) en el noroeste del Chubut, Argentina. *Mastozool. Neotrop.* 11:91–94.

Martin, P. S., C. Gheler-Costa, and L. M. Verdade. 2009. Microestruturas de pêlos de pequenos mamíferos não-voadores: chave para identificação de espécies de agroecossistemas do estado de São Paulo, Brasil. *Biota Neotrop.* 9:233–41.

Martin, R. E. 1970. Cranial and bacular variation in populations of spiny rats of the genus *Proechimys* (Rodentia: Echimyidae) from South America. *Smithsonian Contrib. Zool.* 35:1–19.

Martin, T. 1992. Schmelzmikrostruktur in den Inzisiven alt- und neuweltlicher hystricognather Nagetiere. *Palaeovertebrata, Mémoire extraordinaire*, pp. 1–168.

———. 1994. On the systematic position of *Chaetomys subspinous* (Rodentia: Caviomorpha) based on evidence from the incisor enamel microstructure. *J. Mamm. Evol.* 2:117–23.

———. 2004. Incisor enamel microstructure of South America's earliest rodents: implications for caviomorph origin and diversification. In *The Paleogene mammalian fauna of Santa Rosa, Amazonian Peru*, ed. K. E. Campbell Jr., 131–40. *Sci. Ser. Nat. Hist. Mus. Los Angeles Co.* 40:1–163.

———. 2005. Incisor schmelzmuster diversity in South America's oldest rodent fauna and early Caviomorph history. *J. Mamm. Evol.* 12:405–17.

Martínez, D. R. 1993. Food habits of the rufous-legged owl (*Strix rufipes*) in temperate rainforests of southern Chile. *J. Raptor Res.* 27:214–16.

Martínez, D. R., R. A. Figueroa, C. L. Ocampo, and F. M. Jaksic. 1998. Food habits and hunting ranges of short-eared owls (*Asio flammeus*) in agricultural landscapes of southern Chile. *J. Raptor Res.*32:111–15.

Martínez, D. R., and F. M. Jaksic.1997. Selective predation on scansorial and arboreal mammals by rufous-legged owls (*Strix rufipes*) in southern Chilean rainforest. *J. Raptor Res.*31:370–75.

Martínez, F. A. 1988. Helmintofauna del *Myocastor coypus* y *Lagostomys maximus*. Nematodes. *Vet. Argent.* 5:33–34, 36–37.

Martínez, J. J., and V. Di Cola. 2011. Geographic distribution and phenetic skull variation in two close species of *Graomys* (Rodentia, Cricetidae, Sigmodontinae). *Zool. Anzeiger* 250:175–94.

Martínez, J. J., L. I. Ferro, M. I. Mollerach, and R. M. Barquez. 2012. The phylogenetic relationships of the Andean swamp rat genus *Neotomys* (Rodentia, Cricetidae, Sigmodontinae) based on mitochondrial and nuclear markers. *Acta Theriol.* 57:277–87.

Martínez, J. J., R. E. González-Ittig, G. R. Theiler, R. Ojeda, C. Lanzone, A. Ojeda, and C. R. Gardenal. 2010. Patterns of speciation in two sibling species of *Graomys* (Rodentia, Cricetidae) based on mtDNA sequences. *J. Zool. Syst. Evol. Res.* 48:159–66.

Martínez, J. J., J. M. Krapovickas, and G. R. Theiler. 2010. Morphometrics of *Graomys* (Rodentia, Cricetidae) from Central-Western Argentina. *Mamm. Biol.* 75:180–85.

Martínez, M. M., J. P. Isacch, and F. Donatti. 1996. Aspectos de la distribución y biología reproductiva de *Asioclamator* en la provincia de Buenos Aires, Argentina. *Ornitol. Neotrop.* 7:157–61.

Martínez, R. L., M. E. Bocco, N. Monaco, and J. J. Polop. 1990. Winter diet in Akodon Dolores (Thomas, 1916). *Mammalia* 54:197–205.

Martínez-Crovetto, R. 1963. Esquema fitogeográfico de la Provincia de Misiones (Argentina). *Bonplandia* 1:171–223.

Martino, A. M. G., and M. Aguilera. 1993. Trophic relationships among four cricetid rodents in rice fields. *Rev. Biol. Trop.* 41:131–41.

Martino, A. M. G., and E. Capanna. 2002. Chromosome characterization of an endemic South American rodent, *Calomys hummelincki* (Husson, 1960) (Sigmodontinae, Phyllotini). *Caryologia* 55:331–39.

Martino, A. M. G., E. Capanna, and M. G. Filippucci. 2000. Allozyme variation and divergence in the phyllotine rodent *Calomys hummelincki* (Husson, 1960). *Genetica* 110:163–75.

Martino, A. M. G., M. G. Filippucci, and E. Capanna. 2002. Evolutive pattern of *Calomys hummelicki* (Husson 1960; Rodentia, Sigmodontinae) inferred from cytogenetic and allozymic data. *Mastozool. Neotrop.* 9:187–97.

Martino, P. E., N. Radman, E. Parrado, E. Bautista, C. Cisterna, M. P. Silvestrini, and S. Corba. 2012. Note on the occurrence of parasites of wild nutria (*Myocastorcoypus*, Molina, 1782). *Helminthologia* (Bratislava) 49:164–68.

Martins, A. V., G. Martins, and R. S. Brito. 1955. Reservatórios silvestres de *Schistosoma mansoni* no estado de Minas Gerais. *Rev. Bras. Malariol. Doenças Trop.* 7:259–65.

Martins, R. P., J. G. Ferreira Filho, and V. Zanini. 1980. Dieta de *Tyto alba tuindara* (Gray, 1829) em área de cerrado do sudeste brasileiro. *Resumos VIII Congresso Brasileiro de Zoologia*, Univ. Brasília, pp. 125–26.

Martins, T. F., M. M. Furtado, A. T. de A. Jacomo, L. Silvera, R. Sollmann, N. M. Torres, and M. B. Labruna. 2011. Ticks on free-living wild mammals in Emas National Park, Goias state, central Brazil. *Syst. Appl. Acarol.* 16:201–6.

Martins-Hatano, F., D. Gettinger, and H. G. Bergallo. 2002. Ecology and host specificity of Laelapine mites (Acari: Laelapidae) of small mammals in an Atlantic Forest area of Brazil. *J. Parasitol.* 88:36–40.

Mascarello, J. T., and D. S. Rogers. 1988. Banded chromosomes of *Liomys salvini*, *Heteromys oresterus*, and *H. desmarestianus*. *J. Mamm.* 69:126–30.

Mascheretti, S., P. M. Mirol, M. D. Giménez, C. J. Bidau, J. R. Contreras, and J. B. Searle. 2000. Phylogenetics of the speciose and chromosomally variable genus *Ctenomys* (Ctenomyidae, Octodontoidea) based on mitochondrial cytochrome-b sequence. *Biol. J. Linn. Soc.* 70:361–76.

Mason, M. J. 2004. The middle ear apparatus of the tuco-tuco *Ctenomys sociabilis* (Rodentia, Ctenomyidae). *J. Mamm.* 85:797–805.

Massa, C. 2010. "Descripción de los ensambles de pequeños roedores y su asociación con el paisaje en la Pampa y el Delta e Islas del Paraná en la provincia de Entre Ríos, Argentina." Master's thesis, Universidad de Buenos Aires, Argentina.

Massarini, A. I., M. A. Barros, M. O. Ortells, and O. A. Reig. 1991. Evolutionary biology of fossorial ctenomyine rodents (Caviomorpha, Octodontidae). I. Chromosomal polymorphism and small karyotypic differentiation in central Argentinian populations of tuco-tucos. *Genetica* 83:131–44.

Massarini, A. I., M. A. Barros, V. G. Roig, and O. A. Reig. 1991. Banded karyotypes of *Ctenomys mendocinus* (Rodentia, Octodontidae) from Mendoza, Argentina. *J. Mamm.* 72:194–98.

Massarini, A. I., M. A. Barros, M. O. Ortells, and O. A. Reig. 1995. Variabilidad cromosómica en *Ctenomys talarum* (Rodentia: Octodontidae) de Argentina. *Rev. Chilena Hist. Nat.* 68:207–14.

Massarini, A. I., H. J. Dopazo, J. L. Bouzat, E. Asno, and O. A. Reig. 1992. The population genetic structure of *Ctenomys porteousi* (Rodentia: Octodontidae). *Biochem. Syst. Ecol.* 20:723–34.

Massarini, A. I., F. J. Dyzenchauz, and S. I Tiranti. 1998. Geographic variation of chromosomal polymorphism in nine populations of *Ctenomys azarae*, tuco-tucos of the *Ctenomys mendocinus* group (Rodentia: Octodontidae). *Hereditas* 128:207–11.

Massarini, A. I., and T. R. O. de Freitas. 2005. Morphological and cytogenetics comparison in species of the mendocinus-group (genus *Ctenomys*) with emphasis in *C. australis* and *C. flamarioni* (Rodentia-Ctenomyidae). *Caryologia* 58:21–27.

Massarini, A. I, D. Mizrahi, S. Tiranti, A. Toloza, F. Luna, and C. E. Schleich. 2002. Extensive chromosomal variation in *Ctenomys talarum talarum* from the Atlantic coast of Buenos Aires Province, Argentina (Rodentia: Octodontidae). *Mastozool. Neotrop.* 9:199–207.

Massoia, E. 1961. Notas sobre los cricétidos de la selva marginal de Punta Lara (Mammalia, Rodentia). *Publ. Mus. Mun. Cien. Nat. Trad. Mar del Plata* 1:115–34.

———. 1963a. Sobre la posición sistemática y distribución geográfica de *Akodon* (*Thaptomys*) *nigrita* (Rodentia, Cricetidae). *Physis* 24:73–80.

———. 1963b. *Oxymycterus iheringi* (Rodentia-Cricetidae) nueva especie para la Argentina. *Physis* 24:129–36.

———. 1964. Sistemática, distribución geográfica y rasgos etoecológicos de *Akodon* (*Deltamys*) *kempi* (Rodentia, Cricetidae). *Physis* 24:299–305.

———. 1970. Contribución al conocimiento de los mamíferos de Formosa. *IDIA, Dic.* 1970:55–63.

———. 1971a. Caracteres y rasgos bioecológicos de *Holochilus brasiliensis chacarius* Thomas ("rata nutria") de la provincia de Formosa y comparaciones con *Holochilus brasiliensis vulpinus* (Brants) (Mammalia, Rodentia, Cricetidae). *Rev. Invest. Agropec., INTA,* ser. 1, Biología y Producción Animal 8:13–40.

———. 1971b. Descripción y rasgos bioecológicos de una nueva subespecie de cricétido: *Akodon azarae bibianae* (Mammalia-Rodentia). *Rev. Invest. Agropec. INTA,* ser. 4, Patología Animal 8:131–40.

———. 1971c. *Akodon varius toba* Thomas en la República Argentina (Mammalia-Rodentia-Cricetidae). *Rev. Invest. Agropec. INTA,* ser. 4, Patología Animal 8:123–29.

———. 1973a. Descripción de *Oryzomys fornesi*, nueva especie y nuevos datos sobre algunos especies y subespecies argentinas del subgénero *Oryzomys* (*Oligoryzomys*) (Mammalia - Rodentia - Cricetidae). *Rev. Invest. Agrop. INTA,* ser. 1 Biol. Prod. Anim. *Buenos* Aires 10:21–37.

———. 1973b. Zoogeografía del género *Cavia* en la Argentina con comentarios bioecológicos y systemáticos (Mammalia-Rodentia-Caviidae). *Rev. Invest. Agrop. INTA,* ser. 1 Biol. Prod. Anim. *Buenos* Aires 10:1–11.

———. 1974. Ataques graves de *Holochilus* y otros roedores a cultivos de caña de azúcar. *IDIA* 321(24):1–12.

———. 1976a. Sobre la identidad del holotipo de *Graomys hypogaeus* Cabrera, 1934 (Mammalia-Rodentia-Cricetidae). *Rev. Invest. Agrop. INTA, Buenos* Aires. ser. 5 Patol. Veget. 13:15–20.

———. 1976b. Mammalia. In *Fauna de agua dulce de la República Argentina*, ed. R. A. Ringuelet, 1–44. Buenos Aires, Argentina: Fundación Editorial Ciencia y Cultura.

———. 1977. Mammalia Argentina. II. Los mamíferos de la Provincia de Formosa. *Resúmenes VII Congreso Latinoamericano de Zoología*, Tucumán, Argentina, p. 107.

———. 1978. Descripción de un género y especie nuevos: *Bibimys torresi* (Mammalia-Rodentia-Cricetidae-Sigmodontinae-Scapteromyini). *Resúmenes V Jornadas Argentinas de Zoología*, Villa Giardino, Córdoba, Argentina, p. 56.

———. 1979a. El género *Octodon* en la Argentina. *Neotropica* 25:36.

———. 1979b. Descripción de un genero y especie nuevos: *Bibimys torresi* (Mammalia–Rodentia–Cricetidae–Sigmodontinae–Scapteromyini). *Physis* 38:1–7.

———. 1980a. Nuevos datos sobre *Akodon, Deltamys* y *Cabreramys*, con la descripción de una especie y subespecies nuevas (Mammalia, Rodentia, Cricetidae). *Hist. Nat.* 1:179.

———. 1980b. Mammalia de Argentina. I - Los mamíferos silvestres de la provincia de Misiones. *Iguazú* 1:15–43.

———. 1980c. El estado sistemático de cuatro especies de cricetidos sudamericanos y comentarios sobre otras especies congenericas (Mammalia-Rodentia). *Ameghiniana* 17:280–87.

———. 1981a. Notas sobre los cricétidos mendocinos (Mammalia, Rodentia). *Hist. Nat.* 1:205–8.

———. 1981b. El estado sistemático y zoogeografía de *Mus brasiliensis* Desmarest y *Holochilus sciureus* Wagner (Mammalia–Rodentia–Cricetidae). *Physis Sec. C* 39:31–34.

———. 1982. Diagnosis previa de *Cabreramys temchuki*, nueva especie (Rodentia, Cricetidae). *Hist. Nat.* 2:91–92.

———. 1983. La alimentación de algunas aves del Orden Strigiformes en la Argentina. *El Hornero (Rev. Ornitol. Neotrop.)* 12:24–148.

———. 1985a. Análisis de regurgitados de *Asioflammeus* del arroyo Chasicó. *ACINTACNIA, Instituto Nacional de Tecnología Agropecuaria* 2(15):7–9.

———. 1985b. El estado sistemático de algunos muroideos estudiados por Ameghino en 1889 con la revalidación del género *Necromys* (Mammalia, Rodentia, Myomorpha). *Circ. Inf. Asoc. Paleontol. Argent.* 14:4.

. 1988a. Algunos restos de pequeños roedores y pájaros depredados por aves rapaces en el río Quilquihue, Departamento de Lacar, provincia de neuquén. *Bol. Cien. Asoc. Protec. Nat.* 4:20–23.

———. 1988b. Presas de *Tytoalba* en Saladillo, partido de Saladillo, provincia de Buenos Aires. I. *Bol. Cien. Asoc. Protec. Nat.* 6:10–14.

———. 1988c. Presas de *Tyto alba* en Campo Ramón, departamento Oberá, provincia de Misiones–I. *Bol. Cien. Asoc. Protec. Nat.* 7:4–16.

———. 1988d. Análisis de regurgitados de *Rhinoptynxclamator* del partido de Marcos Paz, provincia de Buenos Aires. *Bol. Cien. Asoc. Protec. Nat.* 9:4–9.

———. 1988e. Pequeños mamíferos depredados por *Geranoaetus melanoleucus* en el Paraje Confluencia, Departamento Collon Cura, provincia de Neuquén. *Bol. Cien. Asoc. Protec. Nat.* 9:13–18.

———. 1989. Nuevos o poco conocidos cráneos de mamíferos vivientes- 1- *Rhipidomys leucodactylus austrinus* de la provincia de Salta, República Argentina. *Bol. Cien. Asoc. Protec. Nat.* 15:14–16.

———. 1990. Nuevos o poco conocidos cráneos de mamíferos vivientes - 2 – *Euryzygomatomys spinosus spinosus* de la provincia de Misiones, República Argentina. *Bol. Cien. Assoc. Protec. Nat.* 17: 9–14.

———. 1993. Los roedores misioneros–1–Lista sistemática comentada y geonemia provincial conocida. *Bol. Cien. Asoc. Protec. Nat.* 25:42–51.

———. 1996. Los roedores con pelaje espinoso de la Argentina (Mammalia). *Bol. Cien. Asoc. Protec. Nat.* 29:26–29.

———. 1998. Roedores vinculados con las virosis humanas en la República Argentina. *2nd Congreso Argentino de Zoonosis y 1st Congreso Argentino y Latinoamericano de Enfermedades Emergentes*, Buenos Aires, pp. 243–46.

Massoia, E., G. Aprile, and B. Lartigau.1995a. Vertebrados depredados por *Tyto alba* en Capitán Solari, partido de Sargento Cabral, provincia de Chaco. *Bol. Cien. Asoc. Protec. Nat.* 27:9–14.

———. 1995b. Análisis de regurgitados de *Tyto alba* de Estación Santa Margarita, Departamento 9 de Julio, Provincia de Santa Fe. *Bol. Cien. Asoc. Protec. Nat.* 27:19–21.

Massoia, E., and J. C. Chebez. 1993. *Mamíferos silvestres del Archipielago Fueguino*. Santa Fe, Argentina: L.O.L.A. (Literature of Latin America) , 200 pp.

Massoia, E., J. C. Chebez, and A. Bosso. 2006. *Los mamíferos silvestres de la provincia de Misiones Argentina*. Buenos Aires: Authors, 512 pp.

Massoia, E., J. C. Chebez, and S. Heinonen Fortabat. 1989a. Segundo análisis de egagrópilas de *Tyto alba tuidara* en el Departamento de Apóstoles, Província de Misiones. *Bol. Cien. Asoc. Protec. Nat.* 13:3–8.

———. 1989b. Análisis de regurgitados de *Tyto alba tuidara* de Los Helechos, Departamento Oberá, Provincia de Misiones. *Bol. Cien. Asoc. Protec. Nat.* 14:16–22.

———. 1989c. Mamíferos e aves depredados por *Tyto alba tuidara* en Bonpland, Departamento Candelaria, Provincia de Misiones. *Bol. Cien. Asoc. Protec. Nat.* 15:19–24.

———. 1989d. Mamíferos e aves depredados por *Tyto alba* en el arroyo Yabebyrí, Departamento de Candelária, Provincia de Misiones. *Bol. Cien. Asoc. Protec. Nat.* 15:8–13.

———. 1991a. Nuevos o poco conocidos cráneos de mamíferos vivientes–3–*Abrawayaomys ruschi* de la provincia de Misiones, República Argentina. *Bol. Cien. Asoc. Protec. Nat.* 19:39–40.

———. 1991b. El estado sistemático de *Thomasomys pictipes* Osgood, 1933 (Rodentia, Cricetidae). *Bol. Cien. Aprona* 19:17–18.

———. 1994a. La depredación de algunos mamíferos por *Bubo virginianus* en el Departamento Malargüe, Mendoza. *Bol. Cien. Asoc. Protec. Nat.* 26:2–5.

———. 1994b. Análisis de regurgitados de *Bubo virginianus* en el lago Cardiel, Departamento Lago Buenos Aires, provincia de Santa Cruz. *Bol. Cien. Asoc. Protec. Nat.* 26:17–21.

Massoia, E., and A. Fornes. 1962. Un cricétido nuevo para la Argentina: *Akodon arviculoides montensis* Thomas (Rodentia). *Physis* 65:185–94.

———. 1964a. Pequeños mamíferos (Marsupialia, Chiroptera y Rodentia) y aves obtenidos en regurgitaciones de lechuzas (Strigiformes) del Delta bonaerense. *Delta Parana Invest. Agric.* 4:27–34.

———. 1964b. Nuevos datos sistemáticos, biológicos y etoecológicos de *Oryzomys (Oligoryzomys) delticola* Thomas (Rodentia, Cricetidae). *Delta Parana Invest. Agric.* 4:35–47.

———. 1964c. Notas sobre el género *Scapteromys* (Rodentia-Cricetidae). I. sistemática, distribución geográfica y rasgos etoecológicos de *Scapteromys tumidus* (Waterhouse). *Physis* 24:279–97.

———. 1964d. Comentarios sobre *Eligmodontia typus typus* (Rodentia, Cricetidae). *Physis* 24:298.

———. 1965a. Notas sobre el género *Scapteromys* (Rodentia–Cricetidae). II. Fundamentos de la identidad específica de *S. principalis* (Lund) y *S. gnambiquarae* (M. Ribeiro). *Neotrópica* 11:1–7.

———. 1965b. Contribución al conocimiento de los roedores miomorfos argentinos vinculados con laFiebre Hemorrágica (Rodentia: Cricetidae y Muridae). *Comisión Nacional Coordinadora para el Estudio y Lucha contrala Fiebre Hemorrágica Argentina, Ministerio de Asistencia Social y Salud Pública*, Buenos Aires, 20 pp.+4 plates.

———. 1967a. El estado sistematico, distribucion geografica y datos etoecologicos de algunos mamiferos neotropicales (Marsupialia y Rodentia) con la descripcion de *Cabreramys*, genero nuevo (Cricetidae). *Acta Zool. Lilloana* 23:407–30.

———. 1967b. Roedores recolectados en la Capital Federal (Caviidae, Cricetidae y Muridae). *INTA, IDIA* 240:47–53.

———. 1969. Caracteres comunes y distintivos de *Oxymycterus nasutus* (Waterhouse) y *Oxymycterus iheringi* Thomas (rodentia, Cricetidae). *Physis* 28:315–21.

Massoia, E., A. Fornes, R. Wainberg, and T. G. Fronza. 1968. Nuevos aportes al conocimiento de las especies bonaerenses del género *Calomys* (Rodentia - Cricetidae). *Rev. Invest. Agropec.* 5:63–92.

Massoia, E., S. Heinonen Fortabat, and A. Diéguez. 1997. Análisis de componentes mastozoológicos y ornitológicos en regurgitados de *Tyto alba* de Estancia Guayacolec, Depto. Pilcomayo, Pcia. de Formosa, República Argentina. *Bol. Cien. Asoc. Protec. Nat.* 32:12–17.

Massoia, E., and U. F. J. Pardiñas. 1988a. Pequeños mamíferos depredados por *Bubo virginianus* en Pampa de Nestares, Departamento Pilcaniyeu, Provincia de Río Negro. *Bol. Cien. Asoc. Protec. Nat.* 3:23–27.

———. 1988b. Presas de *Bubo virginianus* en Cañadón Las Coloradas, Departamento Pilcaniyeu, Provincia de Río Negro. *Bol. Cien. Asoc. Protec. Nat.* 4:14–19.

———. 1993. El estado sistemático de algunos muroideos estudiados por Ameghino en 1889. Revalidación del género *Necromys* (Mammalia, Rodentia, Cricetidae). *Ameghiniana* 30:407–18.

Massoia, E., and H. Pastore. 1997. Análisis de regurgitados de *Bubo virginianus megallanicus* (Lesson, 1828) del Parque Nacional Laguna Blanca, Dpto. Zapala, Pcia. de Neuguén. *Bol. Cien. Asoc. Protec. Nat.* 33:18–19.

Massoia, E., H. Pastore, and J. C. Chebez. 1999. Mamíferos depredados por *Tyto alba* en los departamentos de Gral. Ocampo y Rosario V. Peñaloza, Pcia, de la Rioja. *Bol. Cien. Asoc. Protec. Nat.* 37:17–20.

Massoia, E., Renard of Coquet, S., and J. Fernandez. 1980 (1981). *Lama guanicoe* in the primitive economy, as verified in the archaeological record digging Chenque Haichol, Neuquén. *IDIA* 389–90:79–82.

Massoia, E., and M. J. Silveira. 1996. Los micromamíferos del sitio Alero los Cipreses (Departamento de Los Lagos, provincia de Neuquén, República Argentina). *Rev. Mus. Argentino Cien. Nat. "Bernardino Rivadavia," Cien. Zool.* 131:1–4.

Massoia, E., S. I. Tiranti, and A. J. Diéguez. 1997. Pequeños mamíferos depredados por *Tyto alba* en la provincia de La Pampa, según sucesivas recolecciones. *Bol. Cien. Asoc. Protec. Nat.* 32:19–21.

Massoia, E., S. I. Tiranti, and M. P. Torres. 1989. La depredación de pequenos mamíferos por *Tyto alba* en Canal 6, Partido de Campana, Província de Buenos Aires. *Bol. Cien. Asoc. Protec. Nat.* 13:14–19.

Matamoros, Y. 1981. Anatomia e histology del sistema reproductor del tepezcuintle (*Cuniculus paca*). *Rev. Biol. Trop.* 29:155–64.

Matamoros, Y., and B. Pashov. 1984. Ciclo estral del tepezcuintle (*Cuniculus paca*, Brisson) en cautiverio. *Brenesia* 22:249–60.

Mathis, V. L., M. S. Hafner, D. J. Hafner, and J. W. Demastes. 2013. Resurrection and redescription of the pocket gopher *Thomomys seldoni* from the Sierra Madre Occidental of Mexico. *J. Mamm.* 94:544–60.

Matocq, M .D., J. L. Patton, and M. N. F. da Silva. 2000. Population genetic structure of two ecologically distinct Amazonian spiny rats: separating history and current ecology. *Evolution* 54:1423–32.

Matschie, P. 1898. *Hamburger Magalhaensische Sammelreise, Säugethiere.* Hamburg: L. Friederichsen & Co., 29 pp + 1 plate.

Matson, J. O., and J. P. Abravaya. 1977. Blarinomys breviceps. *Mamm. Species* 74:1–3.

Matson, J. O., and K. A. Shump Jr. 1981. Intrapopulation variation in cranial morphology in the agouti *Dasyprocta punctata* (Dasyproctidae). *Mammalia* 44:559–70. [1980 vol. 44, no. 4: pp. 423ff. "Achevé d'imprimer le 20 février 1981."]

Mattevi, M. S., T. Haag, A. P. Nunes, L. F. B. Oliveira, J. L. P. Cordeiro, and J. Andrades-Miranda. 2002. Karyotypes of Brazilian representatives of genus *Zygodontomys* (Rodentia, Sigmodontinae). *Mastozool. Neotrop.* 9: 33–38.

Mattevi, M. S., T. Haag, L. F. B. Oliveira, and A. R. Langguth. 2005. Chromosome characterization of Brazilian species of *Calomys* waterhouse, 1837 from Amazon, Cerrado and Pampas domains (Rodentia, Sigmodontinae). *Arq. Mus. Nac., Rio de Janeiro* 63:175–81.

Mattevi, M. S., I. J. Sbalqueiro, T. R. O. de Freitas, and L. B. F. Oliveira. 1981. Estudos citotaxonomicos em roedores do extremo sul do Brasil. *Proceedings of the V Congreso Latino Americano de Genética*, Santiago, Chile, p. 67.

Mauffrey, J.-F. 1999. "Rongeurs arboricoles en forêt néotropicale: Première étude de la communauté des Nouragues." Ph.D. diss., Université Montpellier II Sciences et Techniques du Languedoc, France, 30 pp + figures and appendices.

Mauffrey, J.-F., and F. Catzeflis. 2002. Ecological and isotopic discrimination of syntopic rodents in a neotropical rain forest of French Guiana. *J. Trop. Ecol.* 19:209–14.

Mauffrey, J.-F., C. Steiner, and F. M. Catzeflis. 2007. Small-mammal diversity and abundance in a French Guianan rain forest: test of sampling procedures using species rarefaction curves. *J. Trop. Ecol.* 23:419–25.

Mayor, P., R. E. Bodmer, and M. Lopez-Bejar. 2012. Functional anatomy of the female genital organs of the wild black agouti (*Dasyprocta fuliginosa*) female in the Peruvian Amazon. *Anim. Reprod. Sci.* 123:249–57.

Mayr, E. 1963. *Animal species and evolution.* Cambridge, MA: Belknap Press of Harvard University Press, xiv + 797 pp.

McCain, C. M., R. M. Timm, and M. Weksler. 2007. Redescription of the enigmatic long-tailed rat *Sigmodontomys aphrastus* (Cricetidae: Sigmodontinae) with comments on taxonomy and natural history. *Proc. Biol. Soc. Washington* 120:117–36.

McFarlane, D. A., and A. O. Debrot. 2001. A new species of extinct oryzomyine rodent from the Quaternary of Curacao, Netherlands Antilles. *Caribb. J. Sci.* 37:182–84.

McKee, R. C., and G. H. Adler. 2002. Tail autotomy in the Central American spiny rat, *Proechimys semispinosus*. *Stud. Neotrop. Fauna Environ.* 37:181–85.

McKenna, M. C., and S. K. Bell. 1997. *Classification of mammals above the species level.* New York: Columbia University Press, xii + 631 pp.

Medina, A. I., and Bidau, C.J. 2007. Dimorfismo sexual de tamaño y alometría del tamaño testicular en tuco-tucos (*Ctenomys*: Rodentia: Ctenomyidae). *Actas XXI Jornadas Argentinas de Mastozoología*, SAREM, Tafí del Valle, Tucumán.

Medina, A. I., D. A. Martí, and C. J. Bidau. 2007. Subterranean rodents of the genus *Ctenomys* follow the converse to Bergmann's rule. *J. Biogeogr.* 34:1439–54.

Medina, C. E., H. Zeballos, and E. López. 2012. Diversidad de mamíferos en los bosques montanos del valle de Kcosñipata, Cusco, Perú. *Mastozool. Neotrop.* 19:85–104.

Medina, R. A., F. Torres-Pérez, H. Galeno, M. Navarrete, P. A. Vial, R. E. Palma, M. Ferres et al. 2009. Ecology, genetic diversity, and phylogeographic structure of Andes Virus in humans and rodents in Chile. *J. Virol.* 83:2446–59.

Mella, J. E. 2006. Micromamíferos en el Monumento Natural El Morado: abundancia relativa y cambios estacionales. *Notic. Mens. Mus. Nac. Hist. Nat. Chile* 357:10–18.

Mella, J. E., J. A. Simonetti, A. E. Spotorno, and L. C. Contreras. 2002. Mammals of Chile. In *Diversidad y conservación de los mamíferos neotropicales*, ed. G. Ceballos and J. A. Simonetti, 151–83. México, D.F.: CONABIO-UNAM, 582 pp.

Mello, D. A. 1977. Observações preliminares sobre a ecologia de algumas espécies de roedores do Cerrado, município de Formosa, Goiás, Brasil. *Rev. Bras. Pesqui. Med. Biol.* 10:39–44.

———. 1979. *Trypanosoma (Megatrypanum) amileari* n. sp., isolated from *Oryzomys eliurus* (Wagner, 1845) (Rodentia-Cricetidae). *Ann. Parasitol. Hum. Comp.* 54:489–94.

———. 1986. Breeding of wild-caught rodent cricetidae *Holochilus brasiliensis* under laboratory conditions. *Lab. Anim.* 20:195–96.

Mello, D. A., and C. H. Mathias. 1987. Criação de *Akodon arviculoides* (Rodentia, Cricetidae) em laboratório. *Rev. Bras. Biol.* 47:419–23.

Mello, D. A., and M. L. Teixeira. 1977. Nota sobre a infecção natural de *Calomys expulsus*, Lund, 1841 (Cricetidae-Rodentia) pelo *Trypanosoma cruzi*. *Rev. Saide Pub.* 11:561–64.

Melo, F. P. L., and M. Tabarelli. 2003. Seed dispersal and demography of pioneer trees: the case of *Hortia arborea*. *Plant Biol.* 5:359–65.

Melo-Leitão, C. de. 1943. Fauna amazônica. *Rev. Bras. Geogr.* 5:343–70.

Mena, J. L., M. Williams, C, Gazzolo, and F. Montero. 2007. Conservation status of *Melanomys zunigae* (Sanborn 1949) and small mammals in the Lomas of Lima. *Rev. Peru. Biol.* 14:201–7.

Mendes, L. A. F, P. L. B. Rocha, M. F. S. Ribeiro, S. F. Perry, and E. S. Oliveira. 2004. Differences in ingestive balance of two populations of Neotropical *Thrichomys apereoides* (Rodentia, Echimyidae). *Comp. Biochem. Physiol. A Mol. Integr. Physiol.* 138:327–32.

Mendes Pontes, A. R., J. R. Gadelha, E. R. A. Melo, F. B. de Sá, A. C. Loss, V. Caldara Junior, L. P. Costa, and Y. L. R. Leite. 2013. A new species of porcupine, genus *Coendou* (Rodentia: Erethizontidae) from the Atlantic forest of northeastern Brazil. *Zootaxa* 3636:421–438.

Mendez, E. 1969. Four new species of Gyropidae (Mallophaga) from spiny rats in Middle America. *Pac. Insects* 11:497–506.

———. 1971. A new species of the genus *Cummingsia* Ferris from the Republic of Colombia (Mallophaga: Trimenoponidae). *Proc. Entomol. Soc. Wash.* 73:23–27.

———. 1993. *Los roedores de Panama*. Panama, 372 pp [author].

Mendonça, P. P. de, P. Cobra, L. R. Bernardo, and T. Silva-Soares. 2011. Predation of the snake *Spilotes pullatus* (Squamata: Serpentes) upon the rodent *Proechimys gardneri* (Rodentia: Echimyidae) in the Amazon Basin, northwestern Brazil. *Herpetol. Notes* 4:425–27.

Menegaux, M. A. 1902. Catalogue des mammifères rapportés par M. Geay de la Guyane française en 1898 et 1900. *Bull. Mus. Natl. Hist. Nat., Paris* 8:490–96.

Menezes, D. J. A., M. A. M. de Carvalho, A. C. de Assis-Neto, M. F. de Oliveira, E. C. Farias, M. A. Miglino, and G. X. Medeiros. 2003. Morphology of the external male genital organs of agouti (*Dasyprocta aguti*, Linnaeus, 1766). *Braz. J. Vet. Res. Anim. Sci.* 40:148–53.

Menezes, D. J. A., M. A. M. de Carvalho, M. F. Oliveira, and A. C. Assis-Neto. 2001. Morphologic aspects of glans of the penis in agouti, *Dasyprocta* sp, Rodentia. *Rev. Bras. Reprod. Anim.* 25:208–9.

Merani, M. S., M. Lizarralde, D. Olivera, and N. O. Bianchi. 1978. Cytogenetics of South American akodont rodents (Cricetidae). IV. Interspecific crosses between *Akodon dolores × Akodon molinae*. *J. Exp. Zool.* 206:343–46.

Mercado, I., and J. M. Miralles. 1991. Mamíferos. In *Historia natural de un valle en los Andes*, ed. E. Forno and M. Baudoin, 293–343. La Paz: Instituto de Ecológia, 559 pp.

Mercer, J. M., and V. L. Roth. 2003. The effects of Cenozoic global change on squirrel phylogeny. *Science* 299:1568–72.

Merler, J., R. Bó, R. Quintana, and A. Malvarez. 1994. Habitat studies at different spatial scales for multiple conservation goals in the Paraná River delta (Argentina). *Int. J. Ecol. Environ.* 20:149–62.

Merriam, C. H. 1901. Synopsis of the rice rats (genus *Oryzomys*) of the United States and Mexico. *Proc. Wash. Acad. Sci.* 3:273–95.

———. 1892. The geographic distribution of life in North America with special reference to the Mammalia. *Proc. Biol. Soc. Washington* 7:1–64, 1 plate.

———. 1894. A new subfamily of murine rodents—the Neotominae—with description of a new genus and species and a synopsis of the known forms. *Proc. Acad. Nat. Sci. Philadelphia* 1894:225–52, 1 plate.

———. 1895a. Monographic revision of the pocket gophers, family Geomyidae (exclusive of the species of *Thomomys*). *N. Amer. Fauna* 8:1–258.

———. 1895b. Brisson's genera of mammals, 1762. *Science*, n.s. 1(14):375–76.

———. 1898. Descriptions of twenty new species and a new subgenus of *Peromyscus* from Mexico and Guatemala. *Proc. Biol. Soc. Washington* 12:115–25.

———. 1901. Synopsis of the rice rats (genus *Oryzomys*) of the United States and Mexico. *Proc. Washington Acad. Sci.* 3:273–95.

———. 1902. Twenty new pocket mice (*Heteromys* and *Liomys*) from Mexico. *Proc. Biol. Soc. Washington* 15:41–50.

Merrick, M. J., J. L. Koprowski, and R. N. Gwinn. 2012. Sciurus stramineus (Rodentia: Sciuridae). *Mamm. Species* 44(1):44–50.

Merritt, D. A., Jr. 1983. Preliminary observations on reproduction in the Central American agouti, *Dasyprocta punctata. Zoo Biol.* 2:127–31.

———. 1984. The pacarana, *Dinomys branickii.* In *One medicine,* ed. O. A. Ryder and M. L. Byrd, 154–61. New York: Springer-Verlag, 373 pp.

Meserve, P. L. 1981a. Trophic relationships among small mammals in a Chilean semiarid thorn scrub community. *J. Mamm.* 62:304–14.

———. 1981b. Resource partitioning in a Chilean semi-arid small mammal community. *J. Anim. Ecol.* 50:745–57.

———. 1983. Feeding ecology of two Chilean caviormorphs in a central Mediterranean savanna. *J. Mamm.* 64:322–25.

Meserve, P. L., and W. Glanz. 1978. Geographical ecology of small mammals in the northern Chilean arid zone. *J. Biogeogr.* 5:135–48.

Meserve, P. L., J. R. Gutiérrez, and F. M. Jaksic. 1993. Effects of vertebrate predation on a caviomorph rodent (*Octodon degus*) in a semiarid thorn scrub community in Chile. *Oecologia* 94:153–58.

Meserve, P. L., D. A. Kelt, and D. R. Martínez. 1991. Geographical ecology of small mammals in continental Chile Chico, South America. *J. Biogeogr.* 18:179–87.

Meserve, P. L., B. K. Lang, R. Murúa, A. Muñoz-Pedreros, and L. A. González. 1991. Characteristics of a terrestrial small mammal assemblage in a temperate rainforest in Chile. *Rev. Chilena Hist. Nat.* 64:157–69.

Meserve, P. L., B. K. Lang, and B. D. Patterson. 1988. Trophic relationships of small mammals in a Chilean temperate rainforest. *J. Mamm.* 69:721–30.

Meserve, P. L. and E. Le Boulengé. 1987. Population dynamics and ecology of small mammals in the northern Chilean semiarid region. *Fieldiana Zool.,* n.s., 39:413–31.

Meserve, P. L., B. W. Milstead, J. R. Gutierrez, and F. M. Jaksic. 1999. The interplay of biotic and abiotic factors in a semiarid Chilean mammal assemblage: Results of a long-term experiment. *Oikos* 85:364–72.

Meserve, P. L., R. Murúa, O. Lopetegui-N, and J. R. Rau. 1982. Observations in the small mammal fauna of a primary temperate rain forest in southern Chile. *J. Mamm.* 63:315–17.

Meserve, P. L., E. J. Shadrick, and D. A. Kelt. 1987. Diets and selectivity of two Chilean predators in the northern semi-arid zone. *Rev. Chilena Hist. Nat.* 60:93–99.

Meserve, P. L., J. A. Yunger, J. R. Gutiérrez, L C. Contreras, W. B. Milstead, B. K. Lang, K. L. Cramer, et al. 1995. Heterogeneous responses of small mammals to an El Niño Southern Oscillation event in northcentral semiarid Chile and the importance of ecological scale. *J. Mamm.* 76:580–95.

Mesquita, A. de O. 2009. "Comunidades de pequenos mamíferos em fragmentos florestais conectados por corredores de vegetação no sul de Minas Gerais." Master's thesis, Universidade Federal de Lavras, Minas Gerais, Brazil.

Meyen, F. J. F. 1833. Beiträge zur Zoologie, gesammelt auf einer Reise um die Erde. Zweite Abhandlung. Säugethiere. *N. Acta Phys. Med. Acad. Caesar. Leopoldino-Carolinae Nat. Curios.* 16 [for 1832], pt. 2:549–610 + 7 plates.

Meyer, F. 1790. Ueber ein neues Säugethiergeschlecht. *Mag. Thiergesch. Thieranat. Thierarzneykd.* 1:46–55.

Meyerson-McCormick, R., J. C. Cranford, and R. M. Akers. 1990. Milk yield and composition in the punare *Thrichomys apereoides. Comp. Biochem. Physiol. A Mol. Integr. Physiol.* 96:211–14.

Michalski, F., and D. Norris. 2011. Activity pattern of *Cuniculus paca* (Rodentia: Cuniculidae) in relation to lunar illumination and other abiotic variables in the southern Brazilian Amazon. *Zoologia* 28:701–8.

Middleton, J. 2007. "A systematic revision of genus *Isthmomys* (Rodentia: Cricetidae)." Master's thesis, Texas Tech University, Lubbock.

Miglino, M. A., A. M. Carter, C. E. Ambrosio, M. Bonatelli, M. F. de Oliveira, and R. H. dos Santos. 2004. Vascular organization of the hystricomorph placenta: a comparative study in the agouti, capybara, guinea pig, paca and rock cavy. *Placenta* 25:438–48.

Milano, M. Z. 2007. "Ecologia da comunidade de pequenos mamíferos da floresta estacional aluvial da RPPN Cabeceira do Prata, região da Serra da Bodoquena, estado do Mato Grosso do Sul." Master's thesis, Universidade Federal do Paraná, Curitiba, Brazil.

Miles, M. A., A. A. de Souza, and M. M. Póvoa. 1981. Mammal tracking and next location in Brazilian forest with an improved spool-and-line device. *J. Zool.* 195:331–47.

Milishnikov, A. N. 2006. Two types of genetic differentiation in the evolution of sibling species of the spiny rats from genus *Proechimys* native to upper Amazonia. *Dokl. Biol. Sci.* 408:237–41.

Miller, F. W. 1930. Notes on some mammals of southern Matto Grosso, Brazil. *J. Mamm.* 11:10 22, 2 plates.

Miller, G. S., Jr. 1912. List of North American land mammals in the United States National Museum, 1911. *Bull. U.S. Natl. Mus.* 79:1–455.

Miller, G. S., Jr., and J. W. Gidley. 1918. Synopsis of the supergeneric groups of Rodents. *J. Washington Acad. Sci.* 8:431–48.

Miller, G. S., Jr., and R. Kellogg. 1955. List of North American Recent mammals, *Bull. U.S. Natl. Mus.* 205:xii + 1–954.

Miller, G. S., Jr., and J. A. G. Rehn. 1901. Systematic results of the study of North American land mammals to the close of the year 1900. *Proc. Boston Soc. Nat. Hist.* 30:1–352.

Miller, F. W. 1930. Notes on some mammals of southern Matto Grosso, Brazil. *J. Mamm.* 11:10–22, 2 plates.

Miller, L. E. 1918. *In the wilds of South America. Six years of exploration in Colombia, Venezuela, British Guiana, Peru, Bolivia, Argentina, Paraguay, and Brazil.* New York: Charles Scribner's Sons, xiv + 424 pp.

Miller, L. M., and S. Anderson. 1977. Bodily proportions of Uruguayan Myomorph Rodents. *Am. Mus. Novit.* 2615:1–10.

Mills, J. N., B. A. Ellis, J. E. Childs, J. I. Maiztegui, and A. Castro-Vázquez. 1992a. Seasonal changes in mass and reproductive condition of the corn mouse (*Calomys musculinus*) on the Argentine pampa. *J. Mamm.* 73:876–84.

Mills, J. N., B. Ellis, K. McKee Jr., T. Ksiazek, J. Oro, J. Maiztegui, G. Calderon, C.J. Peters, and J. E. Childs. 1991b. Junin virus activity in rodents from endemic and nonendemic loci in central Argentina. *Am. J. Trop. Med. Hyg.* 44:589–97.

Mills, J. N., B. A. Ellis, K. T. McKee, J. I. Maiztegui, and J. E. Childs. 1991a. Habitat associations and relative densities of rodent populations in cultivated areas of central Argentina. *J. Mamm.* 72:470–79.

———. 1992b. Reproductive characteristics of rodent assemblages in cultivated regions of central Argentina. *J. Mamm.* 73:515–26.

Milne-Edwards, A. 1890. *Mammifères. Mission Scientifique du Cap Horn. 1882–1883.* Tome VI. Zoologie. Première partie, pp. 3–32. Paris: Ministères de la Marine et de l'instruction Publique, Gauthier-Villars et fils.

Minding, J. 1829. *Über die geographische Bertheilung der Säugethiere.* Berlin: Enslin'fche Buchhandlung, 103 pp.

Miranda, C. L., R. V. Rossi, T. B. F. Semedo, and T. A. Flores. 2012. New records and geographic distribution extension of *Neusticomys ferreirae* and *N. oyapocki* (Rodentia, Sigmodontinae). *Mammalia* 76:335–39.

Miranda, G. B., L. F. B. Oliveira, J. Andrades-Miranda, A. Langguth, S. M. Callegari-Jacques, and M. S. Mattevi. 2009. Phylogenetic and phylogeographic patterns in sigmodontine rodents of the genus *Oligoryzomys*. *J. Hered.* 100:309–21.

Miranda, J. A., A. P. Nunes, L. F. B. Oliveira, and M. S. Mattevi. 1999. The karyotype of the South American rodent *Kunsia tomentosus* (Lichtenstein, 1830). *Cytobios* 98:137–47.

Miranda-Ribeiro, A. de. 1905. Vertebrados do Itatiaya (Peixes, Serpentes, Sáurios, Aves e Mammíferos). *Arq. Mus. Nac., Rio de Janeiro* 13:163–90 + 3 plates.

———. 1914. *Mammíferos. Cebidae, Hapalidae; Vespertilionidae, Emballonuridae, Phyllostomatidae; Felidae, Mustelidae, Canidae, Procyonidae; Tapyridae; Suidae, Cervidae; Sciuridae, Muridae, Octodontidae, Coenduidae, Dasyproctidae, Caviidae e Leporidae; Platanistidae; Bradypodidae, Myrmecophagidae, Dasypodidae; Didelphyidae.* Comissão de Linhas Telegráficas Estratégicas de Matto-Grosso ao Amazonas, Anexo 5, 49 pp + appendix (3 pp.) + 25 plates.

———. 1918 (1919). *Dinomys pacarana?* Arch. Esc. Super. Agric. Med. Vet. Rio J. 2:13–15.

———. 1936. The new-born of the Brazilian tree-porcupine (*Coendou prehensilis* Linn.) and of the hairy tree-porcupine (*Sphingurus villosus* F. Cuv.). *Proc. Zool. Soc. Lond.* 106:971–74.

———. 1941. Sôbre dois novos sciurideos do Brasil. *O Campo* 139:10–11.

Miranda Ribeiro, P. 1955. Tipos das espécies do Prof. Alípio de Miranda Ribeiro depositados no Museu Nacional (com uma relação dos gêneros, espécies e subespécies descritos). *Arq. Mus. Nac., Rio de Janeiro* 42:389–417.

Mirol, P. M., M. D. Giménez, J. B. Searle, C. J. Bidau, and C. G. Faulkes. 2010. Species and population boundaries in a changing environment in the subterranean South American rodent *Ctenomys*. *Biol. J. Linn. Soc.* 100:369–83.

Miserendino S., R. S., and H. Azurduy F. 2005. Nota sobre el primer specimen de *Myoprocta pratti* (Rodentia, Dasyproctidae) para Bolivia. *Kempffiana* 1:55–57.

Mohr, E. 1937. Vom Pararana (*Dinomys branickii* Peters). *Zool. Gard. (Neue Folge)* 9:204–9.

Molina, C., C. García, and N. Abad. 1995. Notas sobre la distribution de *Dactylomys dactylinus* en Venezuela. *Mem. Soc. Cien. Nat. La Salle* 55:41–45.

Molina, G. I. 1776. *Compendio della storia geografica, natural, e civile del regno del Chile.* Bologna: Stamperia di S. Tommaso d'Aguino, vii, + 245 pp, 10 plates. + 1 map (fold out).

———. 1782. *Saggio sulla storia naturale del Chili,* 1st edition. Bologna: Stamperia di S. Tomnaso d'Aquino, 368 pp. + 1 map. [second edition, published 1810.]

———. 1808. *The geographical, natural and civil history of Chile.* Translated by R. Alsop. Middletown, Connecticut: I. Riley. Originally published two volumes:

Saggio sulla storia naturale del Chili, Bologna, 1782, and *Saggio sulla storia civile del Chili,* Bologna, 1787.

Mollineau, W. M., A. Adogwa, and G. W. Garcia. 2009. The gross and micro anatomy of the accessory sex glands of the male agouti (*Dasyprocta leporina*). *Anatom. Hist. Embryol.* 38:204–7.

Mollineau, W. M., A. Adogwa, N. Jasper, K. Young, and G. Garcia. 2006. The gross anatomy of the male reproductive system of a neotropical rodent: the agouti (*Dasyprota leporina*). *Anatom. Hist. Embryol.* 35:47–52.

Mondolfi, E. 1972. La lapa o paca. *Def. Nat.* 2:4–16.

Mondolfi, E., and B. S. Boher. 1984. Una nueva subespecie de ardilla del grupo *Sciurus granatensis* (Mammalia, Rodentia) en Venezuela. *Acta Cien. Venez.* 35:312–14.

Mones, A. 1968. Proposicion de una nueva terminologia relacionada con el crecimiento de los molars. *Zool. Platense* 1:13–16.

———. 1975. Estudios sobre la familia Hydrochoeridae (Rodentia), VI. Catalogo anotado de los ejemplares-tipo. *Comun. Paleontol. Mus. Hist. Nat. Montevideo* 1:99–130.

———. 1981. Sinopsis sistemática preliminary de la familia Dinomyidae (Mammalia, Rodentia, Caviomorpha). *Anais II Congreso Latino Americano de Paleontología, Porto Alegre, Brazil* 2:605–19.

———. 1986. Paleovertebrata Sudamericana. Catálogo sistemático de los vertebrados fósiles de América del Sur. Parte I. Lista preliminar y Bibliografía. *Courier Forschungsinstit. Senckenberg* 82:1–625.

———. 1991. Monographía de la familia Hydrochoeridae (Mammalia: Rodentia). Sistematica–Paleontologia–Filogenia–Bibliografía. *Courier Forchungsinstit. Senckenberg* 134:1–235.

———. 1995. Estudios sobre la familia Dinomyidae. I. Sobre la nomenclatura y sinonimia familiar (Mammalian: Rodentia). *Comun. Paleontol. Mus. Hist. Nat. Montevideo*28:87–91.

———. 1997. Estudios sobre la familia Dinomyidae, II. Aportes para una osteología comparada de Dinomys branickii peters, 1873 (Mammalia: Rodentia). *Comun. Paleontol. Mus. Hist. Nat. Montevideo* 29:93–132.

———. 2007. *Josephoartigasia,* Nuevo nombre para *Artigasia* Frances & Mones, 1966 (Rodentia, Dinomyidae), non *Artigasia* Christie, 1934 (Nematoda, Thelastomatidae). *Comun. Paleontol. Mus. Hist. Nat. Montevideo* 36:213–14.

Mones A., J. González, R. Praderi, and M. Clara. 2003. Diversidad de la biota uruguaya. Mammalia. *An. Mus. Nac. Hist. Nat. Antropol.,* ser. 2, 10(4):1–28.

Mones, A., and M. A. Klappenbach. 1997. Un ilustrado aragonés en el virreinato del río de La Plata: Félix de Azara 1742 1821). Estudios sobre su vida, su obra, y su pensamiento. *Anal. Mus. Nac. Hist. Nat. Montevideo,* sér. 2, 9:i–viii, 1–231.

Mones, A., and J. Ojasti. 1986. Hydrochoerus hydrochaeris. *Mamm. Species* 264:1–7.

Monjeau, J. A. 1989. "Ecología y distribución geográfica de los pequeños mamíferos del Parque Nacional Nahuel Huapí y áreas adyacentes." Ph.D. diss., Universidad Nacional de La Plata, Argentina.

Monjeau, J. A., R. S. Sikes, E. C. Birney, N. Guthmann, and C. J. Phillips. 1997. Small mammal community composition within the major landscape divisions of Patagonia, southern Argentina. *Mastozool. Neotrop.* 4:113–27.

Monteiro, L. R., L. C. Duarte, and S. F. dos Reis. 2003. Environmental correlates of geographic variation in skull and mandible shape of the punaré rat *Thrichomys apereoides* (Rodentia: Echimyidae). *J. Zool. Lond.* 261:47–57.

Monteiro, L. R., L. G. Lessa, and A. S. Abe. 1999. Ontogenetic variation in skull shape of *Thrichomys apereoides* (Rodentia, Echimyidae). *J. Mamm.* 80:102–11.

Montenegro-Díaz, O., H. López-Arévalo, and A. Cadena. 1991. Aspectos ecológicos del roedor arborícola *Rhipidomys latimanus* Tomes, 1860, (Rodentia: Cricetidae) en el Oriente de Cundinamarca, Colombia. *Caldasia* 16:565–72.

Montes, M. A. 2007. Uma abordagem molecular na análise da filogenia e da filogeografía dos roedores akodontines do neotropico." Ph.D. diss., Universidade Federal do Rio Grande do Sul, Porto Alegre, Brazil.

Montes, M. A., M. D. Giménez, C. J. Bidau, and J. B. Searle. 2001. Chromosomal and molecular delimitation of three species of the *Ctenomys* (Rodentia, Ctenomyidae) from Patagonia (Argentina). *Chromosome Res.* 9 (1, supplement):73–74.

Montes, M. A., L. F. B. Oliveira, S. L. Bonatto, S. M. Callegari-Jacques, and M. S. Mattevi. 2008. DNA sequence analysis and the phylogeographical history of the rodent *Deltamys kempi* (Sigmodontinae, Cricetidae) on the Atlantic Coastal Plain of south of Brazil. *J. Evol. Biol.* 21:1823–35.

Monteverde, M., L. Piudo, K. Hodara, and R. Douglass. 2011. Population ecology of *Eligmodontia morgana* (Rodentia, Cricetidae, Sigmodontinae) in northwestern Argentina. *Ecol. Austral* 21:195–200.

Montgomery, G. G., and Y. D. Lubin. 1978. Movements of *Coendou prehensilis* in the Venezuelan llanos. *J. Mamm.* 59:887–88.

Monticelli, P. F., and C. Ades. 2013. The rich acoustic repertoire of a precious rodent, the wild cavy *Cavia aperea. Bioacoustics* 22:49–66.

Montserin, B. G. 1937. *The Tree Rat as a Pest of Cacao in Trinidad*. Port-of-Spain: Government Printing Office, 12 pp.

Moojen, J. 1942. Sobre os "ciurídeos" das coleções de Museu Nacional, do Departamento de Zoologia de S. Paulo e do Museu Paraense Emilio Goeldi. *Bol. Mus. Nac., Rio de Janeiro*, nova sér, zool., 1:1–56.

———. 1943a. Alguns mamíferos colecionados no nordeste do Brazil com a descrição de duas espécies novas e notas de campo. *Bol. Mus. Nac., Rio de Janeiro*, nova sér, zool., 5:1–14+3 plates.

———. 1943b. Captura e preparação de pequenos mamíferos para coleções de estudo. Rio de Janeiro: Impresna National, 98 pp.

———. 1948. Speciation in the Brazilian spiny rats (genus *Proechimys*, family Echimyidae). *Univ. Kansas Publ., Mus. Nat. Hist.* 1:301–406.

———. 1950. "*Echimys (Phyllomys) kerri*" n. sp. (Echimyidae, Rodenia). *Rev. Bras. Biol.* 10:489–92.

———. 1952a. A new *Clyomys* from Paraguay (Rodentia: Echimyidae). *J. Washington Acad. Sci.* 42:102.

———. 1952b. *Os Roedores do Brasil*. Rio de Janeiro: Instituto Nacional do Livro.

———. 1958. *Sciurus cabrerai*, n. novum (Sciuridae, Rodentia). *An. Acad. Brasil. Ciênc.* 30(4):50–51.

———. 1965. Nôvo gênero de Cricetidae do Brasil Central (Glires, Mammalia). *Rev. Bras. Biol.* 25:281–85.

Moojen, J., M. Locks, A. Langguth. 1997. A new species of *Kerodon* Cuvier, 1825 from the state of Goias, Brazil (Mammalia, Rodentia, Caviidae). *Bol. Mus. Nac., Rio de Janeiro*, nova sér, zool., 377:1–10.

Moojen, J., S. F. dos Reis, and M. V. Dellape. 1988. Quantitative variation in *Thrichomys apereoides* (Lund, 1841) (Rodentia: Echimyidae). *Bol. Mus. Nac., Rio de Janeiro*, nova sér, zool., 316:1–15.

Moore, J. C. 1959. Relationships among living squirrels of the Sciurinae. *Bull. Am. Mus. Nat. Hist.* 118:153–206.

———. 1961a. Geographic variation in some reproductive characteristics of diurnal squirrels. *Bull. Am. Mus. Nat. Hist.* 122:1–32.

———. 1961b. The spread of existing diurnal squirrels across the Bering and Panamanian land bridges. *Am. Mus. Novit.* 2044:1–26.

Mora, M. S., A. P. A. P. Cutrera, E. P. Lessa, A. I. Vassallo, A. D'Anatro, and F. J. Mapelli. 2013. Phylogeography and population genetic structure of the Talas tuco-tuco (*Ctenomys talarum*): integrating demographic and habitat histories. *J. Mamm.* 94:459–76.

Mora, M. S., M. J. Kittlein, A. I. Vassalo, and F. J. Mapelli. 2013. Diferenciación geográfica en caracteres de la morfología craneana en el roedor subterráneo *Cte-*

nomys australis (Rodentia: Ctenomyidae). *Mastozool. Neotrop.* 20:75–96.

Mora, M. S., E. P. Lessa, M. J. Kittlein, and A. I. Vassallo. 2006. Phylogeography of the subterranean rodent *Ctenomys australis* in sand-dune habitats: Evidence of population expansion. *J. Mamm.* 87:1192–203.

Mora, M. S., F. J. Mapelli, O. E. Gaggiotti, M. J. Kittlein, and E. P. Lessa. 2010. Dispersal and population structure at different spatial scales in the subterranean rodent *Ctenomys australis*. *BMC Genet. Online* 11:1–14.

Mora, M. S., A. I. Olivares, and A. I. Vassallo. 2003. Size, shape and structural versatility of the skull of the subterranean rodent *Ctenomys* (Rodentia, Caviomorpha): functional and morphological analysis. *Biol. J. Linn. Soc.* 78:85–95.

Moraes, L. B., D. E. Paolinetti Bossi, and A. X. Linhares. 2003. Siphonaptera parasites of wild rodents and marsupials trapped in three mountain ranges of the Atlantic Forest of southeastern Brazil. *Mem. Inst. Oswaldo Cruz* 98:1071–76.

Moraes-Santos, H. M., C. C. de Sousa de Melo, and P. M. de Toledo. 1999. Ocorrência de *Dactylomys dactylinus* (Caviomorpha, Echimyidae) em material zooarqueológico da Serra dos Carajás, Pará. *Bol. Mus. Paraense Emilio Goeldi*, sér. zool. 15:159–67.

Morales, G. A., and F. Carreno. 1976. The *Proechimys* rat: a potential laboratory host and model for the study of *Trypanosoma evansi*. *Trop. Anim. Health Prod.* 8:122–23.

Morando, M., and J. J. Polop. 1997. Annotated checklist of mammals of Córdoba province. *Mastozool. Neotrop.* 4:129–36.

Moreira, J. C., E. G. Manduca, P. R. Gonçalves, M. M. de Morais Jr., R. F. Pereira, G. Lessa, and J. A. Dergam. 2009. Small mammals from Serra do Brigadeiro state park, Minas Gerais, southeastern Brazil: species composition and elevational distribution. *Arq. Mus. Nac., Rio de Janeiro* 67:103–18.

Moreira, J. C., and J. A. de Oliveira. 2011. Evaluating diversification hypotheses in the South American cricetid *Thaptomys nigrita* (Lichtenstein, 1829) (Rodentia: Sigmodontinae): an appraisal of geographical variation based on different character systems. *J. Mamm. Evol.* 18:201–14.

Moreira, J. R., J. R. Clarke, and D. W. Macdonald. 1997. The testis of capybaras (*Hydrochoerus hydrochaeris*). *J. Mamm.* 78:1096–100.

Moreira, J. R., K. M. P. M. B. Ferraz, E. A. Herrera, and D. W. Macdonald, eds. 2013. *Capybara: Biology, use and conservation of an exceptional Neotropical species*. New York: Springer, i–xvii+1–419 pp.

Moreno, P., and L. Albuja V. 2005. Nuevos registros de *Akodon orophilus* (Rodentia, Muridae) en el Ecuador. *Rev. Politéc.* 21:28–44.

———. 2012. Primer registro de *Thomasomys onkiro* (Rodentia: Cricetidae), para los Andes sur del Ecuador. *Rev. Politéc.* 30:9–17.

Moreno, R. S., R. W. Kays, and R. Samudio Jr. 2006. Competitive release in diets of ocelot (*Leopardus pardalis*) and puma (*Puma concolor*) after jaguar (*Panthera onca*) decline. *J. Mamm.* 87:808–16.

Moreno-Cárdenas, P., and D. G. Tirira. 2011. Ratón Andino del Pichincha *Thomasomys vulcani*. In *Libro rojo de los mamíferos del Ecuador*, 2nd ed., ed. D. G. Tirira, 84. Quito, Ecuador: Fundación Mamíferos y Conservación, Pontificia Universidad Católica del Ecuador y Ministerio del Ambiente del Ecuador. Publicación especial sobre los mamíferos del Ecuador 8, 398 pp.

Moreno Salas, L., D. del C. Castro, J. C. Torres-Mura, and D. Glonzalez-Acuña. 2005 (2006). *Phthiraptera* (Amblycera and Anoplura) parasites of the family Octodontidae, Ctenomyidae and Abrocomidae (Mammalia: Rodentia) from Chile. *Rudolstaedt. Nat. Schr.* 13:115–18.

Moretti, E. R., B. Basso, I. Albesa, A. J. Eraso, and F. O. Kravetz. 1980. Natural infection of *Calomys laucha* by *Trypanosoma cruzi*. *Medicina (Buenos Aires)* 40:181–86.

Morgan, C. C., and A. Alvarez. 2013. The humerus of South American caviomorph rodents: shape, function and size in a phylogenetic context. *J. Zool.* 290:107–16.

Morgan, C. C. and D. H. Verzi. 2011. Carpal-metacarpal specializations for burrowing in South American octodontoid rodents. *J. Anat.* 219:167–75.

Morris, D. 1962. The behavior of the green acouchi (*Myoprocta pratti*) with special reference to scatter hoarding. *Proc. Zool. Soc. Lond.* 139:701–32.

Morrison, P., K. Kerst, C. Reynafarje, and J. Ramos. 1963. Hematocrit and hemoglobin levels in some peruvian rodents from high and low altitude. *Int. J. Biometeorol.* 7:51–58.

Morrison-Scott, T. C. S. 1937. An apparently new form of Cricetinae from Bristish Guiana. *Ann. Mag. Nat. Hist.*, ser. 10, 20:535–38.

Moussy, V. M. 1860. *Description géographique et statistique de la confédéracion Argentine*. Paris: F. Didot, 2:1–671.

Müller, P. L. S. 1776. *Des Ritters Carl von Linné Königlich Schwedischen Leibarztes u. u. vollständiges Natursystem nach der zwölften lateinischen Ausgabe un nach Anleitung des holländischen Houttuynischen Werks mit einer ausführlichen Erklärung*. Supl. Erste Classe Säugende Thiere. Nürenberg: Gabriel Nicolaus Raspe, 62 pp. + 1 unnumbered, 3 plates.

Müller, P. 1973. *The dispersal centres of terrestrial vertebrates in the Neotropical realm*. The Hague: Dr. W. Junk, i–vi + 1–244 pp.

Müller, P., and I. Vesmanis. 1971. Eine neue subspezies von *Sciurus ingrami* (Rodentia: Sciuridae) der Insel von São Sebastião (Staat São Paulo, Brasilien). *Senckenber. Biol.* 52:377–80.

Munari, D. P., C. Keller, and E. M. Venticinque. 2011. An evaluation of field techniques for monitoring terrestrial mammal populations in Amazonia. *Mamm. Biol.* 76:401–8.

Muñoz-Pedreros, A. 2000. Orden Rodentia. In *Mamíferos de Chile*, ed A. Muñoz-Pedreros and J. Yáñez-Valenzuela, 73–106. Valdivia: CEA Ediciones, viii + 463 pp.

Muñoz-Pedreros, A., and R. Murúa. 1990. Control of small mammals in a pine plantation (Central Chile) by modification of the habitat of predators (*Tyto alba*, Strigidae and *Pseudalopex* sp., Canidae). *Acta Oecol.* 11:251–61.

Muñoz-Pedreros, A., R. Murúa, and L. González. 1990. Nicho ecológico de micromamíferos en un agroecosistema forestal de Chile central. *Rev. Chilena Hist. Nat.* 63:267–77.

Muñoz-Pedreros, A., R. Murúa B., and J. Rodríguez P. 1988. Nuevos registros de *Octodon bridgesi* (Waterhouse, 1844) en la costa de la VII y VIII regiones de Chile (Rodentia: Octodontidae). *Medio Ambiente* 9:96–98.

Muñoz-Pedreros, A., and J. Yáñez. 2000. *Mamíferos de Chile*. Valdivia: Ediciones CEA, viii + 463 pp.

Murie, A. 1932. A new *Oryzomys* from a Pine ridge in British Honduras. *Occas. Papers Mus. Zool., Univ. Michigan* 245:1–3.

Murie, J. O. 1977. Cues used for cache-finding by agoutis (*Dasyprocta punctata*). *J. Mamm.* 58:95–96.

Murray, A. 1866. *The geographical distribution of mammals*. London: Day and Son., 420 pp.

Murúa, R., and L. A. González. 1985. A cycling population of *Akodon olivaceus* (Cricetidae) in a temperate rain forest in Chile. *Acta Zool. Fenn.* 173:77–79.

Murúa, R., L. A. González, and M. Briones. 2005. Cambios en el ensamble de micromamíferos durante la sucesíon secundaria en un bosque costero de Valdivia, Chile. In Historia, Biodiversidad y Ecología de los Bosques Costeros de Chile, ed. C. Smith-Ramirez, J. J. Armesto, and C. Valdovinos, 516–31. Santiago, Chile: Editorial Universitaria, 708 pp.

Murúa, R., L. A. González, and P. L. Meserve. 1986. Population ecology of *Oryzomys longicaudatus philippi*

(Rodentia: Cricetidae) in southern Chile. *J. Anim. Ecol.* 55:281–93.

Murúa, R., P. L. Meserve, L. A. González, and C. Jofré. 1987. The small mammal community of a Chilean temperate rain forest: lack of evidence of competition between dominant species. *J. Mamm.* 68:729–38.

Murúa, R., O. Neumann, and I. Dropelmann. 1981. Food habits of *Myocastor coypus* in Chile. *Proceedings of the Worldwide Furbearer Conference* 1:544–58.

Musser, G. G., and M. D. Carleton. 1993. Family Muridae. In *Mammal species of the world,* 2nd ed., ed. D. E. Wilson and D. M. Reeder, 501–755. Washington, DC: Smithsonian Institution Press, xviii + 1,206 pp.

———. Superfamily Muroidea. In *Mammal species of the world,* 3rd ed., ed. D. E. Wilson and D. M. Reeder, 894–1,531. Baltimore, MD: Johns Hopkins Press, 2:xx + 745–2,142.

Musser, G. G., M. D. Carleton, E. Brothers, and A. L. Gardner. 1998. Systematic studies of Oryzomyine rodents (Muridae, Sigmodontinae): diagnoses and distributions of species formerly assigned to *Oryzomys* "*capito.*" *Bull. Am. Mus. Nat. Hist.* 236:1–376.

Musser, G. G., and A. L. Gardner. 1974. A new species of the ichthyomyine *Daptomys* from Peru. *Am. Mus. Novit.* 2537:1–23.

Musser, G. G., and J. L. Patton. Systematic studies of oryzomyine rodents (Muridae): the identity of *Oecomys phelpsi* Tate. *Am. Mus. Novit.* 2961:1–6.

Musser, G. G., and M. M. Williams. 1985. Systematic studies of Oryzomyine rodents (Muridae): definition of *Oryzomys villosus* and *Oryzomys talamancae*. *Am. Mus. Novit.* 2810:1–22.

Myers, P. 1977. A new phyllotine rodent (genus Graomys) from Paraguay. *Occas. Papers Mus. Zool., Univ. Michigan* 676:1–7.

———. 1982. Origins and affinities of the mammal fauna of Paraguay. In *Mammalian biology in South America,* ed. M. A. Mares and H. H. Genoways, 85–93. The Pymatuning Symposia in Ecology 6, Special Publications Series. Pittsburgh: Pymatuning Laboratory of Ecology, University of Pittsburgh, xii + 539.

———. 1990. A preliminary revision of the *varius* group of *Akodon* (*A. dayi, dolores, molinae, neocenus, simulator, toba,* and *varius*). In *Advances in Neotropical mammalogy,* ed. J. F. Eisenberg and K. H. Redford, 5–54. Gainesville: Sandhill Crane Press. [Dated 1989; published 6 February 1990.]

Myers, P., and M. D. Carleton. 1981. The species of *Oryzomys* (*Oligoryzomys*) in Paraguay and the identity of Azara's "rat sixième ou rat à tarse noir." *Misc. Publ. Mus. Zool., Univ. Michigan* 161:1–41.

Myers, P., and R. L. Hansen. 1980. Rediscovery of the Rufous-Faced Crake (*Laterallus xenopterus*). *Auk* 97:901–2.

Myers, P., B. Lundrigan, and P. K. Tucker. 1995. Molecular phylogenetics of oryzomyine rodents: the genus *Oligoryzomys*. *Mol. Phylogenet. Evol.* 4:372–82.

Myers, P., and J. L. Patton. 1989a. A new species of *Akodon* from the cloud forests of eastern Cochabamba Department, Bolivia (Rodentia: Sigmodontinae). *Occas. Papers Mus. Zool., Univ. Michigan* 720:1–28.

———. 1989b. *Akodon* of Peru and Bolivia—revision of the *fumeus* group (Rodentia: Sigmodontinae). *Occas. Papers Mus. Zool., Univ. Michigan* 721:1–35.

Myers, P., J. L. Patton, and M. F. Smith. 1990. A review of the *boliviensis* group of *Akodon* (Rodentia: Sigmodontinae), with emphasis on Peru and Bolivia. *Misc. Publ. Mus. Zool., Univ. Michigan* 177:iv + 1–104..

Myers, P., A. Taber, and I. Gamarra de Fox. 2002. Mamíferos de Paraguay. In *Diversidad y conservación de los mamíferos neotropicales,* ed. G. Ceballos and J. A. Simonetti, 453–502. México, D.F.: CONABIO-UNAM, 582 pp.

Myers, P., and R. M. Wetzel. 1979. New records of mammals from Paraguay. *J. Mamm.* 60:638–41.

Nabte, M. J., A. Andrade, S. L. Saba, and A. Monjeau. 2009. Mammalia, Rodentia, Sigmodontinae, *Akodon molinae* Contreras, 1968: new locality records and filling gaps. *Check List* 5:320–24.

Nabte, M. J., U. F. J. Pardiñas, and S. L. Saba. 2008. The diet of the Burrowing Owl, *Athene cunicularia*, in the arid lands of northeastern Patagonia, Argentina. *J. Arid Environ.* 72:1526–30.

Nabte, M. J., S. L. Saba, and U. F. J. Pardiñas. 2006. Dieta del Búho magallánico (*Bubo magellanicus*) en el Desierto del Monte y la Patagonia Argentina. *Ornitol. Neotrop.* 17:27–38.

Nachman, M. W. 1992a. Geographic patterns of chromosomal variation in South American marsh rats, *Holochilus brasiliensis* and *H. vulpinus*. *Cytogenet. Cell Genet.* 61:10–16.

———. 1992b. Meiotic studies of Robertsonian polymorphisms in the South American marsh rat, *Holochilus brasiliensis*. *Cytogenet. Cell Genet.* 61:17–24.

Nachman, M. W., and P. Myers. 1989. Exceptional chromosomal mutations in a rodent population are not strongly underdominant. *Proc. Natl. Acad. Sci. USA.* 86:6666–70.

Nadler, C. F., and R. S. Hoffmann. 1970. Chromosomes of some Asian and South American squirrels (Rodentia, Sciuridae). *Experientia* 26:1383–86.

Naiff, R. D., W. Y. Mok, and M. F. Naiff. 1985. Distribution of *Histoplasma capsulatum* in Amazonian wildlife. *Micropathologia* 89:165–68.

Napolitano, C., M. Bennett, W. E. Johnson, S. J. O'Brien, P. A. Marquet, I. Barría, E. Poulin, and A. Iriarte. 2008. Ecological and biogeographical inferences on two sympatric and enigmatic Andean cat species using genetic identification of faecal samples. *Mol. Ecol.* 17:678–90.

Nascimento, F. F., A. Lazar, A. N. Menezes, A. da Matta Durans, J. C. Moreira, J. Salazar-Bravo, P. S. D'Andrea, and C. R. Bonvicino. 2013. The role of historical barriers in the diversification process in open vegetation formations during the Miocene/Pliocene using an ancient rodent lineage as a model. *PLos ONE* 8(4):e61924. doi:10.1371/journal.pone.0061924.

Nascimento, F. F., L. G. Pereira, L. Geise, A. M. R. Bezerra, P. S. D'Andrea, and C. R. Bonvicino. 2011. Colonization process of the Brazilian common vesper mouse, *Calomys expulsus* (Cricetidae, Sigmodontinae): a biogeographic hypothesis. *J. Hered.* 102:260–68.

Nasif, N., G. Esteban, and P. E. Ortiz. 2010. Novedoso hallazgo de egagrópilas en el Mioceno tardío, Formación Andalhuala, provincia de Catamarca, Argentina. *Ser. Corr. Geol.* 25:105–14.

Nava, S., M. Lareschi, A. M. Abba, P. M. Beldomenico, J. M. Venzal, A. J. Mangold, and A. A. Guglielmone. 2006. Larvae and nymphs of *Amblyomma tigrinum* Koch, 1844 and *Amblyomma triste* Koch, 1844 (Acari: Ixodidae) naturally parasitizing Sigmodontinae rodents. *Acarologia* 46:135–41.

Nava, S., M. Lareschi, and D. Voglino. 2003. Interrlationship between ectoparasites and wild rodents from northeastern Buenos Aires Province, Argentina. *Mem. Inst. Oswaldo Cruz* 98:45–49.

Nava, S., P. M. Velazco, and A. A. Guglielmone. 2010. First record of *Amblyomma longirostre* (Koch, 1844) (Acari: Ixodidae) from Peru, with a review of this tick's host relationships. *Syst. Appl. Acarol.* 15:21–30.

Navarro, G., and M. Maldonado. 2002. *Geografía ecologica de Bolivia: vegetacion y ambientes acuaticos.* Cochabamba: Editorial del Centro de Ecologia Simon I. Patiño, 719 pp + 197 photos + 18 maps + 34 drawings + 3 satellite images.

Navarro, M. C. 1991. "Ecología de *Akodon molinae* en el Monte Argentino." Master's thesis, Universidad Nacional de Tucumán, Argentina.

Navone, G. T., M. Lareschi, and J. Notarnicola. 2010. Los roedores sigmodontinos y sus parásitos en la región pampeana. In *Biología y ecología de pequeños roedores en la región pampeana de Argentina*, ed.

J. Polop and M. Busch, 217–61. Córdoba, Argentina: Editorial Universidad Nacional de Córdoba, 332 pp.

Navone, G. T., J. Notarnicola, S. Nava, M. R. Robles, C. Galliari and M. Lareschi. 2009. Arthropods and helminthes assemblage in sigmodontine rodents from wetlands of the Río de La Plata, Argentina. *Mastozool. Neotrop.* 16: 121–33.

Naya, D. E., L. A. Ebensperger, P. Sabat, and F. Bozinovic. 2008. Digestive and metabolic flexibility allows female degus to cope with lactation costs. *Physiol. Biochem. Zool.* 81:186–94.

Nehring, A. 1887. Über eine *Ctenomys*-Art aus Rio Grande do Sul (Süd-Brasilien). *Sitzungsberitche Gesellschaft Naturforschung Freunde zu Berlin*, Jahrgang 1887, 4:45–47.

———. 1900a. Über *Ctenomys Pundti* n. sp. und *Ct. minutus* Nhrg. *Zool. Anzeiger* 23:420–25.

———. 1900b. Über *Ctenomys neglectus* n. sp., *Ct. Nattereri* Wagn. und *Ct. lujanensis* Amegh. *Zool. Anzeiger* 23:535–41.

———. 1900c. Uber die Schadel von *Ctenomys minutus* Nhrg., *Ct.* torquatusLicht. und *Ct. pundti* Nhrg. *Sitzungsberichte Gesellschaft Naturforschung Freunde zu Berlin, Jahrgang* 1900, 9:201–10.

Neira, R., X. García, and R. Scheu. 1989. Análisis descriptivo del comportamiento reproductivo y de crecimiento de chinchillas (*Chinchilla laniger* Gray) en confinamiento. *Av. Prod. Anim. Chile* 14:109–19.

Nelson, E. W. 1899a. Descriptions of three new squirrels from South America. *Bull. Am. Mus. Nat. Hist.* 12:77–80.

———. 1899b. Revision of the squirrels of Mexico and Central America. *Proc. Wash. Acad. Sci.* 1:15–110.

Nelson, K., R. J. Baker, H. S. Shelhammer, and R. K. Chesser. 1984. Test of alternative hypotheses concerning the origin of *Reithrodontomys raviventris*: genetic analysis. *J. Mamm.* 65:668–73.

Nelson, T. W., and K. Shump Jr. 1978. Cranial variation and size allometry in *Agouti paca* from Ecuador. *J. Mamm.* 59:387–94.

Neme, G., G. Moreira, A. Atencio, and L. de Santis. 2002. El registro de microvertebrados del sitio arqueológico Arroyo Malo 3 (Provincia de Mendoza, Argentina). *Rev. Chilena Hist. Nat.* 75:409–21.

Neri-Bastos, F. A., D. M. Barros-Battesti, P. M. Linardi, M. Amaku, A. Marcili, S. E. Favorito, and R. Pinto-da-Rocha. 2004. Ectoparasites of wild rodents from Parque Estadual da Cantareira (Pedra Grande Nuclei), São Paulo, Brazil. *Rev. Bras. Parasitol. Vet.* 13:29–35.

Nes, N. 1963. The chromosomes of *Chinchilla laniger*. *Acta Vet. Scand.* 4:128–35.

Neves, A. C. da S. A. 2008. "Descrição de classes etárias para *Thrichomys inermis* (Pictet, 1843); *Thrichomys laurentius* Thomas, 1904 & *Thrichomys pachyurus* (Wagner, 1845) (Rodentia: Echimyidae) e comparação de suas séries ontogenéticas," Graduate monograph, Universidade Federal do Rio de Janeiro, Brazil.

Neves, A. C. da S. A., and L. M. Pessôa. 2011. Morphological distinction of species of *Thrichomys* (Rodentia: Echimyidae) through ontogeny of cranial and dental characters. *Zootaxa* 2804:15–24.

Neveu-Lemaire, M. and G. Grandidier. 1911. *Notes sur les mammifères des hauts plateaux de l'Amerique du Sud*. Paris: Imperimerie Nationale, viii + 1–127 pp.

Nicéforo-María, H. A. 1923. Guagua calalluna, *Dinomys* sp. n. (?). *Bol. Soc. Colomb. Cien. Nat.* 72:317–20.

Nicola, P. A., L. R. Monteiro, L. M. Pessôa, F. J. Von Zuben, F. J., Rohlf, and S. F. dos Reis. 2003. Congruence of hierarchical localized variation in cranial shape and molecular phylogenetic structure in spiny rats, genus *Trinomys* (Rodentia: Echimyidae). *Biol. J. Linn. Soc.* 80:385–96.

Nimer, E. 1979. *Climatologia do Brasil*. Rio de Janeiro: Superintendência de Recursos Naturais e Meio Ambiente, 421 pp.

Nitikman, L. Z. 1985. Sciurus granatensis. *Mamm. Species* 246:1–8 + 3 figs.

Nitikman, L. Z., and M. A. Mares. 1987. Ecology of small mammals in a gallery forest of Central Brazil. *Ann. Carnegie Mus.* 56:75–95.

Noblecilla, M. C., and V. Pacheco. 2012. Dieta de roedores sigmodontinos (Cricetidae) en los bosque montanos tropicales de Huánuco, Perú. *Rev. Peru. Biol.* 19:317–22.

Norris, D., F. Michalski, and C. A. Peres. 2010. Habitat patch size modulates terrestrial mammal activity patterns in Amazonian forest fragments. *J. Mamm.* 91:551–60.

Notarnicola, J., and G. T. Navone. 2011. *Litomosoides pardinasi* n. sp. (Nematoda, Onchocercidae) from two species of cricetid rodents in northern Patagonia, Argentina. *Parasitol. Res.* 108:187–94.

Novack, A. J., M. B. Main, M. E. Sunquist, and R. F. Labisky. 2005. Foraging ecology of jaguar (*Panthera onca*) and puma (*Puma concolor*) in hunted and non-hunted sites within the Maya Biosphere Reserve, Guatemala. *J. Zool. Lond.* 267:167–78.

Novello, A., and C. Altuna. 2002. Cytogenetics and distribution of two new karyomorphs of the *Ctenomys pearsoni* complex (Rodentia, Octodontidae) from southern Uruguay. *Mamm. Biol.* 67:188–92.

Novello, A. F., and E. P. Lessa. 1986. G-band homology in two karyomoprphs of the *Ctenomys pearsoni* complex (Rodentia, Octodntidae). *Z. Säugetierk.* 55:43–48.

Novillo, A., A. Ojeda, and R. Ojeda. 2009. *Loxodontomys pikumche* (Rodentia, Cricetidae) a new species for Argentina. *Mastozool. Neotrop.* 16:239–42.

Nowak, R. M. 1999. *Walker's mammals of the World*. 6th ed. Baltimore: Johns Hopkins University Press, 2:vii-x + 834–1936 pp.

Nunes, A. 2002. First record of *Neusticomys oyapocki* (Muridae: Sigmodontinae) from the Brazilian Amazon. *Mammalia* 66:445–47.

Nutt, K. J. 2007. Socioecology of rock-dwelling rodents. In *Rodent societies: An ecological and evolutionary perspective*, ed. O. J. Wolff. and P. W. Sherman, pp. 416–26. Chicago and London: University of Chicago Press, xv + 610 pp.

Oaks J. R., J. M. Daul, and G. H. Adler. 2008. Life span of a tropical forest rodent, *Proechimys semispinosus*. *J. Mamm.* 89:904–8.

Ocampo-Garces, A., F. Hernandez, W. Mena, and A. G. Palacios. 2005. Wheel-running and rest activity pattern interaction in two octodontids (*Octodon degus*, *Octodon bridgesi*). *Biol. Res.* 38:299–305.

Ochoa G., J., M. Aguilera, V. Pacheco, and P. J. Soriano. 2001. A new species of *Aepeomys* Thomas 1898 (Rodentia: Muridae) from the Andes of Venezuela. *Mamm. Biol.* 66:228–37.

Ochoa G., J., C. Molina, C., and S. Giner. 1993. Inventario y estudio comunitario de los mamíferos del Parque Nacional Canaima, con una lista de las especies registradas para la Guayana Venezolana. *Acta Cient. Venez.* 44:245–62.

Ochoa G., J., J. Sánchez H., M. Bevilacqua, and R. Rivero. 1988. Inventario de los mamíferos de la Reserva Forestal de Ticoporo y la Serranía de los Pijiguaos, Venezuela. *Acta Cient. Venez.* 32:269–80.

Ochoa G., J., and P. Soriano. 1991. A new species of water rat, genus *Neusticomys* Anthony, from the Andes of Venezuela. *J. Mamm.* 72:97–103.

O'Connell, M. A. 1981. "Population ecology of small mammals from northern Venezuela." Ph.D. diss., Texas Tech University, Lubbock.

———. 1982. Population biology of North and South American grassland rodents: a comparative review. In *Mammalian biology in South America*, ed. M. Mares and H. Genoways, 167–85. The Pymatuning Symposia in Ecology 6, Special Publications Series. Pittsburgh: Pymatuning Laboratory of Ecology, University of Pittsburgh, xii + 539.

———. 1989. Population dynamics of Neotropical small mammals in seasonal habitats. *J. Mamm.* 70:532–48.

Oehlmeyer, A., S. J. Narita, F. A. Alves and J. R. V. Lima. 2010. Ocorrência de *Blarinomys breviceps* (Winge, 1888) (Rodentia, Sigmodontinae) em fragmentos florestais da Mata Atlântica no sudeste do Brasil. *V Congresso Brasileiro de Mastozoologia*; Posters, EC117 (available on CD).

Ojasti, J. 1964. Notas sobre el género *Cavia* (Rodentia: Caviidae) en Venezuela con descripción de una nueva subespecie. *Acta Biologia Venezuel.* 4:146–55.

———. 1972. Revisión preliminary de los picures o aguties de Venezuela (Rodentia, Dasyproctidae). *Mem. Soc. Cien. Nat. La Salle* 32:159–204.

———. 1983. Consumo de fauna por una comunidad indígena en el Estado Bolívar, Venezuela. *Symp. Conserv. Manejo Fauna Silvestr. Neotrop.* 9:45–50.

Ojasti, J., R. Guerrero, and O. E. Hernández. 1992. Mamíferos de la expedición de Tapirapeco, estado Amazonas, Venezuela. *Acta Biol. Venez.* 14:27–40.

Ojeda, A. A. 2010. Phylogeography and genetic variation in the South American rodent *Tympanoctomys barrerae* (Rodentia: Octodontidae). *J. Mamm.* 91:302–13.

Ojeda, A. A., G. D'Elía, and R. A. Ojeda. 2005. Taxonomía alfa de *Chelemys* y *Euneomys* (Rodentia, Cricetidae): el número diploide de ejemplares topotípicos de *C. macronyx* y *E. mordax*. *Mastozool. Neotrop.* 12:79–82.

Ojeda, A. A., M. H. Gallardo, F. Mondaca, and R. A. Ojeda. 2007. Nuevos registros de *Tympanoctomys barrerae* (Rodentia, Octodontidae*). Mastozool. Neotrop.* 14:267–70.

Ojeda, A. A., A. Novillo, R. A. Ojeda, and S. Roig-Juñent. 2013. Geographical distribution and ecological diversification of South American octodontid rodents. *J. Zool.* 289:285–93.

Ojeda, A. A., C. A. Ríos, and M. H. Gallardo. 2004. Chromosomal characterization of *Irenomys tarsalis* (Rodentia, Cricetidae, Sigmodontinae). *Mastozool. Neotrop.* 11:95–98.

Ojeda, R. A. 1985. "A biogeographic analysis of the mammals of Salta Province, Argentina: Patterns of community assemblage in the Neotropics." Ph.D. diss., University of Pittsburgh, Pennsylvania.

———. 1989. Small mammal responses to fire in the Monte Desert, Argentina. *J. Mamm.* 70:416–20.

Ojeda, R. A., C. E. Borghi, G. B. Diaz, S. M. Giannoni, M. A. Mares, and J. K. Braun. 1999. Evolutionary convergence of the highly adapted desert rodent *Tympanoctomys barrerae* (Octodontidae). *J. Arid Environ.* 41:443–52.

Ojeda, R. A., C. E. Borghi, and V. G. Roig. 2002. Mamíferos de Argentina. In *Diversidad y conservación de los mamíferos neotropicales*, ed. G. Ceballos and J. A. Simonetti, 23–63. México, D.F.: CONABIO-UNAM, 582 pp.

Ojeda, R. A., and G. B. Diaz. 1997. La categorización de los mamíferos de Argentina. In *Libro rojo de los mamíferos y aves amenazados de Argentina*, ed. F. J. Garcia, R. A. Ojeda, R. M. Fraga, G. B. Diaz, and R. J. Baigún, 73–163. Buenos Aires: Parques Nacionales.

Ojeda, R. A., J. M. Gonnet, C. E. Borghi, S. M. Giannoni, C. M. Campos, and G. B. Diaz. 1996. Ecological observations of the red vizchacha rat, *Tympanoctomys barrerae* in desert habitats of Argentina. *Mastozool. Neotrop.* 3:183–91.

Ojeda, R. A., and M. A. Mares. 1989. *A biogeographic analysis of the mammals of the Salta Province, Argentina: patterns of species assemblage in the Neotropics.* Special Publications of the Museum 27. Lubbock: Texas Tech University Press, 66 pp.

Ojeda, R. A., M. C. Navarro, C. E. Borghi, and A. M. Scollo. 2001. Nuevos registros de *Salinomys* y *Andalgalomys* (Rodentia, Muridae) para la provincia de La Rioja, Argentina. *Mastozool. Neotrop.* 8:69–71.

Ojeda, R. A., V. G. Roig, E. P. Cristaldo, and C. N. Moyano. 1989. A new record of *Tympanoctomys* (Octodontidae) from Mendoza Province, Argentina. *Texas J. Sci.* 41:333–36.

Ojeda, R. A., and S. Tabeni. 2009. The mammals of the Monte Desert revisited. *J. Arid Environ.* 73:173–81.

Ojeda, R. A., S. Tabeni, and V. Corbalán. 2011. Mammals of the Monte Desert: from regional to local assemblages. *J. Mamm.* 92:1236–44.

Oken, L. 1816. *Lehrbuch der Naturgeschichte.* Dritter Theil. Zoologie. Jena: August Schmid und Comp., 3:xvi+1270 pp., 1 table.

———. 1823. Abbildungen zur Naturgeschichte Brasiliens von M. Dr. von Neuwied. *Isis von Oken* 13:1259–60.

Olalla, A. M. 1935. El genero *Sciurillus* representado en la Amazonia y algunos observaciones sobre el mismo. *Rev. Mus. Paulista. São Paulo* 19:425–30.

Olds, N. 1988. "A revision of the genus *Calomys* (Rodentia: Muridae)." Ph.D. diss., City University of New York.

Olds, N., and S. Anderson. 1987. Notes on Bolivian mammals. 2. Taxonomy and distribution of rice rats of the subgenus *Oligoryzomys*. In *Studies in Neotropical mammalogy, essays in honor of Philip Hershkovitz*, ed. B. D. Patterson and R. M. Timm, 261–81. *Fieldiana Zool.* 39:frontispiece, viii + 1–506.

Olds, N., and S. Anderson. 1990. A diagnosis of the tribe Phyllotini (Rodentia, Muridae). In *Advances in Neotropical mammalogy*, ed. K. H. Redford and J. F. Eisenberg, 55–74. Gainesville, FL: Sandhill Crane

Press, x + 614 pp. [Dated 1989; published 6 February 1990.]

Olds, N., S. Anderson, and T. L. Yates. 1987 . Notes on Bolivian mammals 3: a revised diagnosis of *Andalgalomys* (Rodentia, Muridae) and the description of a new subspecies. *Am. Mus. Novit.* 2890:1–17.

Olfers, I. von. 1818. Bemerkungen zu Illiger's Ueberblick der Säugthiere, nach ihrer Vertheilung über die Welttheile, rücksichtig der Südamericanischen Arten (Species). In *Journal von Brasilien, odor vermischte Nachrichten auch Brasilien, auf wissenschaftlichen Reisen gesammelt,* W. L. von Eschwege, 192–237. In *Neue Bibliothek der wichtigsten Reisebeschreibungen zur Erwiterung der Erd-und Völkerkunde;* in Verbindung mit einigen anderen Gelehrten gesmmelt un herausgegben, ed. F. I. Bertuch. Weimar: Verlage des Landes-Industrie-Comptoirs, 15(2):xii + 304 pp., 6 plates.

Olifiers, N., A. Cunha, C. E. V. Grelle, C. R. Bonvicino, L. Geise, L. G. Pereira, L. G., M. V. Vieira, P. D'Andrea, and R. Cerqueira. 2007. Lista de espécies de pequenos mamíferos não-voadores do Parque Nacional da Serra dos Órgãos. In *Ciência e conservação na Serra dos Órgãos,* ed. C. Cronemberger and E. B. V. de Castro, 183–92. Brasília D.F.: IBAMA Press, 296 pp.

Olivares, A. I., D. H. Verzi, and A. I. Vassalo. 2004. Masticatory morphological diversity and chewing modes in octodontid rodents (Rodentia, Octodontidae). *J. Zool.* 263:167–77.

Olivares, A. I., D. H. Verzi, M. G. Vucetich, and C. I. Montalvo. 2012. Phylogenetic affinities of the late Miocene echimyid 133*Pampamys* and the age of *Thrichomys* (Rodentia, Hystricognathi). *J. Mamm.* 93:76–86.

Oliveira, C. G., R. A Martinez, G. A. F. Giné, D. M. Faria, and F. A. Giaotto. 2011. Genetic assessment of the Atlantic Forest bristle porcupine, *Chaetomys subspinosus* (Rodentia: Erethizontidae), an endemic species threatened with extinction. *Genet. Mol. Res.* 10:923–31.

Oliveira, F. F. de, and A. Langguth. 2004. Pequenos mamíferos (Didelphimorphia e Rodentia) de Paraíba e Pernambuco, Brasil. *Rev. Nordest. Biol.* 18:19–86.

Oliveira, F. S., J. C. Canola, P. T. Oliveira, J. D. Pécora, and A. Capelli. 2006. Anatomoradiographic description of the teeth of pacas bred in captivity (*Agouti paca,* Linnaeus, 1766). *Anat. Histol. Embryol.* 35:316–18.

Oliveira, F. S., L. L. Martins, J. C. Canola, P. T. Oliveira, J. D. Pecora, and A. P. Pauloni. 2012. Macroscopic desription of teeth of Azara's agouti (*Dasyprocta azarae*). *Pesqui. Vet. Bras.* 32:93–95.

Oliveira, J. A. de. 1998. "Morphometric assessments of species groups in the South American rodent genus *Oxymycterus* (Sigmodontinae), with taxonomic notes based on the analyses of type material." Ph.D. diss., Texas Tech University, Lubbock.

Oliveira, J. A. de, and C. R. Bonvicino. 2002. A new species of sigmodontine rodent from the Atlantic Forest of eastern Brazil. *Acta Theriol.* 47:307–22.

Oliveira, J. A. de, and C. R. Bonvicino. 2006. Capítulo 12. Ordem Rodentia, in *Mamíferos do Brasil,* ed. N. R. dos Reis, A. L. Petacchi, W. A. Pedro, and I. Passos de Lima, 347–406. Paraná: Londrina, 437 pp.

Oliveira, J. A. de, P. R. Gonçalves, and C. R. Bonvicino. 2003. Mamíferos da Caatinga. In *Ecologia e conservação da Caatinga,* ed. I. R. Leal, M. Tabarelli, and J. M. C. da Silva, 275–335. Recife: Universidade Federal de Pernambuco, xvi + 804 pp.

Oliveira, J. A. de, and L. M. Pessôa. 2005. Mamíferos. In *Flora Acuña Juncá; Lígia Funch; Washington Rocha. (Org.). Biodiversidade e conservação da Chapada Diamantina.* Brasília: Ministério do Meio Ambiente, 1:377–405.

Oliveira, J. A. de, R. S. Strauss, and S. F. dos Reis. 1998. Assessing relative age and age structure in natural populations of *Bolomys lasiurus* (Rodentia: Sigmodontinae) in northeastern Brazil. *J. Mamm.* 79:1170–83.

Oliveira, J. de C., and D. Pattolli. 1964. Contribution to the knowledge of the trypanosomes of rodents, with description of two new species. *Pap. Avulsos Zool.* (São Paulo). 16:217–27.

Oliveira, P. A., R. B. Souto Lima, and A. G. Chiarello. 2011. Home range, movements and diurnal roots of the endangered thin-spined porcupine, *Chaetomys subspinosus* (Rodentia: Erethizontidae), in the Brazilian Atlantic Forest. *Mamm. Biol.* 76:97–107.

Oliveira, P. P. D., and L. Geise. 2006. Notas biogeográficas de Akodon serrensis (Muroidea, Sigmodontnae). *Livro de Resumos do I Congresso Sul-Americano de Mastozoologia, Porto Alegre, UFRGS Gráfica,* 1:93.

Oliveira, R. C., A. Guterres, C. G. Schrago, J. Fernandes, B. R. Teixeira, S. Zeccer, C. R. Bonvicino, et al. 2012. Detection of the first incidence of *Akodon paranaensis* naturally infected with the Jabora virus strain (Hantavirus) in Brazil. *Mem. Inst. Oswaldo Cruz* 107:424–28.

Oliveira, R. C., P. J. Padula, R. Gomes, V. P. Martinez, C. Bellomo, C. R. Bonvicino, D. E. Lima, et al. 2011. Genetic characterization of hantaviruses associated with sigmodontine rodents in an endemic area for Hantavirus Pulmonary Syndrome in Southern Brazil. *Vector Borne Zoonotic Dis* 11:301–14.

Oliveira, T. G. de, R. G. Gerude, and J. S. Silva Jr. 2007. Unexpected mammalian records in the state of

Maranhão. *Bol. Mus. Paraense Emilio Goeldi, Cien. Nat.* 2:23–32.

Oliveira, T. G., and E. R. L. Mesquita. 1998. Notes on the distribution of the white faced tree rat, *Echimys chrysurus* (Rodentia, Echimyidae) in northeastern Brazil. *Mammalia* 62:305–6.

Oliveira-Filho, A. T., and M. A. L. Fontes. 2000. Patterns of floristic differentiation among Atlantic Forests in southeastern Brazil and the influence of climate. *Biotropica* 32:793–810.

Oliveira-Santos, L. G., P. C. Antunes, C. A Zucco, and F. A. S. Fernandez. 2013. Suitable animal movement indexes or just geometric correlations? A comment on Püttker et al. 2012. *J. Mamm.* 94:948–53.

Oliver, W. L .R., and I. B. Santos. 1991. Threatened endemic mammals of the Atlantic Forest region of south-east Brazil. *Wildl. Preserv. Trust Spec. Sci. Rep.* 4:1–125.

Olmos, F. 1992. Observations on the behaviour and population dynamics of some Brazilian Atlantic Forest rodents. *Mammalia* 55:555–65. [1991 vl. 55, no. 4: pp. 489ff. "Achevé d'imprimer le 20 mars 1992."]

———. 1997. The giant Atlantic Forest tree rat *Nelomys thomasi* (Echimyidae): a Brazilian insular endemic. *Mammalia* 61:130–34.

Olmos, F., M. Galetti, M. Paschoal, and S. L. Mendes. 1993. Habits of the southern bamboo rat, *Kannabateomys amblyonyx* (Rodentia, Echimyidae) in southeastern Brazil. *Mammalia* 57:325–35.

Olrog, C. C. 1950. Mamíferos y aves del archipiélago de Cabo de Hornos. *Acta Zool. Lilloana* 9:505–32.

———. 1976. Sobre mamíferos del noroeste argentino. *Acta Zool. Lilloana* 32:5–14.

———. 1979. Los mamíferos de la selva húmeda, Cerro Calilegua, Jujuy. *Acta Zool. Lilloana* 33:9–14.

Olson, S. L. 2008. Falsified data associated with specimens of birds, mammals, and insects from the Veragua Archipelago, Panama, collected by J.H. Batty. *Am. Mus. Novit.* 3620:1–37.

Opazo, J. C. 2005. A molecular timescale for caviomorph rodents (Mammalia, Hystricognathi). *Mol. Phylogenet. Evol.* 37:932–37.

Opazo, J. C., R. E. Palma, M. Melo, and E. P. Lessa. 2005. Adaptive evolution of the insulin gene in Caviomorph rodents. *Mol. Biol. Evol.* 22:1290–98.

Orcés V., G., and L. Albuja V. 2004. Presencia de *Speothos venaticus* (Carnivora: Canidae) en el Ecuador Occidental y nuevo registro de *Coendou rufescens* (Rodentia: Erethizontidae) en el Ecuador. *Politécnica* 25:11–17.

Orlando, L., J.-F. Mauffrey, J. Cuisin, J. L. Patton, C. Hänni, and F. Catzeflis. 2003. Napoleon Bonaparte and the fate of an Amazonian rat: new data on the taxonomy of *Mesomys hispidus* (Rodentia: Echimyidae). *Mol. Phylogenet. Evol.* 27:113–20.

Orr, R. T. 1938. A new rodent of the genus *Nesoryzomys* from the Galápagos Islands. *Proc. California Acad. Sci.*, Ser. 4, 23:303–6.

Ortells, M. O. 1995. Phylogenetic analysis of G-banded karyotypes among the South American Rodents of the genus *Ctenomys* (Caviomorpha: Octodontidae), with special reference to chromosomal evolution and speciation. *Biol. J. Linn. Soc.* 54:43–70.

Ortells, M. O., and G. E. Barrantes. 1994. A study of genetic distances and variability in several species of the genus *Ctenomys* (Rodentia: Octodontidae) with special reference to a probable role of chromosomes in speciation. *Biol. J. Linn. Soc.* 53:189–208.

Ortells, M. O., J. R. Contreras, and O. A. Reig. 1990. New *Ctenomys* karyotypes (Rodentia, Octodontidae) from North-eastern Argentina and from Paraguay confirm the extreme multiformity of the genus. *Genetica* 82:189–201.

Ortells, M. O., O. A. Reig, N. Brum-Zorrilla, and O. A. Scaglia. 1988. Cytogenetics and karyosystematics of phyllotine rodents (Cricetidae, Sigmodontinae). I. Chromosome multiformity and gonosomal-autosomal translocation in *Reithrodon*. *Genetica* 77:53–63.

Ortells, M. O., O. A. Reig, R. L. Wainberg, G. E. Hurtado de Catalfo, and T. M. L. Gentile de Fronza. 1989. Cytogenetics and karyosystematics of phyllotine rodents (Cricetidae, Sigmodontinae). III. Chromosome multiformity and autosomal polymorphism in Eligmodontia. *Z. Säugetierk.* 54:129–40.

Ortiz, J. C., W. Venegas, J. Sandoval, P. Chandía, and F. Torres-Pérez. 2004. *Hantavirus* en roedores de la Octava región de Chile. *Rev. Chilena Hist. Nat.* 77:251–56.

Ortiz, M. I., E. Pinna-Senn, C. Rosa, and J. A. Lisanti. 2007. Localization of telomeric sequences in the chromosomes of three species of *Calomys* (Rodentia, Sigmodontinae). *Cytologia* 72:165–71.

Ortiz, P. E. 2001. "Roedores del Pleistoceno superior del valle de Tafí (Provincia de Tucumán), implicancias paleoambientales y paleobiogeográficas." Ph.D. diss., Universidad Nacional de Tucumán, Argentina, 230 pp.

Ortiz, P. E., S. Cirignoli, D. H. Podesta, and U. F. J. Pardiñas. 2000. New records of sigmodontine rodents (Mammalia: Muridae) from high-andean localities of northwestern Argentina. *Biogeographica* 76:133–40.

Ortiz, P. E., D. A. García López, M. J. Babot, U. F. J. Pardiñas, P. J. AlonsoMuruaga, and J. P. Jayat. 2012. Exceptional Late Pleistocene nicrovertebrate diversity in northwestern Argentina reveals a marked small mammal turnover. *Palaeogeogr. Palaeoclimatol. Palaeoecol.* 361–62:21–37.

Ortiz, P. E., and J. P. Jayat. 2007a. Fossil record of the Andean rat, *Andinomys edax* (Rodentia: Cricetidae), in Argentina. *Mastozool. Neotrop.* 14:77–83.

———. 2007b. Roedores sigmodontinos (Mammalia: Rodentia: Cricetidae) del límite Pleistoceno-Holoceno en el valle de Tafí, provincia de Tucumán (Argentina): tafonomía y significación paleoambiental. *Ameghiniana* 44:641–60.

———. 2012a. Range extension of *Cavia tschudii* Fitzinger, 1867 (Mammalia: Caviidae) and first record in Catamarca, northwestern Argentina. *Check List* 8:782–83.

———. 2012b. The Quaternary record of *Reithrodon auritus* (Rodentia: Cricetidae) in northwestern Argentina and its paleoenvironmental meaning. *Mammalia* 76:455–60.

Ortiz, P. E., J. P. Jayat, N. L. Nasif, P. Teta, and A. Haber. 2012. Late Holocene rodents of the Puna de Atacama, Chico Tebenquiche archaeological site, Catamarca, Argentina. *Archaeofauna* 21:249–66.

Ortiz, P. E., J. P. Jayat, and U. F. J. Pardiñas. 2011a. Roedores y marsupiales en torno al límite Pleistoceno/Holoceno en Catamarca, Argentina: extinciones y evolución ambiental. *Ameghiniana* 48:336–57.

———. 2011b. Fossil sigmodontine rodents of northwestern Argentina: Taxonomy and paleoenvironmental meaning. In *Cenozoic geology of the central Andes of Argentina*, ed. J. A. Salfity and R. A. Marquillas, 301–15. Salta, Argentina: SCS Publisher, 458 pp.

Ortiz, P. E., J. P. Jayat, and S. J. Steppan. 2012. A new fossil phyllotine (Rodentia, Sigmodontinae) fromt the Late Pliocene in the Andes of northern Argentina. *J. Vert. Paleontol.* 32:1429–41.

Ortiz, P. E., M. C. Madozzo Jaen, and J. P. Jayat. 2012. Micromammals and paleoenvironments: climatic oscillations in the Monte desert of Catamarca (Argentina) during the last two millenia. *J. Arid Environ.* 77:103–9.

Ortiz, P. E., and U. J. F. Pardiñas. 2001. Sigmodontinos (Mammalia: Rodentia) del Pleistoceno tardio del valle de Tafí (Tucumãn, Argentina): taxonomia, tafonomía y reconstruccíon paleoambiental. *Ameghiniana* 38:3–26.

Ortiz, P. E., U. F. J. Pardiñas, S. J. and Steppan. 2000. A new fossil phyllotine (Rodentia: Muridae) from northwestern Argentina and relationships of the *Reithrodon* group. *J. Mamm.* 81:37–51.

Osbahr, K. 2010. Evaluación de la tasa de defecación y del uso de letrinas en la guagua loba (*Dinomys brdanickii* Rodentia: Dinomyidae). *Rev. UDCA Actual. Divulg. Cien.* 13(1):57–66.

Osbahr, K., P. Acevedo, A. Villamizar, and D. Espinosa. 2009. Comparación de la estructura y de la función de los miembros anterior y posterior de *Cuniculus taczanowskii* y *Dinomys branickii*. *Rev. UDCA Actual. Divulg. Cien.* 12(1):37–50.

Osbahr, K., and J. L. Azumendi. 2009. Comparación de la cinemática de los miembros de dos especies de roedores histricognatos (*Cuniculus taczanowskii* y *Dinomys branickii*). *Rev. UDCA Actual. Divulg. Cien.* 12(2):39–50.

Osgood, W. H. 1909. Revision of the mice of the American genus *Peromyscus*. *N. Amer. Fauna* 28:1–285.

———. 1910. Mammals from the coast and islands of northern South America. *Field Mus. Nat. Hist.*, zool. ser., 10:23–32.

———. 1912. Mammals from western Venezuela and eastern Colombia. *Field Mus. Nat. Hist.*, zool. ser., 10:33–66+2 plates.

———. 1913. New Peruvian mammals. *Field Mus. Nat. Hist.*, zool. ser., 10:93–100.

———. 1914a. Four new mammals from Venezuela. *Field Mus. Nat. Hist.*, zool. ser., 10:135–41.

———. 1914b. Mammals of an expedition across northern Peru. *Field Mus. Nat. Hist.*, zool. ser., 10:143–85.

———. 1915. New mammals from Brazil and Peru. *Field Mus. Nat. Hist.*, zool. ser., 10:187–98.

———. 1916. Mammals of the Collins-Day South American expedition. *Field Mus. Nat. Hist.*, zool. ser., 10:199–216+2 plates.

———. 1919. Names of some South American mammals. *J. Mamm.* 1:33–36.

———. 1921. Notes on nomenclature of South American mammals. *J. Mamm.* 2:39–40.

———. 1925. The long-clawed South American rodents of the genus *Notiomys*. *Field Mus. Nat. Hist.*, zool. ser., 12:113–25.

———. 1929. A new rodent from the Galapagos Islands. *Field Mus. Nat. Hist.*, zool. ser., 17:21–24.

———. 1933a. The supposed genera *Aepeomys* and *Inomys*. *J. Mamm.* 14:161.

———. 1933b. The generic position of *Mus pyrrhorhinus* Wied. *J. Mamm.* 14:370–71.

———. 1933c. The South American mice referred to Microryzomys and Thallomyscus. *Field Mus. Nat. Hist.*, zool. ser., 20:1–8.

———. 1933d. Two new rodents from Argentina. *Field Mus. Nat. Hist.*, zool. ser., 20:11–14.

———. 1941. The technical name of the chinchilla. *J. Mamm.* 22:407–11.

———. 1943a. A new genus of rodents from Peru. *J. Mamm.* 24:369–71.

———. 1943b. The mammals of Chile. *Field Mus. Nat. Hist.*, zool. ser., 30:1–268.

———. 1944. Nine new South American rodents. *Field Mus. Nat. Hist.*, zool. ser., 29:191–204.

———. 1945. Two new rodents from Mexico. *J. Mamm.* 26:299–301.

———. 1946. A new octodont rodent from the Paraguayan Chaco. *Fieldiana Zool.* 31:47–49.

———. 1947. Cricetine rodents allied to *Phyllotis*. *J. Mamm.* 28:165–74.

Ostojic, H., V. Cifuentes, and C. Monge. 2002. Hemoglobin affinity in Andean rodents. *Biol. Res.* 35:27–30.

Oyarce, C. E., C. M. Campos, G. B. Diaz, and M. A. Dacar. 2003. Hábitos alimentarios de *Octomys mimax* (Rodentia: Octodontidae), en el parque provincial Ischigualasto (San Juan, Argentina). *XVIII Jornadas SAREM*, La Rioja, Argentina, p. 59.

Pacheco, V. 1991. A new species of *Scolomys* (Muridae: Sigmodontinae) from Peru. *Publicaciones del Museo de Historia Natural, UNMSM (A)* 37:1–3.

———. 2002. Mamíferos del Perú. In *Diversidad y conservación de los mamíferos neotropicales*, ed. G. Ceballos and J. A. Simonetti, 503–49. México, D.F.: CONABIO-UNAM, 584 pp.

———. 2003. "Phylogenetic analyses of the Thomasomyini (Muroidea: Sigmodontinae) based on morphological data." Ph.D. diss., City University, New York.

Pacheco, V., R. Cardenillas, E. Salas, C. Tello, and H. Zeballos. 2009. Diversidad y endemismo de los mamíferos del Perú. *Rev. Peru. Biol.* 16:5–32.

Pacheco, V., J. H. Córdova, and M. Velásquez. 2012. Karyotypes of *Akodon orophilus* Osgood 1913 and *Thomasomys* sp. (Rodentia: Sigmodontinae) from Huánuco, Peru. *Rev. Peru. Biol.* 19:107–10.

Pacheco, V., H. de Macedo, E. Vivar, C. Ascorra, R. Arana-Cardó, and S. Solari. 1995. Lista anotada de los mamíferos peruanos. *Occas. Papers Conserv. Biol.* 2:1–35.

Pacheco, V., G. Márquez, E. Salas, and O. Centty. 2012. Diversidad de mamíferos en la cuenca media del río Tambopata, Puno, Perú. *Rev. Peru. Biol.* 18:231–44.

Pacheco, V., B. D. Patterson, J. L. Patton, L. H. Emmons, S. Solari, and C. F. Ascorra. 1993. List of mammal species known to occur in Manu Biosphere Reserve, Peru. *Publ. Mus. Hist. Nat. Univ. Nac. Mayor San Marcos Ser. A Zool.* 44:1–12.

Pacheco, V., and J. L. Patton. 1995. A new species of the Puna mouse, genus *Punomys* Osgood, 1943 (Muridae, Sigmodontinae) from the southeastern Andes of Perú. *Z. Säugetierk.* 60:85–96.

Pacheco, V., and M. Peralta. 2011. Rediscovery of *Rhipidomys ochrogaster* J. A. Allen, 1901 (Cricetidae:

Sigmodontinae) with a redescription of the species. *Zootaxa* 3106:42–59.

Pacheco, V., E. Salas, L. Cairampoma, M. Noblecilla, H. Quintana, F. Ortiz, P. Palermo, and R. Ledesma. 2007. Diversidad y conservación de los mamíferos en la cuenca del río Apurímac, Perú. *Rev. Peru. Biol.* 14:169–80.

Pacheco, V., and J. Ugarte-Núñez. 2011. New records of Stolzmann's fish-eating rat *Ichthyomys stolzmanni* (Cricetidae, Sigmodontinae) in Peru: a rare species becoming a nuisance. *Mamm. Biol.* 76:657–61.

Pacheco, V., and E. Vivar. 1996. Annotated checklist of the non-flying mammals at Pakitza, Manu Reserve Zone, Manu National Park, Peru. In *Manu, the biodiversity of southeastern Peru*, ed. D. E. Wilson and A. Sandoval, 577–92. Washington, DC: Smithsonian Institution Press, 679 pp.

Padula, P., R. Figueroa, M. Navarrete, E. Pizarro, R. Cadiz, C. Bellomo, C. Jofre, et al. 2004. Transmission study of Andes hantavirus infection in wild sigmodontine rodents. *J. Virol.* 78:11972–79.

Paglia, A. P., G. A. B. da Fonseca, A. B. Rylands, G. Herrmann, L. M. S. Aguiar, A. G. Chiarello, Y. L. R. Leite, et al. 2011. Lista Anotado dos Mamíferos do Brasil, 2ª Edição. *Occas. Papers Conserv. Biol.* 6:1–74.

Paglia, A. P., P. De Marco Junior, F. M. Costa, R. F. Pereira, and G. Lessa. 1995. Heterogeneidade estrutural e diversidade de pequenos mamíferos em um fragmento de mata secundária de Minas Gerais, Brasil. *Rev. Bras. Zool.* 12:67–79.

Paglia, A. P., F. A. Perini, M. G. O. Lopes, and C. F. S. Palmuti. 2005. Novo registro de *Blarinomys breviceps* (Winge, 1888) (Rodentia, Sigmodontinae) no estado de Minas Gerais, Brasil. *Lundiana* 6:155–57.

Pagnozzi, J.M., V. Fagundes, F. H. G. Rodrigues, A. A. Bueno, and Y. Yonenanga-Yassuda. 2000. Citogenética comparativa entre *Clyomys bishopi* e *C. laticeps*. *XXIII Congresso Brasileiro de Zoologia*, Cuiabá, MT.

Paise, G., and E. M. Vieira. 2006. Daily activity of a Neotropical rodent (*Oxymycterus nasutus*): seasonal changes and influence of environmental factors. *J. Mamm.* 87:733–39.

Paiva, M. P. 1973. Distribuição e abundância de alguns mamíferos selvagens no Estado do Ceará. *Ciênc. Cult.* 25:442–50.

Pajot, F. X.; F. le Pont, B. Gentile, and R. Besnard. 1982. Epidemiology of leishmaniasis in French Guiana. *Trans. R. Soc. Trop. Med. Hyg.* 76:112–13.

Palacios, A. G. and T. M. Lee. 2013. Husbandry and breeding in the *Octodon degus* (Molina 1782). *Cold Spring Harb. Protoc.* 4:350–53.

Pallas, P. S. 1766. *Miscellanea zoologica quibus novae imprimis atque obscurae animalium species descriuntur et observationibus iconibusque illustrantur.* Hague Comitum: P. van Cleef, xii + 224 pp., 14 plates.

Palma, R. E., D. Boric-Bargetto, F. Torres-Perez, C. E. Hernández, and T.L. Yates. 2012. Glaciation effects on the phylogeographic structure of *Oligoryzomys longicaudatus* (Rodentia: Sigmodontinae) in the southern Andes. *PLoS ONE* 7(3):e32206.

Palma, R. E., R. A. Cancino, and E. Rodríguez-Serrano. 2010. Molecular systematics of *Abrothrix longipilis* (Rodentia: Cricetidae: Sigmodontinae) in Chile. *J. Mamm.* 91:1102–11.

Palma, R. E., P. A. Marquet, and D. Boric-Bargetto. 2005. Inter and intraspecific phylogeography of small mammals in the Atacama Desert and adjacent areas of northern Chile. *J. Biogeogr.* 32:1931–41.

Palma, R. E., E. Rivera-Milla, J. Salazar-Bravo, F. Torres-Pérez, U. F. J. Pardiñas, P. A. Marquet, A. E. Spotorno, et al. 2005. Phylogeography of *Oligoryzomys longicaudatus* (Rodentia: Sigmodontinae) in temperate South America. *J. Mamm.* 86:191–200.

Palma, R. E., E. Rodríguez-Serrano, E. Rivera-Milla, C. E. Hernandez, J. Salazar-Bravo, M. I. Carma, S. Belmar-Lucero, et al. 2010. Phylogenetic relationships of the pygmy rice rats of the genus *Oligoryzomys* Bangs, 1900 (Rodentia, Sigmodontinae). *Zool. J. Linn. Soc.* 160:551–66.

Palmer, T. S. 1897a. The scientific name of the viscacha. *Science* 6, no. 131:21–22.

———. 1897b. A list of the generic and family names of Rodentia. *Proc. Biol. Soc. Washington* 11:241–270.

———. 1903. Some new generic names of mammals. *Science*, n.s., 17:873.

———. 1904. Index Generum Mammalium: A list of the genera and families of mammals. *N. Amer. Fauna* 23:1–984.

Pantaleão. E. 1978. "Caracterização de espécies do gênero *Cavia* por análise de seus cariótipos." Master's thesis, Universidade Federal do Rio Grande do Sul, Porto Alegre, Brazil.

Parada, A, G. D'Elía, C. J. Bidau, and E. P. Lessa. 2011. Species groups and the evolutionary diversification of tuco-tucos, genus *Ctenomys* (Rodentia: Ctenomyidae). *J. Mamm.* 92:671–82.

Parada, A., A. Ojeda, S. Tabeni, and G. D'Elía. 2012. The population of *Ctenomys* from the Ñacuñán Biosphere Reserve (Mendoza, Argentina) belongs to *Ctenomys mendocinus* Philippi, 1869 (Rodentia: Ctenomyidae): molecular and karyotypic evidence. *Zootaxa* 3402:61–68.

Parada, A., U. F. J. Pardiñas, J. Salazar-Bravo, G. D'Elía, and R.E. Palma. 2013. Dating an impressive Neotropical radiation: molecular time estimates for the Sigmodontinae (Rodentia) provide insights into its historical biogeography. *Mol. Phylogenet. Evol.* 66:960–68.

Pardiñas, U. F. J. 1993. El registro más antiguo (Pleistoceno temprano a medio) de *Akodon azarae* (Fischer, 1829) (Mammalia, Rodentia, Cricetidae) en la provincia de Buenos Aires, Argentina. *Ameghiniana* 30:149–53.

———. 1995a. Novedosos cricétidos (Mammalia, Rodentia) en el Holoceno de la región pampeana, Argentina. *Ameghiniana* 32:197–203.

———. 1995b. Capítulo 11. Los roedores cricétidos. In Evolución climática y biológica de la región pampeana durante los últimos cinco millones de años. Un ensayo de correlación con el Mediterráneo occidental, ed. M. T. Alberdi, G. Leone, and E. P. Tonni, 229–56. Madrid: Museo Nacional de Ciencias Naturales, CSIC, Monografías 14:1–423.

———. 1995c. Sobre las vicisitudes de los generos *Bothriomys* Ameghino 1889, *Euneomys* Coues, 1874 y *Graomys* Thomas, 1916 (Mammalia, Rodentia, Cricetidae). *Ameghiniana* 32:173–80.

———. 1996. El registro fósil de *Bibimys* Massoia, 1979 (Rodentia). Consideraciones sobre los Scapteromyini (Cricetidae, Sigmodontinae) y su distribución durante el Plioceno-Holoceno en la región pampeana. *Mastozool. Neotrop.* 3:15–38.

———. 1997. Un nuevo sigmodontino (Mammalia: Rodentia) del Plioceno de Argentina y consideraciones sobre el registro fósil de los Phyllotini. *Rev. Chilena Hist. Nat.* 70:543–55.

———. 1998. Roedores holocénicos del sitio Cerro Casa de Piedra 5 (Santa Cruz, Argentina): tafonomía y paleoambientes. *Palimpsesto* 5:66–90.

———. 1999a."Los roedores muroideos del Pleistoceno tardío-Holoceno en la Región Pampeana (sector Este) y Patagonia (República Argentina): aspectos taxonómicos, importancia bioestratigráfica y significación paleoambiental." Ph.D. diss., Universidad Nacional de La Plata, Argentina, ix + 283 pp.

———. 1999b. Fossil murids: taxonomy, palaeoecology, and palaeoenvironments. *Quat. South Am. Antarct. Penin.* 13:225–54.

———. 2000. Los sigmodontinos (Mammalia, Rodentia) de la coleccion Ameguino (Museo Argentino de Ciências Naturales "Bernardino Rivadavia"): Revision taxonômica. *Rev. Mus. La Plata (N.S.) Paleontol.* 9, 61:247–54.

———. 2001. Condiciones áridas durante el Holoceno Temprano en el sudoeste de la provincia de Buenos

Aires (Argentina): vertebrados y tafonomía. *Ameghiniana* 38:227–36.

———. 2004. Roedores sigmodontinos (Mammalia: Rodentia: Cricetidae) y otros micromamíferos como indicadores de ambientes hacia el Ensenadense cuspidal en el sudeste de la provincia de Buenos Aires. *Ameghiniana* 41:437–50.

———. 2008. A new genus of oryzomyine rodent (Cricetidae: Sigmodontinae) from the Pleistocene of Argentina. *J. Mamm.* 89:1270–78.

———. 2009. El género *Akodon* (Rodentia: Cricetidae) en Patagonia: estado actual de su conocimiento. *Mastozool. Neotrop.* 16:135–52.

———. 2010. Los roedores sigmodontinos más antiguos de América del Sur: nuevos restos fósiles para un viejo problema. *V Congreso Brasileiro de Mastozoologia* (19 a 23 de setembro de 2010, Sao Pedro, Brasil), a construção da Mastozoologia no Brasil.

———. 2013. Localidades típicas de micromamíferos en Patagonia: el viaje de J. Hatcher en las nacientes del Río Chico, Santa Cruz, Argentina. *Mastozool. Neotrop.* 20:413–420.

Pardiñas, U. F. J., A. M. Abba, and M. L. Merino. 2004. Micromamíferos (Didelphimorphia y Rodentia) del sudoeste de la provincia de Buenos Aires (Argentina): taxonomía y distribución. *Mastozool. Neotrop.* 11:211–31.

Pardiñas, U. F. J., A. L. Cione, J. San Cristobal, D. H. Verzi, and E. P. Tonni. 2004. A new last Interglacial continental vertebrate assemblage in central-eastern Argentina. *Curr. Res. Pleistocene* 21:111–12.

Pardiñas, U. F. J., S. Cirignoli, and C. A. Galliari. 2004. Distribution of *Pseudoryzomys simplex* (Rodentia: Cricetidae) in Argentina. *Mastozool. Neotrop.* 11:105–8.

Pardiñas, U. F. J., S. Cirignoli, J. Laborde, and A. Richieri. 2004. Nuevos datos sobre la distribución de *Irenomys tarsalis* (Philippi, 1900) (Rodentia, Sigmodontinae) en Argentina. *Mastozool. Neotrop.* 11:99–104.

Pardiñas, U. F. J., G. D'Elía, and S. Cirignoli. 2003. The genus *Akodon* (Muroidea: Sigmodontinae) in Misiones, Argentina. *Mamm. Biol.* 68:129–43.

Pardiñas, U. F. J., G. D'Elía, S. Cirignoli, and P. Suarez. 2005. A new species of *Akodon* (Rodentia, Cricetidae) from the northern Campos grasslands of Argentina. *J. Mamm.* 86:462–74.

Pardiñas, U. F. J., G. D'Elía, and P. E. Ortiz. 2002. Sigmodontinos fosiles (Rodentia: Muroidea, Sigmodontinae) de America del Sur: estado actual de su conocimiento y prospectiva. *Mastozool. Neotrop.* 9:209–52.

Pardiñas, U. F. J., G. D'Elía, and P. Teta. 2009. Una introducción a los mayores sigmodontinos vivientes: revisión de *Kunsia* Hershkovitz, 1966 y descripción de un nuevo género Rodentia: Cricetidae). *Arq. Mus. Nac., Rio de Janeiro* 66:509–94. [Number 3–4 of the *Arquivos* in which this paper appeared is dated "jul./ des.2008" but did not appear until 2009.]

Pardiñas, U. F. J., G. D'Elía, P. Teta, P. E. Ortiz, J. P. Jayat, and S. Cirignoli. 2006. Akodontini Vorontsov, 1959 (sensu D'Elía, 2003). In *Mamíferos de Argentina: sistemática y distribución*, ed. R. M. Barquez, M. M. Díaz, and R. A. Ojeda, 146–66. Mendoza, Argentina: Sociedad Argentina para el Estudio de los Mamíferos, 359 pp.

Pardiñas, U. F. J., G. D'Elía, P. Teta, and B. D. Patterson. 2008. *Kunsia fronto*. IUCN Red List of Threatened Species, version 2011.1, accessed January 1, 2011, http://www.iucnredlist.org.

Pardiñas, U. F. J., and C. Deschamps. 1996. Sigmodontinos (Mammalia, Rodentia) pleistocénicos del suboeste de la provincial de Buenos Aires (Argentina): aspectos sistemáticos, paleozoogeográficos y paleoambientles. *Estud. Geol.* 52:367–79.

Pardiñas, U. F. J., and C. A. Galliari. 1998a. La distribución del ratón topo *Notiomys edwardsii* (Mammalia: Muridae). *Neotrópica* 44:123–24.

———. 1998b. Comentario sobre el trabajo "Los mamíferos del Parque Biológico Sierra de San Javier, Tucumán, Argentina: observaciones sobre su sistemática y distribución," Capllonch et al., 1997 (*Mastozool. Neotrop.*, 4:49–71). *Mastozool. Neotrop.* 5:61–62.

———. 1998c. Sigmodontinos (Rodentia, Muridae) del Holoceno inferior de Bolivia. *Rev. Esp. Paleontol.* 13:17–25.

———. 1999. La presencia de *Akodon iniscatus* (Mammalia: Rodentia) en la provincia de Buenos Aires (Argentina). *Neotrópica* 45:115–17.

———. 2001. Reithrodon auritus. *Mamm. Species* 664:1–8.

Pardiñas, U. F. J., J. N. Gelfo, J. San Cristobal, A. L. Cione, and E. P. Tonni. 1996. Una asociación de organismos marinos y continentals en el Pleistoceno superior en el sur de la provincial de Buenos Aires, Argentina. *Actas XIII Congreso-Geológico Argentino y III Congreso de Exploración de Hidrocarburos* 5:95–111.

Pardiñas, U. F. J., G. Lessa, P. Teta, J. Salazar-Bravo, and E. M. V. C. Câmara. 2014. A new genus of sigmodontine rodent from eastern Brazil and the origin of the tribe Phyllotini. *J. Mamm.* 95:201–15.

Pardiñas, U. F. J., and M. J. Lezcano. 1995. Cricétidos (Mammalia: Rodentia) del Pleistoceno tardio del nordeste de la provincia de Buenos Aires (Argentina). Aspectos sistemáticos y paleoambientales. *Ameghiniana* 32:249–65.

Pardiñas, U. F. J., and E. Massoia. 1989. Roedores y mar-supialesde Cerro Castillo, Paso Flores, Departamento Pilcaniyeu,provincia de Río Negro. *Bol. Cien. Asoc. Protec. Nat.* 13:9–13.

Pardiñas, U. F. J., G. J. Moreira, C. M. Garcia-Esponda, and L. de Santis. 2000. Deterioro ambiental y micro-mamíferos durante el Holoceno en el nordeste de la estepa patagónica (Argentina). *Rev. Chilena Hist. Nat.* 73:9–21.

Pardiñas, U. F. J., and P. E. Ortiz. 2001. *Neotomys ebrio-sus*, an enigmatic South American rodent (Muridae, Sigmodontinae): its fossil record and present distribu-tion in Argentina. *Mammalia* 65:244–50.

Pardiñas, U. F. J., and P. Ramírez-Llorens. 2005. The ge-nus *Oecomys* (Rodentia, Sigmodontinae) in Argentina. *Mammalia* 69:103–7.

Pardiñas, U. F. J. and P. Teta. 2005. Roedores sigmo-dontinos del Chaco Húmedo de Formosa: aspectos taxonómicos y distribución geográfica. In *Historia natural y paisaje de la Reserva El Bagual, provincia de Formosa, Argentina. Inventario de la fauna de verteb-rados y flora vascular de un área protegida del Chaco Húmedo*, ed. A. G. Di Giacomo and S. F. Krapovickas, 501–17. Temas de Naturaleza y Conservación 4. Bue-nos Aires: Aves Argentinas/Asociación Ornitológica del Plata, 578 pp.

———. 2011a. On the taxonomic status of the Brazilian mouse *Calomys anoblepas* Winge, 1887 (Mammalia, Rodentia, Cricetidae). *Zootaxa* 2788:38–44.

———. 2011b. Fossil history of the marsh rats of the genus *Holochilus* and *Lundomys* (Cricetidae, Sigmo-dontinae) in southern South America. *Estud. Geol.* 61:111–29.

———. 2013a. Taxonomic status of *Mus talpinus* Lund (Rodentia: Sigmodontinae) from the Quaternary deposits of Lagoa Santa, Minas Gerais, Brdazil and its paleoenvironmental meaning. *Mammalia* 77:347–55.

———. 2013b. Holocene stability and recent dramatic changes in micromammalian communities of north-western Patagonia. *Quarternary Int.* 305:127–40.

Pardiñas, U. F. J., P. Teta, J. C. Chebez, J. C., F. Martínez, S. Ocampo, and D. Navas. 2010. Mammalia, Roden-tia, Sigmodontinae, *Euneomys chinchilloides* (Water-house, 1837): range extension. *Check List* 6:167–69.

Pardiñas, U. F. J., P. Teta, S. Cirignoli, and D. H. Podestá. 2003. Micromamíferos (Didelphimorphia y Rodentia) de Norpatagonia Extra Andina, Argentina: taxonomía alfa y biogeografia. *Mastozool. Neotrop.* 10:69–113.

Pardiñas, U. F. J., P. Teta and G. D'Elía. 2009. Taxonomy and distribution of *Abrawayaomys* (Rodentia: Criceti-dae), an Atlantic Forest endemic with the description of a new species. *Zootaxa* 2128: 39–60.

———. 2010. Roedores sigmodontinos de la región pam-peana: historia evolutiva, sistemática y taxonomía. In *Biología y ecología de pequeños roedores en la región pampeana de Argentina. Enfoques y Perspectivas*, ed. J. Polop and M. Busch, 9–36. Córdoba, Argentina: Editorial Universidad Nacional de Córdoba.

Pardiñas, U. F. J., P. Teta, G. D'Elía, S. Cirignoli, and P. E. Ortiz. 2007. Resolution of some problematic type localities for sigmodontine rodents (Cricetidae, Sig-modontinae). In *The quintessential naturalist: Honor-ing the life and legacy of Oliver P. Pearson*, ed. D. A. Kelt, E. P. Lessa, J. Salazar-Bravo, and J. L. Patton, 391–416. *Univ. California Publ. Zool.* 134:v-xii + 1–981.

Pardiñas, U. F. J., P. Teta, G. D'Elía, and G. B. Diaz. 2011. Taxonomic status of *Akodon oenos* (Rodentia, Sigmodontinae), an obscure species from west central Argentina. *Zootaxa* 2749:47–61.

Pardiñas, U. F. J., P. Teta, G. D'Elía, and C. A. Galliari. 2008. Rediscovery of *Juliomys pictipes* (Rodentia: Cri-cetidae) in Argentina: emended diagnosis, geographic distribution, and insights on genetic structure. *Zootaxa* 1758: 29–44.

Pardiñas, U. F. J., P. Teta, G. D'Elía, and E. P. Lessa. 2011. The evolutionary history of sigmodontine rodents in Patagonia and Tierra del Fuego. *Biol. J. Linn. Soc.* 103:495–513.

Pardiñas, U. F. J., P. Teta, A. Formoso, and R. Barberena. 2011. Roedores del extremo austral: tafonomía, diversidad y evolución ambiental duranteel Holoceno tardio. In *Bosques, montañas y cazadores: investiga-ciones arqueológicas en Patagonia Meridional*, ed. L. A. Borrero and K. Borrazzo, 61–84. Buenos Aires: CONICET-IMHICIHU, Editorial Dunken, 239 pp.

Pardiñas, U. F. J., P. Teta, and S. Heinonen Fortabat. 2005. Diet of Barn Owl (*Tyto Alba*) in the largest sub-tropical wetlands of Argentina and eastern Paraguay. *J. Raptor Res.* 39:65–69.

Pardiñas, U. F. J., P. Teta, and D. Udrizar Sauthier. 2008. Mammalia, Didelphimorpia and Rodentia, southwest of the province of Mendoza, Argentina. *Check List* 4:218–25.

Pardiñas, U. F. J., P. Teta, D. Voglino, and F. J. Fernández. 2013. Enlarging rodent diversity in west-central Argen-tina: a new species of the genus *Holochilus*. *J. Mamm.* 94:231–40.

Pardiñas, U. F. J., and E. Tonni. 1998. Procedencia estratigráfica y edad de los más antiguos muroideos (Mammalia, Rodentia) de América del Sur. *Ameghini-ana* 35:473–75.

———. 2000. A giant vampire (Mammalia, Chiroptera) in the Late Holocene from the Argentinean pampas:

paleoenvironmental significance. *Palaeogeogr. Palaeo-climatol. Palaeoecol.* 160:213–21.

Pardiñas, U. F. J., D. E. Udrizar Sauthier, and P. Teta. 2012. Micromammal diversity loss in central-eastern Patagonia over the last 400 years. *J. Arid Environ.* 85:71–75.

Pardiñas U. F. J., D. E. Udrizar Sauthier, P. Teta, and G. D'Elía. 2008. New data on the endemic Patagonian long-clawed Mouse *Notiomys edwardsii* (Rodentia: Cricetidae). *Mammalia* 72:273–85.

Pardini, R. 2004. Effects of forest fragmentation on small mammals in an Atlantic Forest landscape. *Biodiver. Conserv.* 13:2567–86.

Pardini, R., M. de Sousa S, R. Braga-Neto, and J. P. Metzger. 2005. The role of forest structure, fragment size and corridors in maintaining small mammal abundance and diversity in an Atlantic Forest landscape. *Biol. Conserv.* 124:253–66.

Pardini R., and F. Umetsu. 2006. Pequenos mamíferos não-voadores da Reserva Florestal do Morro Grande: distribuição das espécies e da diversidade em uma área de Mata Atlântica. *Biota Neotrop.* 6:1–22.

Pardo, R. 2007. Las culturas fueguinas: Kawéskar, Yagán y Selk-nam en relación con el medio ambiente. *Criíta. cl,* September 9, 2007, http://critica.cl/ciencias-sociales /y-las-culturas-fueguinas-kaweskar-yagan-y-selk-nam -en-relacion-con-el-medio-ambiente.

Paresque, R., A. U. Christoff, and V. Fagundes. 2009. Karyology of the Atlantic Forest rodent *Juliomys* (Cricetidae): a new karyotype from southern Brazil. *Genet. Mol. Biol.* 32:301–5.

Paresque, R., W. P. da Souza, S. L. Mendes, and V. Fagundes. 2004. Composicão cariotípica da fauna de roedores e marsupiais de duas áreas de Mata Atlântica do Espírito Santo state, Brazil. *Bol. Mus. Biol. Mello Leitão, nova serie* 17:5–33.

Parreira, G. G., and F. M. Cardoso. 1993. Seasonal variation of the spermatogenic activity in *Bolomys lasiurus* (Lund, 1841) (Rodentia, Cricetidae) from Southeastern Brazil. *Mammalia* 57:27–34.

Parsons, F. G. 1894. On the morphology of the sciuromorphine and hysrticomorphine rodents. *Proc. Zool. Soc. Lond.* 1894 (part II):251–96.

Paschoal, A. M. O., F. L. Santiago, M. V. Castilho, M. L. L. Perilli, E. V. C. Câmara, and L. C. Oliveira. 2004. Novos registros de *Trinomys moojeni* (Rodentia: Echimyidae) no Parque Nacional da Serra do Cipó, MG. *Resumos do 25° Congresso Brasileiro de Zoologia,* 8–13 Fevereiro de 2004. Brasília, pp. 1080.

Paschoal, M. G., and M. Galetti. 1995. Seasonal food use by the Neotropical squirrel *Sciurus ingrami* in southeastern Brazil. *Biotropica* 27:268–73.

Pascual, R. 2006. Evolution and geography: the biogeographic history of South American land mammals. *Ann. Missouri Bot. Garden* 93:209–30.

Pascual, R., J. Pisano, and E. Ortega Hinojosa. 1965. Un nuevo Octodontidae (Rodentia, Caviomorpha) de la formación Epecuen (Plioceno medio) de Hidalgo (Prov. de la Pampa). *Ameghiniana* 4:19–30.

Passamani, M. 2010. Use of space and activity pattern of *Sphiggurus villosus* (F. Cuvier, 1823) from Brazil (Rodentia, Erethizontidae). *Mamm. Biol.* 75:455–58.

Passamani, M., R. A. S. Cerboncini, and J. E. de Oliveira. 2011. Distribution extension of *Phanenomys ferrugineus* (Thomas, 1894), and new data on *Abrawayaomys ruschii* Cunha and Cruz, 1979 and *Rhagomys rufescens* (Thomas, 1886), three rare species of rodents (Rodentia: Cricetidae) in Minas Gerais, Brazil. *Check List* 7:827–31.

Passamani, M., S. L. Mendes, A. G. Chiarello. 2000. Non volant mammals of the Estação Biológica de Santa Lúcia and adjacent areas of Santa Teresa, Espírito Santo, Brazil. *Bol. Mus. Biol. Mello Leitão* 11/12:201–14.

Pathak, S., T. C. Hsu, L. Shirley, and J. D. Helm, III. 1973. Chromosome homology in the climbing rats, genus *Tylomys* (Rodentia: Cricetidae). *Chromosoma* 42:215–28.

Patterson B., and R. Pascual. 1968a. New echimyid rodents from the Oligocene of Patagonia, and a synopsis of the family. *Breviora* 301:1–14.

———. 1968b. The fossil mammal fauna of South America. *Q. Rev. Biol.* 43:409–51.

———. 1972. The fossil mammal fauna of South America. In *Evolution, mammals, and southern continents,* ed. A. Keast, F. C. Erk, and B. Glass, 247–309. Albany: State University of New York Press, 543.

Patterson, B., and A. E. Wood. 1982. Rodents from the Deaeadan Oligocene of Bolivia and the relationships of the Caviomorpha. *Bull. Mus. Comp. Zool.* 149:371–543.

Patterson, B. D. 1992a. A new genus and species of long-clawed mouse (Rodentia: Muridae) from temperate rainforests of Chile. *Zool. J. Linn. Soc.* 106:127–45.

———. 1992b. Mammals in the Royal Natural History Museum, Stockholm, collected in Brazil and Bolivia by A. M. Olalla during 1934–1938. *Fieldiana Zool.,* n.s., 66:iii + 1–42.

———. 1999. Contingency and deteterminism in mammalian biogeography: the role of history. *J. Mamm.* 80:345–34.

———. 2000. Patterns and trends in the discovery of new Neotropical mammals. *Diversity Distrib.* 6:145–51.

———. 2001. Fathoming tropical biodiversity: the continuing discovery of Neotropical Mammals. *Diversity Distrib.* 7:191–96.

Patterson, B. D., M. H. Gallardo, and K. E. Freas. 1984. Systematics of mice of the subgenus *Akodon* (Rodentia: Cricetidae) in southern South America, with the description of a new species. *Fieldiana Zool.,* n.s., 23:1–16.

Patterson, B. D., P. L. Meserve, and B.K. Lang. 1989. Distribution and abundance of small mammals along an elevational transect in temperate rainforests of Chile. *J. Mamm.* 70:67–78.

———. 1990. Quantitative habitat associations of small mammals along an elevational transect in temperate rain-forests of Chile. *J. Mamm.* 71:620–33.

Patterson, B. D., D. F. Stotz, and S. Solari. 2006. Mammals and birds of the Manu Biosphere Reserve, Peru. *Fieldiana Zool.,* n.s., 110: 1–49.

Patterson, B. D., and P. M. Velazco. 2006. A distinctive new cloud-forest rodent (Hystricognathi: Echimyidae) from the Manu Biosphere Reserve, Peru. *Mastozool. Neotrop.* 13:175–91.

———. 2008. Phylogeny of the rodent genus *Isothrix* (Hystricognathi, Echimyidae) and its diversification in Amazonia and the eastern Andes. *J. Mamm. Evol.* 15:181–201.

Patton, J. L. 1984. Systematic status of the large squirrels (subgenus *Urosciurus*) of the western Amazon Basin. *Stud. Neotrop. Fauna Environ.* 19:53–72.

———. 1986 [1987]. Patrones de distribucion y especiacion de fauna de mamiferos de los bosques nublados andinos del Peru. *An. Mus. Hist. Nat. Valparaiso* 17:87–94.

———. 1987. Species groups of spiny rats genus *Proechimys* (Rodentia: Echimyidae). In *Studies in Neotropical mammalogy, essays in honor of Philip Hershkovitz,* ed. B. D. Patterson and R. M. Timm, 305–45. *Fieldiana Zool.* 39:frontispiece, viii + 1–506.

———. 1993. Family Heteromyidae. In *Mammal species of the world: A taxonomic and geographic reference,* 2nd ed., ed. D. E. Wilson and D. M. Reeder, 477–86. Washington, DC: Smithsonian Institution Press, xviii + 1,206 pp.

———. 2005a. Family Heteromyidae. In *Mammal species of the world,* 3rd ed., D. E. Wilson and D. M. Reeder, 844–58. Baltimore: Johns Hopkins Press.

———. 2005b. Family Geomyidae. In *Mammal species of the world,* 3rd ed., ed. D. E. Wilson and D. M. Reeder, 859–870. Baltimore: Johns Hopkins Press, 2:xvii + 745–2142.

———. 2005c. Species and speciation: changes in a paradigm through the career of a rat trapper. In *Going afield: Lifetime experiences in exploration, science, and the biology of mammals,* ed. C. J. Phillips and C. Jones, 263–76. Special Publications of the Museum 47. Lubbock: Texas Tech University Press, iii + 289 pp.

Patton, J. L., B. Berlin, and E. A. Berlin. 1982. Aboriginal perspectives of a mammal community in Amazonian Peru: knowledge and utilization patterns among the Aguaruna Jívaro. In *Mammalian biology in South America,* ed. M. A. Mares and H. H. Genoways, 111–28. Pymatuning Symposia in Ecology 6, Special Publications Series. Pittsburgh: Pymatuning Laboratory of Ecology, University of Pittsburgh, xii + 539.

Patton, J. L., and L. H. Emmons. 1985. A review of the genus *Isothrix* (Rodentia, Echimyidae). *Am. Mus. Novit.* 2817:1–14.

Patton, J. L., and A. L. Gardner. 1972. Notes on the systematics of *Proechimys* (Rodentia: Echimyidae), with emphasis on Peruvian forms. *Occas. Papers Mus. Zool. Louisiana State Univ.* 44:1–30.

Patton, J. L., and M. S. Hafner. 1983. Biosystematics of the native rodents of the Galapagos Archipelago, Ecuador. In *Patterns of Evolution in Galapagos Organisms,* ed. R. I. Bowman, M. Benson, and E. Leviton, 539–68. San Francisco: American Association for the Advancement of Science, Pacific Division, 568 pp.

Patton, J. L., P. Myers, and M. F. Smith. 1989. Electromorphic variation in selected South American akodontine rodents (Muridae: Sigmodontinae), with comments on systematic implications. *Z. Säugetierk.* 54:347–59.

———. 1990. Vicariant versus gradient model of diversification: the small mammal fauna of eastern Andean slopes of Peru. In *Vertebrates in the Tropics,* ed. G. Peters and R. Hutterer, 355–71. Bonn: Alexander Koenig Zoological Research Institute, 424 pp.

Patton, J. L., and O. A. Reig. 1990. Genetic differentiation among echimyid rodents, with emphasis on spiny rats, genus *Proechimys.* In *Advances in Neotropical mammalogy,* ed. J.F. Eisenberg and K.H. Redford, 75–96. Gainesville: The Sandhill Crane Press. [Dated 1989; published 6 February 1990.]

Patton, J. L., and M. A. Rogers. 1983. Systematic implications on non-geographic variation in the spiny rat genus *Proechimys* (Echimyidae). *Z. Säugetierk.* 48:363–70.

Patton, J. L., and M. N. F. da Silva. 1995. A review of the spiny mouse genus *Scolomys* (Rodentia: Muridae: Sigmodontinae) with the description of a new species from the western Amazon of Brazil. *Proc. Biol. Soc. Washington* 108:319–37.

Patton, J. L., M. N. F. da Silva, and J. R. Malcolm. 1994. Gene geneology and differentiation among arboreal spiny rats (Rodentia: Echimyidae) of the Amazon

Basin: a test of the riverine barrier hypothesis. *Evolution* 48:1314–23.

———. 1996. Hierarchical genetic structure and gene flow in three sympatric species of Amazonian rodents. *Mol. Ecol.* 5:229–38.

———. 2000. Mammals of the Rio Juruá and the evolutionary and ecological diversification of Amazonia. *Bull. Am. Mus. Nat. Hist.* 244:1–306.

Patton, J. L., and M. F. Smith. 1992a. mtDNA plhylogeny of Andean mice: a test of diversification across ecological gradients. *Evolution* 46:174–83.

———. 1992b. Evolution and systematics of akodontine rodents (Muridae: Sigmodontinae) of Peru, with emphasis on the genus *Akodon*. In *Biogeografía, ecología y conservación del bosque montano en el Perú*, ed. K. R. Young, and N. Valencia, 83–103. *Mem. Mus. Hist. Nat. Lima Univ. Nac. Mayor San Marcos* 21:1–227.

———. 1994. Paraphyly, polyphyly, and the nature of species boundaries in pocket gophers (genus *Thomomys*). *Syst. Biol.* 43:11–26.

Paula, A. C. 1983. "Relações espaciais de pequenos mamíferos em uma comunidade de mata de galeria do Parque Nacional de Brasília." Master's thesis, Universidade de Brasília, Distrito Federal, Brazil.

Paula Couto, C. 1950. *Peter Wilhelm Lund, Memórias sôbre a paleontologia Brasileira*. Rio de Janeiro, Brazil: Instituto Nacional do Livro, 1–589 + 1 [unnumbered] + 56 plates.

Paulsen, R. M. M., and J. G. W. Brum. 1999. Nematodes (Nematode) parasites of nutria (*Myocastor coypus*) in a farming exploitation area, county of Rio Grande, RS, Brazil. *Arq. Inst. Biol., São Paulo* 66:15–20.

Paulsen, R. M. M., G. Muller, J. G. W. Brum, and A. L. Sinkoc. 1999. *Hydrochoeristrema massaroi* n. sp. (Trematoda: Paramphistomidae: Cladorchilini) parasite of nutria (*Myocastor coypus* Molina, 1782)(Rodentia, Capromyidae). *Arq. Inst. Biol., São Paulo* 66:135–37.

Pautasso, A. A. 2008. Mamíferos de la provincial de Santa Fe, Argentina. *Commun. Mus. Prov. Cien. Nat. Florentino Ameghino* 13:1–248.

Pavan, S. E., and Y. L. R. Leite. 2011. Morphological diagnosis and geographic distribution of Atlantic Forest red-rumped mice of the genus *Juliomys* (Rodentia: Sigmodontinae). *Zoologia* 28:663–72.

Paynter, R. A., Jr. 1982. *Ornithological gazetteer of Venezuela*. Cambridge: Harvard College Press, iv + 245 pp, 1 map.

———. 1988. *Ornithological gazetteer of Chile*. Cambridge: Harvard College Press, v + 329 pp, 1 map.

———. 1989. *Ornithological gazetteer of Paraguay*. 2nd ed. Cambridge: Harvard College Press, iv + 59 pp, 1 map.

———. 1993. *Ornithological gazetteer of Ecuador*. 2nd ed. Cambridge: Harvard College Press, xii + 247 pp, 1 map.

———. 1994. *Ornithological gazetteer of Uruguay*. 2nd ed. Cambridge: Harvard College Press, vi + 111 pp, 1 map.

———. 1995. *Ornithological gazetteer of Argentina*. 2nd ed. Cambridge: Harvard College Press, ix + 1043 pp, 1 map.

Paynter, R. A., Jr. 1997. *Ornithological gazetteer of Colombia*. 2nd ed. Cambridge: Harvard College Press, ix + 537 pp, 1 map.

Paynter, R. A., Jr., and M. A. Traylor Jr. 1991. *Ornithological gazetteer of Brazil*. Cambridge: Harvard College Press, 1:viii + 1–352; 2:353–788, 1 map.

Pearson, O. P. 1939. Three new small mammals from eastern Panama. *Not. Nat. Acad. Nat. Sci. Philadelphia* 6:1–5.

———. 1948. Life history of mountain viscachas in Peru. *J. Mamm.* 29:345–74.

———. 1949. Reproduction of a South American rodent, the mountain viscacha. *Am. J. Anat.* 84:143–74.

———. 1951. Mammals in the highlands of southern Peru. *Bull. Mus. Comp. Zool.* 106(3):117–74.

———. 1957. Additions to the mammalian fauna of Peru and notes on some other Peruvian mammals. *Brevoria* 73:1–7.

———. 1958. A taxonomic revision of the rodent genus *Phyllotis*. *Univ. Calif. California Zool.* 56:391–496, plates 6–13.

———. 1960 [1959]. Biology of the subterranean rodents, *Ctenomys*, in Peru. *Mem. Mus. Hist. Nat. Javier Prado* 9:1–56. [Dated 1959, but issued April, 1960.]

———. 1967. La estructura por edades y la dinámica reproductiva de una población del roedor de campo *Akodon azarae*. *Physis* 27:53–58.

———. 1972. New information on ranges and relationships within rodent genus *Phylotis* in Peru and Ecuador. *J. Mamm.* 53:677–86.

———. 1976. An outbreak of mice in the coastal desert of Peru. *Mammalia* 39:375–86. [1975 vol. 39, no. 3: pp. 343–522. "Achevé d'imprimer le 31 mars 1976."]

———. 1982. Distribución de pequeños mamíferos en el Altiplano y los desiertos del Perú. In *Zoología Neotropical. Actas del VIII Congreso Latinoamericano de Zoología*, ed. P. Salinas, 263–84. Mérida, Venezuela.

———. 1983. Characteristics of a mammalian fauna from forests in Patagonia, southern Argentina. *J. Mamm.* 64:476–92.

———. 1984. Taxonomy and natural history of some fossorial rodents of Patagonia, southern Argentina. *J. Zool.* 202:225–37.

———. 1987. Mice and the postglacial history of the Traful Valley of Argentina. *J. Mamm.* 68:469–78.

———. 1984. Taxonomy and natural history of some fossorial rodents of Patagonia, southern Argentina. *J. Zool. Lond.* 202:225–37.

———. 1988. Biology and feeding dynamics of a South American herbivorous rodent, *Reithrodon*. *Stud. Neotrop. Fauna Environ.* 23:25–39.

———. 1992. Reproduction in a South American mouse, *Abrothrix longipilis*. *Anat. Rec.* 234:73–88.

———. 1995. Annotated keys for identifying small mammals living in or near Nahuel Huapi National Park or Lanin National Park, southern Argentina. *Mastozool. Neotrop.* 2:99–148.

———. 2002. A perplexing outbreak of mice in Patagonia, Argentina. *Stud. Neotrop. Fauna Environ.* 3 7:187–200.

Pearson, O. P., N. Binsztein, L. Boiry, C. Busch, M. Di Pace, G. Gallopin, P. Penchaszadeh and M. Piantanida. 1968. Estructura social, distribución espacial and composición por edades de una población de tucotucos (*Ctenomys talarum*). *Invest. Zool. Chilenas* 13:47–80.

Pearson, O. P., and M. I. Christie. 1991. Sympatric species of *Euneomys* (Rodentia, Cricetidae). *Stud. Neotrop. Fauna Environ.* 26:121–27.

———. 1985. Los tuco-tucos (género *Ctenomys*) de los Parques Nacionales Lanín y Nahuel Huapi, Argentina. *Hist. Nat.* 5:337–43.

Pearson, O. P., and H. A. Lagiglia. 1992. "Fuerte de San Rafael": una localidad tipo ilusoria. *Rev. Mus. Hist. Nat. San Rafael (Mendoza, Argentina)* 12:35–43.

Pearson, O. P., and J. L. Patton. 1976. Relationships among South American phyllotine rodents based on chromosome analysis. *J. Mamm.* 57:339–50.

Pearson, O. P., and A. K. Pearson. 1982. Ecology and biogeography of the southern rainforests of Argentina. In *Mammalian biology in South America*, ed. M. Mares and H. Genoways, 129–42. The Pymatuning Symposia in Ecology 6, Special Publications Series. Pittsburgh: Pymatuning Laboratory of Ecology, University of Pittsburgh, xii + 539.

Pearson, O. P. and A. K. Pearson. 1993. La fauna de mamíferos pequeños de Cueva Traful I, Argentina: pasado y presente. *Praehistoria* 1:211–24.

Pearson, O. P., and C. P. Ralph. 1978. The dversity and abundance of vertebrates along an altitudinal gradient in Peru. *Mem. Mus. Hist. Nat. Javier Prado* 18:1–97.

Pearson, O. P., S. Martin, and J. Bellati. 1987. Demography and reproduction of the silky desert mouse (*Eligmodontia*) in Argentina. In *Studies in Neotropical mammalogy, essays in honor of Philip Hershkovitz*, ed.

B. D. Patterson and R. M. Timm, 433–46. *Fieldiana Zool.* 39:frontispiece, viii + 1–506.

Pearson, O. P., and M. F. Smith. 1999. Genetic similarity between *Akodon olivaceus* and *Akodon xanthorhinus* (Rodentia: Muridae) in Argentina. *J. Zool.* 247:43–52.

Peceño, M. C. 1983. "Estudio citogenético y genético-evolutivo del chigüire, género *Hydrochaeris*." Undergraduate thesis, Universidad Simón Bolívar, Caracas, Venezuela), 119 pp. [Not seen, cited in Mones and Ojasti 1986.]

Pecnerová, P., and N. Martínková. 2012. Evolutionary history of tree squirrels (Rodentia, Sciurini) based on multilocus phylogeny reconstruction. *Zool. Scr.* 41:211–19.

Pedó, E., T. R. O. de Freitas, and S. M. Hartz. 2010. The influence of fire and livestock grazing on the assemblage of non-flying small mammals in grassland-Araucaria Forest ecotones, southern Brazil. *Zoologia* 27:533–540.

Péfaur, J. E., and A. Díaz de Pascual. 1985. Small mammal species diversity in the Venezuelan Andes. *Acta Zool. Fenn.* 173:57–59.

Peichel, L., A. E. Chavez, A. Ocampo, W. Mena, F. Bozinovic, and A. G. Palacios. 2005. Eye and vision in the subterranean rodent cururo (*Spalacopus cyanus*, Octodontidae). *J. Comp. Neurol.* 486:197–208.

Peláez-Campomanes, P., and R. A. Martin. 2005. The Pliocene and Pleistocene history of cotton rats in the Meade basin of southwestern Kansas. *J. Mamm.* 86:475–94.

Pelzeln, A. von. 1883. *Brasilische Säugethiere. Resultate von Johann Natterer's Reisen in den Jahren 1817 bis 1835. Verhandl. Kaiserl Königl. Zool. Bot. Ges. Wien* 33 (Beiheft [Suppl.]):1–139.

Pennant, T. 1771. *Synopsis of Quadrupeds*. Chester: J. Monk, i–xxv + 1–382, 15 plates.

———. 1781. *History of Quadrupeds*. London: B. White, 2:285–566 + 14 (unnumbered), pls. 32–52.

Pepa, H. M. 2001. Productos naturales utilizados por los aborígenes del fin del mundo. 34° Acta, Congressus Internationalis Historiae Pharmaciae Firenze 1999; S:116–19.

Peppers, L. L., and R. D. Bradley. 2000. Cryptic species in *Sigmodon hispidus*: evidence from DNA sequences. *J. Mamm.* 81:332–43.

Peppers, L. L., D. S. Carroll, and R. D. Bradley. 2002. Molecular systematics of the genus *Sigmodon* (Rodentia: Muridae): evidence from the mitochondrial cytochrome-*b* gene. *J. Mamm.* 83: 396–407.

Percequillo, A. R. 1998."Sistemática de *Oryzomys* Baird, 1858 do leste do Brasil (Muroidea, Sigmodontinae)." Master's thesis, Universidade de São Paulo, Brazil.

———. 2003. "Sistemática de *Oryzomys* Baird, 1858: definição dos grupos de espécies e revisão taxonômica do grupo *albigularis* (Rodentia, Sigmodontinae)." Ph.D. diss., Universidade de São Paulo, Brazil.

———. 2006. Guia para nomenclatura e padronização da descrição da dentição nos roedores sigmodontineos. *Bol. Soc. Bras. Mastozool.* 47:5–11.

Percequillo, A.R., A.P. Carmignotto, and M.J. de Silva. 2005. A new species of *Neusticomys* (Ichthyomyini, Sigmodontini) from central Brazilian Amazonia. *J. Mamm.* 86:873–80.

Percequillo, A. R., P. R. Gonçalves, and J. A. de Oliveira. 2004. The rediscovery of *Rhagomys rufescens* (Thomas, 1886), with a morphological redescription and comments on its systematic relationships based on morphological and molecular (cytochrome-b) characters. *Mamm. Biol.* 69: 238–57.

Percequillo, A. R., E. Hingst-Zaher, and C. R. Bonvicino. 2008. Systematic review of genus *Cerradomys* Weksler, Percequillo and Voss, 2006 (Rodentia: Cricetidae: Sigmodontinae: Oryzomyini), with description of two new species from eastern Brazil. *Am. Mus. Novit.* 3622:1–46.

Percequillo, A. R., F. P. Tirelli, F. Michalski, and E. Eizirik. 2011. The genus *Rhagomys* (Thomas 1917) (Rodentia, Cricetidae, Sigmodontinae) in South America: morphological considerations, geographic distribution and zoogeographic comments. *Mammalia* 75:195–99.

Percequillo, A. R., M. Weksler, and L. P. Costa. 2011. A new genus and species of rodent from the Brazilian Atlantic Forest (Rodentia: Cricetidae: Sigmodontinae: Oryzomyini), with comments on oryzomyine biogeography. *Zool. J. Linn. Soc.* 161:357–90.

Percich, R. E., and V. L. Hodara. 1990. Crecimiento en laboratorio de roedores cricetidos: variables craneanas y peso de cristalino de *Calomys callidus. Physis*, sec. C 48:7–14.

Pereira, C., and W. F. de Almeida. 1942. *Amylophorus rochalimai*, n. g., n. sp., (Ciliata: Butschliidae) do intestino grosso de "*Dasyprocta azarae* Lichtenstein, 1823" (Rodentia: Caviidae). *Arq. Inst. Biol., São Paulo* 13:261–70.

Pereira, J., P. Teta, N. Fracassi, A. Johnson, and P. Moreyra. 2005. Sigmodontinos (Rodentia, Cricetidae) de la Reserva de Vida Silvestre Urugua-í (provincia de Misiones, Argentina), con la confirmación de la presencia de "*Akodon*" *serrensis* en la Argentina. *Mastozool. Neotrop.* 12:83–89.

Pereira, L. A. 1982. "Uso ecológico do espaço de *Zygodontomys lasiurus* (Rodentia, Cricetinae) en habitat natural de Cerrado do Brazil central." Master's thesis, Universidade de Brasília, Brazil.

Pereira, L. G. 2006. "Chapada Diamantina e Vale do Rio Jequitinhonha: composição da mastofauna e estrutura microevolutiva de algumas espécies de pequenos mamíferos." Ph.D. diss., Universidade Federal do Rio de Janeiro, Brazil.

Pereira, L. G., and L. Giese. 2007. Karyotype composition of some rodents and marsupials from Chapada Diamantina (Bahia, Brasil). *Brazil. J. Biol.* 67:509–18.

Pereira, L. G., and L. Giese. 2009. Non-flying mammals of Chapada Diamantina (Bahia, Brazil). *Biota Neotrop.* 9:185–96.

Pereira, L. G., L. Geise, A. A. Cunha, and R. Cerqueira. 2008. *Abrawayaomys ruschii* Cunha & Cruz, 1979 (Rodentia, Cricetidae) no estado do Rio de Janeiro, Brasil. *Pap. Avulsos Zool.* (São Paulo) 48:33–40.

Pereira, T. M., I. S. Pinto, L. P. Costa, and C. Graeff-Teixeira. 2010. Absence of Angiostrongylid nematodes in wild non-flying small mammals in Duas Bocas Biological Reserve, Cariacica, southeastern Brazil. *Rev. Patol. Trop.* 39:145–48.

Pereira, V. B, C. A. C. Braga, and M. R. S. Pires. 2010. Biologia de *Blarinomys breviceps* (Winge, 1887): um roedor fossorial Atlântica. *V Congresso Brasileiro de Mastozoologia*; Posters, ST16 (available on CD).

Pereira Barreto, M., and R. Domingues Ribeiro. 1972. Wild vectors and reservoirs of *Trypanosoma cruzi*. LIII: Natural infection of the rat, *Wiedomys pyrrhorhinus*(Wied, 1821) by *T. cruzi. Rev. Bras. Biol.* 32:595–600.

Peres, C. A., and C. Baider. 1997. Seed dispersal, spatial distribution and population structure of brazilnut trees (*Bertholletia excelsa*) in southeastern Amazonia. *J. Trop. Ecol.* 13:595–616.

Peres, C. A., J. Barlow, and T. Haugaasen. 2003. Vertebrate responses to surface fires in Amazonian forests. *Oryx* 37:97–109.

Pérez, E. M. 1992. Agouti paca. *Mamm. Species* 404:1–7.

Pérez, M. E., and D. Pol. 2012. Major radiations in the evolution of caviid rodents: reconciling fossils, ghost lineages, and relaxed molecular clocks. *PLoS ONE* 7(10):e48380.

Pérez, M. E., and M. G. Vucetich. 2011. A new extinct genus of Cavioidea (Rodentia, Hystricognathi) from the Miocene of Patagonia (Argentina) and the evolution of cavioid mandibular morphology. *J. Mamm. Evol.* 18:163–83.

Perez, S. I., J. A. Felizola Diniz-Filho, F. J. Rohlf, and S. F. dos Reis. 2009. Ecological and evolutionary factors in the morphological diversification of South American spiny rats. *Biol. J. Linn. Soc.* 98:646–60.

Pérez-Zapata, A., M. Aguilera, and O. A. Reig. 1992. An allopatric karyomorph of the *Proechimys guairae* complex (Rodentia: Echimyidae) in eastern Venezuela. *Interciencia* 17:235–40.

Pérez-Zapata, A., D. Lew. M. Aguilera, and O. A. Reig. 1992. New data on the systematics and karyology of *Podoxymys roraimae* (Rodentia, Cricetidae). *Z. Säugetierk.* 57:216–24.

Pérez-Zapata, A., O. A. Reig, M. Aguilera, and A. Ferrer. 1986. Cytogenetics and karyosystematics of South American oryzomyine rodents (Cricetidae: Sigmodontinae). I. A species of *Oryzomys* with a low chromosomal number from northern Venezuela. *Z. Säugetierk.* 51:368–78.

Pérez-Zapata, A., A. D. Vitullo, and O. A. Reig. 1987. Karyotypic and sperm distinction of *Calomys hummelincki* from *Calomys laucha* (Rodentia: Cricetidae). *Acta Cient. Venez.* 38:90–93.

Perissinotti, P. P., C. D. Antinucci, R. Zenuto, and F. Luna. 2009. Effect of diet quality and soil hardness on metabolic rate in the subterranean rodent *Ctenomys talarum*. *Comp. Biochem. Physiol. A Mol. Integr. Physiol.* 154:298–307.

Pessino, M. E. M., J. H. Sarasola, C. Wander, and N. Besoky. 2001. Long-term response of puma (*Puma concolor*) to a population decline of the plains vizcacha (*Lagostomus maximus*) in the Monte desert, Argentina. *Ecol. Austral* 11:61–67.

Pessôa, L. M., M. M. de Oliveira Corrêa, E. Bittencourt, and S. F. dos Reis. 2005. Chromosomal characterization of taxa of the genus *Trinomys* Thomas, 1921, in states of Rio de Janeiro and São Paulo (Rodentia: Echimyidae). *Arq. Mus. Nac., Rio de Janeiro* 63:161–68.

Pessôa, L. M., M. M. de Oliveira Corrêa, J. A. de Oliveira, and M. O. G. Lopes. 2004. Karyological and morphometric variation in the genus *Thrichomys* (Rodentia: Echimyidae). *Mamm. Biol.* 69:258–69.

Pessôa, L. M., J. A. de Oliveira, and F. S. Reis. 1990. Quantitative cranial character variation in selected populations of the guyannensis-group of *Proechimys* (Rodentia: Echimyidae) from Brazil. *Zool. Anzeiger* 225:396–400.

———. 1992. A new species of Spiny rat genus *Proechimys*, subgenus *Trinomys* (Rodentia: Echimyidae). *Z. Säugetierk.* 57:39–46.

Pessôa, L.M., and S. F. dos Reis. 1991a. Cranial Infraspecific differenciation in *Proechimys iheringi* Thomas (Rodentia: Echimyidae). *Z. Säugetierk.* 56:34–40.

———. 1991b. The contribution of indeterminate growth to nongeographic variation in adult *Proechimys albispinus* (Is. Geoffroy) (Rodentia: Echimyidae). *Z. Säugetierk.* 56:220–24.

———. 1992a. Bacular variation in the subgenus *Trinomys*, genus *Proechimys* (Rodentia: Echimyidae). *Z. Säugetierk.* 57:100–2.

———. 1992b. An analysis of morphological discrimination between *Proechimys iheringi* and *Proechimys dimidiatus* (Rodentia: Echimyidae). *Zool. Anzeiger* 228:189–200.

———. 1993a. A new subspecies of *Proechimys iheringi* Thomas (Rodentia: Echimyidae) from the State of Rio de Janeiro, Brazil. *Z. Säugetierk.* 58:181–90.

———. 1993b. Proechimys dimidiatus. *Mamm. Species* 441:1–3.

———. 1994. Systematic implications of craniometric variation in *Proechimys iheringi* Thomas (Rodentia: Echimyidae). *Zool. Anzeiger* 232:181–200.

———. 1995. Coat color variation in *Proechinys albispinus* (Geoffroy, 1838) (Rodentia, Echimyidae). *Bol. Mus. Nac., Rio de Janeiro*, nova sér, zool., 361:1–5.

———. 1996. Proechimys iheringi. *Mamm. Species* 536:1–4.

———. 2002. Proechimys albispinus. *Mamm. Species* 693:1–3.

Pessôa, L. M. S. F. dos Reis, and M. F. Pessôa, M. F. 1996. Bacular variation in subspecies of the Brazilian spiny rat *Proechimys* (*Trinomys*) *iheringi*. *Stud. Neotrop. Fauna Environ.* 31:129–32.

Pessôa, L. M., and R. E. Strauss. 1999. Cranial size and shape variation, pelage and bacular morphology, and subspecific differentiation in spiny rats, *Proechimys albispinus* (Is. Geoffroy, 1838), from northeastern Brazil. *Bonner Zool. Beitr.* 48:231–43.

Pessôa, L. M., F. J. Von Zuben, and F. S. dos Reis. 1998. Morphological affinities of *Proechimys yonenagae* Rocha, 1995 (Rodentia: Echimyidae): evidence from bacular and cranial characters. *Bonner Zool. Beitr.* 48:167–77.

Peterka, H. E. 1937. A study of the myology and osteology of tree squirrels with regard to adaptation to arboreal, glissant and fossorial habits. *Trans. Kansas Acad. Sci.* 39:313–32.

Peters, W. 1861. Über einige merkwürdige Nagethiere (*Spalacomys indicus*, *Mus tomentosus* und *Mus squamipes*) des Königl. zoologischen Museums. *Abh. Konigl. Preuss. Akad. Wiss. Berlin* 1860:139–56, 2 plates.

———. 1864. Machte eine Mittheilung über neue Eichhornarten aus Mexico, Costa Rica und Guiana, so wie uber *Scalops latimanus* Bachmann. *Monatsber. Konigl. Preuss. Akad. Wiss. Berlin* 1863:652–56.

———. 1865. A provisional report about new kinds of mammalian genera (translated from the original German). *Monatsber. Konigl. Preuss. Akad. Wiss. Berlin* 1864:177–81.

———. 1866. Über neue oder ungenügend bekannte Flederthiere (*Vampyrops*, *Uroderma*, *Chiroderma*, *Ametrida*, *Tylostoma*, *Vespertilio*, *Vesperugo*) und Nager

(*Tylomys, Lasiomys*). *Monatsber. Konigl. Preuss. Akad. Wiss. Berlin* 1867:392–411 + 2 plates.

———. 1873a. Über *Dinomys* eine merkwürdige neue Gattung der stachelschweinartigen Nagethiere aus den Hochgebirgen von Peru. *Monatsber. Konigl. Preuss. Akad. Wiss. Berlin* 1873:551–52.

———. 1873b. Ueber *Dinomys*, eine merkwürdige neue Gattumg von Nagethieren aus Peru. *Sitz-Ber. Gesellsch. naturforsch. Freunde Berlin*, pp. 227–34, 4 plates.

Peterson, R. L. 1966. Recent mammals records from the Galápagos Islands. *Mammalia* 30:441–45.

Petter, F. 1973. Les noms de genre *Cercomys, Nelomys, Trichomys* et *Proechimys* (Rongeurs, Echimyides). *Mammalia* 37:422–26.

———. 1978. Epidémiologie de la leishmaniose en Guyane française, en relation avec l'existence d'une espèce nouvelle de Rongeurs Echimyidés, *Proechimys cuvieri* sp. n. *C. R. Seances Acad. Sci. D* 287:261–64.

———. 1979 [1980]. Une nouvelle espèce de rat d'eau de Guyane française, *Nectomys parvipes* sp.nov. (Rongeurs, Cricetidae). *Mammalia* 43:507–10. [1979 vol. 43, no 4: pp. 439ff. "Achevé d'imprimer le 15 avril 1980."]

Petter, F., and H. Cuenca Aguirre. 1982. Un *Isothrix* nouveau de Bolivie (Rongeurs, Echimyidés). La denture des *Isothrix*. *Mammalia* 46:191–03.

Petter, F., and O. Tostain. 1981. Variabilité de la 3ᵉ molair supérieure d'*Holochilus brasiliensis* (Rongeurs, Cricetidae). *Mammalia* 45:257–59.

Peurach Collins, S. 1994. "Karyotypes of a selected group of Phyllotine (Muridae, Sigmodontinae) rodents." Undergraduate honors' thesis, University of New Mexico, Albuquerque.

Philippi, R. A. 1857. Descripcion de una nueva especie de rata por el Señor Don Luis Ladbeck de Valdivia, precedida de algunas observaciones jenerales por el Dr. R. A. Philippi. *An. Univ. Chile* 15:360–62.

———. 1858. Beschreibung einiger neuen chilenischen Mäuse. *Arch. Naturgesch.* 24:77–82.

———. 1860. *Reise durch die Wüste Atacama auf Befehl der chilenischen Regierung im Sommer 1843–1854 unternommen und ausgeführt von R. A. Philippi*. Halle: Eduard Anton, x + 192 + 62 pp.

———. 1869. Ueber einige Thiere von Mendoza. *Arch. Naturgesch.* 35:38–41

———. 1872. Drei neue Nager aus Chile. *Z. Gesam. Nat.* 6:442–47.

———. 1880. *Ctenomys fueguinus* Philippi. *Arch. Naturgesch.* 46:276–79.

———. 1896. Descripción de los Mamíferos traidos del viaje de esploración a Tarapacá hecho por órden del Gobierno en el verano de 1884 a 1885 por Federico Philippi. *An. Mus. Nac. Chile Zool.* 13a:1–24 + 7 plates.

———. 1900. Figuras i descripciones de los murideos de Chile. *An. Mus. Nac. Chile Zool.* 14a:1–70 + 25 plates.

Philippi, R. A., and L. Landbeck. 1858. Beschreibung einiger neuen Chilenischen Mäuse. *Arch. Naturgesch.* 24, 1:77–82.

Pia, M. V., M. S. López, and A. Novaro. 2003. Effects of livestock on the feeding ecology of endemic culpeo foxes (*Pseudalopex culpaeus smithersi*) in central Argentina. *Rev. Chilena Hist. Nat.* 76:313–21.

Piana, R. P. 2007. Anidamiento y dieta de *Harpia harpyja* Linnaeus en la comunidad nativa de Infierno, Madre de Dios, Perú. *Rev. Peru. Biol.* 14:135–38.

Piantanida, M. J. 1987. Distintos aspectos de la reproducción en la naturaleza y en cautiverio del roedor cricétido *Akodon dolores*. *Physis Sec. C* 45:47–58.

———. 1993. Datos sobre algunos aspectos de la reprodución en una colonia del roedor *Holochilus chacarius chacarius* (Massoia, 1974), Rodentia, Cricetidae. *Rev. Mus. Argentino Cien. Nat. "Bernardino Rivadavia," Inst. Nac. Invest. Cien. Nat. Ecol.* 4:39–41.

Piantanida, M. J., and N. Nani. 1993. Reproducción y crecimiento de *Bolomys temchuki* (Massoia, 1980) en cautiverio (Rodentia, Cricetidae). *Rev. Mus. Argentino Cien. Nat. "Bernardino Rivadavia," Inst. Nac. Invest. Cien. Nat. Ecol.* 4:39–50.

Picot, H. 1992. *Holochilus brasiliensis* and *Nectomys squamipes* (Rodentia-Cricetidae) natural hosts of *Schistosoma mansoni*. *Mem. Instit. Oswaldo Cruz* 87:255–60.

Pictet, F.-J. 1841a. Notice sur les animaus nouveaux ou peu connus du Muséé de Genève, ser. 1, Mammifères, livr. 2, Pl. 7 (animal) and Pl. 8, the skull and other parts.

———. 1841b. Premiere notice sur les animaux nouveaux ou peu connus du Musée de Genève. Observations sur quelques rongeurs épineux du Brésil. *Mem. Soc. Phys. Hist. Nat. Genève* 9:143–60, plates 1–5.

———. 1843a. Seconde notice sur les animaux noveaux ou peu connus du Musée de Genève. *Mem. Soc. Phys. Hist. Nat. Genève* 10:201–14, plates 1–5.

———. 1843b. Description d'un nouveau genre de rongeurs de la familie des Histricins. *Rev. Zool. Soc. Cuvier.* 6:225–27.

Pictet, F.-J., and C. Pictet. 1844. Notices sur les animaux nouveaux ou peu connus du Musée de Genève. Description des rats envoyés de Bahia au Musée de Genève. In Pictet, F.-J. 1841–1844, *Notices sur les animaux nouveaux ou peu connus du Musée de Genève*. *I*ʳᵉ série. Mammifères, Genève, pp. 41–82, pl. 12–23.

Pigière, F., W. Van Neer, C. Aniseau, and M. Denis. 2012. New archaeological evidence for the introduction of the guinea pig to Europe. *J. Archaeol. Sci.* 39:1020–24.

Pillado, M. S., and A. Trejo. 2000. Diet of the Barn owl (*Tyto alba tuidara*) in northwestern Argentine Patagonia. *J. Raptor Res.* 34:334–38.

Pimentel, D. S., and M. Tabarelli. 2004. Seed dispersal of the palm *Attalea oleifera* in a remnant of the Brazilian Atlantic Forest. *Biotropica* 36:74–84.

Pine, R. H. 1971. A review of the long-whiskered rice rat, *Oryzomys bombycinus*, Goldman. *J. Mamm.* 52:590–96.

———. 1973. Mammals (exclusive of bats) of Belém, Pará, Brazil. *Acta Amaz.* 3:47–79.

———. 1973. Una nueva especie de *Akodon* (Mammalia: Rodentia: Muridae) de la Isla de Wellington, Magallanes, Chile. *An. Inst. Patagon.* 4:423–26.

———. 1976. A new species of *Akodon* (Mammalia: Rodentia: Muridae: Cricetinae) from Isla de los Estados, Argentina. *Mammalia* 40:63–68.

———. 1980. Notes on rodents of the genera *Wiedomys* and *Thomasomys* (including *Wilfredomys*). *Mammalia* 44:195–202.

Pine, R. H., P. Angle, and D. Bridge. 1978. Mammals from the sea, mainland and islands at the southern tip of South America. *Mammalia* 42:105–14.

Pine, R. H., I. R. Bishop, and R. L. Jackson. 1970. Preliminary list of mammals of the Xavantina-Cachimbo Expedition (Central Brazil). *Trans. Roy. Soc. Trop. Med. Hyg.* 64:668–70.

Pine, R. H., and D. C. Carter. 1970. Distributional notes on the thick-spined rat (*Hoplomys gymnurus*) with the first records from Honduras. *J. Mamm.* 51:804.

Pine, R. H., S. D. Miller, and M. L. Schamberger. 1979. Contributions to the mammalogy of Chile. *Mammalia* 43:339–76.

Pine, R. H., N. E. Pine, and S. D. Bruner. 1981. Mammalia. In *Aquatic biota of Tropical South America. Part 2: Anarthropoda*, ed. S. H. Hurlbert, G. Rodriguez, and N. D. Santos, 267–98. San Diego: San Diego State University, San Diego, xi+298 pp.

Pine, R. H., R. T. Timm, and M. Weksler. 2012. A newly recognized clade of trans-Andean Oryzomyini (Rodentia: Cricetidae), with description of a new genus. *J. Mamm.* 93:851–70.

Pine, R. H., and R. M. Wetzel. 1976. A new subspecies of *Pseudoryzomys wavrini* (Mammalia: Rodentia: Muridae: Cricetinae) from Bolivia. *Mammalia* 39:649–55. [1975 vol. 39, no. 4: pp. 523ff. "Achevé d'imprimer le 30 avril 1976."]

Pinheiro, F. P., R. E. Shope, A. H. P. Andrade, G. Bensabath, G.V. Cacios, and J. Casals. 1966. Amaparí, a new

virus of the Tacaribe group from rodents and mites Amapá Territory, Brazil. *Proc. Soc. Exp. Biol. Med.* 122:531–35.

Pinheiro, P. S., P. A. Hartmann, and L. Geise. 2004. New record of *Rhagomys rufescens* (Thomas, 1886) (Rodentia: Muridae: Sigmodontinae) in the Atlantic Forest of southeastern Brazil. *Zootaxa* 431:1–11.

Pini, N. C., S. Levis, G. Calderón, J. Ramirez, D. Bravo, E. Lozano, C. Ripoll, S. St. Jeor, T. G. Ksiazek, R. M. Barquez, and D. Enría. 2003. Hantavirus Infection in humans and rodents, northwestern Argentina. *Emerg. Infect. Dis.* 9:1070–76.

Pinna-Senn, E., D. D. de Barale, J. Polop, M. Y. Ortiz, M. C. Provensal, and J. A. Lisanti. 1992. Estudio cariotípico y morfométrico en una población de *Akodon* sp. (Rodentia, Cricetidae) de Pampa de Achala. *Mendeliana* 10:59–70.

Pinto, C. M., S. Ocaña-Mayorga, M. S. Lascano, and M. J. Grijalva. 2006. Infection by tripanosomes in marsupials and rodents associated with human dwellings in Ecuador. *J. Parasitol.* 92:1251–5.

Pinto, I. S., J. R. Botelho, L. P. Costa, Y. L. R. Leite, and P. M. Linardi. 2009. Siphonaptera associated with wild mammals from the central Atlantic Forest biodiversity corridor in southeastern Brazil. *J. Med. Entomol.* 46:1146–51.

Pinto, I. S., A. C. C. Loss, A. Falqueto, and Y. L. R. Leite. 2009. Pequenos mamíferos não voadores em fragmentos de Mata Atlântica e áreas agrícolas em Viana, Espírito Santo, Brasil. *Biota Neotrop.* 9:355–60.

Pinto, I. S., C. B. Santos, A. L. Ferreira, and A. Falqueto. 2009. Richness and diversity of sand flies (Diptera, Psychodidae) in an Atlantic rainforest reserve in southeastern Brazil. *J. Vector Ecol.* 35:325–32.

Pinto, O. M. de O. 1931. Ensaio sobre a fauna de sciurideos do Brasil. *Rev. Mus, Paulista, São Paulo* 17:263–319.

Pires, A. S., and M. Galetti. 2012. The agouti *Dasyprocta leporina* (Rodentia: Dasyptoctidae) as seed disperser of the palm *Astrocaryum aculeatissimum*. *Mastozool. Neotrop.* 19:147–53.

Piso, W. 1658. *Gulielmi Pisonis medici Amstelaedamensis De Indiae utriusque re naturali et medica: libri quatuordecim, quorum contenta pagina sequens exhibit.* Amsteldaedami [Amsterdam]: Apud Ludovicum et Danielem Elzevirios, [24], 3–327 [329], [5], 39, 226, [2], p. ill (woodcuts).

Pittier, H., and G. H. H. Tate. 1932. Sobre fauna venezolana. Lista provisional de los mamíferos observados en el país. *Bol. Soc. Venez. Cien. Nat.* 7:249–80.

Pizzimenti, J. J., and R. de Salle. 1980. Dietary and morphometric variation in some Peruvian rodent communities: the effect of feeding strategy on evolution. *Biol. J. Linn. Soc.* 13:263–85.

Poche, F. 1912. Über den Inhalt und die Erscheinungszeiten der einzelnen Teile, Hefte etc. und die verschiedenen Ausgaben des Schreber'schen Säugetierwerkes (1774–1855). *Arch. Naturgesch.* 77(1, suppl. 4):124–83.

Po-Chedley, D. S., and A. R. Shadle. 1955. Pelage of the porcupine, *Erethizon dorsatum. J. Mamm.* 36:84–95.

Pocock, R. I. 1913. Description of a new species of Agouti (*Myoprocta*). *Ann. Mag. Nat. Hist.* ser. 8, 12:110–11.

———. 1914. On the facial vibrissae of Mammalia. *Proc. Zool. Soc. Lond.* 1914:889–912.

———. 1922. On the external characters of some hystricognath rodents. *Proc. Zool. Soc. Lond.* 1922:365–427.

———. 1926. The external characters of a young female *Dinomys branickii* exhibited in the Society's Gardens. *Proc. Zool. Soc. Lond.* 1926 (part I):221–30.

Podestá, D. H., S. Cirignoli, and U. F. J. Pardiñas. 2000. New data on the distribution of *Octodon bridgesii* (Mammalia: Rodentia) in Argentina. *Neotropica* 46:75–77.

Poeppig, E. L. 1835a. *Reise in Chile, Perú und auf dem Amazonenstrome während der Jahre 1827–1832.* Leipzig: F. Fleischer, 1:xii+466 pp.

———. 1835b. Über dem Cucurrito Chile's (*Psammoryctes noctivagus* Poepp.). *Arch. Naturgesch.* 1:252–55.

———. 1836. *Reise in Chile, Peru und auf dem Amazonenstrome während der Jahre 1827–1832.* Leipzig: F. Fleischer, 2:1–464, 14 plates.

Poleo, C., and R. Mendoza. 2001. Determinación de picos poblacionales e índices reproductivos de especies de ratas que causan daños en el cultivo de arroz en el Sistema de riego del Río Guárico. In *Memórias IV Jornadas Técnicas Divulgativas. Guárico-Apure* 2001, 25. Maracay, Venezuela: Editora INIA, Gerencia de Nagaciación Tecnológica.

Polop, J. J. 1989. Distribution and ecological observations of wild rodents in Pampa de Achala, Córdoba, Argentina. *Stud. Neotrop. Fauna Environ.* 24:53–59.

———. 1991. Distribución de cricétidos en las Sierras de Achala (provincia de Córdoba, República Argentina). *Rev. Univ. Nac. Río Cuarto, Zool.* 11:115–21.

Polop, J. J., and M. Busch, eds. 2010. *Biología y ecología de pequeños roedores en la región pampeana de Argentina: enfoques y perspectivas.* Córdoba, Argentina: Universidad Nacional de Córdoba, 332 pp.

Polop, J. J., C. N. Gardenal, and M. S. Sabattini. 1982. Comunidades de roedores de cultivos de sorgo en la Provincia de Córdoba y su posible vinculación con la fiebre hemorragica Argentina. *Ecosur* 9:107–16.

Polop, J. J., R. Martínez, and M. Torres. 1985. Distribución y abundancia de poblaciones de pequeños roedores en la zona de embalse de Río Tercero, Córdoba. *Hist. Nat.* 5:33–44.

Polop, J. J., and M. C. Provensal. 2000. Morphological variation and age determination in *Calomys venustus* (Thomas 1894) (Rodentia, Muridae). *Mastozool. Neotrop.* 7:101–15.

Polop, J. J., M. C. Provensal, and P. Dauría. 2005. Reproductive characteristics of free-living *Calomys venustus* (Rodentia, Muridae). *Acta Theriol.* 50:357–66.

Polop, J. J., and M. S. Sabattini. 1993. Rodent abundance and distribution in habitats of agrocenosis in Argentina. *Stud. Neotrop. Fauna Environ.* 28:39–46.

Ponce-Ulloa, H. E. 1989. Descripción de *Jellisonia amadoi* sp. nov. y *J. mexicana* sp. nov. del estado de Guerrero, Méxiso (Siphonaptera: Ceratophyllidae). *Folia Entomol. Mex.* 28:177–85.

Poole, A. J., and V. S. Schantz. 1942. Catalog of the type specimens of mammals in the United States National Museum, including the Biological Surveys collection. *Bull. U.S. Natl. Mus.* 178:xiii+1–705.

Porteous, I. S., and S. J. Pankhurst. 1998. Social structure of the mara (*Dolichotis patagonum*) as a determinant of gastro-intestinal parasitism. *Parasitology* 116:269–75.

Poulton, S. M. C. 1982. An ecological study of petroleum activities; Aguaro-Guariquito National Park, Venezuela. July-August 1981. *Final report, Cambridge Venezuela Research Group, Cambridge, U.K.,* 155 pp.

Poux, C., P. Chevret, D. Huchon, W. W. de Jong, and E. J. P. Douzery. 2006. Arrival and diversification of caviomorph rodents and platyrrhine primates in South America. *Syst. Biol.* 55:228–44.

Powers, A. M., D .R. Mercer, D. M. Watts, H. Guzman, C. F. Fulhorst, V. L. Popov, and R. B. Tesh. 1999. Isolation and genetic characterization of a hantavirus (Bunyaviridae: *Hantavirus*) from a rodent, *Oligoryzomys microtis* (Muridae), collected in northeastern Peru. *Am. J. Trop. Med. Hyg.* 61:92–98.

Pozo, E. J., G. Troncos C., A. Palacios F, F. Arévalo G, G. Carrión T, and V. A. Laguna-Torres. 2005. Distribución y hospederos de pulgas (Siphonaptera) en la provincia de Ayabaca, Piura-1999. *Rev. Peru. Med. Exp. Salud Pub.* 22:316–20.

Pozo-R., W. E., J. Gómez, and M. Acosta. 2007. Cricétidos de la Finca Ana María, noroccidente de la Provincia de Pichincha, Ecuador. *Bol. Tec. Ser. Zool. Sangolqui* 7:49–52.

Pozo-R., W. E., I. Olmedo, and S. Espinoza. 2006. Diversidad rodentológica en remanentes de bosque nativo y cercas vivas de la hacienda El Prado, serranía ecuatoriana. *Bol. Tec. Ser. Zool. Sangolqui* 6:33–44.

Prado, J. L., J. Chiesa, G. Tognelli, E. Cerdeno, and E. Strasser. 1998. Mammals from the Rio Quinto Formation (Pliocene), San Luis province (Argentina): Biostratigraphic, zoogeographic and paleoenvironmental aspects. *Estud. Geol.* (Madrid) 54:153–60.

Prado, J. R., and A. R. Percequillo. 2013. Geographic distribution of the genera of the tribe Oryzomyini (Rodentia: Cricetidae: Sigmodontinae) in South America: patterns of distribution and diversity. *Arch. Zool.* 44:1–120.

Prell, H. 1934a. Die gegenwärtig bekannten Arten der Gattung *Chinchilla* Bennett, (Beiträge zur Kenntnis der Chinchilla II). *Zool. Anzeiger* 108:97–104.

———. 1934b. Über *Mus laniger* Molina. *Zool. Gart. Leipz.* 7:207–09.

Prevedello, J. A., R. G. Rodrigues, and E. L. de Araujo Monteiro-Filho. 2010. Habitat selection by two species of small mammals in the Atlantic Forest, Brazil: Comparing results from live trapping and spool-and-line tracking. *Mamm. Biol.* 75:106–14.

Prevosti, F. J., and U. F. J. Pardiñas. 2009. Comment on "The oldest South American Cricetidae (Rodentia) and Mustelidae (Carnivora): Late Miocene faunal turnover in central Argentina and the Great American Biotic Interchange" by D. H. Verzi and C. I. Montalvo [*Palaeogeogr. Palaeoclimatol. Palaeoecol.* 267 (2008) 284–91]. *Palaeogeogr. Palaeoclimatol. Palaeoecol.* 280(3–4):543–47.

———. In press. The heralds: carnivores (Carnivora) and sigmodontine rodents (Cricetidae) in the Great American Biotic Interchange. In *Origins and Evolution of Cenozoic South American Mammals*, ed. A. L. Rosenberger and M. F. Tejedor. New York: Springer.

Prieto-Torres, D., A. Belandria-Abad, U. Gómez, and R. Calchi. 2011. Lista Preliminar de mamíferos no voladores en tres localidades de la vertiente suroriental de la Sierra de Perijá, Estado Zulia-Venezuela. *Bol. Centr. Invest. Biol.* 45:21–34.

Price, R. D., and K. C. Emerson. 1971. A revision of the genus *Geomydoecus* (Mallophaga: Trichodectidae) of the New World pocket gophers (Rodentia: Geomyidae). *J. Med. Entomol.* 8:228–57.

———. 1986. New species of *Cummingsia* Ferris (Mallophaga: Trineoponidae) from Peru and Venezuela. *Proc. Biol. Soc. Washington* 94:748–52.

Price, R. D., and R. M. Timm. 2002. Two new species of the chewing louse genus Gliricola Mjöberg (Phthriaptera: Gryopidae) from Peruvian rodents. *Proc. Entomol. Soc. Washington* 104:863–67.

Priotto, J. W., M. Morando, and L. Avila. 1996. Nuevas citas de roedores de los pastizales de altura de la Sierra de Comechingones, Córdoba, Argentina. *Facena* 12:135–38.]

Priotto, J. W., and J. Polop. 1997. Space and time use in syntopic populations of *Akodon azarae* and *Calomys venustus* (Rodentia, Muridae). *Z. Säugetierk.* 62:30–36.

Priotto, J. W., A. Steinmann, and J. J. Polop. 2002. Factors affecting home range size and overlap in *Calomys venustus* (Muridae : Sigmodontinae) in Argentine agroecosystems. *Mamm. Biol.* 67:97–104.

Provensal, M. C., and J. J. Polop. 2008. Inter-annual changes in structure of overwintering population in *Calomys venustus*. *Acta Zool. Sinica* 54:36–43.

———. 1993. Morphometric variation in populations of *Calomys musculinus*. *Stud. Neotrop. Fauna Environ.* 28:95–103.

Pucheran, J. 1845. Description de quelques Mammiferes Américans. *Rev. Zool. Soc. Cuvier.* 8:335–37.

Puerta, H., C. Cantillo, J. Mills, B. Hjelle, J. Salazar-Bravo, and S. Mattar. 2006. Hantavirus del Nuevo Mundo: ecología y epidemiología de un virus emergente en Latinoamérica. *Medicina* 66:343–56.

Puig, S., and N. R. Nani. 1981. Estudio del crecimiento y reproducción de "*Graomys griseoflavus*" (Waterhouse, 1837) en una colonia de laboratorio. *Rev. Mus. Argentino Cien. Nat. "Bernardino Rivadavia," Instit. Nac. Invest. Cienc. Nat. Ecol.* 2(8):129–71.

Puig, S., M. I. Rosi, M. I. Cona, V. G. Roig, and S. A. Monge. 1992. Estudio ecológico del roedor subterráneo *Ctenomys mendocinus* en la precordillera de Mendoza, Argentina: densidad poblacional y uso del espacio. *Rev. Chilena Hist. Nat.* 65:247–54.

———. 1999. Diet of a Piedmont population of *Ctenomys mendocinus* (Rodentia, Ctenomyidae): seasonal patterns according sex and relative age. *Acta Theriol.* 44:15–27.

Puig, S., F. Videla, M. Cona, S. Monge, and V. Roig. 1998. Diet of the vizcacha *Lagostomus maximus* (Rodentia, Chinchillidae), habitat preferences and food availability in northern Patagonia, Argentina. *Mammalia* 62:191–204.

Püttker, T., C. Barros, T. K. Martins, S. Sommer, and R. Pardini. 2012. Suitability of distance metrics as indexes of home-range size in tropical rodent species. *J. Mamm.* 93:115–23.

Püttker, T., A. A. Bueno, C. dos Sandos de Barros, S. Sommer, and R. Pardini. 2013. Habitat specialization with habitat amount to determine dispersal success of rodents in fragmented landscapes. *J. Mamm.* 94:714–26.

Püttker, T., Y. Meyer-Lucht, and S. Sommer. 2006. Movement distances of five rodent and two marsupial species in forest fragments of the costal Atlantic rainforest, Brazil. *Ecotropica* 12:131–39.

———. 2008a. Fragmentation effects on population density of three rodent species in secondary Atlantic Rainforest, Brazil. *Stud. Neotrop. Fauna Environ.* 43:11–18.

———. 2008b. Effects of fragmentation on parasite burden (nematodes) of generalist and specialist small mammal species in secondary forest fragments of the coastal Atlantic Forest, Brazil. *Ecol. Res.* 23:207–15.

Püttker, T., R. Pardini, Y. Meyer-Lucht, and S. Sommer. 2008. Responses of five small mammal species to microscale variations in vegetation structure in secondary Atlantic Forest remnants, Brazil. *BMC Ecol.* 8:9.

Püttker, T., P. I. Prado, C. dos Santod de Barros, and T. K. Martins. 2013. Animal movements and geometry: a response to Oliveira-Santos et al. 2013. *J. Mamm.* 94:954–56.

Queirolo, D., and M. A. M. Granzinolli. 2009. Ecology and natural history of *Akodon lindberghi* (Rodentia, Sigmodontinae) in southeastern Brazil. *Iheringia Ser. Zool.* 99:189–93.

Quentin, J. C. 1967. *Rictularia zygodontomys* n. sp. Nematode nouveau parasite de Rongeurs du Brasil. *Bull. Mus. Natl. Hist. Nat., Paris* 39:740–44.

———. 1969. Etude de nematodes *Syphacia* parasites de rongeurs Cricetidae sud-americains et de leurs correlations biogeographiques avec certaines especes nearctiques. *Bull. Mus. Natl. Hist. Nat., Paris* 41:909–25.

Quentin, J. C., Y. Karimi, and C. Rodríguez de Almeida. 1968. *Protospirura numidica criceticola*, n. subsp. parasite de rongeurs cricetidae du Brésil. Cycle evolutif. *Ann. Parasitol. Paris* 43:583–96.

Quiceno, C. A. 1993. "Sistematica y distribución del género *Akodon* (Rodentia: Cricetidae) en el Valle del Cauca." Undergraduate thesis, Universidad del Valle, Cali, Colombia.

Quintana, C. A. 1996. Diversidad del roedor *Microcavia* (Caviomorpha, Caviidae) de America del sur. *Mastozool. Neotrop.* 3:63–86.

———. 1998. Relaciones filogeneticas de roedores Caviinae (Caviomorpha, Caviidae), de America del Sur. *Bol. R. Soc. Esp. Hist. Nat.* 94:125–34.

———. 2001. *Galea* (Rodentia, Caviidae) del Pleistoceno Superior y Holoceno de las sierras de Tandilia oriental, provincial de Buenos Aires, Argentina. *Amerghiniana* 38:399–407.

———. 2002. Roedores cricétidos del Sanandrense (Plioceno tardio) de la provincia de Buenos Aires. *Mastozool. Neotrop.* 9:263–75.

Quintana, H., V. Pacheco, and E. Salas. 2009. Diversidad y conservación de los mamíferos de Ucayali, Perú. *Ecol. Apl.* 8:91–103.

Quintana, R. D. 2002. Influence of livestock grazing on the capybaras trophic niche and forage preferences. *Acta Theriol.* 47:175–83.

Quintana, V. 2009. Registros de *Dromiciops gliroides* y *Chelemys megalonyx* en bosques nativos de centro-sur de Chile. *Gestión Ambiental* 17:45–55.

Quintana Navarrete, H. L. 2011. "Distribucion y modelamiento por Maxent de los mamíferos endémicos de Perú." Master's thesis, Universidad Nacional Mayor de San Marcos, Lima, Peru.

Quintela, F. M., G. L. Gonçalves, S. L. Althoff, I. J. Sbalqueiro, L. F. B. Oliveira, and T. R. O. de Freitas. 2014. A new species of swamp rat of the genus *Scapteromys* Waterhouse, 1837 (Rodentia: Sigmodontinae) endemic to the *Araucaria angustifolia* forest in southern Brazil. *Zootaxa* 381:207–225.

Quirici, V., R. A. Castro, L. Ortiz-Tolhuysen, A. S. Chesh, J. R. Burger, E. Miranda, A. Cortes, et al. 2010. Seasonal variation in the range areas of the diurnal rodent *Octodon degus*. *J. Mamm.* 91:458–66.

Quirici, V., S. Faugeron, L. D. Hayes, and L. A. Ebensperger. 2011. The influence of group size on natal dispersal in the communally rearing and semifossorial rodent, *Octodon degus*. *Behav. Ecol. Sociobiol.* 65:787–98.

Rafinesque, C. S. 1815. *Analyse de la nature, ou tableau de l'univers et des corps organisés*. Palerme: l'Imprimerie de Jean Barravecchia, 224 pp.

Rageot, R., and L. Albuja V. 1994. Mamíferos de un sector de la alta Amazonía Ecuatoriana: Mera, Provincia de Pastaza. *Rev. Polit.* 19:165–208.

Ramdial, B. 1972. The natural history and ecology of the paca (*Cuniculus paca*). Trinidad: Lands and Survey Department, Mapping and Control Section, 11 pp.

Ramirez, P. B., J. W. Bickham, J. K. Braun, and M. A. Mares. 2001. Geographic variation in genome size of *Graomys griseoflavus* (Rodentia: Muridae). *J. Mamm.* 82:102–8.

Ramirez, S. C., and C. M. S. Vela. 2003. *Plan maestro del Parque Nacional del Río Abiseo*. Lima, Peru: Instituto Nacional de Recursos Naturales–INRENA.

Ramos, R. S. L., W. G. Vale, and F. L. Assis. 2003. Karyotypic analysis in species of the genus *Dasyprocta* (Rodentia: Dasyproctidae) found in Brazilian Amazon. *An. Acad. Brasil. Ciênc.* 75:55–69.

Ramos, V. N., and K. G. Facure. 2009. Ecología alimentar de *Calomys tener* (Rodentia, Cricetidae) em áreas naturais de Cerrado. *Anais do III Congresso Latino Americano de Ecologia, 10 a 13 de Setembro de 2009, Sao Lourenco, Minas Gerais, Brasil Inscrição* 624:resumo 27.

Rau, J. R., and J. E. Jiménez. 2002. Diet of puma (*Puma concolor*, Carnivora: Felidae) in coastal and Andean ranges of southern Chile. *Stud. Neotrop. Fauna Environ.* 37:201–5.

Rau, J. R., D. R. Martínez, J. R. Low, and M. S. Tillería. 1995. Depredación por zorros chillas (*Pseudalopexgriseus*) sobre micromamíferos cursorialesescansoriales y arborícolas en un área silvestreprotegida del sur de Chile. *Rev. Chilena Hist. Nat.* 68:333–40.

Ray, C. E. 1958. Fusion of the cervical vertebrae in the Erethizontidae and Dinomyidae. *Breviora* 97:1–11.

Rebane, K. 2002. "The effects of historic climatic change and anthopogenic disturbance on rodent communities in Patagonia, Argentina." Honors' thesis, Stanford University, Palo Alto, California.

Rebelato, G. S. 2006. "Analise ecomorfológica de quatro espécieś de *Ctenomys* do Sul do Brasil (Ctenomyidae-Rodentia)." Master's thesis, Universidade Federal do Rio Grande do Sul, Porto Alegre, Brazil.

Reca, A. R., C. Ubeda, and D. Grigera. 1996. Prioridades de conservación de los mamíferos de Argentina. *Mastozool. Neotrop.* 3:87–118.

Redford, K. H. 1984. Mammalian predation on termites: tests with the burrowing mouse (*Oxymycterus roberti*) and its prey. *Oecologia* 65:145–52.

Redford, K. H., and J. F. Eisenberg. 1992. *Mammals of the Neotropics: The Southern Cone.* Vol. 2, *Chile, Argentina, Uruguay, Paraguay.* Chicago: University of Chicago Press, x + 430 pp. + 18 pls.

Redford, K. H. and G. A. B. da Fonseca. 1986. The role of gallery forests in the zoogeography of the Cerrado's nonvolant mammalian fauna. *Biotropica* 18:126–35.

Reed, E. C. 1877. Apuntes de la zoología de la hacienda de Cauquenes, provincia de Colcague. *An. Univ. Chile* 49:537–41.

Reed, J. T., and J. M. Brennan. 1975. The subfamily Leeuwenhoekinae in the Neotropics (Acarina: Trombiculidae). *Brigh. Young Univ. Sci. Bull.,* biol. ser. 20:1–42.

Reeder, D. M., K. M. Helgen, and D. E. Wilson. 2007. Global trends and biases in new mammal species discoveries. *Occas. Papers Mus. Texas Tech Univ.* 269:1–35.

Reguero, M. A., A. M. Candela, and R. N. Alonso. 2007. Biochronology and biostratigraphy of the Uquía Formation (Pliocene-Early Pleistocene, NW Argentina) and its significance in the Great American Biotic Interchange. *J. South Am. Earth Sci.* 23:1–16.

Reid, F. A. 1997. *A field guide to the mammals of Central America and southeast Mexico.* New York and Oxford: Oxford University Press, i–xvii + 1–334 + 52 plates.

———. 2009. *A field guide to the mammals of Central America and southeast Mexico.* 2nd ed. New York and Oxford: Oxford University Press, v–vxii + 1–346 + 52 plates.

Reid, J. L., and A. Sanchez-Gutierrez. 2010. Two new vertebrate prey in the diet of the blue-crowned Motmot (*Momotus momota*). *Zeledonia* 14:68–72.

Reig, O. A. 1958. Notas para una actualización del conocimiento de la fauna de la Formación Chapadmalal. I. Lista faunistica preliminary. *Acta Geol. Lilloana* 2:241–53.

———. 1964. Roedores y marsupiales del partido de General Pueyrredón y regiones adyacentes (provincia de Buenos Aires, Argentina). *Publ. Mus. Municip. Cienc. Nat. Mar del Plata Lorenzo Scaglia* 1:203–24.

———. 1965. Datos sobre la comunidad de pequeños mamíferos de la region costera del partido de General Pueyrredon y de los partidos limítrofes (Prov. de Buenos Aires, Argentina). *Physis* 25:205–11.

———. 1970. Ecological notes on the fossorial octodont rodent *Spalacopus cyanus* (Molina). *J. Mamm.* 51:592–601.

———. 1972. "The evolutionary history of the South American cricetid rodents." Ph.D. diss., University College, London.

———. 1977. A proposed unified nomenclature for the enameled components of the molar teeth of the Cricetidae (Rodentia). *J. Zool.* 181:227–41.

———. 1978. Roedores cricétidos del Plioceno superior de la provincia de Buenos Aires (Argentina*). Publ. Mus. Municip. Cienc. Nat. Mar del Plata Lorenzo Scaglia* 2:164–90.

———. 1980. A new fossil genus of South American cricetid rodents allied to *Wiedomys,* with an assessment of the Sigmodontinae. *J. Zool.* 192:257–81.

———. 1981. A refreshed unorthodox view of paleobiogeography of South American mammals (review of G. G. Simpson, 1980. *Splendid isolation: The curious history of South American mammals.* New Haven, CT: Yale University Press). *Evolution* 35:1032–35.

———. 1984. Distribuição geográfica e histórica evolutiva dos roedores muroides sulamericanos (Cricetidae: Sigmodontinae). *Rev. Bras. Genet.* 7:333–65.

———. 1986. Diversity patterns and differentiation of high Andean rodents. In *High Altitude Tropical Biogeography,* ed. F. Vuilleumier and M. Monasterio, 404–40. New York: Oxford University Press.

———. 1987. An assessment of the systematics and evolution of the Akodontini, with the description of new fossil species of *Akodon* (Cricetidae: Sigmodontinae). In *Studies in Neotropical mammalogy, essays in honor of Philip Hershkovitz,* ed. B. D. Patterson and R. M. Timm, 347–99. *Fieldiana Zool.* 39:frontispiece, viii + 1–506.

———. 1989. Karyotypic repatterning as one triggering factor in cases of explosive speciation. In *Evolution-*

ary Biology of Transient Unstable Populations, ed. A. Fontdevilla, 246–89. New York: Springer-Verlag.

———. 1994. New species of akodontine and scapteromyine rodents (Cricetidae) and new records of *Bolomys* (Akodontini) from the Upper Pliocene and Middle Plesistocene of Buenos Aires Province, Argentina. *Ameghiniana* 31:99–113.

Reig, O. A., M. Aguilera, M. A. Barros, and M. Useche. 1980. Chromosomal speciation in a rassenkreis of Venezuelan spiny rats (genus *Proechimys*, Rodentia, Echimyidae). *Genetica* 52/53:291–312.

Reig, O. A., M. Aguilera, and A. Pérez-Zapata. 1990. Cytogenetics and karyostystematics of South American oryzomyine rodents (Cricetidae: Sigmodontinae). II. High numbered karyotypes and chromosomal heterogeneity in Venezuelan *Zygodontomys*. *Z. Säugetierk.* 55:361–70.

Reig, O. A., M. A. Barros, M. Useche, M. Aguilera, and O. J. Linares. 1979. The chromosomes of spiny rats, *Proechimys trinitatis*, from Trinidad and eastern Venezuela (Rodentia, Echimyidae). *Genetica* 51:153–58.

Reig, O. A., J. R. Contreras, and M. Piantanida. 1966. Contribución a la elucidación de la sistemática de las entidades del género *Ctenomys* (Rodentia, Octodontidae). I. Relaciones de parentesco entre muestras de ocho poblaciones de tuco-tucos inferidas del estudio estadístico de variables del fenotipo y su correlación con las características del cariotipo. *Contrib. Cien. Ser. Zool.* 2:297–352.

Reig, O. A., and P. Kiblisky. 1968a. Chromosomes in four species of rodents of the genus *Ctenomys* (Rodentia, Octodontidae) from Argentina. *Experientia* 24:274–76.

———. 1968b. Los cromosomas somáticos de *Akodon urichi* (Rodentia, Cricetidae). *Acta Cien. Venez.* 19:73.

———. 1969. Chromosome multiformity in the genus *Ctenomys* (Rodentia, Octodontidae). *Chromosoma* 28:211–44.

Reig, O. A., P. Kiblisky, and O. J. Linares. 1971. Datos sobre *Akodon urichivenezuelensis* (Rodentia: Cricetidae), un ratón de interés para la investigación citogenética. *Acta Biol. Venez.* 7:157–89.

Reig, O. A., and J. A. Kirsch. 1988. Descubrimiento del segundo ejemplar conocido y de la presencia en Argentina del peculiar sigmodontino (Rodentia: Cricetidae) *Abrawayaomys ruschii* Souza Cuña et Cruz. *4 Jornadas Argentinas de Mastozoología, Resúmenes*, p. 80.

Reig, O. A., N. Olivo, and P. Kiblisky. 1971. The idiogram of the Venezuelan vole mouse *Akodon urichi*

venezuelensis Allen (Rodentia, Cricetidae). *Cytogenetics* 10:99–114.

Reig, O. A., and C. A. Quintana. 1991. A new genus of fossil octodontine rodent from the Early Pleistocene of Argentina. *J. Mamm.* 72:292–99.

———. 1992. Fossil ctenomyine rodents of the genus *Eucelophorus* (Caviomorpha: Octodontidae) from the Pliocene and Early Pleistocene of Argentina. *Ameghiniana* 29:363–80.

Reig, O. A., A. Spotorno, and R. Fernández. 1972. A preliminary survey of chromosomes in populations of the Chilean burrowing octodont rodent *Spalacopus cyanus* Molina (Caviomorpha, Octodontidae). *Biol. J. Linn. Soc.* 4:29–38.

Reig, O. A., M. Tranier, and M. A. Barros. 1980. Sur l'identification chromosomique de *Proechimys guyannensis* (E. Geoffroy, 1803) et de *Proechimys cuvieri* Petter, 1978 (Rodentia, Echimyidae). *Mammalia* 43:501–5. [1979 vol. 43, no. 4: pp. 439ff. "Achevé d'imprimer le 15 avril 1980."]

Reig, O. A., and M. Useche. 1976. Diversidad cariotipica y sistematica en poplaciones Venezolanas de *Proechimys* (Rodentia, Echimyidae), con datos adicionales sobre poblaciones de Peru y Colombia. *Acta Cient. Venez.* 27:132–40.

Reinhardt, J. 1849. Iagttagelser om en besynderlig hyppig, abnorm Haleløshed hos flere brasilianske Pigrotter. *Videnskab. Meddel., Kjöbenhavn* 7:110–15.

———. 1851. Description of *Carterodon sulcidens* Lund (translated by Wallich). *Ann. Mag. Nat. Hist.*, ser. 2, 10: 417–20.

Reis, N. R. dos, A. L. Peracchi, W. A. Pedro, and I P. de Lima, editors. 2006. *Mamíferos do Brasil*. Paraná: Londrina.

Reis, S. F. dos, L. C. Duarte, L. R. Monteiro, and F. J. Von Zuben. 2002a. Geographic variation in cranial morphology in *Thrichomys apereoides*: I. Geometric descriptors and patterns of variation in shape. *J. Mamm.* 83:333–44.

———. 2002b. Geographic variation in cranial morphology in *Thrichomys apereoides*: II. Geographic units, morphological discontinuities, and sampling gaps. *J. Mamm.* 83:345–53.

Reis, S. F. dos, and L. M. Pessôa. 1995. *Proechimys albispinus minor*, a new subspecies from the state of Bahia, northeastern Brazil (Rodentia: Echimyidae). *Z. Säugetierk.* 60:181–90.

———. 2004. Thrichomys apereoides. *Mamm. Species* 74:1–5.

Reis, S. F., L. M. Pessôa, and B. Bordin. 1992. Cranial phenotypic evolution in *Proechimys iheringi* Thomas (Rodentia: Echimyidae). *Zool. Scr.* 21:201–4.

Reis, S. F. dos, J. P. Pombal Jr., J. L. Nessimian, and L. M. Pessôa. 1996. Altitudinal distribution and feeding habits of *Blarinomys breviceps* (Winge, 1888). *Z. Säugetierk.* 61:253–55.

Reise, D., and M. H. Gallardo. 1989. Intraspecific variation in facing-water behavior of *Spalacopus cyanus* (Octodontidae, Rodentia). *Z. Säugetierk.* 54:331–33.

———. 1990. A taxonomic study of the South American genus *Euneomys* (Cricetidae, Rodentia). *Rev. Chilena Hist. Nat.* 63:73–82.

Reise, D., and W. Venegas S. 1974. Observaciones sobre el comportamiento de la fauna de micromamiferos en la region de Puerto Ibañez (Lago General Carrera), Aysen, Chile. *Bol. Soc. Biol. Concepc.* 47:71–85.

———. 1987. Catalogue of records, localities, and biotopes from research work on small mammals in Chile and Argentina. *Gayana, Zoología* 51:103–30.

Rendall, P. 1897. Natural history notes from the West Indies. *The Zoologist* 4:341–45.

Rengger, J. R. 1830. *Naturgeschichte der Säugethiere von Paraguay*. Basel: Schweighauserschen Buchhandlumg, xvi + 394 pp.

Rengifo, E. M., and R. Aquino. 2012. Descripción del nido de *Scolomys melanops* (Rodentia, Cricetidae) y su relación con *Lepidocaryum tenue* (Arecales, Arecaceae). *Rev. Peru. Biol.* 19:213–16.

Reus, M. L., B. Peco, C. de los Ríos, S. M. Giannoni, and C. M. Campos. 2013. Trophic interactions betweentwo medium-sized mammals: the case of the native *Dolichotis patagonum* and the exotic *Lepus europaeus* in a hyper-arid ecosystem. *Acta Theriol.* 58:205–14.

Reynafarje, B., and P. Morrison. 1962. Myoglobin levels in some tissues from wild Peruvian rodents native to high altitude. *J. Biol. Chem.* 237:2861–64.

Rezende, E. L., A. Cortes, L. D. Bacigalupe, R. F. Nespolo, and F. Bozinovic. 2003. Ambient temperature limits above-ground activity of the subterranean rodent *Spalacopus cyanus*. *J. Arid Environ.* 55:63–74.

Ribeiro, A. C., A. Maldonado Junior, P. S. D'Andrea, G. O. Vieira, and L. Rey. 1998. Susceptibility of *Nectomys rattus*(Pelzeln, 1883) to experimental infection with *Schistosoma mansoni*(Sambon, 1907): a potential reservoir in Brazil. *Mem. Instit. Oswaldo Cruz* 93:295–99.

Ribeiro, L. F., L. O. M. Conde, L. C.Guzzo, and R. Papalambropoulos. 2009. Behavioral patterns of *Guerlinguetus ingrami* (Thomas, 1901) from three natural populations in Atlantic forest fragments in Espírito Santo state, southeastern Brazil. *Nat. Online* 7(2):92–96.

Ribeiro, M. de F. S., P. L. B. da Rocha, L. A. F. Mendes, S. F. Perry, and E. S. de Oliveira. 2004. Physiological effects of short-term water deprivation in the South American sigmodontine rice rat *Oligoryzomys nigripes* and water rat *Nectomys squamipes* within a phylogenetic context. *Can. J. Zool.* 82:1326–35.

Ribeiro, N. A. B. 2006. "Análises cromossômicas e filogenia de roedores do gênero *Proechimys* (Echimyidae, Rodentia)." Ph.D. diss., Universidade Federal do Pará, Belém, Brazil, 95 pp.

Ribeiro, R., and J. Marinho-Filho. 2005. Community structure of small mammals (Mammalia, Rodentia) from Estacao Ecologica de Aguas Emendadas, Planaltina, Distrito Federal, Brazil. *Rev. Bras. Zool.* 22:898–907.

Ribeiro, R., C. R. Rocha, and J. Marinho-Filho. 2011. Natural history and demography of *Thalpomys lasiotis* (Thomas, 1916), a rare and endemic species from the Brazilian savanna. *Acta Theriol.* 56:275–82.

Ribeiro, R. D., and M. P. Barretto. 1977. Studies on wild reservoirs and vectors of *Trypanosoma cruzi*. Part 64. Natural infection of the rodent *Dasyprocta agouti agouti* by *Trypanosoma cruzi*. *Rev. Bras. Biol.* 37:233–40.

Richmond, C. W. 1899. On the dates of Lacepèdes 'Tableau.' *The Auk* 16:325–29.

Richter, M. H, J. H Hanson, M. N. Cajimat, M. L. Milazzo, and C. F. Fulhorst. 2010. Geographical range of Rio Mamoré Virus (family Bunyaviridae, genus *Hantavirus*) in association with the Small-Eared Pygmy Rice Rat (*Oligoryzomys microtis*). *Vector Borne Zoonotic. Dis.* 10:613–20.

Rieger, T. T., A. Langguth, and T. A. Waimer. 1995. Allozymic characterization and evolutionary relationships in the Brazilian *Akodon cursor* species group (Rodentia-Cricetidae). *Biochem. Genet.* 33:283–95.

Rinderknecht, A., and R. E. Blanco. 2008. The largest fossil rodent. *Proc. R. Soc. Lond. B Biol. Sci.* 275:923–28.

Rinker, G. C. 1954. The comparative myology of the mammalian genera *Sigmodon*, *Oryzomys*, *Neotoma*, and *Peromyscus* (Cricetinae), with remarks on their intergeneric relationships. *Misc. Publ. Mus. Zool., Univ. Michigan* 83:1–124.

Ríos-Uzeda, B., R. B. Wallace, and J. Vargas. 2004. La Jayupa de la Altura (*Cuniculus taczanowskii*, Rodentia, Cuniculidae), un Nuevo registro de mamífero para la fauna de Bolivia. *Mastozool. Neotrop.* 11:109–14.

Riva, R., O. R. Vidal, and N. L. Baro. 1977. Los cromosomas del genero *Holochilus*. II. El cariotipo de *H. brasiliensis vulpinus*. *Physis* 36:215–18.

Rivas, B. A., and O. J. Linares. 2006. Cambios en la forma de la pata posterior entre roedores sigmodontinos según su locomoción y hábitat. *Mastozool. Neotrop.* 13:205–15.

Rivas, B. A., and J. E. Péfaur. 1999a. Variacion geográfica en poblaciones venezolanas de *Oryzomys albigularis* (Rodentia: Muridae). *Mastozool. Neotrop.* 6:47–59.

———. 1999b. Variacion craneana entre sexo y edad en *Oryzomys albigularis* (Rodentia: Muridae). *Mastozool. Neotrop.* 6:61–70.

Rivas, B. A., and M. A. Salcedo. 2006. Lista actualizada de los mamíferos del Parque Nacional El Ávila, Venezuela. *Mem. Soc. Cien. Nat. La Salle* 164:29–56.

Rivas-Rodríguez, B. A., G. D'Elía, and O. Linares. 2010. Diferenciación morfológica en sigmodontinos (Rodentia: Cricetidae) de las Guayanas venezolanas con relación a su locomoción y habitat. *Mastozool. Neotrop.* 17:97–109.

Rivera P. C., R. E. G. Ittig, H. J. R. Fraire, S. Levis, and C. N. Gardenal. 2007. Molecular identification and phylogenetic relationship among the species of the genus *Oligoryzomys* (Rodentia, Cricetidae) present in Argentina, putative reservoirs of hantaviruses. *Zool. Scr.* 36:231–39.

Riverón, S. 2011. "Estructura poblacional e historia demográfica del "pericote patagónico" *Phyllotis xanthopygus*(Rodentia: Sigmodontinae) en Patagonia Argentina." Master's thesis, Universidad de la República, Montevideo, Uruguay.

Roberts, M., S. Brand, and E. Maliniak. 1985. The biology of captive prehensile-tailed porcupines, *Coendou prehensilis*. *J. Mamm.* 66:476–82.

Roberts, M. S., K. V. Thompson, and J. A. Cranford. 1988. Reproduction and growth in captive punare (*Thrichomys apereoides* Rodentia: Echimyidae) of the Brazilian Caatinga with reference to the reproductive strategies of the echimyidae. *J. Mamm.* 69:542–51.

Robereto, C. 1914. Los estratos araucano y sus fósiles. *An. Mus. Nac. Hist. Nat. Buenos Aires* 25:1–247.

Robinson, H. C., and R. C. Wroughton. 1911. On five new sub-species of oriental squirrels. *J. Fed. Malay States Mus.* 4:233–35.

Robinson, M., F. M. Catzeflis, J. Briolay, and D. Mouchiroud. 1997. Molecular phylogeny of rodents, with special emphasis on murids: evidence from nuclear gene LCAT. *Mol. Phylogenet. Evol.* 8:423–34.

Robinson, W., and M. W. Lyon Jr. 1901. An annotated list of mammals collected in the vicinity of La Guaira, Venezuela. *Proc. U.S. Natl. Mus.* 24:135–62.

Robles, L., and M. A. Ruggero. 2001. Mechanics of the mammalian cochlea. *Physiol. Rev.* 81:1305–52.

Robles, M. del R. 2008. "Nematodes Oxiuridae, Trichuridae y Capillariidae en roedores Akodontini (Cricetidae, Sigmodontinae) de la Cuenca del Plata (Argentina): su importancia en la interpretación de las relaciones parásito-hospedador-ambiente." Ph.D. diss., Universidad Nacional de La Plata, Argentina.

———. 2010. La importancia de los nematodes Syphaciini (Syphaciinae-Oxyuridae) como marcadores específicos de sus hospedadores. *Mastozool. Neotrop.* 17:305–15.

Robles, M. del R., O. Bain, and G. T. Navone. 2012. Description of a new Capillariinae (Nematoda: Thichuridae) from *Scapteromys aquaticus* (Cricetidae: Sigmodontinae) from Buenos Aires, Argentina. *J. Parasitol.* 98:627–39.

Robles, M. del R., C. Galliari, and G. T. Navone. 2012. New records of nematode parasites from *Euryzygomatomys spinosus* (Rodentia, Echimyidae) in Misiones Province, Argentina. *Mastozool. Neotrop.* 19:353–58.

Robles, M. del R., and G. T. Navone. 2007. A new species of *Syphacia* (Nematoda: Oxyuridae) from *Oligoryzomys nigripes* (Rodentia: Cricetidae) in Argentina. *Parasitol. Res.* 101:1069–75.

———. 2010. Redescription of *Syphacia venteli* Travassos 1937 (Nematoda: Oxyuridae) from *Nectomys squamipes* in Argentina and Brazil and description of a new species of *Syphacia* from *Melanomys caliginosus* in Colombia. *Parasitol. Res.* 106:1117–26.

Rocha, C. R., R. Ribeiro, F. S. C. Takahasi, and J. Marinho-Filho. 2011. Microhabitat use by rodent species in a central Brazilian cerrado. *Mamm. Biol.* 76:651–653.

Rocha, P. L. B. 1991. "Ecologia e morfologia de uma nova espécie de *Proechimys* (Rodentia: Echimyidae) das dunas interiores do Rio São Francisco, Bahia." Master's thesis, Universidade de São Paulo, Brazil.

———. 1996. *Proechimys yonenagae*, a new species of spiny rat (Rodentia: Echimyidae) from fossil sand dunes in the Brazilian Caatinga. *Mammalia* 59:537–49. [1995 vol. 59, no. 4: pp. 481ff. "Achevé d'imprimer le 30 avril 1996."]

Rocha, P. L. B., S. Renous, A. Abourachid, and E. Höfling. 2007. Evolution toward asymmetrical gaits in Neotropical spiny rats (Rodentia: Echimyidae): evidences favoring adaptation. *Can. J. Zool.* 85:709–17.

Rocha, R. G., B. M. da A. Costa, and L. P. Costa. 2011. *Rhipidomys ipukensis* n. sp. In Small mammals of the mid-Araguaia River in central Brazil, with the description of a new species of climbing rat, R. G. Rocha et al., 23–27. *Zootaxa* 2789:1–34.

Rocha, R. G., E. Ferreira, B. M. A. Costa, I. C. M. Martins, Y. L. R. Leite, L. P. Costa, and C. Fonseca. 2011. Small mammals of the mid-Araguaia River in central Brazil, with the description of a new species of climbing rat. *Zootaxa* 2789:1–34.

Rocha, R. G., E. Ferreira, Y. L. R. Leite, C. Fonseca, and L. P. Costa. 2011. Small mammals in the diet of Barn owls, *Tytoalba* (Aves: Strigiformes) along the mid-Araguaia River in central Brazil. *Zoologia* 28:709–16.

Rocha, R. G., C. Fonseca, Z. Zhou, Y. L. R. Leite, and L. P. Costa.2012. Taxonomic and conservation status of the elusive *Oecomys cleberi* (Rodentia, Sigmodontinae) from central Brazil. *Mamm. Biol.* 77:414–19.

Rocha-Barbosa, O., J. S. L. Bernardo, M. F. C. Loguercio, T. R. O. Freitas, J. R. Santos-Mallet, and C. J. Bidau. 2013. Penial morphology in three species of Brazilian tuco-tucos, *Ctenomys torquatus*, *C. minutus*, and *C. flamarioni* (Rodentia: Ctenomyidae). *Brazilian J. Zool.* 73:201–9.

Rocha-Barbosa, O., D. Youlatos, J.-P. Gasc, S. and S. Renous. 2002. The clavicular region of some cursorial Cavioidea (Rodentia: Mammalia). *Mammalia* 66:413–21.

Rocha-Mendes, F., S. B. Mikich, J. Quadros, and W. A. Pedro. 2010. Feeding ecology of carnivores (Mammalia, Carnivora) in Atlantic Forest remnants, southern Brazil. *Biota Neotrop.* 10:21–30.

Roda, S. A. 2006. Dieta de *Tyto alba* na Estação Ecológica do Tapacurá, Pernambuco, Brasil. *Rev. Brasil. Ornitol.* 14:449–52.

Rodarte, R., and G. Lessa. 2010. Variação sexual em caracteres morfológicos cranianos de *Akodon serrensis* (Thomas, 1902) (Rodentia: Sigmodontinae) do Parque Estadual da Serra do Brigadeiro, MG. *Resumos, 5*th *Congresso Brasileiro de Mastozoologia.*

Rode, P. 1943. Catalogue des types de mammifères du Muséum National d'Histoire Naturelle. Ordre des rongeurs. I. Sciuromorphes. *Bull. Mus. Natl. Hist. Nat., Paris*, 15, sér. 2 :275–82 and 382–85.

———. 1945. Catalogue des types de mammifères du Muséum National d'Histoire Naturelle. Ordre des rongeurs. *Bull. Mus. Natl. Hist. Nat., Paris*, 17, sér. 2 :24–31, 95–102, 201–8, and 292–300.

Rodrigues, F. H. G., A. Hass, A. C. R. Lacerda, R. L. S. C. Grando, M. A. Bagno, A. M. R. Bezerra, and W. R. Silva. 2007. Feeding habits of the maned wolf (*Chrysocyon brachyurus*) in the Brazilian Cerrado. *Mastozool. Neotrop.* 14:37–51.

Rodrigues, F. H. G., L. Silveira, A. T. A. Jácomo, A. P. Carmignotto, A. M. R. Bezerra, D. Coelho, H. Garbogini, et al. 2002. Composição e caracterização da fauna de mamíferos do Parque Nacional das Emas, Goiás. *Rev. Bras. Zool.*19:589–600.

Rodrigues, R. F., A. M. Carter, C. E. Ambrosio, T. Carlesso dos Santos, and M. A. Miglino. 2006. The subplacenta of the red-rumped agouti (*Dasyprocta leporina* L). *Reprod. Biol. Endocrinol.* 4:31.

Rodrigues, V. L., and A. do N. Ferraz Filho. 1984. *Trypanosoma (Megatrypanum) rochasilvai*, sp. n., encontrada no estado de São Paulo, Brasil, parasitando *Oryzomys laticeps* (Leche, 1886) (Rodentia-Cricetidae). *Rev. Bras. Biol.* 44:299–304.

Rodrigues Roque, A. L., P. S. D'Andrea, G B. de Andrade, and A. M. Jansen. 2005. *Trypanosoma cruzi*: Distinct patterns of infection in the sibling caviomorph rodent species *Thrichomys apereoides laurentius* and *Thrichomys pachyurus* (Rodentia, Echimyidae). *Exp. Parasitol.* 111:37–46.

Rodríguez, D., C. Lanzone, V. Chillo, P. A. Cuello, S. Albanese, A. A. Ojeda, and R. A. Ojeda. 2012. Historia natural de un roedor raro del desierto argentino,*Salinomys delicatus* (Cricetidae: Sigmodontinae). *Rev. Chilena Hist. Nat.* 85:13–27

Rodríguez, D., C. Lanzone, P. Teta, and U. F. J. Pardiñas. 2012. *Salinomys delicatus*. In *Libro Rojo de mamíferos amenazados de la Argentina*, ed. R. A. Ojeda, V. Chillo and G. B. Diaz Isenrath. Buenos Aires: Sociedad Argentina para el Estudio de los Mamíferos, 257 pp.

Rodríguez, M., R. Montoya, and W. Venegas. 1983. Cytogenetic analysis of some Chilean species of the genus *Akodon* Meyen (Rodentia, Cricetidae). *Caryologia* 36:129–38.

Rodríguez, V. A. 2005. Determinación específica y niveles de polimorfismo alozímico de *Graomys* (Rodentia, Muridae) de los alrededores de Comodoro Rivadavia. *Mastozool. Neotrop.* 12:105–7.

Rodríguez, V. A., and G. R. Theiler. 2007. Micromamíferos de la región de Comodoro Rivadavia (Chubut, Argentina). *Mastozool. Neotrop.* 14:97–100.

Rodríguez-Mahecha, J. V., J. I. Hernández-Camacho, T. R. Defler, M. Alberico, R. B. Mast, R. A. Mittermeier, and A. Cadena. 1995. Mamíferos Colombianos: sus nombres communes e indígenas. *Occas. Papers Conserv. Biol.* 3:1–56.

Rodríguez-Serrano, E., R. A. Cancino, and R. E. Palma. 2006. Molecular phylogeography of *Abrothrix olivaceus* (Rodentia: Sigmodontinae) in Chile. *J. Mamm.* 87:971–80.

Rodríguez-Serrano, E., C. E. Hernández, and R. E. Palma. 2008. A new record and an evaluation of the phylogenetic relationships of *Abrothrix olivaceus markhami* (Rodentia: Sigmodontinae). *Mamm. Biol.* 73:307–17.

Rodríguez-Serrano, E., E. R. Palma, and C. E. Hernández. 2008. The evolution of ecomorphological traits within the Abrothrichini (Rodentia: Sigmodontinae): A Bayesian phylogenetics approach. *Mol. Phylogenet. Evol.* 48:473–80.

Roger, O. 1887. Verzeichniss der bisher bekannten fossilen Säugethiere. *Ber. Nat. Sensch. Vereines Schwab. Neuburg* 29:1–162.

Rogers, D. S. 1989. Evolutionary implications of chromosomal variation among spiny pocket mice, genus *Heteromys* (order Rodentia). *Southwest. Nat.* 34:85–100.

———. 1990. Genic evolution, historical biogeography, and systematic relationships among spiny pocket mice (subfamily Heteromyinae). *J. Mamm.* 71:668–85.

Rogers, D. S., E. A. Arenas, F. X. González-Cózatl, D. K. Hardy, J. D. Hanson, and N. Lewis-Rogers. 2009. Molecular phylogenetics of *Oligoryzomys fulvescens* based on cytochrome-*b* gene sequences, with comments on the evolution of the genus *Oligoryzomys*. In *60 años de la Colección Nacional de Mamíferos del Instituto de Biología, UNAM. Aportaciones al conocimiento y conservación de los mamíferos Mexicanos*, ed. F. A. Cervantes, Y. Hortelano Moncada, and J. Vargas Cuenca, 179–92. Mexico, DF: Universidad Autonoma de México, 317 pp.

Rogers, D. S., M. D. Engstrom, and E. Arellano. 2004. Phylogenetic relationships among Peromyscine rodents: allozyme evidence. In *Contribuciones mastozoologicas en homenaje a Bernardo Villa*, ed. V. Sánchez-Cordero and R. A. Medellín, 427–40. Mexico, DF: Instituto de Biología e Instituto de Ecología, UNAM, 500 pp.

Rogers, D. S., and D. J. Schmidly. 1982. Systematics of spiny pocket mice (genus *Heteromys*) of the *desmarestianus* species group from México and northern Central America. *J. Mamm.* 63:375–86.

Rogers, D. S., and V. L. Vance. 2005. Phylogenetics of spiny pocket mice (genus *Liomys*): analysis of cytochrome-b based on multiple heuristic approaches. *J. Mamm.* 86:1085–94.

Roguin, L. 1986. Les mammifères du Paraguay dans le collections du Muséum de Genève. *Rev. Suisse Zool.* 93:1009–22.

Roig, V. G. 1965. Elenco sistemático de los mamíferos y aves de la provincia de Mendoza y notas sobre su distribución geográfica. *Bol. Est. Geogr.* 49:175–227.

Roig, V. G., and O. A. Reig. 1969. Precipitin test relationships among Argentinian species of the genus *Ctenomys* (Rodentia, Octodontidae). *Comp. Biochem. Physiol.* 30:665–72.

Rojas-Robles, R., A. Correa, and E. Serna-Sanchez. 2008. Sombra de semillas, supervivencia de plantulas y distribución especial de la palma *Oenocarpus bataua*, en un bosque de los Andes colombianos. *Actual. Biol.* (Medellín) 30(89):135–50.

Roldán, E. R. S., A. D. Vitullo, M. S. Merani, and I. von Lawzewitsch. 1985. Cross fertilization in vivo and in vitro between 3 species of vesper mice, *Calomys* (Rodentia, Cricetidae). *J. Exp. Zool.* 233:433–42.

Roldán, V. L. 2010. "Ensambles de micromamíferos vivientes (roedores y marsupiales) de la Patagonia austral." Undergraduate thesis, Universidad Nacional de la Patagonia San Juan Bosco, Trelew, Argentina.

Romo, M. C. 1995. Food habits of the Andean fox (*Pseudalopex culpaeus*) and notes on the mountain cat (*Felis colocolo*) and puma (*Felis concolor*) in the Río Abiseo National Park, Perú. *Mammalia* 59:335–44.

Rood, J. P. 1963. Observations on the behavior of the spiny rat *Heteromys melanoleucus* in Venezuela. *Mammalia* 27:186–92.

———. 1970. Ecology and social behavior of the desert cavy (*Microcavia australis*). *Am. Midl. Nat.* 83:415–54.

———. 1972. Ecological and behavioral comparisons of three genera of Argentine cavies. *Anim. Behav. Monogr.* 5:1–83.

Rood, J. P., and F. H. Test. 1968. Ecology of the spiny rat, *Heteromys anomalus*, at Rancho Grande, Venezuela. *Am. Midl. Nat.* 79:89–102.

Roosevelt, T. 1914. *Through the Brazilian wilderness.* New York: Charles Scribner's Sons, x+410 pp.

Rosa, A. O., and E. M. Vieira. 2010. Comparação da diversidade de mamíferos entre áreas de floresta de restinga e áreas plantadas com *Pinus elliottii* (Pinaceae) no sul do Brasil. In *Mamíferos de Restingas e Manguezais do Brasil*, ed. L. M. Pessôa, W. C. Tavares, and S. Sciliano, 225–42. Rio de Janeiro: Sociedade Brasileira de Mastozoologia, 282 pp.

Rosa, C. C., T. Flores, J. C. Pieczarka, R. V. Rossi, M. I. C. Sampaio, J. D. Rissino, P. J. S. Amaral, and C. Y. Nagamachi. 2012. Genetic and morphological variability in South American rodent *Oecomys* (Sigmodontinae, Rodentia): evidence for a complex of species. *J. Genet.* 91:265–77.

Rosa, E. S. T., J. N. Mills, P. J. Padula, M. R. Elkhoury, T. G. Ksiazek, W. S. Mendes, E. D. Santos, et al. 2005. Newly recognized hantaviruses associated with hantavirus pulmonary syndrome in northern Brazil: partial genetic characterization of viruses and serologic implication of likely reservoirs. *Vector Borne Zoonotic Dis* 5:11–19.

Rosa, E. S. T., D. B. A. Medeiros, M .R. T. Nunes, D. B. Simith, A. S. Pereira, M. R. Elkhoury, M. L. Nunes,et al. 2011. Pigmy Rice Rat as potential host of Castelo dos Sonhos hantavirus. *Emerg. Infect. Dis.* 17:1527–30.

Rosenmann, A. M. 1959. *Ctenomys fulvus* Phil., su hábitat (Rodentia, Ctenomyidae). *Invest. Zool. Chilenas* 5:217–20.

Rosi, M. I. 1983. Notas sobre la ecologia, distribucion y sistematica de *Graomys griseoflavus griseoflavus* (Waterhouse, 1837) (Rodentia,Cricetidae) in la Provincia de Mendoza. *Hist. Nat.* 3:161–67.

Rosi, M. I., M. I. Cona, S. Puig, F. Videla, and V. G. Roig. 1996. Size and structure of burrow systems of the fossorial rodent *Ctenomys mendocinus* in the piedmont of Mendoza Province, Argentina. *Z. Säugetierk.* 61:352–64.

Rosi, M. I., M. I. Cona, and V. G. Roig. 2002. Estado actual del conocimiento del roedor fosorial *Ctenomys mendocinus* Philippi 1869 (Rodentia: Ctenomyidae). *Mastozool. Neotrop.* 9:277–95.

Rosi, M. I., M. I. Cona, V. G. Roig, A. I. Massarini, and D. H. Verzi. 2005. Ctenomys mendocinus. *Mamm. Species* 777:1–6.

Rosi, M. I., M. I. Cona, F. Videla, S. Puig, and V. G. Roig. 2000. Architecture of *Ctenomys mendocinus* (Rodentia) burrows from two habitats differing in abundance and complexity of vegetation. *Acta Theriol.* 45:491–505.

Rosi, M. I., S. Puig, M. I. Cona, F. Videla, F. Méndez, and V. G. Roig. 2009. Diet of a fossorial rodent (Octodontidae), above-ground food availability, and changes related to cattle grazing in the central Monte (Argentina). *J. Arid Environ.* 73:273–79.

Rosi, M. I., S. Puig, F. Videla, M. I. Cona, and V. G. Roig. 1996. Ciclo reproductivo y estructura etaria de *Ctenomys mendocinus* (Rodentia, Ctenomyidae) del Piedemonte de Mendoza, Argentina. *Ecol. Austral* 6:87–93.

Rosi, M. I., S. Puig, Videla, L. Madoery, and V. G. Roig. 1992. Estudio ecológico del roedor subterráneo *Ctenomys mendocinus* en la precordillera de Mendoza, Argentina: ciclo reproductivo y estructura etaria. *Rev. Chilena Hist. Nat.* 65:221–33.

Rossi, M. S., O. A. Reig, and J. Zorzopulos. 1990. Evidence for rolling-circle replication in a major satellite DNA from the South American rodents of the genus *Ctenomys. Mol. Biol. Evol.* 7:340–50.

Rossin, M. A. 2004. Study of the interrelations existing between *Ctenomy talarum* (Rodentia: Octodontidae) of the Necochea district and its parasites. *Mastozool. Neotrop.* 11:119.

Roth, V. L. 1996. Cranial integration in the Sciuridae. *Am. Zool.* 36:14–23.

Roth, V. L., and J. M. Mercer. 2008. Differing rates of macroevolutionary diversification in arboreal squirrels. *Curr. Sci.* 95:857–61.

Rouaux, R., C. Giai, N. Fernandez, V. Bianco, and L. J. M. De Santis. 2003. Stomach structure in *Akodon azarae* and *Calomys musculinus* (Rodentia; Muridae). *Mastozool. Neotrop.* 10:115–21.

Rovereto, C. 1914. Los estratos Araucanos y sus fósiles. *An. Mus. Nac. Hist. Nat. Buenos Aires* 25:1–247 (31 plates).

Rowe, D. L., K. A. Duna, R. M. Adkins, and R. L. Honeycutt. 2010. Molecular clocks keep dispersal hypotheses afloat: evidence for trans-Atlantic rafting by rodents. *J. Biogeogr.* 37:305–24.

Rowe, D. L., and R. L. Honeycutt. 2002. Phylogenetic relationships, ecological correlates, and molecular evolution within the Cavioidea (Mammalia, Rodentia). *Mol. Biol. Evol.* 19: 263–77.

Rozas, M., C. Botto-Mahan, X. Coronado, S. Ortiz, P. E. Cattan, and A. Solari. 2005. Short report: *Trypanosoma cruzi* infection in wild mammals from a chagasic area of Chile. *Am. J. Trop. Med. Hyg.* 73:517–19.

Rozas, M., S. De Doncker, V. Adaui, X. Coronado, C. Barnabé, M. Tibyarenc, A. Solari, and J.-C. Dujardin. 2007. Multilocus polymerase chain reaction restriction fragment-length polymorphism genotyping of *Trypanosoma cruzi* (Chagas Disease): taxonomic and clinical applications. *J. Infect. Dis.* 195:1381–88.

Roze, U. 1989. *The North American porcupine.* Washington, DC: Smithsonian Institution Press, 296 pp.

Ruedas, L. A., J. A. Cook, T. L. Yates, and J. W. Bickham. 1993. Conservative genome size and rapid chromosomal evolution in the South American tuco-tucos (Rodentia: Ctenomyidae). *Genome* 36:449–58.

Rumiz, D. W., and C. Eulert. 1996. Mamíferos. In *Diagnóstico del component de fauna Silvestre para el plan de manejo del Parque Nacional Amboro*, ed. D. Rumiz, 2–1 to 2–22. Santa Cruz: Informe a FAN-TNC.

Rusconi, C. 1928. Dispersión geográfica de los tuco-tucos vivientes (*Ctenomys*) en la región neotropical. *An. Soc. Argent. Estud. Geogr. GAEA* 3:235–54.

———. 1931a. Las especies fósiles del género *Ctenomys*, con descripción de nuevas especies. *An. Soc. Cien. Argent.* 112:129–42 (1), 217–36 (2).

———. 1931b. Las especies fósiles del género *Ctenomys. An. Soc. Cien. Argen.* 112:129–63.

———. 1934a. Tercera noticia sobre los vertebrados fósiles de las arenas puelchenses de villa Ballester. *An. Soc. Cien. Argent.* 117:19–37.

———. 1934b. Una nueva subespecie de tucu tuco viviente. *Rev. Chilena Hist. Nat.* 38:108–10.

———. 1935. Tres nuevos espécies de mamíferos del Puelchense de Villa Ballester. *Bol. Paleontol. Buenos Aires* 5:1–4.

———. 1936. Nuevo género de roedores del Puelchense de Villa Ballester. *Bol. Paleontol. Buenos Aires* 7:1–4.

Russell, R. J. 1960. Pleistocene pocket gophers from San Josecito Cave, Nuevo León, México. *Univ. Kansas Publ., Mus. Nat. Hist.* 9:539–48.

———. 1968. Evolution and classification of the pocket gophers of the subfamily Geomyinae. *Univ. Kansas Publ., Mus. Nat. Hist.* 16:473–579.

Saavedra, B. 2003. "Disminución en tamaño poblacional y asimetría fluctuante en *Octodon bridgesi* (Rodentia), taxón especialista de habitat." Ph.D. diss., Universidad de Chile, Santiago.

Saavedra, B., D. Quiroz, and J. Iriarte. 2003. Past and present small mammals of Isla Moche (Chile). *Mamm. Biol.* 68:365–71.

Saavedra, B., and J. Simonetti. 2000. A northern and threatened population of *Irenomys tarsalis* (Mammalia: Rodentia) from central Chile. *Z. Säugetierk.* 64: 243–45.

———. 2001. New records of *Dromiciops gliroides* (Microbiotheria: Microbiotheriidae) and *Geoxus valdivianus* (Rodentia: Muridae) in central Chile: their implications for biogeography and conservation. *Mammalia* 65: 96–100.

———. 2003. Holocene distribution of octodontid rodents in central Chile. *Rev. Chilena Hist. Nat.* 76:383–89.

Saavedra-Rodríguez, C. A., G. H. Kattan, K. Osbahr, and J. G. Hoyos. 2012. Multiscale patterns of habitat and space use by the pacarana *Dinomys branickii*: factors limiting its distribution and abundance. *Endang. Species Res.* 16:273–81.

Sabat, P., and C. Veloso. 2003. Ontogenic development of intestinal disaccharidases in the precocial rodent *Octodon degus* (Octodontidae). *Comp. Biochem. Physiol. A Mol. Integr. Physiol.* 13482:393–97.

Sabrosky, C. W. 1967. Comment on the application to place Fischer's names for D'Azara's rodents on the Official List. Z. N.(S.) 1774. *Bull. Zool. Nomencl.* 24:141.

Sachser, N. 1986. Different forms of social organization at high and low population densities in guinea pigs. *Behaviour* 97:252–72.

———. Sachser, N. 1998. Of domestic and wild guinea pigs: studies in sociophysiology, domestication and social evolution. *Naturwissenschaften* 85:307–17.

Sachser, N., E. Schwart-Weig, A. Keil, and J. T. Epplen. 1999. Behavioural strategies, testes size, and reproductive success in two caviomorph rodents with different mating strategies. *Behaviour* 136:1203–17.

Seamon, J. O., and G. H. Adler. 1999. Short-term use of space by a Neotropical forest rodent, *Proechimys semispinosus*. *J. Mamm.* 80:899–904.

Safford, H. F. 1999. Brazilian Paramos. I. An introduction to the physical environment and vegetation of the campos de altitude. *J. Biogeogr.* 26:696–712.

Sage, R. D., J. R. Contreras, V. G. Roig, and J. L. Patton. 1986. Genetic variation in the South American burrowing rodents of the genus *Ctenomys* (Rodentia: Ctenomyidae). *Z. Säugetierk.* 51:158–72.

Sage, R. D., O. P. Pearson, J. Sanguinetti, and A. K. Pearson. 2007. Ratada 2001: a rodent outbreak following the flowering of bamboo (*Chusquea culeou*) in southwestern Argentina. In *The quintessential naturalist: Honoring the life and legacy of Oliver P. Pearson*, ed. D. A. Kelt, E. P. Lessa, J. Salazar-Bravo, and J. L. Patton, 177–224. *Univ. California Publ. Zool.* 134:v-xii + 1–981.

Sahores, M., and A. Trejo. 2004. Diet shift of barn owls (*Tyto alba*) after natural fires in Patagonia, Argentina. *J. Raptor Res.* 38:174–77.

Salazar, J. A., M. L. Campbell, S. Anderson, S. L. Gardner, and J. L. Dunnum. 1994. New records of Bolivian mammals. *Mammalia* 58:125–30.

Salazar-Bravo, J., J. W. Dragoo, M. D. Bowen, C. J. Peters, T. G. Ksiazek, and T. L. Yates. 2002. Natural nidality in Bolivian hemorrhagic fever and the systematics of the reservoir species. *Infect. Genet. Evol.* 1:191–99.

Salazar-Bravo, J., J. W. Dragoo, D. S. Tinnin, and T. L. Yates. 2001. Phylogeny and evolution of the neotropical rodent genus *Calomys*: inferences from mtDNA sequence data. *Mol. Phylogenet. Evol.* 20:173–84.

Salazar-Bravo, J., and L. Emmons. 2003. Mamíferos. In *Biodiversidad: La riqueza de Bolivia estado de conocimiento y conservación*, ed. P. L. Ibisch and G. Mérida, 146–48. Santa Cruz de La Sierra: Editorial FAN, Santa Cruz de La Sierra, 638 pp.

Salazar-Bravo, J., J. Miralles-Salazar, A. Rico-Cernohorska, and J. Vargas. 2011. First record of *Punomys* (Rodentia: Sigmodontinae) in Bolivia. *Mastozool. Neotrop.* 18:143–46.

Salazar-Bravo, J., U. F. J. Pardiñas, and G. D'Elía. 2013. A phylogenetic appraisal of the Sigmodontinae (Rodentia, Cricetidae) with emphasis on phyllotine genera: systematics and biogeography. *Zool. Scr.* 42:250–61.

Salazar-Bravo, J., L. A. Ruedas, and T. L. Yates. 2002. Mammalian reservoirs of arenaviruses. *Curr. Top. Microbiol. Immunol.* 262:25–64.

Salazar-Bravo, J., T. Tarifa, L. F. Aguirre, E. Yensen, and T. L. Yates. 2003. Revised checklist of Bolivian mammals. *Occas. Papers Mus. Texas Tech Univ.* 220:1–27.

Salazar-Bravo, J., and T. L. Yates. 2007. A new species of *Thomasomys* (Cricetidae: Sigmodontinae) from central Bolivia. In *The quintessential naturalist: Honoring the life and legacy of Oliver P. Pearson*, ed. D. A. Kelt, E. P. Lessa, J. Salazar-Bravo, and J. L. Patton, 747–74. *Univ. California Publ. Zool.* 134:v-xii + 1–981.

Salazar-Bravo, J., E. Yensen, T. Tarifa, and T. L. Yates. 2002. Distributional records of Bolivian mammals. *Mastozool. Neotrop.* 9:70–78.

Sallam, H. M., E. R. Seiffert, M. E. Steiper, and E. L. Simons. 2009. Fossil and molecular evidence constrain scenarios for the early evolution and biogeographic history of hystricognathous rodents. *Proc. Natl. Acad. Sci. USA* 106:16722–27.

Salles, L. O., C. Cartelle, P. G. Guedes, P. C. Boggiani, A. Janoo, and C. A. M. Russo. 2006. Quaternary mammals from Serra da Bodoquena, Mato Grosso do Sul, Brazil. *Bol. Mus. Nac., Rio de Janeiro*, nova sér, zool., 521:1–12.

Salles, L. O., G. S. Carvalho, M. Weksler, F. L. Sicuro, A. R. Camardella, P. G. Guedes, L. S. Avilla, E. A. P. Abrantes, V. Sahate, and I. S. A. Costa. 1999. Fauna de Mamíferos do Quaternário de Serra da Mesa (Goiás, Brasil). *Publ. Avulsas Mus. Nac., Rio de Janeiro* 78:1–15.

Salvador, C. H., and F. A. S. Fernandez. 2008a. Population dynamics and conservation status of the insular cavy *Cavia intermedia* (Rodentia: Caviidae). *J. Mamm.* 89:721–29.

———. 2008b. Reproduction and growth of a rare, island-endemic cavy (*Cavia intermedia*) from southern Brazil. *J. Mamm.* 89:909–15.

Sanborn, C. C. 1929. The land mammals of Uruguay. *Field Mus. Nat. Hist.*, zool. ser., 17:147–65.

———. 1931a. Notes on *Dinomys*. *Field Mus. Nat. Hist.*, zool. ser., 18:149–55.

———. 1931b. A new *Oxymycterus* from Misiones, Argentina. *Proc. Biol. Soc. Washington* 44:1–2.

———. 1947a. The South American rodents of the genus *Neotomys*. *Field. Zool.* 31:51–57.

———. 1947b. Geographical races of the rodent *Akodon jelskii* Thomas. *Fieldiana Zool.* 31:133–42.

———. 1949a. Cavies of southern Peru. *Proc. Biol. Soc. Washington* 63:133–34.

———. 1949b. Mammals from the Rio Ucayali, Peru. *J. Mamm.* 30:277–88.

———. 1949c. The status of *Akodon andinus polius* Osgood. *J. Mamm.* 30: 315.

———. 1949d. A new species of rice rat (*Oryzomys*) from the coast of Peru. *Publ. Mus. Hist. Nat. Javier Prado Ser. A Zool.* 3:1–4.

———. 1950. Small rodents from Peru and Bolivia. *Publ. Mus. Hist. Nat. Javier Prado Ser. A Zool.* 6:1–16.

———. 1951. Mammals from Marcapata, southeastern Peru. *Publ. Mus. Hist. Nat. Javier Prado Ser. A Zool.* 6:1–26.

———. 1953. Mammals from the Departments of Cuzco and Puno, Peru. *Publ. Mus. Hist. Nat. Javier Prado Ser. A Zool.*, 12:1–8.

Sanborn, C. C., and O. P. Pearson. 1947. The tuco-tucos of Peru (genus *Ctenomys*). *Proc. Biol. Soc. Washington* 60:135–38.

Sánchez, A., J. A. Marchal, I. Romero-Fernández, E. Pinna-Senn, M. I. Ortiz, J. L. Bella, and J. A. Lisanti. 2010. No differences in the Sry gene between males and XY females in *Akodon* (Rodentia, Cricetidae). *Sex. Dev.* 4:155–61.

Sánchez, F., P. Sánchez-Palomino, and A. Cadena. 2004. Inventario de mamíferos en un bosque de los Andes centrales de Colombia. *Caldasia* 26:291–309.

Sánchez H. J., J. Ochoa G., and R. S. Voss. 2001. Rediscovery of *Oryzomys gorgasi* (Rodenta: Muridae), with notes on taxonomy and natural hHistory. *Mammalia* 65:205–14.

Sánchez, J. P. 2013. "Sifonápteros parásitos de los roedores sigmodontinos de la Patagonia norte de la Argentina: estudios sistemáticos y ecológicos." Ph.D. diss., Universidad Nacional de La Plata, Argentina.

Sánchez, J. P., S. Nava, M. Lareschi, D. E. Udrizar Sauthier, A. J. Mangold, and A. A. Guglielmone. 2010. Host range and geographical distribution of *Ixodes sigelos* (Acari: Ixodidae). *Exp. Appl. Acarol.* 52:199–205.

Sánchez, J. P., D. E. Udrizar Sauthier, and M. Lareschi. 2009. New records of fleas (Insecta, Siphonaptera) parasites on sigmodontine rodents (Cricetidae) from southern Patagonia, Argentina. *Mastozool. Neotrop.* 16:243–46.

Sánchez-Villagra, M. R., O. Aguilera, and I Horovitz. 2003. The anatomy of the world's largest extinct rodent. *Science* 301:1708–10.

Sandweiss, D. H., and E. S. Wing. 1997. Ritual rodents: the guinae pigs of Chincha, Peru. *J. Field Archaeol.* 24:47–58.

Santos, I. B., W. L. R. Oliver, and A. B. Rylands. 1987. Distribution and status of two species of tree porcupines, *Chaetomys subspinosus* and *Sphiggurus insidiosus*, in south-east Brazil. *Dodo* 24:43–60.

Santos, J. W., and E. A. Lacey. 2011. Burrow sharing in the desert-adapted torch-tail spiny rat, *Trinomys yonenagae*. *J. Mamm.* 92:3–11.

Santos, R. A. L., and R. P. B. Henriques. 2010. Spatial variation and the habitat influence in the structure of communities of small mammals in areas of rocky fields in the Federal District. *Biota Neotrop.* 10:31–38.

Santos, T. G. dos, M. R. Spies, K. Kopp, R. Trevisan, and S. Z. Cechin. 2008. Mamíferos do campus da Universidade Federal de Santa Maria, Rio Grande do Sul, Brasil. *Biota Neotrop.* 8:125–31.

Santos-Filho, M., F. Frieiro-Costa, A. R. Ignácio, and M. N. F. Silva. 2012. Use of habitats by non-volant small mammals in Cerrado in central Brazil. *Braz. J. Biol.* 72:893–902.

Santos-Filho, M. dos, M. N. F. da Silva, and D. J. da Silva. 2001. Ocorrência da espécie *Kunsia tomentosus* (Lichtenstein, 1830), (Mammalia, Rodentia) em Unidade de Conservação. *III Simpósio Sobre Recursos Sócios Econômicos do Pantanal Um Desafio do Novo Milênio.* EMBRAPA, Corumbá, Mato Grosso do Sul, http://www.cpap.embrapa.br/agencia/congresso/Bioticos/SANTOSFILHO-047.pdf.

Santos Pires A. dos, F. A. dos Santos Fernandez, B. R. Feliciano, and D. de Freitas. 2008. Use of space by *Necromys lasiurus* (Rodentia, Sigmodontinae) in a grassland among Atlantic Forest fragments. *Mamm. Biol.* 75:270–76.

Sarich, V. 1985. Rodent macromolecular systematics. In *Evolutionary relationships among rodents: A multidisciplinary analysis*, ed. W. P. Luckett and J.L. Hartenberger, 423–52. Berlin: Springer-Verlag, 721 pp.

Sarmiento, L., M. Tantalean, and A. Huiza. 1999. Nemátodos parásitos del hombre y de los animales en el Perú. *Rev. Peru. Parasitol.* 14:9–65.

Sassi, P. L., M. B. Chiappero, C.Borghi, and C. N. Gardenal. 2011. High genetic differentiation among populations of the small cavy *Microcavia australis* occupying different habitats. *J. Exp. Zool.* 315:337–48.

Saunders, R. C. 1975. Venezuelan Macronyssidae (Acarina: Mesostigmata). *Brigham Young Univ. Sci. Bull.*, biol. ser. 20:75–90.

Sausa, K. S., and A. Bager. 2008. Feeding habitas of Geoffroy's cat (*Leopardus geoffroyi*) in southern Brazil. *Mamm. Biol.* 73:303–8.

Saussure, H. de. 1860. Note sur quelques mammifères du Mexique. *Rev. Mag. Zool. Pure Appl.*, ser. 2, 12:1–11, 53–57, 97–110, 241–54, 271–93, 366–83, 425–31, 479–94, 4 pls. [Also printed as an independently paginated, 82-page separate publication.]

Say, T. 1821. On a quadruped, belonging to the order Rodentia. *J. Acad. Nat. Sci. Philadelphia* 2, part 1:330–31.

Say, T., and G. Ord. 1825. Description of a new species of Mammalia, whcreon a new genus is proposed to be founded. *J. Acad. Nat. Sci. Philadelphia* 4:352–55.

Sbalqueiro, I. J. 1989. "Análises cromossômicas e filogenéticas em algumas espécies de roedores da região sul do Brasil." Ph.D. diss., Universidade Federal do Rio Grande do Sul, Porto Alegre, Brazil.

Sbalqueiro, I. J., M. S Bueno, J. Moreira, A. P. D. Ramos, C. Padovani, A. Xinenez, and J. M. S. Agostini. 1988. Cariótipo com 96 cromosomos em *Echimys dasythrix*, o mais elevado múmero diplóide entre os mamíferos.

Resumos do XV Congresso Brasileiro de Zoologia, Curitiba, p. 532.

Sbalqueiro, I. J., C. T. Cordeiro, G. F. Queluz, C. Borges, C. Zotz, M. M. Los, and J. Y. Matsumoto. 1987. Análises cariotípicas em roedores cricetídeos de duas localidades do Paraná. *Ciênc. Cult.* 39(Suppl.):735.

Sbalqueiro, I. J., M. Kiko, M. Lacerda, D. Elachkar, and L. R. Arnt. 1986. Estudos cromossômicos em roedores da família Cricetidae coletados no Paraná. *Ciênc. Cult.* 38(Suppl.):926.

Sbalqueiro, I. J., M. S. Mattevi, T. R. O. Freitas, and L. F. B. Oliveira. 1982. Variabilidade cromossômica em roedores do Rio Grande do Sul. *Resumo do IX Congresso Brasileiro de Zoologia*, Porto Alegre, p. 29.

Sbalqueiro, I. J., M. S. Mattevi, and L. F. B. Oliveira. 1984. An $X_1X_1X_2X_2/X_1X_2Y$ mechanism of sex determination in a South American rodent, *Deltamys kempi* (Rodentia, Cricetidae). *Cytogenet. Cell Genet.* 38:50–55.

Sbalqueiro, I. J., M. S. Mattevi, L. F. B. Oliveira, and T. R. O. Freitas. 1982. Estudos cromossômicos de espécies de roedores akodontinos do Rio Grande do Sul. *Ciênc. Cult.* 34:750.

Sbalqueiro, I. J., M. S. Mattevi, L. F. B. Oliveira, and M .J. V. Solano. 1991. B chromosome system in populations of *Oryzomys flavescens*(Rodentia, Cricetidae) from southern Brazil. *Acta Theriol.* 36:193–99.

Sbalqueiro, I. J., and A. P. Nascimento. 1996. Occurrence of *Akodon cursor* (Rodentia, Cricetidae) with 14, 15 and 16 chromosome cytotypes in the same geographic area in southern Brazil. *Rev. Bras. Genet.* 19:565–69.

Schaldach, W. J., Jr. 1966. New forms of mammals from southern Oaxaca, Mexico, with notes on some mammals of the Coastal range. *Saugetierkd. Mitt.* 14:286–97.

Schaller, G. B. 1983. Mammals and their biomass on a Brazilian ranch. *Arq. Zool., São Paulo* 31:1–36.

Scheibler, D. R., and A. U. Christoff. 2007. Habitat associations of small mammals in southern Brazil and use of regurgitated pellets of birds of prey for inventorying a local fauna. *Brazil. J. Biol.* 67:619–25.

Schenk, J. J., K. C. Rowe, and S. J. Steppan. 2013. Ecological opportunity and incumbency in the diversification of repeated continental colonizations by muroid rodents. *Syst. Biol.* 62:837–864.

Schetino, M. A. A. 2008. "Código de barras de DNA em espécies de *Proechimys* (Rodentia: Echimyidae) da Amazônia." Master's thesis, Programa Integrado do Pós-Graduão em Biologia Tropical e Recursos Naturais, INPA/UFAM, Manaus, Brazil, 81 pp.

Schinz, H. R. 1821. *Das Thierreich eingetheilt nach dem Bau der Thiere als Grundlage ihrer Naturgeschichte und der vergleichenden Anatomie von dem Herrn Ritter von Cuvier.* Erster band. Säugethiere und Vogél. Stuttgart und Tübingen: J. G. Cotta'schen Buchhandlungh, 1:xxxviii + 894 pp.

———. 1825a. *Das Thierreich eingetheilt nach dem Bau der Thiere als Grundlage ihrer Naturgeschichte und der vergleichenden Anatomie von dem Herrn Ritter von Cuvier.* Vierter band. Zoophyten. Stuttgard und Tübingen: J. G. Gota'schen Buchhandlung, 4:xix + 793 pp.

———. 1825b (or 1826). *Naturgeschichte und Abbildungen der Säugethiere.* Zurich: Brodtmann's lithographischer Kunstanstalt, vi + 417 pp., 8 (unnumbered), 178 pls. [See Palmer 1897:22 and Tate 1935:365 for question on date of publication.]

———. 1840. *Naturgeschichte und Abbildungen der Menschen und der Säugethiere nach den neuesten Entdeckungen und vorzüglichsten Originalien bearbeitet. Zweite Abtheilung: Die Säugethiere.* Zürich: Honeggerschen Lithographischen Anstalt, 1–250 + 135 pls.

———. 1845. *Systematisches Verzeichniss aller bis jetzt bekannten Säugethiere oder Synopsis Mammalium nach dem Cuvier'schen System.* Solothurn: Jent und Gassmann, 2:iv + 1–574.

Schleich, C. E., and C. Busch. 2002a. Acoustic signals of a solitary subterranean rodent *Ctenomys talarum* (Rodentia: Ctenomyidae): physical characteristics and behavioural correlates. *J. Ethol.* 20:123–31.

———. 2002b. Juvenile vocalizations of *Ctenomys talarum* (Rodentia: Octodontidae). *Acta Theriol.* 47:25–33.

———. 2004. Energetic expenditure during vocalization in pups of the subterranean rodent *Ctenomys talarum. Naturwissenschaften* 91:548–51.

Schleich, C. E., S. Veitl, E. Knotkova, and S. Begall. 2007. Acoustic communication in subterranean rodents. In *Subterranean rodents: News from underground,* ed. S. Begall, H. Burda and C. E. Schleich, 113–28. Heidelberg: Springer, i–xviii + 398 pp.

Schliemann, H. 1982. Nachweis wilder Meerschweinchen für Ecuador und Beschreibung von *Cavia aperea patzelti* subsp. nov. *Z. Säugetierk.* 47:79–90.

Schlosser, M. 1884. Die Nager des europäischen Tertiärs nebst Betrachtungen über die Organisation und die geschichtliche Entwicklung der Nager überhaupt. *Palaeontographica* 31:19–162, 8 plates.

Schmid, M., A. Fernández-Badillo, W. Feichtinger, C. Steinlein, and J. I. Roman. 1988. On the highest chromosome number in mammals. *Cytogenet. Cell Genet.* 49:305–8.

Schmid, M., C. Steinlein, W. Feichtinger, M. Schmidt, R. Visbal-García, and A. Fernández-Badillo. 1992. An intriguing Y chromosome in *Heteromys anomalus* (Rodentia, Heteromyidae). *Hereditas* 117:209–14.

Schramm, B. A., and R. E. Lewis. 1987. Four new species of *Plocopsylla* (Siphonaptera: Stephanocircidae) from South America. *J. Med. Entomol.* 24:399–407.

———. 1988. A new species of *Plocopsylla* Jordan, 1931 (Siphonaptera: Stephanocircidae) from Argentina. *J. NY. Entomol. Soc.* 96:465–69.

Schreber, J. C. D. von. 1792. *Die Säugethiere in Abbildungen nach der Natur mit Beschreibungen. Vierter Theil. Das Stachelthier. Die Savie. Der Biber. Die Maus. Das Murmelthier. Das Eichhorn. Der Schläfer. Der Springer. Der Hase. Der Klippschliefer.* Erlangen: Wolfgang Walther, 4:591–936, pls. 166–240. [This part of Volume 4 was published in 1792; see Sherborn 1892 for dates of publication of parts.]

———. 1843. *Die Säugethiere in Abbildungen nach der Natur mit Beschreibungen von Dr. Johann Christian Daniel von Schreber. Supplementbaud.* Leipzig: Commission ber Boss'schen Buchhandlung, xiv + 2 (unnumbered) + 614 pp.

Schwanz, L. E., and E. A. Lacey. 2003. Olfactory discrimination of gender by colonial tuco-tucos (*Ctenomys sociabilis*). *Mamm. Biol.* 68:53–60.

Schweigmann, N. J., A. Alberti, S. Pietrokovsky, O. Conti, A. Riarte, S. Montoya, and C. Wisnivesky-Colli. 1992. A new host of *Trypanosoma cruzi* from Jujuy, Argentina: *Octodontomys gliroides* (P. Gervais and d'Orbigny, 1844) (Rodentia, Octodontidae). *Mem. Inst. Oswaldo Cruz* 87:217–20.

Sclater, P. L., and O. Salvin. 1879. On the birds collected by T. K. Salmon in the State of Antiquia, United States of Colombia. *Proc. Zool. Soc. Lond.* 1879 (part III):486–550.

Scoles, R., and R. Gribel. 2012. The regeneration of Brazil nut trees in relation to nut harvest intensity in the Trombetas River valley of northern Amazonia, Brazil. *For. Ecol. Manag.* 265:71–81.

Seevers, C. H. 1955. A revision of the tribe Amblyopinini: staphylinid beetles parasitic on mammals. *Fieldiana Zool.* 37:211–64.

Serra, M. T. 1979. Composición botánica y variación estacional de la alimentación de *Chinchilla lanigera* en condiciones naturales. *Cien. Forest. Chile* 1:11–18.

Servicio Agrícola Ganadero (SAG). 2000. *Cartilla de Caza.* Santiago, Chile: Ministerio de Agricultura, Servicio Agrícola Ganadero, Departamento de Protección de los Recursos Naturales Renovables, 84 pp.

Severo, J. B. 1998. "Cariótipos dos equimídeos de algumas localidades do Brasil." Master's thesis, Universidade Federal do Rio Grando so Sul, Porto Alegre, Brazil.

Shadle, A. R. 1948. Gestation period in the porcupine, *Erethizon dorsatum dorsatum*. *J. Mamm.* 29:162–64.

Shamel, H. H. 1931. *Akodon chacoensis*, a new cricetine rodent from Argentina. *J. Washington Acad. Sci.* 21:427–29.

Shaw, G. 1800. *General zoology or systematic natural history*. London: G. Kearsley, 1(1 and 2):xv +1–552, 121 plates.

———. 1801. *General zoology. Mammalia. Quadrupeds. Order Glires*. London: Thomas Davison, 2(1).:vi + 1–226 + 44 plates.

Sherborn, C. D. 1892. On the dates of the parts, plates, and text of Schreber's 'Saugthiere'. *Proc. Zool. Soc. Lond.* 1891 (part IV):587–92. [Part IV of the 1891 volume of the *Proceedings* was published in April 1892; see Duncan 1937.]

———. 1897. Note on the dates of "The Zoology of the 'Beagle.'" *Ann. Mag. Nat. Hist.*, ser. 6, 20:483.

———. 1899. Lacépèdés "Tableaux . . . des Mammifères et des Oiseaux," 1799. *Natural Science* 15:406–9.

———. 1914. An attempt at a fixation of the dates of issue of the parts of the publications of the Musée [sic] d'Histoire Naturelle of Paris, 1802–1850. *Ann. Mag. Nat. Hist.*, ser. 8, 13:365–68.

———. 1915. [Untitled note]. In *Catalog of books, manuscripts, maps and drawings in the British Museum (Natural History)*, p. 2082. London: British Museum (Natural History), 5, 4 pp. (unnumbered) + 1957–2403.

———. 1922. *Index animalium sive index nominum quae ab A.D. MDCCLVIII generibus et speiebus animalium imposita sunt*. Part 1. Introduction, bibliography and index A-Aff. London: British Museum (Natural History), cxxxii + 128 pp.

———. 1924a. *Index animalium sive index nominum quae ab A.D. MDCCLVIII generibus et speiebus animalium imposita sunt*. Part IV. Index Bail.-Bryzos. London: British Museum (Natural History), pp. 641–943.

———. 1924b. *Index animalium sive index nominum quae ab A.D. MDCCLVIII generibus et speiebus animalium imposita sunt*. Part V. Index C-Ceyl. London: British Museum (Natural History), pp. 945–1196.

———. 1925a. *Index animalium sive index nominum quae ab A.D. MDCCLVIII generibus et speciebus animalium imposita sunt*. Part VI. Index Ceyl.-Concolor. London: British Museum (Natural History), pp. 1197–452.

———. 1925b. *Index animalium sive index nominum quae ab A.D. MDCCLVIII generibus et speiebus animalium imposita sunt*. Part VII. Index Concolor-Czizeki. London: British Museum (Natural History), pp. 1453–771.

———. 1927a. *Index animalium sive index nominum quae ab A.D. MDCCLVIII generibus et speciebus animalium imposita sunt*. Part XIII. Index *implicatus-laminella*. London: British Museum (Natural History), pp. 3137–392.

———. 1927b. *Index animalium sive index nominum quae ab A.D. MDCCLVIII generibus et speciebus animalium imposita sunt*. Part XIV. Index *laminella-Lyzzia*. London: British Museum (Natural History), pp. 3393–746.

———. 1930. *Index animalium sive index nominum quae ab A.D. MDCCLVIII generibus et speiebus animalium imposita sunt*. Part XXIII. Index S. littera-serratus. London: British Museum (Natural History), pp. 5703–910.

———., and F. J. Griffin. 1934. On the dates of publication of the natural history portions of Alcide d'Orbighy's 'Voyage Amérique méridionale.' *Ann. Mag. Nat. Hist.*, ser. 10, 13:130–34.

———., and B. B. Woodward. 1901. Dates of publication of the zoological and botanical portions of some French voyages.—*Part II*. Ferret and Galinier's 'Voyage en Abyssinie'; Lefebvre's 'Voyage en Abyssinie'; 'Exploration scientifique de l'Algérie'; Castelnau's 'Amérique du Sud'; Dumont d'Urville's 'Voyage de l'Astrolabe'; Laplace's 'Voyage sur la Favorite'; Jacquemont's 'Voyage dans l'Inde'; Tréhouart's 'Commission scientifique d'Islande'; Cailliaud, 'Voyage à Méroé'; 'Expédition scientifique de Morée'; Fabre, 'Commission scientifique du Nord'; Du Petit-Thouars, 'Voyage de la Vénus'; and on the dates of the 'Faune Française.' *Ann. Mag. Nat. Hist.*, ser. 7, 8:161–64, 333–36, and 491–94.

Shufeldt, R. W. 1926. Observações sobre certos peixes e mammiferos do Brasil e mais particularmente sobre sua osteologia. *Rev. Mus. Paulista, São Paulo* 14:503–61.

Siegenthaler, G. B., E. Fiorucci, S. Tiranti, P. Borráz, M. Urioste y A. García. 1990. Informe de avance sobre el Plan de Relevamiento de los Vertebrados de la Provincia de La Pampa. *Rev. Agro. Pampeano* 18:38–48.

Siegenthaler, G. B., S. Tiranti, and C. M. Duco. 1993. Relevamiento de los vertebrados de la provincia de La Pampa. *Terc. Inf. 5 J. Pampeanas Cien. Nat.* 1:139–47.

Sierra, R., ed. 1999. *Propuesta preliminary de un sistema de clasificación de vegetación para el Ecuador continental*. Quito: Proyecto INEFAN/GEFBIRF y EcoSiencia.

Sierra-Cisternas, C. X. 2010. "Filogeografía de *Abrothrix longipilis* (Rodentia: Sigmodontinae)." Seminario de Título presentado a la Facultad de Ciencias Naturales y Oceanográficas para optar al titulo de Biólogo. Universidad de Concepción, Facultad de Ciencias Naturales y Oceanográficos, Concepción, Chile, 56 pp.

Sierra de Soriano, B. 1960. Elementos constitutivos de una habitación de *Myocastor coypus bonariensis* (Geoffroy), ("nutria"). *Rev. Fac. Human. Ciénc. Univ. Repúb. Urug.* 18:257–76.

———. 1969. Algunos caracteres externos de cricétinos y su relación con el grado de adaptación a la vida acuática (Rodentia). *Physis* 28:471–86.

Sikes, R. S., J. A. Monjeau, E. C. Birney, C. J. Phillips, and J. R. Hillyard. 1997. Morphological versus chromosomal and molecular divergence in two species of *Eligmodontia*. *Z. Säugetierk.* 62:265–80.

Silva, A. L. G. da. 2005. "Variação craniana, bacular e citogenética em três populações de *Thrichomys apereoides* (Lund, 1839) do nordeste, centro e sudeste do Brasil." Master's thesis, Museu Nacional and Universidade Federal do Rio de Janeiro, Brazil.

Silva, C. E. F., E. S. Eler, M. N. F. da Silva, and E. Feldberg. 2012. Karyological analysis of *Proechimys cuvieri* and *Proechimys guyannensis* (Rodentia, Echimyidae) from central Amazon. *Genet. Mol. Biol.* 35:88–94.

Silva, C. R., A. R. Percequillo, G. E. Iack-Ximenes, and M. de Vivo. 2003. New distriutional records of *Blarinomys breviceps* (Winge, 1888) (Sigmodontinae, Rodentia). *Mammalia* 67:147–52.

Silva, F. R. da, R. M. Begnini, and B. C. Lopes. 2011. Seed dispersal and predation in the palm Syagrus romanzoffiana on two islands with different faunal richness, southern Brazil. *Stud. Neotrop. Fauna Environ.* 46:163–71.

Silva, G. S., and M. Tabarelli. 2001. Seed dispersal, plant recruitment and spatial distribution of *Bactris acanthocarpa* Martius (Arecaceae) in a remnant of Atlantic Forest in northeast Brazil. *Acta Oecol.* 22:259–68.

Silva, J. da, T. R. O. de Freitas, V. Heuser, J. R. Marinho, F. Bittencourt, C. T. S. Cerski, L. M. Kliemann, and B. Erdtmann. 2000. Effects of chronic exposure to coal in wild rodents (*Ctenomys torquatus*) evaluated by multiple methods and tissues. *Mutat. Res.* 470:39–51.

Silva, J. da, T. R. O. de Freitas, J. R. Marinho, G. Speit, and B. Erdtmann. 2000. An alkaline single-cell gel electrophoresis (comet) assay for environmental biomonitoring with native rodents. *Genet. Mol. Biol.* 23:241–45.

Silva, M. J. de J. 1994. "Estudos cromossômicos e de complexos sinaptonêmicos em roedores brasileiros da tribo Oryzomyini (Cricetidae, Rodentia)." Master's thesis, Universidade de São Paulo, São Paulo, Brazil.

———. 1999. Studies of the karyotypic differentiation process on four genera of Brazilian rodents (Sigmodontinae, Rodentia), based on conventional and molecular cytogenetics. *Genet. Mol. Biol.* 22:614.

Silva, M. J. de J., A. R. Percequillo, and Y. Yonenaga-Yassuda. 2000a. Citogenética de pequenos roedores de Pacoti, Serra de Baturité, Ceará. *XXIII Congresso Brasileiro de Zoologia, 2000, Cuiabá, MT. Resumos*, p. 565.

———. 2000b. Cytogenetics and systematic approach on a new *Oryzomys*, of the *nitidus* group (Sigmodontinae, Rodentia) from northeastern Brazil. *Caryologia* 53:219–26.

Silva, M. J. de J., K. Ventura, C. B. Di Nizo, and Y. Yonenaga-Yassuda. 2009. *Abrawayaomys*: a new member of tribe Akodontini (Sigmodontinae, Rodentia)? *10th International Mammalogical Congress, Mendoza, Argentina, Abstracts*, p. 137.

Silva, M. J. de J., and Y. Yonenaga-Yassuda. 1997. New karyotype of two related species of *Oligoryzomys* genus (Cricetidae, Rdentia) involving centric fusion with loss of NORs and distribution of telomeric (TTAGGG)$_n$ sequences. *Hereditas* 127:217–29.

———. 1998. Karyotype and chromosomal polymorphism of an undescribed *Akodon* from central Brazil, a species with the lowest known diploid chromosome number in rodents. *Cytogenet. Cell Genet.* 81:46–50.

———. 1999. Autosomal and sex chromosomal polymorphisms with multiple rearrangements and a new karyotype in the genus *Rhipidomys* (Sigmodontinae, Rodentia). *Hereditas* 131:211–20.

Silva, M. J. de J., J. L. Patton, and Y. Yonenaga-Yassuda. 2006. Phylogenetic relationships and karyotype evolution in the sigmodontine rodent *Akodon* (2n = 10 and 2n = 16) from Brazil. *Genet. Mol. Biol.* 29:469–74.

Silva, M. N. F. da. 1998. Four new species of spiny rats of the genus *Proechimys* (Rodentia: Echimyidae) from the western Amazon of Brazil. *Proc. Biol. Soc. Washington* 111:436–71.

Silva, M. N. F. da, and J. L. Patton. 1993. Amazonian phylogeography: mtDNA sequence variation in arboreal echimyid rodents (Caviomorpha). *Mol. Phylogenet. Evol.* 2:243–55.

Silva, R. B., E. M. Vieira, and P. Izar. 2008. Social monogamy and biparental care of the neotropical southern bamboo rat (*Kannabateomys amblyonyx*). *J. Mamm.* 89:1464–72.

Silva, S. I. 2005. Posiciones tróficas de pequeños mamíferos en Chile: una revisión. *Rev. Chilena Hist. Nat.* 78:589–99.

Silva Gómez, J. A. 2008. "Sistemática molecular de *Thaptomys* Thomas, 1916 (Rodentia, Cricetidae)." Master's thesis, Universidade Federal do Espírito Santo, Vitória, Brazil.

Silva-Júnior, J. de S., and A. P. Nunes. 2000. An extension of the geographical distribution of *Dactylomys dacty-*

linus Desmarest, 1822 (Rodentia, Echimyidae). *Bol. Mus. Paraense Emilio Goeldi, sér. zool.* 16:65–73.

Silveira, F. T., R. Lainson, J. J. Shaw, R. R. Braga, E. E. A. Ishikawa, and A. A. A. Souza. 1991. Leishmaniose cutanea na Amazonia: isolamento de *Leishmania* (*Viannia*) *lainsoni* do roedor *Agouti paca* (Rodentia: Dasyproctidae), no estado do Para, Brasil. *Rev. Inst. Med. Trop. Sao Paulo* 33:18–22.

Silveira, M., P. Teta, V. Aldazabal, and E. Eugenio. 2010. La fauna menor en la subsistencia de los cazadores recolectores del sitio "El Divisadero Monte 6" (partido de General Lavalle, provincia de Buenos Aires). In *Zooarqueología a principios del siglo XXI. Aportes teóricos, metodológicos y casos de estudio*, ed. M. Gutiérrez, M. De Nigris, P. Fernández, P. Giardina, A. Gil, A. Izeta, G. Neme, and H. Yacobaccio, 575–81. Mendoza (Argentina): Ediciones Libros del Espinillo.

Silvius, K. M. 2002. Spatio-temporal patterns of palm endocarp use by three Amazonian forest mammals: granivory or 'grubivory'? *J. Trop. Ecol.* 18:707–23.

Silvius, K. M., and J. M. V. Fragoso. 2003. Red-rumped agouti (*Dasyprocta leporina*) home range use in an Amazonian forest: implications for the aggregated distribution of forest trees. *Biotropica* 35:74–83.

Simões, R., R. Gentile, V. Rademaker, P. D'Andréa, H. Herrera, T. Freitas, R. Lanfredi, and A. Maldonado Jr. 2010. Variation in the helminth community structure of *Thrichomys pachyurus* (Rodentia: Echimyidae) in two sub-regions of the Brazilian Pantanal: the effects of land use and seasonality. *J. Helminthol.* 84:266–75.

Simonetti, J. A. 1983. Effect of goats upon native rodents and European rabbits in the Chilean matorral. *Rev. Chilena Hist. Nat.* 56:27–30.

———. 1989. Sobre la distribución de Aconaemys Ameghino, 1892 en Chile. *Notic. Mens. Mus. Nac. Hist. Nat. Chile* 315:8–9.

———. 1994. Impoverishment and nestedness in Caviomorph assemblages. *J. Mamm.* 75:979–84.

Simonetti, J. A., and J. R. Rau. 1989. Roedores del Holoceno temprano de la cueva del Milodón, Magallanes, Chile. *Notic. Mens. Mus. Nac. Hist. Nat. Chile* 315:3–5.

Simonetti, J. A., and B. Saavedra. 1998. Holocene variation in the small mammal fauna of central Chile. *Z. Säugetierk.* 63:58–62.

Simonetti, J. A., and A. E. Spotorno. 1980. Posición taxonómica de *Phyllotis micropus* (Rodentia: Cricetidae). *An. Mus. Hist. Nat. Valparaiso* 13:285–97.

Simpson, G. G. 1941. Vernacular names of South American mammals. *J. Mamm.* 22:1–17.

———. 1945. The principles of classification and a classification of mammals. *Bull. Am. Mus. Nat. Hist.* 85:xvi + 350 pp.

———. 1950. History of the fauna of Latin America. *Am. Sci.* 38:361–89.

Sist, P. 1989. Demography of *Astrocaryum sciophillum* and understory palm of French Guiana. *Principes* 33:142–51.

Slamovits, C. H., J. A. Cook, E. P. Lessa, and M. S. Rossi. 2001. Recurrent amplifications and deletions of satellite DNA accompanied chromosomal diversification in South American tuco-tucos (genus *Ctenomys*, Rodentia: Octodontidae): a phylogenetic approach. *Mol. Biol. Evol.* 18:1708–19.

Slaughter, B. H., and J. E. Ubelaker. 1984. Relationships of South American cricetines to rodents of North America and the Old World. *J. Vert. Paleontol.* 42:255–64.

Smit, F. G. A. M. 1968. Siphonaptera taken in formalin-traps in Chile. *Zool. Anzeiger* 180:220–28.

———. 1987. *An illustrated catalogue of the Rothschild collection of fleas (Siphonaptera) in the British Museum (Natural History)*. London: British Museum (Natural History), 7:1–380 + 8 plates.

Smith, A. C. 1999. Potential competitors for exudates eaten by saddleback (*Saguinus fuscicollis*) and moustached (*Saguinus mystax*) tamarins. *Neotropical Primates* 7:73–75.

Smith, C. H. 1842. Mammalia. Introduction to Mammalia. In *The naturalist's library*, vol. 15, ed. W. Jardine, 73–313. London: Chatto & Windus, Piccadilly, 1–313 + 31 plates.

Smith, M. F., D. A. Kelt, and J. L. Patton. 2001. Testing models of diversification in mice in the *Abrothrix olivaceus/xanthorhinus* complex in Chile and Argentina. *Mol. Ecol.* 10:397–405.

Smith, M. F., and J. L. Patton. 1991. Variation in mitochondrial cytochrome-*b* sequence in natural populations of South American akodontine rodents (Muridae: Sigmodontinae). *Mol. Biol. Evol.* 8:85–103.

———. 1993. The diversification of South American murid rodents: evidence from mtDNA sequence data for the akodontine tribe. *Biol. J. Linn. Soc.* 50:149–77.

———. 1999. Phylogenetic relationships and the radiation of sigmodontine rodents in South America: evidence from cytochrome-b. *J. Mamm. Evol.* 6:89–128.

———. 2007. Molecular phylogenetics and diversification of South American grass mice, genus *Akodon*. In *The quintessential naturalist: Honoring the life and legacy of Oliver P. Pearson*, ed. D. A. Kelt, E. P. Lessa, J. Salazar-Bravo, and J. L. Patton, 827–58. *Univ. California Publ. Zool.* 134:v-xii + 1–981.

Smith, V. S., J. E. Light, and L. A. Durden. 2008. Rodent louse diversity, phylogeny, and cospeciation in the Manu Biosphere Reserve, Peru. *Biol. J. Linn. Soc.* 95:598–610.

Smithe, F. B. 1975. *Naturalist's color guide. Atlas* + Supplement. New York: American Museum of Natural History, 229 pp.

Smythe, N. 1970. Relationships between fruiting seasons and seed dispersal methods in a Neotropical forest. *Am. Nat.* 104:25–35.

———. 1978. The natural history of the Central American agouti, *Dasyprocta punctata. Smithsonian Contrib. Zool.* 257:iv + 1–52.

———. 1987. The paca (*Cuniculus paca*) as a domestic source of protein for the Neotropical, humid lowlands. *Appl. Anim. Behav. Sci.* 17:155–70.

———. 1991. Steps toward domesticating the paca (*Agouti = Cuniculus paca*) and prospects for the future. In *Neotropical wildlife use and conservation*, ed. J. G. Robinson and K. H. Redford, 202–16. Chicago: University of Chicago Press, 520 pp.

Smythe, N., W. Glanz, and E. Leigh Jr. 1983. Ecology and population regulation in pacas and agoutis. In *The ecology of a tropical forest: Seasonal rhythms and long-term changes*, ed. E. Leigh Jr., A. Rand, and D. Windsor, 227–38. Washington, DC: Smithsonian Institution Press, 468 pp.

Soares V. B., R. W. Carvalho, O. Fernandes, C. Pirmez, and P. C. Sabroza. 1999. Study of *Leishmania* infection in rodents during an outbreak of American tegumentary leishmaniasis in an endemic area of Rio de Janeiro, Brazil. *Mem. Inst. Oswaldo Cruz* 94 (Suppl. II):abstract B-U–54.

Sobrero, R., V. E. Campos, S. M. Giannoni, and L. A. Ebensperger. 2010. Octomys mimax. *Mamm. Species* 42(1):49–57.

Solari, S. 2007. Trophic relationships within a highland rodent assemblage from Manu National Park, Cusco, Peru. In *The quintessential naturalist: Honoring the life and legacy of Oliver P. Pearson*, ed. D. A. Kelt, E. P. Lessa, J. Salazar-Bravo, and J. L. Patton, 225–40. *Univ. California Publ. Zool.* 134:v-xii + 1–981.

Solari, S., V. Pacheco, L. Luna, P. M. Velazco, and B. D. Patterson. 2006. Mammals of the Manu Biosphere Reserve. In *Mammals and Birds of the Manu Biosphere Reserve, Peru*, ed. B. D. Patterson, D. F. Stotz, and S. Solari, 13–23. *Fieldiana Zool.*, n.s., 110: iii + 1–49.

Solari, S., E. Vivar, P. M. Velazco, J. J. Rodríguez, D. E. Wilson, R. J. Baker, and J. L. Mena. 2001. The small mammal community of the Lower Urubamba Region, Peru. In *Urubamba: The biodiversity of a Peruvian rainforest*, ed. A. Alonso, F. Dallmeier, and P. Campbell, 171–82. SI/MAB Series 7. Washington, DC: Smithsonian Institution Press, x + 204 pp.

Solmsdorff, K., D. Kock, C. Hohoff, and N. Sachser. 2004. Comments on the genus *Galea* Meyen 1833 with description of *Galea monasteriensis* n. sp. from Bolivia (Mammalia, Rodentia, Caviidae). *Senckenb. Biol.* 84:137–56.

Solórzano-Filho, J. A. 2006. Mobbing of *Leopardus wiedii* while hunting by a group of *Sciurus ingrami* in an Araucaria forest of southeast Brazil. *Mammalia* 70:156–57.

Sombra, M. S., and A. M. Mangione. 2005. Obsessed with grasses? The case of mara *Dolichotis patagonum* (Caviidae: Rodentia). *Rev. Chilena Hist. Nat.* 78:401–8.

Soriano, P. J., and F. V. Clulow. 1988. Efecto de las inundaciones estacionales sobre poblaciones de pequeños mamíferos en los llanos altos occidentales de Venezuela. *Ecotrópicos* 1:3–10.

Soriano, P. J., A. Díaz de Pascual, J. Ochoa G., and M. Aguilera. 1999. Biogeographic analysis of the mammal communities in the Venezuelan Andes. *Interciencia* 24:17–25.

Soriano, P. J., and J. Ochoa G. 1997. Lista actualizada de los mamiferos de Venezuela. In *Vertebrados actuales e fósiles de Venezuela, Serie Catálogo Zoológico de Venezuela*, Volume I, ed. E. La Marca. Mérida, Venezuela: Museo de Ciencia y Tecnologia de Mérida, 298 pp.

Soriano, P. J., A. Utrera, and M. Sosa. 1990. Inventario preliminar de los mamíferos del Parque Nacional General Cruz Carrillo (Guaramacal), Estado Trujillo, Venezuela. *Biollania* 7:83–99.

Soukup, J. 1961. Materiales para el catalogo de los mamiferos peruanos. *Biota* 3(26):240–76.

———. 1965. Materiales para el catalogo de los mamiferos peruanos. Primer Suplemento. *Biota* 5(44):341–74.

Sousa, G. B. de, N. de Rosa, and C. N. Gardenal. 1996. Protein polymorphism in *Eligmodontia typus*. Genetic divergence with other phyllotine cricetids. *Genetica* 97:47–53.

Sousa, K. S., and A. Bager. 2008. Feeding habvits of Geoffroy's cat (*Leopardusgeoffroyi*) in southern Brazil. *Mamm. Biol.* 73:303–8.

Sousa, M. A. N. de. 2005. "Pequenos mamíferos (Didelphimorphia, Didelphidae e Rodentia, Sigmodontinae) de algumas áreas da Caatinga, Cerrado, Mata Atlântica e Brejo de Altitude do Brasil: Considerações citogenéticas e geográficas." Ph.D. diss., Universidade de São Paulo, São Paulo, Brazil.

Sousa, M. A. N. de, A. Langguth, and E. do A. Gimenez. 2004. Mamíferos dos brejos de altitude de Paraíba e Pernambuco. In *Brejos de altitude em*

Pernambuco e Paraíba: História natural, ecologia e conservação, ed. K. C. Porto, J. J. P. Cabral, and M. Tabarelli, 229–54. Brasília: Ministério do Meio Ambiente, 324 pp.

Souto Lima, R. B. de, P. A. Oliveira, and A. G. Chiarello. 2008. Diet of the thin-spined porcupine (*Chaetomys subspinosus*), an Atlantic Forest endemic threaned with extinction in southeastern Brazil. *Mamm. Biol.* 75:538–46.

Souza, A. L. G. 2011."Sistemática integrativa e abordagem biogeográfica de linhagens de roedores da Caatinga." Ph.D. diss., Museu Nacional, Universidade Federal do Rio de Janeiro, Brazil.

Souza, A. L. G., M. M. O. Corrêa, C. T. de Aguilar, and L. M. Pessôa. 2011. A new karyotype of *Wiedomys pyrrhorhinus* (Rodentia: Sigmodontinae) from Chapada Diamantina, northeastern Brazil. *Zoologia* 28:92–96.

Souza, A. L G., M. M. O. Corrêa, and L. M. Pessôa. 2006. Morphometric discrimination between *Trinomys albispinus* (Is. Geoffroy, 1838) and *Trinomys minor* (Reis & Pessôa, 1995) from Chapada Diamantina, Bahia, Brazil, and the karyotype of *Trinomys albispinus* (Rodentia, Echimyidae). *Arq. Mus. Nac., Rio de Janeiro* 64:325–32.

———. 2007a. The first description of the karyotype of *Dasyprocta azarae* Lichtenstein, 1823 (Rodentia, Dasyproctidae) from Brazil. *Mastozool. Neotrop.* 14:227–33.

———. 2007b. Variabilidad morfológica y citogenética en el género *Rhipidomys* Tschudi, 1844 (Rodentia, Sigmodontinae) de la Chapada Diamantina, Bahia, Brasil. In *Libro de Resúmenes, Programa, XXI Jornadas Argentinas de Mastozoología, 6 al 9 de noviembre de 2007, Tafí del Valle, Tucumán, Argentina,* 217–18. Tucumán: Sociedad Argentina para el Estudio de los Mamíferos–SAREM.

Souza, A. L. G., L. M. Pessôa, A. N. Menezes, A. M. R. Bezerra, and C.R. Bonvicino. 2011. O rio São Francisco como provável barreira geográfica para as duas espécies do gênero *Wiedomys* (Rodentia). *Rev. Mus. La Plata Zool.* 18(172):163R.

Souza, J. G. R., M. C. Digiani, R. O. Simões, J. L. Luque, R. Rodrigues-Silva, and A. Maldonado Jr. 2009. A new heligmonellid species (Nematoda) from *Oligoryzomys nigripes* (Rodentia: Sigmodontinae) in the Atlantic Forest, Brazil. *J. Parasitol.* 95:734–38.

Souza, L. T. M. D., A. Suzuki, L. E. Pereira, I. B. Ferreira, R. Pereira de Souza, A. S. Cruz, T. I. Ikeda, et al. 2002. Identificação das espécies de roedores reservatórios de hantavírus no sul e sudeste do Brasil. *Inf. Epidemiol. Sus* 11:249–51.

Souza, M. J. 1981. "Caracterização cromossômica em oito espécies de roedores brasileiros das famílias Cricetidae e Echimyidae," Ph.D. diss., Universidade de São Paulo, Brazil.

Souza, M. J., and Y. Yonenaga-Yassuda.1982. Chromosomal variability of sex chromosomes and NOR's in *Thrichomys apereoides* (Rodentia, Echimyidae). *Cytogenet. Cell Genet.* 33:197–203.

———. 1984. G- and C-band patterns and nucleolus organizer regions en somatic chromosomes of *Clyomys laticeps laticeps* (Rodentia, Echimyidae). *Experientia* 40:96–97.

Speed, M. P., and G. D. Ruxton. 2005. Warning displays in spiny animals: one (more) evolutionary route to aposematism. *Evolution* 59:2499–508.

Spix, J. B. von, and L. Agassiz. 1829. Selecta genera et species piscium quos in itinere per Brasiliam annos MDCCCXVII-MDCCCXX jussu et auspiciis Maximiliami Josephi I . . . colleget et pingendso curavit Dr. J. B. de Spix . . . Monachii. Part 1: i–xvi + i–ii + 1–82, 48 plates.

Spotorno, A. E. 1976. Análisis taxonómico de tres especies altiplanicas del género *Phyllotis* (Rodentia,Cricetidae). *An. Mus. Hist. Nat. Valparaiso* 9:141–61.

———. 1979. Contrastación de la macrosistemática de roedores caviomorfos por análisis de la morfologia reproductivo masculina. *Arch. Biol. Med. Exp.* 12:97–106.

———. 1986. "Systematics and evolutionary relationships of Andean phyllotine and akodontine rodents." Ph.D. diss., University of California, Berkeley.

———. 1992. Parallel evolution and ontogeny of simple penis among New World cricetid rodents. *J. Mamm.* 73:504–14.

Spotorno, A. E., H. Cofre, G. M. Manríques, Y. Vilina, P. Marquet, and L. I. Walker. 1998. Nueva especie de mamífero filotino *Loxodontomys* en Chile central. *Rev. Chilena Hist. Nat.* 71:359–74.

Spotorno, A. E., and R. Fernandez. 1976. Chromosome stability in southern *Akodon. Mamm. Chromo. Newslett.* 17:13–14.

Spotorno, A. E., G. Manríquez, A. Fernández L., J. C. Marín, F. González, and J. Wheeler. 2007. Domestication of guinea pigs from a southern Peru-northern Chile wild species and their middle pre-Columbian mummies. In *The quintessential naturalist: Honoring the life and legacy of Oliver P. Pearson,* ed. D. A. Kelt, E. P. Lessa, J. Salazar-Bravo, and J. L. Patton, 367–88. *Univ. California Publ. Zool.* 134:v-xii + 1–981.

Spotorno, A. E., J. C. Marín, G. Manríquez, J. P. Valladares, E. Rico, and C. Rivas. 2006. Ancient and modern steps during the domestication of guinea pigs (*Cavia porcellus* L.). *J. Zool.* 270:57–62.

Spotorno, A. E., E. Palma, and J. P. Valladares. 2000. Biología de roedores reservorios de Hantavirus en Chile. *Rev. Chilena Infectol.* 17:197–210.

Spotorno, A. E., J. Sufan-Catalan, and L. I. Walker. 1994. Cytogenetic diversity and evolution of Andean species of *Eligmodontia* (Rodentia ; Muridae). *Z. Säugetierk.* 59:299–308.

Spotorno, A. E., J. P. Valladares, J. C. Marín, R. E. Palma, and C. Zuleta R. 2004. Molecular divergence and phylogenetic relationships of chinchillids (Rodentia: Chinchillidae). *J. Mamm.* 85:384–88.

Spotorno, A. E., J. P. Valladares, J. C. Marín, and H. Zeballos. 2004. Molecular diversity among domestic guinea-pigs (*Cavia porcellus*) and their close phylogenetic relationship with the Andean wild species *Cavia tschudii*. *Rev. Chilena Hist. Nat.* 77:243–50.

Spotorno, A. E., and L. I. Walker. 1979. Analisis de similitud cromosomica según patrones de bandas G en cuatro especies chilenas de *Phyllotis* (Rodentia, Cricetidae). *Arch. Biol. Med. Exp.* (Santiago) 12:83–90.

———. 1983. Análisis electroforético de dos especies de *Phyllotis* en Chile central y sus híbridos experimentales. *Rev. Chilena Hist. Nat.* 56:51–59.

Spotorno, A. E., L. Walker, L. I., Contreras, J. Pincheira, and R. Fernández-Donoso. 1988. Cromosomas ancestrales en Octodontidae y Abromidae. *Arch. Biol. Med. Exp.* (Santiago) 21:527.

Spotorno, A. E., L. I. Walker, L. C. Contreras, J. C. Torres, R. Fernandez-Donoso, M. Soledad Berrios, and J. Pincheira. 1995. Chromosome divergence of *Octodon lunatus* and *Abrocoma bennetti* and the origins of Octodontoidea (Rodentia: Histricognathi). *Rev. Chilena Hist. Nat.* 68:227–39.

Spotorno, A. E., L. I. Walker, S. V. Flores, M. Yevenes, J. C. Marín, and C. Zuleta. 2001. Evolutión de los filotinos (Rodentia, Muridae) en los Andes del Sur. *Rev. Chilena Hist. Nat.* 74:151–66.

Spotorno, A. E., C. Zuleta, and A. Cortes. 1990. Evolutionary systematics and heterochrony in *Abrothrix* species (Rodentia, Cricetidae). *Evol. Biol.* 4:37–62.

Spotorno, A. E., C. Zuleta, A. Gantz, A., F. Saiz, J. Rau, M. Rosenmann, A. Cortes, G. Ruiz, L. Yates, E. Couve, and J. C. Marín. 1998. Sistemática y adaptación de mamíferos, aves e insectos fitófagos de la Región de Antofagasta, Chile. *Rev. Chilena Hist. Nat.* 71:501–26.

Spotorno, A. E., C. A. Zuleta, J. P. Valladares, A. L. Deane, and J. E. Jiménez. 2004. Chinchilla laniger. *Mamm. Species* 758:1–9.

Spotorno, A. E., C. Zuleta R., L. I. Walker G. Manriquez S., J. P. Valladares, and J. C. Marín. 2013. A small,

new gerbil-mouse *Eligmodontia* (Rodentia: Cricetidae) from dunes at the coasts and deserts of north-central Chile: molecular, chromosomic, and morphological analyses. *Zootaxa* 3683:377–94.

Spradling, T. A., S. V. Brant, M. S. Hafner, and C. J. Dickerson. 2004. DNA data support a rapid radiation of pocket gopher genera (Rodentia: Geomyidae). *J. Mamm. Evol.* 11:105–25.

Stallings, J. R. 1988. "Small mammal communities in an eastern Brazlian park." Ph.D. diss., University of Florida, Gainesville.

———. 1989. Small mammal inventories in an eastern Brazilian park. *Bull. Fla. State Mus. Biol. Sci.* 34:153–200.

Stallings, J. R., M. Kierulff, and L. F. B. M. Silva. 1994. Use of space, and activity patterns of Brazilian bamboo rats (*Kannabateomys amblyonyx*) in exotic habitat. *J. Trop. Ecol.* 10:431–38.

Steadman, D. W., and C. E. Ray. 1982. The relationships of *Megaoryzomys curioi*, an extinct cricetine rodent (Muroidea: Muridae) from the Galápagos Islands, Ecuador. *Smithsonian Contrib. Paleobiol.* 51:iii+1–23.

Steadman, D. W., and S. Zousmer. 1988. *Galápagos: Discovery on Darwin's Islands*. Washingon, DC: Smithsonian Institution Press, 208 pp.

Stein, B. R. 1988. Morphology and allometry in several genera of semiaquatic rodents (*Ondatra*, *Nectomys*, and *Oryzomys*). *J. Mamm.* 69:500–11.

———. 2000. Morphology of subterranean rodents. In *Life underground: The biology of subterranean rodents*, ed. E. A. Lacey, J. L. Patton, and G. N. Cameron, 19–61. Chicago: University of Chicago Press.

Steiner, C., P. Sourrouille, and F. Catzeflis. 2000. Molecular characterization and mitochondrial sequence variation in two sympatric species of *Proechimys* (Rodentia: Echimyidae) in French Guiana. *Biochem. Syst. Ecol.* 28: 963–73.

Steiner-Souza, F., T. R. O. de Freitas, and P. Cordeiro-Estrela. 2010. Inferring adaptation within shape diversity of the humerus of subterranean rodent *Ctenomys*. *Biol. J. Linn. Soc.* 100:353–67.

Stephens, L., and M. A. Traylor Jr. 1983. *Ornithological gazetteer of Peru*. Cambridge: Harvard College Press, vi+273 pp, 1 map.

———. 1985. *Ornithological gazetteer of the Guianas*. Cambridge: Harvard College Press, v+121 pp, 1 map.

Steppan, S. J. 1993. Phylogenetic relationships among the Phyllotini (Rodentia: Sigmodontinae) using morphological characters. *J. Mamm. Evol.* 1:187–213.

———. 1995. Revision of the tribe Phyllotini (Rodentia: Sigmodontinae), with a phylogenetic hypothesis for the Sigmodontinae. *Fieldiana Zool.*, n.s., 80:1–112.

————. 1996. A new species of *Holochilus* (Rodentia: Sigmodontinae) from the Middle Pleistocene of Bolivia and its phylogenetic significance. *J. Vert. Paleontol.* 16:522–30.

————. 1998. Phylogenetic relationships and species limits within *Phyllotis* (Rodentia: Sigmodontinae): concordance between mtDNA sequence and morphology. *J. Mamm.* 79:573–93.

Steppan, S. J., R. M. Adkins, and J. Anderson. 2004. Phylogeny and divergence-date estimates of rapid radiations in muroid rodents based on multiple nuclear genes. *Syst. Biol.* 53:533–53.

Steppan, S. J., and U. F. J. Pardiñas. 1998. Two new fossil muroids (Sigmodontinae: Phyllotini) from the early Pleistocene of Argentina: phylogeny and paleoecology. *J. Vert. Paleontol.* 18:640–49.

Steppan, S. J., O. Ramirez, J. Banbury, D. Huchon, V. Pacheco, L. I. Walker, and A. E. Spotorno. 2007. A molecular reappraisal of the systematics of the leaf-eared mice *Phyllotis* and their relatives. In *The quintessential naturalist: Honoring the life and legacy of Oliver P. Pearson*, ed. D. A. Kelt, E. Lessa, J. Salazar-Bravo, and J. L. Patton, 799–826. *Univ. California Publ. Zool.* 134:v-xii + 1–981.

Steppan, S. J., B. L. Storz, and R. S. Hoffmann. 2004. Nuclear DNA phylogeny of the squirrels (Mammalia: Rodentia) and the evolution of arboreality from c-myc and RAG1. *Mol. Phylogenet. Evol.* 30:703–19.

Steppan, S. J., and J. Sullivan. 2000. The emerging statistical perspective in systematics: a comment on Mares and Braun. *J. Mamm.* 81:260–70.

Stern, J. J., and A. Merari. 1969. The bathing behavior of the chinchilla: effects of deprivation. *Phychonom. Sci.* 14:115–16.

Stevens, S. M., and T. P. Husband. 1998. The influence of edge on small mammals: evidence from Brazilian Atlantic Forest fragments. *Biol. Conserv.* 85:1–8.

Stolzmann, J. 1885. Description d'un nouveau rongeur du genre *Coelogenys*. *Proc. Zool. Soc. Lond.* 1885 (part I):161–67.

Stone, W. 1914. On a collection of mammals from Ecuador. *Proc. Acad. Nat. Sci. Philadelphia* 66:9–19.

Storr, G. K. C. 1780. *Prodromus methodi Mammalium*. Tübingen: Litteris Reissianis, 43 pp., 4 tables.

Streilein, K. E. 1982a. The ecology of small mammals in the semiarid Brazilian Caatinga. I. Climate and faunal composition. *Ann. Carnegie Mus.* 51:79–107.

————. 1982b. The ecology of small mammals in the semiarid Brazilian Caatinga. II. Water relations. *Ann. Carnegie Mus.* 51:109–26.

————. 1982c. The ecology of small mammals in the semiarid Brazilian Caatinga. III. Reproductive biol-

ogy and population ecology. *Ann. Carnegie Mus.* 51:251–69.

————. 1982d. The ecology of small mammals in the semiarid Brazilian Caatinga. IV. Habitat selection. *Ann. Carnegie Mus.* 51:331–43.

————. 1982e. The ecology of small mammals in the semiarid Brazilian Caatinga. V. Agonistic behavior and overview. *Ann. Carnegie Mus.* 51:345–69.

Stunkard, H. W. 1953. *Raillietina demerariensis* (Cestoda), from *Proechimys cayennensis trinitatis* of Venezuela. *J. Parasitol.* 39:272–79.

Suárez, O., and S. Bonaventura. 2001. Habitat use and diet in sympatric species of rodents of the low Parana delta, Argentina. *Mammalia* 65:161–76.

Suárez, O., and F. Kravetz. 1998a. Copulatory pattern and mating system of *Akodon azarae* (Rodentia, Muridae). *Iheringia Ser. Zool.* 84:133–40.

Suárez, O., and F. Kravetz. 1998b. Transmission of food selectivity from mothers to offsprings in *Akodon azarae* (Rodentia, Muridae). *Behaviour* 135:251–59.

Suárez, O., G. Cueto, R. Cavia, I. Gómez Villafañe, D. Bilenca, A. Edelstein, P. Martínez, et al. 2003. Prevalence of infection with hantavirus in rodent populations of central Argentina. *Mem. Inst. Oswaldo Cruz* 98:727–32.

Suárez, O. V. 1994. Diet and habitat selection of *Oxymycterus rutilans* (Rodentia, Cricetidae). *Mammalia* 58:225–34.

Suárez-Villota, E. Y., C. B. Di-Nizo, C. L. Neves, and M. J. de Jesus Silva. 2013. First cytogenetic information for *Drymoreomys albimaculatus* (Rodentia, Cricetidae), a recently described genus from Brazilian Atlantic Forest. *ZooKeys* 303:65–76.

Suárez-Villota, E. Y., R. A. Vargas, C. L. Marchant, J. E. Torres, N. Köhler, J. J. Núñez, R. de la Fuente, J. Page, and M. H. Gallardo. 2012. Distribution of repetitive DNAs and the hybrid origin of the red vizcacha rat (Octodontidae). *Genome* 55:105–17.

Sudman, P. D., and M. S. Hafner. 1992. Phylogenetic relationships among Middle American pocket gophers (genus *Orthogeomys*) based on mtDNA sequences. *Mol. Phylogenet. Evol.* 1:17–25.

Sudman, P. D., J. K. Wickliffe, P. Horner, M. J. Smolen, J. W. Bickham, and R. D. Bradley. 2006. Molecular systematics of pocket gophers of the genus *Geomys*. *J. Mamm.* 87:668–76.

Suriano, D. M., and G. T. Navone. 1994. Three new species of the genus *Trichuris* Roederer, 1761 (Nematoda: Trichuridae) from Cricetidae and Octodontidae rodents in Argentina. *Res. Rev. Parasitol.* 54:39–46.

Sussman, D. R. 2011. The erethizontid fossil from the Uquía formation of Argentina should not be referred to the genus *Erethizon*. *J. South Am. Earth Sci.* 31:475–78.

Sutton, C. A. 1989a. Catálogo preliminar de los helmintos parásitos de roedores neotropicales. *Com. Invest. Cient. Prov. Buenos Aires, Monografía* 12:1–122.

———. 1989b. Contribution to the knowledge of Argentina's parasitological fauna xvii. Spirurida nematoda from neotropical Cricetidae *Physaloptera calnuensis* new-species and *Protospirura numidica criceticola* new record Quentin Karimi and Rodriguez de Almeida. *Bull. Mus. Natl. Hist. Nat. A Zool. Biol. Ecol. Anim.* 11:61–68.

Sutton, C. A., and M. C. Durette-Desset. 1995. A description of *Graphidioides kravetzi* n. sp. and the revision of *Graphidioides* Cameron, 1923 (Nematoda, Trichostrongyloidea), parasites of Neotropical rodents. *Syst. Parasitol.* 31:133–45.

Sutton, C. A., and J. P. Hugot. 1987. Contribution a la connaissance de la faune parasitaire d'Argentine, XVIII. Etude morphologique de *Wellcomia dolichotis* n. sp. (Oxyuridae, Nematoda), parasite de *Dolichotis patagonum*. *Syst. Parasitol.* 10:85–93.

Svartman, M. 1989. "Levantamento cariotípico de roedores da região do Distrito Federal." Master's thesis, Universidade de São Paulo, São Paulo, Brazil.

Svartman, M., and E. J. Cardoso de Almeida. 1993a. Pericentric inversion and X chromosome polymorphism in *Rhipidomys* sp. (Cricetidae, Rodentia) from Brazil. *Caryologia* 46:219–25.

———. 1993b. The karyotype of *Oxymycterus* sp (Cricetidae, Rodentia) from central Brazil. *Experientia* 49:718–20.

———. 1993c. Robertsonian fusion and X chromosome polymorphism in *Zygodontomys* (*Bolomys*) *lasiurus* (Cricetidae, Rodentia) from central Brazil. *Braz. J. Genet.* 16:225–35.

———. 1994. The karyotype of *Akodon lindberghi* Hershkovitz, 1990 (Cricetidae, Rodentia). *Braz. J. Genet.* 17:225–27.

Swier, V. J., R. D. Bradley, W. Rens, F. F .B. Elder, and R. J. Baker. 2009. Patterns of chromosomal evolution in *Sigmodon*, evidence from whole chromosome paints. *Cytogenet. Genome Res.* 125:54–66.

Tabeni, S., N. Marcos, M. I. Rosi, and B. Bender. 2012. Vulnerability of small and medium-sized prey mammals in relation to their habitat preferences, age classes and locomotion types in the temperate Monte Desert, Argentina. *Mamm. Biol.* 77:90–96.

Tabeni, S., and R. A. Ojeda. 2005. Ecology of the Monte Desert small mammals in disturbed and undisturbed habitats. *J. Arid Environ.* 63:244–55.

Taraborelli, P., B, Borruel, and A. Mangeaud. 2009. Ability of murid rodents to find buried seeds in the Monte Desert. *Ethology* 115:201–9.

Taraborelli, P., V. Corbalán, and S. Giannoni. 2003. Locomotion and escape modes in rodents of the Monte Desert (Argentina). *Ethology* 109:475–85.

Taraborelli, P., P. Moreno, P. Sassi, M. A. Dacar, and R. Ojeda. 2011. New eco-morphological-behavioural approach of the chinchilla rats in the pre-Andean foothills of the Monte Desert (Argentina). *J. Nat. Hist.* 45:1745–58.

Taber, A. B. 1987. "The behavioural ecology of the mara *Dolichotis patagonum*." Ph.D. diss., Oxford University, United Kingdom.

Taber, A. B., and D. W. MacDonald. 1992. Communal breeding in the mara, *Dolichotis patagonum*. *J. Zool. Lond.* 227:439–52.

Talamoni, S. A., D. Couto, D. A. Cordeiro Júnior, and F. M. Diniz. 2008. Diet of some species of Neotropical small mammals. *Mamm. Biol.* 73:337–41.

Talamoni, S. A., and M. M. Dias. 1999. Population and community ecology of small mammals in southeastern Brazil. *Mammalia* 63:167–81.

Tamayo, M., and D. Frassinetti. 1980. Catálogo de los mamíferos fósiles y vivientes de Chile. *Bol. Mus. Nac. Hist. Nat. Chile* 37:323–99.

Tamayo, M., H. Núñez, and J. Yáñez. 1987. Lista sistematica actualizada de los mamíferos vivientes en Chile y sus nombres comunes. *Notic. Mens. Mus. Nac. Hist. Nat. Chile* 312:1–14.

Tammone, M. N., E. A. Lacey, and M. A. Relva. 2012. Habitat use by colonial tuco-tucos (*Ctenomys sociabilis*): specialization, variation, and sociality. *J. Mamm.* 93:1409–19.

Tantaleán, M., and J. Chavez. 2004. Wild animals endoparasites (Nemathelminthes and Platyhelminthes) from the Manu Biosphere Reserve, Peru. *Rev. Peru. Biol.* 11:219–22.

Tarifa, T., J. Aparicio and E. Yensen. 2007. Mammals, amphibians, and reptiles of the Bolivian High Andes: an initial comparison of diversity patterns in *Polylepis* woodlands. In *The quintessential naturalist: Honoring the life and legacy of Oliver P. Pearson*, ed. D. A. Kelt, E. P. Lessa, J. Salazar-Bravo, and J. L. Patton, 241–73. *Univ. California Publ. Zool.* 134:v-xii + 1–981.

Tarifa, T., C. Azurduy, R. R. Vargas, N. Huanca, J. Terán, G. Arriaran C. Salazar, and L. Terceros. 2009. Observations on the natural history of *Abrocoma* sp. (Rodentia, Abrocomidae) in a *Polylepis* woodland in Bolivia. *Mastozool. Neotrop.* 16:252–58.

Tapia, P. A. 1995. Activity temporal pattern of *Abrothrix xanthorhinus* (Waterhouse) in the Laguna Amarga sector, Torres del Paine National Park. *An. Inst. Patagon. Ser. Cienc. Nat.* 23:48–50.

Tassino, B. 2006. "Estructura poblacional y biología reproductive del tucu-tucu de Río Negro (*Ctenomys rionegrensis*): relaciones entre el comportamiento y los procesos evlutivos." Ph.D. diss. Universidad de la Repúblia, Montevideo, Uruguay.

Tassino, B., and C. A. Passos. 2010. Reproductive biology of Río Negro tuco-tuco, *Ctenomys rionegrensis* (Rodentia: Octodontidae). *Mamm. Biol.* 75:253–60.

Tassino, B., I. E., R. P. Garbero, P. Altesor, and E. A. Lacey. 2011. Space use by Río Negro tuco-tucos (*Ctenomys rionegrensis*): excursions and spatial overlap. *Mamm. Biol.* 76:143–47.

Tate, G. H. H. 1931. The ascent of Mount Turumiquire. *Nat. Hist.* 31:539–548.

———. 1932a. The taxonomic history of the genus *Reithrodon* Waterhouse (Cricetidae). *Am. Mus. Novit.* 529:1–4.

———. 1932b. The taxonomic history of the South American cricetid genera *Euneomys* (subgenera *Euneomys* and *Galenomys*), *Auliscomys*, *Chelemyscus*, *Chinchillula*, *Phyllotis*, *Paralomys*, *Graomys*, *Eligmodontia*, and *Hesperomys*. *Am. Mus. Novit.* 541:1–21.

———. 1932c. The South American Cricetidæ described by Felix Azara. *Am. Mus. Novit.* 557:1–5.

———. 1932d. The taxonomic history of the Neotropical cricetid genera *Holochilus*, *Nectomys*, *Scapteromys*, *Megalomys*, *Tylomys* and *Ototylomys*. *Am. Mus. Novit.* 562:1–19.

———. 1932e. The taxonomic history of the South and Central American cricetid rodents of the genus *Oryzomys*. Part 1: subgenus *Oryzomys*. *Am. Mus. Novit.* 579:1–18.

———. 1932f. The taxonomic history of the South and Central American cricetid rodents of the genus *Oryzomys*. Part 2: Subgenera *Oligoryzomys*, *Thallomyscus*, and *Melanomys*. *Am. Mus. Novit.* 580:1–16.

———. 1932g. The taxonomic history of the South and Central American oryzomyine genera of rodents (excluding *Oryzomys*): *Nesoryzomys*, *Zygodontomys*, *Chilomys*, *Delomys*, *Phaenomys*, *Rhagomys*, *Rhipidomys*, *Nyctomys*, *Oecomys*, *Thomasomys*, *Inomys*, *Aepeomys*, *Neacomys*, and *Scolomys*. *Am. Mus. Novit.* 581:1–28.

———. 1932h. The taxonomic history of the South and Central American akodont rodent genera: *Thalpomys*, *Deltamys*, *Thaptomys*, *Hypsimys*, *Bolomys*, *Chroeomys*, *Abrothrix*, *Scotinomys*, *Akodon* (*Chalcomys* and *Akodon*), *Microxus*, *Podoxymys*, *Lenoxus*, *Oxymycterus*, *Notiomys*, and *Blarinomys*. *Am. Mus. Novit.* 582:1–32.

———. 1932i. The taxonomic history of certain South and Central American cricetid Rodentia: *Neotomys*, with remarks upon its relationships; the cotton rats (*Sigmodon* and *Sigmomys*); and the "fish-eating" rats (*Ichthyomys*, *Anotomys*, *Rheomys*, *Neusticomys*, and *Daptomys*). *Am. Mus. Novit.* 583:1–10.

———. 1935. The taxonomy of the genera of Neotropical hystricoid rodents. *Bull. Am. Mus. Nat. Hist.* 68:295–447.

———. 1936. Some Muridae of the Indo-Australian region. *Bull. Am. Mus. Nat. Hist.* 72:501–728.

———. 1939. The mammals of the Guiana Region. *Bull. Am. Mus. Nat. Hist.* 76:151–229.

———. 1947. A list of the mammals collected at Rancho Grande, in a montane cloud forest of northern Venezuela. *Zoologica* 32:65–66.

Tavares, W. C. 2007."Variação morfológica e citogenética do rato-de-espinho *Trinomys eliasi* (Pessoa & Reis, 1993), com evidências para a descrição de uma nova espécie de *Trinomys* Thomas, 1921 no Estado do Rio de Janeiro (Echimyidae, Rodentia)." Undergraduate monograph, Universidade Federal do Rio de Janeiro, Rio de Janeiro, Brazil.

Tavares, W. C., L. M. and Pessôa. 2010. Variação morfológica em populações de *Trinomys* (Thomas, 1921) de Restingas e Matas de Baixada no Estado do Rio de Janeiro. In *Mamíferos de Restingas e Manguezais do Brasil*, ed. L. M. Pessôa, W. C. Tavares, and S. Siciliano, 127–54. Rio de Janeiro: Sociedade Brasileira de Mastozoologia, 282 pp.

Tavares, W. C., L. M. Pessôa, and P. R. Gonçalves. 2011. New species of *Cerradomys* from coastal sandy plains of southeastern Brazil (Cricetidae: Sigmodontinae). *J. Mamm.* 92:645–58.

Tauber, A. A. 2005. Fossil mammals and age of the Salicas Formation (Late Miocene), Sierra de Velasco, La Rioja province, Argentina. *Ameghiniana* 42:443–60.

Teixeira, B. R. 2005. "Manejo em cativeiro e biologia reprodutiva de duas espécies de *Thrichomys* (Rodentia: Echimyidae) provenientes do Piauí e Mato Grosso do Sul." Master's thesis, Museu Nacional and Universidade Federal do Rio de Janeiro, Brazil.

Teixeira, B. R., A. L. Roque, S. C. Barreiros-Gómez, P. M. Borodin, A. M. Jansen, and P. S. D'Andrea. 2005. Maintenance and breeding of *Thrichomys* (Trouessart, 1880b) (Rodentia: Echimyidae) in captivity. *Mem. Inst. Oswaldo Cruz* 100:527–30.

Telford, S. R., Jr., A. Herrer, and H. A. Christensen. 1972. Enzootic cutaneous leishmaniasis in eastern Panama. 3. Ecological factors relating to the mammalian hosts. *Ann. Trop. Med. Parasitol.* 66:173–79.

Temminck, C. J. 1827. *Monographies de Mammalogie, ou description de quelques generes de mammifères, don't lest espèces ont été observees dan les différens musées de l'Europe.* Leiden: C. C. Vander Hoek, 1:245–68.

Terán, M. F., J. Ayala, and J. C. Hurtado. 2008. Primer registro de *Kunsia tomentosus* (Rodentia: Cricetidae: Sigmodontinae) en el norte del Departamento de La Paz, Bolivia. *Mastozool. Neotrop.* 15:129–33.

Terborgh, J. W., J. W. Fitzpatrick, and L. H. Emmons. 1984. Annotated checklist of bird and mammal species of Cocha Cashu Biological Station, Manu National Park, Peru. *Fieldiana Zool.*, n.s., 21:1–29.

Tesh, R. B. 1970a. Observations on the natural history of *Diplomys darlingi. J. Mamm.* 51:197–99.

———. 1970b. Notes on the reproduction, growth, and development of echimyid rodents in Panama. *J. Mamm.* 51:199–202.

Tesh, R. B., M. L. Wilson, R. Salas, N. M. C. de Manzione, D. Tovar, T. G. Ksiazek, and C. J. Peters. 1993. Field studies on the epidemiology of Venezuelan hemorrhagic fever: implication of the cotton rat *Sigmodon alstoni* as the probable rodent reservoir. *Am. J. Trop. Med. Hyg.* 49:227–35.

Testoni, A. F., S. L. Althoff, A. P. Nascimento, F. Steiner-Souza, and I. J. Sbalqueiro. 2010. Description of the karyotype of *Rhagomys rufescens* Thomas, 1886 (Rodentia, Sigmodontinae) from southern Brazil Atlantic Forest. *Genet. Mol. Biol.* 33:479–85.

Testoni, A. F., J. Furnis, S. L. Althoff, F. R. Tortato, and J. J. Cherem. 2012. *Akodon serrensis* Thomas, 1902 (Mammalia: Rodentia: Sigmodontinae): records in Santa Catarina state, southern Brazil. *Check List* 8:1344–46.

Teta, P., and A. Andrade. 2002. Micromamíferos depredados por *Tyto alba* (Aves, Tytonidae) en las Sierras de Talagapa (provincia de Chubut, Argentina). *Neotrópica* 48:88–90.

Teta, P., A. Andrade, and U. F. J. Pardiñas. 2002. Novedosos registros de roedores sigmodontinos (Rodentia: Muridae) en la Patagonia central Argentina. *Mastozool. Neotrop.* 9:79–84.

———. 2005. Micromamíferos (Didelphimorphia y Rodentia) y paleoambientes del Holoceno tardía en la Patagonia noroccidental extra-andina (Argentina). *Archaeofauna* 14:183–97.

Teta, P., G. Cueto and O. Suárez. 2007. New data on morphology and natural history of *Deltamys kempi* Thomas, 1919 (Cricetidae, Sigmodontinae) from central-eastern Argentina. *Zootaxa* 1665:43–51.

Teta, P., G. D'Elía, and U. F. J. Pardiñas. 2010. *Graomys hypogaeus* Cabrera, 1834 es un sinónimo de *Eligmo-*
dontia moreni (Thomas, 1896). *Mastozool. Neotrop.* 17:201–5.

Teta, P., G. D'Elía, U. F. J. Pardiñas, P. Jayat, and P. Ortiz. 2011. Phylogenetic position and morphology of *Abrothrix illutea* Thomas, 1925, with comments on the incongruence between gene trees of *Abrothrix* (Rodentia, Crticetidae) and their implications for the delimitation of the genus. *Zoosyst. Evol.* 87:227–41.

Teta, P., C. M. González-Fischer, M. Codesido, and D. N. Bilenca. 2010. A contribution from Barn owl pellets analysis to known micromammalian distributions in Buenos Aires province, Argentina. *Mammalia* 74:97–103.

Teta, P., D. Loponte, and A. Acosta. 2004. Sigmodontinos (Mammalia, Rodentia) del Holoceno tardio del nordeste de la provincia de Buenos Aires (Argentina). *Mastozool. Neotrop.* 11:69–80.

Teta, P., S. Malzof, R. Quintana, and Y. J. Pereira. 2006. Presas del Ñacurutu en el Bajo Delta del río Paraná (Buenos Aires, Argentina). *Ornitol. Neotrop.* 17:441–44.

Teta, P., and P. E. Ortiz. 2002. Micromamíferos andinos holocenicos del sitio arqueológico Inca Cueva 5, Jujuy, Argentina: tafonomia, zoogeograpfia y reconstrucción paleoambiental. *Estud. Geol.* (Madrid) 58:117–35.

Teta, P., C. Panti, A. Andrade, and A. E. Perez. 2001. Amplitud y composición de la dieta de *Bubo virginianus* (Aves: Strigiformes: Strigidae) en la Patagonia noroccidental Argentina. *Bol. Soc. Biol. Concepc.* 72:131–38.

Teta, P., and U. F. J. Pardiñas. 2006. Pleistocene record of marsh rats of the genus *Lundomys* in southern South America: paleoclimatic significance. *Curr. Res. Pleistocene* 23:179–80.

———. 2010. Mammalia, Didelphimorphia and Rodentia, central Santa Fe Province, Argentina. *Check List* 6:552–54.

Teta, P., U. F. J. Pardiñas, A. Andrade, aand S. Cirignoli. 2007. Distribución de los géneros *Euryoryzomys* y *Sooretamys* (Rodentia, Cricetidae) en Argentina. *Mastozool. Neotrop.* 14:279–84.

Teta, P., U. F. J. Pardiñas, and G. D'Elía. 2006. Abrotrichinos. In *Mamíferos de Argentina: Taxonomía y distribución,* ed. R. M. Barques, M. M. Díaz, and R. A. Ojeda, 192–97. Tucumán: Sociedad Argentina para el Estudio de los Mamíferos, 359 pp.

———. 2011. On the composite nature of the holotype of *Loxodontomys pikumche* Spotorno, et al., 1998 (Rodentia, Cricetidae, Sigmodontinae). *Zootaxa* 3135:55–58.

Teta, P., U. F. J. Pardiñas, and P. E. Ortiz. 2012. Revisión del registro fósil Plio-Pleistoceno asignado a la tribu Abrotrichini (Cricetidae, Sigmodontinae). *II*

Congresso Latinoamericano de Mastozoología y XXV Jornadas Argentinas de Mastozoología, Buenos Aires. Resúmenes: 275.

Teta, P., U. F. J. Pardiñas, M. Silveira, V. Aldazabal, and E. Eugenio. 2013. Roedores sigmodontinos del sitio arqueológico "El Divisadero Monte 6" (Holoceno tardío, Buenos Aires, Argentina): taxonomía y reconstrucción ambiental. *Mastozool. Neotrop.* 29:171–77.

Teta, P., U. F. J. Pardiñas, D. E. Udrizar Sauthier, and G. D'Elía. 2009. Loxodontomys micropus (Rodentia: Cricetidae). *Mamm. Species* 837:1–11.

Teta, P., U. F. J. Pardiñas, D. E. Udrizar Sauthier, and M. H. Gallardo. 2014. A new species of the tetraploid vizcacha rat *Tympanoctomys* (Caviomorpha, Octodontidae) from central Patagonia, Argentina. *J. Mamm.* 95:60–71.

Teta, P., J. A. Pereira, N. G. Fracassi, S. B. C. Bisceglia, and S. Heinonen Fortabat. 2009. Micromamíferos (Didelphimorphia y Rodentia) del Parque Nacional Lihue Calel, La Pampa, Argentina. *Mastozool. Neotrop.* 16:183–98.

Teta, P., J. A. Pereira, E. Muschetto, and N. Fracassi. 2009. Mammalia, Didelphimorphia, Chiroptera, and Rodentia, Parque Nacional Chaco and Capitán Solari, province of Chaco, Argentina. *Check List* 5:144–50.

Texera, W. A. 1975. Descripción de una nueva subespecie de *Ctenomys magellanicus* (Mammalia: Rodentia: Ctenomyidae) de Tierra del Fuego, Magallanes, Chile. *An. Inst. Patagonia* 6:163–67.

Thaler, L. 1966. Les rongeurs fossils du Bas-Languedoc dans leur rapports avec l'histoire des faunes et la stratigraphie du Tertiaire d'Eureope. *Mem. Mus. Natl. Hist. Nat.*, ser. C, 17:1–295.

Theiler, G. R., and A. Blanco. 1996a. Patterns of evolution in *Graomys griseoflavus* (Rodentia: Muridae): II. Reproductive isolation between cytotypes. *J. Mamm.* 77:776–84.

———. 1996b. Patterns of evolution in *Graomys griseoflavus* (Rodentia: Muridae): III. Olfactory discrimination as a premating isolation mechanism between cytotypes. *J. Exp. Zool.* 274:346–50.

Theiler, G. R., and C. N. Gardenal. 1994. Patterns of evolution in *Graomys griseoflavus* (Rodentia, Cricetidae). I. Protein polymorphism in populations with different chromosome numbers. *Hereditas* 120:225–29.

Theiler, G. R., C. N. Gardenal, and A. Blanco. 1999. Patterns of evolution in *Graomys griseoflavus* (Rodentia, Muridae). IV. A case of rapid evolution. *J. Evol. Biol.* 12:970–79.

Theiler, G. R., R. H. Ponce, R. E. Fretes, and A. Blanco. 1999. Reproductive barriers between the 2N = 42 and

2N = 36–38 cytotype of *Graomys* (Rodentia, Muridae). *Mastozool. Neotrop.* 6:129–33.

Thomas, O. 1880. On mammals from Ecuador. *Proc. Zool. Soc. Lond.* 1880 (part III):393–403 + 1 plate.

———. 1881a. Description of a new species of *Reithrodon*, with remarks on the other species of the genus. *Proc. Zool. Soc. Lond.* 1880 (part IV):691–96. [Part IV of the 1880 volume of the *Proceedings* was published in April 1881; see Duncan 1937.]

———. 1881b. Account of the zoological collections made during the survey of H. M. S. 'Alert' in the Straits of Magellan and on the Coast of Patagonia. I. Mammalia. *Proc. Zool. Soc. Lond.* 1881 (part I):3–6.

———. 1882. On a colection of rodents from north Peru. *Proc. Zool. Soc. Lond.* 1882 (part I):98–111 + 1 plate.

———. 1884. On a collection of Muridae from central Peru. *Proc. Zool. Soc. Lond.* 1884 (part III):447–58 + 3 plates.

———. 1886a. Description of a new Brazilian species of *Hesperomys. Ann. Mag. Nat. Hist.*, ser. 5, 17: 250–51.

———. 1886b. Notes on *Hesperomys pyrrhorhinus* Pr. Max. *Ann. Mag. Nat. Hist.*, ser. 5, 18:421–23.

———. 1887. On the small mammals collected in Demerara by Mr W. L. Sclater. *Proc. Zool. Soc. Lond.* 1887:150–53 + 1 plate.

———. 1888a. On a new species of *Loncheres* from British Guiana. *Ann. Mag. Nat. Hist.*, ser. 6, 1:326.

———. 1888b. On a new and interesting annectant genus of Muridae, with remarks on the relations of the Old- and New-World members of the family. *Proc. Zool. Soc. Lond.* 1888 (part I):130–35 + 1 plate.

———. 1890. Muridae. In *Mission scientifique du Cap Horn, 1882–1883. 6. Zoologie. Mammiferes*, ed. A. Milne-Edwards, 1–32 + 8 plates. Paris: Gauthier-Villars et Fils, 1–32 + 1–341 + 1–35 + 1–62.

———. 1893a. On some mammals from central Peru. *Proc. Zool. Soc. Lond.* 1893 (part I):333–41, 2 plates.

———. 1893b. Notes on some Mexican *Oryzomys. Ann. Mag. Nat. Hist.*, ser 6, 11:402–5.

———. 1894a. On two new Neotropical mammals. *Ann. Mag. Nat. Hist.*, ser. 6, 13:437–39.

———. 1894b. Descriptions of some new Neotropical Muridae. *Ann. Mag. Nat. Hist.*, ser. 6, 14:346–66.

———. 1895a. An analysis of the mammalian generic names given in Dr. C. W. Gloger's 'Naturgeschichte' (1841). *Ann. Mag. Nat. Hist.*, ser. 6, 15:189–93.

———. 1895b. On small mammals from Nicaragua and Bogota. *Ann. Mag. Nat. Hist.*, ser. 6, 16:55–60.

———. 1895c. Descriptions of four small mammals from South America, including one belonging to the peculiar marsupial genus "*Hyracodon*" Tomes. *Ann. Mag. Nat. Hist.*, ser. 6, 16:367–70.

———. 1896. On new small mammals from the Neotropical region. *Ann. Mag. Nat. Hist.*, ser. 6, 18:301–14.

———. 1897a. On the genera of rodents: an attempt to bring up to date the current arrangement of the Order. *Proc. Zool. Soc. Lond.* 1896 (part IV):1012–28. [Part IV of the 1896 volume of the *Proceedings* was published in April 1897; see Duncan 1937.]

———. 1897b. On a new species of *Lagidium* from the eastern coast of Patagonia. *Ann. Mag. Nat. Hist.*, ser. 6, 19:466–67.

———. 1897c. Notes on some S.-American Muridae. *Ann. Mag. Nat. Hist.*, ser. 6, 19:494–501.

———. 1897d. On some small mammals from Salta, N. Argentina. *Ann. Mag. Nat. Hist.* ser. 6, 20:214–18.

———. 1897e. Descriptions of four new South American mammals. *Ann. Mag. Nat. Hist.*, ser. 6, 20:218–21.

———. 1897f. Descriptions of new bats and rodents from America. *Ann. Mag. Nat. Hist.*, ser. 6, 20:544–53.

———. 1898a. On indigenous Muridae in the West Indies, with the description of a new Mexican *Oryzomys*. *Ann. Mag. Nat. Hist.*, ser 7, 1:176–80.

———. 1898b. Description of a new *Echimys* from the neighbourhood of Bogota. *Ann. Mag. Nat. Hist.*, ser. 7, 1:243–45.

———. 1898c. On some new mammals from the neighbourhood of Mount Sahama, Bolivia. *Ann. Mag. Nat. Hist.*, ser. 7, 1:277–83.

———. 1898d. Descriptions of two new Argentine rodents. *Ann. Mag. Nat. Hist.*, ser. 7, 1:283–86.

———. 1898e. On seven new small mammals from Ecuador and Venezuela. *Ann. Mag. Nat. Hist.*, ser. 7, 1:451–57.

———. 1898f. Descriptions of new mammals from South America. *Ann. Mag. Nat. Hist.*, ser. 7, 2:265–75.

———. 1898g. On the mammals obtained by Mr. A. Whyte in Nyasaland, and presented to the British Museum by Sir H. H. Johnston, K.C.B; being a fifth contribution to the mammal-fauna of Nyasaland. *Proc. Zool. Soc. Lond.* 1897 (part IV):925–39 + 1 plate. [Part IV of the 1897 volume of the *Proceedings* was published in April 1898; see Duncan 1937.]

———. 1898h. On some mammals obtained by the late Mr. Henry Durnford in Chubut, E. Patagonia. *Proc. Zool. Soc. Lond.* 1898 (part II):210–12.

———. 1898i. Viaggio del Dott. Borelli nel Chaco Boliviano e nella Republica Argentina. XII. On the small mammals collected by Dr. Borelli in Bolivia and northern Argentina. *Boll. Mus. Zool. Anat. Comp. Regia Univ. Torino* 13:1–4.

———. 1899a. On some small mammals from the District of Cuzco, Peru. *Ann. Mag. Nat. Hist.* ser. 7, 3:40–44.

———. 1899b. On new small mammals from South America. *Ann. Mag. Nat. Hist.*, ser. 7, 3:152–55.

———. 1899c. Description of new Neotropical mammals. *Ann. Mag. Nat. Hist.*, ser. 7, 4:278–88.

———. 1899d. Descriptions of new rodents from the Orinoco and Ecuador. *Ann. Mag. Nat. Hist.*, ser. 7, 4:378–83.

———. 1900a. New South American mammals. *Ann. Mag. Nat. Hist.*, ser. 7, 5:148–53.

———. 1900b. Descriptions of new Neotropical mammals. *Ann. Mag. Nat. Hist.*, ser. 7, 5:217–22.

———. 1900c. Descriptions of new Neotropical mammals. *Ann. Mag. Nat. Hist.*, ser. 7, 5:269–74.

———. 1900d. Descriptions of two new murines from Peru and a new hare from Venezuela. *Ann. Mag. Nat. Hist.*, ser. 7, 5:354–57.

———. 1900e. On giant squirrels from the Amazonian region. *Ann. Mag. Nat. Hist.*, ser. 7, 6:137–39.

———. 1900f. Descriptions of new rodents from western South America. *Ann. Mag. Nat. Hist.*, ser. 7, 6:294–302.

———. 1900g. Description of new rodents from western South America. *Ann. Mag. Nat. Hist.*, ser. 7, 6: 383–87.

———. 1900h. New Peruvian species of *Conepatus*, *Phyllotis*, and *Akodon*. *Ann. Mag. Nat. Hist.*, ser. 7, 6:466–69.

———. 1901a. A new spiny rat from La Guaira, Venezuela. *Proc. Biol. Soc. Washington* 14:27–28.

———. 1901b. New mammals from Peru and Bolivia, with a list of those recorded from the Inambari river, upper Madre de Dios. *Ann. Mag. Nat. Hist.*, ser. 7, 7:178–90.

———. 1901c. New South American Sciuri, *Heteromys*, *Cavia*, and *Caluromys*. *Ann. Mag. Nat. Hist.*, ser. 7, 7:192–96.

———. 1901d. New species of *Saccopteryx*, *Sciurus*, *Rhipidomys*, and *Tatu* from South America. *Ann. Mag. Nat. Hist.*, ser. 7, 7:366–71.

———. 1901e. On a collection of mammals from the Kanuku Mountains, British Guiana. *Ann. Mag. Nat. Hist.*, ser. 7, 8:139–54.

———. 1901f. New Neotropical mammals, with a note on the species of *Reithrodon*. *Ann. Mag. Nat. Hist.*, ser. 7, 8:246–55.

———. 1901g. On mammals obtained by Mr. Alphonse Robert on the Rio Jordão, S. W. Minas Geraes. *Ann. Mag. Nat. Hist.*, ser. 7, 8:526–36.

———. 1901h. New species of *Oryzomys*, *Proechimys*, *Cavia*, and *Sylvilagus* from South America. *Ann. Mag. Nat. Hist.*, ser. 7, 8:536–39.

———. 1902a. On mammals from the Serra do Mar of Paraná, collected by Mr. Alphonse Robert. *Ann. Mag. Nat. Hist.*, ser. 7, 9:59–64.

———. 1902b. On mammals from Cochabamba, Bolivia and the region north of that place. *Ann. Mag. Nat. Hist.*, ser. 7, 9:125–43.

———. 1902c. On mammals collected by Mr. Perry O. Simons in the southern part of the Bolivian Plateau. *Ann. Mag. Nat. Hist.*, ser. 7, 9:222–30.

———. 1902d. On mammals collected at Cruz del Eje, central Cordova, by Mr. P. O. Simmons. *Ann. Mag. Nat. Hist.*, ser. 7, 9:237–45.

———. 1902e. Diagnosis of a new Central American porcupine. *Ann. Mag. Nat. Hist.*, ser. 7, 10:169.

———. 1902f. New forms of *Saimiri*, *Oryzomys*, *Phyllotis*, *Coendou*, and *Cyclopes*. *Ann. Mag. Nat. Hist.*, ser. 7, 10:246–50.

———. 1902g. On two new genera of rodents from the highlands of Bolivia. *Proc. Zool. Soc. Lond.* 1902 (1):114–17 + 2 plates.

———. 1903a. New species of *Oxymycterus*, *Thrichomys*, and *Ctenomys* from S. America. *Ann. Mag. Nat. Hist.*, ser.7, 11:226–29.

———. 1903b. New forms of *Sciurus*, *Oxymycterus*, *Kannabateomys*, *Proechimys*, *Dasyprocta*, and *Caluromys* from South America. *Ann. Mag. Nat. Hist.*, ser. 7, 11:487–93.

———. 1903c. Notes on Neotropical mammals of the genera *Felis*, *Hapale*, *Oryzomys*, *Akodon*, and *Ctenomys*, with descriptions of new species. *Ann. Mag. Nat. Hist.*, ser. 7, 12:234–43.

———. 1903d. Notes on South-American monkeys, bats, carnivores, and rodents, with descriptions of new species. *Ann. Mag. Nat. Hist.*, ser. 7, 12:455–64.

———. 1904a. On the mammals collected by Mr. A. Robert at Chapada, Mato Grosso (Percy Sladen Expedition to Central Brazil). *Proc. Zool. Soc. Lond.* 1903, 2:232–44. [Usually dated to 1903, but volume 2 of the 1903 *Proceedings* was published on 1 April 1904.]

———. 1904b. Two new mammals from South America. *Ann. Mag. Nat. Hist.*, ser. 7, 13:142–44.

———. 1904c. New forms of *Saimiri*, *Saccopteryx*, *Balantiopteryx* and *Thrichomys* from the Neotropical region. *Ann. Mag. Nat. Hist.*, ser. 7, 13:250–55.

———. 1904d. New *Sciurus*, *Rhipidomys*, *Sylvilagus* and *Caluromys* from Venezuela. *Ann. Mag. Nat. Hist.*, ser. 7, 14:33–37.

———. 1904e. New *Callithrix*, *Midas*, *Felis*, *Rhipidomys* and *Proechimys* from Brazil and Ecuador. *Ann. Mag. Nat. Hist.*, ser. 7, 14:188–96.

———. 1905a. New Neotropical *Molossus*, *Conepatus*, *Nectomys*, *Proechimys*, and *Agouti*, with a note on the genus *Mesomys*. *Ann. Mag. Nat. Hist.*, ser. 7, 15:584–91.

———. 1905b. New Neotropical *Chrotopterus*, *Sciurus*, *Neacomys*, *Coendou*, *Proechimys*, and *Marmosa*. *Ann. Mag. Nat. Hist.*, ser. 7, 16:308–14.

———. 1906a. A new aquatic genus of Muridae discovered by consul L. Söderström in Ecuador. *Ann. Mag. Nat. Hist.*, ser. 7, 17:86–88.

———. 1906b. A third genus of the *Ichthyomys* group. *Ann. Mag. Nat. Hist.*, ser. 7, 17:421–23.

———. 1906c. Notes on South-American rodents. *Ann. Mag. Nat. Hist.*, ser. 7, 18:442–48.

———. 1907a. On a remarkable mountain viscacha from southern Patagonia, with diagnoses of other members of the group. *Ann. Mag. Nat. Hist.*, ser. 7, 19:438–44.

———. 1907b. On Neotropical mammals of the genera *Callicebus*, *Reithrodontomys*, *Ctenomys*, *Dasypus*, and *Marmosa* (also *Grison*). *Ann. Mag. Nat. Hist.*, ser. 7, 20:161–68.

———. 1908. A new *Akodon* from Tierra del Fuego. *Ann. Mag. Nat. Hist.*, ser. 8, 2:496–97.

———. 1909. Notes on some South American mammals, with descriptions of new species. *Ann. Mag. Nat. Hist.*, ser. 8, 4:230–42.

———. 1910a. A collection of mammals from eastern Buenos Ayres, with descriptions of related new mammals from other localities. *Ann. Mag. Nat. Hist.*, ser. 8, 5:239–47.

———. 1910b. Mammals from the river Supinaam, Demerara, presented by Mr. F. V. McConnell to the British Museum. *Ann. Mag. Nat. Hist.*, ser. 8, 6:184–89.

———. 1910c. On mammals collected in Ceará, N. E. Brazil, by Fräulein Dr. Snethlage. *Ann. Mag. Nat. Hist.*, ser. 8, 6:500–3.

———. 1910d. Four new South-American rodents. *Ann. Mag. Nat. Hist.* ser. 8, 6:503–6.

———. 1911a. Three new South American mammals. *Ann. Mag. Nat. Hist.*, ser. 8, 7:113–15.

———. 1911b. Three new mammals from the lower Amazonas. *Ann. Mag. Nat. Hist.*, ser. 8, 7:606–8.

———. 1911c. New rodents from S. America. *Ann. Mag. Nat. Hist.*, ser. 8, 8:250–56.

———. 1911d. The mammals of the tenth edition of Linnaeus; an attemp to fix the types of the genera and the exact bases and localities of species. *Proc. Zool. Soc. Lond.* 1911 (part I):120–58.

———. 1912a. A new genus of opossum and a new tucotuco. *Ann. Mag. Nat. Hist.*, ser. 8, 9: 239–41.

———. 1912b. New bats and rodents from S. America. *Ann. Mag. Nat. Hist.*, ser. 8, 10:403–11.

———. 1912c. New *Centronycteris* and *Ctenomys* from S. America. *Ann. Mag. Nat. Hist.*, ser. 8, 10: 638–40.

———. 1912d. On small mammals from the Lower Amazon. *Ann. Mag. Nat. Hist.*, ser. 8, 11:84–90.

————. 1912e. The generic names *Cercomys* and *Proechimys*. *Proc. Biol. Soc. Washington* 25:115–18.

————. 1913a. On small mammals collected in Jujuy by Señor E. Budin. *Ann. Mag. Nat. Hist.*, ser. 8, 11, 136–43.

————. 1913b. New forms of *Akodon* and *Phyllotis*, and a new genus for "*Akodon*" *teguina*. *Ann. Mag. Nat. Hist.*, ser. 8, 11:404–9.

————. 1913c. New mammals from South America. *Ann. Mag. Nat. Hist.*, ser. 8, 12:567–74.

————. 1914a. On various South-American mammals. *Ann. Mag. Nat. Hist.*, ser. 8, 13:345–63.

————. 1914b. Three new S.-American mammals. *Ann. Mag. Nat. Hist.*, ser. 8, 13:573–75.

————. 1914c. New *Nasua*, *Lutra*, and *Proechimys* from South America. *Ann. Mag. Nat. Hist.*, ser. 8, 14:57–61.

————. 1914d. New South-American rodents. *Ann. Mag. Nat. Hist.*, ser. 8, 14:240–44.

————. 1914e. Four new small mammals from Venezuela. *Ann. Mag. Nat. Hist.*, ser. 8,14:410–14.

————. 1914f. "On a remarkable case of affinity between animals inhabiting Guiana, W. Africa, and the Malay Archipelago." *Abstr. Proc. Zool. Soc. Lond.*, 5 May, No. 133:36 [A preliminary version.] [Paper subsequently published as Thomas, O. 1914. On a remarkable case of affinity between animals inhabiting Guiana, West Africa, and the Malay Archipelago. *Proc. Zool. Soc. Lond.* 1914 (part II):415–17.]

————. 1914g. Nomina conservada in Mammalia. *Zool. Anzeiger* 44:284–86.

————. 1915. A new genus of phyllostome bats and a new *Rhipidomys* from Ecuador. *Ann. Mag. Nat. Hist.*, ser. 8, 16:310–12.

————. 1916a. On the grouping of the South-American Muridae that have been referred to *Phyllotis*, *Euneomys*, and *Eligmodontia*. *Ann. Mag. Nat. Hist.*, ser. 8, 17:139–43.

————. 1916b. Notes on the Argentine, Patagonian and Cape Horn Muridae. *Ann. Mag. Nat. Hist.*, ser. 8, 17:182–87.

————. 1916c. On *Rattus* as a generic name, with a note on the nomenclature of *Echimys* and *Loncheres*. *Ann. Mag. Nat. Hist.*, ser. 8, 18:70–72.

————. 1916d. On the generic names *Rattus* and *Phyllomys*. *Ann. Mag. Nat. Hist.*, ser. 8, 18:240.

————. 1916e. Some notes on the Echimyinae. *Ann. Mag. Nat. Hist.*, ser. 8, 18:294–301.

————. 1916f. On the classification of the cavies. *Ann. Mag. Nat. Hist.*, ser. 8, 18:301–3.

————. 1916g. Two new Argentine rodents, with a new subgenus of *Ctenomys*. *Ann. Mag. Nat. Hist.*, ser. 8, 18:304–6.

————. 1916h. Two new species of *Akodon* from Argentina. *Ann. Mag. Nat. Hist.*, ser. 8, 18:334–36.

————. 1916i. The grouping of the South-American Muridae commonly referred to *Akodon*. *Ann. Mag. Nat. Hist.*, ser. 8, 18:336–40.

————. 1916j. Two new Muridae from South America. *Ann. Mag. Nat. Hist.*, ser. 8, 18:478–80.

————. 1917a. Notes on the species of the genus *Cavia*. *Ann. Mag. Nat. Hist.*, ser. 8, 19:152–60.

————. 1917b. A new species of *Aconaemys* from southern Chili. *Ann. Mag. Nat. Hist.*, ser. 8, 19:281–82.

————. 1917c. On small mammals from the Delta of the Parana. *Ann. Mag. Nat. Hist.*, ser. 8, 20:95–100.

————. 1917d. On the arrangement of the South American rats allied to *Oryzomys* and *Rhipidomys*. *Ann. Mag. Nat. Hist.*, ser. 8, 20:192–98.

————. 1917e. A new agouti from the Moon Mountains, southern British Guiana, with notes on other species. *Ann. Mag. Nat. Hist.*, ser. 8, 20:259–61.

————. 1917f. Notes on agoutis, with descriptions of new forms. *Ann. Mag. Nat. Hist.*, ser. 8, 20:310–13.

————. 1917g. Preliminary diagnoses of new mammals obtained by the Yale-National Geographic Society Peruvian expedition. *Smithson. Misc. Collect.* 68:1–3.

————. 1918a. Two new tuco-tucos from Argentina. *Ann. Mag. Nat. Hist.*, ser. 9, 1:38–40.

————. 1918b. On small mammals from Salta and Jujuy collected by Mr. E. Budin. *Ann. Mag. Nat. Hist.*, ser. 9, 1:186–93.

————. 1918c. A new species of *Eligmodontia* from Catamarca. *Ann. Mag. Nat. Hist.*, ser. 9, 2:482–84.

————. 1919a. On some small mammals from Catamarca. *Ann. Mag. Nat. Hist.*, ser. 9, 3:115–18.

————. 1919b. On mammals collected by Sr. E. Budin in North-western Patagonia. *Ann. Mag. Nat. Hist.*, ser. 9, 3:199–212.

————. 1919c. Two new Argentine species of *Akodon*. *Ann. Mag. Nat. Hist.*, ser. 9, 3:213–14.

————. 1919d. On small mammals from "Otro Cerro", north-eastern Rioja, collected by Sr. E. Budin. *Ann. Mag. Nat. Hist.*, ser. 9, 3:489–500.

————. 1919e. A new species of *Euneomys* from Patagonia. *Ann. Mag. Nat. Hist.*, ser. 9, 4:127–28.

————. 1919f. List of mammals from the highlands of Jujuy, north Argentina. *Ann. Mag. Nat. Hist.*, ser. 9, 4:128–35.

————. 1919g. Two new rodents from Tartagal, Salta, N. Argentina. *Ann. Mag. Nat. Hist.*, ser. 9, 4:154–56.

————. 1919h. The method of taking the incisive index in rodents. *Ann. Mag. Nat. Hist.*, ser. 9, 4:289–90.

————. 1920a. A further collection of mammals from Jujuy. *Ann. Mag. Nat. Hist.*, ser. 9, 5:188–96.

————. 1920b. New species of *Reithrodon, Abrocoma,* and *Scapteromys* from Argentina. A. A further collection from Sr. Budin [pp. 473–77]. B. The *Scapteromys* of the Parana Delta [pp. 477–78]. *Ann. Mag. Nat. Hist.,* ser. 9, 5:473–78.

————. 1920c. A new genus of Echimyinae. *Ann. Mag. Nat. Hist.,* ser. 9, 6:113–15.

————. 1920d. On mammals from near Tinogasta, Catamarca, collected by Sr. Budin. *Ann. Mag. Nat. Hist.,* ser. 9, 6:116–20.

————. 1920e. A new tuco-tuco from Tucumán. *Ann. Mag. Nat. Hist.,* ser. 9, 6:243–44.

————. 1920f. On mammals from the lower Amazonas in the Goeldi Museum, Para. *Ann. Mag. Nat. Hist.,* ser. 9, 6:266–83.

————. 1920g. On small mammals from the Famatina Chain, north-western Rioja. *Ann. Mag. Nat. Hist.,* ser. 9, 6:417–22.

————. 1920h. Report on the Mammalia collected by Mr. Edmund Heller during the Peruvian expedition of 1915 under the auspices of Yale University and the National Geographic Society. *Proc. U.S. Natl. Mus.* 58:217–49.

————. 1921a. A new tuco-tuco from Bolivia. *Ann. Mag. Nat. Hist.,* ser. 9, 7:136–37.

————. 1921b. Two new Muridae discovered in Paraguay by the Marquis de Wavrin. *Ann. Mag. Nat. Hist.,* ser. 9, 7:177–79.

————. 1921c. A new mountain vizcacha (*Lagidium*) from N. W. Patagonia. *Ann. Mag. Nat. Hist.,* ser. 9, 7:179–81.

————. 1921d. New *Rhipidomys, Akodon, Ctenomys* and *Marmosa* from the Sierra Santa Barbara, S.E. Jujuy. *Ann. Mag. Nat. Hist.,* ser. 9, 7:183–87.

————. 1921e. On cavies of the genus *Caviella. Ann. Mag. Nat. Hist.,* ser. 9, 7:445–48.

————. 1921f. New *Sigmodon, Oryzomys,* and *Echimys* from Ecuador. *Ann. Mag. Nat. Hist.,* ser. 9, 7:448–50.

————. 1921g. The tuco-tuco of San Juan, Argentina. *Ann. Mag. Nat. Hist.,* ser. 9, 7:523–24.

————. 1921h. On spiny rats of the *Proechimys* group from southeastern Brazil. *Ann. Mag. Nat. Hist.,* 9, 8:140–43.

————. 1921i. On mammals from the Province of San Juan, western Argentina. *Ann. Mag. Nat. Hist.,* ser. 9, 8:214–21.

————. 1921j. New *Cryptotis, Thomasomys,* and *Oryzomys* from Colombia. *Ann. Mag. Nat. Hist.,* ser. 9, 8:354–57.

————. 1921k. On a further collection of mammals from Jujuy obtained by Sr. Budin. *Ann. Mag. Nat. Hist.,* ser. 9, 8:608–17.

————. 1921l. New *Hesperomys* and *Galea* from Bolivia. *Ann. Mag. Nat. Hist.,* ser. 9, 8:622–24.

————. 1923a. Two new mammals from Marajó Island. *Ann. Mag. Nat. Hist.,* ser. 9, 12:341–42.

————. 1923b. Three new mammals from Peru. *Ann. Mag. Nat. Hist.,* ser. 9, 12:692–94.

————. 1924a. On a new fish-eating rat from Bogota. *Ann. Mag. Nat. Hist.,* ser. 9, 13:164–65.

————. 1924b. Some notes on pacas. *Ann. Mag. Nat. Hist.,* ser. 9, 13:237–39.

————. 1924c. On a collection of mammals made by Mr. Latham Rutter in the Peruvian Andes. *Ann. Mag. Nat. Hist.,* ser. 9, 13:530–38.

————. 1924d. A new fish-eating rat from Ecuador. *Ann. Mag. Nat. Hist.,* ser. 9, 13:541–42.

————. 1924e. The geographical races of *Oryzomys ratticeps. Ann. Mag. Nat. Hist.,* ser. 9, 14:143–44.

————. 1924f. New *Callicebus, Conepatus,* and *Oecomys* from Peru. *Ann. Mag. Nat. Hist.,* ser. 9, 14:286–88.

————. 1924g. Nomina conservada in Mammalia. *Proc. Zool. Soc. Lond.* 1924 (part II):345–48.

————. 1925a. A new genus of cavy from Catamarca. *Ann. Mag. Nat. Hist.,* ser. 9, 15:418–20.

————. 1925b. The Spedan Lewis South American exploration.—I. On mammals from southern Bolivia. *Ann. Mag. Nat. Hist.,* ser. 9, 15:575–82.

————. 1925c. On some Argentine mammals. I. Two new rodents from Tucuman Province. *Ann. Mag. Nat. Hist.,* ser. 9, 15:582–83.

————. 1925d. On some Argentine mammals. II. A special genus for *Euneomys fossor. Ann. Mag. Nat. Hist.,* ser. 9, 15:584–85.

————. 1925e. On some Argentine mammals. III. A second species of *Spalacopus. Ann. Mag. Nat. Hist.,* ser. 9, 15:585–86.

————. 1926a. Two new mammals from North Argentina. *Ann. Mag. Nat. Hist.,* ser. 9, 17:311–13.

————. 1926b. The Godman-Thomas expedition to Peru.—I. On mammals collected by Mr. R. W. Hendee near Lake Junin. *Ann. Mag. Nat. Hist.,* ser. 9, 17:313–18.

————. 1926c. The Spedan Lewis South American exploration.—II. On mammals collected in the Tarija Department, southern Bolivia. *Ann. Mag. Nat. Hist.,* ser. 9, 17:318–28.

————. 1926d. The Spedan Lewis South American exploration.—III. On mammals collected by Sr. Budin in the Province of Tucuman. *Ann. Mag. Nat. Hist.,* ser. 9, 17:602–9.

————. 1926e. The Godman-Thomas expedition to Peru.—II. On mammals collected by Mr. Hendee in north Peru between Pacasmayo and Chachapoyas. *Ann. Mag. Nat. Hist.*, ser. 9, 17:610–16.

———. 1926f. On some mammals from the middle Amazons. *Ann. Mag. Nat. Hist.*, ser. 9, 17:635–39.

———. 1926g. The Godman-Thomas expedition to Peru.—III. On mammals collected by Mr. R. W. Hendee in the Chachapoyas region of north Peru. *Ann. Mag. Nat. Hist.*, ser. 9, 18:156–67.

———. 1926h. The Spedan Lewis South Amerian Exploration. IV. List of mammals obtained by Sr. Budin on the boundary between Jujuy and Bolivia. *Ann. Mag. Nat. Hist.*, ser. 9, 18:193–95.

———. 1926i. The Godman-Thomas expedition to Peru.—IV. On mammals collected by Mr. R. W. Hendee north of Chachapoyas, Province of Amazonas, north Peru. *Ann. Mag. Nat. Hist.*, ser. 9, 18:345–49.

———. 1927a. The Godman-Thomas Expedition to Peru.—V. On mammals collected by Mr. R. W. Hendee in the Province of San Martin, N. Peru, mostly at Yurac Yacu. *Ann. Mag. Nat. Hist.*, 9, 19:361–73.

———. 1927b. A selection of lectotypes of American rodents in the collection of the British Museum. *Ann. Mag. Nat. Hist.*, ser. 9, 19:545–54.

———. 1927c. The *Octodon* of the highlands near Santiago. *Ann. Mag. Nat. Hist.*, ser. 9, 19:556–57.

———. 1927d. On a further collection of mammals collected by Sr. E. Budin in Neuquen, Patagonia. *Ann. Mag. Nat. Hist.*, ser. 9, 19:650–58.

———. 1927e. On further Patagonian mammals from Nuequen and Rio Colorado collected by Señor E. Budin. *Ann. Mag. Nat. Hist.*, ser. 9, 20:199–205.

———. 1927f. The Godman-Thomas expection to Peru.—VI. On mammals from the upper Huallaga and neighbouring highlands. *Ann. Mag. Nat. Hist.*, ser. 9, 20:594–608.

———. 1928a. A new *Thomasomys* from Rio Grande do Sul. *Ann. Mag. Nat. Hist.*, ser. 10, 1:154–55.

———. 1928b. The Godman-Thomas expedition to Peru.—VII. The mammals of the Rio Ucayali. *Ann. Mag. Nat. Hist.*, ser. 10, 2:249–65.

———. 1928c. The Godman-Thomas expedition to Peru.—VIII. On mammals obtained by Mr. Hendee at Pebas and Iquitos, upper Amazons. *Ann. Mag. Nat. Hist.*, ser. 10, 2:285–94.

———. 1928d. A new *Echimys* from eastern Ecuador. *Ann. Mag. Nat. Hist.*, ser. 10, 2:409–10.

———.1928e. Note on *Sciurus splendidus. Ann. Mag. Nat. Hist.*, ser. 10, 2:590.

———. 1929. The mammals of Señor Budin's Patagonian expedition, 1927–28. *Ann. Mag. Nat. Hist.*, ser. 10, 4:35–45.

Thomas, O., and J. S. Leger. 1926a. The Godman-Thomas expedition to Peru.—IV. On mammals collected by Mr. R. W. Hendee north of Chachapoyas, Province of Amazonas, north Peru. *Ann. Mag. Nat. Hist.*, ser. 9, 18:345–49.

———. 1926b. The Spedan Lewis South American exploration.—V. Mammals obtained by Señor E. Budin in Neuquen. *Ann. Mag. Nat. Hist.*, ser. 9, 18:635–41.

Thomazini, N. B. 2009. "Correlação entre estrutura cariotípica e filogenia molecular em *Rhipidomys* (Cricetidae, Rodentia) do leste do Brasil." Master's thesis, Universidade Federal do Espírito Santo, Vitoria, Espírito Santo, Brazil.

Thompson, J. V. 1815. Description of a new species of the genus *Mus*, belonging to the section of pouched rats. *Trans. Linn. Soc. Lond.* 11:161–63 + 1 plate.

Thorington, R. W., Jr., and R. S. Hoffmann. 2005. Family Sciuridae. In *Mammalian species of the world*, 3rd ed., ed. D. E. Wilson and D. M. Reeder, 754–818. Baltimore: Johns Hopkins Press, 2:xx + 745–2,142.

Thorington, R. W., Jr., J. L. Koprowski, M. A. Steele, and J. F. Whatton. 2012. *Squirrels of the world*. Baltimore: Johns Hopkins University Press, 5 unnumbered + 459 pp.

Thornback, J., and M. Jenkins (compilers). 1982. *The IUCN mammal red data book*, Part 1. IUCN, Switzerland: Gland, xl + 515 pp.

Thunberg, C. P. 1823. *Fauna Cayanensis*. Upsaliae: Excuderant Palmelad et C., 12 pp.

Tiedemann, F. 1808. *Zoologie. Zu seinen Vorlesungen entworfen. Allegemeine Zoologiek Mensch und Säugthiere*. Landshut: Webershen Buchhandlung, 1:xvi + 610 pp., + 2 (unnumbered).

Tiepolo, L. M. 2007. "Roedores Sigmodontinae do Brasil meridional: composição taxonômica, distribuição e relações fitogeográficas." Ph.D. diss., Museu Nacional and Universidade Federal do Rio de Janeiro, Brazil.

Timm, R. M., and R. D. Price. 1985. A review of *Cummingsia* Ferris (Mallophaga: Trimenoponidae), with a description of two new species. *Proc. Biol. Soc. Washington* 98:391–402.

———. 1994. Revision of the chewing louse genus *Eutrichophilus* (Phthiraptera: Trichodectidae) from the New World porcupines (Rodentia: Erethizontidae). *Field. Zool.*, n.s., 76:1–35.

Timm, R. M., D. E. Wilson, B. L. Clauson, R. K. LaVal, and C. S. Vaughan. 1989. Mammals of the La Selva–Braulio Carillo complex, Costa Rica. *N. Amer. Fauna* 75:1–162.

Tipton, V. J., R. M. Altman, and C. M. Keenan. 1967. Mites of the subfamily Laelaptinae in Panama (Acarina:

Laelaptidae). In *Ectoparasites of Panama*, ed. R. I. Wenzel and V. J. Tipton, 23–45. Chicago: Field Museum of Natural History, xii+861 pp.

Tipton, V. J., and C. E. Machado-Allison. 1972. Fleas of Venezuela. *Brigham Young Univ. Sci. Bull.*, biol. ser., 17:1–115.

Tirado, C., A. Cortés, E. Miranda-Urbina, and M. A. Carretero. 2012. Trophic preferences in an assemblage of mammal herbivores from Andean Puna (northern Chile). *J. Arid Environ.* 79:8–12.

Tiranti, S. I. 1988. Análisis de regurgitados de *Tyto alba* de la provincia de La Pampa. *Bol. Cien. Asoc. Protec. Nat.* 11:8–12.

———. 1989. Tres roedores cricétidos nuevos para la fauna de la provincia de La Pampa. *Actas J. Nac. Fauna Silvestr.*, pp. 489–94.

———. 1992. Barn owl prey in southern La Pampa, Argentina. *J. Raptor Res.* 26:89–92.

———. 1996a. "Cytogenetics of some mammal species from central Argentina." Master's thesis, Texas Tech University, Lubbock.

———. 1996b. Small mammals from Chos Malal, Neuquén, Argentina, based upon owl predation and trapping. *Texas J. Sci.* 48:303–10.

———. 1997. Cytogenetics of silky desert mice, *Eligmodontia* spp. (Rodentia, Sigmodontinae) in central Argentina. *Z. Säugetierk.* 62:37–42.

———. 1998a. Chromosomal variation in the scrub mouse *Akodon molinae* (Rodentia: Sigmodontinae) in central Argentina. *Texas J. Sci.* 50:223–38.

———. 1998b. Cytogenetics of *Graomys griseoflavus* (Rodentia: Sigmodontinae) in central Argentina. *Z. Säugetierk.* 63:32–36.

———. 1999. Observaciones citogenéticas sobre algunos mamíferos del Centro de Argentina. 7th Jornadas Pampeanas de Ciencias Naturales, pp. 297–312.

Tiranti, S. I., F. J. Dyzenchauz, E. R. Hasson, and A. I. Massarini. 2005. Evolutionary and systematic relationships among tuco-tucos of the *Ctenomys pundti* complex (Rodentia: Octodontidae): a cytogenetic and morphological approach. *Mammalia* 69:69–80.

Tirira, D. 1999. *Mamíferos del Ecuador*. Publicación Especial 2, Museo de Zoología, Quito: Pontificia Universidad Católica del Ecuador.

———. 2001. *Lista roja de los mamíferos del Ecuador*, accessed January 1, 2011, http://www.terraecuador.net/mamíferosdelecuador/listaroja.

———. 2004. *Nombres de los mamíferos del Ecuador*. Ediciones Murciélago Blanco y Museo Ecuatoriano de Ciencias Naturales. Quito: Publicación especial sobre los mamíferos del Ecuador 5.

———. 2007. *Mamíferos del Ecuador. Guía de campo*. Publicación especial sobre los mamíferos del Ecuador 6. Quito: Ediciones Murciélago Blanco.

Tirira, D., and C. Boada. 2008. *Melanomys robustulus*. IUCN Red List of Threatened Species Version 2009, accessed July 25, 2009, http://www.iucnredlist.org.

———. 2009. Diversidad de mamíferos en bosques de Ceja Andina alta del nororiente de la provincial de Carchi, Ecuador. Centro de Investigaciones IASDA. *Bol. Tec. 8 Ser. Zool.* 4–5:1–25.

Tognelli, M. F., C. M. Campos, and R. A. Ojeda. 2001. Microcavia australis. *Mamm. Species* 648:1–4.

Toledo, P. M. de, H. M. Moraes-Santos, and C. C. de S. de Melo. 1999. Levantamento preliminar de mamíferos não-voadores da Serra dos Carajás: grupos silvestres recentes e zooarqueológicos. *Bol. Mus. Paraense Emilio Goeldi,sér.* zool. 15:141–57.

Toloza, A. C., O. Scaglia, and A. I. Massarini. 2004. Cytogenetic analysis of *Ctenomys opimus* (Rodentia, Octodontidae) de la Argentina. *Mastozool. Neotrop.* 11:115–18.

Tomasco, I. H., R. Del Río, R. Iturriaga, and F. Bozinovic. 2010. Comparative respiratory strategies of subterranean and fossorial octodontid rodents to cope with hypoxis and hypercapnic atmospheres. *J. Comp. Physiol. B.* 180:877–84.

Tomasco, I. H., and E. P. Lessa. 2007. Phylogeography of the tuco-tuco *Ctenomys pearsoni*: mtDNA variation and its implication for chromosomal differentiation. In *The quintessential naturalist: Honoring the life and legacy of Oliver P. Pearson*, ed. D. A. Kelt, E. P. Lessa, J. Salazar-Bravo, and J. L. Patton, 859–82. *Univ. California Publ. Zool.* 134:v-xii + 1–981.

Tomblin, D. C., and G. H. Adler. 1998. Differences in habitat use between two morphologicall similar tropical forest rodents. *J. Mamm.* 79:953–61.

Tomes, R. F. 1858. Notes on a collection of Mammals made by Mr. Fraser at Gualaquiza. *Proc. Zool. Soc. Lond.*, part XXVI, 1858:546–49.

———. 1860a. Notes on a second collection of Mammalia made by Mr. Fraser in the Republic of Ecuador. *Proc. Zool. Soc. Lond.* 1860 (part II):211–21.

———. 1860b. Notes on a third collection of Mammalia made by Mr. Fraser in the Republic of Ecuador. *Proc. Zool. Soc. Lond.* 1860 (part II):260–68.

———. 1861 [1862]. Report on a collection of mammals made by Osbert Salvin, Esq., F.Z.S., at Dueñas, Guatemala; with notes on some of the species, by Mr. Fraser. *Proc. Zool. Soc. Lond.* 1861 (part III):278–88 + 1 plate. [Year of publication uncertain as part III of the 1861 volume of the *Proceedings* was issued between August 1860 and March 1861; see Duncan 1937.]

Tonni, E. P. 1984. The occurrence of *Cavia tschudii* (Rodentia, Caviidae) in the Southwest of Salta province, Argentina. *Stud. Neotrop. Fauna Environ.* 19:155–58.

Tonni, E. P., J. L. Prado, F. Fidalgo, and J. H. Laza. 1992. El Piso/Edad Montehermosense (Plioceno) y sus mamíferos. *3° Jornadas Geológicas Bonaerenses, Actas* 113–18, La Plata.

Torres, M. R., C. E. Borghi, S. M. Giannoni, and A. Pattini. 2003. Portal orientation and architecture of burrows in *Tympanoctomys barrerae* (Rodentia, Octodontidae). *J. Mamm.* 84:541–46.

Torres-Contreras, H., and F. Bozinovic. 1997. Food selection in an herbivorous rodent: Balancing nutrition with thermoregulation. *Ecology* 78:2230–37.

Torres-Contreras, H., E. Silva-Aranguiz, and F. M. Jaksic. 1994. Diet and selectivity of *Speotyto cunicularia* in a semi-arid locality of northern Chile throughout seven years (1987–1993). *Rev. Chilena Hist. Nat.* 67:329–40.

Torres-Díaz, C., S. Gomez-González, P. Torres-Morales, and E. Gianoli. 2012. Soil disturbance by a native rodent drives microhabitat expansion of an alien plant. *Biol. Invasion.* 14:1211–20.

Torres-Mura, J. C., and L. C. Contreras. 1989. Ecología trófica de la lechuza blanca (*Tyto alba*) en los Andes de Chile central. *Stud. Neotrop. Fauna Environ.* 24:97–103.

———. 1998. Spalacopus cyanus. *Mamm. Species* 594:1–5.

Torres-Mura, J. C., M. L. Lemus, and L. C. Contreras. 1989. Herbivorous specialization of South American desert rodent *Tympanoctomys barrerae*. *J. Mamm.* 70:646–48.

Tort, J., C. M. Campos, and C. E. Borghi. 2004. Herbivory by tuco-tucos (*Ctenomys mendocinus*) on shrubs in the upper limit of the Monte desert (Argentina). *Mammalia* 68:15–21.

Tovar Narváez, A., C. Tovar Ingar, J. Saito Díaz, A. Soto Hurtado, F. Regal Gastelumendi, Z. Cruz Burga, C., Veliz Rosas, P. Vásquez Ruesta, and G. Rivera Campos. 2010. *Yungas Peruanas–Bosques montanos de la vertiente oriental de los Andes del Perú: una perspective ecorregional de conservación.* Lima: CDC-UNALM, 150 pp.

Traba, J., P. Acebes, V. E. Campos, and S. M. Giannoni. 2010. Habitat selection by two sympatric species in the Monte desert, Argentina. First data for *Eligmodontia moreni* and *Octomys mimax. J. Arid Environ.* 74:179–85.

Tranier, M. 1976. Nouvelles données sur l'évolution non-parallèle du caryotype et de la morphologie chez les phyllotinés (Rongeurs, Cricétidés). *C. R. Séances Acad. Sci. D* 283:1201–3.

Trapido, H., and C. Sanmartín. 1971. Pichindé virus: a new virus of the Tacaribe Group from Colombia. *Am. J. Trop. Med. Hyg.* 20:631–41.

Travaini, A., J. A. Donazar, O. Ceballos, A. Rodríguez, F. Hiraldo, and M. Delibes. 1997. Food habits of common barn-owls along an elevational gradient in Andean Argentine Patagonia. *J. Raptor Res.* 31:59–64.

Travassos da Rosa, A. P. A., R. B. Tesh, F. P. Pinheiro, J. F. S. Travassos da Rosa, and N. E. Peterson. 1983. Characterization of eight new phlebotomus fever serogroup arboviruses (Bunyaviridae: *Phlebovirus*) from the Amazon region of Brazil. *Am. J. Trop. Med. Hyg.* 32:1164–71.

Travi, V. H. 1981. Nota prévia sobre nova espécie do gênero *Ctenomys* Blainville, 1826. *Iheringia Ser. Zool.* 60:123–24.

———. 1983. "Etología de *Ctenomys torquatus* Lichtenstein, 1830 (Rodentia-Ctenomyidae) em Taim, Rio Grande do Sul, Brasil." Master's thesis, Universidade Federal do Rio Grande do Sul, Porto Alegre, Brazil.

Trejo, A., and D. Grigera. 1998. Food habits of the Great horned owl (*Bubo virginianus*) in a Patagonian steppe in Argentina. *J. Raptor Res.* 32:306–11.

Trejo, A., and N. Guthmann. 2003. Owl selection on size and sex classes of rodents: activity and microhabitat use of prey. *J. Mamm.* 84:652–58.

Trejo, A., N. Guthmann, and M. Lozada. 2005. Seasonal selectivity of Magellanic horned owl (*Bubo magellanicus*) on rodents. *Eur. J. Wildl. Res.* 51:185–90.

Trejo, A., and S. Lambertucci. 2007. Feeding habits of Barn owls along a vegetative gradient in northern Patagonia. *J. Raptor Res.* 41:277–87.

Trejo, A., and V. Ojeda. 2004. Diet of Barn owls (*Tyto alba*) in forested habitats of northwestern Argentine Patagonia. *Ornitol. Neotrop.* 15:307–11.

Treviranus, G. R. 1802. *Biologie, Oder Philosophie der lebenden Natur für Naturforscher und Aerzte.* Gottingen: Johann Friedrich Röwer, Zweyter Band, 1, xiv+478 pp.

Tribe, C. J. 1996. "The Neotropical rodent genus *Rhipidomys* (Cricetidae, Sigmodontinae)—a taxonomic revision." Ph.D. diss., University College, London.

———. 2005. A new species of *Rhipidomys* (Rodentia, Muroidea) from north-eastern Brazil. *Arq. Mus. Nac., Rio de Janeiro* 63:131–46.

Tribe, C. J., R. Leher, M. C. O. Doglio, M. I. L. Rebello, D. S. Mello, and E. M. M. Guimarães. 1985. Uma tentative de esclarecimento do uso do espaço e estrutura social na cutia *Dasyprocta aguti aguti* (Rodentia). *Resumos do XII Congresso Brasiliero de Zoologia*, Campinas, São Paulo, p. 296.

Trillmich, F., C. Kraus, J. Künkele, M. Asher, M. Clara, G. Dekomien, J. T. Epplen, A. Saralegui, and N. Sachser. 2004. Species-level differentiation of two cryptic species pairs of wild cavies, genera *Cavia* and *Galea*, with a discussion of the relationship between social systems and phylogeny in the Caviinae. *Can. J. Zool.* 82:516–24.

Trouessart, E.-L. 1880a. Revision du genre Écureuil. *Le Naturaliste* 37:290–93.

———. 1880b. Catalogue des mammíferos vivants et fossils. Ordre des rongeurs. *Bull. Soc. Etud. Sci. Angers* 10(2):58–212.

———. 1897. *Catalogus mammalium tam viventium quam fossilium*. Fasciculus III. Rodentia II (Myomorpha, Histricomorpha, Lagomorpha). Berolini: R. Friedländer & Sohn, 1:vi + 664. [Issued in fascicles; the title page to volume 1 (containing Rodentia) bears the typological date "1898–1899".]

———. 1904. *Catalogus mammalium tam viventium quam fossilium. Quinquennale supplementum anno, 1904*, Fasciculus II. Rodentia. Berolini: R. Friedländer & Sohn, vii + 929 pp.

———. 1920. L'*Echinoprocta rufescens* (Hystricidé) décrit par Gray en 1865 retrouvé en Colombie, près de Bogotà. *Bull. Mus. Hist.Nat. Paris* 1920(6):448–53.

True, F. W. 1884. A provisional list of the mammals of North and Central America, and the West Indian islands. *Proc. U.S. Natl. Mus.* 7:587–611.

———. 1889 [1888]. On the occurrence of *Echinomys semispinosus*, Tomes, in Nicaragua. *Proc. U.S. Natl. Mus.* 11:467–68.

———. 1894 [1893]. On the relationships of Taylor's mouse, *Sitomys taylori*. *Proc. U.S. Natl. Mus.* 16:757–58.

Tschudi, J. J. von. 1843. Mammalium conspectus quae in Republica Peruana reperiuntur et pleraque observata vel collecta sunt in itinere. *Arch. Naturgesch.* 10 (1):244–55. [The year of publication of this issue of the *Archiv* is given as 1844 on the title page, but Sherborn (1922:cxxiv) stated that the issue was released in December 1843.]

———. 1844. *Untersuchungen über die Fauna Peruana*. Therologie, [part 1:1–20; part 2:21–76]. St. Gallen: Scheitlin und Zollikofer. [See Sherborn 1922:cxxiv for dates of publication of parts.]

———. 1845. *Untersuchungen über die Fauna Peruana*. Therologie, [part 3:77–132; aret 4:133–88; part 5:189–244]. St. Gallen: Scheitlin und Zollikofer. [See Sherborn 1922:cxxiv for dates of publication of parts.]

———. 1846. *Peru. Reiseskizzen aus den Jahren 1838–1842*. St. Gallen: Scheitlin und Zollikofer, vol. 1, 345 pp., vol. 2, 402 pp.

———. 1847. *Travels in Peru, during the years 1838–1842, on the coast, in the Sierra, across the Cordilleras and the Andes, into the primeval forests*. London: David Bogue. [This is a slightly abridged translation by Thomasina Ross of Tschudi 1846.]

Tsigalidou, V., A. G. Simotas, and A. Fasoulas. 1966. Chromsomes of the coypu. *Nature* 211:994–95.

Tuck Haugaasen, J. M., T. Haugaasen, C. A. Peres, R. Gribel, and P. Wegge. 2012. Fruit removal and natural seen dispersal of the Brazil nut tree (*Bertholletia excelsa*) in central Amazonia, Brazil. *Biotropica* 44:205–10.

Tullberg, T. 1899. Ueber das System der Nagetiere. *N. Acta Reg. Soc. Sci. Ups.*, ser. 3, 18:i-v + 1–514 + A1–18 + 57 plates.

Túnez, J. I., M. H. Cassini, M. Guichon, and D. Centron. 2005. Variabilidad genética en coipos, *Myocastor coypus*, y su relación con la presión de caza. *Rev. Mus. Argent. Cienc. Nat.*, n.s., 7:1–6.

Túnez, J. I., M. L. Guichon, C. Centron, A. P. Henderson, C. Callahan, and M. H. Cassini. 2009. Relatedness and social organization of coypus in the Argentinean pampas. *Mol. Ecol.* 18:147–55.

Turton, W. 1802. *A general system of nature through the three grand kindoms of animals, vegetables, and minerals; systematically divided into their several classes, orders, general, species and varieties, with their habitations, manners, ecolony, structure, and peculiarities. Translated from Gmelin's last edition of the celebrated Systema naturae, by Sir Charles Linnee. Amended and enlarged by the improvements and discoveries of later naturalists and societies, with appropriate copperplates*. London: Lackington, Allen, and Co., vol. 1:vii + 1–943 + 1 (unnumbered).

Turvey, S. T., S. Bdrace, and M. Weksler. 2012. A new species of recently extinct rice rat (*Megalomys*) from Barbados. *Mamm. Biol.* 77:404–13.

Turvey, S. T., M. Weksler, E. L. Morris, and M. Nokkert. 2010. Taxonomy, phylogeny, and diversity of the extinct Lesser Antillean rice rats (Sigmodontinae: Oryzomyini), with description of a new genus and species. *Zool. J. Linn. Soc.* 160:748–72.

Tuttle, M. D. 1970. Distribution and zoogeography of Peruvian bats, with comments on natural history. *Univ. Kansas Sci. Bull.* 49:45–86.

Twigg, G. I. 1962.Notes on *Holochilus sciureus* in British Guiana. *J. Mamm.*4:369–74.

———. 1965. Studies on *Holochilus sciureus berbicensis* a cricetine rodent from the coastal region of British Guiana. *Proc. Zool. Soc. Lond.* 245:263–83.

Ubilla, M., and M. T. Alberdi. 1990. *Hippidion* sp. (Mammalia, Perissodactyla, Equidae) en sedimentos del Pleistocene Superior del Uruguay (Edad Mamífero Lujanense). *Estud. Geol.* 46:453–64.

Ubilla, M., D. Perea, C. Goso, and N. Lorenzo. 2004. Late Pleistocene vertebrates from northern Uruguay: tools for biostratigraphic, climatic and environmental reconstruction. *Quarternary Int.* 114:129–42.

Ubilla, M., G. Piñeiro, and C. A. Quintana. 1999. A new extinct species of the genus *Microcavia* (Rodentia, Caviidae) from the Upper Pleistocene of the northern basin of Uruguay, with paleobiogeographic and paleoenvironmental comments. *Stud. Neotrop. Fauna Environ.* 34:141–49.

Ubilla, M., and A. Rinderknecht. 2001. About the genus *Galea* Meyen, 1831 (Rodentia, Caviidae), its fossil record in the Pleistocene of Uruguay and description of a new extinct species. *Bol. R. Soc. Esp. Hist. Nat. Sec. Biol.* 96:111–22.

———. 2003. A late Miocene Dolichotinae (Mammalia, Rodentia, Caviidae) from Uruguay, with comments about the relationships of some related fossil species. *Mastozool. Neotrop.* 10:293–302.

Udrizar Sauthier, D. E. 2009. "Los micromamíferos y la evolución ambiental durante el Holoceno en el río Chubut (Chubut, Argentina)." Ph.D. diss., Universidad Nacional de La Plata, Argentina.

Udrizar Sauthier, D. E., A. M. Abba, J. B. Bender, and P. M. Simon. 2008. Mamíferos del Arroyo Perucho Verna, Entre Ríos, Argentina. *Mastozool. Neotrop.* 15:75–86.

Udrizar Sauthier, D. E., A. M. Abba, L. G. Pagano and U. F. J. Pardiñas. 2005. Ingreso de micromamíferos brasílicos en la provincia de Buenos Aires, Argentina. *Mastozool. Neotrop.* 12:91–95.

Udrizar Sauthier, D. E., A. Andrade, and U. F. J. Pardiñas. 2005. Predation of small mammals by Rufous-legged owl, Great horned owl and Barn owl, in Argentinean Patagonia forests. *J. Raptor Res.* 39:163–66.

Udrizar Sauthier, D. E., A. E. Formoso, P. Teta, and U. F. J. Pardiñas. 2011. Enlarging the knowledge on *Graomys griseoflavus* (Rodentia: Sigmodontinae) in Patagonia: distribution and environments. *Mammalia* 75:185–93.

Udrizar Sauthier, D. E., U. F. J. Pardiñas, and E. P. Tonni. 2009. *Tympanoctomys* (Mammalia: Rodentia) en el Holoceno de Patagoina, Argentina. *Ameghiania* 46:203–7.

Udrizar Sauthier, D. E., P. Teta, P. Wallace, and U. F. J. Pardiñas. 2008. Mammalia, Rodentia, Sigmodontinae, *Loxodontomys micropus*: new locality records. *Check List* 4:171–73.

Umetsu, F., L. Naxara, and R. Pardini. 2006. Evaluating the efficiency of pitfall traps for sampling small mammals in the Neotropics. *J. Mamm.* 87:757–65.

Umetsu, F., and R. Pardini. 2007. Small mammals in a mosaic of forest remnants and anthropogenic habitats—evaluating matrix quality in an Atlantic Forest landscape. *Landscape Ecol.* 22:517–30.

Upham, N. S., and B. D. Patterson. 2012. Diversification and biogeography of the Neotropical caviomorph lineage Octodontoidea (Rodentia: Hystricognathi). *Mol. Phylogenet. Evol.* 63:417–29.

Upham, N. S., R. Ojala-Barbour, J. Brito M., P. M. Velazco, and B. D. Patterson. 2013. Transitions between Andean and Amazonian centers of endemism in the radiation of some arboreal rodents. *BMC Evol. Biol.* 13:191.

Urban, M. C., J. J. Tewksbury, and K. S. Sheldon. 2012. On a collision course: competition and dispersal differences create non-analog communities and cause extinctions during climate change. *Proc. R. Soc. Lond. B Biol. Sci.* 279:2072–80.

Urich, F. W. 1911. Rats and other mammals on cacao estates. *Circ. Dep. Agric. Trinidad Tobago* 3:19–22.

Urrejola, D., E. A. Lacey, J. R. Wieczorek, and L. A. Ebensperger. 2005. Daily activity patterns of free-living cururos (*Spalacopus cyanus*). *J. Mamm.* 86:302–8.

Valdez, L., and G. D'Elía. 2013. Differentiation in the Atlantic Forest: phylogeography of *Akodon montensis* (Rodentia: Sigmodontinae) and the Carnaval-Moritz model of Pleistocene refugia. *J. Mamm.* 94:911–22.

Valencia-Pacheco, E., J. Avaria-Llautureo, C. Muñoz-Escobar, D. Boric-Bargetto, and C. E. Hernández. 2011. Patrones de distribución geográfica de la riqueza de especies de roedores de la Tribo Oryzomyini (Rodentia: Sigmodontinae) em Sudamérica: evaluando la importancia de los procesos de colonización y extinción. *Rev. Chilena Hist. Nat.* 84:365–77.

Valim, M. P., and P. M. Linardi. 2008. A taxonomic catalog, including host and geographic distribution, of the species of the genus *Gyropus* Nitzsch (Phthiraptera: Amblycera: Gryopidae). *Zootaxa* 1899:1–24.

Valladares, F., M. Espinosa, M. Torres, E. Díaz, N. Zeller, J. de la Riva, M. Grimberg, and A. Spotorno. 2012. Nuevo registro de *Chinchilla chinchilla* (Rodentia, Chinchillidae) para la Región de Atacama, Chile. Implicancias para su estado de conservación. *Mastozool. Neotrop.* 19:173–78.

Valladares, J. P. and A. E. Spotorno. 2003. Case 3278: *Mus laniger* Molina, 1782 and *Eriomys chinchilla* Lichtenstein,1830 (currently *Chinchilla lanigera* and *C. chinchilla*: Mammalia, Rodentia): proposed conservation of the specific names. *Bull. Zool. Nomencl.* 60(3):177.

Valladares, P., and C. Campos. 2012. New record of *Abrocoma bennetti murrayi* (Rodentia, Abrocomidae) from the Atacama Region. Extension of distribution range in Chile. *Idesia* (Chile) 30:115–18.

Valqui, M. H. 2001. "Mammal diversity and ecology of terrestrial small rodents in western Amazonia." Ph.D. diss., University of Florida, Gainesville.

Van Vuuren, B. J., S. Kinet, J. Chopelet, and F. Catzeflis. 2004. Geographic patterns of genetic variation in four Neotropical rodents: conservation implications for small game mammals in French Guiana. *Biol. J. Linn. Soc.* 81:203–18.

Vanzolini, P. E. 1992. A supplement to the *Ornithological Gazetteer of Brazil*. Museu de Zoologia, Universidade de São Paulo, 252 pp.

Vargas, J., P. Flores, and J. Martínez. 2007. Pequeños mamíferos en dos áreas protegidas de la vertiente oriental boliviana, considerando la variación altitudinal y la formación vegetacional. *Revista Virtual REDESMA* 1(2):http://revistavirtual.redesma.org/vol2/pdf/ambiental/mamiferos.pdf.

Vargas, R. 2005. Heterogeneidad de recapturas de *Akodon subfuscus subfuscus* y *Oxymycterus paramensis paramensis* (Rodentia: Muridae) en fragmentos de bosques altoandinos de Bolivia. *Ecol. Boliv.* 40:25–34.

Vasallo, A. I. 1998. Functional morphology, comparative behaviour, and adaptation in two sympatric subterranean rodents genus *Ctenomys* (Caviomorpha: Octodontidae). *J. Zool.* 244:415–27.

———. 2006. Acquisition of subterranean habits in tucotucos (Rodentia, Caviomorpha, *Ctenomys*): role of social transmission. *J. Mamm.* 87:939–43.

Vassallo, A. I., and A. I. Echeverría. 2009. Evolution of brain size in a highly diversifying lineage of subterranean rodent genus *Ctenomys* (Caviomorpha: Ctenomyidae). *Brain Behav. Evol.* 73:138–49.

Vassallo, A. I., and M. S. Mora. 2007. Interspecific scaling and ontogenetic growth patterns of the skull in living and fossil ctenomyid and octodontid rodents (Caviomorpha: Octodontoidea). In *The quintessential naturalist: Honoring the life and legacy of Oliver P. Pearson*, ed. D. A. Kelt, E. P. Lessa, J. Salazar-Bravo, and J. L. Patton, 945–68. *Univ. California Publ. Zool.* 134:v-xii + 1–981.

Vásquez, R. A. 1997. Vigilance and social foraging in *Octodon degus* (Rodentia: Octodontidae) in central Chile. *Rev. Chilena Hist. Nat.* 70:557–63.

Vásquez, R. A., and J. A. Simonetti. 1999. Life history traits and sensitivity to landscape change: the case of birds and mammals of mediterranean Chile. *Rev. Chilena Hist. Nat.* 72:517–25.

Vaz, S.M. 1983. Contribuição ao estudo da fauna de mamíferos da Reserva Biológica de Poço das Antas. *Bras. Florest.* 55:33–35.

———. 2000. Sobre a distribuição geográfica de *Phaenomys ferrugineus* (Thomas) (Rodentia, Muridae). *Rev. Bras. Zool.* 17:183–86.

———. 2002. Sobre a ocurência de *Callistomys pictus* (Pictet) (Rodentia, Echimyidae). *Rev. Bras. Zool.* 19(3):631–35.

———. 2005. Mamíferos colecionados pelo Serviço de Estudos e Pesquisas sobre a Febre Amarela nos Municípios de Ilhéus e Buerarema, Estado da Bahia, Brasil. *Arq. Mus. Nac., Rio de Janeiro* 63:21–28.

Vaz Ferreira, R. 1960. Nota sobre Cricetinae del Uruguay. *Arch. Soc. Biol. Montev.* 24:66–75.

Vega-Zuniga, T., F. S. Medina, F. Fredes, C. Zuniga, D. Severín, A. G. Palacio, H. J. Karte, and J. Mpodozis. 2013. Does nocturality drive binocular vision? octodontine rodents as a case study. *PLoS ONE* 8(12):e84199.

Veitl, S., S. Begall, and H. Burda. 2000. Ecological determinants of vocalisation parameters: the case of the coruro *Spalacopus cyanus* (Octodontidae), a fossorial social rodent. *Bioacoustics* 11:129–48.

Velandia-Perilla, J. H., and C. A. Saavedra-Rodríguez. 2013. Mammalia, Rodentia, Cricetidae, *Neusticomys monticolus* (Anthony, 1921): noteworthy records of the montane fish-eating rat in Colombia. *Check List* 9:686–88.

Velhagen, W. A., and V. L. Roth. 1997. Scaling of the mandible in squirrels. *J. Morphol.* 232:107–32.

Veloso, C., and F. Bozinovic. 2000. Effect of food quality on the energetics of reproduction in a precocial rodent, *Octodon degus*. *J. Mamm.* 81:971–78.

Venegas, S. W. 1974. A cytogenetic study of *Aconaemys fuscus fuscus* (Rodentia, Octodontidae). *Bol. Soc. Biol. Concepc.* 47:207–14.

———. 1974. Variación cariotípica en *Phyllotis micropus micropus* Waterhouse (Rodentia, Cricetidae). *Bol. Soc. Biol. Concepc.* 48:69–76.

———. 1976. Somatic chromosomes of *Octodon bridgesi* (Rodentia, Octodontidae). *Bol. Soc. Biol. Concepc.* 49:7–16.

Ventura, J., M. J. Lopez-Fuster, M. Salazar, and R. Perez-Hernandez. 2000. Morphometric analysis of some Venezuelan akodontine rodents. *Netherlands J. Zool.* 50:487–501.

Ventura, K. 2009. "Estudos de citogenética e de filogenia molecular em roedores da tribo Akodontini." Ph.D. diss., Universidade de São Paulo, Brazil.

Ventura, K., V. Fagundes, G. D'Elía, A. U. Christoff, and Y. Yonenaga-Yassuda. 2011. A new allopatric lineage of the rodent *Deltamys* (Rodentia: Sigmodontinae)and the chromosomal evolution in *Deltamys kempi* and *Deltamys* sp. *Cytogenet. Genome Res.* 135:126–34.

Ventura, K., P. C. M. O'Brien, Y. Yonenaga-Yassuda, and M. A. Ferguson-Smith. 2009. Chromosome homologies of the highly rearranged karyotypes of four *Akodon* species (Rodentia, Cricetidae) resolved by reciprocal chromosome painting: the evolution of the lowest diploid number in rodents. *Chromosome Res.* 17:1063–78.

Ventura, K., Y. Sato-Kuwabara, V. Fagundes, L. Geise, Y. L. R. Leite, L. P. Costa, M. J. J. Silva, et al. 2012. Phylogeographic structure and karyotypic diversity of the Brazilian shrew mouse (*Blarinomys breviceps*, Sigmodontinae) in the Atlantic Forest. *Cytogenet. Genome Res.* 138:31–35.

Ventura, K., M. J. de J. Silva, V. Fagundes, R. Pardini, and Y. Yonenaga-Yassuda. 2004. An undescribed karyotype of *Thaptomys* (2n = 50) and the mechanism of differentiation from *Thaptomys nigrita* (2n = 52) evidenced by FISH and Ag-NORs. *Caryologia* 57:89–97.

Ventura, K., M. J. de J. Silva, L. Geise, Y. L. R. Leite, U. F. J. Pardiñas, Y. Yonenaga-Yassuda, and G. D'Elía. 2013. The phylogenetic position of the enigmatic Atlantic forest-endemic spiny mouse *Abrawayaomys* (Rodentia: Sigmodontinae). *Zool. Stud.* 52:55, 10 pp.

Ventura, K., M. J. de J. Silva, and Y. Yonenaga-Yassuda. 2010. *Thaptomys* Thomas 1915 (Rodentia, Sigmodontinae, Akodontini) with karyotypes 2n = 50, FN = 48 and 2n = 52, FN = 52: Two monophyletic lineages recovered by molecular phylogeny. *Genet. Mol. Biol.* 33:256–61.

Ventura, K., G. E. I. Ximenes, R. Pardini, M. A. Sousa, Y. Yonenaga-Yassuda, and M. J. J. Silva. 2008. Karyotypic analyses and morphological comments on the endemic and endangered Brazilian painted tree rat *Callistomys pictus* (Rodentia, Echimyidae). *Genet. Mol. Biol.* 31:697–703.

Venzal, J. M., O. Castro, P. Cabrera, C. de Souza, G. Fregueiro, D. M. Barros-Battesti, and J. E. Keirans. 2001. *Ixodes* (*Haemixodes*) *longiscutatum* Boero (new status) and *I.* (*H.*) *uruguayensis* Kohls & Clifford, a new synonym of *I.* (*H.*) *longiscutatum* (Acari : Ixodidae). *Mem. Inst. Oswaldo Cruz* 96:1121–22.

Venzal, J. M., S. Nava, A. J. Mangold, M. Mastropaolo, G. Casas, and A. A. Buglielmone. 2012. *Ornithodoros quilinensis sp nov* (Acari, Argasidae), a new tick species from the Chacoan region in Argentina. *Acta Parasitol.* 57:329–36.

Verzi, D. H. 2001. Phylogenetic position of *Abalosia* and the evolution of the extant Octodontinae (Rodentia, Caviomorpha, Octodontidae). *Acta Theriol.* 46:243–68.

———. 2002. Patrones de evolución morfológica en Ctenomyinae (Rodentia, Octodontidae). *Mastozool. Neotrop.* 9:309–28.

———. 2008. Phylogeny and adaptive diversity of rodents of the family Ctenomyidae (Caviomorpha): delimiting lineages and genera in the fossil record. *J. Zool.* 274:386–94.

Verzi, D. H., and A. Alcover. 1990. *Octodon bridgesi* Waterhouse, 1844 (Rodentia: Octodontidae) in the Argentinian living mammalian fauna. *Mammalia* 54:61–67.

Verzi, D. H., C. M. Deschamps, and E. P. Tonni. 2004. Biostratigraphic and palaeoclimatic meaning of the Middle Pleistocene South American rodent *Ctenomys kraglievichi* (Caviomorpha, Octodontidae). *Palaeoecology* 214:315–29.

Verzi, D. H., C. M. Deschamps, and M. G. Vucetich. 2002. Sistemática y antigüedad de *Paramyocastor diligens* (Ameghino, 1888) (Rodentia, Caviomorpha, Myocastoridae). *Ameghiniana* 39:193–200.

Verzi, D. H., and M. C. Iglesias. 1999. Revision of the affinities of a primitive Octodontidae (Rodentia, Caviomorpha) from the Pliocene of Argentina. *Bol. Real Soc. Esp. Hist. Nat. Sec. Geol.* 94:99–104.

Verzi, D. H., and M. Lezcano. 1996. Status sistemático y antigüedad de *Megactenomys kraglievichi* Rusconi, 1930 (Rodentia, Octodontidae). *Rev. Mus. La Plata Paleontol.* 9:239–46.

Verzi, D. H., and C. I. Montalvo. 2008. The oldest South American Cricetidae (Rodentia) and Mustelidae (Carnivora): late Miocene faunal turnover in central Argentina and the Great American Biotic Interchange. *Palaeogeogr. Palaeoclimatol. Palaeoecol.* 267:284–91.

Verzi, D. H., C. I. Montalvo, and M. G. Vucetich. 1999. Relationships and evolutionary significance of *Neophanomys biplicatus* (Rodentia, Octodontidae) from the Late Miocene–Early Pliocene of Argentina. *Ameghiniana* 36:83–90.

Verzi, D. H., A. I. Olivares, and C. C. Morgan. 2010. The oldest SouthAmerican tuco-tuco (late Pliocene, northwestern Argentina) and the boundaries of the genus *Ctenomys* (Rodentia, Ctenomyidae). *Mamm. Biol.* 75:243–52.

———. 2013. Phylogeny, evolutionary patterns and timescale of South American octodontoid rodents. The importance of recognising morphological differentiation in the fossil record. *Acta Palaeontol. Polon.* Published online March 13, 2013, http://dx.doi.org/10.4202/app .2012.0135.

Verzi, D. H., and C. Quintana. 2005. The caviomorph rodents from the San Andrés Formation, east-central Argentina, and global Late Pliocene climatic change. *Palaeogeogr. Palaeoclimatol. Palaeoecol.* 219:303–20.

Verzi, D. H., E. P. Tonni, O. A. Scaglia, and J. O. San Cristóbal. 2002. The fossil record of the desert-adapted

South American rodent *Tympanoctomys* (Rodentia, Octodontidae). Paleoenvironmental and biogeographic significance. *Palaeography, Palaeoclimatology, and Palaeontology* 179:149–58.

Verzi, D. H., M. G. Vucetich, and C. I. Montalvo. 1995. Un nuevo Eumysopinae (Rodentia: Echimyidae) del Mioceno tardio de la Provincia de La Pampa y consideraciones sobre la historia de la subfamília. *Ameghiniana* 32:191–95.

Vianchá Sánchez, Á. P., J. Y. Cepeda-Gómez, E. C. Muñoz López, Á. M. Hernández Ochoa, and L. Rosero Lasprilla. 2012. Mamíferos pequeños no voladores del Parque Natural Municipal Ranchería, Paipa, Boyacá, Colombia. *Rev. Biodivers. Neotrop.* 2:37–44.

Vianna, M. F., M. C. Mosto, and F. J. Fernández. 2011. Ensamble de micromamíferos (Rodentia: Cricetidae) recuperados de una muestra de egaprópilas de *Tyto alba* del sudoeste de la provincia de San Juan, Argentina. *Rev. Mus. La Plata Zool.* 18(172):169R–172R.

Vidal-Rioja, L., R. Riva and N. I. Baro. 1976. Los cromosomas del genero *Holochilus*. 1. Polimorfismo de *H. chacarius* Thomas (1906). *Physis* 35: 75–85.

Vidal-Rioja, L.,, R. Riva, and S. Spirito. 1973a. Los cromosomas de la *Chinchilla brevicaudata*. Contribución a la sistemática del género *Chinchilla* (Rodentia: Chinchillidae). *Physis* 32:141–50.

———. 1973b. The chromosomes of the South American rodent *Lagostomus maximus*. *Caryologia* 26:77–82.

Vie, J.-C. 1999. Wildlife rescues—the case of the Petit Saut hydroelectric dam in French Guiana. *Oryx* 33:115–26.

Vié, J.-C., V. Volobouev, J. L. Patton, and L. Granjon. 1996. A new species of *Isothrix* (Rodentia: Echimyidae) from French Guiana. *Mammalia* 60:393–406.

Vicente, J. J., D. C. Gomes, and N. A. de Araujo Filho. 1982. Alguns helmintos de marsupiais e roedores da ilha Grande, estado do Rio de Janeiro. *Atas Soc. Biol. Rio J.* 23:3–4.

Vieira, C. O. da C. 1945. Sobre uma coleção de mamíferos de Mato Grosso. *Arq. Zool., São Paulo* 4:396–429.

———. 1951. Notas sobre os mamíferos obtidos pela especião do Instituto Butantã ao rio das Mortes e Serra do Roncador. *Pap. Avulsos Zool.* (São Paulo)10:105–25.

———. 1952. Resultados de uma expedição científica ao território do Acre: Mamíferos. *Pap. Avulsos Zool.* (São Paulo)11:21–32.

———. 1953. Roedores e lagomorfos do Estado de São Paulo. *Arq. Zool., São Paulo* 8:129–68.

———. 1955. Lista remissiva dos mamíferos do Brasil. *Arq. Zool., São Paulo* 8:341–474.

Vieira, E. M. 1993. Occurrence and prevalence of botflies, *Metacuterebra apicalis* (Diptera: Cuterebridae) in rodents of Cerrado from central Brazil. *J. Parasitol.* 79:792–95.

Vieira, E. M., and L. C. Baumgarten. 1995. Daily activity patterns of small mammals in a cerrado area from central Brazil. *J. Trop. Ecol.* 11:255–62.

———. 2009. Daily activity patterns of small mammals in a cerrado area from central Brazil. *J. Trop. Ecol.* 11:255–62.

Vieira, E. M., L. C. Baumgarten, G. Paise, and R. G. Becker. 2010. Seasonal patterns and influence of temperature on the daily activity of the diurnal neotropical rodent *Necromys lasiurus*. *Can. J. Zool.* 88:259–65.

Vieira, E. M., and E. L. A. Monteiro-Filho. 2003. Vertical stratification of small mammals in the Atlantic rain forest of south-eastern Brazil. *J. Trop. Ecol.* 19:501–7.

Vieira, E. M., G. Paise, and P. H. D. Machado. 2006. Feeding of small rodents on seeds and fruits: a comparative analysis of three species of rodents of the *Araucaria* forest, southern Brazil. *Acta Theriol.* 51:311–18.

Vieira, M. V. 1989. "Dinâmica de populações, variação sazonal de nichos e seleção de microhabitats numa comunidade de roedores de Cerrado brasileiro." Master's thesis, Universidade Estadual de Campinas, São Paulo, Brazil.

———. 1997. Dynamics of a rodent assemblage in a Cerrado of southeast Brazil. *Rev. Bras. Biol.* 57:99–107.

Vieira, M. V., N. Olifiers, A. C. Delciellos, V. Z. Antunes, L. R. R. Bernardo, C. E. de V. Grelle, and R. Cerqueira. 2009. Land use vs. fragment size and isolation as determinants of small mammal composition and richness in Atlantic Forest fragments. *Biol. Conserv.* 142:1191–200.

Vieytes, E. C., C. C. Morgan, and D. H. Verzi. 2007. Adaptive diversity of incisor enamel microstructure in South American burrowing rodents (family Ctenomyidae, Caviomorpha). *J. Anat.* 211:296–302.

Vilela, J. F. 2005. "Filogenia molecular de *Brucepattersonius* (Rodentia: Sigmodontinae) com uma análise morfométrica craniana do gênero." Master's thesis, Universidade Federal do Rio de Janeiro, Brazil, xvi+67 pp.

Vilela, J. F., B. Mello, C. M. Voloch, and C. G. Schrago. 2013. Sigmodontine rodents diversified in South America prior to the complete rise of the Panamanian Isthmus. *J. Zool. Syst. Evol. Res.*, published on-line 27 Dec. 2013, doi:10.1111/jzs.12057.

Vilela, J. F., J. A. de Oliveira, and C. R. Bonvicino. 2006. Taxonomic Status of *Brucepattersonius albinasus* (Rodentia: Sigmodontinae). *Zootaxa* 1199:61–68.

Vilela, R. do V. 2005. "Estudos em roedores da família Echimyidae, com abordagens em sistemática molecular, citogenética e biogeografía." Master's thesis, Universidade de São Paulo, Brazil.

Vilela, R. do V., T. Machado, K. Ventura, V. Fagundes, M. de J. Silva, and Y. Yonenaga-Yassuda. 2009. The taxonomic status of the endangered thin-spined porcupine, *Chaetomys subspinosus* (Olfers, 1818), based on molecular and karyologic data. *BMC Evol. Biol.* 9:29, doi:10.1186/1471-2148-9-29.

Villa R., B. 1941. Una nueva rata de campo (*Tylomys gymnurus* sp. nov.). *An. Instit. Biol., México* 12:763–66.

Villafañe, G. de. 1981. Reproducción y crecimiento de *Calomys musculinus murillus* (Thomas, 1916). *Hist. Nat.* 1:237–56.

Villafañe, G. de, J. Merler, R. Quintana, and R. Bó. 1992. Habitat selection in cricetine rodent population on maize field in the Pampa region of Argentina. *Mammalia* 56:215–29.

Villalobos, F., and F. Cervantes-Reza. 2007. Phylogenetics relationships of Mesoamerican species of the genus *Sciurus* (Rodentia: Sciuridae). *Zootaxa* 1525:31–40.

Villalpando, G. J., Vargas, and J. Salazar-Bravo. 2006. Record of *Rhagomys* (Rodentia: Cricetidae) for Bolivia. *Mastozool. Neotrop.* 13:143–49.

Villareal, D., K. L. Clark, L. C. Branch, J. L. Hierro, and M. Machicote. 2008. Alteration of ecosystem structure by a burrowing herbivore, the plains vizcacha (*Lagostomus maximus*). *J. Mamm.* 89:700–11.

Villarroel, C. A. 1975. Dos nuevos Ctenomyinae (Caviomorpha, Rodentia) en los estratos de la Formación Umala (Plioceno Superior) de Vizcachani (Prov. Aroma, Dpto. La Paz, Bolivia). *Act. Primer Congr. Argent. Paleontol. Bioestratigr.* 2:495–503.

Vincent, J. F. V., and P. Owers. 1986. Mechanical design of hedgehog spines and porcupine quills. *J. Zool. Lond.* (A) 210:55–75.

Vitullo, A. D., and J. A. Cook. 1991. The role of sperm morphology in the evolution of tuco-tucos (Rodentia, Ctenomyidae): confirmation of results from Bolivian species. *Z. Säugetierk.* 56:359–64.

Vitullo, A. D., M. B. Espinosa, and M. S. Merani. 1990. Cytogenetics of vesper mice, *Calomys* (Rodentia, Cricetidae): Robertsonian variation between *Calomys callidus* and *Calomys venustus*. *Z. Säugetierk.* 55:99–105.

Vitullo, A. D., A. E. Kajon, R. Percich, G. Zuleta, M. S. Merani, and F. Kravetz. 1984. Caracterización citogenética de tres especies de roedores (Rodentia, Cricetidae) de la república Argentina. *Rev. Mus. Argentino Cien. Nat. "Bernardino Rivadavia," Inst. Nac. Invest. Cien. Nat. Zool.* 13:491–98.

Vitullo, A. D., M. S. Merani, O. A. Reig, A. E. Kajon, O. Scaglia, M. B. Espinosa, and A. Pérez-Zapata. 1986. Cytogenetics of South American akodont rodents (Cricetidae): new karyotypes and chromosomal banding patterns of Argentinean and Uruguayan forms. *J. Mamm.* 67:69–80.

Vitullo, A. D., E. R. S. Roldán, and M. S. Merani. 1988. On the morphology of spermatozoa of tuco-tucos, *Ctenomys*: new data and its implication for the evolution of the genus. *J. Zool.* 215:675–83.

Vivanco, P., V. Ortiz, M. A. Rol, and J. A. Madrid. 2007. Looking for the keys to diurnality downstream from the circadian clock: role of melatonin in a dual-phasing rodent, *Octodon degus. J. Pineal Res.* 42:280–90.

Vivar Pinares, S. E. 2006. "Análisis de distribución altitudinal de mamíferos pequeños en el Parque Nacional Yanachaga Chemillén, Pasco, Perú." Master's thesis, Universidad Nacional Mayor de San Marcos, Lima, Peru.

Vivas, A. M. 1984. "Ecología de poblaciones de los roedores de los llanos de Guárico." Undergraduate thesis, Universidad Simón Bolívar, Caracas, Venezuela.

———. 1986. Population biology of *Sigmodon alstoni* (Rodentia, Cricetidae) in the Venezuelan llanos. *Rev. Chilena Hist. Nat.* 59:179–91.

Vivas, A. M., and A. C. Calero. 1985. Algunos aspectos de la ecología poblacional de los pequeños mamíferos en la Estación Biológica de los Llanos. *Bol. Soc. Venez. Cien. Nat.* 143:79–99.

———. 1988. Patterns in the diet of *Sigmodon hispidus* (Rodentia, Cricetidae) in relation to available resources in a tropical savanna. *Ecotropicos* 1:82–91.

Vivas, A. M., R. Roca, E. Weir, K. Gil, and P. Gutiérrez. 1986. Ritmo de actividad nocturna de *Zygodontomys microtinus, Sigmodon alstoni, y Marmosa robinsoni* en Masaguaral, Estado Guárico. *Acta Cient. Venez.* 37:456–58.

Viveiros, E. de. 1958. *Rondon conta a sua vida*. Rio de Janeiro: São Jose, 638 pp.

Vivo, M. 1991. *Taxonomia de Callithrix Erxleben, 1777 (Callitrichidae, Primates)*. Fundação Biodiversitas: Belo Horizonte, Brazil, 105 pp.

Vogelsang, E.G., and J. Espin. 1949. Dos nuevos huéspedes para *Capillaria hepatica* (Bancroft, 1893) Travassos1915; nutria (*M. coypus*) y el raton ratón mochilero (*Akodon venezuelensis*). *Rev. Med. Vet. Parasitol. Caracas* 8:73–78.

Voglino, D., and U. F. J. Pardiñas. 2005. Roedores sigmodontinos (Mammalia: Rodentia: Cricetidae) y otros micromamíferos pleistocénicos del norte de la provincia de Buenos Aires (Argentina): reconstrucción paleoambiental para el Ensenadense cuspidal. *Ameghiniana* 42:143–58.

Voglino, D., U. F. J. Pardiñas, and P. Teta. 2004. *Holochilus chacarius cha*carius (Rodentia, Cricetidae) en la Provincia de Buenos Aires, Argentina. *Mastozool. Neotrop.* 11:243–47.

Volobouev, V. T., and F. M. Catzeflis. 2000. Chromosome banding analysis (G-, R- and C- bands) of *Rhipidomys nitela* and a review of the cytogenetics of *Rhipidomys* (Rodentia: Sigmodontinae). *Mammalia* 64:353–60.

Voloch, C. M., J. F. Vilela, L. Loss-Oliveira, and C. G. Schrago. 2013. Phylogeny and chronology of the major lineages of New World hystricognath rodents: insights on the biogeography of the Ecocene/Oligocene arrival of mammals in South America. *BMC Res. Note.* 6:160.

Vorontsov, N. N. 1959. The system of hamster (Cricetinae) in the sphere of the world fauna and their phylogenetic relations. *Byull. Mosk. Obschchestva Ispytatelei Prirody Otdel Biol.* 64:134–37. Article in Russian.

———. 1979. Evolution of the alimentary system of myomorph rodents [A translation of N. N. Vorontsov, 1967, Evolystsiya pischchevaritel'noi sistemy gryzunov mysheobraznye]. New Delhi: Indian National Scientific Documentation Centre, 346 pp.

Voss, R. S. 1988. Systematics and ecology of ichthyomyine rodents (Muroidea): Patterns of morphological evolution in a small adaptive radiation. *Bull. Am. Mus. Nat. Hist.* 188:259–493.

———. 1991a. An introduction to the Neotropical muroid rodent genus *Zygodontomys*. *Bull. Am. Mus. Nat. Hist.* 210:1–113.

———. 1991b. On the identity of "*Zygodontomys*" *punctulatus* (Rodentia: Muroidea). *Am. Mus. Novit.* 3026:1–8.

———. 1992. A revision of the South American species of *Sigmodon* (Mammalia: Muridae) with notes on their natural history and biogeography. *Am. Mus. Novit.* 3050:1–56.

———. 1993. A revision of the Brazilian muroid rodent genus *Delomys* with remarks on "thomasomyine" characters. *Am. Mus. Novit.* 3073:1–44.

———. 2003. A new species of *Thomasomys* (Rodentia: Muridae) from eastern Ecuador, with remarks on mammalian diversity and biogeography in the Cordillera Oriental. *Am. Mus. Novit.* 3421:1–47.

———. 2011. Revisionary notes on Neotropical porcupines (Rodentia: Erethizontidae). 3. An annotated checklist of species of *Coendou* Lacépède, 1799. *Am. Mus. Novit.* 3720:1–36.

———. 2014. First South American record of *Isthmomys pirrensis* (Goldman, 1912) (Rodentia: Cricetidae: Neotominae). *Check List.* 10:648–49.

Voss, R. S., and N. I. Abramson. 1999. Case 3121. *Holochilus* Brandt, 1835, *Proechimys* J. A. Allen, 1899 and *Trinomys* Thomas, 1921 (Mammalia, Rodentia): proposed conservation by the designation of *H. sciureus* Wagner, 1842 as the type species of *Holochilus*. *Bull. Zool. Nomencl.* 56:255–61.

Voss, R. S., and R. Angermann. 1997. Revisionary notes on Neotropical porcupines (Rodentia: Erethizontidae). 1. Type material described by Olfers (1818) and Kuhl (1820) in the Berlin Zoological Museum. *Am. Mus. Novit.* 3214:1–44.

Voss, R. S., and M. D. Carleton. 1993. A new genus for *Hesperomys molitor* Winge and *Holochilus magnus* Hershkovitz (Mammalia, Muridae) with an analysis of its phylogenetic relationships. *Am. Mus. Novit.* 3085:1–39.

Voss, R. S., and L. H. Emmons. 1996. Mammalian diversity in Neotropical lowland rainforests: a preliminary assessment. *Bull. Am. Mus. Nat. Hist.* 230:1–115.

Voss, R. S., M. Gómez-Laverde, and V. Pacheco. 2002. A new genus for *Aepeomys fuscatus* Allen, 1912, and *Oryzomys intectus* Thomas, 1921: Enigmatic murid rodents from Andean cloud forests. *Am. Mus. Novit.* 3373:1–42.

Voss, R. S., P. D. Heideman, V. L. Mayer, and T. M. Donnelly. 1992. Husbandry, reproduction and postnatal development of the neotropical muroid rodent *Zygodontomys brevicauda*. *Lab. Animal.* 26:38–46.

Voss, R. S., C. Hubbard, and S. A. Jansa. 2013. Phylogenetic relationships of New World porcupines (Rodentia, Erethizontidae): implications for taxonomy, morphological evolution, and biogeography. *Am. Mus. Novit.* 3769:1–36.

Voss, R. S., and A. V. Linzey. 1981. *Comparative Gross Morphology of Male Accessory Glands among Neotropical Muridae (Mammalia: Rodentia), with Comments on Systematic Implications.* Misc. Publ. Mus. Zool. Univ. Michigan 159:1–41.

Voss, R. S., D. P. Lunde, and N. B. Simmons. 2001. The mammals of Paracou, French Guiana: A Neotropical lowland rainforest fauna. Part 2. Nonvolant species. *Bull. Am. Mus. Nat. Hist.* 263:1–236 pp., 99 figures, 69 tables.

Voss, R. S., and L. Marcus. 1992. Morphological evolution in muroid rodents. II. Craniometric factor divergence in seven Neotropical genera, with experimental results from *Zygodontomys*. *Evolution* 46:1918–34.

Voss, R. S., and P. Myers. 1991. *Pseudoryzomys simplex* (Rodentia: Muridae) and the significance of Lund's collections from the caves of Lagoa Santa, Brazil. In *Contributions to Mammalogy in honor of Karl F. Koopman*, ed. T. A. Griffiths and D. Klingener, 414–32. *Bull. Am. Mus. Nat. Hist.* 206:1–432.

Voss, R. S., J. L. Silva L., and J. A. Valdes L. 1982. Feeding behavior and diets of Neotropical water rats, genus *Ichthyomys* Thomas, 1893. *Z. Säugetierk.* 47:364–69.

Voss, R. S., and M. N. F. da Silva. 2001. Revisionary notes on Neotropical porcupines (Rodentia: Erethizontidae). 2. A review of the *Coendou vestitus* group with descriptions of two new species from Amazonia. *Am. Mus. Novit.* 3351:1–36.

Voss, R. S., and M. Weksler. 2009. On the taxonomic status of *Oryzomys curasoae* McFarlane and Debrot, 2001, (Rodentia: Cricetidae: Sigmodontinae) with remarks on the phylogenetic relationships of *O. gorgasi* Hershkovitz, 1971. *Caribb. J. Sci.* 45:73–79.

Vriesendorp, C., T. S. Schulenberg, W. S. Alverson, D. K. Moskovitz, and J.-I. Rojas Moscoso (editors). 2006. Perú: Sierra del Divisor. *Rapid Biol. Inventor.* 17:1–298, 20 plates.

Vucetich, M. G. 1975. La anatomía del oído medio como indicadora de relaciones sistemáticas y filogenéticas en algunos grupos de roedores. *1st Congreso Argentino de Paleontología y Biostratigrafía, Tucumán, Actas* 2:477–94.

———. 1984. Los roedores de la Edad Fraisense (Mioceno medio) de Patagonia. *Rev. Mus. La Plata*, n.s., 8. *Paleontología* 50:47–126.

Vucetich, M. G., C. M. Deschamps, A. I. Olivares, and M. T. Dozo. 2005. Capybaras, shape, size and time: a model kit. *Acta Paleontol. Pol.* 50:259–72.

Vucetich, M. G., C. M. Deschamps, and M. E. Pérez. 2013. In *Capybara: Biology, use and conservation of an exceptional Neotropical species*, ed. J. R. Moreira, K. M. P. M. B. Ferraz, E. A. Herrera, and D. W. Macdonald, 39–60. New York, Heidelberg, Dordrecht/London: Springer, i-xvii +1–419 pp.

Vucetich, M. G., M. M. Mazzoni, and U. F. J. Pardiñas. 1993. Los roedores de la Formacion Collon Cura (Mioceno medio), y la Ignimbrita Pilcaniyeu. Cañadon del Tordillo, Neuquen. *Ameghiniana* 30:361–81.

Vucetich, M. G., and D. Verzi. 1991. Un nuevo Echimyidae (Rodentia, Histricognathi) de la Edad Colhuehuapense de Patagonia y consideraciones sobre la sistematica de la familia. *Ameghiniana* 28:67–74.

———. 1993. Las homologies en los diseños oclusales de los roedores Caviomorpha: un modelo alternative. *Mastozool. Neotrop.* 1:61–72.

———. 2002. First record of Dasyproctidae (Rodentia) in the Pleistocene of Argentina. Paleoclimatic implication. *Palaeogeogr. Palaeoclimatol. Palaeoecol.* 178:67–73.

Vucetich, M. G., D. H. Verzi, and J.-L. Hartenberger. 1999. Review and analysis of the radiation of the South American Hystricognathi (Mammalia, Rodentia). *C. R. Acad. Sci. Earth Planet. Sci.* 329:763–69.

Wagler, J. 1830. *Natürlisches System der Amphibien, mit vorangehender Classification der Säugthiere und Vögel*. München, Stuttgart und Tübingen: J. G. Cottaschen Buchhandlung, vi+354 pp., 2 foldouts.

———. 1831a. Einige Mitteilungen über Thiere Mexicos. *Isis von Oken* 24(6):510–34.

———. 1831b. Gattungen der Sippe *Lagostomus* Brookes, mit ihrer Synonymie. *Isis von Oken* 24(6):612–17.

———. 1831c. Beiträge von Wagler zur Sippe *Dasyprocta* Illig. *Isis von Oken* 24(6):617–22.

———. 1832. Neue Sippen und Gattungen der Säugethiere und vögel. I. Säugthiere. *Isis von Oken* 25, 11:1218–21.

Wagner, J. A. 1840. Beschreibung einiger neuer Nager, welche auf der Reise des Herrn Hofraths von Shubert gesammelt wurden, mit Bezugnahme auf einige andere werwandte Formen. II. Stachelmäus. *Abhandl. Math.–Physik. König. Bayer. Akad. Wiss. München* 3:191–210.

———. 1841. Gruppirung der Gattungen der Nager in natürlichen Familien, nebst Beschreibung einiger neuen Gattungen und Arten. *Arch. Naturgesch.* 7(1):111–38.

———. 1842a. Beschreibung einiger neuer oder minder bekannter Nager. *Arch. Naturgesch.* 8(1):1–33.

———. 1842b. Nachtrag zu meiner Beschreibung von Habrocoma und Holochilus. *Arch. Naturgesch.* 8(1):288.

———. 1842c. Diagnosen neuer Arten brasilischer Säugthiere. *Arch. Naturgesch.* 8(1):356–62.

———. 1843a. *Die Säugthiere in Abbildungen nach der Natur mit Beschreibungen von Dr. Johann Christian Daniel von Schreber. Supplementband*. Dritter Abtheilung: Die Beutelthiere und Nager (erster Abschnitt). Erlangen: Expedition das Schreber'schen Säugthier- und des Esper'sschen Schmetterlingswerkes, und in Commission der Voss'schen Buchhandlung in Leipzig, 3:xiv+614 pp., pls. 85–165. [See Sherborn (1892) and Poche (1912) for dates of publication.]

———. 1843b. Bemerkungen über die jetztlebende brasilianische Thierwelt in den obigen Abhandlungen. *Isis von Oken* 1843:738–60. [Wagner's translation of selected extracts from P. W. Lund's works.]

———. 1843c: Bericht über die Leistungen in der Naturgeschichte der Säugthiere während des Jahres 1842. *Arch. Naturgesch.* 9(2):1–67.

———. 1844. *Die Säugthiere in Abbildungen nach der Natur mit Beschreibungen von Dr. Johann Christian*

Daniel von Schreber. Supplementband. Vierte Abtheilung: Die Nager (zweiter Abschnitt), Zahnlücker, Einhufer, Dickhäuter und Wiederkäuer. Erlangen: Expedition das Schreber'schen Säugthier- und des Esper'sschen Schmetterlingswerkes, und in Commission der Voss'schen Buchhandlung in Leipsiz, 4:xii + 523 pp., pls. 168–327. [See Poche 1912 and Sherborn 1891 for date of publication.]

————. 1845a. Diagnosen einiger neuen Arten von Nagern und Handflülern. *Arch. Naturgesch.* 11(1):145–49.

————. 1845b. Bericht Uuberdie Leistungen in der Naturgeschichte der Säugthiere während des Jahres 1844. *Arch. Naturgesch.* 11(2):1–43.

————. 1848. Beiträge zur Kentnniss der Arten von *Ctenomys. Arch. Naturgesch.* 14(1):72–78.

————. 1850. Beiträge zur Kenntniss der Säugthiere Amerika's. *Abhandl. Math.–Physik. König. Bayer. Akad. Wiss. München* 5:271–332.

Wahlert, J. H. 1985. Skull morphology and relationships of geomyoid rodents. *Am. Mus. Novit.* 2812:1–20.

————. 1991. The Harrymyinae, a new heteromyid subfamily (Rodentia, Geomorpha), based on cranial and dental morphology of Harrymys Munthe, 1988. *Am. Mus. Novit.* 3013:1–23.

Wainberg, R. L., and T. G. Fronza. 1974. Autosomic polymorphism in *Phyllotis griseoflavus griseoflavus* Waterhouse, 1837 (Rodentia, Cricetidae). *Boll. Zool.* 41:19–24.

Walker, E. P. 1968. *Mammals of the world.* 2nd ed. Baltimore: Johns Hopkins University Press, 2:668–1500.

Walker, L. I., M. Rojas, S. Flores, A. E. Spotorno, and G. Manriquez. 1999. Genomic compatibility between two phyllotine rodent species evaluated through their hybrids. *Hereditas* 131:227–38.

Walker, L. I., and A. E. Spotorno. 1992. Tandem and centric fusions in the chromosome evolution of the South American phyllotines of the genus *Auliscomys* (Rodentia; Cricetidae). *Cytogenet. Cell Genet.* 61:135–40.

Walker, L. I., A. E. Spotorno, and J. Arrau. 1984. Cytogenetic and reproductive studies of two nominal subspecies of *Phyllotis darwini* and their experimental hybrids. *J. Mamm.* 65:220–30.

Walker, R. S., G. Ackermann, J. Schachter-Broide, V. Pancotto, and A. J. Navaro. 2000. Habitat use by mountain vizcachas (*Lagidium viscacia* Molina, 1782) in the Patagonian steppe. *Z. Säugetierk.* 65:293–300.

Walker, R. S. A. J. Navaro, and L. C. Branch. 2003. Effects of patch attributes, barriers, and distance between patches on the distribution of a rock-dwelling rodent (*Lagidium viscacia*). *Landscape Ecol.* 18:185–92.

————. 2007. Functional connectivity defined through cost-distance and genetic analyses: a case study for the rock-dwelling mountain vizcacha (*Lagidium viscacia*) in Patagonia, Argentina. *Landscape Ecol.* 22:1303–14.

Walker, R. S., A. N. Navaro, and P. G. Perovic. 2008. Comparison of two methods for estimation of abundance of mountain vizcachas (*Lagidium viscacia*) based on direct counts. *Mastozool. Neotrop.* 13:271–74.

Walle, P. 1914. *Bolivia, its people and its resources, its railways, mines, and rubber-forests.* Translated by B. Mial. London: T. Fisher Unwin, 407 pp. + 62 plates + 4 maps.

Walton, A. H. 1997. Rodents. In *Vertebrate paleontology in the Neotropics: The Miocene fauna of La Venta, Colombia,* ed. by R. F. Kay, R. H. Madden, R. LO. Cifelli, and J. J. Flynn, 392–409. Washington, DC: Smithsonian Institution Press, xvi + 592 pp.

Waterhouse, G. R. 1837. Characters of new species of the genus *Mus,* from the collection of Mr. Darwin. *Proc. Zool. Soc. Lond.* 1837 (part V):15–21, 27–32.

————. 1838. Mammalia. In *The zoology of the Voyage of H.M.S. Beagle under the command of Captain Firzroy, R.N., during the Years 1832–1836,* ed. C. Darwin, Fascicles 2, 4, and 5 (pages i-vi + 1–48, pls. 1–24, 33). London: Smith, Elder and Co., 2:xii + 97 pp., 35 pls. 1838–1839. [See Sherborn 1897 for dates of publication of parts.].

————. 1839. Mammalia. In *The zoology of the Voyage of the H.M.S. Beagle under the command of Captain FitzRoy, R.N., during the Years 1832–1836,* ed. C. Darwin, Fascicle 10 (pages vii-ix + 49–97, pls. 25–32, 34). London: Smith, Elder and Co., 2:xii + 97 pls, 1838–1839. [See Sherborn 1897 for dates of publication of parts.]

————. 1840. On a new genus of the family Muridae and order Rodentia. *Proc. Zool. Soc. Lond.,* 1840, part VIII:1–3.

————. 1842. Report of the meeting of November 9, 1841. *Proc. Zool. Soc. Lond.* 1841 (part IX):89–92. [Published March 1842; see Duncan 1937.]

————. 1845. On the various skins of Mammalia from Chile, with notes relating to them by Mr. Bridges. *Proc. Zool. Soc. Lond.* 1844 (part XII, no. CXL):153–57. [Part XI, no. CXL of the 1844 volume of the *Proceedings* was published in February 1845; see Duncan 1937.]

————. 1846. Hesperomys Boliviensis. *Proc. Zool. Soc. Lond.* 1846 (part XIV):9–10. [Also published in 1846 the *Ann. Mag. Nat. Hist.,* ser. 1, 17:483–84.]

————. 1848. *A natural history of the Mammalia: Rodentia, or gnawing mammalia.* London: Hippolyte Baillière, 2:1–500, 22 pls. [note: Sherborn 1922, pg.

cxxviii, states that pages 1–304 were published in 1847 and pages 305–500 in 1848. 1848 is herein cited as the year of publication for volume 2, as specified on the title page and as followed by previous authors.]

Webb, R. A. 1991. Chinchillas. In *Manual of exotic pets*, ed. P. H. Beynon and J. E. Cooper, 15–21. Cheltenham, UK: British Small Animal Veterinary Association.

Webb, S. D. 2006. The Great American Biotic Interchange: patterns and processes. *Ann. Missouri Bot. Garden* 93:245–57.

Weir, B. J. 1971a. Some observations on reproduction in the female green acouchi, *Myoprocta pratti. J. Reprod. Fertil.* 24:193–201.

———. 1971b. Some observations on reproduction in the female agouti, *Dasyprocta aguti. J. Reprod. Fertil.* 24:203–11.

———. 1971c. The reproductive physiology of the plains viscacha, *Lagostomus maximus. J. Reprod. Fertil.* 25:355–63.

———. 1971d. The reproductive organs of the female plains viscacha, *Lagostomus maximus. J. Reprod. Fertil.* 25:365–73.

———. 1974a. The tuco-tuco and plains vizcacha. In *The Biology of hystricomorph rodents*, ed. I. W. Rowlands and B. J. Weir, 113–30. *Symp. Zool. Soc. Lond.*, 34:xix + 1–482.

———. 1974b. Reproductive characteristics of hystricomorph rodents. In *The biology of hystricomorph rodents*, ed. I. W. Rowlands and B. J. Weir, 265–301. *Symp. Zool. Soc. Lond.* 34:xix+482.

———. 1974c. Notes on the origin of the domestic Guinea pig. In *The biology of hystricomorph rodents*, ed. I. W. Rowlands and B. J. Weir, 437–46. *Symp. Zool. Soc. Lond.* 34:xix+482.

Weksler, M. 1996. "Revisão sistemática do grupo de espécies nitidus do gênero *Oryzomys* (Rodentia: Sigmodontinae)." Master's thesis, Universidade Federal do Rio de Janeiro, Brazil.

———. 2003. Phylogeny of Neotropical oryzomyine rodents (Muridae: Sigmodontinae) based on the nuclear IRBP exon. *Mol. Phylogenet. Evol.* 29:331–49.

———. 2006. Phylogenetic relationships of oryzomyine rodents (Muridae: Sigmodontinae): separate and combined analyses of morphological and molecular data. *Bull. Am. Mus. Nat. Hist.* 296:1–149.

Weksler, M., and C. R. Bonvicino. 2005. Taxonomy of pigmy rice rats (genus *Oligoryzomys*, Rodentia: Sigmodontinae) of the Brazilian Cerrado, with the description of two new species. *Arq. Mus. Nac., Rio de Janeiro* 63:113–30.

Weksler, M., C. R. Bonvicino, I. B. Otazu, and J. S. Silva Júnior. 2001. Status of *Proechimys roberti* and *P. oris* (Rodentia: Echimyidae) from eastern Amazonia and central Brazil. *J. Mamm.* 82:109–22.

Weksler, M., L. Geise, and R. Cerqueira. 1999. A new species of *Oryzomys* (Rodentia, Sigmodontinae) from southeast Brazil, with comments on the classification of the *O. capito* species group. *Zool. J. Linn. Soc.* 125:445–62.

Weksler, M., and A. R. Percequillo. 2011. Key to the genera of the tribe Oryzomyini (Rodentia: Cricetidae: Sigmodontinae). *Mastozool. Neotrop.* 18:281–92.

Weksler, M., A. R. Percequillo, and R. S. Voss. 2006. Ten new genera of oryzomyine rodents (Cricetidae: Sigmodontinae). *Am. Mus. Novit.* 3537:1–29.

Wenzel, R. L., and V. J. Tipton. 1966. Some relationships between mammal hosts and their ectoparasites. In *Ectoparasites of Panama*, ed. R. L. Wenzel and V. J. Tipton, 677–723. Chicago: Field Museum of Natural History, xii + 861 pp.

Werner, F. A., K. J. Kedesma, and R. Hidalgo B. 2006. Mountain vizcacha (*Lagidium* cf. *peruanum*) in Ecuador—first record of Chinchillidae from the northern Andes. *Mastozool. Neotrop.* 13:217–74.

Wesmael, C. 1841. Zoologie. *Bull. Acad. R. Med. Belge.* 8:59–61.

Wetmore, A. 1926. Observations on the birds of Argentina, Paraguay, Uruguay, and Chile. *Bull. U.S. Natl. Mus.* 133:i–iv + 1–448.

Wetzel, R. M., and J. W. Lovett. 1974. A collection of mammals from the Chaco of Paraguay. *Univ. Connecticut Occas. Papers*, biol. sci. ser., 2:203–16.

Weyenbergh, H. 1877. *Dolichotis centralis* Wehenbergh, een nieve worm der Subungulata uit Zuit-Amerika. *Versl. Kon. Akad. Wetensch. Amsterdam*, ser. 2, xi:247–57.

Whitaker, J. O., Jr., and J. M. Dietz. 1987. Ectoparasites and other associates of some mammals from Minas Gerais, Brazil. *Entomol. News* 98:189–97.

White, T. G., and M. S. Alberico. 1992. Dinomys branickii. *Mamm. Species* 410: 1–5.

Whitehead, P. J. P. 1979. Georg Markgraf and Brazilian zoology. In *Johan Maurits van Nassau-Siegen 1604–1679, A Humanist Prince in Europe and Brazil*, ed. E. van den Boogaart, H. R. Hoetink, and P. J. P. Whitehead, 424–71. The Hague: The Johan Maurits van Nassau Stichting, 538 pp.

Whitney, B. M., D. C. Oren, and D. C. Pimentel Neto. 1996. *Uma lista anotada de aves e mamíferos registrados em 6 sítios de setor norte do Parque Nacional da Sierra do Divisor, Acre, Brasil: uma avaliação ecológica*. Informe não publicado. Brasilia e Rio Branco: Nature Conservancy and S.O.S. Amazônica.

Wied-Neuwied, M. P. zu. 1820. Ueber ein noch unbe-schriebenes Säugethier aus der Familie der Nager. *Isis von Oken* 6:43.

———. 1821. *Reise nach Brasilien in den Jahren; 1815 bis 1817*. Frankfurt a. M.: Heinrich Ludwig Brönner, 2:xviii + 345 pp. + 1 (unnumbered), 16 plates, 1 map.

———. 1823. *Abbildungen zur Naturgeschichte Brasiliens*, lief. 3, plate 2. Weimar: Verlage des Grossherzogi. [For contents see Oken, L. 1823. *Isis von Oken* 13:1259.]

———. 1826. *Beiträge zur Naturgeschichte von Brasilien. Werzeichniss der Amphibien, Säugthiere und Vögel, welche auf einer Reise zwischen dem 13ten und dem 23ten Grade südlicher Breite im östlichen Brasilien beobachtet wurden*. II. Abtheilung. Mammalia. Säugthiere. Weimar: Gr. H. S. priv. Landes-Industrie-Comptoirs, 2:1–622, 5 plates.

———. 1839. Über einige Nagher mit äusseren Backen-taschen aus dem westlichyen Nord-America. I. Über ein paar neue Gattungen de Nagethiere mit äusseren Backentaschen. *N. Acta Phys.-Med. Acad. Caesareae Leopoldino-Carolinae* 19:367–74.

Wiegmann, A. F. A. 1835. Eining Bermerkungen über das Chinchilla, vom Herausgeber. *Arch. Naturgesch.* 2:204–14.

———. 1838. Bericht über die Leistungen in Bearbeitung der übrigen Thierklassen, während des Jahres 1837. *Arch. Naturgesch.* 4(2):309–94.

Wilkins, K. T., and L. L. Cunningham. 1993. Relationship of cranial and dental features to direction of mastica-tion in tuco-tucos (Rodentia: *Ctenomys*). *J. Mamm.* 74:383–90.

Williams, D. F., H. H. Genoways, and J. K. Braun. 1993. Taxonomy. In *Biology of the Heteromyidae*, ed. H. H. Genoways and J. H. Brown, 38–196. Special Publica-tion 10. [Provo, UT]: American Society of Mammalo-gists, v–xii + 1–719 pp.

Williams, D. F., and M. A. Mares. 1978. A new genus and species of phyllotine rodent (Mammalia: Muridae) from northwestern Argentina. *Ann. Carnegie Mus.* 47:193–221.

Williams, S. L., H. H. Genoways, and J. A. Groen. 1983. Results of the Alcoa foundation-Surinam expeditions. 7. Records of mammals from central and southern Surinam. *Ann. Carnegie Mus.* 52:329–36.

Willner, G. R. 1982. Nutria (*Myocastor coypus*). In *Wild Mammals of North America*, ed. J. A. Chapman and G. A. Feldhamer, 1059–76. Baltimore: Johns Hopkins University Press, xiv + 1147 pp.

Willner, G. R., J. A. Chapman, and D. Pursley. 1979. Reproduction, physical responses, food habits, and abundance of nutria in Maryland marshes. *Wildl. Monogr.* 65:1–43.

Wilson, D. E., and F. R. Cole. 2000. *Common names of mammals of the world*. Washington, DC: Smithsonian Institution Press, xiv + 204 pp.

Wilson, D. E., and D. M. Reeder, eds. 2005. *Mammal species of the world*, 3rd ed. Baltimore: Johns Hopkins University Press, 1:xxxvii + 743, 2:xvii + 745–2142.

Wing, E. 1986. Domestication of Andean mammals. In *High altitude tropical biogeography*, ed. F. Vuilleumier and M. Monasterio, 246–64. Oxford: Oxford Univer-sity Press, 649 pp.

Winge, H. 1887 [1888]. Jordfunde og nulevende Gnavere (Rodentia) fra Lagoa Santa, Minas Geraes, Brasilien: med udsigt over gnavernes indbyrdes slaegtskab. *E Museo Lundii, Kjöbenhavn* 1(3):1–178 + 8 pls. [Cited as either 1887 or 1888; handwritten year of publica-tion given as "1887 (88)" in digital copy available on at http://www.biodiversitylibrary.org/.]

———. 1891. *Habrothrix hydrobates* n. sp. en Vandrotte fra Venezuela. *Videnskab. Meddel., Kjöbenhavn* 5(3): 20–27, 1 plate.

———. 1941. *The interrelationships of the mammalian genera*. Vol 2: *Rodentia, Carnivora, Primates*. Trans-lated from Danish by E. Deichmann and G. Allen. Copenhagen: C.A. Reitzels Forlag, 376 pp.

Wittouck, P., E. Pinna Senn, C. A. Soñez, M. C. Provensal, J. J. Polop, and J. A. Lisanti. 1995. Chromosomal and synaptonemal complex analysis of Robertsonian poly-morphisms in *Akodon dolores* and *Akodon molinae* (Rodentia: Cricetidae) and their hybrids. *Cytologia* 60:93–102.

Wlasiuk, G., J. C. Garza, and E. P. Lessa. 2003. Gene-tic and geographic differentiation in the Rio Negro tuco-tuco (*Ctenomys rionegrensis*): inferring the roles of migration and drift from multiple genetic markers. *Evolution* 57:913–26.

Wolffsohn, J. A. 1910. Revision de algunos generos de marsupiales i roedores chilenos del Museo Nacional de Santiago. *Bol. Mus. Nac. Hist. Nat. Chile* 2:83–102.

———. 1913. Reseña de los trabajos publicados desde 1895 por autores nacionales y extranjeros sobre la mamalogía chilena. *Acta. Soc. Cien. Chile* 23:57–79.

———. 1916. Description of a new rodent from central Chile. *Rev. Chilena Hist. Nat.* 20:6–7.

———. 1923. Medidas maximas y minimas de algunos mamíferos chilenos colectados entre los años 1896 y 1917. *Rev. Chilena Hist. Nat.* 27:159–67.

Wolffsohn, J. A., and C. E. Porter. 1908. Catalogo metodico de los mamíferos chinenos existentes en el Museo de Valparaiso en Diciembre de 1905. *Rev. Chilena Hist. Nat.* 12:66–85.

Wood, A. E. 1955. A revised classification of the rodents. *J. Mamm.* 36: 165–87.

———. 1965. Grades and clades among rodents. *Evolution* 19:115–30.

Wood, A. E., and B. Patterson. 1959. The rodents of the Deseadan Oligocene of Patagonia and the beginnings of South American rodent evolution. *Bull. Mus. Comp. Zool.* 120:281–428.

Woodman, N., N. A. Slade, R. M. Timm, and C. A. Schmidt. 1995. Mammalian community structure in lowland, tropical Peru, as determined by removal trapping. *Zool. J. Linn. Soc.* 113:1–20.

Woodman, N., R. M. Timm, R. Arana C., V. Pacheco, V., C. A. Schmidt, E. D. Hooper, and C. Pacheco A. 1991. Annotated checklist of the mammals of Cuzco Amazonico, Peru. *Occas. Papers, Univ. Kansas Mus. Nat. Hist.* 145:1–12.

Woodruff, J. A., E. A. Lacey, and G. Bentley. 2010. Contrasting fecal corticosterone metabolite levels in captive and free-living colonial tuco-tucos (*Ctenomys sociabilis*). *J. Exp. Zool.* 313A:498–507.

Woods, C. A. 1972. Comparative myology of jaw, hyoid, and pectoral appendicular regions of New and Old World hystricomorph rodents. *Bull. Am. Mus. Nat. Hist.* 147:115–98.

———. 1982. The history and classification of South American hystricognath rodents: reflections on the far away and long ago. In *Mammalian biology in South America*, ed. M. A. Mares and H. H. Genoways, 377–92. The Pymatuning Symposia in Ecology 6, Special Publications Series. Pittsburgh: Pymatuning Laboratory of Ecology, University of Pittsburgh, xii + 539.

———. 1984. Hystricognath rodents. In *Recent mammals of the world*, ed. S. Anderson and J. K. Jones Jr., 389–446. New York: John Wiley and Sons.

———. 1989. The biogeography of West Indian rodents. In *Biogeography of the West Indies: Past, present, and future*, ed. C. A. Woods, 741–98. Gainesville, FL: Sandhill Crane Press.

———. 1993. Suborder Hystricognathi. In *Mammal species of the world*, 2nd ed., ed. D. E. Wilson and D. M. Reeder, 771–806. Washington, DC: Smithsonian Institution Press, xviii + 1,206 pp.

Woods, C. A., and D. K. Boraker. 1975. Octodon degus. *Mamm. Species* 67:1–5.

Woods, C. A., L. Contreras, G. Willner-Chapman, and H. P. Whidden. 1992. Myocastor coypus. *Mamm. Species* 398:1–8.

Woods, C. A., and E. B. Howland. 1979. Adaptive radiation of capromyid rodents: anatomy of the masticatory apparatus. *J. Mamm.* 60:95–116.

Woods, C. A., and C. W. Kilpatrick. 2005. Infraorder Hystricognathi. In *Mammal species of the world*, 3rd ed., ed. D. E. Wilson and D. M. Reeder, 1538–600. Baltimore: Johns Hopkins Press, 2:xvii + 745–2142.

Worth, C. B., W. G. Downs, T. H. G. Aitken, and E. S. Tikasingh. 1968. Arbovirus studies in Bush Bush Forest, Trinidad, W. I., September 1959–December 1964. IV. Vertebrate populations. *Am. J. Trop. Med. Hyg.* 17:269–75.

Wright, S. 1969. *Evolution and the genetics of populations.* Vol. 2. *The theory of gene frequencies.* Chicago: University of Chicago Press, vii + 511 pp.

Wurster, D. H., J. R. Snapper, and K. Benirschke. 1971. Unusually large sex chromosomes: new methods of measuring and description of karyotypes of six rodents (Myomorpha and Hystricomorpha) and one lagomorph (Ochotonidae). *Cytogenetics* 10:153–76.

Wyss, A. R., J. J. Flynn, M. A. Norell, C. C. Swisher, III, R. Charrier, M. J. Novacek, and M. C. McKenna. 1993. South America's earliest rodent and recognition of a new interval of mammalian evolution. *Nature* 365: 434–37.

Xavier, S. C. C., V. C. Vaz, P. S. D'Andrea, L. Herrera, L. Emperaire, J. R. Alves, O. Fernandes, L. F. Ferreira, and A. M. Jansen. 2007. Mapping of the distribution of *Trypanosoma cruzi* infection among small wild mammals in a conservation unit and its surroundings (Northeast-Brazil). *Parasitol. Int.* 56:119–28.

Ximénez, A. 1965. *Wiedomys pyrrhorhinos* (Rodentia-Cricetidae), un nuevo mamífero para el Uruguay. *Physis* 25:135–36.

———. 1967. Consideraciones sobre un mamífero nuevo para el Uruguay: *Cavia aperea rosida* Thomas 1917 (Mammalia - Caviidae). *Comun. Zool. Mus. Hist. Nat. Montevideo* 9:1–4 + 1 lámina.

———. 1980. Notas sobre el genéro *Cavia* Pallas con la descripción de *Cavia magna* sp. n. (Mammalia-Caviidae). *Rev. Nordest. Biol.* 3:145–79.

Ximénez, A., and A. Langguth. 1970. *Akodon cursor montensis* en el Uruguay (Mammalia-Cricetinae). *Comun. Zool. Mus. Hist. Nat. Montevideo* 128:1–7.

Ximénez, A., A. Langguth, and R. Praderi. 1972. Lista sistemática de los mamiferos del Uruguay. *Anal. Mus. Hist. Nat. Montevideo*, sér. 2, 7:1–49.

Yahnke, C. J. 2006. Habitat use and natural history of small mammals in the central Paraguayan Chaco. *Mastozool. Neotrop.* 13:103–16.

Yahnke, C. J., E. Gamarra de Fox, and F. Colman. 1998. Mammalian species richness in Paraguay: the effectiveness of national parks in preserving biodiversity. *Biol. Conserv.* 84:263–68.

Yahnke, C. J., P. L. Meserve, T. G. Ksiazek, and J. N. Mills. 2001. Patterns of infection with Laguna Negra Virus in wild populations of *Calomys laucha* in the central Paraguayan Chaco. *Am. J. Trop. Med. Hyg.* 65:768–76.

Yáñez, J. L., W. Sielfeld, J. Valencia, and F. Jaksíc. 1978. Relaciones entre la sistematica y la morfometria del subgenero *Abrothrix* (Rodentia: Cricetidae) en Chile. *An. Inst. Patagonia* 9:185–97.

Yáñez, J. L., J. C. Torres-Mura, J. R. Rau, and L. C. Contreras. 1987. New records and current status of *Euneomys* (Cricetidae) in southern South America. In *Studies in Neotropical mammalogy, essays in honor of Philip Hershkovitz*, ed. B. D. Patterson and R. M. Timm, 283–87. *Fieldiana Zool.* 39:frontispiece, viii + 1–506.

Yáñez, J. L., J. Valencia, and F. Jaksíc. 1979. Morfometría sistemática del subgenero *Akodon* (Rodentia) en Chile. *Arc. Biol. Med. Exp.* 12:197–202.

Yáñez, J. L., and R. Zülch. 1981. Morfometría de tres poblaciones de *Spalacopus cyanus* (Rodentia). *Arch. Biol. Med. Exp.* 14:R304.

Yensen, E., and T. Tarifa. 2002. Mammals of Bolivian *Polylepis* woodlands: guild structure and diversity patterns in the world's highest woodlands. *Ecotropica* 8:145–62.

Yepes, J. 1930.Los roedores octodóntinos con distribución en la zone cordillerana de Chile y Argentina. *Rev. Chilena Hist. Nat.* 34:321–31.

———. 1933. Nuevos roedores para la fauna Argentina. *Rev. Chilena Hist. Nat.* 37:46–49.

———. 1935a. Epítome de la sistemática de los roedores argentinos. *Rev. Inst. Bacteriol.* 7:213–69.

———. 1935b. Consideraciones sobre el genero *Andinomys* (Cricetinae) y descripción de una forma nueva. *An. Mus. Argentino Cien. Nat. "Bernardino Rivadavia"* 38:333–48.

———. 1936. Mamíferos coleccionados en la parte central y occidental de la Provincia de la Rioja. *Physis* 12:31–42, 3 láminas.

———. 1942. Zoogeografía de los roedores octodóntidos de Argentina y descripción de un género nuevo. *Rev. Argent. Zoogeograf.* 2:69–81.

Yoneda, M. 1984. Composición por especies y ciclo reproductor de los roedores de la parte norte de los Andes bolivianos. *Ecol. Boliv.* 5:53–62.

Yonenaga, Y. 1972a. "Polimorfismo cromossômico em roedores brasileiros." Master's thesis, Universidade de São Paulo, Brazil.

———. 1972b. Chromosomal polymorphism in the rodent *Akodon arviculoides* ssp. (2N = 14) resulting from two pericentric inversions. *Cytogentics* 11:488–99.

———. 1975. Karyotypes and chromosome polymorphism in Brazilian rodents. *Caryologia* 28:269–86.

Yonenaga-Yassuda, Y. 1979. New karyopypes and somatic and germ-cell banding in *Akodon arviculoi-*

des (Rodentia, Cricetidae). *Cytogenet. Cell Genet.* 23:241–49.

Yonenaga-Yassuda, Y., O. Frota Pessoa, S. Kasahara, and E. J. Cardoso de Almeida. 1976. Cytogenetic studies on Brazilian rodents. *Ciênc. Cult.* 28:202–11.

Yonenaga-Yassuda, Y., S. Kasahara, E. J. C. Almeida, and A. L. Peracchi. 1975. Chromosomal banding patterns in *Akodon arviculoides* (2n = 14), *Akodon* sp. (2n = 24 and 25), and two male hybrids with 19 chromosomes. *Cytogenet. Cell Genet.* 15:388–99.

Yonenaga-Yassuda, Y., M. J. Souza, S. Kasahara, M. L'abbate, and H. T. Chu. 1985. Supernumerary system in *Proechimys iheringi iheringi*(Rodentia, Echimydae) from the State of São Paulo, Brazil. *Caryologia* 38:179–94.

Yoshida, E. L. A., C. A. C. Cuba, R. S. Pacheco, E. Cupolillo, C. C. Tavares, G. M. C. Machado, H. Monen, and G. Grimaldi. 1993. Description of *Leishmania* (*Leishmania*) *forattinii* sp. n., a new parasite infecting opossums and rodents in Brazil. *Mem. Inst. Oswaldo Cruz* 88:97–406.

Youlatos, D. 1999. Locomotor and postural behavior of *Sciurus igniventris* and *Microsciurus flaviventer* (Rodentia, Sciuridae) in eastern Ecuador. *Mammalia* 63:405–16.

———. 2011. Substrate use and locomotor modes of the Neotropical pygmy squirrel *Sciurillus pusillus* (E. Geoffroy, 1803) in French Guyana. *Zool. Stud.* 50:745–50.

Young, K. R., and B. León. 1999. Peru's humid eastern montane forests: an overview of their physical settings, biological diversity, human use and settlement, and conservation needs. *DIVA Tech. Rep.* 5:1–97.

Yunes, R. M. F., R. A. A. P. Cutrera, and A. Castro-Vázquez. 1991. Nesting and digging behavior in 3 species of *Calomys* (Rodentia; Cricetidae). *Physiol. Behav.* 49:489–92.

Zambelli, A., C. I. Catanesi, and L. Vidal-Rioja. 2003. Autosomal rearrangements in *Graomys griseoflavus* (Rodentia): a model of non-random Robertsonian divergence. *Hereditas* 139:167–73.

Zambelli, A., F. Dyzenchauz, A. Ramos, N. de Rosa, R. Wainberg, and O. A. Reig. 1992. Cytogenetics and karyosystematics of phyllotine rodents (Cricetidae, Sigmodontinae). III. New data on the distribution and variability of karyomorphs of the genus *Eligmodontia*. *Z. Säugetierk.* 57:155–62.

Zambelli, A., and L. Vidal-Rioja. 1995. Molecular analysis of chromosomal polymorphism in the South American cricetid, *Graomys griseoflavus*. *Chromosome Res.* 3:361–67.

Zambelli, A., L. Vidal-Rioja, and R. Wainberg. 1994. Cytogenetic analysis of autosomal polymorphism in

Graomys griseoflavus (Rodentia, Cricetidae*)*. *Z. Säugetierkd.* 59:14–20.

Zanchin, N. I. 1988. "Estudos cromossômicos em orizominos e equimídeos da Mata Atlântica." Master's thesis, Universidade Federal do Rio Grande do Sul, Porto Alegre, Brazil.

Zanchin, N. I. T., A. Langguth, and M. S. Mattevi. 1992. Karyotypes of Brazilian species of *Rhipidomys* (Rodentia, Cricetidae). *J. Mamm.* 73:120–22.

Zanchin, N. I. T., I. J. Sbalqueiro, A. Langguth, R. C. Bossle, E. C. Castro, L. F. B. Oliveira, and M. S. Mattevi. 1992. Karyotype and species diversity of the genus *Delomys* (Rodentia, Cricetidae) in Brazil. *Acta Theriol.* 37:163–69.

Zeballos, H., and R. C. Carrera. 2010. Mamíferos de la Reserva Nacional de Salinas y Aguada Blanca, Arequipa y Moquegua, Suroeste del Perú. In *Diversidad Biológica de La Reserva Nacional de Salinas y Aguada Blanca: Arequipa–Moquegua*, ed. H. Zeballos, J. A. Ochoa, and E. López, 249–59. Lima, Peru: Profonanpe-DESCO, Ministerio del Ambiente, 313 pp.

Zeballos, H., V. Pacheco, and L. Baraybar. 2001. Diversidad y conservación de los mamíferos de Arequipa, Peru. *Rev. Peru. Biol.* 8:94–104.

Zeballos, H., L. Villegas, R. Gutiérrez, K. Caballero, and P. Jimenéz. 2000. Vertebrados de las lomas de Atiquipa y Mejía, sur del Perú. *Rev. Ecol. Lat. Am.* 7:11–18.

Zeballos, H., and E. Vivar. 2008. *Melanomys zunigae. IUCN red list of threatened species*, Version 2009.1, accessed July 26, 2009, http://www.iucnredlist.org.

Zenuto, R. R., and C. Busch. 1995. Influence of the subterranean rodent *Ctenomys australis* (tuco-tuco) in a sand-dune grassland. *Z. Säugetierk.* 60:277–85.

Zenuto, R. R., M. S. Fanjul, and C. Busch. 2004. Use of chemical communication by the subterranean rodent *Ctenomys talarum* (tuco-tuco). *J. Chem. Ecol.* 30:2111–26.

Zenuto, R. R., E. A. Lacey, and C. Busch. 1999. DNA fingerprinting reveals polygyny in the subterranean rodent *Ctenomys talarum. Mol. Ecol.* 8:1529–32.

Zenuto, R. R., A. D. Vitullo, and C. Busch. 2003. Sperm characteristics in two populations of the subterranean rodent *Ctenomys talarum* (Rodentia: Octodontidae). *J. Mamm.* 84:877–85.

Zijlstra, J. S. 2012. A new oryzomyine (Rodentia: Sigmodontinae) from the Quaternary of Curaçao (West Indies). *Zootaxa* 3534:61–68.

Zijlstra, J. S., P. A. Madern, and L. W. van den Hoek Ostende. 2010. New genus and two new species of Pleistocene oryzomyines (Cricetidae: Sigmodontinae) from Bonaire, Netherlands Antilles. *J. Mamm.* 91:860–73.

Zimmerman, D. M., M. Douglas, D. R. Reavill, and E. C. Greiner. 2009. *Echinococcus oligarthrus* cystic hydatidosis in Brazilian agouti (*Dasyprocta leporina*). *J. Zoo Wildl. Med.* 40:551–58.

Zimmermann, E. A. W. 1777. *Specimen zoologiae geographicae, quadrupedum domicilia et migrationes sistens.* Leiden: Lugduni Batavorum, 1–24 (unnumbered) + 1–685 pp.

———. 1780. *Geographische Geschichte der Menschen, und der algemein verbreiteten vierfüssigen Thiere.* Zweiter Band. Leipzig: Wenganschen Buchhandlung, 2:[6] + 1–432.

Zortéa, M., and B. F. A. de Brito. 2010. Diurnal roosts and minimum home range defined by sleeping sites of a thin-spined porcupine *Chaetomys subspinosus* (Rodentia: Erethizontidae). *Zoologia* 27:209–12.

Zuercher, G. L., P. S. Gipson, and O. Camillo. 2005. Diet and habitat associations of bush dogs *Speothos venaticus* in the Interior Atlantic Forest of eastern Paraguay. *Oryx* 39:86–89.

Zülch, R., L. E. Walker, and J. Pincheira. 1982. Comparación de los cariotipos gandeados G y C de las formas costera y andina de *Spalacopus cyanus* y *Octodon degus* (Rodentia, Octodontidae). *Arch. Biol. Med. Exp.* 15:R164.

Zuleta, G., and D. Bilenca. 1992. Seasonal shifts within juvenile recruit sex ratio of pampas mice *Akodon azarae. J. Zool.* 227:397–404.

Zuleta, G., F. O. Kravetz, M. Busch, and R. E. Percich. 1988. Dinámica poblacional del ratón del pastizal pampeano (*Akodon azarae*) en ecosistemas agrarios de Argentina. *Rev. Chilena Hist. Nat.* 61:231–44.

Zúñiga, H. 2000. "Estudio sistemático del género Cavia Pallas, 1766 (Rodentia: Caviidae) en Colombia." Master's thesis, Universidad Nacional de Colombia, Santa Fé de Bogotá.

Zuñiga, H., M. Pinto-Nolla, J. I. Hernández-Camacho, and O. M. Torrez-Martínez. 2002. Revisión taxonómica de las species del género *Cavia* (Rodentia: Caviidae) en Colombia. *Acta Zool. Mex.* 87:111–23.

Zuñiga, H., J. R. Rodríguez, and A. Cadena. 1988. Densidad de población de pequeños mamíferos en dos comunidades del bosque Andino. *Acta Biol. Colomb.* 1:85–93.

Gazetteer of Localities

THIS VOLUME follows the same protocol used by A. L Gardner for volume 1 of this series on South American mammals, namely, localities are arranged alphabetically by country and by major political unit (state, district, province, region, or department) within each country, except for French Guiana and Trinidad and Tobago. Place names are arranged alphabetically according to the name of the reported site of capture or to the nearest place name. Different sites of capture referenced to the same place name are arranged from north to south and from west to east. Variants of place names, including errors and alternative spellings used by the authors cited, are placed in their appropriate alphabetical sequence. Note that new high-level administrative units (departments, provinces, regions, or states) have been added to some countries so that the same locality may be placed in a different unit in volume 1 than here. For example, Peru recently changed its highest level administrative divisions from Departamento to Región plus the Provincia de Lima.

The first volume in this series provided coordinates in degrees and minutes, but we have adopted decimal degree designations, in part because the latter are easier to map using various Global Information Systems software or Google Earth. The coordinates themselves come from a variety of sources, including handheld GPS units now in near-universal use by field investigators. The museum collections participating in the Mammal Networked Information Systems project (MaNIS, http://manisnet.org /; which has now been transitioned to VertNet, http:// vertnet.org/) have retrospectively georeferenced locali-

ties based on guidelines developed for that project (see http://manisnet.org/GeorefGuide.html). These collection databases thus served as the source for many of the coordinates listed herein. Other coordinate sets are those given in the gazetteer of volume 1 but converted to decimal degrees. These are largely based on the U.S. Board on Geographic Names or on information contained in the Ornithological Gazetteers of the Neotropics series available through the Bird Department of the Museum of Comparative Zoology at Harvard University (expanded and corrected for Brazil by Vanzolini 1992). This series of gazetteers is particularly useful for older place names that may no longer be in use. Finally, coordinates were also obtained from a number of different website databases, such as the U.S. Board on Geographic Names (http://geonames.usgs.gov), the Alexander Digital Library (http://middleware.alexandria.ucsb.edu/), Glosk (http:// www.glosk.com/), Google Earth (http://earth.google.com /), and especially Geonames (http://www.geonames.org), or from the gazetteers of many publications or papers citing particular localities. The datum for each coordinate pair is not given, as it is unknown for the majority of coordinates cited. This important piece of information is available for localities georeferenced by the MaNIS protocol. English or metric measures of distances and elevations are given as provided in the original citation or locality designation. The political division wherein a specific locality is placed has been updated to reflect current usage if such has changed since the original citation (specimen label or publication).

Argentina

Buenos Aires (Provincia de)

Abra del Hinojo, Sierras de Curamalal, 750 m, 37.73333°S, 62.26667°W

Abra de la Ventana, Parque Provincial Ernesto Tornquist, 1,243 m, 38.15000°S, 61.98333°W

Álamos, 36.90000°S, 62.35000°W

Algarrobo, 38.88333°S, 63.13333°W

Arroyo Brusquitas, Miramar, 100 m, 38.22439°S, 57.77712°W

Arroyo de las Brusquitas, 38.20140°S, 57.78840°W

Arroyo Chasicó, 38.61528°S, 63.03231°W

Arroyo de los Loros, 38.05000°S, 61.91667°W

Arroyo Las Tijeras, 36.36667°S, 56.83333°W

Arroyo Sauce Chico, 38.78333°S, 62.30000°W

Bahía Blanca, 20 m, 38.72735°S, 62.26914°W

Bahía San Blas, 40.55000°S, 62.21667°W

Balcarce, 37.83333°S, 58.25000°W

Balneario San Antonio, 39.65167°S, 62.11917°W

Bañado de Flores, Buenos Aires, 34.63333°S, 58.46667°W

Berazategui, 36.40000°S, 56.96667°W

Bonifacio [= Estación Bonifacio], 36.80900°S, 62.24500°W

Buenos Aires, 34.60000°S, 58.45000°W

Cabaña San José, 40.96417°S, 62.79944°W

Canal 6 and Paraná de las Palmas, 34.15000°S, 58.95000°W

Carmen de Patagones, 40.78333°S, 62.96667°W

Carmen de Patagones, 8 km N, 40.80000°S, 62.98333°W

Cerro de la Gloria, 35.97486°S, 57.45032°W

Ciudad de Lincoln [= Lincoln], 34.86649°S, 61.53020°W

Ciudad de Pehuajó [= Pehuajó], 35.81077°S, 61.89680°W

Claromecó, 38.85000°S, 60.08333°W

Copetonas, 38.66667°S, 60.41667°W

Cristiano Muerto, 27 m, 38.65778°S, 59.60455°W

Daireaux, 36.60000°S, 61.75000°W

Delta del Paraná, 33.71667°S, 59.25000°W

Diego Gaynor, 34.28333°S, 59.23333°W

D'Orbigny, 37.68333°S, 61.71667°W

Ensenada, La Plata, sea level, 34.85000°S, 57.91667°W

Estancia El Abra, 40.50000°S, 63.37250°W

Estancia La Providencia, 35.09730°S, 623.50490°W

Estación Deferrari, 38.30120°S, 59.38620°W

Estación Experimental INTA Hilario Ascasubi, 39.36667°S, 62.64333°W

Estación Experimental INTA Delta del Paraná, 34.17449°S, 58.86728°W

Estación San José, 38.17740°S, 59.00980°W

Ezeiza, Punta Lara, 34.83306°S, 58.51722°W

Fortín Olavarría, 35.70176°S, 63.02429°W

Fulton, 37.41820°S, 58.80980°W

General Lavalle, 36.41667°S, 56.95000°W

Granja 17 de Abril, 34.74372°S, 58.97972°W

INTA Balcarce, 37.73982°S, 59.25999°W

Isla Ella, delta of Río Paraná, 34.36667°S, 58.63333°W

Isla Martín García, 34.18333°S, 58.23333°W

José C. Paz, 34.51917°S, 58.74944°W

Laguna Alsina [= Estación Bonifacio], 36.80958°S, 62.24671°W

Laguna Chasicó, 25 m, 38.64200°S, 63.10350°W

Laguna Las Encadenadas, 38.04744°S, 62.51640°W

Laguna Mar Chiquita, 37.69580°S, 57.40530°W

La Plata, 34.94194°S, 58.22750°W

Las Talas, 34.86668°S, 57.88333°W

Lobería, 15 km SE, 38.15000°S, 58.78333°W

Mar Chiquita [= Balneario Mar Chiquita], 37.74744°S, 57.42482°W

Mar del Plata, 38.00228°S, 57.55754°W

Mar del Tuyú, General Lavalle, sea level, 36.58032°S, 56.69271°W

Marucha, 35.63030°S, 62.22950°W

Médanos, 38.85000°S, 63.01667°W

Monte Hermoso [= Balneario Monte Hermoso], 38.98740°S, 61.31600°W

Necochea, 38.54722°S, 58.73667°W

Olavarría, 36.89120°S, 60.31530°W

Otamendi, vicinity, 38.11667°S, 57.85000°W

Papín [= Garré], 36.56037°S, 62.60217°W

Partido Balcarce, 37.78140°S, 58.24400°W

Pedro Luro, 20 km S on Hwy 3, 39.69217°S, 62.67351°W

Pehuén-có [= Balneario Pehuén-có], 38.99726°S, 61.55225°W

Pirovano, 36.48333°S, 61.56667°W

Puesto El Chara, 39.45000°S, 62.05000°W

Puesto El Plátano, 35.96528°S, 57.45306°W

Punta Alta, 38.87583°S, 62.07333°W

Punta Blanca, 30 road km SW of La Plata, 35.08333°S, 57.51667°W

Punta de Indio, 35.26667°S, 57.25000°W

Punta Lara, 34.81920°S, 57.97372°W

Punta Médanos [= Punta Sur del Cabo San Antonio], 36.88527°S, 56.66894°W

Punta Negra, Necochea, 38.62190°S, 58.82810°W

Quiroga, 35.29292°S, 61.40672°W

Ramallo, 33.48680°S, 60.00400°W

Reserva Ecológica Costanera Sur, 34.60000°S, 58.45000°W

Reserva El Destino, 35.13333°S, 57.38333°W

Río Quequén Salado, 38.93972°S, 60.51000°W

Saladillo, 35.63675°S, 59.77853°W

San Blas, 40.55000°S, 62.25000°W

San Clemente del Tuyú, 36.35694°S, 56.72351°W

San Fernando, 34.181470°S, 58.69637°W

Santa Clara del Mar, 37.83333°S, 57.50000°W

San Pedro, 33.67103°S, 59.65585°W
Sierra de la Ventana, 38.13807°S, 61.79518°W
Tandil, 178 m, 37.31667°S, 59.15000°W
Valeria del Mar, sea level, 37.13333°S, 56.88333°W
Zelaya, 34.35000°S, 34.86667°W

Catamarca (Provincia de)

Agua de Dionisio, 2,221 m, 27.27697°S, 66.72987°W
Alero Los Viscos, Barranca Larga, 2,400 m, 27.01472°S, 66.74167°W
Andalgalá, 1,060 m, 27.58195°S, 66.31646°W
Andalgalá, 13 km NNW, 27.51919°S, 66.41033°W
Andalgalá, 28 km S and 9.3 km W, 680 m, 27.83417°S, 66.26389°W
Antofagasta de la Sierra, Paycuqui, 3,345–3,664 m, 26.05940°S, 67.40636°W
Barranca Larga, 26.91667°S, 66.83333°W
Barranca Larga, Cueva Los Viscos, 2,400 m, 27.01885°S, 66.74517°W
Belén, 26.53300°S, 66.91600°W
Belén, 4.6 km SW and 0.7 km N, 27.679333°S, 67.58333°W
Bolsón de Pipanaco, 27.82115°S, 66.24303°W
Campo Arenal, Los Nacimientos, 2,139 m, 27.14740°S, 66.67850°W
Catamarca, 28.46667°S, 65.78333°W
Catamarca, access to the town by Rte 38, 28.50389°S, 65.78169°W
Chumbicha, 28.85430°S, 66.23500°W
Chumbicha, 0.5 km E of Rte 38 along Rte 60, 28.86120°S, 66.23810°W
Chumbicha, 45 km W, 3,000 m, 28.66541°S, 66.24708°W
Corral Quemado, 2,070 m, 27.10000°S, 66.93333°W
El Bolsón, 2,309 m, 27.02469°S, 66.76106°W
El Bolsón, 5.2 km S, 2,404 m, 27.96760°S, 65.85401°W
El Desmonte, 21 km SW, 2,228 m, 26.98857°S, 66.26506°W
El Espinillo, Campo del Pucará, Las Estancias, 27.59043°S, 66.14291°W
El Peñón, 1.5 km S, 3,485 m, 26.50000°S, 67.25000°W
Establemimiento Río Blanco, 28 km S and 13.3 km W of Andalgalá, 27.85028°S, 66.30472°W
Estancia El Tapón, 463 m, 33.10333°S, 67.20500°W
Huaico Hondo, ca. 2 km SE on highway 42, E of Portezuelo, 1,992 m, 28.41969°S, 65.54458°W
La Carrera, 28.41667°S, 65.71667°W
Laguna Blanca, 3,243 m, 26.54264°S, 66.993266°W
Laguna del Rosario, 532 m, 32.15489°S, 68.24048°W
La Puntilla, near Tinogasta, 1,000 m, 28.13333°S, 67.50000°W
Las Chacritas, ca. 28 km NNW of Singuil, 1,888 m, 27.70672°S, 65.91128°W

Las Higuerillas, 5 km S on Hwy 9, 1,173 m, 27.95792°S, 65.68486°W
Las Juntas, 28.10944°S, 65.91667°W
Loma Atravesada, ca. 3 km NW of Leandro Vega ranch, NW of Chumbicha, 28.79392°S, 66.30729°W
Minas Capillitas, 27.33333°S, 66.41667°W
Mogote Las Trampas, ca. 15 km NW of Chumbicha, 2,300 m, 28.74185°S, 66.32057°W
Otro Cerro, 3,000 m, 28.75000°S, 66.28333°W
Pastos Largos, 3,250 m, 27.68925°S, 68.16416°W
Pipanaco's salt basin [= Bolsón de Pipanaco], 27.82115°S, 66.24303°W
Portezuelo de Pasto Ventura [= Laguna Blanca], 3,960 m, 26.66919°S, 67.22404°W
Quimilo, 29.91667°S, 65.36667°W
Recreo, 29.36667°S, 65.06667°W
Río Amanao, west bank, about 15 km W (on Rte 62) of Andalgalá, 27.60000°S, 66.46667°W
Santa María, 26.68333°S, 66.03333°W
Sierra del Aconquija, 4,000 m, 27.21667°S, 66.13333°W
Trampasacha, 614 m, 28.81667°S, 66.30000°W
Tinogasta, 28.06667°S, 67.56667°W
Tinogasta, 22 km NE, 27.92629°S, 67.40841°W

Chaco (Provincia del)

Campo Aráos, 26.60000°S, 59.25000°W
Campo Bermejo, 61 m, 27.45140°S, 58.98670°W
Capitán Solari, 26.80000°S, 59.55000°W
Ciervo Petiso, 26.58028°S, 59.63083°W
Colonia Benítez, 27.33083°S, 58.94611°W
Colonia Elisa, 26.80000°S, 59.55000°W
Colonia Río Tragadero, 2 km N of Resistencia, 27.39000°S, 65.91228°W
El Colorado, 26.46667°S, 59.36667°W
Estación Experimental Colonia Beneitez, INTA, 27.33000°S, 58.93000°W
General Pinedo, 27.31667°S, 61.28333°W
Laguna Limpia, 26.49556°S, 59.68083°W
Las Breñas, 27.98944°S, 61.08139°W
Las Palmas, 27.05849°S, 58.69363°W
Pampa Bolsa, 26.42302°S, 61.07883°W
Pampa Chica [= Colonia Pampa Chica], 26.10000°S, 59.88333°W
Parque Nacional Chaco, 26.90796°S, 59.52852°W
Parque Nacional Chaco, 26.83333°S, 59.66667°W
Presidente Roca, 26.14083°S, 59.59528°W
Presidente Roque Sáenz Peña, 26.78333°S, 60.43333°W
Puente General Belgrano, 1 km E, Río Paraná, 27.45000°S, 58.85000°W
Río de Oro, ca. 26.79000°S, 58.95000°W

Roque Saenz Peña, 26.78804°S, 60.43991°W
Selvas del Río de Oro, 26.78333°S, 58.95000°W
Tres Isletas, 26.34056°S, 60.43194°W

Chubut (Provincia del)

Alto Río Senguerr, 41 km W, 44.90572°S, 71.22367°W
Astra, 45.73333°S, 67.48333°W
Buenos Aires Chico, 42.06667°S, 71.20000°W
Cabaña Arroyo Pescado, 43.02528°S, 70.79278°W
Campo de Pichiñán, 43.56389°S, 69.06722°W
Capilla El Triana, 45.58721°S, 71.72853°W
Cerro El Sombrero, 44.13917°S, 68.26333°W
Cholila, 42.71000°S, 71.43000°W
Colonia Nahuel Pan, 42.95000°S, 71.16667°W
Comodoro Rivadavia, 45.86667°S, 67.50000°W
Cushamen, 42.20000°S, 70.83333°W
El Hoyo [= El Hoyo de Epuyén], 42.06667°S, 71.50000°W
El Maitén, 42.05000°S, 71.16000°W
El Maitén, 4 km N, 42.38000°S, 71.11000°W
Escuela N859, Fofo Cahuel, 116 m, 42.40833°S, 70.52944°W
Esquel, 42.90000°S, 71.31667°W
Establecimiento La Ollada, 44.74675°S, 69.61881°W
Estancia El Gauchito, 45.18333°S, 67.18333°W
Estancia El Valle, 8.5 km NW, Valle de Lanzaniyeu, 43.88394°S, 70.67072°W
Estancia Escondida, 45.32333°S, 69.83500°W
Estancia La Irma, 42.26667°S, 63.68333°W
Estancia La Porfia, 43.23100°S, 68.64697°W
Estancia Los Manantiales, 475 m, 45.50556°S, 67.49139°W
Estancia Los Nogales, 42.65167°S, 67.05778°W
Estancia San Pedro, 42.06667°S, 67.56667°W
Estancia Santa María, Puesto El Chango, 45.46431°S, 69.43167°W
Estancia Talagapa, 42.13778°S, 68.25472°W
Estancia Talagapa, 1,411 m, 42.15557°S, 68.26947°W
Estancia Valle Huemules, 45.94800°S, 71.50700°W
Estancia Valle Huemules, Lago Blanco, 593 m, 45.95374°S, 71.52026°W
Isla Escondida, 43.66667°S, 65.33333°W
Istmo Ameghino, 42.43000°S, 64.26467°W
José de San Martín, 30 km W, 2,000 ft, 44.53000°S, 70.40000°W
Lago Blanco, 4 km W, 45.91000°S, 71.30000°W
Lago Colhue-Huapi, 45.50000°S, 68.80000°W
Lago Fontana, 44.98278°S, 71.39214°W
Lago Futalaufquen, 42.88367°S, 71.59817°W
Lago La Plata, west side, 44.82361°S, 72.00722°W
Lago Musters, 265 m, 45.42698°S, 69.17917°W
Laguna La Blanca, 42.80000°S, 65.13333°W

Leleque, 5 km W, 42.38000°S, 71.11000°W
Leleque, 20 km S, 42.58333°S, 71.03333°W
Los Altares, 43.86667°S, 68.40000°W
Los Altares, 36 km W, 43.85000°S, 68.81667°W
Pampa de Agnia, 27 km NW, 43.49567°S, 69.83083°W
Pampa de Agnia, 30 km NW, 43.47967°S, 69.81817°W
Pampa de los Guanacos, 45.36306°S, 68.64122°W
Paraje Fofocahuel, Campo Netchovit, 42.32833°S, 70.55917°W
Parque Nacional Los Alerces, 42.88000°S, 7159000°W
Paso de Indios, 18 km W of junction of Hwys 29 and 27 along Hwy 27, 44.58333°S, 68.03389°W
Paso del Sapo, 42.68333°S, 69.71667°W
Península Valdez [= Valdés], 42.51050°S, 63.99930°W
Pico Salamanca, Comodoro Rivadavia, 100 m, 45.78333°S, 67.45000°W
Puente Picún Leufú, 39.21042°S, 70.05914°W
Puerto Lobos, 42.03333°S, 65.07639°W
Puerto Madryn, 8 km ESE (by road), 42.79423°S, 64.959i70°W
Puerto Piojo, 162 m, 44.88333°S, 65.10228°W
Punta Delgada, 42.76667°S, 63.63333°W
Rawson, 43.30016°S, 65.10228°W
Río Arrayanes, Parque Nacional Los Alerces, 42.75000°S, 71.75000°W
Río Tecka, 6.5 km W of the bridge, on RP N°17, 53.59778°S, 71.10950°W
Salina Grande, vicinity, 42.05389°S, 70.10583°W
Sarmiento, 17 km W, 43.60389°S, 69.27306°W
Sierra Apas, 42.00000°S, 67.63333°W
Sierra de Tepuel, Cañadón de la Madera, 43.85160°S, 70.72790°W
Subida del Naciente, 41.67333°S, 67.15417°W
Tecka, 3 km N along Hwy 4, 650 m, 43.46667°S, 70.84722°W
Tres Banderas, 4 km S on RP 11, 42.80861°S, 68.01556°W
Valle Hermoso, Comodoro Rivadavia, 45.10167°S, 68.50972°W
Valle del Lago Blanco, 45.91667°S, 71.25000°W

Córdoba (Provincia de)

Alejo Ledesma, 33.60643°S, 62.62304°W
Candelaria [= Candelaria Sud], 30.83882°S, 63.79982°W
Candonga, 950 m, 31.10000°S, 64.36667°W
Cerro de Oro, 32.60484°S, 64.89964°W
Chucul, 33.00000°S, 64.16667°W
Colonia Tirolesa, 31.30459°S, 63.95440°W
Cosquín, 31.23330°S, 64.15000°W
Cruz del Eje, 30.72644°S, 64.80387°W
Deán Funes, 30.40000°S, 64.35000°W
Espinillo, Río Cuarto, 200 m, 33.01667°S, 64.36667°W

Estancia San Luis, 2,000 m, 31.40000°S, 64.80000°W
Guanaco Muerto, 30.47889°S, 64.06667°W
La Carlota, 33.41993°S, 63.29769°W
Laguna Larga, 31.76667°S, 63.80000°W
La Nacional, 34.75000°S, 64.90000°W
Las Toscas, 31.86667°S, 65.46667°W
Los Mistoles, 30.62526°S, 63.88623°W
Manantiales, 33.53536°S, 63.31722°W
Mar Chiquita, 30.80000°S, 62.88333°W
Marull, 30.90685°S, 62.82927°W
Noetinger, 112 m, 32.36667°S, 62.31667°W
Pampa de Achala, 2,164 m, 31.61366°S, 64.85321°W
Pampa de Achala, 2,000 m, 31.57426°S, 64.80337°W
Puente Olmos, 32.46667°S, 63.31667°W
Río Ceballos, 31.16500°S, 64.33300°W
Río Cuarto, 33.11667°S, 64.35000°W
Río Cuarto and Mercedes, on road between, near border
  with San Luis, approximately 33.58333°S, 65.00000°W
Salinas Grandes, 30.05000°S, 65.08333°W
Santiago Temple, 10 km N, 31.38580°S, 63.40210°W
Villa Dolores, 529 m, 31.93333°S, 65.20000°W
Villa de María del Río Seco, 29.90000°S, 63.71667°W
Washington, 33.87485°S, 64.68607°W
Yacanto, 32.05290°S, 65.05170°W

Corrientes (Provincia de)

Ahoma Sur, Empedrado, 27.95000°S, 58.80000°W
Arroyo Pehuajó, 27.70000°S, 57.96000°W
Chavarría, 28.95488°S, 58.57277°W
Colonia Brougnes, 28.16667°S, 59.81667°W
Colonia Romero, 71 m, 27.65333°S, 57.53085°W
Colonia 3 de Abril, 28.26667°S, 58.91667°W
Contreras Cué, 28.01667°S, 56.60000°W
Costa Mansión, 28.03333°S, 58.81667°W
Curuzú Laurel, 27.91667°S, 57.50000°W
El Sombrero, 27.70000°S, 58.76000°W
Empedrado, 27.95124°S, 58.80542°W
Esquina, 39 m, 30.01444°S, 59.52719°W
Estancia Rincón de Animas, 20 km NE of Sauce, 29.94778°S,
  58.62306°W
Estancia San Juan Poriahú, 27.70972°S, 57.18889°W
Estancia Tacuarita, 27.98244°S, 56.56242°W
Estancia Yacyretá, 27.48270°S, 56.72500°W
Goya, 29.14558°S, 59.26438°W
Ibahay, 27.51969°S, 57.40745°W
Loma Alta, 28.68285°S, 58.97302°W
Loreto, 27.75000°S, 57.26667°W
Los Angeles [= Estancia Los Angeles], 65 m, 28.75172°S,
  58.95764°W
Manantiales, 27.94259°S, 58.10451°W
Mburucuyá, 28.04539°S, 58.22449°W

Pago Alegre, 28.15747°S, 58.40171°W
Palmar Grande, 63 m, 27.94195°S, 57.90057°W
Paraje Angostura, 27.60000°S 57.96000°W
Paraje Caiman, 28.03333°S, 57.68333°W
Paraje Sarandicito, 30 m, 30.21871°S, 59.56134°W
Paraje Uguay, Esteros del Iberá, 28.65000°S, 57.46667°W
Pilar, 27.46667°S, 58.78333°W
Saladas, 2825384°S, 58.62591°W
San Miguel, 27.98333°S, 57.58333°W
San Miguel, San Juan Poriahú, 27.71667°S, 57.19389°W
San Roque, 12 km S, 28.68333°S, 58.70000°W
Santa Rosa, 8 km NW, 28.18333°S, 58.11667°
Santa Tecla, Ruta Nacional 12, km 1287, 27.63333°S,
  56.36667°W
Santo Tomé, 73 m, 28.54939°S, 56.04077°W
Tres Bocas, 57 m, 30.02251°S, 58.29219°W

Entre Ríos (Provincia de)

Arroyo Brazo Largo and Arroyo Brazo Chico, 33.78333°S,
  58.78333°W
Arroyo Perucho Verne, east of Villa Elisa, 32.17492°S,
  58.35790°W
Colonia Yerua, 31.41710°S, 58.42208°W
Concordia, 31.39296°S, 58.02089°W
Diamante, 32.06667°S, 57.18889°W
Estación Paranacito [= Estación General San Martín],
  33.70881°S, 59.03319°W
Gualeguay, 33.15319°S, 59.33750°W
Gualeguaychú, 100 m, 33.01667°S, 58.51667°W
Isla Chapetón, Río Paraná, 31.55000°S, 60.30000°W
Isla Ibicuy, 33.73333°S, 59.21667°W
Médanos, 5 m, 33.43610°S, 59.06995°W
Paraná, 31.56667°S, 60.53333°W
Paranacito, 33.70934°S, 59.03310°W
Parque Nacional El Palmar, 31.87974°S, 58.28693°W
Parque Nacional El Palmar, 31.84306°S, 58.32250°W
Paso Vera, 32.45000°S, 58.20000°W
Puerto Ruiz, 33.21667°S, 59.36667°W
Río Paraná, islands of lower delta, approximately
  34.18333°S, 58.41667°W
Strobel, 32.01917°S, 60.58636°W
Ubajay, 31.79358°S, 58.31350°W
Victoria, 32.61841°S, 60.15478°W
Villa Ramírez, 32.17573°S, 60.19902°W
Yaqueri, 20 m, 31.38333°S, 58.11667°W

Formosa (Provincia de)

Bañado La Estrella, 24.39918°S, 60.34596°W
Clorinda, 13 km S, 25.40083°S, 57.77189°W
Colonia Villafañe, 17 km W, 26.18333°S, 59.25000°W

Cooperative de Villa Dos Trece [= Kilométyro 213], 26.18333°S, 59.36667°W

Estación Experimental El Colorado, INTA, 26.30000°S, 59.37000°W

Estancia Guayacolec, 25.98333°S, 58.25000°W

Ingeniero Guillermo Juarez, 35 km S and 5 km E, Puesto Divisadero, 23.90060°S, 61.86310°W

La Florenca, 24.20000°S, 24.01667°W

Laguna Blanca, vicinity, 25.13330°S, 58.25000°W

Laguna Naick Nect, INTA IPAF-NEA, 25.20000°S, 58.11667°W

Misión Tacaaglé, 24.96667°S, 58.81667°W

Naineck, 25.21667°S, 58.116667°W

Parque Nacional Río Pilcomayo, 25.13333°S, 58.11667°W

Pozo del Tigre, 24.89667°S, 60.32333°W

Reserva El Bagual, 26.16667°S, 58.93333°W

Riacho Pilagra, 26.18333°S, 58.18333°W

Ruta 81 y 95, 7 km N cruce entre, 25.22210°S, 59.71010°W

San Martín Número Uno, 24.53250°S, 59.90056°W

Jujuy (Provincia de)

Abra Pampa, 22.76800°S, 65.70280°W

Abrapampa [= Abra Pampa], 3,500 m, 22.76800°S, 65.70280°W

Aguas Negras, 600 m, 23.75000°S, 64.93333°W

Bárcena, ca. 3 km S, 1,808 m, 24.00056°S, 65.44767°W

Caimancito, 600 m, 23.74070°S, 64.59370°W

Calilegua, 447 m, 23.77268°S, 64.77002°W

Cerro Calilegua, 1,700 m, 23.58333°S, 64.90000°W

Cerro Casabindo, 4,800 m, 22.93333°S, 66.11667°W

Cerro Hermoso, 2,800 m, 23.56667°S, 64.85000°W

Cerro de La Lagunita, 4,500 m, 23.58333°S, 65.20000°W

Cerro Lagunita, near Maimara, 23.61667°S, 65.4000°W

Cerro Morado, sobre Río Morado, 11 km NW of San Antonio, 24.32000°S, 65.40000°W

Cochinoca, 22.74572°S, 65.89566°W

El Chaghuaral, 500 m, 24.26667°S, 64.80000°W

El Durazno, Cerro Calilegua, 2,600 m, 23.56666°S, 64.88333°W

El Palmar, Sierra de Santa Bárbara, 500 m, 24.11667°S, 64.73333°W

El Simbolar, 25 km SW of Palma Sola, 24.14291°S, 64.40458°W

El Toro, 50 km W of Susques, 23.30000°S, 67.00000°W

El Toro, 55 km W of Susques, 4,137 m, 23.07258°S, 66.71462°W

Finca El Piquete, margin of Río Volcán, ca. 5 km from the junction of Río Tamango and logging road, 973 m, 24.18564°S, 64.55949°W

Finca Las Capillas, 1,200 m, 24.06670°S, 65.13330°W

Guairazul, ca. 3 km S, 4,100 m, 22.94589°S, 66.26958°W

Higuerilla, 2000 ft, 24.21000°S, 65.17000°W

Higuerilla, 1,735 m, 23.53333°S, 65.03333°W

Humahuaca, 7,700 ft, 23.20000°S, 65.35000°W

Hwy 9 on border of Salta, at campground on way to El Carmen, 4,600 ft, 24.46667°S, 65.35000°W

Ingenio La Esperanza, 24.23333°S, 64.86667°W

La Antena, Sierra del Centinela, S of El Fuerte, 2,350 m, 24.29901°S, 64.38591°W

Laguna La Brea, 23.93333°S, 64.46667°W

La Herradura, 12 km SW of El Fuerte on Provincial Rte 6, 1,428 m, 24.30158°S, 64.49406°W

La Poma, 3 km E of La Poma, 3,268 m, 24.71375°S, 66.16625°W

La Quica, 22.11667°S, 65.60000°W

Ledesma, 23.83333°S, 64.76667°W

León, 1,668 m, 24.04216°S, 65.42271°W

León, 1 mi W, 5,800 ft, 24.03333°S, 65.43249°W

Maimará, 2,230 m, 23.61667°S, 65.40000°W

Mesada de las Colmenas, 1,150 m, 23.70681°S, 64.85624°W

Quebrada Alumbriojo, ca. 8 km NE of Santa Ana, Parque Nacional Calilegua, 2,900 m, 23.32111°S, 64.91769°W

Rinconada, 6 km N, on road to Timón Cruz, 4,286 m, 22.41600°S, 66.20000°W

Río Las Capillas, 15 km N of Las Capillas by Provincial Rte 20, 957 m, 24.03333°S, 64.98333°W

Ronquil Angosto, 2 km W on Provincial Rte 16, 3,700 m, 23.70000°S, 65.70000°W

Rte 83, on road to Valle Grande, 9 km N of San Francisco, 1,200 m, 23.58333°S, 64.96667°W

Saladillo, 200–300 m, 23.71667°S, 65.10000°W

Saladillo, 8 km E, 705 m, 24.41770°S, 64.90680°W

Salinas Grandes, 18 km E, 3,980 m, 23.68800°S, 65.70080°W

San Antonio de los Cobres, 12.3 km N and 11.5 km W, 4,336 m, 24.23333°S, 66.68333°W

San Francisco and Pampichuela, on road between them, 23.60660°S, 64.95700°W

San Rafael, 994 m, 24.08966°S, 64.39794°W

San Salvador, 24.20000°S, 65.31667°W

Santa Bárbara, 1,649 m, 24.28333°S, 64.40000°W

Santa Catalina, 21.95000°S, 66.06667°W

Sey, 8.2 km S, 24.01347°S, 66.51467°W

Sierra de Tilcara, 4,500 m, 23.58333°S, 66.20000°W

Sierra de Zenta, 4,500 m, 23.58333°S, 65.20000°W

Sierra de Zenta, 14,800 ft, 23.05000°S, 65.08333°W

Sunchal, 1,300 m, 24.26667°S, 64.45000°W

Susques, 23.60000°S, 66.48333°W

Susques, 26 km W, 4,198 m, 23.38580°S, 66.53770°W

Termas de Reyes, Mirador, on Provincial Rte 4, 1,889 m, 24.16291°S, 65.49958°W

Tilcara, 2,440 m, 23.56667°S, 65.36667°W

Tilcara, 0.5 mi E, 3,145 m, 23.56667°S, 65.56667°W

Tiraxi, 1.5 km E, sobre ruta 29, 23.99351°S, 65.31919°W

Tres Cruces, 3,360 m, 22.92031°S, 65.58136°W
Villa Carolina, Río Lavallén, 500 m, 24.23300°S, 64.65000°W
Yala, mountains W of, 24.11667°S, 65.70795°W
Yavi, 22.13333°S, 65.46667°W
Yavi, near La Quiaca, 3,440 m, 22.11667°S, 66.43333°W
Yuto, 23.63333°S, 64.46667°W
Yuto, Río San Francisco, 342 m, 23.64342°S, 64.47194°W

La Pampa (Provincia de)

Algarrobo del Aguila, 36.39944°S, 67.145600°W
Anzoátegui, 40 km N, 38.70420°S, 63.80661°W
Catriló, 36.40000°S, 63.41667°W
Chamaicó, 15 km SW, Loma Loncovaca, 35.15000°S, 65.08333°W
Colonia 25 de Mayo, 38.11667°S, 67.61667°W
Colonia Gobernador Ayala, 37.58333°S, 68.06667°W
El Guanaco [= Médanos El Guanaco], 36.53333°S, 66.43333°W
Estancia El Puma, Chicalcó, 36.00076°S, 67.24797°W
Estancia La Pastoril [= Colonia La Pastoril], 518 m, 36.37312°S, 66.22771°W
Estancia Los Guadales, 38.60000°S, 64.43333°W
Estancia Los Ranqueles, 36.12750°S, 66.18310°W
Estación Perú, 37.65000°S, 64.13333°W
General Acha, 37.75000°S, 65.00000°W
General Pico, 35.65820°S, 63.75460°W
Gran Salitral, 37.40000°S, 67.20000°W
La Florida, 36.18999°S, 66.42388°W
Laguna El Chañar, Puesto El Chato, 36.95000°S, 65.81667°W
Laguna Guatraché, 37.76299°S, 63.55381°W
La Humada, 36.33333°S, 68.00000°W
Lihue Calel, 37.95000°S, 65.65000°W
Loventué, 308 m, 36.18849°S, 65.28751°W
Luán Toro, 36.20347°S, 65.10084°W
Luán Toro, 20 km SE, Estancia La Florida, 36.37000°S, 65.02000°W
Naicó, 117 m, 36.92613°S, 64.40651°W
Naicó, 12 km NNE, Estancia Los Toros, 36.83000°S, 64.35000°W
Parque Nacional Lihue Calel, 38.03070°S, 65.61610°W
Peru Station [= Estación Perú], 38.69779°S, 64.57091°W
Quehué, Estancia Los Molinos, 37.12000°S, 64.50000°W
Río Salado, 270 m, 36.76187°S, 66.90576°W
Santa Rosa, 176 m, 36.61667°S, 64.28333°W
Santa Rosa, 5 km S, 168 m, 36.46660°S, 64.31667°W
Santa Rosa, 10 km SW, Chacra La Lomita, 36.66667°S, 64.36667°W
Utracán, 37.28250°S, 64.833°W
Victoria, 36.21500°S, 65.43583°W

La Rioja (Provincia de)

Ambil, 31.11667°S, 66.35000°W
Cañón de la Reserva, Talampaya, 29.75000°S, 67.90000°W
Chamical, 30.36667°S, 66.21667°W
Chilecito, 29.16667°S, 67.50000°W
Cuesta La Cébila, 1 km from Chumbicha, on Rte 60, 28.83333°S, 66.40000°W
Guandacol, 29.51667°S, 68.53333°W
Guayapa, Patquía, 30.06540°S, 66.88730°W
El Pesebre, 2,448 m, 28.88003°S, 67.69328°W
Famatina, 28.91667°S, 67.51667°W
La Invernada, ca. 35 km N of Nevada de Famatina, 3,800 m, 28.68676°S, 67.84656°W
La Rioja, 490 m, 29.43333°S, 66.85000°W
Machigasta, 777 m, 28.53957°S, 66.79648°W
Pampa de la Viuda, km 19 on Provincial Rte 73, 2,150 m, 29.27025°S, 67.14297°W
Pastillos, Reserva Provincial "Laguna Brava," 38.40000°S, 69.10000°W
Potrerillo [= El Potrerillo], 2,600 m, 28.39603°S, 67.67661°W
Quimilo, 26 km SW, 581 m, 30.04539°S, 65.52039°W
Salar La Antigua, 45 km NE of Chamical, 30.03333°S, 66.06667°W
Salicas, 28.37323°S, 67.07291°W
Tello [= Desiderio Tello], 31.21667°S, 66.31667°W
Ulapes, 31.58333°S, 66.25000°W
Villa Unión, 29.30000°S, 68.20000°W
Villa Unión, 14.5 km N, 1,253 m, 29.18659°S, 68.24847°W

Mendoza (Provincia de)

25 de Mayo, ca. 850 m, 34.58468°S, 68.54491°W
Caverna de las Brujas, 35.80178°S, 69.81798°W
Cerro Colorado, 35.51458°S, 69.92683°W
Cerro Medio, 35 km WNW of 25 de Mayo, 1,600 m, 34.46000°S, 68.90000°W
Costa de Araujo, 32.75781°S, 68.39740°W
Desaguadero, 457 m, 33.40334°S, 67.15564°W
Desaguadero, 27 km N, 1,670 m, 33.17323°S, 67.18333°W
Divisadero Largo, vicinity, 32.91430°S, 68.88960°W
El Algarrobal, 775 m, 31.81667°S, 68.76667°W
Gendarmería Cruz de Piedra, 2 mi E, 3,000 m, 34.18333°S, 69.67778°W
General Alvear, 34.97694°S, 67.69111°W
Hwy 40, 2 km N of junction Hwy 40 and El Manzano Road, 1,300 m, 36.08333°S, 69.72861°W
La Batra, SW of Laguna Llancanelo, 1,425 m, 35.76666°S, 69.36667°W
Lago [= Laguna del] Diamante, 10.5 km W of old Rte 40, along road to Lago Diamante, 34.25000°S, 69.30399°W

Laguna del Diamante, 34.16660°S, 69.68330°W

Laguna del Diamante, 3 km S, 3,270 m, 34.21919°S, 69.70809°W

Laguna de La Niña Encantada, 1,826 m, 35.16056°S, 69.86917°W

Laguna de La Niña Encantada, Los Molles, 35.18333°S, 69.90000°W

Laguna Llancanelo, 1,335 m, 35.75000°S, 69.13333°W

La Paz, 33.46091°S, 67.54956°W

La Pega, 32.80000°S, 68.80000°W

Las Catitas, 33.30000°S, 68.03333°W

Las Higueras, 1,147 m, 32.48333°S, 68.90000°W

Las Lajas, 36.83333°S, 69.03333°W

Las Leñas, 6 km NW (by road), 35.10000°S, 70.12000°W

Lavalle [= General Lavalle], 626 m, 32.72218°S, 68.59137°W

Malargüe, 1,450 m, 35.47546°S, 69.58427°W

Mendoza, 838 m, 32.89084°S, 68.82717°W

Ñacuñán, 34.04564°S, 67.95372°W

Nihuil Dam, 16 km S and 2 km W of El Nihuil, 1,330 m, 35.17129°S, 68.21333°W

Papagallos [= Arroyo de los Papagayos], 1,070 m, 33.95000°S, 69.06667°W

Paraditas, 35 km S by Hwy 40 and 3 km E, 34.25000°S, 69.08472°W

Puesto El Peralito, Malargüe, 35.46667°S, 69.58333°W

Punta de Vacas, 3,000 m, 32.85000°S, 69.75000°W

Quebrada de la Vena, ca. 7 km SSE of Uspallata, 1,880 m, 32.65675°S, 69.35950°W

Refugio Militar General Alvarado, 3 km W, 2,100 m, 34.25000°S, 69.33889°W

Reserva Ñacuñán, 34.03333°S, 67.96667°W

Reserva Telteca, 520 m, 32.39320°S, 68.04630°W

Rincón de Atuel, 34.750000°S, 68.61667°W

RN 40 footbridge over Río Grande, 36.31278°S, 69.66750°W

Salinas de Diamante, 34.91667°S, 68.88111°W

Seccional Horcones PP Aconcagua, 32.81553°S, 69.94128°W

Tambillos, 2,418 m, 32.37645°S, 69.42790°W

Tupungato, 1,060 m, 33.37146°S, 69.14845°W

Uspallata, 2,060–2,136 m, 32.62267°S, 69i.28592°W

Uspallata, ca. 7 km S, 36.56524°S, 69.34852°W

Valle Hermoso, 28 km E of Volcán Peteroa, 2,460 m, 35.09794°S, 70.10247°W

Valle de Las Cuevas, 35.25000°S, 70.56667°W

Vallecitos, ca. 3 km SSE (by road), 32.99657°S, 69.32145°W

Villavicencio, 2 km S of Rte 32, 32.51667°S, 68.98333°W

Villavicencio, between Villavicencio and Uspallata, 32.91722°S, 69.23525°W

Volcán Carapacho, 35.83018°S, 69.14813°W

Volcán Carapacho, Laguna Llancanelo, 35.60000°S, 69.20000°W

Volcán Peteroa, 35.43333°S, 70.33333°W

Misiones (Provincia de)

Aristóbulo del Valle, Cuña Piru, 26.95000°S, 55.11670°W

Arroyo Itaembé Miní, 27.45500°S, 56.05070°W

Arroyo Itaembé Miní, bridge on RN 12, 27.43333°S, 56.00000°W

Arroyo Paraíso, junction with Hwy 2, 27.16660°S, 54.16667°W

Arroyo Paraíso, 6 km NE of junction with Hwy 2, 27.16660°S, 54.16667°W

Arroyo de Salamanca, Parque Provincial "Ernesto Che Guevara", 147 m, 26.61470°S, 54.78083°W

Balneario de la Reserva Privada de Usos Múltiples de la Universidad Nacional de La Plata "Valle del Arroyo Cuña Piru," ca. 200 m, 27.08330°S, 54.95000°W

Candelaria, Arroyo Yabebyrí, 27.46077°S, 55.74353°W

Caraguatay, Río Paraná, 100 m, 26.61667°S, 54.76667°W

Concepción de la Sierra, 27.98333°S, 55.60000°W

Dos de Mayo, Guarani, 300 m, 27.03333°S, 54.65000°W

El Dorado, 26.40000°S, 54.58333°W

Eldorado [= El Dorado], 26.40000°S, 54.58333°W

Eldorado [= El Dorado], Segunda Iglesia Cuadrangular, Barrio Parque Km 11, 26.40000°S, 54.51667°W

Estancia Santa Inés, 95 m, 27.52555°S, 55.87194°W

Estación Experimental INTA Cuartel Río Victoria, 26.92513°S, 54.42484°W

Iguazú, Arroyo Mbocaí and Rte 12, 25.68012°S, 54.50806°W

Los Helechos, 27.55000°S, 55.05667°W

Lote Alto Paraná S.A., ca. 15 km E of Wanda on Rte 19, 26.00100°S, 54.42400°W

Montecarlo, 26.56667°S, 54.78333°W

Oberá, Campo Ramón, 27.46697°S, 54.99953°W

Oberá, Campo Ramón, 27.41667°S, 55.01667°W

Oberá, Escuela Provincial 639 "Rosario Vera Peñaloza," Lote 92, Sección II de Campo Ramón, 27.41667°S, 55.01667°W

Parada Leis, 92 m, 27.60102°S, 55.84070°W

Parque Nacional Iguazú, 25.68333°S, 54.45000°W

Parque Nacional Iguazú, Sendero Macuco, ca. 200 m, 25.68333°S, 54.43333°W

Parque Provincial Islas Malvinas, 25.83333°S, 51.16667°W

Parque Provincial Moconá, ca. 2 km W, junction Hwy 21 and Arroyo Oveja Negra, 27.16660°S, 54.16667°W

Parque Provincial Uruguá-í, 25.82874°S, 54.12938°W

Posadas, Estancia Santa Inés, 27.31320°S, 55.52190°W

Profundidad, 27.56667°S, 55.71667°W

Puerto Esperanza, 26.02641°S, 54.61244°W
Puerto Gisela, San Ignacio, 50 m, 27.01667°S, 55.45000°W
Reserva Privada UNLP "Valle del Arroyo Cuña Pirú," 27.08333°S, 54.95000°W
Reserva Privada de Vida Silvestre Urugua-í, 25.98333°S, 54.08333°W
R.N.E. San Antonio, 26.04479°S, 53.77789°W
RP 2, 6 km NE of Arroyo Paraíso, 27.21325°S, 54.033331°W
San Ignacio, 27.25902°S, 55.53925°W
San Martín, 27.46347°S, 55.33466°W
Santa Ana, 27.35000°S, 55.60000°W
Santa Inés, 27.56667°S, 55.81667°W
Sierra de la Victoria, 25.91670°S, 54.00000°W
Tobunas, Rte 14, Km 352, 26.46667°S, 53.90000°W

Neuquén (Provincia de)

Aeropuerto Caviahue, 1,050 m, 39.33000°S, 71.32000°W
Arroyito, 3 km S, 39.08333°S, 68.58333°W
Arroyo Covunco and RNN 40, 38.78333°S, 70.18333°W
Barda Negra, Parque Nacional Laguna Blanca, 39.03333°S, 70.38333°W
Barrancas, 36.85794°S, 69.92497°W
Cajón Grande, A° Curi Leuvú, 36.81025°S, 70.40019°W
Cañadón del Tordillo, 40.38333°S, 70.18333°W
Caviahue, 2 km E and 1 km N, 1,650 m, 37.84000°S, 70.96000°W
Chapelco, 1,250 m, 40.23000°S, 71.26000°W
Chos Malal, 847 m, 37.37809°S, 70.27085°W
Chos Malal, 3–5 km upstream along Neuquén River, 37.34421°S, 70.28135√W
Collon-Curá, 16 km SE of La Rinconada, 800 m, 40.11667°S, 70.73333°W
Confluencia, 3 km NW, 40.04000°S, 70.85000°W
Copahue, vicinity, 1,600–2,100 m, 37.83333°S, 71.06667°W
Copahue, 1 km E, Las Maquinitas, 2,050 m, 37.83000°S, 71.10000°W
Copahue, 1.5 km S, ca. 2,100 m, 37.80000°S, 71.10000°W
Cueva Traful I, 40.71556°S, 71.11028°W
Estancia Alicura, 1,800 m, 40.43000°S, 70.73000°W
Estancia Fortín Chacabuco, 3 km S and 2 km W of Cerro Puntudo, 1,075 m, 40.96847°S, 71.19181°W
Estancia la Primavera, 11 km NW of Confluencia, 40.71496°S, 71.10825°W
Estancia La Querencia, 39.11667°, 70.93333°W
Estancia Los Helechos, vicinity, 2.5 km W and 5 km S of Cerro Colorado, 1,000 m, 39.74000°S, 71.47000°W
Estancia Paso Coihue, 4 km N and 4 km E, 1,700 m, 40.89000°S, 71.26000°W
Estancia Rincón Grande, 800 m, 49.03333°S, 71.05000°W

Hwy 60 and Hwy 23, 21 km NW and 1 km N of junction, 900 m, 39.66667°S, 71.20333°W
Lago Caviahue, south margin, 37.89265°S, 71.03583°W
Lago Curruhué, Parque Nacional Lanin, 39.91667°S, 71.41667°W
Lago Curruhué Chico, 3 km E, 1,000 m, 39.83000°S, 761.48000°W
Lago Epulafquen, 1,450 m, 39.80000°S, 71.58000°W
Lago Hui Hui, SW end, 8 km W and 2 km S of Cerro Quillén, 39.36000°S, 71.356000°W
Lago Lolog, NE coast, 5 km W of Río Quilquihue, 40.06000°S, 71.34000°W
Lago Quillén, north shore, 39.45000°S, 71.38000°W
La Rinconada, 2 km SE, 40.01000°S, 70.65000°W
Las Coloradas, 5 km N, 39.50000°S, 70.58000°W
Lonco Luan, 8.8 km S, 1,000 m, 39.01667°S, 71.00000°W
Nahuel Huapi, 11 km NNE, 40.95000°S, 71.10000°W
Pampa de Hui Hui, 4 km W and 2 km S of Cerro Quillén, 39.34000°S, 71.31000°W
Paraje La Querencia, 39.16667°S, 70.93333°W
Parque Nacional Laguna Blanca, 39.03333°S, 70.38333°W
Parque Nacional Laguna Blanca, 31 km SW of Zapala, 39.09000°S, 70.32000°W
Paso Córdoba, 40.60000°S, 71.15000°W
Paso Puyehue, 32 km WNW of Villa La Angostura, 40.70000°S, 71.95000°W
Pilolil, 39.65000°S, 70.95000°W
Pilolil, 4 km S, 39.63333°S, 70.93833°W
Piño Hachado, 1,500 m, 38.56186°S, 69.78337°W
PN Laguna Blanca, 0.58 km W and 4.20 km N of Co. Mellizo Sud, 39.03333°S, 70.30000°W
Puente Carreri, 2.6 km NW, 38.87167°S, 70.45833°W
Rahue, 3 km W, 39.45000°S, 70.95000°W
Refugio, 3.5 km N and 1.5 km E of Estancia Paso Coihue, 40.89000°S, 71.29000°W
Refugio Parque Provincial Tromen, 12 km NE, 37.05893°S, 70.08590°W
Refugio Neumeyer, ridge above, 41.25000°S, 7125000°W
Río Carreri, bridge on RNP°13, 38.88722°S, 70.43556°W
Río Neuquén, 1 km SE of bridge RNN 40, 37.41389°S, 70.22806°W
Río Pichi Traful, 0.8 km E and 3.3 km N of junction with Rte 234, 40.47778°S, 71.58889°W
Río Quilquihue, 750 m, 40.05875°S, 71.06668°W
Risco Alto, Auca Mahuida, 37.75114°S, 68.90328°W
Riscos Bayos, 37.95000°S, 70.78333°W
Ruca Malen, N end Lago Correntoso, 40.82000°S, 71.61000°W
Salar de Añelo, 38.01667°S, 68.88333°W
Sierra del Portezuelo, 38.91667°S, 69.53333°W
Zapala, 1,062 m, 38.75000°S, 70.06667°W

Zapala, 7 km NW, 38.85000°S, 70.12000°W
Zapala, 20 km E, 38.88333°S, 69.81667°W

Río Negro (Provincia de)

Aguada Cecilio, 40.85817°S, 65.80583°W

Arrayanes, 33.23133°S, 58.02283°W

Arroyo Pinturas, 41.70111°S, 66.70361°W

Balneario Las Cañas, 33.18900°S, 58.35550°W

Bariloche [= San Carlos de Bariloche], 19 km SE, 41.25000°S, 71.14000°W

Bariloche [= San Carlos de Bariloche], 43 km SSW, 1,030 m, 41.50000°S, 71.46000°W

Barda Esteban, Pilcaniyeu, 40.60875°S, 70.53108°W

Cabaña Cacique Foyel, 41.58333°S, 71.51667°W

Campana Mahuida, 41.50000°S, 66.41667°W

Campo Anexo Pilcaniyeu, 41.02000°S, 70.47000°W

Campo Viejo, Estancia San Ramón, 800 m, 41.03000°S, 71.44000°W

Cañadón Quetrequile, 41.86105°S, 69.46846°W

Cerro Grande, 5 km W, 41.60000°S, 65.42000°W

Cerro Castillo [= Cerro Guacho], 40.55000°S, 70.63333°W

Cerro Corona, 1,354 m, 41.45258°S, 66.91266°W

Cerro Corona, Meseta de Somuncura, 41.45583°S, 66.91444°W

Cerro Leones, 41.08089°S, 71.14109°W

Cerro Leones, 15 km ENE of Bariloche, 41.08210°S, 71.14401°W

Cerro Ñireco, 7.5 km N and 8 km E, 39.00000°S, 70.40000°W

Cerro Puntudo, 41.30472°S, 66.90611°W

Cerro Somuncurá Chico, 41.53333°S, 66.88333°W

Chimpay, 39.16990°S, 66.14720°W

Choele Choel, 39.26667°S, 65.65000°W

Comallo, 10 km S, 41.09000°S, 70.21000°W

Comallo, 10 km WNW, 40.95000°S, 70.31667°W

Comico, 5 km W, 40.32000°S, 67.33000°W

El Bolsón, 19 km NNE, 41.8000°S, 71.42000°W

El Espigón, 29 km S of El Cóndor on RPN 1, 41.11778°S, 63.00861°W

El Rincón [= Establecimiento El Rincón], 40.98333°S, 66.68333°W

Estancia Calcatreo, entrance to, 41.70396°S, 69.37708°W

Estancia Calcatreo, 41.73333°S, 69.36667°W

Estancia Huanuluan, 41.36640°S, 69.81380°W

Estancia Liempi, 41.65000°S, 66.68333°W

Estancia Maquinchao, Puesto de Hornos, 41.70000°S, 68.65000°W

Estancia Pilcaniyeu, 41.13333°S, 70.66667°W

Huanuluan, 3,500 m, 41.36667°S, 69.86667°W

Lago Steffen, 500 m, 41.48000°S, 71.57000°W

La Guardia, 33.05383°S, 57.72250°W

Laguna del Molino, Ingeniero Jacobacci, 41.75583°S, 69.55999°W

Laguna Tromen, 2,100 m, 37.09910°S, 70.11260°W

Las Grutas, 40.83333°S, 65.11667°W

Las Victorias, 4.2 km E of Bariloche, 41.13000°S, 71.25000°W

Maquinchao, 40 mi S, 39.51910°S, 68.05639°W

Mencué, 15 km NE, 40.36033°S, 69.52650°W

Meseta de Somuncurá, 41.35555°S, 67.92817°W

Nahuel Huapi, east of, ca. 40.96667°S, 71.50000°W

Pichi Leufú, 41.13333°S, 70.85000°W

Pilcaniyeu, 41.11881°S, 70.72914°W

San Carlos, 40.03333°S, 6793333°W

Sierra Grande, 5 km W, 41.60000°S, 65.42000°W

Tembrao, 41.14167°S, 66.30833°W

Valcheta, 45 km SE, 49.98000°S, 65.77000°W

Viedma, 18 km SW, 40.94000°S, 63.02167°W

Salta (Provincia de)

Abra de Ciénaga Negra, ca. 3 km SE, 3,090 m, 23.31990°S, 64.90914°W

Aguaray, 22.24043°S, 63.73490°W

Aguaray, 22.24311°S, 63.73786°W

Aguas Blancas, 27 km W, 22.75000°S, 64.66600°W

Anta, Parque Nacional El Rey, 24.70000°S, 64.63333°W

Arroyo Salado, 7 km E of Rosario de la Frontera, al lado del ACA, 25.83190°S, 64.93825°W

Azul Cuesta, ca. 9 km S of Nazareno, 3,285 m, 22.53275°S, 65.11828°W

Barrancas, 16 km S and 1.8 km W along Río de las Burras, 23.41611°S, 66.02639°W

Cabeza de Buey, Campo La Peña, 24.78350°S, 65.01870°W

Cachi, 17 km NW, 25.01667°S, 66.23333°W

Cachi Adentro, ca. 2 km NNE, on the road to Las Pailas, 2,567 m, 24.88333°S, 65.66667°W

Cachi Adentro, ca. 3 km N, 24.06667°S, 66.21667°W

Cafayate, 1,623 m, 26.07295°S, 65.97614°W

Campo Quijano, ca. 5 km NW, Km 30 on RN 51 [Quebrada del Toro], ca. 1,600 m, 24.88333°S, 65.66667°W

Cañadón Ojo de Agua, 10 km S of Rosario de la Frontera, 979 m, 25.89988°S, 64.96191°W

Cauchari, 4,150 m, 24.09756°S, 66.77003°W

Cauchi, 4,400 m, 24.08333°S, 66.83333°W

Chorillos, 4,100 m, 24.121667°S, 66.45000°W

Corralito, 25.95000°S, 65.86000°W

Dragones, 31 km SSW, along Río Bermejo, 23.49367°S, 63.42867°W

El Breal, 6 km W of Santa Victoria E., 23.23333°S, 62.93333°W

El Corralito, ca. 23 km SW of Campo Quijano, sobre Ruta Nacional No. 51, 24.96666°S, 65.80000°W

El Talar, 23.16667°S, 62.90000°W

Embarcación, 23.21600°S, 64.10000°W

Escoipe, ca. 15 km W on Hwy 13, 2,680 m, 25.17410°S, 65.82536°W

Finca Alto Verde, 670 m, 23.20000°S, 64.45000°W

Fortín Ballivián, Río Pilcomayo, 23.00000°S, 63.83333°W

General Ballivián, 20 km W on Puerto Baulés Road, 22.83333°S, 64.05000°W

Isla de Cañas, road to: 48.8 km W of junction of Rtes 50 and 18 on Rte 18, 22.96744°S, 64.58584°W

Las Lajitas, 24.68333°S, 64.25000°W

La Poma, 3,190 m, 24.72196°S, 66.20106°W

La Poma, 3 km E, 24.71375°S, 66.16625°W

La Viña, 25 km SE, 25.65000°S, 65.48333°W

Los Andes, 24.50000°S, 27.33333°W

Los Cardones, 25.18333°S, 65.85000°W

La Paloma, 3 km E, 24.71375°S, 66.16625°W

La Represa, Metán, 25.48333°S, 64.95000°W

Los Andes, 3,000 m, 24.50000°S, 67.33333°W

Los Toldos, 1.6 km W, 1,621 m, 22.27811°S, 64.71275°W

Los Toldos, 5 km S (on road to Vallecito), 22.30854°S, 64.71083°W

Lower Cachi, 2,340 m, 25.11998°S, 66.1306°W

Macapillo, 25.36667°S, 64.01667°W

Manual Elordi, Vermejo, 2,910 m, 23.25444°S, 64.13489°W

Metán, 25.13333°S, 65.01667°W

Orán, 23.13333°S, 64.31667°W

Pampa Verde, ca. 8 km WSW of Los Toldos and S of Cerro Bravo, 2,400 m, 22.28330°S, 64.80000°W

Parque Nacional Baritú, Finca Jakulica, Los Helechos, 1,200 m, 22.66208°S, 64.53291°W

Parque Nacional Baritú, angosto Río Baritú, 1,600 m, 22.50708°S, 64.75958°W

Piquirenda, 700–800 m, 22.33333°S, 63.78333°W

Pulares, 5 km WSW, 25.12592°S, 65.60404°W

Rearte Norte, Rosario de la Frontera, 25.80000°S, 64.96667°W

Río de las Conchas, 5.7 km W of Metán, 996 m, 25.47000°S, 65.01000°W

Rodeo Pampa, 1 km ENE, km 59 on Provincial Rte 7, 3,080 m, 22.24658°S, 65.05119°W

Salar Pastos Grandes, 3,980 m, 24.62552°S, 66.72059°W

Salta, 24.78333°S, 65.41670°W

"Salta Province," exact locality unknown, placed at 25.08333°S, 66.15000°W

Santa Rosa de Tastil, 24.45580°S, 65.95917°W

Santa Victoria, Parque Nacional Baritú, Arroyo Santa Rosa, 22.61667°S, 64.6000°W

Santa Victoria range, 13 km NW of Lizoite, 4,265 m, 22.20225°S, 65.21113°W

Socompa, 24.55000°S, 68.18333°W

Tartagal, 470 m, 22.551636°S, 63.80313°W

Tartagal, Sierra de Aguaray, 800 m, 22.26667°S, 63.73333°W

Tolar Grande, 3,500 m, 24.58932°S, 67.39355°W

Tolombón, 1,600 m, 26.18636°S, 65.93673°W

Vado de Arrazayal, 20 km NW of Aguas Blancas, 22.48333°S, 64.65000°W

Valle Encantado, 3,000 m, 25.19667°S, 65.84278°W

Vega Cortadera, 2,897 m, 25.11667°S, 67.03333°W

Vega Cortadera, 3,610 m, 25.05000°S, 67.10000°W

San Juan (Provincia de)

Angaco Sud, 31.58333°S, 68.28333°W

Cañada Honda, 500 m, 31.98333°S, 68.55000°W

Cañada Honda, 600 m, 31.98043°S, 68.55694°W

Castaño Nuevo, 9 km NW of Villa Nueva, 31.03889°S, 69.54452°SW

Complejo Astronómico "El Leoncito," 4 km W, 31.78333°S, 69.37557°W

Estancia Leoncito, 1 km W of Observatorio Astronómico, 31.83333°S, 69.25000°W

Ischigualasto, 30.08333°S, 67.93333°W

Ischigualasto, Valle Fértil, 30.00000°S, 68.00000°W

José de Martí, 15 km ESE on road to Chañar Seco, 31.85000°S, 68.04611°W

Los Sombreros [= Puesto de los Sombreros], Sierra Tontal, 2,490 m, 31.91841°S, 69.02416°W

Pampa Larga, 32.06667°S, 69.83333°W

Parque Nacional El Leoncito, 2,100–2,900 m, 31.75667°S. 69.22444°W

Parque Provincial Ishigualasto, 3 km N headquarters, 30.;08333°S, 67.91667°W

Pedernal, 1,080 m, 31.98333°S, 68.73333°W

Reserva Ischigualasto, 30.11667°S, 67.93333°W

Sierra del Tontal, Los Sombreros, 2,700 m, 31.55000°S, 69.18333°W

Tudcum [= Nacedero], 30.18707°S, 69.27800°W

Vega Agua del Godo, 3,300 m, 29.03333°S, 69.40000°W

San Luis (Provincia de)

Arizona, 35.72250°S, 65.31972°W

Buena Esperanza, 11 km W, 34.75000°S, 65.37000°W

Eleodoro Lobos, 33.38997°S, 66.01343°W

Hualtaran, 3 km W, Parque Provincial Sierra de las Quijadas, 32.00000°S, 67.00000°W

Hualtaran, 6 km W, Parque Provincial Sierra de las Quijadas, 2800 ft, 32.55000°S 67.03333°W

La Toija, 1.3 km N and 5.8 km W, Pampas de las Salinas, 32.21517°S, 66.63833°W

Naschel, 32.91753°S, 65.38859°W

Pampa de las Salinas, 365 m, 31.92697°S, 66.68771°W

Pampa de las Salinas, 1.3 km N and 7.4 km W of La Botija, 32.20278°S, 66.65222°W

Pampa de las Salinas, ca. 23 km N Rte 20, Pampa de Salinas, near la Botija, 32.17730°S, 66.59560°W

Papagayos, 32.69390°S, 64.98270°W

Papagayos, camp on left bank Río Papagayos, 30.36667°S, 64.36667°W

Parque Provincial Sierra de la Quijadas, 6 km W of Hualtaran, 32.48639°S, 67.00052°W

Paso del Rey, 9 km N, 1,455 m, 32.94588°S, 66.00268°W

Pedernera, 33.66667°S, 65.25000°W

Salinas de Bebedero, 15 km SSE, 33.58333°S, 66.63333°W

San Francisco del Monte de Oro, 32.60000°S, 66.13333°W

San Luis, 708 m, 33.30000°S, 66.35000°W

Varela, 12 km N by road, 34.00000°S, 66.44056°W

Villa Mercedes, 511 m, 33.667571°S, 65.45783°W

Santa Cruz (Provincia de)

Alero Destacamento Guardaparque, 47.85227°S, 72.03391°W

Arroyo Aikén, 46.66667°S, 70.50000°W

Bahía del Fundo, 46.10000°S, 67.60000°°W

Cabo Vírgenes, 52.35000°S, 68.38333°W

Cañadón Minerales, 46.74984°S, 67.60712°W

Cape Fairweather [= Cabo Buen Tiempo], 51.55000°S, 68.95000°W

Casa de Piedra [= Estación Cada de Piedra], Río Ecker, 739 m, 47.12691°S, 70.86325°W

Cerro Casa de Piedra, 47.88333°S, 72.25000°W

Cerro Comisión, 50.33986°S, 72.47133°W

Cerro Fortaleza, 50.23611°S, 70.88831°W

Cueva de las Manos, 47.18333°S, 70.58333°W

Cerro Ventana, 49.06223°S, 70.24233°W

Cordilleras, west of upper Río Chico, 47.12215°S, 70.86456°W

El Calafate, 52 km WSW, 50.37000°S, 72.85000°W

El Chaltén, 8 km N, 49.26365°S, 72.88756°W

El Encanto Cascades, 14.60000°S, 60.70000°W

Estancia Cerro del Paso, 47.85917°S, 66.44056°W

Estancia Cerro Ventana, 220 m, 48.98972°S, 70.25417°W

Estancia La Anita, 50.45000°S, 72.56667°W

Estancia La Ascención, 49.88639°S, 72.04139°W

Estancia La Aurora, 4.4 km E Los Antiguos, 46.55000°S, 71.55000°W

Estancia La Cantera, 46.90083°S, 70.58306°W

Estancia La Dorita, Cañadón Loreley, 47.80306°S, 69.00500°W

Estancia Laguna Manantiales, 47.53333°S, 68.30000°W

Estancia La Julia, 49.59081°S, 69.59225°W

Estancia La María, 48.41011°S, 68.86667°W

Estancia La Vizcaina, 46.91667°S, 70.83333°W

Estancia Tucu Tucu, 48.46000°S, 71.98000°W

Estancia Tucu Tucu, Río Chico, 48.45600°S, 72.05200°W

Faro de Cabo Vírgenes, 4 km W, 52.33331°S, 68.41303°W

Holdich, 13 km SW, 46.01427°S, 68.32861°W

Lago Cardiel, 48.91667°S, 71.25000°W

Lago Cardiel, La Península, Estancia Las Tunas, 48.84483°S, 71.20651°W

Lago Cardiel, N end and RN 40, 48.90000°S, 781.01667°W

Lago San Martín, 48.86667°S, 72.66667°W

Laguna del Diez, 47.88333°S, 67.85000°W

Laguna del Diez, 40 km SW of Monumento Nacural Bosques Petrificados, 47.50000°S, 67.50000°W

Laguna de los Cisnes, 47.50000°S, 70.61667°W

La Porteña [= Estancia La Porteña], Río Vista, 48.03333°S, 71.93333°W

Meseta del Lago Buenos Aires, nacimiento Río Ecker, 47.15050°S, 71.23430°W

Meseta El Pedrero, 46.77583°S, 69.62650°W

Monumento Natural Bosques Petrificados, 47.67188°S, 68.01994°W

Parador Luz Divina, Río La Leona, 49.83333°S, 72.03333°W

Piedra Clavada, 46.58333°S, 68.56667°W

Piedra Clavada Sur, 46.71111°S, 68.682787°W

Port Desire [= Puerto Deseado], 47.75034°S, 65.89382°W

Puerto Deseado, 47.75000°S, 65.88333°W

Puerto Santa Cruz, 50.01667°S, 68.51667°W

Punta Beagle, 49.93622°S, 68.57022°W

Puesto El Cuero, 48.18367°S, 69.26667°W

Reserva Provincial Geológica Laguna Azul, 52.07470°S, 69.58130°W

Río Chico and Río Belgrano, confluence, 48.26149°S, 71.20270°W

Río Coig [= Coyle], mouth, 50.98333°S, 69.21667°W

Río Deseado, 47.83333°S, 65.90000°W

Río Deseado, Estancia Cerro del Paso, 47.85917°S, 66.43583°W

Río Ecker, 500 m aguas abajo Estancia Casa de Piedra, margen norte, 47.12520°S, 70.86039°W

Río Santa Cruz, 4 km W of Punta Quilla s/RP 288, 50.10000°S, 68.43333°W

Río Tucu Tucu, about 8 km downstream from headwaters, 48.47000°S, 71.87000°W

Santa Cruz, 50.01667°S, 68.51667°W

Santa Cruz, 30 mi S, 52.06901°S, 69.20604°W

Seccional Glaciar Moreno, 50.47000°S, 73.00000°W

Sierra de los Baguales y de las Vizcachas, 50.83333°S, 72.33333°W

Valle del Río Tucu-Tucu, 48.40000°S, 71.81667°W

Santa Fe (Provincia de)

Berna, 29.27400°S, 59.84600°W
Caicique Ariacaiquín, 30.65775°S, 60.23041°W
Coronda, 31.97263°S, 60.91982°W
Esperanza, 31.45000°S, 60.93333°W
Estación Santa Margarita, 28.30000°S, 61.53333°W
Estero La Zulema, 4 km NE of Estación Guaycurú, 29.06667°S, 60.18333°W
Helvecia, 31.09834°S, 60.08830°W
Jacinto L. Aráuz, 30.73361°S, 60.97528°W
La Matilde, 16 km N of Alejandra, 29.75000°S, 59.78333°W
Las Palmas, 30.46667°S, 61.71667°W
Los Palomares, 31.68333°S, 60.73333°W
Pedro Gómez Cello, 30.03333°S, 60.30000°W
Puerto Gaboto, 32.45312°S, 60.84020°W
Puerto Ocampo, 28.51667°S, 59.13333°W
Reconquista, 29.15000°S, 59.65000°W
Recreo, 31.48333°S, 60.73333°W
San Javier, 30.58621°S, 59.93684°W
San José de Rincón, 31.60426°S, 60.56761°W
Santa Margarita, 28.30000°S, 61.55000°W
Tostado, 29.23194°S, 61.76917°W

Santiago del Estero (Provincia de)

Bandera, 28.88333°S, 62.26583°W
Buena Vista, 15 km NE of Villa Ojo de Agua on Hwy 13, 29.40000°S, 63.60000°W
Guasayan Vírgen del Valle picnic area on Hwy 64 between Santa Catalina and La Puerta Chiquita, 701 m, 28.13333°S, 64.78333°W
INTA "La María" Research Station, 2.9 km W of station entrance, 137 m, 28.02819°S, 64.26106°W
Lavelle, 28.20000°S, 65.13333°W
Pampa de los Guanacos, 1 km S and 2 km E, 26.24786°S, 61.81779°W
Pampa de los Guanacos, 6 km S and 2 km E, 26.24000°S, 61.86100°W
Pozo Hondo, ca. 30 km N along route 34, 254 m, 26.90139°S, 64.45861°W
San Antonio, 28.62280°S, 63.19983°W
Sucho Corral, 27.93333°S, 63.45000°W

Tierra del Fuego, Antártida e Islas del Atlántico Sur (Provincia de)

Bahía Capitán Canepa, head, Isla de los Estados, 54.85000°S, 64.45000°W
Bahía del Buen Suceso, 54.78333°S, 65.25000°W
Bahía Lapataia, 54.83333°S, 68.43333°W
Bahía San Sebastián, 53.30000°S, 68.46667°W
Estancia Cullen, 52.90000°S, 68.46667°W
Estancia San Martín, Bahía San Sebastián, 53.12493°S, 68.52727°W
Estancia Viamonte [= Estancia Via Monte], 54.03349°S, 67.36678°W
Estancia Via Monte, 54.03349°S, 67.36678°W
Lago Fagnano [= Lago Kami], 54,63333°S, 67.98000°W
Lago Fagnano, near east end, 54.52469°S, 67.20228°W
Lago Yehuin, 54.41670°S, 67.68330°W
Lake Fagnano, north end, 54.53000°S, 67.21000°W
Lake Fagnano, southeastern end, 54.600090°S, 67.38333°W
Lake Yerwin [= Lago Yehuin], 54.51667°S, 67.68333°W
Lapataia, 54.83333°S, 68.56667°W
Las Cotorras (RN 3, km 3012, 25 km N of Ushuaia), 54.71667°S, 68.03333°W
Strait of Magellan, south shore, near eastern entrance, ca. 52.65000°S, 68.60000°W

Tucumán (Provincia de)

Alberdi [= Villa Alberdi], 333 m, 27.58593°S, 65.62037°W
Alberdi, ca. 10 km S of Hualinchay, 2,300 m, 26.32228°S, 65.61264°W
Burruyacu, 490 m, 26.49918°S, 64.74206°W
Camino Ticucho, 26.61667°S, 65.21667°W
Cerro San Javier, 26.80000°S, 65.38333°W
Concepción, ca. 300 m, 27.34278°S, 65.59726°W
Concepción, 27.35187°S, 65.58833°W
Cumbre de Mala-Mala, Sierra de Tucumán, 26.78333°S, 65.55000°W
Cunbre [= Cumbre] de Mala-Mala, Sierra de Tucumán, 26.78333°S, 65.55000°W
El Cadillal, Estación de Piscicultura, 26.68333°S, 65.26667°W
El Infiermillo [= Abra del Infiernillo], 26.73736°S, 65.78404°W
El Naranjal, 26.66667°S, 65.05000°W
El Papal, Parque Nacional Campo de Los Alisos, 2,175 m, 27.18333°S, 65.95000°W
Faimalla [= Famaillá?], 26.81667°S, 65.36667°W
Gobernador Garmendia, 2 km S, por ruta 34, 26.57660°S, 64.56176°W
Horco Molle, 650 m, 26.79400°S, 65.31600°W
Huaca Huasi, 26.55000°S, 65.71660°W
Hualinchay, 26.30470°S, 65.65570°W
Hualinchay, on road to Cafayate, 26.30591°S, 65.61021°W
Hualinchay, 10 km S, 26.32223°S, 65.61266°W
Hualinchay, 10 km S by road on trail to Lara, 2,316 m, 26.32228°S, 65.61624°W

La Calera, 27.59728°S, 65.73062°W

La Cocha, 411 m, 27.77729°S, 65.57035°W

La Junta, headwaters of Río Pavas, 27.13330°S, 66.01668°W

Lamadrid [= La Madrid], 251 m, 27.64674°S, 65.24685°W

Las Cejas, ca. 4 km NW, 348 m, 26.86056°S, 64.76472°W

Las Paras [= Las Pavas], 911 m, 27.25375°S, 65.87399°W

Los Sarmientos, 27.40000°S, 65.68333°W

Las Tipas, Parque Biológico Sierra de San Javier, 26.66418°S, 65.40520°W

La Tranquera, northern border Los Chorillos farm, on Hwy 205, 1,426 m, 26.27620°S, 64.98361°W

Los Cardones, 26.63333°S, 65.81667°W

Monteagudo, 271 m, 27.51162°S, 65.27656°W

Monteros, 27.16667°S, 65.50000°W

Ñorco, 2,500 m, 26.48333°S, 65.36667°W

Paso El Infiernillo, 26.74375°S, 65.75542°W

Reserva Provincial Los Sosa, Rte 307, km 35, campamento de Lialidad Provincial, 1,234 m, 27.01667°S, 65.95000°W

Río Choromoro, 770 m, 26.41218°S, 65.31836°W

Río Colorado, 27.15000°S, 65.35000°W

San Miguel de Tucumán, 26.82508°S, 65.21276°W

San Pedro de Colalao, 26.24180°S, 65.50660°W

Simoca, 300 m, 27.26722°S, 65.35647°W

Tafí del Valle, vicinity, Km 95 of Hwy 307, 26.43333°S, 65.95000°W

Tapia, 665 m, 26.59308°S, 65.27864°W

Ticucho, 26.50000°S, 65.23333°W

Tucumán [= San Miguel de Tucumán], 26.82414°S, 65.22260°W

Villa San Javier, 1,225 m, 26.78168°S, 65.35844°W

Yerba Buena, 26.81667°S, 65.31667°W

Zanjón de Tafí, 2 km SW of Tafí del Valle, 2,000 m, 26.86194°S, 65.71833°W

Bolivia

Beni (Departamento de El)

Bahía de los Casara, 20 km W of Larangiera, Río Itenez, 13.21700°S, 62.35000°W

Barranca Colorada, Río San Luis, ca. 14.25680°S, 62.60500°W

Baures, 13.58300°S, 63.58300°W

Boca del Río Baures, 12,50000°S, 64.30000°W

Costa Marques [Brazil], opposite, Río Iténez, 12.46667°S, 64.28333°W

Costa Marques [Brazil], 4 km above, Río Iténez, 12.48300°S, 64.25000°W

Curicha, 12.61667°S, 63.51667°W

El Consuelo, 14.33333°S, 67.25000°W

El Triunfo, 13.08333°S, 65.46667°W

Estación Biológica del Beni, 14.63333°S, 66.30000°W

Exaltación, 8 km N, 13.26667°S, 65.25000°W

Guayaramerín [= Puerto Sucre], Río Mamoré, 10.81667°S, 65.41667°W

Guayaramerín, 4 km S, 10.85000°S, 65.41667°W

Lago Victoria, 13.76667°S, 63.50000°W

Las Penas, 12.73333°S, 64.46667°W

Mangal, 25 km NW of Exaltación, 13.10000°S, 65.41667°W

Nueva Calama, 12.80000°S, 64.36667°W

Pampa de Meio, Río Iténez, 12.50000°S, 64.31667°W

Piedras Blancas, 13.25000°S, 64.33333°W

Puerto Caballo, 13.71667°S, 65.35000°W

Puerto Salinas, Río Beni, 226 m, 14.33300°S, 67.55000°W

Riberalta, 10.98333°S, 66.10000°W

Río Iténez, Isla Leamel [= Leomil or Liomil], 12.50000°S, 64.16667°W

Río Iténez, 20 km above mouth, 12.00000°S, 65.03333°W

Río Mamoré, 12.43333°S, 65.38333°W

Río Mamoré, 17 km NNW of Nuevo Berlin, 12.53333°S, 65.15000°W

Río Mamoré, 5 km NE of Río Grande mouth, 10.38333°S, 65.38333°W

Río Tijamuchi, 14.93333°S, 65.15000°W

San Antonio de Lora, 10 km E, 15.11667°S, 64.91667°W

San Borja and Trinidad, camino between, 14.83333°S, 66.35000°W

San Ignacio, 14.88333°S, 65.60000°W

San Joaquín, Monto Río Machupo, 13.06667°S, 64.81667°W

Santa Rosa, 13.01667°S, 65.18333°W

Totaisal, 1 km SW of Estación Biológica del Beni, 300 m, 14.61667°S, 66.31667°W

Yucumo, 15.16667°S, 67.06667°W

Yutiole, 12.25000°S, 64.80000°W

Chuquisaca (Departamento de)

Camargo, 2,590 m, 20.65000°S, 65.21667°W

Carandaytí, 9 km E, 20.76667°S, 63.00000°W

Cerro Bufete, 2,000 m, 20.83000°S, 64.375003°W

Cerro Bufete, 2,050 m, 20.83017°S, 64.30533°W

Chuhuayaco [= Chuhuahacu], 2 km E, 1,200 m, 19.71667°S, 63.85000°W

Chuhuayacu, 2 km E, 1,200 m, 19.71667°S, 63.85000°W

Chuquisaca [= Sucre], 19.03333°S, 65.28333°W

Horcos, 80 km SE of Sucal [= Sucre?], 19.48333°S, 64.55000°W

Jamachuma, 1.3 km W, 17.53330°S, 66.11667°W

Monteagudo, 14 km N, Cañon de Herida, Río Bañado, 19.65000°S, 64.10000°W

Monteagudo, 2 km SE, 19.83333°S, 64.98333°W

Padilla, 9 km (by road) N, 19.30000°S, 64.36670°W

Padilla, 11 km N and 16 km W, 2,050 m, 19.20000°S, 64.45000°W

Porvenir, 675 m, 20.75000°S, 63.21667°W

Porvenir, 1.5 km NW, 20.75000°S, 63.21667°W

Potolo [= Estancia Pupayo], 22 km S and 13 km E of Icla, 3,700 m, 19.56666°S, 64.66666°W

Río Limón, 1,300 m, 19.56324°S, 64.11506°W

Sargento Rodríguez (Paraguay), 1 km W, 20.55000°S, 62.28333°W

Sucre, 19.03330°S, 65.28300°W

Sud Cinti, 15.6 km N of El Palmar, above Río Santa Marta, 20.91667°S, 64.91667°W

Tarabuco, 3,250 m, 19.16667°S, 64.91667°W

Tarabuco, 2 km N, 19.16700°S, 64.933300°W

Tarabuco, 4 km N, 19.08000°S, 64.56000°W

Tarabuco, 12 km N and 11 km E, 2,450 m, 19.06667°S, 64.81667°W

Tarabuquillo, 6 km NW, 19.21667°S, 64.56667°W

Tichucha, Río Capirenda, 20.46700°S, 64.06700°W

Tihumayu, 1,500 m, 19.56667°S, 64.13333°W

Tiquipa, 2 km S and 10 km E, Laguna Palmar, 20.93333°S, 63.35000°W

Tola Orco, 40 km from Padilla, Tomina, 2,100 m, 19.45000°S, 64.11667°W

Tola Orko [= Tola Orco], 2,100 m, 19.45000°S, 64.11667°W

Tomina, 19.18333°S, 64.50000°W

Cochabamba (Departamento de)

Altamachi, 17.03333°S, 66.43333°W

Alto Palmar, Chaparé, 1,600 m, 17.10000°S, 65.48300°W

Arani, 2,750 m, 17.56667°S, 65.76667°W

Ayopaya, 3,500 m, 16.50000°S, 77.58333°S

Chaparé, 2,000 m, 17.15000°S, 65.50000°W

Charuplaya, Río Securé, 1,350 m, 16.60000°S, 66.61667°W

Choquecamata, 4,000 m, 16.91667°S, 66.61667°W

Choquecamata, mountains near, 15,000 ft, 16.91667°S, 66.61667°W

Choro [= El Choro], 3,500 m, 16.9333°S, 66.70000°W

Chuchicancha, 17.35000°S, 65.71667°W

Cocapata, 3,115 m, 1751667°S, 65.28333°W

Cochabamba, 17.24000°S, 66.09000°W

Colomi, 3,800 m, 17.35000°S, 65.86666°W

Colomi, ca. 13 km N, 17.13000°S, 65.54000°W

Colomi, 16.5 km NW, 17.23330°S, 65.95000°W

Comarapa [Santa Cruz], 25 to 28 km W, 2,800 m, 17.85000°S, 64.66667°W

Comarapa [Santa Cruz], 31 km by road W, 2,800 m, 17.84470°S, 64.69230°W

Corani, 2,630 m, 17.21194°S, 65.86917°W

Corani, 3,130 m, 17.20000°S, 65.93333°W

Corani hydroelectric plant, near headquarters, 2,630 m, 17.21194°S, 65.89617°W

Cuesta Cucho, 2,300 m, 17.25000°S, 65.75000°W

El Choro, 3,500 m, 16.93333°S, 66.70000°W

El Palmar, Río Cochi Mayu, 17.10433°S, 65.75317°W

El Palmar, Chaparé, 17.25000°S, 65.38333°W

Epizana, 101 km by road SW, Siberia cloud forest, Cordillera Oriental, 2,989 m, 17.80000°S, 64.75000°W

Incachaca, 2,250–3,000 m, 17.23333°S, 65.68333°W

Misión San Antonio [= San Antonio del Chimoré], Río Chimoré, 400 m, 16.95000°S, 65.40000°W

Mosetenes, upper Río Mamoré, 16.66700°S, 66.05000°W

Palmar, Yungas de Cochabamba, 2,600 m, 16.33333°S, 66.75000°W

Parotani, 6 mi W, 17.56667°S, 66.46667°W

Pocona, 2,700 m, 17.65000°S, 65.40000°W

Peña Blanca, ca. 2 km NW, 18.29000°S, 65.10000°W

Punata, 10 mi E, 10,500 ft, 17.35000°S, 65.55000°W

Río Chapare, 15.96667°S, 64.7000°W

Rodeo, 7.5 km SE, 3,800 m, 17.66667°S, 65.58333°W

Rodeo, 9.5 km SE, then 2.5 km on road to ENTEL antenna, 3,875 m, 17.66667°S, 65.58333°W

San Antonio [= Villa Tunari], 16.95000°S, 65.38333°W

["San Antonio, later called Villa Tunari, or the confluence of the San Antonio and Espiritu Santo rivers, or 0.5 km NE of Villa Tunari." (Anderson, Koopman, and Creighton 1982:18)]

Tablas Monte, 4.4 km N (by road), 1,833 m, 17.06667°S, 65.98333°W

Tapacari, 17.51667°S, 66.60000°W

Tapacari, 15 mi E, 17.51667°S, 66.40000°W

Tiraque, 15 mi ENE, 3,200 m, 17.83333°S, 65.61667°W

Todos Santos, 218 m, 16.80000°S, 65.13333°W

Totora, 2 km N of Cocapata, 17.56667°S, 65.30000°W

Totora, 17 km E, Tinkusiri, 2,950 m, 17.75000°S, 65.03333°W

Totora, 20 mi E, 2,960 m, 17.70000°S, 64.86667°W

Ucho Ucho, 17.16667°S, 66.33300°W

Valle Hermoso, 17.38333°S, 66.15000°W

Villa Tunari, 2 km E, 300 m, 16.95000°S, 65.38333°W

Vinto, 2,621 m, 17.26667°S, 66.31667°W

Yungas, 1,000 m, 17.11330°S, 65.68500°W

Yungas, 2,000 m, 16.33333°S, 66.75000°W

Yungas del Palmar, 1,100 m, 17.13333°S, 65.50000°W

Yungas near Pojo, Carrasco, 1,800 m, 17.75000°S, 64.81667°W

Yungas, N of Locotyal, 1,110 m, 17.00000°S, 65.83333°W

Yuquí, 16.60000°S, 64.95000°W

La Paz (Departamento de)

Abel Iturralde, Pampas del Heath, 183 m, 12.88442°S, 68.63464°W

Achacachi, Lake Titicaca, 16.05000°S, 68.71667°W

Achacachi, 10–15 km N, 15.25000°S, 68.71667°W

Alaska Mine, 4,300 m, 16.28333°S, 69.03333°W

Andes de Bolivia, near La Paz, 16.50000°S, 68.15000°W

Antequilla and Pelechuco, 1 km W of pass between, 14,85000°S, 69.16667°W

Apolo, 17 km N, 14.56667°S, 68.46667°W

Astillero, 2,700 m, 16.46670°S, 67.58330°W

Ayane [= Yani], 3,500 m, 15.60000°S, 68.58333°W

Bellavista, 15.33333°S, 68.71667°W

Border of Chile, Bolivia, and Peru, 17.50000°S, 69.50000°W

Campamento Piara, 3,000 m, 14.80361°S, 69.03278°W

Caracato, 2,900 m, 16.98333°S, 67.81667°W

Caranavi, 8 km SE, 15.86667°S, 67.50000°W

Caranavi, 35 km by road N, Serranía Bellavista, 15.66667°S, 67.58333°W

Charuplaya, 17.20000°S, 66.96667°W

Chijchipani, 850 m, 15.63290°S, 67.58540°W

Chilalaya, 4,000 m, 16.21667°S, 68.45000°W

Chimate, 700 m, 15.41700°S, 68.00000°W

Chimosi [= Chimasi], 15.50000°S, 67.88333°W

Chulumani, 1,810 m, 16.04000°S, 67.51667°W

Chuncani, 14.63149°S, 68.11433°W

Cocopunco, 10,000 ft, 15.50000°S, 68.58333°W

Comauchi [= Comanche], 4,121 m, 16.95000°S, 68.41667°W

Copacabana, 16.16667°S, 69.08333°W

Esperanza, near Mount Sajama, 4,200 m, 17.81667°S, 678.78333°W

2° Estación climatológica, 2,552 m, 16.19960°S, 67.89248°W

Frankfurt, Chile at border with Bolivia and Peru, 17.50000°S, 69.50000°W

Franz Tamayo, Palcabamba, 2,441 m, 14.82444°S, 68.94444°W

Franz Tamayo, Pelechuco, Llamachaqui, on road to Apolo, 3,160 m, 14.80000°S, 69.03333°W

Guanay, 427 m, 15.50000°S, 67.88333°W

Huaraco, 17.16667°S, 67.91667°W

Huaraco-Antipampa, 3,650 m, 17.48333°S, 67.61667°W

Irupana, 16.48333°S, 6746667°W

Ixiamas, 358 m, 13.75000°S, 68.15000°W

Jesús de Machaca, 12 km by road SW, Río Desaguadero, 3,850 m, 16.80000°S, 68.86667°W

Khallutaka [near Laja], 16.50000°S, 68.23333°W

La Cumbre, 4,610 m, 16.36667°S, 68.05000°W

[Laguna Estrellani, ca. 2.5 km NE,] cumbre del camino a Yungas, 4,770 m, 16.33000°S, 68.02000°W

Laguna Viscachani, 3,788 m, 16.20000°S, 68.13333°W

La Paz, 16.50000°S, 68.15000°W

La Paz, 20 km S, 3,962 m, 16.80000°S, 68.15000°W

La Reserva, 840 m, 15.73333°S, 67.51667°W

Limani, 3,788 m, 16.86048°S, 68.09080°W

Mapiri, 800–2,000 m, 15.25000°S, 68.16667°W

Mapiri, Larecaja, 610 m, 15.25000°S, 68.16667°W

Mapiri, upper Río Beni, 1,000 m, 15.25000°S, 68.16667°W

Mecapaca, 16.66667°S, 68.01667°W

Moira camp, Alto Río Madidi, 13.58330°S, 68.76670°W

[Site is a lumber camp or saw mill (abandoned) on the upper Río Madidi. Emmons (1991) called the sawmill "Aserradero Moira," while S. Anderson (1997) used the name "Moire" for this site. Referred to as "Alto Madidi, Río Madidi, 270 m" in A. L. Gardner 2008.]

Mt. Sorata, base, 15.78333°S, 68.666667°W

Murillo, Cuticucho, 16.46667°S, 68.10000°W

Ñequejahuira, Río Unduavi, 8,000 ft, 16.31670°S, 67.86670°W

Okara, 15.68330°S, 67.50000°W

Parque Nacional y Area Natural de Manejo Integrado Cotapata, 1,860 m, 16.21333°S, 67.88667°W

Pelechuco, 3,650 m, 14.80000°S, 69.06667°W

Pitiguaya, Río Unduavi, 1,750 m, 16.35000°, 67.76667°W

Pongo, 3,690 m, 16.33330°S, 67.93333°W

Puerto Linares, 1 mi. W, 350 m, 15.46667°S, 67.55000°W

Puente de Choculo, 16.81667°S, 67.68333°W

Quiswarani, 4,230 m, 14.80018°S, 69.11667°W

Reserva de Fauna Ulla-Ulla, 114.86667°S, 69.26667°W

Río Aceromarca, 3,275 m, 16.30000°S, 67.88333°W

Río Alto Madidi, right bank, 400 m, 13.58333°S, 68.76667°W

Río Beni, 13.45000°S, 67.35000°W

Río Desaguadero, 16.56667°S, 68.08333°W

Río Madidi, 12.71667°S, 67.00000°W

Río Madidi, 8 km from mouth, 240 m, 12.56667°S, 67.00000°W

Río Solocame, 16.35000°S, 67.53333°W

Río Zongo, 16.16667°S, 68.03333°W

Salla, 3,820 m, 17.18333°S, 67.63333°W

San Andrés de Machaca, 16.73333°S, 69.01667°W

San Andrés de Machaca, 8.5 km W, 16.98444°S, 69.01667°W

San Ernesto, near Mapimi, 610–1,000 m, 15.25000°S, 68.16667°W

Sayani, 2,500 m, 16.13333°S, 68.10000°W

Serranía Bella Vista, 1,525 m, 15.69417°S, 67.51639°W

Sorata, 15.78333°S, 68.66667°W

Sorata, 2,650 m, 15.76667°S, 68.63333°W

Sorata, 10 km by road N, Moyabaya, Río Bhallapampa, 2,650 m, 15.71667°S, 68.66667°W

Tacacoma, 3,170 m, 15.58333°S, 68.65000°W

Tacacoma, 2.5 km S, 3,600 m, 15.60000°S, 68.65000°W

Tembladerani, 16.51667°S, 68.15000°W

Ticunhuaya, 15.65000°S, 68.40000°W

Ulla-Ulla, 3,571 m, 15.05000°S, 69.26667°W

Ulla-Ulla, 5 km E, ca. 15.07259°S, 69.22537°W

Unduavi, ca. 15 km (by road) NE, 2,400 m, 16.28360°S, 67.78890°W

Yolocito, 16.23333°S, 67.75000°W

Zongo, 18 km by road N, Cuticucho [= Cutikhuchu], 2,697 m, 16.13333°S, 68.11667°W

Zongo, 30 km N, 2,000 m, 15.9000°S, 67.98300°W

Zongo, 30 km by road N, cement mine, 2,000 m, 15.91667°S, 67.81667°W

Oruro (Departamento de)

Camino del Oruro, 37 km SW, 3.5 km NE of Toledo, 18.15000°S, 67.40000°W

Challapata, 3,710 m, 18.90000°S, 66.78333°W

Condo, 19.05000°S, 66.73333°W

Cruce Ventilla, 7 km S and 4 km E, 3,450 m, 19.13000°S, 66.12000°W

Estancia Agua Rica, 22 km S and 40 km SE of Sajama, 4,010 m, 18.33333°S, 68.56667°W

Finca Santa Helena, approx. 10 km (by road) SW of Pazna, 3,750 m, 16.30000°S, 66.91667°W

Huancaroma, near Eucaliptus, 3,720 m, 17.51667°S, 67.51667°W

Huancaroma, 1 km W, 3,730 m, 17.66667°S, 67.50000°W

Luca, 3,650 m, 19.58333°S, 67.90000°W

Mount Sajama [= pueblo at SW base of Nevada de Sajama], 18.11667°S, 69.00000°W

Oruro, 17.98333°S, 67.15000°W

Oruro, 40 km S, 18.55000°S, 67.15000°W

Pampa Aulliaga [= Pampa Aullagas], 19.18333°S, 67.08333°W

Panza, 1 km SW by road, 18.66700°S, 66.98300°W

Poopo, 18.38333°S, 66.98333°W

Quebrada Kohuiri, 18.13333°S, 68.85000°W

Sajama, 1 km SW, 17.91667°S, 68.68333°W

Sajama, 1.5 km SW, 4,200 m, 18.13333°S, 68.96667°W

Sajama, 30 km S and 25 km E, 18.41667°S, 68.80000°W

Pando (Departamento de)

Bella Vista, 170 m, 11.38333°S, 67.20000°W

Bella Vista, 10 km SSW of Mapiri, 15.33333°S, 68.21667°W

Independencia, 11.43333°S, 67.56667°W

Ingavi, on Río Orton, 10.76670°S, 66.73330°W

La Cruz, 170 m, 11.40000°S, 67.21667°W

Luz de América, Reserva Nacional de Vida Silvestre Amazónica Manuripi, 12.12173°S, 68.61042°W

Remanso, 160 m, 10.93333°S, 66.30000°W

Río Madre de Dios near Santa Rosa, left bank, 180 m, 12.21667°S, 68.40000°W

Río Nareuda, 11.28333°S, 68.91667°W

San Juan de Nuevo Mundo, 18 km N, 170 m, 10.76667°S, 68.73333°W

[A privately owned castaña and rubber property of about 40,000 ha; the site lies on the springs of the headwaters of the Río Negro, close to the Río Orton; the collecting locality is 18 km north of this spot.]

Potosí (Departamento de)

Acacio, 31 km on road to Uncia, 18.10000°S, 66.13333°W

Chocaya, 3,765 m, 20.90000°S, 66.28333°W

Cotagaita, 30 mi WNW, 20.50000°S, 66.10000°W

Kirikari Mountains, 19.65000°S, 65.66667°W

Laguna Colorado, 2 km E of ENDE camp, 4,280 m, 22.16667°S, 67.78333°W

Lagunillas, 3,500 m, 19.58333°S, 65.50000°W

Lipez, 4,500 m, 31.85000°S, 66.85000°W

Livichuco, 4,075 m, 18.88333°S, 66.46667°W

Llica, 19.86667°S, 68.25000°W

Otavi, 10 km SSE, 20.13952°S, 65.30837°W

Pampa de Talapaca, 20.00000°S, 65.36667°W

Pocoata, 3 km SE, 18.73333°S, 66.15000°W

Potosí, 19.58333°S, 65.75000°W

Potosí, 20 mi S, 3,749 m, 19.86667°S, 65.76667°W

Quetena Chica on Río Quetena, 4,200 m, 22.18333°S, 67.33333°W

Río Cachimayo, 19.30000°S, 66.20000°W

Serranía Siberia, 11 km (by road) NW of Torrecillas, 17.81667°S, 64.68333°W

Tupiza, 2,000 m, 21.45000°S, 65.71667°W

Uyuni, 3,670 m, 20.46667°S, 66.83333°W

Uyuni, 1 mi E, 3,658 m, 21.11667°S, 66.48333°W

Uyuni, 4 mi E, 21.11667°S, 66.45000°W

Uyuni, 5 mi [= 8 km] E, 4,000 m, 20.46667°S, 66.75000°W

Villa Alota, 9 km W, 21.40000°S, 67.78333°W

Villazón, 5 mi N, 22.00000°S, 65.75000°W

Yuruma, 3,230 m, 21.71667°S, 65.56667°W

Santa Cruz (Departamento de)

Ascención [= Ascención de Guarayos], 6 km W, 15.71667°S, 63.15000°W

Ascención, 6 km W by road, 240 m, 15.71667°S, 63.15000°W

Ascención de Guarayos, 15.71667°S, 63.15000°W

Aserradero Moira, 45 km E, 14.61667°S, 60.80000°W

Ayacucho, 250 m, 17.85000°S, 63.33300°W

Bañados del Izozog, 18.80412°S, 62.16397°W

Boyuibe, near Camiri, 20.10000°S, 63.53332°W

Boyuibe, 26 km E, 20.43333°S, 63.03333°W

Buena Vista, 450 m, 17.45000°S, 63.66667°W

Buenavista [= Buena Vista], 400 m, 17.45000°S, 63.66667°W

Buen Retiro, 6 km N, 300 m, 17.21667°S, 63.63333°W

Camiri, 5 km S of Choreiti, 1,000 m, 20.10000°S, 63.53300°W

Campamento Huanchaca II, Parque Nacional Noel Kempff Mercado, 14.52361°S, 60.73944°W

Campamento Los Fierros, Parque Nacional Noel Kempff Mercado, 14.56111°S, 60.92778°W

Campo de Guanacos, 19.00000°S, 63.00000°W

Caranda, 2 mi S, 17.55000°S, 63.53333°W

Cerro Amboró, 4.5 km N and 1.5 km E, Río Pitasama, 620 m, 17.75006°S, 63.64692°W

Cerro Colorado, 19.45000°S, 62.35000°W

Cocapata, above on old Cochabamba to Santa Cruz road, 3,200 m, 17.51667°S, 65.28333°W

Comarapa, 2,500 m, 17.90000°S, 64.48333°W

Comarapa, 1 km N and 8 km W, 2,450 m, 17.91667°S, 64.56667°W

Comarapa, ca. 25 km by road. W, 2,800 m, 17.82180°S, 64.66110°W

Comarapa, 28 km by road W, 17.85000°S, 64.66667°W

Com. Guirapembi, 19.43333°S, 62.51667°W

Cotoca, 2 km SE, 17.75000°S, 62.95000°W

El Refugio, Parque Nacional Noel Kempff Mercado, 14.75000°S, 61.00000°W

El Refugio, right bank Río Paragua/Tavo, 14.76694°S, 61.03389°W

El Refugio, 2.5 km NE, Parque Nacional Noel Kempff Mercado, 14.76720°S, 61.03470°W

El Refugio Huanchaca, Parque Nacional Noel Kempff Mercado, 170 m, 14.76700°S, 61.03300°W

El Refugio Pampa, 3 km NE from camp, 14.75000°S, 61.02000°W

Esperanza, 515 m, 16.46667°S, 61.26667°W

Estancia Cachuela Esperanza, 300 m, 16.78333°S, 63.23333°W

Estancia Las Cuevas, 1 km NE, 1,300 m, 18.18333°S, 63.73333°W

Estancia Laja, 17 km S of Quiñe, 2,100 m, 18.20000°S, 64.30000°W

Estancia San Marcos, 6 km W of Ascención, 400 m, 15.88641°S, 63.18555°W

Estación El Pailón, 3.5 km W, 300 m, 17.65000°S, 62.75000°W

Floripondio, 18.06667°S, 64.75000°W

Flor de Oro, Parque Nacional Noel Kempff Mercado, 210 m, 13.55167°S, 61.01417°W

Flor de Oro, Río Iténez, Parque Nacional Noel Kempff Mercado, 13.55170°S, 61.56667°W

Guadalupe, 10 km S of Vallegrande, 18.55000°S, 64.08333°W

Huanchaca II, Parque Nacional Noel Kempff Mercado, 700 m, 14.52361°S, 60.73944°W

Ingeniero Mora, 7 km E and 3 km N, 490–580 m, 18.13333°S, 63.20000°W

La Bélgica, 4 km SW, 17.58333°S, 61.25000°W

Lago Caimán, Parque Nacional Noel Kempff Mercado, 200 m, 13.60000°S, 60.91667°W

La Hoyada, 30 km S of Valle Grande, 18.75000°S, 64.10000°W

Las Cruces, 17.78333°S, 63.36667°W

Las Lomitas, 15 km S of Santa Cruz, 17.88333°S, 63.11667°W

Los Fierros, Parque Nacional Noel Kempff Mercado, 14.56111°S, 60.92778°W

Los Fierros, 6 km S, Parque Nacional Noel Kempff Mercado, 14.59000°S, 60.86000°W

Los Fierros, 17 km S, Parque Nacional Noel Kempff Mercado, 15.55000°S, 60.81667°W

Mangabalito, Parque Nacional Noel Kempff Mercado, 14.56111°S, 60.92778°W

Mataracú, Ichilo, 380–420 m, 17.6000°S, 63.93333°

Palmar de Osorio, 17.88333°S, 63.15000°W

Palmarito, Río San Julián, 400 m, 16.81667°S, 62.37000°W

Pampa de Meio, Río Itenéz, 12.50000°S, 64.31667°W

Parapeti, 20.00000°S, 63.00000°W

Parque Nacional Noel Kempff Mercado, 14.26667°S, 60.86667°W

Puerto Pacay, 1 km SE, 17.20000°S, 62.75000°W

Punta Rieles, 16.58300°S, 64.20000°W

Río Ariruma, 7 km by road SE of Ariruma, 1,750 m, 18.33333°S, 64.21667°W

Río Quiser, 16.61667°S, 62.76667°W

Río Surutó, 17.40000°S, 63.85000°W

Roboré, Río Roboré, 300 m, 18.33333°S, 59.75000°W

Roboré, 29.5 km W, 18.31667°S, 60.03333°W

San Ignacio de Velasco, 300 m, 16.38333°S, 60.98333°W

San José de Chiquitos, 17.85000°S, 60.78333°W

San Miguel Rincón, 300 m, 17.38333°S, 63.53333°W

San Rafael de Amboró, 400 m, 17.60000°S, 63.60000°W

San Ramón, 16.60000°S, 62.70000°W

San Ramón, 10 km N, 250 m, 16.60000°S, 62.70000°W

Santa Ana, 16.61700°S, 60.71700°W

Santa Ana, 17.46667°S, 63.73333°W

Santa Cruz at Brecha 5.5, ca. 55 km SE, 18.08333°S, 62.83333°W

Santa Cruz de la Sierra, 17.78333°S, 63.18333°W

Santa Cruz da la Sierra, Andrés Ibáñez, 430 m, 17.75000°S, 63.25000°W

Santa Rosa de la Roca, 250 m, 15.83333°S, 61.45000°W

Santiago de Chiquitos, 700 m, 18.31667°S, 59.56667°W

Santiago de Chiquitos, 4 km N and 1 km W, Río Pitisama, 400–700 m, 18.30257°S, 59.59725°W

Serranía Siberia, 11 km NW by road Torrecillas, 2,650 m, 17.8333°S, 64.83333°W

Siberia, 25 and 30 km W of Comarapa, 2,800 m, 17.85000°S, 64.7000°W

Tita, 18.41667°S, 62.16667°W

Tita, 8 km SE, 290 m, 18.46667°S, 62.11667°W

Torrecillas, 3 km N [by road], 17.850000°S, 64.63333°W

Urubicha, Río Negro, 14.97583°S, 62.60722°W

Vallegrande, 18.46667°S, 64.13333°W

Velasco, Parque Nacional Noel Kempff Mercado, Campamento Los Fierros, 215 m, 14.56651°S, 64.92678°W

Zanja Honda, 10 km S, 750 m, 18.30000°S, 63.20000°W

Tarija (Departamento de)

Caiza, 21.81667°S, 63.56667°W

Camatindi, 1 km S, 650 m, 21.00000°S, 63.38333°W

Caraparí, 1,000 m, 21.26667°S, 63.21667°W

Caraparí, 3 km WNW, 850 m, 21.80000°S, 63.78333°W

Carlazo, 38 km by road ENE of Tarija, 2,400 m, 21.35000°S, 64.30000°W

Carpirenda, 10 km S, 21.21667°S, 63.00000°W

Cieneguillas, 2 km SE, 21.31667°S, 65.03333°W

Cuyambuyo, 3 km SE, 900 m, 22.28333°S, 64.51667°W

Cuyambuyo, 4 km N by road, Fábrica del Papel, Río Sidras, 980 m, 22.21667°S, 64.60000°W

Cuyambuyo, 8 km by road N, 22.20000°S, 64.60000°W

Entre Rios, 21.53333°S, 64.20000°W

Erquis, 2,618 m, 21.46667°S, 64.85000°W

Fábrica del Papel, 4 km N of Cuyambuyo by road, 980 m, 22.25000°S, 64.58333°W

Iscayachi, 3,450 m, 21.48000°S, 64.95000°W

Iscayachi, 1 km E, Río Tomayapo, 3,416 m, 21.48333°S, 64.95000°W

Iscayachi, 4.5 km E, 3,750 m, 21.48333°S, 64.91667°W

Iscayachi, 12 km NW, 21.33333°S, 65.10000°W

Padcaya, 11.5 km N and 5.5 m E, 21.78333°S, 64.66667°W

Palo Marcado, 342 m, 21.45000°S, 63.11667°W

Palos Blancos, 2 km S and 5 km E, 800 m, 21.43000°S, 63.73000°W

pié Sierra Santa Rosa, Itau, 1,000 m, 21.70000°S, 63.90000°W

Pilcomayo, 21.25000°S, 63.50000°W

Pino [= Pinos], 1,800 m, 21.46667°S, 64.30000°W

Porvenir, 20.75000°S, 63.21667°W

Rancho Tambo, 61 km by road E of Tarija, 2,100 m, 21.45000°S, 64.31667°W

Sama, 4,000 m, 21.48333°S, 65.03333°W

Serranía del Sama, 3,200 m, 21.45000°S, 64.86666°W

Tablada, 21.55000°S, 64.78333°W

Tambo, 21.45000°S, 64.38333°W

Tapecua, 1,500 m, 21.26667°S, 63.91667°W

Tarija, 21.51667°S, 64.75000°W

Tarija, 10 mi NW, 8,200 ft, 21.46667°S, 64.70000°W

Taringuti, 21.46667°S, 63.28333°W

Tucumilla, 1 km E, 21.45000°S, 64.81667°W

Vermejo, 3,500 ft, 22.16667°S, 64.70000°W

Villa Montes, 21.25000°S, 63.50000°W

Villa Montes, 35 km SE by road, Taringuiti, 21.63333°S, 62.56667°W

Villa Montes, 8 km S and 10 km E, 21.31670°S, 63.41670°W

Yacuiba, 22.03333°S, 63.75000°W

Brazil

Acre (Estado do)

Fazenda Santa Fé [= Flora], left bank of Rio Juruá, 08.60000°S, 72.85000°W

Igarapé Porongaba, right bank of Rio Juruá, 400 m, 08.60000°S, 72.85000°W

Igarapé Porongaba, opposite of, left bank of Rio Juruá 08.66667°S, 72.78333°W

Iquiri [= Rio Ituxi, right bank affluent of upper Rio Purus], 09.96667°S, 67.80000°W

Manuel Urbano, Sena Madureira, BR 364 [Brazil Highway 364], Km 8, 08.88333°S, 68.66667°W

Nova Vida, right bank of Rio Juruá, 08.36700°S, 72.81700°W

Oriente [= Seringal Oriente], near Taumaturgo, Rio Juruá, 08.80000°S, 72.76667°W

Plácido de Castro, 10.33333°S, 67.18333°W

Paraná do Natal, Rio Juruá, 08.65909°S, 72.50735°W

Parque Nacional da Serra do Divisor, 07.50611°S, 73.77444°W

Rio Branco, 09.96667°S, 67.80000°W

Sena Madureira, 130 m, 09.06667°S, 68.66667°W

Sena Madureira, Km 8 on BR 364 [Brazil Highway 364] between Sena Madureira and Manuel Urbano, ca. 08.88333°S, 69.30000°W

Seringal do Oriente, Rio Juruá, near Vila Taumataurgo, 08.80000°S, 72.76667°W

Seringal Santo Antônio, 08.88333°S, 69.31667°W

Sobral, left bank of Rio Juruá, 08.36670°S, 72.81670°W

Alagoas (Estado de)

Anadia, 09.66667°S, 36.28333°W

Anadia, Sítio Vale Verde, 09.70000°S, 36.30000°W

Capela, 09.40000°S, 36.06667°W

Delmiro Gouveia, 09.38611°S, 37.99556°W

Fazenda Canoas, 09.53333°S, 35.68333°W

Fazenda do Prata, 13 km SSW of São Miguel dos Campos, 09.88333°S, 36.15000°W

Fazenda Santa Justina, 6 km SSE of Matriz de Camaragibe, 09.20000°S, 35.50000°W

Ibateguara, 08.99500°S, 35.84100°W

Matriz de Camaragibe, 09.15170°S, 35.53330°W

Matriz de Camaragibe, Fazenda Santa Justina, 6 km SSW of Matriz de Camaragibe, 16 m, 09.21667°S, 35.50000°W

Palmeira dos Índios, 09.41667°S, 36.61667°W

Penedo, 10.29028°S, 36.58639°W

Quebrangulo, 411 m, 09.31889°S, 36.47111°W

Santana do Ipanema, Sítio Goiabeira, 09.31667°S, 37.25000°W

São Miguel dos Campos, Mangabeiras, sea level, 09.86667°S, 36.15000°W

Sítio Angelim, Viçosa, 09.40000°S, 36.23334°W

Usina Sinimbú, Mangabeiras [= Manimbu], 09.93333°S, 36.08333°W

Viçosa, vicinity, 09.37139°S, 36.24083°W [coordinates of Viçosa]

Amapá (Estado do)

Amapá, 01.44412°N, 52.02154°W

Amapá, 02.05000°N, 50.80000°W

Amapá, 4 km N, 02.05000°N, 50.80000°W

Cachoeira de Santo Antônio, Rio Jari, 00.66667°S, 52.50000°W

Fazenda Asa Branca, Tartarugalzinho, ca. 01.51667°N, 50.90000°W

Fazenda Itapuã, 02.06667°N, 50.48333°W

Macapá, 00.05000°N, 51.05000°W

Macapá, Rio Amapari, 00.03333°N, 51.05000°W

Mazagão, Boa Fortuna, upper Igarapé Rio Branco, 00.55000°N, 51.20000°W

Mazagão, Rio Vila Nova, 00.11667°S, 51.28333°W

Oiapoque, upper Rio Oiapoque, 03.63885°N, 51.85048°W

Porto Grande, 50 km ESE, 00.60000°N, 51.00000°W

Rio Amapari de Macapá, 01.66700°N, 51.06700°W

Serra do Navio, Rio Amapari, 700–800 m, 00.98330°N, 52.05000°W

Serra do Navio, Macapá, 1–100 m, 00.03333°N, 51.08333°W

Tartarugalzinho, 01.28333°N, 50.80000°W

Terezinha, Rio Amapari, Serra do Navio, 00.96667°N, 52.03333°W

Vila Velha do Cassiporé, Oiapoque, 03.21667°N, 51.23333°W

Amazonas (Estado do)

Acajutuba, lower Rio Negro, 03.08333°S, 60.48333°W

Altamira, right bank of Rio Juruá, 06.58333°S, 68.90000°W

Alto Rio Urucu, 04.85000°S, 65.26670°W

Andirá, near Villa da Imperatriz [= Parintins], 02.60000°S, 56.73333°W

Arquipélago da Anavilhanas, Rio Negro, 02.70000°S, 60.75000°W

Ayapua [= Aiapuá], 04.45000°S, 62.13333°W

Barro Vermelho, left bank of Rio Juruá, 06.46667°S, 68.76667°W

Borba, Rio Madeira, 25 m, 04.38890°S, 59.59360°W

BR 319 [= Brazilian Highway 319], Km 62, 03.61667°S, 60.21667°W

Canabouca [= Lago Canabouca], Paraná do Jacaré, 03.50000°S, 60.68333°W

Carvoeiro, 01.43333°S, 62.01667°W

Castanhal, 01.50944°N, 60.97806°W

Cocuy [= Cucuí], Rio Negro, 01.20000°N, 66.83333°W

Codajás, left bank of Rio Solimões, 03.81667°S, 62.08333°W

Colocação Três Barracas, Igarapé Jauari, Rio Aracá, 00.59675°N, 63.10965°W

Colocação Vira-Volta, left bank of Rio Juruá on Igarapé Arabidi, affluent of Paraná Breu, 03.28333°S, 66.23333°W

Comunidade Bela Vista, left bank of Rio Madeira, 05.24611°S, 60.71389°W

Comunidade Colina, Rio Tiquié, Município de São Gabriel da Cachoeira, ca. 00.12000°N, 69.00670°W

Dejedá, Rio Juruá, 06.50000°S, 69.41667°W

Eirunepé, left bank of Rio Juruá, 06.65000°S, 69.86667°W

Estirão do Equador, right bank of Rio Javari, 04.52542°S, 71.56446°W

Faro, north bank of Rio Amazonas, 02.18333°S, 56.73333°W

Fazenda Dimona, 80 km N of Manaus, 02.41700°S, 60.0000°W

Fortaleza, Paraná do Urariá, 03.50000°S, 58.03333°W

Foz do Rio Castanhas, 07.58333°S, 60.33333°W

Humaitá, Escola Agrotécnica de Humaitá, 60 m, 07.55000°S, 63.06667°W

Hyutanaham [= Huitanaã], upper Purus River, 07.66667°S, 65.76667°W

Igarapé de Alvarães, right bank of Rio Solimões, 03.20000°S, 64.83000°W

Igarapé Araújo, left bank of Rio Preto, 00.06778°N, 64.59472°W

Igarapé Auará, Rio Madeira, 04.55000°S, 59.86667°W

Igarapé Grande, left bank of Rio Juruá, 06.58333°S, 69.83333°W

Igarapé Nova Empressa, left bank of Rio Juruá, ca. 06.80000°S, 70.73333°W

Igarapé Tucunaré, 00.17000°N, 63.52000°W

Ilha das Onças, left bank of Rio Negro, 01.83250°S, 61.38030°W

Ilha Paxiuba, right bank of Rio Juruá, 03.31670°S, 66.00000°W

Itacoatiara, Rio Amazonas, 03.13333°S, 58.41667°W

Jainu, right bank of Rio Juruá, 06.46667°S, 68.76670°W

João Pessoa [= Eirunepé], Rio Juruá, 06.66667°S, 69.86667°W

Lago do Baptista, Ilha de Tupinambarama, right bank of Rio Amazonas, 03.28333°S, 58.26667°W

Lago do Mapixi, east of Rio Purus, 05.71667°S, 63.90000°W

Lago Meduinim, left bank of Rio Negro, 01.79170°S, 61.38610°W

Lago do Serpa, 03.08333°S, 58.50000°W

Lago Três Unidos, Igarapé Arabidi, left band Rio Juruá, 03.26670°S, 66.21670°W

Lago Vai-Quem-Quer, right bank of Rio Juruá, 03.31670°S, 66.01670°W

Macaco, left bank of Rio Jaú, 02.09139°S, 62.13944°W

Manacapuru, north bank of Rio Solimões, 03.29972°S, 60.62056°W

Manaos [= Manaus], Rio Negro, 50 m, 03.13333°S, 60.01667°W

Manaquiri, Rio Solimões, 03.31667°S, 60.35000°W

Manaquiri, lower Rio Solimões, 03.48333°S, 60.51667°W

Manaus, 03.13333°S, 60.01667°W

Manaus, ca. 80 km N, 02.41667°S, 59.83333°W

Manaus-Itacoatiara road, Km 50, 02.66667°S, 59.91667°W

Marabitanas, 01.00000°N, 66.85000°W

Nova Empresa, left bank of Rio Juruá, 06.80000°S, 70.73333°W

Nova Jerusalém, 05.55000°S, 61.11667°W

Parintins, 02.60000°S, 56.73333°W

Paraná do Aiapuá, Rio Purus, 04.47000°S, 62.08000°W

Paraná do Manhãna, between Rio Japurá and Rio Solimões, 01.83000°S, 67.00000°W

PDBFF, 82 km N of Manaus, 02.4000°S, 59.86670°W

Penedo, right bank of Rio Juruá, 06.83333°S, 70.08333°W

Redenção, Rio Purus, 04.96667°S, 62.58333°W

Rio Apuaú, left bank of Rio Negro, 02.58333°S, 60.80000°W

Rio Cuicuriari, below São Gabriel da Cachoeira, 00.23300°S, 66.80000°W

Rio Guariba, mouth, left bank of Rio Aripuanã, 07.66667°S, 60.26667°W

Rio Iquiri [= Rio Ituxi], 07.30000°S, 64.85000°W

Rio Jaú (above mouth on right bank), 01.96500°, 61.48720°W

Rio Jaú, Rio Negro, between Cajutuba and Airão, 01.93333°S, 61.35000°W

Rio Madeira, Rosarinho, Lago Miguel, 05.87528°S, 61.39028°W

Rio Pitinga, left bank of Rio Uatumã, 02.58470°S, 57.85750°W

Rio Purus, right bank [Município de Beruri], 03.90000°S, 61.36670°W

Rio Quichito, affluent from the south of the Rio Javari, 04.36667°S, 70.03333°W

Rio Uaupés, opposite Tahuapunta, 00.61700°N, 69.10000°W

Rio Uatumã, right bank 5 km S mouth of Rio Pitinga, 01.06667°S, 59.60000°W

Sacado, right bank of Rio Juruá, 06.75000°S, 70.85000°W

Santo Antônio do Içá, left bank of Rio Içá, 03.06667°S, 67.93333°W

Santo Isidoro, Tefé, 03.40000°S, 64.63333°W

São Gabriel da Cachoeira, Rio Negro, 00.13330°S, 67.08330°W

São João, Aripuanan [= Aripuanã], 05.48333°S, 60.40000°W

Seringal Condor, left bank of Rio Juruá, 06.75000°S, 70.85000°W

Serra do Aracá, 00.90158°N, 63.43392°W

Serra Cucuhy [= Cucuí], Rio Negro, 76 m, 01.12000°N, 66.83333°W

Tabocal, 00.80000°N, 67.23333°W

Tahuapunta, Rio Uaupés, at the Colombian border, 03.61667°N, 69.10000°W

[See Aellen (1970:6) and Paynter and Traylor (1981:252) for discussions on this locality; other spellings include Tahuapunto and Tauapunto (e.g., Hershkovitz 1987:48), and Tauá.]

Tambor, left bank of Rio Jaú, 02.21670°S, 62.43330°W

Tauá, Rio Uaupés, 03.61667°N, 69.10000°W

Tunuhy [= Tunuí], São Gabriel do Cachoeira, Rio Negro, 01.38333°N, 68.15000°W

Umarituba, Rio Negro, 00.06667°N, 69.10000°W

Urucurituba, Rio Amazonas, 02.76667°S, 57.81667°W

Villa Bella da Imperatriz [= Parintins], Boca Rio Andirá, 02.60000°S, 56.73333°W

Villa Bella da Imperatriz [= Parintins], south bank of Rio Amazonas, 02.60000°S, 56.73333°W

Virgem Guajará, 04.89556°S, 59.70722°W

Yavanari, right bank of Rio Negro, 00.51700°S, 64.83300°W

Bahia (Estado da)

Abaíra, Mata do Tijuquinho, Chapada Diamantina, 1,700 m, 13.28333°S, 41.90000°W

Almada, Rio do Braço, 14.68333°S, 39.25000°W

Andaraí, Fazenda Santa Rita, 399 m, 12.80167°S, 41.26667°W

Araçás, 12.20000°S, 38.20000°W

Barra, 11.08333°S, 43.16667°W

Barreiras, 12.15000°S, 45.98333°W

Bom Jesus da Lapa, 13.25000°S, 43.41667°W

Cachoeira do Ferro Doido, Morro do Chapéu, 11.62500°S, 41.99861°W

Caetité, 14.06667°S, 42.46667°W

Caravelas, 17.75000°S, 39.25000°W

Catolés de Cima, Abaíra, 13.28333°S, 41.90000°W

Chapada Diamantina, 11.23333°S, 41.71667°W

Chapada Diamantina, Morro da Torre da TeleBahia, Lençóis, 12.56000°S, 41.40000°W

Cocos, 12.72377°S, 44.56359°W

Cocos, Fazenda Sertão Formoso, 775 m, 14.71667°S, 45.91667°W

Colônia Leopoldina (now Helvécia), 50 km SW of Caravelas, 17.80833°S, 39.66361°W

Correntina, 13.34333°S, 44.63667°W

Cravolândia, 13.36667°S, 39.85000°W

Cumuruxatiba, Prado, 17.10000°S, 39.18333°W

Curaça, 08.99194°S, 39.90806°W

Estação Ecológica do Pau Brasil, 40 m, 16.36667°S, 39.18333°W

Fazenda Aldeia, 7 km NNW of Valença, 13.42111°S, 39.06139°W

Fazenda Almada, Município de Ilhéus, 14.87100°S, 39.317003°W

Fazenda Bolandeira, 10 km S of Una, 15.35000°S, 39.00000°W

Fazenda Imbaçuaba, 30 km N of Prado, 4 m, 17.34110°S, 39.21667°W

Fazenda Jatobá, Correntina, 13.20000°S, 44.30000°W

Fazenda Jatobá-Floryl, Correntina, 13.95000°S, 45.96667°W

Fazenda Jucuruta [= Fazenda Sertão do Formoso], Jaborandi, 14.16667°S, 44.55000°W

Fazenda Juramento, Morro do Chapéu, 11.46667°S, 41.23333°W

Fazenda Piratiquicê, Ilhéus, 14.83000°S, 39.08333°W

Fazenda Salinas, Morro do Chapéu, 11.55000°S, 41.15611°W

Fazenda Ribeirão da Fortuna, 14.95000°S, 39.31667°W

Fazenda Santa Rita, 8 km E of Andaraí, 12.80167°S, 41.26139°W

Fazenda Sertão do Formoso, Jaborandi, 14.61667°S 45.83333°W

Feira de Santana, 12.26667°S, 38.96667°W

Fazenda Unacau, 8 km SE of São José, 11.10000°S, 39.26667°W

Firmino Alves, 14.98333°S, 39.93333°W

Ibipeba, 11.64083°S, 42.01111°W

Ibiraba, 10.78333°S, 42.81667°W

Ibiraba, Barra, 10.80639°S, 42.83778°W

Ilha Madre de Deus, 12.73333°S, 38.61667°W

Ilhéus, Almada, Rio do Braço, 14.68333°S, 39.25000°W

Ilhéus, 1 m, 14.78889°S, 39.04944°W

Ilhéus, 52 m, 14.78900°S, 39.03333°W

Ilhéus, Rio do Braço, 13.35000°S, 39.16667°W

Itaetê, 12.08583°S, 40.97278°W

Itambé, 15.23333°S, 41.11667°W

Itanhém, 17.16667°S, 40.33333°W

Itirussú, 13.51667°S, 40.15000°W

Jaborandi, 13.61889°S, 44.43278°W

Jacobina, Serra das Almas, 11.18545°S, 40.53608°W

Jequié, 13.85000°S, 40.08333°W

Joazeiro [= Joãzeiro], Rio São Francisco, 371 m, 09.41667°S, 40.50000°W

Lagoa de Itaparica, 11.05000°S, 42.78333°W

Lamarão, 11.79361°S, 38.88167°W

Lençóis, Chapada Diamantina, 600 m, 12.46667°S, 41.38333°W

Lençois, 13.81667°S, 41.71661°W

Macaco Seco, Monte Andaraí, Rio Paraguaçu, 12.80000°S, 41.33333°W

Manague do Caritoti, Caravelas, 17.72500°S, 39.25972°W

Mata do Nono, São Felipe, 12.83333°S, 39.08333°W

Mata Ribeirão da Fortuna, 40 km W of Ilhéus, Itabuna, 14.80667°S, 39.24872°W

Mata do Zé Leandro, 13.00000°S, 41.33333°W

Morro do Chapéu, 1,030 m, 11.59111°S, 41.20778°W

Mucugê, Parque Nacional da Chapada Diamantina, Rio Cumbuca, 19.04472°S, 41.34900°W

Mucujê [= Mucugê], 13.00000°S, 41.38333°W

Nova Viçosa, 17.89194°S, 39.37194°W

Palmeiras, Pai Inácio, 12.46317°S, 41.47467°W

Palmeiras, Chapada Diamantina, Gerais da Cachoeira da Fumaça, 1,200 m, 12.60000°S, 41.46667°W

Parque Zoobotânico CEPLAC, 6 km E of Itabuna, 14.80000°S, 39.33333°W

Pico das Almas, 13.55000°S, 41.93333°W

Ponto do Pitu, 14.78333°S, 39.08333°W

Prado, 17.06667°S, 39.23333°W

Queimadas, 10.96667°S, 39.63333°W

Queimadas, 10.38333°S, 42.50000°W

Reserva Biológica Pau Brasil [= Estação Ecológica Pau Brasil], 15 km NW of Porto Seguro, 20 m, 16.40000°S, 39.18333°W

Reserve Particular do Patrimônio Natural Jequitibá, Serra da Jibóia, 12.85000°S, 38.46667°W

Rio Mucuri, sea level, 18.08333°S, 39.56667°W

Rio Una, 10 km ESE of São José, 15.21667°S, 39.03333°W

Rio Unamirim, 14 km W of Valença, 13.28333°S, 39.21667°W

RPPN da Serra do Teimoso, 250 m, 15.15000°S, 39.51667°W

Salvador, 12.98333°S, 38.51667°W

Santo Amaro, 12.53333°S, 38.771667°W

Santo Inácio, 11.10878°S, 42.72368°W

São Felipe, 13.85000°S, 39.10000°W

São Felipe, 13.86667°S, 41.38333°W

São Gonçalo, 30 km SW of Feira de Santana, 12.41667°S, 38.96667°W

São Marcelo, junction of Rio Preto and Rio Sapão, 11.03333°S, 45.53333°W

São Marcelo, upper Rio Preto, 11.03333°S, 45.53333°W

São Marcello [= São Marcelo], Rio Preto, 11°03333S, 45.53333°W

Seabra, 12.46667°S, 41.76667°W

Senhor do Bonfim [= Villa Nova], 10.46667°S, 41.18333°W

Sento Sé, 09.73333°S, 41.88333°W

Sertão do Famoso, Jaborandi, 13.59160°S, 44.52931°W

Trancoso, 16.63333°S, 39.11667°W

Tremendal, Rio Ressaro, 13.97583°S, 41.41083°W

Três Braçõs, 37 km N and 34 km E of Jequié, 450 m, 13.53333°S, 39.75000°W

Vacaria, 10.58333°S, 42.56667°W

Valença, 13.36667°S, 39.08333°W

Vila Brasil, 15.30000°S, 39.06667°W

Vitória da Conquista, 14.85000°S, 40.85000°W

Una, 11 m, 15.29333°S, 39.07528°W

Ceará (Estado do)

Chapada do Araripe, 7 km SW of Crato, 07.27750°S, 39.45083°W

Crato, 07.23417°S, 39.40944°W

Guaraciaba do Norte, 800 m, 04.16667°S, 40.76667°W

Fortaleza, 03.71722°S, 38.54306°W

Ibiapina, Sítio Pejuaba, 03.91667°S, 40.90000°W

Ipu, 04.32222°S, 40.71083°W

Itapajé, 03.68333°S, 39.56667°W

Jaguaruana, 04.83389°S, 37.78111°W

Juá [near Iguatu], 06.36667°S, 39.30000°W

Lagoa do Catu, 03.92522°S, 38.35997°W

Pacoti, Serra de Baturité, 04.21667°S, 38.93333°W

Parque Nacional de Ubajara, Chapada de Ibiapaba, 04.00000°S, 41.00000°W

Russas, 04.94028°S, 37.97583°W

Santanópole [= Santana do Cariri], 07.18333°S, 39.73333°W

São Benedito, 895 m, 04.03333°S, 40.86667°W

São Paulo, Serra de Ibiapaba, 900 m, 04.00000°S, 41.00000°W

Serra de Baturité, Pacoti, 04.21667°S, 38.91667°W

Serra do Castelo, 03.96667°S, 38.70000°W

Serra Dantas, Jaguaruana, 04.83389°S, 37.78111°W

Serra de Maranguape, 919 m, 03.49000°S, 38.71670°W

Sítio Caiana, Crato, 07.23333°S, 39.38333°W

Sítio Camará, Milagres, 07.28333°S, 38.95000°W

Sítio Friburgo, Serra de Baturité, Pacoti, 04.21694°S, 38,89806°W

Sítio Piraguara, São Benedito, 04.05000°S, 40.833333°W

Trairussu (praia) Aguiraz, 04.22912°S, 38.38860°W

Distrito Federal

Brasília, 15.78333°S, 47.91667°W

Brasília, 20 km S, 15.96700°S, 47.91700°W

Fazenda Água Limpa, Universidade de Brasília, 15.95000°S, 45.90000°W

Fundação Zoobotânica, 15.78333°S, 47.91667°W

Matoso [= Matosa], Parque Nacional de Brasília, 1,100 m, 15.78333°S, 47.91667°W

Parque Nacional de Brasília, 15.68117°S, 47.99466°W

Parque Nacional de Brasília, Brasília, 15.47000°S, 47.56000°W

Parque Nacional de Brasília, ca. 20 km NW of Brasília, about 1,100 m, 15.63000°S, 48.05000°W

Planaltina, Águas Emendadas Ecological Station, 15.57000°S, 47.60000°W

Reserva Biológica de Águas Emendadas, Planaltina, 15.35000°S, 47.36000°W

Sgt. Silvio Delmar Hollembach Brasília Zoological Garden, Brasília, 15.93370°S, 47.88470°W

Espírito Santo (Estado do)

Águia Branca, 18.89000°S, 40.82700°W

Alto Alegre, Reserva Biológica de Duas Bocas, 550 m, 20.28000°S, 40.51000°W

Anchieta, 40.80000°S, 40.61667°W

Aracruz Florestal, Aracruz, 19.83777°S, 40.25047°W

Arrozal, Parque Nacional do Caparaó, 3.0 km N and 0.1 km W of Pico da Bandeira, 2,400 m, 20.36667°S, 41.80000°W

Barra do Jucu, Vila Velha, 20.40867°S, 40.32596°W

Cachoeiro de Itapemirim, 4 mi N of Castelinho, 50 m, 20.85000°S, 41.1000°W

Casa Queimada, Parque Nacional do Caparaó, 2,079 m, 20.46000°S, 41.81000°W

Campinho, Colatina, 19.53333°S, 40.61667°W

Cariacica, 19 m, 20.28539°S, 40.49767°W

Castelinho, 1,200 m, 20.51667°S, 40.98333°W

Castelinho, 4 km N, 20.51667°S, 40.51667°W

Castelo, 3 km NE of Forno Grande, 20.50000°S, 41.10000°W

Castelo, Reserva Biológica de Forno Grande, 20.50000°S, 41.10000°W

Coacas, Viana, 20.35917°S, 40.47278°W

Colatina, right bank of Rio Doce, 19.55000°S, 40.61667°W

Ecoporanga, 18.36667°S, 40.83333°W

Engenheiro Reeve [= Rive], 20.76667°S, 41.46667°W

Fazenda Santa Terezinha, 33 km NE of Linhares, 50 m, 19.13333°S, 39.95000°W

Floresta da Caixa d'Água, Santa Teresa, 19.93333°S, 40.61667°W

Floresta da Capela de São Braz, Santa Teresa, 19.91461°S, 40.36205°W

Hidroelétrica São Pedro, Domingos Martins, 20.40000°S, 41.01667°W

Grota, 12 km E of Aracruz, 40 m, 19.81700°S, 40.18300°W

Hotel Fazenda Monte Verde, 24 km SE of Venda Nova do Imigrante, 20.33333°S, 41.13333°W

Jirau, 18.80000°S, 40.01667°W

Lago de Sete Pontas, Itapemirim, 20.93803°S, 40.81215°W

Linhares, 19.41667°S, 40.06667°W

Nova Lombardia Biological Reserve [= Reserva Riológica Augusto Ruschi], 19.88333°S, 40.53333°W

Pancas, 19.20000°S, 40.78333°W

Parque Estadual da Fonte Grande, 1,200–2,039 m, 20.52000°S, 41.00000°W

Parque Estadual da Fonte Grande, Vitória, 1,200–2,039 m, 20.31667°S, 40.34999°W

Parque Nacional do Caparaó, 20.48333°S, 41.71667°W

Praia das Neves, Presidente Kennedy, 21.23333°S, 40.95000°W

Reserva Florestal Nova Lombardia [= Reserva Riológica Augusto Ruschi], 500 m, 19.83333°S, 40.53000°W

Reserva Florestal Nova Lombardia [= Reserva Riológica Augusto Ruschi], 1,200 m, 19.93560°S, 40.60000°W

Rio Itaúnas, 18.38333°S, 39.73333°W

Rio São José, 19.16667°S, 40.20000°W

Rive [= Engenheiro Reeve], 400–600 m, 20.76667°S, 41.46667°W

Santa Clara do Caparaó, PARNA Caparaó, vertente nordeste, 20.42400°S, 41.80810°W

Santa Teresa, 659 m, 19.91667°S, 40.60000°W

São Mateus, 18.72583°S, 39.92250°W

São Mateus, 1 km S, 18.73333°S, 39.866667°W

Vargem Alta, Hotel Fazenda Monte Verde, 20.46660°S, 41.00260°W

Venda Nova, 20.33717°S, 41.13470°W

Viana, 20.38900°S, 40.46100°W

Goiás (Estado de)

Anápolis, 1,000 m, 16.31667°S, 48.96667°W

Aporé, 19.42889°S, 50.88750°[W

Assentamento Babilônia, alto Rio Araguaia, Santa Rita do Araguaia, 17.42000°S, 53.09000°W

Baliza, 16.36672°S, 52.43611°W

Barra do Rio São Domingos, 13.40000°S, 46.31700°W

Caldas Novas, 17.45000°S, 48.63333°W

Calvacante, 13.79750°S, 47.45830°W

Campo Alegre de Goiás, 17.63889°S, 47.78194°W

Colinas do Sul, 14.15140°S, 48.07830°W

Corumbá de Goiás, 15.92360°S, 48.80860°W

Crixás, 14.53000°S, 49.96667°W

Fazenda Cadoz, Mimoso de Goiás, 15.03944°S, 48.16139°W

Fazenda Cana Brava, Nova Roma, 12.85000°S, 46.95000°W

Fazenda Canadá, 13.07111°S, 46.73889°W

Fazenda Fiandeira, Parque Nacional da Chapada dos Veadeiros, 65 km SSW of Cavalcante, 14.06667°S, 47.75000°W

Fazenda Formiga, 15.55000°S, 49.45000°W

Fazenda Santa Elena, at Rio São Mateo, about 72 km from São Domingos and 60 km from Posse (by road), 13.83333°S, 46.83333°W

Fazenda Vão dos Bois, 13.57472°S, 47.182560°W

Fazenda Vão dos Bois, Teresina de Goiás, 13.77639°S, 47.26472°W

Flores de Goiás, 14.44860°S, 47.05030°W

Ipameri, 17.73333°S, 48.61667°W

Mambaí, 14.48333°S, 46.10000°W

Mambaí, 14.48778°S, 46.26470°W

Mimoso de Goiás, 14.05610°S, 48.16140°W

Mimoso de Goiás, 15.04832°S, 48.34448°W

Minaçu, 13.53310°S, 48.22000°W

Mineiros, Emas National Park, 18.08333°S, 52.91667°W

Mineiros, Parque Nacional das Emas, 750 m, 18.25000°S, 52.88333°W

Moro da Baleia, Alto Paraíso, Parque Nacional da Chapada dos Veadeiros, 14.08333°S, 47.51667°W

Morro do Cabludo, Corumbá de Goiás, 15.91667°S, 48.80000°W

Parque Estadual da Serra de Caldas Novas, Caldas Novas, 17.47000°S, 48.40000°W

Parque Nacional da Chapada dos Veadeiros, near Pouso Alto, 14 km NNW of Alto Paraíso, 14.01667°S, 47.51667°W

Parque Nacional das Emas, 18.25000°S, 52.88333°W

Pico Alto, Parque Nacional da Chapada dos Veadeiros, 1,500 m, 14.13250°S, 47.50600°W

Posse, 14.08333°S, 46.36667°W

Rio das Almas, Jaraguá, 666 m, 15.75000°S, 49.33333°W

Rio São Miguel, 14.30000°S, 47.86667°W

São Domingo, Fazenda Cruzeiro do Sul, 13.65910°S, 46.76260°W

Serra da Mesa, 13.83417°S, 48.30444°W

Serra da Mesa, 40 km NE of Uruaçu, 14.28639°S, 48.91833°W

Serranópolis, 18.30583°S, 51.96194°W

Uruaçu, 14.52470°S, 49.14080°W

Usina Hidroelétrica Serra da Mesa, Minaçu, 13.53000°S, 48.19000°W

Usina Hidroelétrica Serra da Mesa, Rio Bagagen, 13.66667°S, 47.83333°W

Maranhão (Estado do)

Aldeia Gurupiuna, Reserva Indígena Alto Turiaçú, 01.68333°S, 45.35000°W

Alto da Alegria, 01.86000°S, 45.67000°W

Alto Parnahyba, 400–600 m, [= region along the Alto Rio Parnaíba, centered at 09.10000°S, 45.95000°W]

Alto Parnaíba [= Alto Parnahyba, region along the Alto Rio Parnaíba, centered at 09.10000°S, 45.95000°W]

Bacabal, 04.28333°S, 44.78333°W

Estiva, 09.27000°S, 46.420000°W

Estreito, right bank of Rio Tocantins, 06.86667°S, 44.30000°W

Fazenda Lagoa Nova, 04.23333°S, 44.78333°W

Fazenda Lagoa Nova, about 32 km NW of Bacabal, 04.06667°S, 44.96667°W

Fazenda Pé de Coco, Estreito, 06.55000°S, 47.45000°W

Ilha de São Luis do Maranhão, 02.60000°S, 44.23333°W

Miritiba [= Humberto de Campos], 02.61667°S, 43.45000°W

Palmeiral, Cantanhede, 03.63333°S, 44.36667°W

Palmeiral, Matões, 03.66667°S, 44.45000°W

Pedra Chata, Rio Gurupi, Carutapera, 01.18333°S, 46.01667°W

Ribeirãozinho, Imperatriz, 05.53333°S, 47.48333°W

São Bento, 02.70000°S, 44.83333°W

UTE Ponte Madeira, São Luis, 02.60000°S, 44.23333°W

Vargem Grande, 03.50000°S, 43.91667°W

Mato Grosso (Estado do)

Acurizal, 17.75000°S, 57.61667°W

Alta Floresta, 09.88333°S, 56.46667°W

Apiacás, 220 m, 09.56667°S, 57.38333°W

Aripuanã, 09.98330°S, 59.30000°W

Aripuanã, 09.56667°S, 59.45000°W

Aripuanã, Alta do Rio Madeira, 105 m, 10.16667°S, 59.46667°W

Aripuanã, Humboldt Laboratorie, Rio Roosevelt, 09.16700°S, 60.63300°W

Aripuanan [= Aripuanã], 09.16667°S, 60.63333°W

Barão de Melgaço, 16.19444°S, 55.96750°W

Cáceres, 16.07060°S, 57.67400°W

Cachoeira Dardanelos, right bank of Rio Aripuanã, 10.16667°S, 59.45000°W

Caicara [= Caiçara], 16.06700°S, 57.71700°W

Caiê-Malu, Chapada dos Parecis, 11.20000°S, 59.00000°W

Campo Novo de Parecis, 13.67500°S, 57.89194°W

Campos Novos, Serra do Norte, Chapada dos Parecis, ca. 440 m, 12.83333°S, 59.75000°W

Casa de Pedra, Chapada dos Guimarães, 15.41667°S, 55.83333°W

Chapada dos Guimarães [= Serra da Chapada], 15.43333°S, 55.75000°W

Chapada dos Guimarães [= Serra da Chapada], Casa de Pedra, 15.43333°S, 55.75000°W

Cidade Laboratório de Humboldt, Aripuanã, 09.16667°S, 60.63333°W

Cocalinho, 14.38330°S, 50.98330°W

Cláudia, 11.58333°S, 55.13333°W

Cuiabá, 15.60445°S, 56.09714°W

Cuiabá, 15.58333°S, 56.08333°W

Estação Ecológica Serra das Araras, 300–800 m, 15.65000°S, 57.21667°W

Estação Ecológica Serra das Araras, Porto Estrela, 15.39000°S, 57.13000°W

Expedition base camp, 260 km N of Xavantina, Serra do Roncador, ca. 368 m, 12.85000°S, 51.76667°W

Fazenda Noirumbá, 34 km NW of Ribeirão Cascalheira, 279 m, 12.63333°S, 51.93333°W

Fazenda São Luís, 30 km N of Barra do Garças, 15.63330°S, 52.35580°W

Feliz Natal, 12.38583°S, 54.92000°W

Gaúcha do Norte, 13.18333°S, 53.25670°W

Jacaré, alto Rio Xingu, 12.00000°S, 53.40000°W

Juruena, 300 m, 10.30583°S, 58.48944°W

Mato Grosso [= Vila Bela da Santíssima Trindade], 15.08333°S, 59.95000°W

Palmeiras, 15.91667°S, 55.46667°W

Peixoto de Azevedo, 10.22889°S, 54.98278°W

Ponte Branca, 16.65000°S, 53.78333°W

Porto do Braço das Lontras, Rio Comemoração de Floriano, alto Rio Jamari, 11.96722°S, 62.28639°W

Porto Estrela, Estação Ecológica Serra das Araras, 300–800 m, 15.65000°S, 57.21667°W

Posto Jacaré, 12.00000°S, 53.40000°W

Posto Leonardo, 12.18333°S, 53.36667°W

Reserva Ecológica Cristalino, 40 km N of Alta Floresta, 09.559694°S, 55.92833°W

Reserva Particular do Patrimônio Natural do Serviço do Comércio, 16.71719°S, 56.18631°W

Ribeirão Cascalheira, Serra do Roncador, 12.85000°S, 51.76667°W

Rio Saueniná, 12.40000°S, 58.6667°W

Rodovia 7 Placas, Cabeceira do Rio Cuiabá, 14.23333°S, 55.03333°W

Santa Anna de Chapada [= Chapada dos Guimarãos], 800 m, 15.43300°S, 55.75000°W

São Domingos, Rio das Mortes, 13.50000°S, 51.400090°W

São João, cabeceiras do Rio Aripuanã, 12.01667°S, 59.43333°W

São José do Xingu, 10.80440°S, 52.74412°W

São Luiz de Cáceres, 16.06667°S, 57.68333°W

São Manoel [= Barra do São Manoel], Rio Teles Pires, 07.33333°S, 58.03333°W

Serra da Chapada [= Chapada dos Guimarães], 15.43333°S, 55.75000°W

Serra do Roncador, 264 km N of Xavantina, 12.81667°S, 51.76667°W

Taiamã Island, Cáceres, 16.49694°S, 56.42139°W

Tapirapoã [= Tapirapuã], on Sepotuba River, 14.84728°S, 57.76588°W

Tapirapoan [= Tapirapuã], 280 m, 14.85000°S, 57.75000°W

Tucum, 16.46667°S, 57.80000°W

UHE Manso, 200 m, 14.70000°S, 56.26667°W

Usina Hidroelétrica Manso, 100 km N of Cuiabá, 14.62269°S, 55.73522°W

Utiariti, Rio Papagaio, 13.03333°S, 58.28333°W

Via Rica, 09.90778°S, 51.20361°W

Villa Maria, NNW of Cáceres, 16.05000°S, 57.70000°W

Xavantina, 260 km N (by road) of Ribeirão Cascalheira, Serra do Roncador, 12.51000°S, 51.46000°W

Xavantina, 264 km N, Serra do Roncador, 200 m, 12.82000°S, 51.76667°W

Xavantina, 264 km N, Serra do Roncador Base Camp, 200 m, 12.82000°S, 51.76667°W

Mato Grosso do Sul (Estado do)

Aquidauana, 20.47083°S, 55.78694°W

Brasilândia, 21.25000°S, 52.03300°W

Cassilândia, 19.11278°S, 51.73339°W

Corumbá, 92 m, 19.00917°S, 57.65333°W

Corumbá, 10 km NE of Urucum, ca. 18.93329°S, 57.60761°W

Corumbá, 7 km WSW of Urucum, São Marcus road, 19.02513°S, 57.68058°W

Fazenda Alegria, Corumbá, 19.25028°S, 57.02472°W

Fazenda Bonito, Bonito, 21.10000°S, 56.27000°W

Fazenda Califórnia, Morraria do Sul, Bodoquena, 520 m, 20.70000°S, 56.86670°W

Fazenda Maringá, 54 km W of Dourados, 22.21670°S, 54.8000°W

Fazenda Primavera, Bataiporã, 334 m, 22.33333°S, 53.28333°W

Fazenda Princesinha, Bonito, Parque Nacional da Serra da Bodoquena, 550 m, 21.08333°S, 57.58333°W

Fazenda Rio Negro, Aquidauna, 20.47111°S, 55.78722°W

Gruta São Miguel, 21.10000°S, 56.56667°W

Ivinheima [= Ivinhema], 22.30472°S, 53.81528°W

Maracaju, 500 m, 21.63333°S, 55.15000°W

Municipalities of Dois Irmãos de Buirití and Terenos, ca. 20.50000°S, 53.30000°W

Nova Alvorada do Sul, 21.45000°S, 54.45000°W

Porto Faya, 18.46667°S, 57.36667°W

Rio Sucuriú, 20.78333°S, 51.63333°W

Salobra, 125 m, 20.16667°S, 56.51667°W

Sidrolândia, 20.93194°S, 54.96083°W

Urucum [de Corumbá], 18 km SSE of Corumbá, 19.15000°S, 57.63333°W

Urucum [de Corumbá], near Corumbá, 19.15000°S, 57.63333°W

Minas Gerais (Estado de)

Além Paraíba, 140 m, 21.86667°S, 42.68333°W

Almenara, 16.058583°S, 40.65722°W

Alto Caparaó, PARNA Caparaó, vertente SE, 20.43360°S, 41.86777°W

Alto da Consulta, Poços de Caldas, 21.79999°S, 46.56667°W

Araguari, Rio Jordão, 700–900 m, 18.63333°S, 48.18333°W

Arrenegado, about 46 km N of Macacos, 1,140 m, 17.56167°S, 43.74000°W

Bandeira, 15.88333°S, 40.56667°W

Barra do Paraopeba, Rio Extrema, 510 m, 18.83333°S, 45.17333°W

Boca da Mata, 771 m, 19.02861°S, 43.41667°W

Boca da Mata, km 104–105 in the road from Logoa Santa to Conceição do Mato Dentro, 1,200 m, 19.16667°S, 43.55000°W

Bocaiuva, 17.11667°S, 43.83333°W

Brejo da Lapa, 22.35000°S, 44.73000°W

Brumadinho, Rio Manso, 20.14333°S, 44.19972°W

Campos Gerais de San Felipe, east of Januária, 15.41667°S, 42.80000°W

Caparaó National Park, 20.31667°S, 41.71667°W

Capitão Andrade, 19.07639°S, 41.87500°W

Caratinga, 19.43000°S, 42.06000°W

Caratinga, Fazenda Montes Claros, 19.83333°S, 41.83333°W

Casa Alpina Hotel, 1,550 m, 22.36417°S, 44.80806°W

Conceição do Mato Dentro, 19.02000°S, 43.25000°W

Conquista, Usina Hidroelétrica de Igarapava, 20.00000°S, 47.58000°W

Coronel Murta, Ponte do Colatino, left bank of Rio Jequitinhonnha, 322 m, 16.60000°S, 42.20000°W

Diamantina, Conselheiro Mata, 18.24556°S, 43.90056°W

Estação Ecológica de Acauã, 17 km N of Turmalina, 17.13333°S, 42.76667°W

Estação Biológica de Caratinga, Caratinga, 19.66670°S, 41.83333°W

Estação Bológica da Mata do Sossego, Simonésia, 20.16667°S, 42.00000°W

Estação Ecológica da Caratinga, Caratinga, 19.72922°S, 42.56104°W

Estação de Pesquisa e Desenvolvimento Ambiental de Peti, 630–806 m, 19.90000°S, 43.37000°W

Famoso, Parque Nacional Grande Sertão Veredas, 15.26670°S, 45.866670°W

Fazenda do Bené, 4 km SE of Passa Vinte, 22.23333°S, 44.20000°W

Fazenda Cafundó, 13 km NE of Nova Ponte, 19.13100°S, 47.73900°W

Fazenda Canoas, 36 km NE and 12 km W of Montes Claros, Juramento, 16.84806°S, 43.58694°W

Fazenda Cauaia, Matozinhos, 19.56083°S, 44.05139°W

Fazenda Chico Cera, Parque Nacional da Serra da Canastra, São Roque de Minas, 20.21000°S, 46.15000°W

Fazenda Esmeralda, 30 km E and 4 km N (by road) Rio Casca, 19.21667°S, 43.05139°W

Fazenda Esperança, Serro, 940 m, 18.61667°S, 42.38333°W

Fazenda do Itaguaré, Passa Quatro, 22.39030°S, 44.96670°W

Fazenda do Itaguaré, 16 km SW of Passa Quatro, 1,500 m, 22.47000°S, 45.08000°W

Fazenda Jaguara, 19.46667°S, 43.96667°W

Fazenda Lapa Vermelha, Pedro Leopoldo, 19.37500°S, 44.00000°W

Fazenda Montes Claros, 19.83333°S, 41.83333°W

Fazenda Neblina, Parque Estadual da Serra do Brigadeiro, 20 km W of Fervedouro, 1,300 m, 20.71667°S, 42.48333°W

Fazenda da Onça, 13 km SW of Delfim Moreira, 22.60000°S, 45.33333°W

Fazenda Paiol, Floresta Alta, Virgem da Lapa, 16.83333°S, 42.21670°W

Fazenda Santa Cruz, Felixlândia, 18.77056°S, 45.14278°W

Fazenda Triángulo Formoso, Buritizeiros, 17.35000°S, 44.96667°W

Indianápolis, Usina Hidroelétrica de Miranda, 18.91222°S, 48.04139°W

Irapé, 16.78333°S, 42.66667°W

Itamonte, 22.28390°S, 44.87000°W

Itamonte, Brejo da Lapa, 22.35440°S, 44.73170°W

Itapetininga, 647 m, 23.60000°S, 48.05000°W

Jacutinga, 22.28746°S, 46.61354°W

Jaíba, 15.33833°S, 43.67444°W

Jambreiro, Nova Lima, 19.98333°S, 43.83333°W

Jequitinhonha, 16.43333°S, 41.00000°W

Jordânia, 15.90000°S, 40.18333°W

Juiz de Fora, 800 m, 21.68333°S, 43.45000°W

Juramento, 16.84778°S, 43.58694°W

Lagoa Santa, near Rio das Velhas, 19.62694°S, 43.88333°W

Lagoa Santa, 760 m, 19.62694°S, 43.88333°W

Lagoa Santa, 104 km N, 18.61667°S, 43.88333°W

Lapa do Capão Seco, 760 m, 19.62720°S, 43.88333°W

Lapa da Escrivânia Nr. 5, ca. 1963333°S, 43.88333°W

Lapa da Serra das Abelhas, ca. 19.55365°S, 43.987245°W

Lassance, 17.83333°S, 44.56667°S

Macacos, 3.25 km NW by road, Parque Nacional Sempre Vivas, 1,251 m, 17.96389°S, 43.78833°W

Mata do Banco, 20.94111°S, 42.10778°W

Mata do Dr. Daniel, Conceição do Mato Dentro, 19.21667°S, 43.50000°W

Mata do Paraíso, Viçosa, 20.80500°S, 42.85889°W

Mata do Paraíso, Viçosa, 20.75000°S, 42.86667°W

Mata do Paraíso, 20.08333°S, 42.05000°W

Mata do Praúno, 5 km N (by road), Conceição do Mato Dentro, 690 m, 18.98583°S, 43.42306°W

Mata da Prefeitura, 6 km SW of Viçosa, 20.78333°S, 42.91667°W

Mata do Sossego, 20.11667°S, 42.01667°W

Mocambinho, Jaíba, 15.10000°S, 44.04999°W

Mocambinho, Manga, 15.33333°S, 43.66667°W

Mocambinho, 15.85000°S, 43.05000°W

Ouro Branco, 20.51709°S, 43.70005°W

Ouro Preto, 1,061 m, 20.38333°S, 42.88333°W

Parque Estadual do Rio Doce, 19.50000°S, 42.51667°W

Parque Estadual do Rio Preto, 15 km S of São Gonçalo do Rio Preto, 18.15000°S, 43.38333°W

Parque Estadual da Serra do Brigadeiro, 20.68333°S, 42.46667°W

Parque Estadual da Serra do Cipó, 20.01778°S, 43.61556°W

Parque Estadual da Serra do Ouro Branco, 20.51079°S, 43.66325°W

Parque Estadual da Serra do Papagaio, 22.15000°S, 44.71667°W

Parque Nacional do Caparaó, 20.46667°S, 41.80000°W

Parque Nacional Grande Sertão Veredas, 700–800 m, 15.26667°S, 42.86667°W

Parque Nacional Grande Sertão Veredas, 15.16000°S, 44.14000°W

Parque Nacional da Serra da Canastra, 20.15496°S, 46.65375°W

Parque Particular do Patrimônio Natural Santuário do Caraça, 25 km SW of Santa Bárbara, 1,300 m, 20.08333°S, 43.50000°W

Peirópolis, 19.73333°S, 47.75000°W

Pico da Bandeira, Parque Nacional do Caparaó, 2,890 m, 20.43393°S, 41.79706°W

Pirapitanga, 20.13300°S, 43.20000°W

Pirapora, 489 m, 17.33570°S, 44.89810°W

Poços de Caldas, 21.78256°S, 46.56375°W

Posses, 13 km SE, 22.38333°S, 44.85000°W

Prados, 21.05000°S, 44.06667°W

Reserva do Jacob, Nova Ponte, 19.13333°S, 47.68333°W

Reserva Particular do Partimônio Natural do Caraça, 25 km SW of Santa Bárbara, 1,300 m, 20.08000°S, 43.50000°W

Riacho da Cruz, 15.33333°S, 44.30000°W

Riacho Mocambinho, Jaíba, 15.10000°S, 44.05000°W

Rio Doce State Forestry Park, 19.80000°S, 42.63333°W

Rio das Velhas, near Lagoa Santa, 17.21667°S, 44.81667°W

Salinas, 16.16667°S, 42.28333°W

Santa Bárbara, 19.96306°S, 42.65667°W

São Roque de Minas, Serra da Canastra, 20.23696°S, 46.36776°W

Serra Azul (COPASA, Mateus Leme), 20.08000°S, 44.43000°W

Serra das Cabeças, Parque Estadual da Serra do Brigadeiro, 20.68778°S, 42.47111°W

Serra Caparaó, Fazenda Cardoso, 2 leagues NE of Caparaó, 1,020 m, 20.51667°S, 41.90000°W

Serra do Caraça Private Reserve, 20.10194°S, 43.49306°W

Serra do Cipó, 19.23333°S, 43.55000°W

Serra do Papagaio, Itamonte, 22.14167°S, 44.72500°W

Sítio Maglândia, Simão Pereira, 610 m, 21.96361°S, 43.31194°W

Sumidouro, ca. 12 km SW of Lagoa Santa, 19.54111°S, 43.94111°W

Viçosa, 650 m, 20.75390°S, 42.88190°W

Viçosa, Mata do Paraíso, 650 m, 20.75000°S, 42.88333°W

Pará (Estado do)

Agrovila União, 18 km S and 19 km W of Altamira, Rio Xingu, 100 m, 04.68333°S, 52.38333°W

Altamira, 52 km SSW, right bank of Rio Xingu, 03.65000°S, 52.36667°W

Altamira, 54 km S and 150 km W, 200 m, 04.0000°S, 52.75000°W

Alter do Chão, right bank of Rio Tapajós, 02.50361°S, 54.95250°W

Amorim [= Igarapé Amorim], Rio Tapajós, 02.53333°S, 55.78333°W

Aramanaí, Rio Tapajós, 02.75000°S, 55.18333°W

Arampucu, Rio Parú do Oeste [= Leste?], 01.13333°N, 54.61667°W

Baião, Rio Tocantins, 02.68333°S, 49.58333°W

Barreirihhas, Rio Tapajós, 04.41667°S, 56.21667°W

Belém, 01.45000°S, 48.48333°W

Belém, Bosque, 01.45000°S, 48.48333°W

Belém, Utinga, sea level, 01.43000°S, 48.42000°W [Wooded area around waterworks on eastern edge of Belém.]

Boim, Rio Tapajós, 02.81667°S, 55.16667°W

BR 165 [Brazil Highway 165], Km 217, Santarém to Cuiabá, 04.00000°S, 54.66700°W

BR 165, Santarém to Cuiabá, Santarém, 100 m, 02.65000°S, 54.56667°W

Bragança, Santa Maria, Tracuateua [= Tracuatena], 01.08333°S, 46.90000°W

Cachoeira da Porteira, Rio Trombetas, 01.08333°S, 57.03333°W

Cametá, west bank of Rio Tocantins, 02.25000°S, 49.50000°W

Campos de Arimba, Igarapé Jaramacaru, 01.16700°S, 55.90000°W

Capim, 01.68333°S, 47.87333°W

Casa Nova, Rio Arapiuns, 02.33333°S, 55.05000°W

Castanhal, Monte Dourado, Rio Jari, 00.68333°S, 52.81667°W

Castelo dos Sonhos, 08.13333°S, 54.93333°W

Castelo dos Sonhos, Altamira, 03.20000°S, 52.20000°W

Caxiuanã, 01.73333°S, 51.38333°W

Curralinho, Marajó, 01.81361°S, 49.79528°W

Curuá-Una, 02.40000°S, 54.08333°W

Cussary, 01.91667°S, 53.66667°W

Faro [= Fazenda Paraíso], lower Rio Nhamudá, 02.18333°S, 56.73333°W

Fazenda Eco-Búfalos, Ilha de Marajó, 00.25000°S, 48.83333°W

Fazenda Gavinha, 00.71667°S, 48.52333°W

Fazenda Paraíso, lower Rio Nhamudá, 02.18333°S, 56.73333°W

Fazenda Recreio, Ilha Caviana, 00.16667°S, 50.16667°W

Fazenda Santana, Ilha Mexiana, Marajó Archipelago, 00.00375°S, 49.58333°W

Fazenda São Raimundo, km 42 (estreito), São João do Araguaia, Marabá, 77 m, 05.38333°S, 48.76667°W

Fazenda São Pedro, Monte Alegre, 02.00778°S, 54.06917°W

Flexal, Itaituba-Jacareacanga, km 212, left bank of Rio Tapajós, 05.75000°S, 57.38333°W

Floresta Estadual de Trombetas, near Rio Curuá, 00.96277°S, 55.52230°W

Floresta Nacional Tapirapé-Aquiri, 05.80140°S, 50.51500°W

Fordlândia, Rio Tapajós, 03.66667°S, 55.50000°W

Foz do Rio Curuá, 02.40000°S, 54.08333°W

Foz do Río Curuá, right bank of Rio Amazonas, 02.55000°S, 54.08333°W

Gonotire, Rio Fresco at confluence with Rio Xingu, 06.65000°S, 51.98300°W

Gorotire, Rio Xingu, 07.78333°S, 51.13333°W

Gradaús, right bank of Rio Fresco, 07.71667°S, 51.15000°W

Igarapé Açú, left bank of Rio Tapajós, 03.73333°S, 55.51667°W

Igarapé Amorim, Rio Tapajós, 02.53333°S, 55.78333°W

Igarapé Amorin [= Igarapé Amorim], Rio Tapajós, 02.53333°S, 55.78333°W

Igarapé Assu [= Igarapé Açú], 01.11667°S, 47.61667°W

Igarapé Assu [= Igarapé Açú], 03.73333°S, 55.51667°W

Igarapé Brabo, left bank of Rio Tapajós, 02.43333°S, 55.00000°W

Ilha de Arapiranga, 01.33333°S, 48.56667°W

Ilha de Marajó, 01.00000°S, 49.50000°W

Ilha Mexiana, 00.03333°S, 49.58333°W

Ilha do Taiuna [= Ilha do Tayaúna], Rio Tocantins, 02.25000°S, 49.50000°W

Ilha Tocantins, Tucurí, Rio Tocantins, 04.41667°S, 49.53333°W

Insel Mexiana [= Ilha Mexiana], 00.03333°S, 49.58333°W

Iroçanga, Rio Tapajós, 02.50000°S, 55.16667°W

Iroçanga, opposite, left bank of Rio Tapajós, 00.25000°S, 55.16667°W

Itaituba, Rio Tapajós, 04.28333°S, 55.98333°W

Itaituba, 19 km S, 04.28333°S, 55.98333°W

Jacareacanga, Rio Tapajós, 06.21667°S, 57.75000°W

Jacareacanga, Tapajós, Transamazônica Pará-Itaituba, Km 19, 04.48333°S, 56.28333°W

Jatobal [= Puerto Jatobal], [near], 73 km N and 45 km W of Marabá, 100 m, 04.68333°S, 49.53333°W

Lago, Cuiteua, north bank of Rio Amazonas, 01.81667°S, 54.96667°W

Marabá, Serra do Norte, 05.35000°S, 49.11667°W

Marajó, 01.00000°S, 49.50000°W

Marissu, 01.52333°S, 52.58167°W

Mata do Sequeirinho, Distrito de Sossego, Parauapebas, 06.44330°S, 50.06880°W

Mocajuba, 02.58333°S, 49.50000°W

Monte Alegre, Rio Gurupatuba, 10 m, 02.01667°S, 54.06667°W

Monte Cristo, Juçarabena, Rio Tapajós, 04.08333°S, 55.65000°W

Monte Dourado, 00.86667°S, 52.52833°W

Muaná, 01.53333°S, 49.21667°W

Óbidos, north bank of Rio Amazonas, 01.91667°S, 55.51667°W

Oriximiná, Porto Trombetas, Rio Saracazinho, 01.70000°S, 56.38333°W

Paraná do Sumaúma, mouth of Rio Tocantins, 01.85000°S, 49.00000°W

Parauapebas, Fazenda São Luiz, 06.18333°S, 50.65000°W

Parque Nacional da Amazônia, 00.45000°S, 56.25000°W

Peixe-Boi, 01.19222°S, 47.31389°W

Pinkaití Research Station, Kayapó Indigenous Area, 07.77056°S, 51.96194°W

Piquiatuba, Rio Tapajós, 25 m, 03.05000°S, 55.11667°W

Portel, Rio Procupi, 01.95000°S, 50.83333°W

Programa de Assentamento Benfica I, Itupiranga, 05.17000°S, 49.36000°W

Project Pinkaití Research Station, Kayapó Indigenous Reserve, 07.68333°S, 51.86667°W

Puerto Jatobal, 73 km N and 45 km W of Maraba, 04.68333°S, 49.53333°W,

Reserva Biológica de Trombetas, Oriximiná, 01.40000°S, 56.47000°W

Rio Arapiuns, Santarém, left bank of Rio Tapajós, 02.36667°S, 55.06667°W

Rio Bracjá, mouth, right bank of Rio Xingu, 03.41667°S, 51.83333°W

Rio Curuá-Tinga, affluent of Rio Curuá-Una, 03.80000°S, 54.35000°W

Rio Jamanxim, right bank of Rio Tapajós, 04.76667°S, 56.41667°W

Rio Riosinho, left bank of Rio Fresco, 07.11667°S, 51.65000°W

Santa Júlia, Rio Iriri [= Ilha de Santa Júlia], 03.91667°S, 53.11667°W

Santa Rosa, Rio Jamanxim, 05.33300°S, 55.28300°W

Santarém, Rio Tapajós, 02.43333°S, 54.70000°W

São Domingos do Capim, 01.68333°S, 47.78333°W

Serra dos Carajás, platô N4, 06.08750°S, 50.12028°W

Serra do Cachimbo, 08.95000°S, 54.90000°W

Serra Norte, 05.35000°S, 49.11667°W

Serra do Tumucumaque, 02.50000°N, 60.71667°W

Sítio Calandrinho, Tucuruí, Rio Tocantins, 03.76611°S, 49.67250°W

Soure, 00.69610°S, 48.51910°W

Soure, Ilha de Marajó, 01.00000°S, 49.50000°W

Sumaúma, left bank of Rio Tapajós, 03.58333°S, 55.36667°W

Tauari, Rio Tapajós, 03.08333°S, 55.10000°W

Tauary [= Tauari], Rio Tapajós, 03.08333°S, 55.10000°W

Urucurituba, Rio Tapajós, 03.50000°S, 55.50000°W

Urumajó [= Vila Urumajá], Arabóia [= mata do Arabóia], east of Bragança, Estrada de Ferro de Bragança, 01.05000°S, 46.65000°W

Utinga, 10 m, 01.48930°S, 48.51360°W

Vila Brabo, right bank of Rio Tocantins, 05.10000°S, 49.35000°W

Vila Braga, Rio Tapajós, 04.41667°S, 56.28333°W

Vilarinho do Monte, Rio Xingu, 01.61667°S, 52.01667°W

Vista Alegre, Praia do Rio Camará, 00.70000°S, 47.71667°W

Paraíba (Estado da)

Areia, Mata do Pau-Ferro, 06.97000°S, 35.70000°W

Fazenda Alagamar, 9 km S and 6 km E of Mamanguape, 06.83000°S, 35.12000°W

Fazenda Pacatuba, Sapé, Corredor São João, 07.03333°S, 35.15000°W

Fazenda Pacatuba, 10 km NE of Sapé, 07.20000°S, 35.15.0000°W

João Pessoa, 07.11667°S, 34.86667°W

Mamanguape, 06.83861°S, 35.12611°W

Mamanguape, Camaratuba, 06.65000°S, 35.13333°W

Mata do Pau Ferro, 6 km de Areia, 06.96667°S, 35.68333°W

Mulungu, Lagoa do Monteiro, 07.00000°S, 35.33333°W

Pirauá, 07.51860°S, 35.50300°W

Pombal, 06.72028°S, 37.80167°W

Reserva Biológica Guaribas, 06.73333°S, 35.15000°W

Rio Cuiá, bairro Valentina, 07.19170°S, 34.84243°W

Salgado São Felix, 07.35611°S, 35.44056°W

Paraná (Estado do)

Castro, 980 m, 24.78333°S, 50.00000°W
Curitiba, 25.43333°S, 49.26667°W
Estação Ecológica do Canguiri, Piraquara, 905 m, 25.45000°S, 49.06667°W
Estação Ecológica do Canguiri, Piraquara, 25.44167°S, 49.06333°W
Estação de Roça Nova, Município de Piraquara, 25.41667°S, 49.08333°W
Fazenda Monte Alegre, 885 m, 24.20000°S, 50.55000°W
Fazenda Panagro, 25.91667°S, 49.40000°W
Fênix, 23.90000°S, 51.95000°W
Foz do Iguaçu, 200 m, 25.65000°S, 54.46667°W
Guajuvira, 25.60000°S, 49.53333°W
Ilha Rasa, Guaraqueçaba, 25.34042°S, 48.409789°W
Jaguariaíva, 850 m, 24.25000°S, 49.70000°W
Mananciais da Serra, Piraquara, 25.44667°S, 49.06333°W
Mangueirinha, 26.91667°S, 52.18333°W
Monte Alegre, 24.28333°S, 50.60000°W
Nova Tirol, 25.46667°S, 49.05000°W
Ortigueira, 24.20000°S, 50.95000°W
Palmira, 25.70000°S, 50.15000°W
Parque Barigüi, Bairro Mercês, Curitiba, 25.41556°S, 49.30083°W
Piraquara, Mananciais da Serra, 25.44167°S, 49.06333°W
Ponta Grossa, 25.09500°S, 50.16190°W
Represa de Guaricana, 25.47694°S, 48.83444°W
Rio Paracaí, 23.68333°S, 53.95000°W
Rio Paranapanema, Fazenda Coioá, Cambará, 455 m, 22.95000°S, 49.98333°W
Roça Nova, 25.46667°S, 48.96667°W
Roça Nova, Serra do Mar, 25.47194°S, 49.01389°W
Salto Caxias Dam, lower Rio Iguaçu, 25.50000°S, 53.50000°W
Telêmaco Borba, 24.10000°S, 50.51667°W
Tijucas do Sul, 25.95000°S, 49.23333°W
Usina Hidroelétrica de Guaricana, Morretes, 400 m, 25.66667°S, 48.91667°W
Usina Hidroelétrica de Salta Caixas, 25.41670°S, 53.18330°W

Pernambuco (Estado de)

Agrestina, 08.45000°S, 35.93333°W
Bodocó, 07.77833°S, 39.94111°W
Bom Conselho, 09.16972°S, 36.67972°W
Bonito, 08.48333°S, 35.73333°W
Buíque, Parque Nacional do Catimbau, 972 m, 08.58161°S, 37.24144°W
Buíque, Sítio Mata Verde, 08.61667°S, 37.15000°W
Caruaru, 08.28240°S, 35.97490°W

Correntes, 09.12890°S, 36.33030°W
Dois Irmãos, 08.05000°S, 35.90000°W
Exu, 07.51667°S, 39.71667°W
Garanhuns, 853 m, 08.90000°S, 36.48333°W
Garanhuns, Sítio Cavaquinho, 08.90000°S, 36.48333°W
Igarassu, 07.83417°S, 34.90639°W
Macaparana, 07.55470°S, 35.45310°W
Macaparana, Fazenda Água Fria, 45 m, 07.56667°S, 35.45000°W
Mata de Dois Irmãos, 08.00000°S, 34.93333°W
Mata do Macuco, Sítio Bituri, Brejo da Madre de Deus, 762 m, 08.14583°S, 36.37111°W
Mata Tauá, Usina Trapiche, 08.52948°S, 35.16919°W
Mata Xanguá, 08.65000°S, 35.16667°W
Parque Nacional do Catimbau, 08.52361°S, 37.24014°W
Pernambuco [= Recife], 08.05000°S, 34.90000°W
Pesqueira, 08.35778°S, 36.69639°W
Recife, 08.05000°S, 34.90000°W
Recife, 40 km W, 08.05000°S, 34.90000°W
Saltinho, Rio Formoso, 08.75000°S, 35.10000°W
São Lorenço, 28–60 m, 08.00000°S, 35.05000°W
São Lorenço da Mata, 45 m, 08.00352°S, 35.02143°W
São Vicente Ferrer, 07.58333°S, 35.48333°W
Sítio Cavaquinho, Garanhuns, 08.90000°S, 36.48333°W
Vila Feira Nova, 07.51667°S, 35.55738°W

Piauí (Estado do)

Coronel José Dias, 08.81583°S, 42.51278°W
Estação Ecológica de Uruçuí-Una, 08.86667°S, 44.96667°W
Estação Ecológica de Uruçuí-Una, 325 m, 08.85000°S, 45.01667°W
Fazenda Felicidade, Coronel José Dias, 08.81639°S, 42.51250°W
João Costa, 08.05148°S, 42.42026°W
Lagoa Alegre, 04.50000°S, 42.61667°W
São João do Piauí, 08.35806°S, 42.24667°W

Rio de Janeiro (Estado do)

Abraão, Ilha Grande, 23.13528°S, 44.17472°W
Angra dos Reis, Terra Indígena Guarani Sapukai, 22.83333°S, 44.38333°W
Angra dos Reis, sea level, 23.00000°S, 44.30000°W
Barro Branco, Duque de Caxias, 22.62028°S, 43.26361°W
Bemposta, 22.15000°S, 43.11667°W
Cabiúnas, Macaé, 22.28616°S, 41.73680°W
Cambuci, 21.48417°S, 41.87278°W
Campos do Itatiaia, Parque Nacional do Itatiaia, 22.39056°S, 44.67056°W
Carapebus, 9 m, 22.27500°S, 41.66889°W
Casimiro de Abreu, 109 m, 23.48056°S, 42.20417°W

Casimiro de Abreu, Reserva Biológica de Poço das Antas, 22.58333°S, 42.28333°W

Caxias [= Duque de Caxias], 22.47080°S, 43.18420°W

Centro de Primatologia, Magé, 22.58333°S, 42.83333°W

Fazenda Alpina, Teresópolis, 22.41667°S, 42.83333°W

Fazenda Boa Fé, 902 m, 22.43000°S, 42.98000°W

Fazenda Boa Fé, Teresópolis, 22.35750°S, 42.90139°W

Fazenda Carlos Guinle, Estrada Rio-Teresópolis, 22.45000°S, 42.95611°W

Fazenda Comari, Teresópolis, 22.43333°S, 42.98333°W

Fazenda Consorciadas, 22.58000°S, 42.91000°W

Fazenda Marimbondo, 1,570 m, 22.36069°S, 44.56667°W

Fazenda Nova Friburgo, 22.31222°S, 42.51472°W

Fazenda Santo Antônio da Aliança, Valença, 22.37167°S, 43.78972°W

Fazenda São José da Serra, Bonsucesso, Sumidouro, 22.22528°S, 42.70806°W

Fazenda São Pedro, 21.30000°S, 41.11667°W

Fazenda Velha, Tijuca, 300 m, 22.90000°S, 43.23333°W

Fazenda do Tenente, São João de Marcos, 22.76667°S, 44.01667°W

Fazenda União, Casimiro de Abreu, 20 m, 22.48333°S, 42.20000°W

Gávea, 22.96667°S, 43.23333°W

Guapimirim, Garrafão, 700 m, 27.47444°S, 42.98333°W

Ilha Grande, 23.15000°S, 44.23333°W

Itaquaí, 22.85220°S, 43.77530°W

Itatiaya [= Pico das Agulhas Negras], 1,455 m, 22.38333°S, 44.63333°W

Macaé de Cima, Pirineus, 22.43000°S, 42.51670°W

Mambucaba, Angra dos Reis, 23.02139°S, 44.51722°W

Mangaratiba, 22.92000°S, 44.10900°W

Maricá, 5 m, 22.96140°S, 42.86090°W

Mata do Carvão, São Francisco do Itabapoana, 21.39218°S, 41.09196°W

Mata do Mamede, 23.00000°S, 44.32000°W

Mata da Rifa, Parque Estadual do Desengano, 21.87861°S, 41.90389°W

Morro do Itaoca, Campos dos Goytacazes, 21.78899°S, 41.43665°W

Nova Friburgo, 22.16550°S, 42.31520°W

Nova Friburgo, Córrego Grande, 26 km NE of Socavão, 1,080 m, 22.28333°S, 42.68333°W

Parque Estadual dos Três Picos, 700 m, 22.40000°S, 42.58333°W

Parque Nacional do Itatiaia, 22.44000°S, 44.62000°W

Parque Nacional do Itatiaia, Maromba, 22.33306°S, 44.62000°W

Parque Nacional da Restinga de Jurubatiba, Carapebus, 22.25000°S, 41.65000°W

Parque Nacional da Serra dos Órgãos, base da Pedra do Sino, Campo das Antas, 22.45944°S, 43.02778°W

Praia Vermelha, Ilha Grande, 23.116560°S, 44.36000°W

Quissamã, 22.14671°S, 41.44043°W

Reserva Biológica de Poço das Antas, Silva Jardim, 22.53330°S, 42.28330°W

Reserva Biológica União, Rio das Ostras, 22.43531°S, 42.03283°W

Restinga da Barra de Maricá, Maricá, 22.95713°S, 42.88719°W

Restinga do Farolzinho, Farol de São Tomé, Campos dos Goytacazes, 22.00000°S, 40.98333°W

Restinga de Iquipari-Grussaí, Grussaí, São João da Barra, 21.74472°S, 41.04028°W

Rio Bonito, 22.66667°S, 42.61667°W

Rio de Janeiro, sea level, 22.90000°S, 43.23333°W

Sepetiba, 22.96667°S, 43.70000°W

Serra de Macaé, 22.31667°S, 42.33333°W

Silva Jardim, 22.65083°S, 42.39167°W

Sítio Santana, Beira de Lagoa, Quissamã, 22.07056°S, 41.35361°W

Sítio Xitaca, Debossan, Nova Friburgo, 22.43333°S, 42.53333°W

Socavão, near Teresópolis, 22.43333°S, 42.90000°W

Teresópolis, 22.43333°S, 42.98333°W

Teresópolis, 902 m, 22.26667°S, 42.98333°W

Teresópolis, Vale das Antas, 22.45000°S, 44.03330°W

Teresópolis, Vieira, 22.21833°S, 42.71652°W

Tijuca, Trapicheiro, 22.93333°S, 43.28333°W

Vale do Paraíba, Itatiaia, 22.38333°S, 44.6333°W

Venâncio Aires, 29.58333°S, 52.20000°W

Rio Grande do Sul (Estado do)

Aceguá, 8 km E, 31.86300°S, 54.08300°W

Alegrete, 29.78333°S, 55.76667°W

Aratiba, 27.40718°S, 52.29625°W

Bagé, 31.31667°S, 54.10000°W

Banhado do Pontal, Triunfo, 29.93333°S, 51.71667°W

Beco dos Cegos, 5 m, 30.23793°S, 51.09527°W

BR 176 [Brazil Highway 176], Manoel Viana, 29.39300°S, 55.42942°W

Butiá, 30.11972°S, 51.96222°W

Cachoeira do Sul, 30.03917°S, 52.89389°W

Cambará do Sul, 1,000 m, 29.20000°S, 50.25000°W

Cambará do Sul, Parque Nacional da Serra Geral, 29.06973°S, 49.99473°W

Cambará do Sul, Parque Nacional da Serra Geral, 29.06972°S, 49.99472°W

Candiota, 31.55806°S, 53.67250°W

Capão do Leão, Mostardas, 31.10000°S, 50.92000°W

Capão Novo, 4 m, 29.67750°S, 49.96889°W

Charqueadas, 29.95000°S, 51.61667°W

Chico Lomã, 52 m, 29.90444°S, 50.50428°W

Cidreira, 30.18111°S, 50.20556°W

Colônia do Mundo Novo [= Colônia de Santa Maria do Mundo Novo], 29.60000°S, 50.76667°W

Dom Pedrito, 30.98278°S, 54.67306°W

Eldorado do Suy, Charqueadas, 50 m, 29.98333°S, 51.63333°W

Esmeralda, 28.05360°S, 51.19030°W

Estação Ecológica do Taim, sea level, 32.65000°S, 52.60000°W

Faxinal, 30.30000°S, 51.68333°W

Faxinalzinho, 27.33664°S, 52.66969°W

Fazenda Aldo Pinto, São Nicolau, 28.13333°S, 55.26667°W

Fazenda Caçapava, 32.86667°S, 52.53333°W

Fazenda Marcelina, 30.36700°S, 50.38700°W

General Câmara, 29.90500°S, 51.76028°W

Guaíba, 30.10000°S, 51.31667°W

Hermenegildo, 33.65000°S, 53.28333°W

Itapeva, Torres, 29.45083°S, 49.92722°W

Itaqui, 28.12528°S, 56.55306°W

Lagoa do Peixe, 31.30000°S, 51.00000°W

Marinheiros Island, 32.02499°S, 52.17218°W

Morro do Osório, 29.88333°S, 50.26667°W

Mostardas, 31.10694°S, 50.92111°W

Osório, 29.88667°S, 50.26972°W

Osório, Morro Osório, 29.90000°S, 50.26667°W

Palmares do Sul, 13 m, 30.25778°S, 50.50972°W

Palmital, 131 m, 29.66667°S, 52.93333°W

Parque Estadual de Itapuã, 30.82440°S, 51.32480°W

Parque Nacional dos Aparados da Serra, Cambará do Sul, 29.25000°S, 48.83333°W

Parque Estadual do Turvo, 100–400 m, 27.13000°S, 53.80000°W

Parque de Itapuã [= Parque Estadual de Itapuã], 170 m, 30.36123°S, 51.02602°W

Passinhos, 135 m, 30.03000°S, 50.38861°W

Passo do Narciso, Maçambará, 29.12628°S, 55.39989°W

Pelotas, 31.77194°S, 52.34250°W

Pinheiros, Candelária, 29.78333°S, 52.73333°W

Piraju, Manoel Viana, 29.40331°S, 55.58028°W

Pontal do Morro Alto, 29.77080°S, 50.18750°W

Porto Alegre, 30.07917°S, 51.12500°W

Praia do Cassino, 32.20900°S, 52.18700°W

Quaraí, 30.38750°S, 56.45139°W

Quintão, 29.66667°S, 50.20000°W

Reserva Ecológica do Taim, 32.66667°S, 52.58333°W

Rio Grande, 32.03500°S, 52.09861°W

Rio Uruguay, south bank, Município de Aratiba, 27.39427°S, 52.30028°W

Rondinha, Parque Estadual Florestal de Rondinha, 28.73882°S, 50.27794°W

Rota do Sol, Tainhas—Terra de Areia, 29.43333°S, 50.15000°W

Santa Maria, 29.68417°S, 53.80694°W

Sapiranga, Alto Ferrabraz, Morro Ferrabraz, 28.89833°S, 49.54916°W

São Borja, 28.61667°S, 56.01667°W

São Francisco de Paula, 900 m, 29.44806°S, 50.58361°W

São João do Monte Negro, 29.70000°S, 51.46667°W

São Lourenço do Sul, 31.36667°S, 51.96667°W

São Lourenço [= São Lourenço do Sul], sea level, 31.36667°S, 51.96667°W

Sapiranga, 29.63810°S, 51.00690°W

Taim, 32.49028°S, 52.58083°W

Tapes, 30.40240°S, 51.23450°W

Taquara [= Taquara do Mundo Novo], 29.64250°S, 50.79535°W

Tavares, 16 m, 31.28722°S, 51.09361°W

Torres, 29.33450°S, 49.72690°W

Torres, 29.33528°S, 49.72694°W

Torres, Faxinal Norte Lagoa Itapeva, 29.50000°S, 49.91667°W

Tramandaí, 29.98472°S, 50.13361°W

Três Barras, margins of Rio Uruguai, Aratiba, 27.32750°S, 52.23083°W

Tupanciretã, 28.93333°S, 53.66667°W

Uruguaiana, 29.75472°S, 57.08333°W

Usina Hidroelétrica de Itá, right margin of Rio Uruguay, 400 m, 27.33333°S, 52.23333°W

Venâncio Aires, 29.58333°S, 52.20000°W

Victor Graeff, 411 m, 28.56667°S, 52.75000°W

Rondônia (Estado de)

Abunã, left bank of Rio Madeira, 09.53494°S, 65.34383°W

Alto Paraíso, 130 m, 09.71667°S, 63.31667°W

Alvarada d'Oeste, BR 429 [Brazil Highway 429], Km 87, 11.34139°S, 62.28639°W

Barão de Melgaço, Rio da Comemoração, upper Rio Ji-Paraná, 350 m, 11.86700°S, 60.71700°W

Cachoeira Nazaré, left bank of Rio Ji-Paraná, 10.85000°S, 61.93333°W

Cachoeira Nazaré, left bank of Rio Ji-Paraná, 100 m, 09.75000°S, 61.91667°W

Calama, 08.05000°S, 62.88333°W

Campos dos Palmares de Maria de Molina, 12.12000°S, 60.48222°W

Jirau, right bank of Rio Madeira, 40 km SE of Jaci-Paraná, Porto Velho, 09.33444°S, 64.72722°W

João Bonifácio, 499 m, 12.16667°S, 60.08333°W

Ouro Preto d'Oeste, Rio Paraíso, 10.71667°S, 62.2333°W

Pimenta Bueno, 11.65000°S, 61.20000°W

Pista Nova, 8 km N of Porto Velho, 08.76667°S, 63.90000°W

Porto Velho, Rio Madeira, 08.76667°S, 63.90000°W

Porto Velho, 49 km E, 08.75000°S, 63.50000°W

Rio Roosevelt, left bank of upper Rio Aripuanã, 12.01667°S, 60.25000°W

Samuel, 08.75000°S, 63.46667°W

Usina Hidroelétrica Samuel, Rio Jamari, 08.75000°S, 63.46667°W

Vilhena, 12.73343°S, 60.14368°W

Vilhena, Fazenda Planalto, 600 m, 12.53333°S, 60.36667°W

Roraima (Estado de)

BR 174 [Brazil Highway 175], frontier between Brazil and Venezuela, frontier mark 8, 04.47000°N, 62.08000°W

Caracaraí, Parque Nacional do Viruá, 01.81611°N, 61.12806°W

Colônia do Apiaú, margem direita do Igarapé Serrinha, Rio Mucajaí, 02.63333°N, 61.20000°W

Ilha de Maracá, 03.36667°N, 61.43333°W

Parque Nacional do Viruá, 01.48335°N, 61.03339°W

Rio Branco, Conceição, 02.18333°N, 60.96667°W

Rio Catrimani, 00.47000°N, 61.73000°W

Rio Cotinga, Limão, 03.93333°N, 60.50000°W

Rio Mucajaí, lower, south of Boa Vista, 02.42000°N, 60.87000°W

Rio Uraricoera, 03.03300°N, 60.50000°W

Serra da Lua, 02.85000°N, 60.71667°W

Sunumu, 04.16667°N, 60.50000°W

Taparinha [= Lago do Cobra, Caracaraí], 01.81611°N, 61.12806°W

Santa Catarina (Estado de)

Alta da Boa Vista, 1,200 m, 27.70000°S, 49.15000°W

Baragem do rio São Bento, 28.60000°S, 49.55000°W

Caldas da Imperatriz, 27.73298°S, 48.81274°W

Caldas da Imperatriz, Parque Estadual Serra do Tabuleiro, 27.83333°S, 48.78333°W

Colônia Hansa, 26.43333°S, 49.23333°W

Corupá, 26.43333°S, 49.23333°W

Fazenda Gateados, 920–1,000 m, 27.96667°S, 50.81667°W

Florianópolis, 27.58333°S, 48.56667°W

Florianópolis, Rio Tavares, CASAN, Rodovia Estadual SC 405, 27.65000°S, 48.50000°W

Hansa [= Corupá], 26.43333°S, 49.23333°W

Ilha de Santa Catarina, 27.60000°S, 48.50000°W

Ilhas Moleques do Sul, 27.85000°S, 48.43333°W

Itá, 27.35000°S, 52.31667°W

Itapiranga, 27.10100°S, 53.42440°W

Jaborá, 27.17580°S, 51.73360°W

Jaguaruna, 20 m, 28.61500°S, 49.02556°W

Joinville, 26.30000°S, 48.83333°W

Mono, Parque Natural Municipal Nascentes do Garcia, 27.01667°S, 49.23333°W

Morro dos Conventos, 28.95000°S, 49.38333°W

Parque Estadual da Serra do Tabuleiro, 27.73333°S, 48.81667°W

Palhoça, 27.63333°S, 48.66667°W

Paulo Lopes, 27.93333°S, 48.70000°W

Pinheiros, Anitápolis, 520 m, 27.90000°S, 49.13333°W

Rancho Queimado, 27.66667°S, 49.016667°W

Reserva Biológica Estadual do Sassafrás, 950 m, 26.70000°S, 49.66667°W

Santo Amaro da Imperatriz, 27.41170°S, 48.46430°W

Teresópolis, 27.83330°S, 48.66700°W

Urubici, Morro da Igreja, Parque Nacional de São Joaquim, 28.12778°S, 49.47083°W

São Paulo (Estado de)

Alto da Serra [= Paranapiacaba, Serra do Mar], 800 m, 23.78333°S, 46.31667°W

Americana, 22.74028°S, 47.33444°W

Araraquara, 21.79440°S, 48.02560°W

Ariri, Cananéia, 25.21472°S, 48.04250°W

Atibaia, 23.11673°S, 46.55640°W

Avanhandava, Rio Tietê, 21.46667°S, 49.95000°W

Bananal, Reserva Ecológica do Bananal, 23.78333°S, 46.30000°W

Barra do Guaraú, Cananéia, 52 m, 24.73333°S, 48.05000°W

Barra do Icapara, 24.68333°S, 47.43333°W

Barra do Ribeirão Onça Parda, 24.31667°S, 47.85000°W

Biritiba Mirim, 23.95000°S, 46.05000°W

Bonito, 24.25000°S, 48.16667°W

Boracéia, 23.63333°S, 45.86667°W

Butantan, 23.53333°S, 46.61667°W

Caçapava, 23.10000°S, 45.71667°W

Capão Bonito, 24.18250°S, 48.23420°W

Campinas, 22.90083°S, 47.05722°W

Campininha, Campinas, 22.52000°S, 47.04000°W

Caraguatatuba, 23.60000°S, 45.40000°W

Casa Grande, 23.43300°S, 46.93300°W

Caucaia do Alto, 100–1,100 m, 23.66670°S, 47.01667°W

Cotia, 23.61667°S, 46.93333°W

Cotia, Reserva Morro Grande, 23.48333°S, 46.78333°W

Cristo, 800–1,000 m, 23.85000°S, 47.47000°W

Estação Biológica do Bananal, 22.80000°S. 44.36667°W

Estação Biológica de Boracéia, 850 m, 23.65000°S, 45.86667°W

Estação Biológica Jataí, 21.55000°S, 47.75000°W

Estação Biológica de Santa Bárbara, 22.80000°S, 49.23333°W

Estação Ecológica do Bananal, Bananal, 22.80000°S, 44.36000°W

Estação Experimental de Ubatuba, 23.41667°S, 45.16667°W

Estação de Ferro Noroeste, Braúna, 19.67014°S, 50.30002°W

Fazenda Banaura, left margin of Rio Branco, 24.21472°S, 46.78889°W

Fazenda Califórnia, Serra do Bananal, 22.83333°S, 44.46667°W

Fazenda Intervales, 24.33333°S, 48.41667°W

Fazenda Intervales, Capão Bonito, 24.25000°S, 48.16667°W

Fazenda Intervales, Base do Carmo, 24.33333°S, 48.41667°W

Fazenda Intervales (base do Carmo), Capão Bonito, 700 m, 24.33333°S, 48.41667°W

Fazenda João XXIII, 689 m, 23.94361°S, 47.66667°W

Fazenda Passes, Santo Antônio do Aracanguá, 20.93670°S, 50.49560°W

Fazenda Sete Lagoas, Mogi Guaçu, 22.21667°S, 47.06667°W

Floresta Nacional de Ipanema, 20 km NW of Sorocaba, 23.43528°S, 47.62806°W

Guarapiranga Reservoir, Interlagos, 23.71667°S, 46.70000°W

Guaratuba, 23.75000°S, 45.91667°W

Iguapé, sea level, 24.70810°S, 47.55530°W

Ilha do Cardoso, 25.13333°S, 47.96667°W

Ilha do Cardoso, Cananéia, 23.83333°S, 47.96667°W

Ilha do Mar Virado, 23.56667°S, 45.16667°W

Ilha de São Sebastião, 23.81667°S, 45.33333°W

Interlagos, São Paulo, 23.71667°S, 46.70000°W

Ipanema [= Bacaetava], 23.43330°S, 23.43330°W

Iporanga, 24.58556°S, 48.59306°W

Iporanga, Parque Estadual Turístico do Alto Ribeira (PE-TAR), 24.56666°S, 48.633330°W

Itapetininga, 23.55972°S, 48.05310°W

Itapetininga, 647 m, 23.60000°S, 48.05000°W

Itapura, 20.66667°S, 51.51667°W

Itararé, 755 m, 24.11250°S, 49.33167°W

Juréia, 24.50000°S, 47.20000°W

Lins, Campestre, 21.66667°S, 49.75000°W

Morretinho, 24.31667°S, 47.80000°W

Mulheres, 800–1,000 m, 24.05000°S, 48.37000°W

Museros, 800–1,000 m, 24.22000°S, 48.40000°W

Paranapiacaba, 820 m, 23.78333°S, 46.30000°W

Parque Estadual da Cantareira, 23.36667°S, 46.60000°W

Parque Estadual Intervales, Barra Grande, ca. 900 m, 24.28333°S, 48.35000°W

Parque Estadual Intervales, Base do Carmo, 24.33333°S, 48.41667°W

Parque Estadual Morro Grande, Caucaia do Alto, 23.68333°S, 46.96667°W

Parque Estadual da Serra do Mar, Núcleo Santa Virgínia, 10 km NW of Ubatuba, 23.35833°S, 45.12500°W

Parque Estadual Turístico do Alto Ribeira, 81 m, 24.58356°S, 48.58333°W

Parque Nacional da Bocaina, 22.83333°S, 44.68333°W

Pedreira, 22.74190°S, 46.90140°W

Picinguaba, Ubatuba, 23.35556°S, 44.86556°W

Pilar do Sul, 23.81667°S, 47.68333°W

Piquete, 22.60000°S, 45.18333°W

Porto Cabral, Rio Paraná, 22.30000°S, 52.63333°W

Primeiro Morro, 24.36667°S, 47.81667°W

Posto Indígena Icatu, Braúna, 400 m, 22.5000°S, 50.31667°W

Presidente Epitácio, left bank of Rio Paraná, 21.76667°S, 52.11667°W

Primeiro Morro, 24.36667°S, 47.81667°W

Quadro Penteado, 24.38333°S, 47.91667°W

Quilombo, Reserva Florestal do Morro Grande, 800–1,000 m, 23.76000°S, 47.00000°W

Reserva Florestal do Morro Grande, 900 m, 23.65000°S, 47.01667°W

Riacho Grande, 777 m, 23.80000°S, 46.58000°W

Ribeirão Grande, 24.16667°S, 48.35000°W

Ribeirão Preto, campus Universidade de São Paulo, 21.18333°S, 47.80000°W

Rio Feio [= Rio Aguapel], 22.01667°S, 49.65000°W

Rio Guaratuba, Santos, 23.78333°S, 45.91667°W

Salto Grande, 22.90000°S, 49.98333°W

Salto de Pirapora, Bairro da Ilha, 23.71667°, 47.61667°W

Salto de Pirapora, 500 m, 23.60000°S, 47.56667°W

Santos, 23.96667°S, 46.31667°W

São José da Boa Vista, 21.96667°S, 46.78333°W

São José do Rio Preto, 20.80000°S, 49.38333°W

São Luís do Paraitinga, 900 m, 23.22000°S, 45.31000°W

São Sebastião, 23.80000°S, 45.41667°W

São Sebastião, Ilha de São Sebastião, 23.83333°S, 45.33333°W

Serra da Bocaina, 22.70000°S, 44.63333°W

Serra da Fartura, 21.89350°S, 46.75300°W

Serra da Juréia, 24.53333°S, 47.25000°W

Tapiraí, 24.58330°S, 47.80000°W

Teodoro Sampaio, 400 m, 22.51667°S, 52.16667°W

Ubatuba, 23.41667°S, 45.11667°W

Ubatuba, Núcleo Picinguaba, Parque Estadual da Serra do Mar, 23.33333°S, 44.85000°W

UHE de Ourinhos, Rio Ribeira [on border with Paraná state], 501 m, 23.15083°S, 49.96944°W

Vanuire, 21.78333°S, 50.78333°W

Varjão, Bertioga, sea level, 23.75000°S, 45.91667°W

Varjão do Guaratuba, Bertioga, 23.71667°S, 45.88333°W

Ypancma [– Bacaetava], 950 m, 23.43722°S, 47.38333°W

Sergipe (Estado de)

Arauá, 11.23333°S, 37.61667°W

Canindé de São Francisco, 09.64222°S, 37.78861°W

Cristinápolis, 11.47528°S, 37.75556°W

Estância, south of, 11.30000°S, 37.44000°W

Fazenda Capivara, 7 km SE of Brejo Grande, 10.48333°S, 36.43333°W

Fazenda Cedro, near Estância, outside Crasto, 11.33694°S, 37.41472°W

Fazenda Cruzeiro, 13 km SSE of Cristinápolis, 80 m, 11.58333°S, 37.71667°W

Nossa Senhora da Glória, 10.21833°S, 37.42028°W

Pacatuba, 1 km W, 10.50000°S, 37.50000°W

Penedo, 2 km SW, ca. 10.30717°S, 36.58685°W

Tocantins (Estado do)

Aliança do Tocantins, 11.30000°S, 48.93333°W

Araguatins, 05.63333°S, 48.11667°W

Área de Proteção Ambiental de Jalapão, 481 m, 10.60000°S, 46.70000°W

Buriti, 05.31583°S, 48.22889°W

Cana Brava, 12.78333°S, 46.86667°W

Dianópolis, 11.63333°S, 46.83333°W

Fazenda Lago Verde, 180 m, 10.86919°S, 49.69781°W

Fazenda Osara, 05.28333°S, 48.30000°W

Formoso de Araguaia, 11.78333°S, 49.75000°W

Jalapão, 10.05000°S, 46.96667°W

Lagoa da Confusão, 10.79360°S, 49.62361°W

Lajeado, 09.75000°S, 48.35000°W

Paranã, Fazenda São João, 12.91389°S, 47.62000°W

Parque Nacional do Araguaia, 10.45000°S, 50.48333°W

Peixe, 240 m, 12.25000°S, 48.26667°W

Pium, Parque Estadual do Cantão, 09.30200°S, 49.96000°W

Ponte Alta do Tocantins, Estação Ecológica Serra Geral do Tocantins, 533 m, 10.66667°S, 46.86667°W

Porto Nacional, 10.70000°S, 48.41667°W

Rio Santa Teresa, 20 km NW of Peixe, 11.84278°S, 48.63556°W

Rio Sono, 09.35989°S, 47.87533°W

Caribbean Netherlands [= formerly Netherlands Antilles]

Aruba

Sero Blanco, 12.51667°N, 70.01667°W

Curaçao

Klein Santa Maria [= Santa Martabaai], 12.28333°N, 69.13333°W

Chile

Aysén (Región de)

Chacabuco valley, 100 km SSW of Puerto Ibáñez, 47.25000°S, 72.66667°W

Chile Chico, 2 km S and Chile Chico Aeródromo, 1 km W, 46.55000°S, 71.93333°W

Chile Chico, 20 km S, 46.53333°S, 71.66667°W

Chonos Islands [= Archipiélago de los Chonos], 45.00000°S, 74.00000°W

Coyhaique [also spelled Coihaique; = Estancia Coyhaique], 45.56667°S, 72.06667°W

Coyhaique, 3 km N, Reserva Nacional Coyhaique, Laguna Venus, 750 m, 45.53014°S, 72.03328°W

Coyhaique Alto, 1 km E, 45.46667°S, 71.58333°W

El Manzano, 47.15000°S, 72.65000°W

Estancia Chacabuco, Cochrane, 47.26667°S, 72.55000°W

Fin del Fiordo Ofhidro, 48.40021°S, 73.83390°W

Fundo Los Flamencos, 4.5 km SE of Coyhaique Alto, 730 m, 45.41281°S, 71.56044°W

General Carrera, 4 km W, Pto. Ibáñez, 46.30000°S, 71.93333°W

Isla Gunther, 43.93333°S, 73.70000°W

Lago Atravesado, 44.70000°S, 72.28333°W

Melinka, Isla Guaitecas, 43.89654°S, 73.74420°W

Osorno, Paso Puyehue, 40.60000°S, 71.83333°W

Parque Nacional Queulat, 44.56667°S, 72.45000°W

Puerto Aisén, 45.40000°S, 72.70000°W

Puerto Chacahuco, ca. 15 km E, NW end of Lago Reisco, 45.45958°S, 72.72785°W

Puerto Ibáñez, 46.26667°S, 71.96667°W

Puerto Ingeniero Ibáñez, 0.25 km W, 46.27100°S, 71.96800°W

Reserva Nacional Coyhaique, 3 km N of Coyhaique, 45.53905°S, 72.06100°W

Río Aysén [= Aisén], 45.41667°S, 72.80000°W

Río Ñirehuao, 45.23333°S, 71.73333°W

Sector El Manzano, 47.15000°S, 72.65000°W

Antofagasta (Región de)

Aguas Calientes, 23.91667°S, 67.66667°W

Calama, 2,300 m, 22.46667°S, 68.93333°W

Copacoya [= Cerro Copacoya], vicinity, 22.31667°S, 68.03333°W

El Laco, 56 km SE of Socaire, 4,400 m, 23.97607°S, 67.49035°W

Llullaillaco, 24.72273°S, 68.53748°S

Ojos de San Pedro, 55 mi NE of Calama, 12,500 ft, 21.96667°S, 68.33333°W

Paposo, 150 m, 25.01667°S, 70.46667°W

Paposo, vicinity, 648 m, 25.00719°S, 70.451367°W

Pingo-Pingo [= Cerro Pingo-Pingo], 3,305 m, 23.96667°S, 68.46667°W

Salar de Atacama, 2,261 m, 23.50000°S, 68.250000°W

San Pedro de Atacama, 22.91667°S, 68.21667°W

San Pedro de Atacama, 5 mi S, 2,400 m, 22.98933°S, 68.21667°W

San Pedro de Atacama, 20 mi E, 12,000 ft, 21.95000°S, 68.56667°W

San Pedro, 20 mi E, 18.49771°S, 69.99990°W

Talabre, 23.36667°S, 67.85000°W

Tilopozo, 2,275 m, 23.78333°S, 68.25000°W

Toconao, 40 km SE, 23.46000°S, 67.76000°W

Toconce, 60 mi ENE of Calama, 4,240 m, 22.26667°S, 68.18333°W

Araucanía (Region de la)

Angol, 27 km WNW, Parque Nacional Nahuelbuta, 37.81667°S, 72.98333°W

Baños Río Blanco, 4 km W, 38.56667°S, 71.61256°W

Comuna de Victoria, cuenca del Río Quino, near Curacautín, 38.41000°S, 72.04000°W

Comuy, 39.06670°S, 73.00000°W

Galvarino, 38.40000°S, 72.78333°W

Los Álamos, 7 km NNW, 37.55840°S, 73.49701°W

Nahuelbuta, Nahuelbuta National Park, 38.66667°S, 73.16667°W

Parque Nacional Nahuelbuta, 37.81667°S, 72.98333°W

Paso de Las Raíces, 58 km W of Curacautín, 38.43333°S, 71.38333°W

Paso Pino Hachado, 1.7 km W, 38.66667°S, 70.95287°W

Paso Pino Hachado, 3.5 km W, 38.63333°S, 70.93333°W

Paso Pino Hachado, 9.4 km W, 1,555 m, 38.61667°S, 70.98333°W

Pedregoso, 38.53333°S, 71.18333°W

Quetropillán, Villarrica National Park, 39.45000°S, 81.8000°W

Reigolil Pass, 39.11667°S, 71.48333°W

Reserva Nacional Malalcahuello, 38.61361°S, 71.54222°W

Río Colorado, 38.41667°S, 71.50000°W

Río Colorado, Lonquimay, 38.56667°S, 71.76667°W

Temuco, 38.75102°S, 72.66571°W

Temuco, 11 km SE, 38.80340°S, 72.51054°W

Termas de San Luis, 39.36667°S, 71.66667°W

Tirúa, 2 km NE, 38.32059°S, 73.48383°W

Tolhuaca, Tolhuaca National Park, 38.21667°S, 71.81667°W

Villa Ranquil, 11 km N, 38.11757°S, 71.18333°W

Villarica, 39.45000°S, 71.80000°W

Arica y Parinacota (Región de, recently split from Tarapacá)

Arica, 3,650 m, 18.41667°S, 69.55000°W

Caldera, 10 mi N, 50 m, 27.52144°S, 70.83333°W

Caritaya, 75 mi SE of Arica, 3,890 m, 19.00000°S, 69.41667°W

Choquelimpie, 114 km NE of Arica, 4,615 m, 18.28333°S, 69.31667°W

Chungará, 18.25000°S, 69.16667°W

Cuesta Zapahuira, 25 km SSW of Putre, 18.26667°S, 69.58333°W

Guallatiri, 18.48300°S, 69.16700°W

Parinacota, 4,500 m, 18.20000°S, 69.26667°W

Putre, 3,500 m, 18.20000°S, 69.58333°W

Putre, 1 km W, 3,500 m, 18.20000°S, 69.65000°W

Putre, 6 mi E, 4,200 m, 18.16000°S, 69.47500°W

Quebrada de Camarones, 19.18333°S, 70.26667°W

Timar, 2,750 m, 18.75000°S, 69.70000°W

Valle de Azapa, 40 km SE of Arica, 18.48333°S, 70.23333°W

Vallenar, 28.56667°S, 70.75000°W

Atacama (Región de)

Altamira, coastal mountains of Vllenar, 26.45000°S, 70.31667°W

Bahía Salado, 55 km S of Caldera, ca. 27.66369°S, 70.94060°W

Caldera, 27.06667°S, 70.08333°W

Copiapó, 45 km S, ca. 27.74955°S, 70.48043°W

Domeyko, 28.95000°S, 70.90000°W

Laguna Santa Rosa, 27.08333°S, 69.16667°W

Leoncitos [= Vegas del Leoncito], 26.53333°S, 68.95000°W

Piedra Colgada, 27.30000°S, 70.48333°W

Playa Rodillo, 1.5 km N of Caldera, ca. 27.05608°S, 70.80842°W

Quebrada de Pajonales, 22 km N of Caldera, 27.05000°S, 70.66667°W

Quebrada Piedras Lindas, Parque Nacional Nevado Tres Cruces, 27.47333°S. 69.00255°W

Ramadilla [= Ramadillas], 27.31667°S, 70.55000°W

Salar de Pedernales, 26.25000°S, 69.16667°W

Santa Rosa, 27.08333°S, 69.16667°W

Vallenar, 28.58333°S, 70.76667°W

Vallenar, 40 km E, 28.56667°S, 70.34118°W

Biobío (Region de)

Aguas Calientes, 5 km E of Termas Chillán, 2,000 m, 36.90000°S, 71.38333°W

Aserradero, 7 km below Termas Chillán, 36.90000°S, 71.41667°W

Chillán, 36.90000°S, 71.41667°W

Concepción, 36.83333°S, 73.05000°W

Isla Moche [= Isla Mocha], 38.36667°S, 73.91667°W

Laguna del Laja, Río Petronquincs, 37.35000°S, 71.30000°W

Las Quilas, Termas de Chillán, 36.90000°S, 71.51667°W

Moche Island [= Isla Mocha], 38.36667°S, 73.91667°W

Quirihue, Los Remates, 36.26667°S, 72.53333°W

Quirihue, 36.28333°S, 72.53333°W

Recinto, 13 km E, 36.80000°S, 71.73333°W

Reserva Nacional Los Queules, 35.98333°S, 72.68333°W

Sierra de Nahuelbuta, 38.01667°S, 73.21667°W

Sierra de Nahuelbuta, 37.80000°S, 73.06667°W

Termas Chillán, 7 km below, 1,250 m, 36.90000°S, 71.41667°W

Tomé, 36.61667°S, 72.95000°W

Club Andino Alemán, vicinity of refugio, on the W side of the Chillán Volcano, 36.90227°S, 71.41707°W

Coquimbo (Región de)

Cerro Potrerillos, 4 km E of Guanaqueros, 20 m, 30.13333°S, 70.95000°W

Conchali, 31.88333°S, 71.48333°W

Coquimbo, 29.96667°S, 71.35000°W

Huetelauquén, 31.55000°S, 71.50000°W

Illapel, 10.5 km E (between Illapel and Salamanca), 31.63333°S, 71.16667°W

Illapel, 15 km N of Aucó, near Reserva Nacional Las Chinchillas, 31.63333°S, 71.16667°W

La Higuera, 29.45898°S, 71.35603°W

La Higuera, comuna, Sector Playa Los Choros, 15 m, 29.23333°S, 71.30000°W

La Serena, 29.81667°S, 71.26667°W

Las Palmas, 95 km N of Los Vilos, 31.26667°S, 71.56667°W

Paihuano [= Paiguano], 1,000 m, 30.01667°S, 70.53333°W

Parque Nacional Fray Jorge, 30.66667°S, 71.66667°W

Parque Nacional Fray Jorge, 275 m, 30.65000°S, 71.63333°W

Parque Nacional Fray Jorge, Quebrada de Las Vacas, 30.66667°S, 71.66667°W

Pte. Los Molles, 10 km N, 32.16667°S, 71.45000°W

Romero, 29.83333°S, 71.11667°W

Vicuña, 30.03194°S, 70.70806°W

Los Lagos (Región de)

Anticura, Puyehue National Park, 40.65000°S, 72.16667°W

Bahía Mansa, 2 km S, 40.55000°S, 73.76667°W

Bahía San Pedro, 49.90000°S, 73.88330°W

Chepu, 15 km E, Isla de Chiloé, 42.08333°S, 73.83667°W

Cuesta El Moraga, 43.30000°S, 72.45000°W

Isla Grande de Chiloé, Río Inio, 43.14390°S, 73.83226°W

La Picada, 41.10000°S, 72.50000°W

Maicolpue, 2 km S of Bahía Mansa, 40.55000°S, 73.76667°W

Parque Nacional Puyehue, 40.75000°S, 72.15000°W

Petrohué, Lago Todos Santos, 41.13333°S, 72.41667°W

Peulla, Parque Nacional Vicente Pérez Rosales, 41.10000°S, 7203333°W

Puerto Montt, 41.46667°S, 72.95000°W

Puyehue, Parque Nacional Puyehue, 40.70000°S 72.13333°W

Quellón, Chiloé Island [= Isla Grande de Chiloé], 43.35000°S, 74.11667°W

Quellón, Isla Grande de Chiloé, 43.13600°S, 73.66300°W

Quellón, Isla Chiloé, 43.11667°S, 73.61667°W

Río Inio, mouth, Isla de Chiloé, 43.30400°S, 74.14250°W

Valle de La Picada, 41.03333°S, 72.50000°W

Los Ríos (Región de)

Costa Río Caunahue, 40.40500°S, 72.25417°W

Fundo San Juan, near La Unión, Valdivia, 40.29407°S, 73.09268°W

Fundo San Martín, San José, Valdivia, 39.63333°S, 73.11667°W

Fundo Santa Rosa, 39.80000°S, 73.23330°W

Mehuín, 39.43333°S, 73.16667°W

Puerto Fuy, 6 km S, 39.92071°S, 71.90000°W

Valdivia, 39.80000°S, 73.24167°W

Magallanes y Antártica Chilena (Región de)

Bahía Felipe, 52.85000°S, 69.88333°W

Bahía Morris, head, Isla Capitán Aracena, 60 m, 54.23333°S, 71.50000°W

Bahía Parry, 54.61667°S, 69.36667°W

Caleta Lientur, Isla Wollaston, 55.73000°S, 67.32000°W

Caleta Toledo, Isla Deceit, 55.81667°S, 67.10000°W

Cerro Cóndor, top, 5 km W of Lapataia, 54.81667°S, 68.63333°W

Cockle Cove [= Caleta Cockle], Isla Madre de Dios, 50.08333°S, 75.03333°W

Cueva del Milodón, 51.58333°S, 72.58333°W

El Torcido, Tierra del Fuego, 53.53333°S, 69.36667°W

Estancia La Frontera, 54.00000°S, 68.75000°W

Estancia Lago Escondido, 500 m, 53.88333°S, 68.86667°W

Estancia Ponsonby, east end of Riesco Island [= Isla Riesco], 52.83333°S, 71.75000°W

Fuerte Bulnes, 53.62000°S, 70.92000°W

Grevy Island [= Isla Grevy], 55.53333°S, 67.61667°W

Harrison Island, 54.05000°S, 71.21667°W

Isla Grevy, 55.53944°S, 67.65389°W

Isla Hornos, 56.0000°S, 67.0000°W

La Cumbre, Baguales, 50.83333°S, 72.41667°W

Lago Pehoé, Torres del Paine, 51.10000°S, 73.06667°W

Lago Sarmiento, Parque Nacional Torres del Paine, 51.05000°S, 72.75000°W

Laguna Lazo, Lake Sarmiento [= Lago Sarmiento], Última Esperanza, 51.12162°S, 72.82328°W

Lake Sarmiento [= Lago Sarmiento], 51.06667°S, 72.75000°W

Martial Bay, Herschel Island, 55.81667°S, 67.61667°W

Orange Bay, Hoste Island, 55.50000°S, 68.06667°W

Pali Aike, 52.41667°S, 69.70000°W

Palli [= Pali] Aike, 52.41667°S, 69.70000°W

Parque Nacional Torres del Paine, 51.00000°S, 73.00000°W

Península Hardy, Isla Hoste, 55.50000°S, 68.00000°W

Port Gregory [= Bahía San Gregorio], 52.61667°S, 70.13333°W

Puerto Edén, 1.2 km NNW, Isla Wellington, 49.13300°S, 74.45000°W

Puerto del Hambre [= Puerto Famine], 53.60250°S, 70.94056°W

Puerto Natales, 51.72000°S, 72.48000°W

Puerto Natales, Última Esperanza, 51.73333°S, 72.51667°W

Punta Arenas, 53.16400°S, 70.93800°W

Punta Arenas, vicinity, 53.16400°S, 70.93800°W

Punta Arenas, vicinity, 53.13333°S, 70.91667°W

Puerto Henry, vicinity, Isla Riesco, 53.41390°S, 72.59350°W

Reserva Nacional Magallanes, 53.13306°S, 71.02500°W

Riesco Island [= Isla Riesco], east end, 53.00000°S, 72.50000°W

Riesco Island [= Isla Riesco], east end, 52.83333°S, 71.75000°W

Río Verde, east end of Skyring Water, 52.55597°S, 71.97210°W

Santa María, near Porvenir, Tierra del Fuego, 53.40000°S, 70.31667°W

San Martín, 53.33333°S, 68.95000°W

Sector lago Toro, Parque Nacional Torres del Paine, 51.16667°S, 72.95000°W

Seno Monteith, Isla Madre de Dios, 50.40306°S, 75.07028°W

Sierra Baguales, Última Esperanza, 50.63333°S, 72.40000°W

St. Martin's Cove, Hermite Island, 55.85000°S, 67.53333°W

Torres del Paine, 51.11667°S, 73.11667°W

Wollaston Island, 55.60000°S, 67.36667°W

Maule (Región de)

Arroyo del Valle, Río Maule, 35.83333°S, 70.70000°W

Baño San Pedro, Romeral, 35.13945°S, 70.47787°W

Constitución, 35.33333°S, 72.41667°W

Curicó, 34.98333°S, 71.23333°W

Laguna del Maule, 2,140 m, 36.05000°S, 70.50000°W

Lircay, Altos de Lircay National Reserve, 35.60000°S, 71.20000°W

Parque Inglés-Siete Tazas, 35.50000°S, 71.00000°W

Pilén Alto, 1.7 km W of Paso Pino Hachado, 35.95000°S, 72.416671°W

Radal, Reserva Nacional Radal, 36.55000°S, 70.98333°W

Reserva Nacional Los Queules, 35.98373°S, 72.68348°W

Río Maule, 14 km above Curillanque, 35.81667°S, 70.91667°W

Valle de Río Teno, 34.98333°S, 71.38333°W

O'Higgins (Región de)

Baños de Cauquenes [= Termas de Cauquenes], 34.25000°S, 70.56667°W

Baños del Flaco [= Termas de Flaco], 34.95000°S, 70.43333°W

Cachapoal, 34.26667°S, 71.45000°W

Los Cipreses, Rancagua, 34.01667°S, 70.48333°W

Rapel, 33.95000°S, 71.75000°W

Santiago (Región Metropolitana de)

Alicahue, 32.31667°S, 70.48333°W

Angostura [=Angostura Paine], 33.91667°S, 70.73333°W

El Colorado, Farellones, 33.33333°S, 70.33330°W

Farellones, 2,000 m, 33.35000°S, 70.33333°W

Fundo Santa Laura, 10 km W of Til Til, 1,100 ft, 33.26667°S, 70.85000°W

La Parva, 3,000 m, 33.33333°S, 70.28333°W

Las Condes, 700 m, 33.36667°S, 70.51667°W

Lagunillas, 33.60000°S, 70.28333°W

Las Melosas, 33.84469°S, 70.20374°W

Puente Alto, 33.61667°S, 70.65000°W

Rinconada de Maipú, 33.61667°S, 71.65000°W

Santuario de la Naturaleza Yerba Loca, 2,780 m, 33.31667°S, 70.28333°W

Valle de la Junta, Cajón del Río Volcán, 2,400 m, 33.80000°S, 70.01667°W

Villa Alhué, 3 km N of Melipilla, 33.68541°S, 71.20089°W

Tarapacá (Región de)

Canchones, Salar de Pintado, 983 m, 20.45000°S, 69.61667°W

Colchane, Suricayo, 3,712 m, 19.28333°S, 68.63333°W

Enquela, vicinity, 3,850 m, 19.21667°S, 68.80000°W

Miñita, 19.11667°S, 69.58333°W

Pica, vicinity, 1,259 m, 20.50000°S, 69.35000°W

Río Loa, mouth, 21.41667°S, 70.08333°W

Valparaíso (Región de)

Aconcagua [= Plaza Vieja, ca. 5 km W of Los Andes], 32.81667°S, 70.65000°W

Farellones, 51 km E of Santiago, 33.21667°S, 70.20000°W

La Calera, 32.78333°S, 71.20000°W

Lago Quintero, 32.78333°S, 7153333°W

Lake of Quintero [= Lago Quintero], 32.79576°S, 7152482°W

La Ligua, 32.45403°S, 71.23159°W

Limache, 33.01667°S, 71.26667°W

Los Molles, 30 km N of La Ligua, 33.31667°S, 71.55000°W

Los Molles, 3 km N, 33.28962°S, 71.5500°W

Aconcagua [= Plaza Vieja, ca. 5 km W of Los Andes], near old village, 32.81667°S, 70.65000°W

Olmué, 33.00000°S, 71.20000°W

Papudo, La Ligua, 32.45000°S, 71.23333°W

Pte. Los Molles, 3 km N, 32.20000°S, 71.45000°W

Quilpué, 33.05000°S, 71.45000°W

San Carlos de Apoquindo, 33.40000°S, 70.48333°W

Valparaíso, sea level, 33.03333°S, 71.63333°W

Colombia

Amazonas (Comisaría de)

Caserío Kuiru, 15 km below Chorrea, Río Igará-Paran, 00.75000°S, 73.08333°W

Corregimiento Puerto Santander, cerca de la Quebrada Bocaduché, margen sur Río Caquetá, 150 m, 00.66667°S, 72.13333°W

Río Apaporis, Ina Gaje, Río Pacoa, 00.11700°N, 71.21700°W

Antioquia (Departamento de)

Alto Bonito, upper Río Sucio, 400 m, 07.01667°N, 76.28333°W

Alto Bonito, 1,500 ft, 07.20000°N, 76.50000°W

Bellavista, 06.55000°N, 75.30000°W

Bellavista, 4 km NE, above Río Porce, 06.55000°N, 75.30000°W

Cisneros, 06.5500°N, 75.06667°W

Cisneros, 11 km S and 30 km E, 03.78333°N, 76.76667°W

Concordia, 2,020 m, 06.05000°N, 75.91667°W

El Cedro, Caicedo, 06.41667°N, 76.00000°W

Finca Katiri, Município Mutatá, 07.21030°N, 76.45148°W

Finca La Reina, Vereda La Sociedad, Corregimiento Santa Rita, 1,900–2,080 m, 05.61667°N, 75.93333°W

Finca de Margarita Molino, 05.97720°N, 75.36500°W

Frijolera, 07.16667°N, 75.41667°W

Guapantal, 2,200 m, 06.33333°N, 76.41667°W

Hacienda Barro, 12 km S of Caucasia, 07.88333°N, 75.20000°W

La Bodega, N side of Río Negrito (Sonsón to Nariño highway), 5,800 ft, 05.70000°N, 71.11667°W

La Pintada, S of Medellín, 05.73333°N, 75.43333°W

La Tirana, 25 km S and 22 km W of Zaragoza, 520–770 m, 07.35000°N, 75.05000°W

Loma Teguerre, 07.90000°N, 77.00000°W

Medellín, Las Palmas, 2,600 m, 06.25110°N, 75.59580°W

Medellín, vicinity of, 1,500 m, 06.29139°N, 75.53611°W

Mpio. Jardín, Vda. La Linda, 2,615 m, 05.60139°N, 75.82306°W

Mirundó, 06.98000°N, 76.73000°W

Paramillo, W Andes, 12,500 ft, 05.49111°N, 75.88806°W

Paramillo, 06.98333°N, 75.85000°W

Páramo, 7 km E of Sonsón, 2,900 m, 05.71667°N, 75.25000°W

Páramo de Belmira, 3,145 m, 06.63333°N, 75.66667°W

Páramo Frontino, Urrao, 3,200–3,500 m, 06.77778°N, 76.12861°W

Páramo Frontino, Urrao, 3,500 m, 06.46667°N, 76.06667°W

Puerto Berrío "Las Virginias," 06.49440°N, 74.40670°W

Puerto Valdivia, 360 ft, 07.30000°N, 75.38333°W

Purí, above Cáceres, ca. 170 m, 07.41667°N, 75.33333°W

Quebrada del Oro, 07.16667°N, 75.45000°W

Reserva San Sebastián–La Castellana (ca. 30 km SE of Medellín), 06.1000°N, 75.55000°W

Río Currulao, 20 km SE of Turbo, 07.95630°N, 76.60336°W

Río Negrito, 15 km E, 1,850 m, 05.71181°N, 75.06542°W

San Jéronimo, 35 km NE of Medellín, 720–800 m, 06.45000°N, 75.75000°W

San Pedro, 2,650 m, 06.46667°N, 75.55000°W

Santa Bárbara, Río Urrao, 2,700 m, 06.41667°N, 76.25000°W

Santa Elena, 2,620–2,750 m, 06.21667°N, 75.50000°W

Santa Elena, SW of, 9,000 ft, 06.21667°N, 75.50000°W

Sierra Santa Elena, near Medellín, 06.21667°N, 75.16667°W

Sonsón, 05.78333°N, 75.30000°W

Sonsón, 7 km E of Río Negrito, 2,000 m, 05.70000°N, 75.30000°W

Sonsón, 7 km E, 3,050 m, 05.71667°N, 75.25000°W

Turbo, 0810000°N, 76.71667°W

Urabá, Río Cumulao, 50 m, 08.00000°N, 76.73333°W

Urrao, Río Aná, 2,200 m, 06.33333°N, 76.18333°W

Valdivia, 1,200 m, 07.18333°N, 75.45000°W

Valdivia, right bank lower Río Cauca, 360–3,800 ft, 07.18000°N, 75.45000°W

Valdivia, La Cabaña, 07.16667°N, 75.43333°W

Valdivia, La Selva, 1,900 m, 07.18333°N, 75.45000°W

Valdivia, Las Ventanas, 2,000 m, 07.18333°N, 75.45000°W

Valdivia, 9 km S, 1,200 m, 07.15000°N, 75.45000°W

Valdivia, 10 km S, 1,500 m, 07.18333°N, 75.45000°W

Ventanas, 04.63333°N, 75.45000°W

Vda. San Miguel de la Cruz, sitio Alto de San Miguel de la Cruz, 2,480 m, 05.97720°N, 75.36500°W

Ventanas, Valdivia, 3,000 m, 07.08333°N, 75.45000°W

Vereda Puente Peláez, Finca Cañaveral, 06.01667°N, 75.50000°W

Villa Arteaga, 07.33333°N, 76.43333°W

Zaragoza, 24–26 km S and 21–22 km W, near La Tirana, 540–670 m, ca. 07.35000°N, 75.05000°W

Arauca (Departamento de)

Campo Petrolero de Caño Limón, 06.93333°N, 70.95000°W
Fatima, Río Cobaria, 695 m, 06.8000°N, 72.16667°W
Río Arauca, 300 m, ca. 07.08780°N 72.75750°W

Atlántico (Departamento de)

Barranquilla, opposite, lower Río Magdalena, 10.98333°N, 74.8000°W
Ciénaga de Guajaro, Sabana Larga, 10.61667°N, 74.91667°W

Bolívar (Departamento de)

Cartagena, 10.41667°N, 75.50000°W
Corregimiento de Ventura, ESE of Guaranda, 400 m, 08.46670°N, 74.53330°W
Morales, Río Viejo, 08.58333°N, 73.85000°W
Norosí, vicinity, Río San Pedro, 178 m, ca. 08.53333°N, 74.03333°W
San Juan de Nepomuceno, 167 m, 09.95000°N, 75.08333°W
Simití, 15 mi W, 6 mi above Santa Rosa, 2,800 ft, 07.97000°N, 74.05000°W

Boyacá (Departamento de)

Arauca, Río Cavaria [= Río Cobaria], 1,100 ft, 07.00000°N, 72.16700°W
Hacienda La Primavera, East Andesm, E side, 7,000 ft, 07.00000°N, 72.33333°W
La Primavera, 7,000 ft, 07.00000°N, 72.33333°W
Miraflores, 6,200 ft, 05.20000°N, 73.20000°W
Muzo, 850 m, 05.53333°N, 74.10000°W
Muzo, 1,300 m, 05.50000°N, 74.16667°W
Paipa, Parque Natural Ranchería, 2,600–3,500 m, 05.84690°N, 73.11850°W
Páramo de Toquilla, 3,200 m, 05.61667°N, 72.83333°W
Puerto Boyacá, Verede Dos Quebrad, 05.95000°N, 74.58330°W
Río Cobaria Fatima, 07.05000°N, 72.06700°W
Río Lengupa, 04.82000°N, 73.07000°W
Villa de Leyva, sector Chaina, vereda Río Abajo, 3,079 m, 05.69194°N, 73.46481°W
Villa de Leyva, sector Chaina, vereda Río Abajo, 3,090 m, 05.68483°N, 73.12947°W

Caldas (Departamento de)

Bosque de Florencia, ca. 05.36333°N, 75.65222°W [precise locality not found]
Bosque de Florencia, 1,200–1,800 m, 05.61000°N, 75.02000°W
Manizales, Río Termales, 3,200 m, 05.07000°N, 75.52956°W
Reserva Río Blanco, 05.06667°N, 75.53333°W
Samaná, 1,000 m, 05.40000°N, 75.00000°W
Samaná, Río Honda, 2,500 m, 05.25000°N, 75.25000°W

Caquetá (Intendencia del)

Florencia, Río Bodoquera, 300 m, 01.60000°N, 75.60000°W
Florencia, Mantanito, 400 m, 01.40000°N, 75.55000°W
La Morelia [= La Murelia], Río Bodoquera, 1,200 m, 01.50000°N, 74.80000°W
La Murelia, Río Bodoquera, ca. 300 m, 01.51667°N, 75.68333°W
Tres Troncos, La Taqua, Río Caquetá, 185 m, 00.13333°N, 74.68333°W

Casanare (Departamento del)

Finca Balmoral, 04.91667°N, 72.38333°W

Cauca (Departamento del)

Almaguer, 10,300 ft, 01.91667°N, 76.83333°W
Cerro Munchique, 2,540 m, 02.53333°N, 76.95000°W
Cerro Munchiquito, 1,500–2,599 m, 02.53333°N, 76.95000°W
Cocal, 4,000–6,000 ft, 02.51667°N, 76.98333°W
Cononuco, Parcque Nacional Puracé, Río Bedón, Versalles, N° Calma 141-M, 02.33333°N, 76.46667°W
El Roble, 03.23333°N, 76.35000°W
Gorgona Island [= Isla Gorgona], sea level, 03.00000°N, 78.20000°W
Güengüé, 03.23333°N, 76.35000°W
Inza, 02.66667°N, 76.08333°W
Isla Gorgona, sea level, 03.00000°N, 78.20000°W
La Boca, Río Saija, 2,500 m, 02.63740°N, 76.96193°W
La Florida, 2,200–2,400 m, 02.58333°N, 76.91667°W
La Gallera, 1,750 m, 02.58000°N, 76.91700°W
Laguna San Rafael, 0.5–0.8 km E, Páramo de Puracé, Parque Nacional Puracé, 3,310–3,320 m, 02.34806°N, 76.50083°W
Munchique [= Cerro Munchique], near El Tambo, 1,500 m, 02.58000°N, 76.95000°W
Paispamba, 02.28333°N, 76.55000°W
Paletará, Parque Nacional Puracé, 02.16667°N, 76.30000°W
Popayan, coast range to W, 10,340 ft, ca. 02.83333°N, 77.00000°W
Río Guachicono, 650 m, 01.93333°N, 77.13333°W
Río Mechengue, 800 m, 02.66667°N, 77.20000°W
Río Munchique, 1,000 m, 02.58000°N, 77.25000°W

Río Saija, 100 m, 02.86667°N, 77.68333°W
Sabanetas, 1,9000 ft, 02.53333°N, 76.88333°W
Salento, west Quindio Andes, 7,000 ft, 04.63333°N, 75.56667°W
Valle de las Papas, 3,050 m, 01.91700°N, 76.60000°W

Cesar (Departamento de)

Colonia Agrícola de Caracolicito, 100–400 m, 10.20000°N, 73.96667°W
Colonia Agrícola de Caracolicito, 335 m, 10.30000°N, 74.00000°W
El Orinoco, Río Cesar, 10.48333°N, 73.25000°W
Pueblo Viejo, Río San Antonio, 2,440 m, 10.53300°N, 73.41700°W
Río Guaimaral [= Río Garupal], [Municipio de] Valledupar, ca. 09.81670°N, 73.61670°W
San Alberto, 350 m, 07.75000°N, 73.38333°W
San Sebastián, 10.56667°N, 73.60000°W
Valledupar, Río Guaimaral, 10.08300°N, 73.53300°W
Villanueva, Sierra Negra, 10.60000°N, 72.91667°W

Chocó (Departamento de El)

Acandí, 08.53333°N, 77.23333°W
Alto de Barrigonal, Serranía del Darién, 2,400 m, 08.05000°N, 77.25000°W
Bagadó [= Andagada of J. A. Allen (1916c)], Río Andagueda, 05.41667°N, 76.40000°W
Bahía Solano, ca. 7 km S, ca. 100 m, 06.17500°N, 77.41100°W
Baudó [= Serranía de Baudó], 06.00000°N, 77.08333°W
Con limite de Departamento de Valle (carretera vereda Pacífico–Mpo. San José del Palmar), 2,000 m, 04.97420°N, 76.22830°W
Condoto, mouth of Río Condoto, 300 ft, 05.10000°N, 76.61700°W
Gorgas Memorial Laboratory, Teresita, 07.43000°N, 77.11700°W
Juntas de Tamaná, Rio Tamaná, 405–800 ft, 04.98000°N, 76.63000°W
Nóvita, Río Tamaná, 150 ft, 04.92000°N, 76.35000°W
Parque Nacional Natural Utría, Ensenada de Utría, ca. 10 m, 06.04600°N, 77.35100°W
Río Docampado, 75 m, 04.75000°N, 77.30000°W
Río Jurubidá, Baudó Mts., 05.83000°N, 77.28000°W
Río Nuquí, base of Baudó Mts., 05.70000°N, 77.27000°W
Río Salaqui, 07.45000°N, 77.11667°W
Río San Juan, 04.05000°N, 77.45000°W
Río Traundó, 07.43000°N, 77.12000°W
Santa Marta, Río Curiche, ca. 30 m, 07.01200°N, 77.63900°W

Serranía de Baudó, 3,500 ft, 06.0000°N, 77.08300°W
Sipí, Río Sipí, Río San Juan, 150 ft, 04.67000°N, 76.63000°W
Teresita, Río Truando, 07.43333°N, 77.11667°W
Unguía, sea level, 08.01667°N, 77.06667°W
Unguía, upper Río Ipetí, 08.01667°N, 77.11667°W

Córdoba (Departamento de)

Catival, upper Río San Jorge, 120 m, 08.28333°N, 75.68333°W
Montería, 08.76667°N, 75.88333°W
San Juan de Río Seco, 04.85000°N, 74.63333°W
Socorré, upper Río Sinú, 100–250 m, 07.85000°N, 76.28333°W

Cundinamarca (Departamento de)

Anolaima, on branch of Río Bogotá, ca. 1,500 m, 04.76667°N, 74.46667°W
Bogotá, 04.59710°N, 74.07350°W
Bogotá, vicinity, 04.60000°N, 74.08333°W
Chipaque, 8,500 ft, 04.44250°N, 74.04417°W
Choachí, 04.53333°N, 73.93333°W
Cordillera Oriental, western slope, ca. 04.60000°N, 74.08300°W
Cuchillas del Carnicero, near Bogotá, 04.60000°N, 74.08333°W
El Roble, above Fusugasugá, east Andes, 8,000 ft, 04.38333°N, 74.31667°W
El Verjón, 04.53000°N, 74.06667°W
Finca El Soche, 15 km W of Soacha, 2,360 m, 04.58330°N, 74.21670°W
Finca El Tabacal, 3,155 m, 05.20942°N, 73.92239°W
Fómeque, 04.48333°N, 73.90000°W
Fusugasugá, 04.34389°N, 74.36778°W
Guaicáramo, 600–700 m, 04.71667°N, 73.03333°W
Guasca, Río Balcones, 2,720 m, 04.86639°N, 73.86611°W
La Aguadita, 1,800 m, 04.41667°N, 74.33333°W
Laguna Verde, Páramo de Guerrero, 3,470 m, 05.21647°N, 74.00475°W
Laguna Vergón, 04.53333°N, 74.06667°W
La Regadera, Usme, 04.47310°N, 74.11610°W
Mámbita, 1,000 m, 04.76667°N, 73.31667°W
Nariño, near Río Mira, 04.40060°N, 74.83580°W
Paime, 1,038 m, 05.36667°N, 74.16667°W
Panamea, 05.16667°N, 74.21667°W
Páramo de í 04.53111°N, 73.92583°W
Quipile, 04.75000°N, 74.53333°W
Reserva Biológica Carpanta, 2,400 m, 04.56667°N, 73.68333°W
Reserva Forestal, Vda. Quebrada Honda, 3,180 m, 05.11225°N, 74.00628°W

Río Balcones, 04.66667°N, 73.55000°W
Río Magdalena, lowlands near, 04.83333°N, 74.750000°W
Sabana Grande, lowlands near Bogotá, 04.60000°N, 74.10000°W
San Cristóbal, Bogotá, 2,900 m, 04.58333°N, 74.08333°W
San Francisco, Bogotá, 3,500 m, 04.72889°N, 73.84333°W
San Juan de Río Seco, 04.85000°N, 74.63333°W
Usaguen, 04.70000°N, 74.03333°W
Volcanes, Municipio de Caparrapí, 05.45000°N, 74.51667°W

Huila (Departamento del)

Acevedo, 1,700 m, 01.81667°N, 75.86667°W
Andalucia, eastern Andes, 5,000 ft, 01.90000°N, 75.66667°W
Andalucia, 7,000 ft, 01.90333°N, 75.68722°W
Belén, 2,135 m, 02.25000°N, 76.08333°W
Camp Coscorrón, Hacienda San Diego, 17 km SE of Villavieja, 1600 ft, 03.05000°N, 75.18333°W
La Candela, 6,500 ft, 01.83333°N, 76.33333°W
La Palma, 5,500 ft, 01.78333°N, 76.36667°W
Meremberg [= Finca Merenberg], 02.23333°N, 76.13333°W
Río Aguas Claras, Río Sauza, 01.63056°N, 75.99167°W
Río Ovejeras, 01.95000°N, 76.48333°W
San Adolfo, Río Aguas Claras, 1,400 m, 01.63056°N, 76.99167°W
San Adolfo, 1,500 m, 01.61667°N, 75.98333°W
San Adolfo, Acevedo, 1,400 m, 01.61667°N, 75.98333°W
San Adolfo, vicinity, Río Aguas Claras, Río Suaza, 1,400–1,600 m, 01.61667°N, 75.98333°W
San Agustín, 01.87863°N, 76.27557°W
San Agustín, Las Bardas, 3,200 m, 01.88333°N, 76.26667°W
San Agustín, Las Ovejeras, 2,350 m, 01.95000°N, 76.48330°W
San Agustín, Páramo de las Papas, 3,600 m, 01.91667°N, 76.55000°W
San Agustín, Río Magdalena, 2,300 m, 01.83333°N, 76.26667°W
San Agustín, Río Ovejeras, 01.95000°N, 76.48333°W
San Agustín, Santa Marta, 2,700 m, 01.93333°N, 76.53333°W
San Antonio, 2,200 m, 01.95000°N, 76.48333°W
San Antonio, 01.96667°N, 76.58333°W
Valle de Sauza, 01.91667°N, 75.66667°W
Villavieja, 03.21667°N, 75.23333°W
Villavieja, 5 km N, 1,400 ft, 03.31700°N, 75.13300°W
Villavieja, vicinity, 03.33333°N, 75.183333°W

La Guajira (Departamento de)

Cabo de la Vela, Arroyo Cerrejón, Cerro Pilón de Azucar, 12.20917°N, 72.14756°W
Chirua, 10.83000°N, 73.38000°W
Fonseca, Las Marimondas, 1,000 m, 10.86667°N, 72.71667°W
Laguna de Junco, 7,500 ft, 10.48333°N, 72.91667°W
Las Marimondas, Sierra de Perijá, 1,000 m, 10.86667°N, 72.71667°W
[Labels may include the name "Fonseca," but see Hershkovitz (1949:383).]
Puerto Estrella, 12.35000°N, 71.31667°W
San Miguel, 1,700 m, 10.96667°N, 73.48333°W
Villanueva, Sierra Negra, 280 m, 10.61667°N, 72.98333°W

Magdalena (Departamento del)

Bonda, 50 m, 11.23333°N, 74.13333°W
Buritaca, 11.25000°N, 73.76667°W
Cerros San Lorenzo, San Lorenzo, 2,200 m, 11.16667°N, 74.11667°W
Cincinati [= Valparaiso, Hacienda Cincinati], 1,480 m, 11.10000°N, 74.10000°W
Don Dago [= Don Diego], sea level, 11.23333°N, 73.68333°W
Don Diego, sea level, 11.23333°N, 73.68333°W
El Dorado Nature Reserve, Sierra Nevada de Santa Marta, 11.10082°N, 74.07204°W
El Libano plantation, near Bonda, 1,250 m, 11.16700°N, 74.0000°W
El Orinodo, Río Cesar, 09.0000°N, 73.96667°W
Hacienda Cincinati [= Cincinati, Valparaiso], 1,480 m, 11.10000°N, 74.10000°W
Macotama, 2,438 m, 10.91700°N, 73.5000°W
Mamancanaca, near, 3,600 m, 10.71667°N, 73.65000°W
Mamatoca [= Mamatoco], Santa Marta, 11.23333°N, 71.16667°W
Manzanares, 11.23333°N, 74.21667°W
Minca, west slope Sierra Nevada de Santa Marta, 2,000 ft, 11.15000°N, 74.11667°W
Onaca, Santa Marta, 11.18333°N, 74.06667°W
Palomino, 11.03333°N, 73.65000°W
Páramo de Macotama, Sierra Nevada de Santa Marta, 11,000 ft, 10.91667°N, 73.50000°W
Pueblo Viejo, Sierra Nevada de Santa Marta, 10.98333°N, 73.45000°W
Santa Marta, sea level, 11.25000°N, 74.21667°W
Santa Marta, Serranía San Lorenzo, Estación INDERENA, 2,000 m, 11.24720°N, 74.20170°W
Sierra El Libano, 6,000 ft, 11.08333°N, 74.06667°W

Meta (Departamento del)

Barrigona [= Puerto Barrigón], 04.16667°N, 72.16667°W
Buenavista, 4,500 ft, 04.16667°N, 73.68333°W

Cabaña Duda, junction of Río Duda and Río Guayabero, 250 m, 02.55000°N, 74.05000°W

Condominio Camino Real, Municípío de Villavicencio, 1,500 m, 04.15000°N, 73.61667°W

Finca El Buque, Villavicencio, 04.15000°N, 73.61667°W

Fundo "Guami," Piñalto, 450 m, 02.98333°N, 73.66667°W

La Macarena, Río Yerli, 03.23000°N, 73.87000°W

La Macarena Parque, Río Guapaya, 02.75000°N, 73.91667°W

Los Micos, 03.28333°N, 73.88333°W

Los Micos, 18 km S of San Juan de Arama, 1,300 ft, 03.28300°N, 73.88300°W

Macarena Mt. [= Serranía de La Macarena], 1,140 m, 02.75000°N, 73.91667°W

Puerto Lleras, 03.26667°N, 73.38333°W

Puerto López, 05.08333°N, 72.96667°W

Restrepo, ca. 500 m, 04.25000°N, 73.56667°W

San Juan de Arama, Plaza Bonita, 400 m, 03.26667°N, 73.86667°W

Serranía de la Macarena, 02.75000°N, 76.91667°W

Serranía de la Macarena, 1,140 m, 02.90000°N, 73.90000°W

Villavicencio, upper Río Meta, 500 m, 04.15000°N, 73.61667°W

Villavicencio, 7 km NE (of airport), 04.15000°N, 73.61667°W

Nariño (Departamento de)

Barbacoas, on Río Telembi, 23 m, 01.68333°N, 78.15000°W

Buenavista, 350 m, 01.48333°N, 78.08333°W

Candelilla, 01.48333°N, 78.71667°W

El Naranjál, Laguna de la Cocha, 2,700 m, 01.08330°N, 77.15000°W

Galera [= Volcán Galera], 4,090 m, 01.22186°N, 76.46667°W

Guayacana, 200 m, 01.43333°N, 78.45000°W

Guayabetal, near, 01.79861°N, 78.81556°W

Finca Arizona, 4.5 km S of Remolino, 500–1,000 m, 01.67333°N, 77.32913°W

La Florida, 2,320 m, 01.29851°N, 77.40614°W

La Laguna, 01.21667°N, 77.20000°W

Pasto, slopes of Volcán Galeras, 3,650 m, 01.21361°N, 77.28111°W

Reserva Natural La Planada, 01.15000°N, 78.00000°W

Ricuarte, 1,500 m, 01.21667°N, 77.98333°W

Río Rumiyacu, headwaters, 755 m, 00.50000°N, 77.23333°W

San Pablo, 1,500 m, 01.10000°N, 78.01667°W

Norte de Santander (Departamento de)

Alturas de Pamplona, 07.38333°N, 72.65000°W

El Guayabal, 10 mi from San José de Cúcuta, 08.01667°N, 72.48333°W

Finca La Palma, Durania, 07.71667°N, 72.66667°W

Guamalito, near El Carmen, 08.56667°N, 73.48333°W

Páramo de Tamá, 8,000 ft, 07.41667°N, 72.43333°W

Río Tarrá, 200 m, 08.60000°N, 73.01700°W

San Calixto, Río Tarrá, 08.40140°N, 73.20830°W

Tarrá, 08.65000°N, 73.01667°W

Tibú, 91 km N, Río de Oro, 07.16667°N, 73.15000°W

Putumayo (Departamento del)

La Tagua, Tres Troncos, Río Caquetá, 185 m, 00.03000°N, 74.65000°W

Río Mecaya, at Río Caquetá, 185 m, 00.46667°N, 75.33333°W

[P. Hershkovitz locality; see Paynter and Traylor (1981:157).]

Quindío (Departamento del)

La Guneta [= Laguneta], 04.58333°N, 75.50000°W

El Roble, Salento, 2,200 m, 04.68333°N, 75.60000°W

Finca La Cubierta, 6 km N of Salento-Cocora road, 3,670 m, 04.70000°N, 75.46700°W

Salento, 2,135 m, 04.63333°N, 75.56667°W

Risaralda (Departamento de)

Finca Cañón, Papayal, Río San Rafael, Parque Nacional Tatamá, 2,435 m, 05.07530°N, 75.96720°W

La Pastora, Reserva Ucumarí, 04.71667°N, 75.48333°W

La Pastora, PRN Ucumarí, 2,550 m, 04.81330°N, 75.69610°W

Río Termales, 3,200 m, 04.83889°N, 75.54583°W

Siató, Río Siató, San Juan, 5,200 ft, 05.22000°N, 76.12000°W

Vereda Siató, 1,520–1,620 m, 05.23333°N, 76.03333°W

Santander (Departamento de)

Barrancabermeja, 07.06528°N, 73.85472°W

Cachirí, 1,890 m, 07.50000°N, 73.01667°W

Finca El Pajal, Vda. Guarumales, 2,150 m, 07.13725°N, 72.99908°W

Finca San Miguel, San Pedro de la Paz, 06.31670°N, 73.95000°W

Hacienda Montebello, near Cerro San Pablo, 350–500 m, 08.01667°N, 73.41667°W

Meseta de Los Caballeros, 5 km NW of La Albania, San Vicente de Chururí, 200 m, 06.90000°N, 73.67000°W

Santander, 10 km W, 1,005 m, 03.01667°N, 76.46667°W

San Vicente de Chucurí, 06.90000°N, 73.41667°W

Sucre (Departamento de)

Las Campanas, 175–200 m, 09.50000°N, 75.35000°W

Tolima (Departamento del)

Chicoral, 04.21667°N, 74.89333°W

El Edén, E of Quindío Andes, 8,300 ft, 04.50000°N, 75.33333°W

El Triunfo, 05.20000°N, 74.75000°W

Hacienda Indostán. 04.63390°N, 75.09720°W

Mariquita, 535 m, 05.20000°N, 74.90000°W

Nevado del Ruiz, 3,300–4,000 m, 04.90000°N, 75.13300°W

Quebrada Aico, between Coyaima and Chaparral, 03.88333°N, 74.23333°W

Río Chilí, 04.11667°N, 75.26667°W

Río Toche, 04.43333°N, 75.36667°W

Valle del Cauca (Departamento del)

Albán [= Salencio], 7,200 ft, 04.78000°N, 76.18000°W

Atuncela, 700–800 m, 03.76667°N, 76.70000°W

Bahía Málaga, Quebrada Valencia, road to Quebrada Alegría, sea level—60 m, 04.11667°N, 77.23333°W

Bosque San Antonio, Cordillera Occidental, 2,250 m, 03.50010°N, 76.63320°W

Buenaventura, Palmares del Pacifico, 03.88333°N, 77.06667°W

Buenaventura, 6 km N, 03.88300°N, 77.06700°W

Cali, 03.45000°N, 76.52000°W

Correa, Farallones de Cali, 03.52138°N, 76.58585°W

Corrigimiento Baragán, Hacienda La Esperanza, 3,000 m, 04.03220°N, 75.88560°W

El Cairo, Cerro del Inglés, 04.76250°N, 76.30444°W

Finca La Playa, 1,800 m, 03.45000°N, 76.61667°W

Hacienda Formoso, 930 m, 04.80000°N, 75.98333°W

Hacienda la Sirena, carretera Palmira-La Nevera, 2,800 m, 03.56100°N, 76.00740°W

Hacienda Los Aples, 2,460 m, 03.33130°N, 76.08160°W

Laguna Sonso, 940 m, 03.85960°N, 76.35030°W

Las Lomitas [= Lomitas], 1,400 m, 03.63333°N, 76.63333°W

Los Cisneros, 600 ft, 03.78333°N, 76.76667°W

Mechenquito, 02.67000°N, 77.20000°W

Miraflores, near Palmira, 6,200 ft, 03.58333°N, 76.16667°W

Municipio de Roldanillo, 980 m, 04.41540°N, 76.14880°W

Palmira, 03.58300°N, 76.16700°W

Palmira, vicinity, central Andes, 6,000 ft, 03.53333°N, 76.26667°W

Pichindé, 1,800 m, 03.44583°N, 76.62444°W

Reserva Forestal Yotoco, 03.88333°N, 76.46667°W

Río Frio, Río Cauca, 3500 ft, 04.15000°N, 76.30000°W

Río Mechengue, 800 m, 02.50000°N, 77.50000°W

Río Raposo, 03.71667°N, 77.13333°W

San Antonio, 2,134 m, 03.50000°N, 76.63333°W

San Antonio, 4 km NW, 1,980 m, 03.25000°N, 76.46667°W

San Isidro, 1 km from on Río Calima, 03.90000°N, 76.25000°W

San José, 03.85000°N, 76.87000°W

Virology Field Station, Río Raposo, 03.63333°N, 77.08333°W

Zabaletas, 50 m, 03.73333°N, 76.95000°W

Vaupés (Comisaría del)

Iino Goje, Río Apaporis, 01.00000°N, 70.00000°W

La Providencia, Río Apaporis, 00.88333°S, 69.96667°W

Maipures, upper Río Orinoco, 05.18333°N, 67.83333°W

Río Vaupés, right bank, near Mitú, 240 m, 01.11667°N, 70.08333°W

Río Vaupés, in front of Tahuapunto, 00.65000°N, 69.20000°W

Yay Gojes, lower Río Apaporis, 01.00000°N, 70.00000°W

Vichada (Comisaría del)

Maipures, middle Río Orinoco, 05.18333°N, 67.81667°W

Ecuador

Azuay (Provincia de)

Baños, 02.90000°S, 79.06667°W

Bestión, 3,080 m, 03.16667°S, 79.21667°W

Giron, 2,100 m, 03.16667°S, 79.13333°W

Las Cajas, Lake Luspa, 02.83333°S, 79.50000°W

Las Cajas National Park, 02.83333°S, 79.50000°W

Las Cajas Plateau, Torreadora, 02.84610°S, 79.20780°W

Mazán, 2,700 m, 02.87000°S, 79.21667°W

Molleturo, 2,315 m, 02.80000°S, 79.43333°W

Tunguilla Valley, 03.30000°S, 79.30000°W

Valle de Yunguilla, 03.30000°S, 79.30000°W

Yunguilla Valley [= Valle de Yunguilla], 1,500 m, 03.43300°S, 79.18300°W

Bolívar (Provincia de)

Carmen, near Sinche [= Hacienda Sinche], 7,500 ft, 01.53333°S, 78.98333°W

Hacienda Porvenir, W of Hacienda Talahua, 01.35000°S, 79.06700°W

Porvenir, 1,500 m, 01.35000°S, 79.07000°W

Río Chimbo and Río Coco, confluence, 732 m, 02.08333°S, 79.00000°W

Río Tatahuazo, 2.5 km E of Cruz de Lizo, 2,800 m, 01.71667°S, 78.98333°W

Río Tatahuazo, 4 km E of Cruz de Lizo, 3,000 m, 01.71667°S, 78.95000°W

San José, 12 mi SW of Huigra, 02.40680°S, 79.10607°W

Sinche [= Hacienda Sinche], 3,200 m, 01.53333°S, 78.98333°W

Sinche [= Hacienda Sinche], 3,385–4,000 m, 01.53300°S, 78.98300°W

Cañar (Provincia de)

Cañar, 2,600 m, 02.55000°S, 78.93333°W

Chícal, Naupan Mtns., 10,100 ft, 02.40000°S, 78.96700°W

Joyapal, 02.48333°S, 79.16667°W

San Antonio, 02.48333°S, 78.95000°W

Carchi (Provincia de)

Atal, near San Gabriél, Montúfar, 3,757 m, 00.60000°N, 77.81700°W

Atal, 5 mi SE of San Gabriel, 00.55000°N, 77.76667°W

El Pailón, 01.06667°N, 78.26667°W

El Pailón, Parroquia Tobar Donoso, 1,400 m, 01.00000°N, 78.23333°W

Los Encinos, 3,423 m, 00.65894°N, 77.86379°W

Montúfar, Atal, 2,700 m, 00.60000°N, 77.81667°W

Montúfar, near San Gabriél, 2,900 m, 00.60000°N, 77.81700°W

Páramo del Artesón, Comuna La Esperanza, 00.77917°N, 77.90611°W

Potrerillos, Ipuerán, 3,100 m, 00.66667°N, 77.63334°W

Potrerillos, ravine near, Ipuerán, 3,672 m, 00.71667°N, 77.86667°W

Quebrada La Buitrera, Reserva Ecológica del Ángel, 3,943 m, 00.73333°N, 77.91667°W

Reserva Biológica Guandera, 00.58333°N, 77.66667°W

Chimborazo (Provincia de)

Alao [= Hochkordillere von Alao], 01.90000°S, 78.48333°W

Cochaseca, 01.46667°S, 78.80000°W

Chunchi, 02.28333°S, 78.91667°W

Mt. Chimborazo, 13,500–13,800 ft, 01.46667°S, 78.80000°W

Pallatanga, 1,510–1,650 m, 01.98333°S, 78.95000°W

Pauchi, 02.28333°S, 78.98333°W

Puente de Chimbo, Bucay, 365 m, 02.16667°S, 79.16667°W

Ríos Chimbo-Coco, 2,400 ft, 02.10000°S, 78.96700°W

Urbina, 11,400 ft, 01.50000°S, 78.73333°W

Cotopaxi (Provincia de)

Hacienda Sr. Cepeda, 10 km NW of Chugchilán, 00.78194°S, 78.94444°W

Otonga, 00.43333°S, 79.00000°W

Reserva Bosque Nublado Otonga, San Francisco de las Pampas, 1,900–2,000 m, 00.41667°S, 79.00000°W

El Oro (Provincia de)

Arenillas, 60 m, 03.55000°S, 80.06667°W

El Chiral, 1,630 m, 03.63333°S, 79.68333°W

Los Pozos, 90 m, 03.55000°S, 80.06667°W

Pinas, 03.83333°S, 79.75000°W

Portovelo, 03.71667°S, 79.65000°W

Río Pindo, 1,850 ft, 04.01667°S, 79.66667°W

Salvias, 03.78300°S, 79.35000°W

Salvias, above, Cordiller de Chilla, 03.46667°S, 79.58333°W

Santa Rosa, 33 m, 03.48333°S, 79.98333°W

Taraguacocha, Cordillera de Chilla, 2,970 m, 03.66667°S, 79.66667°W

Taraguacocha, Cordillera de Chilla, páramo 1,000 ft above camp, 10,750 ft, 03.66667°S, 79.66667°W

Zaruma, 6,000 ft, 03.70000°S, 79.61667°W

Esmeraldas (Provincia de)

Bayone, 01.10000°N, 78.98333°W

Cachabí, 500 ft, 00.96667°N, 78.80000°

Cachavi [= Cachabí], 00.96667°N, 78.80000°W

Carondelet, 60 ft, 01.10000°N, 78.70000°W

El Salto, estero, 00.98889°N, 79.50000°W

Esmeraldas, sea level, 00.98333°N, 79.06667°W

La Bocana, 01.07500°N, 79.30560°W

Majua, 3 km W, 00.70000°N, 79.55000°W

Mindo, vicinity, Río Blanco, 4,000 ft, 00.03333°N, 79.80000°W

Pamibilar, 00.76667°N, 79.08333°W

Quinindé [= Rosa Zárate], 00.33333°N, 79.46667°W

Rosa Zárate, near Río Quininde, 100 m, 00.33333°N, 79.46667°W

San Javier, Río Cachaví, 01.06667°N, 78.78333°W

Galápagos (Territorio de)

Academy Bay, Isla Santa Cruz [= Indefatigable Island], 00.74145°S, 90.30272°W

Barrington Cove, Isla Santa Fe [= Barrington Island], 00.81667°S, 90.06667°W

Cabo Douglas, Isla Fernandina [= Narborough Island], 00.030139°S, 91.65222°W

Conway Bay, Isla Santa Cruz [= Indefatigable Island], 00.55194°S, 90.51613°W

Fortuna, near Bellavista, Isla Santa Cruz [= Indefatigable Island], 00.69401°S, 90.32480°W

Isla Baltra [= South Seymour Island], 00.44012°S, 90.28059°W

Isla Fernandina [= Narborough Island], caldera rim, 00.35147°S, 91.56385°W

Cerro Brujo, Bahía d'Esteban, Isla San Cristóbal [= Chatham Island], 00.77766°S, 89.46058°W

James Bay, Isla Santiago [= James Island], 00.24171°S, 90.86022°W

La Bomba, Isla Santiago [= James Island], 00.18611°S, 90.69972°W

Mangrove Point [= Punta Mangle], Isla Fernandina [= Narborough Island], 00.45328°S, 91.38909°W

Punta Espinosa, Isla Fernandina [= Narborough Island], 00.26543°S, 91.44919°W

Sullivan Bay, Isla Santiago [= James Island], 00.28102°S, 90.57522°W

Volcán Fernandina, west slope, Isla Fernandina [= Narborough Island], 00.40000°S, 91.60000°W

Guayas (Provincia de)

Bucay, Río Chimbo, 312 m, 02.16667°S, 79.10000°W

Cerro Baja Verde [= Cerro de Bajo Verde], 32 km NW of Guayaquil, 150 m, 01.98333°S, 80.03333°W

Chongón, 40 m, 02.23333°S, 80.06667°W

Chongoncito, 0–100 m, 02.23333°S, 80.08333°W

Dualé, 0–100 m, 01.86667°S, 79.93333°W

Guayaquil, 02.16667°S, 79.9000°W

Huerta Negra, 20 km ESE of Baláo, near Tenque, 03.00000°S, 79.76667°W

Naranjo [= Naranjito], 30 m, 02.21667°S, 79.48333°W

Río Chongón, 1.5 km SE of Chongón, 02.34444°S, 80.06667°W

Imbabura (Provincia de)

Intag, near Peñaherrera, 00.35000°N, 78.53330°W

Lita, 3,000 ft, 00.87000°N, 78.47000°W

"Los Cedros" sector, Manduroyacu, Reserva Ecológica Cotacachi-Cayapas, 00.28333°N, 78.76667°W

Paramba, Río Mira, 900 m, 00.81667°N, 78.35000°W

San Nícolas, near Pimanpiro, Valle de la Chota, 6500 ft, 00.40000°N, 77.96700°W

Santa Rosa, 10 km E, 702 m, 00.33083°N, 78.93194°W

Loja (Provincia de)

Alamor, 04.03333°S, 80.03333°W

Alomar, Puyango, 1,014 m, 04.01667°S, 80.01667°W

Celica, 2,100 m, 04.11667°S, 79.98333°W

Cerro Ahuaca, 2 km from Cariamanga, 2,450 m, 04.30817°S, 79.54644°W

El Porotillo, San José, 04.23333°S, 79.25000°W

Guachanamá, 2,760 m, 04.03333°S, 79.88333°W

Guainche, 3,200 ft, 04.50000°S, 80.41667°W

Laguna Negra, Bosque Protector Colombo-Yacurí, 04.71083°S, 79.43833°W

Las Chinchas, 1 km WSW, 7,340 ft, 03.96667°S, 79.43333°W

Loja, 04.00000°S, 79.21667°W

Loja, N side, Malacatos Divide, 2,300 m, 04.23333°S, 79.25000°W

Loja, 8 km W, 2,200 m, 04.00000°S, 79.28333°W

Los Posos, 04.50000°S, 79.83333°W

Los Pozos, 01.00000°S, 80.61667°W

Malacatos, 1,600 m, 04.23333°S, 79.25000°W

Sabanilla, near Río Destrozo, a small tributary of the Río Zamora, 5,700 ft, 04.03000°S, 79.02000°W

San Bartolo, 04.03333°S, 79.91667°W

Los Ríos (Provincia de)

Estación Biológica Pedro Franco Dávila, Jauneche, 50 m, 01.33333°S, 79.58333°W

Hacienda Pijigual, Vinces, 0–100 m, 01.55000°S, 79.73333°W

Hacienda Santa Teresita (Abras de Mantequyilla), ca. 12 km NE of Vinces, 01.53333°S, 79.75000°W

Pimocha, Río Babahoyo, 01.83333°S, 79.58333°W

San Carlos, vicinity, ca. 01.38333°S, 79.50000°W

Manabí (Provincia de)

Bahía de Caráquez, Río Briseño, 00.60000°S, 80.41667°W

Balzar mountains, upper Río Palenque, 00.91667°S, 79.91667°W

Cerro de Pata de Pajaro, 00.03300°N, 79.98300°W

Chone, 00.68333°S, 80.10000°W

Cuaque [= Coaque], sea level, 00.00000°S, 80.10000°W

Manaví, Río de Oro, 00.46667°S, 79.60000°W

Río de Oro, 02.13333°S, 79.60000°W

Tama [= Jama], sea level, 00.18333°S, 80.26667°W

Morona–Santiago (Provincia de)

Cerro Bosco, forest surrounding Pacifictel Transmitting/ Receiving Antenna, 03.00611°S, 78.50083°W

Chiguaza, 1,100 m, 02.01667°S, 77.96667°W

Domono, 1,170 m, 02.21734°S, 78.12644°W

Gualaquiza, 750 m, 03.40000°S, 78.55000°W

Mendéz Sur, Oriente, 02.71667°S, 78.31667°W

Río San José, 03.28333°S, 78,61667°W

Río Upano, Sangay National Park, 2,962 m, 02.20496°S, 78.45859°W

Tinajillas, 2,300 m, 03.01658°S, 78.58032°W

Napo (Provincia de)

Antisana, 4,115 m, 00.50000°S, 78.13300°W

Archidona, 00.916667°S, 77.80000°W

Archidona, vicinity, 1,067 m, ca. 00.91667°S, 77.80000°W

Baeza, 1,980 m, 00.46460°S, 77.88924°W

Cerro Antisana, oriente, 4,000 m, 00.50000°S, 78.13333°W

Cerro Antisana, 4,100 m, 00.50000°S, 78.13333°W

Cerro Galeras, 00.83333°S, 77.58333°W

Chaco [= El Chaco], Río Oyacachi, 1,615 m, 00.38333°S, 77.81667°W

Cosanga, Río Aliso, 00.56667°S, 77.86667°W

Cuyuco [probably Cyuyja], below Papallacta, 7,000 ft, ca. 00.40000°S, 78.00000°W

Lago Agrio, 12 km NE, 00.19167°N, 76.78333°W

La Selva Jungle Lodge, 00.50000°S, 76.36667°W

Oyacachi, El Chaco County, 3,444 m, 00.21667°S, 78.06667°W

Papallacta, 11,500 ft, 00.36667°S, 78.13333°W

Papallacta, near, 00.36667°S, 78.13333°W

Papallacta, 9 km E, 9,280 ft, 00.38713°S, 78.05460°W

Papallacta, 1.6 km W, 10,500 ft, 00.36408°S, 78.14720°W

Papallacta, 6.2 km W, 11,700 ft, 00.34151°S, 78.18040°W

Papallacta, 6.9 km by road W, 00.38516°S, 78.16868°W

Papallacta, 8.2 km W, 12,200 ft, 00.38104°S, 78.19526°W

Papallacta, 10.6 km W by road, 3,560 m, 00.36667°S, 78.13333°W

Río Ansu, ca. 01.31700°S, 77.88300°W

Río Coca, mouth of, upper Río Napo, 00.48333°S, 76.96667°W

Río Jatun Yacu, 01.08166°S, 77.86911°W

Río Lagarto Cocha, boca [= mouth], 00.65000°S, 75.26667°W

Río Papallacta valley, 11,100 ft, 00.36700°S, 78.13300°W

Río Pucuno, 00.80000°S, 77.26667°W

Río Napo, 2,000 ft, 00.08000°S, 78.00000°W

Río Napo (probably near Tena), 00.98333°S, 77.81667°W

Río Napo (presumably near Puerto Napo), 01.05000°S, 77.78333°W

Río Oyacachi, 00.36667°S, 77.78333°W

San Javier, 00.97273°S, 77.07273°W

San José abajo, 00.41667°S, 77.33333°W

Santa Cecilia, 00.05000°S, 76.70000°W

Sumaco arriba [= Volcán Sumaco], 8,000–9,000 ft, 00.56667°S, 77.63333°W

Sumaco abajo, 00.56667°S, 77.63333°W

Volcán Sumaco, 00.56667°S, 77.63333°W

Orellana (Provincia de)

Llunchi, 00.43905°S, 76.81362°W

["An island about 18 km below mouth of Río Coca." (Hershkovitz 1944:98)]

Parque Nacional Yasuní, 01.08333°S, 75.91667°W

Parque Nacional Yasuní, 38 km W of Pompeya Sur, 00.68333°S, 76.43333°W

Río Cotapino, 00.80000°S, 77.43333°W

Río Suno (abajo), 00.70464°S, 77.24074°W

San José Abajo [= San José Nuevo], 250 m, 00.43333°S, 77.33333°W

San José de Payamino, 00.50000°S, 77.28330°W

Tiputini Biodiversity Reserve, 00.61667°S, 76.16667°W

Pastaza (Provincia de)

Canelos, Río Bobonazo, 01.58333°S, 77.75000°W

Mera, Río Alpuyuca [= Río Alpayacu], 1,140 m, 01.46667°S, 78.13333°W

Montalvo [= Andoas], Río Bobonazo, 250 m, 02.06667°S, 76.96667°W

Puyo, 975 m, 01.46700°S, 77.98300°W

Río Capahuari, 02.51667°S, 76.85000°W

Río Conambo, 01.86667°S, 76.78333°W

Río Pindo Yacu, 01.53333°S, 77.95000°W

Río Tigre, 2,000 ft, 02.11700°S, 76.06700°W

Río Yana Rumi, 01.63333°S, 76.98333°W

Sarayacu, Río Bobonaza, 700 m, 01.73333°S, 77.48333°W

Tiguino, 130 km S of Coca, 300 m, ca. 01.64700°S, 76.98100°W

Pichincha (Provincia de)

Concepción, 00.10000°S, 768.50000°W

Cooperativa Salcedo Lindo, 4 km S of Encampamento de CODESA on road from Pedro Vicente Maldonado, 00.21667°N, 79.08333°W

Cotopaxi, Atacazo, 00.36667°S, 78.61667°W

Gualea, Río Tulipe, 1,200 m, 00.08300°S, 78.78300°W

Gualea, 1,500 m, 00.11667°N, 78.73333°W

Gualea, Ilambo Valley, 1,800 m, 00.11667°N, 78.83333°W

Guápulo, 00.20000°S, 78.48333°W

Guarmai, 00.28333°S, 78.71667°W

Guarumos, 2,000 m, 00.06667°S, 78.63333°W

Hacienda Monjas, 4,500 m, 00.23333°S, 78.61667°W

Huila North, oriente, 1,400 m, 00.05000°N, 78.88333°W

Las Máquinas, on trail from Aloag to Santo Domingo de los Colorados, 2,180 m, 00.43333°S, 78.73333°W

Mindo, Río Blanco, 1,260 m, 00.03333°S, 78.76667°W

Mojanda, 00.13333°N, 78.28333°W

Mount Pichincha [= Volcán Pichincha], 01.00000°S, 78.60000°W

Mount Pichincha, east side, 3,500 ft, 00.43300°S, 78.68300°W

Mt. Pichincha, 12,000 ft, 00.16667°S, 78.55000°W

Nanegal, 00.11667°N, 78.76667°W

Mount Illiniza, vicinity, 00.66667°N, 78.70000°W

Pacto, 1,400 m, 00.20000°N, 78.86700°W

Pichincha, 3,353 m, 00.16667°S, 78.55000°W

Pichincha, 3,300–3,800 m, 00.16667°S, 78.55000°W

Pinantura, 00.41667°S, 78.36667°W

Puembo, 00.18333°S, 78.35000°W

Quito, 00.21667°S, 78.50000°W

Quito, 00.25000°S, 78.58000°W

Quito, 15 mi S, 00.45386°S, 78.57680°W

Río Pita, upper, 11,500 ft, ca. 00.30000°S, 78.46667°W

Río Tulipe, near Gualea, 1,200 m, ca. 00.08333°S, 78.78333°W

Río Verde, 975 m, 00.20000°N, 78.86667°W

Saloya [= misspelled Galaya], 00.30000°S, 78.66667°W

San Ignacio [= Hacienda San Ignacio], 00.38333°S, 78.51667°W

San José, occidente, 2,000 m, 00.18333°N, 78.40000°W

San Tadeo, 00.01667°N, 78.80000°W

Tablón, 00.36667°S, 78.25000°W

Tablón, road to Papallacta, 00.36667°S, 78.25000°W

Tanda [= Hacienda Tanda], ca. 8,000 ft, 00.00001°N, 78.33333°W

Volcán Pichincha, 3,600–3,700 m, 00.16667°S, 78.55000°W

West Mindo, W of Andes, 4,000 ft, ca. 00.03333°S, 78.80000°W

Zapadores, Río Saloya, 6,400 ft, 00.01667°N, 78.95000°W

Santa Elena (Provincia de)

Cerro Manglar Alto [= Cerro Manglaralto, headwaters of Río Manglaralto; also Cerros de Colonche, following Painter 1993], 460 m, 01.78333°S, 80.61667°W

Cerro de Manglaralto, 460 m, 01.78333°S, 80.61667°W

Manglar Alto, vicinity, Cordillera de Colconche, 01.83330°S, 80.73330°W

Santa Domingo de los Tsáchilas (Provincia de)

Ila, 36 mi SW Santo Dominto, 800 ft, 00.58000°S, 79.17000°W

Santo Domingo, 485 m, 00.21667°S, 79.10000°W

Santo Domingo, 00.25000°S, 79.15000°W

Santo Domingo de los Colorados, 00.25000°S, 79.15000°W

Sucumbíos (Provincia de)

Lagarto Yaco [= Lagarto Cocha], 00.65000°S, 75.21700°W

Laguna Grande, Río Cuyabeno, 00.00000°S, 76.18333°W

Limoncocha [= Limon Cocha], Río Napo, 00.41667°S, 76.63333°W

["Missionary station established 1955 on unspoiled lemon-colored lake in tropical forest, 2 km from mouth of Río Jivino, tributary of the Río Napo." (Paynter and Traylor 1977:66)]

Reserve de Producción Faunística Cuyabeno, 220 m, 00.00001°N, 76.20000°W

Santa Cecilia, 340 m, 00.08400°N, 76.98900°W

Tungurahua (Provincia de)

Baños, 1,770 m, 01.40000°S, 78.41667°W

Hacienda San Francisco, east of Ambato, 8,000 ft, 01.01667°S, 78.50000°W

Laguna Pisayambo, 2 km SW, Pisayambo, Pargue Nacional Llanganates, 3,630 m, 01.08333°S, 78.31667°W

Mirador, 1,500 m, 01.43333°S, 78.25000°W

Palmera, 4,000 ft, 01.41667°S, 78.2000°W

San Antonio, 6,727 ft, 01.43333°S, 78.36667°W

San Francisco, E of Ambato, ca. 2,440 m, 01.30000°S, 78.50000°W

Zamora–Chinchipe (Provincia de)

Alto Machinaza, 1,400 m, 03.76606°S, 78.50517°W

Los Encuentros, 4 km ENE, 850 m, 03.75000°S, 78.61667°W

Sabanilla, 4 km E, 1,585 m, 04.03000°S, 79.02000°W

San Antonio, 02.23333°S, 78.41667°W

Zamora, 1,000 m, 04.06667°S, 78.96667°W

French Guiana

Arataye, 2,100 m, 04.03333°N, 52.70000°W

Awala-Yalimapo, 05.74111°N, 53.92778°W

Cacao, 04.58333°N, 52.46667°W

Camopi, 03.21667°N, 52.46667°W

Cayenne, 04.93333°N, 52.333333°W

Cayenne, vicinity, 04.86667°N, 52.31667°W

Inini, River Arataye, 04.00000°N, 52.33333°W

Iracoubo, 05.48333°N, 53.31667°W

Kaw, 100 m, 04.48333°N, 52.03333°W

Kourou, sea level, 05.15000°N, 52.65000°W

Les Nouragues, 120 m, 04.08333°N, 52.66667°W

Mont Saint Marcel, 02.36667°N, 53.01667°W

Nouragues, 04.08300°N, 52.66700°W

Paracou, near Sinnamary, 05.28333°N, 52.91667°W

Piste de Saint Élie, 04.83333°N, 53.28333°W

River Iracoubo, 05.50000°N, 53.23333°W
River Oyapock, 04.13333°N, 51.66667°W
Riviere Approuague, 04.63300°N, 51.96700°W
Saül, 03.61667°N, 53.20000°W
Saut Pararé, Rivière Arataye, 04.05000°N, 52.66667°W
Sinnamary River, right bank 21 km upstream from Petit
    Saut Dam, 04.94670°N, 53.03170°W
St. Eugène, 04.85000°N, 53.06667°W
St.-Laurent du Maroni, 05.50000°N, 54.03333°W
Tamanoir, Mana River, 05.15000°N, 53.75000°W
Trois Sauts, 02.16667°N, 53.18333°W

Guyana
Barima–Waini (District)

Baramita, Old World, 07.36667°N, 60.48333°W
Santa Cruz, 07.66667°N, 59.23333°W

Cuyuni–Mazaruni (District)

Kalakun, 06.40000°N, 58.65800°W
Kartabo [= Kartabu], 06.38333°N, 58.68333°W
Mount Roraima, summit, 05.20222°N, 60.73528°W
Mount Roraima, slopes, 05.23333°N, 60.71667°W
Venamo River, Potaro Highlands, 610 m, 06.00000°N,
    61.33333°W

Demerara–Mahaica (District)

Buxton, 1 mi E, 06.78333°N, 58.03333°W
Demerara [= Georgetown], Supenaam River, 06.80448°N,
    58.15527°W
Demerara River, 06.80000°N, 58.16667°W
Georgetown, sea level, 06.8044°N, 58.15527°W
Hyde Park, Demerara River, 100 m, 06.50000°N,
    58.26667°W
Loo Creek, 68 km by road S of Georgetown, 06.23333°N,
    58.25000°W

East Berbice–Corentyne (District)

Berbice [= New Amsterdam], 06.25000°N, 57.51667°W
Potaro Highlands, 395 m, 04.41667°N, 58.26667°W

Essequibo Islands–West Demerara (District)

Bonasika River, 06.75000°N, 58.50000°W
Essequibo River, 13 mi from mouth, ca. 06.98300°N,
    58.38300°W
Lower Essequibo River, 06.85556°N, 58.50589°W

Mahaica–Berbice (District)

Blairmont Plantation, 06.26667°N, 57.53333°W
Tauraculi, 05.95000°N, 57.80000°W

Pomeroon-Supenaam (District)

Makasima, Pomeroon River, 07.50000°N, 58.71667°W
Supinaam [= Supenaam] River, 06.98333°N, 58.51667°W

Potaro–Siparuni (District)

Kurupukari, 04.66667°N, 58.66667°W
Paramakatoi, 04.71667°N, 59.70000°W
Surama [Upper Takutu–Upper Essequibo District], 40 km
    NE, 04.38300°N, 58.86700°W

Upper Demerara–Berbice (District)

Comackpea [= Comaccka], Rio Cunerara, 05.83300°N,
    58.36700°W
Great Falls of Demerara River [= Ororo Marali], 05.31667°N,
    58.51667°W
Potaro River, 05.36667°N, 58.90000°W

Upper Takutu–Upper Essequibo (District)

Awarawaunowa, vicinity, 02.66667°N, 59.20000°W
Chodikar River, ca. 55 km SW of Wai-Wai village of Gunn's
    Strip, 01.36667°N, 58.76667°W
Dadanawa, 150 m, 02.83333°N, 59.50000°W
Dadanawa, 10 mi E, 02.83333°N, 59.50000°W
Dadanawa, vicinity, 02.83333°N, 59.51667°W
Kanuku Mountains, 600 ft, 03.20000°N, 59.35000°W
Nappi Creek, Kanuku Mountains, 03.38333°N, 59.80000°W
Quarter Mile Landing, 5 km S of Annai, 03.91700°N,
    59.10000°W
Surama, 5 km SE, 04.16670°N, 59.08330°W
Tamton, 02.35000°N, 59.71667°W

Paraguay
Alto Paraguay (Departamento de)

Agua Dulce, 54 km E, 20.01667°S, 59.76667°W
Cerro León, 50 km WNW of Fortín Madrejón, 20.38300°S,
    60.31700°W
Estancia Kambá Aka, Parque Nacional Río Negro,
    19.83333°S, 58.75000°W
Estancia Tres Marias, 21.27867°S, 59.55217°W
Estancia 3 Maria [= Estancia Tres Marias], 70 m, 21.27867°S,
    59.55217°W

Fortín Madrejón, 50 km WNW, 20.46761°S, 60.31908°W
Laguna Placenta, 70 m, 21.14396°S, 5941483°W
Palmar de las Islas, 215 m, 19.62981°S, 60.61250°W
Puerto Casado [= Puerto La Victoria], 22.33333°S, 57.91667°W

Alto Paraná (Departamento de)

Puerto Bertoni, 91 m, 25.63333°S, 54.66667°W
Refugio Biológico Limoy, north of Río Limoy, ca. 270 m, 24.80000°S, 54.45000°W
Reserva Biológica Limoy, 24.73000°S, 54.40000°W

Amambay (Departamento de)

Cerro Corá, 4 km SW by road, 22.61667°S, 55.98333°W
Colonia Sargento Dure, 3 km E of Río Apa, 22.16700°S, 56.45000°W
Parque Nacional Cerro Corá, 22.65000°S, 56.000°W
Parque Nacional Cerro Corá, 22.66667°S, 55.98333°W
Parque Nacional Cerro Corá, 22.50000°S, 56.18000°W
Pedro Juan Caballero, 28 km SW, 22.74549°S, 55.80921°W

Boquerón (Departamento de)

Colonia Fernheim, 112 m, 22.25000°S, 59.83333°W
Estancia Iparoma, 19 km N of Filadelfia, 22.17591°S, 60.02905°W
Filadelfia, 22.34540°S, 60.03150°W
Fortín Guachalla, 22.45000°S, 62.33333°W
Fortín Juan de Zalazar, 23.10000°S, 59.30000°W
Fortín Teniente Pratts Gil, 22.68333°S, 59.43333°W
Fortín Toledo, 22.02230°S, 64.30000°W
Loma Plata, 22.383330°S, 59.85000°W
Parque Nacional Teniente Enciso, 19.84202°S, 58.75332°W
Teniente Enciso, 21.08333°S, 61.76667°W
Villa Hayes, 410 km NW by road, 22.48333°S, 60.43333°W
Villa Hayes, 420 km NW by road, 22.48333°S, 59.98333°W

Caaguazú (Departamento de)

Carayaó, 24 km NNW, Estancia San Ignacio, 25.18333°S, 56.40000°W
Colonia Somerfield, 25.43333°S, 55.71667°W
Coronel Oviedo, 22.5 km N by road, 24.41667°S, 56.45000°W
Estancia San Ignacio, 24 km NNW of Carayaó, 25.08333°S, 56.60000°W
Sommerfield Colony [= Colonia Somerfield], 25.43333°S, 55.71667°W

Caazapá (Departamento de)

Estancia Dos Marías, 26.76667°S, 56.53333°W

Canindeyú (Departamento de)

Curuguaty, 6.3 km N by road, 24.51667°S, 55.70000°W
Curuguaty, 13.3 km N by road, ca. 24.51667°S, 55.70000°W
Estancia Salazar, Río Verde, 24.18333°S, 55.26667°W
Reserva Natural del Bosque Mbaracayú, 24.12000°S, 55.51000°W
Reserva Natural del Bosque Mbaracayú, 200 m, 24.15000°S, 55.31667°W
Reserva Natural del Bosque Mbaracayú, 24.09000°S, 55.17000°W
Reserva Natural del Bosque Mbaracayú, headquarters, 24.16667°S, 56.91667°W
Reserva Natural Privada Morombí, 24.67000°S, 55.38000°W
Reserva Natural Privada Morombí, 24.60000°S, 55.43000°W
Sendero Morotí, Reserva Mbaracayú, 24.13333°S, 55.41667°W

Central (Departamento de)

Asunción, 25.26667°S, 57.66667°W

Concepción (Departamento de)

Concepción, 7 km NE, 23.41667°S, 57.28333°W
Concepción, 8 km E, 23.41670°S, 57.15010°W
Concepción, vicinity, 23.41667°S, 57.28333°W
Horqueta, 23.40000°S, 56.88333°W
Rte. 3, 28 km S junction Rte. 3 and Rte. 5, 22.71667°S, 56.30000°W
Yby-Yaú, 22.93333°S, 56.53333°W

Cordillera (Departamento de)

Atyrá, 22.15000°S, 57.10000°W
Tobatí, 1.6 km S by road, 150 m, 25.25000°S, 57.06667°W
Tobatí, 12 km N by road, 25.14170°S, 57.06670°W

Itapúa (Departamento de)

El Tirol, 27.18333°S, 55.71667°W
Encarnación, 27.33333°S, 55.90000°W
Estación Parabel, 0.3 km E of house, 26.34850°S, 55.51250°W
Parque Nacional San Rafael, 26.50389°S, 55.79222°W
Reserva de Recursos Manejados San Rafael, 26.54000°S, 55.77000°W
San Rafael, 8 km N, 27.06189°S, 54.38393°W

Misiones (Departamento de)

Ayolas, 5 km ENE, 27.34943°S, 56.80331°W
Corate-í, 27.40000°S, 57.01667°W
Isla Yaciretá, 27.40000°S, 56.75000°W
Refugio Faunistico "Atingy," 75 m, 27.35000°S, 56.70000°W
San Francisco, 26.86667°S, 57.05000°W
San Francisco, 36 km NE of San Ignacio, 26.88680°S, 57.03210°W
San Ignacio, 409 km S, 26.86700°S, 57.75000°W
San Pablo, 26.86667°S, 57.05000°W

Ñeembucú (Departamento de)

Desmochado, 27.11667°S, 58.10000°W
Estancia San Felipe, 22.68267°S, 57.35867°W
Estancia Yacaré, 26.58333°S, 58.13333°W
Mayor Martínez, 27.15000°S, 58.21667°W
Paso Pucú, 27.13333°S, 58.51667°W
Pilar, 26.86879°S, 58.29346°W

Paraguarí (Departamento de)

Costa del Río Tebicuary, 26.51667°S, 57.23333°W
Parque Nacional Ybycuí, 26.01667°S, 57.05000°W
Sapucaí, ca. 300 m, 25.66667°S, 56.91667°W
Sapucaí, 25.66000°S, 56.95000°W
Sapucay [= Sapucaí], ca. 300 m, 25.66667°S, 56.91667°W
Ybycuí, 26.01670°S, 57.05000°W

Presidente Hayes (Departamento de)

Chaco-í, 15.5 km by road NNW, 25.16667°S, 57.73333°W
Estancia Juan de Zalazar, 2 km S of headquarters, 23.11280°S, 59.31380°W
Estancia Laguna Porá, 22.33333°S, 59.43333°W
Estancia Samaklay, 23.46667°S, 59.86667°W
Fortín Juan de Zalazar, 23.10000°S, 59.30000°W
Juan de Zalazar [= Fortín Juan de Zalazar], 23.10000°S, 59.30000°W
Juan de Zalazar [=Fortín Juan de Zalazar], 8 km NE, 23.02594°S, 59.17994°W
La Golondrina, 25.10000°S, 57.56667°W
Puerto Falcón, 83 km NW, 24.80000°S, 58.35000°W
Puerto Pinasco, 22.71667°S, 57.83333°W
Riacho Negro, 25.06667°S, 57.91667°W
Villa Hayes, 25.10000°S, 57.56667°W
Villa Hayes, 24 km NE, 24.93333°S, 57.75000°W
Villa Hayes, 24 km W, 25.08333°S, 57.7667°W
Villa Hayes, 69 km NW by road, 24.68875°S, 57.96547°W
Waikthlatingwaialwa [or Waikthlatingwayalwa; see Paynter 1989], 23.10007°S, 59.30036°W

San Pedro (Departamento de)

Aca-Poi, 1 km N, 23.50000°S, 56.70000°W
Ganadera Jejui, 24.09000°S, 56.48000°W
Tacuati, Aca-Poi, 23.45000°S, 56.58333°W

Peru

Amazonas (Regíon de, formerly Departamento de)

ACP Abra Patricia-Alto Nieva, 2,282 m, 05.68733°S, 77.80930°W
ACP Huiquilla El Choctamal, 2,762 m, 06.38039°S, 77.95500°W
Bagua, 8 km WSW, 05.66667°S, 78.51667°W
Balsas, mountains to E, 10,000 ft, 06.83333°S, 77.83333°W
Balsas, mountains to E, 3,048 m, 06.75000°S, 77.83300°W
Chachapoyas, 06.21667°S, 77.85000°W
Communidad Estera, Cerro Calle Nueva, 3,651 m, 12.90521°S, 73.80436°W
Condechaca, Río Utcubamba, 06.35000°S, 77.90000°W
Cordillera Colán, E of La Peca, 3,260 m, ca. 05.50000°S, 78.16700°W
Cordillera Colán, E of La Peca (ridge W of peaks), 10,000 ft, 05.51092°S, 78.33683°W
Cordillera Colán, E of La Peca (ridge W of peaks), 11,100 ft, 05.56550°S, 78.30967°W
Cordillera Colán, NE of La Peca, 10,000 ft, 05.48087°S, 78.28260°W
Cordillera del Cóndor, Valle Río Comaina, Puesta Vigilancia 3, Alfonso Ugarte, 03.90810°S, 78.42110°W
Goncha, 8,500 ft, 05.75565°S, 78.46664°W
Goncha, 2,700 m, 05.75470°S, 78.46670°W
Guayabamba, 06.36667°S, 77.41667°W
Huambo, 1,130 m, 06.36667°S, 77.46667°W
Huampami [= Huampam], vicinity, Río Cenepa, 700 ft, 04.58333°S, 78.20000°W
Huampami [= Huampam], Río Cenepa, 210 m, 04.46670°S, 78.16667°W
Kagka (Aguaruna village), Río Kagka, tributary of Río Comaina, 04.38333°S, 78.19667°W
Lake Pomachcoas, 6 km SW by road, 6,000 ft, 05.86989°S, 77.98731°W
La Peca Nueva, 12 km E by trail, 1,760 m, 05.56667°S, 78.28333°W
La Poza, 170 m, 04.05000°S, 77.76667°W
Leimebamba [= Leymebamba], ca. 20 km by road W, 2,804 m, 06.75000°S, 77.80000°W
Molinopampa, Tambo Ventija [= Ventilla], 2,407 m, 06.18333°S, 77.61666°W
Molinopampa, 10 mi E, 06.18333°S, 77.50000°W

Pomacocha [= Florida], 5 km N and 5 km E, 1,830 m, 05.78812°S, 77.87152°W

Pongo de Rentema, Río Marañón, 05.50109°S, 78.50000°W

Río Kagka, headwaters, tributary of the Río Cenepa, 04.26667°S, 78.15000°W

Río Santiago, mouth at Río Marañón, 04.42727°S, 77.63636°W

Río Utcubamba, about 15 mi S of Chachapoyas, 06.33333°S, 77.86667°W

Río Utcubamba, 15 mi above Chachapoyas, 05.93333°S, 77.98333°W

Río Utcubamba, Uchco, 05.93333°S, 77.98333°W

Rodriguez de Mendoza, Tambo Almirante, near Uchco, 06.18333°S, 77.21667°W

Santa Rosa, 05.40000°S, 78.50000°W

San Pedro, 06.63333°S, 77.70000°W

Tambo Carrizal, mountains E of Balsas, 2,000 m, 06.78333°S, 77.86667°W

Tingo, 30 km S and 41 km E of Bagua, Río Utcabamba, 05.90000°S, 78.20000°W

Tseasim (Aguaruna village), Río Huampani of Río Cenepa, 04.45667°S, 78.15333°W

Uchco [= Uschco or Uscho], Tambo Almirante, 1,525 m, 06.11700°S, 77.33300°W

Uscho [= Uschco or Uchco], about 50 km E of Chachapoyas, 5,000 ft, 06.18333°S, 77.21667°W

Yambrasbamba, 6,500 ft [= 1,970 m], 05.76667°S, 77.90000°W

Ancash (Regíon de, formerly Departamento de)

Callon, 2 km S and 11 km W of Huaras, 09.55142°S, 77.63352°W

Carhuaz, Jangas, Antauran, Cordillera Negra, Quebrada Lancash, 3,000 m, 09.40000°S, 77.58333°W

Casma, 29 km S (by road), 30 m, 09.46667°S, 78.31667°W

Chasquitambo, 4 km by road NE, km 51, 13.80000°S, 73.38333°W

Cuenca del Río Pampas, below Conzuso, 08.17234°S, 77.85613°W

Huaras, 3,052 m, 09.53333°S, 77.58333°W

Huaras, Quilcahuanca, 4,300 m, 09.50000°S, 77.41700°W

Huaras, 25 km S, 3,810 m, 09.86950°S, 77.39767°W

Huaras, 12 km W, 12,680 ft, 09.53333°S, 77.64263°W

Huaras, 3 km S and 12 km W, 12,500 ft, 09.56042°S, 77.654260°W

Huari, Yanacocha, 09.65556°S, 77.13389°W

Macate, 2,712 m, 08.76667°S, 78.08333°W

Pariacoto, 1 km N and 12 km E, ca. 09.50763°S, 77.77404°W

Quebrada Chalhuacocha, Acrana, 3,844 m, 08.24411°S, 77.77064°W

Rachococo, 45 km S of Huaraz, 09.90000°S, 77.40000°W

Recuay, 4 mi S and 8 mi E, 3,810 m, 09.77487°S, 77.34934°W

Ticapampa, 4,200 m, 09.80000°S, 77.70000°W

Uramarca [= Yuramarca], near Paloma, 08.75000°S, 77.90000°W

Apurímac (Regíon de, formerly Departamento de)

Abancay, 10 km N, 13.59405°S, 72.85024°W

Abancay, 28 km NE (by road), 3,620 m, 13.55083°S, 72.75407°W

Abancay, 10 km SSE, 13.66684°S, 72.88131°W

Chalhuanca, 25 km SW, 14,500 ft, 14.53897°S, 73.51213°W

Chalhuanca, 36 km S (by road), 3,510 m, 14.54777°S, 73.30943°W

Grau, Virundo, 14.25022°S, 72.68144°W

Arequipa (Regíon de, formerly Departamento de)

Antiquipa, 15.78333°S, 74.26667°W

Antiquipa, 1 km SW, 325 m, 15.80000°S, 74.36667°W

Antiquipa, 2 km E, 15.80000°S, 74.33680°W

Arequipa, 16.40000°S, 71.55000°W

Arequipa, 12 km E, 3,230 m, 16.70000°S, 71.43749°W

Arequipa, 43 km E, 3,930 m, 16.38333°S, 71.38333°W

Arequipa, ca. 33 rd km E, 16.36600°S, 71.28962°W

Arequipa, ca. 48 km by road E, 16.34967°S, 71.16393°W

Cabrerías, 16.25000°S, 71.48330°W

Cailloma, 5 km W, 4,530 m, 15.18806°S, 71.81875°W

Callalli, 15 km S, 4,150 m, 15.64389°S, 71.44556°W

Camana, 5 mi ENE, 3,000 ft, 16.58900°S, 72.63073°W

Chala, 16 km ESE (by road), 10 ft, 16.07183°S, 74.09667°W

Chavina, on coast near Acari, Lomos River, 15.61667°S, 74.63333°W

Chivay, 1 km N, 3,700 m, 15.63096°S, 71.60000°W

Chucarapi, Tambo Valley, 300 ft, 17.07083°S, 71.72195°W

El Rayo, 15.91667°S, 71.58333°W

Huancarama, Orcopampa, 3,791 m, 15.26000°S, 72.34000°W

Huaylarco, 54–55 mi ENE of Arequipa, 15,300 ft, 16.03333°S, 70.83333°W

Laguna Salinas, 14,100 ft, 16.36667°S, 71.13333°W

Lomas, 7 km E of Matarani, 15.56667°S, 74.83333°W

Salinas, 4,316 m, 16.37840°S, 71.13330°W

Salinas, 22 mi E of Arequipa, 14,200 ft, 16.36667°S, 71.13333°W

San Ignacio, 4,375 m, 15.16100°S, 71.78500°W

San Juan de Tarucani, 4,345 m, 16.18000°S, 71.06000°W

Sumbay, 15.96667°S, 71.40202°W

Sumbay, 2 km W, 4,200 m, 15.96667°S, 71.40202°W

Tambo, 17.06667°S, 71.83333°W

Toccra Pampa, Caylloma, 3,970 m, 15.96840°S, 71.72780°W

Ayacucho (Región de, formerly Departamento de)

Anchihuay, 3,250 m, 13.00598°S, 73.73698°W

Arizona (fish farm), Río Vinchos, 3,397 m, 13.34653°S, 74.43020°W

Chinquintirca, 2,685 m, 13.04309°S, 73.68315°W

El Bagrecito, Río Vinchos, 3,526 m, 13.34476°S, 74.45077°W

Hacienda Luisiana, Río Ayacucho, 12.66667°S, 73.73333°W

Huahuanchayo [= Huanhuachayo], ca. 1,660 m, 12.73333°S, 73.78333°W

[Huanhuachayo is "a clearing along the Andean mule trail connecting Hacienda Luisiana and nearby communities along the Río Apurímac and Río Santa Rosa with the mountain town of Tambo" (A. L. Gardner and Patton 1976:42).]

Huanta, 2 mi SE, 12.95379°S, 74.22915°W

Nazca, 35 km ENE, 3,200 m, 14.74713°S, 75.01044°W

Pampamarca, 4 km W (by road), 4,230 m, 14.56667°S, 73.58712°W

Pampamarca, 23 km (by road) W, 14.56667°S, 73.76343°W

Puncu, 30 km NE of Tambo, 3,370 m, 12.78300°S, 73.81700°W

Puquio, 10 km WNW, 14.64466°S, 74.27059°W

Puquio, 11 km NW, 14.58752°S, 74.24889°W

Puquio, 15 mi WNW, 12,000 ft, 14.61699°S, 74.33921°W

Puquio, 18 km E by road, 3,770 m, 14.58497°S, 74.01515°W

Puquio, 21 km ENE, 14,500 ft, 14.58379°S, 73.84510°W

Río Santa Rosa, 12.70000°S, 73.73333°W

San José, Río Santa Rosa above Hacienda Luisiana, 1,000 m, 12.73333°S, 73.76667°W

San Miguel Tambo, 3,500 m, 13.00000°S, 73.98333°W

Tambo, 6 mi NNE, 12,300 ft, 12.93333°S, 74.01667°W

Tucumachay Cave, 4,250 m, 11.67150°S, 75.04990°W

Vischongo, Río Pomacochas, 2,800 m, 13.58860°S, 73.99610°W

Yanamonte, 2,751 m, 12.78500°S, 73.99500°W

Yuraccyacu, 2,600 m, 12.75000°S, 73.80000°W

Cajamarca (Región de, formerly Departamento de)

Bosque Cachil, between Cascas and Contumaza, 2,500 m, 07.38925°S, 78.78188°W

Cajabamba, 07.61667°S, 78.05000°W

Cajamarca, 2,720 m, 07.16667°S, 78.51667°W

Catamarca, 35 mi WNW, 06.97176°S, 78.98780°W

Celendín, Hacienda Limón, 6,720 ft, 06.83333°S, 78.08333°W

Chaupe, 05.16667°S, 79.16667°W

Chaupe, 4 km W, ca. 05.16667°S, 79.16667°W

Chota, 7 km N and 3 km E, 8,650 ft, 06.48670°S, 78.62287°W

Contumaza, Bosque Cachil, between Cascas y Contumaza, 2,739 m, 07.36667°S, 78.81667°W

Cutervo, San Andrés de Cutervo, 1,998 m, 06.21667°S, 78.6667°W

Cutervo, San Andrés de Cutervo, Cutervo National Park, 100 m over El Tragadero, 3,000 m, 06.24997°S, 78.76653°W

Cutervo, 1 km NW, 8,700 ft, 06.36027°S, 78.85639°W

El Arenal, 1 km S and 6 km W of Pomahuaca, 915 m, 07.12889°S, 79.05772°W

Hacienda Limón, 06.83333°S, 78.08333°W

Hacienda Taulis, 06.90000°S, 79.05000°W

Huarandosa, 05.21667°S, 78.8000°W

Las Ashitas, 4 km W of Pachapiriana, 9,200 ft, 05.65521°S, 79.16010°W

Las Juntas, 3 km W of Pachapiriana, 7,600 ft, 06.65275°S, 79.15160°W

Monte Seco, 2.5 km N, 06.86667°S, 79.08333°W

Perico, 500 m, 05.25000°S, 78.75000°W

Pisit, T 25, 3,399 m, 06.81139°S, 78.87052°W

San Ignacio, 05.13333°S, 78.96333°W

San Ignacio, Tabaconas, Cerro La Viuda (Tabaconas-Namballe National Sanctuary buffer zone), 1,897 m, 05.60139°S, 79.32176°W

San Ignacio, Tabaconas, Piedra Cueva in Cerro Coyona (Tabaconas-Namballe National Sanctuary), 3,343 m, 05.26722°S, 79.27052°W

San Pablo, 4,000 m, 07.11667°S, 78.83333°W

Santa Cruz, Catache, 3.81 km NE from Monteseco, 2150 m, 06.84365°S, 79.08133°W

Tambillo, Río Malleta, 5,800 ft, 06.16667°S, 78.75000°W

Taulís, 2,700 m, 06.90000°S, 79.05000°W

Cusco (Región de, formerly Departamento de)

Amacho, 2,750 m, 13.50000°S, 70.91667°W

Amaibamba [= Amaybamba], Urubamba Valley, 13.06306°S, 72.45370°W

Amaybamba, 13.06306°S, 72.45370°W

Bosque Aputinye above Huyro, 12.96667°S, 72.60000°W

Calca, 55.4 km by road N, 3,560 m, 13.13333°S, 72.00000°W

Ccachubamba, 2,850 m, 13.50000°S, 70.91667°W

Ccolini, near Marcapata, 3,900 m, 13.51667°S, 70.96667°W

Centro de Investigación Wayqecha, 2,550–2,900 m, 13.176488°S, 71.58723°W

Chancarara, Hatunpampa, 3,150 m, 13.62336°S, 70.83411°W

Chirapata, Cossireni Pass, 12.71667°S, 72.23333°W

Chospyoc [= Chospiyoc], Urubamba, 10,000 ft, 13.26667°S, 72.35000°W

Consuelo, 15.9 km SW of Pilcopata, 1,000 m, 13.02362°S, 71.49185°W

Consuelo, W of Pilcopata, 1,000 m, 13.13333°S, 71.25000°W

Cordillera de Sicuani, 14.26670°S, 71.21667°W

Cordillera Vilcabamba, 3,350 m, 11.66000°S, 73.66722°W

Cordillera Vilcabamba, northern end, 3,370 m, 11.66000°S, 73.66722°W

Cuzco, 94 km W on road to Abancay, 13.58333°S, 72.03460°W

Hacienda Cadena, ca. 1,000 m, 13.40000°S, 70.71667°W

Hacienda Villa Carmen, Cosñipata [= Río Cosñipata], 850 m, 12.83300°S, 71.25000°W

Hadquiña, 1,500 m, 13.11667°S, 72.65000°W

Huancarani, 5 km N, 3,140 m, 14.25370°S, 71.22750°W

Idma, Santa Ana Valley, 4,600 ft, 12.88333°S, 72.81667°W

Idma, road, 11,200 ft, 12.88330°S, 72.81700°W

Kiteni, Río Urubamba, 12.33333°S, 72.83333°W

Kiteni, 66 km from Rosalina on road from Quillabamba, 12.33333°S, 72.83333°W

La Convencíon, 12.90000°S, 72.53333°W

Laguna Sibinacocha, 13.84650°S, 72.05720°W

Lauramarca [= Labramarca], 13.70000°S, 71.31667°W

Lucma, Cosireni Pass, Quirapata [= Chirapata], 2,985 m, 13.05111°S, 72.56667°W

Machu Picchu, 1,830 m, 13.16611°S, 72.49833°W

Machu Picchu, 3,900 m, 13.15833°S, 72.53139°W

Machu Picchu, above timberline, Runcaraccay ruins, 14,000 ft, 13.23333°S, 72.48333°W

Machu Picchu, San Miguel bridge, 6,000 ft, 13.10000°S, 72.63333°W

Marcapata, 2,743 m, 13.50000°S, 70.91667°W

Marcapata, 3,900 m, 13.58722°S, 70.95833°W

Marcapata, directly below, 13.50000°S, 70.91667°W

Marcapata, Camante, 2,000 m, 13.50000°S, 70.91667°W

Marcapata, Limacpunco, 2,400 m, 13.40000°S, 70.71667°W

Marcapata, 1 km below, 9,000 ft, 13.59876°S, 70.97222°W

Ocobamba, near Cuzco, 12.83333°S, 74.23333°W

Occobamba [= Ocobamba] Pass, 4,000 m, 12.89000°S, 72.81333°W

Ollantaytambo [= Ollantaitambo], 13.26667°S, 72.26667°W

Paltaybamba [= Paltaibamba], 13.03333°S, 72.08333°W

Paucartambo, 20 km N (by road), km 100, 3,580 m, 13.20269°S, 70.66667°W

Paucartambo, 32 km (by road) NE at km 112, 3,140 m, 13.13333°S, 71.36667°W

Paucartambo, 54 km NE (by road), km 135, 2,190 m, 13.14814°S, 71.58655°W

Pacuartambo, 72 km NE (by road), km 152, 1,460 m, ca. 13.11667°S, 71.28333°W

Paso Ocobamba, 3,800–4,210 m, 12.93333°S, 72.26667°W

Pillahuata, 2,460 m, 13.16218°S, 71.59751°W

Puesto de Vigilancia Acjanacu, 3,350–3,500 m, 13.19639°S, 71.61972°W

Quillabamba, 90 km SE (by road), 3,450 m, 13.11707°S, 7233935°W

Quince Mil [= Quincemil], 650 m, 13.26667°S, 70.63333°W

Quincemil [= Quince Mil], Río Marcapata, 680 m, 13.26700°S, 70.63300°W

Río San Miguel, 4,500 ft, 12.71667°S, 73.23333°W

Río Urubamba, 50 km NE of Cuzco, 13.10135°S, 72.643356°W

San Juan Grande, Quincemil, Quispicanchis, 650 m, 13.21667°S, 70.73333°W

Santa Ana, upper Río Urubamba Valley, 12.86667°S, 72.71667°W

Suecia, km 128 on Carretera Shintuya, 13.10053°S, 71.56875°W

Suecia, below, Manu Biosphere Reserve, 1,900 m, 13.10053°S, 71.56875°W

Tangoshiari, 2 km SW, Río Pagoreni, 11.76670°S, 73.32580°W

Tocapoqueu, Occobamba [= Ocobamba] Valley, 9,100 ft, 12.88333°S, 72.35000°W

Torontoy, 9,500 ft, 13.16700°S, 72.50000°W

Tres Cruces, 18 km N of Paucartambo, 11,900 ft, 13.10000°S, 71.76667°W

Urubamba, 13.30000°S, 72.11667°W

Yauri, 16 km SW, 3,960 m, 14.89364°S, 71.51732°W

Huancavelica (Regíon de, formerly Departamento de)

Hacienda Piso, Lorroja, 3,018 m, 12.73333°S, 74.45000°W

Huancavelica, 3,660 m, 12.76667°S, 75.03333°W

Lircay, 12.98333°S, 74.73333°W

Ticrapo, 2 km S, Pisco Valley, 2,440 m, 13.35000°S, 75.39820°W

Huánuco (Regíon de, formerly Departamento de)

Agua Caliente, Río Pachitea, 08.83333°S, 74.68333°W

Ambo, 2,064 m, 10.11667°S, 76.16667°W

Bosque Zapatagocha, above Acomayo, 9,250 ft, 09.66667°S, 76.05000°W

Campamento Provias, 2,890 m, 08.68828°S, 76.96922°W

Campamento Regional, 3,010 m, 08.65519°S, 77.00231°W

Carpish Pass, below, on trail to Hacienda Paty, 2,165 m, 09.70000°S, 76.15000°W

Caserío de San Pedro de Carpish, 3,015 m, 09.72972°S, 76.10694°W

Cayna, Chiliatuna, 10.18333°S, 76.33333°W

Chinchao, 5,700 ft, 09.63333°S, 76.06667°W

Chinchavito, at the mouth of the Río Chiraco, 09.48333°S, 75.91667°W

Chinchuragra, 3,850 m, 09.45889°S, 76.83611°W

Chinchuagra, Punchao, 09.46556°S, 76.83278°W

Cordillera Carpish, 09.66667°S, 76.15000°W

Cordillera Carpish, east slope, Carretera Central, 2,400 m, 09.70483°S, 76.07400°W

Cullcui, 09.38333°S, 76.70000°W

Dos de Mayo, Cayna, Chiliatuna, 10.18333°S, 76.33333°W

Galloganán, 10.15889°S, 76.13611°W

Hacienda Buena Vista, Río Chinchao, ca. 1,070 m, 09.63333°S, 76.06667°W

Hacienda Éxito, Río Cayumba, 915 m, 09.43333°S, 76.00000°W

Hatoncucho, 10.15444°S, 76.12917°W

Huánuco, 3,200–3.720 m, 09.91700°S, 76.23300°W

Huánuco, mountains 15 mi NE, 3,200 m, 09.88333°S, 76.11667°W

Huánuco, mountains 15 mi NE, 12,200 ft, ca. 09.88300°S, 76.20000°W

Ichocán, 10.17083°S, 76.12000°W

Iscarag, 3,630 m, 09.30472°S, 76.59306°W

Kenqarajra, 3,190 m, 09.08917°S, 76.79583°W

Pampa Hermoza, 3,130 m, 08.89306°S, 76.97222°W

Panao, mountains, 09.83333°S, 76.03333°W

Río Chinchao, 09.50000°S, 75.93333°W

Shogos, 2,986 m, 09.88542°S, 76.43942°W

Tingo Maria, Río Huallaga, 09.28333°S, 75.98333°W

Tingo Maria, 35 km NE on Carretera Central, 09.09795°S, 75.80578°W

Unchog [= Bosque Unchog], pass between Churrubamba and Hacienda Paty, NNW of Acomayo, 3,450 m, 09.68333°S, 76.11667°W

Ica (Regíon de, formerly Departamento de)

Hacienda San Jacinto, Ica, 14.15000°S, 75.75000°W

Ica, 70 mi E of Pisco, 14.06667°S, 75.70000°W

Pisco, 10 km SSE, 200 ft, 13.78351°S, 76.18129°W

Junín (Regíon de, formerly Departamento de)

Acobamba, 45 mi [= 72 km] NE of Cerro de Pasco, 2,440 m, 11.33300°S, 75.41000°W

Amable Maria, 11.16667°S, 75.16667°W

Carhuamayo, 10.92000°S, 76.05000°W

Casapalca, 1 mi E, 14,000 ft, 11.08667°S, 76.57674°W

Chanchamayo, 1,300 m, 11.25000°S, 75.31667°W

Chanchamayo, 1,500 m, 10.91667°S, 75.30000°W

Chanchamayo [= Pueblo Nuevo], 1,000–1,200 m, 11.05000°S, 75.31700°W

[see A. L. Gardner and Patton 1976:42]

Colonia Amable Maria, Montaña de Vitoc, 11.16667°S, 75.31667°W

Concepción, 7 km E, 11.91667°S, 75.21900°W

Cordillera Vilcabamba, 11.55972°S, 73.64111°W

Huancayo, 17 mi WNW, 11.97257°S, 75.46414°W

Incapirca, Zezioro, 11.0000°S, 76.20000°W

Jauja, 10 km W, 11,550 ft, 11.80000°S, 75.15222°W

La Garita del Sol, 1,750 m, 11.28333°S, 75.35000°W

La Oroya, 22 km N, 4,040 m, 11.38267°S, 75.87967°W

Lurin, 15 mi E, 12.31667°S, 76.06613°W

Maraynioc, Vitoc Valley, 3,800 m, 11.36667°S, 75.40000°W

Maraynioc, 45 mi NE of Tarma, 3,655–3,960 m, 11.36667°S, 75.40000°W

Palca, 16 km NNE (by road), 2,540 m, 11.21637°S, 75.46057°W

Perené [= Colonia del Perené], 800–1,200 m, 10.96667°S, 75.21667°W

Pomacocha, Yauli Valley, 14,200 ft, 11.73556°S, 76.13250°W

San Blas, vicinity, 4,250 m, 11.01603°S, 76.19009°W

San Ramón, 2 km NE, 10.96667°S, 75.21667°W

San Ranón, 3 mi SW, ca. 11.16437°S, 75.33998°W

San Ramón, 10 km WSW, 1,275 m, 08.13300°S, 75.33300°W

Satipo, on Río Satipo–Río Tambo drainage, 629 m, 11.26700°S, 74.61700°W

Satipo, Cordillera de Vilcabamba, 2,015–2,050 m, 11.56667°S, 73.63333°W

Tarma, Tiambra, 2,960 m, 11.24000°S, 75.68747°W

Tarma, 2 mi NW of San Ramón, 2,000 ft, 11.41667°S, 75.71667°W

Tarma, 22 mi E, 7,300 ft, 11.24850°S, 75.54420°W

Vitoc Valley, 11.30000°S, 75.33333°W

Yano Mayo, Río Tarma, 2,590 m, 11.41700°S, 75.7000°W

La Libertad (Regíon de, formerly Departamento de)

Cachicadán, 3,100 m, 08.06463°S, 78.17321°W

Hacienda Llagueda, 07.76667°S, 78.58333°W

Huamachuco, 8 mi S, 07.80000°S, 78.06667°W

Huamachuco, south of, 07.81483°S, 78.0503°W

Mashua, east of Tayabamba on trail to Ongón, 3,350 m, 08.20000°S, 77.23333°W

Menocucho, 08.01667°S, 78.83333°W

Otuzco, 5 mi SW, 8,000 ft, 07.95145°S, 78.63493°W

Otuzco, mountains to NE, 3,050 m, 07.90000°S, 78.58333°W

Pacasmayo, 8 m, 07.40000°S, 79.56667°W

Pacasmayo, 5 km NE, 60 m, 07.36640°S, 79.53435°W

Piedra Negra, on trail from Los Alisos to Parque Nacional Río Abiseo, 07.77356°S, 77.54445°W

Sanagorán, 2,700 m, 07.78312°S, 78.14639°W

Santiago de Chuco, 10 mi WNW, 3,960 m, 08.09464°S, 78.31747°W

Trujillo, 08.11667°S, 79.03333°W

Utcubamba, on trail to Ongón, 2,075 m, 08.21667°S, 77.13333°W

Lambayeque (Regíon de, formerly Departamento de)

Bosque de Chiñama, 2,550 m, 06.10000°S, 79.43333°W

Boyovar, 12 km S and 8 km E, 05.95851°S, 80.99443°W

Chongoyape, Cabache, 210 m, 06.64506°S, 79.389170°W

Etén, 20 m, 06.90000°S, 79.96667°W

Morrope, 2 mi SE, 500 ft, 06.57046°S, 79.99621°W

Morrope, 8 km S, 06.55000°S, 80.01667°W

Reserva Ecológica Chaparri, 06.51667°S, 79.45000°W

Seques, 06.90000°S, 79.30000°W

Uyurpampa, 2,827 m, 06.21997°S, 79.36236°W

Lima (Provincia de, formerly Departamento de)

Cañete, 19 km N, 30 m, 13.08333°S, 76.40000°W

Canta, 1 mi W, 11.46722°S, 76.63886°W

Casapalca, 1 mi E, 14,000 ft, 11.08667°S, 76.57674°W

Casapalca, 1.5 mi W, 13,200 ft, 11.08667°S, 76.61337°W

Casapalca, 6.3 mi W, 11.65000°S, 76.23333°W

Cerro Azul, Río Cañete Valley, 100 m from ocean, 13.05000°S, 76.5000°W

Cerro San Jerónimo, 12.00000°S, 77.03333°W

Chancay, 20 km N and 6 km W, 90 m, 11.58333°S, 77.26667°W

Huaros, 12,000 ft, 11.28333°S, 76.53333°W

Huaros, below, 11.28330°S, 76.53333°W

Lomas de Atocongo, 12.13333°S, 76.93333°W

Lomas de Lachay, 12 km N and 11 km W of Cancay, 11.35000°S, 77.38333°W

Matucana, 11.85000°S, 76.40000°W

Naña, 11.98333°S, 76.83333°W

Pacomanta, 3,850 m, 12.18972°S, 76.30361°W

Pacomanta, Km 120 on Lima to Huarochirí highway, 3,875 m, 12.19000°S, 76.30000°W

Pucusana, 4 km ENE, 150 ft, 12.46950°S, 77.76600°W

Pucusana, 10 km ENE, 12.48333°S, 76.80000°W

Rimac Calley, 4,400 ft, 12.03333°S, 77.15000°W

Surco, 1 mi W, 1,830 m, 11.86667°S, 76.46667°W

Yangas, 3,200 ft, 11.69472°S, 76.8456°W

Yauyos, 5 mi E, 9,000 ft, 12.45000°S, 75.87599°W

Yauyos, 8 mi NE, 9,500 ft, 12.36771°S, 75.868626°W

Zarate, 6 mi E of Pueblo San Bartolomé, 2,725 m, 11.88333°S, 76.45000°W

Loreto (Regíon de, formerly Departamento de)

Balsapuerto, 700 ft, 05.83333°, 76.60000°W

El Chino, right bank Río Tahuayo, 04.30140°S, 73.23110°W

Estación Biológica Allpahuayo, 25 km SW of Iquitos, 171 m, 03.96667°S, 73.41667°W

Estación Biológica Quebrada Blanco, right bank Quebrada Blanco, 04.35000°S, 73.15000°W

Genero [= Jenero] Herrera, 04.98330°S, 73.76670°W

Hacienda Santa Elena, ca. 35 km NE of Tingo Maria, 1,000 m, 04.83333°S, 74.21667°W

Iquitos, 03.74806°S, 73.24722°W

Jenaro Herrera, 04.86667°S, 73.65000°W

Jenaro Herrera, 2.8 km E, 135 m, 04.86667°S, 73.65000°W

Jenaro Herrera, 7 km E, right bank Río Ucayali, 04.91667°S, 73.76667°W

Lagunas, 05.23333°S, 75.63333°W

Mishana Allpahuayo, 04.00000°S, 73.25000°W

Nazareth [= Amelia; see Stephens and Traylor 1983], 04.33333°S, 70.08333°W

Nuevo San Juan, Río Gálvez, 150 m, 05.25830°S, 73.16390°W

Orosa, Río Amazonas, 03.43333°S, 72.13333°W

Pebas, Río Amazonas, 300 ft, 03.33333°S, 71.81667°W

Puerto Punga, Río Tapiche, 06.23333°S, 74.03333°W

Quebrada Orán, ca. 5 km N of Río Amazonas, 85 km NE of Iquitos, 110 m, 03.47500°S, 72.51700°W

Quebrada Pshaga, left bank Río Morona, Alto Río Amazonas, 2,200 m, 04.00000°S, 77.33333°W

Reserva Nacional Allpahuayo-Mishana, 28 km SW of Iquitos, 03.96667°S, 73.41667°W

Río Apayacu, 03.31667°S, 72.10000°W

Río Curaray, boca [= mouth], 02.36667°S, 74.08333°W

Río Curaray, vicinity, 02.36667°S, 74.08333°W

Río Peruate boca [= mouth], Río Amazonas, 90–100 m, 03.70000°S, 71.48300°W

Río Pisqui, west tributary of Río Ucayali, 07.75000°S, 75.01667°W

Río Tigre, 1 km below Río Tigrillo, 04.28333°S, 74.31667°W

Río Yaquerana, left bank at mouth, alto Río Yavarí, 05.71667°S, 72.96667°W

San Fernando, left bank Río Yavarí, 100 m, 04.20000°S, 70.23333°W

San Jacinto, 02.31667°S, 74.86667°W

San Jerónimo, west bank Río Ucayali, 305 m, 07.92000°S, 74.91000°W

San Lorenzo, Río Marañón, 04.83333°S, 76.66667°W

San Pedro, right bank Río Blanco, 04.33690°S, 73.19780°W

San Pedro, 80 km NE of Jenaro Herrera, 04.33333°S, 73.20000°W

Santa Cecilia, Río Maniti, 110 m, 03.43333°S, 72.76667°W

Santa Elena, Río Samiria, 130 m, 04.83333°S, 74.21667°W
Santa Luisa, Río Nanay, 03.33333°S, 74.58333°W
Santa Rita, Iquitos, 120 m, 03.75000°S, 73.18333°W
Sarayacu, Río Ucayali, 06.73333°S, 75.10000°W
Teniente Lopez, 1.5 km N, 175 m, 02.85000°S, 76.11000°W
Yurimaguas, Río Huallaga, 05.91667°S, 76.08333°W

Madre de Dios (Región de, formerly Departamento de)

Alberque Cusco Amazónica [= Reserva Cusco Amazónico], Río Madre de Dios, ca. 12 km E of Puerto Maldonado, 200 m, 12.55000°S, 69.05000°W
Altamira, Manu, 400 m, 12.25000°S, 71.00000°W
Cocha Cashu, 45 km NW [= 80 river km upstream] of the mouth of the Río Manu and about 8 km inside the border of Manu National Park, 380 m, 11.85000°S, 71.31667°W
Cocha Cashu Biologial Station, Parque Nacional Manu, 11.90000°S, 71.36667°W
Itahuana, right bank Río Madre de Dios, 12.78333°S, 71.21667°W
Maskoitania, 13.4 km NNW of Atalaya, left bank Río Madre de Dios, 450 m, 12.77167°S, 71.38547°W
Pakitza, Río Manu, 350 m, 11.94960°S, 71.28330°W
Pampas de Heath [= Pampa del Heath], ca. 50 km (by river) S of Puerto Pardo, 160 m, ca. 12.95000°S, 68.95000°W
Pampas del Heath, 12.51786°S, 69.17189°W
Quebrada Aguas Calientes, left bank Río Alto Madre de Dios, 2.75 km E of Shintuya, 12.66833°S, 71.26900°W
Reserva Cuzco Amazónico, north bank of Río Madre de Dios, 14 km E of Puerto Maldonado, ca. 200 m, 12.55000°S, 69.05000°W
Río Colorado, zona boca, 12.65000°S, 70.33333°W
Río Inambari, 2,000 ft, 13.91667°S, 70.25000°W
Río Manu, 57 km above mouth, 12.26667°S, 70.85000°W
Río Tambopata, 30 km above mouth, 12.73333°S, 69.18333°W
Río Tavara, Fila Boca Guacamayo, 13.50500°S, 69.68330°W
Tambopata, 12.40000°S, 69.13333°W

Moquegua (Región de, formerly Departamento de)

Caccachara, 4,630 m, 16.68330°S, 70.06667°W
Toquepala, 5 mi NW, 2,950 m, 17.25053°S, 70.69408°W
Torata, 19 km NE, 3,625 m, 16.94518°S, 70.72839°W

Pasco (Región de, formerly Departamento de)

Cerro de Pasco, restaurant, 4,500 m, 10.66667°S, 76.33333°W
Cerro de Pasco, 10 mi NE, 13,000 ft, 10.58105°S, 76.16325°W
Chipa, 4,170 m, 10.70000°S, 75.94000°W
Chiquirin [= Chigrín?], near La Quinua, 3,360 m, 10.36667°S, 75.95000°W
Cumbre de Ollon, ca. 12 km E of Oxapampa, 10.56667°S, 75.29035°W
Estación Biológica de San Alberto, 2,430 m, 10.55020°S, 75.33120°W
Huariaca, 2,740 m, 10.43278°S, 69.65214°W
La Quinua, mountains N of Cerro de Pasco, 3,536 m, 10.60000°S, 76.16700°W
Millpo, east of Tambo de Vascas on Pozuzo-Chagalla trail, 3,450 m, 10.38000°S, 76.25139°W
Montsinery, 10.45000°S, 74.86667°W
Nevati, 300 m, 10.35000°S, 74.85000°W
["Nevati is a mission station surrounded by a village of about 250 Campa Indian inhabitants. It is located on the north bank of the Río Pichis, about 10 km. SE Puerto Bermudez. The surrounding land is undulating and the virgin evergreen forest is 60–120 feet tall." (Tuttle 1970:81)]
Posuzo [= Pozuzo], upper Río Pachitea (Río Pozuzo), 1,000 m, 10.06667°S, 75.53333°W
Pozuzo, Delfín, 1,300 m, 10.00000°S, 76.00000°W
Puerto Bermudez, 10.33333°S, 74.90000°W
Puerto Bermudez, ca. 10 km N, 10.18333°S, 74.96667°W
Runicruz, 9,700 ft, 10.73333°S, 75.91667°W
San Pablo, 900 ft [= 273 m], 10.45000°S, 74.86667°W
["San Pablo is a Campa Indian village of about 175 inhabitants located in undulating country on the east bank of the Río Azupizu." (Tuttle 1970:81)]
Santa Cruz, ca. 9 km SSE of Oxapampa, 2,050 m, 10.56700°S, 75.16700°W

Piura (Región de, formerly Departamento de)

Asiayacu, 04.61667°S, 79.61667°W
Ayabaca, , 2,709 m, 04.63333°S, 79.71667°W
Ayabaca, bosque de Huamba, 44 km E of Ayabaca, 2,950 m, 04.71872°S, 79.53085°W
Batan, on Zapalache-Carmen trail, 9,000 ft, 05.11667°S, 79.38333°W
Canchaque, 1,230 m, 05.40000°S, 79.60000°W
Canchaque, 6.4 mi E (by road), 1,675 m, 05.31600°S, 79.52617°W
Canchaque, 15 km by road E, 05.40000°S, 79.46466°W
Catacaos, 05.26667°S, 80.68333°W
Cerro Amotape, 190 km N and 40 km W of Sullana, 04.86667°S, 81.16667°W
Cerro Chinguela, 2,900 m, 05.11670°S, 79.38330°W
Cerro Chinguela, 2,990 m, 05.12668°S, 79.39014°W

Cerro Chinguela, Huancabamba, 05.12667°S, 79.39014°W

Cerro Chinguela, ca. 5 km NE of Zapalache, 2,700–2,900 m, 05.11667°S, 79.38333°W

Cerro Prieto, Lancones, Encuentros, 04.25000°S, 80.51667°W

Chulucanas, 45 km W and 15 km N, 330 m, 04.95686°S, 80.56806°W

Hacienda Bigotes, Salitral, 180 m, 05.31667°S, 79.80000°W

Huancabamba, 3,000 m, 05.23300°S, 79.46700°W

Huancabamba, Canchaque, 1,198 m, 05.37597°S, 79.60613°W

Huancabamba, Tambo, 3,000 m, 05.35000°S, 79.55000°W

Huancabamba, 33 road km SW, W slope, 05.45545°S, 79.66216°W

Faique, Los Potreros, 05.40000°S, 80.73333°W

Jilili, 04.58333°S, 79.81667°W

Laguna, 1,150 m, 04.68333°S, 79.83333°W

Le Payta [= Paita], 0–100 m, 05.09580°S, 81.10630°W

Minera Majaz, Campamento Nuevo York, 3,071 m, 04.90456°S, 79.37481°W

Monte Grande, 14 km N and 25 km E of Talara, 04.45889°S, 81.04611°W

Morropón, Chalaco, 161 m, 05.18222°S, 79.96889°W

Palambla, 05.38333°S, 79.61667°W

Pariamarca Alto, 2,990 m, 05.15867°S, 79.54901°W

Pariñas Valley, near Talara, 04.56667°S, 81.28333°W

Piura, 10 km E, 300 ft, 05.20000°S, 80.54313°W

Porculla Pass, 7,000 ft, 05.83333°S, 79.51667°W

Portachuelo, 2,106 m, 05.02422°S, 79.91036°W

Porculla Pass, 2 km W, 6,500 ft, 05.85000°S, 79.53473°W

Quebrada Bandarrango (locality not found)

Quebrada El Gallo, Campamento Bomba Quemada 1, 04.89739°S, 79.73033°W

Reventazon, 64 km S and 19 km W of Sechura, 100 ft, 06.16667°S, 80.96667°W

Sechura, 30 km SSE, 30 ft, 05.80369°S, 80.71667°W

Talara, 10 km ENE, 04.53206°S, 81.20008°W

Tambo, Huancabamba, 05.35000°S, 79.55000°W

Puno (Región de, formerly Departamento de)

Abra Aricoma, 13 mi ENE of Crucero, 15,000 ft, 14.27806°S, 69.82185°W

Agualani, 9 km N of Limbani, 2,840 m, 14.11129°S, 69.69153°W

Aguas Claras camp, Río Heath, 12.95000°S, 68.9000°W

Ananea, 11 km W and 12 km S, 4,200 m, 14.59115°S, 69.65214°W

Ancomarca, 1 mi SW, 17.21022°S, 69.66064°W

Arapa, 3 mi NE, 15.01248°S, 70.08491°W

Asillo, 14.78333°S, 70.35000°W

Bella Pampa, 16 mi N and 2 mi E of Limbani, 13.90061°S, 69.67019°W

Caccachara, 16.68330°S, 69.89963°W

Caccachara, 50 mi [= 80 km] SW of Ilave, 15,200 ft, 16.59453°S, 70.19540°W

Crucero, 13 mi W and 2 mi N, 3,960 m, 14.32108°S, 70.19283°W

Hacienda Calacala, 7 mi SW of Putina, 4,140 m, 15.03333°S, 69.93333°W

Hacienda Collacachi, 3,900 m, 15.75000°S, 70.08333°W

Hacienda Ontave, 12,900 ft, 16.50000°S, 69.65000°W

Hacienda Pairumani, 24 mi S of Ilave, 3,960 m, 16.43034°S, 69.66667°W

Hacienda Pichupichuni, 6 km NW of Huacullani, 3,820 m, 16.58333°S, 69.38333°W

Hacienda Pichupichuni, Río Callacame, 8 km NW of Huacullani, 16.58333°S, 69.38333°W

Huacullani, 3,900 m, 16.63333°S, 69.33333°W

Huancullani, 10 km SW, 12,900 ft, 16.69723°S, 69.39961°W

Inca Mines [= Santo Domingo], Río Inambari, 1,830 m, 13.85000°S, 69.68333°W

Juli, 16.21670°S, 69.45000°W

Juli, 4 km E, 3,870 m, 12,700 ft, 16.20270°S, 69.4226°W

Lago Loriscota, 21 km WSW of Mazocruz, 14,900 ft, 16.86667°S, 70.03333°W

La Pampa, Limbani-Asterillos road, 573 m, 13.65000°S, 69.60000°W

Limbani, 12,000 ft, 14.13333°S, 69.70000°W

Limbani, 1 mi S, 3,485 m, 14.162245°S, 69.68889°W

Limbani, 3 mi N, 14.08996°S, 69.68889°W

Limbani, 4 mi N, 8,800 ft, 14.08996°S, 69.68889°W

Limbani, 6 mi N, 14.06101°S, 69.68889°W

Limbani, 9 km N, 2,840 m, 14.01866°S, 69.70000°W

Limbani, 5 km SSW, 13,800 ft, 14.21458°S, 69.71725°W

Limbani, 8 mi SSW, 15,000 ft, 14.25466°S, 69.73426°W

Macusani, 36 km SE (by road), 4,080 m, 14.26667°S, 70.28750°W

Mazocruz, 3,960 m, 16.75000°S, 69.73333°W

Mazocruz, 5 mi N, 16.67771°S, 69.73333°W

Oconeque, Río Quitún, 10 mi N of Limbani, 2,133 m, 14.00317°S, 69.68889°W

Ollachea, 6.5 km SW, 3,350 m, 13.85821°S, 70.52584°S

Ollachea, 11 km NNE, 1,875 m, 13.72481°S, 70.44440°W

Pairumani, 22 km SW of Llave on Río Huanque, 16.41667°S, 69.75000°W

Pampa de Ancomarca, 123 km W of Ilave, 4,150 m, 16.08333°S, 70.81631°W

Pampa Grande, Sandia, 14.28333°S, 69.43333°W

Pampa Grande, below San Ignacio on the Río Tambopata, 380–420 m, 13.81667°S, 68.98333°W

Pampa de Quellecot, 60 km S of Ilave, 16.62554°S, 69.66667°W

Pauquiplaya, below San Ignacio, Río Tambopata, 3,000 ft, 14.01667°S, 68.95000°W

Pisacoma, 16.90778°S, 69.37250°W

Pisacoma, 25 km SW, 14,000 ft, 17.07640°S, 69.53261°W

Pomata, 4 km NW, 12,500 ft, 16.24111°S, 69.32646°W

Pucara, 6 km S (by road), 3,850 m, 15.10303°S, 70.36467°W

Puno, 3,800 m, 15.23558°S, 70.05703°W

Puno, 5 km W, 3,960 m, 15.83333°S, 70.08001°W

Putina, 6 km N, 3,900 m, 14.86244°S, 69.86667°W

Putino Punco (not located)

Río Ccallacami, near Huacullani, 3,928 m, 16.61707°S, 69.33848°W

Río Huanque, 3,990 m, 16.20000°S, 69.73333°W

Río Inambari, probably near Oroya, 13.88333°S, 696667°W

Río Santa Rosa, 8 mi W of Mazocruz, 13,300 ft, 16.75000°S, 69.85338°W

Sagrario [= Segrario], upper Río Inambarí, 1,030 m, 13.91667°S, 69.68333°W

San Antón, 14.58333°S, 70.31667°W

San Antón, 4.5 km NE (by road), 14.56459°S, 70.29599°W

San Antonio de Esquilache, 16,000 ft, 16.10000°S, 70.18000°W

Sandia, 1,985 m, 14.28333°S, 69.43333°W

San Fermín, 850 m, 13.94360°S, 68.97610°W

San Juan del Oro, 14.18333°S, 69.16667°W

Santa Rosa, 4,195 m, 16.76000°S, 69.85000°W

Santa Rosa (de Ayaviri), 12 km S, 3,960 m, 14.71927°S, 70.78580°W

Santo Domingo, Río Inambari, 1,800 m, ca. 13.85000°S, 69.68333°W

Sillustani, 3,800 m, 17.73000°S, 70.17000°W

Tincopalca, 50 mi W of Puno, 4,115 m, 15.86667°S, 70.75000°W

Vilque, 13 km W, 13,300 ft, 15.76667°S, 70.37132°W

Yanacocha, valley of Río Tambopata, 1,942 m, 14.19375°S, 69.25567°W

Yanahuaya, 14 km W (Abra Marrancunca), 2,210 m, 14.26667°S, 69.32974°W

San Martín (Regíon de, formerly Departamento de)

Área de Conservación Municipal Mishquiyacu–Rumiyacu y Almendra, 06.08333°S, 76.98333°W

La Palmas, Parque Nacional Río Abiseo, 2,100 m, 07.97700°S, 77.35500°W

Las Papayas, ca. 5 km W of Pajaten ruins, 2,550 m, 07.48333°S, 77.37863°W

Laurel, 2,755 m, 06.68736°S, 77.69697°W

Los Chochos, 25 km NNE of Pataz, Parque Nacional Río Abiseo, 07.64486°S, 77.48128°W

Mariscal Caceres Huicungo, Parque National Río Abiseo, 2,650 m, 07.64486°S, 77.48128°W

Moyobamba, 860 m, 06.05000°S, 76.96667°W

Pampa de Cuy, 24 km NE of Pataz, Parque Nacional Río Abiseo, 3,300 m, 07.65000°S, 77.50000°W

Puca Tambo, 1,480 m, 06.16667°S, 77.26667°W

Puerta del Monte, Parque Nacional Río Abiseo, 3,200 m, 07.65778°S, 77.46944°W

Puerta del Monte, ca. 30 km NE of Los Alisos, Parque Nacional Río Abiseo, 3,230 m, 07.53333°S, 77.48333°W

Puerta del Monte, ca. 30 km NE of Los Alisos, Parque Nacional Río Abiseo, 3,250 m, 07.65778°S, 77.46944°W

Río Negro, about 35 mi W of Moyobamba, 2,600 ft, 05.80000°S, 77.31667°W

Rioja, 842 m, 06.08333°S, 77.15000°W

Valle de los Chochos, ca. 25 km NE of Pataz, 3,280 m, 07.64486°S, 77.48128°W

Vilcabamba del Pajate, ca. 31 km NE of Pataz [La Libertad], 2,800 m, 07.58977°S, 77.39415°W

Yurac Yacu, 750 m, 05.95000°S, 77.18300°W

["Yurac Yacu is a rather large village at the junction of the Yurac Yacu stream with the Río Mayo. It is 35 miles east of Jumbilla on the opposite side of the mountains, and a little over 20 miles WNW of Moyobamba." (O. Thomas 1927a:361)]

Tacna (Regíon de, formerly Departamento de)

Challapalca, 0.5 mi W, 17.23333°S, 69.79086°W

Challapalca, 1 mi NE, 17.23111°S, 69.77270°W

Lago Suche, 14,500 ft, 16.93333°S, 70.38333°W

Lago Suche, 5 km N, 4,450 m, 16.93333°S, 70.33639°W

Morro de Sama, 65 km W of Tacna, 100–1,000 ft, 17.98333°S, 70.88333°W

Nevado Livine, 2 km N, 4,640 m, 17.24017°S, 69.80751°W

Nevado Livine, 2 km NW, 15,300 ft, 17.24546°S, 69.82081°W

Pampa de Tetire, 29 km NE of Tarata, 14,600 ft, 17.22916°S, 69.83978°W

Río Tarata, 11,200 ft, 16.44450°S, 70.98433°W

Tarata, 10,100 ft, 17.47444°S, 70.03278°W

Tarata, 1.5 mi N (10 km by road), 11,600 ft, 17.45276°S, 70.03278°W

Tarata, 2.6 mi N (23 km by road), 13,000 ft, 17.43681°S, 70.03278°W

Tarata, 5 km NE, 3,688 m, 17.46667°S, 70.03333°W

Tarata, 6 km NE, 12,900 ft, 17.43611°S, 69.99284°W

Tarata, 12 mi NE, 14,600 ft, 17.39139°S, 69.94623°W

Tarata, 13 km NE, 14,500 ft, 17.39139°S, 69.94623°W

Tumbes (Región de, formerly Departamento de)

Huasimo, 750 m, 03.56667°S, 80.46667°W
Laguna Lamadero, 03.60000°S, 80.20000°W
Matapalo, 03.68333°S, 80.20000°W
Quebrada Los Naranjos, Zarumilla, ca. 03.81107°S, 80.32175°W
Tumbes, 03.56667°S, 80.46667°W

Ucayali (Región de, formerly Departamento de)

Balta, Río Curanja, 300 m, 10.13333°S, 71.21667°W
Cumaría, 30 m, 09.85000°S, 74.01667°W
Largato, upper Río Ucayali, 10.66667°S, 73.90000°W
Pucallpa, 08.63750°S, 74.89683°W
Pucallpa, 59 km SW, 08.58333°S, 74.86667°W
Reserva Nacional Sierra del Divisor, ca. 07.98320°S, 73.89866°W
Río Alto Ucayali, 09.75000°S, 74.13333°W
San José, on Río Santa Rosa, 12.73333°S, 73.76667°W
Sarayacu, Río Ucayali, 125 m, 06.73300°S, 75.10000°W
Sepahua, Río Sepahua, 11.16667°S, 73.06667°W
Suayo, 07.30000°S, 74.91667°W
Tushemo [= Tushma], near Masisea, Río Ucayali, 225 m, 08.61667°S, 74.35000°W
[See O. Thomas (1928a:250) for location; specimens from this place often reported as from Masisea.]
Yarinacocha, Río Ucayali, 08.25000°S, 74.71667°W

Surinam

Brokopondo (District)

Afobaka, 05.00000°N, 54.98333°W
Brownsberg Nature Park, 7 km S and 18.5 km W of Afobakka, 04.91700°N, 55.20000°W

Commewijne (District)

Peninika boarding school, near confluence of Peninika Creek and upper Commewijnw River, 05.59461°N, 54.68104°W

Marowijne (District)

Albina, 3 km SW, 05.48300°N, 54.06700°W
Perica, 05.85000°N, 54.71667°W
Powakka, ca. 50 km S of Paramaribo, 05.58333°N, 54.25000°W
Wiawia Bank, 05.96667°N, 54.35000°W

Nickerie (District)

Kutari River, 02.18333°N, 56.78333°W
Poder, near Nieuw Nickerie, 05.90000°N, 57.00000°W

Para (District)

Carolina Kreek, 05.41700°N, 55.18333°W
Finisanti [= Finisanti Presie], Saramacca River, 05.13333°N, 55.48333°W
Lelydorpplan, 05.61659°N, 55.19996°W
Loksie Hattie [= Loksie Hatti], 05.15000°N, 55.46667°W
Matta, 15 km W of Zanderij airport, 05.46667°N, 55.35000°W
Zanderij, 05.45000°N, 55.20000°W

Paramaribo (District)

Paramaribo, 05.83333°N, 55.16667°W

Saramacca (District)

La Poule, 05.78333°N, 55.41667°W

Sipaliwini (District)

Avanavero, 04.83300°N, 57.23300°W
Avanavero Falls, Kabaleba River, 04.81700°N, 57.40000°W
Emmaketen [= Emma Kenten], 350 m, 04.08333°N, 56.20000°W
Frederik Willem IV Falls, Courantyne River [= Corantijn River], 112 m, 03.46667°N, 57.61667°W
Kutari River, camp 1, ca. 02.17829°N, 56.75788°W
Oelemarie, 03.10000°N, 54.53333°W
Sipaliwini airstrip, 02.03333°N, 56.13333°W
Sipaliwini-savanne-vliegeveld [= Sipaliwini savanna airstrip], 02.08333°N, 56.16667°W
Tafelberg, 700 m, 03.91667°N, 56.16667°W

Wanica (District)

Kwatta, west of Paramaribo, 05.85000°N, 55.30000°W
Santo Boma Locks, about 12 km SW of Paramaribo, 05.78333°N, 55.28333°W

Trinidad and Tobago
Tobago

Little Tobago Island, 11.30000°N, 60.50000°W
Pigeon Peak, Tobago Forest Reserve, 1.5 km SSW of Charlotteville, 550 m, 11.30000°N, 60.55000°W

Pigeon Peak, Tobago Forest Reserve, 2.5 km SSW of Charlotteville, 335–550 m, 11.30000°N, 60.55000°W

Richmond, 11.21667°N, 60.61667°W

Runnemede, 11.25000°N, 60.70000°W

Speyside, 11.30000°N, 60.53333°W

Trinidad

Botanic Gardens, St. Anne's Barracks, 10.66667°N, 61.51667°W

Bush Bush Forest, 10.51667°N, 51.03333°W

Caparo, 10.45000°N, 61.33330°W

Carenage, 10.26000°N, 61.20000°W

Caura, 334 m, 10.71667°N, 61.35000°W

Chaguaramas, 10.68330°N, 61.63330°W

Cumaca, 10.70000°N, 61.15000°W

Diego Martin, 10.72000°N, 61.57000°W

Las Cuevas, Maraba Bay, 10.78333°N, 61.38333°W

Mayaro, 10.16700°N, 61.08330°W

Nariva, Nariva Swamp, 10.41670°N, 61.06670°W

Oropouche Heights, Siparia Ward, 10.76667°N, 61.15000°W

Princes Town, 10.26667°N, 61.38333°W

Princestown [= Princes Town], 10.26667°N, 61.38333°W

Princes Town, 7 mi SE, 10.21667°N, 61.33333°W

Savanna Grande, 10.30000°N, 61.36667°W

Uruguay

Artigas (Departamento de)

Artigas, 30.40000°S, 56.46667°W

La Isleta, Colonia Artigas, 30.40000°S, 56.63333°W

Paso del Campamento, 30.78333°S, 56.78333°W

Rincón de Franquía, 30.41667°S, 57.65000°W

Canelones (Departamento de)

Arroyo Tropa Vieja, sea level, 34.78333°S, 55.86667°W

Bañado de Tropa Vieja, 34.78333°S, 55.86667°W

Cuchilla Alta, 34.78500°S, 55.49861°W

Laguna del Cisne, 34.75250°S, 55.83660°W

Cerro Largo (Departamento de)

Melo, 6 km SE, 92 m, 32.36667°S, 54.18333°W

Paso del Dragón [= Plácido Rosas], 1 km NW, 32.75680°S, 53.72264°W

Colonia (Departamento de)

Arroyo Artilleros, Santa Ana, 34.83333°S, 57.55000°W

Arroyo Limetas [= Arroyo de las Limetas], 34.18333°S, 58.10000°W

Colonia del Sacramento, 34.46667°S, 57.85000°W

Estanzuela, 80 m, 34.30000°S, 57.73333°W

Playa Ferrando, 2 km E of Cnia. Sacramento, 34.41667°S, 57.88333°W

Puente del Arroyo Pereira sobre RN 1, 34.48333°S, 56.85000°W

Durazno (Departamento de)

Arroyo Cordobés y Río Negro, 35 m, 32.50000°S, 55.31667°W

Durazno, 33.08333°S, 56.08333°W

Florida (Departamento de)

Arteaga, Cerro Colorado, 33.50000°S, 55.50000°W

Lavalleja (Departamento de)

Minas, 34.37589°S, 55.23771°W

Paso Averías, Río Cebollatí, 33.66667°S, 54.33333°W

Maldonado (Departamento de)

Barra del Arroyo Maldonado, 34.91667°S, 54.85000°W

Chihuahua, 34.93444°S, 54.94639°W

Hostería La Laguna, N de Maldonado, 34.25000°S, 54.75000°W

José Ignacio, 34.83944°S, 54.64778°W

Las Flores, 50 m, 34.75000°S, 55.33333°W

Maldonado, 34.90386°S, 54.95050°W

Punta del Este, 34.96667°S, 54.95000°W

San Carlos, 15 km N, 100 m, 34.66667°S, 54.91667°W

Solís, 34.78333°S, 55.38333°W

Montevideo (Departamento de)

Carrasco, 34.88333°S, 56.03333°W

Montevideo, 34.83333°S, 56.18333°W

Parque Lecoq, 34.81667°S, 56.35000°W

Villa Colón, 34.80210°S, 56.22420°W

Paysandú (Departamento de)

Guabiyú, 31.73333°S, 58.03333°W

Río Negro (Departamento de)

Fray Bentos, 33.13333°S, 58.30000°W

La Tabaré, 33.35683°S, 58.30950°W

Marfalda, 32.87967°S, 57.97167°W
Paso de las Piedras, 34.00000°S, 54.66667°W
Paysandú, 15 km S, Río Negro, 32.35758°S, 57.45896°W

Rivera (Departamento de)

Cuñapirú [= Represa de Cuñapirú], 31.53333°S, 55.58333°W
Estancia La Quemada, 32.02000°S, 54.57033°W
Minas de Corrales, 31.50000°S, 55.50000°W

Rocha (Departamento de)

Arroyo La Palma, Ruta 15 km 10, La Paloma, 34.58633°S, 54.17850°W
Laguna de Castillos, 34.35000°S, 53.86667°W
Laguna Negra, 34.08333°S, 53.76667°W
Laguna de Rocha, 34.62250°S, 54.25750°W
Lascano, 22 km SE, 50 m, 33.78294°S, 54.05453°W
Parque Nacional Refugio de Fauna Laguna de Castillos, 34.36667°S, 53.49000°W
Parque Nacional Santa Teresa, 34.00000°S, 53.50000°W
Valizas, 34.35167°S, 53.84167°W

Salto (Departamento de)

Salto, 31.41667°S, 57.0000°W

San José (Departamento de)

Arazatí [= Puerto Arazatí], 34.56667°S, 57.00000°W
Bañados del Arazatí, 34.52830°S, 57.05240°W
Barra del Río Santa Lucía, 34.73333°S, 56.40000°W
Kiyú, 34.68333°S, 56.73333°W
Río Santa Lucía (barra), 24 m, 34.78611°S, 56.35833°W

Soriano (Departamento de)

Cardona, 3 km E, 150 m, 33.83333°S, 57.38333°W
Dolores, 15 mi SW, 33.50000°S, 58.21667°W
Nueva Palmira, 1 km N, 33.83333°S, 58.41667°W

Tacuarembó (Departamento de)

Ansina, 31.90000°S, 55.46667°W
Paso Baltasar, Arroyo Tres Cruces, 31.66667°S, 55.83333°W
Pueblo Ansina [= Ansina], 31.90000°S, 55.46667°W
Tacuarembó, 31.70000°S, 55.98333°W

Treinta y Tres (Departamento de)

Arrozal 33, 33.00000°S, 53.66667°W
Boca del Río Tacuari, 32.76667°S, 53.30000°W

Boca del Río Tacuari, 16 km SSW, 32.76667°S, 53.30000°W
Paso Ancho, Cañada de las Piedras and Rte 8 [= 8 km N of Treinta y Tres], 32.71667°S, 55.33333°W
Río Tacuari, 16 km SSW of mouth, 33.86074°S, 53.48953°W
Treinta y Tres, 8 km E, 33.23333°S, 5438333°W

Venezuela

Amazonas (Estado de)

Acanana, 48 km NW of Esmeralda, Río Cuucunuma, 03.53333°N, 65.80000°W
Belén [= Culebra], Río Cunucunuma, 56 km NNW of Esmeralda, 150 m, 03.65000°N, 65.76700°W
Boca Mavaca, 84 km SSE of Esmeralda, Río Manavichi, 138 m, 02.50000°N, 65.21667°W
Boca del Río Ocamo, Río Orinoco, 02.80000°N, 65.23333°W
Caño León, Río Orinoco, Cerro Duida, 03.41667°N, 65.66667°W
Capibara, 106 km SW of Esmeralda, Brazo Casiquiare, 02.61667°N, 66.31667°W
Casiquiare Canal, Capibara, 02.61667°N, 66.31667°W
Cerro Aracamuri, Cumbre Sur, 1,500 m, 01.47770°N, 65.83600°W
Cerro Duida, 03.61667°N, 65.68333°W
Cerro Duida, Agüita Camp, 03.33333°N, 65.53333°W
Cerro Duida, Cabecera del Caño Culebra, 40 km NNW of Esmeralda, 1,400 m, 03.50000°N, 65.71667°W
Cerro Duida, Caño Culebra, 750 m, 03.61667°N, 65.68333°W
Cerro Neblina Base Camp, left (west) bank Río Baria [= Río Mawarinuma], 140 m, 00.83055°N, 66.16111°W
Cerro Neblina, Camp V, 1,800 m, 00.83300°N, 65.98300°W
Cerro de la Neblina, Camp VII, 5.1 km NE of Pico Phelps, 1,800 m, 00.84444°N, 65.96944°W
Cerro Neblina, Camp XI, 00.86667°N, 65.96667°W
Cerro de Tamacuare, 01.27500°N, 64.78333°W
Cerro Yapacana, 03.75000°N, 66.80000°W
Coyowatari, Quebrada Orinoquito, 02.43667°N, 64.26667°W
Esmeralda, 03.16667°N, 65.55000'W
Jawasu, 12 mi W, left bank Río Casiquiare, 01.96700°N, 66.70000°W
Monduapo, upper Orinoco, 04.75000°N, 67.80000°W
Monduapo, right bank upper Río Orinoco, 04.90000°N, 67.80000°W
Nericagua (Caño Usate), Río Orinoco, 04.42000°N, 67.80000°W
Paria Grande, 00.56667°N, 67.58333°W
Ponzón, 50 km NE of Puerto Ayacucho, 06.05000°N, 67.41667°W

Puerto Ayacucho, 18 km SSE, 05.66667°N, 67.58333°W

Puerto Ayacucho, 32 km S, 05.66667°N, 67.63333°W

Raya, 32 km SSE of Puerto Ayacucho, 05.30000°N, 67.53000°W

Río Siapa, 525 m, 01.43333°N, 64.36667°W

San Carlos de Río Negro, ca. 4 km N of Isla Saramá, 01.91667°N, 67.06667°W

San Juan [= Manapiare], Río Manapiare, 163 km ESE of Puerto Ayacucho, 155 m, 05.30000°N, 66.21667°W

San Juan Manapiare, 05.30000°N, 66.21667°W

Sierra Parima, 02.50000°N, 64.00000°W

Tama Tama [= Tamatama], 135 m, 03.16667°N, 65.81667°W

Tamatama, Río Orinoco, 135 m, 03.16667°N, 65.81667°W

Anzoátegui (Estado de)

Cantaura, 09.31667°N, 64.35000°W

Cerro La Laguna, cumbre, 2,200 m, 10.02400°N, 64.13000°W

Cueva de Agua, 8 km SE of Guanta, 10.16667°N, 64.58333°W

El Merey, 09.16667°N, 64.73333°W

Paso Los Cocos, Río Caris, 280 m, 08.60000°N, 64.06667°W

Pekin Abajo, Río Neverí, 100 m, 10.15000°N, 64.51667°W

Río Orituapano, 08.95000°N, 66.21667°W

Apure (Estado de)

Guasdualito, 07.25000°N, 70.73333°W

Hato Caribén, Río Cinaruco, 32 km NE of Puerto Páez, 06.55000°N, 67.21700°W

Hato El Frío, 30 km W (by road) El Samán, 07.71667°N, 68.90000°W

Nulita, Selvas de San Camilo, 29 km SSW of Santo Domingo, 24 m, 07.31667°N, 71.95000°W

Nuitla, 3 km N of Nula, 07.30000°N, 71.88333°W

Puerto Paez, 06.20762°N, 67.45123°W

Puerto Páez, 38 km NNW, Río Cinaruco, 06.55000°N, 67.51667°W

Río Cinaruco, 65 km NW of Puerto Páez, 76 m, 06.55000°N, 67.91667°W

San Fernando de Apure, 07.88782°N, 67.47236°W

Tama, 07.40000°N, 72.40000°W

Aragua (Estado de)

Campamento Rafael Rangel, Loma de Hierro, 1,200–1,260 m, 10.15000°N, 67.15000°W

Camp Rafael Rangel, 09.76700°N, 63.61700°W

Colonia Tovar, 10.41667°N, 67.28333°W

Estación Biológica Rancho Grande, 13 km NW of Maracay, 1,050 m, 10.36667°N, 67.68333°W

Lake [= Lago] Valencia, 10.18333°N, 67.75000°W

Monumento Natural Pico Codazzi, 600–2,429 m, 10.37740°N, 67.29710°W

Ocumare de la Costa, 10.45000°N, 67.76667°W

Rancho Grande, 10.36667°N, 67.68333°W

Rancho Grande Biological Station, 13 km NW of Maracay, 1,050–1,100 m, 10.36667°N, 67.68333°W

Barinas (Estado de)

Altamira, 750 m, 08.8333°N, 70.50000°W

Altamira, vicinity, ca. 08.08333°N, 70.50000°W

Buena Vista, 08.40000°N, 70.08333°W

Hato Corozal, Caño Tutumito, Arismendi, 08.48333°N, 68.35000°W

Reserva Forestal de Ticoporo, 250 m, 07.75000°N, 69.91667°W

Santa Bárbara, 07.78333°N, 71.16667°W

Bolívar (Estado de)

Acopán-tepuí, 1,524 m, 05.21667°N, 62.06667°W

Altagracia, Immataca District, 08.05000°N, 62.43333°W

Arabupo [= Arabopó], Mt. Roraima, 05.10000°N, 60.73333°W

Arabupu [= Arabopó], Mt. Roraima, 05.10000°N, 60.73333°W

Auyán-tepuí, 1,000 m, 05.91667°N, 62.53333°W

Auyán-tepuí, 1,850 m, 05.75000°N, 62.53333°W

Auyán-tepuí, south slope, Río Caroní, 3,500 ft, 05.91700°N, 62.53300°W

Boca de Parguaza, 06.40000°N, 67.20000°W

Caicara, Río Orinoco, 07.64639°N, 66.17222°W

Cerro Auyan-Tepui, 05.91667°N, 62.53333°W

Cuidad Bolívar, 08.12000°N, 63.55000°W

Ciudad Bolívar, Suapure, Río Caura valley, 08.13333°N, 63.55000°W

Ciudad Bolívar, 46 km S and 7 km E, 08.13333°N, 63.55000°W

Churi-tepuí, Camp 5, 1,494 m, 05.21667°N, 61.90000°W

El Dorado, 85 km SSE [= km 125, vicinity of Salto de El Danto], 05.96670°N, 61.41333°W

El Llagual [= El Yagual], lower Río Caura, 07.41667°N, 65.16667°W

El Manaco, 56 to 59 km SE of El Dorado, 150 m, 06.31667°N, 61.31667°W

El Yagual, lower Río Caura, 07.41667°N, 65.16667°W

Embalse Guri, Isla Panorama, Río Caroni, 275 m, 07.34194°N, 62.84722°W

Hato La Florida, 44 km SE of Caicara, 07.50000°N, 65.78000°W

Hato La Florida, 45 km SE of Caicara, 07.50000°N, 65.78000°W

Hato San José, 20 km W of La Paragua, 300 m, 06.81667°N, 63.48333°W

Icabarú, 45 km NE, 04.55000°N, 61.41667°W

Icabarú, 43 to 45 km NE, 851–854 m, 04.58333°N, 61.31667°W

Icabarú, 45 km NE, Santa Lucia de Surukun, 851 m, 04.55000°N, 61.42000°W

Icabarú, 55 km NE, 04.33333°N, 612.75000°W

Imataca, 07.75000°N, 61.00000°W

La Unión, Río Caura, 200 m, 06.91667°N, 64.91667°W

Mount Roraima, Arabopo, 05.20000°N, 60.73333°W

Puerto Cabello del Caura, 3 im E, 07.16667°N, 64.98333°W

Río Cuyuni, 69 km SE by road, 06.15000°N, 61.43333°W

Río Supamo, 50 km SE of El Manteco, 350 m, 07.00000°N, 62.25000°W

Río Yuruaní, 12 km from mouth, 100 m, 06.80000°N, 61.83300°W

Rondon Camp, Mt. Roraima, 05.16667°N, 60.76667°W

Roraima Tepui, summit, 2,050–2,580 m, 05.14400°N, 60.76200°W

San Ignacio de Yuruaní, 850 m, 05.03333°N, 61.13333°W

Serranía de los Pijiguaos, 140 km SW of Caicara, 300 m, 06.48333°N, 66.71667°W

Sipao, 07.44309°N, 65.40335°W

Vetania, 04.33333°N, 61.76667°W

Yagual, 07.41667°N, 65.16667°W

Carabobo (Estado de)

Campamento La Justa, Río Morón, 350 m, 10.38333°N, 6823333°W

El Trompillo, 1,200 ft, 10.06700°N, 67.76700°W

La Copa, 4 km NW of Montalban, 1,537 m, 10.25000°N, 68.35000°W

La Cumbre de Valencia, 1,700 m, 10.33333°N, 68.00000°W

Montalbán, 10.21667°N, 68.03333°W

Patanemo, 10.43333°N, 67.91667°W

San Esteban, 10.43333°N, 68.01667°W

San Esteban, near Venezuela Hills, 2,000 ft, 10.43333°N, 68.01667°W

Urama, E of Yaracuy, 10.45000°N, 68.31667°W

Caracas

Caracas, 10.50000°N, 66.91667°W

Caracas, Phelps Hill, 10.51667°N, 66.95000°W

Cojedes (Estado de)

El Baul, 08.95000°N, 68.30000°W

Hato Itabana, 09.58333°N, 68.61667°W

Delta Amacuro (Estado de)

Caño Araguabisi, 09.22416°N, 61.00444°W

Caño Guiniquina, 09.48333°N, 60.98333°W

La Horquta, Tucupita, 09.06667°N, 62.05000°W

Falcón (Estado de)

Capatárida, 40 m, 11.16700°N, 70.61700°W

Carrizalito, Sierra San Luis, 11.13333°N, 69.75000°W

Cerro La Danta, Parque Nacional J. C. Falcón, 1,300–1,470 m, 11.23333°N, 69.60000°W

Cerro Santa Ana, Peninsula de Paraguaná, 560 m, 11.81667°N, 69.96667°W

Cerro Santa Ana, Peninsula de Paraguaná, 49 km N and 32 km W of Coro, 420–615 m, 11.81667°N, 69.95000°W

Cerro Socopo, 84 km NW of Carora, 1,250 m, 10.46667°N, 70.80000°W

Coro, 11.41667°N, 69.68333°W

Isiro, 11.94929°N, 70.09942°W

La Pastora, 14 km ENE of Mirimire, 190 m, 11.20000°N, 68.61667°W

La Pastora, near, 11 km ENE of Mirimiri, 11.20000°N, 68.60000°W

Mirimire, 11.16667°N, 68.71667°W

Mirimire, near 11.20000°N, 68.61667°W

Reio Cocopo, 80 km NW of Carora, 10.46667°N, 70.80000°W

Sanare, 10.88333°N, 68.38333°W

Serranía de San Luis, Parque Nacional Juan Crisótoma Falcón, Sector Cumbre de Uria, ca. 9 km N of Cabure, 1,320–1,370 m, 11.22795°N, 69.61477°W

Urama, 19 km NW, 10.61667°N, 68.40000°W

Guárico (Estado de)

Calabozo, 08.84528°N, 67.54278°W

Dos Caminos, 09.58333°N, 67.33333°W

Guárico, 08.66667°N, 66.58333°W

Hato Las Palmitas, 09.60000°N, 67.45000°W

Parque Nacional Guatopo, 15 km NW of Altagracia, 720 m, 09.97000°N, 66.42000°W

Parque Nacional Guatopo, 40 km SSE of Caracas, 710 m, 10.05000°N, 66.43333°W

Río Orituco, 10 km W of Chaguaramas, 09.40000°N, 66.46667°W

San José de Tiznados, 52 km NNW of Calabozo, 150 m, 09.38300°N, 67.55000°W

San Juan de los Morros, 09.91667°N, 67.35000°W

Zaraza, 09.35000°N, 65.31667°W

Lara (Estado de)

Antoátequi, W of Guarico, 4,705 ft, 09.60000°N, 69.90000°W

Caserío Boro, 10 km N of El Tocuyo, 520 m, 09.88333°N, 69.78333°W

Caserío Boro, 13 km W of El Tocuyo, 900 m, 09.75000°N, 69.90000°W

Curarigua, 09.97861°N, 69.93694°W

El Blanquito, Parque Nacional Yacambu, 17 km SE of Sanare, 09.66667°N, 69.61667°W

El Tocuyo, vicinity, 09.88333°N, 69.78333°W

Páramo de las Rosas, 09.58833°N, 70.11444°W

Mérida (Estado de)

Bosque San Eusebio, 7 km SSE of La Azulita, 2,500 m, 08.64417°N, 07.147583°W

El Baho, 5 km SW of Santo Domingo, 2,500 m, 08.83752°N, 70.72578°W

Escorial, Sierra de Mérida, 2,500 m, 08.63333°N, 71.08333°W

Fundo Vista Alegre, San Isidro de Bejuquero, 08.80000°N, 70.85000°W

Hacienda La Guapa, 6 km S of Río Chico, 10.28300°N, 65.96700°W

Hacienda Santa Catarina, Río Chama, 1,372 m, 08.53333°N, 71.35000°W

La Azulita, 1,135 m, 08.71667°N, 71.45000°W

La Coromoto, 7 km SE of Tabay, 3,350 m, 08.60000°N, 71.01667°W

Laguna Brava, Páramo de Mariño, 2,090 m, 08.33500°N, 71.78917°W

Laguna Negra, 5.75 km ESE of Apartaderos, 3,500 m, 08.80000°N, 70.76667°W

Laguna Verde, 9 km SE of Tabay, 3,550 m, 08.57440°N, 71.00920°W

La Montana, 4.1 km SE of Mérida, 08.17250°N, 71.59556°W

La Mucuy, 08.63333°N, 71.03333°W

Mérida, 1,475 m, 08.59524°N, 71.14340°W

Mérida, 1,600 m, 08.60000°N, 71.13333°W

Mesa Bolívar, vicinity, ca. 08.43333°N, 71.56667°W

Montes de Los Nevados, 2,500 m, 08.46667°N, 71.06667°W

Monte Zerpa, 6 km N of Mérida, 2,160 m, 08.63778°N, 71.16306°W

Mucugají, 08.78333°N, 70.81667°W

Mucubaji, 3.25 km ESE of Apartaderos, 3,600 m, 08.80611°N, 70.84472°W

Paramito, 3 km W of Timotes, 08.98333°N, 70.76667°W

Paramito, 4 km W of Timotes, 3,265 m, 08.56821°N, 70.76620°W

Páramo de los Conejos, 15 mi N of Mérida, 08.83333°N, 71.256000°W

Páramo de Mariño, Mérida/Táchira border, 2,600 m, 08.25000°N, 71.86667°W

Páramo Tambor, 2,400–2,920 m, 08.60000°N, 71.40000°W

Río Milla, 1,630 m, 08.50000°N, 71.36667°W

Santa Rosa, 2 km N of Mérida, 08.06868°N, 71.14110°W

Sierra de Mérida, 2,800 m, 08.50000°N, 71.05000°W

Sierra de Mérida, 08.66667°N, 71.00000°W

Tabay, 6 km ESE, 2,630 m, 08.61842°N, 71.00850°W

Tabay, 8 km SE, near La Coromoto, 08.60000°N, 71.01667°W

Timotes, 08.98333°S, 70.73333°W

Miranda (Estado de)

Altagracia de Orituco, 25 km (by road) N, 10.03333°N, 66.45000°W

Alto de Nuevo Leon, 33 km W of Caracas, 1,996 m, 10.43300°N, 67.16700°W

Curupao, 10.51667°N, 66.63333°W

Curupao, 19 km E of Caracas, 1,160 m, 10.50000°N, 66.63333°W

Estación Experimental Río Negro, 50–70 m, 10.33333°N, 66.283333°W

Hacienda Las Planadas, 10.53333°N, 66.50000°W

Parque Nacional Guatopo, 10.08300°N, 66.41700°W

Río Chico, 10.31667°N, 65.96667°W

Río Chico, 1 km S, 10.30000°N, 65.96667°W

Monagas (Estado de)

Cachipo, 09.93333°N, 63.13333°W

Cachipo, 6.5 km NE, 09.91667°N, 63.16667°W

Caripe, Cumaná, 800 m, 10.20000°N, 63.48333°W

Caripe, 2 km N and 4 km W, 1,170–1,335 m, 10.18333°N, 63.53333°W

Caripito, 100 m, 10.166678°N, 63.10000°W

Hato Mata de Bejuco, 54–55 km SSE of Maturín, 18 m, 09.31667°N, 62.93333°W

Isla Guara, 09.06667°N, 62.08333°W

Los Barrancos, 8 km NNW, 09.83827°N, 63.38235°W

Río Chiquito, Guanaguana, 10.06667°N, 63.56667°W

San Agustín, near, 2 km N and 4 km E of Caripe, 1,180 m, 10.20000°N, 63.48333°W

San Agustín, 5 km NW of Caripe, 1,340 m, 10.20000°N, 63.53333°W

San Agustín, 3 to 5 km NW of Caripe, 10.20000°N, 63.53333°W

San Antonio, 10.11667°N, 63.71667°W

San Juan, 15 km N of Areo, 09.86667°N, 63.88333°W

Nueva Esparta (Estado de)

Isla Margarita

Cerro Matasiete, 3 km NE of La Asunción, 410 m, 11.04500°N, 63.84500°W

Dos Ríos, 10.25000°N, 63.88333°W

El Valle, 50 m, 10.98333°N, 63.86667°W

El Valle, Margarita, 10.98333°N, 63.86667°W

Lalal, 10.16667°N, 63.91667°W

Península Macanao, 11.00000°N, 64.28333°W

San Juan, Cerro Copey, 800 m, 11.01667°N, 63.90000°W

Portuguesa (Estado de)

Cogollal, near Guanarito, 08.70000°N, 69.21667°W

La Arenosa, 08.68330°N, 69.53333°W

La Hoyada, near Guarnarito, 09.05000°N, 69.75000°W

Piritu, 10.03333°N, 65.03333°W

Sucre (Estado de)

Campo Alegre, 10.16667°N, 63.75000°W

Carapas, 1,707 m, 10.20000°N, 63.53333°W

Cariaco-Chacopata, carretera between Cariaco and Chacopata, 10.65000°N, 63.71667°W

Cerro Negro, 10 km NE of Caripe, 1,630–1,690 m, 10.20000°N, 63.53000°W

Cristóbal Colón [= Macuro], 10.65000°N, 61.93333°W

Cumaná, sea level, 10.46667°N, 64.16667°W

Cumaná, 2 km E, 10.46667°N, 64.13333°W

Cumanacoa, 10.25000°N, 63.91667°W

El Argarrobo, 10.66667°N, 62.80000°W

Ensenada Cauranta, 7 km N and 5 km E of Guira, 4 m, 10.66700°N, 62.25000°W

Guaraúnos, 10.56667°N, 63.13333°W

Hacienda Hunantal, vicinity, 21 km E of Cumaná, 15 m, 10.45000°N, 63.96700°W

Ipure, near Cumaná, 685 m, 10.36667°N, 64.16667°W

Macuro, 50 m, 10.66667°N, 61.95000°W

Manacal, 10.28333°N, 63.05000°W

Manacal, 10.61667°N, 63.01667°W

Manacal, 26 km ESE of Carupano, 190 m, 10.62000°N, 63.02000°W

Nevera, 2,400 ft, 10.10000°N, 64.63333°W

Neverí, 10.25000°N, 63.91667°W

Quebrada Seca [= Villarroel], 200 m, 09.51700°N, 70.66700°W

Táchira (Estado de)

Buena Vista, 07.43333°N, 72.43333°W

Buena Vista, 450 m, 07.89806°N, 72.01667°W

Buena Vista, 41 km SW of Can Cristobal, near Páramo de Tamá, 2,350–2,420 m, 07.45000°N, 72.43333°W

El Caimito, 07.81667°N, 72.35000°W

Estación Experimental, 07.56667°N, 72.08333°W

Las Mesas, 08.15000°N, 72.16667°W

Orope, 08.36667°N, 71.30000°W

Páramo de Tamá, 2,134 m, 07.41667°N, 72.43333°W

Páramo de Tamá, 3,070 m, 07.42639°N, 72.39333°W

Páramo El Zumbador, 8 km SSW of El Cobre, 2,750 m, 07.97111°N, 72.07444°W

Paso Honda, Río Potosí, 07.95000°N, 71.65000°W

Presa La Honda, 10 km SSE of Pregonero, 1,100 m, 07.95000°N, 71.70000°W

San Juan de Colón, 797 m, 08.03333°N, 72.28333°W

Táchira, 07.83333°N, 72.08333°W

Villarroel, Quebrada Seca, 200 m, 10.30000°N, 63.95000°W

Uribante, Río Potosí, 1,050 m, 07.96667°N, 71.68333°W

Trujillo (Estado de)

Agua Santa, 23 km NW of Valera, 90 m, 09.51700°N, 70.66700°W

Hacienda Misisí, 2,210 m, 09.35000°N, 70.30000°W

Hacienda Misisí, 14 km E of Trujillo, 2,215–2,365 m, 09.35833°N, 70.30722°W

Hacienda Misisí, 15 km E of Trujillo, 2,350–2,360 m, 09.35833°N, 70.30722°W

Isnotú, 10 km WNW of Valera, 930 m, 09.36700°N, 70.70000°W

Macizo de Guaramacal, 9 km ENE of Boconó, 3,100 m, ca. 09.23333°N, 70.18472°W

Vargas (Estado de)

Galiparé, Cerro del Ávila, near Caracas [= Galipán, or Pichaco de Galipán], 6,000 ft, 10.56667°N, 66.90000°W

La Guaira, 10.60000°N, 66.93333°W

Los Venados, 4 km NNW of Caracas, 1,465–1,524 m, 10.53333°N, 66.90000°W

Naiguatá, 10.61667°N, 66.73333°W

Pico Ávila, 5 km NNE of Caracas, near Hotel Humboldt, 2,184 m, 10.55000°N, 66.87000°W

Pico Ávila, near Hotel Humboldt, 2,223 m, 10.55000°N, 66.86667°W

San Julián, 8 mi E of La Guaira, sea level, 10.61700°N, 66.83300°W

Naiguatá, los canals de, Parque Nacional El Ávila, 720 m, 10.57000°N, 67.75000°W

Yaracuy (Estado de)

Finca El Jaguar, 12 km NW of Aroa, 10.43333°N, 68.90000°W

Minas de Aroa, 10.41667°N, 68.90000°W

["Aroa is the name now used to designate the district of which Pueblo Nuevo is the center, but it was formerly applied to the copper mine situated about three miles up the gorge of the Aroa River. It is the headquarters of the Bolivar Railway." (Carriker *in* J. A. Allen 1911:241)]

Palmichal, 23 km N of Bejuma, 1,000 m, 10.05000°N, 68.61700°W

Sierra de Aroa, Parque Nacional Yurubí, Sector El Silencio, La Trampa del Tigre, 1,940 m, 10.40306°N, 68.80028°S

Urama, 19 km NW, Km 40, 10.61700°N, 68.40000°W

Urama, vicinity, 10.53333°N, 68.38333°W

Zulia (Estado de)

BMUL:Cerro Azul, vicinity, 10.85000°N, 72.26667°W

Cojoro, near, 34 km NNE of Paraguaipoa, 11.63333°N, 71.83333°W

El Caimito, Refugio de Fauna Silvestre y Reserva de Pesca Ciénaga de Los Olivitos, 40 km NE of Maracaibo, 10.95000°N, 71.38333°W

El Panorama, Río Aurare, 10.66667°N, 71.41667°W

El Rosario, 09.15000°N, 72.60000°W

El Rosario, 45–51 km WNW of Encontrados, 37–50 m, 09.21667°N, 72.56667°W

Empalado Savannas [= Empalado Sabana], 10.66667°N, 71.50000°W

Hacienda El Tigre, 17 km N and 55 km W of Maracai, 80 m, 10.66700°N, 71.61700°W

Hacienda Platanal, 18 km N and 49 km W of Maracaibo, 10.86667°N, 72.05000°W

Kasmera, 21 km SW of Machiques, 270 m, 09.98333°N, 72.71667°W

Los Angeles del Tucuco, 09.80000°N, 72.83333°W

Misión Tukuko, 09.83333°N, 72.86667°W

Misión Tukuko, 60 km SW of Machiques, 200–400 m, 10.06700°N, 72.56700°W

Novito, 3.1 km S and 19 km W of Machiquos, 1,135 m, 10.06667°N, 72.56667°W

Novito, 19 km WSW of Machiques, 1,155 m, 10.03333°N, 72.71667°W

Perijá, Río Cogolla, 09.85000°N, 72.10000°W

Río Cachiri, 10.83333°N, 72.21667°W

Río Cogollo, 10.25000°N, 72.50000°W

Río Lajas, confluence with El Palmar, 10.66215°N, 72.29821°W

Río Limón, 10.46091°N, 71.82469°W

List of Taxa

Genus *Deltamys*
   D. kempi . . . . Delta Mouse
Genus *Gyldenstolpia*
   G. fronto . . . . Fossorial Giant Rat
   G. planaltensis . . . . Cerrado Giant Rat
Genus *Juscelinomys*
   J. candango . . . . Candango Akodont
   J. huanchacae . . . . Huanchaca Akodont
Genus *Kunsia*
   K. tomentosus . . . . Woolly Giant Rat
Genus *Lenoxus*
   L. apicalis . . . . White-tailed Akodont
Genus *Necromys*
   N. amoenus . . . . Pleasant Akodont
   N. lactens . . . . White-chinned Akodont
   N. lasiurus . . . . Hairy-tailed Akodont
   N. lenguarum . . . . Paraguayan Akodont
   N. obscurus . . . . Dark-furred Akodont
   N. punctulatus . . . . Ecuadorean Akodont
   N. urichi . . . . Northern Grass Mouse
Genus *Oxymycterus*
   O. amazonicus . . . . Amazon Hocicudo
   O. caparaoe . . . . Mt. Caparaó Hocicudo
   O. dasytrichus . . . . . Atlantic Forest Hocicudo
   O. delator . . . . Spy Hocicudo
   O. hiska . . . . Small Yungus Hocicudo
   O. huchucha . . . . Quechuan Hocicudo
   O. inca . . . . Incan Hocicudo
   O. josei . . . . Cook's Hocicudo
   O. juliacae . . . . . Upper Yungus Inca Hocicudo
   O. nasutus . . . . Darwin's Hocicudo
   O. nigrifrons . . . . Elfin Forest Hocicudo
   O. paramensis . . . . Páramo Hocicudo
   O. quaestor . . . . Quaestor Hocicudo
   O. rufus . . . . Red Hocicudo
   O. wayku . . . . Ravine Hocicudo
Genus *Podoxymys*
   P. roraimae . . . . Roraima Akodont
Genus *Scapteromys*
   S. aquaticus . . . . Argentinean Swamp Rat
   S. tumidus . . . . Uruguay Swamp Rat
Genus *Thalpomys*
   T. cerradensis . . . . Cerrrado Akodont
   T. lasiotis . . . . Hairy-eared Akodont
Genus *Thaptomys*
   T. nigrita . . . . Ebony Akodont
Tribe Ichthyomyini
Genus *Anotomys*
   A. leander . . . . Earless Water Mouse
Genus *Chibchanomys*
   C. orcesi . . . . Orces's Andean Water Mouse
   C. trichotis . . . . Northern Andean Water Mouse

Genus *Ichthyomys*
   I. hydrobates . . . . . Silver-bellied Ichthyomyine
   I. pitieri . . . . Pittier's Ichthyomyine
   I. stolzmanni . . . . . . Stolzmann's Ichthyomyine
   I. tweedii . . . . Tweedy's Ichthyomyine
Genus *Neusticomys*
   N. ferreirai . . . . Ferreira's Ichthyomyine
   N. monticolus . . . . Montane Ichthyomyine
   N. mussoi . . . . Musso's Ichthyomyine
   N. oyapocki . . . . Guianan Ichthyomyine
   N. peruviensis . . . . Peruvian Ichthyomyine
   N. venezuelae . . . . Venezuelan Ichthyomyine
Tribe Oryzomyini
Genus *Aegialomys*
   A. galapagoensis . . . . Galapagos Aegialomys
   A. xantheolus . . . . Yellowish Aegialomys
Genus *Amphinectomys*
   A. savamis . . . . Ucayali Water Rat
Genus *Cerradomys*
   C. goytaca . . . . Goytacá Cerradomys
   C. langguthi . . . . Langguth's Cerradomys
   C. maracajuensis . . . . . . Maracaju Cerradomys
   C. marinhus . . . . Marinho's Cerradomys
   C. scotti . . . . Lindbergh's Cerradomys
   C subflavus . . . . Flavescent Cerradomys
   C. vivoi . . . . Vivo's Cerradomys
Genus *Drymoreomys*
   D. albimaculatus . . . .
   . . . . White-throated Montane Forest Rat
Genus *Eremoryzomys*
   E. polius . . . . Gray Eremoryzomys
Genus *Euryoryzomys*
   E. emmonsae . . . . Emmons's Euryoryzomys
   E. lamia . . . . Buffy-sided Euryoryzomys
   E. legatus . . . . Tarija Euryoryzomys
   E. macconnelli . . . . Macconnell's Euryoryzomys
   E. nitidus . . . . Elegant Euryoryzomys
   E. russatus . . . . Russet Euryoryzomys
Genus *Handleyomys*
   H. fuscatus . . . .
   . . . . Colombian Western Andes
   . . . . Cloud Forest Mouse
   H. intectus . . . .
   . . . . Colombian Central Andes
   . . . . Cloud Forest Mouse
Genus "*Handleyomys*"
   "*H.*" *alfaroi* . . . . Alfaro's Rice Rat
Genus *Holochilus*
   H. brasiliensis . . . . Brazilian Marsh Rat
   H. chacarius . . . . Chacoan Marsh Rat
   H. sciureus . . . . Amazonian Marsh Rat
   H. venezuelae . . . . Venezuelan Marsh Rat

H. vulpinus . . . . .Crafty Marsh Rat
Genus *Hylaeamys*
    H. acritus . . . . Bolivian Hylaeamys
    H. laticeps . . . . Atlantic Forest Hylaeamys
    H. megacephalus . . . .
    . . . . Azara's Broad-headed Hylaeamys
    H. oniscus . . . .
    . . . . Northern Atlantic Forest Hylaeamys
    H. perenensis . .Western Amazonian Hylaeamys
    H. tatei. . . . . Tate's Hylaeamys
    H. yunganus . . . . Amazonian Hylaeamys
Genus *Lundomys*
    L. molitor. . . . . Lund's Water Rat
Genus *Melanomys*
    M. caliginosus . . . . Dusky Rice Rat
    M. columbianus . . . Colombian Dusky Rice Rat
    M. robustulus. . . . . Robust Dark Rice Rat
    M. zunigae . . . . Zuniga's Dark Rice Rat
Genus *Microakodontomys*
    M. transitorius . . . . Transitional Colilargo
Genus *Microryzomys*
    M. altissimus . . . . .Páramo Colilargo
    M. minutus. . . . . .Montane Colilargo
Genus *Mindomys*
    M. hammondi . . . . Hammond's Mindomys
Genus *Neacomys*
    N. dubosti . . . . Dubost's Spiny Mouse
    N. guianae . . . . Guianan Spiny Mouse
    N. minutus. . . . . Minute Spiny Mouse
    N. musseri . . . . Musser's Spiny Mouse
    N. paracou. . . . . .Paracou Spiny Mouse
    N. spinosus . . . . .Common Spiny Mouse
    N. tenuipes. . . . . . . Slender Foot Spiny Mouse
Genus *Nectomys*
    N. apicalis . . . . Western Amazonian Water Rat
    N. grandis . . . . Magdalena Water Rat
    N. palmipes . . . . Trinidad Water Rat
    N. rattus. . . . . .Common Water Rat
    N. squamipes . . . . Atlantic Water Rat
Genus *Nephelomys*
    N. albigularis . . . . White-throated Nephelomys
    N. auriventer . . . . . Golden-bellied Nephelomys
    N. caracolus . . . Costal Cordilleran Nephelomys
    N. childi. . . . . Child's Nephelomys
    N. keaysi . . . . .Keays's Nephelomys
    N. levipes . . . . Nimble-footed Nephelomys
    N. maculiventer . . . . .Santa Marta Nephelomys
    N. meridensis. . . . . .Mérida Nephelomys
    N. moerex . . . . .Gray-bellied Nephelomys
    N. nimbosus. . . . .
    . . . . .Lesser Golden-bellied Nephelomys
    N. pectoralis. .Western Colombian Nephelomys

Genus *Nesoryzomys*
    N. darwini . . . . .Darwin's Nesoryzomys
    N. fernandinae . . .Small Fernandina Nesoryzomys
    N. indefessus . . . . . . Santa Cruz Nesoryzomys
    N. narboroughi . . . .
    . . . . Large Fernandina Nesoryzomys
    N. swarthi . . . . .Santiago Nesoryzomys
Genus *Oecomys*
    O. auyantepui . . . . Guianan Oecomys
    O. bicolor . . . . White-bellied Oecomys
    O. catherinae . . . . . . Atlantic Forest Oecomys
    O. cleberi . . . . .Cleber's Oecomys
    O. concolor . . . . Natterer's Oecomys
    O. flavicans . . . . .Tawny Oecomys
    O. mamorae. . . . . Mamoré Oecomys
    O. paricola. . . . . Brazilian Oecomys
    O. phaeotis . . . . .Dusky Oecomys
    O. rex . . . . Regal Oecomys
    O. roberti. . . . . Robert's Oecomys
    O. rutilus . . . . Reddish Oecomys
    O. speciosus. . . . . Savanna Oecomys
    O. superans . . . . Large Oecomys
    O. sydandersoni. . . . . . . Anderson's Oecomys
    O. trinitatis . . . . Long-furred Oecomys
Genus *Oligoryzomys*
    O. andinus . . . . .Andean Colilargo
    O. arenalis . . . . Sandy Colilargo
    O. brendae. . . . . .San Javier's Colilargo
    O. chacoensis. . . . . .Chacoan Colilargo
    O. delicatus . . . . Delicate Colilargo
    O. destructor . . . . .Tschudi's Colilargo
    O. flavescens . . . . .Flavescent Colilargo
    O. fornesi. . . . . Forne's Colilargo
    O. griseolus . . . . Grizzled Colilargo
    O. longicaudatus . . . . . . .Long-tailed Colilargo
    O. magellanicus . . . . Patagonian Colilargo
    O. mattogrossae. . . . . . Mato Grosso Colilargo
    O. messorius . . . . Hairy Colilargo
    O. microtis. . . . . .Small-eared Colilargo
    O. moojeni . . . . Moojen's Colilargo
    O. nigripes . . . . .Black-footed Colilargo
    O. rupestris . . . . .Highlands Colilargo
    O. stramineus. . . . . . .Straw-colored Colilargo
    O. utiaritensis . . . . Utiariti Colilargo
Genus *Oreoryzomys*
    O. balneator. . . . . . . Ecuadorean Oreoryzomys
Genus *Oryzomys*
    O. couesi . . . . .Coues's Marsh Rice Rat
    O. gorgasi . . . . Gorgas Marsh Rice Rat
Genus *Pseudoryzmys*
    P. simplex. . . . . False Oryzomys
Genus *Scolomys*

S. melanops . . . . Gray Spiny Mouse
S. ucayalensis . . . . Ucayali Spiny Mouse
Genus *Sigmodontomys*
 S. alfari . . . . Alfaro's Rice Rat
Genus *Sooretamys*
 S. angouya . . . . Angouya Sooretamys
Genus *Tanyuromys*
 T. aphrastus . . . . .Long-tailed Montane Rat
Genus *Transandinomys*
 T. bolivaris . . . . Long-whiskered Oryzomys
 T. talamancae . . . . Transandean Oryzomys
Genus *Zygodontomys*
 Z. brevicauda . . . . .Cane Mouse
 Z. brunneus . . . . Colombian Cane Mouse
Tribe Phyllotini
Genus *Andalgalomys*
 A. olrogi . . . . Olrog's Chaco Mouse
 A. pearsoni . . . . Pearson's Chaco Mouse
 A. roigi . . . . Roig's Chaco Mouse
Genus *Auliscomys*
 A. boliviensis . . . . Bolivian Pericote
 A. pictus . . . . Colorful Pericote
 A. sublimis . . . . .Lofty Pericote
Genus *Calassomys*
 C. apicalis . . . . . . Calaça's White-tailed Mouse
Genus *Calomys*
 C. boliviae . . . . Bolivian Laucha
 C. callidus . . . . Crafty Vesper Mouse
 C. callosus . . . . .Large Vesper Mouse
 C. cerqueirai . . . . . . Cerqueira's Vesper Mouse
 C. expulsus . . . . Caatinga Laucha
 C. hummelincki . . . . Hummelinck's Laucha
 C. laucha . . . . .Small Vesper Mouse
 C. lepidus . . . . Andean Vesper Mouse
 C. musculinus . . . . Drylands Vesper Mouse
 C. sorellus . . . . Peruvian Vesper Mouse
 C. tener . . . . .Delicate Vesper Mouse
 C. tocantinsi . . . . Tocantins Vesper Mouse
 C. venustus . . . . Córdoba Laucha
Genus *Eligmodontia*
 E. bolsonensis . . . . Bolsón Gerbil Mouse
 E. dunaris . . . . Dune Gerbil Mouse
 E. hirtipes . . . . Hairy-footed Gerbil Mouse
 E. moreni . . . . Monte Gerbil Mouse
 E. morgani . . . . Morgan's Gerbil Mouse
 E. puerulus . . . . Andean Gerbil Mouse
 E. typus . . . . Highland Gerbil Mouse
Genus *Galenomys*
 G. garleppi . . . . Garlepp's Pericote
Genus *Graomys*
 G. chacoensis . . . . .Chaco Pericote
 G. domorum . . . . .Pale Pericote

G. edithae . . . . Otro Cerro Pericote
G. griseoflavus . . . . Common Pericote
Genus *Loxodontomys*
 L. micropus . . . . .Southern Pericote
Genus *Phyllotis*
 P. alisosiensis . . . . . Los Alisos Leaf-eared Mouse
 P. amicus . . . . .Friendly Leaf-eared Mouse
 P. andium . . . . Andean Leaf-eared Mouse
 P. anitae . . . . .Anita's Leaf-eared Mouse
 P. bonariensis . . Buenos Aires Leaf-eared Mouse
 P. caprinus . . . . . . Capricorn Leaf-eared Mouse
 P. darwini . . . . Darwin's Leaf-eared Mouse
 P. definitus . . . . Ancash Leaf-eared Mouse
 P. gerbillus . . . . Peruvian Leaf-eared Mouse
 P. haggardi . . . . . . Haggard's Leaf-eared Mouse
 P. limatus . . . . Lima Leaf-eared Mouse
 P. magister . . . . Master Leaf-eared Mouse
 P. osgoodi . . . . .Osgood's Leaf-eared Mouse
 P. osilae . . . . . . Bunch Grass Leaf-eared Mouse
 P. xanthopygus . . . .
 . . . . Yellow-rumped Leaf-eared Mouse
Genus *Salinomys*
 S. delicatus . . . . Delicate Salt Flat Mouse
Genus *Tapecomys*
 T. primus . . . . .Tapecua Leaf-eared Mouse
 T. wolffsohni . . . Wolffsohn's Leaf-eared Mouse
Tribe Reithrodontini
Genus *Reithrodon*
 R. auritus . . . . Hairy-soled Conyrat
 R. typicus . . . . .Naked-soled Conyrat
Tribe Sigmodontini
Genus *Sigmodon*
 S. alstoni . . . . .Groove-toothed Cotton Rat
 S. hirsutus . . . . .Burmeister's Cotton Rat
 S. inopinatus . . . . Ecuadorean Cotton Rat
 S. peruanus . . . . Peruvian Cotton Rat
Tribe Thomasomyini
Genus *Aepeomys*
 A. lugens . . . . Mérida Aepeomys
 A. reigi . . . . Reig's Aepeomys
Genus *Chilomys*
 C. fumeus . . . . Smoky Chilomys
 C. instans . . . . .Andean Chilomys
Genus *Rhagomys*
 R. longilingua . . . . . . .Long-tongued Rhagomys
 R. rufescens . . . . Rufescent Rhagomys
Genus *Rhipidomys*
 R. austrinus . . . . Southern Andean Rhipidomys
 R. cariri . . . . .Caatinga Rhipidomys
 R. caucensis . . . .Lesser Colombian Rhipidomys
 R. couesi . . . . .Coues's Rhipidomys
 R. emiliae . . . . Snethlage's Rhipidomys

R. fulviventer . . . . . .Tawny-bellied Rhipidomys
R. gardneri. . . . . Gardner's Rhipidomys
R. ipukensis . . . . Ipuka Rhipidomys
R. itoan . . . . .Sky Rhipidomys
R. latimanus. . . . . . .Northwestern Rhipidomys
R. leucodactylus. . . . . Great Rhipidomys
R. macconnelli . . . .
. . . . .Macconnell's Tepui Rhipidomys
R. macrurus . . . . Cerrado Rhipidomys
R. mastacalis . . . .
. . . . North Atlantic Forest Rhipidomys
R. modicus. . . . . . Lesser Peruvian Rhipidomys
R. nitela . . . . .Guianan Rhipidomys
R. ochrogaster . . . . . . .Buff-bellied Rhipidomys
R. similis . . . . Greater Colombian Rhipidomys
R. tenuicauda. . . . . . . .Turmiquire Rhipidomys
R. tribei . . . . Tribe's Rhipidomys
R. venezuelae . . . . . . .Venezuelan Rhipidomys
R. venustus. . . . . Mérida Highland Rhipidomys
R. wetzeli . . . . Wetzel's Rhipidomys
Genus *Thomasomys*
T. andersoni . . . . Anderson's Thomasomys
T. apeco . . . . .Apeco Thomasomys
T. aureus . . . . .Golden Thomasomys
T. auricularis . . . . . Red Andean Thomasomys
T. australis . . . . Austral Thomasomys
T. baeops . . . . .Short-faced Thomasomys
T. bombycinus . . . . .Silky Thomasomys
T. caudivarius. . . . . White-tipped Thomasomys
T. cinereiventer. . . . . Ashy-bellied Thomasomys
T. cinereus . . . . Olive-gray Thomasomys
T. cinnameus . . . .
. . . . Cinnamon-colored Thomasomys
T. contradictus . . Central Andean Thomasomys
T. daphne. . . . . Daphne's Thomasomys
T. dispar. . . . . Colombian Thomasomys
T. eleusis. . . . . Peruvian Thomasomys
T. emeritus . . . . Venezuelan Thomasomys
T. erro . . . . Wandering Thomasomys
T. fumeus . . . . Smokey Thomasomys
T. gracilis . . . . Gracile Thomasomys
T. hudsoni . . . . Hudson's Thomasomys
T. hylophilus . . . . . . .Woodland Thomasomys
T. incanus. . . . . Inca Thomasomys
T. ischyrus . . . . .Long-tailed Thomasomys
T. kalinowskii. . . . . .Kalinowski's Thomasomys
T. ladewi . . . . Ladew's Thomasomys
T. laniger . . . . Soft-furred Thomasomys
T. macrotis. . . . . . . .Long-eared Thomasomys
T. monochromos . . . . Unicolored Thomasomys
T. nicefori. . . . . Nicéforo María's Thomasomys
T. niveipes . . . . White-footed Thomasomys

T. notatus. . . . . Dusky-footed Thomasomys
T. onkiro . . . . Ashaninka Thomasomys
T. oreas . . . . Montane Thomasomys
T. paramorum . . . . .Páramo Thomasomys
T. popayanus . . . . Popayán Thomasomys
T. praetor . . . . Cajamarca Thomasomys
T. princeps . . . . .Principal Thomasomys
T. pyrrhonotus . . Reddish-backed Thomasomys
T. rosalinda . . . . . . .Rosalinda's Thomasomys
T. silvestris . . . . Sylvan Thomasomys
T. taczanowskii . . Taczanoswki's Thomasomys
T. ucucha . . . . .Ucucha Thomasomys
T. vestitus . . . . Mérida Thomasomys
T. vulcani . . . . Pichincha Thomasomys
Tribe Wiedomyini
Genus *Wiedomys*
W. cerradensis . . . . Cerrado Wiedomys
W. plyrrhorhinos . . . . . . .Red-nosed Wiedomys
Subfamily Tylomyinae
Genus *Tylomys*
T. mirae . . . . Southern Climbing Rat
Suborder Hystricomorpha
Superfamily Cavioidea
Family Caviidae
Subfamily Caviinae
Genus *Cavia*
C. aperea . . . . .Brazilian Guinea Pig
C. fulgida . . . . Shiny Guinea Pig
C. intermedia . . . . . . .Intermediate Guinea Pig
C. magna . . . . .Greater Guinea Pig
C. patzelti. . . . . Sacha Guinea Pig
C. porcellus . . . . Domestic Guinea Pig
C. tschudii . . . . Montane Guinea Pig
Genus *Galea*
G. comes . . . .
. . . . . .Southern Highland Yellow-toothed Cavy
G. flavidens . . . . .Eastern Yellow-toothed Cavy
G. leucoblephara Lowland Yellow-toothed Cavy
G. musteloides . Highland Yellow-toothed Cavy
G. spixii . . . . Spix's Yellow-toothed Cavy
Genus *Microcavia*
M. australis . . . . . . .Southern Mountain Cavy
M. niata . . . . Northern Mountain Cavy
M. shiptoni. . . . . . . .Shipton's Mountain Cavy
Subfamily Dolichotinae
Genus *Dolichotis*
D. patagonum . . . . Patagonian mara
D. salinicola. . . . . Chacoan mara
Subfamily Hydrochoerinae
Genus *Hydrochoerus*
H. hydrochaeris . . . . .Capybara
H. isthmius. . . . . .Lesser Capybara

Genus *Kerodon*

  K. acrobata . . . . .Acrobatic Moco

  K. rupestris. . . . . .Moco

Family Cuniculidae

  Genus *Cuniculus*

    C. paca. . . . . .Paca

    C. taczanowskii . . . . .Mountain Paca

Family Dasyproctidae

  Genus *Dasyprocta*

    D. azarae . . . . Azara's Agouti

    D. croconota . . . . .Orange Agouti

    D. fuliginosa . . . . Black Agouti

    D. guamara . . . . Guamara Agouti

    D. iacki . . . . Iack-Ximenes's Agouti

    D. kalinowskii . . . . Kalinowski's Agouti

    D. leporina. . . . . Red-rumped Agouti

    D. prymnolopha. . . . . .Black-rumped Agouti

    D. punctata . . . . Central American Agouti

    D. variegata . . . . Brown Agouti

  Genus *Myoprocta*

    M. acouchy . . . . Red Acouchy

    M. pratti. . . . . Green Acouchy

Superfamily Chinchilloidea

Family Chinchillidae

  Subfamily Chinchillinae

    Genus *Chinchilla*

      C. chinchilla. . . . . .Short-tailed Chinchilla

      C. lanigera . . . . Chilean Chinchilla

    Genus *Lagidium*

      L. ahuacaense. . . . . Ecuadorean Viscacha

      L. viscacia . . . . Mountain Viscacha

      L. wolffsohni . .Wolffshon's Mountain Viscacha

  Subfamily Lagostominae

    Genus *Lagostomus*

      L. crassus . . . . Peruvian Plains Viscacha

      L. maximus . . . . .Plains Viscacha

Family Dinomyidae

  Genus *Dinomys*

    D. branickii . . . . Pacarana

Superfamily Erethizontoidea

Family Erethizontidae

  Genus *Chaetomys*

    C. subspinosus . . . . . . .Bristle-spined Porcupine

  Genus *Coendou*

    C. bicolor. . . . . Bicolor-spines Porcupine

    C. ichillus. . . . . .Small Porcupine

    C. insidiosus. . . . . Bahia Hairy Dwarf Porcupine

    C. melanurus . . . . Black-tailed Porcupine

    C. nycthemera . . . . Silver Porcupine

    C. prehensilis . . . . Brazilian Porcupine

    C. pruinosus. . . . . Frosted Porcupine

    C. quichua . . . . Quichua Porcupine

C. roosmalenorum . . . .Roosmalens's Porcupine

C. rufescens . . . . Stump-tailed Porcupine

C. spinosus. . . . .

. . . . .Paraguayan Hairy Dwarf Porcupine

C. vestitus . . . . Brown Hairy Dwarf Porcupine

Superfamily Octodontoidea

Family Abrocomidae

  Genus *Abrocoma*

    A. bennettii . . . . .Bennett's Chinchilla Rat

    A. boliviensis . . . . Bolivian Chinchilla Rat

    A. budini . . . . Budin's Chinchilla Rat

    A. cinerea. . . . . .Ashy Chinchilla Rat

    A. famatina . . . . Famatina Chinchilla Rat

    A. schisticea . . . Sierra del Tontal Chinchilla Rat

    A. uspallata . . . . Uspallata Chinchilla Rat

    A. vaccarum . . . . Mendozan Chinchilla Rat

  Genus *Cuscomys*

    C. ashaninka . . . .

. . . . Ashaninka Arboreal Chinchilla Rat

    C. oblativus . . . .

. . . . Machu Picchu Arboreal Chinchilla Rat

Family Ctenomyidae

  Genus *Ctenomys*

    C. argentinus . . . . .Argentine Tuco-tuco

    C. australis. . . . . .Dune Tuco-tuco

    C. azarae . . . . Azara's Tuco-tuco

    C. bergi . . . . Cordobea Tuco-tuco

    C. bicolor. . . . . .Bicolored Tuco-tuco

    C. boliviensis . . . . .Bolivian Tuco-tuco

    C. bonettoi. . . . . Bonetto's Tuco-tuco

    C. brasiliensis . . . . Brazilian Tuco-tuco

    C. colburni. . . . . . . .White-bellied Tuco-tuco

    C. coludo . . . . Puntilla Tuco-tuco

    C. conoveri. . . . . Chacoan Tuco-tuco

    C. coyhaiquensis . . . . . . .Coyhaique Tuco-tuco

    C. dorbignyi . . . . D'Orbignyi's Tuco-tuco

    C. dorsalis . . . . .Black-backed Tuco-tuco

    C. emilianus . . . . .Emily's Tuco-tuco

    C. famosus . . . . .Famatina Tuco-tuco

    C. flamarioni . . . . Flamarion's Tuco-tuco

    C. fochi . . . . Foch's Tuco-tuco

    C. fodax. . . . . Lago Blanco Tuco-tuco

    C. frater . . . . Forest Tuco-tuco

    C. fulvus. . . . . Long-tailed Tuco-tuco

    C. goodfellowi . . . . . . .Goodfellow's Tuco-tuco

    C. haigi . . . . Patagonian Tuco-tuco

    C. ibicuiensis . . . . .Ibicui Tuco-tuco

    C. johannis. . . . . San Juan Tuco-tuco

    C. juris. . . . . Jujuy Tuco-tuco

    C. knighti. . . . . .Catamarca Tuco-tuco

    C. lami. . . . . Lami Tuco-tuco

    C. latro. . . . . .Mottled Tuco-tuco

C. leucodon . . . . White-toothed Tuco-tuco
C. lewisi . . . . .Lewis's Tuco-tuco
C. magellanicus . . . . Magellan Tuco-tuco
C. "mariafarelli" . . . . Maria Farell's Tuco-tuco
C. maulinus . . . . Maule Tuco-tuco
C. mendocinus . . . . Mendoza Tuco-tuco
C. minutus . . . . Minute Tuco-tuco
C. nattereri . . . . Natterer's Tuco-tuco
C. occultus . . . . .Furtive Tuco-tuco
C. opimus . . . . Andean Tuco-tuco
C. osvaldoreigi . . . . . .Osvaldo Reig's Tuco-tuco
C. paraguayensis . . . . . . Paraguayan Tuco-tuco
C. pearsoni . . . . Pearson's Tuco-tuco
C. perrensi . . . . Perrens's Tuco-tuco
C. peruanus . . . . Peruvian Tuco-tuco
C. pilarensis . . . . .Pilar Tuco-tuco
C. pontifex . . . . Brown Tuco-tuco
C. porteousi . . . . Cinnamon Tuco-tuco
C. pundti . . . . Pundt's Tuco-tuco
C. rionegrensis . . . . Rio Negro Tuco-tuco
C. roigi . . . . Roig's Tuco-tuco
C. rondoni . . . . .Rondon's Tuco-tuco
C. rosendopascuali . . . .
. . . . Rosendo Pascual's Tuco-tuco
C. saltarius . . . . Salta Tuco-tuco
C. scagliai . . . . .Scaglia's Tuco-tuco
C. sericeus . . . . Silky Tuco-tuco
C. sociabilis . . . . Colonial Tuco-tuco
C. steinbachi . . . . .Steinbach's Tuco-tuco
C. talarum . . . . Talas Tuco-tuco
C. torquatus . . . . Collarded Tuco-tuco
C. tuconax . . . . .Robust Tuco-tuco
C. tucumanus . . . . Tucumán Tuco-tuco
C. tulduco . . . . Sierra Tontal Tuco-tuco
C. validus . . . . .Guaymallén Tuco-tuco
C. viperinus . . . . Monte Tuco-tuco
C "yolandae" . . . . .Santa Fe Tuco-tuco
Family Echimyidae
　Subfamily Dactylomyinae
　　Genus *Dactylomys*
　　　D. boliviensis . . . . .Bolivian Bamboo Rat
　　　D. dactylinus . . . . .Amazon Bamboo Rat
　　　D. peruanus . . . . Montane Bamboo Rat
　　Genus *Kannabateomys*
　　　K. amblyonyx . . . . Atlantic Bamboo Rat
　　Genus *Ollamys*
　　　O. albicaudus . . . . White-tailed Ollala Rat
　　　O. edax . . . . Greedy Ollala Rat
　Subfamily Echimyinae
　　Genus *Callistomys*
　　　C. pictus . . . . Painted Tree Rat
　　Genus *Diplomys*

D. caniceps . . . . . Colombian Rufous Tree Rat
D. labilis . . . .
. . . . Central American Rufous Tree Rat
Genus *Echimys*
　E. chrysurus . . . . .White-faced Tree Rat
　E. saturnus . . . . Dark Tree Rat
　E. vieirai . . . . .Vieira's Tree Rat
Genus *Isothrix*
　I. barbarabrownae . . . .
. . . . Barbara Brown's Brush-tailed Rat
　I. bistriata . . .Yellow-crowned Brush-tailed Rat
　I. negrensis . . . . Rio Negro Brush-tailed Rat
　I. orinoci . . . . . .Orinoco Brush-tailed Rat
　I. pagurus . . . . Plain Brush-tailed Rat
　I. sinnamariensis . . Sinnamary Brush-tailed Rat
Genus *Makalata*
　M. didelphoides . .Red-nosed Armored Tree Rat
　M. macrura . . . .Long-tailed Armored Tree Rat
　M. obscura . . . . .Dark Armored Tree Rat
Genus *Pattonomys*
　P. carrikeri . . . . . . Carriker's Speckled Tree Rat
　P. flavidus . . . . .Yellow Speckled Tree Rat
　P. ocasius . . . . Bare-tailed Tree Rat
　P. punctatus . . . . . .Orinocan Speckled Tree Rat
　P. semivillosus . . . .
. . . . Colombian Speckled Tree Rat
Genus *Phyllomys*
　P. blainvilii . . . . Golden Atlantic Tree Rat
　P. brasiliensis . . . . . .Brazilian Atlantic Tree Rat
　P. dasythrix . . . . .Drab Atlantic Tree Rat
　P. kerri . . . . Kerr's Atlantic Tree Rat
　P. lamarum . . . . Pallid Atlantic Tree Rat
　P. lundi . . . . Lund's Atlantic Tree Rat
　P. mantiqueirensis . . . .
. . . . . . Serra da Mantiqueira Atlantic Tree Rat
　P. medius . . . . . . Long-furred Atlantic Tree Rat
　P. nigrispinus . . .Black-spined Atlantic Tree Rat
　P. pattoni . . . . Patton's Atlantic Tree Rat
　P. sulinus . . . . .Southern Atlantic Tree Rat
　P. thomasi . . . . Thomas's Atlantic Tree Rat
　P. unicolor . . . . . . Unicolored Atlantic Tree Rat
Genus *Santamartamys*
　S. rufodorsalis . . . . .Red Crested Tree Rat
Genus *Toromys*
　T. grandis . . . . Black Toro
　T. rhipidurus . . . . Peruvian Toro
Subfamily Eumysopinae
　Genus *Carterodon*
　　C. sulcidens . . . . Groove-toothed Spiny Rat
　Genus *Clyomys*
　　C. laticeps . . . . .Broad-headed Spiny Rat
　Genus *Euryzygomatomys*

E. spinosus. Guiara
Genus *Hoplomys*
   H. gymnurus . . . . Armored Rat
Genus *Lonchothrix*
   L. emiliae. . . . . . . .Tufted-tailed Spiny Tree Rat
Genus *Mesomys*
   M. hispidus . . . . Ferriera's Spiny Tree Rat
   M. leniceps. . . . . . . Long-haired Spiny Tree Rat
   M. occultus . . . . .Hidden Spiny Tree Rat
   M. stimulax . . . . Pará Spiny Tree Rat
Genus *Proechimys*
   P. brevicauda . . . . Short-tailed Spiny Rat
   P. canicollis. . . . . .Colombian Spiny Rat
   P. chrysaeolus. . . . . .Boyacá Spiny Rat
   P. cuvieri. . . . . Cuvier's Spiny Rat
   P. decumanus . . . . Pacific Spiny Rat
   P. echinothrix. . . . . .Stiff-spined Spiny Rat
   P. gardneri . . . . .Gardner's Spiny Rat
   P. goeldii. . . . . Goeldi's Spiny Rat
   P. guairae . . . . La Guaira Spiny Rat
   P. guyannensis . . . . Guyenne Spiny Rat
   P. hoplomyoides. . . . . .Guianan Spiny Rat
   P. kulinae . . . . Kulina Spiny Rat
   P. longicaudatus . . . . . . Long-tailed Spiny Rat
   P. mincae . . . . Minca Spiny Rat
   P. oconnelli. . . . . O'Connell's Spiny Rat
   P. pattoni . . . . Patton's Spiny Rat
   P. quadruplicatus . . . . Napo Spiny Rat
   P. roberti . . . . Robert's Spiny Rat
   P. semispinosus. . . . . Tomes's Spiny Rat
   P. simonsi. . . . . .Simons's Spiny Rat
   P. steerei. . . . . .Steere's Spiny Rat
   P. trinitatis . . . . .Trinidad Spiny Rat
Genus *Thrichomys*
   T. apereoides . . . . Lagoa Santa Punaré
   T. inermis. . . . . .Jacobina Punaré
   T. laurentius . . . . .São Lourenço Punaré
   T. pachyurus. . . . . Paraguayan Punaré

Genus *Trinomys*
   T. albispinus. . . . . White-spined Spiny Rat
   T. dimidiatus . . . . Rio de Janeiro Spiny Rat
   T. eliasi. . . . . .Elias's Spiny Rat
   T. gratiosus. . . . . Gracile Atlantic Spiny Rat
   T. iheringi. . . . . .São Paulo Spiny Rat
   T. mirapitanga . . . . .Pau Brasil Spiny Rat
   T. moojeni . . . . Moojen's Spiny Rat
   T. paratus. . . . . Rigid-spined Atlantic Spiny Rat
   T. setosus . . . .
   . . . . Elegant-spined Atlantic Spiny Rat
   T. yonenagae . . . . Torch-tailed Spiny Rat
Subfamily Myocastorinae
  Genus *Myocastor*
   M. coypus . . . . Coypu
Family Octodontidae
  Genus *Aconaemys*
   A. fuscus . . . . .Chilean Rock Rat
   A. porteri . . . . .Porter's Rock Rat
   A. sagei . . . . Sage's Rock Rat
  Genus *Octodon*
   O. bridgesii . . . . Bridges's Degu
   O. degus. . . . . .Common Degu
   O. lunatus . . . . Coastal Degu
   O. pacificus . . . . Mocha Degu
  Genus *Octodontomys*
   O. gliroides . . . . Mountain Degu
  Genus *Octomys*
   O. mimax. . . . . Viscacha Rat
  Genus *Spalacopus*
   S. cyanus . . . . .Cururo
  Genus *Tympanoctomys*
   T. aureus . . . . Golden Viscacha Rat
   T. barrerae . . . . .Red Viscacha Rat
   T. kirchnerorum . . . . . .Kirchners' Viscacha Rat
   T. loschalchalerosorum. . . . .
   . . . . Chalchalero Viscacha Rat

Contributors

Diego F. Alvarado-Serrano
2095 Museum of Zoology
University of Michigan
1109 Geddes Avenue
Ann Arbor, MI 48109-1079 USA
dalvarad@umich.edu

Sergio Ticul Álvarez-Castañeda
Centro de Investigaciones Biológicas del Noroeste, S.C.
Mar Bermejo 195
La Paz, Baja California Sur, 23009, México
sticul@cibnor.mx

Robert P. Anderson
Department of Biology
City College of New York
City University of New York
New York, NY 10031 USA
anderson@sci.ccny.cuny.edu

Elizabeth Arellano
Centro de Investigqción en Biodiversidad y Conservación
Universidad Autónoma del Estado de Morelos
Av. Universidad 1001 Col. Chamilpa
Cuernavaca, Morelos 62209, México
elisabet@uaem.mx

Rubén M. Barquez
Programa de Investigaciones de Biodiversidad
   Argentina (PIDBA)
Universidad National de Tucumán
Miguel Lillo 255
4000 San Miguel de Tucumán, Argentina
rubenbarquez@arnet.com.ar

Alexandra M. R. Bezerra
Instituto Nacional de Câncer
Divisão de Genética
and
Laboratório de Biologia e Controle da Esquistossomose
Departamento Medicina Tropical
IOC-FIOCRUZ
Rio de Janeiro, Brazil
abezerra@fst.com.br

Claudio J. Bidau
Universidad Nacional de Río Negro
Sede Alto Valle
Subsede Villa Regina
Tacuarí 669, Río Negro, Argentina
bicau47@yahoo.com

Cibele R. Bonvicino
Instituto Nacional de Câncer
Divisão de Genética
and
Laboratório de Biologia e Controle da Esquistossomose
Departamento Medicina Tropical
IOC-FIOCRUZ
Rio de Janeiro, Brazil
cibelerb@inca.gov.br

Janet K. Braun
Department of Mammalogy
Sam Noble Oklahoma Museum of Natural History
2401 Chautauqua Avenue
University of Oklahoma
Norman, OK 73072-7029, USA
jkbraun@ou.edu

Michael D. Carleton
Department of Vertebrate Zoology
Division of Mammals
National Museum of Natural History
Smithsonian Institution
10th and Constitution Avenue NW
Washington, D.C. 20560-0108 USA
carletonm@si.edu

Ana Paula Carmignoto
Universidade Federal de São Carlos
Rodovia João Leme dos Santos, Km 110 – SP-264
Bairro do Itinga, Sorocaba, SP
Brazil – CEP 18052-780
apcarmig@ufscar.br

Guillermo D'Elía
Instituto de Ciencias Ambientales y Evolutivas
Facultad de Ciencias
Universidad Austral de Chile
Campus Isla Teja s/n
Valdivia, Chile
guile.delia@gmail.com

M. Mónica Díaz
Programa de Investigaciones de Biodiversidad Argentina (PIDBA)
Universidad National de Tucumán
Miguel Lillo 255
4000 San Miguel de Tucumán, Argentina
mmonicadiaz@yahoo.com.ar

Robert C. Dowler
Department of Biology

Angelo State University
San Angelo, TX 76909 USA
robert.dowler@angelo.edu

Jonathan L. Dunnum
Division of Mammals
Museum of Southwestern Biology
University of New Mexico
Albuquerque, NM 87131-0001 USA
jldunnum@unm.edu

Louise H. Emmons
Department of Vertebrate Zoology
Division of Mammals
National Museum of Natural History
Smithsonian Institution
10th and Constitution Ave., NW
Washington, D.C. 20560-0108 USA
emmonsl@si.edu

Carlos A. Galliari
Centro de Estudios de Parasitológicos y de Vectores (CEPAVE)
Calle 2 no. 584
La Plata, Buenos Aires, Argentina
cailogalliari@gmail.com

Lena Geise
Departamento de Zoologia – IB
Universidade do Estado do Rio de Janeiro
Rua São Francisco Xavier, 524 Maracanã
Rio de Janeiro, RJ, Brazil
lenageise@gmail.com
geise@uerj.br

Marcela Gómez-Laverde
Fundación Ulamá
Apartado Aéreo 93674
and
Instituto de Ciencias Naturales
Universidad Nacional de Colombia
Bogotá, Colombia
calima@andinet.com

Pablo R. Gonçalves
Programa de Pós-Graduação em Ciências Ambientais e
    Conservação
Universidade Federal do Rio de Janeiro
CP 119331, CEP 27910-970
Macaé, RJ, Brazil
prg@acad.ufrj.br

Enrique M. González
Museo Nacional de Historia Natural
Casilla de Correos 399
11.000 Montevideo, Uruguay
vidasilvestre@interamerica.com.uy

Mark S. Hafner
Museum of Natural Science
119 Foster Hall
Louisiana State University

Baton Rouge, LA 70803 USA
namark@lsu.edu

J. Delton Hanson
Research Testing Laboratory
Lubbock, TX 79416 USA

Jorge Pablo Jayat
Laboratorio de Investigaciones Ecológicas de las Yungas (LIEY)
Universidad Nacional de Tucumán
Casilla de Correo 34
4107 Yerba Buena
Tucumán, Argentina
pjayat@proyungas.com.ar
elijayat@gmail.com

Cecilia Lanzone
Grupo de Investigaciones de la Biodiversidad
IADIZA, CONICET
CCT-Mendoza, CC 507
5500 Mendoza, Argentina
celanzone@mendoza-conicet.gob.ar

Rafael N. Leite
Department of Biolgy
491 WIDB
Brigham Young University
Provo, UT 84602 USA
rnleite@ymail.com

Yuri L. R. Leite
Laboratório de Mastozoologia e Biogeografia
Departamento de Ciências Biológicas
Universidade Federal do Espírito Santo
Avenida Marechal Campos 1468
Maruípe
29043-900, Vitória, ES, Brasil
yleite@gmail.com

Ana Carolina Loss
Laboratório de Mastozoologia e Biogeografia
Departamento de Ciências Biológicas
Universidade Federal do Espírito Santo
Avenida Marechal Campos 1468
Maruípe
29043-900, Vitória, ES, Brasil
carol.loss@gmail.com

Simone Lóss
Sackler Institute for Comparative Genomics
American Museum of Natural History
New York, NY 10024 USA
simoneloss@gmail.com

Lucía Luna
Museum of Zoology, Mammal Division
University of Michigan
Ann Arbor, MI, 48109 USA
1928 Lindsay Lane
Ann Arbor, MI 48104 USA
llunawo@mich.edu

Guy G. Musser
Department of Mammalogy
American Museum of Natural History
Central Park West at 79th Street
New York, NY 10024 USA
holdenmusser@gmail.com

Antonio C. da S. A. Neves
Departamento de Zoologia
Instituto de Biologia
Universidade Federal do Rio de Janeiro
CCS, Room A1-121
Ilha do Fundão
21941-590, Rio de Janeiro, RJ, Brazil
antonio8519@hotmail.com

João A. de Oliveira
Museu Nacional / Universidade Federal
do Rio de Janeiro
Departamento de Vertebrados
Quinta da Boa Vista s/n
20940-040, Rio de Janeiro, RJ, Brazil
jaoliv@mn.ufrj.br

Pablo E. Ortiz
Cátedra de Paleontología
Facultad de Ciencias Naturales e Instituto Miguel Lillo
Universidad Nacional de Tucumán
Miguel Lillo 205
4000 San Miguel de Tucumán, Argentina
peortiz@uslsinectis.com.ar

Víctor Pacheco
Departamento de Mastozoología
Museo de Historia Natural
Universidad Nacional Mayor de San Marcos
Apartado 14-0434
Lima-14, Perú
vpachecot@unmsm.edu.pe

Ulyses F. J. Pardiñas
Unidad de Investigación Diversidad
Sistemática y Evolución
Centro Nacional Patagónico
Casilla de Correo 128
9120 Puerto Madryn
Chubut, Argentina
ulyses@cenpat.edu.ar

Roberta Paresque
Centro Universitário do Norte do Espírito Santo
Universidade Federal do Espírito Santo
Rua Humberto de Almeida Franklin
257 Universitário
CEP 29933-480 São Mateus, ES, Brazil
robertaparesque@ceunes.ufes.br

Bruce D. Patterson
Department of Zoology
Field Museum of Natural History
1400 S. Lake Shore Drive

Chicago, IL 60605-2496 USA
bpatterson@fieldmuseum.org

James L. Patton
Museum of Vertebrate Zoology
3101 Valley Life Sciences Building
University of California
Berkeley, CA 94720 USA
patton@berkeley.edu

Alexandre R. Percequillo
Departamento de Ciências Biológicas
Escola Superior de Agricultura 'Luiz de Queiroz'
Universidade de São Paulo
Avenue Pádua Dias, 11
Caixa Postal 9
Piracicaba, São Paulo, 13418-900, Brazil
percequi@usp.br

Leila M. Pessôa
Departamento de Zoologia
Instituto de Biologia
Universidade Federal do Rio de Janeiro
CCS, Room A1-121
Ilha do Fundão
21941-590, Rio de Janeiro, RJ, Brazil
pessoa@acd.ufrj.br

Oswaldo Ramirez
Departmento de Ciencias Biológicas y Fisiológicas
Facultad de Ciencias y Filosofía
Universidad Peruana Cayetano Heredia
San Martín de Porres, Lima, Perú
oramirez@upch.edu.pe

Jorge Salazar-Bravo
Department of Biological Sciences
Texas Tech University
Lubbock, TX 79409-3131 USA
j.salazar-bravo@ttu.edu

André L. G. da Silva
Departamento de Zoologia
Instituto de Biologia
Universidade Federal do Rio de Janeiro
CCS, room A1-121
Ilha do Fundão
21941-590, Rio de Janeiro, RJ, Brazil
andrenoctivagous@yahoo.com.br

Margaret F. Smith
Museum of Vertebrate Zoology
3101 Valley Life Sciences Building
University of California
Berkeley, CA 94720 USA
margaretfsmith@comcast.net

Angel O. Spotorno
Labratorio de Genómica Evolutiva de Mamíferos
Instituto de Ciencias Biomédicas
Facultad to Medicina

Universidad de Chile
Casilla 70061
Santiago 7, Chile
aspotorn@med.uchile.cl

Scott J. Steppan
Department of Biological Sciences
Florida State University
Tallahassee, FL 32306-1100 USA
steppan@bio.fsu.edu

William C. Tavares
Departamento de Zoologia
Instituto de Biologia
Universidade Federal do Rio de Janeiro
CCS, Room A1-121
Ilha do Fundão
21941-590, Rio de Janeiro, RJ, Brazil
tavares_w@yahoo.com.br

Pablo Teta
Unidad de Investigación Diversidad
Sistemática y Evolución
Central Nacional Patagónico
Casilla de Correo 128
9120 Puerto Madryn
Chubut, Argentina
antheca@yahoo.com.ar

Christopher J. Tribe
University Museum of Zoology Cambridge
Department of Zoology
Downing Street
Cambridge CB2 3EJ, United Kingdom
cjt14@hermes.cam.ac.uk

Michael Valqui
Director Ejecutivo
Centro para la Sostenibilidad Ambiental
Universidad Peruana Cayetano Heredia
Av. Armendáriz 445
MIraflores, Lima 18, Peru
michael.valqui@gmail.com

Diego H. Verzi
Sección Mastozoología
Museo de La Plata
Paseo del Bosque s.n
La Plata 1900, Argentina
dverzi@fcnym.unlp.edu.ar

Júlio Fernando Vilela
Laboratório de Biologia Evolutiva Teórica e Aplicada
Departamento de Genética-Instituto de Biologia
Prédio do CCS, Bloco A, Sala A2-092
Universidade Federal do Rio de Janeiro
Rua Prof. Rodolpho Paulo Rocco, S/N
Cidade Universitaria - Rio de Janeiro, RJ, Brazil
CEP: 21941-617
julio.vilela@gmail.com

Mario de Vivo
Museu de Zoologia
Universitydade de São Paulo
Av. Nazaré, 481 Ipiranga
São Paulo, SP, 04263-000, Brazil
mdvivo@usp.br

Robert S. Voss
Department of Mammalogy
American Museum of Natural History
Central Park West at 79th Street
New York, NY 10024 USA
voss@amnh.org

Marcelo Weksler
Universidade Federal do Estado do Rio de Janeiro (UNIRIO)
Departamento de Zoologia
Instituto de Biociências
Av. Pasteur 458
20290-240, Urca
Rio de Janeiro, RJ, Brazil
marcelo.weksler@gmail.com

Index